ASTRONOMY AND ASTROPHYSICS ABSTRACTS

A Publication of the Astronomisches Rechen-Institut Heidelberg
Member of the International Council
for Scientific and Technical Information
Astronomy and Astrophysics Abstracts is Prepared
Under the Auspices of the International Astronomical Union

Volume 41
Literature 1986, Part 1

Edited by
S. Böhme U. Esser H. Hefele I. Heinrich W. Hofmann
D. Krahn V. R. Matas L. D. Schmadel G. Zech

Springer-Verlag Berlin Heidelberg GmbH 1986

Astronomisches Rechen-Institut, Mönchhofstraße 12–14,
D-6900 Heidelberg 1, F.R. Germany
Telex: 461 336 ARIHD D
Director: Prof. Dr. Roland Wielen

Astronomy and Astrophysics Abstracts
Department head: Dr. Lutz D. Schmadel
Editors-in-Chief: Inge Heinrich, Dr. Lutz D. Schmadel

ISBN ISBN 978-3-662-12357-7 ISBN 978-3-662-12355-3 (eBook)
DOI 10.1007/978-3-662-12355-3

Ursprünglich erschienen bei Springer-Verlag Berlin Heidelberg New York 1986
Softcover reprint of the hardcover 1st edition 1986
Library of Congress Catalog Card Number 72-104650.
Media conversion: Daten- und Lichtsatz-Service, Würzburg.

Preface

Astronomy and Astrophysics Abstracts aims to present a comprehensive documentation of the literature concerning all aspects of astronomy, astrophysics, and their border fields. It is devoted to the recording, summarizing, and indexing of the relevant publications throughout the world. *Astronomy and Astrophysics Abstracts* is prepared by a special department of the *Astronomisches Rechen-Institut* under the auspices of the *International Astronomical Union.*

Volume 41 records literature published in 1986 and received before August 15, 1986. Some older documents which we received late and which are not surveyed in earlier volumes are included too. We acknowledge with thanks contributions of our colleagues all over the world. We also express our gratitude to all organizations, observatories, and publishers which provide us with complimentary copies of their publications.

Starting with Volume 33, all the recording, correction, and data processing work was done by means of computers. The recording was done by our technical staff members Ms. Helga Ballmann, Ms. Beate Göbel-Karl, Ms. Monika Kohl, Ms. Sylvia Matyssek, Ms. Doris Schmitz-Braunstein, Ms. Utta-Barbara Stegemann. Ms. Karin Burkhardt, Mr. Jochen Heidt, Mr. Martin Schlötelburg, and Mr. Stefan Wagner supported our task by careful proof-reading. It is a pleasure to thank them all for their encouragement.

Heidelberg, September 1986 **The Editors**

Contents

Introduction

Astronomical Bibliographies

Astronomy and Astrophysics Abstracts started documentation and abstracting work in 1969 as the direct successor of the Astronomischer Jahresbericht. For information on astronomical literature before this date consultation of one of the following bibliographies is suggested:

(1) J. J. de Lalande, Bibliographie Astronomique, Paris 1803 (this work covers the time from 480 B. C. to the year 1803, VIII + 966 pages).

(2) J. C. Houzeau, A. Lancaster, Bibliographie générale de l'astronomie, Volume I (in two parts), Bruxelles 1887, 1889, Volume II, Bruxelles 1882. The complete title of Volume II is "Bibliographie générale de l'astronomie ou catalogue méthodique des ouvrages, des mémoires et des observations astronomiques, publiés depuis l'origine de l'imprimerie jusqu'en 1880". A new edition of these volumes was prepared by D. W. Dewhirst in 1964.

(3) Bibliography of Astronomy, 1881–1898. The literature of this period was recorded on standard slips by the Observatoire Royal de Belgique. From the material (some 52,000 items) a microfilm version was produced by University Microfilms Limited, Tylers Green, High Wycombe, Buckinghamshire, England, in 1970.

(4) Astronomischer Jahresbericht, 1899 gegründet von Walter Wislicenus, herausgegeben vom Astronomischen Rechen-Institut in Heidelberg (formerly in Berlin), Verlag W. de Gruyter, Berlin. For the period from 1899 to 1968 sixty-eight volumes were published, each of which, in general, covers the literature of one year.

Concept of *Astronomy and Astrophysics Abstracts*

This abstracting service aims to present a comprehensive documentation of the literature in all fields of astronomy and astrophysics and their border fields. It appears in semi-annual volumes. Two of these volumes cover the literature of one calendar year. Every effort will be made to ensure that the average time interval between the receiving date of the original documents and publication of the abstracts will not exceed eight months. This time interval is near to that achieved by monthly abstracting journals, compared to which our system of accumulation of information over six months offers the advantage of greater convenience for the user.

The main characteristics of the concept of *Astronomy and Astrophysics Abstracts* may be summarized as follows:

(1) The subdivision of astronomy and its border fields into subject categories is facilitated by the fact that the astronomical objects appear to be particularly well suited for the formation of categories. It may be assumed that such subdivisions can be maintained for a long period. Experience shows, however, that progress in research might imply minor changes in the classification scheme.

(2) Each paper has been classified into one of 106 numbered subject categories and given a serial number within the category. In this way each item is numbered by six figures: the first three indicate the number of the category, the following three the serial number within the category. Reference to an abstract in Volume 1 is indicated by "01" before the number of the category; for example: 01.074.028, denotes Volume 1, category 074, abstract 028. A paper might be classified into more than one category. In this case, its abstract is placed only in one category, whereas in the other categories only cross references are given. These are listed at the end of each category.

(3) Authors' abstracts are used whenever possible. Popular articles are not abstracted.

(4) If possible, titles of papers and abstracts are given in English. A special reference is made to titles which we have not taken in the original language.

The whole material was recorded by means of modified ITT 3030 microcomputers. All text recording programs and other data processing software were developed by Multicom GmbH, Gröbenzell, F. R. Germany and by our staff members as well. The index computations were carried out on the IBM 3081 D computer of the University of Heidelberg.

Classification Systems

The two most common and widely used classification systems in astronomy and astrophysics are given by Class 9 of the International Classification System for Physics, published by the International Council of Scientific Unions Abstracting Board (Second edition 1978. ICSU-AB, 17 Rue Mirabeau, 75017 Paris, France, ISSN 0305-9618), and the *Astronomy and Astrophysics Abstracts* classification. In order to facilitate literature searches, we introduce a concordance relation between these two very different systems. This solution is only a unilateral one. Starting from the fourth hierarchical level of the ICSU-AB system, the appropriate *Astronomy and Astrophysics Abstracts* chapter numbers are listed. This cannot imply an identical content of the respective chapters in both systems. In many cases there is only a rather partial concordance, and therefore the *Astronomy and Astrophysics Abstracts* numbers are enclosed in parentheses. Considering our objectives, only the astronomical part of Class 9 of the ICSU-AB scheme is covered.

Transliteration Scheme for the Russian Alphabet

The transliteration of the Russian alphabet in use in *Astronomy and Astrophysics Abstracts* is presented here.

А	а	a	П	п	p
Б	б	b	Р	р	r
В	в	v	С	с	s
Г	г	g	Т	т	t
Д	д	d	У	у	u
Е	е	e	Ф	ф	f
Ё	ё	e	Х	х	kh
Ж	ж	zh	Ц	ц	ts
З	з	z	Ч	ч	ch
И	и	i	Ш	ш	sh
Й	й	j	Щ	щ	shch
К	к	k	Ы	ы	y
Л	л	l	Ь	ь	'
М	м	m	Э	э	eh
Н	н	n	Ю	ю	yu
О	о	o	Я	я	ya

This transliteration was recommended by the Abstracting Board of the International Council of Scientific Unions in 1969. It corresponds essentially to the transliteration proposed by the Academy of Sciences, Moscow. In this case the letters can be read and printed by usual data processing machines. If the names of Russian authors in the literature are transliterated in a different scheme, we present the names as they are given in the references cited and in addition in brackets according to our transliteration table.

Sources of Information

The majority of sources of information for this volume is given in category **001 Periodicals** and in category **008 Observatories, Institutes.** It may be noted that the titles of the periodicals are given in the original languages, and that Russian titles have been transliterated applying the transliteration scheme given above. Category 008 records publication series of observatories and astronomical institutes. Titles of the periodicals have been given following the recommendations of the "International List of Periodical Title Word Abbreviations" and its additions (see also **Abbreviations**). In most cases they permit recognition of the full title without recourse to the key in category 001. If other secondary sources have been consulted, we cite these papers and give reference to the respective services.

The total number of papers (some do not give names of authors) recorded in this volume amounts to 10,392.

Author, Subject, and Object Indexes

The subject category and the serial number have been used as a reference in all three indexes. These references are more precise than page references and offer considerable advantages in indexing by means of computers.

The author index of this volume contains 12,223 names.

We consider the subject index as an approximation to an optimal index covering all fields of astronomy and astrophysics. Starting with Volume 18, the subject index was enlarged to a certain extent in order to provide a thesaurus of astronomical and astrophysical terms. At present, the *Astronomy and Astrophysics Abstracts Vocabulary* 10.86, containing 2,305 key words, is in use. This is done not only for the users' convenience, but also with the intention to propose the use of special key words to authors and publishers.

While each volume is scheduled to contain an author index and a subject index, the magnetic tapes containing the index information will be used to produce separate index volumes (authors and subjects) at intervals of five years.

Beginning with Vol. 39 *Astronomy and Astrophysics Abstracts* will additionally contain an object index providing a further key to the documents abstracted. For detailed information concerning selection, standardization, sorting, etc, see introduction to the object index.

The sorting program for the indexes is based on the IBM SORT/MERGE Program. This program sorts blank before hyphen and before letters. Apostrophes are ignored by a special routine.

The users are requested to inform us on spelling errors within the author index in order to assist us in eliminating mistakes in future cumulative indexes.

Astronomy and Astrophysics Abstracts is prepared at the *Astronomisches Rechen-Institut,* Heidelberg under the auspices of the *International Astronomical Union* on a non-profit basis. The editors urge publishers of literature related to astronomy and astrophysics to provide our service in due time with complimentary copies of their material.

Publications should be mailed to:

Astronomy and Astrophysics Abstracts
Astronomisches Rechen-Institut
Moenchhofstrasse 12–14
D 6900 Heidelberg 1
F. R. Germany
Telex: 461 336 ARIHD D

Concordance Relation

between the ICSU-AB International Classification System for Physics
and the *Astronomy and Astrophysics Abstracts* Classification Scheme

ICSU-AB International Classification System for Physics	**Astronomy and Astrophysics Abstracts Classification Scheme**
0 General	
01.10 Announcements, news, and organizational activities	
01.10.C	006 (010)
01.10.F	011
01.10.H	013 (010)
01.30 Physics literature and publications	
01.30.B	012 (014)
01.30.C	012
01.30.E	003
01.30.K	002 (003)
01.30.M	003
01.30.P	003
01.30.R	014
01.30.T	002
01.40 Education	014
01.50 Educational aids	014
01.60 Biographical, historical, and personal notes	004 (005, 006, 007)
01.65 History of science	004
01.90 Other topics of general interest	015
9 Geophysics, Astronomy, and Astrophysics	
91.10 Geodesy and gravity	
91.10.B	045
91.10.N	044
91.10.Q	081

ICSU-AB International Classification System for Physics	Astronomy and Astrophysics Abstracts Classification Scheme
91.25 Geomagnetism and paleomagnetism; geoelectricity	084
91.35 Earth's interior structure and properties	081
91.90 Other topics in solid Earth physics	081
92.60 Meteorology	082
92.65 Atmospheric optics	082
94.10 Physics of the neutral atmosphere	
94.10.B	082
94.10.D	082
94.10.F	082
94.10.G	082 (063)
94.10.H	082
94.10.L	082
94.10.N	106
94.10.Q	082
94.10.S	084
94.20 Physics of the ionosphere	
94.20.B	083
94.20.D	083
94.20.M	083 (084)
94.20.P	083 (084)
94.20.W	083 (062)
94.20.Y	083 (084)
94.30 Physics of the magnetosphere	
94.30.C	084
94.30.D	084
94.30.E	084
94.30.F	084 (062)
94.30.G	084 (062)
94.30.H	084
94.30.L	084
94.30.M	084
94.30.S	084
94.30.V	084 (074)
94.30.W	084 (078, 144)

ICSU-AB International Classification System for Physics	Astronomy and Astrophysics Abstracts Classification Scheme
94.40 Cosmic rays	
94.40.C	144
94.40.E	144 (078, 106)
94.40.H	078
94.40.K	144 (085)
94.40.L	144 (078)
94.40.V	105 (144)
94.60 Interplanetary space	
94.60.D	074
94.60.F	074
94.60.G	074 (062)
94.60.K	106
94.60.M	106
94.60.Q	074 (091, 094)
94.60.R	106 (062)
94.80 Aerospace facilities and techniques, space research	
94.80.P	053 (051)
94.80.R	053 (051)
94.80.W	035
95.10 Fundamental astronomy	
95.10.C	042 (043, 052)
95.10.E	042 (052)
95.10.G	041 (079, 095, 096)
95.10.J	041
95.30 Fundamental aspects of astrophysics	
95.30.C	061 (022)
95.30.E	022
95.30.G	022 (061)
95.30.J	063
95.30.L	062
95.30.Q	062
95.30.S	066 (161)
95.45 Observatories	008 (009)

ICSU-AB International Classification System for Physics	Astronomy and Astrophysics Abstracts Classification Scheme
95.55 Astronomical instruments	
95.55.B	032
95.55.C	032 (031)
95.55.E	032
95.55.J	033
95.55.L	035
95.65 Auxiliary and recording instruments	034
95.70 Other instrumentation and techniques (including clocks, frequency standards, etc.)	034 (036)
95.75 Techniques of observation and reduction	
95.75.D	036
95.75.F	036
95.75.H	036
95.75.K	036
95.75.M	036
95.75.P	036 (021)
95.80 Catalogues, atlases, etc.	002 (046)
96.10 General, solar nebula, and cosmogony	107 (091)
96.20 Moon	
96.20.B	094
96.20.D	094
96.20.J	094
96.30 Planets and satellites (excluding the moon)	
96.30.D	092
96.30.E	093
96.30.G	097
96.30.H	098
96.30.K	099
96.30.M	100
96.30.T	101

ICSU-AB International Classification System for Physics	Astronomy and Astrophysics Abstracts Classification Scheme
96.50 Other objects in the planetary system	
96.50.D	106
96.50.G	102 (103)
96.50.K	104
96.50.M	105
96.60 Solar physics	
96.60.C	080 (075)
96.60.F	071 (080)
96.60.K	080
96.60.M	071
96.60.N	073 (074)
96.60.P	074
96.60.Q	072
96.60.R	073 (076, 077)
96.60.S	073
96.60.V	078 (074)
97.10 Stellar characteristics and properties	
97.10.B	131
97.10.C	065 (061)
97.10.E	064 (063)
97.10.F	112 (064)
97.10.H	112 (064)
97.10.K	116 (065)
97.10.L	116 (065)
97.10.N	115
97.10.Q	115
97.10.R	115 (113, 114)
97.10.T	114
97.10.V	111
97.10.W	111
97.20 Normal stars (by class): general or individual	
97.20.D	121
97.20.R	126
97.30 Variable and peculiar stars (including novae)	
97.30.E	114 (112)
97.30.F	116

ICSU-AB International Classification System for Physics	Astronomy and Astrophysics Abstracts Classification Scheme
97.30.G	122 (123)
97.30.J	122 (123)
97.30.K	122 (123)
97.30.N	122 (123)
97.30.Q	124 (117, 122, 123)
97.30.S	122 (123)
97.60 Late stages of stellar evolution (including black holes)	
97.60.B	125
97.60.G	126 (067)
97.60.J	067
97.60.L	067
97.60.S	067 (065)
97.80 Binary and multiple stars (including extrasolar planetary systems)	
97.80.D	118 (002)
97.80.F	120
97.80.H	119
97.80.J	142 (117)
97.80.K	118
97.80.M	118
98.10 Stellar dynamics	151
98.20 Stellar clusters and associations	
98.20.C	152
98.20.E	153
98.20.H	154
98.40 Interstellar matter and nebulae	
98.40.B	131
98.40.C	131
98.40.F	131
98.40.H	132
98.40.J	133
98.40.K	131
98.40.M	134
98.40.N	125

ICSU-AB International Classification System for Physics	Astronomy and Astrophysics Abstracts Classification Scheme

98.50 The Galaxy; extragalactic objects and systems

98.50.C	157
98.50.E	157 (161)
98.50.H	157 (158, 161)
98.50.K	160
98.50.L	155
98.50.M	160 (156)
98.50.R	158
98.50.T	160

98.70 Other objects and background radiations of unknown origin or distances

98.70.D	141
98.70.J	159
98.70.L	133
98.70.Q	142 (143)
98.70.S	144
98.70.V	161 (142, 143)

98.80 Cosmology

98.80.B	161
98.80.D	161 (066)
98.80.F	061 (161)

Abbreviations

Abbreviations used in *Astronomy and Astrophysics Abstracts* are primarily based on the 'International List of Periodical Title Word Abbreviations', prepared for the UNISIST/ICSU-AB Working Group on Bibliographic Descriptions (1970).

A.A.B.	Associazione Astrofili Bolognesi
Aarg.	Aargang
AAS	American Astronomical Society
AAVSO	American Association of Variable Star Observers
Abh.	Abhandlung –
Abstr.	Abstract –
Abt.	Abteilung
Acad.	Academi –, Academy
Accad.	Accademi –
Act.	Active, Activit –
Adm.	Administr –
Adv.	Advanc –
Aehron.	Aehronomi –
Aeron.	Aeronom –
Aeronaut.	Aeronauti –
Aerosp.	Aerospace
Afr.	Africa –
AG	Astronomische Gesellschaft
AIAA	American Institute of Aeronautics and Astronautics
AJB	Astronomischer Jahresbericht
Akad.	Akadem –
Ala.	Alabama
Alm.	Almanac –
Am.	America –, Amerika –, Amerique –
Amat.	Amateur –
Amst.	Amsterdam
An.	Anais, Anale –, Anali –, Anals
Anal.	Analis –, Analit –, Analys –, Analyt –
Angew.	Angewandt –
Ann.	Annaes, Annal –
Annu.	Annu –
Anst.	Anstalt
Anu.	Anual –, Anuar –
Anz.	Anzeiger
Appl.	Applied
Arb.	Arbeit
Arch.	Archiv –
Årg.	Årgang
Argent	Argentin –
Ariz.	Arizona
Ark.	Arkiv –
Arkh.	Arkhiv –
Artif.	Artifici –
ASA	Astronomical Society of Australia
Asoc.	Asocia –
ASP	Astronomical Society of the Pacific
ASSA	Astronomical Society of Southern Africa
Assem.	Assembl –
Assoc.	Associ –
Assoz.	Assozi –
Astrofis.	Astrofisic –
Astrofiz.	Astrofizi –
Astrometr.	Astrometr –
Astron.	Astronom –
Astronaut.	Astronauti –, Astronauty –
Astrophys.	Astrophys –
ASV	Astronomical Society of Victoria
ASWA	Astronomical Society of Western Australia
At.	Atom –

Atmos.	Atmosf –, Atmosph –
Aust.	Australi –
BAA	British Astronomical Association
Barc.	Barcelona
Bayer.	Bayerisch –
Beitr.	Beitrag, Beiträge
Belg.	Belge –, Belgi –
Beob.	Beobacht –
Beogr.	Beograd –
Ber.	Bericht –
Bibl.	Bibliot –
Bibliogr.	Bibliograf –, Bibliograph –
BIH	Bureau International de l'Heure
Bimest.	Bimestr –
Bl.	Blatt, Blätter
Bol.	Boletin
Boll.	Bolletino
Br.	British
Bras.	Brasil –
Brun.	Brunens –
Bruss.	Brussel, s
Brux.	Bruxelles
Bul.	Buleten –, Buletin –, Bulten
Bulg.	Bulgar –
Bull.	Bulletin –, Bullettino
Bur.	Bureau –
Byul.	Byuleten –, Byuletin –
Byull.	Byulleten –
C. R.	Comptes Rendus
Cah.	Cahier –
Calif.	California
Camb.	Cambridge
Can.	Canadi –, Canada
Carol.	Carolina –
Cas.	Casopis
Cat.	Catalog –
Celest.	Celestial
Cent.	Center, Central, Centrale, Centrally, Centre
Cercet.	Cercetari
Cesk.	Ceskoslov –
Chem.	Chemi –
Chim.	Chimi –
Chin.	Chinese
Chron.	Chronic –, Chronik, Chronique
Chronom.	Chronometr –
Cie.	Compagnie
Cienc.	Ciencia –
Cient.	Cientific –
Circ.	Circolar –, Circolo, Circolaire –, Circular –, Circulo
Cirk.	Cirkulaer –
Cl.	Clasa, Classe –
Co.	Companies, Company
Coll.	College
Collect.	Collect –
Colloq.	Colloqui –
Colo.	Colorado

Comet. Cometary
Commentat. Commentat –
Commun. Communica –
Comput. Computation, Computer –, Computing
Comun. Comunica –
Conf. Conferen –
Congr. Congres –
Conn. Connecticut
Contract. Contract –
Contrib. Contribu –
Copenh. Copenhagen
Cosm. Cosmic –
Cosmochim. Cosmochimi –
COSPAR Committee on Space Research
Crystallogr. Crystallograph –
CSIRO Commonwealth Scientific and Industrial Research Organization
Cult. Cultur –, Cultuur
Curr. Current
Czech. Czechoslovak –

D. C. District of Columbia
DDR Deutsche Demokratische Republik
Del. Delaware
Dep. Departament, Département, Department
Dev. Development –, Développement –
Dig. Digest
Dir. Director –
Diss. Disserta –
Div. Divis –
Doc. Document –
Dok. Dokument –
Dokl. Doklad –
Dom. Dominion
Dtsch. Deutsch

Ed. Edit –
Edinb. Edinburgh
Ehksp. Ehksperiment –
Eidg. Eidgenössisch –
Eksp. Eksperiment –
Electron. Electroni –
Eng. Engineer –
Environ. Environment –
Equip. Equipement, Equipment
Ergeb. Ergebnis –
ESA European Space Agency
ESO European Southern Observatory
Espec. Especial –
ESRO European Space Research Organization
Eur. Europ –
Eval. Evaluation –
Exp. Experiment –
Extraterr. Extraterrestr –

F. R. Germany Federal Republic of Germany
Fac. Facolt –, Faculd –, Facult –
Fak. Fakult
Fasc. Fascicul –
Fenn. Fenni –
Finn. Finni –
Fis. Fisic –, Fisik –
Fiz. Fizic –, Fizik –, Fizyk –
Fla. Florida
Fluid. Fluidi –
Fond. Fondation –, Fondazione
Fortschr. Fortschritt –
Fotogr. Fotograf –
Found. Foundation –
Fr. Français –
Freq. Frequen –
Fundam. Fundamenta –
Fys. Fysik –, Fysisch, Fysisk –
Fyz. Fyzik –

G. Giornale
Ga. Georgia
Gaz. Gazeta, Gazette
Gazz. Gazzetta
Gen. General
Geochem. Geochem –
Geochim. Geochim –
Geod. Geodaes –, Geodaet –, Geodes –, Geodet –, Geodez –
Geofis. Geofis –
Geofiz. Geofiz –
Geofys. Geofys –
Geogr. Geograf –, Geograph –
Geokhim. Geokhim –
Geol. Geolog –, Geolosk –
Geomagn. Geomagneti –
Geophys. Geophys –
Ges. Gesellschaft
Gesch. Geschichte
Gl. Glavno –
Glas. Glasnik
Gos. Gosudarst –
Gov. Government –
Grenzgeb. Grenzgebiet –
GSFC Goddard Space Flight Center

H.M. Her Majesty's, His Majesty's
Hamb. Hamburg
Handb. Handbook, Handbuch
Heidelb. Heidelberg –
Helv. Helveti –
Her. Herald –
Hist. History
Hochsch. Hochschule
Hoegsk. Hoegskol –
HR-diagram Hertzsprung-Russell diagram
Hung. Hungar –
Hydrogr. Hydrograf –, Hydrograph –

IAF International Astronautical Federation
IAU International Astronomical Union
IBM International Business Machines Corporation
ICSTI International Council for Scientific and Technical Information
ICSU International Council of Scientific Unions
ICSU-AB International Council of Scientific Unions – Abstracting Board
IEEE Institute of Electrical and Electronics Engineers
Ill. Illinois
Inc. Incorporated
Ind. Industr –
Inf. Informat –, Informaz –, Informe –
Ing. Ingenieur
INIS International Nuclear Information System
INSPEC International Information Services for the Physics and Engineering Communities
Inst. Institut –, Instytut –
Instrum. Instrument –
Int. International, Internazional –
Intell. Intelligenc –
Inter. Intérieur –, Interior
Interplanet. Interplanetary
Intez. Intezet –
Invest. Investiga –
Ionos. Ionosfer –, Ionospher –
Ir. Irish
Iskusstv. Iskusstvenn –
Isr. Israel –
Issled. Issledovan –
Ist. Istitut
Ital. Itali –
Izd. Izdatel –
Izv. Izvesti –

J.	Joernaal –, Jornal –, Journal –
Jaarb.	Jaarboek –
Jahr.	Jahrbuch, Jahrbücher
Jahresber.	Jahresbericht –
Jahresschr.	Jahresschrift
Jahrg.	Jahrgang
JPL	Jet Propulsion Laboratory
Jpn.	Japan –
K.	Königlich –, Koninkljik –, Kunglig –
Kans.	Kansas
Kartogr.	Kartograf –
Kernforsch.	Kernforschung
Kernphys.	Kernphysik –
Khem.	Khemyi –
Khim.	Khimi –
Kim.	Kimija –, Kimya
Kl.	Klass –
Kolloq.	Kolloquium –
Komet.	Kometnyj
Komm.	Kommission –
Konf.	Konfer –
Kongr.	Kongress
Kosm.	Kosmich –
Kosmog.	Kosmogon –
Kozp.	Kozponti
KPNO	Kitt Peak National Observatory
Ky.	Kentucky
La.	Louisiana
Lab.	Laborato –
LEST	Large European Solar Telescope
Lett.	Letter –, Lettra, Lettre
Libr.	Librair –, Librar –
Madr.	Madrid
Mag.	Magasin, Magazin –
Magn.	Magneti –, Magnitn –
Mar.	Marin –
Mass.	Massachusetts
Mat.	Matemaat –, Matemat –
Mater.	Material –
Math.	Mathemat –
Md.	Maryland
Meas.	Measur –
Mec.	Mecani –
Mech.	Mechani –
Medd.	Meddelande –, Meddelelse
Meded.	Mededeeling, Mededeling –
Mekh.	Mekhani –
Mem.	Memento –, Memoir –, Memori –, Memory –, Memuary
Memo.	Memorand –
Mens.	Mensile, Mensual –, Mensuel –
Messtech.	Messtechni –
Meteorol.	Meteorolog –
Mex.	Mexic –
Mich.	Michigan
Micromec.	Micromecaniq –
Miner.	Mineral, Minerale –, Minerali –
Mineral.	Mineralog –
Minn.	Minnesota
Miss.	Mississippi
MIT	Massachusetts Institute of Technology
Mitt.	Mitteilung –
Mo.	Missouri
Mod.	Modern –
Mol.	Molecul –, Molekul –
Mon.	Monat, Monatlich –, Month –
Monogr.	Monograph –
Mont.	Montana
MPI	Max-Planck-Institut
Mt.	Mount
Muench.	Muenchen
Mus.	Museum
N. C.	North Carolina
N. D.	North Dakota
N. H.	New Hampshire
N. J.	New Jersey
N. M.	New Mexico
N. Y.	New York
N. Z.	New Zealand
Nablyud.	Nablyudeni –
Nac.	Nacion –
Nachr.	Nachricht –
NASA	National Aeronautics and Space Administration
Nat.	Natur –
Natl.	National –
Naturforsch.	Naturforsch –
Naturwiss.	Naturwissenschaft –
Natuurkd.	Natuurkunde
Nauchn.	Nauchny –
Nauk.	Nauka, Naukite, Naukov –, Naukow –
Naut.	Nautic –
Nav.	Naval –
Navig.	Navigat –
Naz.	Nazion –
Nebr.	Nebraska
Ned.	Nederland –
Nev.	Nevada
Newsl.	Newsletter –
Not.	Notationes, Notic –, Notise, Notizi –
Nouv.	Nouveau –, Nouvell –
Nov.	Novoe
Nucl.	Nucléaire –, Nuclear –, Nucl –
Nukl.	Nukle –
Numer.	Numeri –
O-va	Obshchestva
O-vo	Obshchestvo
Obs.	Observ –
Österr.	Österreich –
Off.	Offic –
Okla.	Oklahoma
Opt.	Optic –, Optik –, Optique
Oreg.	Oregon
Oss.	Osserva –
Pa.	Pennsylvania
Pac.	Pacific
Paleontol.	Paleontolog –
Pap.	Paper –, Papier
Part.	Particle
Pekin.	Pekinens –
Perem.	Peremenn –
Period.	Periodi –
Petrol.	Petrolog –
Philos.	Philosoph –
Photogr.	Photograf –, Photograph –
Photogramm.	Photogrammetr –
Photom.	Photometr –
Phys.	Physic –, Physik –, Physique –, Physisch
Pict.	Picture –
Planet.	Planetary
Pol.	Polish, Polon –
Pr.	Prac –
Prelim.	Prelimin –
Prepr.	Preprint
Prib.	Pribor –
Prikl.	Prikladn –
Prilozh.	Prilozhen –
Prir.	Prirodn –
Prirodoved.	Prirodoved –

Probl.	Problem–
Proc.	Proceedings
Prod.	Prodott–, Produc–, Produkt,
Prog.	Progres–
Propag.	Propagation
Prospect.	Prospecting
Prov.	Provinc–, Provints–, Provinz–
Pubbl.	Pubblicazion–
Publ.	Publicac–, Publicas–, Publicat–, Publikas–, Publikat–
Q.	Quarterly
Quant.	Quantit–
R.	Royal
R. I.	Rhode Island
Radiat.	Radiati–
Radioact.	Radioactiv–, Radioaktiv–
Radioisot.	Radioisotop–
Rap.	Raport–
Rapp.	Rapport–
RAS	Royal Astronomical Society
Rec.	Record–
Rech.	Recherche–
Ref.	Referat–, Reference–, Referieren
Relat.	Related, Relation–
Relativ.	Relativit–
Rend.	Rendicont–
Rep.	Report–
Repr.	Reprint–
Repub.	Republi–
Res.	Research–
Result.	Resultad–, Resultat–
Rev.	Review–, Revisio, Revista, Revue–
Rezul't.	Rezul'tat–
Ric.	Ricerca, Ricerche
Riv.	Rivist–
Roum.	Roumain–
Rundsch.	Rundschau
S. Afr.	South Africa
S. C.	South Carolina
S. D.	South Dakota
SAF	Société Astronomique de France
SAI	Società Astronomica Italiana
Samml.	Sammlung–
SAO	Smithsonian Astrophysical Observatory
SAS	Société Astronomique de Suisse
Satell.	Satellite
Sb.	Sbornik–
Scand.	Scandinavi–
Sch.	Schul–
Schr.	Schrift–
Schriftenr.	Schriftenreihe
Schweiz.	Schweizer–
Sci.	Scienc–, Scient–, Scienz–
Scr.	Scripta, Scritt–
Secc.	Seccion–
Sect.	Secti–
Sekc.	Sekci–, Sekcj–
Sekt.	Sektion–, Sektor–
Sekts.	Sektsi–
Sel.	Seleccion–, Select–, Selek–, Selezione
Selsk.	Selskab–, Selskap–
Semin.	Séminair–, Seminar–
Sep.	Separat–
Ser.	Seria–, Serie–, Seriya
Serv.	Servic–, Serviz–
Sess.	Sessi–
Signal.	Signalétique–
Simp.	Simpoz–
Sin.	Sinica
Sitzungsber.	Sitzungsbericht–
Skr.	Skrift–
Smithson.	Smithsonian
Soc.	Sociedad–, Societ–
Sol.	Solar
Soln.	Solnechn–
Sonderdr.	Sonderdruck–
Soobshch.	Soobshchen–
South.	Southern
Spacecr.	Spacecraft
Spat.	Spatial–
Spec.	Special–
Spectrosc.	Spectroscop–
Spectrosk.	Spectroskop–
Spets.	Spetsial–
Spez.	Spezial–, Speziell–
SSR	Sovetskaya Sotsialisticheskaya Respublika
SSSR	Soyuz Sovetskikh Sotsialisticheskikh Respublik
St.	Saint–, Sankt–, Sant–
–St.	–Straße, Street
Stand.	Standard–, Standart–
Sternw.	Sternwarte–
Stiint.	Stiintific–
Stn.	Station, Stazione
Stud.	Studia, Studie–, Studii
Supl.	Suplement–, Supliment–
Suppl.	Supplement–
Surv.	Survey–
Syd.	Sydney
Symp.	Sympos–, Sympoz–
Syst.	System–
Sz.	Szemle
Teach.	Teacher–, Teaching
Tec.	Tecni–
Tech.	Techni–
Technol.	Technolog–
Tecnol.	Tecnolog–
Teh.	Tehnic–, Tehnika, Tehnisk–
Tehnol.	Tehnolog–, Tehnolosk–
Tek.	Tekni–
Tekh.	Tekhni–
Tekhnol.	Tekhnolog–
Teknol.	Teknolog–
Telesc.	Telescop–
Telev.	Television–
Tenn.	Tennessee
Teor.	Teoret–, Teori–
Terr.	Terrestr–
Test.	Testing
Tex.	Texas
TH	Technische Hochschule
Theor.	Theoret–, Theori–
Tidschr.	Tidschrift–
Tidskr.	Tidskrift–
Tidsskr.	Tidsskrift–
Top.	Topic–
Torun.	Torunensis
Tr.	Trudy
Trans.	Transactions, Transazione
Tsentr.	Tsentral–
Tsirk.	Tsirkulyar–
TU	Technical University
Uch.	Uchen–
Uchebn.	Uchebn–
UK	United Kingdom
Umsch.	Umschau
UN	United Nations
Univ.	Universidad–, Universit–, Univerzitet–
Ups.	Upsaliens–

US	United States
USA	United States of America
USSR	Union of Soviet Socialist Republics
Utr.	Utrecht
Va.	Virginia
Var.	Various
Ver.	Verein –, Verenig –
Veränderl.	Veränderlich –
Verh.	Verhandl –
Vermess.	Vermessung –
Vermessungswes.	Vermessungswesen
Veröff.	Veröffentlich –
Vesn.	Vesnik
Vestn.	Vestnik
Vetensk.	Vetenskap –
Vgl.	Vergleich –
Vidensk.	Videnskab –, Videnskap
Vierteljahresschr.	Vierteljahresschrift –
Vierteljahrsschr.	Vierteljahrsschrift
VLB	Very Long Baseline
Volcanol.	Volcanolog –
Vopr.	Vopros –
Vortr.	Vorträge
Vses.	Vsesoyuzn –
Vt.	Vermont
Vyp.	Vypusk –
Vyssh.	Vyssh –
Vyzk.	Vyzkum –
W. Va.	West Virginia
Wash.	Washington
West.	Western
Wet.	Wetenschap –, Wetenskap –
Wis.	Wisconsin
Wiss.	Wissenschaft –
Wyo.	Wyoming
Yad.	Yadern –
Z.	Zeitschrift –
ZA	Zero Age
ZAED	Zentralstelle für Atomkernenergie-Dokumentation
Zap.	Zapisk –, Zapyisk –
Zaved.	Zaveden –
Zent.	Zentral
Zentralbl.	Zentralblatt
Zesz.	Zeszyt
Zh.	Zhurnal –
Zirk.	Zirkular

Periodicals, Proceedings, Books, Activities

001 Periodicals

A.A.O. Newsl.
A.A.O. Newsletter. Anglo–Australian Observatory, PO Box 296, Epping, N.S.W. 2121, Australia.

AAS Photo–Bull.
American Astronomical Society Photo–Bulletin. Working Group on Photographic Materials, 211 Space Sciences Building. University of Florida, Gainesville, Fla. 32611, USA. ISSN 0065–7433.

Abastumanskaya Astrofiz. Obs., Byull.
Abastumanskaya Astrofizicheskaya Observatoriya, Gora Kanobili. Byulleten'. Akademiya Nauk Gruzinskoj SSR. Izdatel'stvo Metsniereba, Tbilisi, USSR. ISSN 0375–6644.

Abh. Akad. Wiss. DDR
Abhandlungen der Akademie der Wissenschaften der DDR. Abteilung Mathematik, Naturwissenschaften, Technik. Akademie–Verlag, Berlin, German Democratic Republic.

Abh. Hamb. Sternw.
Abhandlungen aus der Hamburger Sternwarte. Hamburger Sternwarte, Universität Hamburg, Gojenbergsweg 2, D–2050 Hamburg 80, F.R. Germany. ISSN 0374–1583.

Abstr. Submitted Pap.
Abstracts of Submitted Papers. Institute of Astronomy, The Observatories, Madingley Road, Cambridge CB3 0HA, England.

Acad. R. Belg., Bull. Cl. Sci.
Académie Royale de Belgique, Bulletin de la Classe des Sciences (Koninklijke Academie van België, Mededelingen van de Klasse der Wetenschappen). 5e Série, Palais des Académies, Bruxelles, Belgium.

Acad. R. Belg., Mém. Cl. Sci.
Académie Royale de Belgique, Mémoires de la Classe des Sciences. Collection in 8°, 2e Série, Palais des Académies, Bruxelles, Belgium.

Acad. Sci. Est. SSR, Div. Phys. Math. Tech. Sci., Prepr.
Academy of Sciences of the Estonian SSR, Division of Physical, Mathematical and Technical Sciences, Preprints. Institute of Astrophysics and Atmospheric Physics, Academy of Sciences of the Estonian SSR, 202444 Tartu, Tôravere, Estonian SSR, USSR.

Acta Acust.
Acta Acustica. Science Press, Beijing, People's Republic of China. Subscription address: Guozi Shudian, PO Box 399, Beijing, People's Republic of China.

Acta Astron.
Acta Astronomica. An International Quarterly Journal. Polska Akademia Nauk, Komitet Astronomii, Państwowe Wydawnictwo Naukowe, Warszawa–Kraków, Poland. Subscription address: Ars Polona, 00–068 Warszawa, Krakowskie Przedmieście 7, Poland. ISSN 0001–5237.

Acta Astron. Sin.
Acta Astronomica Sinica. Purple Mountain Observatory, Academia Sinica. Nanjing, People's Republic of China. English translation in Chin. Astron. Astrophys. ISSN 0001–5245.

Acta Astronaut.
Acta Astronautica. Journal of the International Academy of Astronautics. Pergamon Press, Oxford – New York – Toronto – Paris – Frankfurt – Sydney. ISSN 0094–5765.

Acta Astrophys. Sin.
Acta Astrophysica Sinica. Beijing Astronomical Observatory, Academia Sinica, Beijing, People's Republic of China. Subscription address: Guozi Shudian, PO Box 399, Beijing, People's Republic of China. English translation in Chin. Astron. Astrophys.

Acta Cosmologica
Acta Cosmologica. Zeszyty Naukowe Universytetu Jagiellońskiego. Państwowe Wydawnictwo Naukowe, Warszawa–Kraków, Poland. ISSN 0137–2386.

Acta Fac. Rerum Nat. Univ. Comenianae, Astron. Geophys.
Acta Facultatis Rerum Naturalium Universitatis Comenianae, Astronomia et Geophysica. Published for the Komenský University by Slovenské pedagogické nakladatel'stvo, 89112 Bratislava, Czechoslovakia.

Acta Geod. Geophys.
Acta Geodaetica et Geophysica. Institute of Geodesy and Geophysics, Academia Sinica, Beijing, People's Republic of China. Subscription address: Guozi Shudian, PO Box 399, Beijing, People's Republic of China.

Acta Geod. Geophys. Montan.
Acta Geodaetica, Geophysica et Montanistica. Akademiai Kiado, H–1054 Budapest, Alkotmany utca 21, Hungary. ISSN 0374–1842.

Acta Geophys. Pol.
Acta Geophysica Polonica. Subscription address: Ars Polona, 00–068 Warszawa, Krakowskie Przedmieście 7, Poland. ISSN 0001–5725.

Acta Geophys. Sin.
Acta Geophysica Sinica. Department of Geophysical Research, Academia Sinica, Beijing, People's Republic of China. Subscription address: Guozi Shudian, PO Box 399, Beijing, People's Republic of China. ISSN 0001–5733.

Acta Math. Sin.
Acta Mathematica Sinica. Academia Sinica, Beijing, People's Republic of China. Subscription address: Guozi Shudian, PO Box 399, Beijing, People's Republic of China.

Acta Mech. Sin.
Acta Mechanica Sinica. Academia Sinica, Beijing, People's Republic of China. Subscription address: Guozi Shudian, PO Box 399, Beijing, People's Republic of China. ISSN 0254–3060.

Acta Meteorol. Sin.
Acta Meteorologica Sinica. Subscription address: Guozi Shudian, PO Box 399, Beijing, People's Republic of China. ISSN 0577–6619.

Acta Phys. Austriaca
Acta Physica Austriaca. Springer–Verlag, Wien, Austria. ISSN 0001–6713.

Acta Phys. Hung.
Acta Physica Hungarica. Akademiai Kiado, Publishing House of the Hungarian Academy of Sciences, H–1054 Budapest, Alkotmany U. 21, Hungary. ISSN 0231–4428.

Acta Phys. Pol., Ser. A
Acta Physica Polonica, Ser. A. (General Physics, Solid State Physics, Atomic and Molecular Spectroscopy, Applied Physics). Polska Akademia Nauk, Warszawa, Poland. Subscription address: Ars Polona, 00–068 Warszawa, Krakowskie Przedmieście 7, Poland. ISSN 0587–4246.

Acta Phys. Pol., Ser. B
Acta Physica Polonica, Ser. B (Elementary Particle Physics, Nuclear Physics, Theory of Relativity, Field Theory). Polska Akademia Nauk, Warszawa, Poland. Subscription address: Ars Polona, 00–068 Warszawa, Krakowskie Przedmieście 7, Poland. ISSN 0587–4254.

Acta Phys. Sin.
Acta Physica Sinica. Institute of Physics, Academia Sinica, Beijing, People's Republic of China. ISSN 0372–736X.

Acta Phys. Slovaca
Acta Physica Slovaca. VEDA Publishing House of the Slovak Academy of Sciences, 89530 Bratislava, Klemensova 19, Czechoslovakia. ISSN 0323–0465.

Acta Polytech. Scand., Appl. Phys. Ser.
Acta Polytechnica Scandinavia, Applied Physics Series. Finnish Academy of Technical Sciences, Kansakoulukatu 10A, SF–00100 Helsinki, Finland. ISSN 0355–2721.

Acta Sci. Nat. Univ. Pekin.
Acta Scientiarum Naturalium Universitatis Pekinensis. Beijing, People's Republic of China. ISSN 0479–8023.

Acta Sci. Nat. Univ. Sunyatseni
Acta Scientiarum Naturalium Universitatis Sunyatseni (Zhongshandaxue Xuebao). Canton Post Office, Canton, People's Republic of China.

Acta Tech. Acad. Sci. Hung.
Acta Technica Academiae Scientiarum Hungaricae. Akademiai Kiado, Budapest 1363, PO Box 24, Hungary. ISSN 0001–7035.

Acta Tech. ČSAV
Acta Technica Československá akademie věd. Academia, Publishing House of the Czechoslovak Academy of Sciences, Vodičkova 40, 112 29 Praha 1, Czechoslovakia. Subscription address: John Benjamins N.V., Periodical Trade, Warmoesstraat 54, Amsterdam, The Netherlands. ISSN 0001–7043.

Acta Univ. Carol. Math. Phys.
Acta Universitatis Carolinae. Mathematica et Physica. Fakulta matematicko–fyzikální, Karlova universita, Praha, Czechoslovakia. ISSN 0001–7140.

A.D.I.O.N. Bull.
A.D.I.O.N. Bulletin. Association pour le Développement International de l'Observatoire de Nice. Observatoire de Nice, B.P. 252, F–06007 Nice Cedex, France.

Adv. Astronaut. Sci.
Advances in the Astronautical Sciences. American Astronautical Society, Publicatons Office, PO Box 28130, San Diego, Calif. 92128, USA. ISSN 0065–3438.

Adv. Phys.
Advances in Physics. Taylor & Francis Ltd., London, England. ISSN 0001–8732.

Adv. Space Res.
Advances in Space Research. The official journal of the Committee on Space Research (COSPAR). Pergamon Press, Oxford – New York – Toronto – Sydney – Paris – Frankfurt. ISSN 0273–1177.

Aéronaut. Astronaut.
L'Aéronautique et l'Astronautique. Editions Air et Cosmos, 6 Rue Anatole de la Forge, F–75017 Paris, France. ISSN 0001–9275.

AIAA J.
AIAA Journal. American Institute of Aeronautics and Astronautics, 1290 Avenue of the Americas, New York, N.Y. 10019, USA. ISSN 0001–1452.

AIP Conf. Proc.
AIP Conference Proceedings. American Institute of Physics, 335 East 45th Street, New York, N.Y. 10017, USA. ISSN 0094–243X.

Algoritm. Nebesnoj Mekh.
Algoritmy Nebesnoj Mekhaniki (Materialy Matematicheskogo Obespecheniya EhVM). Institut Teoreticheskoj Astronomii Akademii Nauk SSSR. 191187 Leningrad, D–187, nab. Kutuzova, dom–10. USSR.

Alta Freq.
Alta Frequenza. Ufficio Centrale AEI–CEI, Viale Monza 259, I–20126 Milano, Italy. ISSN 0002–6557.

Am. Assoc. Variable Star Obs. Bull.
The American Association of Variable Star Observers Bulletin. The American Association of Variable Star Observers, 187 Concord Avenue, Cambridge, Mass. 02138, USA.

Am. Assoc. Variable Star Obs. Rep.
The American Association of Variable Star Observers Report. The American Association of Variable Star Observers, 187 Concord Avenue, Cambridge, Mass. 02138, USA.

Am. J. Phys.
American Journal of Physics. Published for the American Association of Physics Teachers by the American Institute of Physics, 335 East 45th Street, New York, N.Y. 10017, USA. ISSN 0002–9505.

Am. Mineral.
American Mineralogist. Mineralogical Society of America, 2000 Florida Avenue, N.W., Washington, D.C. 20009, USA. ISSN 0003–004X.

Am. Sci.
American Scientist. Society of Sigma XI, 345 Whitney Avenue, New Haven, Conn. 06510, USA. ISSN 0003–0996.

An. Acad. Bras. Cienc.
Anais da Academia Brasileira de Ciencias. Caixa Postal 229, ZC–00 Rio de Janeiro gb, Brazil. ISSN 0001–3765.

An. Fis., Ser. A
Anales de Fisica, Serie A (Fenomenos e Interacciones). Real Sociedad Espanola de Fisica, Facultades de Ciencias, Ciudad Universitaria, Madrid 3, Spain. ISSN 0211–6243.

An. Fis., Ser. B
Anales de Fisica, Serie B (Aplicaciones, Metodes e Instrumentos). Real Sociedad Espanola de Fisica, Facultades de Ciencias, Ciudad Universitaria, Madrid 3, Spain. ISSN 0211-6251.

An. Obs. Astron. Univ. Coimbra
Anais do Observatório Astronómico da Universidade de Coimbra. Observatório Astronómico da Universidade de Coimbra, Coimbra, Portugal. ISSN 0870-2853.

Anal. Chem.
Analytical Chemistry. American Chemical Society, 1155 16th Street N.W., Washington, D.C. 20036, USA. ISSN 0003-2700.

Anglo-Aust. Obs., Prepr.
Anglo-Australian Observatory, Preprints (AAO PP). Anglo-Australian Observatory, PO Box 296, Epping, N.S.W. 2121, Australia.

Anglo-Aust. Telesc., Annu. Rep.
Anglo-Australian Telescope, Annual Report. Anglo-Australian Observatory, PO Box 296, Epping, N.S.W. 2121, Australia.

Ann. Acad. Sci. Fenn., Ser. A VI
Annales Academiae Scientiarum Fennicae, Series A VI (Physica). Academia Scientiarum Fennica, Snellmaninkato 9-11, 00 170 Helsinki 17, Finland. ISSN 0066-2003.

Ann. Fond. Louis de Broglie
Annales de la Fondation Louis de Broglie. Fondation Louis de Broglie, 1 rue Montgolfier, F-75003 Paris, France.

Ann. Geophys.
Annales Geophysicae. European Geophysical Society. Gauthier-Villars. Subscription address: C.D.R. Centrale des Revues, 11 rue Gossin, F-92543 Montrouge Cedex, France.

Ann. Inst. Henri Poincaré Phys. Théor.
Annales de l'Institut Henri Poincaré Physique Théorique. Gauthier-Villars, C.D.R., Centrale des Revues, 11 rue Gossin, 92543 Montrouge Cedex, France. ISSN 0246-0211.

Ann. Isr. Phys. Soc.
Annals of the Israel Physical Society. c/o Department of Physics, Bar-Ilan University, Ramat-Gan, Israel. Adam Hilger Ltd., Bristol, England. ISSN 0309-8710.

Ann. Nucl. Energy
Annals of Nuclear Energy. Pergamon Press, Oxford - New York - Toronto - Paris - Frankfurt - Sydney. ISSN 0306-4549.

Ann. N.Y. Acad. Sci.
Annals of the New York Academy of Sciences. The New York Academy of Sciences, 2 East 63rd Street, New York, N.Y., USA. ISSN 0077-8923.

Ann. Obs. Astron. Alger
Annales de l'Observatoire Astronomique d'Alger. Observatoire Astronomique de l'Université d'Alger, Algiers, Algeria.

Ann. Phys. (Leipzig)
Annalen der Physik (Leipzig). Johann Ambrosius Barth, Salomonstr. 18B, Leipzig 701, German Democratic Republic. ISSN 0003-3804.

Ann. Phys. (N.Y.)
Annals of Physics (New York). Academic Press Inc., New York - London. ISSN 0003-4916.

Ann. Phys. (Paris)
Annales de Physique (Paris). Masson et Cie S.A., 120 Boulevard Saint-Germain, F-75280 Paris Cedex 06, France. ISSN 0003-4169.

Ann. Sci.
Annals of Science. Taylor & Francis Ltd., London, England. ISSN 0003-3790.

Ann. Sci. Univ. Besançon, Phys.
Annales Scientifiques de l'Université de Besançon, Physique. Institut des Sciences Naturelles, Place Leclerc, F-25000 Besançon, France. ISSN 0365-6543.

Ann. Shanghai Obs., Acad. Sin.
Annals of Shanghai Observatory, Academia Sinica. Shanghai Scientific and Technical Publishers, Shanghai, Rei Jing Er Street 450, People's Republic of China.

Ann. Soc. Sci. Brux., Sér. I
Annales de la Société Scientifique de Bruxelles, Série I (Sciences Mathématiques, Astronomiques et Physiques). Rue de Bruxelles 61, B-5000 Namur, Belgium. ISSN 0037-959X.

Ann. Tokyo Astron. Obs., Second Ser.
Annals of the Tokyo Astronomical Observatory, Second Series. Tokyo Astronomical Observatory, University of Tokyo, Mitaka, Tokyo 181, Japan. ISSN 0082-4704.

Ann. Univ.-Sternw. Wien
Annalen der Universitäts-Sternwarte Wien. Institut für Astronomie der Universität Wien, Türkenschanzstr. 17, A-1180 Wien, Austria. Published by Ferd. Dümmlers Verlag, Bonn, F.R. Germany. ISSN 0342-4030.

Annu. Rep. Astron. Inst. Greece
Annual Reports of the Astronomical Institutes of Greece. Published by the Greek National Committee for Astronomy, Athens, Greece.

Annu. Rep. (B.I.H.)
Annual Report (B.I.H.). Bureau International de l'Heure, 61, avenue de l'Observatoire, F-75014 Paris, France.

Annu. Rep. Dir., Mt. Wilson Las Campanas Obs.
Annual Report of the Director, The Mt. Wilson and Las Campanas Observatories. Mt. Wilson and Las Campanas Observatories of the Carnegie Institution of Washington, 813 Santa Barbara Street, Pasadena, Calif. 91101-1292, USA.

Annu. Rep. (ERP)
Annual Report (ERP). Xu-Jia-Hui Section, Shanghai Observatory, Academia Sinica. Subscription address: Shanghai Scientific and Technical Publishers, Shanghai, Rei Jing Er Street 450, People's Republic of China.

Annu. Rep. Geophys. Obs.
Annual Report of Geophysical Observations. The International Latitude Observatory of Mizusawa, Mizusawa-Shi, Iwate-Ken, Japan. ISSN 0579-5958.

Annu. Rep. Int. Polar Motion Serv.
Annual Report of the International Polar Motion Service. Central Bureau of the International Polar Motion Service, International Latitude Observatory of Mizusawa, Mizusawa-Shi, Iwate-Ken, Japan. ISSN 0074-7432.

Annu. Rep. Meteorol. Obs. Int. Latitude Obs. Mizusawa
Annual Report of the Meteorological Observations made at the International Latitude Observatory of Mizusawa. The International Latitude Observatory of Mizusawa, Mizusawa-Shi, Iwate-Ken, Japan. ISSN 0303-8378.

Annu. Rev. Astron. Astrophys.
Annual Review of Astronomy and Astrophysics. Annual Reviews Inc., 4139 El Camino Way, Palo Alto, Calif. 94306, USA. ISSN 0066–4146.

Annu. Rev. Earth Planet. Sci.
Annual Review of Earth and Planetary Sciences. Annual Reviews Inc., 4139 El Camino Way, Palo Alto, Calif. 94306, USA. ISSN 0084–6597.

Annu. Rev. Fluid Mech.
Annual Review of Fluid Mechanics. Annual Reviews Inc., 4139 El Camino Way, Palo Alto, Calif. 94306, USA. ISSN 0066–4189.

Annu. Rev. Nucl. Part. Sci.
Annual Review of Nuclear and Particle Science. Annual Reviews Inc., 4139 El Camino Way, Palo Alto, Calif. 94306, USA. ISSN 0163–8998.

Annu. Univ. Sofia
Annuaire de l'Université de Sofia. Faculté de Physique. Bibliothèque de l'Université, Sofia, Bulgaria. ISSN 0584–0279.

Antenna
L'Antenna. Via Monte Generoso 6/a, I–20155 Milano, Italy. ISSN 0003–5386.

Anz. Österr. Akad. Wiss., Math.–Naturwiss. Kl.
Anzeiger der Österreichischen Akademie der Wissenschaften, Mathematisch–Naturwissenschaftliche Klasse. Springer–Verlag, Wien, Austria. ISSN 0065–535X.

Appl. Opt.
Applied Optics. A monthly publication of the Optical Society of America. American Institute of Physics, 335 East 45th Street, New York, N.Y. 10017, USA. ISSN 0003–6935.

Appl. Phys., B
Applied Physics, B (Photophysics and Laser Chemistry). Springer–Verlag, Berlin – Heidelberg – New York – Tokyo. ISSN 0721–7269.

Appl. Phys. Lett.
Applied Physics Letters. American Institute of Physics, 335 East 45th Street, New York, N.Y. 10017, USA. ISSN 0003–6951.

Appl. Spectrosc.
Applied Spectroscopy. Society for Applied Spectroscopy, 428 East Preston Street, Baltimore, Md. 21202, USA. ISSN 0003–7028.

Appl. Spectrosc. Rev.
Applied Spectroscopy Reviews. Marcel Dekker Inc., 270 Madison Avenue, New York, N.Y. 10016, USA. ISSN 0570–4928.

Arch. Elektrotech.
Archiv für Elektrotechnik. Springer–Verlag, Berlin – Heidelberg – New York – Tokyo. ISSN 0003–9039.

Arch. Hist. Exact Sci.
Archive for History of Exact Sciences. Springer–Verlag, Berlin – Heidelberg – New York – Tokyo. ISSN 0003–9519.

Arch. Int. Hist. Sci.
Archives Internationales d'Histoire des Sciences. Istituto dell'Enciclopedia Italiana, fondata da Giovanni Treccani, Roma, Italy. ISSN 0003–9810.

Arch. Sci.
Archives des Sciences. Société de Physique et d'Histoire Naturelle de Genève. Subscription address: Librairie Payot, 6 Rue Grenus, CH–1211 Geneva 11, Switzerland. ISSN 0003–9705.

Archaeoastronomy (U.K.)
Archaeoastronomy. Supplement to Journal for the History of Astronomy. Science History Publications Ltd., Halfpenny Furze, Mill Lane, Chalfont St Giles, Bucks. England, HP8 4NR. ISSN 0142–7253.

Archaeoastronomy (U.S.A.)
Archaeoastronomy. The Bulletin of the Center for Archaeoastronomy, Space Sciences Building, University of Maryland, College Park, Md. 20742, USA. ISSN 0190–9940.

Archaeometry
Archaeometry. Research Laboratory for Archaeology and the History of Art, Oxford University, 6 Keble Road, Oxford OX1 3QJ, England. ISSN 0003–813X.

Archenhold–Sternw. Berlin–Treptow, Sonderdr.
Archenhold–Sternwarte Berlin–Treptow, Sonderdruck. Archenhold–Sternwarte, DDR–1193 Berlin, Alt Treptow 1, German Democratic Republic.

Archenhold–Sternw. Berlin–Treptow, Vortr. Schr.
Archenhold–Sternwarte Berlin–Treptow, Vorträge und Schriften. Archenhold–Sternwarte, DDR–1193 Berlin, Alt Treptow 1, German Democratic Republic. ISSN 0570–6262.

Arecibo Obs./NAIC, Newsl.
Arecibo Observatory/NAIC Newsletter. National Astronomy and Ionosphere Center, PO Box 995, Arecibo, P.R. 00613, Puerto Rico.

Ark. Fys. Semin. Trondheim
Arkiv for det Fysiske Seminar i Trondheim. c/o Institutt for Teoretisk Fysikk, Universitetet i Trondheim, NTH, N–7034 Trondheim, Norway. ISSN 0365–2459.

Ark. Mat.
Arkiv för Matematik. Institut Mittag–Leffler, Auravägen 17, S–182 62 Djursholm, Sweden. ISSN 0004–2080.

Armagh Obs., Repr.
Armagh Observatory, Reprints. Armagh Observatory, Armagh BT61 9DG, Northern Ireland.

Artif. Satell.
Artificial Satellites. Polish Academy of Sciences, Space Research Committee. Państwowe Wydawnictwo Naukowe, Warszawa–Lódź, Poland. Subscription address: Ars Polona, 00–068 Warszawa, Krakowskie Przedmieście 7, Poland. ISSN 0571–205X.

Astrofiz. Issled. Izv. Spets. Astrofiz. Obs.
Astrofizicheskie Issledovaniya. Izvestiya Spetsial'noj Astrofizicheskoj Observatorii. Akademiya Nauk SSSR. Izdatel'stvo Nauka, Leningradskoe Otdelenie, Leningrad. 199164 Leningrad, V–164, Mendeleevskaya l., 1, USSR. English translation in Bull. Spec. Astrophys. Obs. – North Caucasus. ISSN 0324–1459.

Astrofizika
Astrofizika. Izdatel'stvo Akademii Nauk Armyanskoj SSR, Erevan, USSR. English translation in Astrophysics. ISSN 0571–7132.

Astrofys. Inst., Vrije Univ. Bruss., Overdruk
Astrofysisch Instituut, Vrije Universiteit Brussel, Overdruk. Astrofysisch Instituut, Vrije Universiteit Brussel, Pleinlaan 2, B–1050 Brussel, Belgium.

Astron. Astrophys.
Astronomy and Astrophysics. A European Journal. Springer–Verlag, Berlin – Heidelberg – New York – Tokyo. ISSN 0004–6361.

Astron. Astrophys., Suppl. Ser.
Astronomy and Astrophysics, Supplement Series. A European Journal. Les Editions de Physique, Z.I. de Courtaboeuf, B.P. 112, F–91944 Les Ulis Cedex, France. ISSN 0365–0138.

Astron. Bull., Carter Obs.
Astronomical Bulletin, Carter Observatory. Carter Observatory, PO Box 2909, Wellington 1, New Zealand. ISSN 0373–7268.

Astron. Circ.
Astronomical Circular. Edited by the Chinese Astronomical Society. Compiled by the Editors of Acta Astronomica Sinica, Purple Mountain Observatory, Academia Sinica, Nanjing, People's Republic of China.

Astron. Contrib. Univ. Manchester, Ser. II: Jodrell Bank Repr.
Astronomical Contributions from the University of Manchester, Series II: Jodrell Bank Reprints. Nuffield Radio Astronomy Laboratories, Jodrell Bank, Macclesfield, Cheshire SK11 9DL, England.

Astron. Contrib. Univ. Manchester, Ser. III
Astronomical Contributions from the University of Manchester, Series III. Department of Astronomy, University of Manchester, Oxford Road, Manchester M13 9PL, England.

Astron. Data Cent. Bull.
AstronomicaL Data Center Bulletin. National Space Science Data Center/World Data Center A for Rockets and Satellites. National Aeronautics and Space Administration, Goddard Space Flight Center, Greenbelt, Md. 20771, USA.

Astron. Express
Astronomy Express. Cambridge University Press, London – New York – New Rochelle – Melbourne – Sydney. ISSN 0265–5365.

Astron. Her.
Astronomical Herald. Astronomical Society of Japan, Tokyo Astronomical Observatory, Oosawa Mitaka, Tokyo, Japan. ISSN 0374–2466.

Astron. Inst. "Anton Pannekoek", Univ. Amst., Repr.
Astronomical Institute "Anton Pannekoek", University of Amsterdam, Reprint.

Astron. Inst. Univ. Brno, Contrib.
Astronomical Institute of the University Brno, Contributions. Astronomical Institute, University Brno, Kotlářská 2, Brno, Czechoslovakia.

Astron. Inst. Univ. Brno, Publ.
Astronomical Institute of the University Brno, Publications. Astronomical Institute, University Brno, Kotlářská 2, Brno, Czechoslovakia.

Astron. J.
The Astronomical Journal. Published for the American Astronomical Society by the American Institute of Physics, 335 East 45th Street, New York, N.Y. 10017, USA. ISSN 0004–6256.

Astron. Mitt. Wien
Astronomische Mitteilungen Wien. Institut für Astronomie der Universität Wien, Türkenschanzstraße 17, A–1180 Wien, Austria.

Astron. Nachr.
Astronomische Nachrichten. Akademie–Verlag, DDR–108 Berlin, Leipziger Straße 3–4, German Democratic Republic. ISSN 0004–6337.

Astron. Obs. Trieste, Publ.
Astronomical Observatory Trieste, Publications. Osservatorio Astronomico di Trieste, Via G.B. Tiepolo 11, I–34131 Trieste, Italy.

Astron. Pap.
Astronomical Papers prepared for the use of the American Ephemeris and Nautical Almanac. Published by the Nautical Almanac Office, U.S. Naval Observatory by direction of the Secretary of the Navy and under the authority of Congress. U.S. Government Printing Office, Washington, D.C., USA.

Astron. Q.
The Astronomy Quarterly. Pachart Publishing House, 1130 San Lucas Circle, Tucson, Ariz. 85704, USA. ISSN 0364–9229.

Astron. Raumfahrt
Astronomie und Raumfahrt. Kulturbund der DDR, Zentrale Kommission Astronomie und Raumfahrt. Redaktionssitz: 9630 Crimmitschau, Pionier– und Jugendsternwarte "Johannes Kepler", Strasse der Jugend 8. Available from Zeitungsvertriebsamt, Abt. Export, 1004 Berlin, Strasse der Pariser Kommune 3–4, German Democratic Republic. ISSN 0587–565X.

Astron. Rechen–Inst. Heidelb., Mitt., Ser. A
Astronomisches Rechen–Institut Heidelberg, Mitteilungen, Serie A. Astronomisches Rechen–Institut, Mönchhofstraße 12–14, D–6900 Heidelberg, F.R. Germany.

Astron. Rechen–Inst. Heidelb., Mitt., Ser. B
Astronomisches Rechen–Institut Heidelberg, Mitteilungen, Serie B. Astronomisches Rechen–Institut, Mönchhofstraße 12–14, D–6900 Heidelberg, F.R. Germany.

Astron. Rep.
The Astronomical Reports. Polskie Towarzystwo Miłośników Astronomii, Polish Amateur Astronomical Society, ul. Ludwika Solskiego 30/8, PL–31–027 Kraków, Poland.

Astron. Sch.
Astronomie in der Schule. Verlag Volk und Wissen, DDR–1086 Berlin, Krausenstraße 50, Postfach 1213, German Democratic Republic. ISSN 0004–6310.

Astron. Tidsskr.
Astronomisk Tidsskrift. Astronomisk Selskab, Kobenhavn; Norsk Astronomisk Selskap, Oslo; Svenska Astronomiska Sällskapet, Stockholm. Subscription address: Svenska Astronomiska Sällskapet, Stockholms Observatorium, S–13300 Saltsjöbaden, Sweden. ISSN 0004–6345.

Astron. Tsirk.
Astronomicheskij Tsirkulyar. Izdavaemyj Byuro Astronomicheskikh Soobshchenij Akademii Nauk SSSR. ISSN 0365–7248.

Astron. Vestn.
Astronomicheskij Vestnik. Izdatel'stvo Nauka, 918, Moskva, K–9, USSR. English translation in Sol. Syst. Res. ISSN 0320–930X.

Astron. Zh.
Astronomicheskij Zhurnal. Akademiya Nauk SSSR. Izdatel'stvo Nauka, Moskva, 103717, GSP, Moskva, K–62, Podsosenskij per., 21, USSR. English translation in Sov. Astron. ISSN 0004–6299.

Astronomia
Astronomia. Periodico trimestrale dell'Unione Astrofili Italiani. Subscription address: L. Baldinelli, C.P. 1630, I–40100 Bologna A.D., Italy. ISSN 0392–2308.

Astronomie
L'Astronomie et Bulletin de la Société Astronomique de France. Société Astronomique de France, 3, Rue Beethoven, F-75016 Paris, France. ISSN 0004-6302.

Astronomy
Astronomy. Astro Media Corp., 625 E. St. Paul Avenue, PO Box 92788 Milwaukee, Wis. 53202, USA. ISSN 0091-6358.

Astrophys. J.
The Astrophysical Journal. Published by The University of Chicago Press for the American Astronomical Society. The University of Chicago Press, 5801 S. Ellis Avenue, Chicago, Ill. 60637, USA. ISSN 0004-637X.

Astrophys. J., Lett. Ed.
The Astrophysical Journal, Letters to the Editor. Published by The University of Chicago Press for the American Astronomical Society. The University of Chicago Press, 5801 S. Ellis Avenue, Chicago, Ill. 60637, USA. ISSN 0571-7248.

Astrophys. J., Suppl. Ser.
The Astrophysical Journal, Supplement Series. Published by The University of Chicago Press for the American Astronomical Society. The University of Chicago Press, 5801 S. Ellis Avenue, Chicago, Ill. 60637, USA. ISSN 0067-0049.

Astrophys. Lett.
Astrophysical Letters. Gordon and Breach Science Publishers Inc., New York – London – Paris. ISSN 0004-6388.

Astrophys. Prepr. Ser.
Astrophysics Preprint Series. Astronomy and Astrophysics Group, Faculty of Physics, Pontificia Universidad Católica de Chile, Casilla 6014, Santiago de Chile, Chile.

Astrophys. Relativ., Prepr. Ser.
Astrophysics and Relativity, Preprint Series. Department of Applied Mathematics and Astronomy, University College, PO Box 78, Cardiff CF1 1XL, England.

Astrophys. Space Phys. Rev.
Astrophysics and Space Physics Reviews. Soviet Scientific Reviews, Section E. Harwood Academic Publishers GmbH, PO Box 786, Cooper Station, New York, N.Y. 10276, USA. ISSN 0143-0432.

Astrophys. Space Sci.
Astrophysics and Space Science. An International Journal of Cosmic Physics. D. Reidel Publishing Company, Dordrecht – Boston. ISSN 0004-640X.

Astrophysics
Astrophysics. A cover-to-cover translation of Astrofizika of the Academy of Sciences of the Armenian SSR. Consultants Bureau, 227 West 17th Street, New York, N.Y. 10011, USA. ISSN 0004-6396.

Atmos. Environ.
Atmospheric Environment. Pergamon Press, Oxford – New York – Toronto – Paris – Frankfurt – Sydney. ISSN 0004-6981.

Atomkernenerg. Kerntech.
Atomkernenergie und Kerntechnik. Verlag Karl-Thiemig AG, Postfach 900740, D-8000 München 90, F.R. Germany. ISSN 0171-5747.

Atti Accad. Naz. Lincei, Ser. Ottava, Rend.
Atti della Accademia Nazionale dei Lincei. Serie Ottava, Rendiconti. Classe di Scienze fisiche, matematiche e naturali. Accademia Nazionale dei Lincei, Roma, Italy. ISSN 0001-4435.

Atti Accad. Sci. Torino I
Atti della Accademia delle Scienze di Torino I. Classe di Scienze fisiche, matematiche e naturali. Via Accademia delle Scienze 6, Via Maria Vittoria 3, Torino (208), Italy. ISSN 0001-4419.

Atti Fond. Giorgio Ronchi
Atti della Fondazione Giorgio Ronchi. Largo Enrico Fermi 1, I-50125 Arcetri-Firenze, Italy. ISSN 0015-606X.

Aust. Comput. J.
Australian Computer Journal. Australian Trade Publications, 28 Chippen Street, Chippendale, N.S.W. 2008, Australia. ISSN 0004-8917.

Aust. J. Astron.
Australian Journal of Astronomy. Astral Press, PO Box 107, Wembley, W.A. 6014, Australia. ISSN 0814-5628.

Aust. J. Geod. Photogramm. Surv.
Australian Journal of Geodesy, Photogrammetry, and Surveying. School of Surveying, University of New South Wales, PO Box 1, Kensington, NSW 2033, Australia. ISSN 0313-9220.

Aust. J. Phys.
Australian Journal of Physics. Commonwealth Scientific and Industrial Research Organization (CSIRO), 314 Albert Street, East Melbourne, Victoria 3002, Australia. ISSN 0004-9506.

Aust. Phys.
Australian Physicist. Australian Institute of Physics, Science Centre, 35-43 Clarence Street, Sydney, NSW 2000, Australia. ISSN 0004-9972.

Austrian Pap. Asteroids
Austrian Papers on Asteroids. Institut für Astronomie, Universitätsplatz 5, A-8010 Graz, Austria.

Automatica
Automatica. Pergamon Press, Oxford – New York – Frankfurt – Paris – Sydney – Toronto – Tokyo. ISSN 0005-1098.

BAV Mitt.
BAV Mitteilungen. Berliner Arbeitsgemeinschaft für Veränderliche Sterne e.V., Sternwarte, Munsterdamm 90, D-1000 Berlin 41, F.R. Germany.

BAV Rundbrief
BAV Rundbrief. Mitteilungsblatt der Berliner Arbeitsgemeinschaft für Veränderliche Sterne. BAV Berliner Arbeitsgemeinschaft für Veränderliche Sterne e.V., Sternwarte, Munsterdamm 90, D-1000 Berlin 41, F.R. Germany. ISSN 0405-5497.

BBSAG Bull.
Bedeckungsveränderlichen Beobachter der Schweizerischen Astronomischen Gesellschaft Bulletin. Available from K. Locher, Rebrain 39, 8624 Grüt, Switzerland.

B.I.H. Circ.
Bureau International de l'Heure (B.I.H.) Circulars A, D, E. 61, avenue de l'Observatoire, F-75014 Paris, France.

Biul. Obs. Astron. Uniw. M. Kopernika Toruniu
Biuletyn Obserwatorium Astronomicznego Uniwersytetu M. Kopernika w Toruniu. Institute of Astronomy, Nicolaus Copernicus University, Chopina 12/18, PL-87-100 Toruń, Poland.

Bol. Acad. Cienc. Fis. Mat. Nat.
Boletin de la Academia de Ciencias Fisicas Matematicas y Naturales. Academia de Ciencias Fisicas Matematicas y Naturales, Apartado de Correo 1421, Caracas, Venezuela. ISSN 0366-1652.

Bol. Asoc. Argent. Astron.
Boletin de la Asociación Argentina de Astronomía, La Plata, Argentina. ISSN 0571–3285.

Bol. Astron.
Boletin Astronômico. Observatório do Capricórnio, Prefeitura Municipal de Campinas–SP, Brazil.

Bol. Astron. Obs. Madr.
Boletín Astronómico del Observatorio de Madrid. Instituto Geografico Nacional, General Ibáñez de Ibero, 3. Madrid 3, Spain. ISSN 0373–7101.

Bol. Inst. Tonantzintla
Boletin del Instituto de Tonantzintla. Instituto Nacional de Astrofisica, Optica y Electronica, Apartados Postales Nos. 216 y 51, Puebla, Pue., Mexico. ISSN 0303–7584.

Bol. Obs. Ebro
Boletín del Observatorio del Ebro. Consejo Superior de Investigaciones Científicas, Observatorio del Ebro, Roquetas (Tarragona), Spain. ISSN 0211–5166.

Boll. Geofis. Teor. Appl.
Bollettino di Geofisica Teorica ed Applicada. Osservatorio Geofisico Sperimentale, I–34123 Trieste, Italy. ISSN 0006–6729.

Boundary–Layer Meteorol.
Boundary–Layer Meteorology. D. Reidel Publishing Company, Dordrecht – Boston – London. ISSN 0006–8314.

Boyden Obs., Occas. Publ.
Boyden Observatory, Occasional Publication. Boyden Observatory, Institute and Department of Astronomy, University of the Orange Free State, PO Box 339, Bloemfontein 9300, South Africa.

Boyden Obs., Repr.
Boyden Observatory, Reprint. Boyden Observatory, Astronomy Department, University of the Orange Free State, PO Box 339, Bloemfontein 9300, South Africa.

Br. Astron. Assoc. Circ.
British Astronomical Association Circular. British Astronomical Association, Burlington House, Piccadilly, London W1V 0NL, England. ISSN 0264–4185.

Br. J. Philos. Sci.
British Journal for the Philosophy of Science. Cambridge University Press, Cambridge – London – New York – New Rochelle – Melbourne – Sydney. ISSN 0007–0882.

Br. J. Photogr.
British Journal of Photography. Henry Greenwood & Co. Ltd., 28 Great James Street, London WC1N 3HL, England. ISSN 0007–1196.

Bulg. J. Phys.
Bulgarian Journal of Physics. Bulgarian Academy of Sciences, Faculty of Physics, 5 Anton Ivanov Blvd., 1126 Sofia, Bulgaria. ISSN 0323–9217.

Bull. Am. Astron. Soc.
Bulletin of the American Astronomical Society. Published by the American Institute of Physics, 335 East 45th Street, New York, N.Y. 10017, USA. ISSN 0002–7537.

Bull. Assoc. Fr. Obs. Etoiles Variables
Bulletin de l'Association Française des Observateurs d'Etoiles Variables. Revue trimestrielle. A.F.O.E.V. Observatoire de Lyon, F–69230 Saint Genis Laval, France. ISSN 0153–9949.

Bull. Astron.
Bulletin Astronomique. Observatoire Royal de Belgique (Astronomisch Bulletin. Koninklijke Sterrenwacht van België). Observatoire Royal de Belgique, 3, avenue Circulaire, Uccle, B–1180 Bruxelles, Belgium.

Bull. Astron. Inst. Czech.
Bulletin of the Astronomical Institutes of Czechoslovakia. Academia, Publishing House of the Czechoslovak Academy of Sciences, Vodičkova 40, 11229 Praha 1, Czechoslovakia. ISSN 0004–6248.

Bull. Astron. Soc. India
Bulletin of the Astronomical Society of India. Astronomical Society of India, Osmania University, Hyderabad 500 007, India. ISSN 0304–9523.

Bull. Aust. Math. Soc.
Bulletin of the Australian Mathematical Society. University of Queensland Press, St. Lucia, Queensland 4067, Australia. ISSN 0004–9727.

Bull. Cl. Sci., Acad. R. Belg.
Bulletin de la Classe des Sciences, Académie Royale de Belgique. Académie Royale des Sciences des Lettres et de Beaux–Arts de Belgique, Brussels, Belgium. ISSN 0001–4141.

Bull. Crimean Astrophys. Obs.
Bulletin of the Crimean Astrophysical Observatory. A cover–to–cover translation of Izv. Krymskoj Astrofiz. Obs. Allerton Press, Inc. 150 5th Avenue, New York, N.Y. 10011, USA.

Bull. Etoiles Tardives Spectre Particulier
Bulletin sur les Etoiles Tardives à Spectre Particulier (Newsletter of Chemically Peculiar Late–type Stars). Observatoire de Strasbourg, 11, rue de l'Université, F–67000 Strasbourg, France.

Bull. Géod.
Bulletin Géodésique. The Journal of the International Association of Geodesy. Bureau Central de l'Association Internationale de Géodésie, 39 rue Gay–Lussac, F–75005 Paris, France. ISSN 0007–4632.

Bull. Geogr. Surv. Inst.
Bulletin of the Geographical Survey Institute. Geographical Survey Institute, Ministry of Construction, Kitasato–1, Yatabe–Machi, Tsukuba–Gun, Ibaraki–ken, Japan. ISSN 0373–7160.

Bull. Geophys.
Bulletin of Geophysics. National Central University, Chung–Li, Taiwan, Republic of China. ISSN 0253–4800.

Bull. Inf. Cent. Données Stellaires
Bulletin d'Information du Centre de Données Stellaires. Observatoire de Strasbourg, 11, rue de l'Université, F–67000 Strasbourg, France. ISSN 0242–6536.

Bull. Inf. Etoiles Be
Bulletin d'Information sur les Etoiles Be (Be Star Newsletter). Observatoire de Strasbourg, 11, rue de l'Université, F–67000 Strasbourg, France.

Bull. Obs. Astron. Belgr.
Bulletin de l'Observatoire Astronomique de Belgrade. Observatoire Astronomique de Belgrade, Beograd, Volgina 7, Yugoslavia. ISSN 0373–3734.

Bull. Res. Inst. Sci. Meas., Tôhoku Univ.
Bulletin of the Research Institute for Scientific Measurements, Tôhoku University, Sendai, Japan. ISSN 0040–8689.

Bull. Soc. R. Sci. Liège
Bulletin de la Société Royale des Sciences de Liège. L'Université, 15 Avenue des Tilleurs, Liège, Belgium. ISSN 0037-9565.

Bull. Spec. Astrophys. Obs. – North Caucasus
Bulletin of the Special Astrophysical Observatory – North Caucasus. A cover-to-cover translation of Astrofiz. Issled. Izv. Spets. Astrofiz. Obs. Allerton Press, Inc., 150 5th Avenue, New York, N.Y. 10011, USA.

Bull., Time Serv. Mizusawa Obs.
Bulletins, Time Service of the Mizusawa Observatory. The International Latitude Observatory of Mizusawa, Mizusawa-Shi, Iwate-Ken, Japan. ISSN 0580-6585.

Bull. Tokyo Gakugei Univ., Ser. IV
Bulletin of Tokyo Gakugei University, Series IV (Mathematics and Natural Sciences) 4-1-1 Nukui-kita-machi, Koganei, Tokyo, Japan. ISSN 0371-6813.

Byull. Inst. Astrofiz.
Byulleten' Instituta Astrofiziki. Akademiya Nauk Tadzhikskoj SSR. Izdatel'stvo Donish, Dushanbe, USSR. ISSN 0568-6865.

Byull. Inst. Teor. Astron.
Byulleten' Instituta Teoreticheskoj Astronomii. Akademiya Nauk SSSR. Leningradskoe Otdelenie, Izdatel'stvo Nauka, Leningrad V-164, Mendeleevskaya l., 1, USSR. ISSN 0002-3302.

C. R. Acad. Sci., Sér. Gén., Vie Sci.
Comptes Rendus de l'Académie des Sciences, Série Générale, La Vie des Sciences. Académie des Sciences, Paris. Subscription address: Gauthier-Villars, C.D.R., Centrale des Revues, 11, rue Gossin, F-92543 Montrouge Cedex, France.

C. R. Acad. Sci., Sér. II
Comptes Rendus de l'Académie des Sciences, Série II: Mécanique, Physique, Chimie, Sciences de l'Univers, Sciences de la Terre. Académie des Sciences, Paris. Subscription address: Gauthier-Villars, C.D.R., Centrale des Revues, 11, rue Gossin, F-92543 Montrouge Cedex, France. ISSN 0249-6305.

Can. J. Earth Sci.
Canadian Journal of Earth Sciences. National Research Council of Canada, Ottawa KIA OR6, Canada. ISSN 0008-4077.

Can. J. Phys.
Canadian Journal of Physics. National Research Council of Canada, Ottawa KIA OR6, Canada. ISSN 0008-4204.

Carter Obs., Repr. Ser.
Carter Observatory, Reprint Series. Carter Observatory, PO Box 2909, Wellington 1, New Zealand.

Cartes Synoptiques
Cartes Synoptiques de la Chromosphère Solaire et Catalogues des Filaments et des Centres d'Activité. Observatoire de Paris, Section d'Astrophysique, F-92190 Meudon, France.

Celest. Mech.
Celestial Mechanics. An International Journal of Space Dynamics. D. Reidel Publishing Company, Dordrecht – Boston. ISSN 0008-8714.

Cent. Astron. Sci. Spat., Obs. Sol.
Centre de l'Astronomie et des Sciences Spatiales, Observations Solaires. Centre de l'Astronomie et des Sciences Spatiales, Academiei Republicii Socialiste România, Bucuresti, Rumania.

Cent. Astrophys., Prepr. Ser.
Center for Astrophysics, Preprint Series. Center for Astrophysics, 60 Garden St., Cambridge, Mass. 02138, USA.

Centaurus
Centaurus. International Magazine of the History of Mathematics, Science, and Technology. Munksgaard Ltd., Copenhagen, Denmark. ISSN 0008-8994.

Česk. Čas. Fyz., Sekce A
Československý časopis pro fyziku, Sekce A. Academia Publishing House of the Czechoslovak Adademy of Sciences, Vodičkova 40, 11229 Praha 1, Czechoslovakia. ISSN 0009-0700.

Chem. Phys.
Chemical Physics. North-Holland Publishing Company, PO Box 211, 1000 AE Amsterdam, The Netherlands. ISSN 0301-0104.

Chem. Phys. Lett.
Chemical Physics Letters. North-Holland Publishing Company, PO Box 211, 1000 AE Amsterdam, The Netherlands. ISSN 0009-2614.

Chin. Astron. Astrophys.
Chinese Astronomy and Astrophysics. A selected translation from the current issues of Acta Astronomica Sinica, Acta Astrophysica Sinica and Chinese Journal of Space Science. Pergamon Press, Oxford – New York – Toronto – Paris – Frankfurt – Sydney. ISSN 0275-1062.

Chin. J. Phys.
Chinese Journal of Physics. Physical Society of the Republic of China, Physics Department, National Taiwan University, Taipei, Taiwan, Republic of China. ISSN 0577-9073.

Chin. J. Space Sci.
Chinese Journal of Space Science. Subscription address: China International Book Trading Corporation, Guozi Shudian, PO Box 2820, Beijing, People's Republic of China.

Chin. Phys.
Chinese Physics. Selected translations from current issues of major Chinese physics and astronomy journals. American Institute of Physics, 335 East 45th Street, New York, N.Y. 10017, USA. ISSN 0273-429X.

Chin. Phys. Lett.
Chinese Physics Letters. Science Press, 137 Chaoyangmennei Street, Beijing, People's Republic of China.

Ciel
Le Ciel. Bulletin de la Société Astronomique de Liège. Société Astronomique de Liège, B-4410 Vottem, Belgium.

Ciel Espace
Ciel et Espace. Association Française d'Astronomie, 115 rue de Charenton, F-75012 Paris, France.

Ciel Terre
Ciel et Terre. Bulletin de la Société Royale Belge d'Astronomie, de Météorologie et de Physique du Globe. Société Royale Belge d'Astronomie, de Météorologie et de Physique du Globe, 3, avenue Circulaire, B-1180 Bruxelles, Belgium. ISSN 0009-6709.

Circ. Czech. Obs., Time Latitude
Circular of the Czechoslovak Observatories, Time and Latitude. Czechoslovak Academy of Sciences, Astronomical Institute, Prague, Czechoslovakia.

Circ. Inf.
Circulaire d'Information. Union Astronomique Internationale. Commission des Etoiles Doubles. Observatoire de Paris, F-92190 Meudon, France.

Circ. Stn. Astron. Int. Latitudine, Carloforte–Cagliari
Circolari della Stazione Astronomica Internazionale di Latitudine, Carloforte–Cagliari. Series A, B, C. Stazione Astronomica Internazionale di Latitudine, Carloforte–Cagliari, Italy.

Circ. Time Latitude Serv.
Circular Time and Latitude Service. Polish Academy of Sciences, Astronomical Latitude Observatory, Borowiec, Poland.

Classical Quantum Gravity
Classical and Quantum Gravity. Institute of Physics, 47 Belgrave Square, London SW1X 8QX, England.

Clim. Change
Climatic Change. D. Reidel Publishing Company, Dordrecht – Boston – London. ISSN 0165–0009.

CODATA Bull.
CODATA Bulletin. Committee on Data for Science and Technology of the International Council of Scientific Unions. Pergamon Press, Oxford – New York – Toronto – Sydney – Paris – Frankfurt. ISSN 0366–757X.

Coelum
Coelum. Periodico bimestrale per la Divulgazione dell'Astronomia. Osservatorio Astronomico Universitario di Bologna, Cassella Postale 596, I–40100 Bologna, Italy. ISSN 0366–7588.

Commentat. Phys.–Math.
Commentationes Physico–Mathematicae. Societas Scientiarum Fennica, Snellmaninkatu 9–11, SF–00170 Helsinki, Finland. ISSN 0069–6609.

Comments Astrophys.
Comments on Astrophysics. A Journal of Critical Discussion of the Current Literature. Comments on Modern Physics: Part C. Gordon and Breach Science Publishers, New York – London. ISSN 0146–2970.

Comments At. Mol. Phys.
Comments on Atomic and Molecular Physics. Gordon and Breach Science Publishers, New York – London. ISSN 0010–2687.

Comments Nucl. Part. Phys.
Comments on Nuclear and Particle Physics. Gordon and Breach Science Publishers, New York – London. ISSN 0010–2709.

Comments Plasma Phys. Controlled Fusion
Comments on Plasma Physics and Controlled Fusion. Gordon and Breach Science Publishers, New York – London. ISSN 0374–2806.

Commun. Astron. Dep., Univ. Ankara
Communications from the Astronomical Department, University of Ankara. Fen Fakültesi, Ankara, Turkey.

Commun. Fac. Sci. Univ. Ankara, Sér. A3
Communications de la Faculté des Sciences, de l'Université d'Ankara, Série A3 (Astronomie). Ankara Üniversitesi, Fen Fakültesi, Besevler–Ankara, Turkey.

Commun. Konkoly Obs.
Communications from the Konkoly Observatory of the Hungarian Academy of Sciences, Budapest, Hungary. ISSN 0324–2234.

Commun. Math. Phys.
Communications in Mathematical Physics. Springer–Verlag, Berlin – Heidelberg – New York – Tokyo. ISSN 0010–3616.

Commun. Theor. Phys.
Communications in Theoretical Physics. Huazhong University of Science and Technology Press, Wuhan, People's Republic of China.

Commun. Univ. Lond. Obs.
Communications from the University of London Observatory. University of London Observatory, Mill Hill Park, London NW7 2QS, England. ISSN 0458–2128.

Commun. Univ. Obs., St. Andrews
Communications from the University Observatory, St. Andrews. University Observatory, Buchanan Gardens, St. Andrews, Fife KY16 9LZ, Scotland.

Comput. Graphics
Computer Graphics. Association for Computing Machinery, 1133 Avenue of the Americas, New York, N.Y. 10036, USA. ISSN 0097–8930.

Comput. Methods Appl. Mech. Eng.
Computer Methods in Applied Mechanics and Engineering. North–Holland Publishing Company, PO Box 211, 1000 AE Amsterdam, The Netherlands. ISSN 0045–7825.

Comput. Phys. Commun.
Computer Physics Communications. North–Holland Publishing Company, Amsterdam, The Netherlands. ISSN 0010–4655.

Comput. Syst.
Computer Systems. Techpress Publishing Company Ltd., Walton House, 93 High Street, Bromley BR1 1JW, England. ISSN 0264–4193.

Computer
Computer. The Institute of Electrical and Electronic Engineers Computer Society, 5855 Naples Plaza, Suite 301, Long Beach, Calif. 90803, USA. ISSN 0018–9162.

Comun. Obs. Astron. Univ. Coimbra
Comunicações do Observatório Astronomico da Universidade de Coimbra, Portugal.

Contemp. Phys.
Contemporary Physics. Taylor and Francis Ltd., 10–14 Macklin Street, London, WC2B 5NF, England. ISSN 0010–7514.

Contrib. Astron. Obs. Skalnaté Pleso
Contributions of the Astronomical Observatory Skalnaté Pleso. VEDA, vydavatel'stvo Slovenskej akadémie vied, Bratislava, Czechoslovakia.

Contrib. Atmos. Phys.
Contributions to Atmospheric Physics. (Beiträge zur Physik der Atmosphäre). Friedrich Vieweg & Sohn Verlagsgesellschaft mbH, Postfach 5829, D–6200 Wiesbaden, F.R. Germany. ISSN 0005–8173.

Contrib. Bosscha Obs.
Contributions from the Bosscha Observatory. Bosscha Observatory, Bandung Institute of Technology, Department of Science, Lembang, Java, Indonesia.

Contrib. Dep. Astron., Univ. Kyoto
Contributions from the Department of Astronomy, University of Kyoto. Department of Astronomy, Faculty of Science, University of Kyoto, Sakyo–ku, Kyoto 608, Japan. ISSN 0388–0230.

Contrib. Dep. Astron., Univ. Tokyo
Contributions from the Department of Astronomy, University of Tokyo. Department of Astronomy, University of Tokyo, Bunko–ku, Tokyo 113, Japan. ISSN 0563–8038.

Contrib. Dep. Geod. Astron., Univ. Thessaloniki
Contributions from the Department of Geodetic Astronomy, University of Thessaloniki. Department of Geodetic Astronomy, University of Thessaloniki, Thessaloniki, Greece.

Contrib. Dom. Astrophys. Obs.
Contributions from the Dominion Astrophysical Observatory. Dominion Astrophysical Observatory, National Research Council of Canada, 5071 West Saanich Road, Victoria, B.C. V8X 4M6, Canada.

Contrib. Dunsink Obs.
Contributions from the Dunsink Observatory. Dunsink Observatory, Castleknock, County Dublin, Republic of Ireland.

Contrib. Geophys. Inst. Slovak Acad. Sci.
Contributions of the Geophysical Institute of the Slovak Academy of Sciences. Slovak Academy of Sciences, Dubravská 9, 84228 Bratislava, Czechoslovakia. ISSN 0586–4607.

Contrib. Inst. Argent. Radioastron.
Contribuciones del Instituto Argentino de Radioastronomia. Instituto Argentino de Radioastronomia, Casilla de Correo No. 5, RA–1894 Villa Elisa, Prov. de Buenos Aires, Argentina.

Contrib. Inst. Astron. Res.
Contributions of the Institute for Astronomical Research. Institute for Astronomical Research, Inc., PO Box 15854, Baton Rouge, La. 70895, USA.

Contrib. Kwasan Hida Obs., Univ. Kyoto
Contributions from the Kwasan and Hida Observatories, University of Kyoto. Kwasan Observatory, University of Kyoto, Yamashina, Kyoto, 607, Japan. Hida Observatory, University of Kyoto, Kamitakara, Gifu–ken, 506–13, Japan. ISSN 0388–2349.

Contrib. La. State Univ. Obs.
Contributions of the Louisiana State University Observatory. Louisiana State University Observatory, Baton Rouge, La. 70803, USA.

Contrib. Lick Obs.
Contributions from the Lick Observatory. Lick Observatory, Board of Studies in Astronomy and Astrophysics, University of California at Santa Cruz, Santa Cruz, Calif. 95064, USA.

Contrib. Nicholas Copernicus Obs. Planetarium Brno
Contributions of the Nicholas Copernicus Observatory and Planetarium in Brno. Nicholas Copernicus Observatory and Planetarium Brno, Czechoslovakia.

Contrib. Nizamiah Japal–Rangapur Obs.
Contributions from Nizamiah and Japal–Rangapur Observatories. Centre of Advanced Study in Astronomy, Osmania University, Hyderabad–500 007, India.

Contrib. Oss. Astron. Torino
Contributi dell'Osservatorio Astronomico di Torino (Pino Torinese). Osservatorio Astronomico di Torino, I–10025 Pino Torinese (Torino), Italy.

Contrib. Van Vleck Obs.
Contributions from the Van Vleck Observatory. Van Vleck Observatory, Middletown, Conn., USA.

Contrib. Wrocław Astron. Obs.
Contributions from the Wrocław Astronomical Observatory. Wrocław University Observatory, Wrocław, Poland.

Copenh. Univ. Obs., Repr.
Copenhagen University Observatory, Reprints. University Observatory, Øster Voldgade 3, DK–1350 Copenhagen K, Denmark.

Cosmic. Res.
Cosmic Research. A cover–to–cover translation of Kosm. Issled. Consultants Bureau, 227 West 17th Street, New York, N.Y. 10011, USA. ISSN 0010–9525.

Cracow Obs. Repr.
Cracow Observatory Reprints. Cracow Observatory, Jagellonian University, Cracow, Poland.

CSELT Rapp. Tec.
CSELT Rapporti Tecnici. Centro Studi e Laboratori Telecomunicazioni, Via Guglielmo Reiss Romoli 274, Torino, Italy. ISSN 0390–1815.

Curr. Sci.
Current Science. Current Science Association, Raman Research Institute, Bangalore 560006, India. ISSN 0011–3891.

Curr. Top. Chin. Sci., Sect. E: Astron.
Current Topics in Chinese Science, Section E: Astronomy. An annual selection of papers published in Scientia Sinica and Kexue Tongbao. Gordon and Breach Science Publishers, Inc., London – New York – Paris. ISSN 0732–4421.

Czech. J. Phys., Sect. B
Czechoslovak Journal of Physics, Section B. Academia Publishing House of the Czechoslovak Academy of Sciences, Vodičkova 40, 11229 Praha 1, Czechoslovakia. ISSN 0011–4626.

Data Rep. Hydrogr. Obs., Ser. Astron. Geod.
Data Report of Hydrographic Observations, Series of Astronomy and Geodesy. Hydrographic Department of Japan, Tsukiji–5, Chuo–ku, Tokyo 104, Japan.

Debrecen Heliophys. Obs. Hung. Acad. Sci., Repr.
Debrecen Heliophysical Observatory of the Hungarian Academy of Sciences, Reprint. Heliophysical Observatory of the Hungarian Academy of Sciences, 4010 Debrecen, Hungary.

Dějiny Věd Tech.
Dějiny Věd a Techniky. Academia Praha, Praha, Czechoslovakia. ISSN 0300–4414.

Dep. Astron. McDonald Obs. Univ. Tex., Repr.
Department of Astronomy and McDonald Observatory of the University of Texas, Reprints. Astronomy Department, R.L.M. 15.220, University of Texas, Austin, Tex. 78712, USA.

Dep. Astrophys., Univ. Oxford, Publ.
Department of Astrophysics, University of Oxford, Publication. Department of Astrophysics, University of Oxford, South Parks Road, Oxford OX1 3RQ, England.

Diss. Abstr. Int., Sect. B
Dissertation Abstracts International – Section B (The Sciences and Engineering). University Microfilms International, Ann Arbor, Michigan 48106, USA. ISSN 0419–4217.

Dokl. Akad. Nauk BSSR
Doklady Akademii Nauk BSSR. Akademiya Nauk, Minsk, Belorussian SSR. ISSN 0002–354X.

Dokl. Akad. Nauk SSSR. Ser. Mat. Fiz.
Doklady Akademii Nauk SSSR. Seriya Matematika, Fizika. Izdatel'stvo Nauka, Moskva. Editorial address: 117874 GSP–7 Moskva, B–485, Profsoyuznaya Ul., 90 Kom. 533, USSR. ISSN 0002–3264.

Dokl. Bolg. Akad. Nauk
Doklady Bolgarskoj Akademii Nauk. Sofiya, Bulgaria. ISSN 0366–8681.

Dtsch. Geod. Komm. Bayer. Akad. Wiss., Reihe A
Deutsche Geodätische Kommission bei der Bayerischen Akademie der Wissenschaften, Reihe A: Theoretische Geodäsie. Deutsche Geodätische Kommission, Marstallplatz 8, D–8000 München 22, F.R. Germany. ISSN 0065–5309.

Dtsch. Geod. Komm. Bayer. Akad. Wiss., Reihe B
Deutsche Geodätische Kommission bei der Bayerischen Akademie der Wissenschaften, Reihe B: Angewandte Geodäsie. Deutsche Geodätische Kommission, Marstallplatz 8, D–8000 München 22, F.R. Germany. ISSN 0065–5317.

Dtsch. Geod. Komm. Bayer. Akad. Wiss., Reihe C
Deutsche Geodätische Kommission bei der Bayerischen Akademie der Wissenschaften, Reihe C: Dissertationen. Deutsche Geodätische Kommission, Marstallplatz 8, D–8000 München 22, F.R. Germany. ISSN 0065–5325.

Dtsch. Geod. Komm. Bayer. Akad. Wiss., Reihe D
Deutsche Geodätische Kommission bei der Bayerischen Akademie der Wissenschaften, Reihe D: Tafelwerke. Deutsche Geodätische Kommission, Marstallplatz 8, D–8000 München 22, F.R. Germany.

Dtsch. Geod. Komm. Bayer. Akad. Wiss., Reihe E
Deutsche Geodätische Kommission bei der Bayerischen Akademie der Wissenschaften, Reihe E: Geschichte und Entwicklung der Geodäsie. Deutsche Geodätische Kommission, Marstallplatz 8, D–8000 München 22, F.R. Germany. ISSN 0065–5341.

Dudley Obs. Rep.
Dudley Observatory Report. The Dudley Observatory, 69 Union Avenue, Schenectady, N.Y. 12308, USA.

Dunsink Obs. Repr.
Dunsink Observatory Reprints. Dunsink Observatory, Castleknock, County Dublin, Republic of Ireland.

Earth, Moon, Planets
Earth, Moon, and Planets. An International Journal of Comparative Planetology. D. Reidel Publishing Company, Dordrecht – Boston. ISSN 0167–9295.

Earth–Oriented Appl. Space Technol.
Earth–Oriented Applications of Space Technology. Pergamon Press, Oxford – New York – Frankfurt. ISSN 0277–4488.

Earth Planet. Sci. Lett.
Earth and Planetary Science Letters. Elsevier Scientific Publishing Company, PO Box 211, 1000 AE Amsterdam, The Netherlands. ISSN 0012–821X.

Earth–Sci. Rev.
Earth–Science Reviews. Elsevier Scientific Publishing Company, PO Box 330, 1000 AH Amsterdam, The Netherlands. ISSN 0012–8252.

Electron. Lett.
Electronics Letters. Institution of Electrical Engineers, Savoy Place, London, WC2R 0BL, England. ISSN 0013–5194.

Electronics
Electronics. McGraw–Hill Publishing Company, 1221 Avenue of the Americas, New York, N.Y. 10020, USA. ISSN 0013–5070.

Elektronik
Elektronik. Franzis–Verlag GmbH, D–8000 München 37, Postfach 3701 20, Karlstraße 37, F.R. Germany. ISSN 0013–5658.

Elettron. Telecomun.
Elettronica e Telecomunicazioni. Via Arsenale 41, I–10121 Torino, Italy. ISSN 0013–6123.

Endeavour New Ser.
Endeavour New Series. Pergamon Press, Oxford – New York – Toronto – Paris – Frankfurt – Sydney. ISSN 0013–7162.

EOS Trans. Am. Geophys. Union
EOS Transactions of the American Geophysical Union. American Geophysical Union, 2000 Florida Avenue N.W., Washington, D.C. 20009, USA. ISSN 0096–3941.

ESA Brochure
ESA Brochure (ESA BR). European Space Agency, Scientific and Technical Publications Branch, ESTEC, Postbus 299, 2200 AG Noordwijk, The Netherlands. ISSN 0250–1589.

ESA Bull.
ESA Bulletin. European Space Agency, Scientific and Technical Publications Branch, ESTEC, Postbus 299, 2200 AG Noordwijk, The Netherlands. ISSN 0376–4265.

ESA IUE Newsl.
ESA IUE Newsletters. The ESA IUE Observatory. Apartado 540 65, Madrid, Spain. Subscription address: European Space Agency, 8–10 rue Mario–Nikis, F–75738 Paris Cedex 15, France.

ESA J.
ESA Journal. European Space Agency, Scientific and Technical Publications Branch, ESTEC, Postbus 299, 2200 AG Noordwijk, The Netherlands. ISSN 0379–2285.

ESA SCI
European Space Agency SCI. European Space Agency, Scientific and Technical Publications Branch, ESTEC, Postbus 299, 2200 AG Noordwijk, The Netherlands.

ESA Sci. Tech. Rep.
ESA Scientific and Technical Reports. European Space Agency, Scientific and Technical Publications Branch, ESTEC, Postbus 299, 2200 AG Noordwijk, The Netherlands.

ESA Spec. Publ.
ESA Special Publication (ESA SP). European Space Agency, Scientific and Technical Publications Branch, ESTEC, Postbus 299, 2200 AG Noordwijk, The Netherlands. ISSN 0379–6566.

ESO Annu. Rep.
ESO Annual Report. European Southern Observatory, Karl–Schwarzschild–Straße 2, D–8046 Garching bei München, F.R. Germany. ISSN 0531–4496.

ESO Sci. Prepr.
ESO Scientific Preprint. European Southern Observatory, Karl–Schwarzschild–Straße 2, D–8046 Garching bei München, F.R. Germany.

ESO Sci. Rep.
ESO Scientific Report. European Southern Observatory, Karl–Schwarzschild–Straße 2, D–8046 Garching bei München, F.R. Germany.

ESO Tech. Rep.
ESO Technical Report. European Southern Observatory, Karl–Schwarzschild–Straße 2, D–8046 Garching bei München, F.R. Germany.

Eur. J. Phys.
European Journal of Physics. Institute of Physics, 47 Belgrave Square, London SW1X 8QX, England. ISSN 0143–0807.

Europhys. News
Europhysics News. European Physical Society, PO Box 69, CH–1213 Petit–Lancy 2, Switzerland. ISSN 0531–7479.

EXOSAT Express
EXOSAT Express. ESA, EXOSAT Observatory, ESOC, Robert-Bosch-Str. 5, D-6100 Darmstadt, F.R. Germany.

Feingerätetechnik
Feingerätetechnik. VEB Verlag Technik, Oranienburger Straße 13/14, DDR-1020 Berlin, German Democratic Republic. ISSN 0014-9683.

Fis. Tecnol.
Fisica e Tecnologia. Societa Italiana di Fisica, Via Loderingo Degli Andalo 2, I-40124 Bologna, Italy. ISSN 0391-9757.

Fiz. Sz.
Fizikai Szemle. Kiadja a Lapkiado Vallalat, Budapest VII, Lenin korut 9-11, Hungary. ISSN 0015-3257.

Fizika
Fizika. Mladost Export-Import, 41000 Zagreb, Ilica 30, Yugoslavia. ISSN 0015-3206.

Folia Fac. Sci. Nat. Univ. Purkynianae Brun., Phys.
Folia Facultatis Scientiarum Naturalium Universitatis Purkynianae Brunensis, Physica. University J. E. Purkyne, 61137 Brno-Kotlarska 2, Czechoslovakia. ISSN 0323-0287.

Fortschr. Phys.
Fortschritte der Physik. Akademie-Verlag, DDR-108 Berlin, Leipziger Straße 3-4, German Democratic Republic. ISSN 0015-8208.

Found. Phys.
Foundations of Physics. Plenum Publishing Corporation, 227 West 17th Street, New York, N.Y. 10011, USA. ISSN 0015-9018.

Fra Fys. Verden
Fra Fysikkens Verden. Fysisk Institutt, Universitetet i Trondheim, Norges Laererhogskole, N-7000 Trondheim, Norway. ISSN 0015-9247.

Fujitsu Sci. Tech. J.
Fujitsu Scientific and Technical Journal. Fujitsu Ltd., 1015 Kamikodanaka, Nakahara-ku, Kawasaki 211, Kanagawa, Japan. ISSN 0016-2523.

Fundam. Cosmic Phys.
Fundamentals of Cosmic Physics. Gordon and Breach Science Publishers, New York - London - Paris. ISSN 0094-5846.

Funkschau
Funkschau. Franzis-Verlag GmbH, Postfach 370120, Karlstr. 37, D-8000 München 2, F.R. Germany. ISSN 0016-2841.

Fys. Tidsskr.
Fysisk Tidsskrift. Subscription address: Jul. Gjellerups Boghandel, Solvgade 87, DK-1307 Kobenhavn, Denmark. ISSN 0016-3392.

G. A.A.B.
Giornale dell'A.A.B. Associazione Astrofili Bolognesi, Casella Postale 1630, I-40100 Bologna A.D., Italy. ISSN 0392-3932.

G. Astron.
Giornale di Astronomia. Società Astronomica Italiana, Largo E. Fermi, 5, I-50125 Firenze, Italy. ISSN 0390-1106.

Gemini
Gemini. Newsletter of the Royal Greenwich Observatory. Royal Greenwich Observatory, Herstmonceux Castle, Hailsham, East Sussex BN27 1RP, England.

Gen. Relativ. Gravitation
General Relativity and Gravitation. Published under the auspices of the International Committee on General Relativity and Gravitation GRG. Plenum Publishing Corporation, 233 Spring Street, New York, N.Y. 10013, USA. ISSN 0001-7701.

Geochim. Cosmochim. Acta
Geochimica et Cosmochimica Acta. Journal of the Geochemical Society and the Meteoritical Society. Pergamon Press, New York - Oxford - Toronto - Paris - Frankfurt - Sydney. ISSN 0016-7037.

Geod.-Geophys. Arb. Schweiz
Geodätisch-Geophysikalische Arbeiten in der Schweiz. Schweizerische Geodätische Kommission, Institut für Geodäsie und Photogrammetrie, ETH-Hönggerberg, CH-8093 Zürich, Switzerland.

Geod. Geophys. Veröff., Reihe II
Geodätische und Geophysikalische Veröffentlichungen, Reihe II (Solar-terrestrische Beziehungen und Physik der Atmosphäre). Nationalkomitee für Geodäsie und Geophysik bei der Akademie der Wissenschaften der Deutschen Demokratischen Republik, DDR-1500 Potsdam, Telegrafenberg, German Democratic Republic.

Geod. Geophys. Veröff., Reihe III
Geodätische und Geophysikalische Veröffentlichungen, Reihe III (Physik der festen Erde). Nationalkomitee für Geodäsie und Geophysik bei der Akademie der Wissenschaften der Deutschen Demokratischen Republik, DDR-1500 Potsdam, Telegrafenberg, German Democratic Republic. ISSN 0435-6187.

Geod. Kartogr.
Geodezja i Kartografia. Komitet Geodezji Polskiej Akademii Nauk. Publisher: Pań1stwowe Wydawnictwo Naukowe, Warszwa, Poland. ISSN 0016-7134.

Geomagn. Aehron.
Geomagnetizm i Aehronomiya. Akademiya Nauk SSSR. Izdatel'stvo Nauka, Moskva. English translation in Geomagn. Aeron. ISSN 0016-7940.

Geomagn. Aeron.
Geomagnetism and Aeronomy. A cover-to-cover translation of Geomagn. Aehron. American Geophysical Union, 2000 Florida Avenue, N.W., Washington, D. C. 20009, USA. ISSN 0016-7932.

Geophys. Astrophys. Fluid Dyn.
Geophysical and Astrophysical Fluid Dynamics. Gordon and Breach Science Publishers, New York - London - Paris. ISSN 0309-1929.

Geophys. J. R. Astron. Soc.
Geophysical Journal of the Royal Astronomical Society. Published for the Royal Astronomical Society by Blackwell Scientific Publications, Oxford - London - Edinburgh - Boston - Melbourne. ISSN 0016-8009.

Geophys. Norv.
Geophysica Norvegica. Universitetsforlaget, PO Box 2959, Toyen, Oslo 6, Norway. ISSN 0332-5903.

Geophys. Prospect.
Geophysical Prospecting. European Association of Exploration Geophysicists, PO Box 162, 2501 AN The Hague, The Netherlands. ISSN 0016-8025.

Geophys. Res. Lett.
Geophysical Research Letters. American Geophysical Union, 2000 Florida Avenue, N.W., Washington, D.C. 20009, USA. ISSN 0094-8276.

Geophysics
Geophysics. Society of Exploration Geophysicists, PO Box 3098, Tulsa, Okla. 74101, USA. ISSN 0016–8033.

GEOS Circ.
GEOS Circular. Series: Cep (Cepheids), EB (eclipsing binaries), RR (RR Lyrae type variables), SA (small–amplitude variables), SR (red variables). GEOS (Groupe d'Etude et d'Observation Stellaire), 12 rue Bezout, F–75014 Paris, France.

Gerlands Beitr. Geophys.
Gerlands Beiträge zur Geophysik. Akademische Verlagsgesellschaft Geest & Portig KG, DDR–7010 Leipzig, Sternwartenstraße 8, German Democratic Republic. ISSN 0016–8696.

Glas. Mat., Ser. III
Glasnik Matematički, Serija III. Društvo matematičara i fizičara SRH, Marulićev trg 19, YU–41001 Zagreb, P.P. 187, Yugoslavia. ISSN 0017–095X.

Greenwich Time Rep.
Greenwich Time Report. Royal Greenwich Observatory, Time and Latitude Service, Herstmonceux Castle, Hailsham, East Sussex BN27 1RP, England. ISSN 0264–4177.

Hadronic J.
Hadronic Journal. Hadronic Press Inc., Nonantum, Mass. 02195, USA. ISSN 0162–5519.

Heavens
The Heavens. The Oriental Astronomical Association, Ōtsu–shi, Shiga–ken, Japan. In Japanese.

Helv. Phys. Acta
Helvetica Physica Acta. Schweizerische Physikalische Gesellschaft. Birkhäuser Verlag, Elisabethenstraße 19, CH–4000 Basel 10, Switzerland. ISSN 0018–0238.

H.M. Naut. Alm. Off., Libr. Repr.
H.M. Nautical Almanac Office, Library Reprint. H.M. Nautical Almanac Office, Royal Greenwich Observatory, Herstmonceux Castle, Hailsham, East Sussex BN27 1RP, England.

Hvar Obs. Bull.
Hvar Observatory Bulletin. Faculty of Geodesy, University of Zagreb, Kačićeva 26, YU–41000 Zagreb, Yugoslavia. ISSN 0351–2651.

I.A.P.P.P. Commun.
International Amateur–Professional Photoelectric Photometry Communication. Fairborn Observatory, 1247 Folk Road, Fairborn, Ohio 45324, USA.

IAU Circ.
International Astronomical Union, Circular. Central Bureau for Astronomical Telegrams, Smithsonian Astrophysical Observatory, 60 Garden Street, Cambridge, Mass. 02138, USA. ISSN 0081–0304.

IAU Inf. Bull.
International Astronomical Union Information Bulletin. IAU–UAI Secrétariat, 61, Avenue de l'Observatoire, F–75014 Paris, France. Published by D. Reidel Publishing Company, Dordrecht – Boston – London.

Icarus
Icarus. International Journal of Solar System Studies. Academic Press Inc., San Diego – Orlando – New York – Austin – London – Montreal – Sydney – Tokyo – Toronto. ISSN 0019–1035.

IEE Proc. F
IEE Proceedings F (Communications, Radar and Signal Processing). Institution of Electrical Engineers, PO Box 8, Southgate House, Stevenage, Herts. SG1 1HQ, England. ISSN 0143–7070.

IEEE J. Solid–State Circuits
IEEE Journal of Solid–State Circuits. The Institute of Electrical and Electronics Engineers, 345 East 47th Street, New York, N.Y. 10017, USA. ISSN 0018–9200.

IEEE Spectrum
IEEE Spectrum. The Institute of Electrical and Electronics Engineers, 345 East 47th Street, New York, N.Y. 10017, USA. ISSN 0018–9235.

IEEE Trans. Aerosp. Electron. Syst.
IEEE Transactions on Aerospace and Electronic Systems. The Institute of Electrical and Electronics Engineers, 345 East 47th Street, New York, N.Y. 10017, USA. ISSN 0018–9251.

IEEE Trans. Antennas Propag.
IEEE Transactions on Antennas and Propagation. The Institute of Electrical and Electronics Engineers, 345 East 47th Street, New York, N.Y. 10017, USA. ISSN 0018–926X.

IEEE Trans. Commun.
IEEE Transactions on Communications. The Institute of Electrical and Electronics Engineers, 345 East 47th Street, New York, N.Y. 10017, USA. ISSN 0090–6778.

IEEE Trans. Geosci. Remote Sensing
IEEE Transactions on Geoscience and Remote Sensing. The Institute of Electrical and Electronics Engineers, 345 East 47th Street, New York, N.Y. 10017, USA. ISSN 0196–2892.

IEEE Trans. Ind. Electron.
IEEE Transactions on Industrial Electronics. The Institute of Electrical and Electronics Engineers, 345 East 47th Street, New York, N.Y. 10017, USA. ISSN 0278–0046.

IEEE Trans. Instrum. Meas.
IEEE Transactions on Instrumentation and Measurement. The Institute of Electrical and Electronics Engineers, 345 East 47th Street, New York, N.Y. 10017, USA. ISSN 0018–9456.

IEEE Trans. Microwave Theory Tech.
IEEE Transactions on Microwave Theory and Techniques. The Institute of Electrical and Electronics Engineers, 345 East 47th Street, New York, N.Y. 10017, USA. ISSN 0018–9480.

IEEE Trans. Nucl. Sci.
IEEE Transactions on Nuclear Science. The Institute of Electrical and Electronics Engineers, 345 East 47th Street, New York, N.Y. 10017, USA. ISSN 0018–9499.

IEEE Trans. Plasma Sci.
IEEE Transactions on Plasma Science. The Institute of Electrical and Electronics Engineers, 345 East 47th Street, New York, N.Y. 10017, USA. ISSN 0093–3813.

IHW Newsl.
The International Halley Watch Newsletter. Jet Propulsion Laboratory, California Institute of Technology, Pasadena, Calif. 91109, USA.

IMA J. Appl. Math.
IMA Journal of Applied Mathematics. Academic Press Inc., London – New York. ISSN 0272–4960.

Indian East. Eng.
Indian and Eastern Engineer. "Piramal Mansion", 235 Dr. D. Naoroji Road, Bombay 400 001, India. ISSN 0019–4352.

Indian Inst. Astrophys., Prepr.
Indian Institute of Astrophysics, Preprints. Indian Institute of Astrophysics, Sarjapur Road, Bangalore – 560 034, India.

Indian J. Cryog.
Indian Journal of Cryogenics. Indian Cryogenics Council, Jadavpur University, Calcutta 700 032, India. ISSN 0379–0479.

Indian J. Hist. Sci.
Indian Journal of History of Science. Indian National Science Academy, Bahadur Shah Zafar Marg, New Delhi 110 002, India.

Indian J. Phys., Part B
Indian Journal of Physics, Part B. Indian Association for the Cultivation of Science, 2 & 3 Raja Subodh Chandra Mallik Road, Calcutta 700 032, India. ISSN 0374–3330.

Indian J. Pure Appl. Math.
Indian Journal of Pure and Applied Mathematics. Indian National Science Academy, Bahadur Shah Zafar Marg, New Delhi 110 002, India. ISSN 0019–5588.

Indian J. Pure Appl. Phys.
Indian Journal of Pure and Applied Physics. Council of Scientific & Industrial Research. Publications & Information Directorate, Hillside Road, New Delhi 110 012, India. ISSN 0019–5596.

Indian J. Radio Space Phys.
Indian Journal of Radio and Space Physics. Council of Scientific & Industrial Research. Publications & Information Directorate, Hillside Road, New Delhi 110 012, India. ISSN 0367–8393.

Indian J. Theor. Phys.
Indian Journal of Theoretical Physics. Institute of Theoretical Physics, Bognan Kutir, 4–1 Mohan Bagan Lane, Calcutta 700 004, India. ISSN 0019–5693.

Inf. Bull. Variable Stars
Information Bulletin on Variable Stars. Commission 27 of the IAU. Konkoly Observatory, Budapest, Hungary. ISSN 0374–0676.

Infrared Phys.
Infrared Physics. An International Research Journal. Pergamon Press, Oxford – New York – Toronto – Sydney – Paris – Frankfurt. ISSN 0020–0891.

Inst. Astron. Astrophys. Tech. Univ. Berlin, Mitt.
Institut für Astronomie und Astrophysik der Technischen Universität Berlin, Mitteilungen. Institut für Astronomie und Astrophysik der Technischen Universität Berlin, Hardenbergstraße 36, D–1000 Berlin 12, F.R. Germany.

Inst. Astron. Fis. Espacio, Tirada Aparte
Instituto de Astronomía y Fisica del Espacio, Tirada Aparte. Instituto de Astronomía y Fisica del Espacio, Casilla de Correo 67, Sucursal 28, 1428 Buenos Aires, Argentina.

Inst. Astron., Univ. Camb., Annu. Rep.
Institute of Astronomy, University of Cambridge, Annual Report. The Observatories, Madingley Road, Cambridge CB3 0HA, England.

Inst. Astrophys. Paris, Pré–Publ.
Institut d'Astrophysique de Paris, Pré–Publication. Institut d'Astrophysique, 98 bis, Boulevard Arago, F–75014 Paris, France.

Inst. Obs. Mar., Bol. Astron.
Instituto y Observatorio de Marina, Boletin Astronomico. Instituto y Observatorio de Marina, San Fernando (Cadiz), Spain.

Inst. Obs. Mar., Bol. Ser. B
Instituto y Observatorio de Marina, Boletin Serie B. (Pequeños Planetas). Instituto y Observatorio de Marina, San Fernando (Cadiz), Spain. ISSN 0558–3993.

Inst. Obs. Mar., Bol. Ser. C
Instituto y Observatorio de Marina, Boletin Serie C. (Rotacion de la Tierra). Instituto y Observatorio de Marina, San Fernando (Cadiz), Spain. ISSN 0210–6485.

Inst. Obs. Mar., Bol. Ser. D
Instituto y Observatorio de Marina, Boletin Serie D. (Ocultaciones). Instituto y Observatorio de Marina, San Fernando (Cadiz), Spain.

Inst. Obs. Mar., Mem. Act.
Instituto y Observatorio de Marina, Memoria de las Actividades. Instituto y Observatorio de Marina, San Fernando (Cadiz), Spain.

Inst. Teor. Astrofys., Blindern–Oslo, Småtrykk
Institutt for Teoretisk Astrofysikk, Blindern–Oslo, Småtrykk. Institute of Theoretical Astrophysics, University of Oslo, PO Box 1029, Blindern, N–0315 Oslo 3, Norway.

Inst. Theor. Astrophys., Blindern–Oslo, Rep.
Institute of Theoretical Astrophysics, Blindern–Oslo, Reports. Institute of Theoretical Astrophysics, University of Oslo, PO Box 1029, Blindern, N–0315 Oslo 3, Norway. ISSN 0078–6780.

Inst. Theor. Astrophys., Blindern–Oslo, Repr.
Institute of Theoretical Astrophysics, Blindern–Oslo, Reprints. Institute of Theoretical Astrophysics, University of Oslo, PO Box 1029, Blindern, N–0315 Oslo 3, Norway.

Inst. Theor. Astrophys., Univ. Oslo, Publ. Ser.
Institute of Theoretical Astrophysics, University of Oslo, Publication Series. Institute of Theoretical Astrophysics, University of Oslo, PO Box 1029, Blindern, N–0315 Oslo 3, Norway. ISSN 0800–6652.

Inst. Theor. Phys. Sternw. Univ. Kiel, Repr.
Institut für Theoretische Physik und Sternwarte der Universität Kiel, Reprints. Institut für Theoretische Physik und Sternwarte der Universität Kiel, D–2300 Kiel, F.R. Germany.

Int. Comet Q.
The International Comet Quarterly. Department of Physics and Astronomy, Appalachian State University, Boone, N.C. 28608, USA. ISSN 0736–6922.

Int. J. Gen. Syst.
International Journal of General Systems. Gordon and Breach Science Publishers Inc., New York – London – Paris. ISSN 0308–1079.

Int. J. Heat Mass Transfer
International Journal of Heat and Mass Transfer. Pergamon Press, Oxford – New York – Toronto – Sydney – Paris – Frankfurt. ISSN 0017–9310.

Int. J. Infrared Millimeter Waves
International Journal of Infrared and Millimeter Waves. Plenum Publishing Corporation, 227 West 17th Street, New York, N.Y. 10011, USA. ISSN 0195–9271.

Int. J. Mass Spectrom. Ion Processes
International Journal of Mass Spectrometry and Ion Processes. Elsevier Scientific Publishing Company, PO Box 330, 1000 AH Amsterdam, The Netherlands. ISSN 0020–7381.

Int. J. Non–Linear Mech.
International Journal of Non–Linear Mechanics. Pergamon Press, Oxford – New York – Toronto – Paris – Frankfurt – Sydney. ISSN 0020–7462.

Int. J. Theor. Phys.
International Journal of Theoretical Physics. Plenum Publishing Corporation, 233 Spring Street, New York, N.Y. 10013, USA. ISSN 0020–7748.

Interdisciplinary Sci. Rev.
Interdisciplinary Science Reviews. Heyden & Son Ltd., Spectrum House, Hillview Gardens, London NW4 2JQ, England. ISSN 0308–0188.

Ir. Astron. J.
The Irish Astronomical Journal. A Half–Yearly Publication under the Auspices of the Observatories of Armagh and Dunsink. Armagh Observatory, Armagh BT61 9DG, Northern Ireland. ISSN 0021–1052.

IRIS Bull. A
IRIS Bulletin A (Earth Orientation). Subcommission International Radio Interferometric Surveying (IRIS) Steering Committee. Available from National Geodetic Survey, National Oceanic and Atmospheric Administration, N/CG114, Rockville, Md. 20852, USA.

ISIS
ISIS. An International Review devoted to the History of Science and its Cultural Influences. Department of History and Sociology of Science, University of Pennsylvania, Philadelphia, Pa. 19104, USA. ISSN 0021–1753.

Issled. Solntsa Krasnykh Zvezd
Issledovaniya Solntsa i Krasnykh Zvezd. (Investigations of the Sun and Red Stars). Akademiya Nauk Latvijskoj SSR, Radioastrofizicheskaya Observatoriya. Izdatel'stvo Zinatne, Riga, USSR. ISSN 0135–1303.

Izv. Akad. Nauk Arm. SSR, Ser. Fiz.
Izvestiya Akademii Nauk Armyanskoj SSR, Seriya Fizika. Akademiya Nauk Armyanskoj SSR, Erevan. 375019 Erevan, ul. Barekamutyan, 24, USSR. ISSN 0002–3035.

Izv. Astron. Ehngel'gardt. Obs.
Izvestiya Astronomicheskoj Ehngel'gardtovskoj Observatorii. Izdatel'stvo Kazanskogo Universiteta, Kazan, ul. Lenina, d. 4/5, USSR.

Izv. Glav. Astron. Obs. Pulkovo
Izvestiya Glavnoj Astronomicheskoj Observatorii v Pulkove. Akademiya Nauk SSSR. Leningradskoe Otdelenie, Izdatel'stvo Nauka, Leningrad. 199164 Leningrad, V–164, Mendeleevskaya l., 1, USSR. ISSN 0367–7966.

Izv. Krymskoj Astrofiz. Obs.
Izvestiya Ordena Trudovogo Krasnogo Znameni Krymskoj Astrofizicheskoj Observatorii. Akademiya Nauk SSSR. Izdatel'stvo 'Nauka', Moskva, USSR. English translation in Bull. Crimean Astrophys. Obs. ISSN 0367–8466.

Izv. Vyssh. Uchebn. Zaved., Radiofiz.
Izvestiya Vysshikh Uchebnikh Zavedenij, Radiofizika. Gor'kovskij Universitet. Gor'kij, ul. Lyadova 25, USSR. English translation in Radiophys. Quantum Electron. ISSN 0021–3462.

J. Am. Assoc. Variable Star Obs.
The Journal of the American Association of Variable Star Observers. The American Association of Variable Star Observers, 187 Concord Avenue, Cambridge, Mass. 02138, USA.

J. Am. Chem. Soc.
Journal of the American Chemical Society. American Chemical Society, 1155 16th Street N.W., Washington, D.C. 20036, USA. ISSN 0002–7863.

J. Appl. Meteorol.
Journal of Applied Meteorology. American Meteorological Society, 45 Beacon Street, Boston, Mass. 02108, USA. ISSN 0021–8952.

J. Appl. Phys.
Journal of Applied Physics. American Institute of Physics, 335 East 45th Street, New York, N.Y. 10017, USA. ISSN 0021–8979.

J. Astron. Soc. Egypt
Journal of the Astronomical Society of Egypt. Astronomical Society of Egypt, Astronomy Department, Faculty of Sciences, Cairo University, Cairo, Egypt.

J. Astron. Soc. West. Aust.
Journal of the Astronomical Society of Western Australia. Astronomical Society of Western Australia, PO Box S1460, Perth, W.A. 6000, Australia.

J. Astronaut. Sci.
Journal of the Astronautical Sciences. American Astronautical Society, 6060 Duke Street, Alexandria, Va. 22304, USA. ISSN 0021–9142.

J. Astrophys. Astron.
Journal of Astrophysics and Astronomy. Indian Academy of Sciences, PO Box 8005, Bangalore 560 080, India. ISSN 0250–6335.

J. Atmos. Sci.
Journal of the Atmospheric Sciences. American Meteorological Society, 45 Beacon Street, Boston, Mass. 02108, USA. ISSN 0022–4928.

J. Atmos. Terr. Phys.
Journal of Atmospheric and Terrestrial Physics. Pergamon Press, Oxford – New York – Frankfurt. ISSN 0021–9169.

J. Aust. Math. Soc., Ser. B
Journal of the Australian Mathematical Society, Series B. (Applied Mathematics). Department of Mathematics, University of Queensland, St. Lucia, QLD 4067, Australia. ISSN 0334–2700.

J. Br. Astron. Assoc.
Journal of the British Astronomical Association. The British Astronomical Association, Burlington House, Piccadilly, London, W1V 0NL, England. ISSN 0007–0297.

J. Br. Interplanet. Soc.
Journal of the British Interplanetary Society. The British Interplanetary Society, 27–29 South Lambeth Road, London, SW8 1SZ, England. ISSN 0007–084X.

J. Chem. Educ.
Journal of Chemical Education. American Chemical Society, 119 West 24th Street, New York, N.Y. 10011, USA. ISSN 0021–9584.

J. Chem. Phys.
Journal of Chemical Physics. American Institute of Physics, 335 East 45th Street, New York, N.Y. 10017, USA. ISSN 0021–9606.

J. Comput. Phys.
Journal of Computational Physics. Academic Press Inc., 111 5th Avenue, New York, N.Y. 10003, USA. ISSN 0021–9991.

J. Differ. Equations
Journal of Differential Equations. Academic Press Inc., New York – London. ISSN 0022–0396.

J. Electron Spectrosc. Relat. Phenom.
Journal of Electron Spectroscopy and Related Phenomena. Elsevier Scientific Publishing Company, PO Box 330, 1000 AH Amsterdam, The Netherlands. ISSN 0368–2048.

J. Electrostat.
Journal of Electrostatics. Elsevier Scientific Publishing Company, PO Box 330, 1000 AH Amsterdam, The Netherlands. ISSN 0304–3886.

J. Environ. Sci.
Journal of Environmental Sciences. Institute of Environmental Sciences, 940 East Northwest Highway, Mt. Prospect, Ill. 60056, USA. ISSN 0022–0906.

J. Fluid Mech.
Journal of Fluid Mechanics. Cambridge University Press, Bentley House, 200 Euston Road, London, NW1 2DB, England. ISSN 0022–1120.

J. Geodyn.
Journal of Geodynamics. Geophysical Press, Brouwersgracht 236, 1013 HE Amsterdam, The Netherlands. ISSN 0264–3707.

J. Geol.
Journal of Geology. The University of Chicago Press, 5801 S. Ellis Avenue, Chicago, Ill. 60637, USA. ISSN 0022–1376.

J. Geomagn. Geoelectr.
Journal of Geomagnetism and Geoelectricity. University of Tokyo Press, c/o Business Centre for Academic Societies, 4–16 Yayoi 2–chome, Bunkyo–ku, Tokyo 113, Japan. ISSN 0022–1392.

J. Geophys.
Journal of Geophysics. Springer–Verlag, Berlin – Heidelberg – New York – Tokyo. ISSN 0340–062X.

J. Geophys. Res.
Journal of Geophysical Research. Section A: Space Physics, Section B: Solid Earth and Planets, Section C: Ocean, Section D: Atmosphere. American Geophysical Union, 2000 Florida Avenue, N.W., Washington, D.C. 20009, USA. ISSN 0148–0227.

J. Guid. Control Dyn.
Journal of Guidance, Control, and Dynamics. American Institute of Aeronautics and Astronautics, 1290 Avenue of the Americas, New York, N.Y. 10104, USA. ISSN 0731–5090.

J. Hist. Astron.
Journal for the History of Astronomy. Published by Science History Publications Ltd., Halfpenny Furze, Mill Lane, Chalfont St Giles, Bucks. HP8 4NR, England. ISSN 0021–8286.

J. Inst. Electron. Commun. Eng. Jpn.
Journal of the Institute of Electronics and Communications Engineers of Japan. Institute of Electronics and Communications Engineers of Japan, Denshi Tsushin Gakkai, Kikai–Shinko–Kaikan, 5–8, Shibakoen 3 Chome Minato–ku, Tokyo 105, Japan. ISSN 0373–6121.

J. Inst. Electron. Telecommun. Eng.
Journal of the Institution of Electronics and Telecommunication Engineers. Institution of Electronics and Telecommunication Engineers, 2 Institutional Area, Lodi Road, New Delhi 110 003, India. ISSN 0377–2063.

J. Korean Astron. Soc.
The Journal of the Korean Astronomical Society. The Korean Astronomical Society, Seoul, Korea.

J. Magn. Magn. Mater.
Journal of Magnetism and Magnetic Materials. North–Holland Publishing Company, PO Box 211, 1000 AE Amsterdam, The Netherlands. ISSN 0304–8853.

J. Math. Phys.
Journal of Mathematical Physics. American Institute of Physics, 335 East 45th Street, New York, N.Y. 10017, USA. ISSN 0022–2488.

J. Math. Phys. Sci.
Journal of Mathematical and Physical Sciences. Indian Institute of Tech., Madras, India. ISSN 0047–2557.

J. Mec. Theor. Appl.
Journal de Mécanique Théorique et Appliquée. Centrale des Revues, 11 rue Gossin, F–92543 Montrouge Cedex, France. ISSN 0750–7240.

J. Mech. Eng. Lab.
Journal of Mechanical Engineering Laboratory. Namiki Sakura–mura, Niihari–gun, Ibaraki, Japan. ISSN 0388–4252.

J. Microcomput. Appl.
Journal of Microcomputer Applications. Academic Press Inc. Ltd., 24 – 28 Oval Road, London NW1 7DX, England. ISSN 0143–3792.

J. Mol. Spectrosc.
Journal of Molecular Spectroscopy. Academic Press Inc., 111 5th Avenue, New York, N.Y. 10003, USA. ISSN 0022–2852.

J. Nanjing Univ.
Journal of Nanjing University. (Natural Sciences Edition). Nanjing University, Nanjing, People's Republic of China.

J. Navig.
The Journal of Navigation. The Royal Institute of Navigation. Cambridge University Press, Cambridge – London – New York – New Rochelle – Melbourne – Sydney. ISSN 0020–3009.

J. Non–Cryst. Solids
Journal of Non–Crystalline Solids. North–Holland Publishing Company, PO Box 211, 1000 AE Amsterdam, The Netherlands. ISSN 0022–3093.

J. Opt. (Calcutta)
Journal of Optics. Optical Society of India, Department of Applied Physics, University of Calcutta, 92 Acharya Prafulla Chandra Road, Calcutta 700 009, India. ISSN 0304–5811.

J. Opt. (Paris)
Journal of Optics. Masson Editeur, 120 Boulevard Saint–Germain, F–75280 Paris Cedex 06, France. ISSN 0150–536X.

J. Opt. Soc. Am. A
Journal of the Optical Society of America A. Optics and Image Science. Published for the Optical Society of America by the American Institute of Physics, 335 East 45th Street, New York, N.Y. 10017, USA. ISSN 0740–3232.

J. Opt. Soc. Am. B
Journal of the Optical Society of America B. Optical Physics. Published for the Optical Society of America by the American Institute of Physics, 335 East 45th Street, New York, N.Y. 10017, USA. ISSN 0740–3224.

J. Phys.
Journal de Physique. Les Editions de Physique, Z. I. de Courtabœuf, B.P. 112, F–91944 Les Ulis Cedex, France. ISSN 0302–0738.

J. Phys. A
Journal of Physics A. (Mathematical and General Physics). Institute of Physics, 47 Belgrave Square, London SW1X 8QX, England. ISSN 0305–4470.

J. Phys. B
Journal of Physics B. (Atomic and Molecular Physics). Institute of Physics, 47 Belgrave Square, London SW1X 8QX, England. ISSN 0022–3700.

J. Phys. Chem.
Journal of Physical Chemistry. American Chemical Society, 1155 16th Street N.W., Washington, D.C. 20036, USA. ISSN 0022–3654.

J. Phys. Chem. Ref. Data
Journal of Physical and Chemical Reference Data. American Chemical Society, 1155 16th Street, N. W., Washington, D. C. 20036, USA. ISSN 0047–2689.

J. Phys. Colloq.
Journal de Physique Colloque. Les Editions de Physique, Z.I. de Courtabœuf, B.P. 112, F–91944 Les Ulis Cedex, France. ISSN 0449–1947.

J. Phys. D
Journal of Physics D. (Applied Physics). Institute of Physics, 47 Belgrave Square, London SW1X 8QX, England. ISSN 0022–3727.

J. Phys. E
Journal of Physics E. (Scientific Instruments). Institute of Physics, 47 Belgrave Square, London SW1X 8QX, England. ISSN 0022–3735.

J. Phys. Earth
Journal of Physics of the Earth. University of Tokyo Press. Subscription address: Japan Publications Trading Company Ltd., C.P.O. 722, Tokyo, Japan. ISSN 0022–3743.

J. Phys. F
Journal of Physics F. (Metal Physics). Institute of Physics, 47 Belgrave Square, London SW1X 8QX, England. ISSN 0305–4608.

J. Phys. G
Journal of Physics G. (Nuclear Physics). Institute of Physics, 47 Belgrave Square, London SW1X 8QX, England. ISSN 0305–4616.

J. Phys. Lett.
Journal de Physique Lettres. Les Editions de Physique, Z.I. de Courtabœuf, B.P. 112, F–91944 Les Ulis Cedex, France. ISSN 0302–072X.

J. Phys. Soc. Jpn.
Journal of the Physical Society of Japan. Room 211, Kikai Shinko Building, 3–5–8 Shiba Koen, Minato–ku, Tokyo 105, Japan. ISSN 0031–9015.

J. Plasma Phys.
Journal of Plasma Physics. Cambridge University Press, Cambridge – London – New York – New Rochelle – Melbourne – Sydney. ISSN 0022–3778.

J. Proc. R. Soc. N.S.W.
Journal and Proceedings of the Royal Society of New South Wales. Science Centre, 35 Clarence Street, Sydney, N.S.W. 2000, Australia. ISSN 0035–9173.

J. Quant. Spectrosc. Radiat. Transfer
Journal of Quantitative Spectroscopy and Radiative Transfer. Pergamon Press, Oxford – New York – Toronto – Paris – Frankfurt – Sydney. ISSN 0022–4073.

J. R. Astron. Soc. Can.
The Journal of the Royal Astronomical Society of Canada. The Royal Astronomical Society of Canada, 124 Merton Street, Toronto, Ont. M4S 2Z2, Canada. ISSN 0035–872X.

J. Radio Res. Lab.
Journal of the Radio Research Laboratories. Radio Research Laboratories, Ministry of Posts & Telecommunications, Nukui–Kitamachi, Konganei–shi, Tokyo 184, Japan. ISSN 0033–8001.

J. Res. Natl. Bur. Stand.
Journal of Research of the National Bureau of Standards. Subscription address: US Government Printing Office, Washington, D.C. 20402, USA. ISSN 0091–0635.

J. Sci. Ind. Res.
Journal of Scientific and Industrial Research. Council of Scientific & Industrial Research, Publications & Information Directorate, Hillside Road, New Delhi 110 012, India. ISSN 0022–4456.

J. Sci. Res. Banaras Hindu Univ.
Journal of Scientific Research of the Banaras Hindu University. Department of Physics, Faculty of Science, PO Banaras Hindu University, Varnasi 221 005, India. ISSN 0447–9483.

J. Soc. Instrum. Control Eng.
Journal of the Society of Instrument and Control Engineers. Society of Instrument and Control Engineers, Hongo, 1–35–28–303, Bunkyo–ku, Tokyo 113, Japan. ISSN 0453–4662.

J. Spacecr. Rockets
Journal of Spacecraft and Rockets. American Institute of Aeronautics and Astronautics, 1290 Avenue of the Americas, New York, N.Y. 10104, USA. ISSN 0022–4650.

J. Spectrosc. Soc. Jpn.
Journal of the Spectroscopical Society of Japan. Spectroscopical Society of Japan, Room 301, Clean Building, 1–13, Kanda–Awaji–cho, Chiyoda–ku, Tokyo 101, Japan. ISSN 0038–7002.

J. Stat. Phys.
Journal of Statistical Physics. Plenum Publishing Corporation, 227 West 17th Street, New York, N.Y. 10011, USA. ISSN 0022–4715.

J. Toyo Univ., Gen. Educ., Nat. Sci.
Journal of the Toyo University, General Education, Natural Science. Toyo University, 28, Hakusan 5–chôme, Bunkyo–ku, Tokyo, Japan.

Jenaer Rundsch.
Jenaer Rundschau. Jenoptik Jena GmbH. VEB Verlag Technik, Oranienburger Str. 13/14, DDR–1020 Berlin, German Democratic Republic. ISSN 0368–203X.

JETP Lett.
JETP Letters. A cover–to–cover translation of Pis'ma v Zhurnal Ehksperimental'noj i Teoreticheskoj Fiziki. American Institute of Physics, 335 East 45th Street, New York, N.Y. 10017, USA. ISSN 0021–3640.

Johns Hopkins APL Tech. Dig.
Johns Hopkins APL Technical Digest. The Johns Hopkins University Applied Physics Laboratory. Johns Hopkins Road, Laurel, Md. 20707, USA. ISSN 0270–5214.

Jpn. J. Appl. Phys., Part 1
Japanese Journal of Applied Physics, Part 1 (Regular Papers and Short Notes). Publication Office, Daini Toyokaiji Building, 24–8, Shinbashi 4–chome, Minato–ku, Tokyo 105, Japan. ISSN 0021–2922.

Jpn. J. Appl. Phys., Part 2
Japanese Journal of Applied Physics, Part 2 (Letters). Publication Office, Daini Toyokaiji Building, 24–8, Shinbashi 4–chome, Minato–ku, Tokyo 105, Japan. ISSN 0021–4922.

Kandilli Obs., Heliophys. Serv. Publ., Second Ser.
Kandilli Observatory, Heliophysics Service Publications, Second Series. Kandilli Observatory, Istanbul, Turkey.

Kapteyn Astron. Inst., Annu. Rep.
Kapteyn Astronomical Institute, Annual Report. Department of Astronomy Rijksuniversiteit Groningen, Groningen, The Netherlands.

KDD Tech. J.
KDD Technical Journal. International Communications Research Institute, Tokyo, Japan. ISSN 0452–3431.

Kexue Tongbao
Kexue Tongbao (Science Bulletin). Science Press, No. 137, Chaoyangmennei Street, Beijing, People's Republic of China. ISSN 0250–7862.

Kinematika Fiz. Nebesn. Tel
Kinematika i Fizika Nebesnykh Tel. Akademiya Nauk Ukrainskoj SSR. Otdelenie Fiziki i Astronomii. Glavnaya Astronomicheskaya Observatoriya Akademii Nauk USSR. Izdatel'stvo Naukova Dumka, 252127 Kiev 127, Goloseevo, USSR. ISSN 0233–7665.

Kodaikanal Obs. Bull., Ser. A
Kodaikanal Observatory Bulletins, Series A. Indian Institute of Astrophysics, Bangalore 560 034, India.

Kodaikanal Obs. Repr.
Kodaikanal Observatory Reprints. Indian Institute of Astrophysics, Bangalore 560 034, India.

Komet. Tsirk.
Kometnyj Tsirkulyar. Gruppa po Issledovaniyu Komet Sektsii 'Solnechnaya Sistema' Astronomicheskogo Soveta AN SSSR. Kievskij Universitet im. T.G. Shevchenko. Glavnaya Astronomicheskaya Observatoriya AN USSR.

Komety Meteory
Komety i Meteory. Akademiya Nauk Tadzhikskoj SSR. Astronomicheskij Sovet Akademii Nauk SSSR. Izdatel'stvo 'Donish', 734029 Dushanbe, ul. Ajni, 121, korp. 2, USSR. ISSN 0568–6199.

Kosm. Issled.
Kosmicheskie Issledovaniya. Akademiya Nauk SSSR. Izdatel'stvo 'Nauka', Moskva. Editorial address: 103717 GSP, Moskva, K–62, Podsosenskij per., 21. USSR. English translation in Cosmic Res., Consultants Bureau, New York, N.Y., USA. ISSN 0023–4206.

Kosm. Luchi
Kosmicheskie Luchi. Rezul'taty Issledovanij po Mezhdunarodnym Geofizicheskim Proektam. Mezhduvedomstvennyj Geofizicheskij Komitet pri Prezidiume Akademii Nauk SSSR. Izdatel'stvo Sovetskoe radio, Moskva, Glavpochtamt, a/ya No. 693. USSR.

Kozmos
Kozmos. Popular Astronomical Journal. Slovenské ústredie amatérskej astronomie, Hurbanovo.

Latitude Circ.
Latitude Circular. Astronomic–Geodetical Observatory at Józefosław. Warsaw Technical University, Warsaw, Poland.

LEST Found., Annu. Rep.
Large Earth–based Solar Telescope Foundation, Annual Report. Institute of Theoretical Astrophysics, University of Oslo, PO Box 1029, Blindern, N–0315 Oslo 3, Norway. ISSN 0800–7799.

LEST Found., Tech. Rep.
Large Earth–based Solar Telescope Foundation, Technical Report. Institute of Theoretical Astrophysics, University of Oslo, PO Box 1029, Blindern, N–0315 Oslo 3, Norway. ISSN 0800–7780.

Lett. Math. Phys.
Letters in Mathematical Physics. D. Reidel Publishing Company, Dordrecht – Boston. ISSN 0377–9017.

Lett. Nuovo Cimento
Lettere al Nuovo Cimento della Società Italiana di Fisica. Società Italiana di Fisica, Editrice Compositori, viale XII Giugno, 1, I–40124 Bologna, Italy. ISSN 0024–1318.

Lick Obs. Bull.
Lick Observatory Bulletin. Lick Observatory, Board of Studies in Astronomy and Astrophysics, University of California at Santa Cruz, Santa Cruz, Calif. 95064, USA.

Lohrmann–Obs., Tech. Univ. Dresden, Zirk.
Lohrmann–Observatorium, Technische Universität Dresden, Zirkular. Lohrmann–Observatorium, Technische Universität Dresden, DDR–8027 Dresden, Mommsenstraße 13, German Democratic Republic.

Lowell Obs. Bull.
Lowell Observatory Bulletin. Lowell Observatory, Flagstaff, Ariz., USA.

Madà
Madà (Science). The Weizmann Science Press of Israel, Jerusalem, Israel. ISSN 0368–833X.

Magy. Geofiz.
Magyar Geofizika. Lapkiado Vallalat, H–1073 Budapest, Lenin korut 9–11, Hungary. ISSN 0025–0120.

Manuscr. Geod.
Manuscripta Geodaetica. Springer–Verlag, Berlin – Heidelberg – New York – London – Paris – Tokyo. ISSN 0340–8825.

Math. Modelling
Mathematical Modelling. Pergamon Press, New York – Oxford – Toronto – Paris – Frankfurt – Sydney. ISSN 0270–0255.

Math. Proc. Camb. Philos. Soc.
Mathematical Proceedings of the Cambridge Philosophical Society. Cambridge University Press, Cambridge – London – New York – New Rochelle – Melbourne – Sydney. ISSN 0305–0041.

Meccanica
Meccanica. Pitagora Editrice, via Zamboni 57, I–40126 Bologna, Italy. ISSN 0025–6455.

Meded. K. Acad. Wet., Lett. Schone Kunsten Belg., Kl. Wet.
Mededelingen van de Koninklijke Academie voor Wetenschappen, Letteren en Schone Kunsten van Belgie, Klasse der Wetenschappen. Koninklijke Academie voor Wetenschappen, Letteren en Schone Kunsten van Belgie. Paleis der Academien, Hertogsstraat 1, B–1000 Bruxelles, Belgium.

Mem. Astron. Soc. India
Memoirs of the Astronomical Society of India. Astronomical Society of India, Department of Astronomy, Osmania University, Hyderabad 500 007, India.

Mem. Fac. Eng., Kyoto Univ.
Memoirs of the Faculty of Engineering, Kyoto University. Faculty of Engineering, Kyoto University, Kyoto, Japan. ISSN 0023–6063.

Mem. Fac. Eng., Osaka City Univ.
Memoirs of the Faculty of Engineering, Osaka City University. Faculty of Engineering, Osaka City University, 459 Sugimoto–cho, Sumi Yoshi–kum, Osaka, Japan. ISSN 0078–6659.

Mem. Fac. Sci., Kyoto Univ., Ser. Phys., Astrophys., Geophys., Chem.
Memoirs of the Faculty of Science, Kyoto University, Series of Physics, Astrophysics, Geophysics, and Chemistry. Faculty of Science, University of Kyoto, Kyoto 608, Japan. ISSN 0368–9689.

Mem. Jpn. Astron. Study Assoc.
Memoirs of the Japan Astronomical Study Association. National Science Museum, Ueno Park, Taito–ku, Tokyo, Japan.

Mem. Soc. Astron. Ital.
Memorie della Società Astronomica Italiana. Società Astronomica Italiana, Largo Fermi, 5, I–50125 Firenze, Italy. ISSN 0037–8720.

Mercury
Mercury. The Journal of the Astronomical Society of the Pacific. The Astronomical Society of the Pacific, 1290 24th Avenue, San Francisco, Calif. 94122, USA. ISSN 0047–6773.

Meres Autom.
Meres es Automatika. Lapkiado Vallalat, Budapest VII, Lenin Korut 9 - 11, Hungary. ISSN 0025–9993.

Messenger
The Messenger – El Mensajero. European Southern Observatory, Karl–Schwarzschild–Straße 2, D–8046 Garching bei München, F.R. Germany. ISSN 0722–6691.

Meteoritics
Meteoritics. The Journal of the Meteoritical Society. Center for Meteorite Studies, Arizona State University, Tempe, Ariz. 85287, USA. ISSN 0026–1114.

Meteoritika
Meteoritika. Akademiya Nauk SSSR. Komitet po Meteoritam. Izdatel'stvo 'Nauka', Moskva. 117864 GSP–7, Moskva V–485, Profsoyuznaya ul., d. 90, USSR. ISSN 0369–2507.

Meteorol. Mag.
Meteorological Magazine. General Meteorological Office, London Road, Bracknell, Berks. RG12 2SZ, England. ISSN 0026–1149.

Metrologia
Metrologia. Springer–Verlag, Berlin – Heidelberg – New York. ISSN 0026–1394.

Micro Syst.
Micro Systèmes. Société Parisienne d'Edition, 2 à 12 rue de Bellevue, F–75940 Paris Cedex 19, France. ISSN 0183–5084.

Mikrocomput. Z.
Mikrocomputer Zeitschrift. Franzis–Verlag GmbH, Karlstr. 37, D–8000 München 2, F.R. Germany. ISSN 0720–4442.

Mikrowellen Mag.
Mikrowellen Magazin. Sprechsaal–Verlag, PO Box 401, D–8630 Coburg, F.R. Germany. ISSN 0722–8244.

Minor Planet Bull.
The Minor Planet Bulletin. Bulletin of the Minor Planets Section of the Association of Lunar and Planetary Observers. Editorial Office: R. P. Binzel, Department of Astronomy, University of Texas, Austin, Tex. 78712, USA.

Minor Planet Circ.
The Minor Planet Circulars/Minor Planets and Comets. Minor Planet Center, Smithsonian Astrophysical Observatory, 60, Garden Street, Cambridge, Mass. 02138, USA. ISSN 0736–6884.

Mitsubishi Denki Giho
Mitsubishi Denki Giho. Mitsubishi Electric Corporation, Mitsubishi Denki Building, Marunouchi, Tokyo 100, Japan. ISSN 0369–2302.

Mitt. Archenhold–Sternw. Berlin–Treptow
Mitteilungen der Archenhold–Sternwarte Berlin–Treptow. Archenhold–Sternwarte, DDR–1193 Berlin, Alt Treptow 1, German Democratic Republic.

Mitt. Astron. Ges.
Mitteilungen der Astronomischen Gesellschaft, Hamburg. Subscription address: Astronomisches Institut der Universität Bochum, Postfach 102148, D–4630 Bochum, F.R. Germany. ISSN 0172–5483.

Mitt. Astrophys. Obs. Potsdam
Mitteilungen des Astrophysikalischen Wissenschaften der DDR, Zentralinstitut für Astrophysik. Astrophysikalisches Observatorium Potsdam, DDR–1500 Potsdam, Telegrafenberg, German Democratic Republic.

Mitt. Geod. Inst. Rheinischen Friedrich–Wilhelms–Univ. Bonn
Mitteilungen aus den Geodätischen Instituten der Rheinischen Friedrich–Wilhelms–Universität Bonn. Geodätische Institute der Rheinischen Friedrich–Wilhelms–Universität Bonn, Nußallee 17, D–5300 Bonn 1, F.R. Germany. ISSN 0723–4325.

Mitt. Inst. Astron. Phys. Geod. Tech. Hochsch. Münch.
Mitteilungen aus dem Institut für Astronomische und Physikalische Geodäsie der Technischen Hochschule München. Institut für Astronomische und Physikalische Geodäsie der Technischen Hochschule München, D–8000 München, F.R. Germany.

Mitt. Karl–Schwarzschild–Obs. Tautenburg
Mitteilungen des Karl–Schwarzschild–Observatoriums Tautenburg. Akademie der Wissenschaften der DDR, Zentralinstitut für Astrophysik. Karl–Schwarzschild–Observatorium Tautenburg, DDR–6901 Tautenburg, German Democratic Republic.

Mitt. Lohrmann–Obs. Tech. Univ. Dresden
Mitteilungen des Lohrmann–Observatoriums der Technischen Universität Dresden. Lohrmann–Observatorium, Technische Universität Dresden, DDR–8027 Dresden, Mommsenstraße 13, German Democratic Republic.

Mitt. Satell.–Beobachtungsstn. Zimmerwald
Mitteilungen der Satelliten–Beobachtungsstation Zimmerwald. Druckerei der Universität Bern, Switzerland.

Mitt. Sonnenobs. Kanzelhöhe
Mitteilungen des Sonnenobservatoriums Kanzelhöhe. Sonnenobservatorium Kanzelhöhe, A–9520 Sattendorf, Austria.

Mitt. Sternw. Münch.
Mitteilungen der Sternwarte München. Institut für Astronomie und Astrophysik der Universität München, Universitäts–Sternwarte, Scheinerstr. 1, D–8000 München 80, F.R. Germany.

Mitt. Sternw. Sonneberg
Mitteilungen der Sternwarte zu Sonneberg. Akademie der Wissenschaften der DDR, Zentralinstitut für Astrophysik. Sternwarte Sonneberg, DDR–6400 Sonneberg, German Democratic Republic.

Mitt. Univ.–Sternw. Innsb.
Mitteilungen der Universitäts–Sternwarte Innsbruck. Institut für Astronomie der Universität Innsbruck, Universitätsstraße 4, A–6020 Innsbruck, Austria.

Mitt. Univ.–Sternw. Jena
Mitteilungen der Universitäts–Sternwarte zu Jena. Universitätssternwarte Jena, Schillergäßchen 2, DDR–6900 Jena, German Democratic Republic.

Mitt. Universitätssternw. Graz
Mitteilungen der Universitätssternwarte Graz. Institut für Astronomie, Universitätsplatz 5, A–8010 Graz, Austria.

Mitt. Veränderliche Sterne
Mitteilungen über Veränderliche Sterne. Akademie der Wissenschaften der DDR, Zentralinstitut für Astrophysik, Sternwarte Sonneberg, DDR–6400 Sonneberg, German Democratic Republic.

Mitt. Zentralinst. Phys. Erde
Mitteilungen des Zentralinstituts für Physik der Erde. Akademie der Wissenschaften der DDR, Zentralinstitut für Physik der Erde, DDR–1500 Potsdam, Telegrafenberg A17, German Democratic Republic.

Mod. Geol.
Modern Geology. Gordon and Breach Science Publishers, New York – London. ISSN 0026–7775.

Mon. Not. R. Astron. Soc.
Monthly Notices of the Royal Astronomical Society. Published for the Royal Astronomical Society by Blackwell Scientific Publications, Oxford – London – Edinburgh – Boston – Melbourne. ISSN 0035–8711.

Mon. Notes Astron. Soc. S. Afr.
Monthly Notes of the Astronomical Society of Southern Africa. South African Astronomical Observatory, PO Box 9, Observatory, 7935 Cape, South Africa. ISSN 0024–8266.

Mon. Notes Int. Polar Motion Serv.
Monthly Notes of the International Polar Motion Service. Central Bureau of the International Polar Motion Service, International Latitude Observatory of Mizusawa, Mizusawa-shi, Iwate–ken, Japan. ISSN 0020–8337.

Mons Astrophys. Pap.
Mons Astrophysical Papers. (Communications du Département d'Astrophysique de la Faculté des Sciences de Mons. Departement d'Astrophysique, Université de Mons, B–7000 Mons, Belgium.

MPA Rep.
Max–Planck–Institut für Physik und Astrophysik, Institut für Astrophysik, Reports. Max–Planck–Institut für Physik und Astrophysik, Institut für Astrophysik, Karl–Schwarzschild–Straße 1, D–8046 Garching bei München, F.R. Germany.

MPE Contrib.
Max–Planck–Institut für Physik und Astrophysik, Institut für Extraterrestrische Physik, Contributions. Max–Planck–Institut für Extraterrestrische Physik, Giessenbachstraße, D–8046 Garching bei München, F.R. Germany.

MPE Intern. Rep.
Max–Planck–Institut für Physik und Astrophysik, Institut für Extraterrestrische Physik, Internal Report. Max–Planck–Institut für Extraterrestrische Physik, Giessenbachstraße, D–8046 Garching bei München, F.R. Germany. ISSN 0178–0719.

MPE Prepr.
Max–Planck–Institut für Physik und Astrophysik, Institut für Extraterrestrische Physik, Preprints. Max–Planck–Institut für Extraterrestrische Physik, Giessenbachstraße, D–8046 Garching bei München, F.R. Germany. ISSN 0340–8922.

MPE Rep.
Max–Planck–Institut für Physik und Astrophysik, Institut für Extraterrestrische Physik, Reports. Max–Planck–Institut für Extraterrestrische Physik, Giessenbachstraße, D–8046 Garching bei München, F.R. Germany. ISSN 0173–699X.

MPE Repr.
Max–Planck–Institut für Physik und Astrophysik, Institut für Extraterrestrische Physik, Reprints. Max–Planck–Institut für Extraterrestrische Physik, Giessenbachstraße, D–8046 Garching bei München, F.R. Germany.

MPG Spiegel
MPG Spiegel. Max–Planck–Gesellschaft zur Förderung der Wissenschaften, Residenzstraße 1a, D–8000 München 2, F.R. Germany. ISSN 0341–7727.

MSN Microwave Syst. News
MSN Microwave Systems News. EW Communications Inc., 1170 East Meadow Drive, Palo Alto, Calif. 94303, USA. ISSN 0164–3371.

Mt. Stromlo Siding Spring Obs., Repr.
Mount Stromlo and Siding Spring Observatories, Reprints. Mount Stromlo and Siding Spring Observatories, Research School of Physical Sciences, The Australian National University, Private Bag, Woden PO, ACT 2606, Australia.

Nablyud. Iskusstv. Nebesn. Tel
Nablyudeniya Iskusstvennykh Nebesnykh Tel. Published by Astronomicheskij Sovet Akademii Nauk SSSR, Moskva. Moskva Zh–17, ul. Pyatnitskaya, 48, Astronomicheskij Sovet AN SSSR.

Nachr. Akad. Wiss. Göttingen. II
Nachrichten der Akademie der Wissenschaften in Göttingen. II. Mathematisch–Physikalische Klasse. Vandenhoeck und Ruprecht, Göttingen, F.R. Germany. ISSN 0065–5295.

Nachr. Karten–Vermessungswes., Reihe I
Nachrichten aus dem Karten– und Vermessungswesen, Reihe I: Orginalbeiträge. Verlag des Instituts für Angewandte Geodäsie, Frankfurt a.M., F.R. Germany. ISSN 0469–4236.

Nachr. Olbers–Ges. Bremen
Nachrichten der Olbers–Gesellschaft Bremen. Dr. Walter Stein, Werderstr. 73, Bremen, F.R. Germany.

Nanjing Univ. Obs., Publ.
Nanjing University Observatory, Publications. Department of Astronomy and Astrophysics, Nanjing University, Nanjing, People's Republic of China.

NASA Conf. Publ.
NASA Conference Publication. National Aeronautics and Space Administration, Scientific and Technical Information Branch, Washington, D.C. 20546, USA. For sale by the National Technical Information Service, Springfield, Va. 22161, USA.

NASA Contract. Rep.
NASA Contractor Report. National Aeronautics and Space Administration, Scientific and Technical Information Branch, Washington, D.C. 20546, USA. For sale by the National Technical Information Service, Springfield, Va. 22161, USA. ISSN 0565–7059.

NASA IUE Newsl.
NASA IUE Newsletter. National Aeronautics and Space Administration, Goddard Space Flight Center, Greenbelt, Maryland 20771, USA. ISSN 0738–2677.

NASA Ref. Publ.
NASA Reference Publication. National Aeronautics and Space Administration, Scientific and Technical Information Branch, Washington, D.C. 20546, USA. For sale by the National Technical Information Service, Springfield, Va. 22161, USA.

NASA Spec. Publ.
NASA Special Publication. National Aeronautics and Space Administration, Scientific and Technical Information Branch, Washington, D.C. 20546, USA. For sale by the National Technical Information Service, Springfield, Va. 22161, USA. ISSN 0091–0805.

NASA Tech. Brief
NASA Technical Brief. NASA Technology Utilization Program, Technology Transfer Division, PO Box 8757, Baltimore/Washington International Airport, Md. 21240, USA. ISSN 0096–7494.

NASA Tech. Memo.
NASA Technical Memorandum. National Aeronautics and Space Administration, Scientific and Technical Information Branch, Washington, D.C. 20546, USA. For sale by the National Technical Information Service, Springfield, Va. 22161, USA. ISSN 0499–9320.

NASA Tech. Note
NASA Technical Note. National Aeronautics and Space Administration, Scientific and Technical Information Branch, Washington, D.C. 20546, USA. For sale by the National Technical Information Service, Springfield, Va. 22161, USA. ISSN 0499–9339.

NASA Tech. Pap.
NASA Technical Paper. National Aeronautics and Space Administration, Scientific and Technical Information Branch, Washington, D.C. 20546, USA. For sale by the National Technical Information Service, Springfield, Va. 22161, USA.

Natl. Acad. Sci. Lett.
National Academy Science Letters. National Academy of Sciences, 5–Lajpat Rai Road, Allahabad 211 002, India. ISSN 0250–541X.

Natl. Astron. Ionos. Cent., Astron. Prepr.
National Astronomy and Ionosphere Center, Astronomy Preprints. National Astronomy and Ionosphere Center, Space Sciences Building, Cornell University, Ithaca, N.Y. 14853, USA.

Natl. Astron. Ionos. Cent., Astron. Publ.
National Astronomy and Ionosphere Center, Astronomy Publications. National Astronomy and Ionosphere Center, Space Sciences Building, Cornell University, Ithaca, N.Y. 14853, USA.

Natl. Electron. Rev.
National Electronics Reviews. National Electronics Council, Abell House, John Islip Street, London SW1P 4LN, England. ISSN 0305–2257.

Natl. Geogr.
National Geographic. National Geographic Society, 17th and M Sts. N.W., Washington, D.C. 20036, USA. ISSN 0027–9358.

Natl. Radio Astron. Obs., Repr., Ser. A
National Radio Astronomy Observatory, Reprints, Series A. National Radio Astronomy Observatory, PO Box 2, Green Bank, W.Va. 24944, USA.

Natl. Radio Astron. Obs., Repr., Ser. B
National Radio Astronomy Observatory, Reprints, Series B. National Radio Astronomy Observatory, PO Box 2, Green Bank, W.Va. 24944, USA.

Nature
Nature. Macmillan Journals Ltd., London, England. Subscription address: Nature, Circulation Dept., Brunel Road, Basingstoke, Hants RG21 2XS, England. ISSN 0028–0836.

Naturwissenschaften
Die Naturwissenschaften. Springer–Verlag, Berlin – Heidelberg – New York – Tokyo. ISSN 0028–1042.

Nauchn. Inf.
Nauchnye Informatsii. Astronomicheskij Sovet Akademii Nauk SSSR, Moskva, USSR. ISSN 0130–9773.

Naučna Misao
Naučna Misao (Scientific Idea). Society for Promotion and Propagation of Science, 41000 Zagreb, Pavleka Miškine 37, Yugoslavia. ISSN 0467–0468.

Navigation (Paris)
Navigation. Institut Française de Navigation, 3, avenue Octave–Greard, F–75340 Paris Cedex 07, France. ISSN 0028–1530.

Navigation (Wash.)
Navigation. Journal of the Institute of Navigation. Institute of Navigation, Suite 832. 815 15th Street, N.W., Washington, D.C. 20005, USA. ISSN 0028–1522.

Ned. Tijdschr. Natuurkd. A
Nederlands Tijdschrift voor Natuurkunde A. Nederlandse Natuurkundige Vereniging, Princetonplein 5, NL–3508 TA Utrecht, The Netherlands. ISSN 0378–6374.

NEOS Earth Orientation Bull.
National Earth Orientation Service Earth Orientation Bulletin. Incorporating U.S. Naval Observatory, Time Service Publications, Series 7. U.S. Naval Observatory, Time Service Division (62C), Washington, D.C. 20390, USA.

New Phys., Korean Phys. Soc.
New Physics, Korean Physical Society. Korea Institute for Industrial Economics & Technology, PO Box 205, Chungryang, Seoul, Republic of Korea. ISSN 0374–4914.

New Sci.
New Scientist. New Science Publications, Commonwealth House, 1–19 New Oxford Street, London WC1A 1NG, England. ISSN 0028–6664.

News Lett. Astron. Soc. N.Y.
News Letter of the Astronomical Society of New York. Astronomial Society of New York, 1125 Oxford Place, Schenectady, N.Y. 12308, USA.

Nizamiah Rangapur Obs. Dep. Astron. Osmania Univ., Repr.
Nizamiah and Rangapur Observatories and Department of Astronomy of Osmania University, Reprint. Centre of Advanced Study in Astronomy, Osmania University, Hyderabad 500 007, India.

Nova Acta Leopoldina
Nova Acta Leopoldina. Abhandlungen der Deutschen Akademie der Naturforscher Leopoldina. Deutsche Akademie der Naturforscher Leopoldina, DDR–4010 Halle (Saale), German Democratic Republic. Johann Ambrosius Barth, DDR–7010 Leipzig. ISSN 0369–5034.

Nova Acta Regiae Soc. Sci. Ups., Ser. V: A
Nova Acta Regiae Societatis Scientiarum Upsaliensis, Ser. V: A (Astronomy and Mathematical Sciences). Royal Society of Sciences of Uppsala. Subscripton address: Almqvist & Wiksell International, Stockholm, Sweden. ISSN 0346–6253.

NSSDC Newsl.
National Space Science Data Center Newsletter. National Space Science Data Center, NASA/Goddard Space Flight Center, Greenbelt, Md. 20771, USA.

NSSDC/WDC–A–R&S, Publ.
NSSDC/WDC–A–R&S, Publication. National Space Science Data Center/World Data Center A for Rockets and Satellites, National Aeronautics and Space Administration, Goddard Space Flight Center, Greenbelt, Md. 20771, USA.

Nucl. Instrum. Methods Phys. Res., Sect. A
Nuclear Instruments and Methods in Physics Research, Section A. Accelerators, Spectrometers, Detectors and Associated Equipment. North–Holland Publishing Company, PO Box 211, 1000 AE Amsterdam, The Netherlands. ISSN 0167–5087.

Nucl. Instrum. Methods Phys. Res., Sect. B
Nuclear Instruments and Methods in Physics Research, Section B. Beam Interactions with Materials and Atoms. North–Holland Publishing Company, PO Box 211, 1000 AE Amsterdam, The Netherlands. ISSN 0168–583X.

Nucl. Phys. A
Nuclear Physics A. North–Holland Publishing Company, PO Box 211, 1000 AE Amsterdam, The Netherlands. ISSN 0029–5582.

Nucl. Phys. B, Part. Phys.
Nuclear Physics B, Particle Physics. North–Holland Publishing Company, PO Box 211, 1000 AE Amsterdam, The Netherlands. ISSN 0550–3213.

Nucl. Sci. Appl., Sect. A
Nuclear Science Applications, Section A. Harwood Academic Publishers GmbH, Poststraße 22, CH–7000 Chur, Switzerland. ISSN 0191–1686.

Nucl. Tracks Radiat. Meas.
Nuclear Tracks and Radiation Measurements. Pergamon Press, Oxford – New York – Toronto – Paris – Frankfurt – Sydney. ISSN 0191–278X.

Numer. Math.
Numerische Mathematik. Springer–Verlag, Berlin – Heidelberg – New York – Tokyo. ISSN 0029–599X.

Nuovo Cimento A
Il Nuovo Cimento della Società Italiana di Fisica A. (Nuclei, particles and fields). Società Italiana di Fisica, Editrice Compositori, viale XII Giugno, 1, I–40124 Bologna, Italy. ISSN 0369–3546.

Nuovo Cimento B
Il Nuovo Cimento della Società Italiana di Fisica B. (General physics, relativity, astronomy and mathematical physics and methods). Società Italiana di Fisica, Editrice Compositori, viale XII Giugno, 1, I–40124 Bologna, Italy. ISSN 0369–3554.

Nuovo Cimento C
Il Nuovo Cimento della Società Italiana di Fisica C. (Geophysics and space physics). Società Italiana di Fisica, Editrice Compositori, viale XII Giugno, 1, I–40124 Bologna, Italy. ISSN 0390–5551.

Nuovo Cimento D
Il Nuovo Cimento della Società Italiana di Fisica D. (Condensed matter, atomic, molecular and chemical physics, biophysics). Società Italiana di Fisica, Editrice Compositori, viale XII Giugno, 1, I–40124 Bologna, Italy. ISSN 0392–6737.

N.Z. J. Sci.
New Zealand Journal of Science. New Zealand Department of Scientific and Industrial Research. Subscription address: Science Information Division, DSIR, PO Box 9741, Wellington, New Zealand. ISSN 0028–8365.

OAA Comput. Sect. Circ.
OAA Computing Section Circular. S. Nakano, PO Box No. 32, Sumoto Post Office, Hyogo–Ken, 656–91, Japan.

Obs. Astron. Antares, Contrib. Cient.
Observatório Astronômico Antares, Contribuição Cientifica. Universidade Estadual de Feira de Santana, Feira de Santana, Brazil.

Obs. Astron. Córdoba, Tirada Aparte
Observatorio Astronomico Córdoba, Tirada Aparte. Observatorio Astronomico, Laprida 854, 5000 Córdoba, Argentina.

Obs. Astron. La Plata, Sep. Astron.
Observatorio Astronómico La Plata, Separata Astronomica. Observatorio Astronómico, Universidad Nacional de La Plata, La Plata, Argentina.

Obs. Astron. Univ. Nac. La Plata, Ser. Astron.
Observatorio Astronómico de la Universidad Nacional de La Plata, Serie Astronómica. Observatorio Astronómico, Universidad Nacional de La Plata, La Plata, Argentina. ISSN 0325–3163.

Obs. Astron. Univ. Nac. La Plata, Ser. Espec.
Observatorio Astronómico de la Universidad Nacional de La Plata, Serie Especial. Observatorio Astronómico, Universidad Nacional de La Plata, La Plata, Argentina. ISSN 0325–3015.

Obs. Astrophys. Lab., Univ. Helsinki, Rep.
Observatory and Astrophysics Laboratory, University of Helsinki, Report. Observatory and Astrophysics Laboratory, University of Helsinki, Tähtitorninmäki, SF–00130 Helsinki 13, Finland. ISSN 0355–9289.

Obs. Haute Provence, Pré–Publ.
Observatoire de Haute Provence, Pré–Publication. Observatoire de Haute Provence, F–04870 St. Michel l'Observatoire, France.

Obs. Lyon Prepr.
Observatoire de Lyon Preprints. Observatoire de Lyon, F–69230 St. Genis Laval, France.

Obs. Lyon Repr.
Observatoire de Lyon Reprint. Observatoire de Lyon, F–69230 St. Genis Laval, France.

Obs. R. Belg., Commun., Sér. A
Observatoire Royal de Belgique, Communications, Série A. (Koninklijke Sterrenwacht van België, Mededelingen, Reeks A). Observatoire Royale de Belgique, 3, avenue Circulaire, Uccle, B-1180 Bruxelles, Belgium.

Obs. R. Belg., Commun., Sér. B
Observatoire Royal de Belgique, Communications, Série B. (Koninklijke Sterrenwacht van België, Mededelingen, Reeks B). Observatoire Royal de Belgique, 3, avenue Circulaire, Uccle, B-1180 Bruxelles, Belgium.

Obs. Satell.
Observations of Satellites. Finnish Meteorological Institute. Subscription address: Government Printing Centre, Marketing Department, PO Box 516, SF-00101 Helsinki 10, Finland. ISSN 0355-2004.

Obs. Variable Stars, Rep.
Observations of Variable Stars, Report. Nederlandse Vereniging voor Weer-en Sterrenkunde. Kapteyn Astronomical Institute, Postbus 800, 9700 AV Groningen, The Netherlands.

Observatory
The Observatory. A Review of Astronomy. Royal Greenwich Observatory, Herstmonceux Castle, Hailsham Sussex, BN27 1RP England, ISSN 0029-7704.

Occas. Rep. R. Obs., Edinb.
Occasional Reports of the Royal Observatory, Edinburgh. Royal Observatory, Blackford Hill, Edinburgh EH9 3HJ, Scotland. ISSN 0309-099X.

Occultation Newsl.
Occultation Newsletter. International Occultation Timing Association (IOTA), PO Box 596, Tinley Park, Ill. 60477, USA.

Österr. Z. Vermessungswes. Photogramm.
Österreichische Zeitschrift für Vermessungswesen und Photogrammetrie. Österreichischer Verein für Vermessungswesen und Photogrammetrie, Friedrich-Schmidt-Platz 3, A-1082 Wien, Austria.

Onsala Space Obs., Prepr.
Onsala Space Observatory, Preprint. Chalmers University of Technology, Onsala Space Observatory, S-43900 Onsala, Sweden.

Opt. Acta
Optica Acta. Taylor and Francis Ltd., 4 John Street, London WC1N 2ET, England. ISSN 0030-3909.

Opt. Commun.
Optics Communications. North-Holland Publishing Company, PO Box 211, 1000 AE Amsterdam, The Netherlands. ISSN 0030-4018.

Opt. Eng.
Optical Engineering. Society of Photo-Optical Instrumentation Engineers (SPIE), PO Box 10, 405 Fieldston Road, Bellingham, Wash. 98225, USA. ISSN 0091-3286.

Opt. Lett.
Optics Letters. Published for the Optical Society of America by the American Institute of Physics, 335 East 45th Street, New York, N.Y. 10017, USA. ISSN 0146-9592.

Opt. Pura Apl.
Optica Pura y Aplicada. Sociedad Española de Optica, Serrano 121, Madrid 6, Spain. ISSN 0030-3917.

Origins Life
Origins of Life and Evolution of the Biosphere. The Journal of the International Society for the Study of the Origin of Life. D. Reidel Publishing Company, Dordrecht - Boston - London. ISSN 0302-1688.

Orion
Orion. Zeitschrift der Schweizerischen Astronomischen Gesellschaft (SAG). Zentralsekretariat, Hirtenhofstraße 9, CH-6005 Luzern, Switzerland. ISSN 0030-557X.

Orione
Orione. Rivista Trimestrale di Astronomia. Rivista Orione, via Roma 6, I-10025 Pino Torinese, Italy.

Oss. Astrofis. Catania, Pubbl.
Osservatorio Astrofisico di Catania, Pubblicazione. Osservatorio Astrofisico, città universitaria, I-95125 Catania, Italy.

Oss. Astron. Milano-Merate, Contrib.
Osservatorio Astronomico Milano-Merate, Contributo. Osservatorio Astronomico, Via Emilio Bianchi 46, I-22055 Merate, Como, Italy.

Oss. Mem. Oss. Astrofis. Arcetri
Osservazioni e Memorie dell'Osservatorio Astrofisico di Arcetri. Università Degli Studi di Firenze, Firenze, Italy.

Oss. Privato, Specola "Ariel", Pubbl.
Osservatorio Privato, Specola "Ariel", Pubblicazione. Osservatorio Privato, Specola "Ariel", Via S.Antonio 7, I-3100 Treviso, Italy.

Oyo Buturi
Oyo Buturi. Japan Society of Applied Physics, Room No. 209-2, Kikai-Shinko Building, 3 Shiba-Koen Minato-ku, Tokyo, Japan. ISSN 0369-8009.

Pascal Explore
Pascal Explore. E48 (Environnement cosmique terrestre, astronomie et géologie extraterrestre). Centre National de la Recherche Scientifique, Centre de Documentation Scientifique et Technique, 26, rue Boyer, F-75971 Paris Cedex 20, France. ISSN 0761-2109.

Patrika
Patrika. Newsletter of the Indian Academy of Sciences. Indian Academy of Sciences, Bangalore 560 080, India.

Perem. Zvezdy
Peremennye Zvezdy (Variable Stars). Sbornik statej, izdavaemyj Astronomicheskim Sovetom Akademii Nauk SSSR. Editorial address: Sternberg State Astronomical Institute of the Moscow University, Universitetskij prospekt, 13, 117234 Moscow, USSR. ISSN 0373-7683.

Perem. Zvezdy, Prilozh.
Peremennye Zvezdy, Prilozhenie (Variable Stars, Supplement). Sbornik statej, izdavaemyj Astronomicheskim Sovetom Akademii Nauk SSSR. Astronomicheskij Sovet Akademii Nauk SSSR, Moskva, USSR.

Perth Obs. Commun.
Perth Observatory Communications. Perth Observatory, Bickley, Australia.

Philos. Trans. R. Soc. London, Ser. A
Philosophical Transactions of the Royal Society of London, Series A: Mathematical and Physical Sciences. The Royal Society, 6 Carlton House Terrace, London SW1Y 5AG, England. ISSN 0080-4614.

Photogr. Sci. Eng.
Photographic Science and Engineering. Society of Photographic Scientists and Engineers, Suite 204, 1330 Massachusetts Avenue N.W., Washington, D.C. 20005, USA. ISSN 0031-8760.

Photogramm. Eng. Remote Sensing
Photogrammetric Engineering and Remote Sensing. American Society of Photogrammetry, 105 North Virginia Avenue, Falls Church, Va. 22046, USA. ISSN 0099–1112.

Photonics Spectra
Photonics Spectra. The Magazine of Optical/Electro–Optical/Laser Technology. Published by the Optical Publishing Co., Inc., PO Box 1146, Pittsfield, Mass. 01201, USA. ISSN 0191–0647.

Photorin
Photorin, Mitteilungen der Lichtenberg–Gesellschaft e.V.

Phys. Abstr.
Physics Abstracts. Science Abstracts Series A. An INSPEC Publication published by the Institution of Electrical Engineers in association with the Institute of Electrical and Electronics Engineers Inc. Subscription address: INSPEC Marketing Department, IEE, Station House, Nightingale Road, Hitchin, Herts. SG5 1RJ, England. ISSN 0036–8091.

Phys. Bl.
Physikalische Blätter. Physik–Verlag GmbH, Postfach 1260/1280, D–6940 Weinheim, F.R. Germany. ISSN 0031–9279.

Phys. Briefs
Physics Briefs. Physikalische Berichte. Edited by Deutsche Physikalische Gesellschaft und Fachinformationszentrum Energie, Physik, Mathematik in cooperation with American Institute of Physics. Physik–Verlag GmbH, Postfach 1260/1280, D–6940 Weinheim, F.R. Germany. ISSN 0170–7434.

Phys. Bull.
Physics Bulletin. Institute of Physics, 47 Belgrave Square, London SW1X 8QX, England. ISSN 0031–9112.

Phys. Chem. Miner.
Physics and Chemistry of Minerals. Springer–Verlag, Berlin – Heidelberg – New York – Tokyo. ISSN 0342–1791.

Phys. Earth Planet. Inter.
Physics of the Earth and Planetary Interiors. Elsevier Scientific Publishing Company, PO Box 211, 1000 AE Amsterdam, The Netherlands. ISSN 0031–9201.

Phys. Educ.
Physics Education. Institute of Physics. 47 Belgrave Square, London SW1X 8QX, England. ISSN 0031–9120.

Phys. Energ. Fortis Phys. Nucl.
Physica Energiae Fortis et Physica Nuclearis. Science Press, Beijing. Subscription address: Guozi Shudian, PO Box 399, Beijing, People's Republic of China. ISSN 0254–3052.

Phys. Fluids
Physics of Fluids. American Institute of Physics, 335 East 45th Street, New York, N.Y. 10017, USA. ISSN 0031–9171.

Phys. Lett. A
Physics Letters A. (General, Atomic, and Solid State Physics). North–Holland Publishing Company, PO Box 211, 1000 AE Amsterdam, The Netherlands. ISSN 0031–9163.

Phys. Lett. B
Physics Letters B. (Nuclear, Elementary Particle, and High–Energy Physics). North–Holland Publishing Company, PO Box 211, 1000 AE Amsterdam, The Netherlands. ISSN 0031–9163.

Phys. Rep.
Physics Reports. A review section of Physics Letters. North–Holland Publishing Company, PO Box 211, 1000 AE Amsterdam, The Netherlands. ISSN 0370–1573.

Phys. Rev. A
Physical Review A. (General Physics). Published for the American Physical Society by the American Institute of Physics, 335 East 45th Street, New York, N.Y. 10017, USA. ISSN 0556–2791.

Phys. Rev. B
Physical Review B. (Condensed Matter). Published for the American Physical Society by the American Institute of Physics, 335 East 45th Street, New York, N.Y. 10017, USA. ISSN 0163–1829.

Phys. Rev. C
Physical Review C. (Nuclear Physics). Published for the American Phyiscal Society by the American Institute of Physics, 335 East 45th Street, New York, N.Y. 10017, USA. ISSN 0556–2813.

Phys. Rev. D
Physical Review D. (Particles and Fields). Published for the American Physical Society by the American Institute of Physics, 335 East 45th Street, New York, N.Y. 10017, USA. ISSN 0556–2821.

Phys. Rev. Lett.
Physical Review Letters. Published for the American Physical Society by the American Institute of Physics, 335 East 45th Street, New York, N.Y. 10017, USA. ISSN 0031–9007.

Phys. Scr.
Physica Scripta. An International Journal for Experimental and Theoretical Physics. The Royal Swedish Academy of Sciences, Publications Department, Box 50005, S–104 05 Stockholm, Sweden. ISSN 0031–8949.

Phys. Teach.
Physics Teacher. American Institute of Physics, 335 East 45th Street, New York, N.Y. 10017, USA. ISSN 0031–921X.

Phys. Technol.
Physics in Technology. Institute of Physics, 47 Belgrave Square, London SW1X 8QX, England. ISSN 0305–4624.

Phys. Today
Physics Today. American Institute of Physics, 335 East 45th Street, New York, N.Y. 10017, USA. ISSN 0031–9228.

Phys. Unserer Zeit
Physik in unserer Zeit. Verlag Chemie GmbH, Postfach 1260/1280, D–6940 Weinheim, F.R. Germany. ISSN 0031–9252.

Physica A
Physica A. North–Holland Publishing Company, PO Box 211, 1000 AE Amsterdam, The Netherlands. ISSN 0378–4371.

Physica B, C
Physica B and C (Low Temperature and Solid State Physics; Atomic, Molecular and Plasma Physics; Optics). North–Holland Publishing Company, PO Box 211, 1000 AE Amsterdam, The Netherlands. ISSN 0378–4363.

Physica D
Physica D. North–Holland Publishing Company, PO Box 211, 1000 AE Amsterdam, The Netherlands. ISSN 0167–2789.

Pis'ma Astron. Zh.
Pis'ma v Astronomicheskij Zhurnal. Akademiya Nauk SSSR. Izdatel'stvo 'Nauka', Moskva. 103717, GSP, Moskva, K-62, Podsosenskij per., 21. USSR. English translation in Soviet Astron. Lett. ISSN 0320-0108.

Pis'ma Zh. Ehksp. Teor. Fiz.
Pis'ma v Zhurnal Ehksperimental'noj i Teoreticheskoj Fiziki. Akademiya Nauk SSSR. Izdatel'stvo Nauka, Moskva, Leninskij Prospekt, 14, USSR. English translation in JETP Lett. ISSN 0370-274X.

Planet. Space Sci.
Planetary and Space Science. Pergamon Press, Oxford - New York - Beijing - Frankfurt - São Paulo - Sydney - Tokyo - Toronto. ISSN 0032-0633.

Plasma Phys. Controlled Fusion
Plasma Physics and Controlled Fusion. Pergamon Press, Oxford - New York - Toronto - Paris - Frankfurt - Sydney. ISSN 0032-1028.

Pokroky Mat., Fyz. Astron.
Pokroky matematiky, fyziky a astronomie. Academia, Praha, Czechoslovakia. ISSN 0032-2423.

Port. Phys.
Portugaliae Physica. Sociedade Portuguesa de Fisica, Avenida de Republica, 37-40, P-1000 Lisboa, Portugal. ISSN 0048-4903.

Postepy Astron.
Postepy Astronomii. Czasopismo Poświecone Upowszechnianiu Wiedzy Astronomicznej. Polskie Towarzystwo Astronomiczne, Warszawa. Printed in Poland by Pánstwowe Wydawnictwo Naukowe, Warszawa-Lódź, Poland. ISSN 0032-5414.

Postepy Fiz.
Postepy Fiziyki. Polskie Towarzystwo Fizykczne, 00-681 Warszawa, ul. Hoza 69, Poland. ISSN 0032-5430.

Pramāna
Pramāna. Indian Academy of Sciences, Bangalore 560 006, India. ISSN 0304-4289.

Prepr. Astron. Inst. Univ. Basel
Preprint of the Astronomical Institute of the University of Basel. Astronomisches Institut der Universität Basel, Venusstraße 7, CH-4102 Binningen, Switzerland.

Prepr. Steward Obs.
Preprints of the Steward Observatory. Steward Observatory, The University of Arizona, Tucson, Ariz. 85721, USA.

Priroda
Priroda. Izdatel'stvo 'Nauka', Moskva. Editorial address: 117049, Moskva, GSP-1, Maronovskij per., 26, USSR. ISSN 0032-874X.

Probl. Kosm. Fiz.
Problemy Kosmicheskoj Fiziki. Respublikanskij Mezhvedomstvennyj Nauchnyj Sbornik. Izdatel'stvo pri Kievskom Gosudarstvennom Universitete Izdatel'skogo Obedineniya 'Vishcha Shkola'. Kiev, USSR. ISSN 0555-2796.

Proc. Astron. Soc. Aust.
Proceedings of the Astronomical Society of Australia. Astronomical Society of Australia, Sydney. Subscription address: Division of Radiophysics, PO Box 76, Epping, N.S.W. 2121, Australia. ISSN 0066-9997.

Proc. Cosmic-Ray Res. Lab. Nagoya Univ.
Proceedings of the Cosmic-Ray Research Laboratory of Nagoya University. Cosmic Ray Research Laboratory and Department of Physics, Nagoya University, Chikusa-ku, Nagoya, Japan. ISSN 0250-5444.

Proc. IEEE
Proceedings of the IEEE. The Institute of Electrical and Electronics Engineers, 345 East 47th Street, New York, N.Y. 10017, USA. ISSN 0018-9219.

Proc. Indian Acad. Sci., Earth Planet. Sci.
Proceedings of the Indian Academy of Sciences, Earth and Planetary Sciences. Indian Academy of Sciences, Bangalore 560 080, India. ISSN 0370-0089.

Proc. Indian Acad. Sci., Eng. Sci.
Proceedings of the Indian Academy of Sciences, Engineering Sciences. Indian Academy of Sciences, Bangalore 560 080, India. ISSN 0253-4096.

Proc. Indian Natl. Sci. Acad., Part A
Proceedings of the Indian National Science Academy, Part A. (Physical Sciences). The Indian National Science Academy, Bahadur Shah Zafar Marg, New Delhi 110 002, India. ISSN 0370-0046.

Proc. Int. Latitude Obs. Mizusawa
Proceedings of the International Latitude Observatory of Mizusawa. The International Latitude Observatory of Mizusawa, Mizusawa-Shi, Iwate-Ken, Japan. ISSN 0536-3403.

Proc. Jpn. Acad., Ser. A
Proceedings of the Japan Academy, Series A. (Mathematical Sciences). Japan Academy, Ueno Park, Tokyo 110, Japan. ISSN 0386-2194.

Proc. Jpn. Acad., Ser. B
Proceedings of the Japan Academy, Series B. (Physical and Biological Sciences). Japan Academy, Ueno Park, Tokyo 110, Japan. ISSN 0386-2208.

Proc. K. Ned. Akad. Wet., Ser. B
Proceedings of the Koninklijke Nederlandse Akademie van Wetenschappen, Series B. (Physical Sciences). North-Holland Publishing Company, PO Box 211, 1000 AE Amsterdam, The Netherlands. ISSN 0023-3366.

Proc. Natl. Acad. Sci. India, Sect. A
Proceedings of the National Academy of Sciences of India, Section A (Physical Sciences). National Academy of Sciences of India, 5 Lajpatrai Road, Allahabad-2, India. ISSN 0369-8203.

Proc. Natl. Acad. Sci. U.S.A.
Proceedings of the National Academy of Sciences of the United States of America. National Academy of Sciences, 2101 Constitution Avenue, Washington, D.C. 20418, USA. ISSN 0027-8424.

Proc. R. Soc. Edinb., Sect. A
Proceedings of the Royal Society of Edinburgh, Section A (Mathematical and Physical Sciences). Royal Society of Edinburgh, 22 George Street, Edinburgh EH2 2PQ, Scotland. ISSN 0308-2105.

Proc. R. Soc. London, Ser. A
Proceedings of the Royal Society of London, Series A. (Mathematical and Physical Sciences). The Royal Society, 6 Carlton House Terrace, London, SW1Y 5AG, England. ISSN 0080-4630.

Proc. SPIE Int. Soc. Opt. Eng.
Proceedings of the SPIE – The International Society for Optical Engineering. The International Society for Optical Engineering, PO Box 10, Bellingham, Wash. 98227, USA. ISSN 0277–786X.

Prog. Part. Nucl. Phys.
Progress in Particle and Nuclear Physics. Pergamon Press, New York – Toronto – Paris – Frankfurt – Sydney. ISSN 0146–6410.

Prog. Theor. Phys.
Progress of Theoretical Physics. Research Institute for Fundamental Physics and the Physical Society of Japan. Subscription address: Yukawa Hall, Kyoto University, 606 Kyoto, Japan. ISSN 0033–068X.

Prog. Theor. Phys., Suppl.
Progress of Theoretical Physics, Supplement. Research Institute for Fundamental Physics and the Physical Society of Japan. Subscription address: Yukawa Hall, Kyoto University, 606 Kyoto, Japan. ISSN 0375–9687.

PTB Mitt.
PTB Mitteilungen. Forschen + Prüfen. Amts- und Mitteilungsblatt der Physikalisch-Technischen Bundesanstalt Braunschweig – Berlin. Friedr. Vieweg & Sohn Verlagsgesellschaft mbH, Faulbrunnenstraße 13, D–6200 Wiesbaden, F.R. Germany. ISSN 0030–834X.

Pubbl. Stn. Astron. Int. Latitudine, Carloforte–Cagliari, Nuov. Ser.
Pubblicazioni della Stazione Astronomica Internazionale di Latitudine, Carloforte–Cagliari, Nuova Serie. Stazione Astronomica Internazionale di Latitudine, Carloforte–Cagliari, Italy.

Pubbl. Varie Fuori Ser. Oss. Astron. Torino (Pino Torinese)
Pubblicazioni Varie Fuori Serie dell'Osservatorio Astronomico di Torino (Pino Torinese). Osservatorio Astronomico di Torino, I–10025 Pino Torinese (Torino), Italy.

Publ. Astron. Dep. Eötvös Univ.
Publications of the Astronomy Department of the Eötvös University. Department of Astronomy, Eötvös Lorand University, Budapest, Hungary.

Publ. Astron. Inst. Czech. Acad. Sci.
Publications of the Astronomical Institute of the Czechoslovak Academy of Sciences. Astronomical Institute, Czechoslovak Academy of Sciences, Budečská 6, 120 23 Praha 2, Czechoslovakia.

Publ. Astron. Obs. Sarajevo
Publications of the Astronomical Observatory of Sarajevo. Astronomical Observatory of Sarajevo, Sarajevo, M. Tita 44, Yugoslavia. ISSN 0351–4587.

Publ. Astron. Opservatorije Beogr.
Publikacija Astronomske Opservatorije u Beogradu (Publications de l'Observatoire Astronomique de Beograd). Astronomska Opservatorija u Beogradu, YU–11050 Beograd, Volgina 7, Yugoslavia. ISSN 0373–3742.

Publ. Astron. Soc. Jpn.
Publications of the Astronomical Society of Japan. Astronomical Society of Japan, Tokyo Astronomical Observatory. Mitaka-shi, Tokyo 181, Japan. ISSN 0004–6264.

Publ. Astron. Soc. Pac.
Publications of the Astronomical Society of the Pacific. The Astronomical Society of the Pacific, 1290 24th Avenue, San Francisco, Calif. 94122, USA. ISSN 0004–6280.

Publ. Astron. Soc. "Rudjer Bošković"
Publication of the Astronomical Society "Rudjer Bošković". Astronomical Society "Rudjer Bošković", Beograd, Yugoslavia. ISSN 0506–4295.

Publ. Astrophys. Obs. Potsdam
Publikationen des Astrophysikalischen Observatoriums zu Potsdam. Akademie der Wissenschaften der DDR, Zentralinstitut für Astrophysik. Astrophysikalisches Observatorium Potsdam, DDR–1500 Potsdam, Telegrafenberg, German Democratic Republic.

Publ. Beijing Astron. Obs.
Publications of the Beijing Astronomical Observatory. Beijing Astronomical Observatory, Academia Sinica, Beijing, People's Republic of China.

Publ. Bosscha Obs.
Publications of the Bosscha Observatory. Bandung Institute of Technology, Department of Science, Lembang, Indonesia.

Publ. Debrecen Heliophys. Obs.
Publications of Debrecen Heliophysical Observatory. Hungarian Academy of Sciences, H–4010 Debrecen, Hungary. ISSN 0209–7567.

Publ. Dep. Astron., Univ. Beogr.
Publications of the Department of Astronomy, University of Beograd. Department of Astronomy, University of Beograd, Beograd, Yugoslavia. ISSN 0350–3283.

Publ. Dep. Astron., Univ. Cape Town
Publications of the Department of Astronomy of the University of Cape Town. Astronomy Department, University of Cape Town, Rondebosch, 7700 Cape, South Africa.

Publ. Dep. Astron., Univ. Chile
Publicaciones Departamento de Astronomía, Universidad de Chile. Observatorio Astronomico Nacional, Cerro Calan, Santiago de Chile, Chile.

Publ. Dep. Geod. Astron., Univ. Thessaloniki
Publications of the Department of Geodetic Astronomy, University of Thessaloniki. Department of Geodetic Astronomy, University of Thessaloniki, Thessaloniki, Greece.

Publ. Dom. Astrophys. Obs.
Publications of the Dominion Astrophysical Observatory. Dominion Astrophysical Observatory, National Research Center of Canada, 5071 West Saanich Road, Victoria, B.C. V8X 4M6, Canada. ISSN 0078–6950.

Publ. Ege Univ. Obs.
Publications of Ege University Observatory. Ege University Observatory, P.K. 21, Bornova–Izmir, Turkey.

Publ. Espec. Obs. Astron. Quito
Publicacion Especial del Observatorio Astronomico de Quito. Observatorio Astronomico de Quito, Escuela Politécnica Nacional, Quito, Ecuador.

Publ. Finn. Geod. Inst.
Publications of the Finnish Geodetic Institute. Finnish Geodetic Institute, Helsinki, Finland. ISSN 0085–6932.

Publ. Inst. Geophys., Pol. Acad. Sci., Ser. F (Planet Geod.)
Publications of the Institute of Geophysics, Polish Academy of Sciences, Series F (Planetary Geodesy). Państwowe Wydawnictwo Naukowe, Warszawa–Łódź, Poland. ISSN 0318–0222.

Publ. Int. Latitude Obs. Mizusawa
Publications of the International Latitude Observatory of Mizusawa. The International Latitude Observatory of Mizusawa, Mizusawa-shi, Iwate-ken, Japan. ISSN 0386–0779.

Publ. Istanbul Univ. Obs.
Publications of the Istanbul University Observatory. Istanbul University Observatory, Istanbul, Turkey.

Publ. Lick Obs.
Publications of the Lick Observatory. Lick Observatory, Board of Studies in Astronomy and Astrophysics, University of California at Santa Cruz, Santa Cruz, Calif. 95064, USA.

Publ. Obs. Astron. Nac., Univ. Colombia
Publicaciones del Observatorio Astronomico Nacional. Observatorio Astronomico Nacional, Universidad Nacional de Colombia, Apartado Aéreo 2584, Bogotá, Colombia.

Publ. Obs. Astron. "Prof. Manuel de Barros", Fac. Cienc. Porto
Publicações do Observatório Astronómico "Prof. Manuel de Barros", da Faculdade de Ciências do Porto. Observatório Astronómico "Prof. Manuel de Barros", Universidade do Porto, Monte da Virgem – Vila Nova de Gaia, Portugal.

Publ. Obs. Astron. Univ. Cluj
Publicatiile Observatorului Astronomic al Universitatii din Cluj. Astronomical Observatory, University Babes–Bolyai, Cluj–Napoca, Romania.

Publ. Obs. Bordeaux, Nouv. Sér.
Publications de l'Observatoire de Bordeaux, Nouvelle Série. Observatoire de l'Université de Bordeaux I, F–33270 Floirac, France.

Publ. Obs. Ebro
Publicaciones del Observatorio del Ebro (Miscelanea). Consejo Superior de Investigaciones Cientificas, Observatorio del Ebro, Roquetas (Tarragona), Spain.

Publ. Obs. Genève, Sér. A
Publications de l'Observatoire de Genève, Série A. Observatoire de Genève, CH–1290 Sauverny, Switzerland.

Publ. Obs. Genève, Sér. B
Publications de l'Observatoire de Genève, Série B. Observatoire de Genève, CH–1290 Sauverny, Switzerland.

Publ. Purple Mt. Obs.
Publications of the Purple Mountain Observatory. Purple Mountain Observatory, Academia Sinica, Nanking, People's Republic of China.

Publ. R. Obs., Edinb.
Publications of the Royal Observatory, Edinburgh. Science Research Council. Royal Observatory, Blackford Hill, Edinburgh EH9 3HJ, Scotland. ISSN 0305–2001.

Publ. Semin. Astron. Geod. Univ. Complutense Madr.
Publicaciones del Seminario de Astronomia y Geodesia de la Universidad Complutense de Madrid. Seminario de Astronomia y Geodesia, Facultad de Ciencias Matemáticas, Universidad Complutense, Madrid, Spain.

Publ. Shaanxi Astron. Obs.
Publications of the Shaanxi Astronomical Observatory. Shaanxi Astronomical Observatory, Academia Sinica, Lintong, Xian, People's Republic of China.

Publ. Spéc. Cent. Données Stellaires
Publication Spéciale du Centre de Données Stellaires. Observatoire de Strasbourg, 11, rue de l'Université, F–67000 Strasbourg, France.

Publ. U.S. Nav. Obs., Second Ser.
Publications of the United States Naval Observatory, Second Series. U.S. Government Printing Office, Washington, D.C. 20402, USA.

Publ. Variable Star Sect., R. Astron. Soc. N.Z.
Publications of Variable Star Section, Royal Astronomical Society of New Zealand. Astronomical Research Ltd., PO Box 3093, Greerton, Tauranaga, New Zealand.

Publ. Warner Swasey Obs.
Publications of the Warner and Swasey Observatory. Warner and Swasey Observatory, Case Western Reserve University, Department of Astronomy, Smith Bldg., Cleveland, Ohio 44106, USA. ISSN 0160–2500.

Publ. Yunnan Obs.
Publications of Yunnan Observatory. Yunnan Observatory, Academia Sinica, PO Box 110, Kunming, Yunnan Province, People's Republic of China.

Pure Appl. Geophys.
Pure and Applied Geophysics. Birkhäuser Boston Inc., 380 Green Street, Cambridge, Mass. 02139, USA. ISSN 0033–4553.

Q. Appl. Math.
Quarterly of Applied Mathematics. American Mathematical Society, PO Box 1571, Providence, R.I. 02901, USA. ISSN 0033–569X.

Q. Bull. Sol. Act.
Quarterly Bulletin on Solar Activity. International Astronomical Union. Published by the Tokyo Astronomical Observatory, University of Tokyo, Mitaka, Tokyo 181, Japan.

Q. J. R. Astron. Soc.
The Quarterly Journal of the Royal Astronomical Society. Published for the Royal Astronomical Society by Blackwell Scientific Publications, Oxford – London – Edinburgh – Boston – Melbourne. ISSN 0035–8738.

Q. J. R. Meteorol. Soc.
Quarterly Journal of the Royal Meteorological Society. Cromwell House, High Street, Bracknell, Berks., England. ISSN 0035–9009.

R. Greenwich Obs., Annu. Rep.
Royal Greenwich Observatory, Annual Report. Royal Greenwich Observatory, Herstmonceux Castle, Hailsham, East Sussex BN27 1RP, England. ISSN 0308–3322.

R. Greenwich Obs., Bull.
Royal Greenwich Observatory, Bulletins. Royal Greenwich Observatory, Herstmonceux Castle, Hailsham, East Sussex BN27 1RP, England. ISSN 0308–5074.

R. Obs. Ann.
Royal Observatory Annals. Royal Greenwich Observatory, Herstmonceux Castle, Hailsham, East Sussex BN27 1RP, England. ISSN 0080–4371.

R. Obs. Edinb., Res. Facilities
Royal Observatory Edinburgh, Research and Facilities. Royal Observatory, Blackford Hill, Edinburgh EH9 3HJ, Scotland.

Radiant
Radiant. Journal of the Dutch Meteor Society. Federation of European Meteor Astronomers (FEMA), Morssingel 35a, 2312 AZ Leiden, The Netherlands.

Radiat. Eff.
Radiation Effects. Gordon and Breach Science Publishers Inc., New York – London – Paris. ISSN 0033–7579.

Radio Sci.
Radio Science. American Geophysical Union, 2000 Florida Avenue, N.W., Washington, D.C. 20009, USA. ISSN 0048–6604.

Radiochem. Radioanal. Lett.
Radiochemical and Radioanalytical Letters. Elsevier Sequoia S.A., PO Box 851, CH–1001 Lausanne, Switzerland. ISSN 0079–9483.

Rapp. Act. Obs. Paris
Rapport d'Activité de l'Observatoire de Paris. Observatoire de Paris, F–92190 Meudon, France.

Rech. Aerosp.
Recherche Aerospatiale. Office National d'Etudes et de Recherches Aerospatiales, 29 Avenue de la Division Leclerc, F–92320 Chatillon, France. ISSN 0034–1223.

Recherche
Recherche. Société d'Editions Scientifiques, 57, rue de Seine, F–75006 Paris, France. ISSN 0029–5671.

Ref. Zh., 51. Astron.
Referativnyj Zhurnal, 51. Astronomiya. Gosudarstvennyj Komitet SSSR po Nauke i Tekhnike. Vsesoyuznyj Institut Nauchnoj i Tekhnicheskoj Informatsii. Akademiya Nauk SSSR. Moskva. 125219, Moskva, A–219, Baltijskaya ul., 14, USSR. ISSN 0486–2236.

Ref. Zh., 52. Geod. Aehrosemka
Referativnyj Zhurnal, 52. Geodeziya i Aehrosemka. Gosudarstvennyj Komitet SSSR po Nauke i Tekhnike. Vsesoyuznyj Institut Nauchnoj i Tekhnicheskoj Informatsii. Akademiya Nauk SSSR. Moskva. 125219, Moskva, A–219, Baltijskaya ul., 14, USSR. ISSN 0375–9717.

Ref. Zh., 62. Issled. Kosm. Prostranstva
Referativnyj Zhurnal, 62. Issledovanie Kosmicheskogo Prostranstva. Gosudarstvennyj Komitet SSSR po Nauke i Tekhnike. Vsesoyuznyj Institut Nauchnoj i Tekhnicheskoj Informatsii. Akademiya Nauk SSSR. Moskva. 125219, Moskva, A–219, Baltijskaya ul., 14, USSR. ISSN 0034–2408.

Rep. Finn. Geod. Inst.
Reports of the Finnish Geodetic Institute. Finnish Geodetic Institute, Helsinki, Finland. ISSN 0355–1962.

Rep. Inst. Phys. Chem. Res.
Reports of the Institute of Physical and Chemical Research. Institute of Physical and Chemical Research, Rikagaku Kenkyushu, Wako–shi, Saitama 351, Japan. ISSN 0020–3084.

Rep. Obs. Lund
Reports from the Observatory of Lund. Institutionen för Astronomi, Lunds Universitet, Box 1107, 22104 Lund, Sweden. ISSN 0349–4217.

Rep. Prog. Phys.
Reports on Progress in Physics. Published by the Institute of Physics, 47 Belgrave Square, London SW1X 8QX, England. ISSN 0034–4885.

Rep. Ser., Dep. Phys. Sci., Univ. Turku
Report Series, Department of Physical Sciences, University of Turku. Department of Physical Sciences, University of Turku, SF–20500 Turku 50, Finland. ISSN 0356–9896.

Repr. Theor. Astrophys. Group, Dep. Math., UMIST
Reprints from the Theoretical Astrophysics Group, Department of Mathematics, UMIST. Department of Mathematics, UMIST, PO Box 88, Manchester M6O 1QD, England.

Res. Bull. Meisei Univ., Phys. Sci. Eng.
Research Bulletin of Meisei University, Physical Sciences and Engineering. Meisei University, Hino–Shi, Tokyo, Japan. ISSN 0388–130X.

Res. Cent. Astron. Appl. Math., Acad. Athens, Contrib. Ser. I
Research Center for Astronomy and Applied Mathematics, Academy of Athens, Contributions Series I (Astronomy). Research Center for Astronomy and Applied Mathematics, Academy of Athens, Athens, Greece.

Res. Lab. Electron. Onsala Space Obs., Res. Rep.
Research Laboratory of Electronics and Onsala Space Observatory, Research Report. Chalmers University of Technology, Research Laboratory of Electronics, Fack, S–40220 Göteborg 5, Sweden.

Rev. Astron.
Revista Astronomica. Organo de la Asociación Argentina Amigos de la Astronomia. Asociación Argentina Amigos de la Astronomia, C. C. 369–C. Central, 1000–Buenos Aires, Argentina. ISSN 0044–9253.

Rev. Bras. Fis.
Revista Brasileira de Fisica. Sociedade Brasileira de Fisica, Caixa Postal 20553, Sao Paulo SP, Brazil. ISSN 0374–4922.

Rev. Fac. Sci. Univ. Istanbul, Ser. C
Review of Faculty of Science University of Instanbul, Serie C. (Üniversitesi fen Fakültesi Mecmuasi, Seri C). Fen Fakültesi Basimevi, Istanbul, Turkey. ISSN 0444–7298.

Rev. Geophys.
Reviews of Geophysics. American Geophysical Union, 2000 Florida Avenue, N.W., Washington, D.C. 20009, USA. ISSN 8755–1209. Formerly: Rev. Geophys. Space Phys.

Rev. Hist. Sci.
Revue d'Histoire des Sciences. Presses Universitaires de France, Paris, France. ISSN 0151–4105.

Rev. Mex. Astron. Astrofis.
Revista Mexicana de Astronomia y Astrofisica. Instituto de Astronomia, Universidad Nacional Autónoma de México, Apartado Postal 70–264, Mexico 20, D.F., Mexico. ISSN 0185–1101.

Rev. Mex. Fis.
Revista Mexicana de Fisica. Sociedad Mexicana de Fisica, Apartado Postal No. 20–364, Mexico 20, D.F., Mexico. ISSN 0035–001X.

Rev. Mod. Phys.
Reviews of Modern Physics. Published for the American Physical Society by the American Institute of Physics, 335 East 45th Street, New York, N.Y. 10017, USA. ISSN 0034–6861.

Rev. Phys. Appl.
Revue de Physique Appliquée. Editions de Physique, Z.I. de Courtabœuf, B.P. 112, F–91944 Les Ulis Cedex, France. ISSN 0035–1687.

Rev. Quest. Sci.
Revue des Questions Scientifiques. Société Scientifique de Bruxelles, 61, rue de Bruxelles, B–5000 Namur, Belgium. ISSN 0035–2160.

Rev. Radio Res. Lab.
Review of the Radio Research Laboratories. Ministry of Posts & Telecommunications, Nukui–Kitamachi, Koganei–shi, Tokyo 184, Japan. ISSN 0033–801X.

Rev. Roum. Phys.
Revue Roumaine de Physique. Academie Republicii Populare Romine, B.P. 134–135, Bucuresti, Rumania. ISSN 0035–4090.

Rev. Roum. Sci. Tech., Sér. Mec. Appl.
Revue Roumaine des Sciences Techniques, Série de Mécanique Appliquée. Academia R.S.R., Str. Constantin Mille 15, Bucuresti, Rumania. ISSN 0035–4074.

Rev. Sci. Instrum.
Review of Scientific Instruments. American Institute of Physics, 335 East 45th Street, New York, N.Y. 10017, USA. ISSN 0034–6748.

Rezul't. Nablyud. Iskusstv. Sputnikov Zemli
Rezul'taty Nablyudenij Iskusstvennykh Sputnikov Zemli. Published by Astronomicheskij Sovet Akademii Nauk SSSR, Ryazanskij Gosudarstvennyj Pedagogicheskij Institut, Ryazan'. 390000 GSP, Ryazan, ul. Svobody, 46, USSR. ISSN 0131–8586.

Ric. Astron.
Ricerche Astronomiche. Specola Vaticana, I–00120 Città del Vaticano.

Rijksuniv. Gent, Sterrenkundig Obs., Meded.
Rijksuniversiteit Gent, Sterrenkundig Observatorium, Mededeling. Astronomical Observatory, State University of Ghent, Krijgslaan 271, B–9000 Ghent, Belgium.

Říše hvězd
Říše hvězd. Czech popular astronomical journal. Panorama, Praha, Czechoslovakia.

Riv. Nuovo Cimento
La Rivista del Nuovo Cimento della Società Italiana di Fisica. Società Italiana di Fisica, Editrice Compositori, viale XII Giugno 1, I–40124 Bologna, Italy. ISSN 0035–5917.

Rutherford Appleton Lab., Rep.
Rutherford Appleton Laboratory, Report. Rutherford Appleton Laboratory, Chilton, Didcot, Oxfordshire OX11 0QX, England.

SAAO Newsl.
SAAO Newsletter. South African Astronomical Observatory, PO Box 9, Observatory 7935, Cape, South Africa.

Sac Peak Update
Sac Peak Update. Sacramento Peak Observatory, Sunspot, N.M. 88349, USA.

S.Afr. Astron. Obs., Annu. Rep.
South African Astronomical Observatory, Annual Report. South African Astronomical Observatory, PO Box 9, Observatory, 7935 Cape, South Africa. ISSN 0250–0671.

S.Afr. Astron. Obs., Circ.
South African Astronomical Observatory, Circulars. South African Astronomical Observatory, PO Box 9, Observatory, 7935 Cape, South Africa.

S.Afr. J. Phys.
South African Journal of Physics. Bureau for Scientific Publications, PO Box 1758, Pretoria 0001, South Africa. ISSN 0379–4377.

Schweiz. Tech. Z.
Schweizerische Technische Zeitschrift. Schweizerischer Technischer Verband, Weinbergstraße 41, CH–8023 Zürich, Switzerland. ISSN 0040–151X.

Sci. Am.
Scientific American. Scientific American, Inc., 415 Madison Avenue, New York, N.Y. 10017, USA. ISSN 0036–8733.

Sci. Atmos. Sin.
Scientia Atmospherica Sinica. Science Press, Peking. Subscription address: Guozi Shudian, PO Box 399, Beijing, People's Republic of China.

Sci. Dimension
Science Dimension. National Research Council of Canada, Ottawa K1A OR6, Canada. ISSN 0036–830X.

Sci. Pap. Inst. Phys. Chem. Res.
Scientific Papers of the Institute of Physical and Chemical Research. Rikagaku Kenkyusho, Wako–shi, Saitama 351, Japan. ISSN 0020–3092.

Sci. Prog.
Science Progress. Blackwell Scientific Publications, Oxford – London – Edinburgh – Boston – Melbourne. ISSN 0036–8504.

Sci. Rep. Tôhoku Univ., Eighth Ser.
The Science Reports of the Tôhoku University, Eighth Series (Physics and Astronomy). Faculty of Science, Tôhoku University, Sendai 980, Japan. ISSN 0388–5607.

Sci. Sin., Ser. A
Scientia Sinica, Series A (Mathematical, Physical, Astronomical and Technical Sciences). Academia Sinica, Beijing. Science Press, No. 137, Chaoyangmennei Street, Beijing, People's Republic of China. Subscription address: Scientific and Technical Books Service Ltd., PO Box 197, London WC2N 4DE, England. ISSN 0253–5831.

Science
Science. American Association for the Advancement of Science, 1515 Massachusetts Avenue, N.W., Washington, D.C. 20005, USA. ISSN 0036–8075.

Scr. Fac. Sci. Nat. Univ. Purkynianae Brun., Phys.
Scripta Facultatis Scientiarum Naturalium Universitatis Purkynianae Brunensis, Physica. University J.E. Purkyne, 61137 Brno–Kotlarská 2, Czechoslovakia. ISSN 0231–6129.

Semin. Astron. Geod., Univ. Madr., Publ.
Seminario de Astronomia y Geodesia, Universidad Complutense, Facultad de Ciencias Matemáticas, Madrid, Publicación. Universidad Complutense, Madrid, Spain.

Sendai Astron. Rap.
Sendai Astronomiaj Raportoj. Astronomical Institute, Tôhoku University, Sendai 980, Japan.

Shaanxi Astron. Obs., Repr.
Shaanxi Astronomical Observatory, Reprints. Shaanxi Astronomical Observatory, Academia Sinica, Lintong, Xian, People's Republic of China.

Shanghai Obs., Acad. Sin., Prepr.
Shanghai Observatory, Academia Sinica, Preprint. Shanghai Observatory, Academia Sinica, 80 Nandan Road, Shanghai, People's Republic of China.

Sitzungsber. Akad. Wiss. DDR, Math. Naturwiss. Tech.
Sitzungsberichte der Akademie der Wissenschaften der DDR, Mathematik – Naturwissenschaften – Technik. Akademie–Verlag, DDR–1086 Berlin, Leipziger Straße 3–4, German Democratic Republic. ISSN 0138–3956.

Sitzungsber. Heidelb. Akad. Wiss.
Sitzungsberichte der Heidelberger Akademie der Wissenschaften. Mathematisch–Naturwissenschaftliche Klasse. Springer–Verlag, Berlin – Heidelberg – New York – Tokyo. ISSN 0371–0165.

Sitzungsber., Österr. Akad. Wiss., Math.–Naturwiss. Kl. Abt. II
Sitzungsberichte, Österreichische Akademie der Wissenschaften, Mathematisch–Naturwissenschaftliche Klasse, Abteilung II (Mathematik, Astronomie, Physik, Meteorologie und Technik). Springer–Verlag, Wien, Austria. ISSN 0029–8816.

Sky Telesc.
Sky and Telescope. Sky Publishing Corporation, 49 Bay State Rd., Cambridge, Mass. 02238–1290, USA. ISSN 0037–6604.

Smithson. Astrophys. Obs., Spec. Rep.
Smithsonian Astrophysical Observatory, Special Report. Available from the Publications Division, Distribution Section, Smithsonian Astrophysical Observatory, Cambridge, Mass. 02138, USA.

Sol. Energy
Solar Energy. Pergamon Press, Oxford – New York – Toronto – Paris – Frankfurt – Sydney. ISSN 0038–092X.

Sol. Maps Act.
Solar Maps and Activity. Manila Observatory, Solar Division, Manila, Philippines.

Sol. Phys.
Solar Physics. A Journal for Solar Research and the Study of Solar Terrestrial Physics. D. Reidel Publishing Company, Dordrecht – Boston – Lancaster – Tokyo. ISSN 0038–0938.

Sol. Radio Data
Solar Radio Data. OSRA, Observatorium für solare Radioastronomie, Zentralinstitut für Astrophysik, Akademie der Wissenschaften der DDR, DDR–1501 Tremsdorf/Potsdam, German Democratic Republic.

Sol. Syst. Res.
Solar System Research. Translation of Astron. Vestn. Consultants Bureau, 227 West 17th Street, New York, N.Y. 10011, USA. ISSN 0038–0946.

Sol. Terr. Environ. Res. Jpn.
Solar Terrestrial Environmental Research in Japan. Institute of Space and Aeronautical Science, University of Tokyo, 4–6–1, Komaba, Meguro–ku, Tokyo 153, Japan. ISSN 0386–5444.

Solid State Phys.
Solid State Physics. Agne Gijutsu Center, Kitamura Building, 5–1–25, Minamiayama, Minatoku, Tokyo, Japan. ISSN 0454–4544.

Soln. Dannye, Byull.
Solnechnye Dannye, Byulleten' (Solar Data). Akademiya Nauk SSSR. Astronomicheskij Sovet i Glavnaya Astronomicheskaya Observatoriya. Izdatel'stvo 'Nauka', Leningradskoe Otdelenie, Leningrad. 199164 Leningrad, V–164, Mendeleevskaya l. 1, USSR. ISSN 0552–5829.

Sonne
Sonne. Mitteilungsblatt der Amateursonnenbeobachter. Wilhelm–Foerster–Sternwarte, Munsterdamm 90, D–1000 Berlin 41, F.R. Germany. ISSN 0721–0094.

Soobshch. Akad. Nauk Gruz. SSR
Soobshcheniya Akademii Nauk Gruzinskoj SSR. Akademiya Nauk Gruzinskoj SSR. ISSN 0002–3167.

Soobshch. Byurakan. Obs.
Soobshcheniya Byurakanskoj Observatorii. Akademiya Nauk Armyanskoj SSR, Erevan. 375019 Erevan, Barekamutyan, 24, USSR. ISSN 0370–8691.

Soobshch. Gos. Astron. Inst. Shternberg
Soobshcheniya Gosudarstvennogo Astronomicheskogo Instituta im. P.K. Shternberga. Moskovskij Gosudarstvennyj Universitet im. M.V. Lomonosova. Izdatel'stvo Moskovskogo Universiteta, Moskva. Moskva, K–9, ul. Gertsena, 5/7, USSR. ISSN 0038–1489.

Soobshch. Spets. Astrofiz. Obs.
Soobshcheniya Spetsial'noj Astrofizicheskoj Observatorii. Akademiya Nauk SSSR. Izdanie Spetsial'noj Astrofizicheskoj Observatorii, USSR.

South. Stars
Southern Stars. Journal of the Royal Astronomical Society of New Zealand (Inc.). Royal Astronomical Society of New Zealand, PO Box 3181, Wellington, New Zealand. ISSN 0049–1640.

Sov. Astron.
Soviet Astronomy. A translation of Astron. Zh. American Institute of Physics, 335 East 45th Street, New York, N.Y. 10017, USA. ISSN 0038–5301.

Sov. Astron. Lett.
Soviet Astronomy Letters. A translation of Pis'ma Astron. Zh. American Institute of Physics, 335 East 45th Street, New York, N.Y. 10017, USA. ISSN 0360–0327.

Space Educ.
Space Education. A Publication of The British Interplanetary Society. The British Interplanetary Society, 27/29 South Lambeth Road, London SW8 1SZ, England. ISSN 0621–1813.

Space Sci. Rev.
Space Science Reviews. D. Reidel Publishing Company, Dordrecht – Boston. ISSN 0038–6308.

Space Telesc. Sci. Inst., Instrum. Sci. Rep.
Space Telescope Science Institute, Instrument Science Report. Space Telescope Science Institute, 3700 San Martin Drive, Homewood Campus, Baltimore, Md. 21218, USA.

Space Telesc. Sci. Inst., Newsl.
Space Telescope Science Institute, Newsletter. Space Telescope Science Institute, 3700 San Martin Drive, Homewood Campus, Baltimore, Md. 21218, USA.

Space Telesc. Sci. Inst., Prepr. Ser.
Space Telescope Science Institute, Preprint Series. Space Telescope Science Institute, 3700 San Martin Drive, Homewood Campus, Baltimore, Md. 21218, USA.

Spaceflight
Spaceflight. A Publication of The British Interplanetary Society. The British Interplanetary Society, 27/29 South Lambeth Road, London SW8 1SZ, England. ISSN 0038–6340.

Spectrosc. Lett.
Spectroscopy Letters. Marcel Dekker Inc., 270 Madison Avenue, New York, N.Y. 10016, USA. ISSN 0038–7010.

Speculations Sci. Technol.
Speculations in Science and Technology. Elsevier Sequoia, S.A., PO Box 851, CH–1001 Lausanne 1, Switzerland. ISSN 0155–7785.

ST–ECF Newsl.
Space Telescope–European Coordinating Facility Newsletter. Space Telescope–European Facility, European Southern Observatory, Karl–Schwarzschild–Straße 2, D–8046 Garching bei München, F.R. Germany.

Stand. Star Newsl.
Standard Star Newsletter. Working Group on Standard Stars (IAU Commissions 29, 30, 45). Dipartimento di Fisica, Università degli Studi di Milano, Milano, Italy.

Sterne
Die Sterne. Zeitschrift für alle Gebiete der Himmelskunde. Johann Ambrosius Barth, DDR–7010 Leipzig, Salomonstraße 18b, German Democratic Republic. ISSN 0039–1255.

Sterne Weltraum
Sterne und Weltraum. Astronomische Monatsschrift. Verlag Sterne und Weltraum Dr. Vehrenberg GmbH, Portiastraße 10, D–8000 München 90, F.R. Germany, ISSN 0039–1263.

Sternenbote
Der Sternenbote. Österreichische Astronomische Monatsschrift. Astronomisches Büro, Hasenwartgasse 32, A–1238 Wien, Austria. ISSN 0039–1271.

Stockholms Obs., Rep.
Stockholms Observatorium, Report. Stockholms Observatorium, S–13300 Saltsjöbaden, Sweden.

Strolling Astron.
The Strolling Astronomer. The Journal of The Association of Lunar and Planetary Observers. The Strolling Astronomer, Box 3 AZ, University Park, N.M. 88003, USA. ISSN 0039–2502.

Stud. Cercet. Fiz.
Studii si Cercetari de Fizica. Academia Republicii Populare Romine, PO Box 134–5, Calca Victoriei 126, Bucuresti, Rumania, ISSN 0039–3940.

Stud. Geophys. Geod.
Studia Geophysica et Geodaetica. Geophysical Institute of the Czechoslovak Academy of Sciences. Published by Academia, Prague, Czechoslovakia.

Stud. Hist. Philos. Sci.
Studies in History and Philosophy of Science. Pergamon Press, Oxford – New York – Toronto – Paris – Frankfurt – Sydney. ISSN 0039–3681.

Stud. Soc. Sci. Torun., Sect. F
Studia Societatis Scientiarum Torunensis, Sectio F (Astronomia). Państwowe Wydawnictwo Naukowe, Warszawa – Poznań – Toruń, Poland. ISSN 0082–5573.

Surv. Geophys.
Surveys in Geophysics. An International Review Journal of Geophysics and Planetary Sciences. D. Reidel Publishing Company, Dordrecht – Boston – Lancaster – Tokyo. ISSN 0169–3298. Formerly entitled Geophys. Surv.

Surv. High Energy Phys.
Surveys in High Energy Physics. Harwood Academic Publishers GmbH, Poststraße 22, CH–7000 Chur, Switzerland. ISSN 0142–2413.

Swarthmore Astron. Repr.
Swarthmore Astronomical Reprints. Sproul Observatory, Swarthmore College, Swarthmore, Pa. 19081, USA.

Syd. Obs. Pap.
Sydney Observatory Papers. School of Physics, University of Sydney, Sydney, N.S.W. 2006, Australia.

Syst. Int.
Systems International. IPC Electrical Electronic Press Ltd., Quadrant House, The Quadrant, Sutton, Surrey SM2 5AS, England. ISSN 0309–1171.

Tartu Astrofüüs. Obs. Publ.
W. Struve nimelise Tartu Astrofüüsika Observatooriumi, Publikatsioonid. Eesti NSV Teaduste Akadeemia, Tartu.

Tartu Astrofüüs. Obs. Teated
Tartu Astrofüüsika Observatoorium Teated. Eesti NSV Teaduste Akadeemia W. Struve nim. Tartu Astrofüüsika Observatoorium. Valgus, Tallinn.

Tech. Mess. – TM
Technisches Messen – TM. Archiv für Technisches Messen. R. Oldenbourg Verlag GmbH, Rosenheimer Str. 145, D–8000 München 80, F.R. Germany. ISSN 0171–8096.

Tech. Mitt. Krupp Forschungsber.
Technische Mitteilungen Krupp Forschungsberichte. Fachbücherei Krupp, Postfach 917, D–4300 Essen 1, F.R. Germany. ISSN 0494–9382.

Tectonophysics
Tectonophysics. International Journal of Geotectonics and the Geology and Physics of the Interior of the Earth. Elsevier Publishing Company, PO Box 211, 1000 AE Amsterdam, The Netherlands. ISSN 0040–1951.

Telecommun. J.
Telecommunication Journal. (English Edition). International Telecommunications Union. Place des Nations, CH–1221 Genève 20, Switzerland. ISSN 0497–137X.

Tellus, Ser. B
Tellus, Series B. (Chemical and Physical Meteorology). Swedish Geophysical Society, Arrhenius Laboratoriet, S–10691 Stockholm, Sweden. ISSN 0280–6509.

Tex. J. Sci.
Texas Journal of Science. The Talley Press, San Angelo, Tex., USA. ISSN 0040–4403.

THEOCHEM
THEOCHEM. Elsevier Scientific Publishing Company, PO Box 330, 1000 AH Amsterdam, The Netherlands. ISSN 0166–1280.

Theor. Pap.
Theoretic Papers. The Blindern Theoretic Research Team, P.B. 1029, Blindern, Oslo 3, Norway.

Time Freq. Serv., Bull.
Time and Frequency Services, Bulletin. Shaanxi Astronomical Observatory, Academia Sinica, Lintong, Xian, People's Republic of China.

Time Serv. Bull.
Time Service Bulletin. Osservatorio Astronomico di Torino, I–10025 Pino Torinese (Torino), Italy.

Tôhoku Geophys. J.
Tôhoku Geophysical Journal. Science Reports of the Tôhoku University, Fifth Series. Faculty of Science, Tôhoku University, Sendai 980, Japan. ISSN 0040–8794.

Tokyo Astron. Bull., Second Ser.
Tokyo Astronomical Bulletin, Second Series. Tokyo Astronomical Observatory, University of Tokyo, Mitaka, Tokyo 181, Japan. ISSN 0082–4690.

Tokyo Astron. Obs., Kiso Inf. Bull.
Tokyo Astronomical Observatory, Kiso Information Bulletin. Tokyo Astronomical Observatory, University of Tokyo, Mitaka, Tokyo 181, Japan. ISSN 0286–1380.

Tokyo Astron. Obs. Rep.
Tokyo Astronomical Observatory Report. Tokyo Astronomical Observatory, University of Tokyo, Mitaka, Tokyo 181, Japan. ISSN 0374–4639.

Tokyo Astron. Obs. Repr.
Tokyo Astronomical Observatory Reprints. Tokyo Astronomical Observatory, University of Tokyo, Mitaka, Tokyo 181, Japan. ISSN 0082–4712.

Tokyo Astron. Obs., Time Latitude Bull.
Tokyo Astronomical Observatory, Time and Latitude Bulletins. Tokyo Astronomical Observatory, University of Tokyo, Mitaka, Tokyo 181, Japan. ISSN 0388–3701.

Top. Astrophys. Astron. Space Sci.
Topics in Astrophysics, Astronomy, and Space Sciences. Centre for Astronomy and Space Sciences, Central Institute of Physics. CIP Press, Cutitul de Argint 5, 75 212 Bucharest, Romania.

Tr. Astrofiz. Inst. Alma–Ata
Trudy Astrofizicheskogo Instituta. Akademiya Nauk Kazakhskoj SSR. Izdatel'stvo Nauka Kazakhskoj SSR, Alma–Ata. 480021, Alma–Ata, ul. Shevchenko, 28, USSR.

Tr. Astron. Obs., Leningrad
Trudy Astronomicheskoj Observatorii. Uchenye Zapiski Gosudarstvennogo Universiteta im. A.A. Zhdanova, Seriya matematicheskikh nauk = Izdatel'stvo Leningradskogo Gosudarstvennogo Universiteta A.A. Zhdanova, Leningrad. 199164, Leningrad V–164, Universitetskij nab., 7/9, USSR. ISSN 0136–8109; ISSN 0136–8141.

Tr. Glav. Astron. Obs. Pulkovo, Ser. 2
Trudy Glavnoj Astronomicheskoj Observatorii v Pulkove, Seriya 2. Akademiya Nauk SSSR. Leningradskoe Otdelenie, Nauka, Leningrad. 199164, Leningrad V–164, Mendeleevskaya l., 1, USSR.

Tr. Gos. Astron. Inst. Shternberg
Trudy Gosudarstvennogo Astronomicheskogo Instituta im. P.K. Shternberga. Izdatel'stvo Moskovskogo Universiteta, Moskva, USSR. ISSN 0371–6791.

Tr. Inst. Teor. Astron., Leningrad
Trudy Instituta Teoreticheskoj Astronomii, Akademiya Nauk SSSR. Izdatel'stvo Nauka, Leningrad. 199164, Leningrad V–164, Mendeleevskaya l., 1, USSR. ISSN 0568–6016.

Tr. Kazan. Gorod. Astron. Obs.
Trudy Kazanskoj Gorodskoj Astronomicheskoj Observatorii. Izdatel'stvo Kazanskogo Universiteta, Kazan. Kazan, ul. Lenina, 2, USSR. ISSN 0371–8247.

Tr. Tashkent. Astron. Obs.
Trudy Ordena Trudovogo Krasnogo Znameni Astronomicheskogo Instituta Akademii Nauk Uzbekskoj SSR. Izdatel'stvo FAN Uzbekskoj SSR, Tashkent. Tashkent, ul. Gololya, 70, USSR.

Trans. Am. Nucl. Soc.
Transactions of the American Nuclear Society. American Nuclear Society, 555 North Kensington Avenue, La Grange Park, Ill. 60525, USA. ISSN 0003–018X.

Trans. Astron. Obs. Yale Univ.
Transactions of the Astronomical Observatory of Yale University. Yale University Observatory, 260 Whitney Avenue, New Haven, Conn. 06511, USA.

Trans. Bose Res. Inst.
Transactions of the Bose Research Institute. Bose Research Institute, 93/1 Acharya Prafulla Chandra Road, Calcutta 9, India. ISSN 0006–7903.

Trans. IAU
Transactions of the International Astronomical Union. Published on behalf of the IAU by D. Reidel Publishing Company, Dordrecht – Boston – London.

Trans. Inst. Electron. Commun. Eng. Jpn., Part C
Transactions of the Institute of Electronic and Communication Engineers of Japan, Part C. Denshi Tsushin Gakkai, Kikai–Shinko–Kaikan, 5–8 Shibakoen 3 Chome, Minato–ku, Tokyo 105, Japan. ISSN 0373–6113.

Transp. Theory Stat. Phys.
Transport Theory and Statistical Physics. Marcel Dekker Inc., 270 Madison Avenue, New York, N.Y. 10016, USA. ISSN 0041–1450.

Tsirk. Astron. Inst., Tashkent
Tsirkulyar Astronomicheskogo Instituta, Tashkent. Akademiya Nauk Uzbekskoj SSR. Izdatel'stvo FAN Uzbekskoj SSR, Tashkent. Tashkent, ul. Gololya, 70, USSR. ISSN 0373–7675.

Tsirk., Astron. Obs. L'vov
Tsirkulyar, Astronomicheskaya Observatoriya L'vov. L'vovskij Ordena Lenina Gosudarstvennyj Universitet im. Ivana Franko. Izdatel'stvo L'vovskogo Universiteta, L'vov. 290005, L'vov–5, ul. Lomonosova, 8, USSR. ISSN 0374–0722.

Tsirk. Shemakh. Astrofiz. Obs.
Tsirkulyar Shemakhinskoj Astrofizicheskoj Observatorii. Akademiya Nauk Azerbajdzhanskoj SSR. Izdatel'stvo Ehlm, Baku. 370143 Baku–143, prospekt Narimanova, 31. ISSN 0135–0420.

Turku Univ. Obs., Informeto
Turku University Observatory, Informeto. Turku University Observatory, Tuorla, SF–2150 Piikkiö, Finland.

Turku Univ. Obs., Informo
Turku University Observatory, Informo. Turku University Observatory, Tuorla, SF–21500 Piikkiö, Finland.

UKIRT Rep.
UKIRT Report. United Kingdom Infrared Telescope, 900 Leilani Street, Hilo, Hawaii 96720, USA. ISSN 0260–9983.

Umschau
Die Umschau. Das Wissenschaftsmagazin. Postfach 110262, Stuttgarter Straße 18–24, D–6000 Frankfurt am Main 11, F.R. Germany. ISSN 0722–8562.

Univ. Barc., Dep. Fis. Tierra Cosmos, Publ.
Universidad de Barcelona, Departamento de Fisica de la Tierra y del Cosmos, Publicacion. Departamento de Fisica de la Tierra y del Cosmos, Universidad de Barcelona, Spain.

Univ. Hannover, Astron. Stn., Veröff.
Universität Hannover, Astronomische Station, Veröffentlichungen. Universität Hannover, Astronomische Station, Hannover, F.R. Germany.

Univ. Obs., St. Andrews, Repr.
University Observatory, St. Andrews, Reprint. University Observatory, Buchanan Gardens, St. Andrews, Fife KY16 9LZ, Scotland.

Univ. Tex., Monogr. Astron.
The University of Texas, Monographs in Astronomy. Department of Astronomy, University of Texas, Austin, Tex. 78712, USA.

Univ. Tex., Publ. Astron.
The University of Texas, Publications in Astronomy. Department of Astronomy, University of Texas, Austin, Tex. 78712, USA. ISSN 0276–1106.

Upps. Astron. Obs. Ann.
Uppsala Astronomiska Observatoriums Annaler. Astronomiska Observatoriet, Box 515, S–75120 Uppsala, Sweden.

Upps. Astron. Obs. Rep.
Uppsala Astronomical Observatory, Report. Astronomiska Observatoriet, Box 515, S–75120 Uppsala, Sweden.

U.S. Nav. Obs., Circ.
U.S. Naval Observatory, Circular. U.S. Naval Observatory, Washington, D.C. 20390, USA.

U.S. Nav. Obs., Time Serv. Publ., Ser. 01
U.S. Naval Observatory, Time Service Publications, Series 1 (Worldwide Primary Time and Frequency VLF and HF Transmissions). U.S. Naval Observatory, Time Service Division (62C), Washington, D.C. 20390, USA.

U.S. Nav. Obs., Time Serv. Publ., Ser. 04
U.S. Naval Observatory, Time Service Publications, Series 4 (Daily Time Differences and Relative Phase Values). U.S. Naval Observatory, Time Service Division (62C), Washington, D.C. 20390, USA.

U.S. Nav. Obs., Time Serv. Publ., Ser. 06
U.S. Naval Observatory, Time Service Publications, Series 6 (A.1–UT1 Data). U.S. Naval Observatory, Time Service Division (62C), Washington, D.C. 20390, USA.

U.S. Nav. Obs., Time Serv. Publ., Ser. 10
U.S. Naval Observatory, Time Service Publications, Series 10 (Astronomical Programs). U.S. Naval Observatory, Time Service Division (62C), Washington, D.C. 20390, USA.

U.S. Nav. Obs., Time Serv. Publ., Ser. 11
U.S. Naval Observatory, Time Service Publications, Series 11 (Time Service Report). U.S. Naval Observatory, Time Service Division (62C), Washington, D.C. 20390, USA.

U.S. Nav. Obs., Time Serv. Publ., Ser. 14
U.S. Naval Observatory, Time Service Publications, Series 14 (Time Service Announcement). U.S. Naval Observatory, Time Service Division (62C), Washington, D.C. 20390, USA.

Utr. Sterrekundige Overdrukken
Utrechtse Sterrekundige Overdrukken. Sonnenborgh Observatory, Zonnenburg 2, NL–3512 Utrecht, The Netherlands.

Uttar Pradesh State Obs., Repr.
Uttar Pradesh State Observatory, Reprint. Uttar Pradesh State Observatory, Manora Peak, Naini Tal 263 139, India.

Vasiona
Vasiona. Revue d'Astronomie. Bulletin de la Société Astronomique 'R. Bošković'. Vasiona, Narodna opservatorija, Kalemegdan, Gornji Grad, Beograd, Yugoslavia. ISSN 0506–4295.

Vatican Obs. Publ.
Vatican Observatory Publications. Specola Vaticana, I–00120 Città del Vaticano.

Vatican Obs. Publ., Spec. Ser., Studi Galileiani
Vatican Observatory Publications, Special Series, Studi Galileiani. Specola Vaticana, I–00120 Città del Vaticano.

Veröff. Archenhold–Sternw. Berlin–Treptow
Veröffentlichungen der Archenhold–Sternwarte Berlin–Treptow. Archenhold–Sternwarte, DDR–1193 Berlin, Alt–Treptow 1, German Democratic Republic.

Veröff. Astron. Inst. Bonn
Veröffentlichungen der Astronomischen Institute Bonn. Ferd. Dümmler Verlag, Kaiserstraße 31–37, D–5300 Bonn 1, F.R. Germany. ISSN 0340–9821.

Veröff. Astron. Rechen–Inst. Heidelb.
Veröffentlichungen des Astronomischen Rechen–Instituts Heidelberg. Verlag G. Braun, Karl–Friedrich–Straße 14–18, D–7500 Karlsruhe 1, F.R. Germany.

Veröff. Bayer. Komm. Int. Erdmessung Bayer. Akad. Wiss., Astron.–Geod. Arb.
Veröffentlichungen der Bayerischen Kommission für die Internationale Erdmessung der Bayerischen Akademie der Wissenschaften, Astronomisch–Geodätische Arbeiten. Bayerische Akademie der Wissenschaften, Marstallplatz 8, D–8000 München 22, F.R. Germany. ISSN 0340–7691.

Veröff. Forschungsinst. Dtsch. Mus. Gesch. Naturwiss. Tech., Reihe A
Veröffentlichungen des Forschungsinstituts des Deutschen Museums für die Geschichte der Naturwissenschaften und der Technik, Reihe A. Forschungsinstitut des Deutschen Museums für die Geschichte der Naturwissenschaften und der Technik, D–8000 München, F.R. Germany. ISSN 0418–9949.

Veröff. Remeis–Sternw. Bamberg
Veröffentlichungen der Remeis–Sternwarte Bamberg. Astronomisches Institut der Universität Erlangen–Nürnberg. Remeis–Sternwarte, Sternwartstraße 7, D–8600 Bamberg, F.R. Germany.

Veröff. Sternw. Pulsnitz
Veröffentlichungen der Sternwarte Pulsnitz. Sternwarte Pulsnitz, DDR–8514 Pulsnitz, German Democratic Republic.

Veröff. Sternw. Sonneberg
Veröffentlichungen der Sternwarte in Sonneberg. Akademie der Wissenschaften der DDR, Zentralinstitut für Astrophysik, Sternwarte Sonneberg, DDR–6400 Sonneberg, German Democratic Republic. ISSN 0373–6199.

Veröff. Wilhelm–Foerster–Sternw.
Veröffentlichungen der Wilhelm–Foerster–Sternwarte. Wilhelm–Foerster–Sternwarte, Munsterdamm 90, D–1000 Berlin 41, F.R. Germany.

Veröff. Zentralinst. Phys. Erde
Veröffentlichungen des Zentralinstituts für Physik der Erde. Akademie der Wissenschaften der DDR, Zentralinstitut für Physik der Erde, DDR–1500 Potsdam, Telegrafenberg A17, German Democratic Republic. ISSN 0514-8790.

Vesmír
Vesmír. Přírodovědecký časopis Čs. akademie věd. Publisher: Academia, Praha, Czechoslovakia.

Vestn. Kiev. Univ. Astron.
Vestnik Kievskogo Universiteta. Astronomiya. Izdatel'stvo pri Kievskom Gosudarstvennom Universitete Izdatel'skogo Obedineniya Vishcha Shkola, Kiev. 252001, Kiev–1, Kreshchatik, 4, USSR. ISSN 0203–7319; ISSN 0321–3927.

Vestn. Leningr. Univ., Mat. Mekh. Astron.
Vestnik Leningradskogo Universiteta Matematika, Mekhanika, Astronomiya. Leningradskij Universitet, Universitetskaya Nab. 7/9 Leningrad B–164, USSR. ISSN 0024–0850.

Vestn. Mosk. Univ., Ser. 3. Fiz. Astron.
Vestnik Moskovskogo Universiteta, Seriya 3. Fizika, Astronomiya. Izdatel'stvo Moskovskogo Universiteta, 103009, Moskva, Ul.Gertsena 5/7, USSR. English translation in Mosc. Univ. Phys. Bull. ISSN 0579–9392.

Villanova Univ. Obs. Contrib.
Villanova University Observatory Contributions. Department of Astronomy, Villanova University, Villanova, Pa. 19085, USA.

Vilniaus Astron. Obs. Biul.
Vilniaus Astronomijos Observatorijos Biuletenis. Vilniaus Valstybinis V. Kapsuko Universitetas, LTSR Mokslu Akademijos Fizikos Institutas, Vilnius. 232031, Vilnius 31, Čiurlionio 29, Lithuania, USSR. ISSN 0136–3697.

Vistas Astron.
Vistas in Astronomy. An International Review Journal. Pergamon Press, Oxford – New York – Frankfurt. ISSN 0083–6656.

Warner Swasey Obs., Repr.
Warner and Swasey Observatory, Reprints. Warner and Swasey Observatory, Case Western Reserve University, Department of Astronomy, Smith Bldg., Cleveland, Ohio 44106, USA.

Weather
Weather. James Glaisher House, Grenville Place, Bracknell, Berks, RG12 1BX, England. ISSN 0043–1656.

Werkgroepnieuws
Werkgroepnieuws. WGN, the international circular for meteor observers. Kontaktblad van meteoorwaarnemers in de Benelux. Tweemaandelijks tijdschrift. P. Roggemans, Dellingstraat 25, B–2800 Mechelen, Belgium.

Wiss. Z. Friedrich–Schiller–Univ. Jena, Math.–Naturwiss. Reihe
Wissenschaftliche Zeitschrift der Friedrich–Schiller–Universität Jena, Mathematisch–Naturwissenschaftliche Reihe. Friedrich–Schiller–Universität Jena. Subscription address: Zeitungsvertriebsamt, Abt. Export, DDR–1004 Berlin, Straße der Pariser Kommune 3/4, German Democratic Republic. ISSN 0043–6836.

Wiss. Z. Humboldt–Univ. Berlin, Math.–Naturwiss. Reihe
Wissenschaftliche Zeitschrift der Humboldt–Universität zu Berlin, Mathematisch–Naturwissenschaftliche Reihe. Humboldt–Universität Berlin. Subscription address: Zeitschriftenvertriebsamt, Abt. Export, DDR–1004 Berlin, Straße der Pariser Kommune 3/4, German Democratic Republic. ISSN 0522–9863.

Wrocław Astron. Obs., Repr.
Wrocław Astronomical Observatory, Reprint. Wrocław University Observatory, Wrocław, Poland.

Wuli
Wuli. Science Press, Beijing. Subscription address: Guozi Shudian, PO Box 399, Beijing, People's Republic of China. ISSN 0379–4148.

Yad. Fiz.
Yadernaya Fizika. Izdatel'stvo Nauka, 103717 GSP Moskva K–62. Podsosenskij Per. 24, SSSR. English translation in Sov. J. Nucl. Phys. ISSN 0044–0027.

Yamamoto Circ.
Yamamoto Circular. Yamamoto Observatory, 289, Kamitanakami–kiryutyo, OTU, Sigaken, 520–21 Japan.

Z. Angew. Math. Mech.
Zeitschrift für angewandte Mathematik und Mechanik. Akademie–Verlag GmbH, DDR–108 Berlin, Leipziger Str. 3–4, German Democratic Republic. ISSN 0044–2267.

Z. Angew. Math. Phys.
Zeitschrift für Angewandte Mathematik und Physik. Verlag Birkhäuser, Postfach 4000, Basel 24, Switzerland. ISSN 0044–2275.

Z. Naturforsch., A
Zeitschrift für Naturforschung, Teil A (Physik, Physikalische Chemie, Kosmophysik). Verlag der Zeitschrift für Naturforschung, PO Box 2645, D–7400 Tübingen, F.R. Germany. ISSN 0340–4811.

Z. Phys., A
Zeitschrift für Physik, A. (Atoms und Nuclei). Springer–Verlag, Berlin – Heidelberg – New York – Tokyo. ISSN 0340–2193.

Z. Phys., B
Zeitschrift für Physik, B (Condensed Matter). Springer–Verlag, Berlin – Heidelberg – New York – Tokyo. ISSN 0722–3277.

Z. Phys., C
Zeitschrift für Physik, C (Particles and Fields). Springer–Verlag, Berlin – Heidelberg – New York – Tokyo. ISSN 0170–9739.

Z. Vermessungswes.
Zeitschrift für Vermessungswesen. Verlag Konrad Wittner KG, Postfach 147, D–7000 Stuttgart 1, F.R. Germany. ISSN 0513–9155.

Zeiss Inf.
Zeiss Information. Carl Zeiss, Oberkochen. F.R. Germany. ISSN 0044–2054.

Zeit–Breitenbestimmungen
Zeit– und Breitenbestimmungen, Empfangszeiten von Zeitsignalen, Präzisionszeitvergleiche. Akademie der Wissenschaften der DDR, Zentralinstitut für Physik der Erde, DDR–1500 Potsdam, Telegrafenberg A 17, German Democratic Republic.

Zemlya Vselennaya
Zemlya i Vselennaya. Astronomiya, Geofizika. Issledovaniya Kosmicheskogo Prostranstva. Nauchno–Populyarnyj Zhurnal Akademii Nauk SSSR. Izdatel'stvo Nauka, Moskva, USSR. ISSN 0044–3948.

Zenit
Zenit. Populair–wetenschappelijk maandblad over sterrenkunde, weerkunde, ruimtevaart, ruimte–onderzoek en aanverwante wetenschappen en technieken. Stichting De Koepel, Nachtegaalstraat 82 bis, Utrecht, The Netherlands. ISSN 0165–0211.

Zentralbl. Math. Grenzgeb. – Math. Abstr.
Zentralblatt für Mathematik und ihre Grenzgebiete – Mathematics Abstracts. Heidelberger Akademie der Wissenschaften und Fachinformationszentrum Energie Physik Mathematik GmbH, Karlsruhe. Springer–Verlag, Berlin – Heidelberg – New York – Tokyo. ISSN 0044–4235.

Zentralinst. Astrophys., Sternw. Babelsberg, Mitt.–Neue Folge
Zentralinstitut für Astrophysik, Sternwarte Babelsberg, Mitteilungen–Neue Folge. Akademie der Wissenschaften der DDR. Sternwarte Babelsberg, DDR–1502 Potsdam–Babelsberg, Rosa–Luxemburg–Straße 17a, German Democratic Republic.

Zh. Ehksp. Teor. Fiz.
Zhurnal Ehksperimental'noj i Teoreticheskoj Fiziki. Akademiya Nauk SSSR, Leninskij Prosp. 14, Moskva, USSR. English translation in Sov. Phys.–JETP. ISSN 0044–4510.

Zvaigžnota Debess
Zvaigžnota Debess. Latvijas PSR Zinātnu Akadēmijas Radioastrofizikas Observatorijas Populārzinatnisks Rakstu Krāgums. Izdevnieciba Zinātne, Riga. 226004 Riga, Vienibas Gatve 11, USSR. ISSN 0135–129X.

Journals Abstracted Completely

A selected number of journals listed in category 001 (periodicals) are central to the subject scope of *Astronomy and Astrophysics Abstracts*. Depending on their relevance, almost all papers of the journals listed below are abstracted in our service.

AAS Photo-Bull.
Acta Astron.
Acta Astron. Sin.
Acta Astrophys. Sin.
Acta Cosmologica
Algoritm. Nebesnoj Mekh.
Am. Assoc. Variable Star Obs. Bull.
Annu. Rev. Astron. Astrophys.
Archaeoastronomy (U.K.)
Archaeoastronomy (U.S.A.)
Astrofizika
Astron. Astrophys.
Astron. Astrophys., Suppl. Ser.
Astron. Circ.
Astron. Data Cent. Bull.
Astron. J.
Astron. Nachr.
Astron. Q.
Astron. Tidsskr.
Astron. Tsirk.
Astron. Vestn.
Astron. Zh.
Astronomia
Astronomie
Astrophys. J.
Astrophys. J., Lett. Ed.
Astrophys. J., Suppl. Ser.
Astrophys. Lett.
Astrophys. Space Sci.
Aust. J. Astron.

BAV Rundbrief
BBSAG Bull.
Bol. Asoc. Argent. Astron.
Bol. Astron.
Br. Astron. Assoc. Circ.
Bull. Am. Astron. Soc.
Bull. Assoc. Fr. Obs. Etoiles Variables
Bull. Astron. Inst. Czech.
Bull. Astron. Soc. India
Bull. Inf. Cent. Données Stellaires

Celest. Mech.
Chin. Astron. Astrophys.
Ciel
Ciel Terre
Circ. Inf.
Coelum
Comments Astrophys.
Curr. Top. Chin. Sci., Sect. E: Astron.

Earth, Moon, Planets

Fundam. Cosmic Phys.

G. Astron.
Gen. Relativ. Gravitation
GEOS Circ.

I.A.P.P.P. Commun.
IAU Circ.
Icarus
IHW Newsl.
Inf. Bull. Variable Stars
Int. Comet Q.
Ir. Astron. J.
Issled. Solntsa Krasnykh Zvezd

J. Am. Assoc. Variable Star Obs.
J. Astrophys. Astron.
J. Br. Astron. Assoc.
J. Hist. Astron.
J. Korean Astron. Soc.
J. R. Astron. Soc. Can.

Kinematika Fiz. Nebesn. Tel
Komet. Tsirk.
Komety Meteory

Mem. Astron. Soc. India
Mem. Soc. Astron. Ital.
Mercury
Messenger
Meteoritics
Meteoritika
Minor Planet Bull.
Minor Planet Circ.
Mitt. Astron. Ges.
Mon. Not. R. Astron. Soc.
Mon. Notes Astron. Soc. S.Afr.

Nablyud. Iskusstv. Nebesn. Tel
Nauchn. Inf.
News Lett. Astron. Soc. N.Y.

Observatory
Occultation Newsl.
Orion
Orione

Perem. Zvezdy
Perem. Zvezdy, Prilozh.
Pis'ma Astron. Zh.
Postępy Astron.
Probl. Kosm. Fiz.
Proc. Astron. Soc. Aust.
Publ. Astron. Soc. Jpn.
Publ. Astron. Soc. Pac.
Publ. Variable Star Sect., R. Astron. Soc. N.Z.

Q. J. R. Astron. Soc.

Rev. Astron.
Rev. Mex. Astron. Astrofis.

Sky Telesc.
Sol. Phys.
Soln. Dannye Byull.
South. Stars
Space Sci. Rev.
Sterne
Sterne Weltraum
Strolling Astron.

Trans. IAU

Vasiona
Vestn. Kiev. Univ. Astron.
Vistas Astron.

Yamamoto Circ.

Zenit
Zvaigžnota Debess

Publications of Observatories and Astronomical Institutes

Reports, communications, publications, numbered series of reprints and preprints of observatories and astronomical institutes (listed in category 001) which are scanned completely in our service are listed below.

A.A.O. Newsl.
Abastumanskaya Astrofiz. Obs., Byull.
Abh. Hamb. Sternw.
Abstr. Submitted Pap.
Acad. Sci. Est. SSR, Div. Phys. Math. Tech. Sci., Prepr.
A.D.I.O.N. Bull.
An. Obs. Astron. Univ. Coimbra
Anglo-Aust. Obs., Prepr.
Anglo-Aust. Telesc., Annu. Rep.
Ann. Obs. Astron. Alger
Ann. Shanghai Obs., Acad. Sin.
Ann. Tokyo Astron. Obs., Second Ser.
Ann. Univ.-Sternw. Wien
Annu. Rep. Astron. Inst. Greece
Annu. Rep. (B.I.H.)
Annu. Rep. Dir., Mt. Wilson Las Campanas Obs.
Annu. Rep. Geophys. Obs.
Annu. Rep. Int. Polar Motion Serv.
Annu. Rep. Meteorol. Obs. Int. Latitude Obs. Mizusawa
Archenhold-Sternw. Berlin-Treptow, Sonderdr.
Archenhold-Sternw. Berlin-Treptow, Vortr. Schr.
Arecibo Obs./NAIC, Newsl.
Armagh Obs., Repr.
Astrofiz. Issled. Izv. Spets. Astrofiz. Obs.
Astrofys. Inst., Vrije Univ. Bruss., Overdruk
Astron. Bull., Carter Obs.
Astron. Contrib. Univ. Manchester, Ser. II: Jodrell Bank Repr.
Astron. Contrib. Univ. Manchester, Ser. III
Astron. Inst., "Anton Pannekoeck", Univ. Amst., Repr.
Astron. Inst. Univ. Brno, Contrib.
Astron. Inst. Univ. Brno, Publ.
Astron. Mitt. Wien
Astron. Obs. Trieste, Publ.
Astron. Pap.
Astron. Rechen-Inst. Heidelb., Mitt., Ser. A
Astron. Rechen-Inst. Heidelb., Mitt., Ser. B
Astron. Zeit-Breitenbestimmungen
Astrophys. Prepr. Ser.
Astrophys. Relativ., Prepr. Ser.
Austrian Pap. Asteroids

BAV Mitt.
B.I.H. Circ.
Biul. Obs. Astron. Uniw. M. Kopernika Toruniu
Bol. Astron. Obs. Madr.
Bol. Inst. Tonantzintla
Bol. Obs. Ebro
Boyden Obs., Occas. Publ.
Boyden Obs., Repr.
Bull. Astron.
Bull. Etoiles Tardives Spectre Particulier
Bull. Inf. Cent. Données Stellaires
Bull. Inf. Etoiles Be
Bull. Obs. Astron. Belgr.
Bull., Time Serv. Mizusawa Obs.
Byull. Inst. Astrofiz.
Byull. Inst. Teor. Astron.

Carter Obs., Repr. Ser.
Cartes Synoptiques
Cent. Astron. Sci. Spat., Obs. Sol.
Cent. Astrophys., Prepr. Ser.
Circ. Czech. Obs., Time Latitude
Circ. Stn. Astron. Int. Latitudine, Carloforte-Cagliari
Circ. Time Latitude Serv.
Commun. Astron. Dep., Univ. Ankara
Commun. Konkoly Obs.
Commun. Univ. Lond. Obs.
Commun. Univ. Obs., St. Andrews
Comun. Obs. Astron. Univ. Coimbra
Contrib. Astron. Obs. Skalnaté Pleso
Contrib. Bosscha Obs.
Contrib. Dep. Astron., Univ. Kyoto
Contrib. Dep. Astron., Univ. Tokyo
Contrib. Dep. Geod. Astron., Univ. Thessaloniki
Contrib. Dom. Astrophys. Obs.
Contrib. Dunsink Obs.
Contrib. Inst. Argent. Radioastron.
Contrib. Inst. Astron. Res.
Contrib. Kwasan Hida Obs., Univ. Kyoto
Contrib. La. State Univ. Obs.
Contrib. Lick Obs.
Contrib. Nicholas Copernicus Obs. Planetarium Brno
Contrib. Nizamiah Japal-Rangapur Obs.
Contrib. Oss. Astron. Torino
Contrib. Van Vleck Obs.
Contrib. Wrocław Astron. Obs.
Copenh. Univ. Obs., Repr.
Cracow Obs. Repr.

Data Rep. Hydrogr. Obs., Ser. Astron. Geod.
Debrecen Heliophys. Obs., Hung. Acad. Sci., Repr.
Dep. Astron. McDonald Obs. Univ. Tex., Repr.
Dep. Astrophys., Univ. Oxford, Publ.
Dudley Obs. Rep.
Dunsink Obs. Repr.

ESO Annu. Rep.
ESO Sci. Prepr.
ESO Sci. Rep.
ESO Tech. Rep.

Gemini
Greenwich Time Rep.

H.M. Naut. Alm. Off., Libr. Repr.
Hvar Obs. Bull.

Inst. Astron. Astrophys. Tech. Univ. Berlin, Mitt.
Inst. Astron. Fis. Espacio, Tirada Aparte
Inst. Astron., Univ. Camb., Annu. Rep.
Inst. Astrophys. Paris, Pré-Publ.
Inst. Obs. Mar., Bol. Astron.
Inst. Obs. Mar., Bol. Ser. B
Inst. Obs. Mar., Bol. Ser. C
Inst. Obs. Mar., Bol. Ser. D
Inst. Obs. Mar., Mem. Act.
Inst. Teor. Astrofys., Blindern-Oslo, Småtrykk
Inst. Theor. Astrophys., Blindern-Oslo, Rep.
Inst. Theor. Astrophys., Blindern-Oslo, Repr.
Inst. Theor. Astrophys., Univ. Oslo, Publ. Ser.
Inst. Theor. Phys. Sternw. Univ. Kiel, Repr.
Izv. Astron. Ehngel'gardt. Obs.
Izv. Glav. Astron. Obs. Pulkovo
Izv. Krymskoj Astrofiz. Obs.

Kandilli Obs., Heliophys. Serv. Publ., Second Ser.
Kapteyn Astron. Inst., Annu. Rep.
Kodaikanal Obs. Bull., Ser. A
Kodaikanal Obs. Repr.

Latitude Circ.
Lick Obs. Bull.

Lohrmann-Obs., Tech. Univ. Dresden, Zirk.
Lowell Obs. Bull.

Messenger
Mitt. Archenhold-Sternw. Berlin-Treptow
Mitt. Astrophys. Obs. Potsdam
Mitt. Geod. Inst. Rheinischen Friedrich-Wilhelms-Univ. Bonn
Mitt. Karl-Schwarzschild Obs. Tautenburg
Mitt. Lohrmann-Obs. Tech. Univ. Dresden
Mitt. Satell.-Beobachtungsstn. Zimmerwald
Mitt. Sonnenobs. Kanzelhöhe
Mitt. Sternw. Münch.
Mitt. Sternw. Sonneberg
Mitt. Universitätssternw. Graz
Mitt. Univ.-Sternw. Innsb.
Mitt. Univ.-Sternw. Jena
Mitt. Veränderliche Sterne
Mitt. Zentralinst. Phys. Erde
Mon. Notes Int. Polar Motion Serv.
Mons Astrophys. Pap.
MPE Contrib.
MPE Intern. Rep.
MPE Prepr.
MPE Rep.
MPE Repr.
Mt. Stromlo Siding Spring Obs., Repr.

Nanjing Univ. Obs., Publ.
NASA IUE Newsl.
Natl. Astron. Ionos. Cent., Astron. Prepr.
Natl. Astron. Ionos. Cent., Astron. Publ.
Natl. Radio Astron. Obs., Repr., Ser. A
Natl. Radio Astron. Obs., Repr., Ser. B
NEOS Earth Orientation Bull.
Nizamiah Rangapur Obs. Dep. Astron. Osmania Univ., Repr.
NSSDC Newsl.

Obs. Astron. Antares, Contrib. Cient.
Obs. Astron. Córdoba, Tirada Aparte
Obs. Astron. La Plata, Sep. Astron.
Obs. Astron. Univ. Nac. La Plata, Ser. Astron.
Obs. Astron. Univ. Nac. La Plata, Ser. Espec.
Obs. Astrophys. Lab., Univ. Helsinki, Rep.
Obs. Haute Provence, Pré-Publ.
Obs. Lyon Prepr.
Obs. Lyon Repr.
Obs. R. Belg., Commun., Sér. A
Obs. R. Belg., Commun., Sér. B
Obs. Variable Stars, Rep.
Occas. Rep. R. Obs., Edinb.
Onsala Space Obs., Prepr.
Oss. Astrofis. Catania, Publ.
Oss. Astron. Milano-Merate, Contrib.

Perth Obs. Commun.
Prepr. Astron. Inst. Univ. Basel
Prepr. Steward Obs.
Proc. Int. Latitude Obs. Mizusawa
Pubbl. Stn. Astron. Int. Latitudine Carloforte-Cagliari, Nuov. Ser.
Pubbl. Varie Fuori Ser. Oss. Astron. Torino (Pino Torinese)
Publ. Astron. Dep. Eötvös Univ.
Publ. Astron. Inst. Czech. Acad. Sci.
Publ. Astron. Obs. Sarajevo
Publ. Astron. Opservatorije Beogr.
Publ. Astrophys. Obs. Potsdam
Publ. Beijng Astron. Obs.
Publ. Bosscha Obs.
Publ. Debrecen Heliophys. Obs.
Publ. Dep. Astron. Univ. Beogr.
Publ. Dep. Astron. Univ. Cape Town
Publ. Dep. Astron., Univ. Chile
Publ. Dep. Geod. Astron., Univ. Thessaloniki
Publ. Dom. Astrophys. Obs.
Publ. Ege Univ. Obs.
Publ. Inst. Geophys., Pol. Acad. Sci., Ser. F (Planet. Geod.)
Publ. Int. Latitude Obs. Mizusawa
Publ. Istanbul Univ. Obs.
Publ. Lick Obs.
Publ. Obs. Astron. Nac., Univ. Colombia
Publ. Obs. Astron. "Prof. Manuel de Barros", Fac. Cienc. Porto
Publ. Obs. Astron. Univ. Cluj
Publ. Obs. Bordeaux, Nouv. Ser.
Publ. Obs. Ebro
Publ. Obs. Genève, Sér. A
Publ. Obs. Genève, Sér. B
Publ. Purple Mt. Obs.
Publ. R. Obs., Edinb.
Publ. Semin. Astron. Geod. Univ. Complutense Madr.
Publ. Shaanxi Astron. Obs.
Publ. Spéc. Cent. Données Stellaires
Publ. U.S. Nav. Obs., Second Ser.
Publ. Warner Swasey Obs.
Publ. Yunnan Obs.

Q. Bull. Sol. Act.

R. Greenwich Obs., Annu. Rep.
R. Greenwich Obs., Bull.
R. Obs. Ann.
R. Obs. Edinb., Res. Facilities
Rapp. Act. Obs. Paris
Rep. Obs. Lund
Rep. Ser., Dep. Phys. Sci., Univ. Turku
Repr. Theor. Astrophys. Group, Dep. Math., UMIST
Res. Cent. Astron. Appl. Math., Acad. Athens, Contrib. Ser. I
Res. Lab. Electron. Onsala Space Obs., Res. Rep.
Ric. Astron.
Rijksuniv. Gent, Sterrenkundig Obs., Meded.
Rutherford Appleton Lab., Rep.

SAAO Newsl.
Sac Peak Update
S.Afr. Astron. Obs., Annu. Rep.
S.Afr. Astron. Obs., Circ.
Sci. Rep. Tôhoku Univ., Eighth Ser.
Semin. Astron. Geod., Univ. Madr., Publ.
Sendai Astron. Rap.
Shaanxi Astron. Obs., Repr.
Shanghai Obs., Acad. Sin., Prepr.
Smithson. Astrophys. Obs., Spec. Rep.
Sol. Maps Act.
Sol. Radio Data
Soobshch. Byurakan. Obs.
Soobshch. Gos. Astron. Inst. Shternberg
Soobshch. Spets. Astrofiz. Obs.
Space Telesc. Sci. Inst., Instrum. Sci. Rep.
Space Telesc. Sci. Inst., Newsl.
Space Telesc. Sci. Inst., Prepr. Ser.
ST-ECF Newsl.
Stockholms Obs., Rep.
Stud. Soc. Sci. Torun., Sect. F
Swarthmore Astron. Repr.
Syd. Obs. Pap.

Tartu Astrofüüs. Obs. Publ.
Tartu Astrofüüs. Obs. Teated
Theor. Pap.
Time Freq. Serv., Bull.
Time Serv. Annu. Rep.
Time Serv. Bull.
Tokyo Astron. Bull., Second Ser.
Tokyo Astron. Obs., Kiso Inf. Bull.
Tokyo Astron. Obs. Rep.
Tokyo Astron. Obs. Repr.
Tokyo Astron. Obs., Time Latitude Bull.
Tr. Astrofiz. Inst. Alma-Ata
Tr. Astron. Obs., Leningrad
Tr. Glav. Astron. Obs. Pulkovo, Ser. 2
Tr. Gos. Astron. Inst. Shternberg
Tr. Inst. Teor. Astron., Leningrad

Tr. Kazan. Gorod. Astron. Obs.
Tr. Tashkent. Astron. Obs.
Trans. Astron. Obs. Yale Univ.
Tsirk. Astron. Inst. Tashkent
Tsirk. Astron. Obs. L'vov
Tsirk. Shemakh. Astrofiz. Obs.
Turku Univ. Obs., Informeto
Turku Univ. Obs., Informo

UKIRT Rep.
Univ. Barc., Dep. Fis. Tierra Cosmos, Publ.
Univ. Hannover, Astron. Stn., Veröff.
Univ. Obs., St. Andrews, Repr.
Univ. Tex., Monogr. Astron.
Univ. Tex., Publ. Astron.
Upps. Astron. Obs. Ann.
Upps. Astron. Obs., Rep.
U.S. Nav. Obs., Circ.
U.S. Nav. Obs., Time Serv. Publ., Ser. 1
U.S. Nav. Obs., Time Serv. Publ., Ser. 4
U.S. Nav. Obs., Time Serv. Publ., Ser. 6
U.S. Nav. Obs., Time Serv. Publ., Ser. 10
U.S. Nav. Obs., Time Serv. Publ., Ser. 11
U.S. Nav. Obs., Time Serv. Publ., Ser. 14
Utr. Sterrekundige Overdrukken
Uttar Pradesh State Obs., Repr.

Vatican Obs. Publ.
Vatican Obs. Publ., Spec. Ser., Studi Galileiani
Veröff. Archenhold-Sternw. Berlin-Treptow
Veröff. Astron. Inst. Bonn
Veröff. Astron. Rechen-Inst. Heidelb.
Veröff. Remeis-Sternw. Bamberg
Veröff. Sternw. Pulsnitz
Veröff. Sternw. Sonneberg
Veröff. Wilhelm-Foerster-Sternw.
Veröff. Zentralinst. Phys. Erde
Villanova Univ. Obs. Contrib.
Vilniaus Astron. Obs. Biul.

Warner Swasey Obs., Repr.
Wrocław Astron. Obs., Repr.

Zentralinst. Astrophys., Sternw. Babelsberg, Mitt.-Neue Folge

002 Bibliographical Publications, Documentation, Catalogues, Atlases

002.001 **Bibliography of publications on white holes.**
A. P. Trofimenko, V. S. Gurin.
Gen. Relativ. Gravitation, Vol. 18, No. 1, p. 53 – 58 (1986).

002.002 **A catalogue of southern dark clouds.**
M. Hartley, R. N. Manchester, R. M. Smith, S. B. Tritton, W. M. Goss.
Astron. Astrophys., Suppl. Ser., Vol. 63, No. 1, p. 27 – 48 (1986).
A catalogue of 1101 dark clouds has been compiled from visual inspection of ESO/SERC Southern *J* survey plates for declinations south of –33°. This catalogue complements that of Lynds (1962) which is based on the Palomar Sky Survey. Equatorial positions of accuracy 10 arcsec or better are listed for each cloud along with size (major axis × minor axis), density class and ESO/SERC field number. Galactic coordinates, derived from the equatorial positions, are used to define a name for each cloud. Complexes, consisting of several separately identifiable concentrations, are listed and identified by a trailing C on the name. The conclusions of Lynds, that clouds are concentrated along the galactic equator and toward the galactic centre, are reinforced. No clear connection with Gould's Belt is evident in the distribution of southern clouds.

002.003 **A collection of Galilean satellite eclipse observations, 1652 – 1983: II.**
J. H. Lieske.
Astron. Astrophys., Suppl. Ser., Vol. 63, No. 1, p. 143 – 202 (1986).
The largest collection of Galilean satellite eclipse observations extent has been gathered from the published literature and from manuscript collections from 1652 to 1983. Many of the observations were thought to have been irretrievably lost for more than a century. The collection will be of great value in evaluating long–term effects on the Galilean satellites such as the possible existence of tidal dissipation on their periods.

002.004 **Erratum: "Catalogue of dark nebulae and globules for galactic longitudes 240 to 360 degrees" [Astron. Astrophys., Suppl. Ser., Vol. 58, No. 2, p. 365 – 386 (1984).**
J. V. Feitzinger, J. A. Stüwe.
Astron. Astrophys., Suppl. Ser., Vol. 63, No. 1, p. 203 (1986). See Abstr. 38.002.034.

002.005 **IDAM: a catalogue of DM, HD(E) and other cross–identifications for stars in open clusters (magnetic tape).**
J. C. Mermilliod.
Astron. Astrophys., Suppl. Ser., Vol. 63, No. 2, p. 293 – 296 (1986).
This catalogue presents cross–identifications between usual numbering systems for stars in open clusters and several astronomical catalogues: BS, HD–HDE, DM, NLS, LSS, GCVS, IDS, ADS, SAO and miscellaneous lists or catalogues. It concerns 345 open clusters, for a total of 9543 entries. Inverse tables are given for each catalogue. The present catalogue completely supersedes the 1976 version.

002.006 **Electronographic *BV* photometry of three distant clusters of galaxies. II. The catalogue.**
O. Le Fèvre, A. Bijaoui, G. Mathez, J. P. Picat, G. Lelièvre.
Astron. Astrophys., Suppl. Ser., Vol. 63, No. 2, p. 297 – 303 (1986).
In this paper the authors give the photometric catalogue of three distant clusters of galaxies, for 223 objects (stars and galaxies) detected in A 992 field, 318 in 3C 299, 295 in 1305 + 295, with 9 parameters, including the rectangular positions, R_{25}, V_{25} and $(B–V)_{25}$.

002.007 **Designation problems in astronomy.**
C. Jaschek.
Q. J. R. Astron. Soc., Vol. 27, No. 1, p. 60 – 63 (1986).
Designation practices in use are discussed and the need for differentiating designation and description is underlined. A series of practical recommendations is proposed to keep the existing confusion down to acceptable levels.

002.008 **Bibliografia su metodologie di osservazione, riduzione dei dati e applicazioni della fotometria stellare multicolore UBV.**
M. Santangelo, E. Mazzoni, U. Mallegni.
Coelum, Vol. 54, N. 1, p. 11 – 15 (1986).

002.009 **Gli ammassi globulari. III.**
R. Monella.
Coelum, Vol. 54, N. 1, p. 17 – 32 (1986).

002.010 **Interstellar travel and communication bibliography – 1985 update.**
Z. Paprotny, J. Lehmann, J. Prytz.
J. Br. Interplanet. Soc., Vol. 39, No. 3, p. 127 – 136 (1986).

002.011 **New Hα emission stars found above 10° galactic latitude.**
C. B. Stephenson.
Astrophys. J., Vol. 300, No. 2, p. 779 – 784 (1986).
Approximately 1300 5°2 × 5°2 red–sensitive objective prism plates have been searched for Hα–emission stars in order to complete the search of the entire sky. The survey here reported represents 59% of the whole sky, defined by the region north of declination –25° and mostly further than 10° from the galactic plane. Two hundred and six stars have been newly found, mostly of unknown spectral type. The stars' latitude distribution suggest that they are mainly moderately high luminosity disk population stars. There are new finds especially in Orion and the Taurus dark clouds (limiting mag ≈13).

002.012 **Third catalogue of Am stars with known spectral types.**
B. Hauck.
Astron. Astrophys., Suppl. Ser., Vol. 64, No. 1, p. 21 – 23 (1986).
A catalogue has been compiled of Am stars with known spectral types with respect to K line, hydrogen–lines and metallic–lines. It contains 2269 determinations for 1805 stars. The catalogue is available on magnetic tape from the Centre de Données Stellaires, Strasbourg.

002.013 **The Case low–dispersion northern sky survey. III.**
P. Pesch, N. Sanduleak.
Astrophys. J., Suppl. Ser., Vol. 60, No. 2, p. 543 – 549 (1986). With plates 50 – 56.
Positions, estimated magnitudes, and finding charts (when needed) are provided for 130 blue and/or emission–line galaxies, probable H II regions in six galaxies, 73 blue and/or emission–line unresolved objects including QSO candidates, and 58 known and probable blue stars contained within the region $08^h05^m < \alpha < 10^h10^m$ and $+33°0 < \delta < +38°0$ (1950). The objects, whose blue magnitudes range from 12 to 18, were identified on low–dispersion, objective–prism plates taken with the Burrell Schmidt telescope at Kitt Peak.

002.014 **The second Byurakan spectral sky survey. IV. The field $\alpha = 12^h22^m$, $\delta = +55°00'$.**
B. E. Markarian, D. A. Stepanyan, L. K. Erastova.
Astrofizika, Tom 23, Vyp. 3, p. 439 – 450 (1985). In Russian. English translation in Astrophysics, Vol. 23, No. 3.
The fourth list of objects of the second Byurakan survey covering a field of 4 × 4 degrees and centred on $\alpha = 12^h22^m$, $\delta = +55°00'$ is presented. This list contains data on 106 objects and galaxies and 12 blue stars. The type distribution of objects is

as follows: 16 candidates of QSO, 29 BSO, 32 galaxies with significant UV continuum, the Seyfert characteristics can be suspected among three of them and 29 emission line galaxies without noticeable UV continuum. The surface density of QSO and Sy objects till 19^m is more than one per square degree.

002.015 **Directory of European Observatories 1986 (DEO 1986).**
P. A. J. Vercoutter.
Philippe's Astronomical Series, Vol. 2. Astronomical Contact Group, Jan Van Eyckdreef 7, B-8900 Ieper, Belgium. 10+193 pp. Price US$ 10.00, BF 500.00 (1986).

002.016 **Atlas of historical eclipse maps. East Asia 1500 BC – AD 1900.**
F. R. Stephenson, M. A. Houlden.
Cambridge University Press, Cambridge – London – New York – New Rochelle – Melbourne – Sydney. 15+431 pp. Price £ 40.00, US$ 79.50 (1986). ISBN 0-521-26723-4.
This book presents a series of computer-drawn maps and tables for all total and annular eclipses of the Sun calculated to have been observable in East Asia in the 3400 years from 1500 BC to AD 1900. An introduction explains the methods used to calculate the parameters and determine the trajectories of each eclipse. Then follow the maps themselves, 862 in all, which plot the path of the umbral region over East Asia and tabulate the longitude, limits of latitude, local time, and mean altitude for each event.

002.017 **On the statistics of papers being published during 1975 – 1983.**
K. Rudnicki.
Postepy Astron., Tom 33, Zesz. 1–2, p. 71 – 72 (1985). In Polish.

002.018 **Statistics of astronomical publications.**
J. Fukue.
Astron. Her., Vol. 79, No. 4, p. 98 – 104 (1986). In Japanese.

002.019 **The designation of radio sources observed with the DRAO synthesis telescope.**
L. A. Higgs.
J. R. Astron. Soc. Can., Vol. 80, No. 1, p. 51 – 54 (1986).
A simple designation scheme for referencing radio sources in lists of objects observed at the Dominion Radio Astrophysical Observatory is introduced. Sources are referenced by a prefix "*nP*" where *n* is the serial number of the list and *P* refers to Penticton, followed by the serial number of the source in the list. Retroactive designations for lists in existence are presented.

002.020 **Nomenclature for objects in the Magellanic Clouds.**
M.-C. Lortet.
Astron. Astrophys., Suppl. Ser., Vol. 64, No. 2, p. 303 – 324 (1986).
The aims and presentation of the present dictionary are similar to that of the First Dictionary of the Nomenclature of Celestial Objects (Fernandez et al. 1983). It is intended to be a basic help for astronomers, librarians, and users of Data Centers.

002.021 **Nomenclature for objects in the galaxy M33.**
M.-C. Lortet.
Astron. Astrophys., Suppl. Ser., Vol. 64, No. 2, p. 325 – 328 (1986).
This is a complement to the First Dictionary of Nomenclature of Celestial Objects (Fernandez et al., 1983). A dictionary of acronyms used for objects in the galaxy M33 is given. Basic charts or catalogues are listed.

002.022 **First Supplement to the First Dictionary of the Nomenclature of Celestial Objects (solar system excluded).**
M.-C. Lortet, F. Spite.
Astron. Astrophys., Suppl. Ser., Vol. 64, No. 2, p. 329 – 389 (1986).

002.023 **Solar Physics. Subject Index and Name Index Volumes 1 – 100 (January 1967 – October 1985).**
C. de Jager, Z. Švestka (Editors).
Sol. Phys., Vol. 101, Nos. 1/2, 7+671 pp. (1985).

002.024 **Catalogue of cometary orbits.**
B. G. Marsden.
Fifth Edition. International Astronomical Union, Minor Planet Center, Smithsonian Astrophysical Observatory, 60 Garden Street, Cambridge, Mass. 02138, USA. 102 pp. Price US$ 12.00 (1986).
Review in IAU Circ., No. 4168 (1986).
The fifth edition contains orbits for 1187 cometary apparitions and is intended to be complete for comets observed up to the end of December 1985. It is arranged in essentially the same manner as the fourth edition. In addition to new discoveries and recoveries, the fifth edition includes new entries 1618 III, 1639, 1908 IV, 1953 VIII, 1967 XV and 1976 XV. Improved orbits are included in the fifth edition for many more of the older comets.

002.025 **Catalogue of right ascensions of 241 near-equatorial stars in the system of the Catalogue of the Time Service.**
I. T. Inogamov.
Tsirk. Astron. Inst., Tashkent, No. 114, p. 14 – 31 (1985). In Russian.

002.026 **Catalogue of proper motions of stars in the region of the galactic star cluster NGC 7243.**
Eh. Rakhmatov, M. Muminov.
Tsirk. Astron. Inst., Tashkent, No. 116, p. 17 – 55 (1985). In Russian.

002.027 **Catalogue of 900 faint-star ultraviolet spectra based on observed data of the Space Observatory Orion-2.**
G. A. Gurzadyan, D. B. Oganesyan, S. S. Rustambekova, R. A. Epremyan, edited by V. A. Ambartsumyan.
Byurakanskaya Astrofizicheskaya Observatoriya. Izdatel'stvo Akademii Nauk Armyanskoj SSR, Erevan. 20+344 pp. Price 6 Rbl. 60 Kop. (1985). In Russian and English.

002.028 **Documentation for the machine-readable version of *The Henry Draper Catalogue*. (Edition 1985).**
N. G. Roman, W. H. Warren Jr.
NSSDC/WDC-A-R&S, Publ., 84-18, 5+27 pp. (1985).
An updated, corrected and extended machine-readable version of the Henry Draper Catalogue is described. Published and unpublished errors discovered in the previous version have been corrected; letters indicating supplemental stars in the BD have been moved to a new byte to distinguish them from double-star components; and the machine-readable portion of The Henry Draper Extension (HDE) has been converted to the same format as the main catalog, with additional data added as necessary.

002.029 **Documentation for the machine-readable version of *Faint Blue Objects at High Galactic Latitude* (Warnock and Usher 1982).**
W. H. Warren Jr.
NSSDC/WDC-A-R&S, Publ., 85-06, 5+9 pp. (1985).
A detailed description of the machine-readable data file is given. The data set is a catalog of faint blue objects selected according to relative ultraviolet excess from *ubv* three-color 1.2-m Palomar Schmidt plates. Five selected area fields centered on SA 28, SA 29, SA 55, SA 57 and SA 94 are included. The data consist of color classifications, B magnitudes, 1950 equatorial coordinates and remarks; the current file contains 3678 objects.

002.030 **The Bedford Catalogue. From "A cycle of celestial objects".**
W. H. Smyth, with a foreword by G. Lovi.
Reprinted from the first edition, London 1844. Willmann-Bell, Inc., Richmond, Va., USA. 28+560 pp. Price US$ 18.95 (1986). ISBN 0-943396-10-7.

002.031 **Death and the astronomer: effect of cessation of personal influence upon rates of citation of astronomical papers.**
V. Trimble.
Bull. Am. Astron. Soc., Vol. 17, No. 4, p. 827 (1985). Abstract. – See Abstr. 010.062.

002.032 **The Voyager catalogue of far and extreme ultraviolet observations.**
T. E. Carone, J. B. Holberg, R. S. Polidan.
Bull. Am. Astron. Soc., Vol. 17, No. 4, p. 838 (1985). Abstract. – See Abstr. 010.062.

002.033 **New carbon stars in the IRAS Point Source Catalogue.**
D. J. MacConnell, R. R. Landis, P. B. Baker.
Bull. Am. Astron. Soc., Vol. 17, No. 4, p. 877 (1985). Abstract. – See Abstr. 010.062.

002.034 **Continuing results from the Michigan Catalogue of MK types for the HD stars (–90° to –12°) and future plans.**
N. Houk, J. Sowell.
Bull. Am. Astron. Soc., Vol. 17, No. 4, p. 878 (1985). Abstract. – See Abstr. 010.062.

002.035 **A spectrophotometric atlas of southern MK standards from 5800 Å to 10,200 Å.**
A. C. Danks, M. Dennefeld.
Bull. Am. Astron. Soc., Vol. 17, No. 4, p. 881 (1985). Abstract. – See Abstr. 010.062.

002.036 **A new catalogue of quasi–stellar objects.**
G. Burbidge, A. Hewitt.
Bull. Am. Astron. Soc., Vol. 17, No. 4, p. 890 (1985). Abstract. – See Abstr. 010.062.

002.037 **Differential catalogue of declinations of 2884 stars of the latitude programs.**
L. R. Tokhtas'eva, A. A. Krasikova, A. I. Nefed'eva, Eh. N. Vorob'eva.
Izv. Astron. Ehngel'gardt. Obs., No. 49, p. 3 – 56 (1985). In Russian.

002.038 **Comparison of the differential catalogue of declinations of 2884 stars of the latitude programs with the AGK3R catalogue.**
L. R. Tokhtas'eva.
Izv. Astron. Ehngel'gardt. Obs., No. 51, p. 50 – 57 (1985). In Russian.

The results of a comparison of the differential catalogue of declinations from observations carried out at the Engelhardt Observatory with the AGK3R catalogue are given. The distribution of the residual differences of the catalogues are discussed.

002.039 **Catalogue of magnitudes and colour indices of stars in the direction of NGC 7062.**
S. S. Tokhtas'ev.
Izv. Astron. Ehngel'gardt. Obs., No. 51, p. 60 – 89 (1985). In Russian.

2035 stars in the field with coordinates $\alpha = 21^h23^m30^s$, $\delta = +46°$ were measured from Schmidt UBV–plates. Charts of the investigated field, numbers of stars, V–magnitudes, (U–B) and (B–V) colour–indices are given.

002.040 **Bibliography of Earth, Moon, and Planets.**
Z. Kopal, M. Moutsoulas, F. B. Waranius.
Earth, Moon, Planets, Vol. 35, No. 2, p. 125 – 200 (1986).

Articles entered into the data base at the Lunar and Planetary Institute Library, May – October, 1984. Concerning: The Moon. Space utilization (including lunar base). Planets: Jupiter, Mars, Mercury, Neptune, Pluto, Saturn, Uranus, Venus. Other objects: asteroids, comets, meteorites, cosmic dust, other particles, etc.

002.041 **A review of problems in the fundamental radio–source catalog.**
K. Nakajima.
Proceedings of the Symposium on Application of Space Techniques to Astronomy and Geophysics, p. 1 – 4 (1986). In Japanese. – See Abstr. 012.027.

Problems which will arise in compiling a fundamental catalog of radio–sources (FRK) are reviewed. Some recent works on them are referred.

002.042 **A catalogue of Seyfert galaxies and related objects.**
N. Kaneko.
Department of Physics, Faculty of Science, Hokkaido University, Sapporo 060, Japan. 85 pp. (1986).

A catalogue of Seyfert galaxies and related objects, which contains 560 objects, is given. The coordinate designation, and the right ascension and declination are given for the epochs 1950.0 and 2000.0. The radial velocity, activity class, visual magnitude, morphological form, remarks for morphology, membership in pairs, groups and clusters, peculiarities, and miscellaneous names are included in the catalogue. The activity class, which is widely applicable to all active objects, is a generalization of the Seyfert class or type. The Markarian Seyfert galaxies are listed in an appendix.

002.043 **Bibliografia di sequenze e cataloghi fotometrici UBV nel cielo boreale.**
M. Santangelo, E. Mazzoni, U. Mallegni, G. Fornaciari, S. Orlandi, F. Banducci.
Coelum, Vol. 54, N. 2, p. 89 – 92 (1986).

002.044 **A comprehensive index to the Minor Planet Bulletin: 1973 – 1985.**
C. Cunningham.
Minor Planet Bull., Vol. 12, No. 4, Part 2, p. 43 – 54 (1985).

002.045 **Alphabetic list of minor planet names and naming references.**
K. Kelly.
Minor Planet Bull., Vol. 13, No. 2, p. 11 – 20 (1986).

An alphabetic list of 2785 asteroid names and their references, complete through December 1985, is presented.

002.046 **Preserving astronomical papers.**
D. Schaumberg, D. Osterbrock.
Mercury, Vol. 15, No. 1, p. 19, 31 (1986).

002.047 **Catalogue of ultraviolet spectra of 900 faint stars. Observational material of the space station Orion 2.**
G. A. Gurzadyan, D. B. Oganesyan, S. S. Rustambekova, R. A. Epremyan.
Izdatel'stvo Akad. Nauk ArmSSR. Erevan. 343 pp. (1985). In Russian.
Review in Ref. Zh., 51. Astron., 5.51.66 (1986).

002.048 **Catalogue of proper motions of PFKSZ–2 stars.**
A. D. Polozhentsev.
Red. Zh., Vestn. Leningr. Univ., Mat. Mekh. Astron., 12 pp. (1985). In Russian. Abstr. in Ref. Zh., 51. Astron., 5.51.145 (1986).

002.049 **Compiled catalogue of fundamental faint stars of the southern sky.**
A. D. Polozhentsev.
Red. Zh., Vestn. Leningr. Univ., Mat. Mekh. Astron., 19 pp. (1985). In Russian. Abstr. in Ref. Zh., 51. Astron., 5.51.146 (1986).

002.050 **General catalogue of ephemerides and apparent orbits of 736 visual binary stars.**
L.–s. Yan, Z.–y. Chu, D.–s. Pan.
Zo–Se Section, Shanghai Observatory, Academia Sinica, Shanghai, The People's Republic of China. 6+313 pp. (1985).

Information on 736 pairs of visual binaries is presented in tabular form. For each pair ephemerides are given for the period (1984 – 2003). A graph indicates relative positions of the components at different epochs for each system. Orbital elements are compiled and a geometrical classification of orbits is given. Finally, the systems are classified according to the physical nature of the component stars.

002.051 **Culgoora catalogue of solar radio noise storms, 1973 to 1984.**
R. T. Stewart, H. M. C. Eason, L. H. Heisler.
Proc. Astron. Soc. Aust., Vol. 6, No. 2, p. 231 – 271 (1985).
A unique data set of 160 MHz solar noise storm positions and polarizations covering a complete sunspot cycle interval from the Skylab period of 1973 through the Solar Maximum intervals of 1980 and 1984 is presented in the form of 27.28–day synoptic plots.

002.052 **A multicolor photometric catalog of galaxies and stars in the field of the rich cluster II Zw 1305.4+2941 at z = 0.24.**
D. C. Koo, R. G. Kron, D. Nanni, D. Trevese, A. Vignato.
Space Telesc. Sci. Inst., Prepr. Ser., No. 82, 40 pp. (1985). Submitted to Astron. J.

002.053 **Quasars to B > 22.5 in Selected Area 57: a catalog of multicolor photometry, variability, and astrometry.**
D. C. Koo, R. G. Kron, K. M. Cudworth.
Space Telesc. Sci. Inst., Prepr. Ser., No. 85, 58 pp. (1985). Submitted to Publ. Astron. Soc. Pac.

002.054 **Catalog of the Vatican Meteorite Collection.**
R. Salvatori, A. Maras, E. A. King.
Vatican Obs. Publ., 75 pp. (1986).

002.055 **Astronomy and Astrophysics Abstracts. Vol. 40. Literature 1985, Part 2.**
S. Böhme, U. Esser, W. Fricke, H. Hefele, I. Heinrich, W. Hofmann, D. Krahn, V. R. Matas, L. D. Schmadel, G. Zech (Editors).
Published for Astronomisches Rechen–Institut by Springer–Verlag, Berlin – Heidelberg – New York – London – Paris – Tokyo. 14+1096 pp. Price DM 198.00, US$ 86.00 [Subscription Price DM 158.40, US$ 71.70] (1986). ISBN 3–540–16655–6, ISBN 0–387–16655–6 (USA).

002.056 **Unified catalog of Washington and Mizusawa PZT stars.**
S. Manabe, S. Sakai, T. Hara, H. Kitago, K. Iwadate.
Publ. Int. Latitude Obs. Mizusawa, Vol. 18, No. 2, p. 63 – 92 (1984).
A unified catalog of Washington and Mizusawa PZT stars is constructed by using recompiled and rereduced data spanning from 1954 to 1983. The reduction is made based on the MERIT Standards. Internal precisions of the right ascensions and declinations at the mean epochs of observations are 0.6 ms and 0.″008, respectively, for the best determined stars. Those of proper motions are 0.08 ms/year and 0.″001/year, respectively.

002.057 **Catalog of O–B stars observed with Tokyo Meridian Circle.**
H. Yasuda, H. Hara, R. Fukaya, H. Ishii.
Ann. Tokyo Astron. Obs., Second Ser., Vol. 21, No. 1, p. 107 – 126 (1986).
A catalog of the O–B stars, selected from "Blaauw–Parenago" list and Rubin's catalog, has been compiled on the FK4 system by the observations made with the Gautier 8–inch Meridian Circle at the Tokyo Astronomical Observatory during the period, 1971 – 1979. It contains 1059 stars and was compiled for the future establishment of high precision proper motions of O–B stars.

002.058 **On indexing papers on astronomy with standard terms.**
T. N. Mandryka.
Vestn. Khar'kov. Univ., No. 278, p. 79 – 85 (1985). In Russian.

002.059 **Catalogue of sunspot groups at heliographic latitudes $B \leqslant 4°$ and $B \geqslant 38°$ from 1874 to 1983.**
G. Lustig, H. Haupt.
Astron. Astrophys., Suppl. Ser., Vol. 64, No. 3, p. 507 – 544 (1986).
For the sake of completeness the authors present a catalogue of sunspots in two different extreme latitude zones, $B \leqslant 4°$ and $B \geqslant 38°$, using the «Photoheliographic Results» from Greenwich for the period 1874 – 1976 and records from the observatories of Greenwich, Istanbul, Kanzelhöhe and Rome for the years 1977 – 1983. The catalogue has been prepared as a homogeneous basis and as a reference–book for statistical and prognostic purposes, Haupt (1983).

002.060 **A multicolor photometric catalog of galaxies and stars in the field of the rich cluster II Zw 1305.4+2941 at z = 0.24.**
D. C. Koo, R. G. Kron, D. Nanni, D. Trevese, A. Vignato.
Astron. J., Vol. 91, No. 3, p. 478 – 493 (1986). With plate 41.
Techniques are described for automatic photometry of faint images on fine–grained plates, and are applied to an area of 350 square arcminutes in SA 57 containing a rich cluster of galaxies at $z = 0.24$. The authors present a photometric catalog based upon 4 m plates in four bandpasses for 738 objects (494 galaxies, 244 stars) brighter than 22.5 mag in the blue or 21.0 mag in the red.

002.061 **The CDS and the SIMBAD data base.**
D. Egret.
Bull. Inf. Cent. Données Stellaires, No. 30, p. 25 – 30 (1986). – See Abstr. 012.049.
The CDS and the SIMBAD data base are presented. Details are given on the present uses of the data base and the expected future developments.

002.062 **Les catalogues d'étoiles variables et le satellite astrométrique HIPPARCOS.**
J. P. Cordoni, M. O. Mennessier.
Bull. Inf. Cent. Données Stellaires, No. 30, p. 55 – 60 (1986).
In order to find and follow a star to be observed, its apparent HIPPARCOS magnitude must be known in advance with an accuracy better than 0.5 mag. Evidently the star must also be brighter than the satellite detection threshold. The variables are classified into 7 groups which are defined in this paper. So each star can be marked by a variability flag in the Input Catalogue. The statistics by flag is given for the stars of the GCVS and the NSV.

002.063 **Finding charts of radio stars for the HIPPARCOS mission.**
R. Hering, H. M. Schwerdtfeger, H. G. Walter.
Bull. Inf. Cent. Données Stellaires, No. 30, p. 61 – 77 (1986).
Currently a total of 275 confirmed, suspected and potential radio stars has been proposed to the HIPPARCOS observing list. A subset of these stars has positions insufficiently accurate for the HIPPARCOS mission. To facilitate reobservation finding charts are provided for this subset consisting of 50 stars with confirmed radio emission; they are produced from the POSS, SRC and ESO(B) atlantes.

002.064 **The Working Group "Designations" of Commission 5 at the New Delhi I.A.U. meeting.**
F. Spite.
Bull. Inf. Cent. Données Stellaires, No. 30, p. 113 – 116 (1986).

002.065 **Progress report on the Bonner Durchmusterung.**
M. J. Wagner.
Bull. Inf. Cent. Données Stellaires, No. 30, p. 117 (1986).

002.066 **On the errata in the "14ème Catalogue Complémentaire de Binaires Spectroscopiques".**
A. Pedoussaut.
Bull. Inf. Cent. Données Stellaires, No. 30, p. 119 (1986). See Abstr. 40.002.103.

002.067 **The Bibliographic Catalogue of Stellar Radial Velocities.**
M. Barbier–Brossat, M. Petit.
Bull. Inf. Cent. Données Stellaires, No. 30, p. 121 (1986).

002.068 **Catalogue of orbital elements of 3143 minor planets numbered by November 1, 1984, on magnetic tape.**
Yu. V. Batrakov, V. A. Shor.
Bull. Inf. Cent. Données Stellaires, No. 30, p. 123 – 124 (1986).
The catalogue on magnetic tape containing the orbital elements of 3143 permanently numbered minor planets and some connected information is described.

002.069 **Catalogue of isolated galaxies.**
V. E. Karachentseva, V. S. Lebedev, A. L. Shcherbanovskij.
Bull. Inf. Cent. Données Stellaires, No. 30, p. 125 – 126 (1986).

002.070 **Machine–readable version of the "14ème Catalogue Complémentaire de l'Observatoire de Toulouse" by A. Pedoussaut, N. Ginestet and J. M. Carquillat.**
J. L. Halbwachs.
Bull. Inf. Cent. Données Stellaires, No. 30, p. 127 (1986).
Machine–readable version of a catalogue published in Astron. Astrophys., Suppl. Ser., Vol. 58, No. 3, p. 601 – 655 (1984). See Abstr. 38.002.037.

002.071 **Index to the first twenty–nine issues of the CDS Bulletin.**
C. Jaschek, M. J. Wagner.
Bull. Inf. Cent. Données Stellaires, No. 30, p. 131 – 163 (1986).

002.072 **Multiple citations of multiple–author publications.**
A. R. Rao, M. N. Vahia.
Publ. Astron. Soc. Pac., Vol. 98, No. 603, p. 511 – 512 (1986).
In continuation of a recent study of Abt (1984) that multiple author papers get higher citations, it is suggested that this may be due to a higher circulation of the preprints/reprints of the publication as well as higher frequency of contact of the authors with the other scientists in the same field. This suggestion is further supported by the authors' earlier study (Rao and Vahia 1984) that joint–author papers tend to be written by comparatively senior scientists who are definitely well established and consequently extensively read.

002.073 **The Royal Astronomical Society instrument collection: 1827 – 1985.**
H. D. Howse.
Q. J. R. Astron. Soc., Vol. 27, No. 2, p. 212 – 236 (1986).

002.074 **The Palomar–Green Catalog of Ultraviolet–Excess Stellar Objects.**
R. F. Green, M. Schmidt, J. Liebert.
Astrophys. J., Suppl. Ser., Vol. 61, No. 2, p. 305 – 352 (1986). With plates 14–34.
The Palomar–Green Catalog consists of positions, photographic *B*–magnitudes, spectral types, and photoelectric measurements when available, for 1874 stellar objects. The objects were chosen for ultraviolet excess on photographs from the Palomar 18 inch (46 cm) Schmidt telescope and then spectroscopically classified. Of these, 1715 comprise a statistically complete sample, covering 10,714 square degrees down to a typical limiting magnitude of $B = 16.1$. Extensive discussion is given of the magnitude calibration and completeness determination. The dominant subgroup is that of the sdB stars, which show a spatial distribution consistent with an old disk population.

002.075 **Star catalogs and files available at the Stellar Data Center.**
Bull. Inf. Cent. Données Stellaires, No. 30, p. 191 – 194 (1986).

002.076 **Catalogs to become available from Centre de Données Stellaires.**
Bull. Inf. Cent. Données Stellaires, No. 30, p. 195 (1986).

002.077 **Catalogs currently available on microfiche from Centre de Données Stellaires.**
Bull. Inf. Cent. Données Stellaires, No. 30, p. 197 – 203 (1986).

002.078 **Bibliography on Be stars.**
D. Ballereau, H. Hubert, A. M. Hubert–Delplace, J. Jugaku, P. Koubský, E. Mendoza, M. Ruusalepp, A. Slettebak.
Bull. Inf. Etoiles Be, No. 13, p. 31 – 34 (1986).
Bibliography of papers on Be stars published 1985/86.

002.079 **The spectral energy distribution of early type stars. I. A catalogue of photometric data of 259 stars from 0.15 to 4.8 μm.**
P. S. Thé, P. R. Wesselius, I. Janssen.
Working Group Stars With Extended Atmospheres, Prepr., No. 76, 27 pp. (1986). To appear in Astron. Astrophys., Suppl. Ser.

002.080 **Death comes as the end–effect of cessation of personal influence upon rates of citation of astronomical papers.**
V. Trimble.
Czech. J. Phys., Sect. B, Vol. B36, No. 1, p. 175 – 179 (1986). Abstr. in Phys. Abstr., Vol. 89, No. 1280, Entry 54070 (1986).

002.081 **Mean light and colour (B–V and U–B) curves of 210 field RR Lyrae type stars.**
N. Nikolov, N. Buchantsova, M. Frolov.
University of Sofia, Sofia, Bulgaria, 424 pp. (1984). Abstr. in Phys. Abstr., Vol. 89, No. 1281, Entry 61199 (1986).

002.082 **List of papers published by Prof. B. V. Sreekantan.**
Cosmic pathways, p. 396 – 406 (1986). – See Abstr. 003.045.

002.083 **Statistics of the papers published in Acta Astronomica during the period 1975 – 1983.**
J. Smak.
Postepy Astron., Tom 32, Zesz. 2, p. 181 (1984). In Polish.

002.084 **The Observatorio Nacional – Center for Astrophysics Redshift Survey. A status report.**
L. N. da Costa, P. S. Pellegrini, C. Willmer, M. A. Nunes.
Rev. Mex. Astron. Astrofis., Vol. 12, Spec. Issue, p. 63 – 72 (1986). – See Abstr. 012.066.
The authors report on the present status of the ongoing redshift survey of the southern sky.

002.085 **The new index catalog of visual double stars – WDS.**
C. E. Worley, G. G. Douglass.
Astrometric techniques, p. 649 – 651 (1986). – See Abstr. 012.068 (IAU Symp. No. 109).
A new index catalog of visual double stars, the Washington Double Star Catalog (WDS), 1984.0, is now available in tape form. All of the double star data has been updated, as well as the notes to the catalog. This report summarizes the contents of the WDS and describes its compilation.

002.086 **The IAU numbering system of radio sources.**
B. Elsmore.
Astrometric techniques, p. 737 – 738 (1986). – See Abstr. 012.068 (IAU Symp. No. 109).
The convention for numbering radio sources adopted by the IAU uses the 1950 coordinates of right ascension in hours and minutes, followed by the declination truncated to tenths of a degree. Now that the current practice is to use the new system of astronomical constants and to adopt the equator and equinox of 2000 as standard, the question must be raised again as to whether the 1950 coordinates should continue to be used in an approved numbering system.

002.087 **The preparation and distribution of machine–readable astrometric data.**
W. H. Warren Jr.
Astrometric techniques, p. 739 – 748 (1986). – See Abstr. 012.068 (IAU Symp. No. 109).
The acquisition, preparation and distribution of machine–readable astrometric data by the Astronomical Data Center are described. Current development work in the areas of astrometric and positional catalogs is mentioned, too.

002.088 **The status of the plates taken for the charts and the catalogue of the Carte du Ciel (Astrographic Catalogue).**
H. Eichhorn, C. Jaschek.
Astrometric techniques, p. 807 – 810 (1986). – See Abstr. 012.068 (IAU Symp. No. 109). Appendix B of IAU Symp. No. 109.

002.089 **An atlas of nova shells.**
W. C. Seitter, H. Duerbeck.
Mitt. Astron. Ges., No. 65, p. 208 (1986). – See Abstr. 012.072.

002.090 **The Astrophysical Journal General Index by authors and by subjects, Volumes 243 – 299 (1 January 1981 – 15 December 1985) and to the Supplement Series, Volumes 45 – 59 (January 1981 – December 1985).**
Published by the University of Chicago Press for the American Astronomical Society. The University of Chicago Press, 5801 S. Ellis Avenue, Chicago, Ill. 60637, USA. 3 + 710 pp. (1986).

002.091 **Data catalog series for space science and applications flight missions. Volume 3B. Descriptions of data sets from low– and medium–altitude scientific spacecraft and investigations.**
J. E. Jackson, R. Horowitz (Editors).
NSSDC/WDC–A–R&S, Publ., No. 86–01, 6 + 102 pp. (1986).

002.092 **A statistical analysis and overview of the IRAS point source catalog.**
T. Chester.
Light on dark matter, p. 3 – 22 (1986). – See Abstr. 012.080.
The IRAS point source catalog can be segregated into stars, "IR galactic components", galaxies, cirrus, and "disk and bulge stars" categories. For each source type, the types and distances of the sources are discussed, and the sky distribution, log N/log S, and identification rate are given.

002.093 **A statistical analysis of the LRS catalog.**
F. M. Olnon.
Light on dark matter, p. 31 – 38 (1986). – See Abstr. 012.080.
The IRAS catalog of low–resolution spectra contains 8 – 22 μm spectra of 5425 relatively bright point sources distributed over the whole sky. The various types of spectra (LRS classes) can readily be associated with well–known groups of objects. About 96% of the sources are stars, most of them with oxygen– or carbon–rich envelopes, five sources are galaxies, and the others are compact H II regions, planetary nebulae, and 'peculiar stars'.

002.094 **Black holes: an annotated bibliography, 1975 – 1983.**
S. I. Danko.
Scarecrow Press, Inc., 52 Liberty St., P.O. Box 656, Metuchen, N.J. 08840, USA. 14 + 282 pp. Price US$ 25.00 cloth (1985). ISBN 0–8108–1836–1.
Review in Space Sci. Rev., Vol. 44, Nos. 1/2, p. 188; 1986 (*V. Verbunt*).

002.095 **Atlas van de sterrenhemel.**
S. Dunlop, W. Tirion.
Uitgeverij Becht, Amsterdam, The Netherlands. 80 pp. Price *f* 34.90 (1985). ISBN 90–230–0584–8.
Review in Zenit, 13. Jaarg., Nr. 5, p. 186; 1986 (*A. Nagel*).

002.096 **The Macmillan Dictionary of Astronomy.**
V. Illingworth.
Second, revised, and extended edition. Macmillan Press, London, England. 5 + 437 pp. Price £ 25.00 cloth (1985). ISBN 0–333–39062–8 cloth, ISBN 0–333–39243–4 paper.
Review in Observatory, Vol. 106, No. 1072, p. 80; 1986 (*C. Jenkins*).

002.097 **Check a possible supernova – Set 2. (*Forty photographs of galaxies*).**
J. Salmi.
JS–Observatory, Juhani Salmi, Versijärvenk, 36 C 40, SF–15140 Lahti, Finland. Available from Webb Society, 9 Fairacres Close, Herne Bay CT6 6LL, England. Axel Brandenburg, Eichholz 54, D–2000 Hamburg 11, F.R. Germany. 6 pp. + 40 cards, Price FM 35.00, £ 3.65, DM 14.00 (1985).
Reviews in Ciel Terre, Vol. 102, No. 1, p. 29, 1986 (*A. Heck*); Sterne Weltraum, 25. Jahrg., Nr. 1, p. 54; 1986 (*M. Möller*).

002.098 **Sarna Deep Sky Atlas.**
T. M. Sarna.
Second edition. Willmann–Bell Inc., P.O. Box 3135, Richmond, Virginia 23235, USA. 110 pp. Price US$ 29.95 paper (1984). ISBN 0–943396–06–9.
Reviews in Mercury, Vol. 15, No. 3, p. 86 (1986); Orion, 44. Jahrg., Nr. 212, p. 34; 1986 (*A. Tarnutzer*).

An assessment of research done at the national optical observatories.
See Abstr. 013.001.

Problems in data compilation.
See Abstr. 036.066.

The Southern Reference Star program: a 20–year international project nearing completion.
See Abstr. 041.021.

The extension of the fundamental system to fainter magnitudes.
See Abstr. 041.022.

Results of a comparison of the Kharkov catalogue of declinations of 1407 nearpolar stars with the catalogue of Fabricius.
See Abstr. 041.039.

On the SRS catalogue.
See Abstr. 041.074.

Presentation of new solar and planetary tables of interest for historical calculations.
See Abstr. 046.005.

An atlas of photospheric magnetic field observations and computed coronal magnetic fields: 1976 – 1985.
See Abstr. 075.050.

The Infrared Astronomical Satellite Asteroid database.
See Abstr. 098.018.

Atlas of photometric asteroid lightcurves.
See Abstr. 098.026.

A minor planet database for your microcomputer.
See Abstr. 098.085.

Statistical catalogue of the parameters of orbits of long–period comets in the Laplace coordinate system.
See Abstr. 102.056.

Presentation of an Atlas of Cometary Spectra.
See Abstr. 103.008.

Stellar motions in the solar neighborhood.
See Abstr. 111.009.

New subdwarfs. V. Radial velocities for 889 high–proper–motion stars measured with the Mount Wilson 100 inch reflector.
See Abstr. 111.016.

Common proper motion stars in the AGK3.
See Abstr. 111.019.

Proper motion survey with the 48–inch Schmidt Telescope. LXVI. Proper motions for 7698 stars in the Hyades region.
See Abstr. 111.027.

Proper motion survey with Schmidt telescopes. LXVII. Proper motions for 210 faint stars.
See Abstr. 111.028.

OH/IR catalogue and correlation with the IRAS data base.
See Abstr. 112.141.

A high resolution atlas of two T Tauri stars.
See Abstr. 121.006.

IAU archives of unpublished photoelectric observations of variable stars (1985/6).
See Abstr. 123.016.

A catalog of the [Hα + N II] emission features in the Kepler SNR.
See Abstr. 125.083.

Dark cloud statistics.
See Abstr. 131.390.

High latitude molecular clouds: completeness of the survey and implications for molecular surveys.
See Abstr. 131.391.

Positions of 127 Hyads and 6–cm observations of 320 Hyads.
See Abstr. 153.023.

Catalogue of stars of the central region of the M13 globular cluster.
See Abstr. 154.011.

Systematic reinvestigation of the radial velocities of the galactic globular clusters: image–tube results.
See Abstr. 154.057.

The Massachusetts–Stony Brook galactic plane CO survey: observations, galactic structure, and cloud identification.
See Abstr. 155.090.

H I–observations of galaxies in the Kraan–Korteweg – Tammann catalogue of nearby galaxies. I. The data.
See Abstr. 157.014.

Static properties of galaxies in the Shapley–Centaurus cluster. I. The catalogue.
See Abstr. 157.139.

A new faint QSO survey.
See Abstr. 159.060.

Quasars to $B > 22.5$ in Selected Area 57: a catalog of multicolor photometry, variability, and astrometry.
See Abstr. 159.086.

Compilation of clusters of galaxies with published redshifts.
See Abstr. 160.116.

The Centaurus cluster of galaxies – I. The data.
See Abstr. 160.118.

Anisotropy of the galaxies detected by *IRAS*.
See Abstr. 161.125.

003 Books

003.001 **Physics of the Sun. Vol. I: The solar interior. Vol. II: The solar atmosphere. Vol. III: Astrophysics and solar–terrestrial relations.**
P. A. Sturrock, T. E. Holzer, D. M. Mihalas, R. K. Ulrich (Editors), with an introduction and summary by P. A. Sturrock. Geophysics and Astrophysics Monographs. D. Reidel Publishing Company, Dordrecht – Boston – Lancaster – Tokyo.
Vol. I: 9 + 257 pp. Price Dfl. 130.00, US$ 44.50, £ 36.50. ISBN 90–277–1860–1.
Vol. II: 11 + 385 pp. Price Dfl. 180.00, US$ 62.00, £ 49.95. ISBN 90–277–1861–X.
Vol. III: 11 + 287 pp. Price Dfl. 140.00, US$ 49.00, £ 38.95. ISBN 90–277–1862–8.
Set: Price Dfl. 360.00, US$ 125.00, £ 99.75 (1986). ISBN 90–277–1823–7.
The individual contributions are included in their corresponding subject categories – see abstracts 064.010, 073.008 – 073.010, 074.007, 077.003, 080.010 – 080.019, 084.007, 085.003, 085.004, 107.004, 116.004.

003.002 **Spinors and space–time. Volume 2: Spinor and twistor methods in space–time geometry.**
R. Penrose, W. Rindler.
Cambridge monographs on mathematical physics. Cambridge University Press, Cambridge – London – New York – New Rochelle – Melbourne – Sydney. 9 + 501 pp. Price £ 45.00, US$ 89.50 (1986). ISBN 0–521–25267–9.
This is a companion volume to the introductory work *Spinors and space–time*, Volume 1: *two–spinor calculus and relativistic fields*. (See Abstr. 39.003.151). The main topics introduced and discussed in the present volume are twistor theory and related matters such as massless fields and the geometry of light rays, energy–momentum and angular momentum, and the conformal structure of infinity. In addition, a detailed classification of space–time curvature by 2–spinor means is provided. The main concern is the classification of the Weyl (or empty–space Riemann) tensor, but a comprehensive treatment of the Ricci tensor is also given, showing how this arises as a particular case of the classification of symmetric spinors generally.

003.003 **History of British space science.**
H. Massey, M. O. Robins.
Cambridge University Press, Cambridge – London – New York – New Rochelle – Melbourne – Sydney. 21 + 514 pp. Price £ 45.00, US$ 89.50 (1986). ISBN 0–521–30783–X.
Contents: The scientific background. The technical background. The initiation of the Skylark rocket programme – the IGY and artificial satellites. Post–IGY developments – NASA – COSPAR – British National Committee for Space Research – British satellite experiments. The Ariel programme. The European Space Research Organization. Commonwealth co–operation in space research. Smaller rockets for scientific purposes – Skua and Petrel. Attitude controlled Skylark rockets. The Trend Committee and the Science Research Council. The

transformation of ESRO into ESA. The Space Science Committee for Europe. Scientific studies by British space scientists – figure of the earth and the neutral atmosphere. Scientific studies by British space scientists – the ionosphere, the magnetosphere and cosmic rays. The contribution from British space scientists to astronomy. Concluding remarks.

003.004 **Photochemistry of the atmospheres of Mars and Venus.**
V. A. Krasnopolsky (*V. A. Krasnopol'skij*),
U. von Zahn (Technical Editor).
Physics and Chemistry in Space, Vol. 13. Springer–Verlag, Berlin – Heidelberg – New York – Tokyo. 9+334 pp. Price DM 248.00 (1986). ISBN 3–540–15336–5 (F.R. Germany), ISBN 0–387–15336–5 (USA). For the Russian original see 33.003.062.

003.005 **Investigations on classical and quantum theory of gravitation.**
Collection of papers of the Dnepropetrovsk University, Dnepropetrovsk. 85 pp. (1985). In Russian.
Review in Ref. Zh., 51. Astron., 3.51.52 (1986).

003.006 **Sothis- und Monddaten. Studien zur astronomischen und technischen Chronologie Altägyptens.**
R. Krauss.
Hildesheimer ägyptologische Beiträge, Vol. 20. Gerstenberg Verlag, Hildesheim, F.R. Germany. 12+219 pp. (1985). ISBN 3–8067–8086–X.

003.007 **Galaxies, axisymmetric systems and relativity. Essays presented to W. B. Bonnor on his 65th birthday.**
M. A. H. MacCallum (Editor).
Cambridge University Press, Cambridge – London – New York – New Rochelle – Melbourne – Sydney. 20+300 pp. Price £ 25.00, US$ 49.50 (1985). ISBN 0–521–30812–7.
Reviews in Astrophys. Space Sci., Vol. 122, No. 1, p. 189 – 190; 1986 (*J. Dyson*); Space Sci. Rev., Vol. 43, Nos. 3/4, p. 393 – 394; 1986 (*V. Icke*).
The individual contributions are included in their corresponding subject categories – see abstracts 005.004, 066.051 – 066.066, 161.049 – 161.052.

003.008 **Long-term evolution of short-period comets.**
A. Carusi, Ľ. Kresák, E. Perozzi, G. B. Valsecchi.
Adam Hilger Ltd, Bristol – Boston. 350 pp Price £ 50.00, US$ 85.00 hardcover ringbound (1985). ISBN 0–85274–489–7.
Reviews in Icarus, Vol. 66, No. 3, p. 632; 1986 (*D. K. Yeomans*); Planet. Space Sci., Vol. 34, No. 6, p. 573; 1986 (*R. Larmor*); Space Sci. Rev., Vol. 44, Nos. 1/2, p. 186; 1986 (*J. M. A. Danby*).
This book contains some results of the Long-Term Evolution Project (LTEP) conducted in the last three years by the Planetology Branch of the Space Astrophysics Institute, CNR, Roma, Italy, and the Interplanetary Matter Department of the Astronomical Institute, SAV, Bratislava, Czechoslovakia. The LTEP consists in the integration of motion of all known 132 comets with revolution periods less than 200 years, over a time span of 821.4 years (exactly: 300,000 days), covering the period from 1585 to 2406 AD. This represents a grand total of 108,421 years, including 11,793 revolutions of different comets. For comparison, observational records cover, as to 1 January 1985, only 556 apparitions of these objects.

003.009 **Eléments d'astronomie fondamentale.**
M. Danloux–Dumesnils.
Librairie Scientifique et Technique, Albert Blanchard, 9, rue de Médicis, Paris, France. 407 pp. Price FF 210.00 (1985). ISBN 2–9500722–0–8.
Review in Ciel Terre, Vol. 102, No. 2, p. 61; 1986 (*L. Louys*).

003.010 **Astronomy with your personal computer.**
P. Duffett-Smith.
Cambridge University Press, Cambridge – London – New York – New Rochelle – Melbourne – Sydney. 10+256 pp. Price £ 25.00, US$ 44.50 cloth, £ 9.95, US$ 14.95 paper (1985). ISBN 0–521–26620–3 cloth, ISBN 0–521–31976–5 paper.

003.011 **Fabry–Perot interferometers.**
G. Hernandez.
Cambridge Studies in Modern Optics, Vol. 3. Cambridge University Press, Cambridge – London – New York – New Rochelle – Melbourne – Sydney. 16+343 pp. Price £ 37.50, US$ 69.50 (1986). ISBN 0–521–32238–3.
Contents: Historical perspective. Mathematical development. Optimum operation. Multiple–etalon devices. Luminosity and resolution considerations. Active media in the etalon spacer. Practical Fabry–Perot interferometers. Non–classical Fabry–Perot devices.

003.012 **Space missions to Halley's comet.**
R. Reinhard, B. Battrick (Editors).
ESA Spec. Publ., ESA SP–1066. 16+253 pp. Price FF 200.00 (1986).
The individual contributions are included in their corresponding subject categories – see abstracts 013.015 – 013.018, 051.011 – 051.021, 103.615 – 103.617.

003.013 **The Giotto mission – its scientific investigations.**
R. Reinhard, B. Battrick (Editors).
ESA Spec. Publ., ESA SP–1077. 25+193 pp. Price FF 200.00 (1986).
The individual contributions are included in their corresponding subject categories – see abstracts 035.005 – 035.015, 051.022, 051.023.

003.014 **From SU(3) to gravity. Festschrift in honor of Yuval Ne'eman.**
E. Gotsman, G. Tauber (Editors).
Cambridge University Press, Cambridge – London – New York – New Rochelle – Melbourne – Sydney. 17+457 pp. Price £ 40.00, US$ 59.50 (1985). ISBN 0–521–30784–8.
The individual contributions within the subject scope of Astronomy and Astrophysics Abstracts are included in their corresponding categories – see abstracts 013.024, 015.022, 061.019, 061.020, 066.077 – 066.082, 067.038, 080.037, 112.042, 151.042.

003.015 **Astronomy & Astrophysics.**
M. S. Roberts (Editor).
AAAS Publication No. 84–5. American Association for the Advancement of Science, 1333 H Street, NW, Washington, D.C. 20005, USA. 7+383 pp. Price US$ 29.95 cloth, US$ 17.95 paper (1985). ISBN 0–87168–311–3 cloth, ISBN 0–87168–275–3 paper.
The material in chapters 1 through 24 has been updated and corrected from the original articles that appeared in Science issues published between 1982 and 1984. For the original articles see abstracts 31.085.30, 31.073.135, 33.144.010, 33.102.012, 33.155.115, 37.115.007, 37.064.021, 37.131.161, 37.131.225, 38.117.018, 38.151.013, 38.116.019, 39.065.080, 31.158.148, 31.151.059, 33,157.246, 37.159.089, 37.161.160, 37.161.290, 38.158.070, 39.061.048, 39.161.223, 32.033.002, 39.051.081.

003.016 **Inverse problems in astronomy. A guide to inversion strategies for remotely sensed data.**
I. J. D. Craig, J. C. Brown.
Adam Hilger Ltd., Bristol – Boston. 9+152 pp. Price £ 19.00, US$ 29.00, DM 80.10 (1986). ISBN 0–85274–369–6.
Contents: Introduction. Some linear integral equations and their occurrence in astronomy: A: Fredholm equations, B: Volterra equations. Mathematical aspects of the inversion problem. Classical methods of numerical inversion. Non–classical inversion and stabilisation techniques. Numerical examples. Discussion, conclusions and recommended strategy.

003.017 **Die Entstehung der Sterne. Interstellare Materie.**
Selected papers from Spektrum der Wissenschaft. With an introduction by J. Krautter.
Spektrum der Wissenschaft Verlagsgesellschaft mbH & Co., Heidelberg, F.R. Germany. 190 pp. Price DM 39.00 (1986). ISBN 3–922508–35–9.

003.018 **Stellar flares. Physics. Cosmogony.**
G. A. Gurzadyan.
Glavnaya Redaktsiya Fiziko–Matematicheskoj Literatury, Nauka, Moskva. 560 pp. Price 6 Rbl. 40 Kop. (1985). In Russian.

003.019 **Sun and earth.**
H. Friedman.
Scientific American Library, No. 15. Scientific American Books, Inc., New York, USA. 12+251 pp. Price £ 15.95 (1986). ISBN 0–7167–5012–0.
Contents: From Stonehenge to Skylab. Inside the sun: core to surface. The solar atmosphere. An inconstant sun: electromagnetic radiation. An inconstant sun: solar particle radiation. In search of a climate connection. Origins and endings.

003.020 **Einstein's legacy. The unity of space and time.**
J. Schwinger.
Scientific American Library, No. 16. Scientific American Books, Inc., New York, USA. 14+250 pp. Price £ 15.95 (1986). ISBN 0–7167–5011–2.
Review in Nature, Vol. 320, No. 6063, p. 662; 1986 (*A. Pais*).

003.021 **Der Himmel bei Tag und Nacht. Erkennen – Beobachten – Fotografieren.**
C. A. Ronan (Editor), translated from the English original by H.–M. Hahn.
Kosmos, Franckh'sche Verlagshandlung, W. Keller & Co., Stuttgart, F.R. Germany. 223 pp. Price DM 39.50 (1986). ISBN 3–440–05627–9. For the English original see 40.003.166.
Contents: Der Taghimmel (*S. Dunlop*). Der Nachthimmel (*C. A. Ronan*). Die Beobachtung und Fotografie des Himmels (*B. Jones*).

003.022 **Die Kosmos–Himmelskunde.**
J. Herrmann.
Kosmos, Franckh'sche Verlagshandlung, W. Keller & Co., Stuttgart, F.R. Germany. 191 pp. Price DM 39.50 (1986). ISBN 3–440–05618–X.
Review in Sternenbote, 29. Jahrg., Nr. 6, p. 122 – 123 (1986).

003.023 **Problems of astronomy.**
Odes. univ. Odessa. 120 pp. (1985). In Russian.
Review in Ref. Zh., 51. Astron., 3.51.39 (1986).
See abstracts 022.057, 036.077, 036.078, 042.049, 062.050, 064.067, 098.068, 117.137 – 117.139, 119.043, 122.080 – 122.083.

003.024 **Astronomical–geodetical investigations.**
Collection of papers of the Ural University, Sverdlovsk. 178 pp. (1985). In Russian.
Review in Ref. Zh., 51. Astron., 4.51.41 (1986).
See abstracts 072.046, 085.014, 113.035, 117.152, 119.045 – 119.047, 151.078 – 151.080, 153.028, 157.175.

003.025 **The ionosphere and solar–terrestrial relations.**
V. I. Drobzhev (Editor).
Nauka, Alma–Ata. 187 pp. (1985). In Russian.
Review in Ref. Zh., 51. Astron., 4.51.49 (1986).

003.026 **Dark companions of stars. Astrometric commentary on the lower end of the main sequence.**
P. van de Kamp, with an introduction by W. B. Burton and contributions by C. W. Tombaugh and R. S. Harrington.
Space Sci. Rev., Vol. 43, Nos. 3/4, p. 211 – 327 (1986). Available also as hardbound edition. Price Dfl. 95.00, US$ 44.50, £ 30.25. ISBN 90–277–2270–6.
Contents: Part I. Our cosmic neighbourhood: The "fixed" stars and the "wandering" planets. Stellar motions. Stellar distances. Stellar parameters. The nearby stars. The lower end of the main sequence. Our solar system. Origins. Part II. Unseen stars and planets: Historical I. Historical II. Perturbations. Perturbations – early mostly photographic studies. Perturbations – current photographic studies. Two cases of serendipity: Ross 614 and VW Cephei. Astrometric study of Barnard's star. Epilogue.

003.027 **Kinetic theory of particles and photons. Theoretical foundations of non–LTE plasma spectroscopy.**
J. Oxenius.
Springer Series in Electrophysics, Vol. 20. Springer Verlag, Berlin – Heidelberg – New York – Tokyo. 12+353 pp. Price DM 120.00 (1986). ISBN 3–540–15809–X (F.R. Germany), ISBN 0–387–15809–X (USA).
Contents: Thermal equilibrium and detailed balance. Kinetic equations of particles. The kinetic equation of photons. Moment equations and fluid description. H theorem for gases and radiation. Energy exchange between matter and radiation. Momentum exchange between matter and radiation.

003.028 **Methods in the qualitative theory of dynamical systems in astrophysics and gas dynamics.**
O. I. Bogoyavlensky (*O. I. Bogoyavlenskij*), translated from the Russian by D. Gokhman.
Springer Series in Soviet Mathematics. Springer Verlag, Berlin – Heidelberg – New York – Tokyo. 9+301 pp. Price US$ 49.00 (1985). ISBN 3–540–13614–2 (F.R. Germany), ISBN 0–387–13614–2 (USA). For the Russian original see 29.003.026.
Contents: Methods of qualitative analysis of multi–dimensional dynamical systems. Qualitative theory of homogeneous cosmological models without the motion of matter. Qualitative theory of homogeneous cosmological models with the motion of matter and electromagnetic fields. Self–similar spherically symmetric solutions for the general theory of relativity. Self–similar motion of self–gravitating gas in stars. Self–similar rotation of an ideal gas. The dynamics of a gaseous ellipsoid. The dynamics of perturbations of the periodic Toda lattice.

003.029 **Astrophysical phenomena and radio carbon.**
Fiz.–tekh. inst. Akad. Nauk SSSR, Leningrad. 199 pp. (1985). In Russian.
Review in Ref. Zh., 51. Astron., 6.51.70 (1986).

003.030 **Gamma–ray astronomy.**
P. V. Ramana Murthy, A. W. Wolfendale.
Cambridge Astrophysics Series. Cambridge University Press, Cambridge – London – New York – New Rochelle – Melbourne – Sydney. 10+248 pp. Price £ 30.00, US$ 49.50 (1986). ISBN 0–521–30528–4.
Contents: (1) Production and absorption mechanisms. (2) Gamma–ray line astronomy. (3) Gamma–ray bursts. (4) Medium energy gamma–rays. (5) Ultra–high energy gamma–rays.

003.031 **Der Halleysche Komet im Jahre 1910. Eine Sammlung zeitgenössischer Berichte (1909 – 1911).**
Mit einem Vorwort von K. Fritze.
Zentralantiquariat der Deutschen Demokratischen Republik, Postfach 1080, DDR–7010 Leipzig, GDR. 15+143 pp. Price DM 15.00 (1985).
This collection of reprints contains among many brief reports F. S. Archenhold's extensive essay "Kometen, Weltuntergangsprophezeiungen und der Halleysche Komet. Zweite Auflage, Berlin–Treptow (1910)."

003.032 **Introduction to gravitation.**
V. de Sabbata, M. Gasperini.
World Scientific Publishing Co. Pte. Ltd., Singapore. 9+346 pp. Price US$ 36.00 cloth (1985). ISBN 9971–50–049–3 cloth, ISBN 9971–50–050–7 paper.
Contents: (1) Geometry and gravitation. (2) The formalism of general relativity. (3) Gravitational field equations. (4) The three classical tests of Einstein's theory. (5) Elements of cosmology. (6) Relativistic cosmological models. (7) Non–static models of the universe. (8) Gravitational waves. (9) Dense and collapsed matter. (10) The Einstein–Cartan theory. (11) The strong gravity theory. (12) Gauge theory of gravity. (13) Supergravity. (14) Gravitational theory in the language of the exterior forms.

003.033 **Kometen.**
D. Möhlmann, K. Sauer, R. Wäsch.
Akademie–Verlag, Berlin, GDR. 119 pp. Price DM 18.00 (1986).
Contents: (1) Einleitung. (2) Aus der Geschichte der Kometenbeobachtung. (3) Himmelsmechanik der Kometen. (4) Der Kometenkern. (5) Chemie und Mineralogie der festen Kometenmaterie. (6) Die Kometenatmosphäre und ihre Wechselwirkung mit dem Sonnenwind. (7) Der Staubschweif. (8) Herkunft und Entstehung der Kometen. (9) Missionen zum Kometen Halley.

003.034 **Physics of comets.**
K. S. Krishna Swamy.
World Scientific Publishing Co. Pte. Ltd., Singapore. 15+273 pp. Price US$ 24.00 (1986). ISBN 9971–978–18–0.
Contents: (1) General introduction. (2) Dynamics. (3) Physical aspects. (4) Spectra. (5) Spectra of coma. (6) Gas–production rates in coma. (7) Dust tail. (8) Light scattering theory. (9) The nature of the dust particles. (10) Ion tails. (11) Nucleus. (12) Origin. (13) Relation to other Solar System studies. (14) Problems and prospects.

003.035 **Photometric and polarimetric investigations of celestial bodies.**
A. V. Morozhenko (Editor).
Akademiya Nauk Ukrainskoj SSR, Glavnaya Astronomicheskaya Observatoriya. Naukova Dumka, Kiev. 200 pp. Price 3 Rb. 20 Kop. (1985). In Russian.
The individual contributions are included in their corresponding subject categories – see abstracts 031.026, 034.104 – 034.108, 035.099, 036.149, 036.150, 036.189, 063.079 – 063.084, 063.108, 063.109, 071.037, 091.043, 100.060, 102.071, 103.202, 103.203, 114.109 – 114.112, 122.133 – 122.135, 131.292, 158.255.

003.036 **Elements of relativity theory.**
D. F. Lawden.
John Wiley & Sons, Chichester – New York – Brisbane – Toronto – Singapore. 10+108 pp. Price £ 7.95 paper (1985). ISBN 0–471–90852–5.
Contents: (1) Special principle of relativity. (2) Lorentz transformation. (3) Velocity transformations. (4) Relativistic mechanics. (5) General theory of relativity.

003.037 **Short guide to modern star names and their derivation.**
P. Kunitzsch, T. Smart.
Otto Harrassowitz, Wiesbaden, F.R. Germany, 5+64 pp. Price DM 39.80 (1986). ISBN 3–447–02580–8.

003.038 **Comet Halley 1909 – 1911. Retrospect of observations.**
D. Möhlmann, K. Sauer (Editors).
Akademie–Verlag, Berlin, GDR. 161 pp. Price DM 26.80 (1986).

003.039 **Interferometry and synthesis in radio astronomy.**
A. R. Thompson, J. M. Moran, G. W. Swenson Jr.
A Wiley–Interscience Publication. John Wiley & Sons, New York – Chichester – Brisbane – Toronto – Singapore. 21+534 pp. Price US$ 59.95 (1986). ISBN 0–471–80614–5.
Contents: (1) Introduction and historical review. (2) Introductory theory of interferometers and correlator arrays. (3) Further theory of the interferometer response. (4) Geometrical relationships and other practical considerations. (5) Design of arrays. (6) Response of the receiving system. (7) Design of the analog receiving system. (8) Digital signal processing. (9) Very–long–baseline interferometry. (10) Calibration and Fourier transformation of visibility data. (11) Image processing and enhancement. (12) Interferometer techniques for astrometry and geodesy. (13) Propagation effects. (14) Radio interference. (15) Related techniques.

003.040 **Physical processes in the circumterrestrial plasma.**
Akademiya Nauk SSSR. Ordena Lenina Sibirskoe Otdelenie Yakutskij Filial, Institut kosmofizicheskikh issledovanij i aehronomii, Yakutsk. 140 pp. Price 1 Rbl. 9 Kop. (1984). In Russian.
The individual contributions within the subject scope of Astronomy and Astrophysics Abstracts are included in their corresponding categories – see abstracts 084.057, 085.024, 085.025, 106.073, 144.100, 144.101.

003.041 **Annual Review of Nuclear and Particle Science. Vol. 35.**
J. D. Jackson, H. E. Gove, R. F. Schwitters (Editors).
Annu. Rev. Nucl. Part. Sci., Vol. 35, 684 pp. (1985). ISBN 0–8243–1535–9.
Review in Phys. Abstr., Vol. 89, No. 1282, Entry 67676 (1986). See abstracts 034.177, 067.139.

003.042 **Annual Review of Earth and Planetary Sciences, Volume 14.**
G. W. Wetherill, A. L. Albee, F. G. Stehli (Editors).
Annu. Rev. Earth Planet. Sci., Vol. 14, 9+593 pp. Price US$ 47.00 (1986). ISBN 0–8243–2014–X.
The individual contributions within the subject scope of Astronomy and Astrophysics Abstracts are included in their corresponding categories – see abstracts 081.047, 081.048.

003.043 **The Venus International Reference Atmosphere.**
A. J. Kliore, V. I. Moroz, G. M. Keating (Editors).
Adv. Space Res., Vol. 5, No. 11, 8+305 pp. (1985). ISBN 0–08–034631–6.
The individual contributions are included in their corresponding subject categories – see abstracts 093.050 – 093.056.

003.044 **The Cambridge photographic atlas of the planets.**
G. Briggs, F. Taylor.
Cambridge University Press, Cambridge – London – New York – New Rochelle – Melbourne – Sydney. 256 pp. Price £ 7.95, US$ 14.95 (1986). ISBN 0–521–31058–X paper.
Paperback edition of the title first published in 1982 as cloth edition (see Abstr. 32.002.144), with an additional chapter (4 pp.) on the Uranian system based on Voyager 2 fly–by results.

003.045 **Cosmic pathways. Contemporary perspectives in physics and astrophysics. Essays written in honour of Prof. B. V. Sreekantan on his 60th birthday.**
R. Cowsik (Editor), with a foreword by M. G. K. Menon.
Tata McGraw–Hill Publishing Company Limited, New Delhi. 17+411 pp. (1986). ISBN 0–07–451567–5.
The individual contributions within the subject scope of Astronomy and Astrophysics Abstracts are included in their corresponding categories – see abstracts 002.082, 004.101, 005.024 – 005.026, 009.060, 013.059 – 013.061, 067.144, 067.145, 117.288, 142.099, 142.100, 143.030, 143.031, 144.110 – 144.119, 161.393, 161.394.

003.046 **"Lilienthal oder die Astronomen". Historische Materialien zu einem Projekt Arno Schmidts.**
J. Drews, H. Schwier (Editors).
edition text + kritik, München, F.R. Germany. 351 pp. Price DM 69.00 (1984). ISBN 3–88377–169–4.
The individual contributions are included in their corresponding subject categories – see abstracts 004.103, 004.104, 005.028 – 005.038.

003.047 **Physik des Planeten Erde. Physik des Erdkörpers, der Hydrosphäre und der Atmosphäre.**
R. Lauterbach (Editor).
2., neu bearbeitete Auflage. Akademie–Verlag, Berlin, GDR. 286 pp. Price DM 51.40. (1985).
The individual contributions within the subject scope of Astronomy and Astrophysics Abstracts are included in their corresponding categories – see abstracts 081.056, 081.057, 082.087, 082.088, 084.068, 094.070, 107.048.

003.048 **Space, time, infinity.**
J. S. Trefil.
Pantheon Books, New York; Smithsonian Books, Washington. 256 pp. Price US$ 29.95 (1985). ISBN 0–394–54843–4.

003.049 **Planetary landscapes.**
R. Greeley.
Allen & Unwin Ltd., London – Boston – Sydney. 9 + 265 pp. Price £ 24.95, US$ 44.95 (1985). ISBN 0–04–551080–6.
Contents: Introduction. Geologic exploration of the solar system. Planetary morphologic processes. The moon. Mercury. Venus. Mars. The Jupiter system. The Saturn system. Epilogue.

003.050 **Galaxies.**
P. W. Hodge.
Harvard University Press, Cambridge, Mass., USA – London, England. 9 + 174 pp. Price US$ 25.95, £ 18.95 (1986). ISBN 0–674–34065–5.
Contents: Galaxies and the Universe. Galactic structure. The formation and evolution of galaxies. Stars and clusters, gas and dust. The missing mass. The Magellanic Clouds. The Local Group. The nearest spirals. Clusters and superclusters of galaxies. The distance scale. Disturbed galaxies. Quasars.

003.051 **Advances in planetary geology. Vol. 1.**
NASA Tech. Memo., NASA TM–88784, Vol. 1, 5 + 572 pp. (1986).
The individual contributions are included in their corresponding subject categories – see abstracts 091.056, 097.036.

003.052 **Giotto special issue.**
ESA Bull., No. 46, 114 pp. (1986).
The individual contributions within the subject scope of Astronomy and Astrophysics Abstracts are included in their corresponding categories – see abstracts 013.051, 035.098, 051.101, 051.102, 052.062, 052.063, 103.722, 103.723.

003.053 **The analysis of extraterrestrial materials.**
I. Adler.
Chemical analysis, Vol. 81. John Wiley & Sons, New York – Chichester – Brisbane – Toronto – Singapore, 21 + 346 pp. Price DM 164.30 (1986). ISBN 0–471–87880–4.
Contents: Part 1: (1) Planetary exploration: its impact on the development of analytical methods. (2) The first ET's, the meteorites. (3) The lunar receiving laboratory. (4) Lunar samples: detailed analysis. (5) Isotopes and cosmochronology. (6) Remote analysis. (7) Remote sensing by reflectance spectroscopy. (8) Surface analysis from orbiting spacecraft. (9) Remote in situ analysis of planetary surfaces. (10) Mars atmopsheric and surface measurements. (11) The Viking X–ray fluorescence experiment. (12) Venus – the chemistry of its atmosphere and surface. (13) A brief look at the future.

003.054 **Halley: the once–in–a–lifetime comet.**
D. Allen, C. Allen.
Allen & Unwin Ltd., London – Boston – Sydney. 101 pp. Price $A 9.95 (1985). ISBN 0–86861–567–6.
Reviews in Mercury, Vol. 15, No. 3,p. 87 (1986); Proc. Astron. Soc. Aust., Vol. 6, No. 2, p. 272 – 273; 1986 (*J. W. O'Byrne*); Sky Telesc., Vol. 71, No. 1, p. 43 (1986).

003.055 **The extraterrestrial encyclopedia. Our search for life in outer space.**
J. A. Angelo Jr.
Facts on File, New York, USA. 10 + 254 pp. Price US$ 24.95 (1985).
From Science, Vol. 231, No. 4743, p. 1318 (1986).

003.056 **The exploding suns: the secrets of the supernovas.**
I. Asimov.
Dutton. 276 pp. Price US$ 18.95 cloth (1985).
Review in Mercury, Vol. 15, No. 3, p. 94 (1986).

003.057 **Guide de la comète de Halley.**
I. Asimov.
Translated from the English edition. Editions du Rocher, Monaco, Monaco. 124 pp. Price FF 78.00 paper (1985). For the English original see 39.003.193.
Review in Ciel Terre, Vol. 102, No. 2, p. 64; 1986 (*J. Sauval*).

003.058 **Halley's comet.**
M. E. Bailey.
Department of Astronomy, The University, Manchester, M13 9PL, England. 90 pp. Price £ 1.50 (1985).
Review in Spaceflight, Vol. 28, No. 3, p. 130 (1986).

003.059 **Die asymmetrische Schöpfung. Ursprung und Ausdehnung des Universums.**
J. D. Barrow, J. Silk, translated from the English by G. Kurz and S. Summerer, with an introduction by R. Kippenhahn.
R. Piper GmbH & Co. KG, München – Zürich. 270 pp. Price DM 44.00 (1986). ISBN 3–492–02940–X. For the English original see Abstr. 37.003.042.

003.060 **The anthropic cosmological principle.**
J. D. Barrow, F. J. Tipler.
Clarendon Press, New York, USA; Oxford University Press, Oxford, England. 22 + 706 pp. Price US$ 29.95, £ 25.00 (1986).
Reviews in Nature, Vol. 320, No. 6060, p. 315 – 316; 1986 (*W. H. Press*); Science, Vol. 232, No. 4753, p. 1036 – 1037; 1986 (*J. Silk*).

003.061 **L'ordre dans le chaos.**
P. Bergé, Y. Pomeau, C. Vidal.
Hermann, Paris, France. 353 pp. Price FF 230.00 (1984).
Review in Phys. Today, Vol. 39, No. 1, p. 85 – 86; 1986 (*J. P. Gollup*).

003.062 **Build your own telescope.**
R. Berry.
Charles Scribner's Sons, New York, USA. 10 + 276 pp. Price cloth US$ 24.95 (1985). ISBN 0–684–184761.
From J. Br. Astron Assoc., Vol. 96, No. 3, p. 186 (1986).

003.063 **The star guide. A unique system for identifying the brightest stars in the night sky.**
S. L. Beyer.
Little, Brown, Boston, USA. 12 + 404 pp. Price US$ 22.95 cloth, US$ 11.95 paper (1986).
From Science, Vol. 232, No. 4754, p. 1156 (1986).

003.064 **Johannes Kepler Gesammelte Werke. XI, Part 1: Ephemerides novae motuum coelestium.**
V. Bialas (Editor).
C. H. Beck, München, F.R. Germany. 597 pp. Price DM 98.00 (1983).
Review in J. Hist. Astron., Vol. 17, Part 1, No. 48, p. 63 – 65; 1986 (*C. Wilson*).

003.065 **Chronology of the Ancient World.**
E. J. Bickerman.
Second Edition. Cornell University Press, Ithaca, NY, USA. 223 pp. Price US$ 18.50 (1980).
Review in Archaeoastronomy (U.S.A.), Vol. 7, Nos. 1 – 4, p. 138 – 139; 1984 (*O. L. Harvey*).

003.066 **Astronomie, een praktische handleiding.**
D. Block.
A. W. Sijthoff, Amsterdam, The Netherlands. 184 pp. Price *f* 24.50 (1984). ISBN 90–218–3350–6.
Review in Zenit, 13. Jaarg., Nr. 1, p. 35; 1986 (*M. Drummen*).

003.067 **A manual on astronomic and grid azimuth.**
R. B. Buckner.
Landmark Enterprises, 10324 Newton Way, Rancho Cordova, Calif. 95670, USA. 255 pp. Price $ 19.00 (1984). ISBN 0–910845–22–0.
Review in Sky Telesc., Vol. 71, No. 3, p. 265 (1986).

003.068 **The propagation of radio waves.**
K. G. Budden.
Cambridge University Press, Cambridge – London – New York – New Rochelle – Melbourne – Sydney. 669 pp. Price £ 60.00, US$ 89.50 cloth (1985).

Reviews in Planet. Space Sci., Vol. 34, No. 6, p. 573; 1986 (*D. R. Bates*); Space Sci. Rev., Vol. 44, Nos. 1/2, p. 184; 1986 (*C. Slottje*).

003.069 **Stargazer.**
F. E. Butler.
Pack contains booklet (Signposts to the stars, 32 pp. paperback), planisphere, chart of the stars, and Halley's comet guide.
Georg Philip London, England. Price £ 4.95 (1985). ISBN 0–540–01096–0.
From J. Br. Astron. Assoc., Vol. 96, No. 3. p. 187 (1986).

003.070 **Recent advances in the aerospace sciences.**
C. Casci (Editor).
Plenum Press, New York, USA. 421 pp. (1985).
Review in Astrophys. Space Sci., Vol. 123, No. 2, p. 415 – 416; 1986 (*B. Ružička*).

003.071 **Histoire de l'astronomie (occidentale).**
L. M. Celnikier.
Collection "Petite Collection d'Histoire des Sciences". Editions Technique et Documentation Lavoisier, Paris, France. 6 + 266 pp. Price FF 125.00 (1986).
Review in Ciel Terre, Vol. 102, No. 3, p. 93; 1986 (*A. Heck*).

003.072 **Mapping the cosmos.**
J. Chance, R. O. Wells Jr. (Editors).
Rice University Press, Houston, Tex., USA. 8 + 175 pp. Price US$ 19.95 (1985).
From Science, Vol. 231, No. 4736, p. 416 (1986).

003.073 **Le château des étoiles.**
P. Chatel.
Editions Liana Lévi–Sylvie Messinger (distribué en Belgique par Hachette). 336 pp. Price FF 89.00 paper (1985).
Review in Ciel Terre, Vol. 102, No. 1, p. 28; 1986 (*A. Heck*).

003.074 **The life and death of stars.**
D. A. Cooke.
Crown Publishers Inc., New York, N.Y., USA. 193 pp. Price US$ 29.95 (1985). ISBN 0–517–55268–X.
Review in Sky Telesc., Vol. 71, No. 4, p. 362 – 363; 1986 (*D. Mihalas*).

003.075 **Imaging Saturn. The Voyager flights to Saturn.**
H. S. F. Cooper Jr.
Reprint of the 1983 edition. Holt, Rinehart and Winston, New York, N.Y., USA. 14 + 210 pp. Price US$ 9.95 (1985).
From Science, Vol. 231, No. 4735, p. 280 (1986).

003.076 **Infinite vistas: new tools for astronomy.**
J. Cornell, J. Carr (Editors).
Charles Scribner's Sons, New York, USA. 302 pp. Price US$ 18.95 cloth (1985). ISBN 0–684–18287–4.
Reviews in Mercury, Vol. 15, No. 3, p. 89 (1986); Sky Telesc., Vol. 71, No. 6, p. 579 (1986).

003.077 **The planets.**
H. Couper, N. Henbest.
Pan Books, Cavaye Place, London SW10 9PG, England. 144 pp. Price £ 10.95 cloth, £ 6.95 paper (1985). ISBN 0–330–29082–7.
Reviews in Phys. Abstr., Vol. 89, No. 1277, Entry 36431 (1986); Spaceflight, Vol. 28, No. 1, p. 47 (1986).

003.078 **The universe.**
H. Couper, D. Pelham.
Random House, Inc., New York, N.Y., USA. 12 pp. Price US$ 19.95 (1985). ISBN 0–394–54691–1.
Review in Sky Telesc., Vol. 71, No. 4, p. 367 (1986).

003.079 **A random walk in relativity and cosmology.**
N. Dadhich, J. V. Narlikar, J. Krishna Rao, C. V. Vishveshwara (Editors).
John Wiley and Sons, Chichester – New York – Brisbane – Toronto – Singapore. 236 pp. Price US$ 34.95 (1985). ISBN 0–470–20198–3.
Review in Sky Telesc., Vol. 71, No. 3, p. 263 (1986).

003.080 **Les planètes artificielles.**
W. Dannau.
Editions Bordas, Paris, France. 192 pp. Price FB 1.413 (1985).
Review in Ciel Terre, Vol. 102, No. 2, p. 61; 1986 (*J. Vercheval*).

003.081 **On the astroclimate in the USSR.**
Sh. P. Darchiya.
Nauka, Moskva. 175 pp. (1985). In Russian.
Review in Ref. Zh., 51. Astron., 3.51.56 (1986).

003.082 **Was sind Quasare?**
G. Dautcourt.
Translation of the German edition. Rad. shk., Kiev. 131 pp. (1985). In Russian.
From Ref. Zh., 51. Astron., 5.51.67 (1986). For the original see 19.003.050.

003.083 **Astronomy and the imagination: a new approach to man's experience of the stars.**
N. Davidson.
Routledge & Kegan Paul, London – Boston – Melbourne – Henley. 237 pp. Price £ 12.95 (1985). ISBN 0–7102–0371–3.
From Nature, Vol. 319, No. 6048, p. 79 (1986).

003.084 **Problems of nonlinear and turbulent processes in physics. Part 2.**
A. S. Davydov, V. M. Chernousenko (Editors).
Naukova dumka, Kiev. 359 pp. (1985). In Russian.
Review in Ref. Zh., 51. Astron., 4.51.48 (1986).

003.085 **Observing Halley's comet.**
G. Dawes et al.
Allen & Unwin Ltd., London – Boston – Sydney. 66 pp. Price $A 8.95 (1985). ISBN 0–86861–920–5.
Reviews in Proc. Astron. Soc. Aust., Vol. 6, No. 2, p. 272; 1985 (*J. W. V. Storey*); Sky Telesc., Vol. 71, No. 1, p. 43 (1986).

003.086 **Halley's comet: a mysterious visitor from outer space.**
T. Dickinson.
Edmund Scientific. 125 pp. (1985).
Review in Mercury, Vol. 15, No. 3, p. 87 (1986).

003.087 **Albert Ducrocq raconte le ciel.**
A. Ducrocq.
Editions Nathan, Paris, France. 62 pp. (1985).
Review in Ciel Terre, Vol. 102, No. 1, p. 27; 1986 (*M. Delcourt–Honorez*).

003.088 **Medieval cosmology. Theories of infinity, place, time, void, and the plurality of worlds.**
P. Duhem, edited and tranlated by R. Ariew.
The University of Chicago Press, Chicago – London. 31 + 601 pp. Price US$ 35.00, £ 24.75 (1985). ISBN 0–226–16922–7.

This book is a selection, in English translation, of Pierre Duhem's *Le système du monde. Histoire des doctrines cosmologiques de Platon à Copernic*, written in the years 1909 – 1916, and originally published in ten volumes, in French, by Hermann, Éditeurs des Sciences et des Arts, Paris.

003.089 **Low light level detectors in astronomy.**
M. J. Eccles, M. E. Sim, K. P. Tritton.
Translated from the English edition. Mir, Moskva. 200 pp. (1986). In Russian. For the English original see Abstr. 34.003.061.
Review in Ref. Zh., 51. Astron., 6.51.74 (1986).

003.090 **Photometry and polarization in remote sensing.**
W. G. Egan.
Elsevier Science, New York – Amsterdam – Oxford. 502 pp. (1985).
Review in Icarus, Vol. 66, No. 3, p. 632 – 634; 1986 (*R. H. Giese*).

003.091 **Alle tiders bog.**
M. Eilertsen (Editor).
Politikens Forlag, København, Denmark. 208 pp. Price D.Kr. 168.00 (1985).
Review in Astron. Tidsskr., Årg. 19, Nr. 2, p. 86 – 87; 1986 (*T. Johnsen*).

003.092 **General theory of the rotation of the earth.**
Zh. S. Erzhanov, A. A. Kalybaev.
Nauka, Moskva. 256 pp. Price 1 Rbl. 60 Kop. (1984). In Russian.
Review in Astron. Zh., Tom 63, Vyp. 2, p. 407 – 409; 1986 (*K. A. Kulikov*).

003.093 **Kalendarium der Jahre 1700 – 2080.**
O. Fambach.
Verlag Ludwig Röhrscheid, Bonn, F.R. Germany. 7 + 253 pp. Price DM 58.00 (1982). ISBN 3–7928–0459–X.
Review in Astron. Nachr., Vol. 307, No. 3, p. 200; 1986 (*H.–J. Felber*).

003.094 **Verdensrom og romfart.**
A. Feldman, E. Tronstad.
Faktum Forlag A/S, Oslo, Norway. 336 pp. Price N.Kr. 320.00 (1985).
Review in Astron. Tidsskr., Årg. 19, Nr. 2, p. 87; 1986 (*Ø. Hauge*).

003.095 **William Whiston: honest Newtonian.**
J. E. Force.
Cambridge University Press, The Edinburgh Building, Shaftesbury Road, Cambridge, CB2 2RU. 208 pp. Price £ 25.00 (1985).
Review in J. Br. Interplanet. Soc., Vol. 39, No. 4, p. 189; 1986 (*L. J. Carter*).

003.096 **Het zonnestelsel.**
K. Frazier.
Time–Life Books, Amsterdam. 176 pp. Price *f* 45.25 (1985). ISBN 90–6182–498–2.
Review in Zenit, 13. Jaarg., Nr. 2, p. 75 – 76; 1986 (*G. Schilling*).

003.097 **A la découverte de l'astronomie.**
P. Gavazzi, translated from the German by P. H. Carlier.
Editions Casterman, Tournai, Belgium. Collection "A la découverte de". 34 pp. paper (1984).
Review in Ciel Terre, Vol. 102, No. 1, p. 27; 1986 (*M. Delcourt–Honorez*).

003.098 **Planispheric astrolabes from the National Museum of American History.**
S. Gibbs, G. Saliba.
Smithsonian Institution Press, Cambridge, Mass., USA. 231 pp. Price US$ 7.50 (1984).
Review in Sky Telesc., Vol. 71, No. 1, p. 43 (1986).

003.099 **Comets, meteors & asteroids – how they affect earth.**
S. Gibilisco.
TAB Books Inc., Blue Ridge Summit, Pennsylvania, USA; John Wiley & Sons Ltd., Chichester, England. 208 pp. Price US$ 12.95, £ 11.45 paper (1985). ISBN 0–8306–1905–4.
Review in Astron. Tidsskr., Årg. 19, Nr. 2, p. 94 (1986).

003.100 **Halley's comet for classrooms.**
W. H. Glenn.
National Science Teachers Association, 1742 Connecticut Ave., NW., Washington D.C. 20009, USA. 16 pp. Price $ 1.50 (1985). ISBN 0–87355–047–1.
Review in Sky Telesc., Vol. 71, No. 3, p. 263 (1986).

003.101 **On some problems of astrophysics.**
A. V. Goncharskij, A. M. Cherepashchuk, A. G. Yagola.
Nauka, Moskva. 250 pp. (1985). In Russian.
Review in Ref. Zh., 51. Astron., 5.51.63 (1986).

003.102 **Planetary landscapes.**
R. Greeley.
Allen & Unwin, Winchester, Mass. – Hemel Hempstead, UK. 265 pp. Price US$ 44.94, £ 24.95 (1985). ISBN 0–04–551080–6.
Reviews in Nature, Vol. 320, No. 6063, p. 664; 1986 (*M. H. Carr*); Sky Telesc., Vol. 71, No. 6, p. 577 (1986).

003.103 **Photons, galaxies and stars. Selected papers of R. Hanbury Brown, Raman Professor, 1974.**
R. Hanbury Brown.
Indian Academy of Sciences, P.B. No. 8005, Bangalore 560 080, India. 9 + 426 pp. Price US$ 30.00 (1985).
This book contains a selection of scientific papers by R. Hanbury Brown published during the period 1952 – 1982. They are arranged in the following chapters: (1) The emission from radio sources. (2) The angular diameters of radio sources. (3) Techniques of measuring angular size. (4) Photon optics. (5) Optical astronomy. (6) Gamma ray astronomy. (7) Lectures: Measuring the size of the stars; Bosons and stars; Concern about the control of science; Science and faith.

003.104 **A short introduction to astronomy.**
L. Harris.
Vantage Press, New York, USA. 165 pp. Price US$ 12.95 (1985).
Review in Observatory, Vol. 106, No. 1070, p. 28; 1986 (*P. W. Hill*).

003.105 **Out of the cradle – exploring the frontiers beyond earth.**
W. K. Hartmann, R. Miller, P. Lee.
Workman, New York, USA. 192 pp. Price US$ 19.95 cloth, US$ 11.95 paper (1984).
Review in Icarus, Vol. 64, No. 3, p. 589 – 590; 1985 (*R. A. Schorn*).

003.106 **Progrès et découvertes en astronomie.**
M. Harwit, translated from the English by F. Heudier and J. L. Heudier.
Editions Masson, Paris – New York – Barcelona – Milan – Mexico – São Paulo. 301 pp. (1984).
Review in Astronomie, Vol. 100, p. 265 – 266; 1986 (*J. L. Masnou*).

003.107 **Kometer och kometskräck.**
B. Hedberg.
Bokförlaget Rubicon, Uppsala, Sweden. 192 pp. Price Sv.kr. 106.00 (1985).
Review in Astron. Tidsskr., Årg. 19, Nr. 1, p. 47 (1986).

003.108 **Dissertatio Philosophica de Orbitis Planetarum. Philosophische Erörterung über die Planetenbahnen (1801).**
G. W. F. Hegel. Übersetzt, eingeleitet und kommentiert von W. Neuser.
Schriften zur Naturphilosophie, Band 2. VCH Verlagsgesellschaft, D–6940 Weinheim, F.R. Germany. 7 + 176 pp. Price DM 64.00 (1986). ISBN 3–527–17547–4.

003.109 **Halley's comet.**
N. Henbest (Editor).
New Science Publications, Holborn Publishing Group. Commonwealth House, 1 – 19 New Oxford St., London WC1A 1NG, England. 68 pp. Price £ 2.25, US$ 3.25 (1985).
Review in Sky Telesc., Vol. 71, No. 4, p. 367 (1986).

003.110 **The new astronomy.**
N. Henbest, M. Martin.
Cambridge University Press, Cambridge – London – New York – New Rochelle – Melbourne – Sydney. 240 pp. Price £ 9.95 (1986). Paper version of 37.003.094.
Reviews in Astron. Tidsskr., Årg. 19, Nr. 2, p. 94 (1986); Spaceflight, Vol. 28, No. 6, p. 269 (1986).

003.111 **The Universe next door: a complete guide exploring the skies and understanding what you see.**
T. Holt.
Charles Scribner's Sons, New York, USA. 6+408 pp. Price US$ 24.95 cloth (1985). ISBN 0–684–18358–7.
From Phys. Today, Vol. 39, No. 2, p. 78 (1986).

003.112 **Isaac Newton: reluctant genius.**
D. C. Ipsen.
Enslow Publishers, P.O. Box 28, Aldershot, Hants, GU12 6BP, England. 93 pp. Price £ 11.95.
Review in Spaceflight, Vol. 28, No. 6, p. 269 (1986).

003.113 **Reversals of the earth's magnetic field.**
J. A. Jacobs.
Adam Hilger Ltd., Bristol, England. 10+240 pp. Price £ 19.95 (1984). ISBN 0–85274–442–0.
Review in Phys. Earth Planet. Inter., Vol. 42, No. 3, p. 197 – 198; 1986 (*K. A. Hoffmann*).

003.114 **Kosmologie. Geschichte und Systematik in philosophischer Perspektive.**
B. Kanitscheider.
Verlag Phillip Reclam Jr., Stuttgart, F.R. Germany. 504 pp. Price DM 13.20 paper, DM 19.80 cloth (1984). ISBN 3–15–008025–8 paper, ISBN 3–15–028025–7 cloth.
Review in Sterne Weltraum, 25. Jahrg., Nr. 1, p. 52 – 53; 1986 (*F. Krojer*).

003.115 **100 billion suns.**
R. Kippenhahn.
Unwin Paperbacks, London, England. 266 pp. Price £ 4.50 (1985). Paperback edition of 34.003.090.
Review in Observatory, Vol. 106, No. 1071, p. 51; 1986 (*A. King*).

003.116 **The solar system: observations and interpretations.**
M. G. Kivelson.
Prentice–Hall, Englewood Cliffs, N.J., USA. 436 pp. Price US$ 43.95 (1986).
From Phys. Today, Vol. 39, No. 6, p. 92 (1986).

003.117 **L'univers et la terre.**
J. Kleczek, P. Jakes.
Editions Cercle d'Art, Encyclopédies et Connaissances, Paris, France. 319 pp.
Review in Astronomie, Vol. 100, p. 319; 1986 (*P. Simon*).

003.118 **Polarized light in nature.**
G. P. Können.
Translated from the Dutch edition 1980. Cambridge University Press, Cambridge – London – New York – New Rochelle – Melbourne – Sydney. 172 pp. Price £ 19.50, US$ 32.50 cloth (1985).
Reviews in Astrophys. Space Sci., Vol. 120, No. 2, p. 307; 1986 (*J. E. Dyson*); Space Sci. Rev., Vol. 44, Nos. 1/2, p. 178; 1986 (*R. Zerull*).

003.119 **The physical universe.**
K. B. Krauskopf, A. Beiser.
5th edition. McGraw–Hill Book Co., New York, USA. 21+760 pp. Price US$ 35.95 (1986). ISBN 0–07–035484–7.

003.120 **Astronomy and sodium lighting.**
W. H. Krase, K. Wolf.
Rand Corp. Publications Dept., 1700 Main St., PO Box 2138, Santa Monica, Calif. 90406, USA. 57 pp. Price US$ 7.50 (1984).
Review in Sky Telesc., Vol. 71, No. 1, p. 41 (1986).

003.121 **Planet Venus.**
L. V. Ksanfomaliti.
Nauka, Moskva. 376 pp. (1985). In Russian.
Review in Ref. Zh., 51. Astron., 5.51.64 (1986).

003.122 **High–energy physics.**
B. Kursunoglu, S. L. Mintz, A. Perlmutter (Editors).
Plenum Press, New York, USA. 10+270 pp. Price US$ 47.50 cloth (1985). ISBN 0–306–42070–8.
From J. Br. Astron. Assoc., Vol. 96, No. 3, p. 186 (1986).

003.123 **The world of science: astronomy.**
P. Lancaster Brown.
Facts on File Publications, 460 Park Ave. S., New York, N.Y. 10016, USA. 64 pp. Price US$ 9.95 (1984). ISBN 0–87196–985–8.
Review in Sky Telesc., Vol. 71, No. 5, p. 473 (1986).

003.124 **Fenêtres sur l'univers. L'histoire des instruments de l'astronomie.**
R. Learner. Translated from the English by H. Biraud, C. Casamatta, F. Combes, R. Monier.
Editions Denoël, Paris, France. 224 pp. Price FF 295.00 (1984).
Review in Astronomie, Vol. 100, p. 268; 1986 (*J. Pernet*).

003.125 **Halley, le roman des comètes.**
A. C. Levasseur–Regourd, P. de La Cotardière.
Editions Denoël, Paris, France. 290 pp. Price FF 128.00.
Review in Astronomie, Vol. 100, p. 159; 1986 (*P. Simon*).

003.126 **Astronomy.**
E. P. Levitan.
Prosveshchenie, Moskva. 175 pp. (1985). In Russian.
Review in Ref. Zh., 51. Astron., 4.51.42 (1986).

003.127 **Physics of shock waves in gases and plasmas.**
M. A. Liberman, A. L. Velikovich, translated into English by A. S. Dobroslavskij.
Springer Series in Electrophysics, Vol. 19. Springer–Verlag, Berlin – Heidelberg – New York – Tokyo. 11+385 pp. Price DM 135.00 (1986). ISBN 3–540–15605–4 (F.R. Germany), ISBN 0–387–15605–4 (USA).

This book deals with the problems of shock structures, stability and dynamics, shock–shaping physical processes and mechanisms. Special attention is paid to the shock phenomena in plasma with magnetic fields, whose effects on the flow dynamics and ionization kinetics lead to particularly unconventional shock–wave behavior.
Contents: (1) Introduction to the theory of gasdynamic shock waves. (2) Gas shock ionization and shock–wave structures in plasmas. (3) Magnetohydrodynamic shock waves in plasmas. (4) Ionizing shock waves in magnetc fields: structures and stability. (5) Dynamics of shock waves in magnetic fields.

003.128 **Astrophotography: a step–by–step approach.**
R. T. Little.
Macmillan Publishing Co., New York, USA. 79 pp. Price US$ 19.95 (1986). ISBN 0–02–948980–6.
Review in Sky Telesc., Vol. 71, No. 5, p. 464 – 465; 1986 (*L. C. Coombs*).

003.129 **The theory of gravitation.**
A. A. Logunov, M. A. Mestvirishvili.
Moskovskij Gosudarstvennyj Universitet, Moskva. 80 pp. (1985). In Russian.
Review in Ref. Zh., 51. Astron. 5.51.68 (1986).

003.130 **A decade of comets.**
D. Machholz.
D. Machholz, 5234 Camden Ave., San Jose, Calif. 95124, USA. 111 pp. Price US$ 8.00 paper (1985).
Reviews in Mercury, Vol. 15, No. 3, p. 88 (1986); Space Educ., Vol. 1, No. 11, p. 527 (1986).

003.131 **Astrophotographie. Ein Leitfaden für den Amateurastronomen.**
P. Martinez, translated from the French by E. Lamla.
Verlag Darmstädter Blätter, Schwarz & Co., Haubachweg 5, D-6100 Darmstadt, F.R. Germany. 375 pp. Price DM 55.00 (1985). ISBN 3-87139-081-X. For the French original see 33.003.069.
Reviews in Orion, 44. Jahrg., Nr. 212, p. 32 - 33; 1986 (*A. von Rotz*); Sterne Weltraum, 25. Jahrg., Nr. 3, p. 174; 1986 (*P. Riepe*).

003.132 **A la poursuite du Soleil. La construction du coronographe d'amateur.**
P. Mazereau, P. Bourge.
Editions Eyrolles, Paris, France. 112 pp. Price FF 140.00 (1985).
Review in Ciel Terre, Vol. 102, No. 2, p. 60; 1986 (*F. Clette*).

003.133 **Instabilities in space and laboratory plasmas.**
D. B. Melrose.
Cambridge University Press, Cambridge - London - New York - New Rochelle - Melbourne - Sydney. 12+280 pp. Price £ 35.00, US$ 49.50 (1986). ISBN 0-521-30541-1.
This book is an introductory account of instabilities in plasmas. It concentrates on laboratory plasmas, such as those encountered in fusion research, and the space plasmas studied in physics of the magnetosphere and solar atmosphere. An important feature is the stress placed on the similarities between astrophysical and laboratory plasmas.
Contents: (1) Introduction. (2) The response of an unmagnetized plasma. (3) Reactive instabilities. (4) Kinetic instabilities. (5) Particle motions in waves. (6) Weak turbulence theory. (7) Nonlinear instabilities and strong turbulence. (8) Magnetohydrodynamics. (9) MHD instabilities. (10) Dispersion in a magnetized plasma. (11) Electron cyclotron maser emission. (12) Instabilities in warm and inhomogeneous plasmas. (13) Instabilities due to anisotropic fast particles.

003.134 **Rendezvous with Halley's comet.**
S. S. Mims.
Space News Publishing Co., PO Box 66521, Baton Rouge, La. 70896, USA. 44 pp. Price US$ 4.95 (1985).
Review in Sky Telesc., Vol. 71, No. 2, p. 161 (1986).

003.135 **Invitation to astronomy.**
S. Mitton, J. Mitton.
Basil Blackwell Inc., Oxford - New York. 210 pp. Price £ 14.50, US$ 24.95 cloth, £ 4.95, US$ 8.95 paper (1986). ISBN 0-631-14699-7 cloth, ISBN 0-631-14695-4 paper.
Review in Sky Telesc., Vol. 71, No. 5, p. 473 (1986).

003.136 **1986 Yearbook of Astronomy.**
P. Moore.
W. W. Norton and Co., 500 Fifth Avenue, New York, N.Y. 10036, USA; Sidgwick & Jackson, London, England. 240 pp. Price US$ 15.95, £ 8.95 cloth, £ 6.50 paper (1985). ISBN 0-393-30254-7.
Review in Observatory, Vol. 106, No. 1072, p. 86; 1986 (*W. Nicholson*).

003.137 **How to make the most of your telescope.**
P. Moore.
Longman, Harlow, England. 80 pp. Price £ 6.95 cloth (1985). ISBN 0-582-23535-9.
From J. Br. Astron. Assoc., Vol. 96, No. 3, p. 186 (1986).

003.138 **Stargazing: astronomy without a telescope.**
P. Moore.
Barron's Educational Series Inc., Woodbury, N.Y., USA. 176 pp. Price US$ 19.95 (1985). ISBN 0-8120-5644-2.
From Nature, Vol. 319, No. 6048, p. 79 (1986).

003.139 **The sky at night.**
P. Moore.
Patrick Stephens, Wellingborough, England. 192 pp. Price £ 6.99 paper (1985). ISBN 0-85059-753-6.
Review in Observatory, Vol. 106, No. 1072, p. 91 (1986).

003.140 **Halley's comet pop-up book.**
P. Moore, H. Couper.
Bonanza Pop-Up Books, 1 Park Ave., New York, N.Y. 10016, USA. (Available from Crown Publishers Inc.). 12 pp. Price US$ 9.98 (1985). ISBN 0-517-48351-3.
Review in Sky Telesc., Vol. 71, No. 2, p. 161 (1986).

003.141 **Halleys komet.**
P. Moore, J. Mason.
Natur och Kultur, Stockholm, Sweden. 159 pp. Price Sv.kr. 90.00 (1985).
Review in Astron. Tidsskr., Årg. 19, Nr. 1, p. 48 (1986).

003.142 **Het heelal.**
P. Moore, I. Nicolson.
Elsevier Scientific Publishing Company, Amsterdam - Bruxelles. 256 pp. Price *f* 69.50 (1985). ISBN 90-10-05447-0. For the English original see 40.003.155.
Review in Zenit, 13. Jaarg., Nr. 2, p. 75; 1986 (*M. Drummen*).

003.143 **Astronomy with a small telescope.**
J. Muirden (star maps by W. Tirion).
George Philip, 12 - 14 Long Acre, London WC2E, England. 224 pp. Price £ 9.95 cloth (1985). ISBN 0-540-01088-X.
From Nature, Vol. 319, No. 6049, p. 161 (1986).

003.144 **Spacefarers of the '80s and '90s. The next thousand people in space.**
A. R. Oberg.
Columbia University Press. 238 pp. Price US$ 24.95 (1985). ISBN 0-231-05906-X.
From Nature, Vol. 319, No. 6048, p. 79 (1986).

003.145 **The Megalithic Odyssey.**
C. O'Brien.
Turnstone Press, Wellingborough, England. 176 pp. Price £ 4.95 (1983).
Review in Archaeoastronomy (U.S.A.), Vol. 7, Nos. 1 - 4, p. 142 - 143; 1984 (*J. Barnatt*).

003.146 **Manuel pratique de l'astronome amateur.**
J. M. Oliver, translated from the Spanish by V. Boro.
Editions de Vecchi, Paris, France. 174 pp. Price FB 706.00 (1984).
Review in Ciel Terre, Vol. 102, No. 1, p. 27; 1986 (*L. Louys*).

003.147 **Transfer processes in an anisotropic circumterrestrial plasma.**
V. N. Oraevskij, Yu. V. Konikov, G. V. Khazanov.
Nauka, Moskva. 173 pp. (1985). In Russian.
Review in Ref.Zh., 62. Issled. Kosm. Prostranstva, 2.62.617 (1986).

003.148 **Elements of modern optical design.**
D. C. O'Shea.
John Wiley & Sons, Chichester - New York - Brisbane - Toronto - Singapore. 402 pp. Price US$ 39.95 (1985). ISBN 0-471-07796-8.
Review in Sky Telesc., Vol. 71, No. 4, p. 367 (1986).

003.149 **A brief view of astronomy.**
J. M. Pasachoff.
Saunders, Philadelphia, USA. 285 pp. Price US$ 19.95 (1985).
From Phys. Today, Vol. 39, No. 6, p. 92 (1986).

003.150 **Capire l'astronomia.**
J.–C. Pecker.
U. Hoepli, Milano, Italy. 295 pp. Price L 25,000 (1986).
Review in Coelum, Vol. 54, N. 2, p. 106 – 107; 1986 (*R. Monella*).

003.151 **Histoire du ciel et de la terre (3 volumes): Les premiers astronomes – L'astronomie moderne – Depuis le big bang.**
A. Perez de Laborda.
Editions Belin, Paris, France. 46+46+38 pp. Price FB 1,140 (1984).
Review in Ciel Terre, Vol. 102, No. 2, p. 60; 1986 (*J. C. Jodogne*).

003.152 **The teaching in astronomy.**
P. Preece, D. Clish.
University of Exeter, Exeter, England. 73 pp. Price £ 2.00 paper (1985). ISBN 5068–072–7.
From J. Br. Astron. Assoc., Vol. 96, No. 3, p. 187 (1986).

003.153 **The planets.**
B. Preiss (Editor).
Bantam, New York, USA. 336 pp. Price US$ 24.95 (1985).
From Phys. Today, Vol. 39, No. 2, p. 78 (1986).

003.154 **Order out of chaos.**
I. Prigogine, I. Stengers, with a foreword by A. Toffler.
Bantam Books, Toronto – New York – London – Sydney. 349 pp. Price US$ 8.95 paper (1984).
Review in Space Sci. Rev., Vol. 44, Nos. 1/2, p. 179 – 180; 1986 (*Y. Néeman*).

003.155 **Halley's comet in South Africa October 1985 – May 1986.**
R. Quarmby.
Delta Books, PO Box 41021, Craighall 2024, South Africa. 55 pp. Price R 8.95 cloth, R 5.95 paper.
Review in Mon. Notes Astron. Soc. S. Afr., Vol. 44, Nos. 11 – 12, p. 121 – 122; 1985 (*I. S. Glass*).

003.156 **The soul of the night: an astronomical pilgrimage.**
C. Raymo.
Prentice–Hall, Inc., Englewood Cliffs, N.J. 07632, USA. 209 pp. Price US$ 15.95. ISBN 0–13–822883–3.
Review in Sky Telesc., Vol. 71, No. 2, p. 161 (1986).

003.157 **De evolutie van het heelal.**
H. Reeves.
Translation from: "Patience dans l'azur, l'Evolution cosmique". Van Gennep, Amsterdam, The Netherlands. 269 pp. Price *f* 58.80 paper (1986). ISBN 90–6012–614–9.
Review in Zenit, 13. Jaarg., Nr. 6, p. 228; 1986 (*G. Schilling*).

003.158 **Bibliotheca Mesopotamica. Vol. 2, Fasc. 1. Babylonian Planetary Omens: Part 1, Enūma Anu Enlil, Tablet 63: The Venus Tablet of Ammisaduqa.**
E. Reiner, in collaboration with D. Pingree.
Undena Publications, Malibu, Calif., USA. 65 pp. Price US$ 15.00 (1975).
Review in Archaeoastronomy (U.S.A.), Vol. 7, Nos. 1 – 4, p. 136 – 137; 1984 (*A. Aaboe*).

003.159 **Bibliotheca Mesopotamica. Vol. 2, Fasc. 2. Babylonian Planetary Omens: Part 2, Enūma Anu Enlil, Tablets 50 – 51.**
E. Reiner in collaboration with D. Pingree.
Undena Publications, Malibu, Calif., USA. 100 pp. Price US$ 12.00 (1981).
Review in Archaeoastronomy (U.S.A.), Vol. 7, Nos. 1 – 4, p. 136 – 137; 1984 (*A. Aaboe*).

003.160 **Secrets of the sky.**
I. Ridpath.
Hamlyn, Twickenham, England. 64 pp. Price £ 3.99 cloth (1985). ISBN 0–600–30946–0.
From J. Br. Astron. Assoc., Vol. 96, No. 3, p. 185 (1986).

003.161 **Collins gem guide to the night sky.**
I. Ridpath, W. Tirion.
Collins, London, England. 240 pp. Price £ 1.95 paper (1985). ISBN 0–00–458817–7.
From J. Br. Astron. Assoc., Vol. 96, No. 3, p. 186 (1986).

003.162 **Universe guide to stars and planets.**
I. Ridpath, W. Tirion.
Universe Books, 381 Park Ave. S., New York. N.Y. 10016, USA. 384 pp. Price US$ 19.95 cloth, US$ 10.95 paper (1985). ISBN 0–87663–366–1 cloth, 0–87663–859–0 paper.
Review in Sky Telesc., Vol. 71, No. 1, p. 35 – 36; 1986 (*G. Lovi*).

003.163 **Capture a comet: an amateur photographer's guide.**
M. Roberts.
Aurorae Publishing. 220 pp. Price US$ 12.95 paper (1985).
Review in Mercury, Vol. 15, No. 3, p. 88 (1986).

003.164 **The cosmic zygote: cosmology in the Amazon basin.**
P. G. Roe.
Rutgers University Press. 384 pp. Price US$ 40.00 (1982).
Review in Archaeoastronomy (U.S.A.), Vol. 7, Nos. 1 – 4, p. 154 – 159; 1984 (*S. Hugh–Jones*).

003.165 **Tools of radio astronomy.**
K. Rohlfs.
Astronomy and Astrophysics Library, Vol. 1. Springer–Verlag, Berlin – Heidelberg – New York – London – Paris – Tokyo. 12+319 pp. Price DM 118.00 (1986). ISBN 3–540–16188–0 (F.R. Germany), ISBN 0–387–16188–0 (USA).
Contents: (1) Radio astronomical fundamentals. (2) Electromagnetic wave propagation fundamentals. (3) Wave polarization. (4) Fundamentals of antenna theory. (5) Filled aperture antennas. (6) Interferometers and aperture synthesis. (7) Receivers. (8) Emission mechanisms of continuous radiation. (9) Some examples of thermal and nonthermal radio sources. (10) Line radiation fundamentals. (11) Line radiation of neutral hydrogen. (12) Recombination lines. (13) Interstellar molecules and their line radiation.

003.166 **A spacefaring people: perspectives on early spaceflight.**
A. Roland.
Scientific and Technical Information Branch, National Aeronautics and Space Administration, Washington, USA. 156 pp. (1985).
From Nature, Vol. 319, No. 6049, p. 161 (1986).

003.167 **Introduzione all'astronomia, esercitazioni e problemi.**
G. Romano.
F. Muzzio & Co., Padua, Italy. 337 pp. Price L. 36,000.
Review in Astronomia, N. 1, p. 13; 1986 (*G. Favero*).

003.168 **Gli astri.**
L. Rosino.
Ed. UTET, Torino, Italy. 569 pp. (1985).
Review in Coelum, Vol. 54, N. 1, p. 45; 1986 (*G. Romano*).

003.169 **Spectral line shapes. Vol. 3.**
F. Rostas (Editor).
Walter de Gruyter, Berlin – New York. 769 pp. Price DM 340.00 (1985).
Review in Space Sci. Rev., Vol. 44, Nos. 1/2, p. 177; 1986 (*E. A. Gurtovenko, R. I. Kostik*).

003.170 **Contact: a novel.**
C. Sagan.
Simon & Schuster, New York, USA. 432 pp. Price US$ 18.95 (1985). ISBN 0–671–43400–4.
From Sky Telesc., Vol. 71, No. 6, p. 579 (1986).

003.171 **Der Komet.**
C. Sagan, A. Druyan.
Buchklub Ex Libris, Zürich, Switzerland. 330 pp. Price SFr. 39.00 (1985). For the English original see 40.003.171.
Review in Orion, 44. Jahrg., Nr. 212, p. 32; 1986 (*A. von Rotz*).

003.172 **Solar flare prediction.**
C. Sawyer, J. W. Warwick, J. T. Dennett.
Colorado Associated University Press, Boulder, Colo., USA. 12+179 pp. Price US$ 25.00 (1986).
From Science, Vol. 232, No. 4758, p. 1656 (1986).

003.173 **The electronic sky. Digital images of the cosmos.**
R. Schild, J. Cornell.
Smithsonian Institution, Photographic Services, MAH CB–054, Washington, D.C. 20560. 19 pp., 41 slides. Price US$ 31.00 (1985).
Review in Sky Telesc., Vol. 71, No. 4, p. 369 (1986).

003.174 **Blik op oneindig.**
G. Schilling.
Unieboek/Romen, Weesp. 100 pp. Price *f* 24.50 (1985). ISBN 90–228–3797–1.
Review in Zenit, 13. Jaarg., Nr. 2, p. 74 – 75; 1986 (*B. Ernst*).

003.175 **Deterministic chaos: an introduction.**
H. G. Schuster.
Physik–Verlag, Weinheim, F.R. Germany (US dist. Verlag Chemie, Deerfield Beach, Florida). 220 pp. Price US$ 43.75 (1984).
Review in Phys. Today, Vol. 39, No. 1, p. 85 – 86; 1986 (*J. P. Gollub*).

003.176 **Superstrings. The first 15 years of superstring theory. Vol. 1 – 2.**
J. H. Schwarz (Editor).
World Scientific Publ. Co., Singapore/John Wiley & Sons Ltd., Chichester. 1141 pp. Price £ 35.75 (1985).
Review in Astron. Tidsskr., Årg. 19, Nr. 2, p. 95 (1985).

003.177 **Light. The mystery of the Universe.**
K. Seyrafi.
Electro–Optical Research Co., Los Angeles, Calif., USA. 12+241 pp. Price US$ 18.95 (1986).
From Science, Vol. 232, No. 4753, p. 1050 (1986).

003.178 **Black holes, white dwarfs and neutron stars. The physics of compact objects.**
S. L. Shapiro, S. A. Teukolsky. Translated from the English edition. Ya. A. Smorodinskij (Editor).
Mir, Moskva. Part 1, translated by A. D. Dolgov. 256 pp. Price 2 Rbl. 60 Kop. Part 2, translated by S. N. Rodionov. 399 pp. Price 3 Rbl. 60 Kop. (1985). In Russian. For the original see 33.003.134.
Review in Priroda, No. 4, p. 123 (1986).

003.179 **Asteroids (or thorny way of investigations).**
A. N. Simonenko.
Nauka, Glavnaya Redaktsiya Fiziko–Matematicheskoj Literatury, Moskva. 208 pp. Price 65 Kop. (1985). In Russian.
Review in Priroda, No. 1, p. 123 (1986).

003.180 **Investigation of the earth atmosphere from the aureols of the sun and moon.**
G. F. Sitnik.
Moskovskij Gosudarstvennyj Universitet, Moskva. 116 pp. (1985). In Russian.
Review in Ref. Zh., 51. Astron., 3.51.49 (1986).

003.181 **To acknowledge the wonder.**
E. Squires.
Adam Hilger Ltd., Accord, Mass. 02018, USA. 199 pp. Price US$ 29.00 (1985). ISBN 0–85274–786–1.
Review in Sky Telesc., Vol. 71, No. 1, p. 41 (1986).

003.182 **Stjernehimmelen.**
R. Stabell, F. Kolberg.
Universitetsforlaget, Oslo – Bergen – Stavanger – Tromsö. 210 pp. Price N.kr. 140.00 (1985).
Review in Astron. Tidsskr., Årg. 19, Nr. 2, p. 95 (1986).

003.183 **The mathematical theory of turbulence.**
M. M. Stanišič.
Springer–Verlag, Berlin – Heidelberg – New York – London – Paris – Tokyo. 16+429 pp. Price DM 87.00 (1985).
Review in Astron. Nachr., Vol. 307, No. 1, p. 68; 1986 (*D. Elstner*).

003.184 **General relativity: an introduction to the theory of the gravitational field.**
H. Stephani.
Cambridge University Press. 298 pp. Price US$ 24.95 (1985). ISBN 0–521–31534–4. Paperback edition of 33.003.120.
Review in Sky Telesc., Vol. 71, No. 6, p. 577 (1986).

003.185 **Collisionless shocks in the heliosphere. A tutorial review.**
R. G. Stone, B. T. Tsurutani (Editors).
Geophysical Monograph 34. American Geophysical Union, Washington, D.C., USA. 8+115 pp. Price US$ 18.00 (1985).
From Science, Vol. 231, No. 4745, p. 1611 (1986).

003.186 **Honneur aux comètes.**
E. Stram.
Numéro 16–17–18 de 1984 de la revue LYRE, édité par VEGA (Villepreux Espace Groupe Astronomique: 20, chemin de Rambouillet, F–78450 Villepreux). 80 pp. Price FF 49.50 paper.
Review in Ciel Terre, Vol. 102, No. 1, p. 30; 1986 (*J. Sauval*).

003.187 **Astronomie sur ZX Spectrum et Spectrum +.**
L. Strebler.
Editions du P.S.I., 77402 Lagny, France. 76 pp. Price FF 85.00 paper (1984).
Review in Ciel Terre, Vol. 102, No. 2, p. 61; 1986 (*P. Demoulin*).

003.188 **Cosmic investigations of planets and satellites.**
Yu. A. Surkov.
Nauka, Moskva. 309 pp. Price 4 Rbl. 40 Kop. (1985). In Russian.
Review in Priroda, No. 4, p. 123 (1986).

003.189 **Autoionization, recent developments, and applications.**
A. Temkin (Editor).
Plenum Press, New York – London. 261 pp. (1985).
Review in Astrophys. Space Sci., Vol. 123, No. 1, p. 216 – 217; 1986 (*Z. J. Horak*).

003.190 **Oculus Enoch... Ein Beitrag zur Entdeckungsgeschichte des Fernrohrs.**
A. Thewes.
Verlag Isensee, Oldenburg, F.R. Germany. 51 pp. Price DM 18.00 (1983). ISBN 3–920557–50–6.
Review in Sterne Weltraum, 25. Jahrg., Nr. 6, p. 353; 1986 (*J. Solf*).

003.191 **De mens in de kosmos.**
C. Titulaer.
Elsevier Scientific Publishing Company, Amsterdam, The Netherlands. 129 pp. Price f 39.90, US$ 17.50 (1985). ISBN 90–10–05560–4.
Reviews in Sky Telesc., Vol. 71, No. 6, p. 577 (1986); Zenit, 13. Jahrg., Nr. 1, p. 34 – 35; 1986 (*G. Schilling*).

003.192 **Het grote hemelboek.**
C. Titulaer.
A. W. Sijthoff, Amsterdam, The Netherlands. 128 pp. Price *f* 24.50 (1985). ISBN 90–218–3481–2.
Review in Zenit, 13. Jaarg., Nr. 1, p. 34 – 35; 1986 (*G. Schilling*).

003.193 **L'univers ou l'astronomie moderne pour tous.**
R. Tocquet, with a preface by L. Leprince-Ringuet.
Editions Godefroy, BP 9, 27760 La Ferrière-sur-Risle. Price F 490.00 (1985).
Review in Astronomie, Vol. 100, p. 212; 1986 (*M. Dumont*).

003.194 **Space, time, infinity.**
J. S. Trefil.
Pantheon Books, New York; Smithsonian Books, Washington. (Available from Sky Publishing Corp.) 255 pp. Price US$ 29.95 cloth (1985). ISBN 0-394-54843-4.
Reviews in Mercury, Vol. 15, No. 3, p. 90 (1986); Sky Telesc., Vol. 71, No. 2, p. 157, 159; 1986 (*H. Shipman*).

003.195 **Here's Halley.**
J. C. Troeger.
Ames Community Schools, 120 South Kellogg, Ames, Iowa 50010, USA. 62 pp. Price US$ 9.50 (1985).
Review in Sky Telesc., Vol. 71, No. 3, p. 263 (1986).

003.196 **Microcomputer control of telescopes.**
M. Trueblood, R. Genet.
Willmann-Bell Inc., P.O. Box 3125, Richmond, Virginia 23235, USA. 377 pp. Price US$ 19.95 (1985). ISBN 0-943396-05-0.
Review in Orion, 44. Jahrg., Nr. 212, p. 33; 1986 (*A. Tarnutzer*).

003.197 **Collisionless shocks in the heliosphere. Review of current research.**
B. T. Tsurutani, R. G. Stone (Editors).
Geophysical Monograph 35. American Geophysical Union, Washington, D.C., USA. 6+303 pp. Price US$ 36.00 (1985).
From Science, Vol. 231, No. 4745, p. 1611 (1986).

003.198 **The cosmic inquirers. Modern telescopes and their makers.**
W. Tucker, K. Tucker.
Harvard University Press, Cambridge, Mass., USA. 12+221 pp. Price US$ 20.00 (1986).
From Science, Vol. 232, No. 4754, p. 1154 (1986).

003.199 **Halley: Komeet op komst.**
P. van Dooren.
Uitgeverij Lannoo, Tiel/Weesp. 80 pp. Price *f* 19.90 (1985). ISBN 90-209-1322-0.
Review in Zenit, 13. Jaarg., Nr. 5, p. 186; 1986 (*G. W. E. Beekman*).

003.200 **Kijk op kometen.**
R. van Helden.
Zomer & Keuning, Ede/Antwerpen, Belgium. 131 pp. Price *f* 19.90 (1985). ISBN 90-210-5102-8.
Review in Zenit., 13. Jaarg., Nr. 5, p. 186; 1986 (*G. Schilling*).

003.201 **A brief history of the Cincinnati Astronomical Society.**
J. Ventre, E. Goodman.
Cincinnati Astron. Soc., 5247 Zion Rd., Cleves, Ohio 45002, USA. 131 pp. Price US$ 6.00 paper (1985).
Review in Mercury, Vol. 15, No. 3, p. 91 (1986).

003.202 **Un ticket pour l'espace.**
M. Vieillefosse, M. Vieillefosse.
Editions Belfond, Collection Belfond/Sciences, Paris, France. 221 pp. Price FF 89.00 paper (1985).
Review in Ciel Terre, Vol. 102, No. 1, p. 27; 1986 (*R. Dejaiffe*).

003.203 **The radiative field in multilayer media.**
T. Viik.
Valgus, Tallin. 90 pp. (1985). In Russian.
Review in Ref. Zh., 51. Astron., 4.51.143 (1986).

003.204 **Laplace.**
B. A. Vorontsov-Vel'yaminov.
2nd revised edition. Nauka, Moskva. 286 pp. (1985). In Russian.
Review in Ref. Zh., 51. Astron., 1.51.37 (1986).

003.205 **Method of teaching astronomy at secondary school. Handbook for teachers.**
B. A. Vorontsov-Vel'yaminov, M. M. Dagaev, A. V. Zasov, Eh. V. Kononovich, K. A. Portsevskij, N. K. Semakin, E. K. Straut, S. V. Shirokov.
Second revised edition. Prosveshchenie, Moskva. 240 pp. (1985). In Russian.
Review in Ref. Zh., 51. Astron., 5.51.91 (1986).

003.206 **An encounter with Halley's comet 1985 – 1986.**
S. Walker, G. Blow.
Wilson & Horton Ltd. 88 pp. Price $ 5.95 (1985).
Review in South. Stars, Vol. 31, No. 4, p. 223 – 224; 1985 (*A. Woodger*).

003.207 **Comet! The story behind Halley's comet.**
G. Walz-Chojnacki.
AstroMedia, PO Box 92788, Milwaukee, Wis. 53202, USA. 64 pp. Price US$ 9.75 (1985). ISBN 0-913135-03-8.
Reviews in Sky Telesc., Vol. 71, No. 2, p. 159 (1986); Spaceflight, Vol. 28, No. 3, p. 129 (1986).

003.208 **Quasar astronomy.**
D. W. Weedman.
Cambridge astrophysics series. Cambridge University Press, Cambridge – London – New York – New Rochelle – Melbourne – Sydney. 13+217 pp. Price £ 25.00, US$ 39.50 (1986). ISBN 0-521-30318-4.
Concentrating on observational astrophysics rather than theory, this book serves as both a summary of what is known about quasars and as a guide to research methods and unanswered questions. After reviewing the cosmological framework and necessary equations, this book discusses all aspects of observed quasar properties. Techniques are summarized for analysing quasar data obtained with various kinds of telescopes and extensive references are given to recent publications. The major original contributions deal with the distribution of quasars in spacetime.
Contents: (1) The technology. (2) Quasar surveys. (3) Cosmology. (4) Quasars as cosmological probes. (5) Luminosity functions. (6) Quasar evolution. (7) Continuous spectra. (8) Emission line spectroscopy. (9) Quasar structure.

003.209 **Beatrice (Hill) Tinsley 1941 – 1981, astronomer.**
S. Whineray.
Massy University and New Zealand Institute of Physics, Palmerston North, New Zealand. 97 pp. (1985).
From Phys. Today, Vol. 39, No. 2, p. 80 (1986).

003.210 **Whitney's star finder. A field guide to the heavens.**
C. A. Whitney.
4th edition. Knopf, New York, USA. 14+107 pp. + star finder. Price US$ 12.95 (1985).
From Science, Vol. 231, No. 4738, p. 628 (1986).

003.211 **Theory and experiment in gravitational physics.**
C. M. Will.
Cambridge University Press, Cambridge – London – New York – New Rochelle – Melbourne – Sydney. 342 pp. Price £ 15.00, US$ 24.95 (1985). Paperback edition of 32.003.126.
Review in Gen. Relativ. Gravitation, Vol. 18, No. 4, p. 433 – 434; 1986 (*F. de Felice*).

003.212 **Comet watch: the return of Halley's comet.**
F. H. Winter.
Lerner Publications Company, 241 First Avenue North, Minneapolis, Minn. 55401, USA. 64 pp. Price US$ 9.95 (1985).
Review in Spaceflight, Vol. 28, No. 3, p. 129 (1986).

003.213 **Selected works. Particles, nuclei and the universe.**
Ya. B. Zel'dovich.
Nauka, Moskva. 463 pp. (1985). In Russian.
Review in Ref. Zh., 51. Astron., 1.51.34 (1986).

003.214 **Guide to observing the moon.**
Edited by British Astronomical Association.
Enslow Publishers, Inc., P.O. Box 38, Aldershot, Hants., GU12 6BP, England. 128 pp. Price £ 9.95, US$ 14.95 (1986).
Review in Sternenbote, 29. Jahrg., Nr. 6, p. 121 – 122 (1986).

003.215 ***Space Station program*, description, applications, and opportunities.**
NASA Space Station Task Force; Noyes Publications, Park Ridge, NJ, USA. 29 + 754 pp. Price US$ 67.00 cloth (1985).
Review in Astrophys. Space Sci., Vol. 122, No. 1, p. 185; 1986 (*J. Kleczek*).

003.216 **Vie et mort des étoiles.**
Editions Belin, Paris, France. 192 pp. Price FF 97.00 (1985).
Review in Ciel Terre, Vol. 102, No. 2, p. 60; 1986 (*M. Gabriel*).

004 History of Astronomy

004.001 **William Herschel and the making of modern astronomy.**
M. Hoskin.
Sci. Am., Vol. 254, No. 2, p. 90 – 96 (1986).

004.002 **Comètes: histoire et histoires.**
P. Véron.
C. R. Acad. Sci., Sér. Gén., Vie Sci., Tome 3, No. 1, p. 39 – 60 (1986).

004.003 **Vad var fysik och astronomi för grekerna? En idéhistorisk diskussion.**
B. Nilsson.
Astron. Tidsskr., Årg. 19, Nr. 1, p. 1 – 8 (1986).

004.004 **Kometen av år 1240 i Håkon Håkonssons saga.**
C. Andersson, N. Bjørgo.
Astron. Tidsskr., Årg. 19, Nr. 1, p. 9 – 15 (1986).

004.005 **Das "Auge von Mainz". Der frühe Plan eines Observatoriums im 17. Jahrhundert.**
A. Thewes.
Sterne Weltraum, 25. Jahrg., Nr. 1, p. 16 – 17 (1986).

004.006 **Über antike und neuzeitliche Fernrohre.**
R. Lukas.
Sterne Weltraum, 25. Jahrg., Nr. 2, p. 96 – 97 (1986).

004.007 **Comet Halley's first expected return: English public apprehensions, 1755 – 58.**
C. B. Waff.
J. Hist. Astron., Vol. 17, Part 1, No. 48, p. 1 – 37 (1986).

004.008 **John of London and his unknown Arabic source.**
P. Kunitzsch.
J. Hist. Astron., Vol. 17, Part 1, No. 48, p. 51 – 57 (1986).

004.009 **No edition of Copernicus in 1640 or 1646.**
E. Rosen.
J. Hist. Astron., Vol. 17, Part 1, No. 48, p. 58 – 59 (1986).

004.010 **Islamic astronomy.**
O. Gingerich.
Sci. Am., Vol. 254, No. 4, p. 68 – 75, 70A, 70D (1986).
While astronomy withered in medieval Europe it flourished in Islam. Renaissance astronomers learned from the texts of Islamic scholars who had preserved and transformed the sicence of the ancient Greeks.

004.011 **Die Sonne im alten Ägypten. Teil III. Sonnenheiligtümer.**
C. Regner.
Sonne, Jahrg. 10, Nr. 37, p. 22 – 23 (1986).

004.012 **Newton and the infinite universe.**
E. Harrison.
Phys. Today, Vol. 39, No. 2, p. 24 – 30, 32 (1986).
Newton said that if the starry heavens were of finite extent they would "fall down to the middle" and there "compose one great spherical mass," yet he avoided calculating the time for gravitational collapse.

004.013 **Un anniversaire: l'expédition Lapérouse.**
S. Grillot.
Astronomie, Vol. 100, p. 33 – 37 (1986).

004.014 **A propos des observations des satellites de Jupiter par Galilée en 1610.**
J. E. Arlot, C. Ruatti, D. T. Vu.
Astronomie, Vol. 100, p. 71 – 76 (1986).

004.015 **A la recherche du coronium.**
S. Laloë, J.-C. Pecker.
Astronomie, Vol. 100, p. 175 – 180 (1986).

004.016 **Light in the temples.**
E. C. Krupp.
Archaeoastronomy (U.S.A.), Vol. 7, Nos. 1 – 4, p. 18 (1984). Abstract. – See Abstr. 010.061.

004.017 **Ancient heliocentrists' unburial & the Almagest crimes' motive.**
D. Rawlins, H. Perry.
Archaeoastronomy (U.S.A.), Vol. 7, Nos. 1 – 4, p. 20 (1984). Abstract. – See Abstr. 010.061.

004.018 **Halley's comet: Canadian observations and reactions, 1835 – 36 and 1910.**
J. A. Smith.
J. R. Astron. Soc. Can., Vol. 80, No. 1, p. 1 – 15 (1986).
The files of old newspapers provide a rich source of Canadian data about the past returns of Halley's comet. A description is presented of the 1835 – 36 and 1910 stories, editorials, advertisements, poems, and columns that are informative.

004.019 **Mannetjes op de maan.**
P. van der Krogt.
Zenit, 13. Jaarg., Nr. 2, p. 54 – 60 (1986).

004.020 **De hemel op aarde. De lange weg van reuzenglobe naar projectieplanetarium.**
G. W. E. Beekman.
Zenit, 13. Jaarg., Nr. 3, p. 84 – 90 (1986).

004.021 **The correlation of dynastic and planetary cycles in Ancient China.**
D. W. Pankenier.
Archaeoastronomy (U.S.A.), Vol. 7, Nos. 1 – 4, p. 29 (1984). Abstract. – See Abstr. 012.012.

004.022 **A research on the lists of corrected values for the fluctuations in the speed of the planets in Ancient China.**
M.–d. Chen.
Archaeoastronomy (U.S.A.), Vol. 7, Nos. 1 – 4, p. 29 (1984). Abstract. – See Abstr. 012.012.

004.023 **Gongsun Qing and the Taichu reforms.**
C. Cullen.
Archaeoastronomy (U.S.A.), Vol. 7, Nos. 1 – 4, p. 29 (1984). Abstract. – See Abstr. 012.012.

004.024 **Philosophy of San–tong–li.**
H. Kawahara.
Archaeoastronomy (U.S.A.), Vol. 7, Nos. 1 – 4, p. 29 (1984). Abstract. – See Abstr. 012.012.

004.025 **The method of eclipse forecast of "Shih–Hsien" calendar in Mongolia and Tibet.**
M.–x. Huang.
Archaeoastronomy (U.S.A.), Vol. 7, Nos. 1 – 4, p. 29 – 30 (1984). Abstract. – See Abstr. 012.012.

004.026 **On the interpretation of some "strange" astronomical phenomena recorded in Chinese sources.**
M. Teboul.
Archaeoastronomy (U.S.A.), Vol. 7, Nos. 1 – 4, p. 30 – 31 (1984). Abstract. – See Abstr. 012.012.

004.027 **Ming astronomical records.**
H. P. Yoke.
Archaeoastronomy (U.S.A.), Vol. 7, Nos. 1 – 4, p. 31 (1984). Abstract. – See Abstr. 012.012.

004.028 **The Hêng Hsing P'ing Chang and Ch'ih Tao Nan Pei Liang Tsung Hsing T'u: the star catalogue and atlas in late Ming China.**
K. Hashimoto.
Archaeoastronomy (U.S.A.), Vol. 7, Nos. 1 – 4, p. 31 (1984). Abstract. – See Abstr. 012.012.

004.029 **Research on armillary sphere made in the Ming Dynasty.**
S.–t. Yi.
Archaeoastronomy (U.S.A.), Vol. 7, Nos. 1 – 4, p. 31 – 32 (1984). Abstract. – See Abstr. 012.012.

004.030 **The first western cosmology introduced into Japan in comparison with the case of China.**
S. Ito.
Archaeoastronomy (U.S.A.), Vol. 7, Nos. 1 – 4, p. 32 (1984). Abstract. – See Abstr. 012.012.

004.031 **Astronomical orientation of tombs in Ancient China.**
V. R. Tiede.
Archaeoastronomy (U.S.A.), Vol. 7, Nos. 1 – 4, p. 32 (1984). Abstract. – See Abstr. 012.012.

004.032 **Solar eclipse in "Shijing" and rotation of the Earth.**
J.–y. Liu.
Archaeoastronomy (U.S.A.), Vol. 7, Nos. 1 – 4, p. 32 (1984). Abstract. – See Abstr. 012.012.

004.033 **Identification of implicit periodicity of historical recorded meteorite falls.**
S.–y. Chang.
Archaeoastronomy (U.S.A.), Vol. 7, Nos. 1 – 4, p. 32 (1984). Abstract. – See Abstr. 012.012.

004.034 **New archaeoastronomical discoveries in China.**
Z.–z. Xi.
Archaeoastronomy (U.S.A.), Vol. 7, Nos. 1 – 4, p. 34 – 45 (1984).

004.035 **The Mesoamerican rabbit in the moon: an influence from Han China?**
C. R. Wicke.
Archaeoastronomy (U.S.A.), Vol. 7, Nos. 1 – 4, p. 46 – 55 (1984).

004.036 **The tomb of Antiochus revisited: planetary alignments and the deification of the king.**
V. S. Tuman.
Archaeoastronomy (U.S.A.), Vol. 7, Nos. 1 – 4, p. 56 – 69 (1984).

004.037 **A possible equinoctial sun–sighting station at Tsiping, New Mexico.**
M. Zeilik.
Archaeoastronomy (U.S.A.), Vol. 7, Nos. 1 – 4, p. 70 – 75 (1984).

004.038 **Summer solstice at Casa Rinconada: calendar, hierophany, or nothing?**
M. Zeilik.
Archaeoastronomy (U.S.A.), Vol. 7, Nos. 1 – 4, p. 76 – 81 (1984).

004.039 **Numerical characteristics of three engraved bison scapulae from the Texas Gulf Coast.**
W. B. Murray.
Archaeoastronomy (U.S.A.), Vol. 7, Nos. 1 – 4, p. 82 – 88 (1984).

004.040 **The origins of astronomy and astrology in Mesopotamia.**
R. Chadwick.
Archaeoastronomy (U.S.A.), Vol. 7, Nos. 1 – 4, p. 89 – 95 (1984).

004.041 **Armenian astronomy in the Bronze Age.**
J. G. Ouzounian.
Archaeoastronomy (U.S.A.), Vol. 7, Nos. 1 – 4, p. 105 – 109 (1984).

004.042 **Astronomy in Taíno mythology.**
S. Robiou–Lamarche.
Archaeoastronomy (U.S.A.), Vol. 7, Nos. 1 – 4, p. 110 – 115 (1984).

004.043 **Scepticism and French interest in Copernicanism to 1630.**
F. J. Baumgartner.
J. Hist. Astron., Vol. 17, Part 2, No. 49, p. 77 – 88 (1986).

004.044 **The star catalogue commonly appended to the Alfonsine Tables.**
P. Kunitzsch.
J. Hist. Astron., Vol. 17, Part 2, No. 49, p. 89 – 98 (1986).

004.045 **Failure and success: two early experiments with concave gratings in stellar spectroscopy.**
D. E. Osterbrock.
J. Hist. Astron., Vol. 17, Part 2, No. 49, p. 119 – 129 (1986).

004.046 **Origins of the spectral sequence.**
J. B. Kaler.
Sky Telesc., Vol. 71, No. 2, p. 129 – 134 (1986).

004.047 **Newton, Halley, and the comet.**
O. Gingerich.
Sky Telesc., Vol. 71, No. 3, p. 230 – 232 (1986).

004.048 **Portrait of a 40–foot giant.**
B. Warner.
Sky Telesc., Vol. 71, No. 3, p. 253 – 254 (1986).

004.049 **The journey of the 200–inch mirror.**
D. di Cicco.
Sky Telesc., Vol. 71, No. 4, p. 347 – 348 (1986).

004.050 **Stonehenge – an observatory of the Neolithic age?**
Z. Alksne.
Zvaigžnota Debess, 1984. Gada Pavasaris, p. 9 – 12 (1984). In Latvian.

004.051 **Dark lunar eclipses in classical antiquity.**
R. B. Stothers.
J. Br. Astron. Assoc., Vol. 96, No. 2, p. 95 – 97 (1986).

004.052 **Shapley's calibration of the P–L curve.**
B. L. Welther.
Bull. Am. Astron. Soc., Vol. 17, No. 4, p. 827 (1985). Abstract. – See Abstr. 010.062.

004.053 **A spectrum of the Sun from a balloon–borne spectrograph in 1899.**
D. H. DeVorkin.
Bull. Am. Astron. Soc., Vol. 17, No. 4, p. 828 (1985). Abstract. – See Abstr. 010.062.

004.054 **The solar–stellar connection in historical perspective.**
K. G. Hufbauer.
Bull. Am. Astron. Soc., Vol. 17, No. 4, p. 828 (1985). Abstract. – See Abstr. 010.062.

004.055 **Cassini, Flamsteed, and Halley on the dimensions of the solar system.**
A. Van Helden.
Bull. Am. Astron. Soc., Vol. 17, No. 4, p. 844 (1985). Abstract. – See Abstr. 010.062.

004.056 **Leonardo da Vinci's drawings of the surface features of the moon.**
G. Reaves.
Bull. Am. Astron. Soc., Vol. 17, No. 4, p. 844 (1985). Abstract. – See Abstr. 010.062.

004.057 **Kepler's Mysterium Cosmographicum: the most fruitful error in the history of astronomy.**
L. C. Hill.
Bull. Am. Astron. Soc., Vol. 17, No. 4, p. 844 (1985). Abstract. – See Abstr. 010.062.

004.058 **Samuel Scott's comet Halley of 1759.**
R. E. Schmidt.
Bull. Am. Astron. Soc., Vol. 17, No. 4, p. 844 (1985). Abstract. – See Abstr. 010.062.

004.059 **Ethnoastronomy and cultural symbolism: the problem of interpretation.**
M. J. Young.
Bull. Am. Astron. Soc., Vol. 17, No. 4, p. 851 (1985). Abstract. – See Abstr. 010.062.

004.060 **Monitoring phases of the moon and eclipse prediction with Mayan time cycles.**
W. T. Murawski.
Bull. Am. Astron. Soc., Vol. 17, No. 4, p. 851 (1985). Abstract. – See Abstr. 010.062.

004.061 **Mt. Meru cosmology in Tibetan Buddhism.**
J. M. Malville.
Bull. Am. Astron. Soc., Vol. 17, No. 4, p. 851 (1985). Abstract. – See Abstr. 010.062.

004.062 **Historic Pueblo moon watching: Shalako at Zuni.**
M. Zeilik.
Bull. Am. Astron. Soc., Vol. 17, No. 4, p. 851 – 852 (1985). Abstract. – See Abstr. 010.062.

004.063 **Seven bridges for seven brothers.**
E. C. Krupp.
Bull. Am. Astron. Soc., Vol. 17, No. 4, p. 852 (1985). Abstract. – See Abstr. 010.062.

004.064 **The mean motion of the planets in the Almagest.**
S. J. Goldstein Jr.
Bull. Am. Astron. Soc., Vol. 17, No. 4, p. 852 (1985). Abstract. – See Abstr. 010.062.

004.065 **"Grimnismál" 10: old Norse astronomy.**
D. O. Comstock.
Bull. Am. Astron. Soc., Vol. 17, No. 4, p. 852 (1985). Abstract. – See Abstr. 010.062.

004.066 **Physkon, Cleopatra, & Antoninus tables: ancient ephemerides' base.**
D. Rawlins.
Bull. Am. Astron. Soc., Vol. 17, No. 4, p. 852 (1985). Abstract. – See Abstr. 010.062.

004.067 **Aristarchos as source of Babylonian & Ptolemaic mean motions.**
D. Rawlins.
Bull. Am. Astron. Soc., Vol. 17, No. 4, p. 901 (1985). Abstract. – See Abstr. 010.062.

004.068 **The discovery of the Saturn system: Cassini's life and activity.**
M. Fulchignoni.
ESA Spec. Publ., ESA SP–241, p. 289 – 294 (1985). – See Abstr. 012.018.

Cassini's life and activity are shortly described. The discovery of Iapetus, Rhea, Thetis and Dione are reported with the words of Cassini himself as well as the description of the first observation of a gap in the Saturn ring (Cassini division).

004.069 **V. P. Engelhardt and the new Engelhardt Observatory of the Kazan University during the transition period of its activity (1906 – 1910).**
S. N. Korytnikov.
Izv. Astron. Ehngel'gardt. Obs., No. 48, p. 38 – 56 (1984). In Russian.

004.070 **V. P. Engelhardt and the new Engelhardt Observatory of the Kazan University during 1910 – 1915.**
S. N. Korytnikov.
Izv. Astron. Ehngel'gardt. Obs., No. 48, p. 57 – 75 (1984). In Russian.

004.071 **Through the infrared with logbook and lantern slides – a history of infrared astronomy from 1868 to 1960.**
W. M. Sinton.
Publ. Astron. Soc. Pac., Vol. 98, No. 600, p. 246 – 251 (1986).

004.072 **The mystery of Bayer's *Uranometria* – Part 3: Letters and signatures.**
L. J. Carter, A. T. Lawton.
J. Br. Interplanet. Soc., Vol. 39, No. 4, p. 177 – 188 (1986).

004.073 **Les instruments des observatoires français au 19e siècle.**
S. Grillot.
Astronomie, Vol. 100, p. 275 – 289 (1986).

004.074 **La relativité générale à l'étiage: 1925 – 1955.**
J. Eisenstaedt.
Arch. Hist. Exact Sci., Vol. 35, No. 2, p. 115 – 185 (1986).

004.075 **Sirius is nooit rood geweest.**
R. H. van Gent.
Zenit, 13. Jaarg., Nr. 6, p. 212 (1986).

004.076 **Ulugh Beg's forgotten observatory.**
E. W. Piini.
Sky Telesc., Vol. 71, No. 6, p. 542 – 544 (1986).

004.077 **On the geometry of Plato's universe.**
S. V. Zhitomirskij.
Vopr. istor. estestvozn. i tekh., No. 4, p. 104 – 107 (1985). In Russian. Abstr. in Ref. Zh., 51. Astron., 5.51.3 (1986).

004.078 **Tycho Brahe – from astrology to astronomy.**
M. Berić.
Vasiona, Année 33, No. 5, p. 93 – 96 (1985). In Croatian.

004.079 **Ruder Bošković and the discovery of Uranus.**
A. Tomić.
Vasiona, Année 34, No. 1 – 2, p. 6 – 8 (1986). In Serbian.

004.080 **The sundials of the monastery Studenica.**
M. Tadić.
Vasiona, Année 34, No. 1 – 2, p. 13 – 16 (1986). In Croatian.

004.081 **Ancient sunspot observations in China and solar variability.**
Z. T. Xu.
Galaxies, quasars and cosmology, p. 201 – 221 (1985). – See Abstr. 012.042.
On the basis of naked–eye sunspot records in ancient China the solar variability over two thousand years may be revealed. The dispute over the Maunder Minimum could lead to a new understanding of solar activity and a revision of the solar dynamo theory.

004.082 **Galileo e il problema della misura del tempo.**
E. Proverbio.
Suppl. Ann. Ist. Mus. Storia Sci., Anno 1983, Fasc. 2, p. 63 – 72 (1983). = Pubbl. Stn. Astron. Int. Latitudine, Carloforte–Cagliari, Nuov. Ser., N. 94.

004.083 **Le applicazioni del telescopio all'astronomia di posizione nel XVII e XVIII secolo.**
E. Proverbio.
Boll. Cent. Int. A. Beltrame Storia Spazio Tempo, N. 3 – 4, p. 23 – 33 (1984). = Pubbl. Stn. Astron. Int. Latitudine, Carloforte–Cagliari, Nuov. Ser., N. 110.

004.084 **L'horloge mécanique en astronomie de position jusqu'à la moitié du XVIII siècle.**
E. Proverbio.
Proceedings of the XIth International Congress of Chronometry, p. 187 – 193 (1984). = Pubbl. Stn. Astron. Int. Latitudine, Carloforte–Cagliari, Nuov. Ser., N. 114.
The introduction of the mechanical clock at about the beginning of the XIV century has played a very important role for the improvement of the astronomical observation in positional astronomy. At the beginning two different periods can be singled out in the astronomical applications of the mechanical clock. The first period, put between the end of the XVI century and the first half of the following century, is characterized by an evolution of the old mechanical clocks. The second on the contrary is distinguished by the application of the theoretic mechanics to the horology technique and leads to considerable improvement in the accuracy of astronomical time–keepers.

004.085 **Historia de la ciencia árabe. Los sistemas astronómicos.**
J. M. Torroja.
Semin. Astron. Geod., Univ. Madr., Publ., No. 122, 16 pp. (1981).

004.086 **La mecánica celeste.**
J. M. Torroja.
Semin. Astron. Geod., Univ. Madr., Publ., No. 132, 20 pp. (1983).

004.087 **Sugli orientamenti di alcune strutture preistoriche.**
G. Romano, M. Tonon.
Riv. Archeol., Anno 9, p. 5 – 8 (1985). = Oss. Privato, Specola "Ariel", Pubbl. No. 91.

004.088 **Alcune osservazioni sull'orientamento delle chiese medioevali e moderne entro la cinta muraria di Treviso.**
G. Romano.
Riv. Archeol., Anno 9, p. 56 – 58 (1985). = Oss. Privato, Specola "Ariel", Pubbl. No. 92.

004.089 **Per un catalogo di motte e castellieri nella pianura tra il Piave ed il Tagliamento e su alcuni loro allineamenti astronomici.**
G. Romano, M. Tonon.
Atti Mem. Ateneo Treviso, Nuova Ser., No. 1, p. 131 – 139 (1985). = Oss. Privato, Specola "Ariel", Pubbl. No. 93.

004.090 **Una raccolta di disegni astronomici tolemaici.**
G. Romano.
Boll. Cent. Int. Storia Spazio Tempo, No. 5, p. 53 – 60 (1986). = Oss. Privato, Specola "Ariel", Pubbl. No. 94.

004.091 **Intimachay: a December solstice observatory at Machu Picchu, Peru.**
D. S. P. Dearborn, K. J. Schreiber, R. E. White.
Prepr. Steward Obs., No. 638, 15 pp. (1986). To appear in Am. Antiquities.

004.092 **Chinese ancient cosmology and astronomical instruments.**
S.–z. Zhang.
Publ. Purple Mt. Obs., Vol. 4, No. 4, p. 57 – 65 (1985). In Chinese.

004.093 **Science and superstition: comets and the French public in the 18th century.**
P. Stewart.
Am. J. Phys., Vol. 54, No. 1, p. 16 – 24 (1986). Abstr. in Phys. Abstr., Vol. 89, No. 1277, Entry 36452 (1986).

004.094 **A star map – origin unknown.**
D. W. Hughes.
J. Br. Astron. Assoc., Vol. 96, No. 4, p. 230 – 231 (1986).

004.095 **Stelle e comete nella astronomia precolombiana.**
G. Romano.
Coelum, Vol. 54, N. 3, p. 113 – 126 (1986).

004.096 **Shapley's standard candles.**
B. L. Welther.
J. Am. Assoc. Variable Star Obs., Vol. 14, No. 2, p. 77 (1985). Abstract. – See Abstr. 010.021.

004.097 **The discovery of Pluto. Some generally unknown aspects of the story.**
C. W. Tombaugh.
Mercury, Vol. 15, No. 3, p. 66 – 72 (1986).

004.098 **Galileo and optical illusion.**
G. D. Parker.
Am. J. Phys., Vol. 54, No. 3, p. 248 – 252 (1986). Abstr. in Phys. Abstr., Vol. 89, No. 1281, Entry 61252 (1986).

004.099 **Waves in plasmas: some historical highlights.**
T. H. Stix.
Proceedings of the 1984 International Conference on Plasma Physics, Vol. 2, p. 735 – 759 (1985). Abstr. in Phys. Abstr., Vol. 89, No. 1281, Entry 61255 (1986). – See Abstr. 012.056.

004.100 **Berkeley, Newton and the stars.**
K. P. Winkler.
Stud. Hist. Philos. Sci., Vol. 17, No. 1, p. 23 – 42 (1986).
In his criticism of Newton, Berkeley makes two claims about the fixed stars. His first is that the stars do not appear to exert attractive power. The second is that the stars can serve as the reference frame against which natural philosophers measure "true motion and true rest". In this paper the author shows how these claims fit together and how they support, and are in turn supported by, Berkeley's more general views on motion and force.

004.101 **Experimental particle physics in the age of innocence. (The decay of mesotrons: 1939 – 1943).**
B. B. Rossi.
Cosmic pathways, p. 331 – 348 (1986). – See Abstr. 003.045.
This paper presents an account of the discovery of cosmic ray mesons.

004.102 **James Bradley and the eighteenth century 'gap' in attempts to measure annual stellar parallax.**
M. Williams.
Notes Rec. R. Soc. London, Vol. 37, No. 1, p. 83 – 100 (1982).

004.103 **Die Himmelskunde in neuerer Zeit. Herschel's und seiner Zeitgenossen Wirksamkeit bis zum Schlusse des 18. Jahrhunderts.**
J. H. Mädler.
Lilienthal oder die Astronomen, p. 47 – 85 (1984). – See Abstr. 003.046.
Reprinted from J. H. Mädler, Geschichte der Himmelskunde von der ältesten bis auf die neueste Zeit. Unveränderter Neudruck der Ausgabe von 1873, Band II, p. 1 – 72 (1973).

004.104 **Die Lilienthaler Sternwarte. Ein Bild aus der Geschichte der Himmelskunde in Deutschland.**
H. A. Schumacher.
Lilienthal oder die Astronomen, p. 86 – 252 (1984). – See Abstr. 003.046.
Reprinted from Abhandlungen, herausgegeben vom naturwissenschaftlichen Vereine zu Bremen, Band XI, p. 39 – 170 (1890).

004.105 **Die Kosmogonie Anton von Zachs.**
P. Brosche.
Berichte zur Wissenschaftsgeschichte, Band 9, p. 89 – 93 (1986).
In his "Cosmogenische Betrachtungen" (1804), Anton von Zach rediscovered – probably independently – some aspects of the theories of Kant and Laplace. More originally, he envisaged also the consequences of an era of heavy impacts in the early history of the Earth.

004.106 **L'astronomia a Napoli nel '700.**
S. Mancuso.
G. Astron., Vol. 11, N. 4, p. 533 – 536 (1985).

004.107 **Sugli orientamenti di alcuni dolmen pugliesi.**
G. Romano, M. Perissinotto.
G. Astron., Vol. 11, N. 4, p. 537 – 544 (1985).

004.108 **L'histoire de l'astronomie et la philosophie.**
A. B. K. Stojković.
Publ. Astron. Soc. "Rudjer Bošković", No. 4, p. 9 – 13 (1985). – See Abstr. 012.078. In Croatian.

004.109 **Muvekithane – first astronomic institutions in Bosnia and Hercegovina.**
J. Mulaomerović.
Publ. Astron. Soc. "Rudjer Bošković", No. 4, p. 25 – 27 (1985). – See Abstr. 012.078. In Croatian.

004.110 **The observatory in Pula.**
J. Bačić.
Publ. Astron. Soc. "Rudjer Bošković", No. 4, p. 29 – 34 (1985). – See Abstr. 012.078. In Croatian.

004.111 **The astronomical observations of the count L. F. Marsigli on the territory of Vojvodina during summer of the year 1696.**
J. Francisty.
Publ. Astron. Soc. "Rudjer Bošković", No. 4, p. 67 – 76 (1985). – See Abstr. 012.078. In Croatian.

004.112 **Astronomy in the works of J. Rajić.**
N. D. Janković.
Publ. Astron. Soc. "Rudjer Bošković", No. 4, p. 77 – 86 (1985). – See Abstr. 012.078. In Serbian.

004.113 **Astronomical texts in national language in Croatia until 18th century.**
Ž. Dadić.
Publ. Astron. Soc. "Rudjer Bošković", No. 4, p. 87 – 89 (1985). – See Abstr. 012.078. In Croatian.

004.114 **On the next transit of Venus across the solar disc by Ivan Milošević.**
E. Stipanić.
Publ. Astron. Soc. "Rudjer Bošković", No. 4, p. 95 – 98 (1985). – See Abstr. 012.078. In Croatian.

004.115 **The Serbs and calendar reforms.**
N. D. Janković.
Publ. Astron. Soc. "Rudjer Bošković", No. 4, p. 103 – 110 (1985). – See Abstr. 012.078. In Serbian.

004.116 **A contribution to the knowledge of astronomical terminology by the end of the XVIII century in slavenoservian language.**
G. M. Popović.
Publ. Astron. Soc. "Rudjer Bošković", No. 4, p. 111 – 113 (1985). – See Abstr. 012.078. In Croatian.

004.117 **Notices on astronomy in Montenegro during the XIX century.**
V. Gledić.
Publ. Astron. Soc. "Rudjer Bošković", No. 4, p. 123 – 125 (1985). – See Abstr. 012.078. In Croatian.

004.118 **A note on Copernicus from 1837.**
V. Trajkovska.
Publ. Astron. Soc. "Rudjer Bošković", No. 4, p. 127 – 128 (1985). – See Abstr. 012.078. In Croatian.

004.119 **The three body problem in the Algol system in Slavko Rozgaj's dissertation.**
G. Ivanišević, K. Pavlovski.
Publ. Astron. Soc. "Rudjer Bošković", No. 4, p. 147 – 148 (1985). – See Abstr. 012.078. In Croatian.

004.120 **Growth of knowledge of astronomical refraction.**
G. Teleki, O. Atanacković.
Publ. Astron. Soc. "Rudjer Bošković", No. 4, p. 151 – 155 (1985). – See Abstr. 012.078. In Croatian.

004.121 **A survey of radio astronomical researches from their earliest beginnings to our days.**
B. M. Ševarlić.
Publ. Astron. Soc. "Rudjer Bošković", No. 4, p. 157 – 159 (1985). – See Abstr. 012.078. In Croatian.

004.122 **Microturbulence and stellar spectral lines.**
V. Kršljanin.
Publ. Astron. Soc. "Rudjer Bošković", No. 4, p. 177 – 179 (1985). – See Abstr. 012.078. In Croatian.

004.123 **The oldest diplomawork on geodetical astronomy in Slovenia.**
B. Kilar.
Publ. Astron. Soc. "Rudjer Bošković", No. 4, p. 189 – 191 (1985). – See Abstr. 012.078. In Croatian.

The Bedford Catalogue. From "A cycle of celestial objects".
See Abstr. 002.030.

History of British space science.
See Abstr. 003.003.

Sothis– und Monddaten. Studien zur astronomischen und technischen Chronologie Altägyptens.
See Abstr. 003.006.

Johannes Kepler Gesammelte Werke. XI, Part 1: Ephemerides novae motuum coelestium.
See Abstr. 003.064.

Histoire de l'astronomie (occidentale).
See Abstr. 003.071.

Dissertatio Philosophica de Orbitis Planetarum. Philosophische Erörterung über die Planetenbahnen (1801).
See Abstr. 003.108.

Kosmologie. Geschichte und Systematik in philosophischer Perspektive.
See Abstr. 003.114.

Fenêtres sur l'univers. L'histoire des instruments de l'astronomie.
See Abstr. 003.124.

The Megalithic Odyssey.
See Abstr. 003.145.

Bibliotheca Mesopotamica. Vol. 2, Fasc. 1. Babylonian Planetary Omens: Part 1, Enūma Anu Enlil, Tablet 63: The Venus Tablet of Ammisaduqa.
See Abstr. 003.158.

Bibliotheca Mesopotamica. Vol. 2, Fasc. 2. Babylonian Planetary Omens: Part 2, Enūma Anu Enlil, Tablets 50 – 51.
See Abstr. 003.159.

The cosmic zygote: cosmology in the Amazon basin.
See Abstr. 003.164.

Edwin Hubble: legal eagle.
See Abstr. 005.001.

Planetariums can be visual computers for ethnoastronomers.
See Abstr. 009.009.

Der Einsteinturm in Potsdam und die Relativitätstheorie.
See Abstr. 009.011.

Arqueoastronomía y etnoastronomía en Mesoamérica. UNAM symposium report, 24 – 29 September 1984.
See Abstr. 011.010.

Prehistoric rock art and archaeoastronomy, Little Rock conference report, 7 – 12 October 1984.
See Abstr. 011.011.

L'Académie royale des Sciences et la figure de la Terre.
See Abstr. 011.038.

Solar studies in China.
See Abstr. 013.003.

The persistence of Lowell's legacy: life, Mars, and the U.S. Space Program 1958 – 1967.
See Abstr. 013.026.

New tools for ancient skies.
See Abstr. 014.011.

The legacy of the 200–inch.
See Abstr. 032.011.

Presentation of new solar and planetary tables of interest for historical calculations.
See Abstr. 046.005.

Coordination of minor planet orbital studies: past, present and future.
See Abstr. 098.065.

The historical records of Halley's comet in China.
See Abstr. 103.611.

Halley's comet in history.
See Abstr. 103.616.

The history of Comet Halley.
See Abstr. 103.724.

005 Biography

005.001 **Edwin Hubble: legal eagle.**
N. Hetherington.
Nature, Vol. 319, No. 6050, p. 189 – 190 (1986).
Edwin Hubble skilfully employed trial tactics and techniques to attain favourable verdicts from the court of science.

005.002 **Gennem Italien i Galileis fodspor.**
P. Gadman.
Astron. Tidsskr., Årg. 19, Nr. 1, p. 16 – 23 (1986).

005.003 **The legacy of Schiaparelli and Lowell.**
F. I. Ordway.
J. Br. Interplanet. Soc., Vol. 39, No. 1, p. 19 – 27 (1986).
A retrospective is presented of Giovanni Virginio Schiaparelli of the Brera Observatory in Milan and his discovery of markings on Mars described as canali, and of Percival Lowell, founder of an observatory in Arizona dedicated to the study of those markings. A summary is given of events from the eruption of the Martian canal debate a century ago, to the establishment of the observatory in 1894 and beyond Lowell's death in 1916.

005.004 **Prof. William Bowen Bonnor: a biographical sketch and a bibliography.**
M. A. H. MacCallum.
Galaxies, axisymmetric systems and relativity, p. XIII – XX (1985). – See Abstr. 003.007.

005.005 **Der Nachlaß von Friedrich Wilhelm Bessel (1784 – 1846) im Zentralen Archiv der Akademie der Wissenschaften der DDR.**
K. Klauss.
Sterne, 62. Band, Heft 1, p. 35 – 39 (1986).

005.006 **A neglected Galilean letter.**
S. Drake.
J. Hist. Astron., Vol. 17, Part 2, No. 49, p. 99 – 108 (1986).

005.007 **Old stars: Ormsby Mitchel.**
K. J. Weddle.
Sky Telesc., Vol. 71, No. 1, p. 14 – 16 (1986).

005.008 **The outstanding Soviet astrophysicist V. Krat (1911 – 1983).**
J. Vitinskis (*Yu. I. Vitinskij*), A. Balklavs.
Zvaigžnota Debess, 1984. Gada Pavasaris, p. 28 – 31 (1984). In Latvian.

005.009 **Simon Newcomb: a new view.**
B. G. Corbin, L. E. Doggett.
Bull. Am. Astron. Soc., Vol. 17, No. 4, p. 827 – 828 (1985). Abstract. – See Abstr. 010.062.

005.010 **Alexander Roberts: distinguished South African amateur astronomer and humanitarian.**
T. R. Williams.
Bull. Am. Astron. Soc., Vol. 17, No. 4, p. 828 (1985). Abstract. – See Abstr. 010.062.

005.011 **Nicholas T. Bobrovnikoff and the scientific study of comet Halley 1910.**
D. E. Osterbrock.
Bull. Am. Astron. Soc., Vol. 17, No. 4, p. 828 (1985). Abstract. – See Abstr. 010.062.

005.012 **Fricke's influence on the world of astronomy.**
G. Westerhout.
Celest. Mech., Vol. 37, No. 3, p. 345 – 348 (1985). – See Abstr. 012.017.

005.013 **Edmond Halley (III).**
K. Städeli.
Orion, 44. Jahrg., Nr. 212, p. 8 – 9 (1986).

005.014 **Bernhardt Schmidt e l'astronomia con i telescopi Schmidt.**
F. Börngen.
Coelum, Vol. 54, N. 2, p. 57 – 72 (1986).

005.015 **Nicholas T. Bobrovnikoff and the scientific study of Comet Halley 1910.**
D. E. Osterbrock.
Mercury, Vol. 15, No. 2, p. 46 – 50, 63 (1986).

005.016 **Edmond Halley.**
A. H. Jarrett.
Boyden Obs., Occas. Publ., No. 3, 2+10 pp. (1986).

005.017 **The phenomenon Shklovskij (18.VI (1.VII) 1916 – 3.III.1985), on the occasion of the 70th anniversary of his birthday.**
N. S. Kardashev, L. S. Marochnik.
Priroda, No. 6, p. 84 – 95 (1986). In Russian.

005.018 **Christopher Hansteen (1784 – 1873). En pioner i norsk universitetsmiljø.**
J. M. Hansteen.
Inst. Teor. Astrofys., Blindern–Oslo, Småtrykk, Nr. 105A, 3 pp. (1984).

005.019 **Anne S. Young and Alice H. Farnsworth: 58 years of astronomy at Mount Holyoke College.**
M. L. Hazen.
J. Am. Assoc. Variable Star Obs., Vol. 14, No. 2, p. 49 – 51 (1985).

005.020 **Louise F. Jenkins, astronomer and missionary in Japan.**
S.–I. Sakuma.
J. Am. Assoc. Variable Star Obs., Vol. 14, No. 2, p. 67 – 68 (1985).

005.021 **Solon I. Bailey and *THE* number of galactic novae.**
P. L. Collins.
J. Am. Assoc. Variable Star Obs., Vol. 14, No. 2, p. 74 (1985). Abstract. – See Abstr. 010.021.

005.022 **Shapley in Washington: a scientist among politicians.**
B. L. Welther.
J. Am. Assoc. Variable Star Obs., Vol. 14, No. 2, p. 76 (1985). Abstract. – See Abstr. 010.021.

005.023 **From the collapse of the atom to the collapse of the universe – Niels Bohr and astrophysics.**
L.–z. Fang.
Chin. Astron. Astrophys., Vol. 10, No. 1, p. 34 – 37 (1986). English translation of Ziran Zazhi, Vol. 8, No. 10, p. 691 – 694 (1986).

005.024 **From Scorpius X–1 to quarks.**
G. Garmire.
Cosmic pathways, p. 361 – 363 (1986). – See Abstr. 003.045.
The author presents a brief personal account of his collaboration with B. V. Sreekantan in the years 1964 – 1966 at MIT.

005.025 **B. V. Sreekantan – profile of a scientist.**
S. Naranan.
Cosmic pathways, p. 387 – 395 (1986). – See Abstr. 003.045.

005.026 **A biographical note on Prof. B. V. Sreekantan.**
Cosmic pathways, p. 407 – 408 (1986). – See Abstr. 003.045.

005.027 **A few flashes on Hans Bethe contributions during the thirties.**
E. Amaldi.
From nuclei to stars, p. 1 – 16 (1986). – See Abstr. 012.070.

005.028 **Der Briefwechsel zwischen J. H. Schroeter und G. Chr. Lichtenberg.**
W. Gresky.
Lilienthal oder die Astronomen, p. 7 – 34 (1984). – See Abstr. 003.046.
Reprinted from Mitt. Gauß Ges., Göttingen, Heft 11, p. 24 – 49 (1974).

005.029 ***Joh. Hieron. Schröter,* als Astronom.**
F. von Zach.
Lilienthal oder die Astronomen, p. 35 – 36 (1984). – See Abstr. 003.046.
Reprinted from Allgemeine Geographische Ephemeriden III (1799).

005.030 **Beitrag zur Geschichte des Lebens und Wirkens des Astronomen Johann Hieronymus Schroeter.**
H. W. Rotermund.
Lilienthal oder die Astronomen, p. 37 – 46 (1984). – See Abstr. 003.046.
Reprinted from Neues Vaterländisches Archiv II, Lüneburg, p. 272 – 291 (1824).

005.031 **Johann Hieronymus Schröter, Schöpfer der Selenographie.**
H.–B. Brenske.
Lilienthal oder die Astronomen, p. 259 – 272 (1984). – See Abstr. 003.046.
Reprinted from Nachr. Olbers–Ges. Bremen, Nr. 60, p. 28 – 35 (1964).

005.032 **Johann Hieronymus Schroeter im Urteil seiner Zeit.**
D. B. Herrmann.
Lilienthal oder die Astronomen, p. 273 – 281 (1984). – See Abstr. 003.046.
Reprinted from Sterne, Jahrg. 41, Heft 7 – 8, p. 135 – 143 (1965).

005.033 **Der Amateur von Lilienthal.**
H.-B. Brenske.
Lilienthal oder die Astronomen, p. 282 – 288 (1984). – See Abstr. 003.046.
Reprinted from Nachr. Olbers-Ges. Bremen, Nr. 67, p. 5 – 8 (1966).

005.034 **Bessels Weggang aus Lilienthal und sein Verhältnis zu Schroeter in späterer Zeit. Unbekannte Dokumente über Bessel, Schroeter und Lilienthal.**
D. B. Herrmann.
Lilienthal oder die Astronomen, p. 289 – 296 (1984). – See Abstr. 003.046.
Reprinted from Nachr. Olbers-Ges. Bremen, Nr. 67, p. 9 – 12 (1966).

005.035 **Die "Biographie Universelle" über Schroeter (1847).**
Lilienthal oder die Astronomen, p. 297 – 303 (1984). – See Abstr. 003.046. In French.

005.036 **Die "Allgemeine deutsche Biographie" über Schroeter (1891).**
Lilienthal oder die Astronomen, p. 304 – 307 (1984). – See Abstr. 003.046.

005.037 **Sternwarte Lilienthal. Von Johann Hieronymus Schroeter, Olbers, Harding und der "Vereinigten Astronomischen Gesellschaft" Lilienthal.**
D. Gerdes.
Lilienthal oder die Astronomen, p. 308 – 319 (1984). – See Abstr. 003.046.
Reprinted from Nachr. Olbers-Ges. Bremen, Nr. 127, p. 2 – 9 (1983).

005.038 **Ein "unerleDichter Fall" oder Arno Schmidt auf Schroeters Spur.**
J. Huerkamp.
Lilienthal oder die Astronomen, p. 320 – 338 (1984). – See Abstr. 003.046.

005.039 **Herschel, músico y astrónomo.**
D. J. J. de Orús Navarro.
Memorias de la Real Academia de Ciencias y Artes de Barcelona, Vol. 46, Núm. 9, p. 227 – 247 (1985).

005.040 **Radovan Danić – a popularizer of astronomy (I): Printed papers.**
M. S. Dimitrijević, A. Tomić.
Publ. Astron. Soc. "Rudjer Bošković", No. 4, p. 41 – 45 (1985). – See Abstr. 012.078. In Croatian.

005.041 **Radovan Danić – a popularizer of astronomy (II): Lectures and courses.**
A. Tomić, M. S. Dimitrijević.
Publ. Astron. Soc. "Rudjer Bošković", No. 4, p. 47 – 49 (1985). – See Abstr. 012.078. In Croatian.

005.042 **Notes of Bošković's lectures on practical astronomy in academical year 1754/55.**
I. Martinović.
Publ. Astron. Soc. "Rudjer Bošković", No. 4, p. 91 – 93 (1985). – See Abstr. 012.078. In Croatian.

005.043 **Life and work of Jovan (Julijan) Čokor (Baja, 21.1.1810 – Sr. Karlovci 10. or 13.6.1871).**
B. D. Jovanović.
Publ. Astron. Soc. "Rudjer Bošković", No. 4, p. 99 – 102 (1985). – See Abstr. 012.078. In Croatian.

005.044 **Vojvodina and men from Vojvodina who popularized astronomy up to 1941.**
B. D. Jovanović.
Publ. Astron. Soc. "Rudjer Bošković", No. 4, p. 117 – 121 (1985). – See Abstr. 012.078. In Croatian.

005.045 **The notions of general, particular, cause and consequence in the work of Prof. M. Milanković.**
B. Popović.
Publ. Astron. Soc. "Rudjer Bošković", No. 4, p. 129 – 130 (1985). – See Abstr. 012.078. In Croatian.

005.046 **Stjepan Mohorovičić (1890 – 1980) and his "private station for cosmic physics".**
G. Ivanišević.
Publ. Astron. Soc. "Rudjer Bošković", No. 4, p. 131 – 134 (1985). – See Abstr. 012.078. In Croatian.

005.047 **Vojislav V. Mišković and his works in history of astronomy.**
M. Mužijević.
Publ. Astron. Soc. "Rudjer Bošković", No. 4, p. 135 – 137 (1985). – See Abstr. 012.078. In Croatian.

005.048 **Johann Kepler resided on Yugoslav territory.**
B. D. Jovanović.
Publ. Astron. Soc. "Rudjer Bošković", No. 4, p. 149 – 150 (1985). – See Abstr. 012.078. In Croatian.

005.049 **Knut Lundmark (1889 – 1958) and some of his contributions to extragalactic astronomy.**
G. Ivanišević.
Publ. Astron. Soc. "Rudjer Bošković", No. 4, p. 161 – 162 (1985). – See Abstr. 012.078. In Croatian.

Isaac Newton: reluctant genius.
See Abstr. 003.112.

Oculus Enoch... Ein Beitrag zur Entdeckungsgeschichte des Fernrohrs.
See Abstr. 003.190.

Beatrice (Hill) Tinsley 1941 – 1981, astronomer.
See Abstr. 003.209.

No edition of Copernicus in 1640 or 1646.
See Abstr. 004.009.

The discovery of the Saturn system: Cassini's life and activity.
See Abstr. 004.068.

Galileo e il problema della misura del tempo.
See Abstr. 004.082.

Elias Loomis and the Loomis Observatory.
See Abstr. 009.058.

The case of Galileo.
See Abstr. 011.039.

006 Personal Notes

006.001 **Robert S. Dietz received the Barringer Medal.**
Meteoritics, Vol. 20, No. 4, p. 597 (1985).

006.002 **Sandra M. Faber received the 1985 Dannie Heineman Prize for Astrophysics of the American Institute of Physics and the American Astronomical Society.**
Phys. Today, Vol. 39, No. 3, p. 119 (1986).

006.003 **Peter Goldreich received the Chapman Medal 1985.**
D. Lynden-Bell.
Q. J. R. Astron. Soc., Vol. 27, No. 1, p. 1 – 2 (1986).

006.004 **Pierre Lacroute received the Prix Janssen.**
C. R. Acad. Sci., Sér. Gén., Vie Sci., Tome 2, No. 6, p. 600 (1986). In French.

006.005 **Eugene M. Shoemaker received the Leonard Medal.**
Meteoritics, Vol. 20, No. 4, p. 595 (1985).

006.006 **Judith S. Young received the Maria Goeppert-Mayer Award.**
Phys. Today, Vol. 39, No. 5, p. 111 (1986).

007 Obituaries

007.001 **Friedrich Becker, 12. Juni 1900 – 25. Dezember 1985.**
W. Priester.
Sterne Weltraum, 25. Jahrg., Nr. 3, p. 126 – 127 (1986).

007.002 **Ludwig Biermann, 13. März 1907 – 12. Januar 1986.**
R. Lüst.
Sterne Weltraum, 25. Jahrg., Nr. 5, p. 250 – 251 (1986).

007.003 **In ricordo di Marcello Ceccarelli.**
A. Braccesi.
G. Astron., Vol. 11, N. 4, p. 547 – 552 (1985).

007.004 **John Paul Cox, 1926 November 4 – 1984 August 19.**
R. H. Garstang.
Q. J. R. Astron. Soc., Vol. 27, No. 2, p. 302 – 305 (1986).

007.005 **La vie et l'œuvre de Paul Dirac, 1902 August 8 – 1984 October 20.**
L. Michel.
C. R. Acad. Sci., Sér. Gén., Vie Sci., Tome 2, No. 6, p. 619 – 622 (1986).

007.006 **Paul Adrien Maurice Dirac, 1902 August 8 – 1984 October 20.**
D. W. Sciama.
Q. J. R. Astron. Soc., Vol. 27, No. 1, p. 124 – 126 (1986).

007.007 **Eric Gray Forbes, 1933 March 30 – 1984 November 21.**
D. Howse.
Q. J. R. Astron. Soc., Vol. 27, No. 1, p. 126 – 128 (1986).

007.008 **Eric Gray Forbes, 30.3.1933 – 21.11.1984.**
D. B. Herrmann.
Archenhold-Sternw. Berlin-Treptow, Sonderdr., Nr. 29, 4 pp. (1985).

007.009 **André Hamon (1900 – 1984).**
R. Sagot, B. Clouet.
Astronomie, Vol. 100, p. 49 – 50 (1986).

007.010 **Caroline Winifred Herschel, 1906 – 1985 May 5.**
C. Stott, T. Fanning.
Q. J. R. Astron. Soc., Vol. 27, No. 1, p. 128 – 129 (1986).

007.011 **B. K. Ioannisiani (*Ioannisyan*), 1911 – 1985, December 9.**
Astrofizika, Tom 24, Vyp. 1, p. 208 (1986). In Russian. English translation in Astrophysics, Vol. 24, No. 1.

007.012 **Lincoln La Paz (1897 – 1985).**
W. H. Haas.
Strolling Astron., Vol. 31, Nos. 7–8, p. 173 (1986).

007.013 **Laurence Patrick Lee (1913 – 1985).**
G. H. Rowe.
South. Stars, Vol. 31, No. 4, p. 221 – 222 (1985).

007.014 **Orren C. Mohler, 29 July 1908 – 17 September 1985.**
F. D. Miller.
Phys. Today, Vol. 39, No. 4, p. 74 (1986).

007.015 **Gordon Allen Newkirk, Jr., 1928 – 1985.**
J. A. Eddy, R. M. MacQueen.
Sol. Phys., Vol. 104, No. 2, p. 256 – 258 (1986).

007.016 **In memoriam: Ernst Julius Öpik, 1893 – 1985.**
G. Wetherill.
Icarus, Vol. 66, No. 2, p. 193 – 194 (1986).

007.017 **The turbulent life of E. J. Öpik.**
P. Moore.
Sky Telesc., Vol. 71, No. 2, p. 149 (1986).

007.018 **K. F. Ogorodnikov, 1900 July 30 – 1985 June 29.**
Zemlya Vselennaya, No. 1, p. 44 – 46 (1986). In Russian.

007.019 **Stefan Piotrowski, 1910 – 1985.**
J. Smak, A. Sołtan.
Postepy Astron., Tom 33, Zesz. 1–2, p. 65 – 68 (1985). In Polish.

007.020 **Professor Harry Pollard, 1919 – 1985.**
D. G. Saari.
Celest. Mech., Vol. 37, No. 4, p. 349 (1985).

007.021 **Beryl H. Potter died 1985 August 10.**
Phys. Today, Vol. 39, No. 2, p. 92 (1986).

007.022 **Dr. Antoni Przybylski died 1984 September 21.**
S. C. B. Gascoigne.
Proc. Astron. Soc. Aust., Vol. 6, No. 2, p. 275 (1985).

007.023 **Edward Rosen (1906 – 1985).**
I. B. Cohen.
J. Hist. Astron., Vol. 17, Part 2, No. 49, p. 150 – 153 (1986).

007.024 **Svein Rosseland, 1894 March 31 – 1985 January 19.**
E. Jensen.
Inst. Teor. Astrofys., Blindern–Oslo, Småtrykk, Nr. 106, 11 pp. (1985).

007.025 **V. P. Shabanskij, 1928 April 13 – 1985 November 16.**
Geomagn. Aehron., Tom 26, No. 1, p. 172 (1986). In Russian. English translation in Geomagn. Aeron.

007.026 **La vie et l'œuvre de Pol Swings, 1906 Septembre 24 – 1983 Octobre 28.**
C. Fehrenbach.
C. R. Acad. Sci., Sér. Gén., Vie Sci., Tome 2, No. 6, p. 657 – 662 (1986).

007.027 **Pol Swings, 1906 September 24 – 1983 October 28.**
R. H. Garstang.
Q. J. R. Astron. Soc., Vol. 27, No. 2, p. 305 – 308 (1986).

007.028 **Pol Swings, 24.9.1906 – 28.10.1983.**
A. Woszczyk.
Postepy Astron., Tom 32, Zesz. 1, p. 87 – 98 (1984). In Polish.

007.029 **Stefan Temesvary, 1915 – 1984 September 13.**
E. Trefftz.
Q. J. R. Astron. Soc., Vol. 27, No. 1, p. 129 – 130 (1986).

007.030 **Alexander Thom, 1894 March 26 – 1985 November 7.**
R. J. C. Atkinson.
J. Hist. Astron., Vol. 17, Part 1, No. 48, p. 73 – 75 (1986).

007.031 **Edna M. Wadsworth died 1985 March 28.**
D. H. Sadler.
Q. J. R. Astron. Soc., Vol. 27, No. 1, p. 131 (1986).

007.032 **Bernhard Wedel, 11. Juli 1936 – 14. März 1986.**
R. Lukas.
Sterne Weltraum, 25. Jahrg., Nr. 6, p. 350 (1986).

007.033 **Thomas Wilson Wormell died 1985 January 3.**
R. Goody.
Q. J. R. Astron. Soc., Vol. 27, No. 1, p. 131 – 133 (1986).

008 Publications of Observatories, Institutes

Reports, communications, and publications of observatories and astronomical institutes are recorded in this section; included are numbered series of reprints. Whenever possible, the numbers of the abstracts referring to the publications are given. The places of observatories and institutes are listed in alphabetical order. If only the formal name of an observatory or institute is known, its place can be found in the following index list.

Aerospace Corporation	**El Segundo,** Calif.
Algonquin Radio Observatory	**Lake Traverse,** Canada
Allegheny Observatory	**Pittsburgh,** Pa.
Anglo-Australian Observatory	**Epping,** Australia
Antares Observatory	**Feira de Santana,** Brazil
Applied Research Corporation	**Landover,** Md.
Archenhold Observatory	**Berlin,** German Democratic Republic
Argentine Institute of Radio Astronomy	**Villa Elisa,** Argentina
Argentine Radio Astronomy Institute	**Pereyra,** Argentina
Arizona State University	**Tempe,** Ariz.
Astronomisches Rechen-Institut	**Heidelberg,** F.R. Germany
Babelsberg Observatory	**Potsdam,** German Democratic Republic
Bartol Research Foundation	**Newark,** Del.
Battelle Memorial Institute	**Richland,** Wash.
Bell Telephone Laboratories	**Holmdel,** N.J.
Bergedorf Observatory	**Hamburg,** F.R. Germany
Blindern Institute of Theoretical Astrophysics	**Oslo,** Norway
Bosscha Observatory	**Lembang,** Indonesia
Bouzareah Observatory	**Algiers,** Algeria
Boyden Observatory	**Bloemfontein,** South Africa
Bureau International de l'Heure	**Paris,** France
C. E. Kenneth Mees Observatory	**Rochester,** N.Y.
California Institute of Technology	**Pasadena,** Calif.
Cantonal Observatory	**Neuchâtel,** Switzerland
Carlsberg Automatic Meridian Circle	**La Palma,** Spain
Carter Observatory	**Wellington,** New Zealand
Center for Astrophysics	**Cambridge,** Mass.
Center for High Angular Resolution Astronomy, Georgia State University	**Atlanta,** Ga.
Centre de Données Stellaires	**Strasbourg,** France
Centro Astronomico Hispano-Aleman	**Calar Alto,** Spain
Centro de Investigación de Astronomía (CIDA)	**Mérida,** Venezuela
Cerro Calan National Astronomical Observatory	**Santiago,** Chile
Cerro Tololo Interamerican Observatory	**La Serena,** Chile
Chamberlin Observatory	**Denver,** Colo.
Charles University Astronomical Institute	**Prague,** Czechoslovakia
Climenhaga Observatory	**Victoria,** Canada
Cointe Observatory	**Liège,** Belgium
Computer Sciences Corporation	**Silver Spring,** Md.
Cornell University	**Ithaca,** N.Y.
Crawford Hill Observatory	**Holmdel,** N.J.
David Dunlop Observatory	**Richmond Hill,** Canada
Dearborn Observatory	**Evanston,** Ill.
Department of Astronomy and Space Sciences, Punjabi University	**Patiala,** India
Department of Mathematics, University of Poona	**Poona,** India
Deutsches Hydrographisches Institut	**Hamburg,** F.R. Germany
Dominion Astrophysical Observatory	**Ottawa,** Canada
Dominion Astrophysical Observatory	**Victoria,** Canada
Dominion Radio Astrophysical Observatory	**Penticton,** Canada
Dudley Observatory	**Schenectady,** N.Y.
Dunsink Observatory	**Dublin,** Ireland
Dyer Observatory	**Nashville,** Tenn.
Ebro Observatory	**Roquetas,** Spain
Effelsberg Radio Observatory	**Bonn,** F.R. Germany
Ege University Observatory	**Izmir,** Turkey
Electronics Research Laboratory	**Göteborg,** Sweden
Engelhardt Observatory	**Kazan,** USSR
Erwin W. Fick Observatory	**Ames,** Iowa
European Southern Observatory	**Garching,** F.R. Germany
European Southern Observatory	**La Silla,** Chile

Fabra Observatory	**Barcelona,** Spain
Felix Aguilar Observatory	**San Juan,** Argentina
Fernbank Science Center	**Atlanta,** Ga.
Figl Observatory	**Vienna,** Austria
Five College Astronomy Department	**Amherst,** Mass.
Floirac Observatory	**Bordeaux,** France
Florida State University Radio Observatory	**Tallahassee,** Fla.
Flower and Cook Observatory	**Malvern,** Pa.
Fort Skala Station	**Cracow,** Poland
Franko State University Astronomical Observatory	**Lvov,** USSR
Georgia State University	**Atlanta,** Ga.
Goddard Space Flight Center	**Greenbelt,** Md.
Goethe Link Observatory	**Bloomington,** Indiana
Hale Observatories	**Pasadena,** Calif.
Hartebeesthoek Radio Astronomy Observatory	**Johannesburg,** South Africa
Haute Provence Observatory	**Saint Michel,** France
Haystack Observatory	**Westford,** Mass.
Heinrich-Hertz-Institut	**Berlin,** German Democratic Republic
Heliophysical Observatory	**Debrecen,** Hungary
Herzberg Institute of Astrophysics	**Ottawa,** Canada
High Altitude Observatory	**Boulder,** Colo.
H.M. Nautical Almanac Office	**Greenwich,** England
Hoher List Observatory	**Bonn,** F.R. Germany
Hopkins Observatory	**Williamstown,** Mass.
Horn d'Arturo Observatory	**Bologna,** Italy
Hvar Observatory	**Zagreb,** Yugoslavia
IBM Thomas J. Watson Research Center	**Yorktown Heights,** N.Y.
Indian Institute of Astrophysics	**Bangalore,** India
Infrared Telescope Facility	**Honolulu,** Hawaii
Institut Astrofiziki	**Dushanbe,** USSR
Instituto Nacíonal de Astrofisica, Optica y Electronica	**Puebla,** Pue., México
International Latitude Observatory	**Carloforte-Cagliari,** Italy
International Latitude Observatory	**Mizusawa,** Japan
IUE Observatory	**Villafranca,** Spain
James Mims Observatory	**Baton Rouge,** La.
Jet Propulsion Laboratory	**Pasadena,** Calif.
Jodrell Bank Radio Observatory	**Manchester,** England
Judson B. Coit Observatory	**Boston,** Mass.
Kandilli Observatory	**Istanbul,** Turkey
Kanzelhöhe Solar Observatory	**Graz,** Austria
Kapteyn Astronomical Laboratory	**Groningen,** The Netherlands
Karl Schwarzschild Observatory	**Tautenburg,** German Democratic Republic
Kiepenheuer Institut	**Freiburg,** F.R. Germany
Kiso Observatory	**Tokyo,** Japan
Kitt Peak National Observatory	**Tucson,** Ariz.
Kodaikanal Observatory	**Bangalore,** India
Königstuhl Observatory	**Heidelberg,** F.R. Germany
Konkoly Observatory	**Budapest,** Hungary
Korean National Astronomical Observatory	**Seoul,** Korea
Kwasan and Hida Observatories	**Kyoto,** Japan
Laboratory for High Energy Astrophysics	**Greenbelt,** Md.
Las Campanas Observatory	**Pasadena,** Calif.
Latitude Station of the Polish Academy of Sciences	**Borowiec,** Poland
Lawrence Livermore National Laboratory	**Livermore,** Calif.
Leander McCormick Observatory	**Charlottesville,** Va.
Lick Observatory	**Santa Cruz,** Calif.
Lindheimer Astronomical Research Center	**Evanston,** Ill.
Llano del Hato Observatory	**Merida,** Venezuela
Lockheed Palo Alto Research Laboratory	**Palo Alto,** Calif.
Lockheed Solar Observatory	**Saugus,** Calif.
Lohrmann Observatory	**Dresden,** German Democratic Republic
Louisiana State University Observatory	**Baton Rouge,** La.
Lowell Observatory	**Flagstaff,** Ariz.
Lunar and Planetary Laboratory	**Tucson,** Ariz.
Main Astronomical Observatory of the USSR Academy of Sciences	**Pulkovo,** USSR
Max-Planck-Institut für Astronomie	**Calar Alto,** Spain

Max-Planck-Institut für Astronomie	**Heidelberg,** F.R. Germany
Max-Planck-Institut für Kernphysik	**Heidelberg,** F.R. Germany
Max-Planck-Institut für Physik und Astrophysik	**Garching,** F.R. Germany
Max-Planck-Institut für Radioastronomie	**Bonn,** F. R. Germany
McDonald Observatory	**Fort Davis,** Tex.
McDonnell Center for the Space Sciences	**St. Louis,** Mo.
Meudon Observatory	**Paris,** France
Michigan State University Observatory	**East Lansing,** Mich.
Millstone Hill Radar Observatory	**Westford,** Mass.
Minor Planet Center	**Cambridge,** Mass.
Molonglo Radio Observatory	**Sydney,** Australia
Monterey Institute for Research in Astronomy	**Carmel Valley,** Calif.
Mount Hamilton	**Santa Cruz,** Calif.
Mount John University Observatory	**Lake Tekapo,** New Zealand
Mount Palomar Observatory	**Pasadena,** Calif.
Mount Stromlo Observatory	**Canberra,** Australia
Mount Wilson Observatory	**Pasadena,** Calif.
Mullard Radio Astronomy Observatory	**Cambridge,** England
Mullard Space Science Laboratory	**London,** England
Multiple Mirror Telescope Observatory	**Tucson,** Ariz.
N. Copernicus Astronomical Center	**Warsaw,** Poland
N. Copernicus University Observatory	**Torun,** Poland
Narrabri Observatory	**Sydney,** Australia
NASA Headquarters	**Washington,** D.C.
National Astronomy and Ionosphere Center	**Ithaca,** N.Y.
National Bureau of Standards	**Washington,** D.C.
National Radio Astronomy Observatory	**Green Bank,** W.Va.
National Radio Astronomy Observatory	**Socorro,** N.M.
National Solar Observatory	**Sunspot,** N.M.
Naval Observatory	**San Fernando** (Cadiz), Spain
New Mexico State University Observatory	**Las Cruces,** N.M.
Nicholas Copernicus Observatory	**Brno,** Czechoslovakia
Nizamiah and Japal-Rangapur Observatories	**Hyderabad,** India
Nuffield Radio Astronomy Laboratories	**Manchester,** England
Oak Ridge Observatory	**Cambridge,** Mass.
Observatorio del Ebro	**Roquetas,** Spain
Ole Roemer Observatory	**Aarhus,** Denmark
Onsala Space Observatory	**Göteborg,** Sweden
Owens Valley Radio Observatory	**Big Pine,** Calif.
Pacific Northwest Laboratory	**Richland,** Wash.
Pennsylvania State University	**University Park,** Pa.
Perkins Observatory	**Delaware,** Ohio
Perth Observatory	**Bickley,** Australia
Pic du Midi Observatory	**Toulouse,** France
Pino Torinese Observatory	**Turin,** Italy
Purple Mountain Observatory	**Nanking,** China
Raman Research Institute	**Bangalore,** India
Rattlesnake Mountain Observatory	**Richland,** Wash.
Remeis Observatory	**Bamberg,** F.R. Germany
Rensselaer Observatory	**Troy,** N.Y.
Ritter Astrophysical Research Center	**Toledo,** Ohio
Rosemary Hill Observatory	**Bronson,** Fla.
Rothney Astrophysical Observatory	**Calgary,** Canada
Royal Observatory	**Edinburgh,** Scotland
Royal Observatory	**Greenwich,** England
Royal Observatory of Belgium	**Uccle,** Belgium
Rutgers University	**Piscataway,** N.J.
Rutherford Appleton Laboratory	**Chilton,** England
Sagamore Hill Radio Observatory	**Hanscom,** Mass.
Saltsjöbaden Observatory	**Stockholm,** Sweden
San Vittore Observatory	**Bologna,** Italy
Satelliten-Beobachtungsstation	**Zimmerwald,** Switzerland
Shaanxi Astronomical Observatory	**Lintong,** China
Smithsonian Astrophysical Observatory	**Cambridge,** Mass.
Sommers-Bausch Observatory	**Boulder,** Colo.
Sonnenborgh Observatory	**Utrecht,** The Netherlands
Sonoma State University	**Rohnert Park,** Calif.
South African Astronomical Observatory	**Cape Town,** South Africa
Space Telescope Science Institute	**Baltimore,** Md.
Special Astrophysical Observatory	**Zelenchukskaya,** USSR

Sproul Observatory	**Swarthmore,** Pa.
Stanford Center for Radar Astronomy	**Menlo Park,** Calif.
Stellar Data Center	**Strasbourg,** France
Sternberg State Astronomical Institute	**Moscow,** USSR
Steward Observatory	**Tucson,** Ariz.
Stockert Radio Observatory	**Bonn,** F.R. Germany
Struve Astrophysical Observatory	**Tartu,** USSR
Tata Institute of Fundamental Research	**Bombay,** India
Tôhoku University Observatory	**Sendai,** Japan
United Kingdom Infrared Telescope (UKIRT)	**Hilo,** Hawaii
University of Sussex	**Brighton,** England
U.S. Naval Observatory	**Washington,** D.C.
UT Radio Astronomy Observatory	**Marfa,** Tex.
Uttar Pradesh State Observatory	**Naini Tal,** India
Van Vleck Observatory	**Middletown,** Conn.
Vatican Observatory	**Castel Gandolfo,** Vatican City
Warner and Swasey Observatory	**Cleveland,** Ohio
Washburn Observatory	**Madison,** Wis.
Wesleyan Radio Observatory	**Delaware,** Ohio
Western Ontario University Observatory	**London,** Canada
Whipple Observatory (MMT)	**Tucson,** Ariz.
Wilhelm Foerster Observatory	**Berlin,** F.R. Germany
Yale University Astronomical Observatory	**New Haven,** Conn.
Yerkes Observatory	**Williams Bay,** Wis.
Yunnang Obs.	**Kunming,** China
Zentralinstitut für Physik der Erde	**Potsdam,** German Democratic Republic

008.021 **Ames, Iowa**

Iowa State University, Erwin W. Fick Observatory, Ames, Iowa 50011. Report for the period 1 September 1984 – 31 August 1985.
Bull. Am. Astron. Soc., Vol. 18, No. 1, p. 200 – 202 (1986).

008.024 **Amherst, Mass.**

Five College Astronomy Department: Amherst College, Amherst, Massachusetts 01002, Hampshire College, Amherst, Massachusetts 01002, Mt. Holyoke College, South Hadley, Massachusetts 01075, Smith College, Northampton, Massachusetts 01063, University of Massachusetts, Amherst, Massachusetts 01003. Report for the academic year 1984 – 1985.
Bull. Am. Astron. Soc., Vol. 18, No. 1, p. 136 – 143 (1986).

008.039 **Arecibo, Puerto Rico**

Arecibo Observatory, Arecibo, Puerto Rico 00613. Report for the period July 1984 – June 1985.
F. Six.
Bull. Am. Astron. Soc., Vol. 18, No. 1, p. 10 – 21 (1986).

008.054 **Austin, Tex.**

University of Texas at Austin, Department of Astronomy, Austin, Texas 78712–1083, McDonald Observatory, Fort Davis, Texas 79737. Report for the period 1 September 1984 – 31 August 1985.
H. J. Smith, F. Bash, J. N. Douglas.
Bull. Am. Astron. Soc., Vol. 18, No. 1, p. 534 – 559 (1986).

008.057 **Baltimore, Md.**

The Johns Hopkins University, Center for Astrophysical Sciences, Department of Physics and Astronomy, Baltimore, Maryland 21218. Report.
A. Davidsen.
Bull. Am. Astron. Soc., Vol. 18, No. 1, p. 203 – 210 (1986).

Space Telescope Science Institute, Baltimore, Maryland 21218. Report for the period 1 October 1984 – 30 September 1985.
Bull. Am. Astron. Soc., Vol. 18, No. 1, p. 498 – 514 (1986).

Space Telescope Science Institute, Preprint Series.
Nos. 80 (41.131.280), 81 (41.156.025), 82 (41.002.052), 83 (41.125.079), 84 (41.131.281), 85 (41.002.053), 86 (41.159.122), 87 (41.160.091), 88 (41.154.040), 89 (41.062.046), 90 (41.157.188), 91 (41.035.057), 92 (41.160.092), 93 (41.036.129), 94 (41.114.107), 95 (41.034.098), 96 (41.122.126), 97 (41.159.123), 98 (41.154.041), 99 (41.035.058), 100 (41.012.040), 101 (41.036.132), 102 (41.160.093), 103 (41.160.124), 104 (41.151.089), 105 (41.064.047), 106 (41.157.189), 107 (41.160.094), 108 (41.159.125), 109 (41.154.042), 110 (41.151.090), 112 (41.124.009), 113 (41.124.010).

Space Telescope Science Institute, Newsletter.
Vol. 3, Nos. 1 – 2 (1986).

008.063 **Bangalore**

Indian Institute of Astrophysics, Preprints.
Nos. 141, 142 (41.075.038), 143 (41.009.047), 144 (41.075.039), 145, 146 (41.077.035), 147 (41.067.104), 148 – 192, 193 (41.034.110), 194 (41.122.136), 195 (41.067.105), 196, 197 (41.122.137), 198 (41.122.138), 199 (41.117.225), 200, 201 (41.122.139), 202 – 207, 208 (41.061.068), 209 (41.066.177), 210, 211 (41.161.219), 212, 213, 214 (41.122.140).

008.072 **Baton Rouge, La.**

Louisiana State University Observatory, Baton Rouge, Louisiana 70803–4001. Report for the period 1 July 1984 – 30 June 1985.
A. U. Landolt.
Bull. Am. Astron. Soc., Vol. 18, No. 1, p. 257 – 259 (1986).

008.078 **Belgrade**

Publications of the Department of Astronomy, University of Beograd.
No. 13 (41.042.068, 41.098.106, 41.072.066, 41.065.098, 41.044.039).

008.081 **Berkeley, Calif.**

University of California at Berkeley, Berkeley Astronomy Department, Berkeley, California 94720. Annual Report 1985.
Bull. Am. Astron. Soc., Vol. 18, No. 1, p. 44 – 52 (1986).

W. M. Keck Observatory, Berkeley, California 94720; Pasadena, California 91106. Report for the period July 1984 – June 1985.
J. E. Nelson, G. M. Smith.
Bull. Am. Astron. Soc., Vol. 18, No. 1, p. 211 (1986).

008.084 **Berlin (East)**

Archenhold–Sternwarte Berlin–Treptow, Sonderdrucke.
Nr. 27 (41.015.038), 29 (41.007.008).

Mitteilungen der Archenhold–Sternwarte Berlin–Treptow.
Nr. 159 (40.003.064), 160 (41.015.039).

Blick in die Sternenwelt 1986.
(41.046.016).

008.087 **Berlin (West)**

Veröffentlichungen der Wilhelm–Foerster–Sternwarte.
Nr. 62 (40.004.049).

008.088 **Bickley**

Perth Observatory Communications.
No. 4 (41.098.105).

008.093 **Bloemfontein**

University of the Orange Free State: Boyden Observatory, Department of Astronomy. Report 1984.
A. H. Jarrett.
Mon. Notes Astron. Soc. S. Afr., Vol. 44, Nos. 3 – 4, p. 30 – 31 (1985).

Boyden Observatory, Occasional Publication.
No. 3 (41.005.016).

Boyden Observatory, Reprint.
No. 58 (41.009.046).

008.096 **Bloomington, Indiana**

Goethe Link Observatory, Indiana University, Bloomington, Indiana 47405. Report for the period September 1984 – August 1985.
R. K. Honeycutt.
Bull. Am. Astron. Soc., Vol. 18, No. 1, p. 144 – 153 (1986).

008.112 **Boone, N.C.**

Appalachian State University, Department of Physics and Astronomy, Boone, North Carolina 28608. Report for the period July 1981 – August 1985.
D. B. Caton, J. T. Pollock.
Bull. Am. Astron. Soc., Vol. 18, No. 1, p. 3 – 5 (1986).

008.117 **Borowiec**

Circular Time and Latitude Service.
Nos. 174 – 175 (41.044.038).

008.120 **Boston, Mass.**

Boston University, Astronomy Department, Boston, Massachusetts 02215. Report for the period September 1984 – August 1985.
Bull. Am. Astron. Soc., Vol. 18, No. 1, p, 40 – 43 (1986).

008.123 **Boulder, Colo.**

High Altitude Observatory, National Center for Atmospheric Research, Boulder, Colorado 80307. Report for the year 1984.
Bull. Am. Astron. Soc., Vol. 18, No. 1, p. 171 – 178 (1986).

008.138 **Bucharest**

Centre de l'Astronomie et des Sciences Spatiales, Observations Solaires.
Rotations 1744 – 1756 (41.072.065).

008.141 **Budapest**

Communications from the Konkoly Observatory.
Nos. 85 (41.155.095), 86 (41.012.039).

008.156 **Cambridge, Engl.**

Institute of Astronomy, University of Cambridge, Annual Report 1984 – 1985.
D. Lynden–Bell.
Inst. Astron., Univ. Camb., Annu. Rep., 44 pp. (1985).

008.159 **Cambridge, Mass.**

Center for Astrophysics, Harvard College Observatory and Smithsonian Astrophysical Observatory, Cambridge, Massachusetts 02138. Report for the period 1 October 1984 – 30 September 1985.
Bull. Am. Astron. Soc., Vol. 18, No. 1, p. 78 – 97 (1986).

The Minor Planet Circulars/Minor Planets and Comets.
Nos. 10329 – 10884 (1986).

IAU Circulars.
Nos. 4157 – 4230 (1986).

008.165 **Cape Town**

The South African Astronomical Observatory.
T. S. Glass, R. M. Catchpole.
South African Astronomical Observatory, PO Box 9, Observatory, 7935 Cape, South Africa. 8 pp. (1986).

South African Astronomical Observatory. Report for the year ending 1984 December 31.
M. W. Feast.
Q.J.R. Astron. Soc., Vol. 27, No. 2, p. 264 – 298 (1986).

SAAO Newsletter.
No. 7 (1986).

University of Cape Town: Department of Astronomy. Report 1984.
B. Warner.
Mon. Notes Astron. Soc. S. Afr., Vol. 44, Nos. 3 – 4, p. 23 – 27 (1985).

University of Cape Town: Department of Applied Mathematics. Report 1984.
G. F. R. Ellis.
Mon. Notes Astron. Soc. S. Afr., Vol. 44, Nos. 3 – 4, p. 22 – 23 (1985).

008.171 **Cardiff**

Astrophysics and Relativity, Preprint Series.
Nos. 122 (41.102.069), 123 (41.102.070), 125 (41.015.036).

008.174 **Carloforte–Cagliari**

Circolari della Stazione Astronomica Internazionale di Latitudine, Carloforte–Cagliari.
N. 20, 26, 27 (41.044.045), 28 (41.044.046), 29 (41.082.070), 30, 31.

Pubblicazioni della Stazione Astronomica Internazionale di Latitudine, Carloforte–Cagliari, Nuova Serie.
N. 94 (41.004.082), 107 (41.044.041), 108 (38.044.001), 110 (41.004.083), 111, 112 (39.004.117), 113 (39.004.142), 114 (41.004.084), 115 (41.044.042), 116 (40.044.070), 117, 122 (41.044.043), 123 (41.044.044), 124.

008.177 **Carmel Valley, Calif.**

The Monterey Institute for Research in Astronomy, Carmel Valley, California 93924. Report for the period 1 September 1984 – 1 September 1985.
W. B. Weaver.
Bull. Am. Astron. Soc., Vol. 18, No. 1, p. 320 (1986).

008.180 **Castel Gandolfo**

Annual Report 1985.
G. V. Coyne.
Vatican. Obs. Publ., Vol. 2, No. 9, p. 113 – 123 (1986).

Vatican Observatory Publications.
Vol. 2, No. 9 (41.008.180).
(41.002.054).

008.189 **Charlottesville, Va.**

National Radio Astronomy Observatory, Charlottesville, Virginia 22903–1797. Report for the period July 1984 – June 1985.
P. A. Vanden Bout, R. J. Havlen.
Bull. Am. Astron. Soc., Vol. 18, No. 1, p. 374 – 424 (1986).

University of Virginia, Department of Astronomy and Leander McCormick Observatory, Charlottesville, Virginia 22903. Report for the period 1 September 1983 – 31 August 1985.
R. Chevalier.
Bull. Am. Astron. Soc., Vol. 18, No. 1, p. 581 – 588 (1986).

008.192 **Chicago, Ill.**

University of Chicago, Department of Astronomy and Astrophysics, Yerkes Observatory, Chicago, Illinois 60637. Report for the period 1 July 1984 – 30 June 1985.
Bull. Am. Astron. Soc., Vol. 18, No. 1, p. 98 – 109 (1986).

008.193 **Chilton**

Rutherford Appleton Laboratory, Report.
RAL–85–086 (41.012.060), 85–094 (41.084.043), 85–097 (41.061.060), 85–106, 85–107 (41.035.060), 85–108 (41.035.061), 85–109 (41.033.021), 85–115 (41.161.190), 86–009 (41.122.127), 86–017 (41.131.282), 86–023 (41.082.066), 86–029 (41.034.099), 86–034 (41.159.126), 86–047 (41.160.111), 86–054.

008.198 **Cleveland, Ohio**

Warner and Swasey Observatory, Case Western Reserve University, Cleveland, Ohio 44106. Report for the period 1 July 1984 – 30 June 1985.
Bull. Am. Astron. Soc., Vol. 18, No. 1, p. 589 – 590 (1986).

008.204 **Coimbra**

Anais do Observatório Astronómico da Universidade de Coimbra.
Fenómenos solares 1979 (41.072.067).

008.207 **College Park, Md.**

University of Maryland, College Park, Maryland 20742. Report for the period 1 October 1984 – 30 September 1985.
Bull. Am. Astron. Soc., Vol. 18, No. 1, p. 285 – 295 (1986).

008.216 **Columbus, Ohio**

The Observatories of the Ohio State and Ohio Wesleyan Universities, Columbus, Ohio 43210 and Delaware, Ohio 43015. Report.
E. R. Capriotti.
Bull. Am. Astron. Soc., Vol 18, No. 1, p. 432 – 437 (1986).

008.219 **Copenhagen**

Copenhagen University Observatory, Reprints.
Nos. 548 (39.131.186), 549 (39.113.005), 550 (39.155.153), 551 (39.122.083), 552 (39.122.084), 553 (39.051.090), 554 (40.041.016), 555 (40.036.067), 556 (40.121.016), 557 (40.114.093), 558, 559 (40.119.003), 560 (28.032.005), 562 (34.041.003), 563 (37.041.002), 564 (40.160.029), 565 (40.117.179), 566 (40.119.041), 567 (40.157.083), 568 (40.156.019), 569 (40.119.023), 570 (41.131.303), 571 (40.111.001), 572 (41.160.060), 573 (41.160.059), 574 (40.119.006), 575 (38.157.237), 576 (40.157.106), 577 (41.131.012), 578 (39.031.029), 579 (41.113.003), 580 (39.041.031), 581 (39.013.084), 582 (40.080.105), 583 (39.063.099), 584 (39.080.070), 585 (38.063.008), 586 (40.071.066), 587 (40.071.067), 588 (40.071.030), 589 (40.075.032), 590 (41.122.016), 591 (41.131.143), 592 (39.113.018), 593 (41.156.004).

008.228 **Crimea**

Chronicle.
Izv. Krymskoj Astrofiz. Obs., Tom 70, p. 172 – 173 (1985). In Russian. English translation in Bull. Crimean Astrophys. Obs., Vol. 70.

Chronicle.
Izv. Krymskoj Astrofiz. Obs., Tom 72, p. 239 – 241 (1985). In Russian. English translation in Bull. Crimean Astrophys. Obs., Vol. 72.

Chronicle.
Izv. Krymskoj Astrofiz. Obs., Tom 74, p. 180 – 181 (1986). In Russian. English translation in Bull. Crimean Astrophys. Obs., Vol. 74.

Izvestiya Ordena Trudovogo Krasnogo Znameni Krymskoj Astrofizicheskoj Observatorii.
Tom 70 (41.073.030, 41.073.031, 41.065.043, 41.091.015, 41.083.002, 41.071.010, 41.022.031, 41.114.033, 41.119.030, 41.112.035, 41.158.072, 41.143.010, 41.031.004, 41.031.005, 41.034.018, 41.008.228),
Tom 72 (41.124.241, 41.112.122, 41.122.031, 41.122.032, 41.114.035, 41.114.036, 41.116.017, 41.126.024, 41.126.025, 41.158.074, 41.073.036, 41.073.037, 41.075.007, 41.072.022, 41.075.008, 41.080.028, 41.091.017, 41.077.011, 41.080.029, 41.036.037, 41.036.038, 41.031.006, 41.008.228),
Tom 73 (41.073.038, 41.072.023, 41.080.030, 41.080.031, 41.076.003, 41.077.012, 41.075.009, 41.073.039, 41.077.013, 41.122.192, 41.114.038, 41.117.097, 41.120.005, 41.153.007, 41.122.033, 41.158.075, 41.141.008, 41.036.039, 41.031.007, 41.033.005),
Tom 74 (41.114.103, 41.122.115, 41.114.104 – 41.114.106, 41.116.076, 41.116.048, 41.022.101, 41.117.210, 41.072.064, 41.073.096, 41.073.097, 41.074.094, 41.080.102, 41.084.042, 41.071.036, 41.034.097, 41.063.074, 41.031.057, 41.036.126, 41.036.127, 41.008.228).

008.234 **Delaware, Ohio**

The Observatories of the Ohio State and Ohio Wesleyan Universities, Columbus, Ohio 43210 and Delaware, Ohio 43015. Report.
E. R. Capriotti.
Bull. Am. Astron. Soc., Vol. 18, No. 1, p. 432 – 437 (1986).

008.240 **Dresden**

Lohrmann–Observatorium, Technische Universität Dresden, Zirkular.
Nr. 112 – 113 (41.044.047).

008.255 **East Lansing, Mich.**

Michigan State University, Department of Physics and Astronomy, East Lansing, Michigan 48824. Report for the period 1 September 1984 – 31 August 1985.
Bull. Am. Astron. Soc., Vol. 18, No. 1, p. 309 – 311 (1986).

008.264 **Epping**

Anglo–Australian Telescope 1984 – 85. Report of the Anglo–Australian Telescope Board, 1 July 1984 to 30 June 1985.
Australian Government Publishing Service, Canberra, Australia. 8 + 56 pp. (1985).

Anglo–Australian Observatory. Report for the year ending 1985 June 30.
D. C. Morton.
Q.J.R. Astron. Soc., Vol. 27, No. 2, p. 237 – 250 (1986).

Anglo–Australian Observatory, Preprints.
Nos. 209 (41.036.136), 210 (41.159.127), 211 (41.158.247), 212 (41.157.195), 213 (41.134.046).

AAO Newsletter.
Nos. 36, 37 (1986).

008.270 **Evanston, Ill.**

Lindheimer Astronomical Research Center and Dearborn Observatory, Northwestern University, Evanston, Illinois 60201. Report for the period 15 October 1984 – 15 October 1985.
Bull. Am. Astron. Soc., Vol. 18, No. 1, p. 228 – 230 (1986).

008.282 **Flagstaff, Ariz.**

Lowell Observatory, Flagstaff, Arizona 86001. Report for the period 1 July 1984 – 30 June 1985.
A. A. Hoag, H. Horstman.
Bull. Am. Astron. Soc., Vol. 18, No. 1, p. 260 – 266 (1986).

008.285 **Fort Davis, Tex.**

University of Texas at Austin, Department of Astronomy, Austin, Texas 78712–1083, McDonald Observatory, Fort Davis, Texas 79737. Report for the period 1 September 1984 – 31 August 1985.
H. J. Smith, F. Bash, J. N. Douglas.
Bull. Am. Astron. Soc., Vol. 18, No. 1, p. 534 – 559 (1986).

008.297 **Garching**

European Southern Observatory. Annual Report 1985.
L. Woltjer.
ESO Annu. Rep., 1985, 91 pp. (1986).

ESO Scientific Preprint.
Nos. 411 (41.125.081), 412 (41.061.067), 413 (41.132.042), 414 (41.112.092), 415 (41.115.019), 416 (41.154.045), 417 (41.159.129), 418 (41.122.131), 419 (41.064.051), 420 (41.160.096), 421 (41.159.130), 422 (41.159.131), 423 (41.158.251), 424 (41.034.102), 425 (41.157.196), 426 (41.122.132), 427 (41.132.043), 428 (41.114.108), 429 (41.157.197), 430 (41.142.089), 431 (41.159.132), 432 (41.042.071), 433 (41.151.094), 434 (41.155.089), 435 (41.158.252), 436 (41.151.095), 437 (41.065.105), 438 (41.117.224), 439 (41.157.198), 440 (41.161.217), 441 (41.155.099), 442 (41.157.199), 443 (41.063.078), 444 (41.158.253), 445 (41.131.291), 446 (41.125.082), 447 (41.161.218), 448 (41.036.137), 449 (41.111.012).

The Messenger – El Mensajero.
Nos. 43 – 44 (1986).

ST–ECF Newsletter.
No. 5 (1986).

MPA Reports.
Nos. 217 (41.064.049), 218 (41.066.167), 219 (41.066.168), 220 (41.065.102), 221 (41.066.169), 222 (41.066.170), 223 (41.066.171), 224 (41.158.249, 41.159.128), 225 (41.065.103), 226 (41.075.037), 227 (41.117.222), 228 (41.067.101), 229 (41.066.172), 230 (41.117.223), 231 (41.161.215), 232 (41.064.050), 233 (41.065.104), 234 (41.066.173), 235 (41.131.288), 236 (41.161.216), 237 (41.066.174), 238 (41.067.102), 239 (41.066.175), 240 (41.067.103).

Max–Planck–Institut für Physik und Astrophysik. Institut für Extraterrestrische Physik – Tätigkeitsbericht 1985.
MPE Rep., No. 194, 6 + 111 pp. (1986).

MPE Reports.
Nos. 193 (41.158.248), 194 (41.008.297), 195 (41.078.016), 197.

MPE Contributions.
Nos. 2294 (40.035.013), 2295, 2298 (39.084.087), 2302, 2312 (40.035.068), 2314 (40.008.297), 2315 – 2317, 2323, 2326 (40.144.028), 2327 (40.155.034), 2328 (40.142.013), 2329 (40.125.044), 2330 (40.131.119), 2331 (40.107.022), 2333 – 2335, 2336 (40.131.208), 2337 (41.142.008), 2338 (41.142.006), 2339 (40.062.146), 2340 (41.125.011), 2341 (41.131.049).

008.309 **Goeteborg**

Onsala Space Observatory, Preprint.
Nos. 85:48 (41.082.071), 85:53 (41.112.091), 86:5 (41.131.289), 86:24 (41.131.290).

008.315 **Montréal**

Université de Montréal, Département de Physique, Montréal, Québec H3C 3J7, Canada. Report for the period 1 September 1984 – 31 August 1985.
F. Wesemael, G. Fontaine.
Bull. Am. Astron. Soc., Vol. 18, No. 1, p. 321 – 328 (1986).

008.321 **Green Bank, W.Va.**

National Radio Astronomy Observatory, Green Bank, West Virginia 24944–0002. Report for the period July 1984 – June 1985.
P. A. Vanden Bout, R. J. Havlen.
Bull. Am. Astron. Soc., Vol. 18, No. 1, p. 374 – 424 (1986).

National Radio Astronomy Observatory, Reprints, Series A.
Nos. 1395 (39.008.189, 39.008.315, 39.008.705, 39.008.780), 1396 (39.033.007), 1397 (39.158.263), 1398 (41.033.024), 1399 (41.033.025), 1400 (40.033.026).

National Radio Astronomy Observatory, Reprints, Series B.
Nos. 541 (41.158.250), 542 (39.155.046), 543 (39.141.039), 544 (40.117.312), 545 (40.117.313), 546 (40.112.114), 547 (40.117.322), 548 (40.036.253).

008.330 **Groningen**

Kapteyn Astronomical Institute. Annual report 1985.
R. Sancísi, P. R. Roelfsema.
Kapteyn Astron. Inst., Annu. Rep., 130 pp. (1986).

008.333 **Hamburg**

Deutsches Hydrographisches Institut Hamburg. 39. Jahresbericht 1984.
G. Zickwolff.
Dtsch. Hydrogr. Inst. Hamb., 117 pp. (1986).

008.345 **Heidelberg**

Astronomisches Rechen–Institut Heidelberg, Mitteilungen, Serie A.
Nr. 171 (40.111.007), 172 (39.151.012), 173 (39.151.129), 174 (41.098.046), 175 (41.098.045), 176 (41.155.100), 177 (40.042.104), 178 (41.041.005), 179 (41.041.071), 180 (41.043.018), 181 (41.043.019), 182 (41.043.028), 183 (41.041.058).

Astronomisches Rechen–Institut Heidelberg, Mitteilungen, Serie B.
Nr. 141 (40.043.008), 142 (40.043.010), 143 (41.041.009), 144 (41.041.042).

Astronomy and Astrophysics Abstracts.
Vol. 40 (41.002.055).

Astronomische Grundlagen für den Kalender 1987.
(41.046.035).

Max–Planck–Institut für Astronomie, D–6900 Heidelberg–Königstuhl, F.R. Germany. Foreign Station: Centro Astronómico Hispano–Aleman, Apartado Correos 511, Almería, Spain. Report for the period 1 January – 31 December 1984.
G. Münch.
Bull. Am. Astron. Soc., Vol. 18, No. 1, p. 296 – 303 (1986).

Max–Planck–Institut für Kernphysik. Jahresbericht 1985.
H. V. Klapdor, E. K. Jessberger.
Max–Planck–Institut für Kernphysik, Heidelberg, F.R. Germany. 201 pp. (1986).

008.354 **Holmdel, N.J.**

AT&T Bell Laboratories, Crawford Hill Laboratory, Holmdel, New Jersey 07733. Report.
Bull. Am. Astron. Soc., Vol. 18, No. 1, p. 27 – 32 (1986).

008.357 **Honolulu, Hawaii**

University of Hawaii, Institute for Astronomy, Honolulu, HI 96822. Report for the period 1 July 1984 – 30 June 1985.
Bull. Am. Astron. Soc., Vol. 18, No. 1, p. 154 – 164 (1986).

The Infrared Telescope Facility, Honolulu, Hawaii 96822. Report for the period 1 July 1984 – 30 June 1985.
Bull. Am. Astron. Soc., Vol. 18, No. 1, p. 196 (1986).

008.360 **Houston, Tex.**

William Marsh Rice University, Houston, Texas 77251. Report for the period July 1984 – July 1985.
Bull. Am. Astron. Soc., Vol. 18, No. 1, p. 488 – 490 (1986).

008.375 **Istanbul**

Publications of the Istanbul University Observatory.
No. 136 (41.073.098), 137 (41.073.099), 138 (41.073.100), 139 (41.072.068).

008.381 **Izmir**

Publications of the Ege University Observatory.
No. 26 (39.119.054, 39.119.011, 39.119.013, 39.117.250, 39.119.085, 40.117.144, 39.119.064, 39.119.065, 39.117.391, 40.117.221, 40.117.222, 40.119.067).

008.387 **Johannesburg**

Hartebeesthoek Radio Astronomy Observatory. Annual report 1984.
G. D. Nicolson.
Mon. Notes Astron. Soc. S. Afr., Vol. 44, Nos. 3 – 4, p. 27 – 29 (1985).

008.390 **Kazan**

Izvestiya Astronomicheskoj Observatorii Ehngel'gardta.
Nos. 48 (41.117.134, 41.119.042, 41.098.067, 41.044.018, 41.004.069, 41.004.070, 41.082.028, 41.100.053), 49 (41.002.037, 41.117.135, 41.117.136, 41.041.024, 41.082.029), 50 (41.091.023, 41.044.019, 41.079.101, 41.104.028),

51 (41.036.070, 41.094.029, 41.104.029, 41.002.038, 41.132.033, 41.002.039).

008.396 **Kiev**

Kinematika i Fizika Nebesnykh Tel.
Tom 2, Nos. 1 – 3 (1986).

008.403 **Kunming, Yunnan**

Publications of Yunnan Observatory.
No. 1 (41.036.138 – 41.036.141, 41.044.048, 41.036.142 – 41.036.147, 41.073.101, 41.072.069, 41.158.254, 41.034.103, 41.033.026 – 41.033.030).

008.408 **La Jolla, Calif.**

University of California at San Diego, La Jolla, California 92093. Report for the period 1 July 1984 – 30 June 1985.
E. M. Burbidge, B. Jones.
Bull. Am. Astron. Soc., Vol. 18, No. 1, p. 59 – 67 (1986).

008.414 **La Serena**

National Optical Astronomy Observatories: Tucson, Arizona 85726, La Serena, Chile, Sunspot, New Mexico 88349. Report for the period 1 October 1983 – 30 September 1984.
J. T. Jefferies.
Bull. Am. Astron. Soc., Vol. 18, No. 1, p. 332 – 373 (1986).

008.420 **Lake Tekapo**

Physics Department and Mt John University Observatory, University of Canterbury. Annual report 1984.
B. G. Wybourne.
Q.J.R. Astron. Soc., Vol. 27, No. 1, p. 120 – 123 (1986).

008.426 **Landover, Md.**

Applied Research Corporation, Landover, Maryland 20785. Report for the period October 1984 – September 1985.
Bull. Am. Astron. Soc., Vol. 18, No. 1, p. 6 – 9 (1986).

008.429 **Las Cruces, N.M.**

New Mexico State University, Department of Astronomy, Las Cruces, New Mexico 88003. Report for the years 1984 – 1985.
Bull. Am. Astron. Soc., Vol. 18, No. 1, p. 427 – 429 (1986).

008.432 **Lawrence, Kans.**

Clyde W. Tombaugh Observatory, University of Kansas, Lawrence, Kansas 66045. Report for the period August 1984 – August 1985.
Bull. Am. Astron. Soc., Vol. 18, No. 1, p. 560 – 563 (1986).

008.446 **Lincoln, Nebr.**

Behlen Observatory, Department of Physics and Astronomy, University of Nebraska, Lincoln, Nebraska 68508–0111. Report for the period January 1984 – July 1985.
E. G. Schmidt.
Bull. Am. Astron. Soc., Vol. 18, No. 1, p. 37 – 39 (1986).

008.447 **Lintong**

Publications of the Shaanxi Astronomical Observatory.
Vol. 7, No. 2 (41.034.111 – 41.034.146, 41.021.019, 41.034.147, 41.082.072, 41.044.062, 41.034.148).

Time and Frequency Services, Bulletin.
Nos. 73 – 78 (41.044.061).

008.450 **Livermore, Calif.**

Lawrence Livermore National Laboratory, University of California, Livermore, California 94550. Report for the period October 1984 – September 1985.
Bull. Am. Astron. Soc., Vol. 18, No. 1, p. 231 – 240 (1986).

008.453 **London, Canada**

The Observatories of the University of Western Ontario, London, Ontario N7A 3K7, Canada. Report for the period 1 July 1984 – 30 June 1985.
W. H. Wehlau.
Bull. Am. Astron. Soc., Vol. 18, No. 1, p. 619 – 620 (1986).

008.456 **London, England**

Mullard Space Science Laboratory, University College, London. Report for the period 1984 November 1 to 1985 October 31.
J. L. Culhane.
Q.J.R. Astron. Soc., Vol. 27, No. 2, p. 251 – 263 (1986).

008.459 **Los Alamos, N.M.**

Los Alamos National Laboratory, Los Alamos, New Mexico 87545. Report for the period 1 July 1984 – 30 June 1985.
A. N. Cox, J. G. Hills.
Bull. Am. Astron. Soc., Vol. 18, No. 1, p. 241 – 256 (1986).

008.462 **Los Angeles, Calif.**

The Aerospace Corporation, Los Angeles, California 90009. Report for the period 1 October 1984 – 30 September 1985.
C. J. Rice.
Bull. Am. Astron. Soc., Vol. 18, No. 1, p. 1 – 2 (1986).

University of California at Los Angeles, Department of Astronomy, Los Angeles, California 90024. Report.
Bull. Am. Astron. Soc., Vol. 18, No. 1, p. 53 – 58 (1986).

University of Southern California, Department of Astronomy, Los Angeles, California 90089. Report for the period September 1984 – August 1985.
Bull. Am. Astron. Soc., Vol. 18, No. 1, p. 496 – 497 (1986).

008.474 **Madison, Wis.**

Washburn Observatory, University of Wisconsin at Madison, Madison, Wisconsin 53706. Report for the period 1 October 1984 – 30 September 1985.
Bull. Am. Astron. Soc., Vol. 18, No. 1, p. 591 – 596 (1986).

008.477 **Madrid**

Boletín Astronómico del Observatorio de Madrid.
Vol. 11, No. 3 (41.121.050, 41.122.141, 41.112.093, 41.112.094, 41.133.012), No. 4 (41.123.006, 41.113.049, 41.117.226, 41.072.071).

Seminario de Astronomia y Geodesia, Publicación.
Nos. 122 (41.004.085), 123 (41.044.055), 124 (41.036.151), 125 (41.081.036), 126 (41.044.056), 127 (41.081.037), 128 (41.043.012), 129 (41.045.019), 130 (41.044.057, 41.044.058, 41.045.020, 41.045.021, 41.043.013), 131 (41.045.022), 132 (41.004.086).

008.486 **Manila**

Solar Maps and Activity.
(41.072.072).

008.498 **Middletown, Conn.**

Van Vleck Observatory, Wesleyan University, Middletown, Connecticut 06457. Report for the period 1 September 1984 – 31 August 1985.
A. R. Upgren.
Bull. Am. Astron. Soc., Vol. 18, No. 1, p. 575 – 577 (1986).

008.504 **Minneapolis, Minn.**

University of Minnesota, School of Physics and Astronomy, Minneapolis, Minnesota 55455. Report for the academic year 1984 – 1985.
Bull. Am. Astron. Soc., Vol. 18, No. 1, p. 312 – 319 (1986).

008.507 **Mizusawa**

Proceedings of the International Latitude Observatory of Mizusawa.
No. 24 (41.032.035, 41.021.018).

Publications of the International Latitude Observatory of Mizusawa.
Vol. 18, No. 2 (41.032.036, 41.002.056, 41.036.152, 41.032.037).

Monthly Notes of the International Polar Motion Service.
Nos. 12 (1985), 1 – 4 (1986) (41.044.060).

Bulletins, Time Service of the Mizusawa Observatory.
Vol. 29 (41.044.059).

008.528 **Naini Tal**

Uttar Pradesh State Observatory, Reprint.
Nos. 230, 231 (34.153.021), 232 (37.073.171), 233 (37.075.032), 234 (37.073.140), 235 (38.034.007), 236 (37.119.062), 237 (37.022.076), 238 (37.117.166), 239 (37.103.161), 240 (37.112.109), 241, 242 (38.071.019), 243 (37.112.111), 244 (37.112.114), 245 (38.153.023), 246 (38.083.047), 247 (38.072.075), 248 (38.119.088), 249 (38.071.034), 250 (38.073.088), 251 (38.101.074), 252 (38.077.042), 253 (38.072.076), 254 (38.117.036), 255 (37.114.146), 256 (38.117.195), 257 (34.117.192), 258 (34.114.197), 259 (38.117.039), 260 (38.096.007), 261 (38.153.023), 262 (38.153.024), 263 (38.073.070), 264 (38.113.020), 266 (38.112.051), 267 (39.074.105), 268 (39.071.031), 269 (39.074.106), 270 (39.073.184), 271 (38.113.042), 272 (39.022.074), 273 (38.112.093), 274 (39.100.003), 275 (39.153.011), 276 (39.112.033), 277 (38.119.085), 279 (40.072.043), 280 (40.077.033), 281 (39.114.071), 282 (39.112.080), 283 (39.119.052), 284 (38.122.125), 285 (38.119.053), 286 (39.112.121), 287 (39.113.051), 288 (39.114.120), 289 (39.119.084), 290 (40.071.027), 291 (40.073.072).

008.531 **Nanking**

A review of the scientific research work of Purple Mountain Observatory in 1985.
Z.–t. Xu, Y.–b. Wang.
Publ. Purple Mountain Obs., Vol. 5, No. 1, p. 71 – 77 (1986).

Fifty years of Purple Mountain Observatory (1934 – 1984).
(41.009.061).

Publications of Purple Mountain Observatory.
Vol. 4, No. 2 (41.045.023, 41.131.313, 41.085.021, 41.072.075, 41.098.109, 41.034.153, 41.032.038),
Vol. 4, No. 3 (41.075.040, 41.085.022, 41.073.110, 41.044.066, 41.098.110, 41.035.071, 41.034.154, 41.021.021, 41.021.014, 41.015.042),
Vol. 4, No. 4 (41.067.116, 41.096.009, 41.041.037, 41.041.038, 41.031.033, 41.004.092, 41.009.049, 41.009.050, 41.009.051),
Vol. 5, No. 1 (41.067.117, 41.073.111, 41.064.053, 41.063.091, 41.098.111, 41.046.017, 41.036.162, 41.033.032, 41.035.072, 41.008.531).

008.534 **Nashville, Tenn.**

Dyer Observatory, Vanderbilt University, Nashville, Tennessee 37235. Report for the period 1 October 1984 – 31 October 1985.
Bull. Am. Astron. Soc., Vol. 18, No. 1, p. 133 – 135 (1986).

008.537 **Neuchâtel**

Rapport de l'Observatoire cantonal de Neuchâtel. Rapport d'activité pour l'Exercice 1985.
J. Bonanomi, G. Fischer.
Observatoire cantonal de Neuchâtel, Neuchâtel, Suisse. 23 pp. (1986).

008.540 **New Haven, Conn.**

Yale University Observatory, New Haven, Connecticut 06511. Report for the period 1 July 1984 – 30 June 1985.
R. B. Larson.
Bull. Am. Astron. Soc., Vol. 18, No. 1, p. 623 – 630 (1986).

008.543 **New York, N.Y.**

Columbia University in the City of New York, Department of Astronomy and Department of Physics, New York, New York 10027. Report for the period September 1984 – August 1985.
Bull. Am. Astron. Soc., Vol. 18, No. 1, p. 114 – 118 (1986).

008.546 **Newark, Del.**

Bartol Research Foundation and University of Delaware, Newark, Delaware 19716. Report for the period September 1984 – September 1985.
Bull. Am. Astron. Soc., Vol. 18, No. 1, p. 33 – 36 (1986).

008.558 **Oslo**

Institutt for teoretisk astrofysikk, Blindern–Oslo, Småtrykk.
Nr. 105 (41.009.048), 105A (41.005.018), 106 (41.007.024).

008.573 **Paris**

Planètes et petit corps du sytème solaire. Rapport thématique 1980 – 1984.
(41.091.049).

Institut d'Astrophysique de Paris, Pré–Publication.
Nos. 123 (41.117.251), 124 (41.103.730), 125 (41.159.156), 126 (41.158.291), 127 (41.079.102), 128 (41.124.011), 129 (41.160.113), 130 (41.134.053), 131 (41.073.120), 132 (41.131.326), 133 (41.065.116), 134 (41.036.194), 135 (41.121.054), 136 (41.121.055), 137 (41.131.327), 138 (41.160.114), 139 (41.160.115), 140 (41.112.099), 141 (41.065.117), 142 (41.131.328), 143 (41.161.315), 144 (41.103.731), 145 (41.159.157), 146 (41.064.054), 147 (41.102.079), 148 (41.103.732), 149 (41.112.100), 150 (41.103.733), 151 (41.115.025).

Annual Report. (B.I.H.).
1985 (41.044.081).

Bureau International de l'Heure (B.I.H.) Circulars.
D230 – D236 (41.044.075).

008.576 **Pasadena, Calif.**

W. M. Keck Observatory, Berkeley, California 94720; Pasadena, California 91106. Report for the period July 1984 – June 1985.
J. E. Nelson, G. M. Smith.
Bull. Am. Astron. Soc., Vol. 18, No. 1, p. 211 (1986).

The Palomar Observatory, Pasadena, California 91125. Annual Report for 1984 – 85.
Bull. Am. Astron. Soc., Vol. 18, No. 1, p. 438 – 460 (1986).

008.585 **Peking**

Publications of the Beijing Astronomical Observatory.
No. 7/1985 (41.033.034 – 41.033.046, 41.034.158, 41.034.159, 41.036.176, 41.033.047, 41.033.048).

008.594 **Philadelphia, Pa.**

University of Pennsylvania, Department of Astronomy, Philadelphia, Pennsylvania 19101. Report.
Bull. Am. Astron. Soc., Vol. 18, No. 1, p. 461 – 466 (1986).

008.606 **Poona**

Department of Mathematics, University of Poona. Report for the period 1984 April 1 – 1985 March 31.
Bull. Astron. Soc. India, Vol. 13, No. 4, p. 399 – 401 (1985).

008.609 **Porto**

Publicacões do Observatório Astronómico "Prof. Manuel de Barros", da Faculdade de Ciências do Porto.
Nos. 35 (41.036.175), 36 (41.052.050).

008.615 **Potsdam**

Zeit- und Breitenbestimmungen, Zeitsysteme, Präzisionszeitvergleiche.
Jahrg. 1983, Nr. 3 – 4 (41.044.072).

Solar Radio Data.
Part I – II, 1986 (41.077.038).

008.618 **Prague**

Circular of the Czechoslovak Observatories, Time and Latitude.
October 1983 – December 1983 (41.044.073), January 1984 – December 1984 (41.044.074).

008.624 **Princeton, N.J.**

Princeton University Observatory, Princeton, New Jersey 08544. Report for the period 1 July 1984 – 30 June 1985.
Bull. Am. Astron. Soc., Vol. 18, No. 1, p. 479 – 485 (1986).

008.645 **Richmond Hill**

David Dunlap Observatory, University of Toronto, Richmond Hill, Ontario L4C 4Y6, Canada. Report for the period 1 July 1984 – 30 June 1985.
Bull. Am. Astron. Soc., Vol. 18, No. 1, p. 125 – 132 (1986).

008.651 **Rochester, N.Y.**

C. E. Kenneth Mees Observatory, University of Rochester, Rochester, New York 14627. Report.
Bull. Am. Astron. Soc., Vol. 18, No. 1, p. 304 – 308 (1986).

008.654 **Rohnert Park, Calif.**

Sonoma State University, Department of Physics and Astronomy, Rohnert Park, California 94928. Report for the period September 1983 – August 1985.
G. G. Spear, J. S. Tenn.
Bull. Am. Astron. Soc., Vol. 18, No. 1, p. 495 (1986).

008.681 **Santa Cruz, Calif.**

University of California at Santa Cruz, Board of Studies in Astronomy and Astrophysics, Santa Cruz, California 95064. Report for 1984 – 1985.
Bull. Am. Astron. Soc., Vol. 18, No. 1, p. 68 – 77 (1986).

Lick Observatory, University of California, Santa Cruz, California 95064. Report for the period 1 July 1984 – 31 June 1985.
Bull. Am. Astron. Soc., Vol. 18, No. 1, p. 212 – 227 (1986).

008.684 **Santiago**

Instituto Isaac Newton, Casilla 8 – 9, Correo 9, Santiago, Chile. Report.
G. Alcaino.
Bull. Am. Astron. Soc., Vol. 18, No. 1, p. 197 – 199 (1986).

Astrophysics Preprint Series.
No. 14 (41.120.023).

008.696 **Seattle, Wash.**

University of Washington, Astronomy Department, Seattle, Washington 98195. Report for the academic year 1984 – 85.
B. Margon.
Bull. Am. Astron. Soc., Vol. 18, No. 1, p. 614 – 618 (1986).

008.699 **Sendai**

Annual report of research activities of the Astronomy Group in 1984/85.
Sci. Rep. Tôhoku Univ., Eigth Ser., Vol. 6, No. 2, p. 149 – 154 (1985).

Sendai Astronomiaj Raportoj.
Nos. 279 (39.064.077), 280 (39.064.079), 281 (39.065.076), 285 (40.134.019), 286 (41.008.699), 287 (40.117.184), 288 (40.151.043), 289 (40.131.246), 296 (40.131.089), 297 (40.117.147).

008.705 **Shanghai**

Annual Report (ERP).
1984 (41.044.071).

008.711 **Silver Spring, Md.**

Computer Sciences Corporation, Silver Spring, Maryland 20910. Report for the period July 1984 – June 1985.
E. H. Scott.
Bull. Am. Astron. Soc., Vol. 18, No. 1, p. 119 – 124 (1986).

008.714 **Skalnaté Pleso**

Contributions of the Astronomical Observatory Skalnaté Pleso.
Vol. 14 (41.074.098, 41.074.099, 41.117.243, 41.034.157, 41.104.068, 41.104.069, 41.080.106, 41.119.065, 41.072.079, 41.106.070).

008.717 **Socorro, N.M.**

National Radio Astronomy Observatory, Socorro, New Mexico 87801–0387. Report for the period July 1984 – June 1985.
P. A. Vanden Bout, R. J. Havlen.
Bull. Am. Astron. Soc., Vol. 18, No. 1, p. 374 – 424 (1986).

Astrophysics Research Center, New Mexico Institute of Mining and Technology, Socorro, New Mexico 87801. Report for the period September 1984 – September 1985.
S. N. Shore.
Bull. Am. Astron. Soc., Vol. 18, No. 1, p. 425 – 426 (1986).

008.720 **Sonneberg**

Mitteilungen über Veränderliche Sterne.
Band 10, Heft 7 (41.123.008, 41.119.066, 41.123.009 – 41.123.011, 41.122.147, 41.124.302, 41.122.148, 41.119.067, 41.123.012, 41.123.013).

008.726 **St. Louis, Mo.**

Washington University, McDonnell Center for the Space Sciences, St. Louis, Missouri 63130. Report for 1984 – 1985.
Bull. Am. Astron. Soc., Vol. 18, No. 1, p. 597 – 613 (1986).

008.729 **St. Michel l'Observatoire**

Observatoire de Haute Provence, Pré–Publication.
Nos. 7 (41.142.091), 8 (41.159.150), 9 (41.142.092).

008.732 **Stanford, Calif.**

Wilcox Solar Observatory, Center for Space Science and Astrophysics, Stanford University, Stanford, California 94305. Report for the period 1983 – 1985.
Bull. Am. Astron. Soc., Vol. 18, No. 1, p. 621 – 622 (1986).

008.738 **Stony Brook, N.Y.**

State University of New York at Stony Brook, Stony Brook, New York 11794. Report for the period 1 September 1984 – 30 September 1985.
D. M. Peterson.
Bull. Am. Astron. Soc., Vol. 18, No. 1, p. 430 – 431 (1986).

008.741 **Strasbourg**

Bulletin d'Information du Centre de Données Stellaires.
No. 30 (1986).

Bulletin d'Information sur les Etoiles Be.
No. 13 (1986).

Bulletin sur les Etoiles Tardives à Spectre Particulier.
No. 3 (1985).

008.744 **Sunspot, N.M.**

National Optical Astronomy Observatories: Tucson, Arizona 85726, La Serena, Chile, Sunspot, New Mexico 88349. Report for the period 1 October 1983 – 30 September 1984.
J. T. Jefferies.
Bull. Am. Astron. Soc., Vol. 18, No. 1, p. 332 – 373 (1986).

008.747 **Swarthmore, Pa.**

Department of Astronomy – Sproul Observatory, Swarthmore College, Swarthmore, Pennsylvania 19081. Report for the period 1 July 1984 – 30 June 1985.
J. E. Gaustad.
Bull. Am. Astron. Soc., Vol. 18, No. 1, p. 515 – 516 (1986).

008.759 **Tartu**

Tartu Astrofüüsika Observatoorium Teated.
Nr. 77 (41.115.009), 78 (41.119.029), 79 (41.063.035), 80 (41.114.030), 81 (41.112.033).

008.762 **Tashkent**

Tsirkulyar Astronomicheskogo Instituta Tashkent.
Nos. 113 (41.044.016, 41.072.030, 41.032.013 – 41.032.015, 41.103.321), 114 (41.002.025, 41.119.037), 115 (41.111.007), 116 (41.002.026).

008.768 **Tempe, Ariz.**

Arizona State University, Astronomy Group, Tempe, Arizona 85287. Report for the period October 1984 – September 1985.
P. A. Wehinger.
Bull. Am. Astron. Soc., Vol. 18, No. 1, p. 22 – 26 (1986).

008.774 **Tokyo**

Annals of the Tokyo Astronomical Observatory, Second Series.
Vol. 21, No. 1 (41.072.073, 41.160.099, 41.151.097, 41.157.200, 41.002.057, 41.157.201).

Tokyo Astronomical Bulletin, Second Series.
Nos. 274 (41.104.066), 275 (41.094.037).

Tokyo Astronomical Observatory Reprints.
Nos. 759 (41.077.008), 760 (41.062.094), 761 (40.161.321), 762 (40.073.126), 763 (40.072.095), 764 (40.073.134), 765 (41.117.077), 766 (40.074.053), 767 (41.074.003), 768 (41.073.004), 769 (41.114.011), 770 (41.116.003), 771 (41.062.004), 772 (40.155.039), 774 (41.158.259), 775 (41.131.294), 776 (41.076.016), 777 (41.074.095), 778 (41.072.074), 779 (41.131.024), 780 (41.080.104), 781 (41.114.022), 782 (41.131.046), 783 (41.131.109), 784 (41.131.212).

Tokyo Astronomical Observatory, Time and Latitude Bulletins.
Vol. 59, No. 4 (41.044.063), Vol. 60, No. 1 (41.044.064).

Quarterly Bulletin on Solar Activity.
Vol. 23, Part V (41.077.036).

008.777 **Toledo, Ohio**

Ritter Astrophysical Research Center, The University of Toledo, Toledo, Ohio 43606. Report for the period 1 July 1984 – 30 June 1985.
Bull. Am. Astron. Soc., Vol. 18, No. 1, p. 491 – 494 (1986).

008.788 **Treviso**

Osservatorio Privato, Specola "Ariel", Pubblicazione.
Nos. 85 (41.015.040), 86 (33.004.075), 87 (41.015.021), 88 – 90, 91 (41.004.087), 92 (41.004.088), 93 (41.004.089), 94 (41.004.090).

008.792 **Troy, N.Y.**

Rensselaer Polytechnic Institute, Troy, New York 12180–3590. Report for the period October 1984 – September 1985.
Bull. Am. Astron. Soc., Vol. 18, No. 1, p. 486 – 487 (1986).

008.795 **Tucson, Ariz.**

University of Arizona, Lunar and Planetary Laboratory, Tucson, Arizona 85721. Report.
Bull. Am. Astron. Soc., Vol. 18, No. 1, p. 267 – 284 (1986).

Multiple Mirror Telescope Observatory, Tucson, Arizona 85721. Report for the period 1 October 1984 – 30 September 1985.
Bull. Am. Astron. Soc., Vol 18, No. 1, p. 329 – 331 (1986).

National Optical Astronomy Observatories: Tucson, Arizona 85726, La Serena, Chile, Sunspot, New Mexico 88349. Report for the period 1 October 1983 – 30 September 1984.
J. T. Jefferies.
Bull. Am. Astron. Soc., Vol. 18, No. 1, p. 332 – 373 (1986).

National Radio Astronomy Observatory, Tucson, Arizona 85721–0655. Report for the period July 1984 – June 1985.
P. A. Vanden Bout, R. J. Havlen.
Bull. Am. Astron. Soc., Vol. 18, No. 1, p. 374 – 424 (1986).

Steward Observatory, University of Arizona, Tucson, Arizona 85721. Report for the period 1 January 1984 – 31 March 1985.
P. A. Strittmatter, W. S. Fitch.
Bull. Am. Astron. Soc., Vol. 18, No. 1, p. 517 – 533 (1986).

Preprints of the Steward Observatory.
Nos. 613 (41.157.202), 614 (41.159.134), 615 (41.157.203), 616 (41.115.020), 617 (41.126.064), 618 (41.035.067), 619 (41.118.028), 620 (41.161.222), 621 (41.154.046), 622 (41.031.028), 623 (41.031.029), 624 (41.031.030), 625 (41.031.031), 626 (41.159.135), 628 (41.066.179), 629 (41.131.295), 630 (41.131.296), 631 (41.131.297 – 41.131.299), 632 (41.036.153), 633 (41.157.204), 634 (41.117.227), 635 (41.122.142), 636 (41.157.205), 637 (41.126.065), 638 (41.004.091), 639 (41.126.066), 640 (41.035.068), 641 (41.131.300), 642 (41.021.020), 643 (41.036.154), 644 (41.133.013), 645 (41.133.014), 646 (41.065.106), 647 (41.131.301), 648 (41.036.155), 649 (41.034.149), 650 (41.126.067), 651 (41.036.156), 652 (41.034.150), 653 (41.099.068), 654 (41.115.021), 655 (41.115.022), 656 (41.036.157), 657 (41.034.151), 658 (41.117.228), 659 (41.103.713), 660 (41.154.047), 661 (41.131.302), 662 (41.161.223), 663 (41.099.069), 664 (41.160.100), 665 (41.157.206), 666 (41.161.224), 668 (41.082.073).

008.801 **Turku**

Turku University Observatory, Informo.
Nos. 96 (41.031.027), 97 (41.155.101), 98 (41.158.260), 99 (41.042.072), 100 (41.151.099), 101 (41.155.102), 102 (41.159.133).

Report Series, Department of Physical Sciences, University of Turku.
FTL–R84 (41.031.027), FTL–R86 (41.155.101), FTL–R89 (41.158.260), FTL–R90 (41.042.072), FTL–R92 (41.151.099), FTL–R94 (41.155.102), FTL–R95 (41.159.133).

008.804 **Uccle**

Observatoire Royal de Belgique, Communications, Série A.
Nos. 78 (39.071.033), 79 (40.103.029, 40.102.087, 40.103.030, 40.004.032), 80 (41.036.148), 81 (40.131.185).

Observatoire Royal de Belgique, Communications, Série B.
Nos. 132 (41.045.018), 133 (39.118.019), 134, 135 (41.098.107), 136 (40.074.039).

008.810 **University Park, Pa.**

The Pennsylvania State University, Department of Astronomy, University Park, Pennsylvania 16802. Report for the period 1 September 1984 – 31 August 1985.
S. Matsushima.
Bull. Am. Astron. Soc., Vol. 18, No. 1, p. 467 – 478 (1986).

008.813 **Uppsala**

Uppsala Astronomical Observatory. Annual Report for 1985.
C.–I. Lagerkvist (Editor).
Upps. Astron. Obs. Rep., No. 39, 31 pp. (1986).

Uppsala Astronomical Observatory, Report.
Nos. 37 (41.098.107), 39 (41.008.813), 40 (41.098.108).

008.816 **Urbana, Ill.**

University of Illinois at Urbana–Champaign, Department of Astronomy, Urbana, Illinois 61801–3000. Report for the period 1 September 1984 – 31 August 1985.
K. M. Yoss, R. F. Webbink.
Bull. Am. Astron. Soc., Vol. 18, No. 1, p. 182 – 195 (1986).

008.825 **Victoria**

Climenhaga Observatory, Physics Department, University of Victoria, Victoria, British Columbia V8W 2Y2, Canada. Report for 1984 – 1985.
Bull. Am. Astron. Soc., Vol. 18, No. 1, p. 110 – 113 (1986).

008.837 **Villanova, Pa.**

Villanova University, Villanova, Pennsylvania 19085. Report for the period September 1984 – September 1985.
Bull. Am. Astron. Soc., Vol. 18, No. 1, p. 578 – 580 (1986).

008.840 **Vilnius**

Chronicle. Astronomy in Lithuania in the years 1982 – 1984.
V. Straižys.
Vilniaus Astron. Obs. Biul., Nr. 69, p. 33 – 39 (1985). In Russian.

Vilniaus Astronomijos Observatorijos Biuletenis.
Nos. 69 (41.113.024, 41.114.063, 41.113.025, 41.008.840), 70 (41.114.064, 41.113.026, 41.117.142, 41.034.047), 71 (41.034.048, 41.034.049), 72 (41.126.036), 73 (41.126.037, 41.126.038).

008.843 **Warsaw**

Latitude Circular.
No. 99 (41.044.049).

008.846 **Washington, D.C.**

U.S. Naval Observatory, Washington, DC 20390–5100. Report for the period 1 July 1984 – 30 June 1985.
G. Westerhout, C. K. Roberts.
Bull. Am. Astron. Soc., Vol. 18, No. 1, p. 564 – 574 (1986).

U.S. Naval Observatory, Time Service Publications, Series 4.
Nos. 988 – 1012 (41.044.050).

Earth Orientation Bulletin.
Vol. 4, Nos. 1 – 13 (41.044.051).

NEOS. Earth Orientation Bulletin.
Vol. 1, Nos. 1 – 13 (41.044.052).

U.S. Naval Observatory, Time Service Announcements.
Nos. 41 – 42 (41.044.053).

U.S. Naval Observatory, Circular.
No. 167, Update No. 1 (41.044.054).

008.849 **Wellington**

Carter Observatory. Report for the year ended 31 March 1985.
R. J. Dodd.
South. Stars, Vol. 31, No. 4, p. 215 – 220 (1985).

008.852 **Westford, Mass.**

Haystack Observatory, Northeast Radio Observatory Corporation (NEROC), Westford, Massachusetts 01886. Report for the period 1 July 1984 – 30 June 1985.
Bull. Am. Astron. Soc., Vol. 18, No. 1, p. 165 – 170 (1986).

008.858 **Williamstown, Mass.**

Hopkins Observatory, Williams College, Williamstown, Massachusetts 01267. Report for the 1984 – 85 academic year.
J. M. Pasachoff.
Bull. Am. Astron. Soc., Vol. 18, No. 1, p. 179 – 180 (1986).

008.864 **Wrocław**

Contributions from the Wrocław Astronomical Observatory.
No. 20 (41.041.036).

Wrocław Astronomical Observatory, Reprint.
Nr. 118 (37.072.003), 119 (38.122.122), 120 (38.122.124), 121 (38.072.049), 122 (39.122.111), 123 (39.114.068), 124 (39.120.009).

008.870 **Yorktown Heights, N.Y.**

IBM Thomas J. Watson Research Center, Yorktown Heights, New York 10598. Report.
B. G. Elmegreen.
Bull. Am. Astron. Soc., Vol. 18, No. 1, p. 181 (1986).

008.873 **Zagreb**

Hvar Observatory Bulletin.
Vol. 9, No. 1 (41.072.070, 41.077.034, 41.032.034, 41.073.102).

008.876 **Zelenchukskaya**

Chronicle.
Astrofiz. Issled. Izv. Spets. Astrofiz. Obs., Tom 22, p. 136 – 137 (1986). In Russian. English translation in Bull. Spec. Astrophys. Obs. – North Caucasus.

Astrofizicheskie Issledovaniya. Izvestiya Spetsial'noj Astrofizicheskoj Observatorii.
Tom 21 (41.114.069 – 41.114.071, 41.036.086, 41.116.075, 41.114.073, 41.131.202, 41.131.203, 41.125.058, 41.033.012, 41.036.087 – 41.036.090, 41.033.013, 41.036.091, 41.033.072, 41.036.093),
Tom 22 (41.114.075, 41.114.076, 41.122.101, 41.116.031, 41.114.077, 41.158.201, 41.114.078, 41.034.065, 41.036.095, 41.036.096, 41.034.066 – 41.034.069, 41.032.024, 41.033.014, 41.033.015, 41.008.876).

Soobshcheniya Spetsial'noj Astrofizicheskoj Observatorii.
Vyp. 48 (41.034.016, 41.158.062, 41.141.005, 41.160.033).

008.882 **Zimmerwald**

Mitteilungen der Satelliten–Beobachtungsstation Zimmerwald.
Nr. 18 (41.052.044, 41.045.016, 41.045.017).

009 Notes on Observatories, Planetaria, Exhibitions

009.001 **The East–Siberian complex of SibIZMIR solar observatories.**
G. Ya. Smolkov (*G. Ya. Smol'kov*), V. E. Stepanov, V. M. Grigoryev (*V. M. Grigor'ev*), V. G. Banin.
Astrophys. Space Sci., Vol. 118, Nos. 1/2, p. 21 – 30 (1986). – See Abstr. 012.001.

The Sayan Solar Observatory, the Tunkinskaya Radiophysical Observatory, and the Baikal Astrophysical Observatory of SibIZMIR have been described. Some results of the investigations carried out over the past years have been expounded. Clarifying solar activity mechanisms and the nature of solar magnetic fields, active regions and flares as well as solar wind are their main problems. In the interest of these investigations the Institute has initiated and developed laboratory–scale simulation of plasma processes.

009.002 **Galactic and extragalactic studies with the KISO Schmidt.**
B. Takase.
Astrophys. Space Sci., Vol. 118, Nos. 1/2, p. 35 – 37 (1986). – See Abstr. 012.001.

On 1 October, 1974 the 105 cm Schmidt–telescope was dedicated at the Kiso Observatory. Since then it has been used for several galactic and extragalactic studies. The main works are described here together with some data on the telescope.

009.003 **The Iraqi National Astronomical Observatory.**
H. M. K. Al–Naimiy.
Astrophys. Space Sci., Vol. 118, Nos. 1/2, p. 51 – 56 (1986). – See Abstr. 012.001.

The current plans of the Space and Astronomy Research Center (SARC) include building a major observatory to work in the optical, IR and radio region of the spectrum. The core of the optical facility will be a 3.5 m optical telescope, together with a 1.25 m telescope designed for efficient performance in the IR. These telescopes will be equipped with instruments for photographic, photometric and spectroscopic observations. A 30 m dish is also being built for millimeter/radio observations.

009.004 **Results from Nobeyama Radio Observatory (NRO) – a progress report.**
M. Morimoto.
Astrophys. Space Sci., Vol. 118, Nos. 1/2, p. 63 – 65 (1986). – See Abstr. 012.001.

The title telescope – in operation since 1982 – gives a surface accuracy of better than 0.2 mm (r.m.s.), with an aperture efficiency of about 25% at 115 GHz. A 5–element interferometer is at final adjustment stage. Observations of proto–stellar objects, extragalactic objects, and spectral lines are briefly described.

009.005 **Radio astronomy at the University of Sydney.**
A. J. Turtle.
Astrophys. Space Sci., Vol. 118, Nos. 1/2, p. 83 – 85 (1986). – See Abstr. 012.001.

The recent improvements to the radio telescopes at the Molonglo (MOST) and Fleurs (FST) Observatories are described along with some of the astronomical projects.

009.006 **Astronomical research activities with the 74 inch telescope at Kottamia Observatory.**
I. A. Issa, A. I. Gamal El Din.
Astrophys. Space Sci., Vol. 118, Nos. 1/2, p. 87 – 93 (1986). – See Abstr. 012.001.

Two topics of research, namely extragalactic and variable star studies represent the main attitudes of astronomical work at the Helwan Observatory. The 74 inch telescope at Kottamia (476 m, 2^h12, $29^\circ93$ N) some 60 km to the east of Cairo is the main instrument. Some other auxiliary equipment (Cassegrain and echelle spectrographs, Westinghouse and Japanese cameras and photoelectric photometer) can be attached at both Cassegrain and Newtonian foci.

009.007 **L'ultimo paradiso degli astronomi.**
P. Bianucci.
Orione, Vol. 6, N. 1, p. 31 – 37 (1986).

009.008 **Das Kölner 3–m–Radioteleskop auf dem Gornergrat.**
G. Winnewisser.
Sterne Weltraum, 25. Jahrg., Nr. 2, p. 77 – 82 (1986).

009.009 **Planetariums can be visual computers for ethnoastronomers.**
V. D. Chamberlain.
Archaeoastronomy (U.S.A.), Vol. 7, Nos. 1 – 4, p. 12 – 15 (1984).

009.010 **Ein neues Planetarium aus Jena.**
L. Meier, J. Reiche.
Sterne, 62. Band, Heft 1, p. 3 – 13 (1986).

009.011 **Der Einsteinturm in Potsdam und die Relativitätstheorie.**
F. W. Jäger.
Sterne, 62. Band, Heft 1, p. 14 – 23 (1986).

009.012 **Popularization of astronomy in Thailand.**
N. Saibejra.
Report on the VIII. International Planetarium Directors Conference, p. 44 – 48 (1985). – See Abstr. 012.014.

009.013 **The Mobile Bochum Observatory – an example for the application of a small planetarium in the field of "spaceology".**
U. Lemmer.
Report on the VIII. International Planetarium Directors Conference, p. 50 – 56 (1985). – See Abstr. 012.014.

009.014 **Active and independent contribution from children – the essential part in the didactical/methodical plan of a planetarium show for children.**
J. Prölss.
Report on the VIII. International Planetarium Directors Conference, p. 58 – 63 (1985). – See Abstr. 012.014.

009.015 **Anticipation and participation as didactic principles of "spaceology".**
T. H. Weyer.
Report on the VIII. International Planetarium Directors Conference, p. 64 – 73 (1985). – See Abstr. 012.014.

009.016 **Götterdämmerung – the withering temples of science.**
S. Wieser.
Report on the VIII. International Planetarium Directors Conference, p. 80 – 82 (1985). – See Abstr. 012.014.

009.017 **The traditional planetarium – alive and well and ready for the years ahead.**
J. M. Chamberlain.
Report on the VIII. International Planetarium Directors Conference, p. 84 – 87 (1985). – See Abstr. 012.014.

009.018 **The new Mannheim Planetarium.**
W. Wacker.
Report on the VIII. International Planetarium Directors Conference, p. 88 – 91 (1985). – See Abstr. 012.014.

009.019 **Didactics value of the planetarium for primary and secondary students.**
O. Vaio.
Report on the VIII. International Planetarium Directors Conference, p. 92 – 94 (1985). – See Abstr. 012.014.

009.020 **A new Convention Center in Morelia Planetarium.**
G. R. Munoz B.
Report on the VIII. International Planetarium Directors Conference, p. 96 – 97 (1985). – See Abstr. 012.014.

009.021 **Planetaria of the USSR and their role in dissemination of information on natural sciences.**
J. K. Fishevskij.
Report on the VIII. International Planetarium Directors Conference, p. 102 – 105 (1985). – See Abstr. 012.014.

009.022 **Recent programs and activities at the "American Museum – Hayden Planetarium" and a call for tapes for the I.P.S. video project.**
W. A. Gutsch.
Report on the VIII. International Planetarium Directors Conference, p. 106 – 107 (1985). – See Abstr. 012.014.

009.023 **The new "La Villette" – planetarium project.**
J.–F. Delorme.
Report on the VIII. International Planetarium Directors Conference, p. 108 – 120 (1985). – See Abstr. 012.014.

009.024 **The Stellarium at "La Villette".**
R. Steinmetz.
Report on the VIII. International Planetarium Directors Conference, p. 122 – 135 (1985). In French and English. – See Abstr. 012.014.

009.025 **The Praha Planetarium.**
A. Rükl.
Report on the VIII. International Planetarium Directors Conference, p. 136 – 138 (1985). – See Abstr. 012.014.

009.026 **Local circumstances of solar and lunar eclipses in five millenia: a new Zeiss eclipse projector and new data.**
H. Mucke.
Report on the VIII. International Planetarium Directors Conference, p. 140 – 145 (1985). – See Abstr. 012.014.

009.027 **Striving for excellence – the OMNIMAX/IMAX–projection system.**
W. Shaw.
Report on the VIII. International Planetarium Directors Conference, p. 146 – 152 (1985). – See Abstr. 012.014.

009.028 **Planetaria from Jena – scope and tendencies.**
P. Köhler.
Report on the VIII. International Planetarium Directors Conference, p. 154 – 157 (1985). – See Abstr. 012.014.

009.029 **New possibilities of fixed star projection using conventional projection bulbs.**
L. Meier.
Report on the VIII. International Planetarium Directors Conference, p. 158 – 161 (1985). – See Abstr. 012.014.

009.030 **Brightness and contrast consideration in projection on large format curved surfaces.**
L. P. Skolnick.
Report on the VIII. International Planetarium Directors Conference, p. 162 – 177 (1985). – See Abstr. 012.014.

009.031 **Trends and developments in Zeiss planetarium instruments.**
H. Raff.
Report on the VIII. International Planetarium Directors Conference, p. 178 – 199 (1985). – See Abstr. 012.014.

009.032 **The history of McDonald Observatory.**
D. S. Evans, J. D. Mulholland.
Bull. Am. Astron. Soc., Vol. 17, No. 4, p. 827 (1985). Abstract. – See Abstr. 010.062.

009.033 **Solar observatories of France.**
M. N. Gnevyshev, V. I. Makarov.
Zemlya Vselennaya, No. 2, p. 48 – 53 (1986). In Russian.

009.034 **The Rutherford Appleton Laboratory's programme in space.**
J. E. Harries.
J. Br. Interplanet. Soc., Vol. 39, No. 5, p. 195 –198 (1986).

009.035 **Palmerston North's first observatory.**
J. B. Seymour.
South. Stars, Vol. 31, No. 4, p. 209 – 214 (1985).

009.036 **L'Osservatorio Astronomico di Parigi.**
L. Candiano.
Coelum, Vol. 54, N. 2, p. 73 – 87 (1986).

009.037 **The Tokyo Schmidt telescope and its programs.**
B. Takase.
Proceedings of the Second Asian–Pacific Regional Meeting on Astronomy, p. 341 – 346 (1984). – See Abstr. 012.031.

009.038 **Recent work with the UK 1.2 m Schmidt Telescope.**
S. B. Tritton.
Proceedings of the Second Asian–Pacific Regional Meeting on Astronomy, p. 358 – 363 (1984). – See Abstr. 012.031.

009.039 **Braeside Observatory 1985.**
R. E. Fried.
I.A.P.P.P. Commun., No. 23, p. 2 (1986). Abstract. – See Abstr. 010.521.

009.040 **Photoelectric photometry in the undergraduate physics program at East Tennessee State University.**
H. D. Powell.
I.A.P.P.P. Commun., No. 23, p. 2 (1986). Abstract. – See Abstr. 010.521.

009.041 **Photometry at Minot State College Observatory.**
D. Martin.
I.A.P.P.P. Commun., No. 23, p. 24 – 25 (1986).

009.042 **Photometry at Lewis and Clark Observatory.**
W. E. Newsome.
I.A.P.P.P. Commun., No. 24, p. 22 – 25 (1986).

009.043 **Planetarium Stuttgart. Jahresbericht 1985.**
H.–U. Keller.
Sterne Weltraum, 25. Jahrg., Nr. 6, p. 351 (1986).

009.044 **Spektroskopische Untersuchungen am Einsteinturm: Zur Physik der Korona und der Sonnenflecken.**
J. Staude.
Sterne, 62. Band, Heft 2, p. 109 – 116 (1986).

009.045 **The story of the Paris Observatory.**
M. S. Dimitrijević.
Vasiona, Année 34, No. 1 – 2, p. 9 – 12 (1986). In Croatian.

009.046 **American observatories in Southern Africa.**
A. H. Jarrett.
Suid–Afr. Tydskr. Wet., Vol. 82, p. 11 – 13 (1986). = Boyden Obs., Repr., No. 58.

009.047 **Indian Institute of Astrophysics – instruments over the years (1786 – 1985).**
J. C. Bhattacharyya, R. K. Kochhar.
Indian Inst. Astrophys., Prepr., No. 143, 8 pp. (1985). To appear in CZ Indian Instrum. Rev.

009.048 **Institutt for teoretisk astrofysikk 50 år.**
Inst. Teor. Astrofys., Blindern–Oslo, Småtrykk, Nr. 105, 12 pp. (1984).
Contents: Svein Rosseland – Norges første astrofysiker (*Ø. Elgarøy*). Universitetet og universet (*Ø. Elgarøy, T. S. Ringnes*).

Astronomi over landegrensene (*T. S. Ringnes*). Jakten på det beste observasjonssted (*E. Jensen*). Astronomi mot år 2000 (*P. Maltby*). Astronomie of romforskning (*O. Kjeldseth Moe*). Solobservasjoner – ny teknologi (*O. Engvold*).

009.049 **Some historical data of the Kunming Astronomical Station and the Qingdao Observatory.**
L. Zhu.
Publ. Purple Mt. Obs., Vol. 4, No. 4, p. 66 – 73 (1985). In Chinese.

009.050 **A short description of the VAX–11/780 computer system of Purple Mountain Observatory.**
Y.–l. Chen, Y.–y. Zhang, H. Fang.
Publ. Purple Mt. Obs., Vol. 4, No. 4, p. 74 – 81 (1985). In Chinese.

The performance and features of VAX–11 series computers are analysed. A brief account of the VAX–11/780 computer system of Purple Mountain Observatory is also given.

009.051 **Foreign affairs of Purple Mountain Observatory during 1985.**
Z.–t. Xu, G.–f. Xie.
Publ. Purple Mt. Obs., Vol. 4, No. 4, p. 82 – 85 (1985). In Chinese.

009.052 **"Save Herstmonceux!"**
N. Henbest.
J. Br. Astron. Assoc., Vol. 96, No. 4, p. 189 (1986).

009.053 **Le Centre de Données Astronomiques de Strasbourg.**
A. Heck.
Ciel Terre, Vol. 102, No. 3, p. 81 – 82 (1986).

009.054 **Una visita all'ESO l'osservatorio australe europeo.**
W. Ferreri.
Orione, Vol. 6, N. 3, p. 30 – 38 (1986).

009.055 **Die Dr. Remeis–Sternwarte in Bamberg und die Veränderlichen Sterne.**
H. Schmidt.
BAV Rundbrief, 35. Jahrg., Nr. 2, p. 64 – 69 (1986).

009.056 **Statistical astronomical work at Padua Astronomical Observatory.**
C. Bonoli.
Bull. Inf. Cent. Données Stellaires, No. 30, p. 168 (1986). = Newsl. Working Group Mod. Astron. Methodol., No. 2.

009.057 **Statistical work at the Astronomical Institutes of the University of Bonn.**
P. Brosche.
Bull. Inf. Cent. Données Stellaires, No. 30, p. 179 – 180 (1986). = Newsl. Working Group Mod. Astron. Methodol., No. 2.

009.058 **Elias Loomis and the Loomis Observatory.**
A. J. Stokes.
J. Am. Assoc. Variable Star Obs., Vol. 14, No. 2, p. 72 (1985).

The history and present state of the Loomis Observatory, built in 1838, is described.

009.059 **Completion near Beijing of China's first over–water solar observing station.**
J.–m. Wang.
Chin. Astron. Astrophys., Vol. 10, No. 1, p. 1 – 2 (1986). English translation of Acta Astrophys. Sin., Vol. 5, No. 4, p. 254 (1985).

009.060 **The story of Ooty Radio Telescope.**
G. Swarup.
Cosmic pathways, p. 349 – 360 (1986). – See Abstr. 003.045.

009.061 **Fifty years of Purple Mountain Observatory (1934 – 1984).**
L. Zhu, P.–y. Zhang (Editors).
Purple Mountain Observatory, Academia Sinica, Nanjing, People's Republic of China. 11 + 134 pp. (1985). In Chinese, with extended English summaries.

This book documents the celebrations on the occasion of the 50th anniversary of Purple Mountain Observatory on September 1, 1984. Furthermore it presents a brief history of the observatory and a collection of photographs of its instruments and facilities. A catalogue of published papers, reports, books and translations from Purple Mountain Observatory is included.

009.062 **Potchefstroomse Universiteit: Departement Wiskunde en Toegepaste Wiskunde.**
J. H. de Klerk.
Mon. Notes Astron. Soc. S. Africa, Vol. 44, Nos. 3 – 4, p. 31 (1985).

009.063 **Rhodes University: Department of Physics and Electronics. Report 1984.**
E. E. Baart.
Mon. Notes Astron. Soc. S. Africa, Vol. 44, Nos. 3 – 4, p. 32 – 33 (1985).

009.064 **University of South Africa: Department of Mathematics, Applied Mathematics and Astronomy.**
W. F. Wargau, J. Wolterbeek.
Mon. Notes Astron. Soc. S. Afr., Vol. 44, Nos. 5–6, p. 42 – 44 (1985).

009.065 **Plans for a fast image recording system at ESO.**
P. J. Grosbøl.
NASA Conf. Publ., NASA CP–2317, p. 317 – 328 (1984). – See Abstr. 012.073.

A 256 diode–array will be installed as detector on the ESO OPTRONICS S–3000 measuring machine in order to increase the acquisition rate. A high intensity LED will be used as light source in a pulse mode. The data will be stored on a random access mass storage device as density values for later reduction. The scanning time for a 30 cm × 30 cm plate with a step size of 10 micron will be less than 10 hours while the dynamic range of the data is expected to be 2.5 density units with an offset of at least 1 unit.

009.066 **A new measuring machine in Paris.**
J. Guibert, P. Charvin.
NASA Conf. Publ., NASA CP–2317, p. 329 – 332 (1984). – See Abstr. 012.073.

A new photographic measuring machine is under construction at Paris Observatory. The amount of transmitted light is measured by a linear array of 1024 photodiodes. Carriage control, data acquisition and on line processing are performed by microprocessors, a S.E.L. 32/27 computer, and an AP 120–B Array Processor. It is expected that a Schmidt telescope plate of size 360 mm square will be scanned in one hour with pixel size of ten microns.

009.067 **The preparation of the measurements at the PDS at the Padova Observatory: reports on an experience with the development of an astronomical data base environment.**
L. Benacchio.
NASA Conf. Publ., NASA CP–2317, p. 361 – 369 (1984). – See Abstr. 012.073.

This paper describes some software facilities used mainly for information, retrieval and analysis at the Padova–Asiago Observatory. These facilities help guest and resident astronomers to make easier the preparation of plate measurements. The problems connected with the creation, use and management of a data base in a scientific (astronomical) environment are reviewed on the basis of the experience gathered during the last three years.

009.068 **Photographic surveys of the southern sky.**
M. E. Sim.
NASA Conf. Publ., NASA CP–2317, p. 379 – 385 (1984). – See Abstr. 012.073.

The Royal Observatory, Edinburgh (ROE) is an establishment of the UK Science and Engineering Research Council, with the primary task of providing high quality research facilities for the UK astronomical community. The UK Schmidt Telescope Unit is responsible for the operation of a 1.2 metre telescope at Siding Spring Observatory, Australia, and for the Plate Library facility and a suite of associated Photographic Laboratories in Edinburgh.

009.069 **Sternentheater Zeiss Planetarium.**
H.–U. Keller.
Dipl.–Ing. Gerd Fassbender, Gevelsberg, F.R. Germany, 70 pp. (1985).

009.070 **The beginning and end of a Jesuit observatory (1841 – 1972).**
F. J. Heyden.
Manila Observatory, UP PO Box 122, UP Post Office, Diliman, Quezon City, Philippines, 24 pp. (1986).

De hemel op aarde. De lange weg van reuzenglobe naar projectieplanetarium.
See Abstr. 004.020.

The observatory in Pula.
See Abstr. 004.110.

The Mt. John 1 metre telescope project.
See Abstr. 032.002.

Dependence of seeing correlation on image separation at the CFH telescope on Mauna Kea.
See Abstr. 082.009.

Extinction variations at La Silla.
See Abstr. 082.075.

Seeing at La Silla: LASSCA 86.
See Abstr. 082.076.

Cometary observations at Onsala Space Observatory since 1980.
See Abstr. 103.012.

Observations of active nuclei at Calar Alto Observatory.
See Abstr. 158.108.

010 Societies, Associations, Organizations

American Association of Variable Star Observers (AAVSO)

010.021 **Abstracts of papers presented at the 74th annual meeting of the AAVSO held in South Hadley, Mass., November 1 – 2, 1985.**
J. Am. Assoc. Variable Star Obs., Vol. 14, No. 2, p. 74 – 77 (1985).

010.022 **Annual report of the director for the fiscal year 1984 – 1985.**
J. A. Mattei.
J. Am. Assoc. Variable Star Obs., Vol. 14, No. 2, p. 82 – 101 (1985).

010.023 **Meetings and activities of the Society, committee reports.**
J. Am. Assoc. Variable Star Obs., Vol. 14, No. 2, p. 78 – 81, 102 – 107 (1985).

010.024 **The Journal of the American Association of Variable Star Observers.**
Vol. 14, No. 2 (1985).

American Astronomical Society (AAS)

010.061 **Historical Astronomy Division, Fourth Annual Meeting, held at Baltimore, Maryland, 9 – 13 June 1984. Abstracts of presented papers.**
Archaeoastronomy (U.S.A.), Vol. 7, Nos. 1 – 4, p. 16 – 20 (1984). First published in Bull. Am. Astron. Soc., Vol. 16, No. 2 (1984).

For the individual contributions see abstracts 37.004.047, 37.004.089 – 37.004.100, 37.004.105, 37.004.106, 37.015.040, 37.015.041, 37.015.043, 41.004.016, 41.004.017.

010.062 **The 167th meeting of the American Astronomical Society, 5 – 9 January 1986, Houston, Texas. Abstracts of presented papers.**
Bull. Am. Astron. Soc., Vol. 17, No. 4, p. 801 – 914 (1985).

010.063 **Late–paper abstracts from the 17th annual meeting of the Division for Planetary Sciences, held 28 October – 1 November 1985 at Baltimore, Maryland.**
Bull. Am. Astron. Soc., Vol. 17, No. 4, p. 915 – 927 (1985).

010.064 **Minutes of the Division for Planetary Sciences annual business meeting 11 October 1984 Kamehameha Hotel, Kailua–Kona, Hawaii.**
Bull. Am. Astron. Soc., Vol. 17, No. 4, p. 928 – 930 (1985).

010.065 **Late paper abstracts of the Solar Physics Division, 13 – 15 May 1985, Tucson, Arizona.**
Bull. Am. Astron. Soc., Vol. 17, No. 4, p. 931, 933 (1985).

010.066 **Late paper abstracts from the 167th meeting of the American Astronomical Society, 5 – 9 January 1986, Houston, Texas.**
Bull. Am. Astron. Soc., Vol. 18, No. 1, p. 631 – 641 (1986).

010.067 **Bulletin of the American Astronomical Society.**
Vol. 17, No. 4 (1985), Vol. 18, No. 1 (1986).

010.068 **AAS Photo–Bulletin.**
No. 39 (1985).

Association Française des Observateurs d'Etoiles Variables

010.101 **Activité de l'A.F.O.E.V. en 1985.**
E. Schweitzer.
Bull. Assoc. Fr. Obs. Etoiles Variables, No. 36, p. 1 – 6 (1986).

Contents: Activité générale. Variables du type Mira et SR. Variables du type Mira (observations en "m_{pg}").

010.102 **L'Association Française des Observateurs d'Etoiles Variables.**
E. Schweitzer.
Bull. Inf. Cent. Données Stellaires, No. 30, p. 85 – 89 (1986).
The methods for entering, handling and stocking variable star observations (long period and cataclysmic variables) made by the AFOEV are discussed. The observations from 1921 to 1979 are being entered; those from 1980–85 are already available on magnetic tape from the CDS.

010.103 **La vie de l'Association.**
E. Schweitzer.
Bull. Assoc. Fr. Obs. Etoiles Variables, No. 35, p. 9, No. 36, p. 9 (1986).

010.104 **Bulletin de l'Association Française des Observateurs d'Etoiles Variables.**
Nos. 35 – 36 (1986).

Association of Lunar and Planetary Observers (A.L.P.O.)

010.121 **The 1985 business meeting of the A.L.P.O.**
J. L. Benton Jr.
Strolling Astron., Vol. 31, Nos. 5–6, p. 125 – 126 (1986).

010.122 **Amateur observing by the A.L.P.O. Minor Planets Section.**
F. Pilcher.
Minor Planet Bull., Vol. 12, No. 4, Part 1, p. 35 – 36 (1985).

010.123 **The Minor Planet Bulletin.**
Vol. 12, Nos. 2 – 4 (1985), Vol. 13, Nos. 1 – 2 (1986).

010.124 **The Strolling Astronomer. The Journal of the Association of Lunar and Planetary Observers.**
Vol. 31, Nos. 5 – 8 (1986).

Astronomical Society of Australia (ASA)

010.161 **Proceedings of the Astronomical Society of Australia.**
Vol. 6, No. 2 (1985).

Astronomical Society of India

010.201 **Bulletin of the Astronomical Society of India.**
Vol. 13, No. 4 (1985).

Astronomical Society of Japan

010.221 **Publications of the Astronomical Society of Japan.**
Vol. 38, Nos. 1 – 3 (1986).

010.222 **The Astronomical Herald.**
Vol. 79, Nos. 1 – 6 (1986). In Japanese.

Astronomical Society of New York

010.241 **Abstracts of papers given at the Fall meeting of the Astronomical Society of New York, held October 25, 1985 at Union College, Schenectady, New York.**
News Lett. Astron. Soc. N.Y., Vol. 2, No. 9, p. 1 – 50 (1986).

010.242 **News Letter of the Astronomical Society of New York.**
Vol. 2, No. 9 (1986).

Astronomical Society of Southern Africa (ASSA)

010.261 **Proceedings of the Annual General Meeting, 1985.**
Mon. Notes Astron. Soc. S. Afr., Vol. 44, Nos. 9–10, p. 69 – 76 (1985).

010.262 **Centre reports: Bloemfontein Centre, Cape Centre, Harare Centre, Natal Centre, Natal Midlands Centre, Pretoria Centre, Transvaal Centre.**
J. J. van Ellinckhuyzen, C. R. G. Turk, R. Fleet,
J. A. da S. Campos, J. W. Watson, J. Wolterbeek, B. Fraser.
Mon. Notes Astron. Soc. S. Afr., Vol. 44, Nos. 11–12, p. 94 – 101 (1985).

010.263 **Section reports.**
Mon. Notes Astron. Soc. S. Afr., Vol. 44, Nos. 9 – 10, p. 77 – 86 (1985).

010.264 **Monthly Notes of the Astronomical Society of Southern Africa.**
Vol. 44, Nos. 3 – 12 (1985).

Astronomical Society of the Pacific (ASP)

010.281 **The Astronomical Society of the Pacific: what it is and how it works.**
A. Fraknoi.
Mercury, Vol. 15, No. 1, p. 28 – 30 (1986).

010.282 **Publications of the Astronomical Society of the Pacific.**
Vol. 97, Nos. 597 – 598 (1985), Vol. 98, Nos. 599 – 604 (1986).

010.283 **Mercury. The Journal of the Astronomical Society of the Pacific.**
Vol. 15, Nos. 1 – 3 (1986).

Astronomical Society of Western Australia (ASWA)

010.301 **Journal of the Society of Western Australia.**
Vol. 36, Nos. 6 – 10 (1986).

Astronomical Society "Rudjer Bošković"

010.321 **Les premiers présidents de la Société Astronomique.**
N. D. Janković.
Publ. Astron. Soc. "Rudjer Bošković", No. 4, p. 37 – 39 (1985).
– See Abstr. 012.078. In Croatian.

010.322 **Publication of the Astronomical Society "Rudjer Bošković".**
No. 4 (1985).

Astronomische Gesellschaft (AG)

010.341 **Mitteilungen der Astronomischen Gesellschaft.**
Nr. 65 (1986).

British Astronomical Association (BAA)

010.401 **British Astronomical Association Circular.**
Nos. 655 – 656 (1986).

010.402 **Meetings and activities of the Association.**
J. Br. Astron. Assoc., Vol. 96, Nos. 2 – 4, p. 108 – 119, 176 – 180, 241 – 246 (1986).

010.403 **Section reports.**
J. Br. Astron. Assoc., Vol. 96, Nos. 2 – 4, p. 68 – 71, 73 – 80, 129 – 130, 148 – 175, 212 – 222 (1986).

010.404 **Journal of the British Astronomical Association.**
Vol. 96, Nos. 2 – 4 (1986).

010.405 **The British Astronomical Association Newsletter.**
Nos. 16 – 18 (1986).

British Interplanetary Society (BIS)

010.421 **JBIS. Journal of the British Interplanetary Society.**
Vol. 39, Nos. 1 – 6 (1986).

010.422 **Spaceflight. A publication of the British Interplanetary Society.**
Vol. 28, Nos. 1 – 6 (1986).

010.423 **Space Education. A publication of the British Interplanetary Society.**
Vol. 1, No. 11 (1986).

European Space Agency (ESA)

010.501 **ESA Bulletin.**
Nos. 45 – 46 (1986).

010.502 **ESA Journal.**
Vol. 9, No. 4 (1985), Vol. 10, No. 1 (1986).

010.503 **ESA IUE Newsletter.**
No. 25 (1986).

010.504 **EXOSAT Express.**
Nos. 15 – 16 (1986).

010.505 **ESA Special Publication.**
ESA SP–226 (012.075), SP–239 (012.019), SP–241 (012.018), SP–246 (012.028).

International Amateur–Professional Photoelectric Photometry

010.521 **Abstracts of papers and papers presented at the 1985 I.A.P.P.P. symposium in Melbourne, Florida.**
I.A.P.P.P. Commun., No. 23, p. 1 – 10, 14 – 17 (1986).

010.522 **Five years of I.A.P.P.P.**
D. S. Hall.
I.A.P.P.P. Commun., No. 23, p. 1 (1986). Abstract. – See Abstr. 010.521.

010.523 **I.A.P.P.P. Photoelectric Photometer Electronics Committee report.**
J. L. Hopkins.
I.A.P.P.P. Commun., No. 23, p. 26 – 28 (1986).

010.524 **I.A.P.P.P. Communication.**
Nos. 23 – 24 (1986).

International Astronomical Union (IAU)

010.561 **IAU Information Bulletin 55.**
J.–P. Swings.
D. Reidel Publishing Company, Dordrecht, The Netherlands, 42 pp. (1986).
Contents: General Assemblies. Executive Committee. Commissions. International Organizations. IAU Symposia and Colloquia. Meetings co–sponsored by the IAU. Other scientific meetings. IAU publications. Other publications. Membership. Other matters. International schools for young astronomers.

010.562 **IAU Commision 5, Newsletter No. 1 (1986A), activities in documentation and astronomical data.**
G. A. Wilkins.
Bull. Inf. Cent. Données Stellaires, No. 30, p. 185 – 190 (1986).

010.563 **Minutes of the meeting of the working group on Be stars of IAU Commissions 29 and 45.**
A. Slettebak.
Bull. Inf. Etoiles Be, No. 13, p. 1 – 10 (1986).

010.564 **Commission 27 of the IAU, Information Bulletin on Variable Stars.**
Nos. 2844 – 2910 (1986).

010.565 **Circulaire d'Information.**
Nos. 98 – 99 (1986).

010.566 **The Minor Planet Circulars/Minor Planets and Comets.**
Nos. 10329 – 10884 (1986).

010.567 **IAU Circulars.**
Nos. 4157 – 4230 (1986).

010.568 **IAU Symposia.**
Nos. 109 (012.068), 114 (012.002), 116 (012.079).

010.569 **Quarterly Bulletin on Solar Activity.**
Vol. 23, Part V (1986).

Korean Astronomical Society

010.621 **The Journal of the Korean Astronomical Society.**
Vol. 18, No. 2 (1985), Vol. 19, No. 1 (1986).

LEST Foundation

010.641 **LEST Foundation. Annual Report 1985.**
Ø. Hauge.
LEST Found., Annu. Rep., 32 pp. (1986).

010.642 **LEST Foundation, Technical Report.**
Nos. 13 – 15 (1985), No. 16 (1986).

010.643 **LEST Large Earth–based Solar Telescope. An overview.**
Institute of Theoretical Astrophysics, University of Oslo, PO Box 1029, Blindern, N–0315 Oslo 3, Norway, 32 pp. (1986).

Meteoritical Society

010.661 **Abstracts of papers presented at the 48th annual meeting of the Meteoritical Society, July 16 – 19, 1985, Bordeaux, France.**
Meteoritics, Vol. 20, No. 4, p. 599 – 799 (1985).

010.662 **Meteoritics. The Journal of the Meteoritical Society.**
Vol. 20, No. 4 (1985), Vol. 21, No. 1 (1986).

Oriental Astronomical Association

010.721 **The Heavens.**
Vol. 67, Nos. 1 – 6 (1986). In Japanese.

Royal Astronomical Society (RAS)

010.741 **X–ray astronomy. Summaries of papers presented at the RAS specialist discussion, held 1985 May 10 at the Scientific Societies' Lecture Theatre, Savile Row.**
Observatory, Vol. 106, No. 1070, p. 5 – 12 (1986).

010.742 **Meetings of the Society.**
Observatory, Vol. 106, Nos. 1070 – 1072, p. 1 – 5, 29 – 35, 53 – 66 (1986).

010.743 **Meetings and activities of the Society.**
Q. J. R. Astron. Soc., Vol. 27, Nos. 1 – 2, p. 134, 311 – 312 (1986).

010.744 **Monthly Notices of the Royal Astronomical Society.**
Vol. 218, Nos. 1 – 4, Vol. 219, Nos. 1 – 4, Vol. 220, Nos. 1 – 4 (1986).

010.745 **Geophysical Journal of the Royal Astronomical Society.**
Vol. 84, Nos. 1 – 3, Vol. 85, Nos. 1 – 3 (1986).

010.746 **The Quarterly Journal of the Royal Astronomical Society.**
Vol. 27, Nos. 1 – 2 (1986).

Royal Astronomical Society of Canada

010.761 **Annual Report 1985.**
R. L. Bishop.
Suppl. J. R. Astron. Soc. Can., 38 pp. (1986).
This Annual Report summarizes the past year of activity of the Society and of its individual Centres.

010.762 **The Journal of the Royal Astronomical Society of Canada.**
Vol. 80, Nos. 1 – 2 (1986).

010.763 **National Newsletter. Supplement to the Journal of the Royal Astronomical Society of Canada.**
Vol. 80, Nos. 1 – 2 (1986).

Royal Astronomical Society of New Zealand

010.781 **Report of the Variable Star Section, Royal Astronomical Society of New Zealand for the year ended 1985 December 31.**
F. M. Bateson.
Publ. Variable Star Sect., R. Astron. Soc. N.Z., No. 13 (C85), p. 85 – 87 (1986).

010.782 **Southern Stars. Journal of the Royal Astronomical Society of New Zealand.**
Vol. 31, No. 4 (1985).

010.783 **Publications of Variable Star Section, Royal Astronomical Society of New Zealand.**
No. 13(C85) (1986).

Schweizerische Astronomische Gesellschaft (SAG)

010.801 **BBSAG Bulletin.**
No. 79 (1986).

010.802 **Mitteilungen.**
Orion, 44. Jahrg., Nr. 212 – 213, p. 17 – 20, 53 – 60 (1986).

010.803 **Orion. Zeitschrift der Schweizerischen Astronomischen Gesellschaft. Revue de la Société Astronomique Suisse.**
44. Jahrg., Nr. 212 – 214 (1986).

Società Astronomica Italiana (S.A.It.)

010.821 **XXIX annual meeting of the Italian Astronomical Society, held at Bologna, Italy, 25 – 27 April 1985. Contributed papers.**
Mem. Soc. Astron. Ital., Vol. 56, No. 4, p. 769 – 822 (1985).

010.822 **Memorie della Società Astronomica Italiana.**
Vol. 56, No. 4 (1985), Vol. 57, No. 1 (1986).

010.823 **Giornale di Astronomia.**
Vol. 11, N. 4 (1985).

Société Astronomique de France

010.841 **Réunion annuelle des commissions du 18 mai 1985.**
Astronomie, Vol. 100, p. 98 – 105 (1986).

010.842 **Séances, commissions, activités de la Société.**
Astronomie, Vol. 100, p. 32, 256 – 258 (1986).

010.843 **L'Astronomie et Bulletin de la Société Astronomique de France.**
Vol. 100, janvier – juin (1986).

Société Astronomique de Liège

010.861 **Le Ciel.**
Vol. 48, janvier – juin, p. 1 – 296 (1986).

Société Royale Belge d'Astronomie

010.881 **Ciel et Terre. Bulletin de la Société Royale Belge d'Astronomie, de Météorologie et de Physique du Globe.**
Vol. 102, Nos. 1 – 3 (1986).

Vereinigung der Sternfreunde e.V. (VdS)

010.921 **Nachrichten der Vereinigung der Sternfreunde e.V.**
Sterne Weltraum, 25. Jahrg., Nr. 1 – 6, p. 48 – 51, 108 – 110, 168 – 171, 228 – 231, 288 – 291, 348 –351 (1986).

010.922 **Sonne. Mitteilungsblatt der Amateursonnenbeobachter.**
Jahrg. 10, Nr. 37 (1986).

011 Reports on Colloquia, Congresses, Meetings, Symposia, Expeditions

011.001 **New results from Exosat.**
A. C. Fabian.
Nature, Vol. 319, No. 6053, p. 451 – 452 (1986).
This note reports on an Exosat discussion meeting, held at the Institute of Astronomy, Cambridge, UK, 5 – 6 November 1985.

011.002 **XIXth General Assembly of the International Astronomical Union.**
Y. Kozai.
Astron. Her., Vol. 79, No. 3, p. 71 – 72 (1986). In Japanese.

011.003 **All–Union conference on "Problems of remote sensing of the moon", held at the Astronomical Observatory of the Kharkov University, 23 – 26 April 1985.**
Yu. G. Shkuratov.
Kinematika Fiz. Nebesn. Tel, Tom 2, No. 1, p. 95 (1986). In Russian. English translation in Kinematics Phys. Celest. Bodies.

011.004 **The 23rd Astrometric Conference of the USSR, Pulkovo, 19 – 22 March 1955.**
D. P. Duma, D. D. Polozhentsev.
Kinematika Fiz. Nebesn. Tel, Tom 2, No. 1, p. 96 (1986). In Russian. English translation in Kinematics Phys. Celest. Bodies.

011.005 **Report on the January, 1985 meeting of the AAS Working Group on Photographic Materials.**
B. Schoening.
AAS Photo–Bull., No. 39, p. 15 (1985).

011.006 **All–Union conference on "Radiation and composition of the sun", held in Pulkovo, 9 – 11 October 1985.**
Eh. A. Gurtovenko, N. G. Shchukina.
Kinematika Fiz. Nebesn. Tel, Tom 2, No. 2, p. 95 – 96 (1986). In Russian. English translation in Kinematics Phys. Celest. Bodies.

011.007 **First All–Union conference on pulsars, held at the Lebedev Physical Institute of the USSR Academy of Sciences in Pushchino, 18 – 19 June 1985.**
I. F. Malov.
Astron. Zh., Tom 63, Vyp. 2, p. 410 – 411 (1986). In Russian. English translation in Sov. Astron., Vol. 30, No. 2.

011.008 **How no(w) brown dwarfs?**
V. Trimble.
Nature, Vol. 320, No. 6060, p. 311 – 312 (1986).
This note reports on the 1985 George Mason workshop on brown dwarfs, held at Fairfax, Virginia, USA, 14 – 18 October 1985.

011.009 **Galaxy distances and deviations from universal expansion.**
J. R. Bond, S. van den Bergh.
Nature, Vol. 320, No. 6062, p. 489 – 490 (1986).
A report on a NATO Advanced Research Workshop, held at Kona, Hawaii, 13 – 17 January 1986, is presented.

011.010 **Arqueoastronomía y etnoastronomía en Mesoamérica. UNAM symposium report, 24 – 29 September 1984.**
Archaeoastronomy (U.S.A.), Vol. 7, Nos. 1 – 4, p. 21 – 23 (1984).

011.011 **Prehistoric rock art and archaeoastronomy, Little Rock conference report, 7 – 12 October 1984.**
Archaeoastronomy (U.S.A.), Vol. 7, Nos. 1 – 4, p. 24 – 27 (1984).

011.012 **Cracow Cosmological Summer School of PAS.**
P. Flin, R. Juszkiewicz.
Postepy Astron., Tom 33, Zesz. 1–2, p. 69 – 70 (1985). In Polish.

011.013 **L'Académie des Sciences et la figure de la Terre.**
H. Lacombe.
C. R. Acad. Sci., Sér. Gén., Vie Sci., Tome 3, No. 2, p. 157 – 180 (1986).
Report on the "1er Colloque national de l'Academie des Sciences", held at Paris, 29 – 31 January 1986.

011.014 **Scientific session of the Department of General Physics and Astronomy and the Department of Nuclear Physics of the USSR Academy of Sciences (24 – 25 April 1985).**
Usp. fiz. nauk, Tom 147, No. 2, p. 421 – 427 (1985). In Russian. From Ref. Zh., 51. Astron., 3.51.23 (1986).

011.015 **Conference on "New investigations of the solar system", held in Houston, USA, March 1985.**
V. V. Shevchenko.
Zemlya Vselennaya, No. 1, p. 53 – 57 (1986). In Russian.

011.016 **All–Union student conference "Instationary stars", held in Odessa, 15 – 17 May 1985.**
V. G. Karetnikov, I. L. Andronov.
Zemlya Vselennaya, No. 2, p. 43 – 44 (1986). In Russian.

011.017 **Scientific session of the Department of General Physics and Astronomy and of the Department of Nuclear Physics of the USSR Academy of Sciences.**
Usp. fiz. nauk, Tom 147, No. 3, p. 621 – 625 (1985). In Russian. Abstr. in Ref. Zh., 51. Astron., 4.51.28 (1986).

011.018 **The XVIth lunar–planetary conference, held in Houston, 11 – 15 March 1985.**
Yu. A. Shchukolyukov.
Vestn. Akad. Nauk SSSR., No. 11, p. 108 – 110 (1985). In Russian. Abstr. in Ref. Zh., 62. Issled. Kosm. Prostranstva, 4.62.5 (1986).

011.019 **Chapman conference on magnetotail physics.**
A. T. Y. Lui.
Johns Hopkins APL Tech. Dig., Vol. 7, No. 1, p. 113 – 116 (1986).
Report on a conference held 1985 October 28 – 31.

011.020 **The physics of accretion onto compact objects. Report on a meeting, held at Tenerife, 21 – 25 April 1986.**
A. Pollock.
EXOSAT Express, No. 16, p. 9 – 17 (1986).

011.021 **Resolutions adopted by the Second Asian–Pacific Regional Meeting on Astronomy (Bandung 1981).**
Proceedings of the Second Asian–Pacific Regional Meeting on Astronomy, p. 567 (1984). – See Abstr. 012.031.

011.022 **Astronautics at Stockholm.**
L. R. Shepherd, L. J. Carter.
Spaceflight, Vol. 28, No. 2, p. 76 – 78 (1986).
Report on the 36th International Astronautical Congress, held at Stockholm, Sweden, October 1985.

011.023 **A working conference on quasars held in Leningrad, 13 – 14 February 1985.**
S. A. Levshakov.
Sov. Astron., Vol. 29, No. 5, p. 604 – 605 (1985). English translation of 40.011.015.

011.024 **The 3rd All–Union conference on coordination of ground–based and space observations of X–ray sources, held in Leningrad, 17 – 20 September 1985.**
T. S. Khruzina.
Astron. Zh., Tom 63, Vyp. 3, p. 617 – 619 (1986). In Russian. English translation in Sov. Astron., Vol. 30, No. 3.

011.025 **Conference on the variability of radiation in active galactic nuclei, held at the Crimean Astrophysical Observatory, 4 – 6 June 1985.**
V. I. Pronik.
Astron. Zh., Tom 63, Vyp. 3, p. 619 – 621 (1986). In Russian. English translation in Sov. Astron., Vol. 30, No. 3.

011.026 **IAU Symposium No. 115, "Star forming regions".**
N. Kaifu.
Astron. Her., Vol. 79, No. 6, p. 154 – 156 (1986). In Japanese.

011.027 **A solution to the solar neutrino puzzle?**
M. M. Waldrop.
Science, Vol. 232, No. 4750, p. 578 (1986).
Report on the Tenth International Workshop on Weak Interactions, Savonlinna, Finland, 16 – 25 June 1985.

011.028 **Venus is looking more like Earth than Mars.**
R. A. Kerr.
Science, Vol. 232, No. 4751, p. 709 – 710 (1986).
Report on the Seventeenth Lunar and Planetary Science Conference, 17 – 21 March 1986, Houston, Texas.

011.029 **Observatories of the future.**
M. G. Edmunds.
Nature, Vol. 321, No. 6068, p. 378 – 379 (1986).
This note reports on a meeting discussing future interferometric gravitational wave detectors. The meeting was held at Cardiff University on 17 February 1986.

011.030 **The double tilt of Uranus.**
F. Bagenal.
Nature, Vol. 321, No. 6073, p. 809 – 810 (1986).
This note summarizes the major scientific results of the Voyager 2 fly–by of Uranus, which were presented at a meeting of the American Geophysical Union, held at Baltimore, Maryland, 19 – 22 May 1986.

011.031 **The VIIIth national conference of Yugoslav astronomers, held in Priština, 23 – 27 September 1985.**
A. Tomić, S. Ninković.
Vasiona, Année 34, No. 1 – 2, p. 16 – 18 (1986). In Croatian.

011.032 **First colloquium of Yugoslav astrophysicists, held in Ljubljana, 13 – 14 February 1986.**
M. S. Dimitrijević.
Vasiona, Année 34, No. 1 – 2, p. 18 – 21 (1986). In Serbian.

011.033 **Spring MIST meeting, 1985.**
P. A. Hadjiry, M. J. Laird, M. Lockwood, R. P. Rijnbeek.
Q. J. R. Astron. Soc., Vol. 27, No. 2, p. 194 – 204 (1986).

011.034 **Summary report on the International Workshop "Catastrophic disruption of asteroids and satellites" (Pisa, Italy, July 30 – August 2, 1985).**
D. R. Davis, P. Farinella, P. Paolicchi, V. Zappalà.
Mem. Soc. Astron. Ital., Vol. 57, No. 1, p. 7 – 10 (1986). – See Abstr. 012.061.

011.035 **Physics of nebulae and interstellar matter. Conference held at the Abastumani Astrophysical Observatory 14 – 16 May 1985.**
N. G. Bochkarev, G. M. Rudnitskij.
Sov. Astron., Vol. 29, No. 6, p. 720 – 722 (1985). English translation of 40.011.038.

011.036 **Council on Training of Personnel in Astronomy, plenum held in Alma–Ata, 27 – 30 May 1985.**
V. A. Hagen–Thorn.
Sov. Astron., Vol. 29, No. 6, p. 723 – 724 (1985). English translation of 40.011.039.

011.037 **Highlights of the 1985 URSI/COSPAR workshop on the IRI.**
Y. V. Ramanamurty.
Adv. Space Res., Vol. 5, No. 10, p. 3 – 7 (1985). – See Abstr. 012.064.
Some of the presentations made at the IRI workshop held at Louvain in 1985 have not appeared in the proceedings. Some discussions which took place during the workshop could be of interest in future work. This article tries to highlight the IRI modelling aspects in the four principal height regions of the ionosphere.

011.038 **L'Académie royale des Sciences et la figure de la Terre.**
J.–J. Levallois.
C. R. Acad. Sci., Sér. Gén., Vie Sci., Tome 3, No. 3, p. 261 – 301 (1986).
Report on the 1er Colloque national de l'Académie des Sciences, held at Paris, 29 – 31 January 1986.

011.039 **The case of Galileo.**
K. Rudnicki.
Postepy Astron., Tom 32, Zesz. 2, p. 175 – 179 (1984). In Polish.
Report on a conference, held at Cracow, May 24 – 26, 1984.

011.040 **Flare Build–up Study Workshop.**
V. Gaizauskas.
Sol. Phys., Vol. 105, No. 1, p. 67 – 72 (1986).
This final Workshop of the Flare Build–up Study (FBS) aimed for an up–to–date synthesis of the preflare state based on new knowledge gained during and since the Solar Maximum Year. More joint discussions were held than was customary at previous FBS Workshops among the study groups. Consensus was possible on some broad issues, but for the thornier aspects of preflare activity the Workshop had to settle for tentative conclusions and to redefine goals for improved future studies. Some of the highlights are summarized.

011.041 **IAU Symposium 109, Gainesville, "Astrometric techniques". Concluding remarks.**
G. Westerhout.
Astrometric techniques, p. 799 – 803 (1986). – See Abstr. 012.068 (IAU Symp. No. 109).

011.042 **Berichte von den Workshops.**
M. Schüssler, J. Trümper, D. Offermann, R. Porcas.
Mitt. Astron. Ges., No. 65, p. 149 – 153 (1986). – See Abstr. 012.072.
Workshop I: Solar/stellar activity and winds (*M. Schüssler*), Workshop II: "ROSAT", Stand des Projekts (*J. Trümper*), Workshop III: Variabilität der mittleren Atmosphäre – Vorbereitung eines Shuttle–Experiments (*D. Offermann*), Workshop IV: VLBI of galactic objects (*R. Porcas*).

012 Proceedings of Colloquia, Congresses, Meetings, Symposia

012.001 **Third Asian–Pacific Regional Meeting of the International Astronomical Union, held at Kyoto, Japan, 30 September – 5 October 1984.**
M. Kitamura, E. Budding (Editors), with remarks by D. C. Morton, welcome addresses by T. Sawada, R. M. West, Y. Kozai, T. Kogure and a brief summary of topics by D. C. Morton, T. Kogure.
Part 1: Astrophys. Space Sci., Vol. 118, Nos. 1/2, 10+537 pp. (1986). Available also as hardbound edition, Price Dfl. 230.00, US$ 98.00, £ 63.75. ISBN 90–277–1867–9.
Part 2: Astrophys. Space Sci., Vol. 119, No. 1, 10+265 pp. (1986). Available also as hardbound edition, Price Dfl. 115.00, US$ 49.00, £ 31.95. ISBN 90–277–2210–2.
Supplement: Astronomical education in the Asian–Pacific region. The Organizing Comittees of the Third Asian–Pacific Regional Meeting of the IAU, Kyoto, Japan. 50 pp. (1986).
The individual contributions are included in their corresponding subject categories – see abstracts 009.001 – 009.006, 013.002 – 013.008, 014.039 – 014.045, 022.006, 032.001 – 032.003, 033.001 – 033.003, 034.001, 035.001, 051.002, 062.004 – 062.006, 063.007, 063.008, 064.006 – 064.009, 065.007 – 065.012, 067.009 – 067.011, 072.003, 073.004 – 073.007, 074.002, 074.003, 075.002, 076.001, 077.001, 077.002, 080.005, 080.006, 106.002, 107.001, 107.002, 112.007 – 112.010, 114.011, 114.012, 115.002, 115.003, 116.003, 117.019 – 117.029, 118.002, 122.007, 122.008, 124.141, 125.004, 126.005, 131.019 – 131.032, 132.004, 132.005, 142.003, 142.004, 151.006 – 151.012, 154.003, 155.012 – 155.020, 156.003, 157.015 – 157.019, 158.018 – 158.022, 159.016 – 159.024, 160.006 – 160.009, 161.018 – 161.021.

012.002 **Relativity in celestial mechanics and astrometry. High precision dynamical theories and observational verifications. Proceedings of the 114th Symposium of the International Astronomical Union, held in Leningrad, USSR, May 28 – 31, 1985.**
J. Kovalevsky, V. A. Brumberg (Editors), with introductory remarks by J. Kovalevsky, and a general discussion by V. A. Brumberg.
D. Reidel Publishing Company, Dordrecht – Boston – Lancaster – Tokyo. 18+426 pp. Price Dfl. 160.00, US$ 64.95, £ 44.50 (1986). ISBN 90–277–2189–0 cloth, ISBN 90–277–2190–4 paper.
The individual contributions are included in their corresponding subject categories – see abstracts 021.001, 036.002, 036.003, 041.002 – 041.007, 042.001 – 042.003, 043.001 – 043.003, 044.001 – 044.003, 046.001, 046.002, 052.001, 052.002, 066.020 – 066.034, 080.007, 081.003, 081.004, 091.004 – 091.006, 094.001, 094.002, 098.001, 099.002 – 099.005, 103.602, 111.001, 143.002.

012.003 **14th International Academy of Astronautics Review Meeting on communication with extraterrestrial intelligence, held in Stockholm, Sweden, October 1985.**
J. Br. Interplanet. Soc., Vol. 39, No. 3, p. 137 – 143 (1986).
The individual contributions are included in their corresponding subject categories – see abstracts 015.004 – 015.016, 131.040.

012.004 **Particles and gravity. Proceedings of the 8th Johns Hopkins Workshop on Current Problems in Particle Theory, held at Baltimore, USA, 20 – 22 June 1984.**
G. Domokos, S. Kovesi–Domokos (Editors).
World Scientific Publishing Co. Pte. Ltd., Singapore. 9+257 pp. Price US$ 38.00 (1984). ISBN 9971–966–90–5.
The individual contributions within the subject scope of Astronomy and Astrophysics Abstracts are included in their corresponding categories – see abstracts 061.004 – 061.007, 066.037 – 066.040, 161.027 – 161.034.

012.005 **Supersymmetry, supergravity and related topics. Proceedings of the XVth GIFT International Seminar on Theoretical Physics, held at Sant Feliu de Guixols, Girona, Spain, 4 – 9 June 1984.**
F. del Aguila, J. A. de Azcárraga, L. E. Ibáñez (Editors).
World Scientific Publishing Co. Pte., Ltd., Singapore. 8+355 pp. Price US$ 52.00 (1985). ISBN 9971–966–79–4.
The individual contributions within the subject scope of Astronomy and Astrophysics Abstracts are included in their corresponding categories – see abstracts 061.008 – 061.011, 066.041, 161.035, 161.036.

012.006 **Cosmic radiation in contemporary astrophysics. Proceedings of a NATO Advanced Study Institute, held at Erice, Sicily, Italy, November 1984.**
M. M. Shapiro (Editor).
NATO Advanced Science Institutes. Series C, Mathematical and Physical Sciences, Vol. 162. D. Reidel Publishing Company, Dordrecht – Boston – Lancaster – Tokyo. 8+274 pp. Price Dfl. 125.00, US$ 49.00, £ 34.75 (1986). ISBN 90–277–2144–0.
The individual contributions are included in their corresponding subject categories – see abstracts 061.012, 067.015, 067.016, 142.006 – 142.008, 143.003, 143.004, 144.012 – 144.020, 159.026.

012.007 **Cosmical gas dynamics. Proceedings of the Manchester Conference, 15 – 19 April 1985.**
F. D. Kahn (Editor).
VNU Science Press, PO Box 2073, 3500 GB Utrecht, The Netherlands. 7+329 pp. Price DM 120.00 (1985). ISBN 90–6764–053–0.
The individual contributions are included in their corresponding subject categories – see abstracts 062.013 – 062.015, 064.011, 065.015, 112.011, 121.007 – 121.010, 124.001, 125.011, 125.012, 131.048 – 131.054, 144.021 – 144.023, 151.015 – 151.017, 155.025 – 155.027, 157.024 – 157.026, 158.031, 158.032.

012.008 **Protostars & planets II. Based on a conference held in Tucson, Arizona, January 1984.**
D. C. Black, M. S. Matthews (Editors), with a foreword by C. J. Lada.
The University of Arizona Press, Tucson, Arizona, USA. 20+1293 pp. Price US$ 45.00 (1985) ISBN 0–8165–0950–6. Additional papers of this conference are published in Icarus, Vol. 61, No. 1, p. 1 – 59 (1985). – See Abstr. 39.012.004.
Reviews in Science, Vol. 232, No. 4753, p. 1037; 1986 (*G. Wynn–Williams*); Sky Telesc., Vol. 71, No. 4, p. 369 (1986).
The individual contributions are included in their corresponding subject categories – see abstracts 065.025, 091.010, 102.008, 105.008 – 105.010, 107.011 – 107.021, 112.019, 121.012 – 121.015, 131.071 – 131.091.

012.009 **The connection between nonradial pulsations and stellar winds in massive stars. Invited papers presented at a workshop held at the University of Colorado, Boulder, Colorado, April 17 – 19, 1985.**
Publ. Astron. Soc. Pac., Vol. 98, No. 599, p. 29 – 55 (1986).
The individual contributions are included in their corresponding subject categories – see abstracts 064.016, 064.017, 065.033, 112.027, 114.027, 116.010, 122.025, 122.026.

012.010 **Asteroids, comets, meteors II. Proceedings of a meeting, held at the Astronomical Observatory of the Uppsala University, Uppsala, Sweden, 3 – 6 June 1985.**
C.-I. Lagerkvist, B. A. Lindblad, H. Lundstedt, H. Rickman (Editors).
Astronomiska Observatoriet, Box 515, S–75120 Uppsala, Sweden. 14+620 pp. Price SEK 400.00. (1986). ISBN 91–506–0492–9.
The individual contributions are included in their corresponding subject categories – see abstracts 022.045 – 022.047, 036.047 – 036.050, 051.037, 074.020, 098.017 – 098.050, 102.018 – 102.037,

103.007 – 103.021, 103.241, 103.261, 103.281, 103.301, 103.631 – 103.638, 104.009 – 104.022, 106.023, 106.024, 107.030 – 107.034.

012.011 **Structure and evolution of active galactic nuclei. Proceedings of an international meeting, held at Trieste, Italy, 10 – 13 April 1985.**
G. Giuricin, F. Mardirossian, M. Mezzetti, M. Ramella (Editors), with welcoming addresses by F. Fusaroli and A. Salam.
Astrophysics and Space Science Library, Vol. 121. D. Reidel Publishing Company, Dordrecht – Boston – Lancaster. 26+772 pp. Price Dfl. 260.00, US$ 99.00, £ 72.25 (1986). ISBN 90–277–2155–6.
The individual contributions are included in their corresponding subject categories – see abstracts 036.051, 061.023, 061.024, 063.045 – 063.047, 067.041 – 067.045, 117.108, 155.046, 158.091 – 158.141, 159.051 – 159.067, 160.043 – 160.045, 161.079, 161.080.

012.012 **The Third International Conference on the History of Chinese Science, held at Beijing, 20 – 25 August 1984. Abstracts of presented papers.**
Z.–z. Xi (Editor).
Archaeoastronomy (U.S.A.), Vol. 7, Nos. 1 – 4, p. 28 – 32 (1984).
See abstracts 004.021 – 004.033.

012.013 **Proceedings of the nineteenth symposium on celestial mechanics, held at Akashi, Japan, 14 – 15 February 1986.**
H. Kinoshita, H. Nakai (Editors).
Tokyo Astronomical Observatory, Tokyo, Japan. 3+112 pp. (1986).
The individual contributions are included in their corresponding subject categories – see abstracts 021.006 – 021.008, 041.020, 042.045, 043.005, 043.006, 044.017, 052.010, 081.019, 098.054 – 098.058, 099.031, 099.032, 100.019, 103.645, 151.051, 151.052.

012.014 **Report on the VIII. International Planetarium Directors Conference, held at Stuttgart, Berlin (West), Hamburg, 28 August – 2 September 1984.**
T. W. Kraupe (Editor).
Carl Zeiss, Oberkochen – Planetarium Stuttgart, Stuttgart, F.R. Germany. 201 pp. (1985).
The individual contributions are included in their corresponding subject categories – see abstracts 009.012 – 009.031, 014.019, 079.002.

012.015 **Spectral evolution of galaxies. Proceedings of the 4th workshop of the Advanced School of Astronomy of the "Ettore Majorana" Centre for Scientific Culture, held at Erice, Italy, 12 – 22 March 1985.**
C. Chiosi, A. Renzini (Editors).
Astrophysics and Space Science Library, Vol. 122. D. Reidel Publishing Company, Dordrecht – Boston – Lancaster – Tokyo. 12+490 pp. Price Dfl. 190.00, US$ 78.00, £ 52.75 (1986). ISBN 90–277–2187–4.
The individual contributions are included in their corresponding subject categories – see abstracts 061.027, 065.050, 131.152, 155.049 – 155.051, 156.014, 157.077 – 157.107.

012.016 **The sun and the heliosphere in three dimensions. Proceedings of the 19th ESLAB Symposium, held at Les Diablerets, Switzerland, 4 – 6 June 1985.**
R. G. Marsden (Editor), with summary remarks by L. A. Fisk.
Astrophysics and Space Science Library, Vol. 123. D. Reidel Publishing Company, Dordrecht – Boston – Lancaster – Tokyo. 12+525 pp. Price Dfl. 195.00, US$ 79.00, £ 54.25. (1986). ISBN 90–277–2198–X.
The individual contributions are included in their corresponding subject categories – see abstracts 051.042, 051.043, 073.045, 074.032 – 074.049, 076.004 – 076.006, 078.006 – 078.009, 106.037 – 106.055, 144.048 – 144.054.

012.017 **Fundamental astronomy and solar system dynamics. Invited papers honoring Prof. Walter Fricke on the occasion of his 70th birthday. A meeting held at Austin, Texas, USA, 27 March 1985.**
R. L. Duncombe, J. H. Lieske, P. K. Seidelmann (Editors), with a welcoming address by J. P. Lamb.
Celest. Mech., Vol. 37, No. 3, p. I – IX, 199 – 348 (1985). Available also as hardbound edition. Price Dfl. 100.00, US$ 47.00, £ 31.95. ISBN 90–277–2268–4.
The individual contributions are included in their corresponding subject categories – see abstracts 005.012, 036.066, 041.021 – 041.023, 043.007 – 043.011, 045.001, 051.051, 098.065, 151.061.

012.018 **The atmospheres of Saturn and Titan. Proceedings of an international workshop held at Alpbach, Austria, 16 – 19 September 1985.**
E. Rolfe, B. Battrick (Editors), with a foreword by S. J. Bauer.
ESA Spec. Publ., ESA SP–241, 7+311 pp. Price Dfl. 55.00, US$ 18.00, DM 50.00. (1985).
The individual contributions are included in their corresponding subject categories – see abstracts 004.068, 015.024, 035.029 – 035.033, 051.052 – 051.056, 052.011, 091.022, 100.030 – 100.052.

012.019 **A cosmic X–ray spectroscopy mission. Proceedings of a workshop held at Lyngby, Denmark, 24 – 26 June 1985.**
N. Longdon, O. Melita (Editors), with an introduction by R. M. Bonnet, and a summary of recommendations for a European mission on high throughput X–ray spectroscopy by J. A. M. Bleeker.
ESA Spec. Publ., ESA SP–239, 6+308 pp. Price FF 150.00 (1985).
The individual contributions are included in their corresponding subject categories – see abstracts 031.009 – 031.011, 035.034 – 035.050, 036.067, 051.057, 067.059 – 067.062, 112.057, 116.027, 117.130 – 117.133, 125.052 – 125.054, 131.187, 142.032 – 142.034, 153.014, 153.015, 156.018, 157.126, 158.175, 160.059, 160.060.

012.020 **Radioastronomical equipment. Proceedings of the 17th All–Union conference, held in Erevan, 10 – 12 October 1985.**
Izdatel'stvo Akad. Nauk ArmSSR, Erevan. 95 pp. (1985). In Russian.
Review in Ref. Zh., 51. Astron., 3.51.55 (1986).

012.021 **Solar flares and coronal physics using P/OF as a research tool. Proceedings of a workshop held at Marshall Space Flight Center, Huntsville, Alabama, USA, 8 – 10 May 1985.**
E. Tandberg–Hanssen, R. M. Wilson, H. S. Hudson (Editors), with an introduction by H. S. Hudson, E. Tandberg–Hanssen, R. M. Wilson, J. R. Dabbs, and a workshop summary by H. S. Hudson, E. Tandberg–Hanssen.
NASA Conf. Publ., NASA CP–2421. 7+303 pp. (1986).
The individual contributions are included in their corresponding subject categories – see abstracts 013.029, 013.030, 035.051, 051.059, 051.060, 071.024, 073.076 – 073.083, 074.065 – 074.070, 076.013, 077.023 – 077.027, 080.048.

012.022 **The Minnesota lectures on active galactic nuclei.**
T. W. Jones.
Publ. Astron. Soc. Pac., Vol. 98, No. 600, p. 129 – 184 (1986).
These reviews are based on a graduate seminar held at the University of Minnesota during the spring quarter 1985. Each of the authors was invited for a one–week visit to Minneapolis to discuss active galactic nuclei with Minnesota students and faculty.
See abstracts 155.065, 157.136, 157.137, 158.179 – 158.184.

012.023 **The Jovian atmospheres. Review and contributed papers presented at a conference held at the Goddard Institute for Space Studies, New York, N.Y., USA, 6 – 8 May 1985.**
With an introduction by J. A. Burns.
Icarus, Vol. 65, Nos. 2/3, p. 159 – 466 (1986).
The individual contributions are included in their corresponding subject categories – see abstracts 062.055, 099.048 – 099.056, 101.045 – 101.047.

012.024 **Dust on Mars. Papers presented at a MECA workshop held at Arizona State University, Tempe, Ariz., USA, in February 1985.**
With an introduction by J. A. Burns.
Icarus, Vol. 66, No. 1, p. 1 – 142 (1986).
The individual contributions are included in their corresponding subject categories – see abstracts 022.062, 022.063, 097.013 – 097.021.

012.025 **Binary and multiple systems. Proceedings of the Platinum Jubilee Symposium of Nizamiah Observatory, held at Hyderabad, India, 22 – 24 November 1983.**
K. D. Abhyankar, G. M. Ballabh (Editors).
Centre of Advanced Study in Astronomy, Osmania University, Hyderabad, India. 8 + 162 pp. (1984).
The individual contributions are included in their corresponding subject categories – see abstracts 098.076, 102.059, 117.155 – 117.157, 118.015, 131.199, 131.200, 155.070, 157.140, 159.087 – 159.089, 161.107.

012.026 **Seismology of the Sun and the distant stars. Proceedings of a NATO Advanced Research Workshop, held at Cambridge, UK, 17 – 21 June 1985.**
D. O. Gough (Editor).
NATO Advanced Science Institutes, Series C, Vol. 169. D. Reidel Publishing Company, Dordrecht – Boston – Lancaster – Tokyo. 12 + 479 pp. Price Dfl. 195.00, US$ 79.00, £ 54.25 (1986). ISBN 90–277–2196–3.
The individual contributions are included in their corresponding subject categories – see abstracts 034.059 – 034.064, 036.083, 036.084, 062.057, 065.067 – 065.069, 072.047, 080.050 – 080.080, 112.065, 115.016, 118.016, 122.096, 122.097, 126.044.

012.027 **Proceedings of the Symposium on Application of Space Techniques to Astronomy and Geophysics, held at Tokyo, Japan, 30 January – 1 February 1986.**
H. Kinoshita, I. Murata, K. Nakajima (Editors).
University of Tokyo Press, Tokyo, Japan. 8 + 202 pp. (1986).
The individual contributions within the subject scope of Astronomy and Astrophysics Abstracts are included in their corresponding categories – see abstracts 002.041, 013.031, 034.193, 036.085, 044.023 – 044.029, 045.002 – 045.010, 051.061, 051.062, 082.043.

012.028 **Re–entry of space debris. Proceedings of an ESA workshop held at the European Space Operations Centre, Darmstadt, F.R. Germany, 24 – 25 September 1985.**
N. Longdon (Editor).
ESA Spec. Publ., ESA SP–246. 3 + 85 pp. Price FF 70.00 (1986).
The individual contributions within the subject scope of Astronomy and Astrophysics Abstracts are included in their corresponding categories – see abstracts 052.012 – 052.018, 082.046 – 082.048.

012.029 **Galactic structure and variable stars. Proceedings of the six–year co–operation in astronomy between Indonesia and Japan (1979 – 1984).**
T. Kogure, B. Hidayat (Editors).
Department of Astronomy, Faculty of Science, Kyoto University, Kyoto 606, Japan. 3 + 117 pp (1985).
The individual contributions are included in their corresponding subject categories – see abstracts 013.037 – 013.040, 034.077, 067.070, 112.071, 114.081, 131.223, 131.224, 151.075, 155.071, 157.155.

012.030 **Interrelationships among circumstellar, interstellar, and interplanetary dust. Proceedings of a workshop held at the Aspen Institute's Wye Plantation Conference Center, Wye, Maryland, USA, 27 February – 1 March 1985.**
J. A. Nuth III, R. E. Stencel (Editors).
NASA Conf. Publ., NASA CP–2403. 14 + 345 pp. (1986).
The individual contributions are included in their corresponding subject categories – see abstracts 022.084 – 022.087, 051.067, 051.068, 064.034, 102.064, 105.195, 106.058, 106.059, 112.072 – 112.077, 131.225 – 131.239, 132.034, 134.035.

012.031 **Proceedings of the Second Asian–Pacific Regional Meeting on Astronomy, held at Bandung, Indonesia, 24 – 29 August 1981.**
B. Hidayat, M. W. Feast (Editors).
Tira Pustaka Publishing House, P.O. Box 3056, Jakarta, Indonesia. 12 + 571 pp. (1984).
The individual contributions are included in their corresponding subject categories – see abstracts 009.037, 009.038, 011.021, 013.041, 013.042, 014.026 – 014.029, 015.030, 015.031, 031.013, 032.026, 036.103, 036.104, 063.064, 064.039, 066.103 – 066.105, 067.071, 071.027, 072.050, 073.089, 082.049, 112.078, 112.079, 116.034, 117.167 – 117.170, 118.021, 118.022, 122.105 – 122.107, 126.047, 131.240 – 131.243, 132.035, 133.011, 142.036, 151.076, 151.077, 154.027, 155.072 – 155.085, 156.022, 157.156 – 157.160, 158.196 – 158.199, 159.094, 159.095, 160.068, 160.069, 161.121 – 161.123.

012.032 **Proceedings of the international conference "Use of observations of artificial earth satellites for geodesy and geophysics", held in Suzdal, 13 – 17 September 1982.**
A. G. Masevich (Editor).
Publications of scientific results of Intercosmos–cooperation. Nablyud. Iskusstv. Sputnikov Zemli, No. 21, Chast' I, 328 pp. (1982) = Nablyud. Iskusstv. Nebesn. Tel, Spets. Vyp. (1984). In Russian and English.
The individual contributions are included in their corresponding subject categories – see abstracts 036.110, 036.112, 045.011 – 045.015, 045.041, 045.042, 052.019 – 052.033, 052.064 – 052.066, 053.013, 081.024 – 081.028, 082.050 – 082.055, 083.013, 085.016, 085.017.

012.033 **X–ray astronomy '84. International symposium on X–ray astronomy, held at Bologna, Italy, 26 – 30 June 1984.**
M. Oda, R. Giacconi (Editors).
Institute of Space and Astronautical Science, Komaba, Meguro–ku, Tokyo 153, Japan. 19 + 506 pp. (1985).
The individual contributions are included in their corresponding subject categories – see abstracts 036.116, 051.084,062.063, 063.066 – 063.069, 067.072, 067.073, 116.037 – 116.041, 117.174 – 117.185, 119.054, 121.045, 122.110, 125.063 – 125.070, 126.049 – 126.051, 134.038, 142.037 – 142.065, 155.089, 157.168 – 157.171, 158.207 – 158.214, 159.098 – 159.102, 160.074 – 160.078, 161.126.

012.034 **Japan – U.S. seminar on galactic and extragalactic compact X–ray sources, held at Tokyo, Japan, January 1985.**
Y. Tanaka, W. H. G. Lewin (Editors).
Institute of Space and Astronautical Science, Komaba, Meguro–ku, Tokyo, Japan. 8 + 325 pp. (1985).
The individual contributions are included in their corresponding subject categories – see abstracts 067.076 – 067.083, 117.187 – 117.191, 126.052, 126.053, 142.067 – 142.084, 158.221 – 158.225, 159.108.

012.035 **Scientific–theoretical conference of young scientists and specialists of the Tadzhik SSR. Department of physical and mathematical sciences.**
A. A. Adkhamov (Editor).
Donish, Dushanbe. 137 pp. (1985). In Russian.
Review in Ref. Zh., 51. Astron., 5.51.48 (1986).

012.036 **Sir Arthur Eddington Centenary Symposium. Vol. 3: Gravitational radiation and relativity. Proceedings of a symposium, held at Nagpur, India, 21 – 27 January 1984.**
J. Weber, T. M. Karade (Editors).
World Scientific Publishing Co. Pte. Ltd., Singapore. 9 + 465 pp. Price US$ 42.00 (1986). ISBN 9971–978–88–1.
The individual contributions are included in their corresponding subject categories – see abstracts 061.036 – 061.044, 066.119 – 066.126, 067.089, 081.032, 161.148 – 161.150.

012.037 **Radioastronomical investigations of the solar system. Abstracts of a conference, held in Odessa in September 1985.**
Kiev. 57 pp. (1985). In Russian.
Review in Ref. Zh., 51. Astron., 6.51.71 (1986).

012.038 **Proceedings of the Third Seminar on Quantum Gravity, held in Moscow, USSR, 23 – 25 October 1984.**
M. A. Markov, V. A. Berezin, V. P. Frolov (Editors).
World Scientific Publishing Co. Pte. Ltd., Singapore, 9 + 705 pp. Price US$ 78.00 (1985). ISBN 9971–978–90–3.
Review in Astron. Tidsskr., Årg. 19, Nr. 2, p. 94 (1986).
The individual contributions are included in their corresponding subject categories – see abstracts 061.051 – 061.059, 066.137 – 066.155, 067.094, 161.181 – 161.189.

012.039 **Eruptive phenomena in stars. Proceedings of a symposium, held at Budapest, 10 – 12 September 1985.**
L. Szabados (Editor).
Commun. Konkoly Obs., No. 86, p. 331 – 446 (1986). ISBN 963–8361–22–0.
The individual contributions are included in their corresponding subject categories – see abstracts 036.128, 064.043 – 064.045, 067.095, 116.049 – 116.051, 117.211 – 117.219, 121.049, 122.117 – 122.125, 123.005, 124.008, 124.243, 124.244, 125.078.

012.040 **Synthetic photometry. Proceedings of the Joint Meeting of the I.A.U. Commissions 25, 29, 36, and 45 held at the XIXth General Assembly of the IAU, New Delhi, India, November 20, 1985.**
R. Buser (Editor).
Space Telesc. Sci. Inst., Prepr. Ser., No. 100, 5 + 45 pp. (1986). To appear in Highlights of Astronomy, Vol. 7.
The individual contributions are included in their corresponding subject categories – see abstracts 035.059, 036.130, 036.131, 064.046, 113.046, 113.047.

012.041 **Supersymmetry and its applications: superstrings, anomalies and supergravity. Proceedings of the Nuffield Workshop, held at Cambridge, U.K., 23 June – 14 July 1985.**
G. W. Gibbons, S. W. Hawking, P. K. Townsend (Editors).
Cambridge University Press, Cambridge – London – New York – New Rochelle – Melbourne – Sydney. 14 + 481 pp. Price £ 27.50, US$ 49.50 (1986). ISBN 0–521–30721–X.
Supersymmetry is a symmetry of gauge invariance which connects bosons (particles with integer spin) with fermions (particles with half integer spin). The study of supersymmetry has led to the construction of supergravity and superstring theories which seem to be the best hope for a complete unified theory of all physical interactions including gravity. The present volume presents a co–ordinated account of the recent advances in supersymmetric theories. The topics of supersymmetry, supergravity and superstrings are introduced in wide ranging review articles, which are supplemented by research papers describing the latest developments (28 contributions).

012.042 **Galaxies, quasars and cosmology. Lectures presented at the First Equatorial School of Relativistic Astrophysics, held at Bogota, Colombia, 13 – 24 February 1984.**
L. Z. Fang, R. Ruffini (Editors).
Advanced Series in Astrophysics and Cosmology, Vol. 2. World Scientific Publishing Co. Pte. Ltd., Singapore. 7 + 221 pp. Price US$ 26.00 cloth (1985). ISBN 9971–978–93–8 cloth, ISBN 9971–50–083–3 paper.
The individual contributions are included in their corresponding subject categories – see abstracts 004.081, 067.097, 151.091, 151.092, 154.044, 157.194, 161.191 – 161.195.

012.043 **Bridge between control science and technology. Proceedings of the Ninth Triennial World Congress of IFAC, held at Budapest, Hungary, 2 – 6 July 1984.**
J. Gertler, L. Keviczky (Editors).
Pergamon Press, Oxford – New York – Frankfurt – Paris – Sydney – Toronto – Tokyo. (1985).
From Phys. Abstr., Vol. 89, No. 1273, Entry 18626 (1986).

012.044 **Geometrical optics. Proceedings of a conference held at Los Angeles, Calif., USA, 23 – 25 January 1985.**
Proc. SPIE Int. Soc. Opt. Eng., Vol. 531 (1985).
Review in Phys. Abstr., Vol. 89, No. 1275, Entry 23940 (1986).

012.045 **Measurement and effects of surface defects and quality of polish. Proceedings of a SPIE conference, held at Los Angeles, Calif., USA, 21 – 22 January 1985.**
Proc. SPIE Int. Soc. Opt. Eng., Vol. 525 (1985).
Review in Phys. Abstr., Vol. 89, No. 1276, Entry 30525 (1986).

012.046 **12th Annual Conference Plasma Physics Group of the Institute of Physics, held at Glasgow, Scotland, 26 – 28 June 1985.**
Plasma Phys. Controlled Fusion, Vol. 27, No. 12A (1985).
Review in Phys. Abstr., Vol. 89, No. 1277, Entry 36407 (1986).
See abstracts 062.100, 084.043.

012.047 **Recent developments in quantum field theory. Proceedings of the Niels Bohr Centennial Conference, held at Copenhagen, Denmark, 6 – 10 May 1985.**
J. Ambjorn, B. J. Durhuus, J. L. Petersen (Editors).
North–Holland Publishing Company, Amsterdam – New York – Oxford – Tokyo. 8 + 298 pp. (1985). ISBN 0–444–86978–6.
Review in Phys. Abstr., Vol. 89, No. 1277, Entry 36419 (1986).
See abstracts 067.118, 161.283.

012.048 **Nuclear structure 1985. Proceedings of the Niels Bohr Centennial Conference, held at Copenhagen, Denmark, 20 – 24 May 1985.**
R. Broglia, G. Hagemann, B. Herskind (Editors).
North–Holland Publishing Company, Amsterdam – New York – Oxford – Tokyo. 18 + 614 pp. (1985). ISBN 0–444–86979–4.
Review in Phys. Abstr., Vol. 89, No. 1277, Entry 36420 (1986).

012.049 **Réseaux informatiques en astronomie. Comptes Rendus de la Réunion du Conseil Scientifique du CDS, held at Strasbourg, France, 7 – 8 November 1985.**
C. Jaschek (Editor), with an opening address by J. Goormaghtigh.
Bull. Inf. Cent. Données Stellaires, No. 30, p. 1 – 53 (1986).
The individual contributions are included in their corresponding subject categories – see abstracts 002.061, 036.177 – 036.183.

012.050 **Space Station planetology experiments (SSPEX). Proceedings of a workshop held at the U.S. Geological Survey, Flagstaff, Arizona, USA, 20 – 22 June 1985.**
R. Greeley, R. J. Williams (Editors).
NASA Conf. Publ., NASA CP 2424, 5 + 91 pp. (1986).
A meeting of 50 planetary scientists considered the uses of the Space Station to support experiments in their various disciplines. Abstracts (28) present various concepts for impact and aeolian processes, particle formation and interaction, and other planetary science experiments. Summaries of the rationale, hardware concepts, accommodations, and recommendations are included.

012.051 **Proceedings of the Southwest Conference on Optics, held at Albuquerque, N.M., USA, 4 – 8 March 1985.**
Proc. SPIE Int. Soc. Opt. Eng., Vol. 540 (1985).
Review in Phys. Abstr., Vol. 89, No. 1278, Entry 42404 (1986).
See abstracts 031.035, 031.040, 031.048.

012.052 **Proceedings of the optical fabrication and testing workshop: large telescope optics, held at Albuquerque, N.M., USA, 4 – 5 March 1985.**
Proc. SPIE Int. Soc. Opt. Eng., Vol. 542 (1985).
Review in Phys. Abstr., Vol. 89, No. 1278, Entry 42405 (1986).
See abstracts 031.036 – 031.039, 031.041 – 031.047.

012.053 **Shergotty–Nakhla–Chassigny Symposia, 16th Lunar and Planetary Science Conference, 1985.**
Geochim. Cosmochim. Acta, Vol. 50, No. 6, p. 875 – 1091 (1986).
The individual contributions are included in their corresponding subject categories – see abstracts 105.219 – 105.238.

012.054 **Italian astronomy and the scientific potential of Space Telescope. Proceedings of a workshop, held in honour of Riccardo Giacconi at Brugine (Padova), Italy, 8 – 9 February 1984.**
F. Bertola (Editor).
Mem. Soc. Astron. Ital., Vol. 56, No. 4, p. 572 – 768 (1985).
The individual contributions are included in their corresponding subject categories – see abstracts 036.193, 041.032, 051.095, 102.078, 114.125, 116.060, 117.250, 118.032, 118.033, 134.052, 141.021, 151.108, 153.036, 153.037, 154.060, 155.109, 157.239, 158.287 – 158.289, 159.154, 160.112, 161.306 – 161.308.

012.055 **Relativity, supersymmetry and cosmology. Proceedings of the 5th Simposio Latino Americano de Relatividad y Gravitacion – SILARG V –, held at Bariloche, Argentina, 4 – 11 January 1985.**
O. J. Bressan, M. Castagnino, V. H. Hamity (Editors).
World Scientific Publishing Co. Pte. Ltd., Singapore. 9 + 348 pp. Price US$ 32.00, £ 31.80 (1985). ISBN 9971–50–003–5.
The individual contributions are included in their corresponding subject categories – see abstracts 061.096 – 061.100, 066.206 – 066.216, 067.129, 161.309 – 161.314.

012.056 **Proceedings of the 1984 International Conference on Plasma Physics, held at Lausanne, Switzerland, 27 June – 3 July 1984.**
M. Q. Tran, R. J. Verbeek (Editors).
EUR 9708 EN. Comm. Eur. Communities, Luxembourg, Luxembourg. 2 Vols., 1155 pp. (1985).
Review in Phys. Abstr., Vol. 89, No. 1280, Entry 54018 (1986).
See abstracts 004.099, 062.108, 062.111, 067.136, 080.112, 106.074, 144.107.

012.057 **Nucleus–nucleus collisions II. Proceedings of the second international conference, held at Visby, Sweden, 10 – 14 June 1985.**
Nucl. Phys. A, Vol. A447 (1986).
Review in Phys. Abstr., Vol. 89, No. 1280, Entry 54964 (1986).
See abstracts 144.095, 144.096, 161.328.

012.058 **X– and gamma–ray sources and applications. Proceedings of the sixth symposium, held at Ann Arbor, Mich., USA, 21 – 23 May 1985.**
Nucl. Instrum. Methods Phys. Res., Sect. A, Vol. A242, No. 3 (1986).
Review in Phys. Abstr., Vol. 89, No. 1280, Entry 55583 (1986).
See abstracts 035.073, 035.076 – 035.078.

012.059 **Sensor design using computer tools II. Proceedings of a conference, held at Arlington, Va., USA, 11 – 12 April 1985.**
Proc. SPIE Int. Soc. Opt. Eng., Vol. 550 (1985).
From Phys. Abstr., Vol. 89, No. 1280, Entry 60646 (1986).
See abstracts 031.049, 035.079.

012.060 **Extragalactic infrared astronomy. Proceedings of the 4th RAL Workshop on Astronomy and Astrophysics, held at The Cosener's House, Abingdon, UK, 20 – 22 May 1985.**
P. M. Gondhalekar (Editor).
Rutherford Appleton Lab., Rep. RAL 85–086, 8 + 181 pp. (1985).
The individual contributions are included in their corresponding subject categories – see abstracts 157.240 – 157.248, 158.293, 158,294, 159.158, 161.339 – 161.341.

012.061 **Catastrophic disruption of asteroids and satellites. An International Workshop held in Pisa, Italy, July 30 – August 2, 1985.**
D. R. Davis, P. Farinella, P. Paolicchi, V. Zappalà (Editors).
Mem. Soc. Astron. Ital., Vol. 57, No. 1, p. 7 – 114 (1986).
The individual contributions are included in their corresponding subject categories – see abstracts 011.034, 022.128, 098.123, 098.124, 107.046, 107.047.

012.062 **Techniques and concepts of high–energy physics III. Proceedings of the Third NATO Advanced Study Institute, held at St. Croix, Virgin Islands, 2 – 13 August 1984.**
T. Ferbel (Editor).
Plenum Publishing Corporation, New York, USA. 7 + 436 pp. (1985). ISBN 0–306–42106–2.
Review in Phys. Abstr., Vol. 89, No. 1282, Entry 67694 (1986).

012.063 **Proceedings of the 1985 International Symposium on Lepton and Photon Interactions at High Energies, held at Kyoto, Japan, 19 – 24 August 1985.**
M. Konuma, K. Takahashi (Editors).
Organ. Committee of the symposium, Kyoto, Japan. 7 + 977 pp. (1986).
Review in Phys. Abstr., Vol. 89, No. 1283, Entry 73456 (1986).

012.064 **International Reference Ionosphere – status 1985/86. Proceedings of the URSI/COSPAR workshop held in Louvain–la–Neuve, Belgium, 28th October – 1st November 1985.**
K. Rawer, Y. V. Ramanamurty (Editors).
Adv. Space Res., Vol. 5, No. 10, 6 + 131 pp. (1985). ISBN 0–08–034026–1.
The individual contributions within the subject scope of Astronomy and Astrophysics Abstracts are included in their corresponding categories – see abstracts 011.037, 083.020 – 083.032.

012.065 **Proceedings of the Fifteenth Lunar and Planetary Science Conference, held at Houston, Texas, USA, 12 – 16 March 1984. Parts 1 + 2.**
G. Ryder, G. Schubert (Editors).
J. Geophys. Res., Vol. 89, Suppl., 364 pp. (1984); Vol. 90, Suppl., 489 pp. (1985). Price US$ 100.00 cloth, US$ 70.00 paper. ISBN 0–87590–236–7.
Review in Geochim. Cosmochim. Acta, Vol. 50, No. 4, p. 643; 1986 *(J. A. O'Keefe)*.
The individual contributions are included in their corresponding subject categories – see abstracts 022.108, 022.109, 036.204 – 036.207, 081.052 – 081.055, 091.052 – 091.054, 093.060, 093.061, 094.043 – 094.069, 097.034, 097.035, 099.085, 100.064, 102.083, 105.247 – 105.270.

012.066 **Memorias de la Cuarta Reunión Regional Latinoamerica de Astronomía, held in Rio de Janeiro, Brasil, November 18 – 23, 1984.**
J. A. de Freitas Pacheco (Editor).
Rev. Mex. Astron. Astrofis., Vol. 12, Spec. Issue, 452 pp. (1986).
The individual contributions are included in their corresponding subject categories – see abstracts 002.084, 013.062, 022.107, 032.045, 032.046, 033.070, 034.186 – 034.188, 036.208 – 036.212, 041.046 – 041.048, 042.089, 044.030, 062.125 – 062.128, 064.069, 067.150, 074.107, 077.043 – 077.045, 080.114, 101.061, 101.062, 103.154, 103.331, 103.737, 106.076, 112.118, 112.119, 113.060, 114.133 – 114.136, 115.027, 116.067 – 116.070, 117.294, 119.084, 120.030, 120.031, 121.059, 121.060, 125.087, 125.088, 126.075 – 126.077, 131.334 – 131.345, 132.050 – 132.056, 133.017, 134.057 – 134.061, 141.022, 151.114 – 151.117, 152.006, 153.043, 153.044, 154.067, 155.115, 115.116, 156.027, 156.028, 157.263 – 157.267, 158.302 – 158.310, 159.168, 159.169, 160.123, 160.124, 161.395 – 161.398.

012.067 **Radio continua during solar flares. Selected contributions to the workshop held at Duino, Italy, 27 – 31 May 1985.**
A. O. Benz (Editor).
Sol. Phys., Vol. 104, No. 1, 8 + 256 pp. (1986). Available also as hardbound edition. Price Dfl. 165.00, US$ 79.00, £ 53.00. ISBN 90–277–2291–9.
The individual contributions are included in their corresponding subject categories – see abstracts 073.135 – 073.142, 074.108, 077.046 – 077.064, 122.182.

012.068 **Astrometric techniques. Proceedings of the 109th Symposium of the International Astronomical Union, held in Gainesville, Florida, USA, 9 – 12 January 1984.**
H. K. Eichhorn, R. J. Leacock (Editors), J. M. Kerrick (Assistant Editor).
D. Reidel Publishing Company, Dordrecht – Boston – Lancaster – Tokyo. 21 + 838 pp. Price Dfl. 308.00, US$ 144.00, £ 101.75 cloth; Dfl. 130.00, US$ 61.50, £ 42.95 paper (1986). ISBN 90–277–2256–0 cloth, ISBN 90–277–2257–9 paper.
The individual contributions are included in their corresponding subject categories – see abstracts 002.085 – 002.088, 011.041, 013.063 – 013.065, 032.049 – 032.072, 034.190 – 034.192, 035.085, 036.213 – 036.230, 041.049 – 041.079, 043.016 – 043.029, 111.024 – 111.026.

012.069 **Nucleosynthesis and its implications on nuclear and particle physics. Proceedings of a NATO Advanced Research Workshop (Fifth Moriond Meeting), held at Les Arcs, France, 17 – 23 March 1985.**
J. Audouze, N. Mathieu (Editors).
NATO Advanced Science Institutes, Ser. C: Mathematical and Physical Sciences, Vol. 163. D. Reidel Publishing Company, Dordrecht – Boston – Lancaster – Tokyo. 17 + 502 pp. Price Dfl. 195.00, US$ 79.00, £ 54.25 (1986). ISBN 90–277–2173–4.
The individual contributions are included in their corresponding subject categories – see abstracts 061.122 – 061.143, 065.128 – 065.146, 080.116, 080.117, 125.089, 125.090, 131.347, 143.032, 161.400 – 161.406.

012.070 **From nuclei to stars. A meeting in nuclear physics and astrophysics exploring the path opened by H. A. Bethe. Proceedings of the International School of Physics, "Enrico Fermi", Course XCI, held at Varenna, Italy, 18 – 23 June 1984.**
A. Molinari, R. A. Ricci (Editors).
Enrico Fermi International Summer School of Physics, Vol. 91. North–Holland Physics Publishing, Amsterdam – Oxford – New York – Tokyo. 14 + 323 pp. Price US$ 70.50, Dfl. 190.00 (1986). ISBN 0–444–86988–3.
The individual contributions within the subject scope of Astronomy and Astrophysics Abstracts are included in their corresponding categories – see abstracts 005.027, 013.066, 061.144 – 061.146, 065.147, 065.148.

012.071 **Plasma penetration into magnetospheres. Proceedings of a conference held at Colymbari (Chania), Crete, Greece, 23 September – 4 October 1985.**
N. Kylafis, J. Papamastorakis, J. Ventura (Editors).
Mediterranean School on Plasma Astrophysics. Crete University Press, P.O. Box 1527, Iraklion, Crete, Greece. 5 + 227 pp. Price US$ 23.00 (1986).
The individual contributions are included in their corresponding subject categories – see abstracts 062.131, 062.132, 067.151 – 067.155, 080.118, 084.074, 117.299 – 117.303, 142.103, 143.033, 143.034.

012.072 **Kosmische Plasmen. Kleine Körper im Sonnensystem. Bericht über die wissenschaftliche Tagung in Göttingen, 7.–11. April 1986.**
Mitt. Astron. Ges., Nr. 65, 254 pp. Price DM 30.00 (1986).
Available from H. U. Keller, Planetarium Stuttgart, Neckarstr. 47, D–7000 Stuttgart, F.R. Germany.
The individual contributions are included in their corresponding subject categories – see abstracts 002.089, 011.042, 036.232, 036.233, 051.098, 062.133 – 062.135, 063.105, 063.106, 065.150 – 065.152, 067.156 – 067.158, 072.097, 075.051 – 075.053, 091.055, 094.071, 098.128, 103.739, 112.120, 116.072, 121.062, 121.063, 124.014, 124.015, 125.091, 131.349, 131.350, 134.063, 134.064, 144.121, 151.040, 151.119, 151.120, 153.047, 155.118, 156.029, 157.269 – 157.273, 158.315 – 158.318, 159.174 – 159.177, 160.125, 161.409.

012.073 **Astronomical microdensitometry conference. Proceedings of a workshop held at Goddard Space Flight Center, Greenbelt, Maryland, USA, 11 – 13 May 1983.**
D. A. Klinglesmith (Editor).
NASA Conf. Publ., NASA CP–2317. 7 + 432 pp. (1984).
The individual contributions are included in their corresponding subject categories – see abstracts 009.065 – 009.068, 034.196 – 034.216, 036.234 – 036.244.

012.074 **Star–forming dwarf galaxies and related objects. Proceedings of a workshop held at Paris, France, 1 – 3 July 1985.**
D. Kunth, T. X. Thuan, J. Tran Thanh Van (Editors), with introductory remarks by J. Lequeux and concluding remarks by J. Audouze.
Editions Frontières, 91190 Gif sur Yvette, France. 13 + 517 pp. Price FF 440.00, US$ 59.00 (1986). ISBN 2–86332–037–8.
The individual contributions are included in their corresponding subject categories – see abstracts 013.068, 065.153, 065.154, 131.351, 131.352, 142.104, 156.030, 157.274 – 157.307, 160.126, 160.127.

012.075 **Proceedings of the colloquium on kilometric optical arrays in space, held at Cargèse, Corsica, France, 23 – 25 October 1984.**
N. Longdon, O. Melita (Editors).
ESA Spec. Publ., ESA SP–226. 5 + 191 pp. Price FF 100.00 (1985).
The individual contributions are included in their corresponding subject categories – see abstracts 013.069, 031.055, 031.056, 032.033, 034.217, 035.086 – 035.097, 036.245 – 036.256, 051.099, 051.100.

012.076 **The evolution of galactic X–ray binaries. Proceedings of a NATO Advanced Research Workshop, held at Ringberg Castle, Rottach–Egern, F.R. Germany, 17 – 21 June 1985.**
J. Truemper, W. H. G. Lewin, W. Brinkmann (Editors).
NATO Advanced Science Institutes, Ser. C: Mathematical and Physical Sciences, Vol. 167. D. Reidel Publishing Company, Dordrecht – Boston – Lancaster – Tokyo. 13 + 309 pp. Price Dfl. 130.00, US$ 54.00, £ 36.25 (1986). ISBN 90–277–2184–X.
The individual contributions are included in their corresponding subject categories – see abstracts 065.155 – 065.159, 067.159 – 067.161, 117.304 – 117.315, 142.105 – 142.109, 151.039.

012.077 **Proceedings of the international conference on Earth rotation and the terrestrial reference frame, held at Columbus, Ohio, USA, 31 July – 2 August 1985.**
I. I. Mueller (Editor), with opening remarks by J. M. Hollander.
Reports on the MERIT–COTES campaign on Earth rotation and reference systems. MERIT–COTES joint working groups, International Astronomical Union, International Association of Geodesy, International Union of Geodesy and Geophysics, Part II. Two–volume set published by and available from Department of Geodetic Science and Surveying, The Ohio State University, 1958 Neil Avenue, Columbus, Ohio 43210–1247, USA. 12 + 771 pp. Price US$ 30.00 (1985).
The individual contributions are included in their corresponding subject categories – see abstracts 013.070, 036.092, 036.111, 036.257, 043.030, 043.031, 044.095 – 044.143, 045.029 – 045.038, 081.058.

012.078 **Proceedings of the VII national conference of Yugoslav astronomers, held at Belgrade, May 9 – 11, 1984.**
N. Janković, Ž. Dadić (Editors).
Publ. Astron. Soc. "Rudjer Bošković", No. 4, 195 pp. (1985).
The individual contributions are included in their corresponding subject categories – see abstracts 38.032.021, 39.010.321, 004.078, 004.108 – 004.123, 005.040 – 005.049, 010.321, 013.071 – 013.073, 015.051, 034.109, 045.040, 046.023 – 046.025, 161.410.

012.079 **Luminous stars and associations in galaxies. Proceedings of the 116th IAU Symposium, held at Porto Heli, Greece, 26 – 31 May 1985.**
C. W. H. de Loore, A. J. Willis, P. Laskarides (Editors), with a summary talk by J. A. Graham and T. E. Armandroff.
D. Reidel Publishing Company, Dordrecht – Boston – Lancaster – Tokyo. 19 + 533 pp. Price Dfl. 200.00, US$ 94.00, £ 62.75 cloth; Dfl. 85.00, US$ 39.50, £ 27.25 paper (1986). ISBN 90–277–2272–2 cloth; ISBN 90–277–2273–0 paper.
The individual contributions are included in their corresponding subject categories – see abstracts 064.071 – 064.073, 065.160 – 065.174, 112.123 – 112.133, 113.015, 113.040, 113.045, 114.034, 114.037, 114.072, 114.138 – 114.143, 115.030 – 115.032, 117.326 – 117.328, 122.193 – 122.196, 131.354 – 131.357, 132.059, 132.060, 152.010 – 152.012, 155.119 – 155.121, 156.031 – 156.041, 157.309 – 157.320, 158.322, 158.323.

012.080 **Light on dark matter. Proceedings of the first IRAS conference, held in Noordwijk, The Netherlands, 10 – 14 June 1985.**
F. P. Israel (Editor).
Astrophysics and Space Science Library, Vol. 124. D. Reidel Publishing Company, Dordrecht – Boston – Lancaster – Tokyo, 23 + 541 pp. Price Dfl. 210.00, US$ 98.00, £ 66.95 (1986). ISBN 90–277–2254–4.
The individual contributions are included in their corresponding subject categories – see abstracts 002.092, 002.093, 013.075, 022.132, 022.133, 035.100, 036.258, 036.259, 051.104, 065.175, 102.084, 106.079, 106.080, 112.134 – 112.147, 113.062 – 113.065, 115.033, 117.329, 121.066, 124.018 – 124.020, 125.094 – 125.096, 131.346, 131.358 – 131.391, 132.061 – 132.063, 133.018, 133.019, 134.065, 134.066, 155.122 – 155.124, 156.042, 157.321 – 157.328, 158.324 – 158.335, 161.411, 161.412.

012.081 **Workshop on the Early Earth: the interval from accretion to the old Archean. Proceedings of a conference, held at Houston, Tex., USA, 23 – 24 April 1984.**
K. Burke, L. D. Ashwal (Editors).
LPI 85–01. Lunar and Planetary Inst., Houston, Tex., USA. 10 + 88 pp. (1985). Papers in summary form only.
Review in Phys. Abstr., Vol. 89, No. 1276, Entry 30546 (1986).

012.082 **Workshop on dust on Mars. Conference held at Tempe, Ariz., USA, 4 – 5 February 1985.**
S. Lee (Editor).
LPI 85–02. Lunar & Planetary Inst., Houston, Tex., USA. 2 + 23 pp. (1985). Papers in summary form only.
Review in Phys. Abstr., Vol. 89, No. 1278, Entry 42421 (1986).

012.083 **Workshop on water on Mars. Conference held at Moffat Field, Calif., USA, 30 November – 1 December 1984.**
S. Clifford (Editor).
LPI 85–03. Lunar & Planetary Inst., Houston, Tex., USA. 105 pp. (1985). Papers in summary form only.
Review in Phys. Abstr., Vol. 89, No. 1278, Entry 42413 (1986).

012.084 **Supersymmetry, supergravity and related topics. Proceedings of the XVth GIFT International Seminar on Theoretical Physics.**
F. del Aguila, J. A. de Azcárraga, L. E. Ibáñez (Editors).
World Scientific Publishing Co. Pte Ltd., Singapore. 355 pp. Price US$ 52.00 cloth, US$ 24.00 paper (1985). ISBN 9971–966–79–4 cloth, ISBN 9971–966–92–1 paper.
Review in Astron. Nachr., Vol. 307, No. 2, p. 136; 1986 (*U. Bleyer*).

012.085 **The polar cusp. Proceedings of a NATO Advanced Workshop on the "Morphology and dynamics of the polar cusp", held at Lillehammer, Norway, 7 – 12 May 1984.**
J. A. Holtet, A. Egeland.
NATO Advanced Science Institutes, Series C. D. Reidel Publishing Company, Dordrecht – Boston – Lancaster – Tokyo. 422 pp. Price Dfl. 160.00, US$ 55.00, £ 40.75 (1985).
Review in Space Sci. Rev., Vol. 43, Nos. 3/4, p. 392; 1986 (*J. Lemaire*).

012.086 **The Tenth UK Geophysical Assembly, held at the University of Glasgow, 7 – 9 April 1986.**
Geophys. J. R. Astron. Soc., Vol. 85, No. 1, p. 243 – 280 (1986).
The individual contributions within the subject scope of Astronomy and Astrophysics Abstracts are included in their corresponding categories – see abstracts 045.043, 045.044, 066.184, 082.089, 084.076, 105.175.

012.087 **Satellites of Jupiter. Part 1.**
D. Morrison (Editor).
Mir, Moskva. 263 pp. (1985). In Russian. For the English original see 32.012.066.
Review in Ref. Zh., 51. Astron., 3.51.48 (1986).

012.088 **Himmelskunde und Kleinrechner. Seminarpapiere vom 13. Sternfreundeseminar 1985.**
H. Mucke (Editor).
Astronomisches Büro, Hasenwartgasse 32, A–1238 Wien, Austria. Price öS 180.00 (1985).
Review in Sterne Weltraum, 25. Jahrg., Nr. 3, p. 174 – 175; 1986 (*A. Kunert*).

012.089 **Proceedings of the Ninth Symposium on Antarctic Meteorites.**
T. Nagata (Editor).
Memoirs of the National Institute of Polar Research. Special Issue No. 35. National Institute of Polar Research, Tokyo, Japan. 326 pp. (1984).
Review in Geochim. Cosmochim. Acta, Vol. 50, No. 1, p. 177; 1986 (*E. Olsen*).

012.090 **Proceedings of the Riverside telescope makers conference, 1985.**
J. Sanford (Editor).
Orange County Astronomers, 2215 Martha Ave., Orange, Calif. 92667, USA. 56 pp. paper (1985).
Review in Mercury, Vol. 15, No. 3, p. 86 (1986).

012.091 **Astrophoto VI Proceedings, March 2, 1985.**
Orange County Astronomers Publication. Available at OCA Publications, 2215 Martha Ave., Orange, Calif. 92667, USA. Price US$ 15.00.
Review in Orion, 44. Jahrg., Nr. 213, p. 60; 1986 (*H. Blikisdorf*).

012.092 **Terrestrial planets: comparative planetology. Conference held at Pasadena, Calif., USA, 5 – 7 June, 1985.**
LPI contribution No. 569. Lunar and Planetary Inst., Houston., Tex., USA. 6 + 23 pp. (1985). Papers in summary form only.
Review in Phys. Abstr., Vol. 89, No. 1274, Entry 19016 (1986).

013 Reports on Astronomy in Various Countries and Particular Fields

013.001 **An assessment of research done at the national optical observatories.**
H. A. Abt.
Publ. Astron. Soc. Pac., Vol. 97, No. 597, p. 1050 – 1052 (1985).
The author considered the papers published in 1980 and 1981 and based on data obtained primarily from four large telescopes, two at national and two at university observatories. The former telescopes produce more papers, but are they as useful to astronomy as judged by citation counts? There is statistically no significant difference in the average citations per paper for the four telescopes and the 5% most–cited papers came from all four telescopes in statistically equal numbers.

013.002 **The Japanese National Large Telescope (JNLT) Project.**
K. Kodaira, T. Kogure.
Astrophys. Space Sci., Vol. 118, Nos. 1/2, p. 5 – 7 (1986). – See Abstr. 012.001.
The astronomical community in Japan is promoting a national project to construct a large optical infrared telescope on Mauna Kea in Hawaii. The scientific objectives and the basic concept of the project are briefly presented, as they stand in September 1984.

013.003 **Solar studies in China.**
S.–h. Ye.
Astrophys. Space Sci., Vol. 118, Nos. 1/2, p. 9 – 14 (1986). – See Abstr. 012.001.
After a brief historical retrospect, this paper describes the present status of solar physics in China. Besides an overall account of the observational network, instrumentation, topics of research, personal and international cooperation, emphases are laid on the works about the maximum year of the 21st cycle of solar activity, as well as observations of the solar total eclipses in 1968, 1980, and 1983. A short description of the trends of further developments is presented.

013.004 **New telescopes in China.**
S.–m. Gong.
Astrophys. Space Sci., Vol. 118, Nos. 1/2, p. 15 – 20 (1986). – See Abstr. 012.001.
This paper covers the main aspects of three new optical telescopes: a 1.26 m aperture one for use in the infrared, a 1.56 m aperture one for astrometry, and a 2.16 m reflector for general astrophysical work. It also briefly mentions the 13.7 m telescope designed for the mm wavelength band, the first VLBI in China and the meter wavelength aperture synthesis telescope. All these telescopes, optical and radio, are now being built in China.

013.005 **Cooperation in astronomy between Indonesia and Japan.**
T. Kogure.
Astrophys. Space Sci., Vol. 118, Nos. 1/2, p. 39 – 43 (1986). – See Abstr. 012.001.
The cooperation programme in astronomy between Indonesia and Japan is presented. This cooperation has started in 1979, in the fields of galactic astronomy and stellar physics.

013.006 **New telescopes in India.**
J. C. Bhattacharyya.
Astrophys. Space Sci., Vol. 118, Nos. 1/2, p. 45 – 49 (1986). – See Abstr. 012.001.
The review covers the description and present status of the two optical/infrared telescopes designed and presently near fabrication in India. The 234 cm aperture optical telescope expected to be installed at Kavalur shortly has several interesting features in its control and data acquisition systems. Design features of the 122 cm Infrared Astronomical Telescope expected to be completed next year are described. The review also covers the Ooty Synthesis Radio Telescope and the Gauribidanur Low Frequency Array operating at 327 and 34 MHz, respectively, and a 10 m aperture millimeter wave telescope being built in Bangalore.

013.007 **Astronomy in Tadjikistan.**
P. B. Babadzhanov.
Astrophys. Space Sci., Vol. 118, Nos. 1/2, p. 95 – 97 (1986). – See Abstr. 012.001.
The astronomical investigations in the present Tadjikistan (USSR) are carried out by the Institute of Astrophysics of the Academy of Sciences of Tadjik SSR. The main scientific fields of investigation are meteors, comets, and variable stars. The Institute has two observational stations: the Gissar Observatory with photographic and radar meteor patrols, a 70 cm telescope, a 40 cm Zeiss astrograph, etc., and the Sanglok Observatory with the Ritchey–Chrétien 1 m telescope.

013.008 **Present and future of astronomical science at Gadjah Mada University, Central Java, Indonesia.**
M. S. A. Sastroamidjojo.
Astrophys. Space Sci., Vol. 118, Nos. 1/2, p. 99 – 100 (1986). – See Abstr. 012.001.
Reasons for the growth of interest in astrophysics and astronomy are discussed. The two main reasons are firstly the solar eclipse of June 1983 and secondly publicity given to the regular N.A.S.A. shuttle flights. However, radioastronomy will develop much faster than optical astronomy because the site requirements are much less stringent than for optical astronomy and the technical infrastructure is readily available.

013.009 **"A.K.V."–cercle d'étude "étoiles variables" de la fédération culturelle de R.D.A.**
H. Busch.
Bull. Assoc. Fr. Obs. Etoiles Variables, No. 35, p. 1 (1986).

013.010 **P/Halley activities in the German Democratic Republic.**
D. Möhlmann, S. Marx.
IHW Newsl., No. 8, p. 18 (1986).

013.011 **The cosmos – as scientists understand it today.**
R. Lüst.
ESA Bull., No. 45, p. 9 – 13 (1986).

013.012 **Zur Forschungsplanung Astronomie für das nächste Jahrzehnt. Gedanken zur Denkschrift.**
P. G. Mezger.
Sterne Weltraum, 25. Jahrg., Nr. 2, p. 83 – 85 (1986).

013.013 **Zur neuen Denkschrift Astronomie.**
H. Elsässer.
Sterne Weltraum, 25. Jahrg., Nr. 3, p. 143 – 145 (1986).

013.014 **Astronomy in the Ukrainian SSR: tendencies of its development and vistas.**
Ya. S. Yatskiv.
Kinematika Fiz. Nebesn. Tel, Tom 2, No. 1, p. 3 – 13 (1986). In Russian. English translation in Kinematics Phys. Celest. Bodies.
The present state and basic tendencies of the development of astronomical investigations in the Ukrainian SSR are considered. New special programs are discussed.

013.015 **The International Halley Watch (IHW) and its observing nets.**
ESA Spec. Publ., ESA SP–1066, p. 161 – 178 (1986). – See Abstr. 003.012.
Preparations for Halley's next visit were initiated already in 1980 with the formation of the International Halley Watch (IHW), six years before the comet's closest approach to the Sun. The IHW is responsible for stimulating, coordinating and archiv-

ing all ground–based observations of Halley's comet throughout the apparition.

013.016 **The Inter–Agency Consultative Group (IACG) and its associated working groups.**
R. Reinhard.
ESA Spec. Publ., ESA SP–1066, p. 199 – 215 (1986). – See Abstr. 003.012.
After 76 years, Halley's comet is again approaching the inner solar system. A fleet of five spacecraft, from three space agencies – Intercosmos, ISAS and ESA – are being prepared to encounter the comet in March 1986. In addition, NASA is participating with a significant programme of space observations. The four space agencies have agreed to coordinate informally all matters related to the space missions to Halley's comet and have formed the Inter–Agency Consultative Group (IACG) for this purpose. Since its formation in 1981, the IACG has demonstrated an ever–growing usefulness for the various flight projects as a focal point for exchange of information, discussion on common problems, and mutual support to enhance the overall scientific return of the space missions to Halley's comet.

013.017 **The Halley dust and gas environment (IACG WG–1).**
R. Reinhard.
ESA Spec. Publ., ESA SP–1066, p. 217 – 231 (1986). – See Abstr. 003.012.

013.018 **Coordinated multispacecraft plasma science in the solar–wind and comet–plasma environments (IACG WG–2).**
A. A. Galeev, F. L. Scarf.
ESA Spec. Publ., ESA SP–1066, p. 233 – 243 (1986). – See Abstr. 003.012.

013.019 **Infrared astronomy after IRAS.**
G. H. Rieke, M. W. Werner, R. I. Thompson, E. E. Becklin, W. F. Hoffmann, J. R. Houck, F. J. Low, W. A. Stein, F. C. Witteborn.
Science, Vol. 231, No. 4740, p. 807 – 814 (1986).

013.020 **On the rotation of the Earth and the terrestrial reference system: Joint summary report of the IAU/IUGG Working Groups MERIT and COTES.**
G. A. Wilkins, I. I. Mueller.
Bull. Géod., Vol. 60, No. 1, p. 85 – 100 (1986).

013.021 **Physics News in 1985. An American Institute of Physics Special Report. Astrophysics Section.**
P. F. Schewe (Editor).
Phys. Today, Vol. 39, No. 1, p. S5 – S12 (1986).
The individual contributions are included in their corresponding subject categories – see abstracts 013.022, 032.005, 103.623, 112.036, 116.016, 118.008, 125.025, 126.021, 144.026.

013.022 **Astrophysics.**
H. Shipman.
Phys. Today, Vol. 39, No. 1, p. S5 – S6 (1986). Short review paper. – See Abstr. 013.021.

013.023 **Gravitation, cosmology and cosmic–ray physics.**
B. Schechter.
Phys. Today, Vol. 39, No. 4, p. 43 – 45 (1986).

013.024 **Observational cosmology, the Space Telescope and science planning.**
E. E. Salpeter.
From SU(3) to gravity, p. 253 – 257 (1985). – See Abstr. 003.014.

013.025 **Unstable stars and their investigation at the National Astronomical Observatory.**
B. Kovachev, M. Tsvetkov.
Spisanie Blg. Akad. Nauk, Vol. 31, No. 3, p. 48 – 57 (1985). In Bulgarian. From Ref. Zh., 51. Astron., 1.51.31 (1986).

013.026 **The persistence of Lowell's legacy: life, Mars, and the U.S. Space Program 1958 – 1967.**
J. N. Tatarewicz.
Bull. Am. Astron. Soc., Vol. 17, No. 4, p. 828 (1985). Abstract. – See Abstr. 010.062.

013.027 **Supercomputers for radio astronomical imaging.**
R. D. Ekers.
Bull. Am. Astron. Soc., Vol. 17, No. 4, p. 896 (1985). Abstract. – See Abstr. 010.062.

013.028 **Supercomputers in astronomy.**
D. S. De Young.
Bull. Am. Astron. Soc., Vol. 17, No. 4, p. 897 (1985). Abstract. – See Abstr. 010.062.

013.029 **X–ray physics subpanel summary.**
R. P. Lin, B. R. Dennis.
NASA Conf. Publ., NASA CP–2421, p. 297 – 298 (1986). – See Abstr. 012.021.
The X–ray subpanel of the P/OF Workshop provided the opportunity to get input from a broad–based scientific community, including SMM experimenters and data analysts, solar flare theorists, and ground–based observers, on scientific and technical trade–offs involved in the X–ray instrumentation. The basic objective was to review the strawman payload set up for the Pinhole/Occulter Facility in the light of scientific developments of the last five years.

013.030 **Coronal physics subpanel summary.**
R. H. Munro.
NASA Conf. Publ., NASA CP–2421, p. 299 – 300 (1986). – See Abstr. 012.021.
The objectives of the coronal physics subpanel's discussions were to clarify the scientific background for the instrumentation of the Pinhole/Occulter Facility.

013.031 **Observing programs planned for the Tokyo photoelectric meridian circle.**
M. Miyamoto, M. Yoshizawa.
Proceedings of the Symposium on Application of Space Techniques to Astronomy and Geophysics, p. 5 – 10 (1986). – See Abstr. 012.027.

013.032 **UK involvement in the ROSAT project.**
G. M. Courtier.
J. Br. Interplanet. Soc., Vol. 39, No. 5, p. 217 – 223 (1986).
The ROSAT X–ray astronomy project is described in general terms from its beginning to the time of writing, about two years before launch. A summary of the mission objectives, the instrumentation, and the planned observations is presented with emphasis on the British telescope, the Wide Field Camera (WFC), which will make the first all–sky survey at XUV wavelengths from 60 to 300 Å.

013.033 **Die Kanarischen Inseln – Zentrum europäischer Astronomie.**
H. Haupt.
Sternenbote, 29. Jahrg., Nr. 6, p. 106 – 110 (1986).

013.034 **La astronomía y la ciencia.**
J. Sahade.
Rev. Astron., Tomo 56, No. 233, p. 11 – 14 (1985).

013.035 **Some work in progress in APL's Space Department.**
G. C. Weiffenbach, E. B. Dobson, S. M. Krimigis.
Johns Hopkins APL Tech. Dig., Vol. 7, No. 1, p. 117 – 121 (1986).
The authors consider the operational longevity of APL Space Department oldest satellites, OSCAR 13, and a status report on the newest, Geosat, together with a listing of current analytic efforts using Geosat data. Finally, they review recent scientific results obtained from Magsat and update current Space Department research activities.

013.036 **Photoelectric studies of asteroids: the amateur contribution.**
K. W. Zeigler.
Minor Planet Bull., Vol. 12, No. 4, Part 1, p. 36 – 37 (1985).
High school students at Globe High School in Globe, Arizona are currently involved in a program of photoelectric photometry of asteroids which has yielded a considerable volume of important information as to the rotational rates and surface properties of 25 bright asteroids.

013.037 **Co–operation in astronomy between Indonesia and Japan.**
T. Kogure, B. Hidayat.
Galactic structure and variable stars, p. 3 – 7 (1985). – See Abstr. 012.029.
The background, form, and activities in the six–year co–operation in astronomy which have been carried out in the fields of galactic astronomy and stellar physics in 1979 – 1984 are briefly summarized, along with an aspect for future co–operation.

013.038 **Astronomy in Indonesia.**
B. Hidayat.
Galactic structure and variable stars, p. 8 – 24 (1985). – See Abstr. 012.029.
The origin of the Bosscha Observatory and the development of astronomy in Indonesia are described. A view relating to education of astronomy is also presented.

013.039 **Observational study of close binary stars.**
M. Kitamura, A. Yamasaki, A. Okazaki, H. L. Malasan.
Galactic structure and variable stars, p. 78 – 87 (1985). – See Abstr. 012.029.

013.040 **Cooperation in instrumentation.**
S. Nishimura, T. Noguchi.
Galactic structure and variable stars, p. 102 – 106 (1985). – See Abstr. 012.029.
Cooperation in astronomical instrumentations between Indonesia and Japan is briefly summarized. Two main items, an electronic circuit for the photoelectric photometer and a sub–beam prism for the Schmidt telescope, are discussed.

013.041 **Proposal for an international Institute of Space Sciences and a Giant Equatorial Radio Telescope as a collaborative effort between the developing countries.**
G. Swarup.
Proceedings of the Second Asian–Pacific Regional Meeting on Astronomy, p. 22 – 31 (1984). – See Abstr. 012.031.
It is proposed to construct a 2 km long and 50 m wide cylindrical radiotelescope at a site close to the earth's equator. The radio telescope would be operated at a few discrete bands protected for radio astronomy in the range of about 38 to 328 MHz. Alongwith a proposed 14 × 12 km synthesis radio telescope, it would become a unique facility for radio astronomy at meter wavelengths.

013.042 **Galactic astronomy at millimeter wavelength in Japan.**
K. Akabane.
Proceedings of the Second Asian–Pacific Regional Meeting on Astronomy, p. 162 – 176 (1984). – See Abstr. 012.031.
Japanese galactic studies in the mm–wave region are reviewed. Galactic mm–wave observation in Japan has been continued in the past 10 years. A brief description on the mm–wave telescope in Japan (6mx1, and 1.5mx2 dishes) is presented. Instrumentations, observations and theoretical studies are summarized.

013.043 **Contributions by R. G. Giovanelli to solar instrumentation.**
J. M. Beckers.
Aust. J. Phys., Vol. 38, No. 6, p. 769 – 773 (1985). Paper presented at the R. G. Giovanelli Commemorative Colloquium, Part I, Sydney, 26 – 29 November 1984.

013.044 **Contributions by R. G. Giovanelli to the study of magnetic field structures.**
W. Livingston.
Aust. J. Phys., Vol. 38, No. 6, p. 775 – 780 (1985). Paper presented at the R. G. Giovanelli Commemorative Colloquium, Part I, Sydney, 26 – 29 November 1984.

013.045 **And the natives are friendly.**
R. E. Fried.
I.A.P.P.P. Commun., No. 23, p. 6 – 10 (1986). – See Abstr. 010.521.
The author discusses the development of amateur photometry.

013.046 **How do I choose a photoelectric observing project?**
D. S. Hall.
I.A.P.P.P. Commun., No. 24, p. 1 – 5 (1986).

013.047 **Die "Denkschrift Astronomie".**
H. J. Völk.
Sterne Weltraum, 25. Jahrg., Nr. 5, p. 258 – 259 (1986).

013.048 **Amateurastronomie in Finnland.**
A. Kellomäki.
Sterne Weltraum, 25. Jahrg., Nr. 5, p. 276 – 278.

013.049 **The large–scale phenomena network of the International Halley Watch.**
J. C. Brandt, M. B. Niedner, J. Rahe.
Int. Comet Q., Vol. 8, No. 1, p. 3 – 7 (1986).

013.050 **Main stages of development of cosmic investigations in the USSR (1957 – 1982).**
V. S. Avduevskij.
Issled. po istor. i teor. razvitiya aviats. i raket.–kosm. nauk. i tekh. Moskva, No. 4, p. 16 – 28 (1985). In Russian. Abstr. in Ref. Zh., 62. Issled. Kosm. Prostranstva, 5.62.8 (1986).

013.051 **The role of the Inter–Agency Consultative Group (IACG) and its associated working groups.**
R. Reinhard.
ESA Bull., No. 46, p. 81 – 94 (1986). – See Abstr. 003.052.

013.052 **È difficile scoprire una cometa?**
V. Zappalà.
Orione, Vol. 6, N. 3, p. 45 – 50 (1986).

013.053 **The Italian contribution to Project MERIT.**
G. Caprioli, F. Carta, F. Chlistovsky, G. Chiumiento, V. Gusai, F. Mazzoleni, G. Picchio, A. Poma, M. Sarasso.
Mem. Soc. Astron. Ital., Vol. 56, No. 4, p. 791 – 794 (1985). – See Abstr. 010.821.

013.054 **Investigations of B and Be stars in the U.S.S.R. in 1985.**
M. Ruusalepp.
Bull. Inf. Etoiles Be, No. 13, p. 19 (1986).

013.055 **Report of activities concerning Be stars from Italy.**
L. Pastori.
Bull. Inf. Etoiles Be, No. 13, p. 20 – 21 (1986).

013.056 **Photometric observing campaign for Be stars. Progress Report No. 9.**
P. Harmanec, J. Horn, P. Koubský.
Bull. Inf. Etoiles Be, No. 13, p. 23 – 29 (1986).

013.057 **Research activity and the state of astrosciences in Turkey.**
A. Uzun.
Czech. J. Phys., Sect. B, Vol. B36, No. 1, p. 185 – 187 (1986). Abstr. in Phys. Abstr., Vol. 89, No. 1280, Entry 54085 (1986).

013.058 **Europe's superior position in infrared astronomy.**
Nachrichtentech. Z., Vol. 39, No. 2, p. 104 – 105 (1986). In German. Abstr. in Phys. Abstr., Vol. 89, No. 1281, Entry 67352 (1986).

013.059 **Old soldier's tale – how the "HAKUCHO" satellite went into orbit.**
M. Oda.
Cosmic pathways, p. 364 – 366 (1986). – See Abstr. 003.045.

013.060 **1962 – 1972 (up through UHURU).**
R. Giocconi.
Cosmic pathways, p. 367 – 384 (1986). – See Abstr. 003.045.
The development of X–ray astronomy and the major scientific discoveries of the UHURU satellite are outlined.

013.061 **Cosmic ray and particle physics deep underground.**
W. R. Kropp, F. Reines, H. W. Sobel.
Cosmic pathways, p. 385 – 386 (1986). – See Abstr. 003.045.

013.062 **Astronomy in Brazil.**
S. Ferraz Mello.
Rev. Mex. Astron. Astrofis., Vol. 12, Spec. Issue, p. 13 – 18 (1986). – See Abstr. 012.066.
Survey of the progresses of Brazilian astronomy from its beginning, around 1960, till nowadays.

013.063 **A photographic astrometric telescope for Brazil.**
L. E. da S. Machado.
Astrometric techniques, p. 229 – 230 (1986). – See Abstr. 012.068 (IAU Symp. No. 109).
This paper describes a short focus astrometric telescope (D = 40 cm; F = 200 cm) at the astronomical station of Campinas, in the State of Sao Paulo.

013.064 **Astrometry in China.**
S.–h. Ye.
Astrometric techniques, p. 729 – 736 (1986). – See Abstr. 012.068 (IAU Symp. No. 109).
Astrometry is one of the important branches of astronomy in China. Six observatories are devoting their efforts to classical astrometry, including star catalogues, positions of major planets and radio stars, and determination of the parameters of the earth's rotation (ERP). Recently, developments of new techniques such as laser ranging and VLBI have pushed astronomers to consider more deeply the frame of reference coordinates.

013.065 **The need for better co–operation and intercomparison in fundamental astrometry.**
G. Teleki.
Astrometric techniques, p. 749 – 755 (1986). – See Abstr. 012.068 (IAU Symp. No. 109).
The present status and practice in fundamental astrometry are analysed and relying upon the findings the author suggests the organization of an international action aimed achieving better co–operation in the star position observations and celestial reference system determinations, as well as performing the intercomparison of the different techniques and methods.

013.066 **Heliosphere, collapsed stars and fusion reactors.**
B. Coppi.
From nuclei to stars, p. 210 – 211 (1986). Abstract. – See Abstr. 012.070.

013.067 **Automatic telescopes large and small.**
L. J. Boyd, R. M. Genet, D. S. Hall.
Publ. Astron. Soc. Pac., Vol. 98, No. 604, p. 618 – 621 (1986).
Completely automatic (unmanned) telescopes of small size are now being routinely used for photometry of brighter stars. The advantages of unmanned operation include lower operating costs and highly efficient operation in well–defined observational programs. While problems may exist in extending unmanned operation to larger ground–based optical telescopes and other sensory modes besides photometry, this extension is already underway.

013.068 **New techniques and prospects.**
L. Vigroux.
Star–forming dwarf galaxies and related objects, p. 489 – 496 (1986). – See Abstr. 012.074.
New techniques and observational facilities will become available soon. In this paper the author describes first the new techniques which will be put on routine service within 5 years, then those which will appear on a longer time scale. Finally, the author has selected some still open problems in BCG and he has tried to see how these new techniques will help to solve them.

013.069 **VLBI at optical and radio wavelengths: congruences and contrasts.**
B. F. Burke.
ESA Spec. Publ., ESA SP–226, p. 177 – 183 (1985). – See Abstr. 012.075.

013.070 **Current work on the Earth's rotation at the United Kingdom Meteorological Office.**
K. D. B. Whysall, R. Hide, M. J. Bell.
Earth rotation and the terrestrial reference frame, p. 417 – 433 (1985). – See Abstr. 012.077.

013.071 **Popularization of astronomy in Vojvodina (1945 – 1984).**
Ž. Ćulum, J. Francisty.
Publ. Astron. Soc. "Rudjer Bošković", No. 4, p. 51 – 56 (1985). – See Abstr. 012.078. In Croatian.

013.072 **The astronomical society "Oton Kučera" 1954 – 1984.**
V. Vlah.
Publ. Astron. Soc. "Rudjer Bošković", No. 4, p. 59 – 60 (1985). – See Abstr. 012.078. In Croatian.

013.073 **The astronomical knowledge of the south–west Macedonia's people.**
G. Cenev.
Publ. Astron. Soc. "Rudjer Bošković", No. 4, p. 139 – 146 (1985). – See Abstr. 012.078. In Serbian.

013.074 **The UK remote and service observing programme.**
M. S. Longair, J. M. Stewart, P. M. Williams.
Q. J. R. Astron. Soc., Vol. 27, No. 2, p. 153 – 165 (1986).
In this review of current progress, the authors describe what has been achieved so far, what can be provided to the remote observer in the future, UK plans to implement these facilities and the main issues as the authors see them.

013.075 **Astronomy after IRAS.**
M. S. Longair.
Light on dark matter, p. 511 – 521 (1986). – See Abstr. 012.080.

A brief history of the Cincinnati Astronomical Society.
See Abstr. 003.201.

The East–Siberian complex of SibIZMIR solar observatories.
See Abstr. 009.001.

The Tokyo Schmidt telescope and its programs.
See Abstr. 009.037.

Recent work with the UK 1.2 m Schmidt Telescope.
See Abstr. 009.038.

American observatories in Southern Africa.
See Abstr. 009.046.

Photographic surveys of the southern sky.
See Abstr. 009.068.

The astrograph of the Valongo Observatory (UFRJ) and the astrometry program at Campinas, São Paulo, Brazil.
See Abstr. 032.046.

HIPPARCOS satellite and the organization of the project.
See Abstr. 041.066.

Preparation of the Hipparcos Input Catalogue. Astrometric programs for Hipparcos. Preliminary astrometric observations.
See Abstr. 041.067.

Astronomical and Doppler observations in Italy during the main MERIT campaign.
See Abstr. 044.044.

Astrometric determinations of UT1 in Italy during the MERIT main campaign.
See Abstr. 044.097.

Ballooning around Venus.
See Abstr. 051.031.

Coordinating work on P/Halley, the International Halley Watch.
See Abstr. 103.021.

014 Teaching in Astronomy

014.001 **Computer recreations: how close encounters with star clusters are achieved with a computer telescope.**
A. K. Dewdney.
Sci. Am., Vol. 254, No. 1, p. 12 – 16 (1986).

014.002 **Die Erde im Weltall. Vier Musterversuche für die Sekundarstufe I, Teil 3.**
W. Schlosser.
Sterne Weltraum, 25. Jahrg., Nr. 1, p. 26 – 28 (1986).

014.003 **Die scheinbare Helligkeit der Sterne. Vorschläge für einen an Versuchen orientierten Unterricht, Teil 1.**
M. Otter.
Sterne Weltraum, 25. Jahrg., Nr. 2, p. 86 – 88 (1986).

014.004 **Die scheinbare Helligkeit der Sterne. Vorschläge für einen an Versuchen orientierten Unterricht, Teil 2.**
M. Otter.
Sterne Weltraum, 25. Jahrg., Nr. 3, p. 146 – 148 (1986).

014.005 **Teaching astronomy at Union College.**
F. Titus.
News Lett. Astron. Soc. N.Y., Vol. 2, No. 9, p. 11 – 12 (1986). Abstract. – See Abstr. 010.241.

014.006 **Comment choisir votre premier télescope? (II).**
S. Harrington.
Astronomie, Vol. 100, p. 84 – 90 (1986).

014.007 **Comment choisir votre premier télescope.**
S. Harrington.
Astronomie, Vol. 100, p. 153 – 158 (1986).

014.008 **Sterrenkunde op de huiscomputer: Eenheid in verscheidenheid.**
J. Loonen.
Zenit, 13. Jaarg., Nr. 3, p. 101 (1986).

014.009 **Sterrenkunde op de huiscomputer: Fotografie van maan en planeten.**
J. Loonen.
Zenit, 13. Jaarg., Nr. 4, p. 141 (1986).

014.010 **Sterrenkunde op de huiscomputer: De tijdvereffening.**
J. Loonen.
Zenit, 13. Jaarg., Nr. 5, p. 177 (1986).

014.011 **New tools for ancient skies.**
D. S. P. Dearborn, B. Bell.
Archaeoastronomy (U.S.A.), Vol. 7, Nos. 1 – 4, p. 96 – 104 (1984).

014.012 **Backyard astronomy: the aurora.**
A. MacRobert.
Sky Telesc., Vol. 71, No. 1, p. 30 – 32 (1986).

014.013 **Astronomical computing: Making slides from computer displays.**
T. B. Hunter, M. L. Brewer.
Sky Telesc., Vol. 71, No. 1, p. 86 – 87 (1986).

014.014 **Discover the southern skies – I.**
G. Cozens.
Sky Telesc., Vol. 71, No. 2, p. 126 – 128 (1986).

014.015 **Astronomical computing: tracking tolerances in astrophotography.**
B. H. Rowe.
Sky Telesc., Vol. 71, No. 2, p. 190 – 191 (1986).

014.016 **Discover the southern skies – II.**
G. Cozens.
Sky Telesc., Vol. 71, No. 3, p. 237 – 239 (1986).

014.017 **Astronomical computing: when is Easter?**
G. H. Horne, W. H. Freeman, R. W. Sinnott.
Sky Telesc., Vol. 71, No. 3, p. 294 – 295 (1986).

014.018 **Astronomical computing: Making your own globular cluster.**
K. Velt.
Sky Telesc., Vol. 71, No. 4, p. 398 – 399 (1986).

014.019 **The 50th anniversary of the Young Astronomers Club at Moscow Planetarium.**
K. A. Porzewskij (*K. A. Portsevskij*).
Report on the VIII. International Planetarium Directors Conference, p. 98 – 100 (1985). – See Abstr. 012.014.

014.020 **Star Date: the Astronomy News Report.**
D. Byrd.
Bull. Am. Astron. Soc., Vol. 17, No. 4, p. 844 (1985). Abstract. – See Abstr. 010.062.

014.021 **Quellen kosmischer Gamma–Strahlungsausbrüche als mögliche Beobachtungsobjekte für Amateurastronomen.**
J. Flohrer.
Astron. Raumfahrt, 24. Jahrg., Heft 2, p. 42 – 45 (1986).

014.022 **Microcomputación y astronomía: Cálculo numérico de órbitas.**
D. H. Zanette.
Rev. Astron., Tomo 56, No. 233, p. 7 – 8 (1985).

014.023 **Microcomputación y astronomía: Programa de los n cuerpos.**
L. Graciotti.
Rev. Astron., Tomo 56, No. 233, p. 9 – 10, 15 – 16 (1985).

014.024 **Astronomy, space flight and education.**
D. Lunan, B. Ramsay.
Space Educ., Vol. 1, No. 11, p. 488 – 489 (1986).

014.025 **Sterrenkunde op de huiscomputer: Commensurabiliteit en kettingbreuken.**
P. L. Cijsouw.
Zenit, 13. Jaarg., Nr. 6, p. 213 (1986).

014.026 **Astronomy education at universities.**
E. A. Muller.
Proceedings of the Second Asian–Pacific Regional Meeting on Astronomy, p. 227 – 230 (1984). – See Abstr. 012.031.

014.027 **Designing astronomy courses.**
R. H. Wilkinson.
Proceedings of the Second Asian–Pacific Regional Meeting on Astronomy, p. 235 – 237 (1984). – See Abstr. 012.031.

014.028 **Comment on education in astronomy.**
B. Rawi.
Proceedings of the Second Asian–Pacific Regional Meeting on Astronomy, p. 246 (1984). – See Abstr. 012.031.

014.029 **Summary of the session on education in astronomy.**
D. G. Wentzel.
Proceedings of the Second Asian–Pacific Regional Meeting on Astronomy, p. 247 (1984). – See Abstr. 012.031.

014.030 **Astronomical computing: Tracking satellites on your home computer.**
R. W. Sinnott.
Sky Telesc., Vol. 71, No. 5, p. 501 – 502 (1986).

014.031 **Astronomical computing: Observing geosynchronous satellites.**
D. L. Welch.
Sky Telesc., Vol. 71, No. 6, p. 606 – 607 (1986).

014.032 **Science Fiction im Astronomieunterricht.**
G. Habersack.
Sterne Weltraum, 25. Jahrg., Nr. 5, p. 264 – 267 (1986).

014.033 **Astronomical resources: interdisciplinary approaches to astronomy.**
A. Fraknoi.
Mercury, Vol. 15, No. 2, p. 59 – 62 (1986).

014.034 **An Apple Pascal simulation of eclipsing binary star systems.**
D. Nagy.
J. Pascal, Ada Modula–2, Vol. 4, No. 1, p. 39 – 50 (1985). Abstr. in Phys. Abstr., Vol. 89, No. 1273, Entry 13318 (1986).

014.035 **The slingshot effect: explanation and analogies.**
A. A. Bartlett, C. W. Hord.
Phys. Teach., Vol. 23, No. 8, p. 466 – 473 (1985). Abstr. in Phys. Abstr., Vol. 89, No. 1276, Entry 30567 (1986).

014.036 **A simple theory of limiting masses.**
P. T. Landsberg, D. F. Falla.
Eur. J. Phys., Vol. 6, No. 4, p. 271 – 274 (1985). Abstr. in Phys. Abstr., Vol. 89, No. 1278, Entry 42463 (1986).

014.037 **The hazards of encountering a black hole. I. Testing the limits of physics.**
D. E. Hall.
Phys. Teach., Vol. 23, No. 9, p. 540 – 544 (1985). Abstr. in Phys. Abstr., Vol. 89, No. 1280, Entry 54062 (1986).

014.038 **Very high–energy gamma–ray sources.**
T. C. Weekes.
Phys. Teach., Vol. 24, No. 1, p. 20 – 28 (1986). Abstr. in Phys. Abstr., Vol. 89, No. 1281, Entry 61213 (1986).

014.039 **Astronomy education in Malaysia.**
M. Othman.
Third Asian–Pacific Regional Meeting of the IAU, Supplement: Astronomical education in the Asian–Pacific region, p. 14 – 15 (1986). Abstract. – See Abstr. 012.001.

014.040 **Astronomy education in Thailand.**
Y. Vanichai.
Third Asian–Pacific Regional Meeting of the IAU, Supplement: Astronomical education in the Asian–Pacific region, p. 16 – 17 (1986). Abstract. – See Abstr. 012.001.

014.041 **The present state of astronomical education in China.**
K.–j. Feng.
Third Asian–Pacific Regional Meeting of the IAU, Supplement: Astronomical education in the Asian–Pacific region, p. 18 – 22 (1986). Abstract. – See Abstr. 012.001.

014.042 **Present situation of astronomical education in Korea.**
J. O. Woo.
Third Asian–Pacific Regional Meeting of the IAU, Supplement: Astronomical education in the Asian–Pacific region, p. 23 – 24 (1986). Abstract. – See Abstr. 012.001.

014.043 **Problems in teaching astronomy in school education.**
N. Owaki.
Third Asian–Pacific Regional Meeting of the IAU, Supplement: Astronomical education in the Asian–Pacific region, p. 25 – 27 (1986). Abstract. – See Abstr. 012.001.

014.044 **Astronomy education in India.**
V. R. Venugopal.
Third Asian–Pacific Regional Meeting of the IAU, Supplement: Astronomical education in the Asian–Pacific region, p. 28 – 31 (1986). Abstract. – See Abstr. 012.001.

014.045 **Colour photography in astronomical education.**
D. F. Malin.
Third Asian–Pacific Regional Meeting of the IAU, Supplement: Astronomical education in the Asian–Pacific region, p. 32 (1986). Abstract. – See Abstr. 012.001.

014.046 **Determination of star masses.**
M. Prosén.
Obzornik Mat. Fiz., Vol. 33, No. 3/4, p. 118 – 120 (1986). In Serbo–Croatian.

A method of star mass determination is described suitable for use in secondary school.

014.047 **Exercises for distance estimates of the Local Group galaxies using Cepheid variables.**
F. Sato.
Educ. Earth Sci., Vol. 37, No. 1, p. 1 – 9 (1984). In Japanese.

This paper presentes exercises for distance estimates of the Local Group galaxies LMC, SMC, M31, and NGC 6822 by use of data of the Cepheid variables in them and those in our own Galaxy taken from various literatures. The exercises are suitable for students of senior high schools and /or of liberal arts course of universities.

014.048 **Photographing slides of stars for teaching stellar brightnesses and colors in lower secondary schools.**
F. Ito, F. Sato.
Educ. Earth Sci., Vol. 37, No. 6, p. 155 – 162 (1984). In Japanese.

014.049 **Misuras dell'intensità relativa dello spettro solare.**
G. Comai, M. Elena, L. Mich, V. Zanetti, A. Zecca.
G. Astron., Vol. 11, N. 4, p. 521 – 531 (1985).

The teaching in astronomy.
See Abstr. 003.152.

Astronomie sur ZX Spectrum et Spectrum +.
See Abstr. 003.187.

Method of teaching astronomy at secondary school. Handbook for teachers.
See Abstr. 003.205.

Comet! The story behind Halley's comet.
See Abstr. 003.207.

Active and independent contribution from children – the essential part in the didactical/methodical plan of a planetarium show for children.
See Abstr. 009.014.

Anticipation and participation as didactic principles of "spaceology".
See Abstr. 009.015.

Götterdämmerung – the withering temples of science.
See Abstr. 009.016.

Didactics value of the planetarium for primary and secondary students.
See Abstr. 009.019.

Photographic photometry of the eclipsing binary U Cephei with small equipment.
See Abstr. 119.086.

015 Miscellanea (Philosophical Aspects, Extraterrestrial Life, etc.)

015.001 **Space science, space technology and the space station.**
J. A. Van Allen.
Sci. Am., Vol. 254, No. 1, p. 22 – 29 (1986).
The space–station program will seriously diminish the opportunities for advancing space science and technology if it proceeds as planned. Most national goals in space are better realized by robot spacecraft.

015.002 **Possible extraterrestrial strategy for Earth.**
J. W. Deardorff.
Q. J. R. Astron. Soc., Vol. 27, No. 1, p. 94 – 101 (1986).

015.003 **The compatibility of the universe to complex order: paradigms and speculations.**
R. D. Meisner.
J. Br. Interplanet. Soc., Vol. 39, No. 3, p. 121 – 126 (1986).
Recent discoveries have indicated that very slight changes in the magnitude of any of the fundamental forces or constants would disallow the existence of such familiar material structures as atoms, stars, and life. Examples of such effects are surveyed, concentrating on those involving the universe's primary macroscopic structures, main–sequence stars. Improper applications of the "anthropic principle" are discussed.

015.004 ***In situ* search for extraterrestrial life.**
H. P. Klein.
J. Br. Interplanet. Soc., Vol. 39, No. 3, p. 137 (1986). Abstract. – See Abstr. 012.003.

015.005 **The evolution of complex life.**
J. Billingham.
J. Br. Interplanet. Soc., Vol. 39, No. 3, p. 137 – 138 (1986). Abstract. – See Abstr. 012.003.

015.006 **Panspermia: unlikely, unsupported, but just possible?**
R. F. Davies.
J. Br. Interplanet. Soc., Vol. 39, No. 3, p. 138 (1986). Abstract. – See Abstr. 012.003.

015.007 **The anthropic principle: probability and the possibility of extraterrestrial life.**
P. J. Hall.
J. Br. Interplanet. Soc., Vol. 39, No. 3, p. 138 – 139 (1986). Abstract. – See Abstr. 012.003.

015.008 **Some diplomatic and journalistic implications of SETI.**
N. N. Nelson.
J. Br. Interplanet. Soc., Vol. 39, No. 3, p. 139 (1986). Abstract. – See Abstr. 012.003.

015.009 **Objectives and first results of the NASA SETI sky field tests at Goldstone, California.**
S. Gulkis.
J. Br. Interplanet. Soc., Vol. 39, No. 3, p. 139 – 140 (1986). Abstract. – See Abstr. 012.003.

015.010 **A directed search for extraterrestrial laser signals.**
A. Betz.
J. Br. Interplanet. Soc., Vol. 39, No. 3, p. 140 (1986). Abstract. – See Abstr. 012.003.

015.011 **Statistics of "excess" observatory noise at the Nançay telescope and elsewhere.**
J. Tarter.
J. Br. Interplanet. Soc., Vol. 39, No. 3, p. 140 (1986). Abstract. – See Abstr. 012.003.

015.012 **SETI and some consequences of the existence of a galactic "belt of intelligent life".**
B. A. Balazs.
J. Br. Interplanet. Soc., Vol. 39, No. 3, p. 141 (1986). Abstract. – See Abstr. 012.003.

015.013 **The inclusion of interaction terms into population dynamics equations of interstellar colonization.**
J. P. Ostriker, E. L. Turner.
J. Br. Interplanet. Soc., Vol. 39, No. 3, p. 141 (1986). Abstract. – See Abstr. 012.003.

015.014 **Toward a multi–path approach for SETI.**
M. D. Papagiannis.
J. Br. Interplanet. Soc., Vol. 39, No. 3, p. 141 – 142 (1986). Abstract. – See Abstr. 012.003.

015.015 **On the possible existence of terrestrial–like scientific–technical civilizations only in the whole Galaxy: one more solution of the Fermi paradox.**
M. Subotowicz.
J. Br. Interplanet. Soc., Vol. 39, No. 3, p. 142 (1986). Abstract. – See Abstr. 012.003.

015.016 **Very low frequency SETI search strategy.**
S. Gorgolewski.
J. Br. Interplanet. Soc., Vol. 39, No. 3, p. 142 – 143 (1986). Abstract. – See Abstr. 012.003.

015.017 **Where does time asymmetry come from?**
G. Vollmer.
Naturwissenschaften, 72. Jahrg., Heft 9, p. 470 – 481 (1985). In German.
All natural processes are going one way: they are irreversible. Surprisingly enough, this "time arrow" is not found in the basic equations of physics, but only in boundary and initial conditions which in turn should be open to explanation. The best candidate for such a prime or master arrow is the cosmic one. Its origin is speculated upon, but might have to wait for a future cosmophysics.

015.018 **Physical cosmology and the anthropic principle.**
B. Kanitscheider.
Naturwissenschaften, 72. Jahrg., Heft 12, p. 613 – 618 (1985). In German.
In the past ten years a new type of argument came to the fore in physical cosmology in which man with his intrinsic properties is used to give explanations for some values of parameters describing the large–scale structure of the universe. Here it is argued that the so–called anthropic explanations are not explanations proper but can be taken as hints that there are more causal connections within the hierarchy of physical systems than hitherto recognized.

015.019 **Between experience and theory.**
P. Brosche.
Naturwissenschaften, 72. Jahrg., Heft 12, p. 668 – 669 (1985). In German.

015.020 **Kometen, Weltbild und weltanschauliche Auseinandersetzung.**
F. Gehlhar.
Astron. Sch., 23. Jahrg., Heft 2, p. 28 – 30 (1986).

015.021 **Orologi e meridiane naturali.**
G. Romano.
Coelum, Vol. 54, N. 1, p. 1 – 9 (1986).

015.022 **Speculation in cosmology.**
F. Hoyle.
From SU(3) to gravity, p. 289 – 295 (1985). – See Abstr. 003.014.

015.023 **How many more discoveries in the Universe?**
M. Harwit, R. Hildebrand.
Nature, Vol. 320, No. 6064, p. 724 – 726 (1986).
It is generally assumed that the potential scope of any scientific discipline is quite incalculable. Yet the discovery of the Earth's major rivers, mountains and seas, an enterprise that may have seemed endless to Ptolemy's contemporaries, is now essentially complete. The same may soon be said about the discovery of new plants and animals on Earth. Harwit has put forward a procedure for estimating the number of observational phenomena remaining to be discovered in astronomy. Here the authors describe a statistical approach which supports the conclusion that the number of phenomena already discovered represents an appreciable fraction of those that can be found.

015.024 **Search for chiral molecules and optical activity in extraterrestrial systems. The Cassini mission opportunity.**
A. Brack, G. Spach.
ESA Spec. Publ., ESA SP–241, p. 189 – 191 (1985). – See Abstr. 012.018.
Search for optical activity arising from chiral molecules on extraterrestrial objects is of great interest for the studies of the origin of life. In the close vicinity of the Earth, two systems call attention in this domain. The planet Mars, for which a form of life may have been granted or is still granted, and the planet–sized moon Titan on which organic chemistry evolution is presently occurring in an abiotic way. Exploration of this last satellite during the Cassini mission offers a unique opportunity to test present day speculations.

015.025 **The viability with respect to temperature, of micro–organisms incident on the Earth's atmosphere.**
F. Hoyle, N. C. Wickramasinghe, S. Al–Mufti.
Earth, Moon, Planets, Vol. 35, No. 1, p. 79 – 84 (1986).
Using laboratory measurements of the resistance of E. coli to flash–heating, it is shown that a large fraction of interplanetary micro–organisms in prograde orbits could be added to the Earth without losing viability due to heating by the atmospheric gases.

015.026 **Martian standard time.**
T. E. Gangale.
J. Br. Interplanet. Soc., Vol. 39, No. 6, p. 282 – 288 (1986).
The factors which influenced the development of the various units of time now used on Earth are briefly considered and compared with conditions on other planets likely to be settled in the 21st Century to show that the practical need for a new system of time measurement will probably exist nowhere but on Mars. A new clock and a new calendar for Mars are proposed.

015.027 **The extraterrestrial life debate – historical perspectives.**
S. J. Dick.
South. Stars, Vol. 31, No. 4, p. 202 – 208 (1985).

015.028 **Absurdity in science fiction.**
A. Satō.
Heavens, Vol. 67, No. 5, p. 147 – 148 (1986). In Japanese.

015.029 **Ein astronomisches Kartenspiel aus dem Jahre 1719.**
C. Zille.
Sterne Weltraum, 25. Jahrg., Nr. 4, p. 226 (1986).

015.030 **The import of science.**
B. Suprapto.
Proceedings of the Second Asian–Pacific Regional Meeting on Astronomy, p. 231 – 234 (1984). – See Abstr. 012.031.

015.031 **Some aspects of science education and the cultural background.**
J. I. G. M. Drost.
Proceedings of the Second Asian–Pacific Regional Meeting on Astronomy, p. 238 – 245 (1984). – See Abstr. 012.031.

015.032 **Night brought to numbers.**
C. Raymo.
Sky Telesc., Vol. 71, No. 6, p. 554 – 556 (1986).

015.033 **Comets and the muse.**
L. A. Marschall.
Mercury, Vol. 15, No. 1, p. 10 – 12 (1986).

015.034 **Visions of space.**
A. C. Clarke.
Spaceflight, Vol. 28, No. 5, p. 201 – 204 (1986).
The author considers some of the possible futures and pitfalls for mankind.

015.035 **On the detectability of antimatter propulsion spacecraft.**
M. J. Harris.
Astrophys. Space Sci., Vol. 123, No. 2, p. 297 – 303 (1986).
It is shown that the NASA Gamma–Ray Observatory will be able to detect large interstellar spacecraft at distances up to ~ 300 pc by the γ–ray emission from the propulsion system alone. The distance limit is set by the possibility of recognizing such objects by their proper motions.

015.036 **On deliberate misreferencing as a tool of science policy.**
F. Hoyle, N. C. Wickramasinghe.
Astrophys. Relativ., Prepr. Ser., No. 125 (1986).

015.037 **When reference means deference.**
J. Maddox.
Nature, Vol. 321, No. 6072, p. 723 (1986).
Hoyle and Wickramasinghe protest that none of the contributors to recent accounts of the encounters with Halley's comet have referred to them. In this note the editor of "Nature" expounds his viewpoint of the matter.

015.038 **Mensch und Kosmos.**
D. B. Herrmann.
Archenhold-Sternw. Berlin-Treptow, Sonderdr., Nr. 27, 11 pp. (1985).

015.039 **Erkenntnisfortschritt und Technik in der modernen Astronomie.**
D. B. Herrmann.
NTM-Schriftenr. Gesch. Naturwiss., Tech., Med., Band 22, Nr. 2, p. 101 - 105 (1985). = Mitt. Archenhold-Sternw. Berlin-Treptow, Nr. 160.

015.040 **Alcune considerazioni sui numeri cosmici.**
G. Romano.
Atti Mem. Accad. Patavina Sci., Lett. Arti, Vol. 91, p. 107 - 122 (1979). = Oss. Privato, Specola "Ariel", Pubbl. No. 85.

015.041 **Signal processing in SETI.**
D. K. Cullers, I. R. Linscott, B. M. Oliver.
Computer, Vol. 18, No. 11, p. 37 - 47 (1985). Abstr. in Phys. Abstr., Vol. 89, No. 1276, Entry 35987 (1986).

015.042 **The vital nature sign and oriental physics.**
D.-l. Zhao.
Publ. Purple Mt. Obs., Vol. 4, No. 3, p. 80 - 87 (1985). In Chinese.
The Yin-Yang theory is an essential principle of the ancient chinese philosophy. This paper discusses some of its properties from the point of view of modern physics and points out its relations to vital bodies.

015.043 **Using the Very Large Array (VLA) and other radio telescopes to perform a parasitic search for extraterrestrial intelligence (SETI).**
J. Tarter.
Acta Astronaut., Vol. 12, No. 11, p. 959 - 962 (1985). Abstr. in Phys. Abstr., Vol. 89, No. 1277, Entry 42186 (1986).

015.044 **The Search for Extraterrestrial Artifacts (SETA).**
R. A. Freitas Jr., F. Valdes.
Acta Astronaut., Vol. 12, No. 12, p. 1027 - 1034 (1985). Abstr. in Phys. Abstr., Vol. 89, No. 1277, Entry 42187 (1986).

015.045 **Signal processing in SETI.**
D. K. Cullers, I. R. Linscott, B. M. Oliver.
Commun. ACM, Vol. 28, No. 11, p. 1151 - 1163 (1985). Abstr. in Phys. Abstr., Vol. 89, No. 1277, Entry 42191 (1986).

015.046 **L'anno di Marte.**
P. Amoroso, M. Arpino.
Orione, Vol. 6, N. 3, p. 56 - 61 (1986).

015.047 **The Moon illusion.**
W. G. Rees.
Q. J. R. Astron. Soc., Vol. 27, No. 2, p. 205 - 211 (1986).
The Moon illusion, or celestial illusion, is the illusion that the Moon near the horizon is larger than the Moon near the zenith, usually by a factor of about 2 in the diameter. The illusion has been known for over 2,000 years, and many explanations have been advanced for it. Four modern theories are discussed in this paper, and new data are presented which tend to confirm the common 'flattened celestial vault' hypothesis.

015.048 **Speaking of God and the universe.**
C. Stuhr.
J. R. Astron. Soc. Can., Vol. 80, No. 2, p. 87 - 90 (1986).

015.049 **The lunacy of it all: lunar phases and human behavior.**
J. Rotton, I. W. Kelly.
Mercury, Vol. 15, No. 3, p. 73 - 75, 95 (1986).

015.050 **Galactic position of our Sun and the optimization of search strategies for detecting extraterrestrial civilizations.**
B. A. Balazs.
Acta Astronaut., Vol. 13, No. 3, p. 123 - 126 (1986). Abstr. in Phys. Abstr., Vol. 89, No. 1283, Entry 78360 (1986).

015.051 **Philosophie - cosmologie - cosmogonie.**
A. B. K. Stojković.
Publ. Astron. Soc. "Rudjer Bošković", No. 4, p. 19 - 23 (1985). - See Abstr. 012.078. In Croatian.

The extraterrestrial encyclopedia. Our search for life in outer space.
See Abstr. 003.055.

Order out of chaos.
See Abstr. 003.154.

Contact: a novel.
See Abstr. 003.170.

Astronomical resources: interdisciplinary approaches to astronomy.
See Abstr. 014.033.

A space telescope for infrared spectroscopy of earthlike planets.
See Abstr. 035.068.

Exobiological aspects in view of Titan.
See Abstr. 100.049.

A search for the tritium hyperfine line from nearby stars.
See Abstr. 114.050.

Diatoms on Earth, comets, Europa and in interstellar space.
See Abstr. 131.196.

Applied Mathematics, Physics

021 Mathematical Papers Related to Astronomy and Astrophysics, Computing

021.001 **Algebraic manipulations on computers for lunar and planetary theories.**
J. Henrard.
Relativity in celestial mechanics and astrometry, p. 59 – 62 (1986). – See Abstr. 012.002 (IAU Symp. No. 114).

A new software to handle the large expressions involved in high accuracy analytical theories has been developed by H. Claes, M. Moons, J. M. Zune and J. Henrard. The main ideas on which this software is based are described here.

021.002 **On the function inverse to the exponential integral function.**
P. Pecina.
Bull. Astron. Inst. Czech., Vol. 37, No. 1, p. 8 – 12 (1986).

The paper deals briefly with one integral of the basic equations of meteoric physics which includes the exponential integral function Ei(x). It is further shown that the approximation of the function inverse to Ei(x), $\mathrm{Ei}^{-1}(x)$, is also required. The construction of this function, using series expansion by means of Chebyshev polynomials, is described. Computer programs, written in Fortran IV G, which serve to compute the values of Ei(x) and $\mathrm{Ei}^{-1}(x)$, accurate to 13 decimal figures, are included.

021.003 **The binning analysis: towards a better significance test.**
E. Gosset, B. Louis.
Astrophys. Space Sci., Vol. 120, No. 2, p. 263 – 306 (1986).

The authors demonstrate that the problems encountered with the binning analysis essentially lie in the significance test which is used. They review and analyse most of the existing significance tests and show that the use of those based on randomization processes considerably improves the situation. For the one–dimensional binning analysis, they apply the '2 within 4' randomization test whereas for the two– and the three–dimensional binning analyses they introduce two innovations: the '4 within 16' randomization test and the '8 within 64' randomization test, respectively.

021.004 **An outline of the practical use of Lie series.**
J. C. van der Ha.
ESA Sci. Tech. Rep., ESA STR–215, 6+30 pp. Price FF 70.00 (1985).

The paper provides an outline of the many practical uses of Lie series. These series are shown to arise naturally in the theory of differential equations when the solution is expanded in a time series. A few properties of Lie series are discussed and illustrated by means of examples. The theory is applied to the Kepler problem, resulting in a universal solution involving Stumpff functions. Finally, the usefulness of a Lie–series approach in modern perturbation theory (particularly in the field of celestial mechanics) is underlined and illustrated. It is demonstrated that the inherent automation of a Lie–series approach facilitates the derivation of perturbation solutions, in particular the higher–order contributions.

021.005 **Application of the distribution of simultaneous events to the investigation of nonlinear relations of temporal processes.**
M. Paupere.
Issled. Solntsa Krasnykh Zvezd, No. 22, p. 16 – 25 (1985). In Russian.

021.006 **Interpolation on a simplex.**
T. Fukushima.
Proceedings of the nineteenth symposium on celestial mechanics, p. 45 – 52 (1986). – See Abstr. 012.013.

The interpolation problem in a general multidimensional space is reduced to that on simplices. A family of smooth interpolation functions on a simplex is constructed from the information on its boundary only.

021.007 **Approximations et erreurs.**
T. Inoue.
Proceedings of the nineteenth symposium on celestial mechanics, p. 59 – 63 (1986). In Japanese. – See Abstr. 012.013.

021.008 **Prep77: a preprocessor of Fortran77.**
T. Fukushima.
Proceedings of the nineteenth symposium on celestial mechanics, p. 93 – 98 (1986). – See Abstr. 012.013.

A preprocessor of Fortran77 named Prep77 has been created. This preprocessor is now being used to create a software system predicting the orbit of artificial satellites at the mainframe computer of the Hydrographic Department, Tokyo.

021.009 **On some possibilities of representing the gravitational potential of an arbitrary body by power series converging everywhere excluding its surface.**
N. A. Chujkova.
Vestn. Mosk. Univ., Ser. 3. Fiz. Astron., Tom 26, No. 5, p. 77 – 81 (1985). In Russian. Abstr. in Ref. Zh., 51. Astron., 3.51.132 (1986).

021.010 **Table of Lamé functions of the first kind.**
N. B. Batueva.
Astron. Tsirk., No. 1393, p. 5 – 7 (1985). In Russian.

021.011 **Simulazione sul calcolatore Cray XMP/12 di un modello numerico per la ricostruzione della sfera celeste.**
B. Bucciarelli, M. Lattanzi.
G. Astron., Vol. 11, N. 4, p. 411 – 418 (1985).

021.012 **Confidence limits for small numbers of events in astrophysical data.**
N. Gehrels.
Astrophys. J., Vol. 303, No. 1, p. 336 – 346 (1986).

Convenient tables and approximate formulae are presented for confidence limits based on Poisson and binomial statistics. Poisson statistics apply when event rates are calculated from small numbers of observed events, and binomial statistics apply when ratios of two different event types are calculated from small numbers of observed events. The limits in the tables are given for all confidence levels commonly used in astrophysics.

021.013 **On the use of quaternions in spherical and positional astronomy.**
M. Firneis, F. Firneis.
Astron. J., Vol. 91, No. 1, p. 177 – 178 (1986).

Quaternions can be used advantageously in order to describe rotations in three–dimensional space without having to resort to the juxtaposition of three Eulerian rotations for this movement.

As half-angle arguments enter into the formulation of the quaternions, a method is at hand to derive systematically all half-angle formulas of spherical trigonometry. This is achieved in two alternative ways which transfer a tangent vector from one side of the spherical triangle to the opposite vertex. Supplementary to all the half-angle formulas, a set of 12 additional half-angle formulas is derived.

021.014 **Magnetic tape read/write operations on PDP-11 series computers.**
C.-j. Shen.
Publ. Purple Mt. Obs., Vol. 4, No. 3, p. 71 – 79 (1985). In Chinese.
The author describes the methods for reading/writing ANSI or non-ANSI magnetic tapes by means of QIO or WTQIO on PDP-11 series computers.

021.015 **Poisson clusters and Poisson voids.**
H. D. Politzer, J. P. Preskill.
Phys. Rev. Lett., Vol. 56, No. 2, p. 99 – 102 (1986).
Expressions are derived for the expected abundance of clusters and voids in a sample of randomly distributed objects.

021.016 **An absolutely stable domain of Hill's equation.**
J.-h. Mai.
Sci. Sin., Ser. A, Vol. 29, No. 5, p. 488 – 501 (1986).

021.017 **Numerical exploration of C^1 billiards constructed with four arcs of circles.**
A. Hayli, T. Dumont.
Celest. Mech., Vol. 38, No. 1, p. 23 – 66 (1986). In French.
The paper deals with some properties of the dynamical system with two degrees of freedom defined by the motion of a particle in a certain type of billiard.

021.018 **Conversion to MELCOM COSMO 900II from Mark III VLBI archive data.**
T. Ishikawa.
Proc. Int. Latitude Obs. Mizusawa, No. 24, p. 9 – 19 (1985).

021.019 **Program design for compiling the Time Bulletin by means of a microcomputer at the Yunnan Observatory.**
J.-g. Zhou.
Publ. Shaanxi Astron. Obs., Vol. 7, No. 2, p. 107 (1984). Abstract. In Chinese and English.

021.020 **A data reduction package for multiple object spectroscopy.**
J. M. Hill, J. D. Eisenhamer, D. R. Silva.
Prepr. Steward Obs., No. 642, 14 pp. (1986). To appear in Proc. SPIE Int. Soc. Opt. Eng., Vol. 672 "Instrumentation in Astronomy IV", March 1, 1986.

021.021 **Analysis of the kernel part of the NRAO stand-alone FORTH system.**
X.-h. Di, F. Han, X.-l. Wang.
Publ. Purple Mt. Obs., Vol. 4, No. 3, p. 52 – 70 (1985). In Chinese.

Astronomy with your personal computer.
See Abstr. 003.010.

Inverse problems in astronomy. A guide to inversion strategies for remotely sensed data.
See Abstr. 003.016.

Himmelskunde und Kleinrechner. Seminarpapiere vom 13. Sternfreundeseminar 1985.
See Abstr. 012.088.

A fast and reliable method for computing free-bound emission coefficients for hydrogenic ions.
See Abstr. 022.088.

Einiges über optisches Rechnen. Teil 2.
See Abstr. 031.014.

A prescription for period analysis of unevenly sampled time series.
See Abstr. 036.068.

Period search using Turbo Pascal.
See Abstr. 036.109.

Programa "ARIES 2 ML" para reducción de posiciones de estrellas.
See Abstr. 036.184.

An analysis of the properties of CLEAN and Smoothness Stabilized CLEAN – some warnings.
See Abstr. 036.198.

Computation of correction values for the determination of planetary right ascensions with the photoelectric transit instrument.
See Abstr. 041.037.

The application of the proper reference framework in astronomy.
See Abstr. 041.038.

Un catálogo provisional FK5 para uso en astronomía geodésica.
See Abstr. 043.013.

A real-time catalogue of the apparent places of stars.
See Abstr. 046.017.

Order-of-scattering and doubling method for the solution of transfer equations.
See Abstr. 063.072.

Eine neue Methode zur Berechnung des Sternaufbaus.
See Abstr. 065.150.

Simulatie meteoorwaarnemingen. De computer als visuele waarnemer.
See Abstr. 104.072.

The formation of planetary systems – a simulation on the microcomputer.
See Abstr. 107.049.

022 Physical Papers Related to Astronomy and Astrophysics

022.001 **Excitation rate coefficients for optically allowed transitions in C–like ions.**
K. M. Aggarwal.
Mon. Not. R. Astron. Soc., Vol. 218, No. 1, p. 123 – 127 (1986).

Based on the **R**–matrix and Coulomb–Born data of collision strengths for carbon–like O, Ne, Mg and Si ions, empirical constants are derived for the optically allowed transitions from which the excitation rate coefficients can be analytically evaluated. The effective collision strengths for these transitions calculated from these constants can be reproduced to within 10 per cent accuracy for the $1s^22s^22p^2 \rightarrow 1s^22s2p^3$ transitions and within 20 per cent accuracy for the $1s^22s2p^3 \rightarrow 1s^22p^4$ transitions in comparison to the numerical results published earlier.

022.002 **Amorphous bronzite – a silicate of astronomical importance.**
J. Dorschner, C. Friedmann, J. Gürtler, T. Henning, H. Wagner.
Mon. Not. R. Astron. Soc., Vol. 218, No. 2, p. 37P – 40P (1986).

Mass absorption coefficients measured for artificially produced submicrometre–sized grains of amorphous bronzite are reported in the wavelength region 5 – 40 µm. Such grains enable an excellent representation of the spectra of T Tauri stars and very young stellar objects in the middle–infrared.

022.003 **Isotope anomalies induced in laboratory distillation.**
T. M. Esat, R. H. Spear, S. R. Taylor.
Nature, Vol. 319, No. 6054, p. 576 – 578 (1986).

The authors have carried out high–temperature distillation experiments at 5×10^{-5}torr on terrestrial pyroxene crystals and measured ^{24}Mg, ^{25}Mg and ^{26}Mg simultaneously using a 61–cm multi–cup mass spectrometer. The raw data were corrected for isotope fractionation using the power law and the Rayleigh law. Regardless of the correction procedure, the sample residues are depleted in ^{26}Mg and the condensates are enriched in ^{26}Mg, relative to non–distilled standards. These effects mimic the "isotope anomalies" in Mg observed in Allende coarse– and fine–grained inclusions, previously attributed to nucleosynthetic processes involving ^{26}Al.

022.004 **Astrophysical implications of amorphous ice – a microporous solid.**
E. Mayer, R. Pletzer.
Nature, Vol. 319, No. 6051, p. 298 – 301 (1986).

Since the observation of the 3–µm band in interstellar infrared sources, vapour–deposited amorphous ice, H_2O(as), has been discussed as a major component of comets, of satellites of the outer planets and of interstellar dust. The authors have reinvestigated adsorption of N_2 on H_2O(as) and find that it is a microporous solid. The differences between adsorption phenomena taking place in micropores and on the surface of intermediate pores or non–porous adsorbents should be of general importance for the various speculations of the role of H_2O(as) in space. In particular, the authors discuss the influence of micropores on the H_2 recombination rate on H_2O(as) in interstellar dust and on adsorption of volatile gases in comets.

022.005 **Einstein *A*–values for rotational transitions in the H_2CO molecule.**
S. Jaruschewski, S. Chandra, D. A. Varshalovich, W. H. Kegel.
Astron. Astrophys., Suppl. Ser., Vol. 63, No. 2, p. 307 – 312 (1986). In French.

Einstein *A*–values have been computed for the electric dipole transitions of the H_2CO molecule between all rotational levels of the vibrational ground state up to 350 cm^{-1}. From these *A*–values the radiative life times have been derived.

022.006 **Grain formation experiments by a plasma jet apparatus.**
T. Tanabé, F. Kamijo, T. Onaka, A. Sakata, S. Wada.
Astrophys. Space Sci., Vol. 119, No. 1, p. 147 – 149 (1986). – See Abstr. 012.001.

022.007 **Differential excitation cross section of atomic oxygen by electron impact: (^{3}P–^{1}D transition).**
T. W. Shyn, W. E. Sharp.
J. Geophys. Res., Vol. 91, No. A2, p. 1691 – 1697 (1986).

Absolute differential excitation cross sections of atomic oxygen (^{3}P–^{1}D transition) by electron impact have been measured by a crossed–beam method. Atomic oxygen was generated by a thermal dissociation method. The energy and angular range covered were from 7 to 30 keV and from 30° to 150°, respectively. A few theoretical results were compared with the present measurements and they are in good agreement with the experimental results within the accuracy of the measurement.

022.008 **New measurements of the absorption cross–sections in the Herzberg continuum of molecular oxygen in the region between 205 and 240 nm.**
A. Jenouvrier, B. Coquart, M. F. Merienne–Lafore.
Planet. Space Sci., Vol. 34, No. 2, p. 253 – 254 (1986).

Laboratory measurements of the absorption of molecular oxygen between 205 and 240 nm have been made with a 50 m multiple reflection cell leading to an absorption path of about 1400 m at pressures between 5 to 100 Torrs.

022.009 **Errata: "Prediction and assignment of the FIR spectrum of hydrogen peroxide" [J. Quant. Spectrosc. Radiat. Transfer, Vol. 32, No. 4, p. 325 – 333 (1984)].**
P. Helminger, J. K. Messer, W. C. Bowman, F. C. De Lucia.
J. Quant. Spectrosc. Radiat. Transfer, Vol. 35, No. 1, p. 79 (1986). See Abstr. 38.022.153.

022.010 **Self, nitrogen and oxygen broadening of the 115–GHz line of carbon monoxide.**
J. M. Colmont, N. Monnanteuil.
J. Quant. Spectrosc. Radiat. Transfer, Vol. 35, No. 2, p. 81 – 85 (1986).

The purpose of the study was measurements of nitrogen–, oxygen– and air–broadened half–widths of the 115–GHz line of carbon monoxide and the determination of the variation of these half–widths with temperature in the temperature range of the earth's middle atmosphere.

022.011 **N_2 and H_2 broadening parameters in the fundamental band of CO.**
M. F. Le Moal, F. Severin.
J. Quant. Spectrosc. Radiat. Transfer, Vol. 35, No. 2, p. 145 – 152 (1986).

Pressure broadening parameters of lines belonging to the (1–0) band of CO in collision with N_2 or H_2 have been measured with a vacuum, high–resolution grating spectrometer. The measurements were carried out in the *R* branch up to $J = 20$ at 298 and 195K. The experimental results are compared with values calculated from two semiclassical theoretical models.

022.012 **Classical binary–encounter collision theory for electron–impact ionization of ions.**
Y. T. Lee.
J. Quant. Spectrosc. Radiat. Transfer, Vol. 35, No. 2, p. 153 – 165 (1986).

The author calculated electron–ion impact ionization using classical binary–encounter collision theory. A differential cross section per unit energy and momentum transfer is derived for two–electron scattering. The author obtains the total ionization cross section by integrating this differential cross section over the appropriate energy and momentum transfer and the bound electron–velocity distribution. The cross sections and rate coefficients, which are sufficiently accurate for many practical applications, are expressed in computationally simple form.

022.013 **Further atomic calculations for Fe XXIII.**
A. K. Bhatia, H. E. Mason.
Astron. Astrophys., Vol. 155, No. 2, p. 413 – 416 (1986).
Electron collision strengths for Fe XXIII have been obtained using the 'distorted wave' approximation. Previous results (Bhatia and Mason, 1981) were obtained including the configurations $2s^2$, $2s2p$, $2p^2$, $2s3s$, $2s3p$ and $2s3d$. The correlation configurations $2p3s$, $2p3p$ and $2p3d$ are also included in the solution of the electron scattering problem.

022.014 **Theoretical intensity ratios for the X–ray lines from Ca XV.**
A. K. Bhatia, H. E. Mason.
Astron. Astrophys., Vol. 155, No. 2, p. 417 – 419 (1986).
Electron collision strengths for Ca XV have been obtained using the 'distorted wave' approximation. Previous results (Mason, 1975; Dere et al., 1979) were obtained for the configurations $2s^22p^2$, $2s2p^3$, $2p^4$. The authors present results for transitions $2s^22p^2 \rightarrow 2s^22p3s$, $2s^22p3d$ which give rise to lines in the X–ray spectra of solar active regions and flares.

022.015 **Production and loss of HC_3N in interstellar clouds: some relevant laboratory measurements.**
J. S. Knight, C. G. Freeman, M. J. McEwan, S. C. Smith, N. G. Adams, D. Smith.
Mon. Not. R. Astron. Soc., Vol. 219, No. 1, p. 89 – 94 (1986).
The results of recent selected ion flow tube (SIFT) experiments on the ion–molecule chemistry of cyanoacetylene are considered in the context of the chemistry of HC_3N in the interstellar environment. Important errors revealed by this SIFT investigation, following an earlier flowing afterglow study in the authors' laboratory, have led to a different perception of the ion–molecule chemistry that HC_3N may undergo in interstellar clouds. It is now evident that insertion and association occur in the reactions of hydrocarbon ions with HC_3N.

022.016 **Temperature and density determination from Si III lines.**
H. Nussbaumer.
Astron. Astrophys., Vol. 155, No. 1, p. 205 – 209 (1986).
It is shown how under conditions of gaseous nebulae, the temperature T_e and the density N_e can be derived from observed Si III fluxes at $\lambda\lambda$ 1206.5, 1301, 1882.7, 1892.0. Transition probabilities among the energetically 17 lowest terms are given and theoretical line intensity ratios are calculated.

022.017 **Optical properties of solid and liquid sulfur at visible and infrared wavelengths.**
R. Sasson, R. Wright, E. T. Arakawa, B. N. Khare, C. Sagan.
Icarus, Vol. 64, No. 3, p. 368 – 374 (1985).
Knowledge of the optical constants of elemental sulfur has potential applications to Venus, Jupiter, Io, Amalthea, and the Earth. The real part, n, of the index of refraction of liquid sulfur (at 133°C) and of solid orthorhombic sulfur (at 25°C) for the wavelength range 0.4 – 2.0 µm were measured ellipsometrically. The imaginary part, k, of the refractive index of liquid sulfur was obtained by transmittance measurements at the same temperature and wavelength range. The reflectance of semi–infinite slabs of solid and liquid sulfur is calculated using the measured n and k values. The authors confirm that sulfur melts on Io would be classified as "black" by the Voyager imaging system.

022.018 **Optical constants of chlorite and serpentine between 2.5 and 50 µm.**
T. Mooney, R. F. Knacke.
Icarus, Vol. 64, No. 3, p. 493 – 502 (1985).
The real and imaginary parts of the 2.5– to 50–µm optical constants of chlorite and serpentine minerals have been determined by a combination of infrared transmission and reflectance spectroscopy.

022.019 **Laboratory measurements of the microwave opacity and vapor pressure of sulfuric acid vapor under simulated conditions for the middle atmosphere of Venus.**
P. G. Steffes.
Icarus, Vol. 64, No. 3, p. 576 – 585 (1985).
Microwave absorption observed in the 35– to 48–km altitude region of the Venus atmosphere has been attributed to the presence of gaseous sulfuric acid (H_2SO_4) in that region. This has motivated the laboratory measurement of the microwave absorption at 13.4– and 3.6–cm wavelengths from gaseous H_2SO_4 in a CO_2 atmosphere under simulated conditions for that region. As part of the same experiments, upper limits on the saturation vapor pressure of gaseous H_2SO_4 have also been determined. The measurements for microwave absorption have been made in the 1– to 6–atm pressure range, with temperatures in the 500 to 575K range.

022.020 **The effect of particle size and porosity on spectral contrast in the mid–infrared.**
J. W. Salisbury, J. W. Eastes.
Icarus, Vol. 64, No. 3, p. 586 – 588 (1985).
The authors find that the decreasing intensity of fundamental molecular vibration bands with decreasing particle size is due primarily to increasing porosity of the finer particle size ranges. This implies that laser reflectance measurements from orbiting spacecraft should avoid loss of spectral contrast for fine particulate surfaces, because such measurements near zero phase angle will benefit from the opposition effect.

022.021 **Dynamical problems of thermoelastic solids.**
A. M. Elnagar, A. M. Abd–Allah.
Earth, Moon, Planets, Vol. 34, No. 3, p. 201 – 211 (1986).
The dynamical problem for thermal stresses in an infinite isotropic elastic cylinder of radius a with its axis along the z–axis, subject to fixed boundary conditions is studied. The Fourier heat conduction equation has been solved applying the Fourier transform and the theory of complex variable. The thermoelastic equation of motion has been separated into two wave equations which can be solved separately. The temperature, the displacement and the stress components have been obtained in analytical form as series involving Bessel function of first kind and of order zero.

022.022 **Restricted quantum–mechanical three–body problems. IV. Asymptotically exact wave functions of a helium–like ion in all singular points of its Schrödinger equation.**
S. Barcza.
Astrophys. Space Sci., Vol. 120, No. 1, p. 61 – 95 (1986).
This paper aims to present some new results on the asymptotic behaviour of the solutions of an infinitely coupled system of second–order ordinary differential equations emerging from a perturbative treatment of the Schrödinger equation of helium–like ions. The essence of the perturbative treatment is that the eigenfunction of a helium–like ion is expanded in the Hilbert space built up from the eigenfunctions of an electron in the field of two fixed centres.

022.023 **Stability of amorphous circumstellar silicate grains.**
T. Henning, J. Svatoš.
Astron. Nachr., Vol. 307, No. 1, p. 49 – 52 (1986). = Mitt. Univ.–Sternw. Jena, Nr. 170.
Within the framework of classical nucleation theory the authors have shown that Mg_2SiO_4 grains (radii of nearly 0.1 µm) which are likely candidates as condensation products in circumstellar shells around late–type stars must condense at about 750K to remain amorphous. If the condensation temperature is about 1000K the newly formed grains must be rather small to rest in the amorphous state during cooling. Submicron sized $MgSiO_3$ grains are entire stable against crystallization for $T_{cond} \leqslant 1000$K.

022.024 **Millimeter wave spectrum of tricarbon oxide sulfide, OCCCS.**
M. Winnewisser, F. Holland.
Astron. Astrophys., Vol. 157, No. 1, p. 19 – 22 (1986).
The millimeter wave rotational transitions of OCCCS (1,2–Propadien–1–one, 3–thioxo–; 2219–62–7) in the vibrational ground state and the first excited vibrational state of the lowest lying doubly degenerate bending mode ν_7 are reported. The lines were measured in the frequency range from 105 to 345 GHz.

022.025 **Plasma densities from the He–like ion Ne IX.**
J. G. Doyle, F. P. Keenan.
Astron. Astrophys., Vol. 157, No. 1, p. 116 – 118 (1986).
The authors present line emissivity data for the three principal lines of the He–like ion Ne IX. This data is compared with high resolution solar flare observations from the Solar Maximum Mission. During the rise phase of flares it is shown that the forbidden line is blended with a high temperature line, as opposed to the intercombination line suggested by McKenzie (1985). It is suggested that the electron density during the rise phase of flares be determined from the intercombination to resonance line ratio which the authors have integrated over a multi–thermal plasma and is shown to be only slightly dependent on the assumed differential emission measure distribution. Scattering of resonance line photons from the line–of–sight is shown to be unimportant.

022.026 **Structure and intensities of microwave lines in the spectra of diatomic molecules.**
J. B. Tatum.
Astrophys. J., Suppl. Ser., Vol. 60, No. 2, p. 433 – 474 (1986).
The structure of microwave lines in the spectrum of a diatomic molecule having zero, one, or two nuclear spins is reviewed. In molecules, such as DCN, with two nuclear spins, three limiting cases of coupling between the nuclear and electronic angular momenta are conceivable. Novel methods of displaying the energy levels and line structures by means of a three–dimensional ternary diagram and a movie film are described. The method of calculation is described in some detail.

022.027 **Millimeter wave spectrum of ^{13}C substitutions of cyclopropenylidene C_3H_2.**
M. Bogey, J. L. Destombes.
Astron. Astrophys., Vol. 159, No. 1/2, p. L8 – L9 (1986).
The authors report the rotational spectrum observation in the laboratory of the two ^{13}C isotopomers of cyclopropenylidene C_3H_2, a molecule recently discovered in abundance in the interstellar medium. The accurate molecular parameters deduced from their analysis are used to predict the frequencies of lines of astrophysical interest.

022.028 **Diffusion and thermal diffusion in partially ionized gases in the atmospheres of the Sun and planets.**
J. Geiss, A. Bürgi.
Astron. Astrophys., Vol. 159, No. 1/2, p. 1 – 15 (1986).
The collision terms of the momentum equations for the components of a partially ionized gas including the effects of ionization, recombination and charge exchange are investigated. The drag and thermal diffusion coefficients are derived for a gas consisting of electrons, one atomic and one ionic species as major components and charged or neutral minor constituents for the case of equal temperatures of atoms and ions. Drag terms and thermal diffusion coefficients of the major and minor components in partially ionized hydrogen are explicitly derived. The results are discussed in regard to solar and stellar surface processes. Finally, the effects of atoms on the diffusion of ions in the upper ionosphere and plasmasphere are discussed.

022.029 **Molecular electrostatic potentials as a first approach to protonated interstellar molecules.**
F. Pauzat, D. Talbi, Y. Ellinger.
Astron. Astrophys., Vol. 159, No. 1/2, p. 246 – 250 (1986).
The applicability of ab–initio molecular electrostatic potentials to examine the protonated adducts of interstellar molecules is tested on the series: CO, CS, SiO, SiS, N_2, H_2 and CO_2. The stable isomers HCO^+ (linear), HCS^+ (linear), $SiOH^+$ (linear), $SiSH^+$ (bent), $HN_2{}^+$ (linear), $H_3{}^+$ (triangular), and $HOCO^+$ (bent) are correctly predicted. The method used provides rapid and inexpensive semi–quantitative results particularly useful for a first selection of the stable protonated entities in chemistry models involving a large number of proton acceptors.

022.030 **The establishment of internally consistent solar scales of oscillator strengths and abundances of chemical elements. Zirconium.**
Eh. A. Gurtovenko, R. I. Kostyk, T. V. Orlova.
Kinematika Fiz. Nebesn. Tel, Tom 2, No. 1, p. 20 – 23 (1986). In Russian. English translation in Kinematics Phys. Celest. Bodies.
Oscillator strengths for 19 Zr I lines (r.m.s. error ± 0.078 dex) and 10 Zr II lines (± 0.081 dex) are determined from observed equivalent widths and central depths of solar Fraunhofer lines. The analysis of the differences of oscillator strengths derived from equivalent widths and from central depths leads to the conclusion that the lines under study have no essential hyperfine structure.

022.031 **Oscillator strengths for neutral iron and its abundance in the solar atmosphere.**
A. A. Boyarchuk, I. S. Savanov.
Izv. Krymskoj Astrofiz. Obs., Tom 70, p. 57 – 128 (1985). In Russian. English translation in Bull. Crimean Astrophys. Obs., Vol. 70.
Published values of the oscillator strengths for neutral iron lines were critically reviewed. Special attention was drawn to the influence of systematic errors on the *gf* values. The established data on the reduced and accepted *gf* values for 3420 lines of neutral iron are presented in a table. Using these established values of oscillator strengths, the authors reviewed the earlier published results on the iron abundance in the atmosphere of the sun. Based only on the weak lines, the uncertainties related with turbulent velocity and damping were reduced. The reviewed results are compiled in a table, and the mean iron abundance in the solar atmosphere was adopted accordingly as equal to $\lg \varepsilon(Fe) = 7.60 \pm 0.10$.

022.032 **La lumière raconte.**
M. Gros.
Astronomie, Vol. 100, p. 187 – 197 (1986).

022.033 **^{53}Mn in the early solar system.**
J. L. Birck, C. J. Allègre.
Meteoritics, Vol. 20, No. 4, p. 609 (1985). Abstract. – See Abstr. 010.661.

022.034 **Gallium and germanium as tracers of formation and differentiation processes in terrestrial planetary objects.**
D. J. Malvin, M. J. Drake.
Meteoritics, Vol. 20, No. 4, p. 704 – 705 (1985). Abstract. – See Abstr. 010.661.

022.035 **On the unification of thin target and thick target approaches to the interpretation of cosmogenic nuclides in meteorites.**
R. Michel.
Meteoritics, Vol. 20, No. 4, p. 711 – 712 (1985). Abstract. – See Abstr. 010.661.

022.036 **A contribution to the low temperature Fe–Ni phase diagram.**
K. B. Reuter, D. B. Williams, J. I. Goldstein.
Meteoritics, Vol. 20, No. 4, p. 742 – 743 (1985). Abstract. – See Abstr. 010.661.

022.037 **Isotopic anomalies scattering processes and the Pauli exclusion principle.**
F. Robert.
Meteoritics, Vol. 20, No. 4, p. 744 – 746 (1985). Abstract. – See Abstr. 010.661.

022.038 **Formation of organic compounds by impact of energetic atoms and ions on ice and frozen volatiles.**
K. Rössler, B. Nebeling, M. C. Batista.
Meteoritics, Vol. 20, No. 4, p. 748 – 749 (1985). Abstract. – See Abstr. 010.661.

022.039 **Systematic studies of ^{10}Be and ^{26}Al production in meteorites: simulation by isotropically irradiated meteorite models.**
S. Theis, P. Englert, R. Michel, G. F. Herzog, T. H. Kruse, J. Klein, R. Middleton.
Meteoritics, Vol. 20, No. 4, p. 772 – 774 (1985). Abstract. – See Abstr. 010.661.

022.040 **Production of mass independently fractionated oxygen isotopic components by ultra–violet light.**
M. H. Thiemens, T. Jackson.
Meteoritics, Vol. 20, No. 4, p. 775 – 776 (1985). Abstract. – See Abstr. 010.661.

022.041 **On the structure of the energy spectrum of Tc I.**
A. S. Bakhtiyarov, P. O. Bogdanovich, G. L. Žukauskas, A. A. Nikitin, Z. B. Rudzikas.
Kinematika Fiz. Nebesn. Tel, Tom 2, No. 2, p. 13 – 20 (1986). In Russian. English translation in Kinematics Phys. Celest. Bodies.
The status of the analysis of technetium lines in the spectra of stars and the sun is considered. The suitability and efficiency of various approaches of the self–consistent Hartree–Fock field for theoretical investigations of the energy spectra of neutral technetium, as well as for the wavelengths and the other characteristics of electronic transitions between its levels are discussed. The $4d^55s^2$, $4d^65s$, $4d^65p$ and $4d^55s5p$ configurations are studied. The results show that the energy spectra of these configurations are strongly overlapped, therefore correlation effects ought to be taken into account.

022.042 **2.5 – 300 μm laboratory observations of submicron SiC particles as cosmic dust candidates.**
A. Borghesi, E. Bussoletti, L. Colangeli, V. Orofino, M. Guido, S. Nunziante–Cesaro.
Infrared Phys., Vol. 26, No 1, p. 37 – 42 (1986).
Three different kinds of α–SiC and one β–SiC sample which have undergone the same treatment procedures have been studied morphologically. Their absorption spectra have also been measured in the range 2.5 – 300 μm. Outside the typical 11.5 μm absorption band, all spectra show an FIR trend which can be fitted by a power law $\lambda^{-\gamma}$. The results of this analysis seem to suggest that α–SiC and β–SiC differ morphologically, while their FIR spectra follow a similar $\lambda^{-1.4}$ trend. Astronomical observations at these long wavelengths do not allow any identification of the actual SiC type existing in space which, on the contrary, can probably be identified by analysing the 11.5 μm band.

022.043 **Oscillator strengths for allowed and intercombination electric dipole transitions in Fe XII.**
S. S. Tayal, R. J. W. Henry.
Astrophys. J., Vol. 302, No. 1, p. 200 – 204 (1986).
The authors have evaluated oscillator strengths for several transitions between the ground $3s^23p^3$ configuration and the excited $3s3p^4$ and $3s^23p^23d$ configurations using fairly large configuration–interaction expansions. The calculations are performed in both *LS* and intermediate couplings. The relativistic effects in intermediate coupling are incorporated by means of the Breit–Pauli Hamiltonian. Comparison with other available results is generally good.

022.044 **Theoretical study of CS–containing molecules I. *Ab initio* Hartree–Fock predictions for C_2S_2.**
J. S. Gómez–Jeria, R. G. E. Morales, L. M. Reyes.
Astrophys. J., Vol. 302, No. 1, p. 488 – 491 (1986).
Calculations at the HF/4–31G level were performed for the C_2S_2 molecule. The ground state is predicted to be $^3\Sigma_g^-$ and is of diradical nature. The lowest lying ionic and excited states were also analyzed. It is concluded that formation of C_2S_2 involving CS species is possible only under the influence of ultraviolet excitation.

022.045 **The production and sputtering of S_2 by keV ion bombardment.**
J. W. Boring, N.–s. Zhao, D. B. Chrisey, D. J. O'Shaughnessy, J. A. Phipps, R. E. Johnson.
Asteroids, comets, meteors II, p. 229 – 234 (1986). – See Abstr. 012.010.
Targets of solid sulfur, H_2S, and CS_2 at 15K were bombarded by keV ions and the ejected neutral molecules were ionized and observed using a quadrupole mass spectrometer. For sulfur targets the dominant species sputtered are S_1 and S_2 for H_2S both H_2S and S_2 dominate the ejecta. For the CS_2 target signals of S, CS, S_2 and CS_2 were observed for short irradiation times. These signals rapidly decreased and a sputter resistant, dark brown residue, stable at room temperature was produced. Residues are produced in the bombardment of all carbon and sulfur containing condensed gases. The authors have considered the implication of these results in terms of the possible production of S_2 by solar and galactic particle bombardment of cometary nuclei and grains.

022.046 **On clathrate hydrates in comets.**
J. Klinger, J. Ocampo, B. Schmitt.
Asteroids, comets, meteors II, p. 241 – 243 (1986). – See Abstr. 012.010.
The decomposition of clathrate hydrates of CO_2 and of krypton have been studied in the laboratory. The experiments showed that the vapor phase over the decomposing clathrate contained a much higher proportion of guest molecules or guest atoms than that contained in the solid. This means that production rates deduced from spectroscopic data of cometary comae cannot inform us whether or not cometary nuclei contain clathrate hydrates.

022.047 **Parents for the cometary CN radical.**
J. Crovisier, D. Bockelée–Morvan.
Asteroids, comets, meteors II, p. 289 – 292 (1986). – See Abstr. 012.010.
The authors suggest that HCN is not the (only) parent for cometary CN radicals. Other candidates are C_2N_2, HC_3N and CH_3CN. The authors discuss the photodissociation properties and excitation conditions of all these parent molecules.

022.048 **Absolute experimental differential and integral electron excitation cross sections for atomic oxygen. 1. The ($^3P \rightarrow {}^3S^0$) transition (1304 Å) at 100 eV.**
J. P. Doering, S. O. Vaughan.
J. Geophys. Res., Vol. 91, No. A3, p. 3279 – 3286 (1986).

022.049 **The laboratory millimeter–wave spectrum of methyl formate in its ground torsional *E* state.**
G. M. Plummer, E. Herbst, F. C. De Lucia, G. A. Blake.
Astrophys. J., Suppl. Ser., Vol. 60, No. 4, p. 949 – 961 (1986).
Over 250 rotational transitions of the internal rotor methyl formate ($HCOOCH_3$) in its ground ($\upsilon_t = 0$) degenerate (*E*) torsional substate have been measured in the millimeter–wave spectral region. The resulting rotational constants allow accurate prediction of the $\upsilon_t = 0$ *E* substate methyl formate spectrum below 300 GHz between states with angular momentum $J \leqslant 30$ and rotational energy $E_{rot} \leqslant 350$ cm^{-1}. The calculated transition frequencies for the *E* state, when combined with the results of a previous analysis of the ground symmetric, nondegenerate (*A*) state, account for over 200 of the emission lines observed toward Orion in a recent survey of the 215 – 265 GHz band.

022.050 **Franck–Condon factors for some molecules of astrophysical interest.**
J.–I. Straume, V. Tsymbal.
Issled. Solntsa Krasnykh Zvezd, No. 20, p. 13 – 25 (1984). In Russian.
Franck–Condon factors with vibration–rotation interaction for Morse potentials are calculated for SiCl, SiN, C_2^-, CN.

022.051 **Resonance effects on excitation rates for highly charged helium-like ions.**
H. Zhang, D. H. Sampson.
Bull. Am. Astron. Soc., Vol. 17, No. 4, p. 832 (1985). Abstract. – See Abstr. 010.062.

022.052 **Band oscillator strengths for TiO.**
J. E. Littleton, S. P. Davis.
Bull. Am. Astron. Soc., Vol. 17, No. 4, p. 847 (1985). Abstract. – See Abstr. 010.062.

022.053 **Statistics of equivalent width data and new oscillator strengths for Si II, Fe II and Mn II.**
D. Van Buren.
Bull. Am. Astron. Soc., Vol. 17, No. 4, p. 847 (1985). Abstract. – See Abstr. 010.062.

022.054 **Atomic data for the S IV 10.5–μm fine–structure transition.**
C. T. Johnson, A. E. Kingston, P. L. Dufton.
Mon. Not. R. Astron. Soc., Vol. 220, No. 1, p. 155 – 163 (1986).
New atomic data relevant to the fine–structure S IV transition at a wavelength of 10.5 μm are presented. The transition probability deduced from configuration interaction wavefunctions agrees with that obtained by Naqvi, but not with the recent result of Bhatia, Doschek & Feldman. Electron impact collision strengths for this transition have been calculated using the R–matrix method. They confirm the important resonance contribution found by Dufton et al. but lead to collision rates which are up to 15 per cent larger. The effect of these data on current models of H II regions is discussed and the authors suggest that the accuracy of such models may be limited by errors in the atomic data for many of the fine–structure lines involved.

022.055 **Splitting of some Fe I spectral lines in a magnetic field in the 8800 – 10800 Å region.**
E. N. Zemanek.
Vestn. Kiev. Univ. Astron., No. 27, p. 26 – 32 (1985). In Russian.

022.056 **Splitting of some ionized iron lines in a magnetic field.**
E. N. Zemanek, A. P. Stefanov.
Vestn. Kiev. Univ. Astron., No. 27, p. 32 – 38 (1985). In Russian.

022.057 **The absorption coefficient in biatomic molecules.**
V. D. Motrich.
Problems of astronomy, p. 60 – 64 (1985). In Russian. Abstr. in Ref. Zh., 51. Astron., 3.51.187 (1986). – See Abstr. 003.023.

022.058 **The influence of heating on the Doppler profiles of spectral lines.**
D. V. Dimitrov, V. G. Shkodrov, B. J. Kovachev.
Dokl. Bolg. Akad. Nauk, Vol. 38, No. 7, p. 811 – 813 (1985). Abstr. in Ref. Zh., 51. Astron., 3.51.659 (1986).

022.059 **On a possibility of experimental investigation of the anomalous ionization phenomenon.**
L. Bankov, A. A. Galeev, Eh. M. Dubinin, I. M. Podgornyj, Yu. N. Potanin.
Kosm. Issled., Tom 24, Vyp. 3, p. 404 – 410 (1986). In Russian. English translation in Cosm. Res.

022.060 **Oscillator strengths of iron lines.**
R. I. Kostyk, T. V. Orlova.
Kinematika Fiz. Nebesn. Tel, Tom 2, No. 3, p. 63 – 67 (1986). In Russian. English translation in Kinematics Phys. Celest. Bodies.
Oscillator strengths of 380 iron lines have been determined from the observed central intensities and equivalent widths of the Fraunhofer solar lines with r.m.s. error not exceeding 12%.

022.061 **Radiative lifetimes of some states of Ne II.**
K. Blagoev, K. Stankova.
J. Quant. Spectrosc. Radiat. Transfer, Vol. 35, No. 6, p. 483 – 486 (1986).
The radiative lifetimes of eight states of Ne II, namely $3d$, $3d'$, $4s$ and $4s'$, are measured. The delay–coincidence method with pulsed electron excitation is employed. A comparison with other results is also reported.

022.062 **Martian dust particles as condensation nuclei: a preliminary assessment of mineralogical factors.**
J. L. Gooding.
Icarus, Vol. 66, No. 1, p. 56 – 74 (1986). – See Abstr. 012.024.
This paper presents a review of mineral properties as they relate to condensate formation under Martian conditions in order to identify key problems that deserve further investigation. New calorimetry data are reported for water–ice formation on candidate Martian materials. It appears that calorimetry data can be resolved with cloud–chamber data and that additional calorimetry experiments on condensate formation involving both H_2O and CO_2 should lead to significantly improve understanding of condensation phenomena on Mars.

022.063 **Properties of filamentary sublimation residues from dispersions of clay in ice.**
R. S. Saunders, F. P. Fanale, T. J. Parker, J. B. Stephens, S. Sutton.
Icarus, Vol. 66, No. 1, p. 94 – 104 (1986). – See Abstr. 012.024.
On simulated Martian polar deposits and cometary dirty ices a series of low–pressure and low–temperature experiments was conducted during which the water ice was sublimed from dilutions, of various ratios, of frozen water–clay dispersions.

022.064 **Spectral reflectance of solid sulfur trioxide (0.25 – 5.2 μm): implications for Jupiter's satellite Io.**
R. M. Nelson, W. D. Smythe.
Icarus, Vol. 66, No. 1, p. 181 – 187 (1986).
The authors present the spectral UV–vis–near–IR reflectance of SO_3 that they obtained in the laboratory and they discuss its significance in the context of sulfur trioxide's other physical properties and the Io environment.

022.065 **Measurements of transition probabilities for doubly ionized vanadium.**
T. Wujec, J. Musielok.
J. Quant. Spectrosc. Radiat. Transfer, Vol. 35, No. 4, p. 239 – 245 (1986).
Transition probabilities of 43 doubly ionized V lines have been measured in the spectral range 2280 – 3600 Å. These lines belong to the $3d^24s - 3d^24p$ transition array. Emission spectroscopy was used to obtain line intensities. The spectra were emitted from a He plasma with added $VOCl_3$ vapour in a wall–stabilized arc.

022.066 **Stark broadening along homologous sequences of singly ionized noble gases.**
T. L. Pittman, N. Konjević.
J. Quant. Spectrosc. Radiat. Transfer, Vol. 35, No. 4, p. 247 – 253 (1986).
The authors report measured Stark widths for eight Ne II, eight Ar II, five Kr II and seven Xe II spectral lines, all belonging to np–nd doublets.

022.067 **Temperature dependence of the far–infrared absorption spectrum of gaseous methane.**
P. Codastefano, P. Dore, L. Nencini.
J. Quant. Spectrosc. Radiat. Transfer, Vol. 35, No. 4, p. 255 – 263 (1986).
The rototranslational absorption spectrum of gaseous methane has been measured at seven different temperatures from 296 to 140 K. The authors have analyzed both the spectral moments and the experimental absorption shapes, assuming that only octupolar and hexadecapolar induction mechanisms contribute to the absorption. This assumption allows to parameterize the temperature dependence of both the intensity and the shape of the absorption band. The results obtained indicate that other contributions to absorption are not negligible.

022.068 **The absorption spectrum of B II, C III, N IV and O V beryllium–like ions.**
J. M. P. Serrão.
J. Quant. Spectrosc. Radiat. Transfer, Vol. 35, No. 4, p. 265 – 276 (1986).
Energy terms, dipole oscillator strengths and photoionization cross sections from the ground state are calculated for beryllium–like ions. Autoionizing state transition energies and line widths for the $^1P^0$ resonances in the continuum are also obtained. The configuration interaction method for initial and final states is used, and the atomic orbitals are generated through angular–momentum–dependent, scaled Thomas–Fermi–Dirac potentials.

022.069 **Absolute transition probabilities in Nb I and Hf I and a solution to the problem of missing infrared branches.**
D. W. Duquette, E. A. Den Hartog, J. E. Lawler.
J. Quant. Spectrosc. Radiat. Transfer, Vol. 35, No. 4, p. 281 – 301 (1986).
Accurate branching ratios and absolute transition probabilities for 320 lines in Nb I and 187 lines in Hf I are reported. A technique is developed for testing the completeness of emission branching ratios and, if necessary, determining the total transition probability of unknown infrared (IR) branches. The technique is based on a comparison of non–IR fluorescence produced by laser excitation to two upper levels from the same lower level. There are no important IR branches beyond 1.0 μm from any of the Nb I levels studied. Several Hf I levels show strong IR branches that account for most of the radiative decay from these levels.

022.070 **Infrared absorption by gas mixtures in the 300 – 850K temperature range I – 4.3 μm and 2.7 μm CO_2 spectra.**
R. Levi Di Leon, J. Taine.
J. Quant. Spectrosc. Radiat. Transfer, Vol. 35, No. 5, p. 337 – 343 (1986).
The spectra have been obtained in a vacuum–tight cell in the 300 – 850K range with a spectral resolution of 0.45 cm^{-1} at 4.3 μm and 0.8 cm^{-1} at 2.7 μm. Experimental details are given and theoretical studies are discussed. Experimental and theoretical spectra are compared.

022.071 **Infrared absorption by gas mixtures in the 500 – 700K temperature range II – 2.7 μm H_2O spectra.**
R. Levi Di Leon, J. Taine, S. Marbot.
J. Quant. Spectrosc. Radiat. Transfer, Vol. 35, No. 5, p. 345 – 347 (1986).
The absorption spectra of water vapor in the 2.7 μm–band have been obtained in an isothermal, vacuum–tight cell with a resolution of 0.8 cm^{-1}. They are compared with a line–by–line calculation using AFGL data and the approximate temperature–dependence of the line–widths. Agreement is very satisfactory in some spectral regions and rather poor in others.

022.072 **Diode–laser measurements and calculations of CO 1–0 P(4) line broadening in the 294– to 765–K temperature range.**
J. M. Hartmann, M. Y. Perrin, J. Taine, L. Rosenmann.
J. Quant. Spectrosc. Radiat. Transfer, Vol. 35, No. 5, p. 357 – 363 (1986).

022.073 **Shock tube measurements of *f*–values for iron and silicon.**
R. E. Pitts.
J. Quant. Spectrosc. Radiat. Transfer, Vol. 35, No. 5, p. 365 – 381 (1986).
Iron–silicon spectra have been photographed in the 3400– to 6200–Å region using a 3.4–m Ebert–type spectrograph. The light source was a He–driven shock tube using a mixture powdered FeSi in an Ar carrier gas. Relative oscillator strengths for 253 Fe I lines and the 3906–Å line of Si I are obtained. These relative oscillator strengths have been placed on an absolute scale using the published gf–values of Blackwell et al. (1979). The published absolute scale of Fe I has been found to be in agreement with those published absolute scales of Si I having a log gf–value of approximately –1.1 for the 3906–Å line.

022.074 **Shock tube measurements of Y I and Y II oscillator strengths.**
R. E. Pitts, G. H. Newsom.
J. Quant. Spectrosc. Radiat. Transfer, Vol. 35, No. 5, p. 383 – 399 (1986).
Yttrium–silicon spectra have been photographed in the 3400– to 6200–Å region using a 3.4–m Ebert–type spectrograph. The light source was a helium–driven shock tube containing a mixture of powdered YSi_2 in an Ar carrier gas. Relative oscillator strengths for 255 Y I lines and 90 Y II lines are obtained. These relative oscillator strengths have been placed on an absolute scale using the published gf–values of Hannaford et al. (1982) and the 3906–Å line of Si I. A comparison of these results with those of previous workers shows general agreement, although significant differences are seen in several cases.

022.075 **The argon–resonance continuum in the far–VUV between 50 and 79 nm.**
G. Trommer.
J. Quant. Spectrosc. Radiat. Transfer, Vol. 35, No. 5, p. 407 – 413 (1986).
The photoionization cross section $\sigma_1(\lambda)$ of the Ar(I) ground state was determined by absolute measurements of the Ar(I) resonance continuum betwen 50 and 79 nm. The vacuum ultraviolet (VUV) radiation originates from the axis of an Ar–He mixture plasma generated in a cascaded arc chamber of 2 mm diameter at atmospheric pressure with arc currents of 80 and 96 A. Good agreement is found between the resulting cross section of this work and data from absorption experiments. Only three theories are confirmed by the experiments. A pure Ar cascade arc is proposed as calculable stationary source of VUV continuum radiation between 55 and 79 nm.

022.076 **Measurement of relative oscillator strengths for Ti I – IV. Transitions from levels of excitation energy between 1.42 and 2.31 eV.**
D. E. Blackwell, A. J. Booth, S. L. R. Menon, A. D. Petford.
Mon. Not. R. Astron. Soc., Vol. 220, No. 2, p. 289 – 302 (1986).
Measurements of the relative oscillator strengths of 52 lines of Ti I with excitation potential between 1.42 eV and 2.31 eV, made to an accuracy of 1 per cent using the Oxford furnace technique, are reported. The absolute scale for Ti I derived in a previous paper in this series for 0 eV lines, is rediscussed using the additional oscillator strengths for high excitation lines contained in this paper. No change is suggested, and all the relative oscillator strengths are placed on this absolute scale. Comparisons are made with the calculated values of Kurucz and Peytremann, the solar values of Kostyk, and with other experimental results. Empirical conversion formulae are given.

022.077 **Measurement of relative oscillator strengths for Cr I lines – II. Measures for transitions from levels with excitation energy 2.54 – 3.46 eV.**
D. E. Blackwell, A. J. Booth, S. L. R. Menon, A. D. Petford.
Mon. Not. R. Astron. Soc., Vol. 220, No. 2, p. 303 – 312 (1986).
Relative oscillator strengths for 53 lines of Cr I, with excitation energies between 2.54 and 3.46 eV, have been measured using the Oxford spectroscopic furnace. The oscillator strengths for lines in the region of 2.5 eV have been placed on an absolute scale by comparison with 1 eV Cr I lines through temperatures measured using photoelectric pyrometry. The lifetime measurements by Kwiatowski et al. have also been used. The two methods agree to better than 1 per cent. This absolute scale has been extended to lines of excitation 3.5 eV using photoelectric pyrometry. The accuracy of relative oscillator strengths within this range of excitation is about 1 per cent, whilst that of the absolute scale is about 7 per cent. Comparison is made with the experimental results and calculations of other authors.

022.078 **Measurement of the oscillator strengths of very weak 1–eV Fe I lines.**
D. E. Blackwell, A. J. Booth, D. J. Haddock, A. D. Petford, S. K. Leggett.
Mon. Not. R. Astron. Soc., Vol. 220, No. 2, p. 549 – 553 (1986).

Relative oscillator strengths are measured for nine very weak 1–eV Fe I lines, using the Oxford spectroscopic furnace. These oscillator strengths are placed on an absolute scale using previous Oxford measurements of stronger 1–eV Fe I lines. The lines are important for the interpretation of cool star spectra because they are the weakest Fe I lines in such spectra whose oscillator strengths are known with good accuracy. It is suggested that the accuracy is 0.015 dex (3.5 per cent).

022.079 **Fine–structure population ratios for the ^2P ground state of C II.**
F. P. Keenan, D. J. Lennon, C. T. Johnson, A. E. Kingston.
Mon. Not. R. Astron. Soc., Vol. 220, No. 2, p. 571 – 576 (1986).
Results of *R*–matrix calculations of electron excitation rates for the $2s^22p^2P_{1/2}$–$2s^22p^2P_{3/2}$ fine–structure transition in C II are used to derive density sensitive population ratios $n(2s^22p^2P_{3/2})/n(2s^22p^2P_{1/2})$ for the ranges of electron and neutral hydrogen densities ($n_e = 10^{-3} - 10^3 \mathrm{cm}^{-3}$; $n_H = 0 - 1000\ \mathrm{cm}^{-3}$) applicable to H I and H II regions. These are found to differ significantly from the earlier calculations of Smeding & Pottasch, and lead to electron density estimates between 40 and 60 per cent smaller.

022.080 **Oscillator strength measurements in the vacuum–ultraviolet. I. The strong 1243, 1493 and 1743 Å multiplets of neutral nitrogen.**
C. Goldbach, M. Martin, G. Nollez, P. Plomdeur, J.–P. Zimmermann, D. Babič.
Astron. Astrophys., Vol. 161, No. 1, p. 47 – 54 (1986).
Oscillator strengths of seven lines (or strongly blended line pairs) of neutral nitrogen, belonging to the strong vacuum–ultraviolet multiplets at 1243, 1493 and 1743 Å, are measured in emission with a wall–stabilized arc. Self–absorption of radiation in the plasma is fully taken into account while boundary layer absorption is carefully eliminated. The measured relative oscillator strengths are normalized to an absolute scale set by lifetime measurements. The use of digital data processing techniques permits the separation of overlapping lines and a precise determination of the extended wing contribution. The uncertainty on the oscillator strength values is between ±12% and ±20%.

022.081 **The reaction of C^+ with HCN: isomeric identity of the C_2N^+ product.**
R. G. Daniel, E. R. Keim, J. M. Farrar.
Astrophys. J., Vol. 303, No. 1, p. 439 – 445 (1986).
The authors present a crossed ion beam – molecular beam study of the reaction of C^+ with HCN at relative energies between 0.62 and 1.57 eV. The large exothermicity of the reaction is shown to be consistent only with the production of the more stable isomer CNC^+. The reaction proceeds through the participation of a long–lived $HCNC^+$ complex which decays in a fraction of a rotational period at the highest collision energy.

022.082 **The rotational spectrum of the C_3H radical.**
C. A. Gottlieb, E. W. Gottlieb, P. Thaddeus, J. M. Vrtilek.
Astrophys. J., Vol. 303, No. 1, p. 446 – 450 (1986).
The rotational spectrum of the recently discovered C_3H radical in both fine–structure ladders of its electronic ground state ($X^2\Pi$) with lambda doubling and hyperfine structure has been determined to high accuracy, and the strongest transitions at frequencies less than 376 GHz are tabulated. The hyperfine constants of the sequence of four hydrocarbon radicales CH, C_2H, C_3H, and C_4H are briefly described.

022.083 **Theoretical collision–induced rototranslational absorption spectra for modeling Titan's atmosphere: H_2–N_2 pairs.**
A. Borysow, L. Frommhold.
Astrophys. J., Vol. 303, No. 1, p. 495 – 510 (1986).
The authors compute from theory the collision–induced rototranslational absorption coefficient of the H_2–N_2 molecular complex at frequencies from 0 to beyond 1000 cm^{-1}, and at temperatures from 50 to 300 K, for which no laboratory measurements exist. The authors account for quadrupolar and hexadecapolar induction of both species, H_2 and N_2, and use a rigorous quantum line shape formalism combined with numerical procedures. The work is of interest for a detailed analysis of the Voyager IRIS spectra of Titan's atmosphere, especially in the regions of the rotational $S_0(0)$ and $S_0(1)$ lines of hydrogen, from ~200 to 700 cm^{-1}.

022.084 **An experimental study of plasma grain interactions.**
R. C. Hazelton, E. J. Yadlowsky.
NASA Conf. Publ., NASA CP–2403, p. A 10 – A 11 (1986). – See Abstr. 012.030.

022.085 **Simulation of cosmic dust spectra.**
J. H. Hecht, R. W. Russell, J. R. Stephens, P. Grieve.
NASA Conf. Publ., NASA CP–2403, p. A 12 (1986). – See Abstr. 012.030.

022.086 **Laboratory studies of the infrared small–particle extinction of amorphous silicates.**
W. Krätschmer.
NASA Conf. Publ., NASA CP–2403, p. A 13 – A 15 (1986). – See Abstr. 012.030.

022.087 **The electrostatics of a dusty plasma.**
E. C. Whipple, D. A. Mendis, T. G. Northrop.
NASA Conf. Publ., NASA CP–2403, p. A 48 (1986). – See Abstr. 012.030.

022.088 **A fast and reliable method for computing free–bound emission coefficients for hydrogenic ions.**
A. Sarmiento, J. Cantó.
Rev. Mex. Astron. Astrofis., Vol. 11, No. 2, p. 61 – 70 (1985). = Contrib. Inst. Astron., Univ. Nac. Autónoma Mex., No. 157.
An approximate formula for the computation of the free–bound emission coefficient for hydrogenic ions is presented. The approximation is obtained through a manipulation of the (free–bound) Gaunt factor which intentionally distinguish the dependence on frequency from the dependence on temperature and ionic composition. Numerical tests indicate that the derived formula is very precise, fast and easy to use, making the calculation of the free–bound contribution from an ionized region of varying temperature and ionic composition a very simple and time saving task.

022.089 **Branching ratios and transition probabilities in rhodium I.**
D. W. Duquette, J. E. Lawler.
J. Opt. Soc. Am. B, Vol. 2, No. 12, p. 1948 – 1952 (1985).

022.090 **$3s^23p^2$–$3s3p^3$, $3s^23p^2$–$3s^23p3d$, and $3s^23p^2$–$3s^23p4s$ transitions in the silicon sequence for elements potassium through titanium.**
E. Biémont.
J. Opt. Soc. Am. B, Vol. 3, No. 2, p. 163 – 176 (1986).

022.091 **Wavelength predictions for the $1s^22s$–$1s2s3p$ satellite transitions.**
M. Mazzoni.
Sol. Phys., Vol. 103, No. 2, p. 231 – 234 (1986).
The experimentally known energies of the $1s^22s \rightarrow 1s2s(^1S,^3S)3p$ transitions are considered for lithium to calcium. Their differences from the corresponding He–like resonances are fitted to the atomic number by means of a polynomical function and, through interpolation, the wavelengths for nitrogen to sodium are calculated.

022.092 **Collision strengths for the intercombination lines of S III.**
Y. K. Ho, R. J. W. Henry.
Astrophys. J., Vol. 303, No. 2, p. 797 – 799 (1986).
Collision strengths for electron impact excitation of doubly ionized sulfur ions (S III) from the ground $3p^2\ ^3P$ state to excited $3s3p^3\ ^1P^0$, $^1D^0$, $3p3d\ ^1P^0$, $^1D^0$ and $3p4s\ ^1P^0$ states are calculated in a two–state close–coupling approximation for the energy range

up to 10^6K. Configuration interaction target wave functions are used in the calculations of the collision strengths.

022.093 **Oscillator strengths of the CN red system.**
S. P. Davis, D. Shortenhaus, G. Stark, R. Engleman Jr., J. G. Phillips, R. P. Hubbard.
Astrophys. J., Vol. 303, No. 2, p. 892 – 896 (1986).
The authors have made direct comparisons of oscillator strengths of bands in the CN red system ($A\ ^2\Pi$–$X\ ^2\Sigma$) with the oscillator strength of the CN violet system ($B\ ^2\Sigma$–$X\ ^2\Sigma$) (0–0) band, by means of absorption measurements in a carbon tube King–type high–temperature furnace. The results show consistency between measured lifetimes in the violet and red systems, and confirm oscillator strengths derived from the solar spectrum of CN.

022.094 **Infrared transition frequencies for methyleneimine, CH_2NH.**
L. Halonen, G. Duxbury.
Astrophys. J., Vol. 303, No. 2, p. 897 – 900 (1986).
Frequencies, intensities, and simulated low–temperature spectra have been calculated for strong mid–infrared (9 – 12 μm) transitions of the CH_2NH molecule. The parameters used are derived from the recent detailed analysis of the high–resolution room temperature Fourier transform spectra of CH_2NH.

022.095 **The reaction $H+C_4H_2$: absolute rate constant measurement and implication for atmospheric modeling of Titan.**
D. F. Nava, M. B. Mitchell, L. J. Stief.
J. Geophys. Res., Vol. 91, No. A4, p. 4585 – 4589 (1986).
The absolute rate constant for the reaction $H+C_4H_2$ has been measured over the temperature interval 210 – 423K, using the technique of flash photolysis–resonance fluorescence. The rate of this reaction, relative to that of the analogous, but slower, reaction of $H+C_2H_2$, appears to make $H+C_4H_2$ a very feasible reaction pathway for effective conversion of H atoms to molecular hydrogen in the stratosphere of Titan.

022.096 **Pressure broadening of ammonia lines in the 6475 Å band at room and low temperatures.**
C. E. Keffer, C. P. Conner, W. H. Smith.
J. Quant. Spectrosc. Radiat. Transfer, Vol. 35, No. 6, p. 487 – 493 (1986).
Measurements of pressure broadened half–widths have been performed for four ammonia vibration–rotation lines in the 6475 Å band. Self–, hydrogen– and helium–broadening have been measured over a temperature range of 175 – 295K. Pressure broadening coefficients and temperature dependence indices have been obtained for each line and broadening gas. A rotational quantum number dependence for the line width has been observed for all broadening gases studied.

022.097 **Proton threshold states in ^{26}Al and their role in astrophysics.**
V. Wijekumar.
Diss. Abstr. Int., Sect. B, Vol. 46, No. 9, p. 3093 (1986). Thesis, The Ohio State University, 114 pp. (1985). Order No. DA8526275.

022.098 **Shock–induced CO_2 loss from $CaCO_3$; implications for early planetary atmospheres.**
M. A. Lange, T. J. Ahrens.
Earth Planet. Sci. Lett., Vol. 77, No. 3/4, p. 409 – 418 (1986).
The authors report new results of shock recovery experiments on single crystal calcite. Recovered samples are subjected to thermogravimetric analysis. This yields the maximum amount of post–shock CO_2, the decarbonization interval and the energy of association (or vaporization), for the removal of remaining CO_2 in shock–loaded calcite. Comparison of post–shock CO_2 with that initially present determines shock–induced CO_2 loss as a function of shock pressure. The experimental results are used to place bonds on models of impact production of CO_2 during accretion of the terrestrial planets.

022.099 **On a comparison of the efficiency of dissociative recombination and dissociation by collisions of electrons with molecular ions.**
V. P. Zhdanov.
Astron. Zh., Tom 63, Vyp. 3, p. 465 – 469 (1986). In Russian. English translation in Sov. Astron., Vol. 30, No. 3.
General relationships are obtained between the cross section of the dissociative recombination with the formation of highly excited atoms and that of the dissociation by collisions of electrons with molecular ions. Some concrete calculations are carried out for the H_2^+ ion. The accuracy of the dipole approximation for calculations of the cross sections of the excitation of ions by collisions with electrons is discussed.

022.100 **Atomic structure calculations and nebular diagnostics.**
S. J. Czyzak, C. D. Keyes, L. H. Aller.
Astrophys. J., Suppl. Ser., Vol. 61, No. 1, p. 159 – 175 (1986).
Considerable differences in predicted transition probabilities and collision strengths can result from differing wave–function choices. For the [O III] λ5007/λ4959 ratio, precise theoretical predictions range from 2.88 to 2.90; the mean photoelectrically observed value (45 nebulae) is 2.98 ± 0.07 p.e. Theory and observation seem to agree for A–value ratios in nebular–type transitions of [N I], [O II], [S II], and possibly [Cl III]. Line–ratio diagrams are presented for various ions of interest in nebular plasma diagnostics.

022.101 **Fe II oscillator strengths.**
A. A. Boyarchuk, I. S. Savanov.
Izv. Krymskoj Astrofiz. Obs., Tom 74, p. 49 – 69 (1986). In Russian. English translation in Bull. Crimean Astrophys. Obs., Vol. 74.
Published values of the oscillator strengths for ionized iron lines are reviewed. The absolute scale of *gf* values is based on life–time measurements. The established data on the reduced and accepted *gf* values for 704 lines of Fe II are presented. A table contains the data on "solar" and "stellar" oscillator strengths for 94 lines. Using the established values of oscillator strengths the authors reviewed the earlier published results on the iron abundance in the atmosphere of the sun. This value is equal to lg ε(Fe) = 7.80 ± 0.10 and differs from the estimate based on neutral iron lines obtained in the same manner.

022.102 **Measurement of wavelengths and Lamb shifts for inner–shell transitions in Fe XVIII – XXIV.**
J. F. Seely, U. Feldman, U. I. Safronova.
Astrophys. J., Vol. 304, No. 2, p. 838 – 848 (1986).
The wavelengths of inner–shell $1s$–$2p$ transitions in the ions Fe XVIII – XXIV have been measured in solar flare spectra recorded by the Naval Research Laboratory crystal spectrometer on the *P78–1* spacecraft. The measurements are compared with previous measurements and with recently calculated wavelengths. It is found that the measured wavelengths are systematically larger than the wavelengths calculated using the *Z*–expansion method by up to 0.65 mÅ. For the more highly charged ions, these differences can be attributed to the QED contributions to the transition energies.

022.103 **Theoretical collision–induced rototranslational absorption spectra for the outer planets: H_2–CH_4 pairs.**
A. Borysow, L. Frommhold.
Astrophys. J., Vol. 304, No. 2, p. 849 – 865 (1986).
Computations of the rototranslational absorption spectra of H_2–CH_4 molecular complexes are presented which are based on the classical multipole expansion; spectral profiles are obtained from an exact quantum formalism. The interaction potential is based on laboratory measurements of H_2–CH_4 pairs at 195 and 297K. The computed spectra provide the most reliable temperature dependence of the absorption coefficient as a function of frequency that can be made under the present circumstances. A theoretical description of the H_2CH_4 dimer features is given in the isotropic potential approximation. This work is significant for the modeling of the far–infrared absorption of the outer planets' atmospheres, where H_2 and CH_4 are present.

022.104 **On a similarity between clusters obtained by SIMS and molecules observed in astrophysics.**
M. Leleyter.
J. Phys. Lett., Vol. 46, No. 19, p. L915 – L922 (1985). Abstr. in Phys. Abstr., Vol. 89, No. 1273, Entry 14785 (1986).

022.105 **Absolute line strengths of $^{74}GeH_4$ near 5 μm.**
R. D. Schaeffer, R. W. Lovejoy.
J. Mol. Spectrosc., Vol. 113, No. 2, p. 310 – 314 (1985). Abstr. in Phys. Abstr., Vol. 89, No. 1274, Entry 20244 (1986).

022.106 **Electron impact excitation of forbidden and allowed transitions in Al II.**
S. S. Tayal, P. G. Burke, A. E. Kingston.
J. Phys. B, Vol. 18, No. 21, p. 4321 – 4325 (1985). Abstr. in Phys. Abstr., Vol. 89, No. 1274, Entry 20441 (1986).

022.107 **Interpolation functions for molecular partition functions.**
S. C. F. Rossi, W. C. Maciel, P. Benevides–Soares.
Rev. Mex. Astron. Astrofis., Vol. 12, Spec. Issue, p. 327 – 329 (1986). Abstract.– See Abstr. 012.066.

022.108 **Grain size evolution and fractionation trends in an experimental regolith.**
F. Hörz, M. J. Cintala, T. H. See, F. Cardenas, T. D. Thompson.
J. Geophys. Res., Vol. 89, Suppl., p. C183 – C196 (1984). – See Abstr. 012.065.
The comminution of blocky planetary surfaces into fine–grained regoliths was simulated by impacting a fragmental gabbro target 200 times with stainless steel projectiles (diameter = 6.35 mm; mass = 1.2 g; V_i = 1.35 km/s).

022.109 **Experimental methods for quenching structures in lunar–analog silicate melts: variations as a function of quench media and composition.**
M. D. Dyar.
J. Geophys. Res., Vol. 89, Suppl., p. C233 – C239 (1984). – See Abstr. 012.065.
Compositions analogous to lunar green, orange, and brown glasses were synthesized under consistent conditions, then quenched into a variety of different media when the samples were removed from the furnace. Iron valence and coordination are a direct function of quench media used. Therefore the remote–sensed spectrum of a glass–bearing regolith on the moon may be influenced by the process by which the glass cooled, and extreme caution must be used when comparing spectra of synthetic glass analogs with real lunar glasses.

022.110 **Rotational analysis of the red electronic emission spectrum of molybdenum nitride (MoN).**
R. C. Carlson, J. K. Bates, T. M. Dunn.
J. Mol. Spectrosc., Vol. 110, No. 2, p. 215 – 241 (1985). Abstr. in Phys. Abstr., Vol. 89, No. 1275, Entry 25537 (1986).

022.111 **Laboratory measurements of physical properties of submicron particles candidates for cosmic dust.**
A. Borghesi, E. Bussoletti, L. Colangeli, C. De Blasi.
Nuovo Cimento C, Vol. 8C, Ser. 1, No. 4, p. 345 – 352 (1985). Abstr. in Phys. Abstr., Vol. 89, No. 1276, Entry 35962 (1986).

022.112 **Collisional excitation of molecules in dense interstellar clouds.**
S. Green.
Nuovo Cimento C, Vol. 8C, Ser. 1, No. 4, p. 435 – 449 (1985). Abstr. in Phys. Abstr., Vol. 89, No. 1276, Entry 35965 (1986).

022.113 **The microwave and far–infrared spectra of the ^{18}OH radical.**
E. R. Comben, J. M. Brown, T. C. Steimle, K. R. Leopold, K. M. Evenson.
Astrophys. J., Vol. 305, No. 1, p. 513 – 517 (1986).
The frequencies, wavelengths, and line strengths for transitions of the ^{18}OH molecule at microwave and far–infrared frequencies have been calculated from an analysis of its far–infrared laser magnetic resonance spectrum.

022.114 **Calculated forbidden rotational spectra of H_3^+.**
F.–S. Pan, T. Oka.
Astrophys. J., Vol. 305, No. 1, p. 518 – 525 (1986).
The forbidden rotational spectra of the H_3^+ molecular ion have been calculated in the ground state and in the υ_2 state, anticipating the observation of such spectra in space and in the laboratory. The small moments of inertia and the large vibrational amplitude of H_3^+ make the forbidden lines unusually strong. The calculated spectral pattern for laboratory observations is given, and the intensities of spontaneous emissions from dense molecular clouds are also discussed.

022.115 **The submillimeter wave spectrum of the protonated and deuterated carbon dioxide.**
M. Bogey, C. Demuynck, J.–L. Destombes.
J. Chem. Phys., Vol. 84, No. 1, p. 10 – 15 (1986). Abstr. in Phys. Abstr., Vol. 89, No. 1277, Entry 37969 (1986).

022.116 **C–N bound formation in the gas phase reaction of C^+ with NH_3.**
R. A. Curtis, J. M. Farrar.
J. Chem. Phys., Vol. 84, No. 1, p. 127 – 134 (1986). Abstr. in Phys. Abstr., Vol. 89, No. 1277, Entry 41127 (1986).

022.117 **Analytical electron microscopy of Mg–SiO smokes: a comparison with infrared and XRD studies.**
F. J. M. Rietmeijer, J. A. Nuth, I. D. R. Mackinnon.
Icarus, Vol. 66, No. 2, p. 211 – 222 (1986).
Experimentally obtained Mg–SiO smokes were studied by analytical electron microscopy using the same samples that had been previously characterized by repeated infrared spectroscopy. The authors suggest that infrared spectroscopy and X–ray diffraction analysis should be complemented by detailed analytical electron microscopy to detect budding crystallinity in vapor phase condensates.

022.118 **Cyanoallene, $CH_2=C=CH–CN$: astrophysical predictions.**
G. Schwahn, G. Winnewisser, W. Joentgen, H.–J. Altenbach, E. Vogel.
Astron. Astrophys., Vol. 161, No. 2, p. 327 – 328 (1986).
As an isomer of methylcyanoacetylene, CH_3CCCN, cyanoallene, $CH_2=C=CH–CN$, is of potential astrophysical interest. Highly precise frequency predictions in the 70 – 100 GHz region are presented, based on microwave and very recent millimeter wave measurements. A possible interstellar formation scheme of the various isomers from $C_4H_4N^{+*}$ is proposed.

022.119 **Atomic partition functions for chromium.**
J. Halenka, B. Grabowski.
Astron. Astrophys., Suppl. Ser., Vol. 64, No. 3, p. 495 – 498 (1986).
More realistic numerical values of the atomic partition functions, $U^{(r)}$, for Cr I, Cr II and Cr III are given for a wide range of the physical conditions of astrophysical and laboratory plasmas.

022.120 **Dielectronic recombination at low temperatures. III. Recombination coefficients for Mg, Al, Si.**
H. Nussbaumer, P. J. Storey.
Astron. Astrophys., Suppl. Ser., Vol. 64, No. 3, p. 545 – 555 (1986).
Effective dielectronic recombination coefficients are calculated for the total recombination, as well as for selected lines and ground and metastable terms of ions of Mg, Al and Si. The authors restrict the calculation to those ions for which the recombining ion has n = 3 valence electrons. Coupling schemes are discussed, and it is concluded that in most cases pure LS coupling is not appropriate for the resonance states. The effective recombination coefficients are fitted to a convenient function of temperature in the range 10^3K to $6x10^4$K.

022.121 **On the Stark broadening within a F I transition array.**
M. S. Dimitrijević.
Astron. Astrophys., Suppl. Ser., Vol. 64, No. 3, p. 591 – 594 (1986).
Using a semiclassical formalism, electron impact widths and shifts within 3s – 3p transition array of F I have been computed. A detailed analysis of Stark broadening parameters as functions of the impact electron angular momentum quantum number and the temperature is carried out. On the basis of the results obtained, the differences between the contribution of elastic, inelastic and strong collisions to the line widths within a transition array are discussed, as well as the Stark shift and width behaviour of the various components.

022.122 **Multi–configuration Hartree–Fock plus Breit–Pauli results for some forbidden transitions.**
C. F. Fischer, M. Godefroid.
J. Phys. B, Vol. 19, No. 2, p. 137 – 148 (1986). Abstr. in Phys. Abstr., Vol. 89, No. 1279, Entry 49396 (1986).

022.123 **Pressure broadening of methane lines in the 6190 Å and 6825 Å bands at room and low temperatures.**
C. E. Keffer, C. P. Conner, W. H. Smith.
J. Quant. Spectrosc. Radiat. Transfer, Vol. 35, No. 6, p. 495 – 499 (1986).
Measurements of pressure broadened–half–widths have been performed for one vibration–rotation line in the 6190 Å methane band and one vibration–rotation line in the 6825 Å methane band. Self–, nitrogen–, hydrogen– and helium–broadening have been measured over a temperature range of 77 – 295K. Pressure broadening coefficients and temperature dependence indices have been obtained for each line and broadening gas.

022.124 **Temperature and pressure dependence of the reaction of He^+ ions with H_2.**
H. Böhringer, F. Arnold.
J. Chem. Phys., Vol. 84, No. 3, p. 1459 – 1462 (1986). Abstr. in Phys. Abstr., Vol. 89, No. 1280, Entry 59517 (1986).

022.125 **Observation of the fourth positive system of CO in dissociative recombination of vibrationally excited CO_2^+.**
F. Vallee, B. R. Rowe, J. C. Gomet, J. L. Queffelec, M. Morlais.
Chem. Phys. Lett., Vol. 124, No. 4, p. 317 – 320 (1986). Abstr. in Phys. Abstr., Vol. 89, No. 1280, Entry 59540 (1986).

022.126 **Asteroidal catastrophic collisions simulated by hypervelocity impact experiments.**
F. Capaccioni, P. Cerroni, M. Coradini, M. Di Martino, P. Farinella, E. Flamini, G. Martelli, P. Paolicchi, P. N. Smith, A. Woodward, V. Zappalà.
Icarus, Vol. 66, No. 3, p. 487 – 514 (1986).
The authors report the results of six impact fragmentation experiments carried out with free–falling macroscopic targets of different compositions and shapes, and with projectile velocities close to 9 km/sec., i.e., significantly higher than the sound velocity in the target materials. The data have been examined by deriving the mass and shape distributions of the fragments, by reconstructing two of the shattered targets in order to study the geometry of the fracture surfaces, and by analyzing the properties of the fine–grained high–velocity ejecta. The fragment mass distributions show clearly that the degree of target fragmentation depends strongly on the impact parameter. While the collisional theory for the origin of families is fully consistent with the experimental results, the elongated shapes of several Apollo–Amor objects are much rarer among the laboratory fragments, and thus appear to require a different explanation.

022.127 **The origin of the Moon and the single–impact hypothesis I.**
W. Benz, W. L. Slattery, A. G. W. Cameron.
Icarus, Vol. 66, No. 3, p. 515 – 535 (1986).
Recently the single–impact hypothesis for forming the Moon has gained some favorable attention. The authors present a series of three–dimensional numerical simulations of an impact between the protoearth and an object about 0.1 of its mass. For computational convenience both objects were assumed to be composed of granite. The authors studied the effects on the outcome of the collision of varying the impact parameter, the initial internal energy, and the relative velocity. The results show that if the impact parameter is large enough so that the center of the impactor approximately grazes the limb of the protoearth, the impactor is not completely destroyed; part of it forms a clump in a large elliptical orbit about the Earth. This clump does not collide with the Earth. However, since the orbit of the clump comes close to the Earth (within the Roche limit) the clump will be destroyed and spread out to form a disk around the Earth. The scenario for making the Moon from a single–impact event is supported by these simulations.

022.128 **Results obtained by laboratory simulations of catastrophic impact.**
A. Fujiwara.
Mem. Soc. Astron. Ital., Vol. 57, No. 1, p. 47 – 64 (1986). – See Abstr. 012.061.
Results obtained in the laboratory impact simulations of catastrophic disruption of minor bodies are reviewed. The present knowledge on the destruction type, mass of the largest fragment, size and shape distribution of the fragments, fracture patterns, ejection velocities and spin periods of the fragments, and energy partition rate are provided with some suggested future works.

022.129 **Changes of chemical compositions of ejecta induced by hypervelocity impacts.**
K. Yamakoshi, A. Fujiwara, N. Asada.
J. Geomagn. Geoelectr., Vol. 37, No. 11, p. 987 – 997 (1985). Abstr. in Phys. Abstr., Vol. 89, No. 1281, Entry 67420 (1986).

022.130 **Astrophysical implications of ice sputtering.**
L. J. Lanzerotti, W. L. Brown, R. E. Johnson.
Nucl. Instrum. Methods Phys. Res., Sect. B, Vol. B14, No. 4 – 6, p. 373 – 377 (1986). Abstr. in Phys. Abstr., Vol. 89, Nr. 1284, Entry 83430 (1986).

022.131 **Stark broadening of singly ionized neon lines.**
N. Konjević, T. L. Pittman.
J. Quant. Spectrosc. Radiat. Transfer, Vol. 35, No. 6, p. 473 – 477 (1986).
Stark profiles of 21 Ne II lines from 10 multiplets were measured in a low pressure, pulsed arc. An electron density of $1.4 \times 10^{23} m^{-3}$ was determined by laser interferometry and an electron temperature of 28000K was measured using relative intensities of O II impurity lines. These experimental data are compared with a previous experiment, with semiclassical theoretical results, and with semiempirical calculations.

022.132 **Ultraviolet photoprocessing and infrared spectroscopy of laboratory simulated grain mantles.**
L. B. d'Hendecourt.
Light on dark matter, p. 247 – 252 (1986). – See Abstr. 012.080.
A series of experiments in which mixtures of ices are irradiated with ultraviolet light have been performed. The evolution of these simulated interstellar grain mantles is followed by infrared spectroscopy. The appearance of new bands has been observed and possible identifications of these bands are discussed. Implications for interstellar chemistry are discussed.

022.133 **Radiation effects on grain materials.**
G. Strazzulla.
Light on dark matter, p. 253 – 260 (1986). – See Abstr. 012.080.
By comparing the carbonaceous and deduced organic components of the Murchison meteorite with MeV proton irradiated solid methane residues the author has tried to suggest a method for making an evolutionary connection between protostellar and interplanetary material.

022.134 **Experimental evidence for the effect of resonance structures on electron excitation rates for Si III.**
T. L. Yu, M. Finkenthal, H. W. Moos.
Astrophys. J., Vol. 305, No. 2, p. 880 – 883 (1986).
Laboratory experimental data are presented which confirm the effect of resonance structures on the electron excitation rates of Si III and, through them, on line ratios used for astrophysical and laboratory diagnostic purposes. In particular, the spectrum of silicon emitted by low–density magnetic mirror plasma was recorded for the 1100 – 1400 Å range, and the relative ratios of Si III lines emitted around 1300 Å were compared with two different sets of theoretical predictions.

022.135 **Direct measurement of the fine–structure interval and g_J factors of singly ionized atomic carbon by laser magnetic resonance.**
A. L. Cooksy, G. A. Blake, R. J. Saykally.
Astrophys. J., Lett. Ed., Vol. 305, No. 2, p. L89 – L92 (1986).
The authors present the results of laser magnetic resonance measurements performed on the ground ^2P state of singly ionized atomic carbon (C II). The $^2P_{3/2} \leftarrow {}^2P_{1/2}$ fine–structure intervals of both $^{12}C^+$ and $^{13}C^+$ have been determined with a precision of approximately 1 ppm, and the g_J factors to approximately one part in 10^4. The highly precise values of the ^{12}C II and ^{13}C II fine–structure intervals should allow the interstellar $^{12}C/^{13}C$ ratio to be unambiguously determined in a number of environments.

Problems of nonlinear and turbulent processes in physics. Part 2.
See Abstr. 003.084.

Polarized light in nature.
See Abstr. 003.118.

Spectral line shapes. Vol. 3.
See Abstr. 003.169.

Autoionization, recent developments, and applications.
See Abstr. 003.189.

Rigorous estimates for the series expansions of Hamiltonian perturbation theory.
See Abstr. 042.039.

Motions in the field of two rotating magnetic dipoles. I. Equilibrium points.
See Abstr. 042.041.

Attractors in a dissipative dynamical system with three dimensions.
See Abstr. 042.043.

Optical properties of simulated astrophysical grains and their dynamics in the near–Earth environment.
See Abstr. 051.104.

An attempt for unification of microcosmos and mega(macro)-cosmos on the base of the order (causality).
See Abstr. 066.129.

S V line ratios in the Sun.
See Abstr. 071.001.

Red/violet contrast reversal on Mars: significance for eolian sediments.
See Abstr. 097.015.

Iron oxide and hydroxide precipitation from ferrous solutions and its relevance to Martian surface mineralogy.
See Abstr. 097.018.

Numerical and laboratory simulations of photometric properties of asteroids.
See Abstr. 098.024.

Peculiar shaped asteroids: some laboratory results.
See Abstr. 098.027.

Chimie organique dans Titan: expériences de simulation en laboratoire et speculations.
See Abstr. 100.050.

Time dependent chemistry in dense molecular clouds. III. Infrared band cross sections of molecules in the solid state at 10K.
See Abstr. 131.318.

Optical luminescence from reflection nebulae?
See Abstr. 131.368.

Formation of organic molecules on interstellar dust particles.
See Abstr. 131.372.

Astronomical Instruments and Techniques

031 Astronomical Optics

031.001 **Optical analysis of thermally cycled 515 mm metallic Al/Al–alloy mirrors.**
L. Noethe, P. Giordano, K. Mischung, F. Franza, R. Wilson.
Astron. Astrophys., Vol. 156, No. 1/2, p. 323 – 336 (1986).

The figure stability of 16 spherical Al or Al–alloy mirrors of 515 mm diameter was tested in a thermal cycling experiment. The optical analysis was done by the Shack–Hartmann method. The general conclusion is that Al or Al–alloys are suitable for the fabrication of telescope mirrors, if, apart from the choice of the material and precautions during the fabrication, the blank is thermally stabilized before the final polishing.

031.002 **A new type of field corrector.**
D.–q. Su.
Astron. Astrophys., Vol. 156, No. 1/2, p. 381 – 385 (1986).

A new type of field corrector, which can correct both the aberrations and the atmospheric dispersion, is proposed. There are two cemented lenses in such a corrector. Each cemented lens is made of two different kinds of glasses, of which the indexes of refraction are near and the dispersions are different. The cemented surface is tilted. These two "lensms" are not only the components for correcting aberrations but also form spectra of objects in the field of view. When the two "lensms" are rotated relatively around the optical axis the dispersion will be changed. By means of this the atmospheric dispersion at different zenith distance can be compensated. A practical example is given and some other methods for correcting the atmospheric dispersion are also mentioned.

031.003 **15 meter multiple mirror telescope design study.**
J. R. P. Angel, N. J. Woolf.
Astrophys. J., Vol. 301, No. 1, p. 478 – 501 (1986).

Taking as a starting point the existing Multiple Mirror Telescope (MMT), a concept for a large and more advanced instrument has been developed. It makes use of four 7.5 m diameter paraboloidal glass primaries of the honeycomb–sandwich type being developed by the University of Arizona. These are mounted quite close together in a square configuration, with their axes coaligned. Two distinct optical configurations are provided. Wide fields of view in the optical and near–infrared are achieved at four separate fast Cassegrain foci (f/4.5). Secondaries of 3 m diameter are used with three–element refractive correctors designed by Epps to give flat 1° fields. When it is necessary to work at a combined focus an alternate set of small Cassegrain secondaries is used. The light is relayed to a coherent central focus, that is made available at several Nasmyth and other instrument stations via steering mirrors. The mechanical and thermal design of the mirrors and the optical designs are aimed at an overall telescope image quality of 0\".25. The use of advanced detectors and instruments now being designed to operate with this telescope is also outlined.

031.004 **Astronomical objective–apochromats of three lenses.**
G. M. Popov, M. B. Popova.
Izv. Krymskoj Astrofiz. Obs., Tom 70, p. 158 – 163 (1985). In Russian. English translation in Bull. Crimean Astrophys. Obs., Vol. 70.

Objective apochromats consisting of three lenses of ordinary glasses and objectives with special glasses OF 3 or OF 4 are discussed. The objectives have focal length of about 1 meter. They are corrected for spherical aberration, secondary spectrum and coma.

031.005 **Laser interferometers for testing astronomical optics.**
G. M. Popov, E. G. Popov.
Izv. Krymskoj Astrofiz. Obs., Tom 70, p. 163 – 168 (1985). In Russian. English translation in Bull. Crimean Astrophys. Obs., Vol. 70.

Three types of laser interferometers are discussed: interferometer for testing fast–speed optics, meniscus interferometer consisting of three cemented meniscus lenses, and a shear interferometer. Shear interferometer consists of a flat glass plate with one semi–transparent mirror surface. All interferometers are compact, stable and solid. Meniscus interferometer and shear interferometer are almost insensitive to vibration.

031.006 **Easily adjusted two–lens null corrector for testing concave astronomical mirrors.**
V. V. Kuznetsov, G. M. Popov, V. I. Pronik.
Izv. Krymskoj Astrofiz. Obs., Tom 72, p. 231 – 238 (1985). In Russian. English translation in Bull. Crimean Astrophys. Obs., Vol. 72.

Consideration is given to the use of a special Offner kind null corrector. The shape of lenses permits their easy adjustment with the null corrector for mirror testing. Some versions of systems to test fast speed *F*/2.5 hyperboloid mirrors of about 4 m diameter are designed.

031.007 **Computation algorithm for investigation of concave astronomical mirrors with the Hartmann method.**
I. V. Il'in.
Izv. Krymskoj Astrofiz. Obs., Tom 73, p. 176 – 189 (1985). In Russian. English translation in Bull. Crimean Astrophys. Obs., Vol. 73.

A modification of the computation algorithm for the Hartmann test is proposed. The logical scheme of the proposed algorithm and principles of its action are described. Illustrative calculations of the proposed program are given.

031.008 **Closure phase in high–resolution optical imaging.**
J. E. Baldwin, C. A. Haniff, C. D. Mackay, P. J. Warner.
Nature, Vol. 320, No. 6063, p. 595 – 597 (1986).

The angular resolution of long–exposure optical images taken with large telescopes at the best sites is limited by atmospheric phase fluctuations to ~0.5 – 1 arc s; that is, ~10 – 50 times worse than the theoretical diffraction limit. In contrast, the measurement of visibilities and closure phases of fringe patterns in short–exposure images, taken through an aperture mask comprising a non–redundant array of three or more holes, offers the prospect of reliable diffraction–limited imaging using methods well established in very long baseline interferometry at radio wavelengths The authors report successful first measurements of fringe visibility and closure phase by this technique, together with evidence that the method is applicable to objects as faint as magnitude 15.

031.009 **Light weight aluminum optics.**
R. C. Catura, J. R. Vieira.
ESA Spec. Publ., ESA SP–239, p. 173 – 175 (1985). – See Abstr. 012.019.

Several light weight mirror blanks have been fabricated by dip–brazing a core of low mass aluminum foam (Duocel) material to thin face sheets of solid aluminum. Optical interferometry was used to assess their dimensional stability over a seven month period. No changes in flatness were observed to the sensitivity of the measurements of a half wavelength of red light.

031.010 **Electroform replication of grazing incidence X–ray optics.**
M. P. Ulmer, W. R. Purcell Jr., D. Bedford, G. R. Simnett.
ESA Spec. Publ., ESA SP–239, p. 195 – 198 (1985). – See Abstr. 012.019.

Work done to produce mirrors via electroform replication is reported. The authors summarize earlier work on small (about 6 cm by 9 cm) cylindrical pieces and new work on substantially larger (40 cm long by 12 cm wide) Wolter I shaped mirrors.

031.011 **Measurements of surface scattering from mirrored surfaces using a triple axis X–ray spectrometer.**
F. E. Christensen, P. L. Jensen, H. W. Schnopper.
ESA Spec. Publ., ESA SP–239, p. 221 – 228 (1985). – See Abstr. 012.019.

In the process of developing a high throughput X–ray telescope for XMM, various techniques have been employed to deposit a thin film of Au on a smooth substrate. A triple axis X–ray spectrometer has been employed to measure the reflectivity and surface scattering of the gold surface or the substrates. For comparison the same kind of measurements have been performed on a high quality Au–coated testflat for EXOSAT and a similar Pt–coated test flat for AXAF.

031.012 **Optical systems of unequal–arm laser interferometers for astronomical–optics testing.**
G. M. Popov, E. G. Popov.
Bull. Crimean Astrophys. Obs., Vol. 69, p. 126 – 131 (1986). English translation of 39.031.005.

031.013 **Notes on the use of Pickering–Racine wedges.**
V. M. Blanco.
Proceedings of the Second Asian–Pacific Regional Meeting on Astronomy, p. 488 – 489 (1984). – See Abstr. 012.031.

031.014 **Einiges über optisches Rechnen. Teil 2.**
G. Traving.
Sterne Weltraum, 25. Jahrg., Nr. 5, p. 274 – 275 (1986).

031.015 **Use of nonrotational aspherical surfaces in optical systems.**
J. G. Baker.
J. Opt. Soc. Am. A, Vol. 2, No. 12, p. P57 (1985). Abstract.

031.016 **Approaches to fabrication and testing of large asymmetric aspherics.**
R. E. Parks.
J. Opt. Soc. Am. A, Vol. 2, No. 12, p. P57 – P58 (1985). Abstract.

031.017 **Poor man's telescope.**
D. Shafer.
J. Opt. Soc. Am. A, Vol. 2, No. 12, p. P65 (1985). Abstract.

This paper describes an unusual telescope design, with a 2–m aperture *f*/2 spherical primary mirror, and a 1/4–m aperture spherical secondary mirror. The resulting *f*/20 Cassegrain system is made aplanatic and diffraction–limited in the visible, over a modest field of view, by the addition of two lenses, each as large as the secondary. These lenses are each flat on one side.

031.018 **Rich man's 2–m telescope.**
D. Shafer.
J. Opt. Soc. Am. A, Vol. 2, No. 12, p. P65 (1985). Abstract.

The author showed how the coma of a classical Cassegrain telescope can be fixed by adding a very weak meniscus lens next to the secondary mirror. This paper is about a simple way to fix both the coma and astigmatism of a Gregorian, again in such a way that the design can be used with or without the "fix" in place. Here, however, the "fix" is very small – a fraction of the secondary diameter. A small single lens is all that is added.

031.019 **Broadband atmospheric dispersion corrector using three materials.**
S. Eisenberg, K.–L. Shu.
J. Opt. Soc. Am. A, Vol. 2, No. 12, p. P65 (1985). Abstract.

031.020 **Very large optics of the future.**
A. B. Meinel, M. P. Meinel.
J. Opt. Soc. Am. A, Vol. 2, No. 12, p. P83 (1985). Abstract.

031.021 **Atmospheric dispersion corrector from the ultraviolet to the near infrared.**
S. Eisenberg, K.–L. Shu.
Appl. Opt., Vol. 24, No. 23, p. 3944 – 3945 (1985).

031.022 **Numerical processing of dynamic subaperture testing measurements.**
C. R. De Hainaut, A. Erteza.
Appl. Opt., Vol. 25, No. 4, p. 503 – 509 (1986).

Subaperture testing is a technique for interferometrically testing optical telescopes with an array of small flats rather than the usual single large flat. Special subaperture testing analysis algorithms are then required to remove the effects of small misalignments of the subaperture flats. This paper covers the case where a series of measurements are taken over a period of time. A dynamic subaperture testing numerical processor is presented which incorporates a discrete Kalman filter to exploit the sequential nature of these measurements. The processor is then simplified so that it may be used when subaperture dynamics are not known. Tuning the Kalman filter and simulating interferometer measurements of known aberrations are discussed.

031.023 **Effect of a small centering error of the Hartmann screen on the computed wave–front aberration.**
J. E. A. Landgrave, J. R. Moya.
Appl. Opt., Vol. 25, No. 4, p. 533 – 536 (1986).

A small displacement of the Hartmann screen from the optic axis of the surface under test may introduce substantial errors on the departures of this surface from the target surface which are found through the test. An expression for the magnitude of these errors shows that they increase with the asphericity of the target surface, being negligible for spherical surfaces which are near completion. This expression was verified by a computer–simulated test of a perfect hyperbolic mirror.

031.024 **Simple numerical measure of the manufacturability of aspheric optical surfaces.**
J. W. Foreman Jr.
Appl. Opt., Vol. 25, No. 6, p. 826 – 827 (1986).

031.025 **Holographic and aspheric Wadsworth mounting in space astronomy.**
M. Duban.
C. R. Acad. Sci., Sér. II, Tome 302, No. 20, p. 1215 – 1218 (1986). In French.

In astrophysics, spectral analysis of a point source can favourable take place, in far ultraviolet, with a Wadsworth mounted spherical grating. Use of holographic ruling on quadrics of revolution improve image quality leading to perfect stigmatism. The author proposes three types of solutions, whose complementary properties make them especially efficient in space astronomy.

031.026 **Multicomponent symmetric achromatic phase plates. I. Broadening of the Pancharatnam system for the case of an arbitrary number of components.**
V. A. Kucherov.
Photometric and polarimetric investigations of celestial bodies, p. 152 – 160 (1985). In Russian. – See Abstr. 003.035.

A procedure is proposed for synthesis of achromatic systems consisting of an arbitrary number of components. The construction contains 2k + 1 cascaded elements made of the same material. All the internal phase plates are half–wave ones. Retardation of external components depends on equivalent retardation of the system. The axes of single plates are placed symmetrically with respect to the central element.

031.027 **Manufacturing optics for the Nordic 2.5 m telescope. Technical report.**
T. Korhonen, S. Haarala, J. Piironen, A. Sillanpää.
Rep. Ser., Dep. Phys. Sci., Univ. Turku, No. FTL–R84, 15 pp. (1985). = Turku Univ. Obs., Informo No. 96.

031.028 **Glass mirrors for space telescopes.**
J. R. P. Angel.
Prepr. Steward Obs., No. 622, 3 pp. (1986). To appear in Proc. SPIE Int. Soc. Opt. Eng., Vol. 542 "Optical Fabrication and Testing Workshop: Large Telescope Optics".

031.029 **Steps toward 8 m honeycomb mirrors. VI. Thermal control.**
J. R. P. Angel, A. Y. S. Cheng, N. J. Woolf.
Prepr. Steward Obs., No. 623, 8 pp. (1986). To appear in Proc. SPIE Int. Soc. Opt. Eng., Vol. 571.

031.030 **Steps toward 8 m honeycomb mirrors. VII. Spin casting an experimental f/1 1.8 m honeycomb blank or borosilicate glass.**
L. Goble, J. R. P. Angel, J. M. Hill.
Prepr. Steward Obs., No. 624, 9 pp. (1986). To appear in Proc. SPIE Int. Soc. Opt. Eng., Vol. 571.

031.031 **Opticians as astronauts.**
J. R. P. Angel.
Prepr. Steward Obs., No. 625, 2 pp. (1986). To appear in Proc. SPIE Int. Soc. Opt. Eng., Vol. 571.

031.032 **Mirror systems: engineering features, benefits, limitations and applications.**
I. R. Abel.
Proc. SPIE Int. Soc. Opt. Eng., Vol. 531, p. 121 – 134 (1985). Abstr. in Phys. Abstr., Vol. 89, No. 1275, Entry 26076 (1986). – See Abstr. 012.044.

031.033 **The process of honeycomb holes of a 1.2 m mirror and the test of an experimental tube for a 1.2 m telescope.**
G.–t. Zheng, G.–x. Huang.
Publ. Purple Mt. Obs., Vol. 4, No. 4, p. 53 – 56 (1985). In Chinese.
In the first part of this paper some experiences of digging 54 honeycomb holes in the back of a 1.2 m mirror are described. The second part deals with a new kind of tube, which supports a secondary mirror through the central hole of the primary mirror. The simulation test manifests the feasibility of such a tube for a f/1.5 primary mirror.

031.034 **Coherence in astrophysical problems.**
A. M. Gretskij.
Vestn. Khar'kov. Univ., No. 278, p. 16 – 18 (1985). In Russian.

031.035 **Optical performance of synthetic aperture telescope configurations.**
J. E. Harvey, P. R. Silverglate, A. B. Wissinger.
Proc. SPIE Int. Soc. Opt. Eng., Vol. 540, p. 110 – 118 (1985). Abstr. in Phys. Abstr., Vol. 89, No. 1278, Entry 44177 (1986). – See Abst. 012.051.

031.036 **An overview of optical telescope technology.**
A. B. Meinel, M. P. Meinel.
Proc. SPIE Int. Soc. Opt. Eng., Vol. 542, p. 35 – 41 (1985). Abstr. in Phys. Abstr., Vol. 89, No. 1278, Entry 44180 (1986). – See Abstr. 012.052.

031.037 **Design aspects of the University of Texas 7.6 m telescope. I.**
F. B. Ray.
Proc. SPIE Int. Soc. Opt. Eng., Vol. 542, p. 42 – 46 (1985). Abstr. in Phys. Abstr., Vol. 89, No. 1278, Entry 44181 (1986). – See Abstr. 012.052.

031.038 **Design aspects of the University of Texas 7.6 m telescope. II.**
F. B. Ray.
Proc. SPIE Int. Soc. Opt. Eng., Vol. 542, p. 47 (1985). Abstr. in Phys. Abstr., Vol. 89, No. 1278, Entry 44182 (1986). – See Abstr. 012.052.

031.039 **Optics for the 15 m National New Technology Telescope.**
L. D. Barr.
Proc. SPIE Int. Soc. Opt. Eng., Vol. 542, p. 68 – 77 (1985). Abstr. in Phys. Abstr., Vol. 89, No. 1278, Entry 44183 (1986). – See Abstr. 012.052.

031.040 **Multilayer coatings for astronomical telescope mirrors.**
D. Y. Song, H. A. Macleod.
Proc. SPIE Int. Soc. Opt. Eng., Vol. 540, p. 156 – 159 (1985). Abstr. in Phys. Abstr., Vol. 89, No. 1278, Entry 44185 (1986). – See Abstr. 012.051.

031.041 **Improved coatings for large astronomical telescope mirrors.**
H. A. Macleod, D. Y. Song.
Proc. SPIE Int. Soc. Opt. Eng., Vol. 542, p. 25 – 27 (1985). Abstr. in Phys. Abstr., Vol. 89, No. 1278, Entry 44186 (1986). – See Abstr. 012.052.

031.042 **Problem areas in large optics fabrication: a project manager's perspective.**
J. W. Mayo III.
Proc. SPIE Int. Soc. Opt. Eng., Vol. 542, p. 1 – 4 (1985). Abstr. in Phys. Abstr., Vol. 89, No. 1278, Entry 44245 (1986). – See Abstr. 012.052.

031.043 **New opportunities from materials selection trade–offs for high precision space mirrors.**
C. J. Pellerin, F. Ayer, Y. Mehrotra, A. K. Hopkins.
Proc. SPIE Int. Soc. Opt. Eng., Vol. 542, p. 5 – 18 (1985). Abstr. in Phys. Abstr., Vol. 89, No. 1278, Entry 44246 (1986). – See Abstr. 012.052.

031.044 **Mirror substrate scalability: fabrication limitations.**
D. J. Martin, B. E. Nelson.
Proc. SPIE Int. Soc. Opt. Eng., Vol. 542, p. 19 – 24 (1985). Abstr. in Phys. Abstr., Vol. 89, No. 1278, Entry 44247 (1986). – See Abstr. 012.052.

031.045 **Glass mirrors for space telescopes.**
J. R. P. Angel.
Proc. SPIE Int. Soc. Opt. Eng., Vol. 542, p. 32 – 34 (1985). Abstr. in Phys. Abstr., Vol. 89, No. 1278, Entry 44249 (1986). – See Abstr. 012.052.

031.046 **Fabrication of the Keck Ten Meter Telescope primary mirror.**
T. S. Mast, J. E. Nelson.
Proc. SPIE Int. Soc. Opt. Eng., Vol. 542, p. 48 – 59 (1985). Abstr. in Phys. Abstr., Vol. 89, No. 1278, Entry 44250 (1986). – See Abstr. 012.052.

031.047 **"Bend–and–Polish" fabrication techniques.**
L. D. Barr, J. H. Richardson, K.–L. Shu.
Proc. SPIE Int. Soc. Opt. Eng., Vol. 542, p. 60 – 67 (1985). Abstr. in Phys. Abstr., Vol. 89, No. 1278, Entry 44251 (1986). – See Abstr. 012.052.

031.048 **Grazing incidence metal optics for the Berkeley Extreme Ultraviolet Explorer satellite: a progress report.**
D. Finley, R. F. Malina, S. Bowyer.
Proc. SPIE Int. Soc. Opt. Eng., Vol. 540, p. 89 – 93 (1985). Abstr. in Phys. Abstr., Vol. 89, No. 1278, Entry 47894 (1986). – See Abstr. 012.051.

031.049 **Adaptive telescope design using computer tools.**
H. Levenstein, G. E. Seibert.
Proc. SPIE Int. Soc. Opt. Eng., Vol. 550, p. 75 – 86 (1985). Abstr. in Phys. Abstr., Vol. 89, No. 1280, Entry 60647 (1986). – See Abstr. 012.059.

031.050 **Atmospheric dispersion correctors at the Cassegrain focus.**
C. G. Wynne, S. P. Worswick.
Mon. Not. R. Astron. Soc., Vol. 220, No. 3, p. 657 – 670 (1986).
The first order aberration theory of an assembly of thin prisms is discussed and the results applied to derive design criteria for atmospheric dispersion correctors located in a converging beam. Two numerical examples are given of Cassegrain focus correctors for a 4.2–m *f*/11 and a 2.5–m *f*/15 telescope giving subarcsec imagery over wide spectral ranges and field angles.

031.051 **Wavefront sensing deconvolution and adaptive optics.**
J. C. Fontanella.
J. Opt. (Paris), Vol. 16, No. 6, p. 257 – 268 (1985). In French. Abstr. in Phys. Abstr., Vol. 89, No. 1282, Entry 73063 (1986).

031.052 **Focal plane optics in far–infrared and submillimeter astronomy.**
R. H. Hildebrand.
Opt. Eng., Vol. 25, No. 2, p. 323 – 330 (1986). Abstr. in Phys. Abstr., Vol. 89, No. 1283, Entry 78280 (1986).

031.053 **Improvement of the image quality of spherical lens correctors for Cassegrain telescopes.**
M.–l. Yi, D.–q. Su.
Acta Astron. Sin., Vol. 27, No. 1, p. 80 – 84 (1986). In Chinese.
The optical design of a two–lens fused silica Cassegrain focus field corrector for Ritchey–Chrétien configurations is presented. The achieved image quality over the spectral range 365 to 1400 nm is well under the 0.″2 limit.

031.054 **Geometrical and diffractional analysis of LEST.**
D. Bertani, M. Cetica, D. Iorio–Fili.
LEST Found., Tech. Rep., No. 13, 29 pp. (1985).
The optical system of LEST is well corrected. The configuration meets its theoretical limit in the sense that the encircled energy diameter of the actual diffractional spot and the theoretical are the same.

031.055 **Future space telescope design concepts.**
A. N. Bunner.
ESA Spec. Publ., ESA SP–226, p. 71 – 77 (1985). – See Abstr. 012.075.
This paper discusses the astronomical goals of an advanced space telescope that might be launched and assembled in space about the year 2005. A fundamental trade–off concerns designs that provide very high angular resolution and designs that provide sensitivity for faint extended source imaging. Some of the advantages and disadvantages of several possible configurations are reviewed.

031.056 **TRIO optics.**
B. Authier, P. Rabout, A. Labeyrie.
ESA Spec. Publ., ESA SP–226, p. 79 – 84 (1985). – See Abstr. 012.075.
Preliminary studies of the optical train for TRIO are reported. Performances and the difficulties expected are outlined. Calculations of the thermal variations and their effect on the Cassegrain telescopes are given. Different designs are considered for the beam recombiner in the central station. Straylight in the coudé beams, generated by structure lighting rather than by meteoritic dust, may affect the limiting magnitude if it is not efficiently baffled.

031.057 **A Ritchey–Chrétien system with prefocal achromatic meniscus for wide spectral region observations.**
G. M. Popov.
Izv. Krymskoj Astrofiz. Obs., Tom 74, p. 161 – 171 (1986). In Russian. English translation in Bull. Crimean Astrophys. Obs., Vol. 74.
A meniscus–type achromatic field corrector for a Ritchey–Chrétien telescope has been considered which is applied in the wide spectral region from far infrared to far ultraviolet.

Elements of modern optical design.
See Abstr. 003.148.

On a method for apodization of telescopes.
See Abstr. 032.004.

IR background speckle noise induced by adaptive optics in astronomical telescopes.
See Abstr. 032.019.

Experience of adjustment of the BTA prime focus optics.
See Abstr. 032.024.

Aperture synthesis of astronomical telescopes.
See Abstr. 032.031.

032 Astronomical Instruments

032.001 **Recent developments with the Anglo–Australian Telescope.**
D. C. Morton.
Astrophys. Space Sci., Vol. 118, Nos. 1/2, p. 31 – 33 (1986). – See Abstr. 012.001.
This report describes the instruments currently available on the 3.9 m Anglo–Australian Telescope, with particular reference to the recent additions of optical fibres for multi–object spectroscopy, a low resolution spectrograph with a red–sensitive charge–coupled device and software for obtaining infrared images by scanning the telescope in synchronism with the recording of data from the InSb photometer.

032.002 **The Mt. John 1 metre telescope project.**
J. B. Hearnshaw.
Astrophys. Space Sci., Vol. 118, Nos. 1/2, p. 79 – 81 (1986). – See Abstr. 012.001.
The Dall–Kirkham 1 m Cassegrain reflecting telescope to be installed at Mt. John University Observatory is described.

032.003 **Photographic and spectrographic observations with a reducing camera on the 61 cm NCU telescope.**
H. H. Wu.
Astrophys. Space Sci., Vol. 118, Nos. 1/2, p. 415 – 418 (1986). – See Abstr. 012.001.
A reducing camera designed by Meinel has been constructed and tested on the 61 cm NCU Cassegrain telescope. The basic principle of RC is to reduce the reflector's *f*/13.5 to the equivalent *f*/2.7 such that the Cassegrain focus image can be reduced on a 35 mm film with almost no loss in sky coverage and in resolution but the exposure time is reduced by a factor of 25. The camera is able to obtain stellar spectra. With a grism and a standard lens the dispersion is 450 Å mm^{-1}, which is comparable to that required for the MK classification. RC constructure and testing results are reported. The method of a wide field spectrograph and that of narrow band differential photography of RC are also discussed.

032.004 **On a method for apodization of telescopes.**
A. V. Lenskij.
Kinematika Fiz. Nebesn. Tel, Tom 2, No. 1, p. 77 – 81 (1986). In Russian. English translation in Kinematics Phys. Celest. Bodies.
A rather simple and sufficiently effective technique of apodization is suggested for telescopes with or without central obscuration. The strip which crosses the pupil parallel to binary star orientation in the image plane serves as an apodizing screen. A comparison with a more complicated technique is made.

032.005 **New telescopes for the 1990s.**
H. Shipman.
Phys. Today, Vol. 39, No. 1, p. S9 – S10 (1986). Short review paper. – See Abstr. 013.021.

032.006 **Amateurs et chambre photographique de Schmidt: Mythe ou réalité? (I).**
M. Cola.
Ciel, Vol. 48, p. 139 – 142 (1986).

032.007 **Amateurs et chambre photographique de Schmidt: Mythe ou réalité? (II).**
M. Cola.
Ciel, Vol. 48, p. 182 – 194 (1986).

032.008 **On the temperature coefficient of the micrometer of the ZTL–180 zenith telescope.**
V. M. Gorban'.
Kinematika Fiz. Nebesn. Tel, Tom 2, No. 2, p. 90 – 92 (1986). In Russian. English translation in Kinematics Phys. Celest. Bodies.
The expression for the scale–temperature coefficient of the zenith telescope ZTL–180 accounting for constructive peculiarities of its micrometer has been obtained. The temperature coefficient is shown to be a variable value depending on the angular distance measured by the eyepiece micrometer.

032.009 **An investigation of the rotation and position–stability characteristics of the prime (vertical) axis of the great azimuth telescope.**
V. Ya. Vainberg (*V. Ya. Vajnberg*).
Bull. Spec. Astrophys. Obs. – North Caucasus, Vol. 19, p. 81 – 86 (1985). English translation of 39.032.011.

032.010 **The influence of ellipticity and nonuniformity of the pivots of a transit instrument on observations for determination of the azimuth according to the direct method.**
Ts. Darakchiev.
Geod., kartogr., zemeustr., Vol. 25, No. 2, p. 16 – 19 (1985). In Bulgarian. Abstr. in Ref. Zh., 52. Geod. Aehrosemka, 2.52.84 (1986).

032.011 **The legacy of the 200–inch.**
R. Learner.
Sky Telesc., Vol. 71, No. 4, p. 349 – 353 (1986).

032.012 **A synchronous telescope and dome rotating system.**
A. Avotinš, J. Andersons.
Issled. Solntsa Krasnykh Zvezd, No. 21, p. 80 – 84 (1984). In Russian.
The azimuthal mounting of a telescope and the synchronous optical system of dome rotation, in which there is no direct mechanical contact between telescope and tracking system are described.

032.013 **Investigation of the system of the Tashkent meridian circle in the 1978 – 1979 period.**
G. G. Khodak.
Tsirk. Astron. Inst., Tashkent, No. 113, p. 22 – 31 (1985). In Russian.

032.014 **Investigation of the influence of the residual coma of the Zeiss double astrograph on the determination of the position of an object.**
Yu. Baltabaev, Yu. M. Ivanov, A. G. Rakhimov, K. Umarova.
Tsirk. Astron. Inst., Tashkent, No. 113, p. 32 – 34 (1985). In Russian.

032.015 **On the scale on the images obtained with the Tashkent normal and the Kitab double astrograph.**
Yu. M. Ivanov, A. G. Rakhimov, K. Umarova.
Tsirk. Astron. Inst., Tashkent, No. 113, p. 35 – 36 (1985). In Russian.

032.016 **Innovative optical telescope design and ideas of new telescopes.**
A. Balklavs.
Zvaigžnota Debess, 1984/85. Gada Ziema, p. 2 – 9 (1984). In Latvian.

032.017 **Preliminary results from the Athens–Hawaii–Purdue–Wisconsin Gamma Ray Telescope.**
A. H. Szentgyorgyi, L. Resvanis, J. G. Learned, V. J. Stenger, D. D. Weeks, J. Gaidos, F. Loeffler, T. R. Palfrey, G. H. Sembroski, C. Wilson, L. K. Rangan, U. Camerini, W. Fry, R. H. March, R. J. Loveless, J. M. Matthews, R. M. Morse, D. D. Reeder, P. Slane, A. Tsirou.
Bull. Am. Astron. Soc., Vol. 17, No. 4, p. 851 (1985). Abstract. – See Abstr. 010.062.

032.018 **The coalignment and cophasing of the National New Technology Telescope.**
J. M. Beckers, K. L. Shu, S. Shaklan.
Bull. Am. Astron. Soc., Vol. 17, No. 4, p. 900 (1985). Abstract. – See Abstr. 010.062.

032.019 **IR background speckle noise induced by adaptive optics in astronomical telescopes.**
P. Eisenhardt, F. Roddier.
Bull. Am. Astron. Soc., Vol. 17, No. 4, p. 901 (1985). Abstract. – See Abstr. 010.062.

032.020 **The astronomical CCD camera Vega–202.**
M. V. Goryachev.
Astron. Tsirk., No. 1392, p. 7 – 8 (1985). In Russian.

032.021 **On the quality of the Moscow Observatory's transit circle GOMZ.**
V. G. Tauber.
Astron. Tsirk., No. 1400, p. 6 – 8 (1985). In Russian.

032.022 **A built–in system for adjustment of a three–mirror telescope.**
N. N. Fashchevskij.
Astron. Tsirk., No. 1402, p. 6 – 8 (1985). In Russian.

032.023 **Giant infrared telescopes for astronomy: a scientific rationale.**
H. M. Dyck, E. J. Kibblewhite.
Publ. Astron. Soc. Pac., Vol. 98, No. 600, p. 260 – 267 (1986).
The authors develop scientific criteria for choosing the dimensions of an interferometric array of telescopes to be used in the infrared. The ideal array consists of small collecting apertures which are either few in number, widely spaced and easily moved to new baselines or many in number, closely spaced and on fixed baselines. The proposed and planned "next–generation" of ground–based telescopes are found to be inadequate to solve the interesting problems of infrared astronomy which require high spatial resolution.

032.024 **Experience of adjustment of the BTA prime focus optics.**
V. Ya. Vajnberg, N. A. Vikul'ev, L. I. Snezhko.
Astrofiz. Issled. Izv. Spets. Astrofiz. Obs., Tom 22, p. 112 – 118 (1986). In Russian. English translation in Bull. Spec. Astrophys. Obs. – North Caucasus.
The use of the Hartmann test in controlling the adjustment of the BTA prime focus optics is described. Methods of the adjustment of the lens field corrector are presented. The current data on the adjustment quality of the BTA prime focus optics are given.

032.025 **Nordisk optisk teleskop. Prosjektgruppen på Risø – Mekanikk og styring.**
J.–E. Solheim.
Astron. Tidsskr., Årg. 19, Nr. 2, p. 49 – 56 (1986).

032.026 **Comparison of a 1.8–meter CCD scannerscope with a 1.2–m Schmidt camera.**
T. Gehrels.
Proceedings of the Second Asian–Pacific Regional Meeting on Astronomy, p. 335 – 340 (1984). – See Abstr. 012.031.
A comparison is made of a Schmidt camera that has a primary mirror of 1.8 m in diameter with a scannerscope that also has a 1.8–m primary, but with charge coupled devices (CCDs) at its prime focus. The comparison comes out close in costs and in area coverage per unit time. The scannerscopy is advantageous in instantaneous data processing, deeper limiting magnitude for moving objects such as comets, and in providing a 1.8–m telescope during moonlight, for instance for infrared astronomy.

032.027 **Valdosta State College's computerized 16–inch telescope.**
M. A. Leake, D. W. Marks, K. S. Rumstay.
I.A.P.P.P. Commun., No. 23, p. 2 – 3 (1986). Abstract. – See Abstr. 010.521.

032.028 **Development of a 2.5–meter mobile observatory.**
W. B. Stanwix–Hay, L. B. Reed, E. F. Strother.
I.A.P.P.P. Commun., No. 23, p. 3 (1986). Abstract. – See Abstr. 010.521.

032.029 **Chopping secondary mirror for a cooled infrared telescope.**
A. Krabbe, D. Lemke, K. Proetel.
Appl. Opt., Vol. 25, No. 4, p. 523 – 527 (1986).
A 10–cm chopping secondary for operation in the cryovacuum has been developed. It is momentum compensated and allows square wave modulation of up to 20 Hz and a beam separation of up to several minutes of arc in the sky. The low heat dissipation of 5 mW at a temperature of 4.2K makes the chopper suitable for a liquid helium–cooled space telescope.

032.030 **Combining beams from separated telescopes.**
W. A. Traub.
Appl. Opt., Vol. 25, No. 4, p. 528 – 532 (1986).
To use separated telescopes to achieve diffraction–limited performance over a finite field of view, the following rule must be obeyed: as viewed from a point in the focal plane, beams from separated telescopes must be recombined so that they appear to be coming directly from a single large telescope which has been masked so as to reproduce exactly the ensemble of collecting telescopes. This condition leads to constraints and tolerances which are of central interest in multiple–beam telescope systems.

032.031 **Aperture synthesis of astronomical telescopes.**
N. A. Dimov.
Opt.–mekh. prom–st', No. 12, p. 32 – 38 (1985). In Russian. Abstr. in Ref. Zh., 51. Astron., 5.51.1024 (1986).

032.032 **Das 3.5–m–Teleskop auf dem Calar Alto.**
K. Bahner.
Sterne Weltraum, 25. Jahrg., Nr. 6, p. 310 – 315 (1986).

032.033 **Results and limitations of ground–based optical arrays.**
A. Labeyrie.
ESA Spec. Publ., ESA SP–226, p. 1 – 6 (1985). – See Abstr. 012.075.
Interference data are produced by ground–based interferometers with baselines reaching 76 meters. Current projects may reach 300 meters, with four 8–meter apertures in the case of the VLT if it is built as an interferometer. In the visible, the limiting magnitude is severely limited by the atmosphere. The possibilities for reconstructing images are restricted, and it is difficult to find mountaintop sites with flat platforms larger than 300 m.

032.034 **The wind protection of the Hvar solar telescope.**
P. Ambrož.
Hvar Obs. Bull., Vol. 9, No. 1, p. 33 – 60 (1985).

032.035 **Microseismicity measurement of the PZT base in the observing house above ground.**
K. Iwadate, H. Kitago.
Proc. Int. Latitude Obs. Mizusawa, No. 24, p. 1 – 8 (1985).

032.036 **Time and latitude observation of the Mizusawa PZT in the observation room above the ground.**
C. Kakuta, T. Hara, H. Kitago, K. Iwadate, S. Takano.
Publ. Int. Latitude Obs. Mizusawa, Vol. 18, No. 2, p. 49 – 62 (1984).
The Mizusawa PZT (Photographic Zenith Tube) was moved from its semibasement location (the second PZT observation room) to a location 6.65 m above ground level (the third PZT observation room) in order to avoid the effects of temperature variations near the ground during observations. Some provisions are made for prevention of heat flow from the ground and for radiation flux. The 3rd observation room shows some satisfactory results on the astronomical observations.

032.037 **Automatic electronic astrolabe at Mizusawa.**
I. Tsubokawa, T. Tsubokawa, K. Asari, C. Kakuta, S. Manabe, S. Sakai.
Publ. Int. Latitude Obs. Mizusawa, Vol. 18, No. 2, p. 105 – 125 (1984).
A Tsubokawa type astrolabe has been developed at the International Latitude Observatory of Mizusawa (ILOM). It works

fully automatically under the control of a computer. Using this instrument, one performs high accuracy measurements of the time (UT0–UTC) and latitude. The accuracy is comparable to that of the PZT of the ILOM. The design, construction and adjustment of the astrolabe are described.

032.038 **The automatic photoelectric transit instrument and its further improvement.**
Z.–l. Zou, J.–s. Yao.
Publ. Purple Mt. Obs., Vol. 4, No. 2, p. 83 – 87 (1985). In Chinese.
The principle and operation of the automatic photoelectric transit instruments are described in this paper. To increase observable magnitude, the authors present a new design about its further improvement. This design is using the photon–counting technique and the automatic track–micrometer.

032.039 **Investigation of the level of the transit instrument of the Kharkov Time Service with the help of an autocollimator.**
S. R. Izmajlov, N. G. Litkevich, T. I. Suchkova.
Vestn. Khar'kov. Univ., No. 278, p. 77 – 78 (1985). In Russian.

032.040 **Very Large Telescope: recent developments.**
D. Enard.
Messenger, No. 44, p. 37 – 40 (1986).

032.041 **On the location and restoration of an 1849 Henry Fitz telescope.**
R. B. Ariail.
J. Am. Assoc. Variable Star Obs., Vol. 14, No. 2, p. 69 – 71 (1985).

032.042 **Meridian telescope for US Naval Observatory, Washington, DC.**
A. Zollner.
Feinwerktech. Messtech., Vol. 94, No. 1, p. 17 – 19 (1986). In German. Abstr. in Phys. Abstr., Vol. 89, Nr. 1284, Entry 83433 (1986).

032.043 **The 1.29 m infrared telescope of Beijing Observatory in pilot operation.**
Z.–y. Qian.
Acta Astrophys. Sin., Vol. 6, No. 2, p. 168 (1986). In Chinese.

032.044 **Principles of a solar magnetic field telescope.**
G. x. Ai, Y.–f. Hu.
Acta Astron. Sin., Vol. 27, No. 2, p. 173 – 180 (1986). In Chinese.
A new solar magnetic field telescope vector video magnetograph has been described. The principle of measurements of solar vector magnetic fields, line–of–sight velocity field and the general quality have been discussed.

032.045 **Pivot errors.**
P. Benevides–Soares, R. Teixeira.
Rev. Mex. Astron. Astrofis., Vol. 12, Spec. Issue, p. 424 – 426 (1986). – See Abstr. 012.066.
The determination of pivot errors for the AM–190 transit circle is described. The method rests on readings of the graduated circles. The measured cross sections are circular within 0.1 μm.

032.046 **The astrograph of the Valongo Observatory (UFRJ) and the astrometry program at Campinas, São Paulo, Brazil.**
L. E. Machado.
Rev. Mex. Astron. Astrofis., Vol. 12, Spec. Issue, p. 441 – 443 (1986). – See Abstr. 012.066.

032.047 **A proposal for the LEST control system.**
O. Engvold, T. E. Andersen, M. Carlsson, J. R. Jensen, K. Klim.
LEST Found., Tech. Rep., No. 15, 55 pp. (1985).
Contents: Trends in development of computers and peripherals. Trends in telescope control and operation. Control system requirements for LEST. Proposed LEST control system. Remote control of LEST. Cost estimate. Estimated total cost of control and data reduction system.

032.048 **Finite element calculation of LEST structure.**
T. E. Andersen, N. C. Jessen.
LEST Found., Tech. Rep., No. 16, 13 pp. (1986).
The structural design of the LEST telscope is presented. Then a computer model for use with the finite element program NASTRAN is described. Finally the results of the computations are presented.

032.049 **Parallaxes with large reflectors.**
R. S. Harrington.
Astrometric techniques, p. 271 – 273 (1986). – See Abstr. 012.068 (IAU Symp. No. 109).
Reflecting telescopes designed to do astrometry can obtain parallaxes essentially free of systematic errors. Improvements in detection and mensuration now make it possible to produce parallaxes of unprecedented accuracy without undue cost or effort.

032.050 **Design characteristics of the 1.56 m astrometric telescope and its usage in astrometry.**
L. Wan, N.–h. Zhu, L.–j. Wang, Z.–h. Yang, Y.–j. Zheng.
Astrometric techniques, p. 275 – 287 (1986). – See Abstr. 012.068 (IAU Symp. No. 109).

032.051 **Astrometric characteristics of the Abastumani Astrophysical Observatory 125 cm reflector (AZT–11) of Ritchey–Chrétien optical system.**
G. N. Salukvadze.
Astrometric techniques, p. 289 – 292 (1986). – See Abstr. 012.068 (IAU Symp. No. 109).

032.052 **Present status and future plans for the two color astrometric interferometer project.**
M. Shao, M. Colavita, D. Staelin, R. Simon, K. Johnston.
Astrometric techniques, p. 331 – 340 (1986). – See Abstr. 012.068 (IAU Symp. No. 109).
A 3–meter two telescope stellar interferometer has been used to observe fringes for the past year in order to study the feasibility of using long baseline interferometers for astrometry. The results of the study will be incorporated into the design of the 20–meter astrometric interferometer. The key to the 20–meter interferometer is the laser system which the authors expect to monitor all the mechanical and thermal imperfections relevant to astrometry at the 10^{-3} to 10^{-4}arcsec level. A slight modification of this system could be used in a space based interferometer for 1 to 10 microarcsecond astrometry of faint objects.

032.053 **A new astrometric system.**
G. Gatewood, J. Stein, J. Kiewiet de Jonge, C. DiFatta, L. Breakiron.
Astrometric techniques, p. 341 – 352 (1986). – See Abstr. 012.068 (IAU Symp. No. 109).
The 30–inch Thaw Refractor of the University of Pittsburgh's Allegheny Observatory has been completely rebuilt. Changes include a new objective lens, a new detector system, and computer control and data acquisition. With an observational accuracy exceeding one–thousandth of an arc second, the new system will greatly expand the domain of astrometric research.

032.054 **Photoelectric astrolabe and astrolabe star catalogues.**
D.–j. Luo, D.–m. Li.
Astrometric techniques, p. 375 – 377 (1986). – See Abstr. 012.068 (IAU Symp. No. 109).

032.055 **The oscillating slit micrometer of the meridian circle PMC 190 Tokyo.**
C. Kühne, M. Miyamoto, M. Yoshizawa.
Astrometric techniques, p. 379 – 388 (1986). – See Abstr. 012.068 (IAU Symp. No. 109).
The meridian circle installed at the Tokyo Astronomical Observatory in 1982/83 is equipped with a photoelectric double slit micrometer. A slit plate is located in the image field of the telescope. It oscillates parallel to right ascension while being guided

at the mean traveling speed of the star. The paper describes the procedure by which the moment of the star passage through the instrument's meridian and declination is determined.

032.056 **A new astrolabe.**
C. Billaud.
Astrometric techniques, p. 389 – 395 (1986). – See Abstr. 012.068 (IAU Symp. No. 109).

To obtain a quiet impersonal astrolabe, a new instrument has been built. The first UT0–UTC results show very good agreement with the BIH scale.

032.057 **Timing and data acquisition system for a field astrolabe.**
C. F. Lukac, P. J. Wheeler, R. E. Keating, R. T. Clarke.
Astrometric techniques, p. 397 – 406 (1986). – See Abstr. 012.068 (IAU Symp. No. 109).

The Danjon Astrolabe at the Naval Observatory has been traditionally used to determine Universal Time and improve the systematic accuracy of star positions. This paper describes the performance of the astrolabe timing and data acquisition system, gives engineering design considerations, and describes the equipment and instrumentation.

032.058 **The photoelectric meridian circle of Pulkovo.**
V. N. Ershov, V. E. Pliss, Yu. S. Streletsky (*Yu. S. Streletskij*).
Astrometric techniques, p. 407 – 411 (1986). – See Abstr. 012.068 (IAU Symp. No. 109).

The paper reports on the new semiautomatical meridian circle of the Pulkovo Observatory. It is equipped with a scanning photoelectric micrometer and a photoelectric circle reading system. The standard errors of a single registration of a star are $0\overset{s}{.}016$ and $0\overset{''}{.}35$ in RA and declination, respectively. The limiting magnitude is $10\overset{m}{.}5$.

032.059 **The glass meridian circle.**
E. Høg.
Astrometric techniques, p. 413 – 419 (1986). – See Abstr. 012.068 (IAU Symp. No. 109).

Some aspects of the design for a glass meridian circle with 24 cm aperture are presented and discussed. The instrument is very compact and portable and it requires only a small dome.

032.060 **Design for a large transit circle with reflecting optics.**
N.-s. Hu.
Astrometric techniques, p. 421 – 426 (1986). – See Abstr. 012.068 (IAU Symp. No. 109).

This paper describes a conceptual design for an automatic transit circle with Ritchey–Chrétien Cassegrain type optics of 40 cm aperture. The instrument takes reversal observation both in the meridian and in the prime vertical. Methods of adopting active optics to achieve a stable boresight of the reflecting telescope and of automatically reversing and leveling the instrument are presented.

032.061 **A proposal for scanning meridian circle.**
A. S. Kharin.
Astrometric techniques, p. 427 (1986). – See Abstr. 012.068 (IAU Symp. No. 109).

Calculations show that in the presence of devices for fast reading of a divided circle and of star positions in the field of telescope, a fast transit circle with a continuous working scanning tube can be created on the basis of TV equipment. When the recording time of a single circle reading is 10^{-4}sec and the rotational speeds do not exceed 7.5 deg/min, the reading error for displacement remains smaller than $0\overset{''}{.}05$. The time required to record star transit in the field of telescope is 2–4 sec.

032.062 **Impersonal techniques of transit circle observations. "The Carlsberg Automatic Meridian Circle".**
L. Helmer.
Astrometric techniques, p. 429 – 433 (1986). – See Abstr. 012.068 (IAU Symp. No. 109).

The conventional reversible meridian circle of Copenhagen University, has over the last years been fully automated, and equipped with a new photoelectrical transit micrometer and a revised photoelectric circle recording system. Setting, observation and data recording is completely controlled by two minicomputers. The entire system has been tested during two years of operation resulting in two differential catalogues. The number of observations included in the two catalogues is about 20,000 and the obtained mean errors are: $\varepsilon_\alpha \cos_\delta = 0\overset{s}{.}014$, $\varepsilon_\delta = 0\overset{''}{.}22$, $\varepsilon_m = 0\overset{m}{.}12$. The faintest star observed was $11\overset{m}{.}98$.

032.063 **Automatic horizontal meridian circle at Pulkovo.**
R. I. Gumerov, V. B. Kapkov, G. I. Pinigin.
Astrometric techniques, p. 459 – 462 (1986). – See Abstr. 012.068 (IAU Symp. No. 109).

Computer control was introduced for the automation of all the major operations of the Pulkovo horizontal meridian circle. This involves an automatic setting system (mirror setting), a circle reading system, two photoelectric eyepiece micrometers, meteorological data sensors and a rotating drive of the pendulum horizon.

032.064 **Automation and software of the Wanschaff vertical circle at Goloseevo.**
A. S. Kharin, L. A. Kukharskij, P. F. Lazorenko, N. F. Minyajlo, M. L. Tsesis.
Astrometric techniques, p. 463 – 464 (1986). – See Abstr. 012.068 (IAU Symp. No. 109).

032.065 **The seven–inch transit circle and its New Zealand program.**
J. A. Hughes, M. D. Robinson, F. S. Gauss, R. C. Stone.
Astrometric techniques, p. 483 – 496 (1986). – See Abstr. 012.068 (IAU Symp. No. 109).

032.066 **The refurbished six–inch transit circle.**
B. L. Klock.
Astrometric techniques, p. 507 – 515 (1986). – See Abstr. 012.068 (IAU Symp. No. 109).

The U.S. Naval Observatory (USNO) six–inch transit circle at Washington underwent extensive refurbishment during the interval between October 1982 and May 1983. Major items refurbished on the instrument included the micrometer, the clamp and slow motion arm, the screen system, and a circle. Support instrumentation refurbished included a new digital motor drive system, a new digital micrometer display system, and new environmental measuring systems.

032.067 **Vacuum meridian marks of the Belgrade large transit instrument.**
L. A. Mitić, I. Pakvor.
Astrometric techniques, p. 525 – 527 (1986). – See Abstr. 012.068 (IAU Symp. No. 109).

More than a dozen years ago vacuum meridian marks were built for the Belgrade large transit instrument with the purpose of completely keeping the air from interfering with the meridian mark readings. Information is presented on the favorable results achieved in regard to both accidental and systematic accuracy.

032.068 **The vacuum mires of the transit instrument at Nikolaev.**
G. M. Petrov, R. T. Fedorova, P. N. Fedorov.
Astrometric techniques, p. 529 – 531 (1986). – See Abstr. 012.068 (IAU Symp. No. 109).

032.069 **Internal refraction in meridian circles.**
E. Høg.
Astrometric techniques, p. 533 – 542 (1986). – See Abstr. 012.068 (IAU Symp. No. 109).

Tube refraction of the order $0\overset{''}{.}3$ inside a meridian circle is a source of systematic errors that has not been noticed before. Natural convection due to cooling at night or heating during day will create primarily vertical temperature gradients in the air. A method of tangential ventilation of the tube was used to eliminate the gradients. The design for a conventional meridian circle and for a glass meridian circle is discussed.

032.070 **Determination of the division corrections of the Bordeaux declination circle by the Benevides–Boczko method.**
Y. Requième, M. Rapaport.
Astrometric techniques, p. 543 – 550 (1986). – See Abstr. 012.068 (IAU Symp. No. 109).
The automatic declination reading system implemented on the Bordeaux meridian circle with a new divided circle is presented. The determination of the division errors by the Benevides–Boczko method was carried out in December 1982 and in March 1983: the standard deviation between the two sets of corrections is about 0."015.

032.071 **A new method of determining absolute azimuth and latitude and suggestion for a new type of meridian circle.**
W. Mao, Z.–m. Li, Y. Fan, S.–s. Hu, M.–h. Du, H. Li.
Astrometric techniques, p. 551 – 555 (1986). – See Abstr. 012.068 (IAU Symp. No. 109).
This article describes a new method by which azimuth and the instantaneous latitude of a meridian–prime vertical–transit instrument can be determined absolutely in mid– and low–latitude areas, and discusses some experimental observations obtained on a remodeled Zeiss transit instrument. Requirements for the development of a new type of transit circle along the lines of this new method are also presented.

032.072 **A new method for zenith distance determination in meridian observations.**
M. Miyamoto, M. Yoshizawa.
Astrometric techniques, p. 557 – 565 (1986). – See Abstr. 012.068 (IAU Symp. No. 109).
In order to obtain higher accuracy in the declination observation, besides a full automation device, the photoelectric meridian circle at Tokyo Astronomical Observatory (Tokyo PMC) is equipped with photoelectric devices for circle reading and a zenith mirror in addition to the conventional mercury horizon. The authors report on a preliminary determination of the horizontal and vertical flexures, and of the division errors of the glass circle.

The Royal Astronomical Society instrument collection: 1827 – 1985.
See Abstr. 002.073.

Infinite vistas: new tools for astronomy.
See Abstr. 003.076.

A la poursuite du Soleil. La construction du coronographe d'amateur.
See Abstr. 003.132.

The cosmic inquirers. Modern telescopes and their makers.
See Abstr. 003.198.

Über antike und neuzeitliche Fernrohre.
See Abstr. 004.006.

Bernhardt Schmidt e l'astronomia con i telescopi Schmidt.
See Abstr. 005.014.

Galactic and extragalactic studies with the KISO Schmidt.
See Abstr. 009.002.

The Iraqi National Astronomical Observatory.
See Abstr. 009.003.

Astronomical research activities with the 74 inch telescope at Kottamia Observatory.
See Abstr. 009.006.

The Tokyo Schmidt telescope and its programs.
See Abstr. 009.037.

Completion near Beijing of China's first over–water solar observing station.
See Abstr. 009.059.

The Japanese National Large Telescope (JNLT) Project.
See Abstr. 013.002.

New telescopes in China.
See Abstr. 013.004.

New telescopes in India.
See Abstr. 013.006.

A photographic astrometric telescope for Brazil.
See Abstr. 013.063.

Automatic telescopes large and small.
See Abstr. 013.067.

A new type of field corrector.
See Abstr. 031.002.

15 meter multiple mirror telescope design study.
See Abstr. 031.003.

Mirror systems: engineering features, benefits, limitations and applications.
See Abstr. 031.032.

An overview of optical telescope technology.
See Abstr. 031.036.

Design aspects of the University of Texas 7.6 m telescope. I.
See Abstr. 031.037.

Design aspects of the University of Texas 7.6 m telescope. II.
See Abstr. 031.038.

Optics for the 15 m National New Technology Telescope.
See Abstr. 031.039.

Multilayer coatings for astronomical telescope mirrors.
See Abstr. 031.040.

Improved coatings for large astronomical telescope mirrors.
See Abstr. 031.041.

Fabrication of the Keck Ten Meter Telescope primary mirror.
See Abstr. 031.046.

Wavefront sensing deconvolution and adaptive optics.
See Abstr. 031.051.

Focal plane optics in far–infrared and submillimeter astronomy.
See Abstr. 031.052.

Geometrical and diffractional analysis of LEST.
See Abstr. 031.054.

A multislit photoelectric star micrometer for the meridian circle of the Nikolayev Astronomical Observatory.
See Abstr. 034.192.

Determining the absolute paramters of a meridian circle by observing the first vertical (V). The effect of the higher power terms of the directional error in the determined values of latitude.
See Abstr. 036.138.

Determining the absolute parameters of a meridian circle by observing the first vertical (VI). Curvature correction to the recording times in the first vertical direction.
See Abstr. 036.139.

Determining the absolute parameters of a meridian circle by observing the first vertical (VII). Curvature correction of the determination value of an apparent zenith distance in the first vertical direction.
See Abstr. 036.140.

Determining the absolute parameters of a meridian circle by observing the first vertical (VIII). The effect of the diurnal variation term of the latitude on the determined values of the azimuth equation and latitude.
See Abstr. 036.141.

Comparison of two methods for the absolute determination of the azimuth of a meridian circle.
See Abstr. 036.142.

Analysis of the effects of errors on the reading of the graduated circle.
See Abstr. 036.143.

Remote control of 2.2–m telescope from Garching.
See Abstr. 036.191.

The design of modern meridian circles for the observation of faint objects.
See Abstr. 036.224.

The use of photographic positions in determining azimuth of a meridian circle.
See Abstr. 036.225.

The use of astrometric instruments in vacuum chambers.
See Abstr. 036.226.

On the implementation of absolute meridian observations in low latitude stations.
See Abstr. 036.227.

An improved high–precision instrument for astrometric observations under lunar conditions.
See Abstr. 041.040.

The Bordeaux automatic transit circle: first catalogues, current and future programs.
See Abstr. 041.061.

Current work with the photoelectric transit instrument at the Observatory of Torino.
See Abstr. 041.062.

Carlsberg Automatic Transit Circle: first two test catalogues and the programme for La Palma.
See Abstr. 041.063.

033 Radio Telescopes and Equipment

033.001 **The Australian radio–telescope.**
B. J. Robinson.
Astrophys. Space Sci., Vol. 118, Nos. 1/2, p. 57 – 61 (1986). – See Abstr. 012.001.

A large synthesis radio telescope is under construction in Australia at a cost of $A 30.7 million. It comprises a 6 km–long "compact array" of 22 m antenna at Culgoora and a "long baseline array" of 319 km, achieved by adding another 22 m antenna near Coonabarabran and the existing 64 m antennae at Parkes. On occasion it will be possible to use this array with one of the deep space network antennas at Tidbinbilla to achieve a baseline of 580 km.

033.002 **A design for digital signal processing in a large field radio patrol camera.**
T. Daishido, K. Asuma, T. Ohkawa, H. Obara, S. Komatsu, K. Nagane.
Astrophys. Space Sci., Vol. 118, Nos. 1/2, p. 467 – 469 (1986). – See Abstr. 012.001.

An FFT processor is being developed for the present pilot system of the large field radio patrol camera. A design of the processor is discussed.

033.003 **The Siberian Solar Radio–Telescope: parameters and principle of operation, objectives and results of first observations of spatio–temporal properties of development of active regions and flares.**
G. Ya. Smolkov (*G. Ya. Smol'kov*), A. A. Pistolkors, T. A. Treskov, B. B. Krissinel (*B. B. Krissinel'*), V. A. Putilov, N. N. Potapov.
Astrophys. Space Sci., Vol. 119, No. 1, p. 1 – 4 (1986). – See Abstr. 012.001.

At the SibIZMIR, the Siberian Solar Radio–Telescope (SSRT) has been devised, built and aimed to diagnose the state of solar activity in the microwave band, and to study the structure and development of active regions and flares in the solar atmosphere with high two–dimensional resolution on a real–time basis.

033.004 **VLBI network at 18 cm wavelength.**
L. I. Matveenko, R. Z. Sagdeev, V. M. Balebanov, V. I. Shevchenko, V. I. Kostenko, V. A. Grishmanovskij, V. E. Velikhov, S. P. Ignatov, B. Z. Kanevskij, L. R. Kogan, A. N. Kozlov, G. D. Kopelyanskij, A. P. Molodyanu, E. P. Molotov, A. Kh. Papatsenko, A. M. Romanov, I. A. Strukov, V. V. Timofeev, A. V. Shevchenko, A. B. Severnyj, I. G. Moiseev, R. L. Sorochenko, A. P. Tsivilev, R. M. Martirosyan, A. M. Aslanyan, A. G. Gulyan, Ya. S. Yatskiv, M. V. Golovnya.
Pis'ma Astron. Zh., Tom 12, No. 1, p. 59 – 65 (1986). In Russian. English translation in Sov. Astron. Lett., Vol. 12.

The main parameters are given of the Soviet 18–cm VLBI network including five elements with interferometric baselines from 100 to 7000 km. Antennas are equipped by hydrogen frequency standards, low–noise amplifiers and 2–MHz bandwidth recording systems. The network is destined for astrophysical observations as well as for astronavigation and geodesy purposes.

033.005 **Systematic pointing errors of the 22–meter radio telescope.**
N. S. Nesterov.
Izv. Krymskoj Astrofiz. Obs., Tom 73, p. 189 – 198 (1985). In Russian. English translation in Bull. Crimean Astrophys. Obs., Vol. 73.

The pointing errors of the 22–m radio telescope of the Crimean Astrophysical Observatory are analysed using data of discrete radio sources observations made in 1974 – 1984. It is shown that the used analytical expressions are adequate for approximation. The vertical axis inclination is measured and the location of the radio telescope is determined more precisely. Some factors leading to pointing instability are established.

033.006 **Certain aspects of the use of radio holography to test the RATAN–600 radio telescope.**
G. A. Pinchuk, A. A. Stotskii (*A. A. Stotskij*).
Bull. Spec. Astrophys. Obs. – North Caucasus, Vol. 19, p. 70 – 74 (1985). English translation of 39.033.009.

033.007 **"Sliding" technique with the RATAN–600 radio telescope.**
M. G. Mingaliev, Z. E. Petrov, V. I. Filipenko, L. N. Cherkov.
Bull. Spec. Astrophys. Obs. – North Caucasus, Vol. 19, p. 75 – 80 (1985). English translation of 39.033.010.

033.008 **Selection of a polarimeter to measure the linear polarization of radio emission with the RATAN–600 radio telescope.**
V. I. Abramov, D. V. Korol'kov.
Bull. Spec. Astrophys. Obs. – North Caucasus, Vol. 19, p. 87 – 91 (1985). English translation of 39.033.011.

033.009 **A study of the instrumental linear–polarization of the south sector of the RATAN–600 radio telescope with flat reflector at 13 cm.**
V. I. Abramov, E. N. Vinyaikin (*E. N. Vinyajkin*).
Bull. Spec. Astrophys. Obs. – North Caucasus, Vol. 19, p. 92 – 99 (1985). English translation of 39.033.012.

033.010 **An experimental study of the directional patterns of the RATAN–600 south sector with periscopic reflector.**
A. V. Temirova.
Bull. Spec. Astrophys. Obs. – North Caucasus, Vol. 19, p. 100 – 107 (1985). English translation of 39.033.013.

033.011 **Phototracking system for the RT–10 radio telescope.**
M. Eliass.
Issled. Solntsa Krasnykh Zvezd, No. 21, p. 16 – 23 (1984). In Russian.
A phototracking system for the radio telescope RT–10 is worked out and tested. The accuracy of tracking of the sun is ±0′.5 in azimuth and ±0′.2 in elevation.

033.012 **Accuracy improvement of the surface of the reflecting elements of the main mirror of the radio telescope RATAN–600.**
A. P. Glumov, S. Ya. Golosova, Yu. K. Zverev, Eh. I. Korkin, A. I. Kopylov, O. N. Shivris.
Astrofiz. Issled. Izv. Spets. Astrofiz. Obs., Tom 21, p. 65 – 72 (1986). In Russian. English translation in Bull. Spec. Astrophys. Obs. – North Caucasus.
Investigations on improving the accuracy of the reflecting elements of the main mirror of RATAN–600 are described. The possibility to get the reflecting surface with r.m.s. deviation less than 0.1 mm is shown. The results obtained with the first series of elements with the improved accuracy of the reflecting surface are presented.

033.013 **The receiving device of a spectrometer at 21 cm wavelength.**
A. P. Venger, L. G. Gassanov, Yu. D. Gudz', G. N. Il'in, N. F. Ryzhkov, A. D. Tuz.
Astrofiz. Issled. Izv. Spets. Astrofiz. Obs., Tom 21, p. 98 – 104 (1986). In Russian. English translation in Bull. Spec. Astrophys. Obs. – North Caucasus.
A description of the receiving device at 21 cm wavelength of the spectrometric complex of the radio telescope RATAN–600 is presented. At the input of this device a transistor amplifier performed in hydrid–integral design and cooled up to the temperature of liquid nitrogen is used. The noise temperature of the receiver does not exceed 40K. Measurement methods with the noise pilot–signal is described.

033.014 **Calculation of the parameters for the circular focusing system of the radio telescope RATAN–600 at observations in the zenith.**
E. K. Majorova, A. A. Stotskij.
Astrofiz. Issled. Izv. Spets. Astrofiz. Obs., Tom 22, p. 119 – 127 (1986). In Russian. English translation in Bull. Spec. Astrophys. Obs. – North Caucasus.
The results of calculation of the effective area and the noise temperature of the radio telescope RATAN–600 at observations in zenith with the use of the circular mirror and special two–mirrors feed are presented. An optimum form of the beam in the receiving horn is discussed.

033.015 **A variant of geodetic adjustment of the radio telescope RATAN–600.**
Yu. K. Zverev.
Astrofiz. Issled. Izv. Spets. Astrofiz. Obs., Tom 22, p. 128 – 135 (1986). In Russian. English translation in Bull. Spec. Astrophys. Obs. – North Caucasus.
The description and the results of practical application of the geodetic angular intersection method for the adjustment of the south sector of the RATAN–600 radio telescope are presented. This method permits to determine the radial position of the elements of the south sector with the rms 0.16 mm; that is fully enough for the provision of radio astronomical observations at wavelengths ⩾2 cm.

033.016 **The James Clerk Maxwell Telescope.**
R. W. Newport.
J. Br. Interplanet. Soc., Vol. 39, No. 5, p. 211 – 216 (1986).
This paper presents a short account of the main features of the James Clerk Maxwell Telescope, a sub–millimetre wave telescope under construction on the island of Hawaii. The technical requirements and design concepts of this instrument are discussed and a summary is given of its present status.

033.017 **Superconducting tunnel detectors in radio astronomy.**
T. G. Phillips, D. B. Rutledge.
Sci. Am., Vol. 254, No. 5, p. 78 – 84 (1986).
A sensitive new radiation detector enables astronomers to explore regions of the universe hidden from optical telescopes, such as sites where stars are now being born.

033.018 **Utrecht ontwerpt nieuwe spiegel voor Hawaii–telescoop.**
H. van de Stadt.
Zenit, 13. Jaarg., Nr. 6, p. 210 – 211 (1986).

033.019 **Determination of the baseline of the Simeiz–Pushchino interferometer.**
V. E. Zharov, L. R. Kogan, E. N. Fedoseev.
Sov. Astron., Vol. 29, No. 5, p. 593 – 595 (1985). English translation of 40.033.025.

033.020 **The Parkes–Tidbinbilla interferometer.**
R. P. Norris, M. J. Batty, M. J. Kesteven, K. J. Wellington.
Proc. Astron. Soc. Aust., Vol. 6, No. 2, p. 137 – 140 (1985).
The authors report the development of a radio–linked interferometer which uses the 64–m telescope at Parkes, NSW, and one of the NASA antennas (64–m or 34–m) at Tidbinbilla, ACT. With a baseline of approximately 275 km, this is the world's longest real–time interferometer; it will be usable at frequencies of 1.6, 2.3, 8.4, and 22 GHz to give angular resolutions of 0.13, 0.09, 0.03, and 0.01 arcsec respectively. The interferometer has already operated successfully in a limited mode and is expected to become fully operational in its initial configuration by September 1985.

033.021 **Millimetre Wave Telescope Project: receiver working group package WP7 – Schottky diode multipliers, final report.**
D. R. Vizard, D. L. John.
Rutherford Appleton Lab., Rep., RAL–85–109, 10+44 pp. (1985).

033.022 **The very long baseline interferometry. Part VI. Methods of spectral line VLBI.**
K. M. Borkowski.
Postepy Astron., Tom 32, Zesz. 3/4, p. 205 – 219 (1984). In Polish.
This part deals with the final analysis of spectral line VLBI data obtained after correlation and postprocessing. Theoretical background for the analysis of differential observables is given. Principles of the fringe rate mapping, multiple–point fringe rate mapping, phase modeling, aperture synthesis and maser cluster

distance determination are discussed. A generalized projection theorem is proved, which relates one–dimensional nonradial Fourier transforms taken on the spatial frequencies plane to these taken on the sky plane.

033.023 **FX: a CMOS–implemented digital spectro–correlator system for radio astronomy.**
T. Nakazuru, K. Miura, Y. Chikada.
Migrating systems to silicon. Proceedings of the 4th Australian Microelectronics Conference, held at Sydney, Australia, 13 – 15 May 1985. Instn. Radio & Electron Eng., Sydney, Australia, p. 10 – 15 (1985). Abstr. in Phys. Abstr., Vol. 89, No. 1273, Entry 18670 (1986).

033.024 **High–performance, 2.5–K cryostat incorporating a 100 – 120–GHz dual polarization receiver.**
J. W. Archer.
Rev. Sci. Instrum., Vol. 56, No. 3, p. 449 – 458 (1985). = Natl. Radio Astron. Obs., Repr., Ser. A, No. 1398.
A hybrid cryostat is described which was designed to cool two SIS mixers in a 100 – 120–GHz dual polarization receiver. A mechanical refrigerator is used to cool radiation shields surrounding a 4.3–1 capacity metal tank containing liquid helium (He4). The temperature of the liquid can be lowered to 2.5K by reducing the helium vapor pressure in the tank with the aid of a mechanical vacuum pump. A liquid hold time, at 2.5K, of longer than five days has been achieved.

033.025 **High–output, single– and dual–diode, millimeter–wave frequency doublers.**
J. W. Archer, M. T. Faber.
IEEE Trans. Microwave Theory Tech., Vol. MTT–33, No. 6, p. 533 – 538 (1985). = Natl. Radio Astron. Obs., Repr., Ser. A, No. 1399.

033.026 **The development of a 10–meter radio telescope.**
Publ. Yunnan Obs., No. 1, p. 104 – 110 (1986). In Chinese.

033.027 **A report on the development of the total flux receiver at 1420 MHz for the 10–meter radio telescope.**
R.–b. Yang.
Publ. Yunnan Obs., No. 1, p. 111 – 125 (1986). In Chinese.

033.028 **Adjustment of the polar axis of the 10–meter radio telescope.**
B.–m. Su, S.–b. Shi, W.–h. Li.
Publ. Yunnan Obs., No. 1, p. 126 – 130 (1986). In Chinese.

033.029 **Determination of the pointing and tracking accuracies of the 10–meter radio telescope.**
B.–m. Su, J.–s. Wang, S.–b. Shi, S.–q. Lu.
Publ. Yunnan Obs., No. 1, p. 131 – 141 (1986). In Chinese.

033.030 **The determination of electrical properties for the 10–meter antenna.**
Q.–z. Shang, H.–m. Hu, W.–h. Li, Y.–f. Gong, H.–r. Yan.
Publ. Yunnan Obs., No. 1, p. 142 – 154 (1986). In Chinese.

033.031 **A heterodyne receiver for the submillimeter wavelength region based on cyclotron resonance in InSb at low temperatures.**
E. R. Brown, J. Keene, T. G. Phillips.
Int. J. Infrared Millimeter Waves, Vol. 6, No. 11, p. 1121 – 1138 (1985). Abstr. in Phys. Abstr., Vol. 89, No. 1276, Entry 30943 (1986).

033.032 **On the sampling method of the JSZ–4 Doppler receiver.**
D.–y. Cha, K.–y. Huang.
Publ. Purple Mt. Obs., Vol. 5, No. 1, p. 57 – 65 (1986). In Chinese.
The authors discuss the properties of the JSZ–4 Doppler receiver and the problem of optimal record. It is shown that the original sampling method losses information. A procedure of improvement is proposed.

033.033 **Microprocessor–controlled stabilised voltage supplies for the "Northern Cross" radiotelescope.**
A. Gallerani, E. Gandolfi.
Elettrotecnica, Vol. 72, No. 9, p. 879 – 887 (1985). In Italian. Abstr. in Phys. Abstr., Vol. 89, No. 1277, Entry 42169 (1986).

033.034 **Design, construction and results of test observations of the Miyun aperture synthesis radio telescope.**
S.–g. Wang.
Publ. Beijing Astron. Obs., No. 7, p. 1 – 10 (1985). In Chinese.

033.035 **An overall introduction to the Mi Yun metre–wave synthesis radio telescope.**
H.–s. Chen, Y.–j. Zheng.
Publ. Beijing Astron. Obs., No. 7, p. 11 – 23 (1985). In Chinese.

033.036 **The overall system adjustment of the Mi Yun synthesis telescope.**
H.–s. Chen, Y.–j. Zheng, Y.–p. Yang, H.–q. Bao.
Publ. Beijing Astron. Obs., No. 7, p. 24 – 26 (1985). In Chinese.

033.037 **The 232 MHz front end of the MSRT (*Mi Yun Synthesis Radio Telescope*).**
T.–y. Piao, G.–q. Zhang.
Publ. Beijing Astron. Obs., No. 7, p. 27 – 35 (1985). In Chinese.

033.038 **The antenna and transmission–line system for the MSRT (*Mi Yun Synthesis Radio Telescope*).**
X.–z. Zhang, F.–y. Liu, L.–s. Kang.
Publ. Beijing Astron. Obs., No. 7, p. 36 – 44 (1985). In Chinese.

033.039 **The automatic tracking system of the meter–wave radiotelescope array.**
T.–l. Jin, L. Pang, H.–q. Bao.
Publ. Beijing Astron. Obs., No. 7, p. 45 – 56 (1985). In Chinese.

033.040 **The constant temperature box.**
T.–y. Piao, H.–p. Zhao.
Publ. Beijing Astron. Obs., No. 7, p. 57 – 61 (1985). In Chinese.
This paper describes a constant temperature system using semiconductor refrigeration. Its operating temperature is +15°C with a control accuracy of ±1°C. The control range of the system is from –25°C to +45°C.

033.041 **The feed and the antenna.**
L.–s. Kang, X.–z. Zhang.
Publ. Beijing Astron. Obs., No. 7, p. 62 – 66 (1985). In Chinese.
Paper concerning the Mi Yun Synthesis Radio Telescope.

033.042 **A digital AGC (*automated gain control*) loop used in the Mi Yun synthesis telescope.**
H. Wang, Y.–p. Yang, X.–m. Wang.
Publ. Beijing Astron. Obs., No. 7, p. 67 – 72 (1985). In Chinese.

033.043 **The digital correlating receiver of the MSRT (*Mi Yun Synthesis Radio Telescope*).**
Y.–p. Yang, H.–q. Bao, H. Wang, G.–q. Chang, H.–p. Zhao.
Publ. Beijing Astron. Obs., No. 7, p. 73 – 84 (1985). In Chinese.

033.044 **The real time acquisition and data handling of the Mi Yun Synthesis Radio Telescope.**
Y.–h. Qiu, Z.–j. Zheng, M.–z. Wei, R.–d. Nan.
Publ. Beijing Astron. Obs., No. 7, p. 85 – 92 (1985). In Chinese.

033.045 **The data acquisition system for the MSRT (*Mi Yun Synthesis Radio Telescope*).**
L. Pang.
Publ. Beijing Astron. Obs., No. 7, p. 93 – 95 (1985). In Chinese.

033.046 **Digital delay system for the MSRT (*Mi Yun Synthesis Radio Telescope*).**
X.–m. Wang, L. Pang, H.–s. Chen.
Publ. Beijing Astron. Obs., No. 7, p. 96 – 102 (1985). In Chinese.

033.047 **The calibration of the Mi Yun aperture synthesis radio telescope by using Cyg A.**
R.–d. Nan.
Publ. Beijing Astron. Obs., No. 7, p. 122 – 126 (1985). In Chinese.

033.048 **Using Cyg A to calibrate the gain of an interferometer.**
R.–d. Nan.
Publ. Beijing Astron. Obs., No. 7, p. 127 – 132 (1985). In Chinese.

033.049 **A very low–noise single–sideband receiver for 200 – 260 GHz.**
N. R. Erickson.
IEEE Trans. Microwave Theory Tech., Vol. MTT–33, No. 11, p. 1179 – 1188 (1985). Abstr. in Phys. Abstr., Vol. 89, No. 1278, Entry 47892 (1986).

033.050 **Radio spectrographs for observing the planet Jupiter.**
R. S. Flagg, W. B. Greenman, P. P Gombola, G. R. Lebo.
Radio Sci., Vol. 20, No. 6, p. 1625 – 1638 (1985). Abstr. in Phys. Abstr., Vol. 89, No. 1278, Entry 47895 (1986).

033.051 **Development of K–3 VLBI automatic operation software.**
A. Kaneko, H. Murakami, S. Kozono, F. Takahashi, T. Yoshino.
J. Radio Res. Lab., Vol. 32, No. 137, p. 155 – 166 (1985). Abstr. in Phys. Abstr., Vol. 89, No. 1279, Entry 48436 (1986).

033.052 **Analog–filter digital–correlator hybrid spectrometer.**
S. Weinreb.
IEEE Trans. Instrum. Meas., Vol. IM–34, No. 4, Part 2, p. 670 – 675 (1985). Abstr. in Phys. Abstr., Vol. 89, No. 1279, Entry 53746 (1986).

033.053 **Outline of the K–3 VLBI data acquisition system.**
N. Kawaguchi.
Rev. Radio Res. Lab., Vol. 30, Spec. Issue, No. 1, p. 43 – 47 (1984). In Japanese. Abstr. in Phys. Abstr., Vol. 89, No. 1279, Entry 53748 (1986).

033.054 **26 m antenna and its feed system.**
N. Kawaguchi, A. Amegaya.
Rev. Radio Res. Lab., Vol. 30, Spec. Issue, No. 1, p. 49 – 54 (1984). In Japanese. Abstr. in Phys. Abstr., Vol. 89, No. 1279, Entry 53749 (1986).

033.055 **26 m antenna drive and pointing program.**
N. Kurihara, H. Murakami.
Rev. Radio Res. Lab., Vol. 30, Spec. Issue, No. 1, p. 55 – 62 (1984). In Japanese. Abstr. in Phys. Abstr., Vol. 89, No. 1279, Entry 53750 (1986).

033.056 **Front end (*for K–3 VLBI 26–m antenna*).**
A. Amegaya, N. Kawaguchi.
Rev. Radio Res. Lab., Vol. 30, Spec. Issue, No. 1, p. 63 – 68 (1984). In Japanese. Abstr. in Phys. Abstr., Vol. 89, No. 1279, Entry 53751 (1986).

033.057 **Back end (*for K–3 VLBI 26–m antenna*).**
H. Kuchi, H. Kuroiwa.
Rev. Radio Res. Lab., Vol. 30, Spec. Issue, No. 1, p. 69 – 85 (1984). In Japanese. Abstr. in Phys. Abstr., Vol. 89, No. 1279, Entry 53752 (1986).

033.058 **Data recorder (*for K–3 VLBI 26–m antenna*).**
S. Hama, A. Amegaya.
Rev. Radio Res. Lab., Vol. 30, Spec. Issue, No. 1, p. 87 – 98 (1984). In Japanese. Abstr. in Phys. Abstr., Vol. 89, No. 1279, Entry 53753 (1986).

033.059 **Atomic hydrogen maser standard.**
H. Morikawa, Y. Ota, H. Kiuchi.
Rev. Radio Res. Lab., Vol. 30, Spec. Issue, No. 1, p. 99 – 107 (1984). In Japanese. Abstr. in Phys. Abstr., Vol. 89, No. 1279, Entry 53754 (1986).

033.060 **Phase fluctuation measurement with HF–band VLBI system and its application to observation of Jupiter's decametric emissions.**
I. Tomizawa.
Rep. Univ. Electro–Commun., Vol. 36, No. 1, p. 99 – 112 (1985). Abstr. in Phys. Abstr., Vol. 89, No. 1280, Entry 60666 (1986).

033.061 **The design of the 64 m antenna system for deep–space projects.**
K. Miura, M. Ichikawa, S. Urusaki, S. Sato, O. Matsumoto.
Mitsubishi Denki Giho, Vol. 59, No. 9, p. 625 – 630 (1985). In Japanese. Abstr. in Phys. Abstr., Vol. 89, No. 1281, Entry 67351 (1986).

033.062 **Phaselock system for millimeter–wave Gunn oscillators used in radio astronomy and laboratory spectroscopy.**
M. Bester.
Int. J. Infrared Millimeter Waves, Vol. 7, No. 1, p. 43 – 63 (1986). Abstr. in Phys. Abstr., Vol. 89, No. 1282, Entry 73050 (1986).

033.063 **Reflector surface measurements of the IRAM 30–m radio telescope.**
A. Greve.
Int. J. Infrared Millimeter Waves, Vol. 7, No. 1, p. 121 – 135 (1986). Abstr. in Phys. Abstr., Vol. 89, No. 1282, Entry 73051 (1986).

033.064 **System delay time calibration equipment (*for K–3 VLBI 26–m antenna*).**
M. Imae, H. Kiuchi.
Rev. Radio Res. Lab., Vol. 30, Spec. Issue, No. 1, p. 109 – 113 (1984). In Japanese. Abstr. in Phys. Abstr., Vol. 89, No. 1282, Entry 73064 (1986).

033.065 **Standard receiving equipment using a delay reference (*for VLBI*).**
N. Kurihara, N. Kawaguchi, A. Amegaya.
Rev. Radio Res. Lab., Vol. 30, Spec. Issue, No. 1, p. 121 – 125 (1984). In Japanese. Abstr. in Phys. Abstr., Vol. 89, No. 1282, Entry 73065 (1986).

033.066 **Low–noise 460 GHz waveguide Schottky mixer radiometer for radioastronomy.**
N. J. Keen, K. D. Mischerikow, G. A. Ediss, E. Perchtold.
Electron. Lett., Vol. 22, No. 7, p. 353 – 355 (1986). Abstr. in Phys. Abstr., Vol. 89, No. 1283, Entry 78276 (1986).

033.067 **Our place in the Universe. II. To the boundaries of the Universe with radioastronomy.**
G. F. Koch.
Nachrichtentech. Z., Vol. 39, No. 3, p. 168 – 170, 172, 174, 176 – 177 (1986). In German. Abstr. in Phys. Abstr., Vol. 89, No. 1283, Entry 78523 (1986).

033.068 **Miyun metre–wave aperture synthesis radio telescope.**
Chin. Astron. Astrophys., Vol. 10, No. 1, p. 3 – 11 (1986). English translation of 40.033.061.

033.069 **The phase calibration of the Miyun aperture synthesis radio telescope with Cyg A.**
R.–d. Nan.
Acta Astrophys. Sin., Vol. 6, No. 1, p. 67 – 71 (1986). In Chinese.
The work on the phase calibration of Miyun system is reported. Processing the data of Cyg A observation, the author ascertains the phase corrections of the interferometers and gives an estimated accuracy of the baselines. The problems in phase–recognition on computer are also discussed.

033.070 **Acousto–optical spectrometers at Itapetinga Radio Observatory.**
E. Scalise Jr., J. L. Monteiro do Vale, Y. Bakor, R. E. Schaal.
Rev. Mex. Astron. Astrofis., Vol. 12, Spec. Issue, p. 409 – 413 (1986). – See Abstr. 012.066.

The characteristics of two low cost acousto–optical spectrometers are presented. They will be used mainly for line work in the frequency range from 20 to 45 GHz. The first instrumental stability tests indicate that integration times lasting several minutes are possible.

033.071 **Dual–shaped offset reflector antennas with circular aperture for realizing low crosspolarization.**
R.–r. Zhang, H.–l. Wang, K.–z. Yang.
Sci. Sin., Ser. A, Vol. 29, No. 6, p. 662 – 672 (1986).

An expression of aperture field crosspolarization is derived for dual shaped offset reflector antennas. The first–order partial differential equation is also presented for the mapping function. The problem of making a search for the mapping function can be simplified as to solve a constant differential equation in terms of an iterative process.

033.072 **Interferometric method of phase synchronization for remote local oscillators.**
A. F. Dravskikh.
Astrofiz. Issled. Izv. Spets. Astrofiz. Obs., Tom 21, p. 113 – 122 (1986). In Russian. English translation in Bull. Spec. Astrophys. Obs. – North Caucasus.

It is shown that time axis positions of VLBI responses for synthesised and not synthesised bandwidths are different. This time difference is proportional to the accumulated phase difference of the local oscillators (frequency standards). This phenomenon opens the possibility to evaluate the phase difference of the local oscillators and to organize their phaselock control by observing only one point wideband radio source.

Interferometry and synthesis in radio astronomy.
See Abstr. 003.039.

Infinite vistas: new tools for astronomy.
See Abstr. 003.076.

Tools of radio astronomy.
See Abstr. 003.165.

The Iraqi National Astronomical Observatory.
See Abstr. 009.003.

Radio astronomy at the University of Sydney.
See Abstr. 009.005.

The story of Ooty Radio Telescope.
See Abstr. 009.060.

Radioastronomical equipment. Proceedings of the 17th All–Union conference, held in Erevan, 10 – 12 October 1985.
See Abstr. 012.020.

New telescopes in China.
See Abstr. 013.004.

New telescopes in India.
See Abstr. 013.006.

Supercomputers for radio astronomical imaging.
See Abstr. 013.027.

Proposal for an international Institute of Space Sciences and a Giant Equatorial Radio Telescope as a collaborative effort between the developing countries.
See Abstr. 013.041.

New telescopes for the 1990s.
See Abstr. 032.005.

Time synchronized systems [*for VLBI*].
See Abstr. 034.176.

Japan–US time synchronization experiments by VLBI.
See Abstr. 034.193.

Estimate of the accuracy of calibration of the "Relict" radiometer from the radio radiation of the moon.
See Abstr. 036.071.

A new method of long–term tracking of sources by the radio telescope RATAN–600.
See Abstr. 036.087.

«Estafette with zoning» method.
See Abstr. 036.088.

Experimental observations with the radio telescope RATAN–600 using the method «estafette with zoning».
See Abstr. 036.089.

Azimuthal aperture synthesis of the sun with the radio telescope RATAN–600 in the regime "South with periscope".
See Abstr. 036.090.

Map analysing and cleaning for the Mi Yun aperture synthesis radio telescope.
See Abstr. 036.176.

Outline of the K–3 data processing system.
See Abstr. 036.203.

The use of the RATAN–600 radio telescope in astrometry.
See Abstr. 041.054.

Angular expansion measurement with the VLA: the distance to NGC 7027.
See Abstr. 134.018.

Determination of the flux densities of radio sources on the set of broadband continuous–spectrum radiometers for the RATAN–600 radio telescope.
See Abstr. 141.011.

Radio galaxies.
See Abstr. 158.200.

034 Auxiliary Instrumentation, Photographic Materials, Clocks

034.001 **A faint object spectrometer for the infrared.**
A. T. Tokunaga, R. G. Smith.
Astrophys. Space Sci., Vol. 118, Nos. 1/2, p. 471 – 472 (1986). – See Abstr. 012.001.
A cryogenic astronomical spectrometer for the NASA Infrared Telescope Facility is described. This spectrometer will employ an array of at least 20 detectors and provide a resolving power of 100 to 1500 at 1 – 5 μm. The resolving power will be adjustable by changing gratings.

034.002 **Gitter mit Blazeeigenschaften in der Stellarspektroskopie.**
H. Lehmann.
Sterne, 62. Band, Heft 1, p. 28 – 34 (1986).

034.003 **Un "camera driver" per pose non guidate.**
L. Balbi.
Orione, Vol. 6, N. 1, p. 52 – 58 (1986).

034.004 **Beobachtungen mit dem 3.5–m–Teleskop auf dem Calar Alto.**
K. Birkle.
Sterne Weltraum, 25. Jahrg., Nr. 1, p. 15 (1986).

034.005 **Ein CCD–Versuch.**
W. Bickel.
Sterne Weltraum, 25. Jahrg., Nr. 1, p. 40 – 41 (1986).

034.006 **Erfahrungen mit Fuji–Farbfilmen in der Astrophotographie.**
G. Jacobs.
Sterne Weltraum, 25. Jahrg., Nr. 2, p. 106 – 107 (1986).

034.007 **Evaluation of the performance of the 576384 Thomson CCD for astronomical use.**
Y. Mellier, M. Cailloux, J. P. Dupin, B. Fort, C. Lours, J. P. Picat, P. Tilloles.
Astron. Astrophys., Vol. 157, No. 1, p. 96 – 100 (1986).
A slow scan CCD camera system was built at the Toulouse Observatory and used to evaluate the performance of the 576384 CCD from Thomson–CSF (new THX 31133) at low temperatures (150K). The authors have emphasized the optimization of the most important parameters for astronomical applications and have compared the Thomson CCD with other CCDs, now currently used in astronomy.

034.008 **Design, construction and initial application of a low–light–level imaging detector.**
A. A. Henden.
Diss. Abstr. Int., Sect. B, Vol. 46, No. 6, p. 1957 (1985). Thesis, Indiana University, 221 pp. (1985). Order No. DA8516616.

034.009 **A new type optical fiber T.V. adaptor–link between telescope and spectrograph.**
S.–d. Wang, S.–y. Jiang, L. Men.
Astrophys. Space Sci., Vol. 121, No. 2, p. 345 – 350 (1986).
A new type of optical fiber T.V. adaptor–link device is developed to bring a star image at the focus of a Schmidt telescope to the slit of a spectrograph located at the observing floor, as well as part of the field to a T.V. monitor for acquisition and guiding purposes. The device is constructed by 60000 thin glass fibers (18 micron in diameter and 1.2 m in length) used as the adaptor, and a single long (20 m) thicker (25 micron core diameter) fused silica fiber used as the link between the telescope and the spectrograph.

034.010 **A high–speed, dual–CCD imaging photometer.**
E. W. Dunham, R. L. Baron, J. L. Elliot, J. V. Vallerga, J. P. Doty, G. R. Ricker.
Publ. Astron. Soc. Pac., Vol. 97, No. 598, p. 1196 – 1204 (1985).
The SNAPSHOT is a high–speed, dual–CCD imaging photometer, with special capability for occultation observations. It is capable of recording occultation light curves in the presence of bright, nonuniform background illumination, such as occultations of stars by Saturn's rings or by comets. Such events are difficult or impossible to observe with conventional photometers. The fundamental idea that makes high time–resolution observations possible is to reduce the data rate by digitizing only those pixels in subframes containing the objects of interest.

034.011 **The Palomar wedge sensitometer.**
J. G. Hoessel, M. S. Jewison, W. R. Oegerle, B. K. Armstrong.
Publ. Astron. Soc. Pac., Vol. 97, No. 598, p. 1205 – 1208 (1985).
A simple sensitometer for use in providing calibration on photographic plates, which has been in use for many years at Palomar Observatory, is described. External comparison between reduced photographic aperture magnitudes for galaxies and CCD photometry for the same objects shows that use of this device results in proper intensity scales. The intensity scale produced using this instrument also agrees well with that produced by the KPNO spot sensitometers.

034.012 **Stellar image profiles from linear detectors and the throughput of astronomical instruments.**
F. Diego.
Publ. Astron. Soc. Pac., Vol. 97, No. 598, p. 1209 – 1214 (1985).
This paper analyzes the spatial energy distribution of stellar images formed by ground–based astronomical telescopes. Lorentzian functions are fitted to intensity profiles of real images obtained with a CCD and the results are used to calculate the energy transmission of astronomical instruments which accept an image with a rectangular or circular stop. The results have been used in the design of the new large echelle spectrographs for the Anglo Australian Telescope and the William Herschel Telescope. A series of figures is presented showing the transmission as a function of the seeing FWHM and the aperture size.

034.013 **A shaped aperture for two–dimensional microdensitometry.**
B. M. Lasker, M. Damashek, P. M. Garnavich.
Publ. Astron. Soc. Pac., Vol. 98, No. 599, p. 123 – 127 (1986).
The desirable profile of an illuminating spot for two–dimensional microdensitometry is often a circularly symmetric bell–shaped function. A useful approximation to this may be achieved by using an in–focus, sharp–edged, cusp–shaped aperture in the measuring microscope, together with a low–pass filter in the output of the photometer.

034.014 **A polarimeter with a new device for eliminating the instrumental phase–plate polarization.**
K. Metz.
Astron. Astrophys., Vol. 159, No. 1/2, p. 333 – 335 (1986).
The European Southern Observatory is providing a new type of a two channel polarimeter which allows to save valuable telescope time by an automatic elimination of the sky background polarization. A simple rotation of the phase–plate system, which is necessary for this elimination, will allow also to avoid the determination of the additional phase–plate polarization. A special technique for manufacturing the compensating phase plate and the diaphragm wheel as well had to be applied in order to manage the correct function of the elimination of phase plate and sky background polarization.

034.015 **In–flight analysis of IUE cross–disperser scattered light.**
P. D. Feldman.
Astron. Astrophys., Vol. 159, No. 1/2, p. 342 – 345 (1986).
The scattered light properties of the IUE cross–disperser gratings are investigated using in–flight data of the emission spectrum of Comet Bradfield (1979 X). The scattering function in the vicinity of H I Lyman–α at 1216 Å is found to be significantly lower

than that derived from either the laboratory measurements of Mount and Fastie (1978) or the analysis by Basri et al. (1985) of cool star observations.

034.016 **Elaboration and investigation of designs for spectral equipment of BTA.**
O. N. Gusev.
Soobshch. Spets. Astrofiz. Obs., Vyp. 48, p. 5 – 26 (1985). In Russian.
The main design principles of the two astronomical spectrographs for the 6–meter telescope: location, and device protection from the outer influence, a device for moving their removable units, mounting and bolting of camera mirrors, and also the spectrum register of the reference emission sources are described.

034.017 **Multi–component symmetric achromatic retarders. II. Composite retarders with small number of elements.**
V. A. Kucherov.
Kinematika Fiz. Nebesn. Tel, Tom 2, No. 1, p. 82 – 87 (1986). In Russian. English translation in Kinematics Phys. Celest. Bodies.
Five–element half– and quarter–wave composite phase plates have been elaborated theoretically. The phase shift deflection for a half–wave retarder achieves less than $\pm 1\overset{\circ}{.}5$ within 0.35 – 1.1 μm. In this case the position of the optical axis remains constant with an accuracy of $\pm 10'$. For a quarter–wave retarder such deflections are 3′ and 10′ respectively in the range of 0.39 – 0.92 μm. Five–element retarders are compared with three–element ones and with other similar systems. The calculations of parameters for three– and five–element retarders are given for arbitrary retardation.

034.018 **Multichannel digital angle converter.**
V. G. Shitov.
Izv. Krymskoj Astrofiz. Obs., Tom 70, p. 168 – 171 (1985). In Russian. English translation in Bull. Crimean Astrophys. Obs., Vol. 70.
The scheme of a multichannel digital angle converter is described for the automatic control system of a gamma telescope. Using the converter error analysis, it has been proposed to construct a multichannel digital angle converter on 14 binary digits on the basis of a sine–cosine rotating transformer.

034.019 **Gas hypersensitization of Kodachrome 64 Film.**
D. Thoeni, A. G. Smith.
AAS Photo–Bull., No. 39, p. 3 – 5 (1985).
Kodachrome 64 Film has been successfully hypersensitized by baking in hydrogen gas. The films were evaluated both by laboratory testing and by exposures on the sky. Speed gains in excess of a factor of 5 were obtained for exposures of astronomical duration.

034.020 **Performance of Kodak Technical Pan Emulsion 2415 on glass plates.**
W. Liller.
AAS Photo–Bull., No. 39, p. 6 – 7 (1985).
Kodak Technical Pan Emulsion 2415 performs as well coated on glass plates as on film base, but requires rather longer baking times to reach its optimum performance level. Eighteen hours in 2% forming gas at 65°C produced a gain of 8.7 with a background fog density of 0.20. A sample of hypersensitized Kodak Spectroscopic Emulsion, Type IIIa–F, had a speed ~1.5 times higher but distinctly coarser grain. At the wavelength of H–alpha, the 2415 emulsion on glass responds excellently.

034.021 **Hypersensitization and astronomical tests of X–ray films.**
A. G. Smith, C. A. Phillips, E. J. Hahn, R. J. Leacock.
AAS Photo–Bull., No. 39, p. 8 – 14 (1985).
For a decade and a half astronomers have been developing new methods of hypersensitizing photographic plates. Application of these techniques to several X–ray films has dramatically reduced low–intensity reciprocity failure during exposures as long as a week. Since laboratory sensitometry and sky tests have shown that one of the films, Kodak SB film, is considerably faster than optimally hypersensitized Kodak spectroscopic plate, type 103a–0, this emulsion might be considered a candidate for astronomical applications.

034.022 **Searching for gravity waves with interferometers.**
B. Schechter.
Phys. Today, Vol. 39, No. 2, p. 17 – 18 (1986).

034.023 **Multicomponent symmetric achromatic retarders. III. Theoretical modelling of the optical properties of a ten–element quartz retardation system.**
V. A. Kucherov.
Kinematika Fiz. Nebesn. Tel, Tom 2, No. 2, p. 59 – 66 (1986). In Russian. English translation in Kinematics Phys. Celest. Bodies.
Formulae are given allowing the calculation of the optical characteristics of a composite retardation system which consists of *N* components. The properties of a ten–element (127°) retarder with accidental deviations of thickness and orientation angle of its elements from calculated values are studied.

034.024 **Frequency and time – national standards.**
A. G. Mungall.
Proc. IEEE, Vol. 74, No. 1, p. 132 – 136 (1986).
The design and performance of the national standards of time and frequency are outlined for countries which maintain primary cesium beam devices. Typical accuracies now attainable are of the order of 10^{-13}, and new developments may improve these significantly within the next few years. Current work is directed toward optical pumping techniques of state selection and to studies of different atoms as the atomic reference.

034.025 **Development and performance of the PTB's CS1 primary clock.**
K. Dorenwendt.
Proc. IEEE, Vol. 74, No. 1, p. 137 – 140 (1986).
The development of the CS1 primary clock is outlined from the first concept to the actual state of the standard. Improvements that have been reached in the past and the actual uncertainty of the standard are described. Finally, some international time–scale comparisons are discussed.

034.026 **Frequency standards based on optically pumped cesium.**
R. E. Drullinger.
Proc. IEEE, Vol. 74, No. 1, p. 140 – 142 (1986).
Optically pumped cesium time and frequency standards using laser diodes offer great potential for improved performance of both laboratory and portable standards. The technology appears to be in hand for the realization of this potential in laboratory based standards but some further laser development is required before the ultimate potential of field standards is reached.

034.027 **Frequency standards based on atomic hydrogen.**
F. L. Walls.
Proc. IEEE, Vol. 74, No. 1, p. 142 – 146 (1986).
The current state of hydrogen maser frequency standards is briefly reviewed. Particular emphasis is placed on the discussion of physical mechanisms which affect long–term stability. While questions concerning absolute accuracy still remain, recent experiments suggest that long–term stability can be at least as good as the best primary standards (1×10^{-13}/year). This long–term stability can be realized by small passive hydrogen maser field standards, a fact which makes these attractive alternatives to cesium field standards.

034.028 **Frequency standards based on stored ions.**
D. J. Wineland.
Proc. IEEE, Vol. 74, No. 1, p. 147 – 150 (1986).
The state of development of frequency standards based on stored ions is reviewed. Several preliminary demonstrations of the concept have already shown a level of performance approaching that of today's cesium–beam standards (accuracy of one part in 10^{13}). The potential for accurately measuring or reducing all known systematic effects suggests that frequency standards based on stored ions with inaccuracies of one part in 10^{15} are obtainable and eventually they could be orders of magnitude better than this.

034.029 **OPTOPUS observations of Halley's comet.**
G. Lund, J. Surdej.
Messenger, No. 43, p. 1 – 5 (1986).

034.030 **First images of globular clusters using a GEC CCD with UV sensitive coating.**
M. Aurière.
Messenger, No. 43, p. 15 – 16 (1986).

034.031 **A new camera and a CCD detector for the Coudé Echelle Spectrograph.**
H. Dekker, B. Delabre, S. D'Odorico, H. Lindgren, F. Maaswinkel, R. Reiss.
Messenger, No. 43, p. 27 – 29 (1986).

034.032 **Contrast met excentrisch diafragma.**
J. Hermsen.
Zenit, 13. Jaarg., Nr. 4, p. 145 – 146 (1986).

034.033 **A polarimeter–magnetometer for hydrogen lines.**
V. T. Schtol (*V. T. Shtol'*), V. D. Bychkov, N. A. Vikul'ev, O. Yu. Georgiev, Yu. V. Glagolevskii (*Yu. V. Glagolevskij*), S. V. Drabek, I. D. Naidenov (*I. D. Najdenov*), I. I. Romanyuk.
Bull. Spec. Astrophys. Obs. – North Caucasus, Vol. 19, p. 66 – 69 (1985). English translation of 39.034.036.

034.034 **Shooting Halley.**
D. Di Cicco.
Sky Telesc., Vol. 71, No. 1, p. 23 – 26 (1986).

034.035 **Photographic astrometry with short focal length objectives.**
A. J. Hollis, D. McAdam, R. Miles.
J. Br. Astron. Assoc., Vol. 96, No. 3, p. 152 – 155 (1986).

Image distortion of a 300mm f/4 photographic lens and Kodak type 2415 35mm roll film has been analysed using a precision X–Y measuring machine. The plate scale between selected pairs of stars was calculated from measured rectangular coordinates of 32 SAO stars. Physical distortion of the film base was found to be negligible compared with off–axis radial distortion of the image. Photographic astrometry of an asteroid (15 Eunomia) was carried out from four exposures made with the 300mm lens using reference stars within one degree of the optical axis. Mean residuals of ± 1.4 arcsec in both RA and Dec were derived from calculated positions of the minor planet. It is concluded that precision of measurements is the major factor limiting arcsecond–accurate astrometry from exposures made with 2415 roll film using short focal length objectives.

034.036 **The rotating full disk Reticon photometer at the San Fernando Observatory.**
A. D. Herzog, G. A. Chapman, J. K. Lawrence, S. Templer.
Bull. Am. Astron. Soc., Vol. 17, No. 4, p. 833 (1985). Abstract. – See Abstr. 010.062.

034.037 **On the angular resolution of air shower array–telescopes.**
J. Linsley.
Bull. Am. Astron. Soc., Vol. 17, No. 4, p. 856 (1985). Abstract. – See Abstr. 010.062.

034.038 **A two–dimensional solar photometer scanned by the Earth's rotation.**
G. A. Chapman, A. D. Herzog, S. Templer.
Bull. Am. Astron. Soc., Vol. 17, No. 4, p. 896 (1985). Abstract. – See Abstr. 010.062.

034.039 **Signal to noise characteristics of data from pulse–counting imaging detectors.**
D. C. Ebbets.
Bull. Am. Astron. Soc., Vol. 17, No. 4, p. 898 – 899 (1985). Abstract. – See Abstr. 010.062.

034.040 **A four–star photoelectric photometer design.**
D. B. Caton, J. T. Pollock.
Bull. Am. Astron. Soc., Vol. 17, No. 4, p. 899 (1985). Abstract. – See Abstr. 010.062.

034.041 **Medium speed photometry with a CCD.**
S. B. Howell, G. H. Jacoby, P. Szkody.
Bull. Am. Astron. Soc., Vol. 17, No. 4, p. 899 (1985). Abstract. – See Abstr. 010.062.

034.042 **A multistep detector for X–ray astronomy.**
B. Ramsey, M. C. Weisskopf, R. F. Elsner.
Bull. Am. Astron. Soc., Vol. 17, No. 4, p. 900 (1985). Abstract. – See Abstr. 010.062.

034.043 **Thermal detectors for X–ray spectroscopy.**
M. Juda, D. McCammon, J. Zhang, R. L. Kelley, S. H. Moseley, A. E. Szymkowiak.
Bull. Am. Astron. Soc., Vol. 17, No. 4, p. 900 (1985). Abstract. – See Abstr. 010.062.

034.044 **A detailed description of proportional counter response below 300 eV.**
K. Jahoda, D. McCammon.
Bull. Am. Astron. Soc., Vol. 17, No. 4, p. 900 (1985). Abstract. – See Abstr. 010.062.

034.045 **As astronomical electro–optical polarimeter for the 2 to 5 micron wavelength region.**
H. B. Ellis Jr., M. W. Werner, M. W. Castelaz, R. W. Capps.
Bull. Am. Astron. Soc., Vol. 17, No. 4, p. 901 (1985). Abstract. – See Abstr. 010.062.

034.046 **Multi–channel electrophotometers for recording night–sky emission.**
V. A. Kravchenko, V. P. Lapchuk, V. N. Ivchenko, G. P. Milinevskij.
Vestn. Kiev. Univ. Astron., No. 27, p. 82 – 87 (1985). In Russian.

034.047 **Design and basic properties of the Vilnius electronographic camera.**
R. Drazdys.
Vilniaus Astron. Obs. Biul., Nr. 70, p. 35 – 42 (1985). In Russian.

Basic properties of existing electronographic cameras (EC) are discussed. Using a detective quantum efficiency criterion the properties of different types of EC are analysed. Main differences between the Vilnius EC and its prototype, Kron's camera, are discussed.

034.048 **The influence of the basic parameters of the photon counter on the quality of measurements.**
R. Kalytis, D. A. Ralys, F. Ralienė.
Vilniaus Astron. Obs. Biul., Nr. 71, p. 3 – 18 (1985). In Russian.

An analysis of the dependence of the sensitivity threshold and the information transmittance of a photon counter on its basic parameters and experiment conditions is given. The considered basic parameters include the counting quantum efficiency, the intensity of the dark signal, the dead–time duration and the non–poissonic factor. The duration of exposure and the background intensity are considered as the experiment conditions. It is shown that the information quality estimation criterion enables to estimate quantitatively the influence of the separate parameters of the photon counter and the conditions of experiment on the efficiency of information transmittance.

034.049 **Investigation of FEU–79 and FEU–106 photomultipliers and their quality estimate.**
R. Kalytis, F. Ralienė, N. Stulpinienė.
Vilniaus Astron. Obs. Biul., Nr. 71, p. 19 – 44 (1985). In Russian.

The basic characteristics and parameters of more than 400 photomultipliers FEU–79 and FEU–106 have been investigated. The counting characteristics for light and dark signals have been measured, and the dependence of the slope of the counting characteristics plateau on the anode voltage as well as the dependence of the dark signal on the temperature have been determined. The counting quantum efficiency at the wavelength 544 nm and the intensity of the dark signal at +20°C have been measured for each of them. The sensitivity threshold values and the quality estimation criterion values at the threshold have been calculated. The dependence of the sensitivity threshold and the quality criterion on the photomultiplier parameters have been determined. The distributions of the basic parameters and information characteristics of the investigated photomultipliers have been obtained.

034.050 **On a possibility of using FEU–115 photomultipliers for recording low–level light streams.**
B. M. Glukhovskoj, N. L. Kozidaeva, A. S. Kutyrev, A. A. Semenikin, A. A. Tokovinin.
Astron. Tsirk., No. 1384, p. 6 – 8 (1985). In Russian.

034.051 **Study of the instrumental profile of a diffraction spectrograph.**
M. I. Divlekeev.
Astron. Tsirk., No. 1385, p. 3 – 5 (1985). In Russian.

034.052 **An efficient method of determining and removing the inclination of microscope–micrometers.**
D. Bozhichkovich.
Astron. Tsirk., No. 1395, p. 4 – 5 (1985). In Russian.

034.053 **Scattered light in a spectrograph and its effect on the measured darkening to the solar limb.**
Eh. A. Gurtovenko, S. N. Osipov.
Kinematika Fiz. Nebesn. Tel, Tom 2, No. 3, p. 59 – 62 (1986). In Russian. English translation in Kinematics Phys. Celest. Bodies.
The procedure of corrections of observed solar limb darkening for the influence of spectrograph scattered light is presented. The measurements of scattered light in the first order of diffraction carried out by three different methods agree well. With the chosen filters the corrections are shown to be insignificant for most of the region of the spectrum.

034.054 **Sensitivity and working temperature of a cooled matrix photosensitive CCD detector.**
V. Yu. Berezin, L. G. Val'skaya, V. I. Gulyaev, S. A. Ioffe, I. N. Surikov, B. N. Formozov, A. T. Khvilivitskij.
Kinematika Fiz. Nebesn. Tel, Tom 2, No. 3, p. 75 – 80 (1986). In Russian. English translation in Kinematics Phys. Celest. Bodies.
The temperature dependences of noise level and threshold irradiation of the polysilicon matrix CCD are considered. It is shown that there is the optimum value of the working temperature (when the threshold irradiation has a minimum value) for specific parameters of CCD and operating conditions.

034.055 **A simple irradiance monitor for testing solar global oscillation network sites.**
G. Fischer, F. Hill, W. Jones, J. Leibacher, W. McCurnin, R. Stebbins, J. Wagner.
Sol. Phys., Vol. 103, No. 1, p. 33 – 39 (1986).
The authors describe a simple irradiance monitor intended for use in assessing the suitability of candidate sites for a worldwide network of small solar telescopes. The network will observe the Sun as continuously as possible in order to provide high quality solar oscillation data with low diurnal sidelobe contamination and high temporal frequency resolution.

034.056 **Efficiency of lased concave diffraction gratings.**
A. V. Bruns.
Bull. Crimean Astrophys. Obs., Vol. 69, p. 132 – 137 (1986). English translation of 39.034.028.

034.057 **Calibration of a photoelectric instrument.**
Yu. V. Khan.
Inst. Akad. Nauk TSSR. Ser. fiz.–tekh., khim. i geol. nauk, No. 4, p. 82 – 83 (1985). In Russian. Abstr. in Ref. Zh., 51. Astron., 4.51.943 (1986).

034.058 **Transmission of light by slit designed for astronomical spectrograph.**
M. A. Ismail.
Astrophys. Space Sci., Vol. 122, No. 1, p. 1 – 32 (1986).
The paper discusses an investigation of linear polarization produced by the transmission of light through an astronomical spectrograph slit. An experimental apparatus was designed and set up to carry out this work. The parallel beam of plane polarized light was rotated in the measuring system, by optically active of half–wave plate. The intensity of normally incident of polarized light of wavelength 0.436 μm transmitted by various slit was measured as a function of slit width.

034.059 **Ground based measurements of global solar intensity oscillations.**
B. N. Andersen.
Seismology of the Sun and the distant stars, p. 339 – 342 (1986). – See Abstr. 012.026.
An instrument designed to observe the low l–value solar intensity oscillations from ground based sites is described. This instrument measures the relative change in the line equivalent width. By taking the ratio between the integrated line intensity and a nearby continuum the effect of the variations in the atmospheric transmission is to a large extent removed. The instrument is expected to have an accuracy of 2 – 5 ppm.

034.060 **A liquid crystal modulator.**
D. G. Eccles, Y. Elsworth, H. B. van der Raay, P. L. Palle, T. Roca Cortes.
Seismology of the Sun and the distant stars, p. 343 – 345 (1986). – See Abstr. 012.026.
The use of a liquid crystal as an optical modulator for use with a resonant scattering spectrometer has been investigated. Preliminary data indicate that using this simple device velocity signals corresponding to 5 minute oscillations have been detected.

034.061 **A Fabry–Perot based stellar seismometer.**
H. R. Butcher, T. R. Hicks.
Seismology of the Sun and the distant stars, p. 347 – 357 (1986). – See Abstr. 012.026.
A simple stellar seismometer for velocity or line strength observations is described, based on a stabilized tunable Fabry–Perot interferometer. Preliminary studies indicate that a stability of near 30 cm/s over many hours should be attainable. The techniques for achieving such stability are described, and a discussion is given of the observational program possible with such an instrument on existing large telescopes.

034.062 **Applications of the magneto–optical filter to stellar pulsation measurements.**
E. J. Rhodes Jr., A. Cacciani, S. Tomczyk.
Seismology of the Sun and the distant stars, p. 359 – 375 (1986). – See Abstr. 012.026.
A proposed method of employing the Cacciani magneto–optical filter for stellar seismology studies is described.

034.063 **A new Fabry–Perot radial velocity spectrometer.**
K. A. R. B. Pietraszewski, N. K. Reay, J. Ring.
Seismology of the Sun and the distant stars, p. 377 – 389 (1986). – See Abstr. 012.026.
The authors have developed a Fabry–Perot interferometric stellar oscillation spectrometer to measure the radial velocity curves of ultra short–period variable stars and to search for solar–type oscillations in stars. The Fabry–Perot etalon is servo–stabilised and the instrument operates under full microprocessor control.

034.064 **Interferometric stellar oscillation spectrometry.**
K. A. R. B. Pietraszewski, J. Ring, A. K. Forrest.
Seismology of the Sun and the distant stars, p. 391 – 402 (1986). – See Abstr. 012.026.
The Michelson Interferometric Stellar Oscillation Spectrometer (ISOS) has been developed and used over the past several years. The authors present here a description of the instrument, the method by which it measures relative radial velocities, a

discussion of some of the data obtained with it and a brief outline of how the accuracy can still be improved.

034.065 **Dual–processor apparatus–program complex – scanner of BTA. I. New possibilities and description of operating.**
S. V. Drabek, I. M. Kopylov, N. N. Somov, T. A. Somova.
Astrofiz. Issled. Izv. Spets. Astrofiz. Obs., Tom 22, p. 64 – 72 (1986). In Russian. English translation in Bull. Spec. Astrophys. Obs. – North Caucasus.

The next stage of development of a television 1000–channel spectrophotometer of BTA is presented. Data acquisition and processing system are carried out on the basis of a measurement–computer complex. The functional scheme of the television–computing complex at whole is described, and the new regimes of apparatus are presented. The properties of application of various regimes for information registration in routine observations are discussed.

034.066 **Efficiency of highly–resolved photographic plates M–300 and 2415 after their hypersensitization in hydrogen.**
N. A. Tikhonov, M. F. Shabanov.
Astrofiz. Issled. Izv. Spets. Astrofiz. Obs., Tom 22, p. 89 – 94 (1986). In Russian. English translation in Bull. Spec. Astrophys. Obs. – North Caucasus.

The efficiency of the photographic emulsions M–300, 2415 and 103a–F is compared before and after the hypersensitization in hydrogen. The spectral sensitivity curves, characteristic curves and the results of noise measurements, signal–to–noise ratios, the number of gradations, quantum efficiency and information capacity for these emulsions are presented.

034.067 **Most effective astronomical photoemulsions after their hypersensitization in hydrogen.**
M. F. Shabanov.
Astrofiz. Issled. Izv. Spets. Astrofiz. Obs., Tom 22, p. 95 – 103 (1986). In Russian. English translation in Bull. Spec. Astrophys. Obs. – North Caucasus.

Nine most effective astronomical photoemulsions are selected, basing on the investigation of more than 300 samples of 15 emulsion types. Their spectral sensitivity curves, the characteristic curves, the number of gradations, the signal–to–noise ratios, the quantum efficiency and information capacity are presented.

034.068 **On spectrophotometry without dispersion elements.**
B. I. Gil'man, V. K. Dubrovich, S. L. Liberman.
Astrofiz. Issled. Izv. Spets. Astrofiz. Obs., Tom 22, p. 104 – 107 (1986). In Russian. English translation in Bull. Spec. Astrophys. Obs. – North Caucasus.

The possibility of constructing a spectrophotometer on the basis of hard–body structures with variable (in coordinate) width of the forbidden zone is considered. It is shown that in such structures due to the selective internal photoeffect a division of the integral emission flux in synonymous correspondence coordinate–wavelength occurs. This gives the possibility to construct spectrophotometers without additional dispersion elements at the input (prisms, gratings, and etc.). Some modifications of the main principle are considered in addition to the analysis of impulse signals with simultaneous determination of their spectra.

034.069 **Contact image transfer from the output screen of an image tube to a photographic film.**
L. V. Gyavgyanen, V. S. Rylov, T. A. Skosyrskaya.
Astrofiz. Issled. Izv. Spets. Astrofiz. Obs., Tom 22, p. 108 – 111 (1986). In Russian. English translation in Bull. Spec. Astrophys. Obs. – North Caucasus.

On the basis of many years experience in application of the contact transfer method to image tubes the design of the last version of a contact transfer unit and its operating conditions are discussed.

034.070 **L'intensificateur de lumière à microcanaux.**
A. Behrend.
Orion, 44. Jahrg., Nr. 212, p. 10 – 12 (1986).

034.071 **Eine Sonnenuhr mit Zeitgleichungskorrektur und Datumanzeige.**
R. Stracke.
Astron. Raumfahrt, 24. Jahrg., Heft 2, p. 58 – 61 (1986).

034.072 **Das Spektrohelioskop – ein klassisches Instrument der Sonnenerkundung.**
A. Grünberg.
Astron. Raumfahrt, 24. Jahrg., Heft 2, p. 61 – 65 (1986).

034.073 **Entraînement horaire type secteur lisse d'un télescope. Contrôle de la précision du mouvement.**
R. Verseau.
Astronomie, Vol. 100, p. 299 – 307 (1986).

034.074 **Astrometry and spider thread.**
R. Fukaya.
Astron. Her., Vol. 79, No. 5, p. 118 – 121 (1986). In Japanese.

034.075 **Probleme der Filmplanlage.**
P. Riepe.
Sterne Weltraum, 25. Jahrg., Nr. 4, p. 216 – 217 (1986).

034.076 **Astrospektrographie mit einfachen Mitteln.**
B. Wagner.
Sterne Weltraum, 25. Jahrg., Nr. 4, p. 218, 220 (1986).

034.077 **Interference filters for Schmidt telescopes.**
S. D. Wiramihardja, T. Tsujimura, T. Sasaki, T. Kogure, B. Hidayat, M. Raharto, M. Ratag.
Galactic structure and variable stars, p. 107 – 117 (1985). – See Abstr. 012.029.

The properties of interference filters currently used in the observational program at the Ouda Schmidt Telescope, Kyoto University and at the Bosscha Observatory, Institute of Technology Bandung are described.

034.078 **New materials applications in solar spectral analysis.**
D. M. Rust.
Aust. J. Phys., Vol. 38, No. 6, p. 781 – 790 (1985). Paper presented at the R. G. Giovanelli Commemorative Colloquium, Part I, Sydney, 26 – 29 November 1984.

The use of lithium niobate and liquid crystals in solar instrumentation designed for automatic measurement of spectral line shifts is described. A solid Fabry–Perot etalon of lithium niobate has an acceptance angle 5.3 times greater than an air–spaced Fabry–Perot filter for the same allowed passband broadening, and the lithium niobate device has no moving parts. The use of liquid crystals in Zeeman–effect analysers is also described. Progress toward implementation of devices with lithium niobate and liquid crystals in a solar telescope is described.

034.079 **Future techniques and instrumentation in solar–stellar physics.**
J. M. Beckers.
Aust. J. Phys., Vol. 38, No. 6, p. 791 – 798 (1985). Paper presented at the R. G. Giovanelli Commemorative Colloquium, Part I, Sydney, 26 – 29 November 1984.

The author reviews the likely instrumental developments for the study of solar–stellar physics. The study of processes known to exist on the sun and occurring on stellar surfaces will require increasingly more sophisticated instrumentation and larger telescopes. The author specifically describes instrumentation required for the study of stellar activity and activity cycles, stellar rotation, stellar surface motions and magnetic fields, stellar pulsations, stellar chromospheres and coronae and stellar winds.

034.080 **Fast timing of air shower fronts.**
R. W. Clay, D. Ciampa.
Aust. J. Phys., Vol. 39, No. 1, p. 93 – 98 (1986).

In order to design a telescope for use in ultra–high energy γ–ray astronomy one needs to carefully consider those effects which determine the system angular resolution. This paper considers detector and discriminator designs for such a system including a consideration of the effect of the spread of particles in the shower front. An angular resolution of $\sim 1°$ at energies of 10^{15}eV can be achieved.

034.081 **A portable occultation photometry system.**
R. S. Ireland, M. Mooney.
I.A.P.P.P. Commun., No. 23, p. 3 (1986). Abstract. – See Abstr. 010.521.

034.082 **Thermal effects on the Starlight–1R photoelectric photometer.**
J. A. Smith, B. J. Smith, E. F. Strother.
I.A.P.P.P. Commun., No. 23, p. 3 – 4 (1986). Abstract. – See Abstr. 010.521.

034.083 **The design of a data logger/telescope controller for photoelectric photometry.**
N. L. Markworth.
I.A.P.P.P. Commun., No. 23, p. 4 (1986). Abstract. – See Abstr. 010.521.

034.084 **The four star photometer.**
E. N. Walker.
I.A.P.P.P. Commun., No. 23, p. 18 – 20 (1986).

034.085 **The Joint European Amatuer Photometer (JEAP).**
E. N. Walker.
I.A.P.P.P. Commun., No. 23, p. 21 – 23 (1986).

034.086 **Ohmic leakage currents and the dark performance of the 1P21 side–window photomultiplier.**
R. Miles.
I.A.P.P.P. Commun., No. 24, p. 6 – 21 (1986).

034.087 **Shearing interferometer for phase shifting interferometry with polarization phase shifter.**
M. P. Kothiyal, C. Delisle.
Appl. Opt., Vol. 24, No. 24, p. 4439 – 4442 (1985).
The use of a cyclic shear interferometer for phase shifting interferometry is discussed. The phase shifting is provided by polarization techniques. Some useful cyclic phase shifting interferometer arrangements are discussed, and the working of a polarization phase shifting cyclic shear interferometer is demonstrated.

034.088 **Laser heterodyne spectrometer for helioseismology.**
D. A. Glenar, D. Deming, F. Espenak, T. Kostiuk, M. J. Mumma.
Appl. Opt., Vol. 25, No. 1, p. 58 – 62 (1986).
The technique of laser heterodyne spectroscopy has been applied to the measurement of solar oscillations. Coherent mixing of solar radiation with the output of a frequency–stabilized CO_2 laser permits the measurement of fully resolved profiles of solar absorption lines with high spectral purity and excellent frequency stability. The authors have used this technique to measure OH pure rotation lines in the infrared (11–μm) solar spectrum. Power spectra of these line frequency measurements show the well–known 5–min oscillations as well as significant velocity power at shorter periods.

034.089 **Subcompact field optics for far–infrared/submillimeter photometers.**
D. Hamilton.
Appl. Opt., Vol. 25, No. 4, p. 537 – 539 (1986).
The severe truncation of light collectors used in compact far–infrared/submillimeter heat trap field optics is discussed. The design of yet more compact heat trap optics and a novel technique for the production of small light collectors are also briefly discussed.

034.090 **Photon noise in photometric instruments at far–infrared and submillimeter wavelengths.**
J. M. Lamarre.
Appl. Opt., Vol. 25, No. 6, p. 870 – 876 (1986).
Photon noise in photometric instruments is computed using two different methods. A first expression, derived from quantum physics, corresponds to the most widely used formulas and is applicable when diffraction phenomena are negligible. The second one is derived from the semiclassical approach of Hanbury Brown and Twiss. Properties of this expression, including consistency with the former one, asymptotical behavior, and scale properties are derived and discussed. The effect of polarization is studied, and practical examples are given.

034.091 **Stressed Ge:Ga infrared detectors: performance and operational parameters.**
D. Lutz, D. Lemke, J. Wolf.
Appl. Opt., Vol. 25, No. 10, p. 1698 – 1700 (1986).
A stressed Ge:Ga infrared detector was tested in various operating conditions. Background–limited performance was demonstrated at photon backgrounds down to $<10^8$photon $s^{-1}cm^{-2}$. The optimal operational parameters were determined.

034.092 **An InSb charge amplifier for use in a spectrometer array.**
D. Hartill, F. Lisi, A. Bettarini.
Appl. Opt., Vol. 25, No. 10, p. 1701 – 1706 (1986).
The authors present a cryogenic charge amplifier system developed for an InSb photodiode array for use in high–resolution infrared spectrometers. The detailed circuit diagram of the charge integrator is presented. The analysis of the expected performance is given.

034.093 **A Cerenkov–ΔE–Cerenkov detector for high energy cosmic ray isotopes and an accelerator study of ^{40}Ar&^{56}Fe fragmentation.**
K. H. Lau.
Diss. Abstr. Int., Sect. B, Vol. 46, No. 8, p. 2693 (1986). Thesis, California Institute of Technology, 222 pp. (1985). Order No. DA8522621.

034.094 **Measurement of sky radiation distribution using a scanner and analyis of circumsolar radiation data.**
Kurisman.
Diss. Abstr. Int., Sect. B, Vol. 46, No. 9, p. 3091 (1986). Thesis, University of New South Wales (1985).

034.095 **Heterodyne spectroscopy of the sun at 12 microns using a diode laser as the local oscillator.**
J. J. Ottusch.
Diss. Abstr. Int., Sect. B, Vol. 46, No. 9, p. 3091 (1986). Thesis, University of California, Berkeley, 152 pp. (1985). Order No. DA8525075.

034.096 **Cryogenic resonant–mass gravitational wave detectors.**
M. Bassan.
Diss. Abstr. Int., Sect. B, Vol. 46, No. 12, p. 4281 (1986). Thesis, Stanford University, 176 pp. (1985). Order No. DA8602453.

034.097 **Experience of the use of the integral spectrometer on the solar tower telescope BST–2 of the Crimean Astrophysical Observatory.**
L. V. Didkovskij, I. E. Kozhevatov, N. N. Stepanyan.
Izv. Krymskoj Astrofiz. Obs., Tom 74, p. 142 – 158 (1986). In Russian. English translation in Bull. Crimean Astrophys. Obs., Vol. 74.
The integral spectrometer was used on the BST–2 to measure the Doppler shift of a spectral line determined by the shift of its center of gravity. The spectrometer is stabilized by the emission from the laser LG–77. It significantly increases the sensitivity of the device.

034.098 **Hybrid analog/digital, large format, photon counting detectors for astronomy.**
J. Crocker, M. Rafal, F. Paresce, A. Hiltner, B. Denman.
Space Telesc. Sci. Inst., Prepr. Ser., No. 95, 8 pp. (1986). To be published in Proc. SPIE, Int. Soc. Opt. Eng. "Instrumentation in Astronomy VI", March 1986.

034.099 **Possible experiments for direct detection of particle candidates for the galactic dark matter.**
P. F. Smith.
Rutherford Appleton Lab., Rep., RAL–86–029, 26 pp. (1986). Invited review presented at 2nd ESO/CERN Symposium on "Cosmology, astronomy, and fundamental physics" (ESO, Garching, March 1986).

034.100 **The Utah Fly's Eye detector.**
R. M. Baltrusaitis, R. Cady, G. L. Cassiday, R. Cooper, J. W. Elbert, P. R. Gerhardy, S. Ko, E. C. Loh, M. Salamon, D. Steck, P. Sokolsky.
Nucl. Instrum. Methods Phys. Res., Sect. A, Vol. A240, No. 2, p. 410 – 428 (1985). Abstr. in Phys. Abstr., Vol. 89, No. 1273, Entry 14387 (1986).

034.101 **Mepsicron and the advancement in high resolution image detection.**
C. Firmani.
Rev. Mex. Fis., Vol. 31, No. 3, p. 503 – 519 (1985). In Spanish. Abstr. in Phys. Abstr., Vol. 89, No. 1274, Entry 20769 (1986).

034.102 **The time–resolved imaging mode (TRIM) of the ESA photon counting detector.**
S. Di Serego Alighieri, M. A. C. Perryman.
ESO Sci. Prepr., No. 424, 5 pp. (1986). Paper presented at the SPIE conference on 'Instrumentation in astronomy VI', Tucson, 3 – 8 March 1986.

034.103 **The microcomputerization of a microphotometer of type MΦ4.**
Z. Li.
Publ. Yunnan Obs., No. 1, p. 94 – 103 (1986). In Chinese.

034.104 **Quantitative two–dimensional spectral classification for the prism spectrograph of the Shemakha Astrophysical Observatory.**
L. V. Timoshenko.
Photometric and polarimetric investigations of celestial bodies, p. 78 – 86 (1985). In Russian. – See Abstr. 003.035.
A system of quantitative two–dimensional spectral classification has been developed for the prism spectrograph of the 2–m reflector of the Shemakha Astrophysical Observatory at the Azerbaidzhan SSR Academy of Sciences. Spectra with dispersion of 9.4 nm/mm near H_γ in the region λλ365 – 500 nm have been obtained for 42 standard stars of I – V luminosity classes and B9 – F8 spectral classes. The system obtained for spectral classification agrees with the MK standard system within ±0.5 of the spectral subclass.

034.105 **Astronomical spectropolarimeter. I. Basic principles of operation.**
O. I. Bugaenko, A. L. Gural'chuk.
Photometric and polarimetric investigations of celestial bodies, p. 160 – 164 (1985). In Russian. – See Abstr. 003.035.
The basic principles of operation of the astronomical spectropolarimeter developed and made at the Main Astronomical Observatory (Kiev) are given. The device consists of a spectrometer, filter photometer and enables all four Stokes vectors to be measured simultaneously. The device analyzer is described. The modulation formula is deduced for determination of the Stokes vector components.

034.106 **Astronomical spectropolarimeter. II. Optical and mechanical block.**
L. A. Bugaenko, M. A. Mel'nikov, L. E. Ragozina, V. S. Samojlov.
Photometric and polarimetric investigations of celestial bodies, p. 164 – 169 (1985). In Russian. – See Abstr. 003.035.
The astronomical spectropolarimeter developed and made at the Main Astronomical Observatory (Kiev) enables the radiation measurement to be carried out under four conditions: photometry, spectrometry, polarimetry, spectropolarimetry as well as a quasi–simultaneous analysis of two neighbouring areas of the celestial sphere.

034.107 **Astronomical spectropolarimeter. III. Information–measuring and recording system.**
L. A. Bugaenko, O. I. Bugaenko, A. L. Gural'chuk, A. S. Delets, I. G. Kesel'man.
Photometric and polarimetric investigations of celestial bodies, p. 169 – 178 (1985). In Russian. – See Abstr. 003.035.
Information–measuring and recording systems of the spectrophotopolarimeter constructed on the basis of a specialized computer are described. The structure of the system is presented, its selection is motivated, algorithms for information processing, control and output are given.

034.108 **Electropolarimeter for the 1–m telescope at the Astrophysical Institute of the Kazakh SSR Academy of Sciences.**
A. V. Kurchakov, F. K. Rspaev.
Photometric and polarimetric investigations of celestial bodies, p. 178 – 181 (1985). In Russian. – See Abstr. 003.035.
An electropolarimeter and methods of multicolour observations of stellar polarization with the 1–m telescope are described.

034.109 **The classical sundials in Jugoslavia. The problems of the exact reconstruction.**
M. Tadić.
Publ. Astron. Soc. "Rudjer Bošković", No. 4, p. 61 – 65 (1985). – See Abstr. 012.078. In Croatian.

034.110 **A single–cell photopolarimeter.**
S. K. Jain, J. C. Bhattacharyya.
Indian Inst. Astrophys., Prepr., No. 193, 9 pp. (1985). To appear in Kodaikanal Obs. Bull.

034.111 **A method of reducing the temperature coefficient of the resonant cavity structure in hydrogen masers.**
Z.–b. Chen.
Publ. Shaanxi Astron. Obs., Vol. 7, No. 2, p. 1 – 7 (1984). In Chinese.

034.112 **Improvement of the tuning cavity of a hydrogen maser.**
G.–m. Chen, R.–q. Tian, W. Shen.
Publ. Shaanxi Astron. Obs., Vol. 7, No. 2, p. 8 – 14 (1984). In Chinese.

034.113 **The use, development, and improvement of the hydrogen maser in Shanghai Observatory.**
W.–w. Kou, Y.–l. Zhang, J.–f. Lu, Y.–q. Peng, H.–x. Huang.
Publ. Shaanxi Astron. Obs., Vol. 7, No. 2, p. 15 – 23 (1984). In Chinese.

034.114 **The characteristics of hydrogen maser receivers.**
W.–f. Li, R.–m. Fan.
Publ. Shaanxi Astron. Obs., Vol. 7, No. 2, p. 24 – 28 (1984). In Chinese.

034.115 **Noise analysis of a beam tube in cesium atomic frequency standards.**
Z. Qin.
Publ. Shaanxi Astron. Obs., Vol. 7, No. 2, p. 29 – 33 (1984). In Chinese.

034.116 **A practical commercial cesium clock.**
B.–y. Huang, C.–h. Wu.
Publ. Shaanxi Astron. Obs., Vol. 7, No. 2, p. 34 – 37 (1984). In Chinese.

034.117 **A new hexapole cesium beam tube.**
B.–y. Chen.
Publ. Shaanxi Astron. Obs., Vol. 7, No. 2, p. 38 – 41 (1984). In Chinese.

034.118 **The shortwave time signals of China in view of the carrier frequencies.**
H.–q. Chen.
Publ. Shaanxi Astron. Obs., Vol. 7, No. 2, p. 42 – 45 (1984). In Chinese.

034.119 **Measurement of frequency stability via correlation technique.**
D.–k. Ma.
Publ. Shaanxi Astron. Obs., Vol. 7, No. 2, p. 46 – 50 (1984). In Chinese.

034.120 **Relativistic effects in the measurement of time and frequency.**
G.–r. Qi.
Publ. Shaanxi Astron. Obs., Vol. 7, No. 2, p. 51 – 57 (1984). In Chinese.

Until recently, the instabilities of clocks and frequency standards used in laboratories were much larger than the corrections required for time measurements carried out with clocks moving about or at rest in locations having different gravitational potential. Therefore, with accuracies and stabilities approaching the order of 10^{-13}, the consideration of relativistic effects is necessary. In order to obtain an idea about the orders of magnitude involved, the author calculates the frequency correction which should be introduced in the clock transport between USNO and CSAO. He obtains $\Delta f/f \approx -4.84 \times 10^{-13}$.

034.121 **A review of one–way satellite time synchronization techniques.**
D.–c. Luo.
Publ. Shaanxi Astron. Obs., Vol. 7, No. 2, p. 58 – 63 (1984). In Chinese.

034.122 **Improvement of a TV comparison circuit.**
M. Fang.
Publ. Shaanxi Astron. Obs., Vol. 7, No. 2, p. 64 – 67 (1984). In Chinese.

034.123 **Analysis of time synchronization accuracy via the Loran–C navigation system.**
J.–t. Zhang.
Publ. Shaanxi Astron. Obs., Vol. 7, No. 2, p. 68 – 70 (1984). In Chinese.

The time synchronization accuracy of Loran–C navigation system has been analysed. The time synchronization accuracy of ground wave signal reaches ± 0.2 μs and that of sky wave signal is ± 8 μs. If the effects of sunsrise and sunset are taken into account, the time synchronization accuracy of the sky wave signal may reach ± 1 μs.

034.124 **Discussion of the accuracy for the measured value of LF propagation delay.**
D.–m. Li.
Publ. Shaanxi Astron. Obs., Vol. 7, No. 2, p. 71 – 77 (1984). In Chinese.

034.125 **A solid state microwave excitation source with phase locked.**
G.–x. Jiang.
Publ. Shaanxi Astron. Obs., Vol. 7, No. 2, p. 78 – 84 (1984). In Chinese.

034.126 **Passive hydrogen maser frequency standard.**
Q.–j. Xu.
Publ. Shaanxi Astron. Obs., Vol. 7, No. 2, p. 85 (1984). Abstract. In Chinese and English.

034.127 **The propagation delay of BPL signal and the ground conductivity along serial paths.**
Z.–h. Liang, H.–Q. Chen, D.–m. Li, F.–c. Wang.
Publ. Shaanxi Astron. Obs., Vol. 7, No. 2, p. 86 (1984). Abstract. In Chinese and English.

034.128 **Progress of Chinese joint atomic time and the problem of rate control.**
Q.–x. Zhuang.
Publ. Shaanxi Astron. Obs., Vol. 7, No. 2, p. 87 (1984). Abstract. In Chinese and English.

034.129 **The experiment of active TV time synchronization.**
H.–q. Zheng, X.–y. Han, S.–y. Lian.
Publ. Shaanxi Astron. Obs., Vol. 7, No. 2, p. 88 (1984). Abstract. In Chinese and English.

034.130 **Measurement on time delay of the longwave reception system for the synthetic atomic time.**
H.–q. Chen, C.–y. He.
Publ. Shaanxi Astron. Obs., Vol. 7, No. 2, p. 89 (1984). Abstract. In Chinese and English.

034.131 **Review of time/frequency research in PTB.**
B.–f. Dai.
Publ. Shaanxi Astron. Obs., Vol. 7, No. 2, p. 90 (1984). Abstract. In Chinese and English.

034.132 **Recent improvement of Shanghai Observatory's atomic time scale.**
Y.–q. Liu, D.–y. Xu, J.–z. Zhang, P.–c. Huang.
Publ. Shaanxi Astron. Obs., Vol. 7, No. 2, p. 91 (1984). Abstract. In Chinese and English.

034.133 **On the clock room's constant temperature.**
K.–y. Chen.
Publ. Shaanxi Astron. Obs., Vol. 7, No. 2, p. 92 (1984). Abstract. In Chinese and English.

034.134 **Introduction of BPM time–coding.**
G.–y. Li.
Publ. Shaanxi Astron. Obs., Vol. 7, No. 2, p. 93 (1984). Abstract. In Chinese and English.

034.135 **Transmitting a coded standard time message via broadcasting networks.**
D.–k. Ma.
Publ. Shaanxi Astron. Obs., Vol. 7, No. 2, p. 94 (1984). Abstract. In Chinese and English.

034.136 **The effects of abnormal phenomena of shortwave time signals on the timing accuracy.**
H.–q. Chen.
Publ. Shaanxi Astron. Obs., Vol. 7, No. 2, p. 95 (1984). Abstract. In Chinese and English.

034.137 **The distinguished period of the LF time signal.**
D.–m. Li.
Publ. Shaanxi Astron. Obs., Vol. 7, No. 2, p. 96 (1984). Abstract. In Chinese and English.

034.138 **Development of a long–wave time service transmitter modulator.**
Z.–l. He.
Publ. Shaanxi Astron. Obs., Vol. 7, No. 2, p. 97 (1984). Abstract. In Chinese and English.

034.139 **The measurement of the absolute time delay of a TV receiving system.**
H.–q. Zheng, X.–y. Han, S.–y. Liang.
Publ. Shaanxi Astron. Obs., Vol. 7, No. 2, p. 98 (1984). Abstract. In Chinese and English.

034.140 **The development of a programmed controller/scanner.**
H.–q. Zheng, S.–y. Liang.
Publ. Shaanxi Astron. Obs., Vol. 7, No. 2, p. 99 (1984). Abstract. In Chinese and English.

034.141 **The precision comparison of the Loran–C method and passive TV synchronization.**
C.–l. Cheng, J.–z. Zheng.
Publ. Shaanxi Astron. Obs., Vol. 7, No. 2, p. 100 (1984). Abstract. In Chinese and English.

034.142 **Determination of the position of a satellite (714 MHz) via three received points and the satellite time comparison.**
J.–a. Song, H.–y. Wang.
Publ. Shaanxi Astron. Obs., Vol. 7, No. 2, p. 101 (1984). Abstract. In Chinese and English.

034.143 **A direct TV satellite receiver for time comparison.**
H.–y. Wang, J.–a. Song.
Publ. Shaanxi Astron. Obs., Vol. 7, No. 2, p. 102 (1984). Abstract. In Chinese and English.

034.144 **Clock/synchronizer.**
G.–r. Ni, A.–h. Liu.
Publ. Shaanxi Astron. Obs., Vol. 7, No. 2, p. 103 (1984). Abstract. In Chinese and English.

034.145 **An automatic multiple switcher constructed by means of a single board microprocessor.**
B.–t. Zhu, R.–l. Sang.
Publ. Shaanxi Astron. Obs., Vol. 7, No. 2, p. 104 (1984). Abstract. In Chinese and English.

034.146 **Preliminary discussion and analysis about V/F converting techniques in frequency measurements.**
J.–x. Peng.
Publ. Shaanxi Astron. Obs., Vol. 7, No. 2, p. 105 – 106 (1984). Abstract. In Chinese and English.

034.147 **Time and frequency transfer via a microwave line.**
J.–c. Li.
Publ. Shaanxi Astron. Obs., Vol. 7, No. 2, p. 108 (1984). Abstract. In Chinese and English.

034.148 **A simulated test of clock travel times via the MX 1502 satellite surveyor.**
Y.–h. Hu, R.–c. Chiao, K.–x. Shen.
Publ. Shaanxi Astron. Obs., Vol. 7, No. 2, p. 113 (1984). Abstract. In Chinese and English.

034.149 **Evaluation of Rockwell HgCdTe arrays for astronomical use.**
M. J. Lebofsky, E. F. Montgomery, W. F. Kailey.
Prepr. Steward Obs., No. 649, 14 pp. (1986).

034.150 **Deployment of the MX spectrometer.**
J. M. Hill, M. P. Lesser.
Prepr. Steward Obs., No. 652, 18 pp. (1986). To appear in Proc. SPIE Int. Soc. Opt. Eng., Vol. 627 "Instrumentation in Astronomy VI", March 31, 1986.

034.151 **Evaluation of dielectric film beamsplitters at cryogenic temperature.**
H. P. Larson.
Prepr. Steward Obs., No. 657, 14+4 pp. (1986). To appear in Appl. Opt.

034.152 **An astronomical heterodyne spectrometer at 10.6 μm.**
Y.–s. Yu.
Chin. J. Infrared Res., Vol. 4, No. 4, p. 261 – 269 (1985). In Chinese. Abstr. in Phys. Abstr., Vol. 89, No. 1276, Entry 35975 (1986).

034.153 **Two–dimensional astronomical photometry techniques.**
C.–j. Wang, J. Lu.
Publ. Purple Mt. Obs., Vol. 4, No. 2, p. 52 – 82 (1985). In Chinese.
Great change has taken place in the observational astronomy with the advent of the new two–dimensional photometry techniques. This article briefly outlines the development of this field in the past ten years. The authors start with a review of the progress in the optical detectors, including the photographic emulsions, the photoconductive devices and the photoemissive devices. Then, an overall description of various types of the two–dimensional astronomical photometry systems is given.

034.154 **System function experiments of an acousto–optical spectrometer by simulating frequency switch.**
L.–b. Pei, X.–l. Wang.
Publ. Purple Mt. Obs., Vol. 4, No. 3, p. 47 – 51 (1985). In Chinese.
The system function experimental project of acousto–optical spectrometer at IF of the receiver by simulating frequency switch is described. The system control, data acquisition and reduction are realized on a micro–computer. Some experimental results are given.

034.155 **Coupling of a high–sensitivity superconducting amplifier to a gravitational–wave antenna.**
P. Carelli, M. G. Castellano, C. Cosmelli, V. Foglietti, I. Modena.
Phys. Rev. A, Vol. 32, No. 6, p. 3258 – 3265 (1985). Abstr. in Phys. Abstr., Vol. 89, No. 1277, Entry 42178 (1986).

034.156 **New photoelectric apparatus of the Kharkov Time Service for recording of star transits.**
V. D. Simonenko, S. A. Stepanov.
Vestn. Khar'kov. Univ., No. 278, p. 74 – 77 (1985). In Russian.

034.157 **Pulse–counting photometer at the Skalnaté Pleso Observatory.**
Ľ. Klocok, J. Zverko, J. Žižňovský.
Contrib. Astron. Obs. Skalnaté Pleso, Vol. 14, p. 97 – 104 (1986).
A digital semi–automatic measuring system of the photoelectric photometer of the 0.6 m telescope at the Skalnaté Pleso Observatory is described.

034.158 **A digital circuit to transform mean solar time into mean sidereal time.**
Y.–j. Zheng, C.–l. Zhang.
Publ. Beijing Astron. Obs., No. 7, p. 103 – 107 (1985). In Chinese.

034.159 **Digital clock.**
C.–l. Zhang, Y.–j. Zheng.
Publ. Beijing Astron. Obs., No. 7, p. 108 – 112 (1985). In Chinese.

034.160 **Use of a birefringent filter for the H_α line on the Pamirs solar telescope.**
A. V. Kründal (*A. V. Kryundal'*).
Soln. Dannye, Byull., 1986, No. 3, p. 85 – 88 (1986). In Russian.
An additional optical channel for the H_α line incorporated by the author on the Pamirs solar telescope of the Pulkovo Observatory is described. The channel enables to take large–scale filtergrams simultaneously with white light photographs and spectrograms.

034.161 **Multianode microchannel array detector systems: performance characteristics.**
J. G. Timothy.
Opt. Eng., Vol. 24, No. 6, p. 1066 – 1071 (1985). Abstr. in Phys. Abstr., Vol. 89, No. 1278, Entry 47896 (1986).

034.162 **A low–noise low–input conductance preamplifier for gravitational research.**
G. V. Pallottino, G. Vannaroni.
IEEE Trans. Instrum. Meas., Vol. IM–34, No. 4, Part 2, p. 676 – 680 (1985). Abstr. in Phys. Abstr., Vol. 89, No. 1279, Entry 48290 (1986).

034.163 **A millimeter continuum astronomical detector system using a 0.1K bolometer cooled by adiabatic demagnetization.**
L. Lesyna.
Bull. Am. Astron. Soc., Vol. 18, No. 1, p. 634 (1986). Abstract. – See Abstr. 010.066.

034.164 **More accurate time from the Heathkit Most Accurate Clock.**
M. V. Tollefson, R. H. Bloomer Jr.
Publ. Astron. Soc. Pac., Vol. 98, No. 603, p. 520 – 523 (1986).
A simple circuit is described which may be added to the Heathkit Most Accurate Clock to provide a computer with more accurate absolute time. The circuit improves attainable accuracy from ±29 ms to ±3 ms.

034.165 **Astronomia al microscopio.**
P. Stroppa.
Orione, Vol. 6, N. 3, p. 39 – 44 (1986).

034.166 **Un "camera driver" per pose non guidate.**
L. Balbi.
Orione, Vol. 6, N. 3, p. 51 – 55 (1986).

034.167 **Besonderheiten der geneigten oder inklinierten Horizontal–Sonnenuhr.**
A. Zenkert.
Astron. Raumfahrt, 24. Jahrg., Heft 3, p. 84 – 86 (1986).

034.168 **Reducing the granularity noise of an image photon counting system by magnetic scanning.**
A. R. Jorden, J. L. A. Fordham.
Q. J. R. Astron. Soc., Vol. 27, No. 2, p. 166 – 181 (1986).
One of the limitations of the performance of an image photon counting system (IPCS), as installed at AAO and La Palma, is the small–scale pixel–to–pixel sensitivity variations of the detector. These variations are difficult to calibrate accurately and can limit the photometric performance by adding granularity noise to the data. A system of magnetic scanning has been implemented to improve this granularity considerably, such that photon noise is the dominant limitation for most observations. This paper describes the mechanisms in the IPCS which cause granularity, following a brief description of the IPCS itself. The results obtained with the system are described, followed by a discussion of the effects of flat–field calibration on signal–to–noise ratio.

034.169 **IRSPEC: ESO's new infrared spectrometer.**
A. Moorwood, P. Biereichel, G. Finger, J.–L. Lizon, M. Meyer, W. Nees, J. Paureau.
Messenger, No. 44, p. 19 – 24 (1986).

034.170 **The universal CCD controller of Asiago Observatory.**
F. Bortoletto, M. D'Alessandro.
Mem. Soc. Astron. Ital., Vol. 56, No. 4, p. 781 – 784 (1985). – See Abstr. 010.821.

034.171 **Multidetector grating spectrometer for infrared astronomy.**
J. Slingerland.
Rev. Sci. Instrum., Vol. 57, No. 2, p. 248 – 252 (1986). Abstr. in Phys. Abstr., Vol. 89, No. 1280, Entry 60650 (1986).

034.172 **Universal CCD–controller system.**
F. Bortoletto, M. D'Alessandro.
Rev. Sci. Instrum., Vol. 57, No. 2, p. 253 – 258 (1986). Abstr. in Phys. Abstr., Vol. 89, No. 1280, Entry 60651 (1986).

034.173 **The high speed photometer at the Toruń Observatory.**
B. Wikierski, Z. Turło.
Acta Astron., Vol. 35, Nos. 3–4, p. 441 – 452 (1985).
The design of the high speed computer–controlled photometer built and operated at the Toruń Observatory is discussed. It is characterized by the time resolution in the microsecond range and a high accuracy absolute timing. Some relevant details of the electronics are also included. Sample observational results of the Crab pulsar are presented.

034.174 **Hypersensitization of astronomical emulsions used at the Byurakan Astrophysical Observatory (Kodak IIIaJ).**
J. A. Stepanyan (*D. A. Stepanyan*).
Sov. Astron., Vol. 29, No. 6, p. 704 – 708 (1985). English translation of 40.034.177.

034.175 **Applications of Fabry–optics in infrared astronomical photometers.**
Z.–y. Qian.
Chin. J. Infrared Res., Vol. 4, No. 5, p. 399 – 401 (1985). In Chinese. Abstr. in Phys. Abstr., Vol. 89, No. 1281, Entry 67358 (1986).

034.176 **Time synchronized systems [*for VLBI*].**
T. Sato, N. Kurihara, M. Aida.
Rev. Radio Res. Lab., Vol. 30, Spec. Issue, No. 1, p. 115 – 119 (1984). In Japanese. Abstr. in Phys. Abstr., Vol. 89, No. 1282, Entry 68061 (1986).

034.177 **Observatory for ultra high–energy processes: the Fly's Eye.**
G. L. Cassiday.
Annu. Rev. Nucl. Part. Sci., Vol. 35, p. 321 – 349 (1985). Abstr. in Phys. Abstr., Vol. 89, No. 1282, Entry 73072 (1986). – See Abstr. 003.041.

034.178 **X–ray films: suppression of reciprocity failure by astronomical techniques.**
A. G. Smith, C. A. Phillips, E. J. Hahn.
J. Imaging Technol., Vol. 11, No. 1, p. 27 – 32 (1985). Abstr. in Phys. Abstr., Vol. 89, No. 1283, Entry 73872 (1986).

034.179 **Detector system for the measurement of the muon array density spectrum in cosmic ray showers.**
M. Jaksic, V. Valkovic.
Nucl. Instrum. Methods Phys. Res., Sect. A, Vol. A244, No. 3, p. 571 – 573 (1986). Abstr. in Phys. Abstr., Vol. 89, No. 1283, Entry 74519 (1986).

034.180 **Experimental test of a prototype gravitational radiation detector employing an active cavity laser sensor.**
A. Abramovici, Z. Vager, M. Weksler.
J. Phys. E, Vol. 19, No. 3, p. 182 – 188 (1986). Abstr. in Phys. Abstr., Vol. 89, No. 1283, Entry 78286 (1986).

034.181 **Data analysis for a gravitational wave antenna with resonant capacitive transducer.**
E. Amaldi, G. Pizzella, P. Rapagnani, F. Ricci, P. Bonifazi, G. Cavallari, E. Coccia, G. V. Pallottino.
Nuovo Cimento C, Vol. 9C, Ser. 1, No. 1, p. 51 – 73 (1986). Abstr. in Phys. Abstr., Vol. 89, Nr. 1284, Entry 78732 (1986).

034.182 **A rigorous demonstration of the conservation of integrated transmission of the tunable Lyot filter.**
D.–q. Su.
Chin. Astron. Astrophys., Vol. 10, No. 1, p. 81 – 85 (1986). English translation of Acta Astrophys. Sin., Vol. 6, No. 1, p. 72 – 79 (1986).
The author derived the transmission function of a Lyot filter. He also proposed a method of measuring the total amount of light outside the main band of filters.

034.183 **A new method for fabrication of diamond–dust blocking filters.**
H. R. Collard, R. C. Hogan.
Infrared Phys., Vol. 26, No 1, p. 51 – 54 (1986).
A new method for making IR blocking filters by coating polyethylene film on both sides with diamond dust is described in detail. A special holding device built to facilitate the coating process is described and the coating procedure explained in a step–by–step manner.

034.184 **Photographier la Lune. Quel temps de pose?**
J. Jonlet.
Ciel, Vol. 48, p. 22 – 23, 26 – 30 (1986).

034.185 **Hypersensibilisation pour l'astrophotographie amateur?**
M. Cola, J. Jonlet.
Ciel, Vol. 48, p. 31 – 36 (1986).

034.186 **The MEPSICRON: a new bidimensional photon counting system.**
C. Firmani, E. Ruíz, J. Bohigas, G. F. Bisiacchi.
Rev. Mex. Astron. Astrofis., Vol. 12, Spec. Issue, p. 379 – 384 (1986). – See Abstr. 012.066.
The authors present the principal characteristics of a new bidimensional photon counting system (MEPSICRON), as well as some astronomical results obtained with it.

034.187 **The spectrophotometer of the National Observatory.**
S. J. C. Landaberry, J. A. de Freitas Pacheco, J. L. Kohl, B. Bazzanella, R. T. Ribeiro.
Rev. Mex. Astron. Astrofis., Vol. 12, Spec. Issue, p. 405 (1986). Abstract. – See Abstr. 012.066.

034.188 **A new data acquisition system for the IAG (Instituto Astronômico e Geofisico) polarimeter.**
W. F. Velloso Jr., A. M. Magalhães, C. Gneiding.
Rev. Mex. Astron. Astrofis., Vol. 12, Spec. Issue, p. 406 (1986). Abstract. – See Abstr. 012.066.

034.189 **Recommendations on super universal multichannel filters for LEST.**
G.–x. Ai, Y.–f. Hu.
LEST Found., Tech. Rep., No. 14, 22 pp. (1985).
The present paper describes a super universal multichannel filter, and proposes optional filter configurations to be considered for LEST.

034.190 **Laser interferometer measurement system on the Yale PDS 2020G.**
J.–F. Lee, W.–S. Tsay, W. van Altena.
Astrometric techniques, p. 237 – 242 (1986). – See Abstr. 012.068 (IAU Symp. No. 109).
The authors have installed a dual–axis laser interferometer system on the Yale PDS 2020G microdensitometer to replace the linear encoders. Various diagnostic tests show that the laser interferometer is a superior encoder system for the PDS with a positional accuracy of about 0.3 micron.

034.191 **Vidicon photometry and astrometry applied to objects of the solar system.**
J.–E. Arlot, W. Thuillot.
Astrometric techniques, p. 369 – 373 (1986). – See Abstr. 012.068 (IAU Symp. No. 109).
A SIT–Vidicon camera has been used for astrometric and photometric observations: phenomena of the Galilean Satellites, occultations of stars by asteroids and comets, closest approaches of fast moving objects. This paper tries to give a preliminary evaluation of the efficiency of these observations.

034.192 **A multislit photoelectric star micrometer for the meridian circle of the Nikolayev Astronomical Observatory.**
V. V. Konin, A. D. Pogonij.
Astrometric techniques, p. 569 – 570 (1986). – See Abstr. 012.068 (IAU Symp. No. 109).

034.193 **Japan–US time synchronization experiments by VLBI.**
S. Hama, T. Yoshino, H. Kiuchi, T. Morikawa, T. Sato, W. J. Klepczynski.
Proceedings of the Symposium on Application of Space Techniques to Astronomy and Geophysics, p. 127 – 131 (1986). In Japanese. – See Abstr. 012.027.
Since 1984, Radio Research Laboratory has conducted VLBI experiments with USA; Richmond and Maryland Pt. for precise time synchronization. The authors achieved a precision of about 0.2 ns.

034.194 **Automatic nodding and the Mk III infrared photometer.**
I. S. Glass.
Mon. Notes Astron. Soc. S. Afr., Vol. 44, Nos. 5–6, p. 45 – 47 (1985).
The automatic nodding system of the 1.9 m telescope at Sutherland and the Mk III infrared photometer are described.

034.195 **A compact photon–counting imaging system.**
P. Hickson.
Publ. Astron. Soc. Pac., Vol. 98, No. 604, p. 622 – 626 (1986).
A compact photon–counting imaging system with a 500x488 pixel format and 33 ms time resolution is described. The detector is a two–stage microchannel intensifier fiber–optically coupled to a CCD camera operating at standard video rate. Photon events are centered in one dimension, digitized, and recorded on videotape. The videotape is later played into the video digitizer input of an image processor and consecutive frames are added to produce the final image.

034.196 **Diagnosing a PDS microdensitometer.**
W. van Altena, J.–F. Lee, A. Wandersee.
NASA Conf. Publ., NASA CP–2317, p. 3 – 18 (1984). – See Abstr. 012.073.
Over the past four years a number of diagnostic tests have been developed for the PDS 2020G microdensitometer to monitor its performance and to isolate various electro–mechanical problems. In this paper the authors describe a number of tests which have helped to diagnose problems with the photometer, positional accuracy and data collection.

034.197 **Performance testing of the High Altitude Observatory PDS microdensitometer.**
A. Poland, R. Munro, D. Friend.
NASA Conf. Publ., NASA CP–2317, p. 19 – 34 (1984). – See Abstr. 012.073.
The High Altitude Observatory PDS microdensitometer, purchased in 1972, is one of the earlier machines produced. Particular features of this machine include: PDP–8 control, 12 bit A/D converter, high density log amplifier, updated locking microscope head, auto–loc, teflon drive screws and nuts, and high speed operation (up to 60 mm/sec). The microdensitometer is generally used with 10 or 50 micron apertures on 35 mm and 70 mm film. The HAO microdensitometer undergoes monthly testing to assure its consistent performance. These tests are designed to check positional and photometric stability at the 10 micron aperture level.

034.198 **Photometric and positional accuracy of the PDS Bonn in view of astronomical requirements.**
H. J. Becker.
NASA Conf. Publ., NASA CP–2317, p. 34 – 43 (1984). – See Abstr. 012.073.
The PDS 1010A of the Astronomical Institutes of the University of Bonn is the PDS version characterized by the 10x10 inch measuring field, density range 0...5, and 4 cm/s maximum scanning speed. The system is controlled by a PDP 8/m computer, which is now being replaced by a link to the VAX computer of the institute with image processing facilities. Until now, PDS measurements were stored on magnetic tape and reduced off–line at the CYBER 172 of the Max–Planck–Institut für Radioastronomie. In 1977 extensive tests of the PDS performance and accuracy were begun. Since then the system has been used in spectroscopic studies, astrometry, and two–dimensional photometry.

034.199 **Positive–feedback photometric drift in the PDS.**
R. H. Cornett, R. C. Bohlin, J. K. Hill, T. P. Stecher.
NASA Conf. Publ., NASA CP–2317, p. 59 – 69 (1984). – See Abstr. 012.073.
The authors have found that digitizing flatfield images produces conditions in the PDS 1010A at the GSFC Laboratory for Astronomy and Solar Physics which cause the measured density to drift by as much as 0.1 DN during a 10 minute interval. The drift occurs when the PDS, set up in equilibrium at fog level, subsequently scans a reasonably dense (>1.2) region for periods of longer than a few minutes. The drift is manifested primarily as a positive shift in density that is approximately the same for all densities.

034.200 **Characteristics of the MSFC, PDS microdensitometer.**
W. F. Fountain, G. A. Gary, H. Oda.
NASA Conf. Publ., NASA CP–2317, p. 71 – 88 (1984). – See Abstr. 012.073.

This report summarizes the results of several parametric studies carried out on the Marshall Space Flight Center (MSFC) PDS–10 microdensitometer for the purposes of documenting and understanding the operation and limitations of the system for inhouse research and to provide a bench mark for comparison with other microdensitometers.

034.201 **Performance of the ESO PDS.**
P. J. Grosbøl.
NASA Conf. Publ., NASA CP–2317, p. 89 – 95 (1984). – See Abstr. 012.073.

The astrometric and photometric performance of the PDS 1010A microdensitometer at ESO is discussed including the tests used for checking it.

034.202 **Experiences with a PDS 2020 GM.**
D. Teuber.
NASA Conf. Publ., NASA CP–2317, p. 97 – 106 (1984). – See Abstr. 012.073.

In April/May 1982 a PDS 2020 GM connected to a Perkin–Elmer PE 3220 computer was installed at the Astronomical Institute of Münster University. This microdensitometer is capable of scanning areas up to 20″ × 20″. The single scan segments are controlled by a microprocessor, while the overall control program resides with the PE 3220. The main part of the PDS is a 52″ × 54″ × 10″ granite block resting in a steel frame. It is the support for the scanning mechanics. Almost all electronics is mounted inside the steel frame.

034.203 **PDS concepts and astronomical applications: an overview.**
B. M. Lasker.
NASA Conf. Publ., NASA CP–2317, p. 109 – 119 (1984). – See Abstr. 012.073.

This paper introduces the topic of PDS status and improvements by reviewing the basic concepts of PDS scanning microdensitometers, including their major electronic, optical and mechanical variations. Certain improvements to the basic PDS design have been found necessary for the effective collection of astronomical data and have been widely implemented in the community; an overview of these improvements is also presented.

034.204 **PDS microdensitometer electronics modifications.**
A. V. Hewitt.
NASA Conf. Publ., NASA CP–2317, p. 121 – 128 (1984). – See Abstr. 012.073.

The design of the detector and front–end electronics in the PDS cause nonlinearity and instability of the system at both low and high densities. Minor and reversible modifications (replacement of the photomultiplier and front–end analog circuitry) have significantly improved the performance. The modified PDS has no transmission mode.

034.205 **The current status and future upgrade of the PDS at the Royal Greenwich Observatory.**
I. G. van Breda, H. E. Davies, A. J. Penny, C. D. Pike.
NASA Conf. Publ., NASA CP–2317, p. 129 – 134 (1984). – See Abstr. 012.073.

The PDS at the Royal Greenwich Observatory is a model 1010A purchased in 1974 with a PDP 11 interface. Comprehensive scanning facilities are available using FORTH software. This paper describes some aspects of the machine's present performance and gives the outline of an upgrade proposal which the authors feel will be necessary in order that their PDS can continue to offer a convenient and reliable service to the astronomical community.

034.206 **Improvements to the photometric response and positional accuracy of the Yale PDS 2020G microdensitometer.**
J.–F. Lee, W. van Altena.
NASA Conf. Publ., NASA CP–2317, p. 151 – 161 (1984). – See Abstr. 012.073.

In the course of optimizing the Yale PDS 2020G microdensitometer for photometric and astrometric research, it has been necessary for the authors to incorporate several hardware changes. In this paper they describe the properties of a new high speed photometer and a positional calibration system.

034.207 **An improved logarithmic amplifier circuit for PDS microdensitometers.**
C. M. Anderson, M. H. Slovak, D. E. Michalski.
NASA Conf. Publ., NASA CP–2317, p. 163 – 173 (1984). – See Abstr. 012.073.

A high speed logarithmic–amplifier circuit for the Perkin–Elmer PDS microdensitometer is discussed. The new amplifier is capable of producing undistorted profiles of rapidly changing, dense (>4) images over the available range of scanning speeds.

034.208 **The PDS–VAX intelligent plate scanning system of Trieste Astronomical Observatory.**
M. Pucillo, G. Sedmak.
NASA Conf. Publ., NASA CP–2317, p. 175 – 187 (1984). – See Abstr. 012.073.

An intelligent subsystem for interfacing a PDS 1010A digital microdensitometer to a DEC DR11W DMA 16 bit UNIBUS port in DEC VAX VMS environment was designed and is presently under test at Trieste Observatory.

034.209 **A microprocessor–based control system for the Vienna PDS microdensitometer.**
H. Jenkner, M. Stoll, J. Hron.
NASA Conf. Publ., NASA CP–2317, p. 197 – 206 (1984). – See Abstr. 012.073.

The aging DEC PDP–12 computer, serving as the control processor for the Vienna PDS 1010 microdensitometer, was replaced by a Motorola Exorset 30 system, based on a Motorola 6809 microprocessor. User communication and instrument control are fully implemented in this system; data transmission to a host computer is provided via standard interfaces. The Vienna PDS System software was developed in BASIC and M6809 assembler. Extensive user information includes continuously updated status and coordinate displays, as well as a real–time graphic display during scanning.

034.210 **Why use a microdensitometer other than a PDS?**
R. S. Stobie.
NASA Conf. Publ., NASA CP–2317, p. 209 – 228 (1984). – See Abstr. 012.073.

The motivations for constructing a special purpose built microdensitometer are explored. A brief description of the salient points of some of these microdensitometers is given stressing in particular the advantages and disadvantages of the system in comparison to a PDS machine.

034.211 **Automated microdensitometer for digitizing astronomical plates.**
J. Angilello, W.–H. Chiang, D. M. Elmegreen, A. Segmüller.
NASA Conf. Publ., NASA CP–2317, p. 229 – 241 (1984). – See Abstr. 012.073.

A precision microdensitometer has been built under control of an IBM S/1 time–sharing computer system. The instrument's spatial resolution is better than 20 μm. A raster scan of an area of 10x10 mm^2 (500x500 raster points) takes 255 minutes. The reproducibility is excellent and the stability is good over a period of 30 hours, which is significantly longer than the time required for most scans. The intrinsic accuracy was found to be better than 3%. The results showed an accuracy excellent for astronomical applications.

034.212 **The Sacramento Peak Fast Microphotometer.**
M. R. Arrambide, R. B. Dunn, A. W. Healy, R. Porter, A. L. Widener, L. J. November, G. E. Spence.
NASA Conf. Publ., NASA CP–2317, p. 243 – 254 (1984). – See Abstr. 012.073.

The Sacramento Peak Observatory Fast Microphotometer translates an optical system that includes a laser and photodiode detector across the film to scan the Y direction. A stepping motor moves the film gate in the X direction. This arrangement affords high positional accuracy, low noise (0.002 RMS density units), modest speed (5000 points/second), large dynamic range (4.5 density units), high stability (0.005 density units), and low scattered light. The Fast Microphotometer is interfaced to the host computer by a 6502 microprocessor.

034.213 **The automated photographic measuring facility at Cambridge.**
E. J. Kibblewhite, M. T. Bridgeland, P. Bunclark, M. Irwin.
NASA Conf. Publ., NASA CP–2317, p. 277 – 288 (1984). – See Abstr. 012.073.

The design and performance automated photographic measuring facility at Cambridge is described. It consists of a precision laser scanning microdensitometer connected to a series of computers that process the data on–line. Plates up to 350 mm square can be measured. The microdensitometer samples the plate to 12 bit accuracy at a speed of 230,000 samples/second. The positional accuracy is better than a micron.

034.214 **A Next Generation Microdensitometer?**
D. G. Monet.
NASA Conf. Publ., NASA CP–2317, p. 291 – 305 (1984). – See Abstr. 012.073.

Motivations for construction of a Next Generation Microdensitometer are presented and their effect on the NGM design is discussed. A prototype of such an engine has been constructed at KPNO. Its design and performance is reviewed.

034.215 **A preliminary analysis of a diode array for densitometry.**
K. A. Janes.
NASA Conf. Publ., NASA CP–2317, p. 307 – 315 (1984). – See Abstr. 012.073.

A diode–array based image digitizer manufactured by the Eikonix Corp. has been tested to see if it can be adapted to the exacting requirements of astronomical densitometry. Preliminary tests indicate that several relatively simple enhancements can improve both the photometric and the positional accuracy of the device.

034.216 **Next generation microdensitometer.**
E. J. Kibblewhite.
NASA Conf. Publ., NASA CP–2317, p. 333 – 341 (1984). – See Abstr. 012.073.

The fundamental limitations of microdensitometers are reviewed and the design of a high speed microdensitometer is described. The system will digitise to 16 bits in transmission at a speed of 100 kHz using a laser beam moving over the emulsion. Other features are automatic platen rotation and autofocus. The cost will be of order $ 200,000.

034.217 **The intensified multi CCD camera.**
A. Blazit.
ESA Spec. Publ., ESA SP–226, p. 155 – 156 (1985). – See Abstr. 012.075.

Interferometers give images with an important number of pixels: in the range of a few millions in space. Much more for ground based ones, because of speckles. For this need the author is building a photon counting TV camera of 40 mm diameter. It is an intensified mosaic of 4 CCD. CCD pixels are subdivised in 4x4 subpixels by a photon centroiding processor, yielding a total of 7000000 image elements.

The Royal Astronomical Society instrument collection: 1827 – 1985.
See Abstr. 002.073.

Fabry–Perot interferometers.
See Abstr. 003.011.

Low light level detectors in astronomy.
See Abstr. 003.089.

Photometry and polarization in remote sensing.
See Abstr. 003.090.

Astrophotography: a step–by–step approach.
See Abstr. 003.128.

Astrophotographie. Ein Leitfaden für den Amateurastronomen.
See Abstr. 003.131.

Capture a comet: an amateur photographer's guide.
See Abstr. 003.163.

L'horloge mécanique en astronomie de position jusqu'à la moitié du XVIII siècle.
See Abstr. 004.084.

Plans for a fast image recording system at ESO.
See Abstr. 009.065.

A new measuring machine in Paris.
See Abstr. 009.066.

Astrophoto VI Proceedings, March 2, 1985.
See Abstr. 012.091.

Cooperation in instrumentation.
See Abstr. 013.040.

Astronomical computing: tracking tolerances in astrophotography.
See Abstr. 014.015.

Notes on the use of Pickering–Racine wedges.
See Abstr. 031.013.

Recent developments with the Anglo–Australian Telescope.
See Abstr. 032.001.

Comparison of a 1.8–meter CCD scannerscope with a 1.2–m Schmidt camera.
See Abstr. 032.026.

Chopping secondary mirror for a cooled infrared telescope.
See Abstr. 032.029.

The constant temperature box.
See Abstr. 033.040.

Das Positionsfadenmikrometer in der Praxis.
See Abstr. 036.011.

Use of a prism on the correction plate of the Tautenburg Schmidt telescope for position determination.
See Abstr. 036.015.

Comparative performances of an objective grating concept in UV spectroscopic observations of solar system objects. II. D/H ratio observations in cometary atmospheres.
See Abstr. 036.023.

Comparative performances of an objective grating concept in UV spectroscopic observations of solar system objects. III. Observations of diffuse emission lines.
See Abstr. 036.025.

Obstacles facing the Venus radar mapper – the implications of gestalt formation in stereo–radargrammetry.
See Abstr. 036.081.

Absolute astronomical accelerometry.
See Abstr. 036.084.

Optical astronomical interferometry: requirements on stellar Michelson interferometry and heterodyne interferometry for high-resolution imaging.
See Abstr. 036.114.

Imaging infrared heterodyne spectrometry.
See Abstr. 036.115.

Thin-film phase filters in coherent-optical systems for image processing.
See Abstr. 036.167.

Astrometry with a scanning CCD.
See Abstr. 036.172.

Sobre a precisão de medidas de chapas num comparador Zeiss-Ascorecord.
See Abstr. 036.175.

Microcomputer controls of spectroscopic instruments.
See Abstr. 036.186.

Technique and advantages of using multiple-zone apertures in Fabry-Pérot spectroscopy.
See Abstr. 036.194.

First results from the new GSU (*Georgia State University*) CCD speckle camera.
See Abstr. 036.221.

Moving-image astrometry with the multi-anode microchannel array (MAMA) detector.
See Abstr. 036.223.

Microdensitometer errors: their effect on photometric data reduction.
See Abstr. 036.234.

The PDS for Space Telescope guide star selection system: status and improvements.
See Abstr. 036.235.

The VAX/VMS device driver and user operation program for the Midwestern Astronomical Data Reduction and Analysis Facility PDS microdensitometer.
See Abstr. 036.236.

Data compression techniques for astronomy.
See Abstr. 036.237.

Techniques of isophotometry.
See Abstr. 036.238.

Crystal gazing. II. Implications of advances in digital data storage technology.
See Abstr. 036.239.

Coordinate time in the vicinity of the earth.
See Abstr. 044.003.

How important is gravitational-wave astronomy? Part I.
See Abstr. 066.102.

Detailed analysis of solar photospheric parameters by quantitative imagery.
See Abstr. 071.033.

Recent results in the spectrophotometry of solar prominences.
See Abstr. 073.089.

Solar vector magnetograms of the Okayama Astrophysical Observatory.
See Abstr. 075.002.

The 1984 solar oscillation program of the Mt. Wilson 60-foot tower.
See Abstr. 080.079.

Submillimetric photometry of the integrated galactic emission.
See Abstr. 155.004.

Brackett-α imaging of the galactic center using a cooled Fabry-Perot array spectrometer.
See Abstr. 155.055.

035 Space Instrumentation

035.001 **STARLAB: an ultraviolet/optical space telescope.**
I. R. Tuohy.
Astrophys. Space Sci., Vol. 118, Nos. 1/2, p. 71 – 77 (1986). – See Abstr. 012.001.

Mount Stromlo and Siding Spring Observatories are actively developing large format photon counting detectors suitable for operation in space over a broad wavelength range. It is intended that these detectors be used in the Scientific Instrument Package of the STARLAB telescope. A description is given of the STARLAB facility, with emphasis on the contribution being made by Australia.

035.002 **Hubble Space Telescope. Dagen d nalkas.**
J. Trojefors, R. Wirestam.
Astron. Tidsskr., Årg. 19, Nr. 1, p. 24 – 31 (1986).

035.003 **LWR camera and ITF performance with 4.5 kV UVC.**
C. L. Imhoff.
ESA IUE Newsl., No. 25, p. 45 – 56 (1986).

035.004 **A rocket experiment to make far-UV images of comet Halley.**
R. P. McCoy, G. R. Carruthers, R. R. Meier, C. B. Opal.
IHW Newsl., No. 8, p. 19 – 24 (1986).

035.005 **The Giotto magnetic-field investigation.**
F. M. Neubauer, M. H. Acuna, L. F. Burlaga, B. Franke, B. Gramkow, F. Mariani, G. Musmann, N. F. Ness, H. U. Schmidt, T. Terenzi, E. Ungstrup, M. Wallis.
ESA Spec. Publ., ESA SP-1077, p. 1 – 14 (1986). – See Abstr. 003.013.

The main objective of the Giotto Magnetometer Experiment is the investigation of the interaction between Comet Halley and the solar wind at a distance of 0.9 AU from the Sun, to within 500 km of the cometary nucleus. A second objective is the study of the interplanetary magnetic field. The instrumentation is described in detail.

035.006 **The Giotto three–dimensional positive ion analyser.**
A. D. Johnstone, J. A. Bowles, A. J. Coates, A. J. Coker, S. J. Kellock, J. Raymont, B. Wilken, W. Studemann, W. Weiss, R. Cerulli–Irelli, V. Formisano, E. de Giorgi, P. Perani, M. de Bernardi, H. Borg, S. Olsen, J. D. Winningham, D. A. Bryant.
ESA Spec. Publ., ESA SP–1077, p. 15 – 32 (1986). – See Abstr. 003.013.
This instrument is designed to measure the three–dimensional energy distribution of positive ions in order to study the interaction between the solar wind and ionised cometary particles. The two sensors measure the distribution from 10 eV to 20 eV once per spacecraft spin and the distribution from 90 eV to 90 keV, with coarse mass discrimination, once every 32 spins.

035.007 **The Giotto RPA–Copernic plasma experiment.**
H. Rème, F. Cotin, A. Cros, J. L. Médale, J. A. Sauvaud, C. d'Uston, A. Korth, A. K. Richter, A. Loidl, K. A. Anderson, C. W. Carlson, D. W. Curtis, R. P. Lin, D. A. Mendis.
ESA Spec. Publ., ESA SP–1077, p. 33 – 52 (1986). – See Abstr. 003.013.
The RPA–Copernic experiment aboard Giotto is designed to measure the three–dimensional distributions of electrons between 10 eV and 30 keV, and the composition and distribution, close to the comet, of thermal positive ions, including clathrate hydrates, in the mass range 10 to 203 amu. It should help in achieving several of the major scientific objectives of the Giotto mission, by determining: (1) the nature of the comet/solar–wind interaction and of comet tails, (2) the chemical and physical nature of the comet's atmosphere and ionosphere, (3) the chemical and physical structure of the comet's nucleus.

035.008 **The Giotto Energetic Particle Experiment.**
S. McKenna–Lawlor, A. Thompson, D. O'Sullivan, E. Kirsch, D. Melrose, K.–P. Wenzel.
ESA Spec. Publ., ESA SP–1077, p. 53 – 65 (1986). – See Abstr. 003.013.
The Energetic Particle Experiment (EPA) onboard Giotto will measure the energy distribution of electrons, protons and heavier nuclei with $E \geqslant 20$ keV during the cruise phase and in the cometary environment during Halley encounter. The detector system and the main scientific objectives of EPA are described.

035.009 **The Giotto Particulate Impact Analyser.**
J. Kissel.
ESA Spec. Publ., ESA SP–1077, p. 67 – 83 (1986). – See Abstr. 003.013.
The Giotto Particulate Impact Analyser (PIA) will provide information on the composition of individual dust particles and the average composition and mass distribution of cometary matter released by Comet Halley. Knowledge of the material's composition and isotopic ratios (e.g. $^7Li/^6Li$, $^{11}B/^{10}B$, and $^{13}C/^{12}C$) will significantly increase our understanding of the current status and the genesis of comets. Variations in particle type or mean composition as a function of distance from the comet will yield information on the release mechanism. The instrument will be sensitive to particles ranging in size from 3×10^{-16} to 5×10^{-10}g (0.1 – 10 μm diameter) with impact rates of up to 100 per second. Depending on the actual impact rate, data on some 1000 – 5000 cometary particles should be gathered by the Analyser.

035.010 **The Giotto Dust Impact Detection System.**
J. A. M. McDonnell, W. M. Alexander, W. M. Burton, E. Bussoletti, D. H. Clark, G. C. Evans, S. T. Evans, J. G. Firth, R. J. L. Grard, E. Grün, M. S. Hanner, D. W. Hughes, E. Igenbergs, H. Kuczera, B. A. Lindblad, J.–C. Mandeville, A. Minafra, D. Reading, A. Ridgeley, G. H. Schwehm, T. J. Stevenson, Z. Sekanina, R. F. Turner, M. K. Wallis, J. C. Zarnecki.
ESA Spec. Publ., ESA SP–1077, p. 85 – 107 (1986). – See Abstr. 003.013.
The Dust Impact Detection System (DIDSY) consists of six independent subsystems with the primary aim of registering the impacts of all particulates of significant mass incident on the probe during the post–perihelion encounter with Comet Halley in 1986. Mounted on Giotto's front dust shield, the detector will determine the mass spectrum of the dust, with a limiting sensitivity of some 10^{-17}g, increasing to the largest grain masses encountered along Giotto's trajectory through the cometary environment with an ultimate spatial resolution of some 70 km. An additional detector is located on the rear shield to monitor those dust particles ($m \gtrsim 5 \times 10^{-7}$g) that are able to penetrate the front dust shield. An ambient plasma monitor is also incorporated into DIDSY to measure the impact plasma generated by both dust and gas impacts on the spacecraft.

035.011 **The Giotto Neutral Mass Spectrometer.**
D. Krankowsky, P. Lämmerzahl, D. Dörflinger, I. Herrwerth, U. Stubbemann, J. Woweries, P. Eberhardt, U. Dolder, J. Fischer, U. Herrmann, H. Hofstetter, M. Jungck, F. O. Meier, W. Schulte, J. J. Berthelier, J. M. Illiano, M. Godefroy, G. Gogly, P. Thévenet, J. H. Hoffman, R. R. Hodges, W. W. Wright.
ESA Spec. Publ., ESA SP–1077, p. 109 – 128 (1986). – See Abstr. 003.013.
Determination of the chemical nature of Comet Halley's volatile components is one of the key objectives of the Giotto mission. It requires measurements of the chemical, elemental, and isotopic composition in the coma, for which the Neutral Mass Spectrometer (NMS) is the primary experiment. This experiment consists of two analysers, a double–focusing mass spectrometer of novel design and an electrostatic energy analyser, covering a mass range up to 36 amu and an energy range up to 2180 eV (corresponding to 86 amu mass equivalent for thermal species), respectively. A detailed instrument description is given.

035.012 **The Giotto Ion Mass Spectrometer.**
H. Balsiger, K. Altwegg, F. Bühler, J. Fischer, J. Geiss, A. Meier, U. Rettenmund, H. Rosenbauer, R. Schwenn, J. Benson, P. Hemmerich, K. Säger, G. Kulzer, M. Neugebauer, B. E. Goldstein, R. Goldstein, E. G. Shelley, T. Sanders, D. Simpson, A. J. Lazarus, D. T. Young.
ESA Spec. Publ., ESA SP–1077, p. 129 – 148 (1986). – See Abstr. 003.013.
The Giotto Ion Mass Spectrometer (IMS) consists of two sensors: one optimised for the outer and the other for the inner coma of Halley's comet. Both sensors feature mass imaging characteristics, thereby permitting simultaneous measurements of several ion species by means of multi–detector arrays. Resultant mass–per–charge resolution is $\geqslant 20$. In addition to mass per charge, the energy per charge, the elevation and azimuth of incident ions are measured.

035.013 **The Giotto Halley Multicolour Camera.**
W. K. H. Schmidt, H. U. Keller, K. Wilhelm, C. Arpigny, C. Barbieri, L. Biermann, R. M. Bonnet, S. Cazes, C. B. Cosmovici, W. A. Delamere, W. F. Huebner, D. W. Hughes, C. Jamar, D. Malaise, H. Reitsema, P. Seige, F. L. Whipple.
ESA Spec. Publ., ESA SP–1077, p. 149 – 172 (1986). – See Abstr. 003.013.
In recent decades, Whipple's "dirty snowball" picture of the cometary nucleus has become widely, but not universally accepted. Proof of the nucleus' existence and study of its properties require pictures in various colour bands, preferably together with other information. Acquisition of these pictures is the task of Halley Multicolour Camera (HMC), which is described in detail.

035.014 **The Giotto Radio–Science Experiment.**
P. Edenhofer, M. K. Bird, H. Buschert, P. B. Esposito, H. Porsche, H. Volland.
ESA Spec. Publ., ESA SP–1077, p. 173 – 186 (1986). – See Abstr. 003.013.
The Giotto spacecraft is equipped with redundant radio transponders capable of downlink transmission at S–band (2.3 GHz) and X–band (8.4 GHz). The goal of the Giotto Radio–Science Experiment is to take measurements of the phase (Doppler) shifts of the downlink carrier signals both at S– and X–band as a

function of time during the Halley encounter. The specific objectives are determination of: (1) the columnar electron content of the ionosphere of Comet Halley, (2) the cometary mass fluence and total mass flow due to atmospheric drag.

035.015 **The Giotto Optical Probe Experiment.**
A. C. Levasseur–Regourd, J. L. Bertaux, R. Dumont, M. Festou, R. H. Giese, F. Giovane, P. Lamy, A. Llebaria, J. L. Weinberg.
ESA Spec. Publ., ESA SP–1077, p. 187 – 193 (1986). – See Abstr. 003.013.
The Halley Optical Probe Experiment (HOPE) has been designed to provide in–situ photopolarimetric data on both the dust cloud and the gaseous atmosphere in Halley's coma. The Optical Probe's concept is presented here, together with a description of the instrumentation and the possibilities for cross–correlation between the HOPE results and those of other space and ground–based–experiments. The instrument was turned on successfully on 13 September 1985.

035.016 **Nouvelles du Télescope Spatial.**
P. Y. Bély.
Astronomie, Vol. 100, p. 43 – 46 (1986).

035.017 **VEGA balloon system and instrumentation.**
R. S. Kremnev, V. M. Linkin, A. N. Lipatov, K. M. Pichkadze (*K. M. Pichkhadze*), A. A. Shurupov, A. V. Terterashvili, R. V. Bakitko, J. E. Blamont, C. Malique, B. Ragent, R. A. Preston, L. S. Elson, D. Crisp.
Science, Vol. 231, No. 4744, p. 1408 – 1411 (1986).
The VEGA Venus balloon radio transmissions received on Earth were used to measure the motion of the balloons and to obtain the data recorded by onboard sensors measuring atmospheric characteristics. Thus the balloons themselves, the gondolas, the onboard sensors, and the radio transmission system were all components of the experiment. A description of these elements is given, and a few details of data sampling and formatting are discussed.

035.018 **An X–ray all–sky monitor for the Space Station.**
S. S. Holt, W. C. Priedhorsky.
Bull. Am. Astron. Soc., Vol. 17, No. 4, p. 848 (1985). Abstract. – See Abstr. 010.062.

035.019 **The Spartan–1 X–ray astronomy experiment.**
G. G. Fritz, R. G. Cruddace.
Bull. Am. Astron. Soc., Vol. 17, No. 4, p. 848 (1985). Abstract. – See Abstr. 010.062.

035.020 **The AXAF CCD imaging spectrometer experiment (ACIS).**
J. A. Nousek, G. P. Garmire, G. R. Ricker, S. A. Collins.
Bull. Am. Astron. Soc., Vol. 17, No. 4, p. 864 – 865 (1985). Abstract. – See Abstr. 010.062.

035.021 **Low energy transmission grating spectrometer (LETGS) for AXAF.**
A. C. Brinkman, J. J. Van Rooijen, J. A. M. Bleeker, J. H. Dijkstra, J. Heise, R. Mewe, F. Paerels, P. A. J. De Korte.
Bull. Am. Astron. Soc., Vol. 17, No. 4, p. 865 (1985). Abstract. – See Abstr. 010.062.

035.022 **Expected AXAF mirror characteristics and their implications for measurements of the Hubble constant using the Sunyaev Zel'dovich effect.**
L. P. Van Speybroeck.
Bull. Am. Astron. Soc., Vol. 17, No. 4, p. 887 – 888 (1985). Abstract. – See Abstr. 010.062.

035.023 **The AXAF High Resolution Camera (HRC) and its use for observations of distant clusters of galaxies.**
S. S. Murray, J. H. Chappell, M. S. Elvis, W. R. Forman, J. E. Grindlay, F. R. Harnden Jr., T. Maccacaro, H. D. Tananbaum, G. S. Vaiana, K. A. Pounds, G. W. Fraser, J. P. Henry.
Bull. Am. Astron. Soc., Vol. 17, No. 4, p. 888 (1985). Abstract. – See Abstr. 010.062.

035.024 **X–ray spectroscopy of AGN with the AXAF "Microcalorimeter".**
S. S. Holt.
Bull. Am. Astron. Soc., Vol. 17, No. 4, p. 888 (1985). Abstract. – See Abstr. 010.062.

035.025 **A high quality movie from the SOUP instrument on SpaceLab 2.**
A. M. Title, T. D. Tarbell.
Bull. Am. Astron. Soc., Vol. 17, No. 4, p. 896 (1985). Abstract. – See Abstr. 010.062.

035.026 **Calibration of the wide field/planetary camera on Space Telescope.**
H. C. Harris, W. A. Baum.
Bull. Am. Astron. Soc., Vol. 17, No. 4, p. 899 (1985). Abstract. – See Abstr. 010.062.

035.027 **The X–ray telescope for the LAMAR Space Shuttle experiment.**
D. Fabricant, M. Conroy, L. Cohen, P. Gorenstein.
Bull. Am. Astron. Soc., Vol. 17, No. 4, p. 899 – 900 (1985). Abstract. – See Abstr. 010.062.

035.028 **The Wisconsin Ultraviolet Photo–Polarimeter Experiment.**
S. A. Stanford, M. A. Murison, B. A. Whitney, G. C. Clayton.
Bull. Am. Astron. Soc., Vol. 17, No. 4, p. 900 (1985). Abstract. – See Abstr. 010.062.

035.029 **The Cassini mission. Complementary observations with the Space Telescope.**
R. Prange.
ESA Spec. Publ., ESA SP–241, p. 209 – 217 (1985). – See Abstr. 012.018.
The Space Telescope and its five focal instruments are presented and their characteristics are summarized. Then a comparison is made, discussing which instrument is best adapted to a given kind of observation. Finally, the last section rapidly reviews the observations which have been up to now considered with the ST on the Saturn and Titan atmospheres.

035.030 **Microwave spectroscopy and radiometry of Saturn and Titan.**
D. O. Muhleman.
ESA Spec. Publ., ESA SP–241, p. 267 – 269 (1985). – See Abstr. 012.018.
Experiments using microwave emission techniques are described for the investigation of the deep atmosphere of Saturn, the Ring system, the entire atmosphere of Titan and its surface. A microwave spectrometer and radiometer system is being designed for the Cassini mission. It would sound the atmosphere of Saturn in the 0.5 to 15 bar pressure range at wavelengths of about 2 and 13 cm. It would map Saturn's ring in thermal emission at wavelengths of about 3 mm and 2 cm and measure the polarization of the scattered radiation at 2 cm. A system consisting of 1 or 2 receivers is discussed to sound the atmosphere of Titan from the cold trap to about 500 km altitude to determine temperature, pressure of the atmosphere of a function of altitude and mixing ratio profiles of certain species. Surface emission of Titan at 2 cm is discussed.

035.031 **Neutral mass spectrometry and gas chromatography.**
H. B. Niemann.
ESA Spec. Publ., ESA SP–241, p. 271 – 279 (1985). – See Abstr. 012.018.
A gas chromatograph–mass spectrometer system which alternatively takes atmospheric gas samples directly into the mass spectrometer, or indirectly through a chromatographic column,

or an aerosol pyrolyzer is proposed for the analysis of the atmosphere of Titan using an entry probe.

035.032 **An aerosol collector pyrolyser for the Cassini Titan probe.**
G. Israel.
ESA Spec. Publ., ESA SP–241, p. 281 – 286 (1985). – See Abstr. 012.018.
During the descent of a Titan probe, as the one proposed in the frame of the Cassini mission, it is very much recommended to undertake aerosols sampling. The preferred analytical method is to use pyrolysis technics in which the gaz products can be identified by the MS and/or the GC–MS instruments which will be part of the probe payload. Sampling aerosols in the region 60 – 30 km seems the best way to determine the nature and importance of the organic polymers (incorporated in the aerosols of the condensed clouds). Another region for sampling is envisaged in the upper part of the atmosphere (200 – 70 km). Preliminary studies on the collecting plates, the pneumatic and mechanic systems are described.

035.033 **Aeronomical measurements by Cassini.**
D. M. Hunten.
ESA Spec. Publ., ESA SP–241, p. 287 – 288 (1985). – See Abstr. 012.018.
The capability of the instruments on the proposed Cassini payload to make in–situ and remote measurements of Titan's upper atmosphere is discussed. The multiple fly–throughs of the Orbiter at heights well below the ionospheric peak offer a truly first–class opportunity for detailed aeronomical studies of this unique atmosphere. They will be supplemented by solar and stellar occultations and measurements of airglow and aurora on both Titan and Saturn.

035.034 **Instrumentation for X–ray spectroscopy.**
H. W. Schnopper.
ESA Spec. Publ., ESA SP–239, p. 77 – 86 (1985). – See Abstr. 012.019.
This discussion will point out the various concentrator, spectroscopic and imaging systems which are likely to be available at the time when payload decisions for ESA's High Throughput X–ray Mission will be made. The most useful parameters, e.g. collecting area, spectral resolution, spatial resolution, and energy range, are summarized and the sensitivity of various combinations suitable for specific observational programs are illustrated.

035.035 **The LAMAR telescope module for high throughput imaging and spectroscopy.**
P. Gorenstein, D. Fabricant, L. Cohen.
ESA Spec. Publ., ESA SP–239, p. 177 – 182 (1985). – See Abstr. 012.019.
The first of eight Kirkpatrick–Baez mirror assemblies is being built for the LAMAR experiment of the Space Shuttle. The aperture is 20 cm × 30 cm and the distance from the front of the mirror to the focal plane is 3.4 m. The reflectors are made from 1.8 mm gold coated float glass. A prototype "brassboard" mirror was constructed and tested.

035.036 **A thin foil high throughput X–ray telescope.**
P. L. Jensen, N. J. Westergaard.
ESA Spec. Publ., ESA SP–239, p. 183 – 187 (1985). – See Abstr. 012.019.
A thin foil reflector X–ray telescope has been proposed for a high throughput spectroscopic satellite (XMM). A prototype of a single quadrant has been built and the precision of a few test mirrors is being measured.

035.037 **Carbon–fibre reinforced epoxy techniques for the X.M.M. LE–telescopes.**
P. A. J. de Korte, J. Klug, G. Moreau, G. P. Whitcomb.
ESA Spec. Publ., ESA SP–239, p. 189 – 193 (1985). – See Abstr. 012.019.
A development study has been carried out in the context of the LE–telescopes for the X–ray Multi–Mirror mission (X.M.M.). In that study an X–ray paraboloidal mirror has been fabricated by means of the replica method developed for EXOSAT. Instead of the use of a beryllium carrier the authors have investigated the use of a carbon–fibre reinforced epoxy carrier, which would have a big advantage for the mass production envisaged for X.M.M.

035.038 **XMM imaging Bragg spectrometer with a doubly curved crystal.**
B. Byrnak, F. E. Christensen, N. J. Westergaard, H. W. Schnopper.
ESA Spec. Publ., ESA SP–239, p. 209 – 212 (1985). – See Abstr. 012.019.
This spectrometer is sensitive in the energy range from 0.5 to 7 keV. The Bragg reflected rays from a doubly bent crystal positioned down–stream of the focal plane of a grazing–incidence concentrator are focused along the axis of a position sensitive detector. The main advantages are: simultaneous (scanning free) coverage of the spectral range of interest, nearly stigmatic imaging, and crystal limited resolution.

035.039 **Measurements of intrinsic width and mosaicity of test crystals with a novel four–crystal X–ray spectrometer.**
F. E. Christensen, H. W. Schnopper, E. H. Silver, N. J. Westergaard.
ESA Spec. Publ., ESA SP–239, p. 213 – 220 (1985). – See Abstr. 012.019.
The high resolution dispersive X–ray spectrometer for the XMM mission proposed in part by the Danish Space Research Institute, will use a variety of curved crystals with varying lattice spacings as the energy dispersive element. This paper reports on an X–ray study of a number of test crystals for which the mosaicity and the perfection of the microcrystallites were measured.

035.040 **Bolometers as spectrometers for X–ray astronomy.**
N. Coron, G. Artzner, G. Dambier, G. Jegoudez, J. Leblanc, J. Y. Deschamps, R. Rocchia, A. Tarrius, O. Testard, H. R. Ravn, H. H. Stroke, B. Jonson.
ESA Spec. Publ., ESA SP–239, p. 229 – 231 (1985). – See Abstr. 012.019.
The authors have investigated the possibility to use composite diamond–germanium bolometers, originally developed for I.R. astronomy, for the individual detection of X–ray photons.

035.041 **Room temperature solid state spectrometer for X–ray astronomy.**
G. Di. Cocco, W. Dusi, C. Labanti, G. F. Spada, G. Ventura, P. Siffert.
ESA Spec. Publ., ESA SP–239, p. 233 – 236 (1985). – See Abstr. 012.019.
The purpose of this paper is to propose the use of room temperature HgI_2 solid state X–ray spectrometers in space borne experiments. The main factors which characterize the behaviour of HgI_2 detectors and their related electronics are presented. The performance of a position sensitive spectrometer, based on available experimental results on HgI_2 detectors, is described and its possible application as a focal plane detector for X–ray imaging concentrators is suggested.

035.042 **The application of reflection gratings to a large area X–ray spectroscopy mission: a discussion of scientific requirements and various design options.**
S. M. Kahn, M. C. Hettrick.
ESA Spec. Publ., ESA SP–239, p. 237 – 244 (1985). – See Abstr. 012.019.

035.043 **A multiwire proportional counter for low energy X–ray imaging.**
M. M. Madsen, P. Jonasson, P. L. Jensen, H. E. Rasmussen, P. Ørup, H. W. Schnopper.
ESA Spec. Publ., ESA SP–239, p. 245 – 254 (1985). – See Abstr. 012.019.
Development of a detector for a high spectral resolution imaging Bragg spectrometer experiment is described, including the test set–up and the read–out electronics. Examples of the position resolution at X–ray energies from aluminum and manganese K radiation (1.49 and 5.9 keV) are presented.

035.044 **An imaging optical/UV monitor for XMM.**
K. O. Mason, J. Allington–Smith, J. L. Culhane, F. A. Cordova, W. C. Priedhorsky, P. G. Murdin, R. Bingham, B. Margon, J. B. Hutchings.
ESA Spec. Publ., ESA SP–239, p. 255 – 263 (1985). – See Abstr. 012.019.

The authors propose that a sensitive instrument affording photometry and spectroscopy at optical and ultraviolet wavelengths could be added, with minimal expenditure, to the complement of AXAF detectors. An AXAF Imaging Optical/UV Monitor would enable simultaneous X–ray, optical, and ultraviolet data to be obtained for every AXAF field.

035.045 **X–ray polarimetry on XMM.**
R. Novick, G. Chanan, D. J. Helfand.
ESA Spec. Publ., ESA SP–239, p. 265 – 275 (1985). – See Abstr. 012.019.

The authors propose that a single XMM Mirror Module be devoted to X–ray polarimetry. The instrument would exploit the polarization dependence of Bragg reflection from a variety of crystals and Thomson scattering from a lithium target. It would be sensitive over the full bandwidth of the mirror and could be used for unique and important studies of as many as 50 different astrophysical objects.

035.046 **A multiplex coding imaging spectrometer for X–ray astronomy.**
R. Rocchia, J. Y. Deschamps, L. Koch–Miramond, A. Tarrius.
ESA Spec. Publ., ESA SP–239, p. 277 – 281 (1985). – See Abstr. 012.019.

The paper describes a multiplex coding system associated with a solid state spectrometer Si(Li) designed to be placed at the focus of a grazing incidence telescope. Simulations and results obtained with a laboratory model are presented.

035.047 **Objective crystal spectrometer (OXS) for imaging extended cosmic X–ray sources.**
H. W. Schnopper, B. Byrnak.
ESA Spec. Publ., ESA SP–239, p. 283 – 286 (1985). – See Abstr. 012.019.

Results from Einstein and EXOSAT have demonstrated that there is a need for a high resolution ($E/\Delta E$ 500 – 1000), imaging (5×5 arcmin pixel), spectrometer which is capable of studying sources with appreciable angular extent (up to 1 – 2 deg). The instrument of choice for diffuse sources has a flat Bragg crystal located ahead of the telescope. As an example of its performance, a source emitting O VII with a surface brightness ~ 10 photons $cm^{-2}s^{-1}ster^{-1}$ will be detected in 7.8×7.8 arcmin pixels over a 2 deg radius field of view. A total observation of $\sim 3 \times 10^5$s will map the source in each of the three O VII helium–like emission lines with a spectral resolution $E/\Delta E$ in excess of 600.

035.048 **The Penning Gas Imager in a space application.**
H. E. Schwarz.
ESA Spec. Publ., ESA SP–239, p. 287 – 290 (1985). – See Abstr. 012.019.

The prototype Penning Gas Imager or PGI(1) and its principles of operation are briefly described. The author presents data taken with laboratory detectors and describes the space version of the PGI. This 120 mm detector will be flown as a focal plane imager on the Soft X–ray Telescope (SXT), an objective grating spectrometer. A novel background rejection scheme is discussed and preliminary experimental data obtained in the laboratory are presented.

035.049 **X–ray applications of deep–depletion CCDs.**
D. M. Walton, J. L. Culhane, R. A. Stern, R. C. Catura.
ESA Spec. Publ., ESA SP–239, p. 291 – 295 (1985). – See Abstr. 012.019.

Results are presented demonstrating the successful use of deep–depletion CCDs as single photon X–ray detectors. CCDs as focal plane X–ray detectors for X–ray astronomy offer high spatial resolution combined with good energy resolution. The importance of deep–depletion CCDs is a significantly higher X–ray quantum efficiency than that of conventional CCDs. The application of CCDs to an X–ray astronomy mission such as XMM is described.

035.050 **X–ray imaging spectroscopy with CCDs.**
A. Wells, D. Lumb, R. Willingale.
ESA Spec. Publ., ESA SP–239, p. 297 – 302 (1985). – See Abstr. 012.019.

A future ESA spectroscopy mission in X–ray astronomy requires a high sensitivity broad band spectrometer operating in the energy band 1 – 10 keV. The detector system, based on a CCD imager, offers spectral resolution comparable with solid state Si(Li) detectors but with spatial resolution matched to the telescope thus allowing simultaneous mapping of extended sources and discrimination between point sources in crowded fields.

035.051 **The concept of the Pinhole/Occulter Facility.**
H. S. Hudson.
NASA Conf. Publ., NASA CP–2421, p. 5 – 13 (1986). – See Abstr. 012.021.

The Pinhole/Occulter Facility is based upon a simple idea for obtaining high angular resolution in astronomical X–ray observations, for example for solar flares at energies >10 keV. The scheme uses a coded aperture device (multiple pinhole camera) with a large separation between the aperture encoder and the detector. Such an imaging device can have an angular resolution much better than 1 arc s if desired. A large structure would also make it possible to have a large external occulter, which would have powerful applications, notably for high–sensitivity observations of the corona in EUV and white light. This capability leads to the definition of the Pinhole/Occulter Facility, which combines both X–ray and coronal observations. The present concept is based upon a 35–m deployable boom, erected in the cargo bay of the Space Shuttle and pointed in the solar direction by the Instrument Pointing System of Spacelab.

035.052 **Mass spectrometers for investigation of the elementary composition of the dust component in the Venus–Halley program.**
R. Z. Sagdeev, S. R. Tabaldyev, V. B. Leonas.
Mass–spectrometriya i khim. kinet. Moskva, p. 209 – 216 (1985). In Russian. Abstr. in Ref. Zh., 62. Issled. Kosm. Prostranstva, 4.62.105 (1986).

035.053 **Miniature Stirling cycle refrigerators for space use.**
T. W. Bradshaw.
J. Br. Interplanet. Soc., Vol. 39, No. 5, p. 224 – 227 (1986).

The Rutherford Appleton Laboratory is involved in two space projects that require cooled infrared detectors. This paper describes the development, operating principles and performance of the coolers to be used with these devices. Further developments in cooler technology are discussed.

035.054 **The Infrared Space Observatory (ISO).**
R. J. Emery.
J. Br. Interplanet. Soc., Vol. 39, No. 5, p. 235 – 240 (1986).

This paper gives some technical background to the ISO project, outlines a range of astronomical topics which could be addressed with ISO, and then describes the instrumental payload and some aspects of its performance.

035.055 **The night eye in the sky.**
H. W. Brandli.
I.A.P.P.P. Commun., No. 23, p. 4 (1986). Abstract. – See Abstr. 010.521.

035.056 **A high–resolution Bragg spectrometer for the observation of the hot interstellar medium X–ray emission.**
J. Y. Deschamps, R. Rocchia, A. Tarrius.
Astrophys. Space Sci., Vol. 121, No. 2, p. 321 – 332 (1986).

The authors present a high–resolution Bragg spectrometer designed for the observation of the soft X–ray cosmic diffuse background. The instrument concept is derived from the de Broglie geometry for the study of extended sources. The estimated sensitivity and the simulations of actual space observations show that the instrument is capable to separate the strongest lines emitted by the most abundant ions (O VII, O VIII, Fe XVII, Ne IX, etc.) and to "map" the whole sky during a six month mission.

035.057 **Synthetic photometry and the calibration of the Hubble Space Telescope.**
J. Koornneef, R. Bohlin, R. Buser, K. Horne, D. Turnshek.
Space Telesc. Sci. Inst., Prepr. Ser., No. 91, 11 pp. (1986). To appear in Highlights in Astronomy, Vol. 7.

035.058 **The ultraviolet calibration of the Hubble Space Telescope: 1. Secondary standards of absolute ultraviolet flux and the recalibration of IUE.**
R. C. Bohlin.
Space Telesc. Sci. Inst., Prepr. Ser., No. 99, 33 pp. (1986). To appear in Astrophys. J.

035.059 **Synthetic photometry and the calibration of the Hubble Space Telescope.**
J. Koornneef, R. Bohlin, R. Buser, K. Horne, D. Turnshek.
Space Telesc. Sci. Inst., Prepr. Ser., No. 100, p. 35 – 45 (1986). – See Abstr. 012.040.

035.060 **Miniature Stirling cycle refrigerators for space use.**
T. W. Bradshaw.
Rutherford Appleton Lab., Rep., RAL–85–107, 14 pp. (1985). See Abstr. 035.053.

035.061 **A report on the mapping of the DIDSY (*Dust Impact Detection System*) sensor responses as a function of micro–meteorite impact position on the Giotto bumper shield.**
A. Ridgeley.
Rutherford Appleton Lab., Rep., RAL–85–108, 44 pp. (1985).
A realistic mock–up of the GIOTTO bumper shield has been used to calibrate the response of the flight bumper shield to micro–meteorite bombardment. A pulsed laser was used to simulate particle impacts, and a detailed mapping of the shield was made in order to determine the response of the DID sensors in terms of azimuthal and radial impact position.

035.062 **The reflector system for the German X–ray satellite ROSAT.**
D. Reinhardt.
Feinwerktech. Messtech., Vol. 93, No. 5, p. 251 – 253 (1985). In German. Abstr. in Phys. Abstr., Vol. 89, No. 1273, Entry 18623 (1986).

035.063 **Space Telescope.**
F. S. Wojtalik.
Bridge between control science and technology, Vol. 5, p. 2909 – 2913 (1985). Abstr. in Phys. Abstr., Vol. 89, No. 1273, Entry 18661 (1986). – See Abstr. 012.043.

035.064 **High energy gamma ray observatories for the study of cosmic ray electrons above 10^{14}eV.**
S. A. Stephens, V. K. Balasubrahmanyan.
Nucl. Instrum. Methods Phys. Res., Sect. A, Vol. A241, No. 1, p. 257 – 264 (1985). Abstr. in Phys. Abstr., Vol. 89, No. 1274, Entry 20058 (1986).

035.065 **Response of high resolution drift tubes to relativistic heavy ions.**
A. Tomasch, S. P. Ahlen, C. Bower, P. B. Price, M. H. Salamon, G. Tarle, J. Musser, H. Crawford.
Nucl. Instrum. Methods Phys. Res., Sect. A, Vol. A241, No. 1, p. 265 – 274 (1985). Abstr. in Phys. Abstr., Vol. 89, No. 1274, Entry 20059 (1986).

035.066 **Proposal for investigation of minor bodies of the Solar System using remote sensing of electron beam induced X–ray fluorescence.**
G. Hrehuss, T. I. Gombosi, I. Naday, L. Pogany, K. Szego.
Acta Phys. Hung., Vol. 58, No. 1 – 2, p. 83 – 100 (1985). Abstr. in Phys. Abstr., Vol. 89, No. 1274, Entry 23751 (1986).

035.067 **MIPS: the Multiband Imaging Photometer for SIRTF.**
G. H. Rieke, C. Lada, M. Lebofsky, F. Low, P. Strittmatter, E. Young, J. Arens, E. Haller, P. Richards, C. Beichman, T. N. Gautier, J. Mould, G. Neugebauer, S. Gaalema, M. Werner.
Prepr. Steward Obs., No. 618, 7 pp. (1986).

035.068 **A space telescope for infrared spectroscopy of earthlike planets.**
J. R. P. Angel, A. Y. S. Cheng, N. J. Woolf.
Prepr. Steward Obs., No. 640, 9 pp. (1986). To appear in Nature.

035.069 **Space Telescope: a better window to the Universe.**
T. A. Nobbe.
Mach. Des., Vol. 57, No. 23, p. 22 – 24, 26 (1985). Abstr. in Phys. Abstr., Vol. 89, No. 1275, Entry 30163 (1986).

035.070 **A study of surface particulate contamination on the primary mirror of the Hubble Space Telescope.**
T. A. Facey.
Proc. SPIE Int. Soc. Opt. Eng., Vol. 525, p. 140 – 146 (1985). Abstr. in Phys. Abstr., Vol. 89, No. 1276, Entry 35973 (1986). – See Abstr. 012.045.

035.071 **A beryllium proportional counter for observing cosmic X–ray sources.**
H.–r. Hang, Q.–x. Li.
Publ. Purple Mt. Obs., Vol. 4, No. 3, p. 41 – 46 (1985). In Chinese.
A braze welded proportional counter with the effective window area of 81 cm^2 and its electronics are described, which were made for monitoring and survey of X–ray binaries of our Galaxy and were used as a balloon–borne detector to measure the cosmic ray intensity.

035.072 **A pulse shape discriminator and its efficiency.**
Q.–x. Li.
Publ. Purple Mt. Obs., Vol. 5, No. 1, p. 66 – 70 (1986). In Chinese.
A pulse shape discriminator has been developed for an X–ray proportional counter to reduce the background. Test measurements in laboratory have shown its efficiency on reducing the background in the energy range 2 – 12 keV to be at least 95%. About 78% – 80% of background can be rejected on balloon–borne observations from the Earth's surface to 30 km above the Earth.

035.073 **Possible planetary XRF instrument with X–ray generator.**
L. I Yin, J. I. Trombka.
Nucl. Instrum. Methods Phys. Res., Sect. A, Vol. A242, No. 3, p. 544 – 549 (1986). Abstr. in Phys. Abstr., Vol. 89, No. 1279, Entry 53723 (1986). – See Abstr. 012.058.

035.074 **The background counting rates in a balloon borne hard X–ray telescope.**
A. J. Dean, N. A. Dipper, R. A. Lewis.
Nucl. Instrum. Methods Phys. Res., Sect. A, Vol. A242, No. 2, p. 342 – 345 (1986). Abstr. in Phys. Abstr., Vol. 89, No. 1279, Entry 53747 (1986).

035.075 **Second generation spectrograph for the Hubble Space Telescope.**
S. P. Maran, B. E. Woodgate, A. Boggess, T. R. Gull, S. R. Heap, R. Melcher, R. F. Green, S. C. Wolff, J. B. Hutchings, E. B. Jenkins, J. L. Linsky, H. W. Moos, F. Roesler, R. A. Shine, J. G. Timothy, D. E. Weistrop.
Bull. Am. Astron. Soc., Vol. 18, No. 1, p. 636 (1986). Abstract. – See Abstr. 010.066.

035.076 **Calibration of Si(Li) detectors for X–ray astronomy.**
J. C. Lochner, E. A. Boldt.
Nucl. Instrum. Methods Phys. Res., Sect. A, Vol. A242, No. 3, p. 382 – 386 (1986). Abstr. in Phys. Abstr., Vol. 89, No. 1280, Entry 55586 (1986). – See Abstr. 012.058.

035.077 **The position sensitive proportional counter (PSPC) of the Rosat telescope.**
U. G. Briel, E. Pfeffermann.
Nucl. Instrum. Methods Phys. Res., Sect. A, Vol. A242, No. 3, p. 376 – 381 (1986). Abstr. in Phys. Abstr., Vol. 89, No. 1280, Entry 60581 (1986). – See Abstr. 012.058.

035.078 **Data acquisition and -evaluation electronics for the Rosat PSPCs.**
G. Metzner.
Nucl. Instrum. Methods Phys. Res., Sect. A, Vol. A242, No. 3, p. 493 – 500 (1986). Abstr. in Phys. Abstr., Vol. 89, No. 1280, Entry 60582 (1986). – See Abstr. 012.058.

035.079 **Precise pointing of Space Telescope using a quadrant detector.**
T. E. Strikwerda, K. Strohbehn, G. A. Heyler.
Proc. SPIE Int. Soc. Opt. Eng., Vol. 550, p. 51 – 57 (1985). Abstr. in Phys. Abstr., Vol. 89, No. 1280, Entry 60646 (1986). – See Abstr. 012.059.

035.080 **Technology for photographing Halley's comet.**
J. R. Kramin, H. J. Meyer.
Elektronik, Vol. 35, No. 5, p. 91 – 96 (1986). In German. Abstr. in Phys. Abstr., Vol. 89, No. 1283, Entry 78277 (1986).

035.081 **Design of a precision pointing control system for the Space Infrared Telescope Facility.**
B. Sridhar, J.-N. Aubrun, K. R. Lorell.
IEEE Control Syst. Mag., Vol. 6, No. 1, p. 28 – 34 (1986). Abstr. in Phys. Abstr., Vol. 89, No. 1283, Entry 78278 (1986).

035.082 **Response of a 7.6 cm × 7.6 cm bismuth germanate spectrometer for solar gamma ray observations.**
M. Yoshimori, H. Watanabe, F. Shiraishi.
Nucl. Instrum. Methods Phys. Res., Sect. A, Vol. A245, No. 1, p. 191 – 198 (1986). Abstr. in Phys. Abstr., Vol. 89, Nr. 1284, Entry 83439 (1986).

035.083 **A Doppler filter technique to measure the hydrogen content of planetary surfaces.**
W. C. Feldman, D. M. Drake.
Nucl. Instrum. Methods Phys. Res., Sect. A, Vol. A245, No. 1, p. 182 – 190 (1986). Abstr. in Phys. Abstr., Vol. 89, Nr. 1284, Entry 83444 (1986).

035.084 **HIPPARCOS: the instrument.**
J. Schrijver.
Manuscr. Geod., Vol. 11, No. 2, p. 97 – 102 (1986).
After a brief overview of ESA's mission for global space astrometry HIPPARCOS, the main parts of the instrument are described in more detail with emphasis on the aspects that influence the quality of the data to be obtained.

035.085 **Microarcsecond astrometric interferometry.**
R. D. Reasenberg.
Astrometric techniques, p. 321 – 330 (1986). – See Abstr. 012.068 (IAU Symp. No. 109).
A small astrometric optical interferometer, Mini-POINTS, could measure the angle between stars about 90° apart. For 10th magnitude stars, an observation time of about 20 minutes would yield a measurement uncertainty of 5 microarcseconds.

035.086 **TRIO: a kilometric array stabilized by solar sails.**
A. Labeyrie, B. Authier, T. De Graauw, E. Kibblewhite, G. Weigelt.
ESA Spec. Publ., ESA SP-226, p. 27 – 33 (1985). – See Abstr. 012.075.
For kilometric baselines and variable aperture patterns allowing image reconstructions TRIO has free-flying telescopes and a focal station. Small solar sails serve to point and stabilize the system. Electrostatic mounts support the telescopes. 2 to 4 telescopes will be part of the first-generation instrument, but many apertures can be utilized in later versions.

035.087 **SAMSI: an orbiting spatial interferometer for micro-arcsecond astronomical observations.**
R. V. Stachnik, D. Y. Gezari.
ESA Spec. Publ., ESA SP-226, p. 35 – 42 (1985). – See Abstr. 012.075.
The authors describe the concept and predicted performance of SAMSI, Spacecraft Array for Michelson Spatial Interferometry, an orbiting spatial interferometer comprised of three free-flying spacecraft, two collector telescopes and a central mixing station.

035.088 **COSMIC: a high resolution, large collecting area telescope.**
W. A. Traub, N. P. Carleton.
ESA Spec. Publ., ESA SP-226, p. 43 – 48 (1985). – See Abstr. 012.075.
The key advantages of COSMIC (Coherent Optical System of Modular Imaging Collectors) are high angular resolution and the ability to produce images of complex, low-surface-brightness objects such as distant galaxies. If configured as a 36 m filled linear array, COSMIC can have 15 times better angular resolution and 10 times greater collecting area than ST. The authors review the considerations which have led them to the current design concept, they list a number of scientific goals, and they point out the potentially useful role of a Space Station for assembly.

035.089 **A fiber-linked version of project TRIO.**
P. Connes, C. Froehly, P. Facq.
ESA Spec. Publ., ESA SP-226, p. 49 – 61 (1985). – See Abstr. 012.075.
The authors show that an appreciable simplification of Project TRIO may be obtained by the use of two equal-length single-mode fibers connecting the two telescopes to the central station.

035.090 **OASIS: optical aperture synthesis in space.**
J. E. Noordam, P. D. Atherton, A. H. Greenaway.
ESA Spec. Publ., ESA SP-226, p. 63 – 69 (1985). – See Abstr. 012.075.
The authors present some of the techniques necessary to accomplish optical aperture synthesis in space. The main purpose is to obtain reliable, high resolution images of astronomical sources, in emission lines, absorption lines and the continuum, with high dynamic range. The technique of redundant spacing calibration allows the identification of phase and amplitude errors associated with a particular telescope in an array. This requires the separate availability of the visibility phase and amplitude for all interferometers (telescope pairs). The authors describe this technique and discuss the various issues which determine the design and its feasibility.

035.091 **Propulsion solaire pour l'interféromètre TRIO.**
J. L. Boit.
ESA Spec. Publ., ESA SP-226, p. 85 – 90 (1985). – See Abstr. 012.075.
Large displacement and high accuracy positioning of each of the three satellites of the TRIO interferometer require a long-service-life, clean, modulated propulsion system. A solar sailing device is described. The propulsion is obtained from the deviation of concentrated solar beams. The problems related to mechanics, thermal behaviour, celestial mechanics, attitude correction strategy, positioning error detection, are then approached.

035.092 **Preliminary studies of a spinning tether-connected TRIO concept.**
E. B. Crellin.
ESA Spec. Publ., ESA SP-226, p. 91 – 94 (1985). – See Abstr. 012.075.
Long baseline interferometry in space may be possible using a spinning central station to which two telescopes are attached by long tethers. This paper outlines the potential advantages of such a system and presents the results from some preliminary feasibility studies.

035.093 **TRIO interferometer positioning.**
B. Authier.
ESA Spec. Publ., ESA SP–226, p. 95 – 97 (1985). – See Abstr. 012.075.
The high accuracy positioning of each of the three satellites of the TRIO interferometer, with respect to the others and to the reference stars, is a new problem whose strategy is approached in this paper.

035.094 **TRIANGLE: a different design for a three satellite interferometer in space.**
F. Vakili.
ESA Spec. Publ., ESA SP–226, p. 107 – 111 (1985). – See Abstr. 012.075.
TRIANGLE consists of three identical satellites; each one acts as a telescope and a central station. These satellites are maintained on the summits of an equilateral triangle and provide a triple baseline interferometer. The configuration of the system is presented. Some of the problems concerning the optics, imaging efficiency and stabilization of the geometry are discussed.

035.095 **The OASIS correlator.**
P. D. Atherton, A. H. Greenaway, J. E. Noordam.
ESA Spec. Publ., ESA SP–226, p. 141 – 144 (1985). – See Abstr. 012.075.
The authors describe the principle of non–redundant recombination of a redundant array of telescopes, the design of the OASIS correlator and the advantages of pupil plane over image plane interference.

035.096 **Space antenna for gravitational wave astronomy.**
J. E. Faller, P. L. Bender, J. L. Hall, D. Hils, M. A. Vincent.
ESA Spec. Publ., ESA SP–226, p. 157 – 163 (1985). – See Abstr. 012.075.
The authors are investigating possible designs for a laser gravitational wave antenna in space using free test masses and heterodyne (interferometric) detection. One possibility is to use baselines about 10^6km long between three spacecraft in nearly circular one–year orbits about the sun. If the orbit elements are chosen properly, the distances between the spacecraft can be kept constant to roughly 1 part in 10^3 without orbit corrections. With milliwatt–transmitted laser power levels and 50 cm diameter optics, a strain sensitivity of $10^{19}/\sqrt{}$Hz over at least the period range from 10 to 10^4 seconds appears feasible. The primary goal of the measurements is to observe gravitational radiation associated with present or past interactions of super–massive objects. A number of binary sources can, however, also be studied.

035.097 **DUO: a far infrared heterodyne concept for space interferometry.**
J. Gay, Y. Rabbia.
ESA Spec. Publ., ESA SP–226, p. 171 – 176 (1985). – See Abstr. 012.075.
The authors propose some outlines of what could be a far infrared heterodyne concept for space interferometry, after a reminding of the advantages offered by this technics, the most interesting being the need for only two satellites instead of three.

035.098 **The Giotto experiments.**
R. Reinhard.
ESA Bull., No. 46, p. 41 – 51 (1986). – See Abstr. 003.052.
The scientific payload of Giotto was made up of ten experiments: a camera for imaging the comet nucleus, three mass spectrometers for analysing the elemental and isotopic composition of the cometary dust and gas environment, various dust impact detectors, a photopolarimeter for measuring the coma's brightness, and a set of plasma instruments for studying the solar–wind/comet interaction.

035.099 **A television sensor device for the international project "Venus–Halley".**
B. N. Formozov.
Photometric and polarimetric investigations of celestial bodies, p. 181 – 184 (1985). In Russian. – See Abstr. 003.035.
A television sensor to obtain the image of comet Halley is described. It includes the 576 × 512 pixels CCD–matrix and provides cooling the matrix down to the temperature 230 – 235K. All the parameters of the unit and its appearance are presented as well as possible ways of its using.

035.100 **Far–infrared (100 – 200 μm) photometry of H II regions with a 1 m balloon borne telescope.**
R. R. Daniel, S. K. Ghosh, K. V. K. Iyengar, T. N. Rengarajan, S. N. Tandon, R. P. Verma.
Light on dark matter, p. 305 – 306 (1986). – See Abstr. 012.080.
A 1 m balloon–borne telescope has been developed at TIFR for far–infrared astronomical observations. The telescope was successfully flown on March 2, 1985 from Hyderabad, India. During the flight RCW 57, RCW 108, RCW 122, W 31, G 351.6–1.3 and Carina nebula were observed. This paper describes the telescope, its performance and very preliminary results from this flight.

X–ray physics subpanel summary.
See Abstr. 013.029.

Coronal physics subpanel summary.
See Abstr. 013.030.

Light weight aluminum optics.
See Abstr. 031.009.

Electroform replication of grazing incidence X–ray optics.
See Abstr. 031.010.

Glass mirrors for space telescopes.
See Abstr. 031.028.

Mirror systems: engineering features, benefits, limitations and applications.
See Abstr. 031.032.

New opportunities from materials selection trade–offs for high precision space mirrors.
See Abstr. 031.043.

Glass mirrors for space telescopes.
See Abstr. 031.045.

Grazing incidence metal optics for the Berkeley Extreme Ultraviolet Explorer satellite: a progress report.
See Abstr. 031.048.

Adaptive telescope design using computer tools.
See Abstr. 031.049.

Focal plane optics in far–infrared and submillimeter astronomy.
See Abstr. 031.052.

Future space telescope design concepts.
See Abstr. 031.055.

TRIO optics.
See Abstr. 031.056.

Present status and future plans for the two color astrometric interferometer project.
See Abstr. 032.052.

Multichannel digital angle converter.
See Abstr. 034.018.

A multistep detector for X–ray astronomy.
See Abstr. 034.042.

Thermal detectors for X–ray spectroscopy.
See Abstr. 034.043.

A detailed description of proportional counter response below 300 eV.
See Abstr. 034.044.

Evaluation of dielectric film beamsplitters at cryogenic temperature.
See Abstr. 034.151.

Multipurpose electrical ground–support equipment for the Giotto Halley Multicolour Camera.
See Abstr. 036.008.

Stellar interferometry in space: fringe detection.
See Abstr. 036.202.

The PDS for Space Telescope guide star selection system: status and improvements.
See Abstr. 036.235.

Extragalactic observations with optical space arrays.
See Abstr. 036.245.

Astronomical criteria for interferometric arrays.
See Abstr. 036.247.

High angular resolution in the infrared: prospects for space observations.
See Abstr. 036.248.

HIPPARCOS satellite and the organization of the project.
See Abstr. 041.066.

Preparing for the future. The ESA technological research and development programme 1986 – 1988. Executive summary.
See Abstr. 051.010.

The Giotto project.
See Abstr. 051.011.

The Vega missions.
See Abstr. 051.012.

The Suisei/Sakigake (Planet–A/MS–T5) missions.
See Abstr. 051.013.

The Astro–1 mission.
See Abstr. 051.016.

The Giotto project.
See Abstr. 051.023.

The MIT X–ray spectroscopy investigation of AXAF.
See Abstr. 051.046.

Prospects for astrometry with the Hubble Space Telescope.
See Abstr. 051.051.

Earth–orbiting telescopes in the Cassini era.
See Abstr. 051.052.

The Cassini mission to Saturn and Titan. An overview.
See Abstr. 051.053.

Infrared spectroscopy aboard Cassini.
See Abstr. 051.056.

The X–Ray Timing Explorer.
See Abstr. 051.084.

An Australian broad–band satellite for X–ray astronomy.
See Abstr. 051.087.

The Astro Observatory guest observer program.
See Abstr. 051.094.

Die Ladungszustände solarer energetischer Teilchen.
See Abstr. 078.016.

Asteroids and the Space Telescope.
See Abstr. 098.020.

Analysis of imaging data from a balloon–borne coded aperture gamma–ray telescope.
See Abstr. 143.013.

^{26}Al experimental results.
See Abstr. 143.032.

The X–ray structure of the 3C 295 cluster: a cooling flow at a redshift of 0.5.
See Abstr. 160.035.

The spectrum of the cosmic background radiation from 2.3 cm^{-1} to 11.0 cm^{-1}.
See Abstr. 161.138.

036 Methods of Observation and Reduction, Data Processing

036.001 **Calibration of the Lick red CCD bandpasses.**
S. Djorgovski.
Publ. Astron. Soc. Pac., Vol. 97, No. 597, p. 1119 – 1126 (1985).

Red bandpasses indigenous to the Lick Observatory CCD imaging and photometry are defined and described. A set of primary calibration standards which define the magnitude zero points is presented. Color equations relating these bandpasses to the other, commonly used photometric systems are evaluated by numerical integration of stellar spectra, and compared to the real data. It is estimated that the present calibration is accurate and consistent to about 2 – 3% as a magnitude predictor, if the primary standards are used; this accuracy is only about 3 – 5% if *VR* photometry is used for calibration, and only about 5 – 8% if *BV* photometry is used. This is sufficient for most scientific applications of data taken in these bandpasses.

036.002 **Relativistic effects in earth based and cosmic long baseline interferometry.**
W. H. Cannon, D. Lisewski, A. M. Finkelstein (*A. M. Finkel'shtejn*), V. Ya Kreinovich (*V. Ya. Krejnovich*).
Relativity in celestial mechanics and astrometry, p. 255 – 268 (1986). – See Abstr. 012.002 (IAU Symp. No. 114).

This paper presents an analysis of relativistic effects on VLBI measurements with an accuracy of 0.0001 arc seconds.

036.003 **Relativistic effects in geodetic VLBI measurements.**
M. Soffel, H. Ruder, M. Schneider, J. Campbell, H. Schuh.
Relativity in celestial mechanics and astrometry, p. 277 – 282 (1986). – See Abstr. 012.002 (IAU Symp. No. 114).

The various relativistic effects occurring in VLBI measurements are discussed. A concrete example showing the influence of the gravitational delay due to the Sun upon the delay residuals from fit in a Mark III geodetic VLBI experiment is given. It is argued that regular geodetic VLBI observations might provide the best test of Einstein's theory of gravity on the post-Newtonian level in the near future.

036.004 **Initiation à la photographie astronomique.**
F. Clette.
Ciel Terre, Vol. 102, No. 1, p. 3 – 6 (1986).

036.005 **L'astrospectrographie à la portée de l'amateur grâce à l'utilisation d'un prisme–objectif.**
P. F. Janssens.
Ciel Terre, Vol. 102, No. 1, p. 11 – 14 (1986).

036.006 **On a possibility of correct extrapolation for a linear part of calibration curve in a stellar photographic photometry.**
S. B. Vladimirov.
Astron. Rep., Vol. 6, p. 5 – 18 (1981). In Russian.
This paper contains some theoretical considerations on a possibility of correct extrapolation for a linear part of calibration curve in a stellar photometry. A method for determination of limits of the linear part of a calibration curve is discussed together with an example for practical solution of this problem.

036.007 **Observer effects by inspection of galaxies on 125 cm Palomar Schmidt Plates.**
M. Burzyński, B. Cader, J. Papaj, E. Rachwalska, S. Ratajczyk, P. Rudawy, E. Szuszkiewicz, S. Szutowicz.
Astron. Rep., Vol. 6, p. 19 – 23 (1981).
Random errors of individual estimates of parameters of galaxies by visual inspection of plates are discussed.

036.008 **Multipurpose electrical ground–support equipment for the Giotto Halley Multicolour Camera.**
J. Turner, G. Schwarz.
ESA Bull., No. 45, p. 68 – 73 (1986).
This article describes the concept and design of the EGSE and its application in the various development and operational phases of the camera.

036.009 **Algorithmen zur automatischen Bildverarbeitung.**
R. Krätschmar.
Sterne Weltraum, 25. Jahrg., Nr. 2, p. 100 – 101 (1986).

036.010 **Ein persönliches Farbempfindlichkeits–Diagramm.**
R. Geckeler.
Sterne Weltraum, 25. Jahrg., Nr. 3, p. 155 – 156 (1986).

036.011 **Das Positionsfadenmikrometer in der Praxis.**
M. Möller.
Sterne Weltraum, 25. Jahrg., Nr. 3, p. 157 – 161 (1986).

036.012 **Noise and astrometric precision. I. Detection noises.**
J. Gay, J. M. Pochet.
Astron. Astrophys., Vol. 155, No. 2, p. 256 – 264 (1986). In French.
The influence of detection noise on the accuracy of stellar position measurements is studied. For this purpose the astrometric instrument can be reduced to an objective forming, in its focal plane, an image which, because of the diurnal motion, crosses the optical field in front of a grid. The radiation transmitted by the grid is collected by a light detector which delivers a signal whose temporal evolution permits the instant that the star passes through the centre of the optical field to be determined. The accuracy of this determination is limited by the noise affecting the recorded signal. In this first part the influence of additive noise resulting from photonic fluctuations of the sky background and the photon noise of the star itself are studied. It is shown that for each type of noise source there is an optimal numerical process which makes the best use of the recorded information. The optimisation of the focal grid for the photoelectric astrolabe, as well as for HIPPARCOS, can be studied directly from the expression proposed here.

036.013 **Noise and astrometric precision. II. Atmospheric effects.**
J. Gay, J. M. Pochet.
Astron. Astrophys., Vol. 155, No. 2, p. 265 – 277 (1986). In French.
The authors propose a study of atmospheric turbulence deterioration of the signal. The analytic difficulties are circumvented by the use of some approximations which do not affect the physical significance of the results. In the case of weak turbulence, the authors separate the effects of scintillation and those, notably more important, of agitation. The other extreme, very strong turbulence is also discussed. The theory developed explains the homogeneity of results noted by observers when this situation exists. There is no manageable analytic theory for the intermediate case. Finally, the authors deduce the criteria which should allow to make the best choice for the focal grid of an astrometric instrument.

036.014 **Intermediate mass stars undergoing a very hot phase: can we measure their temperatures?**
G. Stasińska, R. Tylenda.
Astron. Astrophys., Vol. 155, No. 1, p. 137 – 144 (1986).
The theory of stellar evolution predicts the existence of very hot phases for stars of intermediate masses. However, effective temperatures exceeding 2×10^5K have never been measured (if one leaves aside a few rather suspect results). After having analysed in detail all the methods which have been used to determine effective temperatures, the authors have come to the conclusion that none of them is adequate to reveal the presence of very hot stars. The advent of the Space Telescope should change the situation, at least for the case of very hot nuclei of planetary nebulae.

036.015 **Use of a prism on the correction plate of the Tautenburg Schmidt telescope for position determination.**
S. Hirte, E. Schilbach.
Astron. Nachr., Vol. 307, No. 1, p. 55 – 59 (1986). In German.
To overcome a possible magnitude equation a weakly refracting prism fitted on the correction plate of the Tautenburg Schmidt telescope was used, forming a secondary image of every object 3.8 mag weaker than the primary one. Because the distances between these two images are not constant, a polynomial was searched which could reflect these variation in dependence on the magnitude of a star and its position on the plate. For each of the 11 plates tested the authors have got another polynomial, therefore this way to tide over large magnitude differences is unefficient for astrometric investigations.

036.016 **Determination of the magnitude equation on Tautenburg Schmidt plates.**
E. Schilbach.
Astron. Nachr., Vol. 307, No. 1, p. 61 – 63 (1986). In German.
By measurement of 3 pairs of plates taken in the B system with the Tautenburg Schmidt telescope the coefficients d_x and d_y of the reduction model were derived for each pair in dependence on the magnitude of the stars. It was found that proper motions of stars with a magnitude difference more than 6 – 7 mag with respect to the reference points can be affected by systematic errors larger than the errors of measurement. These errors are due to the magnitude equation and vary from one plate pair to another.

036.017 **Machine measurements of the objective–prism spectra of faint galaxies – I. Plate measurements and initial data reduction.**
J. A. Cooke, S. M. Beard, D. Emerson, B. D. Kelly, H. T. MacGillivray.
Mon. Not. R. Astron. Soc., Vol. 219, No. 2, p. 241 – 249 (1986).
See also 160.019.
The use of the COSMOS measuring machine and related software to extract the spectra of objects on UK Schmidt Telescope objective–prism plates is described. The spectra produced are suitable for use in a wide range of research programmes such as the determination of the redshifts of faint galaxies, the automated detection of quasars and the classification of stars by spectral type.

036.018 **Biases of digital VLBI observables due to small signal approximation and imperfect fringe rotation.**
K. M. Borkowski.
Astron. Astrophys., Vol. 157, No. 1, p. 91 – 95 (1986).
The analysis presented in this paper shows that the usual small signal approximation in VLBI is acceptable in the majority of applications. However, it leads to nonlinear systematic overestimation of the correlation coefficient, which in some cases cannot be considered negligible. Formulae suitable for accurate conversion of measured amplitudes into the correlation coefficient for a few useful fringe rotation schemes are given. The analysis also brings evidence for a small bias in the phase measurements made with present day VLBI correlators, which is entirely due to imperfect fringe rotation.

036.019 **Visibility and phase analysis for image and pupil plane interferometry at optical wavelengths.**
A. Chelli, J. M. Mariotti.
Astron. Astrophys., Vol. 157, No. 2, p. 372 – 382 (1986).
Comparing the radio domain arrays and the future optical (visible/infrared) ones, it has often been argued that the latter are only able to measure the modulus of the source's spatial spectrum, and therefore are poorly suited for imaging. This paper proposes a method for reconstruction of the phase of the spectrum, by continuity in the spatial frequency space. The basic idea is that the size of the support of the transfer function enables, through techniques similar to those developed for speckle interferometry, to reconstruct the phase inside a circle of radius D/λ in the u–v plane (D being the diameter of each telescope). The authors examine the situation for interference in the image plane or in the pupil plane. The latter is favourable, as already known, for visibility determination. But, the former is better for phase reconstruction according to the method described in this paper.

036.020 **A proposed method for the determination of the bivariate luminosity function utilizing an incomplete sample with an application to Seyfert galaxies.**
M. A. Arakelian (*M. A. Arakelyan*).
Astrophys. J., Vol. 301, No. 1, p. 92 – 97 (1986).
The radio and X–ray surveys of cataloged objects are made with fixed limiting flux densities; however, the distribution of optical magnitudes in these listings is often arbitrary. Thus the subsamples with detected radio or X–ray energy are not complete because the original optically surveyed samples are incomplete. The present paper shows that the results of such surveys can be used for the determination of the bivariate luminosity functions, provided that both the distribution of luminosities of objects in the detected subsample, as well as the distribution of apparent magnitudes in the surveyed sample are taken into account. The proposed method is illustrated in the determination of the bivariate optical and radio luminosity function of Seyfert galaxies.

036.021 **Transformation equations and other aids for *VRI* photometry.**
B. J. Taylor.
Astrophys. J., Suppl. Ser., Vol. 60, No. 2, p. 577 – 599 (1986).
This paper describes two new devices for transforming among *VRI* systems, and presents a large number of newly calculated transformations for these systems. Values of $E(V–R)/E(B–V)$ and $E(R–I)/E(B–V)$ are also presented for the Cousins system. Extensive tables of standard–star and other data obtained from the new transformations are given. The use of some of these data for Johnson photometry and techniques for reproducing the Johnson system correctly are discussed.

036.022 **On the determination of the corrections to fundamental star catalogues using observations of artificial earth satellites with inaccurate moments of time.**
R. A. Zejnalov.
Astron. Zh., Tom 63, Vyp. 1, p. 191 – 196 (1986). In Russian. English translation in Sov. Astron., Vol. 30, No. 1.
A possibility of the determination of zero points and systematic errors of star catalogues from optical observations of artificial earth satellites with inaccurate moments of time in the case when these errors are presented by expansions on spherical functions are investigated. The results of model calculations and their estimates are given.

036.023 **Comparative performances of an objective grating concept in UV spectroscopic observations of solar system objects. II. D/H ratio observations in cometary atmospheres.**
P. Bruston.
Astron. Astrophys., Vol. 158, No. 1/2, p. 97 – 101 (1986).
The determination of the D/H ratio in comets is an exciting aim all the more because no evaluation or significant upper limit has been established up to now. The authors discuss the comparative performances of future UV instrumentation to perform this objective. An objective grating instrument has been demonstrated in a previous paper. It is shown that for being able to separate the contribution of the D Lyman α emission to the image obtained on the detector, either a sharp variation in the hydrogen distribution or a clear spatial differentiation between the H and D emission patterns is needed. If achieved, a careful analysis of the data will allow to study the D/H emission ratio and the comparative spatial distributions which should lead to the determination of the D/H abundance ratio.

036.024 **Narrow–band photometry and the evolution of galaxies.**
N. Fiala, K. D. Rakos, A. Stockton.
Publ. Astron. Soc. Pac., Vol. 98, No. 599, p. 70 – 80 (1986). With a correction in Vol. 98, No. 600, p. 268 (1986).
A new approach to the use of narrow–band filters in extragalactic research is described. The photometric system *vz*, *bz*, *yz* essentially imitates, at the redshift of the cluster of galaxies in question, the Strömgren *v*, *b*, *y* passbands. This system has some obvious advantages: the uncertainties associated with *k*–corretions are bypassed, and the passbands avoid the stronger emission lines appearing in the spectra of some normal and most active galaxies. It is necessary to acquire filters for each redshift value and to have accurate spectrophotometry for at least one galaxy at the redshift in order to calibrate the photometry. It is shown that the *vz*, *bz*, *yz* filter system is especially suitable for testing the evolutionary synthesis models. For that purpose 48 CCD frames of three clusters of galaxies (Abell 665, the small cluster around 3C 234, and Cl 1021.1+0427) were obtained. For galaxies, metric magnitudes were derived corresponding to a linear physical standard diameter of 32 kpc (H_0 = 60 km s^{-1}Mpc^{-1}, q_0 = 0.5). Galaxies showing a strong color gradient along their images (disk galaxies) are distinguished from galaxies with no, or very small, color gradient along their images (elliptical and lenticulars).

036.025 **Comparative performances of an objective grating concept in UV spectroscopic observations of solar system objects. III. Observations of diffuse emission lines.**
P. Bruston, C. Emerich.
Astron. Astrophys., Vol. 159, No. 1/2, p. 101 – 107 (1986).
The observational capabilities of an objective grating instrument are fully described – when extended to the particular case of observations of diffuse emission lines – and compared with the usual techniques. It is shown that higher spectral resolutions can be obtained in particular conditions, using occultation by the moon for example. This original method is suggested to be applied to measure both the H Lyα 1216 Å and the He I 584 Å line profiles emitted through resonance scattering of solar photons, by the local interstellar medium.

036.026 **On the accuracy of photographic observations of the moon on the star background.**
R. L. Semerenko.
Kinematika Fiz. Nebesn. Tel, Tom 2, No. 1, p. 55 – 59 (1986). In Russian. English translation in Kinematics Phys. Celest. Bodies.
The errors of the method of snap exposures (effect of atmospheric turbulence, guiding of the telescope and non–uniformity of the images of different lunar zones) have been considered. These errors do not exceed the measurement accuracy (0″.2). It has been determined the optimum number of plates for one observational night and the maximum acceptable value of the exposure in the method. Comparison of SAO and AGK3 catalogues

shows that the main error of the method is due to the reference catalogue.

036.027 **On the allowance for an excess of error distribution when comparing accuracy of different series of astrometric observations.**
L. V. Dzhun'.
Kinematika Fiz. Nebesn. Tel, Tom 2, No. 1, p. 88 – 94 (1986). In Russian. English translation in Kinematics Phys. Celest. Bodies.
The paper deals with the distribution of observational errors assuming that the excess ε_i is not equal to zero.

036.028 **Elemente–Ableitungsmöglichkeiten am Beispiel EZ Lyr.**
W. Braune.
BAV Rundbrief, 35. Jahrg., Nr. 1, p. 14 – 19 (1986).

036.029 **Observing reappearances.**
D. Büttner.
Occultation Newsl., Vol. 3, No. 15, p. 329 – 330 (1986).

036.030 **Etude du spectre solaire à l'aide d'un réseau utilisé en incidence rasante.**
M. Chapelet.
Astronomie, Vol. 100, p. 23 – 32 (1986).

036.031 **System of KONAB programmes for organization of a satellite observation bank.**
N. V. Emel'yanov.
Nauchn. Inf., Vyp. 53, p. 3 – 54 (1982). In Russian.
The KONAB system is described. This system is designed to operate on results of satellite observations by a computer for the organization of a data bank. Classifying, regularization, selection and several forms of printing are possible. Transition between some data forms is also possible.

036.032 **Input and output of satellite observations in foreign formats when handling the KONAB system.**
O. M. Bulygina, R. V. Gabbasov, N. V. Emel'yanov.
Nauchn. Inf., Vyp. 53, p. 55 – 67 (1982). In Russian.
A possibility of operations on satellite observations recorded in different formats using the KONAB system is described. SAO format, its description and listing of necessary programmes are given as an example.

036.033 **A catalogue of results of automatic reduction of photographic satellite tracking data.**
N. D. Kostyuk.
Nauchn. Inf., Vyp. 53, p. 68 – 71 (1982). In Russian.
A catalogue of results of automatic reduction of photographic satellite tracking data obtained with the computer EC–1033 is described. The records are made in Standard Universal Format.

036.034 **The special network for telemetry reception and interferometric measurements of the Vega balloon experiment.**
N. A. Armand, A. F. Bogomolov, V. I. Shevchenko, V. V. Kerzhanovich, Yu. N. Aleksandrov, V. I. Altunin, R. A. Andreev, A. S. Vyshlov, Yu. N. Gorshenkov, V. M. Dubrovin, A. L. Zajtsev, O. G. Zoteev, S. P. Ignatov, I. A. Knorin, A. N. Kozlov, B. Z. Kanevskij, V. I. Kostenko, K. K. Lubny–Gerzyk, L. I. Matveenko, E. P. Molotov, S. V. Pogrebenko, V. I. Rogal'skij, V. A. Rudakov, I. A. Strukov, B. P. Trusov, G. V. Turusin.
Pis'ma Astron. Zh., Tom 12, No. 2, p. 131 – 136 (1986). In Russian. English translation in Sov. Astron. Lett., Vol. 12.
An special receiving system was developed to optimize signal–to–noise ratio for the Vega balloon experiment.

036.035 **Interferometric measurements and telemetry acquisition in the special network of the Vega balloon experiment.**
Yu. N. Aleksandrov, N. A. Armand, A. A. Krymov, R. A. Andreev, V. V. Kerzhanovich, S. Yu. Sila–Novitskij, Ya. D. Khatskelevich, V. I. Efremov, V. A. Prasolov, F. I. Ryukin.
Pis'ma Astron. Zh., Tom 12, No. 2, p. 137 – 142 (1986). In Russian. English translation in Sov. Astron. Lett., Vol. 12.
Optimal computer algorithms of digital signal processing of telemetry acquisition and interferometric measurements in the Vega balloon experiment are described.

036.036 **Interferometric processing of the data from the Vega balloon probe and flyby station.**
L. R. Kogan, E. N. Fedoseev.
Pis'ma Astron. Zh., Tom 12, No. 2, p. 143 – 148 (1986). In Russian. English translation in Sov. Astron. Lett., Vol. 12.
Processing of the data received by the interferometric network from the balloon probe and flyby station of the Vega mission is described. Requirements to the accuracy of the baseline and to the clock synchronization accuracy are formulated.

036.037 **Device for measurements of solar brightness variations based on an integral photodetector matrix.**
L. V. Didkovskij.
Izv. Krymskoj Astrofiz. Obs., Tom 72, p. 217 – 225 (1985). In Russian. English translation in Bull. Crimean Astrophys. Obs., Vol. 72.
An instrument for observations of the brightness variations of the sun was developed on the basis of the integral array detector matrix. Four differential (center–to–limb) brightness signals corresponding to limb zones N, S, E, W as well as the intensity in the center of the solar disk are registered with time resolution 24 ms. Results of the laboratory tests of the instrument are described. The observations performed at the Crimean solar tower telescope using an effective wavelength of 0.85 μm showed that the daily trend of the differential signal is about 2×10^{-3} in units of the center disk intensity.

036.038 **A system for scanning and preliminary processing of spectra with the micro–computer ISKRA–1256.**
V. P. Malanushenko, A. G. Shcherbakov, V. I. Shuteev.
Izv. Krymskoj Astrofiz. Obs., Tom 72, p. 226 – 231 (1985). In Russian. English translation in Bull. Crimean Astrophys. Obs., Vol. 72.
An automatic system using the microphotometer GII (DDR) and the micro–computer «ISKRA–1256» has been developed at the Crimean Astrophysical Observatory to convert densities into intensities and residual intensities, noise clearing and cross–correlation of numbers of spectra. The operational performances of «ISKRA–1256» and the PDS microdensitometer (USA) are compared.

036.039 **Determination of the light and energy distribution of a star image when diffraction and aberrations are taken into account.**
G. M. Popov, M. B. Popova.
Izv. Krymskoj Astrofiz. Obs., Tom 73, p. 165 – 175 (1985). In Russian. English translation in Bull. Crimean Astrophys. Obs., Vol. 73.
The principles of an algorithm for calculation of the light and energy distribution in the image of a star are described.

036.040 **Determination of the radiation field in the chromosphere using the method of observed line profile inversion.**
S. A. Grigor'eva, R. B. Teplitskaya.
Kinematika Fiz. Nebesn. Tel, Tom 2, No. 2, p. 67 – 74 (1986). In Russian. English translation in Kinematics Phys. Celest. Bodies.
The inversion method is modified using the local second–degree B–splines, cubic fundamental splines and polynomials. The technique is generalized for the case of an arbitrary non–thermal velocity field in the solar atmosphere.

036.041 **Isolation of the mean latitude using Orlov and Sakharov filters.**
B. S. Dubik.
Kinematika Fiz. Nebesn. Tel, Tom 2, No. 2, p. 85 – 86 (1986). In Russian. English translation in Kinematics Phys. Celest. Bodies.

Corrected filters for isolation the mean latitude are given. The filters exclude the Chandler component having the velocity 30°16′.

036.042 **A high–precision determination of angular coordinates of geostationary artificial satellites with a wide–angle astrograph with special equipment.**
D. P. Duma, Yu. N. Ivashchenko, Yu. A. Shokin.
Kinematika Fiz. Nebesn. Tel, Tom 2, No. 2, p. 86 – 88 (1986). In Russian. English translation in Kinematics Phys. Celest. Bodies.
Using 8 plates with 32 images of stationary artificial satellites obtained on a wide–angle astrograph with special equipment 14 angular coordinates of satellites are calculated with accuracy of 0″.3 – 0″.6.

036.043 **The potential precision of speckle interferometric measurements. Stellar angular diameters and limb darkening.**
S. G. Kuz'menkov.
Astron. Zh., Tom 63, Vyp. 2, p. 389 – 398 (1986). In Russian. English translation in Sov. Astron., Vol. 30, No. 2.
The potentialities are studied of a joint estimation of several unknown parameters of an arbitrary object from a speckle image power spectrum. A typical problem of joint estimation of the stellar angular diameter and the limb–darkening coefficient for the linear limb–darkening law is considered. The optimum strategy for the solution of the stated problem is suggested, accounting for invariance of the mean number of detected photons per speckle for different models describing the brightness distribution over the stellar disc. The results obtained allow to plan an experiment for the speckle interferometric measurements of stellar angular diameters and limb darkening.

036.044 **On the optimum nonlinear filtering of radio astronomical observation data.**
A. G. Gorshkov, I. K. Rozgacheva.
Astron. Zh., Tom 63, Vyp. 2, p. 399 – 401 (1986). In Russian. English translation in Sov. Astron., Vol. 30, No. 2.
A method is suggested for reduction of radio astronomical observations which allows to obtain an optimum estimate for the sources' flux density distribution function. To describe the random signal statistics, the Kolmogorov–Feller equation is used.

036.045 **Performance tests of DAOPHOT/INVENTORY photometry programmes in dense stellar fields.**
S. Ortolani.
Messenger, No. 43, p. 23 – 25 (1986).

036.046 **On the MIDAS reduction package for CASPEC spectra.**
D. Ponz, E. Brinks.
Messenger, No. 43, p. 31 – 34 (1986).

036.047 **CCD scanning.**
T. Gehrels.
Asteroids, comets, meteors II, p. 19 – 20 (1986). – See Abstr. 012.010.
The sky is being scanned for comets and asteroids with a charge–coupled device (CCD) on a Newtonian 0.9–m reflector, the "Spacewatch Camera." A new method in astrometry is described which scans across the images of comets and asteroids and of SAO standards as well. It has some of the advantages of a transit technique and may be of interest to other observatories.

036.048 **A review of some digital image processing in cometary research.**
S. M. Larson.
Asteroids, comets, meteors II, p. 449 – 459 (1986). – See Abstr. 012.010.
The development of electronic digitizers, digital detector arrays and modern high speed computer processing has led to more efficient, quantitative methods of studying the spatial, temporal and photometric properties of cometary phenomena. Digital image processing techniques are being used and further developed to reduce two dimensional data, to enhance the visibility of cometary features, and to quantify spatial and temporal changes. Some of these methods are reviewed, and their merits and limitations are discussed.

036.049 **On the possibility of observing reflection phenomena in meteor trails.**
Y. Öhman.
Asteroids, comets, meteors II, p. 599 – 602 (1986). – See Abstr. 012.010.
The possibility is considered, that effects due to selective reflection at grazing incidence may be observed in meteor trails. Dark structures may be observed, if a meteor happens to cross the solar disk, and particularly near the very limb and if the observations are made by means of solar filtergrams in some of the strong meteor spectrum lines. A similar reflection effect, where light is lost, may appear at the border of the meteor trail itself. Because of the anomalous dispersion near strong spectrum lines and the condition for total reflection, a loss of light in the red wing of such spectrum lines should be expected at the very border of the meteor trail. As a consequence high dispersion spectra of meteor trails should be expected to show a small displacement towards the violet of some of the strong meteor spectrum lines.

036.050 **Meteor photography limitations and the beginning heights of meteors.**
R. Malmström.
Asteroids, comets, meteors II, p. 609 – 614 (1986). – See Abstr. 012.010.
The observed discrete levels appearing in meteor beginning–height/velocity diagrams are shown to be photographic system (film and optics) generated patterns. A photographic–system–independent way to separate different meteor materials/structures is presented.

036.051 **Evolution of the luminosity function of extragalactic objects.**
V. Petrosian.
Structure and evolution of active galactic nuclei, p. 353 – 381 (1986). – See Abstr. 012.011.
The author describes the steps and procedure required for determination of the evolution of the luminosity function of extragalactic objects and applies the results to quasars and, in particular, addresses the question of the evolution at redshifts higher than three. There are four distinct steps involved in this procedure. These are: selection of complete samples with known observational selection effects. Choice of a cosmological model. Statistical analysis of the sample and determination of the evolution of the luminosity function. Use of the luminosity function for determination of the formation rate and the physical evolution of the objects.

036.052 **On the construction of digital filters for processing of latitude observational data.**
B. S. Dubik.
Red. Zh., Kinematika Fiz. Nebesn. Tel, 8 pp. (1985). In Russian. Abstr. in Ref. Zh., 51. Astron., 1.51.124 (1986).

036.053 **On optimum arrangement of measurements of radial accelerations of an artificial lunar satellite for determination of the zonal harmonics of the potential.**
N. G. Babaeva.
Ivz. Akad. Nauk AzSSR. Ser. fiz.–tekh. i mat. nauk, Tom 5, No. 6, p. 142 – 145 (1984). In Russian. Abstr. in Ref. Zh., 52. Geod. Aehrosemka, 1.52.198 (1986).

036.054 **Reduction of nonrectilinear stellar spectra.**
I. I. Nazarenko, V. S. Shergin.
Bull. Spec. Astrophys. Obs. – North Caucasus, Vol. 19, p. 108 – 114 (1985). English translation of 39.036.070.

036.055 **Taking into account the contribution of scattered atmospheric radiation in photoelectric measurements of stellar brightness.**
Yu. N. Kozulin, V. G. Sekrieru, V. A. Chernobaj.
Izv. Akad. Nauk MSSR. Ser. fiz.-tekh. i mat. nauk, No. 2, p. 54 – 56 (1985). In Russian. Abstr. in Ref. Zh., 51. Astron., 2.51.1069 (1986).

036.056 **Results of an investigation of the symmetry of micrometric measurements from observations of zenith stars at the ZTL–180 zenith telescope in Poltava.**
A. V. Gozhij, V. M. Gorban', L. S. Otkidach.
Prilivy i vrashchenie Zemli, p. 100 – 104 (1985). In Russian. Abstr. in Ref. Zh., 51. Astron., 2.51.1080 (1986).

036.057 **A method for interpolation of the angles of observation of artificial earth satellites.**
V. V. Lavrov.
Kosm. Issled., Tom 24, Vyp. 1, p. 138 – 139 (1986). In Russian. English translation in Cosm. Res.

036.058 **Possibilities of the scintillation method for investigation of the velocity of circumsolar plasma.**
D. Blums, N. Lotova.
Issled. Solntsa Krasnykh Zvezd, No. 21, p. 5 – 16 (1984). In Russian.
The possibilities of solar wind velocity determination by the scintillation method are discussed. It is shown that one-site Fourier or Bessel temporal scintillation spectra can be used to derive the solar wind velocity close to the sun.

036.059 **On a possibility to detect neutral points of a coronal magnetic field.**
B. I. Ryabov.
Issled. Solntsa Krasnykh Zvezd, No. 22, p. 5 – 15 (1985). In Russian.
One of the available radio methods for diagnostics of peculiarities in coronal magnetic structures is described. The method based on an expected polarization effect would permit to localize a neutral point of the coronal magnetic field and to evaluate the electron density in its vicinity.

036.060 **Visual spectroscopy.**
R. J. Livesey.
J. Br. Astron. Assoc., Vol. 96, No. 2, p. 84 – 87 (1986).

036.061 **A graphics tablet and its use in solar disc reduction.**
L. M. Dougherty, D. Fickling, A. W. Perry.
J. Br. Astron. Assoc., Vol. 96, No. 2, p. 88 – 94 (1986).

036.062 **Image restoration and superresolution in the far infrared.**
D. F. Lester, P. M. Harvey, M. Joy.
Bull. Am. Astron. Soc., Vol. 17, No. 4, p. 863 (1985). Abstract. – See Abstr. 010.062.

036.063 **A new abundance calibration for low resolution spectra.**
K. Ratnatunga.
Bull. Am. Astron. Soc., Vol. 17, No. 4, p. 877 – 878 (1985). Abstract. – See Abstr. 010.062.

036.064 **A new hybrid mapping method for VLBI data.**
R. P. Linfield.
Bull. Am. Astron. Soc., Vol. 17, No. 4, p. 901 (1985). Abstract. – See Abstr. 010.062.

036.065 **Inverting remote sensing data by simulated annealing.**
W. Jeffrey, R. Rosner.
Bull. Am. Astron. Soc., Vol. 17, No. 4, p. 901 (1985). Abstract. – See Abstr. 010.062.

036.066 **Problems in data compilation.**
H. Eichhorn, C. S. Cole.
Celest. Mech., Vol. 37, No. 3, p. 263 – 275 (1985). – See Abstr. 012.017.
Concepts involved in the estimation of target quantities and other adjustment parameters are critically discussed. The authors introduce a measure for the efficiency of a set of adjustment parameters such that adjustments carried out using different precepts can be objectively compared. Finally, having applied their suggestions to a specific problem, the authors show that they have obtained estimates of a set of target quantities (in their case, star positions and proper motions) which have smaller formal errors than estimates following traditional procedures.

036.067 **Dispersive X-ray spectroscopy with transmission gratings.**
F. Paerels, J. A. M. Bleeker, A. C. Brinkman, J. H. Dijkstra, R. Mewe, C. J. Schrijver.
ESA Spec. Publ., ESA SP-239, p. 205 – 208 (1985). – See Abstr. 012.019.
To illustrate the potential of a Transmission Grating Spectrometer on future X-ray observatories such as AXAF and XMM the authors present some simulated X-ray spectra of both galactic and extragalactic sources, in comparison with spectra obtained with Einstein and EXOSAT.

036.068 **A prescription for period analysis of unevenly sampled time series.**
J. H. Horne, S. L. Baliunas.
Astrophys. J., Vol. 302, No. 2, p. 757 – 763 (1986).
The authors present a technique for detecting the presence and significance of a period in unequally sampled time series data. The authors review the calculation of the modified periodogram for unevenly sampled data and clarify the proper definition of the variance that is used to normalize the power of the modified periodogram. It is demonstrated that the probability that a peak in the periodogram is noise or signal can be easily assessed by the method given here only when the *total* variance of the data is used to normalize the periodogram power.

036.069 **Application of the Fourier transform for solving and improving spectroscopic binary star orbits.**
A. Abad, A. Elipe.
Astrophys. J., Vol. 302, No. 2, p. 764 – 766 (1986).
A new method of improving a preliminary orbit is found by means of the corrections of the Fourier coefficients. An application of the method is made for the systems HD 149240 and 105 Herculis.

036.070 **Possibilities and restrictions of software of modern computers in regression analysis.**
S. G. Valeev.
Izv. Astron. Ehngel'gardt. Obs., No. 51, p. 3 – 16 (1985). In Russian.

036.071 **Estimate of the accuracy of calibration of the "Relict" radiometer from the radio radiation of the moon.**
L. S. Gurin, V. D. Maslov, I. A. Petryakova, D. P. Skulachev.
Inst. kosm. issled. Akad. Nauk SSSR, Prepr. No. 1019, 14 pp. (1985). In Russian. Abstr. in Ref. Zh., 62. Issled. Kosm. Prostranstva, 3.62.90 (1986).

036.072 **A new method for fast median filtering.**
P. P. Korsun.
Kinematika Fiz. Nebesn. Tel, Tom 2, No. 3, p. 81 – 84 (1986). In Russian. English translation in Kinematics Phys. Celest. Bodies.
A new method for a real-time median filtering is suggested. The saving of computer time amounts to ~ 32 for a filter with aperture of 81 elements.

036.073 **On convergence of measurements in laser observations of artificial earth satellites.**
G. S. Kurbasova, L. V. Rykhlova, G. N. Shlikar'.
Kinematika Fiz. Nebesn. Tel, Tom 2, No. 3, p. 85 – 90 (1986). In Russian. English translation in Kinematics Phys. Celest. Bodies.

Two approaches for the improvement of convergence in the least–squares method when applied to laser observations of artificial earth satellites are discussed. The first approach is the truncation of outstanding measurements and subsequent implication of usual criteria and routines. The second approach is the strong routine of the least–squares method. For the improvement of convergence in the strong routine the method for determination of the initial parameters by the spline approximation is suggested.

036.074 **Statistical methods of stellar polarimetry.**
D. Clarke, B. G. Stewart.
Vistas Astron., Vol. 29, Part 1, p. 27 – 51 (1986).

In principle, optical polarimetry can provide direct information about the geometries of stellar envelopes and binary systems, about stellar magnetic fields, about the properties of scattering particles (electrons, dust) within the stellar environments and about the dust in the interstellar medium. In this paper the statistics associated with polarization measurements will be presented in a deeper and more unified way than has been done previously, some of the material being presented for the first time.

036.075 **Method of observation of meteor streams on adjustable antennae following the daily motion of the radiant.**
A. E. Epishova, Z. M. Ioffe, L. N. Rubtsov.
Komety Meteory, No. 36, p. 3 – 7 (1984). In Russian.

A method of meteor swarms observation on an adjustable antenna with narrow diagram of direction is considered. This method is based on the calculation of the angle of orientation of the maximum line of radiation depending on the place of the radiant swarms on the celestial sphere.

036.076 **On methods of radar measurements of the drift of meteor trains.**
V. M. Kolmakov.
Komety Meteory, No. 37, p. 34 – 39 (1985). In Russian.

Some peculiarities of simple statistical methods of measurements of meteor trains drift without determination of coordinates are discussed.

036.077 **Use of microprocessor techniques for control of astronomical instruments.**
A. F. Pereverzentsev, M. G. Arkhipov, V. N. Ivanov.
Problems of astronomy, p. 93 – 95 (1985). In Russian. Abstr. in Ref. Zh., 51. Astron., 4.51.931 (1986). – See Abstr. 003.023.

036.078 **On a project for automation of observations with the meridian circle of the Astronomical Observatory of the Odessa University.**
I. I. Genovskij.
Problems of astronomy, p. 80 – 87 (1985). In Russian. Abstr. in Ref. Zh., 51. Astron., 4.51.934 (1986). – See Abstr. 003.023.

036.079 **Programs for observations of asteroids.**
A. V. Kuz'min, A. V. Yakimenko.
Chisl. anal.: metody, algoritmy, pril. Moskva, p. 19 – 23 (1985). In Russian. Abstr. in Ref. Zh., 51. Astron., 4.51.984 (1986).

036.080 **Automation of investigations of extended radio sources with the radio telescope of the Institute of Physics of the Academy of Sciences.**
A. V. Kovalenko, S. M. Kutuzov, V. A. Udal'tsov.
Fiz. inst. Akad. Nauk SSSR. Prepr., No. 312, 20 pp. (1985). In Russian. Abstr. in Ref. Zh., 51. Astron., 4.51.990 (1986).

036.081 **Obstacles facing the Venus radar mapper – the implications of gestalt formation in stereo–radargrammetry.**
R. L. Wildey.
Earth, Moon, Planets, Vol. 35, No. 1, p. 47 – 54 (1986).

The question of adapting to radar images the existing hardware that form topographic maps through stereo–photogrammetric models, is examined in principle.

036.082 **A new photographic method for mapping lunar eclipse shadow.**
T. Nakamura, T. Hirayama, M. Noguchi.
Earth, Moon, Planets, Vol. 35, No. 1, p. 55 – 71 (1986).

A new photographic method suitable for digital processing of a two–dimensional density distribution of lunar eclipse shadow has been developed and applied to the observation of the eclipse of December 30, 1982.

036.083 **Coments on techniques for spectral deconvolution.**
P. H. Scherrer.
Seismology of the Sun and the distant stars, p. 117 – 120 (1986). – See Abstr. 012.026.

Current observational questions in asteroseismology require high spectral resolution that can only be obtained with observations spanning many days or months. The primary constraint in the full utilization of single mid–latitude observing sites is the presence of diurnal data gaps. Several methods for removing the effect of these gaps in the spectra obtained from velocity observations have been suggested. The limitations of these methods and their applicability to the helioseismology problem is discussed.

036.084 **Absolute astronomical accelerometry.**
P. Connes.
Seismology of the Sun and the distant stars, p. 403 – 404 (1986). Abstract. – See Abstr. 012.026.

036.085 **The distance determination of celestial objects by new technologies.**
H. Hirabayashi.
Proceedings of the Symposium on Application of Space Techniques to Astronomy and Geophysics, p. 17 – 26 (1986). – See Abstr. 012.027.

Recent techniques such as VLBI make it possible to determine the distance to the celestial objects more directly than the conventional method. Such topics and future possibilities are presented.

036.086 **System of photon counting.**
V. L. Afanas'ev, A. V. Zhuravkov, A. A. Pimonov.
Astrofiz. Issled. Izv. Spets. Astrofiz. Obs., Tom 21, p. 16 – 20 (1986). In Russian. English translation in Bull. Spec. Astrophys. Obs. – North Caucasus.

Some ideas and methods of investigation of statistical properties of the registered photon flux are presented. The system of registering the light outburst of an image tube based on the photon counting method is suggested. The results of experimental investigations of this system are given.

036.087 **A new method of long–term tracking of sources by the radio telescope RATAN–600.**
N. L. Kajdanovskij.
Astrofiz. Issled. Izv. Spets. Astrofiz. Obs., Tom 21, p. 73 – 74 (1986). In Russian. English translation in Bull. Spec. Astrophys. Obs. – North Caucasus.

036.088 **«Estafette with zoning» method.**
O. A. Golubchina.
Astrofiz. Issled. Izv. Spets. Astrofiz. Obs., Tom 21, p. 75 – 84 (1986). In Russian. English translation in Bull. Spec. Astrophys. Obs. – North Caucasus.

The paper is devoted to the elobaration of the new regime of observations at the radio telescope RATAN–600, to the method «estafette with zoning». Geometrical areas of the working part of the antenna, angular resolutions in horizontal direction, possible frequency bands of radiometers and the sensitivity variations of the system radio telescope–radiometer at the transition from the usual regime of observations with the RATAN–600 to the observations by «estafette with zoning» method are estimated.

036.089 **Experimental observations with the radio telescope RATAN–600 using the method «estafette with zoning».**
O. A. Golubchina.
Astrofiz. Issled. Izv. Spets. Astrofiz. Obs., Tom 21, p. 85 – 90 (1986). In Russian. English translation in Bull. Spec. Astrophys. Obs. – North Caucasus.
The results of the first experimental observations of W49, Cygnus A and the sun with the radio telescope RATAN–600 by the method of «estafette with zoning» are discussed. A comparison of theoretically expectable antenna temperatures of the radio emission sources with those obtained from the observations using the method «estafette with zoning» is presented. It is shown that theoretically expectable values of signals from the observable radio emission sources are in agreement with the results obtained after the experiment.

036.090 **Azimuthal aperture synthesis of the sun with the radio telescope RATAN–600 in the regime "South with periscope".**
B. S. Minchenko.
Astrofiz. Issled. Izv. Spets. Astrofiz. Obs., Tom 21, p. 91 – 97 (1986). In Russian. English translation in Bull. Spec. Astrophys. Obs. – North Caucasus.

036.091 **Investigation of the possibility of using acousto–optical correlators for radio interferometry.**
N. A. Esepkina, A. P. Lavrov, S. V. Pruss–Zhukovskij, P. A. Fridman, M. N. Anan'ev.
Astrofiz. Issled. Izv. Spets. Astrofiz. Obs., Tom 21, p. 105 – 112 (1986). In Russian. English translation in Bull. Spec. Astrophys. Obs. – North Caucasus.
The possibility of using the optical reduction for calculation of the reciprocal correlation function of signals obtained with the long–base–line radio interferometer is discussed. The optical correlator consists of a laser diode, an acousto–optical modulator and a photoreceiver made on the basis of the device with charge communication. For modulation of the diode emission and of the acousto–optical modulator carrier the noise signals limited by the zero level are used. A desciption of the experimental device is presented.

036.092 **The correlations between VLBI observables and its effects for the determinations of ERP.**
Z.–h. Qian.
Earth rotation and the terrestrial reference frame, p. 360 – 365 (1985). – See Abstr. 012.077.

036.093 **Presentation of time delay and fringe frequency in coordinate–independent form through measureable values.**
S. V. Reznikov.
Astrofiz. Issled. Izv. Spets. Astrofiz. Obs., Tom 21, p. 123 – 129 (1986). In Russian. English translation in Bull. Spec. Astrophys. Obs. – North Caucasus.
The relativistic coordinate independent expressions for time delay and fringe frequency through measurable values are obtained. The relativistic expression for the rotating base of an interferometer is obtained. The role of geodetic precession of the interferometer base which is the consequence of transformation from the geocentric reference system to the heliocentric one for the procedure of invariant relativistic reductions is studied.

036.094 **Chromatic position difference: a technique for studying double stars.**
L. Yu. Sorokin, A. A. Tokovinin.
Sov. Astron. Lett., Vol. 11, No. 4, p. 226 – 232 (1985). English translation of 40.036.008.

036.095 **Dual–processor apparatus–program complex – scanner of BTA. II. SIPRAO – specialized language of programming.**
N. N. Somov.
Astrofiz. Issled. Izv. Spets. Astrofiz. Obs., Tom 22, p. 73 – 76 (1986). In Russian. English translation in Bull. Spec. Astrophys. Obs. – North Caucasus.
A description of the specialized language of programming of SIPRAO (specialized interpretator for programming and reduction of astronomical observations) which frees an astronomer from non–astronomical aspects of computer programming is presented. The command processor is described and the functional facilities of SIPRAO are enumerated. The possibility of using SIPRAO in related fields of scientific investigations is pointed out.

036.096 **Dual–processor apparatus–programm complex – scanner of BTA. III. Automatized express processing of stellar spectra.**
I. M. Kopylov, N. N. Somov, T. A. Somova.
Astrofiz. Issled. Izv. Spets. Astrofiz. Obs., Tom 22, p. 77 – 88 (1986). In Russian. English translation in Bull. Spec. Astrophys. Obs. – North Caucasus.
Programs of rapid processing of stellar stectra obtained with the help of the television 1000–channel spectrophotometer of BTA are discussed. The program testings carried out during the processing of SS433 spectra showed a good agreement of «computer» and «hand» parameters of spectral lines. Measurement accuracies of these parameters are given.

036.097 **Cómo observar cometas.**
J. E. Bortle.
Rev. Astron., Tomo 56, No. 233, p. 2 – 6 (1985).

036.098 **Waarnemers en integralen. Van cosinus tot kansfunktie.**
P. van der Veen.
Radiant, Jaarg. 8, Nr. 1, p. 1 – 5 (1986).

036.099 **Alle waarnemers en hun lijnen. Een onderzoek naar individuele magnitudenschalen.**
P. Jenniskens, W. Nobel.
Radiant, Jaarg. 8, Nr. 1, p. 6 – 12 (1986).
This article intends to point at severe errors in the meteor magnitude classifications and shows, to a certain level, an increasing alertness for meteors by observers with increasing experience. As a result of teaching and interaction during the observing sessions the systematic effects that are found in the magnitude distributions tend to be strongly group dependent. In order to arrive at observer independent stream parameters, some knowledge of a personal probability function is necessary.

036.100 **Limiting magnitude correction.**
J. Wood.
Werkgroepnieuws, Vol. 14, Nr. 1, p. 20 – 33 (1986).

036.101 **France, Puimichel = Perseids 1986 (I). Personal errors on visual magnitude estimates.**
P. Roggemans.
Werkgroepnieuws, Vol. 14, Nr. 3, p. 93 – 102 (1986).

036.102 **Revised Me difference spectra.**
A. N. Parmar, C. Izzo.
EXOSAT Express, No. 16, p. 21 – 26 (1986).

036.103 **Special photographic techniques applied to Schmidt plates.**
D. F. Malin.
Proceedings of the Second Asian–Pacific Regional Meeting on Astronomy, p. 398 – 401 (1984). – See Abstr. 012.031.

036.104 **Lunar occultation techniques in investigations of binary stars.**
J. C. Bhattacharyya.
Proceedings of the Second Asian–Pacific Regional Meeting on Astronomy, p. 467 – 472 (1984). – See Abstr. 012.031.
The paper discusses the possibility of direct measurement of angular separations and luminosity ratios of components of unresolved binary systems through observations of lunar occultations. A current program at Kavalur Observatory is described and some results presented.

036.105 **A computer approach to handling extinction.**
J. D. Fernie.
I.A.P.P.P. Commun., No. 23, p. 1 (1986). Abstract. – See Abstr. 010.521.

036.106 **Personal computer use of SAO Star Catalog.**
D. M. Kornbluh.
I.A.P.P.P. Commun., No. 23, p. 1 – 2 (1986). Abstract. – See Abstr. 010.521.

036.107 **Apple interface for photometric measurements.**
J. R. Miller.
I.A.P.P.P. Commun., No. 23, p. 11 – 13 (1986).

036.108 **Microcomputer reduction of digital photometric occultation data.**
T. D. Oswalt.
I.A.P.P.P. Commun., No. 23, p. 14 – 17 (1986). – See Abstr. 010.521.

036.109 **Period search using Turbo Pascal.**
D. L. DuPuy.
I.A.P.P.P. Commun., No. 24, p. 26 – 28 (1986).

036.110 **Software development for Doppler data at the Potsdam Station.**
F. Barthelmes, R. Dietrich, G. Gendt.
Nablyud. Iskusstv. Sputnikov Zemli, No. 21, Chast' I, p. 6 – 11 (1982). – See Abstr. 012.032.
The preparation program and the enlarged adjustment program connected with the orbital program POTSDAM–4 have been developed. These software parts are described and first numerical results are discussed. It is shown that this software complex is able to compute long orbital arcs over several days as well as to solve the problems of regional network determination based on Doppler observations.

036.111 **MERIT data processing at Nottingham.**
V. Ashkenazi, T. Moore, P. D. Howard.
Earth rotation and the terrestrial reference frame, p. 399 – 416 (1985). – See Abstr. 012.077.
Details are given of VLBI and SLR (Satellite Laser Ranging) software developed at the University of Nottingham, and the determination of earth rotation parameters from the analysis of data collected during the short and main MERIT campaigns.

036.112 **Precise satellite positioning by the photographic method.**
D. P. Duma.
Nablyud. Iskusstv. Sputnikov Zemli, No. 21, Chast' I, p. 177 – 181 (1982). In Russian.– See Abstr. 012.032.
In order to establish a connection between radio sources and stellar coordinate systems, to study systematic errors of stellar catalogues, to calculate the earth's geodetic parameters as well as to solve other problems, it is reasonable to improve the determination accuracy of satellite angle coordinates of a remote satellite. With this point of view the author proposes to use an astrograph (f = 2 – 3 m) with equipment for exposure time recording with the accuracy up to 10^{-3}s as well as with special cassettes for obtaining point images of stars and satellites.

036.113 **Signal–to–noise ratio in differential speckle interferometry.**
R. Petrov, F. Roddier, C. Aime.
J. Opt. Soc. Am. A, Vol. 3, No. 5, p. 634 – 644 (1986).
Expressions are given for the signal–to–noise ratio of the amplitude and for the uncertainty of the phase of the cross spectrum of two speckle images in the presence of photon shot noise. Applications to astronomical observations are discussed.

036.114 **Optical astronomical interferometry: requirements on stellar Michelson interferometry and heterodyne interferometry for high–resolution imaging.**
P. R. Silverglate.
Appl. Opt., Vol. 25, No. 2, p. 204 – 208 (1986).
The requirements for optical path compensation on space–based (no atmospheric–induced disturbance) heterodyne interferometers and generalized Michelson stellar interferometers are developed. It is demonstrated that heterodyne interferometers require optical path compensation to well within the coherence length dictated by its bandwidth, while Michelson stellar interferometers require compensation to a fraction of their utilized wavelength.

036.115 **Imaging infrared heterodyne spectrometry.**
H. Rothermel, U. Schrey.
Appl. Opt., Vol. 25, No. 4, p. 474 – 477 (1986).

036.116 **Optimized variability analysis in non periodic sources: application to X–ray emission from normal stars.**
A. Collura, S. Sciortino, A. Maggio, S. Serio, R. Rosner.
X–ray astronomy '84, p. 47 – 50 (1985). – See Abstr. 012.033.
The authors discuss the application of a χ^2 method for variability analysis of aperiodic sources. By binning photon arrival times, they consider the deviation of the values of χ^2 from those expected under the hypothesis of a constant source with Poisson noise, and its dependence on the binning rate and phase. They determine from this dependence an "effective" fractional variability and the dominant time scale of the variable component (if any). They use Monte Carlo simulations to check the validity of the method employed. They show applications of this method to X–ray emission from normal stars.

036.117 **System of exact time and frequency for lunar laser ranging.**
Yu. L. Kokurin, V. V. Kurbasov, V. F. Lobanov, N. M. Lypkan', M. A. Ovsyankin.
Fiz. inst. Akad. Nauk SSSR, Prepr., No. 294, 11 pp. (1985). In Russian. Abstr. in Ref. Zh., 51. Astron., 5.51.1042 (1986).

036.118 **Automatic reduction of spectrograms of stars with digital recording.**
Ya. A. Bellas.
Sbornik tr. Mezhdunar. konf. na temu: Avtomatiz. nauchn. issled., Provdiv, 15 – 20 okt., 1984. Tom 1. Sofiya, p. 180 – 186 (1985). In Russian. Abstr. in Ref. Zh., 51. Astron., 5.51.1081 (1986).

036.119 **Special instrument for calculation of the coordinates of the moon.**
O. N. Panasyuk.
Mater. 12 konf. mol. uchenykh Fiz.–mekh. inst. Akad. Nauk, USSR. Sets. otbora i peredachi inf., L'vov, 23 – 25 okt., 1985. Fiz.–mekh. inst. USSR. L'vov, p. 116 – 119 (1986). In Russian. Abstr. in Ref. Zh., 62. Issled. Kosm. Prostranstva, 5.62.389 (1986).

036.120 **Choice of optimum conditions for absolute determination of star declinations with astrolabes.**
E. I. Krejnin.
Astron. Zh., Tom 63, Vyp. 3, p. 589 – 595 (1986). In Russian. English translation in Sov. Astron., Vol. 30, No. 3.
The method suggested by the author in 1968 is further developed. A combination of two astrolabes with prism angles 30° and 45° is found to be sufficiently effective. The joint reduction of all observations when compiling a general catalogue is proved to be

preferable to the consecutive adjustment of coordinates obtained with a single astrolabe to the system of the primary one. Two formulae are derived for the ratio of the declination weights reduced to the system of the astrolabe with $z = 30°$ to those reduced to the system of the astrolabe with $z = 45°$.

036.121 **The potential precision of speckle interferometric measurements. Binary stars.**
S. G. Kuz'menkov.
Astron. Zh., Tom 63, Vyp. 3, p. 596 – 605 (1986). In Russian. English translation in Sov. Astron., Vol. 30, No. 3.
The limiting possibilities of speckle interferometric measurements are studied and the factors limiting these are analysed. The potential precision of the estimate of binary star angular separation, position angle and magnitude difference under separate and joint estimation of these parameters is studied. The limiting magnitude for given observation circumstances is defined in connection with the solution of the concrete measurement problem. The results obtained allow to design the experiment of speckle interferometric measurements of binary stars.

036.122 **Comparison of methods for azimuth determination from photographic observations of the sun.**
A. N. Denisov.
L'vov. politekh. inst. L'vov, 7 pp. (1986). In Russian. Abstr. in Ref. Zh., 51. Astron., 6.51.162 (1986).

036.123 **A machine–readable star catalogue for the automatic processing of measurements with astrographs.**
V. N. Susnin.
Ryazan. radiotekh. inst. Ryazan', 8 pp. (1986). In Russian. Abstr. in Ref. Zh., 51. Astron., 6.51.1004 (1986).

036.124 **Some problems of increasing the accuracy of photoelectric observations in geodetic astronomy.**
V. N. Baranov, R. A. Aleksankina.
Izv. vuzov. Geod. Aehrofotosemka, No. 6, p. 68 – 79 (1985). In Russian. Abstr. in Ref. Zh., 52. Geod. Aerosemka, 6.52.133 (1986).

036.125 **Speckle observations with a photon counting detector.**
P. R. Wood.
Proc. Astron. Soc. Aust., Vol. 6, No. 2, p. 120 – 126 (1985).
A description of the production of astronomical speckles is given in terms of elementary physical principles. Using simple probability arguments, estimates are made of signal–to–noise requirements and magnitude limits for detection; data reduction techniques are discussed. A speckle system used on the AAT (Anglo–Australian Telescope) with the IPCS as detector is described.

036.126 **A system for acquisition and processing of radio astronomical data.**
I. G. Moiseev, N. S. Nesterov, P. S. Nikitin, P. N. Stezhka.
Izv. Krymskoj Astrofiz. Obs., Tom 74, p. 171 – 176 (1986). In Russian. English translation in Bull. Crimean Astrophys. Obs., Vol. 74.
A description of a double processor system intended for automation of observations, sampling and processing of observational data is given. The structure of programs used in the system, observational techniques realized with help of the system and investigations made using it have been described briefly.

036.127 **An automatic system for multichannel radio astronomical observations.**
V. M. Demkin, I. I. Zinchenko, N. S. Nesterov, V. N. Shanin.
Izv. Krymskoj Astrofiz. Obs., Tom 74, p. 176 – 179 (1986). In Russian. English translation in Bull. Crimean Astrophys. Obs., Vol. 74.
An automatic system based on the computer "Electronica–60" and intended for spectral line and continuum observations in the 2.6 – 4 mm wave range has been realized. It makes data sampling and processing in real time and control of the regime of the work of the radio telescope with a distance up to 150 m.

036.128 **A system for recording fast variations of stellar brightness.**
V. P. Zalinian (*V. P. Zalinyan*), H. M. Tovmassian (*G. M. Tovmasyan*).
Eruptive phenomena in stars, p. 435 – 436 (1986). – See Abstr. 012.039.

036.129 **Principles and scope of synthetic photometry.**
R. Buser.
Space Telesc. Sci. Inst., Prepr. Ser., No. 93, 7 pp. (1986). To appear in Highlights of Astronomy, Vol. 7.

036.130 **Principles and scope of synthetic photometry.**
R. Buser.
Space Telesc. Sci. Inst., Prepr. Ser., No. 100, p. 1 – 6 (1986). – See Abstr. 012.040.

036.131 **The possibilities of synthetic photometry.**
B. Gustafsson.
Space Telesc. Sci. Inst., Prepr. Ser., No. 100, p. 7 – 14 (1986). – See Abstr. 012.040.

036.132 **An optimal extraction algorithm for CCD spectroscopy.**
K. Horne.
Space Telesc. Sci. Inst., Prepr. Ser., No. 101, 26 pp. (1986). To appear in Publ. Astron. Soc. Pac.

036.133 **Repositioning the stars in Starfinder.**
M. O. Duke.
Program. J., Vol. 3, No. 3, p. 33 – 36 (1985). Abstr. in Phys. Abstr., Vol. 89, No. 1273, Entry 18667 (1986).

036.134 **Phase closure in optical astronomy.**
N. R. Arnot, P. D. Atherton, A. H. Greenaway.
Trait. Signal, Vol. 2, No. 2, p. 129 – 136 (1985). Abstr. in Phys. Abstr., Vol. 89, No. 1273, Entry 18669 (1986).

036.135 **Design of an expert system for object classification through an application to the classification of galaxies.**
M. Thonnat, C. Granger, M. Berthod.
Proceedings CVPR '85: IEEE Computer Society Conference on Computer Vision and Pattern Recognition, held at San Francisco, Calif., USA, 19 – 23 June 1985. IEEE Comput. Soc. Press, Silver Spring, Md., USA, p. 206 – 208 (1985). Abstr. in Phys. Abstr., Vol. 89, No. 1273, Entry 18671 (1986).

036.136 **Optimal extraction of single–object spectra from observations with two–dimensional detectors.**
J. G. Robertson.
Anglo–Aust. Obs., Prepr., No. 209, 36 pp. (1986). Submitted to Publ. Astron. Soc. Pac.

036.137 **Clustering techniques and their applications.**
F. Murtagh.
ESO Sci. Prepr., No. 448, 14 pp. (1986). To appear in 'Data analysis and astronomy'.

036.138 **Determining the absolute paramters of a meridian circle by observing the first vertical (V). The effect of the higher power terms of the directional error in the determined values of latitude.**
X.–c. Hu.
Publ. Yunnan Obs., No. 1, p. 1 – 9 (1986). In Chinese.
The effects of the higher power terms of the directional error of the instrument, the diurnal aberration and the atmospheric refraction on the results are discussed in the process of the determination of the instantaneous latitude of an observation station by means of the first vertical observation method. The results show that the effect of the higher power terms of the directional error of the instrument is comparatively great within the observable range of the zenith distance, and correction to the results must be made.

036.139 **Determining the absolute parameters of a meridian circle by observing the first vertical (VI). Curvature correction to the recording times in the first vertical direction.**
M.–h. Du, W. Mao.
Publ. Yunnan Obs., No. 1, p. 10 – 17 (1986). In Chinese.
The expression of the curvature correction of the recording times in the first vertical direction is derived by means of the method of the solution for a spherical triangle in combination with the plane projection of the field of view.

036.140 **Determining the absolute parameters of a meridian circle by observing the first vertical (VII). Curvature correction of the determination value of an apparent zenith distance in the first vertical direction.**
M.–h. Du.
Publ. Yunnan Obs., No. 1, p. 18 – 24 (1986). In Chinese.
The expression, accurate to the quadratic terms, of the curvature correction of the determination value of a zenith distance in the first vertical direction is strictly derived by means of the expansion of trigonometric series.

036.141 **Determining the absolute parameters of a meridian circle by observing the first vertical (VIII). The effect of the diurnal variation term of the latitude on the determined values of the azimuth equation and latitude.**
M.–h. Du.
Publ. Yunnan Obs., No. 1, p. 25 – 30 (1986). In Chinese.
An expression for the effect of the diurnal variation term of the latitude on the values of the azimuth equation and the instantaneous latitude which are absolutely determined by observing a star alternately in the first vertical and meridian directions is derived. There is an effect of the same order of magnitude as the variation in the determined values of the azimuth equation, while in the determined values of the instantaneous latitude, the effect is greater than the magnitude of the variation of the observation at the mean time interval, and therefore the effect can be determined from this observation itself.

036.142 **Comparison of two methods for the absolute determination of the azimuth of a meridian circle.**
Y. Fan, W. Mao.
Publ. Yunnan Obs., No. 1, p. 36 – 40 (1986). In Chinese.
The conventional circumpolar star – azimuth mark observation method is compared to the newly proposed first vertical meridian circle observation method. The results show that much more accurate determined values can be obtained if the new method is used.

036.143 **Analysis of the effects of errors on the reading of the graduated circle.**
Y. Fau, W. Mao.
Publ. Yunnan Obs., No. 1, p. 41 – 46 (1986). In Chinese.
The authors propose a "one–shot determination method" for fixing the angular distance of the reading microscope. The effects of the eccentricity error of the circle, the tilt of the circle and the reading microscopes and the flexure of the circle on the circle reading are discussed. It is proved that the arbitrary distribution of microscopes by means of the method for fixing the angular distance will not affect the final accuracy of the circle reading.

036.144 **Construction of an inertial coordinate system with a CCD. (I). Measuring the relative position of an extragalactic system with respect to the reference stars.**
W. Mao, X.–j. Guo.
Publ. Yunnan Obs., No. 1, p. 47 – 54 (1986). In Chinese.
In this article, a method is described for measuring the relative position of an extragalactic system with respect to the reference stars by using a CCD. Since the exposure time of a CCD is short, the successive overlap exposure method can be used in order to expand the observable range of the sky and to overcome the shortage that the measurable field of view is too small. Only two reference stars are needed to determine the relative position of the extragalactic system with respect to the reference stars in the whole range to be measured by the overlap exposure.

036.145 **Construction of an inertial coordinate system with a CCD. (II). The determination of the scales and the correction for the inclination of the measured coordinate system.**
X.–j. Guo, W. Mao.
Publ. Yunnan Obs., No. 1, p. 55 – 61 (1986). In Chinese.
A method for measuring the scales of the two coordinate axes of a CCD plate and the non–perpendicularity of the axes is described. The correction formulae for inclination of the axes and plane of the measured coordinates of each plate with respect to the ideal coordinate system in the course of the overlap exposure are also derived.

036.146 **Construction of an inertial coordinate system with a CCD. (III). Some astronomical corrections of the measured coordinates.**
X.–j. Guo, W. Mao.
Publ. Yunnan Obs., No. 1, p. 62 – 67 (1986). In Chinese.
The correction for the differential refraction and differential aberration of the measured coordinates on each plate during the overlap exposure measurement carried out with a CCD is described, and the correction formulae for the measured coordinates are als derived.

036.147 **Construction of an inertial coordinate system with a CCD. (IV). The transition to the inertial coordinate system.**
W. Mao, X.–j. Guo.
Publ. Yunnan Obs., No. 1, p. 68 – 73 (1986). In Chinese.
Kinematic and dynamic methods for the construction of an inertial coordinate system are described. The geometrical method using extragalactic systems as reference points is described. The precessional rotation of the fundamental catalogue system is derived through the determination of the absolute proper motions of some stars. The CCD measurements are also compared with photographic measurements, and the conclusion is that the inertial coordinate system can be set up by the overlap exposure measurements obtained with a CCD with higher accuracy.

036.148 **The interest of simultaneous spectral and spatial high resolution spectroscopy in the infrared.**
L. Delbouille, N. Grevesse, A. J. Sauval.
Obs. R. Belg., Commun., Sér. A, No. 80, 5 pp. (1985).
After having recalled some of the main advantages of infrared observations, the authors give a few samples of high spectral resolution/low noise solar spectra now routinely obtained in this spectral range by using Fourier transform spectrometers. They then show that such high resolution spectra could be obtained in a short period of time and for small–scale solar structures, using a LEST type collector.

036.149 **On a possibility of realizing the UBVRI system band I using the photomultiplier FEhU–79.**
G. U. Koval'chuk.
Photometric and polarimetric investigations of celestial bodies, p. 188 – 190 (1985). In Russian. – See Abstr. 003.035.
A possibility to expand spectral intervals of photoelectric observations into the infrared region is shown by means of realization of the UBVRI system band I using the photomultiplier FEhU–79.

036.150 **Errors in photometry of artificial celestial bodies.**
M. V. Bratijchuk, V. G. Vygon, S. N. Malikov, Ya. M. Motrunich.
Photometric and polarimetric investigations of celestial bodies, p. 190 – 195 (1985). In Russian. – See Abstr. 003.035.
The radiation effect of stars in the photometer field of view on the photometry of moving objects has been estimated. It is shown that the use of just a two–channel photometer reduces the error resulting from the random number of stars occurring in the photometer field of view. The results of calculations are compared with mathematical modelling data.

036.151 **Aplicación del método de colocación a la reducción de placas fotográficas de estrellas.**
M. J. Sevilla, P. Romero.
Téc. Topográfica, Vol. 10, No. 45, 16 pp. (1982). = Semin. Astron. Geod., Univ. Madr., Publ., No. 124.
In this paper the least-squares collocation method is applied to the reduction of photografic plates of stars, and the definitive formulae for the model that considers all the stochastic processes are derived.

036.152 **Physical models adopted in KAPRI.**
S. Manabe, Y. Takahashi, H. Hanada, T. Ishikawa, M. Fujishita, T. Sato, K. Koike, T. Yoshino.
Publ. Int. Latitude Obs. Mizusawa, Vol. 18, No. 2, p. 93 – 104 (1984).
Descriptions are given for the astronomical, geophysical and geometrical models and formulae adopted in KAPRI, a program for computing the theoretical delay and delay–rate of geodetic VLBI observations and their partial derivatives with respect to physical parameters of interest. The MERIT standards are used as a general reference. However, more sophisticated theories are adopted for the tidal and relativistic effects.

036.153 **Images from astronomical speckle data: weighted shift–and–add analysis.**
J. C. Christou, E. Ribak, E. K. Hege, J. D. Freeman.
Prepr. Steward Obs., No. 632, 27 pp. (1986). To appear in Opt. Eng.

036.154 **Comparative study of deconvolution techniques for infrared speckle interferometry.**
M. L. Cobb, D. W. McCarthy Jr.
Prepr. Steward Obs., No. 643, 8 pp. (1986). To appear in Proc. SPIE Int. Soc. Opt. Eng., Vol. 627 "Instrumentation in Astronomy VI", March 31, 1986.

036.155 **Isoplanicity measurements for calibration of speckle holography amplitudes.**
J. C. Christou, E. K. Hege.
Prepr. Steward Obs., No. 648, 20 pp. (1986). To appear in Opt. Commun.

036.156 **Application of image sharpness criteria in infrared speckle interferometry.**
D. W. McCarthy Jr., M. L. Cobb.
Prepr. Steward Obs., No. 651, 7 pp. (1986). To appear in Proc. SPIE Int. Soc. Opt. Eng., Vol. 627 "Instrumentation in Astronomy VI", March 31, 1986.

036.157 **Maximum entropy image restoration in astronomy.**
R. Narayan, R. Nityananda.
Prepr. Steward Obs., No. 656, 81 pp. (1986). To appear in Annu. Rev. Astron. Astrophys.

036.158 **A calibration method for a cometary plate using field stars.**
Z.–w. Hu, L. Chen, M.–x. Bao.
Kexue Tongbao, Vol. 30, No. 9, p. 1190 – 1195 (1985).
A calibration method for a cometary plate using brightness profiles of the stars is described. The method is divided into two steps. In the first step, the characteristic curves of the plate were given from the magnitudes of the standard stars and their photographic densities; furthermore the brightness profiles of the stars were established. In the second step, the transformation coefficient from the relative brightness to the absolute one can be found by comparing the integrated relative brightness with its standard magnitude of the chosen stars. The principle and its result of the calibration are described.

036.159 **A maximum flatness method for noise reduction in blurred images.**
J. L. Jonas.
S.Afr. J. Phys., Vol. 8, No. 3, p. 67 – 72 (1985). Abstr. in Phys. Abstr., Vol. 89, No. 1275, Entry 30172 (1986).

036.160 **Minimum information method: image sharpening by minimization of structural information.**
J. Pfleiderer.
S.Afr. J. Phys., Vol. 8, No. 3, p. 72 – 78 (1985). Abstr. in Phys. Abstr., Vol. 89, No. 1275, Entry 30173 (1986).

036.161 **An automated astronomical image processor.**
M. P. Grishin, Sh. M. Kurbanov, S. V. Pakhomov, N. E. Piskunov, N. L. Svyatoskavskij.
Sov. Astron. Lett., Vol. 11, No. 5, p. 335 – 338 (1985). English translation of 40.036.082.

036.162 **X–ray imaging techniques.**
D.–x. Fan.
Publ. Purple Mt. Obs., Vol. 5, No. 1, p. 47 – 56 (1986). In Chinese.

036.163 **Effect of amplitude and phase scintillations on radio astronomical observations at decimeter wavelengths.**
T. A. T. Spoelstra.
Tijdschr. Ned. Elektron. Radiogenoot., Vol. 50, No. 5, p. 173 – 179 (1985). In Dutch. Abstr. in Phys. Abstr., Vol. 89, No. 1277, Entry 42183 (1986).

036.164 **Possibilities of improving spatial resolution.**
V. N. Dudinov.
Vestn. Khar'kov. Univ., No. 278, p. 3 – 15 (1985). In Russian.

036.165 **Restoration of images recorded with photon counters.**
Yu. V. Kornienko.
Vestn. Khar'kov. Univ., No. 278, p. 18 – 29 (1985). In Russian.

036.166 **Interferometric synthesis with parallel recording: restoration of the amplitude.**
V. I. Bychinskij, V. N. Uvarov.
Vestn. Khar'kov. Univ., No. 278, p. 29 – 35 (1985). In Russian.

036.167 **Thin–film phase filters in coherent–optical systems for image processing.**
V. G. Vakulik, G. N. Pavlov, S. K. Skorik.
Vestn. Khar'kov. Univ., No. 278, p. 35 – 38 (1985). In Russian.

036.168 **Photometry of astronomical images exposed to linear filtration.**
V. S. Tsvetkova, V. G. Chernyj.
Vestn. Khar'kov. Univ., No. 278, p. 38 – 43 (1985). In Russian.

036.169 **A photometric method for determining the relief of a homogeneous surface with isotropic reflection law.**
N. N. Evsyukov, E. M. Borodenko.
Vestn. Khar'kov. Univ., No. 278, p. 43 – 51 (1985). In Russian.

036.170 **On the use of zenith symmetry in meridian observations.**
N. G. Zuev.
Vestn. Khar'kov. Univ., No. 278, p. 65 – 68 (1985). In Russian.

036.171 **Stellar photometry on glass copies of Schmidt photographic plates.**
M. Pierre.
Astron. Astrophys., Vol. 161, No. 2, p. 408 – 411 (1986).
The author discusses a method for stellar photometry on glass copies of Schmidt photographic plates. Since most of the stars present a saturated profile on the copies, this method is based on the measurement of the area of the stellar images at a given density. Using photoelectric sequences calibration curves giving the magnitude as a function of the square root of this area are built up. These curves are very close to straight lines. It is possible to reach a very good accuracy, around one tenth of magnitude, for a broad magnitude range (about 6 mag), if the field is not too crowded. This is almost as good as what can be achieved on the original plates using conventional methods.

036.172 **Astrometry with a scanning CCD.**
T. Gehrels, B. G. Marsden, R. S. McMillan, J. V. Scotti.
Astron. J., Vol. 91, No. 5, p. 1242 – 1243 (1986).
The authors use a scanning CCD for acquisition and astrometry of new and recovered comets, asteroids, and other solar system objects. The Newtonian focus of a 91 cm telescope has been modified from f/5 to f/3.85 with a relay lens to give a plate scale of 1.73 arcsec per pixel, a scale chosen for efficient coverage of sky. A typical scan will cover 30 minutes of time in right ascension by 0.156 deg in declination and is stored as a digital array of 14848x320 CCD pixel signals. Several potential reference (SAO) stars are observed in one such scan. The reduction from pixel coordinates to apparent topocentric equatorial coordinates of date involves only three free parameters.

036.173 **A pattern–matching algorithm for two–dimensional coordinate lists.**
E. J. Groth.
Astron. J., Vol. 91, No. 5, p. 1244 – 1248 (1986).
A pattern–matching algorithm for two–dimensional coordinate lists is described. The algorithm matches pairs of coordinates in two lists based on the triangles that can be formed from triplets of points in each list. The algorithm is insensitive to coordinate translation, rotation, magnification, or inversion and can tolerate random errors or distortions.

036.174 **Determining dependences between directional quantities and position on a sphere.**
M. F. Bietenholz.
Astron. J., Vol. 91, No. 5, p. 1249 – 1252 (1986).
Two astronomically useful statistical tests to detect dependences of a circular quantity on a spherical one are presented: a nonparametric test and one, developed by Jupp and Mardia, which is based on multivariate statistics. A preliminary investigation of some of the statistical properties of the nonparametric test is undertaken. These tests are applied to a set of radio data to show that neither the integrated polarization vector nor the alignment of the source axes show any dependence on the source's position in the sky. This applies to a global sample, as well as to subsets in redshift, rotation measure, resolution of the radio map, etc.

036.175 **Sobre a precisão de medidas de chapas num comparador Zeiss–Ascorecord.**
L. Bastos.
Publ. Obs. Astron. "Prof. Manuel de Barros", Fac. Cienc. Porto, No. 35, 23 pp. (1983).

036.176 **Map analysing and cleaning for the Mi Yun aperture synthesis radio telescope.**
M.–z. Wei.
Publ. Beijing Astron. Obs., No. 7, p. 113 – 121 (1985). In Chinese.

036.177 **Astronet.**
G. Sedmak.
Bull. Inf. Cent. Données Stellaires, No. 30, p. 7 – 12 (1986). – See Abstr. 012.049.

036.178 **The Starlink astronomical network.**
M. D. Lawden.
Bull. Inf. Cent. Données Stellaires, No. 30, p. 13 – 24 (1986). – See Abstr. 012.049.

036.179 **Network activities at the European Space Agency and at the Space Telescope European Coordinating Facility.**
P. Benvenuti.
Bull. Inf. Cent. Données Stellaires, No. 30, p. 31 – 33 (1986). – See Abstr. 012.049.

036.180 **Networks – definition and clarifications.**
R. Albrecht.
Bull. Inf. Cent. Données Stellaires, No. 30, p. 35 – 36 (1986). – See Abstr. 012.049.

036.181 **The DAN (*Deutsches Astro–Netz*) proposal.**
U. Mebold.
Bull. Inf. Cent. Données Stellaires, No. 30, p. 37 – 40 (1986). – See Abstr. 012.049.

036.182 **The European Academic and Research Network.**
J. L. Delhaye.
Bull. Inf. Cent. Données Stellaires, No. 30, p. 41 – 46 (1986). – See Abstr. 012.049.

036.183 **Coping with IUE and IRAS data from the user's viewpoint.**
J. Mead.
Bull. Inf. Cent. Données Stellaires, No. 30, p. 47 – 48 (1986). – See Abstr. 012.049.

036.184 **Programa "ARIES 2 ML" para reducción de posiciones de estrellas.**
M. J. Sevilla.
Semin. Astron. Geod., Univ. Madr., Publ., No. 127, p. 715 – 741 (1983).

036.185 **A statistical analysis of errors in the 21–cm line redshifts of galaxies.**
G. G. C. Palumbo, G. C. Basiesi–Pillastrini.
Bull. Am. Astron. Soc., Vol. 18, No. 1, p. 639 – 640 (1986). Abstract. – See Abstr. 010.066.

036.186 **Microcomputer controls of spectroscopic instruments.**
G. S. Rossano.
Publ. Astron. Soc. Pac., Vol. 98, No. 603, p. 513 – 519 (1986).
A small, versatile microcomputer–based data–acquisition system has been developed for astronomical observing in the infrared. The hardware is a 16–bit 8086/8087 based system. The hardware is operated using interactive software which permits high–speed data acquisition and real–time analysis of spectroscopic data. The combined hardware and software system facilitates flexible, efficient, and uniform operation of a variety of spectroscopic instruments.

036.187 **A microcomputer–assisted photoelectric photometry system.**
P. Ells.
J. Br. Astron. Assoc., Vol. 96, No. 4, p. 204 – 209 (1986).
This paper describes the use of a microcomputer in a photoelectric photometry system for data collection, reduction and analysis. The results of two evenings' observations of eclipsing binary stars are used for illustration.

036.188 **Experiments with solid–state imaging devices and image processing in amateur astronomy.**
A. C. Cook, J. D. Cook.
J. Br. Astron. Assoc., Vol. 96, No. 4, p. 223 – 229 (1986).
The authors have investigated two types of electronic imaging device in making astronomical observations and in enhancing deep–sky photographs. Software has been written to interface these devices to a microcomputer, so that the images may be integrated for a present time from 1/40 second to 20 seconds, and then stored on floppy disk. Images recorded on disk can subsequently be loaded back into the microcomputer where they undergo processing to enhance certain features while suppressing others.

036.189 **On the use of instrumental photometric standards in stellar TV photometry.**
A. N. Abramenko, E. P. Pavlenko, V. V. Prokof'eva.
Photometric and polarimetric investigations of celestial bodies, p. 184 – 187 (1985). In Russian. – See Abstr. 003.035.
Images of the instrumental standards (star–like images) are projected on the photocathode of a TV system in the neighbourhood of the stars under investigation. Light diodes are used to illuminate a plate with small holes covered with neutral filters. The dependence of the standard brightness on the diode current was measured in the range of 10^5. Standard stellar magnitudes are estimated with an accurcy of $\pm 0\overset{m}{.}05$.

036.190 **Is robust estimation useful for astronomical data reduction?**
R. L. Branham Jr.
Q. J. R. Astron. Soc., Vol. 27, No. 2, p. 182 – 193 (1986).
Robust estimation is a technique for data reduction that is insensitive to departures from assumptions, particularly the assumption that the observational errors follow the normal curve. The commonly used procedure of outlier rejection can be considered robust estimation. This investigation uses 21,365 observations of five minor planets to study two methods of robust estimation.

036.191 **Remote control of 2.2–m telescope from Garching.**
G. Raffi, M. Ziebell.
Messenger, No. 44, p. 26 – 30 (1986).

036.192 **Computer aided design of printed circuit boards at ESO.**
H. Kasten.
Messenger, No. 44, p. 30 – 31 (1986).

036.193 **Advances in information analysis expected for Space Telescope data.**
G. Sedmak.
Mem. Soc. Astron. Ital., Vol. 56, No. 4, p. 691 – 698 (1985). – See Abstr. 012.054.
This paper reviews recent advances in image processing techniques adequate for handling the large amount of two–dimensional images expected from the Space Telescope.

036.194 **Technique and advantages of using multiple–zone apertures in Fabry–Pérot spectroscopy.**
K. C. Sahu, J. N. Desai, M. Srinivasan, R. Gupta.
Inst. Astrophys. Paris, Pré–Publ., No. 134, 8 pp. (1986). To appear in Proc. SPIE Int. Soc. Opt. Eng., Vol. 627.

036.195 **Reports on statistical and pattern recognition software.**
F. Murtagh.
Bull. Inf. Cent. Données Stellaires, No. 30, p. 169 – 170 (1986). = Newsl. Working Group Mod. Astron. Methodol., No. 2.

036.196 **The software library of the Space Telescope – European Coordinating Facility.**
F. Murtagh.
Bull. Inf. Cent. Données Stellaires, No. 30, p. 174 (1986). = Newsl. Working Group Mod. Astron. Methodol., No. 2.

036.197 **Phase–gradient stellar image reconstruction.**
G. J. M. Aitken, R. Johnson, R. Houtman.
Opt. Commun., Vol. 56, No. 6, p. 379 – 383 (1986). Abstr. in Phys. Abstr., Vol. 89, No. 1280, Entry 60664 (1986).

036.198 **An analysis of the properties of CLEAN and Smoothness Stabilized CLEAN – some warnings.**
S. M. Tan.
Mon. Not. R. Astron. Soc., Vol. 220, No. 4, p. 971 – 1001 (1986).
The CLEAN algorithm is widely used in the deconvolution of radio astronomical images. This paper reviews the theoretical basis of the method and shows that the progress of the algorithm may be analysed in terms of a transient phase followed by an asymptotic phase. The convergence result by Schwarz refers to the asymptotic phase, and this result is extended to show how and why the algorithm can diverge under certain conditions. It is shown theoretically and by a one–dimensional example that the unique solution to which this algorithm converges does *not* have the desirable properties of sidelobe suppression usually attributed to it.

036.199 **Potential accuracy of speckle–interferometric measurements. Angular diameters of stars.**
S. G. Kuz'menkov.
Sov. Astron., Vol. 29, No. 6, p. 699 – 704 (1985). English translation of 40.036.173.

036.200 **A new algorithm for signal and image recovery from triple correlations.**
A. W. Lohmann.
Optik, Vol. 72, No. 4, p. 171 – 172 (1986). Abstr. in Phys. Abstr., Vol. 89, No. 1281, Entry 62852 (1986).

036.201 **Radarclinometry for the Venus Radar Mapper.**
R. L. Wildey.
Photogramm. Eng. Remote Sensing, Vol. 52, No. 1, p. 41 – 50 (1986). Abstr. in Phys. Abstr., Vol. 89, No. 1281, Entry 67363 (1986).

036.202 **Stellar interferometry in space: fringe detection.**
L. Koechlin.
J. Opt. (Paris), Vol. 16, No. 6, p. 269 – 276 (1985). In French. Abstr. in Phys. Abstr., Vol. 89, No. 1282, Entry 73081 (1986).

036.203 **Outline of the K–3 data processing system.**
H. Sugimoto.
Rev. Radio Res. Lab., Vol. 30, Spec. Issue, No. 1, p. 159 – 161 (1984). In Japanese. Abstr. in Phys. Abstr., Vol. 89, No. 1282, Entry 73083 (1986).

036.204 **Quantitative determination of mineral types and abundances from reflectance spectra using principal components analysis.**
M. O. Smith, P. E. Johnson, J. B. Adams.
J. Geophys. Res., Vol. 90, Suppl., p. C797 – C804 (1985). – See Abstr. 012.065.
A procedure was developed for analyzing remote reflectance spectra, including multispectral images, that quantifies parameters such as types of mineral mixtures, the abundances of mixed minerals, and particle sizes. The approach is applicable to remote sensing data of planetary surfaces for quantitative determinations of mineral abundances.

036.205 **Quantitative analysis of planetary reflectance spectra with principal components analysis.**
P. E. Johnson, M. O. Smith, J. B. Adams.
J. Geophys. Res., Vol. 90, Suppl., p. C805 – C810 (1985). – See Abstr. 012.065.
A technique is presented for quantitative analysis of planetary reflectance spectra as mixtures of particles on microscopic and macroscopic scales using principal components analysis. This technique allows for determination of the endmembers being mixed, their abundance, and the scale of mixing, as well as other physical parameters. Eighteen lunar telescopic reflectance spectra of the Copernicus crater region, from 600 nm to 1800 nm in wavelength, are modeled in terms of five likely endmembers: mare basalt, mature mare soil, anorthosite, mature highland soil, and clinopyroxene.

036.206 **A Monte Carlo study of parameters affecting computer simulations of crater saturation density.**
A. Woronow.
J. Geophys. Res., Vol. 90, Suppl., p. C817 – C824 (1985). – See Abstr. 012.065.
Computer models of cratered surfaces often use inputs of uncertain nature and importance. This work evaluates the sensitivity of the resulting crater–saturation estimates to the input parameters, principally applicable to the study of craters upward from 8 km diameter.

036.207 **Recovering crater production functions from densely cratered surfaces.**
A. Woronow.
J. Geophys. Res., Vol. 90, Suppl., p. C825 – C827 (1985). – See Abstr. 012.065.
Two techniques proposed for the recovery of the crater–production functions on densely cratered surfaces have been appraised. The first technique classifies craters according to their degree of rim preservation, then interprets the statistics of the freshest craters as closely mimicking the full population's production function. Although commonly practiced, in fact, this method is demonstrably misleading for densely cratered surfaces. The

second technique uses a weighting scheme to recover the production function. Each crater diameter measured is weighted by the fraction of the crater that remains unoverlapped by subsequent impacts. While this method is theoretically sound even at saturation crater densities, it too fails at high densities due to practical difficulties.

036.208 **The Observatório Nacional image processing system.**
L. N. da Costa, R. R. de Carvalho, P. S. Pellegrini, R. Chan, P. Scheid, C. Rite, M. A. Nunes, D. Nascimento.
Rev. Mex. Astron. Astrofis., Vol. 12, Spec. Issue, p. 385 – 392 (1986). – See Abstr. 012.066.

The authors describe the image processing system developed at the Observatório Nacional. The integrated hardware and software system is the result of an ongoing effort to provide adequate facilities for astronomical data analysis.

036.209 **Automatic identification of faint images on the ESO/SRC Sky Atlas.**
R. Chan, P. S. Pellegrini, L. N. da Costa.
Rev. Mex. Astron. Astrofis., Vol. 12, Spec. Issue, p. 393 – 398 (1986). – See Abstr. 012.066.

This paper describes the automatic image classification system developed at the Observatório Nacional. The system is currently being used in photometric studies of moderately distant clusters of galaxies utilizing the on–film copies of the ESO/SRC Sky Atlas.

036.210 **Image processing techniques for position measurements of Uranus' satellites.**
R. Vieira–Martins, R. R. de Carvalho, C. H. Veiga.
Rev. Mex. Astron. Astrofis., Vol. 12, Spec. Issue, p. 399 – 404 (1986). – See Abstr. 012.066.

The authors describe a new procedure to determine the coordinates of the center of Uranus' satellites. Two different methods were used depending on how much the satellite light distribution is affected by that of the planet. Comparison with the results obtained from visual measurements suggests that the procedure proposed does indeed provide much more accurate position measurements for all satellites.

036.211 **Calibration methods in millimeter–wave radioastronomy.**
Z. Abraham, L. C. L. Botti, L. F. del Ciampo.
Rev. Mex. Astron. Astrofis., Vol. 12, Spec. Issue, p. 414 – 418 (1986). – See Abstr. 012.066.

The authors discuss a calibration method for millimeter wave observations that automatically compensates the atmospheric attenuation, even in the presence of a radome, or when the mean kinetic temperature of the sky is smaller than the ambient temperature.

036.212 **Global reduction of fundamental astrometric data: computation of star catalogues.**
P. Benevides–Soares, L. B. F. Clauzet.
Rev. Mex. Astron. Astrofis., Vol. 12, Spec. Issue, p. 447 (1986). Abstract. – See Abstr. 012.066.

036.213 **Statistical analysis of proper motion surveys.**
R. B. Hanson.
Astrometric techniques, p. 43 – 45 (1986). – See Abstr. 012.068 (IAU Symp. No. 109).

Proper motion surveys offer a great deal of data bearing on important astronomical problems such as stellar kinematics and the luminosity function in the solar neighborhood. Major obstacles to the full use of proper motions have long been posed by: (1) incompleteness of proper motion surveys, (2) proper motion bias in kinematic studies, and (3) the indirect approaches and kinematical assumptions needed in traditional luminosity studies. These obstacles can be largely overcome by a new approach (Hanson 1983) using multivariate and conditional distribution statistics (Hanson, Lutz, and Murray 1984) to model the proper motion survey data.

036.214 **Statistical problems encountered in using trigonometric parallaxes.**
T. E. Lutz.
Astrometric techniques, p. 47 – 52 (1986). – See Abstr. 012.068 (IAU Symp. No. 109).

The importance of understanding the properties of a data sample which is to be used for calibrating luminosities can hardly be overemphasized. In an earlier paper (Lutz, 1983) the author points out the methods for dealing with two well–defined cases. In this paper he presents some further results on the topic.

036.215 **Application de la méthode par déconvolution intégrée au cas des occultations d'étoiles doubles.**
M. Froeschlé, C. Meyer.
Astrometric techniques, p. 113 – 121 (1986). – See Abstr. 012.068 (IAU Symp. No. 109).

The most usual method for the determination of the angular separation ϱ and the magnitude difference Δm of the two components is the direct comparison of the observed diffraction pattern with a theoretical model. However, it is possible to use a deconvolution process involving both the derivative of the Fresnel diffraction function for a point–like source, and that of the observed intensity. The authors describe a new method, using the integrated deconvolution process (Froeschlé, Meyer, 1983), in which they determine the variations of the apparent surface of the two sources of a double star while it is being occulted.

036.216 **The astrometric possibilities of very–long–baseline interferometry.**
D. S. Robertson.
Astrometric techniques, p. 143 – 155 (1986). – See Abstr. 012.068 (IAU Symp. No. 109).

The significance of the VLBI observations for astrometry results from the fact that the delays can be determined with a precision of about 0.1 nanosecond, equivalent to a few centimeters. For continental–scale baselines this translates into an angular precision for the resulting source coordinate determinations of the order of milliseconds of arc.

036.217 **Astrometry with the Lowell PDS.**
L. Wasserman, E. Bowell.
Astrometric techniques, p. 231 – 236 (1986). – See Abstr. 012.068 (IAU Symp. No. 109).

The Lowell Observatory PDS is currently used in several projects which necessitate astrometric reduction of photographic plates. To meet this need, the authors have developed a series of computer programs for astrometric data reduction which enable the user to obtain maximum flexibility from the PDS while at the same time retaining ease of use.

036.218 **Instrument and method for determination of high–precision coordinates of geostationary artificial satellites.**
D. P. Duma, L. N. Kizjun (*L. N. Kizyun*), N. I. Laptienko, M. A. Mel'nikov, Yu. I. Safronov.
Astrometric techniques, p. 243 – 245 (1986). – See Abstr. 012.068 (IAU Symp. No. 109).

An experiment for determining precise positions of artificial satellites has been carried out at the Main Astronomical Observatory of the Academy of Sciences of the Ukrainian SSR with the help of the Carl–Zeiss double wide–angle astrograph (2 m focal length, 40 cm objective diameter). Estimates of the precision of the satellite position obtained from 9 plates are given. The standard errors are in the range $0.48'' - 1.67''$.

036.219 **Astrometry with objective prism.**
J. Stock.
Astrometric techniques, p. 253 – 257 (1986). – See Abstr. 012.068 (IAU Symp. No. 109).

A number of years ago the author (Stock, 1978) showed that astrometric positions free of magnitude– or color–dependent errors may be obtained from pairs of objective prism plates with opposite dispersion, since these allow monochromatic measurements as well as a restriction of the photographic density range without restricting the range of magnitudes that can be covered. Results from a series of plates taken originally for a radial velocity survey are compared with two new catalogues of positions from meridian circle positions, namely the Perth 70 (Høg 1976) and the Santiago 67 (Carrasco 1982) catalogues.

036.220 **Speckle interferometry in astrometry.**
H. A. McAlister.
Astrometric techniques, p. 293 – 300 (1986). – See Abstr. 012.068 (IAU Symp. No. 109).
Speckle interferometry, which is characterised by diffraction limited resolution and enhanced accuracy, offers a wide range of possibilities for astrometry. Speckle applications are limited by the constraint of isoplanicity with the result that only differential rather than absolute astrometry can be carried out with the technique. Nevertheless potentially important applications exist in solar system and stellar astrometry.

036.221 **First results from the new GSU (*Georgia State University*) CCD speckle camera.**
W. I. Hartkopf.
Astrometric techniques, p. 301 – 307 (1986). – See Abstr. 012.068 (IAU Symp. No. 109).
The CCD speckle camera now in use at Georgia State University has brought about considerable improvement in limiting magnitude and photometric accuracy over the old photographic system, but has also necessitated considerable changes in data storage and reduction. These new data handling techniques are outlined.

036.222 **Infrared speckle interferometry: a sensitive technique for physical measurements of unseen companions to nearby stars.**
D. W. McCarthy Jr.
Astrometric techniques, p. 309 – 319 (1986). – See Abstr. 012.068 (IAU Symp. No. 109).
Infrared speckle interferometry combines the full resolving power of large telescopes with high photometric sensitivity over the wavelength range 2.2 to 12 microns. Seeing fluctuations limit measurement precision. Astrometric companions have been detected with angular separations $\geqslant 0.1$ arcsec and magnitude differences $\leqslant 3.7$ mag. In some cases, orbital motion is detected. A method of "shift–and–add" enables detection of substellar (0.04 to 0.08 $M_{\odot}$) companions. Future improvements involving detector arrays and seeing monitors are discussed.

036.223 **Moving–image astrometry with the multi–anode micro-channel array (MAMA) detector.**
J. D. Scargle.
Astrometric techniques, p. 353 – 368 (1986). – See Abstr. 012.068 (IAU Symp. No. 109).
If the image of a field of stars is moved past a high–speed detector array at a precisely known velocity, relative positons can be determined by measuring time intervals. The only available detector with spatial and temporal resolution and stability adequate for this technique is the MAMA. A rudimentary observational test with this detector has yielded promising results. Further observations are planned to determine whether milliarcsecond accuracy can be obtained.

036.224 **The design of modern meridian circles for the observation of faint objects.**
M. Yoshizawa, M. Miyamoto.
Astrometric techniques, p. 449 – 458 (1986). – See Abstr. 012.068 (IAU Symp. No. 109).
Definitions of instantaneous instrumental coordinates and local horizontal coordinates of the photoelectric meridian circle of Tokyo Astronomical Observatory are given, and a method to determine the positions of stars by using a scanning double slits micrometer are presented. It is found that stars up to 12.2 mag are observable by making a composite photon intensity. Discussions on the application of observations of faint objects by meridian circles are presented.

036.225 **The use of photographic positions in determining azimuth of a meridian circle.**
C. Fabricius.
Astrometric techniques, p. 465 – 468 (1986). – See Abstr. 012.068 (IAU Symp. No. 109).
The azimuth of a meridian circle may be determined using stars within 1° – 2° of the pole combining observations in upper and lower culmination of stars which have well known right ascensions. The right ascensions are determined photographically using only declinations in the reduction, thus making the azimuth determination independent of the right ascension system. The declinations of the pole near stars are determined with the meridian circle.

036.226 **The use of astrometric instruments in vacuum chambers.**
N.–s. Hu.
Astrometric techniques, p. 469 – 481 (1986). – See Abstr. 012.068 (IAU Symp. No. 109).
This paper discusses the effect of using a vacuum chamber with a horizontal entrance window to eliminate astronomical refraction, dispersion and irregular refraction in a telescope tube. Some engineering problems arising in connection with the entrance window are dealt with. The actual results obtained by vacuum photoelectric astrolabes and a vacuum photographic zenith tube are presented.

036.227 **On the implementation of absolute meridian observations in low latitude stations.**
D.–m. Li.
Astrometric techniques, p. 567 – 568 (1986). – See Abstr. 012.068 (IAU Symp. No. 109).

036.228 **HIPPARCOS data reductions.**
L. Lindegren.
Astrometric techniques, p. 593 – 604 (1986). – See Abstr. 012.068 (IAU Symp. No. 109).
The raw data produced by HIPPARCOS – mainly photon counts from the primary detector and star mappers – will be analysed in full by two independent reduction consortia. The aim is to generate, by 1993, a single catalogue of positions, proper motions, and parallaxes for the 10^5 programme stars, with additional information on photometry, multiple stars, and minor planets. This review outlines some practical and theoretical features of the data reductions.

036.229 **Software for Space Telescope astrometry.**
W. H. Jefferys, J. Fco.
Astrometric techniques, p. 637 – 641 (1986). – See Abstr. 012.068 (IAU Symp. No. 109).
A general overlapping–plate reduction program, REDUCE, has been written for the reduction of astrometric data from the Space Telescope. This paper describes the main features of the program.

036.230 **A fully–automated system of astrometric data collection and processing.**
V. I. Sergienko, A. G. Radchuk, B. A. Pavlov, V. S. Kudeeva.
Astrometric techniques, p. 795 – 798 (1986). – See Abstr. 012.068 (IAU Symp. No. 109).
The automated astronomical system "Astrometrist" for the determination of the parameters of terrestrial rotation is intended for astrometric and meteorological data collection and on–line processing and feedback to the user by a computer.

036.231 **An optimal extraction algorithm for CCD spectroscopy.**
K. Horne.
Publ. Astron. Soc. Pac., Vol. 98, No. 604, p. 609 – 617 (1986).
The author describes an optimal spectrum extraction algorithm and illustrates its performance with spectrum data obtained with a CCD. The main advantage offered by this optimal extraction method in comparison with conventional extraction procedures is improved signal–to–noise ratio, which can be the equivalent of a 70% increase in the exposure time.

036.232 **Automatische Klassifikation von Spektren geringer Dispersion.**
P. Schücker.
Mitt. Astron. Ges., No. 65, p. 192 – 195 (1986). – See Abstr. 012.072.

036.233 **A statistical search for intergalactic dust.**
H. Horstmann.
Mitt. Astron. Ges., No. 65, p. 237 – 238 (1986). – See Abstr. 012.072.

036.234 **Microdensitometer errors: their effect on photometric data reduction.**
E. P. Bozyan, C. B. Opal.
NASA Conf. Publ., NASA CP–2317, p. 45 – 58 (1984). – See Abstr. 012.073.
The authors analyze the performance of densitometers used for photometric data reduction of high dynamic range electrographic plate material. They test densitometer repeatability by comparing two scans of one plate and examine internal densitometer errors by constructing histograms of digitized densities and find inoperative bits and differential non–linearity in the analog–to–digital converter. Such problems appear common to the four densitometers used in this investigation and introduce systematic algorithm dependent errors in the results. Strategies are suggested to improve densitometer performance.

036.235 **The PDS for Space Telescope guide star selection system: status and improvements.**
J. H. Kinsey.
NASA Conf. Publ., NASA CP–2317, p. 135 – 149 (1984). – See Abstr. 012.073.

036.236 **The VAX/VMS device driver and user operation program for the Midwestern Astronomical Data Reduction and Analysis Facility PDS microdensitometer.**
C. M. Anderson, T. W. Lynch, M. P. Dearing.
NASA Conf. Publ., NASA CP–2317, p. 189 – 195 (1984). – See Abstr. 012.073.
The Midwestern Astronomical Data Reduction and Analysis Facility PDS microdensitometer was the first astronomical PDS which was provided with a Motorola 6800 microprocessor control/interface and the first which was to be interfaced to a 32–bit instruction computer operating in a multiuser environment. Command and response hand shaking are handled on an RS 232 serial line while data transfer occurs on a dedicated Direct Memory Access line. This set of conditions necessitated the creation of a specialized device driver for the system. The general outline of this software is discussed.

036.237 **Data compression techniques for astronomy.**
R. Landau, F. D. Ghigo.
NASA Conf. Publ., NASA CP–2317, p. 255 – 265 (1984). – See Abstr. 012.073.
The photometric and astrometric accuracy of the Minnesota automated dual–plate scanner is summarized.

036.238 **Techniques of isophotometry.**
F. D. Ghigo, R. Landau.
NASA Conf. Publ., NASA CP–2317, p. 267 – 276 (1984). – See Abstr. 012.073.
The authors discuss the expected performance of the Minnesota Automated Dual–Plate Scanner with regard to photometry, position measurement accuracy, and ability to classify images. Since the current rebirth of the machine is not complete, the authors have not yet done extensive tests with real plate material, and so some of the following discussion is more theoretical than one might wish.

036.239 **Crystal gazing. II. Implications of advances in digital data storage technology.**
D. C. Wells.
NASA Conf. Publ., NASA CP–2317, p. 345 – 359 (1984). – See Abstr. 012.073.
During the next 5 – 10 years it is likely that the bit density available in digital mass storage systems (magnetic tapes, optical and magnetic disks) will be increased to such an extent that it will greatly exceed that of the conventional photographic emulsions like IIIaJ which are used in astronomy. The implications for designers of future microdensitometers are: (1) there will be a continuing need for precision, digitization of large–format photographic imagery, and (2) that the need for real–time analysis of the output of microdensitometers will decrease.

036.240 **Storage, retrieval, and analysis of ST data.**
R. Albrecht.
NASA Conf. Publ., NASA CP–2317, p. 371 – 378 (1984). – See Abstr. 012.073.
This paper presents an overview of the ST Science Ground System between carrying out the observations and the interactive analysis of preprocessed data. The ground system elements used in data archival and retrieval are described and operational procedures are discussed.

036.241 **Photographic memory – the storage and retrieval of data.**
J. Horton.
NASA Conf. Publ., NASA CP–2317, p. 387 – 413 (1984). – See Abstr. 012.073.
The concept of density encoding digital data in a mass–storage computer peripheral is proposed. This concept requires that digital data be encoded as Distinguishable Density Levels of the film to be used as the storage medium. These DDL's are then recorded on the film in relatively large pixels. Retrieval of the data would be accomplished by scanning the photographic record using a relatively small aperture. Multiplexing of the pixels is used to store data of a range greater than the number of DDL's supportable by the film in question.

036.242 **The role of the automated plate measuring (APM) machine in developing a program to discover QSO's automatically.**
D. Turnshek, F. Briggs, C. Hazard, R. McMahon, R. Terlevich, E. Kibblewhite, M. Irwin.
NASA Conf. Publ., NASA CP–2317, p. 417 – 418 (1984). – See Abstr. 012.073.
During the past several years Hazard, Terlevich and collaborators have visually assessed a wide variety of photographic 1.2–m UK Schmidt data (primarily IIIaJ objective prism plates) in an effort to discover QSOs. The authors are now beginning to develop automatic discovery procedures. They discuss the general role of the APM type machine (developed by Kibblewhite at Cambridge University) in a program to discover QSOs automatically. In general, the methods employed can be used to study stars and galaxies as well.

036.243 **RIMSTAR: rapid image scanning technology and archiving.**
E. R. Craine, J. S. Scott.
NASA Conf. Publ., NASA CP–2317, p. 419 – 427 (1984). – See Abstr. 012.073.
The authors are presently exploring designs for linking one– or two–dimensional detector arrays in large mosaics without gaps between elements of the detectors. These techniques essentially allow construction of arbitrarily large detector mosaics, for which high speed is preserved because individual array components can be read out in parallel. As an example, one design envisions a 1000 × 1000 element array with a frame rate of one per millisecond, 300 times faster than for comparable conventional monolithic detectors. Clear applications for this technology are recognized in the following areas: large scale, real time digital imagery, e.g. surveillance, guidance, etc., and rapid digitization and archiving of documents, e.g. maps, recon photographs, text, etc.

036.244 **Computer simulation of the effects of transmission–averaging in microdensitometry.**
H. M. Heckathorn.
NASA Conf. Publ., NASA CP–2317, p. 429 – 431 (1984). – See Abstr. 012.073.

036.245 **Extragalactic observations with optical space arrays.**
F. Bertola.
ESA Spec. Publ., ESA SP–226, p. 7 – 9 (1985). – See Abstr. 012.075.
The author reviews a few extragalactic topics which could be the subject of investigation for a space interferometer allowing, in addition to high resolution imaging, the possibility of precise measurements of proper motions and parallaxes in nearby galaxies.

036.246 **Surface and circumstellar stellar imaging.**
A. K. Dupree.
ESA Spec. Publ., ESA SP–226, p. 11 (1985). Abstract. – See Abstr. 012.075.

036.247 **Astronomical criteria for interferometric arrays.**
E. Kibblewhite.
ESA Spec. Publ., ESA SP–226, p. 13 – 16 (1985). – See Abstr. 012.075.
The design of optical interferometric array depends on the type of astronomical object one wishes to observe. This paper discusses the key features of certain objects as they affect spacecraft design. Magnitude limits and measuring strategies are discussed for a TRIO–type configuration.

036.248 **High angular resolution in the infrared: prospects for space observations.**
P. Léna.
ESA Spec. Publ., ESA SP–226, p. 17 – 21 (1985). – See Abstr. 012.075.
The author stresses the growth of high angular resolution (sub arc second) programs in the infrared, from 1 to 1000 μm. Ground based programs (speckle or two–telescopes interferometry) are currently producing interesting results. He discusses the future of these ground programs and shows that interferometry in space will improve the sensitivity, extend the wavelength coverage and open a new field in astrophysics, even with the current detectors and system performances.

036.249 **Fringe drift compensation in computer memory.**
L. Kœchlin.
ESA Spec. Publ., ESA SP–226, p. 99 – 105 (1985). – See Abstr. 012.075.
This paper first presents fringe detection and contrast measurement techniques presently used on a ground based 6 to 67 meters baseline Michelson interferometer at CERGA. A proposal follows for very faint object interferometric observation with spacecraft such as Trio using developments of presently tested methods.

036.250 **Problems in image reconstruction with optical arrays.**
A. Labeyrie.
ESA Spec. Publ., ESA SP–226, p. 117 – 119 (1985). – See Abstr. 012.075.
Different ways of reconstructing images with space arrays are discussed. With two apertures, the 0.1″ field makes it difficult to obtain phase information, but the readily available "multicolor phases" can help phaseless reconstruction algorithms. With three or more apertures, phases are obtained through phase–closure algorithms.

036.251 **Numerical experiments in image reconstruction.**
M. Faucherre, R. V. Stachnik, S. Aram.
ESA Spec. Publ., ESA SP–226, p. 121 – 127 (1985). – See Abstr. 012.075.
The authors describe an approach to Fourier transform phase recovery which can work for two–element interferometers and does not require a reference star. They present the results of a series of numerical experiments in which they (1) illustrate the Fourier amplitude and phase structure associated with a number of realistic sources, (2) produce reconstructions by simple Fourier inversion for sources of varying brightness, (3) simulate degradation in the recovered image due to reference phase errors, and (4) show the consequence of invoking the simplest of a number of more sophisticated image recovery algorithms, phase array relaxation.

036.252 **Radio astronomy experience in image reconstruction. A history lesson for optical astronomers.**
P. J. Warner.
ESA Spec. Publ., ESA SP–226, p. 129 – 132 (1985). – See Abstr. 012.075.
The difficulties faced by radio astronomers in reconstructing images without phase information are reviewed.

036.253 **COSMIC image reconstruction.**
M. G. Lacasse, W. A. Traub.
ESA Spec. Publ., ESA SP–226, p. 133 – 140 (1985). – See Abstr. 012.075.
The authors present numerical as well as experimental demonstrations of aperture synthesis in the optical regime, developed for the proposed COSMIC telescope array (Coherent Optical System of Modular Imaging Collectors).

036.254 **Imaging Speckle Interferometer (ISI) in space: digital simulations of image reconstruction and photon noise.**
K.–H. Hofmann, G. Weigelt.
ESA Spec. Publ., ESA SP–226, p. 145 – 151 (1985). – See Abstr. 012.075.
The authors show digital simulations of image reconstruction from ISI data. In these experiments they have simulated various multiple mirror pupil functions and various amounts of photon noise. The authors also discuss speckle spectroscopy with ISI and image reconstruction from 2–telescope interferograms.

036.255 **On line astronomical image processing using microcomputers.**
C. Thom.
ESA Spec. Publ., ESA SP–226, p. 153 – 154 (1985). – See Abstr. 012.075.
Microprocessors as fast as the MC 68000 make it now possible to process images in some circumstances in real time, using only software algorithms. The author built on these principles a system intended for reducing images on the big interferometer at CERGA. This device is now operational as autocorrelator.

036.256 **Aperture synthesis with or without phase information for kilometric arrays. Interpolation algorithms for stabilized image reconstruction.**
A. Lannes.
ESA Spec. Publ., ESA SP–226, p. 165 – 169 (1985). – See Abstr. 012.075.
At the cost of some compromise between resolution and robustness, the problem of reconstructing high–resolution images from a sparsely sampled 2D–frequency plane has to be stated in terms of interpolation, extrapolation being forbidden. Using the method of steepest descent the author introduces a very simple and powerful algorithm for interpolation with or without phase data. In particular, the constraint of nonnegativity on some limited field is taken into consideration. The Fienup algorithm can be retrieved as a special case and improved by modifying its constraint operator and the implicit choice of its relaxation parameter.

036.257 **Stable method for estimation of laser ranging.**
G. S. Kurbasova, L. V. Rykhlova.
Earth rotation and the terrestrial reference frame, p. 213 – 216 (1985). – See Abstr. 012.077.
The authors propose to use a noise–immunity variant of the least squares method in the preliminary analysis of experimental data. The advantage of such a method is shown on the example of Lageos observations.

036.258 **Semi–automated identification of IRAS point sources using UKST plates and the COSMOS measuring machine.**
A. Savage, R. G. Clowes, M. Kalafi, S. K. Leggett, H. T. MacGillivray, R. D. Wolstencroft.
Light on dark matter, p. 23 – 28 (1986). – See Abstr. 012.080.
The authors have initiated a large–scale programme to identify IRAS point sources, and now report on their initial study covering three hundred square degrees, centred on the South Galactic Pole and containing 300 IRAS point sources.

036.259 **Spectral decomposition of IRAS maps.**
R. Braun, R. G. Strom, H. van der Laan, H. Greidanus.
Light on dark matter, p. 47 – 48 (1986). – See Abstr. 012.080.

The CDS and the SIMBAD data base.
See Abstr. 002.061.

Interferometry and synthesis in radio astronomy.
See Abstr. 003.039.

Infinite vistas: new tools for astronomy.
See Abstr. 003.076.

Photometry and polarization in remote sensing.
See Abstr. 003.090.

Photons, galaxies and stars. Selected papers of R. Hanbury Brown, Raman Professor, 1974.
See Abstr. 003.103.

Microcomputer control of telescopes.
See Abstr. 003.196.

Guide to observing the moon.
See Abstr. 003.214.

The preparation of the measurements at the PDS at the Padova Observatory: reports on an experience with the development of an astronomical data base environment.
See Abstr. 009.067.

Supercomputers for radio astronomical imaging.
See Abstr. 013.027.

Supercomputers in astronomy.
See Abstr. 013.028.

The UK remote and service observing programme.
See Abstr. 013.074.

The binning analysis: towards a better significance test.
See Abstr. 021.003.

Confidence limits for small numbers of events in astrophysical data.
See Abstr. 021.012.

Magnetic tape read/write operations on PDP–11 series computers.
See Abstr. 021.014.

A data reduction package for multiple object spectroscopy.
See Abstr. 021.020.

Analysis of the kernel part of the NRAO stand–alone FORTH system.
See Abstr. 021.021.

Photographic and spectrographic observations with a reducing camera on the 61 cm NCU telescope.
See Abstr. 032.003.

Nordisk optisk teleskop. Prosjektgruppen på Risø – Mekanikk og styring.
See Abstr. 032.025.

Comparison of a 1.8–meter CCD scannerscope with a 1.2–m Schmidt camera.
See Abstr. 032.026.

The oscillating slit micrometer of the meridian circle PMC 190 Tokyo.
See Abstr. 032.055.

A proposal for scanning meridian circle.
See Abstr. 032.061.

Impersonal techniques of transit circle observations. "The Carlsberg Automatic Meridian Circle".
See Abstr. 032.062.

Internal refraction in meridian circles.
See Abstr. 032.069.

A new method of determining absolute azimuth and latitude and suggestion for a new type of meridian circle.
See Abstr. 032.071.

A new method for zenith distance determination in meridian observations.
See Abstr. 032.072.

The very long baseline interferometry. Part VI. Methods of spectral line VLBI.
See Abstr. 033.022.

The real time acquisition and data handling of the Mi Yun Synthesis Radio Telescope.
See Abstr. 033.044.

The data acquisition system for the MSRT (*Mi Yun Synthesis Radio Telescope*).
See Abstr. 033.045.

Evaluation of the performance of the 576384 Thomson CCD for astronomical use.
See Abstr. 034.007.

Photographic astrometry with short focal length objectives.
See Abstr. 034.035.

Interferometric stellar oscillation spectrometry.
See Abstr. 034.064.

Future techniques and instrumentation in solar–stellar physics.
See Abstr. 034.079.

Laser heterodyne spectrometer for helioseismology.
See Abstr. 034.088.

Mepsicron and the advancement in high resolution image detection.
See Abstr. 034.101.

Two–dimensional astronomical photometry techniques.
See Abstr. 034.153.

Vidicon photometry and astrometry applied to objects of the solar system.
See Abstr. 034.191.

PDS concepts and astronomical applications: an overview.
See Abstr. 034.203.

A microprocessor–based control system for the Vienna PDS microdensitometer.
See Abstr. 034.209.

Why use a microdensitometer other than a PDS?
See Abstr. 034.210.

The automated photographic measuring facility at Cambridge.
See Abstr. 034.213.

A preliminary analysis of a diode array for densitometry.
See Abstr. 034.215.

Microarcsecond astrometric interferometry.
See Abstr. 035.085.

Techniques for dealing with discordant observations.
See Abstr. 041.006.

A review of astrometric observing.
See Abstr. 041.023.

Determination of the stellar right ascension and photoelectric measurement of the time of transit.
See Abstr. 041.036.

Brosche's method for representing systematic differences in positions and proper motions of stars.
See Abstr. 041.051.

Global reduction of fundamental astrometric data.
See Abstr. 041.053.

Trigonometric parallaxes obtained with the UK Schmidt Telescope.
See Abstr. 041.057.

Space Telescope motion limitations for Fine Guidance Sensor astrometry.
See Abstr. 041.070.

Objets accessibles aux astrolabes de haute précision: étoiles et radioétoiles brillantes, planètes, soleil.
See Abstr. 041.076.

The conversion from the B1950 FK4–based position to the J2000 position of celestial objects.
See Abstr. 043.020.

The precision of J2000.0 radio source positions from Mark III VLBI.
See Abstr. 043.022.

Comparison of two methods for the absolute determination of latitude.
See Abstr. 044.048.

The effect of abnormal atmospheric refraction on the precision of time determination.
See Abstr. 044.066.

Research on the fine structures in polar motion with multi–stage filter.
See Abstr. 044.101.

On the use of the collocation method for estimation of the Earth rotation parameters by satellite laser ranging.
See Abstr. 044.105.

Determination of ERP with lunar laser ranging and discussion on the influence of the adopted parameters.
See Abstr. 044.112.

Earth orientation determinations from VLBI observations.
See Abstr. 044.113.

Earth rotation parameters (ERP) from VLBI.
See Abstr. 044.115.

The U.S. Naval Observatory C.O.R.E. (Combined solution for the orientation of the Earth) solution during project MERIT.
See Abstr. 044.118.

Combination of recent polar motion observations.
See Abstr. 044.119.

Combined determination of the ERP from the results of different techniques, Jan. 1, 1980 – Oct. 31, 1984.
See Abstr. 044.120.

Analysis of short periodical variations of pole coordinates determined by different techniques in the MERIT campaign.
See Abstr. 044.127.

A modern Earth orientation monitoring service: functions, goals, and methods of observation.
See Abstr. 044.129.

Intercomparison of UT1 measurements during the MERIT campaign period.
See Abstr. 044.132.

Intercomparison of polar motion measurements during the MERIT period.
See Abstr. 044.133.

A comparison of Earth orientation data obtained by different space techniques.
See Abstr. 044.134.

Comparison and comment on the techniques for determining ERP (Earth Rotation Parameters).
See Abstr. 044.135.

Agreement in polar motion measurements during the MERIT campaign.
See Abstr. 044.136.

MERIT main campaign: reference frame intercomparisons.
See Abstr. 044.137.

Sensitivity of Lageos Earth orientation results to changes in the MERIT standard model.
See Abstr. 044.138.

SLR (Satellite Laser Ranging) and VLBI technology trends.
See Abstr. 044.139.

Measuring Earth rate perturbations with a large passive ring laser gyro.
See Abstr. 044.140.

On the optimal conditions of determining ERP (Earth Rotation Parameters).
See Abstr. 044.141.

Predicting Earth rotation parameters using a multi–channel Wiener filter.
See Abstr. 044.143.

Actualización de los procesos de reducción de observaciones en astronomía geodésica.
See Abstr. 045.020.

GINFEST: geodetic intercomparison network for evaluating space techniques.
See Abstr. 045.036.

High resolution spectroscopy with ST.
See Abstr. 051.095.

Relativistic effects in astronomical timing measurements.
See Abstr. 066.196.

The correction of observed limb darkening for instrumental and atmospheric stray light.
See Abstr. 071.008.

Polish Spot Numbers.
See Abstr. 072.004.

Application of complete linearization method to the analysis of the emission lines of solar prominences.
See Abstr. 073.129.

Numerical image processing applied to the solar corona.
See Abstr. 074.084.

Sky brightness and atmospheric extinction at A.N.R.A.O., Parkes and Siding Spring Observatory.
See Abstr. 082.063.

Spectropolarimetry of planets. I. The procedure of observations and reduction of data.
See Abstr. 091.027.

Infrared heterodyne spectroscopy of planetary atmospheres.
See Abstr. 091.037.

Observing Mars in the 1980s: reference maps for the visual observer.
See Abstr. 097.010.

Identifications of minor planets.
See Abstr. 098.017.

Toward high–precision observation of Galilean satellites.
See Abstr. 099.031.

Surface brightness of a cometary dust tail. I. Calculation technique.
See Abstr. 102.049.

Analysis of meteor data on a microcomputer system.
See Abstr. 104.071.

Meteoorwaarnemingen bij totale maansverduistering.
See Abstr. 104.077.

Statistical "Z" parallaxes.
See Abstr. 111.002.

Trigonometric parallaxes determined with the Yerkes Observatory 40 inch refractor. VI. Measurements made with PDS microdensitometers.
See Abstr. 111.018.

On a correct algorithm for determining photographic relative proper motions of open clusters.
See Abstr. 111.023.

***UBVRI* photometry of red stars.**
See Abstr. 113.016.

On the Johnson *U* passband.
See Abstr. 113.034.

Some recent results on high dispersion spectroscopy.
See Abstr. 114.125.

How well do we know stellar parameters?
See Abstr. 115.016.

Two dimensional images of Alpha Orionis.
See Abstr. 115.022.

Angular expansion measurement with the VLA: the distance to NGC 7027.
See Abstr. 134.018.

Magnesium gradients in elliptical galaxies. I. Method and initial results.
See Abstr. 157.042.

Photometric redshifts of galaxies.
See Abstr. 157.152.

Beobachtung von Centaurus A im MeV–Gammabereich.
See Abstr. 158.248.

Quasar research with Schmidt telescopes.
See Abstr. 159.095.

Electronographic BV photometry of three distant clusters of galaxies. I. Observations and reduction techniques.
See Abstr. 160.003.

Machine measurements of the objectve–prism spectra of faint galaxies – II. Interactive redshift measurements.
See Abstr. 160.019.

Automated data retrieval in observational cosmology.
See Abstr. 161.155.

Positional Astronomy, Celestial Mechanics

041 Astrometry

041.001 **Erratum: "Astrolabe observations of radio stars at Valinhos" [Astron. Astrophys., Suppl. Ser., Vol. 61, No. 2, 211 – 212 (1985)].**
L. B. F. Clauzet, R. Atalla, S. Débarbat.
Astron. Astrophys., Suppl. Ser., Vol. 63, No. 1, p. 204 (1986). See Abstr. 40.041.001.

041.002 **Relativity in astrometry.**
C. A. Murray.
Relativity in celestial mechanics and astrometry, p. 169 – 175 (1986). – See Abstr. 012.002 (IAU Symp. No. 114).
The coordinate dependence of the definition of light deflexion is discussed and the concepts of "natural" and "proper" reference frames are introduced in order to derive astrometric observables from coordinate quantities. Confusion in published formulations of VLBI data analysis is discussed in the context of the IAU definition of the coordinate time scale TDB. It is pointed out that with this definition, the unit of TDB differs from that of TDT and it is suggested that the speed of light in barycentric units be adopted as a defining constant in the system of astronomical constants.

041.003 **Relativistic interpretation of astrometric data.**
N. V. Pavlov.
Relativity in celestial mechanics and astrometry, p. 191 – 192 (1986). Abstract. – See Abstr. 012.002 (IAU Symp. No. 114).

041.004 **A formal approach to stellar positions in solar barycentric coordinates.**
J. F. Lestrade.
Relativity in celestial mechanics and astrometry, p. 213 (1986). Abstract. – See Abstr. 012.002 (IAU Symp. No. 114).

041.005 **Relativistic effects included in the apparent positions of fundamental stars (APFS).**
T. Lederle, H. Schwan.
Relativity in celestial mechanics and astrometry, p. 223 – 227 (1986). – See Abstr. 012.002 (IAU Symp. No. 114).
The relativistic effects included in the APFS from 1984 onwards are described. They are discussed in comparison with the accuracy available with modern techniques of observation.

041.006 **Techniques for dealing with discordant observations.**
R. L. Branham Jr.
Relativity in celestial mechanics and astrometry, p. 229 – 231 (1986). – See Abstr. 012.002 (IAU Symp. No. 114).

041.007 **Space astrometry prospects and limitations.**
J. Kovalevsky, F. Mignard, M. Froeschlé.
Relativity in celestial mechanics and astrometry, p. 369 – 382 (1986). – See Abstr. 012.002 (IAU Symp. No. 114).
Among the few parameters that describe the generalized space–time metrics, astrometric techniques are essentially sensitive to the displacement of the apparent positions of celestial bodies. This includes the relativistic light deflection and aberration. The possibilities of small field and wide field astrometry in measuring these effects are described. The case of the second order aberration terms is considered with some detail from the theoretical point of view, both for stellar and planetary aberration. New results are presented in the latter case. A section is devoted to a description of the existing space astrometry projects among which Space Telescope and HIPPARCOS are approved but will not contribute significantly to relativistic studies.

041.008 **Cometary coordinate systems for spacecraft data interpretation.**
N. Divine.
IHW Newsl., No. 8, p. 11 – 17 (1986).
Vector quantities will play important roles in the interpretation of data from the spacecraft encounters with Comet P/Halley in March 1986. To facilitate the comparison of results from the several on–board experiments, it is suggested that components of such vector quantities be expressed in one or more of the five candidate coordinate systems defined and illustrated in the paper.

041.009 **Optical astrometry of extragalactic radio sources.**
H. G. Walter, R. M. West.
Astron. Astrophys., Vol. 156, No. 1/2, p. 1 – 7 (1986).
Precise optical positions of 50 extragalactic radio sources between –45° and +5° declination have been uniformly derived from Schmidt plates relative to reference stars of the Perth 70 Catalogue. The mean internal error is 0″.19 in right ascension and 0″.21 in declination. The external error is estimated at 0″.23 to 0″.37 in both coordinates by accounting for errors contributed by the reference stars, errors of the systematic differences between the Perth 70 and FK4 catalogues and errors of the FK4 system itself. It is concluded that the radio survey is well tied to the FK4 system. Three new identifications of optical counterparts are reported, others are confirmed.

041.010 **Expansion of annual aberration into trigonometric series.**
C. Ron, J. Vondrák.
Bull. Astron. Inst. Czech., Vol. 37, No. 2, p. 96 – 103 (1986).
The vector of the Earth's instantaneous velocity with respect to the center of mass of the solar system, needed for the precise reduction of annual aberration, is expanded into a series of trigonometric terms whose arguments are the mean longitudes of the planets and the Moon, and Delaunay arguments, *l, l′, F, D*. The development is based on Bretagnon's planetary theory *VSOP 82* for the heliocentric motion of the planets and Chapront–Touzé's and Chapront's theory *ELP 2000/82* of the geocentric motion of the Moon.

041.011 **Un modello per la ricostruzione della sfera celeste.**
I. Galligani.
G. Astron., Vol. 11, N. 4, p. 419 – 425 (1985).

041.012 **Basis of the international program IRS (program of observations, selection of the reference system).**
E. V. Khrutskaya.
Glav. astron. obs. Akad. Nauk SSSR. Leningrad, 13 pp. (1985). In Russian. Abstr. in Ref. Zh., 51. Astron., 1.51.107 (1986).

041.013 **Right ascensions of the Sun, Mercury and Venus observed with the transit instrument in Nikolaev in 1981.**
G. M. Petrov, R. T. Fedorova, P. N. Fedorov.
Glav. Astron. Obs. Akad. Nauk SSSR. Leningrad, 6 pp. (1985). In Russian. Abstr. in Ref. Zh., 51. Astron., 1.51.111 (1986).

041.014 **Right ascensions of the Sun, Mercury and Venus observed with the transit instrument in Nikolaev in 1983.**
G. M. Petrov, R. T. Fedorova, P. N. Fedorov.
Glav. Astron. Obs. Akad. Nauk SSSR. Leningrad, 10 pp. (1985). In Russian. Abstr. in Ref. Zh., 51. Astron., 1.51.112 (1986).

041.015 **Right ascensions of the Sun, Mercury and Venus observed with the transit instrument in Nikolaev in 1984.**
G. M. Petrov, R. T. Fedorova, P. N. Fedorov.
Glav. Astron. Obs. Akad. Nauk SSSR. Leningrad, 10 pp. (1985). In Russian. Abstr. in Ref. Zh., 51. Astron., 1.51.113 (1986).

041.016 **Declinations of the Sun, Mercury and Venus obtained from observations with the vertical circle of the Nikolaev Observatory in 1983.**
P. V. Majgurov, V. P. Sibilev.
Glav. Astron. Obs. Akad. Nauk SSSR. Leningrad, 8 pp. (1985). In Russian. Abstr. in Ref. Zh., 51. Astron., 1.51.114 (1986).

041.017 **Declinations of the Sun, Mercury and Venus obtained from observations with the vertical circle of the Nikolaev Observatory in 1984.**
N. V. Kostrubina, P. V. Majgurov, V. P. Sibilev.
Glav. Astron. Obs. Akad. Nauk SSSR. Leningrad, 11 pp. (1985). In Russian. Abstr. in Ref. Zh., 51. Astron., 1.51.115 (1986).

041.018 **Relativistic effects in variations of right ascensions of stars and clock corrections.**
R. K. Kadyev.
Issled. po teor. i yader. fiz. Samarkand, p. 4 – 8 (1985). In Russian. Abstr. in Ref. Zh., 51. Astron., 2.51.133; 52. Geod. Aehrosemka, 5.52.99 (1986).

041.019 **Use of Chebyshev polynomials for computing the declinations of the sun and the equation of time.**
I. V. On'kov, S. N. Loginovskij, A. S. Lebedev.
Geod., kartogr., No. 8, p. 34 – 36 (1985). In Russian. Abstr. in Ref. Zh., 52. Geod. Aehrosemka, 2.52.85 (1986).

041.020 **Numerical integration of precession and nutation.**
Y. Kubo.
Proceedings of the nineteenth symposium on celestial mechanics, p. 76 – 81 (1986). In Japanese. – See Abstr. 012.013.

041.021 **The Southern Reference Star program: a 20–year international project nearing completion.**
C. Smith, E. Jackson.
Celest. Mech., Vol. 37, No. 3, p. 277 – 284 (1985). – See Abstr. 012.017.

The present paper describes progress in the compilation of observations of the Southern Reference Star program which were made from 1961 to 1973 into a reference catalog of mean positions at the mean epoch of observation referred to the system of the FK4 catalog.

041.022 **The extension of the fundamental system to fainter magnitudes.**
T. E. Corbin.
Celest. Mech., Vol. 37, No. 3, p. 285 – 298 (1985). – See Abstr. 012.017.

During the past 100 years the fundamental system has developed from a small list of 623 bright stars to one of 1535 stars. The compilation of the next fundamental catalog, the FK5, will involve the addition of some 3200 stars and will extend the system to magnitude 9.0. A number of criteria were used for selecting the 2000 stars in the seventh to ninth magnitude portion of the extension, the Faint Fundamentals. These criteria include distribution in magnitude and spectral type, quality of the observational histories, mean errors of the proper motions and distribution over the sky. These criteria have proven difficult to satisfy, especially at the faint end of the magnitude range and in the Southern Hemisphere.

041.023 **A review of astrometric observing.**
R. S. Harrington.
Celest. Mech., Vol. 37, No. 3, p. 325 – 328 (1985). – See Abstr. 012.017.

041.024 **Results of a determination of absolute declinations of stars with the Sanders–Raimond method from Gordon's observations.**
A. A. Krasikova.
Izv. Astron. Ehngel'gardt. Obs., No. 49, p. 75 – 81 (1985). In Russian.

Absolute declinations of 119 stars obtained from Gordon's observations in Pulkovo during 1939 – 1941 are presented. The results of a comparison of Gordon's catalogue with the AGK3 and GC are given.

041.025 **Investigation of the accuracy of the azimuthal angle–measuring system in the near zenith zone.**
V. N. Barabanov, R. A. Aleksankina.
Izv. vuzov. Geod. Aehrofotosemka, No. 5, p. 88 – 94 (1985). In Russian. Abstr. in Ref. Zh., 52. Geod. Aehrosemka, 3.52.87 (1986).

041.026 **Calculation of the equatorial coordinates of the sun.**
N. V. Ryabov.
Geod. kartogr., No. 11, p. 18 – 19 (1985). In Russian. Abstr. in Ref. Zh., 52. Geod. Aehrosemka, 3.52.88 (1986).

041.027 **Results of first observations with a meridian circle under mountain conditions. II. Processing of right ascensions.**
V. G. Shamaev, N. N. Simonov, T. L. Sorokina.
Astron. Tsirk., No. 1385, p. 5 – 7 (1985). In Russian.

041.028 **Expansion of the AGK3 system to the positions and proper motions of bright stars in the vicinity of galaxies.**
S. P. Rybka.
Kinematika Fiz. Nebesn. Tel, Tom 2, No. 3, p. 30 – 33 (1986). In Russian. English translation in Kinematics Phys. Celest. Bodies.

Accurate positions and proper motions of 592 bright stars ($m < 12^m$) in the vicinity of 15 areas with galaxies are determined. The accuracy of star positions is $\varepsilon_\alpha \cos\delta = \pm 0\farcs20$, $\varepsilon_\delta = \pm 0\farcs18$ (1982.1), that of proper motions is $\varepsilon_{\mu_\alpha} \cos\delta = \pm 0\farcs0106$ per year and $\varepsilon_{\mu_\delta} = \pm 0\farcs0095$ per year.

041.029 **A method for combined determination of astronomical coordinates from observations of star pairs at equal heights.**
Yu. N. Agafonov.
Geod., kartogr., No. 10, p. 13 – 16 (1985). In Russian. Abstr. in Ref. Zh., 52. Geod. Aehrosemka, 4.52.58 (1986).

041.030 **Determination of zero points and systematic errors of fundamental catalogs from observations of earth satellites with imprecise times.**
R. A. Zeinalov (*R. A. Zejnalov*).
Sov. Astron., Vol. 29, No. 4, p. 470 – 474 (1985). English translation of 40.041.008.

041.031 **Measurements of pulsar parallaxes by VLBI.**
C. R. Gwinn, J. H. Taylor, J. M. Weisberg, L. A. Rawley.
Astron. J., Vol. 91, No. 2, p. 338 – 342 (1986). With a correction in No. 4, p. 974.

The authors present results of a program of astrometric observations of pulsars by very long baseline interferometry, including two measurements of annual parallaxes. The authors also observe the absence of relative motion of two quasar pairs. Observations were made at 1.66 GHz over a period of 18 months. For pulsars PSR 0823+26 and PSR 0950+08 parallaxes of 2.8 ± 0.6 and 7.9 ± 0.8 mas, respectively, are obtained. In combination with published measurements of pulse dispersion, the results yield average free–electron densities of 0.054 ± 0.011 and $0.023 \pm 0.002\ \mathrm{cm}^{-3}$ along the lines of sight to these pulsars.

041.032 **Astrometry with ST.**
M. G. Fracastoro.
Mem. Soc. Astron. Ital., Vol. 56, No. 4, p. 579 – 582 (1985). – See Abstr. 012.054.

041.033 **Recalculation of the magnitude equation of bright stars and of Newcomb's precession corrections for Fatchikhin's catalogue of proper motions of 14600 stars.**
V. V. Bobylev.
Astron. Zh., Tom 63, Vyp. 3, p. 584 – 588 (1986). In Russian. English translation in Sov. Astron., Vol. 30, No. 3.
The magnitude equation was recalculated for the bright (8.0 to 10.5 photographic magnitude) stars from Fatchikhin's catalogue. It was found to be: $\Delta m_x = -0''.0088 \pm 0''.0016$, $\Delta m_y = +0''.0115 \pm 0''.0024$ (mean errors). With this value of the magnitude equation corrections of Newcomb's constant of precession were computed for the 768 stars common with the AGK3 catalogue. These are equal to (for 100 years): $\Delta k = -0''.17 \pm 0''.10$, $\Delta n = +0''.29 \pm 0''.20$ (probable errors).

041.034 **Right ascensions of the Sun, Mercury and Venus observed with the transit instrument in Nikolaev in 1980 – 1984 with corrections for lateral refraction.**
P. N. Fedorov.
Glav. Astron. Obs. Akad. Nauk SSSR. Leningrad, 13 pp. (1986). In Russian. Abstr. in Ref. Zh., 51. Astron., 6.51.148 (1986).

041.035 **Right ascensions of the Sun, Mercury and Venus observed in 1981 – 1982 near Kislovodsk with corrections for lateral refraction.**
P. N. Fedorov.
Glav. Astron. Obs. Akad. Nauk SSSR. Leningrad, 9 pp. (1986). In Russian. Abstr. in Ref. Zh., 51. Astron., 6.51.149 (1986).

041.036 **Determination of the stellar right ascension and photoelectric measurement of the time of transit.**
J. Bem.
Acta Universitatis Wratislaviensis, No. 641, Matematyka, Fizyka, Astronomia, XLII, 55 pp. Price zł 35.00 (1983). = Contrib. Wrocław Astron. Obs., No. 20.
The first part of the paper contains a conception of immediate calculation of correction to the right ascension of stars using the atomic time scale and independent information about irregularities of the earth's rotation. The second part describes a method and arrangement of the measurement of the time of transit of the image of a star across the slit system.

041.037 **Computation of correction values for the determination of planetary right ascensions with the photoelectric transit instrument.**
C.–a. Sun, B.–y. Yang.
Publ. Purple Mt. Obs., Vol. 4, No. 4, p. 23 – 32 (1985). In Chinese.

041.038 **The application of the proper reference framework in astronomy.**
L.–h. Li.
Publ. Purple Mt. Obs., Vol. 4, No. 4, p. 33 – 52 (1985). In Chinese.
The application of the proper reference framework to coordinate transformations is discussed, and the formulae and procedures, which are used to calculate the relativistic effects of coordinate transformations between popular astronomical reference frameworks, are given.

041.039 **Results of a comparison of the Kharkov catalogue of declinations of 1407 nearpolar stars with the catalogue of Fabricius.**
L. S. Pavlenko.
Vestn. Khar'kov. Univ., No. 278, p. 69 – 72 (1985). In Russian.

041.040 **An improved high–precision instrument for astrometric observations under lunar conditions.**
A. A. Gurshtein (A. A. Gurshtejn).
Sov. Astron., Vol. 29, No. 6, p. 708 – 713 (1985). English translation of 40.041.041.

041.041 **A method of correction for the determination of planetary right ascensions by photoelectric transit instruments.**
B.–y. Yang, C.–a. Sun.
Acta Astron. Sin., Vol. 27, No. 1, p. 23 – 29 (1986). In Chinese.
This paper deals with the problem of how the displacement between the photocenter and the true center of planetary surface varies with the phase angle of the planet. It leads to a method of correction for the determination of planetary right ascensions by photoelectric transit instruments.

041.042 **Apparent and geometric star positions for the HIPPARCOS mission.**
H. G. Walter, F. Mignard, R. Hering, M. Froeschlé, J. L. Falin.
Manuscr. Geod., Vol. 11, No. 2, p. 103 – 114 (1986).
Geometric and apparent star positions are defined, and the mode of application to the HIPPARCOS data reduction is described. Two independent approaches are treated, the one using ecliptic, the other equatorial coordinates. Both methods agree within 0.5×10^{-3}mas for the geometrical positions and 1×10^{-3}mas for the apparent positions. On the basis of the uncertainties of the earth's and spacecraft's positions the external error of the apparent positions is estimated at 0.2 mas.

041.043 **The sphere reconstitution problem in the HIPPARCOS project.**
I. Galligani, B. Betti, P. L. Bernacca.
Manuscr. Geod., Vol. 11, No. 2, p. 124 – 132 (1986).

041.044 **The continuous analog of the sphere reconstruction in HIPPARCOS project.**
B. Betti, F. Sansò.
Manuscr. Geod., Vol. 11, No. 2, p. 133 – 145 (1986).

041.045 **Geodetic, geometric and computational aspects of HIPPARCOS.**
D. T. van Daalen, H. van der Marel.
Manuscr. Geod., Vol. 11, No. 2, p. 146 – 166 (1986).

041.046 **Absolute azimuth at low latitude with a polarissima.**
R. Teixeira, P. Benevides–Soares.
Rev. Mex. Astron. Astrofis., Vol. 12, Spec. Issue, p. 421 – 423 (1986). – See Abstr. 012.066.
A series of observations of the polarissima star FK4Suppl 3985 ($\delta = -89°42'$, m = 6.6) is discussed. The absolute azimuth of the meridian marks is obtained with the accuracy of 0''.85 for a single observation.

041.047 **Astrolabe observation at Natal, Brazil.**
S. Takagi, J. F. da Silva.
Rev. Mex. Astron. Astrofis., Vol. 12, Spec. Issue, p. 438 – 440 (1986). – See Abstr. 012.066.

041.048 **Optical observations of radio stars with the astrolabe at Valinhos.**
L. B. F. Clauzet, R. M. Atalla.
Rev. Mex. Astron. Astrofis., Vol. 12, Spec. Issue, p. 446 (1986). Abstract. – See Abstr. 012.066.

041.049 **Relativistic reduction of astrometric observations.**
V. A. Brumberg.
Astrometric techniques, p. 19 – 42 (1986). – See Abstr. 012.068 (IAU Symp. No. 109).
The nonuniqueness of the quasi–Galilean coordinates of general relativity leads to the emergence of unmeasurable coordinate–dependent quantities in astronomical practice. One may offer three possible ways to overcome the related difficulties: (1) developing theoretical conclusions only in terms of measur-

able quantities, (2) using arbitrary coordinates and developing an unambiguous procedure for comparing measurable and calculated quantities, and (3) agreement to utilize one and only one coordinate system. The author prefers the second way. For illustration, the relativistic reduction of stellar parallaxes, Doppler satellite observations, navigation measurements with the aid of satellites and radiointerferometric measurements are presented in detail.

041.050 **Computation of compilation catalogs.**
T. E. Corbin.
Astrometric techniques, p. 75 – 86 (1986). – See Abstr. 012.068 (IAU Symp. No. 109).

The computation of mean positions and proper motions for the International Reference Stars (IRS) is hampered by large variations in the observational histories of the stars and lack of overlap between the magnitudes of the IRS and of the FK4. The much–needed extension of the fundamental system to the ninth magnitude will be made in the FK5. The Faint Fundamental Extension is currently being selected at the U.S. Naval Observatory. A proposed list of 1030 faint fundamental stars has been prepared for the northern hemisphere, and work has begun on the selection in the southern hemisphere.

041.051 **Brosche's method for representing systematic differences in positions and proper motions of stars.**
V. V. Vityazev.
Astrometric techniques, p. 87 – 94 (1986). – See Abstr. 012.068 (IAU Symp. No. 109).

A variant of Brosche's well–known method is proposed which allows one to represent systematic differences many times faster than with the original method by the use of spherical harmonics. The proposed method was tested on models and used for representing systematic corrections to the catalogues GC and N30.

041.052 **Some effects of different sources of variation of latitude data on meridian circle catalogues.**
T. J. Rafferty.
Astrometric techniques, p. 95 – 102 (1986). – See Abstr. 012.068 (IAU Symp. No. 109).

Fundamental meridian circle observations require the variation of latitude to be applied before the observations are combined to form a catalogue. In the past 80 years, different observatories have used various determinations of polar motion to reduce their observations. This study shows that the use of different determinations of the latitude variation significantly affects the individual observations and can introduce large zonal errors of the level of 0.05 seconds of arc into the catalogue's declination system.

041.053 **Global reduction of fundamental astrometric data.**
P. Benevides–Soares, L. B. F. Clauzet.
Astrometric techniques, p. 103 – 111 (1986). – See Abstr. 012.068 (IAU Symp. No. 109).

Fundamental observational programs are often lengthened because of the sparsity of data. The whole set of observational material is treated as a single least squares problem, whose unknowns include corrections to the star positions. The problem is usually of huge dimensions, but the authors show that it can be reduced to quite tractable sizes. As an example, the method is applied to a two year series of astrolabe observations. The time and latitude curves are solved for under the form of cubic spline functions.

041.054 **The use of the RATAN–600 radio telescope in astrometry.**
P. M. Afanasieva (*P. M. Afanas'eva*), V. A. Fomin,
Yu. K. Zverev, M. G. Mingaliev, V. N. L'vov, A. A. Pozhalov.
Astrometric techniques, p. 169 – 171 (1986). – See Abstr. 012.068 (IAU Symp. No. 109).

The RATAN–600 radio telescope was designed for solving various problems of astrophysics and radiophysics. Because of its high resolving power and large collecting area, this instrument can be used to measure accurate positions of radio sources. The precision of differential positional observations at wavelength 2.08 cm at the azimuths $\pm 60°$ at this time corresponds to standard errors of 0.″3 to 0.″6 in declination and 0.″5 in the right ascension for declinations between +44° and –10°. Further south, down to declinations of about –25°, observations at $\pm 30°$ azimuths will result in standard errors of a single declination determination of about 1″.

041.055 **Preliminary precise radio/optical positions of selected stars.**
D. R. Florkowski, K. J. Johnston, C. de Vegt, C. M. Wade.
Astrometric techniques, p. 179 – 181 (1986). – See Abstr. 012.068 (IAU Symp. No. 109).

Positions of the radio emission associated with stars were determined along with precise optical positions of the stars. The high precision obtained shows that radio star observations are a practical means of determining the relationship between the optical and radio reference frames.

041.056 **Plans for the second epoch of the Southern Proper Motion Program.**
C. E. López, J.–F. Lee, W. van Altena.
Astrometric techniques, p. 209 – 211 (1986). – See Abstr. 012.068 (IAU Symp. No. 109).

The first photographs for the Yale–San Juan Southern Proper Motion program with respect to faint galaxies were taken with the collaboration of Columbia University in 1965. The first epoch photography was essentially completed in 1974 and plans are now under way to begin the second epoch observations in 1986.

041.057 **Trigonometric parallaxes obtained with the UK Schmidt Telescope.**
C. A. Murray, P. M. Corben, R. W. Argyle.
Astrometric techniques, p. 213 – 221 (1986). – See Abstr. 012.068 (IAU Symp. No. 109).

In continuation of a long term programme of measurement of parallaxes and proper motions in the South Galactic Cap from plates taken with the UK Schmidt Telescope and measured in GALAXY at RGO, data for some 800 red stars brighter than B = 17.5 have been derived from two independent plate series. The combined trigonometric parallaxes show very satisfactory agreement with photometric parallaxes, and the average r.m.s. error of a combined parallax is found to be about $\pm$0.″012 both from internal and external evidence.

041.058 **Optical counterparts of extragalactic radio sources in the southern sky: positions obtained from Schmidt plates.**
H. G. Walter.
Astrometric techniques, p. 223 – 228 (1986). – See Abstr. 012.068 (IAU Symp. No. 109).

The usefulness of Schmidt plates for measuring astrometric positions of extragalactic radio sources with an average uncertainty of less than 0.3 arcseconds is emphasized in the discussion of the results on 50 optical counterparts and by the comparison with corresponding positions obtained through radio interferometry.

041.059 **Variations in external parallax errors.**
A. R. Upgren.
Astrometric techniques, p. 247 – 251 (1986). – See Abstr. 012.068 (IAU Symp. No. 109).

The present astrometric program of the Van Vleck Observatory began in 1967. An analysis was made from parallax and proper motion data from 70 stars in the spectral range dK3 – M2 (Upgren 1973) which found the average external mean error in parallax to be 8.0 ± 1.7 mas (milliarcseconds). Hanson and Lutz (1983) confirm this result using the parallaxes of 14 members of the Hyades cluster also determined at Van Vleck. They find a mean external parallax error of 9.4 ± 1.8 mas and suggest that the proximity of the two determinations may be evidence that Van Vleck parallaxes may be characterized by a single external error.

041.060 **Testing Schmidt plates for astrometric purposes.**
J. L. Russell, C. A. Williams.
Astrometric techniques, p. 259 – 270 (1986). – See Abstr. 012.068 (IAU Symp. No. 109).

The Guide Star Selection System for the Space Telescope is being developed to find positions (±0.25 arcsec) and magnitudes (±0.4 mag) for guide star pairs for every possible ST target. The data available for the astrometric portion of this task are measurements (±1 μm) of plate surveys from the Palomar and UK Schmidt telescopes and positional data from the AGK3 and the SAOC. Described here are the test procedures derived to measure the effect of the astrometric accuracy of the guide star positions on ST pointing success.

041.061 **The Bordeaux automatic transit circle: first catalogues, current and future programs.**
Y. Requième, J. M. Mazurier.
Astrometric techniques, p. 435 – 442 (1986). – See Abstr. 012.068 (IAU Symp. No. 109).

The Bordeaux automatic meridian circle has been used for routine differential observations for three years. Current programs concern NPZT stars, faint nearby stars from the Gliese catalogue, Mars, Uranus, Neptune, Jupiter and Saturn satellites, and minor planets. From 1984 on, 5000 faint stars from the HIPPARCOS Input Catalogue will be observed.

041.062 **Current work with the photoelectric transit instrument at the Observatory of Torino.**
G. Chiumiento, M. G. Fracastoro, M. Sarasso.
Astrometric techniques, p. 443 – 447 (1986). – See Abstr. 012.068 (IAU Symp. No. 109).

041.063 **Carlsberg Automatic Transit Circle: first two test catalogues and the programme for La Palma.**
L. V. Morrison, P. Gibbs.
Astrometric techniques, p. 497 – 505 (1986). – See Abstr. 012.068 (IAU Symp. No. 109).

The first two catalogues produced from test observations made with the Carlsberg Automatic Transit Circle (CATC) at the Copenhagen University Observatory in the period 1981/83 have internal mean errors of 0".21 in right ascension and 0".22 in declination. In 1983 the CATC was moved to the international Observatorio del Roque de los Muchachos where it will begin a programme of differential and fundamental astrometry in 1984.

041.064 **The transit instrument under optimum conditions.**
G. M. Petrov.
Astrometric techniques, p. 517 – 524 (1986). – See Abstr. 012.068 (IAU Symp. No. 109).

At high geographical latitudes the polar night conditions are nearly optimal for absolute determinations of stellar coordinates; high altitude sites near the equator are best for day time observations. It is suggested that observatories be built for absolute position determinations on Spitzbergen, and on McMurdo on the Antarctic Continent. The link between observations in the northern and the southern hemispheres can be established through observations at a high–altitude equatorial observatory, where the Sun and the planets can also be observed.

041.065 **Comparison of Tokyo PZT catalogues with AGK3 and with three other independent catalogues.**
S. Sadžakov, M. Dačić, V. A. Fomin.
Astrometric techniques, p. 571 – 580 (1986). – See Abstr. 012.068 (IAU Symp. No. 109).

Results are presented of the comparison of five Tokyo PZT catalogues with the AGK3 and three other independent catalogues obtained with meridian circles.

041.066 **HIPPARCOS satellite and the organization of the project.**
J. Kovalevsky.
Astrometric techniques, p. 581 – 591 (1986). – See Abstr. 012.068 (IAU Symp. No. 109).

A sketch of the past and future development of the project is given. The resulting astrometric parameters and magnitudes of 100000 stars should be available circa 1993. The main features of the satellite are presented. The main limitations that have been identified are discussed: basic angle unstability, mechanical jitter, grid irregularity, diffraction chromatism, shape of the image dissector sensitivity profile. All these effects are taken into account when assessing a final accuracy of 0".002 for the astrometric parameters of a 9–th magnitude star.

041.067 **Preparation of the Hipparcos Input Catalogue. Astrometric programs for Hipparcos. Preliminary astrometric observations.**
C. Turon–Lacarrieu, Y. Réquième.
Astrometric techniques, p. 605 – 612 (1986). – See Abstr. 012.068 (IAU Symp. No. 109).

The various aspects of the preparation of the Hipparcos Input Catalogue are presented. The astrometric parts of the work are pointed out in more details, especially the new observations and measurements already undertaken. The astrometric proposals submitted for observation with Hipparcos are also presented, as they will allow the link between Hipparcos, Space Telescope and ground–based astrometric programs.

041.068 **The use of Space Telescope to tie the HIPPARCOS reference frame to an extragalactic reference frame.**
P. D. Hemenway, R. L. Duncombe.
Astrometric techniques, p. 613 – 624 (1986). – See Abstr. 012.068 (IAU Symp. No. 109).

414 stars, mostly in the magnitude range 9–11, in the vicinity of 175 radio sources (mainly QSOs and BL Lac objects) have been submitted to the HIPPARCOS project. A subset of these objects will be observed with the NASA/ESA Space Telescope's Fine Guidance Sensors to determine the positions and motions of the extragalactic objects in the HIPPARCOS instrumental system, and thence to derive the overall rotation of the HIPPARCOS instrumental system with respect to the extragalactic system.

041.069 **TYCHO astrometry and photometry.**
E. Høg.
Astrometric techniques, p. 625 – 635 (1986). – See Abstr. 012.068 (IAU Symp. No. 109).

The HIPPARCOS Star Mappers were designed to determine the satellite attitude by observation of stars with known positions. ESA has now enhanced them so that the continuous photon counts from two photomultipliers with B and V response will be recorded. The records will be used to derive an astrometric and photometric catalogue, the TYCHO Catalogue, complete to about B = 11 mag and thus containing about 500000 stars. The positions, annual proper motions and parallaxes will have an accuracy of about 0.03 arcsec and the magnitudes one of 0.03 mag. The instrument is described and some aspects of the data reduction are discussed.

041.070 **Space Telescope motion limitations for Fine Guidance Sensor astrometry.**
A. Fresneau.
Astrometric techniques, p. 643 – 647 (1986). – See Abstr. 012.068 (IAU Symp. No. 109).

Basic Fine Guidance Sensor data have been simulated to correlate the power spectrum densities of the Space Telescope motion disturbances with the "seeing" of an astrometric target. The goal of this study is to describe whether a complete identification of the jitter of the line of sight during an astrometric observation is required to improve the precision of this observation. The present preliminary results indicate that the performance in astrometry are close to the predictions.

041.071 **Astrometric desiderata for nearby stars.**
W. Gliese.
Astrometric techniques, p. 653 – 667 (1986). – See Abstr. 012.068 (IAU Symp. No. 109).

Astrometric desiderata for nearby stars concern, in first line, the parallax measurements, avoidance and elimination of effects of bias in trigonometric parallax programs: bias towards large proper motions, preference of parallaxes with positive accidental errors, the Malmquist bias. For some of the nearby stars one does not yet know accurate positions, for some other objects no reliable proper motions were measured. Finally, for calibrating color–luminosity relations the necessity to observe further precise trigonometric parallaxes is demonstrated and emphasized.

041.072 **On desiderata for star catalogs for the remainder of the twentieth century: a report on catalog work now in progress at the U.S. Naval Observatory.**
C. Smith.
Astrometric techniques, p. 669 – 676 (1986). – See Abstr. 012.068 (IAU Symp. No. 109).

The results of observations of Southern Reference Stars (SRS) of all participants in the SRS program (1961 – 1973) have been received at the U.S. Naval Observatory. It is anticipated that compilation of the final SRS catalog of mean positions will be completed by the end of 1984. A supplementary list to the International Reference Star (IRS) catalog is described. A comparison of solutions for equator and equinox corrections and corrections to the assumed elements of the earth's orbit from daytime observations of the Sun, Mercury, and Venus and nighttime observations of Ceres, Pallas, Juno, and Vesta from Six–inch Transit Circle observations made during the W5–50 program (1963 – 1971) shows that a very significant part of the systematic daytime error has been removed.

041.073 **Comparison of instruments and methods of positional observations of the Sun and major planets.**
A. S. Kharin.
Astrometric techniques, p. 685 – 689 (1986). – See Abstr. 012.068 (IAU Symp. No. 109).

During the time interval from 1960 to 1980, altogether 19 meridian instruments, 12 astrographs and 5 astrolabes took part in observations of the Sun and the major planets. More than 40,000 observations of single coordinates – right ascension and declinations have been obtained. The meridian method was used for the largest number of observations accounting for about 80% of all observations. The smallest contributions, less than 2% of the total number, was made by observations with astrolabes.

041.074 **On the SRS catalogue.**
M. Z. Zverev, D. D. Polozhentsev, E. A. Stepanova, E. V. Khrutskaya, L. I. Yagudin, A. D. Polozhentsev.
Astrometric techniques, p. 691 – 696 (1986). – See Abstr. 012.068 (IAU Symp. No. 109).

The completion of the SRS general catalogue is a result of an important international undertaking where 130 observers and 21 observatories took part. Due to the errors of proper motions the precision of the compiled catalogue will deteriorate with time. The approximate accuracy of the SRS catalogue for the epoch 1970 is $\pm 0^{s}.005$ and $\pm 0''.12$, respectively.

041.075 **The establishment of an astrometric standard region – a description of the method with reference to the astrometric standard region in Praesepe (M44).**
J. L. Russell.
Astrometric techniques, p. 697 – 703 (1986). – See Abstr. 012.068 (IAU Symp. No. 109).

An astrometric standard catalog has been derived for the region centered on the Praesepe open star cluster. This catalog includes positions and proper motions of 408 objects on the system of the FK4 reduced from 9 catalogs and 145 plates from 15 different telescopes and spanning 110 years. The accuracy of the positions ranges from 0.01 to 0.15 arcsec and for the proper motions from 0.0002 to 0.004 arcsec/yr.

041.076 **Objets accessibles aux astrolabes de haute précision: étoiles et radioétoiles brillantes, planètes, soleil.**
F. Chollet, S. Débarbat.
Astrometric techniques, p. 705 – 713 (1986). – See Abstr. 012.068 (IAU Symp. No. 109).

Capabilities of the Danjon astrolabe and its transformation are examined regarding both fields of instrumental modifications and improvement of the reduction techniques. For bright objects (stars and radiostars, planets, Galilean satellites, Sun) these instruments will insure a valuable contribution to results obtained by other techniques.

041.077 **Astrolabe observations of radio stars at the southern hemisphere.**
F. Noël.
Astrometric techniques, p. 715 – 719 (1986). – See Abstr. 012.068 (IAU Symp. No. 109).

A brief description is given of the program of radio star observations with the Danjon Astrolabe of Santiago.

041.078 **The Guide Star Selection System and the Guide Star Catalog for Space Telescope.**
J. L. Russell.
Astrometric techniques, p. 721 – 727 (1986). – See Abstr. 012.068 (IAU Symp. No. 109).

The Guide Star Selection System (GSSS) will provide relative positions of two guide stars and the target for each observation of the Space Telescope as well as photometry for the guide stars. The GSSS is a production system which is based on PDS measurements of Schmidt survey plates. The specified accuracy for the system is ± 0.33 arcsec for the separation of the guide stars. The Guide Star Catalog will include the list of all possible guide stars and brighter, positions (at least ± 1 arcsec) and magnitudes (at least ± 0.4 mag) for essentially the whole sky complete to visual magnitude 15.

041.079 **Results of VLBI observations of radio stars and their potential for linking the HIPPARCOS and extragalactic reference frames.**
J.–F. Lestrade, R. A. Preston, A. E. Niell, R. L. Mutel, R. B. Phillips.
Astrometric techniques, p. 779 – 788 (1986). – See Abstr. 012.068 (IAU Symp. No. 109).

Through VLBI observations of a sample of 20 known radio stars the authors have identified 11 stars that should be appropriate for both astrometric VLBI and HIPPARCOS observations. The measurements indicate that the angular extent of their radio emitting regions is small, i.e. < 3 mas for 7 of them. Most of these radio stars belong to the RS CVn class of binary systems.

Catalogue of right ascensions of 241 near–equatorial stars in the system of the Catalogue of the Time Service.
See Abstr. 002.025.

Differential catalogue of declinations of 2884 stars of the latitude programs.
See Abstr. 002.037.

Comparison of the differential catalogue of declinations of 2884 stars of the latitude programs with the AGK3R catalogue.
See Abstr. 002.038.

Quasars to B > 22.5 in Selected Area 57: a catalog of multicolor photometry, variability, and astrometry.
See Abstr. 002.053.

Unified catalog of Washington and Mizusawa PZT stars.
See Abstr. 002.056.

Catalog of O–B stars observed with Tokyo Meridian Circle.
See Abstr. 002.057.

Les catalogues d'étoiles variables et le satellite astrométrique HIPPARCOS.
See Abstr. 002.062.

Finding charts of radio stars for the HIPPARCOS mission.
See Abstr. 002.063.

Progress report on the Bonner Durchmusterung.
See Abstr. 002.065.

The preparation and distribution of machine–readable astrometric data.
See Abstr. 002.087.

The status of the plates taken for the charts and the catalogue of the Carte du Ciel (Astrographic Catalogue).
See Abstr. 002.088.

IAU Symposium 109, Gainesville, "Astrometric techniques". Concluding remarks.
See Abstr. 011.041.

Observing programs planned for the Tokyo photoelectric meridian circle.
See Abstr. 013.031.

Astrometry in China.
See Abstr. 013.064.

The need for better co–operation and intercomparison in fundamental astrometry.
See Abstr. 013.065.

Conversion to MELCOM COSMO 900II from Mark III VLBI archive data.
See Abstr. 021.018.

Investigation of the influence of the residual coma of the Zeiss double astrograph on the determination of the position of an object.
See Abstr. 032.014.

Parallaxes with large reflectors.
See Abstr. 032.049.

Design characteristics of the 1.56 m astrometric telescope and its usage in astrometry.
See Abstr. 032.050.

Astrometric characteristics of the Abastumani Astrophysical Observatory 125 cm reflector (AZT–11) of Ritchey–Chrétien optical system.
See Abstr. 032.051.

Present status and future plans for the two color astrometric interferometer project.
See Abstr. 032.052.

A new astrometric system.
See Abstr. 032.053.

Photoelectric astrolabe and astrolabe star catalogues.
See Abstr. 032.054.

The oscillating slit micrometer of the meridian circle PMC 190 Tokyo.
See Abstr. 032.055.

A new astrolabe.
See Abstr. 032.056.

Timing and data acquisition system for a field astrolabe.
See Abstr. 032.057.

The photoelectric meridian circle of Pulkovo.
See Abstr. 032.058.

A proposal for scanning meridian circle.
See Abstr. 032.061.

Impersonal techniques of transit circle observations. "The Carlsberg Automatic Meridian Circle".
See Abstr. 032.062.

The seven–inch transit circle and its New Zealand program.
See Abstr. 032.065.

Vacuum meridian marks of the Belgrade large transit instrument.
See Abstr. 032.067.

The vacuum mires of the transit instrument at Nikolaev.
See Abstr. 032.068.

Internal refraction in meridian circles.
See Abstr. 032.069.

Determination of the division corrections of the Bordeaux declination circle by the Benevides–Boczko method.
See Abstr. 032.070.

A new method of determining absolute azimuth and latitude and suggestion for a new type of meridian circle.
See Abstr. 032.071.

A new method for zenith distance determination in meridian observations.
See Abstr. 032.072.

Vidicon photometry and astrometry applied to objects of the solar system.
See Abstr. 034.191.

A multislit photoelectric star micrometer for the meridian circle of the Nikolayev Astronomical Observatory.
See Abstr. 034.192.

HIPPARCOS: the instrument.
See Abstr. 035.084.

Microarcsecond astrometric interferometry.
See Abstr. 035.085.

Noise and astrometric precision. I. Detection noises.
See Abstr. 036.012.

On the determination of the corrections to fundamental star catalogues using observations of artificial earth satellites with inaccurate moments of time.
See Abstr. 036.022.

On the accuracy of photographic observations of the moon on the star background.
See Abstr. 036.026.

Problems in data compilation.
See Abstr. 036.066.

Choice of optimum conditions for absolute determination of star declinations with astrolabes.
See Abstr. 036.120.

Construction of an inertial coordinate system with a CCD. (I). Measuring the relative position of an extragalactic system with respect to the reference stars.
See Abstr. 036.144.

Construction of an inertial coordinate system with a CCD. (II). The determination of the scales and the correction for the inclination of the measured coordinate system.
See Abstr. 036.145.

Construction of an inertial coordinate system with a CCD. (III). Some astronomical corrections of the measured coordinates.
See Abstr. 036.146.

Construction of an inertial coordinate system with a CCD. (IV). The transition to the inertial coordinate system.
See Abstr. 036.147.

Astrometry with a scanning CCD.
See Abstr. 036.172.

Global reduction of fundamental astrometric data: computation of star catalogues.
See Abstr. 036.212.

The astrometric possibilities of very–long–baseline interferometry.
See Abstr. 036.216.

Astrometry with the Lowell PDS.
See Abstr. 036.217.

Astrometry with objective prism.
See Abstr. 036.219.

Speckle interferometry in astrometry.
See Abstr. 036.220.

Moving–image astrometry with the multi–anode microchannel array (MAMA) detector.
See Abstr. 036.223.

The design of modern meridian circles for the observation of faint objects.
See Abstr. 036.224.

On the implementation of absolute meridian observations in low latitude stations.
See Abstr. 036.227.

HIPPARCOS data reductions.
See Abstr. 036.228.

Software for Space Telescope astrometry.
See Abstr. 036.229.

A fully–automated system of astrometric data collection and processing.
See Abstr. 036.230.

Coordinate systems in the general relativistic framework.
See Abstr. 043.001.

Fundamental reference systems; past, present, and future.
See Abstr. 043.007.

The evolution of adopted values for precession.
See Abstr. 043.008.

On the orientation of ephemeris reference frames.
See Abstr. 043.009.

Fundamental coordinate ties using laser ranging data.
See Abstr. 043.010.

Precise position measurements of Jupiter, Saturn and Uranus systems with the Very Large Array.
See Abstr. 043.011.

El sistema IAU (1976) de constantes astronómicas y su repercusión en la reducción de posiciones de estrellas.
See Abstr. 043.012.

Un catálogo provisional FK5 para uso en astronomía geodésica.
See Abstr. 043.013.

The conversion from the B1950 FK4–based position to the J2000 position of celestial objects.
See Abstr. 043.020.

The precision of J2000.0 radio source positions from Mark III VLBI.
See Abstr. 043.022.

The JPL/DSN (*Jet Propulsion Laboratory/Deep Space Network*) J2000 radio reference frame.
See Abstr. 043.023.

The extragalactc optical/radio reference frame. A progress report, IAU Comm. 24 Working Group.
See Abstr. 043.024.

On the reference frame of the planetary ephemerides.
See Abstr. 043.025.

The reconciliation of optical and radio positions.
See Abstr. 043.026.

On the feasibility of a star coordinate determination in the radio astrometry reference system.
See Abstr. 043.027.

Connection between the Hipparcos catalogue and the FK5.
See Abstr. 043.028.

Relating the JPL VLBI reference frame and the planetary ephemerides.
See Abstr. 043.029.

Actualización de los procesos de reducción de observaciones en astronomía geodésica.
See Abstr. 045.020.

Prospects for astrometry with the Hubble Space Telescope.
See Abstr. 051.051.

Progetto HIPPARCOS.
See Abstr. 051.086.

A program for observations of the nearest eclipsing binaries from the HIPPARCOS satellite.
See Abstr. 051.091.

The project HIPPARCOS.
See Abstr. 051.097.

On observations in Schwarzschild background.
See Abstr. 066.027.

Accuracy estimates for the determination of the solar space–time metric by HIPPARCOS.
See Abstr. 066.032.

An astrometric program to define coordinate frames for general relativity experiments.
See Abstr. 066.091.

Meridian circle observations of major planets at the Engelhardt Astronomical Observatory in 1973 – 1974.
See Abstr. 091.023.

Relativistic and perspective effects in proper motions and radial velocities of stars.
See Abstr. 111.001.

Stellar motions in the solar neighborhood.
See Abstr. 111.009.

Proper motions with respect to galaxies.
See Abstr. 111.025.

Astrometry of star clusters: problems, techniques, and opportunities.
See Abstr. 111.026.

Coordonnées équatoriales précises de composantes connues ou nouvelles d'une quatrième série de quelques 200 systèmes doubles et multiples.
See Abstr. 118.029.

Coordonnées équatoriales précises de composantes connues ou nouvelles d'une cinquième série de quelque 200 systèmes doubles et multiples.
See Abstr. 118.030.

Optical observations of the radio star β Persei at the Paris Observatory astrolabe from 1980 to 1983.
See Abstr. 119.002.

Precise positions for two supernovae.
See Abstr. 125.085.

The relativistic effects in localization of gamma–burst sources.
See Abstr. 143.002.

Positions and magnitudes in the area of the open cluster NGC 752.
See Abstr. 153.026.

The optical–radio structure of the southwest hotspot of 3C 33: astrometry.
See Abstr. 158.203.

VLBI limits on the proper motion of the 'core' of the superluminal quasar 3C345.
See Abstr. 159.038.

Positions of extragalactic objects in the vicinity of M3 and M31 (5C3 radio field) and in the 5C1 radio field.
See Abstr. 159.159.

042 Celestial Mechanics, Figures of Celestial Bodies

042.001 **Construction of a planetary solution with the help of an n–body program and analytical complements.**
P. Bretagnon.
Relativity in celestial mechanics and astrometry, p. 69 (1986). Abstract. – See Abstr. 012.002 (IAU Symp. No. 114).

042.002 **Polynomial expansion of the planetary secular terms: relativistic and lunar perturbations.**
J. Laskar.
Relativity in celestial mechanics and astrometry, p. 85 – 92 (1986). – See Abstr. 012.002 (IAU Symp. No. 114).
The relativistic and lunar perturbations must be included in a realistic theory of the secular evolution of the planetary elements. In a general theory, the author includes the first order of these perturbations. Comparison with more elaborated studies shows that it is sufficient with respect to the accuracy of the theory.

042.003 **Local frames.**
B. Bertotti.
Relativity in celestial mechanics and astrometry, p. 233 – 240 (1986). – See Abstr. 012.002 (IAU Symp. No. 114).
It is shown how the Fermi construction of local inertial frames can be generalized, in the slow motion approximation, to the neighbourhood of the earth. This allows a clear and simple description of the relativistic effects for the motion of an earth satellite. A proposal by I. Ciufolini to measure the relativistic precession is reviewed.

042.004 **Approximation of Keplerian motion by epicycle or eccentric motion. (I).**
A. Satō.
Heavens, Vol. 67, No. 1, p. 3 – 7 (1986). In Japanese.

042.005 **Approximation of Keplerian motion by epicycle or eccentric motion. (II).**
A. Satō.
Heavens, Vol. 67, No. 2, p. 44 – 47 (1986). In Japanese.

042.006 **Conformal regularization of the Kepler problem.**
B. Cordani.
Commun. Math. Phys., Vol. 103, No. 3, p. 403 – 413 (1986).

042.007 **Sundman's inequality and zero velocity hypersurfaces in the general *N*–body problem.**
J. M. Muller.
Astron. Astrophys., Vol. 155, No. 1, p. L1 – L2 (1986).
As some confusion has arisen concerning the derivation of expressions for the zero velocity hypersurfaces in many body problems, it is shown that these expressions follow directly from Sundman's generalized inequality, a result which is not widely known.

042.008 **Validation of Sagnier's theory by its application to the major planets.**
R. da Silva Gomes.
Earth, Moon, Planets, Vol. 34, No. 2, p. 103 – 116 (1986).
In this paper, the theory of Sagnier was applied to the major planets, in order to have a first evaluation of this theory which was built, primarily, to study the Galilean satellites of Jupiter. In this application the author included the first order of the intermediary solution and the solution of the linear variational equations. Results from other theories, mainly those of Simon and Bretagnon, have been used as basis for comparisons.

042.009 **Secular terms of classical planetary theories using the results of general theory.**
J. Laskar.
Astron. Astrophys., Vol. 157, No. 1, p. 59 – 70 (1986).
The author has analytically computed the differential system giving the secular variations of the orbital elements for the 8 major planets, at the order 2 with respect to the masses and up to degree 5 in the eccentricity–inclination variables with a relative precision of 10^{-6}. Relativistic and lunar perturbations are included. The entire system is integrated numerically over 10000 years and then developed in Taylor expansion around J2000. The author has obtained new polynomial secular terms for the inner planets up to the power 10 of the time. Comparisons are made with Bretagnon's theory VSOP82 and with the numerically integrated JPL ephemeris, DE102 (Newhall et al., 1983). The global accuracy is approximately 0″.02/1000 yr for the inclination of the Earth. Using the theory of the rotation of the rigid Earth of Kinoshita (1977), the author has derived new formulas for the precessional quantities, up to t^{10}, and valid over 10000 years.

042.010 **On the restricted three rigid bodies problem near the Lagrangian point L_4.**
A. Elipe, S. Ferrer.
Bull. Astron. Inst. Czech., Vol. 37, No. 1, p. 12 – 18 (1986).
The restricted problem of three rigid bodies, when the two primaries are spheres, is analyzed assuming that the satellite is moving near the libration point L_4. Moreover, the authors suppose that the differential equations of the orbital motion can be separated from the rotational motion: thus, the formal solution for the orbital motion given by Cid et al. (1985) is used. Then, a first–order solution for the rotational motion of an axisymmetric body is obtained by employing Deprit's method in the non–resonant case. It is shown that the resonant cases of the rotational motion can be reduced to a generalized Ideal Resonance Problem, in which the coefficients are functions of time.

042.011 **An expansion of the solution of the generalized two–fixed centers problem in Poisson series with the help of a computer.**
Nguyen Dinh Huan, N. V. Emel'yanov.
Astron. Zh., Tom 63, Vyp. 1, p. 197 – 198 (1986). In Russian. English translation in Sov. Astron., Vol. 30, No. 1.
A method and the algorithm of obtaining Poisson series which represent the solution of the generalized two–fixed centers problem have been constructed, developments in powers of orbital eccentricity having been used. By means of a computer the above algorithm allows to construct in literal form the series for values contained in the perturbing function expression.

042.012 **Gravitational splitting of Quadrantid–like meteor streams in resonance with Jupiter.**
C. Froeschlé, H. Scholl.
Astron. Astrophys., Vol. 158, No. 1/2, p. 259 – 265 (1986).
The authors investigate numerically the dynamical evolution of meteor streams with the same orbital elements as the Quadrantids except for a slightly different range of semimajor axes ($3.24 < a < 3.36$ AU). This range covers the whole width of the observed 2/1 Kirkwood gap in the asteroidal belt. They find a peculiar behaviour for those streams in resonant motion with Jupiter: The streams break up into arcs of different sizes with distinctly different dynamical evolutions due to possible modes for resonant motion. Large, isolated arcs with differing orbital parameters are formed. The authors conjecture that such a gravitational splitting which causes the formation of isolated stream arcs with slow dispersion rates might occur in various ring systems in mean motion resonance with a perturbing body.

042.013 **Regularization of the equatorial magnetic–binary problem.**
A. G. Mavraganis, T. Kalvouridis.
Astrophys. Space Sci., Vol. 120, No. 2, p. 205 – 210 (1986).
The differential equations which describe the equatorial motion of a particle in a magnetic–binary system are regularized by a transformation of the dependent and independent variables. The new equations of motion, which are simpler than the original ones, with only linear velocity and acceleration terms occurring, show the invariant form of the law of motion. Finally, the distinction between the zero–velocity curves and the isotachs in the transformed space is also discussed.

042.014 **Bifurcations, stability and universality of families of periodic orbits in the restricted three–body problem.**
A. D. Pinotsis.
Astron. Astrophys., Vol. 159, No. 1/2, p. 231 – 238 (1986).
The author studies some bifurcations of the main families of simple periodic orbits a, b, c, f, g, h, i, k and l of the restricted three–body problem in the case of equal masses of the primaries. There are several cases of infinite pitchfork period–doubling bifurcations, in which the bifurcation ratio is consistent with the universal ratio $\delta \cong 8.72$. But the author found also cases of bifurcations of unstable families of periodic orbits, which do not present infinite period doubling bifurcations. He followed the evolution of the families of double periodic orbits bifurcating from the above families.

042.015 **On the irrotational and the zero angular momentum ellipsoids in the Dirichlet problem.**
B. P. Kondrat'ev.
Astrofizika, Tom 23, Vyp. 3, p. 487 – 502 (1985). In Russian. English translation in Astrophysics, Vol. 23, No. 3.
Two new exact solutions have been found in the Dirichlet problem for the oscillations of self–gravitating incompressible (compressible) homogeneous ellipsoidal configurations. These solutions describe the ellipsoids with zero vorticity and the ellipsoids with zero angular momentum. Eight (seven) integrals of motion exist. It has been shown that for the case of the ellipsoidal figures of equilibrium (the Riemann ellipsoids) with oblique rotation these two types of ellipsoids also exist.

042.016 **Adelphic potentials.**
G. Bozis.
Astron. Astrophys., Vol. 160, No. 1, p. 107 – 110 (1986).
The present paper examines the conditions under which two given potentials $U = A(x,y)$, and $U = B(x,y)$ have in common a monoparametric family of orbits $f(x,y) = c$, not given in advance. Such potentials are defined as adelphic. It is shown that if the given functions A and B are functionally independent and if they satisfy a certain condition, then they are adelphic and the corresponding family $f(x,y) = c$ is, in general, determined uniquely. If A and B are functionally dependent and if A satisfies a certain condition, then so does B and the potentials are adelphic, having in common a family $f(x,y) = c$ which again is determined uniquely. In this latter case, $f(x,y) = c$ is a family of isotach orbits. A partial differential equation giving all possible planar isotach orbits is also derived.

042.017 **Approximation of Keplerian motion by epicycle or eccentric motion (III).**
A. Satō.
Heavens, Vol. 67, No. 3, p. 76 – 79 (1986). In Japanese.

042.018 **On retrograde resonant rotation of natural celestial bodies.**
A. A. Khentov.
Astron. Zh., Tom 63, Vyp. 2, p. 349 – 355 (1986). In Russian. English translation in Sov. AstrOn., Vol. 30, No. 2.
The instability is proved of retrograde rotational motion of a natural celestial body synchronized with its orbital motion under the action of gravitation torque and tidal friction whose values are comparable.

042.019 **Improvement of convergence of planetary potential expansion. General ideas.**
M. S. Petrovskaya.
Astron. Zh., Tom 63, Vyp. 2, p. 356 – 364 (1986). In Russian. English translation in Sov. Astron., Vol. 30, No. 2.
General ideas of the methods of improvement of convergence of the planetary external potential in solid spherical functions are outlined. One of these is based on the application of a series in Chebyshev polynomials as an intermediate one, and the other deals with a modification of Lanczos' method of "economization" of power series. Different forms of generalized potentials are discussed to choose the most appropriate ones for the procedure of improvement of convergence of the planetary potentials.

042.020 **The Hamiltonian equations of the planetary N–body problem in the post–Newtonian approximation.**
A. V. Krivov.
Astron. Zh., Tom 63, Vyp. 2, p. 371 – 380 (1986). In Russian. English translation in Sov. Astron., Vol. 30, No. 2.
The canonical equations in the planetary problem of N bodies having a spherical symmetry in the post–Newtonian approximation of the general relativity theory are obtained. It is ascertained that the influence of the rotational motion of planets in the solar system on the translational one has the same order as the primary components of relativistic effects. The influence of the post–Newtonian term in the Hamiltonian function on the secular changes of the orbital elements is considered. The relativistic correction is shown to have a value and an analytical structure such that it is possible to apply the KAM theory conclusions on the existence of nearly–periodic solutions of the planetary N–body problem quite as much in the post–Newtonian case as in the classical one.

042.021 **Analytical continuation of expansions of the external potentials of ellipsoidal bodies.**
M. V. Belikov.
Astron. Zh., Tom 63, Vyp. 2, p. 381 – 388 (1986). In Russian. English translation in Sov. Astron., Vol. 30, No. 2.
Gravitational external potential expansions of both level and homogeneous ellipsoids of revolution with large flattenings, as well as that of a circular disc with the surface density distribution $\mu(r) = \mu(0)\cdot\sqrt{(1-r^2/a^2)}$ have been constructed, where a means

the radius of the disc and r is the distance from its centre to a point on its surface. In distinction to the earlier constructed series in spherical functions, each of these expansions converges in the whole exterior space down to the surface of the ellipsoid (or the disc) and it can be differentiated term by term an arbitrary number of times.

042.022 **An integral of the Jacobi type in the circular restricted problem of three rigid bodies.**
L. Yu. Khorseva.
Pis'ma Astron. Zh., Tom 12, No. 5, p. 398 – 401 (1986). In Russian. English translation in Sov. Astron. Lett., Vol. 12.
A first integral has been obtained in the circular restricted problem of the translatory–rotational motion of three rigid bodies. It is supposed that the passive body has arbitrary form and structure and the main bodies permit circular motion of the mass centers under definite orientation of the bodies in a rotating system of coordinates.

042.023 **Methods of computation of positions of the inner planets.**
B. Todorovic–Juchniewicz.
Postepy Astron., Tom 33, Zesz. 1–2, p. 59 – 64 (1985). In Polish.
The paper gives a short description of analytical methods of computation of positions of the inner planets which are based on Newcomb's theory of planetary motion. An estimation of their accuracy within 1900 – 2000 was made.

042.024 **Nearly periodic circular orbits in the restricted three–body problem.**
V. P. Evteev, B. M. Nagorev.
Dokl. Akad. Nauk TadzhSSR, Tom 28, No. 3, p. 144 – 147 (1985). In Russian. Abstr. in Ref. Zh., 51. Astron., 1.51.79 (1986).

042.025 **A family of nearly circular orbits for the external variant of the averaged restricted elliptical three–body problem.**
B. M. Nagorev.
Dokl. Akad. Nauk TadzhSSR, Tom 28, No. 4, p. 208 – 211 (1985). In Russian. Abstr. in Ref. Zh., 51. Astron., 1.51.80 (1986).

042.026 **On some problems of analytical mechanics.**
V. V. Rumyantsev.
Probl. analit. mekh. i upr. dvizheniem. Moskva, p. 20 – 36 (1985). In Russian. Abstr. in Ref. Zh., 51. Astron., 1.51.83 (1986).

042.027 **Application of the Lagrange–Jacobi equation to the analytical theory of absolute stability according to Hill.**
V. G. Golubev, T. S. Golubeva.
Inst. teor. ehksp. fiz. Moskva, Prepr., No. 90, 22 pp. (1985). In Russian. Abstr. in Ref. Zh., 51. Astron., 1.51.84 (1986).

042.028 **Additional results in the theory of absolute stability according to Hill resulting from the Lagrange–Jacobi equation.**
V. G. Golubev, T. S. Golubeva.
Inst. teor. ehksp. fiz. Moskva, Prepr., No. 91, 22 pp. (1985). In Russian. Abstr. in Ref. Zh., 51. Astron., 1.51.85 (1986).

042.029 **Stability of motion of an equatorial satellite of an axisymmetric planet.**
M. Kh. Khasanova, M. Eh. Ehrgasheva.
Izv. Akad. Nauk TadzhSSR. Otd–nie fiz.–mat., khim. i geol. nauk, No. 1, p. 77 – 79 (1985). In Russian. Abstr. in Ref. Zh., 51. Astron., 1.51.87 (1986).

042.030 **On the stability of collinear libration points of the photogravitational three–body problem.**
A. L. Kunitsyn, A. T. Tureshbaev.
Nekotor. zadachi i metody issled. dinam. mekh. sistem, Moskva, p. 26 – 31 (1985). In Russian. Abstr. in Ref. Zh., 62. Issled. Kosm. Prostranstva, 1.62.215 (1986).

042.031 **On a family of nearly circular orbits in the internal variant of the circular three–body problem.**
V. P. Evteev.
Dokl. Akad. Nauk TadzhSSR, Tom 28, No. 2, p. 83 – 87 (1985). In Russian. Abstr. in Ref. Zh., 62. Issled. Kosm. Prostranstva, 1.62.216 (1986).

042.032 **On the shift of the triangular libration points for bodies of arbitrary structure.**
N. I. Gamarnik.
Differents. uravneniya i ikh pril., Dnepropetrovsk, p. 105 – 108 (1985). In Russian. Abstr. in Ref. Zh., 62. Issled. Kosm. Prostranstva, 1.62.217 (1986).

042.033 **The stability of satellite orbits in the planar restricted four–body problem.**
S. G. Orudzhev.
Shemakhin. astrofiz. obs. Baku, 8 pp. (1985). In Russian. Abstr. in Ref. Zh., 62. Issled. Kosm. Prostranstva, 1.62.218 (1986).

042.034 **On universal Keplerian algorithms.**
A. V. Krivov.
Vestn. Leningr. Univ., Mat. Mekh. Astron., No. 22, p. 73 – 81 (1985). In Russian. Abstr. in Ref. Zh., 51. Astron., 2.51.87 (1986).

042.035 **Periodic orbits of the two–body problem in a rotating coordinate system.**
V. P. Semenko.
Razvitie idej K. Eh. Tsiolkovskogo v mekh. kosm. poleta. Tr. 18 Chtenij, posvyashch. razrab. nauchn. naslediya i razvitiyu idej K. Eh. Tsiolkovskogo, Kaluga 16 – 19 sent., 1983 Sekts. mekh. kosm. poleta. Moskva, p. 102 – 109 (1984). In Russian. Abstr. in Ref. Zh., 51. Astron., 2.51.88; 62. Issled. Kosm. Prostranstva, 2.62.159 (1986).

042.036 **New stationary solutions in the problem of motion of a satellite in a central field.**
Yu. V. Barkin, Yu. M. Gubarev.
Razvitie idej K. Eh. Tsiolkovskogo v mekh. kosm. poleta. Tr. 18 Chtenij, posvyashch. razrab. nauchn. naslediya i razvitiyu idej K. Eh. Tsiolkovskogo, Kaluga, 16 – 19 sent. 1983. Sekts. Mekh. kosm. poleta. Moskva, p. 82 – 90 (1984). In Russian. Abstr. in Ref. Zh., 62. Issled. Kosm. Prostranstva 2.62.177 (1986).

042.037 **Periodic resonance motions of an axisymmetric satellite in an elliptical orbit.**
T. N. Chekhovskaya.
Kosm. Issled., Tom 24, Vyp. 1, p. 15 – 23 (1986). In Russian. English translation in Cosm. Res.

042.038 **Libration motion of an equatorial satellite in the gravitational field of a rotating nonspherical planet.**
V. A. Kitova.
Kosm. Issled., Tom 24, Vyp. 1, p. 24 – 29 (1986). In Russian. English translation in Cosm. Res.

042.039 **Rigorous estimates for the series expansions of Hamiltonian perturbation theory.**
A. Giorgilli, L. Galgani.
Celest. Mech., Vol. 37, No. 2, p. 95 – 112 (1985).
The authors prove a theorem giving rigorous estimates in the problem of bringing to normal form a nearly integrable Hamiltonian system, using methods of classical perturbation theory. As an application, they consider the case of weakly coupled harmonic oscillators with highly nonresonant frequencies.

042.040 **On the motion of three rigid bodies: central configurations.**
R. Cid, A. Elipe.
Celest. Mech., Vol. 37, No. 2, p. 113 – 126 (1985).
The motion of three rigid bodies is considered. With a set of new variables, and the 10 first integrals of the motion, the problem is reduced to a system of order 25 and one quadrature. The plane motions are characterized, and finally, an equation for the existence of central configurations (in particular, Lagrangian and

Eulerian solutions) has been found. Besides, the case of three axisymmetric ellipsoids is studied.

042.041 **Motions in the field of two rotating magnetic dipoles. I. Equilibrium points.**
C. L. Goudas, M. Leftaki, E. G. Petsagourakis.
Celest. Mech., Vol. 37, No. 2, p. 127 – 148 (1985).
The equilibrium points of charged particles moving under the Lorentz forces of two parallel or anti–parallel magnetic dipoles located at the two primaries of the restricted three–body problem are calculated. The configuration studied simulates the case of close–binary stars with dominant zero–order harmonics in the magnetic field of each star. Depending on the mass parameter μ and the magnetic moment ratio λ of the two dipoles, it is found that there exist from one to nine equilibrium points in the plane of motion of the primaries.

042.042 **Areas of equatorial motion in the magnetic–binary problem.**
T. Kalvouridis, A. Mavraganis, C. Pangalos, C. Zagouras.
Celest. Mech., Vol. 37, No. 2, p. 161 – 170 (1985).
In the process of modelling the magnetic–binary systems, the authors deal with the permissible areas of motion in the equatorial planes of the representative model. The zero velocity curves, derived by the well known steepest descent method, provide the means for determining these areas. Further these curves reveal all the equilibrium points located on the above planes.

042.043 **Attractors in a dissipative dynamical system with three dimensions.**
Y.–s. Sun.
Celest. Mech., Vol. 37, No. 2, p. 171 – 181 (1985).
The author studies an extension of the Hénon mapping to a dissipative dynamical system with three–dimensions and discusses the behavior of the attractors of the Hénon mapping in the extended mapping. The results show that the strange attractor is destroyed by perturbed extension more easily than the trivial attractor and the invariant manifold of the conservative dynamical system.

042.044 **The ideal resonance problem, a comparison of two formal solutions. II.**
A. H. Jupp, A. Y. Abdulla.
Celest. Mech., Vol. 37, No. 2, p. 183 – 197 (1985).
This paper is a sequel to an earlier article of the same title. The two formal analytical solutions of the ideal resonance problem developed respectively by Garfinkel and Jupp are here compared, at second–order in the appropriate small parameter, with numerical integrations; the second–order circulation solution for Jupp's theory being presented for the first time. It transpires that throughout most of the deep resonance regime the second–mentioned solution provides greater accuracy. In addition, it is demonstrated that the first solution is not appropriate when general initial values of the variables are prescribed.

042.045 **Secular perturbation in planetary motion.**
G. Hori.
Proceedings of the nineteenth symposium on celestial mechanics, p. 64 – 66 (1986). In Japanese. – See Abstr. 012.013.

042.046 **A general behaviour of disks in resonance with a satellite.**
B. Sicardy, N. Meyer–Vernet.
Bull. Am. Astron. Soc., Vol. 17, No. 4, p. 924 (1985). Abstract. – See Abstr. 010.063.

042.047 **Osculating orbital elements of a natural satellite of the earth in the function of the phase angle.**
A. A. Aryutkina.
Arzamas. gos. ped. inst. Arzamas, 25 pp. (1985). In Russian. Abstr. in Ref. Zh., 51. Astron., 3.51.146 (1986).

042.048 **On libration points of the photogravitational restricted elliptical three–body problem.**
A. T. Tureshbaev.
Mosk. aviats. inst. Moskva, 10 pp. (1985). In Russian. Abstr. in Ref. Zh., 51. Astron., 3.51.681 (1986).

042.049 **Determination of the semi–major axis, of the eccentricity of the orbit and of the true anomaly of a celestial body from three and more positions.**
N. G. Pal'tsev.
Problems of astronomy, p. 76 – 79 (1985). In Russian. Abstr. in Ref. Zh., 62. Issled. Kosm. Prostranstva, 3.62.172 (1986). – See Abstr. 003.023.

042.050 **Stationary solutions of the restricted three–body problem in the satellite variant.**
S. G. Zhuravlev, V. V. Anikovskii (*V. V. Anikovskij*).
Sov. Astron., Vol. 29, No. 4, p. 465 – 469 (1985). English translation of 40.042.015.

042.051 **On single resonance in the problem of the translational–rotational motion of three rigid bodies.**
Zh. B. Rakishev.
Sov. Astron., Vol. 29, No. 4, p. 484 – 485 (1985). English translation of 40.042.016.

042.052 **Periodic Poincaré solutions for a canonical system with one degree of freedom.**
A. P. Markeev, N. I. Churkina.
Sov. Astron. Lett., Vol. 11, No. 4, p. 267 – 269 (1985). English translation of 40.042.005.

042.053 **Periodic solutions of Hamiltonian systems and their application to satellite dynamics. I.**
Yu. V. Barkin, A. A. Pankratov.
Kosm. Issled., Tom 24, Vyp. 3, p. 345 – 357 (1986). In Russian. English translation in Cosm. Res.

042.054 **On the stability of periodic motions of a satellite around the triangular libration point of the restricted elliptical three–body problem.**
S. N. Lelyavin.
Kosm. Issled., Tom 24, Vyp. 3, p. 358 – 362 (1986). In Russian. English translation in Cosm. Res.

042.055 **Analytical algorithm of the Hori method in the problem of rotation of a celestial body around the mass center.**
A. V. Rodnikov, I. I. Kosenko.
Kosm. Issled., Tom 24, Vyp. 3, p. 363 – 368 (1986). In Russian. English translation in Cosm. Res.

042.056 **A family of nearly circular orbits of arbitrary inclination in the external variant of the restricted circular three–body problem.**
B. M. Nagorev.
Dokl. Akad. Nauk TadzhSSR, Tom 28, No. 5, p. 273 – 277 (1985). In Russian. Abstr. in Ref. Zh., 51. Astron., 5.51.96 (1986).

042.057 **Quaternion methods in problems of the perturbed central motion of a material point. Part 1. General theory. Application to the problem of regularization and to the problem of motion of artificial earth satellites.**
Yu. N. Chelnokov.
Saratov. politekh. inst. Saratov, 36 pp. (1985). In Russian. Abstr. in Ref. Zh., 51. Astron., 5.51.107 (1986).

042.058 **Quaternion methods in problems of the perturbed central motion of a material point. Part 2. The spatial problem of nonperturbed central motion. The problem with initial conditions.**
Yu. N. Chelnokov.
Saratov. politekh. inst. Saratov, 18 pp. (1985). In Russian. Abstr. in Ref. Zh., 51. Astron., 5.51.108 (1986).

042.059 **Algorithms of some problems of celestial mechanics.**
V. G. Shkodrov, Kh. N. Kazasov, V. G. Ivanova.
Sbornik tr. Mezhdunar. konf. na temu: Avtomatiz. nauchn. issled., Plovdiv, 15 – 20 okt., 1984, Tom 2. Sofiya, p. 526 – 529. In Russian. Abstr. in Ref. Zh., 51. Astron., 5.51.121 (1986).

042.060 **Using the logarithm principle for the analysis of perihelion displacements of planetary orbits.**
Ya. G. Klyushin.
Red. Zh., Vestn. Leningr. Univ., Mat. Mekh. Astron., 10 pp. (1986). In Russian. Abstr. in Ref. Zh., 51. Astron., 5.51.124 (1986).

042.061 **Trigonometric expansions of the coordinates in the generalized problem of two fixed centers.**
Nguyen Dinh Huan.
Sov. Astron., Vol. 29, No. 5, p. 584 – 588 (1985). English translation of 40.042.058.

042.062 **Motion near the triangular libration points with variable mass ratio of the components.**
B. B. D'yakov, B. I. Reznikov.
Astron. Zh., Tom 63, Vyp. 3, p. 574 – 583 (1986). In Russian. English translation in Sov. Astron., Vol. 30, No. 3.

Trojan–type orbits are investigated in the system with a variable mass ratio of the components within the framework of the restricted circular three–body problem. For the negligible mass change during a single libration period the increase of the relative mass μ of the smaller component beyond a certain limit leads to the end of the librations and to the entry of the librator into Hill's sphere of this component. This limit is lower than the critical value $\mu_* = 0.0385$ determining the stable equilibrium and depends only on the initial displacement from the libration point. The region of the conditional stability of the libration motion with different mass ratios is established by a numerical method. The results are applied to estimate the hypothesis of Trojans' origin as past satellites of Jupiter.

042.063 **Investigation of the efficiency of algorithms of numerical forecast of the motion of celestial bodies in the framework of numerical experiment of the USSR.**
T. V. Bordovitsyna.
Prikl. nebes. mekh. i upr. dvizheniem. Tr. 7 Obed. nauchn. chtenij po kosmonavt., posvyashch. pamyati vydayushch. sov. uchenykh–pionerov osvoeniya kosm. prostranstva, Moskva, 24 – 28 yanv., 1983. Moskva, p. 49 – 58 (1983). In Russian. Abstr. in Ref. Zh., 51. Astron., 6.51.108 (1986).

042.064 **Invariant transformation of the two–body problem associated with eccentricity.**
C. Edelman.
Celest. Mech., Vol. 37, No. 4, p. 351 – 358 (1985). In French.

For the non–circular two–body problem, having non–zero angular momentum a parametric transformation may be defined changing only the eccentricity and the time. This transformation is of homographic type, and it is an analytical expression depends on the value of the eccentricity with respect to unity. Therefore an hyperbolic or parabolic solution may be changed into an elliptic solution. The application and usefulness of this transformation is concerned with the capture of comets.

042.065 **Second order adiabatic invariants associated with the two–body problem with slowly varying mass.**
M. Salmassi.
Celest. Mech., Vol. 37, No. 4, p. 359 – 369 (1985). In French.

An adiabatic invariant is characterized by the property that its derivative is small and oscillatory. Therefore, assuming that such a quantity is constant does not lead to a cumulative error as $t \to \infty$. In this paper, using action and angle variables, adiabatic invariants to $0(1)$ and $0(\varepsilon)$ are found for the two–body problem with a slowly varying gravitational constant.

042.066 **Series expansions for encounter–type solutions of Hill's problem.**
M. Hénon, J.–M. Petit.
Celest. Mech., Vol. 38, No. 1, p. 67 – 100 (1986).

Hill's problem is defined as the limiting case of the planar three–body problem when two of the masses are very small. This paper describes analytic developments for encounter–type solutions, in which the two small bodies approach each other from an initially large distance, interact for a while, and separate.

042.067 **Numerical continuation of periodic orbits from the restricted to the general 3–dimensional 3–body problem.**
K. Katopodis.
Astrophys. Space Sci., Vol. 123, No. 2, p. 335 – 349 (1986).

In this paper several monoparametric families of periodic orbits of the 3–dimensional general 3–body problem are presented.

042.068 **Change in the mutual distance of celestial bodies resulting from changes in some of their orbital elements.**
J. Lazović.
Publ. Dep. Astron., Univ. Beogr., No. 13, p. 5 – 11 (1985).

A formula is developed for the distance change of two celestial bodies, resulting from the changes in the perihelia arguments, longitudes of the ascending nodes and the inclinations of their orbits. For illustration, the formula is applied to a pair of minor planets at their closest approach.

042.069 **On the dynamical evolution of binary systems.**
Z.–t. Zheng, C.–x. Ni.
Kexue Tongbao, Vol. 30, No. 4, p. 482 – 487 (1985). Abstr. in Phys. Abstr., Vol. 89, No. 1273, Entry 18642 (1986).

042.070 **On the solvability of an equation describing the periodic motions of a satellite in its elliptical orbit.**
W. V. Petryshyn, Z. S. Yu.
Nonlinear Anal. Theory, Methods Appl., Vol. 9, No. 9, p. 969 – 975 (1985). Abstr. in Phys. Abstr., Vol. 89, No. 1273, Entry 18643 (1986).

042.071 **Bifurcations in systems of 3 degrees of freedom.**
G. Contopoulos.
ESO Sci. Prepr., No. 432, 48 pp. (1986). To appear in Celest. Mech.

042.072 **The effect of total angular momentum on the break up of a three–body system.**
S. Mikkola, M. J. Valtonen.
Rep. Ser., Dep. Phys. Sci., Univ. Turku, No. FTL–R90, 17 pp. (1986). = Turku Univ. Obs., Informo No. 99.

The dependence of the break–up energy of initially bound three–body systems on the angular momentum have been studied by means of numerical experiments. The authors derive fitting formulae for the energy distribution as a function of the angular momentum. An approximation for the escape probability of a given mass in case of unequal masses has also been derived.

042.073 **General relativistic celestial mechanics of binary systems. I. The post–Newtonian motion.**
T. Damour, N. Deruelle.
Ann. Inst. Henri Poincaré Phys. Théor., Vol. 43, No. 1, p. 107 – 132 (1985). Abstr. in Phys. Abstr., Vol. 89, No. 1275, Entry 30146 (1986).

042.074 **On the inverse Kepler problem.**
J. Sivardiere.
Eur. J. Phys., Vol. 6, No. 4, p. 245 – 248 (1985). Abstr. in Phys. Abstr., Vol. 89, No. 1278, Entry 42458 (1986).

042.075 **Bounds on the solution to Kepler's equation.**
R. A. Serafin.
Celest. Mech., Vol. 38, No. 2, p. 111 – 121 (1986).

For Kepler's equation two general linear methods of the bounds determination for E_0 root are presented. The methods based on elementary properties of convex functions allow an approach to E_0 root arbitrarily close.

042.076 **Secular variations of major planets' orbital elements.**
Z. Knežević.
Celest. Mech., Vol. 38, No. 2, p. 123 – 138 (1986).
The secular variations of the orbital elements of principal planets are calculated by means of classical Lagrange's method. The terms of the second order with respect to mass, introduced by Hill (1897) and Brouwer and van Woerkom (1950), have been taken into account as well. The best contemporary values of planetary masses and mean elements (Bretagnon, 1982) served as the starting data set for this calculation. Considerable differences with respect to previous solutions of the same type (Brouwer and van Woerkom, 1950: Sharaf and Boudnikova, 1967) were found. Results are compared with some higher order/higher degree solutions and their accuracy discussed.

042.077 **On the occurrence of near syzygies in the solar system.**
A. J. C. McDonald.
Celest. Mech., Vol. 38, No. 2, p. 139 – 153 (1986).
This paper derives the average period of occurrence of syzygies (conjunctions and oppositions) of p planets in unperturbed, coplanar, circular orbits about the sun. Numerical results indicate the reliability of the theory and applications are suggested. For example, the theory is relevant to the search for resonances and near mirror configurations.

042.078 **Canonical solution of the two critical argument problem.**
J. Wisdom.
Celest. Mech., Vol. 38, No. 2, p. 175 – 180 (1986).
The solution by Sessin and Ferraz–Mello (1984) of the Hori auxiliary system for the motion of two planets with periods nearly commensurate in the ratio 2:1 is considerably simplified by the introduction of canonical variables. An analogous canonical transformation simplifies the elliptic restricted problem.

042.079 **Zero velocity hypersurfaces for the general three–dimensional three–body problem.**
R. Sergysels.
Celest. Mech., Vol. 38, No. 3, p. 207 – 214 (1986).
The equation of zero velocity surfaces for the general three–body problem can be derived from Sundman's inequality. The geometry of those surfaces was studied by Bozis in the planar case and by Marchal and Saari in the three–dimensional case. More recently, Saari, using a geometrical approach, has found an inequality stronger than Sundman's. Using Bozis' algebraic method, and a rotating frame which does not take into account entirely the rotation of the three–body system, the author also finds an inequality stronger than Sundman's. The comparison with Saari's inequality is more difficult.

042.080 **On the refutation of the Brown conjecture.**
B. Garfinkel.
Celest. Mech., Vol. 38, No. 3, p. 231 (1986).

042.081 **A new integrated, general time transformation in the Kepler problem.**
J. M. Ferrándiz, S. Ferrer.
Bull. Astron. Inst. Czech., Vol. 37, No. 4, p. 226 – 229 (1986).
A one–parametric family of time transformations is introduced. The corresponding anomalies for Keplerian motions are intermediate between elliptic and true ones, which appear as particular cases. The canonical Delaunay–similar variables and time elements may then be obtained directly.

042.082 **The determination of the gravitational potential in spherical coordinates. III: Radial recursive integration of the Poisson integral.**
R. Tschaepe.
Astron. Nachr., Vol. 307, No. 3, p. 217 – 219 (1986).
Radial recursive integration improves the Legendre representation of the gravitational potential of axisymmetric mass distribution and reduces the number of operations from nN^2M to nNM.

042.083 **Satellite encouters.**
J.–M. Petit, M. Henon.
Icarus, Vol. 66, No. 3, p. 536 – 555 (1986).
The authors study numerically the interaction of two small satellites, initially on circular orbits with slightly different radii. They show first that by going to Hill's limit of vanishing masses, one can reduce the problem to a simpler form in which only one dimensionless parameter remains: the reduced impact parameter. The authors present then a detailed study of the family obtained when this parameter is varied. They compute the minimal distance of approach of the two satellites. In the limiting cases of large and small impact parameters, the results are compared with the predictions of perturbation theories. Finally they study the "transitions", which are apparent discontinuities of the family with a sudden change of the direction of departure.

042.084 **Th problem of two bodies with variable masses: classification of different cases.**
E. P. Razbitnaya.
Sov. Astron., Vol. 29, No. 6, p. 684 – 687 (1985). English translation of 40.042.111.

042.085 **Expansion of the external potential of a homogeneous ellipsoid of revolution with a large flattening.**
M. V. Belikov.
Sov. Astron., Vol. 29, No. 6, p. 688 – 693 (1985). English translation of 40.042.112.

042.086 **Nearly parabolic motions in the plane case of Fatou's problem.**
I. A. Gerasimov.
Sov. Astron., Vol. 29, No. 6, p. 718 (1985). English translation of 40.042.113.

042.087 **A group theoretical approach to a certain class of perturbations of the Kepler problem.**
M. Kummer.
Arch. Ration. Mech. Anal., Vol. 91, No. 1, p. 55 – 82 (1985). Abstr. in Phys. Abstr., Vol. 89, No. 1281, Entry 61288 (1986).

042.088 **The existence of global solution of the N–body problem.**
Q.–d. Wang.
Chin. Astron. Astrophys., Vol. 10, No. 2, p. 135 – 142 (1986). English translation of 40.042.109.

042.089 **Validation of Sagnier's theory by its application to the major planets.**
R. da Silva Gomes.
Rev. Mex. Astron. Astrofis., Vol. 12, Spec. Issue, p. 435 – 437 (1986). – See Abstr. 012.066.
The theory of Sagnier is applied to the major planets, in order to have a first evaluation of this theory which was built, primarily, to study the Galilean satellites of Jupiter. In this application the author includes the first order of the intermediary solution and the solution of the linear variational equations. Results from other theories have been used as basis for comparisons.

042.090 **Stability of the 6th and the 7th libration point in the photo–gravitational restricted circular three body problem.**
S. Vijay Kumar, R. K. Choudhury.
J. Bihar Math. Soc., Vol. 9, p. 48 – 53 (1985).
The Liapunov instability of the sixth and the seventh libration point in the photo–gravitational circular restricted three body problem has been proved for all the planets of the solar system.

042.091 **Errata: "The Theory of the Trojan asteroids. Part V" [Celest. Mech., Vol. 36, No. 1, p. 19 – 45 (1985)].**
B. Garfinkel.
Celest. Mech., Vol. 38, No. 3, p. 232 (1986). – See Abstr. 40.042.009.

042.092 **Idee novelle a proposito della soluzione di un vecchio problema.**
P. Sconzo.
G. Astron., Vol. 11, N. 4, p. 427 – 440 (1985).

042.093 **On a possibility of using the classical representation of the gravitational potential near the surface of a planet.**
V. A. Antonov, K. V. Kholshevnikov.
Nauchn. probl. aviatsii, kosmonavt.: Istor,. sovremennost'. Moskva, p. 162 – 165 (1985). In Russian. Abstr. in Ref. Zh., 51. Astron., 4.51.98 (1986).

042.094 **Expansion of the potentials of some rotating bodies converging at the outer boundaries.**
M. V. Belikov.
Inst. teor. astron. Akad. Nauk SSSR. Leningrad, 47 pp. (1985). In Russian. Abstr. in Ref. Zh., 51. Astron., 5.51.137 (1986).

La mecánica celeste.
See Abstr. 004.086.

Algebraic manipulations on computers for lunar and planetary theories.
See Abstr. 021.001.

An outline of the practical use of Lie series.
See Abstr. 021.004.

On some possibilities of representing the gravitational potential of an arbitrary body by power series converging everywhere excluding its surface.
See Abstr. 021.009.

An absolutely stable domain of Hill's equation.
See Abstr. 021.016.

Expansion of annual aberration into trigonometric series.
See Abstr. 041.010.

Coordinate systems in the general relativistic framework.
See Abstr. 043.001.

Construction of a planetary solution with the help of an *N*–body program, and analytical complements.
See Abstr. 046.021.

Earth gravitational perturbation sensitivity and orbital inaccuracy for GEO–IK orbit.
See Abstr. 052.004.

Two locations, two times, and the element set.
See Abstr. 052.009.

Motion of a lunar orbiter.
See Abstr. 052.010.

Periodic rotational motions of space vehicles around the libration point of the restricted three–body problem.
See Abstr. 052.037.

Quasi–analytical solutions for the dynamics of a class of tethered satellites with Danby's aerodynamical drag.
See Abstr. 052.040.

Nonsingular elements in description of the motion of small eccentricity and inclination satellites.
See Abstr. 052.052.

Classes of orbits in the main problem of satellite theory.
See Abstr. 052.053.

Present problems in relativistic celestial mechanics.
See Abstr. 066.020.

Equations of motion for isolated bodies with relativistic corrections including the radiation reaction force.
See Abstr. 066.021.

Post–Newtonian treatise on the rotational motion of a finite body.
See Abstr. 066.022.

Planetary perturbation equations based on relativistic Keplerian motion.
See Abstr. 066.023.

On the dominant relativistic terms in the lunar theory.
See Abstr. 066.024.

Relativistic Kepler's third law.
See Abstr. 066.235.

High speed evaluation of the geopotential.
See Abstr. 081.019.

Testing solar system stability.
See Abstr. 091.002.

Long–term changes in the semimajor axes of the outer planets.
See Abstr. 091.003.

A relativistic theory of motion of the inner planets.
See Abstr. 091.004.

Account of solar perturbations in the motion of distant planetary satellites.
See Abstr. 091.026.

The dominant relativistic terms in the lunar theory.
See Abstr. 094.009.

Comparison of the asteroid proper elements obtained from various theories.
See Abstr. 098.041.

Orbital evolution at secular resonances in the asteroidal belt.
See Abstr. 098.042.

Secular perturbations of asteroids in secular resonance.
See Abstr. 098.043.

The Hilda group and the Hecuba gap.
See Abstr. 098.044.

On the computation of characteristic orbital elements for the Trojan group of asteroids.
See Abstr. 098.045.

Two Trojans near a secular resonance: (1173) Anchises and (1871) Astyanax.
See Abstr. 098.046.

The orbits of Trojan asteroids.
See Abstr. 098.047.

Orbital studies of 1982 YA, 1983 VA, and 1984 BC.
See Abstr. 098.048.

Subtle gravitational effects in the motion of small bodies in the solar system.
See Abstr. 098.050.

Structure of the 2:1 and 3:2 jovian resonances.
See Abstr. 098.052.

On the accuracy of orbital elements for unnumbered minor planets.
See Abstr. 098.057.

On the theory of motion of the Trojan asteroids.
See Abstr. 098.071.

Methods for the investigation of the evolution of asteroid orbits and crossing conditions with the orbit of the perturbing body.
See Abstr. 098.072.

Perturbing action of Ceres, Pallas, Juno and Vesta at their proximities with other selected minor planets.
See Abstr. 098.106.

The collision lifetimes of four asteroids.
See Abstr. 098.111.

Period analysis of Laplacean resonance amongst Uranian inner satellites.
See Abstr. 101.062.

Some remarks on cometary outbursts.
See Abstr. 103.020.

On the capture origin of the satellites of the solar system.
See Abstr. 107.030.

A comparison between the Roche model and the model of triaxial ellipsoids in close binary systems.
See Abstr. 117.271.

Solutions for the modified Newtonian dynamics field equation.
See Abstr. 151.064.

Simple three-dimensional periodic orbits in a galactic-type potential.
See Abstr. 151.094.

A method for obtaining Hamiltonians with simple periodic solutions.
See Abstr. 151.105.

043 Astronomical Constants, Reference Systems

043.001 **Coordinate systems in the general relativistic framework.**
T. Fukushima, M.-F. Fujimoto, H. Kinoshita, S. Aoki.
Relativity in celestial mechanics and astrometry, p. 145 – 168 (1986). – See Abstr. 012.002 (IAU Symp. No. 114).

The treatment of the coordinate systems is briefly reviewed in the Newtonian mechanics, in the special theory of relativity, and in the general relativistic theory, respectively. Some reference frames and coordinate systems proposed within the general relativistic framework are introduced. With use of the ideas on which these coordinate systems are based, the proper reference frame comoving with a system of mass-points is defined as a general relativistic extension of the relative coordinate system in the Newtonian mechanics. The coordinate transformation connecting this and the background coordinate systems is presented explicitly in the post-Newtonian formalism. The conversion formulas of some physical quantities caused by this coordinate transformation are discussed. The concept of the rotating coordinate system is reexamined within the relativistic framework. A modification of the introduced proper reference frame is proposed as the basic coordinate system in astrometry. The relation between the solar system barycentric coordinate system and the terrestrial coordinate system is given explicitly.

043.002 **System of astronomical constants in the relativistic framework.**
T. Fukushima, M. K. Fujimoto, H. Kinoshita, S. Aoki.
Relativity in celestial mechanics and astrometry, p. 187 – 188 (1986). Abstract. The full paper will be published in Celestial Mechanics. – See Abstr. 012.002 (IAU Symp. No. 114).

043.003 **Numerical effects of gravitational light deflection on the determination of the equinox and equator.**
J. A. Hughes, D. K. Scott, C. A. Smith.
Relativity in celestial mechanics and astrometry, p. 205 – 211 (1986). – See Abstr. 012.002 (IAU Symp. No. 114).

043.004 **Hypothesis of variable *G* and the secular accelerations of the Sun and the Moon.**
S. Yabushita.
Earth, Moon, Planets, Vol. 34, No. 2, p. 139 – 148 (1986).

It is known that the observed secular accelerations of the Sun and Moon are not consistent with the tidal interactions of the Earth with the Sun and Moon. Following Dicke, the hypothesis of variable constant of gravity is adopted and expressions for the accelerations are derived. It is shown that if the theoretical ratio of the acceleration is equated the observed one, a unique value for $-\dot{G}/G$ can be calculated. Adopting the accelerations obtained by Fotheringham, Newton, Muller and Stephenson, and Stephenson, it is found that $-\dot{G}/G$ ranges from 1.4×10^{-11} to $3.3 \times 10^{-11} \mathrm{yr}^{-1}$. This estimate is consistent with the one based upon the comparison of the lunar accelerations measured with respect to atomic and ephemeris times.

043.005 **Reference frame determination using minor planets observed through Tokyo PMC.**
M. Sôma.
Proceedings of the nineteenth symposium on celestial mechanics, p. 24 – 28 (1986). In Japanese. – See Abstr. 012.013.

043.006 **Relativistic coordinates and time scales.**
M.-K. Fujimoto.
Proceedings of the nineteenth symposium on celestial mechanics, p. 82 – 86 (1986). In Japanese. – See Abstr. 012.013.

043.007 **Fundamental reference systems; past, present, and future.**
P. K. Seidelmann.
Celest. Mech., Vol. 37, No. 3, p. 199 – 207 (1985). – See Abstr. 012.017.

The present fundamental reference system is defined by the FK5 Fundamental Reference Catalog and the associated IAU Resolutions of 1976 through 1982. The past is defined to be what preceded the present fundamental reference system and future is defined to be what ever will come after it. The resulting characteristics of the different fundamental reference systems are considered from the aspects of the observational data, the computational methods, and the underlying theories. Having developed an understanding of these operational fundamental reference systems, the desired properties of the ideal fundamental reference system are presented. Finally, some questions concerning the future fundamental reference system are presented.

043.008 **The evolution of adopted values for precession.**
J. H. Lieske.
Celest. Mech., Vol. 37, No. 3, p. 209 – 238 (1985). – See Abstr. 012.017.

A survey of values of the speed of general precession in longitude is presented from the time of Bessel to the adoption of the new value by the IAU in 1976 for use in the FK5 and the J2000 reference system. The important contributions of Walter Fricke of this area of fundamental astronomy are outlined.

043.009 **On the orientation of ephemeris reference frames.**
E. M. Standish Jr.
Celest. Mech., Vol. 37, No. 3, p. 239 – 242 (1985). – See Abstr. 012.017.

This paper presents estimates of the equinox offset and equinox motion of the FK4 as determined by the ephemeris fitting process and compares them with corresponding determinations by Fricke. No significant differences are found. Further, it is indicated how one may also estimate a value for precession and the value of the obliquity from the ephemerides. These values are also compared wih the presently adopted ones.

043.010 **Fundamental coordinate ties using laser ranging data.**
P. J. Shelus.
Celest. Mech., Vol. 37, No. 3, p. 243 – 246 (1985). – See Abstr. 012.017.

One of the main goals of fundamental astronomy is to allow us to locate ourselves accurately in space and time. Each observational technique cannot supply all of the answers all of the time, and the joining of results from different techniques requires a true understanding of the particular reference frame which underlies each one. In this paper the author tries to state the present situation in the attempts to tie artificial satellite and lunar laser ranging data through the construction of stations designed to make both kinds of observations. This type of development strengthens both data types and eases their combination with other data types.

043.011 **Precise position measurements of Jupiter, Saturn and Uranus systems with the Very Large Array.**
D. O. Muhleman, G. L. Berge, D. J. Rudy, A. E. Niell, R. P. Linfield, E. M. Standish.
Celest. Mech., Vol. 37, No. 3, p. 329 – 337 (1985). – See Abstr. 012.017.

Radio tracking measurements of spacecraft associated with the inner planets, such as the Viking Orbiters around Mars, combined with VLBI measurements indicate that the frames of the inner planets (i.e. their ephemerides), stellar system, and the VLBI reference sources are consistent to better than $\pm 0\farcs3$ (Newhall et al., 1984). In the work reported here, the authors have attempted to utilize the outer planets for the same purpose with the additional goal of improving the outer–planet ephemerides. The results for the three planets are very similar and suggest that the ephemeris reference frame for the outer planets is in error by roughly –0.2 arc sec in right ascension.

043.012 **El sistema IAU (1976) de constantes astronómicas y su repercusión en la reducción de posiciones de estrellas.**
M. J. Sevilla.
Bol. Inf. Serv. Geogr. Ejército, No. 57, p. 1 – 48 (1983). = Semin. Astron. Geod., Univ. Madr., Publ., No. 128.

043.013 **Un catálogo provisional FK5 para uso en astronomía geodésica.**
P. Romero.
Semin. Astron. Geod., Univ. Madr., Publ., No. 130, p. 55 – 73 (1983).

The author describes a procedure to obtain from the FK4 a provisional catalog in accordance with the IAU (1976) resolutions.

043.014 **Celestial reference coordinate systems: submilliarcsecond precision demonstrated with VLBI observations.**
D. S. Robertson, F. W. Fallon, W. E. Carter.
Astron. J., Vol. 91, No. 6, p. 1456 – 1462 (1986).

VLBI observations from the POLARIS and IRIS projects have been used to determine positions for 26 extragalactic radio sources. Nearly all the formal errors of the estimated coordinates are less than one millisecond of arc. Repeatability tests using subsets of the total data set indicate that systematic errors in the source coordinates are of the order of one half millisecond of arc. In order to attain submillisecond repeatability it was necessary to correct millisecond–level errors in the adopted IAU 1980 nutation model.

043.015 **A system of astronomical constants in the relativistic framework.**
T. Fukushima, M.–K. Fujimoto, H. Kinoshita, S. Aoki.
Celest. Mech., Vol. 38, No. 3, p. 215 – 230 (1986).

The relation between the units and readings of time and space coordinates of terrestrial and barycentric reference frames is discussed from the viewpoint of general relativity. Attention is paid to the unit of space coordinates since the International Astronomical Union (IAU) regulates only the unit of time in the above two frames. Two definitions of unit of length are examined and their effects on the numerical expression of coordinate transformation, equations of planetary motions, and those for light propagation time are discussed. A clear conflict is found between the IAU (1976) recommendation on the definition of the time–scales in different frames of reference and the statement that all constants in the IAU (1976) new system of astronomical constants are defined in terms of the International System of units (SI units).

043.016 **Concepts of reference systems.**
B. Guinot.
Astrometric techniques, p. 1 – 12 (1986). – See Abstr. 012.068 (IAU Symp. No. 109).

With the atomic clocks and VLBI catalogues of extragalactic sources, the theory–independent references for both space and time will be quasi ideal in their underlying principles (non–rotation of very distant objects, constancy of the frequencies associated to atomic transitions). The paper is an attempt to clarify the role of these natural references with respect to those which are implicitly defined by the dynamical theories.

043.017 **Discussion of meaning and definition of UT.**
B. X. Xu, S. Y. Zhu, H. Zhang.
Astrometric techniques, p. 13 – 17 (1986). – See Abstr. 012.068 (IAU Symp. No. 109).

The new definition of UT1 adopted by the IAU is useful but for many reasons not quite satisfactory. It depends e.g., on the approximate values of some astronomical constants, and is therefore subject to revision in the future. Since it is used for the FK5–based astronomical reference system, its eventual usefulness for space techniques is questioned. Although the new and old UT1 merge continuously at a chosen epoch, they do not form a homogeneous series of data. Neither the new definition, nor the way to convert the old to the new one is based on simple concepts. A conceptual definition of UT1 is suggested in order to correct this situation.

043.018 **The search for unwanted signals: systematic trends in proper motions of fundamental stars.**
P. Brosche, H. Schwan.
Astrometric techniques, p. 53 – 62 (1986). – See Abstr. 012.068 (IAU Symp. No. 109).

An investigation of the N30 along the same lines as that of the FK4 (Brosche and Schwan, 1981) reveals quite similar non–standard systematic motions. Some consequences are indicated for the determination of cluster parallaxes, of secular parallaxes and of the galactic rotation. In general, the authors stress the necessity of always searching for every signal in the data – including those ones which are not understood and hence not "wanted".

043.019 **A new technique for the analytical determination of a fundamental system of positions and proper motions.**
H. Schwan.
Astrometric techniques, p. 63 – 74 (1986). – See Abstr. 012.068 (IAU Symp. No. 109).

A survey of the various tasks involved in the construction of the Fifth Fundamental Catalogue (FK5) is given. The basic characteristics of the methods are described, and information on the most significant errors in the FK4 system is given.

043.020 **The conversion from the B1950 FK4–based position to the J2000 position of celestial objects.**
S. Aoki, M. Sôma, K. Nakajima, Y. Niimi, M. Fujishita, Y. Takahashi.
Astrometric techniques, p. 123 – 131 (1986). – See Abstr. 012.068 (IAU Symp. No. 109).
The authors present the fundamental idea of the conversion method between old and new reference frameworks. Some practical applications are made for the optical observations for Tokyo PZT.

043.021 **The observations of the earth rotation and the stellar system.**
K. Sato, K. Yokoyama, K. Tanikawa, Y. Goto.
Astrometric techniques, p. 133 – 141 (1986). – See Abstr. 012.068 (IAU Symp. No. 109).
One of the main objectives of this paper is to show how the errors in the reference coordinate system, especially the errors in the nutation series, affect the determination of the earth orientation parameters. Another is to estimate the effect of the nutation errors on the corrections to stellar positions in source catalogs.

043.022 **The precision of J2000.0 radio source positions from Mark III VLBI.**
C. Ma.
Astrometric techniques, p. 157 – 162 (1986). – See Abstr. 012.068 (IAU Symp. No. 109).
A celestial coordinate frame defined by extragalactic radio sources has been developed from Mark III VLBI data. 33000 delay and delay rate pairs acquired between August 1979 and December 1982 have been analyzed to give positions for 82 radio sources. Standard J2000.0 astronomical models were applied. 90% of the formal errors for arc lengths between sources are less than 0".003. Arc lengths estimated from separate one–year data sets are consistent at the 0".001 to 0".004 level.

043.023 **The JPL/DSN (*Jet Propulsion Laboratory/Deep Space Network*) J2000 radio reference frame.**
A. E. Niell, J. L. Fanselow, K. M. Liewer, O. J. Sovers, J. B. Thomas, R. N. Treuhaft, K. S. Wallace.
Astrometric techniques, p. 163 – 167 (1986). – See Abstr. 012.068 (IAU Symp. No. 109).
Development of a radio reference frame of very high accuracy has become necessary in order to obtain the required accuracy in spacecraft navigation for current and future planetary missions. The antennas of the NASA Deep Space Network are being utilized on a regular basis to obtain the positions of over 100 compact extragalactic radio sources distributed over the sky down to a declination of –45 degrees with uncertainties of less than 0.005 arcseconds. The authors report results based on data taken from 1978 September to 1983 May, describe the differences from other radio astrometric programs, and give a comparison with other VLBI results.

043.024 **The extragalactc optical/radio reference frame. A progress report, IAU Comm. 24 Working Group.**
C. de Vegt.
Astrometric techniques, p. 173 – 177 (1986). – See Abstr. 012.068 (IAU Symp. No. 109).
A catalog of 233 strong compact extragalactic radio sources which display optical counterparts has been prepared by the working group as a basis for the establishment of an almost inertial extragalactic reference frame. Optical properties of the catalog sources are summarized and further requirements for a refinement and extension of the present data are discussed.

043.025 **On the reference frame of the planetary ephemerides.**
E. M. Standish Jr.
Astrometric techniques, p. 677 – 683 (1986). – See Abstr. 012.068 (IAU Symp. No. 109).
The planetary and lunar ephemerides have their own inherent reference frame which is one of the key frames to be considered in the unification of the many present day frames, both celestial and terrestrial. The use of this frame requires an understanding of its own accuracies, both internal (i.e., relative) and external (e.g., w.r.t. the FK4 system or w.r.t. the dynamical equinox). Such accuracies are discussed in this paper.

043.026 **The reconciliation of optical and radio positions.**
A. N. Argue.
Astrometric techniques, p. 757 – 763 (1986). – See Abstr. 012.068 (IAU Symp. No. 109).
The optical and radio positional reference systems are reviewed. A preliminary comparison indicates that the two do not at present agree to better than 0".2 or 0".3.

043.027 **On the feasibility of a star coordinate determination in the radio astrometry reference system.**
V. S. Gubanov.
Astrometric techniques, p. 765 – 771 (1986). – See Abstr. 012.068 (IAU Symp. No. 109).
At present it is evident that the stability of a celestial coordinate system with an accuracy of at least 0".01 per century, with respect to an internal frame can be derived only from VLBI observations of compact extragalactic radio sources. The author discussed the question whether one can use the high precision and stability of the radio astrometric reference frame to stabilize and improve the reference frame given by star catalogues.

043.028 **Connection between the Hipparcos catalogue and the FK5.**
S. Röser.
Astrometric techniques, p. 773 – 778 (1986). – See Abstr. 012.068 (IAU Symp. No. 109).
Two methods are presented which can be applied to determine the rotation of the Hipparcos system of stars with respect to the FK5. These methods are presently tested in numerical simulation runs.

043.029 **Relating the JPL VLBI reference frame and the planetary ephemerides.**
X X Newhall, R. A. Preston, P. B. Esposito.
Astrometric techniques, p. 789 – 794 (1986). – See Abstr. 012.068 (IAU Symp. No. 109).
A series of VLBI observations of planet–orbiting spacecraft and angularly nearby quasars has provided an estimate of the relative orientation of the JPL VLBI reference frame and the ephemeris frame of the inner planets. The authors find the two frames to be coincident in both right ascension and declination to less than 20 mas.

043.030 **Geodesy by radio interferometry: corrections to the IAU 1980 nutation series.**
T. A. Herring, C. R. Gwinn, I. I. Shapiro.
Earth rotation and the terrestrial reference frame, p. 307 – 325 (1985). – See Abstr. 012.077.
The analysis of VLBI data has yielded corrections to the coefficients of the IAU 1980 nutation series with standard deviations between 0.10 and 0.17 mas (68% confidence intervals). The authors have estimated corrections to the fourteen terms in the nutation series that have the largest coefficients and that could be separated spectrally using the 4.9 years of VLBI data available. They also observe long period drifts which probably arise from errors in the precession constant, or in the 18.6–year–period nutations, or in both. The corrections to the amplitudes of the annual and semiannual terms, and about 60% of the drift in the nutation in obliquity, can be explained by the earth's fluid core being more strongly coupled to the mantle than in the model used for the construction of the IAU 1980 nutation series.

043.031 **An analysis and intercomparison of VLBI nutation estimates.**
T. M. Eubanks, J. A. Steppe, O. J. Sovers.
Earth rotation and the terrestrial reference frame, p. 326 – 340 (1985). – See Abstr. 012.077.
The authors intercompared nutation estimates from long duration VLBI experiments with similar estimates from the IRIS/POLARIS data. The two series have a rms difference of 1.6 mas or less. After removal of the IAU 1980 theory of nutation, there

exist large annual and smaller semiannual oscillations, as well as linear trends, in both obliquity and longitude. The authors regard the existence of these features as well established, and have developed a polynomial and sinusoidal model of corrections to the IAU 1980 nutation theory. Changing the resonance period to 431.5 solar days can account for most of the observed errors in the IAU nutation theory.

IAU Symposium 109, Gainesville, "Astrometric techniques". Concluding remarks.
See Abstr. 011.041.

On the rotation of the Earth and the terrestrial reference system: Joint summary report of the IAU/IUGG Working Groups MERIT and COTES.
See Abstr. 013.020.

Construction of an inertial coordinate system with a CCD. (I). Measuring the relative position of an extragalactic system with respect to the reference stars.
See Abstr. 036.144.

Construction of an inertial coordinate system with a CCD. (II). The determination of the scales and the correction for the inclination of the measured coordinate system.
See Abstr. 036.145.

Construction of an inertial coordinate system with a CCD. (III). Some astronomical corrections of the measured coordinates.
See Abstr. 036.146.

Construction of an inertial coordinate system with a CCD. (IV). The transition to the inertial coordinate system.
See Abstr. 036.147.

Relativity in astrometry.
See Abstr. 041.002.

Numerical integration of precession and nutation.
See Abstr. 041.020.

The Southern Reference Star program: a 20–year international project nearing completion.
See Abstr. 041.021.

The extension of the fundamental system to fainter magnitudes.
See Abstr. 041.022.

Recalculation of the magnitude equation of bright stars and of Newcomb's precession corrections for Fatchikhin's catalogue of proper motions of 14600 stars.
See Abstr. 041.033.

Computation of compilation catalogs.
See Abstr. 041.050.

Brosche's method for representing systematic differences in positions and proper motions of stars.
See Abstr. 041.051.

Some effects of different sources of variation of latitude data on meridian circle catalogues.
See Abstr. 041.052.

The use of Space Telescope to tie the HIPPARCOS reference frame to an extragalactic reference frame.
See Abstr. 041.068.

On desiderata for star catalogs for the remainder of the twentieth century: a report on catalog work now in progress at the U.S. Naval Observatory.
See Abstr. 041.072.

Results of VLBI observations of radio stars and their potential for linking the HIPPARCOS and extragalactic reference frames.
See Abstr. 041.079.

Rediscussion on the effect of precession correction on UT1 and the definition of Universal Time.
See Abstr. 044.062.

Systematic differences between planetary observations and ephemerides.
See Abstr. 046.002.

Reference frames and gravitational effects in the general theory of relativity.
See Abstr. 066.025.

Relativistic effects in astronomical timing measurements.
See Abstr. 066.196.

044 Time and Latitude Determination, Earth Rotation, Polar Motion

044.001 **Relativity effects in the rotation of the Earth and the motion of atmospheric masses.**
V. G. Shkodrov, V. G. Ivanova.
Relativity in celestial mechanics and astrometry, p. 289 – 292 (1986). – See Abstr. 012.002 (IAU Symp. No. 114).

On the basis of observational data on atmospheric pressure (1963 – 1967), the variation of the moment of inertia, and, with certain restrictions, the changes in the angular velocity of the Earth are obtained. The numerical results derived are compared to the relativity effects in Earth rotation. The comparison shows that both effects are equal in periods and very close in amplitudes.

044.002 **The International Atomic Time, definition, realization.**
B. Guinot.
Relativity in celestial mechanics and astrometry, p. 297 (1986). Abstract. The complete paper has been submitted for publication to "Celestial Mechanics". – See Abstr. 012.002 (IAU Symp. No. 114).

044.003 **Coordinate time in the vicinity of the earth.**
D. W. Allan, N. Ashby.
Relativity in celestial mechanics and astrometry, p. 299 – 313 (1986). – See Abstr. 012.002 (IAU Symp. No. 114).

Atomic clock accuracies continue to improve rapidly, requiring the inclusion of general relativity for unambiguous time and frequency clock comparisons. Atomic clocks are now placed on space vehicles and there are many new applications of time and frequency metrology. This paper addresses theoretical and practical limitations in the accuracy of atomic clock comparisons arising from relativity, and demonstrates that accuracies of time and frequency comparison can approach a few picoseconds and a few parts in 10^{16}, respectively.

044.004 **The Earth's differential rotation.**
R. Hide.
Q. J. R. Astron. Soc., Vol. 27, No. 1, p. 3 – 20 (1986).

044.005 **Modelling the pole tide and its effect on the Earth's rotation.**
J. A. Carton, J. M. Wahr.
Geophys. J. R. Astron. Soc., Vol. 84, No. 1, p. 121 – 137 (1986).
The authors construct both analytical and numerical models for the pole tide. The analytical models consider the tide first in a global ocean and then in an enclosed basin on a beta–plane. The results are found to approach equilibrium linearly with decreasing frequency and inversely with increasing basin depth. The numerical models solve Laplace's tidal equations over the world's oceans using realistic continental boundaries and bottom topography. The results indicate that the effects of the non–equilibrium portion of the deep ocean tide on the Chandler wobble period and damping are negligible.

044.006 **Cosmic radiation and the earth's rotation.**
G. P. Pil'nik.
Astron. Zh., Tom 63, Vyp. 1, p. 184 – 190 (1986). In Russian. English translation in Sov. Astron., Vol. 30, No. 1.
On the basis of classical astronomical observations of time waves of nonuniformity in the earth's rotation with periods less than a day were found. The wave with the period of $159^{m}.566$ is very close to the period of global oscillations of the sun's surface $160^{m}.01$ and to the period of the Geminga gamma–ray radiation $159^{m}.96$.

044.007 **Determination of the Earth rotation parameters from optical astrometry observations, 1962.0 – 1982.0.**
Z.–x. Li, M. Feissel.
Bull. Géod., Vol. 60, No. 1, p. 15 – 28 (1986).
The optical astrometric data of the years 1962 – 1982 have been reduced once again at the Bureau International de l'Heure (BIH) in order to redetermine the Earth Rotation Parameters (ERP). This new reduction is based on series largely revised by the stations since their use in the operational work of the BIH, and on some series which were not available previously. A new approach is developed: the catalog and local errors are analysed and corrected as group unknowns, which values are adjusted together with the main unknowns. The results obtained in the new reduction are compared to other series obtained by astrometry and space geodesy.

044.008 **Checking of the double–frequency feature of Chandler main peak for different periods.**
H.–z. Zhang, Y.–b. Han, Z.–a. Li.
Acta Geophys. Sin., Vol. 29, No. 1, p. 16 – 27 (1986). In Chinese.
By use of ILS data, the "double–frequency" feature of the Chandler peak has been checked for different intervals of period from 1900.0 to 1979.0. The conclusion is that the "double–frequency" of Chandler peak is only a statistic characteristic for longer intervals.

044.009 **Pourquoi la Terre tourne–t–elle sur elle–même?**
C. Denis.
Ciel, Vol. 48, p. 110 – 116 (1986).

044.010 **Comparison of the earth rotation parameters determined by different methods during the international experiment MERIT.**
V. D. Krajchev, V. V. Nesterov.
Pis'ma Astron. Zh., Tom 12, No. 3, p. 253 – 254 (1986). In Russian. Englisch translation in Sov. Astron. Lett., Vol. 12.
Preliminary results of the main campaign of the MERIT experiment are analysed. It is concluded that the external accuracy of the new methods of determination of the earth rotation parameters is not so high, as is sometimes believed, and the duration of the MERIT experiments is unsatisfactory to solve the proper problems.

044.011 **Changes at USNO in global timekeeping.**
G. M. R. Winkler.
Proc. IEEE, Vol. 74, No. 1, p. 151 – 155 (1986).
Changes have taken place and new tendencies are now apparent in the practice of time keeping and time service for large geographical areas. This is due to an increased use of high–performance clocks in a variety of electronic systems. The new methods provide increased precision which is also useful in determining earth rotation parameters and other astronomical data. In the following, a brief account of these changes and methods is given as they have taken place in the Time Service of the U.S. Naval Observatory (USNO).

044.012 **Recent trends in NBS time and frequency distribution services.**
R. E. Beehler, D. W. Allan.
Proc. IEEE, Vol. 74, No. 1, p. 155 – 157 (1986).
Since 1967 the National Bureau of Standards has improved its traditional radio broadcast services from radio stations WWV, WWVH, and WWVB and introduced several new services to meet changing needs. The new services are described briefly, including the GOES satellite time code, the Frequency Measurement Service using Loran–C and WWVB, and the Global Time Service based on the GPS satellite common–view technique.

044.013 **Research activities on time and frequency transfer using space links.**
K. Yoshimura, M. Imae, M. Urazuka, T. Morikawa, T. Yoshino, S. Kobayashi, T. Igarashi.
Proc. IEEE, Vol. 74, No. 1, p. 157 – 160 (1986).
Time and Frequency (T/F) transfer systems using space links developed at the Radio Research Laboratories (RRL) are described. T/F transfer experiments using a VLBI system developed at RRL have been conducted on a routine basis about once a month with the U.S. Naval Observatory since April of 1985. The present goal of accuracy is 10 ns with 10–min or longer observations.

044.014 **Time transfer via geostationary satellites.**
C. C. Costain.
Proc. IEEE, Vol. 74, No. 1, p. 161 – 162 (1986).
Numerous experiments using geostationary satellites for two–way time transfer have demonstrated that this method can achieve subnanosecond precision and nanosecond accuracy. Efforts are now directed towards establishing operational networks.

044.015 **The influence of geophysical processes on the Earth's rotation. Part I.**
J. Nastula.
Postepy Astron., Tom 33, Zesz. 1–2, p. 3 – 27 (1985). In Polish.
The problem of the influence of geophysical processes on the Earth's rotation is presented. The role of these processes in the variations of the length of day is described in this part.

044.016 **Determination of time (UT1–UTC). 1983 January – 1983 July.**
O. S. Tursunov.
Tsirk. Astron. Inst., Tashkent, Nos. 113, 114, 116 (1985). In Russian.

044.017 **Analytical expression of TDB–TDT and its accuracy.**
T. Hirayama, H. Kinoshita.
Proceedings of the nineteenth symposium on celestial mechanics, p. 87 – 88 (1986). In Japanese. – See Abstr. 012.013.

044.018 **History of long–term observations of latitude variations and tidal deviations of the vertical in Kazan.**
I. A. Urasina.
Izv. Astron. Ehngel'gardt. Obs., No. 48, p. 30 – 37 (1984). In Russian.

044.019 **High-accuracy reduction of 19 years (1957 – 1976) latitude observations at the Engelhardt Astronomical Observatory.**
V. V. Lapaeva, I. A. Urasina, N. G. Sokolova, N. N. Chudinov.
Izv. Astron. Ehngel'gardt. Obs., No. 50, p. 111 – 197 (1985). In Russian.

044.020 **Vergleichende Analyse der Ergebnisse zweier astronomischer Beobachtungsreihen und einer Dopplermeßreihe zur Bestimmung von Breiten- und Längenvariationen über eine Chandlerperiode.**
K. Kaniuth, W. Schlüter, G. Soltau, W. Wende.
Veröff. Bayer. Komm. Int. Erdmessung Bayer. Akad. Wiss., Astron.-Geod. Arb., Heft Nr. 46, 45 pp. (1985).
During the period 1975.4 – 1976.9 at the satellite observatory Wettzell astronomical observations using two precision astrolabes were carried out parallel to the Navy Navigation Satellite System Doppler measurements permanently performed. This paper investigates the latitudes and longitudes resulting from all three observation series in regard to systematic effects, systematic differences between both independent techniques as well as temporal and, in particular, periodical variations.

044.021 **Deceleration of the earth rotation under the action of the gravitational field of the moon and its density model.**
E. M. Filippov.
Mor. gidrofiz. inst. Akad. Nauk USSR. Sevastopol', 16 pp. (1985). In Russian. Abstr. in Ref. Zh., 51. Astron., 4.51.114 (1986).

044.022 **Chandler wobble parameters from BIH and ILS data.**
H. Lenhardt, E. Groten.
Manuscr. Geod., Vol. 10, No. 4, p. 296 – 305 (1985).
In order to overcome some uncertainties in parametrizing the Chandler wobble BIH and ILS data are evaluated by various methods. A comparison of the corresponding spectra shows no evidence of a several component wobble in BIH spectra. Multiple Chandler spectral peaks usually obtained from ILS data are probably a misleading result caused by irregularities some decades ago. From BIH data a Chandler wobble period of about 431 days is obtained and the quality factor Q is 24, which is an unexpected low value.

044.023 **Short term fluctuations of earth rotation.**
T. Hara, K. Yokoyama.
Proceedings of the Symposium on Application of Space Techniques to Astronomy and Geophysics, p. 11 (1986). – See Abstr. 012.027.

044.024 **Local variations of time observations derived from optical astrometry and motions in the earth's interior.**
C. Kakuta.
Proceedings of the Symposium on Application of Space Techniques to Astronomy and Geophysics, p. 12 – 16 (1986). – See Abstr. 012.027.

044.025 **The analyses of the earth orientation parameters by using the Japan-USA joint VLBI experiments and IRIS data (Part 1).**
Y. Takahashi, K. Heki.
Proceedings of the Symposium on Application of Space Techniques to Astronomy and Geophysics, p. 108 – 113 (1986). In Japanese. – See Abstr. 012.027.

044.026 **The new international earth rotation service.**
K. Yokoyama.
Proceedings of the Symposium on Application of Space Techniques to Astronomy and Geophysics, p. 155 – 159 (1986). In Japanese. – See Abstr. 012.027.
A short review is given about the IAU resolutions and the MERIT/COTES recommendations concerning the new international earth rotation service to be started from January 1, 1988, replacing both the IPMS and the BIH. The most suitable role for Japan in the new service is to act as an analysis center of the VLBI, in addition to the role of observatories.

044.027 **An outline of the project of the VLBI for the earth rotation study and astrometry (VERA).**
T. Hara.
Proceedings of the Symposium on Application of Space Techniques to Astronomy and Geophysics, p. 160 – 163 (1986). – See Abstr. 012.027.

044.028 **QSO's in the post-fit residual of the JPL VLBI data.**
M. Fujishita.
Proceedings of the Symposium on Application of Space Techniques to Astronomy and Geophysics, p. 164 – 168 (1986). In Japanese. – See Abstr. 012.027.
The effects of radio source structure and its time variation were looked for in the post-fit residual of VLBI group delay data observed by the JPL. The estimated values of the effects were calculated from the VLBI map data observed between 1 GHz and 10 GHz. Almost all effects caused by the sources listed in the JPL 1984-2 catalogue of the radio source positions are less than 0.05 nano-second except the singular points in the U – V plane. It seems that there are no systematic errors in the sidereal time – post-fit residual graphs.

044.029 **Nutations derived from Polaris-IRIS VLBI observations and the earth's interior.**
T. Sasao.
Proceedings of the Symposium on Application of Space Techniques to Astronomy and Geophysics, p. 169 – 174 (1986). In Japanese. – See Abstr. 012.027.
A recent analysis of the forced nutations of the earth by Herring, Gwinn and Shapiro based on Polaris-IRIS VLBI observations is reviewed. Possible geophysical implications of their results are discussed.

044.030 **Astrolabe 45° zenith distance observations: preliminary results in time and latitude.**
L. B. F. Clauzet, N. V. Leister, W. Monteiro.
Rev. Mex. Astron. Astrofis., Vol. 12, Spec. Issue, p. 445 (1986). Abstract. – See Abstr. 012.066.

044.031 **Measurements of the parameters of the earth's rotation and the Love numbers *h*, *l*, and *k* with an astrometric very-long-baseline radio interferometer.**
B. N. Lipatov, A. S. Sizov.
Sov. Astron., Vol. 29, No. 4, p. 478 – 483 (1985). English translation of 40.044.008.

044.032 **Zur Zeithaltung auf der Fundamentalstation Wettzell.**
W. Schlüter, D. Feil.
Nachr. Karten-Vermessungswes., Reihe I, Heft Nr. 96, p. 25 – 35 (1985).
The timing system of the Fundamentalstation Wettzell is described. The connection of the local timescale UTC (Wettzell) to the international Atomic Time UTC (BIH) is explained and the offsets are published.

044.033 **The earth's inconstant rotation.**
J. Wahr.
Sky Telesc., Vol. 71, No. 6, p. 545 – 549 (1986).

044.034 **Rotation of the earth in the gravitational field of the moon and sun.**
V. P. Dolgachev.
Sov. Astron., Vol. 29, No. 5, p. 581 – 583 (1985). English translation of 40.044.017.

044.035 **On separation of the semi-annual component from data of latitude observations.**
B. S. Dubik.
Red. Zh., Kinematika Fiz. Nebesn. Tel, 4 pp. (1986). In Russian. Abstr. in Ref. Zh., 51. Astron., 6.51.157 (1986).

044.036 **New techniques in earth rotation study.**
S. Manabe.
Astron. Her., Vol. 79, No. 6, p. 146 – 149 (1986). In Japanese.

044.037 **Mass extinction, Earth's rotation and periodic cratering.**
D.–h. Chen, L.–z. Liu, J.–q. Zheng.
Sci. Sin., Ser. A, Vol. 29, No. 5, p. 531 – 539 (1986).
This paper discusses the relations between the periodic cratering events in the geologic past and the mass extinction and the secular variations of the Earth's rotation. The authors find a statistical relation between the extinction intensity of major extinction and the crater's diameter. The paper presents an analysis of the data of paleontologic clocks within the past 100 Myr. After eliminating the tidal evolutional trend, the analytical results indicate that the secular variation of the Earth's rotation has a 30 Myr cyclicity.

044.038 **Time and Latitude Service, 1985 April – September.**
Circ. Time Latitude Serv., Nos. 174 – 175 (1985).

044.039 **The stability of the amplitudes and phases of the Chandler's, annual and semi–annual term in the polar motion.**
N. Pejović.
Publ. Dep. Astron., Univ. Beogr., No. 13, p. 31 – 39 (1985).
Using the unsmoothed ILS coordinates of the instantaneous pole (Yumi and Yokoyama, 1980) the stability of the amplitudes and phases of the main harmonic terms in the polar motion are investigated. It has been established that the amplitudes and the phases were not subject to systematic fluctuations. As for the amplitude and the phase of the Chandler nutation, it seems that the amplitude forms a periodic function of time.

044.040 **On tidal friction and the decelerations of the Earth's rotation and Moon's revolution.**
E. W. Schwiderski.
Mar. Geod., Vol. 9, No. 4, p. 399 – 450 (1985). Abstr. in Phys. Abstr., Vol. 89, No. 1274, Entry 23408 (1986).

044.041 **Variazioni di latitudine e tempo determinate con l'astrolabio Danjon di Cagliari. Analisi preliminare.**
A. Poma.
Rend. Semin. Fac. Sci., Univ. Cagliari, Vol. 53, Fasc. 2, p. 51 – 62 (1983). = Pubbl. Stn. Astron. Int. Latitudine, Carloforte–Cagliari, Nuov. Ser., N. 107.
Variations of the latitude and astronomical time obtained with the Danjon astrolabe at Cagliari Observatory in the years 1982 – 83 are given. A comparison with the BIH results shows the existence of interesting seasonal variations for both time and latitude. Preliminary analysis and possible explanations are also presented.

044.042 **Dinamica della Terre e telemetria laser lunare.**
L. Cugusi.
Mem. Soc. Astron. Ital., Suppl., Vol. 55, No. 1, p. 37 – 58 (1984). = Pubbl. Stn. Astron. Int. Latitudine, Carloforte–Cagliari, Nuov. Ser., N. 115.

044.043 **Secular polar motion from classical and modern observations.**
A. Poma, E. Proverbio.
Comptes rendus sur les Journées Luxembourgeoises de Géodynamique, 3 pp. (1985). = Pubbl. Stn. Astron. Int. Latitudine, Carloforte–Cagliari, Nuov. Ser., N. 122.

044.044 **Astronomical and Doppler observations in Italy during the main MERIT campaign.**
A. Poma, L. Mureddu, F. Mazzoleni, F. Chlistovsky, G. Chiumiento, G. Picchio, M. Sarasso, G. Caprioli.
Comptes rendus sur les Journées Luxembourgeoises de Géodynamique, 2 pp. (1985). = Pubbl. Stn. Astron. Int. Latitudine, Carloforte–Cagliari, Nuov. Ser., N. 123.

044.045 **Astrolabe observations of time and latitude at Cagliari Observatory during the MERIT main campaign.**
A. Poma, V. Gusai.
Circ. Stn. Astron. Int. Latitudine, Carloforte–Cagliari, N. 27 (Ser. B(12)), 17 pp. (1985).
Time and latitude results of observations made at Cagliari Observatory with the Astrolabe Danjon during the main campaign of Project MERIT (1983 September 1 – 1984 October 31) are given. The results are consistent with the 1976 IAU system of astronomical constants.

044.046 **Time service for the year 1984.**
L. Mureddu.
Circ. Stn. Astron. Int. Latitudine, Carloforte–Cagliari, N. 28 (Ser. A(11)), 33 pp. (1985).

044.047 **Breitenbestimmungen. Juli – Dezember 1985.**
Lohrmann–Obs., Tech. Univ. Dresden, Zirk., Nos. 112 – 113 (1985).

044.048 **Comparison of two methods for the absolute determination of latitude.**
X.–c. Hu, W. Mao.
Publ. Yunnan Obs., No. 1, p. 31 – 35 (1986). In Chinese.
The alternate first vertical meridian circle observation method is compared to the traditional meridian circle method for the absolute latitude determination. It is suggested that, in principle, the latitude values determined by means of the former method are absolute, but those determined by the latter method are not really absolute.

044.049 **Results of determination of latitude in Józefosław, October – December 1985.**
L. Pieczyński.
Latitude Circ., No. 99, 6 pp. (1986).

044.050 **Daily time differences and phase values. 1986 January – June.**
U.S. Nav. Obs., Time Serv. Publ., Ser. 04, Nos. 988 – 1012 (1986).

044.051 **Earth Orientation. 1986 January – March.**
Earth Orientation Bull., Vol. 4, Nos. 1 – 13 (1986).

044.052 **NEOS. Earth Orientation. 1986 April – June.**
NEOS Earth Orientation Bull., Vol. 1, Nos. 1 – 13 (1986).

044.053 **Time Service Announcement.**
G. M. R. Winkler.
U.S. Nav. Obs., Time Serv. Publ., Ser. 14, Nos. 41 – 42 (1986).

044.054 **Project MERIT standards. Update No. 1 (Dec. 85).**
U.S. Nav. Obs., Circ., No. 167, 13 pp. (1985).
The paper contains an updating of U.S. Nav. Obs., Circ., No. 167 – see Abstr. 37.044.055.

044.055 **Modificaciones del movimiento del eje de rotación en la tierra al considerar modelos deformables.**
M. J. Sevilla.
Rev. Real Acad. Cienc., Tomo 76, p. 177 – 180 (1982). = Semin. Astron. Geod., Univ. Madr., Publ., No. 123.
Some results obtained for the polar motion of the rotation axis of the earth under the assumptions of a rigid model and an elastic one, are summarized.

044.056 **Obtención de las medidas de la precisión en la determinación de la latitud y la longitud por fotografías cenitales de estrellas.**
M. J. Sevilla, P. Romero.
Téc. Topográfica, Vol. 10, No. 48, 10 pp. (1982). = Semin. Astron. Geod., Univ. Madr., Publ., No. 126.
By application of a differential method, the variance–covariance matrix for the longitude and latitude determination by zenith photographs is obtained.

044.057 **Movimiento solo forzado del polo para una tierra elástica.**
M. J. Sevilla, P. Romero.
Semin. Astron. Geod., Univ. Madr., Publ., No. 130, p. 1 – 19 (1983).
The purpose of this paper is to obtain the motion of the Earth's axis of rotation in an Earth's fixed frame. The authors consider only the effect of the perturbations due to the external torque and also the elastic deformations produced by the lunisolar tidal force.

044.058 **Autofrecuencias de las ecuaciones de Helmholtz y Liouville para un modelo de tierra tipo Jeffreys simplificado.**
M. J. Sevilla, A. González–Camacho.
Semin. Astron. Geod., Univ. Madr., Publ., No. 130, p. 21 – 36 (1983).
The authors obtain expressions for the free frequencies of polar motion for an ellipsoidal, rotating and perturbed earth model constituted by an elastic mantle with an homogeneous liquid core of additional simple motion.

044.059 **Bulletins, Time Service of the Mizusawa Observatory in 1984.**
I. Tsubokawa.
Bull., Time Serv. Mizusawa Obs., Vol. 29, 2 + 36 pp. (1984/85).

044.060 **Monthly Notes of the International Polar Motion Service.**
Mon. Notes Int. Polar Motion Serv., Nos. 12, 1 – 4, p. 187 – 201, 1 – 71 (1985/86).

044.061 **Time and Frequency Services Bulletin, 1985 October – 1986 March.**
Time Freq. Serv., Bull., Nos. 73 – 78, 36 pp. (1985/86).

044.062 **Rediscussion on the effect of precession correction on UT1 and the definition of Universal Time.**
B.–x. Xu, S.–y. Zhu, H. Zhang.
Publ. Shaanxi Astron. Obs., Vol. 7, No. 2, p. 111 – 112 (1984). Abstract. In Chinese and English.

044.063 **Time and Latitude Bulletins. 1985 October – December.**
Tokyo Astron. Obs., Time Latitude Bull., Vol. 59, No. 4, p. 63 – 90 (1985).

044.064 **Time and Latitude Bulletins. 1986 January – March.**
Tokyo Astron. Obs., Time Latitude Bull., Vol. 60, No. 1, p. 1 – 26 (1986).

044.065 **Comparative stability of pole coordinate systems.**
A. A. Korsun'.
Sov. Astron. Lett., Vol. 11, No. 5, p. 333 – 334 (1985). English translation of 40.044.018.

044.066 **The effect of abnormal atmospheric refraction on the precision of time determination.**
B.–y. Yang.
Publ. Purple Mt. Obs., Vol. 4, No. 3, p. 25 – 28 (1985). In Chinese.
The results of time determination are analysed. It is shown that the precision of time determination with the transit is degraded by the abnormal atmospheric refraction that is caused by the structure and site of the building.

044.067 **The IRIS Earth orientation parameters.**
IRIS Bull. A, Nos. 22 – 28 (1985/86).

044.068 **Rotation of a deformable Earth with dynamical superficial fluid layer and liquid core. I. Fundamental equations.**
H. Legros, M. Amalvict.
Ann. Geophys., Vol. 3, No. 5, p. 655 – 670 (1985). Abstr. in Phys. Abstr., Vol. 89, No. 1277, Entry 41621 (1986).

044.069 **Polar motion and European baselines determined by analysis of satellite laser ranging data.**
A. Caporali, F. Palutan, A. Cenci, S. Casotto.
Lett. Nuovo Cimento, Vol. 44, Ser. 2, No. 7, p. 513 – 518 (1985). Abstr. in Phys. Abstr., Vol. 89, No. 1277, Entry 41635 (1986).

044.070 **The Green Bank interferometer as a tool for the measurement of Earth orientation parameters.**
D. N. Matsakis, F. J. Josties, P. E. Angerhofer,
D. R. Florkowski, D. D. McCarthy, J.–y. Xu, Y.–l. Peng.
Astron. J., Vol. 91, No. 6, p. 1463 – 1473 (1986).
Comparison of UT0 as measured by the connected–element interferometer at Green Bank, West Virginia with similar data derived from VLBI and the BIH indicates that UT0 can be obtained from the longest CEI baseline to an accuracy of 0.8 ms. A variety of improvements are discussed, including a new model atmosphere, a new method of extracting UT0 from the authors' data, and an optimal source observation schedule.

044.071 **Annual Report (ERP).**
Annu. Rep. (ERP), 148 pp. (1986).

044.072 **Zeit– und Breitenbestimmungen, Zeitsysteme, Präzisionszeitvergleiche. 1983 Juli – Dezember.**
Zeit–Breitenbestimmungen, Jahrg. 1983, Nr. 3 – 4, 18 + 16 pp. (1985).

044.073 **Time and Latitude. October – December 1983.**
V. Ptáček, J. Vondrák, R. Weber.
Circ. Czech. Obs., Time Latitude, 23 pp. (1984).

044.074 **Time and Latitude. January – December 1984.**
V. Ptáček, J. Vondrák, R. Weber.
Circ. Czech. Obs., Time Latitude, 25 + 25 + 25 + 20 pp. (1985).

044.075 **Bureau International de l'Heure. (B.I.H.). Circular D.**
B.I.H. Circ., D230 – D236 (1986).
Contents: UTC, TAI, UT and coordinates of the pole, rotation of the Earth. Data for 1985 November – 1986 May.

044.076 **Comparación de series de nutación.**
M. J. Sevilla, P. Romero.
Semin. Astron. Geod., Univ. Madr., Publ., No. 127, p. 975 – 1028 (1983).
The authors compare the nutation series of Woolard, the IAU–1979 and the IAU–1980 ones in terms of the 106 principal waves and in the computed values of nutation in longitude and obliquity.

044.077 **Models for true polar wander from palaeomagnetism, geoid and hot–spot studies.**
P. Marton.
Tectonophysics, Vol. 121, No. 2 – 4, p. 331 – 344 (1986). Abstr. in Phys. Abstr., Vol. 89, No. 1278, Entry 47417 (1986).

044.078 **Pulsar time scale–PT.**
V. G. Il'in, L. K. Isaev, S. B. Pushkin, G. N. Palii,
Yu. P. Ilyasov, A. D. Kuz'min, T. V. Shabanova, Yu. P. Shitov.
Metrologia, Vol. 22, No. 1, p. 65 – 67 (1986). Abstr. in Phys. Abstr., Vol. 89, No. 1279, Entry 48367 (1986).

044.079 **Real polar wandering.**
P. Marton.
Magy. Geofiz., Vol. 26, No. 5 – 6, p. 217 – 225 (1985). In Hungarian. Abstr. in Phys. Abstr., Vol. 89, No. 1279, Entry 53093 (1986).

044.080 **The interpretation of cyclic geophysical processes by the inverse solution of the differential equation of the Earth's rotation.**
H. Jochmann.
Geod. Geophys. Veröff., Reihe III, Heft 53, p. 132 – 139 (1985). = Mitt. Zentralinst. Phys. Erde, Nr. 1285.

044.081 **Bureau International de l'Heure. Annual Report for 1985.**
Annu. Rep. (B.I.H.), 6+A24+B62+C11+D102 pp. (1986).
Contents: Part A: Explanation of tables and other information. Part B: Tables and figures. Part C: Time signals. Part D: Earth rotation and related reference systems.

044.082 **Velocity–dependent inertial induction and secular retardation of the Earth's rotation.**
A. Ghosh.
Pramāna, Vol. 26, No. 1, p. P217 (1986). Abstr. in Phys. Abstr., Vol. 89, No. 1280, Entry 60141 (1986).

044.083 **Is the International Atomic Time TAI a coordinate time or a proper time?**
G. Guinot.
Celest. Mech., Vol. 38, No. 2, p. 155 – 161 (1986).
The International Atomic Time TAI is a physically realized time scale which is ultimately used for comparisons between observations and dynamical theories. Its definition should tell unambiguously what an ideal TAI should be. For terrestrial applications, TAI has been defined as a geocentric coordinate time. In solar system dynamics, a barycentric coordinate time is needed. In general, it is not possible to convert a coordinate time into another coordinate time. But a specific clock synchronized on TAI in the terrestrial system can be considered as reading a "modified" proper time $[TAI]_i$, which can be converted into a barycentric coordinate time.

044.084 **The residual vector in polar motion as a random vector.**
R. A. Serafin, W. Jakś.
Acta Astron., Vol. 35, Nos. 3–4, p. 421 – 432 (1985).
The authors consider the hypothesis that the residual vector in polar motion is the random vector on the R^2 plane. The empirical distribution functions of $X(t)$ and $Y(t)$ pole coordinates, after the separation of linear and periodic components obtained by different observational data, have been tested. The performed test investigations allow the following hypothesis: the residual vector $r(t) = (x(t), y(t))$ in polar motion is the random vector with the uniform distribution of the polar angle and its coordinate values have normal distribution.

044.085 **Polar motion estimates from linar combinations of independent series.**
R. O. Vicente, C. R. Wilson.
Astron. Nachr., Vol. 307, No. 2, p. 137 – 147 (1986).
This paper describes methods for combining independent observations of polar motion into a single time series, and applies these methods to satellite laser ranging, Doppler, and astrometric polar motion data taken over the period 1981 – 1983. The results of analyzing three years of observations show little difference between the Allan and conventional variance. Present day Doppler determinations do not show marked improvement compared to classical astrometric results. Thus, the astrometric observations are still useful.

044.086 **On the geophysical explanation of some irregular variations of the mean latitudes of latitude stations.**
L. Czechowski, B. Kolaczek.
Pure Appl. Geophys., Vol. 123, No. 3, p. 398 – 406 (1985). Abstr. in Phys. Abstr., Vol. 89, No. 1281, Entry 66854 (1986).

044.087 **Measurement of the quality factor Q of the Chandler wobble.**
D.–n. Dong.
Acta Astron. Sin., Vol. 27, No. 1, p. 16 – 22 (1986). In Chinese.
The data of the polar coordinates of ILS(H), IPMS, BIH are analysed using O–K and ARH methods. More accordant results of the quality factor Q of the Chandler wobble are obtained.

044.088 **Acceleration of the earth's rotation obtained from records on occultation of stars by the moon before 600 A.D. in China.**
C.–y. Liu.
Acta Astron. Sin., Vol. 27, No. 1, p. 69 – 79 (1986). In Chinese.
From 83 carefully selected records of occultations of stars by the moon (186 – 567 A.D.), the author obtained the acceleration $\dot{\omega} = -49.7 \pm 4.2$ sec/century^2 or $\Delta t = 5.42 \times 10^3$sec for the weighted mean epoch 423 A.D. He adopts lunar tidal acceleration $\dot{n} = -22''\!.4$/century^2.

044.089 **Determination of ERP with the data of LLR.**
W.–j. Jin, Q.–g. Wang.
Acta Astron. Sin., Vol. 27, No. 2, p. 85 – 94 (1986). In Chinese.

044.090 **On the optimal conditions for determination of Earth rotation parameters.**
M. Zhao, S.–y. Zhu.
Acta Astron. Sin., Vol. 27, No. 2, p. 96 – 105 (1986). In Chinese.
There are various techniques used for determination Earth rotation parameters (ERP). In this paper the observation equation suitable for all techniques is given, and the most favorable situation to determine generalized zenith distance is discussed. Under the assumption that all the equivalent observables have the same weight, the optimal conditions for determining ERPs are reduced from the general observation principle. The result looks very simple, yet, it can be widely used for all techniques. It is also shown, that to obtain the optimal ERPs, the distribution of the source vector is as important as that of the station vector.

044.091 **The influence of the atmospheric zonal tides on the Earth's rotation rate.**
Y.–z. Zhu.
Acta Astron. Sin., Vol. 27, No. 2, p. 106 – 112 (1986). In Chinese.

044.092 **Influence of non–dipole field on determination of Plio–Pleistocene true polar wander.**
D. A. Schneider, D. V. Kent.
Geophys. Res. Lett., Vol. 13, No. 5, p. 471 – 474 (1986).
The authors show that much of the polar offset determined from the paleomagnetic data may be due to deviations in pole position caused by persistent non–dipole zonal components of the geomagnetic field. A correction of paleomagnetic poles for the long–term non–dipole field reduces the polar offset and thus suggests a slower or shorter episode of true polar wander over the past 5 million years.

044.093 **Excitation of the earth's rotational axis by recent glacial discharges.**
P. Gasperini, R. Sabadini, D. A. Yuen.
Geophys. Res. Lett., Vol. 13, No. 6, p. 533 – 536 (1986).
The authors study the effects of present–day glacial discharges and the growth of the Antarctic ice sheet on exciting the earth's rotational axis. Glacial forcing could cause a maximum change in J_2 of about one–third of the observed amount, for the Maxwell rheology and for Burgers' body models with a long–term, lower–mantle viscosity greater than about 10^{23}P. For transient rheologies the amount of excitation due to glacial melting decreases. Polar wander is not much excited by recent glacial melting for the various types of rheologies examined.

044.094 **Length of day and atmospheric angular momentum: a comparison for 1981 – 1983.**
P. J. Morgan, R. W. King, I. I. Shapiro.
J. Geophys. Res., Vol. 90, No. B14, p. 12645 – 12652 (1985).
The authors combined measurements from laser ranging to the moon, laser ranging to the LAGEOS satellite, and very long baseline interferometric observations of extragalactic radio sources, to produce a series of universal time (UT1) and length–of–day (LOD) variations for 1981 – 1983. They first describe the observations and computation procedures used in their analysis. The authors then examine both seasonal and short–period differences between the observed (UT1) and inferred atmospheric angular momentum changes in LOD.

044.095 **Precision and accuracy of Earth rotation determinations derived at BIH from optical astrometry.**
M. Feissel, Z.-x. Li.
Earth rotation and the terrestrial reference frame, p. 3 – 14 (1985). – See Abstr. 012.077.

Several solutions for the Earth Rotation Parameters have been derived at BIH from the measurements of latitude and universal time obtained from 1962 to 1984 by the optical astrometry network. The series obtained are analysed and compared with space geodesy series.

044.096 **IPMS Earth orientation parameters by optical astrometry.**
K. Yokoyama, K. Tanikawa, S. Manabe.
Earth rotation and the terrestrial reference frame, p. 15 (1985). Abstract. – See Abstr. 012.077.

044.097 **Astrometric determinations of UT1 in Italy during the MERIT main campaign.**
G. Chiumiento, G. Caprioli, F. Chlistovsky, F. Mazzoleni, G. Picchio, A. Poma, M. Sarasso.
Earth rotation and the terrestrial reference frame, p. 16 – 28 (1985). – See Abstr. 012.077.

This paper gives a brief description on the activities in Italy during the MERIT main campaign in the field of optical astrometry. The collaborating stations, the instruments, the reductions in the MERIT Standards as well as a preliminary analysis of astrometric determinations of UT1 are here presented.

044.098 **The circumzenithal in the project MERIT – some results.**
G. Karský, J. Kostelecký.
Earth rotation and the terrestrial reference frame, p. 29 – 33 (1985). – See Abstr. 012.077.

The results of the first analysis of observations with a circumzenithal at the Geodetical Observatory Pecný in the main campaign of the MERIT project are presented. The analysis is aimed at the detection of systematic deviations of observational data from the BIH data as well as at the comparison of these results with a detailed analysis for the period 1970 – 1983.

044.099 **Geodetic–astronomical results of the Potsdam Observatory during the MERIT main campaign.**
M. Meinig.
Earth rotation and the terrestrial reference frame, p. 34 – 41 (1985). – See Abstr. 012.077.

Results for latitude variations, UT1–UTC and length of day are presented and quality comparisons are carried out with other stations.

044.100 **A sequence of astronomical ERP during the MERIT main campaign.**
M. Zhao, X.-q. He, D.-w. Zheng.
Earth rotation and the terrestrial reference frame, p. 42 – 54 (1985). – See Abstr. 012.077.

The purpose of this work is to present an ERP sequence, which is based on the data by only a few instruments and whose precision can be comparable to that of the BIH astronomical solution.

044.101 **Research on the fine structures in polar motion with multi–stage filter.**
D.-w. Zheng, D.-n. Dong.
Earth rotation and the terrestrial reference frame, p. 55 – 63 (1985). – See Abstr. 012.077.

A new multi–stage filtering technique to perform narrow band filtering on signals by the Vondrak filters is suggested. This idea using the narrow band filter is proved to be correct by a digital simulated test. The annual, Chandler and long period terms in polar motion are successfully separated for homogeneous series of polar coordinates (1899.85 – 1978.95) of ILS and the instantaneous series of each term in 0.05 years are also available.

044.102 **Investigation of polar motion and the length of day by means of SLR (Satellite Laser Ranging) data of the MERIT campaign.**
H. Montag, G. Gendt, R. Dietrich, K. Kurth.
Earth rotation and the terrestrial reference frame, p. 71 – 87 (1985). – See Abstr. 012.077.

The LAGEOS satellite laser ranging data of the MERIT campaign were analysed. The internal precision of the station position is about ± 2 cm. The pole coordinates and length of day (LOD) were estimated for overlapping 1–, 2–, 3– and 5–day intervals. For the 5– and 3–day resolution the precision is nearly the same, about $\pm 0''.0005$ (± 1.5 cm) for the pole coordinates and ± 0.02 ms for the LOD in average.

044.103 **The SLR operational center series ERP(CSR)84L02.**
B. E. Schutz, R. J. Eanes, B. D. Tapley.
Earth rotation and the terrestrial reference frame, p. 104 – 110 (1985). – See Abstr. 012.077.

Results from the operational center located at the Center for Space Research (CSR), The University of Texas at Austin, were computed and reported weekly during the main MERIT campaign. This report summarizes the computational cycle, the techniques and the results obtained using quick–look data.

044.104 **UT/CSR analysis of Earth rotation from Lageos SLR (Satellite Laser Ranging) data.**
B. D. Tapley, R. J. Eanes, B. E. Schutz.
Earth rotation and the terrestrial reference frame, p. 111 – 125 (1985). – See Abstr. 012.077.

The data set collected during the MERIT campaign in 1983 and 1984 has been used to generate solutions routinely at three–day intervals with an accuracy of <2 mas for polar motion and <0.15 ms for the short period (<60 days) fluctuations in UT1 or, equivalently, <0.07 ms for the length of day obtained by differencing successive UT1 determinations.

044.105 **On the use of the collocation method for estimation of the Earth rotation parameters by satellite laser ranging.**
N. A. Sorokin, E. A. Samus', S. K. Tatevian (*S. K. Tatevyan*).
Earth rotation and the terrestrial reference frame, p. 129 – 137 (1985). – See Abstr. 012.077.

An application of the least square collocation for the estimation of the Earth's rotation parameters: pole's positions and variations of the day length – is considered. The values of the pole positions and length of the day are compared with the same data obtained by other authors. The final accuracy of the estimated parameters was $0''.004$, $0''.003$ and $0^s.00029$ respectively.

044.106 **Polar motion and length of day determination from satellite laser ranging.**
D. E. Smith, D. C. Christodoulidis, M. H. Torrence, S. M. Klosko, P. J. Dunn.
Earth rotation and the terrestrial reference frame, p. 142 – 171 (1985). – See Abstr. 012.077.

Laser ranging data to LAGEOS have been analyzed at the Goddard Space Flight Center for the period September 1983 through October 1984, the main MERIT campaign. Comparison of the polar motion time series with BIH shows a 4.6 mas rms about a mean offset of 1 mas in the x component and 2.9 mas rms about a 21 mas offset in the y component. The precision of the GSFC earth orientation is estimated to be better than 1 mas for polar motion and 0.1 msec for excess length–of–day.

044.107 **Pole position from laser ranging to LAGEOS. University of Padova/Telespazio Report.**
A. Caporali, F. Palutan, A. Cenci, S. Casotto.
Earth rotation and the terrestrial reference frame, p. 172 – 189 (1985). – See Abstr. 012.077.

044.108 **Earth rotation solution from Starlette orbit analysis during the MERIT campaign.**
M. K. Cheng, R. J. Eanes, C. K. Shum, B. E. Schutz, B. D. Tapley.
Earth rotation and the terrestrial reference frame, p. 190 – 204 (1985). – See Abstr. 012.077.

The solution for the Earth Rotation Parameters (ERP), which include the polar motion coordinates, x_p and y_p, and the length of the day, have been derived from the analysis of the satellite laser range (SLR) data to Starlette during the MERIT campaign from 30 August 1983 to 2 December 1984. The Starlette ERP solutions have been compared with the solutions derived from Lageos SLR and the weighted rms about the mean of the difference between these solutions are 17.7 mas, 13.3 mas and 0.65 ms, respectively.

044.109 **A series of ERP (Earth Rotation Parameters) from LAGEOS during the MERIT campaign.**
M.-f. He, C. Huang, C.-g. Fong, S.-y. Zhu.
Earth rotation and the terrestrial reference frame, p. 205 – 212 (1985). – See Abstr. 012.077.

044.110 **Earth rotation, station coordinate and orbit solutions from Lageos during the MERIT campaign.**
J. M. Dow, L. G. Agrotis.
Earth rotation and the terrestrial reference frame, p. 217 – 235 (1985). – See Abstr. 012.077.

A solution for the coordinates of 18 laser stations and a 5 day resolution ERP series for the pole, UT1R and length of day have been derived from 6 months of Lageos ranging data taken during the MERIT campaign. A high correlation with features of VLBI derived UT1R has been observed. The feasibility of higher time resolution is briefly considered.

044.111 **Determination of polar motion and Earth rotation from laser tracking of Lageos.**
M. Murata, H. Shingu.
Earth rotation and the terrestrial reference frame, p. 236 – 245 (1985). – See Abstr. 012.077.

044.112 **Determination of ERP with lunar laser ranging and discussion on the influence of the adopted parameters.**
W.-j. Jin, Q.-g. Wang.
Earth rotation and the terrestrial reference frame, p. 287 – 295 (1985). – See Abstr. 012.077.

044.113 **Earth orientation determinations from VLBI observations.**
D. S. Robertson, W. E. Carter.
Earth rotation and the terrestrial reference frame, p. 296 – 306 (1985). – See Abstr. 012.077.

The intercomparison of VLBI and satellite laser ranging determinations of pole position has established absolute accuracy levels of 2 milliseconds of arc or less for both techniques for time scales of five days or less. Intercomparison of different VLBI observing sessions has demonstrated accuracies of 0.1 milliseconds of time for UT1 determinations.

044.114 **Connected-Element Interferometer results in project MERIT.**
D. D. McCarthy, P. E. Angerhofer, D. R. Florkowski, F. J. Josties, D. N. Matsakis.
Earth rotation and the terrestrial reference frame, p. 341 – 350 (1985). – See Abstr. 012.077.

During project MERIT the Connected Element Interferometer at Green Bank, West Virginia continued in routine daily operation. One antenna was added to the array to provide Earth orientation data from three baselines each of which is about 35 km in length. The precision, accuracy, reliability and timeliness of the interferometer are discussed.

044.115 **Earth rotation parameters (ERP) from VLBI.**
S.-f. Luo, S.-h. Wang.
Earth rotation and the terrestrial reference frame, p. 351 – 359 (1985). – See Abstr. 012.077.

A series of ERP are obtained with formal errors of $\pm 0.''001 - \pm 0.''005$ for x, y; and ± 0.07 ms – 0.5 ms for UT1. The noise level in UT1 determined regularly (every 5 day) is estimated as ± 0.2 ms.

044.116 **The determinations of Earth rotation parameters with SHVGP (Shanghai VLBI Geodynamics computer Program).**
Z.-h. Qian.
Earth rotation and the terrestrial reference frame, p. 366 – 370 (1985). – See Abstr. 012.077.

This paper gives Earth rotation parameter results derived from IRIS data of the MERIT campaign. The formal standard errors of 5-day interval results typically are ± 1 to 2 mas for x and y components of polar motion, and ± 0.05 to 0.10 ms for UT1 in most cases.

044.117 **Accuracy and stability of time series of Earth rotation parameters obtained from space geodesy and astrometric observations.**
M. Feissel, R. Jung.
Earth rotation and the terrestrial reference frame, p. 371 (1985). Abstract. – See Abstr. 012.077.

044.118 **The U.S. Naval Observatory C.O.R.E. (Combined solution for the orientation of the Earth) solution during project MERIT.**
D. D. McCarthy, A. K. Babcock.
Earth rotation and the terrestrial reference frame, p. 372 – 380 (1985). – See Abstr. 012.077.

The U.S. Naval Observatory combination solution for Earth orientation parameters utilizes data from a number of techniques including very long baseline interferometry (POLARIS/IRIS), connected-element interferometry, lunar laser ranging, satellite laser ranging, satellite Doppler ranging, and optical astrometry. The reliability and timeliness of each contributed data type are assessed for its usefulness as a contributor to a service which must provide accurate and timely data to a number of users. All of these aspects of the CORE solution are discussed, along with the accuracy of the results and of the predictions based on these results.

044.119 **Combination of recent polar motion observations.**
R. O. Vicente, C. R. Wilson.
Earth rotation and the terrestrial reference frame, p. 381 – 382 (1985). – See Abstr. 012.077.

This summary describes methods for combining independent polar motion series and the application of those methods to satellite laser ranging, Doppler, and astrometric series for the period 1981 – 1983. The relative weighting given to each series is found to be fairly insensitive to the way in which the noise level is estimated from the data.

044.120 **Combined determination of the ERP from the results of different techniques, Jan. 1, 1980 – Oct. 31, 1984.**
Z.-x. Li, S.-h. Ye.
Earth rotation and the terrestrial reference frame, p. 383 – 398 (1985). – See Abstr. 012.077.

The results of the different networks of six techniques (Astr., Doppler, SLR, LLR, VLBI, and CEI) have been reduced in order to give a combined determination of the ERP. The accuracies and the systematic errors of the 15 networks of six techniques in this solution are also given.

044.121 **A computation of atmospheric excitation functions for the Earth's rotation based on JMA (Japan Meteorological Agency) global analysis data.**
I. Naito, K. Yokoyama.
Earth rotation and the terrestrial reference frame, p. 434 – 439 (1985). – See Abstr. 012.077.

044.122 **Computations of atmospheric angular momentum, with emphasis on the MERIT period.**
D. A. Salstein, R. D. Rosen.
Earth rotation and the terrestrial reference frame, p. 440 – 449 (1985). – See Abstr. 012.077.

044.123 **Variations in LOD (length of day) and pole position as inferred from atmospheric angular momentum.**
C. E. Dutton, F. W. Fallon.
Earth rotation and the terrestrial reference frame, p. 450 – 468 (1985). – See Abstr. 012.077.

044.124 **The atmospheric excitation of Earth orientation changes during MERIT.**
T. M. Eubanks, J. A. Steppe, J. O. Dickey.
Earth rotation and the terrestrial reference frame, p. 469 – 483 (1985). – See Abstr. 012.077.

044.125 **Water storage effects on the Earth's rotation.**
C. R. Wilson, L. Hinnov.
Earth rotation and the terrestrial reference frame, p. 484 – 489 (1985). – See Abstr. 012.077.

044.126 **A synoptic analysis of short period fluctuations of UT1.**
T. Hara, K. Yokoyama.
Earth rotation and the terrestrial reference frame, p. 490 – 504 (1985). – See Abstr. 012.077.

044.127 **Analysis of short periodical variations of pole coordinates determined by different techniques in the MERIT campaign.**
B. Kołaczek, W. Kosek.
Earth rotation and the terrestrial reference frame, p. 505 – 523 (1985). – See Abstr. 012.077.
Pole coordinates determined by different techniques during the MERIT campaign have been analysed by the maximum entropy spectral analysis and the Ormsby filter in order to investigate their short periodical variations with periods ranging from 10 – 50 days. Several periodical terms have been detected in each set of pole coordinates.

044.128 **Irregular variations of the pole motions during the MERIT campaign and in several previous years.**
B. Kołaczek, J. Nastula.
Earth rotation and the terrestrial reference frame, p. 524 – 535 (1985). – See Abstr. 012.077.
In order to detect irregular variations of the pole motion, its velocity on the Earth surface has been computed by Lagrange formula as well as from the first order differences of pole coordinates on the basis of the MERIT data. Correlations of irregular variations of the pole motion with some geophysical phenomena are investigated.

044.129 **A modern Earth orientation monitoring service: functions, goals, and methods of observation.**
W. E. Carter, D. S. Robertson.
Earth rotation and the terrestrial reference frame, p. 536 – 550 (1985). – See Abstr. 012.077.
Three new methods of observation, satellite laser ranging, lunar laser ranging, and very long baseline interferometry, are achieving the required accuracies and resolutions in determining one or more of the quantities needed to form the first true Earth orientation monitoring service, i.e., polar motion, UT1, precession and nutation, and the time–varying positions of the observatories caused by deformations and motions of the tectonic plates. The three techniques are examined.

044.130 **Towards an improved realization of the BIH terrestrial frame.**
C. Boucher, Z. Altamimi.
Earth rotation and the terrestrial reference frame, p. 551 – 564 (1985). – See Abstr. 012.077.
In order to redefine the BIH terrestrial system from 1984 onwards (to be known as BTS), a global adjustment has been carried out using, in addition to the ERP series, sets of coordinates of sites where space geodesy stations are operated. The model used in the analysis is recalled and the calculation realized for the BIH Annual Report for 1984 is described. The future maintenance as well as improvements of the BTS are also outlined.

044.131 **Definition of reference frames and computation of station coordinates and Earth rotation parameters from MERIT laser ranging observations.**
H. Hauck, D. Ehlert, D. Lelgemann.
Earth rotation and the terrestrial reference frame, p. 565 – 583 (1985). – See Abstr. 012.077.
For coordinate and baseline determination using laser ranging data to LAGEOS an analysis was made to get appropriate definitions for a conventional terrestrial reference system which can be realized by the use of satellite laser ranging data. From bimonthly coordinate solutions variations of the origin (center of mass of the Earth) and scale are derived. Two appropriate definitions for the directions of the coordinate axis have been analysed.

044.132 **Intercomparison of UT1 measurements during the MERIT campaign period.**
M. A. Spieth, T. M. Eubanks, J. A. Steppe.
Earth rotation and the terrestrial reference frame, p. 609 – 621 (1985). – See Abstr. 012.077.
The authors have intercompared measurements of UT1 obtained from two space–geodetic techniques: very long baseline interferometry (VLBI); and lunar laser ranging (LLR). Analysis of data series in both the frequency and time domains shows agreement of the various data types to within the formal errors.

044.133 **Intercomparison of polar motion measurements during the MERIT period.**
J. A. Steppe, T. M. Eubanks, M. A. Spieth.
Earth rotation and the terrestrial reference frame, p. 622 – 636 (1985). – See Abstr. 012.077.
The authors intercompared polar motion estimates for the MERIT period produced by satellite laser ranging and by very long baseline interferometry from the International Radio Interferometric Surveying (IRIS) project and from the Deep Space Network (DSN). All three services agree at least as well as is expected from the measurement formal errors. By contrast, the two "classical" techniques of optical astrometry and Doppler tracking of satellites produce results with residual scatters of 10 to 20 mas, with definite systematic errors.

044.134 **A comparison of Earth orientation data obtained by different space techniques.**
A. Mallama.
Earth rotation and the terrestrial reference frame, p. 637 – 647 (1985). – See Abstr. 012.077.

044.135 **Comparison and comment on the techniques for determining ERP (Earth Rotation Parameters).**
M. Zhao, Z.–n. Gu.
Earth rotation and the terrestrial reference frame, p. 648 – 655 (1985). – See Abstr. 012.077.

044.136 **Agreement in polar motion measurements during the MERIT campaign.**
P. Pâquet, D. Djurovic, C. Techy.
Earth rotation and the terrestrial reference frame, p. 656 – 666 (1985). – See Abstr. 012.077.
From the original polar motion (PM) measurements performed during the MERIT campaign, the Chandler and the annual components are removed. The analysis of the residuals shows a high level of significant correlation between the various techniques mainly for phenomena ranging from 30 days to a few months.

044.137 **MERIT main campaign: reference frame intercomparisons.**
I. I. Mueller, Z. Wei.
Earth rotation and the terrestrial reference frame, p. 667 – 672 (1985). – See Abstr. 012.077.
The polar coordinates obtained in the MERIT main campaign were selectively compared and analyzed. The preliminary results demonstrate that the polhode centers obtained from astrometry (BIH), satellite laser ranging (Lageos) and very long baseline interferometry are within an area 6 marcsec (19 cm) across, with satellite Doppler polhode center more than 10 marcsec (32 cm)

away, and that the equatorial planes of the inertial systems inherent in all techniques mentioned above are aligned with one another on the order of milliarcseconds.

044.138 **Sensitivity of Lageos Earth orientation results to changes in the MERIT standard model.**
R. Eanes, B. Schutz, B. Tapley.
Earth rotation and the terrestrial reference frame, p. 673 – 699 (1985). – See Abstr. 012.077.
To facilitate the intercomparison of new techniques for determination of the Earth orientation parameters (EOP), a set of standards has been adopted for project MERIT. These standards have been evaluated by comparing EOP series computed with the MERIT standards to series computed with various model changes. The model changes investigated include the geopotential, solar and earth radiation pressure, nutation, tides, ocean loading station displacements and tectonic plate motion. The sensitivity of the EOP series to these factors is also assessed and different techniques are compared.

044.139 **SLR (Satellite Laser Ranging) and VLBI technology trends.**
R. J. Coates, T. A. Clark, T. K. Varghese.
Earth rotation and the terrestrial reference frame, p. 700 – 718 (1985). – See Abstr. 012.077.
SLR and VLBI technology improvements are continuing, aimed at millimeter system performance. New SLR receiver components have achieved single shot rms precision of 1 cm on LAGEOS. Recent VLBI phase delay observations have achieved sub–cm repeatability. These improvements will improve the accuracy of future polar motion measurements.

044.140 **Measuring Earth rate perturbations with a large passive ring laser gyro.**
J. R. Rotge, G. L. Shaw, H. W. Emrick.
Earth rotation and the terrestrial reference frame, p. 719 – 729 (1985). – See Abstr. 012.077.

044.141 **On the optimal conditions of determining ERP (Earth Rotation Parameters).**
M. Zhao, S.–y. Zhu.
Earth rotation and the terrestrial reference frame, p. 730 – 735 (1985). – See Abstr. 012.077.

044.142 **Polar motion–induced gravity.**
J. M. Wahr.
Earth rotation and the terrestrial reference frame, p. 736 – 741 (1985). – See Abstr. 012.077.
Variations in the geocentric position of the earth's rotation axis (polar motion) cause deformation within the earth. The author estimates the effects of this deformation on surface gravity, and on radial and horizontal positions of points on the earth's surface. The effects of the oceans and of the earth's inelasticity on this deformation are found to be negligible. Peak–to–peak variations in surface gravity of 10 μ Gals or more, and in radial motion of 1 – 2 cm are possible over six months or so.

044.143 **Predicting Earth rotation parameters using a multi–channel Wiener filter.**
P. Morgan, X. Chen.
Earth rotation and the terrestrial reference frame, p. 749 – 764 (1985). – See Abstr. 012.077.

044.144 **Die gesetzliche Zeit im Großherzogtum Luxemburg. Nach welcher Zeit gingen und gehen unsere Uhren? Zeitzonen–Tabelle.**
J. E. Weiland.
Sonderdruck aus dem Kalender "Je maintiendrai" 1986, Luxemburg, 11 pp. (1986).

044.145 **The influence of earthquakes on the Chandler wobble during 1977 – 1983.**
R. S. Gross.
Geophys. J. R. Astron. Soc., Vol. 85, No. 1, p. 161 – 177 (1986).
A direct calculation of the effect on the Chandler wobble of 1287 earthquakes that occurred during 1977 – 1983 is made by resorting to neither an empirical moment–magnitude relationship nor statistical arguments. This is made possible by the recent availability of a set of moment tensors representing the source mechanisms of 1287 earthquakes that occurred during 1977 – 1983. This moment tensor data set is described along with the theory used to calculate the expected change in the Chandler wobble's excitation function. The author presents the outcome of this calculation and compares the resulting theoretical earthquake excitation function to an "observed" excitation function that is derived from observations of polar motion.

General theory of the rotation of the earth.
See Abstr. 003.092.

Solar eclipse in "Shijing" and rotation of the Earth.
See Abstr. 004.032.

Galileo e il problema della misura del tempo.
See Abstr. 004.082.

On the rotation of the Earth and the terrestrial reference system: Joint summary report of the IAU/IUGG Working Groups MERIT and COTES.
See Abstr. 013.020.

The Italian contribution to Project MERIT.
See Abstr. 013.053.

Current work on the Earth's rotation at the United Kingdom Meteorological Office.
See Abstr. 013.070.

Program design for compiling the Time Bulletin by means of a microcomputer at the Yunnan Observatory.
See Abstr. 021.019.

Time and latitude observation of the Mizusawa PZT in the observation room above the ground.
See Abstr. 032.036.

Automatic electronic astrolabe at Mizusawa.
See Abstr. 032.037.

Relativistic effects in the measurement of time and frequency.
See Abstr. 034.120.

A review of one–way satellite time synchronization techniques.
See Abstr. 034.121.

Isolation of the mean latitude using Orlov and Sakharov filters.
See Abstr. 036.041.

On the construction of digital filters for processing of latitude observational data.
See Abstr. 036.052.

Results of an investigation of the symmetry of micrometric measurements from observations of zenith stars at the ZTL–180 zenith telescope in Poltava.
See Abstr. 036.056.

The correlations between VLBI observables and its effects for the determinations of ERP.
See Abstr. 036.092.

MERIT data processing at Nottingham.
See Abstr. 036.111.

Determining the absolute paramters of a meridian circle by observing the first vertical (V). The effect of the higher power terms of the directional error in the determined values of latitude.
See Abstr. 036.138.

Numerical integration of precession and nutation.
See Abstr. 041.020.

Some effects of different sources of variation of latitude data on meridian circle catalogues.
See Abstr. 041.052.

Relativistic coordinates and time scales.
See Abstr. 043.006.

Fundamental coordinate ties using laser ranging data.
See Abstr. 043.010.

Concepts of reference systems.
See Abstr. 043.016.

Discussion of meaning and definition of UT.
See Abstr. 043.017.

The observations of the earth rotation and the stellar system.
See Abstr. 043.021.

Geodesy by radio interferometry: corrections to the IAU 1980 nutation series.
See Abstr. 043.030.

An analysis and intercomparison of VLBI nutation estimates.
See Abstr. 043.031.

Satellite laser ranging and its applications.
See Abstr. 045.001.

Geodesy by radio interferometry: determinations of baseline vector, earth rotation, and solid earth tide parameters with the Mark I very long baseline radio interferometry system.
See Abstr. 045.026.

Geodesy by radio interferometry: studies of the forced nutations of the earth. 1. Data analysis.
See Abstr. 045.027.

Geodesy by radio interferometry: studies of the forced nutations of the earth. 2. Interpretation.
See Abstr. 045.028.

MERIT Lageos laser ranging data analysis at SFB78/DGFI.
See Abstr. 045.030.

LAGEOS observations and precision estimation of Shanghai SLR (Satellite Laser Ranging) station during the main MERIT campaign.
See Abstr. 045.031.

On the use of the Kiev geodynamics program complex for analyzing the MERIT laser ranging data.
See Abstr. 045.032.

Report from Orroral.
See Abstr. 045.034.

Multi–station lunar laser ranging: an analysis of data quality and Earth rotation results.
See Abstr. 045.035.

GINFEST: geodetic intercomparison network for evaluating space techniques.
See Abstr. 045.036.

The coordinate frame of the lunar laser ranging network.
See Abstr. 045.037.

Station coordinates and earth rotation parameters determined from satellite laser ranging during the MERIT campaign.
See Abstr. 045.043.

Tracking station coordinates and earth rotation parameters from satellite laser ranging to LAGEOS during the main MERIT campaign.
See Abstr. 045.044.

Relativistic effects in astronomical timing measurements.
See Abstr. 066.196.

Geodynamic implications of short time scale changes in the geomagnetic dynamo.
See Abstr. 081.029.

Deformación rotacional de una tierra elástica.
See Abstr. 081.036.

Relationship between 20–year geomagnetic variations and polar changes of the latitude.
See Abstr. 084.024.

Variations in the Moon's mean motion due to the Earth's tides.
See Abstr. 094.010.

045 Astronomical Geodesy, Satellite Geodesy, Navigation

045.001 **Satellite laser ranging and its applications.**
B. D. Tapley, B. E. Schutz, R. J. Eanes.
Celest. Mech., Vol. 37, No. 3, p. 247 – 261 (1985). – See Abstr. 012.017.

Satellite laser ranging (SLR) provides an important capability for precise orbit determination and for geophysical parameter estimation to support a number of contemporary geodynamic and oceanographic investigations. The precision of the SLR measurement has improved from the early meter–level systems to the current capabilities of a few centimeters for the best systems. The accuracy of the orbits and geophysical parameter recovery have shown an associated improvement. Polar motion with accuracies of 2 mas, station coordinates better than 10 cm, and interstation baseline rates indicative of tectonic motion are determined routinely with the current set of global SLR data. This discussion reviews the SLR measurement, analysis approach, and some of the recent results derived from the current SLR data set.

045.002 **NNSS Doppler tracking at the Geographical Survey Institute.**
T. Wada.
Proceedings of the Symposium on Application of Space Techniques to Astronomy and Geophysics, p. 36 – 38 (1986). In Japanese. – See Abstr. 012.027.

045.003 **Geodetic results from Starlette laser ranging data.**
K. Kobayashi, H. Inaba, M. Murata.
Proceedings of the Symposium on Application of Space Techniques to Astronomy and Geophysics, p. 39 – 45 (1986). In Japanese. – See Abstr. 012.027.

The station coordinates and earth rotation parameters have been determined using Starlette laser ranging data from 22 stations participated in the MERIT Main Campaign (1983.9 – 1984.10).

045.004 **A survey of radio navigation satellite system alternatives to the GPS.**
M. Murata.
Proceedings of the Symposium on Application of Space Techniques to Astronomy and Geophysics, p. 46 – 51 (1986). In Japanese. – See Abstr. 012.027.

045.005 **Activities for introducing geodetic GPS systems.**
T. Kato.
Proceedings of the Symposium on Application of Space Techniques to Astronomy and Geophysics, p. 52 – 56 (1986). In Japanese. – See Abstr. 012.027.

Recent Japanese activities for introducing GPS system for precise relative positioning are briefly reported.

045.006 **The high accuracy GPS system with differential range technique.**
H. Kiuchi, Y. Sugimoto, N. Kurihara, A. Kaneko, M. Imae, N. Kawano.
Proceedings of the Symposium on Application of Space Techniques to Astronomy and Geophysics, p. 57 – 62 (1986). In Japanese. – See Abstr. 012.027.

045.007 **Effects of the ionosphere on differential GPS.**
T. Kato, I. Murata, A. Tsuchiya.
Proceedings of the Symposium on Application of Space Techniques to Astronomy and Geophysics, p. 63 – 67 (1986). In Japanese. – See Abstr. 012.027.

045.008 **The Japan–China preliminary VLBI experiment. I. Observation system.**
J. Amagai, N. Kawaguchi, A. Kaneko, K. Yamamoto, T.–s. Wan, H.–w. Wu, Z.–h. Qian.
Proceedings of the Symposium on Application of Space Techniques to Astronomy and Geophysics, p. 114 – 119 (1986). In Japanese. – See Abstr. 012.027.

The first joint VLBI experiment was successfully carried out in September 1985. The main purpose of this experiment was to determine the baseline vector between Shanghai Observatory, China and Kashima Space Research Center, Japan and to test the possibility of monitoring the plate motion and the crustal deformation in the eastern Asia. In this paper, the observation systems used at Shanghai station are described.

045.009 **The Japan–China preliminary VLBI experiment. II. Data analysis.**
N. Kawaguchi, J. Amagai, H. Kuroiwa, F. Takahashi.
Proceedings of the Symposium on Application of Space Techniques to Astronomy and Geophysics, p. 120 – 126 (1986). – See Abstr. 012.027.

The first Japan–China VLBI experiment was carried out in September, 1985. From good fringes obtained in two 24–hour observations, the authors determined the baseline length between Kashima and Shanghai with an error less than 4 centimeters.

045.010 **The geodetic results of VLBI experiment on Kashima–Nobeyama baseline.**
H. Kunimori, N. Kawaguchi, J. Amagai, Y. Takahashi, H. Hirabayashi, M. Inoue, M. Tsuboi, H. Iwashita, M. Morimoto.
Proceedings of the Symposium on Application of Space Techniques to Astronomy and Geophysics, p. 138 – 145 (1986). – See Abstr. 012.027.

X–band in both–hand circular polarization VLBI have been made on the baseline between Kashima and Nobeyama in March 1985. The VLBI coordinates at Nobeyama are determined for the first time with a precision of about 6 cm.

045.011 **The influence of errors of orbital elements of satellites on the accuracy of determination of the coordinates of observational stations.**
J. Lamparski.
Nablyud. Iskusstv. Sputnikov Zemli, No. 21, Chast' I, p. 86 – 96 (1982). In Russian.– See Abstr. 012.032.

The paper presents an algorithm of computation of the influence of systematic and accidental errors of the satellite orbit elements on the coordinates of the observational station.

045.012 **The accuracy of station positioning based on photodoppler data.**
J. Lamparski, A. Protnitçka, K. Sviontek.
Nablyud. Iskusstv. Sputnikov Zemli, No. 21, Chast' I, p. 97 – 105 (1982). In Russian.– See Abstr. 012.032.

An accuracy analysis of determination of station coordinates based on photographic observations of the TRANSIT satellites is presented.

045.013 **Effectiveness of computational algorithms of the dynamical method for determination of geodetic parameters.**
Yu. V. Surnin, S. V. Kuzhelev, V. I. Dudarev.
Nablyud. Iskusstv. Sputnikov Zemli, No. 21, Chast' I, p. 106 – 115 (1982). In Russian.– See Abstr. 012.032.

Criteria of the effectiveness of the computational algorithms of the dynamical method (the estimation of state parameters and the prediction of satellite trajectories) are determined. On the basis of the computer–generated data algorithms optimizing the corresponding criteria have been chosen.

045.014 **Combination of normal equations of a Doppler net and normal equations of a terrestrial net.**
I. Gajderowicz.
Nablyud. Iskusstv. Sputnikov Zemli, No. 21, Chast' I, p. 153 – 163 (1982). – See Abstr. 012.032.

The paper presents the way of substituting the unknowns dX, dY, dZ for the unknowns dB, dL, dE in normal equations of a given net and 7–parameter transformation of normal equations between reference systems.

045.015 **Use of the optimization method for transformation of coordinate systems.**
M. Mojzes.
Nablyud. Iskusstv. Sputnikov Zemli, No. 21, Chast' I, p. 164 – 170 (1982). In Russian.– See Abstr. 012.032.
The paper deals with the D–optimization method for the determination of the transformation parameters between two reference coordinate systems. The method is used in the case of correlated and uncorrelated observations of coordinates of a single satellite tracking station.

045.016 **Evaluation of GPS carrier difference observations: the Bernese second generation software package.**
W. Gurtner, G. Beutler, I. Bauersima, T. Schildknecht.
Mitt. Satell.–Beobachtungsstn. Zimmerwald, Nr. 18, p. 17 – 28 (1985). Paper presented at the First International Symposium on Precise Positioning with the Global Positioning System 1985. Preprint.

045.017 **Test of GPS on the CERN–LEP Control Network.**
J. Gervaise, M. Mayoud, G. Beutler, W. Gurtner.
Mitt. Satell.–Beobachtungsstn. Zimmerwald, Nr. 18, p. 29 – 52 (1985). Paper presented at the Joint Meeting FIG–Study Groups 5B and 5C on Inertial, Doppler and GPS measurements for national and engineering surveys, held at Munich 1985, July 1 – 3. Preprint.

045.018 **Les systèmes TRANSIT et GPS navigation et localisation de précision.**
M. P. Paquet.
Obs. R. Belg., Commun., Sér. B, No. 132, 30 pp. (1984).

045.019 **Geodesia por satélites y navegación.**
M. J. Sevilla.
Téc. Topográfica, Vol. 12, No. 54, 28 pp. (1983). = Semin. Astron. Geod., Univ. Madr., Publ., No. 129.

045.020 **Actualización de los procesos de reducción de observaciones en astronomía geodésica.**
M. J. Sevilla.
Semin. Astron. Geod., Univ. Madr., Publ., No. 130, p. 37 – 38 (1983).

045.021 **Coeficientes de la ecuación de Guier y su aplicación al filtrado de datos Doppler.**
M. J. Sevilla, L. Garcia Asensio.
Semin. Astron. Geod., Univ. Madr., Publ., No. 130, p. 39 – 53 (1983).
The authors obtain the Guier equation by formulating the range–rate error as a function of the receiver and satellite position errors and orbital parameters. In addition, they derive the equation of filtering the Doppler counts in the Guier plane.

045.022 **Anomalias de la gravedad basadas en el sistema geodésico de referencia 1980.**
M. J. Sevilla.
Téc. Topográfica, Vol. 11, Nos. 51 – 52, 11 pp. (1983). = Semin. Astron. Geod., Univ. Madr., Publ., No. 131.
The computing formulae for the determination of gravity anomalies in the 1980 Geodetic Reference System are presented, as well as the transformation formulae from the 1967 G.R.S. to the 1980 one. An application is made.

045.023 **A method for computing perturbations by means of tesseral harmonics.**
L.–d. Wu, T.–x. Weng.
Publ. Purple Mt. Obs., Vol. 4, No. 2, p. 1 – 6 (1985). In Chinese.

045.024 **Modern methods in geodesic position–finding.**
H. Wolf.
Naturwissenschaften, 73. Jahrg., Heft 6, p. 291 – 298 (1986). In German.
The main objectives of operational geodesy, being the determination of positions and heights, have been greatly affected by recent technological developments. Electronics, satellite methods, and computer facilities have opened new possibilities. Present problems and links with related sciences are considered.

045.025 **The chord and relative height estimation from single passes of satellite laser ranging data.**
M.–f. He, C.–g. Feng, C. Huang.
Acta Astron. Sin., Vol. 27, No. 2, p. 113 – 120 (1986). In Chinese.
The application of the multiple very short arc technique to determine the chord and relative height between two laser tracking stations from single passes of satellite laser ranging data is described. It is shown that this technique for processing laser ranging data could be regarded as an important supplement to the conventional dynamic technique.

045.026 **Geodesy by radio interferometry: determinations of baseline vector, earth rotation, and solid earth tide parameters with the Mark I very long baseline radio interferometry system.**
J. W. Ryan, T. A. Clark, R. J. Coates, C. Ma, W. T. Wildes, C. R. Gwinn, T. A. Herring, I. I. Shapiro, B. E. Corey, C. C. Counselman, H. F. Hinteregger, A. E. E. Rogers, A. R. Whitney, C. A. Knight, N. R. Vandenberg, J. C. Pigg, B. R. Schupler, B. O. Rönnäng.
J. Geophys. Res., Vol. 91, No. B2, p. 1935 – 1946 (1986).
The authors analyzed 37 very long baseline radio interferometry experiments performed between 1972 and 1978 and derived estimates of baseline vectors between six sites, five in the continental United States and one in Europe. They found no evidence of significant changes in baseline length. For example, with a statistical level of confidence of approximately 85%, upper bounds on such changes within the United States ranged from a low of 10 mm/yr for the 850 km baseline between Westford, Massachusetts, and Green Bank, West Virginia, to a high of 90 mm/yr for the nearly 4000 km baseline between Westford and Goldstone, California. The authors also obtained estimates for universal time and for the x component of the position of the earth's pole.

045.027 **Geodesy by radio interferometry: studies of the forced nutations of the earth. 1. Data analysis.**
T. A. Herring, C. R. Gwinn, I. I. Shapiro.
J. Geophys. Res., Vol. 91, No. B5, p. 4745 – 4754 (1986).
The analysis of 235 very long baseline interferometry observing sessions carried out between July 1980 and December 1984 has yielded estimates of corrections to the seven largest nutation series coefficients (with periods $\leqslant 1$ year) in the Wahr nutation series, the most complete such model available at this time. The largest correction is to the amplitude of the retrograde annual nutation: $-1.80 \pm 0.18 - i(0.42 \pm 0.18)$ milliarc seconds (mas), where the estimated standard deviations contain contributions from the effects of systematic as well as random errors. The authors interpret this correction as being due to an incompleteness in the model of the core–mantle coupling used by Wahr. They have also examined their results for evidence of any "free core nutation" and can place an upper bound of 1.0 mas (99.5% confidence interval) on its amplitude.

045.028 **Geodesy by radio interferometry: studies of the forced nutations of the earth. 2. Interpretation.**
C. R. Gwinn, T. A. Herring, I. I. Shapiro.
J. Geophys. Res., Vol. 91, No. B5, p. 4755 – 4765 (1986).
The authors' nutation measurements imply a deviation of the amplitude of the retrograde annual nutation from its value as calculated by Wahr and by Sasao et al. If this deviation is interpreted as the effect of a departure of the core–mantle boundary from its hydrostatic figure, then the observed amplitude is consistent with a core–mantle boundary that has a second zonal harmonic deviation from the hydrostatic equilibrium figure, with the peak–to–valley deviation being 490 ± 110 m. The part of the retrograde annual nutation out of phase with the driving torques yields an upper limit on the kinematic viscosity at the surface of the fluid core of 0.54 m^2/s (99.5% confidence limit).

045.029 **Performance of NASA laser ranging systems during MERIT.**
R. J. Coates.
Earth rotation and the terrestrial reference frame, p. 64 – 70 (1985). – See Abstr. 012.077.

Fourteen NASA satellite laser ranging systems participated in the MERIT campaign. Laser upgrades beginning in the last half of 1983 produced significant improvements in ranging accuracy and an order of magnitude increase in data points. The details of these improvements in performance are presented.

045.030 **MERIT Lageos laser ranging data analysis at SFB78/DGFI.**
C. Reigber, H. Müller, P. Schwintzer, F. H. Massmann, E. C. Pavlis.
Earth rotation and the terrestrial reference frame, p. 88 – 103 (1985). – See Abstr. 012.077.

The authors present the results for station coordinates and earth orientation parameters based on satellite laser ranging data collected on Lageos during the MERIT campaign. Intercomparison with results of other centers indicates an rms agreement of 5 cm in station coordinates, 1 mas in x_p and y_p, and 0.16 ms for the excess length of day.

045.031 **LAGEOS observations and precision estimation of Shanghai SLR (Satellite Laser Ranging) station during the main MERIT campaign.**
F.–m. Yang, D.–t. Tan, C.–k. Xiao, Z.–y. Li, W.–h. Lu.
Earth rotation and the terrestrial reference frame, p. 126 – 128 (1985). – See Abstr. 012.077.

045.032 **On the use of the Kiev geodynamics program complex for analyzing the MERIT laser ranging data.**
Ya. Yatskiv, V. Taradij, N. Mironov, M. Tsesis, K. Nurutdinov.
Earth rotation and the terrestrial reference frame, p. 138 – 141 (1985). – See Abstr. 012.077.

045.033 **Laser network designed for the Moon and artificial Earth satellite ranging.**
N. K. Abalakin, M. K. Abele, Yu. N. Artyukh, N. G. Basov, V. N. Bronnikov, Yu. V. Ignatenko, Yu. L. Kokurin, V. V. Kurbasov, K. K. Lapushka, V. F. Lobanov, A. G. Masevich, A. N. Sukhanovsky (*A. N. Sukhanovskij*), Yu. P. Shilokhvost, S. G. Shubin, Ya. S. Yatskiv.
Earth rotation and the terrestrial reference frame, p. 246 – 256 (1985). – See Abstr. 012.077.

045.034 **Report from Orroral.**
J. Mck. Luck, B. A. Greene.
Earth rotation and the terrestrial reference frame, p. 257 – 273 (1985). – See Abstr. 012.077.

The Natmap Laser Ranging System at Orroral is now producing results from LAGEOS and the moon. It is well connected to Doppler, VLBI and astrometric sites nearby, and is being considered as a base site for GPS (Global Positioning System) programs.

045.035 **Multi–station lunar laser ranging: an analysis of data quality and Earth rotation results.**
J. O. Dickey, X X Newhall, J. G. Williams.
Earth rotation and the terrestrial reference frame, p. 274 – 286 (1985). – See Abstr. 012.077.

Sixty–five Earth rotation values (UT0) were derived from LLR (Lunar Laser Ranging) data during the main MERIT campaign with the best accuracies during this period being 0.25 ms. During the post–MERIT period (November 1, 1984, through August 12, 1985), 115 determinations of UT0 were calculated with best inherent accuracies of about 0.1 ms. Intercomparisons with determinations from other techniques are discussed.

045.036 **GINFEST: geodetic intercomparison network for evaluating space techniques.**
V. Ashkenazi.
Earth rotation and the terrestrial reference frame, p. 584 – 589 (1985). – See Abstr. 012.077.

Details are given of a geodetic network connecting the major radio telescopes and SLR facilities in Western and Central Europe, which is to be used in a co–location exercise involving VLBI, CERI, SLR and GPS observations, with the aim of evaluating the relative accuracies and system biases of these geodetic space observation techniques.

045.037 **The coordinate frame of the lunar laser ranging network.**
J. G. Williams, X X Newhall, J. O. Dickey.
Earth rotation and the terrestrial reference frame, p. 590 – 600 (1985). – See Abstr. 012.077.

Fits of lunar laser ranges have produced geocentric coordinates for four instruments at three observatories: the 2.7 m and MLRS telescopes at McDonald Observatory, the Haleakala site, and the CERGA site near Grasse, France. These coordinates are compared with coordinates determined from tracking the Lageos satellite. The rms coordinate difference is 18 cm. Tables are given for the expected station motion due to continental drift. The Haleakala site is projected to move 10 cm/yr and the Orroral station 7 cm/yr.

045.038 **Creation of a global geodetic network using Mark III VLBI.**
C. Ma, T. A. Clark, J. W. Ryan.
Earth rotation and the terrestrial reference frame, p. 601 – 608 (1985). – See Abstr. 012.077.

The positions of 15 permanent VLBI stations have been determined using Mark III with one–sigma uncertainties of less than 5 cm. 46070 delay/delay rate observations acquired by the Crustal Dynamics Project and POLARIS/IRIS from 1980 – 84 were included in a least squares solution to estimate the station positions, 44 radio source positions, and earth orientation parameters.

045.039 **GPS Research 1985 at the Institute of Astronomical and Physical Geodesy.**
H. Landau, B. Eissfeller, G. W. Hein.
Schriftenreihe Universitärer Studiengang Vermessungswesen, Universität der Bundeswehr München, F.R. Germany, Heft 19, 208 pp (1986).

The Institute of Astronomical and Physical Geodesy decided 1984 to concentrate its satellite research on the topic of GPS (Global Positioning System). The present report summarizes the investigations and achievements in 1985 in the form of seven papers. Starting with basic physics and mathematics behind the analysis of GPS data the work covers aspects of GPS satellite selection in the field work, the incorporation of GPS baseline vectors in network adjustments with special emphasis of the determination of orthometric heights by combination with gravity data, and a description of one of the GPS analysis software products of IAPG, the phase–difference processing program. Finally, the required force–modelling for the orbit determination of GPS satellites is outlined, as it is used in the program system.

045.040 **Astronomische Messungen auf Laplace– und Geoid–Punkten in SFRJ 1954 – 1973.**
D. Štemberger.
Publ. Astron. Soc. "Rudjer Bošković", No. 4, p. 183 – 188 (1985). – See Abstr. 012.078. In Croatian.

045.041 **Covariance analysis of satellite laser data.**
S. K. Tatevyan, O. A. Petrova, V. V. Kirichuk, O. A. Abrikosov, A. N. Marchenko.
Nablyud. Iskusstv. Sputnikov Zemli, No. 21, Chast' I, p. 32 – 37 (1982). In Russian. – See Abstr. 012.032.

The foundation of the application of the first model least–squares collocation for the preliminary reduction of satellite laser ranging is given. The statistical properties of the laser range to "Lageos" in the MERIT program are considered. The elimination trend problem up to and after correction of the initial conditions is investigated. It is shown that the polynomial elimination of the trend is equivalent to the differential correction of the initial conditions. A method of construction of the empirical covariance functions of the laser range satellite is developed. An optimal model covariance function is proposed.

045.042 **Processing of the INTERCOSMOS – BULGARIA 1300 laser ranging data.**
O. M. Koroleva, R. R. Nazirov, A. A. Tikhonov, P. E. Ehl'yasberg.
Nablyud. Iskusstv. Sputnikov Zemli, No. 21, Chast' I, p. 190 – 197 (1982). In Russian.– See Abstr. 012.032.
Methods and several results of processing of the INTERCOSMOS – BULGARIA 1300 laser ranging data obtained at Riga, Potsdam, Simeiz and Zvenigorod are presented.

045.043 **Station coordinates and earth rotation parameters determined from satellite laser ranging during the MERIT campaign.**
G. M. Appleby, A. T. Sinclair, J. Xia.
Geophys. J. R. Astron. Soc., Vol. 85, No. 1, p. 253 (1986). Abstract. – See Abstr. 012.086.

045.044 **Tracking station coordinates and earth rotation parameters from satellite laser ranging to LAGEOS during the main MERIT campaign.**
T. Moore, C. J. Hill, V. Ashkenazi.
Geophys. J. R. Astron. Soc., Vol. 85, No. 1, p. 254 (1986). Abstract. – See Abstr. 012.086.

Relativistic effects in geodetic VLBI measurements.
See Abstr. 036.003.

MERIT data processing at Nottingham.
See Abstr. 036.111.

Some problems of increasing the accuracy of photoelectric observations in geodetic astronomy.
See Abstr. 036.124.

Physical models adopted in KAPRI.
See Abstr. 036.152.

Stable method for estimation of laser ranging.
See Abstr. 036.257.

Geodetic, geometric and computational aspects of HIPPARCOS.
See Abstr. 041.045.

Un catálogo provisional FK5 para uso en astronomía geodésica.
See Abstr. 043.013.

Dinamica della Terre e telemetria laser lunare.
See Abstr. 044.042.

Polar motion and European baselines determined by analysis of satellite laser ranging data.
See Abstr. 044.069.

Polar motion estimates from linar combinations of independent series.
See Abstr. 044.085.

Precision and accuracy of Earth rotation determinations derived at BIH from optical astrometry.
See Abstr. 044.095.

Investigation of polar motion and the length of day by means of SLR (Satellite Laser Ranging) data of the MERIT campaign.
See Abstr. 044.102.

UT/CSR analysis of Earth rotation from Lageos SLR (Satellite Laser Ranging) data.
See Abstr. 044.104.

On the use of the collocation method for estimation of the Earth rotation parameters by satellite laser ranging.
See Abstr. 044.105.

Polar motion and length of day determination from satellite laser ranging.
See Abstr. 044.106.

Pole position from laser ranging to LAGEOS. University of Padova/Telespazio Report.
See Abstr. 044.107.

Earth rotation solution from Starlette orbit analysis during the MERIT campaign.
See Abstr. 044.108.

A series of ERP (Earth Rotation Parameters) from LAGEOS during the MERIT campaign.
See Abstr. 044.109.

Earth rotation, station coordinate and orbit solutions from Lageos during the MERIT campaign.
See Abstr. 044.110.

Determination of polar motion and Earth rotation from laser tracking of Lageos.
See Abstr. 044.111.

Determination of ERP with lunar laser ranging and discussion on the influence of the adopted parameters.
See Abstr. 044.112.

Connected–Element Interferometer results in project MERIT.
See Abstr. 044.114.

Accuracy and stability of time series of Earth rotation parameters obtained from space geodesy and astrometric observations.
See Abstr. 044.117.

The U.S. Naval Observatory C.O.R.E. (Combined solution for the orientation of the Earth) solution during project MERIT.
See Abstr. 044.118.

Combination of recent polar motion observations.
See Abstr. 044.119.

Combined determination of the ERP from the results of different techniques, Jan. 1, 1980 – Oct. 31, 1984.
See Abstr. 044.120.

A modern Earth orientation monitoring service: functions, goals, and methods of observation.
See Abstr. 044.129.

Towards an improved realization of the BIH terrestrial frame.
See Abstr. 044.130.

Definition of reference frames and computation of station coordinates and Earth rotation parameters from MERIT laser ranging observations.
See Abstr. 044.131.

Intercomparison of UT1 measurements during the MERIT campaign period.
See Abstr. 044.132.

Intercomparison of polar motion measurements during the MERIT period.
See Abstr. 044.133.

A comparison of Earth orientation data obtained by different space techniques.
See Abstr. 044.134.

Comparison and comment on the techniques for determining ERP (Earth Rotation Parameters).
See Abstr. 044.135.

MERIT main campaign: reference frame intercomparisons.
See Abstr. 044.137.

SLR (Satellite Laser Ranging) and VLBI technology trends.
See Abstr. 044.139.

Observation project of Japanese Geodetic Satellite GS–1.
See Abstr. 051.061.

Modelling and estimating the orbits of GPS satellites.
See Abstr. 052.044.

Spacetime coordinates in the geocentric reference frame.
See Abstr. 066.028.

Measurement of the Lense–Thirring drag on high–altitude, laser–ranged artificial satellites.
See Abstr. 066.117.

Relativistic effects in geodynamics.
See Abstr. 081.003.

Methods to correct for the tropospheric delay in satellite–earth range measurements.
See Abstr. 082.071.

Observations of the total electron content (TEC) of the ionosphere in Neustrelitz (GDR) in the years 1976 – 1980 and possibilities of using these data for geodetic purposes.
See Abstr. 083.013.

046 Ephemerides, Almanacs, Calendars, Chronology

046.001 **Numerical planetary and lunar ephemerides: present status, precision and accuracies.**
E. M. Standish Jr.
Relativity in celestial mechanics and astrometry, p. 71 – 83 (1986). – See Abstr. 012.002 (IAU Symp. No. 114).
The Ephemeris Development Program has been in existence for nearly 20 years at JPL. This paper gives an outline of the least–squares adjustment of the ephemerides to the observations, the present physical (dynamical) model, the present observations to which the ephemerides are fit, the expected accuracies of various ephemeris elements, recent and future observations and features of the solar system which are poorly determined (and thereby place limits upon the accuracies). The export procedure, by which an outside user may obtain and use the JPL ephemerides, is described.

046.002 **Systematic differences between planetary observations and ephemerides.**
P. K. Seidelmann, E. J. Santoro, K. F. Pulkkinen.
Relativity in celestial mechanics and astrometry, p. 99 – 103 (1986). – See Abstr. 012.002 (IAU Symp. No. 114).
This paper is primarily concerned with the problem of making the theoretical computations of celestial mechanics agree with the observational data of the real world.

046.003 **One more proposition of the calendar reform.**
J. Konowski.
Astron. Rep., Vol. 7, p. 49 – 54 (1982).

046.004 **Wochentagsberechnungen – im Kopf!**
M. Gossler.
Sternenbote, 29. Jahrg., Nr. 1, p. 2 – 5 (1986).

046.005 **Presentation of new solar and planetary tables of interest for historical calculations.**
P. Bretagnon, J. L. Simon, J. Laskar.
J. Hist. Astron., Vol. 17, Part 1, No. 48, p. 39 – 50 (1986).
For about twenty years, historians of astronomy have been using the tables of B. Tuckerman which give the geocentric positions of the Sun, the Moon and the five bright planets (Mercury, Venus, Mars, Jupiter and Saturn) over the period -600 to +1649 (601 B.C. to A.D. 1649). Recently, the authors have extended these theories for a period of 6000 years and they have constructed tables that give the positions to the Sun and the five bright planets with good precision over the period -4000, +2000. The authors now describe these theories and present the tables they plan to establish.

046.006 **Efemérides Astronômicas 1986.**
CNPq Conselho Nacional de Desenvolvimento Científico e Tecnológico Observatório Nacional, Rio de Janeiro, Brazil. 18 + 534 pp. (1985).

046.007 **Tables of sunrise, sunset, twilight, moonrise, & moonset 1986.**
Prepared by the Astronomical Observation Division, National Geophysical and Astronomical Office.
Philippine Atmospheric, Geophysical and Astronomical Services Administration, Quezon City, Philippines, 14 + 57 pp. (1985). ISSN 0115–3307.

046.008 **Connaissance des Temps. Ephémérides Astronomiques pour l'An 1987.**
Bureau des Longitudes, 77 Avenue Denfert Rochereau, F–75014 Paris, France. 54 + 124 pp. Price FF 170.00 (1986). ISBN 2–11–080482–3, ISSN 0181–3048.

046.009 **Astronomical Calendar of the Sofia Observatory for the year 1986.**
V. Shkodrov D. Rajkova, V. Ivanova, V. Umlenski, A. Antov; edited by B. Kovachev.
Izdatelstvo na Blgarskata Akademiya na Naukite, Sofiya, Bulgaria. 123 pp. Price 1.22 Lv. (1985). In Bulgarian.

046.010 **Tartu Tähetorni Kalender 1986. aastaks: LXII aastakäik.**
Compiled by M. Jõeveer.
Eesti NSV Teaduste Akadeemia, Astrofüüsika ja Atmosfäärifüüsika Instituut, W. Struve nimeline Tartu Astrofüüsika Observatoorium. Publishing House Valgus, Tallinn. 91 pp. Price 35 Kop. (1985). ISSN 0202–2214.

046.011 **The length of the year in the original proposal for the Gregorian Calendar.**
N. M. Swerdlow.
J. Hist. Astron., Vol. 17, Part 2, No. 49, p. 109 – 118 (1986).

046.012 **Anuario del Observatorio Astronómico de Madrid para 1986.**
Instituto Geográfico Nacional, Madrid, Spain. 421 pp. Price Pts. 450.00 (1985). ISBN 84–505–2848–8, ISSN 0373–5125.

046.013 **The Star Almanac for Land Surveyors for the year 1987.**
Prepared by H. M. Nautical Almanac Office.
Her Majesty's Stationery Office, London, England. 16+80 pp. Price £ 2.50 (1986). ISBN 0-11-886928-0.

046.014 **1987 Japanese Ephemeris. Pub. No. 684.**
Hydrographic Department, Maritime Safety Agency, Tokyo, Japan. 6+499 pp. (1986). ISSN 0373-3696.

046.015 **Astronomical ephemeris for the year 1986.**
N. Čabrić.
Vasiona, Année 33, No. 4, p. 65 - 83 (1985). In Croatian.

046.016 **Blick in die Sternenwelt 1986. Astronomischer Kalender der Archenhold-Sternwarte.**
E. Rothenberg.
Archenhold-Sternw. Berlin-Treptow, 48 pp. (1985).

046.017 **A real-time catalogue of the apparent places of stars.**
Z.-x. Li.
Publ. Purple Mt. Obs., Vol. 5, No. 1, p. 41 - 46 (1986). In Chinese.
A real-time catalogue of the apparent places of stars is given. It is a software package set up in a microcomputer.

046.018 **The Astronomical Almanac for the year 1987.**
Data for astronomy, space sciences, geodesy, surveying, navigation and other applications.
Issued by the Nautical Almanac Office, United States Naval Observatory (Washington) and Her Majesty's Nautical Almanac Office, Royal Greenwich Observatory (London).
US Edition: Superintendent of Documents, US Government Printing Office, Washington, D.C. 20402, USA. UK Edition: Her Majesty's Stationery Office, London, England. 9+536 pp. Price £ 18.00 (1986). ISBN 0-11-886925-6.

046.019 **Efemérides Astronómicas para o ano de 1986.**
Observatório Astronómico da Universidade de Coimbra, Coimbra, Portugal. 17+252 pp. (1986). ISSN 0870-1199.

046.020 **Calendar problem.**
M. L. Chaudhury.
Sci. Cult., Vol. 50, No. 10, p. 276 - 283 (1984). Abstr. in Phys. Abstr., Vol. 89, No. 1280, Entry 54067 (1986).

046.021 **Construction of a planetary solution with the help of an *N*-body program, and analytical complements.**
P. Bretagnon.
Celest. Mech., Vol. 38, No. 2, p. 181 - 190 (1986).
The author deals with a solution for the motion of all planets, that is being constructed with numerical integration of the *N*-body Newtonian problem and analytical complements. He has compared the thus obtained solution to the JPL DE200 ephemeris (Standish, "DE200", magnetic tape, 1982) over the interval 1900 - 2000 in order to test its accuracy.

046.022 **ALPL-1 Newtonian ephemeris of the planetary system spanning 4000 years.**
P. A. Dybczyński, T. Jopek.
Acta Astron., Vol. 36, No. 1, p. 153 - 164 (1986).
Numerical integration of the equations of motion of the solar system planets (except Mercury) is described. The continuous ephemeris obtained in result, spanning the time interval from 1918 BC to 2092 AD is presented.

046.023 **The Nautical Almanac 1934 - 1984.**
B. Franušić.
Publ. Astron. Soc. "Rudjer Bošković", No. 4, p. 163 - 165 (1985). - See Abstr. 012.078. In Croatian.

046.024 **70 solar calendars.**
D. V. Slavić.
Publ. Astron. Soc. "Rudjer Bošković", No. 4, p. 167 - 171 (1985). - See Abstr. 012.078. In Croatian.

046.025 **66 lunar calendars.**
D. V. Slavić.
Publ. Astron. Soc. "Rudjer Bošković", No. 4, p. 173 - 175 (1985). - See Abstr. 012.078. In Croatian.

046.026 **Astronomical ephemerides - their evolution and role as foundations of realizing space experiments.**
V. K. Abalakin.
Prikl. nebes. mekh. i upr. dvizheniem. Tr. 7 Obed. nauchn. chtenij po kosmonavt., posvyashch. pamyati vydayushch. sov. uchenykh-pionerov osvoeniya kosm. prostranstva, Moskva, 24 - 28 yanv., 1983. Moskva, p. 28 - 34 (1983). In Russian. Abstr. in Ref. Zh., 51. Astron., 6.51.136 (1986).

046.027 **Stjärnhimlen 1986.**
P. Ahlin, T. Jürisoo, M. Malmort, P. Schlyter.
Bokförlaget INOVA, Stockholm. 112 pp. Price Sv.kr. 50.00 (1985).
Review in Astron. Tidsskr., Årg. 19, Nr. 2, p. 83; 1986 (*S. Söderhjelm*).

046.028 **Almanacco Astronomico 1986 - Effemeridi e fenomeni del Sistema Solare.**
S. de Meis.
Ed. Hoepli. 98 pp. Price L. 12,000 (1986).
Review in Astronomia, N. 1, p. 13; 1986 (*L. Baldinelli*).

046.029 **Astronomisk årsbok 1986.**
A. Larsson, J. Schildt (Editors).
Bokförlaget INOVA, Stockholm, Sweden. 96 pp. Price Sv.kr. 35.00 (1985).
Review in Astron. Tidsskr., Årg. 19, Nr. 2, p. 94 (1986).

046.030 **Astronomical calendar 1986.**
G. Ottewell.
Furman University Physics Dept., Greenville, S.C. 29613, USA. Available from Sky Publishing Corporation. 66 pp. Price US$ 10.00 (1986). ISBN 0-934546-14-2.
Review in Sky Telesc., Vol. 71, No. 5, p. 471 (1986).

046.031 **The A.L.P.O. solar system ephemeris: 1986.**
J. E. Westfall (Editor).
Association of Lunar and Planetary Observers. J. E. Westfall, P.O. Box 16131, San Francisco, Calif. 94116, USA. 83 pp. Price US$ 6.00 (1985).
Review in Sky Telesc., Vol. 71, No. 4, p. 367 (1986).

046.032 **Compact data for navigation and astronomy for the years 1986 - 1990.**
B. D. Yallop, C. Y. Hohenkerk.
HMSO, London, England. 70+27 pp. Price £ 4.95 paper (1985). ISBN 0-11-886926-4.
From J. Br. Astron. Assoc., Vol. 96, No. 3, p. 187 (1986).

046.033 **Anuário Astronômico 1986.**
Instituto Astronômico e Geofísico, Universidade de São Paulo, Caixa Postal 30627, 01051 São Paulo, SP Brasil. 12+279 pp. (1985). ISSN 0080-6412.

046.034 **Das Himmelsjahr 1987. Sonne, Mond und Sterne im Jahreslauf.**
Herausgegeben von H.-U. Keller unter Mitarbeit von E. Karkoschka.
Astro-Kosmos, Franckh'sche Verlagshandlung, W. Keller & Co., Stuttgart, F.R. Germany. 173 pp. (1986). ISBN 3-440-05591-4, ISSN 0439-1551.

046.035 **Astronomische Grundlagen für den Kalender 1988.**
Compiled by T. Lederle, edited by Astronomisches Rechen-Institut, Heidelberg.
Verlag G. Braun, Karl-Friedrich-Straße 14 - 18, D-7500 Karlsruhe 1, F.R. Germany. 84 pp. Price DM 59.00 (1986). ISBN 3-7650-0187-2, ISSN 0067-0014.

Chronology of the Ancient World.
See Abstr. 003.065.

Kalendarium der Jahre 1700 – 2080.
See Abstr. 003.093.

The Serbs and calendar reforms.
See Abstr. 004.115.

Relativistic effects included in the apparent positions of fundamental stars (APFS).
See Abstr. 041.005.

Methods of computation of positions of the inner planets.
See Abstr. 042.023.

Relativistic effects from planetary and lunar observations of the XVIII – XX centuries.
See Abstr. 091.006.

Ephemerides of minor planets for 1987.
See Abstr. 098.103.

The precision of modern theories of Galilean satellites as derived by their comparison with photographic observations made with the 26″ refractor at Pulkovo Observatory.
See Abstr. 099.004.

The accuracy of space ephemerides for Halley's comet.
See Abstr. 103.633.

Space Research

051 Extraterrestrial Research Related to Astronomy and Astrophysics

051.001 **First results from Spacelab 2.**
E. W. Urban.
Nature, Vol. 319, No. 6054, p. 540 – 542 (1986).

051.002 **Space astronomy in Japan.**
M. Oda.
Astrophys. Space Sci., Vol. 118, Nos. 1/2, p. 67 – 70 (1986). – See Abstr. 012.001.
Japan's two X–ray astronomy satellites, HAKUCHO and TENMA, have been actively in operation leading to deeper understanding of the neutron star physics. Other satellite and balloon projects have been performed under international collaboration.

051.003 **Rendezvous mit Giacobini–Zinner.**
R. Froböse.
Umschau, 86. Jahrg., Nr. 2, p. 116 – 118 (1986).

051.004 **Das Phobos–Projekt.**
D. Möhlmann.
Astron. Raumfahrt, 24. Jahrg., Heft 1, p. 5 – 7 (1986).

051.005 **The Soviet Mars programme.**
P. S. Clark.
J. Br. Interplanet. Soc., Vol. 39, No. 1, p. 2 – 18 (1986).
Beginning in 1960, the Soviet Union launched a number of probes to explore the planet Mars. This paper reviews the Mars programme, looking at the launch and arrival conditions that have applied and how the spacecraft have performed. Details will be given of future launch opportunities and a possible sample return mission will be examined.

051.006 **The Chinese space year of 1984.**
P. S. Clark.
J. Br. Interplanet. Soc., Vol. 39, No. 1, p. 28 – 35 (1986).
During 1984 there were three successful Chinese satellite launches to Earth orbit. This paper will review the developments of 1984 and look forward to possible future space activities.

051.007 **Astronomy from the Space Station.**
J. K. Davies.
J. Br. Interplanet. Soc., Vol. 39, No. 2, p. 51 – 56 (1986).
The suitability of the basic Space Station for astronomical missions is reviewed. The requirements for free flying astronomical platforms are discussed. The astronomical uses to which space platforms and the associated space infrastructure can be put are given, and a number of large space astronomy missions are described.

051.008 **Heading for the Giotto encounter with comet Halley.**
R. E. Münch.
ESA Bull., No. 45, p. 14 – 20 (1986).
This article concentrates on the flight–dynamics efforts supporting the historic rendezvous of the Giotto space probe with comet Halley.

051.009 **Cos–B – a mission fully accomplished.**
K. Bennett.
ESA Bull., No. 45, p. 40 – 43 (1986).
On 27 September 1985 the scientific database of the gamma–ray astronomy satellite Cos–B was formally released to the scientific community.

051.010 **Preparing for the future. The ESA technological research and development programme 1986 – 1988. Executive summary.**
N. Longdon (Editor).
ESA Spec. Publ., ESA SP–1079, 54 pp. (1985).

051.011 **The Giotto project.**
R. Reinhard.
ESA Spec. Publ., ESA SP–1066, p. 25 – 48 (1986). – See Abstr. 003.012.
ESA's Giotto Mission to Halley's comet is a fast flyby at 00 UT on 14 March 1986, about four weeks after the comet's perihelion passage, when it will have been at its most active. The scientific payload is made up of ten experiments with a total mass of approximately 60 kg: a camera for imaging the comet nucleus, three mass spectrometers for analysing the elemental and isotopic composition of the cometary dust and gas environment, various dust impact detectors, a photopolarimeter for measuring the coma's brightness, and a set of plasma instruments for studying the solar–wind/comet interaction. The Giotto spacecraft will be targetted to pass the nucleus at a distance of 500 km on the sunward side.

051.012 **The Vega missions.**
R. Grard, T. I. Gombosi, R. Z. Sagdeev.
ESA Spec. Publ., ESA SP–1066, p. 49 – 70 (1986). – See Abstr. 003.012.
A unique opportunity to combine missions to Venus and comet Halley is available in 1985 – 1986 by employing a two–element space vehicle consisting of a Venus lander and a Halley flyby probe. This mission is being conducted by the USSR with the cooperation of a number of other countries.

051.013 **The Suisei/Sakigake (Planet–A/MS–T5) missions.**
K. Hirao.
ESA Spec. Publ., ESA SP–1066, p. 71 – 84 (1986). – See Abstr. 003.012.
The Japanese mission to comet Halley, called the "Planet–A" mission, involved the launch of the "Suisei", or Planet–A, spacecraft on 19 August 1985, to encounter comet Halley in March 1986. Additionally, a test spacecraft called "Sakigake" (i.e. Pioneer), or MS–T5, was launched on 8 January 1985. Hence, the Planet–A project actually involves two spacecraft, Planet–A and MS–T5. Planet–A has two major scientific objectives: (1) to study the growth and the decay of the comet's hydrogen corona and to determine the total hydrogen production rate by taking UV images of the corona over an extended period, (2) to study the interaction of the solar wind with the cometary ionosphere by measuring the three–dimensional distributions of ions and electrons (energy range 30 eV – 16 keV). MS–T5 has as its scientific objective: to study the solar–wind plasma and, if possible, deter-

mine the signatures of the solar–wind/comet interaction at large distances from the comet. The scientific experiments carried by Planet–A and MS–T5 are described.

051.014 **Impact–induced spacecraft charging and related effects during the Halley encounters.**
D. T. Young.
ESA Spec. Publ., ESA SP–1066, p. 85 – 98 (1986). – See Abstr. 003.012.

051.015 **The International Cometary Explorer (ICE) mission to comets Giacobini–Zinner and Halley.**
J. C. Brandt.
ESA Spec. Publ., ESA SP–1066, p. 99 – 115 (1986). – See Abstr. 003.012.

051.016 **The Astro–1 mission.**
J. C. Brandt, M. B. Niedner.
ESA Spec. Publ., ESA SP–1066, p. 117 – 127 (1986). – See Abstr. 003.012.
The Astro–1 mission is scheduled to be launched in March 1986 for an effective duration of 7 days, specifically chosen to overlap two spacecraft encounters with Halley's comet. These have been tentatively identified as Vega–2 (on 9 March) and Giotto (on 13/14 March). The Astro–1 payload will be launched by the Space Shuttle and observations carried out by the Shuttle crew under the direction of instrument Principal Investigators and comet scientists on the ground.

051.017 **International Ultraviolet Explorer (IUE) observation of Halley's comet.**
Y. Kondo, J. C. Brandt.
ESA Spec. Publ., ESA SP–1066, p. 129 – 133 (1986). – See Abstr. 003.012.

051.018 **Observations of Halley's comet by the Solar Maximum Mission (SMM).**
M. B. Niedner.
ESA Spec. Publ., ESA SP–1066, p. 135 – 143 (1986). – See Abstr. 003.012.

051.019 **The Spartan mission.**
C. A. Barth, S. A. Stern, A. I. F. Stewart.
ESA Spec. Publ., ESA SP–1066, p. 145 – 150 (1986). – See Abstr. 003.012.
The Spartan Halley experiment was aboard Space Shuttle Challenger when it exploded shortly after take–off on 26 January 1986. The article has been retained for the sake of completeness of reporting of the 1986 Halley endeavours and the appreciation of the authors' efforts.

051.020 **NASA–supported smaller programmes to observe comet Halley.**
W. Brunk.
ESA Spec. Publ., ESA SP–1066, p. 151 – 159 (1986). – See Abstr. 003.012.
NASA is supporting several smaller programmes of space observations of comet Halley. They include two rocket flights, the use of an instrument on Pioneer–7, and a portable 35–mm camera system carried as an astronaut option on the Space Shuttle.

051.021 **The Pathfinder Concept.**
R. E. Münch.
ESA Spec. Publ., ESA SP–1066, p. 245 – 253 (1986). – See Abstr. 003.012.
The Pathfinder Concept allows Giotto to be targeted more precisely by using comet Halley nucleus position data provided by the Vega Project of Intercosmos. NASA supports this activity by determining the positions of the two Vega spacecraft very accurately using their Deep–Space Network and Very–Long–Baseline Interferometry techniques.

051.022 **Giotto mission background.**
R. Reinhard.
ESA Spec. Publ., ESA SP–1077, p. vii – x (1986). – See Abstr. 003.013.

051.023 **The Giotto project.**
R. Reinhard.
ESA Spec. Publ., ESA SP–1077, p. xi – xxv (1986). – See Abstr. 003.013.
ESA's Giotto mission to Halley's Comet is a fast flyby in March 1986, about four weeks after the comet's perihelion passage, when it is most active. The scientific payload consists of 10 experiments with a total mass of about 60 kg: a camera for imaging the comet nucleus, three mass spectrometers for analysis of the elemental and isotopic composition of the cometary gas and dust environment, various dust impact detectors, a photopolarimeter for measurements of the coma brightness, and a set of plasma instruments for studies of the solar–wind/comet interaction. The mission scientific objectives, the spacecraft, and the mission are described.

051.024 **The first stage of the Vega project.**
R. Z. Sagdeev, V. I. Moroz.
Pis'ma Astron. Zh., Tom 12, No. 1, p. 5 – 9 (1986). In Russian. English translation in Sov. Astron. Lett., Vol. 12.
Some general data related to the Venus part of the Vega mission are presented and scientific experiments with landers and planetary balloons are listed.

051.025 **Vega balloon experiment.**
R. Z. Sagdeev, R. S. Kremnev, V. M. Linkin, J. Blamont, R. Preston, A. S. Selivanov.
Pis'ma Astron. Zh., Tom 12, No. 1, p. 10 – 15 (1986). In Russian. English translation in Sov. Astron. Lett., Vol. 12.
A short history of development of Vega balloons, basic data on operations and experiment itself, on–board equipment and ground network are presented.

051.026 **Vega balloon experiment: the ground network.**
R. Z. Sagdeev, L. I. Matveenko, V. M. Linkin, V. V. Kerzhanovich.
Pis'ma Astron. Zh., Tom 12, No. 1, p. 16 – 18 (1986). In Russian. English translation in Sov. Astron. Lett., Vol. 12.
A special ground network was created to provide Vega balloon telemetry acquisition and very–long–base interferometry. Practically all the largest radio telescopes of the world were involved and they formed a unique global interferometric network at 18 cm wavelength.

051.027 **Vega balloon as a tool to study the dynamics of the Venus atmosphere.**
R. S. Kremnev, A. S. Selivanov, V. M. Linkin, J. Blamont, R. V. Bakit'ko, A. N. Lipatov, I. Ya. Tarnoruder, V. I. Puchkov, V. D. Kustodiev, A. A. Shurupov, S. P. Ignatova, V. V. Kerzhanovich, L. I. Khlyustova, G. A. Frank, A. I. Turkova, V. P. Karyagin, A. V. Kostin, V. I. Mashkov, K. M. Pichkhadze, A. V. Terterashvili, C. Malique, B. Ragent, R. Preston.
Pis'ma Astron. Zh., Tom 12, No. 1, p. 19 – 24 (1986). In Russian. English translation in Sov. Astron. Lett., Vol. 12.
The balloon and meteorological sensors of the Vega stations are described. The cyclogram and characteristics of measurements, transformation and transmission of meteorological data are given.

051.028 **Vega balloon experiment: the global network of radio telescopes and first results.**
R. Preston, R. Z. Sagdeev, J. Balmont, L. I. Matveenko, V. M. Linkin, V. V. Kerzhanovich, A. B. Severnyj, G. Laurans, C. Hildebrand, G. Purcell, S. Finley, C. Stelzried, J. Ellis,

G. Petit, L. Boloh, A. Ortega–Molina, L. Rosolen, A. Boichot, F. Biraud, D. Collin.
Pis'ma Astron. Zh., Tom 12, No. 1, p. 25 – 29 (1986). In Russian. English translation in Sov. Astron. Lett., Vol. 12.
A unique network of 20 radio telescopes was formed to provide 24–hour telemetry acquisition and trajectory measurements of Vega balloons. Differential very–long–base radio interferometry was used to determine trajectory of the balloons. First results from Doppler data allowed to estimate the mean zonal wind velocity and to find an anomaly in Vega 2 trajectory when it flied over the mountain in Aphrodite Terra.

051.029 **Meteorological measurements of Vega 1 and Vega 2 balloons along their trajectories.**
R. Z. Sagdeev, V. M. Linkin, J. Blamont, R. Preston, V. V. Kerzhanovich, A. N. Lipatov, A. A. Shurupov, A. Ingersoll, D. Crisp, A. V. Terterashvili, N. A. Armand, R. V. Bakit'ko, A. S. Selivanov, B. Ragent, C. Malique, A. Seiff, Yu. N. Aleksandrov, L. Elson, J. Urech, J. Morales, R. Young.
Pis'ma Astron. Zh., Tom 12, No. 1, p. 30 – 35 (1986). In Russian. English translation in Sov. Astron. Lett., Vol. 12.
Direct measuremets of pressure, temperature, vertical wind velocity relative to the balloons, cloud particle backscattering and ambient light level were made with Vega balloons. Estimates of velocities of atmospheric motions are obtained from Doppler data.

051.030 **Dust and minor bodies – the picture from IRAS.**
A. J. Meadows.
Meteoritics, Vol. 20, No. 4, p. 711 (1985). Abstract. – See Abstr. 010.661.

051.031 **Ballooning around Venus.**
D. E. Koshland Jr.
Science, Vol. 231, No. 4744, p. 1349 (1986).

051.032 **Weather balloons at Venus.**
R. A. Kerr.
Science, Vol. 231, No. 4744, p. 1369 (1986).

051.033 **The VEGA Venus balloon experiment.**
R. Z. Sagdeev, V. M. Linkin, J. E. Blamont, R. A. Preston.
Science, Vol. 231, No. 4744, p. 1407 – 1408 (1986).
In June 1985, two instrumental baloons were placed in the atmosphere of Venus as part of the VEGA mission. In situ sensors measured pressure, temperature, vertical wind velocity, cloud particle backscatter, ambient light level, and frequency of lightning. The history, organization, and principal characteristics of this international balloon experiment are described.

051.034 **Quasars and galaxies observable with the Space Telescope.**
C. J. Krieger.
J. R. Astron. Soc. Can., Vol. 80, No. 1, p. 56 – 58 (1986).

051.035 **Alle ogen zijn gericht op komeet Halley.**
G. Schilling.
Zenit, 13. Jaarg., Nr. 3, p. 96 – 99 (1986).

051.036 **Ruimte–onderzoek: een toekomstblik.**
W. de Graaff.
Zenit, 13. Jaarg., Nr. 5, p. 164 – 172 (1986).

051.037 **The International Cometary Explorer mission to comets Giacobini–Zinner and Halley: an update.**
J. C. Brandt.
Asteroids, comets, meteors II, p. 433 – 442 (1986). – See Abstr. 012.010.
On September 11, 1985 the in situ exploration of comets will begin with the pass of the International Cometary Explorer (ICE) spacecraft through the plasma tail of Comet Giacobini–Zinner (G/Z). The primary scientific goal of the ICE mission is to study the interaction of the solar wind with comets by penetrating the downstream plasma environment. The ICE studies of Comet Giacobini–Zinner are complementary to the direct exploration of Halley's Comet in March 1986, and the ICE encounter provides a first test of spacecraft systems in a cometary environment. ICE will also serve as an upstream monitor of solar–wind conditions for Comet Halley during two time periods. The near–radial alignments occur on October 31, 1985 and March 28, 1986 when ICE will be 0.93 and 0.21 AU upstream, respectively.

051.038 **Catching a volatile Halley before it's gone.**
R. A. Kerr.
Science, Vol. 232, No. 4748, p. 320 – 323 (1986).
Although a disappointing sight to many, those with a good view are finding Comet Halley to be bubbling with activity and new insights.

051.039 **The International Cometary Explorer mission to comet Giacobini–Zinner.**
T. T. von Rosenvinge, J. C. Brandt, R. W. Farquhar.
Science, Vol. 232, No. 4748, p. 353 – 356 (1986).
The International Cometary Explorer (ICE) became the first spacecraft ever to encounter a comet when it passed through the tail of comet Giacobini–Zinner. An overview of this encounter is presented, including highlights of the results.

051.040 **The high road to Halley.**
J. K. Beatty.
Sky Telesc., Vol. 71, No. 3, p. 244 – 245 (1986).

051.041 **Giotto and Vega results on Halley's comet.**
N. Henbest.
J. Br. Astron. Assoc., Vol. 96, No. 3, p. 125 – 127 (1986).

051.042 **Coronal spectroscopy and imaging from Spartan during the polar passage of Ulysses.**
J. L. Kohl, H. Weiser, G. L. Withbroe, G. Noci, R. H. Munro.
The sun and the heliosphere in three dimensions, p. 39 – 43 (1986). – See Abstr. 012.016.
A joint payload consisting of a UV coronal spectrometer and a white light coronagraph is being developed for a series of flights on NASA's Spartan carrier. Spartan, which is deployed and retrieved by Shuttle, is intended to provide at least 27 orbits of solar observations per mission. The first flight of this payload is planned for October 1986, and it is anticipated that a subsequent flight will occur at each polar passage of Ulysses.

051.043 **The Ulysses mission.**
R. G. Marsden, K.–P. Wenzel, E. J. Smith.
The sun and the heliosphere in three dimensions, p. 477 – 490 (1986). – See Abstr. 012.016.
The path followed by Ulysses will enable us, for the first time, to explore the heliosphere within a few astronomical units of the Sun over the full range of heliographic latitudes. Instrumentation carried on board the spacecraft is designed to measure the properties of the solar wind, the Sun/wind interface, the heliospheric magnetic field, solar radio bursts and plasma waves, solar X–rays, solar and galactic cosmic rays, and interplanetary/interstellar neutral gas and dust. Ulysses will also be used to detect cosmic gamma–ray bursts and search for gravitational waves.

051.044 **The Advanced X–ray Astrophysics Facility (AXAF): an overview.**
M. C. Weisskopf.
Bull. Am. Astron. Soc., Vol. 17, No. 4, p. 864 (1985). Abstract. – See Abstr. 010.062.

051.045 **Energy balance in stellar coronae.**
J. L. Linsky.
Bull. Am. Astron. Soc., Vol. 17, No. 4, p. 864 (1985). Abstract. – See Abstr. 010.062.

051.046 **The MIT X–ray spectroscopy investigation of AXAF.**
C. R. Canizares.
Bull. Am. Astron. Soc., Vol. 17, No. 4, p. 864 (1985). Abstract. – See Abstr. 010.062.

051.047 **Studies of isolated neutron stars, pulsars and pulsar–driven nebulae with the AXAF.**
A. S. Wilson.
Bull. Am. Astron. Soc., Vol. 17, No. 4, p. 864 (1985). Abstract. – See Abstr. 010.062.

051.048 **Low–frequency variability and VLBI from space.**
B. Dennison, R. S. Booth.
Bull. Am. Astron. Soc., Vol. 17, No. 4, p. 890 (1985). Abstract. – See Abstr. 010.062.

051.049 **The science mission of AXAF.**
M. S. Davis.
Bull. Am. Astron. Soc., Vol. 17, No. 4, p. 899 (1985). Abstract. – See Abstr. 010.062.

051.050 **Science operations and data analysis for the Extreme Ultraviolet Explorer.**
L. R. Cominsky, H. L. Marshall, C. A. Dobson.
Bull. Am. Astron. Soc., Vol. 17, No. 4, p. 900 (1985). Abstract. – See Abstr. 010.062.

051.051 **Prospects for astrometry with the Hubble Space Telescope.**
W. H. Jefferys, G. F. Benedict, P. D. Hemenway, P. J. Shelus, R. L. Duncombe.
Celest. Mech., Vol. 37, No. 3, p. 299 – 305 (1985). – See Abstr. 012.017.

The Hubble Space Telescope (HST), a large optical telescope having an aperture of 2.4 meters and a length of 8.8 meters, is being developed by the National Aeronautics and Space Administration. This telescope will be placed into earth orbit by the space shuttle. Astrometric observations with the HST are made using a Fine Guidance Sensor which is capable of measuring the position of one object relative to another with an accuracy of ± 0.002 arcseconds. The astrometric user of HST will be provided with an Astrometric Data Reduction Software package (ADRS). The variety of astrometric problems to be investigated with HST is discussed.

051.052 **Earth–orbiting telescopes in the Cassini era.**
J. Caldwell.
ESA Spec. Publ., ESA SP–241, p. 219 – 222 (1985). – See Abstr. 012.018.

This paper describes two satellites that are expected to provide important observational capability for planetary studies before and during the operational phase of the Cassini project. These are: the Hubble Space Telescope, which is scheduled to be launched in 1986; and the Far Ultraviolet Spectroscopic Explorer, which is in a preliminary definition phase that is very similar to the current status of Cassini. For the former, the following are discussed: initial complement of instruments; initial observing plans and prospects for second generation instruments. For the latter telescope, a list of desired features, as determined by various advisory groups, is given.

051.053 **The Cassini mission to Saturn and Titan. An overview.**
J.–P. Lebreton.
ESA Spec. Publ., ESA SP–241, p. 225 – 232 (1985). – See Abstr. 012.018.

The feasibility study of Cassini, a Saturn Orbiter and Titan Probe mission, has been jointly conducted by ESA and NASA in 1984/1985. In this paper the author gives an overview of the mission.

051.054 **Cassini – Titan probe concept.**
G. E. N. Scoon.
ESA Spec. Publ., ESA SP–241, p. 245 – 260 (1985). – See Abstr. 012.018.

Cassini target launch date is in May 1994 using the Space Shuttle/Centaur G Prime launch vehicle. The combined Orbiter and Probe Spacecraft is delivered to the Saturn system via a delta–VEGA trajectory, which requires a flight time of about 7.7 years. Opportunities for asteroid flybys will occur as the Spacecraft traverses the asteroid belt, which happens twice during the interplanetary cruise to Saturn. Arrival at Saturn is predicted for January 2002, when the Orbiter and Probe combination will be placed into an initial three–month orbit. Near the end of this first orbit, the Orbiter targets the Probe for separation and delivery to Titan.

051.055 **Cassini – the imaging experiment.**
T. V. Johnson.
ESA Spec. Publ., ESA SP–241, p. 261 – 265 (1985). – See Abstr. 012.018.

The Cassini mission offers many opportunities for imaging science. Just as with the Galileo mission, the combination of improved imaging detectors, multiple close flybys, and a long stay in the system allows significant advances over the reconnaissance performed by the two Voyager spacecraft in 1981. The Cassini mission objectives fall into roughly four catagories, Saturn atmosphere, Titan atmosphere, Satellites and Rings, and Magnetosphere. High resolution imaging and synoptic observations of dynamic features throughout the system will be major contributers to achievements in each of these areas.

051.056 **Infrared spectroscopy aboard Cassini.**
D. Gautier.
ESA Spec. Publ., ESA SP–241, p. 297– 305 (1985). – See Abstr. 012.018.

The information on thermal structures and compositions of Saturn and Titan atmospheres which can be retrieved from spectroscopy of the far infrared, submillimeter, and microwave spectra emitted by these objects is analyzed. The corresponding instrumentation proposed aboard the Orbiter of the Cassini mission is briefly described.

051.057 **Capabilities of spectroscopy in the context of X–ray astronomy in the 1990's.**
J. L. Culhane.
ESA Spec. Publ., ESA SP–239, p. 5 – 14 (1985). – See Abstr. 012.019.

A brief comparison of the mission philosophies and parameters of the proposed NASA AXAF and ESA XMM observatories is presented. In particular, from the way in which mission philosophy has evolved in X–ray astronomy, the role of a mission such as the XMM is explained. The scientific topics that require an observatory of the XMM class are outlined and the relevance of angular and spectral resolution to these aims is discussed.

051.058 **Project INTERSHOCK – investigation of the fine structure of shock waves in cosmic plasma – aims, problems, methods.**
A. A. Galeev, V. Bumba, O. L. Vajsberg, S. Fischer, G. N. Zastenker.
Kosm. Issled., Tom 24, Vyp. 2, p. 147 – 150 (1986). In Russian. English translation in Cosm. Res.

051.059 **The role of high resolution observations in determining energy release and transport processes.**
D. F. Smith.
NASA Conf. Publ., NASA CP–2421, p. 139 – 141 (1986). – See Abstr. 012.021.

With high resolutions one shall be able to determine where the energy release occurs and, with comparable optical and UV data to ascertain the preflare energy storage, to determine whether coronal storage is sufficient or energy must be transported from the photosphere. The role of the interaction of loops in this release and in particular acceleration should also become clear.

051.060 **The Pinhole/Occulter as a tool for X–ray astronomy.**
K. S. Wood.
NASA Conf. Publ., NASA CP–2421, p. 277 – 294 (1986). – See Abstr. 012.021.

The two X–ray transform telescopes on the Pinhole/Occulter Facility (P/OF) will provide sub–arcsecond angular resolution, high sensitivity imaging in hard X–rays for the first time, as well as improved capabilities for timing studies and spectroscopy. Although the design was driven by solar observing requirements, the result is an instrument of major importance for astronomical as well as solar observations. Fundamental astrophysical measurements could be made on supernova remnants, clusters of galaxies, active galaxy cores, accreting binary systems, and active coronae of nearby stars. Ways in which astronomical observations will benefit solar research are discussed.

051.061 **Observation project of Japanese Geodetic Satellite GS–1.**
M. Sasaki.
Proceedings of the Symposium on Application of Space Techniques to Astronomy and Geophysics, p. 27 – 31 (1986). – See Abstr. 012.027.

The Japanese Geodetic Satellite, GS–1, will be launched in August or September 1986. The observation project of GS–1 which will be conducted by the Hydrographic Department of Japan is presented.

051.062 **Japanese space VLBI mission.**
T. Nishimura, T. Takano, T. Yamada, H. Hirabayashi.
Proceedings of the Symposium on Application of Space Techniques to Astronomy and Geophysics, p. 152 – 154 (1986). – See Abstr. 012.027.

A space VLBI mission is being studied by the ISAS–NRO–RRL joint working group. This report summarizes the fundamental requirements on the mission and some of the technical proposals studied by the working group.

051.063 **Comet rendezvous: the next stage in cometary exploration.**
D. H. Collins, S. L. Miller.
J. Br. Interplanet. Soc., Vol. 39, No. 6, p. 263 – 272 (1986).

As a number of spacecraft carried out flybys of Halley's comet in March 1986, plans were underway for the next stage of cometary exploration: the first comet rendezvous. Selection is scheduled for August 1986 of the scientific payload destined for the first sustained, close–up observations of a comet. The auhors, from NASA's Comet Rendezvous Asteroid Flyby Project at JPL, describe the planned mission and spacecraft designs.

051.064 **IRAS: Le cosmos infrarouge.**
G. Burki.
Orion, 44. Jahrg., Nr. 212, p. 13 – 16, 21 (1986).

051.065 **European satellite studies of minor planets. II.**
C. J. Cunningham.
Minor Planet Bull., Vol. 12, No. 3, p. 29 – 30 (1985).

A review is presented of European spacecraft that will carry the study of asteroids into the 21st century.

051.066 **On the use of a sunward libration–point–orbiting spacecraft as an interplanetary magnetic field monitor for magnetospheric studies.**
T. J. Kelly, N. U. Crooker, G. L. Siscoe, C. T. Russell, E. J. Smith.
J. Geophys. Res., Vol. 91, No. A5, p. 5629 – 5636 (1986).

051.067 **Preparation, analysis, and release of simulated interplanetary grains into low earth orbit.**
J. R. Stephens, I. B. Strong, T. D. Kunkle.
NASA Conf. Publ., NASA CP–2403, p. A 34 – A 39 (1986). – See Abstr. 012.030.

The authors discuss the formation and optical characterization of simulated interstellar and interplanetary dust with particular emphasis on studying the properties on irregularly shaped particles. They also discuss efforts to develop the techniques to allow dust experiments to be carried out in low–earth orbit, thus extending the conditions under which dust experiments may be performed.

051.068 **Ultraviolet spectroscopy of meteoric debris of comets.**
T. J. Wdowiak, W. R. Kubinec, J. A. Nuth III.
NASA Conf. Publ., NASA CP–2403, p. A 45 – A 47 (1986). – See Abstr. 012.030.

051.069 **IRAS und der neue Infrarothimmel. Teil 1: Die technische Meisterleistung.**
D. Lemke, U. Klaas.
Sterne Weltraum, 25. Jahrg., Nr. 5, p. 260 – 263 (1986).

051.070 **The birth of X–ray astronomy, Part II.**
W. Tucker, R. Giacconi.
Mercury, Vol. 15, No. 1, p. 13 – 18 (1986).

051.071 **Space Report.**
Spaceflight, Vol. 28, No. 1, p. 6 – 8 (1986).

1986 January review of space news and events.

051.072 **Towards Neptune.**
C. E. Kohlhase, R. V. Frampton, J. W. Gerschultz.
Spaceflight, Vol. 28, No. 1, p. 10 – 15 (1986).

A new chapter in the remarkable journey of Voyager 2 is about to be written during the forthcoming flyby of Uranus and in three years' time with an encounter of Neptune.

051.073 **Space at JPL.**
W. McLaughlin.
Spaceflight, Vol. 28, No. 1, p. 20 – 23 (1986).

Contents: Mars Aeronomy Orbiter. Keck telescope groundbreaking. ICE bisects Giacobini–Zinner. Life in the cosmos.

051.074 **International Space Report.**
Spaceflight, Vol. 28, No. 2, p. 54 – 55 (1986).

1986 February review of space news and events.

051.075 **Space at JPL.**
W. McLaughlin.
Spaceflight, Vol. 28, No. 2, p. 68 – 71, 94 (1986).

Contents: Galileo flight engineering. Unusual asteroid discovery. Future mission technologies. A cometary tale. Voyager returns first Uranus picture.

051.076 **International Space Report.**
Spaceflight, Vol. 28, No. 3, p. 117 (1986).

1986 March review of space news and events.

051.077 **Space at JPL.**
W. McLaughlin.
Spaceflight, Vol. 28, No. 3, p. 122 – 126 (1986).

Contents: Uranus discoveries thrill. A Voyager diary.

051.078 **International Space Report.**
Spaceflight, Vol. 28, No. 4, p. 164 (1986).

1986 April review of space news and events.

051.079 **Space at JPL.**
W. McLaughlin.
Spaceflight, Vol. 28, No. 4, p. 176 – 178 (1986).

Contents: Ploughmen to the search. Joint plan for Saturn Orbiter. Space Station boosts robotics.

051.080 **International Space Report.**
Spaceflight, Vol. 28, No. 5, p. 214 – 215 (1986).

1986 May review of space news and events.

051.081 **Space at JPL.**
W. McLaughlin.
Spaceflight, Vol. 28, No. 5, p. 232 – 236 (1986).

Contents: Satellite discovery bonus for navigation optics. Galileo window opens June 1987. Where are they?

051.082 **International Space Report.**
Spaceflight, Vol. 28, No. 6, p. 260 – 263 (1986).
1986 June review of space news and events.

051.083 **Space at JPL.**
W. McLaughlin.
Spaceflight, Vol. 28, No. 6, p. 270 – 274 (1986).
Contents: Locating comet Halley. Protecting Voyager. Radio science sheds light on Uranus.

051.084 **The X–Ray Timing Explorer.**
D. E. Gruber.
X–ray astronomy '84, p. 503 – 506 (1985). – See Abstr. 012.033.
The X–Ray Timing Explorer, whose launch is currently anticipated in 1991, is the next U.S. satellite mission dedicated to X–ray astronomy. The major scientific goals are reviewed, short descriptions are given of the detector and information systems, and the spacecraft and planned mission are briefly described.

051.085 **More missions to explore the solar system.**
J. R. French.
Spaceflight, Vol. 28, No. 5, p. 207 – 210 (1986).

051.086 **Progetto HIPPARCOS.**
P. L. Bernacca.
G. Astron., Vol. 11, N. 4, p. 375 – 409 (1985).

051.087 **An Australian broad–band satellite for X–ray astronomy.**
J. G. Greenhill, K. B. Fenton, R. K. Sood, I. R. Tuohy.
Proc. Astron. Soc. Aust., Vol. 6, No. 2, p. 186 – 194 (1985).
A broad–band (2 – 190 keV) Australian X–ray satellite could provide a spectral sensitivity substantially better than HEAO–1 or any presently approved spacecraft. It would be virtually unique by providing simultaneously data over a wide energy range with high sensitivity and energy resolution in the little measured region above 30 keV. Details of the proposed spacecraft and scientific objectives are given.

051.088 **Comet nucleus sample return missions with electrically propelled spacecraft.**
E. Stuhlinger, H. Fechtig, E. Igenbergs, H. Loeb.
J. Br. Interplanet. Soc., Vol. 39, No. 6, p. 273 – 281 (1986).
Analysis of pristine cometary material in laboratories would decisively improve our understanding of cosmological evolution. Electric propulsion systems provide the propulsive capabilities which make such missions practical. The present study describes a mission with an electrically propelled spacecraft, launched by the Shuttle with the Centaur G' as upper stage. The spacecraft will travel to the close vicinity of a comet, make a number of scientific observations, collect several drill core samples from the nucleus, and return those samples to Earth. An almost ideal mission opportunity with a rendezvous early in 1997 will be offered by comet P/Wild 2.

051.089 **Communicating from the edge of the solar system.**
E. C. Posner, R. L. Horttor, T. L. Grant.
IEEE Spectrum, Vol. 23, No. 6, p. 60 – 65 (1986).
Centimeter–band transmissions, three–dimensional data compression, and laser–based systems may launch unmanned missions to deep space.

051.090 **IRAS und der neue Infrarothimmel. Teil 2: Aufregende Entdeckungen.**
U. Klaas, D. Lemke.
Sterne Weltraum, 25. Jahrg., Nr. 6, p. 321 – 325 (1986).

051.091 **A program for observations of the nearest eclipsing binaries from the HIPPARCOS satellite.**
T. Z. Dworak, E. Oblak.
Postepy Astron., Tom 32, Zesz. 3/4, p. 187 – 203 (1984). In Polish.
The paper gives a review of arrangements of the astrometric HIPPARCOS satellite mission. Technical data as well as programs of observations are briefly described. The authors' program of observations of the nearest eclipsing binaries (and also bright systems) is extensively presented. These observations will serve for the determination of heliocentric parallaxes of these stars as well as for the determination of their geometric and physical parameters.

051.092 **Space mission "Vega": encounter with comet Halley.**
V. M. Balebanov, V. I. Moroz.
Priroda, No. 5, p. 3 – 17 (1986). In Russian.

051.093 **On the feasibility of observing small asteroids with Galileo, Venera, and Comet–Rendezvous – Asteroid–Flyby missions.**
T. Gehrels.
Icarus, Vol. 66, No. 2, p. 288 – 296 (1986).
Galileo, Venera, and CRAF flyby are being planned for known asteroids that are in size range of tens or hundreds of kilometers. This paper examines the feasibility of observing, as targets of opportunity, asteroids that are smaller. They are different objects, collisional debris most likely. The spacecraft will encounter thousands of small asteroids, within reach of the imaging cameras on board. The magnitude–frequency relation can be reconfirmed, and extended to meteorite sizes. Possibilities are also discussed for observing a few suitable candidates from the earth, in support of the missions so that they can obtain data on composition.

051.094 **The Astro Observatory guest observer program.**
K. H. Nordsieck, T. Gull.
Bull. Am. Astron. Soc., Vol. 18, No. 1, p. 636 (1986). Abstract. – See Abstr. 010.066.

051.095 **High resolution spectroscopy with ST.**
M. Hack.
Mem. Soc. Astron. Ital., Vol. 56, No. 4, p. 681 – 686 (1985). – See Abstr. 012.054.

051.096 **A look at the primeval explosion.**
J. Mather.
New Sci., Vol. 109, No. 1491, p. 48 – 50 (1986). Abstr. in Phys. Abstr., Vol. 89, No. 1280, Entry 60580 (1986).

051.097 **The project HIPPARCOS.**
J. Kovalevsky.
Manuscr. Geod., Vol. 11, No. 2, p. 89 – 96 (1986).
The scientific background of the project HIPPARCOS is given first. Some of the astronomical objectives are shortly described. Then, the main features of the satellite and its mode of operation are presented. A global description of the organization of the project is given.

051.098 **Observations of artificially induced suprathermal electron fluxes on board Spacelab 1.**
J. Watermann, K. Wilhelm, K. M. Torkar, W. Riedler.
Mitt. Astron. Ges., No. 65, p. 166 – 169 (1986). – See Abstr. 012.072.

051.099 **Strategy for the development of space arrays.**
A. Labeyrie.
ESA Spec. Publ., ESA SP–226, p. 23 – 25 (1985). – See Abstr. 012.075.

051.100 **Mission analysis aspects.**
E. A. Roth.
ESA Spec. Publ., ESA SP–226, p. 113 – 116 (1985). – See Abstr. 012.075.
In this contribution only those projects are considered where the baseline of the interferometer is represented by two (or more) free–flying satellites. This leads to difficult but challenging problems of astrodynamics which are still awaiting solutions. Orbits which are likely candidates are discussed: high terrestrial orbits, quasi–periodic orbits around the triangular libration points (L_4, L_5) in the Earth–Moon system or orbits near these points in the Earth–Sun system. Some of the problems arising from such missions are briefly mentioned.

051.101 **A brief history of the Giotto mission.**
R. Reinhard.
ESA Bull., No. 46, p. 19 – 21 (1986). – See Abstr. 003.052.

051.102 **The Giotto project: from early concepts to flight model.**
D. Dale, F. Felici, P. L. Galbo.
ESA Bull., No. 46, p. 22 – 33 (1986). – See Abstr. 003.052.
An attempt will be made to provide a summary of the major decisions that had to be taken and the milestones that had to be met, the technology that was used, the inherent mission risks, and the difficulties that were encountered along the way.

051.103 **Programmes under development and operations/Programmes en cours de réalisation et d'exploitation.**
ESA Bull., No. 46, p. 95 – 104 (1986).

051.104 **Optical properties of simulated astrophysical grains and their dynamics in the near–Earth environment.**
J. R. Stephens, T. D. Kunkle, I. B. Strong.
Light on dark matter, p. 245 – 246 (1986). – See Abstr. 012.080.
Plans are underway to prepare simulated interplanetary grains and study their behavior when released into low–Earth orbit. The program is twofold: preparing and characterizing, in the laboratory, grains which simulate interstellar and interplanetary grains, and monitoring the optical properties of grains released into low–Earth orbit to study their dynamical behavior and interaction with the fields and plasma near the Earth. Such phenomena as grain coagulation and alignment and also the effect of grains on the Earth's plasma environment will be studied.

Data catalog series for space science and applications flight missions. Volume 3B. Descriptions of data sets from low– and medium–altitude scientific spacecraft and investigations.
See Abstr. 002.091.

History of British space science.
See Abstr. 003.003.

Les planètes artificielles.
See Abstr. 003.080.

***Space Station program*, description, applications, and opportunities.**
See Abstr. 003.215.

Space Station planetology experiments (SSPEX). Proceedings of a workshop held at the U.S. Geological Survey, Flagstaff, Arizona, USA, 20 – 22 June 1985.
See Abstr. 012.050.

The Inter–Agency Consultative Group (IACG) and its associated working groups.
See Abstr. 013.016.

The Halley dust and gas environment (IACG WG–1).
See Abstr. 013.017.

Coordinated multispacecraft plasma science in the solar–wind and comet–plasma environments (IACG WG–2).
See Abstr. 013.018.

Infrared astronomy after IRAS.
See Abstr. 013.019.

UK involvement in the ROSAT project.
See Abstr. 013.032.

Astronomy after IRAS.
See Abstr. 013.075.

Search for chiral molecules and optical activity in extraterrestrial systems. The Cassini mission opportunity.
See Abstr. 015.024.

Future space telescope design concepts.
See Abstr. 031.055.

TRIO optics.
See Abstr. 031.056.

The AXAF CCD imaging spectrometer experiment (ACIS).
See Abstr. 035.020.

Low energy transmission grating spectrometer (LETGS) for AXAF.
See Abstr. 035.021.

The Cassini mission. Complementary observations with the Space Telescope.
See Abstr. 035.029.

Microwave spectroscopy and radiometry of Saturn and Titan.
See Abstr. 035.030.

A report on the mapping of the DIDSY (*Dust Impact Detection System*) sensor responses as a function of micro–meteorite impact position on the Giotto bumper shield.
See Abstr. 035.061.

TRIO: a kilometric array stabilized by solar sails.
See Abstr. 035.086.

SAMSI: an orbiting spatial interferometer for micro–arcsecond astronomical observations.
See Abstr. 035.087.

COSMIC: a high resolution, large collecting area telescope.
See Abstr. 035.088.

A fiber–linked version of project TRIO.
See Abstr. 035.089.

OASIS: optical aperture synthesis in space.
See Abstr. 035.090.

Propulsion solaire pour l'interféromètre TRIO.
See Abstr. 035.091.

Preliminary studies of a spinning tether–connected TRIO concept.
See Abstr. 035.092.

TRIANGLE: a different design for a three satellite interferometer in space.
See Abstr. 035.094.

The OASIS correlator.
See Abstr. 035.095.

Space antenna for gravitational wave astronomy.
See Abstr. 035.096.

DUO: a far infrared heterodyne concept for space interferometry.
See Abstr. 035.097.

The special network for telemetry reception and interferometric measurements of the Vega balloon experiment.
See Abstr. 036.034.

Interferometric processing of the data from the Vega balloon probe and flyby station.
See Abstr. 036.036.

Advances in information analysis expected for Space Telescope data.
See Abstr. 036.193.

Storage, retrieval, and analysis of ST data.
See Abstr. 036.240.

Astrometry with ST.
See Abstr. 041.032.

Satellite tour design for the Cassini mission.
See Abstr. 052.011.

Coronal transients at high heliographic latitudes.
See Abstr. 074.038.

Interplanetary energetic particles, coronal flares, and hard X–ray microflares.
See Abstr. 076.013.

Ion release experiments in the solar wind.
See Abstr. 084.043.

Asteroid 29 Amphitrite: surface characteristics and prospects for the Galileo flyby.
See Abstr. 098.014.

29 Amphitrite: the first asteroid to be visited by a space mission.
See Abstr. 098.034.

Interior of Saturn.
See Abstr. 100.030.

Thermal structure of Saturn's atmosphere.
See Abstr. 100.032.

Theories for the general circulation of Saturn's atmosphere.
See Abstr. 100.034.

Titan's surface: nature and implications for Cassini.
See Abstr. 100.039.

Thermal structure and meteorology of Titan.
See Abstr. 100.040.

Photochemistry of Titan.
See Abstr. 100.047.

VLA–observations of Saturn – can we do better from space?
See Abstr. 100.052.

Voyager 2 swings by Uranus.
See Abstr. 101.001.

Cometary observations from space.
See Abstr. 102.078.

ICE: mission to a comet.
See Abstr. 103.129.

Halley's comet as an outstanding target.
See Abstr. 103.615.

Debris from comet Halley and the risk for space probes.
See Abstr. 103.638.

The Planet–A Halley encounters.
See Abstr. 103.676.

The Giotto encounter with comet Halley.
See Abstr. 103.682.

First journey to a comet.
See Abstr. 103.699.

Stellar activity in the evolutionary framework: the Space Telescope perspective.
See Abstr. 116.060.

New insight on the evolution of binary X–ray sources from the X–ray observations of nearby spiral galaxies.
See Abstr. 142.084.

ST to the discovery of the galactic halo.
See Abstr. 154.060.

052 Astrodynamics, Navigation of Space Vehicles

052.001 **On calculation of relativistic effects in numerical prediction of artificial satellites' motion.**
T. V. Bordovitsyna, N. A. Sharkovsky (*N. A. Sharkovskij*).
Relativity in celestial mechanics and astrometry, p. 283 – 288 (1986). – See Abstr. 012.002 (IAU Symp. No. 114).

This paper discusses the algorithm of the calculation of relativistic effects in numerical prediction of the artificial satellites' motion. An attempt to estimate the values of relativistic effects in the motion of two artificial satellites Navstar and Lageos is made. These values are compared with the quantities of other weak perturbations and the errors of the calculated position of the satellite due to the inaccurate approximation of perturbations.

052.002 **The motion of a geosynchronous satellite.**
K. B. Bhatnagar.
Relativity in celestial mechanics and astrometry, p. 293 – 295 (1986). – See Abstr. 012.002 (IAU Symp. No. 114).

This paper presents the results of an investigation done on the motion of a geosynchronous satellite under the gravitational forces of the earth (including the ellipticity of the earth equator) the sun, the moon and the solar radiation pressure. The study presented here can be applied to satellites in near circular orbits at any inclination and with orbital radii less than ten times the earth's radius.

052.003 **Orbit determination and analysis for 1970–97B at 14th–order resonance.**
A. N. Winterbottom, D. G. King–Hele.
Planet. Space Sci., Vol. 34, No. 1, p. 47 – 57 (1986).

The orbit of the satellite 1970–97B, the rocket of Cosmos 378, at inclination 74°, has been determined at 18 epochs between April and August 1971, when the effects of 14th–order resonance with the Earth's gravitational field were appreciable.

052.004 **Earth gravitational perturbation sensitivity and orbital inaccuracy for GEO–IK orbit.**
J. Klokočník, J. Kostelecký.
Bull. Astron. Inst. Czech., Vol. 37, No. 2, p. 84 – 90 (1986).

The sensitivity of the orbit of the proposed GEO–Intercosmos satellite to the Earth's gravity field perturbations is studied as well as its radial orbital inaccuracy as a function of the errors in the harmonic coefficients of the geopotential. Numerical results are presented.

052.005 **On Tisserand's polynomials and inclination functions in the theory of motion of artificial earth satellites.**
E. P. Aksenov.
Astron. Zh., Tom 63, Vyp. 2, p. 365 – 370 (1986). In Russian. English translation in Sov. Astron., Vol. 30, No. 2.

The connection of Tisserand's polynomials with inclination functions in the theory of motion of artificial earth satellites is established. The most important properties of these special functions of celestial mechanics are presented. The problem of expansion of the perturbaton function in satellite problems is considered.

052.006 **Investigation of the motion of a space apparatus in the vicinity of the collinear libration center.**
P. E. Ehl'yasberg, V. M. Sinitsin, T. A. Timokhova.
Inst. kosm. issled. Akad. Nauk SSSR. Prepr., No. 995, 59 pp. (1985). In Russian. Abstr. in Ref. Zh., 51. Astron., 1.51.100; 62. Issled. Kosm. Prostranstva, 1.62.241 (1986).

052.007 **Bifurcation of resonance curves of the equation of satellite motion in an elliptical orbit.**
Z. S. Batalova, N. A. Ezhevskaya.
Dinam. sistem: Dinam. upr. Gor'kij, p. 78 – 93 (1984). In Russian. Abstr. in Ref. Zh., 62. Issled. Kosm. Prostranstva 2.62.178 (1986).

052.008 **Periodic orbits around the moon and their use in super-long–baseline radio interferometry.**
A. Yu. Kogan.
Kosm. Issled., Tom 24, Vyp. 1, p. 52 – 57 (1986). In Russian. English translation in Cosm. Res.

052.009 **Two locations, two times, and the element set.**
L. G. Taff, P. M. S. Randall.
Celest. Mech., Vol. 37, No. 2, p. 149 – 159 (1985).
The authors have delved into that aspect of classical initial orbit determination wherein one transforms two locations at two times into an orbital element set. The particular realm of application they have in mind is artificial satellites whose locations are uncertain to $\cong 1\ km^3$, the two locations being separated in time by no more than 30 min. Not only have they investigated Gauss's method, iteration on the semi–latus rectum, iteration on the true anomaly, and the Lambert–Euler technique, they have made substantial practical progress with the latter two algorithms.

052.010 **Motion of a lunar orbiter.**
H. Kinoshita.
Proceedings of the nineteenth symposium on celestial mechanics, p. 53 – 58 (1986). In Japanese. – See Abstr. 012.013.

052.011 **Satellite tour design for the Cassini mission.**
R. E. Diehl.
ESA Spec. Publ., ESA SP–241, p. 233 – 243 (1985). – See Abstr. 012.018.
The mission design for the Project Cassini satellite tour of Saturn is described. The use of Titan encounters for gravity–assisted trajectory shaping to satisfy atmospheres, magnetosphere and satellite science requirements is discussed. Three candidate satellite tours are presented that illustrate the trade–offs between different strategies to satisfy the science objectives.

052.012 **Detection of the malfunction of Cosmos 954 and Cosmos 1402 and the subsequent predictions of their re–entries.**
G. E. Perry.
ESA Spec. Publ., ESA SP–246, p. 25 – 28 (1986). – See Abstr. 012.028.

052.013 **Lifetime predictions for the RAE Table of Satellites.**
D. M. C. Walker, D. G. King–Hele.
ESA Spec. Publ., ESA SP–246, p. 29 – 37 (1986). – See Abstr. 012.028.
After studies of Earth satellites in 1955 – 56, the authors operated the UK satellite prediction service in 1958 and began a table of satellites in response to requests. This has now grown to the 770–page RAE Table of Earth Satellites 1957 – 1982, which includes lifetime estimates for all satellites still in orbit, excluding fragments. The authors have made about 10000 such lifetime predictions and in this paper they describe the methods of calculation.

052.014 **Long–term analytical orbit decay and reentry prediction.**
H. H. Klinkrad.
ESA Spec. Publ., ESA SP–246, p. 39 – 48 (1986). – See Abstr. 012.028.
The paper outlines the main features of a new analytical approach to the prediction of close–earth satellite orbits of moderate eccentricities.

052.015 **Cosmos 1402 re–entry predictions: a retrospective analysis.**
L. Anselmo.
ESA Spec. Publ., ESA SP–246, p. 49 – 54 (1986). – See Abstr. 012.028.

052.016 **Methods for orbit decay predictions used at the FGAN–FHP.**
H. G. Peters.
ESA Spec. Publ., ESA SP–246, p. 61 – 62 (1986). – See Abstr. 012.028.
This article deals with two methods for satellite lifetime prediction used at the FHP. One of these methods is the well–known procedure due to King–Hele. It is described how the second method originated as by–product of the effort to overcome certain difficulties in the supply of the input parameters for King–Hele's method.

052.017 **Organization of the orbit and re–entry prediction computations at CNES for Skylab and Cosmos 1402.**
F. Nouel.
ESA Spec. Publ., ESA SP–246, p. 63 – 73 (1986). – See Abstr. 012.028.

052.018 **Decay of ESA space objects.**
G. Janin, J. de Leeuw.
ESA Spec. Publ., ESA SP–246, p. 75 – 79 (1986). – See Abstr. 012.028.
Methods for orbital lifetime predictions are reviewed and the set up of the orbit generation software used at ESOC for lifetime estimation is outlined. After a discussion of the limitations in accuracy of lifetime predictions, a list of all ESRO/ESA objects launched up to now is given, together with their actual or predicted deday dates.

052.019 **Application of the resonance theory to transformation of the elements of a resonance orbit of close artificial earth satellites.**
S. N. Yashkin.
Nablyud. Iskusstv. Sputnikov Zemli, No. 21, Chast' I, p. 12 – 20 (1982). In Russian.– See Abstr. 012.032.
The motion of a close artificial earth satellite in the libration region is studied by the method of the ideal resonance problem. The accuracy of solution is up to $J_2{}^2$.

052.020 **Development of a semi–analytical theory of motion of resonant satellites in the gravitational field of the earth.**
I. V. Tupikova.
Nablyud. Iskusstv. Sputnikov Zemli, No. 21, Chast' I, p. 21 – 24 (1982). In Russian.– See Abstr. 012.032.
A third–order semi–analytical theory has been developed for the motion of resonant satellites in the gravitational field of the earth. The short–period terms have been removed from the Hamiltonian by the Hori–Deprit method. The equations of motion with the averaged Hamiltonian have been integrated numerically. The comparison of the results of the semi–analytical theory with those of numerical integration of the equations of motion in osculating elements showed coincidence within several meters over the period of one month.

052.021 **On the computation of perturbations in the elements of 24–hours nearly circular satellite orbits.**
A. S. Sochilina.
Nablyud. Iskusstv. Sputnikov Zemli, No. 21, Chast' I, p. 25 – 31 (1982). In Russian.– See Abstr. 012.032.
The perturbation theory based on Krylov's, Bogolyubov's and Mitropolsky's method of averaging is used to obtain the long–

period solution for longitude λ_N of a 24–h satellite, as well as for $h = \sin i \sin \Omega$ and $k = \sin i \cos \Omega$, where i is the inclination of the orbit to the equator, Ω is the right ascension of the ascending node.

052.022 **Calculation of satellite positions for nearly circular orbits.**
M. N. Boyarskij, R. R. Nazirov.
Nablyud. Iskusstv. Sputnikov Zemli, No. 21, Chast' I, p. 48 – 54 (1982). In Russian.– See Abstr. 012.032.
An analytic technique of rectangular coordinates calculation of a satellite moving in a roughly circular orbit with short–period disturbances caused by the earth's oblateness is described. The osculating orbit elements are assumed to be known exactly at each time of crossing the node. The results obtained by the proposed method are compared with those by numerical integration for INTERCOSMOS–BULGARIA 1300.

052.023 **Processing of LAGEOS observation series using the analytical theory.**
I. S. Gayazov.
Nablyud. Iskusstv. Sputnikov Zemli, No. 21, Chast' I, p. 38 – 41 (1982). In Russian.– See Abstr. 012.032.
The analysis of LAGEOS observation series, selected for comparison of various satellite orbit determination programs, has been undertaken using the analytical theory. A short discussion both of the data processing and the results obtained is given.

052.024 **Evolution of artificial earth satellite orbits under the influence of gravitational perturbations.**
F. Veres.
Nablyud. Iskusstv. Sputnikov Zemli, No. 21, Chast' I, p. 42 – 47 (1982). In Russian.– See Abstr. 012.032.
The evolution of AES orbits is studied. The effects of the moon and the sun and the deviation of the earth's figure from spherical symmetry are taken into account.

052.025 **On the effectiveness of numerical integration methods depending on the form of equations of motion.**
E. N. Makarova.
Nablyud. Iskusstv. Sputnikov Zemli, No. 21, Chast' I, p. 55 – 58 (1982). In Russian.– See Abstr. 012.032.
The effectiveness of the numerical integration methods for two types of the equations of motion (in rectangular coordinates and in perturbed elements) has been compared and some conclusions have been drawn.

052.026 **Improvement of the initial orbital elements of AES from observations at short arcs.**
N. Petrova, I. Trenkov.
Nablyud. Iskusstv. Sputnikov Zemli, No. 21, Chast' I, p. 59 – 66 (1982). In Russian.– See Abstr. 012.032.
A method is suggested for improved least–squares estimates and respective stochastic characteristics of initial orbital conditions of artificial Earth satellites based on data drawn out from new visual, photographic and laser observations of short arcs without the necessity of synchronization. New data for the orbit of the satellite can be included – orbital elements in the Greenwich coordinate system. A flattened spheroidal coordinate system is applied and development of coordinates in power series at regularized time and the respective coefficients are computed using recurrent formulae.

052.027 **Results of application of the AMP–method to orbital determination of earth satellites.**
O. P. Bykov, S. A. Solov'ev.
Nablyud. Iskusstv. Sputnikov Zemli, No. 21, Chast' I, p. 67 – 74 (1982). In Russian.– See Abstr. 012.032.
The method of apparent motion parameters (the AMP–method) proposed by Kisilev and developed at the Pulkovo Observatory was applied to the determination of AES orbits under conditions of fictitious positional AES observations on a short topocentric arc observed from a single tracking station. It is shown that the AMP–method solves the problem of AES–orbit determination from the positional data more reliable, accurate and effective than the so–called "exact method". Examples of the AMP–method application to AES–orbit determination under conditions of fictitious positional observations with their different random errors are given.

052.028 **Electrodynamic effects in the problem of precise ephemeris computation of artificial earth satellites.**
S. A. Vavilov, V. G. Degtyarev, E. K. Kolesnikov, Yu. M. Ehventave.
Nablyud. Iskusstv. Sputnikov Zemli, No. 21, Chast' I, p. 75 – 77 (1982). In Russian.– See Abstr. 012.032.
The problem of the influence of quantum reactive forces due to presence of direct powerful radiowave systems on the navigation of artificial satellites and the influence of forces caused by electrodynamic effects of interaction of an artificial satellite with the surrounding cosmic plasma on the motion of the artificial satellite are considered. It is ascertained that these forces should be under consideration for the long–term prognosis of artificial satellite motion.

052.029 **Geocentric coordinates of perigee motion.**
B. B. Baghos, Kh. I. Khalil.
Nablyud. Iskusstv. Sputnikov Zemli, No. 21, Chast' I, p. 78 – 85 (1982). – See Abstr. 012.032.
The mean daily geocentric position, the longitude and the latitude of the perigee and its daily variation for five geodetic satellites are computed by using SAO orbital ephemeris data.

052.030 **Application of point mass models to orbital determination.**
N. A. Sorokin, O. A. Abrikosov, A. N. Marchenko.
Nablyud. Iskusstv. Sputnikov Zemli, No. 21, Chast' I, p. 116 – 123 (1982). In Russian.– See Abstr. 012.032.
A comparison of rangings of Lageos being calculated as a result of its orbit prediction during a 2.5–day interval on the basis of the point mass model of the geopotential M3 with laser ranging data is presented. The (O–C) values, determined with the use of M3 and GEM–10, were compared and a conclusion about the possibility of utilization of the multipoint models M3 and M4 for orbital predictions during several days intervals was drawn. Estimates of satellite positions defined by different geopotential models were obtained.

052.031 **Motion of a satellite in the earth's atmosphere based on spherical harmonics representation.**
M. Vykutilova.
Nablyud. Iskusstv. Sputnikov Zemli, No. 21, Chast' I, p. 240 – 248 (1982). In Russian.– See Abstr. 012.032.
An atmospheric density model based on spherical harmonic analysis is proposed. The expression for perturbations in the orbital elements of a satellite moving in the earth's atmosphere is obtained. The method is applied to computing of the semimajor–axis perturbations of the Intercosmos 17 satellite.

052.032 **Influence of the atmospheric drag on orbits of geodetic satellites.**
T. V. Kasimenko, N. A. Sorokin.
Nablyud. Iskusstv. Sputnikov Zemli, No. 21, Chast' I, p. 258 – 263 (1982). In Russian.– See Abstr. 012.032.
The influence of atmospheric drag on satellite orbits of up to 1000 km is discussed. It is shown that the accuracy of orbital determination is to a great extent limited by errors of the models used (in particular, that of the atmosphere).

052.033 **Determination of the periods of rotation of different satellites by two methods using observations from one station.**
M. V. Bratijchuk, A. G. Kirichenko.
Nablyud. Iskusstv. Sputnikov Zemli, No. 21, Chast' I, p. 302 – 307 (1982). In Russian. See Abstr. 012.032.
Periods of rotation of the satellites Geos–B, Geos–C, Intercosmos 19, Bulgaria 1300 determined by two methods using results of observations by one station are presented. The atmospheric density has been calculated from periods of Geos–C and its dependences on solar activity is shown.

052.034 **Bifurcations of resonance curves of the equation of motion of a satellite in an elliptical orbit.**
Z. S. Batalova, N. A. Ezhevskaya.
Din. sistem: Din. upr., Gor'kij, p. 78 – 93 (1984). In Russian. Abstr. in Ref. Zh., 51. Astron., 5.51.135 (1986).

052.035 **Stationary motions of gravitating bodies.**
S. Ya. Stepanov.
Zadachi issled. ustojchivosti i stabiliz. dvizheniya. Moskva, p. 85 – 89 (1985). In Russian. Abstr. in Ref. Zh., 62. Issled. Kosm. Prostranstva, 5.62.208 (1986).

052.036 **Numerical analysis of the orbits of high–apogee artificial earth satellites.**
A. I. Averbukh, B. V. Girshovich, A. I. Shejkhet.
Prikl. nebes. mekh. i upr. dvizheniem. Tr. 7 Obed. nauchn. chtenij po kosmonavt., posvyashch. pamyati vydayushch. sov. uchenykh–pionerov osvoeniya kosm. prostranstva, Moskva, 24 – 28 yanv., 1983. Moskva, p. 5 – 11 (1983). In Russian. Abstr. in Ref. Zh., 51. Astron., 6.51.134 (1986).

052.037 **Periodic rotational motions of space vehicles around the libration point of the restricted three–body problem.**
Yu. V. Barkin, S. N. Lelyavin.
Prikl. nebesn. mekh. i upr. dvizheniem. Tr. 7 Obed. nauchn. chtenij po kosmonavt., posvyashch. pamyati vydayushch. sov. uchenykh–pionerov osvoeniya kosm. prostranstva, Moskva, 24 – 28 yanv., 1983. Moskva, p. 133 – 124 (1983). In Russian. Abstr. in Ref. Zh., 62. Issled. Kosm. Prostranstva, 6.62.180 (1986).

052.038 **On a method of calculation of the motion of a 12–hour artificial earth satellite.**
M. A. Vashkov'yak.
Prikl. nebesn. mekh. i upr. dvizheniem. Tr. 7 Obed. nauchn. chtenij po kosmonavt., posvyashch. pamyati vydayushch. sov. uchenykh–pionerov osvoeniya kosm. prostranstva, Moskva, 24 – 28 yanv., 1983. Moskva, p. 12 – 27 (1983). In Russian. Abstr. in Ref. Zh., 62. Issled. Kosm. Prostranstva, 6.62.181 (1986).

052.039 **The stability of nonlinear oscillations of a rotating solid body relative to the mass center in an elliptical orbit.**
V. I. Gulyaev, P. P. Lizunov.
Prikl. nebesn. mekh. i upr. dvizheniem. Tr. 7 Obed. nauchn. chtenij po kosmonavt., posvyashch. pamyati vydayushch. sov. uchenykh–pionerov osvoeniya kosm. prostranstva, Moskva, 24 – 28 yanv., 1983. Moskva, p. 144 – 153 (1983). In Russian. Abstr. in Ref. Zh., 62. Issled. Kosm. Prostranstva, 6.62.190 (1986).

052.040 **Quasi–analytical solutions for the dynamics of a class of tethered satellites with Danby's aerodynamical drag.**
I. Bonzani, M. G. Zavattaro Chiado Piat.
Celest. Mech., Vol. 37, No. 4, p. 371 – 385 (1985). In French.
An analytic solution, in terms of expansion of a small parameter, for the classical two–body tethered satellite with a specific drag model is obtained. The time of flight as well as the solution regarding the state variable of the system are expressed in terms of quadratures of the anomaly.

052.041 **HIPPARCOS astrometry satellite: closed–loop digital attitude control using gas jets.**
D. Carlucci, F. Donati.
Bridge between control science and technology, Vol. 5, p. 2921 – 2926 (1985). Abstr. in Phys. Abstr., Vol. 89, No. 1273, Entry 18626 (1986). – See Abstr. 012.043.

052.042 **High accuracy on–ground attitude reconstitution for the ESA astrometry mission HIPPARCOS.**
F. Donati, P. Belforte, E. Canuto, D. Carlucci, A. Villa.
Bridge between control science and technology, Vol. 5, p. 2927 – 2932 (1985). Abstr. in Phys. Abstr., Vol. 89, No. 1273, Entry 18627 (1986). – See Abstr. 012.043.

052.043 **Determination of satellite orbits and the global positioning system.**
L. G. Agrotis.
Thesis, Univ. Nottingham, England (1984). Abstr. in Phys. Abstr., Vol. 89, No. 1274, Entry 23737 (1986).

052.044 **Modelling and estimating the orbits of GPS satellites.**
G. Beutler, W. Gurtner, I. Bauersima, R. Langley.
Mitt. Satell.–Beobachtungsstn. Zimmerwald, Nr. 18, p. 1 – 15 (1985). Paper presented at the First International Symposium on Precise Positioning with the Global Positioning System 1985. Preprint.

052.045 **Trajectories and orbital maneuvers for the ISEE–3/ICE comet mission.**
R. Farquhar, D. Muhonen, L. C. Church.
J. Astronaut. Sci., Vol. 33, No. 3, p. 235 – 254 (1985). Abstr. in Phys. Abstr., Vol. 89, No. 1276, Entry 35914 (1986).

052.046 **Alternative gravity–assist sequences for the ISEE–3 escape trajectory.**
D. Muhonen, S. Davis, D. Dunham.
J. Astronaut. Sci., Vol. 33, No. 3, p. 255 – 273 (1985). Abstr. in Phys. Abstr., Vol. 89, No. 1276, Entry 35915 (1986).

052.047 **Optimization of a multiple lunar–swingby trajectory sequence.**
D. W. Dunham, S. A. Davis.
J. Astronaut. Sci., Vol. 33, No. 3, p. 275 – 278 (1985). Abstr. in Phys. Abstr., Vol. 89, No. 1276, Entry 35916 (1986).

052.048 **Accelerometer–enhanced trajectory control for the ISEE–3 halo orbit.**
D. Muhonen, D. Folta.
J. Astronaut. Sci., Vol. 33, No. 3, p. 289 – 300 (1985). Abstr. in Phys. Abstr., Vol. 89, No. 1276, Entry 35917 (1986).

052.049 **ISEE–3/ICE navigation analysis.**
L. Efron, D. K. Yeomans, A. F. Schanzle.
J. Astronaut. Sci., Vol. 33, No. 3, p. 301 – 323 (1985). Abstr. in Phys. Abstr., Vol. 89, No. 1276, Entry 35918 (1986).

052.050 **Determinação de órbitas de satélites terrestres.**
I. Vergueiro Pereira do Couto.
Publ. Obs. Astron. "Prof. Manuel de Barros", Fac. Cienc. Porto, No. 36, 111 pp. (1984).

052.051 **Computational code of interplanetary trajectory of a spacecraft. Application to Galileo mission.**
A. Mizuta, K. Ide, T. Fujiwara.
Mem. Fac. Eng. Nagoya Univ., Vol. 37, No. 1, p. 105 – 116 (1985). Abstr. in Phys. Abstr., Vol. 89, No. 1280, Entry 60578 (1986).

052.052 **Nonsingular elements in description of the motion of small eccentricity and inclination satellites.**
I. Wytrzyszczak.
Celest. Mech., Vol. 38, No. 2, p. 101 – 109 (1986).
Nonsingular elements are used for description of the motion of small eccentricity and inclination satellites. The theory described in the Nacozy and Dallas's (1977) paper was further developed.

052.053 **Classes of orbits in the main problem of satellite theory.**
J. Martínez Alfaro, V. Reglero.
Celest. Mech., Vol. 38, No. 2, p. 163 – 173 (1986).
The authors consider the main problem in satellite theory restricted to the polar plane. For suitable values of the energy the system has two unstable periodic orbits. The authors classify the trajectories in terms of their ultimate behavior with respect to these periodic orbits in: oscillating, asymptotic and capture orbits.

052.054 **On the radial intermediaries and the time transformation in satellite theory.**
R. Cid, S. Ferrer, M. L. Sein–Echaluce.
Celest. Mech., Vol. 38, No. 2, p. 191 – 205 (1986).
Taking advantage of the radial intermediaries and the regularization and linearization methods, the zonal Earth satellite theory is studied in the polar nodal canonical set of variables.

052.055 **Earth albedo and the orbit of Lageos.**
D. P. Rubincam.
Celest. Mech., Vol. 38, No. 3, p. 233 – 296 (1986).
This paper is concerned with an analytical treatment of what is commonly called "the albedo problem", and how it relates to the orbit of the Lageos satellite. The long–period perturbations in the orbit of Lageos satellite due to the earth's albedo have been found using a new analytical formalism. The earth is assumed to be a sphere whose surface diffusely reflects sunlight according to Lambert's law. The formalism is based on spherical harmonics; it produces equations which hold regardless of whether the terminator is seen by the satellite or not. Specializing in the case of a realistic zonal albedo shows that Lageos' orbital semimajor axis changes periodically by only about a centimeter and the eccentricity by two parts in 10^5.

052.056 **Expansion of the perturbation function due to the influence of lunar and solar gravitation on the motion of an artificial earth satellite.**
N. V. Emel'yanov.
Sov. Astron., Vol. 29, No. 6, p. 681 – 684 (1985). English translation of 40.052.060.

052.057 **Improvement of the elements of satellite orbits jointly with refinement of the observation times.**
R. A. Zeinalov (*R. A. Zejnalov*).
Sov. Astron., Vol. 29, No. 6, p. 694 – 698 (1985). English translation of 40.052.061.

052.058 **A design technique for trajectory correction maneuvers.**
G. R. Hintz, C. Chadwick.
J. Astronaut. Sci., Vol. 33, No. 4, p. 429 – 443 (1985). Abstr. in Phys. Abstr., Vol. 89, No. 1281, Entry 67318 (1986).

052.059 **On the problem of the critical inclination and commensurability in the motion of artificial satellites.**
L. Liu, K. A. Innanen.
Acta Astron. Sin., Vol. 27, No. 1, p. 1 – 8 (1986). In Chinese.
The resonant problem resulting from the critical inclination and commensurability in the motion of an artificial satellite is discussed. Considering the non–spherical perturbation of the earth for the averaged model, the quality of the equilibrium solution and the corresponding libration region are given. Finally, the authors compare this solution with the actual motion of an artificial satellite. They find that the resonance plays a stabilizing role in the motion and that the luni–solar perturbations have a significant effect on the orbital resonance of 24–h synchronous satellites.

052.060 **High accuracy on–ground attitude reconstitution for the ESA astrometry HIPPARCOS mission.**
F. Donati, E. Canuto, B. Fassino, P. Belforte.
Manuscr. Geod., Vol. 11, No. 2, p. 115 – 123 (1986).
This paper is concerned with the attitude reconstitution procedure that will be used by the Scientific Consortium FAST for the reduction of the HIPPARCOS data.

052.061 **Effect of the Earth's shadow on the motion of a system of two satellites connected by extensible cable under the influence of solar radiation pressure.**
A. K. P. Singh, H. S. P. Singh.
J. Bihar. Math. Soc., Vol. 9, p. 97 – 105 (1985).

052.062 **The Giotto encounter and post–encounter operations.**
D. E. B. Wilkins, A. Parkes, H. Nye.
ESA Bull., No. 46, p. 66 – 70 (1986). – See Abstr. 003.052.
The Giotto spacecraft is now progressing around the Sun to meet up with the Earth in 1990. The spacecraft has survived the encounter with less damage than had once been feared, but we will have to wait the test of time to see if it can be reactivated on schedule in 1990.

052.063 **The Pathfinder Concept, its implementation and its results.**
R. E. Münch.
ESA Bull., No. 46, p. 75 – 79 (1986). – See Abstr. 003.052.
Thanks in large part to a historic example of interagency collaboration, Giotto was able to achieve a final cometary flyby distance of 600 km on the sunward side, with an uncertainty of just 40 km.

052.064 **Determination of the orbital elements of the INTERCOSMOS – BULGARIA 1300 satellite from laser observations.**
N. V. Emel'yanov.
Nablyud. Iskusstv. Sputnikov Zemli, No. 21, Chast' I, p. 198 – 204 (1982). In Russian.– See Abstr. 012.032.
Orbital parameters of the satellite "Intercosmos – Bulgaria 1300" from laser observations obtained at 4 stations during a 3–month time interval have been determined. The observations were derived in series approximately on 5–day intervals and from each individual series orbital parameters have been derived. The mean deviations of observational values of the satellite's distance from the theoretical ones were about 5 meters. The theoretical accuracy of the obtained orbital parameters is approximately 0.3 second of arc. The evolution of orbital parameters during a 3–month interval is represented on graphs. An analytical theory of the satellite's motion based on the generalized problem of two fixed centers has been applied.

052.065 **Methods and results of ballistic and navigational supply of the INTERCOSMOS – BULGARIA 1300 satellite.**
V. D. Yastrebov, V. D. Glotov, R. A. Dzesov, V. N. Zhukov, V. P. Semenko.
Nablyud. Iskusstv. Sputnikov Zemli, No. 21, Chast' I, p. 205 – 212 (1982). In Russian.– See Abstr. 012.032.
Methods of processing of trajectory measurements of the INTERCOSMOS – BULGARIA 1300 satellite, determination and prediction of its orbit and timing of scientific information obtained from the satellite are discussed.

052.066 **A posteriori estimate of the accuracy of prediction of the INTERCOSMOS – BULGARIA 1300 motion.**
R. R. Nazirov, A. A. Tikhonov.
Nablyud. Iskusstv. Sputnikov Zemli, No. 21, Chast' I, p. 213 – 219 (1982). In Russian.– See Abstr. 012.032.

The slingshot effect: explanation and analogies.
See Abstr. 014.035.

Prep77: a preprocessor of Fortran77.
See Abstr. 021.008.

TRIO interferometer positioning.
See Abstr. 035.093.

Local frames.
See Abstr. 042.003.

New stationary solutions in the problem of motion of a satellite in a central field.
See Abstr. 042.036.

Quaternion methods in problems of the perturbed central motion of a material point. Part 1. General theory. Application to the problem of regularization and to the problem of motion of artificial earth satellites.
See Abstr. 042.057.

Investigation of the efficiency of algorithms of numerical forecast of the motion of celestial bodies in the framework of numerical experiment of the USSR.
See Abstr. 042.063.

The JPL/DSN (*Jet Propulsion Laboratory/Deep Space Network*) J2000 radio reference frame.
See Abstr. 043.023.

Satellite laser ranging and its applications.
See Abstr. 045.001.

On the use of the Kiev geodynamics program complex for analyzing the MERIT laser ranging data.
See Abstr. 045.032.

Astronomical ephemerides – their evolution and role as foundations of realizing space experiments.
See Abstr. 046.026.

Determination of microaccelerations in the orbital stations Salyut 6 and Salyut 7.
See Abstr. 053.014.

14th–order harmonics in the geopotential from analysis of satellite orbits at resonance.
See Abstr. 081.005.

Further comparisons of Earth models by means of lumped coefficients. Part II.
See Abstr. 081.009.

A simple model for the gravitational field of the earth.
See Abstr. 081.027.

Relative motion of the geodynamic twin–satellite.
See Abstr. 081.041.

Anomalous perigee shift and eccentricity variation due to air drag in the reentry phase.
See Abstr. 082.048.

Dynamical parameters of the upper atmosphere from an orbital analysis of the satellites INTERCOSMOS 18 and MAGION.
See Abstr. 082.052.

Several results of drag determination for EXPLORER 39 by the parallactic method.
See Abstr. 082.053.

053 Artificial Satellites, Space Probes

053.001 **Photoelectric observations of the Intercosmos–Bulgaria–1300 satellite.**
M. V. Bratijchuk, A. B. Gvardionov, V. P. Epishev, I. F. Najbauehr, O. E. Starodubtseva.
Kinematika Fiz. Nebesn. Tel, Tom 2, No. 1, p. 60 – 65 (1986). In Russian. English translation in Kinematics Phys. Celest. Bodies.
From the results of complex observations of the satellite Intercosmos–Bulgaria–1300 the character of light scattering by different parts of its surface has been determined. The analytical expressions of phase functions for observations obtained from March 1982 to October 1983 are given.

053.002 **Visual observations, horizontal and equatorial (1950.0) coordinates of Bolgariya 1300.**
Rezul't. Nablyud. Iskusstv. Sputnikov Zemli, Vyp. 107 (247), 70 pp. (1985). In Russian.
Concerning Bolgariya 1300–1981–75–1. January – November 1983.

053.003 **Visual observations, equatorial (1950.0) and horizontal coordinates of Bolgariya 1300.**
Rezul't. Nablyud. Iskusstv. Sputnikov Zemli, No. 108 (248), 84 pp. (1985). In Russian.
Concerning Bolgariya 1300–1981–75–1. August 1983 – October 1984.

053.004 **Geostationary satellites.**
G. E. Taylor.
Sky Telesc., Vol. 71, No. 6, p. 557 (1986).

053.005 **Observing artificial satellites.**
D. King–Hele, R. Eberst.
Sky Telesc., Vol. 71, No. 5, p. 457 – 463 (1986).

053.006 **Photographing Earth satellites.**
P. D. Maley.
Sky Telesc., Vol. 71, No. 6, p. 563 – 567 (1986).

053.007 **Satellite Digest–188.**
R. D. Christy.
Spaceflight, Vol. 28, No. 1, p. 41 (1986).
1986 January listing of satellite and spacecraft launches.

053.008 **Satellite Digest–189.**
R. D. Christy.
Spaceflight, Vol. 28, No. 2, p. 56 – 57 (1986).
1986 February listing of satellite and spacecraft launches.

053.009 **Satellite Digest–190.**
R. D. Christy.
Spaceflight, Vol. 28, No. 3, p. 118 – 119 (1986).
1986 March listing of satellite and spacecraft launches.

053.010 **Satellite Digest–191.**
R. D. Christy.
Spaceflight, Vol. 28, No. 4, p. 165 (1986).
1986 April listing of satellite and spacecraft launches.

053.011 **Satellite Digest–192.**
R. D. Christy.
Spaceflight, Vol. 28, No. 5, p. 216 (1986).
1986 May listing of satellite and spacecraft launches.

053.012 **Satellite Digest–193.**
R. D. Christy.
Spaceflight, Vol. 28, No. 6, p. 262 (1986).
1986 June listing of satellite and spacecraft launches.

053.013 **Results of the observational programme "INTERCOSMOS – BULGARIA 1300".**
N. I. Georgiev, Ch. B. Dokev, Ya. G. Chapanov, S. K. Tatevyan, P. E. Ehl'yasberg, V. I. Adas'ko.
Nablyud. Iskusstv. Sputnikov Zemli, No. 21, Chast' I, p. 183 – 189 (1982). In Russian.– See Abstr. 012.032.
The results obtained by the programme of observations of the artificial Earth satellite "INTERCOSMOS – BULGARIA 1300" are discussed. Information about the avail-

able observations in the data base of the Central Laboratory for Geodesy of the Bulgarian Academy of Sciences and of the Astronomical Council of the Academy of Sciences of the USSR is included.

053.014 **Determination of microaccelerations in the orbital stations Salyut 6 and Salyut 7.**
V. A. Sarychev, M. Yu. Belyaev, V. V. Sazonov, T. N. Tyan.
Kosm. Issled., Tom 24, Vyp. 3, p. 337 – 344 (1986). In Russian. English translation in Cosm. Res.

053.015 **The structure of Planet–A, a Halley's comet explorer.**
K. Miura, H. Hasegawa, K. Matsubara, K. Kawakami.
Mitsubishi Denki Giho, Vol. 59, No. 9, p. 642 – 644 (1985). In Japanese. Abstr. in Phys. Abstr., Vol. 89, No. 1281, Entry 67317 (1986).

***Space Station program*, description, applications, and opportunities.**
See Abstr. 003.215.

The Giotto experiments.
See Abstr. 035.098.

A high–precision determination of angular coordinates of geostationary artificial satellites with a wide–angle astrograph with special equipment.
See Abstr. 036.042.

A method for interpolation of the angles of observation of artificial earth satellites.
See Abstr. 036.057.

Instrument and method for determination of high–precision coordinates of geostationary artificial satellites.
See Abstr. 036.218.

Space astronomy in Japan.
See Abstr. 051.002.

The International Cometary Explorer mission to comets Giacobini–Zinner and Halley: an update.
See Abstr. 051.037.

The Giotto project: from early concepts to flight model.
See Abstr. 051.102.

Programmes under development and operations/Programmes en cours de réalisation et d'exploitation.
See Abstr. 051.103.

Theoretical Astrophysics

061 General Aspects (Nucleosynthesis, Elementary Particles, Neutrino Astronomy, etc.)

061.001 **^{148,150}Sm: a test for *s*–process nucleosynthesis.**
R. R. Winters, F. Käppeler, K. Wisshak,
A. Mengoni, G. Reffo.
Astrophys. J., Vol. 300, No. 1, p. 41 – 55 (1986).
The authors have measured to a precision of $\sim$4.5% the neutron capture cross sections of ^{148,149,150}Sm over the neutron energy range $4 < E < 250$ keV. These data, in conjunction with calculated cross sections for ^{147}Nd and ^{147,148}Pm, are used to establish a set of Maxwellian–averaged cross sections useful for investigation of branchings in the *s*–process of stellar nucleosynthesis.

061.002 **Neutrino capture in ^{69,71}Ga and the detection of solar and galactic neutrinos.**
K. Grotz, H. V. Klapdor, J. Metzinger.
Astron. Astrophys., Vol. 154, No. 1/2, p. L1 – L4 (1986).
Calculations of the neutrino–capture cross sections of ^{69,71}Ga based on a microscopic treatment of the Gamow Teller matrix elements are presented. A strong enhancement of the cross section for high energetic neutrinos is found compared to previous phenomenological estimates. Implications for the detection of solar and galactic neutrinos by a gallium neutrino detector are discussed.

061.003 **The screening of photodisintegration reactions.**
R. Mochkovitch, K. Nomoto.
Astron. Astrophys., Vol. 154, No. 1/2, p. 115 – 118 (1986). With a correction in Vol. 157, No. 2, p. 403 (1986).
The authors calculate the rates of photodisintegration reactions in a strongly coupled plasma for the non–resonant and resonant cases. The screening produces a decrease of the potential barrier between two interacting nuclei together with an increase of the energy threshold of the reaction. The authors show that these two effects nearly cancel out, the final result being that the photodisintegration rates are slightly reduced by the screening effects. This is briefly discussed with a special attention to the photodisintegration of ^{20}Ne in the degenerate O + Ne + Mg core of evolved 8 – 10 $M_\odot$ stars.

061.004 **Gauged $N = 8$ supergravity in $d = 11$ dimensions.**
B. de Wit.
Particles and gravity, p. 11 – 22 (1984). – See Abstr. 012.004.
The embedding of pure gauged $N = 8$ supergravity in $d = 11$ supergravity is discussed. Explicit results are presented for two SO(7) invariant field configurations. These configurations contain the supersymmetric SO(8) invariant solution and two SO(7) invariant solutions of $d = 11$ supergravity.

061.005 **A relation between $N = 2$ supergravities in 10 and 4 dimensions.**
G. Chapline, R. Slansky.
Particles and gravity, p. 43 – 49 (1984). – See Abstr. 012.004.
The authors list the 4–dimensional supergravities that are derived from the $N = 2$, 10–dimensional supergravities using the method of coset–space dimensional reduction. The spectrum of each 4–dimensional theory consists of those supermultiplets that are invariant under the symmetry transformations of the coset space.

061.006 **Isometries as probes of the extra dimensions.**
F. Mansouri, L. Witten.
Particles and gravity, p. 51 – 66 (1984). – See Abstr. 012.004.
It is suggested that the consequences of dependence on the extra dimensions in Kaluza–Klein theories can best be studied in terms of isometries and their spontaneous symmetry breaking. The authors show explicitly how a dimensional reduction can be carried out in which isometries remove the traces of the extra dimensions from the reduced theory.

061.007 **Towards unification of elementary particle physics and cosmology in 10–dimensions.**
G. Chapline, G. Gibbons.
Particles and gravity, p. 121 – 128 (1984). – See Abstr. 012.004.
Ten–dimensions seems to be a unique setting for unifying at the classical level cosmology and elementary particle physics. Some interesting results along these lines are obtained starting with a Yang–Mills coupled to supergravity theory in 10–dimensions. However, further progress will require finding an underlying quantum theory.

061.008 **The group manifold approach to unified gravity.**
T. Regge.
Supersymmetry, supergravity and related topics, p. 73 – 108 (1985). – See Abstr. 012.005.
This paper discusses the application of the concept of superforms, originating from the theory of Lie groups, in constructing unified theories of gravity.

061.009 **Aspects of supergravity theories.**
G. W. Gibbons.
Supersymmetry, supergravity and related topics, p. 123 – 181 (1985). – See Abstr. 012.005.
Contents: (1) An introduction to anti–de Sitter space. (2) Black hole solitons in ungauged extended supergravity. (3) Supersymmetric monopoles in Kaluza–Klein theory.

061.010 **$d = 4$ supergravity from $d = 11$: conjectures revisited.**
M. J. Duff.
Supersymmetry, supergravity and related topics, p. 182 – 213 (1985). – See Abstr. 012.005.
This paper examines the relationship between the massless sector of the theory obtained by Kaluza–Klein compactification of $d = 11$ supergravity on the round S^7 and the $N = 8$ gauged $SO(8)$ supergravity in $d = 4$.

061.011 **Vacuum stability in supergravity.**
D. Z. Freedman.
Supersymmetry, supergravity and related topics, p. 214 – 225 (1985). – See Abstr. 012.005.
A discussion of the vacuum stability problem in supergravity theories is given. The definition of energy in a gravitational field theory, and the small fluctuation stability argument for anti de Sitter space are outlined. The notion of Killing spinors and the related proof of global stability are discussed briefly.

061.012 **Particle acceleration by magnetic reconnection.**
W. H. Sorrell.
Cosmic radiation in contemporary astrophysics, p. 79 – 85 (1986). – See Abstr. 012.006.

Dissipation of magnetic energy during field–line reconnection could lead to runaway acceleration of charged particles to high energies. The particle energy gain is directly proportional to the product of the strength of the pre–merging magnetic field and the plasma resistivity. Particle acceleration to relativistic energies is possible provided that plasma turbulence or some form of instability leads to a drastic reduction of the classical (Coulomb) conductivity.

061.013 **The *s*–process branching at ^{85}Kr.**
G. Walter, H. Beer, F. Käppeler, R.–D. Penzhorn.
Astron. Astrophys., Vol. 155, No. 2, p. 247 – 255 (1986).

The Maxwellian average neutron capture cross sections of ^{86}Kr, ^{85}Rb and ^{87}Rb have been measured at $kT = 25$ keV by activation. These data together with recently published neutron capture cross sections of Kr and Sr isotopes allow for a determination of the neutron density of the weak *s*–process component in the framework of the classical *s*–process model with two independent fluence distributions. The *s*–process abundances from this study are compared with the pure *s*–process krypton discovered in primitive meteorites.

061.014 **Limits on mass of missing mass.**
J. Maddox.
Nature, Vol. 319, No. 6056, p. 717 (1986).

Of the particles whose mass may account for the missing mass of the universe, those known as axions are the most shadowy. What is known of stellar evolution surprisingly helps to define their properties.

061.015 **Inflation and shadow matter.**
L. M. Krauss, A. H. Guth, D. N. Spergel,
G. B. Field, W. H. Press.
Nature, Vol. 319, No. 6056, p. 748 – 751 (1986).

Recent work on superstring theories has prompted interest in 'shadow matter', exotic matter which interacts only gravitationally with normal matter. Gross et al. have constructed a superstring theory resulting in a gauge group $E_8 \times E_8$. Such a theory could result, at low energies, in the existence of two sectors: an 'observed' sector associated with all familiar particles and interactions, and another 'hidden' sector, previously discussed in various low–energy supersymmetry models. The authors demonstrate that if, in the early Universe, an inflationary phase is associated with the breaking of one of the symmetries in the $E_8 \times E_8$ theory, this strongly constrains the physics of both sectors if shadow matter is to be the missing mass in the Universe.

061.016 **Direct laboratory determination of the ^{187}Re half–life.**
M. Lindner, D. A. Leich, R. J. Borg, G. P. Russ,
J. M. Bazan, D. S. Simons, A. R. Date.
Nature, Vol. 320, No. 6059, p. 246 – 248 (1986).

The long–lived, naturally occurring radionuclide ^{187}Re is important in geochemistry and cosmology as a nucleochronometer. The authors report here a half–life of $(4.35 \pm 0.13) \times 10^{10}$yr, based on the growth of ^{187}Os over a 4–yr period into a large source of osmium–free rhenium. Since this result agrees with the best geochemically determined values, no significant revision of the present galactic age limits based on the geochemical values is necessary. The agreement at the 5% level of the decay rates determined on laboratory and meteorite–age timescales places a tight constraint on cosmological models in which fundamental "constants" are allowed to vary.

061.017 **A new estimate of the ^{19}Ne(p,γ)^{20}Na and ^{15}O(α,γ)^{19}Ne reaction rates at stellar energies.**
K. Langanke, M. Wiescher, W. A. Fowler, J. Görres.
Astrophys. J., Vol. 301, No. 2, p. 629 – 633 (1986).

The authors have improved the Wallace and Woosley estimates of the ^{19}Ne(p,γ)^{20}Na and ^{15}O(α,γ)^{19}Ne reaction rates at stellar energies by using the most recent experimental and theoretical data and a firmer theoretical background. They find a considerable increase in the ^{19}Ne(p,γ)^{20}Na reaction rate at temperatures $T_9 \leqslant 0.3$, which may lead to a greater production of intermediate mass elements in the rp–process in stellar evolution.

061.018 **On the nucleosynthesis beyond the iron peak.**
D. A. Ptitsyn, V. M. Chechetkin.
Nauchn. Inf., Vyp. 56, p. 38 – 41 (1984). In Russian.

A process of synthesis of nuclei beyond the iron peak at high temperature, density and neutron excess is proposed. It is shown that there exist two groups of nuclei as a result of the breakdown of the detailed balance between (p,γ) and (γ,p) reactions for high–Z elements. Relative abundances of the main group are described by the nuclear statistical equilibrium, whereas for heavier nuclei a kinetic equilibrium of (γ,p) reactions and β–processes turns out to occur.

061.019 **Constraints in conformal simple supergravity.**
P. van Nieuwenhuizen.
From SU(3) to gravity, p. 369 – 382 (1985). – See Abstr. 003.014.

All constraints of conformal simple ($N = 1$) supergravity are derived from closure of the gauge algebra alone. In particular, closure on the dilaton field replaces previous derivations of the constraint which involves the Lorentz curvature.

061.020 **York's cosmic time versus proper time as relevant to changes in the dimensionless "constants", *K*–meson decay, and the unity of black hole and big crunch.**
A. Qadir, J. A. Wheeler.
From SU(3) to gravity, p. 383 – 394 (1985). – See Abstr. 003.014.

Contents: (1) What measure of time governs change? (2) Change of "constants" with time? (3) "Collapse syndrome" puts a limit of the order of 5×10^{-37} per year on any rate of change of electron properties with *proper* time. (4) *K*–meson decay. (5) Conclusion.

061.021 **Superstrings and supersymmetry.**
A. De Rújula.
Nature, Vol. 320, No. 6064, p. 678 (1986).

061.022 **Quantum tunnelling towards an exploding universe?**
M. J. Perry.
Nature, Vol. 320, No. 6064, p. 679 (1986).

061.023 **The magnetic monopoles content of galactic nuclei, quasars, stars and planets.**
Q.–h. Peng, Z.–y. Li, D.–y. Wang.
Structure and evolution of active galactic nuclei, p. 659 – 661 (1986). – See Abstr. 012.011.

There are two possible sources of monopoles in celestial bodies: (1) They may be preserved in the collapse of the objects from original clouds or nebulae. (2) They may be captured in the time after formation of the objects. The main goal of this work is to estimate the efficiency of the first process.

061.024 **A monopole model for galactic nuclei.**
Q.–h. Peng, D.–y. Wang, Z.–y. Li.
Structure and evolution of active galactic nuclei, p. 663 – 667 (1986). – See Abstr. 012.011.

The aim of this text is to discuss the possibility for setting up a model of galactic nuclei and quasars not based upon black holes.

061.025 **Conditions for the dynamical *r*–process.**
O. N. Sinyukova, S. S. Filippov, V. M. Chechetkin.
Inst. prikl. mat. Akad. Nauk SSSR. Prepr., No. 67, 22 pp. (1985). In Russian. Abstr. in Ref. Zh., 51. Astron., 1.51.321 (1986).

061.026 **The concept of matter in contemporary science.**
A. Balklavs.
Zvaigžnota Debess, 1984. Gada Pavasaris, p. 4 – 8, Gada Vasara, p. 2 – 6 (1984). In Latvian.

061.027 **Effects of metal-dependent stellar models on the yield of nitrogen.**
C. Forieri.
Spectral evolution of galaxies, p. 473 – 475 (1986). – See Abstr. 012.015.
The relation between nitrogen and oxygen is studied in the case of variable yields. The influence of the chemical composition on stellar evolution is taken into account. The author found that new models mimick the primary origin of nitrogen.

061.028 **Using an X-ray color-color diagram to search for hot neutron stars and set a limit on the magnetic monopole flux.**
F. A. Córdova, T. W. Bell, E. W. Kolb.
Bull. Am. Astron. Soc., Vol. 17, No. 4, p. 906 (1985). Abstract. – See Abstr. 010.062.

061.029 **A grand superspace of unified field theories.**
A. E. Fischer.
Gen. Relativ. Gravitation, Vol. 18, No. 6, p. 597 – 608 (1986).
A grand superspace is proposed as the phase space for gauge field theories with a fixed structure group G over a fixed space-time manifold M. This superspace incorporates all principal fiber bundles with these data. This phase space is the space of isomorphism classes of all connections on all G-principal fiber bundles over M (fixed G and M). The justification for choosing this grand superspace for the phase space is that the space-time and the structure group are determinants of the physical theory, but the principal fiber bundle with the given G and M is not.

061.030 **Isotopic compositions of bismuth, lead, thallium, and mercury from mini *r*-processing.**
D. Heymann, K. Liffman.
Meteoritics, Vol. 21, No. 1, p. 95 – 108 (1986).
Meteorites contain isotopically strange Xe and Kr from a "mini *r*-process" in stars. There can be no doubt that mercury is associated with these inert gases at their source. The most probable isotopic composition of this Hg has not been calculated. Such calculations are reported in this paper along with calculations for the neighboring elements Tl and Pb. Among the principal objectives of the present work is the comparison of the deduced compositions of these elements from the mini *r*-process with their normal solar-system composition to learn whether the two are sufficiently different for successful detection of the mini *r*-components.

061.031 **Thermal diffusion in ionized gases and planetary and solar processes.**
L. Monchick.
Johns Hopkins APL Tech. Dig., Vol. 7, No. 1, p. 39 – 41 (1986).
Thermal diffusion, which has been presumed to be an important process in ionized gases such as those that occur in planetary ionospheres and in stellar interiors, is recalculated, taking into account two effects that have previously been ignored. The first effect is that quantum mechanics must be used to calculate collision cross sections at high temperatures as well as at low. The second effect is that the sea of charges surrounding two moving charges does not instantaneously follow the relative motion of the two charges. In ionized gases that are characteristic of stellar interiors, the newer estimates of thermal diffusion differ substantially from the estimates produced by the older theory.

061.032 **How does the SU(2) ⊗ U(1) symmetry break in the early universe?**
K. Takahashi.
Phys. Rev. Lett., Vol. 56, No. 1, p. 7 – 10 (1986).

061.033 **Astrophysical constraints on the couplings of axions, majorons, and familons.**
D. S. P. Dearborn, D. N. Schramm, G. Steigman.
Phys. Rev. Lett., Vol. 56, No. 1, p. 26 – 29 (1986).
The spontaneous breaking of a global symmetry leads to the existence of Nambu-Goldstone bosons, which through their coupling to electrons and/or photons, can transport energy from the cores of stars and affect significantly the course of stellar evolution. The authors find by following in detail the evolution of stars that if the couplings to electrons and/or photons is too strong, helium never ignites – in contradiction with the observational evidence.

061.034 **Cosmic quarkonium: a probe of dark matter.**
M. Srednicki, S. Theisen, J. Silk.
Phys. Rev. Lett., Vol. 56, No. 3, p. 263 – 265 (1986).
If the galactic halo is composed of heavy (several gigaelectronvolts), weakly interacting particles, pair annihilation into a heavy-antiquark bound state plus a monochromatic photon can produce potentially observable sharp peaks in the diffuse cosmic-γ-ray background.

061.035 **Toward a proof of entropy increase in the presence of quantum black holes.**
R. D. Sorkin.
Phys. Rev. Lett., Vol. 56, No. 18, p. 1885 – 1888 (1986).

061.036 **A model-theory for tachyons in two dimensions.**
E. Recami, W. A. Rodrigues Jr.
Sir Arthur Eddington Centenary Symposium, Vol. 3: Gravitational radiation and relativity, p. 151 – 203 (1986). – See Abstr. 012.036.
This paper first presents a formulation of special relativity describing both particles and their antiparticles and then discusses an "extended relativity" theory of tachyons in two dimensions.

061.037 **Physical symmetry groups and associated bundles in field theory.**
A. Crumeyrolle.
Sir Arthur Eddington Centenary Symposium, Vol. 3: Gravitational radiation and relativity, p. 204 – 223 (1986). – See Abstr. 012.036.

061.038 **Quantum equivalence principle.**
A. V. Osipova.
Sir Arthur Eddington Centenary Symposium, Vol. 3: Gravitational radiation and relativity, p. 236 – 244 (1986). – See Abstr. 012.036.
The geometrical structure of physical objects in general relativity and gauge theories is discussed in order to formulate the relation of internal symmetry quantum numbers to the geometry of the space-time manifold. Some arguments are given for the use of the Borel-Hirzebruch description of characteristic classes for vector bundles.

061.039 **Eddington's numbers and quantum gravity.**
H.-H. von Borzeszkowski, H.-J. Treder.
Sir Arthur Eddington Centenary Symposium, Vol. 3: Gravitational radiation and relativity, p. 245 – 270 (1986). – See Abstr. 012.036.
The Einstein-Maxwell theory does not unify gravitation and electromagnetism; it harmonizes however the two phenomena. Following Eddington's discussion, one must incorporate cosmological features into local physics to arrive at a unification. Reviving this discussion, one finds that the difficulties arising for a unified gravito-electromagnetic theory are the same which appear on the way to quantum gravity.

061.040 **Principle-theoretic approach of Kondo and construction-theoretic formalism of gauge theories.**
L. C. Jain.
Sir Arthur Eddington Centenary Symposium, Vol. 3: Gravitational radiation and relativity, p. 271 – 282 (1986). – See Abstr. 012.036.

061.041 **Dominance of B^{10} or C^{12} nuclear production during gravitational collapse in the stellar interior?**
K. L. Narayana.
Sir Arthur Eddington Centenary Symposium, Vol. 3: Gravitational radiation and relativity, p. 298 – 314 (1986). – See Abstr. 012.036.

061.042 **Physical processes of gravitons and gravy quarks in the interior of cosmod and supergravity cosmic stellar entities.**
K. L. Narayana.
Sir Arthur Eddington Centenary Symposium, Vol. 3: Gravitational radiation and relativity, p. 315 – 325 (1986). – See Abstr. 012.036.

061.043 **Tachyonic dyons.**
B. S. Rajput.
Sir Arthur Eddington Centenary Symposium, Vol. 3: Gravitational radiation and relativity, p. 326 – 374 (1986). – See Abstr. 012.036.
The author has constructed a self consistent Lorentz covariant quantum field theory for spin–1 and spin–1/2 particles carrying electric and magnetic charges without using controvertial string variables. He derives the gauge invariant properties of monopoles by generalizing classical electrodynamics to include Abelian and non–Abelian gauge groups.

061.044 **Does a free tachyon exist?**
T. Chang.
Sir Arthur Eddington Centenary Symposium, Vol. 3: Gravitational radiation and relativity, p. 431 – 436 (1986). – See Abstr. 012.036.
According to a new space–time theory with a privileged frame, the introduction of superluminous particles is natural. The key problem is how to discover a tachyon in experiment. After studying the properties of the neutrino, it is suggested that the neutrino is a probable candidate for a free tachyon.

061.045 **Is nature supersymmetric?**
H. E. Haber, G. L. Kane.
Sci. Am., Vol. 254, No. 6, p. 42 – 50 (1986).
Supersymmetry could represent the next step in the quest for a few simple laws that explain the nature of matter. Physicists have been seeking evidence to test the theory.

061.046 **Radiation and energy loss by monopole in cosmic plasma.**
X.–q. Li.
Astrophys. Space Sci., Vol. 123, No. 1, p. 125 – 135 (1986).
Radiation and energy loss by monopole in cosmic plasma are investigated in detail. The general formulae for monopole radiation are derived; the Čerenkov power and spectra radiation differing from those of charged particles are found; and its energy loss in cosmic plasma and metal is obtained. It is pointed out that it is possible to examine monopole in virtue of the Čerenkov radiation in ordinary medium.

061.047 **Indirect experimental evidence against the tachyons.**
A. Mészáros.
Astrophys. Space Sci., Vol. 123, No. 2, p. 409 – 410 (1986).
It is noted that the empirically confirmed existence of the massive $W^{\pm}$ and Z^0 particles is in fact an indirect experimental evidence against the existence of tachyons.

061.048 **Numerical nucleosynthesis results for Bianchi type–V cosmological models.**
M. S. Madsen, D. R. Matravers, D. Milun.
Astrophys. Space Sci., Vol. 123, No. 2, p. 411 – 414 (1986).
Results of numerical calculations of helium–4 and deuterium abundances in a wide range of Bianchi type–V locally rotationally–symmetric cosmolgical models are presented. The tabulated element abundances show that these models generically yield helium fractions which are unacceptably large compared with the available observational data.

061.049 **Observation of an unexplained event from a magnetic monopole detector.**
A. D. Caplin, M. Hardiman, M. Koratzinos, J. C. Schouten.
Nature, Vol. 321, No. 6068, p. 402 – 406 (1986).
A single event was recorded by a 0.18 m^2 inductive magnetic monopole detector, in a total effective observation time of 8,242 hours. None of the monitoring channels that record vibration, radiofrequency interference, and so on, indicated any extraneous cause for this event. An astrophysical estimate of the maximum number flux of monopoles suggests that the chances of a monopole having caused the event are less than 1 in 10^3. Alternative explanations for the candidate are considered, but non of them seem likely.

061.050 **Towards the geometrization of physics.**
G. A. J. Sparling.
Nature, Vol. 321, No. 6068, p. 417 – 419 (1986).
The concept of symmetry breaking may be expressed mathematically by the reduction of a structure group, for example, choosing a point on a sphere reduces its symmetry group from the group of all rotations to the subgroup of rotations fixing that point. A systematic analysis of Penrose's twistor equation has revealed that twistor theory embodies a truth of nature: that interactive gravitational field theories in space–time correspond to such a symmetry breakdown – a breakdown that occurs, not in space–time, but in twistor space – the space of null geodesics of space–time. In particular, twistor theory lies at the heart of supergravity.

061.051 **On the many–worlds interpretation of quantum theory.**
V. F. Mukhanov.
Proceedings of the Third Seminar on Quantum Gravity, p. 16 – 38 (1985). – See Abstr. 012.038.

061.052 **Supergrand unified composite model in pregeometry.**
H. Terazawa.
Proceedings of the Third Seminar on Quantum Gravity, p. 236 – 245 (1985). – See Abstr. 012.038.
The supergrand unified theory of composite particles and fields in pregeometry is reviewed. It effectively reproduces not only unified gauge theories for the strong and electroweak forces of leptons and quarks but also general relativity for gravity at low energies.

061.053 **Present state of quantum supergravity.**
E. S. Fradkin, A. A. Tseytlin.
Proceedings of the Third Seminar on Quantum Gravity, p. 303 – 307 (1985). – See Abstr. 012.038.
It is outlined that the most interesting candidates for a fundamental theory consistently incorporating quantized gravity are at present the finite conformal supergravities in four dimensions and superstring theories in ten dimensions.

061.054 **Geometrical origin of new unconstrained superfields.**
A. A. Rosly, A. S. Schwarz.
Proceedings of the Third Seminar on Quantum Gravity, p. 308 – 324 (1985). – See Abstr. 012.038.
The authors show how the unconstrained formulations of $N = 2$ and $N = 3$ super gauge theories can be obtained from geometrical considerations. An analogous formulation of supersymmetric hyper–Kähler sigma model is given.

061.055 **Geometry of 11–dimensional supergravity.**
R. E. Kallosh.
Proceedings of the Third Seminar on Quantum Gravity, p. 325 – 339 (1985). – See Abstr. 012.038.
A new geometrical form of a $d = 11$ supergravity Lagrangian is suggested. The 3–index gauge potential in this theory is a supercovariant torsion tensor.

061.056 **De Sitter supergravity and generalized Lie superalgebras.**
M. A. Vasiliev (*M. A. Vasil'ev*).
Proceedings of the Third Seminar on Quantum Gravity, p. 340 – 362 (1985). – See Abstr. 012.038.

061.057 **Higher conservation laws for supersymmetric gauge theories.**
I. Ya. Aref'eva, I. V. Volovich.
Proceedings of the Third Seminar on Quantum Gravity, p. 363 – 375 (1985). – See Abstr. 012.038.

061.058 **Kaluza–Klein theories and spontaneous compactification mechanisms of extra space dimensions.**
D. P. Sorokin, V. I. Tkach, D. V. Volkov.
Proceedings of the Third Seminar on Quantum Gravity, p. 376 – 392 (1985). – See Abstr. 012.038.

061.059 **The chiral anomaly in conformal and ordinary simple supergravity in Fujikawa's approach.**
P. H. Frampton, D. R. T. Jones, P. van Nieuwenhuizen, S. C. Zhang.
Proceedings of the Third Seminar on Quantum Gravity, p. 393 – 409 (1985). – See Abstr. 012.038.
In this contribution the authors reobtain the chiral anomaly of simple ordinary supergravity by means of Fujikawa's method as well as by the Pauli–Villars method. Then they present, as a new result, the axial gravitational spin 3/2 anomaly for simple conformal supergravity.

061.060 **New heterotic string theories in uncompactified dimensions < 10.**
K. S. Narain.
Rutherford Appleton Lab., Rep., RAL–85–097, 12 pp. (1985).

061.061 **Bose condensation of particle–antiparticle systems.**
K. P. Sinha.
Pramāna, Vol. 25, No. 4, p. 467 – 472 (1985). Abstr. in Phys. Abstr., Vol. 89, No. 1273, Entry 13517 (1986).

061.062 **The axion problem.**
K. Lanius.
Ann. Phys. (Leipzig), Vol. 42, No. 4 – 6, p. 476 – 484 (1985). In German. Abstr. in Phys. Abstr., Vol. 89, No. 1273, Entry 13776 (1986).

061.063 **Neutron capture process in astrophysics.**
G. J. Mathews, R. A. Ward.
Rep. Prog. Phys., Vol. 48, No. 10, p. 1371 – 1418 (1985). Abstr. in Phys. Abstr., Vol. 89, No. 1273, Entry 18655 (1986).

061.064 **Saving the axions in superstring models.**
K. Yamamoto.
Phys. Lett. B, Vol. 161B, No. 4 – 6, p. 289 – 293 (1985). Abstr. in Phys. Abstr., Vol. 89, No. 1274, Entry 19442 (1986).

061.065 **Quantum fluctuation effects due to baryon resonances in nuclear mean field theory.**
T. Nakai, S. Takagi.
Prog. Theor. Phys., Vol. 74, No. 2, p. 315 – 326 (1985). Abstr. in Phys. Abstr., Vol. 89, No. 1274, Entry 19662 (1986).

061.066 **The fate of relic neutrinos.**
J. Balog, G. Marx.
Acta Phys. Hung., Vol. 58, No. 1 – 2, p. 35 – 38 (1985). Abstr. in Phys. Abstr., Vol. 89, No. 1274, Entry 23739 (1986).

061.067 **Some considerations on the origin of nitrogen.**
F. Matteucci.
ESO Sci. Prepr., No. 412, 25 pp. (1986). Submitted to Mon. Not. R. Astron. Soc.

061.068 **At what energy does gravity unite with grand unified theories?**
C. Sivaram.
Indian Inst. Astrophys., Prepr., No. 208, 3 pp. (1986).

061.069 **A boundary layer model for the Kaluza–Klein monopole.**
G. Kunstatter, D. E. Vincent.
Phys. Lett. B, Vol. 163B, No. 1 – 4, p. 96 – 100 (1985). Abstr. in Phys. Abstr., Vol. 89, No. 1275, Entry 24198 (1986).

061.070 **Bianchi–Kantowski–Sachs solutions of $N = 1$ supergravity in $d = 11$ dimensions for a torus space.**
D. Lorenz–Petzold.
Classical Quantum Gravity, Vol. 2, No. 6, p. 829 – 839 (1985). Abstr. in Phys. Abstr., Vol. 89, No. 1275, Entry 24223 (1986).

061.071 **Cosmological instability in ten–dimensional supergravity.**
D. Bailin, A. Love.
Phys. Lett. B, Vol. 163B, No. 1 – 4, p. 135 – 139 (1985). Abstr. in Phys. Abstr., Vol. 89, No. 1275, Entry 24227 (1986).

061.072 **Geometrical aspects of magnetic monopoles.**
P. Bandyopadhyay, S. Roy.
Int. J. Theor. Phys., Vol. 24, No. 10, p. 987 – 1000 (1985). Abstr. in Phys. Abstr., Vol. 89, No. 1275, Entry 24598 (1986).

061.073 **Wave equation for a magnetic monopole.**
G. Lochak.
Int. J. Theor. Phys., Vol. 24, No. 10, p. 1019 – 1050 (1985). Abstr. in Phys. Abstr., Vol. 89, No. 1275, Entry 24599 (1986).

061.074 **Monopole abundance in the Solar System and the intrinsic heat in the Jovian planets.**
J. Arafune, M. Fukugita, S. Yanagita.
Phys. Rev. D, Vol. 32, No. 10, p. 2586 – 2590 (1985). Abstr. in Phys. Abstr., Vol. 89, No. 1275, Entry 24612 (1986).

061.075 **Cosmological constraints on heavy unstable particles.**
K. Choi, K. Kang, J. E. Kim.
Phys. Rev. D, Vol. 32, No. 10, p. 2822 – 2827 (1985). Abstr. in Phys. Abstr., Vol. 89, No. 1275, Entry 24613 (1986).

061.076 **Beta–decay half–lives of new neutron–rich chromium–to–nickel isotopes and their consequences for the astrophysical r–process.**
U. Bosch, W.–D. Schmidt–Ott, P. Tidemand–Petersson, E. Runte, W. Hillebrandt, M. Lechle, F.–K. Thielemann, R. Kirchner, O. Klepper, E. Roeckl, K. Rykaczewski, D. Schardt, N. Kaffrell, M. Bernas, P. Dessagne, W. Kurcewicz.
Phys. Lett. B, Vol. 164B, No. 1 – 3, p. 22 – 26 (1985). Abstr. in Phys. Abstr., Vol. 89, No. 1275, Entry 24840 (1986).

061.077 **A monopole model for galactic nuclei.**
Q.–h. Peng, D.–y. Wang, Z.–y. Li.
Kexue Tongbao, Vol. 30, No. 8, p. 1056 – 1061 (1985). Abstr. in Phys. Abstr., Vol. 89, No. 1275, Entry 30429 (1986).

061.078 **Thermonuclear rates of the $^{28}Si(p,\gamma)$ reaction.**
M. Kicinska–Habior, T. Matulewicz.
J. Phys., Vol. 46, No. 11, p. 1881 – 1885 (1985). Abstr. in Phys. Abstr., Vol. 89, No. 1276, Entry 31289 (1986).

061.079 **Constraints to the radiative lifetime of a light neutral fermion in the galactic halo from IUE.**
G. Auriemma, F. Bochicchio, A. Frenkel, F. Massa.
Nuovo Cimento C, Vol. 8C, Ser. 1, No. 4, p. 398 – 409 (1985). Abstr. in Phys. Abstr., Vol. 89, No. 1276, Entry 35963 (1986).

061.080 **And if neutrinos do have mass!**
J. J. Simpson.
Recherche, Vol. 16, No. 171, p. 1408 – 1410 (1985). In French. Abstr. in Phys. Abstr., Vol. 89, No. 1276, Entry 36388 (1986).

061.081 **Massive–neutrino decays and the cosmological nucleosynthesis.**
B. V. Vajner.
Sov. Astron. Lett., Vol. 11, No. 5, p. 275 – 277 (1985). English translation of 40.061.022.

061.082 **Fermion masses and cosmology in multidimensional unified theories.**
R. Bergamini, P. Giacconi, G. Venturi.
Nuovo Cimento A, Vol. 90A, Ser. 11, No. 2, p. 223 – 231 (1985). Abstr. in Phys. Abstr., Vol. 89, No. 1277, Entry 36614 (1986).

061.083 **Supergravity limit of superstring theory and Friedmann–Robertson–Walker cosmology.**
D. Balin, A. Love, D. Wong.
Phys. Lett. B, Vol. 165B, No. 4 – 6, p. 270 – 274 (1985). Abstr. in Phys. Abstr., Vol. 89, No. 1277, Entry 36620 (1986).

061.084 **Primordial two–component maximally symmetric inflation.**
K. Enqvist, D. V. Nanopoulos, M. Quiros, C. Kounnas.
Nucl. Phys. B, Part. Phys., Vol. B262, No. 3, p. 538 – 555 (1985). Abstr. in Phys. Abstr., Vol. 89, No. 1277, Entry 36635 (1986).

061.085 **Non–minimal supergravity at finite temperatures and cosmological applications.**
K. Enqvist, D. V. Nanopoulos, M. Quiros, C. Kounnas.
Nucl. Phys. B, Part. Phys., Vol. B262, No. 3, p. 556 – 574 (1985). Abstr. in Phys. Abstr., Vol. 89, No. 1277, Entry 36636 (1986).

061.086 **Constraints on Hamiltonian lattice formulations of field theories in an expanding universe.**
N. Weiss.
Phys. Rev. D, Vol. 32, No. 12, p. 3228 – 3232 (1985). Abstr. in Phys. Abstr., Vol. 89, No. 1277, Entry 36988 (1986).

061.087 **Magnetic monopoles and strange matter.**
J. Sanudo, A. Segui.
Phys. Lett. B, Vol. 166B, No. 1, p. 45 – 48 (1986). Abstr. in Phys. Abstr., Vol. 89, No. 1277, Entry 37047 (1986).

061.088 **Search for possible superheavy particles in sodium nuclei.**
W. J. Dick, G. W. Greenlees, S. L. Kaufman.
Phys. Rev. D, Vol. 33, No. 1, p. 32 – 38 (1986). Abstr. in Phys. Abstr., Vol. 89, No. 1277, Entry 37054 (1986).

061.089 **Stellar energy loss through scalar–neutrino production.**
A. Halprin.
Phys. Rev. D, Vol. 32, No. 12, p. 3081 – 3083 (1985). Abstr. in Phys. Abstr., Vol. 89, No. 1277, Entry 42244 (1986).

061.090 **Tuning the cosmological constant in $N = 1$ supergravity.**
M. Visser.
Phys. Lett. B, Vol. 165B, No. 4 – 6, p. 289 – 291 (1985). Abstr. in Phys. Abstr., Vol. 89, No. 1277, Entry 42369 (1986).

061.091 **Effect of the axion halo on bound electrons.**
J. C. Slonczewski.
Phys. Rev. D, Vol. 32, No. 12, p. 3338 – 3341 (1985). Abstr. in Phys. Abstr., Vol. 89, No. 1277, Entry 42380 (1986).

061.092 **Production of massive spin–1/2 particles in anisotropic spacetimes.**
K. H. Lotze.
Classical Quantum Gravity, Vol. 3, No. 1, p. 81 – 95 (1986). Abstr. in Phys. Abstr., Vol. 89, No. 1278, Entry 42617 (1986).

061.093 **Auger–like formation of monopolium.**
L. Bracci, G. Fiorentini, M. Lissia, P. Quarati.
Phys. Lett. B, Vol. 165B, No. 4 – 6, p. 425 – 428 (1985). Abstr. in Phys. Abstr., Vol. 89, No. 1278, Entry 42985 (1986).

061.094 **Detection rates for "invisible"–axion searches.**
P. Sikivie.
Phys. Rev. D, Vol. 32, No. 11, p. 2988 – 2991 (1985). Abstr. in Phys. Abstr., Vol. 89, No. 1278, Entry 48083 (1986).

061.095 **Global existence of time–dependent Yang–Mills–Higgs monopoles.**
J. Burzlaff, N. O'Murchadha.
Commun. Math. Phys., Vol. 105, No. 1, p. 85 – 98 (1986).
The authors study the Cauchy problem for non–abelian Yang–Mills–Higgs theory in (3 + 1)–dimensional Minkowski spacetime. With suitable conditions on the background fields and a suitable choice of a Sobolev space for the subtracted gauge potentials and the Higgs field, the authors establish local existence. They then prove global existence by showing that an appropriate norm of the solutions cannot blow up in a finite time.

061.096 **Quantum field theory of mutually interacting particles in expanding universes.**
J. Audretsch.
Relativity, supersymmetry and cosmology, p. 152 – 180 (1985). – See Abstr. 012.055.
The author discusses the quantum field theory of mutually interacting particles in a classical curved space–time (external gravitational field) using the *S*–matrix approach in the interaction picture. Because the external field is capable of producing pairs out of the vacuum, the specification of the outcome of the pure mutual interaction is based on the registration of the participating particles which show no vacuum creation. A detailed analysis of the corresponding added–up probability is given.

061.097 **Supersymmetric Klein–Gordon equation in *D*–dimensions.**
C. G. Bollini, J. J. Giambiagi.
Relativity, supersymmetry and cosmology, p. 215 – 224 (1985). – See Abstr. 012.055.

061.098 **Scalar particle production in anisotropic metrics.**
G. Bender, L. P. Chimento.
Relativity, supersymmetry and cosmology, p. 224 – 231 (1985). – See Abstr. 012.055.
The authors study anisotropy damping due to scalar particle creation. The field is real, massless and conformally coupled. Using a phenomenological model it is shown that a Bianchi type I metric become rapidly isotropic, which evolves as a radiation dominated Friedmann Universe.

061.099 **Susy effective potential using superfields.**
P. P. Srivastava.
Relativity, supersymmetry and cosmology, p. 308 – 317 (1985). – See Abstr. 012.055.
The author illustrates the use of superfields for computing the effective potential in supersymmetric theories by applying the superfield tadpole and vacuum bubble methods to the general renormalizable action of interacting chiral superfields.

061.100 **Some remarks on the structure of $N = 1$ supersymmetric quantum mechanics.**
J. C. D'Olivo, L. F. Urrutia, F. Zertuche.
Relativity, supersymmetry and cosmology, p. 318 – 323 (1985). – See Abstr. 012.055.

061.101 **World–sheet instantons and the Peccei–Quinn symmetry.**
X.–G. Wen, E. Witten.
Phys. Lett. B, Vol. 166B, No. 4, p. 397 – 401 (1986). Abstr. in Phys. Abstr., Vol. 89, No. 1280, Entry 54274 (1986).

061.102 **Quantum instabilities and the cosmological constant.**
W. A. Hiscock.
Phys. Lett. B, Vol. 166B, No. 3, p. 285 – 288 (1986). Abstr. in Phys. Abstr., Vol. 89, No. 1280, Entry 54285 (1986).

061.103 **The Maxwell–Boltzmannian approach to the nuclear reaction rate theory.**
H. J. Haubold, A. M. Mathai.
Fortschr. Phys., Vol. 33, No. 11 – 12, p. 623 – 644 (1985). Abstr. in Phys. Abstr., Vol. 89, No. 1280, Entry 54961 (1986).

061.104 **Axion constraints from white dwarf cooling times.**
G. G. Raffelt.
Phys. Lett. B, Vol. 166B, No. 4, p. 402 – 406 (1986). Abstr. in Phys. Abstr., Vol. 89, No. 1280, Entry 61143 (1986).

061.105 **Implications of a non–zero cosmological constant on self–gravitating fermionic systems.**
A. F. Pacheco, J. Sanudo, A. Segui.
Phys. Lett. B, Vol. 168B, No. 3, p. 223 – 225 (1986). Abstr. in Phys. Abstr., Vol. 89, No. 1281, Entry 61398 (1986).

061.106 **$N = 1$ supergravity grand unified theories with noncanonical kinetc energy terms.**
M. Drees.
Phys. Rev. D, Vol. 33, No. 5, p. 1468 – 1475 (1986). Abstr. in Phys. Abstr., Vol. 89, No. 1281, Entry 61444 (1986).

061.107 **The electron neutrino mass double beta decay and cosmology.**
P. Langacker, B. Sathiapalan.
Nucl. Phys. B, Part. Phys., Vol. B266, No. 3 – 4, p. 669 – 686 (1986). Abstr. in Phys. Abstr., Vol. 89, No. 1281, Entry 61796 (1986).

061.108 **Searching for supersymmetry in high–energy cosmic–ray interactions.**
R. W. Robinett.
Phys. Rev. D, Vol. 33, No. 5, p. 1239 – 1246 (1986). Abstr. in Phys. Abstr., Vol. 89, No. 1281, Entry 61949 (1986).

061.109 **Can scalar neutrinos or masive Dirac neutrinos be the missing mass?**
K. Freese.
Phys. Lett. B, Vol. 167B, No. 3, p. 295 – 300 (1986). Abstr. in Phys. Abstr., Vol. 89, No. 1281, Entry 67336 (1986).

061.110 **Astrophysical axion bounds diminished by screening effects.**
G. G. Raffelt.
Phys. Rev. D, Vol. 33, No. 4, p. 897 – 909 (1986). Abstr. in Phys. Abstr., Vol. 89, No. 1281, Entry 67649 (1986).

061.111 **Cosmology and the causal interpretation of quantum mechanics.**
C. Dewdney, P. R. Holland, A. Kyprianidis, J. P. Vigier.
Phys. Lett. A, Vol. 114A, No. 7, p. 365 – 370 (1986). Abstr. in Phys. Abstr., Vol. 89, No. 1282, Entry 67833 (1986).

061.112 **The invisible axion in a SUSY GUT SO(10) model.**
G. Minggao, L. Jizong.
J. Phys. G, Vol. 12, No. 3, p. 161 – 165 (1986). Abstr. in Phys. Abstr., Vol. 89, No. 1282, Entry 67899 (1986).

061.113 **Tunneling without barriers.**
K. Lee, E. J. Weinberg.
Nucl. Phys. B, Part. Phys., Vol. B267, No. 1, p. 181 – 202 (1986). Abstr. in Phys. Abstr., Vol. 89, No. 1282, Entry 68246 (1986).

061.114 **Grand unification of fermion generations: neutrino masses and cosmological constraints.**
D. Chang, A. Kumar, R. N. Mohapatra.
Phys. Rev. D, Vol. 33, No. 6, p. 1777 – 1784 (1986). Abstr. in Phys. Abstr., Vol. 89, No. 1282, Entry 68305 (1986).

061.115 **Strange matter.**
E. Farhi.
Nucl. Phys. A, Vol. A450, p. 505c – 506c (1986). Abstr. in Phys. Abstr., Vol. 89, No. 1282, Entry 68354 (1986).

061.116 **Destruction of ^{26}Al in explosive nucleosynthesis.**
H. P. Trautvetter, H. W. Becker, U. Heinemann, L. Buchmann, C. Rolfs, F. Käppeler, M. Baumann, H. Freiesleben, H.–J. Lutke–Stetzkamp, P. Geltenbort, F. Gonnenwein.
Z. Phys., A, Vol. 323, No. 1, p. 1 – 11 (1986). Abstr. in Phys. Abstr., Vol. 89, No. 1282, Entry 68513 (1986).

061.117 **Proposed test of a strange matter pulsar by observation of the neutrino flux from Cygnus X–3.**
G. L. Shaw, G. Benford, D. J. Silverman.
Phys. Lett. B, Vol. 169B, No. 2 – 3, p. 275 – 279 (1986). Abstr. in Phys. Abstr., Vol. 89, No. 1283, Entry 78440 (1986).

061.118 **On the use of SSC as a detector of certain dark matter candidates.**
A. K. Drukier.
Acta Phys. Pol., Ser. B, Vol. B17, No. 3, p. 229 – 238 (1986). Abstr. in Phys. Abstr., Vol. 89, No. 1283, Entry 78538 (1986).

061.119 **Direct–annihilation fraction of positrons in a neutral low–density galactic environment.**
B. L. Brown, M. Leventhal, A. P. Mills Jr.
Phys. Rev. A, Vol. 33, No. 4, p. 2281 – 2283 (1986). Abstr. in Phys. Abstr., Vol. 89, Nr. 1284, Entry 79123 (1986).

061.120 **Resonant amplification of ν oscillations in matter and solar–neutrino spectroscopy.**
S. P. Mikheyev (*S. P. Mikheev*), A. Yu. Smirnov.
Nuovo Cimento C, Vol. 9C, Ser. 1, No. 1, p. 17 – 26 (1986). Abstr. in Phys. Abstr., Vol. 89, Nr. 1284, Entry 83395 (1986).

061.121 **The effect of the new $^{12}C(\alpha,\gamma)^{16}O$ rate on the chemical evolution of the solar neighborhood.**
F. Matteucci.
Astrophys. J., Lett. Ed., Vol. 305, No. 2, p. L81 – L84 (1986).
New models of chemical evolution of the solar neighborhood have been computed by taking into account the effect of the revised rate of the $^{12}C(\alpha,\gamma)^{16}O$ reaction on the chemical yields from massive stars, together with the yields from low– and intermediate–mass stars which also include those from Type I supernova explosions. In particular, the evolution of ^{12}C, ^{14}N, ^{16}O, ^{20}Ne, ^{24}Mg, ^{28}Si, and ^{56}Fe has been followed in detail, and their predicted solar absolute abundances as well as their relative ratios, both in the Sun and in metal–poor stars, have been compared with the observed ones.

061.122 **The puzzle of lithium in evolved stars.**
H. Reeves.
Nucleosynthesis and its implications on nuclear and particle physics, p. 13 – 21 (1986). – See Abstr. 012.069.
The following observations are reported for some evolved stars: (1) CNO abundances and isotopes, implying CNO cycle alteration; (2) Li abundances reaching, but not above, the interstellar value ($Li/H \leqslant 10^{-9}$); (3) Lithium isotopic ratio as in the solar system ($^6Li/^7Li \sim 0.1$); (4) X–ray fluxes of 10^{29}erg/sec. It is shown here that neither spallation processes in the stellar surfaces, nor ($He^3 + He^4$) reaction after completion of H burning, nor mixing with interstellar material can account for all these features at the same time.

061.123 **The survival of helium–3 in stars.**
G. Steigman, D. S. P. Dearborn, D. N. Schramm.
Nucleosynthesis and its implications on nuclear and particle physics, p. 37 – 44 (1986). – See Abstr. 012.069.
If the observed abundance of ^{3}He is to be used – in concert with the predictions of primordial nucleosynthesis – to bound from below the universal density of nucleons, the survival of ^{3}He in the material processed through stars must be estimated. The results of detailed stellar evolution calculations which follow the destruction of ^{3}He are outlined. The survival of ^{3}He as a function of stellar mass, composition and mass loss is presented. The cosmological consequences are summarized.

061.124 **How degenerate can we be?**
G. Steigman.
Nucleosynthesis and its implications on nuclear and particle physics, p. 45 – 55 (1986). – See Abstr. 012.069.
Since relic neutrinos are virtually unobservable, a large – but undetectable – lepton asymmetry is possible. Cosmological constraints can provide bounds to such a universal neutrino degeneracy. The most stringent bounds to neutrino degeneracy follow from comparisons of the predictions of big bang nucleosynthesis

with the observed abundances of the light elements. It is shown that neutrino degeneracy will *not* permit the universe to be "closed" by nucleons.

061.125 **s–process nucleosynthesis – stellar aspects and the classical model.**
F. Käppeler.
Nucleosynthesis and its implications on nuclear and particle physics, p. 253 – 262 (1986). – See Abstr. 012.069.
The sensitivity and the limits of *s*–process analyses with the classical model are discussed with respect to the interpretation of abundance patterns in the atmospheres of red giant stars. In particular, the effects of neutron capture cross sections, mean neutron exposure, and neutron density on the resulting *s*–process overabundances are considered.

061.126 **s–process nucleosynthesis below A = 90.**
H. Beer.
Nucleosynthesis and its implications on nuclear and particle physics, p. 263 – 270 (1986). – See Abstr. 012.069.
A complete *s*–process analysis of heavy elements is carried out in the frame of the classical model with three exponential neutron exposure distributions. Special emphasis is placed on the mass region below A = 90 where the main and weak *s*–process component are effective. The derived astrophysical parameters of this model are able to reproduce heavy *s*–process element abundances correctly.

061.127 **Pulsed–neutron–source models for the astrophysical s–process.**
W. M. Howard, G. J. Mathews, K. Takahashi, R. A. Ward.
Nucleosynthesis and its implications on nuclear and particle physics, p. 271 – 275 (1986). – See Abstr. 012.069.
The astrophysical *s*–process is a sequence of neutron–capture and beta–decay reactions on a slow time scale compared to beta–decay lifetimes near the line of stability. The authors study this detailed sequence of neutron capture, continuum and bound–state beta decay, positron decay, and electron–capture reactions that comprise the *s*–process for a broad range of astrophysical environments. The results are compared with the solar–system abundances of heavy elements.

061.128 **A parametrized study of the $^{13}C(\alpha,n)^{16}O$ neutron source.**
A. Jorissen, M. Arnould.
Nucleosynthesis and its implications on nuclear and particle physics, p. 303 – 311 (1986). – See Abstr. 012.069.
After briefly reviewing the astrophysical sites where protons could be mixed into He– and C–rich layers, the authors present the main features of the neutron processing resulting from that mixing. These are used as a guideline for an analytical investigation of the conditions required for $^{13}C(\alpha,n)^{16}O$ to be an efficient neutron source when $T \lesssim 1.5 \times 10^8$K.

061.129 **Magnesium isotopes and galactic evolution.**
B. Barbuy.
Nucleosynthesis and its implications on nuclear and particle physics, p. 325 – 329 (1986). – See Abstr. 012.069.
The carbon–burning products $^{24,25,26}Mg$ in halo stars and super–metal–rich stars are studied. The isotopic composition as a function of metallicity is derived and a comparison of the results with nucleosynthesis theories is presented.

061.130 **^{26}Al–destruction in neutron rich environments.**
H. P. Trautvetter.
Nucleosynthesis and its implications on nuclear and particle physics, p. 347 – 354 (1986). – See Abstr. 012.069.
There is strong observational evidence for the existence of ^{26}Al in galactic interstellar matter. Models for the abundance of this isotope need accurate nuclear reaction rates for its production as well as for its destruction. This paper presents experimental results for the destruction of ^{26}Al by neutrons.

061.131 **Certainties and uncertainties in long–lived chronometers.**
B. S. Meyer, D. N. Schramm.
Nucleosynthesis and its implications on nuclear and particle physics, p. 355 – 362 (1986). – See Abstr. 012.069.
Evidence is provided for the confirmation of the long–held suspicion that $^{232}Th/^{238}U$ is not a very long–lived chronometric pair. It is argued, however, that it nevertheless provides a firm lower limit to the age of the Galaxy. It is also shown that $^{187}Re/^{187}Os$ is a very long–lived chronometric pair; thus, the Re/Os pair gives an upper limit to T_{Gal}, the Galaxy's age. From current data, limits give a model independent range for T_{Gal} of 6.4 Gyr $\lesssim T_{Gal} \lesssim$ 22.6 Gyr with a best fit model of $T_{Gal} \approx$ 15 Gyr.

061.132 **More about nucleocosmochronology: the reliability of the long–lived chronometers, and the production of extinct radioactivities.**
M. Arnould, N. Prantzos.
Nucleosynthesis and its implications on nuclear and particle physics, p. 363 – 372 (1986). – See Abstr. 012.069.
After reassessing the opinion that the $^{232}Th - ^{238}U$, $^{235}U - ^{238}U$ and $^{187}Re - ^{187}Os$ pairs do not appear to be very reliable galactic age indicators, the authors examine a model for the origin of extinct radionuclides calling for massive mass losing Of and WR stars. It is shown that these stars may be responsible for the inferred presence of ^{26}Al, ^{107}Pd and possibly other short–lived species in certain meteoritic condensates.

061.133 **Chronometer studies with initial galactic enrichment.**
F.–K. Thielemann, J. W. Truran.
Nucleosynthesis and its implications on nuclear and particle physics, p. 373 – 388 (1986). – See Abstr. 012.069.
The long–lived actinide chronometers $^{232}Th/^{238}U$ and $^{235}U/^{238}U$ are used to determine the galactic age. Emphasis is put on constraints for the nucleosynthesis production function from galactic evolution models and astronomical observation. The results are strongly dependent on the amount of initial enrichment and vary betwen 12.6 and 24.6 billion years. The inclusion of the $^{187}Re/^{187}Os$ pair within the same framework shows the existence of consistent solutions, but the intrinsic uncertainties in the Re/Os pair cannot lead to a more precise age determination. Thus, at present, nucleocosmochronology cannot provide smaller uncertainties than the analysis of globular cluster ages or determinations of the Hubble constant.

061.134 **A possible unified interpretation of the solar $^{48}Ca/^{46}Ca$ abundance ratio and Ca – Ti isotopic anomalies in meteorites.**
K.–L. Kratz, W. Hillebrandt, J. Krumlinde, P. Möller, F.–K. Thielemann, M. Wiescher, W. Ziegert.
Nucleosynthesis and its implications on nuclear and particle physics, p. 389 – 396 (1986). – See Abstr. 012.069.
When taking into account nuclear structure effects in β–decay properties of neutron–rich S to K isotopes, neutron exposures similar to those occurring in explosive He–burning can explain the solar abundance ratio $^{48}Ca/^{46}Ca$ as well as the correlated Ca – Ti isotopic anomalies observed in the EK–1–4–1 Allende inclusion, assuming both to be the result of the same nucleosynthesis process.

061.135 **The screening of photodisintegration reactions.**
R. Mochkovitch, K. Nomoto.
Nucleosynthesis and its implications on nuclear and particle physics, p. 399 – 405 (1986). – See Abstr. 012.069.
The authors calculate the rates of photodisintegration reactions in a strongly coupled plasma for the non–resonant and resonant cases. The screening produces a decrease of the Coulomb barrier between two interacting nuclei together with an increase of the energy threshold of the reaction. The authors show that these two effects nearly cancel out so that the photodisintegration rates differ little from their values in a dilute medium. Some astrophysical consequences of this result are briefly discussed.

061.136 **Electron polarization in nuclear reactions at high density.**
R. Mochkovitch, M. Hernanz.
Nucleosynthesis and its implications on nuclear and particle physics, p. 407 – 411 (1986). – See Abstr. 012.069.
In most calculations of the enhancement factor of nuclear reactions at high density, the degenerate electrons have been treated as a uniform non-polarized background. The authors check the validity of this assumption by computing the correction for finite electron polarizability. It is shown that electron polarization can be safely neglected for example in deflagration scenarios for Type I supernovae or in X–γ–ray burst models.

061.137 **Large-basis shell-model technology in nucleosynthesis and cosmology.**
G. J. Mathews, S. D. Bloom, K. Takahashi, G. M. Fuller, R. F. Hausman Jr.
Nucleosynthesis and its implications on nuclear and particle physics, p. 413 – 421 (1986). – See Abstr. 012.069.
The authors discuss various applications of the Lanczos method to describe properties of many-body microscopic systems in nucleosynthesis and cosmology. These calculations include: solar neutrino detectors; beta-decay of excited nuclear states; electron-capture rates during a core-bounce supernova; exotic quarked nuclei as a catalyst for hydrogen burning; and the quark-hadron phase transition during the early universe.

061.138 **Properties of anomalous nuclei and their possible effects on stellar burning.**
R. N. Boyd.
Nucleosynthesis and its implications on nuclear and particle physics, p. 423 – 430 (1986). – See Abstr. 012.069.
The effects of the possible existence of anomalous nuclei, Q-nuclei, on stellar burning are investigated. In particular, they could catalyze a hydrogen-burning cycle, similar to the CNO cycle, which could contribute appreciably to solar energy generation with an abundance of about 1×10^{-15} Q-nuclei per normal nucleus. Furthermore, they are found to be able to solve the solar neutrino problem.

061.139 **Stellar reaction rate of $^{14}N(p,\gamma)^{15}O^{+}$.**
U. Schröder, H. W. Becker, J. Görres, C. Rolfs, H. P. Trautvetter, R. E. Azuma, J. King.
Nucleosynthesis and its implications on nuclear and particle physics, p. 431 – 434 (1986). – See Abstr. 012.069.
The $^{14}N(p,\gamma)^{15}O$ reaction is a member of the chain of reactions in the hydrogen burning CNO cycles. It is the slowest reaction in the main CN cycle and thus controls the energy generation in the CNO cycles. This paper results from a new experimental study of this reaction, with proton energies ranging from 170 keV to 3.5 MeV.

061.140 **Recent progress in experimental determination of the $^{12}C(\alpha,\gamma)^{16}O$-reaction rate.**
H. P. Trautvetter, A. Redder, C. Rolfs.
Nucleosynthesis and its implications on nuclear and particle physics, p. 435 – 442 (1986). – See Abstr. 012.069.
The $^{12}C(\alpha,\gamma)^{16}O$ reaction rate is important for nucleosynthesis during hydrostatic shell burning. This paper reviews first existing experimental data for this important reaction and then presents new experimental results. The implications of these new data for theoretical extrapolations to stellar energies are discussed.

061.141 **Radioactive ion beams: research motivation and methods of production.**
R. N. Boyd.
Nucleosynthesis and its implications on nuclear and particle physics, p. 443 – 450 (1986). – See Abstr. 012.069.
Radioactive ion beams can provide research opportunities not available with ordinary ion beams. In particular, radioactive beams allow investigation of nuclear reactions important to the stellar burning and nucleosynthesis which occur in high temperature and/or density environments in stars. General design considerations are presented which apply to any such facility. The facility presently being built at Ohio State University is described.

061.142 **Can neutrinos from Cygnus X–3 be seen by proton decay detectors?**
A. Dar.
Nucleosynthesis and its implications on nuclear and particle physics, p. 477 – 486 (1986). – See Abstr. 012.069.
The flux of high energy neutrinos from Cygnus X-3 is estimated directly from the observed light curve of high energy γ-rays from this source. These neutrinos induce an underground muon flux which can be distinguished by its characteristic direction and duration from the muon background induced by atmospheric neutrinos. For zenith angles $\gtrsim 84°$ the expected rate of muon events in the IMB detector due to Cygnus X-3 is between 1 to 10 events/year, depending on the actual duration of the γ-ray pulses from this source.

061.143 **Detection of magnetic monopoles with metastable type I superconductors.**
L. Gonzalez-Mestres, D. Perret-Gallix.
Nucleosynthesis and its implications on nuclear and particle physics, p. 487 – 494 (1986). – See Abstr. 012.069.
Metastable type I superconductors provide a natural way to detect magnetic monopoles. The method is independent of the speed of the monopole, and provides tracking the timing information. The detector consists of a colloid of superheated tin microspheres (30 to 100 μm diameter). Time resolution is in the range 100 nsec to 1 μsec and space resolution is ~1 cm. Therefore, the speed and direction of a monopole can be determined with good accuracy. The present status of the project is discussed.

061.144 **Nuclear matter under extreme conditions and the early universe.**
L. Van Hove.
From nuclei to stars, p. 170 – 180 (1986). – See Abstr. 012.070.
Contents: (1) Introduction. (2) QCD strings and the deconfinement transition. (3) Quark-gluon plasma droplets in high energy collisions. (4) The hadronic phase transition in the very early universe.

061.145 **Direct neutron capture and element synthesis.**
H. Morinaga.
From nuclei to stars, p. 212 – 221 (1986). – See Abstr. 012.070.
Contents: (1) Introduction. (2) Physics of the direct neutron capture. (3) Multiple neutron capture and direct neutron capture. (4) Direct neutron capture and the *r*-process path.

061.146 **Observational evidence of element formation processes in stars.**
M. Hack.
From nuclei to stars, p. 241 – 256 (1986). – See Abstr. 012.070.
Contents: (1) Introduction. (2) Chemical composition of unevolved stars. (3) The problem of the missing zero-metal population. (4) Chemical composition of the interstellar medium. (5) The abundance of deuterium. (6) ^{3}He abundance. (7) Li and Be abundances. (8) Evolved stars. (9) Conclusion.

061.147 **Field of an accelerating celestial body with a large number of magnetic monopoles.**
Y.-j. Wang, Z.-m. Tang.
Sci. Sin., Ser. A, Vol. 29, No. 6, p. 639 – 649 (1986).
Solving Einstein-Maxwell equations with the cosmological constant, the authors have reached a new metric and the Maxwell tensor, and discussed the effect of magnetic charge and acceleration.

061.148 **Thermonuclear reaction rates from (p, n) reactions on targets with $A = 92 - 122$.**
S. Kailas.
J. Astrophys. Astron., Vol. 7, No. 2, p. 53 – 57 (1986).
From the experimentally measured (p, n) cross-sections for ^{92,94}Zr, ^{93}Nb, ^{95,96,98}Mo, ^{103}Rh, ^{107,109}Ag, ^{110}Pd, ^{115}In, ^{117,122}Sn nuclides, for proton energies below 7 MeV, thermonuclear reaction rates in the temperature range $1 \leqslant T_9 \leqslant 5$ ($T_9 = 10^9$K) have been extracted. These reaction rates have been fitted to a three parameter empirical expression proposed by Fowler.

Supersymmetry and its applications: superstrings, anomalies and supergravity. Proceedings of the Nuffield Workshop, held at Cambridge, U.K., 23 June – 14 July 1985.
See Abstr. 012.041.

Heliosphere, collapsed stars and fusion reactors.
See Abstr. 013.066.

Possible experiments for direct detection of particle candidates for the galactic dark matter.
See Abstr. 034.099.

An analytic approach to the connection between stellar structure parameters and the neutrino emission rate in a simple stellar model.
See Abstr. 065.022.

Supernova nucleosynthesis in low–metallicity populations.
See Abstr. 065.039.

Explosive hydrogen burning in novae.
See Abstr. 065.040.

Destruction of ^{3}He in stars.
See Abstr. 065.048.

Stellar technetium and niobium abundances as a measure of the lifetime of AGB stars in the third dredge–up phase.
See Abstr. 065.049.

Nucleosynthesis accompanying classical nova outbursts.
See Abstr. 065.128.

Reaction rates in the *rp*–process and nucleosynthesis in novae.
See Abstr. 065.129.

Nucleosynthesis and Type I supernovae.
See Abstr. 065.130.

Nucleosynthesis and massive star evolution.
See Abstr. 065.135.

Evolution and nucleosynthesis of massive stars with extended mixing.
See Abstr. 065.136.

Nucleosynthesis in massive, mass losing, stars.
See Abstr. 065.137.

Nucleosynthesis in massive stars: winds from WR stars and isotopic anomalies in cosmic rays.
See Abstr. 065.138.

Stellar *s*–process diagnostics.
See Abstr. 065.142.

Production and survival of ^{99}Tc in He–shell recurrent thermal pulses.
See Abstr. 065.143.

Neutron capture nucleosynthesis in massive stars.
See Abstr. 065.144.

Type I SNe from binary systems: consequences on galactic chemical evolution.
See Abstr. 065.145.

^{26}Al produced by Wolf–Rayet stars and the 1.8 MeV line emission of the Galaxy.
See Abstr. 065.146.

Supernova theory.
See Abstr. 065.147.

On the production of heavy neutron–rich isotopes during He and C burning in massive stars.
See Abstr. 065.148.

Luminous content of galaxies: inferences from supernovae.
See Abstr. 065.173.

The gravitational acceleration of antiprotons.
See Abstr. 066.003.

Gauge theory in curved space.
See Abstr. 066.037.

Spontaneous quantum compactification in seventeen dimensions.
See Abstr. 066.038.

Geometry as particles and particles as geometry.
See Abstr. 066.039.

Gravity on a lattice.
See Abstr. 066.041.

On the Anandan–Stodolsky effect.
See Abstr. 066.046.

Classical unified fields: physical space.
See Abstr. 066.047.

On topologically massive hypergravity.
See Abstr. 066.077.

Discrete Yang–Mills theories.
See Abstr. 066.079.

An alternative to general relativity.
See Abstr. 066.082.

Model systems of fields in the theory of gravitation and the problems of cosmology.
See Abstr. 066.125.

Superluminal separation caused by gravitational lens' effect of neutrino objects.
See Abstr. 066.131.

Quantized strings and curved space–time.
See Abstr. 066.138.

Quantum gravity effects in particle physics and cosmology.
See Abstr. 066.148.

An inertial interpretation of acceleration radiation.
See Abstr. 066.152.

Scalar fields and the cosmological constant.
See Abstr. 066.207.

Gravitational collapse and the cosmic antineutrino background.
See Abstr. 067.035.

A relativistic rotating massive object with saturation of magnetic monopoles.
See Abstr. 067.061.

Forces exerting on a test particle outside a relativistic rotating massive object with saturation of magnetic monopoles and jets from the object.
See Abstr. 067.062.

Does dark matter affect the solar neutrino flux?
See Abstr. 080.025.

The solar neutrino problem: astronomy or physics?
See Abstr. 080.037.

acceleration by shock waves. If low-energy cosmic rays are to be accelerated by the Fermi process at a shock front, then either the wave turbulence ahead of the shock must have large amplitude or hydromagnetic turbulence must be generated behind the shock front. The role of firehose instabilities in generating such post-shock turbulence is briefly discussed.

062.004 **Formation of astrophysical jets by a contracting magnetic accretion disk.**
K. Shibata, Y. Uchida.
Astrophys. Space Sci., Vol. 118, Nos. 1/2, p. 443 – 447 (1986). – See Abstr. 012.001.
The authors discuss an MHD model for the formation of astrophysical jets, in which the directed flows are ejected along the rotation axis of an accretion disk formed from a cloud having a large scale magnetic field parallel to the angular momentum axis of the disk. The acceleration of jets is due to the $\mathbf{j} \times \mathbf{B}$ force in the relaxing magnetic twist which is produced by the rotation of the disk.

062.005 **Effect of magnetic field on a shock-induced thermal instability.**
S. S. Hong, B. C. Koo.
Astrophys. Space Sci., Vol. 119, No. 1, p. 141 – 142 (1986). – See Abstr. 012.001.
The effect of a magnetic field on a thermal instability has been studied in a radiatively cooling region behind an interstellar shock of moderate propagation velocity (~ 10 km s^{-1}). It is shown that the presence of a magnetic field of a few microgauss is very effective in preventing the thermal instability from building-up density concentrations. In the absence of the magnetic field, the shock-induced thermal instability will amplify a pre-shock density inhomogeneity by more than an order of magnitude. However, in the field's presence, the amplified density contrast is shown to be only a factor 2. Implications for the "trace of a sweeping broom" in the Pleiades nebula are discussed.

062.006 **Formation of collimated beams.**
M. Yokosawa.
Astrophys. Space Sci., Vol. 119, No. 1, p. 213 – 215 (1986). – See Abstr. 012.001.
A mechanism is proposed for the formation of collimated beams in radio galaxies. Collimated flows are considered to be non-thermally driven by high energy particles and magneto-hydrodynamic (MHD) waves. The galactic nucleus is regarded as being surrounded by a cool gas. The cool gas accretes onto the nucleus, and then the high energy particles are completely locked to the MHD waves. When a quasi-radial magnetic field is embedded in the accretion flow, the resulting MHD wave packets are collimated into the direction of the symmetry axis of the galactic nuclear disc. The fluid around the nucleus is considered to be accelerated and heated by these MHD waves. The fluid beam is ejected along the symmetry axis.

062.007 **Free-convection and mass transfer effects on the flow past a vertical plate.**
C. Perdikis.
Astrophys. Space Sci., Vol. 119, No. 2, p. 295 – 303 (1986).
An analysis of the laminar free-convection and mass transfer effects of a viscous incompressible heat generating fluid-flow past an impulsively vertical plate, with constant suction and heat flux at the plate, in the presence of free stream oscillations is considered. Approximate solutions of the coupled, nonlinear equations relevant to the problem are obtained and analytical expressions for the velocity field, temperature field, skin friction and the rate of heat transfer are presented. The influence on the various parameters entering into the problem is discussed.

062.008 **Unsteady magnetohydrodynamic free-convection flows in a rotating fluid.**
J. N. Tokis.
Astrophys. Space Sci., Vol. 119, No. 2, p. 305 – 313 (1986).
The three-dimensional free-convection flow near an infinite vertical plate moving in a rotating fluid in the presence of a transverse magnetic field is studied in the case when the plate temperature undergoes a thermal transient. An exact solution has been obtained by defining a complex velocity with the help of the Laplace-transform technique, when the plate is moving with a velocity which is an arbitrary function of time. Three special cases of physical interest are also discussed.

062.009 **Strong plane shock waves in exponential and optically-thin atmospheres, II.**
J. B. Singh.
Astrophys. Space Sci., Vol. 119, No. 2, p. 373 – 379 (1986).
Non-similarity solutions for the propagation of strong plane shock waves in an optically-thin grey atmosphere are investigated. The density increases exponentially under low pressure. The shock moves with variable velocity and the total energy of the wave is not constant. Planck's diffusion approximation has been taken into account in the present problem.

062.010 **Fast and optimal solution to the "Rankine-Hugoniot problem".**
A. F. Viñas, J. D. Scudder.
J. Geophys. Res., Vol. 91, No. A1, p. 39 – 58 (1986).
A new, definitive, reliable and fast iterative method is described for determining the geometrical properties of a shock (i.e., θ_{Bn}, $\mathbf{n}$, V_S and M_A), the conservation constants and the self-consistent asymptotic magnetofluid variables using the Rankine-Hugoniot conservation equations. The technique uses the three dimensional magnetic field and plasma observations. The method is well conditioned and reliable at all θ_{Bn} angles regardless of the shock strength or geometry. Explicit proof of "uniqueness" of the shock geometry solution by either analytical or graphical methods is given. The method is applied to synthetic and real interplanetary shocks, including a bow shock event and the results are then compared with those determined by preaveraging methods and other iterative schemes. A complete analysis of the confidence region and error bounds of the solution is also presented.

062.011 **A parametric study of slow shock Rankine-Hugoniot solutions and critical Mach numbers.**
J. P. Edmiston, C. F. Kennel.
J. Geophys. Res., Vol. 91, No. A2, p. 1361 – 1372 (1986).
The authors present a parametric survey of the Rankine-Hugoniot solutions for slow MHD shock waves. They also calculate the parametric dependences of the critical Mach numbers above which resistivity or thermal conduction alone can no longer provide all the shock dissipation, as well as the Mach number above which the flux of shock-catching ions can be significant. Slow shocks rival fast shocks in the degree to which their properties might depend upon upstream parameters.

062.012 **Laboratory observations of ion cyclotron waves associated with a double layer in an inhomogeneous magnetic field.**
M. J. Alport, S. L. Cartier, R. L. Merlino.
J. Geophys. Res., Vol. 91, No. A2, p. 1599 – 1608 (1986).

062.013 **Non-steady radiative shocks.**
D. E. Innes, J. R. Giddings, S. A. E. G. Falle.
Cosmical gas dynamics, p. 143 – 148 (1985). – See Abstr. 012.007.
Earlier work has shown that supernova shocks entering the radiative phase go through a period of unstable (catastrophic) cooling, in which one or more secondary shocks are produced. The authors have demonstrated that the cooling instability is fundamental and not related to the way in which the shocks are formed. This has been done by means of detailed hydrodynamical and ionisation calculations for a "steady" 200 km/s shock moving into a medium containing a small (but plausible) density perturbation.

062.014 **Numerical gasdynamics.**
S. A. E. G. Falle.
Cosmical gas dynamics, p. 149 – 162 (1985). – See Abstr. 012.007.
The equations of inviscid gasdynamics are a set of hyperbolic conservation laws which determine the state of the gas at any instant, given an initial state. These equations have wave-type solutions and also solutions with discontinuities, either in the

conserved quantities or their derivatives. Any numerical approximation to these equations must be able to correctly model this behaviour.

062.015 **The reconfinement of jets.**
M. J. Wilson, S. A. E. G. Falle.
Cosmical gas dynamics, p. 259 – 265 (1985). – See Abstr. 012.007.
The authors give the conditions under which shocks will be formed within a steady jet surrounded by a spatially varying external pressure; what is required is that a non–equilibrium jet be reconfined by the ambient medium. The flow in a reconfining gas jet is investigated by a numerical calculation. The authors also present a model for the M87 jet based upon the idea that it is being reconfined by the surrounding medium.

062.016 **Weak discontinuities in relativistic MHD.**
A. V. Gopalakrishna, M. S. Ganagi.
Astrophys. Space Sci., Vol. 120, No. 1, p. 139 – 149 (1986).
By using singular surface theory and ray theory the speeds of propagation of fast and slow waves, propagating into a medium in arbitrary motion, have been obtained in relativistic magnetohydrodynamics. The differential equation governing the growth of these waves along the rays has been derived and the solution has been presented in integral form.

062.017 **The relativistic short wavelength instabilities in astrophysical jets. II. The case of magnetofluid–dynamics.**
L. Zaninetti.
Astron. Astrophys., Vol. 156, No. 1/2, p. 194 – 198 (1986).
The Kelvin–Helmholtz instabilities in an axisymmetric configuration with a magnetic field produce a dispersion relation that is solved using a numerical routine when the wavelengths are shorter than the jet radius.

062.018 **Magnetic non–equilibrium in a stratified atmosphere.**
S. I. Vainshtein (*S. I. Vajnshtejn*).
Astron. Astrophys., Vol. 156, No. 1/2, p. 295 – 300 (1986).
A study has been made of equilibrium conditions of the magnetic field in a stratified medium influenced by a gravity force. Two classes of equilibrium solutions arising in the presence of a neutral line of the magnetic field, are derived. When the topology of an initial magnetic field does not correspond to these two classes of solutions, reconnection is taking place and current sheet is produced. Some examples of magnetic non–equilibrium are given. The topology of the lines of force for these configurations is not equivalent to the final equilibrium state and the formation of current sheets with a subsequent dissipation of the magnetic field is inevitable. The situations under examination are all readily realized in the solar corona. Magnetic non–equilibrium may represent one of the reconnnection of lines of force mechanisms in the solar corona.

062.019 **Oscillatory motions in intense flux tubes.**
S. S. Hasan.
Mon. Not. R. Astron. Soc., Vol. 219, No. 2, p. 357 – 372 (1986).
The purpose of the present investigation is to examine the detailed nature of oscillatory behaviour that can occur in intense flux tubes. In order to keep the analysis sufficiently general the author treated first a polytropic atmosphere under certain approximations for which an analytic solution was possible. He then solved the problem for a realistic stratification using model opacities so as to obtain results applicable to the Sun.

062.020 **On the behavior of hydromagnetic surface waves.**
M. A. Lee, B. Roberts.
Astrophys. J., Vol. 301, No. 1, p. 430 – 439 (1986).
The behavior of velocity, magnetic field, and pressure perturbations about a continuously varying interface in pressure equilibrium is investigated in detail within ideal incompressible magnetohydrodynamics. A specific initial value problem is solved in quadrature for a thin interface and compared with the solution for a discontinuous interface. The unattenuated surface wave about a discontinuous interface is replaced at a thin interface by a collective surface disturbance which decays, with the associated energy density flowing into local oscillations within the interface. Implications for the propagation and dissipation of surface waves in the solar corona are discussed.

062.021 **Quasistatic broadening of hydrogen spectral lines in plasmas.**
S. A. Gulyaev, G. V. Sholin.
Astron. Zh., Tom 63, Vyp. 1, p. 50 – 61 (1986). In Russian. English translation in Sov. Astron., Vol. 30, No. 1.
Calculations of quasistatic Stark profiles of hydrogen and hydrogen–like ion lines forming in plasmas are presented. The difference between the exact profiles and approximate ones (which are usually used) is discussed. In a set of important cases, under $n \gg 1$, this difference may be very large. Impact profiles of central (unshifted) Stark components, which are not subject to quasistatic broadening, are calculated. The criterion of applicability of quasistatic broadening is specified and the range of validity of various mechanisms of broadening for different transitions in hydrogen–like atoms and ions is determined. The results may be used for interpretation of observations of radio recombination lines of high orders in nebulae and in the interstellar medium or in problems connected with diagnostics of laboratory plasma in a wide interval of densities and temperatures.

062.022 **Space charge sheath in plasma–neutral gas interaction.**
N. Venkataramani, S. K. Mattoo.
Astrophys. Space Sci., Vol. 121, No. 1, p. 83 – 103 (1986).
A space charge sheath is found to be formed whenever a high–velocity magnetized plasma stream penetrates a gas cloud. The sheath is always located at the head of the plasma stream, and its thickness is very small compared to the length of the plasma stream. Soon after the sheath is formed it quickly slows down to the Alfvén critical velocity. Presently available models of the sheath may explain the initiation of the sheath formation. Physical processes like heating of the electrons and ionization of the gas cloud which come into play at a later stage of the interaction are not included in these models. These processes considerably alter the potential structure in the sheath region. A schematic model of the observed sheath is presented here.

062.023 **Cylindrical oscillations of force–free electromagnetic fields.**
V. A. Osherovich.
Astrophys. Space Sci., Vol. 121, No. 2, p. 255 – 264 (1986).
The force–free electromagnetic field represents a natural generalization of the well–known force–free magnetic field model and allows the magnetic field to maintain electric charge separation. The basic equation for the cylindrical oscillations of the force–free electromagnetic field is obtained and solved for a linear case. The spectrum of possible resonances in a magnetized atmosphere is discussed.

062.024 **Equations of motion and the equilibrium conditions for an ideal rotating liquid in parametrized post–Newtonian magnetohydrodynamics.**
N. P. Bondarenko.
Astrofizika, Tom 23, Vyp. 3, p. 595 – 603 (1985). In Russian. English translation in Astrophysics, Vol. 23, No. 3.
Magnetohydrodynamics is involved in the parametrized post–Newtonian formalism. The equations of motion are obtained as well as the conservation laws and the equilibrium conditions of an ideal rotating liquid.

062.025 **Numerical solutions for time–dependent hydrodynamic flow with mass loading.**
N. I. Kömle, H. I. M. Lichtenegger.
Planet. Space Sci., Vol. 34, No. 4, p. 403 – 408 (1986).
The flow of a magnetized collisionless plasma can to a large extent be treated in terms of hydrodynamics. If in addition to the plasma neutral particles are present which are added to the flow by some ionization process, a "mass loading" of the plasma takes place. The authors investigate with the aid of an appropriate numerical code the time–dependent solutions of this problem in a one–dimensional geometry. Extensive computer experiments reveal that, depending on the initial conditions, there exists a

variety of qualitatively different solutions. Only in certain cases the solutions approach a steady state.

062.026 **The relativistic maximum instability in astrophysical jets.**
L. Zaninetti.
Astron. Astrophys., Vol. 160, No. 1, p. 135 – 140 (1986).
The study of the Kelvin–Helmholtz instabilities in cylindrical boundary layer flows leads to a series of modes, one of which grows faster than the surrounding ones and becomes dominant. The author studies the relativistic effects on the *e*–folding length, the wavelength and the time–scale of the mode of this maximum instability. These deduced quantities could explain some observed parameters of astrophysical jets such as blobs, oscillations and typical lengths.

062.027 **On the Kelvin–Helmholtz instabilities of high–velocity magnetized shear layers.**
S. Roy Choudhury, R. V. E. Lovelace.
Astrophys. J., Vol. 302, No. 1, p. 188 – 199 (1986).
This paper investigates the Kelvin–Helmholtz instability of a finite–width, ideal magnetohydrodynamic shear layer with a "linear" velocity profile in the transition region. The magnetic field is assumed constant and parallel to the direction of the streaming. Also, the density and temperature of the plasma are assumed constant. The scalar perturbation quantities have the form $f(x)\exp[i(k_y y+k_z z-\omega t)]$. For astrophysical plasma flows such as those thought to exist in extragalactic objects with radio jets, the present results indicate the nature of the most rapidly growing perturbations as a function of the shear–layer properties. An illustrative application of the present analysis is given for the Earth's magnetosphere.

062.028 **Finite parallel resistivity and uniform plasma motion across magnetic fields.**
J. L. Sperling.
J. Geophys. Res., Vol. 91, No. A3, p. 3254 – 3262 (1986).

062.029 **Decay instability of finite–amplitude circularly polarized Alfvén waves: a numerical simulation of stimulated Brillouin scattering.**
T. Terasawa, M. Hoshino, J.–I. Sakai, T. Hada.
J. Geophys. Res., Vol. 91, No. A4, p. 4171 – 4187 (1986).
By means of a numerical simulation, nonlinear evolution of large amplitude dispersive Alfvén waves is studied. An energy transfer from the parent wave to two daughter Alfvén–like waves and a soundlike wave is observed (a stimulated Brillouin scattering process). The observed growth rates and propagation characteristics of these daughter waves agree with the analytical results, which the authors obtain by extending the previous treatments by Goldstein, Derby, Sakai, and Sonnerup.

062.030 **Shock wave detection by analyzing the variation of energy spectra.**
M. Ya. Natenzon, I. Yu. Serova.
Inst. kosm. issled. Akad. Nauk SSSR. Prepr., No. 979, 22 pp. (1985). In Russian. Abstr. in Ref. Zh., 62. Issled. Kosm. Prostranstva, 1.62.133 (1986).

062.031 **Stratification of emission from interstellar shock waves.**
K. V. Bychkov.
Bull. Spec. Astrophys. Obs. – North Caucasus, Vol. 19, p. 40 – 52 (1985). English translation of 39.062.073.

062.032 **Internal gravitational waves in a plasma sheet.**
V. V. Safargaleev, Yu. P. Mal'tsev.
Geomagn. Aehron., Tom 26, No. 2, p. 270 – 274 (1986). In Russian. English translation in Geomagn. Aeron.

062.033 **On methods for calculating heat transfer in a flare plasma.**
B. Sermulina, A. Spektor.
Issled. Solntsa Krasnykh Zvezd, No. 20, p. 5 – 12 (1984). In Russian.
Different methods of flare plasma heat transfer simulation are considered for conditions when classical (Coulomb) heat transfer is not valid. The comparative analysis of methods is executed. The method of real heat flux approximation by heat–transfer coefficient variation is found.

062.034 **Poloidal and toroidal fields in geomagnetic field modeling.**
G. Backus.
Rev. Geophys., Vol. 24, No. 1, p. 75 – 109 (1986).

062.035 **Numerical modeling of magnetized jets.**
K. R. Lind.
Bull. Am. Astron. Soc., Vol. 17, No. 4, p. 832 (1985). Abstract. – See Abstr. 010.062.

062.036 **Evolution of magnetized jets in power–law potentials.**
J. Siah, P. J. Wiita.
Bull. Am. Astron. Soc., Vol. 17, No. 4, p. 832 (1985). Abstract. – See Abstr. 010.062.

062.037 **Standing waves in stratified finite–radius magnetic flux tubes.**
Y. Q. Lou, R. Rosner.
Bull. Am. Astron. Soc., Vol. 17, No. 4, p. 833 (1985). Abstract. – See Abstr. 010.062.

062.038 **Nonlinear numerical simulation of planar thermal instability.**
R. B. Dahlburg, C. R. DeVore, J. M. Picone, J. T. Karpen, J. T. Mariska.
Bull. Am. Astron. Soc., Vol. 17, No. 4, p. 833 (1985). Abstract. – See Abstr. 010.062.

062.039 **The Bondi–Hoyle accretion problem revisited.**
R. E. Taam, B. A. Fryxell.
Bull. Am. Astron. Soc., Vol. 17, No. 4, p. 856 (1985). Abstract. – See Abstr. 010.062.

062.040 **On the physics of thick disks and inner jets.**
S. K. Chakrabarti.
Bull. Am. Astron. Soc., Vol. 17, No. 4, p. 869 (1985). Abstract. – See Abstr. 010.062.

062.041 **The formation of supersonic jets in axisymmetric accretion flows.**
B. A. Fryxell, R. E. Taam.
Bull. Am. Astron. Soc., Vol. 17, No. 4, p. 869 (1985). Abstract. – See Abstr. 010.062.

062.042 **Instabilities of thick accretion tori and redistribution of angular momentum.**
W. Benz, W. H. Zurek.
Bull. Am. Astron. Soc., Vol. 17, No. 4, p. 869 – 870 (1985). Abstract. – See Abstr. 010.062.

062.043 **Multicomponent, reacting MHD flows.**
B. T. Draine.
Mon. Not. R. Astron. Soc., Vol. 220, No. 1, p. 133 – 148 (1986).
The equations describing multicomponent, reacting MHD flows are derived. Arbitrarily oriented magnetic fields are considered. The differential equations are obtained for: (1) general flows, and (2) the special case of plane–parallel, steady flows. The effects of reactive processes, such as recombination, photoionization, photodissociation, charge exchange, and reactive scattering are discussed. Departures from Maxwellian velocity distributions are considered.

062.044 **Multicomponent reacting MHD flows – a reply.**
D. R. Flower, G. Pineau des Forêts.
Mon. Not. R. Astron. Soc., Vol. 220, No. 1, p. 149 – 153 (1986).
In an accompanying paper, Draine considers the formulation of the problem of multicomponent reacting MHD flows. Draine notes certain discrepancies with earlier work on this same problem by Flower, Pineau des Forêts & Hartquist. In this paper, the

authors consider the quantitative significance of these discrepancies in the context of shocks propagating in interstellar clouds.

062.045 **On the stability of incompressible constant angular momentum cylinders.**
O. M. Blaes, W. Glatzel.
Mon. Not. R. Astron. Soc., Vol. 220, No. 1, p. 253 – 258 (1986).
It is shown that incompressible cylindrical shells rotating in an external potential and having constant angular momentum distributions can be unstable with respect to non–axisymmetric perturbations. The instability occurs on dynamical time–scales and is due to the interaction of two free boundaries. The growth rates depending on the azimuthal wavenumber and the thickness of the shell are given in analytical form.

062.046 **On pumping astronomical masers.**
N. D. Kylafis, C. Norman.
Space Telesc. Sci. Inst., Prepr. Ser., No. 89, 12 pp. (1986). To appear in Astrophys. J., Lett. Ed.

062.047 **The similarity solutions for the kinematic force–free magnetic field.**
W.–r. Hu, N.–h. Xu.
Kexue Tongbao, Vol. 30, No. 4, p. 457 – 462 (1985). Abstr. in Phys. Abstr., Vol. 89, No. 1273, Entry 14789 (1986).

062.048 **A reconsideration of thermal diffusion in ionized gases: quantal and dynamic shielding effects.**
L. Monchick, E. A. Mason.
Phys. Fluids, Vol. 28, No. 11, p. 3341 – 3348 (1985). Abstr. in Phys. Abstr., Vol. 89, No. 1274, Entry 21241 (1986).

062.049 **Derivation of equations of hydrodynamics with source of mass, momentum and energy.**
M. V. Konyukov.
Fiz. inst. Akad. Nauk SSSR. Prepr., No. 144 (1985). In Russian. Abstr. in Ref. Zh., 51. Astron., 3.51.192 (1986).

062.050 **On a mechanism of instability of an accreting column.**
I. L. Andronov.
Problems of astronomy, p. 113 – 118 (1985). In Russian. Abstr. in Ref. Zh., 51. Astron., 3.51.198 (1986). – See Abstr. 003.023.

062.051 **Calculation of effects of generation and transfer of magnetic fields for a rotating convective envelope.**
L. L. Kichatinov, N. A. Plieva.
Issled. Geomagn., Aehron. Fiz. Solntsa, Moskva, No. 73, p. 162 – 167 (1985). In Russian. Abstr. in Ref. Zh., 51. Astron., 3.51.415 (1986).

062.052 **Nonlinear rapid symmetric waves in a magnetic layer.**
E. G. Merzlyakov, M. S. Ruderman.
Inst. kosm. issled. Akad. Nauk SSSR. Prepr., No. 990, 43 pp. (1984). In Russian. Abstr. in Ref. Zh., 51. Astron., 3.51.427 (1986).

062.053 **On the important role of relativistic effects in the electrodynamics of plasma.**
A. I. Laptukhov.
Fiz. protsessy v ionos. i magnitosfere. Moskva, p. 131 – 146 (1984). In Russian. Abstr. in Ref. Zh., 51. Astron., 3.51.429 (1986).

062.054 **On the existence of the continuous spectrum of ideal MHD in a 2D magnetostatic equilibrium.**
M. Goossens, S. Poedts, D. Hermans.
Sol. Phys., Vol. 102, Nos. 1/2, p. 51 – 66 (1985).
The continuous spectrum of a 2D magnetostatic equilibrium with y–invariance is derived. It is shown that the continuous spectrum is given by an eigenvalue problem on each magnetic surface and is related to the different behaviour of the equilibrium quantities in different magnetic surfaces. The special case of a uniform poloidal magnetic field in a 1D equilibrium that is stratified with height, has been considered in detail and it is found that there is no continuous spectrum. It is shown that this result relies completely on the artificial property that the behaviour of the equilibrium quantities along a magnetic field line is independent of the field line considered. As a consequence the non–existence of a continuous spectrum in a 1D equilibrium with a uniform magnetic field cannot be used to argue that the continuous spectrum has no physical relevance.

062.055 **Stable, baroclinic eddies on Jupiter and Saturn: a laboratory analog and some observational tests.**
P. L. Read.
Icarus, Vol. 65, Nos. 2/3, p. 304 – 334 (1986). – See Abstr. 012.023.
Laboratory experiments (and corresponding numerical simulations) on free thermal convection in a rotating fluid, subject to horizontal differential heating and cooling, are examined in the light of recent observations of the longest–lived eddies on Jupiter and Saturn (including Jupiter's Great Red Spot and White Ovals). Both laboratory and atmospheric systems are shown to be capable of satisfying scaling requirements for mutual dynamical similarity, within the uncertainties of the Jovian observations.

062.056 **Propagation of spherical magnetogasdynamic shock waves.**
R. C. Srivastava, D. Leutloff, J. P. Vishwakarma, A. Kumar.
Astrophys. Space Sci., Vol. 122, No. 1, p. 33 – 40 (1986).
Magnetogasdynamic shock waves propagating in a medium of increasing density are discussed. The shock travels in a dense atmosphere. The authors have used the Runge–Kutta method to obtain a numerical solution of the problem. The distribution of flow variables behind the shock are shown by graphs.

062.057 **On wave reflection or absorption at hydromagnetic critical levels.**
L. M. B. C. Campos, J. C. G. C. Leitão.
Seismology of the Sun and the distant stars, p. 281 – 292 (1986). – See Abstr. 012.026.
The authors consider the effects of magnetic field on acoustic waves, first in a homogeneous medium and then in an atmospheric slab. They choose the case of an horizontal magnetic field.

062.058 **Helical twisting on an adiabatically expanding jet. II. Spatial growth.**
P. E. Hardee.
Astrophys. J., Vol. 303, No. 1, p. 111 – 118 (1986).
The effect of adiabatic jet expansion on helical twisting has been investigated by studying spatial growth of the fundamental helical mode. One result of the spatial analysis is that harmonic components propagate with invariant $\omega R/u$ and can remain at nearly the instantaneous fastest spatially growing frequency. Comparison between spatial growth and convected temporal growth reveals that the fastest growing wavelength and the maximum growth rate are similar but not equal in the two approaches.

062.059 **Acceleration of charged particles in magnetic reconnection: solar flares, the magnetosphere, and solar wind.**
M. L. Goldstein, W. H. Matthaeus, J. J. Ambrosiano.
Geophys. Res. Lett., Vol. 13, No. 3, p. 205 – 208 (1986).
Recent simulations using test particles suggests that reconnection may efficiently accelerate particles to the maximum energies that are observed in several astrophysical contexts. A simple analytic formula is used in conjunction with the simulation results to predict the maximum energy achievable in a particular plasma environment with the result that in solar flares reconnection is capable of accelerating particles to several GeV. In magnetospheric substorms the predicted maximum can reach several hundred keV, and near magnetic sector crossings in the solar wind the maximum energy can approach 100 keV.

062.060 **Dynamic time dependent hexagonal magnetoconvection.**
J. M. Lopez, J. O. Murphy.
Aust. J. Phys., Vol. 38, No. 6, p. 885 – 910 (1985). Paper presented at the R. G. Giovanelli Commemorative Colloquium, Part II, Tucson, Arizona, 17 – 18 January 1985.

The time dependence of the single mode hexagonal magneto-convective system has been investigated numerically at high Rayleigh number. It is established that, in certain parameter ranges, the system has oscillatory solutions which not only have a periodic nature, but also develop into chaotic and intermittent solutions. Further, the system generates nonzero mean kinetic and magnetic helicity together with substantial magnetic field amplification. These features are shown to be maintained in time without any externally imposed rotation of the system.

062.061 **The possibility of a photospheric dynamo.**
P. R. Wilson.
Aust. J. Phys., Vol. 38, No. 6, p. 911 – 918 (1985). Paper presented at the R. G. Giovanelli Commemorative Colloquium, Part II, Tucson, Arizona, 17 – 18 January 1985.

Several recent observations are discussed which suggest that the current model for the emergence and decay of photospheric flux does not provide a complete account of all the processes involved. An elementary two–dimensional dynamo is discussed and it is shown how this may be adapted to photospheric conditions in order to provide a plausible kinematic account of these observations.

062.062 **The eigenvalue approach in modelling solar magnetic structures.**
V. A. Osherovich.
Aust. J. Phys., Vol. 38, No. 6, p. 975 – 980 (1985). Paper presented at the R. G. Giovanelli Commemorative Colloquium, Part II, Tucson, Arizona, 17 – 18 January 1985.

A requirement for continuous magnetic fields and finite magnetic energy leads to the eigenvalue approach in magnetohydrodynamics. Concrete models are developed for sunspots (return–flux sunspot model), prominences, coronal loops and transients. This approach allows a classification of a variety of magnetic topologies and reveals the relations between magnetic and thermodynamic structures. The eigenvalue approach is extended to three–dimensional modelling of the global solar magnetic field.

062.063 **On the hydrodynamical equations of astrophysical jets.**
M. Calvani, L. Nobili, R. Turolla.
X–ray astronomy '84, p. 491 – 493 (1985). – See Abstr. 012.033.

The authors present a model for jet acceleration that takes into account full hydrodynamical treatment of the flow together with a careful analysis of the matter–radiation coupling.

062.064 **Long nonlinear waves in a compressible magnetically structured atmosphere. II. Slow body waves in a magnetic slab.**
E. G. Merzljakov (*E. G. Merzlyakov*), M. S. Ruderman.
Sol. Phys., Vol. 103, No. 2, p. 259 – 276 (1986).

An approximate equation describing long nonlinear slow body wave propagation in a magnetic slab in magnetic environments is derived. It is shown that this equation has no solutions in the form of solitary waves. Second–harmonic generation by a sinusoidal wave is studied. Results of numerical computations are presented. They show that a sinusoidal disturbance overturns, i.e., infinite gradients arise. An approximate formula for the overturning time is derived. The application of these results to the problem of coronal heating is discussed.

062.065 **Leaky and non–leaky oscillations in magnetic flux tubes.**
P. S. Cally.
Sol. Phys., Vol. 103, No. 2, p. 277 – 298 (1986).

An extensive analyis, both analytic and numerical, of waves in flux tubes imbedded in (possibly) magnetic surroundings is given. It is shown that any wave confined to the tube and its neighbourhood can be put into one of seven categories. Simple criteria for deciding the existence of each typ in any particular case are derived. Many other (leaky) modes are found which excite waves in the external medium and thereby lose energy to the surroundings. A number of asymptotic analyses allow much information to be gained about these without the need for numerical solution of the complicated equations involved. Three particular cases, pertaining to photospheric flux tubes, Hα fibrils, and coronal loops, are considered in detail.

062.066 **Local dynamic thermal instability.**
S. A. Balbus.
Astrophys. J., Lett. Ed., Vol. 303, No. 2, p. L79 – L82 (1986).

A very simple but general criterion for local thermal instability in dynamical systems is derived. The background flow may be time–dependent. The instability criterion is very similar to the classical Field result, except that the cooling function L is replaced by L/T, where T is the temperature. A third–order differential equation for the evolution of small–wavelength radial perturbations in spherical systems is derived. With self–gravity, it is the exact equation that would result for the evolution of perturbations in simple matter–dominated Friedmann cosmologies subject to external heating and cooling. The equation is solved by WKBJ techniques, allowing acoustical and condensational modes to be treated on an equal footing.

062.067 **Collision of charged particles with an MHD–shock front of finite depth.**
A. A. Rumyantsev, M. B. Kerimbekov.
Inst. fiz. Akad. Nauk AzSSR., Prepr. No. 159, 15 pp. (1985). In Russian. Abstr. in Ref. Zh., 51. Astron., 5.51.198 (1986).

062.068 **Radiation resonance interactions in a plasma (generation of fast particle streams, spectra of cosmic radiation, radiation corrections to the Landau damping).**
V. N. Tsytovich.
Probl. nelinejn. i turbulent. protsessov v fiz. Tr. 2 Mezhdunar. rab. gruppy, Kiev, 1983. Chast' 2. Kiev, p. 334 – 346 (1985). In Russian. Abstr. in Ref. Zh., 51. Astron., 5.51.207 (1986).

062.069 **Some astrophysical applications of MHD instabilities of differentially rotating fluids.**
B. D. Popp.
Diss. Abstr. Int., Sect. B, Vol. 46, No. 7, p. 2346 – 2347 (1986). Thesis, Harvard University, 115 pp. (1985). Order No. DA8520271.

062.070 **Computational hydrodynamics in astrophysical studies: highly shocked flows in three dimensions.**
C. N. Arnold.
Diss. Abstr. Int., Sect. B, Vol. 46, No. 11, p. 3881 (1986). Thesis, The University of Michigan, 217 pp. (1985). Order No. DA8600402.

062.071 **The onset of Alfvénic turbulence.**
S. Ghosh.
Diss. Abstr. Int., Sect. B, Vol. 46, No. 12, p. 4282 (1986). Thesis, University of Maryland, 220 pp. (1985). Order No. DA8528974.

062.072 **Interaction of two collisionless shocks.**
P. J. Cargill, C. C. Goodrich, K. Papadopoulos.
Phys. Rev. Lett., Vol. 56, No. 18, p. 1988 – 1991 (1986).

Kinetic simulations of the interaction between two collisionless shocks are presented. During the collision of two perpendicular shocks, the shock electromagnetic field structures pass through each other, while the previously shocked ions are kept separate by the electric fields arising in the collision. When two supercritical shocks collide, a fraction of ions are accelerated up to an order of magnitude in energy by first being reflected at one shock, then interacting with the electric fields of the other shock.

062.073 **Numerical investigation of magnetohydrodynamical models of cosmic plasma.**
V. V. Denisenko, N. V. Erkaev, V. G. Pivovarov.
Primenenie EhVM v modelir. zadach mat. fiz. Krasnoyarsk, p. 40 – 51 (1985). In Russian. Abstr. in Ref. Zh., 62. Issled. Kosm. Prostranstva, 6.62.469 (1986).

062.074 **Non neutral plasmas in the laboratory and astrophysics.**
F. C. Michel.
Proc. Astron. Soc. Aust., Vol. 6, No. 2, p. 127 – 129 (1985).
Recent laboratory experiments have confirmed theoretical expectations that the aligned rotator model (Goldreich and Julian 1969) does not function in the way originally expected, if at all. These experiments confirm that the nonneutral (completely charge–separated) plasma is isolated into finite regions, which in the case of the magnetosphere about an aligned rotator means in general that there is no plasma from the neutron star to be found at the light–cylinder.

062.075 **A method of calculating radiative heat diffusion in particle simulations.**
L. Brookshaw.
Proc. Astron. Soc. Aust., Vol. 6, No. 2, p. 207 – 210 (1985).
A new method for solving heat diffusion in three dimensional particle simulations is described. The difficulties encountered by other authors is discussed, in particular the difficulty of including boundary conditions in particle simulations. One and three dimensional tests of the method are described.

062.076 **The role of kinetic boundary conditions in generating type II solutions for Rayleigh–Benard convection.**
J. O. Murphy, N. Yannios.
Proc. Astron. Soc. Aust., Vol. 6, No. 2, p. 216 – 219 (1985).

062.077 **Time dependent multimode hexagonal magnetoconvection.**
J. M. Lopez, J. O. Murphy.
Proc. Astron. Soc. Aust., Vol. 6, No. 2, p. 227 – 231 (1985).
The time dependent multimode equations for magneto-convection have been solved numerically for a range of parameter values and cell sizes. The solutions demonstrate the nonlinear interactions between the different scales of motion and how the induced magnetic field is concentrated into the smaller length scale. The generation of mean kinetic and magnetic helicity occurs at low Prandtl number which has significant physical interpretations in the solar situation.

062.078 **Turbulent regeneration of magnetic fields.**
J. Carvalho, N. Pires.
Astrophys. Space Sci., Vol. 122, No. 2, p. 193 – 199 (1986).
The authors calculate an exact form for the regeneration term of the turbulent dynamo equation which is valid for arbitrary values of magnetic Reynolds number. It is shown that finite conductivity can change the regenerative character of the dynamo, depending on the geometry and character of the fluid motion in the turbulent eddies.

062.079 **Force–free equilibrium at magnetic neutral points.**
N. Seehafer.
Astrophys. Space Sci., Vol. 122, No. 2, p. 247 – 255 (1986).
It is shown that, at neutral points of force–free magnetic fields, the electric current density must vanish. This property is independent of whether the neutral points are isolated or (e.g.) fill lines or surfaces. One implication is the fact that in a cold pressureless plasma the formation of neutral current sheets cannot be adiabatically slow. The field–line topology in the neighbourhood of neutral points is discussed.

062.080 **A numerical solution of MHD free–convection flow in the general nonlinear Stokes problem by the finite–difference method.**
N. G. Kafousias, J. Daskalakis.
Astrophys. Space Sci., Vol. 123, No. 1, p. 193 – 203 (1986).
With viscous dissipation and Joule heating taking into account a numerical solution of magneto–hydrodynamic free convection flow, in the Stokes problem, is obtained for different values of Prandtl number P. The fluid is viscous, incompressible, and electrically conducting and the magnetic lines of force are assumed to be fixed relative to the plate which is started moving impulsively in its own plane or it is uniformly accelerated. The solution is obtained with an implicit second–order method, for $P = 0.71$ (air) and $P = 7$ (water).

062.081 **Mass and particle "loading" in plasma–neutral gas interaction.**
M. Danz, T. Roatsch.
Astrophys. Space Sci., Vol. 123, No. 2, p. 221 – 235 (1986).
Qualitative considerations of mass and particle "loading" in plasma–neutral gas interaction are made by the help of a one–dimensional steady model. In the case of particle "loading" the plasma flow behind a strong shock is accelerated as in the case of mass "loading" regarded by Biermann et al. (1967) whereas behind a weak shock the flow is decelerated. As an example, one–dimensional time–dependent flows with ionization of the neutral gas by Alfvén's critical velocity effect are calculated. Because of the acceleration of the subsonic flow by "loading" processes, the existence of a tangential discontinuity in the flow around comets seems questionable.

062.082 **Gravitational instability of thermally anisotropic plasma.**
B. Singh, G. L. Kalra.
Astrophys. J., Vol. 304, No. 1, p. 6 – 10 (1986).
The equations of Chew, Goldberger, and Low (1956) modified to include the heat flux vector and self–gravitation are used to study the gravitational instability of unbounded plasma placed in a uniform static magnetic field. The linear stability analysis shows the importance of some of the additional terms which arise as a result of higher moments. The influence of these terms on the gravitational instability is examined.

062.083 **On the structure and stability of radiative shock waves.**
E. Bertschinger.
Astrophys. J., Vol. 304, No. 1, p. 154 – 177 (1986).
The structure of a spherical, radiative, interstellar shock wave (e.g., a supernova remnant or a stellar wind bubble) is derived under the assumption that the cooling rate per unit volume has the form $\Lambda \propto \varrho^{\alpha} T^{\beta}$ for T above a cutoff temperature T_c and $\Lambda = 0$ for T below T_c. In the limit of short cooling time and large isothermal cooling Mach number, the flow has an exact analytic solution. Three–dimensional linear stability analyses of the analytic solutions have been performed. The results demonstrate that the radiative phase of shock evolution cannot be adequately modeled as being spherically symmetric and that high spatial resolution is necessary to follow the growth of instabilities.

062.084 **Viscous normal modes on coronal inhomogeneities and their role as a heating mechanism.**
R. S. Steinolfson, E. R. Priest, S. Poedts, L. Nocera, M. Goossens.
Astrophys. J., Vol. 304, No. 1, p. 526 – 531 (1986).
Viscous damping of Alfvén surface waves is examined both analytically and numerically using incompressible MHD. The waves experience negligible decay below the transition zone. High–frequency waves damp just above the transition region, while those of lower frequency lose energy further out. A comparison of dissipative decay rates shows that wave damping by viscosity proceeds approximately two orders of magnitude faster than by resistivity.

062.085 **Condensation modes in sheared magnetic fields.**
C.–H. An.
Astrophys. J., Vol. 304, No. 1, p. 532 – 541 (1986).
This paper studies the condensation modes in coronal cylindrical plasmas, concentrating on how magnetic shear affects the stability. It is found that magnetic shear does not have important effects on initiating condensation for low–β coronal plasmas. The effects of magnetic field (shear, twist, and strength) on the condensation modes are investigated. The stability of condensation modes strongly depends on the choice of equilibrium temperature and density profiles. The result demonstrates the necessity of knowing the ambient heating function, which determines temperature and density profiles, for better understanding prominence formation and stability.

062.086 **Magnetized thermal conduction fronts.**
S. A. Balbus.
Astrophys. J., Vol. 304, No. 2, p. 787 – 798 (1986).
The evolution of planar thermal conduction fronts in the presence of a dynamically weak, but otherwise self–consistent, magnetic field is considered. The field is assumed to be connected and untangled. In the diffusion limit for the thermal conductivity, these fronts exhibit self–similar behavior, even in the presence of a field. The role of the field is restricted to channeling the heat flux along its lines of force. The large–scale effects of a magnetic field on cloud evaporation in the interstellar medium are briefly discussed, and it is suggested that these fields preclude the presence of time–independent evaporative solutions.

062.087 **Magnetic nonequilibrium and current sheet formation.**
S. I. Vainshtein (*S. I. Vajnshtejn*), E. N. Parker.
Astrophys. J., Vol. 304, No. 2, p. 821 – 827 (1986).
This paper investigates the equilibrium of a cluster of long, straight, twisted magnetic flux tubes extending in the z–direction in a highly conducting fluid, subject to the boundary condition that the cluster is surrounded by a uniform pressure P. It is shown that there is equilibrium only for axial symmetry, i.e., only for a single twisted flux tube. Any more complicated cluster of twisted tubes is subject to nonequilibrium reconnection of the transverse component of the field.

062.088 **Solitary waves in a magnetic flux tube.**
B. Roberts.
Phys. Fluids, Vol. 28, No. 11, p. 3280 – 3286 (1985). Abstr. in Phys. Abstr., Vol. 89, No. 1274, Entry 21259 (1986).

062.089 **Kelvin–Helmholtz instabilities of high–velocity magnetized, anisotropic shear layers.**
S. Roy Choudhury, V. L. Patel.
Phys. Fluids, Vol. 28, No. 11, p. 3292 – 3301 (1985). Abstr. in Phys. Abstr., Vol. 89, No. 1274, Entry 21267 (1986).

062.090 **The stability of dissipative magnetohydrodynamic shear flow in a parallel magnetic field.**
J. Lerner, E. Knobloch.
Geophys. Astrophys. Fluid Dyn., Vol. 33, No. 3 – 4, p. 295 – 314 (1985). Abstr. in Phys. Abstr., Vol. 89, No. 1274, Entry 23745 (1986).

062.091 **Gravitating configurations with a magnetic field. I. The Dirichlet problem.**
O. V. Kravtsov.
Astrofizika, Tom 24, Vyp. 3, p. 603 – 619 (1986). In Russian. English translation in Astrophysics, Vol. 24, No. 3.
It has been proved that the Dirichlet problem is in principle solvable and the conditions of existence of conjugated configurations have been determined (in terms of the Dedekind theorem) for objects with a magnetic field. The author has investigated the role of the surface current and found expressions for the magnetic fields within and outside the configuration which satisfy the conditions of the Dirichlet problem.

062.092 **Optically thin, adiabatic electron–positron pair wind.**
J. Fukue.
Publ. Astron. Soc. Jpn., Vol. 38, No. 2, p. 167 – 182 (1986).
Steady spherically symmetric e^+e^- pure pair plasma wind is investigated by taking into account the pair creation and annihilation and by adopting the equations of state which are valid in the full range of temperature. The effects of equations of state adopted are demonstrated. Electrons and positrons, ejected from a relativistic plasma which is confined near the center of active galactic nuclei, may be kept to be almost decoupled.

062.093 **Reaction instability in electron–positron pair equilibrium plasmas.**
J. Fukue.
Publ. Astron. Soc. Jpn., Vol. 38, No. 2, p. 183 – 198 (1986).
Stability of an electron–positron pair equilibrium plasma is investigated assuming the polytropic relation. The author obtains three kinds of instability criteria analogous to the case of thermal instability (Field 1965). The growth rates of perturbation in several limiting cases are obtained. Application is made to the e^+e^- pair equilibrium plasmas studied previously.

062.094 **Critical–level crossing of Rossby–gravity wave–packets in a linear–shear, stratified flow.**
H. Ando.
Geophys. Astrophys. Fluid Dyn., Vol. 35, p. 71 – 92 (1986). = Tokyo Astron. Obs. Repr., No. 760.

062.095 **Magnetostatic equilibrium in a nonuniform gravity field.**
S. Wang, A.–h. Wang.
Kexue Tongbao, Vol. 30, No. 8, p. 1031 – 1035 (1985). Abstr. in Phys. Abstr., Vol. 89, No. 1275, Entry 30154 (1986).

062.096 **Magnetohydrodynamic solitons and radio knots in jets.**
R. Fiedler.
Astrophys. J., Vol. 305, No. 1, p. 100 – 108 (1986).
Weakly nonlinear surface waves are examined in the context of the beam model for jetlike radio sources. By introducing a finite scale length, viz. the beam radius, geometrical dispersion can act to balance nonlinear wave growth and thereby produce solitons, localized wave packets of stable waveform. A method for obtaining a soliton equation from the MHD equations is presented and then applied to radio knots in jets.

062.097 **Ion–sound turbulence due to shock gradients in collisionless plasmas.**
N. Gopalswamy, G. Thejappa.
Pramāna, Vol. 25, No. 5, p. 575 – 585 (1985). Abstr. in Phys. Abstr., Vol. 89, No. 1277, Entry 39022 (1986).

062.098 **Magnetic buoyancy instabilities incorporating rotation.**
D. W. Hughes.
Geophys. Astrophys. Fluid Dyn., Vol. 34, No. 1 – 2, p. 99 – 142 (1985). Abstr. in Phys. Abstr., Vol. 89, No. 1277, Entry 42162 (1986).

062.099 **Macrodynamics of α^2 dynamos.**
G. R. Ierley.
Geophys. Astrophys. Fluid Dyn., Vol. 34, No. 1 – 2, p. 143 – 173 (1985). Abstr. in Phys. Abstr., Vol. 89, No. 1277, Entry 42163 (1986).

062.100 **Beams and jets in astrophysical plasmas.**
J. C. Brown.
Plasma Phys. Controlled Fusion, Vol. 27, No. 12A, p. 1359 – 1368 (1985). Abstr. in Phys. Abstr., Vol. 89, No. 1277, Entry 42164 (1986). – See Abstr. 012.046.

062.101 **Numerical results on instabilities of "top hat" jets.**
L. Zaninetti.
Phys. Fluids, Vol. 29, No. 1, p. 332 – 333 (1986). Abstr. in Phys. Abstr., Vol. 89, No. 1279, Entry 50235 (1986).

062.102 **Topological invariants of field lines rooted to planes.**
M. A. Berger.
Geophys. Astrophys. Fluid Dyn., Vol. 34, No. 4, p. 265 – 281 (1986). Abstr. in Phys. Abstr., Vol. 89, No. 1280, Entry 60616 (1986).

062.103 **Equilibrium of magnetic fields with arbitrary interweaving of the lines of force. I. Discontinuities in the torsion.**
E. N. Parker.
Geophys. Astrophys. Fluid Dyn., Vol. 34, No. 4, p. 243 – 264 (1986). Abstr. in Phys. Abstr., Vol. 89, No. 1280, Entry 60825 (1986).

062.104 **Investigations of spherical kinematic mean–field dynamo models.**
K.–H. Rädler.
Astron. Nachr., Vol. 307, No. 2, p. 89 – 113 (1986).
The paper supplements an earlier one on the mean–field approach to spherical kinematic dynamo models (Rädler 1980, 27.062.129) by results of numerical investigations.

062.105 **Large–scale magnetic fields in discs with strong differential rotation.**
G. D. Chagelishvili, J. G. Lominadze (*D. G. Lominadze*), Z. A. Sokhadze.
Astron. Nachr., Vol. 307, No. 2, p. 115 – 118 (1986).
The evolution of a large–scale magnetic field is considered in the non–linear dynamo situation in discs with α–effect due to non–mirror–symmetric turbulence and with strong differential rotation. The maximal value of the field strength is estimated by using an idealized dependence of the α–coefficient on the field strength.

062.106 **About the third integral of charged particle motion in strongly curved magnetic fields.**
J. Büchner.
Astron. Nachr., Vol. 307, No. 3, p. 191 – 199 (1986).
Charged particle motion is studied in magnetic fields with an inversion of field direction and a strong curvature in the reversal region. Investigations are performed in a typical field model. Due to its symmetry and stationarity two of the integrals of motion of the resulting Hamiltonian immediately arise. The question is analyzed whether the system is integrable (this means regular solutions exist) and of what kind the third integral is. Two different third integrals are found, which are valid in different parts of the phase space.

062.107 **Theoretical studies of interstellar molecular shocks – III. The formation of CH^+ in diffuse clouds.**
G. Pineau des Forêts, D. R. Flower, T. W. Hartquist, A. Dalgarno.
Mon. Not. R. Astron. Soc., Vol. 220, No. 4, p. 801 – 824 (1986).
The authors present the results of computations of both one–fluid hydrodynamic and three–fluid magnetohydrodynamic models of interstellar molecular shocks. A network of over 150 chemical reactions was incorporated in the models with a view to correctly predicting the variation of the fractional ionization in the precursors of MHD shocks and the column densities of a number of chemical species, particularly CH^+, which may be selectively formed in shocks. They find that one–fluid hydrodynamic models predict column densities of CH^+ which are too small to be considered compatible with observations. The MHD models are more successful in this respect.

062.108 **Theory of galactic disks and its relevance to the solar wind magnetic configuration.**
G. Bertin, B. Coppi.
Proceedings of the 1984 International Conference on Plasma Physics, Vol. 2, p. 1077 – 1084 (1985). Abstr. in Phys. Abstr., Vol. 89, No. 1281, Entry 67307 (1986). – See Abstr. 012.056.

062.109 **Super–rotation and diffusion of axial angular momentum. I. "Speed limits" for axisymmetric flow in a rotating cylindrical fluid annulus.**
P. L. Read.
Q. J. R. Meteorol. Soc., Vol. 112, No. 471, p. 231 – 251 (1986). Abstr. in Phys. Abstr., Vol. 89, No. 1281, Entry 67337 (1986).

062.110 **Super–rotation and diffusion of axial angular momentum. II. A review of quasi–axisymmetric models of planetary atmospheres.**
P. L. Read.
Q. J. R. Meteorol. Soc., Vol. 112, No. 471, p. 253 – 272 (1986). Abstr. in Phys. Abstr., Vol. 89, No. 1281, Entry 67338 (1986).

062.111 **The slide–away regime in space and laboratory plasmas.**
B. Coppi.
Proceedings of the 1984 International Conference on Plasma Physics, Vol. 2, p. 1039 – 1042 (1985). Abstr. in Phys. Abstr., Vol. 89, No. 1281, Entry 67341 (1986). – See Abstr. 012.056.

062.112 **The Benjamin–Ono–Burgers equation: an application in solar physics.**
P. M. Edwin, B. Roberts.
Wave Motion, Vol. 8, No. 2, p. 151 – 158 (1986). Abstr. in Phys. Abstr., Vol. 89, No. 1282, Entry 73039 (1986).

062.113 **Steepening of waves in radiative magnetohydrodynamics.**
R. Shyam, L. P. Singh, V. D. Sharma.
Acta Astronaut., Vol. 13, No. 3, p. 95 – 100 (1986). Abstr. in Phys. Abstr., Vol. 89, No. 1283, Entry 75699 (1986).

062.114 **Lower–hybrid instabilities driven by an ion velocity ring.**
K. Akimoto, K. Papadopoulos, D. Winske.
J. Plasma Phys., Vol. 34, Part 3, p. 445 – 465 (1985). Abstr. in Phys. Abstr., Vol. 89, No. 1283, Entry 75752 (1986).

062.115 **Propagating double–layer solution in two–beam plasma systems.**
L.–t. Song.
Chin. Astron. Astrophys., Vol. 10, No. 1, p. 67 – 72 (1986). English translation of 40.062.034.

062.116 **Multiple Compton scattering in strongly magnetized plasma.**
Y.–m. Xu, Q.–y. Qu.
Chin. Astron. Astrophys., Vol. 10, No. 2, p. 88 – 93 (1986). English translation of Acta Astrophys. Sin., Vol. 6, No. 1, p. 58 – 65 (1986).
The authors discussed the effect of multiple scattering on the annihilation line photons and γ–ray continuum. In their calculations they assumed the scattering layer to be a non–degenerate plasma with a temperature between (+7)K and (+8)K, and a strong magnetic field of 4.414(+12)G. The effect of the strong field will help to gain a deeper understanding of the γ–ray burst spectrum.

062.117 **Self–excitation effect in the scattering of electromagnetic radiation in a strong magnetic field.**
Z.–w. Liu.
Chin. Astron. Astrophys., Vol. 10, No. 2, p. 154 – 161 (1986). English translation of Chin. J. Space Sci., Vol. 6, p. 33 – 42 (1986).
It is pointed out that the electron scattering of incident electromagnetic radiation of frequency ω_i includes also a component with the cyclotron frequency ω_L. Except in the limit $\omega_i \gg \omega_L$, this "self–excited effect" modifies appreciably the cross section. The modified cross section for a classical non–relativistic electron is given in this paper for both a single electron and a cold plasma. The spectral distribution of electron scattering is discussed and the Thomson absorption coefficient of a magnetized thermal equilibrium plasma is obtained for some typical cases. The results obtained here may be relevant to astrophysical processes, such as the spectrum formation of X–ray pulsars.

062.118 **Quasielectrostatic wave propagation in a hot anisotropic plasma.**
S. S. Sazhin.
Planet. Space Sci., Vol. 34, No. 6, p. 497 – 509 (1986).

062.119 **Spherically symmetric accretion: an improved gas–dynamic solution.**
E. Ya. Gidalevich, A. B. Egorov.
Sov. Astron., Vol. 29, No. 6, p. 628 – 629 (1985). English translation of 40.062.148.

062.120 **Is magnetic energy the source of synchrotron radiation in jets?**
L. Turner.
Astrophys. J., Vol. 305, No. 2, p. 668 (1986).
It has been suggested recently that the source of synchrotron radiation in certain radio jets, such as NGC 6251, may originate in the dissipation of magnetic energy necessary to maintain the helical magnetic field configuration in a state of minimum energy. Upon reexamination, however, it is demonstrated here that the assumed model leaves no residue of magnetic energy for fueling jet synchrotron radiation.

062.121 **Magnetic structure of a flux rope.**
T. Yeh.
Astrophys. J., Vol. 305, No. 2, p. 884 – 891 (1986).
A class of mathematical solutions is presented for the magnetic structure of a flux rope. These solutions in an orthogonal system of tubal coordinates describe the solenoidal distributions of the magnetic field and electric current in a flux rope. The model structure is useful in the studies of the dynamics of looplike coronal transients and the equilibria of solar prominences. It shows that the Lorentz force affects mainly the lateral motion or equilibrium of mass elements in a flux rope relative to the axis.

062.122 **Erratum: "Magnetohydrodynamic stability of an axisymmetric, line–tied, diamagnetic plasmoid embedded in a uniform magnetic field" [Astrophys. J., Vol. 288, No. 2, p. 672 – 678 (1985)].**
T. J. Bogdan.
Astrophys. J., Vol. 305, No. 2, p. 954 (1986). See Abstr. 39.062.061.

062.123 **Erratum: "Force–free equilibria of magnetized jets" [Astrophys. J., Vol. 289, No. 1, p. 173 – 187 (1985)].**
A. Königl, A. R. Choudhuri.
Astrophys. J., Vol. 305, No. 2, p. 954 (1986). See Abstr. 39.062.066.

062.124 **Three–dimensional structure and dynamics of a supersonic jet.**
C. N. Arnold, W. D. Arnett.
Astrophys. J., Lett. Ed., Vol. 305, No. 2, p. L57 – L60 (1986).
Using a different numerical technique, the authors have reproduced in two dimensions and three dimensions the 1982 results of Norman and colleagues for an axisymmetric supersonic jet. Supersonic, hot, pressure–matched jets ($M = 6.0$) have cocoons with a rich structure of cells that have complicated dynamics including vortical motion in all three degrees of freedom. These cocoon cells shield the beam at least partially from the three–dimensional fluting instability.

062.125 **Magnetohydrodynamic effects in radio jets.**
E. M. de Gouveia Dal Pino, R. Opher.
Rev. Mex. Astron. Astrofis., Vol. 12, Spec. Issue, p. 143 – 152 (1986). – See Abstr. 012.066.
The authors study the dynamic effects of magnetic fields on jets. For a quantitative analysis of the problem, they consider a continuous jet with negligible viscous and thermal conductive effects and infinite electrical conductivity. They show that solutions of the magnetohydrodynamic equations give a natural explanation for the observed slow and rapid changes of opening angle. The authors show also that the observed opening angle and its rate of change put limits on the possible physical parameters of the jet.

062.126 **Sound and thermal waves in a fluid with an arbitrary heat–loss function.**
M. H. Ibáñez S.
Rev. Mex. Astron. Astrofis., Vol. 12, Spec. Issue, p. 339 (1986). Abstract. – See Abstr. 012.066.

062.127 **On the propagation of small disturbances in a radiating fluid in which a general reaction proceeds. I. Eddington approximation.**
M. H. Ibáñez S.
Rev. Mex. Astron. Astrofis., Vol. 12, Spec. Issue, p. 340 (1986). Abstract. – See Abstr. 012.066.

062.128 **Thermochemical instabilities in reacting gases. II. Plasma of hydrogen in presence of a magnetic field.**
M. H. Ibáñez S., E. Sira.
Rev. Mex. Astron. Astrofis., Vol. 12, Spec. Issue, p. 341 (1986). Abstract. – See Abstr. 012.066.

062.129 **A model of hydromagnetic turbulence in differentially rotating disks.**
S. Kato, T. Horiuchi.
Publ. Astron. Soc. Jpn., Vol. 38, No. 3, p. 313 – 333 (1986).
The structure of hydromagnetic turbulence in differentially rotating disks is examined in order to determine the so–called α–value in accretion disks. This examination is made by considering energy interactions among the turbulent gas, turbulent magnetic field, and thermal and kinetic states of differentially rotating disks.

062.130 **Modulation instability of circularly polarized Alfvén waves.**
M. Longtin, B. U. Ö. Sonnerup.
J. Geophys. Res., Vol. 91, No. A6, p. 6816 – 6824 (1986).
The stability of a finite amplitude circularly polarized Alfvén wave of wave number k_0 is studied by using the two–fluid isentropic equations. Linear perturbation analysis, involving two sideband transverse waves having wave numbers $k_0 \pm k$ and a longitudinal wave with wave number k, is used to find the exact sixth–order dispersion relation. The analysis is then limited to the case where $k \ll k_0$. The resulting fourth–order dispersion relation is examined analytically and numerically, and a surface is found that separates stable and unstable regions in parameter space. This surface describes the boundary between stable and unstable regions not only for $k \ll k_0$ but for the entire branch of the dispersion relation which extends to $k = 0$.

062.131 **A simple theory of turbulent transport coefficients.**
D. Montgomery.
Plasma penetration into magnetospheres, p. 15 – 36 (1986). – See Abstr. 012.071.
A coefficient of turbulent diffusion was shown long ago to be expressible in terms of the kinetic energy spectrum of the diffusing turbulence, which is the harder quantity to calculate. By making a few reasonable assumptions, it is possible to obtain the kinetic energy spectrum from an eddy–viscosity model which generalizes well–known results of "alpha–effect" and "beta–effect" dynamo theory. The Kolmogoroff spectrum is recoverable as a special case.

062.132 **Magnetospheric interchange instability: a tutorial lecture.**
A. F. Cheng.
Plasma penetration into magnetospheres, p. 37 – 52 (1986). – See Abstr. 012.071.
The paper reviews the possible significance of magnetospheric interchange instabilities. Flux tube interchanges may be an important mechanism for radial transport of plasma in planetary magnetospheres. The conventional theory of interchange instability in which the instability criterion is derived from a simple energy principle is reviewed. The author derives the interchange instability criterion using a simple dynamical argument in the case of straight parallel field lines.

062.133 **Accretion discs.**
G. E. Morfill, G. Hasinger.
Mitt. Astron. Ges., No. 65, p. 53 – 78 (1986). – See Abstr. 012.072.
In this review the authors discuss the physics of accretion discs and their mathematical description using macroscopic transport coefficients. Examples of accretion discs include the presolar/

preplanetary nebula, Saturn's rings and, of course, binary systems containing a compact object. The compact object – a white dwarf, neutron star, or perhaps even a black hole – accretes matter from its companion, which in most cases is a normal star. The great variety of physical processes found in the different accretion disc systems has given rise to a large number of publications. The authors discuss some of this work and reference additional material.

062.134 **Bremsstrahlung in hot, strongly magnetized plasmas.**
U. Bock, H. Herold, H. Ruder, G. Wunner.
Mitt. Astron. Ges., No. 65, p. 217 – 218 (1986). – See Abstr. 012.072.

062.135 **One–photon annihilation in strongly magnetized relativistic electron–positron pair plasmas.**
G. Wunner, J. Paez, H. Ruder, H. Herold.
Mitt. Astron. Ges., No. 65, p. 218 – 219 (1986). – See Abstr. 012.072.

062.136 **Magnetic fields of cosmic bodies.**
B. V. Vasil'ev.
Obed. inst. yader. issled. Dubna. Prepr., No. R17–65–255, 6 pp. (1985). In Russian. Abstr. in Ref. Zh., 51. Astron., 4.51.149 (1986).

Kinetic theory of particles and photons. Theoretical foundations of non–LTE plasma spectroscopy.
See Abstr. 003.027.

Methods in the qualitative theory of dynamical systems in astrophysics and gas dynamics.
See Abstr. 003.028.

Physics of shock waves in gases and plasmas.
See Abstr. 003.127.

Instabilities in space and laboratory plasmas.
See Abstr. 003.133.

Transfer processes in an anisotropic circumterrestrial plasma.
See Abstr. 003.147.

The mathematical theory of turbulence.
See Abstr. 003.183.

On a possibility of experimental investigation of the anomalous ionization phenomenon.
See Abstr. 022.059.

Atomic partition functions for chromium.
See Abstr. 022.119.

The determination of the gravitational potential in spherical coordinates. III: Radial recursive integration of the Poisson integral.
See Abstr. 042.082.

One–photon pair annihilation in magnetized relativistic plasmas.
See Abstr. 063.001.

Electron–positron pair production by ultrarelativistic electrons in a soft photon field.
See Abstr. 063.002.

A method for solving one–dimensional plane flow problems for inviscid radiating gases using characteristic methods.
See Abstr. 063.020.

Saha decrements and collisional–radiative recombination and ionization coefficients for a nonequilibrium nitrogen plasma.
See Abstr. 063.023.

On the theory of strong shock waves propagation in inhomogeneous gravitating media. II. Radiative cooling phase.
See Abstr. 063.056.

Non–unique solutions to the stellar atmosphere problem.
See Abstr. 064.012.

Steady hydromagnetic flows in open magnetic fields. I. A class of analytic solutions.
See Abstr. 064.027.

The equilibrium structure of thin magnetic flux tubes. II.
See Abstr. 064.042.

A non–local theory of time–dependent convection for numerical application to the radial pulsation of stars.
See Abstr. 065.026.

Gravitational radiation from the collapse and explosion of rapidly rotating, homogeneous, gaseous ellipsoids imbedded in halos.
See Abstr. 066.018.

Gravitational instability in the presence of suspended particles and variable magnetic field.
See Abstr. 066.133.

On the acceleration of charged particles at relativistic shock–fronts.
See Abstr. 066.175.

Relativistic neutral superfluids. Exact solutions.
See Abstr. 066.218.

Electron–positron pairs in a mildly relativistic plasma in active galactic nuclei.
See Abstr. 067.011.

Wave propagation in pulsar magnetospheres: dispersion relations and normal modes of plasmas in superstrong magnetic fields.
See Abstr. 067.036.

Wave propagation in pulsar magnetospheres: refraction of rays in the open flux zone.
See Abstr. 067.037.

On the Papaloizou and Pringle instability.
See Abstr. 067.041.

Plasma instabilities in the magnetosphere of pulsars.
See Abstr. 067.136.

Hydrodynamics of accretion columns.
See Abstr. 067.156.

The dynamics of magnetically trapped fluids. I. Implications for umbral dots and penumbral grains.
See Abstr. 072.037.

Heat transport in steep temperature gradients. I. Small flaring solar loops.
See Abstr. 073.057.

A shock condensation mechanism for loop prominences.
See Abstr. 073.058.

Fast–shock formation in line–tied magnetic reconnection models of solar flares.
See Abstr. 073.106.

Oscillations trapped in a chromosphere with a horizontal magnetic field.
See Abstr. 073.107.

Momentum deposition by electron beams in solar flares.
See Abstr. 073.118.

Modeling of reconnection processes in ribbon flares.
See Abstr. 073.119.

Particle acceleration in solar flares.
See Abstr. 073.131.

Plasma transport across the heliopause.
See Abstr. 074.073.

Dynamic heating of stationary coronal loops.
See Abstr. 074.102.

Numerical solution of the magnetostatic equations for thick flux tubes, with application to sunspots, pores, and related structures.
See Abstr. 075.011.

Magnetic reconnection.
See Abstr. 080.118.

Nonlinear propagation of whistlers in the ionosphere.
See Abstr. 083.017.

The stochastic excitation of reversals in simple dynamos.
See Abstr. 084.041.

A theoretical interpretation of the Σ–D relation for supernova remnants.
See Abstr. 125.020.

The hydrodynamics of clouds overtaken by supernova remnants.
See Abstr. 125.087.

The hydrodynamics of clouds overtaken by supernova remnants. I. Cloud crushing phenomena.
See Abstr. 131.063.

Secondary antiproton production in relativistic plasmas.
See Abstr. 144.005.

Cosmic–ray shock acceleration in oblique MHD shocks.
See Abstr. 144.029.

Computation of the quasi–steady gas flow in a spiral galaxy by means of a multigrid method.
See Abstr. 151.021.

Global simulations of gas flow in disc galaxies – I. Response to a spiral density wave.
See Abstr. 151.062.

Construction of a global gas–dynamical model for our Galaxy.
See Abstr. 155.031.

On the stability of the galactic magnetic field in the presence of a magnetic monopole halo.
See Abstr. 155.096.

On wind–type flows in astrophysical jets. II. Propagation outside the nucleus and the case of M87.
See Abstr. 158.024.

A model for the surface brightness of a turbulent, low Mach number jet. II. The global energy budget and radiative losses.
See Abstr. 158.025.

Beam models for radio sources. V. Collimation in more realistic galactic potentials.
See Abstr. 158.026.

Morphology and asymmetry of active galaxies from cavity stability.
See Abstr. 158.050.

A model for the surface brightness of a turbulent low Mach number jet. III. Adiabatic jets of arbitrary density ratio: application to NGC 315.
See Abstr. 158.261.

Structure and stability of quasar clouds.
See Abstr. 159.140.

Thermal–chemical instability in a pre–galactic gas cloud.
See Abstr. 161.019.

063 Radiative Transfer, Scattering

063.001 **One–photon pair annihilation in magnetized relativistic plasmas.**
A. K. Harding.
Astrophys. J., Vol. 300, No. 1, p. 167 – 177 (1986). With a correction on p. 462.

In superstrong magnetic fields, electron–positron pairs may annihilate into single photons producing spectral features above 1 MeV. The author calculates the exact one–photon annihilation rate in the general case where pairs may annihilate from excited Landau states, extending the previous studies which were restricted to pairs in the ground state. She also derives asymptotic expressions for annihilation spectra and rates in the limit of large pair quantum numbers. It is shown that one–photon annihilation is competitive with the two–photon process at typical neutron star field strengths. The one–photon rates are evaluated for both thermal and nonthermal pair distributions and are compared to the corresponding two–photon annihilation rate.

063.002 **Electron–positron pair production by ultrarelativistic electrons in a soft photon field.**
A. Mastichiadis, A. P. Marscher, K. Brecher.
Astrophys. J., Vol. 300, No. 1, p. 178 – 189 (1986).

The fully differential cross section for photon–electron pair production is integrated numerically over phase space. Results are obtained for the astrophysically interesting case in which the interaction between an ultrarelativistic electron and a soft photon results in $e^{\pm}$ pair production. The positron spectrum is a function of the energies of both the photon and the electron, as well as the angle of interaction. The energy at which the positron distribution peaks is inversely proportional to the photon energy and independent of the electron energy.

063.003 **Circular polarization of astrophysical masers due to overlap of Zeeman components.**
S. Deguchi, W. D. Watson.
Astrophys. J., Lett. Ed., Vol. 300, No. 1, p. L15 – L18 (1986).

Calculational methods are presented and utilized in an idealized case to evaluate the polarization due to the overlap of Zeeman components that arises in the radiation from astrophysical masers. The essential simplicity of this previously unrecognized effect is a compelling argument in its favor as the basic cause for the strong, net circular polarization of the OH masers. Observations can then yield an estimate for the strength of the magnetic fields in the masing regions. Magnetic fields of approximately a milligauss or greater are inferred for the masing gas associated with regions of star formation.

063.004 **The statistical equilibrium of hydrogen and helium in a radiation field, with an application to interpreting Wolf–Rayet spectra.**
A. K. Bhatia, A. B. Underhill.
Astrophys. J., Suppl. Ser., Vol. 60, No. 1, p. 323 – 356 (1986).
A 70–level model helium atom which includes singly and doubly excited states of He I and a 17–level model hydrogen atom are used to solve the equations of statistical equilibrium for helium and hydrogen in a radiation field. The plasma has a velocity gradient, and densities and temperatures typical for the mantles of Wolf–Rayet stars. The emission lines of the Wolf–Rayet stars HD 50896, HD 191765, HD 192103, and HD 192163 are analyzed, as well as those of ζ Pup (O4ef) and α Cam (O9.5 Ia). It is argued that the apparent decrement of intensity along the Pickering series of He II is an ambiguous indicator of the composition of the atmospheres of Wolf–Rayet stars. The electron temperatures in the atmospheres of Wolf–Rayet stars probably exceed 10^5K.

063.005 **Classification of pump mechanisms for astronomical OH masers and a maser model for the H_2O photodissociation pump mechanism.**
P. Andresen.
Astron. Astrophys., Vol. 154, No. 1/2, p. 42 – 54 (1986).
The pump mechanism of astronomical OH masers is discussed in view of two new experimental results on the photodissociation of H_2O in the VUV and on the inelastic collisions of OH with H_2. The experimental results rule out the most favored collisional pump mechanisms for astronomical OH masers and yield two new pump mechanisms. A comparison of features of stellar and interstellar masers leads to the suggestion that most interstellar masers in regions of star formation are pumped by H_2O photodissociation, whereas stellar masers may be pumped by inelastic OH–H_2 collisions. For the H_2O pump mechanism a specific maser model including H_2O and OH photodissociation as well as IR relaxation is presented and simple results are discussed.

063.006 **Compton heating and cooling in thermal electron gases.**
P. W. Guilbert.
Mon. Not. R. Astron. Soc., Vol. 218, No. 1, p. 171 – 175 (1986).
Exact results are derived for calculating Compton heating and cooling in optically thin gases. For electron temperatures less than 50 keV an asymptotic expansion is provided which is free of integrals and special functions which can be used to calculate Compton heating and cooling for photon spectra extending up to 5 MeV with an accuracy of better than 3 per cent.

063.007 **Plasma and radiation processes.**
R. G. Hewitt, D. B. Melrose.
Astrophys. Space Sci., Vol. 118, Nos. 1/2, p. 183 – 193 (1986). – See Abstr. 012.001.
The physical processes leading to metre–wave and microwave emissions from the solar corona are reviewed and the possibility of observing analogous phenomena on other stars is discussed. Particular emphasis is placed on three alternative processes for the emission from flare stars: cyclotron maser emission, second harmonic plasma emission, and gyro–synchrotron emission.

063.008 **Moment method for the inverse radiative transfer in inhomogeneous media.**
S. J. Wilson, K. K. Sen.
Astrophys. Space Sci., Vol. 119, No. 1, p. 93 – 96 (1986). – See Abstr. 012.001.
An inverse problem of radiative transfer in an inhomogeneous plane medium of scattering albedo $\omega = \omega_0 \exp(-\tau/s)$ is solved by a moment method. The results are compared with those obtained by Dunn (1983) using Monte Carlo method.

063.009 **Comments on "The angle–dependent redistribution functions R_{III} and R_{IV}" by S. J. McKenna.**
I. Hubený, P. Heinzel.
Astrophys. Space Sci., Vol. 119, No. 2, p. 409 – 413 (1986). – See Abstr. 38.063.060.
It is shown that the complicated form of the line absorption probability function for scattering in subordinate lines, derived by McKenna (1984), is an artifact of adopting a mathematically inconvenient expression for the corresponding redistribution function. It is demonstrated that the absorption probability must be given by the Voigt function.

063.010 **Matrix formulations for the transfer of solar radiation in a plane–parallel scattering atmosphere.**
T. Nakajima, M. Tanaka.
J. Quant. Spectrosc. Radiat. Transfer, Vol. 35, No. 1, p. 13 – 21 (1986).
Matrix formulations are presented for the discrete–ordinate and the matrix–operator methods of solving the transfer of solar radiation in a plane–parallel scattering atmosphere. Eigenspace transformations of symmetric matrices are introduced into the method of Stamnes and Swanson instead of using the decomposition of an asymmetric matrix. The computational stability is considerably improved by this algorithm, especially for single–precision calculations. Representations of the reflection and transmission matrices in the matrix–operator method are also given, in terms of the indicated formulations, by considering a boundary–value problem of the discrete–ordinate method. The solutions of the discrete–ordinate method for inhomogeneous atmospheres are given by combining discrete–ordinate solutions for respective homogeneous sublayers through the addition technique of the matrix–operator method.

063.011 **Effect of quasi–bound states on collisional redistribution.**
M. Belsley, W. J. Alford, K. Burnett, J. Cooper.
J. Quant. Spectrosc. Radiat. Transfer, Vol. 35, No. 1, p. 53 – 65 (1986).
The authors derive an expression for the collisional redistribution of radiation due to states trapped by an excited state potential barrier in the limit of no subsequent collisions during the excited state lifetime. By considering two simple models, the authors suggest three possible methods of qualitatively determining excited state well depths when no detailed knowledge of the ground or excited state interaction potentials exists a priori.

063.012 **Numerical evaluation of time–dependent reflected intensity from an anisotropically scattering semi–infinite atmosphere.**
B. D. Ganapol, M. Matsumoto.
J. Quant. Spectrosc. Radiat. Transfer, Vol. 35, No. 1, p. 71 – 78 (1986).
A numerical scheme, designed to provide four or five digit accuracy for the reflected photon intensity from an anisotropically scattering semi–infinite medium, is presented. The atmosphere is characterized by both a mean free time between collisions and a mean time of temporal capture. Tables of reflected intensities for a test–phase function are included for use as benchmarks to which other numerical solutions can be compared.

063.013 **Photon path length analysis of radiative transfer in planar layers with single internal sources.**
R. D. Skocypec, R. O. Buckius.
J. Quant. Spectrosc. Radiat. Transfer, Vol. 35, No. 2, p. 87 – 108 (1986).
A solution technique is presented which successfully predicts the radiative intensity and flux path length distributions for a scattering medium with an arbitrary internal source. All scattering effects are included through the optical path length concept. The extension of the optical path length technique to incorporate internal sources permits relatively simple modeling of scattering layers with any source distribution (e.g. temperature) and any absorption feature using a data base of responses (path length distributions) to arbitrarily located internal single sources. Path length distributions are presented for a single source in the layer and exhibit trends similar to those seen in the boundary source

problem. The effects of orders of scattering are demonstrated. Although the technique is directed to radiative heat transfer analyses, the source need not be a thermal source and the solutions are not restricted to thermal applications.

063.014 **On pumping astronomical masers.**
N. D. Kylafis, C. Norman.
Astrophys. J., Lett. Ed., Vol. 300, No. 2, p. L73 – L77 (1986).
The authors have examined the possibility that astronomical masers are pumped collisionally in regions where the charged particle temperature is larger than that of the neutral particles. Such conditions occur naturally in the magnetic precursors of MHD shocks. The authors have derived analytically a condition for masing which depends on the collisional rate coefficients and the fractional ionization only. An important advantage of this pumping mechanism over other pumps is that non–LTE collisions can produce greater power than ordinary pumping schemes for similar conditions. An application of this pumping mechanism to H_2O masers is presented.

063.015 **Magnetization effects in an incoherent synchrotron source.**
A. K. Singal.
Astron. Astrophys., Vol. 155, No. 2, p. 242 – 246 (1986).
The author has derived a lower limit to the magnetic field energy density in terms of the energy density of the relativistic electrons within the radiating regions of an incoherent synchrotron source. This limit, imposed by the catastrophic diamagnetic effects, in turn implies that the maximum brightness temperatures in such sources cannot be higher than about $10^{11.5}$K, which is consistent with the VLBI observations of compact radio sources. It follows further that in such sources, total power radiated by self–Compton emission process cannot be much higher than that radiated by synchrotron mechanism itself.

063.016 **Influence of ion–atom collisions on the absorption of radiation.**
A. A. Mihajlov, M. S. Dimitrijević.
Astron. Astrophys., Vol. 155, No. 2, p. 319 – 322 (1986).
It is shown that for the interpretation of the continuum absorption spectra, of A type stars, the processes $H_2^+ + \hbar\omega \rightarrow H + H^+$ and $H + H^+ + \hbar\omega \rightarrow H + H^+$ must be considered together since their contribution is comparable. The corresponding absorption coefficients are also determined.

063.017 **Analytical expressions for the temporal and spatial evolution of an energetic electron population in an inhomogeneous medium.**
N. Vilmer, G. Trottet, A. L. MacKinnon.
Astron. Astrophys., Vol. 156, No. 1/2, p. 64 – 72 (1986).
In the context of non thermal hard X–ray models, the temporal evolution of the instantaneous X–ray spectrum is related to the temporal evolution of the non thermal electron distribution function in the source. As a necessary preliminary to detailed source modelling, the authors present here time dependent solutions of the electron continuity equation. This equation describes the temporal and spatial evolution of an electron population continuously injected during a finite time and over a finite extent in a stationary and inhomogeneous medium. The solutions obtained for some simple situations are examined and compared to previous results. Potential applications to solar hard X–ray observations are discussed.

063.018 **Two–photon emission in X–ray pulsars. I. Basic formulas.**
D. B. Melrose, J. G. Kirk.
Astron. Astrophys., Vol. 156, No. 1/2, p. 268 – 276 (1986).
It is suggested that the following processes may determine the spectra of X–ray pulsars such as Her X–1. (1) Nonradiative collisional excitation of electrons to their first Landau level, followed by radiative decay, is the only important source of photons. (2) These cyclotron photons are transferred through the source by being absorbed and re–emitted. (3) Non–cyclotron photons below the cyclotron frequency can be produced by the re–emission process. (4) Non–cyclotron photons above the cyclotron frequency can be produced by a related process. Specific formulas describing the two–photon emission process are derived and expressed in the form of an averaged emission probability. Compound probabilities are defined for absorption plus re–emission and for non–radiative collisional excitation plus cyclotron emission. Interpolation formulas relate these compound probabilities to the probabilities for resonant scattering and resonant bremsstrahlung respectively.

063.019 **Two–photon emission in X–ray pulsars. II. Radiation transfer.**
J. G. Kirk, D. B. Melrose.
Astron. Astrophys., Vol. 156, No. 1/2, p. 277 – 282 (1986).
A transport equation governing the transfer of radiation in the accretion columns of strongly magnetized X–ray pulsars is derived using a generalization of a method due to Melrose (1981). Processes included are cyclotron absorption and emission, Compton scattering, bremsstrahlung and double Compton scattering. A prescription is derived to deal with terms which scatter particles from the ground state to the first excited state. The equation is used to demonstrate that the resonant part of double Compton scattering may be expected to dominate bremsstrahlung as a source of photons over a wide parameter range of interest. Current models which neglect this effect are inconsistent in their calculation of continuum spectra.

063.020 **A method for solving one–dimensional plane flow problems for inviscid radiating gases using characteristic methods.**
F. Schmitz, B. E. Wolf.
Astron. Astrophys., Vol. 156, No. 1/2, p. 289 – 294 (1986).
The authors describe a method for the numerical treatment of one–dimensional plane problems of radiation hydrodynamics relating to inviscid fluids subject to selfgravitation and stationary radiation in LTE. The adopted equation of state describes a gas consisting of hydrogen and helium, including effects of ionization and dissociation under LTE conditions. For plane geometry and a Lagrangian description with appropriate variables, the characteristic form of the hydrodynamic equations is very short and compact. The authors also discuss the simple case of the optically thin approximation, a special form of the two–ray approximation and the diffusion approximation. They briefly consider some numerical details of two codes, namely a second order scheme written for a universal computer and a first order scheme written for a vector computer.

063.021 **Shape of spectral lines emitted from a gas in the presence of a condensed phase.**
I. A. Vasilieva (*I. A. Vasil'eva*).
J. Quant. Spectrosc. Radiat. Transfer, Vol. 35, No. 3, p. 171 – 178 (1986).
The shape of emission lines from a spatially inhomogeneous gas containing particles of a condensed phase is discussed. Monochromatic scattering from particles and the walls limiting the infinite slab was taken into account, as well as emission and absorption by atoms, particles and walls. The Sobolev probabilistic method was used to obtain an exact solution of the equation of transfer. The intensity of the radiation emitted from the slab is expressed in terms of the fundamental functions. The final expressions may be used for calculations of emission line shapes.

063.022 **Radiative transport in a planar medium exposed to azimuthally unsymmetric incident radiation.**
S. Kumar, J. D. Felske.
J. Quant. Spectrosc. Radiat. Transfer, Vol. 35, No. 3, p. 187 – 212 (1986).
An analytical solution is developed for the radiation field engendered by illuminating an absorbing/anisotropically scattering planar medium with an azimuthally unsymmetric incident flux. The boundaries of the medium are allowed to possess both specular and diffuse reflection characteristics. The problem is mathematically reduced to solving a system of azimuthally independent equations which may be analyzed by many existing techniques. A solution scheme is formulated in terms of singular eigenfunctions and solved using a modified F_N method. Computations are per-

formed for the azimuthally symmetric case of irradiation by parallel rays at oblique incidence. Several different phase functions are considered and results are presented for the discrete eigenvalues, the heat flux distributions within the medium, the reflectance and transmittance of the slab and the angular distribution of the intensities at the boundaries.

063.023 **Saha decrements and collisional–radiative recombination and ionization coefficients for a nonequilibrium nitrogen plasma.**
R. D. Taylor, A. W. Ali.
J. Quant. Spectrosc. Radiat. Transfer, Vol. 35, No. 3, p. 213 – 230 (1986).
Using a time–dependent approach, a collisional–radiative model is developed for a nitrogen plasma. Effective recombination and ionization coefficients are calculated for optically thin and optically thick cases in bound–bound u.v. radiation. The steady–state population densities for neutral and ionic states are presented in terms of their deviations from Saha equilibrium predictions. These results show the approach of the plasma to local thermodynamic equilibrium as a function of electron temperature and density. The computations focus on temperatures between 1.0 and 3.0 eV and electron densities between 10^{16} and $10^{19}\mathrm{cm}^{-3}$.

063.024 **Conditions for the elements of the scattering matrix.**
J. W. Hovenier, H. C. van de Hulst, C. V. M. van der Mee.
Astron. Astrophys., Vol. 157, No. 2, p. 301 – 310 (1986).
A study is made of general conditions for the 16 elements of the scattering matrix that transforms the Stokes parameters of the incident wave into those of the scattered wave. A number of relations is derived for scattering by one particle of arbitrary size, shape and composition. In this case a complete set of nine independent equations is found. Several methods are described to obtain inequalities for the elements of the scattering matrix for an assembly of particles. A selection of these inequalities is explicitly derived and discussed.

063.025 **The radiative acceleration of astrophysical jets: line locking in SS 433.**
P. R. Shapiro, M. Milgrom, M. J. Rees.
Astrophys. J., Suppl. Ser., Vol. 60, No. 2, p. 393 – 431 (1986).
A radiative acceleration mechanism known as "line locking", suggested earlier to explain both the value and the remarkable invariance in space and time of the jet outflow velocity in SS 433, is examined here in detail. Special relativistic flow equations are derived. Detailed, numerical solutions of these equations are presented, including matter–radiation coupling, ionization balance, and the atomic level population equilibrium, for a number of cases, including line locking by H atoms and by the hydrogenic ions of heavier elements, such as He, C, and Fe. Approximate analytical solutions are also derived, including a fundamental constraint on the efficiency of the radiative acceleration mechanism.

063.026 **A numerical investigation of approximation procedures for optically thick resonance lines.**
M. Faurobert.
Astron. Astrophys., Vol. 158, No. 1/2, p. 191 – 199 (1986).
This paper concerns approximation procedures which can replace the full transfer equation with R_{II} partial frequency redistribution. Existing asymptotic analyses of the formation of optically thick resonance lines in dilute media show that the interior radiation field (away from boundaries) can be approximated by the solution of a space and frequency diffusion equation and provide an expression of the emergent radiation field, away from line center, in terms of the interior solution. Here an approximation for the angle–averaged intensity, to be used at all optical depths, is constructed by adding the solution of the interior diffusion equation to the surface value of the angle–averaged intensity. The validity of this approximation is tested by comparison with the solution of the full transfer equation, in the case where the probability of collisional deexcitation and the Doppler width of the line are uniform. The diffusion equation has then an analytical solution. The tests are performed first for a two–level atom, considering a wide range of optical thicknesses and densities, and also for the Lyα line under physical conditions typical of the quasar broad emission line regions. In the latter case, the validity of a mean escape probability approximation has also been examined.

063.027 **The formation of the continuum spectrum in X–ray pulsars.**
J. G. Kirk.
Astron. Astrophys., Vol. 158, No. 1/2, p. 305 – 309 (1986).
It is suggested that the flat spectrum (in intensity) observed as the phase averaged emission in the range 2 – 20 keV of some X–ray pulsars – notably Her X–1 – is just the optically thin emission from resonant double Compton scattering. Production of photons in this range also involves the participation of two cyclotron photons whose occupation numbers are parameterized by a Bose–Einstein distribution. Estimates are presented in which the required flux is emitted from a hot spot covering a few per cent of the neutron star's surface with a thickness of about one Thomson depth and a temperature of 10 keV.

063.028 **Brightness distribution of resonant line photons in spherical envelopes.**
A. Natta, S. Beckwith.
Astron. Astrophys., Vol. 158, No. 1/2, p. 310 – 324 (1986).
The scattering of resonant line photons in spherically symmetric, expanding envelopes with turbulence is studied numerically using a Monte–Carlo scattering code. The envelopes have different radial density distributions for scatterers, various ratios of systematic to random velocity, and different line center optical depths. These calculations produce both emerging line profiles and spatial brightness distributions for scattered photons as they appear to a distant observer. The line shapes and brightness distributions are weak functions of line center optical depth and the form of the density distribution, being virtually indistinguishable for a wide range of parameters. Differing ratios of systematic to random velocity have a stronger effect on the emerging photons. The authors argue that although it may be difficult to determine densities and temperatures in circumstellar envelopes from spectral line data, it will be easier to detect velocity gradients, especially when brightness data become available.

063.029 **Radiative transfer in spectral lines by noncoherent scattering. I. Isotropically–scattering medium.**
J. S. Aliyev.
Astrophys. Space Sci., Vol. 121, No. 1, p. 45 – 60 (1986).
A functional analytic method of solution of the operator equations is used to obtain a solution of the radiative transfer equation in spectral lines. A problem of scattering in the spectral line with frequency redistribution in an isotropic scattering medium is considered.

063.030 **The comoving–frame equation of radiative transfer in relativistic flows.**
K. Morita, N. Kaneko.
Astrophys. Space Sci., Vol. 121, No. 1, p. 105 – 125 (1986).
The comoving–frame equations of radiative transfer and moment equations to accurate terms of all orders in v/c are derived in the modified Lagrangian form. The equations exactly describe the interaction of radiation with matter in a relativistically moving medium in flat or curved spacetime. Two specialized sets of equations are presented: (1) the equation of radiative transfer and moment equations accurate to terms of second order (v^2/c^2), and (2) the transfer equation and moment equations for a radial flow in curved spacetime with the Schwarzschild–type metric.

063.031 **Angular distribution of radiation in an isotropically scattering slab.**
M. S. Abdel Krim, S. A. El Wakil.
Astrophys. Space Sci., Vol. 121, No. 1, p. 137 – 145 (1986).
The equation of radiative transfer in an isotropically scattering slab subject to general boundary conditions is solved. The Padé approximation technique is used to calculate the reflected and transmitted angular distributions. Numerical results for angular

distributions through and at the boundaries of a finite slab are calculated by the Padé approximation technique. The results for a Padé approximation of [0/1] are compared with results obtained by the Galerkin method.

063.032 **Radiative transfer in spectral lines by noncoherent scattering. II. Anisotropic scattering medium.**
J. S. Aliyev.
Astrophys. Space Sci., Vol. 121, No. 2, p. 283 – 300 (1986).
The functional analytic method of solution is applied to investigation of the radiative transfer equation in spectral lines. A problem of scattering in the spectral line with the frequency redistribution in anisotropic–scattering infinite and semi–infinite media is considered. Continuum absorption in the lines is also taken into account. The albedo problem with the determined expansion coefficients and the intensity of the emergent radiation is given as an example.

063.033 **Modified spherical harmonic method in radiative transfer in linearly anisotropic scattering planar medium.**
F. S. Wan, S. J. Wilson, K. K. Sen.
Astrophys. Space Sci., Vol. 121, No. 2, p. 309 – 315 (1986).
Mengüç and Viskanta (1983) examined the scope of some approximate methods for solving transfer problems in plane medium scattering anisotropically. Their aim was to focus attention on methods capable of extension to multidimensional geometry. In the present paper, it is demonstrated that modified double–interval spherical harmonic method admirably suits that role. The transfer problem of Mengüç and Viskanta's model have been solved by this method. The results computed are found to be in good agreement with those obtained by other methods.

063.034 **Stokes parameters for Thomson scattering in a strong magnetic field.**
C. K. Chou.
Astrophys. Space Sci., Vol. 121, No. 2, p. 333 – 344 (1986).
The effect of a strong magnetic field on neutron stars or white dwarfs is calculated for Thomson scattering. The Stokes parameters for the scattered radiation are computed explicitly in terms of the state of polarization of the incident wave, the electron–cyclotron frequency, the angle of incidence, and the angle of scattering. The effect of the magnetic field on these parameters is considered in detail for incoming radiation propagating along or across the external magnetic field. The criterion for the magnetic field to substantially affect the Stokes parameters is that the photon frequency be less than the electron–cyclotron frequency.

063.035 **The radiation field in multilayer media.**
T. Viik.
Tartu Astrofüüs. Obs. Teated, Nr. 79, 90 pp. (1985). In Russian.
In the atmospheres of planets and stars there may exist sufficiently thick plane–parallel layers in which the optical properties of the medium, e.g. the albedo of single scattering, the phase function etc., are nearly constant. If we use finite difference, differential or integral methods to find the radiation field in such a layer, we have to integrate the radiative transfer equation in short steps due to the inner nature of these methods.

063.036 **Asymptotic scaling in the problems of resonance radiation transfer in linearly expanding media. II. Solutions for infinite and semi–finite media.**
S. I. Grachev.
Astrofizika, Tom 23, Vyp. 3, p. 551 – 568 (1985). In Russian. English translation in Astrophysics, Vol. 23, No. 3.
Resonance radiation transfer is considered for infinite and semi–infinite media expanding isotropically or plane–parallelly with depth independent dimensionless velocity gradient γ. Complete frequency redistribution is assumed for a scattering in a comoving frame of reference. The source function and radiation intensity are shown in the limit of small γ to be expressed through functions of a less number of arguments which are the combinations of old arguments (including parameters γ and single scattering albedo λ among them). For several main problems these functions are found explicitly for the cases of Doppler and power profiles of the absorption coefficient. They are expressed in terms of elementary or known special functions in the case of Doppler absorption coefficient in a one–dimensional medium with $\lambda = 1$. It is shown that the expansion of a medium can lead to the formation of narrow (with a width less than a thermal one) intense features in spectral line profiles.

063.037 **On the nonlinear problem of noncoherent scattering in a one–dimensional medium.**
A. Kh. Khachatryan, A. A. Akopyan.
Astrofizika, Tom 23, Vyp. 3, p. 569 – 581 (1985). In Russian. English translation in Astrophysics, Vol. 23, No. 3.
The paper considers the nonlinear transfer problem within the spectral line in the layer of finite thickness in one–dimensional approximation at general frequency redistribution laws. Numerical calculations in the 4th and 8th approximations are performed for the Doppler law of redistribution. The dependence of the real optical thickness on the finite one is presented in tables.

063.038 **A method of calculation of radiation fields in spectral lines.**
V. M. Serbin.
Astrofizika, Tom 23, Vyp. 3, p. 583 – 594 (1985). In Russian. English translation in Astrophysics, Vol. 23, No. 3.
An iterative method of calculation of the radiation field in a spectral line in plane–parallel moving atmospheres with complete frequency redistribution is proposed.

063.039 **Radiative transfer in a sphere illuminated by a parallel beam: an integral equation approach.**
R.–L. Shia, Y. L. Yung.
Astrophys. J., Vol. 301, No. 2, p. 554 – 570 (1986). = Contrib. No. 4084, Div. Geol. Planet. Sci., Calif. Inst. Technol., Pasadena, Calif. 91125, USA.
The problem of multiple scattering of nonpolarized light in a planetary body of arbitrary shape illuminated by a parallel beam is formulated using the integral equation approach. There exists a simple functional whose stationarity condition is equivalent to solving the equation of radiative transfer and whose value at the stationary point is proportional to the differential cross section. This analysis reveals a direct relation between the microscopic symmetry of the phase function for each scattering event and the macroscopic symmetry of the differential cross section for the entire planetary body, and the inter–connection of these symmetry relations and the variational principle. Illustrative examples of applications to planetary problems are provided.

063.040 **Line formation in the comoving frame: accurate solution using an approximate Newton–Raphson operator.**
K. Hempe, K. Schönberg.
Astron. Astrophys., Vol. 160, No. 1, p. 141 – 147 (1986).
An approximate Newton–Raphson operator has been derived from the comoving frame equation of radiative transfer. This operator has been applied to the two–level line formation problem in spherical expanding atmospheres. Due to its simple structure the operator can be calculated very easily and its application is not restricted to the comoving frame method. Test calculations have been carried out for atmospheric structures, which are typical for early and late type stars. The convergence of the method has been found to be fast, stable, and linear.

063.041 **Fast approximations for the R_{II-A} redistribution function.**
P. Gouttebroze.
Astron. Astrophys., Vol. 160, No. 1, p. 195 – 198 (1986).
Approximations are proposed for the computation of the R_{II-A} redistribution function (which describes the redistribution of frequencies for photons scattered coherently in the atom's frame). These approximations are particularly efficient on vector computers. Precision and timing properties of the method are examined for a typical line formation problem (Lyα in the solar atmosphere).

063.042 **The statistical description of a radiation field on the basis of the invariance principle. III. The mean time of photon travel in a scattering medium.**
A. G. Nikogosyan.
Astrofizika, Tom 24, Vyp. 1, p. 149 – 166 (1986). In Russian. English translation in Astrophysics, Vol. 24, No. 1.
The method suggested in the previous papers, which is based on the invariance principle and the applications of generating and characteristic functions, is used to determine the mean time of photon travel in a three–dimensional semi–infinite medium. Under general assumptions concerning the elementary act of scattering new equations and formulae are obtained to find out the dependence of the mentioned mean quantity on the characteristics of the original photon. It has been shown that the mean time of travel for photons escaping the medium can be found by formal differentiation in respect to β of the appropriate equations and formulae for photon exit probability. The new relation between the mean number of scatterings and the mean time of photon travel in the medium has been found.

063.043 **Astrophysical masers with magnetic fields and velocity gradients.**
S. Deguchi, W. D. Watson, L. R. Western.
Astrophys. J., Vol. 302, No. 1, p. 108 – 119 (1986).
The transfer of polarized radiation in saturated astrophysical masers is treated in the regime in which fractional linear polarizations Q/I greater than one–third can be produced by the influence of magnetic fields. A velocity gradient is assumed to be present, and the LVG (or Sobolev) approximation is utilized. Spontaneous emission, which has been ignored in previous studies, tends to reduce the maximum linear polarizations that can be attained. Parameters relevant to the circumstellar SiO masers are emphasized.

063.044 **Low–energy electron–atom bremsstrahlung.**
R. J. Gould.
Astrophys. J., Vol. 302, No. 1, p. 205 – 207 (1986).
This paper extends recent work providing an elementary calculation of bremsstrahlung and opacity associated with the scattering of low–energy electrons by neutral atoms and molecules. The method applies when the scattering potential is "hard" or when the collision time is short, applying the classical soft–photon emission probability formula for arbitrary $\hbar\omega$. The opacity is evaluated for electron scattering by H, He, and H_2 using experimentally determined elastic scattering cross sections, and is compared with more elaborate calculations.

063.045 **Relativistic particle streams in AGN.**
C. S. Coleman.
Structure and evolution of active galactic nuclei, p. 521 – 523 (1986). – See Abstr. 012.011.
Under certain conditions a relativistic particle stream is stable against disruption by self–generated turbulence. Such streams could persist and propagate at relativistic velocity in regions of strongly magnetized plasma near the cores of active galactic nuclei (AGN). The anomalously high brightness temperatures and superluminal motions observed in many AGN can be understood in terms of synchrotron radiating streams of electrons and/or positrons, without requiring unrealistic magnetic field configurations or extreme degrees of alignment relative to the observer.

063.046 **Cerenkov line radiation – a new interpretation of the emission lines of quasars.**
T. Kiang, J. H. You.
Structure and evolution of active galactic nuclei, p. 583 – 586 (1986). – See Abstr. 012.011.
An electron moving at relativistic speed $\beta = v/c$ in a medium of refractive index n will emit Cerenkov radiation if $\beta n > 1$, and for a given β, this condition may be satisfied over some small wavelength range next to atomic lines, under favourable physical conditions. The radiation so produced will than look like broad lines, asymmetrical in shape and characteristically displaced from the atomic line. For this reason, J. H. You called this radiation the "Cerenkov line radiation". The necessary physical conditions are that there should be a sufficient number of the atom in question and of the relativistic electrons.

063.047 **Distance of the far galaxies and scattering of the photons in the space.**
M. Missana.
Structure and evolution of active galactic nuclei, p. 645 – 647 (1986). – See Abstr. 012.011.
From the theory of the radiative transfer, including the scattering of the light and the Compton effect, it can be deduced that the intensity of the far astrophysical objects decreases approximatively as $1/R^3$.

063.048 **Line formation in expanding atmospheres: multi–level calculations using approximate lambda operators.**
W.–R. Hamann.
Astron. Astrophys., Vol. 160, No. 2, p. 347 – 351 (1986).
The multi–level non–LTE spectral formation in spherically expanding atmospheres is calculated by means of a perturbation technique. The method uses approximate lambda operators which act only locally. Adequate operators are taken for continua and lines, the latter being constructed from the viewpoint of the comoving frame. One obtains modified, non–linear equations for the statistical equilibrium, which are solved by linearization. Test calculations are presented for a typical Wolf–Rayet star atmosphere of pure helium, accounting for 12 atomic levels and 45 He II lines. The method seems to be capable for handling additionally He I and CNO as complex ions.

063.049 **Radiative transfer in photohydrodynamical shocks.**
P. A. Becker, M. C. Begelman.
Bull. Am. Astron. Soc., Vol. 17, No. 4, p. 832 (1985). Abstract. – See Abstr. 010.062.

063.050 **Variable optical depth radiative transport.**
A. Fu, W. D. Arnett.
Bull. Am. Astron. Soc., Vol. 17, No. 4, p. 832 (1985). Abstract. – See Abstr. 010.062.

063.051 **Transport of continuum radiation in disk geometry.**
G. F. Spagna Jr.
Bull. Am. Astron. Soc., Vol. 17, No. 4, p. 847 (1985). Abstract. – See Abstr. 010.062.

063.052 **Confinement of radiation in expanding ejecta.**
A. G. Petschek, S. A. Colgate.
Bull. Am. Astron. Soc., Vol. 17, No. 4, p. 848 (1985). Abstract. – See Abstr. 010.062.

063.053 **Compton scattering in strong magnetic fields.**
A. K. Harding, J. K. Daugherty.
Bull. Am. Astron. Soc., Vol. 17, No. 4, p. 857 (1985). Abstract. – See Abstr. 010.062.

063.054 **Quantum corrections to relativistic synchrotron emission.**
J. J. Brainerd, V. Petrosian.
Bull. Am. Astron. Soc., Vol. 17, No. 4, p. 872 (1985). Abstract. – See Abstr. 010.062.

063.055 **Astrophysical water masers in the presence of magnetic fields.**
S. Deguchi, W. D. Watson.
Astrophys. J., Vol. 302, No. 2, p. 750 – 756 (1986).
The fractional linear polarization Q/I caused by the presence of magnetic fields is calculated for the 22 GHz water maser as a function of the intensity of radiation within the maser. Complications arise due to the high angular momentum and hyperfine structure of the states (6_{16}–5_{23}) that are involved. The large–velocity–gradient approximation is utilized to treat the radiative transfer. Calculations are performed for the regime in which $Q/I \geqslant 1/3$ can be attained. It is shown that $Q/I \geqslant 0.5$ requires magnetic fields of at least 0.1 – 1 G in the masing cloud.

063.056 **On the theory of strong shock waves propagation in inhomogeneous gravitating media. II. Radiative cooling phase.**
V. P. Pas'ko, S. A. Silich.
Kinematika Fiz. Nebesn. Tel, Tom 2, No. 3, p. 15 – 21 (1986). In Russian. English translation in Kinematics Phys. Celest. Bodies.
The effect of the gravitational field of stars on the propagation of strong radiative shock waves is investigated in the infinitely thin layer approximation. The equations of motion for shock waves and their analytical solutions are found. Combining the previously obtained solution for the adiabatic phase with the present one, the complete description of the shock wave propagation is obtained.

063.057 **Polarization in spectral lines. IV: Resonance polarization in the Hanle effect, collisionless regime.**
E. Landi Degl'Innocenti.
Sol. Phys., Vol. 102, Nos. 1/2, p. 1 – 20 (1985).
The general formalism (Landi Degli'Innocenti, 1983) is particularized to deduce the radiative transfer equations for polarized radiation and the statistical equilibrium equations for a multi-level atom in the Hanle effect, collisionless regime. The formulae are developed both in the standard representation and in the representation of the statistical tensors. For resonance scattering in a two-level atom in the presence of a weak magnetic field, in the limiting case of complete depolarization of the ground level, the authors derive with the classical results of the Hanle effect the expression of the phase matrix in terms of ordinary rotation matrices. A magnetic kernel is introduced in the formalism and its main properties are analyzed. Finally, the law of scattering is generalized to take properly into account the influence of the anisotropy of the radiation field on the atomic polarization of the ground level (depopulation pumping in the Hanle effect).

063.058 **On scattering of solar radiation in an one-dimensional atmosphere.**
R. G. Indzhgia.
Tr. Inst. prikl. mat. Tbilis. univ., Tom 15, p. 232 – 241 (1985). In Russian. Abstr. in Ref. Zh., 62. Issled. Kosm. Prostranstva, 4.62.641 (1986).

063.059 **Resonance line transfer and transport of excited atoms – I. Basic relations and limiting cases.**
J. Borsenberger, J. Oxenius, E. Simonneau.
J. Quant. Spectrosc. Radiat. Transfer, Vol. 35, No. 4, p. 303 – 328 (1986).
The classic problem of non-(LTE) spectral line formation by two-level atoms in a self-excited gas is reconsidered. In contrast to previous investigations, the authors take the convective transport of the excited atoms into account through the streaming term of the corresponding kinetic equation. They establish the governing equations and show that non-LTE line formation is, in general, characterized by three dimensionless parameters ε, ζ, η rather than by just one parameter ε. The authors consider some limiting cases that are useful in interpreting physically the numerical solutions of the two coupled kinetic equations of photons and excited atoms, respectively. The main topic of the paper is the investigation of the direct effect of a boundary surface on the distribution function of the excited atoms, as opposed to the indirect effect of the boundary on the distribution function via the radiation field in the spectral line. The authors also study the influence of the corresponding boundary layer on the spectral line emitted by the system. They assume pure Doppler broadening of the spectral line throughout.

063.060 **Chebyshev collocation method for solving the radiative transfer equation.**
K. Kamiuto.
J. Quant. Spectrosc. Radiat. Transfer, Vol. 35, No. 4, p. 329 – 336 (1986).
A Chebyshev collocation method is presented for solving the spherical harmonics approximation to the equation of radiative transfer in a plane-parallel, homogeneous medium. As a result of test computations performed for Rayleigh and Henyey-Greenstein phase functions, it was found that the proposed method can be used to solve transfer problems stably and accurately, even for an optically thick case, and that the accuracy of computations is greatly improved by increasing the order of Chebyshev polynomials in expansion.

063.061 **Radiation transfer due to a point source in an isotropically scattering, inhomogeneous solid sphere.**
S. T. Thynell, M. N. Özisik.
J. Quant. Spectrosc. Radiat. Transfer, Vol. 35, No. 5, p. 349 – 356 (1986).
A method of analysis is presented for solving radiation transfer due to a point source in an absorbing, isotropically scattering, inhomogeneous sphere, which is subjected to a diffusely reflecting boundary. The effects of various functional forms of albedo on the incident radiation, angular distribution of radiation intensity and emissivity are examined.

063.062 **Monte Carlo spectra from pair-equilibrium, weakly magnetized thermal plasmas.**
A. A. Zdziarski.
Astrophys. J., Vol. 303, No. 1, p. 94 – 100 (1986).
The author reports the results of Monte Carlo simultations of weakly magnetized, relativistic thermal plasmas. Thermal synchrotron emission and absorption, bremsstrahlung, and Compton scattering are included, as well as pair annihilation and pair production. The plasmas are assumed to be in pair equilibrium. The resulting spectra are of power-law form, and can be well described by a set of simple analytical formulae. Neither a pair annihilation feature nor a Wien hump is seen in the spectra. The possible relevance of these results to active galactic nuclei and γ-bursts is discussed.

063.063 **Comment on "Nonexistence of two forms of turbulent bremsstrahlung" by J. Kuijpers and D. B. Melrose.**
M. Nambu.
Astrophys. J., Vol. 303, No. 1, p. 347 – 348 (1986).
Kuijpers and Melrose make an error in imposing the symmetry $-k \leftrightarrow k_2$. The "response" and "Lagrangian" formalisms with the correct second symmetry $-k \leftrightarrow k_2$ agree. The most dominant plasma maser effect comes from the polarization term, except for the electrostatic waves not having an applied static magnetic field. (See Abstr. 40.063.009).

063.064 **The moment method for radiative transfer problems in spherical atmospheres with a core.**
S. J. Wilson, F. S. Wan.
Proceedings of the Second Asian-Pacific Regional Meeting on Astronomy, p. 536 – 540 (1984). – See Abstr. 012.031.

063.065 **On the nature of the opposition effect of brightness and negative light polarization of solid cosmic surfaces.**
Yu. G. Shkuratov.
Astron. Tsirk., No. 1400, p. 3 – 6 (1985). In Russian.

063.066 **Theoretical X-ray spectra of accreting magnetized neutron stars and quasars.**
P. Mészáros, W. Nagel, R. Bussard.
X-ray astronomy '84, p. 181 – 186 (1985). – See Abstr. 012.033.
The authors discuss their recent numerical calculations of the radiative transfer in slabs and cylinders with $B \cong 10^{12}$Gauss, as a model for X-ray pulsars, and show the phase-dependent spectra and pulse shapes. For quasars, spectra are calculated in both the standard thin disk and two-temperature thick disk models.

063.067 **Conflicting guiding center motions in cyclotron radiation.**
D. A. Leahy, R. Lieu, A. J. Evans.
X-ray astronomy '84, p. 255 – 258 (1985). – See Abstr. 012.033.
The first order radiative transition of electrons in strong magnetic fields, of relevance to pulsars, has been studied.

063.068 **Recombination and ionisation in strongly magnetized thermal plasmas of accreting neutron stars.**
G. Wunner, F. Geyer, H. Herold, W. Rösner, H. Ruder, A. Steitz.
X–ray astronomy '84, p. 263 – 266 (1985). – See Abstr. 012.033.
For magnetic field strengths $B \gtrsim 10^7T$, which are typical of the hot X–ray emitting plasmas in the vicinity of accreting neutron stars, the authors have rederived the pertaining formulae for the cross sections of the elementary processes of photo-recombination and three–body recombination of protons and electrons, as well as the cross sections of the time–reversed processes (photoionisation and impact ionisation), and have evaluated them numerically for $B = 2.35 \times 10^7T$ and 4.7×10^8T. For the densities and temperatures prevailing in accretion plasmas it turns out that recombination via emission of photons is more frequent by several orders of magnitude than recombination via collisions.

063.069 **Radiative transfer in the accretion column of X–ray pulsars: effects from the hot spot.**
M. Soffel, H. Herold, H. Ruder, J. Ventura.
X–ray astronomy '84, p. 271 – 274 (1985). – See Abstr. 012.033.
The influence of the hot spot that very likely forms at the base of the accretion column of X–ray pulsars with X–ray luminosities not much larger than $\sim 10^{30}W$ onto emitted spectra and angle dependent fluxes is discussed.

063.070 **Erratum: "Bowen fluorescence mechanisms in X–ray binaries" [Astrophys. J., Vol. 291, No. 2, p. 492 – 504 (1985)].**
S. Deguchi.
Astrophys. J., Vol. 303, No. 2, p. 901 (1986). See Abstr. 39.063.046.

063.071 **Some problems of the nonlinear theory of optimization of statistical modelling of radiative transfer in a layer of matter with anisotropic scattering.**
A. V. Starkov.
VTs SO Akad. Nauk SSSR., Prepr. No. 597, 14 pp. (1985). In Russian. Abstr. in Ref. Zh., 51. Astron., 5.51.193 (1986).

063.072 **Order–of–scattering and doubling method for the solution of transfer equations.**
A. P. Wang.
Astrophys. Space Sci., Vol. 122, No. 2, p. 281 – 286 (1986).
A method to obtain scattering and transmission functions for computational purposes has been developed in this paper. This method can be easily extended to the scattering atmosphere involving internal energy and may be to the time–dependent case. Special cases such as the homogeneous process, the Helmholtz's principle of reciprocity and the self–adjoint are presented.

063.073 **Polarization of Lyman–alpha lines from hydrogen–like ions.**
S. Chandra.
Astrophys. Space Sci., Vol. 123, No. 1, p. 205 – 208 (1986).
The Oppenheimer–Penney theory to calculate the polarization of $L\alpha$ lines from hydrogen–like ions, when the impact electrons are distributed such that their probability is more in the regions close to the magnetic field ($f(\alpha) \sim \cos^n\alpha$), is applied by Chandra and Joshi (1984). The work of Chandra and Joshi (1984) has been reinvestigated for the pitch–angle distribution $f(\alpha) \sim \sin^n\alpha$. The degrees of polarization are still found to be independent of the atomic number of a hydrogen–like ion.

063.074 **Numerical algorithm for solution of the radiative transfer equations taking into account the influence of a magnetic field.**
D. N. Rachkovskij.
Izv. Krymskoj Astrofiz. Obs., Tom 74, p. 158 – 160 (1986). In Russian. English translation in Bull. Crimean Astrophys. Obs., Vol. 74.
The main equations and comments to them are presented to compute the radiative transfer equations taking into account the presence of a magnetic field. The LTE condition is assumed and only a true absorption is being accounted. The magnetic field and the velocity gradient are considered as the function of depth of the photosphere.

063.075 **The analysis of ensembles of moderately saturated interstellar lines.**
E. B. Jenkins.
Astrophys. J., Vol. 304, No. 2, p. 739 – 766 (1986).
This paper shows that large populations of interstellar lines can be analyzed collectively using the standard, single–component Gaussian curve of growth. Global abundance determinations can be derived straightforwardly from the sums of equivalent widths taken from surveys which include many lines of sight, provided one is sure that very strongly saturated lines are not present. This simplified approach also applies to any single path with a large number of very different velocity components, either well separated or clumped and blended. In either case, the method enables one to derive column densities without a special need for high signal quality, fine wavelength resolution, or elaborate analysis techniques. Some examples illustrate the use of the method.

063.076 **Relativistic radiation transport in dispersive media.**
S. Kichenassamy, R. A. Krikorian.
Phys. Rev. D, Vol. 32, No. 8, p. 1866 – 1870 (1985). Abstr. in Phys. Abstr., Vol. 89, No. 1273, Entry 13468 (1986).

063.077 **About explanations for X–ray bursts with double peaks and precursors.**
M. A. Schweizer.
Can. J. Phys., Vol. 63, No. 7, p. 956 – 961 (1985). Abstr. in Phys. Abstr., Vol. 89, No. 1273, Entry 18840 (1986).

063.078 **Contribution functions and the depths of formation of spectral lines.**
P. Magain.
ESO Sci. Prepr., No. 443, 17 pp. (1986). To appear in Astron. Astrophys.

063.079 **On correlations between certain photometric and polarimetric characteristics of light scattered by surfaces of atmosphereless celestial bodies.**
L. O. Kolokolova.
Photometric and polarimetric investigations of celestial bodies, p. 38 – 42 (1985). In Russian. – See Abstr. 003.035.
Possible interrelations between the characteristic features of phase relations of brightness and polarization of light are analyzed for atmosphereless celestial bodies. Some correlations are marked between the parameters of the opposition effect and the negative polarization branch. In particular, there are correlations between the opposition effect amplitude and the angle of minimum and inversion angle of the negative polarization branch.

063.080 **Multiple scattering of polarized light in spherically symmetric media.**
J. Freimanis.
Photometric and polarimetric investigations of celestial bodies, p. 128 – 133 (1985). In Russian. – See Abstr. 003.035.
Three different methods for treatment of radiative transfer problems in spherically symmetric media including polarization effects have been developed. The problem for a geometrically thin spherical envelope with Rayleigh scattering surrounding an isotropic unpolarized point source has been reduced to the plane–parallel case; Stokes parameters for the outcoming radiation have been calculated. Linear integral equations are obtained for source functions inside a homogeneous Rayleigh–scattering sphere. A general method has been developed for analytical solution of the problem when an isotropic unpolarized point source is immersed into an infinite homogeneous isotropic medium; closed–form expressions are written for Rayleigh scattering.

063.081 **Complex "RADUGA-4" for a numerical solution of the transfer equation of polarized radiation in axially symmetric media.**
T. A. Germogenova, L. P. Bass, N. V. Konovalov.
Photometric and polarimetric investigations of celestial bodies, p. 133 – 136 (1985). In Russian. – See Abstr. 003.035.
An algorithm is briefly described for solving the transfer equation of polarized radiation in media with axial symmetry by the method of discrete ordinates. It has been shown that the complex "RADUGA-4" gives quite acceptable results, except for the case of small angular distances from the direction to the source.

063.082 **On the physical correctness of the scattering matrix for polarized radiation.**
N. V. Konovalov.
Photometric and polarimetric investigations of celestial bodies, p. 136 – 140 (1985). In Russian. – See Abstr. 003.035.
The conception of physical correctness of the scattering matrix is introduced. Some signs of physical correctness (non-negative) of some specific types of scattering matrices as well as the method of correction of incorrect matrices are proposed.

063.083 **Computation of the nonstationary radiation field in an optically thick planetary atmosphere.**
N. N. Fomin.
Photometric and polarimetric investigations of celestial bodies, p. 140 – 146 (1985). In Russian. – See Abstr. 003.035.
The paper deals with certain functions of the nonstationary radiation transfer theory in a plane parallel homogeneous anisotropically scattering optically thick planetary atmosphere overlying a surface with albedo <1.

063.084 **Intensity and polarization of radiation scattered by spheroidal particles.**
N. V. Voshchinnikov, V. G. Farafonov.
Photometric and polarimetric investigations of celestial bodies, p. 146 – 152 (1985). In Russian. – See Abstr. 003.035.
The intensity functions i_1 and i_2 and polarization degree P are calculated for prolate and oblate spheroids when the direction of incident radiation is parallel to the rotation axis. Spheroidal particles with refractive indices $m = 1.2, 1.33, 1.5$ and 1.65 and the semiaxis ratio $a/b = 1.5 - 10.0$ are considered. Effects of the particle size, shape, refractive index on i_1, i_2 and P are investigated. It is shown that the characteristics of the radiation scattered by spheroids greatly depend on the particle shape, index m, and particle prolateness or oblateness.

063.085 **Light scattering in multi-layer atmospheres. II. The Milne problem.**
Eh. G. Yanovitskij.
Astrofizika, Tom 24, Vyp. 3, p. 535 – 548 (1986). In Russian. English translation in Astrophysics, Vol. 24, No. 3.
A plane semi-infinite atmosphere consisting of an arbitrary number of homogeneous layers with known optical thicknesses is considered. An effective algorithm of the radiative field determination for the Milne problem is given. The case of layers of large optical thickness is studied in detail. Simple asymptotic formulae are found for this case.

063.086 **A point source in an infinite stochastic medium.**
R. S. Vardanyan.
Astrofizika, Tom 24, Vyp. 3, p. 549 – 559 (1986). In Russian. English translation in Astrophysics, Vol. 24, No. 3.
The paper considers the radiation transfer problem in an infinite three-dimensional field when the quantum survival probability λ in the case of an elementary act of scattering is presented by a statistically homogeneous and isotropic Gaussian field with an exponential correlation function. The diagram method is applied to the solution. The first moment of the source function and the average number of quantum scatterings are found in the Bourret approximation. The strongly correlated field is considered as a particular case.

063.087 **Radiation pressure and emission clouds around active galactic nuclei.**
M. Elitzur, G. J. Ferland.
Astrophys. J., Vol. 305, No. 1, p. 35 – 44 (1986).
Detailed calculations of the contribution of the trapped line photons to the overall pressure in line-emitting regions around active galactic nuclei are presented. They lead to the conclusion that radiation pressure is a significant factor in determining the stability of these clouds.

063.088 **On the origin of the infrared and X-ray continua of active galactic nuclei.**
A. A. Zdziarski.
Astrophys. J., Vol. 305, No. 1, p. 45 – 56 (1986).
A model is proposed to explain the overall continuum spectra of radio-quiet active galactic nuclei, steeper in the near-infrared ($\alpha \approx 1.2$), and flatter in the X-ray regime ($\alpha \approx 0.7$). Such spectra can be produced in a synchrotron self-Compton model with a broken power-law electron distribution that is flat at low γ and steep at large γ (γ is the Lorentz factor). The steeper part produces the infrared spectrum via synchrotron process, while both parts together produce the flatter X-ray spectrum via Compton scattering. A broken electron distribution can result, for example, from the Fermi shock-acceleration process, in which accelerated protons outnumber by a large factor accelerated electrons.

063.089 **New results in the optics of grains with resonant impurities.**
P. R. Shapiro, K. A. Holcomb.
Astrophys. J., Vol. 305, No. 1, p. 433 – 454 (1986).
The scattering and absorption cross sections and extinction profiles of grains of arbitrary size compared with the wavelength of incident light are calculated for grains containing resonantly absorbing impurity atoms or molecules. Such grains are candidates for explaining the diffuse interstellar bands. The embedded cavity model of Purcell and Shapiro is used to derive the bulk dielectric constants of the composite grain material. Cross sections are then derived from Mie theory for homogeneous spherical grains and from the solution of Güttler for spherical core-mantle grains. The authors investigate in detail the effect on the grain extinction profiles of variations in such parameters as particle size, host grain composition, impurity oscillator strength, impurity concentration, and the confinement of the resonant impurities to a grain mantle or core.

063.090 **A cloud collision model for water maser excitation.**
J. C. Tarter, W. J. Welch.
Astrophys. J., Vol. 305, No. 1, p. 467 – 483 (1986).
High-velocity collisions between small, dense, neutral clouds or between a dense cloud and a dense shell can provide the energy source required to excite H_2O maser emission. The radiative precursor from the surface of the collisional shock front rapidly diffuses through the cloud, heating the dust grain but leaving the H_2 molecules cool. Transient maser emission occurs as the conditions for the Goldreich and Kwan "hot-dust cold-gas" maser pump scheme are realized locally within the cloud. In time the local maser action quenches due to the heating of the H_2 molecules by collisions against the grains. This model is compared with maser sources in Orion, W51 and W49.

063.091 **The calculation of the Voigt function H(a,v).**
J.-h. Jin.
Publ. Purple Mt. Obs., Vol. 5, No. 1, p. 31 – 35 (1986). In Chinese.

063.092 **Benchmark results in radiative transfer.**
R. D. M. Garcia, C. E. Siewert.
Transp. Theory Stat. Phys., Vol. 14, No. 4, p. 437 – 483 (1985). Abstr. in Phys. Abstr., Vol. 89, No. 1278, Entry 42705 (1986).

063.093 **On the radiative–radiative pumping of circumstellar SiO masers.**
G. T. Bolgova.
Nauchn. Inf., Vyp. 57, p. 39 – 50 (1984). In Russian.
Deguchi and Iguchi's radiative–radiative (RR_υ) pumping model of SiO masers in circumstellar envelopes is extended for more numbers of transitions. A more realistic model of collisions has been used. It is shown that the RR_υ–pumping can in principle invert all observed maser transitions (except the $J = 2$–1 transition in the $\upsilon = 0$ state) and number of nonobserved transitions in Mira–type stars and supergiants. The cycles of the net population transfer leading to the inversion of these transitions are identified from qualitative considerations. The shortcomings of the model are discussed.

063.094 **A method of analysis of the pumping mechanisms of cosmic masers.**
A. M. Sobolev.
Nauchn. Inf., Vyp. 57, p. 51 – 54 (1984). In Russian.
A method of analysis of the pumping mechanisms on the basis of investigation of the cyclic processes of population transfer between energy levels is described.

063.095 **Vector algorithms for H–equations arising in radiative transfer through inhomogeneous media.**
S. L. Hollis, C. T. Kelley.
Transp. Theory Stat. Phys., Vol. 15, No. 1 – 2, p. 33 – 48 (1986). Abstr. in Phys. Abstr., Vol. 89, No. 1282, Entry 68038 (1986).

063.096 **Compton scattering in strong magnetic field.**
Y.–m. Xu.
Chin. Astron. Astrophys., Vol. 10, No. 2, p. 121 – 128 (1986). English translation of 40.063.095.

063.097 **The polarization properties of inverse Compton scattering in strong magnetic field.**
X.–y. Xia, G.–j. Qiao, X.–j. Wu.
Acta Astrophys. Sin., Vol. 6, No. 2, p. 160 – 167 (1986). In Chinese.
The authors discuss the polarization properties of inverse Compton scattering in a strong magnetic field (both relativistic and non–relativistic cases). Some interesting results are obtained and compared with field–free condition.

063.098 **A rapidly convergent iterative solution of the non–LTE radiation transfer problem.**
G. L. Olson, L. H. Auer, J. R. Buchler.
J. Quant. Spectrosc. Radiat. Transfer, Vol. 35, No. 6, p. 431 – 442 (1986).
An iterative scheme has been developed for the solution of the non–LTE line radiation transfer problem. The method uses an approximate operator that is deliberately chosen to be local so that it can be easily extended to multidimensional geometry. The difference between the formal and approximate solutions is used as a driving term for the iterations. In one–dimensional, semi–infinite and free–standing slabs, the technique is found to be very fast, robust, and applicable to a large class of problems.

063.099 **Generalized Eddington approximation method for radiative transfer problems in slab medium.**
S. J. Wilson, K. K. Sen.
J. Quant. Spectrosc. Radiat. Transfer, Vol. 35, No. 6, p. 467 – 472 (1986).
A generalized Eddington approximation method is proposed to solve radiative transfer problems in an anisotropic scattering slab medium. The method, which was developed earlier to tackle transfer problems with asymmetric radiation fields in spherical geometry, proves to be equally useful in a planar geometry. The main feature of the method is to divide the radiation field into subregions identified by some angular parameters which, in turn, are determined from the transfer equation. The results are compared with existing data.

063.100 **The theory of multiple scattering of radiation in plane parallel atmospheres.**
K. Stamnes.
Rev. Geophys., Vol. 24, No. 2, p. 299 – 310 (1986).
A review is provided of multiple scattering theory associated with radiative transfer problems in planetary atmospheres. The radiative transfer equation pertaining to a scattering, absorbing, and emitting medium is considered. Analytic solutions are presented and computational aspects are discussed for plane parallel atmospheres. These computationally efficient and reliable solutions can be used to attack a multitude of radiative transfer problems in realistic atmospheres. Computational illustration is provided, a discussion of certain applications of the theory is presented, and some remaining theoretical problems are pointed out.

063.101 **Radiation of ultrarelativistic electrons moving through a dense medium in the presence of a strong magnetic field.**
Yu. D. Kotov, S. R. Kel'ner, S. V. Bogovalov.
Pis'ma Astron. Zh., Tom 12, No. 5, p. 402 – 409 (1986). In Russian. English translation in Sov. Astron. Lett., Vol. 12.
The synchrotron radiation of electrons with zero pitch angle before entering the matter is studied. Pitch–angle distribution occurs due to multiple scattering. It is shown that in the magnetic fields that are characteristic of pulsars the mechanism considered makes a main contribution to the electron energy losses in a wide range of the initial energies of the particles and concentrations of the nuclei.

063.102 **On the correlation between X–ray flux and hydrogen line intensities in quasars and active galactic nuclei.**
J. Kwan.
Astrophys. J., Vol. 305, No. 2, p. 679 – 683 (1986).
Hydrogen line intensities in quasars and active galactic nuclei are calculated as a function of the UV to X–ray luminosity ratio. The dependence of Lyα/Hα (or Lyα/Hβ) on L_{UV}/L_x is found to be weak and not easily discernible in a large data sample if variations in the cloud column and nucleon density are large. The causes for this weak dependence are examined.

063.103 **Redistribution of radiation in the presence of velocity–changing collisions.**
I. Hubený, J. Cooper.
Astrophys. J., Vol. 305, No. 2, p. 852 – 862 (1986).
The authors present a suitable form of the normalized redistribution function relevant for a treatment of redistribution in the presence of velocity–changing collisions. Attention is devoted to provide practical estimates of the ratio between the usual line–broadening parameters and the velocity–changing rate. It is shown that for virtually all cases of astrophysical interest the effect of velocity–changing collisions is quite negligible.

063.104 **On photoproduction of ultraviolet emission lines.**
G. J. Ferland.
Publ. Astron. Soc. Pac., Vol. 98, No. 604, p. 549 – 551 (1986).
The point of this note has been to draw attention to a physical process of possible importance in some circumstances. Photoproduction of lines such as the C III] λ1909 doublet can be important in low–temperature, low–ionization nebulae and should be included in detailed models.

063.105 **Can velocity gradients explain the observed Stokes *V* asymmetry in the absence of large zero–crossing shifts?**
K. D. Pahlke, S. K. Solanki.
Mitt. Astron. Ges., No. 65, p. 162 – 165 (1986). – See Abstr. 012.072.

063.106 **Strahlungstransport in den Außenschichten weißer Zwerge mit Linienabsorption unter Einwirkung starker Magnetfelder.**
I. Bues.
Mitt. Astron. Ges., No. 65, p. 205 – 206 (1986). – See Abstr. 012.072.

063.107 **The physical meaning of contribution functions and depth of formation of absorption lines.**
A. P. Sarychev.
Astron. Zh., Tom 63, Vyp. 3, p. 556 – 562 (1986). In Russian. English translation in Sov. Astron., Vol. 30, No. 3.
Mathematically exact solution of the problem of the depth of formation of a Fraunhofer line is proposed. As an example the author analyzes the formation of the weak–line centre $\lambda = 23118.28$ Å of the CO molecule.

063.108 **Experimental modeling of polarization properties of atmosphereless cosmic bodies at small phase angles.**
L. A. Akimov, V. P. Tishkovets, Yu. G. Shkuratov.
Photometric and polarimetric investigations of celestial bodies, p. 42 – 47 (1985). In Russian. – See Abstr. 003.035.
The results of polarimetry of terrestrial samples with a complex surface structure are presented. Great attention was paid to a dependence of polarization upon structure formative processes (pressing, crushing) and surface albedo. The authors come to the conclusion that a negative polarization theory for atmosphereless cosmic bodies must be based on a light waves scattering from nearby particle systems.

063.109 **Polarization of radiation in spectral lines.**
D. I. Nagirner.
Photometric and polarimetric investigations of celestial bodies, p. 118 – 128 (1985). In Russian. – See Abstr. 003.035.
A review is given of the theory of transport of polarized radiation in atomic spectral lines. The resonance polarization and polarization arising in a magnetic field for lines formed in LTE and by scattering processes as well as several more complicated cases (microturbulent field and oriented atoms) are considered. The publications reviewed are those devoted to: (1) absorption and scattering matrices; (2) analytical and numerical solutions of transfer equations; (3) applications of the theory to solar phenomena (atmosphere, sunspots, prominences, corona etc.), magnetic stars and some other objects.

Kinetic theory of particles and photons. Theoretical foundations of non–LTE plasma spectroscopy.
See Abstr. 003.027.

The radiative field in multilayer media.
See Abstr. 003.203.

Diffusion and thermal diffusion in partially ionized gases in the atmospheres of the Sun and planets.
See Abstr. 022.028.

Atomic structure calculations and nebular diagnostics.
See Abstr. 022.100.

Strong plane shock waves in exponential and optically–thin atmospheres, II.
See Abstr. 062.009.

Quasistatic broadening of hydrogen spectral lines in plasmas.
See Abstr. 062.021.

Multiple Compton scattering in strongly magnetized plasma.
See Abstr. 062.116.

Self–excitation effect in the scattering of electromagnetic radiation in a strong magnetic field.
See Abstr. 062.117.

Bremsstrahlung in hot, strongly magnetized plasmas.
See Abstr. 062.134.

Multilevel line formation in the expanding envelope of ζ Aur/ VV Cep systems.
See Abstr. 064.001.

Emission line formation in accretion discs.
See Abstr. 064.005.

Spectral flux from low–density photospheres: numerical results.
See Abstr. 064.014.

The vertical temperature stratification and corona formation of accretion disc atmospheres.
See Abstr. 064.018.

Construction of non–LTE model atmospheres using approximate lambda operators.
See Abstr. 064.036.

Radiative transfer in circumstellar 1612 MHz OH masers.
See Abstr. 064.068.

Asymptotics and quantum chaos in stellar oscillations.
See Abstr. 065.067.

Electron–positron pairs in a mildly relativistic plasma in active galactic nuclei.
See Abstr. 067.011.

A flare–induced cascade model of gamma–ray bursts.
See Abstr. 067.052.

How do X–ray pulsars make their X–rays?
See Abstr. 067.077.

QED and X–ray pulsars.
See Abstr. 067.091.

Optical reprocessing of gamma–ray bursts.
See Abstr. 067.148.

Emission characteristics of neutron star polar caps.
See Abstr. 067.155.

Directionality of bremsstrahlung from relativistic electrons in solar flares.
See Abstr. 073.026.

Electron–cyclotron maser emission from the Sun and stars: variations with plasma temperature and density.
See Abstr. 077.053.

Gyrosynchrotron emission of solar flares.
See Abstr. 077.062.

Radiation output.
See Abstr. 080.016.

Attenuation of solar radiation in a homogeneous atmosphere.
See Abstr. 080.095.

Two–scale solution for atmospheric scintillation.
See Abstr. 082.057.

Radiative transfer in the 4.3 μ CO_2 and 4.7 μ CO bands in the non–LTE Venus and Mars atmospheres: transformation of absorbed solar energy.
See Abstr. 093.043.

Mode decoupling during retrorefraction as an explanation for bizarre radar echoes from icy moons.
See Abstr. 099.020.

Constraints on the NH_3 and PH_3 distributions in the Great Red Spot.
See Abstr. 099.057.

Estimates of the bolometric albedos and radiation balance of Uranus and Neptune.
See Abstr. 101.047.

IR heating of the cometary atmosphere.
See Abstr. 102.034.

Reflection effect in close binaries.
See Abstr. 117.157.

Expansion degradation of ultraviolet photons in supernova ejecta.
See Abstr. 125.040.

Temperature fluctuations and infrared emission from dust particles in a hot gas.
See Abstr. 131.134.

Measurements of X–ray scattering from interstellar grains.
See Abstr. 131.135.

Models for IRAS observations of galactic H II regions.
See Abstr. 132.062.

Observations of the Bowen fluorescent mechanism in planetary nebulae.
See Abstr. 134.012.

Pulse phase spectroscopy of Her X–1 with EXOSAT.
See Abstr. 142.050.

The infrared response to electron–positron pair annihilation at the Galactic center.
See Abstr. 155.039.

Active galactic nuclei and electron–positron pairs.
See Abstr. 158.224.

Some comments on iron emission line from NGC 4151: beam model.
See Abstr. 158.225.

Radiatively–driven shock waves in quasar envelopes.
See Abstr. 159.028.

064 Stellar Atmospheres, Stellar Envelopes, Mass Loss, Accretion

064.001 **Multilevel line formation in the expanding envelope of ζ Aur/VV Cep systems.**
R. Baade.
Astron. Astrophys., Vol. 154, No. 1/2, p. 145 – 153 (1986).
This paper describes a multilevel formalism for UV line profiles in ζ Aur systems on the basis of the Sobolev method. With atomic models for Si II and for Fe II UV multiplet 9 and 191 the author studies some problems implicated in the two–level approximation.

064.002 **Conservation of mass and angular momentum in narrow transition fronts in accretion discs.**
F. Meyer.
Mon. Not. R. Astron. Soc., Vol. 218, No. 1, p. 7P – 11P (1986).
The author shows the internal consistency of the quasi–stationary approximation for narrow transition fronts in accretion discs with respect to ordering by the smallness parameter: width of the transition over disc radius. Claims of inconsistency are shown to result only from neglecting terms proportional to the front width in an equation which relates these terms to another term of the same order of magnitude. It is further pointed out that artificial smoothing in transition fronts can lead to unphysical values for front velocities. Instead, transition fronts must be numerically resolved for reliable results in accretion disc problems.

064.003 **Magnetically controlled stellar winds in close binary systems.**
D. Moss.
Mon. Not. R. Astron. Soc., Vol. 218, No. 2, p. 247 – 263 (1986).
Calculations are presented for phenomenological models of magnetically controlled stellar winds in close binary systems of unit mass ratio, in circular orbits with synchronization of orbital and rotational motion. The magnetic field is taken as the sum of that of two equal dipoles placed at the stellar centres, aligned parallel to the rotation axis. Two different assumptions are made about the extent of the "dead zones" on the stellar surface, from which wind does not flow. Case A is the most conservative, and almost certainly exaggerates the size of these regions. This has the consequence that the angular momentum loss rate does not necessarily increase with increasing field strength. Case B gives a smaller dead zone and reduces the importance of this saturation mechanism. Some comparisons are made with other discussions of the binary wind problem.

064.004 **On spectral line formation in winds from extended continuum sources.**
J. E. Drew.
Mon. Not. R. Astron. Soc., Vol. 218, No. 3, p. 41P – 47P (1986).
By means of schematic line profile calculations based upon the Sobolev approximation (Castor, 1970), it is shown how spatial extension of the embedded, illuminating continuum source alters the character of resonance line profiles formed in a high–speed wind. The intended application for these calculations is to the winds of cataclysmic variables. The assumed wind geometry is spherically symmetric, but this does not determine the qualitative nature of the results obtained. An important finding is that, even when the wind is extremely opaque in the line of interest, the depth of minimum absorption need not be any greater than about half of the adjacent, unabsorbed continuum level, if the size of the continuum source is at least comparable with the wind's acceleration length scale.

064.005 **Emission line formation in accretion discs.**
K. Horne, T. R. Marsh.
Mon. Not. R. Astron. Soc., Vol. 218, No. 4, p. 761 – 773 (1986).
The authors examine the formation of emission lines from accretion discs and develop simple expressions for computing the profiles of both optically thin and optically thick lines.

064.006 **Analysis of a magnetohydrodynamic stellar wind.**
Y. Saito, M. Saitō.
Astrophys. Space Sci., Vol. 119, No. 1, p. 33 – 35 (1986). – See Abstr. 012.001.
Characteristic features of solutions are analysed around the Alfvénic point. There are three types of solutions through the point.

064.007 **Supercritical winds from cool "canonical" stars caused by evolution on the main sequence.**
P. Venkatakrishnan.
Astrophys. Space Sci., Vol. 119, No. 1, p. 51 – 55 (1986). – See Abstr. 012.001.
Even the very slow expansion of a star's radius due to evolution on the main sequence is shown to be supercritical for cool stars without coronae. Since steady spherically–symmetric supercritical solutions are theoretically impossible, unsteady supercritical solutions are studied. It is seen that smooth sonic transi-

tions are possible in the unsteady case, but they are accompanied by enhancement of pressure over the critical values.

064.008 **Steady mass–loss from supermassive stars.**
M. Kato.
Astrophys. Space Sci., Vol. 119, No. 1, p. 57 – 59 (1986). – See Abstr. 012.001.
Structures of Newtonian supermassive stars are calculated with the opacity for Compton effect $\varkappa_0/(1+\alpha T)$, where $\varkappa_0 = 0.2(1+X)$ and $\alpha = 2.2\times10^{-9}K^{-1}$. The track of the main–sequence is turned right in upper part of the HR diagram. Mass loss will occur in a main–sequence stage for a star with mass larger than a critical mass. The cause of mass loss and the expansion of the radius is continuum radiation pressure. The critical mass for mass loss is $1.02\times10^6 M_\odot$ for a Population I star, and $1.23\times10^5 M_\odot$ for Population III star. Mass loss rates expected in these stars are 3.3×10^{-3} and $4.0\times10^{-3} M_\odot yr^{-1}$, respectively.

064.009 **Photometric determination of variations in the surface conditions for pulsating stars.**
W. J. Couch, H. J. Trodahl.
Astrophys. Space Sci., Vol. 119, No. 1, p. 97 – 99 (1986). – See Abstr. 012.001.
Two sets of model atmosphere spectra have been used to determine variations of effective temperature, surface gravity, and radius from multicolour indices measured for pulsating stars. The two model atmosphere sets led to surface parameters that differ by considerably more than the measurement uncertainties.

064.010 **Stellar chromospheres, coronae, and winds.**
J. P. Cassinelli, K. B. MacGregor.
Physics of the Sun, Vol. III, p. 47 – 123 (1986). – See Abstr. 003.001.
Contents: Late–type stars: observational evidence for the presence of chromospheres in late–type stellar atmospheres (spectral diagnostics and line formation, observational summary and location in the H–R diagram, the Wilson–Bappu effect), observational evidence for the presence of regions and coronae in late–type stellar atmospheres, chromospheric and coronal heating mechanisms, observational evidence for mass loss from late–type stars (main sequence stars, circumstellar absorption lines, chromospheric emission line asymmetries, circumstellar dust shells), mass loss mechanisms for late–type giants and supergiants. The winds and coronae of early–type stars: the velocity and mass loss rates derived from line and continuum observations (the formation of P Cygni profiles, the free–free continuum energy distribution of hot stars), coronal gas in early–type stars (superionization of the winds, X–ray observation of early–type stars), wind dynamics (radiation forces on line opacity: momentum deposition considerations, radiative acceleration, instability of line driven winds and the consequences, hybrid models with a base coronal zone, magnetically driven winds and magnetically dominated coronae).

064.011 **Propagation of radiation damped acoustic waves in stellar atmospheres.**
P. Ulmschneider.
Cosmical gas dynamics, p. 237 – 242 (1985). – See Abstr. 012.007.
The propagation of radiatively damped acoustic waves along magnetic flux tubes and its implication for the unknown chromospheric and coronal heating mechanism is discussed. It is shown that the behaviour of the radiative cooling strongly determines the structure of the chromosphere and the transition–layer.

064.012 **Non–unique solutions to the stellar atmosphere problem.**
D. Muchmore.
Astron. Astrophys., Vol. 155, No. 1, p. 172 – 174 (1986).
The temperature sensitivity of CO molecular opacity can lead to a bifurcation in the solutions to the stellar atmosphere problem. Models produced with 2–frequency radiative hydrodynamic calculations show that for solar gravity and $T_{eff} = 5800$ to 5890K, an atmosphere in radiative equilibrium can exist either in a warm phase with a structure like a grey atmosphere or in a distinctly different cool phase. The difference in surface temperature between two such cases at the same T_{eff} amounts to 1000 to 1500K. These temperature differences occur only at elevations some 500 km or more above $\tau_{Ross} = 1$.

064.013 **Limits of interacting stellar winds model of planetary nebulae. A numerical study.**
R. Bedogni, A. D'Ercole.
Astron. Astrophys., Vol. 157, No. 1, p. 101 – 112 (1986).
The dynamical evolution of planetary nebulae (PNe) was calculated numerically in the scenario of the interacting stellar wind model (ISWM). Radiative losses and heat conduction were included in the calculations together with the SLIC method to follow the surface separating the winds. The general trend of PN evolution is discussed and comparison with some observations are made. Limitations to the validity of ISWM are presented, both based on dynamical and observational grounds.

064.014 **Spectral flux from low–density photospheres: numerical results.**
S. Hershkowitz, E. Linder, R. V. Wagoner.
Astrophys. J., Vol. 301, No. 1, p. 220 – 229 (1986).
Radiative transfer through sharp, quasi–static atmospheres whose opacity is dominated by hydrogen is considered at densities low enough that scattering usually dominates absorption and radiative excitations usually dominate collisional excitations. Numerical results for the continuum spectral flux are obtained for effective temperatures $T_e = 6000 - 16{,}000K$ and scale heights $\Delta R = 10^{10} - 10^{14}cm$. Comparison with observations of the Type II supernova 1980k tends to increase the value of the Hubble constant previously obtained by the Baade method.

064.015 **A simple explanation for the Linsky–Haisch boundary line for transition layers.**
E. Böhm–Vitense.
Astrophys. J., Vol. 301, No. 1, p. 297 – 301 (1986).
For stars with low gravitational acceleration, transition layers between chromosphere and corona cannot form, because the chromospheres are so extended that the mechanical energy flux decreases faster than the square of the electron density. There is not enough flux left to lead to a steep temperature increase in the transition layer and corona. If the dissipation length λ for the mechanical energy flux is the same for all stars, the boundary line for transition layers would be expected to coincide with a line of g = const in the H–R diagram. The numerical value for this g depends on the dissipation length.

064.016 **Atmospheric structure and mass loss from pulsation with speculative application to B and Be stars.**
L. A. Willson.
Publ. Astron. Soc. Pac., Vol. 98, No. 599, p. 37 – 40 (1986). – See Abstr. 012.009.
The author summarizes some very general results obtained from analyzing models calculated by Bowen (1985) for the atmospheres of radially pulsating cool stars. He then discusses qualitatively some possible implications of the nonradial pulsations observed in B and Be stars for the structure of their atmospheres.

064.017 **Summary comments on the state of stellar–wind theory as it relates to nonradial pulsations.**
J. I. Castor.
Publ. Astron. Soc. Pac., Vol. 98, No. 599, p. 52 – 55 (1986). – See Abstr. 012.009.

064.018 **The vertical temperature stratification and corona formation of accretion disc atmospheres.**
G. Shaviv, R. Wehrse.
Astron. Astrophys., Vol. 159, No. 1/2, p. L5 – L7 (1986).
The authors formulate the radiative transfer problem for an accretion disc and solve for the vertical stratification and radiation field consistently. The method developed can treat any form of energy generation law. The authors show that if for example, the alpha–model is assumed throughout the photosphere a thermal instability develops even for discs which are optically thick. They hypothesise that this instability leads to the formation of a corona. The authors find also that the temperature distribution is very flat even in the optically thick regions and the disc height

($z(\tau = 1)$) is significantly greater than the classical value one obtains from global considerations.

064.019 **Bombardment solutions to the "soft X-ray puzzle" in radial white dwarf accretion.**
A. M. Thompson, J. C. Brown, J. Kuijpers.
Astron. Astrophys., Vol. 159, No. 1/2, p. 202 – 208 (1986).
The problem of the observed low ratio of hard to soft X-rays in accreting white dwarfs, with low accretion rates, (e.g. AM Her) is discussed in terms of models where the accreting matter is treated as a non-thermal stream bombarding a static atmosphere cooled by optically thin radiation.

064.020 **Bipolar hydromagnetic winds from disks around protostellar objects.**
R. E. Pudritz, C. A. Norman.
Astrophys. J., Vol. 301, No. 2, p. 571 – 586 (1986).
The bipolar outflows associated with protostars are modeled as two-component, hydromagnetic winds which originate in the envelopes of rotating, magnetized molecular disks. The authors show that the torques exerted upon disks by centrifugally driven winds can account for both the observed linear momentum and energy transport rates associated with the molecular outflows. The prediction is that the luminosity associated with protostellar objects during the molecular outflow epoch is due to accretion from these disks. These outflows can solve the angular-momentum problem of star formation and drive accretion onto the protostellar cores. The molecular outflow produces a magnetohydrodynamic shock as it collides with ambient molecular cloud gas with characteristic H_2 emission produced in the pre-shock region. Calculations are presented for the masses, rotation rates, and densities, and outflow characteristics for disks and their outflows, for the sources HH 7-11 (SVS 13), Orion-KL (IRc2), L1551 (IRS 5), and B335.

064.021 **Models of transition regions in hybrid stars.**
J. W. Brosius, D. J. Mullan.
Astrophys. J., Vol. 301, No. 2, p. 650 – 663 (1986).
The authors present models of the transition regions of six "hybrid" atmosphere stars. The models include mass loss and are based on an analysis of IUE data. The equations describing the mass loss are used to evaluate the momentum and energy deposition rates which are required at each level of the atmosphere. The final models for all six stars have mass-loss rates which lie below the current VLA upper limits by factors of 2 – 10. The model densities agree with those derived by density-sensitive line ratios. These transition region models are rather extended ($0.1 - 0.5\ R_*$). The source of mechanical energy is identified with Alfvén waves and wave amplitudes and magnetic field strengths are derived.

064.022 **Red giant model atmospheres. IV. Colors of K and M giant models.**
T. Y. Steiman-Cameron, H. R. Johnson.
Astrophys. J., Vol. 301, No. 2, p. 868 – 876 (1986).
Photometric parameters in the Johnson broad-band and Wing narrow-band photometric systems have been computed for 24 stellar atmospheric models of K and M giants of Johnson, Bernat, and Krupp (1980). Color temperatures, color indices, and certain molecular indices predicted by these models are compared with observations of M giant stars. Good agreement is obtained for $T_{eff} \gtrsim 3500$K. For models with T_{eff} below 3500K, the TiO bands are systematically too weak relative to observations. Possible reasons for this discrepancy are discussed. The effects of limb darkening and possible variation of photospheric radius with wavelength on predicted broad-band models colors are discussed.

064.023 **Time-dependent gas motions in the outer regions of late-type giant stars.**
V. G. Gorbatskij.
Nauchn. Inf., Vyp. 56, p. 9 – 16 (1984). In Russian.
A review is given of the observations and theories related to motion of gas in the outer regions of long-period variables and other late-type giant stars. Propagation of shock waves in the star atmosphere and other proposed mechanisms of mass loss from the star are briefly considered. Attention is paid to possible instabilites in the envelopes of stars and to some consequences of such instabilites.

064.024 **Radiation pressure on molecules as a mechanism of mass loss by red giants. 1. Qualitative analysis.**
V. S. Strel'nitskij, A. V. Fedorova, O. V. Fedorova, I. K. Shmeld (*I. Šmelds*).
Nauchn. Inf., Vyp. 57, p. 30 – 38 (1984). In Russian.
A qualitative analysis of the mechanism of mass loss of luminous cool stars due to radiation pressure on molecules is presented. It is shown that the previous conclusions of inadequateness of this mechanism in producing the necessary gas acceleration or mass-loss rate are not well grounded, because they are based on the poorly known elemental abundances in the atmospheres and gas densities on the basis of the expanding envelope. Rough estimates show that this mechanism can provide the preliminary gas accelerations in oxygen-rich giants of solar mass. It must be particularly efficient in carbon-rich giants due to high abundances of carbon-bearing molecules in atmospheres with augmented C/H ratios. Here the radiation pressure on molecules can provide gas outflows in stars of any mass.

064.025 **An investigation of the atmospheres of O-type main sequence stars by non-LTE.**
L. I. Solov'eva.
Astron. Zh., Tom 63, Vyp. 2, p. 310 – 316 (1986). In Russian. English translation in Sov. Astron., Vol. 30, No. 2.
Calculations of equivalent widths of N III multiplets λ4640 Å, λ4097 Å, λ4379 Å have been performed on the basis of a detailed solution of the coupled statistical-equilibrium and radiative-transfer equations for a multiline, multilevel, multi-ion ensemble and for non-LTE model atmospheres with $T_{eff} = 32500 - 55000$K and $\lg g = 3.3, 3.5, 4.0, 4.5$. The comparison of the observed equivalent widths with the calculated ones by non-LTE of H, He I, He II, C III, C IV, N III, N IV lines in the visual spectra of O-type stars permitted to establish that the model atmospheres with T_{eff} and $\lg g$ determined from H and He lines satisfactorily predict also the behaviour of absorption and emission lines of light ions for 8 stars from the 10 investigated ones. The normal abundances of carbon and nitrogen have been used and the microturbulent velocity ξ_t was found to be 0 – 5 km/s.

064.026 **Mass loss by stars on the asymptotic giant branch.**
Yu. L. Frantsman.
Astrofizika, Tom 24, Vyp. 1, p. 131 – 147 (1986). In Russian. English translation in Astrophysics, Vol. 24, No. 1.
The theoretical populations of white dwarfs and carbon stars were generated for Salpeter's initial mass function and constant stellar birthrate history. The effect of very strong mass loss on the mass distribution of white dwarfs and the luminosity distribution of carbon stars is discussed and the results are compared with observations. This comparison suggests that a significant mass loss by stars on the asymptotic giant branch occurs besides stellar wind and planetary nebulae ejection. Thus it is possible to explain the absence of carbon stars with $M_{bol} < -6^m$ in the Magellanic Clouds, very narrow mass distribution of white dwarfs and very small number of white dwarfs with $M > 1.0\ M_\odot$. The luminosity of asymptotic giant branch stars in the globular clusters of the Magellanic Clouds appears to be a very good indicator of the age.

064.027 **Steady hydromagnetic flows in open magnetic fields. I. A class of analytic solutions.**
B. C. Low, K. Tsinganos.
Astrophys. J., Vol. 302, No. 1, p. 163 – 187 (1986).
This paper introduces a class of axisymmetric steady hydromagnetic flows along open stellar magnetic fields. The full set of hydromagnetic equations that describes these flows in spherical coordinates is reduced to a set of ordinary differential equations in the radial distance from the origin. A judicious choice of free functions admits solutions for a spherically symmetric density in an otherwise nonspherically symmetric flow. In general, the flow requires a formal, steady energy addition whose amount and spatial distribution are calculated self-consistently to conserve

energy everywhere. These solutions provide a rare opportunity to study the basic physics of hydromagnetic stellar winds in terms of exact relationships. The rich variety of solutions presented include flows along dipolar magnetic fields that are purely radial and those that, in addition, have θ–components.

064.028 **Konsistente sphärische NLTE–Modelle für Objekte mit Leuchtkräften entsprechend B–Sternen.**
P. Höflich.
Diss. Nat.–Math. Gesamtfak. Ruprecht–Karls–Univ. Heidelberg, Heidelberg, F.R. Germany, 8+245 pp. (1986).

064.029 **Why do you enjoy studying equatorial disk oscillations?**
A. T. Okazaki.
Astron. Her., Vol. 79, No. 4, p. 88 – 91 (1986). In Japanese.

064.030 **The influence of the dust condensation on the abundances in the envelopes of carbon stars.**
I. Šhmelds.
Issled. Solntsa Krasnykh Zvezd, No. 21, p. 23 – 28 (1984). In Russian.

Calculations of chemical equilibrium in the envelope of a carbon star with allowing of C and SiC dust condensation are carried out. The molecular abundance in that case may not strongly differ from that in the non–dust condensation case. Condensation SiC allows to explain observable deficiency of some silicon containing molecules if the process finishes when Si/S $\approx$ 1, but it is in discrepancy with observable abundance of SiS and absence of H_2S.

064.031 **Stars in a gas–dust envelope.**
I. Šmelds.
Zvaigžnota Debess, 1984. Gada Rudens, p. 12 – 15 (1984). In Latvian.

064.032 **Theory of the transition layer emission measures.**
E. Böhm–Vitense.
Bull. Am. Astron. Soc., Vol. 17, No. 4, p. 847 (1985). Abstract. – See Abstr. 010.062.

064.033 **On molecular absorption in the atmospheres of late–type giants.**
A. V. Shavrina.
Kinematika Fiz. Nebesn. Tel, Tom 2, No. 3, p. 92 – 94 (1986). In Russian. English translation in Kinematics Phys. Celest. Bodies.

The synthetic profile of the (0,1) band of the $B^1\Pi - X^1\Sigma^+$ system of ZrO ($\lambda \sim 693$ nm) has been calculated using model atmospheres for S–type stars. The absorption of six bands of the γ–system of ZrO was considered as "quasi–continuous" one and was included in the continuous opacity. This procedure allows matching the observed band profile to the synthetic spectrum.

064.034 **Chromospheric dust formation, stellar masers and mass loss.**
R. E. Stencel.
NASA Conf. Publ., NASA CP–2403, p. A 29 – A 33 (1986). – See Abstr. 012.030.

The author outlines a multi–step scenario which describes a plausible mass loss mechanism associated with red giant and related stars. The process involves triggering a condensation instability in an extended chromosphere, leading to the formation of cool, dense clouds which are conducive to the formation of molecules and dust grains. Once formed, the dust can be driven away from the star by radiation pressure. Consistency with various observed phenomena is discussed.

064.035 **Structure of magneto–Keplerian discs: an analytical solution for thin accretion discs threaded by stellar magnetic fields.**
O. Kaburaki.
Mon. Not. R. Astron. Soc., Vol. 220, No. 2, p. 321 – 338 (1986).

The structure of Keplerian discs which are threaded by the magnetic field lines of each central star are considered. Under the assumptions of thin discs, stationarity and of axisymmetry, a fairly simple analytical solution is found.

064.036 **Construction of non–LTE model atmospheres using approximate lambda operators.**
K. Werner.
Astron. Astrophys., Vol. 161, No. 1, p. 177 – 182 (1986).

The author introduces a new method of calculating model atmospheres for hot stars subject to constraints of radiative, hydrostatic and statistical equilibrium using approximate lambda operators, while the exact solution of the transfer problem is obtained by combination with a perturbation technique. Model atmospheres with up to 100 atomic levels in non–LTE can be constructed, exceeding remarkably the capacities of the classical linearization approach as presented by Auer and Mihalas (1969).

064.037 **The maximum possible mass loss rate for dust–driven winds.**
H.–P. Gail, E. Sedlmayr.
Astron. Astrophys., Vol. 161, No. 1, p. 201 – 202 (1986).

It is shown that L_*/cv_∞ is not the maximum possible mass loss rate in case of dust driven winds, contrary to what has often been assumed. (L_* is the stellar luminosity, v_∞ the terminal velocity of the wind.) On the basis of a simple analytical approximation a relation is derived which shows that for heavily obscured IR–objects the mass loss rate has to exceed the above "limit value".

064.038 **Spectral flux from low–density photospheres: approximate results.**
S. Hershkowitz, E. Linder, R. V. Wagoner.
Astrophys. J., Vol. 303, No. 2, p. 800 – 809 (1986).

Radiative transfer through sharp, quasi–static atmospheres containing mostly hydrogen is considered at densities low enough that scattering dominates absorption and collisional excitations are negligible. It is found that the balance between photoionization and radiative recombination which governs the atomic level populations greatly reduces the effective absorptive opacity. Approximate results for the emergent continuum spectral flux are obtained, significantly different than if LTE level populations were assumed. The results are applied to spectra of Type II supernovae.

064.039 **On the dimensions of accretion disks in cataclysmic variables.**
M. Kitamura.
Proceedings of the Second Asian–Pacific Regional Meeting on Astronomy, p. 55 – 58 (1984). – See Abstr. 012.031.

The problem of rotating gaseous disks or accretion disks in cataclysmic binary systems is discussed from a simple hydrodynamic point of view for the inviscid gas flow with the aid of the Roche coordinates.

064.040 **Vistas on stellar–solar atmospheres.**
J.–C. Pecker.
Aust. J. Phys., Vol. 38, No. 6, p. 799 – 809 (1985). Paper presented at the R. G. Giovanelli Commemorative Colloquium, Part I, Sydney, 26 – 29 November 1984.

The evolution of the modelling of the stellar atmosphere towards greater physical consistency is described briefly. The progressive relaxation of stringent hypotheses (such as RE, LTE, HE, plane–parallel geometry) in all layers characterizes this evolution. The author discusses roughness (i.e. departures from spherical geometry), the origin of the concept, and how its recent application to Skylab and OSO–8 UV data shows its importance, difficult to take into account accurately, but not to be overlooked.

064.041 **A study of hybrid atmosphere stars: transition region modeling and chromospheric emission variability.**
J. W. Brosius.
Diss. Abstr. Int., Sect. B, Vol. 46, No. 9, p. 3089 (1986). Thesis, University of Delaware, 148 pp. (1985). Order No. DA8525286.

064.042 **The equilibrium structure of thin magnetic flux tubes. II.**
W. Kalkofen, R. Rosner, A. Ferrari, S. Massaglia.
Astrophys. J., Vol. 304, No. 1, p. 519 – 525 (1986).

The thermal structure of the medium inside thin, vertical, magnetic flux tubes embedded in a given external atmosphere is investigated, assuming cylindrical symmetry and a depth–independent plasma β. The authors compute the variation with tube radius of the temperature on the tube axis and estimate the temperature on the tube wall.

064.043 **Ejection of matter by massive stars.**
T. A. Lozinskaya.
Eruptive phenomena in stars, p. 379 – 380 (1986). – See Abstr. 012.039.

064.044 **Shell phenomenon in Be stars.**
P. Harmanec.
Eruptive phenomena in stars, p. 385 – 392 (1986). – See Abstr. 012.039.

064.045 **The common nature of flaring processes on the sun and red dwarfs.**
M. M. Katsova, M. A. Livshits.
Eruptive phenomena in stars, p. 437 – 438 (1986). – See Abstr. 012.039.

064.046 **Theoretical stellar energy distributions.**
R. L. Kurucz.
Space Telesc. Sci. Inst., Prepr. Ser., No. 100, p. 29 – 33 (1986). – See Abstr. 012.040.

064.047 **Stellar winds from young stars.**
N. Panagia.
Space Telesc. Sci. Inst., Prepr. Ser., No. 105, 20 pp. (1986). To appear in Phys. Scr.

064.048 **Continuum spectra of collimated, ionized stellar winds.**
S. P. Reynolds.
Astrophys. J., Vol. 304, No. 2, p. 713 – 720 (1986).
For some pre–main–sequence stars, highly collimated, ionized gas jets have been detected. The author calculates the thermal continuum emission from collimated, ionized winds in the presence of gradients in jet width, velocity, ionization, and temperature. Unresolved sources can exhibit partially opaque spectra with spectral indices anywhere between 2 and –0.1. Well–collimated jets are much more efficient than spherical winds at producing radio flux. Relations for the collimation distance and mass–loss rate for jets as a function of observable parameters are given. The results are applied to continuum sources in NGC 7538 and L1551.

064.049 **Baroclinic waves in a vertically stratified thin accretion disk.**
E. Knobloch, H. C. Spruit.
MPA Rep., No. 217, 27 pp. (1986). Submitted to Astron. Astrophys.

064.050 **Stationary shocks in accretion disks.**
H. C. Spruit.
MPA Rep., No. 232, 36 pp. (1986). Submitted to Astron. Astrophys.

064.051 **Radiatively–driven stellar winds.**
L. B. Lucy.
ESO Sci. Prepr., No. 419, 11 pp. (1986). Paper presented at IAU Colloq. No. 89.

064.052 **A pulsed jet as the source of momentum in bipolar molecular outflows.**
S.–U. Choe, R. N. Henriksen.
Astrophys. J., Vol. 305, No. 1, p. 131 – 135 (1986).
The authors propose that freely falling material inflates and squeezes a hot cavity in the center of an accretion disk, thereby creating a momentum reservoir at the free–fall pressure. The heated, light, material rises through the more dense confining axial accretion, becoming a stellar jet by the usual nozzle mechanism. It is argued that this can solve the momentum problem in the bipolar molecular outflows because the free–fall pressure "head" can drive a momentum flux much in excess of the radiative momentum flux. The model suggests a pulsed jet associated with the larger scale bipolar molecular outflow due to a relaxation oscillation with a 50% duty cycle.

064.053 **The continuous absorption coefficient in the atmospheres of solar–type stars.**
J.–h. Jin.
Publ. Purple Mt. Obs., Vol. 5, No. 1, p. 12 – 30 (1986). In Chinese.
The construction of tables of the continuous absorption coefficient of stellar atmospheres and their applicability are discussed. The continuous absorption coefficient in one gram of stellar material is calculated for atmospheres of solar–type stars with T = 3000 – 30000K (using 47 temperature values). For every temperature the author has chosen 30 electron pressures.

064.054 **Effects of a supercritical wind on nova observations.**
M. Friedjung.
Inst. Astrophys. Paris, Pré–Publ., No. 146, 9 pp. (1986). To appear in "Compte–rendus de la VIe Rencontre de Moriond en Astrophysique", March 1986.

064.055 **Two–dimensional models of stellar wind bubbles. V. The efficiency of momentum transfer between winds and dense clumps in free wind regions.**
M. Różyczka, G. Tenorio–Tagle.
Acta Astron., Vol. 35, Nos. 3–4, p. 213 – 227 (1985).
Hydrodynamical models of dense circumstellar clumps accelerated by supersonic stellar winds are obtained. Dimensions, densities, and velocities of the models are representative of conditions that are likely to be met by Herbig–Haro objects. It is shown that the existing analytical estimates of the acceleration efficiency are too optimistic, what has its source in neglected hydrodynamical effects.

064.056 **Magnetic fields in starspots on late–type giants.**
K. Jahn.
Acta Astron., Vol. 35, Nos. 3–4, p. 261 – 278 (1985).
Computations of models of magnetic starspots on cool active giants show that the value of the magnetic intensity in spots is generally of the order of one kilogauss, although in larger spots the field can be as weak as a few hundred gauss. It is also argued, that spots on giants qualitatively differ from those on late–type dwarfs, since they cannot be too large. λ And is discussed as an example. The comparison of models of spots computed for that star with photometric observations suggest, that a dark region on λ And consists of hundreds of small spots in which the magnetic intensity cannot exceed about 900 gauss.

064.057 **Alfvén wave driven wind in late type giants and supergiants. I. Critical solutions.**
J. Sikorski, E. Bielicz, M. Krogulec.
Acta Astron., Vol. 35, Nos. 3–4, p. 279 – 289 (1985).
The conditions at the critical point for Alfvén wave driven wind theory are investigated. Influence of the mass loss rate, damping length, Alfvén wave energy flux and magnetic field strength on physical conditions at the critical point is analysed.

064.058 **Structure of the accretion disks in cataclysmic variables.**
G. Pojmański.
Acta Astron., Vol. 36, No. 1, p. 69 – 77 (1986).
Models of the vertical structure of the accretion disks are calculated for various metal and hydrogen abundances, opacity tables, values of α and efficiency of the convection. Families of models with α from 0.01 to 1.0, with Cox and Stewart, Cox and Tabor and Alexander opacity tables, with helium and hydrogen content for Population I and II and for different mixing length values are presented in diagrams.

064.059 **Mass loss from very young massive stars.**
T. Henning.
Astron. Nachr., Vol. 307, No. 2, p. 119 – 127 (1986). = Mitt. Univ.–Sternw. Jena, Nr. 172.
A very important property of very young and massive stars (BN objects) is their intensive mass loss. The author describes the main methods to derive the mass loss rates. Available observa-

tions are used to characterize the ionized stellar winds and the CO flows. The results are confronted with theories describing the anisotropic mass loss.

064.060 **Stellar magnetic activity and long term luminosity variations of late type stars. I. On generalization of Öpik's convection theory to a rotating medium.**
C. B. Park, H. S. Yun, J. W. Lee.
J. Korean Astron. Soc., Vol. 18, No. 2, p. 86 – 99 (1985).
A generalization of the original Öpik's cellular convection theory has been made to accomodate a rotating convective medium. With the use of the formulation, a set of rotating model envelopes of the sun and late–type main–sequence stars have been constructed under three different rotation periods. Their thermal structures are presented and characteristics of their convection are discussed in the context of stellar dynamo.

064.061 **Influence of convection on line asymmetries.**
Y. S. Park, H. S. Yun.
J. Korean Astron. Soc., Vol. 19, No. 1, p. 15 – 31 (1986).
The authors have analyzed Gray's observed mean line bisectors of F5, G0, G2, and G5 normal dwarf stars and interpreted them by computing theoretical line bisectors based on a two stream model. A set of perturbed models has been derived, and their detailed structures on temperature fluctuations and velocity fields are presented as a function of depth, which account for the observed bisectors. It is found that the degree of stellar convective overshootings and temperature fluctuations in the upper atmospheres increases towards earlier spectral types.

064.062 **The structure and stability of the accretion disc boundary layer.**
J. C. B. Papaloizou, G. Q. G. Stanley.
Mon. Not. R. Astron. Soc., Vol. 220, No. 3, p. 593 – 610 (1986).
An axially–symmetric accretion disc is treated both analytically and numerically, with special attention paid to the boundary layer. A modified form of the standard viscosity model is used. The authors show the mechanism by which overstable oscillations can evolve there and how they dissipate.

064.063 **Mode identification from line profile variations – II. A quantitative least–squares algorithm.**
L. A. Balona.
Mon. Not. R. Astron. Soc., Vol. 220, No. 3, p. 647 – 656 (1986).
The author presents an algorithm which allows a completely objective determination of all the parameters for line profile variations in a non–radially oscillating star. The method is applicable when the projected rotational velocity is considerably larger than the pulsational velocity. The first few moments of the profile are used to compute a criterion which is a function of the spherical harmonic order and angle of inclination. The most likely spherical harmonic order and angle of inclination is found when the criterion reaches a minimum at which point least–squares estimates and confidence intervals for the remaining pulsational parameters can be obtained.

064.064 **Oscillations of thick accretion discs.**
M. Jaroszyński.
Mon. Not. R. Astron. Soc., Vol. 220, No. 4, p. 869 – 881 (1986).
The non–axisymmetric oscillations of polytropic, ideal fluid tori rotating in the external gravitational field of a point mass are investigated both numerically and analytically. Normal modes of oscillations are found.

064.065 **Accretion of noninteracting particles.**
A. F. Illarionov.
Sov. Astron., Vol. 29, No. 6, p. 630 – 633 (1985). English translation of 40.064.059.

064.066 **Acoustic instability in hot–star subphotospheres.**
N. S. Petrukhin.
Sov. Astron., Vol. 29, No. 6, p. 633 – 638 (1985). English translation of 40.064.060.

064.067 **Coagulation and multiplication of dust particles in media with carbon excess.**
D. N. Dojkov.
Problems of astronomy, p. 56 – 59 (1985). In Russian. Abstr. in Ref. Zh., 51. Astron., 3.51.759 (1986). – See Abstr. 003.023.

064.068 **Radiative transfer in circumstellar 1612 MHz OH masers.**
C. Alcock, R. R. Ross.
Astrophys. J., Vol. 305, No. 2, p. 837 – 851 (1986).
The authors have made a detailed investigation of radiative transfer in the 1612 MHz OH maser in the expanding envelopes around evolved stars, in the context of the standard model in which the mass outflow is windlike. It is shown that neither fully nor partially saturated models are compatible with the observations. The authors propose a model for the mass outflow from these stars in which all mass is lost by the ejection of blobs of material. These blobs, acted on by radiation pressure on grains, evolve into platelike structures in the OH maser region.

064.069 **The reflection effect in model stellar atmospheres. I. Grey atmospheres with convection.**
L. P. R. Vaz, Å. Nordlund.
Rev. Mex. Astron. Astrofis., Vol. 12, Spec. Issue, p. 190 (1986). Abstract. – See Abstr. 012.066.

064.070 **Accretion disks.**
L. Filippov.
Astrofiz. issled., Tom 4, p. 68 – 79 (1985). In Russian. Abstr. in Ref. Zh., 51. Astron., 6.51.197 (1986).

064.071 **Intrinsic parameters of hot blue stars.**
R. P. Kudritzki, D. G. Hummer.
Luminous stars and associations in galaxies, p. 3 – 18 (1986). – See Abstr. 012.079. (IAU Symp. No. 116).
Contents: Introduction. The photosphere of hot stars. Wind blanketing. Metal opactiy and non–LTE line blanketing. Radiation–driven winds. The H–R diagrams of the most massive galactic O–stars. Abundances in massive O–stars. Abundance gradients in the galactic disk from young cluster B–stars. Future developments.

064.072 **Two possible wind models for Wolf–Rayet stars.**
T. Nugis.
Luminous stars and associations in galaxies, p. 223 – 224 (1986). – See Abstr. 012.079. (IAU Symp. No. 116).
The author's purpose was to find out the empirical models of the envelopes of WN5 and WN6 stars which are not in conflict with the continuum runs in the IR spectral region and with the emission line intensities and profiles.

064.073 **Stellar atmospheric instability in the upper part of the Hertzsprung–Russell diagram.**
C. de Jager, H. Nieuwenhuijzen.
Luminous stars and associations in galaxies, p. 255 – 256 (1986). – See Abstr. 012.079. (IAU Symp. No. 116).
The authors list evidences supporting the validity of their instability criterion and the consequent relations (cf. de Jager 1978, 1980, 1984).

Strong plane shock waves in exponential and optically–thin atmospheres, II.
See Abstr. 062.009.

Some astrophysical applications of MHD instabilities of differentially rotating fluids.
See Abstr. 062.069.

A model of hydromagnetic turbulence in differentially rotating disks.
See Abstr. 062.129.

Accretion discs.
See Abstr. 062.133.

The statistical equilibrium of hydrogen and helium in a radiation field, with an application to interpreting Wolf–Rayet spectra.
See Abstr. 063.004.

Brightness distribution of resonant line photons in spherical envelopes.
See Abstr. 063.028.

Line formation in the comoving frame: accurate solution using an approximate Newton–Raphson operator.
See Abstr. 063.040.

Line formation in expanding atmospheres: multi–level calculations using approximate lambda operators.
See Abstr. 063.048.

Transport of continuum radiation in disk geometry.
See Abstr. 063.051.

The moment method for radiative transfer problems in spherical atmospheres with a core.
See Abstr. 063.064.

On the radiative–radiative pumping of circumstellar SiO masers.
See Abstr. 063.093.

Strahlungstransport in den Außenschichten weißer Zwerge mit Linienabsorption unter Einwirkung starker Magnetfelder.
See Abstr. 063.106.

Hayashi limits for carbon stars and the onset of dust–driven winds.
See Abstr. 065.006.

Mass loss and the evolution of supergiants.
See Abstr. 065.046.

Stellar oscillations in the presence of a magnetic field.
See Abstr. 065.120.

Nucleosynthesis in massive, mass losing, stars.
See Abstr. 065.137.

Nucleosynthesis in massive stars: winds from WR stars and isotopic anomalies in cosmic rays.
See Abstr. 065.138.

Instability in massive stars: an overview.
See Abstr. 065.163.

From Miras to planetary nebulae: a model of mass loss.
See Abstr. 065.175.

Boundary behaviour of accretion disks determined by gravity induced cooling near the black hole.
See Abstr. 067.018.

A critical examination of thick disks with equatorial accretion.
See Abstr. 067.034.

On hydrodynamics of radiatively driven winds.
See Abstr. 067.074.

On the stream–disk interaction in accreting compact objects.
See Abstr. 067.114.

Transonic accretion flow in a thin disk around a black hole. III. Analytic considerations.
See Abstr. 067.133.

Interaction of the accretion disk with the magnetosphere of a rotating neutron star.
See Abstr. 067.151.

Structure of the magnetospheres of accreting magnetic compact objects.
See Abstr. 067.153.

Emission characteristics of neutron star polar caps.
See Abstr. 067.155.

Solar granulation.
See Abstr. 071.027.

MHD model for long distance coronal loops.
See Abstr. 074.107.

Heating and mass transfer in solar and stellar atmospheres by turbulent Alfvén waves.
See Abstr. 080.114.

Luminosity and colour variations of 88 Her through phase changes from the far UV to the visual spectral regions. II. Empirical atmospheric modelling: an association between photospheric and envelope variations.
See Abstr. 112.029.

Long–wavelength emission lines from stellar winds.
See Abstr. 112.033.

Infrared recombination line profiles: a diagnostic probe of the velocity field in the S 106 stellar wind.
See Abstr. 112.113.

IRAS observations of cool excess around main sequence stars.
See Abstr. 112.134.

Mass loss by cool stars.
See Abstr. 112.139.

The internal radius of CS shells around cool, oxygen–rich stars.
See Abstr. 112.144.

The flux distribution of Vega for 10 μm < λ < 100 μm, and the calibration of IRAS at 12 μm and 25 μm.
See Abstr. 113.062.

The formation of nitrogen and carbon lines in HD 50896 (WN5).
See Abstr. 114.034.

An investigation of the atmosphere of the supergiant δ CMa (F8 Ia) and determination of its evolutionary parameters.
See Abstr. 114.035.

A model atmosphere analysis of α Lyr and γ Gem.
See Abstr. 114.036.

A model atmosphere analysis of three A–type stars.
See Abstr. 114.038.

Non–LTE analysis of massive stars in the Magellanic Clouds.
See Abstr. 114.072.

Interpretation of $\Delta \upsilon = 2$ carbon monoxide lines in the spectrum of α Her.
See Abstr. 114.080.

Steps to a new calibration of the spectral type – effective temperature relationship for early–type stars.
See Abstr. 114.140.

Photospheric and "chromospheric" C IV lines in B–main–sequence stars.
See Abstr. 114.141.

A light synthesis program for binary stars. II. Light curve and color curve effects in a contact system.
See Abstr. 117.002.

Dwarf nova burst asymmetry and the physics of accretion disks.
See Abstr. 117.086.

Polarization in massive X-ray binaries. I. A low-inclination model for Cygnus X-1.
See Abstr. 117.165.

Models and theoretical spectra of accretion discs in dwarf novae.
See Abstr. 117.204.

The optical spectrum of HDE 226868 = Cygnus X-1. III. A focused stellar wind model for He II λ4686 emission.
See Abstr. 117.208.

Numerical studies of wind accretion.
See Abstr. 117.223.

Monochromatic reflection effect and temperature distributions on the distorted surfaces of close binary systems.
See Abstr. 117.295.

Accretion disks in mass-exchange binaries.
See Abstr. 117.299.

Accretion disks in binary stars.
See Abstr. 117.300.

Accretion onto magnetic white dwarfs.
See Abstr. 117.301.

An accretion disc surrounding a component of Epsilon Aurigae.
See Abstr. 119.023.

Molecular outflows and mass loss in pre-main-sequence stars.
See Abstr. 121.047.

Atmospheres of Delta Scuti pulsating variables. II. The analysis of 44 Tau and V644 Her spectra.
See Abstr. 122.032.

Nonradial modes in line-profile variable stars.
See Abstr. 122.097.

Atmospheres of Delta Scuti-type pulsating variables. III. An analysis of the 28 And spectrum.
See Abstr. 122.192.

CD −24°9052: a new bright and helium-rich subdwarf O star.
See Abstr. 126.011.

A test of the dipole model for the rotating magnetic white dwarf Feige 7.
See Abstr. 126.013.

Model atmospheres of DA white dwarfs. II. Catalogue.
See Abstr. 126.036.

Model atmospheres of DQ white dwarfs. I. Details of computation.
See Abstr. 126.037.

Model atmospheres of DQ white dwarfs. II. Catalogue.
See Abstr. 126.038.

Modelling for O and N infrared line observations of H II blisters.
See Abstr. 132.044.

065 Stellar Structure and Evolution

065.001 **Neutrino diffusion in stellar collapse.**
L. J. van den Horn, J. Cooperstein.
Astrophys. J., Vol. 300, No. 1, p. 142 – 150 (1986).

The authors consider the diffusion flows of neutrinos in stellar collapse. The neutrino energy and number fluxes are driven by the gradients of temperature and chemical potential, and are characterized by four Rosseland mean free paths. The authors derive exact expressions for these Rosseland means, taking scattering and stimulated absorption of the neutrinos into account separately as well as in combination.

065.002 **Relation of chromospheric activity to convection, rotation, and pre-main-sequence evolution.**
R. L. Gilliland.
Astrophys. J., Vol. 300, No. 1, p. 339 – 347 (1986).

Pre-main-sequence, or T Tauri, stars are characterized by much larger fluxes of nonradiative origin than their main-sequence counterparts. The recent success of correlating nonradiative fluxes to the Rossby number, Ro = P_{rot}/τ_{conv}, a central parameter of simple dynamo theories of magnetic field generation, has led to the suggestion that the same relation might be of use in explaining the pre-main-sequence stars. This study shows that this dependence alone cannot account for the high observed nonradiative fluxes. When the role of evolving acoustic flux is taken into account, however, consistency with the observations can be obtained.

065.003 **Evolution of white dwarfs.**
D. Koester, D. Schönberner.
Astron. Astrophys., Vol. 154, No. 1/2, p. 125 – 134 (1986).

The evolution of white dwarfs with 0.598 and 0.546 $M_\odot$ is followed from the stage of a central star of planetary nebulae to very low luminosities using a stellar evolution code, realistic initial models and equation of state. Since the authors' model end up with a somewhat smaller hydrogen envelope mass than found in a recent paper by Iben and Tutukov (1984) they do not find a late phase where hydrogen burning is the dominant energy source. The reason for this difference and implications concerning the amount of hydrogen present in cool white dwarfs are discussed. The main emphasis is, however, on the deviations of radii and surface gravities from the Hamada and Salpeter (1961) zero temperature models. These deviations are large at effective temperatures above 40000K and have to be taken into account in the determination of masses and space densities for white dwarfs.

065.004 **An exact solution for a rotating selfgravitating gas mass with oscillatory density.**
F. Schmitz, R. Ebert.
Astron. Astrophys., Vol. 154, No. 1/2, p. 214 – 218 (1986).

An oscillatory solution of the equations of structure of isothermal selfgravitating gas masses with stationary differential rotation is given. Both the density and the rotational velocity show oscillations. The solution is an axially symmetric analogue to a solution found by Schmid-Burgk (1967) which shows a periodic structure depending on two cartesian coordinates. It falls into two families with distinct density structures but a common rotation law. The physical meaning of the solutions is discussed. In order to draw conclusions about other configurations with differential rotation the authors consider the influence of an applied external gravitational field.

065.005 **Boundary layer theory and its application to Algol systems: a model for the pseudo–photosphere stars.**
S. N. Shore, A. R. King.
Astron. Astrophys., Vol. 154, No. 1/2, p. 263 – 266 (1986).
Mass accretion onto main sequence stars in Algol systems must take place in a shear–dominated boundary layer. This study shows that many of the gross observational properties of these systems can be understood to arise from such regions, in which turbulent coupling to the underlying star is important. Specific systems are discussed, and comparison is also made to previous numerical hydrodynamic simulations.

065.006 **Hayashi limits for carbon stars and the onset of dust–driven winds.**
L. B. Lucy, J. A. Robertson, C. M. Sharp.
Astron. Astrophys., Vol. 154, No. 1/2, p. 267 – 273 (1986).
Hayashi limits are computed for carbon stars on the asymptotic giant branch. With previously published treatments of the opacity of cool, carbon–enriched matter, the derived limit is such that the predicted extreme redward evolution of these stars falls far short of observed effective temperatures as well as of those required for grain condensation in the surface layers. However, with new opacity calculations – which include the isotopic variants of important diatomic molecules and adopt the observationally–justified microturbulent velocity of 5 km s^{-1} – the Hayashi limit shifts to substantially cooler effective temperatures and becomes just consistent with the belief that N–type carbon stars subsequently evolve into dust–enshrouded configurations.

065.007 **Nonlinear hydrodynamical models of stellar convective zones.**
W. Unno, M.–A. Kondo.
Astrophys. Space Sci., Vol. 118, Nos. 1/2, p. 223 – 225 (1986). – See Abstr. 012.001.
Large–eddy simulation (LES) of turbulent convection is discussed in various versions (mixing–length theory, modal theory and spectral theory) in respect to the application to stellar convective zones. The eddy diffusivities are essential in these theories, and the nonlinear treatment of convection consistent with turbulent diffusivities (of effective Reynolds number of about 20 and Prandtl number of 0.4) offers a simple method of constructing stellar models without the use of the mixing–length.

065.008 **The effect of back pressure on the contact evolution of a close binary system.**
M. Nakamura, Y. Nakamura.
Astrophys. Space Sci., Vol. 118, Nos. 1/2, p. 283 – 286 (1986). – See Abstr. 012.001.
The contact condition which includes the effect of pressure in common envelope is derived and employed in the evolutionary calculation of a close binary system in contact phase. The result indicates that the effect is negligibly small and the equality of surface potentials of both components is a sufficiently good approximation for the contact condition.

065.009 **Carbon deflagration models for type–I supernovae and theoretical optical spectra.**
K. Nomoto, F.–K. Thielemann, K. Yokoi, D. Branch.
Astrophys. Space Sci., Vol. 118, Nos. 1/2, p. 305 – 309 (1986). – See Abstr. 012.001.
A carbon deflagration model in the accreting C+O white dwarf is presented as a plausible model for type–I supernovae. The evolution of the white dwarf is calculated from the beginning of accretion through complete disruption. The synthetic spectrum at maximum light for this carbon deflagration model is calculated and compared to the observed spectrum of SN 1981b in NGC 4536 at maximum light. If the radial stratification of composition in the outer layers of the model is removed by mixing, the theoretical spectrum is in good agreement with the observed spectrum.

065.010 **Study of time–evolving hydrodynamic cepheid models.**
M. Takeuti.
Astrophys. Space Sci., Vol. 119, No. 1, p. 37 – 40 (1986). – See Abstr. 012.001.
For studying modal coupling, the author ran the hydrodynamic models of cepheids and observed the growth and disappearance of various modes in their pulsation. The existence of modal coupling has been confirmed.

065.011 **The phase transitions of superdense matter and supernova explosions.**
M. Takahara, K. Sato.
Astrophys. Space Sci., Vol. 119, No. 1, p. 45 – 49 (1986). – See Abstr. 012.001.
Effects of the phase transitions of superdense matter on supernova explosions are investigated with the aid of an idealized equation of state on the assumptions of adiabatic collapse. It is found that in the case of strong phase transitions explosions become weaker, while in the case of weak phase transitions explosions become stronger. However, the increment of the ejected energy is not so large as suggested by Migdal et al. (1979).

065.012 **One–armed oscillations of a non–selfgravitating polytropic disk.**
A. T. Okazaki, S. Kato.
Astrophys. Space Sci., Vol. 119, No. 1, p. 109 – 110 (1986). – See Abstr. 012.001.
One–armed global oscillations in a non–selfgravitating polytropic disk rotating around a star are investigated. The unperturbed disk is axisymmetric, geometrically thin, and extends infinitely in the radial direction keeping its thickness constant. Perturbations considered are inviscid and adiabatic. It is found that there are one–armed retrograde wave modes which are trapped in an inner region of the disk.

065.013 **Radial and nonradial pulsation of a rotating partially degenerate standard model.**
M. K. Das.
Astrophys. Space Sci., Vol. 119, No. 2, p. 263 – 277 (1986).
The effect of slow uniform rotation on the radial and nonradial modes of partially degenerate standard models has been investigated. For the case of the radial mode, it is shown that the destabilizing effect of slow uniform rotation reduces with increased central condensation of the model. However, it is found that the models with a strongly degenerate interior, become more unstable dynamically as a result of slow uniform rotation. Further, it is noted that the frequency of the nonradial modes of oscillation (Kelvin mode) increases due to the presence of rotation. Thus the period of radial modes of oscillation of a slowly uniformly rotating partially degenerate model are much larger than the corresponding periods of nonradial (Kelvin) modes.

065.014 **Speculations on the fizzled collapse of a massive star.**
A. Burrows.
Astrophys. J., Vol. 300, No. 2, p. 488 – 491 (1986).
This paper studies the physics of the long–term accretion phase of a stellar collapse that does not lead to a supernova. A trapped surface should form early on in this accretion phase. It is shown that, even when nuclear burning and supersonic turbulence are included, the envelope, where most of the progenitor's mass resides, can never achieve temperatures sufficient to radiate more than a negligible fraction of its rest mass in neutrinos before it is swallowed by the black hole. The neutrino emissions during the pre–black hole, Kelvin–Helmholtz phase of the inner core should dominate the total integrated neutrino flux even when the supernova fizzles.

065.015 **Collapse of an isothermal protostar: the effect of initial and boundary conditions.**
R. E. Dentus.
Cosmical gas dynamics, p. 243 – 250 (1985). – See Abstr. 012.007.
Results are presented from a 1 dimensional model of the isothermal hydrodynamic collapse phase of a spherically symmetric protostar. These show that the initial conditions imposed affect the final appearance of the protostar.

065.016 **Rotational and tidal perturbations of linear isentropic oscillations of a gaseous star.**
L. Martens, P. Smeyers.
Astron. Astrophys., Vol. 155, No. 2, p. 211 – 226 (1986).
A previous investigation on the rotational and tidal perturbations of linear isentropic oscillations of a gaseous star has been continued (31.117.029). First– and second–order corrections of eigenfrequencies of rotationally and tidally distorted polytropic models with indexes $n = 3$ and $n = 4$ are determined for various modes. The most striking result is that the relative first–order corrections of the eigenfrequencies of the p–modes decrease when the radial order of these modes increases. Because of this result, the authors have developed an asymptotic approximation for p–modes of high radial orders in a rotationally and tidally distorted star.

065.017 **Mode identification from line profile variations.**
L. A. Balona.
Mon. Not. R. Astron. Soc., Vol. 219, No. 1, p. 111 – 129 (1986).
The author proposes a simple method of analysing spectral line profiles in non–radially oscillating, rotating stars in order to determine the spherical harmonic order of the oscillation. Unlike the usual method, profile modelling is not required. Moreover, the proposed method automatically gives the periods of oscillation and provides information on the projected rotational velocity and angle of inclination.

065.018 **Theoretical Stokes profiles for magnetic stars – II. An appraisal of statistical methods for magnetic field determination.**
M. J. Stift.
Mon. Not. R. Astron. Soc., Vol. 219, No. 1, p. 203 – 216 (1986).
Synthetic spectra in the Stokes I and V parameters calculated within the framework of the rigid rotator model are used in a discussion of the respective performances of three different magnetic field determination methods. Starting with the mathematical formulation of Babcock's method and of two statistical methods, viz. a template method developed by Borra, Fletcher & Poeckert and a correlation method proposed by Weiss, Jenkner & Wood, the authors then analyse the basic differences between these techniques. They conclude that while both statistical methods are to a limited extent suitable for survey purposes they do not provide a sound basis for detailed modelling of stellar magnetic field geometries.

065.019 **Observational implications of the explosion of white dwarfs.**
R. López, J. Isern, R. Canal, J. Labay.
Astron. Astrophys., Vol. 155, No. 1, p. 1 – 5 (1986).
The authors present models for Type I supernova light curves, based on the partial explosion (leaving a bound remnant) of a mass–accreting white dwarf in a close binary system. These models would correspond to partially solid white dwarfs and they can reproduce the Pskovskii–Branch effect, in contrast with those based on the more usual assumption of total disruption of the white dwarf. A value of $H_0 = 100\ \mathrm{km\ s^{-1} Mpc^{-1}}$ seems to be excluded by the authors' models. Remaining uncertainties are discussed.

065.020 **Protostellar angular momentum transport by spiral density waves.**
C. Yuan, P. Cassen.
Icarus, Vol. 64, No. 3, p. 435 – 447 (1985).
The gravitational collapse of molecular clouds or cloud cores is expected to lead to the formation of stars that begin their lives in a state of rapid rotation. It is known that, in at least some specific cases, rapidly rotating, self–gravitating bodies are subject to instabilities that cause them to assume ellipsoidal shapes. In this paper the authors investigate the consequences of such instabilities on the angular momentum evolution of a star in the process of formation from a collapsing cloud, and surrounded by a protostellar disk, with a view toward applications to the formation of the Solar System.

065.021 **Stellar models under homoaxial rotation: Eulerian equations and simulation of non–uniformly rotating stars.**
V. S. Geroyannis.
Astrophys. Space Sci., Vol. 120, No. 1, p. 43 – 59 (1986).
The Eulerian equations are set up for a model subject to homoaxial rotation and suitable for simulation of a non–uniformly rotating star. These equations are formulated in a non–inertial frame of reference, rotating uniformly, (i.e., rigidly) with respect to the inertial common frame.

065.022 **An analytic approach to the connection between stellar structure parameters and the neutrino emission rate in a simple stellar model.**
H. J. Haubold, A. M. Mathai.
Astron. Nachr., Vol. 307, No. 1, p. 1 – 7 (1986).
In order to obtain the internal structure of a main–sequence star such as the Sun usually one has to solve the detailed structure equations numerically. This paper is an attempt to construct analytic models for the stellar nuclear energy generation. The authors give closed–form analytic results for the stellar luminosity and stellar neutrino emission rate when the radial matter density of the spherical star under consideration is linear. For the numerical estimation of the neutrino flux of a specified stellar nuclear reaction they take into account parameters of the standard solar model. The present paper gives for the first time the connection between stellar structure parameters and neutrino fluxes in an analytic stellar model.

065.023 **Analytic solution to the problem of nuclear energy generation rate in a simple stellar model.**
H. J. Haubold, A. M. Mathai.
Astron. Nachr., Vol. 307, No. 1, p. 9 – 12 (1986).
This paper deals with complete analytic solution to the problem of evaluating the total net rate of nuclear energy generation and the basic integral of the luminosity of a purely gaseous star in hydrostatic equilibrium. It is assumed that the total energy production is proportional to certain powers of the density and temperature and that the density inside the star is a nonlinear function varying from the centre to the surface.

065.024 **The formation of red giants.**
M. Beech.
Astron. Astrophys., Vol. 156, No. 1/2, p. 391 – 392 (1986).
The author considers the processes that may explain why dwarf stars become red giants after they complete central hydrogen burning. It seems clear that it is the development of a molecular weight gradient and an increased gravitational field at the central core boundary that conspire to transform a star into a red giant. The author indicates here that both of these effects can have an explanation in terms of simple polytropes; nuclear evolution and the change to shell hydrogen burning both act in the sense of increasing the star averaged polytropic index $\langle n \rangle$ towards its maximum value of five.

065.025 **Fission of rapidly rotating fluid systems.**
R. H. Durisen, J. E. Tohline.
Protostars & planets II, p. 534 – 575 (1985). – See Abstr. 012.008.
Sufficiently rapid rotation can cause breakup of an equilibrium fluid system through the growth of dynamic distortional instabilities. This process is called fission and has been invoked to explain the formation of close binary and multiple systems of celestial bodies. Both classic and modern results of fission theory are reviewed in this paper.

065.026 **A non–local theory of time–dependent convection for numerical application to the radial pulsation of stars.**
G. Gonczi.
Astron. Astrophys., Vol. 157, No. 1, p. 133 – 138 (1986).
A linear perturbation of Ulrich's (1976) non–local theory of convection in non–pulsating stars is used to build up a non–local theory for the influence of the convective heat flux on the pulsation of cepheid type stars. The unphysical oscillations observed in eigenfunctions given by local theories are completely eliminated. It is shown that, in regions where the time scale of convection is smaller than that of pulsation, the convective heat flux drives the

pulsation. This driving must be added to the damping due to turbulent viscosity which may be included in the theory.

065.027 The effects of diffusion due to gravity and due to composition gradients on the rate of hydrogen burning in a cooling degenerate dwarf. II. Dependence on initial metallicity and on buffer mass.
I. Iben Jr., J. MacDonald.
Astrophys. J., Vol. 301, No. 1, p. 164 – 176 (1986).
This paper addresses the degree to which diffusion and nuclear burning depend on the assumed metallicity of an AGB precursor star and on the precise point in the thermal pulse cycle of this precursor at which planetary ejection is assumed to occur. If planetary nebula ejection occurs during the quiescent hydrogen-burning phase, then the mass ΔM_{BUF} of the helium–nitrogen zone which separates matter which has been processed by partial helium burning during a preceding convective helium shell flash from matter in and above the hydrogen–burning shell can have a value which varies from nearly 0 to 1 times the mass ΔM_H processed by hydrogen burning between flashes. It is found that, if $\Delta M_{BUF}/\Delta M_H = 0.1$, the abundance of CNO elements in the burning region is enhanced by chemical diffusion to a sufficient degree that a hydrogen shell flash is initiated. The effect of the flash is to first mix helium to the surface and to then cause the star to expand to giant dimensions. The authors speculate further that additional flashes may occur and that wind mass loss may ultimately convert stars which experience a diffusion induced shell flash into non–DA white dwarfs. This latter possibility thus provides a channel whereby non–DA dwarfs may be formed if planetary nebula ejection occurs during either the helium shell flash on the AGB or during the subsequent quiescent helium–burning phase.

065.028 Normal modes of oscillation for rotating stars. III. Variational calculations with an improved set of basis vectors.
M. J. Clement.
Astrophys. J., Vol. 301, No. 1, p. 185 – 203 (1986).
Two types of basis vectors are incorporated into a variational principle with the aim of computing gravity modes which are as accurate as the pressure modes of corresponding order. Both types of vectors are derivable from a scalar potential. One generates longitudinal waves and dominates the p–mode eigenfunctions, and the other generates isobaric waves and dominates the g–modes. In the final analysis, the g–mode eigenfrequencies are evaluated to the same level of accuracy as those of the p–modes. Numerical results based on fast Fourier transform techniques are presented for 48 axisymmetric normal modes. The equilibrium states are rigidly rotating sequences of $n = 2$ and $n = 3$ polytropes.

065.029 Brunt–Väisälä frequency and semiconvection.
W. D. Pesnell.
Astrophys. J., Vol. 301, No. 1, p. 204 – 206 (1986).
A derivation of the Brunt–Väisälä frequency, valid in regions of a star where both partial ionization and a varying nuclear composition are present, is given. The final result shows that even in regions where the total ionization approximation is valid, a reduction of the Brunt–Väisälä when compared to the Ledoux criterion is necessary. Areas where this function is important are discussed, and examples are presented using different forms of equations of state.

065.030 Explosive nucleosynthesis in carbon deflagration models of Type I supernovae.
F.–K. Thielemann, K. Nomoto, K. Yokoi.
Astron. Astrophys., Vol. 158, No. 1/2, p. 17 – 33 (1986).
There is increasing evidence that Type I supernovae (SN I) are the main producers of iron–peak elements in the Galaxy. In addition observations of SN I also indicate the existence of appreciable amounts of intermediate elements like O, Mg, Si, S, and Ca in the outer layers of the exploding star. Explosive nucleosynthesis results of carbon deflagration supernovae are presented in detail. Special emphasis is given to the discussion of burning conditions and corresponding nucleosynthesis products, and the nuclear uncertainties which govern the amount of electron captures in the central part of the core. The latter gives constraints on the isotopic composition of the iron peak elements. The overproduction of $^{54}Fe + ^{58}Ni$ is still existing. Assuming that SN I which contributed to the abundances in the solar system, originated from white dwarfs with a metallicity range $0.1 < Z/Z_\odot < 1$, might remove this overproduction. This would allow for SN I to be the major contributors of Fe–group nuclei in galactic nucleosynthesis.

065.031 Further studies of a fragmentation problem.
J. J. Monaghan, J. C. Lattanzio.
Astron. Astrophys., Vol. 158, No. 1/2, p. 207 – 211 (1986).
The authors have re–examined the fragmentation of a rotating isothermal cloud using a particle method with a resolution approximately twice as fine as that used previously. The maximum densities at a given time exceed those found in simulations of the same system using donor cell methods, but the overall evolution is similar to that found in early particle simulations. In particular there is no indication of a run–away collapse of entire fragments. The calculations using ~ 7000 particles were followed for the longest time and they show that gravitational interactions between the fragments eventually become important.

065.032 A theorem concerning nonradial oscillations.
M. Gabriel.
Astron. Astrophys., Vol. 158, No. 1/2, p. 395 (1986).
The author shows that, in Cowling's approximation, the modulus of the eigenvalues of nonradial oscillations is an increasing function of the degree l.

065.033 Nonradial pulsation theory of massive stars.
Y. Osaki.
Publ. Astron. Soc. Pac., Vol. 98, No. 599, p. 30 – 32 (1986). – See Abstr. 012.009.
The present theory of nonradial pulsation of massive stars is examined, particularly on excitation mechanisms on nonaxisymmetric nonradial pulsation waves. A possible mechanism is suggested for the episodic mass loss from the equator in Be stars by nonradial pulsations.

065.034 The evolution of massive close binary systems with mass loss and overshooting. II. Case A of mass transfer in systems with high mass ratios.
C. H. B. Sybesma.
Astron. Astrophys., Vol. 159, No. 1/2, p. 108 – 116 (1986).
The case A of mass transfer is studied for systems with an initial primary mass of 20 $M_\odot$ with initial mass ratios of 0.8 and 0.9. Simultaneous evolutionary calculations are presented, including the effects of overshooting from the convective core and mass loss through a stellar wind, for different initial periods. Two modes of evolution are found: a sequential mode in which the primary evolves into a WR star before the secondary fills its critical lobe and a parallel mode in which the evolution of the secondary drives it past its own critical radius resulting in long lived contact or near contact systems that cannot evolve conservatively. Possible fates of the two modes are discussed.

065.035 Amplitude instability in pulsating stars: a one–zone model.
M. Auvergne.
Astron. Astrophys., Vol. 159, No. 1/2, p. 197 – 201 (1986).
The author proposes to interpret the amplitude modulation in HR 7308 and also in RR Lyrae stars as an instability appearing at large amplitude during the star evolution. He numerically investigates this instability on a one zone model submitted to a variable external flux. The qualitative behavior of the modulation is in agreement with the observed one in HR 7308.

065.036 Explosive helium burning: I. The kinetics of burning at constant temperature and density.
A. M. Khokhlov, E. V. Ergma.
Astrofizika, Tom 23, Vyp. 3, p. 605 – 619 (1985). In Russian. English translation in Astrophysics, Vol. 23, No. 3.
The kinetics of helium burning at constant temperature ($T \geqslant 10^9$K) and density ($\varrho \geqslant 10^5 g/cm^3$) are investigated. In the

(ϱ, T) plane the region of iron group and lighter element production is located. The mean atomic number of synthesized nuclei for various temperatures and densities is determined. For temperatures $T > 4 \times 10^9$K the "neutron flash" effect is found, which may result in changing the isotope abundance of s– and r–process elements. The range of ϱ and T where the over–abundance (in comparison with NSE) of elements beyond the iron group is possible is found.

065.037 **Initial mass function for zero–metal stars.**
Y. Yoshii, H. Saio.
Astrophys. J., Vol. 301, No. 2, p. 587 – 600 (1986).

The initial mass function (IMF) for zero–metal stars is derived based on the opacity–limited fragmentation theory. If a pregalactic cloud is born just after the decoupling epoch, the fragments evolve isothermally due to the two–photon cooling of atomic hydrogen. The IMF formed in such a cloud has a peak around intermediate mass. The slope toward the massive part of the IMF is steeper than a Salpeter–like IMF. In a cloud which is born in the low–temperature region, the fragments evolve isothermally due to the cooling by molecular hydrogen. Here, fragments with $\sim 10\ M_\odot$ become bound by self–gravitation before the opacity eventually quenches the instability. If the radiation of protostars thus formed inhibits H_2 formation, the resulting IMF would be similar to the case without molecular hydrogen.

065.038 **Models for Type I supernova. I. Detonations in white dwarfs.**
S. E. Woosley, R. E. Taam, T. A. Weaver.
Astrophys. J., Vol. 301, No. 2, p. 601 – 623 (1986). = Lick Obs. Bull., No. 1022.

In a series of papers, the properties of white dwarfs provoked into nuclear instability by accretion will be explored. The results bear many close similarities to the observed properties of Type I supernovae, especially their energetics, light curves, nucleosynthesis, and spectra. In this paper a class of "detonating white dwarfs" is explored. The final evolution of carbon–oxygen white dwarfs in binary systems undergoing mass exchange at rates in the range 10^{-9} to $10^{-8} M_\odot \mathrm{yr}^{-1}$ is considered. Three models are constructed having initial carbon–oxygen core masses 0.5, 0.8, and 1.2 $M_\odot$ and accreting helium at rates of 10^{-8}, 2×10^{-9}, and $10^{-9} M_\odot \mathrm{yr}^{-1}$, respectively. In all cases continued evolution leads to the accumulation of a thick helium layer surrounding the core, degenerate helium ignition, detonation and mass ejection. Detailed isotopic nucleosynthesis has also been computed for these models and implications for Galactic chemical evolution and γ–line astronomy are discussed. At present, the class of detonating models does not agree well with spectroscopic constraints upon common Type I events.

065.039 **Supernova nucleosynthesis in low–metallicity populations.**
M. Jura.
Astrophys. J., Vol. 301, No. 2, p. 624 – 628 (1986).

If radiation pressure on dust is important in deriving the mass loss from red giants, and if these stars do not dredge up large amounts of processed material during this evolution, then the total amount of mass lost by Population II stars with low metallicity is small. Consequently, even intermediate–mass ($\gtrsim 1.5\ M_\odot$) Population II stars become supernovae instead of white dwarfs. As a result, the metal production rate of an ensemble of low–Z stars may be much greater than for high–Z stars, and this can explain why there are relatively few low–metallicity stars in the Galaxy.

065.040 **Explosive hydrogen burning in novae.**
M. Wiescher, J. Görres, F.–K. Thielemann, H. Ritter.
Astron. Astrophys., Vol. 160, No. 1, p. 56 – 72 (1986).

In the present study the authors included all available nuclear information on proton–rich unstable nuclei, to update thermonuclear reaction rates in explosive hydrogen burning. They are applied in a systematic analysis of explosive hydrogen burning for a variety of temperature conditions, appropriate to nova explosions.

065.041 **A model for time–dependent turbulent convection.**
R. Kuhfuß.
Astron. Astrophys., Vol. 160, No. 1, p. 116 – 120 (1986).

A new formulation for time–dependent convection is derived which allows for a description of turbulent convection together with a consistent derivation of transport equations for convective mixing. Convective overshooting and the coupling between radial motions and convection are inherent to the model. The theory is derived from the exact hydrodynamic equations mostly by means of the anelastic approximation and diffusion approximations. It should be applicable to stellar evolution simulations.

065.042 **Exact solutions of the Lane–Emden equation in *N*–dimensional space.**
G. P. Horedt.
Astron. Astrophys., Vol. 160, No. 1, p. 148 – 156 (1986).

The author presents the general solution in terms of Bessel functions for the polytrope of index $n = 1$, and a straigthforward method to obtain the Schuster–Emden integral in N–dimensional space. He presents a compilation of all exact analytic solutions of the Lane–Emden equation, so far found. The behaviour of Lane–Emden functions is studied numerically also for the imaginary sequence, and an exact seven digit tabulation for the boundary values of polytropic sheets, cylinders and spheres is provided for the polytropic indexes of principal interest. Finally, the author quotes an improved determination of the Seidov–Kuzakhmedov minimum mass occurring for spherical polytropes at polytropic index $n \cong 4.823$.

065.043 **On the excitation of oscillations of stars at their encounters.**
A. G. Kosovichev, A. B. Severnyj.
Izv. Krymskoj Astrofiz. Obs., Tom 70, p. 14 – 37 (1985). In Russian. English translation in Bull. Crimean Astrophys. Obs., Vol. 70.

The oscillations arising at close encounters of stars are considered with the aid of numerical methods in linear adiabatic approximation. It is shown that two types of oscillatory motions appear due to the action of tidal force. The first one is a normal mode described by a spherical harmonic of degree $l = 2$ and order $m = 0$. The second one, the more important type of oscillation of several times higher amplitude ($l = 2$, $m = +2$) is a running wave propagating in the direction of the parabolic motion of the disturbing star. The spectral distribution of amplitudes is investigated for polytropic stellar models with the index of polytropes $N = 2, 2.5, 3, 3.25, 3.5$ and for different periastron distances. It is suggested that the 160–min oscillations of the sun can be a relict phenomenon of encounter of the sun with a star, or a cloud, of small (1/5 of solar) mass at the earlier stages of solar evolution when it was among the stars of an early–type star cluster. It is also pointed out that the oscillations arising at close encounters are very similar to those of β Cep stars.

065.044 **FG Sagittae and the stellar evolution theory.**
Yu. A. Fadeev.
Nauchn. Inf., Vyp. 56, p. 3 – 8 (1984). In Russian.

The FG Sge problem is discussed. The supposition that FG Sge is undergoing a thermal flash in the helium burning shell gives the best explanation of FG Sge's observable properties. The relations of core mass–luminosity and core mass–interflash period for double shell source stars as well as analysis of FG Sge's pulsations confirm the connection between planetary nebula formation processes, stellar pulsations and thermal instabilities in the helium burning shell.

065.045 **Study of low–mass stellar models with chemically inhomogeneous zones of neutral stability.**
Yu. V. Vandakurov.
Nauchn. Inf., Vyp. 56, p. 42 – 45 (1984). In Russian.

Convection in a chemically inhomogeneous medium redistributes the matter, thus a neutrally stable radiative zone can be formed. Here a model of low–mass star with convective envelope and neutrally stable core of the type studied earlier by the author is considered.

065.046 **Mass loss and the evolution of supergiants.**
Yu. L. Frantsman.
Nauchn. Inf., Vyp. 56, p. 46 – 48 (1984). In Russian.
Some aspects of the evolution of massive stars on the He–burning stage are considered. The effect of mass loss during the evolution increases the time spent by the star in the red supergiant region, and then increases the number of red supergiants relative to the blue, especially for larger mass stars.

065.047 **Chemically peculiar stars in groups of different age.**
V. G. Klochkova, I. M. Kopylov.
Astron. Zh., Tom 63, Vyp. 2, p. 240 – 245 (1986). In Russian. English translation in Sov. Astron., Vol. 30, No. 2.
The parameters of the atmospheres for 91 Bp and Ap stars in the age range lg $t = 6.4 - 8.7$ have been obtained. A conclusion is made that during the main–sequence phase of evolution the degree of peculiarity of the stars does not depend on their age. It is suggested that the chemically peculiar stars obtain the observed peculiarities at the earliest stages of star formation, in the pre–main–sequence phase.

065.048 **Destruction of ^{3}He in stars.**
D. S. P. Dearborn, D. N. Schramm, G. Steigman.
Astrophys. J., Vol. 302, No. 1, p. 35 – 38 (1986).
The observed abundance of ^{3}He can be used, in conjunction with big bang nucleosynthesis, to set a *lower* bound to the density of nucleons in the universe. Critical to this approach is an estimate of the destruction of ^{3}He in stars. Detailed stellar evolution calculations which explicitly examine ^{3}He destruction are presented. The effect of stellar mass, composition, and mass–loss rate on ^{3}He destruction is studied. Limits on ^{3}He destruction appropriate for big bang nucleosynthesis and for understanding present interstellar ^{3}He abundances are discussed.

065.049 **Stellar technetium and niobium abundances as a measure of the lifetime of AGB stars in the third dredge–up phase.**
G. J. Mathews, K. Takahashi, R. A. Ward, W. M. Howard.
Astrophys. J., Vol. 302, No. 1, p. 410 – 414 (1986).
Recent measurements of quantitative stellar abundances for Tc, Nb, and elements with neighboring atomic numbers are discussed in the framework of detailed network calculations of the production and destruction of Tc during thermal pulses operating in AGB stars. It is found that Tc may not be a good indicator of the *s*–process temperature since the thermally enhanced beta–decay rate for ^{99}Tc appears to be compensated by the increased neutron production in a ^{22}Ne(α,n)^{25}Mg s–process. The paper also investigates the time dependence of the surface abundances of Tc and Nb and other elements during the third dredge–up phase of AGB stars.

065.050 **Advancements in the stellar evolution theory: the role of convective overshooting all across the HR diagram.**
C. Chiosi.
Spectral evolution of galaxies, p. 237 – 262 (1986). – See Abstr. 012.015.
The author presents recent calculations of stellar structure in which the effects of convective overshooting are taken into account during the core H– and He–burning phases of stars in the mass range 1.3 $M_\odot$ to 100 $M_\odot$. In addition to this, the author briefly reviews the effects of mass loss in luminous stars of all spectral types, and in red giant and asymptotic giant branch stars. Finally, the effects of the novel cross section for the ^{12}C(α,γ)^{16}O nuclear reaction are also illustrated.

065.051 **Adiabatic mass–radius relations for polytropic stars.**
M. S. Hjellming, R. F. Webbink.
Bull. Am. Astron. Soc., Vol. 17, No. 4, p. 840 (1985). Abstract. – See Abstr. 010.062.

065.052 **Nonaxisymmetric normal modes of rotating stars.**
R. A. Managan.
Bull. Am. Astron. Soc., Vol. 17, No. 4, p. 840 (1985). Abstract. – See Abstr. 010.062.

065.053 **Massive stars in galaxies.**
R. M. Humphreys.
Bull. Am. Astron. Soc., Vol. 17, No. 4, p. 875 (1985). Abstract. – See Abstr. 010.062.

065.054 **Mixing at the carbon–hydrogen discontinuity in low mass AGB stars.**
D. E. Hollowell, I. Iben Jr.
Bull. Am. Astron. Soc., Vol. 17, No. 4, p. 877 (1985). Abstract. – See Abstr. 010.062.

065.055 **The transformation from planetary nebula nucleus to white dwarf: a seismological study of stellar metamorphosis.**
S. D. Kawaler.
Bull. Am. Astron. Soc., Vol. 17, No. 4, p. 894 (1985). Abstract. – See Abstr. 010.062.

065.056 **A simple model for coupled convection/pulsation.**
R. F. Stellingwerf.
Bull. Am. Astron. Soc., Vol. 17, No. 4, p. 894 (1985). Abstract. – See Abstr. 010.062.

065.057 **Ionization driving mechanisms in pulsating stars.**
W. D. Pesnell.
Bull. Am. Astron. Soc., Vol. 17, No. 4, p. 894 (1985). Abstract. – See Abstr. 010.062.

065.058 **Pulsational mass loss and cepheid evolution.**
W. M. Brunish, L. A. Willson, S. A. Becker.
Bull. Am. Astron. Soc., Vol. 17, No. 4, p. 894 (1985). Abstract. – See Abstr. 010.062.

065.059 **Evolution of massive spinning stars.**
S. R. Sreenivasan, D. Narasimha.
Bull. Am. Astron. Soc., Vol. 17, No. 4, p. 895 (1985). Abstract. – See Abstr. 010.062.

065.060 **Cross sections for tidal capture binary formation and stellar merger.**
H. M. Lee, J. P. Ostriker.
Bull. Am. Astron. Soc., Vol. 17, No. 4, p. 895 (1985). Abstract. – See Abstr. 010.062.

065.061 **The solar–stellar connection: I. Solar studies.**
S. Sofia.
Bull. Am. Astron. Soc., Vol. 17, No. 4, p. 898 (1985). Abstract. – See Abstr. 010.062.

065.062 **The solar–stellar connection: II. Stellar studies.**
P. Demarque.
Bull. Am. Astron. Soc., Vol. 17, No. 4, p. 898 (1985). Abstract. – See Abstr. 010.062.

065.063 **Evolutionary period changes in variable helium–rich white dwarfs.**
S. D. Kawaler, D. E. Winget, I. Iben Jr., C. J. Hansen.
Astrophys. J., Vol. 302, No. 2, p. 530 – 535 (1986).
The authors investigate the rates of pulsation period change in realistic, theoretical evolutionary models of pulsating DB white dwarf (DBV) stars and their presumed progenitors, the pulsating PG 1159 (DOV) stars. The expected rates of period change are computed by solving the equations of linear, adiabatic, nonradial oscillation for nonrotating models of hydrogen–deficient white dwarfs. The value of $d(\ln P)/dt$ varies over the range $(3-5)\times10^{-14}s^{-1}$ for models of DOV stars. Values for DBV stars range from $3\times10^{-16}s^{-1}$ at $T_{eff} = 30{,}000$K to $5\times10^{-17}s^{-1}$ at $T_{eff} = 20{,}000$K. Period changes of this order should be measurable in DBV stars over a few observing seasons. The effects of varying stellar mass and chemical composition on the numerical results are also examined.

065.064 **Primordial stellar evolution: the protostar phase.**
S. W. Stahler, F. Palla, E. E. Salpeter.
Astrophys. J., Vol. 302, No. 2, p. 590 – 605 (1986).
The authors calculate the structure and evolution of a protostar forming from a cloud composed of pure hydrogen and helium gas. The collapse is described numerically as a sequence of steady state accretion flows onto the hydrostatic core, which grows from an initial mass of 0.01 $M_\odot$ to 10.5 $M_\odot$. Because of the high accretion luminosity, the core is surrounded by an optically thick radiative precursor for most of its evolution. The core radius is much larger than for present–day protostars, reaching 47 $R_\odot$ when the mass is 1 $M_\odot$. For sufficiently massive cores the deep interior contracts strongly, driving out a "luminosity wave". This wave reaches the surface when the mass is 8 $M_\odot$, resulting in a large increase in core radius and the disappearance of the radiative precursor. In the H–R diagram, the protostar evolves upward along a nearly vertical track, with effective temperature near 5000K.

065.065 **Charged particles in the magnetosphere of a collapsing star.**
V. G. Krivdik.
Kinematika Fiz. Nebesn. Tel, Tom 2, No. 3, p. 8 – 14 (1986). In Russian. English translation in Kinematics Phys. Celest. Bodies.
The motion of charged particles in the external electromagnetic field of a collapsing star is investigated in drift approximation. The particles' motion trajectories, their drift velocity and energy change in the course of collapse are calculated.

065.066 **Numerical analysis of nonlinear radial pulsations of stars.**
A. G. Kosovichev.
Bull. Crimean Astrophys. Obs., Vol. 69, p. 101 – 115 (1986). English translation of 39.065.029.

065.067 **Asymptotics and quantum chaos in stellar oscillations.**
J. Perdang.
Seismology of the Sun and the distant stars, p. 141 – 171 (1986). – See Abstr. 012.026.
Ray chaos in stars, in the asymptotic formalism of geometric acoustics, is shown to arise under perturbations breaking the spherical symmetry of the star. If mild enough, ray chaos merely induces a random scatter of the actual eigenfrequencies of the perturbed star around the standard asymptotic frequencies $\omega \propto (n+1/2l+\varphi)$ of the radially symmetric model. The statistical propertes of this intrinsic scatter are found to differ from the statistics of a scatter due to observational noise.

065.068 **Finite amplitude limit of the ^{3}He instability.**
I. W. Roxburgh.
Seismology of the Sun and the distant stars, p. 265 – 269 (1986). – See Abstr. 012.026.
The models of solar type stars are unstable to non–radial oscillations driven by burning ^{3}He. These oscillations modify the distribution of ^{3}He leading to finite amplitude oscillations with typical temperature perturbations $\delta T/T \sim 0.1$.

065.069 **Asteroseismology for certain Ap stars.**
W. Dziembowski, P. R. Goode.
Seismology of the Sun and the distant stars, p. 441 – 451 (1986). – See Abstr. 012.026.
The discovery by Kurtz of phase coherent, rapid oscillations in certain Ap stars holds great promise for the new field of asteroseismology. The authors discuss a generalized oblique pulsator model for these stars which allows to compare the effects of rotation and magnetism. A central role for mode coupling and amplitude limitation in the observed period doubling is suggested.

065.070 **More about collapse of a radiating star.**
A. K. G. de Oliveira, J. A. de Freitas Pacheco, N. O. Santos.
Mon. Not. R. Astron. Soc., Vol. 220, No. 2, p. 405 – 412 (1986).
Further results are obtained concerning the model proposed by de Oliveira, Santos and Kolassis (1985). The authors predict the properties of the radiation pulse in the optical, X–ray and γ–ray domains, resulting from collapse of neutron stars and white dwarfs.

065.071 **Extremely metal–deficient stars –II. Evolution of intermediate–mass stars up to carbon ignition or core degeneracy.**
A. Tornambè, A. Chieffi.
Mon. Not. R. Astron. Soc., Vol. 220, No. 2, p. 529 – 547 (1986).
The evolution of stellar models in the mass range 2.5 – 8 $M_\odot$ and log Z = –8, –6 and –4 has been computed from the main sequence up to off–centre carbon ignition or CO core degeneracy, taking into account semi–convection and breathing pulses during the central He–burning phase. The mass limit between carbon ignition and core degeneracy is presented as a function of Z. Evolutionary results have also been interpreted in terms of production rates of Type I supernovae during the halo phase of our Galaxy.

065.072 **The evolution of WR + O binaries into massive X–ray binaries.**
P. Hellings, C. de Loore.
Astron. Astrophys., Vol. 161, No. 1, p. 75 – 84 (1986).
The transition of WR + O binaries into compact + O binaries, and later into massive X–ray binaries is studied by evolutionary computations of models of presently observed WR + O binaries. The final products of the systems after the supernova explosion of the primary are computed, including the possibility of an asymmetric explosion.

065.073 **Non–radial instability of WR models.**
A. Noels, R. Scuflaire.
Astron. Astrophys., Vol. 161, No. 1, p. 125 – 129 (1986).
The development of a H burning shell in a massive star evolving with mass loss triggers a phase of vibrational instability towards g^+ modes of non radial oscillations. The *e*–folding time is of the order of 1000 yr but the unstable phase lasts only about 5 times more. Although the periods seem in accordance with those observed in some WR stars, it is still not clear that such a short unstable phase could be responsible for the observed variability.

065.074 **A simple model for coupled convection and pulsation.**
R. F. Stellingwerf.
Astrophys. J., Vol. 303, No. 1, p. 119 – 129 (1986).
A generalization of Baker's one–zone pulsation model is derived that includes a simple treatment of time–dependent convection. Both nonlinear and linear effects are discussed. The convective flux tends to stabilize the pulsation, most effectively for red models. The turbulent pressure is always destabilizing, with the most effective phasing for "instability strip" models. It is also shown that this model is subject to a number of dynamic instabilities.

065.075 **On the evolutionary state of cataclysmic variable white dwarfs and their recurrence times.**
E. M. Sion, S. G. Starrfield.
Astrophys. J., Vol. 303, No. 1, p. 130 – 135 (1986).
The authors have calculated hydrostatic evolutionary sequences for massive white dwarfs (M > 1.0 $M_\odot$) accreting with a solar mixture of elements. The interflash period depends on both the mass accretion rate and the luminosity of the white dwarf. The authors combine their calculations with earlier results and are able to predict that the classical nova outburst must occur on low–luminosity, massive white dwarfs which are accreting mass at low rates.

065.076 **Stellar evolution in the upper Hertzsprung–Russell diagram.**
C. Doom, J. P. de Greve, C. de Loore.
Astrophys. J., Vol. 303, No. 1, p. 136 – 145 (1986).
Evolutionary computations for massive stars, including mass loss and overshooting from convective cores, are presented. Using these computations, a coherent picture of the evolution of massive stars in the upper H–R diagram is given. Different for-

mation mechanisms for various groups of luminous stars are discussed. These groups include η Car, the Hubble–Sandage variables and the progenitors of W–R systems.

065.077 **Hydrodynamic models for novae with ejecta rich in oxygen, neon, and magnesium.**
S. Starrfield, W. M. Sparks, J. W. Truran.
Astrophys. J., Lett. Ed., Vol. 303, No. 1, p. L5 – L9 (1986).
This letter seeks to explain the characteristics of a new class of novae. This class consists of those objects that have been observed to eject material rich in oxygen, neon, magnesium, and aluminum at high velocities. The authors propose that for this class of novae the outburst is occurring not on a carbon–oxygen white dwarf but on an oxygen–neon–magnesium white dwarf which has evolved from a star which has a main–sequence mass of $\sim 8\,M_{\odot}$ to $\sim 12\,M_{\odot}$. The authors have simulated such an outburst by evolving $1.25\,M_{\odot}$ white dwarfs accreting hydrogen–rich material at various rates. The resulting evolutionary sequences are the most violent that the authors have yet studied and can eject the entire accreted envelope plus core material at high velocities.

065.078 **Finite element calculations for non–radial oscillations of stars.**
R. A. S. Fiedler, A. L. Andrew.
Aust. J. Phys., Vol. 39, No. 1, p. 99 – 102 (1986).
Application of the finite element method to a variational principle of Chandrasekhar, although not suitable for computing stable g modes of non–radial oscillation of stars, is found to be satisfactory for unstable g modes as well as f and p modes.

065.079 **Mixing between the helium core and the hydrogen shell in the core helium flash.**
R. G. Deupree.
Astrophys. J., Vol. 303, No. 2, p. 649 – 658 (1986).
The hydrogen shell has been included in two–dimensional hydrodynamic simulations of the core helium flash. The formulation of this treatment is outlined, and then it is shown that the inclusion of the shell and exterior regions does not affect the early hydrodynamic phase. A buoyant bubble of partially nuclearly processed material is formed in the helium core, and the hydrodynamic interaction of this bubble with hydrogen shell is followed. It is found that the rise of the bubble substantially distorts the hydrogen shell location and that mixing of bubble material into the hydrogen shell can occur.

065.080 **The structure of the thermal modes in pulsating stars.**
W. D. Pesnell, J. R. Buchler.
Astrophys. J., Vol. 303, No. 2, p. 740 – 748 (1986).
The eigenvectors and eigenvalues of the linearized thermal diffusion operator are studied and their effect on the thermal structure of radial stellar pulsations is examined. It is shown on the basis of two specific cases, that the difference between these eigenvectors may explain the lack of pulsational instability of radial modes in the β Cephei stars.

065.081 **The effects of a 2:1 resonance in nonlinear radial stellar pulsations.**
J. R. Buchler, G. Kovács.
Astrophys. J., Vol. 303, No. 2, p. 749 – 765 (1986).
A single set of amplitude equations is derived which applies to and describes the finite–amplitude behavior of a resonant ($2\omega_{\alpha} \approx \omega_{\beta}$) pulsator, both in and out of resonance. A detailed analysis of these equations and a discussion of their relevance for stellar pulsators are presented. The formalism is able to describe the systematics of the observed behavior of the (singly periodic) Cepheid and RR Lyrae stars. It shows a very sharp break (bifurcation) in the vicinity of the resonance.

065.082 **Supermassive star formation – the early phases.**
M. P. Merilan.
Diss. Abstr. Int., Sect. B, Vol. 46, No. 6, p. 1958 (1985). Thesis, The Ohio State University, 219 pp. (1985). Order No. DA8518990.

065.083 **Q–nucleosynthesis: implications for stellar evolution.**
C. L. Joseph.
Diss. Abstr. Int., Sect. B, Vol. 46, No. 12, p. 4282 (1986). Thesis, The Ohio State University, 186 pp. (1985). Order No. DA8518969.

065.084 **Rapidly rotating core–collapse models.**
M. C. Thompson.
Proc. Astron. Soc. Aust., Vol. 6, No. 2, p. 214 – 216 (1985).
A sequence of core collapse models with decreasing rotation rate is considered. These models were calculated using a three dimensional, post–Newtonian, hydrodynamical, numerical code. The results show that for high precollapse rotational energies the core can become unstable resulting in the formation of what resemble 'spiral arms'.

065.085 **Physical structure of a sequence of two zone polytropic stellar models.**
J. O. Murphy, R. Fiedler.
Proc. Astron. Soc. Aust., Vol. 6, No. 2, p. 219 – 222 (1985).
Expressions for the physical structure of a sequence of two zone polytropic stellar models, based on composite analytical solutions of the Lane–Emden equation with indices $n = 1$ and $n = 5$, have been determined. The coefficients of vibrational stability for radial oscillations of this sequence of models have also been calculated and it is found that increasing the extent of the $n = 5$ outer zone has a stabilizing effect.

065.086 **Radial pulsations and vibrational stability of a sequence of two zone polytropic stellar models.**
J. O. Murphy, R. Fiedler.
Proc. Astron. Soc. Aust., Vol. 6, No. 2, p. 222 – 226 (1985).
Characteristic frequencies of some modes of radial pulsations and their relative displacements have been determined for a sequence of composite polytropic stellar models having an inner zone of index $n = 1$ and an envelope $n = 5$. It is demonstrated that vibrational stability increases with the degree of central condensation of the individual two zone models.

065.087 **The analysis of modal coupling in a hydrodynamic model of Cepheids by the maximum entropy method.**
K.–i. Uji–Iye.
Sci. Rep. Tôhoku Univ., Eighth Ser., Vol. 6, No. 3, p. 173 – 179 (1986).
The author constructs a hydrodynamic model of classical Cepheids and tries in confirming modal coupling, which is assumed in theoretical papers. He shows the transient enhancement of a weaker mode at the stage when the amplitude of the main mode is increasing. He finds the decrease of the period–ratio at the doubly periodic stage. The results suggest that the non–linear studies are important on the problem of double–mode Cepheids.

065.088 **Permanently homologous stars.**
T. R. Carson.
Observatory, Vol. 106, No. 1072, p. 71 – 72 (1986).

065.089 **Precursor heating and acceleration by Langmuir waves: the type II supernova case.**
M. Contini, R. Opher.
Astrophys. Space Sci., Vol. 122, No. 2, p. 355 – 370 (1986).
The effect of shock wave propagation is investigated with respect to precursor heating and acceleration, upstream, with the aim of explaining supernova (SN) observations. A model is presented, where two different sources of Langmuir waves produce upstream heating and acceleration: (1) Langmuir waves excited by resonance beam particles, that are accelerated through the shock front by Bell's mechanism; and (2) Langmuir waves, created in the post–shock turbulent zone. The calculations are applied to type II SN.

065.090 **Numerical implementation of a perturbation theory up to the third order for rotating polytropic stars: application to a solar model.**
V. S. Geroyannis, F. N. Valvi.
Astrophys. Space Sci., Vol. 123, No. 2, p. 273 – 295 (1986).
Numerical treatment of a perturbation theory up to the third order yields quantitative estimates for an interesting polytropic solar model possessing differential rotation of Clement's type. A suitably defined "reduction factor", otherwise referred to as "strength", moderates the effects of the differential rotation and the resulting limiting state represents efficiently uniform rotation of the model. The numerical results obtained are compared to those derived from an original numerical implementation of a second–order perturbation theory.

065.091 **On the evolution of binary components which first fill their Roche lobes after the exhaustion of central helium.**
I. Iben Jr.
Astrophys. J., Vol. 304, No. 1, p. 201 – 216 (1986).
The evolution of model close binary components of initial mass 3 – 7 $M_\odot$ and of Population I composition is followed from the beginning of the core helium–burning phase. The primary objective of this paper is to establish the transformation between initial mass and final degenerate dwarf mass of components which do not fill their Roche lobe for the first time until after exhausting helium at the center but before reigniting hydrogen in the shell. It is found that, for a model which would not have experienced a "second–dredge–up" episode as a single star, the mass of the final CO dwarf is essentially the same as the mass of the CO core of the single star analog at the beginning of the thermally pulsing asymptotic giant branch (TP–AGB) phase. For models which would have experienced a second dredge–up as single stars, the final degenerate dwarf mass is somewhat larger than the mass of the CO core of a single star at the beginning of the TP–AGB phase.

065.092 **Relaxation oscillations of a helium–burning star: helium shell flashes not triggered by the accumulation of a critical mass.**
I. Iben Jr., M. Y. Fujimoto, D. Sugimoto, S. Miyaji.
Astrophys. J., Vol. 304, No. 1, p. 217 – 230 (1986).
This paper studies the shell helium–burning phase of a close binary component which has lost all or almost all of its hydrogen–containing fuel by mass loss in an episode of Roche–lobe overflow. It is demonstrated how a thermal instability arises in a shell helium–burning star, not because of compression in a helium zone whose mass is growing due to processing through a hydrogen–burning zone, but because of the compression which follows naturally as the star relaxes following a preceding helium shell flash. By an analysis of the thermodynamic behavior of the relaxation oscillations one gains insight into the structure of the instability region in the gravothermal phase plane. The core helium and shell helium–burning phases of these stars fall in that region of the HR diagram occupied by the subdwarf O and B stars. The authors explore the possibility that a substantial fraction of all sdO and sdB stars are in fact components of close binaries.

065.093 **The evolution of ultrashort period binary systems.**
L. A. Nelson, S. A. Rappaport, P. C. Joss.
Astrophys. J., Vol. 304, No. 1, p. 231 – 240 (1986).
The authors have carried out detailed evolutionary calculations wherein a very low mass, hydrogen–depleted, semi–detached binary containing a collapsed object can attain an exceptionally short orbital period while maintaining a relatively high rate of mass tranfer. It is found that the observed properties of systems like 4U 1626–67, 4U 1916–05 and GP Com can be understood if one assumes that they contain moderately ($X \approx 0.1$) to severely ($X \approx 0.001$) hydrogen–deficient secondary stars that are neither fully degenerate nor burning helium. For $X \approx 5 \times 10^{-4}$ minimum orbital periods as small as $\sim$35 min are attained.

065.094 **Hydrodynamic models of RR*ab* variables: comparison of observed and theoretical light curves.**
N. R. Simon, T Aikawa.
Astrophys. J., Vol. 304, No. 1, p. 249 – 255 (1986).
Eight fundamental–mode RR Lyrae models are integrated using a new code, TGRID, with dynamic zoning in the hydrogen ionization region. When the light curves from these models are Fourier decomposed, it is found that the Fourier phases ϕ_{21} and ϕ_{31} disagree with the values obtained from the observed light curves of RR*ab* pulsators. Other published calculations for these stars show a similar discrepancy. Evidence is presented which indicates that this disagreement may be due to the use of the diffusion approximation and the neglect of atmospheric shocks.

065.095 **Erratum: "Neutrino energy loss in stellar interiors" [Astrophys. J., Vol. 296, No. 1, p. 197 – 203 (1985)].**
H. Munakata, Y. Kohyama, N. Itoh.
Astrophys. J., Vol. 304, No. 1, p. 580 (1986). See Abstr. 40.065.036.

065.096 **Diffusion coefficients for stellar plasmas.**
C. Paquette, C. Pelletier, G. Fontaine, G. Michaud.
Astrophys. J., Suppl. Ser., Vol. 61, No. 1, p. 177 – 195 (1986).
An approximate method of computing diffusion coefficients for the relatively dense plasmas characteristic of white dwarf envelopes is presented. The method is based on the numerical evaluation of collision integrals for a screened Coulomb potential. The main results are presented in the form of high–accuracy analytic fits for the collision integrals. The fits can be used within the framework of Chapman–Enskog's theory or Burgers's method of solution of the Boltzmann equation.

065.097 **Diffusion in white dwarfs: new results and comparative study.**
C. Paquette, C. Pelletier, G. Fontaine, G. Michaud.
Astrophys. J., Suppl. Ser., Vol. 61, No. 1, p. 197 – 217 (1986).
New estimates of diffusion time scales of heavy elements in white dwarf envelopes are presented. The calculations explicitly take into account screening effects in the evaluation of the transport properties of white dwarf plasmas. Several new results are presented and compared in detail with the results of previous investigations. The improved diffusion time scales are still much smaller than evolutionary time scales, which confirms the need for mechanisms competing agains settling to account for the presence of metal traces in the spectra of several cool white dwarfs. A consistent picture of the accretion–diffusion model is shown to emerge.

065.098 **On the alpha–model of turbulence.**
T. Angelov.
Publ. Dep. Astron., Univ. Beogr., No. 13, p. 27 – 30 (1985).
The author considers the structure of stellar discs with an α–model of turbulence. One obtains the equation of the energetic balance for turbulent pulsations and shows that $\alpha(r)$ is, in the central and peripheral parts of the disc, a monotonous, slowly increasing and bounded function. Quick changes occur in the narrow band of the disc near the star's surface.

065.099 **Nucleosynthesis and evolution of massive stars with mass loss and overshooting.**
N. Prantzos, C. Doom, M. Arnould, C. de Loore.
Astrophys. J., Vol. 304, No. 2, p. 695 – 712 (1986).
Evolutionary computations for the H– and He–burning phases of population I stars in the mass range $50 \leqslant M_{ZAMS}/M_\odot \leqslant 100$ are presented. The stellar models include a semiempirical mass loss formalism, the Roxburgh criterion for convection, and a detailed nuclear reaction network for the treatment of energy production and nucleosynthesis (including nuclei up to ^{30}Si and neutron captures). A detailed discussion is presented of the evolutionary tracks, of the abundances in the interior and at the surface of the model stars, and of the composition of the ejecta for individual stars. The implications of the results for cosmic–ray and solar system isotopic anomalies, γ–ray line astronomy and the chemical evolution of the Galaxy are discussed.

065.100 **Effect of rapid mass accretion onto the main sequence stars.**
M. Sarna.
Postepy Astron., Tom 32, Zesz. 3/4, p. 221 – 248 (1984). In Polish.

During the evolution of a close binary system there is a phase of rapid mass exchange between its components. The effect of rapid mass inflow on the internal structure of main sequence stars is studied.

065.101 **How stars explode.**
J. C. Wheeler, K. Nomoto.
Am. Sci., Vol. 73, No. 3, p. 240 – 247 (1985). Abstr. in Phys. Abstr., Vol. 89, No. 1274, Entry 23852 (1986).

065.102 **Nuclear–reaction networks and stellar evolution codes: on the coupling of composition changes and energy release in explosive nuclear burning.**
E. Müller.
MPA Rep., No. 220, 16 pp. (1986). Submitted to Astron. Astrophys.

065.103 **Colors and luminosities of stars with spots.**
H. C. Spruit, A. Weiss.
MPA Rep., No. 225, 29 pp. (1986). Submitted to Astron. Astrophys.

065.104 **Nucleosynthesis in He–flashes on accreting white dwarfs.**
W. Hillebrandt, M. Lechle, W. Ziegert, K. Nomoto, F. K. Thielemann.
MPA Rep., No. 233, 12 pp. (1986). To appear in accretion processes in astrophysics.

065.105 **On the advanced evolution of intermediate mass stars: induced semiconvection, breathing pulses and the upper mass limit for carbon deflagration.**
V. Castellani, A. Chieffi, L. Pulone, A. Tornambè.
ESO Sci. Prepr., No. 437, 57 pp. (1986). To appear in Astrophys. J.

065.106 **The origin and evolution of helium–rich white dwarfs.**
J. Liebert.
Prepr. Steward Obs., No. 646, 15 pp. (1986). To appear in IAU Colloq. No. 87.

065.107 **On the behavior of double degenerate binaries associated with Type I supernovae.**
A. G. W. Cameron, I. Iben Jr.
Astrophys. J., Vol. 305, No. 1, p. 228 – 234 (1986).

The authors present some simple consequences of the hypothesis that double degenerate dwarf binary systems are the precursors to Type I supernova explosions. They consider systems consisting either of two carbon–oxygen dwarfs or of an oxygen–neon–magnesium dwarf and carbon–oxygen dwarf which spiral together until the secondary of lower mass fills its Roche lobe. It is assumed that the former systems, if sufficiently massive, evolve into normal Type I supernovae, while the latter evolve into Type I peculiar supernovae. Processes leading to the supernova explosion are outlined. Evolutionary scenarios for a supernova remnant system are developed depending on whether or not a neutron star has formed during the explosion.

065.108 **Dynamic fission instabilities in rapidly rotating $n = 3/2$ polytropes: a comparison of results from finite–difference and smoothed particle hydrodynamics codes.**
R. H. Durisen, R. A. Gingold, J. E. Tohline, A. P. Boss.
Astrophys. J., Vol. 305, No. 1, p. 281 – 308 (1986).

The authors use three different three–dimensional hydrodynamics codes to study fission instabilities in rapidly rotating, polytropic equilibrium models and to follow the growth of nonaxisymmetric structure in finitely unstable axisymmetric equilibrium states. For the two models considered, the polytropic index $n = 3/2$, and the ratios β of total rotational kinetic energy to gravitational energy are 0.33 and 0.38. Two of the hydrodynamics codes are based on finite–difference, donor–cell techniques, and the third is a smoothed particle hydrodynamics code. All codes agree that production of a detached binary or multiple system is not the direct outcome of dynamic bar mode instability for $n = 3/2$ polytropes. Instead, mass and angular momentum are shed in the form of two trailing spiral arms. The ejected material forms a ring or disk around a central nonaxisymmetric remnant.

065.109 **Supernovae and nuclear structure: electron capture and the nuclear incompressibility.**
J. Cooperstein.
Nuclear structure 1985, p. 299 – 310 (1985). Abstr. in Phys. Abstr., Vol. 89, No. 1277, Entry 42165 (1986). – See Abstr. 012.048.

065.110 **Supernova matter: a semiclassical approach.**
M. Pi, X. Viñas, M. Barranco, A. Perez–Canyellas, A. Polls.
Astron. Astrophys., Suppl. Ser., Vol. 64, No. 3, p. 439 – 451 (1986).

The authors have used a well established semiclassical method to study systematically the properties of hot dense matter in the conditions prevailing during the gravitational collapse of massive stars. Different semiclassical kinetic energy and entropy densities are discussed. Detailed results for different nuclear forces having different nucleon effective mass and nuclear incompressibility are presented. Plasma effects are considered for a typical adiabat. The results here presented cover a wide range in entropy per baryon and electron concentration per baryon for baryon densities ranging from $10^{-3}fm^{-3}$ to $0.30\ fm^{-3}$.

065.111 **Blue–to–red ratios as tests for theoretical models of massive stars.**
W. M. Brunish, J. S. Gallagher, J. W. Truran.
Astron. J., Vol. 91, No. 3, p. 598 – 601 (1986).

The problem of matching observed number ratios of blue to red supergiants (B/R) is re–examined on the basis of single–star evolution models which undergo helium ignition in the blue and self–consistently include mass loss. The theoretical and observed B/R ratios can be considered to agree to within expected errors, but more sensitive techniques are needed to provide sharp tests of stellar evolution theory.

065.112 **Evolution of intermediate and low mass stars.**
C. Chiosi, G. Bertelli, A. Bressan, E. Nasi.
Bull. Am. Astron. Soc., Vol. 18, No. 1, p. 635 (1986). Abstract. – See Abstr. 010.066.

065.113 **Presupernova evolution of 10 – 12 $M_{\odot}$ stars.**
K. Nomoto, M. Hashimoto.
Bull. Am. Astron. Soc., Vol. 18, No. 1, p. 635 (1986). Abstract. – See Abstr. 010.066.

065.114 **Carbon pollution in helium–rich white dwarfs: time–dependent calculations.**
G. Fontaine, C. Pelletier, F. Wesemael, G. Michaud, G. Wegner.
Bull. Am. Astron. Soc., Vol. 18, No. 1, p. 636 – 637 (1986). Abstract. – See Abstr. 010.066.

065.115 **Lower bounds on the masses of rapidly rotating white dwarfs.**
M. Rao, G. Chanmugam, J. E. Tohline.
Bull. Am. Astron. Soc., Vol. 18, No. 1, p. 637 (1986). Abstract. – See Abstr. 010.066.

065.116 **On the evolution of very luminous stars. I. Presupernova evolution.**
N. Langer, M. F. El Eid.
Inst. Astrophys. Paris, Pré–Publ., No. 133, 27 pp. (1986). To appear in Astron. Astrophys.

065.117 **On the evolution of very luminous stars. II. Pair creation supernova in massive Wolf–Rayet stars.**
M. F. El Eid, N. Langer.
Inst. Astrophys. Paris, Pré–Publ., No. 141, 26 pp. (1986). To appear in Astron. Astrophys.

065.118 **Modulation of amplitudes in oscillating stars due to resonant mode coupling.**
P. Moskalik.
Acta Astron., Vol. 35, Nos. 3–4, p. 229 – 254 (1985).
Resonant coupling of an unstable mode to two lower frequency stable modes is studied. It is shown, that under certain conditions (which are not very restrictive) such interaction leads to a periodic amplitude modulation.

065.119 **Pulsational masses of B – G supergiants.**
K. Gesicki, J. Zalewski.
Acta Astron., Vol. 35, Nos. 3–4, p. 255 – 260 (1985).
Periods of the fundamental radial mode are computed for B – G supergiants. The results are used to determine pulsational masses of the 32 known pulsating supergiants. The discrepancy between evolutionary and pulsational masses is discussed.

065.120 **Stellar oscillations in the presence of a magnetic field.**
C. G. Campbell, J. C. B. Papaloizou.
Mon. Not. R. Astron. Soc., Vol. 220, No. 3, p. 577 – 591 (1986).
The authors investigate the problem of stellar pulsation in the presence of a magnetic field, for arbitrary frequencies and horizontal wavenumbers. Magnetic effects are strong in a boundary layer close to the stellar surface, but it is possible to identify independent Alfvénic–type modes at sufficient stellar depth. They show how the boundary layer solutions can be matched to the interior solutions for a range of horizontal wavenumbers, this range increasing with frequency. The authors calculate the energy flux carried by the Alfvénic modes and show that the loss of pulsational energy via these waves is significant, this being especially so for high horizontal wavenumbers. Their results are of particular interest to observations of pulsating magnetic Ap stars.

065.121 **The low–luminosity J–type carbon stars.**
T. Lloyd Evans.
Mon. Not. R. Astron. Soc., Vol. 220, No. 3, p. 723 – 735 (1986).
The He shell flash mechanism does not account for the ^{13}C–rich carbon stars (J stars) found on the lower fringe of the luminosity distribution of carbon stars in Magellanic Cloud clusters. Most cluster carbon stars have weak ^{13}C, like most of the N–type carbon stars in the Galaxy. Galactic R stars almost all have enhanced ^{13}C. They have normal abundances of s–process elements, whereas N stars have large overabundances. These differences suggest a real physical distinction. The faint J stars in the Magellanic Clouds fall on the upper part of the asymptotic giant branch. Enhanced ^{13}C is also expected in the most of the N–type carbon stars in the galaxy. Galactic R stars almost all have enhanced ^{13}C.

065.122 **A further uncertainty in the mixing length theory of convection in the structure of late–type stars.**
R. J. Tayler.
Mon. Not. R. Astron. Soc., Vol. 220, No. 3, p. 793 – 797 (1986).
The key parameter in the mixing length theory of convection is the ratio of the mixing length to the pressure scale height. It is suggested that surface magnetic fields and rotation both affect the energy transport by convection in such a way that the mixing length is replaced in the expression for the convective flux by a smaller quantity, which will be called the effective mixing length. The effective mixing length in a late–type main–sequence star may increase as its rotation rate and magnetic activity decrease.

065.123 **Our place in the universe. III. The lifetime of the stars.**
G. F. Koch.
Nachrichtentech. Z., Vol. 39, No. 4, p. 254 – 256 (1986). In German. Abstr. in Phys. Abstr., Vol. 89, Nr. 1284, Entry 83521 (1986).

065.124 **Peculiar supernova of type I as a result of explosive helium burning in an accreting white dwarf.**
A. M. Khokhlov, E. V. Ergma.
Pis'ma Astron. Zh., Tom 12, No. 5, p. 366 – 372 (1986). In Russian. English translation in Sov. Astron. Lett., Vol. 12.
It is supposed that a peculiar supernova of type I (SN Ip) may result from explosive helium burning in small mass ($M_{WD} < 1\ M_{\odot}$) helium or carbon–oxygen–helium white dwarfs. Incomplete helium burning may explain as well small velocities and lower luminosities of SN Ip. Differences between SN Ip and SN I spectra may be due to particular properties of explosive helium burning.

065.125 **Formation of detonation in degenerate stars.**
S. I. Blinnikov, A. M. Khokhlov.
Pis'ma Astron. Zh., Tom 12, No. 4, p. 318 – 324 (1986). In Russian. English translation in Sov. Astron. Lett., Vol. 12.
Formation of a thermonuclear burning front is considered under the explosion of a degenerate CO star as a supernova. It is shown that assuming an isentropic initial temperature profile in the central region the thermal instability results in a thermonuclear detonation (rather than the deflagration regime commonly believed).

065.126 **Use of variational approach in determining the periods of small adiabatic oscillations of rotationally and tidally distorted polytropic gaseous spheres.**
C. Mohan, R. M. Saxena.
Indian J. Pure Appl. Math., Vol. 17, No. 3, p. 414 – 426 (1986).
The authors consider the feasibility of using the variational approach to determine the eigenfrequencies of pseudoradial and nonradial modes of oscillations of rotationally and tidally distorted gaseous spheres whose undistorted equilibrium structures are polytropic spheres. The variational method developed for this purpose is based on the variational principle of Chandrasekhar and Lebovitz for undistorted gaseous spheres and uses the averaging technique of Kippenhahn and Thomas to incorporate the effects of distortions caused by rotation and tidal forces.

065.127 **On the immediate neighbourhood of the origin of polytropic configurations.**
G. J. Srivastava.
Indian J. Math., Vol. 26, p. 101 – 104 (1984).

065.128 **Nucleosynthesis accompanying classical nova outbursts.**
J. W. Truran.
Nucleosynthesis and its implications on nuclear and particle physics, p. 97 – 103 (1986). – See Abstr. 012.069.
Theoretical studies of thermonuclear runaways in degenerate hydrogen–rich matter leading to classical nova outbursts predict that nova ejecta can be characterized by substantial overabundances of the elements carbon, nitrogen, and oxygen as well as of other interesting nuclear species. This paper reviews the abundance patterns observed in novae and discusses these within the context of the theoretical models. It then examines the possible implications of these abundance enrichments for galactic nucleosynthesis. The possible role of nova in the production of ^{22}Na and ^{26}Al is also discussed.

065.129 **Reaction rates in the *rp*–process and nucleosynthesis in novae.**
M. Wiescher, J. Görres, F.–K. Thielemann, H. Ritter.
Nucleosynthesis and its implications on nuclear and particle physics, p. 105 – 112 (1986). – See Abstr. 012.069.
Nuclear structure information available on proton rich unstable nuclei were used to evaluate thermonuclear reaction rates in the *rp*–process in explosive hydrogen burning. These rates were applied in a systematic analysis for a variety of temperature conditions, appropriate to nova explosions, to study nucleosynthesis for isotopes of Ne, Na, Mg, Al, Si. The results are discussed in comparison with recent observations of elemental abundances in nova ejecta.

065.130 **Nucleosynthesis and Type I supernovae.**
R. Canal, J. Isern, J. Labay, R. López.
Nucleosynthesis and its implications on nuclear and particle physics, p. 121 – 130 (1986). – See Abstr. 012.069.

The white dwarf model of Type I supernovae is discussed with particular emphasis on the effects of crystallization upon thermonuclear ignition and burning propagation. Preliminary hydrodynamic results are presented for different assumptions concerning the outcome of phase transitions in cooling white dwarfs. Light curves associated with these models are shown to reproduce the "Pskovskii–Branch effect".

065.131 **Explosive nucleosynthesis in carbon deflagration models of Type I supernovae.**
F.-K. Thielemann, K. Nomoto, K. Yokoi.
Nucleosynthesis and its implications on nuclear and particle physics, p. 131 – 142 (1986). – See Abstr. 012.069.

There is increasing evidence that Type I supernovae are the main producers of iron–peak elements in the Galaxy. These elements can be produced by carbon deflagration models of accreting carbon–oxygen white dwarfs in binary systems or stars on the asymptotic giant branch which ignite central carbon burning explosively. The present study discusses the nucleosynthesis results of those carbon deflagration supernovae in detail. Special emphasis is given to the discussion of burning conditions and corresponding nucleosynthesis products.

065.132 **Differentially rotating equilibrium models and the collapse of rotating degenerate configurations.**
E. Müller, Y. Eriguchi.
Nucleosynthesis and its implications on nuclear and particle physics, p. 143 – 144 (1986). Abstract. – See Abstr. 012.069.

065.133 **Theoretical models for Type I and Type II supernovae.**
S. E. Woosley, T. A. Weaver.
Nucleosynthesis and its implications on nuclear and particle physics, p. 145 – 166 (1986). – See Abstr. 012.069.

Recent theoretical progress in understanding the origin and nature of Type I and Type II supernovae is discussed. New Type II presupernova models characterized by a variety of iron core masses at the time of collapse are presented. Stars heavier than about 20 $M_{\odot}$ must explode by a "delayed" mechanism not directly related to the hydrodynamical core bounce and a subset is likely to leave black hole remnants. The isotopic nucleosynthesis expected from these massive stellar explosions is in striking agreement with the sun. Type I supernovae result when an accreting white dwarf undergoes a thermonuclear explosion. The critical role of the velocity of the deflagration front in determining the light curve, spectrum, and, especially, isotopic nucleosynthesis in these models is explored.

065.134 **Pair creation supernovae with rotation.**
W. Glatzel, M. F. El Eid, K. J. Fricke.
Nucleosynthesis and its implications on nuclear and particle physics, p. 167 – 175 (1986). – See Abstr. 012.069.

The authors have followed the evolution of rotating massive C–O cores assuming Maclaurin spheroids for the description of their structure. This simplified treatment allows to establish the marginally unstable initial configurations depending on the geometry and the thermodynamic state of the cores. These are then used as initial models for the nonspherical core collapse calculations.

065.135 **Nucleosynthesis and massive star evolution.**
N. Langer, M. F. El Eid, K. J. Fricke.
Nucleosynthesis and its implications on nuclear and particle physics, p. 177 – 187 (1986). – See Abstr. 012.069.

The uncertainty of the theoretical evolution of massive stars due to incomplete knowledge about internal mixing processes is examined, and consequences for the chemical evolution of galaxies are drawn. In contrast to earlier work, the authors find semiconvection to be very important in the post main sequence evolution of stars in the mass range 10 – 50 $M_{\odot}$, when the mixing timescale of semiconvection is accurately taken into account. For more massive stars semiconvection is less important.

065.136 **Evolution and nucleosynthesis of massive stars with extended mixing.**
C. de Loore, N. Prantzos, M. Arnould, C. Doom.
Nucleosynthesis and its implications on nuclear and particle physics, p. 189 – 196 (1986). – See Abstr. 012.069.

The authors discuss the evolution of massive stars ($M_i \geqslant 50\,M_{\odot}$) with extended mixing. The computations take into account detailed nucleosynthesis of 28 isotopes from H to ^{30}Si, and all relevant reactions for hydrogen and helium burning, including the production of neutrons. The principal difference with previous models is a smaller hydrogen burning lifetime.

065.137 **Nucleosynthesis in massive, mass losing, stars.**
N. Prantzos, C. de Loore, C. Doom, M. Arnould.
Nucleosynthesis and its implications on nuclear and particle physics, p. 197 – 206 (1986). – See Abstr. 012.069.

The authors have made detailed nucleosynthesis computations for the H- and He burning phases in massive stars with mass loss and overshooting. They present some results concerning the evolution of core and surface abundances, as well as the composition of the ejecta of these stars.

065.138 **Nucleosynthesis in massive stars: winds from WR stars and isotopic anomalies in cosmic rays.**
A. Maeder.
Nucleosynthesis and its implications on nuclear and particle physics, p. 207 – 214 (1986). – See Abstr. 012.069.

A short account of the main effects of mass loss on stellar nucleosynthesis is given. The problem of the isotopic anomalies of the galactic cosmic rays is also considered. The observed anomalies indicate that these cosmic rays originate from inner galactic regions and that in these regions the relative contribution of WC stars to the galactic enrichment is larger than in the solar neighbourhood.

065.139 **Explosive disruption of stars by big black holes.**
J.-P. Luminet.
Nucleosynthesis and its implications on nuclear and particle physics, p. 215 – 221 (1986). – See Abstr. 012.069.

The fate of a star deeply plunging in a strong external gravitational field may be described in terms of the "pancake" compression process during which explosive nucleosynthesis may be triggered in the core of the star. Consequences on the dynamics of stellar disruption, fuelling of active galactic nuclei and production of heavy elements are briefly examined.

065.140 **Nucleosynthesis in pancake stars.**
B. Pichon.
Nucleosynthesis and its implications on nuclear and particle physics, p. 223 – 231 (1986). – See Abstr. 012.069.

When a star penetrates deeply into the tidal radius of a black hole, it undergoes a violent and large compression and heating, with severe consequences for nucleosynthesis in the stellar core. In this article, on the basis of the affine star model of Carter and Luminet, taking into account the variation of entropy, with the use of a new nucleosynthesis code including a relativistic treatment of the external gravitational field, it is demonstrated that the tidal disruption of stars provides indeed a new site of nucleosynthesis in galactic centers.

065.141 **Light curves of exploding Wolf–Rayet stars.**
S. Cahen, R. Schaeffer, M. Cassé.
Nucleosynthesis and its implications on nuclear and particle physics, p. 243 – 249 (1986). – See Abstr. 012.069.

The removal of the extended hydrogen envelope surrounding the dense core of preexplosive stars by stellar winds or mass transfer onto a companion star has consequences on the optical manifestation of their final outbursts. The stripped models lead to faster and dimmer optical displays. The effective temperature near maximum light is however predicted to be rather similar to that of classical Type II supernovae. The calculations reported here show that in principle it may be possible to distinguish Wolf–Rayet supernovae triggered by core collapse from those induced by pair production.

065.142 **Stellar *s*–process diagnostics.**
G. J. Mathews, R. A. Ward, K. Takahashi, W. M. Howard.
Nucleosynthesis and its implications on nuclear and particle physics, p. 277 – 284 (1986). – See Abstr. 012.069.
It is argued that the solar–system σN curve can be best understood if the *s*–process is largely produced by stars in the mass range of $M \sim 2-4\ M_{\odot}$. Several observations are then studied as indicators of the validity of this hypothesis. It is found that isotopic Zr abundances and elemental (Ba/Sr) vs. (Ba/Nd) ratios are consistent with these conditions.

065.143 **Production and survival of ^{99}Tc in He–shell recurrent thermal pulses.**
K. Takahashi, G. J. Mathews, R. A. Ward, S. A. Becker.
Nucleosynthesis and its implications on nuclear and particle physics, p. 285 – 292 (1986). – See Abstr. 012.069.
The authors report recent work on the long–standing ^{99}Tc problem. Having combined a detailed study of the recurrent He–shell thermal–pulse, third dredge–up episodes in a 2.25 $M_{\odot}$ star and an *s*–process network calculation, they show that a substantial amount of ^{99}Tc can be produced by the *s*–process and can survive to be dredged up to the stellar surface. The calculated surface abundances of ^{99}Tc and elements with neighboring atomic numbers are compared with observations.

065.144 **Neutron capture nucleosynthesis in massive stars.**
N. Prantzos, J.–P. Arcoragi, M. Arnould.
Nucleosynthesis and its implications on nuclear and particle physics, p. 293 – 301 (1986). – See Abstr. 012.069.
Neutron–capture nucleosynthesis during core He burning in massive (50 – 100 $M_{\odot}$) mass losing stars is studied in the framework of new stellar models based on the latest available nuclear data. The resulting stellar surface abundances as well as the composition of the stellar winds corresponding to the WC phase of Wolf–Rayet stars are presented. Some implications for the isotopic composition of galactic cosmic rays are also discussed.

065.145 **Type I SNe from binary systems: consequences on galactic chemical evolution.**
F. Matteucci, L. Greggio.
Nucleosynthesis and its implications on nuclear and particle physics, p. 315 – 323 (1986). – See Abstr. 012.069.
The binary model for Type I SN precursors is assumed, and the abundances relative to the Sun of several chemical elements (^{12}C, ^{16}O, ^{24}Mg, ^{28}Si, ^{56}Fe) are computed by means of a chemical evolution model for the solar vicinity. The contribution to the galactic enrichment by Type II SNe is also taken into account. It is concluded that the binary model for Type I SNe, together with the available nucleosynthesis computations and an initial mass function constant in space and time, can account for the observed pattern of most of the studied elements.

065.146 **^{26}Al produced by Wolf–Rayet stars and the 1.8 MeV line emission of the Galaxy.**
M. Cassé, N. Prantzos.
Nucleosynthesis and its implications on nuclear and particle physics, p. 339 – 346 (1986). – See Abstr. 012.069.
Massive stars (50 to 100 $M_{\odot}$) both synthesize and eject significant quantities of ^{26}Al in their Of and WR stages. The authors estimate the collective contribution of these extreme Population I objects to the 1.8 MeV line emission of the galactic plane.

065.147 **Supernova theory.**
H. A. Bethe.
From nuclei to stars, p. 181 – 209 (1986). – See Abstr. 012.070.
Contents: (1) Introduction. (2) The result of the supernovae. (3) Pre–SN configuration. (4) Entropy. (5) Start of collapse. (6) Electron capture. (7) Neutrino trapping. (8) Energy and entropy. (9) Dynamics. (10) Equation of state. (11) Start of the shock. (12) Cooperstein's shock calculations. (13) Success and failure. (14) Net ram pressure. (15) Low–mass supernovae. (16) Larger–mass supernovae. (17) Revival of shock by neutrino heating. (18) Conclusions.

065.148 **On the production of heavy neutron–rich isotopes during He and C burning in massive stars.**
R. Gallino, M. Busso.
From nuclei to stars, p. 309 – 323 (1986). – See Abstr. 012.070.
The main purpose of this lecture is a re–examination of the yields of *s*–elements by massive short–lived stars. He–burning conditions are reviewed and results are presented for the nucleosynthesis in this phase. The authors briefly analyse the possibility of n–captures during carbon burning and discuss the contribution of massive stars to the enrichment of the interstellar medium in *s*–elements.

065.149 **Modal analysis of rotating polytropic cylinders. II. Gravitationally unstable mode and its relation to r–mode.**
S. Ishibashi, H. Ando.
Publ. Astron. Soc. Jpn., Vol. 38, No. 3, p. 295 – 311 (1986).
The authors analyze the normal modes of adiabatic oscillations in rotating homentropic cylinders ranging from the hot to cold cases, and investigate the physical properties of the r–mode and the gravitationally unstable mode. For both modes the restoring force is derived from the conservation of the total specific vorticity of a fluid element. The prograde nature of the r–mode and the retrograde nature of the gravitationally unstable mode are interpreted by this conservation relation, and the essential difference between these two modes comes from the importance of the self–gravitational force for the latter mode.

065.150 **Eine neue Methode zur Berechnung des Sternaufbaus.**
R. Bulirsch, J. Reiter.
Mitt. Astron. Ges., No. 65, p. 181 – 185 (1986). – See Abstr. 012.072.

065.151 **Synthetische Lichtkurven und Rotationsprofile für Sterne mit Sternflecken.**
K. G. Strassmeier.
Mitt. Astron. Ges., No. 65, p. 199 (1986). – See Abstr. 012.072.

065.152 **Das Schicksal eines 100 $M_{\odot}$–Sternes.**
N. Langer, M. F. El Eid.
Mitt. Astron. Ges., No. 65, p. 212 – 213 (1986). – See Abstr. 012.072.

065.153 **Integrated spectral energy distribution and colors of young stellar populations.**
I. Tarrab, G. Bruzual.
Star–forming dwarf galaxies and related objects, p. 439 – 448 (1986). – See Abstr. 012.074.
The authors construct evolutionary synthesis models of young stellar populations using evolutionary tracks for high mass stars that include stellar mass loss. The spectrum of the resulting population is computed and the derived photometric properties are compared with existing data. The agreement between the predictions and the observations is good for reasonable values of the age of the stellar population.

065.154 **Integrated colours and synthetic HR diagrams for LMC clusters.**
C. Chiosi, G. Bertelli, A. Bressan, E. Nasi, L. Pigatto.
Star–forming dwarf galaxies and related objects, p. 449 – 463 (1986). – See Abstr. 012.074.
Integrated $(B–V)_0$ and $(U–B)_0$ colours and synthetic HR diagrams of stellar clusters as a function of the age are discussed in the light of stellar models which incorporate the effects of convective overshooting all over the major evolutionary phases. Effects of stochastic nature in the initial mass function caused by the finite number of stars per cluster, and effects of dispersion in the stellar ages are also taken into account. The resulting theoretical calibrating relationships are presented and some implications for real clusters of different age are outlined.

065.155 **Common envelope evolution.**
P. Bodenheimer, R. E. Taam.
The evolution of galactic X–ray binaries, p. 13 – 24 (1986). – See Abstr. 012.076.

The short period binary systems are thought to result from a prior phase of common envelope evolution. Recently, improvements in theoretical models have resulted by extension of the numerical simulations to two space dimensions. These calculations now indicate that ejection of much of the common envelope is probable, but that the eventual formation of a short period binary is still uncertain.

065.156 **Wind fed X–ray binaries.**
P. Hellings, C. de Loore.
The evolution of galactic X–ray binaries, p. 51 – 62 (1986). – See Abstr. 012.076.

The effect of a supernova explosion on actually observed WR + O binaries is studied, and the transition of the resulting systems into X–ray sources is followed. In most cases the system remains bound and becomes a compact + O binary with a runaway velocity of the order of 10 km s^{-1}, later evolving into a massive X–ray binary. It is shown that the duration of the X–ray lifetime for all massive X–ray binaries is about 10^5years.

065.157 **From wide to close binaries?**
P. P. Eggleton.
The evolution of galactic X–ray binaries, p. 87 – 105 (1986). – See Abstr. 012.076.

Processes are reviewed which can modify the period of moderately wide binaries ($\sim$10 – 3000 d), principally those of low mass ($\lesssim$8 $M_\odot$). Some processes may shorten, but others may lengthen, the period. Possible end–products are described. The role of stellar winds, either with or without magnetic linkage to one of the components, is emphasised. Finally, the evolution of triple–star systems is mentioned in connection with low mass X–ray binaries, and with SS433.

065.158 **Evolution of close binaries and the formation of neutron stars.**
E. P. J. van den Heuvel.
The evolution of galactic X–ray binaries, p. 107 – 127 (1986). – See Abstr. 012.076.

Mechanisms for the formation of neutron stars in binary systems are discussed. Stellar evolution calculations together with the observed characteristics of massive X–ray binaries indicate that close binary systems stars in the mass range 10 to >40 $M_\odot$ leave neutron stars as remnants. In wide binaries and for single stars the lower limit is reduced to 7 – 8 $M_\odot$. Accretion–induced collapse (AIC) of a white dwarf of suitable mass and chemical composition is a second viable way for the formation of neutron stars in binary systems. Evidence is summarized indicating that both in the wide radio pulsar binaries and in the bright galactic bulge X–ray sources the neutron stars were formed by AIC.

065.159 **The duration of the detached phase in low–mass X–ray binaries in which the neutron star was formed by accretion–induced collapse of a white dwarf.**
W. Sutantyo, E. P. J. van den Heuvel.
The evolution of galactic X–ray binaries, p. 183 – 186 (1986). – See Abstr. 012.076.

065.160 **The cooler supergiants (A to M): crucial signposts in the lifecycles of massive stars.**
R. M. Humphreys.
Luminous stars and associations in galaxies, p. 45 – 59 (1986). – See Abstr. 012.079. (IAU Symp. No. 116).

The intermediate and late–type supergiants are the visually brightest stars. They are among the first stellar objects observed in other galaxies and provide our first clues to the conditions of massive star evolution in galaxies of different types. These supergiants are usually considered to be post hydrogen burning stars, and their relative numbers in the HR diagram provide essential comparisons with models for the later stages of massive star evolution. Most importantly, the most luminous cooler supergiants define the stability limit for massive stars in the HR diagram.

065.161 **Simulations of the brightest stars in galaxies as distance indicators.**
L. Greggio.
Luminous stars and associations in galaxies, p. 105 – 107 (1986). – See Abstr. 012.079. (IAU Symp. No. 116).

The brightest stars in galaxies have been often considered useful as distance indicators. To investigate on this problem the author has performed a number of numerical simulations of the absolute magnitude of the brightest objects in a population of stars more massive than 15 $M_\odot$, in four different bands: two UV bands, centered at 1400 Å and 1750 Å, the B and the V bands.

065.162 **Synthetic HR diagrams for luminous stars: a test of stellar evolution theory.**
E. Nasi.
Luminous stars and associations in galaxies, p. 121 – 122 (1986). – See Abstr. 012.079. (IAU Symp. No. 116).

Synthetic HR diagrams for massive stars were constructed with the aim of reproducing the observed features for galactic luminous stars. By means of a Monte Carlo technique, stars were randomly distributed in the M_{Bol} – Log T_{eff} plane, weighted on the Salpeter's initial mass function and with a constant star formation rate.

065.163 **Instability in massive stars: an overview.**
I. Appenzeller.
Luminous stars and associations in galaxies, p. 139 – 149 (1986). – See Abstr. 012.079. (IAU Symp. No. 116).

Dynamical, vibrational, and thermal instabilities of massive blue stars are discussed as possible mechanisms for the observed brightness variations of such objects. Relaxation oscillations (on local thermal time scales) due to dynamical instabilities of the stellar wind flows appear to be the most likely mechanism, at least for the S Dor variables. Very massive main–sequence stars with $M > 10^3 M_\odot$ should be violently vibrationally unstable and therefore should differ significantly from stable main–sequence stars of lower mass.

065.164 **On the occurrence of WR + O binaries.**
J. P. De Greve, P. Hellings, E. P. J. van den Heuvel.
Luminous stars and associations in galaxies, p. 253 (1986). – See Abstr. 012.079. (IAU Symp. No. 116).

065.165 **How to form a 200 $M_\odot$ star.**
H. Zinnecker.
Luminous stars and associations in galaxies, p. 271 – 273 (1986). – See Abstr. 012.079. (IAU Symp. No. 116).

Stellar coalescence is suggested as a possible mechanism for doubling the upper stellar mass limit from $\sim$100 $M_\odot$ to $\sim$200 $M_\odot$ in a moderately dense cluster of a few hundred young massive stars ($\sim 10^5 M_\odot pc^{-3}$). The merger will be between the two components of the dominant central tight binary formed in the core of the cluster by the N–body evolution. This process may occur in some giant extragalactic H II regions.

065.166 **Massive star evolution: mass loss and mixing.**
A. Maeder.
Luminous stars and associations in galaxies, p. 287 – 300 (1986). – See Abstr. 012.079. (IAU Symp. No. 116).

At first it may be surprising that mass loss, overshooting and mixing, which are indeed very different physical processes, have similar consequences on stellar evolution. These various processes may increase the main–sequence (MS) lifetime, extend the width of the MS, bring CNO–processed materials to stellar surfaces and, in extreme cases, lead to quasi–homogeneous evolution. The present status of the knowledge on these effects is examined in detail.

065.167 **Effects of convective overshooting, mass loss (and chemical composition) across the HR diagram.**
C. Chiosi.
Luminous stars and associations in galaxies, p. 317 – 338 (1986). – See Abstr. 012.079. (IAU Symp. No. 116).
The far reaching consequences of convective overshooting during the core H and He–burning phases of stars in the mass range 1.3 $M_{\odot}$ to 100 $M_{\odot}$ are discussed. In addition, the effects of mass loss in luminous stars of all spectral types, and in the red giant and asymptotic giant branch stars are briefly outlined. Furthermore, the effects of the novel $^{12}C(\alpha\ \gamma)^{16}O$ reaction rate are also illustrated.

065.168 **Binary evolution in the upper HRD.**
J. P. De Greve, C. de Loore.
Luminous stars and associations in galaxies, p. 339 – 352 (1986). – See Abstr. 012.079. (IAU Symp. No. 116).
The importance of close binaries in the domain of luminous objects is examined. The group of unevolved massive stars of O– and B–type, the groups of Wolf–Rayet binaries and X–ray binaries are discussed, and limits on the mass ratios, masses and periods are listed, as well as the average values. The connections between these various groups are investigated, and relations among subgroups are established.

065.169 **The average X–ray lifetime of massive X–ray binaries.**
P. Hellings.
Luminous stars and associations in galaxies, p. 383 – 384 (1986). – See Abstr. 012.079. (IAU Symp. No. 116).
In this study the author computes the transition from compact + O systems into X–ray binaries, with emphasis on the wind powered stage.

065.170 **Consequences of rotational mixing in late type massive stars.**
S. R. Sreenivasan, W. J. F. Wilson.
Luminous stars and associations in galaxies, p. 385 – 387 (1986). – See Abstr. 012.079. (IAU Symp. No. 116).
The authors have examined consequences of mixing induced by rotation in massive stars (Pop I composition: X = 0.7; Y = 0.27) in the mass range 40 – 100 $M_{\odot}$.

065.171 **Convective overshooting: the upper mass limit for stars undergoing core He–flash.**
G. Bertelli, A. Bressan.
Luminous stars and associations in galaxies, p. 421 – 422 (1986). – See Abstr. 012.079. (IAU Symp. No. 116).
The authors have computed evolutionary sequences of 1.4 $M_{\odot}$, 1.5 $M_{\odot}$ and 1.6 $M_{\odot}$ with chemical composition X = 0.700 and Z = 0.020 adopting the description of convective overshooting formulated by Bressan et al. (1981).

065.172 **The final fate of massive stars.**
H. Schild.
Luminous stars and associations in galaxies, p. 423 – 424 (1986). – See Abstr. 012.079. (IAU Symp. No. 116).
The cluster membership of Wolf–Rayet stars, supernova remnants and pulsars is used to study the late stages of stellar evolution of massive stars.

065.173 **Luminous content of galaxies: inferences from supernovae.**
J. C. Wheeler.
Luminous stars and associations in galaxies, p. 467 – 476 (1986). – See Abstr. 012.079. (IAU Symp. No. 116).
Some luminous stars undoubtedly explode as supernovae, but it is not yet certain that fate awaits them all. The connection between luminous stars and supernovae is reviewed in terms of the statistical rates of supernovae and pulsars, the constraints of nucleosynthesis, and the various classifications of supernovae by their spectra and light curves, including a newly confirmed class of peculiar Type I supernovae.

065.174 **Convective overshooting: new integrated colours vs age relations for star clusters.**
G. Bertelli, A. G. Bressan, C. Chiosi, E. Nasi, L. Pigatto.
Luminous stars and associations in galaxies, p. 511 – 512 (1986). – See Abstr. 012.079. (IAU Symp. No. 116).
The aim of this preliminary investigation is to study the effects of convective overshooting on the integrated colours of clusters whose turnoff mass is in that range in which convective overshooting is effective.

065.175 **From Miras to planetary nebulae: a model of mass loss.**
P. J. Bedijn.
Light on dark matter, p. 119 – 126 (1986). – See Abstr. 012.080.
Stellar pulsation in combination with radiation pressure on dust is proposed as the mechanism causing mass loss in Mira variables and an expression for the mass loss rate in terms of the basic stellar parameters mass and photospheric radius is derived which leads to an acceleration of the mass loss rate in time as suggested by observations. Upper Asymptotic Giant Branch evolution with such a mass loss is presented. A model period distribution for classical Miras as well as a model 1612 MHz OH luminosity function for (OH/IR) Miras in the solar neighbourhood are fitted to observed ones.

Methods in the qualitative theory of dynamical systems in astrophysics and gas dynamics.
See Abstr. 003.028.

The exploding suns: the secrets of the supernovas.
See Abstr. 003.056.

How no(w) brown dwarfs?
See Abstr. 011.008.

A simple theory of limiting masses.
See Abstr. 014.036.

The screening of photodisintegration reactions.
See Abstr. 061.003.

Limits on mass of missing mass.
See Abstr. 061.014.

A new estimate of the $^{19}Ne(p,\gamma)^{20}Na$ and $^{15}O(\alpha,\gamma)^{19}Ne$ reaction rates at stellar energies.
See Abstr. 061.017.

Effects of metal–dependent stellar models on the yield of nitrogen.
See Abstr. 061.027.

Thermal diffusion in ionized gases and planetary and solar processes.
See Abstr. 061.031.

Astrophysical constraints on the couplings of axions, majorons, and familons.
See Abstr. 061.033.

Dominance of B^{10} or C^{12} nuclear production during gravitational collapse in the stellar interior?
See Abstr. 061.041.

Neutron capture process in astrophysics.
See Abstr. 061.063.

Beta–decay half–lives of new neutron–rich chromium–to–nickel isotopes and their consequences for the astrophysical r–process.
See Abstr. 061.076.

Stellar energy loss through scalar–neutrino production.
See Abstr. 061.089.

Axion constraints from white dwarf cooling times.
See Abstr. 061.104.

The effect of the new $^{12}C(\alpha,\gamma)^{16}O$ rate on the chemical evolution of the solar neighborhood.
See Abstr. 061.121.

The survival of helium–3 in stars.
See Abstr. 061.123.

***s*–process nucleosynthesis – stellar aspects and the classical model.**
See Abstr. 061.125.

***s*–process nucleosynthesis below A = 90.**
See Abstr. 061.126.

Pulsed–neutron–source models for the astrophysical *s*–process.
See Abstr. 061.127.

A parametrized study of the $^{13}C(\alpha,n)^{16}O$ neutron source.
See Abstr. 061.128.

Magnesium isotopes and galactic evolution.
See Abstr. 061.129.

More about nucleocosmochronology: the reliability of the long–lived chronometers, and the production of extinct radioactivities.
See Abstr. 061.132.

Observational evidence of element formation processes in stars.
See Abstr. 061.146.

Equilibrium of magnetic fields with arbitrary interweaving of the lines of force. I. Discontinuities in the torsion.
See Abstr. 062.103.

Supercritical winds from cool "canonical" stars caused by evolution on the main sequence.
See Abstr. 064.007.

Steady mass–loss from supermassive stars.
See Abstr. 064.008.

Stellar chromospheres, coronae, and winds.
See Abstr. 064.010.

Stellar magnetic activity and long term luminosity variations of late type stars. I. On generalization of Öpik's convection theory to a rotating medium.
See Abstr. 064.060.

Gravitational collapse and the cosmic antineutrino background.
See Abstr. 067.035.

Gravitational collapse and rotation: a quandary for the black–holes.
See Abstr. 067.089.

Magnetic field decay and the origin of neutron star binaries.
See Abstr. 067.112.

Dynamics of an affine star model in a black hole tidal field.
See Abstr. 067.128.

Mass shedding and partition of the *a*/*m* ratio between core and envelope in gravitational collapse.
See Abstr. 067.134.

Theoretical aspects of helio– and asteroseismology.
See Abstr. 080.051.

Asymptotic sound–speed inversions.
See Abstr. 080.062.

An improved asymptotic formula for solar gravity–mode periods.
See Abstr. 080.063.

Solar g modes: a method to find the depth of the convective envelope.
See Abstr. 080.064.

The influence of convection theory uncertainties on the deduction of the speed of sound in the solar interior.
See Abstr. 080.065.

Mass loss from cool stars.
See Abstr. 112.042.

IRAS observations of the cool galactic hypergiants.
See Abstr. 112.124.

IRAS observations of cool excess around main sequence stars.
See Abstr. 112.134.

An investigation of the atmosphere of the supergiant δ CMa (F8 Ia) and determination of its evolutionary parameters.
See Abstr. 114.035.

Non–LTE analysis of massive stars in the Magellanic Clouds.
See Abstr. 114.072.

The formation and evolution of carbon stars.
See Abstr. 115.018.

Solar and stellar magnetic activity.
See Abstr. 116.004.

The stellar seismology of ε Eridani.
See Abstr. 116.007.

Tidal spin–up and magnetic braking in cataclysmic variables.
See Abstr. 117.013.

Massive X–ray binaries: their physics and evolution.
See Abstr. 117.020.

The minimum orbital period for ultra–compact binaries with helium burning secondaries.
See Abstr. 117.039.

On the problem of the gap of the orbital periods in cataclysmic variables.
See Abstr. 117.098.

Evolutionary status of W UMa–type binaries.
See Abstr. 117.099.

Evolution of low–mass close binaries with orbital momentum loss.
See Abstr. 117.102.

Evolution of close binaries under the assumption of momentum loss by magnetic stellar wind.
See Abstr. 117.112.

Non–conservative evolution of close binaries.
See Abstr. 117.161.

On the masses of Wolf–Rayet components in close binary system.
See Abstr. 117.168.

A common origin for Type I supernovae and low–mass binary X–ray sources.
See Abstr. 117.185.

Postthermonuclear runaway angular momentum loss in cataclysmic binaries.
See Abstr. 117.229.

The use of ST in research about evolutionary connections among various types of close binary systems.
See Abstr. 117.250.

Origin and evolution of X–ray binaries in globular clusters.
See Abstr. 117.304.

A note on the nature of the secondaries in ultra–compact binaries.
See Abstr. 117.305.

The evolution of low mass binary systems.
See Abstr. 117.309.

On the evolution of Her X–1.
See Abstr. 117.313.

Secular evolution of cataclysmic binaries.
See Abstr. 117.314.

The evolution of magnetic cataclysmic variables.
See Abstr. 117.315.

Bolometric corrections and magnitudes of WR + O stars.
See Abstr. 117.326.

The evolution of massive close binaries with mass loss and overshooting. An application to V729 Cyg (= BD +40°4220).
See Abstr. 117.328.

The chemical composition of Algol systems – III. Beta Lyrae–nucleosynthesis revealed.
See Abstr. 119.015.

Evolutionary status of V356 Sgr.
See Abstr. 119.080.

On the spectroscopic orbit of the bright WC binary γ^2Velorum, with implications for the evolutionary scenario of Wolf–Rayet stars.
See Abstr. 120.021.

Young stellar objects and their circumstellar dust: an overview.
See Abstr. 121.012.

Models of the active stage of R Coronae Borealis–type stars.
See Abstr. 122.054.

SU Cas: an overtone pulsator.
See Abstr. 122.073.

Nonradial modes in line–profile variable stars.
See Abstr. 122.097.

Flare stars – physics and evolution.
See Abstr. 122.117.

Rapid accretion and hibernation in the pre–outburst history of classical novae.
See Abstr. 124.010.

Type I supernovae.
See Abstr. 125.089.

Feasibility of a low–mass binary source progenitor for PSR1937 + 214.
See Abstr. 126.003.

The outcome of mass transfer in a carbon–oxygen white dwarf binary system.
See Abstr. 126.070.

Analytical, time–dependent mass function and fragmentation rate due to gravitational instability.
See Abstr. 131.114.

Star formation in colliding gas flows.
See Abstr. 131.304.

Electrodynamic forces in stellar formation.
See Abstr. 131.345.

Physical processes in massive star formation.
See Abstr. 131.354.

Oscillating star formation.
See Abstr. 131.355.

Hollow H II regions. I. Effects of stellar evolution and wind energy dissipation.
See Abstr. 132.021.

Outer haloes of planetary nebulae as probes of a fast luminosity decline in their nuclei.
See Abstr. 134.005.

Origin of planetary nebulae: morphology, carbon–to–oxygen abundance ratios, and central star multiplicity.
See Abstr. 134.011.

Breaking through the wall.
See Abstr. 134.019.

Detection of the very hot central star in NGC 2440.
See Abstr. 134.020.

The evolution of planetary nebula nuclei.
See Abstr. 134.040.

Angular momentum distributions and energies of rotating selfgravitating discs.
See Abstr. 151.026.

Dynamical effects of stellar evolution in globular clusters.
See Abstr. 151.029.

A method for integrating the equations that govern linear isentropic oscillations of an incompressible Maclaurin spheroid embedded in a spheroidal halo.
See Abstr. 151.044.

Evolution of massive stars: comparison of cluster sequences and models with mass loss.
See Abstr. 153.005.

The lithium abundance gap in the Hyades F stars: the signature of diffusion.
See Abstr. 153.016.

Problems in the evolution of young stars in open clusters and associations. Perspectives of future work.
See Abstr. 153.036.

Star formation and some aspects of the chemical evolution of the Galaxy.
See Abstr. 155.008.

The effect of the new $^{12}C(\alpha,\gamma)^{16}O$ rate on the chemical evolution of the solar neighbourhood.
See Abstr. 155.098.

Observational constraints to the chemical evolution of the galactic halo from the Space Telescope.
See Abstr. 155.109.

The cool–giants populations in the Magellanic Clouds.
See Abstr. 156.022.

Evolutionary population synthesis and the Magellanic Cloud clusters.
See Abstr. 156.026.

Bimodal star formation and remnant–dominated galactic models.
See Abstr. 157.010.

Metal–enhanced galactic winds. I.
See Abstr. 157.258.

The age of the universe, primordial galaxies, and Space Telescope.
See Abstr. 161.306.

Standard big bang nucleosynthesis and chemical evolution of D and ^{3}He.
See Abstr. 161.401.

Nucleosynthetic consequences of Population III stars.
See Abstr. 161.406.

066 Relativistic Astrophysics, Gravitation Theory

066.001 **Projective and volume–preserving bundle structures involved in the formulation of $A(4)$ gauge theories.**
R. O. Fulp, W. R. Davis, L. K. Norris.
Gen. Relativ. Gravitation, Vol. 18, No. 1, p. 1 – 29 (1986).
The bundle structures required by volume–preserving and related projective properties are developed and discussed in the context of $A(4)$ gauge theories which may be taken as the proper framework for Poincaré gauge theories. The results of this paper include methods for extending both tensors and connections to a principal fiber bundle having $G1(4,R) \times G1(4,R)$ as its structure group. This bundle structure is shown to be a natural arena for the generalized covariant differentiation utilized by Einstein for his extended gravitational theories involving nonsymmetric connections.

066.002 **A magnetic–monopole–type solution in the Poincaré gauge field theory of gravity.**
P. Baekler.
Gen. Relativ. Gravitation, Vol. 18, No. 1, p. 31 – 43 (1986).
In a microscopical theory of gravity the coupling of internal gauge degrees of freedom to those of space–time are studied. A magnetic–monopole–type solution for the coupled $SO(3)$ Yang–Mills–Higgs system in a space–time with curvature and torsion is derived. The coupling constant of the Lorentz gauge bosons can be related directly to the (constant) Higgs fields and to the "cosmological constant" which is induced by the quadratic curvature terms in the Lagrangian.

066.003 **The gravitational acceleration of antiprotons.**
T. Goldman, M. V. Hynes, M. M. Nieto.
Gen. Relativ. Gravitation, Vol. 18, No. 1, p. 67 – 70 (1986).
The authors review the theoretical arguments and experimental evidence (both direct and indirect) for the conventional view that the gravitational acceleration of antimatter must be the same as that of matter. They find there is no compelling support for such a belief. The authors propose to measure the gravitational acceleration of antiprotons using a beam extracted from the low–energy antiproton ring (LEAR) at CERN.

066.004 **Lorentzian framed structures in general relativity.**
K. L. Duggal, R. Sharma.
Gen. Relativ. Gravitation, Vol. 18, No. 1, p. 71 – 77 (1986).
The authors study a class of four–dimensional real Lorentzian manifolds generated by a global operator (satisfying a cubic equation). They characterize empty, Einstein, conformally flat spaces and construct mathematical models of two physical space–times. Locally, the solutions presented have been identified with known solutions. Kruskal space–time is generated by using the warped product technique.

066.005 **A new class of interior solutions for a stress–energy tensor of Segré characteristic [111,1].**
T. Biech, A. Das.
Gen. Relativ. Gravitation, Vol. 18, No. 1, p. 93 – 103 (1986).
The interior solutions of (the tetrad versions of) Einstein's field equations with T_{AB} having Segré characteristic [111,1] (which has all four eigenvalues distinct), are investigated. For this purpose a *mixed method*, which combines Synge's g method and T method, is introduced. Some of the tetrad equations are solved for the metric functions while the remaining equations are used to define the corresponding components of T_{AB}.

066.006 **Complex relativity and real solutions. III. Real type–N solutions from complex $N \otimes N$ ones.**
M. S. Hickman, C. B. G. McIntosh.
Gen. Relativ. Gravitation, Vol. 18, No. 2, p. 107 – 136 (1986).
Einstein's vacuum field equations are integrated in complex relativity in a major subcase of the class whose Weyl tensor is of the type $N \otimes N$, i.e., when the left and right Weyl spinors Ψ and $\tilde{\Psi}$ are each of type N. Before the equations are integrated, the relevant general theory and equations are developed in a tetrad frame which is well suited to the discussion of these and a wider class of complex solutions and is called a left quarter flat frame. The relationship between this frame and the coordinates used and some other frames and coordinates, including the complexified version of the frame often used for real type–N metrics, is discussed.

066.007 **Stationary solutions in five–dimensional general relativity.**
G. Clément.
Gen. Relativ. Gravitation, Vol. 18, No. 2, p. 137 – 160 (1986).
Asymptotically Minkowskian stationary axisymmetric solutions of the five–dimensional Einstein equations are generated from particular classes of stationary axisymmetric solutions of the four–dimensional Einstein–Maxwell equations. First, the five–dimensional electrostatic and magnetostatic solutions are generated from the four–dimensional electrostatic solutions and a harmonic scalar field. Then, a new class of five–dimensional nonstatic solutions are generated from the four–dimensional class $E = +1$.

066.008 **Shell sources of stationary axisymmetric gravitational fields.**
D. Babala.
Gen. Relativ. Gravitation, Vol. 18, No. 2, p. 173 – 191 (1986).
A number of solutions for material shell sources of stationary axisymmetric gravitational fields are presented. Explicit solutions are found for shells lying on equipotential hypersurfaces (g_{tt} = const) and generating static "monopole" fields in prolate and oblate spheroidal coordinates. Numerical solutions are found for shells lying on hypersurfaces of constant $g_{\varphi t}/g_{\varphi\varphi}$ and generating Kerr– and Tomimatsu–Sato fields. The shells have minimum areas allowed by the energy conditions of Hawking and Ellis.

066.009 **A note on the integrability conditions for the existence of rational first integrals of the geodesic equations in a Riemannian space.**
C. D. Collinson.
Gen. Relativ. Gravitation, Vol. 18, No. 2, p. 207 – 214 (1986).
Rational first integrals of the equation determining the geodesics of a given Riemannian space are investigated. It is shown that integrability conditions can be obtained for the existence of such integrals.

066.010 **On the assumptions made in treating the gravitational wave problem by the high–frequency approximation.**
M. Evangelista de Araujo.
Gen. Relativ. Gravitation, Vol. 18, No. 3, p. 219 – 233 (1986).
The assumptions that enter into the formulation of the gravitational wave problem in the limit of high frequency are discussed. It is shown that depending on the relation between the amplitude parameter η and the frequency parameter ε the concept of "back–reaction" can have different physical interpretations. It is also shown that as a direct consequence of these assumptions high–frequency gravitational waves do not disperse in a vacuum.

066.011 **Gravitational radiation reaction in quasistatic, axially symmetric systems with possibly strong fields.**
R. E. Kates.
Gen. Relativ. Gravitation, Vol. 18, No. 3, p. 235 – 253 (1986).
The leading radiation reaction in quasistatic, axially symmetric systems is calculated by matched asymptotic expansions. Earlier results on inductive transfer of near–zone "energy" are combined via matching with a wave–zone expansion in the Regge–Wheeler gauge. Two independent measures of the damping are computed: (1) the change of the $1/R$ coefficient in a Weyl–Levi–Cività type of multipole expansion in the near zone and (2) the change in Bondi energy at future null infinity.

066.012 **The ADM Hamiltonian at the postlinear approximation.**
G. Schäfer.
Gen. Relativ. Gravitation, Vol. 18, No. 3, p. 255 – 270 (1986).
The ADM Hamiltonian for a many–particle system is calculated up to the postlinear approximation, i.e., to the approximation that both the equations of motion for the particles and the equations of motion for the gravitational field in case of no–incoming radiation correctly result up to the postlinear approximation. The relation of this Hamiltonian to the ADM Hamiltonian obtained by a post–Newtonian approximation scheme which was applied up to the first radiation–reaction and radiation levels is discussed.

066.013 **Precession of angular momentum vector in a slowly rotating Kerr metric.**
A. M. Faridi.
Gen. Relativ. Gravitation, Vol. 18, No. 3, p. 272 – 286 (1986).
Specializing Penrose and Floyd's result $F_{ab;c} = F_{[ab;c]}$ to the Kerr metric, the author explicitly constructs the skew symmetric tensor F_{ab} and Carter's quadratic integral of geodesic motion. F_{ab} is then shown to be closely related to the orbital angular momentum encountered in Newtonian mechanics.

066.014 **The behavior of the gravitational field near the initial singularity.**
G. Francisco.
Gen. Relativ. Gravitation, Vol. 18, No. 3, p. 287 – 308 (1986).
The ultralocal representation of the canonically quantized gravitational field is used to obtain the evolution of coherent states in the immediate neighborhood of the singularity. It is shown that smearing functions play the role of classical fields since they correspond to cosmological solutions around the singularity. A special class of ultralocal coherent states is shown to contain the essential aspects of the dynamics of the system.

066.015 **Causally discontinuous space–times.**
U. D. Vyas, G. M. Akolia.
Gen. Relativ. Gravitation, Vol. 18, No. 3, p. 309 – 314 (1986).
Along with certain properties of the region where distinguishingness is violated, it is proved that the violation of distinguishingness and reflectingness cannot occur individually.

066.016 **Characterization of locally rotationally symmetric space–times.**
S. W. Goode, J. Wainwright.
Gen. Relativ. Gravitation, Vol. 18, No. 3, p. 315 – 331 (1986).
The authors give a simple characterization of locally rotationally symmetric space–times in terms of the existence of a canonical null tetrad or canonical orthonormal tetrad. The result is applied to space–times which satisfy the Einstein field equations with a perfect fluid or electromagnetic field as source.

066.017 **Newtonian gravitation corrected.**
J. Maddox.
Nature, Vol. 319, No. 6050, p. 173 (1986).
Suspicions that the inverse square law for the gravitational force between two objects may be incorrect at short distances appear to have been confirmed.

066.018 **Gravitational radiation from the collapse and explosion of rapidly rotating, homogeneous, gaseous ellipsoids imbedded in halos.**
H. Beltrami, W. Y. Chau.
Astrophys. Space Sci., Vol. 119, No. 2, p. 353 – 372 (1986).
The nonlinear motion (collapse/explosion) of a homogeneous, rapidly rotating, gaseous ellipsoid has been studied with the effects of an external gravitational field incorporated. The non–axisymmetric motion has been followed using an efficient numerical code based on certain properties of the potential functions. The gravitational radiation associated with the phenomenon was calculated, and the wave forms studied to extract information on the dynamics of the source. Possible implications of the results for various astrophysical processes are discussed.

066.019 **Friedmann–Robertson–Walker vacuum solutions in the scale covariant theory.**
A. Beesham.
Astrophys. Space Sci., Vol. 119, No. 2, p. 415 – 420 (1986).
The author derives Friedmann–Robertson–Walker vacuum solutions in the scale covariant theory and discusses some of their properties.

066.020 **Present problems in relativistic celestial mechanics.**
V. A. Brumberg.
Relativity in celestial mechanics and astrometry, p. 5 – 17 (1986). – See Abstr. 012.002 (IAU Symp. No. 114)
Review of the present problems of relativistic celestial mechanics. Advantage is taken of the method suggested earlier by the author and based on using quasi–Galilean coordinates with arbitrary coordinate functions or parameters. As compared with the previous papers the new elements are post–post–Newtonian approximation for the circular motion in the Schwarzschild problem and reduction of the artificial satellite problem including the main solar perturbations to the Schwarzschild problem. Some current questions of time scales definitions, reference frames and reduction of observations are briefly discussed.

066.021 **Equations of motion for isolated bodies with relativistic corrections including the radiation reaction force.**
L. P. Grishchuk, S. M. Kopejkin.
Relativity in celestial mechanics and astrometry, p. 19 – 34 (1986). – See Abstr. 012.002 (IAU Symp. No. 114).
The authors have derived in an explicit form the equation of motion for two spherically–symmetric non rotating bodies in the slow motion approximation. The equations include relativistic corrections of order $(v/c)^2$, $(v/c)^4$ and $(v/c)^5$ to the newtonian equations of motion. It is shown that the equations depend on the only parameter characterizing each body, namely on its relativistic mass, regardless of its internal structure and degree of compactness. This means that the equations can also be applied to bodies with a strong internal gravity, such as neutron stars and black holes.

066.022 **Post–Newtonian treatise on the rotational motion of a finite body.**
T. Fukushima.
Relativity in celestial mechanics and astrometry, p. 35 – 40 (1986). – See Abstr. 012.002 (IAU Symp. No. 114).

The theory on the motion of bodies in the gravitational field is not completed until it is formulated in the framework of the general relativity. The post–Newtonian formalism (Misner et al., 1970; Will, 1981) is a suitable approach to deal with slow motions in a weak gravitational field such as the rotation of the Earth. In this paper, the author constructs a post–Newtonian theory of the rotational motion of a finite body with use of the concept of the proper reference frame of a massive body (Fukushima et al., 1986) and a definition of the rigid body in the post–Newtonian framework.

066.023 **Planetary perturbation equations based on relativistic Keplerian motion.**
N. Ashby.
Relativity in celestial mechanics and astrometry, p. 41 – 52 (1986). – See Abstr. 012.002 (IAU Symp. No. 114).

Solutions of the geodesic equations for bound test particle motion in a Schwarzschild field are expressed using Jacobian elliptic functions. Keplerian orbital elements are identified and related to a set of canonical constants. Relativistic Lagrange planetary perturbation equations are derived.

066.024 **On the dominant relativistic terms in the lunar theory.**
M. Soffel, H. Ruder, M. Schneider.
Relativity in celestial mechanics and astrometry, p. 53 – 57 (1986). – See Abstr. 012.002 (IAU Symp. No. 114).

For a simplified 3–body (Earth, Moon, Sun) problem it is shown how the usual Einstein–Infeld–Hoffmann equations for the lunar motion reduce to the Jacobi–equations after the transformation to the proper reference frame. The dominant relativistic contributions to the lunar laser ranging observables are then obtained in a Hill–Brown calculation. It is argued that in the proper reference frame all post–Newtonian variational terms are proportional to $m = n'/(n-n')$ [$n(n')$ = mean motion of Moon (Sun)].

066.025 **Reference frames and gravitational effects in the general theory of relativity.**
O. S. Ivanitskaya, N. V. Mitskiévič, Yu. S. Vladimirov.
Relativity in celestial mechanics and astrometry, p. 177 – 186 (1986). – See Abstr. 012.002 (IAU Symp. No. 114).

The authors summarize the reference frame theory in general relativity (in the monad as well as in the tetrad representation) and give examples of specific gravitational effects.

066.026 **Local measurements and cosmological background.**
P. Teyssandier, P. Tourrenc.
Relativity in celestial mechanics and astrometry, p. 189 (1986). Abstract. – See Abstr. 012.002 (IAU Symp. No. 114).

066.027 **On observations in Schwarzschild background.**
S. V. Khmil.
Relativity in celestial mechanics and astrometry, p. 199 – 203 (1986). – See Abstr. 012.002 (IAU Symp. No. 114).

In the geometrical optics approximation a uniform approach to the description of the influence of strong Schwarzschild field on astronomical observations is given.

066.028 **Spacetime coordinates in the geocentric reference frame.**
M. Fujimoto, E. Grafarend.
Relativity in celestial mechanics and astrometry, p. 269 – 276 (1986). – See Abstr. 012.002 (IAU Symp. No. 114).

A geocentric relativistic reference frame is established which is close to the conventional non–relativistic equatorial frame of reference. Points on the earth surface and near the earth–space are coordinated making use of the Frobenius matrix of integrating factors which connects the geocentric orthonormal tetrad with the tangent spacetime of relativistic pseudo–Riemann geometry. The gravity field of the earth and its relative velocity with respect to the solar system barycentre cause coordinate effects of the order of 10 cm for topocentric point positioning.

066.029 **Application of the Deep Space Network (DSN) to the testing of general relativity.**
J. D. Anderson, G. S. Levy, N. A. Renzetti.
Relativity in celestial mechanics and astrometry, p. 329 – 344 (1986). – See Abstr. 012.002 (IAU Symp. No. 114).

Over the past two decades, radar and radio observations of planets and spacecraft have been made by stations of the Deep Space Network (DSN), which by the unprecedented nature of their accuracy have produced the most accurate tests of general relativity available. The authors review the history of the instrumentation and data analysis of the first spacecraft test as well as tests of the metric nature of gravity using radar, optical observations, and radio astrometry of the planets and the limits placed on the variability of the gravitational constant G. Finally, the authors discuss the prospects for improved accuracy through ongoing upgrades of DSN instrumentation and show the results of covariance analyses for possible future NASA missions.

066.030 **General relativity confronts experiment.**
C. M. Will.
Relativity in celestial mechanics and astrometry, p. 355 – 367 (1986). – See Abstr. 012.002 (IAU Symp. No. 114).

The author reviews the status of experimental tests of general relativity. These include tests of the Einstein equivalence principle, which requires that gravitation be described by a curved–spacetime, "metric" theory of gravity. General relativity is consistent with all tests to date, including the "classical tests": light deflection using radio interferometers, radar time delay using Viking Mars landers, and the perihelion shift of Mercury; and tests of the strong equivalence principle, such as lunar laser ranging tests of the "Nordtvedt effect", and tests for variations in G. The author also reviews ten years of observations of the binary pulsar PSR 1913 + 16, in which the first evidence for gravitational radiation has been found.

066.031 **Prospects for observations of relativistic effects in the solar system.**
R. D. Reasenberg, I. I. Shapiro.
Relativity in celestial mechanics and astrometry, p. 383 – 391 (1986). – See Abstr. 012.002 (IAU Symp. No. 114).

The solar system is the traditional laboratory for testing theories of gravitation. To enhance significantly the accuracy of such tests, one must investigate novel techniques. In this paper the authors concentrate on an experiment that promises a dramatic improvement in a classical test of general relativity – the deflection of light by solar gravity. The goal is to measure the post–post–Newtonian constribution of nearly 11 microarcseconds to this deflection. The technique proposed is based on use of an astrometric optical interferometer, POINTS, which could be operated from the bay of the Space Shuttle, mounted on the proposed Space Station, or supported by an independent spacecraft. POINTS should be able to measure the separation of stars about 90° apart with an uncertainty of only a few microarcseconds.

066.032 **Accuracy estimates for the determination of the solar space–time metric by HIPPARCOS.**
S. Söderhjelm, L. Lindegren.
Relativity in celestial mechanics and astrometry, p. 393 – 399 (1986). – See Abstr. 012.002 (IAU Symp. No. 114).

Parallel to the determination of astrometric parameters for some 100,000 stars, the HIPPARCOS data may also be used to study the light deflection produced by the sun. Cowling (1983, thesis) has calculated the light deflection observable by HIPPARCOS for a general space–time metric, and the authors have included his list of fifteen metric coefficients as "global" parameters in simulated HIPPARCOS reductions. The estimated covariance matrices show that only 7 – 9 linear combinations of the Cowling parameters may be determined with some accuracy ($\sim 10^{-10}$). The value for the standard GR light deflection may be determined if the thermal effects on the satellite are small, but not to better than about 0.5%.

066.033 **Gravitational time delay of signals in the Kerr metric.**
I. G. Dymnikova.
Relativity in celestial mechanics and astrometry, p. 411 – 416 (1986). – See Abstr. 012.002 (IAU Symp. No. 114).
The gravitational time delay of signals in a gravitational field of a rotating massive body is considered both in a weak field approximation and in a strong field caused by a rotating black hole. The expressions describing the time of propagation of signals are obtained by integrating the light geodesics of the Kerr metric in a reference frame of a distant observer using the Boyer–Lindquist coordinates and assuming that the wavelength of radiation is much less than the characteristic scale of the field. The existence of the asymmetry in the time delay is shown depending on the mutual orientation of a photon propagation direction and of the rotation axis. As a result of this asymmetry, the effects of relative time delay are predicted and calculated for the signals focused by a rotating gravitational lens.

066.034 **Does the energy density of the vacuum influence planetary motions?**
R. Shubert.
Relativity in celestial mechanics and astrometry, p. 417 – 418 (1986). – See Abstr. 012.002 (IAU Symp. No. 114).
New cosmological and quantum mechanical evidence for the existence of an ether has made it appropriate to investigate again the effect of such a medium upon the motions of planets and satellites within the solar system. This paper discusses the pertinent consequences of associating the effects of the space–time continuum with the properties of a perfet fluid which can exert measurable forces on planetary bodies.

066.035 **On purely radiative space–times.**
H. Friedrich.
Commun. Math. Phys., Vol. 103, No. 1, p. 35 – 65 (1986).
The existence of space–times representing pure gravitational radiation which comes in from infinity and interacts with itself is discussed. They are characterized as solutions of Einstein's vacuum field equations possessing a smooth structure at past null infinity which forms the "future null cone at past timelike infinity with complete generators". The "pure radiation problem" is analysed where "free initial data" for Einstein's field equations are prescribed on the null cone at past time–like infinity.

066.036 **Vector two–point functions in maximally symmetric spaces.**
B. Allen, T. Jacobson.
Commun. Math. Phys., Vol. 103, No. 4, p. 669 – 692 (1986).
The authors obtain massive and massless vector two–point functions in maximally symmetric spaces (and vacua) of any number of dimensions. These include de Sitter space and anti–de Sitter space, and their Euclidean analogs S^n and H^n.

066.037 **Gauge theory in curved space.**
Y. Hosotani.
Particles and gravity, p. 67 – 80 (1984). – See Abstr. 012.004.
Non–Abelian gauge theory in curved space is analysed from two standpoints. In the first half various exact classical solutions are presented which solve matter field equations and the Einstein equations simultaneously. Their implications to the very early univere are discussed. In the second half the problem of dynamical mass generation by finite curvature is investigated to show the importance of quantum effects to phase transitions in curved space.

066.038 **Spontaneous quantum compactification in seventeen dimensions.**
A. Chodos, E. Myers.
Particles and gravity, p. 81 – 100 (1984). – See Abstr. 012.004.
The authors consider general relativity with a cosmological term in $D > 4$ space–time dimensions. They examine the possibility that the inclusion of one–loop quantum effects will allow compactification on a manifold of the form $M^4 \times S^N$ (N must be odd for technical reasons). The first candidate for such a phenomenon occurs for $N = 13$.

066.039 **Geometry as particles and particles as geometry.**
G. W. Gibbons.
Particles and gravity, p. 101 – 108 (1984). – See Abstr. 012.004.
This paper describes some recent progress in implementing Einstein's dream of constructing models for particles constructed solely from the geometry of spacetime, without singularities. It also mentions the speculative possibility that there may be, at a fundamental level, no real distinction between "particles" and "geometry".

066.040 **Gravitational and Lorentz anomalies.**
F. Langouche, T. Schücker, R. Stora.
Particles and gravity, p. 109 – 118 (1984). – See Abstr. 012.004.
The authors define the anomalies for the symmetries of general relativity and describe the Adler–Bardeen class of solutions of the corresponding consistency equations in even dimensions by a purely algebraic algorithm.

066.041 **Gravity on a lattice.**
T. Regge.
Supersymmetry, supergravity and related topics, p. 109 – 122 (1985). – See Abstr. 012.005.
The current revival of interest on field theories on a lattice has focused the attention on the problems presented by gravity on piecewise linear spaces. The metric tensor, the curvature, the Bianchi identities and the Einstein–Hilbert action can be all given a natural definition on a gravitational lattice. The use of this technique in quantum gravity is outlined in the present paper.

066.042 **On local conditions for multiple imaging by bounded, smooth gravitational lenses.**
K. Subramanian, S. A. Cowling.
Mon. Not. R. Astron. Soc., Vol. 219, No. 2, p. 333 – 346 (1986).
The conditions on the local value of the projected surface density, Σ, of a smooth and bounded gravitational lens, which determine whether such a lens is capable of producing multiple images are derived. It is shown that if Σ exceeds a critical value Σ_c at some point on such a lens then the lens can produce multiple images. The value of Σ_c depends only on fundamental constants and the distance scale of the lensing. The necessary conditions for lenses which have spherical symmetry or elliptical symmetry are given. It is also shown that there is no general lower bound on the value of Σ necessary for multiple imaging.

066.043 **Gravitational lenses: model fitting, time delay and Hubble parameter.**
R. Kayser.
Astron. Astrophys., Vol. 157, No. 1, p. 204 – 206 (1986).
Various formulations have been introduced for the calculations of the time delay between gravitational lensed images. The author shows here, that the concept of the cosmological correction function, originally introduced in the wavefront method, is generally applicable. Expressions are given for the time delay in all formulations.

066.044 **Fokker's type action at a distance theory of gravitation.**
A. Yu. Turygin.
Gen. Relativ. Gravitation, Vol. 18, No. 4, p. 333 – 348 (1986).
The theory of direct interparticle gravitational interaction is constructed, based on an action principle of the Fokker type, including many–particle interactions. The action is defined in a background space–time implying the requirement of a possibility to represent the equations of motion of a test particle in the form of geodesic equations in an effective Riemannian metric in any order of the gravitational constant k. In the first approximation the absorber theory of gravitational radiation is proved.

066.045 **Two–body solution of the Newman–Penrose asymptotic equations.**
A. Cresswell, R. L. Zimmerman.
Gen. Relativ. Gravitation, Vol. 18, No. 4, p. 349 – 366 (1986).
The Newman–Penrose nonlinear asymptotic field equations are separated in terms of spin weight spherical harmonics. As an example, the results are used to study the radiation effects on a two–body system. The presence of radiation is manifest through

the nonlinear terms in the asymptotic equations. If these terms are assumed to be small, the asymptotic equations can be formally solved by an iteration procedure. For the above example the first step of the iteration procedure is implemented to an accuracy that includes the effects of radiation up to octopole order.

066.046 **On the Anandan–Stodolsky effect.**
W. G. Unruh.
Gen. Relativ. Gravitation, Vol. 18, No. 4, p. 367 – 373 (1986).
By placing a force field in the vertical arms of the Colella–Overhauser–Werner interferometer, an additional phase shift, of order the mean potential for this force over mc^2 times the COW phase shift, can be obtained, which comes about because of the different velocity dependences of the gravitational and the extra forces.

066.047 **Classical unified fields: physical space.**
O. J. Bressan.
Gen. Relativ. Gravitation, Vol. 18, No. 4, p. 375 – 388 (1986).
This paper develops a general unified theory of classical mechanics and classical electromagnetism in a gravitational field on Friedman–Schöuten space–time (FSS). In this formalism (1) local equations of a charged fluid in an electromagnetic field are the same as in classical mechanics, (2) local equations for a moving charged fluid are the same as in electromagnetism, (3) the path of a charged particle under gravity and electromagnetism is a geodesic of the four–dimensional FSS, and (4) the strong equivalence principle and a nonzero torsion coexist without conflict.

066.048 **Algebraically special, nonaligned, non–null Einstein–Maxwell fields.**
J. B. Griffiths.
Gen. Relativ. Gravitation, Vol. 18, No. 4, p. 389 – 393 (1986).
A class of algebraically special non–null electrovac solutions is presented, in which the repeated principal null congruence is shear free, twist free, and geodesic, and is not aligned with a principal null direction of the electromagnetic field. This class includes all previously known solutions of this type as special cases. It can also be considered as a generalization of the Robinson–Trautman family of solutions.

066.049 **New exact solutions for a charged fluid sphere in general relativity.**
J. Hajj–Boutros, J. Sfeila.
Gen. Relativ. Gravitation, Vol. 18, No. 4, p. 395 – 410 (1986).
A new generation technique is elaborated in the case of static spherically symmetric distribution of a charged fluid. The technique deals only with a charged perfect fluid verifying a barytropic equation of state, i.e., $P = (\gamma-1)\varrho$. Many new exact solutions are generated from those of Pant and Sah, Banerjee and Santos, Humi and Mansour. Their physical properties are then studied in some detail.

066.050 **A partly reduced system of Hamiltonian equations and the initial value problem in the ECSK theory of gravity.**
K. Frackiewicz, W. Szczyrba.
Gen. Relativ. Gravitation, Vol. 18, No. 4, p. 411 – 431 (1986).
This paper discusses the initial value problem in the ECSK theory of gravity coupled to matter tensor fields. The analysis is performed for the phenomenological ECSK theory and for the ECSK theory coupled to: a covector matter field, the generalized Maxwell electrodynamics, and the generalized Fermi–Dirac electrodynamics. The Poisson brackets of the seven Hamiltonian constraints, which have to be satisfied by the canonical variables, are found. It is proved that they are of the first class.

066.051 **On exact radiative solutions representing finite sources.**
J. Bičák.
Galaxies, axisymmetric systems and relativity, p. 99 – 114 (1985). – See Abstr. 003.007.
The exact radiative solutions of Einstein's vacuum field equations representing finite sources and admitting global null infinity are reviewed. Robinson–Trautman solutions are first briefly mentioned. The boost–rotation symmetric solutions, representing uniformly accelerated objects, are then treated and their radiative properties are discussed.

066.052 **Proof of a generalized Geroch conjecture.**
I. Hauser, F. J Ernst.
Galaxies, axisymmetric systems and relativity, p. 115 – 125 (1985). – See Abstr. 003.007.
The authors extend to the case of N Abelian gauge fields interacting with gravitation an earlier proof that essentially all stationary axisymmetric vacuum fields can be obtained from the Minkowski space solution by means of Kinnersley–Chitre transformations.

066.053 **Limits of the double Kerr solution.**
C. Hoenselaers.
Galaxies, axisymmetric systems and relativity, p. 126 – 137 (1985). – See Abstr. 003.007.
It is shown that the so–called double Kerr solution contains the rotating version of the C metric as a particular limit. Another limiting process yields a generalization of the Kinnersley–Chitre solution.

066.054 **Non–inheritance of static symmetry by Maxwell fields.**
M. A. H. MacCallum, N. Van den Bergh.
Galaxies, axisymmetric systems and relativity, p. 138 – 148 (1985). – See Abstr. 003.007.
The authors present a class of cylindrically symmetric static metrics which satisfy the Einstein–Maxwell equations for a source–free electromagnetic field. The Maxwell field in these solutions has a sinusoidal time dependence. The solutions are given up to a set of three coupled ordinary differential equations of first order, which for a special case reduce to one second–order equation. The existence of these solutions is a counter–example to the conjecture that static electrovac solutions must contain static Maxwell fields.

066.055 **Stationary axisymmetric electrovacuum fields in general relativity.**
G. Neugebauer, D. Kramer.
Galaxies, axisymmetric systems and relativity, p. 149 – 165 (1985). – See Abstr. 003.007.
The SU(2,1) symmetry of the field equations gives rise to a linear problem for the stationary axisymmetric Einstein–Maxwell fields. The linear problems of three subcases (vacuum, electrostatics, pure electromagnetic fields) are studied. The solution generating technique is applied to endow the Kerr solution with a magnetic dipole moment.

066.056 **An almost conformal approach to axial symmetry.**
Z. Perjés.
Galaxies, axisymmetric systems and relativity, p. 166 – 181 (1985). – See Abstr. 003.007.
A new approach to stationary and axially symmetric gravitational fields in general relativity is outlined. It is based on the conformal properties of the 3–space of Killing trajectories. No smooth limit into Weyl's static metrics is allowed. Two degenerate classes of fields are excluded, and the Ernst potential with its complex conjugate are introduced as new coordinates. Two special cases, characterized by conformal flatness, and by Simon flatness, are solved explicitly, and a conformal potential for the general problem is shown to exist.

066.057 **Conformally stationary axisymmetric space–times.**
J. Winicour.
Galaxies, axisymmetric systems and relativity, p. 182 – 190 (1985). – See Abstr. 003.007.
This paper considers the possibility of conformally stationary axisymmetric vacuum space–times. Assuming that the symmetries are orthogonally transitive and excluding the type N case, it is shown that the space–time must be conformally static. The resulting equations are highly overdetermined and offer little hope of solutions useful for the description of radiation from isolated systems.

066.058 **A family of conformally flat space–times having the same curvature tensor in a given co–ordinate frame.**
C. D. Collinson.
Galaxies, axisymmetric systems and relativity, p. 193 – 198 (1985). – See Abstr. 003.007.
The algebraic consequences on the possible metric tensors of space–time with a curvature tensor, whose components are specified in some co–ordinate frame, are well known. In this paper an indication of the importance of the analytic conditions to be derived from the Bianchi identity is given and a family of conformally flat space–times is presented, each member of which has the same curvature tensor components.

066.059 **On the Bell–Szekeres solution for colliding electromagnetic waves.**
J. B. Griffiths.
Galaxies, axisymmetric systems and relativity, p. 199 – 207 (1985). – See Abstr. 003.007.
The Bell–Szekeres solution for the collision of plane electromagnetic waves is described and its structure is analysed. The solution is also extended to the case in which the polarization of the approaching electromagnetic waves is nonaligned.

066.060 **A remark on the Hauser metric.**
A. Held.
Galaxies, axisymmetric systems and relativity, p. 208 – 213 (1985). – See Abstr. 003.007.
In 1974, I. Hauser published a solution to the Einstein vacuum equations which was type N with a twisting principal null congruence. He began by making a general simplifying assumption and then found a particular case, leaving open the question as to whether or not there existed other solutions obeying his original restriction. It is shown that given that assumption, the Hauser solution is the only one possible.

066.061 **Numerical relativity by power series.**
R. Penrose.
Galaxies, axisymmetric systems and relativity, p. 214 – 217 (1985). – See Abstr. 003.007.

066.062 **Projective relativity and the equation of motion.**
E. Schmutzer.
Galaxies, axisymmetric systems and relativity, p. 218 – 236 (1985). – See Abstr. 003.007.
Starting from a new 5–dimensional unified projective field theory, the author investigates the equation of motion within the framework of this theory. He derives the explicit form of the 5–dimensional energy projector of an electrically charged perfect fluid and discusses the equation of motion of an electrically charged test particle.

066.063 **On generalised equations of geodesic deviation.**
B. F. Schutz.
Galaxies, axisymmetric systems and relativity, p. 237 – 246 (1985). – See Abstr. 003.007.
Motivated by applications to the motion of particles in gravitational waves, the author adopts a definition of geodesic deviation based upon a geodesic curve leaving a reference geodesic orthogonally and connecting it to a target geodesic. Working in Riemann normal coordinates to higher order, the author solves the geodesic equation and constructs the appropriate geodesics. In the case of nearly parallel geodesics the resulting deviation equation to second order is derived and discussed.

066.064 **Lobatchevski plane gravitational waves.**
S. T. C. Siklos.
Galaxies, axisymmetric systems and relativity, p. 247 – 274 (1985). – See Abstr. 003.007.
Space–times which admit a *Killing spinor* are investigated. It is found that all such space–times are conformal to *pp*–waves and are characterised by one arbitrary function of three variables. The Weyl tensor is type {4} and the Ricci tensor represents a negative cosmological constant plus a pure radiation field. When the radiation field vanishes, the resulting Einstein spaces are the *only* Einstein spaces (other than *pp*–waves) which are conformal to *pp*–waves.

066.065 **Perfect fluid and vacuum solutions of Einstein's field equations with flat 3–dimensional slices.**
H. Stephani, T. Wolf.
Galaxies, axisymmetric systems and relativity, p. 275 – 287 (1985). – See Abstr. 003.007.
A program for finding perfect fluid and vacuum solutions is proposed. It starts from the assumption that space–time admits a family of flat 3–slices, with special properties of their extrinsic curvature. The solutions which have an extrinsic curvature tensor proportional to the 3–dimensional metric tensor are characterized, and most of them are explicitly given. They contain metrics with three (or more) Killing vectors as well as solutions without any symmetry.

066.066 **Self–similar solutions of Einstein's equations.**
J. Wainwright.
Galaxies, axisymmetric systems and relativity, p. 288 – 300 (1985). – See Abstr. 003.007.
This paper considers self–similar perfect fluid solutions whose similarity group acts transitively on space–time. While exact self–similarity of this type is very restrictive, it is shown here that approximate or asymptotic self–similarity occurs in a variety of different situations. This means that the study of self–similar solutions can give insight into more general solutions.

066.067 **The real value of Mercury's perihelion advance.**
A. M. Nobili, C. M. Will.
Nature, Vol. 320, No. 6057, p. 39 – 41 (1986).
The perihelion advance of the orbit of Mercury has long been one of the observational cornerstones of general relativity. Radar determinations of Mercury's motion since the mid–1960s have produced agreement with general relativity at the level of 0.5%, or to ± 0.2 arc s per 100 yr. The authors have noted an apparent uncertainty in the value of the theoretical prediction for the advance within general relativity, as cited in various sources. The origin of this discrepancy is a 1947 article by Clemence, who used unconventional values for the astronomical unit and the speed of light in his calculation of the predicted advance. Here the authors derive the accurate value and uncover the source of Clemence's value.

066.068 **Gravitational microlensing at large optical depth.**
B. Paczyński.
Astrophys. J., Vol. 301, No. 2, p. 503 – 516 (1986).
A large number of numerical models of gravitational microlensing by stars in the lensing galaxy has been calculated, and properties of the models are described. The expected light intensity variations are more rapid when optical depth (i.e., the surface mass density expressed in proper units) to microlensing is large, but the time scale is a few years in the best cases, and much longer in a typical case. However, microlensing introduces considerable scatter, up to 2 or 3 orders of magnitude, to the intensity of macroimages expected at any given time, and this may considerably complicate the analysis of the observed lenses. It is shown that the macroimage is surrounded with a faint "halo" made of a large number of microimages from individual stars with average surface brightness falling off as R^{-4}.

066.069 **On the local geometry of rotating matter.**
R. J. Petti.
Gen. Relativ. Gravitation, Vol. 18, No. 5, p. 441 – 460 (1986).
This paper shows that general relativity and ordinary continuum models of matter imply the presence of Cartan torsion. The key concept is that torsion can be viewed as translational holonomy per unit area, in the limit of very small areas. Translational holonomy is introduced as the nonclosure of the "development" of a space–time loop into a flat space–time. The translational holonomy around a charged rotating black hole is calculated.

066.070 **Static spherically symmetric fields in the scalar–tetradic theory A.**
D. Sáez.
Gen. Relativ. Gravitation, Vol. 18, No. 5, p. 479 – 496 (1986).
The scalar–tetradic theory A has been recently formulated. In this paper it is proved that, for a suitable tetrad field and any given value of the coupling constant W of the theory, there is a unique static spherically symmetric solution of the field equations describing the outer field of a standard perfect fluid sphere. On account of the Birkhoff theorem this solution appears to have considerable physical interest.

066.071 **Two Kerr–NUT constituents in equilibrium.**
D. Kramer.
Gen. Relativ. Gravitation, Vol. 18, No. 5, p. 497 – 509 (1986).
The equilibrium conditions for the superposition of two Kerr–NUT solutions are revisited. The new derivation of these conditions leads to formulas which include also the hyperextreme case. In the symmetric model of two hyperextreme constituents the surface $f = 0$ of infinite red shift is investigated.

066.072 **Stationary vacuum fields with a conformally flat three–space. II. Proof of axial symmetry.**
Z. Perjés.
Gen. Relativ. Gravitation, Vol. 18, No. 5, p. 511 – 530 (1986).
The conjecture is proved that stationary vacuum space–times having a conformally flat three–space are axially symmetric. The proof uses the Ernst potential and the complex conjugate potential as independent coordinates. Two field equations: a combination of the Einstein equations and an integrability condition are algebraic in one of the field variables. Their coefficients, computed by employing a REDUCE program, separately vanish unless axial symmetry holds. Solution of the coefficient equations yields the proof of axial symmetry.

066.073 **Stationary vacuum fields with a conformally flat three–space. III. Complete solution.**
Z. Perjés.
Gen. Relativ. Gravitation, Vol. 18, No. 5, p. 531 – 547 (1986).
It is shown that stationary vacuum space–times having a conformally flat 3–space are always characterized by a functional relationship between the real and imaginary parts of the Ernst potential. This result is based on a new formulation of Einstein's equation, and it completes the investigation of vacuum space–times with conformally flat spaces.

066.074 **Geometrization of spin and the Weyssenhoff fluid conjecture.**
L. L. Smalley, J. R. Ray.
Gen. Relativ. Gravitation, Vol. 18, No. 5, p. 549 – 553 (1986).
It is shown that the logical treatment of spinning perfect fluids occurs in a metric–torsion space–time with or without mass conservation. The consequence is the geometrization of spin; that is the Weyssenhoff form relating spin and the trace–free torsion is obtained without any ad hoc assumption.

066.075 **Einstein gravity as a low–energy effective theory: comparison with weak and strong interactions.**
C. Sivaram.
Bull. Astron. Soc. India, Vol. 13, No. 4, p. 339 – 345 (1985).
Einstein's theory which correctly describes gravity at long distances (low energies) is first compared with Fermi's theory, which describes weak interactions at low energies. Analogous to strong interactions at high energies being described by the gauge invariant finite theory of quantum chromodynamics (QCD), gravity at high energies would be described by an asymptotically free–scale invariant theory. Then just as an effective theory of pions, describing low energy strong interactions, emerges from QCD at low energies, Einstein's theory would be the low energy effective counterpart of this gauge invariant high energy theory.

066.076 **On the distance measure for gravitational lenses.**
C. Alcock, N. Anderson.
Astrophys. J., Vol. 302, No. 1, p. 43 – 51 (1986).
The authors discuss some of the physical elements that are important for geometric optics in curved space–time, with a view toward interpreting data on gravitational lenses. The propagation of photons along long arcs is highly sensitive to the mean density in the beam, averaged along the arc. Fluctuations about this mean are not important unless their scale is comparable to that of the universe. Tidal distortion is important if only a few lumps are involved. The variation in the geometrical optics is considerably greater if there is a non–zero cosmological constant.

066.077 **On topologically massive hypergravity.**
C. Aragone, S. Deser.
From SU(3) to gravity, p. 327 – 339 (1985). – See Abstr. 003.014.
In three spacetime dimensions, Einstein gravity can be consistently coupled to a massless gauge spin–5/2 field; this hypergravity model has, however, neither spin nor dynamical content. Both are restored by addition of topological mass terms. It is shown here that the resulting coupled system is "almost" but not quite consistent: the Bianchi constraints on the spin 5/2 equations only involve lower rank components of the fields, unlike the $D = 4$ case.

066.078 **On the dynamics of the torsion of spacetime: exact solutions in a gauge theoretical model of gravity.**
P. Baekler, F. W. Hehl.
From SU(3) to gravity, p. 341 – 359 (1985). – See Abstr. 003.014.
The gauging of the Poincaré group yields a spacetime with both torsion and curvature. For illustrating the effects of a propagating torsion field, the authors pick a generic model Lagrangian of the type $(\text{torsion})^2/l^2 + (\text{curvature})^2/\varkappa$. Here l denotes the Planck length and $\varkappa$ the coupling constant of the hypothetical Lorentz gauge potential. For the corresponding field equations, exact classical solutions with electric charge are derived.

066.079 **Discrete Yang–Mills theories.**
T. Regge.
From SU(3) to gravity, p. 395 – 407 (1985). – See Abstr. 003.014.

066.080 **On plane waves and nullicles.**
I. Robinson.
From SU(3) to gravity, p. 409 – 422 (1985). – See Abstr. 003.014.
This paper discusses gravitational plane waves in an attempt to determine the gravitational field of a particle moving at the speed of light.

066.081 **Gravitational energy and bimetric general relativity.**
N. Rosen.
From SU(3) to gravity, p. 423 – 432 (1985). – See Abstr. 003.014.
With the help of the bimetric formalism one can obtain an energy–momentum density tensor for gravitation in general relativity.

066.082 **An alternative to general relativity.**
S. Weinberg.
From SU(3) to gravity, p. 441 – 452 (1985). – See Abstr. 003.014.
The author discusses the possibility that the theory of gravitation in $d > 4$ dimensions is based on some "tangent–space group" G_T of local transformations on tensor and spinor indices. This work is motivated partly by the problems encountered by more conventional "Kaluza–Klein" theories but even more by an old fascination with the question of whether special relativity is really an indispensable part of general relativity. This paper presents a Lagrangian for one particular alternative to general relativity in d dimensions.

066.083 **'Missing image' in gravitational lens systems?**
D. Narasimha, K. Subramanian, S. M. Chitre.
Nature, Vol. 321, No. 6065, p. 45 – 46 (1986).
Gravitational lens models involving extended mass distributions generally predict an odd number of images, with one of the images close to the centre of the principal lens galaxy. In all the lens systems observed up to now, only an even number of images

have been unambiguously detected. The authors suggest here that the presence of a compact nucleus with mass in the range 5×10^9 to $5 \times 10^{10} M_{\odot}$, contained within a region of size $\lesssim 50$ pc at the centre of a lens galaxy, would dim the odd image significantly without affecting the rest of the image configuration.

066.084 **Central charges in the canonical realization of asymptotic symmetries: an example from three dimensional gravity.**
J. D. Brown, M. Henneaux.
Commun. Math. Phys., Vol. 104, No. 2, p. 207 – 226 (1986).
It is shown that the global charges of a gauge theory may yield a nontrivial central extension of the asymptotic symmetry algebra already at the classical level. This is done by studying three dimensional gravity with a negative cosmological constant. The asymptotic symmetry group in that case is either $R \times SO(2)$ or the pseudo–conformal group in two dimensions, depending on the boundary conditions adopted at spatial infinity. In the latter situation, a nontrivial central charge appears in the algebra of the canonical generators, which turns out to be just the Virasoro central charge.

066.085 **On the stabilizing role of the envelope of a relativistic star.**
A. A. Mikhal'chuk.
Red. zh. Izv. vuzov. Fiz. Tomsk, 15 pp. (1985). In Russian. Abstr. in Ref. Zh., 51. Astron., 2.51.193 (1986).

066.086 **Internal gravitational stability of cool relativistic stars in variational approach.**
A. A. Mikhal'chuk.
Red. zh. Izv. vuzov. Fiz. Tomsk, 11 pp. (1985). In Russian. Abstr. in Ref. Zh., 51. Astron., 2.51.194 (1986).

066.087 **Stability of nuclei of relativistic stars and selection of the function of the radial shift.**
A. A. Mikhal'chuk.
Red. zh. Izv. vuzov. Fiz. Tomsk, 13 pp. (1985). In Russian. Abstr. in Ref. Zh., 51. Astron., 2.51.195 (1986).

066.088 **Testing relativity with DI Herculis.**
D. H. Smith.
Sky Telesc., Vol. 71, No. 3, p. 236 (1986).

066.089 **On the theory of gravitation taking into account the role of vacuum. I.**
L. Sh. Grigoryan, G. S. Saakyan.
Astrofizika, Tom 24, Vyp. 2, p. 387 – 395 (1986). In Russian. English translation in Astrophysics, Vol. 24, No. 2.
The paper develops the idea of a definite field of vacuum tensions appearing due to the violation of its symmetry properties caused by gravitation. The gravitational "constant" is assumed to be defined by the local state of vacuum. Equations describing the gravitational field and the mass distribution are deduced.

066.090 **A search for sinusoidal and burst VLF gravitational waves using Pioneer 10 and 11 spacecraft tracking data.**
J. W. Armstrong, F. B. Estabrook, H. D. Wahlquist.
Bull. Am. Astron. Soc., Vol. 17, No. 4, p. 872 (1985). Abstract. – See Abstr. 010.062.

066.091 **An astrometric program to define coordinate frames for general relativity experiments.**
A. Buffington.
Bull. Am. Astron. Soc., Vol. 17, No. 4, p. 898 (1985). Abstract. – See Abstr. 010.062.

066.092 **Diffraction in gravitational lensing for compact objects of low mass.**
S. Deguchi, W. D. Watson.
Bull. Am. Astron. Soc., Vol. 17, No. 4, p. 907 (1985). Abstract. – See Abstr. 010.062.

066.093 **The Curzon singularity. I: Spatial sections.**
S. M. Scott, P. Szekeres.
Gen. Relativ. Gravitation, Vol. 18, No. 6, p. 557 – 570 (1986).
Some new properties of geodesics and other curves lying in the spatial sections t = const. of the Curzon solution are derived. These are shown to allow one to build up a new coordinate system in which the singularity appears unambiguously as a ring.

066.094 **The Curzon singularity. II: Global picture.**
S. M. Scott, P. Szekeres.
Gen. Relativ. Gravitation, Vol. 18, No. 6, p. 571 – 583 (1986).
A new coordinate system is presented for the Curzon metric such that a class of incomplete geodesics are extendible both in the past and the future into regions of flat space. All other incomplete geodesics terminate at the (infinite curvature) singularity, which appears as a ring with finite radius and is naked.

066.095 **Note on equivalence of cosmological space–times.**
R. J. Torrence, W. E. Couch.
Gen. Relativ. Gravitation, Vol. 18, No. 6, p. 585 – 589 (1986).
It is shown that only de Sitter space and Minkowski space can be expressed in more than one of the standard forms for Robertson–Walker space–times.

066.096 **A charged generalization of Florides' interior Schwarzschild solution.**
Ø. Grøn.
Gen. Relativ. Gravitation, Vol. 18, No. 6, p. 591 – 596 (1986).
The new class of interior Schwarzschild solutions found by Florides is generalized to the charged case. A particular solution within this class is found, which represents an electromagnetic mass–model of a neutral spherically symmetric system. The pressure is isotropic, decreasing monotonously with increasing radius and vanishes at the surface of the matter distribution.

066.097 **The gravitational anomaly: an elementary, coordinate–space approach.**
S. A. Fulling.
Gen. Relativ. Gravitation, Vol. 18, No. 6, p. 609 – 615 (1986).
Alvarez–Gaumé and Witten have shown that energy–momentum conservation must be violated in certain parity–violating quantum field theories involving gravity. In two dimensions this effect can be studied without the aid of Feynman diagrams or calculations in momentum space. It is shown that if matter is created at all by the gravitational field, it must appear in both left–moving and right–moving modes if the conservation law is to hold always.

066.098 **Stationary axisymmetric solutions of three–dimensional Einstein gravity.**
S. Deser, B. Laurent.
Gen. Relativ. Gravitation, Vol. 18, No. 6, p. 617 – 632 (1986).
Einstein theory in three space–time dimensions is studied in the stationary, rotationally symmetric regime. Explicit interior solutions for various types of matter sources are found and their properties are discussed.

066.099 **"Infrared" Maxwell fields.**
S. Chakravarty, J. Ivancovich, E. T. Newman.
Gen. Relativ. Gravitation, Vol. 18, No. 6, p. 633 – 640 (1986).
This paper studies the analytic behavior of a class, the so–called infrared sector, of globally regular vacuum Maxwell fields in the neighborhood of the boundary of compactified Minkowski space.

066.100 **Nonstatic analogues of Kohler–Chao solution of imbedding class one.**
Y. K. Gupta, R. S. Gupta.
Gen. Relativ. Gravitation, Vol. 18, No. 6, p. 641 – 648 (1986).
The nonstatic analogues of the Kohler–Chao perfect fluid solution have been derived starting with a spherically symmetric flat V_5. It is also established that all the Petrov type D perfect fluid solutions of imbedding class one do not possess a pressure–free surface at a finite radius and therefore cannot be fitted to the external Schwarzschild's model.

066.101 **Irrotational and conformally Ricci–flat perfect fluids.**
N. Van den Bergh.
Gen. Relativ. Gravitation, Vol. 18, No. 6, p. 649 – 668 (1986).
This paper investigates irrotational perfect fluids in a conformally Ricci–flat space–time. Three cases are analyzed separately, according to whether the gradient of the conformal scalar field is (1) aligned with, (2) orthogonal to the fluid velocity, or (3) forms an arbitrary (and position–dependent) angle with the latter.

066.102 **How important is gravitational–wave astronomy? Part I.**
P. Kafka.
Naturwissenschaften, 73. Jahrg., Heft 5, p. 248 – 257 (1986). In German.
The gravitational–wave research group at the Max–Planck–Institute for Quantum Optics in Garching has presented a detailed proposal for a laser interferometer with 3 km arm–length in order to measure relative changes of length of about 10^{-21} and perhaps detect cosmic collapse events from a few thousand neighbouring galaxies. The paper reviews the physics and astrophysics of gravitational radiation in the context of the proposed big antennae. Part I treats general properties of the waves and the antenna sensitivity. For Part II see Abstr. 066.205.

066.103 **Timing observations of the pulsar signal propagations in the galactic gravitational field as precision tests of the Einstein Equivalence Principle.**
W.–T. Ni.
Proceedings of the Second Asian–Pacific Regional Meeting on Astronomy, p. 441 – 448 (1984). – See Abstr. 012.031.
If Einstein Equivalence Principle (EEP) is observed, photons with different polarizations as test particles shall follow identical trajectories in a gravitational field. In pulsar observations, the pulses and micropulses with different polarizations are correlated in general structure and timing. Due to precise timing and rich polarization data, pulsar signal propagations in galactic gravitational field verify this aspect of EEP to an accuracy of 10^{-8} to 10^{-10}. The significance of these timing tests and their relation to other test of EEP are studied in a general phenomenological framework.

066.104 **Linking Maxwell and Einstein through the Large Numbers Hypothesis.**
W. Davidson.
Proceedings of the Second Asian–Pacific Regional Meeting on Astronomy, p. 460 – 466 (1984). – See Abstr. 012.031.
Dirac has deduced an 'atomic' space–time metric A for the large scale universe in which conventional atomic/electromagnetic theory holds but gravity weakens, so that the gravitational constant $G \propto t^{-1}$, where t is atomic time. He suggests that conventional (Einsteinian) gravitational theory (in which G is constant) holds in another space–time metric E, in which the time $\tau \propto t^2$. In this paper the A and E metrics are deduced directly from the Large Number Hypothesis, and several implications of the theory that may be tested in principle are indicated.

066.105 **On the Lichnerowicz theory of gravitational shock waves.**
J. Ibrahim, K. Supelli.
Proceedings of the Second Asian–Pacific Regional Meeting on Astronomy, p. 482 (1984). Abstract. – See Abstr. 012.031.

066.106 **Mikro–Gravitationslinsen und ihre Bedeutung für die Astrophysik.**
R. Kayser.
Sterne Weltraum, 25. Jahrg., Nr. 5, p. 252 – 254 (1986).

066.107 **Gravitational light deflection effects in eclipsing binary pulsars.**
P. W. Gorham.
Astrophys. J., Vol. 303, No. 2, p. 601 – 604 (1986).
A study of the general relativistic deflection of light in an idealized eclipsing binary pulsar system is performed, which indicates that the pulse train of the pulsar is differentially delayed through the course of the orbit in a way distinct from other observed delays. The effect is calculated for the Hercules X–1 – HZ Herculis system. It is shown to be marginally detectable at present for Her X–1, and may possibly be more evident in other eclipsing X–ray binary pulsars.

066.108 **A class of internal spherically–symmetrical statistical solutions of gravity equations. I. Construction of an analytical solution.**
A. M. Baranov, Yu. N. Vlasenko.
Krasnoyar. univ. Krasnoyarsk, 10 pp. (1985). In Russian. Abstr. in Ref. Zh., 51. Astron., 5.51.214 (1986).

066.109 **New soliton–like exact solutions of general relativity equations modelling rotating supermassive astrophysical objects.**
V. G. Pisarenko, A. N. Krishtal', Yu. A. Selivanov.
Probl. nelinejn. i turbulent. protsessov v fiz. Tr. 2 Mezhdunar. rab. gruppy, Kiev, 1983. Chast' 2. Kiev, p. 320 – 322 (1985). In Russian. Abstr. in Ref. Zh., 51. Astron., 5.51.219 (1986).

066.110 **The time of revolution of a satellite around a planet in the relativistic theory of gravitation and in general relativity and the possibility of experimental testing of delay due to gravitational interaction.**
A. A. Logunov, Yu. M. Loskutov.
Inst. fiz. vysok. ehnergij. Serpukhov., Prepr. No. 187, 9 pp. (1985). In Russian. Abstr. in Ref. Zh., 51. Astron., 5.51.230 (1986).

066.111 **Phenomenological formalism in the relativistic theory of gravitation and observations.**
V. I. Zhdanov.
Metrologiya, No. 10, p. 3 – 14 (1985). In Russian. Abstr. in Ref. Zh., 51. Astron., 5.51.231 (1986).

066.112 **Observations of new gravitational effects.**
G. I. Kuz'menko.
Odes. inst. inzh. mor. flota. Odessa, 9 pp. (1985). In Russian. Abstr. in Ref. Zh., 51. Astron., 5.51.232 (1986).

066.113 **A two–dimensional model for relativistic gas jets.**
K. W. Thompson.
Diss. Abstr. Int., Sect. B, Vol. 46, No. 7, p. 2347 (1986). Thesis, Princeton University, 233 pp. (1985). Order No. DA8520752.

066.114 **The equation of state and general relativity in supernovae.**
E. A. Baron.
Diss. Abstr. Int., Sect. B, Vol. 46, No. 10, p. 3480 (1986). Thesis, State University of New York at Stony Brook, 127 pp. (1985). Order No. DA8527931.

066.115 **Electron–positron pair production by ultrarelativistic electrons in a soft photon field.**
A. Mastichiadis.
Diss. Abstr. Int., Sect. B, Vol. 46, No. 11, p. 3882 (1986). Thesis, Boston University, 92 pp. (1985). Order No. DA8601362.

066.116 **Erratum: "Gravitational–wave emission from rotating gravitational collapse" [Phys. Rev. Lett., Vol. 55, No. 8, p. 891 – 894 (1985)].**
R. F. Stark, T. Piran.
Phys. Rev. Lett., Vol. 56, No. 1, p. 97 (1986). See Abstr. 40.066.125.

066.117 **Measurement of the Lense–Thirring drag on high–altitude, laser–ranged artificial satellites.**
I. Ciufolini.
Phys. Rev. Lett., Vol. 56, No. 4, p. 278 – 281 (1986).
The author describes a new method of measuring the Lense–Thirring relativistic nodal drag using LAGEOS together with another similar high–altitude, laser–ranged satellite with appropriately chosen orbital parameters. He proposes, for this purpose, that a future satellite such as LAGEOS II have an inclination supplementary to that of LAGEOS. The experiment

proposed would provide a method for experimental verification of the general relativistic formulation of Mach's principle and measurement of the gravitomagnetic field.

066.118 **Must nonspherical collapse produce black holes? A gravitational confinement theorem.**
W. Israel.
Phys. Rev. Lett., Vol. 56, No. 8, p. 789 – 791 (1986).
It is shown that a trapped surface can be extended, in a locally area–preserving fashion, to a three–cylinder that is everywhere spacelike, provided it encounters no singularities in its development. Thus the interior of a regular, area–preserving, initially trapped surface is permanently sealed off from causal influence on the exterior world.

066.119 **Gravitational antennas and the search for gravitational radiation.**
J. Weber.
Sir Arthur Eddington Centenary Symposium, Vol. 3: Gravitational radiation and relativity, p. 1 – 77 (1986). – See Abstr. 012.036.
Methods are described and compared for observing gravitational radiation. The gravitational radiation antenna measures the spacetime curvature. Classical and quantum theories of the antenna are presented. Antennas in thermal equilibrium exchange energy with gravitational radiation at a rate proportional to the square of the Riemann tensor. It is shown that the preparation of antennas in correlated quantum states results in energy exchange linear in the Riemann tensor. A review of all experiments attempting to observe gravitational radiation is presented.

066.120 **Nonstatic massless scalar field in the theory of gravitation.**
V. S. Dorodnykh.
Sir Arthur Eddington Centenary Symposium, Vol. 3: Gravitational radiation and relativity, p. 136 – 138 (1986). – See Abstr. 012.036.
Two classes of exact solutions for self–consistent system of equations describing spherically symmetric gravitation and massless scalar field are obtained. The solutions of the first class correspond to the case, when only massless scalar matter is present. The solutions of the other class take into account the sources of scalar field on a spherical shell.

066.121 **Estimation of the gravitational radiation (GR) from the pulsar PSR 1937 + 214.**
R.–x. Qin, D.–n. Mao.
Sir Arthur Eddington Centenary Symposium, Vol. 3: Gravitational radiation and relativity, p. 292 – 297 (1986). – See Abstr. 012.036.
An analysis of the pulsar PSR 1937 + 214 using the tilted rotator model is presented. It is estimated that the gravitational radiation power of the pulsar is $\sim 1.5 \times 10^{36}$erg s^{-1} and that the dimensionless stress in a GR detector on Earth is of the order 10^{-29} to 10^{-27}.

066.122 **On the correlation between six–dimensional and four–dimensional coordinate transformations.**
M. T. Teli.
Sir Arthur Eddington Centenary Symposium, Vol. 3: Gravitational radiation and relativity, p. 375 – 385 (1986). – See Abstr. 012.036.
It is shown that Cole's approach of parallel time vectors in a frame is not sufficient to reproduce Lorentz transformations from six–dimensional coordinate transformations. The author obtains this correlation by introducing the subspace concept in which the time direction of moving frame is chosen as the preferred time direction.

066.123 **On the electromagnetic and gravitational radiation in general relativity.**
Z. Ahsan.
Sir Arthur Eddington Centenary Symposium, Vol. 3: Gravitational radiation and relativity, p. 395 – 400 (1986). – See Abstr. 012.036.

066.124 **The consistency of theory of developed Newtonian gravitation (TDNG) partially alias TETG in respect of the attenuation.**
I. A. Adămuti.
Sir Arthur Eddington Centenary Symposium, Vol. 3: Gravitational radiation and relativity, p. 401 – 417 (1986). – See Abstr. 012.036.

066.125 **Model systems of fields in the theory of gravitation and the problems of cosmology.**
V. N. Mel'nikov, K. P. Stanukovich (*K. P. Stanyukovich*).
Sir Arthur Eddington Centenary Symposium, Vol. 3: Gravitational radiation and relativity, p. 418 – 430 (1986). – See Abstr. 012.036.
Exact solutions of classical and quantum fields in GR and quantum cosmology are obtained and applied to the solution of fundamental problems of cosmology. The problem of variations of fundamental constants and some relations among basic parameters of fundamental interactions are also discussed.

066.126 **Some structural considerations on the theory of gravitational field in Finsler spaces – II.**
S. Ikeda.
Sir Arthur Eddington Centenary Symposium, Vol. 3: Gravitational radiation and relativity, p. 437 – 451 (1986). – See Abstr. 012.036.
New field equations are proposed for a Finslerian gravitational field with a non–Cartan connection and the intrinsic behaviour of the internal vector is discussed from the standpoint of the theory of tangent bundles.

066.127 **General–relativistic equations of binary motion for extended bodies, with conservative corrections and radiation damping.**
S. M. Kopejkin.
Sov. Astron., Vol. 29, No. 5, p. 516 – 524 (1985). English translation of 40.066.087.

066.128 **Doppler shifting of a distant light source in a Schwarzschild gravitational field.**
M. D. Kislik.
Sov. Astron., Vol. 29, No. 5, p. 524 – 528 (1985). English translation of 40.066.088.

066.129 **An attempt for unification of microcosmos and mega-(macro)cosmos on the base of the order (causality).**
A. Kartsaklis.
Astrophys. Space Sci., Vol. 122, No. 2, p. 201 – 211 (1986).
The author gives an axiomatic system of causal space and its topological structure, that is convenient for many kinds of spacetime frame reference. He gives also a topology of the pure states, according to axiomatic foundation of quantum mechanics by J. Jauch. Finally, he attempts to unify, from a causality point of view, the microcosmos and mega(macro)cosmos supported by the transition probability.

066.130 **Comment on the GRT–LRS Bianchi type V solutions.**
S. R. Roy, J. P. Singh.
Astrophys. Space Sci., Vol. 122, No. 2, p. 395 – 396 (1986).
The points raised by Lorenz–Petzold (1985) on the authors' paper (1983) are discussed. It is shown that the conclusions drawn by him are not substantiated.

066.131 **Superluminal separation caused by gravitational lens' effect of neutrino objects.**
S.–p. Xiang, R. Ruffini.
Astrophys. Space Sci., Vol. 123, No. 1, p. 1 – 4 (1986).
Neutrino objects can cause the apparent superluminal separation in quasars as gravitational lenses. The authors analyse the possibility and calculate the differential bending for different equilibrium configurations of neutrino objects.

066.132 **Astronomical tests in a Kerr field.**
K. D. Krori, M. Barua.
Astrophys. Space Sci., Vol. 123, No. 1, p. 21 – 29 (1986).
The authors calculate the astronomical effects that accrue in the equatorial plane of a stellar object due to rotation according to Kerr theory.

066.133 **Gravitational instability in the presence of suspended particles and variable magnetic field.**
R. C. Sharma, J. N. Misra.
Astrophys. Space Sci., Vol. 123, No. 2, p. 375 – 380 (1986).
The gravitational instability of an infinite homogeneous self–gravitating and infinitely conducting gas–particle medium is considered in the presence of suspended particles and a variable horizontal magnetic field varying in vertical direction. It is found that the Jeans's criterion of instability remains unaffected even if the effects due to suspended particles and variable horizontal magnetic field are included.

066.134 **Random gravitational lensing.**
R. J. Nemiroff.
Astrophys. Space Sci., Vol. 123, No. 2, p. 381 – 387 (1986).
A simple three–dimensional approach to calculate the probability of observing gravitational lensing, caused by a chance superposition of lens and source, is presented. The formalism is then used to estimate the lensing probabilities in a number of astrophysical situations. As indicated in earlier works, the author finds that the chance of random gravitational lensing being detected locally is small, and the chance at extra–galactic distances is large.

066.135 **Comment on the big–bang singularity in the scale–covariant theory.**
A. Beesham.
Astrophys. Space Sci., Vol. 123, No. 2, p. 405 – 407 (1986).
In a recently published article, Rahman and Banerji (1985) have shown that certain Friedman–Lemaître–Robertson–Walker cosmological models with zero cosmological constant in the scale covariant theory of Canuto et al. (1977) do not have an observable singularity of infinite density in the beginning. This is in contrast to the corresponding general relativistic models, each of which has a big–bang singularity. In this note, the author shows that the cosmological models in the scale–covariant theory which have been considered as possible viable models of the Universe by Canuto and his co–workers each possesses a singularity where the energy density diverges.

066.136 **Gravitational microlensing by the galactic halo.**
B. Paczyński.
Astrophys. J., Vol. 304, No. 1, p. 1 – 5 (1986).
The massive halo of our Galaxy has an optical depth to gravitational microlensing $\tau \approx 10^{-6}$. If the halo is made of objects more massive than $\sim 10^{-8} M_{\odot}$, then any star in a nearby galaxy has a probability of 10^{-6} to be strongly microlensed at any time. Monitoring the brightness of a few million stars in the Magellanic Clouds over a time scale between 2 hr and 2 yr may lead to a discovery of "dark halo" objects in the mass range $10^{-6} - 10^{2} M_{\odot}$ or it may put strong upper limits on the number of such objects.

066.137 **Relativistic theory of gravitation.**
A. A. Logunov, M. A. Mestvirishvili.
Proceedings of the Third Seminar on Quantum Gravity, p. 39 – 71 (1985). – See Abstr. 012.038.
A relativistic theory of gravitation (RTG) is constructed in a unique way on the basis of special relativity and a geometrization principle. In this, a gravitational field is treated as the Faraday–Maxwell spin–2 and spin–0 physical field possessing energy and momentum. The source of a gravitational field is the total conserved energy–momentum tensor of matter and of a gravitational field in Minkowski space. RTG leads to a strong prediction: the universe is not closed but just "flat".

066.138 **Quantized strings and curved space–time.**
E. S. Fradkin, A. A. Tseytlin.
Proceedings of the Third Seminar on Quantum Gravity, p. 72 – 102 (1985). – See Abstr. 012.038.
The authors develop a field–theoretic description of closed Bose string theory giving a covariant definition of the effective action Γ for fields corresponding to different string "excitations". The formalism makes possible a self–consistent study of the ground state problem (including the problem of space–time compactification) in the quantum string theory. The authors suggest a solution to the old "tachyon problem" based on generation of non–trivial vacuum values for the lowest–mass "excitations".

066.139 **Dynamical suppression of topological change.**
B. DeWitt.
Proceedings of the Third Seminar on Quantum Gravity, p. 103 – 122 (1985). – See Abstr. 012.038.
This paper reports on some work concerning spontaneous quantum fluctuations of 3–space topology.

066.140 **Simplicial quantum gravity.**
J. B. Hartle.
Proceedings of the Third Seminar on Quantum Gravity, p. 123 – 140 (1985). – See Abstr. 012.038.
Simplicial approximation and the ideas associated with the Regge calculus provide a concrete way of implementing a sum over histories formulation of quantum gravity. A simplicial geometry is made up to flat simplices joined together in a prescribed way together with an assignment of lengths to their edges. The construction of the simplicial Euclidean action for this approach to quantum general relativity is illustrated.

066.141 **The generalized Schwinger–DeWitt technique and the unique effective action in quantum gravity.**
A. O. Barvinsky (*A. O. Barvinskij*), G. A. Vilkovisky (*Eh. Ya. Vil'koviskij*).
Proceedings of the Third Seminar on Quantum Gravity, p. 141 – 160 (1985). – See Abstr. 012.038.

066.142 **Conformal anomaly for gravitons and the coupled gravity–matter excitations.**
L. P. Grishchuk, A. D. Popova.
Proceedings of the Third Seminar on Quantum Gravity, p. 161 – 164 (1985). – See Abstr. 012.038.
The conformal anomaly for gravity and the coupled gravity–matter excitations is computed in general form for an arbitrary background space–time.

066.143 **Anomalies, cohomology and generalized secondary classes.**
H.–y. Guo, K. Wu, B.–y. Hou, S.–k. Wang.
Proceedings of the Third Seminar on Quantum Gravity, p. 165 – 187 (1985). – See Abstr. 012.038.
The authors discuss recent work on the generalized secondary characteristic classes, the cohomologies of gauge groups realized upon these classes and the degenerate forms of these classes, as well as their applications to the analysis on both gauge and gravitational anomalies in spacetimes of different dimensions.

066.144 **On quantum theory of measurements of gravitational field.**
M. B. Mensky.
Proceedings of the Third Seminar on Quantum Gravity, p. 188 – 204 (1985). – See Abstr. 012.038.
Quantum measurement theory for the gravitational field is formulated on the basis of Feynman integrals in finite limits. An application of this approach to the local measurement of curvature gives absolute limits on its precision and localization.

066.145 **The concept of gravitons and the measurement of effects of quantum gravity.**
H.–H. von Borzeszkowski.
Proceedings of the Third Seminar on Quantum Gravity, p. 205 – 222 (1985). – See Abstr. 012.038.

It is demonstrated, first, that there arise limitations in quantum gravity which show that general relativity theory (GRT) is essentially classical and, second, that the graviton language is accordingly only useful in certain approximations to GRT, in order to harmonise GRT and the quantum theory of matter.

066.146 **Classical and quantum pregeometry.**
D.-E. Liebscher.
Proceedings of the Third Seminar on Quantum Gravity, p. 223 – 235 (1985). – See Abstr. 012.038.
The program for theories explaining the metric structure of space–time as dynamical consequence of a pregeometric structure is considered. It is shown that an affine space allows the definition of a metric by dynamics. The character of small deviations from a classical metric structure is explained.

066.147 **Thermodynamics and field statistics of a relativistic superfluid.**
W. Israel.
Proceedings of the Third Seminar on Quantum Gravity, p. 246 – 269 (1985). – See Abstr. 012.038.
This paper outlines the covariant (nonequilibrium) phenomenological theory of a relativistic superfluid, and analyzes its microscopic justification by studying the statistical thermodynamics of the superfluid in the case of thermal equilibrium, using a Euclidean path–integral formalism.

066.148 **Quantum gravity effects in particle physics and cosmology.**
V. A. Kuz'min.
Proceedings of the Third Seminar on Quantum Gravity, p. 270 – 288 (1985). – See Abstr. 012.038.
The author estimates possible violations, at Planck scales, of baryon number conservation, CPT and Pauli principle. It is found that in some cases these violations, although small, are within the possibility of observation with present experimental techniques. It is shown that the observed baryon asymmetry of the universe could be explained by the possible CPT violation due to quantum gravity.

066.149 **The use of finite energy sum rules for the calculation of the induced gravitational constant.**
N. V. Krasnikov, A. A. Pivovarov.
Proceedings of the Third Seminar on Quantum Gravity, p. 289 – 299 (1985). – See Abstr. 012.038.
The authors use the finite energy sum rules for the calculation of the induced gravitational constant in gauge theories of the QCD type. The induced gravitational constant is determined by the gluon vacuum condensate.

066.150 **Improved characterization of the Kerr metric.**
Z. Perjés.
Proceedings of the Third Seminar on Quantum Gravity, p. 446 – 461 (1985). – See Abstr. 012.038.
The condition of asymptotic flatness is removed from Simon's characterization of the Kerr metric by vanishing of the complexified Bach tensor. The solution of the stationary vacuum equations of relativity is given for a vanishing Simon tensor.

066.151 **Casimir effect in space–times with non–euclidean topology.**
S. G. Mamayev (*S. G. Mamaev*), V. M. Mostepanenko.
Proceedings of the Third Seminar on Quantum Gravity, p. 462 – 478 (1985). – See Abstr. 012.038.
The non–trivial topology of space–time leads to the vacuum polarization of quantized fields characterized by non–zero vacuum expectation values of the stress–energy tensor. The authors present an approach to its computation which allows an analytical investigation of the dependence of the effect upon topology and geometrical parameters of the manifold. A new interpretation for the effective vacuum temperature is also given.

066.152 **An inertial interpretation of acceleration radiation.**
R. M. Wald.
Proceedings of the Third Seminar on Quantum Gravity, p. 479 – 485 (1985). – See Abstr. 012.038.
As shown by Unruh in 1976, if a quantum field is in its vacuum state in Minkowski spacetime, an accelerating observer would feel himself immersed in a thermal bath of quanta of the field and a particle detector he carries would get excited. How does an inertial observer interpret this "detection" process, in view of the fact that according to him, no quanta of the field were present initially? It is shown that for a simple model detector the absorption by the detector of a "Rindler particle" corresponds in the inertial viewpoint to emission by the detector of an ordinary Minkowski particle.

066.153 **Quantum field theory for a general class of accelerated observers.**
N. Sanchez.
Proceedings of the Third Seminar on Quantum Gravity, p. 486 – 495 (1985). – See Abstr. 012.038.
The author formulates QFT in a wide class of accelerated coordinates in four dimensions. This generalizes the approach given previously by the author in terms of holomorphic (and/or anti–holomorphic) mappings. A characterization of global and asymptotic thermal equilibrium situations is given, and the quantum detection and the isotropy of thermal radiation are discussed.

066.154 **On the vacuum definition in curved space–time.**
M. Castagnino.
Proceedings of the Third Seminar on Quantum Gravity, p. 496 – 506 (1985). – See Abstr. 012.038.
Two new definitions (strong and minimal vacua) for the vacuum in Robertson–Walker universe are introduced and their physical consequences are given.

066.155 **Tunneling transitions with gravitation: breaking of the quasiclassical approximation.**
G. V. Lavrelashvili, V. A. Rubakov, P. G. Tinyakov.
Proceedings of the Third Seminar on Quantum Gravity, p. 628 – 661 (1985). – See Abstr. 012.038.
The authors discuss the problem of the validity of the quasiclassical approximation for the description of tunneling transitions with gravitational effects taken into account. They present two model examples in which quasiclassics is not applicable. The first example is the closed homogeneous and isotropic universe with positive cosmological constant undergoing the transition from the Friedmann regime to de Sitter–like one. The second example is the transition leading to the materialization of the bubble of a new phase in the metastable vacuum.

066.156 **An elementary course on general relativity.**
T. Regge.
Vasiona, Année 33, No. 3, p. 41 – 62 (1985). In Croatian.
Contents: Historical introduction. The equivalence principle. The field equations and the gravitational potential. The classical tests for general relativity. Cosmology. Beyond relativity.

066.157 **The new formula for gravitational radiation energy loss: applications to the axisymmetric two–body problem and the binary pulsar.**
F. I. Cooperstock, P. H. Lim.
Astrophys. J., Vol. 304, No. 2, p. 671 – 681 (1986).
By accounting for nonlinearities and applying the Gauss theorem for retarded functions, the authors develop the formula for gravitational radiation energy loss in general relativity. This replaces the quadrupole formula, while clarifying the evident minimal domain of validity of the quadrupole formula. The new formula is considered in detail for the axisymmetric two–body problem with conventional equation of state, $\varepsilon = \varepsilon(P)$. An order–of–magnitude analysis is performed for the binary pulsar.

066.158 **Considerations regarding a Space–Shuttle measurement of the gravitational constant.**
Y. Avron, M. Livio.
Astrophys. J., Lett. Ed., Vol. 304, No. 2, p. L61 – L64 (1986).
The authors propose to measure the gravitational constant G by observing, in space, the orbit of two solid balls of known masses. Such an experiment can be made on board the Space Shuttle. It is a relatively simple experiment which has the potential for high accuracy. The use of differential materials can also serve to test the recently proposed existence of an intermediate–range coupling to hypercharge or baryon number.

066.159 **Gravitational radiation from dust.**
R. A. Isaacson.
J. Math. Phys., Vol. 26, No. 11, p. 2859 – 2868 (1985). Abstr. in Phys. Abstr., Vol. 89, No. 1273, Entry 13473 (1986).

066.160 **The gravitational arrow of time.**
W. B. Bonnor.
Phys. Lett. A, Vol. 112A, No. 1 – 2, p. 26 – 28 (1985). Abstr. in Phys. Abstr., Vol. 89, No. 1273, Entry 13488 (1986).

066.161 **Symmetry breaking, phase transition and gravity.**
K. P. Sinha.
Pramāna, Vol. 25, No. 4, p. 343 – 347 (1985). Abstr. in Phys. Abstr., Vol. 89, No. 1273, Entry 13489 (1986).

066.162 **Considerations about the apparent "superluminal expansions" in astrophysics.**
E. Recami, A. Castellino, G. D. Maccarrone, M. Rodonò.
Report INFN/AE–85/3, Ist. Naz. Fis. Nucl., Catania, Italy, 21 pp. (1985). Abstr. in Phys. Abstr., Vol. 89, No. 1273, Entry 18658 (1986).

066.163 **Relation between the gravitational and the cosmological constant.**
F. Tomasek.
Lett. Nuovo Cimento, Vol. 44, Ser. 2, No. 4, p. 241 – 245 (1985). Abstr. in Phys. Abstr., Vol. 89, No. 1273, Entry 18971 (1986).

066.164 **The search for gravitational waves.**
S. L. Shapiro, R. F. Stark, S. A. Teukolsky.
Am. Sci., Vol. 73, No. 3, p. 248 – 257 (1985). Abstr. in Phys. Abstr., Vol. 89, No. 1274, Entry 19142 (1986).

066.165 **On the collision of impulsive gravitational waves when coupled with fluid motions.**
S. Chandrasekhar, B. C. Xanthopoulos.
Proc. R. Soc. London, Ser. A, Vol. 402, No. 1822, p. 37 – 65 (1985). Abstr. in Phys. Abstr., Vol. 89, No. 1274, Entry 19143 (1986).

066.166 **Consistency of the Kaluza–Klein ansatz for cosmological solutions.**
Y. Fujii, K. Yamagishi.
Phys. Lett. B, Vol. 162B, No. 4 – 6, p. 317 – 321 (1985). Abstr. in Phys. Abstr., Vol. 89, No. 1274, Entry 19144 (1986).

066.167 **Statistical gravitational lensing: influence of compact objects on the observed quasar luminosity function.**
P. Schneider.
MPA Rep., No. 218, 48 pp. (1986). Submitted to Astron. Astrophys.

066.168 **On charge integrals of the Riemann tensor and their relation to the notion of angular momentum at future null infinity.**
O. M. Moreschi.
MPA Rep., No. 219, 53 pp. (1986).
Charge integrals of the Riemann tensor are studied. A new definition of angular momentum at future null infinity is proposed which is free of supertranslation ambiguities. A prescription is given to single out a unique Bondi system.

066.169 **Numerical relativity on null cones.**
R. Gomez, R. A. Isaacson, J. S. Welling, J. Winicour.
MPA Rep., No. 221, 18 pp. (1986).
The aim of the authors has been to develop numerical methods for exploring general relativity as a theory, rather than to provide realistic simulations of complex astrophysical systems. They have concentrated on the gravitational degrees of freedom, and maintained a strict integrity with the geometrical nature of key theoretical concepts such as Bondi mass, energy flux, asymptotic flatness and null infinity. While they had originally hoped to tackle applications to existing theoretical problems such as the effect of supermomentum on angular momentum loss, the work has not progressed this far. Rather, it has led the authors to totally new theoretical tools and problems.

066.170 **The Newtonian limit of Penrose's quasi–local mass.**
B. J. Jeffryes.
MPA Rep., No. 222, 24 pp. (1986). Submitted to Classical Quantum Gravity.

066.171 **Propagating modes in gauge field theories of gravity.**
R. Kuhfuß, J. Nitsch.
MPA Rep., No. 223, 22 pp. (1986). Submitted to Gen. Relativ. Gravitation.

066.172 **Gravitational radiation near spatial and null infinity.**
B. G. Schmidt.
MPA Rep., No. 229, 12 pp. (1986).

066.173 **A gravitational lens origin of AGN–variability? Consequences of micro–lensing.**
P. Schneider, A. Weiß.
MPA Rep., No. 234, 57 pp. (1986). Submitted to Astron. Astrophys.

066.174 **Multiple scattering gravitational lens theory and general properties of gravitational light deflection.**
P. Schneider, U. Borgeest.
MPA Rep., No. 237, 53 pp. (1986). Submitted to Astron. Astrophys.

066.175 **On the acceleration of charged particles at relativistic shock–fronts.**
J. G. Kirk, P. Schneider.
MPA Rep., No. 239, 22 pp. (1986).
The authors consider the process of particle acceleration at relativistic shocks, assuming that the particles undergo only pitch–angle scattering on each side of the shock, in which case the acceleration proceeds via the first order Fermi process. The solutions of the particle transport equation on each side of the shock are obtained, satisfying the boundary conditions far from the shock. The authors' method converges well for all values of the shock velocity they consider. They conclude that it provides accurate solutions of the particle transport equation.

066.176 **"Skyhook" – a detector of gravitational waves.**
A. G. Polnarev.
Priroda, No. 6, p. 108 – 109 (1986). In Russian.

066.177 **Constraints on the constant in Rosen's bimetric general relativity theory.**
C. Sivaram.
Indian Inst. Astrophys., Prepr., No. 209, 9 pp. (1986). Submitted to Found. Phys.

066.178 **On the theory of gravitation taking into account the role of vacuum. II.**
L. Sh. Grigoryan, G. S. Saakyan.
Astrofizika, Tom 24, Vyp. 3, p. 591 – 601 (1986). In Russian. English translation in Astrophysics, Vol. 24, No. 3.
The idea concerning the definite field of tensions appearing in a vacuum in the presence of gravity and on the gravitational "constant" defined by the local state of the vacuum suggested in Part I of the paper (1984) is further developed. The static and spherically symmetric gravitational field has been investigated.

The post–Newtonian limit of the theory has been studied and the integration procedure of field equations has been suggested.

066.179 **Gravitational lensing – models.**
R. Narayan.
Prepr. Steward Obs., No. 628, 8 pp. (1986). To appear in IAU Symp. No. 119.

066.180 **Twistors, minimal surfaces and strings.**
W. T. Shaw.
Classical Quantum Gravity, Vol. 2, No. 6, p. L113 – L119 (1985). Abstr. in Phys. Abstr., Vol. 89, No. 1275, Entry 24168 (1986).

066.181 **Spacetimes with a transitive similarity group.**
K. Rosquist, R. T. Jantzen.
Classical Quantum Gravity, Vol. 2, No. 6, p. L129 – L133 (1985). Abstr. in Phys. Abstr., Vol. 89, No. 1275, Entry 24171 (1986).

066.182 **Induction of correct centrifugal force in a rotating mass shell.**
H. Pfister, K. H. Braun.
Classical Quantum Gravity, Vol. 2, No. 6, p. 909 – 918 (1985). Abstr. in Phys. Abstr., Vol. 89, No. 1275, Entry 24175 (1986).

066.183 **Point particle limit and the far–zone quadrupole formula in general relativity.**
T. Futamase.
Phys. Rev. D, Vol. 32, No. 10, p. 2566 – 2574 (1985). Abstr. in Phys. Abstr., Vol. 89, No. 1275, Entry 24187 (1986).

066.184 **The search for gravitational waves.**
J. Hough.
Geophys. J. R. Astron. Soc., Vol. 85, No. 1, p. 245 (1986). Abstract. – See Abstr. 012.086.

066.185 **Spontaneous compactification with quadratic and cubic curvature terms.**
F. Müller–Hoissen.
Phys. Lett. B, Vol. 163B, No. 1 – 4, p. 106 – 110 (1985). Abstr. in Phys. Abstr., Vol. 89, No. 1275, Entry 24199 (1986).

066.186 **Some exact inhomogeneous solutions of Einstein's equations with symmetries on the hypersurfaces t = const.**
F. Argueso, J. L. Sanz.
J. Math. Phys., Vol. 26, No. 12, p. 3118 – 3124 (1985). Abstr. in Phys. Abstr., Vol. 89, No. 1275, Entry 30499 (1986).

066.187 **4–geodesy.**
W. A. Miller, J. A. Wheeler.
Nuovo Cimento C, Vol. 8C, Ser. 1, No. 4, p. 418 – 434 (1985). Abstr. in Phys. Abstr., Vol. 89, No. 1276, Entry 35964 (1986).

066.188 **Relativistic shock propagation in nonuniform media.**
B. I. Gnatyk.
Sov. Astron. Lett., Vol. 11, No. 5, p. 331 – 332 (1985). English translation of 40.066.090.

066.189 **Two mass solutions of Einstein's vacuum equations: the double Kerr solution.**
W. Dietz, C. Hoenselaers.
Ann. Phys. (N.Y.), Vol. 165, No. 2, p. 319 – 383 (1985). Abstr. in Phys. Abstr., Vol. 89, No. 1277, Entry 36596 (1986).

066.190 **Static planar domain wall in general relativity with a cosmological constant.**
B. Linet.
Int. J. Theor. Phys., Vol. 24, No. 12, p. 1159 – 1164 (1985). Abstr. in Phys. Abstr., Vol. 89, No. 1277, Entry 36599 (1986).

066.191 **A class of inhomogeneous Gödel–type models.**
M. J. Reboucas, J. Tiomno.
Nuovo Cimento B, Vol. 90B, Ser. 11, No. 2, p. 204 – 210 (1985). Abstr. in Phys. Abstr., Vol. 89, No. 1277, Entry 36603 (1986).

066.192 **Regular reduction of relativistic theories of gravitation with a quadratic Lagrangian.**
L. Bel, H. S. Sirousse–Zia.
Phys. Rev. D, Vol. 32, No. 12, p. 3128 – 3135 (1985). Abstr. in Phys. Abstr., Vol. 89, No. 1277, Entry 36604 (1986).

066.193 **Isometries of homogeneous Gödel–type spacetimes.**
A. F. F. Teixeira, M. J. Reboucas, J. E. Aman.
Phys. Rev. D, Vol. 32, No. 12, p. 3309 – 3311 (1985). Abstr. in Phys. Abstr., Vol. 89, No. 1277, Entry 36605 (1986).

066.194 **Canonical quantum Kaluza–Klein theory.**
M. I. Beciu.
Nuovo Cimento B, Vol. 90B, Ser. 11, No. 2, p. 223 – 231 (1985). Abstr. in Phys. Abstr., Vol. 89, No. 1277, Entry 36615 (1986).

066.195 **Kaluza–Klein solutions with non–compact internal spaces.**
S. Randjbar–Daemi, C. Wetterich.
Phys. Lett. B, Vol. 166B, No. 1, p. 65 – 68 (1986). Abstr. in Phys. Abstr., Vol. 89, No. 1277, Entry 36624 (1986).

066.196 **Relativistic effects in astronomical timing measurements.**
R. W. Hellings.
Astron. J., Vol. 91, No. 3, p. 650 – 659 (1986).

The author derives the relativistic transformations that are necessary to transform the results of Earth–based astronomical timing observations from the terrestrial frame in which the observations are made to the proper frame of the astronomical object being observed. Two coordinate systems are considered, the first being the parametrized post–Newtonian coordinate system and the second being a coordinate system based on the use of temps dynamique barycentrique as a time coordinate. The results are applied to three modern accurate data types – spacecraft ranging, pulsar timing, and VLBI.

066.197 **Morphological aberration in gravitational–lens images.**
J. M. Barnothy, M. F. Barnothy.
Astron. J., Vol. 91, No. 4, p. 755 – 757 (1986).

The Baldwin effect, the BL Lac phenomenon, the color changes, and brightness fluctuations in QSOs may arise from morphological aberrations in images, produced by gravitational lenses. As an example, the behavior of BL Lac object 0846 + 51W1 is discussed.

066.198 **Some exact solutions of gravitational waves coupled with fluid motions.**
S. Chandrasekhar, B. C. Xanthopoulos.
Proc. R. Soc. London, Ser. A, Vol. 402, No. 1823, p. 205 – 224 (1985). Abstr. in Phys. Abstr., Vol. 89, No. 1278, Entry 42594 (1986).

066.199 **Spontaneous compactification in quantum Kaluza–Klein theories.**
M. H. Sarmadi.
Nucl. Phys. B, Part. Phys., Vol. B263, No. 1, p. 187 – 206 (1986). Abstr. in Phys. Abstr., Vol. 89, No. 1278, Entry 42603 (1986).

066.200 **The special theory of relativity on a gravitational field.**
M. Note.
Int. J. Fusion Energy, Vol. 3, No. 2, p. 23 – 31 (1985). Abstr. in Phys. Abstr., Vol. 89, No. 1279, Entry 48280 (1986).

066.201 **A simple coupling with cosmological implications.**
D. Saez, V. J. Ballester.
Phys. Lett. A, Vol. 113A, No. 9, p. 467 – 470 (1986). Abstr. in Phys. Abstr., Vol. 89, No. 1279, Entry 48283 (1986).

066.202 **Is the number of spatial dimensions an integer?**
C. Jarlskog, F. J. Yndurain.
Europhys. Lett., Vol. 1, No. 2, p. 51 – 53 (1986). Abstr. in Phys. Abstr., Vol. 89, No. 1279, Entry 53726 (1986).

066.203 **The unnecessarily mystical motion.**
G. F. Sanger.
Bull. Am. Astron. Soc., Vol. 18, No. 1, p. 641 (1986). Abstract. – See Abstr. 010.066.

066.204 **The problem of a self–gravitating scalar field.**
D. Christodoulou.
Commun. Math. Phys., Vol. 105, No. 3, p. 337 – 361 (1986).
In this paper the author begins the study of the global initial value problem for Einstein's equations in the spherically symmetric case with a massless scalar field as the material model. He reduces the problem to a single nonlinear evolution equation. Taking as initial hypersurface a future light cone with vertex at the center of symmetry, he proves the local, in retarded time, existence and global uniqueness of classical solutions. The author also proves that if the initial data is sufficiently small there exists a global classical solution which disperses in the infinite future.

066.205 **How important is grvitational–wave astronomy? Part II.**
P. Kafka.
Naturwissenschaften, 73. Jahrg., Heft 6, p. 305 – 313 (1986). In German.
Part II reviews the expected sources of gravitational radiation and confronts them with the planned sensitivity of the antennae. For Part I see Abstr. 066.102.

066.206 **Motion and radiation in general relativity.**
B. F. Schutz.
Relativity, supersymmetry and cosmology, p. 3 – 80 (1985). – See Abstr. 012.055.
This paper reviews recent progress on the connected problems of the motion of bodies in general relativity, the gravitational radiation they generate, and the reaction effects of this radiation on the motion. It treats three separate problems in a unified manner. The first is linearized theory and the post–linear approximations, including a slow–motion expansion. The second is the Newtonian limit, in which self–gravity is weak but not negligible. The third is the strong–field point–particle limit, in which bodies with strong internal gravity interact by their "Newtonian" gravitational field. In each case, the so–called quadrupole formula for gravitational radiation is verified.

066.207 **Scalar fields and the cosmological constant.**
L. F. Abbott.
Relativity, supersymmetry and cosmology, p. 130 – 151 (1985). – See Abstr. 012.055.
This paper explores the possibility of introducing a compensating scalar field which dynamically adjusts its vacuum energy in order to produce, in a natural way, an acceptably small cosmological constant. The necessary dynamical mechanisms are developed in a detailed discussion of symmetry breaking and phase transitions in de Sitter and anti–de Sitter spacetimes.

066.208 **Unified classical physics on a metric space with torsion.**
O. J. Bressan.
Relativity, supersymmetry and cosmology, p. 232 – 237 (1985). – See Abstr. 012.055.
Classical mechanics and electromagnetism are unified in a metric space with semi–symmetrical connection. The path of a charged trial particle under gravity and electromagnetic field is a geodesic of the four dimensional space.

066.209 **The renormalization group and the singularity.**
F. E. Figueirido, M. Castagnino.
Relativity, supersymmetry and cosmology, p. 253 – 258 (1985). – See Abstr. 012.055.

066.210 **The geometric structure of a class of 2*n*–solitons solutions of Einstein's equations.**
M. C. Díaz, R. J. Gleiser.
Relativity, supersymmetry and cosmology, p. 259 – 266 (1985). – See Abstr. 012.055.
An analysis of the geometric structure of a class of $2n$–soliton solutions of Einstein's equations, obtained from a flat background by the inverse scattering method of Belinsky and Zakharov, is presented.

066.211 **The zero–point field in non–inertial frames.**
S. Hacyan.
Relativity, supersymmetry and cosmology, p. 267 – 270 (1985). – See Abstr. 012.055.
The energy spectrum of the zero–point field as seen in non–inertial frames is investigated. Uniformly accelerated frames and black holes are considered. It is suggested that the radiation produced by black holes or acceleration is a manifestation of the zero–point field and of the same nature.

066.212 **Spherically symmetric perfect fluid solutions in relativistic astrophysics.**
D. García Lambas, W. Lamberti, V. H. Hamity.
Relativity, supersymmetry and cosmology, p. 271 – 278 (1985). – See Abstr. 012.055.
The general problem of matching two solutions of the Einstein equations through a regular or singular hypersurface is reviewed. The particular case of matching two spherically symmetric expanding perfect fluid solutions is discussed and its possible application to some interesting problems like a non–static spherically symmetric star model and an inhomogeneous cosmological model is briefly presented.

066.213 **The local term of the energy momentum tensor for a quantum scalar field in a curved background.**
D. Harari, C. Núñez, M. Castagnino.
Relativity, supersymmetry and cosmology, p. 279 – 285 (1985). – See Abstr. 012.055.
The local part of the renormalized energy–momentum tensor of a quantum scalar free field in a curved background is assumed to be computable from Hadamard–type kernels constructed in terms of geometric objects. A comparison of the results with vacuum polarization in Robertson–Walker metrics is presented.

066.214 **On torsion detection using post–Newtonian approximation.**
M. Levinas, M. Castagnino.
Relativity, supersymmetry and cosmology, p. 286 – 291 (1985). – See Abstr. 012.055.
The authors study the post–Newtonian approximation of the ECSK theory and analyze the torsion detection when the matter Lagrangian is minimally and non–minimally coupled.

066.215 **The quantization of matter and gravitational fields – the cosmological aspect.**
E. Nowotny.
Relativity, supersymmetry and cosmology, p. 292 – 295 (1985). – See Abstr. 012.055.
Reducing the metric to the conformal flat space–time, it is shown that this degree of freedom of the metric coupled to a spin zero matter field of type ϕ^4 leads to a renormalizable quantum model if some conditions between the coupling parameter of the nonlinear matter field, the mass and cosmological constant are satisfied.

066.216 **Canonical quantum gravity.**
J. Zanelli.
Relativity, supersymmetry and cosmology, p. 324 – 339 (1985). – See Abstr. 012.055.
The author presents an ansatz for carrying out in part the program of quantization of gravity in canonical form. The idea is to exploit the covariance of any Lagrangian theory under changes of generalized coordinates, in order to select a particular form for the operator constraints of quantum gravity. The resulting commutator algebra of these operators is found to close, as required by consistency.

066.217 **Simulating gravitational lenses.**
E. E. Falco.
Bull. Inf. Cent. Données Stellaires, No. 30, p. 175 – 178 (1986). = Newsl. Working Group Mod. Astron. Methodol., No. 2.

066.218 **Relativistic neutral superfluids. Exact solutions.**
G. Papini, M. Weiss.
Nuovo Cimento B, Vol. 91B, Ser. 11, No. 1, p. 31 – 56 (1986). Abstr. in Phys. Abstr., Vol. 89, No. 1280, Entry 54256 (1986).

066.219 **On the collision of impulsive gravitational waves when coupled with null dust.**
S. Chandrasekhar, B. C. Xanthopoulos.
Proc. R. Soc. London, Ser. A, Vol. 403, No. 1825, p. 189 – 198 (1986). Abstr. in Phys. Abstr., Vol. 89, No. 1280, Entry 54260 (1986).

066.220 **Gravitational lenses. I. Deflection of light by an opaque and spherical mass or a point mass.**
R. Bergh, E. Ostgaard.
Ark. Fys. Semin. Trondheim, No. 17, p. 1 – 19 (1985). Abstr. in Phys. Abstr., Vol. 89, No. 1280, Entry 60595 (1986).

066.221 **On analytical continuation of the Kerr–Newman metric with cosmological constant.**
V. S. Gurin.
Acta Phys. Pol., Ser. B, Vol. B17, No. 1, p. 3 – 10 (1986). In Russian. Abstr. in Phys. Abstr., Vol. 89, No. 1280, Entry 61129 (1986).

066.222 **Eigenschaften singularitätenfreier, statischer, zentralsymmetrischer Lösungen gewisser Gravitationsfeldgleichungen 4. Ordnung.**
B. Fiedler, M. Günther.
Astron. Nachr., Vol. 307, No. 2, p. 129 – 135 (1986).
Fiedler and Schimming (1983) proved that the fourth order gravitational field equations with a linear combination of Bach's and Einstein's tensors on the left–hand side, which were proposed by Treder, admit static centrally symmetric solutions which are analytical and non–flat in some neighbourhood of the centre of symmetry. The existence of these solutions, known at first only in a small neighbourhood of $r = 0$ (r radius), can now be extended to intervals $0 \leqslant r \leqslant \alpha$ with arbitrarily large α.

066.223 **Accretion disks in bimetric theory of gravitation. A strong field test of general relativity?**
H. G. Paul.
Astron. Nachr., Vol. 307, No. 3, p. 183 – 189 (1986).
The sensitivity of accretion disk models with respect to its inner boundary conditions, being located in the strong gravity region of a highly compact central body, is used to discuss a possibility of strong field tests of gravitation via compact source observations. Within the bimetric theory of gravitation the optically thin bremsstrahlung model is calculated as an example for this possibility. Using this model to describe the innermost region of the "bimetric disk", significant differences with respect to the corresponding results in Einstein's theory were found.

066.224 **Astigmatic property of an n–body gravitational lens.**
K. Chang.
J. Korean Astron. Soc., Vol. 19, No. 1, p. 11 – 14 (1986).
It is shown that the astigmatic property of a single gravitational lens in a static bounded gravitational field can be retained, if an n–gravitating body as a whole acts simultaneously as gravitational lens.

066.225 **Mercury's precession according to special relativity.**
T. E. Phipps Jr.
Am. J. Phys., Vol. 54, No. 3, p. 245 – 247 (1986). Abstr. in Phys. Abstr., Vol. 89, No. 1281, Entry 61220 (1986).

066.226 **Dimensional reduction caused by a cosmological constant.**
E. J. Squires.
Phys. Lett. B, Vol. 167B, No. 3, p. 286 – 288 (1986). Abstr. in Phys. Abstr., Vol. 89, No. 1281, Entry 61390 (1986).

066.227 **Gravitational radiation in electromagnetic universes.**
V. I. Khlebnikov.
Classical Quantum Gravity, Vol. 3, No. 2, p. 169 – 173 (1986). Abstr. in Phys. Abstr., Vol. 89, No. 1281, Entry 61401 (1986).

066.228 **On the vanishing of the cosmological constant.**
N. Deruelle, J. Madore.
Phys. Lett. A, Vol. 114A, No. 4, p. 185 – 190 (1986). Abstr. in Phys. Abstr., Vol. 89, No. 1281, Entry 61409 (1986).

066.229 **Dimensionality of space and the pulsating Universe.**
F. R. Tangherlini.
Nuovo Cimento B, Vol. 91B, Ser. 11, No. 2, p. 209 – 217 (1986). Abstr. in Phys. Abstr., Vol. 89, No. 1282, Entry 67867 (1986).

066.230 **Thermodynamic instability of de Sitter space.**
E. Mottola.
Phys. Rev. D, Vol. 33, No. 6, p. 1616 – 1621 (1986). Abstr. in Phys. Abstr., Vol. 89, No. 1282, Entry 67895 (1986).

066.231 **Gravitational lenses. II. Deflection of light by galaxies with spheroidal volume mass densities and elliptical surface mass densities.**
R. Bergh, E. Ostgaard.
Ark. Fys. Semin. Trondheim, No. 5, p. 1 – 64 (1986). Abstr. in Phys. Abstr., Vol. 89, No. 1283, Entry 78262 (1986).

066.232 **Birkhoff–type theorem in self–creation cosmology.**
T. Singh.
J. Math. Phys., Vol. 27, No. 4, p. 1074 – 1075 (1986). Abstr. in Phys. Abstr., Vol. 89, No. 1283, Entry 78545 (1986).

066.233 **Symmetric solutions to the Gauss–Bonnet extended Einstein equations.**
J. T. Wheeler.
Nucl. Phys. B, Part. Phys., Vol. B268, No. 3 – 4, p. 737 – 746 (1986). Abstr. in Phys. Abstr., Vol. 89, Nr. 1284, Entry 78735 (1986).

066.234 **An approximate correction to the quadrupole radiation of rotating stars.**
M. q. Zhong.
Chin. Astron. Astrophys., Vol. 10, No. 1, p. 77 – 80 (1986). English translation of 40.066.188.

066.235 **Relativistic Kepler's third law.**
B. M. Barker, G. G. Byrd, R. F. O'Connell.
Astrophys. J., Vol. 305, No. 2, p. 623 – 633 (1986).
The authors give a post–Newtonian generalization of Kepler's third law that includes spin and quadrupole moment effects, parameterized post–Newtonian parameters γ and β, and the Nordtvedt effect. The time from periastron to apastron is also found, and it turns out, in general, not to be equal to half the period. The results are given in a very general coordinate system which is specified by three arbitrary dimensionless parameters.

066.236 **Does the gravitational constant change?**
V. K. Milyukov.
Priroda, No. 6, p. 96 – 104 (1986). In Russian.

066.237 **Static solutions and equilibrium distributions of a gravitating liquid with plane symmetry in the general relativity theory.**
V. A. Ruban.
Astron. Zh., Tom 63, Vyp. 2, p. 233 – 239 (1986). In Russian. English translation in Sov. Astron., Vol. 30, No. 2.
Static V_4 fields and equilibrium distributions of a plane–symmetric gravitating liquid are studied in the framework of the general relativity theory, on the basis of the obtained analytic solutions, for polytropic relativistic equations of state. $P = n\varepsilon$ $(0 < n \leqslant 1)$, as well as for a plane homogeneous layer of a non–compressible medium with $\varepsilon_0 =$ const. It is shown that plane–symmetric static solutions of the Einstein equations always contain a physical vacuum singularity, a singular repulsing plane with negative gravitational mass, and are thus principally differ-

ent from an infinite plane equilibrium layer of liquid in the Newtonian theory.

066.238 **Regularities of irregular variation of the gravitational constant.**
Yu. P. Kozlov.
Kazan. khim.–tekhnol. inst. Kazan', 12 pp. (1985). In Russian. Abstr. in Ref. Zh., 51. Astron., 5.51.138 (1986).

Spinors and space–time. Volume 2: Spinor and twistor methods in space–time geometry.
See Abstr. 003.002.

Investigations on classical and quantum theory of gravitation.
See Abstr. 003.005.

Einstein's legacy. The unity of space and time.
See Abstr. 003.020.

Methods in the qualitative theory of dynamical systems in astrophysics and gas dynamics.
See Abstr. 003.028.

Introduction to gravitation.
See Abstr. 003.032.

Elements of relativity theory.
See Abstr. 003.036.

A random walk in relativity and cosmology.
See Abstr. 003.079.

The theory of gravitation.
See Abstr. 003.129.

Theory and experiment in gravitational physics.
See Abstr. 003.211.

La relativité générale à l'étiage: 1925 – 1955.
See Abstr. 004.074.

Observatories of the future.
See Abstr. 011.029.

Supersymmetry and its applications: superstrings, anomalies and supergravity. Proceedings of the Nuffield Workshop, held at Cambridge, U.K., 23 June – 14 July 1985.
See Abstr. 012.041.

Supersymmetry, supergravity and related topics. Proceedings of the XVth GIFT International Seminar on Theoretical Physics.
See Abstr. 012.084.

Gravitation, cosmology and cosmic–ray physics.
See Abstr. 013.023.

Searching for gravity waves with interferometers.
See Abstr. 034.022.

Cryogenic resonant–mass gravitational wave detectors.
See Abstr. 034.096.

Relativistic effects in the measurement of time and frequency.
See Abstr. 034.120.

Coupling of a high–sensitivity superconducting amplifier to a gravitational–wave antenna.
See Abstr. 034.155.

A low–noise low–input conductance preamplifier for gravitational research.
See Abstr. 034.162.

Experimental test of a prototype gravitational radiation detector employing an active cavity laser sensor.
See Abstr. 034.180.

Data analysis for a gravitational wave antenna with resonant capacitive transducer.
See Abstr. 034.181.

Relativistic effects in earth based and cosmic long baseline interferometry.
See Abstr. 036.002.

Relativistic effects in geodetic VLBI measurements.
See Abstr. 036.003.

Relativity in astrometry.
See Abstr. 041.002.

Relativistic interpretation of astrometric data.
See Abstr. 041.003.

Relativistic effects included in the apparent positions of fundamental stars (APFS).
See Abstr. 041.005.

Relativistic effects in variations of right ascensions of stars and clock corrections.
See Abstr. 041.018.

Relativistic reduction of astrometric observations.
See Abstr. 041.049.

Polynomial expansion of the planetary secular terms: relativistic and lunar perturbations.
See Abstr. 042.002.

Local frames.
See Abstr. 042.003.

The Hamiltonian equations of the planetary *N*–body problem in the post–Newtonian approximation.
See Abstr. 042.020.

General relativistic celestial mechanics of binary systems. I. The post–Newtonian motion.
See Abstr. 042.073.

Coordinate systems in the general relativistic framework.
See Abstr. 043.001.

System of astronomical constants in the relativistic framework.
See Abstr. 043.002.

Numerical effects of gravitational light deflection on the determination of the equinox and equator.
See Abstr. 043.003.

Hypothesis of variable *G* and the secular accelerations of the Sun and the Moon.
See Abstr. 043.004.

Relativistic coordinates and time scales.
See Abstr. 043.006.

Relativity effects in the rotation of the Earth and the motion of atmospheric masses.
See Abstr. 044.001.

Coordinate time in the vicinity of the earth.
See Abstr. 044.003.

On calculation of relativistic effects in numerical prediction of artificial satellites' motion.
See Abstr. 052.001.

Gauged $N = 8$ supergravity in $d = 11$ dimensions.
See Abstr. 061.004.

A relation between $N = 2$ supergravities in 10 and 4 dimensions.
See Abstr. 061.005.

Isometries as probes of the extra dimensions.
See Abstr. 061.006.

The group manifold approach to unified gravity.
See Abstr. 061.008.

Aspects of supergravity theories.
See Abstr. 061.009.

$d = 4$ supergravity from $d = 11$: conjectures revisited.
See Abstr. 061.010.

Vacuum stability in supergravity.
See Abstr. 061.011.

Constraints in conformal simple supergravity.
See Abstr. 061.019.

York's cosmic time versus proper time as relevant to changes in the dimensionless "constants", *K*–meson decay, and the unity of black hole and big crunch.
See Abstr. 061.020.

Superstrings and supersymmetry.
See Abstr. 061.021.

Quantum tunnelling towards an exploding universe?
See Abstr. 061.022.

A grand superspace of unified field theories.
See Abstr. 061.029.

A model–theory for tachyons in two dimensions.
See Abstr. 061.036.

Quantum equivalence principle.
See Abstr. 061.038.

Eddington's numbers and quantum gravity.
See Abstr. 061.039.

Physical processes of gravitons and gravy quarks in the interior of cosmod and supergravity cosmic stellar entities.
See Abstr. 061.042.

Towards the geometrization of physics.
See Abstr. 061.050.

Supergrand unified composite model in pregeometry.
See Abstr. 061.052.

Present state of quantum supergravity.
See Abstr. 061.053.

At what energy does gravity unite with grand unified theories?
See Abstr. 061.068.

Global existence of time–dependent Yang–Mills–Higgs monopoles.
See Abstr. 061.095.

Quantum field theory of mutually interacting particles in expanding universes.
See Abstr. 061.096.

Scalar particle production in anisotropic metrics.
See Abstr. 061.098.

Weak discontinuities in relativistic MHD.
See Abstr. 062.016.

The comoving–frame equation of radiative transfer in relativistic flows.
See Abstr. 063.030.

Relativistic radiation transport in dispersive media.
See Abstr. 063.076.

Hot spots on neutron stars. II. The observer's sky.
See Abstr. 067.001.

On hydrodynamics of radiatively driven winds.
See Abstr. 067.074.

Effect of vacuum polarization near black holes.
See Abstr. 067.094.

Relativistic effects in geodynamics.
See Abstr. 081.003.

Proposals for an experiment to detect the earth's gravitomagnetic field.
See Abstr. 081.004.

A relativistic theory of motion of the inner planets.
See Abstr. 091.004.

Simulation of general relativistic corrections in long term numerical integrations of planetary orbits.
See Abstr. 091.005.

Relativistic effects from planetary and lunar observations of the XVIII – XX centuries.
See Abstr. 091.006.

New relativistic effects in the motion of the Moon.
See Abstr. 094.002.

The dominant relativistic terms in the lunar theory.
See Abstr. 094.009.

Interplanetary plasma turbulence and the Doppler detection of a gravitational wave background.
See Abstr. 106.007.

Relativistic and perspective effects in proper motions and radial velocities of stars.
See Abstr. 111.001.

An analysis of pulsar timing irregularities: is it induced by a source of gravitational waves in the Galaxy?
See Abstr. 126.068.

Can superconducting cosmic strings explain anomalous radio sources?
See Abstr. 141.012.

The relativistic effects in localization of gamma–burst sources.
See Abstr. 143.002.

A new approach to a relativistic model of galaxies.
See Abstr. 151.014.

The third image, the redshift of the lens, and new components of the gravitational lens 2016 + 112.
See Abstr. 158.274.

CCD camera brightness monitoring of Q0957 + 561 A, B.
See Abstr. 159.002.

Motion of quasar images by gravitational lens galaxies.
See Abstr. 159.020.

Apparent superluminal expansion in a quasar neutrino halo.
See Abstr. 159.035.

Les mirages gravitationnels: un outil de choix pour sonder l'univers.
See Abstr. 159.043.

A test for the gravitational lens hypothesis in the most luminous quasar: S5 0014+81.
See Abstr. 159.065.

High-resolution imaging from Mauna Kea: the triple quasar in 0.3-arc s seeing.
See Abstr. 159.084.

An apparent gravitational lens with an image separation of 2.6 arc min.
See Abstr. 159.085.

Gravitational lenses and multiple quasars.
See Abstr. 159.087.

Detection of the lensing galaxy in PG 1115+08.
See Abstr. 159.103.

Is there a black hole in the sky?
See Abstr. 159.118.

Is QSO 1146+111 B, C due to lensing by a cosmic string?
See Abstr. 159.119.

Q1146+111B, C quasar pair: illusion or delusion?
See Abstr. 159.120.

Test of the gravitational lens hypothesis for the quasar pair 1146+111B, C.
See Abstr. 159.121.

Deep CCD images of 2345+007: lensing by dark matter.
See Abstr. 159.149.

A cosmic string in Leo?
See Abstr. 159.152.

Snarls from the cosmic string in Leo.
See Abstr. 159.153.

VV 172: an effect of gravitational amplification by a massive halo?
See Abstr. 160.014.

Superstring gravity and the early universe.
See Abstr. 161.004.

Bianchi type-V perfect fluid models.
See Abstr. 161.005.

Gravitational radiation from cosmological phase transitions.
See Abstr. 161.013.

Comments on the BDT-FRW perfect fluid solutions.
See Abstr. 161.023.

Comments on the BDT-FRW solutions with a cosmological constant.
See Abstr. 161.024.

Problems of dimensional reduction and inflationary cosmology.
See Abstr. 161.034.

Supersymmetry and cosmology.
See Abstr. 161.035.

Statistical gravitational lensing and quasar-galaxy associations.
See Abstr. 161.038.

Friedmann-Robertson-Walker stiff matter solutions in the scale covariant theory.
See Abstr. 161.039.

The automorphism group and field equations for Bianchi universes.
See Abstr. 161.051.

Production of massive spin-1/2 particles in Robertson-Walker universes with external electromagnetic fields.
See Abstr. 161.054.

Bianchi type-II cosmological model with viscous fluid.
See Abstr. 161.065.

Comment on stiff matter solutions in cosmological models based on Lyra's manifold.
See Abstr. 161.106.

Improved bounds on the dimension of space-time.
See Abstr. 161.142.

An upper limit to the density inhomogeneity in the Universe.
See Abstr. 161.164.

On the causal geometry of spacetime: some applications to cosmology.
See Abstr. 161.165.

Bianchi type-I Lyttleton-Bondi universe.
See Abstr. 161.166.

Robertson-Walker type Lyttleton-Bondi universe.
See Abstr. 161.167.

An analysis of the derivation of Friedmann's equation.
See Abstr. 161.168.

Thermodynamics of the vacuum.
See Abstr. 161.169.

A plane symmetric cosmological model in Lyra manifold.
See Abstr. 161.170.

The Bianchi type-V vacuum cosmological model in the scale covariant theory.
See Abstr. 161.173.

Entropy in an oscillating universe in the assumption of universe "splitting" into numerous smaller "daughter" universes.
See Abstr. 161.181.

Quantum fluctuations as the cause of inhomogeneity in the universe.
See Abstr. 161.182.

Inflationary stages in cosmological models with a scalar field.
See Abstr. 161.183.

Dynamics of inflating bubbles in the early universe.
See Abstr. 161.185.

Primordial black holes and observational restrictions on quantum gravity.
See Abstr. 161.189.

Hawking's theory of quantum cosmology.
See Abstr. 161.194.

Inflationary cosmology from a classical gauge field.
See Abstr. 161.195.

Entropy production due to gravitational-wave viscosity in a Kaluza-Klein inflationary universe.
See Abstr. 161.205.

On Komar integrals in asymptotically anti-de Sitter space-times.
See Abstr. 161.238.

Anisotropic cosmological model in Nordtvedt's scalar–tensor theory of gravitation.
See Abstr. 161.239.

Bianchi type–I model with cosmological constant in a generalized scalar–tensor theory of gravitation.
See Abstr. 161.275.

On the uniqueness of the Einstein–de Sitter cosmology in spinor–tetrad theory.
See Abstr. 161.285.

Behaviour of homogeneous five–dimensional space–times.
See Abstr. 161.290.

Supersymmetry and cosmology.
See Abstr. 161.309.

Self–consistent cosmogenesis.
See Abstr. 161.310.

De Sitter self consistent cosmology with adiabatic regularization.
See Abstr. 161.312.

On the meaning of quantum cosmology.
See Abstr. 161.313.

Weyl (integrable) spacetimes as a model for our cosmos.
See Abstr. 161.314.

Cosmic Lorentz transformation.
See Abstr. 161.317.

Comment on horizons in Dirac cosmology.
See Abstr. 161.346.

A cosmological model of matter in the framework of special relativity.
See Abstr. 161.378.

Cosmology with time–varying *G*.
See Abstr. 161.385.

067 Astrophysics of Compact Objects (Neutron Stars, Black Holes)

067.001 **Hot spots on neutron stars. II. The observer's sky.**
C. Ftaclas, M. W. Kearney, K. Pechenick.
Astrophys. J., Vol. 300, No. 1, p. 203 – 208 (1986).
The authors explore the gravitational lensing properties of neutron stars and apply the results to X–ray emission from radio and X–ray binary pulsars. A simple method is described and illustrated for calculating the apparent brightness and appearance of the observer's sky for any model of emission from the stellar surface. The authors examine the case where the emission arises solely from two antipodal hot spots. For typical neutron star parameters the inclusion of gravitational effects makes it 3 – 10 times less likely, on average, that a given observer will see only one hot spot of the star, assuming no bias in the angle between the magnetic and spin axes.

067.002 **Sound horizon of accretion onto a Kerr black hole.**
J. F. Lu.
Gen. Relativ. Gravitation, Vol. 18, No. 1, p. 45 – 51 (1986).
Two accretion modes (disclike and Bondi type), based on a discontinuous change in location of the sound horizon, exist generally for accretion onto a Kerr black hole, except for the case of a corotating accretion flow onto a very rapidly rotating hole, in which only Bondi–type mode is possible.

067.003 **Properties of condensed matter in very strong magnetic fields.**
P. B. Jones.
Mon. Not. R. Astron. Soc., Vol. 218, No. 3, p. 477 – 485 (1986).
The ground–state energies of rhombohedral and body–centred tetragonal atomic lattices in very strong magnetic fields have been calculated by means of density functionals. The exchange–correlation functional is that adopted in previous calculations of atomic and molecular chain ground–state energies. Approximate electronic eigenfunctions and eigenvalues of the lattice Khon–Sham potential have been obtained by a local approximation in one angular variable. Atomic cohesive energies and electron chemical potentials have been derived for atomic numbers and magnetic flux densities relevant to the stellar surface in radio pulsars. The cohesive energies are small and, though subject to errors caused by the local approximation, show unambiguously that the electric field boundary condition is $\mathbf{E}\cdot\mathbf{B} = 0$ at the surface.

067.004 **On accretion of angular momentum from an inhomogeneous medium.**
M. Livio, N. Soker, M. de Kool, G. J. Savonije.
Mon. Not. R. Astron. Soc., Vol. 218, No. 4, p. 593 – 604 (1986).
The problem of a compact object accreting from an inhomogeneous medium has been studied, using a three–dimensional numerical scheme. When pressure effects are neglected, it has been shown that the mass accretion rate is given by the Bondi–Hoyle value. Not more than a few per cent of the angular momentum deposited into the accretion cylinder are accreted by the compact object. Some possible consequences for the case of neutron stars and white dwarfs accreting from a stellar wind are discussed.

067.005 **Origin of ultra–high–energy γ–rays from Cygnus X–3 and related sources.**
D. Kazanas, D. C. Ellison.
Nature, Vol. 319, No. 6052, p. 380 – 382 (1986).
Ultra–high–energy (UHE) γ–ray emission from Cygnus X–3 and the several other binary X–ray sources has been observed at γ–ray energies $\geqslant 10^{15}$eV. Also, collisionless shocks are expected to form in accretion flows onto neutron stars or black holes. The authors consider the diffusive shock acceleration of ions as the mechanism responsible for the γ–ray emission observed. The shock can accelerate ions to energies near 10^{16}eV. The subsequent proton–proton (p–p) collisions and photodissociation of ^{4}He can produce a flux of neutrons that escapes from the acceleration site despite the high magnetic fields ($\sim 10^8$G). These neutrons, by interacting with the binary companion, produce the observed UHE radiation.

067.006 **Quasi–periodic oscillations from accretion disk coronas.**
C. B. Boyle, A. C. Fabian, P. W. Guilbert.
Nature, Vol. 319, No. 6055, p. 648 – 649 (1986).
Recent observations of galactic bulge X–ray sources have revealed in some cases the existence of a hot accretion disk corona and in others quasi–periodic oscillations (QPOs) in the X–ray flux. The authors suggest that these properties are directly related and that the oscillations are produced in the innermost parts of the corona. The observed variations in the centroid frequency of the QPOs with source intensity are explained and further variations with the photon spectrum predicted. Unlike some alternative models, it is not required that the central object is a rapidly spinning weakly magnetized neutron star.

067.007 **Pair production instabilities as a source of X-ray flares from accreting black holes.**
P. Moskalik, M. Sikora.
Nature, Vol. 319, No. 6055, p. 649 – 652 (1986).
Several active galactic nuclei emit most of their energy at $h\nu > 100$ keV. Observations suggest that their high-energy spectra are strongly variable and are cut off or broken at ~ 1 MeV. A possible explanation of these properties is that the high-energy radiation is produced in the central region (radius $R < 10\ R_s$, where R_s is the Schwarzschild radius) of the two-temperature plasma accreting onto a supermassive black hole. There, the e_+–e_- pair creation can be very efficient and, at a certain luminosity range around $L_{cr} = 0.01\ L_{Edd}$ (where L_{Edd} is the Eddington luminosity) it gives rise to a plasma instability. It is shown that this pair-production instability leads to cyclic variations of the accretion flow, during which high-energy flares are produced.

067.008 **A qualitative model for GX5–1.**
N. M. Berman, G. M. Stollman.
Astron. Astrophys., Vol. 154, No. 1/2, p. L23 – L25 (1986).
Recent observations of the galactic bright bulge source GX5–1 have shown quasi-periodic oscillations in the X-ray flux (1 – 18 keV), which appear as a broad peak in the power spectrum with a centroid frequency in the range 20 – 40 Hz. In addition there is a red noise component in the spectrum, the slope of which, along with the centroid frequency and width of the quasi-oscillation peak, is correlated with the intensity of the source. The authors present a qualitative model to explain the essential features of GX5–1 described above.

067.009 **Estimation of some parameters of pulsars and their applications.**
X.-j. Wu, G.-j. Qiao, X.-y. Xia, F. Li.
Astrophys. Space Sci., Vol. 119, No. 1, p. 101 – 104 (1986). – See Abstr. 012.001.
The estimation of some important parameters of pulsars is discussed. The results have been compared with the RS model and applied to the estimation of radio luminosity and the beaming factor.

067.010 **Pulsar magnetosphere with conspicuous trans-field flow.**
S. Shibata.
Astrophys. Space Sci., Vol. 119, No. 1, p. 105 – 107 (1986). – See Abstr. 012.001.
It is shown that (1) in the pulsar magnetosphere the violation of the ideal-MHD condition, $E + v \times B \neq 0$, (i.e., conspicuous trans-field motion and non-zero field aligned electric field $E_{\parallel} \neq 0$) appears owing to relativistic large inertia, and (2) an axisymmetric numerical model with tenuous plasma suggests that in the region of the trans-field flow a vacuum-like electric field and a closed current circuit develop.

067.011 **Electron–positron pairs in a mildly relativistic plasma in active galactic nuclei.**
F. Takahara, M. Kusunose.
Astrophys. Space Sci., Vol. 119, No. 1, p. 217 – 219 (1986). – See Abstr. 012.001.
The authors investigate the electron–positron pair concentration in an optically thin mildly relativistic plasma which is supposed to exist in active galactic nuclei. Firstly the equilibrium concentration is calculated when copious soft photons are supplied through the cyclotron higher harmonics. Secondly, the authors examine the pair production in a hot accretion plasma around a massive black hole, comparing relevant time scales.

067.012 **Charged particle trajectories in the field of a charge near a Schwarzschild black hole.**
V. P. Sonar, S. V. Dhurandhar, N. Dadhich.
Bull. Astron. Soc. India, Vol. 13, No. 4, p. 372 – 386 (1985).
The authors consider the motion of charged particles in the field of a charge placed near a Schwarzschild black hole. The electromagnetic field gets modified owing to gravitational field of the black hole. The system, charge plus the hole, is axisymmetric (no longer spherically symmetric) which poses difficulty in obtaining analytic solutions of equations of motion. However, motion along the axis and circular orbits about the axis of symmetry are discussed.

067.013 **Energetic radiation from rapidly spinning pulsars. I. Outer magnetosphere gaps.**
K. S. Cheng, C. Ho, M. Ruderman.
Astrophys. J., Vol. 300, No. 2, p. 500 – 521 (1986).
Very rapidly spinning magnetized neutron stars such as the Crab and Vela pulsars are strong γ-ray sources with photon energies extending to 10^{12}eV and possibly beyond. These special pulsars must contain extremely ultrarelativistic particle accelerators. In this paper the authors consider the origin and structure of particle accelerators in the outer magnetosphere of pulsars characterized by large spindown energy loss rates. The discussion is concerned with the class of (mainly young) pulsars which can sustain pair production in otherwise charge-depleted regions of the outer magnetosphere. The resulting electrostatic accelerating region ("gap") is expected to exist far from the neutron star, where a corotating magnetosphere would be rotating at speeds a significant fraction of c. The size and location of this gap supports the production of energetic radiation beams of quite different character from the usual coherent microwave emission of radio pulsars.

067.014 **Energetic radiation from rapidly spinning pulsars. II. Vela and Crab.**
K. S. Cheng, C. Ho, M. Ruderman.
Astrophys. J., Vol. 300, No. 2, p. 522 – 539 (1986).
The growth of charge-depleted regions (gaps) in the outer magnetosphere of rapidly spinning magnetized neutron stars is limited by $e^{\pm}$ production. This paper considers details of the outer magnetosphere accelerator pair production and injection processes. The resulting radiation intensities and spectra are calculated and observable consequences for the Vela and the Crab pulsars are derived.

067.015 **Cosmic ray acceleration by binary neutron stars.**
W. Kundt.
Cosmic radiation in contemporary astrophysics, p. 57 – 65 (1986). – See Abstr. 012.006.
Young binary neutron stars – the elder brothers of pulsars – are proposed as the boosters of the ionic component of cosmic rays. Their rotational energy can be converted into beams of cosmic rays if there is enough coupling between the corotating magnetosphere and the impinging plasma, in a manner similar to the sparking of a grindstone. Power-law spectra in energy are obtained from a power-law dependence of the accelerating fields. The upper cutoff energy should not greatly exceed 10^{20}eV. The observed ionic cosmic-ray spectrum would result from a superposition of the injection by $\lesssim 10^6$ young binary neutron stars.

067.016 **Particle acceleration by pulsars.**
W. Kundt.
Cosmic radiation in contemporary astrophysics, p. 67 – 78 (1986). – See Abstr. 012.006.
It is argued that the winds from fast-rotating magnets – like pulsars – consist primarily of (relativistic) electrons and positrons with Lorentz factors in excess of 10^2. The particle energies are likely distributed according to a power law, with Lorentz factors reaching and exceeding $\gamma = 10^9$.

067.017 **The neutron-star oscillations model for gamma-ray bursts.**
A. G. Muslimov, A. I. Tsygan.
Astrophys. Space Sci., Vol. 120, No. 1, p. 27 – 41 (1986).
Exact analytic expressions for the vacuum electromagnetic fields produced by an oscillating magnetized sphere are obtained. The solutions are analyzed for various modes of pulsation and for low-order multipole magnetic moments. Within the context of neutron star oscillations, the possibility of gamma-ray generation is discussed. It is shown that the radial pulsations provide an efficient mechanism for generation of gamma-radiation and electron–positron pairs in some regions around the neutron star.

067.018 **Boundary behaviour of accretion disks determined by gravity induced cooling near the black hole.**
H. G. Paul.
Astron. Nachr., Vol. 307, No. 1, p. 41 – 47 (1986).
In the framework of thin, rigorous relativistic accretion disks the inner boundary behaviour is studied taking into consideration the internal energy of the disk plasma. Consuming the internal energy stored in the accretion flow the disk forms a cusplike wall in the effective gravitational potential; the disk matter is raised on higher levels of the angular momentum by cooling simultaneously. This gravity induced cooling leads to a rigorous braking of the radial accretion flow. There is a steep, inner boundary with a very small gap very close at the orbit of last bound motion of test matter. Especially natural boundary conditions can be imposed in contrast to the boundary value problem in the divergent, relativistic standard models.

067.019 **The lower mass limit of the compact object in LMC X–3.**
T. Mazeh, J. van Paradijs, E. P. J. van den Heuvel, G. J. Savonije.
Astron. Astrophys., Vol. 157, No. 1, p. 113 – 115 (1986).
The discovery of the binary period and the high radial velocity amplitude of the optical counterpart of the X–ray source LMC X–3 (Cowley et al., 1983) provided strong evidence for the existence of a black hole in this system (Cowley et al., 1983; Paczynski, 1983). In this paper the authors comment on some further observational data that are needed to establish the black hole assumption for LMC X–3.

067.020 **Can comet clouds around neutron stars explain gamma–ray bursts?**
S. Tremaine, A. N. Żytkow.
Astrophys. J., Vol. 301, No. 1, p. 155 – 163 (1986).
The authors rediscuss the proposal of Harwit and Salpeter that gamma–ray bursts are due to impacts of comets onto neutron stars. It is assumed that most stars are formed with comet clouds similar to the Oort comet cloud which surrounds the Sun. Possible models for the formation of a neutron star which retains its comet cloud are investigated. The authors then estimate the rate of impacts onto the neutron star by comets from the cloud and show that (1) the recurrence rate of impacts can be as large as the value $\sim 1\ \mathrm{yr}^{-1}$ implied by observations; and (2) the observed log N–log S relation for gamma–ray bursts, interpreted in the context of the model, implies that the number density of burst sources is $\sim 6 \times 10^{-3} \mathrm{pc}^{-3}$, although the uncertainty in this value is large.

067.021 **The stability of radiation pressure–dominated accretion discs.**
M. Camenzind, F. Demole, N. Straumann.
Astron. Astrophys., Vol. 158, No. 1/2, p. 212 – 216 (1986).
The stability of thin accretion discs against axisymmetric perturbations is analyzed for polytropic equations of state. The authors derive a general dispersion relation as a function of the polytropic index N and the ratio between the radiation pressure and the total pressure in the disc. As in the case of constant density models, standard alpha–discs are found to be unstable when radiation pressure dominates. Models based on a magnetic viscosity are, however, always stable against axisymmetric perturbations.

067.022 **Episodic accretion onto massive black holes.**
R. Bottema, R. H. Sanders.
Astron. Astrophys., Vol. 158, No. 1/2, p. 297 – 304 (1986).
Episodic activity in galactic nuclei may be the result of an encounter between an interstellar cloud and a massive black hole. The present paper describes two–dimensional hydrodynamical calculations of cloud–hole encounters in the limits of clouds larger than and comparable to the accretion radius. For large clouds the total mass accreted and the accretion time scale is shown to be simply related to initial density and velocity gradients in the cloud. Seyfert luminosities ($>10^{43}$erg/s) can be achieved for time scales of 10^5 to 10^7yr. For small clouds, the captured material develops a complex structure on the scale of the accretion radius. This structure bears a resemblance to the morphology of the ionized gas in the central two parsecs of the Galaxy; however, the gas velocity field does not reproduce in detail the observed kinematics of the ionized gas in this region.

067.023 **Supermassive stars in galactic nuclei.**
E. Krügel, A. Tutukov.
Astron. Astrophys., Vol. 158, No. 1/2, p. 367 – 370 (1986).
The authors discuss in the framework of a simple analytical model supermassive stars (SMS) in the mass range between 10^3 to $\sim 5 \times 10^5 M_\odot$. An SMS is expected to possess a strong stellar wind and, if it is located in the potential well of a galactic nucleus, a high accretion rate fed by mass loss from the old stars in the galactic bulge is also very likely. The coexistence of accretion and a strong stellar wind can theoretically lead to an "eternal" SMS, where accretion occurs from a disk and mass loss into all other directions. The authors also present the IR spectra of an SMS with luminosity $3 \times 10^8 L_\odot$ buried in a dust medium closely resembling the inner 300 pc of our galactic center. From comparison of these spectra with observations of our galactic center it is concluded that an SMS of such a luminosity cannot be responsible for either the ionizing flux or for the IR emission. However, a less luminous SMS ($\leqslant 10^8 L_\odot$) cannot be excluded as an additional source of energy and ionization.

067.024 **Production of ultra–high–energy γ–rays by accreting neutron stars.**
Y.–M. Wang.
Astrophys. Space Sci., Vol. 121, No. 1, p. 193 – 200 (1986).
Several well–known binary X–ray sources have been reported to emit copious γ–radiation at energies up to and exceeding 10^{15}eV. It is proposed here that the observed events occur during episodes of non–steady accretion onto neutron stars, when MHD instabilities give rise to vortex motions on very large scales deep inside the magnetosphere. The magnetic lines of force are strongly distorted and reconnect in neutral sheets, along which extremely high voltage drops are maintained and a small fraction of the particles are accelerated to ultra–relativistic energies. The γ–rays are produced in nuclear collisions undergone by runaway ions traversing regions of high–density, diamagnetic plasma in the accretion flow.

067.025 **Erratum: "Revival of the Penrose process for astrophysical applications" [Astrophys. J., Vol. 290, No. 1, p. 12 – 14 (1985)].**
S. M. Wagh, S. V. Dhurandhar, N. Dadhich.
Astrophys. J., Vol. 301, No. 2, p. 1018 (1986). See Abstr. 039.067.066.

067.026 **Models of magnetic white dwarfs.**
D. Adam.
Astron. Astrophys., Vol. 160, No. 1, p. 95 – 106 (1986).
This paper is devoted to the effects brought by the presence of a strong internal magnetic field on the structure of white dwarfs. The spherical equations of hydrostatic equilibrium are solved by using certain approximations. Comparisons with the model of Ostriker and Hartwick (1968) are made and show an excellent agreement. The effects of electrostatic corrections to cold electron pressure, of β–inverse decays and of pycno–nuclear reactions have also been considered. Mass–radius relationships are given for various values of the ratio of the magnetic energy to the gravitational energy of the white dwarf, and for various values of the highest excited Landau level.

067.027 **Strange quark stars.**
P. Haensel, J. L. Zdunik, R. Schaeffer.
Astron. Astrophys., Vol. 160, No. 1, p. 121 – 128 (1986).
The existence of self–bound, or slightly unbound strange quark matter might be highly relevant for physics of neutron stars. In the present paper the authors construct neutron star models. They derive the equation of state of three flavor quark matter. Using two sets of the bag model parameters, the authors calculate two equations of state corresponding to bound and slightly unbound strange quark matter. The strange quark star models, calculated for these equations of state are presented, and their possible astronomical implications are discussed.

067.028 **Generation of neutron star magnetic fields by thermomagnetic effects.**
V. A. Urpin, S. A. Levshakov, D. G. Yakovlev.
Mon. Not. R. Astron. Soc., Vol. 219, No. 4, p. 703 – 717 (1986).
This paper considers a linear stage of the magnetic field generation in outer ($\varrho \lesssim 10^{11} \mathrm{g\,cm^{-3}}$) layers of rather hot rotating neutron stars by thermomagnetic phenomena which accompany the heat transport from the stellar core.

067.029 **X–ray bursts from soft X–ray transients.**
T. Hanawa, M. Y. Fujimoto.
Publ. Astron. Soc. Jpn., Vol. 38, No. 1, p. 13 – 28 (1986).
The thermal evolution of a neutron star in a soft X–ray transient is computed from an off–accretion phase through the first shell flash after an onset of the accretion. The structure and the evolution of the whole neutron star are solved by taking account of general relativity properly. The results are compared with observations of the short–lived event detected about two days before the 1969 outburst of Cen X–4.

067.030 **Steady mass loss from X–ray bursting neutron stars with H–rich envelope.**
M. Kato.
Publ. Astron. Soc. Jpn., Vol. 38, No. 1, p. 29 – 45 (1986).
The structure of a spherically symmetric, steady mass–losing Newtonian He shell with an H–rich envelope on a neutron star is calculated. Along an evolutional path of an X–ray burst, sequences of static and steady mass–loss solutions with and without an H–rich layer on the He layer are constructed. Solutions in the presence of the H–rich layer are presented and discussed.

067.031 **The ratio of the luminosities of the boundary layer and the disk in the case of disk accretion onto a neutron star with weak magnetic field.**
R. A. Syunyaev, N. I. Shakura.
Pis'ma Astron. Zh., Tom 12, No. 4, p. 286 – 292 (1986). In Russian. English translation in Sov. Astron. Lett., Vol. 12.
It is shown that for disk accretion onto a neutron star with weak magnetic field the luminosity of the boundary layer exceeds remarkably the luminosity of the extended accretion disk. In the extreme case of a fast rotating star this ratio decreases.

067.032 **Two–dimensional numerical model of a rotating neutron star with pion condensation.**
Yu. A. Berezin, B. G. Mukanova.
Pis'ma Astron. Zh., Tom 12, No. 5, p. 383 – 387 (1986). In Russian. English translation in Sov. Astron. Lett., Vol. 12.
Numerical solutions of the axisymmetrical equations of gravitational gas dynamics for neutron stars with non–monotonous equation of state associated with pion condensation are studied. The phase transition gives rise to formation of a shock wave and a discontinuity of the rotation velocity at the boundary between two phases with different densities.

067.033 **Models of extended relativistic envelopes of neutron stars.**
B. Paczyński, N. Anderson.
Astrophys. J., Vol. 302, No. 1, p. 1 – 10 (1986).
Models of static, extended, neutron star envelopes are calculated numerically, and a simple analytical approximation of their structure is given. Such envelopes may exist over a range of neutron star luminosities $0.7 < L_\infty/L_E < 1$ if the effects of general relativity and temperature dependence of electron scattering are taken into account. Both effects are shown to be essential for the structure of extended envelopes. Most of the envelope mass is confined to a layer only a few meters thick, while there is very little mass in the extended part, which may be over 100 km thick. These models may be applicable to some X–ray bursts which show evidence for radius expansion.

067.034 **A critical examination of thick disks with equatorial accretion.**
J. H. Hunter Jr., R. E. Wilson.
Astrophys. J., Vol. 302, No. 1, p. 11 – 18 (1986). = Contrib. Dep. Astron., Univ. Florida, No. 84.
The authors develop a method for direct integration of the structural differential equations of the thick disk models with equatorial accretion by Paczyński and Abramowicz. They discuss their method of finding consistent solutions and describe several numerical difficulties. Quantitatively, the authors find differences from Paczyński and Abramowicz, for which they suggest several possible explanations. The structure at small radii is extremely sensitive to the conditions at large radii. The authors outline a procedure for computing equipressure contours in the disk interior, and find them to have a simple form.

067.035 **Gravitational collapse and the cosmic antineutrino background.**
S. E. Woosley, J. R. Wilson, R. Mayle.
Astrophys. J., Vol. 302, No. 1, p. 19 – 34 (1986). = Lick Obs. Bull., No. 1023.
During their collapse to neutron stars or black holes, stellar cores emit an intense flux of electron antineutrinos produced primarily by positron capture and pair annihilation. Providing that the typical energy of these neutrinos exceeds $\sim$10 MeV, it has been suggested that the composite signal of all gravitational collapses that have ever occurred might be visible to detectors currently contemplated for the study of proton decay. Estimates of the antineutrino flux and spectra from Type II supernovae and the gravitational collapse of still more massive stars to black holes are obtained from numerical and semianalytic calculations of models in the mass range $10 - 5\mathrm{x}10^5 M_\odot$. The determined fluxes and spectra probably lie outside the range of detectability, at least in the near future.

067.036 **Wave propagation in pulsar magnetospheres: dispersion relations and normal modes of plasmas in superstrong magnetic fields.**
J. Arons, J. J. Barnard.
Astrophys. J., Vol. 302, No. 1, p. 120 – 137 (1986).
The authors derive the dispersion relations and polarization characteristics of the normal modes of radiation in superstrong magnetic fields, with particular attention to those attributes of importance to the transfer of radiation in the relativistic electron–positron plasmas expected to occur in the magnetospheres of radio pulsars. The distinction between superluminous and subluminous modes is made in the superstrong regime, where drift motions across the field are negligible, and useful formulae for the Landau damping of the subluminous branch of ordinary mode (the Alfvén wave) are derived. These are used to set observational constraints on the geometry of the emission zone in radio pulsars, if the emission mechanism generates radiation in the form of subluminous waves.

067.037 **Wave propagation in pulsar magnetospheres: refraction of rays in the open flux zone.**
J. J. Barnard, J. Arons.
Astrophys. J., Vol. 302, No. 1, p. 138 – 162 (1986).
The authors investigate the propagation of waves through a relativistically outflowing electron–positron plasma in a very strong dipolar magnetic field, conditions expected in pulsar magnetospheres. They derive Hamilton's equations for the propagation of rays through a plasma which is inhomogeneous in density, magnetic field direction, and Lorentz factor. These equations are numerically and analytically solved for rays propagating through the plasmas outflowing along the "open" dipolar field lines in which the density decreases inversely as the radius cubed and in the case where gradients transverse to the radial direction exist.

067.038 **Cosmic superfluidity: the evidence for superfluidity in neutron stars.**
D. Pines.
From SU(3) to gravity, p. 297 – 320 (1985). – See Abstr. 003.014.
Neutron stars are expected to be superfluid on quite general grounds. Following a brief historical review of theoretical calculations which have suggested the existence of at least three distinct superfluids inside neutron stars, the astrophysical evidence for their existence, which comes from the agreement between theory and observation for the observed relaxation of the rotation and spindown rates of the Vela and Crab pulsars, and of

PSR 0525+21, following sudden jumps (glitches) in these quantities, is reviewed. The prospects for observing similar consequences of the pinned crustal neutron superfluid in the recently-discovered millisecond pulsars are examined.

067.039 **A flare-induced cascade model of γ-ray bursts.**
P. A. Sturrock.
Nature, Vol. 321, No. 6065, p. 47 – 49 (1986).
The author proposes that the basic energy-release mechanism of γ-ray bursts is a flare in the magnetosphere of a neutron star. This involves reconnection that leads to an electric field parallel to the magnetic field. This accelerates an electron along the magnetic field, producing high-energy γ rays (by curvature radiation) that promptly annihilate in the magnetic field, resulting in an electron-positron cascade as in radio pulsars. This model offers an explanation of the continuum spectrum, of the 511-keV annihilation line, and possibly of the optical radiation that seems to accompany some γ-ray bursts.

067.040 **X-ray spectra and atmospheric structures of bursting neutron stars.**
T. Ebisuzaki.
Astron. Her., Vol. 79, No. 4, p. 84 – 87 (1986). In Japanese.

067.041 **On the Papaloizou and Pringle instability.**
M. A. Abramowicz, O. M. Blaes, J. Lu.
Structure and evolution of active galactic nuclei, p. 113 – 128 (1986). – See Abstr. 012.011.
The recent discovery of Papaloizou and Pringle that fluid rings are dynamically unstable against global, nonaxisymmetric perturbations is of great importance to the theory of accretion disks. However, before these disks can be proved to be stable or unstable, substantial theoretical progress along the lines sketched in this lecture must be achieved.

067.042 **The birth of AGNs and quasars via the collapse of dense star clusters to supermassive black holes.**
S. L. Shapiro.
Structure and evolution of active galactic nuclei, p. 129 – 147 (1986). – See Abstr. 012.011.
The author considers the fate of a dense cluster of compact stars – neutron stars or stellar-mass black holes – embedded in the nucleus of an active galaxy. The combined effects of two-body dynamical relaxation (the "gravothermal catastrophe") and star-star collisions and coalescence can drive such a cluster to a relativistic state in a Hubble time. When the core becomes sufficiently relativistic, the cluster inevitably undergoes catastrophic gravitational collapse to a supermassive black hole. This process leads naturally to the birth of supermassive black holes of the "right size" to explain quasars and AGNs: $10^6 \lesssim M/M_\odot \lesssim 10^9$.

067.043 **An idealized model for dust accretion in active galactic nuclei.**
J. Buitrago, E. Mediavilla.
Structure and evolution of active galactic nuclei, p. 503 – 507 (1986). – See Abstr. 012.011.
With an adequate supply of matter, the inwards drifting of dust particles induced by the Poynting-Robertson effect can lead to the formation of a shell or disk-like structure around the source of radiation. In a steady state, the Poynting-Robertson effect produce a spatial density of dust going like 1/r which in turn gives rise to a mid-infrared spectrum $F_\nu \propto 1/\nu$. The authors think that this short report could be of interest for the interpretation of the infrared excess found in many extragalactic sources.

067.044 **Effect of dynamical friction on the escape of a supermassive black hole ejected from the center of a galaxy.**
R. C. Kapoor.
Structure and evolution of active galactic nuclei, p. 573 – 577 (1986). – See Abstr. 012.011.
The author has used the impulsive approximation technique to numerically estimate the effect of dynamical friction on the motion of a supermassive black hole (mass $\cong 10^9 M_\odot$) through a galaxy (mass $\cong 10^{11} M_\odot$) which has recoiled from the center of the latter as a result of rocket effect (anisotropic emission of gravitational radiation or plasma emission etc.)

067.045 **Floppy discs: a recipe for the "observers' dream" model.**
M. D. Smith, D. J. Raine.
Structure and evolution of active galactic nuclei, p. 717 – 721 (1986). – See Abstr. 012.011.
Compton-heating of a disc's photosphere by a hard central X-ray source leads to a back-pressure on the disc surface. This pressure dominates low mass disc (LMD) systems, leading to disc asymmetries and, as a result, large asymmetries in the corona/wind in the nuclear environment. Present knowledge of accretion discs indicate that LMD systems may exist in radio galaxies. The authors propose that the environment controls jet visibility, thus producing 'one-sided' jet structures. Bridle's (1984) "Menu for an All-Purpose Source Model" can then be interpreted.

067.046 **Changes in stellar parameters implied by a neutron star corequake.**
P. Haensel, J. L. Zdunik, R. Schaeffer.
Astron. Astrophys., Vol. 160, No. 2, p. 251 – 258 (1986).
First order phase transitions at supranuclear densities may lead to neutron star corequakes. Universal formulae for the changes of the stellar parameters implied by the appearance of a small condensed core of the new phase are given. The only relevant property of the new phase is the relative density jump at the transition pressure. The parameters determining the linear response of the stellar radius, surface redshift, moment of inertia, and energy release are calculated for severel realistic equations of state of the ordinary phase of cold dense matter, in a broad interval of central densities of neutron stars. The results are given in the form of plots suitable for practical use in actual calculations. The linear response parameters represent an additional characteristics of realistic neutron star models.

067.047 **Photodissociation of molecules in the magnetic field of a neutron star.**
V. K. Khersonskii (*V. K. Khersonskij*).
Bull. Spec. Astrophys. Obs. – North Caucasus, Vol. 19, p. 53 – 58 (1985). English translation of 39.067.065.

067.048 **The magnetic moments of neutron stars with various equations of state.**
D. M. Sedrakyan, A. G. Movsisyan.
Astrofizika, Tom 24, Vyp. 2, p. 279 – 286 (1986). In Russian. English translation in Astrophysics, Vol. 24, No. 2.
The magnetic moment generated by superfluid protons dragged by superfluid neutrons is calculated for various stars and various equations of state. The results are of the order of $10^{29} - 10^{31}$ Gauss cm^3.

067.049 **About observational manifestation of pion stars.**
G. S. Adzhyan, Yu. L. Vartanyan.
Astrofizika, Tom 24, Vyp. 2, p. 301 – 306 (1986). In Russian. English translation in Astrophysics, Vol. 24, No. 2.
It has been shown that the Kelvin time of cooling of the nucleus-like superdense cosmic objects (pion stars) is conditioned only by the photon canal of energy loss and is much less than that of ordinary neutron stars. The course of the specific nuclear processes on the surface of such objects in the time of matter accretion can lead to the emergence of a γ-pulsar.

067.050 **Maximum mass of pion stars.**
Yu. L. Vartanyan, G. B. Alaverdyan.
Astrofizika, Tom 24, Vyp. 2, p. 420 – 423 (1986). In Russian. English translation in Astrophysics, Vol. 24, No. 2.
To check the existence of superdense nucleus-like self-confined cosmic objects (pion stars) in the case of a highly hard equation of state ($\partial P/\partial \varrho = 1$) the authors have calculated the maximum values of their masses for various densities at the surface. For such configurations the values of the radius and the relativistic moment of inertia are also calculated.

067.051 **Nonradial g–mode oscillations in X–ray bursting neutron stars.**
P. N. McDermott, R. E. Taam.
Bull. Am. Astron. Soc., Vol. 17, No. 4, p. 849 (1985). Abstract. – See Abstr. 010.062.

067.052 **A flare–induced cascade model of gamma–ray bursts.**
P. A. Sturrock.
Bull. Am. Astron. Soc., Vol. 17, No. 4, p. 850 (1985). Abstract. – See Abstr. 010.062.

067.053 **A model of electromagnetic damping mechanisms for neutron star oscillations.**
B. W. Carroll, E. G. Zweibel, C. J. Hansen, P. N. McDermott, M. P. Savedoff, J. H. Thomas, H. M. Van Horn.
Bull. Am. Astron. Soc., Vol. 17, No. 4, p. 855 (1985). Abstract. – See Abstr. 010.062.

067.054 **Detailed numerical calculations of nucleosynthesis in thick accretion disks around black holes.**
L. Jin, W. D. Arnett, S. K. Chakrabarti.
Bull. Am. Astron. Soc., Vol. 17, No. 4, p. 856 (1985). Abstract. – See Abstr. 010.062.

067.055 **Hydrodynamics near black holes.**
J. F. Hawley.
Bull. Am. Astron. Soc., Vol. 17, No. 4, p. 857 (1985). Abstract. – See Abstr. 010.062.

067.056 **Neutron star evolution with internal energy dissipation by superfluid vortex creep.**
N. Shibazaki, F. K. Lamb.
Bull. Am. Astron. Soc., Vol. 17, No. 4, p. 908 (1985). Abstract. – See Abstr. 010.062.

067.057 **Hypercritical spherical accretion onto compact objects.**
J. M. Blondin.
Bull. Am. Astron. Soc., Vol. 17, No. 4, p. 908 (1985). Abstract. – See Abstr. 010.062.

067.058 **Coherent pulsar radiation from nonneutral plasma.**
F. C. Michel.
Bull. Am. Astron. Soc., Vol. 17, No. 4, p. 908 (1985). Abstract. – See Abstr. 010.062.

067.059 **Accreting neutron stars in close binary systems.**
S. A. Ilovaisky.
ESA Spec. Publ., ESA SP–239, p. 49 – 58 (1985). – See Abstr. 012.019.

The author briefly reviews the current status of our knowledge about accretion phenomena on neutron stars in X–ray binary systems and highlights the role played by X–ray spectroscopic measurements with particular emphasis on recent results from the Tenma and EXOSAT X–ray observatories.

067.060 **X–ray spectra of bursting neutron stars: effects of Comptonization and bound–free opacity.**
T. Ebisuzaki, K. Nomoto.
ESA Spec. Publ., ESA SP–239, p. 111 – 116 (1985). – See Abstr. 012.019.

The authors calculate atmospheric models and emitted X–ray spectra for X–ray bursting neutron stars taking into account the effects of Comptonization and free–free and bound–free emission and absorption. The color temperatures obtained from the calculated X–ray spectra are significantly higher than the effective temperatures. The theoretical color temperature to luminosity diagram is in good agreement with the observational diagram of MXB 1636–536 which was observed with both TENMA and EXOSAT.

067.061 **A relativistic rotating massive object with saturation of magnetic monopoles.**
Q. Peng.
ESA Spec. Publ., ESA SP–239, p. 127 – 128 (1985). – See Abstr. 012.019.

067.062 **Forces exerting on a test particle outside a relativistic rotating massive object with saturation of magnetic monopoles and jets from the object.**
Q. Peng.
ESA Spec. Publ., ESA SP–239, p. 129 – 130 (1985). – See Abstr. 012.019.

067.063 **Unravelling fates of black holes.**
D. N. Page.
Nature, Vol. 321, No. 6066, p. 111 (1986).

067.064 **Models of radiation–driven winds from general relativistic neutron stars.**
B. Paczyński, M. Prószyński.
Astrophys. J., Vol. 302, No. 2, p. 519 – 529 (1986).

Fully relativistic models of winds driven by the super–Eddington luminosity of X–ray bursting neutron stars are presented. It is found that all the super–Eddington energy flux is used to gently blow off matter with a terminal velocity of the order of 0.01 c, and that the radiative luminosity never exceeds the Eddington limit by more than 1%. The photospheric radius of an outflowing envelope is always at least 10 times larger than the radius of a neutron star. There is an almost continuous transition between static extended envelopes of neutron stars that have a slightly sub–Eddington luminosity and outflowing envelopes that are slightly super–Eddington and have a low mass outflow rate.

067.065 **Black hole as thermal machine and quantum source of energy.**
I. D. Novikov.
Zemlya Vselennaya, No. 2, p. 19 – 24 (1986). In Russian.

067.066 **Black hole stability.**
C. C. Leiby Jr.
Astrophys. Space Sci., Vol. 122, No. 1, p. 157 – 161 (1986).

A vector theory of electromagnetism and gravitation has indicated a possible equivalence of gravitational energy and electric charge. If true, the resulting electromagnetic and gravitational forces within a stationary, gravitationally collapsed body of radius $R = GM/c^2$ are everywhere in balance within the body, if it has a radial mass density distribution proportional to $1/r^2$. In addition, radial perturbation of such a body will result in a force imbalance which is restorative. Hence, the equilibrium is stable.

067.067 **On gravitational field energy of a charged black hole.**
N. Dadhich, V. Chellathurai.
Mon. Not. R. Astron. Soc., Vol. 220, No. 2, p. 555 – 558 (1986).

By using the Lynden–Bell & Katz invariant definition for gravitational field energy for static spherical systems, the authors compute the field energy density for a charged sphere. It turns out that the field energy gets decreased by the presence of charge and in particular for a maximally charged ($m = e$) black hole it is equal to $1/2\ Mc^2$. Unlike the Schwarzschild black hole, wherein the entire energy lies outside the hole, in this case it is shared by both interior and exterior of the hole, and for $m = e$ it is shared equally. The authors also establish a complementary relationship between the Penrose and the Lynden–Bell & Katz definitions for energy inside and outside of a charged black hole.

067.068 **Estimates of the maximum angular velocity of rotating neutron stars.**
L. Lindblom.
Astrophys. J., Vol. 303, No. 1, p. 146 – 153 (1986).

The instabilities that limit the rotation rate of neutron stars are driven by gravitational radiation reaction and moderated by viscosity. This paper estimates the angular velocities where these instabilities set in. These estimates improve on previous work in two ways: (1) by approximating the influence of the dissipative mechanisms on the locations of the critical angular velocities, and (2) by using fully relativistic calculations of some of the relevant frequencies and time scales. The calculations show that the critical rotation periods are very insensitive to the value of the viscosity but depend strongly on the equation of state of the nuclear matter and on the mass of the neutron star.

067.069 **Probing the magnetic field of radio pulsars: a reexamination of polarization position angle swings.**
J. J. Barnard.
Astrophys. J., Vol. 303, No. 1, p. 280 – 291 (1986).
The author calculates the polarization position angle as a function of pulse longitude and polarization limiting radius in radio pulsars assuming a vacuum magnetic field. He also estimates the polarization radius r_{pl}, based on the calculations of Cheng and Ruderman and Stinebring. It is found that for rotation periods $P < 0.06$ s this radius is $r_{pl} \approx 2 \times 10^5 (P/1\ s)^{-2}$cm and for $P > 0.06$ s it is $r_{pl} \approx 9 \times 10^8 (P/1\ s)^{0.4}$cm for a surface magnetic field strength of 10^{12}G and radio frequency of 10^9Hz. For short rotation periods, r_{pl} approaches the light cylinder radius r_{lc}.

067.070 **The evolution of neutron stars.**
W. Sutantyo.
Galactic structure and variable stars, p. 88 – 94 (1985). – See Abstr. 012.029.
The author discusses the evolution of neutron stars in the BP (magnetic field – pulse period) diagram in the way described by Radhakrishnan and Srinivasan (1981). Neutron stars in close binary systems will follow different evolutionary tracks from those of single neutron stars. The author suggests that the fast X–ray pulsar 4U 0115+63 and A 0538–66 are young neutron stars, while Cen X–3 and SMC X–1 are "recycled" pulsars.

067.071 **The origin of high velocity pulsars.**
W. Sutantyo.
Proceedings of the Second Asian–Pacific Regional Meeting on Astronomy, p. 432 – 440 (1984). – See Abstr. 012.031.
Theories of the origin of high velocity pulsars are critically reviewed. The author argues that if observations confirm the alignment between the velocity vectors and the spin axis, mostly, only neutron stars which originate from single stars can be detected as runaway radio pulsars. Those only form some 1/4 of all neutron stars. On the other hand, if the velocity vectors are proven to be perpendicular to the spin axis, high velocity pulsars may have originated from both, single stars and binary systems, or only from binary systems, depending on the nature of the explosion.

067.072 **Internal torques in neutron stars.**
M. A. Alpar.
X–ray astronomy '84, p. 239 – 242 (1985). – See Abstr. 012.033.
An understanding of the dynamics of neutron star interiors is important for studies of X–ray sources containing neutron stars, in order to separate possible effects of internal torques from those of accretion torques. A summary of recent results on the dynamical properties of the neutron star superfluids are given.

067.073 **A pulsar magnetosphere model with streaming.**
D. A. Leahy.
X–ray astronomy '84, p. 251 – 254 (1985). – See Abstr. 012.033.
A fully coordinate invariant formalism is used to express the physical laws of the magnetosphere in a rotating frame. This allows time–independence to be applied for any arbitrary pulsar surface field. Here several stages in the magnetosphere model are outlined, starting with the force free case, then adding particle inertia and streaming.

067.074 **On hydrodynamics of radiatively driven winds.**
R. Turolla, L. Nobili, M. Calvani.
Astrophys. J., Vol. 303, No. 2, p. 573 – 581 (1986).
The problem of radiation–driven spherically symmetric winds is reexamined. The relativistic hydrodynamical equations are derived both for an optically thick and for an optically thin flow in a Schwarzschild background. A detailed analysis clarifies the relevance of critical and subcritical points in the quest for a regular solution. Numerical models are constructed, and their connection with astrophysical phenomena is briefly discussed.

067.075 **Rotating wind solution and the acceleration of the cosmic radio jets.**
S. K. Chakrabarti.
Astrophys. J., Vol. 303, No. 2, p. 582 – 588 (1986).
A wind–type solution is obtained in an axisymmetric spacetime in the case when matter has angular momentum. An application of the solution in understanding the acceleration of matter in radio jets as it flows along the funnel of a thick accretion disk is suggested. It is observed that the location of the critical point depends strongly upon the angular momentum of the wind. Apart from the usual acceleration of matter due to the expansion of the flow tube, the energy due to the rotational motion is found to contribute to the poloidal motion, hence enhancing the flow speed.

067.076 **Pulse frequency variations in pulsars.**
F. K. Lamb.
Japan – U.S. seminar on galactic and extragalactic compact X–ray sources, p. 19 – 32 (1985). – See Abstr. 012.034.
Recent theoretical work on pulsars, coupled with extremely precise measurements of pulse frequency variations, has led to fundamental changes in our understanding of the dynamical properties of neutron stars. This research also indicates that new work is needed on the physics of accretion flows in massive X–ray binaries.

067.077 **How do X–ray pulsars make their X–rays?**
J. H. Krolik.
Japan – U.S. seminar on galactic and extragalactic compact X–ray sources, p. 33 – 39 (1985). – See Abstr. 012.034.
Although a general consensus exists that the X–rays from binary X–ray pulsars come from accretion hot spots on the surface of rotating neutron stars, there is very little knowledge of how the kinetic energy of accreting matter is transformed into X–rays. Qualitative arguments point to the outgoing radiation itself as the primary agent decelerating the accretion flow, but real calculations are made extremely difficult by the complicated character of the photon–electron cross section in a strong magnetic field.

067.078 **X–ray spectra of bursting neutron stars and color temperature – luminosity diagram.**
T. Ebisuzaki, K. Nomoto.
Japan – U.S. seminar on galactic and extragalactic compact X–ray sources, p. 101 – 107 (1985). – See Abstr. 012.034.
The authors calculate atmospheric models and emitted X–ray spectra for X–ray bursting neutron stars taking into account the effects of Comptonization and free–free and bound–free emission and absorption. The color temperatures obtained from the calculated X–ray spectra are higher than the effective temperatures. The theoretical color–temperature to luminosity diagram is in good agreement with the observational diagram of MXB 1636–536 which was observed with Tenma satellite.

067.079 **Accretion flow.**
S. Hayakawa.
Japan – U.S. seminar on galactic and extragalactic compact X–ray sources, p. 165 – 173 (1985). – See Abstr. 012.034.
A scenario of accretion flow from a companion star to a compact object is described to give an overview together with salient points of some papers only orally presented at the workshop.

067.080 **Neutron star formation in close binaries: collapse of accreting white dwarfs.**
K. Nomoto.
Japan – U.S. seminar on galactic and extragalactic compact X–ray sources, p. 227 – 232 (1985). – See Abstr. 012.034.
The evolutionary origin of low mass X–ray binaries, i.e., the origin of a neutron star in close binaries is discussed. In particular the condition for accreting white dwarfs to collapse is examined. The O + Ne + Mg white dwarf with a subgiant companion is the most likely candidate of the neutron star progenitor of the low mass X–ray binaries. Also demonstrated is the evolution of merging double white dwarfs to form a single neutron star. This could be the origin of a single millisecond pulsar.

067.081 **Some comments on neutron star formation.**
S. Tsuruta, K. Nomoto.
Japan – U.S. seminar on galactic and extragalactic compact X–ray sources, p. 245 – 249 (1985). – See Abstr. 012.034.

Comparing the results from the latest neutron star cooling calculations with the observational data, it is suggested that many neutron stars may be formed through processes other than supernova explosions.

067.082 **The physics of black holes.**
R. H. Price.
Japan – U.S. seminar on galactic and extragalactic compact X–ray sources, p. 251 – 260 (1985). – See Abstr. 012.034.

A new formalism is described for conceptually and calculationally dealing with black holes in astrophysical contexts. Though the formalism is not an approximation it allows the application of non–general–relativistic astrophysical insight and intuition. As an illustration the formalism is used to elucidate the nature of the Blandford–Znajek energy mechanism for active galactic nuclei.

067.083 **Cosmic X–ray jets.**
E. D. Feigelson.
Japan – U.S. seminar on galactic and extragalactic compact X–ray sources, p. 301 – 309 (1985). – See Abstr. 012.034.

Jets, or material ejected by compact objects in collimated beams, have been found in a wide variety of astrophysical circumstances. They can occur in radio galaxies, quasars and other active galaxies; in protostars and T Tauri stars; in X–ray binaries; and in pulsars associated with supernova remnants. A few jets have been detected in the X–ray band.

067.084 **Thermal convection and generation of a large–scale magnetic field in the inner region of the accreting disc of black holes.**
D. G. Lominadze, G. D. Chagelishvili.
Probl. nelinejn. i turbulent. protsessov v fiz. Tr. 2 Mezhdunar. rab. gruppy, Kiev, 1983. Chast' 2. Kiev, p. 311 – 318 (1985). In Russian. Abstr. in Ref. Zh., 51. Astron., 5.51.205 (1986).

067.085 **Electromagnetic torques and alignment of neutron stars.**
K. K. Ng.
Diss. Abstr. Int., Sect. B, Vol. 46, No. 6, p. 1958 – 1959 (1985). Thesis, State University of New York at Stony Brook, 93 pp. (1985). Order No. DA8518373.

067.086 **Nonradial oscillations of neutron stars.**
P. N. McDermott.
Diss. Abstr. Int., Sect. B, Vol. 46, No. 8, p. 2693 (1986). Thesis, The University of Rochester, 223 pp. (1985). Order No. DA8516481.

067.087 **Buoyancy limits on magnetic viscosity stress law scalings in quasi–stellar object accretion disk models.**
P. J. Sakimoto.
Diss. Abstr. Int., Sect. B, Vol. 46, No. 9, p. 3092 (1986). Thesis, University of California, Los Angeles, 270 pp. (1985). Order No. DA8525879.

067.088 **The quantum ergosphere and dynamical temperature of the Reissner–Nordström black hole.**
B. R. Smith Jr.
Diss. Abstr. Int., Sect. B, Vol. 46, No. 11, p. 3883 (1986). Thesis, The University of North Carolina at Chapel Hill, 84 pp. (1985). Order No. DA8527323.

067.089 **Gravitational collapse and rotation: a quandary for the black–holes.**
F. de Felice.
Sir Arthur Eddington Centenary Symposium, Vol. 3: Gravitational radiation and relativity, p. 224 – 235 (1986). – See Abstr. 012.036.

A necessary condition that a rotating black hole forms in the course of gravitational collapse, is that $a/m < 1$, where a is the specific angular momentum of the collapsing body and m its total mass (in geometrized units). In this paper the values of (a/m) are estimated at different stages of the stellar evolution with the purpose of outlining an evolutionary track for that ratio, from the pre–main–sequence to the core collapse phase, and to see whether there exist conditions which might realistically inhibit formation of the black holes.

067.090 **Cross section for the capture of test particles by a charged collapsar.**
A. B. Gaina.
Sov. Astron., Vol. 29, No. 5, p. 589 – 591 (1985). English translation of 40.067.133.

067.091 **QED and X–ray pulsars.**
M. C. Allen, D. B. Melrose, A. J. Parle.
Proc. Astron. Soc. Aust., Vol. 6, No. 2, p. 211 – 214 (1985).

Techniques in QED (quantum electrodynamics) have been developed previously allowing one to treat electron–photon and photon–photon interactions exactly in the magnetized vacuum and allowing one to include the effects of a medium. These techniques are extended to include particle–particle interactions. Exact cross–sections for electron–electron collisions are derived and compared with known expressions. Such calculations have application in studies of the formation and transfer of radiation in the atmospheres surrounding neutron stars.

067.092 **Rapidly rotating neutron star models.**
J. L. Friedman, J. R. Ipser, L. Parker.
Astrophys. J., Vol. 304, No. 1, p. 115 – 139 (1986).

Results are reported from a numerical investigation of the structure of rapidly rotating relativistic models, based on equations of state proposed for neutron star matter. Sequences of rotating stars with baryon mass approximately equal to 1.4 $M_\odot$ were constructed using 10 equations of state (EOSs). Along each sequence the angular velocity increases from zero to the Keplerian frequency that marks the termination point. For the four stiffest EOSs, the upper limit on rotation set by gravitational instability is consistent with the observed frequency of the fast pulsar. Upper limits on mass, baryon mass, moment of inertia, red– and blueshifts, equatorial velocity, and on the parameter cJ/GM^2 are found for representative EOSs.

067.093 **The central engine of quasars and active galactic nuclei: hadronic interactions of shock–accelerated relativistic protons.**
D. Kazanas, D. C. Ellison.
Astrophys. J., Vol. 304, No. 1, p. 178 – 187 (1986).

The authors consider spherical accretion on a massive black hole as the energy source for active galactic nuclei and quasars. In this model the kinetic energy of the infalling material is randomized by a collisionless shock around the black hole. Relativistic protons are produced with high efficiency via first–order Fermi shock acceleration and can provide sufficient pressure to support the shock. The model can naturally produce ultrarelativistic electrons with flat spectra, a feature which, when coupled with e^+–e^- pair production, can account for the observed continuum radiation from AGNs.

067.094 **Effect of vacuum polarization near black holes.**
V. P. Frolov, A. I. Zel'nikov.
Proceedings of the Third Seminar on Quantum Gravity, p. 413 – 445 (1985). – See Abstr. 012.038.

This paper presents the main formulas for the Green's functions in the spacetime of a black hole. These formulas are then used for investigating the contribution of massless fields to the vacuum polarization near black holes. The contribution of massive fields is also discussed.

067.095 **Runaway instability in the inner cooling region of optically thin, bremsstrahlung disks around Kerr black holes.**
H. G. Paul.
Eruptive phenomena in stars, p. 383 – 384 (1986). – See Abstr. 012.039.

067.096 **Gamma–ray burst annihilation lines and neutron star structure.**
E. P. Liang.
Astrophys. J., Vol. 304, No. 2, p. 682 – 687 (1986).
The author considers the neutron star redshift data base provided by the recent measurements of gamma–ray burst pair annihilation lines in the range 300 – 511 keV. Despite the controversies surrounding their reality and interpretations, it is shown that there is tentative evidence to support a neutron star redshift range of 0.2 – 0.5 with the highest concentration in the narrow range 0.25 – 0.35. Combined with mass data from X–ray and binary pulsars, these data favor the softer equation of state neutron star models. Prospects of detecting gravitational waves from events like that of 1979 March 5 are also discussed.

067.097 **Black hole models for quasar–like objects: some selected topics.**
S. Tsuruta.
Galaxies, quasars and cosmology, p. 61 – 94 (1985). – See Abstr. 012.042.
After summarizing the characteristic properties of "quasar–like objects" this paper reviews supermassive black hole models for these sources and examines in particular emission mechanisms operating in accretion disks.

067.098 **Absence of super–radiance of the Dirac particles in the Kerr–Newman geometry and the weak positive–energy condition.**
S. M. Wagh, N. Dadhich.
Phys. Rev. D, Vol. 32, No. 8, p. 1863 – 1865 (1985). Abstr. in Phys. Abstr., Vol. 89, No. 1273, Entry 18829 (1986).

067.099 **On the energy emission by a Kerr black hole in the superradiant range.**
A. Curir.
Phys. Lett. B, Vol. 161B, No. 4 – 6, p. 310 – 312 (1985). Abstr. in Phys. Abstr., Vol. 89, No. 1274, Entry 23859 (1986).

067.100 **Gamma ray bursts.**
J. M. Hameury.
Ann. Phys. (Paris), Vol. 10, No. 5, p. 369 – 413 (1985). In French. Abstr. in Phys. Abstr., Vol. 89, No. 1274, Entry 23913 (1986).

067.101 **On the interpretation of multi–peak structure observed in some X–ray bursts from 4U/MXB 1636–53.**
M. Y. Fujimoto, M. Sztajno, W. H. G. Lewin, J. van Paradijs.
MPA Rep., No. 228, 12 pp. (1986). Submitted to Astron. Astrophys.

067.102 **Thermal evolution of accreting neutron stars. II. Long X–ray bursts as a probe into the interior.**
M. Y. Fujimoto, T. Hanawa, I. Iben Jr., M. B. Richardson.
MPA Rep., No. 238, 30 pp. (1986). Submitted to Astrophys. J.

067.103 **Two photon emission in X–ray pulsars: models of the phase dependent spectra and the pulse–shapes.**
J. G. Kirk, W. Nagel, M. C. Storey.
MPA Rep., No. 240, 19 pp. (1986). Submitted to Astron. Astrophys.

067.104 **General relativistic decollimation of a jet.**
R. C. Kapoor.
Indian Inst. Astrophys., Prepr., No. 147, 13 pp. (1985). Submitted to Astrophys. Lett.

067.105 **Central engine or locomotive?**
R. C. Kapoor.
Indian Inst. Astrophys., Prepr., No. 195, 16 pp. (1985).
The astrophysical aspects of proposals of escape of supermassive black holes from galactic nuclei and their implications in relation to quasars are discussed. It is concluded that high velocity recoil of the central engine can at best be considered as an exception rather than a rule since it requires a violent release of mass energy in an asymmetrical manner, spectacular in nature.

067.106 **An upper limit on the mean mass of black holes in the galactic halo.**
H. Kamahori, M. Fujimoto.
Publ. Astron. Soc. Jpn., Vol. 38, No. 2, p. 151 – 159 (1986).
The possible existence of black holes in the galactic halo and their dynamical influence on the structure of the Galaxy are discussed. Massive black holes would scatter disk stars gravitationally to make both their velocity dispersion and the disk thickness very large. The authors find that when the whole halo mass is referred to the black holes, a mean mass of the halo black holes must be smaller than $2 \times 10^6 M_\odot$ in order for the thin disk of our Galaxy to have remained in the present structure for 2×10^{10}yr without further destruction.

067.107 **Discussion about pulsar's braking index n.**
G.–j. Qiao, X.–j. Wu, H. Zhang, X.–y. Xia.
Kexue Tongbao, Vol. 30, No. 8, p. 1062 – 1067 (1985).
The pulsar's braking index $n = \ddot{\Omega}\Omega/\dot{\Omega}^2$ (Ω is the angular rate of a pulsar) can be completely determined through observation. Because n is independent of the parameters of the neutron star and a direct connection between n and theoretical models can be built, this braking index, n, has got much attention. The authors test the present pulsar theories by comparing theoretical n values with observational ones.

067.108 **Effect of two superposed Schwarzschild black holes' field on the proper area of the horizon.**
Y.–c. Wang.
Kexue Tongbao, Vol. 30, No. 9, p. 1272 – 1273 (1985). Abstr. in Phys. Abstr., Vol. 89, No. 1275, Entry 24194 (1986).

067.109 **Statistical entropy of a Schwarzschild black hole.**
M. Alexanian.
J. Stat. Phys., Vol. 41, No. 3 – 4, p. 709 – 717 (1985). Abstr. in Phys. Abstr., Vol. 89, No. 1275, Entry 30298 (1986).

067.110 **Quantum state for a black hole in a de Sitter universe.**
R. Balbinot, R. Bergamini.
Lett. Nuovo Cimento, Vol. 44, Ser. 2, No. 6, p. 423 – 428 (1985). Abstr. in Phys. Abstr., Vol. 89, No. 1276, Entry 36211 (1986).

067.111 **Accretion disks and periodic outbursts of active galactic nuclei.**
D. N. C. Lin, G. A. Shields.
Astrophys. J., Vol. 305, No. 1, p. 28 – 34 (1986). = Lick. Obs. Bull., No. 1032.
The authors examine the local thermal stability of accretion disks around supermassive black holes in active galactic nuclei. Such disks are unstable at radii where the surface temperature is several thousand degrees. Supermassive disks therefore should undergo limit–cycle outbursts similar to those believed to occur in dwarf novae. Operating on a time scale of $\sim 10^4 - 10^7$yr this mechanism will result in alternating periods of higher and lower accretion rate onto the black hole and, consequently, higher and lower luminosity. Quasi–periodic outbursts on this time scale may be recorded in the structure of extended radio sources, a possible example being 4C 29.47.

067.112 **Magnetic field decay and the origin of neutron star binaries.**
R. E. Taam, E. P. J. van den Heuvel.
Astrophys. J., Vol. 305, No. 1, p. 235 – 245 (1986).
The authors show that the origin and evolution of magnetized neutron stars in binary systems can be consistently interpreted within the framework of the surface magnetic field decay model for radio pulsars with a decay time scale of order 5×10^6yr. The progenitors of these magnetized neutron stars appear to fall into two classes. Either they are the helium core remnants resulting from a case B mass exchange phase, in which case the neutron star forms in the core collapse following the Fe photodisintegration (or electron capture) phase, or they are massive white dwarfs which have been driven over the Chandrasekhar mass limit by accretion. The immediate progenitors of the binary radio pulsars PSR 0655+64 and PSR 1913+16 fall into the

former category, and those of PSR 0820+02 and PSR 1953+29 fall into the latter category.

067.113 **The mass and radius of a neutron star in X–ray burster MXB 1636–536.**
M. Y. Fujimoto, R. E. Taam.
Astrophys. J., Vol. 305, No. 1, p. 246 – 250 (1986).
The observational data of the X–ray burst source MXB 1636–536, obtained from the Tenma satellite, have been used to place constraints on the mass and radius of a neutron star. The results of theoretical models for the emergent spectral energy distribution for bursting neutron stars are combined with the transverse Doppler and gravitational redshift interpretation of the 4.1 keV absorption line. The authors find a neutron star mass and radius of 1.45 $M_{\odot}$ and 10.3 km.

067.114 **On the stream–disk interaction in accreting compact objects.**
M. Livio, N. Soker, R. Dgani.
Astrophys. J., Vol. 305, No. 1, p. 267 – 280 (1986).
The interaction between the stream of material originating from the inner Lagrangian point L_1 and the accretion disk is studied in the framework of an idealized picture of hypersonic flow. When strong simplifying assumptions are made, fully analytic solutions are found for the physical quantities of the shocked matter flowing above and below the disk surface. The authors then solve the *three–dimensional* problem numerically, using a pseudoparticle method. The results are compared with observations of compact X–ray sources and cataclysmic variables.

067.115 **Confronting X–ray observations of young supernova remnants with neutron star cooling models.**
K. Nomoto, S. Tsuruta.
Astrophys. J., Lett. Ed., Vol. 305, No. 1, p. L19 – L22 (1986).
"Standard" cooling models of neutron stars are obtained by solving exactly the fully general relativistic version of stellar structure–evolution equations, with the best physical input currently available. It is found that for medium to stiff equations of state, it takes $\sim 10^{3-5}$yr to reach the isothermal state. These time scales would be considerably longer than the ages of most young supernova remnants. The authors report results obtained by comparing these theoretical models with the Einstein data of young supernova remnants.

067.116 **The distribution of hyper–relativistic matter around a black hole.**
L.–z. Liu, Q.–h. Peng.
Publ. Purple Mt. Obs., Vol. 4, No. 4, p. 1 – 4 (1985). In Chinese.
The distribution of hyper–relativistic matter around a black hole is discussed. The authors assume that the black hole and the matter around it are in the static state with spherical symmetry. The measurement is non–static, because the black hole is accumulating. The authors do not consider the variation of measurement with time.

067.117 **Maser instability theory of emission lines from Her X–1.**
D.–y. Wang, D.–y. Mao.
Publ. Purple Mt. Obs., Vol. 5, No. 1, p. 1 – 6 (1986). In Chinese.

067.118 **Black holes and quantum mechanics.**
G. t'Hooft.
Recent developments in quantum field theory, p. 237 – 249 (1985). Abstr. in Phys. Abstr., Vol. 89, No. 1277, Entry 36608 (1986). – See Abstr. 012.047.

067.119 **Vacuum polarization of the electromagnetic field near a rotating black hole.**
V. P. Frolov, A. I. Zel'nikov.
Phys. Rev. D, Vol. 32, No. 12, p. 3150 – 3163 (1985). Abstr. in Phys. Abstr., Vol. 89, No. 1277, Entry 36611 (1986).

067.120 **The globally singularity–free space–time of Schwarzschild black hole.**
S.–c. Zhu, J.–m. Zhu, D.–x. Tang.
Chin. Phys. Lett., Vol. 2, No. 10, p. 445 – 448 (1985). Abstr. in Phys. Abstr., Vol. 89, No. 1277, Entry 42296 (1986).

067.121 **Kaluza–Klein black hole in a six–dimensional Poincaré gauge theory of gravitation.**
R.–R. Hsu, W. B. Yeung.
Phys. Lett. A, Vol. 113A, No. 4, p. 197 – 199 (1985). Abstr. in Phys. Abstr., Vol. 89, No. 1277, Entry 42297 (1986).

067.122 **Black holes in two spacetime dimensions.**
J. D. Brown, M. Henneaux, C. Teitelboim.
Phys. Rev. D, Vol. 33, No. 2, p. 319 – 323 (1986). Abstr. in Phys. Abstr., Vol. 89, No. 1278, Entry 42597 (1986).

067.123 **An analytic representation for the quasi–normal modes of Kerr black holes.**
E. W. Leaver.
Proc. R. Soc. London, Ser. A, Vol. 402, No. 1823, p. 285 – 298 (1985). Abstr. in Phys. Abstr., Vol. 89, No. 1278, Entry 42598 (1986).

067.124 **Exterior solutions for electric and magnetic stars in 2+1 dimensions.**
M. A. Melvin.
Classical Quantum Gravity, Vol. 3, No. 1, p. 117 – 131 (1986). Abstr. in Phys. Abstr., Vol. 89, No. 1278, Entry 42601 (1986).

067.125 **Modelling of gas dynamical processes with nonmonotonic equations of state.**
Yu. A. Berezin, O. E. Dmitrieva, N. N. Yanenko.
Math. Modelling, Vol. 6, No. 5, p. 413 – 419 (1985). Abstr. in Phys. Abstr., Vol. 89, No. 1278, Entry 47882 (1986).

067.126 **γ–positronium in strong magnetic fields.**
L. B. Leinson, V. N. Oraevsky (*V. N. Oraevskij*).
Phys. Lett. B, Vol. 165B, No. 4 – 6, p. 422 – 424 (1985). Abstr. in Phys. Abstr., Vol. 89, No. 1278, Entry 47885 (1986).

067.127 **Accretion disks and periodic outbursts of AGN.**
G. A. Shields, D. N. C. Lin.
Bull. Am. Astron. Soc., Vol. 18, No. 1, p. 633 (1986). Abstract. – See Abstr. 010.066.

067.128 **Dynamics of an affine star model in a black hole tidal field.**
J.–P. Luminet, B. Carter.
Astrophys. J., Suppl. Ser., Vol. 61, No. 2, p. 219 – 248 (1986).
Tidal deformations of a body moving in the field of a large black hole are analyzed within the framework of an affine star model. A number of results are given, including study of the incompressible case which rounds off the classical work on the subject, detailed calculation of the tidal disruption limit for parabolic orbits of polytropes, and discussion of the effects of radiation pressure when the star is highly compressed by the tidal forces.

067.129 **Classical and relativistic models of X–ray bursters.**
R. Aquilano, M. Castagnino.
Relativity, supersymmetry and cosmology, p. 248 – 252 (1985). – See Abstr. 012.055.

067.130 **An optically thin inner accretion torus around a weakly magnetized neutron star.**
I. Hishikawa, R. Hoshi.
Prog. Theor. Phys., Vol. 74, No. 5, p. 1158 – 1160 (1985). Abstr. in Phys. Abstr., Vol. 89, No. 1280, Entry 60828 (1986).

067.131 **Gravitational repulsion by a Reissner–Nordström black hole.**
M. Barua.
Indian J. Pure Appl. Phys., Vol. 23, No. 12, p. 634 – 635 (1985). Abstr. in Phys. Abstr., Vol. 89, No. 1280, Entry 60934 (1986).

067.132 **Are black holes black bodies?**
T. D. Lee.
Nucl. Phys. B, Part. Phys., Vol. B264, No. 2 – 3, p. 437 – 486 (1986). Abstr. in Phys. Abstr., Vol. 89, No. 1280, Entry 60936 (1986).

067.133 **Transonic accretion flow in a thin disk around a black hole. III. Analytic considerations.**
B. Muchotrzeb–Czerny.
Acta Astron., Vol. 36, No. 1, p. 1 – 17 (1986).
Analytic considerations presented in this paper help to understand in physical terms the difference between the solutions describing the disk accretion in the cases of high and low viscosity. The author discusses the connection between viscosity and the type of critical point of the transonic solutions describing the inflow of the matter. The analytic results are presented for the isothermal flow. The character of solutions and some considerations of the stability of isothermal flow give an argument that the accretion with high viscosity may be nonstationary.

067.134 **Mass shedding and partition of the *a/m* ratio between core and envelope in gravitational collapse.**
F. de Felice, Y.–q. Yu.
Mon. Not. R. Astron. Soc., Vol. 220, No. 3, p. 737 – 743 (1986).
The authors show that, even taking into account redistribution of angular momentum, the ratio (a/m) ($a/m = cJ/GM^2$, where J and M are the total angular momentum and gravitational mass) of a collapsing and rotating body varies slowly with the mass, when mass shedding takes place. Thus formation of an extended structure outside a collapsing body, like rings, discs or diffuse matter, is not in general a guarantee that the ratio (a/m) of the inner object is decreased appreciably from its initial value.

067.135 **A kind of supermassive dense celestial body (not black hole).**
Q.–h. Peng, Y.–j. Wang.
Kexue Tongbao, Vol. 30, No. 12, p. 1586 – 1590 (1985). Abstr. in Phys. Abstr., Vol. 89, No. 1281, Entry 67509 (1986).

067.136 **Plasma instabilities in the magnetosphere of pulsars.**
J. G. Lominadze, G. I. Melikidze, A. D. Pataraya.
Proceedings of the 1984 International Conference on Plasma Physics, Vol. 2, p. 1043 – 1066 (1985). Abstr. in Phys. Abstr., Vol. 89, No. 1281, Entry 67521 (1986). – See Abstr. 012.056.

067.137 **Membrane viewpoint on black holes: properties and evolution of the stretched horizon.**
R. H. Price, K. S. Thorne.
Phys. Rev. D, Vol. 33, No. 4, p. 915 – 941 (1986). Abstr. in Phys. Abstr., Vol. 89, No. 1281, Entry 67522 (1986).

067.138 **Nucleonic superfluidity in neutron stars: $^1S_0$ neutron pairing in the inner crust.**
J. M. C. Chen, J. W. Clark, E. Krotscheck, R. A. Smith.
Nucl. Phys. A, Vol. A451, No. 3, p. 509 – 540 (1986). Abstr. in Phys. Abstr., Vol. 89, No. 1282, Entry 73276 (1986).

067.139 **Nuclear processes at neutron star surfaces.**
R. E. Taam.
Annu. Rev. Nucl. Part. Sci., Vol. 35, p. 1 – 23 (1985). Abstr. in Phys. Abstr., Vol. 89, No. 1282, Entry 73277 (1986). – See Abstr. 003.041.

067.140 **The Hawking radiation of charged Dirac particles in Kerr–Newman–de Sitter spacetime.**
D.–y. Xu, H.–y. Wang.
Acta Sci. Nat. Univ. Pekin., No. 5, p. 63 – 74 (1985). In Chinese. Abstr. in Phys. Abstr., Vol. 89, No. 1282, Entry 73432 (1986).

067.141 **Particle acceleration in the "unfavorable" zone of a pulsar.**
L.–f. Chen, S. r. Liang.
Chin. Phys. Lett., Vol. 3, No. 3, p. 137 – 140 (1986). Abstr. in Phys. Abstr., Vol. 89, Nr. 1284, Entry 83591 (1986).

067.142 **Interaction of fermions with black holes.**
T. O. Jones III.
Phys. Rev. D, Vol. 33, No. 8, p. 2494 – 2495 (1986). Abstr. in Phys. Abstr., Vol. 89, Nr. 1284, Entry 83596 (1986).

067.143 **Effect of hyperons on the physical properties of neutron stars.**
Z.–c. Liu, S.–r. Liang, S.–h. Gao.
Chin. Astron. Astrophys., Vol. 10, No. 1, p. 30 – 33 (1986). English translation of 40.067.127.

067.144 **Accretion disk instabilities in the Rapid Burster.**
S. Hayakawa.
Cosmic pathways, p. 243 – 248 (1986). – See Abstr. 003.045.
Rapidly repetitive X–ray bursts from MXB 1730–335 (Rapid Burster) are considered to be due to accretion instabilities. The present paper shows that a simple disk model can explain essential properties of the rapidly repetitive bursts.

067.145 **Cyclotron emission near stellar mass black holes.**
K. M. V. Appa Rao (*K. M. V. Apparao*).
Cosmic pathways, p. 249 – 254 (1986). – See Abstr. 003.045.
Cyclotron emission in the inner regions of an accretion disk around a matter accreting black hole can be appreciable. In the case of the X–ray source Cyg X–1, cyclotron emission may provide the soft photons needed for "Comptonization" to produce high–energy X–rays. The inverse correlation between the fluxes of high–energy and low–energy X–rays during the "high" and "low" states of Cyg X–1, may be understood as a result of the variation of the rate of accretion and the Compton scattering of the cyclotron photons.

067.146 **Accretion onto neutron stars with the presence of a double layer.**
A. C. Williams, M. C. Weisskopf, R. F. Elsner, W. Darbro, P. G. Sutherland.
Astrophys. J., Vol. 305, No. 2, p. 759 – 766 (1986).
It is known, from laboratory experiments, that double layers will form in plasmas, usually in the presence of an electric current. In this paper the authors argue that a double layer may be present in the accretion column of a neutron star in a binary system. It is suggested that the double layer may be the predominant deceleration mechanism for the accreting ions, especially for sources with X–ray luminosities of less than $\sim 10^{37}$ergs s^{-1}.

067.147 **Oscillation spectra of neutron stars with strong magnetic fields.**
B. W. Carroll, E. G. Zweibel, C. J. Hansen, P. N. McDermott, M. P. Savedoff, J. H. Thomas, H. M. Van Horn.
Astrophys. J., Vol. 305, No. 2, p. 767 – 783 (1986).
The authors have investigated the effects of a strong vertical magnetic field on the oscillation spectrum of a cylindrical slab model for the surface layers of a neutron star. In particular, they have considered the effects on those modes known to be concentrated in the surface layers: the pseudo–toroidal *t*– and *a*–modes and the pseudo–spheroidal *s*–, *i*–, and *m*/*g*–modes. The authors have derived and solved the Newtonian pulsation equations for the cylindrical model. The electromagnetic boundary conditions at the neutron star surface match the magnetohydrodynamic motions within the star to outgoing electromagnetic radiation in the evacuated waveguide above the surface; pseudo–toroidal modes couple to TM electromagnetic modes in the waveguide, while pseudo–spheroidal modes match to TE radiation. The periods of the *t*– and *s*–modes are similar to the quasi–periodicities observed in some pulsars.

067.148 **Optical reprocessing of gamma–ray bursts.**
F. Melia, S. Rappaport, P. C. Joss.
Astrophys. J., Lett. Ed., Vol. 305, No. 2, p. L51 – L55 (1986).
One model for the optical flashes associated with three cosmic γ–ray burst sources invokes the reprocessing of some of the γ–radiation emitted by a hypothesized collapsed object in the surface layers of a nearby companion star. The authors have investigated this model by carrying out detailed, fully hydrodynamical calculations of such reprocessing in the surface layers of

very low mass stars. It is found that, at most, ~7% of the γ-ray fluence incident on the companion star is reprocessed into the blue band; the time scale for this reprocessing is typically ~100 s.

067.149 **X-ray spectra of bursting neutron stars and the color temperature versus luminosity diagram.**
T. Ebisuzaki, K. Nomoto.
Astrophys. J., Lett. Ed., Vol. 305, No. 2, p. L67 – L70 (1986).
The authors calculate atmospheric models and emitted X-ray spectra for X-ray bursting neutron stars taking into account the effects of Comptonization and free-free and bound-free emission/absorption. The theoretical color temperature versus luminosity diagram is in good agreement with the observational diagram for MXB 1636-536. This is used to infer the distances to the burster and the galactic center, and the mass and radius of the bursting neutron star.

067.150 **On the stability of accretion disks.**
C. M. Filho.
Rev. Mex. Astron. Astrofis., Vol. 12, Spec. Issue, p. 319 – 326 (1986). – See Abstr. 012.066.
The stability of standard α accretion disks against perturbations in the radial direction is taken in the light of linear perturbation theory. This analysis starts from slightly different, but correct, continuity and momentum equations, besides imposing null boundary conditions on the perturbations and continuity along the flow. With such a treatment, one does obtain more stable disks, leading to some modifications in the results earlier obtained by Shakura and Sunyaev (1975).

067.151 **Interaction of the accretion disk with the magnetosphere of a rotating neutron star.**
U. Anzer, G. Börner.
Plasma penetration into magnetospheres, p. 101 – 114 (1986). – See Abstr. 012.071.

067.152 **Centrifugally driven winds from X-ray pulsars.**
J. Arons.
Plasma penetration into magnetospheres, p. 115 – 124 (1986). – See Abstr. 012.071.
The author claims that magnetized neutron stars accreting plasma from a disk can strip gas off the disk surfaces and throw the plasma away at high speed ($\sim 10^4$km s^{-1}), in the form of a centrifugally driven wind, an idea previously outlined by Anzer and Börner (1980) using a rather different geometry and constituent physics. The author has reached a number of conclusions.

067.153 **Structure of the magnetospheres of accreting magnetic compact objects.**
J. J. Aly.
Plasma penetration into magnetospheres, p. 125 – 140 (1986). – See Abstr. 012.071.
The author reviews some of the work which has been recently done to try to understand the interaction between the magnetic field of a strongly magnetized compact object and either a thin keplerian accretion disk, or a spherically symmetric accretion flow. In each case, he pays particular attention to the following problems: (1) how the external plasma interacting with the magnetosphere can get mixed with the field; (2) by which mechanism does the magnetic field control the transfer of mass-momentum-energy to the compact object. In conclusion, the author compares the magnetosphere of an accreting compact object with that one of a planet.

067.154 **Physics of polar cap accretion onto neutron stars.**
J. Arons.
Plasma penetration into magnetospheres, p. 141 – 142 (1986). Abstract. – See Abstr. 012.071.

067.155 **Emission characteristics of neutron star polar caps.**
P. Mészáros.
Plasma penetration into magnetospheres, p. 143 – 152 (1986). – See Abstr. 012.071.
Accretion powered X-ray pulsars show phase dependent spectral effects which give clues for determining the geometry and physical conditions of the emitting regions. Recent model calculations are discussed, which indicate that the behaviour of the cyclotron line features, the continuum spectrum and the pulse shapes can be reproduced with a pencil beam model, with magnetic field perpendicular to the emitting surface.

067.156 **Hydrodynamics of accretion columns.**
H. Herold, K. Wolf, H. Ruder.
Mitt. Astron. Ges., No. 65, p. 214 – 215 (1986). – See Abstr. 012.072.

067.157 **Polarized radiation transport in accretion columns.**
H. Herold, T. Maile, H. Ruder.
Mitt. Astron. Ges., No. 65, p. 215 – 217 (1986). – See Abstr. 012.072.

067.158 **Photoionization of positronium in strong magnetic fields.**
F. Geyer, H. Herold, H. Ruder, G. Wunner.
Mitt. Astron. Ges., No. 65, p. 220 – 221 (1986). – See Abstr. 012.072.

067.159 **Quasi-periodic oscillations as a probe of galactic-bulge X-ray sources.**
F. K. Lamb.
The evolution of galactic X-ray binaries, p. 151 – 171 (1986). – See Abstr. 012.076.
Quasi-periodic oscillations with frequencies in the range 5 – 50 Hz have recently been discovered in X-rays from two bright galactic-bulge sources and from Sco X-1. A shot-noise model that provides a mathematical framework for analyzing the constraints imposed on physical models by the observed power density spectra is described. A variety of physical models are then examined briefly. One of the most promising, the beat-frequency-modulated accretion model, is adopted as a working hypothesis for interpreting the galactic-bulge sources.

067.160 **On the nature of the quasiperiodic oscillations of low mass X-ray binaries.**
G. E. Morfill, J. Truemper.
The evolution of galactic X-ray binaries, p. 173 – 181 (1986). – See Abstr. 012.076.
Physical arguments, based on the available observations, are used to determine where the quasiperiodic emission, found in some low mass X-ray binaries, originates. A model is presented, where the accretion rate of the neutron star is close to saturation, and the stellar emission is heavily blanketed. The quasiperiodic emission originates in the inner edge of the accretion disc as the result of a beating between density inhomogeneities (plasmoids) in essentially Keplerian rotation, and the stirring of the disc by the stellar magnetic field.

067.161 **Do neutron star magnetic fields decay?**
W. Kundt.
The evolution of galactic X-ray binaries, p. 263 – 269 (1986). – See Abstr. 012.076.
It is argued that neutron-star magnetic dipole moments may not decay within 10^{10}yr. Dying pulsars may align in a quasi-oscillatory manner. In "cool" binary system, accretion of matter from the inner edge of an accretion disk onto the surface of the neutron star may occur in the form of discrete clumps of metallic hydrogen, taking the shape of thin blades.

067.162 **Inversion symmetric initial data for *N* charged black holes.**
J. M. Bowen.
Ann. Phys. (N.Y.), Vol. 165, No. 1, p. 17 – 37 (1985). Abstr. in Phys. Abstr., Vol. 89, No. 1275, Entry 24189 (1986).

Bibliography of publications on white holes.
See Abstr. 002.001.

Black holes: an annotated bibliography, 1975 – 1983.
See Abstr. 002.094.

Black holes, white dwarfs and neutron stars. The physics of compact objects.
See Abstr. 003.178.

The physics of accretion onto compact objects. Report on a meeting, held at Tenerife, 21 – 25 April 1986.
See Abstr. 011.020.

The hazards of encountering a black hole. I. Testing the limits of physics.
See Abstr. 014.037.

Aspects of supergravity theories.
See Abstr. 061.009.

York's cosmic time versus proper time as relevant to changes in the dimensionless "constants", *K*–meson decay, and the unity of black hole and big crunch.
See Abstr. 061.020.

Using an X–ray color–color diagram to search for hot neutron stars and set a limit on the magnetic monopole flux.
See Abstr. 061.028.

Toward a proof of entropy increase in the presence of quantum black holes.
See Abstr. 061.035.

Dominance of B^{10} or C^{12} nuclear production during gravitational collapse in the stellar interior?
See Abstr. 061.041.

Quantum fluctuation effects due to baryon resonances in nuclear mean field theory.
See Abstr. 061.065.

Proposed test of a strange matter pulsar by observation of the neutrino flux from Cygnus X–3.
See Abstr. 061.117.

Can neutrinos from Cygnus X–3 be seen by proton decay detectors?
See Abstr. 061.142.

Formation of collimated beams.
See Abstr. 062.006.

Non neutral plasmas in the laboratory and astrophysics.
See Abstr. 062.071.

Accretion discs.
See Abstr. 062.133.

Bremsstrahlung in hot, strongly magnetized plasmas.
See Abstr. 062.134.

One–photon annihilation in strongly magnetized relativistic electron–positron pair plasmas.
See Abstr. 062.135.

One–photon pair annihilation in magnetized relativistic plasmas.
See Abstr. 063.001.

The radiative acceleration of astrophysical jets: line locking in SS 433.
See Abstr. 063.025.

Stokes parameters for Thomson scattering in a strong magnetic field.
See Abstr. 063.034.

Quantum corrections to relativistic synchrotron emission.
See Abstr. 063.054.

Conflicting guiding center motions in cyclotron radiation.
See Abstr. 063.067.

Recombination and ionisation in strongly magnetized thermal plasmas of accreting neutron stars.
See Abstr. 063.068.

Radiative transfer in the accretion column of X–ray pulsars: effects from the hot spot.
See Abstr. 063.069.

Radiation of ultrarelativistic electrons moving through a dense medium in the presence of a strong magnetic field.
See Abstr. 063.101.

Accretion disks.
See Abstr. 064.070.

Neutrino diffusion in stellar collapse.
See Abstr. 065.001.

The phase transitions of superdense matter and supernova explosions.
See Abstr. 065.011.

Speculations on the fizzled collapse of a massive star.
See Abstr. 065.014.

On the behavior of double degenerate binaries associated with Type I supernovae.
See Abstr. 065.107.

Explosive disruption of stars by big black holes.
See Abstr. 065.139.

Nucleosynthesis in pancake stars.
See Abstr. 065.140.

Evolution of close binaries and the formation of neutron stars.
See Abstr. 065.158.

The duration of the detached phase in low–mass X–ray binaries in which the neutron star was formed by accretion–induced collapse of a white dwarf.
See Abstr. 065.159.

Precession of angular momentum vector in a slowly rotating Kerr metric.
See Abstr. 066.013.

On the local geometry of rotating matter.
See Abstr. 066.069.

Diffraction in gravitational lensing for compact objects of low mass.
See Abstr. 066.092.

Gravitational light deflection effects in eclipsing binary pulsars.
See Abstr. 066.107.

New soliton–like exact solutions of general relativity equations modelling rotating supermassive astrophysical objects.
See Abstr. 066.109.

Electron–positron pair production by ultrarelativistic electrons in a soft photon field.
See Abstr. 066.115.

Must nonspherical collapse produce black holes? A gravitational confinement theorem.
See Abstr. 066.118.

Estimation of the gravitational radiation (GR) from the pulsar PSR 1937 + 214.
See Abstr. 066.121.

Improved characterization of the Kerr metric.
See Abstr. 066.150.

The new formula for gravitational radiation energy loss: applications to the axisymmetric two–body problem and the binary pulsar.
See Abstr. 066.157.

Point particle limit and the far–zone quadrupole formula in general relativity.
See Abstr. 066.183.

Two mass solutions of Einstein's vacuum equations: the double Kerr solution.
See Abstr. 066.189.

The zero–point field in non–inertial frames.
See Abstr. 066.211.

Tenma observation of the X–ray pulsar 4U1626–67.
See Abstr. 117.027.

X–ray emissions from Vela X–1 during its eclipsing period.
See Abstr. 117.028.

Periodic optical line intensity variations in X Persei.
See Abstr. 117.036.

The most reliable black hole yet.
See Abstr. 117.105.

The black–hole candidate A0620–00.
See Abstr. 117.116.

Inverse Compton models of SS 433 and radio quiet AGN.
See Abstr. 117.118.

Triple star evolution and the formation of short–period, low–mass X–ray binaries.
See Abstr. 117.159.

On the dust jet model of SS 433.
See Abstr. 117.220.

***V/R* variability in X–ray emitting Be star binaries.**
See Abstr. 117.234.

The single–line Wolf–Rayet star EZ CMa revisited: additional evidence for a 3.8 day binary period.
See Abstr. 117.238.

Neutron star evolution in eccentric–orbit binaries.
See Abstr. 117.278.

Black holes in massive close binaries: observational data and evolutionary status.
See Abstr. 117.280.

What do we know about Cyg X–1.
See Abstr. 117.288.

***IRAS* observations of binaries with compact objects.**
See Abstr. 117.297.

A note on the nature of the secondaries in ultra–compact binaries.
See Abstr. 117.305.

EXOSAT observations of the 35–day cycle of Her X–1; evidence for neutron star precession.
See Abstr. 117.312.

On the evolution of Her X–1.
See Abstr. 117.313.

Rapid ultraviolet spectral variations in HD 50896 (WN5 + ?).
See Abstr. 117.327.

Spectroscopic and polarimetric parameters of the runaway WN7 binary system HD 197406: is the secondary an X–ray–quiet black hole?
See Abstr. 120.018.

Neutron stars in supernova remnants.
See Abstr. 125.038.

Feasibility of a low–mass binary source progenitor for PSR1937 + 214.
See Abstr. 126.003.

The radio luminosity and magnetic field of pulsars.
See Abstr. 126.012.

Toward an empirical theory of pulsar emission. III. Mode changing, drifting subpulses, and pulse nulling.
See Abstr. 126.020.

Longitudinal plasma waves generated in the atmosphere of pulsars.
See Abstr. 126.028.

Pulse timing and neutron star structure.
See Abstr. 126.055.

Are pulsars born as slow rotators?
See Abstr. 126.058.

EXOSAT measurements of the quiescent X–ray spectrum of the burster 2S 1636–536: evidence for Comptonization in a hot corona.
See Abstr. 142.001.

The interpretation of blackbody spectra and radii during X–ray bursts.
See Abstr. 142.009.

Angular distribution and polarization of X–ray burster radiation.
See Abstr. 142.011.

Quasi–periodic oscillations in galactic X–ray sources.
See Abstr. 142.027.

The creation of compact objects in the Local Group.
See Abstr. 142.082.

Variability of apparent radii in the decay part of type I X–ray bursts.
See Abstr. 142.109.

Constraints on neutron star models of gamma–burst sources from the Einstein Observatory.
See Abstr. 143.009.

Les sursauts gamma cosmiques.
See Abstr. 143.012.

Cosmic gamma ray bursts.
See Abstr. 143.033.

Gamma ray bursts.
See Abstr. 143.034.

Is Cygnus X–3 a quark star?
See Abstr. 144.025.

Enhancement of high–energy cosmic–ray spectrum by type–II supernovae.
See Abstr. 144.075.

Effects of viscous drag on a slowly rotating supermassive star due to an embedded stellar system.
See Abstr. 151.038.

Black holes and local dark matter.
See Abstr. 155.022.

Infrared emission from the galactic center.
See Abstr. 155.075.

On wind–type flows in astrophysical jets. II. Propagation outside the nucleus and the case of M87.
See Abstr. 158.024.

Gas in galactic centres.
See Abstr. 158.031.

Internal dynamics and formation of emission clouds in active galactic nuclei.
See Abstr. 158.094.

Theory of active galactic nuclei.
See Abstr. 158.179.

Active galactic nuclei. I. Central source as a cluster of accreting black holes.
See Abstr. 158.193.

Possible evidence for a massive black hole in a quasar with double redshift systems.
See Abstr. 159.115.

Is there a black hole in the sky?
See Abstr. 159.118.

Primordial black holes and observational restrictions on quantum gravity.
See Abstr. 161.189.

Does thermodynamics require a new expansion after the "big crunch" of our cosmos?
See Abstr. 161.296.

Formation of black holes in quantum cosmology.
See Abstr. 161.390.

Sun

071 Photosphere, Spectrum

071.001 **S V line ratios in the Sun.**
P. L. Dufton, A. Hibbert, F. P. Keenan, A. E. Kingston, G. A. Doschek.
Astrophys. J., Vol. 300, No. 1, p. 448 – 452 (1986).
New atomic data for S V have been used to predict level populations and emission–line intensity ratios for electron densities and temperatures appropriate to the Sun. The electron impact collision rates for spin–forbidden transitions and the intercombination transition spontaneous radiative rate are substantially larger than those previously published. The calculated intensity ratios show correspondingly significant changes. Line ratios deduced from observations obtained with the NRL slit spectrograph aboard Skylab are in good agreement with the new theoretical values.

071.002 **Ändert sich die Struktur der Granulation mit dem solaren Aktivitätszyklus?**
P. N. Brandt.
Sterne Weltraum, 25. Jahrg., Nr. 3, p. 128 – 134 (1986).

071.003 **Short term profile variations of three Fe I photospheric lines in solar quiet regions at the disk center.**
F. Cavallini, G. Ceppatelli, A. Righini, N. Alamanni.
Astron. Astrophys., Vol. 156, No. 1/2, p. 310 – 314 (1986).
Recent measurements of the 7699 Å K I line by other authors have shown that the line asymmetry, during the 5–min oscillations, varies in antiphase with the line shift. The authors have measured, at the center of the solar disk, the bisector variations of three Fe I photospheric lines at 6297.8 Å, 6301.5 Å and 6302.5 Å. The oscillating power of the line bisectors and of the line flanks is distributed in several modes, the highest one is at 3.45 mHz. The power decreases along the bisector from the line core to the continuum, confirming previous results on the 7699 Å K I line. The oscillating power, at equal intensity levels in all the three observed lines, is generally different in the two line flanks, and it decreases differently from the line core to the continuum.

071.004 **Solar Fe II line asymmetries and wavelength shifts.**
D. Dravins, B. Larsson, Å. Nordlund.
Astron. Astrophys., Vol. 158, No. 1/2, p. 83 – 88 (1986).
Convective motions of solar granulation are manifest in the spatially unresolved spectrum as slight asymmetries and wavelength shifts of photospheric spectral lines. In a previous paper (Dravins et al., 1981) that dependence for Fe I lines with line strength, excitation potential and wavelength region was analyzed. This paper extends that work to Fe II lines, examining bisector shapes and wavelength shifts of "unblended" Fe II lines both at disk center and in integrated sunlight. Fe II lines form predominantly in the hotter and denser regions of the deep photosphere, and these different line formation conditions for Fe II manifest themselves in well–defined differences from Fe I: the average Fe II bisectors show a more articulated curvature and a larger convective blueshift. Synthetic spectral lines, computed from a three–dimensional time–dependent hydrodynamic simulation of solar photospheric convection confirm the observed behavior.

071.005 **Long–term width and asymmetry variation of some Fe I photospheric lines in solar quiet regions at the disk center.**
F. Cavallini, G. Ceppatelli, A. Righini.
Astron. Astrophys., Vol. 158, No. 1/2, p. 275 – 278 (1986).
Observations obtained in the years 1983 and 1984 at the Donati Solar Tower in Arcetri show that at the disk center the asymmetry and width of three Fe I photospheric lines show a long term variation. These results are in agreement with previous full disk observations of secular trends in asymmetry and equivalent width.

071.006 **The absolute energy distributions of the Sun, of the 'solar analogs' 16 Cyg B, Hyades VB 64, 16 Cyg A, and of the standard stars α Lyr and 29 Psc.**
H. Neckel.
Astron. Astrophys., Vol. 159, No. 1/2, p. 175 – 188 (1986).
The absolute energy distributions of the solar analog 16 Cyg B, which have been published by Hardorp (1980) and Taylor (1984), are compared with the Sun's absolute energy distribution by Neckel and Labs (1984). The magnitude differences star minus Sun reveal an overall, $1/\lambda$–proportional gradient corresponding to $\Delta\Theta = \Delta(5040/T) \approx 0.009 \pm 0.002$, on which is superimposed a wavy pattern with amplitudes being generally near $0^m.02$. Regarding also the differences (1) between observed and model–predicted energy distributions, (2) between 'old' and 'new' energy distribution of α Lyr, and (3) between the energy distributions of different solar analogs, it seems that most if not all of the wavy pattern is caused by systematic, wavelength–dependent errors originating somewhere in the long sequence of observations rather than by genuine characteristics of the stellar or solar radiation.

071.007 **The effect of radiation on the population of oxygen levels in the solar atmosphere. III. The radiation field of O I bound–free continua.**
N. G. Shchukina.
Kinematika Fiz. Nebesn. Tel, Tom 2, No. 1, p. 14 – 19 (1986). In Russian. English translation in Kinematics Phys. Celest. Bodies.
The effect of solar radiation in the bound–free continuum of neutral oxygen on the population of high–excited O I levels and lower O II levels is studied. The Lyman continuum radiation field in the upper chromosphere is shown to rise the pumping of high–excited levels of neutral oxygen. Peculiarities of departure of the oxygen ionization equilibrium from LTE are caused by deviation of the mean bound–free intensity from the Planck function. In the absence of the chromosphere oxygen over–ionization occurs in the upper layers of the solar atmosphere.

071.008 **The correction of observed limb darkening for instrumental and atmospheric stray light.**
S. N. Osipov.
Kinematika Fiz. Nebesn. Tel, Tom 2, No. 1, p. 24 – 27 (1986). In Russian. English translation in Kinematics Phys. Celest. Bodies.
The method of accounting for instrumental and atmospheric stray light is analysed. The concrete recommendations to the application of the method are given on the basis of comparison of computed and observed intensity distribution in the halo and at the solar limb.

071.009 **Continuous and weak–line absorption by neutral iron in the solar atmosphere.**
M. Rudkjøbing.
Astron. Astrophys., Vol. 160, No. 1, p. 132 – 134 (1986).
An investigation carried out by Gustafsson and Bell (1979) shows the ratio of observed to calculated spectral intensity of light from the centre of the solar disc to decrease from about unity in the blue to 0.9 at about the Balmer limit and further to 0.8 at about 3300 Å. It is the aim of the present paper to estimate the order of magnitude of the total absorption in those weak Fe I lines and continua that cannot be dealt with separately in stellar–atmosphere calculations, in order to see whether the intensity discrepancy of the solar spectrum might be due to this part of the iron absorption.

071.010 **The magnetic split factors for selected solar spectral lines in the wavelength range λ4400 – 6750 Å calculated by the experimental values of Landé factors combination terms.**
M. D. Gusejnov.
Izv. Krymskoj Astrofiz. Obs., Tom 70, p. 51 – 57 (1985). In Russian. English translation in Bull. Crimean Astrophys. Obs., Vol. 70.
Using the experimental values of Landé factors combination terms g_{ex}, which were estimated with high precision the author has computed the Zeeman splitting factors $\bar{g}_{ex}$ for 57 Fraunhofer lines of the solar spectrum in the wavelength range λ4400 – 6750 Å. The comparison with the theoretical values $\bar{g}_{theor}$ (for the case of L–S coupling) for the same spectral lines showed that the values $\bar{g}_{ex}$ and $\bar{g}_{theor}$ are principally in good agreement, though, however, in some cases they are markedly different.

071.011 **An investigation of the CH molecular line profiles in the solar photosphere. III. Determination of the CH molecular line formation depths.**
N. A. Drake, N. V. Ehzerin'sh.
Soln. Dannye, Byull., 1985, No. 11, p. 72 – 76 (1985). In Russian.
The formation depths of a number of CH molecular lines in the solar photosphere were calculated by the method of weight functions, the HSRA model of the solar atmosphere has been used. The centre–to–limb dependence of the effective formation depths of these lines was obtained. The change of the formation depths was investigated in passing from the centre of the lines towards the wing. The CH molecular lines are shown to be formed in the photospheric layer located much lower than the temperature minimum region.

071.012 **Flare–related mass motion in the photosphere.**
T. I. Redyuk.
Soln. Dannye, Byull., 1985, No. 11, p. 87 – 94 (1985). In Russian.
Using spectral observations line–of–sight velocities are derived at several photospheric heights in a flare active region at various moments with respect to a flare. Prior to the flare, some lines reveal strengthened motion to the observer; in the flare course the velocities are mostly negative and increased, while after it they can change their direction or decrease.

071.013 **On the role of the collisional processes in the formation of the O I solar resonance doublet 1355–8 Å.**
N. G. Shchukina.
Soln. Dannye, Byull., 1985, No. 12, p. 49 – 51 (1986). In Russian.
The importance of collisional excitation of the $^5S^0$–level and collisional coupling between higher triplet–quintets levels for formation of solar oxygen doublet 1355–8 Å is considered. The collisional coupling of these levels causes a considerable dependence of doublet intensity on Lyman pumping of the $^3D^0$–level O I.

071.014 **On the transfer equation for a depression.**
B. T. Babij, V. B. Kobylyanskij.
Vestn. L'vov. univ. Ser. astron., No. 59, p. 3 – 4 (1985). In Russian. Abstr. in Ref. Zh., 51. Astron., 2.51.413 (1986).

071.015 **On the behaviour of the carbon line C I 538.0322 nm on the solar disk.**
B. T. Babij, I. V. Ozhiivskij, R. E. Rykalyuk.
Vestn. L'vov. univ. Ser. astron., No. 59, p. 5 – 10 (1985). In Russian. Abstr. in Ref. Zh., 51. Astron., 2.51.414 (1986).

071.016 **Application of a two–dimensional model to the interpretation of the equivalent widths of absorption lines of solar granulation.**
M. M. Koval'chuk, P. A. Olijnyk.
Vestn. L'vov. univ. Ser. astron., No. 59, p. 11 – 14 (1985). In Russian. Abstr. in Ref. Zh., 51. Astron., 2.51.415 (1986).

071.017 **Intensity variations of solar short–wave radiation in the solar activity cycle.**
A. A. Nusinov.
Tr. Inst. prikl. geofiz. Gos. kom SSSR po gidrometeorol. i kontrolyu prirod. sredy, No. 65, p. 25 – 36 (1985). In Russian. Abstr. in Ref. Zh., 51. Astron., 2.51.481 (1986).

071.018 **An initial study of granulation using data from the SOUP (Solar Optical Universal Polarimeter) instrument on Spacelab 2.**
T. D. Tarbell, A. M. Title.
Bull. Am. Astron. Soc., Vol. 17, No. 4, p. 833 (1985). Abstract. – See Abstr. 010.062.

071.019 **Association of small, bright photospheric features with magnetic fields.**
K. Topka, R. Shine.
Bull. Am. Astron. Soc., Vol. 17, No. 4, p. 834 (1985). Abstract. – See Abstr. 010.062.

071.020 **Two–dimensional differential photometry of low contrast white light solar structures.**
E. J. Seykora.
Bull. Am. Astron. Soc., Vol. 17, No. 4, p. 834 (1985). Abstract. – See Abstr. 010.062.

071.021 **Observations of the Ca II K spectral line in He 10830 Å dark points.**
R. Holt, N. Fortune, D. Braun, D. Mullan.
Bull. Am. Astron. Soc., Vol. 17, No. 4, p. 933 (1985). Abstract. – See Abstr. 010.065.

071.022 **Energy distribution in the spectrum of the sun as a star and UBV photometry.**
E. A. Makarova, A. V. Kharitonov.
Astron. Tsirk., No. 1406, p. 5 – 8 (1985). In Russian.

071.023 **The center–to–limb wavelength shift of a number of Fraunhofer lines.**
J. C. LoPresto, A. K. Pierce.
Sol. Phys., Vol. 102, Nos. 1/2, p. 21 – 27 (1985).
The center–to–limb wavelength shifts of the cores of faint, medium and strong Fraunhofer lines are presented: Fe I 3767, Ca II 3968, Ce II 4222, Ca I 4227, Na I 5896, Ba II 6497, and K I 7699.

071.024 **Optical imaging spectroscopy.**
R. C. Canfield.
NASA Conf. Publ., NASA CP–2421, p. 120 – 131 (1986). – See Abstr. 012.021.
The author reviews the technological and operational advances, some of the physical understanding that has come about as a result, and the possibilities for the future.

071.025 **An approximate calculation of the effect of opacity on Si III emission line ratios near the solar limb.**
F. P. Keenan, A. E. Kingston.
Mon. Not. R. Astron. Soc., Vol. 220, No. 2, p. 493 – 500 (1986).
The authors analyse the Si III spectrum observed by the Naval Research Laboratory's spectrograph (S082–B) aboard Skylab as the slit was stepped across the disc and out above the limb. They interpret the ratio between a pair of these lines as due to the effect

of opacity, and show that its variation is consistent with a simple model for the atmosphere. In this way the authors are able to calculate the optical depths of several transitions in Si III, both at the solar limb and disc centre, which are found to compare favourably with those estimated from solar model D of Gabriel.

071.026 **Evolution of "exploding granules".**
O. Namba.
Astron. Astrophys., Vol. 161, No. 1, p. 31 – 38 (1986).
The author presents the results of a morphological study on the evolution of "exploding granules" on the surface of the sun, on the basis of high–resolution photographs from the Stratoscope–I film B. A case study has been done, including microphotometry, for a typical exploding granule. The size and the intensity depression of the central darkening are found to evolve in parallel with the expansion of the granule itself in the rapidly growing phase.

071.027 **Solar granulation.**
R. Van der Borght, P. A. Fox.
Proceedings of the Second Asian–Pacific Regional Meeting on Astronomy, p. 483 – 487 (1984). – See Abstr. 012.031.
The present model of solar granulation, in which the parameters vary across the layer, is based on the anelastic approximation and assumes that the convection does not alter appreciably the mean density and temperature distributions in the layer.

071.028 **Mg VII and Si IX line ratios in the sun.**
F. P. Keenan, A. E. Kingston, K. M. Aggarwal, K. G. Widing.
Sol. Phys., Vol. 103, No. 2, p. 225 – 230 (1986).
Recent *R*–matrix calculations of electron excitation rates for Mg VII and Si IX are used to determine the theoretical density sensitive emission line ratios R_1 and R_2. These are found to be quite similar to the earlier results of Mason and Bhatia. Electron densities derived using observed R_1 and R_2 ratios from Skylab NRL XUV spectra of solar flares and active regions are in good agreement, and compare favourably with those deduced from ions formed at similar electron temperatures to Mg VII and Si IX.

071.029 **On the difference in the depths of formation of the centers of weak Fraunhofer lines and the far wings of strong ones.**
Eh. A. Gurtovenko, V. A. Sheminova.
Sov. Astron., Vol. 29, No. 5, p. 572 – 574 (1985). English translation of 40.071.015.

071.030 **Determination of turbulent velocities in the solar photosphere from weak cyanogen lines in the spectral range 4179 – 4213 Å.**
G. A. Porfir'eva.
Astron. Zh., Tom 63, Vyp. 3, p. 550 – 555 (1986). In Russian. English translation in Sov. Astron., Vol. 30, No. 3.
Weak cyanogen–line profiles (the violet system of bands with $\upsilon' - \upsilon'' = -1$) are analyzed with the aim of redetermination of turbulent velocities in the upper layers of the solar photosphere. From the Liège Solar Atlas some spectral regions of the width 0.3 – 0.6 Å were examined. For the solar disk center resulting Doppler widths, central depths and improved central wavelengths of the elemental components of the cyanogen blends have been obtained. The determined vertical component of the total photospheric velocity field $\xi = 2.1 \text{ km s}^{-1}$ has been found to be appreciably smaller than the value $\xi = 3.5 \text{ km s}^{-1}$ obtained earlier from the analysis of mean–resolution spectra of CN.

071.031 **Granulation deformation near and in sunspot regions.**
M. Collados, E. Marco, J. C. del Toro, M. Vázquez.
Sol. Phys., Vol. 105, No. 1, p. 17 – 25 (1986).
High–resolution white–light pictures are analyzed to study the differences between the granular size near sunspot penumbrae and in light–bridges presenting granular structure and that of the quiet photosphere. No difference is found between the mean granular diameter in light–bridges and the quiet photosphere. The dispersion found in the results corresponding to different zones around the sunspots indicates that the size of the granulation may vary from place to place near the sunspots, its mean value not differing significantly from that of the quiet photosphere. A possible systematic bias in the selection of the granules by Macris (1979) is found.

071.032 **Abundance of rare–earth elements in the solar photosphere.**
V. N. Gorshkov, V. A. Komarovskij.
Astron. Zh., Tom 63, Vyp. 3, p. 563 – 566 (1986). In Russian. English translation in Sov. Astron., Vol. 30, No. 3.
Lifetimes for excited states of the first ions of rare–earth elements obtained are used to correct the *gf* values and the abundances of rare–earths in the solar photosphere determined earlier in other papers. Data for the abundances of the rare–earth elements in the solar photosphere are compared with their abundances in chondrite meteorites. The lifetimes for 4 levels in Pr II, 5 levels in Sm II and 2 levels in Gd II are measured to calculate the *gf* values of the spectral lines.

071.033 **Detailed analysis of solar photospheric parameters by quantitative imagery.**
A. Dollfus, D. Crussaire, E. Pernod, A. Lioure.
C. R. Acad. Sci., Sér. II, Tome 303, No. 2, p. 153 – 158 (1986). In French.
The solar refractor at Observatoire de Meudon produces images of the solar surface through a birefringent filter of high spectral selectivity, combined with a spectral profile modulator which yields several transmission band configurations. The images are calibrated, they are obtained without slits, holes or scans and do not discretize the field in pixels, but are made with short exposure pictures which preserve the telescopic image sharpness. For observations at Meudon Observatory with a 28 cm diameter solar refractor, the velocity and magnetic images exhibit 1,000 km size structures at the solar surface. On September 2, 1984, the spots of a same group were separated by a shear area with sharp gradients reaching 450 *G* and 300 m s^{-1} over 1,000 km, which produced a flare.

071.034 **Electrical conductivity in the solar photosphere and chromosphere.**
J. Kubát, M. Karlický.
Bull. Astron. Inst. Czech., Vol. 37, No. 3, p. 155 – 163 (1986).
The method of computing of isotropic, Pedersen, Hall and Cowling conductivities for the model of a quiet photosphere and chromosphere (C–model of VAL) is developed. Numerical results and graphs are calculated for magnetic fields of 10^{-2}T and 10^{-4}T using new values of the proton–hydrogen scattering cross–section. In addition, the graph of conductivities in the solenoid magnetic field model is presented. A table of the Braginski coefficients calculated from VAL is also presented.

071.035 **Fourier transform spectrometer observations of solar carbon monoxide. II. Simultaneous cospatial measurements of the fundamental and first–overtone bands, and Ca II K, in quiet and active regions.**
T. R. Ayres, L. Testerman, J. W. Brault.
Astrophys. J., Vol. 304, No. 1, p. 542 – 559 (1986).
The authors have used the 1 m Fourier transform spectrometer (FTS) of the McMath telescope to obtain simultaneous cospatial measurements of important diagnostics of thermal structure in the high photosphere and low chromosphere of the Sun: the 2.3 μm first–overtone vibration–rotation bands of carbon monoxide and the 0.3934 μm resonance line of Ca II during one observing run; and the 4.7 μm fundamental and 2.3 μm first–overtone bands of CO during a second run. Spectra were recorded in regions strongly affected by magnetic activity and in quiet areas at similar heliocentric angles. An analysis of the "anomalous" behavior of the CO bands with respect to Ca II K indicates the existence of a highly structured, thermally bifurcated outer atmosphere on the Sun.

071.036 **Observations of 5-minute oscillations of the photospheric brightness of the sun.**
L. V. Didkovskij, V. A. Kotov.
Izv. Krymskoj Astrofiz. Obs., Tom 74, p. 132 – 142 (1986). In Russian. English translation in Bull. Crimean Astrophys. Obs., Vol. 74.

The results of ground-based observations of temporal variations of the solar brightness at 0.82±0.13 μm are presented. The authors find a statistically significant enhancement of the spectral power for frequencies between 2.6 and 3.4 mHz. These variations are attributed to global *p*-mode for solar intensity oscillations because 5-minute peaks in the power spectrum reveal a discrete pattern with the well-known equidistant spacing of 68 microHz. The frequencies agree well with the values of 5-minute oscillations obtained from the total solar irradiance measurements made on board the SMM satellite. Five minute oscillations are seen in both "direct" (for the central area of the solar disk) and "differential" (center-to-limb) brightness signals and are thought to be associated with acoustic compression waves in the photosphere.

071.037 **On measurements of the solar spectral energy transmitted through narrow-band filters.**
Eh. A. Gurtovenko, V. I. Troyan, N. G. Shchukina.
Photometric and polarimetric investigations of celestial bodies, p. 47 – 52 (1985). In Russian. – See Abstr. 003.035.

The calculations have been carried out and practical recommendations were given to construct a photometric model of solar radiation in absolute energy units for the wavelength range of 0.32 – 1.2 μm.

071.038 **Infrared spectrograms of the sun, 11600 – 12800 Å.**
Z. A. Shcherbakova, A. G. Shcherbakov, B. I. Shapiro.
Sov. Astron. Lett., Vol. 11, No. 5, p. 327 – 328 (1985). English translation of 40.071.016.

071.039 **Search for lines of neutral dysprosium in the Fraunhofer solar spectrum.**
A. P. Kucherov, A. V. Perekhod.
Soln. Dannye, Byull., 1986, No. 3, p. 79 – 85 (1986). In Russian.

On the basis of experimental values of the oscillator strengths and the adopted value of dysprosium abundance the expected equivalent widths of the lines of this element were estimated. An attempt to identify at least a single line of the Fraunhofer spectrum as a Dy I line failed.

071.040 **The effects of acoustic-gravity waves on the K I 7699 line.**
G. Severino, G. Roberti, C. Marmolino, M. T. Gomez.
Sol. Phys., Vol. 104, No. 2, p. 259 – 272 (1986).

The authors examine the effects of acoustic-gravity waves with long and short periods on the solar profile of the K I 7699 line using a dynamic model of line formation.

A spectrum of the Sun from a balloon-borne spectrograph in 1899.
See Abstr. 004.053.

The establishment of internally consistent solar scales of oscillator strengths and abundances of chemical elements. Zirconium.
See Abstr. 022.030.

On the structure of the energy spectrum of Tc I.
See Abstr. 022.041.

Oscillator strengths of iron lines.
See Abstr. 022.060.

Measurement of relative oscillator strengths for Ti I – IV. Transitions from levels of excitation energy between 1.42 and 2.31 eV.
See Abstr. 022.076.

Measurement of relative oscillator strengths for Cr I lines – II. Measures for transitions from levels with excitation energy 2.54 – 3.46 eV.
See Abstr. 022.077.

Wavelength predictions for the $1s^2 2s$–$1s2s3p$ satellite transitions.
See Abstr. 022.091.

Fe II oscillator strengths.
See Abstr. 022.101.

Scattered light in a spectrograph and its effect on the measured darkening to the solar limb.
See Abstr. 034.053.

New materials applications in solar spectral analysis.
See Abstr. 034.078.

A high quality movie from the SOUP instrument on SpaceLab 2.
See Abstr. 035.025.

Etude du spectre solaire à l'aide d'un réseau utilisé en incidence rasante.
See Abstr. 036.030.

Device for measurements of solar brightness variations based on an integral photodetector matrix.
See Abstr. 036.037.

Coments on techniques for spectral deconvolution.
See Abstr. 036.083.

The interest of simultaneous spectral and spatial high resolution spectroscopy in the infrared.
See Abstr. 036.148.

The possibility of a photospheric dynamo.
See Abstr. 062.061.

Polarization in spectral lines. IV: Resonance polarization in the Hanle effect, collisionless regime.
See Abstr. 063.057.

Numerical algorithm for solution of the radiative transfer equations taking into account the influence of a magnetic field.
See Abstr. 063.074.

Oszillationen in Sonnenflecken-Umbren, Penumbren und der benachbarten Photosphäre.
See Abstr. 072.097.

Energy deposit at temperature minimum level and white light flares.
See Abstr. 073.015.

The relationship between coronal bright points as seen in He I λ10830 and the evolution of the photospheric network magnetic fields.
See Abstr. 074.078.

Magnetic fields and vertical motions in supergranules.
See Abstr. 075.009.

On the magnetic field of solar granulation.
See Abstr. 075.017.

Simulations of magnetic-flux transport in solar active regions.
See Abstr. 075.020.

Recent studies of magnetic canopies.
See Abstr. 075.026.

An atlas of photospheric magnetic field observations and computed coronal magnetic fields: 1976 – 1985.
See Abstr. 075.050.

Solar photospheric and coronal abundances from solar energetic particle measurements.
See Abstr. 078.015.

Radiation output.
See Abstr. 080.016.

Solar Doppler shifts: sources of continuous spectra.
See Abstr. 080.060.

The K 769.9 nm line profile.
See Abstr. 080.080.

Limb darkening fluctuations: the discrepancy with Doppler data and the 160–min oscillation.
See Abstr. 080.081.

On the surface response of solar giant cells.
See Abstr. 080.103.

The temperatures of Alpha Centauri A and B.
See Abstr. 114.025.

On the synthetic spectra computation for stars and the sun.
See Abstr. 114.103.

072 Sunspots, Faculae, Activity Cycles, Solar Patrol

072.001 **A comparison of photospheric electric current and ultraviolet and X–ray emission in a solar active region.**
B. M. Haisch, M. E. Bruner, M. J. Hagyard, R. M. Bonnet.
Astrophys. J., Vol. 300, No. 1, p. 428 – 437 (1986). With plates 1 – 2.

Coordinated observations of Active Region 2684 were carried out on 1980 September 23. Photospheric magnetic fields were measured by a vector magnetograph; these vector magnetic data were used to infer the distribution of vertical electric current density (J_z) in the active region. High–resolution ultraviolet filtergrams in Lyman–alpha and in the 1600 Å ultraviolet continuum have been compared to J_z maps. Transition–region spectroheliograms obtained by the Solar Maximum Mission (SMM) in C IV, Si IV, and O IV, and soft X–ray raster maps of the active region obtained by the SMM in coronal lines of O VIII, Ne IX, and Mg XI were also compared to J_z maps.

072.002 **Meridional motions of sunspots from 1947.9 – 1985.0. I. Latitude drift at the different solar–cycles.**
A. Hanslmeier, G. Lustig.
Astron. Astrophys., Vol. 154, No. 1/2, p. 227 – 230 (1986).

The authors examine the meridional motion of sunspots from 1947.9 – 1985.0 with the data and records that have been made at the Solar–Observatory Kanzelhöhe. For the first time also cycle 21 is analysed.

072.003 **On the Spörer Minimum.**
Y.–t. Jiang, Z.–t. Xu.
Astrophys. Space Sci., Vol. 118, Nos. 1/2, p. 159 – 162 (1986). – See Abstr. 012.001.

The authors have examined the real behaviour of solar activity during the period AD 1400 – 1600. The results are as follows: (1) the distributions of the 20 naked–eye sunspot records are inhomogeneous. There are 2 sightings in the 15th century and 18 sightings in the 16th century; (2) the distributions of auroral records are similar to sunspot. There are 33 records in the 15th century and 315 records in the 16th century; (3) the climatic fluctuations in China show that the period AD 1430 – 1520 was cold while the period AD 1520 – 1620 was warm. These facts clearly demonstrate that the Spörer Minimum, if it extended from AD 1460 to 1550, could be a specious result and it, if its extent was AD 1400 – 1510, is a real feature of solar variability in that time.

072.004 **Polish Spot Numbers.**
W. S. Szymański.
Astron. Rep., Vol. 7, p. 29 – 48 (1982).

The paper presents definition and description of the so–called Polish Spot Numbers which are suggested in exchange for Wolf's numbers. An instruction of methods for observations of sunspots is given together with some examples of reductions of observations.

072.005 **Sonnenfleckenrelativzahlen des SONNE–Netzes und des S.I.D.C. für November 1985.**
Sterne Weltraum, 25. Jahrg., Nr. 1, p. 32 (1986).

072.006 **Sonnenfleckenrelativzahlen des SONNE–Netzes und des S.I.D.C. für Dezember 1985.**
Sterne Weltraum, 25. Jahrg., Nr. 2, p. 90 (1986).

072.007 **Sonnenfleckenrelativzahlen des SONNE–Netzes und des S.I.D.C. für Januar 1986.**
Sterne Weltraum, 25. Jahrg., Nr. 3, p. 152 (1986).

072.008 **The sharp decrease of Evershed effect and magnetic field at the outer sunspot border.**
E. Wiehr, G. Stellmacher, M. Knölker, H. Grosser.
Astron. Astrophys., Vol. 155, No. 2, p. 402 – 406 (1986).

Spatially high–resolved spectra of a small split line ($g = 0.5$) in four sunspots at different heliographic angles show that the combined effect of line core velocities and line asymmetries (defined as Evershed effect) is strongly limited to the sunspot area. Occasionally observed "adjoining velocity maxima" just outside the sunspot edges are not related to line asymmetries, thus indicating a different origin than that of the classical Evershed effect. A similar concentration of the very spot area is found for the magnetic line broadening of two large split lines. Both results indicate that magnetic field and velocity field are sharply restricted to the visible sunspot, the often observed smooth discontinuity of both quantities are thus considered to be due to seeing.

072.009 **Size and temperature of umbral dots.**
U. Grossmann–Doerth, W. Schmidt, E. H. Schröter.
Astron. Astrophys., Vol. 156, No. 1/2, p. 347 – 353 (1986). = Mitt. Kiepenheuer–Inst. Sonnenphys., Nr. 262.

In 12 white light broad band pictures of 6 different sunspots, taken in 1980 and 1981 with the 40 cm Newtonian Vacuum Telescope operated at the Observatorio del Teide (Iazaña), the authors identified nearly 100 umbral dots and attempted to derive their temperature and diameter. They found strong evidence that the diameter of the dots is in the range 0".4 to 0".9 and the temperature is several 100K up to more than 1000K cooler than the photosphere.

072.010 **Structure of a solar active region from RATAN 600 and Very Large Array observations.**
Sh. B. Akhmedov, V. N. Borovik, G. B. Gelfreikh (*G. B. Gel'frejkh*), V. M. Bogod, A. N. Korzhavin, Z. E. Petrov, V. N. Dikij, K. R. Lang, R. F. Willson.
Astrophys. J., Vol. 301, No. 1, p. 460 – 464 (1986).
Solar active region AR 3804 was observed simultaneously with the RATAN 600 and the VLA in 1982 July. The emission at wavelengths between 2 and 4 cm consisted of narrow ($\phi < 40''$), bright ($T_B \approx 0.2$ to 6×10^6K) core sources surrounded by a weaker ($T_B \approx 10^4 - 10^5$K), extended ($\phi \approx 200''$) halo. The brightest core sources are associated with sunspots and are interpreted in terms of the gyroradiation of thermal electrons at the second and third harmonics of the gyrofrequency. Two of the core sources were associated with a filament that lies above the magnetic neutral line in the photosphere. Magnetic field strengths are derived for these sources.

072.011 **Energy transport in sunspot penumbrae.**
H. U. Schmidt, H. C. Spruit, N. O. Weiss.
Astron. Astrophys., Vol. 158, No. 1/2, p. 351 – 360 (1986).
It is proposed that the magnetic field in the outer penumbra of a sunspot is almost horizontal and that the penumbra itself is very shallow, with a sharp lower boundary. A simple model of energy transport in the outer penumbra predicts that there is a Wilson depression of about 100 km, below which the penumbra extends for only 80 km. A two–component model with bright and dark filaments suggests that the associated differences in observed field strength and corrugations of the visible surface of the penumbra will be small. The authors argue that flows along the field are needed to explain the existence of bright and dark filaments while the Evershed flow is driven by pressure differences along flux tubes.

072.012 **Photoelectric observations of chromospheric sunspot oscillations. III. Spatial distribution of power and frequency in umbrae.**
B. W. Lites.
Astrophys. J., Vol. 301, No. 2, p. 992 – 1004 (1986).
An analysis of the spatial distribution of power and frequency of oscillation is presented for selected data sets of Paper II (Lites). The following conclusions are drawn: (1) the multiple peaks frequently observed in average power spectra are due to the presence of cospatial modes with different frequencies; (2) changes in the frequencies and amplitudes of these modes with time and place within an umbra appear to arise from an interference of modes in the driving source in the photosphere; (3) regions of high oscillatory power in the 3 min band are uncorrelated with regions of high oscillatory power in the 5 min band within umbrae, thereby suggesting that the 5 min oscillations in the photosphere do not drive the 3 min chromospheric oscillations.

072.013 **Photoelectric observations of chromospheric sunspot oscillations. IV. The Ca II H line and He I λ10830.**
B. W. Lites.
Astrophys. J., Vol. 301, No. 2, p. 1005 – 1017 (1986).
Simultaneous observations of profiles of He I λ10830 and the Ca II H line in sunspot umbrae have been made with the Tower telescope, echelle spectrograph, and multidiode array at the National Solar Observatory/Sunspot. These observations reveal the nature of fluid motions in the upper chromosphere during the cycle of umbral oscillations. Time sequences of profiles show that the chromospheric He I line is in weak absorption during all phases of the umbral oscillations, and that it undergoes periodic Doppler shifts of high amplitude (up to 11 km s^{-1}). The amplitudes and waveforms of the Doppler shifts in the He I line indicate that the oscillations are frequently nonlinear (i.e., forming shocks) and therefore contribute to the heating of the upper umbral chromosphere.

072.014 **Lifetimes of umbral dots in sunspots.**
U. Kusoffsky, H. Lundstedt.
Astron. Astrophys., Vol. 160, No. 1, p. 51 – 55 (1986).
A 143 min time sequence of video recordings of a sunspot with high spatial resolution was used to study the lifetime of umbral dots. Video recordings and white light photographs of many types of sunspot were used to study the morphology of dots. The half–life time of "normal" umbral dots is found to be about 60 min, longer by at least a factor 2 than earlier believed. The umbral dots are found to be most often aligned. There exists also umbral dots with lifetimes longer than 140 min which are closely related to the formation of light bridges. Generally the findings are in accordance with the theory of the origin of umbral dots, Parker (1979).

072.015 **Vergleich der Sonnenaktivität im Zyklus 18, 19 und 20.**
J. Jahn.
Sonne, Jahrg. 10, Nr. 37, p. 17 – 21 (1986).

072.016 **On the "ruptures" of solar activity.**
V. F. Chistyakov.
Soln. Dannye, Byull., 1985, No. 11, p. 51 – 57 (1985). In Russian.
Series of time parameters of 11–year solar cycles and the time–frequency spectrum of Wolf numbers (1700 – 1982) manifest "ruptures" or "jumps" of solar activity. These had place at the 18, 19 and 20 centuries. In about 1976 a new rupture started in the behaviour of solar activity. The duration of ruptures is about 20 years. The series of 11–year cycles of sunspots is presented during the ruptures.

072.017 **Classification of cycle pairs on the basis of maximum height – ascending branch length relationship types of pair members.**
G. V. Kuklin.
Soln. Dannye, Byull., 1985, No. 11, p. 57 – 60 (1985). In Russian.
A classification of cycle pairs (including cycles 20 – 21) is made on the basis of $W_M(T_A)$ relationship types to which cycles of pairs belong. A rule for a pair classification using ascending branch lengths of cycle composing this pair is constructed. A relation between the types of pair cycles and the pair property "low–high" or "high–low" is established.

072.018 **On the annual variation of solar activity and the influence of the external factor.**
R. N. Ikhsanov, E. V. Miletskij.
Soln. Dannye, Byull., 1985, No. 12, p. 60 – 67 (1986). In Russian.
The problem of annual variation in solar activity has been studied by means of a statistic analysis using available Wolf numbers, total sunspot areas and solar flares. A conclusion is made that in most cases the annual variation, if any, is not essential at a 0.1% significance level. In the cases of the presence of the variation it can be explained with intersolar processes, mostly with solar magnetic field properties.

072.019 **Prediction of monthly indices of solar activity $F_{10.7}$ on the basis of a multiplicative autoregression model.**
A. V. Mordvinov.
Soln. Dannye, Byull., 1985, No. 12, p. 67 – 73 (1986). In Russian.
A new autoregression method for forecasting monthly values of the solar radio emission flux density at the wavelength 10.7 cm $F_{10.7}$ is proposed. A forecasting model is built on the basis of an analysis of a centered series of logarithms of the $F_{10.7}$ index. A forecast a month in advance can be made using a high order model. The filter parameters of the forecast are found using the recursive Burg algorithm. The length of the optimum filter is determined from the condition of the minimum of the variation coefficient. Using as examples retrospective forecasts basic properties of the forecasting model are studied. A comparison is made between the forecasts made by this method and by Mayot's method.

072.020 **Anomalies of solar activity.**
V. F. Chistyakov.
Soln. Dannye, Byull., 1985, No. 12, p. 73 – 79 (1986). In Russian.
It is shown that solar activity is a sum of continuous and discrete processes. The discrete processes have cycles of about 90 years. A significant phase between the cycle curves of the northern and southern hemispheres of the sun has place at the moments of the "delta–function" inclusion. Moreover, at the mo-

ments many anomalies of solar active processes occur. Such anomalies have been very frequent since 1971.

072.021 **A relationship between sunspot numbers and active region parameters.**
N. N. Kontor, T. G. Khotilovskaya.
Soln. Dannye, Byull., 1985, No. 12, p. 80 – 84 (1986). In Russian.
Annual average sunspot numbers are shown to estimate the parameters of the long–lived active region maximum area distribution, i.e. annual numbers of active region and their average area. This allows to predict other solar activity parameters taking into account both the sunspot forecast and the "active region syndrome".

072.022 **The evolution of an active region and its large–scale current system.**
S. I. Gopasyuk, B. Kalman, V. A. Romanov.
Izv. Krymskoj Astrofiz. Obs., Tom 72, p. 171 – 183 (1985). In Russian. English translation in Bull. Crimean Astrophys. Obs., Vol. 72.
The distribution of the vector **H** computed for the magnetic field in an active region using the observations from 08.06 to 14.06 1969 has been compared with the structure of a potential field calculated by the observed vertical component $\mathbf{H}_z$. It is shown that the method of deviation of vector azimuths of the transverse magnetic field $\mathbf{H}\perp$, computed both for a potential model and the observational data, is extremely sensitive to detect electric currents. The results of computation allowed to distinguish the regions of strong current flows and to trace the evolution of large–scale current systems.

072.023 **On the possible existence of torsional oscillations with a period close to 40 min in sunspots.**
S. I. Gopasyuk.
Izv. Krymskoj Astrofiz. Obs., Tom 73, p. 9 – 15 (1985). In Russian. English translation in Bull. Crimean Astrophys. Obs., Vol. 73.
The gas rotation in two single sunspots has been studied using the observational data of the velocity field and brightness obtained with a double magnetograph in λ5253.5 Å Fe I. One of these spots was located in the northern and the other one in the southern hemisphere of the sun. Two components of the velocity of gas rotation in sunspots were found. One of them corresponds to oscillations with the period close to 40 minutes. It is superposed over the other (main) component slowly changing with time.

072.024 **On solar faculae heating by electric currents.**
Eh. V. Kononovich, I. V. Mironova, B. E. Serebryakov.
Pis'ma Astron. Zh., Tom 12, No. 5, p. 393 – 397 (1986). In Russian. English translation in Sov. Astron. Lett., Vol. 12.
A solar facula model is considered describing the facula as a magnetic tube with radius of 100 – 200 km magnetic field strengths of 1500 – 2000 Gauss and solenoid–type configuration of the magnetic field. The circle electric currents related to the magnetic field are located at the boundary of the tube in a thin transition layer. The heating of a facula is due to Joule dissipation. The corresponding thermal conductivity equation is solved. It is shown that the Ohmic heating is significant, and may essentially contribute to the increase of the facula effective temperature when the magnetic field gradient in a transition layer is $dB/dr \sim (3-7)$Gauss/cm.

072.025 **On the penumbra/umbra ratio of sunspot groups.**
J. Mergentaler.
Postepy Astron., Tom 33, Zesz. 1–2, p. 45 – 50 (1985). In Polish.
The ratio of the area of penumbra to the area of umbra for 2000 sunspot groups are calculated. The mean values of this ratio for all types of groups and its variation with time are given. The area ratio is correlated with total area for the I type groups, but there exists no similar correlation for the E groups.

072.026 **Sunspot numbers: October 1985.**
Sky Telesc., Vol. 71, No. 1, p. 117 (1986).

072.027 **Sunspot numbers: November 1985.**
Sky Telesc., Vol. 71, No. 2, p. 220 (1986).

072.028 **Sunspot numbers: December 1985.**
Sky Telesc., Vol. 71, No. 3, p. 323 (1986).

072.029 **Sunspot numbers: January 1986.**
Sky Telesc., Vol. 71, No. 4, p. 428 (1986).

072.030 **Solar activity. 1983 January – 1983 July.**
F. G. Mustaeva.
Tsirk. Astron. Inst., Tashkent, Nos. 113, 114, 116 (1985). In Russian.

072.031 **Nombres de Wolf pour l'année 1984.**
A. Koeckelenbergh.
Astronomie, Vol. 100, p. 32 (1986).

072.032 **A.L.P.O.S.S. solar observations for rotations 1739 – 1743.**
R. E. Hill.
Strolling Astron., Vol. 31, Nos. 5–6, p. 126 – 131 (1986).

072.033 **A.L.P.O.S.S. solar observations for rotation 1744.**
R. E. Hill.
Strolling Astron., Vol. 31, Nos. 7–8, p. 167 – 171 (1986).

072.034 **Young active regions.**
D. Y. Chou.
Bull. Am. Astron. Soc., Vol. 17, No. 4, p. 895 (1985). Abstract. – See Abstr. 010.062.

072.035 **Sunspot blocking and energy storage in the convection zone.**
A. H. Nye, B. J. LaBonte, D. H. Bruning.
Bull. Am. Astron. Soc., Vol. 17, No. 4, p. 895 (1985). Abstract. – See Abstr. 010.062.

072.036 **Ratio of calcium plage to sunspot area of solar active regions.**
J. K. Lawrence.
Bull. Am. Astron. Soc., Vol. 17, No. 4, p. 895 – 896 (1985). Abstract. – See Abstr. 010.062.

072.037 **The dynamics of magnetically trapped fluids. I. Implications for umbral dots and penumbral grains.**
A. R. Choudhuri.
Astrophys. J., Vol. 302, No. 2, p. 809 – 825 (1986).
The author presents a study of the magnetohydrodynamic system in which a nonmagnetized fluid in a gravitational field is surrounded by a fluid carrying a vertical magnetic field. This study can throw some light on the fine–structural features of a sunspot. The equilibrium configuration of the field–free fluid is a tapering column ending at an apex. A scheme developed to treat the apex indicates that, just below the apex, the radius of the tapering column opens up with a 3/2 power dependence on the depth below the apex. If the internal pressure of the field–free fluid is increased, the apex rises, and a static equilibrium may not be possible. Under conditions inside a sunspot, a column of field–free gas is found to rise with a velocity of about 100 km hr^{-1}. If umbral dots and penumbral grains are interpreted as regions where the field–free gas ultimately emerges, a natural explanation of most of their observed properties is obtained.

072.038 **East–western asymmetry of the Wilson effect.**
G. V. Kuklin.
Issled. Geomagn., Aehron. Fiz. Solntsa, Moskva, No. 73, p. 52 – 60 (1985). In Russian. Abstr. in Ref. Zh., 51. Astron., 3.51.430 (1986).

072.039 **Solar charts obtained by observations in the He I 10830 Å line with Kaliniak's infrared spectrophotometer.**
Eh. V. Kandrashov.
Soln. Dannye, Byull. 1986, No. 1, p. 74 – 82 (1986). In Russian.
The first in the USSR He I 10830 Å line charts of the sun taken by the author with A. A. Kaliniak's IR photoelectric low–noise spectrophotometer are discussed. Some methods of identification of the observed features as active regions, filaments and coronal holes against the "quiet sun" background are considered.

072.040 **Longitudinal distribution of active regions and solar cycles.**
Yu. I. Vitinskij.
Soln. Dannye, Byull. 1986, No. 2, p. 67 – 72 (1986). In Russian.
Peculiarities of the longitudinal distribution in a number of active regions during 1974 – 1981 are given. It is shown that the behavior of the longitudinal distribution varies in an 11–year solar cycle and from cycle to cycle. An accidental longitudinal distribution of active regions is more characteristic of the solar cycle 21 than that of the solar cycle 20. This is interpreted as a manifestation of the secular solar cycle. Active longitudes of active regions for 1976 – 1981 are analysed. Their instability is noted in the northern hemisphere of the sun in comparison with the southern hemisphere for the period 1961 – 1981.

072.041 **Coordinates of sunspots and pores during the ring–like solar eclipse of May 30, 1984.**
Yu. A. Nagovitsyn, E. Yu. Nagovitsyna.
Soln. Dannye, Byull. 1986, No. 2, p. 80 – 83 (1986). In Russian.

072.042 **Some results of a morphological study of the evolution of the active region of July 1982.**
M. B. Ogir', L. D. Parfinenko, M. N. Stoyanova.
Sol. Phys., Vol. 102, Nos. 1/2, p. 67 – 78 (1985).
The photospheric, chromospheric, and magnetic field structure of large active region No. 18474 of July 1982 are studied. Various morphological features observed are described and their possible role in the evolution of sunspots group discussed. The abundance of different light bridges is a characteristic feature for this group. It is shown that the light bridges in the photosphere coincide with the location and direction of arch filaments or fibril streams in the chromosphere, and appear in the region of a developing rotating magnetic field.

072.043 **A detection of the Faraday rotation by the solar vector magnetograph.**
M. Makita.
Sol. Phys., Vol. 103, No. 1, p. 1 – 10 (1986).
The Faraday rotation in the sunspot atmosphere is statistically detected by examining directions of the linear polarization obtained with the vector magnetograph of the Okayama Astrophysical Observatory. It is very effective near the spectral line center and the azimuth of the linear polarization deviates greatly from the magnetic field azimuth. In the case of the iron line, 5250 Å, the magnetic field azimuth will be obtained with an accuracy better than 15°, if observed in the line wing from 27 to 80 mÅ relative to the line center.

072.044 **Solar irradiance variations from photometry of active regions.**
G. A. Chapman, A. D. Meyer.
Sol. Phys., Vol. 103, No. 1, p. 21 – 31 (1986).
The Extreme Limb Photometer (ELP) has been used to measure the irradiance fluctuation of the Sun due to selected active regions. Forty–five active regions that were completely scanned at various disk positions are included in the analysis. The contribution of these active regions to a global solar irradiance fluctuation has been correlated with photometric sunspot and facular indices (PSI and PFI) using published values of sunspot and calcium plage areas. The measured ELP fluctuations are converted to a global brightness fluctuation, $\Delta B/B$. The sunspot component of $\Delta B/B$ correlates with PSI with $r = 0.95$. The facular component of $\Delta B/B$ correlates with PFI with $r = 0.72$. The expression for PFI is important to the question of energy balance between sunspots and faculae and the results presented here are not incompatible with energy balance between the two phenomena; that is the energy deficit of sunspots may be balanced by the energy excess of faculae.

072.045 **Delta–configurations: flare activity and magnetic–field structure.**
S. R. Patty, M. J. Hagyard.
Sol. Phys., Vol. 103, No. 1, p. 111 – 128 (1986).
Complex sunspots in four active regions of April and May 1980, all exhibiting regions of magnetic classification delta, were studied using data from the NASA Marshall Space Flight Center vector magnetograph. The vector magnetic field structure in the vicinity of each delta was determined, and the location of the deltas in each active region was correlated with the locations and types of flare activity for the regions. Two types of delta–configuration were found to exist, active and inactive, as defined by the relationships between magnetic field structure and activity.

072.046 **Investigation of magnetic field strengths of sunspots from spectral lines with different excitation.**
G. V. Lyamova.
Astronomical–geodetical investigations, p. 134 – 146 (1985). In Russian. Abstr. in Ref. Zh., 51. Astron., 4.51.291 (1986). – See Abstr. 003.024.

072.047 **On umbral oscillations as a sunspot diagnostic.**
L. M. B. C. Campos.
Seismology of the Sun and the distant stars, p. 293 – 301 (1986). – See Abstr. 012.026.
The observation that p–modes are attenuated in sunspots is explained by the transfer of energy to motions along the magnetic field lines. The latter appear as Alfvén waves reflected between the atmosphere and anchoring points in the convection zone, and have periods and amplitudes consistent with observations of umbral oscillations. The depth of generation of these waves is about 10 Mm. The oscillation model is used to calibrate a general dimensional law for the intensity of radiation of MHD waves, which gives reasonable results for the Sun, and may apply to other stars.

072.048 **Rotation and expansion within sunspot groups.**
P. A. Gilman, R. Howard.
Astrophys. J., Vol. 303, No. 1, p. 480 – 485 (1986).
By superposing data for many sunspot groups measured on the Mount Wilson white–light plate collection, the authors demonstrate that differential rotation, about equal to the ambient rate, occurs between sunspots within the group. It is also shown that the relative motions of leader and follower sunspots can be characterized primarily as a simple expansion of the group along its major axis, with very little rotation of the pattern about group center.

072.049 **Sonnenfleckenrelativzahlen des SONNE–Netzes und des S.I.D.C. für Februar 1986.**
Sterne Weltraum, 25. Jahrg., Nr. 4, p. 214 (1986).

072.050 **Solar activity.**
F. J. Heyden.
Proceedings of the Second Asian–Pacific Regional Meeting on Astronomy, p. 473 – 474 (1984). – See Abstr. 012.031.

072.051 **Oscillations in sunspots.**
J. H. Thomas.
Aust. J. Phys., Vol. 38, No. 6, p. 811 – 824 (1985). Paper presented at the R. G. Giovanelli Commemorative Colloquium, Part I, Sydney, 26 – 29 November 1984.
Recent observational and theoretical work on oscillations in sunspots is reviewed. The characteristic 3–minute umbral oscillations and flashes are resonant modes of the sunspot itself, whereas the 5–minute oscillations in the umbra are a passive response to forcing by p modes in the surrounding convection zone. The observational evidence suggests that the fundamental cause of the 3–minute oscillations is the photospheric fast–mode resonance, with chromospheric slow–mode resonances perhaps producing additional oscillation frequencies in the chromo-

sphere. Observations and theoretical models of the interaction of 5-minute p-mode oscillations with a sunspot offer a means of probing the structure of a sunspot magnetic flux tube beneath the solar surface.

072.052 **Sunspot number series envelope and phase.**
R. N. Bracewell.
Aust. J. Phys., Vol. 38, No. 6, p. 1009 – 1025 (1985). Paper presented at the R. G. Giovanelli Commemorative Colloquium, Part II, Tucson, Arizona, 17 – 18 January 1985.

This paper presents a method of analysis of annual time series containing solar cycles. It is applicable both to sunspot number counts and to varve thickness measurements.

072.053 **The sunspot cycle and solar magnetic fields. I. The mechanism as inferred from observation.**
R. G. Giovanelli.
Aust. J. Phys., Vol. 38, No. 6, p. 1045 – 1066 (1985).

The author describes how surface observations of motions and magnetic fields combine to provide a fairly unique model of the sunspot and magnetic cycle phenomena. Flux released from sunspots is dispersed by supergranule motions, and carried polewards by a meridional flow of $\sim 20\ m\ s^{-1}$. Because of the tilt of spot-group axes, flux from follower spots arrives in polar zones two weeks or so before leader flux until sunspot maximum. This results in the arrival of more follower than leader flux and the consequent reversal of the polar field. Later in the cycle, when some sunspots appear close to the equator, flux can be dispersed into the other hemisphere by supergranule motions, more from the leader than the follower. Calculations show that this plus the complementary flow of flux from the opposite hemisphere is of the correct magnitude to account for the observed net change in polar flux during the cycle.

072.054 **The sunspot cycle and solar magnetic fields. II. The interaction of flux tubes with the convection zone.**
R. G. Giovanelli.
Aust. J. Phys., Vol. 38, No. 6, p. 1067 – 1089 (1985).

Mechanisms of interaction between flux tubes or ropes and the convection zone are examined insofar as they are relevant to the sunspot cycle. These include floating, transport, and the penetration of gas from outside the tubes. It is found that all previous studies contain one or more major errors of physics which render their conclusions invalid. Some of the results presented here are explanations of (1) the transport of flux tubes by the slow observed poleward motions and the even slower systems; (2) their magnetic field strengths ($\sim 10^4$G at a depth 10 Mm to $(6-12)$x10^4G at ~ 150 Mm); and (3) the amplitudes of the torsional oscillation.

072.055 **Sunspot numbers: February 1986.**
Sky Telesc., Vol. 71, No. 5, p. 532 (1986).

072.056 **Sunspot numbers: March 1986.**
Sky Telesc., Vol. 71, No. 6, p. 635 (1986).

072.057 **Sonnenfleckenrelativzahlen des SONNE-Netzes und des S.I.D.C. für März 1986.**
Sterne Weltraum, 25. Jahrg., Nr. 5, p. 271 (1986).

072.058 **On the initial orientation of emerging active regions.**
J. I. Garcia de la Rosa.
Sol. Phys., Vol. 103, No. 2, p. 249 – 257 (1986).

The initial orientation of 74 active regions is measured in high resolution pictures. The sample has been devided into two classes, whose dividing line is roughly set at 5×10^{21}Mx in the magnetic flux. It is observed that only the large regions tend to display the correct inclination (*p*-part at lower latitude). Their evolution towards a more parallel inclination to the equator is also observed. The first finding is interpreted in terms of the primary nature of the large ARs which are supposed to be directly linked to the basic mechanism of solar activity. The evolution is interpreted as a reaction of the lower layers of the tube to the tilt introduced by the Coriolis effect.

072.059 **Spectral analysis of variations of the solar activity and cosmic ray intensity at the decay phase of the 19th solar cycle.**
E. S. Vernova, V. I. Pochtarev, N. G. Ptitsyna, M. I. Tyasto.
Geomagn. Aehron., Tom 26, No. 3, p. 493 – 496 (1986). In Russian. English translation in Geomagn. Aeron.

072.060 **Semiempirical models of sunspots.**
M. Sobotka.
Sov. Astron., Vol. 29, No. 5, p. 576 – 580 (1985). English translation of 40.072.034.

072.061 **Sonnenfleckenrelativzahlen des SONNE-Netzes und des S.I.D.C. für April 1986.**
Sterne Weltraum, 25. Jahrg., Nr. 6, p. 335 (1986).

072.062 **Eigenbewegungen einer Sonnenfleckengruppe.**
C.-H. Jahn.
Sterne Weltraum, 25. Jahrg., Nr. 6, p. 340 – 342, 344 (1986).

072.063 **The visibility function and its effect on the observed characteristics of sunspot groups. 4. The effect of the shape of the curve of development of sunspot group area on the observational conditions of sunspot groups.**
M. Kopecký.
Bull. Astron. Inst. Czech., Vol. 37, No. 3, p. 163 – 175 (1986).

Using diagrams of the observational conditions of sunspot groups, it is demonstrated how the shape of the curve of development of the sunspot group area affects the frequency distribution of sunspot groups with regard to their apparent lifetimes and east-west asymmetry of appearance and disappearance of groups. It is also shown that, if the curves of development of the sunspot group area have particular shapes, apparent "interrupted" groups may be generated as a result of the existence of the visibility function and solar rotation.

072.064 **Velocity field and flare activity in a sunspot group.**
S. I. Gopasyuk, L. F. Lazareva.
Izv. Krymskoj Astrofiz. Obs., Tom 74, p. 84 – 92 (1986). In Russian. English translation in Bull. Crimean Astrophys. Obs., Vol. 74.

The effect of the photospheric velocity field of the active region on a flare appearance has been studied. Simultaneous observations of the radial velocities and the longitudinal component of the magnetic field in the active region CMP on 12 June 1963 were carried out in the λ5250.2 Å Fe I line. The previously found correlation between flare activity in a sunspot group and sunspot rotation as well as the tendency of flares to appear along the zero line of radial velocities have been confirmed. It has been found that radial velocities and gas fluxes increase at both sides of this zero line prior to the flare appearance.

072.065 **Observations solaires. Rotations 1744 – 1756, 9 janvier – 28 décembre 1984.**
E. Tifrea, A. Dimitriu, S. Dinulescu, G. Maris, A. Oncică, C. Dumitrache.
Cent. Astron. Sci. Spat., Obs. Sol., 48 pp., Price Lei 7.00 (1985).

072.066 **Short-period cyclic variations of solar activity and their geophysical consequences.**
D. Djurović, D. Stajić.
Publ. Dep. Astron., Univ. Beogr., No. 13, p. 19 – 25 (1985).

Spectral analysis of the geomagnetic index and Wolf number has been made. The authors assume that solar activity and geomagnetic field have the common cyclic variations of 2, 3.5 and 5.5 year.

072.067 **Fenómenos solares 1979.**
An. Obs. Astron. Univ. Coimbra, 147 pp. (1986).

072.068 **Sunspot observations made in 1982.**
O. Gölbasi.
Rev. Fac. Sci. Univ. Istanbul, Ser. C, Vol. 48, p. 131 – 157 (1983). = Publ. Istanbul Univ. Obs., No. 139.

072.069 **The spatial distribution of larger sunspot active regions in the 21st cycle.**
B.–s. Tang.
Publ. Yunnan Obs., No. 1, p. 84 – 87 (1986). In Chinese.
The author analyses statistically the spatial distribution of the major sunspots during the period June 1976 – June 1984 of the 21st cycle. He obtains the following conclusions: (1) The sunspots cluster in certain zones of heliographic longitude, (2) the distributions of the sunspots in the northern and southern hemispheres are asymmetric, (3) the distribution of spatial spectra of the sunspots were derived by using the method of autocorrelation analysis. The main result is: the autocorrelation function $R(\tau)$ has its maximum for $\tau = 18$.

072.070 **Observations of an active region with successive flares.**
P. Kotrč, J. Suda.
Hvar Obs. Bull., Vol. 9, No. 1, p. 1 – 23 (1985).
The development of an active region with large flares observed at the Ondrejov and Hvar observatories is described. A connection between penumbral fibrils orientation and the orientation of the filaments in relation to the flares is pointed out. The active region configuration and appropriate flare model are discussed with respect to the occurrence of homologous flares.

072.071 **Actividad solar en 1983: I. Números relativos de Wolf. II. Estadística de manchas y superficie de las mismas.**
M. Lopez Arroyo.
Bol. Astron. Obs. Madr., Vol. 11, No. 4, p. 25 – 65 (1985).

072.072 **Manila Observatory, Solar Division. Solar Maps and Activity. 1985 November – 1986 April.**
F. J. Heyden.
Sol. Maps Act. (1985/86).

072.073 **The broad–band polarization of sunspots observed from 8 February 1981 to 30 August 1982.**
M. Makita, Y. Ohki.
Ann. Tokyo Astron. Obs., Second Ser., Vol. 21, No. 1, p. 1 – 30 (1986).
The broad–band polarization of sunspots was measured with the polarimeter fed to the solar coudé telescope of the Okayama Astrophysical Observatory. 113 polarization maps of 41 sunspot groups were obtained during this period. The observed circular polarization is strong in the limb of sunspots, roughly in proportion to the linear polarization. Its polarity is negative or positive according as the magnetic polarity is north or south. This polarity rule is found to be the same as that in the previous solar cycle.

072.074 **A study of line asymmetry in unipolar sunspots.**
M. Makita, H. Kawakami.
Publ. Astron. Soc. Jpn., Vol. 38, No. 2, p. 257 – 265 (1986).
The asymmetry of the line profile of Fe I 6301.5 Å in the spectrum of unipolar sunspots was analyzed. Most bisectors of the line profile could be approximated by straight lines and interpreted in terms of the variation of the line–of–sight velocity. The analysis of these bisectors, under the assumption of cylindrical symmetry, gives a sunspot velocity structure consistent with that deduced from the Evershed outflow and inflow.

072.075 **Analysis of an active region with high productivity of flare during November, 1979.**
Z.–y. Zhu.
Publ. Purple Mt. Obs., Vol. 4, No. 2, p. 40 – 45 (1985). In Chinese.
This paper analyses the evolution of the region Hale 16413 during Nov. 1979, presenting its features and the distribution of flares at the stages of progress and decay and the characteristics of the locations of flares.

072.076 **The fuzzy classification of the solar cycle and the prediction for the 22nd solar cycle.**
F.–m. Hu, M.–h. Jiang.
Chin. J. Space Sci., Vol. 5, No. 4, p. 237 – 244 (1985). In Chinese.
Abstr. in Phys. Abstr., Vol. 89, No. 1277, Entry 42235 (1986).

072.077 **Effect of local rigid rotation on sunspots.**
M. J. Martres, Z. Mouradian, I. Soru–Escaut.
Astron. Astrophys., Vol. 161, No. 2, p. 376 – 380 (1986).
The parasitic polarities that appear during the life of a sunspot are always signalled in preceding rotations by a filament pointing to the region. These filaments indicate a rigid local rotation during at least one solar rotation. The spot remains stable during the life of the parasitic polarity and begins to decrease as soon as the latter disappears. Parasitic polarities provide an opportunity to show that an emergence of flux is associated with a region with zero differential rotation.

072.078 **Correlation of the velocity and magnetic field in a solar active region.**
R. Berton.
Astron. Astrophys., Vol. 161, No. 2, p. 381 – 386 (1986).
In this paper it is shown how a simple statistical analysis of the line of sight component of the steady velocity fields, and of its spatial correlation and cross correlation with the magnetic field, may be interpreted by a theoretical model discussed by Berton (1985). The steady velocity field has been obtained after separation of the fluctuating part.

072.079 **The rate of areal decrease of long–lived sunspots.**
A. Antalová, R. Mačura.
Contrib. Astron. Obs. Skalnaté Pleso, Vol. 14, p. 163 – 181 (1986).
The purpose of this paper is to compare the rates of decrease of the areas of two groups of unipolar isolated sunspots: the "naked" spots and H–type spots according to the Zürich classification. The Zürich type H spots, which were analyzed in this paper, were observed only during one solar rotation.

072.080 **On principal phases of the solar cycle.**
Yu. I. Vitinskij, G. V. Kuklin, V. N. Obridko.
Soln. Dannye, Byull., 1986, No. 3, p. 53 – 56 (1986). In Russian.
The necessity of revision in traditional concepts of the solar cycle is argued. A new description of principal phases of the solar cycle using a set of different observational data is given. An approach to a further detailed analysis of this problem is determined.

072.081 **Wolf numbers in solar cycle No. 21.**
M. N. Gnevyshev, Yu. A. Nagovitsyn, E. Yu. Nagovitsyna.
Soln. Dannye, Byull., 1986, No. 3, p. 57 – 62 (1986). In Russian.
Brussels international, the American and Kislovodsk systems of Wolf numbers are studied on the descending branch of solar cycle 21. Using smoothed monthly values of Wolf numbers R_s it is shown that the Kislovodsk reduced series R_{KZ} preserves the Zürich system while the Brussels and American systems R_I and R_A differ significantly on the descending branch of solar cycle 21. Principal characteristics of the present cycle and monthly Wolf numbers are given using the reduced Kislovodsk series.

072.082 **On peculiarities of the fine structure of polar faculae.**
M. N. Stoyanova.
Soln. Dannye, Byull., 1986, No. 3, p. 72 – 75 (1986). In Russian.
A study of the northern polar region of the sun is made on the basis of high–resolution photographs. Contrasts and dimensions of separate facular granules are determined. A characteristic structure of polar faculae is revealed. It consists of a pair of facular elements of 1".2 and approximate distance between the components of 3".

072.083 **Relations between the angular resolution of photographs and observed faculae contrast variation over the solar disk.**
T. M. Minasyants.
Soln. Dannye, Byull., 1986, No. 3, p. 88 – 92 (1986). In Russian.
The dependence of the observed faculae contrast, the position of the contrast maximum and that of the nearest to the limb faculae on the spatial resolution is investigated. On the basis of model calculations of faculae using the derived relation and

knowing the photograph resolution it is possible to predict the behaviour of the center–to–limb variation curve.

072.084 **Daily maps of the sun and of magnetic fields of sunspots.**
Soln. Dannye, Byull., 1985, 1986, Nos. 11 – 12, 1 – 4 (1985/1986). In Russian.

072.085 **Temporal variation of brightness and area of sunspots and flocculi in the active region SD N 135.**
N. B. Ograpishvili.
Soln. Dannye, Byull., 1986, No. 4, p. 88 – 94 (1986). In Russian.
The active region SD N 135 was observed on 12 – 29 June 1984. Characteristics of the active region development according to the data obtained at the Abastumani Astrophysical Observatory are given. The data include K–spectroheliograms and photographs in white light. The analysis of the observational material allows to make the conclusion that the variations of the spots are monotonous at first, pulse–like on the following days. The observational results showed that the life–time of the flocculus knots is longer than 4 – 5 days, and the peripherical knots dissipated much faster than central ones.

072.086 **Periodicities in sunspot numbers.**
R. P. Kane, N. B. Trivedi.
J. Geomagn. Geoelectr., Vol. 37, No. 12, 1071 – 1085 (1985). Abstr. in Phys. Abstr., Vol. 89, No. 1281, Entry 67430 (1986).

072.087 **A study of velocity fields of flare regions.**
J.–y. Xuan, A.–a. Xu, Y.–h. Tang.
Chin. Astron. Astrophys., Vol. 10, No. 1, p. 73 – 76 (1986). English translation of 40.072.093.

072.088 **Magneto–optical effect and structure of sunspot magnetic field.**
S.–h. Ye, J.–h. Jin.
Chin. Astron. Astrophys., Vol. 10, No. 2, p. 103 – 107 (1986). English translation of Acta Astrophys. Sin., Vol. 6, No. 1, p. 35 – 41 (1986).
The authors calculated monochromatic images of unipolar sunspots in the linear polarization signal $\dot{U}$ and compared them with the observed data obtained at Marshall Space Flight Center, U.S.A. The results show that their radial fan–shaped model of sunspot magnetic field give $\dot{U}$ images that are similar to the observed ones while a spiral model gives rather different pictures.

072.089 **Solar cycle variation of heliospheric quantities.**
G.–l. Zhang, Y.–f. Xu.
Acta Astrophys. Sin., Vol. 6, No. 1, p. 51 – 57 (1986). In Chinese.
This paper makes a statistical analysis on the solar cycle variation of the heliospheric quantities observed at 1 AU. Two kinds of solar cycle variation with different characters have been identified, i.e. the sunspot and coronal–hole cycles. The implication of the results is that the solar winds observed during the sunspot maximum and the coronal–hole maximum years differ basically in their characters.

072.090 **Secular cycles of solar activity and predicted characteristic values of solar cycles 22 and 23.**
S.–l. Dong.
Acta Astron. Sin., Vol. 27, No. 1, p. 53 – 58 (1986). In Chinese.
This paper briefly reviews some of the recent typical results about the predicted characteristic values of solar cycles 22 and 23. The average features of secular cycles of solar activity are statistically analysed.

072.091 **A simple method of solar activity forecasting.**
Y.–h. Fan.
Acta Astron. Sin., Vol. 27, No. 1, p. 59 – 60 (1986). In Chinese.

072.092 **The spot groups of abnormal magnetic structures accompanied by quickly variation and SWF flares.**
C.–l. Zhu, G.–q. Zhan.
Acta Astron. Sin., Vol. 27, No. 2, p. 171 – 172 (1986). In Chinese.
The relation between the spot groups of abnormal magnetic structures accompanied with quickly variation and SWF flares occurring in the 21th solar cycle (1976.6 – 1980.12) is discussed. The results indicate that the spot groups of abnormal magnetic structures accompanied by rapid variations are important conditions causing SWF flares.

072.093 **Monochromatic images in Stokes parameters and the structure of magnetic fields in sunspots.**
S.–h. Ye, J.–h. Jin.
Sol. Phys., Vol. 104, No. 2, p. 273 – 285 (1986).
Taking into account magneto–optical effects, the authors have obtained numerical solutions of the transfer equations for the Stokes parameters, calculated the linearly polarized intensity ($\tilde{U}$) and constructed its monochromatic images of unipolar sunspots. By comparison with the observational material of the vector magnetograph of the Marshall Space Flight Center, Huntsville (Alabama), the authors have found that the model of radial magnetic fields may give rise to $\tilde{U}$ monochromatic images close to those observed. Moreover, the authors have shown that the model of spiral magnetic fields leads to results in contrast with observations.

072.094 **Photospheric and chromospheric umbral dots in a decaying sunspot.**
R. Kitai.
Sol. Phys., Vol. 104, No. 2, p. 287 – 301 (1986). = Contrib. Kwasan Hida Obs., Univ. Kyoto, No. 266.
Time–sequenced Hα filtergrams and narrow–band blue filtergrams (λ_0 = 4308 Å, $\Delta\lambda$ = 10 Å) of umbral dots in a decaying sunspot were studied. Lifetimes, proper motions, and the correlation between photospheric and chromospheric umbral dots are described and discussed.

072.095 **Large–scale structure of a sunspot and its surrounding photosphere.**
H. I. Abdussamatov (*Kh. I. Abdusamatov*).
Sol. Phys., Vol. 105, No. 1, p. 27 – 33 (1986).
A preliminary study of photometric properties is given, in particular, of the large–scale ($\approx 3''$) structure of a sunspot and the surrounding photosphere which is practically free from the influence of the Earth's atmospheric agitation. Moreover, the author studies their behaviour (their relationships and mutual influence) with time.

072.096 **On the latitude migration of polar faculae in the solar activity cycle. Period: 1970 – 1978.**
V. I. Makarov, V. V. Makarova.
J. Astrophys. Astron., Vol. 7, No. 2, p. 113 – 118 (1986).
The present paper reports the data on the latitude distribution of polar faculae in the period between the reversal of the magnetic field of the sun in the 20th and 21st solar cycles (1970 – 1978).

072.097 **Oszillationen in Sonnenflecken–Umbren, Penumbren und der benachbarten Photosphäre.**
H. Balthasar, E. Fangmeier, G. Küveler, E. Wiehr.
Mitt. Astron. Ges., No. 65, p. 155 (1986). Abstract. – See Abstr. 012.072.

072.098 **Provisional sunspot numbers for December, 1985 – June, 1986.**
Yamamoto Circ., Nos. 2052, 2054, 2056, 2058, 2061 – 2063 (1986). In Japanese.

072.099 **L'activité solaire.**
M.–J. Martres, G. Zlicaric.
Astronomie, Vol. 100, p. 52 – 53, 108 – 109, 161, 213, 269, 321 (1986).

Catalogue of sunspot groups at heliographic latitudes $B \leqslant 4°$ and $B \geqslant 38°$ from 1874 to 1983.
See Abstr. 002.059.

Ancient sunspot observations in China and solar variability.
See Abstr. 004.081.

Solar studies in China.
See Abstr. 013.003.

The Siberian Solar Radio–Telescope: parameters and principle of operation, objectives and results of first observations of spatio–temporal properties of development of active regions and flares.
See Abstr. 033.003.

A simple irradiance monitor for testing solar global oscillation network sites.
See Abstr. 034.055.

A high quality movie from the SOUP instrument on SpaceLab 2.
See Abstr. 035.025.

Ändert sich die Struktur der Granulation mit dem solaren Aktivitätszyklus?
See Abstr. 071.002.

Intensity variations of solar short–wave radiation in the solar activity cycle.
See Abstr. 071.017.

Granulation deformation near and in sunspot regions.
See Abstr. 071.031.

Fourier transform spectrometer observations of solar carbon monoxide. II. Simultaneous cospatial measurements of the fundamental and first–overtone bands, and Ca II K, in quiet and active regions.
See Abstr. 071.035.

Reference models of sunspot chromospheres.
See Abstr. 073.006.

Interpretation of Hα contrast profiles of active region loops.
See Abstr. 073.011.

The sudden decrease of the large flare occurrence and its manifestation in interplanetary space after the maximum of the solar cycle.
See Abstr. 073.024.

On spectral observations of the chromospheric fine structure in an active region.
See Abstr. 073.039.

Investigation of the information content of a system of parameters for forecast of large flares in sunspot groups.
See Abstr. 073.062.

Small solar flares in an activity complex.
See Abstr. 073.063.

The observed oscillation mode of the chromosphere above sunspots.
See Abstr. 073.068.

Flare prediction as a source of information on processes in an active region.
See Abstr. 073.113.

The observed mode of oscillations in the chromosphere above sunspots. II Running penumbral waves.
See Abstr. 073.114.

Estimation of coronal magnetic fields using Type–I emission.
See Abstr. 074.011.

Changes in the geometry of the solar wind transition region depending on solar activity level.
See Abstr. 074.028.

The solar cycle dependence of coronal mass ejections.
See Abstr. 074.039.

The relationship between coronal bright points as seen in He I λ10830 and the evolution of the photospheric network magnetic fields.
See Abstr. 074.078.

The geometry of the solar wind transition region during the 11–year solar cycle.
See Abstr. 074.092.

Coronal interconnection of two active regions observed in 3.5 – 8.0 keV X–rays.
See Abstr. 074.109.

Numerical solution of the magnetostatic equations for thick flux tubes, with application to sunspots, pores, and related structures.
See Abstr. 075.011.

Quasi–periodic variations of the magnetic field of active regions on the sun and their connection with flare activity.
See Abstr. 075.016.

A determination of the structure of the magnetic field of active regions using orientation of small fibrils.
See Abstr. 075.018.

The cancellation of magnetic flux. II. In a decaying active region.
See Abstr. 075.027.

The boundary value problem of the solar force–free magnetic field with constant α and its analytical solution.
See Abstr. 075.031.

Magnetic shear. III. Hale region 17255.
See Abstr. 075.032.

Magnetic shear. IV. Hale regions 16740, 16815, and 16850.
See Abstr. 075.033.

Methods of measurement of the magnetic field vectors of sunspots.
See Abstr. 075.040.

Magnetic field structures of solar active regions obtained by polarization mapping observation at 32 GHz.
See Abstr. 075.043.

Temporal variations of solar EUV, UV, and 10,830–Å radiations.
See Abstr. 076.015.

HRTS Spacelab 2 observations of the temporal development of the C IV transition zone.
See Abstr. 076.018.

Height measurements of *S*–components.
See Abstr. 077.001.

Variations in quiet Sun radiation at centimetre wavelengths during solar maximum period.
See Abstr. 077.002.

Application of distribution of the number of events to diagnostics of great chromospheric bursts.
See Abstr. 077.015.

High–spectral resolution solar microwave observations.
See Abstr. 077.024.

Unstable chromospheric dark features and type–III bursts association.
See Abstr. 077.029.

A study of RATAN–600 observations of solar *S*–component sources.
See Abstr. 077.066.

What causes the solar cycle?
See Abstr. 080.002.

Time–integrated energy budget of a solar activity complex.
See Abstr. 080.004.

Solar activity and the coupling of *g*–mode oscillations.
See Abstr. 080.020.

Differential rotation of sunspot groups in the period from 1874 through 1976 and changes of the rotation velocity within the solar cycle.
See Abstr. 080.021.

Some properties of the dynamics of the solar convective zone and their consequences on the activity cycle.
See Abstr. 080.026.

The solar rotation elements i and Ω derived from sunspot groups.
See Abstr. 080.039.

The influence of faculae on total solar irradiance and luminosity.
See Abstr. 080.043.

Solar irradiance variations derived from magnetograms.
See Abstr. 080.044.

First evidence of a large–scale circulation in the solar convection zone.
See Abstr. 080.059.

Solar Doppler shifts: sources of continuous spectra.
See Abstr. 080.060.

Change of solar oscillation eigenfrequencies with the solar cycle.
See Abstr. 080.077.

Long period variations in the solar diameter.
See Abstr. 080.078.

Global oscillations and active regions.
See Abstr. 080.086.

Umbral oscillations as a probe of sunspot structure.
See Abstr. 080.092.

Comments on the heating of solar active regions by magnetic energy dissipation.
See Abstr. 080.099.

The latitudinal distribution of sunspot areas and variations of the differential rotation of the Sun during the period 1921 – 1971.
See Abstr. 080.106.

Earth's equatorial ionosphere and the heliospheric current sheet.
See Abstr. 083.012.

Geomagnetic and solar data. September 1985.
See Abstr. 084.003.

Geomagnetic and solar data. October 1985.
See Abstr. 084.004.

Dependence of the aurora borealis occurrences on the solar–terrestrial parameters.
See Abstr. 084.008.

Geomagnetic and solar data. November 1985.
See Abstr. 084.017.

Geomagnetic and solar data. December 1985.
See Abstr. 084.018.

Geomagnetic and solar data. January, February 1986.
See Abstr. 084.073.

Some solar cycle phenomena related to the geomagnetic activity from 1868 to 1980. II. High velocity wind streams and cyclical behaviour of poloidal field.
See Abstr. 085.005.

The heliospheric current sheet: 3–dimensional structure and solar cycle changes.
See Abstr. 106.046.

The effect of the heliospheric current sheet on cosmic ray intensities at solar maximum: two alternative hypotheses.
See Abstr. 144.035.

Cosmic ray intensity variations with periods close to the 27–day periods in the 19th and 20th solar cycles.
See Abstr. 144.056.

073 Chromosphere, Flares, Prominences

073.001 **The role of nonclassical electron transport in the lower solar transition region.**
S. P. Owocki, R. C. Canfield.
Astrophys. J., Vol. 300, No. 1, p. 420 – 427 (1986).

Using a linear BGK method to study electron transport in one–dimensional, planar models of the solar transition region, the authors investigate the dependence of nonclassical electron transport effects on the assumed transition region temperature structure, with particular focus on how these nonclassical effects might influence line emission from the lower transition region.

073.002 **Two–frequency imaging of microwave impulsive flares near the solar limb.**
G. A. Dulk, T. S. Bastian, S. R. Kane.
Astrophys. J., Vol. 300, No. 1, p. 438 – 447 (1986).

Using the VLA, the authors observed two impulsive microwave and hard X–ray flares that occurred close to the solar limb on 1981 November 21 and 22. Images were obtained simultaneously at 4.9 and 15 GHz every 10 s during the 5 minutes durations of the flares. The apparently simple impulsive flares were unexpectedly complicated when imaged with arcsecond resolution: the 4.9 and 15 GHz radiation came from spatially separated sources in one flare, and there was no 4.9 GHz radiation at all in the other. The authors interpret the observations in terms of an inhomogeneous flare volume with the magnetic field strength and orientation varying with position, both transverse to and along the line of sight. The 4.9 GHz radiation of the November 22 flare probably arose mainly from thermal electrons with temperatures between 5 and 20×10^6K. The 15 GHz radiation of the flares on both days probably arose from electrons of $E \approx 300$ keV in a weak, nonthermal tail.

073.003 **Soft X–ray observations of high–velocity features in the 29 June 1980 flares.**
R. D. Bentley, J. R. Lemen, K. J. H. Phillips, J. L. Culhane.
Astron. Astrophys., Vol. 154, No. 1/2, p. 255 – 262 (1986).

During the impulsive phase of two flares on 29 June 1980, short lived emission line features have been observed in soft X–rays, near the resonance lines of Fe XXV and Ca XIX, by the Bent Crystal Spectrometer on the Solar Maximum Mission. They are coincident with hard X–ray bursts and with the onset of Hα sprays. They appear to result from a different type of phenomenon from the unresolved blue–shifted component from the resonance lines reported by Feldman et al. (1980) and Antonucci et al. (1982). The authors believe that these discrete line features are due to Doppler–shifted resonance line emission from well collimated, moving plasma with large line–of–sight velocities.

073.004 **High–energy observations of solar flares.**
K. Tanaka.
Astrophys. Space Sci., Vol. 118, Nos. 1/2, p. 101 – 113 (1986). – See Abstr. 012.001.

Extensive observations of solar flares made in high energy bands during the maximum of the present solar cycle are discussed with a special reference to the results from HINOTORI. The hard X–ray (HXR) images from HINOTORI showed mostly coronal emission at 20 – 25 keV suggesting that the HXR is emitted from multiple coronal loops, consistent with the non–thermal electron beam model in a high density corona. The thermal HXR model seems to be inconsistent with some observations. A summary is given for the characteristics of the energy release in an impulsive burst and an empirical model is described.

073.005 **High resolution observation of Hα solar flares and temporal relation between Hα and X–ray, microwave emission.**
H. Kurokawa, T. Kitahara, Y. Nakai, Y. Funakoshi, K. Ichimoto.
Astrophys. Space Sci., Vol. 118, Nos. 1/2, p. 149 – 152 (1986). – See Abstr. 012.001.

The authors studied the evolutional characteristics of fine structures in Hα flare emitting regions and their relation to X–ray and microwave emissions for selected events observed with the 60 cm Domeless Solar Telescope at Hida Observatory, University of Kyoto. The principal conclusions of this investigation are: (1) Hα kernel consists of some finer bright points or Hα flare points whose individual size is less than 1 arc sec. (2) Impulsive brightenings of Hα flare points occurred simultaneously with the spikes of the hard X–ray and microwave bursts within the time resolution of the authors' Hα observations which varied from 1 to 10 s. (3) It is concluded that fast electron beams must be the principal mechanism of heating Hα flares during the impulsive phase of a flare.

073.006 **Reference models of sunspot chromospheres.**
H. S. Yun, H. A. Beebe.
Astrophys. Space Sci., Vol. 118, Nos. 1/2, p. 173 – 175 (1986). – See Abstr. 012.001.

Recent observational and theoretical studies of the structure of umbral chromospheres have led the authors to propose new models of the region. They combine space and ground based determinations of the density–temperature structure of the transition region with existing Ca II H, K, and IR triplet spectral data to establish relatively high density, "gradient" models of the sunspot chromosphere.

073.007 **High energy observations of June 1980 solar flares.**
B. Lokanadham, P. K. Subramanian, A. L. Kiplinger, B. R. Dennis.
Astrophys. Space Sci., Vol. 119, No. 1, p. 5 – 8 (1986). – See Abstr. 012.001.

The paper presents a detailed study of the high energy X–ray observations of the most unusual solar events observed on 4 and 7 June, 1980 with the Hard X–Ray Burst Spectrometer on Solar Maximum Mission satellite. The hard X–ray data of the events are also compared with the radio microwave fluxes. The X–ray time profiles of these flares are characterized by the occurrence of impulsive phase superposed with a number of narrow spikes before the occurrence of the main energetic events. Various models for explaining the evolution of the events are considered and the sequential firing loop model seems to be consistent with the observations of the events.

073.008 **Chromospheric fine structure.**
R. G. Athay.
Physics of the Sun, Vol. II, p. 51 – 69 (1986). – See Abstr. 003.001.

Contents: Properties of chromospheric structure (network and supergranule cells, network coarse and fine structure, supergranule cell structure, active region structure). Velocity and magnetic structure. Magnetohydrodynamic structure (a new look at fine structure, lifting forces). Major problems (energy balance, momentum balance, the Sun as a star).

073.009 **The acceleration and propagation of solar flare energetic particles.**
M. A. Forman, R. Ramaty, E. G. Zweibel.
Physics of the Sun, Vol. II, p. 249 – 289 (1986). – See Abstr. 003.001.

Contents: Energetic particles in solar flares (electromagnetic radiations: radio emissions, hard X–rays, gamma rays; energetic particles: energy spectra and electron–proton correlations, chemical compositions, isotopic and ionic compositions). Mechanisms of solar flare particle acceleration (stochastic acceleration, shock acceleration, acceleration in direct electric fields). Solar flare particle spectra in interplanetary space. Summary and outlook.

073.010 **Nuclear processes in solar flares.**
R. Ramaty.
Physics of the Sun, Vol. II, p. 291 – 323 (1986). – See Abstr. 003.001.

Contents: Nuclear reactions in solar flares (interaction models and properties of the energetic particles, neutron and 2.223 MeV photon production, positron and 0.511 MeV photon production, prompt de–excitation line production). Implications of gamma–ray observations (interaction model, energetic particle spectrum, number and energy content, time dependence, the photospheric ^3He abundance, beaming of the energetic particles).

073.011 **Interpretation of Hα contrast profiles of active region loops.**
L. E. Cram.
Astrophys. J., Vol. 300, No. 2, p. 830 – 835 (1986).

It has been shown recently that the observed Hα contrast profiles of an active region loop can be matched by theoretical contrast profiles based on a "cloud" model with four free parameters: optical depth, Doppler width, source function, and line–of–sight velocity. On the other hand, the non–LTE theory of the formation of Hα by noncoherent scattering implies that one of these parameters, the source function, can be deduced from the others. This paper examines the consistency between values of the source function inferred by an application of the cloud model and those calculated using non–LTE theory.

073.012 **Rapid acceleration of energetic particles in the 1982 February 8 solar flare.**
S. R. Kane, E. L. Chupp, D. J. Forrest, G. H. Share, E. Rieger.
Astrophys. J., Lett. Ed., Vol. 300, No. 2, p. L95 – L98 (1986).

Hard X–ray and gamma–ray measurements of the 1982 February 8 solar flare made with spectrometers aboard the ISEE 3 and SMM spacecraft show that bursts of photons from $\sim$40 keV to $\sim$40 MeV are coincident to within $\pm$1 s. This indicates that the acceleration of particles to relativistic velocities can occur promptly (within $\sim$1 s). As far as the energetic particles interacting at the Sun are concerned, the "two–phase" process, which requires several minutes for particle acceleration to relativistic velocities, cannot account for the present gamma–ray observations.

073.013 **Broadening of soft X–ray lines during the impulsive phase of solar flares: random or directed mass motions?**
J. G. Doyle, R. D. Bentley.
Astron. Astrophys., Vol. 155, No. 2, p. 278 – 282 (1986).

The authors present spectroscopic data for three flares obtained with the Bent Crystal Spectrometer flown in the Solar

Maximum Mission in 1980. This data is concerned with the structure of the Ca XIX resonance line at 3.176 Å during the impulsive phase of flares. On the basis of high time resolution (6s.) data, the authors suggest that the previous published results concerned with the excess broadening of the resonance line being due to bulk random mass motions may give an over–simplified picture. Instead the authors suggest that during this stage of the flare, the resonance line consists of many discrete features, which are interpreted as mass flows.

073.014 **Oscillations of the Sun's chromosphere. IV. Temporal evolution of Hα profile.**
F. Kneer, M. von Uexküll.
Astron. Astrophys., Vol. 155, No. 1, p. 178 – 184 (1986). = Mitt. Kiepenheuer–Inst., Nr.261.
The authors present Hα observations of the quiet solar chromosphere at disk centre. The very different dynamical behaviour of the interior and boundary of the network cells is emphasized. The dominant processes in the interior are waves in the 5 min period range and with shorter periods. The energy flux in these waves is estimated to be too small to account for the radiative losses from standard chromospheric models. In the cell boundaries, the material flows down, on time average, near the centres of rosettes. The short time fluctuations are mainly stochastic while the oscillations are reduced. This suggests that the energy balance in the cell boundaries is determined by MHD turbulence.

073.015 **Energy deposit at temperature minimum level and white light flares.**
J. Aboudarham, J. C. Hénoux.
Astron. Astrophys., Vol. 156, No. 1/2, p. 73 – 78 (1986).
Empirical atmospheric models are used to estimate the energy deposit required to heat the temperature minimum region of the quiet sun to flare temperature. The calculations show that the necessary energy deposit in the temperature minimum region can be accounted for by electron bombardment. Electron beam heating of the temperature minimum and chromospheric regions produces white light emission with an energy output of about $5 \times 10^9 \mathrm{erg\,cm^{-2}s^{-1}}$. Also the upper photosphere is heated to about 30° above its quiet sun value by the downward radiation field. All layers from the chromosphere to the photosphere contribute to the white light emission process. The emission mechanisms are H^- and hydrogen bound–free emission. The authors conclude, that no in–situ energy release is necessary to explain the heating of the temperature minimum region. An electron bombardment can increase the temperature of this region, and subsequently of the photosphere via radiative back heating, hence producing a white light flare.

073.016 **The linear polarization of hydrogen Hβ radiation and the joint diagnostic of magnetic field vector and electron density in quiescent prominences. I. The magnetic field.**
V. Bommier, J. L. Leroy, S. Sahal–Bréchot.
Astron. Astrophys., Vol. 156, No. 1/2, p. 79 – 89 (1986).
Quasi–simultaneous measurements of the linear polarization of the helium D_3 and hydrogen Hβ lines have been achieved at the Pic–du–Midi coronagraph–polarimeter for fourteen quiescent prominences. Both lines are linearly polarized by resonant scattering of the anisotropic photospheric and chromospheric radiation field, and depolarized by the local magnetic field (i.e. the Hanle effect). The measurements show that another depolarizing mechanism has to be taken into account for the Hβ line: this is confirmed by the theory which predicts that Hβ is also depolarized by collisions with the electrons and protons of the medium. Therefore polarization measurements in those two lines provide a method for determining both the magnetic field **B** and the electron density N_e in prominences.

073.017 **The linear polarization of hydrogen Hβ radiation and the joint diagnostic of magnetic field vector and electron density in quiescent prominences. II. The electron density.**
V. Bommier, J. L. Leroy, S. Sahal–Bréchot.
Astron. Astrophys., Vol. 156, No. 1/2, p. 90 – 94 (1986).
This paper is devoted to the analysis of the results concerning the electron density N_e (See 073.016).

073.018 **Narrowband dm–spikes, intermediate drift bursts and pulsations in the solar flare of August 19, 1981.**
M. Karlický.
Bull. Astron. Inst. Czech., Vol. 37, No. 1, p. 1 – 7 (1986). With plate 1.
In the initial phase (12.51–12.53 UT) of the flare of August 19, 1981, an interesting group of narrowband dm–spikes, intermediate drift bursts and pulsations was observed. This paper tries to explain this group of bursts by a uniform model. The same "period" of intermediate drift bursts and pulsations has been explained. In the following time the flare loop under study explodes and a shock wave (type II radio burst) is generated at the relatively high altitude of ≈100000 km above the photosphere. This process was connected with the radio flux decrease on 10 cm.

073.019 **Two classes of solar energetic particle events associated with impulsive and long–duration soft X–ray flares.**
H. V. Cane, R. E. McGuire, T. T. von Rosenvinge.
Astrophys. J., Vol. 301, No. 1, p. 448 – 459 (1986).
For the period 1978 September to 1983 December the authors have identified 67 solar particle events for which their instruments detected electrons above 3 MeV and for which there are soft X–ray observations. On the basis of their signature in soft X–rays, the authors have divided the events into two classes, impulsive and long duration, and find that they have different properties. A detailed account of these differing properties of the X–ray, radio and energetic particle characteristics of the two classes of flares is presented. It is shown that the two classes of particle events arise because different acceleration processes operate preferentially for the two classes of flares.

073.020 **Possible manifestations of anomalous scattering of the radio emission of solar flares by plasmons.**
L. A. Pustil'nik, Del Rio Henaro.
Astron. Zh., Tom 63, Vyp. 1, p. 155 – 161 (1986). In Russian. English translation in Sov. Astron., Vol. 30, No. 1.
Manifestations of anomalous scattering of flare radio emission by regions of plasma turbulence in solar flares are analysed. It is shown that anomalous scattering must produce effectively distortions in the spectrum of transformed emission (with forming a break or a maximum or an exponential cut–off at low frequencies), and also leads to blurring of time and frequency fine structure. Estimates are made of the expected effects for the main plasma turbulence regions. Characteristic frequencies of spectral distortions in transformed emission and the effects of fine–structure blurring are estimated for each of these regions, and possible observational tests are suggested. Comparison with the existing observational data has shown that the predicted effects seem to have been already observed in a number of flare radio phenomena.

073.021 **Structure and activization of solar prominences in the radio and optical ranges.**
G. P. Apushkinskij.
Astron. Zh., Tom 63, Vyp. 1, p. 162 – 165 (1986). In Russian. English translation in Sov. Astron., Vol. 30, No. 1.
The radio investigations of prominences during solar eclipses showed bright knots and holes without radio emission in them. Prominence activization by flares observed at radio waves goes along the chromosphere and then from the lower to the upper level of the prominence. The radio brightness distribution along the prominence coincides with UV data.

073.022 **Stability of cool flux tubes in the solar chromosphere. Linear analysis.**
S. S. Hasan, F. Kneer.
Astron. Astrophys., Vol. 158, No. 1/2, p. 288 – 296 (1986).
The authors examine the linear stability of cool flux tubes in the solar chromosphere which are initially in radiative equilibrium. Owing to the presence of carbon monoxide, there exists a narrow region near the temperature minimum where the temperature gradient becomes steep enough to drive a convective instability. The authors use the thin flux tube equations and include in a simple manner radiative heat exchange with the ambient

medium. Initial states of constant β (where β is the ratio of gas to magnetic pressure) are considered. They present results for various values of β and calculate growth rates, eigen–vectors of the fundamental modes and examine phase relationships. It is suggested that overstable oscillations should invariably be associated with cool flux tubes. These oscillations transport energy and can thus change the thermodynamic structure of flux tubes.

073.023 **The duration of flares and their radio emission.**
A. Antalová.
Bull. Astron. Inst. Czech., Vol. 37, No. 2, p. 107 – 111 (1986).
The mean duration of 906 H–alpha flares were calculated in dependence on their radio, X–ray and areal characteristics. A similar calculation was carried out of the durations of soft X–ray flares (215 flares). A statistical dependence was found between the duration of H–alpha as well as soft X–ray flares and the type of radio emission accompanying the flare.

073.024 **The sudden decrease of the large flare occurrence and its manifestation in interplanetary space after the maximum of the solar cycle.**
L. Křivský, A. Krüger, B. Růžičková–Topolová.
Bull. Astron. Inst. Czech., Vol. 37, No. 2, p. 111 – 114 (1986).
The effect is documented on the sudden decrease of the annual occurrence of large flares and flare–associated manifestations in interplanetary space (type II and IV radio events, and Forbush decreases) during the 11–year cycles of solar activity. Possible explanations can be found, on the one hand, in the limited interaction of two assumed magnetically counteracting systems below the photosphere after the secondary cycle maximum and, on the other, due to the lower values of the differential rotation near the solar equator.

073.025 **A mechanism for deep chromospheric heating during solar flares.**
M. E. Machado, A. G. Emslie, P. J. Mauas.
Astron. Astrophys., Vol. 159, No. 1/2, p. 33 – 39 (1986).
The authors reexamine the role of the negative hydrogen ion, H^-, in the energy balance of the deep solar chromosphere and find, in contrast with earlier authors, that H^- is a source of heating in these levels. The response of this region to an ionizing flux of flare–associated UV radiation ($1500 \lesssim \lambda \lesssim 1900$ Å) is then addressed: the authors find that the excess ionization of Si to Si^+ increases the local electron number density considerably, since most species are largely neutral at deep chromospheric levels. This in turn increases the electron–hydrogen atom association rate, the H^- abundance, and the rate of absorption of photospheric radiation by this ion. The authors find that the excess absorption by this process may lead to a substantial temperature enhancement at temperature minimum levels during flares.

073.026 **Directionality of bremsstrahlung from relativistic electrons in solar flares.**
C. D. Dermer, R. Ramaty.
Astrophys. J., Vol. 301, No. 2, p. 962 – 974 (1986).
The authors have calculated detailed angular and energy spectra of bremsstrahlung from anisotropic electron distributions in solar flares. The results are compared to observations of gamma–ray limb brightening and to data on the variation of the gamma–ray spectrum with flare position on the Sun.

073.027 **On magnetic field stochasticity and nonthermal line broadening in solar flares.**
E. Antonucci, R. Rosner, K. Tsinganos.
Astrophys. J., Vol. 301, No. 2, p. 975 – 980 (1986).
The authors discuss observations of nonthermal line broadening seen in solar flares by the SMM satellite in light of recent results on the generation of magnetic field stochasticity. A consistent model for the data can be constructed by assuming that the observations signal the destruction of an ambient magnetohydrodynamic equilibrium.

073.028 **Electron density diagnostics in the 10 – 100 Å interval for a solar flare.**
W. A. Brown, M. E. Bruner, L. W. Acton, H. E. Mason.
Astrophys. J., Vol. 301, No. 2, p. 981 – 988 (1986).
Electron density measurements from spectral–line diagnostics are reported for a solar flare on 1982 July 13 1627 UT. The spectrogram, covering the 10 – 95 Å interval, contained usable lines of helium–like ions C V, N VI, O VII, and Ne IX which are formed over the temperature interval $0.7 - 3.5 \times 10^6$K. In addition, spectral–line ratios of Si IX, Fe XIV, and Ca XV were compared with new theoretical estimates of their electron density sensitivity to obtain additional electron density diagnostics. An electron density of $3 \times 10^{10} cm^{-3}$ was obtained.

073.029 **High–resolution photography of the solar chromosphere. XXIII. Direct measurement of the optical thickness of Hα solar loops.**
R. J. Bray, R. E. Loughhead.
Astrophys. J., Vol. 301, No. 2, p. 989 – 991 (1986). With plates 9 – 10.
A method has been devised which, under appropriate observational circumstances, enables a direct determination to be made of the optical thickness, t_L of high–rising Hα solar loops observed on the disk. It is applicable to both bright and dark loops and requires no knowledge of the source function in the loop; it is best adapted to optically thin ones. The technique has been successfully applied to four loops observed under various circumstances, giving values of t_L ranging from 0.06 to 1.3.

073.030 **On the fine structure of polarized flare elements.**
A. N. Babin, A. N. Koval'.
Izv. Krymskoj Astrofiz. Obs., Tom 70, p. 3 – 8 (1985). In Russian. English translation in Bull. Crimean Astrophys. Obs., Vol. 70.
The investigation of linear polarization of solar emission features is being continued. From the spectral observations in the Hα line obtained with high spatial resolution it was found that in a distance scale of some arcsecond in some flare knots not only the value but also the sign of the polarization determined by the authors 1983 is changed. In the flare region some emission knots like moustaches or grains of continuous emission show strong polarization.

073.031 **On quasi–periodicity of the flare activity of the sun.**
V. I. Abramenko, M. B. Ogir'.
Izv. Krymskoj Astrofiz. Obs., Tom 70, p. 8 – 14 (1985). In Russian. English translation in Bull. Crimean Astrophys. Obs., Vol. 70.
On the basis of flare activity observations all over the solar disk carried out in summer of 1980 and 1981 the time distribution of flare appearances on the surface and their energies were investigated by the Fourier–analysis method. The following groups of authentic quasi–periods surpassing the 3σ level were discovered: $109 - 166^m$, $68 - 78^m$, $25 - 46^m$. The quasi–periods $68 - 78^m$ and $25 - 46^m$ characterize the repetition of flare groupings, and quasi–periods $109 - 166^m$ are manifested as modulations of the flare energy of the groupings.

073.032 **Solar flare of 1982 April 14.**
S. P. Bagare.
Bull. Astron. Soc. India, Vol. 13, No. 4, p. 394 – 398 (1985).
The author has obtained a good time sequence of spectroheliograms and spatial sequence spectra in the K–line of Ca II using the Kodaikanal tower telescope. He has derived the two dimensional pattern of the mass motions of the plasma associated with the flare at this level in the solar atmosphere. He finds that significant mass motions are practically confined to the two ribbons that developed soon after the flaring event.

073.033 **Investigation of the inhomogeneity of quiescent prominences with the help of the method of halfwidth curves. II.**
A. K. Kirillov.
Soln. Dannye, Byull., 1985, No. 11, p. 68 – 71 (1985). In Russian.

A discussion of the results of investigation of the macroscopic motions in quiescent prominences on the basis of the phenomenological theory of spectral–line formation in a turbulent medium is presented. The largest scale of turbulence in prominences is found; it varies from 2×10^8cm to 2×10^9cm.

073.034 **Neighbouring areas and their influence on the behaviour of flare activity of the solar surface area in question.**
E. V. Miletskij.
Soln. Dannye, Byull., 1985, No. 11, p. 76 – 81 (1985). In Russian.
On the basis of a regional approach to the analysis of solar activity equations connecting values are formed of the Dodson–Hedeman flare index in separate areas of the solar surface with the values of other indices of solar activity taken with account for neighbouring areas. It is shown that the analysis of neighbouring areas allows for better modelling. The conclusion is made that the value of the flare index for an area in question is determined by the preceding activity of a neighbouring area.

073.035 **A spectral investigation of spicules in the K Ca II line.**
N. I. Pishkalo, V. I. Ivanchuk.
Soln. Dannye, Byull., 1985, No. 11, p. 82 – 87 (1985). In Russian.
Some properties of solar chromospherical spicules are studied as observed spectrally in the K Ca II line at the N–limb. Spicules in the K line are wider and more diffusive than those in the Hα line. The average lifetime of K spicules is 10 – 12 minutes. Radial velocities, halfwidths and equivalent widths are also found.

073.036 **Investigation of the linear polarization and Hα–line profiles during the impulsive phase of two solar flares.**
A. N. Babin, A. N. Koval'.
Izv. Krymskoj Astrofiz. Obs., Tom 72, p. 142 – 153 (1985). In Russian. English translation in Bull. Crimean Astrophys. Obs., Vol. 72.
The linear polarization and Hα–line profiles in importance 2B and 1B flares have been studied. Three types of the emission line profiles were observed simultaneously: nearly symmetrical profiles with and without absorption in the line center and asymmetrical profiles. Two components of the line profiles can be distinguished: I – emission contributing mainly in the center of the line; and II – emission contributing to the extended wings. The linear polarization is revealed in six of twenty emission details for an importance 2B flare and in one of five ones for importance 1B. The polarization is observed either over the whole profile or in the central part of the line only.

073.037 **The energy balance and plasma density in solar flares.**
Eh. A. Baranovskij.
Izv. Krymskoj Astrofiz. Obs., Tom 72, p. 154 – 159 (1985). In Russian. English translation in Bull. Crimean Astrophys. Obs., Vol. 72.
On the basis of energy balance calculations for the flare chromosphere irradiated by the X–ray and far EUV flux the density in the region of formation of Lyman and Balmer lines is estimated. From the comparison of radiative losses for Balmer lines and for negative hydrogen ions it is concluded that the flare regions with density $n_H > 4 \times 10^{15}\mathrm{cm}^{-3}$ will not radiate in Balmer lines, for the absorbed energy will be carried away by the radiation of negative ions at comparatively low temperature. It is shown that a significant part of the incident X–ray and EUV flux penetrates through the region of hydrogen lines formation to deeper levels, thus permitting to explain the white light flares by the influence of the beforesaid flux.

073.038 **On the polarization of emission of moustaches observed near the solar limb.**
A. N. Babin, A. N. Koval'.
Izv. Krymskoj Astrofiz. Obs., Tom 73, p. 3 – 8 (1985). In Russian. English translation in Bull. Crimean Astrophys. Obs., Vol. 73.
H_α spectral observations of moustaches using a linear polarization analyser have been carried out. Four cases of moustaches displaying significant differences in the two orthogonal polarized spectra are studied. The differences are clearly seen in either the structure (double emission features in light of one polarization) or brightness. In all cases major differences are observed in the red wing of the H_α line. It indicates that the polarization structure is strongly asymmetric, however this conclusion needs further confirmation. The degree of polarization of moustache emission is found to be high (up to 24%), exceeding considerably theoretical values for the resonance scattering process. In the case of close pairs of moustaches only the weak component is polarized. The authors suggest that moustaches or their fine structure elements are polarized during the brightening phase only.

073.039 **On spectral observations of the chromospheric fine structure in an active region.**
L. G. Kartashova.
Izv. Krymskoj Astrofiz. Obs., Tom 73, p. 70 – 75 (1985). In Russian. English translation in Bull. Crimean Astrophys. Obs., Vol. 73.
Fourty H_α spectrograms of regular fibriles systems and flocculi mottles have been studied. It is shown that these systems have two–component structure, consisting of intermittent bright and dark features. These elements have different H_α line central intensities, half–widths and line–of–sight velocities.

073.040 **Structurization of a flare–active region and soliton model of the initial energy and matter source in solar flares.**
Eh. I. Mogilevskij.
Kinematika Fiz. Nebesn. Tel, Tom 2, No. 2, p. 75 – 82 (1986). In Russian. English translation in Kinematics Phys. Celest. Bodies.
Noting certain fundamental difficulties of the accepted theoretical models of solar flares, the author suggests a possible model in which the initial source of the flare energy and matter is associated with MHD–solitons. The generation of the latter is possible when a certain general (or local) helical structure exists in the distribution of magnetic fields and motions in the photosphere.

073.041 **A method of calculation of the velocity field and the spatial azimuth of a loop prominence.**
G.–p. Ji, L.–s. Cui, X.–b. Ni.
J. Nanjing Univ., Vol. 21, No. 4, p. 609 – 616 (1985). In Chinese.

073.042 **On a connection of the flare activity of the sun with the characteristics of radio radiation of active regions on it.**
S. I. Avdyushin, A. F. Bogomolov, E. A. Borisova,
E. I. Zajtsev, S. P. Leonenko, B. Ya. Losovbskij,
S. V. Makagonov, A. P. Molchanov, B. A. Poperechenko,
N. G. Franchuk, L. V. Yasnov.
Dokl. Akad. Nauk SSSR, Tom 283, No. 1, p. 67 – 70 (1985). In Russian. Abstr. in Ref. Zh., 51. Astron., 1.51.289 (1986).

073.043 **Observations of the sun at the Astronomical Observatory of the L'vov University in 1983.**
M. B. Girnyak, M. M. Koval'chuk, I. S. Laba, P. A. Olijnyk.
Vestn. L'vov. univ. Ser. astron., No. 59, p. 24 – 26 (1985). In Russian. Abstr. in Ref. Zh., 51. Astron., 2.51.485 (1986).

073.044 **Radioastronomical observations of an interplanetary shock wave from the solar flare of August 18, 1979.**
V. I. Vlasov.
Geomagn. Aehron., Tom 26, No. 2, p. 182 – 186 (1986). In Russian. English translation in Geomagn. Aeron.

073.045 **Three–dimensional reconnection after a prominence eruption.**
R. A. Kopp, G. Poletto.
The sun and the heliosphere in three dimensions, p. 65 – 70 (1986). – See Abstr. 012.016.
A currently widely held explanation of the diverse phenomena following a prominence eruption is that these are the direct result of the ensuing reconnection of open magnetic field lines created by the eruption. Kopp and Pneuman (1976) and Kopp and Poletto (1984) developed a 2–D model wherein an analytical time–dependent magnetic field geometry was used to account quantitatively for the major characteristics of many of these post–flare effects. In the present paper, after mentioning a few situations for which a 2–D treatment would nevertheless clearly

be inadequate, the authors extend to three dimensions their earlier model.

073.046 **Oxygen λ7774 emission in solar prominences.**
N. S. Brickhouse, D. A. Landman.
Bull. Am. Astron. Soc., Vol. 17, No. 4, p. 834 (1985). Abstract. – See Abstr. 010.062.

073.047 **HRTS (*High Resolution Telescope and Spectrograph*) Spacelab 2 observations of spicular emission at the solar limb.**
J. W. Cook, J.–D. F. Bartoe, G. E. Brueckner, K. P. Dere, D. G. Socker.
Bull. Am. Astron. Soc., Vol. 17, No. 4, p. 834 (1985). Abstract. – See Abstr. 010.062.

073.048 **Numerical simulation of a siphon mechanism for quiescent prominences.**
J. T. Mariska, A. I. Poland.
Bull. Am. Astron. Soc., Vol. 17, No. 4, p. 842 (1985). Abstract. – See Abstr. 010.062.

073.049 **MHD simulation of active prominence magnetic evolution.**
C.–H. An, S. T. Wu, J. J. Bao.
Bull. Am. Astron. Soc., Vol. 17, No. 4, p. 842 (1985). Abstract. – See Abstr. 010.062.

073.050 **Submillimeter diagnostics of the response of the solar chromosphere to compressional waves.**
C. Lindsey, T. Roellig.
Bull. Am. Astron. Soc., Vol. 17, No. 4, p. 896 (1985). Abstract. – See Abstr. 010.062.

073.051 **Hole boring by electron beams in solar flares.**
B. Woodgate, L. Orwig.
Bull. Am. Astron. Soc., Vol. 17, No. 4, p. 905 (1985). Abstract. – See Abstr. 010.062.

073.052 **Correlation of hard X–ray and soft X–ray observations in solar flares.**
A. G. Emslie.
Bull. Am. Astron. Soc., Vol. 17, No. 4, p. 905 (1985). Abstract. – See Abstr. 010.062.

073.053 **Plasma parameters from the helium–like ion Mg XI during a small impulsive flare.**
G. A. Linford, C. J. Wolfson.
Bull. Am. Astron. Soc., Vol. 17, No. 4, p. 905 (1985). Abstract. – See Abstr. 010.062.

073.054 **Filament eruptions in the impulsive phase of solar flares.**
R. L. Moore, S. W. Kahler, S. R. Kane, H. Zirin.
Bull. Am. Astron. Soc., Vol. 17, No. 4, p. 905 (1985). Abstract. – See Abstr. 010.062.

073.055 **Location of flares on the sun.**
T. Bai.
Bull. Am. Astron. Soc., Vol. 17, No. 4, p. 905 – 906 (1985). Abstract. – See Abstr. 010.062.

073.056 **The heating of the thermal plasma with energetic electrons in small solar flares.**
H.–A. Lin, R. P. Lin, S. R. Kane.
Bull. Am. Astron. Soc., Vol. 17, No. 4, p. 933 (1985). Abstract. – See Abstr. 010.065.

073.057 **Heat transport in steep temperature gradients. I. Small flaring solar loops.**
D. F. Smith.
Astrophys. J., Vol. 302, No. 2, p. 836 – 848 (1986).
Results on nonlocal heat transport which properly takes into account the presence of fast electrons with mean free paths much longer than the temperature scale height L are reviewed. The heat flux in the hot part of the gradient is reduced relative to the Spitzer–Härm value q_{SH} which does not take into account the heat carried away by the fast electrons. The heat flux in the cold part of the gradient is enhanced relative to the value q_{SH} which does not take into account the heat deposited by the fast electrons. These are applied to the problem of thermal hard X–ray burst models.

073.058 **A shock condensation mechanism for loop prominences.**
T. G. Forbes, J. M. Malherbe.
Astrophys. J., Lett. Ed., Vol. 302, No. 2, p. L67 – L70 (1986).
Self–consistent, numerical solutions of the resistive MHD equations in two dimensions show that a quasi–stationary, fast–mode shock is a characteristic feature of the reconnection dynamics of the Kopp–Pneuman model of two–ribbon flares. A preliminary analysis of the effects of radiative cooling and thermal conduction suggests that the fast shock can help trigger a thermal condensation (i.e., a loop prominence) if the reconnecting magnetic fields are sufficiently strong.

073.059 **Dynamics of chromospheric spicules. Radial, tangential and apparent velocities (Review).**
V. I. Ivanchuk, N. I. Pishkalo.
Vestn. Kiev. Univ. Astron., No. 27, p. 7 – 23 (1985). In Russian.

073.060 **Photometric investigation of the echelle spectrogram of the 15. VII. 1981 solar flare.**
U. M. Lejko, V. A. Ostapenko.
Vestn. Kiev. Univ. Astron., No. 27, p. 43 – 46 (1985). In Russian.

073.061 **On an investigation of conic limb flares.**
V. A. Ostapenko, A. A. Rozhilo.
Vestn. Kiev. Univ. Astron., No. 27, p. 46 – 50 (1985). In Russian.

073.062 **Investigation of the information content of a system of parameters for forecast of large flares in sunspot groups.**
V. V. Tel'nyuk–Adamchuk, S. V. Pasechnik, P. F. Kravchuk.
Vestn. Kiev. Univ. Astron., No. 27, p. 54 – 59 (1985). In Russian.

073.063 **Small solar flares in an activity complex.**
A. V. Borovik.
Issled. Geomagn., Aehron. Fiz. Solntsa, Moskva, No. 73, p. 28 – 37 (1985). In Russian. Abstr. in Ref. Zh., 51. Astron., 3.51.437 (1986).

073.064 **Radial velocities of solar active prominences.**
A. I. Kiryukhina.
Astron. Tsirk., No. 1389, p. 4 – 6 (1985). In Russian.

073.065 **Hα line profiles in limb flares.**
A. B. Delone, E. A. Makarova, G. V. Yakunina.
Astron. Tsirk., No. 1399, p. 4 – 6 (1985). In Russian.

073.066 **Blue asymmetry of the Hα profile in the flare of June 4, 1979.**
A. B. Delone, E. A. Makarova, T. S. Flerova, G. V. Yakunina.
Astron. Tsirk., No. 1402, p. 1 – 3 (1985). In Russian.

073.067 **The solar flare of May 16, 1981.**
I. F. Nikulin, K. V. Alikaeva, P. N. Polupan.
Soln. Dannye, Byull. 1986, No. 1, p. 89 – 92 (1986). In Russian.
The structure and dynamics of the development of the flare of May 16, 1981 are considered on the basis of filter and spectrographic observations. It is established that the tilt with respect to the dispersion direction in the Hα spectrum is due to the downward motion of gas towards the loops foot but not to the rotation.

073.068 **The observed oscillation mode of the chromosphere above sunspots.**
V. S. Loskutnikov.
Soln. Dannye, Byull. 1986, No. 2, p. 61 – 67 (1986). In Russian.
The dynamics of 3–min oscillations in two sunspots of a bipolar group is studied. It has been shown that the oscillation mode in the umbra is essentially unstable, however lasting about 1.5 hours the variations are insignificant. An estimation of the chromospheric height above sunspots is made.

073.069 **An application of the cluster analysis to the study of small area flares.**
I. Ferro Ramos, J. Alfonso Sosa.
Soln. Dannye, Byull. 1986, No. 2, p. 73 – 80 (1986). In Russian.
One of the methods of a cluster analysis is described. Peculiarities of its application to a study of properties of multi–dimensional classification of small area flares are analysed. The conventional empirical classification of flares is approved on the basis of the data for 1972 with the use of the method. Some new parameters are suggested to be introduced in this classification for its improvement.

073.070 **Prominence hydrogen lines at 10 – 20 microns.**
J. B. Zirker.
Sol. Phys., Vol. 102, Nos. 1/2, p. 33 – 40 (1985).
An analysis of prominence hydrogen lines is presented in the spectral band 10 – 20 μ. The results are consistent with earlier work.

073.071 **Further analysis of temperature and emission measure during the decay phase of solar flares derived from soft X–ray light curves.**
P. L. Bornmann.
Sol. Phys., Vol. 102, Nos. 1/2, p. 111 – 130 (1985).
The light curves of soft X–ray lines, observed by the Flat Crystal Spectrometer on Solar Maximum Mission during eight solar flares are modeled to determine the plasma temperature and emission measure as functions of time using the method first presented by Bornmann (1985), but modified to include a χ^2search routine. With this modification the technique becomes more general, more accurate, and applicable throughout the gradual phase of the flare. The model reproduces the light curves of the soft X–ray lines throughout these flares. Model fits were repeated for each flare using five different sets of published line emissivity calculations. The emissivities of Mewe and Gronenschild (1981) consistently gave the best fits to the observed light curves for each flare.

073.072 **Flare morphologies and coronal field configurations.**
F. Tang.
Sol. Phys., Vol. 102, Nos. 1/2, p. 131 – 145 (1985).
Chromospheric flares are the footpoints of closed coronal field lines. The author presents different flare morphologies from observations and examines the implied coronal field configurations above the flaring region. Flares are grouped according to the number of ribbons, from unresolved compact point–like flare to four–ribbon flares. Quiet region flares having characteristics all their own are also presented.

073.073 **An investigation of the linear polarization in emission features of active regions.**
N. M. Firstova.
Sol. Phys., Vol. 103, No. 1, p. 11 – 20 (1986).
A study has been made of linear polarization in "moustaches" using the Hα and Hβ lines. During the first three minutes of the existence of the moustaches, the two lines revealed a linear polarization of about 7% located at the center of the lines. No polarization was found during the later stage of the moustaches' lifetimes. Whenever the linear polarization is present, there is a tendency for the polarization to decrease rapidly, in accompaniment of a turning of the azimuth. Several mechanisms for interpreting the polarization observed in the moustaches are examined. Some suggestions are made in support of the excitation of hydrogen atoms in the moustaches initiated by a high–energy electron flux or vertical heat conduction.

073.074 **Investigation of the Mg XII 8.42 Å doublet in solar flare spectra.**
B. Sylwester, P. Faucher, J. Jakimiec, V. V. Krutov, R. W. P. McWhirter, J. Sylwester, M. Tomczak, S. Volonté, I. A. Zhitnik, V. V. Korneev.
Sol. Phys., Vol. 103, No. 1, p. 67 – 87 (1986).
The intensity ratio of the components of the Mg XII 8.42 Å ($1s^2S_{1/2}$–$2p^2P_{1/2,3/2}$) doublet in solar flare spectra has been investigated using observations recorded from the Intercosmos 7 satellite. The observed values of the ratio fall within the interval 0.38 – 0.66 and have been compared with recent theoretical predictions based on an optically thin collisional–radiative model.

073.075 **The structure of high–temperature flare plasma. II. Comparison with observations.**
A. G. Emslie.
Sol. Phys., Vol. 103, No. 1, p. 103 – 110 (1986).
In Paper I (1985) the author derived "scaling laws" which relate peak loop temperature to injected non–thermal electron flux in electron–heated models of the flare corona. In this paper, he compares these predicted relationships with recent coordinated observations in hard X–rays and soft X–rays. Satisfactory agreement is found for large events, while for smaller events the plasma electron temperature determined by soft X–ray spectral fitting is too high. A possible resolution of this apparent discrepancy is discussed through careful examination of the method used to determine the temperature of the soft X–ray emitting plasma. It is concluded that temperatures determined by spectral fitting over a series of lines are not necessarily representative of the true temperature of the plasma.

073.076 **The flare kernel in the impulsive phase.**
C. de Jager.
NASA Conf. Publ., NASA CP–2421, p. 29 – 40 (1986). – See Abstr. 012.021.
The impulsive phase of a flare is characterized by impulsive bursts of X–ray and microwave radiation, related to impulsive footpoint heating up to 50 or 60 MK, by upward gas velocities (150 to 400 km s^{-1}) and by a gradual increase of the flare's thermal energy content. A hole is "burned" into the chromosphere; at the end of the impulsive phase of an average flare the lower part of that "hole" is situated about 1800 km above the photosphere. Hα and other optical and UV line emission is radiated by a thin layer (≈20 km) at the bottom of the flare kernel. The single flux tube model of a flare does not fit with these observations; instead the spaghetti–bundle model is proposed. Microwave and gamma–ray observations suggest the occurrence of dense flare knots of ~800 km diameter, and of high temperature. Future observations should concentrate on locating the microwave/gamma–ray sources, and on determining the kernel's fine structure and the related multi–loop structure of the flaring area.

073.077 **Results from the Solar Maximum Mission.**
B. R. Dennis.
NASA Conf. Publ., NASA CP–2421, p. 67 – 97 (1986). – See Abstr. 012.021.
The major results from SMM are presented as they relate to our understanding of the energy release and particle transportation processes that led to the high energy X–ray aspects of solar flares. Evidence is reviewed for a 152– to 158–day periodicity in various aspects of solar activity including the rate of occurrence of hard X–ray and gamme–ray flares. The statistical properties of over 7000 hard X–ray flares detected with the Hard X–Ray Burst Spectrometer are presented including the spectrum of peak rates and the distribution of the photon number spectrum. A flare classification scheme introduced by Tanaka is used to divide flares into three different types. Three different types of flares are discussed and new results are presented.

073.078 **Theoretical motivation for high spatial resolution, hard X–ray observations during solar flares.**
A. G. Emslie.
NASA Conf. Publ., NASA CP–2421, p. 132 – 138 (1986). – See Abstr. 012.021.
The author reviews the important role played by hard X–ray radiation as a diagnostic of impulsive phase energy transport mechanisms. It is argued that the sub–arc second resolution offered by an instrument such as the Pinhole/Occulter Facility (P/OF) can greatly increase our understanding of such mechanisms.

073.079 **Why P/OF should look for evidences of over-dense structures in solar flare hard X-ray sources.**
D. F. Neidig, S. R. Kane, J. J. Love, E. W. Cliver.
NASA Conf. Publ., NASA CP-2421, p. 142 – 145 (1986). – See Abstr. 012.021.
White-light and hard X-ray (HXR) observations of two white-light flares (WLFs) show that *if* the radiative losses in the optical continuum are powered by fast electrons directly heating the WLF source, then the column density constraints imposed by the finite range of the electrons requires that the WLF consist of an *over-dense region* in the chromosphere, with density exceeding $10^{14} cm^{-3}$. Thus, the authors recommend that P/OF search for evidences of over-dense structures in HXR images obtained simultaneously with optical observations of flares.

073.080 **The heating of the thermal plasma with energetic electrons in small solar flares.**
H.-a. Lin, R. P. Lin.
NASA Conf. Publ., NASA CP-2421, p. 146 – 149 (1986). – See Abstr. 012.021.
From an analysis of 20 small events, the authors conclude that if the volume rather than the density of the thermal plasma region is assumed constant in the flare process, energetic electrons could heat the soft X-ray plasma.

073.081 **Hard X-ray imaging and the relative contribution of thermal and nonthermal emission in flares.**
G. D. Holman.
NASA Conf. Publ., NASA CP-2421, p. 150 – 152 (1986). – See Abstr. 012.021.
The use of hard X-ray imaging to determine the thermal/nonthermal character of the emission in flares is discussed.

073.082 **Super active regions and production of major solar flares.**
T. Bai.
NASA Conf. Publ., NASA CP-2421, p. 153 – 156 (1986). – See Abstr. 012.021.
The success of imaging detectors with small fields of view such as HXIS or P/OF (Pinhole/Occulter Facility) depends heavily on pointing to the right place at the right time. During the solar maximum years many active regions coexist on the solar disk. Therefore, in order to point the imaging detector to the right place, it is important to know which active region is most likely to produce major flares. As a first step toward this goal, the author has identified active regions which produced major flares observed by HXRBS (Hard X-Ray Burst Spectrometer) on SMM during February 1980 through December 1983.

073.083 **The solar radio corona: manifestations of energetic electrons.**
M. Pick.
NASA Conf. Publ., NASA CP-2421, p. 159 – 177 (1986). – See Abstr. 012.021.
Most of the topics discussed concern the acceleration of particles during solar flares. The energy transferred to particles is in fact a great fraction of the energy released during a flare; hence, the problems related to the acceleration/injection process are of crucial importance for build-up and evolution of the flare. The first chapter of this paper deals with the solar flare development: pre-impulsive and impulsive phases, gradual and post gradual phases. A few points related to electron acceleration outside of flares will be also briefly discussed. The second chapter concerns a few problems related to the radio emission processes.

073.084 **Model of the upper chromosphere and the energy balance in the region of a solar flare.**
E. A. Baranovskii (*Eh. A. Baranovskij*).
Bull. Crimean Astrophys. Obs., Vol. 69, p. 93 – 100 (1986). English translation of 39.073.092.

073.085 **Chromospheric lines in a transversal magnetic field.**
G. M. Nikol'skij, L. B. Demkina.
Inst. zem. magn., ionos. i rasprostr. radiovoln Akad. Nauk SSSR. Prepr., No. 25/558, 9 pp. (1985). In Russian. Abstr. in Ref. Zh., 51. Astron., 4.51.300 (1986).

073.086 **Radio observations of the setting of a quiescent prominence behind the solar limb and the interpretation of results on the eclipse of a prominence on July 31, 1981.**
G. P. Apushkinskii (*G. P. Apushkinskij*), N. A. Topchilo.
Sov. Astron., Vol. 29, No. 4, p. 454 – 456 (1985). English translation of 40.073.014.

073.087 **Reflection of the structure of the chromosphere and corona in the solar wind and interplanetary magnetic field.**
G. P. Lyubimov, N. V. Pereslegina.
Sov. Astron., Vol. 29, No. 4, p. 457 – 462 (1985). English translation of 40.073.015.

073.088 **Het begin van een zonnevlam.**
C. de Jager.
Zenit, 13. Jaarg., Nr. 6, p. 214 – 217 (1986).

073.089 **Recent results in the spectrophotometry of solar prominences.**
D. A. Landman.
Proceedings of the Second Asian-Pacific Regional Meeting on Astronomy, p. 512 – 522 (1984). – See Abstr. 012.031.
Some recent results of the solar prominence spectrophotometry program at Haleakala are presented. The observations were made at the Mees Solar Observatory using specialized instrumentation that is also briefly described.

073.090 **Properties of solar flare plasmas derived from soft X-ray line emission.**
P. L. Bornmann.
Diss. Abstr. Int., Sect. B, Vol. 46, No. 8, p. 2692 (1986). Thesis, University of Colorado at Boulder, 153 pp. (1985). Order No. DA8522647.

073.091 **Impulsive brightenings in quiescent solar prominences.**
G. D. Toot.
Diss. Abstr. Int., Sect. B, Vol. 46, No. 10, p. 3482 – 3483 (1986). Thesis, University of Colorado at Boulder, 250 pp. (1985). Order No. DA8528529.

073.092 **Relation between γ-ray emission, radio bursts, and proton fluxes from solar flares.**
V. V. Fomichev, I. M. Chertok.
Sov. Astron., Vol. 29, No. 5, p. 554 – 559 (1985). English translation of 40.073.033.

073.093 **Dynamics of an active prominence of June 13, 1978.**
V. Yu. Klepikov, Yu. V. Platov.
Sov. Astron., Vol. 29, No. 5, p. 569 – 571 (1985). English translation of 40.073.034.

073.094 **Density and temperature of quiescent solar prominences from observations in the radio range.**
G. P. Apushkinskii (*G. P. Apushkinskij*).
Sov. Astron., Vol. 29, No. 5, p. 574 – 575 (1985). English translation of 40.073.035.

073.095 **Transient magnetic disturbances from solar flare explosions and associated pulsations of type IV radio bursts.**
S. Pintér.
Astrophys. Space Sci., Vol. 123, No. 1, p. 209 – 211 (1986).
The author has proposed a mechanism of arise of transient magnetic disturbances from solar flare explosion which can lead to understanding of observed pulsations of type IV radio emission with period of 0.3 – 3.0 s. According to the proposed mechanism the pulsation activity of the radio emission results from MHD waves accompanying the expanding diamagnetic plasma produced by the explosive flare material.

073.096 **On the question of time-dependent flare grouping on the sun.**
M. B. Ogir'.
Izv. Krymskoj Astrofiz. Obs., Tom 74, p. 92 – 101 (1986). In Russian. English translation in Bull. Crimean Astrophys. Obs., Vol. 74.

The investigation of flare activity all over the sun using H_α films obtained in 1980 – 1982 showed that flare groupings on the sun are formed by such a correlation of flares when each of the following flare appears before the end of the preceding one; but simultaneous appearance of flares is also possible. H_α films were compared with the soft X-ray emission (λ0.5 – 4 Å) that showed that flare groupings are accompanied by complex X-ray bursts overlapping some individual bursts. Spectral analysis of three chosen samples of X-ray observations covering 9 – 17 days revealed the presence of some authentic periods from 20 to 578 minutes. The periods of 20 – 30 min correspond to the appearance of individual flares whereas the periods from 40 to 578 min characterize the flare grouping repetition pattern.

073.097 **Peculiar time-dependent changes in the structure of some chromospheric formations.**
L. G. Kartashova.
Izv. Krymskoj Astrofiz. Obs., Tom 74, p. 101 – 112 (1986). In Russian. English translation in Bull. Crimean Astrophys. Obs., Vol. 74.

Changes in structures of 55 fibrils and arch filament systems are being studied as flare dependet. The evolution of loop systems on the limb in the H_α line has been traced. It has been found that the width of dark and light fibrils and arch filaments that compose the two-component systems in question decreases during the flare and afterwards it increases in average by 1.86 ± 0.02 times. The number of elements in the system is increasing and then consequently decreasing, too, by the same number of times. Such a reconstruction is probably caused by the transformation of elements in the system from a single-component formation into a two-component one and vice versa.

073.098 **Über die Eruption vom 18. Oktober 1983.**
F. Esin.
Rev. Fac. Sci. Univ. Istanbul, Ser. C, Vol. 48, p. 33 – 36 (1983). = Publ. Istanbul Univ. Obs., No. 136.

073.099 **Observations of the solar chromosphere on the descending branch of 20th cycle.**
T. Saygaç, H. Gülseçen, M. T. Özkan, A. Ökten.
Rev. Fac. Sci. Univ. Istanbul, Ser. C, Vol. 48, p. 45 – 83 (1983). = Publ. Istanbul Univ. Obs., No. 137.

073.100 **Observations of the solar chromosphere on the ascending branch of 21st cycle.**
H. Gülseçen, M. T. Özkan, A. Ökten, T. Saygaç.
Rev. Fac. Sci. Univ. Istanbul, Ser. C, Vol. 48, p. 85 – 130 (1983). = Publ. Istanbul Univ. Obs., No. 138.

073.101 **Optical and X-ray features of the 4N-flare loops on the solar surface on April 1, 1981.**
J.-y. Xuan, S.-c. Li, D. Luan.
Publ. Yunnan Obs., No. 1, p. 74 – 83 (1986). In Chinese.

073.102 **Helical prominences. III. The prominence of July 29, 1980.**
B. Vršnak.
Hvar Obs. Bull., Vol. 9, No. 1, p. 61 – 71 (1985).

A prominence, observed at Hvar Observatory on July 29, 1980, is described and interpreted. The lower limit of the magnetic field strength is estimated to 10 G, while lower limits of the total internal current and current density are estimated as 10^{11}A and 8×10^{-5}A/m^2, respectively.

073.103 **High-resolution observations of X-ray transitions in Fe XXV – XXIII.**
J. F. Seely, U. Feldman.
J. Phys. B, Vol. 18, No. 23, p. L797 – L801 (1985). Abstr. in Phys. Abstr., Vol. 89, No. 1276, Entry 36096 (1986).

073.104 **Observations of solar flare photon spectra from 20 keV to 7 MeV.**
M. Yoshimori, H. Watanabe, N. Nitta.
J. Phys. Soc. Jpn., Vol. 54, No. 11, p. 4462 – 4467 (1985). Abstr. in Phys. Abstr., Vol. 89, No. 1276, Entry 36097 (1986).

073.105 **Physical conditions in the cool parts of prominences and spicules: the effects of model atom level truncation on the derived plasma parameters.**
D. A. Landman.
Astrophys. J., Vol. 305, No. 1, p. 546 – 552 (1986).

The effects on calculated lower level population densities of the truncation of Na and Sr^+ model atoms are determined in the context of the author's spectral diagnostic scheme for solar prominences and spicules. It is shown that neglect of the upper atomic levels in Na, in particular, leads to overestimates in electron density and gas pressure by factors of ~2 and ~4, respectively, and to underestimates in the degree of hydrogen ionization and in the line-of-sight thickness of emitting material again by factors of ~2 and ~4, respectively.

073.106 **Fast-shock formation in line-tied magnetic reconnection models of solar flares.**
T. G. Forbes.
Astrophys. J., Vol. 305, No. 1, p. 553 – 563 (1986).

At present magnetic reconnection is the most favored mechanism to explain the energy release observed in solar flares. In the numerical experiment presented here line-tied reconnection is modeled by a configuration which produces two supermagnetosonic outflow jets – one directed upward, away from the photosphere, and one directed downward, toward an acarde of closed magnetic loops tied to the photosphere. The fast shock occurs when the downward-directed jet encounters the obstacle formed by the closed loops. Immediately downstream of the shock there exists a deflection sheath in which the submagnetosonic flow coming out of the shock is diverted around the region of static closed loops. It is suggested that such fast shocks may contribute to particle acceleration and thermal condensation in flares.

073.107 **Oscillations trapped in a chromosphere with a horizontal magnetic field.**
Yu. D. Zhugzhda, N. S. Dzhalilov.
Sov. Astron. Lett., Vol. 11, No. 5, p. 300 – 301 (1985). English translation of 40.073.020.

073.108 **Directionality of solar flare hard X rays: Venera 13 observations.**
S. V. Bogovalov, Yu. D. Kotov, V. M. Zenchenko, G. Vedrenne, M. Niel, C. Barat, G. Chambon, R. Talon.
Sov. Astron. Lett., Vol. 11, No. 5, p. 322 – 324 (1985). English translation of 40.073.037.

073.109 **Dense chromospheric plasma layers, and the dynamics of prominences above active regions.**
L. G. Genkin, L. M. Erukhimov.
Sov. Astron. Lett., Vol. 11, No. 5, p. 324 – 326 (1985). English translation of 40.073.038.

073.110 **The short-term physical forecasts of flares.**
D.-x. Fan.
Publ. Purple Mt. Obs., Vol. 4, No. 3, p. 16 – 24 (1985). In Chinese.

This paper presents a review of some problems about the short-term physical forecasts of flares. These problems are described separately (1) the study of flare mechanisms, (2) the short-term physical forecasts of flares. Finally, the author briefly describes the ideas about the further study.

073.111 **Velocity–field gradients of solar prominences and flares and their spectral–line widths.**
S.–h. Ye, J.–h. Jin.
Publ. Purple Mt. Obs., Vol. 5, No. 1, p. 7 – 11 (1986). In Chinese.
This paper presents a development of the concept of the broadening of spectral lines caused by the gradients of velocity fields. It also gives an improvement of the authors' method of the derivation of velocity–field gradients by the half–widths of spectral lines. This method has been applied to the observations of the prominence of March 30, 1959.

073.112 **Solar flares and particle acceleration.**
C. de Jager.
Space Sci. Rev., Vol. 44, Nos. 1/2, p. 43 – 90 (1986).
Contents: The X–ray component of an average flare. Flare energies and numbers of particles involved. Observational indications on the site and mechanism of the energization. Very high–energy phenomena. Pre– and post–flare energizations: homologous flares. Acceleration in electric fields. Reconnection and energization. Observational verification and constraints. Escape into interplanetary space.

073.113 **Flare prediction as a source of information on processes in an active region.**
V. M. Efimenko, V. V. Tel'nyuk–Adamchuk.
Soln. Dannye, Byull., 1986, No. 3, p. 63 – 68 (1986). In Russian.
A study of flare prediction may be a source of data, which needs to be interpreted physically. These data are: the optimal prehistory interval, the contribution of neighbouring sunspot groups to the flare activity, the existence of several independent groups of parameters, the optimal discretization of predictors and predictants, the level of stochastic noise in the flare activity.

073.114 **The observed mode of oscillations in the chromosphere above sunspots. II Running penumbral waves.**
V. S. Loskutnikov.
Soln. Dannye, Byull., 1986, No. 4, p. 58 – 62 (1986). In Russian.
The H_α brightness oscillations at two points of the penumbra at the distance of 2″ and 6″.6 from the umbrae are studied. The oscillatons are coherent and the period is variable. The velocity of the running penumbral waves is found which is used for determination of the Evershed velocity.

073.115 **The radial velocity structure of an H_α quiescent filament on the solar disc.**
V. G. Utrobin, N. S. Shilova.
Soln. Dannye, Byull., 1986, No. 4, p. 75 – 80 (1986). In Russian.
Doppler velocity values in an H_α quiescent filament are determined. The greater part of the filament is rising matter with the velocity $|v| < 3.5$ km/s. The part of the filament with lowering matter increases with the filament age. Applicability of the "cloud" model and "syphon" model is discussed. The Doppler velocities distribution in the neighbourhood of the filament shows that the matter may flow into the filament both from above and from below.

073.116 **Chromospheric line intensities from observations at Pulkovo in 1982 and variations in an 11–year solar cycle.**
V. M. Sobolev, E. P. Bojtsova.
Soln. Dannye, Byull., 1986, No. 4, p. 81 – 88 (1986). In Russian.
Equivalent widths and intensities of the H_α, H_β, H_ε, H_8 hydrogen D_3 and λ3889 helium lines for altitudes 1100 – 5500 km are determined on the basis of spectral observations of the chromosphere in 1982. A graph of the dependence of mean intensities of the lines on the year of observations (1969 – 1982) is plotted which is different for hydrogen and helium lines.

073.117 **Rapid fluctuations in solar flares observed by the Soft X–ray Polychromator.**
D. M. Zarro, K. T. Strong, J. L. R. Saba.
Bull. Am. Astron. Soc., Vol. 18, No. 1, p. 638 (1986). Abstract. – See Abstr. 010.066.

073.118 **Momentum deposition by electron beams in solar flares.**
F. Reale, G. Peres, S. Serio.
Mem. Soc. Astron. Ital., Vol. 56, No. 4, p. 801 – 804 (1985). – See Abstr. 010.821.
Energy deposition by precipitating electron beams, accelerated near the top of loop structures, is considered one of the most plausible heating mechanisms for solar flares. The authors have calculated in detail the effects of electron beams momentum deposition, and have studied the differences in the evolution of a model flare when this effect is not taken into account.

073.119 **Modeling of reconnection processes in ribbon flares.**
R. A. Kopp, G. Poletto.
Mem. Soc. Astron. Ital., Vol. 56, No. 4, p. 809 – 812 (1985). – See Abstr. 010.821.
After showing that the morphological properties of the decay phase of two–ribbon solar flares may be interpreted in terms of reconnection, the authors briefly describe a 2–D analytical model capable of reproducing the large–scale behavior of observed two–ribbon flares.

073.120 **Structures magnétiques à grande échelle dans les protubérances.**
G. Stellmacher, I. S. Kim, V. Yu. Klepikov, S. Koutchmy, A. I. Stepanov.
Inst. Astrophys. Paris, Pré–Publ., No. 131, 7 pp. (1986). To appear in "Coronographie d'ici à l'an 2000".

073.121 **Formation of the May 1981 solar flare complex and the global magnetic field of the sun (H.R. 17644).**
V. Bumba.
Bull. Astron. Inst. Czech., Vol. 37, No. 4, p. 210 – 218 (1986).
The author demonstrates to what degree the development of the studied flare complex is related to the global changes of the solar background magnetic fields. It is connected with the action of magnetic active longitudes, it lasts about 10 rotations and, in its last phase, it forms a coronal hole.

073.122 **Kinematics of the May 1981 flare complex's magnetic field decrease (H.R. 17644).**
V. Bumba, B. Kálmán, M. Klvaňa, J. Suda.
Bull. Astron. Inst. Czech., Vol. 37, No. 4, p. 219 – 226 (1986). With plates 1 – 6.
The authors investigate some characteristic details of the post–maximum dissolution process of the May 1981 flare complex magnetic and photospheric patterns during about one week of the fine–scale observations. Magnetic flux and area changes in the course of field topology simplification are estimated. In the photosphere only five features with a more stable field configuration survived this evolutionary stage.

073.123 **Measurement of magnetic fields in solar prominences.**
G. M. Nikol'skii (*G. M. Nikol'skij*), I. S. Kim, S. Koutchmy, A. I. Stepanov, G. Stellmacher.
Sov. Astron., Vol. 29, No. 6, p. 669 – 672 (1985). English translation of 40.073.111.

073.124 **Continuous and Hα emission of a flare of July 4, 1974.**
A. N. Babin, N. P. Dyatel, M. A. Livshits.
Sov. Astron., Vol. 29, No. 6, p. 676 – 681 (1985). English translation of 40.073.112.

073.125 **X–ray, H–alpha and radio observational study of the 4N/X2.3 flare of 1981 April 1.**
J.–l. Wang, C.–s. Han, J.–y. Xuan, S.–c. Li, R.–y. Zhao.
Chin. Astron. Astrophys., Vol. 10, No. 2, p. 108 – 114 (1986). English translation of Acta Astrophys. Sin., Vol. 6, No. 1, p. 26 – 34 (1986).

073.126 **Magneto–hydrodynamic simulation of prominence formation.**
S. Wang, A.–h. Wang.
Chin. Astron. Astrophys., Vol. 10, No. 2, p. 129 – 134 (1986). English translation of 40.073.110.

073.127 **Hα post–flare loops on May 16, 1981.**
Q.–r. Su, F.–m. Hu.
Acta Astrophys. Sin., Vol. 6, No. 1, p. 20 – 25 (1986). In Chinese.

073.128 **The magnetostatic equilibrium of polytropic state for a horizontal prominence in closed ringlike configuration.**
H. Long, W.–r. Hu.
Acta Astrophys. Sin., Vol. 6, No. 1, p. 42 – 50 (1986). In Chinese.
The authors study the magnetostatic equilibrium of the prominence with the simplification of polytropic law in place of the energy equation. The partial differential equation for the axisymmetric funtion of the magnetic surface is obtained and solved for typical boundary conditions, and the distribution of thermodynamical quantities can be deduced from the polytropic assumption.

073.129 **Application of complete linearization method to the analysis of the emission lines of solar prominences.**
Q.–z. Zhang, C. Fang.
Acta Astron. Sin., Vol. 27, No. 1, p. 30 – 37 (1986). In Chinese.

073.130 **Numerical modeling of the arch eruptive prominence on Dec. 20, 1982.**
Y.–q. Hu, S.–p. Jin, X.–b. Ni, L.–s. Cui, G.–p. Ji, C. Fang.
Acta Astron. Sin., Vol. 27, No. 1, p. 47 – 52 (1986). In Chinese.
The present paper presents a numerical modeling of the eruptive prominence on Dec. 20, 1982, based on one–dimensional hydrodynamical equations. The effect of various factors, such as front deceleration, thermal conduction, radiation loss and heating, on the formation of the eruptive prominence is preliminarily discussed. The result shows that arch eruptive prominences have completely different dynamic characters from coronal arches.

073.131 **Particle acceleration in solar flares.**
W. Dröge, R. Schlickeiser.
Astrophys. J., Vol. 305, No. 2, p. 909 – 919 (1986).
The authors propose a model for second–phase acceleration of electrons and protons associated with solar flares. Particles are accelerated by stochastic resonant scattering off hydromagnetic waves and by shock waves generated in the impulsive phase of the flare. The solutions allow first– and second–order Fermi acceleration processes and arbitrary power–law momentum dependences of the momentum diffusion coefficient. It is shown that the acceleration time scale to a characteristic energy ~100 MeV for protons can be as short as 3 s.

073.132 **Solar gradual hard X–ray bursts and associated phenomena.**
E. W. Cliver, B. R. Dennis, A. L. Kiplinger, S. R. Kane, D. F. Neidig, N. R. Sheeley Jr., M. J. Koomen.
Astrophys. J., Vol. 305, No. 2, p. 920 – 935 (1986). With plate 14.
Results from an examination of the white–light coronagraph, Hα, and radio (microwave, decimetric, and metric) data associated with a sample of 10 gradual hard X–ray bursts (GHBs) observed with *SMM* and *ISEE 3* are presented. The combined data argue against the type II shock hypothesis for GHBs and are consistent with the acceleration and trapping of the radiating electrons in postflare loop systems following coronal mass ejections.

073.133 **The solar flare extreme ultraviolet to hard X–ray ratio.**
A. N. McClymont, R. C. Canfield.
Astrophys. J., Vol. 305, No. 2, p. 936 – 946 (1986).
The authors have examined a data set consisting of simultaneous measurements of the peak 10 – 1030 Å extreme ultraviolet (EUV) flux enhancement and more than 10 keV hard X–ray (HXR) peak flux of many solar flare bursts, ranging over ~4 orders of magnitude in HXR intensity, from studies by Kane et al. It is shown that the EUV/HXR ratio is not constant, but that the data are well fitted by EUV $\propto$ HXR$^{0.60 \pm 0.05}$. These data are compared with the results of a simple analytic model of impulsive emission generated by a beam of nonthermal electrons accelerated near the top of a coronal loop, with power–law energy distribution. An alternative, nonthermal, thick–target model is also considered.

073.134 **High–resolution telescope and spectrograph images of the solar chromosphere and transition zone.**
K. P. Dere, J.–D. F. Bartoe, G. E. Brueckner.
Astrophys. J., Vol. 305, No. 2, p. 947 – 953 (1986). With plates 15 – 16.
Rastered stigmatic UV slit spectra of the Sun obtained by the NRL High Resolution Telescope and Spectrograph are reduced to form two–dimensional images of intensity, Doppler shift, and line width. The spectral region observed around 1550 Å contains strong lines of chromospheric ions, such as C I and Fe II, and of the transition zone ion C IV. The solar region observed is 10″ wide by 800″ long, from the south solar pole to near Sun center and contains mostly features characteristic of the quiet Sun in addition to a polar coronal hole, an active region plage, a filament and a filament channel. A detailed discussion of the velocity fields in the quiet chromosphere is presented.

073.135 **Magnetohydrodynamic theories of solar flares.**
E. R. Priest.
Sol. Phys., Vol. 104, No. 1, p. 1 – 18 (1986). – See Abstr. 012.067.
Our current understanding of the MHD of the flare process is summarised, with some emphasis on processes which produce strong impulsive electric fields and current filamentation. As an introduction, a description of the two main types of flare (i.e, simple–loop and two–ribbon) is given, together with an account of the two branches of reconnection theory (tearing modes and the Petschek–Sonnerup mechanism). Modern numerical experiments of reconnection suggest impulsive bursty acceleration of particles in many small regions of width a hundred kilometers or less. This is followed by a discussion of the eruptive instability thought to initiate a large flare and of the reconnection process of energy release. Finally, the role of emerging flux and horizontally moving satellite sunspots is discussed briefly.

073.136 **Relative timing of hard X–rays and radio emissions during the different phases of solar flares: consequences for the electron acceleration.**
G. Trottet.
Sol. Phys., Vol. 104, No. 1, p. 145 – 163 (1986). – See Abstr. 012.067.
Observations relevant to the relative timing of hard X–ray, microwave and lower frequency radio bursts in different phases of flare are reviewed. It is shown that such timing comparisons give important information concerning the electron acceleration/injection process, the magnetic field topology at the acceleration site and the flare development itself. In particular it is shown that acceleration begins before the flash phase of flares and that it keeps going on continuously during the entire duration of a flare. Moreover, despite their wide separation in altitude, hard X–ray, microwave and lower frequency sources appear to arise from a common injection of electrons going on continuously through the different phases of flare. In situ acceleration by shock waves giving rise to type II radio emission is briefly discussed. As an alternative interactions between small and large scale magnetic structures are proposed.

073.137 **Radio observations of the solar neutron flare of 3 June, 1982.**
M. Karlický, H. Aurass, G. Mann.
Sol. Phys., Vol. 104, No. 1, p. 165 – 168 (1986). – See Abstr. 012.067.
In this paper the solar neutron and white–light flare is studied on the basis of radioastronomical observations. It is shown that the 3 June, 1982 flare had an impulsive character. A strong shock wave ($M_A \geqslant 2.9$) was observed unusually soon after the impulsive phase of the flare. The radio spectrum of this event shows that the primary acceleration process probably occurred in the region with an electron density of $n_e = 4.4 \times 10^{15} \mathrm{m}^{-3}$. The pulsations of the type IV radio burst, observed 15 min after the mass ejection process, manifest the reconnection process in the post–flare stage.

073.138 **Behaviour of the polarization at dm – m wavelengths during the evolution of five two–ribbon flares.**
V. Ruždjak, B. Vršnak, P. Zlobec, A. Schroll.
Sol. Phys., Vol. 104, No. 1, p. 169 – 174 (1986). – See Abstr. 012.067.
The general behaviour of the circular polarization at dm – m wavelengths during the evolution of five two–ribbon flares is investigated. The changes of polarization, if present, occurred 10 to 20 min after the impulsive phases. Increases of the radio and X–ray fluxes occurred at the moments when the Hα ribbons started to extend over spot umbrae.

073.139 **Electron acceleration in flares inferred from radio and hard X–ray emissions.**
A. O. Benz, S. R. Kane.
Sol. Phys., Vol. 104, No. 1, p. 179 – 185 (1986). – See Abstr. 012.067.
Properties of electron acceleration in flares, especially the density structure in the acceleration region, are deduced from a correlation study between decimetric type III, spike, and hard X–ray (HXR) bursts. The high association rate found (71%) strongly suggests that spikes also originate from energetic electrons. Spikes and type III bursts have been found to be easily identified by their different polarizations. A peak–to–peak correlation does not always exist between type III, spike and HXR bursts. This discrepancy can be interpreted in terms of the different source conditions and propagation properties.

073.140 **Solar VLBI of compact transient sources.**
K. F. Tapping.
Sol. Phys., Vol. 104, No. 1, p. 199 – 206 (1986). – See Abstr. 012.067.
Very long baseline interferometry offers the only technique for resolving the transient compact sources which, according to an increasing weight of observational and theoretical evidence, occur during solar flares. The author discusses the problems of applying VLBI to observing the sun, the results obtained so far and how the technique may be effectively employed in future studies of compact emission structures in solar flares.

073.141 **Can observed hard X–ray and microwave flux from solar flares be explained by a single electron population?**
K. Kai.
Sol. Phys., Vol. 104, No. 1, p. 235 – 241 (1986). – See Abstr. 012.067.
In attempting to explain observed hard X–ray and microwave flux from solar flares by a single population of energetic electrons, one has met a serious discrepancy of the order of 10^3 – 10^5 between the calculated and observed microwave flux. In this paper it is shown that this discrepancy can be removed for impulsive flares by the assumption of a precipitation model for both X–ray and microwave sources and that the magnetic fields of 500 – 1000 G is required in the microwave emitting region. The precipitation model is consistent with the rapid time variation exhibited in both hard X–rays and microwaves.

073.142 **Microwave diagnostics of energetic electrons in flares.**
K.–L. Klein, G. Trottet, A. Magun.
Sol. Phys., Vol. 104, No. 1, p. 243 – 252 (1986). – See Abstr. 012.067.
Electrons accelerated during solar flares are revealed by their electromagnetic radiations in different spectral ranges, emitted at different heights in the solar atmosphere. The observational analysis points to a common and continuous injection of particles. Based on this result, a quantitative investigation of the hard X–ray and microwave emissions observed during the 29 June, 1980 flare at 11:40 UT has been performed. First results demonstrate that during the most energetic phase of the event both hard X–rays and microwaves are described by electron populations resulting from the same injection function, and that the total numbers of electrons required for both emissions are compatible.

073.143 **A siphon mechanism for supplying prominence mass.**
A. I. Poland, J. T. Mariska.
Sol. Phys., Vol. 104, No. 2, p. 303 – 312 (1986).
The authors examine a siphon–like mechanism for moving mass from the chromosphere to a gravitational well at the top of a magnetic loop to form a prominence. The calculations assume no a priori flow velocity at the loop base. Instead heating in the loop legs drives the flow. The prominence formation process requires two steps. First, the background heating rate must be reduced to on the order of 1% of the initial heating rate required to maintain the coronal loop. This forms an initial condensation at the top of the loop. Second, the heating must take place only in the loop legs in order to produce a pressure differential which drives mass up into the well at the top of the loop. The heating rate in the loop must be increased once the prominence has begun to form or full prominence densities can not be achieved in a reasonable time.

073.144 **Periodic oscillations found in the velocity field of a quiescent prominence.**
T. Tsubaki, A. Takeuchi.
Sol. Phys., Vol. 104, No. 2, p. 313 – 320 (1986).
With the purpose of detecting periodic oscillations or waves in a quiescent prominence, temporal variations of a Ca II K line profile have been studied. The most conspicuous phenomenon found here is the fact that the edge of the prominence showed, over some 20000 km along the spectrograph–slit, periodic velocity fluctuations of nearly the same phase with periods of 210 – 240 s and with an amplitude of up to ± 2 km s^{-1}. At other portions, several different periods of peaks (160 – 400 s) can also be seen in the power spectra, but less distinctly. As to the intensity and the line width, however, no periodic variations have been detected.

073.145 **Theoretical Ne V emission line ratios compared to solar observations.**
F. P. Keenan, K. M. Aggarwal, K. G. Widing.
Sol. Phys., Vol. 105, No. 1, p. 47 – 50 (1986).
The recent level population calculations for Ne V by Aggarwal are used to determine theoretical emission line ratios R_1 and R_2. A comparison of these with observational data for a solar flare and erupting prominence obtained with the NRL *XUV* spectrograph on board *Skylab* reveals that R_1 and R_2 are in their predicted high density limits. It is shown that they are temperature sensitive and hence may be employed to determine the electron temperatures of Ne V line emitting regions.

073.146 **Chromospheric "active region loops".**
C. Fang, M. J. Martres.
Sol. Phys., Vol. 105, No. 1, p. 51 – 65 (1986).
Archshaped structures above or around sunspot groups are considered as tracers of the magnetic lines of force. A study of the chromospheric contribution to the 3D general pattern is necessary to quantify this relationship. The emissive features detected in nine different active regions and observed on the disk at different levels in the chromosphere have been analysed. Eleven archshaped structures may be easily interpreted as loops. There are no great morphological differences between the "slowly–varying" arches and the flaring ones. However, a new relation is found between these two kinds of chromospheric features: at the maximum of flares, the flaring arch has one of its footpoints in common with a closer stable, pre–existing arch.

073.147 **Energy of microwave–emitting electrons and hard X–ray/microwave source model in solar flares.**
N. Nitta, T. Kosugi.
Sol. Phys., Vol. 105, No. 1, p. 73 – 85 (1986).
The authors present a new method of estimating the energy of microwave–emitting electrons from the observed rate of increase of the microwave flux relative to the hard X–ray flux measured at various energies during the rising phase of solar flares. A total of 22 flares observed simultaneously in hard X–rays (20 – 400 keV) and in microwaves (17 GHz) were analyzed in this way and the results presented.

073.148 **A new microwave/X–ray diagnostic for the thermal phase of solar flares.**
E. J. Schmahl, M. R. Kundu, F. T. Erskine.
Sol. Phys., Vol. 105, No. 1, p. 87 – 99 (1986).
The authors have observed 10 solar bursts during the thermal phase using the Haystack radio telescope at 22 GHz. They show that these high frequency flux observations, when compared with soft X–ray band fluxes, give useful information about the temperature profile in the flare loops. The microwave and X–ray band fluxes provide determinations of the maximum loop temperature, the total emission measure, and the index δ of the differential emission measure. The results imply that, during the late phases of flares, condensation cooling competes with radiative cooling. The results are in good agreement with previous deductions from XUV rocket spectra.

073.149 **Energy interconnection between the plasma cloud ejected from the active region and the shock wave.**
I. N. Garczyńska, B. Rompolt.
Sol. Phys., Vol. 105, No. 1, p. 101 – 109 (1986).
The relation between the ejected plasma cloud and the shock wave propagating ahead of it is examined for 27 pairs of such events. The flare sprays and the eruptive prominences observed in Hα line as well as the fast–moving sources of type–IV radio bursts have been considered as such ejected plasma clouds. Propagation of the shock wave in the solar corona has been examined from the observations of type–II radio bursts. Using the Parker model of the propagation of a shock wave, the shock wave velocity has been compared with the plasma cloud velocity. Energy interconnection between these two events has been studied. In the majority of the investigated cases there exists an energetic interconnection between the plasma cloud and the shock wave.

073.150 **Physical conditions in eruptive prominences at several solar radii.**
R. M. E. Illing, R. G. Athay.
Sol. Phys., Vol. 105, No. 1, p. 173 – 190 (1986).
SMM data from the Coronagraph/Polarimeter experiment giving intensities of Hα and continuum emission in eight erupting prominences are analyzed to obtain the physical conditions in the regions of Hα emission. Since the Hα intensity depends upon three unknowns whereas only two independent observations are available, it is necessary to assume one additional condition in order to obtain unique solutions. Solutions are chosen that give the maximum expansion of the prominence volume as reflected by minimum values of the electron density. These solutions correspond closely with those giving the best agreement between the gas pressure in the prominence and the ambient coronal pressure. Electron densities are found to be of the order of $10^8 cm^{-3}$ at temperatures near 2×10^4K.

073.151 **The rotation of calcium plages in the years 1967 – 1970.**
M. Ternullo.
Sol. Phys., Vol. 105, No. 1, p. 197 – 204 (1986).
An analysis of Ca II spectroheliograms obtained at Catania Observatory in the period 1967 – 1970 shows that plages rotate – in their first four days of lifetime – with a latitudinal differential rotation profile steeper than for older objects. A closer inspection reveals that plages slow down with age in some latitude strips, and accelerate in some others. That makes the profiles of old and young objects highly different from each other. A close correspondence has been found between the latitudes at which plages accelerate (slow down) and the latitudes of westward (eastward) streams (Howard and LaBonte, 1980). Some implications of these results, in the light of the theory of a direct coupling of magnetic features to deep layers, are discussed. In the years concerned, Ca–plages of whatever age show differential rotation profiles steeper than in other phases of the solar cycle.

Solar flare prediction.
See Abstr. 003.172.

Flare Build–up Study Workshop.
See Abstr. 011.040.

Plasma densities from the He–like ion Ne IX.
See Abstr. 022.025.

Diffusion and thermal diffusion in partially ionized gases in the atmospheres of the Sun and planets.
See Abstr. 022.028.

Measurement of wavelengths and Lamb shifts for inner–shell transitions in Fe XVIII – XXIV.
See Abstr. 022.102.

The Siberian Solar Radio–Telescope: parameters and principle of operation, objectives and results of first observations of spatio–temporal properties of development of active regions and flares.
See Abstr. 033.003.

Determination of the radiation field in the chromosphere using the method of observed line profile inversion.
See Abstr. 036.040.

Ideal condensations due to perpendicular thermal conduction in a sheared magnetic field.
See Abstr. 062.002.

Nonlinear rapid symmetric waves in a magnetic layer.
See Abstr. 062.052.

Acceleration of charged particles in magnetic reconnection: solar flares, the magnetosphere, and solar wind.
See Abstr. 062.059.

Beams and jets in astrophysical plasmas.
See Abstr. 062.100.

Lower–hybrid instabilities driven by an ion velocity ring.
See Abstr. 062.114.

Analytical expressions for the temporal and spatial evolution of an energetic electron population in an inhomogeneous medium.
See Abstr. 063.017.

The common nature of flaring processes on the sun and red dwarfs.
See Abstr. 064.045.

The effect of radiation on the population of oxygen levels in the solar atmosphere. III. The radiation field of O I bound–free continua.
See Abstr. 071.007.

Optical imaging spectroscopy.
See Abstr. 071.024.

Electrical conductivity in the solar photosphere and chromosphere.
See Abstr. 071.034.

Fourier transform spectrometer observations of solar carbon monoxide. II. Simultaneous cospatial measurements of the fundamental and first–overtone bands, and Ca II K, in quiet and active regions.
See Abstr. 071.035.

Photoelectric observations of chromospheric sunspot oscillations. III. Spatial distribution of power and frequency in umbrae.
See Abstr. 072.012.

Photoelectric observations of chromospheric sunspot oscillations. IV. The Ca II H line and He I λ10830.
See Abstr. 072.013.

Young active regions.
See Abstr. 072.034.

Ratio of calcium plage to sunspot area of solar active regions.
See Abstr. 072.036.

Solar irradiance variations from photometry of active regions.
See Abstr. 072.044.

Observations of an active region with successive flares.
See Abstr. 072.070.

Analysis of an active region with high productivity of flare during November, 1979.
See Abstr. 072.075.

Effect of local rigid rotation on sunspots.
See Abstr. 072.077.

Temporal variation of brightness and area of sunspots and flocculi in the active region SD N 135.
See Abstr. 072.085.

A study of velocity fields of flare regions.
See Abstr. 072.087.

The spot groups of abnormal magnetic structures accompanied by quickly variation and SWF flares.
See Abstr. 072.092.

Dynamical processes in the solar corona and interplanetary space.
See Abstr. 074.002.

A comparison of structures of the solar corona and a prominence during flare activity.
See Abstr. 074.017.

On thermal instability connected with transformation of Alfvén waves in the solar corona and on formation of prominences.
See Abstr. 074.026.

X–ray imaging of flare loops and coronal arches.
See Abstr. 074.065.

On the formation of coronal cavities.
See Abstr. 074.070.

The derivation of solar coronal electron temperature from the 1^1S–$n^1P/1^1S$–2^1P emission–line ratios in Mg XI.
See Abstr. 074.074.

Electric fields and plasma structure in coronal magnetic loops.
See Abstr. 074.086.

Coronal transport of solar flare particles.
See Abstr. 074.100.

Comparison of magnetographic and polarimetric measurements of the magnetic field in a polar prominence.
See Abstr. 075.015.

Magnetic reconnection in a high–temperature plasma of solar flares. II. Effects caused by transverse and longitudinal magnetic fields.
See Abstr. 075.021.

Solar wind, non–force–free magnetic field and two–ribbon flares.
See Abstr. 075.022.

Recent studies of magnetic canopies.
See Abstr. 075.026.

A probable initial configuration of a flaring loop.
See Abstr. 075.045.

On the magnetic structure of the quiet transition region.
See Abstr. 075.049.

Magnetic reconnection on the Sun.
See Abstr. 075.052.

Continuum absorption in the solar EUV spectra.
See Abstr. 076.001.

A search for 160–min period in the sequences of X–ray bursts and chromospheric flares on the sun.
See Abstr. 076.003.

Influence of solar flares on the X–ray corona.
See Abstr. 076.006.

Solar–flare gamma–ray line spectroscopy.
See Abstr. 076.008.

The excitation of the iron Kα feature in solar flares.
See Abstr. 076.012.

Solar flare X–ray spectra of Fe XXVI and Fe XXV from the Hinotori satellite.
See Abstr. 076.016.

EUV observations of solar flares based on satellite data.
See Abstr. 076.017.

Variations in quiet Sun radiation at centimetre wavelengths during solar maximum period.
See Abstr. 077.002.

Radio imaging of flares.
See Abstr. 077.023.

Unstable chromospheric dark features and type–III bursts association.
See Abstr. 077.029.

Peculiarities of the outburst of kilometer radio radiation during the event on May 13, 1981.
See Abstr. 077.030.

Relation between solar narrow–band decimetre–wave bursts and associated X–ray bursts.
See Abstr. 077.033.

Microwave emission from the limb of the Sun: the events of September 8, 1980 and May 10, 1981.
See Abstr. 077.034.

Gyrosynchrotron emission of solar flares.
See Abstr. 077.062.

Solar flare neon composition and solar cosmic–ray exposure ages based on lunar mineral separates.
See Abstr. 078.001.

The composition of solar energetic particles.
See Abstr. 078.002.

Solar filament eruptions and energetic particle events.
See Abstr. 078.003.

Separation and analysis of temporal and spatial variations in the 10 April 1969 solar flare particle event.
See Abstr. 078.006.

Three–dimensional gradients of solar particles inside 5 AU.
See Abstr. 078.007.

Time–dependent coronal shock acceleration of energetic solar flare particles.
See Abstr. 078.012.

The heavy–ion compositional signature in ^{3}He–rich solar particle events.
See Abstr. 078.013.

Die Ladungszustände solarer energetischer Teilchen.
See Abstr. 078.016.

The application of fuzzy clustering to the short-term forecasts of the solar proton flares.
See Abstr. 078.018.

First evidence of a large-scale circulation in the solar convection zone.
See Abstr. 080.059.

Sweeping-magnetic-twist mechanism for the acceleration of jets in the solar atmosphere.
See Abstr. 080.091.

Energy loss of electrons in solar atmosphere taking many-body interactions into account.
See Abstr. 080.096.

Observation of bremsstrahlung from solar electrons in the earth's atmosphere.
See Abstr. 082.081.

Three-dimensional properties of interplanetary disturbances in 1978 – 1981.
See Abstr. 106.002.

The three-dimensional extent at 1 AU of interplanetary disturbances associated with disappearing solar filaments.
See Abstr. 106.038.

Three-dimensional, time-dependent, MHD model of a solar flare-generated interplanetary shock wave.
See Abstr. 106.039.

Flare-associated shock wave propagation at large heliocentric distances.
See Abstr. 106.066.

IPS observations of transient interplanetary phenomena associated with solar filament activity in late August 1978.
See Abstr. 106.071.

Flares on dMe stars: IUE and optical observations of AT Mic, and comparison of far-ultraviolet stellar and solar flares.
See Abstr. 114.130.

Magnetodynamic phenomena in the solar and stellar outer atmospheres.
See Abstr. 116.003.

Flares on dMe stars: IUE and optical observations of AT Mic, and comparison of far-ultraviolet stellar and solar flares.
See Abstr. 122.127.

Two-dimensional numerical modeling of the cosmic ray storm.
See Abstr. 144.007.

074 Corona, Solar Wind

074.001 **Emission of Lyman α radiation by solar coronal loops. I. General synopsis.**
P. Gouttebroze, J. C. Vial, G. Tsiropoula.
Astron. Astrophys., Vol. 154, No. 1/2, p. 154 – 170 (1986).
The processes of emission of Lyman α radiation by loop-like structures embedded in the solar corona are investigated, for a large range of physical conditions within these objects. The coupled set of equations for radiative transfer, and hydrogen atom level populations, ionization and pressure equilibria is solved to obtain the emitted Lα intensities as functions of pressure, temperature and size.

074.002 **Dynamical processes in the solar corona and interplanetary space.**
R. T. Stewart.
Astrophys. Space Sci., Vol. 118, Nos. 1/2, p. 115 – 125 (1986). – See Abstr. 012.001.
Observations of the solar corona collected over the past decade are discussed from the point of view of short-term (< 1 day) and long-term (> 1 year) effects. Various phenomena are described, including coronal mass ejections, shock waves, and magnetic sector structures. It is argued that emerging magnetic flux is probably the prime cause of these phenomena.

074.003 **Coronal structure observed at the total solar eclipse of 11 June, 1983 in Indonesia.**
E. Hiei, Y. Shimizu, H. Miyazaki, H. Imai, K. Sato, S. Kuji, W. Sinambela.
Astrophys. Space Sci., Vol. 119, No. 1, p. 9 – 15 (1986). – See Abstr. 012.001.
From the photographs taken at the total solar eclipse of 11 June 1983, the authors derived the electron density for the north polar rays and for the thread-like fine structures above the active region, which are 10^8 at 1.4 solar radii and 3×10^9 at 1.15 solar radii, respectively. The brightness distributions of the corona at the polar region and above the active region, and the flattening index were also derived.

074.004 **The propagation of coronal mass ejection transients.**
R. M. MacQueen, A. J. Hundhausen, C. W. Conover.
J. Geophys. Res., Vol. 91, No. A1, p. 31 – 38 (1986).
Measurements of the direction of propagation of 29 coronal mass ejection events observed during the Skylab epoch (1973 – 1974) and 19 events observed during the SMM epoch (1980) reveal that the former undergo an average 2.2° equatorward deflection, while the latter do not deviate significantly from radial motion. No differences between eruptive prominence-associated or flare-associated events can be detected for either epoch. The results suggest that coronal mass ejection events are influenced by the background coronal magnetic and flow patterns; the nonradial forces affecting the Skylab epoch mass ejections arise from the large-scale dipolar magnetic field and flow configuration present at that time.

074.005 **Solar wind proton temperature-velocity relationship.**
R. E. Lopez, J. W. Freeman.
J. Geophys. Res., Vol. 91, No. A2, p. 1701 – 1705 (1986).
Data from Helios 1 are examined to find an experimental fit for the T(V) relationship at 1 AU. The data show a distinct change in the proton temperature-velocity slope at $V \sim 500\ \mathrm{km\,s^{-1}}$. The results are compared to the results of Burlaga and Ogilvie (1973) and show excellent agreement in the $V < 500\ \mathrm{km\,s^{-1}}$ range. In addition, the secular variation of T(V) is examined.

074.006 **Radial variation of interplanetary Alfvénic fluctuations: Pioneer 10 and 11 observations between 1 and 5 AU.**
B. Bavassano, E. J. Smith.
J. Geophys. Res., Vol. 91, No. A2, p. 1706 – 1710 (1986).
The radial evolution of the power spectra of Alfvénic fluctuations observed within the trailing edge of high speed streams in the solar wind has been investigated using magnetic field measurements by Pioneer 10 and Pioneer 11 at heliocentric distances between 1 and 5 AU. In the analyzed frequency range (2.8×10^{-4} to 8.3×10^{-3}Hz) the spectral index of the computed spectra does not depend on frequency and is close to the value predicted for hydrodynamic turbulence. The radial gradient of the power density is independent of frequency and only slightly steeper than that expected for undamped propagation of Alfvénic fluctuations in a rarefaction region. The relation between the present results and those previously obtained by Helios 1 and 2 inside 1 AU is discussed.

074.007 **Physical processes in the solar corona.**
R. Rosner, B. C. Low, T. E. Holzer.
Physics of the Sun, Vol. II, p. 135 – 180 (1986). – See Abstr. 003.001.
Contents: Transport theory (fundamental parameter regimes, the "ideal" problem, viscosity, parallel thermal energy transport, perpendicular transport, some comments on model building). Magnetohydrodynamic processes in the corona (equilibrium magnetic fields, linear stability and nonequilibrium, time–dependent phenomena). Energy and momentum balance of open and closed coronal structures (coronal holes and high–speed streams, Alfvén waves in the lower solar atmosphere, energy supply to magnetically closed coronal regions).

074.008 **Driving mechanisms for the solar wind.**
G. W. Pneuman.
Space Sci. Rev., Vol. 43, Nos. 1/2, p. 105 – 138 (1986).
In this review, the author considers the central physical aspects pertinent to the acceleration of the solar wind. Special importance is placed on the high–speed streams since the properties of these structures seem to strain the various theoretical explanations the most. Heavy emphasis is also given to the observations – particularly as to what constraints they place on the theories. The author also discusses certain sporadic events such as spicules, macrospicules, X–ray bright points, and outflows seen in the EUV associated with the explosive events, jets, and coronal bullets which could be of relevance to this problem. Three theoretical concepts pertaining to the solar wind acceleration process are examined – purely thermal acceleration with and without extended heating, acceleration due to Alfvén wave pressure, and diamagnetic acceleration. Emphasis is given to how well these theories meet the contraints imposed by the observations.

074.009 **The structure of the static corona and transition region.**
S. K. Antiochos, G. Noci.
Astrophys. J., Vol. 301, No. 1, p. 440 – 447 (1986).
The authors investigate static models of coronal loops. For loops that are low–lying it is shown that a new type of solution appears to the static equations, in addition to the well–known coronal loop solution. The new solution is characterized by a maximum plasma temperature $\lesssim 10^5$K. The structure and properties of these cool solutions are discussed. The authors calculate the differential emission measure $Q(T)$ expected for a magnetic arcade. It is shown that the cool loops have a dramatic effect on the form of $Q(T)$ in the lower transition region. In particular, they can account for the observed rise in Q at low T, which has long been thought to be incompatible with the static–loop model.

074.010 **Coronal heating in closely–packed flux tubes: a Taylor–Heyvaerts relaxation theory.**
P. K. Browning, T. Sakurai, E. R. Priest.
Astron. Astrophys., Vol. 158, No. 1/2, p. 217 – 227 (1986).
The aim of this paper is to take a more quantitative and detailed look at dissipation in an array of closely–packed flux tubes. An initially potential coronal loop is investigated, whose footpoints are twisted up by cellular photospheric motions, forming a network of twisted flux tubes. The motions are assumed to be slow compared with the reconnection time–scale, so that the stressed field reconnects and dissipates some of its energy as heat. The generalised Taylor's hypothesis is used to investigate the effects of reconnection on the flux tubes and to determine the efficiency of the dissipation. A basic mathematical model is set up and the procedure for calculating the evolution is outlined. The authors investigate the response of the field to the footpoint motions and evaluate the heating produced. The results are discussed, applications to the coronal heating problem are considered, and the predictions are compared with the known heating requirements of the corona.

074.011 **Estimation of coronal magnetic fields using Type–I emission.**
N. Gopalswamy, G. Thejappa, C. V. Sastry, A. Tlamicha.
Bull. Astron. Inst. Czech., Vol. 37, No. 2, p. 115 – 120 (1986).
The radial dependence of the coronal magnetic field above active regions is calculated using Type–I chain data existing in the literature. Assuming that Type–I emission is due to shock waves, the upstream shock velocity and the density jump across the shock are obtained respectively from the drift rate and the bandwidth of the Type–I chains. Making use of the shock velocity and the density jump in the Rankine–Hugoniot relation, the Alfvén velocity and hence the magnetic field in the corona is calculated. The results are compared with existing estimates.

074.012 **Development of a real–time algorithm for detection of solar wind discontinuities.**
D. Vandev, K. Danov, P. Mateev, P. Petrov, M. Kartalev, N. Trendafilov, Z. K. Smith, M. Dryer.
Astrophys. Space Sci., Vol. 120, No. 2, p. 211 – 221 (1986).
A study is presented of an algorithm, based on the statistical analysis of interplanetary magnetic field data, for the real–time detection of discontinuities in the solar wind. The analysis is based on the application of a "sliding" or "searching" algorithm together with a minimum variance treatment. A demonstration is given for a Pioneer data set from 29 August 1966. For this particular data set the authors identify the original shock with excellent agreement with the earlier studies of its normal vector, as well as an additional shock and tangential discontinuity, neither of which had been identified previously.

074.013 **Heating of coronal arcades by magnetic tearing turbulence, using the Taylor Heyvaerts hypothesis.**
P. K. Browning, E. R. Priest.
Astron. Astrophys., Vol. 159, No. 1/2, p. 129 – 141 (1986).
The heating of the solar corona by direct currents, which are dissipated by magnetic reconnection, is studied in this paper. The coronal field responds to slow photospheric footpoint motions by evolving through a series of equilibria, which may be unstable to resistive modes. According to a generalisation of Taylor's hypothesis (Heyvaerts and Priest, 1984), the field reconnects and relaxes to a linear force–free state (satisfying $\mathbf{\nabla} \times \mathbf{B} = \alpha\mathbf{B}$), where the parameter α is uniform. During the relaxation process, the field reconnects and dissipates some magnetic energy as heat. The value of α at each time and the energy released during relaxation may be determined from the evolution of magnetic helicity. Two theorems concerning this method are proved. The general conclusion is that tearing turbulence is a viable heating mechanism for the solar corona.

074.014 **Line emission from the quiet sun and the coronal hole regions.**
B. N. Dwivedi, P. K. Raju.
Bull. Astron. Soc. India, Vol. 13, No. 4, p. 387 – 393 (1985).
Using the Kopp & Orrall (1976) model for the electron density and temperature for the quiet sun and the coronal hole regions, the authors have computed emission line intensities from Mg VIII, Si X (boron–like); Mg VII, Si IX (carbon–like) and Si VII, S IX (oxygen–like) ions and compared them with available observational data.

074.015 **Eruption of new magnetic fluxes from the solar surface and the associated loop coronal transients.**
Y.–q. Hu.
Sci. Sin., Ser. A, Vol. 29, No. 2, p. 199 – 209 (1986).
The paper demonstrates the properties of neutral current sheet evolution, and points out that the expanding current sheet has a similar shape as the high–density region of a loop coronal transient and it may be considered as a main mechanism for causing such a transient.

074.016 **Evidence for a possible secular decrease in solar wind intensity.**
J. F. Kerridge, R. H. Becker, R. O. Pepin.
Meteoritics, Vol. 20, No. 4, p. 682 – 683 (1985). Abstract. – See Abstr. 010.661.

074.017 **A comparison of structures of the solar corona and a prominence during flare activity.**
Kim Gun–der.
Soln. Dannye, Byull., 1985, No. 11, p. 48 – 51 (1985). In Russian.
The behaviour of structures of the corona in the $\lambda = 5303$ Å line and a prominence in Hα during flare activity is given in figures. From a comparison the conclusion is made that the coronal structure is more stationary than the prominence structure. The corona increased in height and width due to new loops. The prominence decreased due to weakening or disappearance of some loops. The coronal brightness did not change for the 5.5 hours of observation. The prominence brightness significantly decreased. The coronal structure during the flare activity is similar in the 5303 Å line, soft X–rays and UV emission.

074.018 **On the possibility of a radio echo at plasma frequency in the solar corona.**
A. B. Eremin.
Pis'ma Astron. Zh., Tom 12, No. 5, p. 388 – 392 (1986). In Russian. English translation in Sov. Astron. Lett., Vol. 12.
In the frame of the mechanism of generation of the fundamental mode of type III solar radio bursts suggested by Eremin and Zajtsev (1985) the formation of an echo event in the corona at plasma frequency is shown to be possible. Examples of events are given which were observed during the type III_b–III radio storm in July, 1974 and may be identified as radio echos. A regular "violet" (in comparison with the primary burst) frequency shift of the echo burst has been detected that results from the radiation reflection from moving inhomogeneities of the solar wind. An estimate of the mean velocity of the solar wind of $V_{SW} \cong 10^7$cm/s at the distance $R_\odot$ from the photosphere is obtained.

074.019 **The structure of the coronal solar active region. Part I: X–ray observations of the solar active regions.**
G. Bromboszcz.
Postepy Astron., Tom 33, Zesz. 1–2, p. 29 – 44 (1985). In Polish.
In this paper the results of a study of the solar coronal active regions are presented. The structure, temporal variations and stability of the solar X–ray loops are discussed.

074.020 **Interaction of a solar wind magnetic cloud with comet 1975h.**
H. Lundstedt.
Asteroids, comets, meteors II, p. 419 – 424 (1986). – See Abstr. 012.010.
Magnetic loops in the solar corona are expanded into the interplanetary medium at times of coronal mass ejections. A case of interaction between a magnetic loop and a comet is suggested. The author proposes that the interaction results in a fork structure in the plasma tail. The interaction is explained either by a slipping over of the magnetic field lines or by magnetic reconnection.

074.021 **A two–fluid solar wind model with Alfvén waves: parameter study and application to observations.**
R. Esser, E. Leer, S. R. Habbal, G. L. Withbroe.
J. Geophys. Res., Vol. 91, No. A3, p. 2950 – 2960 (1986).
The effects of Alfvén waves from the inner corona on the solar wind density profile, flow velocity and on the random motion of protons are studied. Different base densities, temperatures, and wave velocity amplitudes, as well as different flow geometries, are considered. The model calculations are compared to simultaneous observations of the electron density profile and the resonantly scattered Lyman alpha line. Present observations, out to 4 solar radii, can be used to place limits on the coronal base density and temperature, and put an upper limit on the wave amplitude. It is pointed out that future observations of the electron density and the Lyman alpha line, out to larger heliocentric distances, and of lines from heavier elements, should be used to place more stringent constraints on the amplitudes of MHD waves in the corona.

074.022 **Ion thermal noise in the solar wind: interpretation of the "excess" electric noise on ISEE 3.**
N. Meyer–Vernet, P. Couturier, S. Hoang, J. L. Steinberg, R. D. Zwickl.
J. Geophys. Res., Vol. 91, No. A3, p. 3294 – 3298 (1986).
The excess electric noise component with a ω^{-1} power spectrum, previously identified on ISEE 3 below the plasma frequency, is quantitatively interpreted. The theory is based on quasi–thermal electron noise to which two effects are added: (1) ion quasi–thermal noise Doppler–shifted by the solar wind velocity and (2) less important, particle impacts on and emission from the antennas. The theory is in excellent agreement with the observations for the long *S* wire antenna ($L = 45$ m) on which the impact noise is negligible. For the *Z* antenna which has a larger surface area, the contributions of both effects are necessary to interpret the measurements. These results could be used to give an independent measurement of the solar wind velocity magnitude and direction whenever the plasma frequency exceeds the lowest frequency channel (30 kHz) of the receiver.

074.023 **Transition region, corona, and solar wind in coronal holes.**
J. V. Hollweg.
J. Geophys. Res., Vol. 91, No. A4, p. 4111 – 4125 (1986).
Previous wave–driven solar wind models (Hollweg, 1978) have been extended by including a new hypothesis for the nonlinear wave dissipation.

074.024 **Viscosity in the solar wind.**
T. E. Holzer, E. Leer, X.–p. Zhao.
J. Geophys. Res., Vol. 91, No. A4, p. 4126 – 4132 (1986).
The effects of viscosity on a steady, radial, spherically symmetric solar wind with an embedded, non–radial magnetic field are reconsidered. The correct expression for the classical viscosity in the presence of a non–radial magnetic field is shown to be different from that used in the past, and a means of describing non–classical viscosity is presented. A physical interpretation of the classical and non–classical descriptions of viscosity is provided, and observational inferences are used in discussing the nature and degree of viscous effects in the solar wind.

074.025 **Solar wind Fe and CNO measurements in high–speed flows.**
F. M. Ipavich, A. B. Galvin, G. Gloeckler, D. Hovestadt, S. J. Bame, B. Klecker, M. Scholer, L. A. Fisk, C. Y. Fan.
J. Geophys. Res., Vol. 91, No. A4, p. 4133 – 4141 (1986).
The authors present solar wind Fe charge state and density measurements from two time periods of high–speed solar wind occurring during a flare–associated driver plasma and two time periods during coronal hole–associated high–speed streams. The observations, made with the ultralow energy charge analyzer (ULECA) sensor on ISEE 3, indicate that the iron in the driver plasma solar wind was predominately of charge state 15 or 16, indicative of an unusually high coronal temperature ($\sim 4 \times 10^6$K). The authors also present the charge state distribution of the CNO group in one coronal hole–associated stream and find an ionization temperature of $(1.3 \pm 0.3) \times 10^6$K. Combining the ULECA measurements with proton observations made with the Los Alamos National Laboratory solar wind instrument on ISEE 3, they find that the speed of solar wind iron is equal to that of

hydrogen in the driver plasma but is substantially higher in the coronal hole flows. The abundance ratio Fe/H, relative to the nominal interstream value, is found to be enhanced by a factor of ~2 in the driver plasma and depressed by a factor of ~2 in the coronal hole–associated solar wind.

074.026 **On thermal instability connected with transformation of Alfvén waves in the solar corona and on formation of prominences.**
L. G. Genkin, L. M. Erukhimov, D. L. Smirnov.
Izv. Vyssh. Uchebn. Zaved., Radiofiz., Tom 28, No. 8, p. 953 – 959 (1985). In Russian. Abstr. in Ref. Zh., 51. Astron., 2.51.438 (1986).

074.027 **The solar wind current system: results of a numerical calculation.**
Yu. V. Pisanko.
Geomagn. Aehron., Tom 26, No. 1, p. 1 – 5 (1986). In Russian. English translation in Geomagn. Aeron.

074.028 **Changes in the geometry of the solar wind transition region depending on solar activity level.**
N. A. Lotova, D. F. Blums.
Geomagn. Aehron., Tom 26, No. 1, p. 6 – 9 (1986). In Russian. English translation in Geomagn. Aeron.

074.029 **The effect of hydromagnetic turbulence on the transport of particles in the magnetic field of the solar wind.**
V. N. Lomonosov.
Geomagn. Aehron., Tom 26, No. 2, p. 177 – 181 (1986). In Russian. English translation in Geomagn. Aeron.

074.030 **Azimuthal propagation and escape of accelerated particles from the solar corona.**
G. A. Bazilevskaya, A. I. Sladkova.
Geomagn. Aehron., Tom 26, No. 2, p. 187 – 190 (1986). In Russian. English translation in Geomagn. Aeron.

074.031 **Acceleration of charged particles by the solar wind. Method of generations.**
S. I. Petukhov, A. A. Turpanov, V. S. Nikolaev.
Geomagn. Aehron., Tom 26, No. 2, p. 197 – 203 (1986). In Russian. English translation in Geomagn. Aeron.

074.032 **Coronal magnetic fields – a mini survey.**
R. M. MacQueen.
The sun and the heliosphere in three dimensions, p. 5 – 18 (1986). – See Abstr. 012.016.
Some recent progress in understanding the nature of the evolution of the global coronal magnetic field is reviewed. Particularly, the efforts of Hoeksema (1984) in defining the character of the evolving modes of the potential coronal field are compared with the currently–known evidence for evolution of coronal white light structures. Recent work in examining the soft X–ray intensity of coronal holes over a major portion of the solar cycle is noted, as are two new studies investigating the relation of coronal mass ejections to the ambient global coronal magnetic field.

074.033 **Origins of the solar wind in the corona.**
G. L. Withbroe.
The sun and the heliosphere in three dimensions, p. 19 – 32 (1986). – See Abstr. 012.016.
The author reviews current knowledge concerning the coronal origins of the solar wind. He considers sources whose role in the generation of the solar wind appears to be well established, such as coronal holes and sources whose role is speculative at the present time. The author also discusses some of the techniques which should lead to improvements in our knowledge concerning this subject.

074.034 **The heliospheric energy source.**
E. N. Parker.
The sun and the heliosphere in three dimensions, p. 33 – 38 (1986). – See Abstr. 012.016.
The solar wind and the heliosphere exist as a consequence of the heat input to the corona, particularly the coronal holes. The necessary energy input to coronal holes has been estimated to be 10^6ergs/cm^{2}sec, requiring Alfvén waves with rms fluid velocities of 10^2km/sec. Observational upper limits on coronal fluid velocities are of the order of 25 km/sec, which may not apply to the transparent coronal hole. Alternatively it has been suggested that coronal holes may be heated by agitation from neighboring active regions, suggesting that the vigor of a coronal hole depends upon its location.

074.035 **The structure and rotation of the solar corona: implications for the heliosphere.**
D. G. Sime.
The sun and the heliosphere in three dimensions, p. 45 – 51 (1986). – See Abstr. 012.016.
Observations of the solar corona and measurements of its rotation are reviewed in order to suggest what rotational characteristics of the solar wind might be observed from the Ulysses mission. Results from 20 years of data from the HAO K–coronameters on Mauna Loa, Hawaii, are described. They show that the corona rotates, on average, rather more rigidly than do any of the usual photospheric or chromospheric tracers on the sun, displaying an average synodic rotation rate of 14.2 degrees per day. The suggestion is made that a similar rotation character will be observed by Ulysses, and some evidence to support that from interplanetary observations is discussed.

074.036 **O VI diagnostics of solar wind generation.**
G. Noci, J. L. Kohl, G. L. Withbroe.
The sun and the heliosphere in three dimensions, p. 53 – 58 (1986). – See Abstr. 012.016.
The O VI resonance doublet is partly collisionally and partly radiatively excited in the solar corona. In the solar wind the O VI ions can attain sufficient outflow speed to cause excitation of the $2P_{1/2}$ level by the chromospheric C II λ1037.0 line. The authors show that this extends the diagnostic possibilities of the O VI resonance doublet. In particular, the determination of the intensity ratio of the doublet lines at several heights can be sufficient to yield the solar wind velocity at those heights and hence information on the mechanisms of solar wind acceleration.

074.037 **Solar wind observations near the sun.**
J. W. Armstrong, W. A. Coles, M. Kojima, B. J. Rickett.
The sun and the heliosphere in three dimensions, p. 59 – 64 (1986). See Abstr. 012.016.
The first observations of interplanetary scintillations with the VLA telescope are reported. The solar wind in the accelerating region from 3 to 12 solar radii was observed by scintillation of the radio source 3C 279. The results obtained outside of 7 solar radii showed good agreement with previous work but observations between 3 and 4.5 solar radii were new and unexpected.

074.038 **Coronal transients at high heliographic latitudes.**
D. G. Sime.
The sun and the heliosphere in three dimensions, p. 101 – 106 (1986). – See Abstr. 012.016.
In this paper, some aspects of our knowledge of the properties of coronal transients (mass ejections) are reviewed in order to anticipate some of the opportunities for the study of coronal transients which may come from the Ulysses observations.

074.039 **The solar cycle dependence of coronal mass ejections.**
R. A. Howard, N. R. Sheeley Jr., D. J. Michels, M. J. Koomen.
The sun and the heliosphere in three dimensions, p. 107 – 111 (1986). – See Abstr. 012.016.
The Solwind white light coronagraph on P78–1 has been making routine observations of the solar corona since 28 March 1979. The authors have now analyzed data from June 1, 1984 – May 1, 1985. They find that the sun is producing far fewer coronal mass ejections than during the 1979 – 1981 period and that their properties are quite different.

074.040 **HELIOS images of coronal mass ejections.**
B. V. Jackson.
The sun and the heliosphere in three dimensions, p. 113 – 118 (1986). – See Abstr. 012.016.
The zodiacal light photometers on board the HELIOS spacecraft can be used to form images of coronal mass ejections in the interplanetary medium. Several aspects of these data are unique: they trace coronal mass ejections using Thomson scattering techniques to distances from the Sun greater than 0.5 AU; their perspective from the HELIOS orbits allow information to be gained about the three–dimensional shapes of specific mass ejections viewed both by coronagraphs and HELIOS; the global view afforded by the spacecraft photometers can image the mass ejection from within and thus relate in situ measurements to the shape of the whole structure.

074.041 **Relationship of coronal transients to interplanetary shocks: 3D aspects.**
R. Schwenn.
The sun and the heliosphere in three dimensions, p. 119 – 121 (1986). – See Abstr. 012.016.

074.042 **Interplanetary scintillation observations of the solar wind at high latitudes.**
W. A. Coles, B. J. Rickett.
The sun and the heliosphere in three dimensions, p. 143 – 151 (1986). – See Abstr. 012.016.
The Ulysses spacecraft will provide the first direct measurements of the polar solar wind. Thus the present knowledge of this region is derived from various indirect methods. These include radio propagation observations, cometary observations and optical observations. The purpose of this paper is to review the radio propagation measurements, particularly their solar cycle dependence; and to discuss the analysis of these observations, the underlying assumptions in the models used, and their potential weaknesses.

074.043 **Evolution of turbulence and waves in the solar wind in radius and latitude.**
A. Barnes.
The sun and the heliosphere in three dimensions, p. 159 – 166 (1986). – See Abstr. 012.016.
The author reviews what is known about the evolution of the hydromagnetic fluctuations in the interplanetary plasma with heliocentric distance, and discusses how these observations bear on the questions of the nature and origin of the fluctuations.

074.044 **A computer simulation study of the microscopic structure of a typical current sheet in the solar wind.**
M. Roth.
The sun and the heliosphere in three dimensions, p. 167 – 171 (1986). – See Abstr. 012.016.
The author gives numerical results for the internal structure of a typical current sheet in the solar wind. The time resolution of plasma measurements across such a layer is discussed in the frame of an Interdisciplinary Study of Directional Discontinuities in the Solar Wind with the Ulysses Mission.

074.045 **Solar wind composition and what we expect to learn from out–of–ecliptic measurements.**
J. Geiss, P. Bochsler.
The sun and the heliosphere in three dimensions, p. 173 – 186 (1986). – See Abstr. 012.016.
Elemental abundances in the solar wind are fractionated relative to the solar abundances by atom–ion separation in the upper chromosphere and by ion–ion separation in the corona, where also the charge states of the ions are frozen–in. Thus solar wind composition and charge states of the elements can be used to study conditions and processes at the solar surface and in the corona. The scan over virtually all solar latitudes effected by Ulysses ought to give new insight into processes and conditions in the solar wind source regions, and it will provide data for a 3–dimensional picture of solar wind expansion and heliospheric processes, which is also important for studying the interactions of the galactic cosmic rays and the interstellar gas with the heliosphere.

074.046 **Comets and three–dimensional solar wind structure.**
J. C. Brandt.
The sun and the heliosphere in three dimensions, p. 187 – 189 (1986). – See Abstr. 012.016.
The principal ways in which the study of comets can contribute to knowledge of three–dimensional solar wind structure are briefly reviewed. Calibration of the cometary results should be considered an important scientific goal of space probes sent to high heliocentric latitudes.

074.047 **Structure and dynamics of corotating and transient streams in three dimensions.**
L. F. Burlaga.
The sun and the heliosphere in three dimensions, p. 191 – 204 (1986). – See Abstr. 012.016.
This paper reviews current ideas concerning the sources of interplanetary flows and their dynamical evolution, with emphasis on results that are relevant to measurements which will be made by Ulysses.

074.048 **Propagation of solar wind features: a model comparison using Voyager data.**
N. I. Kömle, H. I. M. Lichtenegger, H. O. Rucker.
The sun and the heliosphere in three dimensions, p. 205 – 210 (1986). – See Abstr. 012.016.
For a three–dimensional description of the heliosphere it is necessary to know the conditions and time variations of the solar wind at remote positions not directly measured by in–situ observations from a spacecraft. In the present paper the authors compare the predictions of ballistic and hydrodynamic models with the actually observed solar wind parameters. A correlation analysis between the parameter profiles from the model and the Voyager 1 observations shows that the hydrodynamic propagation model yields essentially better results than the ballistic propagation model.

074.049 **Solar wind speed azimuthal variation along the heliospheric current sheet.**
S. T. Suess, P. H. Scherrer, J. T. Hoeksema.
The sun and the heliosphere in three dimensions, p. 275 – 280 (1986). – See Abstr. 012.016.
The authors report on analysis of the speeds measured by Voyager 1 and 2 while skimming along a horizontal (east–west) portion of the current sheet over several days in 1977. The results demonstrate that in this case speed variations exist and are large enough to significantly deform the sheet within a few AU or less if the current sheet were anything but perfectly horizontal. The spatial scale of the speed variation ranges from the smallest measurable scale using 1 hour averaged data up to tens of degrees in longitude. A deformation example is given under the assumption that the observed velocity variation exists on a current sheet that is initially perpendicular to the heliographic equator.

074.050 **The relationship of emission–line transients in the low solar corona to Hα activity.**
R. C. Altrock, R. C. Smith.
Bull. Am. Astron. Soc., Vol. 17, No. 4, p. 842 – 843 (1985). Abstract. – See Abstr. 010.062.

074.051 **Determination of coronal velocities using EUV emission lines.**
M. D. Jones, G. J. Rottman, F. Q. Orrall.
Bull. Am. Astron. Soc., Vol. 17, No. 4, p. 843 (1985). Abstract. – See Abstr. 010.062.

074.052 **High–speed solar wind heating.**
J. W. Freeman, R. Lopez.
Bull. Am. Astron. Soc., Vol. 17, No. 4, p. 843 (1985). Abstract. – See Abstr. 010.062.

074.053 **A Phlegethon theory for coronal heating.**
K. H. Schatten, H. G. Mayr.
Bull. Am. Astron. Soc., Vol. 17, No. 4, p. 843 (1985). Abstract. – See Abstr. 010.062.

074.054 **Experiment on determination of the solar wind ion composition with the complex BIFRAM.**
Yu. I. Ermolaev, G. N. Zastenker, V. T. Kogan, G. E. Kocharov, B. V. Koshevenko, A. V. Lejbov, Z. Nemeček, A. K. Pavlov, A. O. Fedorov, A. A. Kharchenko, Yu. V. Chichagov, I. Šafránková.
Kosm. Issled., Tom 24, Vyp. 2, p. 192 – 199 (1986). In Russian. English translation in Cosm. Res.

074.055 **On the colour of the inner solar corona.**
N. I. Dzyubenko.
Vestn. Kiev. Univ. Astron., No. 27, p. 51 – 54 (1985). In Russian.

074.056 **Geometry of solar plasma flow and the magnetic field in the polar corona.**
V. A. Kovalenko, V. I. Mordvinov.
Issled. Geomagn., Aehron. Fiz. Solntsa, Moskva, No. 73, p. 74 – 80 (1985). In Russian. Abstr. in Ref. Zh., 51. Astron., 3.51.428 (1986).

074.057 **Polarization of the white emission of the corona; relevance to its hydrodynamical expansion.**
O. G. Badalyan.
Astron. Tsirk., No. 1400, p. 1 – 3 (1985). In Russian.

074.058 **On the reliability of observations of coronal holes in the He I 10830 Å line using Kaliniak's infrared spectrophotometer.**
I. I. Sokolova, Eh. V. Kandrashov.
Soln. Dannye, Byull. 1986, No. 1, p. 82 – 88 (1986). In Russian.
Peculiarities of observations of coronal holes on solar charts in the He I 10830 Å line taken with the use of Kaliniak's IR spectrophotometer at Pulkovo are considered. Some problems of identification of coronal holes among the low absorption features and the validity of this identification are discussed.

074.059 **The stationary post–flare arch of May 21/22, 1980.**
P. Hick, Z. Švestka.
Sol. Phys., Vol. 102, Nos. 1/2, p. 147 – 158 (1985).
On May 21/22, 1980 the Hard X–Ray Imaging Spectrometer aboard the SMM imaged an extensive coronal structure after the occurrence of a two–ribbon flare on May 21, 20:50 UT. The structure was observed from 22:20 UT on May 21 until its disappearance at 09:00 UT on May 22.

074.060 **On the formation of coronal cavities.**
C.–H. An, S. T. Suess, E. Tandberg–Hanssen, R. S. Steinolfson.
Sol. Phys., Vol. 102, Nos. 1/2, p. 165 – 176 (1985).
The authors present a theoretical study of the formation of a coronal cavity and its relation to a quiescent prominence. They argue that the formation of a coronal cavity is initiated by the condensation of plasma which is trapped by the coronal magnetic field in a closed streamer and which then flows down to the chromosphere along the field lines due to lack of stable magnetic support against gravity. The existence of a coronal cavity depends on the coronal magnetic field strength; with low strength, the plasma density is not high enough for condensation to occur. Furthermore, the authors suggest that prominence and cavity material is supplied from the chromospheric level.

074.061 **The cyclical variation of energy flux and photospheric magnetic field strength from coronal holes.**
D. F. Webb, J. M. Davis.
Sol. Phys., Vol. 102, Nos. 1/2, p. 177 – 190 (1985).
The authors measured the average soft X–ray emission from coronal holes observed on images obtained during AS&E rocket flights from 1974 to 1981. The variation of this emission over the solar cycle was then compared with photospheric magnetic flux measurements within coronal holes over the same period. The authors found that coronal hole soft X–ray emission could be detected and that this emission appeared to increase with the rise of the sunspot cycle from activity minimum to maximum. Quantitative results confirmed previous suggestions that the coronal brightness contrast between holes and large–scale structure decreased during this period of the cycle. Gas pressures at the hole base were estimated for assumed temperatures and found to vary from about 0.03 dyne cm^{-2} in 1974 to 0.35 dyne cm^{-2} in 1981. The increase in coronal hole X–ray emission was accompanied by a similar trend in the surface magnetic flux of near–equatorial holes between 1975 and 1980 (Harvey et al., 1982).

074.062 **Analysis of an elongated narrow coronal streamer observed at the 1973 solar eclipse.**
H. E. Trigoso, T. Tsubaki.
Sol. Phys., Vol. 103, No. 1, p. 41 – 46 (1986).
A detailed photometric analysis has been made of a narrow elongated coronal streamer observed at the 1973 total eclipse. After deriving real intensity distributions, the electron density has been deduced under the assumption that the distribution is axially symmetric and that the decrease in density from the central axis follows the gaussian law. The results show that the gradient of the electron density is much steeper at the lower part, nearest to the solar limb, than those reported previously for larger scale streamers.

074.063 **A numerical hydrodynamic model of a heated coronal loop.**
P. MacNeice.
Sol. Phys., Vol. 103, No. 1, p. 47 – 66 (1986).
The fluid equations describing a fully ionized single temperature (i.e., electron and proton temperatures assumed identical) hydrogen plasma in a coronal loop subject to a transient heating pulse (2×10^9 ergs $cm^{-2}s^{-1}$) centred about the loop apex have been solved numerically. An adaptive regridding scheme was used to ensure adequate spatial resolution throughout the transition region, and due regard paid to the numerical time constants. Because of the fine gridding made possible by this scheme these results represent the first reliable simulation of the impact of a downward propagating conduction front on the transition region, and the early stages of the development of the downward moving compression and upward ablation. Intensities in the O V (1371 Å) transition region line were calculated from the model results.

074.064 **Abundances of carbon, oxygen, and neon in the solar wind during the period from August 1978 to June 1982.**
P. Bochsler, J. Geiss, S. Kunz.
Sol. Phys., Vol. 103, No. 1, p. 177 – 201 (1986).
From four years of data provided by the Ion Composition Instrument on ISEE–3, the authors have derived flux ratios of minor elements in the solar wind and found He/O = 75 ± 20 and Ne/O = 0.17 ± 0.02. These results are compared with recent solar energetic particle composition data and photospheric values, and they are discussed in the light of theoretical models of ionization and acceleration of heavy ions in the solar chromosphere and corona.

074.065 **X–ray imaging of flare loops and coronal arches.**
Z. Svestka.
NASA Conf. Publ., NASA CP–2421, p. 41 – 66 (1986). – See Abstr. 012.021.
This paper summarizes HXIS results related to preflare, flare–decay, and post–flare phases in the solar corona. Particularly discussed are miniflares, long–lived X–ray enhancements along filament channels, flare precursors (some signifying the flare position, others possibly related to the onset of mass ejections), growth of post–flare loops, images of sites of field–line reconnections late in the flare development, giant post–flare coronal arches with energy contents comparable to coronal transients, and hot loops interconnecting active regions. The extremely low HXIS background, in particular, made it possible to discover many new, previously unknown features in the solar corona. Also emphasized are some problems that have appeared during the

evaluation of HXIS data, for the benefit of those who plan future hard X-ray imaging experiments.

074.066 **Corona and solar wind.**
G. L. Withbroe.
NASA Conf. Publ., NASA CP-2421, p. 221 – 240 (1986). – See Abstr. 012.021.

The purpose of this paper is to briefly discuss the role that the Pinhole/Occulter Facility (P/OF) can play in addressing a few scientific problems in the above areas. Spectroscopic plasma diagnostics and coronal physical processes, the structure and physics of streamers and coronal holes and coronal transients and mass ejections are considered.

074.067 **Solwind observations of coronal mass ejections during 1979 – 1985.**
N. R. Sheeley Jr., R. A. Howard, M. J. Koomen, D. J. Michels.
NASA Conf. Publ., NASA CP-2421, p. 241 – 256 (1986). – See Abstr. 012.021.

Around sunspot maximum, coronal mass ejections (CMEs) occurred at the rate of approximately 2 per day, and had a wide range of physical and morphological properties. During the recent years of relatively low sunspot number, CMEs occurred at the rate of only 0.2 per day, and were dominated by the class of so-called "streamer blowouts." These special CMEs maintained a nearly constant occurrence rate of roughly 0.1 per day during the entire interval.

074.068 **Interpretation of coronal synoptic observations.**
R. H. Munro, R. R. Fisher.
NASA Conf. Publ., NASA CP-2421, p. 257 – 261 (1986). – See Abstr. 012.021.

Three-dimensional reconstruction techniques used to determine coronal density distributions from synoptic data are complicated and time consuming to employ. Current techniques also assume time invariant structures and thus mix both temporal and spatial variations present in the coronal data. The observed distribution of polarized brightness, and brightness of coronal features observed either at eclipses or with coronagraphs depends upon both the three-dimensional distribution of electron density within the structure and the location of the feature with respect to the plane-of-the-sky. By theoretically studying the signature of various coronal structures as they would appear during a limb transit, it is possible to recognize these patterns in real synoptic data as well as estimate temporal evolutionary effects.

074.069 **Wave speeds in the corona and the dynamics of mass ejections.**
S. T. Suess, R. L. Moore.
NASA Conf. Publ., NASA CP-2421, p. 262 – 267 (1986). – See Abstr. 012.021.

A disturbance or coronal mass ejection being advected by the solar wind will expand at the fastest local characteristic speed – typically approximately the fast-mode speed. To estimate this characteristic wave speed and the velocity field in the ambient corona, it is necessary to know the magnetic field, temperature, and density. Only the density is known from coronal observations. The authors estimate the magnetic field, solar wind velocity, and characteristic speeds using an MHD model of coronal expansion between 1 and 5 $R_\odot$ with a dipole magnetic field at the base. This model suggests that the outer edge of a mass ejection will appear to move at a nearly constant rate of 400 to 500 km/s between 2 and 5 $R_\odot$. This result is in agreement with the observations by MacQueen and Fisher (1983) and implies that the acceleration mechanism for coronal mass ejections is other than simple entrainment in the solar wind.

074.070 **On the formation of coronal cavities.**
C.-h. An, S. T. Suess, E. Tandberg-Hanssen, R. S. Steinolfson.
NASA Conf. Publ., NASA CP-2421, p. 268 – 273 (1986). – See Abstr. 012.021.

The authors present a theoretical study of the formation of a coronal cavity and its relation to a quiescent prominence. They argue that the formation of a cavity is initiated by the condensation of plasma which is trapped by the coronal magnetic field in a closed streamer and which then flows down to the chromosphere along the field lines due to lack of stable magnetic support against gravity. The existence of a coronal cavity depends on the coronal magnetic field strength; with low strength, the plasma density is not high enough for condensation to occur. Furthermore, the authors suggest that prominence and cavity material is supplied from the chromospheric level. Whether a coronal cavity and a prominence coexist depends on the magnetic field configuration; a prominence requires stable magnetic support.

074.071 **Polarization of the continuous optical radiation of the corona with hydrostatic density distribution.**
O. G. Badalyan.
Inst. zem. magn., ionos. i rasprostr. radiovoln Akad. Nauk SSSR. Prepr., No. 39/572, 24 pp. (1985). In Russian. Abstr. in Ref. Zh., 51. Astron., 4.51.287 (1986).

074.072 **Thermal conditions of the corona and formation of the solar wind.**
I. V. Chashej, V. I. Shishov.
Fiz. inst. Akad. Nauk SSSR. Prepr., No. 287, 54 pp. (1985). In Russian. Abstr. in Ref. Zh., 51. Astron., 4.51.323 (1986).

074.073 **Plasma transport across the heliopause.**
H. J. Fahr, W. Neutsch, S. Grzedzielski, W. Macek, R. Ratkiewicz-Landowska.
Space Sci. Rev., Vol. 43, Nos. 3/4, p. 329 – 381 (1986).

Existing heliopause models are critically rediscussed under the new aspect of possible plasma mixing between the solar wind and the ambient ionized component of the local interstellar medium (LISM). Based on current kinetic plasma theories, effective diffusion rates across the heliopause are evaluated for several models with turbulence caused by electrostatic or electromagnetic interactions that could be envisaged in this context. Some specific cases that may lead to high diffusion rates are investigated, especially in regard to their LISM magnetic field dependence.

074.074 **The derivation of solar coronal electron temperature from the 1^1S–$n^1P/1^1S$–2^1P emission-line ratios in Mg XI.**
F. P. Keenan, A. E Kingston, D. L. McKenzie.
Astrophys. J., Vol. 303, No. 1, p. 486 – 488 (1986).

Recent *R*-matrix calculations of electron excitation rates for transitions in Mg XI are used to determine the theoretical emission-line ratios $R_1 = I(1s^2\ {}^1S\text{–}1s3p\ {}^1P)/\ I(1s^2\ {}^1S\text{–}1s2p\ {}^1P)$ and $R_2 = I(1s^2\ {}^1S\text{–}1s4p\ {}^1P)/\ I(1s^2\ {}^1S\text{–}1s2p\ {}^1P)$ as a function of electron temperature. Electron temperatures deduced from R_1 and R_2 ratios observed by the P78–I satellite for solar flares and active regions are in excellent agreement with those previously determined by other methods.

074.075 **Electron heating in the solar wind: results from AMPTE.**
D. A. Bryant, C. P. Chaloner, D. S. Hall, D. R. Lepine.
J. Br. Interplanet. Soc., Vol. 39, No. 5, p. 199 – 206 (1986).

An account is given of a series of studies of electron acceleration, or heating, in the solar wind performed within the AMPTE three-spacecraft mission. Discussion centres on measurements made by the electron spectrometer carried on the AMPTE-UKS spacecraft and includes investigations of the Earth's bow shock, ion releases made in the solar wind, natural cavities of "holes" found in the interplanetary magnetic field, and the region known as the magnetopause where solar and terrestrial plasmas meet.

074.076 **Early results from the CHASE experiment flown on Spacelab 2.**
A. H. Gabriel, B. E. Patchett, J. Lang, J. L. Culhane, K. Norman, J. H. Parkinson.
J. Br. Interplanet. Soc., Vol. 39, No. 5, p. 207 – 210 (1986).

The Coronal Helium Abundance Spacelab Experiment (CHASE) was flown on the Space Shuttle as part of the Spacelab 2 payload on July 29, 1985. The primary objective of CHASE was to measure the abundance of helium relative to hydrogen to an accuracy of around 10 per cent, which would be a substantial improvement on previous estimates. If helium abundance is characteristic of the value in the Sun generally then this mixture of about 10 per cent by a number of the heavier helium ions is vitally important to all the physical properties of the solar plasma including all energy transport quantities.

074.077 **A la recherche du coronium. Seconde partie.**
S. Laloë, J.-C. Pecker.
Astronomie, Vol. 100, p. 227 – 239 (1986).

074.078 **The relationship between coronal bright points as seen in He I λ10830 and the evolution of the photospheric network magnetic fields.**
K. L. Harvey.
Aust. J. Phys., Vol. 38, No. 6, p. 875 – 883 (1985). Paper presented at the R. G. Giovanelli Commemorative Colloquium, Part II, Tucson, Arizona, 17 – 18 January 1985.

Transient 'dark points' in He I λ10830 are found to be associated with small magnetic bipoles. The number of these dark points varies inversely with the sunspot number. Sampled over a solar cycle, about one-third of the dark points are associated with erupting magnetic flux (ephemeral regions) while the other two-thirds are associated with chance encounters of opposite magnetic polarity features. Since coronal bright points are associated with He I dark points, it is suggested that the inverse correlation of both of these events with the sunspot number results from the higher probability of chance encounters between magnetic network of opposite polarity (larger areas of mixed magnetic polarity) during low levels of solar activity.

074.079 **Thermal regime of the corona with concentrated heating source.**
I. V. Chashej, V. I. Shishov.
Astron. Tsirk., No. 1392, p. 2 – 4 (1985). In Russian.

The influence of the gravitational energy flow H_{g*} on the thermal characteristics of coronal plasma is considered. If the heating source is sufficiently strong ($H_{g*} \sim H_*$) the thermal regime of the corona is affected by the mass flow which, in turn, is controlled by the coronal structure. The regulation of the mass flow takes place near the transition zone, $r \cong r_*$, if $r_m > 2r_*$, and in the outer region $r > r_m$, if $r_m < 2r_*$ (r_m is the heliocentric distance for the temperature maximum).

074.080 **The determination of coronal potential magnetic fields using line-of-sight boundary conditions.**
T. J. Bogdan.
Sol. Phys., Vol. 103, No. 2, p. 311 – 315 (1986).

Previous efforts to construct solar coronal fields using surface magnetograph data have generally employed a least squares minimization technique in order to determine the spherical harmonic expansion coefficients of the magnetic scalar potential. Provided there is no source surface high up in the corona, the author shows that knowlege of the line-of-sight component of the surface magnetic field, $B_l = B_r \sin\theta + B_\theta \cos\theta$, is sufficient to uniquely determine the potential coronal magnetic field by an explicit construction of the magnetic scalar potential for an arbitrary $B_l(\theta,\phi)$.

074.081 **Ballooning instabilities in the solar corona: conditions for stability.**
A. W. Hood.
Sol. Phys., Vol. 103, No. 2, p. 329 – 345 (1986).

A method for analysing the stability of cylindrical loops and arcades to ballooning modes is presented. Using a WKB method, simple tests for instability (or stability to localised modes) are obtained for the two commonly used, line tying, conditions. In addition, when instabilities are present, the physical growth rate and the radial structure of the least stable can be computed from the solution of a 4th order system of ordinary differential equations. The detailed analysis for a particular field is presented.

074.082 **Helium and minor ions in the corona and solar wind: dynamics and charge states.**
A. Bürgi, J. Geiss.
Sol. Phys., Vol. 103, No. 2, p. 347 – 383 (1986).

A theoretical model of the acceleration region of the solar wind with major species (p, α, e) and minor ions (e.g., ^{3}He, C, O, Mg, Si) is presented. Observed n_e-profiles and the equations of continuity and momentum are used to calculate profiles of T, n, and u for all species, as well as charge states of minor ions. The disagreement of the results of a pure p–e model with observations is discussed in some detail. Finally, a model for a coronal hole is investigated and it is found that wave pressure is necessary to model successfully the observed solar wind speeds and abundances.

074.083 **The *K*-corona under hydrostatic density distribution: relevance to solar wind.**
O. G. Badalyan, M. A. Livshits.
Sol. Phys., Vol. 103, No. 2, p. 385 – 392 (1986).

An analytical expression is obtained for the *K*-corona brightness as a function of the distance from the solar limb. The agreement of the theory with the numerous observational data indicates that the density distribution at the levels located at $1.2 - 2.5\ R_\odot$ heliocentric distances may be treated as hydrostatic with $T =$ const. The deviation of the observed *K*-corona brightness from the expression obtained may be indicative of supersonic fluxes of matter. These concepts may be used to verify the various theoretical models for solar wind source.

074.084 **Numerical image processing applied to the solar corona.**
J. M. Malherbe, J. C. Noëns, T. Roudier.
Sol. Phys., Vol. 103, No. 2, p. 393 – 398 (1986).

Numerical data processing is applied to the high-resolution images of the solar corona obtained with the 20 cm coronagraph of the Pic du Midi observatory. Two complementary methods are proposed to solve some classical difficulties usually met in the morphological analysis of the solar corona, namely the brightness gradient in the inner and medium corona, the low contrast of numerous emissive regions and the superimposition along the line of sight of different structures.

074.085 **Coronal mass ejection associated with the stationary post-flare arch of 21 – 22 May 1980.**
M. K. McCabe, Z. F. Švestka, R. A. Howard, B. V. Jackson, N. R. Sheeley Jr.
Sol. Phys., Vol. 103, No. 2, p. 399 – 408 (1986).

By using a combination of X-ray, Hα, white-light corona, and zodiacal light images on 21 – 22 May, 1980 the authors demonstrate, and try to explain, the co-existence of a coronal mass ejection with a stationary post-flare coronal arch. The mass ejection was seen in prolongation of the path of a powerful spray, whereas the active region filament did not erupt.

074.086 **Electric fields and plasma structure in coronal magnetic loops.**
P. Foukal, C. Hoyt, L. Gilliam.
Astrophys. J., Vol. 303, No. 2, p. 861 – 876 (1986).

The authors have analyzed limb spectra of five postflare loops and three active prominences to determine whether the Stark effect observed in their Balmer line emission can be explained as pressure broadening or whether it might imply macroscopic electric fields. The main data set consists of 43 spectra covering the interval between ~3500 and 6000 Å, obtained at the Sacramento Peak 40 cm coronograph between 1970 and 1983. Some spectra were obtained through a rotatable polarizing prism to measure the dependence of line width upon the transmitted plane of polarization. These data suggest a DC electric field of order 10^2V cm^{-1}, whose projection in the sky plane lies transverse to the loop magnetic field vector. Such a field might arise from reconnection in postflare loops.

074.087 **Coronal holes and properties of solar wind.**
A. A. Florinskij.
Magnitosfer. issled. Moskva, No. 6, p. 5 – 10 (1986). In Russian. Abstr. in Ref. Zh., 62. Issled. Kosm. Prostranstva, 5.62.467 (1986).

074.088 **VLBI experiment on radio occultation of near–solar plasma from signals of Venera 15.**
V. A. Alekseev, V. I. Altunin, A. V. Biryukov, Eh. D. Gatehlyuk, A. F. Dement'ev, I. A. Knorin, N. A. Knyazev, A. E. Kryukov, B. N. Lipatov, V. A. Rudakov, A. S. Sizov.
Pis'ma Astron. Zh., Tom 12, No. 6, p. 486 – 492 (1986). In Russian. English translation in Sov. Astron. Lett., Vol. 12.

The VLBI experiment of radio occultation of the near–solar turbulent medium by Venera 15 signals in the wavelength ~30 cm and ~1200 km base is described. The middle–shape spectral line of the interference fringe and its broadening that contains information about the turbulent plasma are determined. The VLBI observation results and possibilities of investigation of the near–solar turbulent plasma by the radio occultation method are discussed.

074.089 **Transition region of the solar wind.**
N. A. Lotova.
Geomagn. Aehron., Tom 26, No. 3, p. 369 – 372 (1986). In Russian. English translation in Geomagn. Aeron.

074.090 **Instability of MHD waves in the solar wind.**
P. P. Malovichko, Yu. M. Vojtenko.
Geomagn. Aehron., Tom 26, No. 3, p. 373 – 377 (1986). In Russian. English translation in Geomagn. Aeron.

074.091 **On the character of dependence of solar wind modulating properties on the distance to the sun.**
L. I. Dorman, V. Kh. Shogenov.
Geomagn. Aehron., Tom 26, No. 3, p. 378 – 383 (1986). In Russian. English translation in Geomagn. Aeron.

074.092 **The geometry of the solar wind transition region during the 11–year solar cycle.**
N. A. Lotova, D. F. Blums.
Geomagn. Aehron., Tom 26, No. 3, p. 488 – 490 (1986). In Russian. English translation in Geomagn. Aeron.

074.093 **On the origin of low–frequency Alfvén waves in the solar wind.**
I. V. Chashej, V. I. Shishov.
Astron. Zh., Tom 63, Vyp. 3, p. 542 – 549 (1986). In Russian. English translation in Sov. Astron., Vol. 30, No. 3.

The problem is considered of the localization of the source of Alfvén waves with periods $\gtrsim 5$ min observed in the solar wind. The conclusion is drawn that the waves are generated in the region near the transition layer (upper chromosphere and lower corona). A comparison is carried out of the models of the solar atmosphere with a homogeneous and a strongly inhomogeneous (tubes) magnetic field. The form of the power spectrum of the outgoing Alfvén waves is discussed.

074.094 **Coronal millimeter wave sources.**
I. G. Moiseev, N. S. Nesterov.
Izv. Krymskoj Astrofiz. Obs., Tom 74, p. 112 – 117 (1986). In Russian. English translation in Bull. Crimean Astrophys. Obs., Vol. 74.

Radio sources located in the solar corona at heights of $(200 - 300) \times 10^3$km above the photosphere have been detected in the course of observations at 8.2 and 13.5 mm wavelengths. The detected sources were optically thin and had effective temperatures of several units of 10^3K.

074.095 **Observational study for detecting periodic oscillations in the inner solar corona.**
T. Tsubaki, Y. Saito, Y. Suematsu, Y. Nakagomi.
Publ. Astron. Soc. Jpn., Vol. 38, No. 2, p. 251 – 256 (1986).

For the purpose of detecting periodic oscillations or waves in the inner corona, time series observations of Fe XIV λ5303 emission–line spectra were made at the Norikura Solar Observatory. Detailed line–profile analyses followed by power spectral analyses have provided the conclusion that periodic intensity fluctuations of nearly the same phase existed over some 100,000 km along the spectrograph slit with the same period of 235 s, and with the wavelength of 30,000 – 40,000 km, while no periodicity was detected both for the line width and the line–of–sight Doppler velocity.

074.096 **The mechanism for accelerating and heating the high speed solar wind between 0.3 AU and 1 AU.**
Z.–r. Zheng, X.–p. Zhao.
Chin. J. Space Sci., Vol. 5, No. 2, p. 92 – 100 (1985). In Chinese. Abstr. in Phys. Abstr., Vol. 89, No. 1276, Entry 35913 (1986).

074.097 **Coulomb self–collision frequencies for non–thermal velocity distributions in the solar wind.**
E. Marsch, S. Livi.
Ann. Geophys., Vol. 3, No. 5, p. 545 – 556 (1985). Abstr. in Phys. Abstr., Vol. 89, No. 1277, Entry 42135 (1986).

074.098 **Coronal index of solar activity IV a, 1976.**
M. Rybanský.
Contrib. Astron. Obs. Skalnaté Pleso, Vol. 14, p. 7 – 80 (1986).

This paper contains tables of values which were obtained from measurements of the emission line 530.3 nm by all corona stations.

074.099 **On the possibility of forecasting coronal holes.**
V. Rušin, M. Rybanský.
Contrib. Astron. Obs. Skalnaté Pleso, Vol. 14, p. 81 – 85 (1986).

The possibility is studied of forecasting coronal holes from ground–based observations of the corona or radio emissions (2800 MHz) above the eastern limb of the Sun. The analysis indicates that it is possible to forecast coronal holes on the basis of coronal observations, but not on the basis of radio emissions.

074.100 **Coronal transport of solar flare particles.**
J. Perez–Peraza.
Space Sci. Rev., Vol. 44, Nos. 1/2, p. 91 – 138 (1986).

The author has reviewed the observational features supporting the view that azimuthal transport of solar flare particles takes place at the level of the solar corona, and not in the interplanetary medium. A brief description is given of models regarding longitudinal particle transport, in terms of their ability to explain the corresponding observational phenomena. Finally, a model with multiple particle emission phases is proposed, in terms of two coronal azimuthal transport steps, on the basis of the dynamical and static features of the coronal magnetic field, respectively.

074.101 **Relationship of coronal transients to interplanetary shocks: 3D aspects.**
R. Schwenn.
Space Sci. Rev., Vol. 44, Nos. 1/2, p. 139 – 168 (1986).

This review intends to cover most of the relevant issues concerning the three–dimensional structure of coronal mass ejections (CMEs) and interplanetary shocks. The author gives a brief phenomenological survey of CMEs, interplanetary shocks, and the associations between them. He deals with detailed inspection of the topology of both CMEs and shocks and with some recent theoretical models. The global aspects of the shocks' propagation through the heliosphere are reviewed.

074.102 **Dynamic heating of stationary coronal loops.**
F. Reale, G. Peres, S. Serio.
Mem. Soc. Astron. Ital., Vol. 56, No. 4, p. 805 – 808 (1985). – See Abstr. 010.821.

074.103 **Radio emission from a shock wave in the solar wind.**
V. G. Ledenev.
Sov. Astron., Vol. 29, No. 6, p. 673 – 676 (1985). English translation of 40.074.080.

074.104 **Stability of siphon flow in coronal magnetic loops.**
F. Xu, S. Rong.
Chin. Astron. Astrophys., Vol. 10, No. 1, p. 17 – 21 (1986). English translation of 40.074.079.

074.105 **Re–examination of the viscous effect in the solar wind.**
X.–f. Zhao.
Chin. Astron. Astrophys., Vol. 10, No. 2, p. 148 – 153 (1986). English translation of Chin. J. Space Sci., Vol. 6, p. 1 – 7 (1986).
The Helios observations of the plasma and magnetic field in the high–speed solar wind streams between 1 AU and 0.3 AU are used to examine the radial distribution of the various forces exerted on the streams. The radial dependence of the real viscous force in the streams is deduced and the corresponding viscosity is derived. It is found that the ratio of the real to the classical viscosity at 1 AU should be equal to or greater than one tenth.

074.106 **Morphology of low–frequency waves in the solar wind and their relation to ground pulsations.**
T. J. Odera, W. F. Stuart.
J. Atmos. Terr. Phys., Vol. 48, No. 1, p. 83 – 95 (1986).
Three classes of low frequency waves (period range 20 – 80 s) were identified using data from the UCLA fluxgate magnetometer experiment on board the ISEE 2 spacecraft. These are continuous pulsations similar in type to Pc 3, band–limited oscillations distinguished by mixed period fluctuations, and relatively isolated wave bundles. The waves were preferentially observed when the interplanetary magnetic field (IMF) direction was sunward and were most common when the cone angle, i.e. the angle between IMF and the sun–earth line was often between 15° and 45°. Their frequency is proportional to the IMF magnitude. Comparison between the waves observed on board the ISEE 2 spacecraft and the Pc 3 – 4 recorded simultaneously at a mid–latitude ground station, Oulu ($L = 4.5$), showed that similarity of spectra of the waves in the spacecraft and on the ground was very rare and that correspondence between the events in space and on the ground was extremely low.

074.107 **MHD model for long distance coronal loops.**
O. T. Matsuura, P. Boscolo.
Rev. Mex. Astron. Astrofis., Vol. 12, Spec. Issue, p. 365 – 369 (1986). – See Abstr. 012.066.
Horizontal parts of long distance coronal loops are studied. Under currently verifiable hypotheses, the conclusion is drawn that the matter is in isothermal and static state. Therefore the heating has only to balance the radiative loss. Applying this result to different types of steady state loops, the heating can be related to the loops' sizes and ages.

074.108 **Images of post–flare coronal structures in X–rays.**
Z. Švestka, P. Hick.
Sol. Phys., Vol. 104, No. 1, p. 187 – 190 (1986). – See Abstr. 012.067.
This is an extended abstract of several papers describing extensive coronal structures related to radio continua and imaged in > 3.5 keV X–rays.

074.109 **Coronal interconnection of two active regions observed in 3.5 – 8.0 keV X–rays.**
F. Fárník, H. F. van Beek, Z. Švestka.
Sol. Phys., Vol. 104, No. 2, p. 321 – 332 (1986).
Using HXIS data, the authors have studied the further development of the coronal arch extending towards the SE above active region No. 17255 in November 1980. The arch disappeared on 10 November, but since 9 November HXIS revealed another arch–like structure towards the SW. The authors have studied the development of this new structure which appeared to be most likely an arch interconnecting AR 17255 with AR 17251. They have estimated the physical characteristics in this coronal structure and compared them with observations of interconnecting loops made at lower energies by Skylab in 1973.

074.110 **Lifetime, age, and rotation of coronal magnetic fields.**
G. D. Parker.
Sol. Phys., Vol. 104, No. 2, p. 333 – 345 (1986).
The rotation of the solar electron corona is determined for intervals when nearly periodic variations dominated the polarization brightness record during 1964 – 1976. Coronal rotation rates derived for 765 intervals vary with height, latitude, and interval length. These rotation rates show a decrease of differential rotation with height and support earlier rotation studies which included much less stationary data. Analyses of the selected intervals and autocorrelation of the complete *K*–coronameter data set give quantitative estimates of the rotational effects of magnetic tracer age and lifetime. The principal effects detected are a relatively fast rotation of very long–lived tracers at high latitude and a relatively fast rotation of very short–lived tracers at low latitudes. The observations indicate that high–to–low latitude magnetic connections extending through the corona speed up rotation at high latitudes and retard it at low latitudes.

074.111 **Dynamics of electron beams in a coronal loop and the hard X–ray burst.**
T. Takakura.
Sol. Phys., Vol. 104, No. 2, p. 363 – 389 (1986).
A numerical simulation has been made for the dynamics of non–thermal electrons (> 10 keV) injected with spatial, temporal and velocity distributions into a model coronal loop. The time variations of the spatial intensity distribution and the spectrum for the expected hard X–rays are computed for many models in order to find the important physical parameters for those characteristics. The most important one is the column density of plasma along the loop.

074.112 **In–situ observations of the latitudinal gradients of the solar wind parameters during 1976 and 1977.**
R. Bruno, U. Villante, B. Bavassano, R. Schwenn, F. Mariani.
Sol. Phys., Vol. 104, No. 2, p. 431 – 445 (1986).
Interplanetary observations from Helios 1, Helios 2, and IMP–8 spacecraft during 1976 and 1977, namely the early portion of solar cycle 21, have been used to investigate the latitudinal gradients of the solar wind parameters. A latitudinal belt of ±25 deg around the current sheet has been investigated. Large gradients for solar wind flow speed, proton density and temperature have been found. Smoother gradients where also found for particle flux, kinetic, gravitational and thermal energy density flux. The momentum flux density and the total energy flux density which other authors found to be independent of any longitudinal stream structure were also found to be independent of any latitudinal structure. Moreover, these two parameters did not show any temporal variation.

074.113 **The post flare loops observed at the total eclipse of February 16, 1980.**
Y. Hanaoka, H. Kurokawa, S. Saito.
Sol. Phys., Vol. 105, No. 1, p. 133 – 148 (1986). = Contrib. Kwasan Hida Obs., Univ. Kyoto, No. 267.
A post flare loop system was observed on the west limb at the total solar eclipse of February 16, 1980 in Kenya. Analyzing the monochromatic images and the flash spectra, the authors obtained results for the temperatures, diameters, and electron densities. The observational results led to construct an empirical model of the post flare loop system which is consistent with the reconnection model of Kopp and Pneuman (1976).

074.114 **Solar wind expansion in a polar coronal hole: inferences from coronal white light and interplanetary Lyman alpha observations.**
R. Lallement, T. E. Holzer, R. H. Munro.
J. Geophys. Res., Vol. 91, No. A6, p. 6751 – 6759 (1986).
Analysis of white light coronagraph observations of a polar coronal hole led Munro and Jackson (1977) to infer large solar wind speeds in the coronal hole and a consequent need for substantial energy addition to the outflow well above the coronal base. Recent interplanetary Lyman α observations by Kumar and Broadfoot (1979) and Lallement et al. (1985) indicate that the solar wind mass flux from well–developed polar coronal holes may be considerably less than that typically observed in situ near the ecliptic plane. A reexamination of the Munro and Jackson analysis, with attention given both to the inferences drawn from the interplanetary Lyman α observations and to the uncertainties inherent in the coronal white light observations, leads to conclusions significantly different from those drawn by Munro and Jackson.

A la recherche du coronium.
See Abstr. 004.015.

Possibilities of the scintillation method for investigation of the velocity of circumsolar plasma.
See Abstr. 036.058.

On a possibility to detect neutral points of a coronal magnetic field.
See Abstr. 036.059.

The International Cometary Explorer mission to comets Giacobini–Zinner and Halley: an update.
See Abstr. 051.037.

Coronal spectroscopy and imaging from Spartan during the polar passage of Ulysses.
See Abstr. 051.042.

Magnetic non–equilibrium in a stratified atmosphere.
See Abstr. 062.018.

On the behavior of hydromagnetic surface waves.
See Abstr. 062.020.

Nonlinear numerical simulation of planar thermal instability.
See Abstr. 062.038.

A reconsideration of thermal diffusion in ionized gases: quantal and dynamic shielding effects.
See Abstr. 062.048.

Nonlinear rapid symmetric waves in a magnetic layer.
See Abstr. 062.052.

Acceleration of charged particles in magnetic reconnection: solar flares, the magnetosphere, and solar wind.
See Abstr. 062.059.

Long nonlinear waves in a compressible magnetically structured atmosphere. II. Slow body waves in a magnetic slab.
See Abstr. 062.064.

Viscous normal modes on coronal inhomogeneities and their role as a heating mechanism.
See Abstr. 062.084.

Condensation modes in sheared magnetic fields.
See Abstr. 062.085.

Theory of galactic disks and its relevance to the solar wind magnetic configuration.
See Abstr. 062.108.

Plasma and radiation processes.
See Abstr. 063.007.

Steady hydromagnetic flows in open magnetic fields. I. A class of analytic solutions.
See Abstr. 064.027.

The role of nonclassical electron transport in the lower solar transition region.
See Abstr. 073.001.

High–energy observations of solar flares.
See Abstr. 073.004.

Electron density diagnostics in the 10 – 100 Å interval for a solar flare.
See Abstr. 073.028.

The solar radio corona: manifestations of energetic electrons.
See Abstr. 073.083.

Reflection of the structure of the chromosphere and corona in the solar wind and interplanetary magnetic field.
See Abstr. 073.087.

High–resolution telescope and spectrograph images of the solar chromosphere and transition zone.
See Abstr. 073.134.

Chromospheric "active region loops".
See Abstr. 073.146.

Physical conditions in eruptive prominences at several solar radii.
See Abstr. 073.150.

On the magnetic structure of the quiet transition region.
See Abstr. 075.049.

An atlas of photospheric magnetic field observations and computed coronal magnetic fields: 1976 – 1985.
See Abstr. 075.050.

Magnetic reconnection on the Sun.
See Abstr. 075.052.

Das mittlere koronale Magnetfeld.
See Abstr. 075.053.

Continuum absorption in the solar EUV spectra.
See Abstr. 076.001.

Influence of solar flares on the X–ray corona.
See Abstr. 076.006.

Associations of compact C IV events, He I 10830 Å dark points, and magnetic structures.
See Abstr. 076.007.

Interplanetary energetic particles, coronal flares, and hard X–ray microflares.
See Abstr. 076.013.

Variations in quiet Sun radiation at centimetre wavelengths during solar maximum period.
See Abstr. 077.002.

A test for large–angle radio scattering in the solar corona.
See Abstr. 077.007.

Polar–cap and coronal–hole–associated brightenings of the Sun at millimeter wavelengths.
See Abstr. 077.008.

Solar radio signature suggestive of proton beams.
See Abstr. 077.014.

Radio observation of the dynamics of high energy electrons in coronal magnetic loops.
See Abstr. 077.016.

Type I solar radio bursts.
See Abstr. 077.035.

The relation between flare–related metric continuum bursts and coronal mass ejections.
See Abstr. 077.047.

Type IV bursts and coronal mass ejections.
See Abstr. 077.060.

Properties of metre–wavelength solar bursts associated with coronal mass ejections.
See Abstr. 077.068.

Solar filament eruptions and energetic particle events.
See Abstr. 078.003.

Time–dependent coronal shock acceleration of energetic solar flare particles.
See Abstr. 078.012.

Solar photospheric and coronal abundances from solar energetic particle measurements.
See Abstr. 078.015.

Structure, dynamics, and heating of the solar atmosphere.
See Abstr. 080.017.

Dynamics of the F2 region as a function of disturbances of the magnetosphere and solar wind.
See Abstr. 083.014.

Relationship of the interplanetary electric field to the high–latitude ionospheric electric field and currents: observations and model simulation.
See Abstr. 083.033.

The highest geoactivity in 1981 and its source on the sun.
See Abstr. 084.011.

Ion release experiments in the solar wind.
See Abstr. 084.043.

Three–dimensional axisymmetric magnetosphere in pressure balance with the solar wind.
See Abstr. 084.069.

A three–dimensional MHD simulation of the interaction of the solar wind with the earth's magnetosphere: the generation of field–aligned currents.
See Abstr. 084.070.

Some solar cycle phenomena related to the geomagnetic activity from 1868 to 1980. II. High velocity wind streams and cyclical behaviour of poloidal field.
See Abstr. 085.005.

Disappearing solar filaments and geomagnetic activity.
See Abstr. 085.030.

Model of cometary atmosphere – solar wind interaction.
See Abstr. 102.009.

Non–stationary solar wind – comet interaction.
See Abstr. 102.036.

The effect of heavy ions on the formation and structure of cometary bow shocks.
See Abstr. 102.057.

MHD turbulences in solar wind – comet interaction region.
See Abstr. 102.075.

Effect of mass loading outside cometary bow shock for the plasma and wave measurements in the coming cometary missions.
See Abstr. 102.076.

Simulation of ICE–Giacobini–Zinner and VEGA/GIOTTO–Halley encounters.
See Abstr. 103.030.

MHD turbulence in the solar wind–comet interaction region.
See Abstr. 103.102.

The brightness variations of comet Halley at large heliocentric distances.
See Abstr. 103.634.

Nitrogen isotopic systematics in meteorites and in the solar wind.
See Abstr. 105.111.

Interplanetary shock collisions: forward with reverse shocks.
See Abstr. 106.003.

The AMPTE artificial comet experiments.
See Abstr. 106.017.

Optical observations of the AMPTE artificial comet from the Northern Hemisphere.
See Abstr. 106.018.

***In situ* magnetic field observations of the AMPTE artificial comet.**
See Abstr. 106.019.

UKS plasma measurements near the AMPTE artificial comet.
See Abstr. 106.020.

Plasma waves and wave–particle interactions seen at the UKS spacecraft during the AMPTE artificial comet experiment.
See Abstr. 106.021.

Dynamics of the AMPTE artificial comet.
See Abstr. 106.022.

A gas dynamic magnetosheath field model for unsteady interplanetary fields: application to the solar wind interaction with Venus.
See Abstr. 106.026.

Solar wind controlled pulsations: a review.
See Abstr. 106.036.

Three–dimensional configurations of interplanetary disturbances associated with coronal mass ejections.
See Abstr. 106.037.

3–D coronal and heliospheric structure from radio observations.
See Abstr. 106.041.

Interstellar gas parameters and solar wind anisotropies deduced from H and He observations in the solar system.
See Abstr. 106.053.

Superposed epoch analysis of corotating interaction regions at 0.3 and 1.0 AU: a comparative study.
See Abstr. 106.057.

Flare stars and solar bursts: high resolution in time and frequency.
See Abstr. 122.182.

Concerning the problem of collisional heating of the interstellar helium flow by solar wind protons.
See Abstr. 131.118.

A perturbation approach to cosmic ray transients in interplanetary space.
See Abstr. 144.037.

Particle acceleration at a termination shock. 1. Application to the solar wind and the anomalous component.
See Abstr. 144.040.

Pick–up ions in the solar wind as a source of suprathermal particles.
See Abstr. 144.054.

Modulation of the anisotropy of pressure of galactic cosmic rays by the solar wind.
See Abstr. 144.101.

075 Magnetic Fields

075.001 **Structure and merging of solar magnetic fluxtubes.**
G. W. Pneuman, S. K. Solanki, J. O. Stenflo.
Astron. Astrophys., Vol. 154, No. 1/2, p. 231 – 242 (1986).
An expansion technique is employed to derive a model of magnetic fluxtubes in the solar atmosphere which takes into account the effects of field line curvature, internal structural variations, and the merging of the tube with its immediate neighbours as it expands.

075.002 **Solar vector magnetograms of the Okayama Astrophysical Observatory.**
M. Makita, S. Hamana, K. Nishi, M. Shimizu, T. Sakurai, K. Shibasaki.
Astrophys. Space Sci., Vol. 118, Nos. 1/2, p. 163 – 167 (1986). – See Abstr. 012.001.
The title instrument is mounted on the 65 cm solar coudé telescope at the Okayama Observatory. Observation is usually of the Fe I 5250 Å line. The data obtained are briefly described.

075.003 **Small–scale structure of solar magnetic fields.**
V. G. Lozitskij.
Kinematika Fiz. Nebesn. Tel, Tom 2, No. 1, p. 28 – 35 (1986). In Russian. English translation in Kinematics Phys. Celest. Bodies.
Zeeman spectrograms of photospheric magnetic fields of three active regions and magnetic fields in a 2B flare were analyzed. The non–sunspot magnetic fields are shown to have two spatially unresolved components: a quasihomogeneous vertical one with strengths $\lesssim 1$ T and a multipolar one with strengths $\lesssim 0.1$ T. The total magnetic fluxes of these components are approximately the same.

075.004 **Structure of the magnetic field and plasma motions in the elements of the solar magnetic field.**
S. I. Gopasyuk, L. B. Demkina.
Soln. Dannye, Byull., 1985, No. 11, p. 61 – 68 (1985). In Russian.
The structure of the solar magnetic field elements has been investigated on the basis of the observational data obtained for several regions of the quiet photosphere. Simultaneous measurements of the longitudinal magnetic field and radial velocity were carried out in the Fe I λ5250.2 Å line in 1975. On the basis of the observations all the components of the magnetic field and velocity vectors were calculated. The vortex structure of the magnetic field and velocity is revealed, a correlation of their values in the elements is found. A suggestion on the probable existence of torsional oscillations in the elements of the solar magnetic field is made.

075.005 **Longitudinal distribution of magnetic neutral lines. I.**
V. I. Makarov, K. S. Tavastsherna.
Soln. Dannye, Byull., 1985, No. 12, p. 51 – 59 (1986). In Russian.
Half–annual values of the number of neutral lines of the solar large–scale magnetic field (1945 – 1974), calculated on the basis of Hα charts, are given.

075.006 **Dynamics of magnetic flux tubes in the convection zone of the sun.**
A. E. Dudorov, A. K. Kirillov.
Soln. Dannye, Byull., 1985, No. 12, p. 85 – 93 (1986). In Russian.
The motion of horizontal isothermal ($T = T_e$) magnetic flux tubes in the convection zone of the sun is considered. The time of the rise of the tubes from the bottom of the convective zone to the photosphere is determined from numerical modelling, taking into account different resistance factors. A possibility of explaining some solar features using the model is discussed.

075.007 **Measurements of solar magnetic fields outside spots using different–strength lines.**
S. I. Gopasyuk.
Izv. Krymskoj Astrofiz. Obs., Tom 72, p. 159 – 171 (1985). In Russian. English translation in Bull. Crimean Astrophys. Obs., Vol. 72.
The magnetic field in the quiet solar atmosphere has been studied. The spectral lines belonged to different chemical elements, had different potentials of excitation on the lower level and different magnetic splitting. The analysis of the observational data showed that the strength of the field in the magnetic elements as well as the magnetic flux increases with height. The maximum of the magnetic field strength coincides with the level where the central parts of λ5324.19 Å Fe I and λ5269.54 Å Fe I strong–line profiles are formed. Then, the magnetic field strength decreases.

075.008 **Distribution of the magnetic field strength across a sunspot umbra base on Fe I, Zr I and Y I absorption lines.**
M. D. Gusejnov.
Izv. Krymskoj Astrofiz. Obs., Tom 72, p. 184 – 187 (1985). In Russian. English translation in Bull. Crimean Astrophys. Obs., Vol. 72.
The magnetic field strengths H across the sunspot umbra N 361 (numbering according to «Solar Data») are measured observing five different temperature–sensitive absorption lines. Zeeman spectrograms in two wavelength ranges (λ6125 – 6175 Å and λ6410 – 6460 Å) were obtained in October 30, 31 and November 1, 1972. The distributions of H across the sunspot umbra determined with simple Zeeman triplets Zr I λ6127.5 Å, λ6143.2 Å, Fe I λ6173.3 Å, λ6420.0 Å. and Y I λ6435.0 Å are compared, showing in the umbra spectrum approximately the same Rowland intensity.

075.009 **Magnetic fields and vertical motions in supergranules.**
T. T. Tsap, I. S. Laba.
Izv. Krymskoj Astrofiz. Obs., Tom 73, p. 62 – 70 (1985). In Russian. English translation in Bull. Crimean Astrophys. Obs., Vol. 73.
The relation between the magnetic fields and radial velocities at the center of the solar disk is studied. Observations of the magnetic fields and line–of–sight velocities were made in the Fe I λ4808 Å, Ca I λ6103 Å, Fe I λ5233 Å, and Mg I λ5184 Å lines.

075.010 **Radio method of mapping large–scale coronal magnetic fields on active stars.**
R. T. Stewart.
Nature, Vol. 320, No. 6062, p. 509 – 510 (1986).
The author describes a radio method which measures the longitudinal magnetic sector structure of the solar corona by observing the Sun 'as a star' and monitoring the sense of circular polarization of metre–wavelength emission as a function of rotational phase. The method is tested by comparing synoptic plots of solar radio–noise–storm polarities derived from high– and low–resolution observations over a period of four years at Culgoora. This method may be applied to the measurement of large–scale coronal fields on other active stars.

075.011 **Numerical solution of the magnetostatic equations for thick flux tubes, with application to sunspots, pores, and related structures.**
V. J. Pizzo.
Astrophys. J., Vol. 302, No. 2, p. 785 – 808 (1986).
Computational difficulties have long hampered the study of the structure of vertical flux tubes near the solar surface. The author develops an iterative numerical technique that permits solution, in great generality, of the nonlinear boundary value problem associated with two–dimensional magnetostatic equilibrium. It is based on the formalism of Low, in which the hydrostatic atmosphere along each field line is arbitrarily specified in lieu of self–consistent treatment of the energy balance. Smooth, continuous, cylindrically axisymmetric magnetic configurations are considered, with no thin flux tube assumption being made. The technique is applied to generate a model sunspot. Its structure and photometric signature are compared with previous models. The magnetic configuration of vertically oriented flux tubes in the neighborhood of the solar surface is also examined.

075.012 **On photospheric magnetic fields in flare regions.**
K. V. Alikaeva, P. N. Polupan, S. I. Gandzha, N. N. Kondrashova, T. I. Redyuk.
Vestn. Kiev. Univ. Astron., No. 27, p. 3 – 7 (1985). In Russian.

075.013 **Comparison of the results of solar magnetic field measurements obtained with different methods.**
V. G. Lozitskij, N. I. Lozitskaya, V. I. Vlasenko.
Vestn. Kiev. Univ. Astron., No. 27, p. 59 – 66 (1985). In Russian.

075.014 **On periodic components of the photospheric and interplanetary magnetic fields.**
L. A. Plyusnina.
Issled. Geomagn., Aehron. Fiz. Solntsa, Moskva, No. 73, p. 80 – 87 (1985). In Russian. Abstr. in Ref. Zh., 51. Astron., 3.51.422 (1986).

075.015 **Comparison of magnetographic and polarimetric measurements of the magnetic field in a polar prominence.**
V. S. Bashkirtsev, G. P. Mashnich.
Issled. Geomagn., Aehron. Fiz. Solntsa, Moskva, No. 73, p. 22 – 28 (1985). In Russian. Abstr. in Ref. Zh., 51. Astron., 3.51.436 (1986).

075.016 **Quasi–periodic variations of the magnetic field of active regions on the sun and their connection with flare activity.**
N. V. Klochek.
Issled. Geomagn., Aehron. Fiz. Solntsa, Moskva, No. 73, p. 37 – 45 (1985). In Russian. Abstr. in Ref. Zh., 51. Astron., 3.51.441 (1986).

075.017 **On the magnetic field of solar granulation.**
G. F. Vyal'shin.
Astron. Tsirk., No. 1405, p. 7 – 8 (1985). In Russian.

075.018 **A determination of the structure of the magnetic field of active regions using orientation of small fibrils.**
G. N. Kulikova, M. M. Molodenskij, B. P. Filippov.
Soln. Dannye, Byull. 1986, No. 1, p. 57 – 63 (1986). In Russian.
The fibril structure in active regions gives information on the direction of the tangential magnetic field. This information is sufficient for solving a two–dimensional problem of field structure determination, the field being assumed as potential. If information on the spectrum of spatial frequencies of the magnetic field in active regions be added to the problem, a three–dimensional reverse problem can be solved.

075.019 **Longitudinal distribution of neutral magnetic lines. II.**
V. I. Makarov, K. S. Tavastsherna.
Soln. Dannye, Byull. 1986, No. 1, p. 63 – 73 (1986). In Russian.
On the basis of the longitudinal distribution of neutral magnetic lines of the large–scale magnetic solar field for the period 1945 – 1974 three types of their activity are detected: isolated islands of a unipolar field, a family of neutral magnetic lines, complexes of neutral magnetic lines. No significant deviation from the random distribution of neutral lines has been found.

075.020 **Simulations of magnetic–flux transport in solar active regions.**
C. R. DeVore, N. R. Sheeley Jr., J. P. Boris, T. R. Young Jr., K. L. Harvey.
Sol. Phys., Vol. 102, Nos. 1/2, p. 41 – 49 (1985).
The authors simulate the evolution of several observed solar active regions by solving a transport equation for magnetic flux at the photosphere. The rates of rotation, meridional flow, and diffusion of the flux are determined self–consistently in the calculations. The findings are in good quantitative agreement with previous measures of the rotation rate and diffusion constant associated with photospheric magnetic fields.

075.021 **Magnetic reconnection in a high–temperature plasma of solar flares. II. Effects caused by transverse and longitudinal magnetic fields.**
B. V. Somov, V. S. Titov.
Sol. Phys., Vol. 102, Nos. 1/2, p. 79 – 96 (1985).
Simple self–consistent models for 'non–neutral' current sheets are considered. Characteristics of high–temperature turbulent current sheets (HTCS) with a small transverse component of magnetic field are determined for conditions in the solar corona. The energy output of such an HTCS is much larger than that of a neutral sheet. This makes it possible to consider the HTCS as an energy source not only in long–lived X–ray loops but also in flaring loops during the 'not' or 'main' phase of a flare. In this case, the magnetic reconnection velocity agrees with the observed velocity of the loop rise. Thus, these phenomena can be interpreted as a result of magnetic reconnection, for example, between new flux emerging from under the photosphere and an 'old' magnetic field. The role of a longitudinal magnetic field in a current sheet is less important for HTCS.

075.022 **Solar wind, non–force–free magnetic field and two–ribbon flares.**
Q.–r. Su.
Sol. Phys., Vol. 102, Nos. 1/2, p. 159 – 164 (1985).
With the assumption that the magnetic field lines are radial at some quite high level in the solar corona, a non–constant shearing magnetic field is introduced into the magnetohydrostatic equations. It is found that a same critical amount of shearing a magnetic island is formed and then breaks out to form an open magnetic configuration in which resistive tearing–mode instability may occur, and may initiate a two–ribbon flare. In addition, high shearing magnetic fields are investigated. It is shown that high shearing magnetic configurations are weak two–dimensional neutral sheets.

075.023 **Periodic variations in the solar magnetic field.**
V. A. Kotov, L. S. Levitskii (*L. S. Levitskij*).
Bull. Crimean Astrophys. Obs., Vol. 69, p. 84 – 92 (1986). English translation of 39.075.014.

075.024 **Measurement of magnetic fields in flocculi from the polarization of solar radio emission.**
A. S. Grebinskii (*A. S. Grebinskij*).
Sov. Astron., Vol. 29, No. 4, p. 449 – 454 (1985). English translation of 40.075.016.

075.025 **The cancellation of magnetic flux. I. On the quiet sun.**
S. H. B. Livi, J. Wang, S. F. Martin.
Aust. J. Phys., Vol. 38, No. 6, p. 855 – 873 (1985). Paper presented at the R. G. Giovanelli Commemorative Colloquium, Part II, Tucson, Arizona, 17 – 18 January 1985.
The primary emphasis in the present paper is on presenting new quantitative measures of the line–of–sight component of the magnetic flux for examples of disappearing magnetic features recorded both photographically and digitally on 9 July 1984. The examples include the partial disappearance of the new magnetic flux from ephemeral regions. As a background the authors first present recent photographic images illustrating the appearance, disappearance, and interactions of magnetic features on the quiet sun.

075.026 **Recent studies of magnetic canopies.**
H. P. Jones.
Aust. J. Phys., Vol. 38, No. 6, p. 919 – 928 (1985). Paper presented at the R. G. Giovanelli Commemorative Colloquium, Part II, Tucson, Arizona, 17 – 18 January 1985.
Two current studies are described which stem from Giovanelli's seminal studies of the spreading of chromospheric fields near active regions and active–region network. First, improved observational techniques are described for obtaining magnetograms in the Ca II 8542 Å, Fe I 8688 Å, and C I 9111 Å lines which at least in principle allow for more accurate treatment of instrumental noise and allow better inference of field orientation. Second, a generalized response function is developed for calculating theoretical magnetograph signals from arbitrary line–of–sight varia-

tions of magnetic field, and initial applications to two–dimensional, potential–field models of network fields are described.

075.027 **The cancellation of magnetic flux. II. In a decaying active region.**
S. F. Martin, S. H. B. Livi, J. Wang.
Aust. J. Phys., Vol. 38, No. 6, p. 929 – 959 (1985). Paper presented at the R. G. Giovanelli Commemorative Colloquium, Part II, Tucson, Arizona, 17 – 18 January 1985.
An active region was studied in detail during its period of decay from 3 to 8 August 1984 using Hα filtergrams and videomagnetograms acquired at the Big Bear Solar Observatory. The decay was initiated by a process of fragmentation in which very small knots of magnetic flux separated from larger concentrations of flux. The disappearance of magnetic flux was always observed when the small fragments of flux encountered other small fragments or concentrations of flux of opposite polarity. This type of disappearance of magnetic flux, called 'cancellation', is shared by both polarities of magnetic field. Cancellation was the only observed means of major loss of flux in the photospheric magnetic fields of the active region. Approaching fragments of opposite polarity flux always collided and, after apparent collision, permanent loss of magnetic flux was subsequently and invariably observed. Thus, cancellation is a highly predictable phenomenon.

075.028 **Evolution of weak solar magnetic fields.**
H. Zirin.
Aust. J. Phys., Vol. 38, No. 6, p. 961 – 969 (1985). Paper presented at the R. G. Giovanelli Commemorative Colloquium, Part II, Tucson, Arizona, 17 – 18 January 1985.
The author reviews studies of the evolution of weak solar magnetic fields with the Big Bear videomagnetograph. To all detectable limits, the field is clumped in small elements. The intranetwork elements show a shorter lifetime and much more rapid motion than the network elements. In some cases they stream into existing network elements and may merge to form new elements, but many show no preferential motion to the network edges.

075.029 **Numerical simulations of large–scale solar magnetic fields.**
C. R. DeVore, N. R. Sheeley Jr., J. P. Boris, T. R. Young Jr., K. L. Harvey.
Aust. J. Phys., Vol. 38, No. 6, p. 999 – 1007 (1985). Paper presented at the R. G. Giovanelli Commemorative Colloquium, Part II, Tucson, Arizona, 17 – 18 January 1985. With a correction in Vol. 39, No. 1, p. 115 (1986).
The authors have solved numerically a transport equation which describes the evolution of the large–scale magnetic field of the Sun. Data derived from solar magnetic observations are used to initialize the computations and to account for the emergence of new magnetic flux during the sunspot cycle. The authors' objective is to assess the ability of the model to reproduce the observed evolution of the field patterns. They discuss recent results from simulations of individual active regions over a few solar rotations and of the magnetic field of the Sun over sunspot cycle 21.

075.030 **The decay of the mean solar magnetic field.**
N. R. Sheeley Jr., C. R. DeVore.
Sol. Phys., Vol. 103, No. 2, p. 203 – 224 (1986).
The authors have analyzed the effects that differential rotation and a hypothetical meridional flow would have on the evolution of the Sun's mean line–of–sight magnetic field as seen from Earth. By winding the large–scale field into strips of alternating positive and negative polarity, differential rotation causes the mean–field amplitude to decay and the mean–field rotation period to acquire the value corresponding to the latitude of the surviving unwound magnetic flux.

075.031 **The boundary value problem of the solar force–free magnetic field with constant α and its analytical solution.**
Z.–c. Chen, J.–x. Wang.
Sol. Phys., Vol. 103, No. 2, p. 317 – 328 (1986).
The authors present a physical model which uses boundary conditions which seem to correspond more appropriately to actual situations. A boundary value problem of solar force–free magnetic field with constant α has been specified to represent the discretely concentrated characteristics of the longitudinal magnetic field on the photosphere. The solution enables to make an analytical description of the configuration of the magnetic field in the chromosphere and corona.

075.032 **Magnetic shear. III. Hale region 17255.**
R. G. Athay, H. P. Jones, H. Zirin.
Astrophys. J., Vol. 303, No. 2, p. 877 – 883 (1986). With plates 7 – 9.
Hale active region 17255, which in many respects was the most vigorous active region observed during the first operational period of SMM, appeared to lie between two large areas of flow (observed in C IV) converging toward the major axis of the region. In the 6 day period from 1980 November 6 – 12, the major axis of the region rotated by ~25°. Evidence is presented which suggests that magnetic shear was widespread in this region. It is concluded that magnetic shear is often associated with flaring activity but is not a sufficient condition for flaring to occur.

075.033 **Magnetic shear. IV. Hale regions 16740, 16815, and 16850.**
R. G. Athay, J. A. Klimchuk, H. P. Jones, H. Zirin.
Astrophys. J., Vol. 303, No. 2, p. 884 – 891 (1986). With plates 10 – 12.
Dopplergrams made in C IV (λ1548) are studied for evidence of velocity shear near Hα dark filaments and for large–scale flow convergent on active regions. The three regions studied support earlier conclusions that shear is a common property of active regions and that active regions may be the foci of converging plasma flow. Flow patterns near filaments show divergence or convergence as well as shear. Also the sense of the shear can be either cyclonic or anticyclonic. No preference is noted for convergence or divergence or for a particular sense of shear.

075.034 **Effect of a longitudinal magnetic field in current sheets on the sun.**
B. V. Somov, V. S. Titov.
Sov. Astron., Vol. 29, No. 5, p. 559 – 563 (1985). English translation of 40.075.021.

075.035 **Magnetic neutral lines of the large–scale magnetic field and solar activity.**
V. I. Makarov, K. S. Tavastsherna, K. R. Sivaraman.
Astron. Zh., Tom 63, Vyp. 3, p. 534 – 541 (1986). In Russian. English translation in Sov. Astron., Vol. 30, No. 3.
On the basis of Hα synoptic charts for the period 1955 – 1981 two components in the distribution of the scales of structures of the large–scale magnetic field are found. It is shown that the general magnetic field of the sun has a latitudinal zonal structure which is formed from the elements of the 1st and 2nd distributions. The boundaries of latitude zones of the magnetic field either oscillate near average latitudes or have only poleward migration. Three types of quasi–periodic oscillations are noted, their periods being different in the S and N hemispheres. The most probable period of oscillations is found in the southern hemisphere and it equals to about 20 solar rotations.

075.036 **Geschwindigkeits–Oszillationen in solaren Magnetfeld–Gebieten.**
E. Wiehr.
Sterne Weltraum, 25. Jahrg., Nr. 6, p. 316 – 320 (1986).

075.037 **Non linear time–evolution of kink–unstable magnetic flux tubes in the convection zone of the Sun.**
F. Moreno–Insertis.
MPA Rep., No. 226, 48 pp. (1986). Submitted to Astron. Astrophys.

075.038 **Measurements of solar magnetic fields.**
J. C. Bhattacharyya, K. S. Balasubramaniam.
Indian Inst. Astrophys., Prepr., No. 142, 12 pp. (1985). Paper presented at the national workshop on solar physics, Kodaikanal, 4 – 6 September 1985.

075.039 **Search for mechanisms of solar magnetism.**
M. H. Gokhale.
Indian Inst. Astrophys., Prepr., No. 144, 23 pp. (1985). Paper presented at the national workshop on solar physics, Kodaikanal, 4 – 6 September 1985.
Study of "solar magnetism" (totality of solar magnetic phenomena on different scales) and its "mechanisms" (physical processes responsible for the phenomena) is obviously essential for the advancement of solar and solar–planetary physics. Apart from that, it has presently acquired immense importance in stellar physics as a part of the study of non–radiative modes of energy storage and transport.

075.040 **Methods of measurement of the magnetic field vectors of sunspots.**
S.–h. Ye, J.–h. Jin.
Publ. Purple Mt. Obs., Vol. 4, No. 3, p. 1 – 7 (1985). In Chinese.
This paper presents a brief review of the methods of measurement of the vector magnetic fields of sunspots with the profiles of Stokes parameters and also a comparison of the measurements of the strengths of magnetic fields of unipolar spots of May 15 and September 1, 1979. Although one and the same observational material of the Okayama Astrophysical Observatory, the two methods give rise to results which significantly differ with each other.

075.041 **Emerging flux regions with rapid evolution. I. The "magnetic flux–lifetime" diagram.**
A. A. Golovko.
Soln. Dannye, Byull., 1986, No. 4, p. 48 – 53 (1986). In Russian.
The values of the maximal magnetic flux Φ_m and lifetime τ of separate emerging flux regions were estimated using photospheric magnetic field observational data for 1973 – 1982. Analysis of these data using the (Φ_m, τ) diagram leads to the conclusion on the existence of a special kind of the emerging flux regions with rapid evolution. They emerge in complex compact flare–active regions with large magnetic field gradients.

075.042 **Latitude structure of the large–scale solar magnetic field in separate longitudinal intervals. I Solar cycle No. 20.**
V. I. Makarov, M. V. Kushnir.
Soln. Dannye, Byull., 1986, No. 4, p. 63 – 68 (1986). In Russian.
An analysis of the distribution of background magnetic fields in the 20th solar cycle using H_α charts is made. The peculiarities of the time–latitude distribution of the field at all the latitudes are determined predominantly by the poleward migration of magnetic fields of active regions. On the latitude versus time diagram only few cases are observed of equatorial fluxes of the field at some longitudes. They are due to the latitude–longitude structure of separate unipolar regions on H_α charts.

075.043 **Magnetic field structures of solar active regions obtained by polarization mapping observation at 32 GHz.**
H. Kumagai.
J. Radio Res. Lab., Vol. 32, No. 137, p. 167 – 174 (1985). Abstr. in Phys. Abstr., Vol. 89, No. 1279, Entry 53842 (1986).

075.044 **On the formulae for calculating solar force–free magnetic fields with constant α and their applications.**
Z.–c. Chen, R.–y. Zhao.
Kexue Tongbao, Vol. 30, No. 12, p. 1601 – 1606 (1985).
Using the formulae for force–free magnetic fields and taking the observed longitudinal field on the photosphere and the force–free factor $\alpha = 0.3$ as the input data, the authors have calculated the magnetic field configuration in the solar atmosphere for the active region occurring on May 16, 1981.

075.045 **A probable initial configuration of a flaring loop.**
V. Krishan.
Plasma Phys. Controlled Fusion, Vol. 28, No. 2, p. 509 – 510 (1986). Abstr. in Phys. Abstr., Vol. 89, No. 1282, Entry 73187 (1986).

075.046 **Nonlinear response of slender magnetic flux tubes to external pressure fluctuations.**
P. Venkatakrishnan.
Sol. Phys., Vol. 104, No. 2, p. 347 – 357 (1986).
The effect of applying external pressure fluctuations on slender flux tubes is studied as a nonlinear initial value problem. Large amplitude velocity oscillations are seen to be produced when the frequency of the imposed fluctuations matches the natural frequency of the tube. Radiative cooling does not significantly damp these resonantly built–up oscillations. The absence of observational evidence for such a resonant response of the tubes is used to put a constraint on the length of tubes.

075.047 **Discussion of a possible magnetic transient in UVSP observations.**
W. Henze Jr.
Sol. Phys., Vol. 104, No. 2, p. 359 – 362 (1986).
As part of a search for possible magnetic transients in the transition region, polarization observations were made in the C IV line at 1548 Å during a flare on 1980, July 13. In contradiction to earlier informal reports, it is not possible to state that magnetic transients of several thousand gauss have really been detected.

075.048 **The origin of the 28– to 29–day recurrent patterns of the solar magnetic field.**
N. R. Sheeley Jr., C. R. DeVore.
Sol. Phys., Vol. 104, No. 2, p. 425 – 429 (1986).
Numerical simulations of the Sun's mean line–of–sight magnetic field suggest an origin for the 28– to 29–day recurrent patterns of the field and its associated interplanetary phenomena. The patterns are caused by longitudinal fluctuations in the eruption of new magnetic flux, the transport of this flux to mid latitudes by supergranular diffusion and meridional flow, and the slow rotation of the resulting flux distributions at the 28– to 29–day periods characteristic of those latitudes.

075.049 **On the magnetic structure of the quiet transition region.**
J. F. Dowdy Jr., D. Rabin, R. L. Moore.
Sol. Phys., Vol. 105, No. 1, p. 35 – 45 (1986).
The authors show that the observed fine–scale structure of the photospheric magnetic network suggests a two–component picture in which magnetic funnels that open into the corona emerge from only a fraction of the network. The gas that makes up the hotter transition region is mostly contained within these funnels. The remainder of the network is occupied by a population of low–lying loops with lengths $\lesssim 10^4$km. The authors propose that the cooler transition region is mainly located within such loops, which are magnetically insulated from the corona and must, therefore, be heated internally. The fine–scale structure of ultraviolet spectroheliograms is consistent with this proposal, and theoretical models of internally heated loops can explain the behavior of the emission measure below $T \approx 10^5$K.

075.050 **An atlas of photospheric magnetic field observations and computed coronal magnetic fields: 1976 – 1985.**
J. T. Hoeksema, P. H. Scherrer.
Sol. Phys., Vol. 105, No. 1, p. 205 – 211 (1986).
Daily magnetogram observations of the large–scale photospheric magnetic field have been made at the John M. Wilcox Solar Observatory at Stanford since May of 1976. This paper announces the UAG Report No. 94, an Atlas containing the complete set of synoptic charts of the measured photospheric magnetic field, the computed field at the source surface, and the coefficients of the multipole expansion of the coronal field. The general underlying structures of the solar and heliospheric fields, which determine the environment for solar–terrestrial relations and provide the context within which solar activity related events occur, can be approximated from these data.

075.051 **The magnetic field of the sun.**
J. O. Stenflo.
Mitt. Astron. Ges., No. 65, p. 25 – 40 (1986). – See Abstr. 012.072.

To understand the nature of the sun's magnetic field one needs to explore both its small–scale and large–scale (global) properties. On a small scale, the field is extremely intermittent, the basic units (the fluxtubes) being much smaller than the resolving power of present telescopes. Using a Fourier transform spectrometer as a polarimeter the properties of these basic units are being systematically explored. The global field pattern can be decomposed in its spherical harmonics coefficients. Power spectrum analysis of these coefficients has revealed a resonant structure in the global evolution of the fields. The resonances are governed by a parity selection rule.

075.052 **Magnetic reconnection on the Sun.**
E. R. Priest.
Mitt. Astron. Ges., No. 65, p. 41 – 51 (1986). – See Abstr. 012.072.

A brief review is given of the process of magnetic reconnection on the Sun, where it can convert magnetic energy into heat and other forms and where it is probably responsible for solar flares and coronal heating. Recent numerical experiments are linking the classical theory of resistive instability and fast steady–state reconnection and are revealing a regime of impulsive bursty reconnection. The current discoveries about plasma behaviour on the Sun are potentially of great value for the rest of astrophysics.

075.053 **Das mittlere koronale Magnetfeld.**
M. Pätzold, M. K. Bird, H. Volland.
Mitt. Astron. Ges., No. 65, p. 156 – 161 (1986). – See Abstr. 012.072.

Principles of a solar magnetic field telescope.
See Abstr. 032.044.

The eigenvalue approach in modelling solar magnetic structures.
See Abstr. 062.062.

Leaky and non–leaky oscillations in magnetic flux tubes.
See Abstr. 062.065.

Time dependent multimode hexagonal magnetoconvection.
See Abstr. 062.077.

Solitary waves in a magnetic flux tube.
See Abstr. 062.088.

Magnetic buoyancy instabilities incorporating rotation.
See Abstr. 062.098.

Topological invariants of field lines rooted to planes.
See Abstr. 062.102.

The Benjamin–Ono–Burgers equation: an application in solar physics.
See Abstr. 062.112.

Magnetic structure of a flux rope.
See Abstr. 062.121.

Numerical algorithm for solution of the radiative transfer equations taking into account the influence of a magnetic field.
See Abstr. 063.074.

Association of small, bright photospheric features with magnetic fields.
See Abstr. 071.019.

Electrical conductivity in the solar photosphere and chromosphere.
See Abstr. 071.034.

A comparison of photospheric electric current and ultraviolet and X–ray emission in a solar active region.
See Abstr. 072.001.

Structure of a solar active region from RATAN 600 and Very Large Array observations.
See Abstr. 072.010.

The dynamics of magnetically trapped fluids. I. Implications for umbral dots and penumbral grains.
See Abstr. 072.037.

Investigation of magnetic field strengths of sunspots from spectral lines with different excitation.
See Abstr. 072.046.

The sunspot cycle and solar magnetic fields. I. The mechanism as inferred from observation.
See Abstr. 072.053.

The sunspot cycle and solar magnetic fields. II. The interaction of flux tubes with the convection zone.
See Abstr. 072.054.

Effect of local rigid rotation on sunspots.
See Abstr. 072.077.

Correlation of the velocity and magnetic field in a solar active region.
See Abstr. 072.078.

On magnetic field stochasticity and nonthermal line broadening in solar flares.
See Abstr. 073.027.

Three–dimensional reconnection after a prominence eruption.
See Abstr. 073.045.

Chromospheric lines in a transversal magnetic field.
See Abstr. 073.085.

Transient magnetic disturbances from solar flare explosions and associated pulsations of type IV radio bursts.
See Abstr. 073.095.

Formation of the May 1981 solar flare complex and the global magnetic field of the sun (H.R. 17644).
See Abstr. 073.121.

Kinematics of the May 1981 flare complex's magnetic field decrease (H.R. 17644).
See Abstr. 073.122.

Measurement of magnetic fields in solar prominences.
See Abstr. 073.123.

Chromospheric "active region loops".
See Abstr. 073.146.

Dynamical processes in the solar corona and interplanetary space.
See Abstr. 074.002.

Estimation of coronal magnetic fields using Type–I emission.
See Abstr. 074.011.

Coronal magnetic fields – a mini survey.
See Abstr. 074.032.

Geometry of solar plasma flow and the magnetic field in the polar corona.
See Abstr. 074.056.

The relationship between coronal bright points as seen in He I λ10830 and the evolution of the photospheric network magnetic fields.
See Abstr. 074.078.

The determination of coronal potential magnetic fields using line-of-sight boundary conditions.
See Abstr. 074.080.

Electric fields and plasma structure in coronal magnetic loops.
See Abstr. 074.086.

Lifetime, age, and rotation of coronal magnetic fields.
See Abstr. 074.110.

A metric type III burst asymmetry relative to simple bipolar active regions.
See Abstr. 077.067.

What causes the solar cycle?
See Abstr. 080.002.

Global resonances in the evolution of solar magnetic fields.
See Abstr. 080.003.

The solar dynamo: observations and theories of solar convection, global circulation, and magnetic fields.
See Abstr. 080.013.

Magnetic energy storage and conversion in the solar atmosphere.
See Abstr. 080.018.

Solar irradiance variations derived from magnetograms.
See Abstr. 080.044.

Magnetic torques and differential rotation.
See Abstr. 080.073.

Internal rotation of the Sun as a consequence of instabilities in the radiative interior.
See Abstr. 080.074.

Is the existence of a strong magnetic field in the Sun's core plausible?
See Abstr. 080.075.

The influence of magnetic fields on convective motions in the outer layers of the sun.
See Abstr. 080.087.

On the differential rotation of solar background magnetic fields. 1. On the possible existence of torsional waves with the wave number k = 1/hemisphere.
See Abstr. 080.097.

On the differential rotation of solar background magnetic fields. 2. Are torsional waves with the wave number k = 1/hemisphere real?
See Abstr. 080.098.

Magnetic reconnection.
See Abstr. 080.118.

Magnetic reconnection in the Earth's magnetotail.
See Abstr. 084.031.

The relationship of the large-scale solar field to the interplanetary magnetic field: what will Ulysses find?
See Abstr. 106.044.

The large-scale structure of the heliospheric magnetic field.
See Abstr. 106.045.

The relationship between the interplanetary magnetic field inhomogeneous structure and the distribution of large-scale magnetic fields in the photosphere (1969 – 1975).
See Abstr. 106.056.

Magnetodynamic phenomena in the solar and stellar outer atmospheres.
See Abstr. 116.003.

Solar and stellar magnetic activity.
See Abstr. 116.004.

076 UV, X, Gamma Radiation

076.001 **Continuum absorption in the solar EUV spectra.**
T. Nishikawa.
Astrophys. Space Sci., Vol. 119, No. 1, p. 17 – 19 (1986). – See Abstr. 012.001.

The author calculated the weakening of 11 ions by solving the rate equations (Dufton, 1977) in order to examine the wavelength-dependence of the weakening, because some authors did not find it (e.g., Doschek and Feldman, 1982).

076.002 **La phase initiale de l'éruption solaire.**
C. de Jager.
Astronomie, Vol. 100, p. 167 – 173 (1986).

076.003 **A search for 160-min period in the sequences of X-ray bursts and chromospheric flares on the sun.**
V. A. Kotov, V. V. Kasinskij, L. S. Levitskij, V. P. Fomin.
Izv. Krymskoj Astrofiz. Obs., Tom 73, p. 43 – 52 (1985). In Russian. English translation in Bull. Crimean Astrophys. Obs., Vol. 73.

A statistical analysis of 533 hard X-ray bursts of the sun detected in 1981 by the Japanese satellite Hinotori in the energy range 17 – 40 keV shows the presence of 160-min modulation with 98% confidence. The result is reinforced by analysis of the SMM satellite data, where analogous periodicity is found at about 93% confidence level.

076.004 **Stereoscopic measurements of hard solar X-rays, and related topics.**
K. Hurley.
The sun and the heliosphere in three dimensions, p. 73 – 86 (1986). – See Abstr. 012.016.

Solar X-ray observations in the >20 keV range are discussed, with emphasis on their implications for electron acceleration. It is shown how measurements of fine time structure, directivity, and the production height of X-radiation may be interpreted in the contexts of thermal and non-thermal electron populations. Recent measurements are reviewed, and the expected contribution of Ulysses is discussed.

076.005 **Temporal evolution of an energetic electron population in an inhomogeneous medium. Application to X-ray bursts.**
N. Vilmer, A. L. MacKinnon, G. Trottet.
The sun and the heliosphere in three dimensions, p. 87 – 92 (1986). – See Abstr. 012.016.

The interpretation of spatially resolved hard X–ray data implies the use of models where the spectral, spatial and temporal evolution of the X–ray flux can be studied. As a first step to construct detailed hard X–ray source models, the authors present time–dependent solutions of the electron continuity equation in an inhomogeneous medium and they illustrate the evolution of the electron population and of the X–ray flux for simple situations. Applications of the results to stereoscopic X–ray observations is briefly discussed.

076.006 **Influence of solar flares on the X–ray corona.**
D. M. Rust, D. A. Batchelor.
The sun and the heliosphere in three dimensions, p. 93 – 98 (1986). – See Abstr. 012.016.

Sequences of X–ray images of solar flares, obtained with the Hard X–ray Imaging Spectrometer (HXIS) on the SMM spacecraft, reveal many dynamical phenomena. Movies of 20 flares recorded with 6–sec time resolution were examined. The authors present a preliminary analysis of the events as a group, and discuss some new aspects of the well–studied 1980 May 21 flare and a 1980 November 6 flare.

076.007 **Associations of compact C IV events, He I 10830 Å dark points, and magnetic structures.**
J. G. Porter, E. J. Reichmann, R. L. Moore, K. L. Harvey.
Bull. Am. Astron. Soc., Vol. 17, No. 4, p. 842 (1985). Abstract. – See Abstr. 010.062.

076.008 **Solar–flare gamma–ray line spectroscopy.**
R. J. Murphy, R. Ramaty, D. J. Forrest, B. Kozlovsky.
Bull. Am. Astron. Soc., Vol. 17, No. 4, p. 904 – 905 (1985). Abstract. – See Abstr. 010.062.

076.009 **Internal gravitational waves, fluctuations of the ultraviolet solar radiation and of the ionospheric parameters.**
R. S. Steblova, I. A. Frejzon, V. V. Kol'tsov, E. I. Chertok, Yu. V. Gorokhov, V. A. Vazherkin, I. N. Panova, M. G. Kogan.
Dokl. Akad. Nauk SSSR. Ser. Mat. Fiz., Tom 283, No. 2, p. 353 – 357 (1985). In Russian. Abstr. in Ref. Zh., 51. Astron., 3.51.486 (1986).

076.010 **Solar EUV variations obtained from experimental data of Prognoz 7 and model calculations.**
T. V. Kazachevskaya, G. S. Ivanov–Kholodnyj, E. A. Bruevich, D. A. Gonyukh, A. A. Nusinov.
Soln. Dannye, Byull. 1986, No. 2, p. 89 – 94 (1986). In Russian.

Experimental data on extreme ultraviolet radiation ($\lambda \leqslant 130$ nm) are compared with the results of two–component model calculations. The Prognoz–7 experimental data on the daily integral EUV flux were obtained in 1978 – 1979. The two–component model makes it possible to compute the solar EUV flux at the 10 – 105 nm band for every particular day using the daily and monthly mean values of solar radio flux at 10.7 cm. The comparison of observational data with model calculations shows an accordance of day–to–day EUV flux variations. The EUV flux increase during chromospheric flares is estimated.

076.011 **The blanketing in the rocket ultraviolet from 225.2 to 319.6 nm.**
S. N. Osipov.
Sol. Phys., Vol. 102, Nos. 1/2, p. 29 – 32 (1985).

The blanketing effect η_λ for the solar disc center and its value near the limb ($\cos\theta = 0.23$) are obtained for the rocket ultraviolet region using the Harvard center and limb spectra. The value of η_λ for the disc center appears to be 0.67 (on average) relative to the quasi–continuum. The essential decrease of η_λ towards the limb is also revealed.

076.012 **The excitation of the iron Kα feature in solar flares.**
A. G. Emslie, K. J. H. Phillips, B. R. Dennis.
Sol. Phys., Vol. 103, No. 1, p. 89 – 102 (1986).

The authors evaluate the relationship between the hard X–ray photon spectrum and the flux of iron Kα emission in a thick–target electron bombardment model. Results are presented for various power–law hard X–ray spectra. These results are applied to two events observed with the Hard X–Ray Burst Spectrometer and the Kα channel of the X–Ray Polychromator Bent Crystal Spectrometer on the Solar Maximum Mission satellite.

076.013 **Interplanetary energetic particles, coronal flares, and hard X–ray microflares.**
R. P. Lin.
NASA Conf. Publ., NASA CP–2421, p. 98 – 119 (1986). – See Abstr. 012.021.

The author reviews solar electron phenomena which can produce low levels of hard X–ray emission at the Sun. The Pinhole/Occulter Facility X–ray instrumentation will provide several orders of magnitude improvement in sénsitivity as well as vastly improved spatial resolution over present–day instrumentation. The leap in sensitivity makes it possible for P/OF to study coronal phenomena in hard X–rays and thus to provide the connection between phenomena observed in the interplanetary medium and at the Sun, and also to study the characteristics of hard X–ray microflares and extend their distribution to even lower flux levels. In addition it may well be possible to observe directly the region of acceleration of the electrons in flares, which presumably occurs in the low density corona. The author discusses the various interplanetary and coronal phenomena of interest, as well as microflares and flare–related phenomena.

076.014 **Polarization and directivity of hard X–ray bremsstrahlung in solar flares.**
S. R. Kel'ner, Yu. I. Skrynnikov.
Sov. Astron., Vol. 29, No. 4, p. 445 – 449 (1985). English translation of 40.076.007.

076.015 **Temporal variations of solar EUV, UV, and 10,830–Å radiations.**
R. F. Donnelly, H. E. Hinteregger, D. F. Heath.
J. Geophys. Res., Vol. 91, No. A5, p. 5567 – 5578 (1986).

The similarities and differences of the temporal variations of the various solar EUV fluxes are examined for their three intrinsic time scales: (1) the short–term variations (days, weeks) caused by solar rotation and the evolution (birth, growth, peak, and decay) of individual active regions, (2) the intermediate–term variations (months) associated with episodes of major groups of active regions, and (3) the long–term solar cycle variations (the rise of solar cycle 21).

076.016 **Solar flare X–ray spectra of Fe XXVI and Fe XXV from the Hinotori satellite.**
K. Tanaka.
Publ. Astron. Soc. Jpn., Vol. 38, No. 2, p. 225 – 249 (1986).

High–resolution X–ray spectra of 13 large flares observed by the Bragg spectrometers aboard the Hinotori satellite are discussed. Plasma properties are derived from theoretical fits of the two spectral features in the regions 1.77 – 1.80 Å (Fe XXVI spectrum) and 1.85 – 1.87 Å (Fe XXV spectrum).

076.017 **EUV observations of solar flares based on satellite data.**
T. V. Kazachevskaya, E. A. Bruevich, G. S. Ivanov–Kholodnyj.
Soln. Dannye, Byull., 1986, No. 3, p. 68 – 72 (1986). In Russian.

The results of EUV ($\lambda \leqslant 1300$ Å) measurements carried out on board the Prognoz–7 satellite using the thermo–luminescent method are presented. In the period 10 – 13 December 1978 enhanced flare–generated EUV radiation was observed. The enhancement mostly related to H_α flares of importance 1 and 2. 10 cases of intensity increase were analysed.

076.018 **HRTS Spacelab 2 observations of the temporal development of the C IV transition zone.**
K. P. Dere, J.–D. F. Bartoe, G. E. Brueckner, J. W. Cook, P. Lund, D. G. Socker.
Bull. Am. Astron. Soc., Vol. 18, No. 1, p. 633 (1986). Abstract. – See Abstr. 010.066.

076.019 **Solar absolute spectral irradiance 118 – 300 nm: July 25, 1983.**
G. H. Mount, G. J. Rottman.
J. Geophys. Res., Vol. 90, No. D7, p. 13031 – 13036 (1985).
The full–disk solar absolute spectral irradiance in the spectral range 118 – 300 nm was obtained from a rocket observation above White Sands Missile Range, New Mexico, on July 25, 1983, halfway in time between solar maximum and solar minimum. Comparison with measurements made in May 1982 shows no changes within the absolute errors of the experiment. Comparison with measurements made during solar maximum in 1979 and 1980 indicates a large decrease in the absolute solar irradiance at wavelengths below 190 nm. The absolute calibration of the instruments for this flight was accomplished at the National Bureau of Standards Synchrotron Radiation Facility.

076.020 **Absolute solar flux measurement shortward of 575 Å.**
H. S. Ogawa, D. L. Judge.
J. Geophys. Res., Vol. 91, No. A6, p. 7089 – 7092 (1986).
A rare–gas ionization chamber was used to obtain an accurate measurement of the absolute solar EUV flux in the 50– to 575–Å region. The instrument, operating in total and near–total absorption, was flown on a solar–pointing sounding rocket on August 16, 1983. For the day of the flight the solar activity indices were $F_{10.7} = 132.1$ and $R_I = 80$, and the integrated solar irradiance at the earth, corrected for atmospheric absorption, was found to be $4.31 \pm 0.31 \times 10^{10}$ photons $cm^{-2}s^{-1}$. Almost exactly a year earlier (August 10, 1982) the same instrument measured an integrated solar flux of $5.71 \pm 0.42 \times 10^{10}$ photons $cm^{-2}s^{-1}$ during a time of enhanced solar activity ($F_{10.7} = 209.5$ and $R_I = 155$).

Measurement of wavelengths and Lamb shifts for inner–shell transitions in Fe XVIII – XXIV.
See Abstr. 022.102.

A comparison of photospheric electric current and ultraviolet and X–ray emission in a solar active region.
See Abstr. 072.001.

Soft X–ray observations of high–velocity features in the 29 June 1980 flares.
See Abstr. 073.003.

High–energy observations of solar flares.
See Abstr. 073.004.

High resolution observation of Hα solar flares and temporal relation between Hα and X–ray, microwave emission.
See Abstr. 073.005.

High energy observations of June 1980 solar flares.
See Abstr. 073.007.

Nuclear processes in solar flares.
See Abstr. 073.010.

Rapid acceleration of energetic particles in the 1982 February 8 solar flare.
See Abstr. 073.012.

Broadening of soft X–ray lines during the impulsive phase of solar flares: random or directed mass motions?
See Abstr. 073.013.

Two classes of solar energetic particle events associated with impulsive and long–duration soft X–ray flares.
See Abstr. 073.019.

The duration of flares and their radio emission.
See Abstr. 073.023.

Directionality of bremsstrahlung from relativistic electrons in solar flares.
See Abstr. 073.026.

Correlation of hard X–ray and soft X–ray observations in solar flares.
See Abstr. 073.052.

The heating of the thermal plasma with energetic electrons in small solar flares.
See Abstr. 073.056.

Heat transport in steep temperature gradients. I. Small flaring solar loops.
See Abstr. 073.057.

Further analysis of temperature and emission measure during the decay phase of solar flares derived from soft X–ray light curves.
See Abstr. 073.071.

Investigation of the Mg XII 8.42 Å doublet in solar flare spectra.
See Abstr. 073.074.

Results from the Solar Maximum Mission.
See Abstr. 073.077.

Theoretical motivation for high spatial resolution, hard X–ray observations during solar flares.
See Abstr. 073.078.

Why P/OF should look for evidences of over–dense structures in solar flare hard X–ray sources.
See Abstr. 073.079.

The heating of the thermal plasma with energetic electrons in small solar flares.
See Abstr. 073.080.

Hard X–ray imaging and the relative contribution of thermal and nonthermal emission in flares.
See Abstr. 073.081.

Het begin van een zonnevlam.
See Abstr. 073.088.

Properties of solar flare plasmas derived from soft X–ray line emission.
See Abstr. 073.090.

Relation between γ–ray emission, radio bursts, and proton fluxes from solar flares.
See Abstr. 073.092.

High–resolution observations of X–ray transitions in Fe XXV – XXIII.
See Abstr. 073.103.

Observations of solar flare photon spectra from 20 keV to 7 MeV.
See Abstr. 073.104.

Directionality of solar flare hard X rays: Venera 13 observations.
See Abstr. 073.108.

Solar flares and particle acceleration.
See Abstr. 073.112.

Rapid fluctuations in solar flares observed by the Soft X–ray Polychromator.
See Abstr. 073.117.

Solar gradual hard X–ray bursts and associated phenomena.
See Abstr. 073.132.

The solar flare extreme ultraviolet to hard X–ray ratio.
See Abstr. 073.133.

Relative timing of hard X–rays and radio emissions during the different phases of solar flares: consequences for the electron acceleration.
See Abstr. 073.136.

Can observed hard X–ray and microwave flux from solar flares be explained by a single electron population?
See Abstr. 073.141.

Microwave diagnostics of energetic electrons in flares.
See Abstr. 073.142.

Energy of microwave–emitting electrons and hard X–ray/microwave source model in solar flares.
See Abstr. 073.147.

A new microwave/X–ray diagnostic for the thermal phase of solar flares.
See Abstr. 073.148.

The structure of the coronal solar active region. Part I: X–ray observations of the solar active regions.
See Abstr. 074.019.

The stationary post–flare arch of May 21/22, 1980.
See Abstr. 074.059.

The cyclical variation of energy flux and photospheric magnetic field strength from coronal holes.
See Abstr. 074.061.

X–ray imaging of flare loops and coronal arches.
See Abstr. 074.065.

Images of post–flare coronal structures in X–rays.
See Abstr. 074.108.

Coronal interconnection of two active regions observed in 3.5 – 8.0 keV X–rays.
See Abstr. 074.109.

Dynamics of electron beams in a coronal loop and the hard X–ray burst.
See Abstr. 074.111.

A synchrotron/inverse Compton interpretation of a solar burst producing fast pulses at $\lambda < 3$ mm and hard X rays.
See Abstr. 077.004.

Comparable energy release rates at the origin of small and large solar bursts.
See Abstr. 077.018.

A study of hard X–ray associated meter–decameter bursts observed on December 19, 1979.
See Abstr. 077.022.

Short–lived solar burst spectral component at f $\gtrsim$ 100 GHz.
See Abstr. 077.026.

Relation between solar narrow–band decimetre–wave bursts and associated X–ray bursts.
See Abstr. 077.033.

Microwaves and hard X–rays ultrafast time structures at the impulsive phase of solar bursts.
See Abstr. 077.043.

A synchrotron/inverse Compton interpretation of a solar burst producing fast pulses at 3 mm and hard X–rays.
See Abstr. 077.045.

Interpretation of temporal features in an unusual X–ray and microwave burst.
See Abstr. 077.061.

Comments on fast pulses in mm–wave bursts.
See Abstr. 077.064.

Deduction of solar neutron fluences from large gamma–ray flares.
See Abstr. 078.017.

Energy loss of electrons in solar atmosphere taking many–body interactions into account.
See Abstr. 080.096.

Some aspects of the physics of stellar X–ray emission.
See Abstr. 116.072.

077 Radio, Infrared Radiation

077.001 **Height measurements of *S*–components.**
K. Shibasaki.
Astrophys. Space Sci., Vol. 119, No. 1, p. 21 – 25 (1986). – See Abstr. 012.001.

The direct measurement of the height of the radio source above a solar active region was done by the Westerbork Synthesis Radio Telescope when the source crossed the west limb. The height of the brightest part was 12000 km above the limb. The result of the disk observation is also presented and the emission mechanisms of the observed sources are discussed.

077.002 **Variations in quiet Sun radiation at centimetre wavelengths during solar maximum period.**
B. Lokanadham, P. K. Subramanian.
Astrophys. Space Sci., Vol. 119, No. 1, p. 27 – 32 (1986). – See Abstr. 012.001.

The quiet Sun radiations at cm wavelengths are found to vary considerably during solar maximum period and are mainly influenced by the various time–varying active centres present at the upper chromosphere–corona transition region, revealing the inhomogeneous structure of the transition and upper chromospheric levels of the solar atmosphere.

077.003 **Solar radio emission.**
M. V. Goldman, D. F. Smith.
Physics of the Sun, Vol. II, p. 325 – 376 (1986). – See Abstr. 003.001.

Contents: Observational results (Type III bursts: ground–based observations (above ~8 MHz), spacecraft observations (below 1 MHz), Langmuir waves and electron streams; microwave bursts; Type II bursts; moving Type IV bursts; Type I noise storms). Theory of Type III radio bursts (quasilinear theory, induced scatter off ions, wave–wave effects of the nonlinear refractive and self–focusing variety, second harmonic emission from Langmuir waves, fundamental emission from Langmuir waves, density irregularities and ion–accoustic waves). Radio emission from shock waves and current sheets (shock configuration, generation of electron streams by shock waves, radiation mechanisms). Radio emission from moving plasmoids and other traps (plasmoid and other trapping configurations, sources of electrons in plasmoids and traps, loss cone and collisional electron losses, radiation mechanisms). Conclusions and recommendations.

077.004 **A synchrotron/inverse Compton interpretation of a solar burst producing fast pulses at $\lambda < 3$ mm and hard X rays.**
P. Kaufmann, E. Correia, J. E. R. Costa, A. M. Zodi Vaz.
Astron. Astrophys., Vol. 157, No. 1, p. 11 – 18 (1986).
The recently discovered new class of solar burst emission component exhibiting very fast pulses (durations $\approx$ 60 ms) at mm–waves only ($\lambda \lesssim 3$ mm) and at hard X rays, poses serious constraints for interpretations. In the present interpretation the authors associate the mm–wave emission with a synchrotron component and hard X rays with the inverse Compton process.

077.005 **Far–IR solar emission lines from high *n* states of hydrogen.**
R. T. Boreiko, T. A. Clark.
Astron. Astrophys., Vol. 157, No. 2, p. 353 – 356 (1986).
Emission lines arising from high *n* transitions (16–15 and 14–13) in neutral hydrogen have been detected in high resolution solar spectra obtained with a Michelson interferometer on a balloon–borne telescope. Absolute intensities are close to the predicted values of Hoang–Binh when pressure–broadening is included, with observed line intensities of up to 10% above the solar continuum. Equivalent emission from heavier elements, such as Mg I and Si I, which is dominant in the near infrared at lower n values, may be present but is significantly weaker than that from H I transitions.

077.006 **The frequency–time drift of pulsations.**
M. J. Aschwanden, A. O. Benz.
Astron. Astrophys., Vol. 158, No. 1/2, p. 102 – 112 (1986).
A set of 1270 digitally recorded type III bursts and pulsations in the frequency range of 100 – 1000 MHz was analyzed by a structure recognizing program. Numerical fitting methods determined the drift rate with much higher accuracy than previous analysis of analog data. The analysis of the drift of solar radio bursts in frequency vs. time provides information not only about the dynamical parameters of the source but also reveals atmospherical properties. Type III events and pulsations are clearly distinguished in their drift distribution. Pulsations statistically exhibit 3 times higher drift rates than type III events, thus requiring considerably shorter local scale heights. The source direction of the type III bursts is preferentially upwards oriented, the pulsations show a downwards oriented anisotropy. Evidence is given that the pulsations have a different origin than the common type III burst, concerning the coronal environment as well as the emission mechanism. Outlines to construct models of decimetric pulsations are discussed.

077.007 **A test for large–angle radio scattering in the solar corona.**
D. G. Wentzel, P. Zlobec, M. Messerotti.
Astron. Astrophys., Vol. 159, No. 1/2, p. 40 – 48 (1986).
A significant fraction of all solar type–I radio bursts are only partially or negligibly polarized. The polarization during such bursts is shown to remain so constant that the delay during coronal propagation of one circularly polarized mode relative to the other is at most 4 milliseconds. The authors deduce (1) that the radiation is emitted fully polarized, because partially polarized emission anywhere near the plasma level would yield vastly longer delays, and (2) that the depolarization occurs by large–angle scattering at least as high in the corona as the harmonic level. The authors discuss one model of highly directional scattering in which the radiation is "reflected" by the magnetic field, much like from a half–silvered mirror. Bursts within a group arise from a single source.

077.008 **Polar–cap and coronal–hole–associated brightenings of the Sun at millimeter wavelengths.**
T. Kosugi, M. Ishiguro, K. Shibasaki.
Publ. Astron. Soc. Jpn., Vol. 38, No. 1, p. 1 – 11 (1986).
Observations of the Sun at 36 GHz (8.3 mm) and 98 GHz (3.1 mm) were made on July 16 to 22, 1984. Maps taken at 36 GHz provided evidence for 3 – 7% brightness enhancement near the poles. Also a ~5% brightness enhancement was observed at 36 GHz in at least one equatorial coronal hole region. On the other hand, no such brightenings were detected at 98 GHz. It is suggested that the two types of brightenings at 36 GHz are two aspects of one phenomenon, associated with polar and equatorial coronal holes, respectively. Also, it is suggested that the temperature and density structure of the upper chromosphere in coronal holes differs from that outside holes.

077.009 **Do local sources of a slowly varying component of decameter–band solar radio emission exist?**
L. L. Bazelyan.
Pis'ma Astron. Zh., Tom 12, No. 4, p. 270 – 274 (1986). In Russian. English translation in Sov. Astron. Lett., Vol. 12.
Solar 25 MHz observations are presented that contradict the common opinion about the existence of local sources responsible for a slowly varying component in the decametre radio emission of the sun.

077.010 **On the origin of solar *s*–type radio burst trains.**
V. V. Zajtsev, E. Ya. Zlotnik.
Pis'ma Astron. Zh., Tom 12, No. 4, p. 311 – 317 (1986). In Russian. English translation in Sov. Astron. Lett., Vol. 12.
An interpretation of the millisecond structure of meter wavelength solar radio bursts is suggested to be similar to the interpretation of the Jovian *s*–bursts. The source is supposed to be a weakly anisotropic plasma with quasi–homogeneous magnetic field where upper–hybrid plasma waves are excited by hot electrons and converted into ordinary and extraordinary waves with refractive index $n_{0,e} \ll 1$ due to scattering by thermal ions. The negative frequency drift is caused by the group delay of electromagnetic waves from the point $n_{0,e} = 0$, till they escape from the source. Quasi–periodic trains of *s*–bursts arise from the pulsed conversion of plasma waves into ordinary and extraordinary waves.

077.011 **On a possible interpretation of the origin of the variable component of solar noise storms.**
Yu. F. Yurovskij.
Izv. Krymskoj Astrofiz. Obs., Tom 72, p. 208 – 210 (1985). In Russian. English translation in Bull. Crimean Astrophys. Obs., Vol. 72.
The author shows that the observed linear dependence between the «event amplitude» and the solar storm background component can be naturally explained, if the variable component is considered as a result of random background component modulation. In this case one need not to account on a possibility of radio bursts overlapping during the emission.

077.012 **3–minute oscillations of the polarized radio emission of local sources on the sun.**
V. I. Abramenko, L. I. Tsvetkov.
Izv. Krymskoj Astrofiz. Obs., Tom 73, p. 53 – 62 (1985). In Russian. English translation in Bull. Crimean Astrophys. Obs., Vol. 73.
3–minute oscillations of the circularly polarized radio emission of the local source associated with the sunspot group SD N 401 are considered. The observations were made at the wavelength 2.25 cm. The flux density, degrees of polarization and the angular size of the source have been estimated. Dynamics of power spectra of the local source oscillation during the time of passage of the sunspot group across the solar disk has been analyzed. It is shown that the oscillation processes in the microwave emission are developed and faded at the time of about 0.5 – 1.0 day. The appearance of the second modulation of oscillations with the period about 2 – 6 min is interpreted by the passage of an MHD wave packet across the coronal condensation.

077.013 **Pulsation of the radio emission of the sun observed as a star at 10 cm wavelength with period of 160 minutes.**
N. S. Nesterov, Yu. F. Yurovskij.
Izv. Krymskoj Astrofiz. Obs., Tom 73, p. 75 – 82 (1985). In Russian. English translation in Bull. Crimean Astrophys. Obs., Vol. 73.

The oscillation of solar flux density at 10 cm wavelength with period close to the photosphere oscillation period of 160.01 min is detected. Its relative amplitude is about 1.5×10^{-3}, the flux maximum is delayed by 26 min concerning the maximum brightness of the sun at 13.5 mm wavelength. The amplitude and phase relations between 10 cm and 13.5 mm brightness pulsations evidence that they are caused by density disturbances stimulated by photosphere oscillation and moving through the solar atmosphere with sound velocity.

077.014 **Solar radio signature suggestive of proton beams.**
A. O. Benz, G. M. Simnett.
Nature, Vol. 320, No. 6062, p. 508 – 509 (1986).

Electron beams propagating in the solar corona excite the ambient plasma to emit radio waves at the local plasma frequency and/or its harmonic. This general interpretation of type III radio bursts is well confirmed by *in situ* measurements. The rate at which peak flux drifts in frequency and time is a measure of the velocity of the exciter. The authors report a class of radio bursts with a significantly lower drift than normal. Possible exciters are slowly moving beams of protons.

077.015 **Application of distribution of the number of events to diagnostics of great chromospheric bursts.**
M. Paupere.
Issled. Solntsa Krasnykh Zvezd, No. 22, p. 26 – 32 (1985). In Russian.

The relation between solar activity and 755 MHz solar radio–emission event number distribution is found. For the quiet sun the event number of radio–emission is close to the normal random process distribution. Changes in the event number distribution take place 1 – 1.5 h before great chromospheric bursts. The radio–emission event number distribution coincides with normal random process event distribution before chromospheric bursts.

077.016 **Radio observation of the dynamics of high energy electrons in coronal magnetic loops.**
M. Poquérusse.
Bull. Am. Astron. Soc., Vol. 17, No. 4, p. 933 (1985). Abstract. – See Abstr. 010.065.

077.017 **Investigation of long–wave cosmic radio radiation aboard Prognoz 10–Intercosmos.**
V. P. Grigor'eva, V. N. Kuril'chik, S. Pinter, A. Tirpak, O. B. Likin, L. Fischer, A. Yaroshevich, A. B. Delone, E. A. Makarova.
Kosm. Issled., Tom 24, Vyp. 2, p. 228 – 234 (1986). In Russian. English translation in Cosm. Res.

077.018 **Comparable energy release rates at the origin of small and large solar bursts.**
P. Kaufmann.
Sol. Phys., Vol. 102, Nos. 1/2, p. 97 – 98 (1985).

The discovery that the simplest impulsive solar bursts at microwaves and hard X–rays have time scales in the range of 10 – 100 ms suggest that the energy release rates in such small events are comparable to rates attributed to large events. This condition may be reconciled to the concept that observed burst fluxes in solar events are the convolution of many fast primary energetic injections, at various repetition rates.

077.019 **On the origin of type III fundamental.**
A. B. Eremin, V. V. Zaitsev (*V. V. Zajtsev*).
Sol. Phys., Vol. 102, Nos. 1/2, p. 99 – 110 (1985).

A two–component scheme for the generation of type III fundamental radiation is proposed. The first component of the fundamental arises at a plasma level $\omega_L \cong \omega_t$ because of the Rayleigh scattering of the plasma waves into electromagnetic radiation. The other component arises at $\omega'_L \cong \omega_t/2$ because of the decay of the first component into plasma waves and the subsequent rescattering of the plasma waves into electromagnetic radiation $\omega_t \cong 2(\omega_t/2)$. By its properties (location, directivity, polarization) the second component is essentially the same as the second harmonic radiation produced by a stream of fast electrons at $\omega'_L \cong (\omega_t/2)$. This scheme is used to solve the main problems (localization and directivity of the source, polarization of type III fundamental) of the harmonic theory of type III solar bursts.

077.020 **Positional drift within groups of Type I solar radio bursts.**
J. Van Nieuwkoop.
Sol. Phys., Vol. 103, No. 1, p. 129 – 140 (1986).

In the dynamic spectrograms of the 60–channel solar radiospectrograph at Dwingeloo coherent groups of Type I bursts have been selected. At 243 MHz the positions of the bursts were measured. Within many groups the bursts showed a systematic angular displacement with time. Movements with apparent velocities up to 15000 km s^{-1} have been observed. The burst positions have been obtained with a two–element interferometer. A special data reduction method used the amplitude variation of the Type I bursts as a parameter for selection against the background radiation.

077.021 **A theory for the solar type–I radio continuum.**
D. G. Wentzel.
Sol. Phys., Vol. 103, No. 1, p. 141 – 151 (1986).

The type–I radio continuum may arise from the combination of two electrostatic waves, both directed nearly normal to the magnetic field. One wave, near the upper–hybrid frequency, is generated by gyroresonance with superthermal electrons and comes into equilibrium with these electrons. The other wave, at the lower–hybrid frequency, is generated by the loss–cone instability of trapped superthermal protons in those wave directions for which the lower–hybrid frequency is an exact multiple of the proton gyrofrequency. The brightness temperature of the continuum indicates both the energy of the superthermal electrons and the existence of at least a small number of superthermal protons.

077.022 **A study of hard X–ray associated meter–decameter bursts observed on December 19, 1979.**
M. R. Kundu, T. E. Gergely, S. R. Kane, H. S. Sawant.
Sol. Phys., Vol. 103, No. 1, p. 153 – 164 (1986).

The authors present the results of a study of the relationship of a complex meter–decameter wavelength radio burst observed with the Clark Lake E–W and N–S interferometers, with a hard X–ray burst observed with the X–ray spectrometer aboard ISEE–3. The radio burst consisted of several type III's, reverse drift type III's, a U burst, and type II and type IV bursts. The X–ray emission was also complex. The radio as well as hard X–ray emissions were observed before the flash phase of the flare. A detailed correspondence between individual features of the radio and hard X–ray burst emissions could not be found. The type IV burst started after all hard X–ray emissions ceased.

077.023 **Radio imaging of flares.**
G. A. Dulk.
NASA Conf. Publ., NASA CP–2421, p. 178 – 190 (1986). – See Abstr. 012.021.

High resolution (arcseconds to tens of arcseconds) mapping of solar microwave bursts has recently been achieved. The author reviews the properties of the radio images that have been reported to now and, so far as possible, the relation of the radio sources to X–ray observations of the same flares.

077.024 **High–spectral resolution solar microwave observations.**
G. J. Hurford.
NASA Conf. Publ., NASA CP–2421, p. 191 – 200 (1986). – See Abstr. 012.021.

The application of high–spectral resolution microwave observations to the study of solar activity is discussed with particular emphasis on the frequency dependence of microwave emission from solar active regions. A "shell model" of gyroresonance emission from active regions is described which suggests that high–spectral resolution, spatially–resolved observations can provide quantitative information about the magnetic field distribution at the base of the corona. Corresponding observations of a single sunspot with the Owens Valley frequency–agile interferometer at 56 frequencies between 1.2 and 14 GHz are presented. The merits and weaknesses of microwave spectroscopy as a technique for measuring magnetic fields in the solar corona are briefly discussed.

077.025 **Single–dish high sensitivity determination of solar limb emission at 22 and 44 GHz.**
J. E. R. Costa, J. L. Homor, P. Kaufmann.
NASA Conf. Publ., NASA CP–2421, p. 201 – 207 (1986). – See Abstr. 012.021.

A large number of solar maps were obtained with the use of the Itapetinga 14–m (45–ft) antenna at 22 GHz and 44 GHz in the 1978 – 1982 period. A statistical study of these maps, reduced using original techniques, permitted the establishment of the solar radius with great accuracy (<1.5 arc s r.m.s.) at the two frequencies. It has been found that 22 GHz and 44 GHz radiation originates at 16000 km and 12500 km, respectively, above the photosphere. Excess emission due to active regions was clearly identified at lower solar latitudes above and below the equator, extending up to 26000 km and 16500 km above the photosphere at 22 GHz and 44 GHz, respectively.

077.026 **Short–lived solar burst spectral component at $f \gtrsim 100$ GHz.**
P. Kaufmann, E. Correia, J. E. R. Costa, A. M. Zodi Vaz.
NASA Conf. Publ., NASA CP–2421, p. 208 – 214 (1986). – See Abstr. 012.021.

A new kind of burst emission component was discovered, exhibiting fast and distinct pulses (~ 60 ms durations), with spectral peak emission at $f \gtrsim 100$ GHz, and onset time coincident to hard X–rays to within $\lesssim 128$ ms. These features pose serious constraints for the interpretation using current models. One suggestion assumes the $f \gtrsim 100$ GHz pulse emission by synchrotron mechanism of electrons accelerated to ultrarelativistic energies. The hard X–rays originate from inverse–Compton scattering of the electrons on the synchrotron photons. Several crucial observational tests are needed for the understanding of the phenomenon, requiring high sensitivity and high–time resolution (~ 1 ms) simultaneous to high–spatial resolution (<0.1 arc s) at $f \gtrsim 100$ GHz and hard X–rays.

077.027 **Polar–cap and coronal–hole–associated brightenings of the sun at millimeter wavelengths.**
T. Kosugi, M. Ishiguro, K. Shibasaki.
NASA Conf. Publ., NASA CP–2421, p. 215 – 217 (1986). Abstract. – See Abstr. 012.021. For the full paper see Abstr. 077.008.

077.028 **Fluctuations in the degree of circular polarization of the radio emission of an active region on the sun.**
V. I. Abramenko, L. I. Tsvetkov.
Bull. Crimean Astrophys. Obs., Vol. 69, p. 116 – 122 (1986). English translation of 39.077.020.

077.029 **Unstable chromospheric dark features and type–III bursts association.**
F. Chiuderi–Drago, N. Mein, M. Pick.
Sol. Phys., Vol. 103, No. 2, p. 235 – 247 (1986).

The temporal association between the kinematic parameters of chromospheric dark features (DF) and the production of radio type–III bursts is investigated during a period of five months. It appears that 23 out of 24 bursts take place when the DF is totally or predominantly blue–shifted. In 18 out of 21 cases, a maximum of the outward velocity is observed in the optical image closest in time to the radio burst. The following peculiarities are also shown by the analyzed DF: (a) All of them present a lengthened shape, in most cases pointing toward a sunspot: a bright region coinciding with a parasitic polarity is observed in between. (b) Horizontal velocities along the DF major axis are often observed, always in a direction opposite to the sunspots. This association is interpreted by assuming that a shock wave, generated by the moving chromospheric material, is able to accelerate coronal electrons thus producing type–III bursts.

077.030 **Peculiarities of the outburst of kilometer radio radiation during the event on May 13, 1981.**
V. P. Grigor'eva, V. S. Prokudina.
Kosm. Issled., Tom 24, Vyp. 3, p. 411 – 417 (1986). In Russian. English translation in Cosm. Res.

077.031 **Autonomous method for measuring the shift of the solar radio radiation center.**
S. I. Avdyushin, G. M. Gavrilov, A. Kurbanov, L. V. Yasnov.
Izv. Akad. Nauk TadzhSSR. Otd–nie fiz.–mat., khim. i geol. nauk, No. 2, p. 74 – 76 (1985). In Russian. Abstr. in Ref. Zh., 51. Astron., 5.51.433 (1986).

077.032 **Broad–band solar bursts of the "spike" type.**
L. M. Bakunin, G. P. Chernov.
Sov. Astron., Vol. 29, No. 5, p. 564 – 568 (1985). English translation of 40.077.026.

077.033 **Relation between solar narrow–band decimetre–wave bursts and associated X–ray bursts.**
S. Enome, L. E. Orwig.
Nature, Vol. 321, No. 6068, p. 421 – 423 (1986).

The authors compare solar narrow–band decimetre–wave spike bursts with corresponding X–ray events for 25 solar flares. They find that decimetre spike bursts are $10^2 - 10^3$ times stronger than normal impulsive microwave bursts for the same observed amount of hard X–ray emission. This feature, along with the observation of 100% circular polarization, strongly supports a maser origin for the decimetre spike bursts.

077.034 **Microwave emission from the limb of the Sun: the events of September 8, 1980 and May 10, 1981.**
S. Urpo, H. Teräsranta, V. Ruždjak, B. Vršnak, B. Rompolt, G. Kren.
Hvar Obs. Bull., Vol. 9, No. 1, p. 25 – 32 (1985).

Solar microwave maps at 8 mm performed at the Metsähovi Radio Research Station in the period 1978 – 1982 revealed some radiation sources located up to 50000 km outside the optical Sun. The authors describe two characteristic events which were associated with large flares occurring close to the solar limb.

077.035 **Type I solar radio bursts.**
N. Gopalswamy.
Indian Inst. Astrophys., Prepr., No. 146, 23 pp. (1985). Paper presented at the national workshop on solar physics, Kodaikanal, 4 – 6 September 1985.

A brief review of the recent work done on type I radio emission in the Indian Institute of Astrophysics is presented. The most plausible low frequency turbulence needed to generate the type I bursts is shown to be ion–sound turbulence. A method is presented to evaluate the coronal magnetic field using type I radio emission data.

077.036 **Solar radio emission, January – December 1981.**
Q. Bull. Sol. Act., Vol. 23, Part V, p. 1 – 123 (1986).

077.037 **New discoveries with radio telescopes.**
J. Schmidt.
Schweiz. Tech. Z., Vol. 82, No. 19, p. 11 – 13 (1985). In German. Abstr. in Phys. Abstr., Vol. 89, No. 1276, Entry 36101 (1986).

077.038 **Solar Radio Data. 1986 January – June.**
Sol. Radio Data, Parts I – II (1986). In microfiche form.

077.039 **Cyclic variations of the solar radio flux at 10.7 and 3.2 cm and their prognostic implications.**
B. V. Danil'chev, L. P. Morozova, V. N. Obridko.
Soln. Dannye, Byull., 1986, No. 4, p. 69 – 74 (1986). In Russian.

Cyclic variations of the solar radio flux at 3 and 10 cm wavelengths have been investigated. A phenomenon of pre–minimum has been discovered to take place in 1971, when a number of indices indicated the sun–earth conditions similar to those that are usually registered at the minimum of Wolf numbers. Prognostic implications of the deviation from a regression dependence between the 3 and 10 cm fluxes are discussed.

077.040 **Polarization and brightness temperature of "spike" bursts related to metric type IV solar radio events.**
M. Messerotti, M. Nonino, P. Zlobec.
Mem. Soc. Astron. Ital., Vol. 56, No. 4, p. 795 – 799 (1985). – See Abstr. 010.821.
The authors analyze the appearance of narrow bandwidth spikes in 16 solar radio bursts of type IV–dm. It is found that spikes are characterized by extremely high brightness temperatures ($12 < \log T_B < 15$) and nearly constant polarization for the complete group of spikes associated with a single radio burst.

077.041 **Electron – cyclotron maser and solar microwave millisecond spike emission.**
H.–w. Li, C.–s. Li, Q.–j. Fu.
Chin. Astron. Astrophys., Vol. 10, No. 1, p. 22 – 29 (1986). English translation of 40.077.046.

077.042 **The thermal mechanism of the microwave bursts of May 13, 1981.**
J.–x. Yao.
Acta Astron. Sin., Vol. 27, No. 1, p. 38 – 46 (1986). In Chinese.
The parts of impulsive phase in the microwave bursts of May 13, 1981 at the different frequencies are similar to each other. Each time–profile of bursts is characterized by an approximately symmetrical structure around the peak. The durations of impulsive phase at different frequencies are nearly the same (about 25 min.). The total duration of each burst at different waves is also approximately the same. The middle part of the peak flux density spectrum is very flat. This is probably a new kind of bursts – a large gradual burst. The author suggests that the mechanism of the gradual bursts is caused by a thermal process.

077.043 **Microwaves and hard X–rays ultrafast time structures at the impulsive phase of solar bursts.**
P. Kaufmann.
Rev. Mex. Astron. Astrofis., Vol. 12, Spec. Issue, p. 347 – 354 (1986). – See Abstr. 012.066.
Recent results on ultrafast time structures observed at microwaves and hard X–rays in the impulsive phase of solar bursts are briefly reviewed in terms of diagnostics and implication on interpretation of accelerating mechanisms.

077.044 **Solar burst with millimetre–wave emission at high frequency.**
P. Kaufmann, E. Correia, J. E. R. Costa, A. M. Z. Vaz, B. R. Dennis.
Rev. Mex. Astron. Astrofis., Vol. 12, Spec. Issue, p. 370 (1986). Abstract. – See Abstr. 012.066.
For the full paper see Abstr. 39.077.005.

077.045 **A synchrotron/inverse Compton interpretation of a solar burst producing fast pulses at 3 mm and hard X–rays.**
P. Kaufmann, E. Correia, J. E. R. Costa, A. M. Z. Vaz.
Rev. Mex. Astron. Astrofis., Vol. 12, Spec. Issue, p. 374 (1986). Abstract. – See Abstr. 012.066.

077.046 **Observations of radio continua and terminology.**
M. Pick.
Sol. Phys., Vol. 104, No. 1, p. 19 – 32 (1986). – See Abstr. 012.067.
Different components and successive stages of type IV bursts are reviewed. Some simplifications in the schematic representation of well developed type IV bursts are proposed. The existence of two physically distinct acceleration/injection mechanisms is discussed. Suggestions for further work are proposed.

077.047 **The relation between flare–related metric continuum bursts and coronal mass ejections.**
R. D. Robinson.
Sol. Phys., Vol. 104, No. 1, p. 33 – 39 (1986). – See Abstr. 012.067.
The author presents the results of a study to determine the relationship between flare–related metre wavelength continuum outbursts and coronal mass ejection events. Using radio data obtained with the Culgoora radioheliograph and optical data from the SOLWIND experiment aboard the P–78 satellite a good temporal and positional overlap between the two types of events is found. Comparing properties for the events it was found that the continuum radio bursts were preferentially associated with the broader, more rapidly moving CME transients.

077.048 **Discussion on the classification of fine structures in continua.**
Ø. Elgarøy.
Sol. Phys., Vol. 104, No. 1, p. 41 – 42 (1986). – See Abstr. 012.067.

077.049 **On fast pulsations in flare continua.**
Ø. Elgarøy.
Sol. Phys., Vol. 104, No. 1, p. 43 – 49 (1986). – See Abstr. 012.067.
Pulsations with periods down to 0.05 s have been observed in continuum emission following flares. In pulsating events several frequencies may be excited, simultaneously or in time succession. Pulsations occur both in classical type IV events and in short–lasting continuum events. The power spectrum shows no large variation during one or two hours. Some implications of fast pulsations for the generation mechanism are discussed.

077.050 **Exponential decay and exciter profile of fast pulses in type IV events.**
A. Abrami, M. Messerotti, P. Zlobec.
Sol. Phys., Vol. 104, No. 1, p. 51 – 55 (1986). – See Abstr. 012.067.
Some fast bursts occurring at 237 MHz during type IV events in association with sawtooth pulsations have been studied. These bursts, when occurring sufficiently isolated from the adjacent pulsating activity, appear to have an exponential decay phase similar to that of fast type III bursts.

077.051 **The polarization of decimetric pulsations.**
M. J. Aschwanden.
Sol. Phys., Vol. 104, No. 1, p. 57 – 65 (1986). – See Abstr. 012.067.
A sample of 10 decimetric broadband pulsations were observed in 1980 – 1983 and analyzed in polarization. Half of the data set was 85 – 100% circularly polarized, the other half showed a mild polarization of 15 – 55%. The polarization is constant in time and frequency for the strongly polarized group. All the mildly polarized bursts originate from near the limb; the lower degree of circular polarization is likely to be caused by depolarization due to propagation effects. The degree of polarization is constant throughout the event, but varies in frequency for the mild polarized group. Implications for pulsation models are discussed.

077.052 **Interplanetary phenomena and solar radio bursts.**
G. M. Simnett.
Sol. Phys., Vol. 104, No. 1, p. 67 – 91 (1986). – See Abstr. 012.067.
The relationship between solar radio emissions and transient interplanetary phenomena is reviewed. It is believed that the most significant advance in recent years has come from coordinated studies of coronal mass ejections and moving type IV bursts, where the evidence appears to favour the Langmuir wave hypothesis as the emission mechanism. Type II bursts are not generally a signature of the main energetic particle acceleration in flares. They do, however, occasionally propagate to 1 AU, and beyond, where they are normally accompanied by protons in the 20 MeV region. Apart from the impulsive microwave burst, there is no reliable radio signature associated with energetic particle acceleration in flares.

077.053 **Electron–cyclotron maser emission from the Sun and stars: variations with plasma temperature and density.**
R. M. Winglee, G. A. Dulk.
Sol. Phys., Vol. 104, No. 1, p. 93 – 97 (1986). – See Abstr. 012.067.

Very bright and highly circularly polarized radio bursts from the Sun, the planets, flare stars, and close binary stars are attributed to the electron–cyclotron maser instability. The mode and frequency of the dominant radiation from the maser instability is shown to be dependent on the plasma temperature and the ratio ω_p/Ω_e of the plasma frequency to the electron–cyclotron frequency. For the emission from the Sun ω_p/Ω_e is probably greater than 0.3 and for $0.3 < \omega_p/\Omega_e < \sqrt{2}$ the emission can be either in the *x*–mode at the second harmonic or the *o*–and/or *z*–modes at the fundamental. For higher ω_p/Ω_e, the emission moves to higher harmonics of Ω_e with the emission being predominately in the *z*–mode when $\omega_p/\Omega_e \gtrsim \sqrt{3}$.

077.054 **Millisecond radio spikes.**
A. O. Benz.
Sol. Phys., Vol. 104, No. 1, p. 99 – 110 (1986). – See Abstr. 012.067.

Millisecond spikes of the solar radio emission are known for more than two decades. They have recently seen a surge in interest of theoreticians who are fascinated by their high brightness temperature of up to 10^{15}K, their association with hard X–ray bursts, and a possibly very intimate relation to electron acceleration. This review is intended to bridge the gap that presently seems to separate theory and observations. The wide range of spike observations is summarized and brought into the perspective of recent models. It is concluded that spikes yield a considerable potential for the diagnostics of energetic particles, their origin, and history in astrophysical plasmas.

077.055 **The characteristics of type IV–associated spikes at metric wavelengths.**
M. Nonino, A. Abrami, M. Comari, M. Messerotti, P. Zlobec.
Sol. Phys., Vol. 104, No. 1, p. 111 – 116 (1986). – See Abstr. 012.067.

Out of 120 solar type IV events recorded at the Trieste Astronomical Observatory, the authors have selected 15 groups of spike bursts. They analyze their properties, concentrating on the characteristics of their polarization. They find that the polarization of the spikes varies over a wide range, but that within a particular group of spikes it remains almost constant. Sometimes groups of spikes with different degrees of polarization occur almost contemporaneously, probably indicating that more than one source is active at nearly the same time. Occasionally spike bursts accompany type III bursts.

077.056 **The microwave spectrum of solar millisecond spikes.**
M. Stähli, A. Magun.
Sol. Phys., Vol. 104, No. 1, p. 117 – 123 (1986). – See Abstr. 012.067.

The microwave radiation from solar flares sometimes shows short and intensive spikes which are superimposed on the burst continuum. In order to determine the upper frequency limit of their occurrence and the circular polarization, a statistical analysis has been performed on digital microwave observations from 3.2 to 92.5 GHz. Additionally, fine structures have been investigated. About 10% of the bursts show fine structures at 3.2 and 5.2 GHz, whereas none occurred above 8.4 GHz. Most of the observed spikes were very short ($\leqslant$10 ms) and their bandwith varied from below 0.5 MHz to more than 200 MHz. Simultaneous observations at two further frequencies showed no coincident spikes at the second and third harmonic.

077.057 **Occurrence of solar decimetric spike bursts over a period of 4 years during the 21st cycle.**
H. W. Urbarz.
Sol. Phys., Vol. 104, No. 1, p. 125 – 129 (1986). – See Abstr. 012.067.

A total number of 460 DCIM events, single or groups, were observed during the period of October 1980 to December 1984 with the Weissenau spectrograph as dynamic film spectra. Several parameters describing these data were evaluated statistically and the main features were discussed in context with associated burst types.

077.058 **Hollow beam distribution of energetic electrons and higher harmonics of electron cyclotron maser.**
H.–w. Li.
Sol. Phys., Vol. 104, No. 1, p. 131 – 136 (1986). – See Abstr. 012.067.

The variations of the growth rates of ECM at first four harmonics in *X*–, *Z*–, and *O*–modes excited by a hollow beam distribution of weakly relativistic electrons with a parameter $\xi \equiv \omega_p/\Omega_e$ are presented in this paper. It is suggested that the ECM emission at these harmonics in *X*–mode is a possible mechanism to produce solar spike emissions with high brightness temperature at shorter and longer decimetric wavelengths.

077.059 **A model of ultra–fast fine structures of microwave bursts.**
R.–y. Zhao, J.–k. Shi.
Sol. Phys., Vol. 104, No. 1, p. 137 – 144 (1986). – See Abstr. 012.067.

A model is presented using the electron cyclotron maser instability driven by a loss–cone distribution (taking account of the relativistic effects for electron cyclotron frequency and electron velocity as well as Landau damping and cyclotron damping) to excite the UFFS (Ultra–Fast Fine Structures) of microwave bursts and adopting a nonlinear density wave as a trigger mechanism to explain the millisecond time structure of the UFFS.

077.060 **Type IV bursts and coronal mass ejections.**
T. E. Gergely.
Sol. Phys., Vol. 104, No. 1, p. 175 – 178 (1986). – See Abstr. 012.067.

The author discusses the statistical association of moving type IV bursts and coronal mass ejections (CMEs), and finds that about one–third to one–half of the IVs occur in association with CMEs, while only about 5% of the CMEs are accompanied by moving type IVs. The mean speed of the moving IVs is smaller than the mean speed of CMEs, so the author concludes that the type IVs move out with the bulk of the ejecta.

077.061 **Interpretation of temporal features in an unusual X–ray and microwave burst.**
A. L. MacKinnon, J. E. R. Costa, P. Kaufmann, B. R. Dennis.
Sol. Phys., Vol. 104, No. 1, p. 191 – 198 (1986). – See Abstr. 012.067.

Observations are briefly discussed on an event in which microwave and hard X–ray emissions were not correlated in the accepted way. Two impulsive peaks of roughly equal intensity were observed at three different microwave frequencies. The hard X–ray peaks accompanying these, however, differ in intensity by almost two orders of magnitude. Various possible interpretations of this burst are discussed.

077.062 **Gyrosynchrotron emission of solar flares.**
C. A. Alissandrakis.
Sol. Phys., Vol. 104, No. 1, p. 207 – 221 (1986). – See Abstr. 012.067.

The current status of our knowledge on the theory of radio emission from mildly relativistic electrons and its application in the interpretation of solar radio bursts are reviewed. The limitation of the observations and the theory (particularly the effect of mode coupling on the observed polarisation) are pointed out and the potential of the gyrosynchrotron process as a diagnostic of the physical conditions is discussed. This will help us to obtain quantitative information about the changes of the magnetic field and the acceleration of particles in solar flares.

077.063 **Decimeter continuum radio emission from a post–flare loop.**
M. R. Kundu.
Sol. Phys., Vol. 104, No. 1, p. 223 – 226 (1986). – See Abstr. 012.067.

Radio observations offer an important means for providing estimates of magnetic fields in post–flare loops.

077.064 **Comments on fast pulses in mm–wave bursts.**
J. E. R. Costa, P. Kaufmann.
Sol. Phys., Vol. 104, No. 1, p. 253 – 256 (1986). – See Abstr. 012.067.

The first observation of a solar bursts with millisecond time resolution at $\lambda = 3.3$ mm was obtained on 21 May, 1984. The time–scale of the millimetric pulses durations was of about 60 ms. The possible strong correlation between pulsed emission at 3.3 mm and hard X–rays is discussed.

077.065 **Solar microwave bursts recorded at 2.84 GHz with millisecond time resolution.**
S.–z. Jin, R.–y. Zhao, Q.–j. Fu.
Sol. Phys., Vol. 104, No. 2, p. 391 – 413 (1986).

A millisecond recorder for solar observations at 2.84 GHz was put to work in 1981. From April 1981 to September 1982 it recorded 250 "spike pulse events", which are here listed together with information on the associated solar flares, hard X–ray bursts and radio bursts. The spike pulses have shorter duration and higher flux density than were previously known, and the basic units of the spike pulse events are single spike pulses crowded together forming separate clusters. Many spike pulses observed are not yet resolved at 1 ms. Fast spike pulse events are closely correlated with complex magnetic field regions, and are often accompanied by hard X–ray bursts and fast drifting radio bursts. Some theoretical investigations of the spike phenomena have been carried out.

077.066 **A study of RATAN–600 observations of solar *S*–component sources.**
A. Krüger, J. Hildebrandt, V. M. Bogod, A. N. Korzhavin, Sh. B. Akhmedov, G. B. Gelfreikh (*G. B. Gel'frejkh*).
Sol. Phys., Vol. 105, No. 1, p. 111 – 121 (1986).

A sample of 36 *S*–component sources observed by the radio telescope RATAN–600 was compared with calculations of gyromagnetic emission and bremsstrahlung based on recent sunspot models. The diagnostic possibilities of the spectral distributions in the radio flux, the degree of polarization, and the source sizes for the estimation of magnetic scale heights and other source parameters were checked by different methods. Depending on the magnetic field structure, the observations show different types of polarization spectra. The flare production rate of active regions appears to be related to their *S*–component flux and magnetic scale heights.

077.067 **A metric type III burst asymmetry relative to simple bipolar active regions.**
B. V. Jackson.
Sol. Phys., Vol. 105, No. 1, p. 123 – 132 (1986).

Metric type III solar radio burst positions are compared spatially and temporally to underlying active region geometry. The positions of these radio bursts have an asymmetric location distribution relative to simple bipolar regions. The type III bursts show a tendency to occur nearer the leading active region – an association shown before from type III burst and magnetic field polarity measurements. The type III bursts also generally occur to the left of the outward to inward directed magnetic field. The asymmetry relative to the outward directed magnetic field has a sense that is consistent with a mechanism of type III burst production that involves a pre–existing coronal current system situated between expanding closed and open magnetic field lines.

077.068 **Properties of metre–wavelength solar bursts associated with coronal mass ejections.**
R. D. Robinson, R. T. Stewart, N. R. Sheeley Jr., R. A. Howard, J. Koomen, D. J. Michels.
Sol. Phys., Vol. 105, No. 1, p. 149 – 171 (1986).

An investigation is made to determine the relationship between a coronal mass ejection (CME) and the characteristics of associated metre–wave activity. It is found that (1) the CME width and leading edge velocity can be highly influential in determining the intensity, spectral complexity and frequency coverage of both type II and continuum bursts; (2) the presence of a CME is possibly a necessary condition for the production of a metric continuum event and (3) metric continuum bursts as well as intense, complex type II events are preferentially associated with strong, long lasting soft X–ray events.

Culgoora catalogue of solar radio noise storms, 1973 to 1984.
See Abstr. 002.051.

The Siberian Solar Radio–Telescope: parameters and principle of operation, objectives and results of first observations of spatio–temporal properties of development of active regions and flares.
See Abstr. 033.003.

Heterodyne spectroscopy of the sun at 12 microns using a diode laser as the local oscillator.
See Abstr. 034.095.

Experimental observations with the radio telescope RATAN–600 using the method «estafette with zoning».
See Abstr. 036.089.

Azimuthal aperture synthesis of the sun with the radio telescope RATAN–600 in the regime "South with periscope".
See Abstr. 036.090.

The interest of simultaneous spectral and spatial high resolution spectroscopy in the infrared.
See Abstr. 036.148.

Ion–sound turbulence due to shock gradients in collisionless plasmas.
See Abstr. 062.097.

Plasma and radiation processes.
See Abstr. 063.007.

Intensity variations of solar short–wave radiation in the solar activity cycle.
See Abstr. 071.017.

Structure of a solar active region from RATAN 600 and Very Large Array observations.
See Abstr. 072.010.

Prediction of monthly indices of solar activity $F_{10.7}$ on the basis of a multiplicative autoregression model.
See Abstr. 072.019.

Two–frequency imaging of microwave impulsive flares near the solar limb.
See Abstr. 073.002.

High resolution observation of Hα solar flares and temporal relation between Hα and X–ray, microwave emission.
See Abstr. 073.005.

High energy observations of June 1980 solar flares.
See Abstr. 073.007.

Narrowband dm–spikes, intermediate drift bursts and pulsations in the solar flare of August 19, 1981.
See Abstr. 073.018.

Two classes of solar energetic particle events associated with impulsive and long–duration soft X–ray flares.
See Abstr. 073.019.

Possible manifestations of anomalous scattering of the radio emission of solar flares by plasmons.
See Abstr. 073.020.

The duration of flares and their radio emission.
See Abstr. 073.023.

The sudden decrease of the large flare occurrence and its manifestation in interplanetary space after the maximum of the solar cycle.
See Abstr. 073.024.

On a connection of the flare activity of the sun with the characteristics of radio radiation of active regions on it.
See Abstr. 073.042.

Prominence hydrogen lines at 10 – 20 microns.
See Abstr. 073.070.

The solar radio corona: manifestations of energetic electrons.
See Abstr. 073.083.

Relation between γ–ray emission, radio bursts, and proton fluxes from solar flares.
See Abstr. 073.092.

Transient magnetic disturbances from solar flare explosions and associated pulsations of type IV radio bursts.
See Abstr. 073.095.

Solar flares and particle acceleration.
See Abstr. 073.112.

Solar gradual hard X–ray bursts and associated phenomena.
See Abstr. 073.132.

Relative timing of hard X–rays and radio emissions during the different phases of solar flares: consequences for the electron acceleration.
See Abstr. 073.136.

Radio observations of the solar neutron flare of 3 June, 1982.
See Abstr. 073.137.

Behaviour of the polarization at dm – m wavelengths during the evolution of five two–ribbon flares.
See Abstr. 073.138.

Electron acceleration in flares inferred from radio and hard X–ray emissions.
See Abstr. 073.139.

Can observed hard X–ray and microwave flux from solar flares be explained by a single electron population?
See Abstr. 073.141.

Microwave diagnostics of energetic electrons in flares.
See Abstr. 073.142.

Energy of microwave–emitting electrons and hard X–ray/microwave source model in solar flares.
See Abstr. 073.147.

A new microwave/X–ray diagnostic for the thermal phase of solar flares.
See Abstr. 073.148.

Energy interconnection between the plasma cloud ejected from the active region and the shock wave.
See Abstr. 073.149.

Estimation of coronal magnetic fields using Type–I emission.
See Abstr. 074.011.

On the possibility of a radio echo at plasma frequency in the solar corona.
See Abstr. 074.018.

Coronal millimeter wave sources.
See Abstr. 074.094.

Radio method of mapping large–scale coronal magnetic fields on active stars.
See Abstr. 075.010.

Magnetic field structures of solar active regions obtained by polarization mapping observation at 32 GHz.
See Abstr. 075.043.

Radio radius of the sun at 2.25 cm in the direction of the polar regions at solar activity maximum.
See Abstr. 080.049.

On a method of determination of the solar radio radius by scanning technique.
See Abstr. 080.102.

Correlation between solar radio bursts and sudden ionospheric disturbances.
See Abstr. 085.021.

3–D coronal and heliospheric structure from radio observations.
See Abstr. 106.041.

Latitude distribution of interplanetary magnetic field lines rooted in active regions.
See Abstr. 106.042.

Heliospheric structure and multispacecraft observations of Type III radio bursts.
See Abstr. 106.043.

Flare stars and solar bursts: high resolution in time and frequency.
See Abstr. 122.182.

078 Cosmic Radiation

078.001 **Solar flare neon composition and solar cosmic–ray exposure ages based on lunar mineral separates.**
C. M. Nautiyal, J. T. Padia, M. N. Rao, T. R. Venkatesan.
Astrophys. J., Vol. 301, No. 1, p. 465 – 470 (1986).

The authors analyze etched pyroxene grain–size separates from lunar soils 14148 and 24087 and etched feldspar grain–size separates from soil 61221 for neon isotopic and elemental composition and discuss procedures to isolate the implanted solar flare (SF) neon and solar and Galactic cosmic–ray (SCR and GCR)–produced Ne components in these samples. These results indicate that the SF neon composition is $^{20}Ne/^{22}Ne = 11.6 \pm 0.2$. The surface exposure ages are determined from the resolved SCR–produced Ne contents.

078.002 **The composition of solar energetic particles.**
R. E. McGuire, T. T. von Rosenvinge, F. B. McDonald.
Astrophys. J., Vol. 301, No. 2, p. 938 – 961 (1986).

The authors present new measurements of elemental abundances in 15 large solar energetic particle events. These data confirm the existence of two major effects previously reported: (1) There are systematic differences between these solar energetic particle abundances and photospheric abundances that are approximately correlated with first ionization potential. This pattern implies that the solar corona is the likely source population for these particles. (2) In some events, an enhancement of heavy ion abundances is seen relative to a "baseline" of solar energetic particle abundances, with a magnitude of enhancement that increases with increasing atomic number. The present data confirm that the above effects are not the result of time or energy variations within individual events.

078.003 **Solar filament eruptions and energetic particle events.**
S. W. Kahler, E. W. Cliver, H. V. Cane, R. E. McGuire, R. G. Stone, N. R. Sheeley Jr.
Astrophys. J., Vol. 302, No. 1, p. 504 – 510 (1986). With plates 15 – 17.

The authors discuss the 1981 December 5 solar filament eruption that they associate with a solar energetic ($E > 50$ MeV) particle (SEP) event observed at 1 AU. The eruption was photographed in Hα, and was observed by the coronagraph on *P78–1*. It occurred well away from any solar active region and was not associated with an impulsive microwave burst, but an associated interplanetary type II burst was detected by *ISEE 3*. The December 5 SEP event had an unusually steep energy spectrum ($\gamma > 3.5$). The implications of this event for SEP production are examined.

078.004 **Increase of low–energy proton fluxes before solar cosmic ray bursts with $E_p > 30$ MeV.**
G. M. Blokh, B. M. Kuzhevskij.
Geomagn. Aehron., Tom 26, No. 1, p. 135 – 137 (1986). In Russian. English translation in Geomagn. Aeron.

078.005 **The intensity distribution of solar proton events taking into account background radiation fluctuations.**
Yu. P. Ochelkov.
Geomagn. Aehron., Tom 26, No. 1, p. 137 – 139 (1986). In Russian. English translation in Geomagn. Aeron.

078.006 **Separation and analysis of temporal and spatial variations in the 10 April 1969 solar flare particle event.**
R. Reinhard, E. C. Roelof, R. E. Gold.
The sun and the heliosphere in three dimensions, p. 297 – 304 (1986). – See Abstr. 012.016.

The authors analysed the 10 April 1969 energetic particle event with five spacecraft separated over 180° in azimuth. The particle fluxes are mapped back to the coronal injection longitudes which are estimated from the instantaneous solar wind velocities and connected at the same time of observation to give the longitudinal particle gradients. Cutting these gradients at representative longitudes gives the temporal variations at fixed coronal longitudes which turn out to be power–law decays. This shows that the exponential rise and decay observed by spacecraft is due to the spatial gradient sampled by the spacecraft as the connection longitude changes due to the Sun's rotation.

078.007 **Three–dimensional gradients of solar particles inside 5 AU.**
E. C. Roelof.
The sun and the heliosphere in three dimensions, p. 331 – 340 (1986). – See Abstr. 012.016.

The spatial gradients (field–aligned, longitudinal and latitudinal) in solar flare energetic particle events are described in terms of their occurrence in the three phases of the event (rise, maximum and decay). The information that can be derived from these phases is discussed in the context of the Ulysses mission.

078.008 **A spatially confined, long–lived stream of solar particles.**
K. A. Anderson, W. M. Dougherty.
The sun and the heliosphere in three dimensions, p. 341 – 348 (1986). – See Abstr. 012.016.

Studies of low energy electrons and ions of solar origin lead the authors to conclude: (1) That on one occasion in 1979 interplanetary field lines connected Earth at 2°S heliographic latitude to a solar active region situated at 16°N latitude. The width of this field line bundle was 2.5×10^6km at 1 AU. A linear extrapolation gives a 12,000 km width at the Sun. (2) That electrons and ions were continuously injected onto this bundle of field lines but not outside it for at least 6 hours and probably more than 20 hours. The authors hypothesize that flare accelerated particles are trapped in a magnetic structure high in the corona and escape over periods of time on the order of one day.

078.009 **Super–events in the inner solar system and their relation to the solar cycle.**
R. Müller–Mellin, K. Röhrs, G. Wibberenz.
The sun and the heliosphere in three dimensions, p. 349 – 354 (1986). – See Abstr. 012.016.

The authors define a super–event in the energetic particle population by the following characteristics: (1) Its onset and decay is much longer than for usual flare–associated solar particle events, the intensity may remain above background for a period of the order of 40 days. (2) The intensity variation with heliolongitude is small. (3) The events are observed in nucleons and electrons simultaneously. As one hypothesis for the origin and storage mechanism for these events the authors suggest shock waves which accelerate particles also at heliolongitudes far away from the flare site, and act as a barrier for galactic cosmic rays as well as for particles generated in the inner solar system.

078.010 **A spatially confined long–lived stream of solar particles.**
K. A. Anderson, W. M. Dougherty.
Sol. Phys., Vol. 103, No. 1, p. 165 – 175 (1986).

The intensities of low–energy solar–interplanetary electrons and ions at 1 AU occasionally change in a "square–wave" fashion. The changes may be increases or decreases and they have duration of a few hours. In one such example following a solar flare, particles flow away from the Sun in a well–defined channel 2.5×10^6km in width for twenty hours or longer. The authors believe that the interplanetary magnetic lines defined by this channel connect to an active region at 16°N solar latitude. At this time the Earth was located at a solar latitude of 2°S. Evidently the particle channel connects to a region of the solar atmosphere which supplies particles over these long times either via storage of the flare accelerated particles or else by continuous acceleration. Arguments are given against the latter possibility. The authors discuss a model for coronal storage which is consistent with the observations.

078.011 **Peculiarities of penetration of solar protons into the earth's magnetosphere during the event of 24 – 25 April 1979.**
G. A. Glukhov, Yu. P. Kratenko, Yu. V. Mineev.
Kosm. Issled., Tom 24, Vyp. 3, p. 468 – 471 (1986). In Russian. English translation in Cosm. Res.

078.012 **Time-dependent coronal shock acceleration of energetic solar flare particles.**
M. A. Lee, J. M. Ryan.
Astrophys. J., Vol. 303, No. 2, p. 829 – 842 (1986).
The time-dependent collisionless acceleration and propagation of energetic particles at an evolving coronal shock is investigated analytically. The shock is taken to be a spherically symmetric infinite-strength, self-similar critical-case blast wave, for which the cold plasma mass density ahead of the shock decreases as r^{-2}. The particles are injected monoenergetically with spherical symmetry. The model includes several essential features of coronal shock acceleration: acceleration via compression at the shock, adiabatic deceleration in the downstream flow, an evolving shock front, and a three-dimensional geometry which allows diffusive escape of particles upstream.

078.013 **The heavy-ion compositional signature in ^{3}He-rich solar particle events.**
G. M. Mason, D. V. Reames, B. Klecker, D. Hovestadt, T. T. von Rosenvinge.
Astrophys. J., Vol. 303, No. 2, p. 849 – 860 (1986).
A survey of the ~ 1 MeV per nucleon heavy-ion abundances in 66 ^{3}He-rich solar particle events has been performed using instruments on the ISEE 3 spacecraft. The observations were carried out in interplanetary space over the period 1978 October – 1982 June. The data confirm earlier observations which show an enrichment of heavy ions in ^{2}He-rich events, relative to the average solar energetic particle composition in large particle events. Surprising new results emerging from the present broad survey are (1) the heavy-ion enrichment pattern is the same within a factor of ~ 2 for almost all cases, and (2) the degree of heavy-ion enrichment is uncorrelated with the ^{3}He enrichment.

078.014 **Estimate of the spectra of a solar cosmic ray source for a number of flares in the 19th and 20th solar activity cycles.**
V. V. Bengin, N. E. Zhukova, I. A. Muratova, V. M. Petrov.
Geomagn. Aehron., Tom 26, No. 3, p. 490 – 493 (1986). In Russian. English translation in Geomagn. Aeron.

078.015 **Solar photospheric and coronal abundances from solar energetic particle measurements.**
H. H. Breneman.
Diss. Abstr. Int., Sect. B, Vol. 46, No. 8, p. 2692 (1986). Thesis, California Institute of Technology, 178 pp. (1985). Order No. DA8522605.

078.016 **Die Ladungszustände solarer energetischer Teilchen.**
A. M. Luhn.
MPE Rep., No. 195, 9+132 pp. (1986).

078.017 **Deduction of solar neutron fluences from large gamma-ray flares.**
M. Yoshimori, H. Watanabe, K. Takahashi.
J. Phys. Soc. Jpn., Vol. 55, No.2, p. 471 – 474 (1986). Abstr. in Phys. Abstr., Vol. 89, No. 1281, Entry 67431 (1986).

078.018 **The application of fuzzy clustering to the short-term forecasts of the solar proton flares.**
Q. Zhang, Q.-f. Hong.
Acta Astron. Sin., Vol. 27, No. 2, p. 164 – 170 (1986). In Chinese.
The short-term forecasts of solar proton flares were explored by the fuzzy clustering method. The short-term forecasts of the proton flares occurred during 21st solar cycle have been examined by this method. The historical forecast examination of the proton flares in January to June 1984 and in January to May 1985 had been made day by day. The results show that this method is feasible.

078.019 **Solar neutrino flux, cosmic rays, and the solar activity cycle.**
P. Raychaudhuri.
Sol. Phys., Vol. 104, No. 2, p. 415 – 424 (1986).
It is suggested that the experimental data on the solar neutrino flux as measured by Davis and his collaborators from 1970 to 1982 vary with the solar activity cycle to a very high level of statistical significance for all the available tests of the hypothesis when the solar neutrino flux data are computed from the weighted moving averages of order 5. The tests have also been applied to the data that have been generated by the Monte Carlo simulation. It is shown that the Monte Carlo simulated data do not indicate a variation within the solar cycle. Thus the moving average data strongly favours the variation within the solar activity cycle.

The acceleration and propagation of solar flare energetic particles.
See Abstr. 073.009.

Two classes of solar energetic particle events associated with impulsive and long-duration soft X-ray flares.
See Abstr. 073.019.

The solar radio corona: manifestations of energetic electrons.
See Abstr. 073.083.

Relation between γ-ray emission, radio bursts, and proton fluxes from solar flares.
See Abstr. 073.092.

Particle acceleration in solar flares.
See Abstr. 073.131.

Ancient sun was highly active.
See Abstr. 105.214.

Shock acceleration of nucleons at $\geqslant 16°$ solar latitude associated with interplanetary corotating interaction regions.
See Abstr. 106.050.

Latitude dependence of co-rotating shock acceleration in the outer heliosphere.
See Abstr. 106.051.

Interplanetary cosmic-ray radial and latitudinal gradients derived in 1984 using *IMP 8*, *Voyager*, and *Pioneer* data.
See Abstr. 144.028.

079 Solar Eclipses

079.001 **Das Perlschnurphänomen und seine Beobachtung.**
P. D. Maley.
Sternenbote, 29. Jahrg., Nr. 3, p. 42 – 55 (1986).

079.002 **The three total solar eclipses of this decade.**
R. Subramanian.
Report on the VIII. International Planetarium Directors Conference, p. 74 – 79 (1985). – See Abstr. 012.014.

079.003 **Eclipse news.**
D. W. Dunham.
Occultation Newsl., Vol. 3, No. 16, p. 345 (1986).

Solar eclipse 1981 July 31

079.101 **Photoelectric observations of the solar eclipse of 1981 July 31.**
V. B. Kapkov, R. I. Gumerov, F. G. Ayupov.
Izv. Astron. Ehngel'gardt. Obs., No. 50, p. 198 – 199 (1985). In Russian.

The block–diagram of the photometer used for the observations of the total solar eclipse of July 31, 1981 is described. The moment of the total phase of the eclipse is $3^h10^m2^s.986 \pm 0.004$ UT.

079.102 **The 1981 total solar eclipse. III. Photometric study of the prominence remnant in the reversing south polar field.**
G. Stellmacher, S. Koutchmy, C. Lebecq.
Inst. Astrophys. Paris, Pré–Publ., No. 127, 19 pp. (1986). To appear in Astron. Astrophys.

Solar eclipse 1986 October 3

079.121 **De merkwaardige zonsverduistering van 3 oktober 1986.**
Zenit, 13. Jaarg., Nr. 6, p. 205 – 209 (1986).

Solar eclipse 1984 May 30

079.141 **Radio observation of the partial solar eclipse of 1984 May 30.**
J. R. Smith.
J. Br. Astron. Assoc., Vol. 96, No. 4, p. 201 – 203 (1986).

The occultation of solar radio noise and bursts at 375 MHz by the Moon was observed using a 9.15 m (30–foot) diameter dish aerial located near Cranleigh, Surrey, during a partial solar eclipse. These observations have been used to relate the positions of the solar radio sources to the visual sunspots.

Atlas of historical eclipse maps. East Asia 1500 BC – AD 1900.
See Abstr. 002.016.

Solar eclipse in "Shijing" and rotation of the Earth.
See Abstr. 004.032.

Local circumstances of solar and lunar eclipses in five millenia: a new Zeiss eclipse projector and new data.
See Abstr. 009.026.

Solar studies in China.
See Abstr. 013.003.

Coronal structure observed at the total solar eclipse of 11 June, 1983 in Indonesia.
See Abstr. 074.003.

The post flare loops observed at the total eclipse of February 16, 1980.
See Abstr. 074.113.

On the influence of the earth atmosphere on conditions for observing a solar eclipse in the radio range.
See Abstr. 082.040.

The highest geoactivity in 1981 and its source on the sun.
See Abstr. 084.011.

Observations of the May 30, 1984, annular solar eclipse at Millstone Hill.
See Abstr. 085.001.

Ionospheric and thermospheric response over Millstone Hill to the May 30, 1984, annular solar eclipse.
See Abstr. 085.002.

080 Atmosphere, Figure, Internal Constitution, Neutrinos, Rotation, etc.

080.001 **Solar ellipticity fluctuations yield no evidence of *g*–modes.**
J. R. Kuhn, K. G. Libbrecht, R. H. Dicke.
Nature, Vol. 319, No. 6049, p. 128 – 131 (1986).
Although there have been several claims for the detection of solar *g*–modes with periods between 2 and 10 hours and although the present sensitivity of the Princeton Solar Distortion Telescope should allow these low frequency modes to be observed, solar oblateness data from the summers of 1983 and 1984 show no evidence of such oscillations with periods between 1 and 5 hours. In about 250 days (nearly 1,000 hours) of observations, the authors find no evidence for significant spectral power associated with *g*–modes. In particular, there is no evidence of a 160.01–min period solar oscillation.

080.002 **What causes the solar cycle?**
D. Gough.
Nature, Vol. 319, No. 6051, p. 263 – 264 (1986).

080.003 **Global resonances in the evolution of solar magnetic fields.**
J. O. Stenflo, M. Vogel.
Nature, Vol. 319, No. 6051, p. 285 – 290 (1986).
Decomposition of the pattern of solar magnetic fields in spherical harmonics for a data set of 25 years and power spectrum analysis of the harmonic coefficients reveals a strikingly resonant modal structure. The resonance frequencies contain information on the structure of the magnetic fields in the Sun's interior.

080.004 **Time–integrated energy budget of a solar activity complex.**
G. A. Chapman, A. D. Herzog, J. K. Lawrence.
Nature, Vol. 319, No. 6055, p. 654 – 655 (1986).
The authors examine the sunspot and facular contributions to luminosity fluctuations due to a solar activity complex over its lifetime from June to November 1982. Both direct, photometric observations of irradiance fluctuations and modelled 'proxy' fluctuations based on published sunspot and calcium plage areas are used. It is found that the total facular energy excess is between 70 and 120% of the sunspot deficit of $\sim 10^{37}$erg. Thus, at a minimum, a major portion of the missing sunspot flux is radiated by faculae, and energy balance or even an excess is possible.

080.005 **Solar five–minute oscillations of low, intermediate, and high degree.**
H. M. Antia, S. M. Chitre, D. Narasimha.
Astrophys. Space Sci., Vol. 118, Nos. 1/2, p. 169 – 172 (1986). – See Abstr. 012.001.
The overstability of acoustic modes trapped inside the Sun is studied with mechanical and thermal effects of turbulence included in an approximate manner through the eddy diffusivities. Many of the acoustic modes are found to be overstable with the most rapidly growing modes occupying a region centred around 3.3 mHz and spread over a wide range of length–scales. The numerical results turn out to be in reasonable accord with the observed power–spectrum of the five–minute oscillations of arbitrary degree. It is demonstrated that these oscillations are most likely to be driven by a simultaneous operation of the $\varkappa$–mechanism and the convective Cowling mechanism, the dominant contribution to the generation of self–excited acoustic waves arising from the turbulent diffusion.

080.006 **Resonant excitation of the solar *g*–modes through coupling of 5–min oscillations.**
H. Ando.
Astrophys. Space Sci., Vol. 118, Nos. 1/2, p. 177 – 181 (1986). – See Abstr. 012.001.
It is proposed that the interacting solar *g*–modes are driven by the beat forcing of *g*–modes by two 5–min oscillations, excited by a $\varkappa$–mechanism. The observational implications of the proposed mechanism are discussed.

080.007 **The Sun's gravitational quadrupole moment inferred from the fine structure of the acoustic and gravity normal mode spectra of the Sun.**
H. A. Hill, G. R. Rabaey, R. D. Rosenwald.
Relativity in celestial mechanics and astrometry, p. 345 – 354 (1986). – See Abstr. 012.002 (IAU Symp. No. 114).
The fine structure of the acoustic and gravity mode multiplets of the Sun have been analyzed to infer the internal rotation of the Sun and upper limits of the internal magnetic field. Observed fine structure for 137 multiplets has been obtained and the fine structure has been examined for dependence on the angular order, m, of the modes. The inferred angular velocity distribution, together with the estimated upper limits on the internal magnetic fields, yields a gravitational quadrupole moment, J_2, of $\approx 7.7 \times 10^{-6}$. This result is consistent with the result obtained by Hill, Bos and Goode (1982) and has important implications for planetary tests of theories of gravitation.

080.008 **The Sun is fading.**
R. A. Kerr.
Science, Vol. 231, No. 4736, p. 339 (1986).
The decrease in solar brightness is small but possibly climatically significant; a turnaround may be in the offing.

080.009 **Perché si studia il Sole? Motivazioni culturali, pratiche e scientifiche che hanno sollecitato e stanno sollecitando uno studio sempre più approfondito della nostra stella.**
G. Godoli.
Orione, Vol. 6, N. 1, p. 16 – 23 (1986).

080.010 **Thermonuclear reactions in the solar interior.**
P. D. MacD. Parker.
Physics of the Sun, Vol. I, p. 15 – 32 (1986). – See Abstr. 003.001.
Contents: Reaction rate formalism. Carbon–nitrogen cycle. Proton–proton chain.

080.011 **Atomic and radiative processes in the solar interior.**
W. F. Huebner.
Physics of the Sun, Vol. I, p. 33 – 75 (1986). – See Abstr. 003.001.
Contents: Calculation of atomic structure (the explicit and the mean ion model). Calculation of opacity (bound–bound processes, bound–free processes, free–free processes, scattering processes, combining processes to obtain the opacity). Results. Uncertainties in the opacity.

080.012 **Hydrodynamic and hydromagnetic phenomena in the deep solar interior.**
W. H. Press.
Physics of the Sun, Vol. I, p. 77 – 95 (1986). – See Abstr. 003.001.
Contents: Troublesome observations (light element depletion, low ^{37}Cl neutrino counts, possibility of rapidly rotating or non-axisymmetric core, 160–minute oscillations). Theoretical topics (nucleo–thermal destabilizations of *g*–modes, depth of the convective zone).

080.013 **The solar dynamo: observations and theories of solar convection, global circulation, and magnetic fields.**
P. A. Gilman.
Physics of the Sun, Vol. I, p. 95 – 160 (1986). – See Abstr. 003.001.
Contents: Convective zone characteristics (basic structure of depth, oblateness and latitudinal temperature differences, luminosity and radius changes). Observations and theory of solar convection zone motions (granulation, supergranulation and mesogranulation, related theories of convection, global circulation). Phenomenology of the Sun's magnetic field and related

features (flux tubes and network; sunspots, active regions, and global patterns; the solar cycle: manifestations in sunspots, magnetic fields, and other properties). Theories of the solar magnetic field (flux tubes and network, active regions and sunspots, solar dynamo theory).

080.014 **Solar internal stresses: rotation and magnetic fields.**
R. K. Ulrich.
Physics of the Sun, Vol. I, p. 161 – 175 (1986). – See Abstr. 003.001.
Contents: Theory (Eddington–Sweet circulation, rotationally driver instabilities, an internal magnetic field). Observations.

080.015 **Solar waves and oscillations.**
T. M. Brown, B. W. Mihalas, E. J. Rhodes Jr.
Physics of the Sun, Vol. I, p. 177 – 247 (1986). – See Abstr. 003.001.
Contents: Theory of waves and oscillations (basic equations: full hydromagnetic equations, linearized equations; waves: nonmagnetic waves, magnetohydrodynamic waves; general properties of solar oscillations: equations and spheroidal mode solutions, Cowling approximation, asymptotic behavior of *p*– and *g*–mode frequencies, radial oscillations, properties of nonadiabatic solutions, toroidal oscillations; excitation and damping of solar pulsations: excitation and damping mechanisms, mode lifetimes, stability of solar pulsation modes; detailed solutions for frequencies and frequency splitting: effects of structure on unperturbed frequencies, effects of rotation, effects of internal magnetic fields; future theoretical needs). Observations (observational techniques: diagnostics of spectrum lines, techniques for observing oscillations and trapped waves; oscillations observed: observations of 5 minute period *p*–mode oscillations, the 160 minute oscillation, torsional oscillations, localized brightness oscillations; wave observations; waves and oscillations in sunspots; future observational needs). Oscillations as probes of the solar interior (direct method: technique and results; inversion methods: technique and applications).

080.016 **Radiation output.**
R. G. Athay.
Physics of the Sun, Vol. II, p. 1 – 50 (1986). – See Abstr. 003.001.
Contents: Basic concepts and definitions (formation of spectra, energy balance, notation, escape probability and scattering depth, escape coefficient, creation and destruction probabilities, thermalization and degradation lengths, intra–atmosphere exchange probability, the source function, the transfer equation). Spectral diagnostics (temperature, density, velocity, magnetic field, and abundance diagnostics). The role of radiation in determining atmospheric properties (photosphere, line blanketing and cooling, the chromosphere, characteristics of chromosphere radiation loss, temperature minimum, the first and second temperature plateau, the transition region, the corona). Nonradial structure (fluid motions and magnetic fields, photosphere and temperature minimum region, chromosphere and transition region, corona). Temporal fluctuations. Challenges for the future (radiation diagnostics, influence of radiation on atmospheric properties, observations).

080.017 **Structure, dynamics, and heating of the solar atmosphere.**
G. W. Pneuman, F. Q. Orrall.
Physics of the Sun, Vol. II, p. 71 – 134 (1986). – See Abstr. 003.001.
Contents: Basic theory and processes. Overall atmospheric structure. The chromosphere–corona transition region and the base of the inner corona (empirical models of the mean temperature structure, inhomogeneous and dynamic structure of the transition shell). Closed coronal regions (basic structure, phenomenology of coronal loops, quasistatic loop models in energy balance, evolution and stability, systematic flow, heating). Open coronal regions and the coronal expansion into interplanetary space (the solar wind, magnetic fields in the solar wind, coronal holes, coronal–interplanetary modeling). Coronal streamers: gas–magnetic field interactions in the solar corona. Coronal activity (the prominence phenomena, solar flares, impulsive flows, coronal transients).

080.018 **Magnetic energy storage and conversion in the solar atmosphere.**
D. S. Spicer, J. T. Mariska, J. P. Boris.
Physics of the Sun, Vol. II, p. 181 – 248 (1986). – See Abstr. 003.001.
Contents: Fundamental concepts (magnetic energy generation, ideal MHD theory, non–MHD properties, the concept of "anomalous" resistivity, global electrodynamic coupling). Magnetically controlled energy conversion (magnetic modifications of plasma transport, transition region structure and flows, channeling and acceleration of plasma, channeling and dissipation of MHD waves, anomalous dissipation of field–aligned currents). Magnetohydrodynamic energy conversion (magnetic flux tube emergence, geometric rearrangements, reconnection and magnetic tearing, particle acceleration at a neutral sheet, flare trigger mechanisms). Outstanding questions.

080.019 **The solar neutrino problem: gadfly for solar evolution theory.**
M. J. Newman.
Physics of the Sun, Vol. III, p. 33 – 45 (1986). – See Abstr. 003.001.
Contents: Standard theory of solar evolution. The missing solar neutrinos. Have we left something out? (Microscopic physics, rotation, magnetic fields, accretion, star formation). The exotic models (mixing, varying *G*, quark catalysis, depleted Maxwell tail, immiscible H–He, the central black hole, nonconventional energy transport).

080.020 **Solar activity and the coupling of *g*–mode oscillations.**
D. G. Wentzel.
Astrophys. J., Vol. 300, No. 2, p. 824 – 829 (1986).
The solar *g*–mode oscillations depend strongly on the difference between actual and adiabatic temperature gradients in the solar interior. Solar activity anchored at the base of the convection zone disturbs this difference and couples modes with similar eigenfrequencies. The author computes this coupling and discusses the possible observational effects at the photosphere. These include a scatter in the frequencies deduced for weakly coupled modes and the apparent temporal behavior of strongly coupled modes.

080.021 **Differential rotation of sunspot groups in the period from 1874 through 1976 and changes of the rotation velocity within the solar cycle.**
H. Balthasar, M. Vázquez, H. Wöhl.
Astron. Astrophys., Vol. 155, No. 1, p. 87 – 98 (1986). = Mitt. Kiepenheuer–Inst., Nr. 263.
The complete sample of the Greenwich Photoheliographic Results (GPR) of the years 1874 – 1976 has been used to determine the differential rotation of the sun. The rotation depends strongly on the type of the sunspots. *C*–groups show high velocities, and *H*– and *J*–spots have the lowest velocities as already found for the period 1940 – 1968. Using the new fit formula the changes of the rotation velocity of sunspots during the solar cycle are proved also for the GPR. The 22–year cycle in the longitudinal distribution of the sunspots is confirmed.

080.022 **The role of inhomogeneous heat transport and anisotropic momentum exchange in the dynamics of stellar convection zones: application to models of the Sun's differential rotation.**
R. M. Pidatella, M. Stix, G. Belvedere, L. Paternò.
Astron. Astrophys., Vol. 156, No. 1/2, p. 22 – 32 (1986). = Mitt. Kiepenheuer–Inst., Nr. 246.
The authors propose mean field models of the Sun's differential rotation which fully include the mixing–length formalism, ionization and inertial terms due to rotation. They assume that differential rotation can be generated by two distinct mechanisms: (1) latitude–dependent heat transport, HT, modelled through the perturbation of the mixing–length; (2) anisotropy of viscosity, AV, modelled through a special form of the viscous

tensor. They test the response of the models to changes of the input parameters, as the depth of the convection zone, the kinematic viscosity, and the boundary conditions imposed at the lower edge of the convection zone. One of the most interesting results is that the AV mechanism is more effective than the HT mechanism, especially for deep convection zones. The influence of rotation on the latitude–dependent energy transport depends strongly on the choice of the lower boundary conditions. There is a general good agreement of the models with the observable quantities.

080.023 **Convective overshoot at the base of the Sun's convection zone.**
R. M. Pidatella, M. Stix.
Astron. Astrophys., Vol. 157, No. 2, p. 338 – 340 (1986). = Mitt. Kiepenheuer–Inst., No. 253.
The non–local form of mixing length theory, as proposed by Shaviv and Salpeter (1973), is applied to the deeper part of the solar convection zone. The result is a layer of convective overshoot (with inward convective heat transport) which extends over a few tenths of a scale height, and has a boundary layer at its base where the temperature gradient suddenly changes from nearly adiabatic to the radiative value. The overshoot layer is part of a (slightly) sub–adiabatic region, of approximately one scale height thickness. The authors argue that this region should be capable to store sufficient magnetic flux, in the form of thin toroidal flux tubes, to account for the flux observed at the solar surface.

080.024 **Is there an unusual solar core?**
K. G. Libbrecht.
Nature, Vol. 319, No. 6056, p. 753 – 755 (1986).
Various attempts have been made in the past two decades to determine the properties of the core of the Sun, by measurements of solar neutrino radiation, the distortion of the solar gravitational potential and frequency splittings of solar oscillation modes. In the latter category, the two best measurements have been made by Duvall and Harvey and by Brown; both showed splitting roughly independent of spherical harmonic degree l, and both had a peculiar peak in the measured splitting at $l = 11$. The author presents new results, based on the analysis of 6,656 individual oscillation modes for $5 \leqslant l \leqslant 20$. These data yield a splitting spectrum which is consistent with previous measurements, but without the unusual peak at $l = 11$, thus suggesting that a simple standard model for the solar core is essentially correct.

080.025 **Does dark matter affect the solar neutrino flux?**
A. De Rújula, S. L. Glashow, L. Hall.
Nature, Vol. 320, No. 6057, p. 38 – 39 (1986).
An observed deficit of high–energy (^{8}B) solar neutrinos and the galactic dark–mass problem have both been explained in terms of hypothetical particles: 'cosmions' must have masses m_c in the range 5 – 50 AMU, a mean elastic scattering cross–section per hydrogen nucleus in the Sun of $<\sigma>_s \sim 4 \times 10^{-36}$cm^2, local mass density $\varrho \sim 1$ AMU cm^{-3}, and cosmological velocity $v \sim 10^{-3}c$. Accumulated in the Sun, they enhance its thermal conductivity so as to yield the measured solar luminosity with a lower central temperature than in standard models. The authors show that such particles, should they exist, could be detected and studied in the laboratory by bolometric procedures proposed originally for the purpose of measuring solar neutrinos.

080.026 **Some properties of the dynamics of the solar convective zone and their consequences on the activity cycle.**
É. Ribes.
C. R. Acad. Sci., Sér. II, Tome 302, No. 14, p. 871 – 873, 875 – 876 (1986). In French.
A large–scale circulation pattern has been detected in the solar convective zone, for the first time. Tracers used to probe the convective layers are newly–born sunspots and long–lived Hα filaments. Young sunspots trace a roughly axisymmetric circulation in the form of four zonal bands, with a velocity amplitude ranging from zero to 100 m s^{-1}. The coincidence of the zonal meridional circulation with the magnetic pattern drawn by the Hα filaments suggests that we are looking at rotating rolls. The direction of rotation is associated with the magnetic polarity. Another result concerns the rotational rate of the deep convective layers. Young spots seem to rotate more rigidly, in contrast with older spots which exhibit a differential rotation similar to the surface rotation. However, the rotational rate exhibits a large dispersion, partly due to young spots located at the site of converging of diverging rolls. This is the consequence of the Coriolis force which decelerates or accelerates the plasma. These results shed a completely new light on problems dealing with the differential rotation, the transport of angular momentum and the dynamo action.

080.027 **On the interpretation of the 160–min oscillation of the sun.**
A. G. Kosovichev, A. B. Severnyj.
Pis'ma Astron. Zh., Tom 12, No. 3, p. 238 – 246 (1986). In Russian. Englisch translation in Sov. Astron. Lett., Vol. 12.
Identification of solar oscillation with the 160–min period as a non–radial mode of the sun has been considered on the basis of the observational data recently available from several observatories. The amplitudes and phases observed allow the most probable identification of such an oscillation as a mode with degree $l = 3$ of a spherical harmonic. On the other hand however, the observed rotational splitting of the eigenfrequency due to solar rotation gives $l = 1$. New methods for future observations are considered to permit more reliable identification of the 160–min oscillations.

080.028 **Influence of the chemical composition on the stability of solar gravity mode oscillations.**
A. G. Kosovichev, A. B. Severnyj.
Izv. Krymskoj Astrofiz. Obs., Tom 72, p. 188 – 198 (1985). In Russian. English translation in Bull. Crimean Astrophys. Obs., Vol. 72.
Solar models with low heavy elements abundance interior (model C) and with mixing in the core are calculated. It is suggested that mixing leads to more homogeneous distribution of hydrogen concentration in the core compared with the standard model. The non–radial oscillations stability of these models are investigated by quasi–adiabatic approximation. For models C g_1– and g_2–modes belonging to dipole ($l = 1$) oscillations are found unstable due to the ε–mechanism. It means that the mechanism of nonlinear resonance excitation of long–period (over the range 120 – 200^m) solar oscillations for this model is confirmed. The instability of a gravity mode is also revealed in the mixing models when the ratio of hydrogen abundance at the centre and the surface equals approximately to 0.7 – 0.8. It is pointed out that only the models with low Z are in agreement with the solar neutrino experiment and the observed solar gravity mode oscillations.

080.029 **Observations of the moon for more accurate definition of the radio radius of the sun.**
A. F. Bachurin.
Izv. Krymskoj Astrofiz. Obs., Tom 72, p. 211 – 217 (1985). In Russian. English translation in Bull. Crimean Astrophys. Obs., Vol. 72.
The method for measuring the solar radio radius at wavelength 2.25 cm is defined more precisely by using the moon as a standard source. Radio measurements of the lunar radius show agreement with the optical one within r.m.s. error limits of $\pm 1''$.The brightness temperature distribution over the lunar disk is given.

080.030 **On the influence of gravitational radiation on the 160–min solar oscillation.**
A. G. Kosovichev.
Izv. Krymskoj Astrofiz. Obs., Tom 73, p. 15 – 26 (1985). In Russian. English translation in Bull. Crimean Astrophys. Obs., Vol. 73.
The resonance amplitude of solar oscillations forced by gravitational waves is calculated in connection with the recent idea, that the 160–min oscillation might be caused by gravitational waves from the nearby binary system Geminga. It is found that the maximum possible amplitude of the oscillation is at least 10^3 times smaller than the observed value. The author concludes that the 160–min oscillation in the sun cannot be excited by gravita-

tional waves emitted from the binary system Geminga but by any other source. He calculates the gravitational wave luminosity of the sun due to the 160–min oscillation and finds that the influence of the gravitational radiation on the damping of the oscillation is negligible.

080.031 **Global oscillations of the sun observed in the variations of differential brightness of the solar photosphere seen in the near infrared.**
V. A. Kotov, S. Koutchmy, Eh. V. Kononovich, N. N. Ryzhikova, T. T. Tsap.
Izv. Krymskoj Astrofiz. Obs., Tom 73, p. 26 – 43 (1985). In Russian. English translation in Bull. Crimean Astrophys. Obs., Vol. 73.
The instrument designed for monitoring of relative center–to–limb brightness fluctuations of the photosphere in near IR spectral range ($\lambda = 1.65\ \mu m$) is described. Analysis of 67 days of observations carried out in 1981 – 1982 shows clearly the presence of 160–min variations of the differential, center–to–limb intensity with an average amplitude of about 3×10^{-5} in units of the «average solar» brightness near 1.65 μm or 0.2K in terms of the temperature of the photosphere.

080.032 **Internal structure and pulsations of solar models with low opacity.**
P. A. Kuzurman.
Nauchn. Inf., Vyp. 56, p. 17 – 37 (1984). In Russian.
Evolutionary models of the present sun with standard and artifically low opacity of stellar matter are obtained and adiabatic nonradial oscillations of these models are computed. The effect of the change of the different parameters (such as the values of solar age and luminosity, the level of the outer boundary conditions in the atmosphere, the value of the parameter of the mixing–length theory) on adiabatic nonradial oscillations of the standard and nonstandard models is investigated.

080.033 **The effects of different parameters on the structure and neutrino radiation of the sun.**
A. V. Fedorova, S. V. Imshennik.
Nauchn. Inf., Vyp. 57, p. 80 – 84 (1984). In Russian.
Several standard models of the sun for two different chemical compositions are derived. The dependence of structure parameters and neutrino flux of the sun on the equation of state and initial ^3He concentration (X_3) is investigated. The results of calculations show that the application of a more appropriate equation of state leads to neutrino flux decrease, while two–fold X_3 variation does not effect the present sun's parameters. The derivatives of the parameters of the sun by X, Z, X_3, α are also obtained.

080.034 **Two nonstandard solar models: with mixed interiors and with enhanced heavy–element opacity.**
A. A. Pamyatnykh, I. K. Sapsaj.
Nauchn. Inf., Vyp. 57, p. 85 – 96 (1984). In Russian.
A standard and two series of nonstandard solar models have been computed with Paczyński's stellar evolution code. The basic features of the models with slow mixing in interiors, occurring in the time scale of order of $10^7 - 10^8$ years, are in agreement with the results of another similar investigation: namely, the models have lower central temperature and lower central mass concentration as compared with the standard model.

080.035 **5–minute oscillations of a solar model with semi–empirical atmosphere.**
P. A. Kuzurman.
Nauchn. Inf., Vyp. 57, p. 97 – 100 (1984). In Russian.
A model of the present sun ($Z = 0.02$, $X_0 = 0.745$, $\alpha = 1.29$) with semi–empirical atmosphere is obtained, and adiabatic nonradial 5–min oscillations of that model are computed. The model has approximately the standard mass concentration and the standard characteristics of the convection zone. The spectrum of adiabatic nonradial oscillations is similar to that of the standard models.

080.036 **The problem of the differential rotation of the sun: inhomogeneity instead of anisotropy.**
L. L. Kichatinov.
Pis'ma Astron. Zh., Tom 12, No. 5, p. 410 – 416 (1986). In Russian. English translation in Sov. Astron. Lett., Vol. 12.
It is shown that convection in a density–stratified rotating medium, being cyclonic, effectively transports the angular momentum and makes rotation to be differential. The derived dependence of angular velocity on the coordinates qualitatively agrees with the observed rotation of the sun.

080.037 **The solar neutrino problem: astronomy or physics?**
J. N. Bahcall.
From SU(3) to gravity, p. 267 – 278 (1985). – See Abstr. 003.014.
A review of current theoretical and observational attempts to resolve the solar neutrino dilemma is presented.

080.038 **Hans Bethe on solar neutrinos.**
J. Maddox.
Nature, Vol. 320, No. 6064, p. 677 (1986).
This note comments on a recent attempt by Bethe (Phys. Rev. Lett., Vol. 56, p. 1305; 1986) to solve the solar neutrino dilemma.

080.039 **The solar rotation elements *i* and Ω derived from sunspot groups.**
H. Balthasar, G. Lustig, D. Stark, H. Wöhl.
Astron. Astrophys., Vol. 160, No. 2, p. 277 – 279 (1986). = Mitt. Kiepenheuer–Inst., Nr. 265.
Using the independent data of sunspot group observations obtained at Greenwich 1874 to 1976 and at the Kanzelhöhe 1947 to 1984 the solar rotation elements were determined. Ω (1850) is determined as (73.75 ± 0.15) deg and i is (7.137 ± 0.017) deg. A significant difference to the well known Carrington rotation elements is found for i.

080.040 **Inertial oscillations and the rotation rate profile of the sun.**
D. B. Guenther.
Bull. Am. Astron. Soc., Vol. 17, No. 4, p. 833 (1985). Abstract. – See Abstr. 010.062.

080.041 **Two–dimensional nonlinear numerical simulations of thermal instability in the solar atmosphere.**
J. M. Picone, R. B. Dahlburg, C. R. DeVore, J. T. Karpen, J. T. Mariska.
Bull. Am. Astron. Soc., Vol. 17, No. 4, p. 843 (1985). Abstract. – See Abstr. 010.062.

080.042 **Solar oscillations; a method for deriving nonlinear effects.**
D. G. Wentzel.
Bull. Am. Astron. Soc., Vol. 17, No. 4, p. 896 (1985). Abstract. – See Abstr. 010.062.

080.043 **The influence of faculae on total solar irradiance and luminosity.**
P. Foukal, J. Lean.
Astrophys. J., Vol. 302, No. 2, p. 826 – 835 (1986).
The authors investigate the facular contribution to the total solar irradiance, using the daily active cavity radiometer irradiance monitor (ACRIM) radiometry for 1980 – 1982 and the Earth Radiation Budget (ERB) radiometry for 1978 – 1982. A cross–correlation analysis of the total irradiance, the UV flux measured by Nimbus 7 and calculated contributions from observed sunspot and facular areas is presented. It is found that the solar irradiance records, after subtraction of the calculated sunspot–blocking contribution, exhibit a short–term modulation that is better explained by faculae than by errors in the sunspot–blocking function. The data for 1980 indicate that the facular contribution was comparable to that of spots over time scales of active region evolution in that year.

080.044 **Solar irradiance variations derived from magnetograms.**
G. A. Chapman, J. E. Boyden.
Astrophys. J., Lett. Ed., Vol. 302, No. 2, p. L71 – L73 (1986).

Pseudo–irradiance fluctuations have been derived from parameterized magnetogram signals. These pseudo–irradiance fluctuations are calibrated by least–squares comparison with ACRIM/*SMM* data. The resulting Mount Wilson irradiance can be examined for a complete sunspot cycle. The results do not appear to be overly sensitive to the mathematical form of the parameterization.

080.045 **Global oscillations of the sun determined from flares in the past (1859 – 1956).**
V. V. Kasinskij.
Issled. Geomagn., Aehron. Fiz. Solntsa, Moskva, No. 73, p. 45 – 51 (1985). In Russian. Abstr. in Ref. Zh., 51. Astron., 3.51.411 (1986).

080.046 **Results of measurements of circumsolar scattered radiation and estimate of their accuracy.**
Yu. I. Brichkov, N. B. Golub', S. V. Gritsenko, T. I. Kudryashova, Yu. A. Sklyarov.
Vopr. klimata i pogody Nizh. Povolzh'ya. Saratov, p. 15 – 31 (1983). In Russian. Abstr. in Ref. Zh., 51. Astron., 3.51.991 (1986).

080.047 **On short–period brightness oscillations of the sun.**
N. N. Ryzhikova.
Astron. Tsirk., No. 1404, p. 6 – 8 (1985). In Russian.

080.048 **The sun at high spatial resolution: the physics of small spatial structures in a magnetized medium.**
R. T. Rosner.
NASA Conf. Publ., NASA CP–2421, p. 14 – 26 (1986). – See Abstr. 012.021.

The author reviews some of the reasons for thinking that there do exist structures in the solar outer atmosphere with spatial scales smaller than those typically accessible to present–day instrumentation.

080.049 **Radio radius of the sun at 2.25 cm in the direction of the polar regions at solar activity maximum.**
A. F. Bachurin.
Bull. Crimean Astrophys. Obs., Vol. 69, p. 123 – 125 (1986). English translation of 39.080.028.

080.050 **Properties of solar oscillations.**
J. Toomre.
Seismology of the Sun and the distant stars, p. 1 – 22 (1986). – See Abstr. 012.026.

Many of the oscillations that can be observed in the atmosphere of the Sun are resonant acoustic or gravity modes of the interior. Accurate measurement of the frequencies of these *p* and *g* modes permits deductions about the internal structure and dynamics of this star. Some of the methods of interpretation, involving a close interplay between observation and theory, can be carried over to the study of more distant stars.

080.051 **Theoretical aspects of helio– and asteroseismology.**
J. Christensen–Dalsgaard.
Seismology of the Sun and the distant stars, p. 23 – 53 (1986). – See Abstr. 012.026.

The dependence of adiabatic oscillations, particularly their frequencies, on the structure of the Sun or other stars is analyzed, on the basis of asymptotic analysis and numerical computations. As a particular example the author considers the effects of mixing in the solar core; mixing sufficiently strong to bring the computed neutrino flux into agreement with the observations appears to be inconsistent with the oscillation observations. The author also discusses the effects of modifications to the upper, superadiabatic part of the convection zone.

080.052 **Observations of low–degree p–mode oscillations in 1984.**
H. M. Henning, P. H. Scherrer.
Seismology of the Sun and the distant stars, p. 55 – 62 (1986). – See Abstr. 012.026.

Analysis of Stanford differential velocity observations has been extended through the 1984 observing season. Excellent quality observations were obtained in 1984 on 38 days in a 49 day interval from June 20th through August 7th. The power spectrum of this data has been examined and improved frequency determinations have been made for p–modes of degree 2 through 5 and order 5 through 34. Of special interest are the modes of the lower orders, n ranging from 5 to 10, which have not been identified previously.

080.053 **Another reason to search for solar g–modes and new limits from solar ellipticity measurements.**
J. R. Kuhn.
Seismology of the Sun and the distant stars, p. 63 – 72 (1986). – See Abstr. 012.026.

Observations of solar g–modes will teach us some gravitational physics. The present sensitivity of the Princeton Solar Distortion Telescope and other recent claims of solar g–modes suggest that these low frequency modes should be observable in shape measurements. From about 250 days and nearly 1000 hours of observations the author finds no evidence for significant spectral power that can be associated with g–modes and no sign of the elusive 160.01 min period solar oscillation.

080.054 **Phase studies of solar 5 min oscillations.**
S. Frandsen.
Seismology of the Sun and the distant stars, p. 73 – 80 (1986). – See Abstr. 012.026.

Observations of time series of spectral lines contain information about the phase (and the amplitude) of the evanescent waves at many heights in the solar atmosphere. The calculated phases turn out to agree with the observed phases in the middle part of the atmosphere, but in the chromosphere, where the theory is only a crude description of the actual environment, differences occur. More surprisingly the calculations predict wrong phases at the deeper layers around unit optical depth. The reason is not known, but the coupling between convection and pulsation, which is ignored in the model used, might be the responsible agent.

080.055 **Observations of gravity waves in the solar atmosphere.**
F.–L. Deubner.
Seismology of the Sun and the distant stars, p. 81 – 82 (1986). Abstract. – See Abstr. 012.026.

080.056 **Is there an anisotropy of p–modes?**
F.–L. Deubner.
Seismology of the Sun and the distant stars, p. 83 – 84 (1986). Abstract. – See Abstr. 012.026.

080.057 **Influence of spatial filtering on possible anisotropies in solar oscillations.**
F. Hill, D. A. Haber, J. Toomre, L. J. November.
Seismology of the Sun and the distant stars, p. 85 – 92 (1986). – See Abstr. 012.026.

The authors have used full disk Doppler observations of solar oscillations to compare the amplitudes of sectoral modes propagating along the equator with those of similar modes propagating along a great circle aligned with the poles. They find that the amplitudes are generally not equal for the two classes of modes, but the results are sensitive to analysis procedures attempting to isolate the different modes of oscillation. Spatial filtering of the data using spherical harmonics suggests that greater amplitudes are associated with "polar" sectoral modes than with "equatorial" sectoral modes.

080.058 **The detection of global convective wave flows on the Sun.**
P. H. Scherrer, J. T. Hoeksema, H. Yoshimura.
Seismology of the Sun and the distant stars, p. 93 – 102 (1986). – See Abstr. 012.026.

Global convective flows in the solar convection zone have been predictd by theoretical interpretations of the global–scale ordering of magnetic fields and activity centers and by theoretical analyses of rotating convection zones. The signatures of such flows have now been detected by analyzing the daily series of low–resoluton Dopplergrams obtained at the Wilcox Solar Observatory at Stanford University. The signatures are patterns of alternating east and west flows with amplitudes on the order of 25 m/s and longitudinal extent of about 30 degrees.

080.059 **First evidence of a large–scale circulation in the solar convection zone.**
E. Ribes, P. Mein.
Seismology of the Sun and the distant stars, p. 103 (1986). Abstract. – See Abstr. 012.026.

080.060 **Solar Doppler shifts: sources of continuous spectra.**
T. L. Duvall Jr., J. W. Harvey.
Seismology of the Sun and the distant stars, p. 105 – 116 (1986). – See Abstr. 012.026.

Oscillation observations can be used to study non–oscillatory solar phenomena that exhibit Doppler shifts. The authors discuss several effects of these phenomena and their associated temporal and spatial power spectra: (1) They limit the signal–to–noise ratio and sometimes detectability of oscillation modes. (2) There is the potential for better understanding and/or detection of solar phenomena: surface rotation, supergranulation, granulation, active regions, giant cells, and mesogranulation. (3) Large–scale convection may spatially modulate oscillation modes, leading to a continuous background spectrum. (4) In regions of the spectrum where the resolution to separate modes is lacked, it is possible to determine upper limits for the integrated effects of modes.

080.061 **Sound speed in the interior of solar models.**
I. W. Roxburgh.
Seismology of the Sun and the distant stars, p. 121 – 123 (1986). – See Abstr. 012.026.

080.062 **Asymptotic sound–speed inversions.**
D. O. Gough.
Seismology of the Sun and the distant stars, p. 125 – 140 (1986). – See Abstr. 012.026.

An asymptotic analysis of stellar p modes in the short–wavelength limit is summarized. The formula can be used as a means of estimating the sound speed in the solar interior from a knowledge of identified eigenfrequencies. The accuracy of the procedure is discussed, and it is shown how the results can be used to measure properties such as the location of the base of the convection zone.

080.063 **An improved asymptotic formula for solar gravity–mode periods.**
A. N. Ellis.
Seismology of the Sun and the distant stars, p. 173 – 175 (1986). – See Abstr. 012.026.

080.064 **Solar g modes: a method to find the depth of the convective envelope.**
M. Gabriel.
Seismology of the Sun and the distant stars, p. 177 – 186 (1986). – See Abstr. 012.026.

It is shown that the phase ε in the first–order asymptotic formula for g^+ modes of models with a radiative core and a convective envelope is not a constant, but is rather a function of σ^2 and l whose limit is 5/12 when σ^2 goes to zero. Application to the Sun shows that the law $\varepsilon(\sigma^2,l)$ is very sensitive to the depth of the convective envelope. Accurate measurements of periods for g modes will give information on the depth of the solar convective envelope. A fitting method to obtain $\varepsilon(\sigma^2,l)$ is given.

080.065 **The influence of convection theory uncertainties on the deduction of the speed of sound in the solar interior.**
R. K. Ulrich.
Seismology of the Sun and the distant stars, p. 187 – 198 (1986). – See Abstr. 012.026.

Two models differing only in the surface layers are studied to determine how the derived frequencies are changed. These frequency changes can only be partially treated as being caused by a surface phase shift. Some portion of the uncompensated frequency change may be misinterpreted as a result of an internal difference between the models.

080.066 **Measuring the Sun's internal rotation using solar p–mode oscillations.**
T. M. Brown.
Seismology of the Sun and the distant stars, p. 199 – 214 (1986). – See Abstr. 012.026.

2–dimensional velocity images of the Sun were obtained for 5 days in June, 1984. This time series of images has been analyzed to obtain frequencies of solar p–mode oscillations with degrees between 8 and 50, with all azimuthal orders for each degree. The principal results of the analysis are measurements of the frequency splitting between modes with the same degree and radial order; these are related to the latitudinal variation of solar rotation. The observed splittings suggest that for $0.3\,R_\odot \leqslant r \leqslant 0.7\,R_\odot$, the solar latitudinal differential rotation is much smaller than at the surface, and moreover than the rotation rate is close to the surface equatorial value.

080.067 **Rotational splitting of $l = 1$ p modes.**
H. B. van der Raay, P. L. Palle, T. Roca Cortes.
Seismology of the Sun and the distant stars, p. 215 – 221 (1986). – See Abstr. 012.026.

An analysis of the amplitude modulation of specific $l = 1$ p modes, has revealed a mean splitting of 0.74 μHz over the frequency range 2.4 to 4.0 mHz. The effects of large scale surface phenomena rotating at the ~27 day surface rotation rate are also detected. The lifetime of the $l = 1$ modes is seen to vary from >120 days at 2425.1 μHz to ~50 days at 3912.2 μHz.

080.068 **Linewidth and rotational splitting of low degree modes in the 5 minute region.**
G. R. Isaak.
Seismology of the Sun and the distant stars, p. 223 – 228 (1986). – See Abstr. 012.026.

The linewidth of low degree modes in the 5 minute region of the Sun's acoustic spectrum increases dramatically with frequency over the range of 2 mHz to some 5 mHz, in approximate agreement with theory. Below some 2.5 mHz the linewidth is narrow enough to allow rotational splitting of $l = 1$ modes to be measured. These measurements seem to confirm earlier results.

080.069 **Solar core rotation: a down to Earth view.**
P. Connes.
Seismology of the Sun and the distant stars, p. 229 – 235 (1986). – See Abstr. 012.026.

The 13–days periodicity solar oscillation discovered in potassium resonance–cell data has been shown to preserve coherence over several years; the proposed explanation was a fast rotating solar core. The author shows here that this coherence is probably an observational artifact due to a very small non–linear interaction between the actual short–coherence solar phenomenon (arising from rotation of surface features) and the much larger annual and periodic radial velocity change of the observer.

080.070 **The internal rotation of the Sun.**
P. R. Goode.
Seismology of the Sun and the distant stars, p. 237 – 241 (1986). – See Abstr. 012.026.

The internal rotation rate of the Sun has been calculated by Duvall, et al. from the solar oscillation data of Duvall and Harvey. The origin and implications of this result are reviewed and compared with those of an earlier, less reliable determination.

080.071 **On the solar rotation.**
M. Gabriel, F. Nemry.
Seismology of the Sun and the distant stars, p. 243 – 247 (1986). – See Abstr. 012.026.

It is pointed out that the solar rotation curve published by Duvall et al. has been obtained without considering the accuracy of the observations. When the limited accuracy, of the order of 10^{-2}, is taken into account, it is found that the data are compatible with a nearly solid rotation and that they do not contain enough information to give any indication on the rotation rate near the centre. Two rotation laws are considered; the rotation is supposed to be a function of the distance to the centre or to the rotation axis. Both laws lead to the same conclusion.

080.072 **The internal rotation of the Sun.**
I. W. Roxburgh.
Seismology of the Sun and the distant stars, p. 249 – 252 (1986). – See Abstr. 012.026.

Recent results on the internal rotation of the Sun are difficult to explain. These problems are reviewed and it is conjectured that a ^{3}He driven mixed shell near 0.3 $R_\odot$ magnetically separates the inner core from the outer regions. Such a shell may rotate differentially and may separate a more rapidly spinning core from the rest of the Sun.

080.073 **Magnetic torques and differential rotation.**
N. O. Weiss.
Seismology of the Sun and the distant stars, p. 253 – 256 (1986). – See Abstr. 012.026.

The rotation profile deduced from the measured splitting of solar oscillations raises several problems. The slight decrease in angular velocity within the convective zone can be explained and the transition from differential to uniform rotation on spherical surfaces must cause a further reduction in the radiative zone. The bump, associated with harmonics of degree 11, seems implausible and it is hard to reconcile a rapidly rotating core with the presence of a significant magnetic field. A possible explanation is that this core has been magnetically decoupled from outer regions since the Sun evolved on to the Hayashi track.

080.074 **Internal rotation of the Sun as a consequence of instabilities in the radiative interior.**
W. A. Dziembowski, L. Paternò, R. Ventura.
Seismology of the Sun and the distant stars, p. 257 – 263 (1986). – See Abstr. 012.026.

The authors consider the effect of the molecular weight gradient in the Sun's radiative interior on the Goldreich–Schubert–Fricke instability, and deduce the marginal gradient of angular velocity which can be tolerated. The same gradient is also calculated for the axisymmetric baroclinic diffusive instability which has the lowest threshold of the hydrodynamical instabilities. The curves obtained for the angular velocity contradict those inferred from the observed splittings of p–mode oscillations. The authors suggest that angular momentum can be transferred from interior owing to some instability caused by the presence of a global magnetic field in the Sun's core.

080.075 **Is the existence of a strong magnetic field in the Sun's core plausible?**
W. A. Dziembowski, L. Paternò, R. Ventura.
Seismology of the Sun and the distant stars, p. 271 – 274 (1986). – See Abstr. 012.026.

The authors estimate the energies of the identified solar gravity modes from the observed radial velocity amplitudes, and deduce that the energies of some modes are of the same order of the total energy of the convective motions in the solar envelope. This fact rules out the possibility of direct or indirect excitation of these modes by convection. It is suggested that the most plausible excitation mechanism relies upon a magnetic torque caused by the presence of a global magnetic field of the order of a megagauss in the Sun's core.

080.076 **Progress report on helium abundance determination.**
W. Däppen, D. O. Gough.
Seismology of the Sun and the distant stars, p. 275 – 280 (1986). – See Abstr. 012.026.

Intermediate results from a programme to determine the helium abundance of the solar convection zone are presented. The method uses the influence of helium ionization on the local sound speed, which can be gauged from inversions of solar oscillation frequencies. The resolution of the principal diagnostic function by the data justifies optimism for an abundance determination in a future step.

080.077 **Change of solar oscillation eigenfrequencies with the solar cycle.**
M. F. Woodard, R. W. Noyes.
Seismology of the Sun and the distant stars, p. 303 – 304 (1986). Abstract. – See Abstr. 012.026.

080.078 **Long period variations in the solar diameter.**
P. Delache.
Seismology of the Sun and the distant stars, p. 305 – 308 (1986). – See Abstr. 012.026.

Observations of the solar diameter with a Danjon–type visual astrolabe have been recorded at the C.E.R.G.A. Observatory since 1975. An harmonic analysis of these measurements shows the existence of very long solar periods. In order to assess the significance of peaks in the power spectrum the author presents four different methods. A few long periodicities in the solar diameter fluctuations have been identified. Most of the periods appear to be correlated with the solar activity monitored by the Zürich sunspot number. A discussion of these correlations is presented.

080.079 **The 1984 solar oscillation program of the Mt. Wilson 60–foot tower.**
E. J. Rhodes Jr., A. Cacciani, S. Tomczyk, R. K. Ulrich.
Seismology of the Sun and the distant stars, p. 309 – 332 (1986). – See Abstr. 012.026.

The authors describe the instrumentation, data, and preliminary results from the summer, 1984, solar oscillation observing program which was carried out using the 60–foot tower telescope of the Mt. Wilson Observatory.

080.080 **The K 769.9 nm line profile.**
H. B. van der Raay, P. L. Palle, T. Roca Cortes.
Seismology of the Sun and the distant stars, p. 333 – 337 (1986). – See Abstr. 012.026.

The observation of a solar absorption line by means of resonant scattering over several months in the study of solar oscillations, yields data which may be analysed to provide details of the line shape over the region being studied. The asymmetry and nonlinearity of the K 769.9 nm line are found and the effect on the overall response of the resonant scattering spectrometer is illustrated. Finally a simple second order correction term to the response function of the system is empirically found which greatly improves overall linearity.

080.081 **Limb darkening fluctuations: the discrepancy with Doppler data and the 160–min oscillation.**
R. Yerle.
Astron. Astrophys., Vol. 161, No. 1, p. L5 – L8 (1986).

The amplitude discrepancy between solar limb darkening and Doppler measurements in the five–minute range, suggests the possible existence of Earth atmosphere oscillations. Recent aeronomical observations add strong evidence to this point of view. A possible explanation of the mysterious 160–min period is presented as a direct consequence. Short period oscillations are another one. Slight discrepancies between Doppler data and between those with theoretical calculations could be also explained in the same way.

080.082 **The effect of solar oscillations on the brightness of the limb.**
B. Schmieder, P. Mein.
Astron. Astrophys., Vol. 161, No. 1, p. 7 – 11 (1986).
The intensity profile of the solar limb has been calculated assuming a spherical geometry based on the VALC model. Hydrodynamic waves in the atmosphere are simulated by introducing temperature and density fluctuations in the model calculation. It is shown that 5 min oscillations lead to variations in the brightness gradient of less than 2%. The limb brightness fluctuation amplitudes observed in that frequency range, with the heliometer at Pic du Midi, (Yerle, 1981) show peaks of 10%. They are difficult to explain by *p* mode waves.

080.083 **The evolution and helium content of the sun.**
Y. Lebreton, A. Maeder.
Astron. Astrophys., Vol. 161, No. 1, p. 119 – 124 (1986).
The constraint that a solar evolutionary model with an age of 4.6×10^9yr must just have the present solar luminosity, implies an initial helium content of Y = 0.279 for models constructed with Los Alamos opacities including molecular absorption at low temperatures. Some interesting consequences have arisen from the models: (1) The neutrino flux is 74% higher than the generally accepted value found by Bahcall et al. (1982). (2) The value of the period spacing of low–degree gravity modes in the solar oscillations is slightly closer to the observed one. (3) As expected, the value Y = 0.279 is larger than that of Y = 0.24 or 0.25 predicted by Big Bang nucleosynthesis.

080.084 **Comparison of differential infrared and differential radius observations of the solar 160 minute period oscillation.**
H. A. Hill.
Astrophys. J., Vol. 303, No. 1, p. 489 – 494 (1986).
The differential infrared observations of the Sun obtained at the Crimean Astrophysical Observatory are compared with the differential radius observations for the 160 minute period solar oscillation. The observed relative phases restrict $|m|$ to 2. The observed relative amplitudes also give information which is consistent with theoretical predictions on the radial dependence of the temperature eigenfunction at the bottom of the photosphere.

080.085 **Magnetohydrodynamic tube waves: waves in fibrils.**
P. S. Cally.
Aust. J. Phys., Vol. 38, No. 6, p. 825 – 837 (1985). Paper presented at the R. G. Giovanelli Commemorative Colloquium, Part I, Sydney, 26 – 29 November 1984.
A discussion of waves in magnetic flux tubes imbedded in magnetic unstratified surroundings is given. Seven types of non–leaky wave are identified. Many more leaky waves, tube oscillations which drive waves in the external medium and thereby lose energy to it, are also found. The particular example of longitudinal and transverse oscillations in chromospheric fibrils is examined in detail.

080.086 **Global oscillations and active regions.**
C. J. Durrant.
Aust. J. Phys., Vol. 38, No. 6, p. 839 – 844 (1985). Paper presented at the R. G. Giovanelli Commemorative Colloquium, Part I, Sydney, 26 – 29 November 1984.
The author presents further estimates of the amplitude of the modulation of the solar global velocity signal caused by the passage of active regions across the solar disc. Using measurements of the profile of the K I λ769.9 nm line in the quiet sun and in plages he finds a global velocity variation of ~ 2 m s^{-1} during the transit of a typical active region of area 3300 millionths of the hemisphere. However, during the period in which a velocity amplitude of 6 m s^{-1} was reported by Claverie et al. (1982), the sunspot areas were exceptionally large and the author confirms Schröter's (1984) result that the combination of spot and plage contributions is sufficient to account for the observed signal. The velocity modulation is thus attributable to surface inhomogeneities, not to the structure of the solar core.

080.087 **The influence of magnetic fields on convective motions in the outer layers of the sun.**
R. Van der Borght, P. Fox.
Aust. J. Phys., Vol. 38, No. 6, p. 845 – 854 (1985). Paper presented at the R. G. Giovanelli Commemorative Colloquium, Part I, Sydney, 26 – 29 November 1984.
In this paper the authors make a preliminary investigation of the nonlinear equations of compressible convection under the influence of solar–type magnetic fields. A polytropic model of the basic structure is used and, although the model is somewhat restrictive, good agreement is obtained with general observations in both strong and weak field cases. The value and influence of the turbulent magnetic resistivity is investigated and the depth dependence of the vertical velocity within a given period is used to study the way in which the overstable oscillations change their direction of flow from positive to negative.

080.088 **The iron sun.**
J. R. Graham.
Sky Telesc., Vol. 71, No. 5, p. 437 (1986).

080.089 **Weakly interacting massive particles and solar oscillations.**
J. Faulkner, D. O. Gough, M. N. Vahia.
Nature, Vol. 321, No. 6067, p. 226 – 229 (1986).
If the Sun were to contain even a minute mass fraction of weakly interacting massive particles (WIMPs), there could be a significant influence on its central thermal structure. In particular, a relative concentration as small as $\sim 10^{-11}$ by number may lower the central temperature sufficiently to bring the predicted electron neutrino detection rate into agreement with observation. This letter shows that observations of solar oscillations provide a means for testing the validity of this proposed resolution of the solar neutrino problem.

080.090 **Weakly interacting massive particles, solar neutrinos, and solar oscillations.**
W. Däppen, R. L. Gilliland, J. Christensen–Dalsgaard.
Nature, Vol. 321, No. 6067, p. 229 – 231 (1986).
Weakly interacting, massive particles (WIMPs) have recently been proposed as a solution to the solar neutrino problem. Solar models which incorporate a relative number of 10^{-11}WIMPs with appropriate scattering cross–section would reduce the predicted neutrino detections by ^{37}Cl to the observed value, without significant changes in the solar structure outside the central region. The authors have subjected these models to an observational test of *p*–mode oscillation frequencies by computing frequency differences of low–degree, high–order oscillation frequencies. Although standard solar models also pass this test, WIMP models provide a better fit.

080.091 **Sweeping–magnetic–twist mechanism for the acceleration of jets in the solar atmosphere.**
K. Shibata, Y. Uchida.
Sol. Phys., Vol. 103, No. 2, p. 299 – 310 (1986).
A magnetodynamic mechanism for the acceleration of jets in the solar atmosphere is proposed, and a 2.5–dimensional MHD simulation is performed to show how this mechanism operates in the situation of the chromosphere–corona region of the solar atmosphere. One of the characteristic properties of the jet in this model is that the jet, consisting of hot core and cool sheath, has a helical velocity field in it, explaining the thus–far unexplained observed feature.

080.092 **Umbral oscillations as a probe of sunspot structure.**
T. E.–H. Abdelatif.
Diss. Abstr. Int., Sect. B, Vol. 46, No. 7, p. 2345 (1986). Thesis, The University of Rochester, 229 pp. (1985). Order No. DA8516458.

080.093 **Possible explanation of the solar–neutrino puzzle.**
H. A. Bethe.
Phys. Rev. Lett., Vol. 56, No. 12, p. 1305 – 1309 (1986).
Mikheyev and Smirnov have shown that electron neutrinos above a certain minimum energy E_m may all be converted into

μ neutrinos on their way out through the sun. The author assumes that this is the reason why Davis and collaborators, in their experiments, find may fewer solar neutrinos than predicted. The minimum energy E_m is found to be about 6 MeV, the mass m of the μ neutrino must be greater than that of the electron neutrino, $m_2{}^2 - m_1{}^2 = 6 \times 10^{-5} eV^2$, and there is a very minor restriction on the neutrino mixing angle.

080.094 **Numerical simulation of the surging arch in an elementary eruptive phenomenon in the solar atmosphere.**
V. N. Dermendjiev.
Sbornik Tr. Mezhdunar. konf. na temu: Avtomatiz. nauchn. issled., Plovdiv, 15 – 20 okt. 1984, Tom 2, Sofiya, p. 540 – 548. Abstr. in Ref. Zh., 51. Astron., 6.51.395 (1986).

080.095 **Attenuation of solar radiation in a homogeneous atmosphere.**
R. G. Indzhgia.
Soobshch. Akad. Nauk GSSR, Tom 119, No. 3, p. 513 – 515 (1985). In Russian. Abstr. in Ref. Zh., 51. Astron., 6.51.939 (1986).

080.096 **Energy loss of electrons in solar atmosphere taking many–body interactions into account.**
K. L. Moza, P. K. Koul, R. R. Rausaria, P. N. Khosa.
J. Astrophys. Astron., Vol. 7, No. 1, p. 39 – 43 (1986).
The role of many–body interactions on the energy loss of electrons accelerated at neutral point during solar flares has been studied. Energy loss with and without many–body interactions has been computed for different electron–density models as function of height. The energy loss increases by a factor of two by inclusion of many–body interactions for incident electron energies greater than 10 keV.

080.097 **On the differential rotation of solar background magnetic fields. 1. On the possible existence of torsional waves with the wave number k = 1/hemisphere.**
L. Hejna.
Bull. Astron. Inst. Czech., Vol. 37, No. 3, p. 175 – 179 (1986).
Some of the questions concerning the possible existence of torsional waves with the wave number k = 1/hemisphere are discussed with regard to the differential rotation of solar background magnetic fields studied with the use of the time series derived from synoptic Hα maps published by McIntosh.

080.098 **On the differential rotation of solar background magnetic fields. 2. Are torsional waves with the wave number k = 1/hemisphere real?**
L. Hejna.
Bull. Astron. Inst. Czech., Vol. 37, No. 3, p. 179 – 187 (1986).
Questions concerning the statistical significance of the changes of smoothed time series of the latitude gradients B/A and C/A, determined by Hejna (1985), (A, B and C being the coefficients which occur in the relation for differential rotation), are discussed. A model computer experiment is used to prove that these changes can be considered statistically significant.

080.099 **Comments on the heating of solar active regions by magnetic energy dissipation.**
S. Chandra.
Astrophys. Space Sci., Vol. 123, No. 2, p. 403 – 404 (1986).
The method of Kumar and Narain (1985) to calculate the twisting velocity of magnetic field appears to be erroneous. Hence, their conclusion about Tucker's proposal for the heating of active regions by magnetic energy dissipation may not be correct.

080.100 **Interpretation and implications of diameter and differential radius observations of the 160 minute period solar oscillation.**
H. A. Hill, J. Tash, C. Padin.
Astrophys. J., Vol. 304, No. 1, p. 560 – 578 (1986).
An oscillation is observed with a period of 160.06 ± 0.08 minutes in the 1978 diameter observations, and with a period of 159.98 ± 0.03 minutes in the 1979 differential radius observations. This oscillation is classified as the well–known 160 minute period oscillation. This paper examines the hypothesis that the primary physical mechanism by which this long–period oscillation is detected via Doppler shifts involves temperature perturbations conjoined with solar surface rotation, and not physical displacement of the solar surface. This hypothesis is found to be viable for observations on the 160 minute oscillation. The formalism is developed for computing the apparent Doppler shifts from the temperature perturbations obtained in diameter and differential radius observations. Based on these calculations the 160 minute period is classified as an $l = 2, m = 2$ mode.

080.101 **Latitude and depth variation of solar rotation.**
T. L. Duvall Jr., J. W. Harvey, M. A. Pomerantz.
Nature, Vol. 321, No. 6069, p. 500 – 501 (1986).
Measurements of the frequencies of various modes of trapped acoustic waves provide information about rotation and structure within the Sun. The authors present new results for modes with spherical harmonic degrees up to 98 which allow the convective envelope to be isolated. For degrees between 20 and 98 there is no evidence that internal rotation differs significantly with depth or latitude from the rotation of surface magnetic field patterns. Modes covering a wide latitude range have systematically lower frequencies than those confined near the equator, indicating the existence of a structural asymmetry within the Sun.

080.102 **On a method of determination of the solar radio radius by scanning technique.**
A. F. Bachurin.
Izv. Krymskoj Astrofiz. Obs., Tom 74, p. 117 – 121 (1986). In Russian. English translation in Bull. Crimean Astrophys. Obs., Vol. 74.
The expression of antenna response in the equatorial coordinates α and δ to the radio emission from the round disk with uniform brightness distribution is derived. It is shown that the solar radio radius at 2.25 cm wavelength calculated in these coordinates differs by no more than 0".15 from the value obtained by using approximate formula in rectangular coordinates.

080.103 **On the surface response of solar giant cells.**
A. A. van Ballegooijen.
Astrophys. J., Vol. 304, No. 2, p. 828 – 837 (1986).
This paper discusses the discrepancy that exists between theories of giant cell convection, which predict typical flow velocities of order 100 m s^{-1}, and observations, which put an upper limit of about 10 m s^{-1} on the east–west velocity at the solar surface. To explain this discrepancy, the author considers a model of inviscid flows within a stratified, rotating convection zone and solves the vorticity equation. The results suggest that the discrepancy can be resolved if the giant cells have a longitudinal wavelength $\lambda < 180$ Mm. Their typical size may be 100 – 150 Mm, which would make them virtually unobservable at the solar surface.

080.104 **Detection of solar five–minute oscillations through white–light intensity.**
J. Nishikawa, S. Hamana, K. Mizugaki, T. Hirayama.
Publ. Astron. Soc. Jpn., Vol. 38, No. 2, p. 277 – 283 (1986).
The authors report the detection of the solar 5–min oscillations in white–light intensity for spherical harmonic degrees $l = 40 - 600$, where they utilized only an objective lens, a broad–band filter, and a CCD camera. The authors find the rms intensity amplitude corresponding to a temperature difference of about 2.6K from four wavelength bands, and a slightly lower temporal frequency than in the theoretical expectation in high l's.

080.105 **From solar dynamo to terrestrial climate.**
S. Sofia, P. Demarque, A. Endal.
Am. Sci., Vol. 73, No. 4, p. 326 – 333 (1985). Abstr. in Phys. Abstr., Vol. 89, No. 1275, Entry 30216 (1986).

080.106 **The latitudinal distribution of sunspot areas and variations of the differential rotation of the Sun during the period 1921 – 1971.**
A. Antalová.
Contrib. Astron. Obs. Skalnaté Pleso, Vol. 14, p. 121 – 144 (1986).
The distribution of sunspot areas is compared with the distribution of variations of the differential rotation in solar activity cycles Nos. 16 – 20 using butterfly diagrams. The analysis indicates that the zones on the Sun in which there is a large accumulation of sunspot areas display a smaller velocity than the long-term average zonal velocity.

080.107 **Instability of differential rotation and turbulence of the solar core.**
A. I. Laptukhov.
Soln. Dannye, Byull., 1986, No. 4, p. 54 – 58 (1986). In Russian.
The stability criterion of the nonviscous fluid rotation in Goldreich–Schubert–Fricke's form should be substituted with a more rigorous one: $\Delta\Omega = 0$ after account for axial asymmetrical disturbances. As $\Delta\Omega \neq 0$ in the solar core, the development of differential rotation instability causes a turbulent state of the core. It is shown that the entropy is nearly maximum with correct corresponding constraints.

080.108 **The enigma of solar neutrinos: a solution in view.**
M. Spiro.
Recherche, Vol. 16, No. 172, p. 1537 – 1539 (1985). In French. Abstr. in Phys. Abstr., Vol. 89, No. 1278, Entry 47997 (1986).

080.109 **Multiplet classification of intermediate–degree solar f–modes and associated rotational splitting.**
H. A. Hill, R. D. Rosenwald.
Bull. Am. Astron. Soc., Vol. 18, No. 1, p. 638 (1986). Abstract. – See Abstr. 010.066.

080.110 **Variation of the solar constant during 1978 – 79 and 1981.**
J. Pap.
Bull. Astron. Inst. Czech., Vol. 37, No. 4, p. 202 – 210 (1986).
This paper investigates the connections between the solar constant variations and the different solar activity parameters, such as the projected areas of the sunspot groups, the 2800 MHz flux intensity, the Zürich sunspot number and the soft X–ray data during 1979 and 1981.

080.111 **Conversion in matter may account for missing solar neutrinos.**
B. Schwarzschild.
Phys. Today, Vol. 39, No. 6, p. 17 – 20 (1986).

080.112 **Solar physics: "progress in solar hydromagnetics".**
M. Kuperus.
Proceedings of the 1984 International Conference on Plasma Physics, Vol. 2, p. 997 – 1017 (1985). Abstr. in Phys. Abstr., Vol. 89, No. 1281, Entry 67459 (1986). – See Abstr. 012.056.

080.113 **Solar structure without computers.**
D. D. Clayton.
Am. J. Phys., Vol. 54, No. 4, p. 354 – 362 (1986). Abstr. in Phys. Abstr., Vol. 89, Nr. 1284, Entry 78612 (1986).

080.114 **Heating and mass transfer in solar and stellar atmospheres by turbulent Alfvén waves.**
R. Opher, V. J. S. Pereira.
Rev. Mex. Astron. Astrofis., Vol. 12, Spec. Issue, p. 355 – 364 (1986). – See Abstr. 012.066.
The authors examine the possibility that turbulent Alfvén waves, created in the upper transition region or lower corona, can be the principal source of the heating of solar and stellar coronae, and the origin of solar and stellar winds.

080.115 **Properties of *r*–modes in the Sun.**
C. L. Wolff, J. B. Blizard.
Sol. Phys., Vol. 105, No. 1, p. 1 – 15 (1986).
In this paper, the authors collect and compute various properties of *r*–modes, concentrating on those that would occupy the solar convective envelope. Several ways one might try to detect the modes are briefly indicated.

080.116 **The gallium solar neutrino experiment GALLEX.**
D. Vignaud.
Nucleosynthesis and its implications on nuclear and particle physics, p. 453 – 461 (1986). – See Abstr. 012.069.
The low threshold (233 keV) of the neutrino capture reaction $^{71}Ga(\nu_e, e^-)^{71}Ge$ is a real chance for the detection of solar neutrinos, particularly those coming from the pp reaction. The GALLEX experiment will measure the solar neutrino flux by counting the ^{71}Ge atoms produced in a 30 ton gallium target in the form of a $GaCl_3$ solution. A calibration will be done with a ^{53}Cr source which emits 751 keV neutrinos. The detector will be set up in the Gran Sasso Underground Laboratory (Italy).

080.117 **Real–time detection of low energy solar neutrinos with indium and superheated superconductivity.**
G. Waysand.
Nucleosynthesis and its implications on nuclear and particle physics, p. 463 – 475 (1986). – See Abstr. 012.069.
The principle of detection in real–time of low energy solar neutrinos with indium is recalled. Because four tons of indium are necessary to get one event per day, an elegant way to realize this experiment is to use indium as the target *and* the detector of solar neutrinos. Superheated superconductivity provides such an opportunity at the expense of the division of the metal in microspheres embedded in a dielectric.

080.118 **Magnetic reconnection.**
E. R. Priest.
Plasma penetration into magnetospheres, p. 1 – 14 (1986). – See Abstr. 012.071.
A brief review is given of reconnection on the sun, where it is probably responsible for a wide variety of phenomena including coronal heating, solar flares and the newly discovered cancelling magnetic features.

Sun and earth.
See Abstr. 003.019.

A solution to the solar neutrino puzzle?
See Abstr. 011.027.

Oscillator strengths for neutral iron and its abundance in the solar atmosphere.
See Abstr. 022.031.

Ground based measurements of global solar intensity oscillations.
See Abstr. 034.059.

A liquid crystal modulator.
See Abstr. 034.060.

Laser heterodyne spectrometer for helioseismology.
See Abstr. 034.088.

Neutrino capture in $^{69,71}Ga$ and the detection of solar and galactic neutrinos.
See Abstr. 061.002.

Thermal diffusion in ionized gases and planetary and solar processes.
See Abstr. 061.031.

Wave equation for a magnetic monopole.
See Abstr. 061.073.

Detection rates for "invisible"–axion searches.
See Abstr. 061.094.

On the use of SSC as a detector of certain dark matter candidates.
See Abstr. 061.118.

Resonant amplification of ν oscillations in matter and solar–neutrino spectroscopy.
See Abstr. 061.120.

A possible unified interpretation of the solar $^{48}Ca/^{46}Ca$ abundance ratio and Ca – Ti isotopic anomalies in meteorites.
See Abstr. 061.134.

Large–basis shell–model technology in nucleosynthesis and cosmology.
See Abstr. 061.137.

Properties of anomalous nuclei and their possible effects on stellar burning.
See Abstr. 061.138.

Oscillatory motions in intense flux tubes.
See Abstr. 062.019.

On the important role of relativistic effects in the electrodynamics of plasma.
See Abstr. 062.053.

Some astrophysical applications of MHD instabilities of differentially rotating fluids.
See Abstr. 062.069.

Magnetostatic equilibrium in a nonuniform gravity field.
See Abstr. 062.095.

Fast approximations for the R_{II-A} redistribution function.
See Abstr. 063.041.

Theory of the transition layer emission measures.
See Abstr. 064.032.

Vistas on stellar–solar atmospheres.
See Abstr. 064.040.

An analytic approach to the connection between stellar structure parameters and the neutrino emission rate in a simple stellar model.
See Abstr. 065.022.

The solar–stellar connection: I. Solar studies.
See Abstr. 065.061.

The solar–stellar connection: II. Stellar studies.
See Abstr. 065.062.

Finite amplitude limit of the ^3He instability.
See Abstr. 065.068.

Numerical implementation of a perturbation theory up to the third order for rotating polytropic stars: application to a solar model.
See Abstr. 065.090.

The special theory of relativity on a gravitational field.
See Abstr. 066.200.

Observations of 5–minute oscillations of the photospheric brightness of the sun.
See Abstr. 071.036.

Meridional motions of sunspots from 1947.9 – 1985.0. I. Latitude drift at the different solar–cycles.
See Abstr. 072.002.

Photoelectric observations of chromospheric sunspot oscillations. III. Spatial distribution of power and frequency in umbrae.
See Abstr. 072.012.

Photoelectric observations of chromospheric sunspot oscillations. IV. The Ca II H line and He I λ10830.
See Abstr. 072.013.

On umbral oscillations as a sunspot diagnostic.
See Abstr. 072.047.

Effect of local rigid rotation on sunspots.
See Abstr. 072.077.

Correlation of the velocity and magnetic field in a solar active region.
See Abstr. 072.078.

Oscillations of the Sun's chromosphere. IV. Temporal evolution of Hα profile.
See Abstr. 073.014.

Dynamics of magnetic flux tubes in the convection zone of the sun.
See Abstr. 075.006.

Geschwindigkeits–Oszillationen in solaren Magnetfeld–Gebieten.
See Abstr. 075.036.

Magnetic reconnection on the Sun.
See Abstr. 075.052.

Spectrum of variations of the geomagnetic AE–index in the period range 130 – 200 min as possible reflection of global oscillations of the sun.
See Abstr. 084.042.

Period of 160 minutes in the solar system: solar pulsation and spin rates of planets and asteroids.
See Abstr. 091.015.

On a possible relation of the law of planetary distances to the 160–minute pulsation of the sun.
See Abstr. 091.017.

Solar and stellar magnetic activity.
See Abstr. 116.004.

Astroseismology.
See Abstr. 116.016.

Solar oscillations and rotation of close binary systems in the Galaxy: 160 minute period.
See Abstr. 117.210.

Oscillations and pulsations in the sun and stars.
See Abstr. 122.007.

The effect of the rotation of the sun in cosmic ray anisotropy during sporadic and recurrent electromagnetic perturbations in the interplanetary space.
See Abstr. 144.057.

Earth

081 Structure, Figure, Gravity, Orbit, etc.

081.001 **Evolution of an impact–induced atmosphere and magma ocean on the accreting Earth.**
T. Matsui, Y. Abe.
Nature, Vol. 319, No. 6051, p. 303 – 305 (1986).
Early rapid formation of the atmosphere and hydrosphere on the terrestrial planets has recently been proposed. The authors present a quantitative study of this process during accretion by planetesimal impacts. These impacts increase the surface temperature and thus affect the formation of either a proto–atmosphere or a proto–hydrosphere by degassing of volatiles. It is shown that an impact–induced H_2O atmosphere increases the surface temperature of the Earth to a stage where a magma ocean is possible, with the total amount of H_2O in the proto–atmosphere clustering around 10^{21}kg. The apparent coincidence of the H_2O abundance in the proto–atmosphere with the present mass ($\sim 1.4 \times 10^{21}$kg) of the ocean may be evidence for an impact origin for the atmosphere and hydrosphere.

081.002 **Chemical effects of large impacts on the Earth's primitive atmosphere.**
B. Fegley Jr., R. G. Prinn, H. Hartman, G. H. Watkins.
Nature, Vol. 319, No. 6051, p. 305 – 308 (1986).
Intense bombardment of the moon and terrestrial planets $\sim 3.9 - 4.0 \times 10^9$years ago could have caused the chemical reprocessing of the Earth's primitive atmosphere. In particular, the shock heating and rapid quenching caused by the impact of large bodies into the atmosphere could produce molecules such as HCN and H_2CO which are important precursors for the abiotic synthesis of complex organic molecules. The authors model the production of HCN and H_2CO by thermochemical equilibrium and chemical kinetic calculations of the composition of shocked air parcels for a wide range of temperatures, pressures and initial compositions.

081.003 **Relativistic effects in geodynamics.**
C. Boucher.
Relativity in celestial mechanics and astrometry, p. 241 – 253 (1986). – See Abstr. 012.002 (IAU Symp. No. 114).

081.004 **Proposals for an experiment to detect the earth's gravitomagnetic field.**
A. G. Polnarev.
Relativity in celestial mechanics and astrometry, p. 401 – 405 (1986). – See Abstr. 012.002 (IAU Symp. No. 114).
Some experiments are discussed for measuring the earth's gravitomagnetic field, using a pendulum or an oscillator rather than a gyroscope.

081.005 **14th–order harmonics in the geopotential from analysis of satellite orbits at resonance.**
D. G. King–Hele, D. M. C. Walker.
Planet. Space Sci., Vol. 34, No. 2, p. 183 – 195 (1986).

081.006 **Mantle convection under simulated plates: effects of heating modes and ridge and trench migration, and implications for the core–mantle boundary, bathymetry, the geoid and Benioff zones.**
G. F. Davies.
Geophys. J. R. Astron. Soc., Vol. 84, No. 1, p. 153 – 183 (1986).

081.007 **Properties of iron at the Earth's core conditions.**
O. L. Anderson.
Geophys. J. R. Astron. Soc., Vol. 84, No. 3, p. 561 – 579 (1986).
The phase diagram of iron up to 330 GPa is solved using the experimental data of static high pressure (up to 11 GPa) and the experimental data of shock wave data (up to 250 GPa). A solution for the highest triple point is found (P = 280 GPa and T = 5760K) by imposing the thermodynamic constraints of triple points. This pressure of the triple point is less than the pressure of the inner core – outer core boundary of the Earth. These results indicate that the density of iron at the inner core – outer core boundary pressure is close to 13 g cm^{-3}, which lies close to the seismic solutions of the Earth at that pressure. It is thus concluded that the Earth's inner core is very likely to be virtually pure iron in its hexagonal close packed phase.

081.008 **Thermoelastic deformations of the Earth's lithosphere: a mathematical model.**
P. Lanzano.
Earth, Moon, Planets, Vol. 34, No. 3, p. 283 – 304 (1986).
The author examines the problem of the thermoelastic deformation of a spherical Earth with constant elastic parameters heated from within by the spontaneous decay of radiogenic elements. The problem consists of the simultaneous solution of the Navier–Stokes equation and the heat conduction equation. The author obtains analytic solutions for the temperature distribution and radial deformation as infinite series of functions of the radial distance and time, depending also on a sequence of eigenvalues. He believes that his analytic results are applicable to the study of the oceanic lithosphere deformations.

081.009 **Further comparisons of Earth models by means of lumped coefficients. Part II.**
J. Klokočník.
Bull. Astron. Inst. Czech., Vol. 37, No. 1, p. 26 – 42 (1986). With plates 2 – 4.
This paper concludes the series of the author's papers on the intercomparisons and calibrations of the Earth gravity field models (*EMs*) by means of lumped geopotential coefficients. The Earth models are arranged in series which have been worked out in the individual centers (SAO, GSFC NASA, DGFI/GRGS, OSU) and the evolution of the quality of the *EMs*, from the oldest to the most recent, is assessed. Significantly erroneous or suspicious values of the lumped coefficients of these *EMs* have been compiled for the individual orders of the harmonics.

081.010 **An analysis of variance–covariance matrix of harmonic geopotential coefficients of GEM–L 2.**
J. Klokočník.
Bull. Astron. Inst. Czech., Vol. 37, No. 2, p. 91 – 95 (1986). With plates 1, 2.
The errors of the lumped geopotential coefficents of the Earth gravity field model GEM–L 2 are computed twice: (1) just from the variances of its harmonic coefficients, (2) accounting the full variance–covariance matrix. The differences between results (1) and (2) are not negligible. The impact of the differences on the accuracy assessment of Earth models is pursued.

081.011 **Ein Beitrag zur harmonischen Darstellung des gezeitenerzeugenden Potentials.**
F.–J. Büllesfeld.
Dtsch. Geod. Komm. Bayer. Akad. Wiss., Reihe C, Heft Nr. 314, 103 pp. (1985). ISBN 3–7696–9364–7.

081.012 **Astronomisch–geodätische Methoden zur Bestimmung des Schwerefeldes der Erde.**
E. Erker.
Sternenbote, 29. Jahrg., Nr. 5, p. 86 – 97 (1986).

081.013 **Axial melting in central peaks of complex impact structures.**
M. R. Dence.
Meteoritics, Vol. 20, No. 4, p. 635 – 636 (1985). Abstract. – See Abstr. 010.661.

081.014 **On computation of derivatives of the earth's gravitational potential for satellite geodesy and geodynamics.**
O. A. Abrikosov.
Kinematika Fiz. Nebesn. Tel, Tom 2, No. 2, p. 51 – 58 (1986). In Russian. English translation in Kinematics Phys. Celest. Bodies.
The problem of computation of derivatives of the earth's gravitational potential is considered. The derivatives have been represented by series of normalized solid spherical harmonics. Their coefficients have been obtained in form of linear combinations of normalized Stokes constants. On this ground an effective algorithm for computation of the geopotential derivatives in rectangular coordinates has been developed.

081.015 **Transition region of the Earth's upper mantle.**
D. L. Anderson, J. D. Bass.
Nature, Vol. 320, No. 6060, p. 321 – 328 (1986).
The transition region of the Earth's upper mantle, 400 – 650 km deep, appears to be mineralogically and chemically distinct from both the shallow mantle and lower mantle. It contains most of the basalt fraction of the Earth's mantle.

081.016 **Cosmogenic helium in a terrestrial igneous rock.**
M. D. Kurz.
Nature, Vol. 320, No. 6061, p. 435 – 439 (1986).
High $^3He/^4He$ ratios in terrestrial samples are generally interpreted as indicating the presence of primitive or primordial gas. The author presents new helium isotopic measurements on samples from the Kula formation of Haleakala volcano (Hawaii) that are best explained by an *in situ* cosmogenic origin for a significant fraction of the ^3He. Results from crushing and stepwise heating experiments, and consideration of the exposure age of the sample at the surface and the cosmic–ray fluxes strongly support this hypothesis.

081.017 **Precision of the geoid, computed from terrestrial gravity measurements.**
G. Strang van Hees.
Manuscr. Geod., Vol. 11, No. 1, p. 1 – 14 (1986).
A method is developed to compute the precision of the geoid and of geoid differences from gravity measurements. The gravity values may have a variable quality and the density of the gravity networks in different areas may also be variable. This meets the reality. The method can also be applied for planning a gravity survey. Given the needed precision of geoid differences, the density of the gravity points can be estimated.

081.018 **Differential geometry of the gravity field.**
E. W. Grafarend.
Manuscr. Geod., Vol. 11, No. 1, p. 29 – 37 (1986).
The review of the differential geometry of the gravity field is based on field equations for Newton gravity in an earth–fixed rotating frame for a deformable body, for Newton–Cartan gravitation and Einstein gravitation. The differential geometry of equipotential surfaces and their orthogonal trajectories, also called plumblines, is discussed with respect to seven examples, mainly motivating the Marussi parameterization for local problems and the Gauß parameterization in terms of a spherical harmonic series representation of the gravity potential for global problems. The computation of three–dimensional plumblines down to the geoid is outlined.

081.019 **High speed evaluation of the geopotential.**
A. Sengoku.
Proceedings of the nineteenth symposium on celestial mechanics, p. 107 – 112 (1986). In Japanese. – See Abstr. 012.013.

081.020 **Is there coherence between Earth tides and earthquakes?**
P. W. Burton.
Nature, Vol. 321, No. 6066, p. 115 (1986).

081.021 **The relationship between spherical and ellipsoidal gravity coefficients of the geopotential.**
N. B. Batueva.
Astron. Tsirk., No. 1393, p. 3 – 5 (1985). In Russian.

081.022 **Numerical theory of motion of the earth and Venus from radar and optical observational data and from observations of motion of the Venera 9 and Venera 10 artificial earth satellites.**
Eh. L. Akim, V. A. Stepan'yants.
Nauchn. probl. aviatsii i kosmonavt.: Istor. i sovremennost'. Moskva, p. 132 – 135 (1985). In Russian. Abstr. in Ref. Zh., 62. Issled. Kosm. Prostranstva, 4.62.178 (1986).

081.023 **A conceptual model of core dynamics and the earth's magnetic field.**
C. B. Officer.
J. Geophys., Vol. 59, No. 2, p. 89 – 97 (1986).
A conceptual model of core dynamics and the earth's magnetic field is presented. It differs from previous investigations in the use of an estimated core viscosity of $2 \times 10^7 cm^2 s^{-1}$. The simplified derivations predict the correct order of magnitude for the external magnetic field and for the westward drift of the non–dipole field.

081.024 **Optimum approximation of the gravity field by a minimum number of point masses.**
F. Barthelmes.
Nablyud. Iskusstv. Sputnikov Zemli, No. 21, Chast' I, p. 124 – 130 (1982). In Russian.– See Abstr. 012.032.
The problem of the best approximation of the gravity field by a given number of parameters leads in the case of the point masses model to the determination of the optimal coordinates of these masses. The algorithm which solves this problem numerically for the given field is described.

081.025 **On determination of the physical surface of the earth using geopotential parameters determined by methods of satellite geodesy.**
G. A. Meshcheryakov.
Nablyud. Iskusstv. Sputnikov Zemli, No. 21, Chast' I, p. 131 – 137 (1982). In Russian.– See Abstr. 012.032.
The problem of determination of the physical surface of the earth is considered. It can be solved by means of the parameters of the gravitational field, known from satellite geodesy, and the traditional information given by astronomical, geodetical and gravimetrical measurements. The corresponding inverse problem of the potential theory is discussed and its uncorrect character is taken into account.

081.026 **On calculation of the parameters of the earth's ellipsoid from satellite data using the method of L_p–estimates.**
G. A. Meshcheryakov, S. D. Volzhanin.
Nablyud. Iskusstv. Sputnikov Zemli, No. 21, Chast' I, p. 138 – 144 (1982). In Russian.– See Abstr. 012.032.
The determination of the parameters of the earth's ellipsoid is described. The L_p estimation generalizing the well–known methods of least–squares estimation, the method of least modules and Chebyshev approximation are used. The results obtained with the use of geopotential models SAO–5, SAO–6, GEM–10B are given.

081.027 **A simple model for the gravitational field of the earth.**
A. N. Marchenko.
Nablyud. Iskusstv. Sputnikov Zemli, No. 21, Chast' I, p. 145 – 152 (1982). In Russian.– See Abstr. 012.032.
The choice of the normal gravity field of the earth in the form of a five–point masses potential is made for the construction of mathematically homogeneous point–masses models of the geopotential. A comparison of the orbits of 11 satellites obtained by means of numerical integration using harmonic coefficients of GEM 9 and GEM 10B models and the five–point masses model is performed. It is shown that the proposed model represents the satellites orbits with the same accuracy as the sets of modern harmonic coefficients of the 2nd order.

081.028 **On the possibility of a representation of the geopotential by means of a point masses system on the basis of satellite observations.**
V. A. Antonov, E. I. Timoshkova, K. V. Kholshevnikov.
Nablyud. Iskusstv. Sputnikov Zemli, No. 21, Chast' I, p. 171 – 176 (1982). In Russian.– See Abstr. 012.032.
A point–masses model approximating the geopotential is considered. Algorithms of its construction using well known expansion in terms of spherical harmonics are examined. An algorithm using direct satellite observations is proposed.

081.029 **Geodynamic implications of short time scale changes in the geomagnetic dynamo.**
S. K. Runcorn.
Phys. Earth Planet. Inter., Vol. 41, Nos. 2 – 3, p. 73 – 77 (1985).

081.030 **Thermal state and thermal regime of the earth's interior.**
V. N. Zharkov.
Phys. Earth Planet. Inter., Vol. 41, Nos. 2 – 3, p. 133 – 137 (1985).
A simplified version of the thermal state and thermal regime of the Earth's mantle is given. It is supposed that the large–scale convection in the mantle is double–layered. Convection in the upper mantle is determined by heat flux from below, and in the lower mantle by internal heating.

081.031 **Detection of inertial gravity oscillations in the Earth's core with a superconducting gravimeter at Brussels.**
P. Melchior, B. Ducarme.
Phys. Earth Planet. Inter., Vol. 42, No. 3, p. 129 – 134 (1986).
As stated by Crossley and Rochester the discovery and identification of even a few undertone periods would provide valuable gross constraints on the stability profile and hence, the thermal regime of the core. Three years of continuous measurements of gravity variations with a superconducting gravimeter at Brussels offer a possibility to detect such oscillations in that part of the spectrum where tidal contributions can be eliminated. Spectral peaks of 10 – 15 ngal amplitude have been detected around a period of 13.9 h.

081.032 **The structure of the terrestrial planets and tidal friction theory.**
R. A. Lyttleton.
Sir Arthur Eddington Centenary Symposium, Vol. 3: Gravitational radiation and relativity, p. 79 – 135 (1986). – See Abstr. 012.036.
This paper reviews theoretical attempts to derive the internal structure of the Earth and its evolution history from a plausible scenario for the initial protoplanetary dust cloud surrounding the Sun. Implications for the other terrestrial planets are also discussed.

081.033 **Milankovitch climatic origin of mid–Cretaceous black shale rhythms in central Italy.**
T. D. Herbert, A. G. Fischer.
Nature, Vol. 321, No. 6072, p. 739 – 743 (1986).
The Earth's orbital variations, reflected in Pleistocene ice volume, were also recorded in non–glacial times. Carbonate production in pelagic mid–Cretaceous sediments, quantified by calcium carbonate and optical densitometry time series, reflects the orbital eccentricity and precessional cycles. Minimal eccentricity brought deep–sea anoxia. Most of the sedimentary variability of this 100–Myr–old sequence lies within the Milankovitch (orbital) frequency band.

081.034 **Expanding Earth hypotheses in the light of palaeogeophysical results.**
R. Lauterbach.
Ann. Phys. (Leipzig), Vol. 42, No. 4 – 6, p. 489 – 495 (1985). In German. Abstr. in Phys. Abstr., Vol. 89, No. 1273, Entry 18203 (1986).

081.035 **On the coupling of fluid dynamics and electromagnetism at the top of the Earth's core.**
E. R. Benton.
Geophys. Astrophys. Fluid Dyn., Vol. 33, No. 3 – 4, p. 315 – 330 (1985). Abstr. in Phys. Abstr., Vol. 89, No. 1274, Entry 23412 (1986).

081.036 **Deformación rotacional de una tierra elástica.**
M. J. Sevilla, A. González–Camacho.
Rev. Geofis., Vol. 38, p. 179 – 187 (1982). = Semin. Astron. Geod., Univ. Madr., Publ., No. 125.
The changes in the Earth's inertial tensor due to the rotational deformation are studied and some equations for these changes are obtained.

081.037 **Deformación de una tierra elástica por efecto de marea.**
M. J. Sevilla, A. González–Camacho.
Semin. Astron. Geod., Univ. Madr., Publ., No. 127, p. 967 – 973 (1983).
The changes in the elastic earth's inertial tensor due to the deformation produced by the earth–tide generating potential are studied and some equations for these changes are obtained.

081.038 **A sensitive radar search for small natural satellites of the earth.**
M. J. Longo, R. Morris.
Astron. J., Vol. 91, No. 5, p. 1238 – 1241 (1986).
The authors have used radar tracking data from NORAD to perform a sensitive search for small natural satellites of the Earth. This search would be sensitive to satellites ranging from 5 cm in diameter with perigee heights $\lesssim 400$ km to about 40 cm with perigees $\lesssim 10000$ km. While a few unidentified objects were found, these were all near transfer orbits likely to contain debris from satellite launches. No objects that are unlikely to be man-made were seen.

081.039 **Accretion rate of extraterrestrial matter: iridium deposited 33 to 67 million years ago.**
F. T. Kyte, J. T. Wasson.
Science, Vol. 232, No. 4755, p. 1225 – 1229 (1986).
Iridium measured in 149 samples of a continuous 9–meter section of Pacific abyssal clay covering the time span 33 to 67 million years ago shows a well–defined peak only at the Cretaceous/Tertiary boundary. In the rest of the section iridium ranges from a minimum concentration near 0.35 nanograms per gram in the Paleocene to a maximum near 1.7 in the Eocene. Correction for terrestrial iridium leads to an extraterrestrical flux of 9 ± 3 nanograms of iridium per square centimeter per million years, and an estimated annual global influx of 78 billion grams of chondritic matter, consistent with recent estimates of the influx of dust, meteorites, and crater–producing bodies with masses ranging from 10^{-13} to 10^{18}grams. Periodic comet showers should increase the cometary iridium flux by a factor of 200 to 600 on a time scale of 1 to 3 million years; the predicted iridium maxima (more than 30 times background) are not present in the intervals associated with the Cretaceous/Tertiary boundary or the tektite–producing late Eocene events.

081.040 **"Frequency windows" from Earth gravity models and resonant solutions – test of the accuracy.**
Z. Šima, J. Klokočnik.
Bull. Astron. Inst. Czech., Vol. 37, No. 4, p. 229 – 239 (1986). With plates 7 – 8.
The authors suggest and also use a test for evaluating the accuracy of some of the harmonic (Stokes) coefficients in the

Earth gravity field models (EMs). The results of resonant phenomena serve as a standard for testing the EMs. The main result of the test is that a good consistency in shape and phase shifts exists for recent EMs, but even for orders $12 \leqslant m \leqslant 15$, where recent EMs are believed to be very reliable, some increase of the original accuracy estimates given by the authors of the EMs is needed. In general the accuracy has to be increased 2 – 3 times to achieve the real accuracy of the EMs.

081.041 **Relative motion of the geodynamic twin–satellite.**
P. Lála, J. Klokočník.
Bull. Astron. Inst. Czech., Vol. 37, No. 4, p. 239 – 246 (1986).
A new method has been suggested for precise mapping of the Earth gravitational field: a passive laser geodynamic twin–probe for further improvements in the lower degree and order harmonic coefficients. This concept has already been worked out, but many questions still remain to be answered, namely the requirements in terms of the separation impulse and relative motion components of the twins. This motion in the field of various gravitational and non–gravitational perturbing forces is studied and the numerical results are presented.

081.042 **The Earth's gravity.**
A. Cazenave, G. Balmino.
Recherche, Vol. 17, No. 176, p. 438 – 447 (1986). In French. Abstr. in Phys. Abstr., Vol. 89, No. 1283, Entry 77875 (1986).

081.043 **Lead isotope systematics and the evolution of the core, mantle, crust and atmosphere.**
I. J. Duncan.
J. Geodyn., Vol. 2, No. 1, p. 1 – 21 (1985). Abstr. in Phys. Abstr., Vol. 89, Nr. 1284, Entry 83116 (1986).

081.044 **Latent–heat driven convection in the Earth's core: numerical results.**
U. Schmit.
J. Geodyn., Vol. 2, No. 1, p. 23 – 34 (1985). Abstr. in Phys. Abstr., Vol. 89, Nr. 1284, Entry 83117 (1986).

081.045 **Isotopic modeling of the evolution of the mantle and crust.**
D. L. Turcotte, L. H. Kellogg.
Rev. Geophys., Vol. 24, No. 2, p. 311 – 328 (1986).
The geochemical reservoirs and cycles of the earth are discussed. Models involving the isotopic evolution of the enriched continental crustal reservoir and the complementary depleted mantle reservoir are considered. Analytic solutions for the fractionation factors f and isotope ratios ε are obtained using consistent approximations. The results are applied to data on the samarium–neodymium, rubidium–strontium, and thorium–uranium–lead systems. Using a model for instantaneous crustal differentiation, data on young rocks provide constraints on the mean age of the continental crust and on the mass of the depleted mantle reservoir.

081.046 **The deep electrical structure of the Earth.**
R. G. Roberts.
Geophys. J. R. Astron. Soc., Vol. 85, No. 3, p. 583 – 600 (1986).
Long–period estimates of the electromagnetic response of the Earth, which contain information about the variation of electrical conductivity at depth, are inverted to estimate the variation with depth of conductivity at depths between 300 and 1000 km beneath 17 locations which are scattered world–wide but concentrated in Europe. Two different one–dimensional inversion techniques are used. The first is rather approximate but allows evaluation of the reliability of individual response estimates. The second is more mathematically rigorous, and allows estimation of conductivity depth–resolutions. The results from the two schemes are found to be qualitatively in agreement.

081.047 **Temperature distribution in the crust and mantle.**
R. Jeanloz, S. Morris.
Annu. Rev. Earth Planet. Sci., Vol. 14, p. 377 – 415 (1986). – See Abstr. 003.042.
Contents: Introduction. Experimental constraints on the geotherm in the crust and mantle. Heat transfer mechanisms. Geodynamics.

081.048 **Chemical geodynamics.**
A. Zindler, S. Hart.
Annu. Rev. Earth Planet. Sci., Vol. 14, p. 493 – 571 (1986). – See Abstr. 003.042.
Contents: Introduction. Mantle chemical heterogeneity: a matter of scale. Isotopic characterization of geochemically distinct mantle components. The chemical and physical structure of the mantle. Summary and conclusions.

081.049 **Asymptotic normal modes of a laterally heterogeneous earth.**
F. A. Dahlen, I. H. Henson.
J. Geophys. Res., Vol. 90, No. B14, p. 12653 – 12681 (1985).
By combining numerical first–order splitting calculations with WKBJ asymptotic analysis, the authors have arrived at a reasonably complete understanding of the normal mode eigenvalue problem for isolated surface wave equivalent multiplets on a rotating, slightly aspherical earth. The principal limitation of this study is the restriction to isolated multiplets. Both rotational coupling as well as coupling due to the earth's lateral heterogeneity have been ignored.

081.050 **Accuracy estimate of geoid and ocean topography recovered jointly from satellite altimetry.**
C. A. Wagner.
J. Geophys. Res., Vol. 91, No. B1, p. 453 – 461 (1986).
Joint determination of the geoid and permanent dynamic sea topography is demonstrated by simulation from direct use of satellite altimetry. The separation is feasible because of the disparities between the error spectra of the geoid and its unique perturbative effects as seen on the orbit. The solution for the components of the permanent topography from 14 cm Seasat altimetry promises to yield accurate results down to half wavelengths of about 1000 km. Present discrimination of this surface from disjoint satellite altimeter results appears to be at a scale near 4000 km. The radial effects of uncertainties in the M2 ocean tide on the Seasat orbit and altimeter measurements will probably be only a minor contaminant of the solution.

081.051 **Coupled free oscillations of an aspherical, dissipative, rotating earth: Galerkin theory.**
J. Park, F. Gilbert.
J. Geophys. Res., Vol. 91, No. B7, p. 7241 – 7260 (1986).
The goal of this study is to show how free oscillation coupling calculations can be incorporated into the search for global lateral structure in the deep earth. A discussion of the methods of performing coupling calculations is given, followed by recipes and examples of the construction of synthetic seismograms.

081.052 **Thermal structure of the accreting earth.**
D. L. Turcotte, J. C. Pflugrath.
J. Geophys. Res., Vol. 90, Suppl., p. C541 – C544 (1985). – See Abstr. 012.065.
The energy associated with the accretion of the earth and the segregation of the core is more than sufficient to melt the entire earth. In order to understand the thermal evolution of the early earth it is necessary to study the relevant heat transfer mechanisms. In this paper the authors postulate the existence of a global magma ocean and carry out calculations of the heat flux through it in order to determine its depth. In the solid mantle heat is transferred by the upward migration of magma. This magma supplies the magma ocean. The increase in the mantle liquidus with depth (pressure) is the dominant effect influencing heat transfer through the magma ocean. The authors find that a magma ocean with a depth of the order of 20 km would have existed as the earth accreted. They conclude that the core segregated and an atmosphere was formed during accretion.

081.053 **The formation of an impact–generated H_2O atmosphere and its implications for the early thermal history of the earth.**
Y. Abe, T. Matsui.
J. Geophys. Res., Vol. 90, Suppl., p. C545 – C559 (1985). – See Abstr. 012.065.

During accretion of the earth by planetesimal impacts, shock degassing of volatile materials occurs, and a proto–atmosphere is formed. Once such an impact–generated atmosphere is formed, the surface temperature of the earth growing by planetesimal impact increases due to the blanketing effect of the atmosphere. The authors show that the surface of the accreting earth was covered entirely by a "magma ocean". This suggests that the differentiation of the earth proceeds under very low pressure and high temperature conditions during its accretion stage. This places a severe constraint on the amount of volatiles in the mantle, and the formation time of the atmosphere and ocean of the earth.

081.054 **Crustal radiogenic heat production and the selective survival of ancient continental crust.**
P. Morgan.
J. Geophys. Res., Vol. 90, Suppl., p. C561 – C570 (1985). – See Abstr. 012.065.

081.055 **The parental magma for some rocks from the Norite 1 subzone of the Stillwater Complex: a lunar analog study.**
G. Ryder, B. Spettel.
J. Geophys. Res., Vol. 90, Suppl., p. C591 – C600 (1985). – See Abstr. 012.065.

The interpretation of single lunar highlands igneous samples is hindered not only by their brecciation but also by a lack of similar intensive work on terrestrial plutonic rocks. Terrestrial studies of cumulate rocks are more extensive and have a well–developed stratigraphic context. In an attempt to establish how much can be learned about the parent magma from single plutonic hand samples, single samples from three successive homogeneous but contrasting layers (anorthosite, anorthositic norite, norite) in the Stillwater Complex Norite 1 subzone were subjected to detailed petrographic and major and trace element chemical analyses.

081.056 **Paläogeophysik.**
R. Lauterbach.
Physik des Planeten Erde, p. 48 – 69 (1985). – See Abstr. 003.047.

081.057 **Physik und Chemie des Erdinneren.**
H. Kautzleben, H. Stiller.
Physik des Planeten Erde, p. 187 – 211 (1985). – See Abstr. 003.047.

081.058 **A program for the computation of oceanic tidal loading effects 'GOTIC'.**
T. Sato, H. Hanada.
Earth rotation and the terrestrial reference frame, p. 742 – 748 (1985). – See Abstr. 012.077.

L'Académie royale des Sciences et la figure de la Terre.
See Abstr. 011.038.

Mass extinction, Earth's rotation and periodic cratering.
See Abstr. 044.037.

Movimiento solo forzado del polo para una tierra elástica.
See Abstr. 044.057.

Autofrecuencias de las ecuaciones de Helmholtz y Liouville para un modelo de tierra tipo Jeffreys simplificado.
See Abstr. 044.058.

Models for true polar wander from palaeomagnetism, geoid and hot–spot studies.
See Abstr. 044.077.

Real polar wandering.
See Abstr. 044.079.

Polar motion–induced gravity.
See Abstr. 044.142.

Satellite laser ranging and its applications.
See Abstr. 045.001.

Geodesy by radio interferometry: studies of the forced nutations of the earth. 2. Interpretation.
See Abstr. 045.028.

Orbit determination and analysis for 1970–97B at 14th–order resonance.
See Abstr. 052.003.

Earth gravitational perturbation sensitivity and orbital inaccuracy for GEO–IK orbit.
See Abstr. 052.004.

Evolution of the atmosphere and oceans.
See Abstr. 082.011.

Geomagnetic reversal spurts and episodes of extraterrestrial catastrophism.
See Abstr. 084.013.

Transitional field behaviour from Southern Hemisphere lavas: evidence for two–stage reversals of the geodynamo.
See Abstr. 084.014.

The stochastic excitation of reversals in simple dynamos.
See Abstr. 084.041.

Theoretical considerations of the magnetic field in a spherical Earth model.
See Abstr. 084.056.

A perturbed disk dynamo model and polarity reversals of the Earth's magnetic field.
See Abstr. 084.060.

A harmonic and statistical analysis of the topography of the Earth, Moon and Mars.
See Abstr. 091.056.

Gravity field of Venus at constant altitude and comparison with earth.
See Abstr. 093.061.

Two–stage models for lunar and terrestrial anorthosites: petrogenesis without a magma ocean.
See Abstr. 094.068.

Mechanism of the early differentiation of the earth.
See Abstr. 107.038.

Phase modulation of the Sun's *z* oscillations.
See Abstr. 155.034.

082 Atmosphere (Refraction, Scintillation, Extinction, Airglow, Site Testing)

082.001 **Detection of stratospheric N_2O_5 by infrared remote sounding.**
G. C. Toon, C. B. Farmer, R. H. Norton.
Nature, Vol. 319, No. 6054, p. 570 – 571 (1986).
The existence of stratospheric N_2O_5, a temporary reservoir species whose photolysis products catalyse ozone destruction, has long been predicted but has not previously been detected. The authors report the identification of N_2O_5 absorption between 1,230 and 1,260 cm^{-1} in infrared spectra measured by the Atmospheric Trace Molecule Spectroscopy experiment from orbit during the Spacelab 3 mission. Preliminary analysis of spectra recorded at sunrise on 1 May 1985 indicates a broad concentration peak of 4×10^8molecules cm^{-3} between 25 and 31 km.

082.002 **Mid-winter intensities of the night airglow O_2 (0–1) atmospheric band emission at high latitudes.**
H. K. Myrabø, C. S. Deehr, G. J. Romick, K. Henriksen.
J. Geophys. Res., Vol. 91, No. A2, p. 1684 – 1690 (1986).
The purpose of this paper is to report and discuss recent observations of the O_2 (0–1) atmospheric band night airglow emission. These observations were made near Longyearbyen, Svalbard at 78°N during 38 days around the 1982 winter solstice.

082.003 **A theoretical and empirical study of the response of the high latitude thermosphere to the sense of the "Y" component of the interplanetary magnetic field.**
D. Rees, T. J. Fuller–Rowell, R. Gordon, M. F. Smith, N. C. Maynard, J. P. Heppner, N. W. Spencer, L. Wharton, P. B. Hays, T. L. Killeen.
Planet. Space Sci., Vol. 34, No. 1, p. 1 – 40 (1986).

082.004 **The role of nitric oxide on the zonally averaged structure of the thermosphere: solstice conditions for solar cycle minimum.**
J.–C. Gérard, R. G. Roble.
Planet. Space Sci., Vol. 34, No. 2, p. 131 – 144 (1986).
A zonally averaged chemical–dynamical model of the Earth's thermosphere is used to investigate the importance of nitric oxide 5.3 μm cooling in controlling the dynamic structure of the thermosphere. The model calculates the zonally averaged circulation, temperature, major atmospheric constituents, and the odd nitrogen distributions self–consistently for solstice conditions during solar cycle minimum.

082.005 **Meteorologische Optik. Halos und andere Lichtbögen am Himmel, Teil 1.**
C. Leinert.
Sterne Weltraum, 25. Jahrg., Nr. 1, p. 18 – 22 (1986).

082.006 **Meteorologische Optik. Halos und andere Lichtbögen am Himmel, Teil 2.**
C. Leinert.
Sterne Weltraum, 25. Jahrg., Nr. 3, p. 136 – 142 (1986).

082.007 **Does the Moon eat the colours?**
J.–P. Parisot.
Earth, Moon, Planets, Vol. 34, No. 3, p. 273 – 279 (1986). In French.
The hydrogen radicals play an important role in the photochemistry of the troposphere of the Earth. The chemistry of OH, HO_2 and H_2O_2 is linked directly to the photodissociation of O_3 through the production of $O(^1D)$. Gaseous H_2O_2 may be removed by heterogeneous reactions involving aerosols and liquid water. During the day and the night the solubility of ambient H_2O_2 in water is estimated and the oxidant capacity of H_2O_2 may explain the bleaching properties of the dew used in the past. This phenomenon may also explain some old maxims concerning the properties of the Moon's light to corrupt colors.

082.008 **Airglow $O_2(^1\Sigma)$ atmospheric band at 8645 Å and the rotational temperature observed at 23°S.**
H. Takahashi, Y. Sahai, P. P. Batista.
Planet. Space Sci., Vol. 34, No. 3, p. 301 – 306 (1986).
Regular observations of the $O_2(^1\Sigma,0\text{–}1)$ atmospheric band at 8645 Å [$O_2A(0,1)$] and the rotational temperature, together with the OH(9,4) band and O I 5577 Å airglow emissions, using multichannel tilting filter type photometers, have been carried out at Cachoeira Paulista (22.7°S, 45.0°W), Brazil, since February 1983. The $O_2A(0,1)$ band intensities occasionally vary from 200 to 1000 R during a night. Covariations in nocturnal and seasonal variation with the O I 5577 Å emission were observed. The temperatures determined from the P branch of the $O_2A(0,1)$ band vary between 180 and 230K. The amplitude of the nocturnal temperature variation is sometimes larger than that determined from the OH emission, and the phase of the variation, on some occasions, leads that of the OH.

082.009 **Dependence of seeing correlation on image separation at the CFH telescope on Mauna Kea.**
C. Christian, R. Racine.
Publ. Astron. Soc. Pac., Vol. 97, No. 598, p. 1215 – 1226 (1985).
Trailed images of optical pairs of stars obtained with the Canada–France–Hawaii 3.6–m telescope on Mauna Kea were analyzed to study seeing fluctuations. The data show that seeing fluctuations, evaluated from the "macroscopic" properties of a guide star image, i.e., image centroid and image width, are correlated for separations as large as 1 arc min. Seeing fluctuations as frequencies <5 Hz were correlated for most image pairs in this particular data set. Higher frequency variations were correlated only for pairs wih separations <8 arc sec. In a simulation of very high resolution (VHR) imagery using an offset guide star, the quality of detected program images were improved by as much as 70%. Guide decision times ranged from 10 to 50 msec, depending on the observational circumstances.

082.010 **The hydrogen escape and the earth's thermospheric temperature.**
A. I. Semenov, O. T. Yurchenko.
Astron. Vestn., Tom 20, No. 1, p. 15 – 19 (1986). In Russian. English translation in Sol. Syst. Res.
The hydrogen escape fluxes from the earth's atmosphere are estimated on the basis of the experimental data on the atomic hydrogen emission H_α and thermospheric temperature. The time–lag variations of the emission intensity and escape fluxes upon the temperature are determined. The mean rate of the escape is about 3×10^8 hydrogen atoms $cm^{-2}s^{-1}$ for $T \sim 1000K$.

082.011 **Evolution of the atmosphere and oceans.**
H. D. Holland, B. Lazar, M. McCaffrey.
Nature, Vol. 320, No. 6057, p. 27 – 33 (1986).
The residence times of most constituents of the atmosphere and oceans are small fractions of the age of the earth and, in general, their rate of output has been nearly equal to their rate of input. The mineralogy of marine evaporites rules out drastic changes in the composition of sea water during the last 900 Myr. The chemistry of soils formed more than 1,000 Myr ago suggests that the atmosphere then contained significantly more CO_2 and less O_2 than at present. Hydrogen peroxide may well have been the principal oxidant and formaldehyde the main reductant in rain water between 3,000 and 1,000 Myr ago. Major changes in atmospheric chemistry since that time are almost certainly related to the evolution of the biosphere.

082.012 **Corpuscular–tropospheric effect and zonal circulation of the earth atmosphere.**
N. B. Mulyukova, Eh. R. Mustel', V. E. Chertoprud.
Astron. Zh., Tom 63, Vyp. 2, p. 326 – 334 (1986). In Russian. English translation in Sov. Astron., Vol. 30, No. 2.
The study is continued of the atmospheric circulation variations of the earth's northern hemisphere after the earth's entry into solar corpuscular flows (SCF). Daily maps of the surface atmospheric pressure for the years 1880 – 1974 served as the data base. In particular, the zonal circulation of the earth atmosphere was considered.

082.013 **Simultaneous H_α line profile and radar measurements at Arecibo.**
R. B. Kerr, S. Atreya, J. W. Meriwether Jr., C. A. Tepley, R. G. Burnside.
J. Geophys. Res., Vol. 91, No. A4, p. 4491 – 4512 (1986).
High resolution measurements of the geocoronal Balmer alpha (H_α) emission were made with a Fabry–Perot interferometer at Arecibo, Puerto Rico, in six observing campaigns between March 1980 and November 1984. Four of these included simultaneous radar measurements of the topside *F* region. The results are presented and discussed in detail.

082.014 **The O_2 atmospheric dayglow in the thermosphere.**
M. R. Torr, B. Y. Welsh, D. G. Torr.
J. Geophys. Res., Vol. 91, No. A4, p. 4561 – 4566 (1986).
The authors report spectral measurements ($\Delta\lambda$ = 8 Å) from Spacelab 1 of the O_2 atmospheric bands in the dayglow at thermospheric altitudes for a tangent ray height of 150 km. Vibrational levels up to $\upsilon' = 4$ are found in the data, requiring a source in addition to the energy transfer from $O(^1D)$ to O_2. The authors suggest that this source is the collisional deactivation of $N(^2D)$ by O_2.

082.015 **Correction to "The O_2 atmospheric 0–0 band and related emissions at night from Spacelab 1" [J. Geophys. Res., Vol. 90, No. A9, p. 8525 – 8538 (1985)].**
M. R. Torr, D. G. Torr, R. R. Laher.
J. Geophys. Res., Vol. 91, No. A4, p. 4607 (1986). See Abstr. 40.082.045.

082.016 **Cloud dispersal to increase observing time for astronomical instruments.**
I. V. Litvinov, V. G. Tsverava.
Bull. Spec. Astrophys. Obs. – North Caucasus, Vol. 19, p. 115 – 116 (1985). English translation of 39.082.062.

082.017 **Measurement of the total ozone content in Nizhneangarsk during the total solar eclipse of July 31, 1981.**
A. M. Shalamyanskij, N. D. Ionina.
Tr. Glav. geofiz. obs., No. 487, p. 120 – 122 (1985). In Russian. Abstr. in Ref. Zh., 51. Astron., 2.51.31 (1986).

082.018 **On the variation of the total content of water vapour in the atmosphere during a solar eclipse.**
N. I. Nikitinskaya, V. N. Tsvetkova.
Tr. Glav. geofiz. obs., No. 487, p. 86 – 92 (1985). In Russian. Abstr. in Ref. Zh., 51. Astron., 2.51.32 (1986).

082.019 **Brightness of night sky background during full moon and twilight.**
Kh. Mammedov, G. A. Nasyrov, Yu. V. Khan.
Izv. Akad. Nauk TSSR. Ser. fiz.–tekh., khim. i geol. nauk, No. 3, p. 80 – 82 (1985). In Russian. Abstr. in Ref. Zh., 51. Astron., 2.51.1070 (1986).

082.020 **La protection des sites astronomiques.**
J. Kovalevsky.
Astronomie, Vol. 100, p. 219 – 226 (1986).

082.021 **Annual variations of cosmic rays in the lower atmosphere.**
V. P. Okhlopkov, L. S. Okhlopkova, T. N. Charakhch'yan.
Geomagn. Aehron., Tom 26, No. 1, p. 24 – 29 (1986). In Russian. English translation in Geomagn. Aeron.

082.022 **Temporal dependences of the short–range cosmic ray component in the stratosphere.**
T. N. Charakhch'yan.
Geomagn. Aehron., Tom 26, No. 1, p. 30 – 38 (1986). In Russian. English translation in Geomagn. Aeron.

082.023 **The near–equatorial thermosphere and F1 layer around the winter solstice period.**
L. P. Goncharov, G. P. Kushnarenko, L. A. Shchepkin.
Geomagn. Aehron., Tom 26, No. 1, p. 39 – 42 (1986). In Russian. English translation in Geomagn. Aeron.

082.024 **On oscillations of the parameters of the upper atmosphere with a period of about 5.5 years given by twilight and night–sky emission data.**
T. G. Megrelishvili, L. M. Fishkova.
Geomagn. Aehron., Tom 26, No. 1, p. 154 – 156 (1986). In Russian. English translation in Geomagn. Aeron.

082.025 **How good is your observing site?**
M. F. Walker.
Sky Telesc., Vol. 71, No. 2, p. 139 – 143 (1986).

082.026 **Atmospheric CH_4, CO and OH from 1860 to 1985.**
A. M. Thompson, R. J. Cicerone.
Nature, Vol. 321, No. 6066, p. 148 – 150 (1986).
It is known that global atmospheric methane concentrations are increasing; less definite data suggest that carbon monoxide is also increasing. The authors use a photochemical model to simulate the trace–gas composition of the unpolluted troposphere at the start of the industrial era (taken as 1860), and at intervals up to 1985. They find that the OH concentration in the background troposphere has decreased significantly and O_3 has increased due to increases of CH_4 and CO; calculated changes depend on temporal trends of $NO_x(NO_x = NO + NO_2)$, for which no historical data are available.

082.027 **On the yearly cycle of the ozone content.**
M. A. Nuzhdina.
Vestn. Kiev. Univ. Astron., No. 27, p. 38 – 42 (1985). In Russian.

082.028 **Second variant of reductions in astronomical refraction for temperature and pressure.**
A. I. Nefed'eva.
Izv. Astron. Ehngel'gardt. Obs., No. 48, p. 76 – 95 (1984). In Russian.

082.029 **Comparison of the values of refraction calculated from the Pulkovo tables, new tables and tables of Harzer.**
A. I. Nefed'eva.
Izv. Astron. Ehngel'gardt. Obs., No. 49, p. 81 – 105 (1985). In Russian.
Results of a comparison of the new tables of astronomical refraction calculated at the Engelhardt Astronomical Observatory with the Pulkovo tables and with the tables of Harzer are given. It is found that there are differences in the tables between reductions for vapour pressure and for temperature. Corrections for the change of pressure and for the variations of the acceleration of gravity are almost identical.

082.030 **Noctilucent clouds.**
V. A. Bronshtehn.
Zemlya Vselennaya, No. 1, p. 38 – 43 (1986). In Russian.

082.031 **Study of the intensity of the atmospheric water vapor line 10832.15 Å in Moscow and Alma–Ata.**
M. I. Divlekeev.
Astron. Tsirk., No. 1385, p. 1 – 3 (1985). In Russian.

082.032 **Sodar measurements of the turbulent optical factor in the atmospheric boundary layer near the 6–m telescope of the Academy of Sciences of the USSR.**
M. A. Kallistratova, G. A. Karyukin, I. V. Petenko, N. S. Thieme, E. A. Shurygin.
Astron. Tsirk., No. 1388, p. 5 – 7 (1985). In Russian.

082.033 **On night–time image quality at different heights above the ground.**
P. V. Sheglov, A. Eh. Gur'yanov.
Astron. Tsirk., No. 1397, p. 4 – 8 (1985). In Russian. Remarks to the paper of V. V. Kravtsov et al. published in Astron. Tsirk., No. 1376, 1985, see 40.082.103.

082.034 **Study of nightglow in the Hβ region.**
A. S. Kutyrev.
Astron. Tsirk., No. 1403, p. 4 – 5 (1985). In Russian.

082.035 **Meteorological conditions affecting observing quality on Mauna Kea.**
D. A. Erasmus.
Publ. Astron. Soc. Pac., Vol. 98, No. 600, p. 254 – 259 (1986).
The meteorological conditions that affect astronomical observing quality on Mauna Kea are discussed. Prevailing weather systems produce temporal and spatial variations in the atmospheric conditions that influence the observing quality. Cirrus–cloud generation and turbulence in the free atmosphere are linked to synoptic scale weather phenomena. The structure of turbulence in the surface boundary layer is related to mesoscale circulation features and local effects.

082.036 **Model for artificial night–sky illumination.**
R. H. Garstang.
Publ. Astron. Soc. Pac., Vol. 98, No. 601, p. 364 – 375 (1986).
A model has been constructed to allow calculation of the night–sky brightness caused by a city at its center and outside the city, and at arbitrary zenith distances. A circular city of uniform brightness is assumed, with the total brightness proportional to the population. Molecular scattering and aerosol scattering are included, with the amount of aerosols being an adjustable parameter, and different scale heights being adopted for molecules and aerosols. The reflectivity of the ground and the fraction of light radiated above the horizontal are taken as parameters. Applications are given to several cities, to the general population–distance relations, to brightness–distance relations, and to the city center brightness–population relations.

082.037 **The observation, calculation, and possible forecasting of astronomical seeing.**
C. E. Coulman, J.–C. André, P. Lacarrère, P. R. Gillingham.
Publ. Astron. Soc. Pac., Vol. 98, No. 601, p. 376 – 387 (1986).
The possibility of forecasting seeing quality from routine meteorological data used as input to a numerical model of the boundary layer is discussed after demonstrating examples of reasonably good agreement between observed and calculated seeing on three nights when observed image diameter did not exceed 2 arc sec. Under poor seeing conditions (5 arc sec) agreement was not achieved. A numerical model appears useful to predict C_n^2 profiles of the atmosphere but cannot handle dome seeing effects, which often constitute a significant contribution to overall image quality.

082.038 **Deviation of the mean daily values of wind velocity from the mean seasonal ones in the lower thermosphere above Tadzhikistan.**
Kh. N. Nabotov, K. A. Karimov.
Komety Meteory, No. 36, p. 42 – 44 (1984). In Russian.

082.039 **Quasi–periodic variations of gamma–ray intensity in the atmosphere.**
A. M. Gal'per, L. P. Gorbachev, L. V. Kurnosova, N. G. Lejkov, B. I. Luchkov, G. S. Gokhale, S. V. Damle, P. K. Kunte, B. V. Sreekantan.
Tr. Fiz. inst. Akad. Nauk SSSR, Tom 162, p. 55 – 59 (1985). In Russian. Abstr. in Ref. Zh., 51. Astron., 4.51.363 (1986).

082.040 **On the influence of the earth atmosphere on conditions for observing a solar eclipse in the radio range.**
Sh. Sh. Gusejnov, B. Barri, N. Magassuba, A. P. Molchanov, I. E. Pogodin, V. M. Somsikov.
The ionosphere and solar–terrestrial relations, p. 85 – 92 (1985). In Russian. Abstr. in Ref. Zh., 51. Astron., 4.51.923 (1986). – See Abstr. 003.025.

082.041 **Scintillations of stars from observations with the Salyut 7 orbital station.**
A. S. Gurvich, I. Zakharov, V. Kan, V. V. Lebedev, A. A. Nesterenko, L. Neuzhil, A. I. Pakhomov, S. A. Savchenko.
Izv. Akad. Nauk SSSR. Fiz. atmos. okeana, Tom 21, No. 12, p. 1235 – 1241 (1985). In Russian. Abstr. in Ref. Zh., 62. Issled. Kosm. Prostranstva, 4.62.627 (1986).

082.042 **A comparative analysis of the volcanic impact on the climates of the Earth and Mars.**
K. Ya. Kondratyev (*K. Ya. Kondrat'ev*), N. I. Moskalenko, S. N. Parzhin.
Earth, Moon, Planets, Vol. 35, No. 1, p. 13 – 18 (1986).
The climatic impact of volcanic activity on the Earth and Mars is considered and a comparative analysis is made. The effect of gaseous components and of aerosols, formed in the process of the evolution of volcanic cloud, on the greenhouse effect and climate on the Earth and Mars is analyzed. A numerical modelling has shown that the climatic impact of volcanic activity depends on the optical properties. In connection with sporadic variations in the intensity of volcanic activity, a temporal modulation of climate takes place with alternating periods of warming and cooling. An important role of SO_2 and volcanic aerosols in the formation of climate in the presence of evolution of the Earth and Mars is revealed.

082.043 **Water vapor radiometer observation at GSI (*Geographical Survey Institute*).**
T. Saito.
Proceedings of the Symposium on Application of Space Techniques to Astronomy and Geophysics, p. 83 – 87 (1986). In Japanese. – See Abstr. 012.027.

082.044 **Astroclimatic research in Mongolia.**
D. Haltar, N. Tugjsuren.
Sov. Astron., Vol. 29, No. 4, p. 474 – 477 (1985). English translation of 40.082.036.

082.045 **Solar absorption microwave measurement of upper atmospheric water vapor.**
J. J. Olivero, J. J. Tsou, C. L. Croskey, L. C. Hale, R. G. Joiner.
Geophys. Res. Lett., Vol. 13, No. 3, p. 197 – 200 (1986).
Ground–based microwave measurements of upper atmospheric water vapor were made during late 1981 at Penn State University using the spectral line decomposition of solar microwave (22.235 GHz) radiation transmitted through the atmosphere.

082.046 **Comparison of recent empirical models of the thermosphere.**
G. W. Prölss, P. W. Blum.
ESA Spec. Publ., ESA SP–246, p. 3 – 11 (1986). – See Abstr. 012.028.
To evaluate the accuracy of present–day thermospheric models, predictions of the C and MSIS–83 models were compared.

082.047 **Comparison of the ESRO 4 thermosphere model with ESRO 4 data during magnetically quiet conditions.**
G. W. Prölss, M. Roemer.
ESA Spec. Publ., ESA SP–246, p. 13 – 17 (1986). – See Abstr. 012.028.

082.048 **Anomalous perigee shift and eccentricity variation due to air drag in the reentry phase.**
K. G. H. Schuchardt, W. Priester, P. W. Blum, H. G. Peters.
ESA Spec. Publ., ESA SP–246, p. 55 – 59 (1986). – See Abstr. 012.028.
With the help of a new computational scheme, which is based on fundamental equations of satellite drag analysis, the amplitudes and phases of global density variations are derived.

082.049 **Atmospheric turbidity estimations from photometric extinction measurements.**
M. Othman.
Proceedings of the Second Asian–Pacific Regional Meeting on Astronomy, p. 541 – 550 (1984). – See Abstr. 012.031.
The author shows that a routine astronomical procedure such as measuring extinction coefficients can be used for monitoring atmospheric pollution.

082.050 **Upper atmosphere density distribution from a drag analysis of INTERCOSMOS satellites.**
L. Sehnal.
Nablyud. Iskusstv. Sputnikov Zemli, No. 21, Chast' I, p. 230 – 239 (1982). – See Abstr. 012.032.
The density distribution for several altitudes is derived by drag analysis of satellites INTERCOSMOS 10, 11, 14, 17 and 18 and models of distribution are proposed. The comparison to densities computed from models DTM, C and MSIS concentrates on the semi–annual and diurnal effects.

082.051 **Structural variations of neutral gas density during the D–region winter anomaly derived from satellite position measurements.**
K.–H. Ohle, G. Sonnemann, R. Reimer, D. Felske.
Nablyud. Iskusstv. Sputnikov Zemli, No. 21, Chast' I, p. 275 – 283 (1982). – See Abstr. 012.032.

082.052 **Dynamical parameters of the upper atmosphere from an orbital analysis of the satellites INTERCOSMOS 18 and MAGION.**
P. Lala, L. Sehnal.
Nablyud. Iskusstv. Sputnikov Zemli, No. 21, Chast' I, p. 284 – 294 (1982). – See Abstr. 012.032.
Results of a long–time analysis of the relative motion of the first Czechoslovak–made satellite MAGION and its mother satellite INTERCOSMOS 18 are presented. The variations of the mean daily motion are used to compare the computed density values with those derived from the existing models of the earth atmosphere based on spherical harmonics representation. An attempt is made to improve the preliminary estimate of the physical parameters of both bodies.

082.053 **Several results of drag determination for EXPLORER 39 by the parallactic method.**
Yu. V. Surnin, Yu. V. Dement'ev.
Nablyud. Iskusstv. Sputnikov Zemli, No. 21, Chast' I, p. 295 – 301 (1982). In Russian.– See Abstr. 012.032.
The efficiency of optical observations and a parallactic method which is a further development of Zhongolovich's and Lozinsky's techniques for the computation of the quasi–draconic period of a satellite has been tested on the example of EXPLORER 39. The drag determination has been performed by using results of one photographic observation campaign of the "Atmosphera" program. The mean error of the draconic period as determined by the parallactic method is 5 – 7 millisecond. The relatively uniform and dense distribution of period determinations along a monthly arc of EXPLORER 39 allowed for the first time to determine the drag with a rather high accuracy from photographic observations obtained at a small number of stations.

082.054 **Correlation between upper atmospheric density and intensity of galactic cosmic radiation.**
E. Ellés–Almàr.
Nablyud. Iskusstv. Sputnikov Zemli, No. 21, Chast' I, p. 308 – 312 (1982). – See Abstr. 012.032.
It has been suggested earlier that galactic cosmic radiation count rate C_{dr} can be used as a new index of density variations in the upper atmosphere of the earth. Based on drag data of 4 satellites in 1966 – 67 – perigee heights between 200 and 270 km – the authors determined the partial correlation coefficients between the measured density and C_{dr} for these four satellites separately. After removing by a linear regression model other effects the following partial correlation coefficients have been derived: –0.32, –0.36, –0.18 and –0.39. The correlations are significant.

082.055 **Upper atmosphere density determinations from orbital acceleration of PAGEOS 1.**
I. Predeanu.
Nablyud. Iskusstv. Sputnikov Zemli, No. 21, Chast' I, p. 313 – 322 (1982). – See Abstr. 012.032.
The atmospheric densities determined at heights above 2000 km for the interval MJD 39308 – 39976 by applying an analytical method are analyzed and compared with the corresponding ones computed according to the CIRA 1972 atmospheric model. Densities resulting from observed variations of the orbital period of the satellite PAGEOS 1 are generally higher than those calculated theoretically.

082.056 **Scintillations and random refraction during occultations by terrestrial atmosphere.**
G. M. Grechko, A. S. Gurvich, V. Kan, S. V. Sokolovskij, S. A. Savchenko.
J. Opt. Soc. Am. A, Vol. 2, No. 12, p. 2120 – 2123 (1985).
The authors report the result of observations of atmospheric refraction fluctuations and scintillations performed aboard the Salyut–6 orbiter. Bodies of various angular diameters (sun, planets, and stars) were used for the observations. It was ascertained that random variations in refraction owing to the horizontally layered structure of the refractive index take place almost everywhere in the upper troposphere and lower stratosphere. Observations of refraction made it possible to obtain estimates of the characteristic parameters of its horizontally layered structure. A spectrum of glitter was calculated on the basis of observed scintillations of Venus.

082.057 **Two–scale solution for atmospheric scintillation.**
A. M. Whitman, M. J. Beran.
J. Opt. Soc. Am. A, Vol. 2, No. 12, p. 2133 – 2143 (1985).
The authors use a modified Kolmogorov spectrum for the index–of–refraction fluctuations and present results in two dimensions for comparison with numerical solutions. They also present results in three dimensions that should represent a quantitatively accurate, theoretical description of the atmospheric–scintillation problem.

082.058 **Modelling the ion composition of the thermosphere on the basis of mass spectrometer measurements aboard the rocket Vertical 6.**
O. P. Krasitskij, V. A. Ershova, V. A. Kochnev.
Kosm. Issled., Tom 24, Vyp. 3, p. 440 – 447 (1986). In Russian. English translation in Cosm. Res.

082.059 **A method for computing the astronomical refraction from aerological data and compilation of refraction tables.**
I. S. Guseva.
Glav. Astron. Obs. Akad. Nauk SSSR, Leningrad, 45 pp. (1985). In Russian. Abstr. in Ref. Zh., 51. Astron., 5.51.173 (1986).

082.060 **Some results of an analysis of new refraction tables.**
I. S. Guseva.
Glav. Astron. Obs. Akad. Nauk SSSR, Leningrad, 10 pp. (1985). In Russian. Abstr. in Ref. Zh., 51. Astron., 5.51.174 (1986).

082.061 **Anomalies of astronomical refraction – methods of detection and investigation.**
V. V. Kirichuk, A. V. Arkhangel'skij.
L'vov. politekh. inst. L'vov, 12 pp. (1985). In Russian. Abstr. in Ref. Zh., 51. Astron., 5.51.175 (1986).

082.062 **Estimate of the amplitude of scintillation of star images and the contribution of atmospheric layers.**
E. K. Bibikov, V. U. Klimik, I. V. Shvalagin.
Uzhgorod. univ. Uzhgorod, 10 pp. (1985). In Russian. Abstr. in Ref. Zh., 51. Astron., 5.51.176 (1986).

082.063 **Sky brightness and atmospheric extinction at A.N.R.A.O., Parkes and Siding Spring Observatory.**
P. Payne.
Proc. Astron. Soc. Aust., Vol. 6, No. 2, p. 182 – 186 (1985).
In May 1984 site testing began for the Automated Patrol Telescope. A series of atmospheric extinction and night sky brightness measurements were made over the following year at Parkes and Siding Spring Mountain using B and V filters. The paper discusses the procedures used for obtaining the results and their limitations.

082.064 **Gaseous ammonia and ammonium ions in the free troposphere.**
H. Ziereis, F. Arnold.
Nature, Vol. 321, No. 6069, p. 503 – 505 (1986).
Ammonia, the most important alkaline compound commonly found in the atmosphere, has a key role in aerosol chemistry. The authors report on the first measurements of NH_4^+ ions in the free atmosphere at altitudes from 4,000 to 8,000 m. Ammonia vapour abundances inferred from these data are very low suggesting efficient heterogeneous ammonia removal and very low ammonia saturation pressures over aerosols.

082.065 **Acetone measurements in the upper troposphere and lower stratosphere – implications for hydroxyl radical abundances.**
F. Arnold, G. Knop, H. Ziereis.
Nature, Vol. 321, No. 6069, p. 505 – 507 (1986).
The authors report on the first measurements of atmospheric acetone abundances over an extended altitude range (5,900 – 11,300 m) covering the lowermost part of the stratosphere and the upper troposphere. The measured upper tropospheric abundances (average 120 p.p.t.v.) are substantially lower than the average values (470 p.p.t.v.) measured at ground level. Above the tropopause, a marked decrease of the acetone mixing ratio was observed.

082.066 **Solar occultation study. Final report.**
P. H. G. Dickinson, R. Holdaway, J. A. Pyle, J. Ballard, G. Spalding, H. K. Roscoe.
Rutherford Appleton Lab., Rep., RAL–86–023, 175 & Draft 17 pp. (1986).
This study examines the usefulness of solar occultation spectroscopy from space for investigation of the composition of the atmosphere and the usefulness of potential host mission orbits for this purpose.

082.067 **Comparative estimates of climatic consequences of Martian dust storms and of possible nuclear war.**
G. S. Golitsyn, A. S. Ginsburg (*A. S. Ginzburg*).
Tellus, Ser. B, Vol. 37B, No. 3, p. 173 – 181 (1985). Abstr. in Phys. Abstr., Vol. 89, No. 1273, Entry 18452 (1986).

082.068 **Polarization effect on radiative transfer in Chandrasekhar's planetary problem.**
T. Takashima.
Appl. Math. Comput., Vol. 17, No. 3, p. 185 – 227 (1985). Abstr. in Phys. Abstr., Vol. 89, No. 1274, Entry 23660 (1986).

082.069 **Structure of the primordial atmosphere surrounding the early–Earth.**
K. Nakazawa, H. Mizuno, M. Sekiya, C. Hayashi.
J. Geomagn. Geoelectr., Vol. 37, No. 8, p. 781 – 799 (1985). Abstr. in Phys. Abstr., Vol. 89, No. 1274, Entry 23670 (1986).

082.070 **Turbidimetric observations carried out at the Astronomical Observatory of Cagliari in 1982.**
V. Quesada.
Circ. Stn. Astron. Int. Latitudine, Carloforte–Cagliari, N. 29 (Ser. C(6)) (1985).
The author reports on the results of turbidimetric parameters during the first measurement campaign in the interval between June and December 1982 which took place at the Cagliari Observatory (Punta Sa Menta–Capoterra).

082.071 **Methods to correct for the tropospheric delay in satellite–earth range measurements.**
G. Elgered, J. Johansson, B. Rönnäng.
Onsala Space Obs., Prepr., No. 85:48, 17 pp. (1985).

082.072 **On the sensitive measurement of horizontal air temperature gradients near an astrometric instrument for correcting anomalous refraction.**
N.–s. Hu, Z.–h. Wang, X.–r. Jiang.
Publ. Shaanxi Astron. Obs., Vol. 7, No. 2, p. 109 – 110 (1984). Abstract. In Chinese and English.

082.073 **Out of site – out of mind, the commonsense of site selection.**
N. J. Woolf.
Prepr. Steward Obs., No. 668, 8 pp. (1986).

082.074 **Noctilucent clouds over western Europe during 1984.**
D. M. Gavine.
Meteorol. Mag., Vol. 114, No. 1360, p. 349 – 356 (1985). Abstr. in Phys. Abstr., Vol. 89, No. 1277, Entry 42026 (1986).

082.075 **Extinction variations at La Silla.**
F. Rufener.
Messenger, No. 44, p. 32 – 35 (1986).

082.076 **Seeing at La Silla: LASSCA 86.**
M. Sarazin.
Messenger, No. 44, p. 36 – 37 (1986).

082.077 **The semi–annual variation of the upper atmosphere density.**
L. Sehnal, Y. E. Helali, M. Y. Tawadrous, B. B. Baghos.
Bull. Astron. Inst. Czech., Vol. 37, No. 4, p. 196 – 202 (1986).

082.078 **Oxygen emissions in the night airglow of the Earth, Venus and Mars.**
V. A. Krasnopolsky (*V. A. Krasnopol'skij*).
Planet. Space Sci., Vol. 34, No. 6, p. 511 – 518 (1986).
Airglow data are analyzed to obtain a set of excitation, energy transfer and quenching reactions and their rate coefficients. To reduce the number of reactions a restriction is imposed on the reactions proceeding via planar collisional complexes with all reactants and products in Σ or S states. The airglow altitude profiles measured can be explained by the excitation of five metastable O_2 states and $O(^1S)$ via a common precursor having the ratio of the rate coefficient of quenching by O to that by O_2 equal to 5. Only $O_2(^2\pi_g)$ satisfies all constraints imposed on this precursor.

082.079 **Comparison of radiative transfer calculated and measured in the lower stratosphere in the far–infrared.**
P. Rabache, B. Rebours.
Infrared Phys., Vol. 26, No 2, p. 123 – 130 (1986).
The spectra calculated and measured in the FIR during an observation of the terrestrial limb of the lower stratosphere are compared. This comparison makes it quite clear what the current limits on measurements and the difficulties in calibrating the spectra are. In the lower stratosphere, where temperature increases with altitude, the continuum of the spectrum measured in the absolute rotation range is difficult to reproduce. The Lorentz line profile must be corrected as a funtion of wavenumber and temperature.

082.080 **A technique for the analysis of two–dimensional data: application to O I 5577 Å airglow.**
R. Elphinstone, J. S. Murphree, L. L. Cogger.
J. Atmos. Terr. Phys., Vol. 48, No. 3, p. 213 – 220 (1986).
A method to determine whether a two–dimensional data set can be represented by two independent separable functions is demonstrated using both synthetic data and satellite observations of O I 5577 Å airglow. For such cases when the functions are separable it is possible to identify the function in the presence of noise using an iterative procedure based on the properties of an equitable matrix. In the specific case of the airglow, the latitudinal and temporal variations are separable for the data set studied. The spatial structure shows a minimum near the equator, with maxima at 35° in the winter hemisphere and 25° in the summer hemisphere. The long–term temporal function shows maxima after the equinoxes, with minima near the solstice. A 12 month component has a maximum near day 300.

082.081 **Observation of bremsstrahlung from solar electrons in the earth's atmosphere.**
J. R. Barcus, K. D. Hudnut, P. Stauning, I. B. Iversen, J. Phillips.
J. Atmos. Terr. Phys., Vol. 48, No. 4, p. 375 – 384 (1986).
Energetic photons (23 – 142 keV) were detected by a balloon–borne scintillator at 74.8° invariant latitude during the solar particle event originating in a flare observed at 0507 UT on 14 August 1982. An abrupt onset of excess radiation was recorded at 0511 UT both in the earth's atmosphere by the balloon instrumentation at 34 km altitude and also by ground–based riometers located in the central polar caps (> 75° inv. lat.). Solar protons did not appear in the inner magnetosphere until after 06 UT. Comparison of the balloon measurements with bremsstrahlung calculations for solar electron impact has confirmed the solitary role of solar electrons during the early part of the event. Later on, when solar protons were also present in the magnetosphere, energetic photons arising from nuclear reactions of protons with atmospheric nuclei also contributed.

082.082 **Atmospheric X–rays at 11°S geomagnetic latitude.**
O. Pinto Jr., W. D. Gonzalez.
J. Atmos. Terr. Phys., Vol. 48, No. 5, p. 485 – 489 (1986).
X–ray measurements at balloon altitudes were made at São José dos Campos, Brasil. Atmospheric X–rays, namely secondary X–radiation from cosmic rays were measured for the energy interval 30 – 155 keV and up to an atmospheric depth of 5.5 g cm^{-2}. A comparison of the flux measured at the Pfotzer maximum during these measurements with those obtained previously by several research groups at other latitudes and with a similar technique has also been made. Finally, a comparison of the atmospheric component with that attributed to the diffuse component is also presented and it is concluded that both components are of about the same magnitude at ~5 g cm^{-2} and at ~11°S geomagnetic latitude.

082.083 **Radiative cooling of the 30 – 110 km atmospheric layer.**
V. I. Fomichev, G. M. Shved, A. A. Kutepov.
J. Atmos. Terr. Phys., Vol. 48, No. 6, p. 529 – 544 (1986).
Near the solstice and equinox, the altitude–latitude distributions of cooling rate in the 15 μm CO_2 and 9.6 μm O_3 bands and in the rotational H_2O band have been calculated. The atmosphere near 75 km in the vicinity of the equator is heated due to thermal radiation transfer in all seasons. The atmosphere is also heated around the pole near the mesopause in spring, and in autumn it is slightly cooled. Estimates are given of the influence of tropospheric clouds on radiative cooling above 25 km and of additional cooling of the lower thermosphere caused by the propagation of gravity waves and tidal modes. Conclusions have been drawn on the dynamics of the middle atmosphere and its contribution to the thermal regime.

082.084 **Accurate cooling rates of the 15 μm CO_2 band – comparison with recent parameterizations.**
R. Haus.
J. Atmos. Terr. Phys., Vol. 48, No. 6, p. 559 – 561 (1986).
Atmospheric cooling rates of the 15 μm CO_2 band in the height range 30 – 110 km are calculated on the basis of spectral highly resolved transmittances (line–by–line). Simultaneous use of some approximations makes it possible to determine non–LTE cooling rates without knowledge of non–equilibrium source functions. Recent parameterizations produce relatively high accuracies and, therefore, are very useful for applications to dynamical and climate models.

082.085 **Identification of ^{18}O–isotopic lines of ozone in infrared ground–based solar absorption spectra.**
C. P. Rinsland, V. M. Devi, J.–M. Flaud, C. Camy–Peyret, M. A. H. Smith, G. M. Stokes.
J. Geophys. Res., Vol. 90, No. D6, p. 10719 – 10725 (1985).
A number of relatively well–isolated lines of the v_3 bands of $^{16}O^{18}O^{16}O$ and $^{16}O^{16}O^{18}O$ have been identified in high–resolution solar absorption spectra recorded with the Fourier transform interferometer at the National Solar Observatory on Kitt Peak (elevation 2095 m, latitude 31.9°N). Based on spectroscopic parameters derived from ozone laboratory spectra recorded with the same instrument, the heavy ozone content of the total column of ozone has been determined from measurements of the equivalent widths of $^{16}O^{18}O^{16}O$, $^{16}O^{16}O^{18}O$, and $^{16}O_3$ lines in three solar absorption spectra. Relative to the concentrations in a natural sample of ozone, the analysis indicates $^{16}O^{18}O^{16}O$ and $^{16}O^{16}O^{18}O$ isotopic abundances of 1.05 ± 0.07 and 1.11 ± 0.11 in the total column of atmospheric ozone.

082.086 **Sodium nightglow: chemically independent of sodium content.**
W. Swider.
J. Geophys. Res., Vol. 91, No. D6, p. 6742 – 6746 (1986).
Observations that both mesospheric ozone and the sodium nightglow attain maxima during the equinoxes is in qualitative accord with the Chapman mechanism as the dominant source of the sodium nightglow at 5893 Å.

082.087 **Physik der Hochatmosphäre.**
J. Taubenheim.
Physik des Planeten Erde, p. 97 – 127 (1985). – See Abstr. 003.047.

082.088 **Physik der unteren Atmosphäre.**
C. Hänsel.
Physik des Planeten Erde, p. 128 – 148 (1985). – See Abstr. 003.047.

082.089 **Atmospheric refraction: its effects on geodetic levelling.**
A. H. Dodson, M. C. Davidson.
Geophys. J. R. Astron. Soc., Vol. 85, No. 1, p. 252 (1986). Abstract. – See Abstr. 012.086.

On the astroclimate in the USSR.
See Abstr. 003.081.

Transfer processes in an anisotropic circumterrestrial plasma.
See Abstr. 003.147.

Investigation of the earth atmosphere from the aureols of the sun and moon.
See Abstr. 003.180.

Growth of knowledge of astronomical refraction.
See Abstr. 004.120.

The viability with respect to temperature, of micro–organisms incident on the Earth's atmosphere.
See Abstr. 015.025.

Self, nitrogen and oxygen broadening of the 115–GHz line of carbon monoxide.
See Abstr. 022.010.

Shock–induced CO_2 loss from $CaCO_3$; implications for early planetary atmospheres.
See Abstr. 022.098.

Stellar image profiles from linear detectors and the throughput of astronomical instruments.
See Abstr. 034.012.

Noise and astrometric precision. II. Atmospheric effects.
See Abstr. 036.013.

A computer approach to handling extinction.
See Abstr. 036.105.

Effect of amplitude and phase scintillations on radio astronomical observations at decimeter wavelengths.
See Abstr. 036.163.

The effect of abnormal atmospheric refraction on the precision of time determination.
See Abstr. 044.066.

The influence of the atmospheric zonal tides on the Earth's rotation rate.
See Abstr. 044.091.

Motion of a satellite in the earth's atmosphere based on spherical harmonics representation.
See Abstr. 052.031.

Determination of the periods of rotation of different satellites by two methods using observations from one station.
See Abstr. 052.033.

Limb darkening fluctuations: the discrepancy with Doppler data and the 160–min oscillation.
See Abstr. 080.081.

Evolution of an impact–induced atmosphere and magma ocean on the accreting Earth.
See Abstr. 081.001.

Chemical effects of large impacts on the Earth's primitive atmosphere.
See Abstr. 081.002.

The formation of an impact–generated H_2O atmosphere and its implications for the early thermal history of the earth.
See Abstr. 081.053.

Tropical convection, ionospheric potentials and global circuit variation.
See Abstr. 083.005.

A rocket measurement of the $O_2(b^1\Sigma_g{}^+ - X^3\Sigma_g{}^-)$ (0–0) atmospheric band in a pulsating aurora.
See Abstr. 084.051.

Effects of solar electromagnetic radiation on the terrestrial environment.
See Abstr. 085.003.

Density variations in the Earth's atmosphere connected with the 11–year solar cycle.
See Abstr. 085.016.

Stochastic simulation of interaction between solar radio radiation and rarefied gas of the earth's upper atmosphere.
See Abstr. 085.017.

A collisional kinetic theory of a plane parallel evaporating planetary atmosphere.
See Abstr. 091.009.

New evidence for the rôle of photoelectrons in the H_2–dayglow of the giant planets.
See Abstr. 091.032.

Review of relative and absolute abundances of atmophilic gases in glass from the EETA 79001 shergottite, and comparison with the martian and terrestrial atmospheres.
See Abstr. 105.124.

Absolute *UBV* photometry at the Zacatecas Observatory.
See Abstr. 113.038.

083 Ionosphere

083.001 **Polar ionospheric E–region plasma wave stabilization and electron heating by wave–induced enhancement of the electron collision frequency.**
F. Primdahl.
Phys. Scr., Vol. 33, No. 2, p. 187 – 191 (1986).

Comparative measurements indicate that the phase velocity of low frequency waves in the ionospheric E–region is often lower than the drift speed of the electrons that drive the waves unstable. In order to understand these results, the linear instability theory is modified to conform with zero growth rate even above threshold.

083.002 **On the presence of periodic variations of the electron concentration in the mean–latitude E ionosphere with a period close to the solar 160–min oscillations.**
V. P. Bobova, B. M. Vladimirskij, A. M. Lyatskaya.
Izv. Krymskoj Astrofiz. Obs., Tom 70, p. 46 – 51 (1985). In Russian. English translation in Bull. Crimean Astrophys. Obs., Vol. 70.

The critical frequencies f_0E and f_0F_2 derived from ionograms obtained in 1974, 1976 and 1980 were analysed by the cosinor method to search for the period of 160–min coinciding with the period of solar oscillation. It has been revealed that the period which is close to $160^{m}.010$ is present in f_0E at quiet conditions.

083.003 **About the C layer in the D region of the ionosphere.**
K.–j. Yang.
Acta Geophys. Sin., Vol. 29, No. 1, p. 1 – 7 (1986). In Chinese.

083.004 **Experimental observation of ultra–low–frequency waves generated in the ionosphere.**
S. Ganguly.
Nature, Vol. 320, No. 6062, p. 511 – 513 (1986).

Nonlinear properties of the ionosphere can be used to produce electromagnetic waves of ultra–low frequencies (ULF, 0 – 300 Hz). The author reports observations of such ULF signals in two experiments. In the first, four HR transmitters were split into two pairs and operated with a frequency separation Δf between each of the pairs. In the second experiment, all of the transmitters were operated at the same high frequency and were pulsed. The ULF signals at 3, 5 and 6.25 Hz and with intensities of $\sim 150-350\times 10^{-11}$G $Hz^{-1/2}$ were received on the island of Mona, which is ~150 km west of the transmitting site.

083.005 **Tropical convection, ionospheric potentials and global circuit variation.**
R. Markson.
Nature, Vol. 320, No. 6063, p. 588 – 594 (1986).
The ionospheric potential (V_I), a measure of the Earth's overall electric field intensity, correlates with the classic "Carnegie curve" diurnal variation in electric field caused by the global distribution of thunderstorms. A comparison of V_I variation with satellite images of high tropical cloud fields indicates that equatorial thunderstorms over the continents are dominant in maintaining the Earth's electric field.

083.006 **On the interconnection of wave processes in the F region and mesopause.**
V. V. Kol'tsov, I. A. Frejzon, Yu. V. Gorokhov,
V. A. Vazherkin, E. I. Chertok, B. P. Potapov,
V. A. Sukhodoev.
Geomagn. Aehron., Tom 26, No. 1, p. 43 – 46 (1986). In Russian. English translation in Geomagn. Aeron.

083.007 **Transient electrodynamical characteristics of the high-latitude ionosphere.**
G. V. Popov.
Geomagn. Aehron., Tom 26, No. 1, p. 52 – 56 (1986). In Russian. English translation in Geomagn. Aeron.

083.008 **An operative method for solving the diffusion equation for hydrogen ions in the ionosphere.**
M. A. Koen, A. D. Kazmirov.
Geomagn. Aehron., Tom 26, No. 1, p. 57 – 62 (1986). In Russian. English translation in Geomagn. Aeron.

083.009 **Determination of the structure of ionospheric irregularities.**
V. E. Kunitsyn.
Geomagn. Aehron., Tom 26, No. 1, p. 75 – 81 (1986). In Russian. English translation in Geomagn. Aeron.

083.010 **Control of a mathematical model of the ionosphere by current experimental data.**
A. G. Kolesnik, V. A. Belov.
Geomagn. Aehron., Tom 26, No. 2, p. 238 – 242 (1986). In Russian. English translation in Geomagn. Aeron.

083.011 **Wave motions in an anisotropic ionosphere.**
A. I. Gvelesiani, G. V. Dzhandieri.
Geomagn. Aehron., Tom 26, No. 2, p. 325 – 327 (1986). In Russian. English translation in Geomagn. Aeron.

083.012 **Earth's equatorial ionosphere and the heliospheric current sheet.**
J. H. Sastri.
Earth, Moon, Planets, Vol. 35, No. 1, p. 85 – 91 (1986).
Evidence is presented to show that during epochs of high sunspot activity, the duration of manifestation of equatorial spread–F (ESF) irregularities in the Earth's equatorial ionosphere undergoes a systematic modulation around the times of crossing of the heliospheric current sheet by the Earth.

083.013 **Observations of the total electron content (TEC) of the ionosphere in Neustrelitz (GDR) in the years 1976 – 1980 and possibilities of using these data for geodetic purposes.**
N. Jakowski, E. Paasch.
Nablyud. Iskusstv. Sputnikov Zemli, No. 21, Chast' I, p. 264 – 274 (1982). – See Abstr. 012.032.
Observations of the total electron content (TEC) have been performed by differential Doppler and Farady rotation methods in Neustrelitz/GDR in the years 1976 – 1980, a phase of ascending solar activity. Expressions are derived to demonstrate the importance of TEC data for ionospheric correction in geodetic applications. The average level of TEC shows a linear relationship with the smoothed Zurich sunspot numbers $\bar{R}_Z$. The percentage standard deviation of TEC amounts to about 20–25% and is not correlated with monthly R_Z and A_P values. The interrelation between TEC observations and geodetic measurements on the basis of the Doppler effect of satellite signals is briefly discussed.

083.014 **Dynamics of the F2 region as a function of disturbances of the magnetosphere and solar wind.**
R. A. Zevakina, M. V. Kiseleva.
Geomagn. Aehron., Tom 26, No. 3, p. 391 – 395 (1986). In Russian. English translation in Geomagn. Aeron.

083.015 **The mechanism of VLF emission of photoelectrons in the outer ionosphere.**
G. D. Gefan, A. A. Trukhan, G. V. Khazanov.
Geomagn. Aehron., Tom 26, No. 3, p. 416 – 420 (1986). In Russian. English translation in Geomagn. Aeron.

083.016 **Planetary distribution of the amplitude of the 27–day variation of the electron density in the E– and D–layers of the ionosphere.**
A. M. Chkhetiya.
Geomagn. Aehron., Tom 26, No. 3, p. 497 – 499 (1986). In Russian. English translation in Geomagn. Aeron.

083.017 **Nonlinear propagation of whistlers in the ionosphere.**
B. Chakraborty, G. C. Das, A. K. Sur, S. N. Paul.
Astrophys. Space Sci., Vol. 123, No. 2, p. 259 – 271 (1986).
The nonlinear dispersion relation for whistlers in the ionosphere has been derived and then the group travel time for an ion–cyclotron whistler from its source to an observer at the satellite has been theoretically calculated. It is seen that the nonlinear effect has some important contribution in the expression of group travel time. From numerical estimations, it is found that the group travel time of whistlers may be changed reasonably due to nonlinear interaction of the wave and the plasma of the ionosphere.

083.018 **Ion compositions in the auroral ionosphere as observed by EISCAT.**
C. Lathuillere, A. Brekke.
Ann. Geophys., Vol. 3, No. 5, p. 557 – 568 (1985). Abstr. in Phys. Abstr., Vol. 89, No. 1277, Entry 42034 (1986).

083.019 **Dynamical plasmapause positions during the July 29 – 31, 1977, storm period: a comparison of observations and time–dependent model calculations.**
Y. Corcuff, P. Corcuff, J. Lemaire.
Ann. Geophys., Vol. 3, No. 5, p. 569 – 580 (1985). Abstr. in Phys. Abstr., Vol. 89, No. 1277, Entry 42035 (1986).

083.020 **Modelling the plasmasphere for the International Reference Ionosphere.**
M. J. Rycroft, I. R. Jones.
Adv. Space Res., Vol. 5, No. 10, p. 21 – 27 (1985). – See Abstr. 012.064.
The paper discusses how profiles of electron and/or ion distributions that are produced by two different computer models can be smoothly coupled together. The first of these models is the empirical International Reference Ionosphere which produces a vertical profile of ionospheric parameters up to an altitude of 1000 km. The second is a physically–based, diffusive equilibrium model of the plasmasphere based upon the theoretical work of Angerami and Thomas (1964), in which plasma is constrained to move parallel to the earth's magnetic field lines. Some problems associated with this work are considered, as are some initial results.

083.021 **Latitudinal asymmetry in electron and ion density distribution in southern and northern hemispheres.**
Adv. Space Res., Vol. 5, No. 10, p. 33 – 36 (1985). – See Abstr. 012.064.
With data of satellite Intercosmos–Bulgaria–1300 northern and corresponding southern hemisphere plasma densities have been compared. Southern densities are greater in the –90 to –180 and +30 to +120° ranges. The opposite is true for –60 to +30°.

No asymmetry has been observed during daytime. These results are explained by the variations of the magnetic declination.

083.022 **On the relative abundance of helium ions in the topside ionosphere.**
R. Koleva, I. S. Kutiev.
Adv. Space Res., Vol. 5, No. 10, p. 37 – 40 (1985). – See Abstr. 012.064.
Based on data from the satellite Intercosmos–Bulgaria–1300, the latitudinal distribution of oxygen and helium ions in the topside ionosphere is discussed for night–time equinox at high solar activity. A comparison with the corresponding IRI–79 distribution is made. The vertical IRI ion composition profile is checked with measurements made with the Vertical–10 rocket. Some recommendations are made in order to improve the IRI–modelling of the ion composition in the topside ionosphere.

083.023 **Determining electron density profiles for the middle ionosphere.**
K. Rawer.
Adv. Space Res., Vol. 5, No. 10, p. 43 – 49 (1985). – See Abstr. 012.064.
Using daytime numerical ionospheric profiles from W. Becker's mid–latitude collection, the geometric parameters of 3 or 4 LAY–functions were determined by best fit while all amplitudes were redetermined step by step with a least squares criterion. It appeared that the transition height and scale of the main function are interrelated while all other geometric parameters are independent. Median values for a spring and a summer period are found, and relations with the peak altitude and half–density thickness of the input profile are established.

083.024 **Refinement in the diurnal variation of IRI–79 electron density distribution.**
Y. V. Ramanamurty.
Adv. Space Res., Vol. 5, No. 10, p. 73 – 75 (1985). – See Abstr. 012.064.
While briefly touching the interpolation technique, used in the IRI–79 description and those used in other ionospheric (prediction) models, the use of solar zenith angle rather than solar time is recommended in the empirical modelling of the peak parameters of the F1 and F2 regions for incorporation into the contemplated new version of the IRI for the middle ionosphere.

083.025 **Input from station–oriented observations and its assimilation into the new formula for the International Reference Lower Ionosphere (IRLI).**
Y. V. Ramanamurty.
Adv. Space Res., Vol. 5, No. 10, p. 79 – 88 (1985). – See Abstr. 012.064.

083.026 **Report on the discussions on modelling the lower ionosphere.**
Y. V. Ramanamurty.
Adv. Space Res., Vol. 5, No. 10, p. 89 – 90 (1985). – See Abstr. 012.064.

083.027 **Electron density models for the lower ionosphere.**
W. C. Bain.
Adv. Space Res., Vol. 5, No. 10, p. 91 – 94 (1985). – See Abstr. 012.064.
Probably the only reliable method of checking an electron density model below 70 km is to calculate from it what would be obtained by VLF or LF propagation over certain paths, and to compare the results with actual observations. This has been done for the IRI at various frequencies from 16 to 70 kHz; the results agree in places but differ substantially elsewhere. Previous models described by the author give satisfactory results and it is suggested that certain features of them might be incorporated with advantage in the IRI. In particular, it is impossible to get agreement with VLF propagation in all seasons by means of a model varying only with solar zenith angle, such as the IRI from 50 – 90 km.

083.028 **Comparison between D– and lower E–region electron density profiles and IRI–79.**
G. A. Moraitis.
Adv. Space Res., Vol. 5, No. 10, p. 95 – 98 (1985). – See Abstr. 012.064.
The D and E–region electron density profiles obtained by different techniques are compared with the IRI–79 model to see how they fit. The rocket data showed good agreement. However discrepancies between the observed and model values were found especially for solar zenith angle greater than 50 degrees.

083.029 **A simplified D–region ion chemistry scheme and its possible use for IRI lower ionosphere modelling.**
J. Taubenheim, B. S. N. Prasad.
Adv. Space Res., Vol. 5, No. 10, p. 107 – 110 (1985). – See Abstr. 012.064.

083.030 **Normalized electron production rate profiles as a result of penetration of high–energy solar particles into the lower ionosphere.**
P. I. Velinov, M. Gerdjikova.
Adv. Space Res., Vol. 5, No. 10, p. 111 – 114 (1985). – See Abstr. 012.064.
Present aim of the International Reference Ionosphere (IRI) is the modelling of the ionospheric plasma parameters under quiet conditions. Disturbed conditions, though not yet described in IRI, are of some interest for users. This paper deals with solar particle effects, which happen to occur by day and night while the solar UV– and X–ray effects are restricted to daytime only.

083.031 **Implementation of the new electron temperature model in IRI.**
D. Bilitza.
Adv. Space Res., Vol. 5, No. 10, p. 117 – 121 (1985). – See Abstr. 012.064.
Since the first edition of IRI in 1978, considerable efforts have been directed towards an improved and refined electron temperature description. Finally, a new empirical model was presented at the COSPAR meeting at Graz, Austria, in 1984. Based mainly on satellite data it highlights a more detailed diurnal, seasonal and altitudinal variation. The author describes the implementation of this model into the IRI framework. A first comparison with incoherent scatter measurements indicates the improved diurnal and seasonal performance of the model.

083.032 **Heat balance of the ionosphere: implications for the International Reference Ionosphere.**
D. Bilitza.
Adv. Space Res., Vol. 5, No. 10, p. 123 – 130 (1985). – See Abstr. 012.064.
Theoretical considerations can be helpful tools in modelling ionospheric parameters in regions and for times where not enough experimental data are available. This study asks whether results of heat balance calculations should be introduced to supplement the data base for the International Reference Ionosphere (IRI). The author discusses the present status of the theoretical understanding and examines the influence of the following unresolved or neglected terms: (1) electron heating rate, (2) electron cooling by fine structure excitation of atomic oxygen, and (3) height–dependent Coulomb Logarithm.

083.033 **Relationship of the interplanetary electric field to the high–latitude ionospheric electric field and currents: observations and model simulation.**
C. R. Clauer, P. M. Banks.
J. Geophys. Res., Vol. 91, No. A6, p. 6959 – 6971 (1986).

083.034 **Comparison of simultaneous Chatanika and Millstone Hill observations with ionospheric model predictions.**
C. E. Rasmussen, R. W. Schunk, J. J. Sojka, V. B. Wickwar, O. de la Beaujardiere, J. Foster, J. Holt, D. S. Evans, E. Nielsen.
J. Geophys. Res., Vol. 91, No. A6, p. 6986 – 6998 (1986).
As part of the MITHRAS program, the Chatanika and Millstone Hill incoherent–scatter radars made coordinated ob-

servations of the polar ionosphere on June 27 and 28, 1981. The authors compare these data with predictions made by a high–latitude ionospheric model.

083.035 **Effects of different convection models upon the high–latitude ionosphere.**
C. E. Rasmussen, R. W. Schunk, J. J. Sojka.
J. Geophys. Res., Vol. 91, No. A6, p. 6999 – 7005 (1986).
Two empirical models of plasma convection that have been proposed for use in studying the ionosphere are the Volland and Heelis models. Both of these models provide a similar description of two–celled ionospheric convection, but they differ in several ways, in particular, in the manner in which plasma flows over the central polar cap and near the polar cap boundary. In order to obtain a better understanding of the way in which these two models affect the ionosphere, two separate runs of the authors' high–latitude, time–dependent ionospheric model were made, with only the convection models distinguishing the two runs.

The propagation of radio waves.
See Abstr. 003.068.

Highlights of the 1985 URSI/COSPAR workshop on the IRI.
See Abstr. 011.037.

The polar cusp. Proceedings of a NATO Advanced Workshop on the "Morphology and dynamics of the polar cusp", held at Lillehammer, Norway, 7 – 12 May 1984.
See Abstr. 012.085.

Diffusion and thermal diffusion in partially ionized gases in the atmospheres of the Sun and planets.
See Abstr. 022.028.

Internal gravitational waves, fluctuations of the ultraviolet solar radiation and of the ionospheric parameters.
See Abstr. 076.009.

The near–equatorial thermosphere and F1 layer around the winter solstice period.
See Abstr. 082.023.

The highest geoactivity in 1981 and its source on the sun.
See Abstr. 084.011.

Interconnection of auroral electron fluxes and longitudinal currents with optical emissions in the high–latitude ionosphere from Intercosmos–Bulgaria 1300 observations.
See Abstr. 084.021.

A numerical model of magnetosphere–ionosphere coupling: preliminary results.
See Abstr. 084.071.

Electric and electrothermal conductivity of the ionospheres of planets. II. The Earth, Mars, Venus.
See Abstr. 091.018.

084 Aurorae, Geomagnetic Field, Magnetosphere

084.001 **Dependence of auroral middle UV emissions on the incident electron spectrum and neutral atmosphere.**
R. E. Daniell Jr., D. J. Strickland.
J. Geophys. Res., Vol. 91, No. A1, p. 321 – 327 (1986).
The authors examine the relationship between three important middle UV auroral emissions, the electron energy spectrum, and the neutral atmosphere. They display calculated emission rates for the Vegard–Kaplan and second positive band systems of N_2 and the O I($^1S \rightarrow {}^3P$) 2972 Å line.

084.002 **Auroral morphology of the midday oval.**
C.–I. Meng, R. Lundin.
J. Geophys. Res., Vol. 91, No. A2, p. 1572 – 1584 (1986).
In this paper, the auroral displays in the noon sector were examined to determine the morphology; their relationship with dayside magnetospheric structures are discussed.

084.003 **Geomagnetic and solar data. September 1985.**
H. E. Coffey.
J. Geophys. Res., Vol. 91, No. A2, p. 1737 (1986).

084.004 **Geomagnetic and solar data. October 1985.**
H. E. Coffey.
J. Geophys. Res., Vol. 91, No. A2, p. 1738 (1986).

084.005 **A first generation numerical geomagnetic storm prediction scheme.**
S.–I. Akasofu, C. F. Fry.
Planet. Space Sci., Vol. 34, No. 1, p. 77 – 92 (1986).

084.006 **Geomagnetic field analysis. IV. Testing the frozen–flux hypothesis.**
J. Bloxham, D. Gubbins.
Geophys. J. R. Astron. Soc., Vol. 84, No. 1, p. 139 – 152 (1986).
The authors present a model of the magnetic field at the core–mantle boundary, for epoch 1959.5, based on a large set of observatory and survey measurements.

084.007 **The effect of the solar wind on the terrestrial environment.**
N. U. Crooker, G. L. Siscoe.
Physics of the Sun, Vol. III, p. 193 – 249 (1986). – See Abstr. 003.001.
Contents: General morphology. Solar wind and geomagnetic activity (solar wind streams, geomagnetic response to streams, periodic geomagnetic activity). Transfer mechanisms at the magnetopause (magnetic merging, other mechanisms, composite model). Magnetospheric convection. Magnetosphere effects on the ionosphere and thermosphere. Middle and lower atmosphere.

084.008 **Dependence of the aurora borealis occurrences on the solar–terrestrial parameters.**
Y. Liritzis, B. Petropoulos.
Earth, Moon, Planets, Vol. 34, No. 1, p. 65 – 75 (1986).
In the present investigation, the authors establish that, for the 20th solar cycle, the occurrence of visual discrete aurorae A, deduced from a complete set of data, is significantly related to the sunspot numbers R_z, the number of flares F (of importance $\geqslant 1$) the solar wind streams derived from solar coronal holes H, and the geomagnetic index A_p. The A–R_z relationship was investigated in particular for the period 1897 – 1951. For this period spectrum analysis of A annual values revealed the existence of 3 – 4 yr and 8 – 10 yr periodicities of significances 95% and 99% respectively.

084.009 **Maximum entropy and power spectrum analysis of geomagnetic intensity variations from archaeomagnetic data: emphasis on the 200–year period.**
Y. Liritzis.
Earth, Moon, Planets, Vol. 34, No. 3, p. 235 – 249 (1986).
Statistical analyses on available archaeomagnetic intensity data mainly from the Aegean area and supplemented by some from Bulgaria and Czechoslovakia have revealed the existence of periodicities. The power spectrum analysis showed the existence of periods of 100 – 120, 200, 250, 540, 800 and 1600 years with an overall significance level of around 95%. Maximum entropy method showed the existence of significant periods of 100 – 130, 200, 240, 400 – 450 and 1200 – 1350 years. Emphasis is given to the 200–year period in relation to similar periods of other solar–terrestrial phenomena as well as to theoretical calculations.

084.010 **Implications of a steady–state magnetospheric convection.**
A. T. Y. Lui, A. Hasegawa.
Planet. Space Sci., Vol. 34, No. 3, p. 315 – 318 (1986).
Fundamental to the development of many magnetospheric theories is the assumption of a steady–state magnetospheric convection. It is shown for the first time that this basic assumption is implicitly related to the source of magnetospheric plasma. The existence of a steady–state electric field is consistent with plasma entry into the magnetosphere from the North–South direction. If plasma entry is through the flanks of the magnetosphere, the large–scale magnetospheric electric field is intrinsically time–dependent. Thus, magnetospheric theories with the underlying assumption that the large–scale electric field is derivable from a scalar potential cannot deal with such a situation. Therefore, the origin of magnetospheric plasma and the time–variability of magnetospheric convection are directly related.

084.011 **The highest geoactivity in 1981 and its source on the sun.**
J. Sýkora, J. Halenka, J. Laštovička, P. Perglerová.
Bull. Astron. Inst. Czech., Vol. 37, No. 2, p. 120 – 126 (1986). With plates 3, 4.
The last decade of July 1981 was characterized by exceedingly high geoactivity, which the authors present in terms of the activity in the geomagnetic field and in the lower and upper ionosphere. Some interesting deviations from the common behaviour of the ionosphere during the highest geomagnetic storms are described. The studied geoactivity originated during an evidently inactive period on the sun. Based on the monochromatic corona observations carried out during the total solar eclipse of July 31, 1981 the authors conclude that a sporadic, low–latitude coronal hole could have been the source of this exceptionally high geoactivity.

084.012 **Auroral $N_2(c_4'^1\Sigma_u^+ \rightarrow a^1\Pi_g)$ emission, II.**
T. G. Slanger.
Planet. Space Sci., Vol. 34, No. 4, p. 399 – 401 (1986).

084.013 **Geomagnetic reversal spurts and episodes of extraterrestrial catastrophism.**
P. C. Pal, K. M. Creer.
Nature, Vol. 320, No. 6058, p. 148 – 150 (1986).
The possibility that the observed 26 – 32 Myr periodicity in the records of extincion events was caused by periodic episodes of physical bombardment of the planetary surface has raised the likelihood that such an astronomical clock also affects geomagnetic reversals. The authors examine the temporal relation between geomagnetic reversals and those catastrophic episodes which occurred during periods of frequent reversals. The rationale is that the geomagnetic field should be most susceptible to catastrophism during its relatively unstable periods of frequent reversals. The authors find spurts in the frequency of geomagnetic reversals at approximately 30–Myr intervals, correlated with catastrophic episodes. These spurts can be seen as the geomagnetic signatures of terrestrial catastrophism, and they are tentatively ascribed to enhanced core turbulence during episodes of bombardment.

084.014 **Transitional field behaviour from Southern Hemisphere lavas: evidence for two–stage reversals of the geodynamo.**
K. A. Hoffman.
Nature, Vol. 320, No. 6059, p. 228 – 232 (1986).
Palaeomagnetic records from Australasian basalts suggest that a geomagnetic reversal is accomplished in two steps, separated by an interval of relative dynamo quiescence. A potentially long–lived field configuration, developed early, is observed to be the 'stepping stone' from which both successful and unsuccessful reversal attempts are made. Several aborted attempts involving the same intermediate field geometry may precede an actual polarity transition.

084.015 **The earth's magnetotail.**
E. W. Hones Jr.
Sci. Am., Vol. 254, No. 3, p. 32 – 39 (1986).
The solar wind sweeps the earth's magnetic field into a vast tail. Disruptions of the tail generate bright auroras at the earth and propel great bodies of magnetized gas into interplanetary space.

084.016 **The theta aurora.**
L. A. Frank, J. D. Craven, D. A. Gurnett, S. D. Shawhan, D. R. Weimer, J. L. Burch, J. D. Winningham, C. R. Chappell, J. H. Waite, R. A. Heelis, N. C. Maynard, M. Sugiura, W. K. Peterson, E. G. Sheeley.
J. Geophys. Res., Vol. 91, No. A3, p. 3177 – 3224 (1986).
The theta aurora is a remarkable configuration of auroral and polar cap luminosities for which a generally sun–aligned transpolar arc extends contiguously from the dayside to nightside sectors of the auroral oval. Four individual occurrences of theta aurora over earth's northern hemisphere are examined in detail with the global auroral imaging instrumentation on board the high–altitude, polar–orbiting spacecraft DE 1. Simultaneous measurements of fields and plasmas with this high–altitude spacecraft and its low–altitude, polar–orbiting companion, DE 2, are examined in order to establish an overview of auroral and polar cap phenomena associated with the appearance of the theta aurora.

084.017 **Geomagnetic and solar data. November 1985.**
H. E. Coffey.
J. Geophys. Res., Vol. 91, No. A3, p. 3328 (1986).

084.018 **Geomagnetic and solar data. December 1985.**
H. E. Coffey.
J. Geophys. Res., Vol. 91, No. A4, p. 4609 (1986).

084.019 **The ring current dynamics as given by data of direct measurements and by data on solar cosmic rays in the magnetosphere.**
Eh. N. Sosnovets, L. V. Tverskaya.
Geomagn. Aehron., Tom 26, No. 1, p. 107 – 113 (1986). In Russian. English translation in Geomagn. Aeron.

084.020 **Passage of protons precipitating from the radiation belt through the atmosphere.**
A. F. Iyudin, V. G. Kirillov–Ugryumov, Yu. D. Kotov, V. A. Shurshakov.
Geomagn. Aehron., Tom 26, No. 2, p. 214 – 217 (1986). In Russian. English translation in Geomagn. Aeron.

084.021 **Interconnection of auroral electron fluxes and longitudinal currents with optical emissions in the high–latitude ionosphere from Intercosmos–Bulgaria 1300 observations.**
A. K. Kuz'min, V. I. Lazarev, M. V. Tel'tsov, S. I. Shkol'nikova.
Geomagn. Aehron., Tom 26, No. 2, p. 218 – 224 (1986). In Russian. English translation in Geomagn. Aeron.

084.022 **Auroral particles associated with non–linear Alfvén waves.**
V. M. Chmyrev, S. V. Bilichenko, Yu. B. Kazanskaya, V. M. Kostin, V. I. Lazarev, M. V. Tel'tsov.
Geomagn. Aehron., Tom 26, No. 2, p. 275 – 282 (1986). In Russian. English translation in Geomagn. Aeron.

084.023 **Energy spectra of protons with energy T > 150 MeV precipitating from the radiation belt as a function of the height above sea level.**
A. F. Iyudin, V. G. Kirillov–Ugryumov, Yu. D. Kotov, V. A. Shurshakov.
Geomagn. Aehron., Tom 26, No. 2, p. 316 – 318 (1986). In Russian. English translation in Geomagn. Aeron.

084.024 **Relationship between 20–year geomagnetic variations and polar changes of the latitude.**
Yu. D. Kalinin, G. V. Fedyaeva.
Geomagn. Aehron., Tom 26, No. 2, p. 352 (1986). In Russian. English translation in Geomagn. Aeron.

084.025 **Ion distribution near the front of the earth's bow shock (11.V.1985).**
O. L. Vajsberg, G. N. Zastenker, V. N. Smirnov, Z. Nemeček, I. Šafránková, L. A. Avanov, E. I. Kolesnikova.
Kosm. Issled., Tom 24, Vyp. 2, p. 166 – 176 (1986). In Russian. English translation in Cosm. Res.

084.026 **On the microstructure on the earth's bow shock.**
O. L. Vajsberg, V. N. Smirnov, L. A. Avanov.
Kosm. Issled., Tom 24, Vyp. 2, p. 235 – 245 (1986). In Russian. English translation in Cosm. Res.

084.027 **Natural oscillations of the neutral layer of the earth's magnetosphere.**
A. I. Gvelesiani, V. G. Kirtskhaliya.
Tr. Inst. geofiz. Akad. Nauk GSSR, Tom 53, p. 20 – 22 (1985). In Russian. Abstr. in Ref. Zh., 62. Issled. Kosm. Prostranstva, 4.62.555 (1986).

084.028 **Geomagnetic disturbances and their connection with the parameters of the interplanetary magnetic field.**
I. M. Zaalishvili.
Tr. Inst. geofiz. Akad. Nauk GSSR, Tom 53, p. 30 – 36 (1985). In Russian. Abstr. in Ref. Zh., 62. Issled. Kosm. Prostranstva, 4.62.569 (1986).

084.029 **The altitude region sampled by ground–based Doppler temperature measurements of the O I 15867 K emission line in aurorae.**
R. J. Sica, M. H. Rees, R. G. Roble, G. Hernandez, G. J. Romick.
Planet. Space Sci., Vol. 34, No. 5, p. 483 – 488 (1986).

084.030 **The role of energetic O^+ precipitation in a mid–latitude aurora.**
M. Ishimoto, M. R. Torr, P. G. Richards, D. G. Torr.
J. Geophys. Res., Vol. 91, No. A5, p. 5793 – 5802 (1986).
The authors show that flux of precipitating energetic O^+ that have been observed by satellites in the topside ionosphere can explain the magnitude of the N_2^+(1 N) (first negative) 3914–Å and N_2(2 P) (second positive) 3371–Å emission rate observed during a mid–latitude aurora over Logan, Utah (41°N, 111°W), on September 21 – 22, 1982).

084.031 **Magnetic reconnection in the Earth's magnetotail.**
E. W. Hones Jr.
Aust. J. Phys., Vol. 38, No. 6, p. 981 – 997 (1985). Paper presented at the R. G. Giovanelli Commemorative Colloquium, Part II, Tucson, Arizona, 17 – 18 January 1985.
In this paper the author discusses briefly evidence for magnetic reconnection and plasmoid formation on the Sun (in flares and possibly at other times as well) and, at greater length, the evidence that these processes occur in the Earth's magnetotail during substorms. The evidence for their occurrence on the Sun is substantial; that for their occurrence in the magnetotail is even stronger. There are obvious analogies between the reconnection scenario for the Sun and that for the magnetotail.

084.032 **The influence of the magnetospheric electric field on the plasmasphere's structure.**
E. V. Voronov, I. A. Krinberg.
Kosm. Issled., Tom 24, Vyp. 3, p. 418 – 427 (1986). In Russian. English translation in Cosm. Res.

084.033 **Height distribution of luminosity of the molecular nitrogen bands in polar aurorae.**
O. I. Yagodkina, V. G. Vorob'ev, T. N. Kolosova.
Geomagn. Aehron., Tom 26, No. 3, p. 421 – 425 (1986). In Russian. English translation in Geomagn. Aeron.

084.034 **VLF diagnostics of the electron concentration in the earth's magnetosphere.**
V. V. Krechetov.
Geomagn. Aehron., Tom 26, No. 3, p. 437 – 440 (1986). In Russian. English translation in Geomagn. Aeron.

084.035 **The relation of magnetospheric disturbances with the parameters of the interplanetary medium.**
O. V. Khorosheva.
Geomagn. Aehron., Tom 26, No. 3, p. 447 – 453 (1986). In Russian. English translation in Geomagn. Aeron.

084.036 **The decay parameter of the ring current.**
V. Yu. Pisarskij, N. M. Rudneva, Ya. I. Fel'dshtejn, A. Grafe.
Geomagn. Aehron., Tom 26, No. 3, p. 454 – 460 (1986). In Russian. English translation in Geomagn. Aeron.

084.037 **Modelling the geomagnetic effect of the ring current based on the hourly mean parameters of the interplanetary medium.**
V. Yu. Pisarskij, N. M. Rudneva, Ya. I. Fel'dshtejn, A. Grafe.
Geomagn. Aehron., Tom 26, No. 3, p. 461 – 466 (1986). In Russian. English translation in Geomagn. Aeron.

084.038 **Paleoinclination of the magnetic field at Kamchatka for 8500 years.**
N. P. Ben'kova, V. V. Kochegura, V. P. Nazarets.
Geomagn. Aehron., Tom 26, No. 3, p. 480 – 485 (1986). In Russian. English translation in Geomagn. Aeron.

084.039 **The normal geomagnetic field of the USSR as given by aeromagnetic survey data.**
R. T. Vasil'ev, I. P. Zherebchenko, D. A. Zherebchenko.
Geomagn. Aehron., Tom 26, No. 3, p. 519 – 523 (1986). In Russian. English translation in Geomagn. Aeron.

084.040 **The international analytic field and its secular variation for the period of 1980 – 1990.**
V. P. Golovkov, G. I. Kolomijtseva.
Geomagn. Aehron., Tom 26, No. 3, p. 523 – 525 (1986). In Russian. English translation in Geomagn. Aeron.

084.041 **The stochastic excitation of reversals in simple dynamos.**
D. Crossley, O. Jensen.
Phys. Earth Planet. Inter., Vol. 42, No. 3, p. 143 – 153 (1986).
Several variations of the αω disc dynamo models are known to exhibit chaotic magnetic field behaviour typical of non–linear, dissipative systems. Although these models demonstrate that geomagnetic reversals can be generated by simplified dynamo equations, the behaviour of the magnetic field itself is generally too simple, showing especially an absence of long polarity epochs in most of the models. The authors show that the addition of three varieties of stochastic processes (Gaussian, flicker and Brownian noise) enriches the field evolution and can lead to realistic palaeomagnetic behaviour.

084.042 **Spectrum of variations of the geomagnetic AE–index in the period range 130 – 200 min as possible reflection of global oscillations of the sun.**
V. P. Bobova, B. M. Vladimirskij, M. I. Pudovkin.
Izv. Krymskoj Astrofiz. Obs., Tom 74, p. 121 – 131 (1986). In Russian. English translation in Bull. Crimean Astrophys. Obs., Vol. 74.

084.043 **Ion release experiments in the solar wind.**
D. A. Bryant.
Plasma Phys. Controlled Fusion, Vol. 27, No. 12A, p. 1369 – 1378 (1985). = Rutherford Appleton Lab., Rep., RAL–85–094. – See Abstr. 012.046.

Lithium and barium plasmas were injected into the solar wind approximately 110,000 kms above the earth's surface. The effects, monitored in situ by two spacecraft of the AMPTE (Active Magnetospheric Particle Tracer Explorers) mission, included the formation of diamagnetic cavities, as slowing down locally of the solar wind, generation of plasma waves and heating of solar–wind electrons. These comet–like interactions between the solar wind and obstacles presented by the injected plasmas are discussed and compared with the effect of the natural obstacle of the earth's magnetosphere first encountered by the solar wind at the bow shock.

084.044 **On fitting a function to the local magnetic field of the earth's surface.**
W.–j. Wu.
Kexue Tongbao, Vol. 30, No. 4, p. 493 – 497 (1985).

In the study on the time–space variations of a local magnetic field of the earth's surface, the method of fitting a function is often used. In application, the choice of the fitting function is of primary importance and the accuracy of fitting and stability of solution are two basic requirements. The mutual consistency of solutions of various geomagnetic components should be considered in the fitting problems.

084.045 **A rocket measurement of $O(^1S)$ and N_2^+ emissions in a pulsating aurora.**
I. C. McDade, E. J. Llewellyn, B. H. Solheim.
Can. J. Phys., Vol. 63, No. 7, p. 983 – 987 (1985). Abstr. in Phys. Abstr., Vol. 89, No. 1273, Entry 18540 (1986).

084.046 **Finite parallel conductivity in the open magnetosphere.**
M. A. Volkov, Yu. P. Mal'tsev.
J. Geophys., Vol. 57, No. 2, p. 125 – 129 (1985). Abstr. in Phys. Abstr., Vol. 89, No. 1273, Entry 18580 (1986).

084.047 **Gradient and curvature drift of auroral electrons in a dipole geomagnetic field.**
J. B. Clarke, J. P. S. Rash, M. W. J. Scourfield.
S.Afr. J. Phys., Vol. 8, No. 3, p. 62 – 66 (1985). Abstr. in Phys. Abstr., Vol. 89, No. 1275, Entry 30044 (1986).

084.048 **Stabilities of magnetohydrodynamic rotational discontinuities.**
S. Wang, L. Zhu.
Kexue Tongbao, Vol. 30, No. 9, p. 1166 – 1170 (1985). Abstr. in Phys. Abstr., Vol. 89, No. 1275, Entry 30122 (1986).

084.049 **Guided modes in the vicinity of the Earth's geomagnetic current sheet.**
K. I. Hopcraft, P. R. Smith.
Phys. Lett. A, Vol. 113A, No. 2, p. 75 – 78 (1985). Abstr. in Phys. Abstr., Vol. 89, No. 1275, Entry 30123 (1986).

084.050 **On the origin of the cusp field–aligned currents.**
L. C. Lee, J. R. Kan, S.–I. Akasofu.
J. Geophys., Vol. 57, No. 3, p. 217 – 221 (1985). Abstr. in Phys. Abstr., Vol. 89, No. 1276, Entry 35901 (1986).

084.051 **A rocket measurement of the $O_2(b^1\Sigma_g^+–X^3\Sigma_g^-)$ (0–0) atmospheric band in a pulsating aurora.**
I. C. McDade, E. J. Llewellyn, F. R. Harris.
Can. J. Phys., Vol. 63, No. 10, 1322 – 1329 (1985). Abstr. in Phys. Abstr., Vol. 89, No. 1277, Entry 42032 (1986).

084.052 **Field–aligned currents distribution deduced from the global electric field model.**
X.–t. Song, L. Wu, G.–k. Zhu.
Chin. J. Space Sci., Vol. 5, No. 4, p. 245 – 253 (1985). In Chinese. Abstr. in Phys. Abstr., Vol. 89, No. 1277, Entry 42073 (1986).

084.053 **Magnetopause stability threshold for patchy reconnection.**
A. A. Galeev, M. M. Kuznetsova, L. M. Zelenyj.
Space Sci. Rev., Vol. 44, Nos. 1/2, p. 1 – 41 (1986).

Contents: Introduction. The model of the magnetopause boundary layer. The set of equations for electromagnetic perturbations. Dispersion equation in the vicinity of the drift tearing mode stability threshold. General criterion of the stabilization of tearing perturbations due to their coupling with ion–sound oscillations. Marginal magnetopause thickness and spatial scale of the reconnection patchies. Dependence on the angle of magnetic field rotation θ_0. The influence of the convective draping of magnetic flux tubes along the magnetopause on the reconnection dynamics. The limitations of the theory in cases of small ($\theta_0 \cong 0°$) and large ($\theta_0 \cong 180°$) magnetopause rotation angles. Macroscopic pictures of spontaneous magnetopause reconnection. Discussion and conclusions.

084.054 **On analyzing the time series of the recent geomagnetic field.**
W. Webers.
Geod. Geophys. Veröff., Reihe III, Heft 53, p. 120 – 131 (1985). = Mitt. Zentralinst. Phys. Erde, Nr. 1333.

084.055 **The Earth's magnetic field yields its secrets.**
A. Coates, M. Smith.
New Sci., Vol. 108, No. 1486, p. 32 – 34 (1985). Abstr. in Phys. Abstr., Vol. 89, No. 1280, Entry 60563 (1986).

084.056 **Theoretical considerations of the magnetic field in a spherical Earth model.**
H. Greiner–Mai.
Astron. Nachr., Vol. 307, No. 3, p. 201 – 208 (1986). In German.

The theoretical treatment of several geophysical problems presupposes the solution of field equations of the magnetic field in the Earth's mantle. The field equations are given in a scalar form for a spherical model of the Earth. It will be shown that analytical solutions are possible in all cases.

084.057 **Investigation of the quasi–stationary spectrum of turbulence of the auroral electrojet.**
A. V. Volosevich, A. V. Kustov.
Physical processes in the circumterrestrial plasma, p. 69 – 76 (1984). In Russian. – See Abstr. 003.040.

084.058 **Reflections on geomagnetic reversals.**
L. R. Alldredge.
J. Geomagn. Geoelectr., Vol. 37, No. 11, p. 1017 – 1033 (1985). Abstr. in Phys. Abstr., Vol. 89, No. 1281, Entry 66870 (1986).

084.059 **The crustal contribution to the spatial spectra of the geomagnetic secular variation.**
J. Meyer, J.–H. Hufen, M. Sibert, A. Hahn.
J. Geomagn. Geoelectr., Vol. 37, No. 12, p. 1141 – 1145 (1985). In German. Abstr. in Phys. Abstr., Vol. 89, No. 1281, Entry 66874 (1986).

084.060 **A perturbed disk dynamo model and polarity reversals of the Earth's magnetic field.**
Y. Honkura, M. Shimizu.
J. Geomagn. Geoelectr., Vol. 37, No. 12, p. 1147 – 1155 (1985). Abstr. in Phys. Abstr., Vol. 89, No. 1281, Entry 66875 (1986).

084.061 **Internal Geomagnetic Reference Field revision 1985.**
D. R. Barraclough, W. Mundt, F. S. Barker, V. P. Golovkov, P. J. Hood, F. J. Lowes, N. W. Peddie, G.–Z. Qi, S. P. Srivastava, R. Whitworth, D. E. Winch, T. Yukutake, D. P. Zidarov, C.–c. An, E. R. Benton, R. H. Estes, D. J. Kerridge, G. I. Kolomijtseva, G. J. Kuhn, R. A. Langel, J. M. Quinn, M. A. Shea, C. F. Voorhies, A. K. Zunde, M. W. McElhinny.
J. Geomagn. Geoelectr., Vol. 37, No. 12, p. 1157 – 1163 (1985). Abstr. in Phys. Abstr., Vol. 89, No. 1281, Entry 66876 (1986).

084.062 **Study of hydrogen Balmer emissions in proton aurora.**
V. Srivastava, V. Singh.
Indian J. Radio Space Phys., Vol. 14, No. 6, p. 147 – 150 (1985). Abstr. in Phys. Abstr., Vol. 89, No. 1281, Entry 67235 (1986).

084.063 **On the current–voltage relationship of the magnetospheric generator at intermediate spatial scales.**
J. F. Vickrey, R. C. Livingston, N. B. Walker, T. A. Potemra, R. A. Heelis, M. C. Kelley, F. J. Rich.
Geophys. Res. Lett., Vol. 13, No. 6, p. 495 – 498 (1986).
Using data from the drift meter and magnetometer on board the HILAT satellite, the authors have examined fluctuations in high latitude electric and magnetic fields at scale sizes between 80 km and 3 km. A comparison of data from summer and winter allows them to assess the impact of changing ionospheric conductivity on the magnetospheric generator. The authors find that, at these scale sizes, the magnetosphere tends to behave as a constant current source that is independent of ionospheric conductivity. This characteristic was noted on both open and closed field lines. The electric field pattern, on the other hand, is much more highly structured in the winter than in the summer. This behavior implies scale size dependent potential drops on closed field lines.

084.064 **Correction to "The auroral luminosity structure in the high–latitude upper atmosphere: its dynamics and relationship to the large–scale structure of the earth's magnetosphere" [Rev. Geophys., Vol. 23, No. 3, p. 217 – 275 (1985)].**
Y. I. Feldstein (*Ya. I. Fel'dshtejn*), Yu. I. Galperin (*Yu. I. Gal'perin*).
Rev. Geophys., Vol. 24, No. 2, p. 349 (1986). See Abstr. 40.084.041.

084.065 **Hook–shaped arcs in dayside polar cap and their relation to the IMF.**
M. G. Gusev, O. A. Troshichev.
Planet. Space Sci., Vol. 34, No. 6, p. 489 – 496 (1986).

084.066 **Modulation of the solar daily geomagnetic variation.**
F. De Meyer.
J. Atmos. Terr. Phys., Vol. 48, No. 2, p. 115 – 130 (1986).
A least squares spectral analyis is used to investigate cyclic and seasonal changes of the harmonic coefficients of the solar daily and semi–diurnal variation over the 24 year interval from 1 January 1960 to 31 December 1983 at a single magnetic observatory (Dourbes, Belgium). As a preparation for the treatment of other long runs of observations a statistical method is presented to combine the output of the spectral and harmonic analyses of a group of stations. The annual means of the Fourier coefficients are significantly correlated with the solar cycle. The spectra show peaks at periods of 11 years, 1 year, 6 months and 27 days, which entails an important amplitude and phase modulation of the daily and semi–diurnal variation. No simple relationship between the peaks in the broad solar rotation band can be proposed.

084.067 **Long–term variations of the external and internal components of the earth's magnetic field.**
J. Gavoret, D. Gibert, M. Menvielle, J. L. Le Mouël.
J. Geophys. Res., Vol. 91, No. B5, p. 4787 – 4796 (1986).
The present study applies classical techniques of time series analysis to the separation of the external and internal signals from the geomagnetic field.

084.068 **Physik der Magnetosphäre.**
C.–U. Wagner.
Physik des Planeten Erde, p. 70 – 96 (1985). – See Abstr. 003.047.

084.069 **Three–dimensional axisymmetric magnetosphere in pressure balance with the solar wind.**
G. K. Yates, M. Heinemann.
J. Geophys. Res., Vol. 91, No. A6, p. 6777 – 6790 (1986).
A computer model of a magnetosphere in which the magnetic dipole axis points into the solar wind has been developed. The axial symmetry implied by this physical situation admits a two–dimensional treatment. The model features (1) a magnetopause in pressure balance with the solar wind, (2) an analytic asymptotic tail model to account for the distant magnetotail, and (3) a full accounting of the magnetohydrodynamic pressure in the magnetosphere. The principal physical results is that with an increase in the plasma pressure from zero the magnetic field configuration changes slowly at first. As the pressure increases further, the field configuration changes more rapidly until X and O lines are formed.

084.070 **A three–dimensional MHD simulation of the interaction of the solar wind with the earth's magnetosphere: the generation of field–aligned currents.**
T. Ogino.
J. Geophys. Res., Vol. 91, No. A6, p. 6791 – 6806 (1986).
A global computer simulation of the interaction of the solar wind with the earth's magnetosphere was executed by using a three–dimensional magnetohydrodynamic model. As a result, the author was able to reproduce quasi–steady–state magnetospheric configurations and a Birkeland field–aligned current system which depend on the polarity of the z component of the interplanetary magnetic field.

084.071 **A numerical model of magnetosphere–ionosphere coupling: preliminary results.**
K. Watanabe, M. Ashour–Abdalla, T. Sato.
J. Geophys. Res., Vol. 91, No. A6, p. 6973 – 6978 (1986).
A three–dimensional simulation model was developed to study magnetosphere–ionosphere coupling in the auroral region. The authors adopted one–fluid MHD equations to model the magnetosphere, and current density continuity equations were solved consistently to model the ionosphere.

084.072 **A simple model of auroral electrodynamics compared with HILAT measurements.**
M. A. Doyle, W. J. Burke, D. A. Hardy, P. F. Bythrow, F. J. Rich, T. A. Potemra.
J. Geophys. Res., Vol. 91, No. A6, p. 6979 – 6985 (1986).
Simultaneous measurements of cold plasma drift velocities, magnetic field perturbations, and precipitating electron fluxes with instrumentation on board the HILAT satellite provide an opportunity to correlate Birkeland current densities, electric fields, and ionospheric conductivities in the region of an auroral form.

084.073 **Geomagnetic and solar data. January, February 1986.**
H. E. Coffey.
J. Geophys. Res., Vol. 91, No. A6, p. 7161 – 7162 (1986).

084.074 **Kelvin–Helmholtz instability at the magnetopause.**
A. Miura.
Plasma penetration into magnetospheres, p. 53 – 69 (1986). – See Abstr. 012.071.
Linear and nonlinear characteristics of Kelvin–Helmholtz (KH) instability are reviewed in the framework of hydrodynamic and magnetohydrodynamic approximations with an emphasis on basic physical processes of the KH instability at the terrestrial magnetopause boundary and their implications in the "viscous interaction" hypothesis of the magnetospheric plasma circulation (convection).

084.075 **International Geomagnetic Reference Field revision 1985.**
Geophys. J. R. Astron. Soc., Vol. 85, No. 1, p. 217 – 220 (1986).

084.076 **The geomagnetic field during the upper Miocene.**
G. J. Sherwood, J. Shaw.
Geophys. J. R. Astron. Soc., Vol. 85, No. 1, p. 259 (1986). Abstract. – See Abstr. 012.086.

The propagation of radio waves.
See Abstr. 003.068.

Reversals of the earth's magnetic field.
See Abstr. 003.113.

Chapman conference on magnetotail physics.
See Abstr. 011.019.

The polar cusp. Proceedings of a NATO Advanced Workshop on the "Morphology and dynamics of the polar cusp", held at Lillehammer, Norway, 7 – 12 May 1984.
See Abstr. 012.085.

Backyard astronomy: the aurora.
See Abstr. 014.012.

Influence of non–dipole field on determination of Plio–Pleistocene true polar wander.
See Abstr. 044.092.

On the use of a sunward libration–point–orbiting spacecraft as an interplanetary magnetic field monitor for magnetospheric studies.
See Abstr. 051.066.

Poloidal and toroidal fields in geomagnetic field modeling.
See Abstr. 062.034.

Acceleration of charged particles in magnetic reconnection: solar flares, the magnetosphere, and solar wind.
See Abstr. 062.059.

Lower–hybrid instabilities driven by an ion velocity ring.
See Abstr. 062.114.

Short–period cyclic variations of solar activity and their geophysical consequences.
See Abstr. 072.066.

Morphology of low–frequency waves in the solar wind and their relation to ground pulsations.
See Abstr. 074.106.

Peculiarities of penetration of solar protons into the earth's magnetosphere during the event of 24 – 25 April 1979.
See Abstr. 078.011.

A conceptual model of core dynamics and the earth's magnetic field.
See Abstr. 081.023.

Geodynamic implications of short time scale changes in the geomagnetic dynamo.
See Abstr. 081.029.

On the coupling of fluid dynamics and electromagnetism at the top of the Earth's core.
See Abstr. 081.035.

Dynamics of the F2 region as a function of disturbances of the magnetosphere and solar wind.
See Abstr. 083.014.

Dynamical plasmapause positions during the July 29 – 31, 1977, storm period: a comparison of observations and time–dependent model calculations.
See Abstr. 083.019.

Relationship of the interplanetary electric field to the high–latitude ionospheric electric field and currents: observations and model simulation.
See Abstr. 083.033.

A preliminary study of the tail of comet Halley and the geomagnetic storm.
See Abstr. 103.711.

The AMPTE artificial comet experiments.
See Abstr. 106.017.

Optical observations of the AMPTE artificial comet from the Northern Hemisphere.
See Abstr. 106.018.

***In situ* magnetic field observations of the AMPTE artificial comet.**
See Abstr. 106.019.

UKS plasma measurements near the AMPTE artificial comet.
See Abstr. 106.020.

Plasma waves and wave–particle interactions seen at the UKS spacecraft during the AMPTE artificial comet experiment.
See Abstr. 106.021.

Dynamics of the AMPTE artificial comet.
See Abstr. 106.022.

Cyclic variations of a connection between the interplanetary magnetic field sector structure, high–velocity streams and geomagnetic disturbances.
See Abstr. 106.030.

The effect of the sector structure of the interplanetary magnetic field on the intensity of auroral electrojets.
See Abstr. 106.031.

Sector structure of the interplanetary magnetic field and the ring current in the magnetosphere.
See Abstr. 106.073.

085 Solar-terrestrial Relations

085.001 **Observations of the May 30, 1984, annular solar eclipse at Millstone Hill.**
J. E. Salah, W. L. Oliver, J. C. Foster, J. M. Holt, B. A. Emery, R. G. Roble.
J. Geophys. Res., Vol. 91, No. A2, p. 1651 – 1660 (1986).
The Millstone Hill incoherent scatter radar was operated on May 30, 1984, to gather data on the effects of the annular solar eclipse on the structure and dynamics of the earth's ionosphere and thermosphere.

085.002 **Ionospheric and thermospheric response over Millstone Hill to the May 30, 1984, annular solar eclipse.**
R. G. Roble, B. A. Emery, E. C. Ridley.
J. Geophys. Res., Vol. 91, No. A2, p. 1661 – 1670 (1986).
The National Center for Atmospheric Research thermospheric general circulation model is used to calculate the time–dependent thermospheric response to the May 30, 1984, annular solar eclipse.

085.003 **Effects of solar electromagnetic radiation on the terrestrial environment.**
R. E. Dickinson.
Physics of the Sun, Vol. III, p. 155 – 191 (1986). – See Abstr. 003.001.
Contents: Atmospheric structure and composition (thermosphere, stratosphere and mesosphere structure and chemistry, tropospheric chemistry). The climate system (current questions, introduction to simple climate models, trapping of thermal radiation by atmospheric constituents, thermal feedback by clouds and water vapor, anthropogenic modulation of trace gases important for climate, atmospheric and oceanic circulation and the seasons, primitive climate, the carbon cycle and the faint–early–Sun). Solar radiation drives the biosphere (origins of photosynthesis, photosynthesis in action, harvesting the sunlight, net primary productivity).

085.004 **Solar energetic particles and their effects on the terrestrial environment.**
G. C. Reid.
Physics of the Sun, Vol. III, p. 251 – 278 (1986). – See Abstr. 003.001.
Contents: Solar energetic particles and the magnetosphere. Energy loss processes (polar cap absorption, ion chemistry of the middle atmosphere: influence of solar energetic particles, polar glow aurora: optical effects of solar energetic particle precipitation). Atmospheric alterations and nuclear interactions. Effects of solar particle events (effects on radio communication and navigation, effects on global atmospheric electricity, potential impact on climate, solar energetic particles and the evolution of the atmosphere and biosphere).

085.005 **Some solar cycle phenomena related to the geomagnetic activity from 1868 to 1980. II. High velocity wind streams and cyclical behaviour of poloidal field.**
P. A. Simon, J.–P. Legrand.
Astron. Astrophys., Vol. 155, No. 2, p. 227 – 236 (1986).
The authors discuss the geomagnetic activity related to the distribution around the sun of the corotating sources themselves of solar wind. This "wind–activity" contributes to 85% of the activity level and concerns 91.5% of the days. In this paper a first category of wind–activity is studied, the so–called "recurrent storms", occurring during the declining phase of sunspot cycle. Finally the authors raise the new problem of the modulation of the cosmic ray flux. This is interesting if, in place of the limited ecliptic plane wind flow, one also considers the three dimensional properties of the polar wind stream and its cyclical behaviour.

085.006 **Peculiarities of helio–geophysical activity during November – December 1982.**
A. V. Belov, N. P. Ben'kova, V. N. Ishkov, N. S. Kaminer, E. V. Lavrova, E. V. Nepomnyashchaya, N. K. Pereyaslova, Yu. P. Sizov, L. A. Kolesnikova.
Geomagn. Aehron., Tom 26, No. 1, p. 18 – 23 (1986). In Russian. English translation in Geomagn. Aeron.

085.007 **Comparison of variations of solar and geomagnetic activity indices at different frequencies within the range of $10 \leqslant T \leqslant 100$ years.**
Yu. R. Rivin.
Geomagn. Aehron., Tom 26, No. 2, p. 353 – 356 (1986). In Russian. English translation in Geomagn. Aeron.

085.008 **1984 status report on continuous solar–terrestrial monitoring programs.**
H. E. Coffey, M. A. Shea.
Bull. Am. Astron. Soc., Vol. 17, No. 4, p. 933 (1985). Abstract. – See Abstr. 010.065.

085.009 **Solar activity and the Caspian sea level.**
P. R. Romanchuk.
Vestn. Kiev. Univ. Astron., No. 27, p. 23 – 26 (1985). In Russian.

085.010 **Effects of the earth atmosphere and 160–min solar brightness oscillations.**
N. N. Ryzhikova.
Astron. Tsirk., No. 1404, p. 4 – 6 (1985). In Russian.

085.011 **Solar activity and tree ring widths.**
G. E. Kocharov, A. N. Konstantinov, V. M. Ostryakov, A. V. Stupneva.
Soln. Dannye, Byull. 1986, No. 2, p. 84 – 89 (1986). In Russian.
The data on radial tree–ring increments from different regions are considered. A correlation of this parameter with solar activity and radiocarbon data is obtained. The authors believe that the tree rings are a good indicator of the solar activity level and the climate in the past.

085.012 **Radiocarbon cycles of geomagnetic and solar origin.**
N. Tujchiev.
Izv. Akad. Nauk TadzhSSR. Otd–nie fiz.–mat., khim. i geol. nauk, No. 2, p. 29 – 33 (1985). In Russian. Abstr. in Ref. Zh., 51. Astron., 4.51.351 (1986).

085.013 **Solar activity and seismicity of the earth.**
O. M. Barsukov.
Geofiz. polya Prikasp. regiona (Makhachkala), No. 2, p. 123 – 130 (1984). In Russian. Abstr. in Ref. Zh., 51. Astron., 4.51.365 (1986).

085.014 **Analysis of temporal series in solar–terrestrial relations.**
A. K. Kirillov, N. Ya. Semovskikh.
Astronomical–geodetical investigations, p. 157 – 167 (1985). In Russian. Abstr. in Ref. Zh., 51. Astron., 4.51.368 (1986). – See Abstr. 003.024.

085.015 **Solar affinity of sedimentary cycles in the late Precambrian Elatina Formation.**
G. E. Williams.
Aust. J. Phys., Vol. 38, No. 6, p. 1027 – 1043 (1985). Paper presented at the R. G. Giovanelli Commemorative Colloquium, Part II, Tucson, Arizona, 17 – 18 January 1985.

085.016 **Density variations in the Earth's atmosphere connected with the 11–year solar cycle.**
M. J. Ill.
Nablyud. Iskusstv. Sputnikov Zemli, No. 21, Chast' I, p. 221 – 229 (1982). – See Abstr. 012.032.

085.017 **Stochastic simulation of interaction between solar radio radiation and rarefied gas of the earth's upper atmosphere.**
M. Ya. Marov, V. I. Shematovich, G. I. Zmievskaya.
Nablyud. Iskusstv. Sputnikov Zemli, No. 21, Chast' I, p. 249 – 257 (1982). – See Abstr. 012.032.
A numerical model is considered for the interaction between the shortwave solar radiation and the earth's upper atmosphere, which results in photoexcitation, photoionization and photodissociation processes violating thermodynamical equilibrium in a medium. The field of radiation is assumed to be equivalent to a set of photons with energies corresponding to the considered spectrum of radiation. An analogy of collisional processes in a partially ionized rarefied gas is used that leads to photochemical reactions.

085.018 **Variations of the circulation and the electric field of the atmosphere preceding solar flares.**
V. G. Moiseev, N. G. Skryabin, S. N. Samsonov, O. Ya. Kalinina.
Fiz. protsessy v polyar. ionos. 1 Vses. shk. Polyar. ionos. i magnitosfer.–ionos. svyazi. Yakutsk, p. 52 – 57 (1985). In Russian. Abstr. in Ref. Zh., 51. Astron., 5.51.489 (1986).

085.019 **On a dissipative hydrodynamical model of interaction of the solar wind with the earth's magnetic field.**
V. N. Zhigulev, A. V. Kitaev.
Chisl. metod. mekh. splosh. sredy, Novosibirsk, Tom 16, No. 6, p. 66 – 76 (1985). In Russian. Abstr. in Ref. Zh., 62. Issled. Kosm. Prostranstva, 6.62.470 (1986).

085.020 **The effect of a solar perturbation on a global climate model.**
W. D. Sellers.
J. Clim. Appl. Meteorol., Vol. 24, No. 8, p. 770 – 776 (1985). Abstr. in Phys. Abstr., Vol. 89, No. 1273, Entry 18446 (1986).

085.021 **Correlation between solar radio bursts and sudden ionospheric disturbances.**
S.–r. Zhou, J.–z. Sun.
Publ. Purple Mt. Obs., Vol. 4, No. 2, p. 31 – 39 (1985). In Chinese.
This paper describes the correlation between solar radio bursts and sudden ionospheric disturbances (SID). The authors find that (1) when the relative increase of peak flux density of impulsive bursts is greater than 50%, then about 85% – 90% of the bursts at 3.2 cm and 10 cm will correspond to the SID importance 2 (or greater) respectively, and (2) when the peak flux density of gradual bursts is more than 50 s.f.u and 25 s.f.u or its duration is more than 30^m and 25^m at 3.2 cm and 10 cm respectively, then about 93% – 94% of bursts at 3.2 cm and 10 cm will correspond to the SID importance 2 (or greater), respectively.

085.022 **The sun and the ice ages of the earth.**
Y.–t. Jiang, Z.–t. Xu.
Publ. Purple Mt. Obs., Vol. 4, No. 3, p. 8 – 15 (1985). In Chinese.
The authors have reviewed the recent development of physical explanations of the ice ages based on solar flux changes. There are two kinds of physical theories. According to one of them the sun moved around the centre of our galaxy and interacted with interstellar matter. Another kind is concentrated on solar interior and then compared theoretical neutriono flux with ice ages. It shows that cooperations between astronomers and geologists may be an efficient way to resolve the enigma of the ice ages at the end.

085.023 **A relationship between the atmospheric pressure during the warm semi–year in Sofia and the Hale solar cycle.**
D. T. Valev.
Soln. Dannye, Byull., 1986, No. 3, p. 75 – 79 (1986). In Russian.
A statistical relationship is shown to exist between the atmospheric pressure during the warm semi–year in Sofia and "Hale's" Wolf numbers. It is noted that Hale's Wolf numbers show a lag in the phase with respect to the summer time atmospheric pressure and statistical confidence of the correlation found is higher 99%. The suggestion is made that the summer time atmospheric pressure in Sofia changed in phase with the global magnetic solar field.

085.024 **Interaction between the magnetized solar wind and the magnetosphere in the case of finite conductivity.**
P. F. Krymskij.
Physical processes in the circumterrestrial plasma, p. 16 – 28 (1984). In Russian. – See Abstr. 003.040.

085.025 **Empirical dependence of the density of a relativistic electron flux in a geosynchronous orbit on solar wind velocity.**
I. P. Bezrodnykh.
Physical processes in the circumterrestrial plasma, p. 46 – 49 (1984). In Russian. – See Abstr. 003.040.

085.026 **Mesospheric ozone changes associated with 27 day solar ultraviolet flux variations.**
A. C. Aikin, H. J. P. Smith.
Geophys. Res. Lett., Vol. 13, No. 5, p. 427 – 430 (1986).
Solar ultraviolet flux changes associated with the 27–day solar rotational period cause corresponding variations in mesospheric ozone near the maximum of the 11–year sunspot cycle. This statement is based on a correlation and spectral analysis of ozone mixing ratios, deduced from Solar Mesospheric Explorer satellite–based measurements of 1.27–μm O_2 airglow emission and solar flux observations made from the same spacecraft in 1982.

085.027 **Selective factors in sun–weather research.**
H. A. Taylor Jr.
Rev. Geophys., Vol. 24, No. 2, p. 329 – 348 (1986).
The extensive body of investigations of apparent correlations between solar wind/IMF disturbances and subsequent changes in the vorticity of the winter troposphere, is reviewed as representative of the search for sun–weather relationships.

085.028 **Response of ozone to short–term variations in the solar ultraviolet irradiance. 1. A theoretical model.**
R. S. Eckman.
J. Geophys. Res., Vol. 91, No. D6, p. 6695 – 6704 (1986).
The response of atmospheric ozone and temperature to variations in the solar ultraviolet irradiance over time scales corresponding to the solar rotation period is examined using a one dimensional, time–dependent radiative–photochemical model of the upper stratosphere and lower mesosphere.

085.029 **Response of ozone to short–term variations in the solar ultraviolet irradiance. 2. Observations and interpretation.**
R. S. Eckman.
J. Geophys. Res., Vol. 91, No. D6, p. 6705 – 6721 (1986).
An analysis of the response of middle atmospheric ozone to short–term variations in the solar ultraviolet irradiance is presented. Measurements of ozone from the Solar Mesosphere Explorer (SME) ultraviolet spectrometer (UVS) are compared to calculations of a one–dimensional, radiative–photochemical model of the upper stratosphere and lower mesosphere.

085.030 **Disappearing solar filaments and geomagnetic activity.**
J. H. Sastri.
Sol. Phys., Vol. 105, No. 1, p. 191 – 196 (1986).
The author reports that the recurrent high speed streams in the solar wind (Sastri et al., 1985) were consistently preceded by a stream interface, as expected of corotating streams. This observation substantiates the role of corotating streams of coronal hole origin in the apparent link between "disparition brusques" (DBs) and geomagnetic activity, and strengthens the view that DBs are not a unique source of geomagnetic activity.

Sun and earth.
See Abstr. 003.019.

From solar dynamo to terrestrial climate.
See Abstr. 080.105.

Earth's equatorial ionosphere and the heliospheric current sheet.
See Abstr. 083.012.

The effect of the solar wind on the terrestrial environment.
See Abstr. 084.007.

Dependence of the aurora borealis occurrences on the solar–terrestrial parameters.
See Abstr. 084.008.

Planetary System

091 Physics and Dynamics of the Planetary System

091.001 **Planetary surface texture and albedo from parameter plots of optical polarization data.**
J. E. Geake, A. Dollfus.
Mon. Not. R. Astron. Soc., Vol. 218, No. 1, p. 75 – 91 (1986).
Optical reflectance polarization measurements obtained with the Meudon Observatory polarimeter, in the laboratory and on the telescope, are collected together for the first time; surfaces investigated include rock and powder samples, meteorites, lunar samples and atmosphereless planetary bodies. Interpretation of the data involves plotting polarization curve parameters and albedo, two or three at a time; all the types of plot that have been found useful are reviewed, including a new three–dimensional model. The empirical relationships found from these plots, calibrated by using data for known surfaces, are then of use for the remote determination of planetary surface properties, such as surface texture and albedo; a new method of estimating grain size is described.

091.002 **Testing solar system stability.**
P. J. Message.
Nature, Vol. 319, No. 6052, p. 359 – 360 (1986).
This note discusses some recent results from numerical integrations of the motions of the major outer planets extending over several million years. These results are compared with predictions from analytical perturbation theory.

091.003 **Long–term changes in the semimajor axes of the outer planets.**
A. Milani, A. M. Nobili, K. Fox, M. Carpino.
Nature, Vol. 319, No. 6052, p. 386 – 388 (1986).
One of the oldest problems of celestial mechanics is that of the long–term behaviour of the semimajor axes *a* of the planetary orbits. The authors have performed a 9.3–Myr numerical integration of the orbits of the outer planets, using a pure newtonian point mass model. They are able to detect high–order variations in the energies, and hence also in *a*, with periods ranging from tens of thousand to millions of years. The most interesting feature is an energy exchange between Uranus and Neptune with a period of 1,119,000 years, the same as the period of the libration between the perihelia of Jupiter and Uranus. The mechanism involves Jupiter and also Saturn. The energy of Pluto changes mostly with periods close to that of the 3:2 libration in mean motion with Neptune.

091.004 **A relativistic theory of motion of the inner planets.**
Eh. L. Akim, V. A. Brumberg, M. D. Kislik, Ju. F Koljuka (*Yu. F. Kolyuka*), G. A. Krasinsky (*G. A. Krasinskij*), E. V. Pitjeva (*E. V. Pit'eva*), V. A. Shishov, V. A. Stepanianz (*V. A. Stepan'yants*), M. L. Sveshnikov, V. F. Tikhonov.
Relativity in celestial mechanics and astrometry, p. 63 – 68 (1986). – See Abstr. 012.002 (IAU Symp. No. 114).
The theory of the motion of Mercury, Venus, Earth and Mars is constructed by numerical integration. The theory takes into account relativistic corrections in the frame of Schwarzschild's space–time metrics. The theory presented is an ephemeris base for deep space experiments.

091.005 **Simulation of general relativistic corrections in long term numerical integrations of planetary orbits.**
A. M. Nobili, I. W. Roxburgh.
Relativity in celestial mechanics and astrometry, p. 105 – 111 (1986). – See Abstr. 012.002 (IAU Symp. No. 114).
Long term numerical integrations of planetary orbits designed to study the stability of the Solar System over timescales comparable to its age have become very promising thanks to the availability of very powerful computers. The stability of such numerical integrations relies on the ability to control all possible sources of error. Among the errors is the fact that Newton's theory of gravitation is used instead of general relativity. The authors show that the secular advance of perihelia predicted by general relativity can be simulated exactly by a $1/r^2$ perturbing potential with almost negligible additional cost in computer time.

091.006 **Relativistic effects from planetary and lunar observations of the XVIII – XX centuries.**
G. A. Krasinsky (*G. A. Krasinskij*), E. Yu. Aleshkina, E. V. Pitjeva (*E. V. Pit'eva*), M. L. Sveshnikov.
Relativity in celestial mechanics and astrometry, p. 315 – 328 (1986). – See Abstr. 012.002 (IAU Symp. No. 114).
Lunar and planetary observations of different types are discussed for the time span 1717 – 1982. The modern ranging observations and the historical ones (mainly transits of Mercury and Venus, solar eclipses and occultations of the inner planets by the Moon) are treated separately and some attempts to detect relativistic effects are carried out.

091.007 **Landschaftsbezeichnungen auf Planeten und Monden.**
P. M. Millman.
Astron. Raumfahrt, 24. Jahrg., Heft 1, p. 7 – 8 (1986).

091.008 **Preferential perihelion and aphelion distances, and planetary formation.**
N. A. Barricelli.
Earth, Moon, Planets, Vol. 34, No. 1, p. 1 – 33 (1986).
An interpretation of a series of precise relationships (Barricelli, 1972) between planetary perihelion and aphelion distances, leading to the definition of preferential perihelion and aphelion distances, designated as meeting distances is outlined. The interpretation is based on Alfvén's (1969) jet stream theory and the hypothesis of harmonic resonance between intersecting jet streams.

091.009 **A collisional kinetic theory of a plane parallel evaporating planetary atmosphere.**
B. Shizgal, R. Blackmore.
Planet. Space Sci., Vol. 34, No. 3, p. 279 – 291 (1986).
The departure from equilibrium in the exosphere of a planetary atmosphere owing to the loss of energetic atoms is studied with solutions of the Boltzmann equation. A rigorous collisional solution is obtained for a plane parallel model with a newly developed discrete ordinate method, and the decrease in the excape flux relative to Jeans flux is calculated together with density and temperature profiles. The escape of hydrogen and helium atoms from Earth and the escape of hydrogen atoms from Mars is considered. The results are compared with previous studies of the same effect. The departures from equilibrium are largest for the

escape of a light species from an atmosphere with a heavy background gas such as for the escape of hydrogen atoms from Mars.

091.010 **What have we learned from modeling giant planet interiors?**
M. Podolak, R. T. Reynolds.
Protostars & planets II, p. 847 – 872 (1985). – See Abstr. 012.008.
Models of the giant planets are reviewed. The theoretical techniques used in computing the models are described, and the observational and experimental inputs are summarized. Special emphasis is placed on uncertainties in these input data. The models are then examined and the results of various authors presented. It is demonstrated that all the planets have heavy–element enhancements of between 10 and 40 $M_{\oplus}$, with a large fraction of this material residing in the core. It is also shown that the ratio of ice to rock in Uranus and Neptune is on the order of three. The implications of these results for theories of the origin of the solar system are discussed.

091.011 **Semiannual oscillations in planetary atmospheres.**
A. M. Krigel'.
Astron. Zh., Tom 63, Vyp. 1, p. 166 – 169 (1986). In Russian. English translation in Sov. Astron., Vol. 30, No. 1.
The semiannual cyclicity of the wind velocity and of the temperature in the earth's equatorial stratosphere indicates the possible existence of similar oscillations on other planets. A comparison is made of published results of observations of the atmospheric circulation oscillations on Mars, Jupiter, and Saturn. The observational data lend support to the hypothesis of the existence of a semiannual cycle in the equatorial belts of planetary atmospheres, with the maximum of the rotational velocity coinciding with the equinoxes.

091.012 **On the connection of the spherical albedo of a planet with its observable characteristics.**
N. N. Fomin, Eh. G. Yanovitskij.
Astron. Vestn., Tom 20, No. 1, p. 5 – 14 (1986). In Russian. English translation in Sol. Syst. Res.
Half–empirical dependences of the geometrical albedo and the reflection coefficient of the center of a planetary disk on the spherical albedo are found. Nonstationary analogies of these values are studied. They can be used when considering the radiative transfer in molecular absorption bands.

091.013 **Determination of spectral distributions and instrumental magnitudes of planets.**
J. W. Hovenier, G. Cornet, P. Stammes.
ESA Sci. Tech. Rep., ESA STR–214, 6+108 pp. Price FF 100.00 (1985).
An algorithm is presented for calculating instrumental magnitudes in the 300 – 1100 nm range for the Moon and all planets except Pluto. An extensive survey and compilation of the relevant literature has been made to produce best estimates for the spectral distributions of the light reflected by each object. The results, together with estimates of the errors, are presented in a number of tables. The spectral distributions form the basis for the algorithm which provides magnitudes for the disk–integrated light, as a function of the object selected, phase angle and distances to the Sun and observer. In addition, information is provided to enable the calculation of the output signal of an ideal slit sensor scanning the disk of an object.

091.014 **Deuterium in the outer Solar System: evidence for two distinct reservoirs.**
T. Owen, B. L. Lutz, C. de Bergh.
Nature, Vol. 320, No. 6059, p. 244 – 246 (1986).
The authors have completed a series of determinations of the CH_3D/CH_4 ratio in the atmospheres of Saturn, Titan and Uranus. These results, coupled with the work of other investigators, suggest that the Solar System contains at least two distinctly different primordial reservoirs of deuterium: that contained in gaseous hydrogen and that contained in the volatiles which have been maintained at low temperatures or isolated from hydrogen; for example, trapped in cold, solid material. Both of these reservoirs were established before the formation of the Solar System.

091.015 **Period of 160 minutes in the solar system: solar pulsation and spin rates of planets and asteroids.**
V. A. Kotov, S. Koutchmy.
Izv. Krymskoj Astrofiz. Obs., Tom 70, p. 38 – 46 (1985). In Russian. English translation in Bull. Crimean Astrophys. Obs., Vol. 70.
It is suggested that the coincidence of the period of the solar global pulsation $P_0 \approx 160$ min with the 9th and 4th harmonics of Earth's and Saturnian days respectively, might not be a chance coincidence. To test this hypothesis a special commensurability function was constructed and the level of uncertainty was emphasized. The analysis of the commensurability with P_0 was extended to all the available data on spin rates of planets and asteroids (in all 182 objects with known spin periods in the solar system) and it was found that the 160–min commensurability is indeed dominant and remarkably significant.

091.016 **Size–compositional relationship in planetary regoliths and its role in interpreting remote sensing data.**
A. Basu.
Meteoritics, Vol. 20, No. 4, p. 605 – 606 (1985). Abstract. – See Abstr. 010.661.

091.017 **On a possible relation of the law of planetary distances to the 160–minute pulsation of the sun.**
V. A. Kotov, S. Koutchmy.
Izv. Krymskoj Astrofiz. Obs., Tom 72, p. 199 – 208 (1985). In Russian. English translation in Bull. Crimean Astrophys. Obs., Vol. 72.
The discovery of the global solar pulsation with the period of $P_0 = 160^m$ enables to introduce characteristic wavelength $L = c \times P_0 = 19.24$ A.U. for the solar system, where c is the speed of light. A corresponding analysis of planetary distances reveals the existence of a statistically significant quasi–commensurability between L and $2\pi a_i$ for inner planets and between $2a_i$ and L for outer ones (a_i being the major semi–axis of a planetary orbit). It is suggested that this L–commensurability can offer new approach to an explanation of the well–known Titius–Bode law for planetary distances. The physical mechanism enforcing the L–commensurability in the solar system is presumably related with 160^m–gravitational waves from some external source of unknown nature.

091.018 **Electric and electrothermal conductivity of the ionospheres of planets. II. The Earth, Mars, Venus.**
A. V. Pavlov.
Kosm. Issled., Tom 24, Vyp. 1, p. 122 – 127 (1986). In Russian. English translation in Cosm. Res.

091.019 **Systems of planetary rings.**
E. Mūkins.
Zvaigžnota Debess, 1984. Gada Vasara, p. 7 – 16 (1984). In Latvian.

091.020 **The dynamics of dense particle disks.**
S. Araki, S. Tremaine.
Icarus, Vol. 65, No. 1, p. 83 – 109 (1986).
The authors investigate the mechanical equilibrium and collisional transport processes in differentially rotating dense particle disks in which the filling factor is not small, so that the ordinary Boltzmann kinetic theory is not accurate. The treatment is based on the Enskog theory of dense hard sphere gases, except that the spheres are inelastic. It is shown that the viscous instability which has been suggested as a source of the structure in Saturn's B ring does not arise in the authors' models. However, the ring may be subject to a phase transition similar to the liquid–solid transition seen in molecular dynamics simulations of elastic hard spheres; in this case the ring could have alternating zero–shear ("solid") and high–shear ("liquid") zones.

091.021 **Reflectance measurements of planetary surface materials at small phase angles.**
B. J. Buratti, W. D. Smythe, R. M. Nelson, V. Charakhani, B. Hapke.
Bull. Am. Astron. Soc., Vol. 17, No. 4, p. 918 (1985). Abstract. – See Abstr. 010.063.

091.022 **Hot atom chemistry and radiolysis of gaseous and solid components of planetary atmospheres.**
K. Rössler.
ESA Spec. Publ., ESA SP–241, p. 175 – 188 (1985). – See Abstr. 012.018.
Particles with kinetic energies exceeding a few eV are relatively frequent in space. They induce chemical reactions different from those considered in cosmic chemistry hitherto and include endothermic reactions with high activation energies and atom–molecule interactions. This "hot atom chemistry" leads to the formation of molecules in rings and icy surfaces as well as in the upper atmosphere of planets and satellites. An important contribution comes from secondary particles generated by cosmic rays in the deeper layers of the atmospheres and the oceans. Laboratory simulation is performed by ion implantation and optical spectroscopy, chemical sputtering and mass spectroscopy, as well as by nuclear recoil techniques and radiochromatography. The variation of the chemical products with the radiation dose gives information on radiation chemistry in space.

091.023 **Meridian circle observations of major planets at the Engelhardt Astronomical Observatory in 1973 – 1974.**
A. Yu. Yatsenko.
Izv. Astron. Ehngel'gardt. Obs., No. 50, p. 106 – 110 (1985). In Russian.
The observed declinations of Venus, Mercury, Mars, Jupiter, Saturn, Uranus are given. The method and reduction of observations are described.

091.024 **Minimum interorbital distances of planets and long–period comets.**
V. P. Tomanov, A. E. Rozaev, G. V. Rozaeva.
Vologod. gos. ped. inst. Vologda, 26 pp. (1985). In Russian. Abstr. in Ref. Zh., 51. Astron., 3.51.137 (1986).

091.025 **On the causes of circumplanetary rings: auxiliary mechanisms which could lead to tidal destruction of some former satellites.**
V. D. Davydov.
Kosm. Issled., Tom 24, Vyp. 2, p. 298 – 306 (1986). In Russian. English translation in Cosm. Res.

091.026 **Account of solar perturbations in the motion of distant planetary satellites.**
O. N. Baturina.
Astron. Tsirk., No. 1403, p. 5 – 8 (1985). In Russian.
With the help of the Universal Poissonian Processor realized on the computer in Fortran the fourth– and fifth–order approximations to the theory of motion of far planetary satellites have been obtained.

091.027 **Spectropolarimetry of planets. I. The procedure of observations and reduction of data.**
A. L. Gural'chuk, V. A. Kucherov, A. V. Morozhenko.
Kinematika Fiz. Nebesn. Tel, Tom 2, No. 3, p. 41 – 47 (1986). In Russian. English translation in Kinematics Phys. Celest. Bodies.
The procedure of observations with the spectropolarimeter equipped with a 127° achromatic waveplate is briefly described. The algorithm for the data reduction is given.

091.028 **Electrostatic forces and shadow effects in planetary dust rings.**
O. Havnes.
Astrophys. Space Sci., Vol. 122, No. 1, p. 97 – 107 (1986).
The vertical dust distribution of dust clouds around planets, resulting from electrostatic forces, is calculated as a function of dust and plasma parameters. Photoelectric charging is included and differences between clouds on the illuminated side and in the shadow zone are examined. The author computes ring structures for conditions which may apply in the spoke–forming regions and studies at what dust and plasma conditions the shadow has a significant effect on the vertical dust cloud structure.

091.029 **Regression modelling in selenodesy.**
S. G. Valeev.
Earth, Moon, Planets, Vol. 35, No. 1, p. 1 – 5 (1986).
The existing models of different orders that describe surface and gravitational fields of planets contain a redundant number of members with no useful information, which makes it difficult to describe physical phenomena. The possibility of using statistical (regressive) modelling as an effective means of obtaining adequate relief and gravitational field models is considered. A concrete numerical example shows that the use of regression modelling in the task of describing the Moon surface by expansion on spherical functions permit to obtain approximate models with the harmonics less than a half in number compared to the traditional approach; the precision of "shortened" model approximations is higher or, at least, not lower than the precision description for the usual number of harmonics that correspond to the expansion order taken.

091.030 **Structure and evolution of the terrestrial planets.**
P. Janle, R. Meissner.
Surv. Geophys., Vol. 8, No. 2, p. 107 – 186 (1986). = Contrib. No. 309, Inst. Geophys. Univ., Kiel, F.R. Germany.
Since 1959 more than 100 successful spacecraft missions to our planetary neighbours have been carried out. The authors give an overview of the results obtained during this era and restricted their study to the terrestrial planets. A summary of the main physical properties of the whole planetary system is presented. Considering the terrestrial planets, detailed knowledge of the surface structures of Moon and Mars is available, whereas for Mercury and Venus only more general overviews exist. Geophysical measurements permit calculations of the interior structures of all terrestrial planets. Except for Mercury, detailed studies of the lithospheres are possible. The comparison of the results from various disciplines provide the key for the tectonic evolution of the terrestrial planets.

091.031 **A normal space gravity field (with an application to tidal effects).**
M. Caputo.
Manuscr. Geod., Vol. 10, No. 4, p. 245 – 254 (1985).
The author presents a model for the gravitational field of an oblate planet above its surface by giving simple formulae for the potential, the gravitational vector and the tidal tensor fields. The equipotential surfaces are ellipsoids of revolution, the lines of force and the isozenitals are hyperbolae. The tidal field caused by the second order zonal harmonic of the gravitational field of a planet is discussed according to the "Nahewirkungsgesetz" (principle of local action) of Weyl (1921).

091.032 **New evidence for the rôle of photoelectrons in the H_2–dayglow of the giant planets.**
R. Prange.
Astron. Astrophys., Vol. 161, No. 1, p. L1 – L4 (1986).
The dayside, planetwide H_2 Rydberg emissions in the EUV range observed by the UVS experiment on the Voyager spacecraft have often been interpreted as the consequence of electron collisions from precipitating electrons. A critical review is made of the main characteristics of these emissions and it is pointed out that their morphological features very closely agree with a photoelectron distribution one's. For this purpose use is done of the terrestrial example.

091.033 **Origin of the gap between planetary ring and satellite systems, and the size and structure of the ring particles.**
N. N. Gor'kavyj, A. M. Fridman.
Sov. Astron. Lett., Vol. 11, No. 4, p. 264 – 266 (1985). English translation of 40.091.006.

091.034 **On the dynamics of an electron beam in the photoelectron layer of an atmosphereless celestial body.**
E. K. Kolesnikov, A. S. Manujlov.
Red. zh. Vestn. Leningr. Univ., Mat., Mekh. Astron., 8 pp. (1985). In Russian. Abstr. in Ref. Zh., 51. Astron., 2.51.295 (1986).

091.035 **Ringsysteme um Planeten in unserem Sonnensystem (I).**
H. Tiersch, D. Stoll.
Astron. Raumfahrt, 24. Jahrg., Heft 2, p. 45 – 51 (1986).

091.036 **Plasma–Vorgänge in den Ringen der Planeten.**
W.–H. Ip.
Sterne Weltraum, 25. Jahrg., Nr. 4, p. 188 – 194 (1986).

091.037 **Infrared heterodyne spectroscopy of planetary atmospheres.**
M. J. Mumma.
J. Opt. Soc. Am. A, Vol. 2, No. 12, p. P83 (1985). Abstract.

091.038 **On the accuracy of theoretical models of concentrations of the neutral components of planetary exospheres.**
A. V. Pavlov.
Kosm. Issled., Tom 24, Vyp. 3, p. 476 – 479 (1986). In Russian. English translation in Cosm. Res.

091.039 **Convection in the interior of planets with different scales of the processes according to the layers.**
A. V. Kobozev.
VTs Dal'nevost. nauchn. tsentra Akad. Nauk SSSR. Khabarovsk, 29 pp. (1986). In Russian. Abstr. in Ref. Zh., 51. Astron., 6.51.241 (1986).

091.040 **Ice covers on the planets of the solar system.**
M. S. Krass.
Mater. glyatsiol. issled. Moskva, No. 54, p. 24 – 29 (1985). In Russian. Abstr. in Ref. Zh., 51. Astron., 6.51.243 (1986).

091.041 **Tidal contribution of planets to removing angular momentum from the Sun.**
M. Burša.
Bull. Astron. Inst. Czech., Vol. 37, No. 3, p. 142 – 146 (1986).
The angle between the solar equatorial plane and the Laplace invariable plane of the solar system has been computed (5.84°). This phenomenon is to be considered as not accidental. The well known idea arises about a fast–spinning early Sun, the angular momentum of which was mostly shared by the surrounding close disc which gave rise to the planets. The tidal removing of the angular momentum from the Sun by Jupiter, Venus and Mercury has been estimated. It has been concluded that the explanation of the present small angular momentum of the Sun would require the planetary orbits to have been much closer to the Sun during their past evolution.

091.042 **A model of thermal convection in the major planets with a strongly density–stratified atmospheric layer.**
J.–I. Yano.
Geophys. Astrophys. Fluid Dyn., Vol. 33, No. 3 – 4, p. 277 – 293 (1985). Abstr. in Phys. Abstr., Vol. 89, No. 1274, Entry 23788 (1986).

091.043 **Polarimetric investigations of planetary atmospheres.**
A. V. Morozhenko.
Photometric and polarimetric investigations of celestial bodies, p. 9 – 13 (1985). In Russian. – See Abstr. 003.035.
Main polarization properties of light diffusely reflected from planetary atmospheres as well as physical characteristics of cloud particles are given. Prospects are discussed for application of polarimetric methods to the study of planets.

091.044 **Micrometeoroid impact on planetary satellites as a magnetospheric mass source.**
P. K. Haff, A. Eviatar.
Icarus, Vol. 66, No. 2, p. 258 – 269 (1986).
Planetary satellites are an important source of mass for planetary magnetospheres. Meteoroid impact vaporization is a supply mechanism which can potentially compete with charged–particle sputtering. Recent estimates of impact fluxes in the outer solar system vary by several orders of magnitude. For the larger flux values impact vaporization will play a role both at Jupiter and Saturn, although for the most part it will not dominate sputtering. At the small end of the flux range, sputtering dominates magnetospheric mass–loading everywhere.

091.045 **On the sputter alteration of regoliths of outer solar system bodies.**
B. Hapke.
Icarus, Vol. 66, No. 2, p. 270 – 279 (1986).
This paper discusses several processes that are expected to occur when the porous regoliths on bodies without atmospheres in the outer solar system are subjected to energetic ion bombardement.

091.046 **Theory of density waves in narrow planetary rings.**
E. Willerding.
Astron. Astrophys., Vol. 161, No. 2, p. 403 – 407 (1986).
This paper discusses the dynamics of narrow planetary rings in a general central potential. To avoid the instability of a self–gravitating fluid cylinder the concept of a string, which consists of equidistant spherical agglomerations of small particles surrounded by a dust envelope, is introduced. Three different kinds of waves can propagate through the string; two of them have the character of density waves. The most important result of the theory is the stability of the string, if the volume mass density lies below a critical value. For the *F*–ring of Saturn and the ε–ring of Uranus numerical values for the critical densities are given. With the proposed density wave theory the occurrence of clumps in the ring and the sinusoidal distortions of the whole ring in two dimensions can be explained.

091.047 **Evolution der Planetenatmosphären.**
K. Schäfer.
Astron. Sch., 23. Jahrg., Heft 3, p. 52 – 56 (1986).

091.048 **Ringsysteme um Planeten in unserem Sonnensystem (II).**
D. Stoll, H. Tiersch.
Astron. Raumfahrt, 24. Jahrg., Heft 3, p. 73 – 77 (1986).

091.049 **Planètes et petits corps du système solaire. Rapport thématique 1980 – 1984.**
Observatoire de Paris, 61, avenue de l'Observatoire, F–75014 Paris, France. 85 pp. (1986).

091.050 **Planetary rings.**
A. Brahic.
Recherche, Vol. 17, No. 173, p. 40 – 51 (1986). In French. Abstr. in Phys. Abstr., Vol. 89, No. 1280, Entry 60679 (1986).

091.051 **Phase transitions and convection in icy satellites.**
D. Bercovici, G. Schubert, R. T. Reynolds.
Geophys. Res. Lett., Vol. 13, No. 5, p. 448 – 451 (1986).
The effects of solid–solid phase changes of subsolidus convection in the large icy moons of the outer solar system are considered. Phase transitions affect convection via processes that distort the phase change boundary and/or influence buoyancy through thermal expansion. Linear stability analyses are carried out for the ice layers with a phase change at the midplane. Two exothermic phase transitions (ice I – ice II, ice VI – ice VIII) and two endothermic transitions (ice I – ice III, ice II – ice V) are considered.

091.052 **Xenolith digestion in large magma bodies.**
D. Walker, W. S. Kiefer.
J. Geophys. Res., Vol. 90, Suppl., p. C585 – C590 (1985). – See Abstr. 012.065.

091.053 **Sulfide and sulfate saturation in hydrous silicate melts.**
M. R. Carroll, M. J. Rutherford.
J. Geophys. Res., Vol. 90, Suppl., p. C601 – C612 (1985). – See Abstr. 012.065.

Hydrothermal experiments have been conducted on a sulfur saturated dacitic melt over a range of pressure, temperature, oxygen fugacity, and melt FeO content in order to examine the effects of these variables on sulfur solubility in fractionated melts. The results are of basic importance in understanding volatile transport and volcanic degasing processes on planets such as the earth, Mars and Venus.

091.054 **The scaling of complex craters.**
S. K. Croft.
J. Geophys. Res., Vol. 90, Suppl., p. C828 – C842 (1985). – See Abstr. 012.065.

The empirical relation between the transient crater diameter (D_g) and final crater diameter (D_r) of complex craters and basins is estimated using cumulative terrace widths, central uplift diameters, continuous ejecta radii, and transient crater reconstructions determined from lunar and terrestrial impact structures. The continuity of the D_g/D_r relation over the entire spectrum of morphologic types from complex craters to multiring basins implies that the rims of all these structures form in the same tectonic environment despite morphologic differences.

091.055 **Die chemische Zusammensetzung und Bildung der terrestrischen Planeten.**
H. Wänke, G. Dreibus.
Mitt. Astron. Ges., No. 65, p. 9 – 24 (1986). – See Abstr. 012.072.

Compaction of the dust of the solar nebula led to larger and larger planetesimals which finally formed the terrestrial planets. Their different densities indicate compositional differencies. Venus, Earth and Mars can be modelled by assuming a C 1–composition (equal to solar abundances for all "condensable" elements), except for the most volatile elements and a varying degree of oxidation. The K/U–ratios of the Earth, Venus and Mars indicate severe, but varying depletions of K relative to U and the C 1–ratio. The only plausible explanation is depletion of K due to its higher volatility; the temperature required is about 1000K.

091.056 **A harmonic and statistical analysis of the topography of the Earth, Moon and Mars.**
B. G. Bills.
NASA Tech. Memo., NASA TM–88784, Vol. 1, p. 1 – 270 (1986). = Thesis, California Institute of Technology, Pasadena, Calif., USA (1978). – See Abstr. 003.051.

In this thesis first a global lunar topographic map is derived from Earth–based and orbital observations supplemented in areas without data by a linear autocovariance predictor. It is confirmed that the low–degree gravity harmonics are caused primarily by surface height variations and only secondarily by lateral density variations. Second, a series of models of the lunar interior is derived from topographic, gravitational, librational and seismic data. Third, a model of martian global topography is obtained by fitting a spherical harmonic series of degree 16 to occultation, radar, spectral and photogrammetric measurements. It is confirmed that, similar to the Moon, the martian low–degree gravity harmonics are produced primarily by surface height variations and only secondly by lateral density variations. Maps of the global topography and Bouguer gravity are presented. Finally, it is shown that the topographic variance spectra of the Earth, Moon, Mars and Venus are all very similar. Implications of this result for "tectonic" and erosive processes on the planetary surfaces are outlined.

Dark companions of stars. Astrometric commentary on the lower end of the main sequence.
See Abstr. 003.026.

The Cambridge photographic atlas of the planets.
See Abstr. 003.044.

Planetary landscapes.
See Abstr. 003.049.

The planets.
See Abstr. 003.077.

Planetary landscapes.
See Abstr. 003.102.

The solar system: observations and interpretations.
See Abstr. 003.116.

Cosmic investigations of planets and satellites.
See Abstr. 003.188.

Cassini, Flamsteed, and Halley on the dimensions of the solar system.
See Abstr. 004.055.

Space Station planetology experiments (SSPEX). Proceedings of a workshop held at the U.S. Geological Survey, Flagstaff, Arizona, USA, 20 – 22 June 1985.
See Abstr. 012.050.

Terrestrial planets: comparative planetology. Conference held at Pasadena, Calif., USA, 5 – 7 June, 1985.
See Abstr. 012.092.

Optical properties of solid and liquid sulfur at visible and infrared wavelengths.
See Abstr. 022.017.

Absolute experimental differential and integral electron excitation cross sections for atomic oxygen. 1. The ($^3P \rightarrow {}^3S^0$) transition (1304 Å) at 100 eV.
See Abstr. 022.048.

Shock–induced CO_2 loss from $CaCO_3$; implications for early planetary atmospheres.
See Abstr. 022.098.

Theoretical collision–induced rototranslational absorption spectra for the outer planets: H_2–CH_4 pairs.
See Abstr. 022.103.

Grain size evolution and fractionation trends in an experimental regolith.
See Abstr. 022.108.

Observation of the fourth positive system of CO in dissociative recombination of vibrationally excited CO_2^+.
See Abstr. 022.125.

Astrophysical implications of ice sputtering.
See Abstr. 022.130.

Neutral mass spectrometry and gas chromatography.
See Abstr. 035.031.

Proposal for investigation of minor bodies of the Solar System using remote sensing of electron beam induced X–ray fluorescence.
See Abstr. 035.066.

A Doppler filter technique to measure the hydrogen content of planetary surfaces.
See Abstr. 035.083.

Quantitative determination of mineral types and abundances from reflectance spectra using principal components analysis.
See Abstr. 036.204.

A Monte Carlo study of parameters affecting computer simulations of crater saturation density.
See Abstr. 036.206.

Recovering crater production functions from densely cratered surfaces.
See Abstr. 036.207.

Polynomial expansion of the planetary secular terms: relativistic and lunar perturbations.
See Abstr. 042.002.

Secular perturbation in planetary motion.
See Abstr. 042.045.

Secular variations of major planets' orbital elements.
See Abstr. 042.076.

On the occurrence of near syzygies in the solar system.
See Abstr. 042.077.

Construction of a planetary solution with the help of an *N*–body program, and analytical complements.
See Abstr. 046.021.

ALPL–1 Newtonian ephemeris of the planetary system spanning 4000 years.
See Abstr. 046.022.

Earth–orbiting telescopes in the Cassini era.
See Abstr. 051.052.

Communicating from the edge of the solar system.
See Abstr. 051.089.

Thermal diffusion in ionized gases and planetary and solar processes.
See Abstr. 061.031.

A reconsideration of thermal diffusion in ionized gases: quantal and dynamic shielding effects.
See Abstr. 062.048.

On the important role of relativistic effects in the electrodynamics of plasma.
See Abstr. 062.053.

Super–rotation and diffusion of axial angular momentum. II. A review of quasi–axisymmetric models of planetary atmospheres.
See Abstr. 062.110.

Magnetospheric interchange instability: a tutorial lecture.
See Abstr. 062.132.

Radiative transfer in a sphere illuminated by a parallel beam: an integral equation approach.
See Abstr. 063.039.

On correlations between certain photometric and polarimetric characteristics of light scattered by surfaces of atmosphereless celestial bodies.
See Abstr. 063.079.

Computation of the nonstationary radiation field in an optically thick planetary atmosphere.
See Abstr. 063.083.

The theory of multiple scattering of radiation in plane parallel atmospheres.
See Abstr. 063.100.

Experimental modeling of polarization properties of atmosphereless cosmic bodies at small phase angles.
See Abstr. 063.108.

Does the energy density of the vacuum influence planetary motions?
See Abstr. 066.034.

The structure of the terrestrial planets and tidal friction theory.
See Abstr. 081.032.

General circulation model simulations of superrotation in slowly rotating atmospheres: implications for Venus.
See Abstr. 093.046.

Geochemical constraints on the origin of the Moon.
See Abstr. 094.016.

Palaeomagnetism of the moon and the problem of dynamo–fields of planets.
See Abstr. 094.024.

Clouds, aerosols, and photochemistry in the Jovian atmosphere.
See Abstr. 099.048.

Episodic volcanism of tidally heated satellites with application to Io.
See Abstr. 099.070.

Catalogue of minimum distances between the orbits of comets and major planets.
See Abstr. 102.067.

The solar–system abundances of Nb, Ta, and Y, and the relative abundances of refractory lithophile elements in differentiated planetary bodies.
See Abstr. 105.239.

The structure of Phaethon and detonation of its icy envelope.
See Abstr. 107.009.

Upper limits to the masses of objects in the solar comet cloud.
See Abstr. 107.031.

Numerical realisation of a three–dimensional convective model of evolution of the relief of planets.
See Abstr. 107.039.

A numerical simulation of the formation of the terrestrial planets.
See Abstr. 107.042.

Amorphous ice in astrophysics.
See Abstr. 131.329.

Calibration of radio sources in the millimeter wave range.
See Abstr. 141.008.

092 Mercury

092.001 **Observer Mercure au moyen de jumelles!**
J. Jonlet.
Ciel, Vol. 48, p. 78 – 90 (1986).

092.002 **Beobachtungen von Merkur und Venus 1984/1985.**
E. Freydank, H. Freydank.
Orion, 44. Jahrg., Nr. 212, p. 24 – 25 (1986).

092.003 **The sodium exosphere and magnetosphere of Mercury.**
W.–H. Ip.
Geophys. Res. Lett., Vol. 13, No. 5, p. 423 – 426 (1986).

Following the recent optical discovery of intense sodium D–line emission from Mercury, the author explores the scenario of an extended exosphere of sodium and other metallic atoms.

092.004 **Radar altimetry of Mercury: a preliminary analysis.**
J. K. Harmon, D. B. Campbell, D. L. Bindschadler, J. W. Head, I. I. Shapiro.
J. Geophys. Res., Vol. 91, No. B1, p. 385 – 401 (1986).

The Arecibo radar observations of Mercury provide information on the morphology of surface features with horizontal dimensions ranging from the 50 km scales characteristic of scarps, ridges and moderate–sized craters, to the 1000 - 2000 km scales of large impact basins and major upland regions. When used in tandem with Mariner 10–derived images and maps, radar altimetry has allowed the authors to better constrain and more closely examine previous interpretations of the geology of the planet.

Right ascensions of the Sun, Mercury and Venus observed with the transit instrument in Nikolaev in 1981.
See Abstr. 041.013.

Right ascensions of the Sun, Mercury and Venus observed with the transit instrument in Nikolaev in 1983.
See Abstr. 041.014.

Right ascensions of the Sun, Mercury and Venus observed with the transit instrument in Nikolaev in 1984.
See Abstr. 041.015.

Declinations of the Sun, Mercury and Venus obtained from observations with the vertical circle of the Nikolaev Observatory in 1983.
See Abstr. 041.016.

Declinations of the Sun, Mercury and Venus obtained from observations with the vertical circle of the Nikolaev Observatory in 1984.
See Abstr. 041.017.

Right ascensions of the Sun, Mercury and Venus observed with the transit instrument in Nikolaev in 1980 – 1984 with corrections for lateral refraction.
See Abstr. 041.034.

Right ascensions of the Sun, Mercury and Venus observed in 1981 – 1982 near Kislovodsk with corrections for lateral refraction.
See Abstr. 041.035.

The real value of Mercury's perihelion advance.
See Abstr. 066.067.

The special theory of relativity on a gravitational field.
See Abstr. 066.200.

Mercury's precession according to special relativity.
See Abstr. 066.225.

093 Venus

093.001 **Is there lightning on Venus?**
J. G. Luhmann, A. F. Nagy.
Nature, Vol. 319, No. 6051, p. 266 (1986).

093.002 **Venus mesosphere and thermosphere. 1. Heat budget and thermal structure.**
R. E. Dickinson, S. W. Bougher.
J. Geophys. Res., Vol. 91, No. A1, p. 70 – 80 (1986).

The calculation of a reasonable global average temperature structure is the first step in the simulation of the dynamics and observed characteristics of Venus's upper mesosphere and thermosphere. Various physical processes that influence Venus temperatures are examined using updated inputs and new parameterizations.

093.003 **Thermal structure of the Venus ionosphere: a two–dimensional model study.**
R. C. Whitten, R. P. Singhal, W. C. Knudsen.
Geophys. Res. Lett., Vol. 13, No. 1, p. 10 – 13 (1986).

A simple, two–dimensional, spectral model of heat transport in the Venus ionosphere has been constructed. Numerical experiments with the model suggest that a nocturnal heat source about an order of magnitude smaller than the day–time source is required to simulate the observed ion and electron temperatures. The solar zenith angle dependence of the high–altitude ion temperature appears to be caused by expansion followed by compression of the plasma as it flows from the dayside to the nightside. Thermal–flux saturation appears to be significant on the nightside.

093.004 **A dynamic model of Venus's gravity field.**
W. S. Kiefer, M. A. Richards, B. H. Hager, B. G. Bills.
Geophys. Res. Lett., Vol. 13, No. 1, p. 14 – 17 (1986).

Unlike on Earth, long–wavelength gravity anomalies and topography correlate well on Venus. Venus's admittance curve from spherical harmonic degree 2 to 18 is inconsistent with either Airy or Pratt isostasy but is consistent with dynamic support from mantle convection. A model using whole mantle flow and a high viscosity near–surface layer overlying a constant viscosity mantle reproduces this admittance curve.

093.005 **Thermal structure of the Venus atmosphere in the middle cloud layer.**
V. M. Linkin, J. Blamont, A. N. Lipatov, A. A. Shurupov, C. Malique, S. P. Ignatova, G. A. Frank, L. I. Khlyustova, A. V. Terterashvili, A. Seiff, V. V. Kerzhanovich, B. Ragent, R. Young, L. Elson, R. Preston, A. Ingersoll, D. Crisp.
Pis'ma Astron. Zh., Tom 12, No. 1, p. 36 – 40 (1986). In Russian. English translation in Sov. Astron. Lett., Vol. 12.

Thermal structure measurements obtained by the Vega balloons show a generally adiabatic state, but suggest the existence of separate air masses with slightly different potential temperature and entropy. At a given pressure the temperatures measured by Vega 2 are 6.5K cooler than by Vega 1. Data taken by the Vega 2 lander probe agree with those of Vega 2 balloon.

093.006 **Vega balloon experiment: mean zonal wind velocity in the atmosphere of Venus from Doppler measurements of the balloons.**
R. A. Andreev, V. I. Altunin, N. A. Armand, Eh. L. Akim, R. V. Bakit'ko, J. Blamont, L. Boloh, A. S. Vyshlov, Yu. N. Gorshenkov, N. M. Ivanov, V. V. Kerzhanovich, V. I. Kostenko, L. R. Kogan, V. D. Kustodiev, V. M. Linkin, G. Laurans, L. I. Matveenko, E. P. Molotov, J. Morales, R. Preston, G. Petit, S. V. Pogrebenko, V. I. Puchkov, A. S. Selivanov, C. Stelzried, I. Ya. Tarnoruder, V. F. Tikhonov, C. Hildebrand.
Pis'ma Astron. Zh., Tom 12, No. 1, p. 41 – 45 (1986). In Russian. English translation in Sov. Astron. Lett., Vol. 12.

Doppler measurements of Vega balloons provided the estimates of mean zonal wind velocity in the atmosphere of Venus at (53 – 54) km altitude of (69 ± 1) m/s for Vega 1 and (66 ± 1) m/s for Vega 2; the wind has westward direction. A solar–tied perturbation is probably seen in the data on wind velocity.

093.007 **Vega balloon experiment: small–scale turbulence in the middle cloud layer of Venus.**
V. V. Kerzhanovich, Yu. N. Aleksandrov, R. A. Andreev, N. A. Armand, R. V. Bakit'ko, J. Blamont, L. Boloh, V. A. Vorontsov, A. S. Vyshlov, S. P. Ignatov, A. Ingersoll, A. L. Zajtsev, V. P. Lysov, B. I. Mottsulev, K. M. Pichkhadze, R. Preston, C. Hildebrand, G. Petit, R. Young.
Pis'ma Astron. Zh., Tom 12, No. 1, p. 46 – 51 (1986). In Russian. English translation in Sov. Astron. Lett., Vol. 12.

From measurements of variations of Doppler velocity of Vega balloons strong turbulence was found on both the dayside and nightside Venus cloud layer. The amplitude of wind velocity variations with time scale of 30 – 100 s is up to 2 m/s.

093.008 **Vega balloon experiment: preliminary analysis of measurements in application to the dynamics of the atmosphere of Venus.**
J. Blamont, R. Z. Sagdeev, V. M. Linkin, G. S. Golitsyn, V. N. Ivanov, A. Ingersoll, V. V. Kerzhanovich, D. Crisp, R. Preston, B. Ragent, A. Seiff, L. Elson, R. Young.
Pis'ma Astron. Zh., Tom 12, No. 1, p. 52 – 58 (1986). In Russian. English translation in Sov. Astron. Lett., Vol. 12.

Typical vertical wind velocities of (1 – 2) m/s measured by the Vega balloons appear to be caused by thermal convection. The observed temperature difference of 6.5K between Vega 1 and Vega 2 data can be caused by the synoptic or planetary scale disturbances. The mean wind velocities obtained from Doppler measurements exceed those of previous experiments. A solar–tied feature in the wind field is possible. Surface topography appears to have influenced the motion of the Vega 2 balloon.

093.009 **Study of the Venus rock composition in the northern part of Land Aphrodite on the lander of Vega 2.**
Yu. A. Surkov, L. P. Moskaleva, O. P. Shcheglov, A. D. Dudin, V. P. Kharyukova, O. S. Manvelyan, G. G. Smirnov.
Pis'ma Astron. Zh., Tom 12, No. 1, p. 66 – 72 (1986). In Russian. English translation in Sov. Astron. Lett., Vol. 12.

A brief description is given of experiment on X–ray radiometric determination of the elemental composition of Venusian rock carried out with the help of the lander of spacecraft Vega 2. Preliminary data on the rock composition in the northern part of the Land Aphrodite and geochemical interpretation of the studied rock character are given.

093.010 **Water vapor concentration in the Venus atmosphere from Vega 1 and Vega 2 data.**
Yu. A. Surkov, O. P. Shcheglov, M. L. Ryvkin, N. A. Davydov, D. M. Shejnin, D. E. Zhitnikov.
Pis'ma Astron. Zh., Tom 12, No. 1, p. 73 – 77 (1986). In Russian. English translation in Sov. Astron. Lett., Vol. 12.

Results of determination of water vapor concentration in the Venus atmosphere from Vega 1, Vega 2 are described. The measurements were carried out at the altitudes of 62 – 25 km above the surface of the planet. Two vertical water vapor concentration profiles were measured in the Venus night atmosphere.

093.011 **Study of the ultraviolet absorption in the atmosphere of Venus by means of the active spectrometry method.**
J.–L. Bertaux, A. P. Ehkonomov, B. Mege, V. I. Moroz, V. I. Gnedykh, A. V. Grigor'ev, A. Abergel, A. Hauchecorne, J.–P. Pommereau, P. Rigaud, B. E. Moshkin, S. B. Sergeeva.
Pis'ma Astron. Zh., Tom 12, No. 1, p. 78 – 84 (1986). In Russian. English translation in Sov. Astron. Lett., Vol. 12.

The absorption spectrum of the atmosphere of Venus in the range (2300 – 4000) Å was measured on board the landing modules of Vega 1 and Vega 2 stations. The optical path was ~ 170 cm. A broad absorption band without fine structure was found. Spectral and altitude dependence of the absorption makes it possible to suppose that the main absorbent is sulfur S_8 in the gas phase with the mixing ratio (5 – 25) ppm (at the heights 25 – 40 km respectively).

093.012 **Preliminary results of the Vega 1 and Vega 2 optical investigations of aerosol in the atmosphere of Venus at 30 – 60 km.**
B. E. Moshkin, V. I. Moroz, V. I. Gnedykh, A. V. Grigor'ev, L. V. Zasova, A. P. Ehkonomov.
Pis'ma Astron. Zh., Tom 12, No. 1, p. 85 – 93 (1986). In Russian. English translation in Sov. Astron. Lett., Vol. 12.

Profiles of the aerosol particles number density above the landing sites of Vega 1 and Vega 2 landers have been measured. Approximately the same structure as obtained in previous experiments is found: 3–layers main cloud cover, intermediate region and undercloud haze. But there are large differences in number densities of particles and especially in their size distributions. Tropospheric nightglow was discovered at wavelength about 1 μm.

093.013 **La haute atmosphère de Vénus. Tentative d'explication de sa rotation.**
C. Boyer.
Astronomie, Vol. 100, p. 77 – 83 (1986).

093.014 **Vertical thermal structure of the Venus atmosphere from temperature and pressure measurements with Vega 2. Preliminary results.**
V. M. Linkin, J. Blamont, A. N. Lipatov, S. I. Devyatkin, A. V. D'yachkov, S. P. Ignatova, V. V. Kerzhanovich, C. Malique, B. I. Stadnyk, Ya. V. Sanotskij, P. G. Stolarchuk, A. V. Terterashvili, A. A. Shurupov, L. I. Khlyustova.
Pis'ma Astron. Zh., Tom 12, No. 2, p. 100 – 105 (1986). In Russian. English translation in Sov. Astron. Lett., Vol. 12.

Accurate temperature and pressure measurements were accomplished during the whole descent of the Vega 2 lander. Temperature and pressure near the surface are 733K and 89.3 bar respectively. Strong temperature inversion in the upper troposphere is found. A few layers with different static stability are observed in the atmosphere.

093.015 **Chemical analysis of aerosol in the Venusian cloud layer by a reaction gas chromatography on board the Vega landers.**
B. G. Gel'man, Yu. V. Drozdov, V. V. Mel'nikov, V. A. Rotin, V. N. Khokhlov, V. B. Bondarev, G. G. Dol'nikov, A. V. D'yachkov, D. F. Nenarokov, L. M. Mukhin, N. V. Porshnev, A. A. Fursov.
Pis'ma Astron. Zh., Tom 12, No. 2, p. 106 – 109 (1986). In Russian. English translation in Sov. Astron. Lett., Vol. 12.

The experiment on sulphuric acid aerosol determination in the Venusian cloud layer on board the Vega landers is described. The average content of sulphuric acid of ~ 1 mg/m^3 was found for the sample taken from the atmosphere at heights from 63 to 48 km. Sulphur dioxide SO_2 was revealed in the gaseous sample at the height of 48 km. From experimental results and blank run measurements the suggestion is made that the Venusian cloud layer aerosol consists of more complicated particles than the sulphuric acid water solution does.

093.016 **Measurements of the composition of the aerosol component of the Venusian atmosphere with the Vega 1 station. Preliminary data.**
Yu. A. Surkov, V. F. Ivanova, A. N. Pudov, V. P. Volkov, Eh. P. Sheretov, B. I. Kolotilin, M. P. Safonov, R. Thomas, J. Lespagnol, A. Hauser, G. Israel, D. Imbault, D. Caramelle.
Pis'ma Astron. Zh., Tom 12, No. 2, p. 110 – 113 (1986). In Russian. English translation in Sov. Astron. Lett., Vol. 12.

A preliminary investigation of mass spectra of gaseous products of pyrolyzed Venusian cloud particles on board the Vega 1 lander revealed the presence of heavy particles in the upper cloud layer. Sulfur is the main chemical component of cloud particles. An estimate of the lower limit of the sulfuric acid aerosol content at 62 – 54 km heights of ~ 2.0 mg/m^3 is obtained. The chlorine line is also present in the mass spectrum with the lower limit of chlorine concentration of ~ 0.3 mg/m^3.

093.017 **Content of uranium, thorium and potassium in Venusian rocks at the landing sites of Vega 1 and Vega 2.**
Yu. A. Surkov, F. F. Kirnozov, O. P. Sobornov, V. N. Glazov, A. G. Dunchenko, L. P. Tatsij.
Pis'ma Astron. Zh., Tom 12, No. 2, p. 114 – 119 (1986). In Russian. English translation in Sov. Astron. Lett., Vol. 12.

The results of a determination of natural radioactive elements in Venusian rocks at the landing sites of Vega 1 and Vega 2 show that their chemical composition is close to tholeitic basalts and gabbro of the earth crust. A short description of equipment and the gamma–spectrometry method of the determination of uranium, thorium and potassium concentration in rocks are given.

093.018 **Preliminary results of a determination of chemical element concentrations in the aerosol of Venus clouds.**
B. M. Andrejchikov, L. M. Mukhin, B. N. Korchuganov, I. K. Akhmetshin, E. N. Tokarev, A. V. Medvedev, M. N. Gol'dfel'd, V. M. Fajnbojm, A. V. Kalyuzhgnyj, I. V. Petryanov, B. I. Ogorodnikov, V. I. Skitovich.
Pis'ma Astron. Zh., Tom 12, No. 2, p. 120 – 122 (1986). In Russian. English translation in Sov. Astron. Lett., Vol. 12.

An X–ray radiometric experiment is described along with the results of measurements of the elemental composition of aerosol in Venus clouds. A preliminary analysis of the data showed that sulfur is present in the range of heights 63 – 47 km with mean content of 5.8 mg/m^3, and chlorine in the height range 61 – 52 km with the mean content of 4.1 mg/m^3. The results of measurements in the range 52 – 47 km may come to an agreement if phosphorus is present in the aerosol with the mean concentration of 7.7 mg/m^3.

093.019 **Preliminary results of measurements of particle concentrations in clouds of Venus at heights of 47 – 63 km by Vega 1 and Vega 2.**
Yu. V. Zhulanov, L. M. Mukhin, D. F. Nenarokov.
Pis'ma Astron. Zh., Tom 12, No. 2, p. 123 – 130 (1986). In Russian. English translation in Sov. Astron. Lett., Vol. 12.

Results of the preliminary processing of the Vega 1, Vega 2 descenders data on the cloud–layer structure of the Venus atmosphere are discussed. A photoelectric counter for aerosol particles is described together with its optical and pneumatic circuits and operation algorithm. Vertical profiles of concentration of particles with a diameter of 0.4 μm agree qualitatively with the Pioneer–Venus and Venera 9, Venera 10 data. Concentrations of these particles are: in the B layer, up to 1.9×10^2cm^{-3}; in the C layer, up to 10 cm^{-3}, in the D layer, up to 1.3×10^2cm^{-3}. Layers have sharp boundaries with significant vertical inhomogeneity of the aerosol concentration field inside them.

093.020 **Venus: detailed mapping of Maxwell Montes region.**
Yu. N. Alexandrov (*Yu. N. Aleksandrov*), A. A. Crymov (*A. A. Krymov*), V. A. Kotelnikov (*V. A. Kotel'nikov*), G. M. Petrov, O. N. Rzhiga, A. I. Sidorenko, V. P. Sinilo, A. I. Zakharov, E. L. Akim (*Eh. L. Akim*), A. T. Basilevski (*A. T. Bazilevskij*), S. A. Kadnichanski (*S. A. Kadnichanskij*), Yu. S. Tjuflin (*Yu. S. Tyuflin*).
Science, Vol. 231, No. 4743, p. 1271 – 1273 (1986).

From October 1983 to July 1984, the north hemisphere of Venus, from latitude 30° to latitude 90°, was mapped by means of the radar imagers and altimeters of the spacecraft Venera 15 and Venera 16. This report presents the results of the radar mapping of the Maxwell Montes region, one of the most interesting features of Venus' surface. A radar mosaic map and contour map have been compiled.

093.021 **Overview of VEGA Venus balloon in situ meteorological measurements.**
R. Z. Sagdeev, V. M. Linkin, V. V. Kerzhanovich, A. N. Lipatov, A. A. Shurupov, J. E. Blamont, D. Crisp, A. P. Ingersoll, L. S. Elson, R. A. Preston, C. E. Hildebrand, B. Ragent, A. Seiff, R. E. Young, G. Petit, L. Boloh, Yu. N. Alexandrov (*Yu. N. Aleksandrov*), N. A. Armand, R. V. Bakitko, A. S. Selivanov.
Science, Vol. 231, No. 4744, p. 1411 – 1414 (1986).

The VEGA balloons made in situ measurements of pressure, temperature, vertical wind velocity, ambient light, frequency of lightning, and cloud particle backscatter. Both balloons encountered highly variable atmospheric conditions, with periods of intense vertical winds occurring sporadically throughout their flights. Large variations in pressure, temperature, ambient light level, and cloud particle backscatter (VEGA–1 only) correlated well during these excursions, indicating that these properties were strong functions of altitude in those parts of the middle cloud layer sampled by the balloons.

093.022 **Determination of Venus winds by ground–based radio tracking of the VEGA balloons.**
R. A. Preston, C. E. Hildebrand, G. H. Purcell Jr., J. Ellis, C. T. Stelzried, S. G. Finley, R. Z. Sagdeev, V. M. Linkin, V. V. Kerzhanovich, V. I. Altunin, L. R. Kogan, V. I. Kostenko, L. I. Matveenko, S. V. Pogrebenko, I. A. Strukov, E. L. Akim, Yu. N. Alexandrov (*Yu. N. Aleksandrov*), N. A. Armand, R. V. Bakitko, A. S. Vyshlov, A. F. Bogomolov, Yu. N. Gorchankov, A. S. Selivanov, N. M. Ivanov, V. F. Tichonov, J. E. Blamont, L. Boloh, G. Laurans, A. Boischot, F. Biraud, A. Ortega–Molina, C. Rosolen, G. Petit.
Science, Vol. 231, No. 4744, p. 1414 – 1416 (1986).

A global array of 20 radio observatories was used to measure the three–dimensional position and velocity of the two meteorological balloons that were injected into the equatorial region of the Venus atmosphere near Venus midnight by the VEGA spacecraft on 11 and 15 June 1985. Initial analysis of only radial velocities indicates that each balloon was blown westward about 11,500 kilometers (8,000 kilometers on the night side) by zonal winds with a mean speed of about 70 meters per second. Excursions of the data from a model of constant zonal velocity were generally less than 3 meters per second; however, a much larger variation was evident near the end of the flight of the second balloon. Consistent systematic trends in the residuals for both balloons indicate the possibility of a solar–fixed atmospheric feature.

093.023 **VEGA balloon dynamics and vertical winds in the Venus middle cloud region.**
V. M. Linkin, V. V. Kerzhanovich, A. N. Lipatov, K. M. Pichkadze (*K. M. Pichkhadze*), A. A. Shurupov, A. V. Terterashvili, A. P. Ingersoll, D. Crisp, A. W. Grossman, R. E. Young, A. Seiff, B. Ragent, J. E. Blamont, L. S. Elson, R. A. Preston.
Science, Vol. 231, No. 4744, p. 1417 – 1419 (1986).

The VEGA balloons provided a long–term record of vertical wind fluctuations in a planetary atmosphere other than Earth's. The vertical winds were calculated from the observed displacement of the balloon relative to its equilibrium float altitude. The winds were intermittent; a large burst lasted several hours, and the peak velocity was 3 meters per second.

093.024 **Thermal structure of the Venus atmosphere in the middle cloud layer.**
V. M. Linkin, V. V. Kerzhanovich, A. N. Lipatov, A. A. Shurupov, A. Seiff, B. Ragent, R. E. Young, A. P. Ingersoll, D. Crisp, L. S. Elson, R. A. Preston, J. E. Blamont.
Science, Vol. 231, No. 4744, p. 1420 – 1422 (1986).
Thermal structure measurements obtained by the two VEGA balloons show the Venus middle cloud layer to be generally adiabatic. Temperatures measured by the two balloons at locations roughly symmetric about the equator differed by about 6.5 kelvins at a given pressure. The VEGA–2 temperatures were about 2.5 kelvins cooler and those of VEGA–1 about 4 kelvins warmer than temperatures measured by the Pioneer Venus Large Probe at these levels. Data taken by the VEGA–2 lander as it passed through the middle cloud agreed with those of the VEGA–2 balloon.

093.025 **Implications of the VEGA balloon results for Venus atmospheric dynamics.**
J. E. Blamont, R. E. Young, A. Seiff, B. Ragent, R. Z. Sagdeev, V. M. Linkin, V. V. Kerzhanovich, A. P. Ingersoll, D. Crisp, L. S. Elson, R. A. Preston, G. S. Golitsyn, V. N. Ivanov.
Science, Vol. 231, No. 4744, p. 1422 – 1425 (1986).
Both VEGA balloons encountered vertical winds with typical velocities of 1 to 2 meters per second. These values are consistent with those estimated from mixing length theory of thermal convection. However, small–scale temperature fluctuations for each balloon were sometimes larger than predicted. The approximate 6.5–kelvin difference in temperature consistently seen between VEGA–1 and VEGA–2 is probably due to synoptic or planetary–scale nonaxisymmetric disturbances that propagate westward with respect to the planet. There is also evidence from Doppler data for the existence of solar–fixed nonaxisymmetric motions that may be thermal tides.

093.026 **Comment on "Venus nightside ionospheric troughs: implications for evidence of lightning and volcanism" by H. A. Taylor Jr., J. M. Grebowsky, and P. A. Cloutier.**
F. L. Scarf.
J. Geophys. Res., Vol. 91, No. A4, p. 4594 – 4598 (1986). See Abstr. 40.093.036.

093.027 **Reply to the comment on "Venus nightside ionospheric troughs: implications for evidence of lightning and volcanism" by F. L. Scarf.**
H. A. Taylor Jr., J. M. Grebowsky, P. A. Cloutier.
J. Geophys. Res., Vol. 91, No. A4, p. 4599 – 4605 (1986). See Abstr. 093.026.

093.028 **Construction of large–scale photomaps of the Venus surface from data of a radar survey made with the automatic interplanetary stations Venera 15 and Venera 16.**
Yu. N. Aleksandrov, A. I. Zakharov, A. A. Krymov, S. A. Kadnichanskij, L. S. Ledovskaya, M. V. Ostrovskij, G. M. Petrov, O. N. Rzhiga, A. I. Sidorenko, V. P. Sinilo, Yu. S. Tyuflin.
Geod. kartogr., No. 9, p. 41 – 48 (1985). In Russian. Abstr. in Ref. Zh., 52. Geod. Aehrosemka, 1.52.205 (1986).

093.029 **Geometric principles for constructing radar panoramas of the Venus surface.**
O. N. Rzhiga, Yu. S. Tyuflin, E. G. Belen'kij.
Geod. kartogr., No. 9, p. 48 – 50 (1985). In Russian. Abstr. in Ref. Zh., 52. Geod. Aehrosemka, 1.52.206 (1986).

093.030 **On properties of the Venus ionosphere and its sources.**
T. K. Breus, M. I. Verigin, K. I. Gringauz.
Kosm. Issled., Tom 24, Vyp. 1, p. 106 – 121 (1986). In Russian. English translation in Cosm. Res.

093.031 **Spectroscopic limits of the NO_2, Br_2, Cl_2 and SO content in the upper atmosphere of Venus.**
V. A. Krasnopol'skij.
Kosm. Issled., Tom 24, Vyp. 2, p. 311 – 314 (1986). In Russian. English translation in Cosm. Res.

093.032 **Atlas of the planet Venus.**
O. N. Rzhiga.
Zemlya Vselennaya, No. 1, p. 13 – 20 (1986). In Russian.

093.033 **News about the Venus atmosphere.**
V. I. Moroz.
Zemlya Vselennaya, No. 3, p. 19 – 30 (1986). In Russian.

093.034 **Geological–morphological description of the Lakshmi Planum (Photomap of the Venusian surface, Sheet B–4).**
A. A. Pronin, A. L. Sukhanov, Yu. S. Tyuflin, S. A. Kadnichanskij, V. A. Kotel'nikov, O. N. Rzhiga, G. I. Petrov, A. I. Sidorenko, Yu. N. Aleksandrov, A. P. Krivtsov, V. P. Sinilo, G. A. Burba, N. N. Bobina.
Astron. Vestn., Tom 20, No. 2, p. 83 – 98 (1986). In Russian. English translation in Sol. Syst. Res.
A morphological description of Lakshmi Planum and its surroundings is given: they are parts of a single structure. It is suggested that the mechanism of matter ascent from planetary interiors and its horizontal spreading is responsible for the structure formation. Folding and/or imbricating are the result of the process. Thus the Lakshmi structure can be considered as a local center of radial spreading. The structure dimensions suggest that asthenospheric currents have been involved in its formation.

093.035 **Geological–morphological description of the Ishtar Terra (Photomap of the Venusian surface, Sheet B–5).**
A. L. Sukhanov, A. A. Pronin, Yu. S. Tyuflin, M. V. Ostrovskij, V. A. Kotel'nikov, O. N. Rzhiga, G. I. Petrov, A. I. Sidorenko, Yu. N. Aleksandrov, A. I. Zakharov, A. A. Krymov, N. N. Bobina.
Astron. Vestn., Tom 20, No. 2, p. 99 – 111 (1986). In Russian. English translation in Sol. Syst. Res.
The main part of Ishtar Terra surface to the east of Maxwell Montes is covered with areal dislocations of several directions, called "parquet". They can be divided by their structural patterns as follows: 1) central stable block; 2) lesser outlying blocks separated by faults and grabens from the central one; 3) mobilized zones where material moved downslope away from the central block as viscous flows; 4) partly "parqueted" lava sheets. The Maxwell Montes have been formed as a result of collision between the central parquet block and Lakshmi Planum.

093.036 **Mapping the north–eastern region of the western hemisphere of the Venus surface.**
A. F. Bogomolov, Yu. S. Tyuflin, G. I. Skrypnik, L. V. Kudrin, M. V. Ostrovskij.
Geod., kartogr., No. 10, p. 42 – 47 (1985). In Russian. Abstr. in Ref. Zh., 52. Geod. Aehrosemka, 4.52.219 (1986).

093.037 **Radiation transfer in the 4.3–μ CO_2 band and the 4.7–μ CO band in the atmospheres of Venus and Mars with violation of LTE. Populations of vibrational states.**
G. I. Stepanova, G. M. Shved.
Sov. Astron., Vol. 29, No. 4, p. 422 – 428 (1985). English translation of 40.093.033.

093.038 **Average dimension and magnetic structure of the distant Venus magnetotail.**
M. A. Saunders, C. T. Russell.
J. Geophys. Res., Vol. 91, No. A5, p. 5589 – 5604 (1986).
The authors report the first major statistical investigation of the far wake of an unmagnetized object embedded in the solar wind. The investigation is based on Pioneer Venus Orbiter magnetometer data from 70 crossings of the Venus wake at altitudes between 5 and 11 Venus radii during reasonably steady IMF conditions. The authors find that Venus has a well–developed tail, flaring with altitude and possibly broader in the direction parallel to the IMF cross–flow component. Tail lobe field po-

larities and the direction of the cross–tail field are consistent with tail accretion from the solar wind. Average values for the cross–tail field (2 nT) and the distant tail flux (3 MWb) indicate that most distant tail field lines close across the center of the tail and are not rooted in the Venus ionosphere. The authors illustrate their findings in a three–dimensional schematic.

093.039 **Reanalysis of Pioneer Orbiter Ultraviolet Spectrometer data: O I 1304 intensities and atomic oxygen densities.**
L. J. Paxton, R. R. Meier.
Geophys. Res. Lett., Vol. 13, No. 3, p. 229 – 232 (1986).
The Pioneer Venus Orbiter Ultraviolet Spectrometer measured the O I 1304 Å intensity over a range of spacecraft altitudes (150 to 300 km at periapsis). The atomic oxygen density is determined from the dayglow by comparing the data to the intensities predicted in a Monte Carlo model which correctly simulates the partial frequency redistribution of photons in a non–isothermal, absorbing atmosphere. The analysis yields atomic oxygen densities consistent with the in situ determinations of the PV Bus Neutral Mass Spectrometer and PVO Neutral Mass Spectrometer and supports recent laboratory measurements of the electron impact excitation cross section of atomic oxygen.

093.040 **Venus spacecraft infrared radiance spectra and some aspects of their interpretation.**
V. I. Moroz, D. Spänkuch, V. M. Linkin, W. Döhler, I. A. Matsygorin, K. Schäfer, L. V. Zasova, D. Oertel, A. V. Dyachkov (*A. V. D'yachkov*), R. Schuster, V. V. Kerzhanovich, H. Becker–Ross, E. A. Ustinov, W. Stadthaus.
Appl. Opt., Vol. 25, No. 10, p. 1710 – 1719 (1986).
Fourier spectrometers aboard the Venera 15 and 16 orbiters measured the outgoing spectral radiance of Venus in the 250 – 1600–cm^{-1} (Venera 15) and 400 – 1600–cm^{-1} (Venera 16) regions with a spectral resolution of 5 – 7 cm^{-1}. This paper gives a short description of the experimental design and discusses those aspects that can be suitably tackled by wide range spectral measurements. Estimates of H_2O and SO_2 abundances were obtained (~30 ppm at 58 km and ~2 ppm at 65 km, respectively). SO_2 is first identified in the infrared by the measurements.

093.041 **The polar ionosphere of Venus near the terminator of the planet from radio occultation data of the Venera 15, 16 satellites.**
N. A. Savich, V. E. Andreev, A. S. Vyshlov, A. L. Gavrik, V. V. Klimov, V. P. Konofalov, N. V. Laptev, V. A. Marmulev, A. P. Mestehrton, A. S. Nabatov, I. K. Osmolovskij, L. N. Samoznaev, D. Ya. Shtern.
Kosm. Issled., Tom 24, Vyp. 3, p. 448 – 458 (1986). In Russian. English translation in Cosm. Res.

093.042 **Results of observations of the Venus occultation by the sun of June 15, 1984.**
V. P. Mikhajlutsa.
Pis'ma Astron. Zh., Tom 12, No. 6, p. 493 – 496 (1986). In Russian. English translation in Sov. Astron. Lett., Vol. 12.
It is shown that the observational track of Venus on small angles from the sun's surface corresponds to DE 200 and VSOP 82 theories, taking into account the deflection of light in the gravity field of the sun. Without taking into account the General Relativity Theory effect the difference between theoretical and observed tracks exeeds the observational errors. Newcomb's theory of the Venus track in any case is in disagreement with observations.

093.043 **Radiative transfer in the 4.3 μ CO_2 and 4.7 μ CO bands in the non–LTE Venus and Mars atmospheres: transformation of absorbed solar energy.**
G. I. Stepanova, G. M. Shved.
Sov. Astron., Vol. 29, No. 5, p. 528 – 531 (1985). English translation of 40.093.042.

093.044 **Geology of Venus.**
V. L. Barsukov, A. T. Bazilevskij.
Priroda, No. 6, p. 24 – 35 (1986). In Russian.

093.045 **Zonal winds and the angular momentum balance of Venus' atmosphere within and above the clouds.**
R. L. Walterscheid, G. Schubert, M. Newman, A. J. Kliore.
J. Atmos. Sci., Vol. 42, No. 19, p. 1982 – 1990 (1985). Abstr. in Phys. Abstr., Vol. 89, No. 1276, Entry 35998 (1986).

093.046 **General circulation model simulations of superrotation in slowly rotating atmospheres: implications for Venus.**
C. Covey, E. J. Pitcher, J. P. Brown.
Icarus, Vol. 66, No. 2, p. 380 – 396 (1986).
A series of experiments with a three dimensional general circulation model developed to simulate Earth's atmosphere is run with planetary rotation rates varying between 1 and 1/64 times Earth's rotation rate and diurnally averaged thermal forcing. Results are used to evaluate theories of Venus' atmospheric superrotation which invoke upward transport of angular momentum from the solid planet by the zonal mean (i.e., axisymmetric) circulation.

093.047 **Lightning and volcanism at Venus?**
H. A. Taylor Jr., P. A. Cloutier.
Bull. Am. Astron. Soc., Vol. 18, No. 1, p. 634 (1986). Abstract. – See Abstr. 010.066.

093.048 **Venus: eastern and western elongations 1981 – 82.**
J. McCue, J. Nichol.
J. Br. Astron. Assoc., Vol. 96, No. 4, p. 212 – 216 (1986).
This report is for a complete synodic period. It is based on drawings, photographs, descriptions and phase estimates for the interval 1981 April to 1982 November. This report also includes a brief account of experiments on the naked–eye visibility of the planet's phases, and a note about the occultation of Sigma Sagittarii by Venus on 1981 November 17.

093.049 **The results of investigation of some regions of Venus with bistatic radiolocation.**
A. G. Pavelyev (*A. G. Pavel'ev*), A. I. Yakovlev, O. E. Milekhin, A. I. Kucherjavenkov (*A. I. Kucheryavenkov*).
Acta Astronaut., Vol. 13, No. 1, p. 39 – 45 (1986). Abstr. in Phys. Abstr., Vol. 89, No. 1280, Entry 60681 (1986).

093.050 **Models of the structure of the atmosphere of Venus from the surface to 100 kilometers altitude.**
A. Seiff, J. T. Schofield, A. J. Kliore, F. W. Taylor, S. S. Limaye, H. E. Revercomb, L. A. Sromovsky, V. V. Kerzhanovich, V. I. Moroz, M. Ya. Marov.
Adv. Space Res., Vol. 5, No. 11, p. 3 – 58 (1985). – See Abstr. 003.043.
From a critical comparison and synthesis of data from the four Pioneer Venus Probes, the Pioneer Venus Orbiter, and the Venera 10, 12, and 13 landers, models of the lower and middle atmosphere of Venus are derived. The models are consistent with the data sets within the measurement uncertainties and established variability of the atmosphere. The models represent the observed variations of state properties with latitude, and preserve the observed static stability. The rationale and the approach used to derive the models are discussed, and the remaining uncertainties are estimated.

093.051 **Circulation of the atmosphere from the surface to 100 km.**
V. V. Kerzhanovich, S. S. Limaye.
Adv. Space Res., Vol. 5, No. 11, p. 59 – 83 (1985). – See Abstr. 003.043.
This paper presents a model of the gross characteristics of the Venus three–dimensional atmospheric circulation that appear relatively constant over long periods. The authors begin by reviewing the historical observations of the atmospheric circulation and then discuss the observational data that were used in constructing the reference model. The model itself is described for the zonal (east–west), meridional (north–south), and vertical

components of motion in terms of tables and graphs where appropriate. A brief discussion of observed turbulence in the Venus atmosphere is also presented.

093.052 **Particulate matter in the Venus atmosphere.**
B. Ragent, L. W. Esposito, M. G. Tomasko, M. Ya. Marov, V. P. Shari, V. N. Lebedev.
Adv. Space Res., Vol. 5, No. 11, p. 85 – 115 (1985). – See Abstr. 003.043.

The authors present a summary of the data currently available (June 1984) describing the planet–enshrouding particulate matter in the Venus atmosphere. A description and discussion of the state of knowledge of the Venus clouds and hazes precedes the tables and plots. The tabular material includes a precis of upper haze and cloud–top properties, parameters for model–size distributions for particles and particulate layers, and columnar masses and mass loadings.

093.053 **Models of Venus neutral upper atmosphere: structure and composition.**
G. M. Keating, J. L. Bertaux, S. W. Bougher, T. E. Cravens, R. E. Dickinson, A. E. Hedin, V. A. Krasnopolsky (*V. A. Krasnopol'skij*), A. F. Nagy, J. Y. Nicholson III, L. J. Paxton, U. von Zahn.
Adv. Space Res., Vol. 5, No. 11, p. 117 – 171 (1985). – See Abstr. 003.043.

Models of the Venus neutral upper atmosphere, based on both in–situ and remote sensing measurements, are provided for the height interval from 100 to 3500 km. The general approach in model formulation was to divide the atmosphere into three regions: 100 to 150 km, 150 to 250 km, and 250 to 3500 km. Available observations are reviewed. Tables are provided for density, temperature, composition, derived quantities, and day–to–day variability as a function of solar zenith angle on the day– and nightsides. Other tables provide corrections for solar activity effects on temperature, composition, and density. For the exosphere, information is provided on the vertical distribution of normal thermal components as well as the hot components on the day– and nightsides.

093.054 **Composition of the Venus atmosphere below 100 km altitude.**
U. von Zahn, V. I. Moroz.
Adv. Space Res., Vol. 5, No. 11, p. 173 – 195 (1985). – See Abstr. 003.043.

The current knowledge on the composition of the Venus atmosphere in the altitude range from the surface to 100 km is compiled. A listing of the isotopic abundances of elements measured in the Venus atmosphere is also presented.

093.055 **Solar and thermal radiation in the Venus atmosphere.**
V. I. Moroz, A. P. Ekonomov, B. E. Moshkin, H. E. Revercomb, L. A. Sromovsky, J. T. Schofield, D. Spänkuch, F. W. Taylor, M. G. Tomasko.
Adv. Space Res., Vol. 5, No. 11, p. 197 – 232 (1985). – See Abstr. 003.043.

Much has been learned about the solar and thermal radiation fields of Venus as a result of the spacecraft missions of the past decade. Measurements from atmospheric probes, flyby and orbiter spacecraft, as well as ground–based instruments have provided a reasonably complete picture of the nature of the radiation fluxes above the atmosphere and from the surface up to 100 km. This paper is a synthesis of results which, with a few exceptions, have been published elsewhere.

093.056 **The Venus ionosphere.**
S. J. Bauer, L. M. Brace, H. A. Taylor Jr., T. K. Breus, A. J. Kliore, W. C. Knudsen, A. F. Nagy, C. T. Russell, N. A. Savich.
Adv. Space Res., Vol. 5, No. 11, p. 233 – 267 (1985). – See Abstr. 003.043.

Physical properties of the Venus ionosphere obtained by experiments on the US Pioneer Venus and the Soviet Venera missions are presented in the form of models suitable for inclusion in the Venus International Reference Atmosphere. The models comprise electron density (from 120 km), electron and ion temperatures, and relative ion abundance in the altitude range from 150 km to 1000 km for solar zenith angles from 0° to 180°. In addition, information on ion transport velocities, ionopause altitudes, and magnetic field characteristics of the Venus ionosphere, are presented in tabular or graphical form. Also discussed is the solar control of the physical properties of the Venus ionosphere.

093.057 **Support of long–wavelength loads on Venus and implications for internal structure.**
W. B. Banerdt.
J. Geophys. Res., Vol. 91, No. B1, p. 403 – 419 (1986).

A thin–shell formulation for computing stress distributions on the surfaces of one–plate planets is developed. The lithospheric model includes laterally varying density anomalies at two depths, corresponding to undulations on a crust–mantle boundary and variations in upper mantle density. For a given set of model parameters, values for the vertical displacement, crustal thickness anomaly, and mantle density anomaly required to satisfy the topographic and gravitational boundary conditions are computed, along with the resulting stress field. The computed stresses can then be compared to stress directions inferred from observed tectonic features. This formulation is used to study loads on Venus's lithosphere.

093.058 **Global radar units on Venus derived from statistical analysis of Pioneer Venus orbiter radar data.**
P. A. Davis, R. C. Kozak, G. G. Schaber.
J. Geophys. Res., Vol. 91, No. B5, p. 4979 – 4992 (1986). With a correction in Vol. 91, No. B7, p. 7571 – 7572 (1986).

A global radar classification data base for Venus has been produced by unsupervised cluster analysis of Pioneer Venus radar data. This radar classification represents a statistical approach to planetary data analysis that incorporates the radar information from two different data bases and represents an unbiased data base that will aid preliminary geomorphologic/geologic analysis. Of the 15 radar units that resulted from the statistical classification, seven are spatially associated with sites whose geology has been previously inferred from Venera landing site data and Arecibo radar data. If the inferences are correct, these seven units represent materials of basaltic composition that differ only in their surface morphologies, which include banded terrain, rift valley terrain, and gently rolling to flat–lying bedrock terrain; the materials contain variable amounts of "soils".

093.059 **A comparison of the regional slope characteristics of Venus and earth: implications for geologic processes on Venus.**
V. L. Sharpton, J. W. Head III.
J. Geophys. Res., Vol. 91, No. B7, p. 7545 – 7554 (1986).

The authors analyze the surface distribution of regional slope values, examine the large–scale textural variations of Venus and earth topography, and discuss their implications regarding the nature of the geological processes which have shaped the surface of Venus.

093.060 **Venus gravity: a harmonic analysis and geophysical implications.**
N. A. Mottinger, W. L. Sjogren, B. G. Bills.
J. Geophys. Res., Vol. 90, Suppl., p. C739 – C756 (1985). – See Abstr. 012.065.

An improved model of Venusian global gravity has been obtained by fitting a tenth degree and order spherical harmonic series to 78 orbital arcs of Doppler tracking data from Pioneer Venus Orbiter. Maps of the free–air anomaly and its formal error are presented, as are "geoid" and isostatic anomaly maps.

093.061 **Gravity field of Venus at constant altitude and comparison with earth.**
C. Bowin, G. Abers, L. Shure.
J. Geophys. Res., Vol. 90, Suppl., p. C757 – C770 (1985). – See Abstr. 012.065.

This paper presents harmonic spline results of the gravity field of Venus as a series of maps of Venus at spacecraft and constant altitudes. Global (except for polar regions) and local relations of gravity to topography are described.

093.062 **Fluid dynamic constraints of the Venus ionospheric flow.**
H. Pérez–de–Tejada.
J. Geophys. Res., Vol. 91, No. A6, p. 6765 – 6770 (1986).

A critical analysis of the current interpretations of the antisolar motion of the Venus ionosphere is presented. It is argued that the pressure gradient forces present across the terminator are not sufficient to produce the observed acceleration of the plasma. A balance condition between the height–integrated momentum flux of the ionospheric flow and the height–integrated deficiency of momentum flux of the shocked solar wind in the vicinity of the terminator is formulated in terms of measurable quantities. It is found that the observed 2 – 4 km/s flow velocities of the upper ionosphere can be readily inferred from such a relation.

Photochemistry of the atmospheres of Mars and Venus.
See Abstr. 003.004.

The analysis of extraterrestrial materials.
See Abstr. 003.053.

Planet Venus.
See Abstr. 003.121.

Venus is looking more like Earth than Mars.
See Abstr. 011.028.

Laboratory measurements of the microwave opacity and vapor pressure of sulfuric acid vapor under simulated conditions for the middle atmosphere of Venus.
See Abstr. 022.019.

Shock–induced CO_2 loss from $CaCO_3$; implications for early planetary atmospheres.
See Abstr. 022.098.

VEGA balloon system and instrumentation.
See Abstr. 035.017.

Obstacles facing the Venus radar mapper – the implications of gestalt formation in stereo–radargrammetry.
See Abstr. 036.081.

Radarclinometry for the Venus Radar Mapper.
See Abstr. 036.201.

Right ascensions of the Sun, Mercury and Venus observed with the transit instrument in Nikolaev in 1981.
See Abstr. 041.013.

Right ascensions of the Sun, Mercury and Venus observed with the transit instrument in Nikolaev in 1983.
See Abstr. 041.014.

Right ascensions of the Sun, Mercury and Venus observed with the transit instrument in Nikolaev in 1984.
See Abstr. 041.015.

Declinations of the Sun, Mercury and Venus obtained from observations with the vertical circle of the Nikolaev Observatory in 1983.
See Abstr. 041.016.

Declinations of the Sun, Mercury and Venus obtained from observations with the vertical circle of the Nikolaev Observatory in 1984.
See Abstr. 041.017.

Right ascensions of the Sun, Mercury and Venus observed with the transit instrument in Nikolaev in 1980 – 1984 with corrections for lateral refraction.
See Abstr. 041.034.

Right ascensions of the Sun, Mercury and Venus observed in 1981 – 1982 near Kislovodsk with corrections for lateral refraction.
See Abstr. 041.035.

The Vega missions.
See Abstr. 051.012.

The first stage of the Vega project.
See Abstr. 051.024.

Vega balloon experiment.
See Abstr. 051.025.

Vega balloon experiment: the ground network.
See Abstr. 051.026.

Vega balloon as a tool to study the dynamics of the Venus atmosphere.
See Abstr. 051.027.

Vega balloon experiment: the global network of radio telescopes and first results.
See Abstr. 051.028.

Meteorological measurements of Vega 1 and Vega 2 balloons along their trajectories.
See Abstr. 051.029.

Weather balloons at Venus.
See Abstr. 051.032.

The VEGA Venus balloon experiment.
See Abstr. 051.033.

Super–rotation and diffusion of axial angular momentum. I. "Speed limits" for axisymmetric flow in a rotating cylindrical fluid annulus.
See Abstr. 062.109.

Numerical theory of motion of the earth and Venus from radar and optical observational data and from observations of motion of the Venera 9 and Venera 10 artificial earth satellites.
See Abstr. 081.022.

Oxygen emissions in the night airglow of the Earth, Venus and Mars.
See Abstr. 082.078.

Electric and electrothermal conductivity of the ionospheres of planets. II. The Earth, Mars, Venus.
See Abstr. 091.018.

Beobachtungen von Merkur und Venus 1984/1985.
See Abstr. 092.002.

Zonal wind in the stratosphere of Titan.
See Abstr. 100.041.

A summary of extremes of isotopic variations in extra–terrestrial materials.
See Abstr. 105.012.

A gas dynamic magnetosheath field model for unsteady interplanetary fields: application to the solar wind interaction with Venus.
See Abstr. 106.026.

094 Moon

094.001 **Libration of the Moon: shape of the Earth and motion of the ecliptic plane.**
M. Moons.
Relativity in celestial mechanics and astrometry, p. 141 – 144 (1986). – See Abstr. 012.002 (IAU Symp. No. 114).
The author presents the perturbations due to the shape of the Earth and the motion of the ecliptic plane on a theory which already contains planetary perturbations. This theory is completely analytical with respect to the harmonic coefficients of the lunar gravity field which is expanded in spherical harmonics up to the fourth order.

094.002 **New relativistic effects in the motion of the Moon.**
B. Mashhoon.
Relativity in celestial mechanics and astrometry, p. 407 – 410 (1986). – See Abstr. 012.002 (IAU Symp. No. 114).
A summary of the main relativistic effects in the motion of the Moon is presented. The results are based on the application of a novel approach to the restricted three–body problem in general relativity to the lunar motion. It is shown that the rotation of the Sun causes a secular acceleration in the relative Earth–Moon motion. This might appear to be due to a temporal "variation" of the gravitational constant.

094.003 **The origin of the moon.**
A. P. Boss.
Science, Vol. 231, No. 4736, p. 341 – 345 (1986).
The origin of the moon is considered within the theory of formation of the terrestrial planets by accumulation of planetesimals. The theory predicts the occurrence of giant impacts, suggesting that the moon formed after a roughly Mars–sized body impacted on the protoearth. The impact blasted portions of the protoearth and the impacting body into geocentric orbit, forming a prelunar disk from which the moon later accreted.

094.004 **Origin of the Earth's moon: constraints from alkali volatile trace elements.**
M. E. Kreutzberger, M. J. Drake, J. H. Jones.
Geochim. Cosmochim. Acta, Vol. 50, No. 1, p. 91 – 98 (1986).
It has been known conclusively since the return of the Apollo 11 samples that the Moon is impoverished in volatile elements compared to the Earth. This depletion is well illustrated by the ratios of volatile to refractory incompatible elements. This points to high temperature processes being involved in the formation of lunar precursor material and/or in lunar assembly. Most modern hypotheses recognize this fact. One therefore might expect the abundances of volatile elements in the Moon to be inversely related to their volatilities. A detailed examination of alkali metal systematics suggests that this deduction is too simplistic, however.

094.005 **Lunar sample 14425: not a lunar tektite.**
B. P. Glass.
Geochim. Cosmochim. Acta, Vol. 50, No. 1, p. 111 – 113 (1986).
Energy–dispersive X–ray analysis of a polished section of lunar sample 14425 shows, in contradiction to a previous report, that it has a composition similar to Apollo 14 breccias, but not to high–magnesium microtektites. Lunar sample 14425 is a large (8.006 ± 0.006 mm dia.) glass bead. Sample 14425 is one of the largest glass beads returned from the moon. Semiquantitative analysis of the bead indicated that it has a composition unlike other lunar samples, but similar to high–magnesium microtektites (O'Keefe and Glass, 1985). A polished section was prepared at the Lunar Receiving Laboratory, Johnson Space Center. Quantitative analysis of this section shows that the spherule is extremely homogenous in composition and that the composition is similar to Apollo 14 impact breccias and not to any known terrestrial tektite. It appears that the glass bead was formed by impact melting and that it is not a lunar tektite.

094.006 **Names of lunar formations.**
A. Satō.
Heavens, Vol. 67, No. 2, p. 50 – 52 (1986). In Japanese.

094.007 **Hugoniot equation of state of anorthite glass and lunar anorthosite.**
M. B. Boslough, S. M. Rigden, T. J. Ahrens.
Geophys. J. R. Astron. Soc., Vol. 84, No. 3, p. 455 – 473 (1986).
Twenty–one Hugoniot experiments were conducted on an amorphous material of anorthite composition, in the pressure range 8 – 120 GPa, using both routine and new methods. Two Hugoniot measurements at about 120 GPa were made on lunar gabbroic anorthosite (Apollo 15, 418). Theoretical Hugoniots are constructed for both materials assuming they are disproportionate to their component oxides. These accurately predict the P–ϱ behaviour of the lunar anorthosite Hugoniot at 120 GPa and the anorthite glass Hugoniot above 50 GPa, but overestimate the shock temperatures of anorthite glass. The mixed oxide model fails to predict the release paths of either material.

094.008 **Coordinates of the Moon reverse side sector objects.**
S. G. Valeev.
Earth, Moon, Planets, Vol. 34, No. 3, p. 251 – 271 (1986).
The coordinates of 171 objects of the reverse side and marginal zone, and 42 objects of the visible side of the Moon's western hemisphere in survey zone from the automatical interplanetary stations "Zond 6 and 8" type are presented. While dealing with the measuring information, systematical errors arising from the breakdown of orthogonality in coordinate transformation were taken into account, as well as original geometry of the Moon's limb as measured from the space.

094.009 **The dominant relativistic terms in the lunar theory.**
M. Soffel, H. Ruder, M. Schneider.
Astron. Astrophys., Vol. 157, No. 2, p. 357 – 364 (1986).
In the present paper the transition from the frame induced by the Einstein–Infeld–Hoffmann (EIH) coordinate system to a proper reference frame moving with the Earth–Moon barycenter is discussed in a simplified 3–body (Earth, Moon, Sun) problem. It is explicitly demonstrated how for this case the EIH–equations of motion reduce to the Jacobi–equations (equations of geodesic deviation) after the transformation to the proper reference frame. The authors then proceed with a Hill–Brown calculation for the dominant relativistic terms in the lunar laser ranging observables in the proper reference frame. For this simplified problem it is shown that in this frame all post–Newtonian terms are proportional to $m = n'/(n-n')$ [$n(n')$ = mean motion of Moon (Sun)] describing the solar influence upon the coupling strength of the Earth–Moon system and consequently all m–independent terms in the lunar EIH–coordinate motion have to be considered as relics from the heliocentric notion of simultaneity used in this coordinate picture.

094.010 **Variations in the Moon's mean motion due to the Earth's tides.**
M. Burša.
Bull. Astron. Inst. Czech., Vol. 37, No. 2, p. 80 – 84 (1986).
The variations in the Moon's mean motion due to the Earth's tides have been derived under the supposition that the Earth is perfectly elastic. The corresponding variations in the orbital angular momentum of the Earth–Moon system have been computed and it has been demonstrated that the tidal variations in the maximum principal moment of the Earth's inertia should essentially be compensated by the tidal variations in the angular velocity of the Earth's rotation.

094.011 **Be–10 contents of Apollo 17 double drive tube 74002/1.**
D. Aylmer, G. F. Herzog, C. Tuniz, R. K. Moniot, T. H. Kruse, J. Klein, R. Middleton.
Meteoritics, Vol. 20, No. 4, p. 604 – 605 (1985). Abstract. – See Abstr. 010.661.

094.012 **Nitrogen content and isotopic composition of lunar breccia 79035: a high resolution study.**
L. P. Carr, I. P. Wright, C. T. Pillinger.
Meteoritics, Vol. 20, No. 4, p. 622 – 623 (1985). Abstract. – See Abstr. 010.661.

094.013 **Constraints on the Collisional Ejection Hypothesis for the origin of the Moon imposed by noble siderophile elements in the earth's mantle.**
M. J. Drake.
Meteoritics, Vol. 20, No. 4, p. 638 – 639 (1985). Abstract. – See Abstr. 010.661.

094.014 **Cosmic ray exposure age and terrestrial age of the lunar meteorite ALHA 81005.**
O. Eugster, S. Niedermann, U. Krähenbühl.
Meteoritics, Vol. 20, No. 4, p. 642 – 643 (1985). Abstract. – See Abstr. 010.661.

094.015 **Return to Fra Mauro: ^{40}Ar–^{39}Ar ages.**
E. Heusser, N. Müller, E. K. Jessberger.
Meteoritics, Vol. 20, No. 4, p. 660 – 661 (1985). Abstract. – See Abstr. 010.661.

094.016 **Geochemical constraints on the origin of the Moon.**
A. S. P. Rao.
Meteoritics, Vol. 20, No. 4, p. 736 – 738 (1985). Abstract. – See Abstr. 010.661.

094.017 **The significance of agglutinate chemistry.**
O. D. Rode, M. Csajka.
Meteoritics, Vol. 20, No. 4, p. 746 – 747 (1985). Abstract. – See Abstr. 010.661.

094.018 **Lunar palaeomagnetism and primeval lunar satellites.**
S. K. Runcorn.
Meteoritics, Vol. 20, No. 4, p. 751 (1985). Abstract. – See Abstr. 010.661.

094.019 **Luna 16 spinels: revisited.**
P. Steinke, A. El Goresy.
Meteoritics, Vol. 20, No. 4, p. 764 (1985). Abstract. – See Abstr. 010.661.

094.020 **Lunar origins and the Mars–sized impactor model: cutting the Gordian knot?**
S. R. Taylor.
Meteoritics, Vol. 20, No. 4, p. 770 (1985). Abstract. – See Abstr. 010.661.

094.021 **Multifrequency polarization observations of the moon with RATAN–600.**
M. N. Naugol'naya, N. S. Soboleva.
Pis'ma Astron. Zh., Tom 12, No. 3, p. 247 – 252 (1986). In Russian. Englisch translation in Sov. Astron. Lett., Vol. 12.

As a result of polarization measurements of the moon's radio emission at 6 frequencies (from 1.38 to 31 cm) the dielectrical permittivity–wavelength dependence was refined. It is shown that diminishing of ε to shorter wavelengths is connected with density changes in the surface layer of 4 – 6 cm in depth. The obtained value of the mean surface density coincides with the Apollo 12 data to within 10%. A conclusion in made about nonvariability of dielectrical permittivity within temperature variations from 90 to 360K.

094.022 **Banded iron formations.**
K. Lambeck.
Nature, Vol. 320, No. 6063, p. 574 (1986).

This note comments on the determination, by Walker and Zahnle, of the lunar distance in the Precambrian from an analysis of the effects of the lunar nodal tide in an Australian banded iron formation (see Abstr. 094.023).

094.023 **Lunar nodal tide and distance to the Moon during the Precambrian.**
J. C. G. Walker, K. J. Zahnle.
Nature, Vol. 320, No. 6063, p. 600 – 602 (1986).

The pace of tidal evolution for the past ~450 Myr implies an Earth/Moon collision some 1,500 – 2,000 Myr BP. The authors present the first direct determination of the lunar distance in the Precambrian. They interpret a 23.3 ± 0.3–yr periodicity preserved in a 2,500 Myr BP Australian banded iron formation as reflecting the climatic influence of the lunar nodal tide, which has been detected with its modern 18.6–yr periodicity in some modern climate records. The lunar distance at 2,500 Myr BP would then have been about 52 Earth radii. The implied history of Precambrian tidal friction is in accord with both the more recent palaeontological evidence and the long–term stability of the lunar orbit.

094.024 **Palaeomagnetism of the moon and the problem of dynamo–fields of planets.**
Sh. Sh. Dolginov.
Kosm. Issled., Tom 24, Vyp. 1, p. 128 – 137 (1986). In Russian. English translation in Cosm. Res.

094.025 **The lunar crater Plato.**
K. P. Marshall, M. P. Mobberley.
J. Br. Astron. Assoc., Vol. 96, No. 3, p. 156 – 165 (1986).

The crater Plato has been a prime target for lunar observers since the seventeenth century, because of its conspicuous dark floor and large size. Abnormal appearances of the crater have been reported since the 1940s, and these are frequently associated with the variable brightness of the craterlets and markings on the floor. This paper describes the topography of the region, and the current and historical observations of the crater by BAA members.

094.026 **Luna incognita: 1984/85 report and 1986 observing schedule.**
J. E. Westfall.
Strolling Astron., Vol. 31, Nos. 5–6, p. 119 – 124 (1986).

094.027 **Evidence for recent lunar volcanism.**
M. T. Kitt.
Strolling Astron., Vol. 31, Nos. 7–8, p. 141 – 145 (1986).

This paper presents data for fresh craters located on the lunar farside, and confirms the findings of the nearside study. In addition, the distribution of fresh craters within the maria themselves has been investigated and is demonstrated to be non–random. These data can be interpreted as evidence for recent (post–Imbrian) volcanism.

094.028 **Spectral reflectance studies of the Nectaris Basin region.**
B. R. Hawke, P. Lucey, J. F. Bell, C. Coombs, P. D. Spudis.
Bull. Am. Astron. Soc., Vol. 17, No. 4, p. 924 (1985). Abstract. – See Abstr. 010.063.

094.029 **Coordinates of objects on the far side of the moon.**
S. G. Valeev.
Izv. Astron. Ehngel'gardt. Obs., No. 51, p. 17 – 40 (1985). In Russian.

094.030 **The parameter I from observations of the moon in a system of stars at a heliometer.**
Yu. A. Nefed'ev, R. R. Shajmukhametov.
Astron. Tsirk., No. 1391, p. 7 – 8 (1985). In Russian.

094.031 **Derivation of the parameter *r* from the second series of Bel'kovich.**
Yu. A. Nefed'ev.
Astron. Tsirk., No. 1396, p. 5 (1985). In Russian.

094.032 **The inclination of the lunar equator to the ecliptic from large–scale star–calibrated photographs of the moon.**
Yu. A. Nefed'ev, L. I. Rakhimov.
Astron. Tsirk., No. 1396, p. 6 (1985). In Russian.

094.033 **Problems of automation in mapping selenodetic reference nets.**
L. A. Karasev.
Visn. Kiiv. univ. Geogr., No. 27, p. 53 – 55 (1985). In Ukrainian. Abstr. in Ref. Zh., 52. Geod. Aehrosemka, 4.52.220 (1986).

094.034 **Names of lunar formations.**
A. Satō.
Heavens, Vol. 67, No. 5, p. 138 – 140 (1986). In Japanese.

094.035 **Shallow moonquakes and large shallow earthquakes: a temporal correlation.**
J. H. Shirley.
Earth Planet. Sci. Lett., Vol. 76, No. 3/4, p. 241 – 253 (1985/86).
The structure and seismicity of the Moon is very different from that of the Earth. One point of similarity is that statistics of seismic energy release in both bodies are dominated by the largest shallow–focus events. This study presents evidence of a relationship linking the timing of the 28 known shallow moonquakes with the timing of the 26 largest shallow earthquakes occurring in the same years (1971 – 1976).

094.036 **Electrical properties of regolith from the lunar samples 24.101.1; 24.159.1 and 24.191.1.**
R. Wäsch, A. Zedler, I. Faber.
Geokhimiya, No. 3, p. 423 – 430 (1985). In Russian. Abstr. in Ref. Zh., 51. Astron., 6.51.286 (1986).

094.037 **Photometric and polarimetric observations of the Moon's surface during 1984 and 1985.**
N. Sekiguchi, T. Yamaguchi, H. Shibasaki.
Tokyo Astron. Bull., Second Ser., No. 275, p. 3163 – 3184 (1986).
Photometric and polarimetric observations of the Moon's surface were carried out at the Dodaira Station of the Tokyo Astronomical Observatory using the 91 cm reflector equipped with the multichannel polarimeter, and the results are presented in tabular form.

094.038 **Horizons and opportunities in lunar sample science.**
Report LPI–85–04, Lunar Planet. Inst., Houston, Tex., USA, 32 pp. (1985). Abstr. in Phys. Abstr., Vol. 89, No. 1275, Entry 30177 (1986).

094.039 **Observations of contact moments of the earth's shadow using lunar surface features during the total lunar eclipse of January 9, 1982.**
K. N. Derkach, V. M. Kirpatovskij.
Vestn. Khar'kov. Univ., No. 278, p. 87 – 88 (1985). In Russian.

094.040 **The fate of the Moon rocks.**
B. M. Register.
Astronomy, Vol. 13, No. 12, p. 14 – 22 (1985). Abstr. in Phys. Abstr., Vol. 89, No. 1279, Entry 53765 (1986).

094.041 **The nature of crater rays: the Copernicus example.**
C. M. Pieters, J. B. Adams, P. J. Mouginis–Mark, S. H. Zisk, M. O. Smith, J. W. Head, T. B. McCord.
J. Geophys. Res., Vol. 90, No. B14, p. 12393 – 12413 (1985).
Crater rays are formed during a cratering event as target material is ballistically ejected to distances of many crater radii forming narrow, generally high albedo, approximately linear features extending outward from the crater. The nature of crater rays was examined for the lunar crater Copernicus using new information on the composition of surface material (from near–IR reflectance measurements), surface roughness (from radar backscatter measurements), and photogeologic data (from available images).

094.042 **The evolution of impact basins: cooling, subsidence, and thermal stress.**
S. R. Bratt, S. C. Solomon, J. W. Head.
J. Geophys. Res., Vol. 90, No. B14, p. 12415 – 12433 (1985).
The authors assess the contribution of thermal contraction and thermal stress to the topography and tectonics of large lunar impact basins. Exploratory models are developed for the temperature structure following basin formation, for the subsequent cooling of the basin region, and for the resulting thermal displacements and stresses as functions of time. The subsidence and stress at the surface are compared with topography and tectonic features in the comparatively well–preserved Orientale basin. On the basis of these comparisons the authors derive approximate constraints on the quantity and distribution of heat implanted during the basin–formation process.

094.043 **Early differentiation of the moon: evidence from trace elements in plagioclase.**
H. Palme, B. Spettel, H. Wänke, A. Bischoff, D. Stöffler.
J. Geophys. Res., Vol. 89, Suppl., p. C3 – C15 (1984). – See Abstr. 012.065.
The purpose of the present study is to estimate the contents of several elements such as Sc, Cr, and Co in the parent magma of the anorthosites and compare these data with those obtained from the parent magma of the Mg–suite rocks, which are early formed plutonic rocks (norites and troctolites) that are different from the anorthosites. The results have some bearing on the models for the early evolution of the lunar crust.

094.044 **Pristine rocks (8th foray): "plagiophile" element ratios, crustal genesis, and the bulk composition of the moon.**
P. H. Warren, G. W. Kallemeyn.
J. Geophys. Res., Vol. 89, Suppl., p. C16 – C24 (1984). – See Abstr. 012.065.
The authors examine Eu/Al, Sr/Al, Eu/Sr, and similar ratios among pristine nonmare lithologies, with implications for nonmare petrogenesis and for the bulk composition of the moon.

094.045 **Pristine highland clasts in consortium breccia 14305: petrology and geochemistry.**
J. W. Shervais, L. A. Taylor, J. C. Laul, M. R. Smith.
J. Geophys. Res., Vol. 89, Suppl., p. C25 – C40 (1984). – See Abstr. 012.065. With a correction in Vol. 90, Suppl., p. C844 (1985).
The authors characterize six small (< 1 cm) samples of pristine highland crust that occur as clasts in polymict lunar breccia 14305. One is an alkali anorthosite; the others include troctolite, norite, and anorthosite of the Mg–suite. These suites represent separate parent magmas, each of which may have been modified by mixing with KREEP.

094.046 **Magnesian anorthosites and associated troctolites and dunite in Apollo 14 breccias.**
M. M. Lindstrom, S. A. Knapp, J. W. Shervais, L. A. Taylor.
J. Geophys. Res., Vol. 89, Suppl., p. C41 – C49 (1984). With a correction in Vol. 90, Suppl., p. C845 (1985).– See Abstr. 012.065.
Magnesian anorthosite, a new type of pristine lunar highlands rock, has been found in Apollo 14 breccias. It has primitive (high Ca and Mg) silicate mineral compositions, and high and variable REE concentrations. The paper discusses the relationship of this rock and associated mafic rocks to other highlands rock types.

094.047 **Alkali gabbronorite, ultra–KREEPy melt rock and the diverse suite of clasts in North Ray crater feldspathic fragmental breccia 67975.**
M. M. Lindstrom.
J. Geophys. Res., Vol. 89, Suppl., p. C50 – C62 (1984). – See Abstr. 012.065.
Lunar breccia 67975 is a feldspathic fragmental breccia from North Ray crater, Apollo 16. It contains clasts of alkali gabbronorite and ultra–KREEPy mafic fragment–laden melt breccias, which are unique among Apollo 16 samples. Both are alkali– and iron–rich rocks with moderate to high REE concentrations. The nature of these two unusual lithologies, the relationship of 67975 clasts to other feldspathic fragmental breccias, and implications for the history of the North Ray crater source region are the major subjects of this paper.

094.048 **Petrology and geochemistry of lunar dimict breccia 61015.**
O. B. James, M. K. Flohr, M. M. Lindstrom.
J. Geophys. Res., Vol. 89, Suppl., p. C63 – C86 (1984). – See Abstr. 012.065.
Dimict breccia 61015 consists dominantly of granulated anorthosite and impact melt rock of VHA (very–high–alumina)–basalt composition. The anorthosite is a typical Apollo 16 ferroan anorthosite, except that it is slightly more augite–rich than most other ferroan anorthosites. The breccia problably formed in the floor of an expanding crater cavity by the following process: Impact melt generated within the cavity was injected into underlying anorthositic bedrock and solidified rapidly, and the anorthosite and melt rock were then fragmented by deformation in the later stages of the impact.

094.049 **Apollo 16 site geology and impact melts: implications for the geologic history of the lunar highlands.**
P. D. Spudis.
J. Geophys. Res., Vol. 89, Suppl., p. C95 – C107 (1984). – See Abstr. 012.065.
The author reconsiders the geology of the Apollo 16 site, the provenance of collected samples, and implications regarding the general geologic history of the lunar highlands.

094.050 **Petrology of Apollo 11 regolith breccias.**
S. B. Simon, J. J. Papike, C. K. Shearer.
J. Geophys. Res., Vol. 89, Suppl., p. C109 – C132 (1984). – See Abstr. 012.065.
Petrographic and mineral chemical data for 16 Apollo 11 regolith breccias show that: (1) the regolith breccias differ from soil 10084 with respect to (*a*) agglutinate content, (*b*) glass population, (*c*) plagioclase compositions, and (*d*) proportions of high–K mare and low–K mare basalt components; (2) the A–11 breccias and soil have highland components that are similar both in abundance and petrology; and (3) lunar regolith breccias provide a better comparison with howardites than do lunar soils. The data and observations are consistent with formation of the regolith breccias from immature soil. It appears that little or no highland material has been added to the Tranquilitatis regolith since the formation of the breccias.

094.051 **Petrologic profile of Apollo 16 regolith at station 4.**
A. Basu, D. S. McKay.
J. Geophys. Res., Vol. 89, Suppl., p. C133 – C142 (1984). – See Abstr. 012.065.
The Apollo16 double drive tube core 64001/2 (total depth ~60 cm) consists of 4 petrologic units based on the modal abundance of lithic, monomineralic, and other particles in the 20 – 500 µm size range of samples from 12 levels. The authors briefly describe and discuss the petrologic characteristics of the core soil 64001/2, and present new modal data on six size fractions of six samples from different levels of the lower part of the core (64001). From this and other previous data the authors infer the evolution of the regolith at this site, which seems to reveal two new aspects of regolith evolution and lateral transport of "exotic" rocks.

094.052 **Stratigraphy and geochemistry of the Stone mountain core (64001/2).**
R. L. Korotev, R. V. Morris, H. V. Lauer Jr.
J. Geophys. Res., Vol. 89, Suppl., p. C143 – C160 (1984). – See Abstr. 012.065.
The double drive tube core 64001/2 collected on Stone mountain, station 4, Appolo 16 was expected to contain the best samples of material representative of the Descartes formation. Based on the concentrations of 24 elements in one sample from each of the 68 half–centimeter dissection intervals of the lower section of the core tube (64001, 26.4 – 60.5 cm depth), no layer of soil enriched in materials thought to be representative Descartes material was found. However, all samples between 26 and 48 cm contain material of mare origin.

094.053 **Agglutinates as recorders of regolith evolution: application to the Apollo 17 drill core.**
J. C. Laul, M. R. Smith, J. J. Papike, S. B. Simon.
J. Geophys. Res., Vol. 89, Suppl., p. C161 – C170 (1984). – See Abstr. 012.065.
The authors report chemical data (27 major, minor, and trace elements by INAA) for agglutinates from 26 depth intervals of the Apollo 17 deep drill core (sections 70009–70001) and compare their compositions with those of the soils in which they occur.

094.054 **Cosmic ray exposure histories and ^{235}U–^{136}Xe dating of Apollo 11, Apollo 12, and Apollo 17 mare basalts.**
O. Eugster, P. Eberhardt, J. Geiss, N. Grögler, H. Schwaller.
J. Geophys. Res., Vol. 89, Suppl., p. C171 – C181 (1984). – See Abstr. 012.065.
The history of exposure to cosmic rays was studied for nine basaltic rocks based on the cosmic–ray–produced noble gas isotopes. The results for the five basalts 10003, 10047, 12004, 12051, and 77135 are consistent with a single–stage exposure. Four basalts, 10057, 10069, 12040, and 12053, experienced a multistage exposure. Using the ^{235}U–^{136}Xe dating method, the time of their early exposure to cosmic rays on the lunar surface was dated to have begun 3 – 4 aeons ago, immediately after rock formation as defined by their crystallization ages. The durations and shielding depths for the proposed two–stage exposure models of the four basalts were calculated. From the shielding at the early stage the authors estimate a lower limit of 50 cm for the thickness of the lava flows from which the investigated basalts originate.

094.055 **Composition of Orientale basin deposits and implications for the lunar basin–forming process.**
P. D. Spudis, B. R. Hawke, P. Lucey.
J. Geophys. Res., Vol. 89, Suppl., p. C197 – C210 (1984). – See Abstr. 012.065.
The Orientale basin, on the lunar western limb, has served as the prototype lunar multiring basin for many years. In order to constrain possible models of the formation and excavation of basin ejecta, the authors have studied the composition of Orientale basin deposits, using both earth–based spectral reflectance and Apollo orbital geochemical data.

094.056 **Magnetic field amplification and generation in hypervelocity meteoroid impacts with application to lunar paleomagnetism.**
L. L. Hood, A. Vickery.
J. Geophys. Res., Vol. 89, Suppl., p. C211 – C223 (1984). – See Abstr. 012.065.
A continuing difficulty with the core dynamo hypothesis for the origin of lunar paleomagnetic fields is the small maximum size of a present–day iron core and correspondingly small estimates for maximum lunar surface paleofields based on available models for planetary magnetic field generation. This paper explores the possibility that magnetic fields may be significantly amplified or generated in large–scale meteoroid impacts and that these fields may have contributed substantially to the observed magnetization of lunar materials.

094.057 **Formation of Apollo 15 green glass beads.**
J. Arndt, W. v. Engelhardt, I. Gonzalez–Cabeza, B. Meier.
J. Geophys. Res., Vol. 89, Suppl., p. C225 – C232 (1984). – See Abstr. 012.065.
By petrographic investigation of green glass beads in breccia 15427 and cooling and heating experiments with a synthetic green glass melt, the authors have tried to clarify the thermal history of Apollo 15 green glasses. From the results new suggestions are deduced on the process of their formation.

094.058 **Lunar luminescence and the filling–in of Fraunhofer lines in moonlight.**
A. E. Potter, W. Mendell, T. Morgan.
J. Geophys. Res., Vol. 89, Suppl., p. C240 – C244 (1984). – See Abstr. 012.065.

The filling–in of Fraunhofer lines in moonlight has been attributed to lunar luminescence. In order to test mechanisms proposed for this effect, measurements were made of the filling–in of the Hα line, the Na D doublet, and a number of lines near the doublet. The degree of filling–in was not correlated with wavelength in a way expected from a broadband luminescence, such as thermoluminescence. However, it was correlated with equivalent widths of the Fraunhofer line, such that it increased as equivalent width decreased. This suggests that the Fraunhofer line filling at the moon's surface is caused by inelastic scattering of sunlight with a small wavelength shift, as for example, photon–phonon scattering.

094.059 **Apollo 14 aluminous mare basalts and their possible relationship to KREEP.**
T. Dickinson, G. J. Taylor, K. Keil, R. A. Schmitt, S. S. Hughes, M. R. Smith.
J. Geophys. Res., Vol. 90, Suppl., p. C365 – C374 (1985). – See Abstr. 012.065.

The authors studied 36 mare basalt fragments from the processing fines of breccia 14321 in order to determine how many individual flows were represented, to what extent each flow was differentiated, and the nature and origin of their source regions in the lunar mantle.

094.060 **Apollo 14 mare basalts: petrology and geochemistry of clasts from consortium breccia 14321.**
J. W. Shervais, L. A. Taylor, M. M. Lindstrom.
J. Geophys. Res., Vol. 90, Suppl., p. C375 – C395 (1985). – See Abstr. 012.065.

Eleven mare basalt clasts have been extracted from polymict lunar breccia 14321 for a combined petrologic and geochemical study. Ten of these clasts are high–Al, low–Ti mare basalts similar to those discovered previously at Apollo 14. The other is a highly evolved tridymite ferrobasalt with low Al and intermediate Ti contents. A hight–Al, ilmenite ferrobasalt from lunar breccia 14305 is the most evolved high–Al basalt described from the Apollo 14 site. The major and trace element characteristics of the basalts studied here allow them to be grouped into five compositional types that range from LREE–depleted to LREE–enriched.

094.061 **The depths of the mare basalt source region.**
A. B. Binder.
J. Geophys. Res., Vol. 90, Suppl., p. C396 – C404 (1985). – See Abstr. 012.065.

Chemical data for the parental magmas of the nine known VLT Array I pyroclastic glasses show statistically significant trends due to olivine ($\sim Fo_{83}$) control during their formation by partial melting. The compositional scatter is largely due to compositional variations in the source regions on the $\pm 1\%$ level. This compositional scatter is small when one considers that the scale of the source region is up to 1000 km, but is sufficient to make positive identification of the residual phase(s) in the source regions difficult. Nevertheless, when the effects of the scatter are properly modeled, it is relatively clear that olivine is the residual phase in the source region. Hence these data and additional constraints indicate that the source regions are at shallow depths in the moon.

094.062 **Origin of olivine at Copernicus.**
C. M. Pieters, D. E. Wilhelms.
J. Geophys. Res., Vol. 90, Suppl., p. C415 – C420 (1985). – See Abstr. 012.065.

The central peaks of Copernicus are among the few lunar areas where near–infrared telescopic reflectance spectra indicate extensive exposures of olivine. Other parts of Copernicus crater and ejecta, which were derived from highland units in the upper parts of the target site, contain only low–Ca pyroxene as a mafic mineral. The exposure of compositionally distinct layers including the presence of extensive olivine may result from penetration to an anomalously deep layer of the crust or to the lunar mantle. The authors suggest that the Procellarum basin and the younger, superposed Insularum basin have provided access to these normally deep–seated crustal or mantle materials by thinning the upper crustal material early in lunar history. The occurrences of olivine in portions of the compositionally heterogeneous Aristarchus Region, in a related geologic setting, may be due to the same sequence of early events.

094.063 **A transient heating event in the history of a highlands troctolite from Apollo 12 soil 12033.**
U. B. Marvin, D. Walker.
J. Geophys. Res., Vol. 90, Suppl., p. C421 – C429 (1985). – See Abstr. 012.065.

A 1.5 mm clast of a highlands troctolite attached to a glassy, alkali–rich soil breccia was identified in a thin section of 1 – 2 mm particles from Apollo 12 soil sample 12033. The troctolite consists of euhedral olivines in a groundmass of granulitic plagioclase; each olivine grain is rimmed by a thin selvage of olivine–plagioclase glass containing vesicles and minute crystallites of spinel, olivine, and plagioclase. This unique texture records a complex history, beginning with the formation of the troctolite as a plutonic cumulate, followed by subsolidus metamorphism, partial remelting, vesculation, crystallite growth, and quenching.

094.064 **Isotope analysis of crystalline impact melt rocks from Apollo 16 stations 11 and 13, North Ray crater.**
W. U. Reimold, L. E. Nyquist, B. M. Bansal, J. L. Wooden, C.–Y. Shih, H. Weismann, I. D. R. Mackinnon.
J. Geophys. Res., Vol. 90, Suppl., p. C431 – C448 (1985). – See Abstr. 012.065.

Rubidium–strontium and Sm–Nd isotopic data were obtained for lunar impact melt rocks from Apollo 16 stations 11 and 13. Current hypotheses for the site geology, stratigraphy, and radiometric age data lead the authors to suggest that the Nectaris basin and the Descartes formation were formed about 4.1 b.y. ago, and the Imbrium basin and the Cayley formation were formed about 3.81 – 3.86 b.y. ago. The ages of some individual melt rocks probably date "local" cratering events, however.

094.065 **Composition and evolution of the lunar crust in the Descartes highlands, Apollo 16.**
D. Stöffler, A. Bischoff, R. Borchardt, A. Burghele, A. Deutsch, E. K. Jessberger, R. Ostertag, H. Palme, B. Spettel, W. U. Reimold, K. Wacker, H. Wänke.
J. Geophys. Res., Vol. 90, Suppl., p. C449 – C506 (1985). – See Abstr. 012.065.

This study is directed toward an understanding of the North Ray target composition, the cratering process, and the analysis of the pre–North Ray history of the crater basement and its relationship to the major selenologic units, the Descartes and Cayley formations. Specific aims may be summarized as follows: (1) interpretation of the nature and provenance of the constituents of polymict breccias, (2) recognition of pristine igneous lithologies and their bearing on the composition of the magma ocean and the primordial crust, (3) assessment of lunar thermal metamorphism, (4) derivation of the structure, stratigraphy, and composition of the subregolith basement from the properties of breccias and young impact melts, and (5) identification and characterization of the Descartes and Cayley formations and their relationships to the Nectaris and Imbrium basin impacts.

094.066 **Chemistry and petrology of Apollo 12 drive tube 12027.**
M. R. Smith, J. C. Laul, S. B. Simon, J. J. Papike.
J. Geophys. Res., Vol. 90, Suppl., p. C507 – C516 (1985). – See Abstr. 012.065.

The authors report new petrologic and chemical data for drive tube 12027 and compare the findings with bulk and size fraction studies of other Apollo 12 soils. The authors also compare chemical mixing calculations with petrologic observations for four separate depth intervals from the 17–cm core.

094.067 **Noble gas component organization in Apollo 14 breccia 14318: ^{129}I and ^{244}Pu regolith chronology.**
T. D. Swindle, M. W. Caffee, C. M. Hohenberg, G. B. Hudson, J. C. Laul, S. B. Simon, J. J. Papike.
J. Geophys. Res., Vol. 90, Suppl., p. C517 – C539 (1985). – See Abstr. 012.065.

Noble gas, petrological, and chemical studies made on grain–size separates from lunar regolith breccia 14318 demonstrate that the noble gases are organized into two functional components, volume–correlated and surface–correlated.

094.068 **Two–stage models for lunar and terrestrial anorthosites: petrogenesis without a magma ocean.**
J. Longhi, L. D. Ashwal.
J. Geophys. Res., Vol. 90, Suppl., p. C571 – C584 (1985). – See Abstr. 012.065.
Two–stage diapiric models for lunar ferroan anorthosites and terrestrial massif anorthosites are examined. The lunar model is developed to explain early lunar differentiation in the absence of a magma ocean. If correct, the terrestrial model serves as an analog for the development of lunar anorthositic diapirs.

094.069 **A new Monte Carlo model of lunar megaregolith development.**
J. Cashore, A. Woronow.
J. Geophys. Res., Vol. 90, Suppl., p. C811 – C815 (1985). – See Abstr. 012.065.
A new Monte Carlo model is used to examine the development of the lunar megaregolith through impact cratering processes. This improved model predicts greater minimum megaregolith thickness than did the earlier models. If the 25 km seismic horizon on the moon is attributable to the bottom of the crater–excavated material, then the highlands appear to have suffered a bombardment of two to five times the number of craters presently preserved on the surface.

094.070 **Der Mond. Struktur, Entwicklung and Erdbeziehung.**
M. Reichstein.
Physik des Planeten Erde, p. 35 – 47 (1985). – See Abstr. 003.047.

094.071 **Anwendung der Spektrophotometrie in der Erkundung der lunaren Oberfläche.**
R. Jaumann, G. Enderlein, G. Neukum.
Mitt. Astron. Ges., No. 65, p. 170 – 177 (1986). – See Abstr. 012.072.

The analysis of extraterrestrial materials.
See Abstr. 003.053.

Mannetjes op de maan.
See Abstr. 004.019.

Leonardo da Vinci's drawings of the surface features of the moon.
See Abstr. 004.056.

Gallium and germanium as tracers of formation and differentiation processes in terrestrial planetary objects.
See Abstr. 022.034.

Experimental methods for quenching structures in lunar–analog silicate melts: variations as a function of quench media and composition.
See Abstr. 022.109.

The origin of the Moon and the single–impact hypothesis I.
See Abstr. 022.127.

Photographier la Lune. Quel temps de pose?
See Abstr. 034.184.

On the accuracy of photographic observations of the moon on the star background.
See Abstr. 036.026.

On optimum arrangement of measurements of radial accelerations of an artificial lunar satellite for determination of the zonal harmonics of the potential.
See Abstr. 036.053.

Estimate of the accuracy of calibration of the "Relict" radiometer from the radio radiation of the moon.
See Abstr. 036.071.

System of exact time and frequency for lunar laser ranging.
See Abstr. 036.117.

Special instrument for calculation of the coordinates of the moon.
See Abstr. 036.119.

Quantitative analysis of planetary reflectance spectra with principal components analysis.
See Abstr. 036.205.

Hypothesis of variable *G* and the secular accelerations of the Sun and the Moon.
See Abstr. 043.004.

On tidal friction and the decelerations of the Earth's rotation and Moon's revolution.
See Abstr. 044.040.

Periodic orbits around the moon and their use in superlong–baseline radio interferometry.
See Abstr. 052.008.

Motion of a lunar orbiter.
See Abstr. 052.010.

On the dominant relativistic terms in the lunar theory.
See Abstr. 066.024.

Evidence for a possible secular decrease in solar wind intensity.
See Abstr. 074.016.

Solar flare neon composition and solar cosmic–ray exposure ages based on lunar mineral separates.
See Abstr. 078.001.

Observations of the moon for more accurate definition of the radio radius of the sun.
See Abstr. 080.029.

The parental magma for some rocks from the Norite 1 subzone of the Stillwater Complex: a lunar analog study.
See Abstr. 081.055.

Planetary surface texture and albedo from parameter plots of optical polarization data.
See Abstr. 091.001.

Regression modelling in selenodesy.
See Abstr. 091.029.

A harmonic and statistical analysis of the topography of the Earth, Moon and Mars.
See Abstr. 091.056.

Tectonic tests of proposed polar wander paths for Mars and the Moon.
See Abstr. 097.009.

A summary of extremes of isotopic variations in extra–terrestrial materials.
See Abstr. 105.012.

Yamato–82193 meteorite: the third lunar meteorite collected at the Yamato Mountains, Antarctica.
See Abstr. 105.169.

Noble gas isotopic composition, cosmic ray exposure history, and terrestrial age of the meteorite Allan Hills A81005 from the Moon.
See Abstr. 105.205.

The wavelength dependence of phase coefficients.
See Abstr. 105.242.

Petrology and chemistry of hyperferroan anorthosites and other clasts from lunar meteorite ALHA81005.
See Abstr. 105.247.

An apatite–rich, ferroan, mafic lithology from lunar meteorite ALHA81005.
See Abstr. 105.256.

Stability and evolution of primeval lunar satellite orbits.
See Abstr. 107.044.

Protoearth mass shedding and the origin of the Moon.
See Abstr. 107.045.

095 Lunar Eclipses

095.001 **Fotometria fotografica dell'eclisse lunare di penombra dell' 8 Novembre 1984.**
F. Foresta Martin, P. Lisca, M. Missori, M. Pigliucci, D. Ricci, F. Sabatini, P. Tosi.
Astronomia, N. 1, p. 2 – 5 (1986).
This work was aimed to test a home–made photodiode densitometer used to measure densities in the photographic negatives of the Moon taken during the November 8, 1984 eclipse, in order to study ordinary and anomalous variations of brightness. Instrumentation and methods of reduction are described.

095.002 **Lunar eclipses return: April 24, 1986.**
J. E. Westfall.
Strolling Astron., Vol. 31, Nos. 7–8, p. 146 – 149 (1986).

095.003 **On prediction of the brightness of lunar eclipses.**
A. A. Khekalo, V. M. Chernov.
Astron. Vestn., Tom 20, No. 2, p. 152 – 155 (1986). In Russian. English translation in Sol. Syst. Res.
The dependence of lunar eclipses brightness according to the Danjon scale on the solar cycle phase is presented by an analytic equation. A less complicated method of foreseeing the brightness of future lunar eclipses using five equations is proposed as well as a table of supposed lunar eclipses brightness from 1982 through 2000. The influence of the products of terrestrial volcanic activity on lunar eclipses brightness in 1982 – 1983, in particular is also considered.

095.004 **Total eclipse of the moon on 4 May 1985.**
T. P. Cooper.
Mon. Notes Astron. Soc. S. Afr., Vol. 44, Nos. 9–10, p. 87 – 90 (1985).

Dark lunar eclipses in classical antiquity.
See Abstr. 004.051.

A new photographic method for mapping lunar eclipse shadow.
See Abstr. 036.082.

Eclipse news.
See Abstr. 079.003.

Observations of contact moments of the earth's shadow using lunar surface features during the total lunar eclipse of January 9, 1982.
See Abstr. 094.039.

Occultations during the total eclipse of 1985 05 04.
See Abstr. 096.013.

096 Lunar and Planetary Occultations

096.001 ***JHK* photometry of candidate stars for occultation by Neptune.**
F. Vilas, D. J. Mink.
Publ. Astron. Soc. Pac., Vol. 98, No. 599, p. 116 – 117 (1986).
JHK photometry is presented of six stars which will be occulted by Neptune during 1986 and 1987. Two stars are possible candidates for infrared photometric observations of occultations.

096.002 **Where are the good Antares grazes during 1986?**
D. W. Dunham.
Occultation Newsl., Vol. 3, No. 15, p. 319 – 321 (1986).

096.003 **Grazing occultations.**
D. Stockbauer.
Occultation Newsl., Vol. 3, No. 15, p. 324 – 327 (1986).

096.004 **Solar system occultations during 1986.**
D. W. Dunham.
Occultation Newsl., Vol. 3, No. 15, p. 330 – 338 (1986).

096.005 **Observations d'occultations rasantes en 1985.**
J. Bourgeois, J. Schwaenen.
Ciel Terre, Vol. 102, No. 2, p. 55 – 57 (1986).

096.006 **Occultations of stars by the moon in Potava in 1983.**
B. F. Sincheskul.
Prilivy i vrashchenie Zemli, Kiev, p. 112 – 122 (1985). In Russian. Abstr. in Ref. Zh., 51. Astron., 2.51.121 (1986).

096.007 **Observations of occultations of stars by the moon at the Astronomical Observatory of the Kiev University in the years 1981 – 1982.**
V. I. Mazur, A. K. Osipov.
Vestn. Kiev. Univ. Astron., No. 27, p. 78 – 80 (1985). In Russian.

096.008 **Observations of a grazing occultation of λ Sagittarius by the moon.**
A. K. Osipov, V. I. Mazur, A. A. Zhitetskij.
Vestn. Kiev. Univ. Astron., No. 27, p. 81 – 82 (1985). In Russian.

096.009 **Lunar occultations of bright planets and bright stars in China, 1986 – 2000.**
B.-l. Liu.
Publ. Purple Mt. Obs., Vol. 4, No. 4, p. 5 – 22 (1985). In Chinese.
This paper gives the circumstances of lunar occultations of Venus, Mars, Jupiter, Saturn, Spica, Antares, Aldebaran, and Regulus for 32 main cities of China in the period 1986 – 2000.

096.010 **Observations of occultations of stars by the moon at the Astronomical Observatory of the Kharkov University in 1982 and 1983.**
K. N. Derkach, V. M. Kirpatovskij, P. P. Pavlenko.
Vestn. Khar'kov. Univ., No. 278, p. 85 – 87 (1985). In Russian.

096.011 **Grazing occultations.**
D. Stockbauer.
Occultation Newsl., Vol. 3, No. 16, p. 342 – 343 (1986).

096.012 **Past observations and future plans for 1986 grazing occultations of Antares.**
D. W. Dunham.
Occultation Newsl., Vol. 3, No. 16, p. 351 – 352 (1986).

096.013 **Occultations during the total eclipse of 1985 05 04.**
M. D. Overbeek.
Mon. Notes Astron. Soc. S. Afr., Vol. 44, Nos. 11–12, p. 102 – 113 (1985).

096.014 **Occultations by Pluto and its satellite.**
IAU Circ., Nos. 4203, 4206 (1986).

096.015 **Planetary occultations.**
IAU Circ., No. 4207 (1986).

Observing reappearances.
See Abstr. 036.029.

Lunar occultation techniques in investigations of binary stars.
See Abstr. 036.104.

Microcomputer reduction of digital photometric occultation data.
See Abstr. 036.108.

Application de la méthode par déconvolution intégrée au cas des occultations d'étoiles doubles.
See Abstr. 036.215.

Venus: eastern and western elongations 1981 82.
See Abstr. 093.048.

Predictions of occultations by minor planets as a test of accuracy.
See Abstr. 098.001.

Analysis of two positive asteroidal occultation reports.
See Abstr. 098.010.

The 13 November 1984 occultation of BD +8°471 by (1) Ceres.
See Abstr. 098.036.

The analysis of the lunar occultation of Vesta (1984).
See Abstr. 098.058.

Reports of asteroidal appulses and occultations.
See Abstr. 098.121.

Astrometric updates for asteroidal occultations in the U.S.S.R. during 1984.
See Abstr. 098.122.

Predictions of stellar occultations by asteroids.
See Abstr. 098.127.

Observation and preliminary analysis of a recent Jovian satellite phenomenon.
See Abstr. 099.058.

Occultation detection of a neptunian ring-like arc.
See Abstr. 101.004.

New evidence for a ring around Neptune.
See Abstr. 101.013.

Re-analysis of the July 1984 star occultation by a Neptunian ring-like arc.
See Abstr. 101.038.

The 1985 August 20 Neptune occultation observed from CTIO.
See Abstr. 101.039.

The 20 August 1985 stellar occultation by Neptune and its arc system.
See Abstr. 101.040.

Signal-to-noise ratios for possible stellar occultations by Pluto.
See Abstr. 101.057.

JHK photometry of Uranus and Neptune occultation candidate stars: 1986 – 1990.
See Abstr. 101.058.

Occultations and appulses by Halley's comet.
See Abstr. 103.004.

Angular diameters by the lunar occultation technique. VII.
See Abstr. 115.024.

The spectroscopic orbits of HD 44780 and 65 Geminorum.
See Abstr. 120.025.

Sigma Scorpii, a bright close binary.
See Abstr. 122.174.

Observations of the lunar occultations of the sources 4C 24.11 and 4C 17.43 in the decameter radio wave range.
See Abstr. 158.057.

097 Mars, Mars Satellites

097.001 **Geomorphic evidence for the distribution of ground ice on Mars.**
S. W. Squyres, M. H. Carr.
Science, Vol. 231, No. 4735, p. 249 – 252 (1986).
High–resolution Viking orbiter images show evidence for quasi–viscous relaxation of topography. The relaxation is believed to be due to creep deformation of ice in near–surface materials. The global distribution of the inferred ground ice shows a pronounced latitudinal dependence. The equatorial regions of Mars appear to be ice–poor, while the heavily cratered terrain poleward of $\pm 30°$ latitude appears to be ice–rich. The style of creep poleward of $\pm 30°$ varies with latitude, possibly due to variations in ice rheology with temperature. The distribution suggests that ice at low latitudes, which is not in equilibrium with the present atmosphere, has been lost via sublimation and diffusion through the regolith, thereby causing a net poleward transport of ice over martian history.

097.002 **Are Martian clouds fractals?**
A. Henderson–Sellers.
Q. J. R. Astron. Soc., Vol. 27, No. 1, p. 90 – 93 (1986).
It has been proposed by Lovejoy that terrestrial clouds and rain areas are fractals: shapes with structure at all scales and no characteristic length. If this feature is the result of the physical and chemical controls upon atmospheric dynamics, then it should also occur on other planets. Examinations of Viking images of Mars lend some support to the theory that all clouds are fractals.

097.003 **On the maximum elevation obtained by aerosols in the martian atmosphere.**
F. Jaquin, P. Gierasch, R. Kahn.
News Lett. Astron. Soc. N.Y., Vol. 2, No. 9, p. 9 (1986). Abstract. – See Abstr. 010.241.

097.004 **Determination of the $^{12}C/^{13}C$ and $^{16}O/^{18}O$ ratio in the Martian atmosphere by 10 micron heterodyne spectroscopy.**
U. Schrey, H. Rothermel, H. U. Käufl, S. Drapatz.
Astron. Astrophys., Vol. 155, No. 1, p. 200 – 204 (1986).
During the opposition of Mars in May 1984, the planetary atmosphere was studied using a 10 μm heterodyne receiver mounted at the Cassegrain focus of the 3 m IRTF telescope on Mauna Kea/Hawaii. Using a model for the Martian atmosphere, the following isotopic ratios have been obtained: $^{12}C/^{13}C = 96 \pm 6$ and $^{16}O/^{18}O = 518 \pm 64$. They are compared with the results of the in situ measurements made by the Viking spaceprobes and terrestrial values and agree within the statistical errors.

097.005 **Comparison of knobs on Mars to isolated hills in eolian, fluvial and glacial environments.**
L. S. Manent, F. El–Baz.
Earth, Moon, Planets, Vol. 34, No. 2, p. 149 – 167 (1986).
Positive isolated features or knobs have been observed on Mars since Mariner 9 first photographed the planet in 1972. With the use of Viking photomosaics, a systematic search for knobs was completed. The knobs were characterized by length, width, geographic location, proximity to streaks and geologic surroundings. Similar isolated features on Earth eroded by fluvial, glacial, and eolian processes were studied and measured. Comparison of length–to–width ratios of Martian knobs to isolated hills on Earth indicate that the Martian knobs are most similar to the isolated hills formed in a hyper–arid environment.

097.006 **Martian fluidized crater distribution: tectonic implications.**
P. G. Thomas, P. L. Masson.
Earth, Moon, Planets, Vol. 34, No. 2, p. 169 – 176 (1986).
Little scaled studies on the entire planet Mars, and large scale studies of a local area (3600 km in diameter) seem to indicate geographic relations between martian fluidized craters and ridges: all the intensely ridged areas exhibit a lot of fluidized craters. Because the presence of fluidized craters indicates low viscosity states of the martian surface, this relation would indicate that the compressive stresses only induced ridges when they occurred at times and places of low viscosity states of the martian surface.

097.007 **Photopolarimetry of Martian aerosols. II. Limb and terminator measurements.**
R. Santer, M. Deschamps, L. V. Ksanfomaliti, A. Dollfus.
Astron. Astrophys., Vol. 158, No. 1/2, p. 247 – 258 (1986).
The two photopolarimeters VPM placed on board the Soviet spaceprobe MARS–5 which orbited planet Mars in February 1974 recorded a number of intensity and polarization scans from limb to terminator over the planet. Ground properties of the planetary surface and scans analysis over scarce clouds and dust raised features have already been published. Here the authors analyse the permanent aerosols in the free atmosphere. They appear to be mostly concentrated at high altitude between 40 to 60 km above the surface. Polarization indicates sub–micron sizes and a refraction index around 1.55 reminiscent of the terrestrial tropospheric aerosols above continents. The optical depth did not exceed much $\tau = 0.1$ at 592 nm during the measurements which corresponded to an overall period of particularly clean atmosphere.

097.008 **Terrestrial analogues of the surface rocks of Mars?**
R. G. Burns.
Nature, Vol. 320, No. 6057, p. 55 – 56 (1986).
Terrestrial rocks and minerals are continually being sought which might serve as analogues for martian materials. Evidence presented here suggests that hisingerite, iddingsite and several ferric sulphate minerals derived from deuteric alteration of iron–rich basalts may provide clues to the evolutionary history of the surface of Mars.

097.009 **Tectonic tests of proposed polar wander paths for Mars and the Moon.**
R. E. Grimm, S. C. Solomon.
Icarus, Vol. 65, No. 1, p. 110 – 121 (1986).
The authors have tested the polar wander paths recently proposed for Mars by Schultz and Lutz–Garihan and for the Moon by Runcorn through a comparison of the lithospheric stress field predicted for rapid global reorientations against observed tectonic features. They have employed the theory of Vening Meinesz and of Melosh to calculate the reorientation stresses, and they argue that the formation of normal faults or graben in broad regions surrounding the former rotation poles should be the minimum tectonic signature of a reorientation that generates lithospheric stresses in excess of the extensional strength of near–surface material. It is concluded that polar wander of the magnitude and timing proposed for Mars by Schultz and Lutz–Garihan did not occur and that the tectonic test of polar wander on the Moon is inconclusive.

097.010 **Observing Mars in the 1980s: reference maps for the visual observer.**
R. McKim.
J. Br. Astron. Assoc., Vol. 96, No. 3, p. 166 – 169 (1986).

097.011 **Calculating Martian polar cap latitudes.**
J. D. Beish, D. C. Parker, C. F. Capen.
Strolling Astron., Vol. 31, Nos. 7–8, p. 137 – 140 (1986).
In order to understand a planet's geology and climate, one must obtain precise positions of its surface and atmospheric features. These positions are usually expressed in degrees of latitude and longitude. For over 150 years the filar micrometer has been used for these measurements. More recently, earth–based and spacecraft photographs have been measured with measuring engines. In addition, longitudes may be determined accurately by

means of central meridian timings. These methods are summarized.

097.012 **Estimate of the mechanical properties of regolith of Phobos.**
A. V. Kozenko.
Astron. Vestn., Tom 20, No. 2, p. 155 – 157 (1986). In Russian. English translation in Sol. Syst. Res.
Slopes and heights in craters and grooves which appear to be overlain with regolith permit to set lower limits to its strength.

097.013 **Thermal infrared opacity of the Mars atmosphere.**
T. Z. Martin.
Icarus, Vol. 66, No. 1, p. 2 – 21 (1986). – See Abstr. 012.024.
A refined technique is presented for deriving 9–μm extinction opacities of the Mars atmosphere from brightness temperature measurements made by the Viking Infrared Thermal Mapper. Improvements include modeling of the vertical temperature profile, a surface–atmosphere temperature discontinuity, and the effects of surface emissivity and particle scattering. The routine is applied to the Mars Average Data Set to yield zonal mean opacities for more than one Mars year. The mean 9–μm opacity for one year is 0.51.

097.014 **Loss of fine–scale surface texture in Viking Orbiter images and implications for the inferred distribution of debris mantles.**
R. Kahn, E. Guinness, R. Arvidson.
Icarus, Vol. 66, No. 1, p. 22 – 38 (1986). – See Abstr. 012.024.
The authors present a self–consistent model that quantitatively accounts for the observed loss of fine–scale features in two overlapping sets of Viking Orbiter images. They then examine the effects of varying atmospheric conditions on inferences about the Martian surface from Orbiter images, and discuss how one may account for atmospheric effects in future photogeologic interpretations. The authors conclude by making a preliminary reassessment of the evidence constraining the latitudinal distribution of debris deposits on Mars, based upon crater size–frequency data.

097.015 **Red/violet contrast reversal on Mars: significance for eolian sediments.**
P. Thomas, J. Veverka.
Icarus, Vol. 66, No. 1, p. 39 – 55 (1986). – See Abstr. 012.024.
Albedo markings on Mars can exhibit reversed contrast with their surroundings when imaged in "red" and "violet" light. The authors use the Viking Orbiter images to establish an association of contrast reversal with particular surface features on Mars. Laboratory reflectance data on candidate Mars materials are also used to seek plausible explanations for the observed Martian color and contrast patterns.

097.016 **Martian dust storms at the early stage of their evolution.**
S. Ebisawa, A. Dollfus.
Icarus, Vol. 66, No. 1, p. 75 – 82 (1986). – See Abstr. 012.024.
The authors discuss some problems concerning the exact behavior of Martian dust storms at their first occurrence and also concerning their emergence sources, on the basis of Earth–based observations.

097.017 **Surface properties of the Pettit wind streak on Mars: implications for sediment transport.**
J. R. Zimbelman.
Icarus, Vol. 66, No. 1, p. 83 – 93 (1986). = Lunar Planet. Inst., Houston, Tex., USA, Contrib., No. 579. – See Abstr. 012.024.
Pettit crater (12°N, 174°W) is associated with a complex wind streak consisting of both Type I (bright) streak and Type II (dark) streak materials. The dark streak material is consistent with models involving sand saltation and transport but the bright streak thermal inertia is not consistent with a volumetrically significant enhancement of unconsolidated dust in the streak.

097.018 **Iron oxide and hydroxide precipitation from ferrous solutions and its relevance to Martian surface mineralogy.**
J. Posey–Dowty, B. Moskowitz, D. Crerar, R. Hargraves, L. Tanenbaum, E. Dowty.
Icarus, Vol. 66, No. 1, p. 105 – 116 (1986). – See Abstr. 012.024.
This paper investigates one of the possibilities for the formation of magnetic phases on the Martian surface, that of ferric iron precipitation in aqueous solutions under oxidizing conditions at neutral pH. Several key questions are addressed here: (1) What are the major controls on the precipitation of iron oxyhydroxides and oxides? (2) In the presence of "candidate" Martian basaltic minerals, which iron phases precipitate? (3) Are there major differences in the iron phases expected on Earth and on Mars?

097.019 **On the thermal properties of Martian fines.**
B. M. Jakosky.
Icarus, Vol. 66, No. 1, p. 117 – 124 (1986). – See Abstr. 012.024.
A physical interpretation of remotely sensed values of the thermal inertia of the Mars surface depends in part on the relationship between regolith particle size and thermal conductivity. Laboratory measurements are available only for particles within the size range of 50 – 800 μm, yet abundant evidence suggests the importance of particles as small as 1 μm and as large as a centimeter. Theoretical arguments are developed here to extrapolate to particle sizes outside of the range measured.

097.020 **Are the Viking Lander sites representative of the surface of Mars?**
B. M. Jakosky, P. R. Christensen.
Icarus, Vol. 66, No. 1, p. 125 – 133 (1986). – See Abstr. 012.024.
Global remote–sensing observations of Mars are compared with remote–sensing observations of the two Viking Lander site regions and with orbiter and lander imaging of the sites. The lander sites do not fit most of the global trends of remote–sensing data. The lander sites are intermediate in structure between classical bright and dark regions, and are distinctive from most of the rest of the planet.

097.021 **Composition and stability of the condensate observed at the Viking Lander 2 site on Mars.**
H. M. Hart, B. M. Jakosky.
Icarus, Vol. 66, No. 1, p. 134 – 142 (1986). – See Abstr. 012.024.
The authors show via an equilibrium surface energy balance calculation that CO_2 frost is not stable at the Viking Lander 2 site during the period of time condensate was observed on the ground at the site. The authors estimate the stability of H_2O frost based on sublimation rates controlled by the surface temperature; water frost was very stable during this period of time, but was not stable after the condensate disappeared. Using spectral evidence, the thickness of the condensate is estimated to be about 10 μm.

097.022 **Stabilization of atmospheric pressure and seasonal changes in the polar caps in a model with a chemically inhomogeneous Martian atmosphere.**
V. I. Aleshin.
Sov. Astron., Vol. 29, No. 4, p. 462 – 464 (1985). English translation of 40.097.015.

097.023 **The climate of Mars.**
R. M. Haberle.
Sci. Am., Vol. 254, No. 5, p. 42 – 50 (1986).
It started out much like the earth's early climate, but it evolved differently. Once warm enough to support flowing water, Mars is now so cold that carbon dioxide freezes at the poles every winter.

097.024 **The question of life on Mars.**
B. Adelman.
J. Br. Interplanet. Soc., Vol. 39, No. 6, p. 256 – 262 (1986).
The problem of life on Mars has been one of the main Mars investigations since the Mariner IV satellite examined the planet in 1965. It is far from settled and will surely continue to be a driving force in the exploration of Mars. The discoveries that have been made are described here and analysed with particular attention to the Viking Lander biology experiments.

097.025 **Martian south polar cap 1973.**
K. Iwasaki, Y. Saito, T. Akabane.
Publ. Astron. Soc. Jpn., Vol. 38, No. 2, p. 267 – 275 (1986).
The regression of the Martian south polar cap for the 1973 apparition is examined with the measurements of the photographs of Mars obtained at the Hida Observatory. The 1973 regression curve is compared to the 1971 and 1977 regression curves and to the mean regression curve extracted from the Lowell data for the period 1907 – 1956. The 1973 regression appears to be faster than the mean regression.

097.026 **Polar warming in the middle atmosphere of Mars.**
D. Deming, M. J. Mumma, F. Espenak, T. Kostiuk, D. Zipoy.
Icarus, Vol. 66, No. 2, p. 366 – 379 (1986).
The authors report ground–based laser heterodyne spectroscopy of non–thermal emission in the cores of the 10.33–μm $R(8)$ and 10.72–μm $P(32)$ lines of $^{12}C^{16}O_2$, obtained at 23 locations on the disk of Mars during the 1984 opposition, at $L_s = 130°$. They find that the temperature of the middle Martian atmosphere varies with latitude. A comparison with 2–D circulation models shows that the presence of atmospheric dust may enhance dynamical heating at high winter latitudes, and may also account for heating at high latitudes in the summer hemisphere.

097.027 **Geology of the planet Mars.**
P. Masson.
Recherche, Vol. 16, No. 172, p. 1482 – 1491 (1985). In French. Abstr. in Phys. Abstr., Vol. 89, No. 1278, Entry 47923 (1986).

097.028 **I canali di Marte e le illusioni ottiche.**
E. Moltisanti.
Orione, Vol. 6, N. 3, p. 14 – 29 (1986).

097.029 **Die Marsopposition 1983/84.**
E. Freydank, H. Freydank.
Orion, 44. Jahrg., Nr. 214, p. 96 – 98 (1986). In German and French.

097.030 **Density of Martian north polar layered deposits: implications for composition.**
M. C. Malin.
Geophys. Res. Lett., Vol. 13, No. 5, p. 444 – 447 (1986).
Mariner 9 and Viking Orbiter observations of the topography and mass of layered deposits in the north polar region of Mars suggest a density of about 1 g/cm³. Consideration of the self–compaction of these deposits further suggests that this low density cannot be explained by a high porosity sediment. Rather, a significant component of ice is probably required to account for the low value. The density further implies that the silicate fraction, regardless of its density (i.e., its composition), represents a volumetrically almost negligible component of the deposit. The most likely values, 5% silicate dust and 95% ice, agree well with a theoretical model for the formation of the deposits through cyclic incorporation of the perennial polar ice cap.

097.031 **Regional dust deposits on Mars: physical properties, age, and history.**
P. R. Christensen.
J. Geophys. Res., Vol. 91, No. B3, p. 3533 – 3545 (1986).
The purpose of this paper is to describe the use of Viking infrared thermal mapper, earth–based radar, and visual observations to study the existence of regional dust deposits on Mars. These observations provide estimates of particle size, rock abundance, surface texture, thermal emissivity, and albedo that can be used to characterize surface deposits and to determine the degree of surface mantling. The questions to be addressed include (1) where are dust deposits located, (2) what are their thicknesses, (3) what are their ages, (4) how are they formed, (5) what does their presence imply about the surface history, (6) is there evidence for cyclic changes in their location?

097.032 **Global duricrust on Mars: analysis of remote–sensing data.**
B. M. Jakosky, P. R. Christensen.
J. Geophys. Res., Vol. 91, No. B3, p. 3547 – 3559 (1986).
Global remote–sensing data for Mars are analyzed to obtain a simple, self–consistent model for the surface layer. The data sets discussed include radar cross–section measurements at several wavelengths, radio whole–disk thermal emission observations at two wavelengths, the global distribution of thermal inertia, deviations of diurnal temperatures from those of a homogeneous model, and thermal spectral estimates of surface rock abundance and of the thermal inertia of the nonrocky component of the surface.

097.033 **Small Martian valleys: pristine and degraded morphology.**
V. R. Baker, J. B. Partridge.
J. Geophys. Res., Vol. 91, No. B3, p. 3561 – 3572 (1986).
This study attempts a detailed investigation of the small valley networks on Mars. Particular attention is paid to the dual nature of many valley systems: a relatively fresh–appearing network portion versus an apparent larger, less distinct network system. These separate network entities are referred to in this paper as pristine and degraded. Detailed mapping of 25 valley network systems provided data for crater counts and morphometric analysis.

097.034 **Quantification of waste morphology in Martian fretted terrain.**
R. C. Kochel, R. T. Peake.
J. Geophys. Res., Vol. 89, Suppl., p. C336 – C350 (1984). – See Abstr. 012.065.
Extensive geomorphic mapping in the northern Martian fretted terrain of Deuteronilus and Protonilus Mensae shows details of the waste–dominated landform association in this region. Wasting appears to have occured along the cratered terrain boundary at the expense of old cratered terrain. North of the boundary remnants of old cratered terrain occur as island mesas surrounded by extensive debris aprons. Quantitative study using principal components analysis reveals distinct spatial variations in the morphology of waste forms in the region and supports the qualitative interpretations made during geomorphic mapping.

097.035 **Iron–montmorillonite: a spectral analog of Martian soil.**
A. Banin, L. Margulies, Y. Chen.
J. Geophys. Res., Vol. 90, Suppl., p. C771 – C774 (1985). – See Abstr. 012.065.
Light absorption and reflectance by smectite clays containing various absorbed ions were measured in the UV, VIS, and NIR ranges and compared to Martian dust and surface soil spectra.

097.036 **Geologic interpretation of remote sensing data for the Martian volcano Ascraeus Mons.**
J. R. Zimbelman.
NASA Tech. Memo., NASA TM–88784, Vol. 1, p. 271 – 572 (1986). = Thesis, Arizona State University, Tempe, Ariz., USA (1984). – See Abstr. 003.051.
Photographs and other remote sensing data of Ascraeus Mons were analyzed in order to relate the remotely determined surface properties to the geology of the volcano. Photogeologic analysis indicates that Ascraeus Mons is similar to terrestrial basaltic shield volcancoes in both surface morphology and estimated rheologic properties. High spatial resolution infrared sequences are consistent with dust deposits at least 2 cm thick on all terrains.

Photochemistry of the atmospheres of Mars and Venus.
See Abstr. 003.004.

The analysis of extraterrestrial materials.
See Abstr. 003.053.

Workshop on dust on Mars. Conference held at Tempe, Ariz., USA, 4 – 5 February 1985.
See Abstr. 012.082.

Workshop on water on Mars. Conference held at Moffat Field, Calif., USA, 30 November – 1 December 1984.
See Abstr. 012.083.

Search for chiral molecules and optical activity in extraterrestrial systems. The Cassini mission opportunity.
See Abstr. 015.024.

Martian dust particles as condensation nuclei: a preliminary assessment of mineralogical factors.
See Abstr. 022.062.

Properties of filamentary sublimation residues from dispersions of clay in ice.
See Abstr. 022.063.

Shock–induced CO_2 loss from $CaCO_3$; implications for early planetary atmospheres.
See Abstr. 022.098.

The Soviet Mars programme.
See Abstr. 051.005.

Axial melting in central peaks of complex impact structures.
See Abstr. 081.013.

A comparative analysis of the volcanic impact on the climates of the Earth and Mars.
See Abstr. 082.042.

Comparative estimates of climatic consequences of Martian dust storms and of possible nuclear war.
See Abstr. 082.067.

Oxygen emissions in the night airglow of the Earth, Venus and Mars.
See Abstr. 082.078.

A collisional kinetic theory of a plane parallel evaporating planetary atmosphere.
See Abstr. 091.009.

Electric and electrothermal conductivity of the ionospheres of planets. II. The Earth, Mars, Venus.
See Abstr. 091.018.

A harmonic and statistical analysis of the topography of the Earth, Moon and Mars.
See Abstr. 091.056.

Radiation transfer in the 4.3–μ CO_2 band and the 4.7–μ CO band in the atmospheres of Venus and Mars with violation of LTE. Populations of vibrational states.
See Abstr. 093.037.

Radiative transfer in the 4.3 μ CO_2 and 4.7 μ CO bands in the non–LTE Venus and Mars atmospheres: transformation of absorbed solar energy.
See Abstr. 093.043.

A summary of extremes of isotopic variations in extra–terrestrial materials.
See Abstr. 105.012.

Martian meteorites: are they (all) from Mars: evidence from trapped noble gases.
See Abstr. 105.117.

Review of relative and absolute abundances of atmophilic gases in glass from the EETA 79001 shergottite, and comparison with the martian and terrestrial atmospheres.
See Abstr. 105.124.

The case for a martian origin of the shergottites, II. Trapped and indigenous gas components in EETA 79001 glass.
See Abstr. 105.199.

Magnetic properties of Antarctic shergottite meteorites EETA 79001 and ALHA 77005: possible relevance to a Martian magnetic field.
See Abstr. 105.200.

The Shergotty Consortium and SNC meteorites: an overview.
See Abstr. 105.219.

Shock metamorphism and petrography of the Shergotty achondrite.
See Abstr. 105.220.

X–ray investigations related to the shock history of the Shergotty achondrite.
See Abstr. 105.221.

Chemical systematics of the Shergotty meteorite and the composition of its parent body (Mars).
See Abstr. 105.222.

Sr and Nd isotopic systematics of Shergotty meteorite.
See Abstr. 105.224.

Formation ages and evolution of Shergotty and its parent planet from U–Th–Pb systematics.
See Abstr. 105.225.

A discussion of isotopic systematics and mineral zoning in the shergottites: evidence for a 180 m.y. igneous crystallization age.
See Abstr. 105.226.

Oxygen isotopes in Shergotty.
See Abstr. 105.227.

Carbon abundance and isotopic studies of Shergotty and other shergottite meteorites.
See Abstr. 105.228.

Nitrogen and light noble gases in Shergotty.
See Abstr. 105.229.

Xenon and other noble gases in shergottites.
See Abstr. 105.230.

Exposure history of shergottites.
See Abstr. 105.231.

Cosmogenic effects in shergottites.
See Abstr. 105.232.

Thermoluminescence and the shock and reheating history of meteorites – III. The shergottites.
See Abstr. 105.233.

Nuclear tracks, Sm isotopes and neutron capture effects in the Elephant Moraine shergottite.
See Abstr. 105.234.

Magnetic studies on Shergotty and other SNC meteorites.
See Abstr. 105.235.

Martian volatiles in shergottite EETA 79001: new evidence from oxidized sulfur and sulfur–rich aluminosilicates.
See Abstr. 105.236.

The parental magma of the Nakhla achondrite: ultrabasic volcanism on the shergottite parent body.
See Abstr. 105.237.

Core formation in the Earth and shergottite parent body (SPB): chemical evidence from basalts.
See Abstr. 105.238.

098 Minor Planets

098.001 **Predictions of occultations by minor planets as a test of accuracy.**
L. K. Kristensen.
Relativity in celestial mechanics and astrometry, p. 113 – 116 (1986). – See Abstr. 012.002 (IAU Symp. No. 114).
Predictions of tracks of occultations of stars by minor planets give a strong test of the accuracy of observations, star places, and dynamical theories. A critical discussion of the 1983 September 11 occultation of 14 Piscium by (51) Nemausa shows that the systematic regional errors of fundamental stars introduce errors in the long–term predictions of occultations that are larger than relativistic effects.

098.002 **Die Durchmesser der kleinen Planeten.**
H. Haupt.
Sterne Weltraum, 25. Jahrg., Nr. 2, p. 70 – 73 (1986).

098.003 **Reflectance spectrophotometry (~ 0.5 – 1.0 μm) of outer–belt asteroids: implications for primitive, organic Solar System material.**
F. Vilas, B. A. Smith.
Icarus, Vol. 64, No. 3, p. 503 – 516 (1985).
Narrowband spectrophotometric observations of outer–belt asteroids have been used to investigate the surface compositions of the D– and P–class asteroids. Spectra of 19 outer–belt asteroids and 2 main–belt D–class asteroids were obtained using two charge–coupled device spectrographs. The spectra can be divided into four groups based upon four discrete slopes among the spectra. All of the outer–belt asteroids are locked into orbits controlled by Jupiter's gravitational attraction. The distinct slope changes suggest that these asteroids are the remnants of a gradation in composition of planetesimals in the outer Solar System.

098.004 **The detection of fast–moving asteroids and comets by IRAS.**
S. F. Green, J. K. Davies, N. Eaton, B. C. Stewart, A. J. Meadows.
Icarus, Vol. 64, No. 3, p. 517 – 527 (1985).
The Infrared Astronomical Satellite (IRAS) was used to search for fast–moving objects in the inner Solar System. The selection effects and limitations of the search are examined with particular reference to the objects which may be present in the moving–object data base at JPL. It is shown that the two new Apollo asteroid discoveries are in fair agreement with the discovery rate by ground–based telescopes. IRAS discovered six comets which is also consistent with the discovery rate for ground–based searches. It is estimated that there are 20 – 30 comets but no Apollo asteroids in the hours–confirmed moving–object data base established by the IRAS project.

098.005 **Photoelectric lightcurve and phase relation of the asteroid 505 Cava.**
J. W. Young, A. W. Harris.
Icarus, Vol. 64, No. 3, p. 528 – 530 (1985).
Photoelectric observations of 505 Cava taken on 8 nights at the Table Mountain Observatory in the fall of 1982 give a period of rotation of 8.1796 ± 0.0010 hr with an amplitude of $0^{m}.25$. The absolute magnitude vs solar phase angle (phase relation) was measured over the range of phase angle of 10 to 25°. The resulting very steep phase relation is suggestive of a dark object (cf. K. Lumme and E. Bowell, 1981). This is consistent with the "FC" classification suggested by D. Tholen (1984) based on eight color photometry.

098.006 **Some peculiarities in the Asteroidal Belt.**
R. R. Cordeiro, R. J. Quiroga.
Earth, Moon, Planets, Vol. 34, No. 2, p. 117 – 132 (1986).
The analysis of the fine structure of the Asteroidal Belt evidences a group of asteroids next to the resonance 4/9 with Jupiter. In this group and in other groups associated to the Hirayama families there are indications that their orbital parameters can be represented by quantum numbers as defined here and in two previous works. Together with this the distribution of the eccentricities and inclinations of the orbital planes of short period comets and diverse type of asteroids indicates that they can be classified as objects with $e \gtrsim \sin i$ and objects with $e \lesssim \sin i$ with a limit $e = \sin i$ which determinates geometrical properties of the orbits related with discrete states in the solar system.

098.007 **The angular momentum of the asteroids.**
C. Sivaram.
Earth, Moon, Planets, Vol. 34, No. 3, p. 231 – 233 (1986).
Consideration of the basic physics involved in the structure of the object are used to obtain relationships for the radius, period, angular momentum, etc. of a typical asteroid. The mass–angular momentum relation for asteroids would tend to favour the fragmentation hypothesis.

098.008 **Observations of minor planets. V.**
F. Börngen, K. Kirsch.
Astron. Nachr., Vol. 307, No. 1, p. 65 – 67 (1986). In German.
A summary is given about the minor planet survey performed in 1984 on Tautenburg Schmidt plates. The authors discovered 84 planets and calculated 205 positions for them. There are 17 numbered planets among them, 16 already earlier observed and 51 new planets with provisional designations. Tautenburg observations could give a tribute to seven planets numbered in the period of this report. The planet 1964 EC discovered in Tautenburg has received the permanent number (3181). In honor of Dr. Paul Ahnert (*22.11.1897) this planet obtained the name Ahnert.

098.009 **Mapping for 5/2 asteroidal commensurability.**
M. Šidlichovský, B. Melendo.
Bull. Astron. Inst. Czech., Vol. 37, No. 2, p. 65 – 80 (1986).
The method of Wisdom (1982) is applied to the 5/2 Kirkwood gap. A transformation for the asteroidal elements which is 1000 times faster than numerical integration is presented. For the planar case, the calculations strongly support the collisional hypothesis. Using Lindstedt's method, improvement of the theory in the order of the eccentricity is suggested. For completeness also the full mapping is given in which the slow evolution of Jupiter's elements due to planetary perturbations is taken into account.

098.010 **Analysis of two positive asteroidal occultation reports.**
R. Boninsegna.
Occultation Newsl., Vol. 3, No. 15, p. 327 (1986).

098.011 **Astrometric updates for asteroidal occultations.**
D. W. Dunham.
Occultation Newsl., Vol. 3, No. 15, p. 327 – 329 (1986).

098.012 **Nouvelles des Earth–grazers – 6.**
M.–A. Combes, J. Meeus.
Astronomie, Vol. 100, p. 137 – 146 (1986).

098.013 **Nouvelles des Earth–grazers – 7.**
M.–A. Combes, J. Meeus.
Astronomie, Vol. 100, p. 181 – 186 (1986).

098.014 **Asteroid 29 Amphitrite: surface characteristics and prospects for the Galileo flyby.**
J. F. Bell, J. C. Gradie, B. R. Hawke, T. B. McCord, M. J. Gaffey.
Meteoritics, Vol. 20, No. 4, p. 606 – 607 (1985). Abstract. – See Abstr. 010.661.

098.015 **Chemical analysis of planetoidal surfaces by measurement of neutron–induced prompt gamma rays.**
P. Englert, J. Brückner, H. Wänke.
Meteoritics, Vol. 20, No. 4, p. 640 – 642 (1985). Abstract. – See Abstr. 010.661.

098.016 **Die Entdeckung des kleinen Planeten (3181) Ahnert.**
F. Börngen.
Sterne, 62. Band, Heft 1, p. 24 – 27 (1986).

098.017 **Identifications of minor planets.**
B. G. Marsden.
Asteroids, comets, meteors II, p. 3 – 12 (1986). – See Abstr. 012.010.

Photographic plates obtained at numerous observatories around the world during the past century are known to contain more than a third of a million images of minor planets. Other plates must surely contain several times this number of unknown images of minor planets. For the purposes of this discussion, an observation consists of the time of mid–exposure, the observed topocentric right ascension and declination of the image, and the location of the observatory. There may also be an indication of brightness and/or motion. The complete identification process consists of uniquely ascribing all known observations to specific individual minor planets. Methods for the partial accomplishment of this task are reviewed, and suggestions are made for future work.

098.018 **The Infrared Astronomical Satellite Asteroid database.**
E. F. Tedesco.
Asteroids, comets, meteors II, p. 13 – 17 (1986). – See Abstr. 012.010.

The Infrared Astronomical Satellite (IRAS) obtained sky survey observations between 9 February 1983 and 22 November 1983 at wavelengths of 12, 25, 60, and 100 microns (Neugebauer et al. 1984). During the course of this survey numerous asteroids were observed, most of which had never before been seen. The asteroid data obtained are being reduced at the Jet Propulsion Laboratory. A database of ground–based observations was created. This database, together with that produced from the IRAS asteroid data, will constitute the final data products of the IRAS Asteroid Task. This paper describes the creation, maintenance, and distribution of the ground–based portion of this all inclusive asteroid database.

098.019 **International collaboration leads to new Aten asteroid discovery, 1984 QA.**
E. F. Helin, R. S. Dunbar, M. A. Barucci.
Asteroids, comets, meteors II, p. 21 – 24 (1986). – See Abstr. 012.010.

A new Earth–crossing asteroid of the Aten type was discovered by the authors on 30 August 1984 at Palomar. The asteroid, designated 1984 QA, is the fourth known member of the Aten type, and the first discovered since 1978. The discovery of this object was the highlight of a collaboration between the JPL Asteroid Search Team and an ESA scientist, and demonstrates one aspect of an on–going joint search effort with other Schmidt observers.

098.020 **Asteroids and the Space Telescope.**
R. Albrecht, H. J. Schober.
Asteroids, comets, meteors II, p. 25 – 27 (1986). – See Abstr. 012.010.

The Space Telescope will hopefully come to operation in the near future and will open a new field of observations also for asteroids. The program is described, which was proposed and accepted to test the imaging system of the Faint Object Camera FOC with respect to asteroids. This program also represents a kind of pilot study for future applications for observing time at the ST, as far as asteroids are concerned. The authors also hope to get a first image of one of the large asteroids Ceres, Pallas, Juno or Vesta.

098.021 **Centennial ephemerides for Aten–Apollo–Amor objects.**
G. Hahn.
Asteroids, comets, meteors II, p. 29 (1986). Abstract. – See Abstr. 012.010.

098.022 **How many Earth–approaching asteroids are there?**
L. G. Taff.
Asteroids, comets, meteors II, p. 31 – 34 (1986). – See Abstr. 012.010.

The author presents a method, based upon the standard urn models of probability theory, which attempts to calculate the number of Earth–approaching asteroids. Earth–approaching minor planets are defined to include all asteroids with perihelion distances less than that of Mars's. The critical assumption herein is that the number of such objects is not changing on a short (i.e., decades) timescale. The attempt fails for lack of sufficient data.

098.023 **Asteroid lightcurve studies.**
A. W. Harris.
Asteroids, comets, meteors II, p. 35 – 44 (1986). – See Abstr. 012.010.

The number of asteroids with known rotational properties has grown approximately exponentially with a doubling time of about 4 years to a present count of about 400. Statistics from the present data set reveal differences in mean rotation rate among different taxonomic classes, and as a function of size, and also demonstrate that the dispersion of rotation rates about the mean is broad among asteroids smaller than about 100 km, indicating the presence of sub–groups of fast and slow rotators. In the last few years, the abundance of observations has become sufficient for serious studies of shapes, pole orientations, and phase relations.

098.024 **Numerical and laboratory simulations of photometric properties of asteroids.**
M. A. Barucci, M. Fulchignoni.
Asteroids, comets, meteors II, p. 45 – 54 (1986). – See Abstr. 012.010.

A review of the results of analytical/numerical simulations of asteroid rotation in order to obtain information from the lightcurve behaviour varying the orientation parameters is presented. Some numerical and experimental data on the influence of the scattering are discussed in comparison with the results obtained assuming a geometrical scattering. The effects of shape, orientation, surface micro– and macromorphology and albedo variations, are presented on the basis of laboratory simulations of asteroid lightcurves obtained with various regular and irregular shaped asteroid models. Results of high– and hyper–velocity impact experiments on rocky models are discussed in connection with the collisional history of the asteroidal bodies and the evolution of their population.

098.025 **Inversion of asteroid lightcurves using spherical harmonics.**
K. Lumme, H. Karttunen, E. Bowell, M. Poutanen.
Asteroids, comets, meteors II, p. 55 – 59 (1986). – See Abstr. 012.010.

If asteroid lightcurve data, expressed two dimensionally in terms of rotational phase and aspect angles, are expanded in spherical harmonics, then infomation on the pole orientation, rotation period, possible albedo variations, and body shape may be derived separately from the coefficients. By minimizing the odd harmonics of order $\geqslant 3$, the pole orientation and rotation rate may be determined quite generally, without invoking a particular model for the shape. The first–order harmonics yield information on albedo variations over the asteroid's surface, whereas the even harmonics can be related to its shape.

098.026 **Atlas of photometric asteroid lightcurves.**
C.–I. Lagerkvist, V. Zappalà, S. Argentini,
M. A. Barucci, M. T. Capria, M. Fulchignoni, L. Guerriero, E. Perozzi, M. Viaggiu.
Asteroids, comets, meteors II, p. 61 – 62 (1986). – See Abstr. 012.010.

A catalogue of asteroid lightcurves has been compiled. The intention is that the catalogue shall include all observations published up to 1984. So far the atlas consists of two parts, the first dealing with the observing data, aspect data, and references, the second dealing with all the lightcurves. A final version of the catalogue, containing data up to the end of 1985, will be published in 1986.

098.027 **Peculiar shaped asteroids: some laboratory results.**
S. Argentini, M. A. Barucci, M. Fulchignoni, L. Guerriero, R. Salvatori.
Asteroids, comets, meteors II, p. 63 – 66 (1986). – See Abstr. 012.010.

Some asteroids exhibiting an irregular lightcurve have been interpreted as peculiar shaped objects. The data obtained by laboratory simulations on various models and some comparisons with those presented in the literature are discussed.

098.028 **Physical studies of asteroids. XII. UBVRI observations of M and CMEU asteroids.**
C.–I. Lagerkvist, G. Hahn, P. Magnusson, H. Rickman, G. Hammarbäck.
Asteroids, comets, meteors II, p. 67 – 72 (1986). – See Abstr. 012.010.

The 50 cm and 1 m telescopes at ESO, Chile, were used during March 1983 and 1984 for UBVRI observations of the asteroids 76, 125, 135, 250, 369, and 382. Photoelectric lightcurves, UBVRI colours, rotation periods and amplitudes are presented.

098.029 **Is 369 Aeria a slowly rotating asteroid?**
H. J. Schober.
Asteroids, comets, meteors II, p. 73 (1986). – See Abstr. 012.010.

From sporadic day–to–day photoelectric observations of 369 Aeria it was concluded that this asteroid should be one of the very rare cases where no light variation was found at all. But extended measurements of 369 Aeria show that it is probably a slowly rotating asteroid.

098.030 **105 Artemis – a new puzzle among asteroids?**
H. J. Schober, A. Schroll.
Asteroids, comets, meteors II, p. 75 – 76 (1986). – See Abstr. 012.010.

The asteroid 105 Artemis is one of the few low–numbered objects, for which up to now no resulting complete lightcurve data are known. The authors have observed this asteroid during five nights quite extensively and though variations are found, they have doubts about the interpretation. 105 Artemis seems to be another puzzling asteroid.

098.031 **Fine structure lightcurve of (51) Nemausa.**
P. Gammelgaard, L. K. Kristensen.
Asteroids, comets, meteors II, p. 77 – 80 (1986). – See Abstr. 012.010.

A few results based on the 15131 BV observations from 1982, 1983, and 1985 are presented.

098.032 **A photoelectric program for small and unusual asteroids.**
M. Di Martino.
Asteroids, comets, meteors II, p. 81 – 84 (1986). – See Abstr. 012.010.

A program for determining the rotational parameters of small and unusual taxonomic type asteroids has been started at the Astronomical Observatory of Torino. In this paper the author presents a part of the results, obtained during an observing campaign carried out at La Silla Observatory (ESO) in August 1984, regarding the asteroids: 382 Dodona, 800 Kressmannia, 916 America, and 1210 Morosovia.

098.033 **The rotational period and phase coefficient of 344 Desiderata.**
C. J. Cunningham.
Asteroids, comets, meteors II, p. 85 – 87 (1986). – See Abstr. 012.010.

Photoelectric photometry conducted with a 36.2 cm Cassegrain in Waterloo, Ontario, Canada in 1983 revealed the rotational period of 344 Desiderata to be 10.7 ± 0.1 hours. This is in good agreement with an unpublished result from the same opposition by Piironen (10.75 hours) and slightly longer than the 10.53 hours determined by Barucci et al. (1985).

098.034 **29 Amphitrite: the first asteroid to be visited by a space mission.**
M. A. Barucci, M. Fulchignoni.
Asteroids, comets, meteors II, p. 89 – 92 (1986). – See Abstr. 012.010.

The Galileo mission may include a fly–by of the asteroid 29 Amphitrite on December, 1986. Awaiting the Galileo mission data the authors try to interpret the available data on the basis of laboratory simulations carried out by the System for Asteroid Models described by Barucci et al. (1982). They performed two experiments using: (1) a biaxial ellipsoidal model with different albedo spots using the method suggested by Lupishko et al. (1983) to separate the effects of surface heterogeneities from the variations due to shape in a lightcurve; (2) a pyramid model as suggested by Van Houten–Groeneveld et al. (1979).

098.035 **Physical studies of asteroids. XIII. IDS spectra of selected asteroids.**
G. Hahn, C.–I. Lagerkvist, P. Magnusson, H. Rickman.
Asteroids, comets, meteors II, p. 93 – 97 (1986). – See Abstr. 012.010.

IDS spectra of asteroids 29 Amphitrite, 32 Pomona and 51 Nemausa in the wavelength range 3800 – 8700 Å are presented. The spectral slope for 29 Amphitrite and 51 Nemausa changes with rotational phase.

098.036 **The 13 November 1984 occultation of BD + 8°471 by (1) Ceres.**
L. H. Wasserman, R. L. Millis, O. G. Franz, M. F. A'Hearn, W. Osborn, A. Klemola.
Asteroids, comets, meteors II, p. 99 – 103 (1986). – See Abstr. 012.010.

The 13 November 1984 occultation of BD + 8°471 was discovered during a photographic search carried out with the 0.5–meter Carnegie Double Astrograph at Lick Observatory and the Lowell Observatory PDS microdensitometer (Millis et al. 1984). The occultation on 13 November is a particularly good event. The star is 10^m3 in V, yielding a predicted drop at occultation of about 10%. Such a drop can be detected by small telescopes equipped with photoeletric photometers. The track was predicted to cross the Caribbean, Florida, southern Texas, and Mexico. Based on this prediction, the authors prepared to observe the event in Mexico using four portable occultation data systems.

098.037 **Collisional evolution in the asteroid belt: an observational survey.**
R. P. Binzel.
Asteroids, comets, meteors II, p. 105 – 107 (1986). – See Abstr. 012.010.

An observational survey has yielded photoelectric lightcurves for over 100 small (diameter < 30 km) asteroids in the main belt. These asteroids were selected from six distinct dynamical regions. The resulting rotational frequencies and lightcurve amplitudes will be used to constrain models for asteroid collisional evolution.

098.038 **Asteroid families.**
P. Farinella, P. Paolicchi, V. Zappalà.
Asteroids, comets, meteors II, p. 109 – 119 (1986). – See Abstr. 012.010.

The dynamical and physical studies of asteroid families are summarized, pointing out the most relevant difficulties against a satisfactory settlement of the whole subject. A few well–established recent findings, mainly regarding the collisional origin and the effects of self–gravitation on the outcomes of asteroidal impacts, are discussed.

098.039 **Collisional evolution of asteroids constrained by their present rotational properties.**
P. Farinella, P. Paolicchi, V. Zappalà.
Asteroids, comets, meteors II, p. 121 – 124 (1986). – See Abstr. 012.010.

The abundance of asteroids having unusually large angular momenta of rotation in the diameter range 200 – 300 km implies that, if these angular momenta have a collisional origin, the asteroid population was of the order of a few times the present one at the time when the orbits were stirred up and the disruptive collisional process begun. The meaning of this result and the issue of the collisional vs. "primordial" origin of the currently observed asteroid rotations are shortly discussed.

098.040 **Pole and shape determinations for 20 large asteroids.**
P. Magnusson.
Asteroids, comets, meteors II, p. 125 – 128 (1986). – See Abstr. 012.010.
Spin axis orientation and sense of rotation have been computed for 20 large asteroids. The spin axes are quite isotropically distributed with a small and statistically uncertain preference for prograde rotation. Two fundamentally different pole determination methods were combined in order to get as reliable results as possible. The first one is an amplitude–magnitude method which is based on tri–axial ellipsoidal models. The second method uses the variation of the observed synodic period of rotation to derive the axis and sense of rotation.

098.041 **Comparison of the asteroid proper elements obtained from various theories.**
Z. Knežević.
Asteroids, comets, meteors II, p. 129 – 134 (1986). – See Abstr. 012.010.
The proper elements of asteroids are calculated by means of the second order – fourth degree theory of Yuasa (1973). Results are compared with those coming from classical Lagrange procedure (Brouwer, 1951), and with those determined by Williams (1979). The differences are discussed in terms of the accuracy of the theory, and special attention has been paid to the objects of high eccentricity and/or inclination.

098.042 **Orbital evolution at secular resonances in the asteroidal belt.**
H. Scholl, C. Froeschlé.
Asteroids, comets, meteors II, p. 135 – 143 (1986). – See Abstr. 012.010.
According to analytic theories, a secular resonance occurs if the precessional rate of an asteroidal orbit is equal to one of the eigenvalues of the system of differential equations for the orbital precessions of the major planets. The authors review the theoretical evidence for secular resonances. They present recent results obtained from numerical calculations for orbital evolution at the three major secular resonances in the asteroidal belt. Strong variations in eccentricity and inclination are obtained in some cases. According to these numerical results, secular resonances have to be considered as sources of meteorites as conjectured by Williams and Wetherill.

098.043 **Secular perturbations of asteroids in secular resonance.**
H. Nakai, H. Kinoshita.
Asteroids, comets, meteors II, p. 145 (1986). Abstract. – See Abstr. 012.010.

098.044 **The Hilda group and the Hecuba gap.**
A. Milani, C. D. Murray, A. M. Nobili.
Asteroids, comets, meteors II, p. 147 – 151 (1986). – See Abstr. 012.010.
The problem of the Hilda group as opposed to the Hecuba gap is investigated in the pure gravitational hypothesis within the planar elliptic restricted 3–body model. The topology of the level manifolds of the Hamiltonian around the fixed points shows a different structure in the two resonances which might explain the observational evidence.

098.045 **On the computation of characteristic orbital elements for the Trojan group of asteroids.**
J. Schubart, R. Bien.
Asteroids, comets, meteors II, p. 153 – 156 (1986). – See Abstr. 012.010.
In analogy to the proper elements of the regular case of asteroidal motion, the authors propose in the Trojan case characteristic quantities which are expected to be constant over some 10^5 years. The construction of these quantities is made possible by the application of Labrouste's method which allows to isolate basic frequencies from a complex quasi–periodic process. In the present paper the authors describe the computation of the characteristic elements and give first results.

098.046 **Two Trojans near a secular resonance: (1173) Anchises and (1871) Astyanax.**
R. Bien, J. Schubart.
Asteroids, comets, meteors II, p. 157 – 160 (1986). – See Abstr. 012.010.
Using the invariable plane as a reference plane, one finds that (1173) Anchises and (1871) Astyanax have mean nodal rates which are near Jupiter's mean nodal rate. In other words: both asteroids are not far from a secular resonance. In the model Sun–Jupiter–Saturn–Uranus–Neptune, the authors investigate the motion of the two asteroids and study variations of the actual orbits which are closer to the resonance.

098.047 **The orbits of Trojan asteroids.**
G. Barber.
Asteroids, comets, meteors II, p. 161 – 164 (1986). – See Abstr. 012.010.
The long term stability of hypothetical Trojan asteroids was examined in an attempt to explain the difference in the number of observed Trojans in the preceding (L_4) and following (L_5) Lagrangian stability points. A possible differentiation caused by long term perturbations by Saturn was found. The validity of this effect, together with its implications on the primeval orbital elements of the planets proto–Jupiter and proto–Saturn is discussed.

098.048 **Orbital studies of 1982 YA, 1983 VA, and 1984 BC.**
G. Hahn.
Asteroids, comets, meteors II, p. 175 – 180 (1986). – See Abstr. 012.010.
Results from orbital integrations over some 2200 years into the future for the asteroids 1982 YA, 1983 VA, and 1984 BC are presented. Two different integrators were used in the elliptic restricted three–body model and in an n–body model. The results confirm the findings of Hahn and Rickman (1985), showing chaotic evolutions for all three orbits, but with large variations, depending on the solar system model and integration method used. The short–periodic ω libration for 1983 VA found by these authors is not reproduced, leading to the conclusion that this rapid orbital transformation occurs only in a very narrow 'window' in element space.

098.049 **Spectroscopy of asteroids in unusual orbits.**
W. D. Cochran, A. L. Cochran, E. S. Barker.
Asteroids, comets, meteors II, p. 181 – 185 (1986). – See Abstr. 012.010.
The authors have obtained medium resolution spectroscopy of a collection of non–main belt asteroids in order to search for possible comet–like spectral features. The asteroids include 9 Earth approachers, 2 Trojans, and the unusual object 2060 Chiron. All spectra were obtained and reduced in the same manner as comet data in the McDonald Observatory Faint Comet Survey. No indications of cometary activity were found in any of the asteroids observed.

098.050 **Subtle gravitational effects in the motion of small bodies in the solar system.**
B. Todorovic–Juchniewicz.
Asteroids, comets, meteors II, p. 465 – 466 (1986). – See Abstr. 012.010.
The influence of some subtle dynamical effects on the motion of a comet or asteroid was studied numerically. The following problems were considered: (1) the perturbations due to the Earth and Moon which were treated either as separate objects or as a single barycentric system; (2) the disturbing effect of Pluto in the cases of the "old" and "new" values of Pluto's mass.

098.051 **Asteroid strokes against the earth and their possible planetary geophysical consequences.**
Yu. D. Kalinin.
Geomagn. Aehron., Tom 26, No. 2, p. 358 – 359 (1986). In Russian. English translation in Geomagn. Aeron.

098.052 **Structure of the 2:1 and 3:2 jovian resonances.**
C. D. Murray.
Icarus, Vol. 65, No. 1, p. 70 – 82 (1986).
The chaotic regions of the phase space in the vicinity of the 2:1 and 3:2 jovian resonances are identified by using a mapping technique derived from a second–order expansion of the disturbing function for the planar elliptical restricted three–body problem. It is shown that both resonances have extensive chaotic regions which in some cases can lead to large changes in the eccentricity of asteroid orbits. Although the 3:2 resonance is shown to be more chaotic than the 2:1 resonance, the existence of the Hilda group of asteroids and the Hecuba gap may be explained by distinct differences in the location of the high–eccentricity regions at each resonance. The problem of the convergence of the expansion of the disturbing function in the outer asteroid belt is also discussed.

098.053 **Pole coordinates of the asteroid 511 Davida as determined via the Amplitude–Magnitude method.**
V. Zappalà, Z. Knežević.
Icarus, Vol. 65, No. 1, p. 122 – 128 (1986).
The Amplitude–Magnitude method is used for the pole determination of the asteroid 511 Davida, using observations from six oppositions. The possible north poles are found to be $\lambda_1 = 92° \pm 7°$; $\beta_1 = 33° \pm 6°$, and $\lambda_2 = 303° \pm 4°$; $\beta_2 = 34° \pm 5°$, when scattering effect is not taken into account. When scattering is accounted for, solutions not significantly different from (λ_1, β_1) and (λ_2, β_2) are obtained. The moderately eccentric and inclined orbit of 511 Davida does not allow the authors to distinguish between the two pole solutions.

098.054 **The motions of numbered asteroids in resonance regions.**
H. Nakai, H. Kinoshita.
Proceedings of the nineteenth symposium on celestial mechanics, p. 1 – 9 (1986). In Japanese. – See Abstr. 012.013.

098.055 **Kirkwood gaps.**
Y. Kozai.
Proceedings of the nineteenth symposium on celestial mechanics, p. 10 – 13 (1986). – See Abstr. 012.013.

098.056 **Secular resonances and family of asteroids.**
M. Yoshikawa.
Proceedings of the nineteenth symposium on celestial mechanics, p. 14 – 18 (1986). In Japanese. – See Abstr. 012.013.

098.057 **On the accuracy of orbital elements for unnumbered minor planets.**
K. Hurukawa.
Proceedings of the nineteenth symposium on celestial mechanics, p. 19 – 23 (1986). In Japanese. – See Abstr. 012.013.

098.058 **The analysis of the lunar occultation of Vesta (1984).**
A. Sengoku.
Proceedings of the nineteenth symposium on celestial mechanics, p. 29 – 34 (1986). – See Abstr. 012.013.
The lunar occultation of Vesta (Oct. 21, 1984, reappearance) observed at Bisei Hydrographic Observatory is analyzed. The residual of the position of Vesta is 0.″19 for the new Japanese Ephemeris and is 0.″36 for the old one. These residuals are within the original error of the ephemeris of Duncombe (1969) and the error of the limb correction of the moon. The diameter of Vesta is 450 km if the uniform disk model is used and is 530 km if the complete limb darkening model is used.

098.059 **Favourable near–Earth asteroid approaches: 1986 – 2000.**
C. Townsend, J. Rogers.
J. Br. Astron. Assoc., Vol. 96, No. 2, p. 106 – 107 (1986).
The authors have numerically integrated 72 special asteroid orbits forward in 40–day increments to the year 2000. Gravitational perturbations by all nine known planets were taken into account. This article presents classical orbital elements for the 20 asteroids that the authors found would approach the Earth to within 0.20 AU during this period.

098.060 **Collisional evolution in the asteroid belt: new results from observed distributions.**
R. P. Binzel.
Bull. Am. Astron. Soc., Vol. 17, No. 4, p. 872 – 873 (1985). Abstract. – See Abstr. 010.062.

098.061 **The coronagraphic search for material around Amphitrite.**
R. J. Terrile, B. A. Smith.
Bull. Am. Astron. Soc., Vol. 17, No. 4, p. 918 (1985). Abstract. – See Abstr. 010.063.

098.062 **Photometric geodesy of asteroids: the data base.**
D. R. Davis, C. R. Chapman, S. J. Weidenschilling, R. Greenberg, D. Levy.
Bull. Am. Astron. Soc., Vol. 17, No. 4, p. 919 (1985). Abstract. – See Abstr. 010.063.

098.063 **Search for correlations between amplitudes and orbital elements of asteroids.**
C. J. Cunningham.
Bull. Am. Astron. Soc., Vol. 17, No. 4, p. 919 (1985). Abstract. – See Abstr. 010.063.

098.064 **The shape of asteroid 44 Nysa.**
J. Lambert, E. F. Tedesco.
Bull. Am. Astron. Soc., Vol. 17, No. 4, p. 919 (1985). Abstract. – See Abstr. 010.063.

098.065 **Coordination of minor planet orbital studies: past, present and future.**
B. G. Marsden.
Celest. Mech., Vol. 37, No. 3, p. 339 – 344 (1985). – See Abstr. 012.017.
A brief account is given of the coordination of astrometric observations and orbital studies of minor planets, particularly with regard to the role played in the past by the Astronomisches Rechen–Institut. The current situation is surveyed and thoughts are presented on future requirements.

098.066 **Interconnection between small bodies of the solar system.**
L. M. Sherbaum, A. M. Kazantsev.
Vestn. Kiev. Univ. Astron., No. 27, p. 66 – 73 (1985). In Russian.

098.067 **Determination of the positions of celestial bodies from photographic observations. Part 4. Iris and Hebe.**
M. I. Kibardina, L. A. Urasin.
Izv. Astron. Ehngel'gardt. Obs., No. 48, p. 27 – 29 (1984). In Russian.

098.068 **On a possibility of determining the form of an asteroid from the light curve.**
N. I. Koshkin.
Problems of astronomy, p. 69 – 75 (1985). In Russian. Abstr. in Ref. Zh., 51. Astron., 3.51.261 (1986). – See Abstr. 003.023.

098.069 **New names of asteroids.**
L. I. Chernykh.
Zemlya Vselennaya, No. 3, p. 87 – 91 (1986). In Russian.

098.070 **Observation of the occultation of the star BD + 10°2203 by the asteroid (53) Kalypso.**
N. I. Dorokhov, A. I. Movchan.
Astron. Tsirk., No. 1399, p. 8 (1985). In Russian.

098.071 **On the theory of motion of the Trojan asteroids.**
R. V. Zagretdinov.
Kinematika Fiz. Nebesn. Tel, Tom 2, No. 3, p. 68 – 74 (1986). In Russian. English translation in Kinematics Phys. Celest. Bodies.
An analytical theory of the Trojan asteroids motion is considered in the framework of the circular restricted three–body problem. The theory generalizes Garfinkel's results for the case of non–zero inclinations of orbits. Comparison of the theory with the results of numerical integration of equations of motion shows that it provides the accuracy of the order $O(m^{3/2})$.

098.072 **Methods for the investigation of the evolution of asteroid orbits and crossing conditions with the orbit of the perturbing body.**
M. L. Lidov, M. A. Vashkov'yak.
Komety Meteory, No. 36, p. 8 – 17 (1984). In Russian.
A survey of the authors' works, where the evolution of orbits is investigated in an averaged restricted circular N–planetary problem, is given. The limit cases are considered, when an averaged perturbing potential in elementary functions can be found. The numerical–analytical method is described to investigate the averaged problem for arbitrary values of the parameters: the semi–major axis of the asteroid's orbit and the projection of its kinetic moments on the normal to the motion plane of the perturbing bodies. In all the cases the qualitative peculiarities of the behaviour of integral curves are analysed in the plane «argument of pericenter–eccentricity». The extreme characteristics of evolutionary orbits are calculated. The method is applied to the investigation of the evolution of real asteroid orbits.

098.073 **On rotational motion of asteroids.**
Yu. V. Barkin.
Komety Meteory, No. 36, p. 18 – 27 (1984). In Russian.

098.074 **Positions of selected minor planets in the years 1971 – 1982.**
S. V. Kaltygina, S. P. Major, G. V. Moroz, E. M. Sereda, S. V. Shatokhina.
Red. Zh., Kinematika Fiz. Nebesn. Tel, 34 pp. (1985). In Russian. Abstr. in Ref. Zh., 51. Astron., 4.51.110 (1986).

098.075 **Positions of selected minor planets in the years 1976 – 1977 from observations made at the Main Astronomical Observatory of the Ukrainian Academy of Sciences with the double astrograph 400/2000.**
V. V. Golovnya, A. A. Zhitetskij, E. M. Izhakevich, S. P. Major, G. V. Moroz.
Red. Zh., Kinematika Fiz. Nebesn. Tel, 29 pp. (1985). In Russian. Abstr. in Ref. Zh., 51. Astron., 4.51.111 (1986).

098.076 **Minor planets and planetary rings.**
J. C. Bhattacharyya.
Binary and multiple systems, p. 95 – 104 (1984). – See Abstr. 012.025.
Circumstances of the discovery of the first asteroid and the characteristics of various orbits followed by them have been discussed. Their shapes, compositions and evolutionary histories so far known have also been described. The review also covers the present status of knowledge about the rings surrounding 3 giant planets.

098.077 **A propos des noms des petites planètes.**
J. Meeus.
Astronomie, Vol. 100, p. 260 – 261 (1986).

098.078 **Asteroid pole positions: a survey.**
C. Cunningham.
Minor Planet Bull., Vol. 12, No. 2, p. 13 – 16 (1985).
The first comprehensive compilation of asteroid coordinates is presented. A brief comparison with four previous compilations is also given.

098.079 **Photoelectric photometry of asteroids 23 Thalia and 37 Fides.**
K. W. Zeigler, W. B. Florence.
Minor Planet Bull., Vol. 12, No. 3, p. 21 – 23 (1985).
Photoelectric observations of asteroids 23 Thalia and 37 Fides made from Gila Observatory during the period October 18 to November 3, 1984 are described.

098.080 **General report of position observations by the ALPO Minor Planets Section for the year 1984.**
F. Pilcher.
Minor Planet Bull., Vol. 12, No. 3, p. 23 – 29 (1985).
Observations of positions of minor planets by members of the Minor Planets Section in calendar year 1984 are summarized. Some magnitude discrepancies from the predicted values are described.

098.081 **Photoelectric photometry of asteroid 4 Vesta.**
F. J. Melillo.
Minor Planet Bull., Vol. 12, No. 4, Part 1, p. 38 (1985).

098.082 **Minor planets at unusuallly favorable oppositions in 1986.**
F. Pilcher.
Minor Planet Bull., Vol. 13, No. 1, p. 1 – 3 (1986).
A list is presented of minor planets which are much brighter than usual at their 1986 oppositions. Included are two numbered Earth–approachers, 3103 1982 BB and 3199 1982 RA, and two other numbered Mars–crossers, 1011 Laodamia and 1139 Atami.

098.083 **Close approaches of minor planets to naked eye stars in 1986.**
E. Goffin.
Minor Planet Bull., Vol. 13, No. 1, p. 3 – 5 (1986).

098.084 **Close mutual approaches of minor planets in 1986.**
E. Goffin.
Minor Planet Bull., Vol. 13,, No. 1, p. 6 – 7 (1986).

098.085 **A minor planet database for your microcomputer.**
K. Kelly.
Minor Planet Bull., Vol. 13, No. 1, p. 7 – 9 (1986).
General principles are discussed for developing a minor planet database. Three major files are described.

098.086 **Photoelectric photometry of asteroid 15 Eunomia.**
F. J. Melillo.
Minor Planet Bull., Vol. 13, No. 2, p. 21 – 22 (1986).

098.087 **The prediction, refinement, observation, and analysis of stellar occultations by asteroids.**
L. H. Wasserman.
I.A.P.P.P. Commun., No. 23, p. 3 (1986). Abstract. – See Abstr. 010.521.

098.088 **Positions of known and discovered asteroids obtained in La Silla, Chile, at the European Southern Observatory, September 1984.**
H. Debehogne, L. E. Machado.
Rev. Mex. Astron. Astrofis., Vol. 11, No. 2, p. 127 – 137 (1985).
234 positions of asteroids are given. Of these positions, 75 refer to catalogued asteroids in the Ephemeridi Malik Planet (1984) and 159 do not correspond to catalogued planetoids, but possibly to discoveries. The methods of least squares and dependences were used for reductions.

098.089 **Apparent close encounters of minor planets.**
D. A. Pierce, A. L. Whipple.
Astron. J., Vol. 91, No. 2, p. 432 – 437 (1986).
During the period 1985.0 – 2001.4, the 34 minor planets of the Texas Minor Planet Project experience 18 apparent close encoun-

ters, where two of the minor planets appear within one degree of each other. The dynamical importance of these close encounters is discussed. Observations are solicited, and the information needed for the observation of these rare events is illustrated and tabulated.

098.090 **Investigation of the evolution of some asteroid orbits.**
M. A. Vashkov'yak.
Kosm. Issled., Tom 24, Vyp. 3, p. 323 – 336 (1986). In Russian. English translation in Cosm. Res.

098.091 **Occultation and lightcurve analysis: the figure of 2 Pallas.**
J. V. Lambert.
Diss. Abstr. Int., Sect. B, Vol. 46, No. 7, p. 2346 (1986). Thesis, New Mexico State University, 142 pp. (1985). Order No. DA8519956.

098.092 **Orbital elements of numbered minor planets.**
Minor Planet Circ., Nos. 10329 – 10884 (1986).
The minor planets are listed according to their definitive number. Newly numbered objects are indicated by an asterisk. The names of the orbit computers are given behind the respective M.P.C. numbers:
(46) 10752 D. K. Yeomans; (203) 10817, (280), (288), (385), (437) 10818 D. W. E. Green; (727) 10753 L. D. Schmadel; (781) 10818, (1220) 10818–10819, (1481), (1593) 10819 D. W. E. Green; (1627) 10752 D. K. Yeomans; (1814) 10753 L. D. Schmadel; (3358)*–(3362)* 10377–10379 B. G. Marsden; (3363)*–(3366)* 10388–10389 C. M. Bardwell; (3367)* 10392 D. W. E. Green; (3368)*–(3369)* 10392–10393 L. D. Schmadel; (3370)*–(3377)* 10393–10395 K. Hurukawa; (3378)*–(3390)* 10396–10400 S. Nakano; (3391)*–(3393)* 10513–10514 K. Hurukawa; (3394)* 10518 T. Urata; (3395)* 10518–10519 T. Kobayashi; (3396)*–(3403)* 10524–10527 B. G. Marsden; (3404)*–(3414)* 10531–10535 C. M. Bardwell; (3415)*–(3420)* 10610–10612 B. G. Marsden; (3421)*–(3424)* 10626–10627 C. M. Bardwell; (3425)*–(3431)* 10628–10630 S. Nakano; (3432)* 10633 H. Oishi; (3433)*–(3435)* 10753–10754 H. Oishi; (3436)* (3437)* 10755–10756 S. Nakano; (3438)*–(3439)* 10759–10760 B. G. Marsden; (3440)*–(3445)* 10764–10766, (3446)*–(3451)* 10825–10827 C. M. Bardwell; (3452)* 10829 B. G. Marsden; (3453)*–(3458)* 10833–10836 L. D. Schmadel; (3459)* 10837 L. K. Kristensen; (3460)*–(3465)* 10837–10839 K. Hurukawa; (3466)*–(3467)* 10842–10843 H. Oishi; (3468)* 10843 T. Urata.

098.093 **Orbital elements of unnumbered minor planets.**
Minor Planet Circ., Nos. 10329 – 10884 (1986).
The unnumbered minor planets are sorted by their provisionary designation. The names of the orbit computers are given behind the respective M.P.C. numbers:
[1927 UE] 10756 S. Nakano; [1931 TC$_1$] 10841 T. Kobayashi; [1931 UE] 10829 B. G.Marsden; [1934 CC] 10402–10403 T. Urata; [1936 UB] 10401, [1939 SF] 10839, [1940 YE], [1941 SW] 10401, [1955 BG] 10402 S. Nakano; [1956 SC] 10535, [1963 RH] 10535–10536 C. M. Bardwell; [1964 CG] 10522–10523 S. Nakano; [1964 TR$_1$] 10391, [1964 TG$_2$] 10391–10392, [1965 UZ] 10536 C. M. Bardwell; [1967 GF$_1$] 10840 S. Nakano; [1967 UV] 10841 T. Kobayashi; [1968 FJ] 10612 B. G. Marsden; [1970 WC] 10630 S. Nakano; [1971 QU] 10760 B. G. Marsden; [1971 UJ] 10536 C. M. Bardwell; [1973 QG$_2$] 10829, [1973 SO] 10379–10380, [1973 UF$_5$] 10380 B. G. Marsden; [1974 FV$_1$] 10843 H. Oishi; [1974 OE] 10612 B. G. Marsden; [1974 QM$_2$] 10773, [1974 SB$_3$] 10634 T. Kobayashi; [1974 SB$_5$] 10380, [1975 AN] 10527 B. G. Marsden; [1975 BF] 10756 S. Nakano; [1975 VZ] 10829–10830, [1975 VK$_2$] 10761, [1976 GD$_2$] 10830, [1976 GV$_3$], [1976 GM$_7$] 10613 B. G. Marsden; [1976 HQ] 10843 H. Oishi; [1976 SN$_3$] 10527 B. G. Marsden; [1976 SZ$_3$] 10756–10757 S. Nakano; [1977 CC] 10390, [1977 QJ$_2$] 10766, [1977 RE$_2$] 10627 C. M. Bardwell; [1978 PV$_2$] 10630 S. Nakano; [1978 RS] 10390, [1978 SQ$_2$] 10627 C. M. Bardwell; [1978 SE$_3$] 10516 H. Oishi; [1978 SU$_5$] 10536 C. M. Bardwell; [1978 SC$_6$] 10630 S. Nakano; [1978 UO$_2$] 10516 H. Oishi; [1978 VZ$_7$] 10613, [1979 FE] 10527 B. G. Marsden; [1979 GE] 10630–10631, [1979 KH] 10631 S. Nakano; [1979 OB$_9$] 10633 H. Oishi; [1979 QB] 10390 C. M. Bardwell; [1979 QL$_8$] 10631 S. Nakano; [1979 SP$_9$] 10830, [1979 SU$_9$] 10761 B. G. Marsden; [1979 SJ$_{11}$] 10627 C. M. Bardwell; [1979 SO$_{11}$] 10830, [1979 SQ$_{11}$] 10761 B. G. Marsden; [1979 VN] 10516 H. Oishi; [1979 XG], [1979 YM$_8$] 10631, [1979 YN$_8$], [1979 YV$_8$] 10632 S. Nakano; [1980 DE$_1$] 10613, [1980 FB] 10830–10831 B. G. Marsden; [1980 FR$_1$] 10757 S. Nakano; [1980 RZ$_2$] 10537 C. M. Bardwell; [1980 TN$_4$] 10517 H. Oishi; [1980 TL$_{15}$] 10757 S. Nakano; [1981 DM] 10537 C. M. Bardwell; [1981 DZ] 10819 D. W. E. Green; [1981 DZ$_1$] 10614 B. G. Marsden; [1981 DK$_3$] 10514–10515 K. Hurukawa; [1981 EF] 10528 B. G. Marsden; [1981 EN] 10768, [1981 EE$_1$] 10820 D. W. E. Green; [1981 EN$_2$] 10537 C. M. Bardwell; [1981 EH$_3$] 10380 B. G. Marsden; [1981 EX$_3$] 10820, [1981 EH$_4$] 10768 D. W. E. Green; [1981 EA$_5$] 10537 C. M. Bardwell; [1981 EF$_5$], [1981 ED$_6$] 10769 D. W. E. Green; [1981 EA$_7$] 10380–10381, [1981 ET$_7$] 10381, [1981 EV$_7$], [1981 EO$_8$] 10614 B. G. Marsden; [1981 ET$_8$] 10769, [1981 EA$_9$] 10820 D. W. E. Green; [1981 EE$_9$], [1981 EH$_9$] 10381, [1981 EQ$_9$] 10614–10615, [1981 ET$_9$] 10382 B. G. Marsden; [1981 EW$_9$] 10538 C. M. Bardwell; [1981 EP$_{10}$] 10820–10821, [1981 ER$_{10}$] 10769–10770 D. W. E. Green; [1981 ES$_{10}$], [1981 EV$_{10}$] 10382, [1981 EX$_{10}$], [1981 EZ$_{10}$], [1981 ED$_{11}$] 10615, [1981 EE$_{11}$] 10382, [1981 EO$_{11}$] 10761–10762 B. G. Marsden; [1981 ER$_{11}$], [1981 EF$_{12}$] 10538 C. M. Bardwell; [1981 EL$_{12}$] 10383 B. G. Marsden; [1981 EN$_{12}$] 10770, [1981 EQ$_{12}$] 10821, [1981 EH$_{13}$] 10770 D. W. E. Green; [1981 ET$_{13}$] 10538 C. M. Bardwell; [1981 EW$_{13}$] 10770, [1981 EX$_{13}$] 10771 D. W. E. Green; [1981 EC$_{14}$] 10383 B. G. Marsden; [1981 ED$_{14}$] 10539 C. M. Bardwell; [1981 EE$_{14}$] 10771 D. W. E. Green; [1981 EF$_{14}$] 10539 C. M. Bardwell; [1981 ER$_{14}$] 10821 D. W. E. Green.
For a continuation of this list see Abstract 098.094.

098.094 **Orbital elements of unnumbered minor planets.**
Minor Planet Circ., Nos. 10329 – 10884 (1986).
This list is a continuation of Abstract 098.093. The unnumbered minor planets are sorted by their provisional designation. The names of the orbit computers are given behind the respective M.P.C. numbers:
[1981 ES$_{14}$], [1981 ET$_{14}$], [1981 EX$_{14}$] 10616, [1981 EY$_{14}$] 10383 B. G. Marsden; [1981 EZ$_{14}$] 10539 C. M. Bardwell; [1981 EJ$_{15}$] 10616–10617 B. G. Marsden; [1981 EO$_{15}$] 10821 D. W. E. Green; [1981 ER$_{15}$] 10539–10540 C. M. Bardwell; [1981 EU$_{15}$] 10383 B. G. Marsden; [1981 EN$_{16}$] 10822 D. W. E. Green; [1981 ET$_{16}$] 10384, [1981 EJ$_{17}$] 10617 B. G. Marsden; [1981 EN$_{17}$] 10771 D. W. E. Green; [1981 ER$_{17}$] 10617 B. G. Marsden; [1981 EV$_{17}$] 10822 D. W. E. Green; [1981 EE$_{18}$] 10540 C. M. Bardwell; [1981 EV$_{18}$] 10617, [1981 EJ$_{19}$] 10384, [1981 EO$_{19}$] 10618 B. G. Marsden; [1981 EQ$_{19}$], [1981 ET$_{19}$] 10822 D. W. E. Green; [1981 EV$_{19}$], [1981 EV$_{20}$] 10618, [1981 EY$_{20}$] 10384 B. G. Marsden; [1981 EB$_{21}$] 10540 C. M. Bardwell; [1981 EO$_{21}$] 10823 D. W. E. Green; [1981 EA$_{22}$] 10618–10619, [1981 EE$_{22}$] 10384–10385, [1981 EJ$_{22}$] 10619 B. G. Marsden; [1981 EO$_{22}$] 10823 D. W. E. Green; [1981 EZ$_{22}$] 10540 C. M. Bardwell; [1981 EH$_{23}$] 10385 B. G. Marsden; [1981 EJ$_{23}$] 10541 C. M. Bardwell; [1981 EK$_{23}$] 10515 K. Hurukawa; [1981 EB$_{24}$] 10385, [1981 EG$_{24}$] 10619 B. G. Marsden; [1981 ER$_{24}$] 10771–10772 D. W. E. Green; [1981 EC$_{25}$] 10541 C. M. Bardwell; [1981 EF$_{25}$] 10823 D. W. E. Green; [1981 EP$_{25}$] 10619 B. G. Marsden; [1981 EC$_{26}$] 10823 D. W. E. Green; [1981 EN$_{26}$] 10619–10620 B. G. Marsden; [1981 ET$_{26}$] 10541 C. M. Bardwell; [1981 EG$_{27}$] 10620, [1981 ET$_{27}$] 10385, [1981 EV$_{27}$], [1981 EP$_{28}$] 10620, [1981 EV$_{28}$] 10620–10621 B. G. Marsden; [1981 EA$_{29}$] 10772 D. W. E. Green; [1981 EC$_{29}$] 10621, [1981 EE$_{29}$] 10386, [1981 ET$_{29}$] 10621 B. G. Marsden; [1981 EX$_{30}$], [1981 ER$_{31}$] 10824 D. W. E. Green; [1981 EO$_{32}$] 10621 B. G. Marsden; [1981 EW$_{32}$] 10515 K. Hurukawa; [1981 EZ$_{32}$] 10621 B. G. Marsden; [1981 EQ$_{33}$] 10772 D. W. E. Green; [1981 EZ$_{33}$], [1981 ED$_{34}$] 10622 B. G. Marsden; [1981 ED$_{35}$] 10824, [1981 EE$_{35}$] 10772 D. W. E. Green; [1981 EO$_{35}$] 10541–10542, [1981 EY$_{35}$] 10542 C. M. Bardwell; [1981 EB$_{36}$], [1981 EG$_{36}$] 10622, [1981 EW$_{36}$] 10623 B. G. Marsden; [1981 EP$_{37}$], [1981 EU$_{37}$] 10542 C. M. Bardwell; [1981 EZ$_{37}$], [1981 EE$_{38}$] 10386 B. G. Marsden; [1981 EM$_{38}$] 10542–10543 C. M. Bardwell; [1981 EP$_{38}$] 10623 B. G. Marsden; [1981 EY$_{38}$] 10515 K. Hurukawa; [1981 ES$_{39}$] 10543 C. M. Bardwell; [1981 EO$_{40}$],

[1981 EY$_{40}$] 10623, [1981 EA$_{41}$] 10386 B. G. Marsden; [1981 EL$_{41}$] 10632 S. Nakano; [1981 EV$_{41}$] 10543 C. M. Bardwell; [1981 EF$_{42}$] 10623–10624 B. G. Marsden; [1981 EO$_{42}$] 10543 C. M. Bardwell; [1981 EY$_{42}$] 10824, [1981 EA$_{43}$], [1981 EQ$_{43}$] 10825 D. W. E. Green; [1981 ER$_{43}$] 10624, [1981 ET$_{43}$] 10387 B. G. Marsden; [1981 EF$_{45}$] 10772–10773 D. W. E. Green; [1981 EM$_{45}$], [1981 EY$_{45}$] 10624 B. G. Marsden; [1981 EV$_{46}$], [1981 EF$_{47}$] 10544 C. M. Bardwell; [1981 ES$_{47}$] 10773, [1981 FP] 10825 D. W. E. Green; [1981 GC] 10831 B. G. Marsden; [1981 GG], [1981 JQ] 10544 C. M. Bardwell; [1981 JJ$_{2}$] 10831, [1981 QN], [1981 QD$_{2}$], [1981 SU$_{2}$] 10528 B. G. Marsden; [1981 SC$_{7}$] 10836 L. D. Schmadel; [1981 UN] 10624 B. G. Marsden; [1981 UC$_{1}$], [1981 UT$_{15}$] 10757 S. Nakano; [1981 VC$_{1}$] 10831, [1981 WQ] 10762 B. G. Marsden; [1981 YX$_{1}$], [1981 YY$_{1}$] 10758 S. Nakano; [1982 BJ] 10828, [1982 BQ] 10766 C. M. Bardwell.
For a continuation of this list see Abstract 098.095.

098.095 **Orbital elements of unnumbered minor planets.**
Minor Planet Circ., Nos. 10329 – 10884 (1986).

This list is a continuation of Abstract 098.094. The unnumbered minor planets are sorted by their provisional designation. The names of the orbit computers are given behind the respective M.P.C. numbers:
[1982 BS], [1982 BE$_{1}$] 10529, [1982 BS$_{1}$] 10832, [1982 BU$_{1}$] 10625 B. G. Marsden; [1982 DK] 10828 C. M. Bardwell; [1982 DN] 10832, [1982 DQ$_{6}$], [1982 DS$_{6}$] 10387, [1982 FN] 10762 B. G. Marsden; [1982 FH$_{3}$] 10844 T. Urata; [1982 HF$_{1}$] 10625 B. G. Marsden; [1982 KC$_{1}$] 10767, [1982 KN$_{1}$] 10828 C. M. Bardwell; [1982 UV$_{1}$] 10758 S. Nakano; [1982 UM$_{7}$] 10529, [1982 UO$_{7}$] 19762, [1982 VT] 10388 B. G. Marsden; [1982 VR$_{4}$] 10516 K. Hurukawa; [1982 VJ$_{11}$] 10758 S. Nakano; [1982 WB] 10625 B. G. Marsden; [1982 WK] 10403 T. Urata; [1982 XV$_{1}$] 10387 B. G. Marsden; [1983 AE$_{1}$] 10758–10759 S. Nakano; [1983 NT] 10841 T. Kobayashi; [1983 SC] 10517 H. Oishi; [1983 TR$_{2}$] 10529 B. G. Marsden; [1983 VP$_{7}$] 10840, [1983 XU] 10759 S. Nakano; [1984 DS] 10763, [1984 FO] 10530, [1984 HR$_{1}$] 10763 B. G. Marsden; [1984 HZ$_{1}$] 10754 K. Hurukawa; [1984 QN] 10517 H. Oishi; [1984 QO] 10767 C. M. Bardwell; [1984 SM] 10513 L. K. Kristensen; [1984 SX] 10518 H. Oishi; [1984 SE$_{3}$] 10755 K. Hurukawa; [1984 SH$_{5}$] 10530 B. G. Marsden; [1984 UX$_{1}$] 10841 T. Kobayashi; [1984 WX] 10767, [1984 YC] 10390 C. M. Bardwell; [1984 YY] 10842 T. Kobayashi; [1985 BB] 10844 T. Urata; [1985 HV$_{1}$] 10395–10396 K. Hurukawa; [1985 JA] 10530 B. G. Marsden; [1985 JF] 10403 T. Urata; [1985 NE] 10530, [1985 PA] 10531 B. G. Marsden; [1985 PE$_{1}$] 10545 C. M. Bardwell; [1985 QR], [1985 QX] 10403 T. Urata; [1985 RV], [1985 RW] 10531 B. G. Marsden; [1985 RB$_{3}$] 10836, [1985 RG$_{4}$] 10837 L. D. Schmadel; [1985 SA] 10773–10774 L. K. Kristensen; [1985 SE$_{1}$] 10390 C. M. Bardwell; [1985 TB] 10388, 10531 B. G. Marsden; [1985 TC] 10402 T. Urata; [1985 TT] 10634 K. Hurukawa; [1985 TX] 10391, [1985 TC$_{1}$] 10545, [1985 TE$_{1}$], [1985 TF$_{1}$] 10391 C. M. Bardwell; [1985 VS] 10531 B. G. Marsden; [1985 WA] 10767–10768 C. M. Bardwell; [1985 XB] 10763, [1985 YP] 10763–10764 B. G. Marsden; [1986 AL] 10523 S. Nakano; [1986 DA] 10545, 10628, [1986 EB] 10545 C. M. Bardwell; [1986 EB] 10625–10626 B. G. Marsden; [1986 EO] 10768 C. M. Bardwell; [1986 EL$_{1}$] 10755 K. Hurukawa; [1986 EM$_{1}$] 10840 S. Nakano; [1986 FA] 10633 T. Urata; [1986 GC] 10840 S. Nakano; [1986 JK] 10764 B. G. Marsden; [1986 JK] 10833 D. K. Yeomans; [1986 LA], [6541 P–L] 10832, [9602 P–L] 10833 B. G. Marsden.

098.096 **Ephemerides of minor planets and comets.**
Minor Planet Circ., Nos. 10404 – 10456, 10550 – 10586, 10634 – 10668, 10774 – 10780, 10849 – 10884 (1986).

098.097 **Orbital elements of one–opposition minor planets.**
Minor Planet Circ., Nos. 10375 – 10376, 10512 – 10513, 10609 – 10610, 10750 – 10752, 10816 – 10817 (1986).

098.098 **New names of minor planets.**
Minor Planet Circ., Nos. 10545 – 10550, 10844 – 10849 (1986).

098.099 **Critical list of minor planets.**
Minor Planet Circ., No. 10781 (1986).

098.100 **Observations of minor planets.**
Minor Planet Circ., Nos. 10360 – 10375, 10478 – 10512, 10592 – 10609, 10677 – 10750, 10795 – 10816 (1986).

Observations made at the following stations are published: Abastuman, Asiago, Brorfelde, Bulgarian Natl. Obs., Burlington N.J., Calar Alto, Caussols, Cavriana, Centro Astron. Yebes, Crimea, El Leoncito, Eldagsen, ESO, Geisei, Goethe Link Obs., Karasuyama, Kitt Peak, Kleť, Lincoln Lab., Lowell Obs., Lowell Obs. Anderson Mesa Stn., Mauna Kea, Mt. John Obs., Mt. Palomar, Nagatoro, Novi Ligure, Oak Ridge Obs., Peking Obs. Xinglong Stn., Perth Obs., Piszkéstető, Poznan, Purple Mountain Obs., Quonochontaug, Reintal, S. Vittore (Bologna), Seewalchen, Sendai Obs., Sendai Obs. Ayashi Stn., Shizuoka, Tartu, Tautenburg, Toyota, U.S. Naval Obs. Flagstaff Stn., Uenohara, Victoria, Wiesbaden, Zimmerwald.

098.101 **Evolution of the orbits of asteroids in the case of commensurabilities of mean motions. Fundamental equations.**
I. A. Gerasimov.
Astron. Zh., Tom 63, Vyp. 3, p. 567 – 573 (1986). In Russian. English translation in Sov. Astron., Vol. 30, No. 3.

The paper describes the application of the von Zeipel approach to the inner circular restricted three–body problem. Equations of resonant asteroids'motion are obtained. In the case of low commensurabilities a simplified system of equations is obtained.

098.102 **Photographic observations of asteroids made at the Zelenchukskaya Astronomical Station of the Kazan University in 1983 – 1984.**
V. N. Kitkin, I. E. Tselishchev.
Kazan. univ. Kazan', 7 pp. (1986). In Russian. Abstr. in Ref. Zh., 51. Astron., 6.51.150 (1986).

098.103 **Ephemerides of minor planets for 1987.**
Yu. V. Batrakov (Editor).
Published by Institut Teoreticheskoj Astronomii Akademii Nauk SSSR. Izdatel'stvo Nauka, Leningradskoe Otdelenie, Leningrad. 354 pp. Price 4 Rbl. 10 Kop. (1986). In Russian and English.

Contents: Introduction, information on new elements of planets (1) – (3330), lost minor planets, osculating elements of perturbing planets, opposition dates, ephemerides of bright planets, ephemerides of some unusual planets, critical list.

098.104 **Identifications and identification changes of minor planets.**
Minor Planet Circ., Nos. 10457, 10588, 10669, 10782 (1986).

098.105 **Photographic positions of comets, minor planets and Pluto observed during 1975 – 1980.**
M. P. Candy, J. Johnston, P. Jekabsons.
Perth Obs. Commun., No. 4, p. 67 – 197 (1985).

098.106 **Perturbing action of Ceres, Pallas, Juno and Vesta at their proximities with other selected minor planets.**
J. Lazović, M. Kuzmanoski.
Publ. Dep. Astron., Univ. Beogr., No. 13, p. 13 – 18 (1985).

Perturbing effects of the first four asteroids during their proximities with other numbered minor planets are examined for cases of proximity distances being less than 0.001 AU. From among 32 asteroid pairs examined, there were 13 pairs with perturbing effects on orbital elements larger than 10″.

098.107 **Positions of asteroids obtained with the GPO telescope at ESO, Chile, during September and October 1984.**
H. Debehogne, C.–I. Lagerkvist.
Upps. Astron. Obs. Rep., No. 37, 12 pp. (1985). = Obs. R. Belg., Commun. Sér. B, No. 135.

098.108 **Physical studies of asteroids. XIV: Phase relations of the asteroids 1 Ceres, 2 Pallas, 6 Hebe, 7 Iris, 8 Flora, 10 Hygiea, 15 Eunomia, 16 Psyche, 18 Melpomene and 40 Harmonia.**
C.-I. Lagerkvist, I. P. Williams.
Upps. Astron. Obs. Rep., No. 40, 13 pp. (1986).
Multiple scattering factors and absolute magnitudes have been derived for ten asteroids, six of type S, two of type C and one M asteroid. The mean multiple-scattering factors Q were found to be 0.136 and 0.043 for the S and C asteroids, respectively. For the asteroids 2 Pallas, 7 Iris, 15 Eunomia, 16 Psyche and 18 Melpomene composite lightcurves are presented.

098.109 **A discussion on determinations of the pole coordinates for (16) Psyche.**
X.-h. Zhou.
Publ. Purple Mt. Obs., Vol. 4, No. 2, p. 46 – 51 (1985). In Chinese.
A test was made for four determinations of the pole coordinates of (16) Psyche. It is concluded that the two nearly same results obtained by the author and D. F. Lupishko and I. N. Belskaya (1983) are reasonable, because the light maxima can be predicted well from them.

098.110 **Photographic observations of positions of minor planets (1983).**
J.-x. Yang, S.-l. Wei, Q. Wang, Y.-l. Ge, T.-w. Lu.
Publ. Purple Mt. Obs., Vol. 4, No. 3, p. 29 – 40 (1985). In Chinese.

098.111 **The collision lifetimes of four asteroids.**
G.-y. Li.
Publ. Purple Mt. Obs., Vol. 5, No. 1, p. 36 – 40 (1986). In Chinese.
An improved deduction of Kessler's formulae on the collision frequency of orbital objects is given. Futhermore, the collison frequencies between the four asteroids 2263, 2355, 2387 and 2425, were computed with those formulae. Results show that the collision lifetimes are more than 4.5×10^{13} years.

098.112 **On the direction of rotation of asteroids.**
D. F. Lupishko, I. N. Bel'skaya, F. P. Velichko.
Vestn. Khar'kov. Univ., No. 278, p. 51 – 56 (1985). In Russian.

098.113 **Photographic position observations of Pallas in 1982 at the Astronomical Observatory of the Kharkov State University.**
P. P. Pavlenko.
Vestn. Khar'kov. Univ., No. 278, p. 72 – 74 (1985). In Russian.

098.114 **Unidentified moving object.**
R. N. Inasaridze, R. I. Kiladze.
Komet. Tsirk., No. 349 (1986). In Russian.

098.115 **Observations of asteroid (4) Vesta in Tartu.**
H. Raudsaar.
Komet. Tsirk., No. 351 (1986). In Russian.

098.116 **Astrometric positions of minor planets in February 1984. Six discoveries, ESO, La Silla, Chile.**
H. Debehogne, G. Vieira.
Acta Astron., Vol. 35, Nos. 3–4, p. 453 – 456 (1985).
138 astronomical positions of asteroids, residuals and dependences are given as obtained from photographic observations made at the GPO, ESO, La Silla, Chile, during February – March 1984.

098.117 **275 precise astrometric positions of minor planets obtained at the GPO Telescope of ESO, La Silla.**
H. Debehogne, V. Protitch-Benishek (*V. Protic-Benisek*), D. Olević.
Acta Astron., Vol. 35, Nos. 3–4, p. 457 – 472 (1985).
Precise astrometric positions of minor planets: 120 Lachesis, 366 Vincentina, 659 Nestor, 883 Matterania, 1254 Erfordia, 1273 Helma, 1280 Baillauda, 2759 Idomeneus, 1984 SN_4, 1984 SO_4, 1984 SP_4 and 1984 SQ_4 obtained at the GPO of ESO, La Silla, Chile during September–October 1984 are presented.

098.118 **Positions of selected minor planets (1983 – 1984).**
G. de Sanctis, W. Ferreri.
Acta Astron., Vol. 36, No. 1, p. 165 – 168 (1986).
207 positions are presented for 24 asteroids observed from September 1983 to October 1984 at the Observatory of Torino. Reductions were made using the dependence method with 6 reference stars taken from the AGK3 and SAO catalogues.

098.119 **Positions of asteroids observed in April 1983 and September 1984 with the GPO telescope at ESO, Chile.**
H. Debehogne, G. de Sanctis.
Acta Astron., Vol. 36, No. 1, p. 169 – 173 (1986).
294 accurate positions of asteroids are presented as obtained from observations performed during April 1983 and September 1984 at the GPO telescope of ESO, La Silla, Chile.

098.120 **Observations of minor planets.**
M. Geffert, H.-J. Tucholke, R.-J. Dettmar.
Astron. Nachr., Vol. 307, No. 3, p. 213 – 215 (1986).

098.121 **Reports of asteroidal appulses and occultations.**
J. Stamm.
Occultation Newsl., Vol. 3, No. 16, p. 343 (1986).

098.122 **Astrometric updates for asteroidal occultations in the U.S.S.R. during 1984.**
D. W. Dunham, I. S. Balinsky.
Occultation Newsl., Vol. 3, No. 16, p. 351 (1986).

098.123 **Scaling laws for the catastrophic collisions of asteroids.**
K. A. Holsapple, K. R. Housen.
Mem. Soc. Astron. Ital., Vol. 57, No. 1, p. 65 – 85 (1986). – See Abstr. 012.061.
Collisions of asteroids have traditionally been studied through laboratory experiments involving targets with masses some 15 to 20 orders of magnitude less than the bodies they are intended to simulate. The authors consider the problem of extrapolation of experimental results up to the size regimes of interest.

098.124 **Implications of the inferred compositions of asteroids for their collisional evolution.**
C. R. Chapman.
Mem. Soc. Astron. Ital., Vol. 57, No. 1, p. 103 – 114 (1986). – See Abstr. 012.061.
The inferred compositions of mainbelt asteroids set constraints on the nature and degree of collisional evolution of the asteroids. Studies of the compositions of members of dynamical families suggest that most supposed families are either not real, or at least are not genetically related fragments of a cosmochemically reasonable precursor body. The mutual incompatibilities among various "accepted" traits of the asteroids deserve further study in order to develop a self-consistent scenario for the chemical and physical evolution of the asteroids.

098.125 **Recent developments in asteroid science.**
N. Eaton.
Phys. Rep., Vol. 132, No. 5, p. 261 – 276 (1986). Abstr. in Phys. Abstr., Vol. 89, No. 1281, Entry 67385 (1986).

098.126 **Astrometria fotografica di pianetini e comete. Relazione annuale sull'attività svolta dagli astrofili dell'Osservatorio S. Vittore (Bologna) nel corso del 1984.**
E. Colombini.
Astronomia, N. 2, p. 5 – 7 (1986).
The author reports the results of the astrometric observations of minor planets and comets made during the program for 1984. The minor planet 1984 FA was discovered by amateur astronomers of Osservatorio S. Vittore (Bologna).

098.127 **Predictions of stellar occultations by asteroids.**
E. Goffin.
Separate print edited by GEOS, 7 pp. (1986).
Available from the International Occultation Timing Association (I.O.T.A., European Section), Paris – Meudon Observatory, F–92195 Meudon Principal Cedex, France.

098.128 **Statistische Untersuchung der Bahnelemente erdnaher kleiner Planeten.**
M. Hoffmann.
Mitt. Astron. Ges., No. 65, p. 180 (1986). Abstract. – See Abstr. 012.072.

098.129 **1985 TB.**
IAU Circ., No. 4160 (1986).

098.130 **1985 PA.**
IAU Circ., Nos. 4163, 4185 (1986).

098.131 **1986 DA.**
IAU Circ., Nos. 4181, 4186, 4203, 4205 (1986).

098.132 **(3361) 1982 HR.**
IAU Circ., No. 4182 (1986).

098.133 **1986 EB.**
IAU Circ., Nos. 4191, 4192, 4199, 4205 (1986).

098.134 **1986 JK.**
IAU Circ., Nos. 4215, 4218, 4220, 4227 (1986).

098.135 **1986 LA.**
IAU Circ., No. 4228 (1986).

098.136 **Fast–moving asteroid 1986 EB.**
Yamamoto Circ., Nos. 2055, 2056, 2059 (1986). In Japanese.

098.137 **1986 DA.**
Yamamoto Circ., Nos. 2055, 2059 (1986). In Japanese.

098.138 **Object Shoemaker, 1986 JK.**
Yamamoto Circ., Nos. 2060, 2061 (1986). In Japanese.

098.139 **1986 LA.**
Yamamoto Circ., No. 2062 (1986). In Japanese.

A comprehensive index to the Minor Planet Bulletin: 1973 – 1985.
See Abstr. 002.044.

Alphabetic list of minor planet names and naming references.
See Abstr. 002.045.

Catalogue of orbital elements of 3143 minor planets numbered by November 1, 1984, on magnetic tape.
See Abstr. 002.068.

Comets, meteors & asteroids – how they affect earth.
See Abstr. 003.099.

Asteroids (or thorny way of investigations).
See Abstr. 003.179.

Summary report on the International Workshop "Catastrophic disruption of asteroids and satellites" (Pisa, Italy, July 30 – August 2, 1985).
See Abstr. 011.034.

Photoelectric studies of asteroids: the amateur contribution.
See Abstr. 013.036.

Asteroidal catastrophic collisions simulated by hypervelocity impact experiments.
See Abstr. 022.126.

Results obtained by laboratory simulations of catastrophic impact.
See Abstr. 022.128.

CCD scanning.
See Abstr. 036.047.

Programs for observations of asteroids.
See Abstr. 036.079.

Motion near the triangular libration points with variable mass ratio of the components.
See Abstr. 042.062.

Change in the mutual distance of celestial bodies resulting from changes in some of their orbital elements.
See Abstr. 042.068.

On the refutation of the Brown conjecture.
See Abstr. 042.080.

Errata: "The Theory of the Trojan asteroids. Part V" [Celest. Mech., Vol. 36, No. 1, p. 19 – 45 (1985)].
See Abstr. 042.091.

Reference frame determination using minor planets observed through Tokyo PMC.
See Abstr. 043.005.

Dust and minor bodies – the picture from IRAS.
See Abstr. 051.030.

Cassini – Titan probe concept.
See Abstr. 051.054.

European satellite studies of minor planets. II.
See Abstr. 051.065.

On the feasibility of observing small asteroids with Galileo, Venera, and Comet–Rendezvous – Asteroid–Flyby missions.
See Abstr. 051.093.

Period of 160 minutes in the solar system: solar pulsation and spin rates of planets and asteroids.
See Abstr. 091.015.

Planètes et petits corps du système solaire. Rapport thématique 1980 – 1984.
See Abstr. 091.049.

Planetary occultations.
See Abstr. 096.015.

Jupiter und seine Begleiter (II).
See Abstr. 099.021.

Cometary nuclei and asteroids: has a link been found?
See Abstr. 102.018.

Are cometary nuclei like asteroids?
See Abstr. 103.007.

Meteoroids inside the earth's orbit: the Eccentrids system.
See Abstr. 104.007.

Asteroidal origin for the Geminid meteor stream.
See Abstr. 104.016.

Quantitative calculation of meteorite flux from various regions of the inner asteroid belt.
See Abstr. 105.163.

Exposure history of shergottites.
See Abstr. 105.231.

The wavelength dependence of phase coefficients.
See Abstr. 105.242.

The spectral and physical properties of metal in meteorite assemblages: implications for asteroid surface materials.
See Abstr. 105.243.

Preliminary analysis of the IRAS solar system dust data.
See Abstr. 106.023.

The formation and origin of the IRAS zodiacal dust bands as a consequence of single collisions between asteroids.
See Abstr. 106.034.

The structure of Phaethon and detonation of its icy envelope.
See Abstr. 107.009.

Simulation of catastrophic fragmentation events: a review of the accelerating techniques and a resumé of the experiments performed until now.
See Abstr. 107.046.

Catastrophic disruption in the solar system: asteroid collisional history, origin of Hirayama families and disruption of small satellites.
See Abstr. 107.047.

099 Jupiter, Jupiter Satellites

099.001 **A collection of Galilean satellite eclipse observations, 1652 – 1983: part I.**
J. H. Lieske.
Astron. Astrophys., Vol. 154, No. 1/2, p. 61 – 76 (1986).

The most complete collection of extant Galilean satellite eclipse observations since 1652 has been assembled. Since many of the old data exist only in manuscript form or in archaic forms (e.g. apparent time, local time, sidereal time, Julian calendar, etc.), they have been reduced to a modern proleptic Universal Time system on the Gregorian calendar. Many of the data had been presumed to be lost for more than a century and since they are very valuable for discussion of long–term effects on the satellites, the author presents them for present as well as future generations of astronomers.

099.002 **The possibility of estimating tidal perturbations of Jupiter's Galilean satellites.**
J. H. Lieske.
Relativity in celestial mechanics and astrometry, p. 117 – 125 (1986). – See Abstr. 012.002 (IAU Symp. No. 114).

Recent measurements in the infra–red of the energy emitted by Io suggest that if the observed emission is due to tidal dissipation in Io, then one should observe a very large secular acceleration of Io in its orbit. The purpose of this paper is to investigate the observational evidence which might be examined to verify that hypothesis. Preliminary results suggest that large tidal accelerations of Jupiter's satellites do not exist at the suggested magnitude and hence the heat flux measurements probably represent an episodic occurrence.

099.003 **Theoretical and observational efforts in order to obtain high precision positions of the Galilean satellites of Jupiter.**
W. Thuillot, D. T. Vu, J. E. Arlot.
Relativity in celestial mechanics and astrometry, p. 127 (1986). Abstract. – See Abstr. 012.002 (IAU Symp. No. 114).

099.004 **The precision of modern theories of Galilean satellites as derived by their comparison with photographic observations made with the 26″ refractor at Pulkovo Observatory.**
T. P. Kisseleva (*T. P. Kiseleva*), N. I. Glebova, A. G. Malkova.
Relativity in celestial mechanics and astrometry, p. 129 – 134 (1986). – See Abstr. 012.002 (IAU Symp. No. 114).

The regular photographic observations of Galilean satellites has been made with 26″ refractor at Pulkovo since 1975. During 1975 – 1979 over 600 observations of Galilean satellites were carried out. The observed positions were compared with the ephemeris based on Sampson's and Lieske's E–1 theories. The comparison of Sampson's theory with observations showed some systematic error which can be considered as an error in longitude of about 1 min. Lieske's theory E–1 showed a better agreement with the observations, the absolute deviations not exceeding 0″.3.

099.005 **Developing high accuracy numerical theories of motion of outer satellites of Jupiter and Saturn.**
L. E. Bykova, V. V. Shikhalev.
Relativity in celestial mechanics and astrometry, p. 135 – 140 (1986). – See Abstr. 012.002 (IAU Symp. No. 114).

The effect of different perturbing factors on the rectangular coordinates of outer satellites of Jupiter (VI–XII) and Saturn (VI–IX) is investigated by the numerical integration of the equations of motion. The effect of relativistic terms in the equations of motion is considered among other perturbations. The effect of weak perturbations is compared with the accuracy of modern observations of satellites.

099.006 **Coupled hydromagnetic wave excitation and ion acceleration upstream of the Jovian bow shock.**
C. W. Smith, M. A. Lee.
J. Geophys. Res., Vol. 91, No. A1, p. 81 – 90 (1986).

The authors extend the Lee (1982) self–consistent theory of upstream wave excitation and particle energization to address observations by Voyager 2 of sunward propagating MHD waves and diffuse suprathermal particle populations upstream of the Jovian bow shock. Two new ideas are incorporated into the theory. First, the interplanetary seed wave population is taken to be an equal admixture of waves propagating both toward and away from the shock parallel to the interplanetary magnetic field. Second, finite connection times are incorporated approximately into the theory in an effort to understand whether the particle spectra at high energy are limited by particle escape or finite connection time.

099.007 **Is Jupiter's ionosphere a significant plasma source for its magnetosphere?**
A. F. Nagy, A. R. Barakat, R. W. Schunk.
J. Geophys. Res., Vol. 91, No. A1, p. 351 – 354 (1986).

A semikinetic model was used to study the steady state, collisionless, polar wind outflow from the Jovian polar caps. H^+ escape fluxes and energies were calculated for a range of conditions, including several values of the ambient electron temperature, different hot electron populations, and both with and without the effects of the centrifugal force. The calculations indicate that if hot electron populations exist over the Jovian polar caps, as they do on earth, polar wind escape fluxes of the order of $10^8 cm^{-2} s^{-1}$ are possible. When integrated over the polar cap area, escape fluxes of this order of magnitude imply an ionospheric source strength of 2×10^{28}ions/s, which is comparable to the present estimate of the total magnetospheric plasma source population. Therefore, the ionosphere may play an important role in populating the Jovian magnetosphere, specifically the "hidden", low energy, light ion component of the population.

099.008 **Monte Carlo simulation of charged particle impact on the satellites of Jupiter and Saturn.**
E. V. Bell II, T. P. Armstrong.
J. Geophys. Res., Vol. 91, No. A2, p. 1397 – 1403 (1986).
Ensemble-averaged lifetimes of Jovian and Saturnian trapped charged particles against loss by impact on natural satellites are presented. Precise values of these loss rates can be obtained from the known geometry of the satellite orbits, magnetic fields, and particle trajectory. These loss rates are needed as an ingredient of transport theory which is usually formulated in terms of the first and second adiabatic invariants and radial distance. The effects of satellite orbital inclination and eccentricity, ion charge to mass ratio and energy, equatorial pitch, and distance from the planet on the particle lifetime are discussed.

099.009 **On Io's control of Jovian decametric radio emissions.**
R. M. Winglee.
J. Geophys. Res., Vol. 91, No. A2, p. 1405 – 1416 (1986).
Io's control of Jovian decametric radio emission (DAM) has been attributed to Io distorting the electron distribution in the inner Jovian magnetosphere. Observations of Faraday rotation in DAM are used to determine the properties of the electron distribution before and after its interaction with Io. It is shown that there is an enhancement in the density of the energetic component in the Io plasma torus correlated with certain Jovian longitude. Io's interaction with this energetic component can produce heating of this component. The Io-controlled emission is attributed to enhanced emission from the heated electrons moving down the field lines to Jupiter.

099.010 **Spatial and diurnal features of the Jovian equatorial anomaly.**
A. Tan.
Planet. Space Sci., Vol. 34, No. 1, p. 117 – 124 (1986).
The author investigates the possibility of formation of equatorial anomaly in the Jovian ionosphere by mechanisms similar to those which produce equatorial anomaly in the terrestrial *F*-region. He does this by studying the effects of electrodynamic drifts in a time-dependent model of the Jovian *F*-region. In particular, the latitudinal and diurnal features of the resulting anomaly and their similarities and differences with their terrestrial counterparts are examined. Two sinusoidal drift velocity models are considered – one, similar to that of the terrestrial *F*-region and the other having opposite phase.

099.011 **Physics and chemistry of sulfur lakes on Io.**
J. I. Lunine, D. J. Stevenson.
Icarus, Vol. 64, No. 3, p. 345 – 367 (1985).
A model for a convecting sulfur lake, heated from below by a silicate magma chamber, is constructed and applied to major hot spot regions on Jupiter's satellite Io. The authors use a two-layer parametrized convection scheme for sulfur and silicates based on a local boundary layer analysis to calculate temperature profiles in the system and the maximum flux which can be extracted from the silicate magma in steady state. The results indicate that the highest-component temperature of some observed hot spots (J. S. Pearl and W. M. Sinton, 1982) is consistent with a convecting molten sulfur system, and the total flux from the most energetic spot, Loki Patera, is close to the maximum which can be extracted from molten silicates by convection.

099.012 **Supersonic meteorology of Io: sublimation-driven flow of SO_2.**
A. P. Ingersoll, M. E. Summers, S. G. Schlipf.
Icarus, Vol. 64, No. 3, p. 375 – 390 (1985).
This paper describes the flow of condensable gas from day side to night side of a frost-covered planet. The gas is SO_2 and the planet is Io, but the model could apply to other planets with thin atmospheres like Triton and Pluto. The authors are interested in the global resurfacing rate and the distributions of surface pressure, atmospheric temperature, and flow speed.

099.013 **Tidally forced viscous heating in a partially molten Io.**
M. N. Ross, G. Schubert.
Icarus, Vol. 64, No. 3, p. 391 – 400 (1985).
The authors investigate tidal dissipative heating in two different models of Io. The partially molten asthenosphere model consists of a rigid inner core and a thin (less than 400 km thick) partially molten "decoupling" layer (asthenosphere) surrounded by an elastic lithosphere. In the partially molten interior model the interior beneath the lithosphere is partially molten throughout. The partially molten region in each model is assumed to possess negligible shear strength and to be characterized by a Newtonian viscosity. Tidal deformation and dissipation in the core of the thin asthenosphere model are assumed negligible.

099.014 **Vertical cloud structure of Jupiter's Equatorial Plumes.**
C. R. Stoker, C. W. Hord.
Icarus, Vol. 64, No. 3, p. 557 – 575 (1985).
A model for the vertical cloud structure of Jupiter's Equatorial Plumes is deduced based on an analysis of Voyager images of the equatorial region in the 6190-Å methane band and the 6000-Å continuum, and ground-based 8900-Å methane band images of Jupiter. A computer code that represents scattering and absorption from aerosol and gas layers was applied to a hierarchy of increasingly complex model aerosol structures to match the observations in the three wavelengths. The observations are consistent with a model for the vertical cloud structure of the equatorial region that consists of four aerosol layers.

099.015 **Geology of large impact craters on Ganymede: implications on thermal and tectonic histories.**
P. G. Thomas, O. P. Forni, P. L. Masson.
Earth, Moon, Planets, Vol. 34, No. 1, p. 35 – 53 (1986).
The purpose of this paper is to study the geology and the geometry of four large impact craters on Ganymede which are not too much modified by topographic relaxation: the Gilgamesh basin, the Western Equatorial basin, the East Hathor basin and the system of furrows of Galileo Regio which are supposed to be a part of a multi-ring basin (Smith et al., 1979).

099.016 **Beaming of Jupiter's decametric emission.**
J. J. Riihimaa.
Earth, Moon, Planets, Vol. 34, No. 2, p. 133 – 138 (1986).
Dynamic spectra of a Jovian non-Io-A storm recorded simultaneously by the Voyager 1 spacecraft and by the Kiiminki radio spectrograph are compared. It seems that the emission beam of the storm co-rotates with the planet and has a sloped leading edge, in accordance with the result of Maeda and Carr (1984).

099.017 **Ganymedean pedestal craters distribution: implications on thermal and tectonic histories.**
O. Forni, P. G. Thomas, P. Masson.
Earth, Moon, Planets, Vol. 34, No. 2, p. 177 – 188 (1986).
The distribution analysis of the ganymedean pedestal craters shows a very good correlation between them and the grooved terrains. Assuming that a lowering of the terrain's viscosity is the determining factor for the presence of pedestal craters, the authors conclude that at the time of their formation, the grooved terrains have a lower viscosity than the other terrains. Moreover, using the density ratio between non-pedestal craters and pedestal craters, a relative low viscosity state's duration time is calculated. Based on the density ratio between pedestal craters and tectonized pedestal craters, a grooved terrains formation's time is also calculated.

099.018 **Decametric radio emission of Jupiter. II. Localization of *S*-component sources.**
B. P. Ryabov.
Astron. Vestn., Tom 20, No. 1, p. 20 – 34 (1986). In Russian. English translation in Sol. Syst. Res.
New results of high-sensitivity observations of the *S*-componets in Jupiter's sporadic radio emission obtained in the frequency range of 10 to 23 MHz are discussed. The correlation of *S*-bursts parameters (power, frequency, drift rate, upper frequencies) with magnetic field strength, angular and linear velocities of Io flux tube foot are studied vs Io flux tube joviographic

meridian longitudes. The author supposes that the transverse velocity of the southern end of Io's flux tube with respect to magnetospheric plasma is among the principal factors controlling the generation and parameters of *S* burst storms.

099.019 **Some dynamical parameters of Jupiter's atmosphere.**
A. P. Vid'machenko.
Kinematika Fiz. Nebesn. Tel, Tom 2, No. 1, p. 48 – 51 (1986). In Russian. English translation in Kinematics Phys. Celest. Bodies.
The first instability of the Taylor–vortex type in the Couette spherical flow is supposed to be responsible for the observed zonal structure of Jupiter. On the basis of this assumption the depth of the planetary atmosphere ($h_J \cong 16271$ km) and the coefficient of kinematic viscosity for large–scale circulation ($K_J \leqslant 1.9 \times 10^{11} cm^2 sec^{-1}$) are obtained. The adiabatic temperature variation with the height is confirmed independently.

099.020 **Mode decoupling during retrorefraction as an explanation for bizarre radar echoes from icy moons.**
V. R. Eshleman.
Nature, Vol. 319, No. 6056, p. 755 – 757 (1986).
Radar echoes from Europa, Ganymede and Callisto, large ice–covered moons of Jupiter, have characteristics that differ markedly from those of other natural targets. A decoupling of two characteristic modes of propagation appears to be the key to an understanding of the anomalous distribution of power among the polarized components of the echoes. Various concepts of mode decoupling and retrorefraction are combined here to explain the strength, distribution and other features of the echoes. Causes of the decoupling are proposed and their morphological implications are introduced.

099.021 **Jupiter und seine Begleiter (II).**
M. Reichstein.
Astron. Sch., 23. Jahrg., Heft 2, p. 30, 43 – 44 (1986).

099.022 **La planète Jupiter en 1983.**
R. Néel.
Astronomie, Vol. 100, p. 117 – 136 (1986).

099.023 **Optical polarimetry of sulphur and the surface microstructure of Io.**
G. A. Steigmann.
Mon. Not. R. Astron. Soc., Vol. 219, No. 4, p. 823 – 833 (1986).
Polarimetric and albedo measurements of various sulphur samples are reported and compared with data for Io. The most likely form of sulphur covering the surface of Io is the form of μm–sized grains, having been rapidly condensed from the vapour phase in volcanic eruptions. However, polarimetric and albedo measurements of such samples strongly suggest that the sulphur regolith on Io is contaminated with silicate minerals. Good agreement with data for Io is obtained from a thick roughened layer of sublimed sulphur contaminated with a silicate slag of albedo 0.13 covering about 19 per cent of the surface area.

099.024 **Tidal reorientation and the fracturing of Jupiter's moon Europa.**
A. S. McEwen.
Nature, Vol. 321, No. 6065, p. 49 – 51 (1986).
The most striking characteristic of Europa is the network of long linear albedo markings over the surface, suggestive of global–scale tectonic processes. The author presents a global map of lineaments with long arc lengths (> 20° or 550 km), and compares the lineament of orientations to the tensile stress trajectories due to tidal distortions (changes in the lenght of three principal semiaxes) and to nonsynchronous rotation (longitudinal reorientation of two of the principal semiaxes). An excellent orthogonal fit to the lineaments is achieved by the stresses due to nonsynchronous rotation with the axis radial to Jupiter located 25° east of its present position. This fit suggests that nonsynchronous rotation occurred at some time in Europa's history.

099.025 **In situ measurements of the plasma bulk velocity near the Io flux tube.**
A. Barnett.
J. Geophys. Res., Vol. 91, No. A3, p. 3011 – 3019 (1986).
The authors study the flow around the Io flux tube by analyzing the eleven spectra taken by the Voyager 1 Plasma Science (PLS) experiment in its vicinity. They determine the bulk plasma parameters using a procedure that uses the full response function of the instrument and the data in all four PLS sensors.

099.026 **A self–consistent model of Jupiter's magnetodisc including the effects of centrifugal force and pressure.**
G. Caudal.
J. Geophys. Res., Vol. 91, No. A4, p. 4201 – 4221 (1986).
This paper describes a self–consistent axisymmetric stationary MHD model of the Jovian magnetosphere, in which both the centrifugal force and the pressure gradients exerted on the plasma are taken into account. The self–consistent solution to the model, with input parameters suitable for the Voyager 1 encounter, is presented and compared to observations.

099.027 **Energetic neutral particles from Jupiter and Saturn.**
A. F. Cheng.
J. Geophys. Res., Vol. 91, No. A4, p. 4524 – 4530 (1986).
The Voyager 1 spacecraft has observed energetic neutral particle fluxes created by charge exchanges in the magnetospheres of Jupiter and Saturn. Analysis of these observations yields significant new information on the energetic ion and ambient neutral populations there.

099.028 **Jovian VLF chorus and Io torus aurora.**
U. S. Inan.
J. Geophys. Res., Vol. 91, No. A4, p. 4543 – 4550 (1986).
A test particle model of the cyclotron resonance interaction of waves and trapped radiation belt particles is used to estimate the energetic electron fluxes precipitated by Jovian VLF chorus waves observed on the Voyager 1 and 2 spacecraft near the Io torus. The precipitation fluxes induced by 1–s–long chorus wave packets at $L \cong 7.6$ and 8.6 are estimated to be bursts of ~5 s duration with a peak of 0.3 – 3 and 0.7 – 7 ergs/cm²s that consist of electrons of ~5 – 100 keV energy and that arrives at the ionosphere ~15 s after the generation of the chorus wave at the equatorial plane. The effects in the Jovian ionosphere of the chorus–induced precipitation are estimated using existing ionospheric models. A possible experiment for measuring Jovian chorus–induced aurora is proposed and discussed.

099.029 **Correction to "Nonisotropic coronal atmosphere on Io" [J. Geophys. Res., Vol. 90, No. A6, p. 5327 – 5331 (1985)].**
E. M. Sieveka, R. E. Johnson.
J. Geophys. Res., Vol. 91, No. A4, p. 4608 (1986). See Abstr. 40.099.076.

099.030 **Europa: characterization and interpretation of global spectral surface units.**
M. L. Nelson, T. B. McCord, R. N. Clark, T. V. Johnson, D. L. Matson, J. A. Mosher, L. A. Soderblom.
Icarus, Vol. 65, No. 1, p. 129 – 151 (1986). = Planet. Geosci. Div. Publ. 343.
The Voyager global multispectral mosaic for the Galilean satellite Europa (T. V. Johnson et al., 1983) was analyzed to map surface units with similar optical properties (T. B. McCord et al., 1982). Color assignments in the unit map are indicative of the spectral nature of the unit. The unit maps make it possible to infer extensions of the geologic units mapped by B. K. Lucchitta and L. A. Soderblom (1982) beyond the region covered in the high–resolution imagery. The most striking feature in the unit maps is a strong hemispheric asymmetry. It is seen most clearly in the ultraviolet/violet albedo ratio image. It appears as if the surface has been darkened, essentially as the cosine of the angle from the antapex of motion. The cosine pattern suggests that the darkening is exogenic in origin and is interpreted as evidence of alteration of the surface by ion bombardment from the Jovian magnetosphere.

099.031 **Toward high-precision observation of Galilean satellites.**
T. Nakamura, M. Soma.
Proceedings of the nineteenth symposium on celestial mechanics, p. 67 – 69 (1986). In Japanese. – See Abstr. 012.013.

099.032 **Preliminary analysis of mutual phenomena of Galilean satellites.**
A. Sengoku.
Proceedings of the nineteenth symposium on celestial mechanics, p. 70 – 75 (1986). In Japanese. – See Abstr. 012.013.

099.033 **The 1983 apparition of Jupiter: rotation periods.**
P. W. Budine.
Strolling Astron., Vol. 31, Nos. 5–6, p. 96 – 105 (1986).

099.034 **Galilean satellite eclipse timings: 1982/1983 report.**
J. E. Westfall.
Strolling Astron., Vol. 31, Nos. 5–6, p. 105 – 119 (1986).

099.035 **Origin of Jovian cosmic rays.**
K. Brecher.
Bull. Am. Astron. Soc., Vol. 17, No. 4, p. 848 (1985). Abstract. – See Abstr. 010.062.

099.036 **The contraction of Io's orbit.**
S. J. Goldstein Jr., K. C. Jacobs.
Bull. Am. Astron. Soc., Vol. 17, No. 4, p. 873 (1985). Abstract. – See Abstr. 010.062.

099.037 **A relook at the zonal distribution of hydrogen in the Jovian atmosphere.**
R. M. Killen, J. W. Chamberlain.
Bull. Am. Astron. Soc., Vol. 17, No. 4, p. 917 (1985). Abstract. – See Abstr. 010.063.

099.038 **Spectral reflectance of quenched sulfur glasses: implications for Io.**
W. D. Smythe, R. M. Nelson.
Bull. Am. Astron. Soc., Vol. 17, No. 4, p. 920 (1985). Abstract. – See Abstr. 010.063.

099.039 **Spectral reflectance of sulfur trioxide (0.25 – 5.2 microns): implication for resurfacing of Jupiter's satellite Io.**
R. M. Nelson, W. D. Smythe.
Bull. Am. Astron. Soc., Vol. 17, No. 4, p. 920 – 921 (1985). Abstract. – See Abstr. 010.063.

099.040 **Proton irradiation of sulfur: sputtering yields, spectral reflectance changes, and applications to Io.**
D. Nash.
Bull. Am. Astron. Soc., Vol. 17, No. 4, p. 921 (1985). Abstract. – See Abstr. 010.063.

099.041 **Volatiles on icy satellites: Europa's trailing side.**
R. H. Brown, D. P. Cruikshank, R. G. Smith, A. T. Tokunaga, R. N. Clark.
Bull. Am. Astron. Soc., Vol. 17, No. 4, p. 921 (1985). Abstract. – See Abstr. 010.063.

099.042 **Trajectories of weakly-charged Jovian dust.**
L. Schaffer, J. A. Burns.
Bull. Am. Astron. Soc., Vol. 17, No. 4, p. 921 (1985). Abstract. – See Abstr. 010.063.

099.043 **Evolution paths for the Galilean satellites.**
R. Greenberg.
Bull. Am. Astron. Soc., Vol. 17, No. 4, p. 921 – 922 (1985). Abstract. – See Abstr. 010.063.

099.044 **Jupiter and Saturn's magnetic differential rotation and expected periods for magnetic field reversals.**
D. H. Hathaway, A. J. Dessler.
Bull. Am. Astron. Soc., Vol. 17, No. 4, p. 925 (1985). Abstract. – See Abstr. 010.063.

099.045 **On gravitational perturbations of resonances.**
G. V. Stolyarov.
Astron. Tsirk., No. 1386, p. 4 – 6 (1985). In Russian.

099.046 **Optical characteristics of the Galilean satellites of Jupiter (0.347 – 0.792 μm). II. The dependence of the reflectivity on the orbital phase angle for $\alpha = 6°$.**
A. N. Dovgopol, V. I. Shavlovskij.
Kinematika Fiz. Nebesn. Tel, Tom 2, No. 3, p. 48 – 51 (1986). In Russian. English translation in Kinematics Phys. Celest. Bodies.

Photoelectric observations of the Galilean satellites have been carried out in the spectral region 0.347 – 0.792 μm using 11 interference filters. The dependence of the reflectivity on the orbital phase angle was obtained for the solar phase angle $\alpha = 6°$. The most significant differences between the orbital curves for $\alpha = 0°$ and $\alpha = 6°$ were observed for Callisto in the ultraviolet region ($\lambda = 0.347$ μm). The smallest differences were observed for Europa in the whole spectral region.

099.047 **The dependence of the reflectivity of the Galilean satellites of Jupiter on the orbital phase angle with a solar phase angle of 6°.**
A. N. Dovgopol, V. I. Shavlovskij.
Red. Zh., Kinematika Fiz. Nebesn. Tel, 14 pp. (1985). In Russian. Abstr. in Ref. Zh., 51. Astron., 4.51.195 (1986).

099.048 **Clouds, aerosols, and photochemistry in the Jovian atmosphere.**
R. A. West, D. F. Strobel, M. G. Tomasko.
Icarus, Vol. 65, Nos. 2/3, p. 161 – 217 (1986). – See Abstr. 012.023.

The authors review current ideas about the composition, horizontal and vertical distribution, and microphysical properties of clouds and aerosols in Jupiter's upper troposphere and stratosphere. They also discuss several key photochemical species, their relation to aerosol formation, and their implications for transport processes. They treat photochemistry in the context of comparative planetology and point out important similarities and differences among the outer planet atmospheres.

099.049 **Observations of the limb darkening of Jupiter at ultraviolet wavelengths and constraints on the properties and distribution of stratospheric aerosols.**
M. G. Tomasko, E. Karkoschka, S. Martinek.
Icarus, Vol. 65, Nos. 2/3, p. 218 – 243 (1986). – See Abstr. 012.023.

This paper discusses two sets of west–east scans of Jupiter and one south–north scan made by the International Ultraviolet Explorer (IUE) satellite at four wavelengths between 0.22 and 0.33 μm. Three of the wavelength bands studied are between 0.22 and 0.25 μm where the stratospheric aerosols should have their strongest effect on the observations. Details of the observational procedures, the photometric reductions, and the determination of the local scattering geometries of the data are given. The observations are compared to aerosol models. A final section discusses the models in the context of other observations which bear on the properties of the aerosols, and summarizes the authors' conclusions.

099.050 **Submillimeter and millimeter observations of Jupiter.**
M. J. Griffin, P. A. R. Ade, G. S. Orton, E. I. Robson, W. K. Gear, I. G. Nolt, J. V. Radostitz.
Icarus, Vol. 65, Nos. 2/3, p. 244 – 256 (1986). – See Abstr. 012.023.

The authors report narrowband photometry of the Jovian disk in 10 passbands covering the range from 0.35 to 3.3 mm wavelength. The derived brightness temperature spectrum is analyzed in the context of existing constraints on the atmospheric temperature structure and composition from ground-based studies at shorter wavelengths and from various spacecraft measurements. The results for wavelengths between 0.35 and 0.45 mm suggest that the radiances can be matched by models which include NH_3 ice particles which are between 30 and 100 μm in size, regardless of the scale height characterizing the cloud.

099.051 **The abundances of ethane and acetylene in the atmospheres of Jupiter and Saturn.**
K. S. Noll, R. F. Knacke, A. T. Tokunaga, J. H. Lacy, S. Beck, E. Serabyn.
Icarus, Vol. 65, Nos. 2/3, p. 257 – 263 (1986). – See Abstr. 012.023.

Infrared spectra near 780 cm^{-1} of Jupiter and Saturn have been obtained to determine the stratospheric abundances of ethane (C_2H_6) and acetylene (C_2H_2). Atmospheric models using Voyager thermal profiles and density profiles with constant mixing ratios result in the mixing ratios, $X(C_2H_2) = 1.0(\pm 0.3) \times 10^{-7}$ and $X(C_2H_6) = 5.5(\pm 1.5) \times 10^{-6}$ for Jupiter. The results for Saturn are $X(C_2H_2) = 3.0(\pm 1.0) \times 10^{-7}$ and $X(C_2H_6) = 7.0(\pm 1.5) \times 10^{-6}$. The ratio of ethane to acetylene, $n[C_2H_6]/n[C_2H_2]$, is found to be insensitive to model atmosphere assumptions. The ratio is 55 ± 31 for Jupiter and 23 ± 12 for Saturn from models with uniform mixing ratios.

099.052 **The vertical structure of the Jovian atmosphere.**
P. H. Smith.
Icarus, Vol. 65, Nos. 2/3, p. 264 – 279 (1986). – See Abstr. 012.023.

An inhomogeneous atmospheric model, which includes both polarizing aerosols and methane gas, has been developed for the equatorial and tropical regions of Jupiter consistent with the Pioneer polarimetry and photometry data as well as the methane–band data of West (1979). A two–cloud model with an overlying stratospheric haze has been used to explain the relative strengths of weak and strong methane bands.

099.053 **Global dynamics and thermal structure of Jupiter's atmosphere.**
F. M. Flasar.
Icarus, Vol. 65, Nos. 2/3, p. 280 – 303 (1986). – See Abstr. 012.023.

This review is addressed to the general student of planetary atmospheres. The plan of the paper is as follows: (1) Current ideas on the thermal structure of the deep atmosphere and interior are briefly summarized. (2) The mean zonal circulation that has been inferred from studies of cloudtop motions is discussed. (3) The thermal structure of the upper troposphere and stratosphere is treated. Evidence for temporal variability in the zonally averaged thermal structure and for dynamical forcing of the tropopause is presented. (4) Some key observations that are needed to elucidate the many unresolved issues concerning the atmospheric dynamics of Jupiter and Saturn are listed.

099.054 **Jupiter: new estimates of the mean zonal flow at the cloud level.**
S. S. Limaye.
Icarus, Vol. 65, Nos. 2/3, p. 335 – 352 (1986). – See Abstr. 012.023.

This paper presents a new high–resolution latitudinal profile of the zonal mean east–west component of motion averaged over 2 month of the clouds seen in the Voyager images. The results and the technique with which they were obtained are presented along with a comparison with the other Voyager results after a brief background.

099.055 **Merging of vortices in the atmosphere of Jupiter: an analysis of Voyager images.**
M.–M. Mac Low, A. P. Ingersoll.
Icarus, Vol. 65, Nos. 2/3, p. 353 – 369 (1986). = Contrib. Div. Geol. Planet. Sci., Calif. Inst. Technol., Pasadena, Calif., USA, No. 4213. – See Abstr. 012.023.

The authors have studied interactions between stable oval structures (spots) using the Voyager 2 cylindrical projection mosaics. In contrast with the solitary wave type of interaction, collisions between spots are irreversible. Most interactions (23 out of 27 cases) lead to merging of the two original spots. The other type of interaction (4 out of 27 cases) is simply a near miss – the spots pass around each other. Interactions of spots with filamentary regions, which are actively changing and more amorphous than spots, usually lead to the disappearance of the spot. Filamentary regions are also the major source of spots.

099.056 **Motions in the interiors and atmospheres of Jupiter and Saturn. 2. Barotropic instabilities and normal modes of an adiabatic planet.**
A. P. Ingersoll, R. L. Miller.
Icarus, Vol. 65, Nos. 2/3, p. 370 – 382 (1986). = Contrib. Div. Geol. Planet. Sci., Calif. Inst. Technol., Pasadena, Calif., USA, No. 4217. – See Abstr. 012.023.

This is the second paper in a series dealing with possible large–scale motions in the giant planets' fluid interiors. The first paper (Ingersoll and Pollard, 1982, IP) presented a scale analysis and a set of equations describing motions in a rotating sphere of variable density when the eddy viscosity and superadiabaticity are small. The present paper extends the analysis of IP to a more general class of flows.

099.057 **Constraints on the NH_3 and PH_3 distributions in the Great Red Spot.**
R. Wagener, J. Caldwell, T. Owen.
Icarus, Vol. 66, No. 1, p. 188 – 191 (1986).

IUE observations of the Great Red Spot and the South Tropical Zone in the wavelength region of the NH_3 predissociation bands between 1900 and 2200 Å show slightly stronger absorption in the Great Red Spot than in the South Tropical Zone. Vertically inhomogeneous Rayleigh scattering radiative transfer models find an enhanced $[NH_3]/[H_2]$ mixing ratio at the 80– to 125–mbar pressure level in the Great Red Spot of a factor of 3 to 10 with respect to the South Tropical Zone. Upper limits on the mixing ratio of PH_3 and the eddy diffusion coefficient above the Great Red Spot are considerably lower than earlier predictions.

099.058 **Observation and preliminary analysis of a recent Jovian satellite phenomenon.**
W. H. Allen, E. Budding.
Earth, Moon, Planets, Vol. 35, No. 1, p. 73 – 77 (1986).

The occultation of Io by Ganymede as observed on 10 June 1985 is reported. The middle of the occultation minimum was found to occur at $14^h14^m5.7^s$ UT. In the plane perpendicular to the line of sight the centres of the two satellite disks passed to within ~2530 km of each other, at a relative velocity, in this plane, of about 11 km s^{-1}. The values of these last two quantities, however, depend on what assumptions are made about the light distribution over Io's disk.

099.059 **A theory of Jovian shadow bursts.**
N. Gopalswamy.
Earth, Moon, Planets, Vol. 35, No. 2, p. 93 – 115 (1986).

The shadow events in the dynamic spectra of Jovian decametric emission are explained as the result of interaction between electron bunches responsible for S and L emissions. The relevant dispersion relation is derived for the fast extraordinary mode in the cold magnetospheric plasma in the presence of S and L electron bunches. The growth rate of the synchrotron maser instability is studied in the presence and absence of S–electrons. It is shown that the synchrotron maser instability responsible for L–emission can be temporarily quenched by the invasion of S–electrons, thereby stopping the L–emission. The theory accounts for various observed features of the shadow events.

099.060 **Medium–energy electrons and heavy ions in Jupiter's magnetosphere: effects of lower hybrid wave–particle interactions.**
D. D. Barbosa.
J. Geophys. Res., Vol. 91, No. A5, p. 5605 – 5615 (1986).

A theory of medium–energy (~keV) electrons and heavy ions in Jupiter's magnetosphere is presented. Lower hybrid waves are generated by the combined effects of a ring instability of neutral wind pickup ions and the modified two–stream instability associated with transport of cool Iogenic plasma. The quasi–linear energy diffusion coefficient for lower hybrid wave–particle interactions is evaluated, and several solutions to the diffusion equation are given. Calculations based on measured wave properties show that the noise modifies substantially the particle distribution functions. The effects are to accelerate superthermal ions and electrons to keV energies and to thermalize the pickup ions on time scales comparable to the particle residence time.

099.061 **Calculated occultation profiles of Io and the hot spots.**
A. S. McEwen, L. A. Soderblom, D. L. Matson, T. V. Johnson, J. I. Lunine.
Geophys. Res. Lett., Vol. 13, No. 3, p. 201 – 204 (1986).
Occultations of Io by other Galilean satellites in 1985 provide a means to locate volcanic hot spots and to model their temperatures. The authors have computed the expected time variations in the integral reflected and emitted radiation of the occultations as a function of wavelength (visual to 8.7 μm). The best current ephemerides were used to calculate the geometry of each event as viewed from Earth. Visual reflectances were modeled from global mosaics of Io. Thermal emission from the hot spots was calculated from Voyager 1 IRIS observations and, for regions unobserved by IRIS, from a model based on the distribution of low–albedo features.

099.062 **Numerical and analytical theories of motion of the outer satellites of Jupiter.**
T. V. Bordovitsyna, T. S. Boronenko, L. E. Bykova, A. M. Chernitsov.
Nauchn. probl. aviatsii i kosmonavt.: Istor. i sovremennost'. Moskva, p. 165 – 169 (1985). In Russian. Abstr. in Ref. Zh., 51. Astron., 5.51.128 (1986).

099.063 **Decameter radio radiation of Jupiter. Localizing the region of generation of S–radiation.**
B. P. Ryabov.
Inst. radiofiz. i ehlektron. Akad. Nauk USSR, Prepr., No. 280, 20 pp. (1985). In Russian. Abstr. in Ref. Zh., 51. Astron., 5.51.279 (1986).

099.064 **Electropolarimetric observations of bright satellites of Jupiter.**
R. A. Chigladze.
Soobshch. Akad. Nauk GruzSSR, Tom 118, No. 3, p. 513 – 516 (1985). In Russian. Abstr. in Ref. Zh., 51. Astron., 5.51.281 (1986).

099.065 **The gas composition and vertical cloud structure of Jupiter's troposphere derived from five micron spectroscopic observations.**
G. L. Bjoraker.
Diss. Abstr. Int., Sect. B, Vol. 46, No. 5, p. 1588 – 1589 (1985). Thesis, The University of Arizona, 239 pp. (1985). Order No. DA8514901.

099.066 **An observationally compatible model for Jupiter's Io–related decametric emission.**
W.–x. Wang.
Diss. Abstr. Int., Sect. B, Vol. 46, No. 9, p. 3092 – 3093 (1986). Thesis, The University of Florida, 158 pp. (1985). Order No. DA8523894.

099.067 **Decameter radio radiation of Jupiter. Morphology of S–storms.**
B. P. Ryabov, A. V. Arkhipov, V. A. Shevchenko.
Inst. radiofiz. i ehlektron. Akad. Nauk USSR, Prepr., No. 279, 30 pp. (1985). In Russian. Abstr. in Ref. Zh., 51. Astron., 6.51.262 (1986).

099.068 **The gas composition of Jupiter derived from 5 micron airborne spectroscopic observations.**
G. L. Bjoraker, H. P. Larson, V. G. Kunde.
Prepr. Steward Obs., No. 653, 75 pp. (1986). To appear in Icarus.

099.069 **The abundance and distribution of water vapor in Jupiter's atmosphere.**
G. L. Bjoraker, H. P. Larson, V. G. Kunde.
Prepr. Steward Obs., No. 663, 68 pp. (1986). To appear in Astrophys. J.

099.070 **Episodic volcanism of tidally heated satellites with application to Io.**
G. W. Ojakangas, D. J. Stevenson.
Icarus, Vol. 66, No. 2, p. 341 – 358 (1986).
A simple model of the coupled thermal and orbital evolution of a tidally heated satellite in an orbital resonance is presented and applied specifically to Io. The model predicts that Io's mean motion may be currently increasing, a possibility suggested by recent estimates of $\dot{n}_1$ from eclipse data. Since Europa's eccentricity mimics that of Io, the model also implies that the tidal stresses in Europa's ice shell may have recently been large enough to produce the observed fracturing. The episodic heating mechanism may be responsible for the resurfacing of Enceladus $<10^9$years ago.

099.071 **Seasonal north–south asymmetry in solar radiation incident on Jupiter's atmosphere.**
R. F. Beebe, R. M. Suggs, T. Little.
Icarus, Vol. 66, No. 2, p. 359 – 365 (1986).
Although the orbital eccentricity and axial tilt are small, the near–temporal coincidence of perihelion and the northernmost excursion of the subsolar point produce asymmetries in the solar radiation incident on Jupiter's atmosphere. Calculations of the incident radiation and the latitudinal gradient of the insolation are presented. North–south asymmetries in the zonal wind and morphology of the large–scale cloud systems observed by Voyagers 1 and 2 are cited. In the absence of an identified internal mechanism capable of generating the observed asymmetries, seasonal forcing of this magnitude should be considered.

099.072 **Detectability of HD and non–equilibrium species in the upper atmospheres of the giant planets from their submillimeter spectrum.**
B. Bézard, D. Gautier, A. Marten.
Astron. Astrophys., Vol. 161, No. 2, p. 387 – 402 (1986).
The submillimeter spectra of Jupiter, Saturn, Uranus and Neptune were modelled in the 10– to 200–cm^{-1} range (50 μm – 1mm) in view of forthcoming observations from space at high spectral resolution ($\lesssim 0.1$ cm^{-1}). Results are presented for the detectability of HD, photolysed compounds and inorganic trace species possibly quenched from hot deep levels by vertical transport. Frequencies and intensities of the rotation lines were generated, when not available, from rotational constants and dipole moment values.

099.073 **Dawn–dusk asymmetry and spongy nature of the Jovian magnetosphere.**
T. Aoyama, H. Oya.
J. Geomagn. Geoelectr., Vol. 37, No. 9, p. 839 – 867 (1985). Abstr. in Phys. Abstr., Vol. 89, No. 1278, Entry 47936 (1986).

099.074 **Observations of mutual phenomena of the Galilean satellites in 1985 – 1986.**
R. L. Marcialis.
Bull. Am. Astron. Soc., Vol. 18, No. 1, p. 639 (1986). Abstract. – See Abstr. 010.066.

099.075 **Disk–resolved photometry of Io. I. Near–opposition limb darkening.**
D. P. Simonelli, J. Veverka.
Icarus, Vol. 66, No. 3, p. 403 – 427 (1986).
To compare Voyager photometric measurements of individual areas on Io to laboratory data, the two sets of information must be reduced to a comparable format. For spacecraft data, this requires knowledge of the near–opposition limb darkening and phase behaviour of the region under observation. In this work the authors derive more accurate normal reflectance spectra for regions on Io by a two–step process. They obtain the near–opposition limb–darkening characteristics for areas on the sur-

face of Io. Combination of the phase and limb–darkening data leads to improved normal reflectance spectra for regions on Io, which can be compared readily with laboratory spectra of candidate materials.

099.076 **Disk–resolved photometry of Io. II. Opposition surges and normal reflectances.**
D. P. Simonelli, J. Veverka.
Icarus, Vol. 66, No. 3, p. 428 – 454 (1986).
Near–opposition ($\alpha = 2-8°$) phase curves have been derived for individual regions on Io in three color classes. Accurate normal reflectances derived for three color classes are generally consistent with the disk–resolved Io spectra determined by previous researchers, and have been compared with laboratory spectra of SO_2 frost and various sulfurs processed to be directly comparable to the Io data. In agreement with the suggestions of previous researchers, the authors conclude that most of the features in the disk–resolved Io spectra can be explained by mixtures of sulfurs and SO_2 frost, but stress that this consistency need not imply that all or most of the Ionian surface layer is made of sulfurous materials.

099.077 **The gas composition of Jupiter derived from 5–μm airborne spectroscopic observations.**
G. L. Bjoraker, H. P. Larson, V. G. Kunde.
Icarus, Vol. 66, No. 3, p. 579 – 609 (1986).
The atmospheric transmission window between 1850 and 2250 cm^{-1} in Jupiter's atmosphere was observed at a spectral resolution of 0.5 cm^{-1} from the Kuiper Airborne Observatory. The mole fractions of NH_3, PH_3, CH_4, CH_3D, CO, and GeH_4 were derived for the 1– to 6–bar portion of Jupiter's troposphere using a spectrum synthesis program. Knowledge of the abundances of these gases below the visible clouds is necessary to calculate the global inventory of nitrogen, phosphorus, carbon, and deuterium, which, in turn, may constrain models of Jupiter's formation.

099.078 **Enhanced acetylene emission near the north pole of Jupiter.**
P. Drossart, B. Bézard, S. Atreya, J. Lacy, E. Serabyn, A. Tokunaga, T. Encrenaz.
Icarus, Vol. 66, No. 3, p. 610 – 618 (1986).
The observation of a polar bright spot in the atmosphere of Jupiter is characterized by an enhancement in the individual lines of C_2H_2 which can be interpreted as an enhancement in the acetylene abundance, although a purely thermal effect, or non–LTE phenomena cannot be excluded. The intensity of the observed hot band lines are also consistent with either hypothesis. An unambiguous interpretation would require observing other molecules to discriminate between photochemical and thermal effects.

099.079 **Photoelectron excitation of H_2 due to solar EUV interactions in the Jovian atmosphere.**
S. A. Haider, R. Shanker, O. N. Singh.
Indian J. Radio Space Phys., Vol. 15, No. 1, p. 6 – 9 (1986). Abstr. in Phys. Abstr., Vol. 89, No. 1283, Entry 78311 (1986).

099.080 **Radial diffusion and ion partitioning in the Io torus.**
A. F. Cheng.
Geophys. Res. Lett., Vol. 13, No. 6, p. 517 – 520 (1986).
A model is presented for radial diffusion and charge state partitioning of sulfur and oxygen ions in the Io torus, including effects of electron impact and charge exchange.

099.081 **Superrotation in the pre–dawn Jovian magnetosphere: evidence for corotating convection.**
M. R. Hairston, T. W. Hill.
Geophys. Res. Lett., Vol. 13, No. 6, p. 521 – 524 (1986).
The authors report evidence for superrotation of plasma in the pre–dawn magnetosphere of Jupiter during the outbound segments of the Voyager 1 and 2 encounters. The evidence consists of a forward spiral of magnetic field lines crossing the equatorial plasma sheet, as distinct from the more typical backward spiral associated with subcorotational flow. The limited available statistics indicate that superrotational flow is restricted in local time to the night side, and also restricted in system III longitude to the hemisphere opposite the active sector of the magnetic–anomaly model. The results are compatible with the planetary–wind concept in general and with the corotating convection model in particular.

099.082 **Volcanic hot spots on Io: correlation with low–albedo calderas.**
A. S. McEwen, D. L. Matson, T. V. Johnson, L. A. Soderblom.
J. Geophys. Res., Vol. 90, No. B14, p. 12345 – 12379 (1985).
Comparison of Voyager imaging data with Voyager 1 infrared observations (IRIS) and with ground–based telescopic infrared data demonstrates a strong correlation between hot spots and low–albedo features (reflectivity $\leqslant 0.3$ in all visual wavelengths) on Io. Most of these low–albedo areas occur on the floors of volcanic calderas, and many are associated with active or recent volcanism. The authors use this correlation to identifiy additional hot spots in regions not observed by IRIS and to model their temperatures.

099.083 **Silicate volcanism on Io.**
M. H. Carr.
J. Geophys. Res., Vol. 91, No. B3, p. 3521 – 3532 (1986).
This paper reexamines several aspects of silicate volcanism on Io, such as how and in what quantities silicate magma could be generated, how it might reach the surface, and how it might manifest itself there. Of special interest are the surface temperatures that would result as a consequence of silicate eruptions, since the temperatures measured have been regarded as a strong evidence against silicate eruptions and strong support for the presence of low–temperature melts at the surface. The general thrust of the paper is that silicates are the principal means by which the satellite dissipates its heat and that they play a prominent role in the evolution of the surface.

099.084 **On the colors of Jupiter's satellite Io: irradiation of solid sulfur at 77K.**
R. Steudel, G. Holdt, A. T. Young.
J. Geophys. Res., Vol. 91, No. B5, p. 4971 – 4977 (1986).
The present work is concerned with the irradiation of various solid sulfur allotropes at 77K to study the physical and chemical transformations that elemental sulfur possibly present on the surface of Io may undergo when exposed to sunlight.

099.085 **Dark halo craters and the thickness of grooved terrain on Ganymede.**
P. M. Schenk, W. B. McKinnon.
J. Geophys. Res., Vol. 90, Suppl., p. C775 – C783 (1985). – See Abstr. 012.065.
Voyager images of the icy Galilean satellite Ganymede reveal that the surface is approximately equally divided into two general terrain types: a relatively dark, ancient, heavily cratered terrain, and a brighter, less heavily cratered terrain characterized by a smooth surface or abundant parallel groove sets. The authors present evidence that indicates dark cratered terrain underlies the present surface of grooved terrain at relatively shallow depth and has not been absorbed into the interior or displaced laterally to a significant degree.

099.086 **Heating by Alfvén wave dissipation in the Jovian thermosphere.**
J. Kar, K. K. Mahajan.
J. Geophys. Res., Vol. 91, No. A6, p. 7101 – 7104 (1986).
An estimate of heating rates by the dissipation of Io–generated Alfvén waves by ion–neutral collisions in the Jovian thermosphere is made. Heating rates at the foot of the Io flux tube are found to be significant.

099.087 **Comment on "Magnetic field properties of Jupiter's tail at distances from 80 to 7500 Jovian radii" by M. L. Goldstein, R. P. Lepping, and E. C. Sittler, Jr. [J. Geophys. Res., Vol. 90, No. A9, p. 8223 – 8239 (1985)].**
T. W. Hill, A. J. Dessler.
J. Geophys. Res., Vol. 91, No. A6, p. 7131 – 7132 (1986). – See Abstr. 40.099.025.

099.088 **Reply to the comment on "Magnetic field properties ..." by T. W. Hill and A. J. Dessler.**
M. L. Goldstein, R. P. Lepping, E. C. Sittler Jr.
J. Geophys. Res., Vol. 91, No. A6, p. 7133 (1986). – See Abstr. 099.087.

A collection of Galilean satellite eclipse observations, 1652 – 1983: II.
See Abstr. 002.003.

A propos des observations des satellites de Jupiter par Galilée en 1610.
See Abstr. 004.014.

Satellites of Jupiter. Part 1.
See Abstr. 012.087.

The production and sputtering of S_2 by keV ion bombardment.
See Abstr. 022.045.

Spectral reflectance of solid sulfur trioxide (0.25 – 5.2 μm): implications for Jupiter's satellite Io.
See Abstr. 022.064.

Absolute line strengths of $^{74}GeH_4$ near 5 μm.
See Abstr. 022.105.

Radio spectrographs for observing the planet Jupiter.
See Abstr. 033.050.

Phase fluctuation measurement with HF–band VLBI system and its application to observation of Jupiter's decametric emissions.
See Abstr. 033.060.

Precise position measurements of Jupiter, Saturn and Uranus systems with the Very Large Array.
See Abstr. 043.011.

Stable, baroclinic eddies on Jupiter and Saturn: a laboratory analog and some observational tests.
See Abstr. 062.055.

Micrometeoroid impact on planetary satellites as a magnetospheric mass source.
See Abstr. 091.044.

Planètes et petits corps du système solaire. Rapport thématique 1980 – 1984.
See Abstr. 091.049.

Kirkwood gaps.
See Abstr. 098.055.

Minor planets and planetary rings.
See Abstr. 098.076.

Interior of Saturn.
See Abstr. 100.030.

The meteorology of Saturn.
See Abstr. 100.033.

The wavelength dependence of phase coefficients.
See Abstr. 105.242.

Observational constraints on models for giant planet formation.
See Abstr. 107.015.

Evolution of the giant planets.
See Abstr. 107.016.

Rings and moons: clues to understanding the solar nebula.
See Abstr. 107.017.

Catastrophic disruption in the solar system: asteroid collisional history, origin of Hirayama families and disruption of small satellites.
See Abstr. 107.047.

Diatoms on Earth, comets, Europa and in interstellar space.
See Abstr. 131.196.

100 Saturn, Saturn Satellites

100.001 **An upper limit to X–ray emission from Saturn.**
D. A. Gilman, K. C. Hurley, F. D. Seward, H. W. Schnopper, J. D. Sullivan, A. E. Metzger.
Astrophys. J., Vol. 300, No. 1, p. 453 – 455 (1986).
X–rays are produced in auroral discharges. The existence of auroral activity on Saturn was suggested by the observation of a magnetosphere by Pioneer 11 and confirmed by UV measurements during the Voyager encounters. The detection of X–rays from Jupiter with the Einstein Observatory provided the impetus for a subsequent observation of Saturn. No emission was detected. This article presents the upper limit established by the observation and compares it with theoretical expectations.

100.002 **D to H ratio and the origin and evolution of Titan's atmosphere.**
J. P. Pinto, J. I. Lunine, S.–J. Kim, Y. L. Yung.
Nature, Vol. 319, No. 6052, p. 388 – 390 (1986).
Observations indicate that Titan's atmosphere is enriched in deuterium by a factor of $\geqslant 3$ relative to Jupiter and Saturn. Potential causative factors examined here for this enrichment are condensation to form tropospheric methane clouds, fractionation occurring over a hypothetical CH_4–C_2H_6 ocean and between the ocean and the clathrate crust beneath, fractionation which occurred during the formation of Titan and fractionation occurring as a result of the evolution of Titan's atmosphere. It is concluded that the greater part of the observed fractionation is probably derived from the formation of Titan and the subsequent evolution of Titan's atmosphere driven by photochemistry.

100.003 **New models of Saturn's magnetic field using Pioneer 11 Vector Helium Magnetometer data.**
L. Davis Jr., E. J. Smith.
J. Geophys. Res., Vol. 91, No. A2, p. 1373 – 1380 (1986).
In a reanalyisis of the Vector Helium Magnetometer data taken by Pioneer 11 during its Saturn encounter in 1979, using improvements in the data set and in the procedures, studies are made of a variety of models. The best is the $P_{11}84$ model, an

axisymmetric spherical harmonic model of Saturn's magnetic field within 8 Saturn radii of the planet.

100.004 **Thermal ions at Saturn: plasma parameters and implications.**
J. D. Richardson.
J. Geophys. Res., Vol. 91, No. A2, p. 1381 – 1389 (1986).
Plasma data from the PLS experiment on Voyagers 1 and 2 are analyzed at the Saturn encounters. Measured ion currents are simulated using the full response function of the PLS instrument assuming plasma distributions are isotropic Maxwellians. The plasma velocity, densities, and temperatures which best simulate the observations are determined.

100.005 **Limits on ion radial diffusion coefficients in Saturn's inner magnetosphere.**
M. Paonessa, A. F. Cheng.
J. Geophys. Res., Vol. 91, No. A2, p. 1391 – 1396 (1986).
Voyager low energy charged particle (LECP) ion phase space densities at constant first and second adiabatic invariants have been used to place limits on the rate of radial diffusion of energetic ions (30 keV to 1 MeV) in Saturn's inner magnetosphere. Upper and lower limits to the radial diffusion coefficient, D_{LL}, are deduced from physical requirements on the rates of diffusion and loss. If D_{LL} is near the lower limit found in this work, then satellite sweeping accounts for a large fraction of the total ion losses. If D_{LL} is near the upper limit, then ion losses can approach 10% of the strong diffusion rate. In this case, ion losses are dominated by wave–particle interactions, and sweeping losses are relatively unimportant.

100.006 **Voyager Saturnian ring measurements and the early history of the solar system.**
H. Alfvén, I. Axnäs, N. Brenning, P.–A. Lindqvist.
Planet. Space Sci., Vol. 34, No. 2, p. 145 – 154 (1986).
The mass distribution in the Saturnian ring system is investigated and compared with predictions from the plasma cosmogony. According to this theory, the matter in the rings has once been in the form of a magnetized plasma, in which the gravitation is balanced partly by the centrifugal force and partly by the electromagnetic forces. As the plasma is neutralized, the electromagnetic forces disappear and the matter can be shown to fall in to 2/3 of the original saturnocentric distance. This causes the so called "cosmogonic shadow effect", which has been demonstrated earlier for the asteroidal belt and in the large scale structure of the Saturnian ring system. The relevance of the cosmogonic shadow effect is investigated for parts of the fine structures of the Saturnian ring system. It is shown that many structures of the present ring system can be understood as shadows and antishadows of cosmogonic origin. These appear in the form of double rings centered around a position a factor 0.64 (slightly less than 2/3) closer to Saturn than the causing feature. Voyager data agree with an accuracy better than 1%.

100.007 **Resurfacing Saturn's satellites: models of partial differentiation and expansion.**
G. J. Consolmagno.
Icarus, Vol. 64, No. 3, p. 401 – 413 (1985).
The faulting seen on the surfaces of Saturn's icy moons may have been caused either by external events, such as large impacts, or internal stresses caused by the expansion of the moons as long–lived radionuclides produced internal heating and phase changes. The author estimates the stress as a function of radius expansion to be $\sigma = 44\,(\Delta r/r)$ kbar. The extensional stress needed for fracture is probably something less than 40 bar. From his calculations the author concludes that the most reasonable cause for rifting on Dione and Rhea is the refreezing of an ammonia–water eutectic melt inside these moons roughly two billion years after their formation, while the rift on Tethys was caused by a large impact, and little rifting should be expected on Iapetus.

100.008 **Hyperion: analysis of Voyager observations.**
P. Thomas, J. Veverka.
Icarus, Vol. 64, No. 3, p. 414 – 424 (1985).
Saturn's satellite Hyperion has attracted recent attention because of its unusual spin state and also because some explanations of the albedo dichotomy on Iapetus have implications for the state of Hyperion's surface. The authors review published telescopic observations of this +14–mag object and then present new data derived from Voyager observations made in 1980 and 1981.

100.009 **Evolution of the Janus–Epimetheus coorbital resonance due to torques from Saturn's rings.**
J. J. Lissauer, P. Goldreich, S. Tremaine.
Icarus, Vol. 64, No. 3, p. 425 – 434 (1985).
The authors analyze the interactions between Saturn's coorbital satellites, Janus and Epimetheus, and the outer edge of the A ring, which is presumably maintained by these moons at their 7:6 resonance. It is shown that ring torques are driving these satellites into a tighter lock. Unless there is a counterbalancing force which the authors have neglected, their orbital configuration will evolve from the current horseshoe–type lock to one of tadpole orbits around a single Lagrange point in $\sim$20 myr.

100.010 **Saturn's rings: particle size distributions for thin layer models.**
H. A. Zebker, E. A. Marouf, G. L. Tyler.
Icarus, Vol. 64, No. 3, p. 531 – 548 (1985).
In this paper the authors develop a new scattering model that allows the vertical ring profile to vary continously from a near monolayer to a many–particle–thick system. This model is then used to improve upon the results of Marouf et al. (1983) and to give new results for additional locations within the rings.

100.011 **Observations of the $J = 10$ manifold of the pure rotational band of phosphine on Saturn.**
M. R. Haas, E. F. Erickson, D. Goorvitch, D. D. McKibbin, D. M. Rank.
Icarus, Vol. 64, No. 3, p. 549 – 556 (1985).
Saturn was observed in the vicinity of the $J = 10$ manifold of the pure rotational band of phosphine on 1984 July 10 and 12 from NASA's Kuiper Airborne Observatory with the facility far–infrared cooled grating spectrometer. It is found that PH_3 must be strongly depleted above the thermal inversion ($\sim$70 mbar). The best fitting models consistent with other observational constraints indicate that PH_3 is significantly depleted at even deeper atmospheric levels ($\leqslant$500 mbar), implying an eddy diffusion coefficient for Saturn of $\sim 10^4 cm^2 sec^{-1}$.

100.012 **Titan's latitudinal temperature distribution and seasonal cycle.**
D. J. Stevenson, B. E. Potter.
Geophys. Res. Lett., Vol. 13, No. 2, p. 93 – 96 (1986).
Voyager IRIS brightness temperature measurements of Titan at a wavelength of 530 cm^{-1} are crudely indicative of ground or lower tropospheric temperatures and indicate 93K for the equator and 91K for both northern and southern high latitudes. The authors show that this near–symmetry can arise naturally in a model where the poles are "pinned" year–round at the dew point of CH_4–N_2 lakes or, more probably, a CH_4–N_2 rich surface layer on a deep ethane–rich ocean. For a polar temperature of 91K, the model implies that the atmosphere contains somewhat less than 8% mole fraction of CH_4.

100.013 **Electromagnetic angular momentum transport in Saturn's rings.**
C. K. Goertz, G. E. Morfill, W.–H. Ip, E. Grün, O. Havnes.
Nature, Vol. 320, No. 6058, p. 141 – 143 (1986).
The observed 'spokes' in Saturn's rings have been interpreted as consisting of elevated, sub–micrometre sized dust particles. The authors show that submicrometre dust particles sporadically elevated above the ring are subject to electromagnetic forces which will reduce their angular momentum inside synchronous orbit and increase it outside. When the dust is reabsorbed by the ring the angular momentum of the ring is decreased (increased)

inside (outside) of synchronous orbit. For the case of the spokes in Saturn's B–ring it is estimated that the timescale for transporting ring material due to this angular momentum coupling effect is comparable to the viscous transport time or even smaller.

100.014 **Plasmatization and recondensation of the saturnian rings.**
W.–H. Ip.
Nature, Vol. 320, No. 6058, p. 143 – 145 (1986).
One major puzzle from the ground–based and Voyager observations of the saturnian system concerns the presence of the E ring and its possible relationship to the icy satellite, Enceladus. Several issues concerning the surface property and interior condition of Enceladus remain unresolved if it is the main supplier of the E–ring particulate matter. The author explores the intriguing alternative that the mass supply and orbital configuration of the E ring may be derived from a mass and momentum coupling between the A ring and the tenuous E ring a process of plasmatization and recondensation of the A–ring icy material.

100.015 **Chaotic rotation of Hyperion?**
R. P. Binzel, J. R. Green, C. B. Opal.
Nature, Vol. 320, No. 6062, p. 511 (1986).
Saturn's satellite Hyperion has been predicted to be in a chaotic rotation state as a result of its irregular shape and tidal interactions. However, an analysis of Voyager 2 images by Thomas et al. indicated a coherent (non–chaotic) rotation period of 13.1 days. The authors report here new observations which were obtained with a higher sampling frequency. These data conclusively show that the 13.1 day period found by Thomas et al. was not due to coherent rotation.

100.016 **Astrometric observations of Saturn's satellites from ESO during the 1981 opposition. Comparison with theory.**
G. Dourneau, C. Veillet, M. R. Dulou, J. F. Le Campion.
Astron. Astrophys., Vol. 160, No. 2, p. 280 – 286 (1986). In French.
First absolute astrometric positions of satellites I–VIII of Saturn obtained from 55 plates taken at ESO with the 1.50 m Danish reflector of the La Silla Observatory, Chile, during the 1981 opposition are presented. Observations have been reduced by means of secondary reference stars. The positions of satellites are presented both in absolute celestial coordinates – with a r.m.s. error of 0".39 in $\alpha \cos \delta$ and 0".29 in δ – and in intersatellite relative coordinates – with a r.m.s. error better than 0".05 as well in $\alpha \cos \delta$ as in δ. Residuals computed using Kozai's theory improved by Rapaport are given.

100.017 **Long–term changes in Saturn's atmospheric belts and zones.**
A. Sanchez–Lavega, E. Battaner.
Astron. Astrophys., Suppl. Ser., Vol. 64, No. 2, p. 287 – 301 (1986).
An analysis of more than 200 photographs of Saturn taken in three spectral ranges from 1890 to 1984, has been performed in order to detect possible seasonal changes in the latitudinal position of the atmospheric belts and in their morphological structure. Annual saturnigraphic latitudes of the belts do not show a systematic trend associated to the seasonal forcing. Structural changes seem to be sporadic and more common to the equatorial belts and zone, a region occupied by a high speed zonal jet. In some occasions they appeared related to activity in form of white clouds (probably of convective origin).

100.018 **Corotation of the Kronian magnetosphere.**
A. Eviatar, J. D. Richardson.
J. Geophys. Res., Vol. 91, No. A3, p. 3299 – 3303 (1986).
Observations of the radial variation of azimuthal plasma velocity in the inner part of the magnetosphere of Saturn show deviations from corotation occur as far in as $L = 4$. Major deviations from rigid corotation occur near the orbits of Rhea and Dione. The authors use Voyager–derived neutral and plasma density and temperature vs. altitude profiles in the upper atmosphere to verify that the atmospheric torque needed to drive corotation through neutral–ion coupling is available. They find that the observed azimuthal velocities and calculated Pedersen conductance are consistent with one another, but that the conductance is much less than previous values estimated for Saturn and Jupiter. The authors attribute the lower than anticipated conductivity to the large heliocentric distance, to the absence of an active satellite (such as Jovian Io) and to a possible role of ring material in depleting the ionosphere.

100.019 **Improvements of the orbital elements of three Saturnian satellites for the opposition of 1971.**
Y. Hatanaka.
Proceedings of the nineteenth symposium on celestial mechanics, p. 35 – 44 (1986). In Japanese. – See Abstr. 012.013.

100.020 **Osservazioni visuali di Saturno nel 1985.**
G. Adamoli.
Astronomia, N. 1, p. 10 – 11 (1986).

100.021 **Saturn, 1984.**
A. W. Heath.
J. Br. Astron. Assoc., Vol. 96, No. 3, p. 170 – 175 (1986).

100.022 **The 1982 – 83 and 1983 – 84 apparitions of the planet Saturn: visual and photographic observations.**
J. L. Benton Jr.
Strolling Astron., Vol. 31, Nos. 7–8, p. 150 – 164 (1986).

100.023 **Visual observations of Saturn in 1984.**
G. Adamoli.
Strolling Astron., Vol. 31, Nos. 7–8, p. 164 – 167 (1986).

100.024 **Observations of Saturn's rings at a wavelength of 350 μm.**
T. L. Roellig, M. W. Werner, J. Estes, E. E. Becklin.
Bull. Am. Astron. Soc., Vol. 17, No. 4, p. 873 (1985). Abstract. – See Abstr. 010.062.

100.025 **An influx of water from Saturn's rings.**
J. E. P. Connerney.
Bull. Am. Astron. Soc., Vol. 17, No. 4, p. 917 (1985). Abstract. – See Abstr. 010.063.

100.026 **Phantoms of Saturn's F ring.**
J. A. Burns, J. N. Cuzzi.
Bull. Am. Astron. Soc., Vol. 17, No. 4, p. 922 (1985). Abstract. – See Abstr. 010.063.

100.027 **Very large impacts on intermediate–sized icy satellites – Saturnian examples and Uranian prospects.**
J. M. Moore, W. B. McKinnon.
Bull. Am. Astron. Soc., Vol. 17, No. 4, p. 922 (1985). Abstract. – See Abstr. 010.063.

100.028 **Origin of Ithaca Chasma, Tethys.**
W. B. McKinnon.
Bull. Am. Astron. Soc., Vol. 17, No. 4, p. 922 (1985). Abstract. – See Abstr. 010.063.

100.029 **CCD photometry of Hyperion: chaotic rotation revisited.**
R. P. Binzel, J. R. Green, C. B. Opal.
Bull. Am. Astron. Soc., Vol. 17, No. 4, p. 924 (1985). Abstract. – See Abstr. 010.063.

100.030 **Interior of Saturn.**
J. I. Lunine.
ESA Spec. Publ., ESA SP–241, p. 3 – 11 (1985). – See Abstr. 012.018.
Models of the interior of Saturn are currently limited by the low accuracy of data on the gravitational moments of the planet and large uncertainties in atmospheric composition. This is in contrast to Jupiter, for which the gravitational moments are better known and thus the behavior of hydrogen and helium at high pressure is a driving uncertainty. The Galileo probe will provide

highly accurate data on the abundances of helium and minor constituents in Jupiter's troposhere. Cassini will accurately determine the tropospheric helium abundance and improve the precision of the gravitational moment data, and hence permit interior models of Saturn of the same resolution as those of Jupiter.

100.031 **The composition of Saturn's atmosphere.**
R. Courtin.
ESA Spec. Publ., ESA SP–241, p. 13 – 20 (1985). – See Abstr. 012.018.

The abundance determinations for the gaseous constituents of Saturn's atmosphere are reviewed. They show that helium is significantly depleted in the outer atmosphere. It appears also that Saturn's atmosphere is enriched in volatiles CH_4, NH_3, PH_3. With the exception of PH_3, little information is available on species considered to be diagnostic of disequilibrium processes (GeH_4, HCN, CO). Determinations of C_2H_2 and C_2H_6 mixing ratios and of their variations with latitude/altitude yield constraints for the modelling of processes in the stratosphere. The deuterium abundance derived from CH_3D lines is close to the value obtained for the protosolar nebula. Implications of Saturn's composition for the origin and evolution of the planet are outlined.

100.032 **Thermal structure of Saturn's atmosphere.**
B. Bézard.
ESA Spec. Publ., ESA SP–241, p. 21 – 31 (1985). – See Abstr. 012.018.

Current knowledge of the thermal structure of Saturn is surveyed. At mid–latitudes, the radiative–convective boundary is located very likely around 700 mbar. At top of the convective zone, the temperature lapse rate indicates incomplete equilibration of ortho and para H_2. Seasonal radiative models give a good representation of the stratospheric part of the temperature profiles and of the large–scale hemispheric asymmetry observed at tropopause levels at the Voyager encounter time. Small–scale meridional gradients imply a decrease of the zonal winds with height above the 300–mbar level. Observational prospects with the Cassini mission are discussed.

100.033 **The meteorology of Saturn.**
G. E. Hunt.
ESA Spec. Publ., ESA SP–241, p. 33 – 40 (1985). – See Abstr. 012.018.

There has been a rapid increase in our knowledge of the atmosphere of Jupiter and Saturn from the recent spacecraft missions, and the corresponding and complementary ground–based telescopic observations. In this paper the author summarises the knowledge of these planetary atmospheres and reviews the current understanding of their meteorologies.

100.034 **Theories for the general circulation of Saturn's atmosphere.**
M. Allison.
ESA Spec. Publ., ESA SP–241, p. 43 – 52 (1985). – See Abstr. 012.018.

The observed circulation of Saturn's cloud–top winds poses a formidable problem for theoretical understanding. According to one class of theories the multiple jet streams are confined to a shallow geostrophic wind layer supported by horizontal gradients in temperature and composition. According to an alternative view the jets extend all the way through the molecular fluid envelope on convectively driven cylinders of motion concentric with the planetary spin axis. Both types of theories require as a starting point some assumption about the buoyancy contrast and eddy fluxes at levels thus far inaccessible to observation. The questions confronted by current theories suggest several important reconnaissance objectives for the proposed Cassini mission to Saturn.

100.035 **Clouds and aerosols on Saturn.**
M. G. Tomasko, L. R. Doose.
ESA Spec. Publ., ESA SP–241, p. 53 – 61 (1985). – See Abstr. 012.018.

Available information on the distribution and optical properties of cloud and aerosol particles in Saturn's atmosphere is summarized.

100.036 **Upper atmosphere and ionosphere of Saturn.**
S. K. Atreya.
ESA Spec. Publ., ESA SP–241, p. 63 – 68 (1985). – See Abstr. 012.018.

The significant aspects of the upper atmosphere and ionosphere of Saturn are summarized with a view toward resolving the outstanding issues from Cassini.

100.037 **Lightning on Saturn.**
D. M. Hunten.
ESA Spec. Publ., ESA SP–241, p. 69 – 71 (1985). – See Abstr. 012.018.

A survey of the existing observational evidence and theoretical ideas about lightning on Saturn is given. The expected water and ice cloud layer in the 15 – 20 bar region is a suitable medium, especially by analogy with Jupiter where lightning has been unequivocally observed. The Saturn electrostatic discharges observed by Voyager are most naturally explained as due to lightning.

100.038 **The composition of the Titan atmosphere.**
D. Gautier.
ESA Spec. Publ., ESA SP–241, p. 75 – 81 (1985). – See Abstr. 012.018.

The present knowledge of the composition of the Titan atmosphere is reviewed. The author especially discusses uncertainties on abundances of major components as well as vertical and horizontal distributions of stratospheric compounds. Implications of the presence of oxygen compounds are considered. The monodeuterated methane is the only detected isotope up to now. The possibility, based on laboratory simulations, of the formation of complex organics in the stratosphere, is also briefly discussed.

100.039 **Titan's surface: nature and implications for Cassini.**
J. I. Lunine.
ESA Spec. Publ., ESA SP–241, p. 83 – 88 (1985). – See Abstr. 012.018.

A model for a mixed ethane–methane–molecular nitrogen surface ocean was recently proposed to reconcile Voyager data on the lower atmosphere of Titan with current understanding of the photolytic destruction of methane in the stratosphere. Such a reservoir would strongly affect the atmosphere and underlying water ice "bedrock" of Titan, and has profound implications for the evolution of Saturn's giant moon. Understanding of the interior and atmosphere of Titan depends upon the capability of Cassini to determine the nature of the surface on a global scale.

100.040 **Thermal structure and meteorology of Titan.**
B. J. Conrath.
ESA Spec. Publ., ESA SP–241, p. 89 – 94 (1985). – See Abstr. 012.018.

Current knowledge of the thermal structure and meteorology of Titan are briefly reviewed. Atmospheric temperature determinations, from Voyager radio occultation and infrared spectroscopy measurements, provide the principal observational constraints on the atmospheric dynamic regime. Titan is the second known example, along with Venus, of cyclostrophic zonal wind balance with upper stratospheric winds a factor of ~ 10 greater than the equatorial surface velocity. Possible contributions of the Cassini Mission to the understanding of Titan's meteorology are discussed.

100.041 **Zonal wind in the stratosphere of Titan.**
C. B. Leovy.
ESA Spec. Publ., ESA SP–241, p. 95 – 98 (1985). – See Abstr. 012.018.

Possible analogies between the zonal wind systems in the stratospheres of Venus and Titan are examined. Venus' super–rotation wind speed is close to the value that maximizes zonal momentum pumping by the semi–diurnal thermal tide. If Titan's zonal wind also maximizes pumping by this tide, the maximum equatorial super–rotation wind speed will be approximately 100 m s^{-1}. On Venus a zonal jet structure is superimposed on the background super–rotation. It is shown that a simple scaling relation describes well the speed and latitude of low latitude jets on planets from Venus to Saturn. Applied to Titan, this scaling predicts a jet speed of 10 – 20 m s^{-1} in the latitude belt 30° – 40° superimposed on the background super–rotation.

100.042 **Clouds and aerosols of Titan's atmosphere.**
R. E. Samuelson.
ESA Spec. Publ., ESA SP–241, p. 99 – 107 (1985). – See Abstr. 012.018.

Titan's cloud structure appears to consist of three basic components: (1) a photochemical aerosol pervading the stratosphere between altitudes of ~40 and 350 km, (2) a mixture of condensed hydrocarbons and nitriles in the lower stratosphere, probably thickest between 50 and 80 km, and (3) methane clouds, thought to exist between 10 and 35 km. Polarimetry, photometry, and infrared spectroscopy provide the basic data for inferring aerosol particle sizes and vertical structure. Properties of the condensates are inferred from infrared spectroscopy and radio occultation data.

100.043 **Phase equilibria in N_2–hydrocarbon systems: applications to Titan.**
W. R. Thompson.
ESA Spec. Publ., ESA SP–241, p. 109 – 119 (1985). – See Abstr. 012.018.

Phase equilibria in the N_2–CH_4–C_2H_6–C_3H_8 system are modeled using an expression for activity coefficients in multicomponent systems. The Titan cloud droplets are liquid CH_4 with 15 – 25% dissolved N_2 (increasing with altitude), up to the (13 km, 0.7 bar, 80°K) level, and solid with 13 – 26% N_2 above. The distribution of CH_4 and N_2 between atmosphere and ocean is computed: $\gtrsim$2.5% CH_4 in the troposphere is necessary to condense any cloud, and implies an ocean composition of $\gtrsim$16 mol % CH_4 and $\lesssim$75 mol % C_2H_6, of depth $\gtrsim$920 m.

100.044 **New observations of D/H in the atmospheres of Saturn and Titan.**
T. Owen, B. L. Lutz, C. de Bergh.
ESA Spec. Publ., ESA SP–241, p. 123 (1985). Abstract. – See Abstr. 012.018.

100.045 **The upper atmosphere of Titan.**
S. J. Bauer.
ESA Spec. Publ., ESA SP–241, p. 125 – 128 (1985). – See Abstr. 012.018.

Voyager 1 observations have determined the predominance of N_2 in Titan's upper atmosphere, its exospheric temperature and its minor species. Electron impact on N_2 within Saturn's magnetosphere appears to be the main source of the airglow emissions and of a yet un–detected ionosphere. Escape of hydrogen from Titan seems to supply the hydrogen torus in the Saturn system, while non–thermal escape of nitrogen can account for some mass loss of Titan's atmosphere since the origin of the solar system.

100.046 **Titan's hydrogen torus.**
W.–H. Ip.
ESA Spec. Publ., ESA SP–241, p. 129 – 141 (1985). – See Abstr. 012.018.

The author reviews the observational results and theoretical models concerning the configuration and dynamics of the Titan hydrogen torus. Emphasis is given on the formation of several other neutral gas clouds (H_2, O, and N) and their physical interrelations. The exosphere of Titan as an interface between the hydrogen torus and the ionosphere/upper atmosphere driving the synthesis of complex nitriles important to prebiotic chemistry is outlined.

100.047 **Photochemistry of Titan.**
D. F. Strobel.
ESA Spec. Publ., ESA SP–241, p. 145 – 148 (1985). – See Abstr. 012.018.

The photochemistry of Titan's atmosphere is reviewed in the light of the scientific findings from the Voyager Mission. The escape rate of hydrogen from Titan is controlled by photochemical production from hydrocarbons. CH_4 is irreversibly converted to less saturated, complex hydrocarbons which accumulate over geologic time on the surface to form an C_2H_6 ocean with dissolved CH_4, N_2, and trace organic molecules and solid C_2H_2 residing at the bottom of the ocean. Challenging scientific problems which the Cassini Mission can address are discussed.

100.048 **Telescopic studies of the satellites of Saturn.**
D. P. Cruikshank.
ESA Spec. Publ., ESA SP–241, p. 149 – 156 (1985). – See Abstr. 012.018.

Decades of telescopic studies and three spacecraft fly–bys have not solved basic problems of the nature, origin, and evolution of the satellites of Saturn. Most of the satellites are icy objects with nearly pure frozen water on their surfaces. Notable exceptions are the dark hemisphere of Iapetus and the captured object Phoebe. Small amounts of non–water contaminants may occur in the surface ices of several of the larger satellites, and Hyperion may contain a large amount of contaminant. Many unresolved questions about the nature of the Saturn satellite system remain for in situ studies by an orbiting spacecraft.

100.049 **Exobiological aspects in view of Titan.**
G. Horneck.
ESA Spec. Publ., ESA SP–241, p. 157 – 159 (1985). – See Abstr. 012.018.

Comparative studies on the planets and moons of our solar system represent one of the tools to reach a more universal understanding of the cosmic history of the biogenic elements and compounds and of the role the cosmic compounds play in the evolution chain towards origin of life. The Cassini mission represents an excellent candidate in this respect. It is of special exobiological interest to understand the organic chemistry processes occurring in Titan's atmosphere and probably on its surface.

100.050 **Chimie organique dans Titan: expériences de simulation en laboratoire et speculations.**
F. Raulin.
ESA Spec. Publ., ESA SP–241, p. 161 – 173 (1985). – See Abstr. 012.018.

100.051 **Ground–based high resolution spectroscopy of Saturn and Titan.**
C. de Bergh.
ESA Spec. Publ., ESA SP–241, p. 195 – 202 (1985). – See Abstr. 012.018.

This paper is devoted to the most recent results of ground–based studies of Saturn and Titan at high spectral resolution. In spite of the restrictions inherent to ground–based observations, a number of very interesting results have been obtained. These results are due not only to the very high quality of the Saturn and Titan spectra which have been recorded but also, in many cases, to the laboratory work which has been done in parallel to help interpreting the planetary data.

100.052 **VLA–observations of Saturn – can we do better from space?**
I. de Pater.
ESA Spec. Publ., ESA SP–241, p. 203 – 208 (1985). – See Abstr. 012.018.

High resolution radio maps of the planet Saturn, obtained with the VLA over the past few years, are discussed. Emphasis is placed upon the shortcomings and inaccuracies of the data, and how the results may be improved by using a small antenna on the SOTP (Saturn Orbiter and Titan Probe).

100.053 **Photographic observations of the Saturn satellites at the Engelhardt Astronomical Observatory in January – May 1980.**
V. Kitkin, I. G. Chugunov.
Izv. Astron. Ehngel'gardt. Obs., No. 48, p. 96 – 100 (1984). In Russian.
A list of 202 positions of 6 Saturn satellites in the system of the AGK3 catalogue is given.

100.054 **The equatorial diameter of Saturn according to measurements of the Engelhardt Astronomical Observatory.**
Yu. A. Nefed'ev.
Astron. Tsirk., No. 1388, p. 8 (1985). In Russian.

100.055 **On the accuracy of a new theory of motion of the Saturn satellites.**
I. G. Chugunov.
Astron. Tsirk., No. 1390, p. 3 – 5 (1985). In Russian.

100.056 **Occurrence of central peak craters on the Saturnian satellites: implications for surface structure.**
K. D. Trego.
Earth, Moon, Planets, Vol. 35, No. 2, p. 117 – 118 (1986).
The presence of craters with central peaks on the ice satellites of Saturn implies that their surface elastic strength is comparable to that of the Moon, Mars, and Mercury which have central peak craters, rather than that of the Jovian ice satellites Ganymede and Callisto which do not have central peak craters.

100.057 **The albedo of Titan.**
G. W. Lockwood, B. L. Lutz, D. T. Thompson, E. S. Bus.
Astrophys. J., Vol. 303, No. 1, p. 511 – 520 (1986).
Photometric observations of Titan since 1972 show a cyclical variation of $\sim 10\%$. A minimum value of brightness and albedo apparently occurred in 1984. Spectrophotometric observations, made annually since 1980 at 8 Å resolution, 3295 – 8880 Å, were used to derive the value $p^* = 0.156 \pm 0.010$ for integrated geometric albedo in 1984. Variations of the equivalent widths of spectral features were not seen.

100.058 **Observations of Saturn satellites made at the Engelhardt Astronomical Observatory and the Zelenchukskaya Astronomical Station in 1982 – 1984.**
V. N. Kitkin.
Kazan. univ. Kazan', 10 pp. (1985). In Russian. Abstr. in Ref. Zh., 51. Astron., 6.51.151 (1986).

100.059 **Outer rings of Saturn.**
J. C. Bhattacharyya, R. Vasundhara.
Curr. Sci., Vol. 54, No. 13, p. 60 – 65 (1985). Abstr. in Phys. Abstr., Vol. 89, No. 1273, Entry 18715 (1986).

100.060 **Results of Titan observations. II. Polarimetry, spectroscopy, radiometry.**
L. R. Lisina.
Photometric and polarimetric investigations of celestial bodies, p. 26 – 37 (1985). In Russian. – See Abstr. 003.035.
The paper reviews the observational results of polarimetry, spectroscopy and radiometry of Titan by ground–based and space facilities.

100.061 **Satellite "wakes" and the orbit of the Encke Gap moonlet.**
M. R. Showalter, J. N. Cuzzi, E. A. Marouf, L. W. Esposito.
Icarus, Vol. 66, No. 2, p. 297 – 323 (1986).
Quasiperiodic optical depth variations have been observed in the Voyager stellar (PPS) and radio occultation profiles near the Encke Gap of Saturn's rings, and have also been detected in one Voyager image. These fluctuations are believed to be the gravitational "wakes" of a moonlet orbiting within the gap. The authors develop a general model for these wakes, and use the results to estimate the moonlet's orbit and mass from the occultation data; this model may have broader applications to planetary rings.

100.062 **Organic material from Phoebe to Iapetus.**
G. Strazzulla.
Icarus, Vol. 66, No. 2, p. 397 – 400 (1986).
It is suggested that only the outermost layers of Phoebe's crust are carbonized by cosmic protons. Meteorite impact on Phoebe leads to the ejection of materials which coming from deeper layers are different in color than the carbonized ones on the surface. The average material coming at Iapetus from Phoebe has then different spectral properties than that on the surface of Phoebe.

100.063 **Saturn's ionosphere: some evidence on equatorial anomaly.**
K. K. Mahajan, J. Kar, M. V. Srilakshmi.
Indian J. Radio Space Phys., Vol. 14, No. 6, p. 143 – 146 (1985). Abstr. in Phys. Abstr., Vol. 89, No. 1281, Entry 67390 (1986).

100.064 **The geomorphology of Rhea: implications for geologic history and surface processes.**
J. M. Moore, V. M. Horner, R. Greeley.
J. Geophys. Res., Vol. 90, Suppl., p. C785 – C795 (1985). – See Abstr. 012.065.
Analyses of landforms on Rhea are used to define three physiographic provinces (cratered terrain 1 undifferentiated, cratered terrain 1 lineated, cratered terrain 2) for which stratigraphic relationships are derived and examined in relation to major impact basins and tectonic features.

100.065 **Satellites of Saturn and Pluto.**
Br. Astron. Assoc. Circ., No. 655 – 656 (1986).

100.066 **Satellites of Saturn and Pluto.**
IAU Circ., No. 4157 (1986).

100.067 **Satellites of Saturn.**
IAU Circ., No. 4178 (1986).

100.068 **Satellites of Saturn and Pluto.**
Yamamoto Circ., No. 2053 (1986). In Japanese.

Imaging Saturn. The Voyager flights to Saturn.
See Abstr. 003.075.

The discovery of the Saturn system: Cassini's life and activity.
See Abstr. 004.068.

Search for chiral molecules and optical activity in extraterrestrial systems. The Cassini mission opportunity.
See Abstr. 015.024.

Theoretical collision–induced rototranslational absorption spectra for modeling Titan's atmosphere: H_2–N_2 pairs.
See Abstr. 022.083.

The reaction $H + C_4H_2$: absolute rate constant measurement and implication for atmospheric modeling of Titan.
See Abstr. 022.095.

The Cassini mission. Complementary observations with the Space Telescope.
See Abstr. 035.029.

Microwave spectroscopy and radiometry of Saturn and Titan.
See Abstr. 035.030.

Neutral mass spectrometry and gas chromatography.
See Abstr. 035.031.

An aerosol collector pyrolyser for the Cassini Titan probe.
See Abstr. 035.032.

Aeronomical measurements by Cassini.
See Abstr. 035.033.

Precise position measurements of Jupiter, Saturn and Uranus systems with the Very Large Array.
See Abstr. 043.011.

Cassini – Titan probe concept.
See Abstr. 051.054.

Cassini – the imaging experiment.
See Abstr. 051.055.

Infrared spectroscopy aboard Cassini.
See Abstr. 051.056.

Stable, baroclinic eddies on Jupiter and Saturn: a laboratory analog and some observational tests.
See Abstr. 062.055.

Accretion discs.
See Abstr. 062.133.

The dynamics of dense particle disks.
See Abstr. 091.020.

Origin of the gap between planetary ring and satellite systems, and the size and structure of the ring particles.
See Abstr. 091.033.

Micrometeoroid impact on planetary satellites as a magnetospheric mass source.
See Abstr. 091.044.

Theory of density waves in narrow planetary rings.
See Abstr. 091.046.

Planètes et petits corps du système solaire. Rapport thématique 1980 – 1984.
See Abstr. 091.049.

Kirkwood gaps.
See Abstr. 098.055.

Minor planets and planetary rings.
See Abstr. 098.076.

Developing high accuracy numerical theories of motion of outer satellites of Jupiter and Saturn.
See Abstr. 099.005.

Monte Carlo simulation of charged particle impact on the satellites of Jupiter and Saturn.
See Abstr. 099.008.

Energetic neutral particles from Jupiter and Saturn.
See Abstr. 099.027.

Jupiter and Saturn's magnetic differential rotation and expected periods for magnetic field reversals.
See Abstr. 099.044.

On gravitational perturbations of resonances.
See Abstr. 099.045.

The abundances of ethane and acetylene in the atmospheres of Jupiter and Saturn.
See Abstr. 099.051.

Motions in the interiors and atmospheres of Jupiter and Saturn. 2. Barotropic instabilities and normal modes of an adiabatic planet.
See Abstr. 099.056.

Episodic volcanism of tidally heated satellites with application to Io.
See Abstr. 099.070.

Detectability of HD and non–equilibrium species in the upper atmospheres of the giant planets from their submillimeter spectrum.
See Abstr. 099.072.

Observational constraints on models for giant planet formation.
See Abstr. 107.015.

Evolution of the giant planets.
See Abstr. 107.016.

Rings and moons: clues to understanding the solar nebula.
See Abstr. 107.017.

Catastrophic disruption in the solar system: asteroid collisional history, origin of Hirayama families and disruption of small satellites.
See Abstr. 107.047.

Strong nonlinear waves in gaseous disks. I. Stationary periodic waves.
See Abstr. 151.041.

101 Uranus, Neptune, Pluto, Transplutonian Planets

101.001 **Voyager 2 swings by Uranus.**
J. Maddox.
Nature, Vol. 319, No. 6049, p. 95 (1986).
The US spacecraft launched towards Jupiter 8.5 years ago is about to encounter Uranus, a planet that has completed only two of its orbits since it was first discovered.

101.002 **Masses of the satellites of Uranus.**
S. F. Dermott, P. D. Nicholson.
Nature, Vol. 319, No. 6049, p. 115 – 120 (1986).
The dynamical theory used to obtain the masses of the uranian satellites from their orbital precession rates contains a fundamental error. The masses can be derived from the precession rates, but a correct analysis of the available data must allow for the fact that the eccentricities, inclinations and precession rates all vary considerably with time, due to mutual secular perturbations.

101.003 **Does Uranus have a magnetic field?**
A. J. Dessler.
Nature, Vol. 319, No. 6050, p. 174 – 175 (1986).

101.004 **Occultation detection of a neptunian ring–like arc.**
W. B. Hubbard, A. Brahic, B. Sicardy, L.–R. Elicer, F. Roques, F. Vilas.
Nature, Vol. 319, No. 6055, p. 636 – 640 (1986).
The apparent closest approach of the star SAO 186001 to Neptune was observed photoelectrically on 22 July 1984 at Cerro Tololo Inter–American Observatory. A 32% signal drop lasting about 1.2 s was probably caused by a partially transparent arc of material at a distance of 67,000 km from Neptune. Neptune's arc(s) do not vary smoothly with azimuth, unlike the rings of other jovian planets.

101.005 **La scoperta di Urano. Voyager 2 all'appuntamento.**
B. Cester.
Orione, Vol. 6, N. 1, p. 24 – 30 (1986).

101.006 **The implication for the presence of a magnetosphere on Uranus in the relationship of EUV and radio emission.**
D. E. Shemansky, G. R. Smith.
Geophys. Res. Lett., Vol. 13, No. 1, p. 2 – 5 (1986).
Two apparent contradictory sets of facts exist concerning the presence of a magnetosphere at Uranus. The Voyager planetary radio astronomy experiment has not detected a Uranus signal at a range <0.7 AU, whereas IUE satellite observations show relatively strong emission indicating the presence of substantial particle excitation of the atmosphere. The character of the EUV emission implies the presence of an ionosphere, and massloading of the extended system comparable to that of Saturn. If the ingredient for production of an active magnetosphere is present, the non detection of radio emission then suggests that Uranus has a very weak or non existent magnetosphere.

101.007 **The ionosphere of Uranus: a myriad of possibilities.**
M. O. Chandler, J. H. Waite Jr.
Geophys. Res. Lett., Vol. 13, No. 1, p. 6 – 9 (1986).
A one–dimensional model has been used to study the effects of exospheric temperature, methane and water influx, ionospheric outflow, and electron precipitation on the composition and structure of the ionosphere of Uranus. The results of the model are discussed in detail.

101.008 **Voyager 2 beim Uranus – erste Ergebnisse.**
H. Mucke.
Sternenbote, 29. Jahrg., Nr. 2, p. 22 – 30 (1986).

101.009 **Predicted satellite plasma tori in the magnetosphere of Uranus.**
A. Eviatar, J. D. Richardson.
Astrophys. J., Lett. Ed., Vol. 300, No. 2, p. L99 – L102 (1986).
The authors have formulated the rate equations for twelve species: water, hydroxyl, molecular and atomic oxygen, molecular and atomic hydrogen, and their respective first ions for conditions expected to hold in the magnetosphere of Uranus. The authors have solved these equations numerically for maximal and minimal source strengths of the five known satellites and have calculated the expected neutral and plasma number densities in the tori predicted to be associated with them.

101.010 **An accurate representation of the motion of Pluto.**
E. Goffin, J. Meeus, C. Steyaert.
Astron. Astrophys., Vol. 155, No. 2, p. 323 – 325 (1986).
The authors present three series of periodic terms, which allow to calculate the heliocentric coordinates of Pluto (longitude, latitude, radius vector) during a time interval of more than two centuries. The terms and coefficients have been derived indirectly by least–square approximation of a numerical integration of the motion of Pluto. For the years 1885 to 2099, the maximum error is $0\farcs5$ in longitude, $0\farcs1$ in latitude, and 0.00002 a.u. in radius vector as compared to the numerical integration.

101.011 **Microwave measurements of the outer planets.**
R. G. Conway, R. J. Davis, S. Padin.
Mon. Not. R. Astron. Soc., Vol. 219, No. 1, p. 31P – 33P (1986).
In 1985 February the flux densities at $\lambda 6$ cm of Uranus and Neptune were measured to be 45.9 ± 2.2 and 14.7 ± 1.4 mJy, corresponding to disc temperatures of 251 and 225K respectively. No emission (<0.8 mJy) was detected from Pluto. The microwave emission from Uranus varies with time, possibly as a seasonal effect. Since 1970 there has been an increase in disc temperature at $\lambda 6$ cm, as at other wavelengths, but between 1978 and 1985 the increase at $\lambda 6$ cm appears to have stopped.

101.012 **On the origin of Triton.**
V. Čelebonović.
Earth, Moon, Planets, Vol. 34, No. 1, p. 59 – 63 (1986).
On dynamical grounds Neptune's satellite Triton is believed to be a captured body. The author presents nondynamical arguments in favour of this idea.

101.013 **New evidence for a ring around Neptune.**
J. Manfroid, R. Haefner, P. Bouchet.
Astron. Astrophys., Vol. 157, No. 1, p. L3 – L5 (1986).
This paper presents an account of observations of the 1984 July 22 appulse of Neptune to SAO 186001. An occultation of 0.8 s duration (FWHM) was detected from La Silla at 05:40:08.6 UTC by two telescopes operating in I and K bands. This, combined with earlier observations led to the recognition of the existence of an irregular ring around Neptune.

101.014 **On the nature of unidentified features in the spectra of Uranus and Neptune.**
M. S. Dement'ev.
Kinematika Fiz. Nebesn. Tel, Tom 2, No. 1, p. 52 – 54 (1986). In Russian. English translation in Kinematics Phys. Celest. Bodies.
Analysis of the spectral observations of Uranus and Neptune ($\lambda\lambda 390 - 840$ nm) shows that the majority of unidentified weak features in the spectra of these planets are in good agreement with the calculated positions for simultaneous transitions in methane and hydrogen molecules.

101.015 **Uranian H Ly–α emission: the interstellar wind source.**
R. V. Yelle, B. R. Sandel.
Geophys. Res. Lett., Vol. 13, No. 2, p. 89 – 92 (1986).
The authors demonstrate that resonant scattering of interstellar wind H Ly–α is an important component of Uranian H Ly–α emissions. For plausible values of the H column abundance this

source contributes 100 to 500 R to the Uranian H Ly–α brightness. This is a significant fraction of the observed signal and is much greater than the solar scattered component.

101.016 **Voyager finds Uranian shepherds and a well–behaved flock of rings.**
R. A. Kerr.
Science, Vol. 231, No. 4740, p. 793 – 796 (1986).

101.017 **Thermal spectrum of Uranus: implications for large helium abundance.**
G. S. Orton.
Science, Vol. 231, No. 4740, p. 836 – 840 (1986).
An analysis of the infrared spectrum of Uranus' disk between 7 micrometers and 3 millimeters suggests a volume mixing ratio for helium in the atmosphere of 40 ± 20 percent, more than for the sun, Jupiter, or Saturn. Alternative explanations require even more extreme assumptions regarding gas abundances or aerosol vertical distribution and spectral properties. The most serious difficulty with a model containing large amounts of helium is devising a credible evolutionary or chemical model explaining the absence or segregation of so much hydrogen.

101.018 **Voyage to a blue planet.**
M. M. Waldrop.
Science, Vol. 231, No. 4741, p. 916 – 918 (1986).
As in its previous encounters with Jupiter and Saturn, Voyager 2 provided an abundance of data at Uranus – and a generous helping of surprise, puzzlement, and wonder.

101.019 **Uranus. Une autre planète bleue.**
J. Manfroid.
Ciel, Vol. 48, p. 164 – 171, 174 – 180 (1986).

101.020 **More dark matter around Uranus and Neptune?**
B. Sicardy, F. Roques, A. Brahic, P. Bouchet, J. P. Maillard, C. Perrier.
Nature, Vol. 320, No. 6064, p. 729 – 731 (1986).
Recent analyses of isolated events observed around Neptune indicate the existence of an incomplete or at least a highly azimuthally variable ring or 'arc' around Neptune. Similar phenomena may be detectable around Uranus. The authors report here an investigation of the surroundings of Uranus and Neptune using stellar occultations observed at the European Southern Observatory on 22 April 1982 and at the Canada France Hawaii Telescope on 15 June 1983, respectively. The high signal–to–noise ratio of these observations sets stringent limits on the presence of matter around these planets.

101.021 **1981N1: a Neptune arc?**
W. B. Hubbard.
Science, Vol. 231, No. 4743, p. 1276 – 1278 (1986).
An object in the vicinity of Neptune detected in 1981 by simultaneous stellar occultation measurements at observatories near Tucson, Arizona, was interpreted as a new Neptune satellite. A reinterpretation suggests that it may have instead been a Neptune arc similar to one observed in 1984. The 1981 object, however, did not occult the star during simultaneous observations at Flagstaff, Arizona. This result constrains possible arc geometries.

101.022 **Voyager 2 à la découverte d'Uranus.**
J. Manfroid.
Ciel Terre, Vol. 102, No. 2, p. 39 – 44 (1986).

101.023 **Voyager 2 op bezoek bij omgevallen planeet.**
G. Schilling, O. Namba.
Zenit, 13. Jaarg., Nr. 1, p. 4 – 13 (1986).

101.024 **Voyager 2 bezocht mysterie planeet Uranus.**
G. Schilling.
Zenit, 13. Jaarg., Nr. 4, p. 124 – 140 (1986).

101.025 **Uranus' last stand.**
J. K. Beatty.
Sky Telesc., Vol. 71, No. 1, p. 11 – 12 (1986).

101.026 **A place called Uranus.**
J. K. Beatty.
Sky Telesc., Vol. 71, No. 4, p. 333 – 337 (1986).

101.027 **Voyager among the ice worlds.**
A. Chaikin.
Sky Telesc., Vol. 71, No. 4, p. 338 – 343 (1986).

101.028 **Planet X and the origins of the shower and steady state flux of short–period comets.**
J. J. Matese, D. P. Whitmire.
Icarus, Vol. 65, No. 1, p. 37 – 50 (1986).
In the planet X model periodic comet showers are associated with passages of the planet's perihelion and aphelion points through a primordial disk of comets believed to lie beyond the orbit of Neptune. The authors find that the absolute numbers of shower and steady state comets are comparable and compatible with the known terrestrial cratering rate, assuming the existence of long–lived extinct comet cores. Canonical planet X model parameters, deduced in part from the scattering dynamics analysis, are: semimajor axis ≈ 80 AU, eccentricity ≈ 0.3, inclination $\approx 45°$, and mass $\approx 5\ m_{\oplus}$. An analysis is given which suggests that planet X, in its present orbit, can create the requisite density gradient of comets near perihelion and aphelion during the lifetime of the Solar System. Independent of the question of periodic comet showers, the existence of planet X and the comet disk can readily explain the origin of the steady state flux of short–period comets over a wide range of parameters.

101.029 **Qu'y a–t–il autour de Neptune?**
J. Manfroid, R. Haefner, P. Bouchet.
Ciel, Vol. 48, p. 212 – 214 (1986).

101.030 **Uranus: the view from Earth.**
R. M. Baum, A. J. Hollis.
J. Br. Astron. Assoc., Vol. 96, No. 2, p. 65 – 67 (1986).

101.031 **Uranus: the view from Voyager.**
P. Moore, N. Henbest.
J. Br. Astron. Assoc., Vol. 96, No. 3, p. 131 – 137 (1986).

101.032 **Accretion heating of Neptune.**
A. A. Jackson IV, J. J. Matese, D. P. Whitmire.
Bull. Am. Astron. Soc., Vol. 17, No. 4, p. 873 (1985). Abstract. – See Abstr. 010.062.

101.033 **CCD astrometry of Uranian moons.**
R. E. Schmidt, D. Pascu, P. K. Seidelmann, E. J. Santoro, J. L. Hershey.
Bull. Am. Astron. Soc., Vol. 17, No. 4, p. 898 (1985). Abstract. – See Abstr. 010.062.

101.034 **Spatial variability of aerosols in the atmosphere of Uranus.**
K. H. Baines, R. J. Terrile, B. A. Smith, K. R. Stapelfeldt.
Bull. Am. Astron. Soc., Vol. 17, No. 4, p. 917 (1985). Abstract. – See Abstr. 010.063.

101.035 **An overview of four years of observations of the H Lyα emission from Uranus and tests for correlations with the solar wind density.**
J. T. Clarke, S. Durrance, J. Murthy, W. Moos, J. Belcher, A. Pradhan, T. Skinner, S. Atreya.
Bull. Am. Astron. Soc., Vol. 17, No. 4, p. 917 – 918 (1985). Abstract. – See Abstr. 010.063.

101.036 **Why no orbital resonances among the satellites of Uranus?**
S. J. Peale.
Bull. Am. Astron. Soc., Vol. 17, No. 4, p. 922 – 923 (1985). Abstract. – See Abstr. 010.063.

101.037 **Are Uranus' rings black because they are in the solar wind?**
S. A. Curtis, N. F. Ness.
Bull. Am. Astron. Soc., Vol. 17, No. 4, p. 923 (1985). Abstract. – See Abstr. 010.063.

101.038 **Re–analysis of the July 1984 star occultation by a Neptunian ringlike arc.**
B. Sicardy, W. B. Hubbard, A. Brahic, F. Roques.
Bull. Am. Astron. Soc., Vol. 17, No. 4, p. 923 (1985). Abstract. – See Abstr. 010.063.

101.039 **The 1985 August 20 Neptune occultation observed from CTIO.**
F. Vilas, W. B. Hubbard.
Bull. Am. Astron. Soc., Vol. 17, No. 4, p. 923 (1985). Abstract. – See Abstr. 010.063.

101.040 **The 20 August 1985 stellar occultation by Neptune and its arc system.**
B. Sicardy, P. Bouchet, A. Brahic, R. McLaren, C. Perrier, F. Roques.
Bull. Am. Astron. Soc., Vol. 17, No. 4, p. 923 – 924 (1985). Abstract. – See Abstr. 010.063.

101.041 **Predicted chemistry of the deep atmosphere of Uranus prior to the Voyager 2 encounter.**
B. Fegley, R. G. Prinn.
Bull. Am. Astron. Soc., Vol. 17, No. 4, p. 924 – 925 (1985). Abstract. – See Abstr. 010.063.

101.042 **On the resonance nature of the Uranus rings determined by its undiscovered satellites.**
N. N. Gor'kavyj, A. M. Fridman.
Astron. Tsirk., No. 1391, p. 1 – 2 (1985). In Russian.

101.043 **On the possibility of detecting some undiscovered Uranus satellites by terrestrial telescopes.**
V. L. Afanas'ev, N. N. Gor'kavyj, M. A. Smirnov, A. M. Fridman.
Astron. Tsirk., No. 1391, p. 3 – 4 (1985). In Russian.

101.044 **Absorption bands in Neptune's spectrum in the λλ4000 – 5000 Å region.**
A. A. Atai, F. N. Kyrkhlarov.
Astron. Tsirk., No. 1391, p. 5 – 7 (1985). In Russian.

101.045 **Radiative–convective equilibrium models of Uranus and Neptune.**
J. F. Appleby.
Icarus, Vol. 65, Nos. 2/3, p. 383 – 405 (1986). – See Abstr. 012.023.

A study of radiative–convective equilibrium models for Uranus and Neptune is presented, with particular emphasis on the stratospheric energy balance, including the influence of aerosol heating and convective penetration. A straightforward numerical method is employed (J. F. Appleby and J. S. Hogan 1984) along with standard opacity formulations and the assumption of local thermodynamic equilibrium.

101.046 **The structure of the Uranian atmosphere: constraints from the geometric albedo spectrum and H_2 and CH_4 line profiles.**
K. H. Baines, J. T. Bergstralh.
Icarus, Vol. 65, Nos. 2/3, p. 406 – 441 (1986). – See Abstr. 012.023.

Constraints on the atmospheric structure of Uranus are derived from recently acquired, high–quality spectral observations. The analysis, based on detailed modeling of a broadband (7 Å) geometric albedo spectrum from 3500 to 10,500 Å, and high–resolution (30 and 100 mÅ, respectively) observations of $H_2$4–0 quadrupole and 6818.9–Å CH_4 features, yields a family of models which parameterizes an upper tropospheric haze layer, a lower optically infinite cloud, the cloud–level methane molar fraction and the mean ortho/para ratio in the visible atmosphere.

101.047 **Estimates of the bolometric albedos and radiation balance of Uranus and Neptune.**
J. B. Pollack, K. Rages, K. H. Baines, J. T. Bergstralh, D. Wenkert, G. E. Danielson.
Icarus, Vol. 65, Nos. 2/3, p. 442 – 466 (1986). – See Abstr. 012.023.

The authors have attempted to bound the wavelength–averaged phase integrals and bolometric albedos of Uranus and Neptune by fitting a wide range of aerosol model atmospheres to their observed geometric albedo spectra. These models are characterized by an upper haze layer of finite optical depth and a lower cloud layer of infinite optical depth at discrete altitudes. Alternative models differ in the assumed value of the particles' single scattering phase function and the wavelength dependence of the haze optical depth. Phase functions ranging from isotropic to those characteristic of particles in the atmospheres of Titan, Jupiter, and Saturn are considered.

101.048 **First direct measurements of the diameters of the large satellites of Uranus and Neptune.**
D. Bonneau, R. Foy.
Astron. Astrophys., Vol. 161, No. 1, p. L12 – L13 (1986).

The authors report the first direct measurements of the diameter of the satellites of Uranus, Ariel, Umbriel, Titania and Oberon, and of the satellite of Neptune, Triton, utilizing the techniques of speckle interferometry at the CFH Telescope. In the case of Uranus, these direct measurements are discussed with respect to the preliminary values from Voyager 2. For Triton, the authors have measured a diameter markedly smaller than indirect values, $2074 < D < 2500$ km depending upon the assumed limb darkening; consequently the Triton albedo is higher: $1.0 > p_v > 0.7$.

101.049 **Voyager 2–Vorbeiflug an Uranus ein großer Erfolg. Erste Ergebnisübersicht.**
H. W. Köhler.
Sterne Weltraum, 25. Jahrg., Nr. 4, p. 196 – 202 (1986).

101.050 **Uranus.**
M. S. Dimitrijević.
Vasiona, Année 33, No. 5, p. 87 – 90 (1985). In Serbian.

101.051 **First news from Uranus.**
M. Klarić.
Vasiona, Année 34, No. 1 – 2, p. 4 – 6 (1986). In Croatian.

101.052 **The rings of Uranus as resonances with unseen satellites.**
N. N. Gor'kavyj, A. M. Fridman.
Sov. Astron. Lett., Vol. 11, No. 5, p. 302 – 303 (1985). English translation of 40.101.006.

101.053 **A forecast for Uranus.**
I. N. James.
Weather, Vol. 41, No. 1, p. 21 – 28 (1986). Abstr. in Phys. Abstr., Vol. 89, No. 1279, Entry 53799 (1986).

101.054 **Formation of the Uranus system: pre–Voyager 2 predictions.**
A. J. R. Prentice.
Bull. Am. Astron. Soc., Vol. 18, No. 1, p. 634 (1986). Abstract. – See Abstr. 010.066.

101.055 **Blaugrüne Murmel [*Uranus*] enthüllt ihre Geheimnisse.**
M. J. Schmidt.
Orion, 44. Jahrg., Nr. 214, p. 80 – 87 (1986). In German and French.

101.056 **On the satellite system of Uranus.**
K. P. Butusov.
Komet. Tsirk., No. 353 (1986). In Russian.

101.057 **Signal–to–noise ratios for possible stellar occultations by Pluto.**
A. S. Bosh, J. L. Elliot, S. E. Kruse, R. L. Baron, E. W. Dunham, L. M. French.
Icarus, Vol. 66, No. 3, p. 556 – 560 (1986).
Signal–to–noise ratios and magnitudes in the Johnson BVR system are presented for nine stars that might be occulted by Pluto during the period 1985 – 1990. From these calculations of the signal–to–noise ratio that could be achieved with a 1–m telescope, the authors find that each star (if occulted) is sufficiently bright to give useful information about a possible atmosphere of Pluto.

101.058 **JHK photometry of Uranus and Neptune occultation candidate stars: 1986 – 1990.**
C. E. Covault, L. M. French.
Icarus, Vol. 66, No. 3, p. 630 – 631 (1986).
JHK photometry is presented for 12 stars to be occulted by Neptune during the period 1986 – 1990, and for 31 stars to be occulted by Uranus during the period 1987 – 1990. Occultation of brightest stars takes place on 8 July 1989 for Neptune and on 16 April 1987, 4 June 1987, and 21 June 1990 for Uranus.

101.059 **Uranus: predicted origin and composition of its atmosphere, moons and rings.**
A. J. R. Prentice.
Phys. Lett. A, Vol. 114A, No. 4, p. 211 – 216 (1986). Abstr. in Phys. Abstr., Vol. 89, No. 1281, Entry 67401 (1986).

101.060 **The collisionless rings of Uranus.**
F. C. Michel.
Geophys. Res. Lett., Vol. 13, No. 5, p. 442 – 443 (1986).
In light of the Voyager 2 mission, the ringlets of Uranus seem unlikely to be accompanied by the swarm of 10 to 18 "shepherding" satellites generally supposed. Consequently, coherent motion (i.e., bulk circulation) in the ringlets again deserves serious consideration. Moreover the radio occultation data show the rings to be composed of larger boulders but few of the small particles one would except if these boulders bumped and ground against one another as they would have to if the motions were incoherent. Recent evidence that Neptune's ringlets may be incomplete arcs further points in this direction.

101.061 **Astrometric observations of Miranda.**
R. Vieira–Martins, C. H. Veiga, D. Lazzaro.
Rev. Mex. Astron. Astrofis., Vol. 12, Spec. Issue, p. 427 – 429 (1986). – See Abstr. 012.066.
Almost 200 plates of Uranus system were obtained during the years of 1982, 83 and 84. They were measured on a Zeiss machine. Miranda positions obtained by an inter–satellite reduction have been compared with theoretical positions.

101.062 **Period analysis of Laplacean resonance amongst Uranian inner satellites.**
D. Lazzaro, R. Vieira–Martins.
Rev. Mex. Astron. Astrofis., Vol. 12, Spec. Issue, p. 430 – 434 (1986). – See Abstr. 012.066.
The study of the inequalities in the mean longitudes of Uranian satellites is used to determine optimized initial conditions for a numerical integration of resonance equations.

101.063 **Pluto.**
IAU Circ., Nos. 4158, 4173 (1986).

101.064 **Satellites of Uranus and Neptune.**
IAU Circ., Nos. 4159, 4164, 4165 (1986).

101.065 **Satellites and rings of Uranus.**
IAU Circ., No. 4168 (1986).

101.066 **Satellites and rings of Uranus.**
Yamamoto Circ., Nos. 2052, 2053, 2054 (1986). In Japanese.

Ruder Bošković and the discovery of Uranus.
See Abstr. 004.079.

The discovery of Pluto. Some generally unknown aspects of the story.
See Abstr. 004.097.

The double tilt of Uranus.
See Abstr. 011.030.

Image processing techniques for position measurements of Uranus' satellites.
See Abstr. 036.210.

Precise position measurements of Jupiter, Saturn and Uranus systems with the Very Large Array.
See Abstr. 043.011.

Towards Neptune.
See Abstr. 051.072.

Micrometeoroid impact on planetary satellites as a magnetospheric mass source.
See Abstr. 091.044.

Theory of density waves in narrow planetary rings.
See Abstr. 091.046.

***JHK* photometry of candidate stars for occultation by Neptune.**
See Abstr. 096.001.

Occultations by Pluto and its satellite.
See Abstr. 096.014.

Planetary occultations.
See Abstr. 096.015.

Minor planets and planetary rings.
See Abstr. 098.076.

Photographic positions of comets, minor planets and Pluto observed during 1975 – 1980.
See Abstr. 098.105.

Supersonic meteorology of Io: sublimation–driven flow of SO_2.
See Abstr. 099.012.

Detectability of HD and non–equilibrium species in the upper atmospheres of the giant planets from their submillimeter spectrum.
See Abstr. 099.072.

Very large impacts on intermediate–sized icy satellites – Saturnian examples and Uranian prospects.
See Abstr. 100.027.

Satellites of Saturn and Pluto.
See Abstr. 100.065.

Satellites of Saturn and Pluto.
See Abstr. 100.066.

Satellites of Saturn and Pluto.
See Abstr. 100.068.

Observational constraints on models for giant planet formation.
See Abstr. 107.015.

Rings and moons: clues to understanding the solar nebula.
See Abstr. 107.017.

102 Comets (Origin, Structure, Atmospheres, Dynamics)

102.001 **Photochemical heating of cometary comae: the effect of icy water droplet recondensation.**
W.-H. Ip.
Astrophys. J., Vol. 300, No. 1, p. 456 – 460 (1986).

The recent suggestion of recondensation of icy water droplets in the vicinity of cometary nuclei is followed up by investigation of the thermodynamics and hydrodynamics in the inner comae. It is found that, if the IR radiative cooling effect is negligible, the heating effect due to resublimation of H_2O molecules from the icy droplets could maintain the gas temperature ($T \approx 75K$) relatively higher than the values previously estimated ($T \approx 10K$) in the inner coma. Furthermore, the mass-loading effect of the H_2O gas injected in the icy droplet zone would produce a minimum in the radial velocity. This could have very significant effects on the atmospheric expansion of cometary comae.

102.002 **The mean energy transfer rate to comets in the Oort cloud and implications for cometary origins.**
M. E. Bailey.
Mon. Not. R. Astron. Soc., Vol. 218, No. 1, p. 1 – 30 (1986).

The mean energy transfer rate to comets in the Oort cloud is calculated including both stellar and giant molecular cloud perturbations. For representative values of the parameters the clouds dominate the evolution for semi-major axes $a \gtrsim 2 \times 10^4$AU, while stars dominate if $a \lesssim 2 \times 10^4$AU. The resulting unbinding time-scales and half-lives are calculated and compared with the results of Bahcall et al. and Hut & Tremaine. Taking the stars and clouds together the author estimates the critical semi-major axis $a_{1/2}$, beyond which the majority of primordial comets are removed during the age of the solar system, is of order 10^4AU. The implications of this work for the existence of the hypothetical solar companion star Nemesis suggests that in its present form the idea should be rejected. Finally the evolution of the Oort cloud and implications for the origins of comets are considered.

102.003 **The structure of comet tails.**
J. C. Brandt, M. B. Niedner Jr.
Sci. Am., Vol. 254, No. 1, p. 38 – 46 (1986).

The plasma tail forms and disconnects from the comet in response to the solar wind and its magnetic field. Observations of comets Giacobini-Zinner and Halley may help to clarify such phenomena.

102.004 **Near-tail reconnection as the cause of cometary tail disconnections.**
C. T. Russell, M. A. Saunders, J. L. Phillips, J. A. Fedder.
J. Geophys. Res., Vol. 91, No. A2, p. 1417 – 1423 (1986).

In a cometary tail disconnection event the plasma tail appears to separate from the coma and to accelerate away from it. As this occurs a new tail begins to form. The authors propose that these disconnections arise in a manner analogous to geomagnetic substorms, i.e., by the formation of a strongly reconnecting region in the near tail that forms a magnetic island in the coma and ejects the plasma tail by strengthening the magnetic "slingshot" within the tail. This reconnection process may be triggered by several different processes, such as interplanetary shocks or variations in the Alfvén Mach number.

102.005 **Kometenkerne.**
H. Kohl.
Sterne Weltraum, 25. Jahrg., Nr. 1, p. 10 – 14 (1986).

102.006 **The relative importance of optical and chemical pumping of cometary OH.**
G. Piehler, W. H. Kegel.
Astron. Astrophys., Vol. 155, No. 2, p. L13 – L16 (1986).

Recently, Andresen (1985) suggested that the relative population of the levels of cometary OH molecules may be dominated by the process of photodissociation of H_2O molecules, a process which by itself may lead to population inversion in the rotational ground state. The authors studied optical pumping in the UV lines and photochemical reactions. According to their results the occupation numbers are changed by the photochemical reactions under stationary conditions by less than a few per cent. They also calculated for comet Halley the intensity ratios in the FIR doublets (purely rotational transitions), the measurement of which would be a direct and decisive test of the suggestion of Andresen.

102.007 **Thermal conduction in cometary nuclei and its effect on cometary activity.**
S. Yabushita, N. Hatta.
Mon. Not. R. Astron. Soc., Vol. 219, No. 1, p. 197 – 202 (1986).

An investigation is made of heat conduction in a cometary nucleus as it orbits about the Sun. Even if the nucleus is made of crystalline ice which has a higher conductivity than amorphous ice, the surface temperature can penetrate only up to a few tens of metres below the surface. The difference between the surface temperature which a comet without heat conduction would have and the surface temperature allowing for heat conduction is less than 20K. The cometary nucleus with heat conduction gives asymmetry in the heat available for sublimation before and after perihelion, and the result obtained here is in accord with the observed asymmetry in the coma brightness of P/Halley at the 1910 appearance.

102.008 **The origin of comets: implications for planetary formation.**
P. R. Weissman.
Protostars & planets II, p. 895 – 919 (1985). – See Abstr. 012.008.

Theories of cometary origin fall into two groups: primordial theories in which the comets formed concurrently with the Sun and planets and have been stored in the Oort cloud over the history of the solar system, and episodic theories in which comets either formed in distant interstellar clouds and were subsequently captured by the solar system or comets formed as a result of catastrophic, recent, and possibly repeated events in the planetary region. A brief description of the Oort cloud based on observational evidence from long-period comets and dynamical modeling through computer simulation techniques is presented in this paper. Then, both the primordial and episodic theories of cometary origin are discussed.

102.009 **Model of cometary atmosphere – solar wind interaction.**
V. B. Baranov, N. A. Zajtsev, M. G. Lebedev.
Astron. Zh., Tom 63, Vyp. 1, p. 170 – 180 (1986). In Russian. English translation in Sov. Astron., Vol. 30, No. 1.

A mathematical model of solar wind – cometary atmosphere interaction containing three strong discontinuity surfaces defined from the solutions (bow shock in the solar wind, inner shock in the comet's gas and contact discontinuity) is formulated. Effects of solar wind "mass-loading" by means of photoionization of cometary neutrals and resonance charge exchange of cometary ions are considered. The domination of these effects for many comets is shown. A restricted problem of flow of cometary gas and solar wind around a given contact surface was solved.

102.010 **New applications of Tisserand's criterion.**
V. V. Radzievskij, V. P. Tomanov.
Astron. Zh., Tom 63, Vyp. 1, p. 198 – 200 (1986). In Russian. English translation in Sov. Astron., Vol. 30, No. 1.

The radius of the planetary orbit that caused perturbations of 76 short-period comets is calculated according to Radzievkij's formula. It is shown that Jupiter has the dominating influence on these cometary perturbations. The possibility of discovering some non-gravitational effects with the help of the formula is discussed.

102.011 **The isotopes of CN in comets.**
J. M. Zucconi, M. C. Festou.
Astron. Astrophys., Vol. 158, No. 1/2, p. 382 – 388 (1986).

The synthetic spectrum of CN isotope violet bands is computed: it appears that several lines of ^{13}CN and $C^{15}N$ are potentially detectable in spectra obtained with a spectral resolution of 0.1 Å or better. The fluorescence efficiencies of those lines is given as a function of both the radial velocity and the heliocentric distance of the comet. The Swings effect is shown to enhance the intensity of the isotope lines relative to that of the most common lines at specific radial velocities and hence to favour their observation.

102.012 **On the theory of the "spotty" model of a cometary nucleus.**
M. Z. Markovich.
Kinematika Fiz. Nebesn. Tel, Tom 2, No. 1, p. 70 – 76 (1986). In Russian. English translation in Kinematics Phys. Celest. Bodies.

It is shown that the photometric parameter may be expressed as $n = n_1 + n_2$, where n_1 characterizes the variation of gas emission intensity from a free ice surface of the nucleus, and n_2 is the change of the area of this surface with heliocentric distance. The expressions connecting the values n_1 and n_2 with the physical and geometrical charactristics of ice and dust components of the nucleus have been obtained.

102.013 **Are cometary nuclei primordial rubble piles?**
P. R. Weissman.
Nature, Vol. 320, No. 6059, p. 242 – 244 (1986).

Whipple's icy conglomerate model for the cometary nucleus has had considerable success in explaining a variety of cometary phenomena such as gas production rates and nongravitational forces. However, as discussed here, both observational evidence and theoretical considerations suggest that the cometary nucleus may not be a well–consolidated single body, but may instead be a loosely bound agglomeration of smaller fragments, weakly bonded and subject to occasional or even frequent disruptive events. The proposed model is analogous to the "rubble pile" model suggested for the larger main–belt asteroids, although the larger cometary fragments are expected to be primordial condensations rather than collisionally derived debris as in the asteroid case.

102.014 **Size distribution of dust inclusions and the secular brightness decrease of periodic comets.**
O. V. Dobrovol'skij.
Kinematika Fiz. Nebesn. Tel, Tom 2, No. 2, p. 35 – 38 (1986). In Russian. English translation in Kinematics Phys. Celest. Bodies.

The known bifurcated shape of the secular brightness decrease – perihelion distance relation is interpreted as due to a division of comets into two separate groups with different size distribution of dust inclusions.

102.015 **Physical characteristics of comets 1980.**
D. A. Andrienko, A. V. Karpenko.
Kinematika Fiz. Nebesn. Tel, Tom 2, No. 2, p. 39 – 44 (1986). In Russian. English translation in Kinematics Phys. Celest. Bodies.

Orbital elements, description and information on 15 comets in 1980 are given. The absolute magnitudes and other physical characteristics are determined.

102.016 **On the halo of icy grains as a possible source of brightness outbursts of comets. Mathematical model of an outburst.**
D. A. Andrienko, I. I. Mishchishina.
Astron. Zh., Tom 63, Vyp. 2, p. 335 – 348 (1986). In Russian. English translation in Sov. Astron., Vol. 30, No. 2.

A mathematical model is proposed of the mechanism of brightness outbursts of comets based on the process of fragmentation of unstable icy bodies contained in their atmosphere under the effect of high–velocity corpuscular stream particles of the solar wind. The model accounts well for the phenomena observed during cometary brightness outbursts.

102.017 **CO^+ fluorescence in comets.**
L. Magnani, M. F. A'Hearn.
Astrophys. J., Vol. 302, No. 1, p. 477 – 487 (1986).

Calculations of the fluorescent equilibrium of the CO^+ molecule to determine the Swings and Greenstein effects as a function of heliocentric radial velocity have been carried out. The previously ignored Baldet–Johnson (*B–A*) bands were included in addition to the comet tail (*A–X*) and first negative (*B–X*) bands, thus allowing a determination of the absolute intensities of all bands. The flux variations for velocities in the range –50 to +350 km s^{-1} have been calculated and may be used to map the acceleration of CO^+ ions into the tail, even with low–dispersion spectra.

102.018 **Cometary nuclei and asteroids: has a link been found?**
W. K. Hartmann.
Asteroids, comets, meteors II, p. 191 – 193 (1986). – See Abstr. 012.010.

New observations of comets by several groups show that comet dust materials are very low in albedo, and typically reddish in color from 0.4 to 2.6 μm. The results are consistent with the view that comet nuclei are more like Trojan asteroids than like other asteroid spectral classes. Outer solar system planetesimals, including comets and Trojans may be colored by a range of types of reddish–black carbonaceous dust containing reddish organic compounds.

102.019 **The Oort cloud in transition.**
P. R. Weissman.
Asteroids, comets, meteors II, p. 197 – 206 (1986). – See Abstr. 012.010.

The outer Oort cloud which is actively being perturbed by random stars, interstellar clouds, and the galactic field, extends from about 10^4 to 10^5AU and contains some $2x10^{12}$ comets with a mass of 7 to 8 Earth masses. Recent studies have suggested the existence of a massive inner cloud of up to 10^{14} comets, extending from just beyond the orbit of Neptune to 10^4AU or more. The inner cloud serves as a replenishment source for the outer cloud as comets there are stripped away by stellar and GMC perturbations. The possible existence of Oort clouds around other stars in the solar neighborhood has been shown by the IRAS detection of dust shells around many main–sequence stars, and the visible light photograph of the accretion disc around Beta Pictoris.

102.020 **A note on the mean energy transfer rate by point–mass perturbers.**
M. E. Bailey.
Asteroids, comets, meteors II, p. 207 – 210 (1986). – See Abstr. 012.010.

An accurate expression for the mean point–mass energy transfer rate to comets in the Oort Cloud is evaluated from a simple analytic approximation to the size of the mean relative velocity change as a function of the cometary semi–major axis and impact parameter of the perturber.

102.021 **Fluffy comets.**
J. M. Greenberg.
Asteroids, comets, meteors II, p. 221 – 223 (1986). – See Abstr. 012.010.

Comparison of the densities of aggregates of interstellar dust with meteors implies that the original comet nucleus is a very loosely packed birdsnest–like structure with probably 60% of its volume being empty space and even more if meteors have densities less than 0.2 g cm^{-3}. The derived mean comet density is less than about 0.5 g cm^{-3}.

102.022 **Interstellar S_2 in comets.**
J. M. Greenberg, R. Grim, L. van Ijzendoorn.
Asteroids, comets, meteors II, p. 225 – 227 (1986). – See Abstr. 012.010.

The presence of S_2 in comet nuclei is shown to imply that: (1) Comets are formed directly from interstellar dust; (2) The interstellar dust probably coagulated in a region whose temperature was <15K and the grains collided at speeds which did not raise the grains to temperatures $\gtrsim 25$K; (3) Internal heating of the comet by the Sun must remain low even after the comet has entered its periodic phase. The low internal heating implies a very low conductivity characteristic of a fluffy or low density porous nucleus.

102.023 **Elemental, isotopic and molecular abundances in comets.**
A. H. Delsemme.
Asteroids, comets, meteors II, p. 245 – 257 (1986). – See Abstr. 012.010.

The cometary nucleus is an icy conglomerate made up of a volatile and of a refractory fraction. The refractory–to–volatile mass ratio is probably in the general range of 0.5 to 1.0 for pristine material. The refractory dust is quasi–chondritic in nature. The volatile fraction contains mostly molecules made up of H, C, N, O, S atoms, where N, O and S are close to cosmic abundances, whereas H is depleted by a factor of 1000 and C by a factor of 4. The only isotope ratio ever measured in comets is the C^{12}/C^{13}. A balance sheet of the atomic versus molecular abundances suggests that fully hydrogenated compounds like methane and ammonia cannot exist in large quantities, whereas oxides or aldehydes like CO, CO_2 and H_2CO may be somewhat more abundant.

102.024 **Modelling the excitation conditions of the water molecule in comets.**
D. Bockelée–Morvan, J. Crovisier.
Asteroids, comets, meteors II, p. 279 – 282 (1986). – See Abstr. 012.010.

Strong evidences indicate that the water molecule is the dominant volatile constituent of cometary nuclei. Since H_2O has no spectral signature at visible wavelengths it should be searched for in the medium infrared or in the microwave spectral region. The authors present here an attempt to model the excitation of H_2O including radiation trapping of the rotational lines, collisional excitation, and infrared vibrational pumping by the solar flux. From this model they determine which H_2O rotational and vibrational transitions should be observable in the near future and estimate the coma cooling rate by rotational emission of the water molecule.

102.025 **The fluorescence spectrum of the CN radical in comets.**
J. M. Zucconi, M. C. Festou.
Asteroids, comets, meteors II, p. 293 – 297 (1986). – See Abstr. 012.010.

It is shown that the CN spectrum in comets can be explained by the fluorescence of solar light. Although the constants of the CN radical have been measured in the laboratory, it is shown that the study of the rotational structure of the violet $\Delta\upsilon = 0$ sequence offers an alternate solution for evaluating some of them, namely the lifetime of the A–state and the vibrational transition rate $A^{VIB}1 - 0$. The lifetime of the A–state is then measured to be twice as large as the experimental value. As a by–product of this study, the Greenstein effect is quantitatively studied.

102.026 **The derivation of OH gas production rates from visual magnitudes of comets.**
M. C. Festou.
Asteroids, comets, meteors II, p. 299 – 303 (1986). – See Abstr. 012.010.

The analysis of IUE observations of the A–X transition of OH at 3064 Å allowed the author to derive a self–consistent set of OH production rates for more than 20 comets observed in the 0.7 – 1.75 A.U. range of heliocentric distances. If one assumes that the productions of C_2 and OH radicals are in a constant ratio whatever is the heliocentric distance and that they do not vary from comet to comet, then, heliocentric magnitudes and OH production rates should be strongly correlated. This is shown to hold over a very large range of values for those two parameters. Very important and straightforward consequences of that empirical law are presented.

102.027 **Brightness curve and porosity of cometary nuclei.**
R. Smoluchowski.
Asteroids, comets, meteors II, p. 305 – 315 (1986). – See Abstr. 012.010.

The brightness curve of a comet is analysed in terms of porosity and nature of volatiles in its nucleus. Mobility of pores, densification, phase changes, differential thermal expansion and heat flow into and out of the nucleus are discussed leading to a method for estimating the effective heat conductivity and porosity of the nucleus. The method is applied to comets P/Giacobini–Zinner, P/Halley, P/Kopff, P/Encke, P/Tempel 2 and P/d'Arrest.

102.028 **Dust coma onset at large heliocentric distances.**
M. Šolc, V. Vanýsek, M. Wolf.
Asteroids, comets, meteors II, p. 327 – 330 (1986). – See Abstr. 012.010.

The photometry of comets at $r_H > 2.5$ AU indicates the presence of a diffuse coma that consists of solid particles. The expansion energy of the coma may be released by sublimating volatiles (CO_2) concentrated in "basins" on the cometary nucleus surface below a thin dust/ice crust. The elevation and following disintegration of the crust due to vapour pressure below it may lead to a brightness burst as shown in P/Halley at $\cong 8$ AU.

102.029 **Chemical differentiation of the cometary nucleus.**
H. L. F. Houpis, D. A. Mendis, W.–H. Ip.
Asteroids, comets, meteors II, p. 341 – 343 (1986). – See Abstr. 012.010.

Recently the authors discussed the possibility that a cometary nucleus is comprised of three components: dust, clathrate and a separate volatile ice (e.g. CO_2) (1985). Here they present the latest improvements in their model. This includes the comet latitudinal variation of gas and dust emission, the effects of a nuclear spin axis tilt and energy and momentum equations which more properly account for gas diffusion and temperature accomodation through the outer mantles of the nucleus.

102.030 **Comet modeling.**
F. P. Fanale, J. R. Salvail.
Asteroids, comets, meteors II, p. 345 – 348 (1986). – See Abstr. 012.010.

Results of three idealized comet models suggest the following: (1) Along with orbital parameters, rotation rate, spin axis orientation and heat conduction into the nucleus are all critical factors in controlling H_2O flux history, dust flux history and mantle evolution. (2) Particles ejected from the light side of high obliquity comets rarely contribute to mantle formation on the dark side, and (3) for the previous apparition of Halley's comet, the magnitude of the observed postperihelion brightness asymmetry is compatible with a fairly high value for the thermal inertia of the outer nucleus and the radius of Halley's comet is between 1.9 and 2.8 km.

102.031 **Gas production rates from pure CO_2, CO and NH_3 ice nuclei.**
M. Wolf, M. Šolc, V. Vanýsek, P. Žáček.
Asteroids, comets, meteors II, p. 355 – 357 (1986). – See Abstr. 012.010.

An energy balance at the chemically homogeneous surface of a fast rotating spherical cometary nucleus is calculated in order to obtain the production rates for three types of highly volatile species. Resulting values are plotted together with the maximum sizes of particles that can be elevated by the gas flow from the surface.

102.032 **Diatoms in comets.**
R. Hoover, F. Hoyle, M. K. Wallis, N. C. Wickramasinghe.
Asteroids, comets, meteors II, p. 359 – 362 (1986). – See Abstr. 012.010.

The fossil record of the microscopic algae classified as diatoms suggests they were injected to Earth at the Cretaceous boundary. Not only could diatoms remain viable in the cometary environment, but also many species might replicate in illuminated surface layers or early interior layers of cometary ice. Presumably they reached the solar system on an interstellar comet as an already-evolved assemblage of organisms. Diatoms might cause colour changes to comet nuclei while their outgassing decays and revives around highly elliptical orbits. Just as for interstellar absorption, high resolution IR observations are capable of distinguishing whether the 10 μm feature arises from siliceous diatom material or mineral silicates. The 10 – 30 μm band and the UV 2200 Å region can also provide evidence of biological material.

102.033 **Photolytic heating of the H_2O coma.**
M. K. Wallis.
Asteroids, comets, meteors II, p. 363 – 368 (1986). – See Abstr. 012.010.

Energetic H–atoms of about 2 eV released in H_2O–photodissociation are potentially a strong heat source. Other dissociative and ionization processes contribute approximately a further 1 eV per H_2O molecule. The heat conversion is inefficient because vibrational and rovibrational H_2O modes are excited. However, optically thick effects with collisional reabsorption ensure 90% conversion inside 100 km, decreasing to 60 – 80% efficiency by 1000 km radius. Solar IR heating and radiative cooling rates can also have similar magnitudes, but overall, photolytic–type heating at a roughly constant rate per unit mass would overcome adiabatic cooling in $\lesssim 1000$ km in the expanding coma.

102.034 **IR heating of the cometary atmosphere.**
M. L. Marconi, D. A. Mendis.
Asteroids, comets, meteors II, p. 369 – 372 (1986). – See Abstr. 012.010.

In order to investigate the transfer of radiation (solar and dust IR) in the cometary atmosphere a 6–band radiative transfer model was developed. Five bands are chosen to coincide with the strongest IR bands of H_2O, while the sixth corresponds to the bulk of solar energy, to which the H_2O is largely transparent. The mean radiation field for each band was self–consistently calculated as a function of the cometo–centric distance using the three–stream approximation (Wilson and Sen, 1980; see Marconi and Mendis, 1984).

102.035 **A cometary dust environment model for the generation of synthetic images.**
L. Massonne, J. Fertig, E. Grün, G. Schwehm.
Asteroids, comets, meteors II, p. 407 – 410 (1986). – See Abstr. 012.010.

In order to interpret the evolution of the cometary dust coma and its structure, a new fast method of generating dust environment models was developed. This method facilitates the computer generation of synthetic images of the coma. The number density of dust is derived as a function of particle size, time and position in a nucleus fixed coordinate system.

102.036 **Non–stationary solar wind – comet interaction.**
T. Roatsch, K. Sauer, K. Baumgärtel.
Asteroids, comets, meteors II, p. 425 – 427 (1986). – See Abstr. 012.010.

The interaction of the solar wind with cometary neutral gas is studied on the basis of the non–stationary gas–dynamic model of Biermann et al. (1967) without an a priori assumption of any discontinuity. The build–up of a weak shock front is confirmed and a reacceleration of the plasma flow after leaving the sunward side of the comet is indicated.

102.037 **Decay and disappearence of the cometary bow shock.**
M. K. Wallis, M. Dryer.
Asteroids, comets, meteors II, p. 429 – 432 (1986). – See Abstr. 012.010.

While the outer coma primarily adds mass in the form of cometary ions to the supersonic solar wind, the intermediate coma by contrast tends to soak up momentum and energy of accelerated cometary ions via charge exchange reactions. The corresponding contraction of the flow would generate expansion waves, which cause weakening and even complete decay of the flanks of the cometary bow shock.

102.038 **The UV bright spot of water vapor in comets.**
J. L. Bertaux.
Astron. Astrophys., Vol. 160, No. 2, p. L7 – L10 (1986).

Though water ice is thought to be a major component of a cometary nucleus, water vapor is quite difficult to identify directly in the gaseous coma. It is suggested that the cloud of H_2O evaporating from the nucleus could reveal itself through an emissive photodissociation mechanism, in which solar Lyman α photons would dissociate H_2O molecules and create a number of OH radicals in the excited state $A^2\Sigma^+$, immediately de–excited to $X^2\pi$ state by spontaneous emission around 310 nm of wavelength. Within a sphere of radius r = 33 km, this emission should be brighter than direct resonance fluorescence of OH, allowing to map directly the H_2O distribution.

102.039 **On the structure of the Oort cometary cloud and possibilities of its observation in the far infrared region.**
L. S. Marochnik, G. B. Sholomitskij.
Inst. kosm. issled. Akad. Nauk SSSR. Prepr., No. 942, 60 pp. (1985). In Russian. Abstr. in Ref. Zh., 51. Astron., 1.51.167 (1986).

102.040 **Cometary particles: thin sectioning and electron beam analysis.**
J. P. Bradley, D. E. Brownlee.
Science, Vol. 231, No. 4745, p. 1542 – 1544 (1986).

Thin sections (500 to 1000 angstroms thick) of individual micrometeorites (5 to 15 micrometers) have been prepared with ultramicrotome equipped with a diamond knife. Electron microscopic examination of these sections has revealed the internal structures of chondritic micrometeorites, and a subset of highly porous, fragile particles has been identified. Delicate meteoritic materials such as these are characteristic of debris from cometary meteors.

102.041 **Tidal gravitational forces: the infall of "new" comets and comet showers.**
D. E. Morris, R. A. Muller.
Icarus, Vol. 65, No. 1, p. 1 – 12 (1986).

The tidal gravitational field of the Galaxy changes the angular momentum of comets in the Oort cloud. For comet orbits with semimajor axis greater than 2×10^4AU, the change is sufficient to bring comets from the Oort cloud into the visible region, causing the infall of "new" comets. The production of intense comet showers by the tidal field of a solar companion or of an interstellar cloud is considered. It is shown that the direction of a solar companion cannot be found from the present distribution of observable comets. The tidal fields of interstellar clouds are not strong enough to cause comet showers of sufficient intensity to result in Earth impacts.

102.042 **The influence of the Galactic tidal field on the Oort comet cloud.**
J. Heisler, S. Tremaine.
Icarus, Vol. 65, No. 1, p. 13 – 26 (1986).

The authors examine the effect of the Galactic disk potential on the Oort cloud. The tidal force from the disk produces a slow drift of a comet's orbital angular momentum, *J*. If *J* is reduced below a critical value, J_*, then the comet enters the planetary system. Once this occurs the comet is lost from the Oort cloud, either to an escape orbit or a much more tightly bound orbit, within a few orbital periods. The authors determine the rate of orbit evolution due to the Galactic tide by analyzing the orbit averaged Hamiltonian. They calculate the loss rate due to the Galactic tide and compare this with the loss rate due to stellar encounters. The Galactic tide appears to be the dominant cause of orbital evolution in the Oort cloud.

102.043 **Effects of the interaction between plasma and neutrals on the stability of the cometary ionopause.**
A. I. Ershkovich, D. A. Mendis.
Astrophys. J., Vol. 302, No. 2, p. 849 – 852 (1986).

Plasma in the cometary ionosphere is collisionally coupled to neutrals that flow out from the nucleus. The authors consider the effects of this coupling on the stability of the cometary ionopause. It is shown that while this coupling causes damping of waves that are excited in the cometary ionopause, it alone cannot quench the Kelvin–Helmholtz (K–H) instability. Despite the plasma–neutral drag, the entire cometary ionopause can be subjected to the K–H instability, which might be responsible for the penetration of the interplanetary magnetic field into the cometary ionosphere. Furthermore, it is found that the neutrals tend to destabilize rather than stabilize the ionopause.

102.044 **Ionization state of plasmoids generated by collisions of dust particles, and the ion charge of plasma of cometary atmospheres.**
S. Ibadov.
Dokl. Akad. Nauk TadzhSSR, Tom 28, No. 7, p. 389 – 392 (1985). In Russian. Abstr. in Ref. Zh., 51. Astron., 3.51.289 (1986).

102.045 **Experimental modelling of cometary phenomena and criteria of similarity.**
E. A. Kajmakov, Yu. I. Svetov.
Astron. Vestn., Tom 20, No. 2, p. 112 – 115 (1986). In Russian. English translation in Sol. Syst. Res.
The modelling of processes occurring on the surfaces of cometary nuclei and the relation of experimental data with similarity criteria are discussed. A new method of complex modelling of cometary phenomena in laboratory and in space based on formation of a cloud composed by micron–size ice particles is described.

102.046 **Multi–charged ions in comets.**
S. Ibadov.
Astron. Tsirk., No. 1384, p. 1 – 3 (1985). In Russian.

102.047 **Ionization of cometary atmospheres by the zodiacal dust cloud.**
S. Ibadov.
Astron. Tsirk., No. 1402, p. 3 – 5 (1985). In Russian.

102.048 **The dynamical evolution of an initial ensemble of cometary bodies in the solar system.**
F. A. Tsitsin, A. S. Rastorguev, V. M. Chepurova.
Astron. Tsirk., No. 1408, p. 5 – 8 (1985). In Russian.
The dynamical evolution of a ring of cometary bodies from an outer part of a protoplanetary cloud to the present relict comet belt between planets is examined. A general course of this evolution is described as a cosmogonic long dissipation in the Galaxy.

102.049 **Surface brightness of a cometary dust tail. I. Calculation technique.**
G. F. Chernyj, Yu. V. Sizonenko.
Kinematika Fiz. Nebesn. Tel, Tom 2, No. 3, p. 52 – 58 (1986). In Russian. English translation in Kinematics Phys. Celest. Bodies.
A calculation technique of theoretical brightness distribution in a cometary dust tail is described for axisymmetric outflow of the dust from the core.

102.050 **Synthetic rotational spectra of C_2 Swan bands in comets.**
K. S. Krishna Swamy.
Publ. Astron. Soc. Pac., Vol. 98, No. 600, p. 252 – 253 (1986).
The author has calculated the synthetic spectra of the $\Delta\upsilon = +1$ sequence of the C_2 Swan system based on resonance–fluorescence statistical–equilibrium calculations. The synthetic profile is shown to be in good agreement with the high–resolution spectra of Lambert and Danks on comet West.

102.051 **The origin of comets.**
M. E. Bailey, S. V. M. Clube, W. M. Napier.
Vistas Astron., Vol. 29, Part 1, p. 53 – 112 (1986).
In this review the authors have traced developments in the understanding of cometary origins from the earliest days right up to the present, highlighting the important links with other branches of astronomy and the contemporary cosmology of the day.

102.052 **Photometric and dynamical characteristics of the cometary families of Jupiter and Saturn.**
V. P. Konopleva.
Komety Meteory, No. 36, p. 28 – 34 (1984). In Russian.

102.053 **Laboratory modelling of the disintegration of the conglomerate H_2O, CO_2 and NH_3 ice nucleus of a comet.**
Kh. Ibadinov, S. Aliev.
Komety Meteory, No. 36, p. 35 – 37 (1984). In Russian.

102.054 **Anisotropic ejection of matter from cometary nuclei and the phenomenon of anomalous intensity distribution of metallic atom lines in cometary heads.**
S. Ibadov.
Komety Meteory, No. 37, p. 8 – 12 (1985). In Russian.
Anisotropic flow of gas–dust matter in the circumnuclear region of comets is shown to expand the zone of gas influence on the cometary dust temperature. The phenomenon of anomalous intensity distribution of Na I lines across the coma of comet Mrkos 1957d (and possibly other comets) is explained by the cooling effect of gas on cometary dust. The method is proposed for determining the gas to dust ratio in the matter ejected from the cometary nucleus.

102.055 **Mathematical model of the process of brightness change in comets.**
M. Z. Markovich.
Komety Meteory, No. 37, p. 13 – 19 (1985). In Russian.
Mathematical modelling is usually applied to processes having incompletely known mechanisms. This is just the case when studying the brightness changes of comets.

102.056 **Statistical catalogue of the parameters of orbits of long–period comets in the Laplace coordinate system.**
V. V. Radzievskij, V. P. Tomanov.
Vsesoyuznoe astronomo–geodezicheskoe obshchestvo, 132 pp. (1985). In Russian. Abstr. in Ref. Zh., 51. Astron., 4.51.47 (1986).

102.057 **The effect of heavy ions on the formation and structure of cometary bow shocks.**
N. Omidi, D. Winske, C. S. Wu.
Icarus, Vol. 66, No. 1, p. 165 – 180 (1986).
The authors investigate the interaction of heavy cometary ions with the solar wind and the formation of a bow shock in front of a comet by means of a hybrid (particle ion, fluid electron) simulation code that solves self–consistently for the electromagnetic fields and the motion of the charged particles. This kinetic treatment of the solar wind protons and the heavy cometary ions allows to examine two important issues. One is the effect of the velocity distribution function of the heavy ions on the shock formation and structure, and the other is the degree of coupling between the two ion species.

102.058 **Detection of submillimeter size particles in the inner coma of a comet through their forward scattering.**
D. B. Vaidya, J. N. Desai.
Earth, Moon, Planets, Vol. 35, No. 1, p. 7 – 11 (1986).
During the occultation of a star by the inner coma region of a comet, specific forward–scattering effects could be observable if submillimeter size particles contribute significantly to the net extinction of star light. In this paper the authors investigate the possibility of detecting the signature of such particles by observing the dependence of extinction on the angular size of the acceptance aperture used at the focal plane of the telescope. Calculations based on a simple model assuming spherically–symmetric and homogeneous coma are presented.

102.059 **Nature of comets.**
K. S. Krishna Swamy.
Binary and multiple systems, p. 105 – 118 (1984). – See Abstr. 012.025.
The author has discussed in brief the present status of our understanding about the origin, nature and composition of comets.

102.060 **Axisymmetric flow of reacting gas from a comet nucleus.**
M. A. Vergazov, K. V. Krasnobaev.
Sov. Astron. Lett., Vol. 11, No. 4, p. 232 – 235 (1985). English translation of 40.102.001.

102.061 **Comet–solar wind interaction: dynamical length scales and models.**
D. A. Mendis, E. J. Smith, B. T. Tsurutani, J. A. Slavin, D. E. Jones, G. L. Siscoe.
Geophys. Res. Lett., Vol. 13, No. 3, p. 239 – 242 (1986).
While the ICE magnetometer measurements at Comet Giacobini–Zinner led to a number of unexpected discoveries such as the intense hydromagnetic turbulence, the large–scale structure of the induced cometary magnetosphere that it delineated was close to theoretical predictions. The authors briefly outline these theoretical models, discuss the relevant observations, and draw inferences.

102.062 **Plasma flow and critical velocity ionization in cometary comae.**
G. Haerendel.
Geophys. Res. Lett., Vol. 13, No. 3, p. 255 – 258 (1986).
The plasma flow in the cometary coma is described by a simple relation which results from the balance of magnetic stresses and drag forces, the latter being mainly due to ion mass loading. Inertial and pressure effects are not explicitly considered, but lumped into a correcting factor of order unity. The relation is used to infer the ionization rate due to the critical velocity effect. The high plasma density found by ICE at the closest approach to comet Giacobini–Zinner is interpreted as a result of the critical velocity effect.

102.063 **Modeling of dust halo formation following comet outbursts: preliminary results.**
T. I. Gombosi, M. Horányi.
Geophys. Res. Lett., Vol. 13, No. 3, p. 299 – 301 (1986).
Evolution of gas and dust distributions following a spatially and temporally localized comet outburst was calculated using a hybrid kinetic – hydrodynamic method. It was found that a comet outburst resulted in significantly increased dust terminal velocities. As these higher terminal velocity values result in larger apex distances, therefore the outburst generates distinct dust envelopes in front of the regular dust coma. Such envelopes were observed at several comets (cf. comet Donati).

102.064 **The nature of cometary grains from remote sensing.**
M. S. Hanner.
NASA Conf. Publ., NASA CP 2403, p. A 9 (1986). – See Abstr. 012.030.

102.065 **Flying sandbanks versus dirty snowballs: discovering the nature of comets.**
F. L. Whipple.
Mercury, Vol. 15, No. 1, p. 2 – 9 (1986).

102.066 **The 18 cm OH lines in comets: preparation for Halley.**
L. E. Snyder.
Astron. J., Vol. 91, No. 1, p. 163 – 170 (1986).
The astrophysical significance of radio observations of cometary OH is discussed. The interesting solar ultraviolet OH pumping mechanism which profoundly affects the radio observations is reviewed. The radio OH observations currently considered to be important for comet Halley are described.

102.067 **Catalogue of minimum distances between the orbits of comets and major planets.**
Yu. G. Babenko, V. P. Konopleva.
Red. Zh., Kinematika Fiz. Nebesn. Tel, 115 pp. (1986). In Russian. Abstr. in Ref. Zh., 51. Astron., 6.51.300 (1986).

102.068 **Metallic ions in cometary comae and plasma tails.**
W.–H. Ip, W. I. Axford.
Nature, Vol. 321, No. 6071, p. 682 – 684 (1986).
Ion composition measurements made in the environment of the comet Giacobini–Zinner by the ICE spacecraft have detected ions of mass 23 – 24 AMU with a relatively high abundance. According to the experimenters, these ions may be either Na^+ or C_2^+, if not both. The authors suggest that the ions detected may indeed be in part Na^+ and/or Mg^+ and that these and other metallic ions (especially Si^+ and Fe^+) may be an important component of the cometary ionosphere and central plasma tail.

102.069 **Organic grains in comets.**
F. Hoyle, N. C. Wickramasinghe.
Astrophys. Relativ., Prepr. Ser., No. 122, 88 pp. (1986).
A collection of papers concerning this topic (see abstracts 13.102.021, 21.015.020, 26.131.148, 30.015.020, 31.015.014, 32.131.171, 40.103.003, 41.103.710.

102.070 **The mystery of the missing shorter period comets.**
F. Hoyle.
Astrophys. Relativ., Prepr. Ser., No. 123, 90 pp. (1986).

102.071 **Polarimetry of comets. Review.**
O. V. Dobrovol'skij, N. N. Kiselev, G. P. Chernova.
Photometric and polarimetric investigations of celestial bodies, p. 3 – 9 (1985). In Russian. – See Abstr. 003.035.
Nearly all polarimetric observations and their interpretations are analyzed. The phase dependence of polarization obtained for a wide range of phase angles α is given. The mean phase curve for the dust atmosphere of the comet has a maximum when $\alpha = 90°$, passes through zero when $\alpha \sim 20°$, and then transfers into the negative branch when $\alpha < 20°$. As polarimetric data show, the comets are clearly divided into dust and gaseous ones. The wavelength dependence of the polarization degree is discussed. Results of detailed polarimetry are compared with the negative experience, when determining the elliptical polarization.

102.072 **Jet formation in comets.**
R. Z. Sagdeev, V. D. Shapiro, V. I. Shevchenko, K. Szego.
Report KFKI–1985–89, Hungarian Acad. Sci., Budapest, Hungary, 5 pp. (1985). Abstr. in Phys. Abstr., Vol. 89, No. 1276, Entry 36048 (1986).

102.073 **Possible nature of cometary atmosphere particles.**
A. V. Morozhenko, L. O. Kolokolova, E. A. Kajmakov, I. S. Lyzunkova.
Icarus, Vol. 66, No. 2, p. 223 – 229 (1986).
The polarimetric and spectrophotometric data of observations, the results of laboratory simulations, and theoretical calculations are considered as evidence in favor of the presence of large irregular particles in cometary atmospheres. The attempt is made to define more precisely the particle parameters.

102.074 **Axisymmetric dusty gas jet in the inner coma of a comet.**
Y. Kitamura.
Icarus, Vol. 66, No. 2, p. 241 – 257 (1986).
Being interested in the jet phenomena in cometary comae such as observed in Halley's comet, the author has investigated the expansion of pure and dusty gas jets by solving the axisymmetric, time–dependent, coupled hydrodynamic equations for the gas and dust.

102.075 **MHD turbulences in solar wind – comet interaction region.**
R. Z. Sagdeev, V. D. Shapiro, V. I. Shevchenko, K. Szego.
Report KFKI–1985–99, Hungarian Acad. Sci., Budapest, Hungary, 7 pp. (1985). Abstr. in Phys. Abstr., Vol. 89, No. 1278, Entry 47969 (1986).

102.076 **Effect of mass loading outside cometary bow shock for the plasma and wave measurements in the coming cometary missions.**
R. Z. Sagdeev, V. D. Shapiro, V. I. Shevchenko, K. Szego.
Report KFKI–1985–100, Hungarian Acad. Sci., Budapest, Hungary, 31 pp. (1985). Abstr. in Phys. Abstr., Vol. 89, No. 1278, Entry 47970 (1986).

102.077 **The discovery of dust trails in the orbits of periodic comets.**
M. V. Sykes, L. A. Lebofsky, D. M. Hunten, F. Low.
Science, Vol. 232, No. 4754, p. 1115 – 1117 (1986).

Analysis of data from the Infrared Astronomical Satellite has yielded evidence for narrow trails of dust coincident with the orbits of periodic comets Tempel 2, Encke, and Gunn. Dust was found both ahead of and behind the orbital positions of these comets. This dust was produced by the low–velocity ejection of large particles during perihelion passage. More than 100 additional dust trails are suggested by the data.

102.078 **Cometary observations from space.**
C. B. Cosmovici.
Mem. Soc. Astron. Ital., Vol. 56, No. 4, p. 587 – 598 (1985). – See Abstr. 012.054.

102.079 **L'origine des comètes.**
M. Festou.
Inst. Astrophys. Paris, Pré–Publ., No. 147, 9 pp. (1986). Cours présenté à l'école de Goutelas de 1985 organisée sur le thème "Physico–chimie de la matière primitive dans le système solaire".

102.080 **Methods for computing comet core temperatures.**
C. P. McKay, S. W. Squyres, R. T. Reynolds.
Icarus, Vol. 66, No. 3, p. 625 – 629 (1986).

The study of many aspects of cometary behavior requires knowledge of the nucleus core temperature. The authors have developed a formalization which allows the core temperature to be determined from numerical models of the surface temperature. Since thermal conduction into the deep interior has only a minor influence on the heat budget of the surface, surface temperatures can be calculated accurately without detailed models of the deep structure of the comet.

102.081 **Cometary dust grains.**
T. Mukai.
Solid State Phys., Vol. 21, No. 2, p. 55 – 61 (1986). In Japanese. Abstr. in Phys. Abstr., Vol. 89, No. 1283, Entry 78330 (1986).

102.082 **Temperatures within comet nuclei.**
S. W. Squyres, C. P. McKay, R. T. Reynolds.
J. Geophys. Res., Vol. 90, No. B14, p. 12381 – 12392 (1985).

The authors developed a model of heat transport within comet nuclei that allows them to calculate temperature profiles beneath the surface of the nucleus and to investigate how such profiles depend on nucleus properties. Calculations are performed for a comet with an orbit like those of periodic comets Wild 2 and Kopff.

102.083 **An improved thermal model for cometary nuclei.**
P. R. Weissman, H. H. Kieffer.
J. Geophys. Res., Vol. 89, Suppl., p. C358 – C364 (1984). – See Abstr. 012.065.

The authors have modified their earlier thermal model of cometary nuclei by incorporating a more accurate calculation of the opacity of the cometary dust coma. The new model uses a modified Sekanina–Miller dust particle size distribution and a Delsemme–Miller gas outflow dynamics model. The improved model has been applied to the case of the 1986 apparition of Halley's comet, also incorporating a revised radius and albedo for the Halley nucleus.

102.084 **S_2 formation in interstellar dust; a diagnostic of the maximum aggregation temperature for a comet.**
R. J. A. Grim, J. M. Greenberg, L. J. van Ijzendoorn.
Light on dark matter, p. 225 – 228 (1986). – See Abstr. 012.080.

Ultraviolet photolysis of interstellar grain types of dirty ices containing H_2O, CO, CH_4 and H_2S produces S_2. At low temperature (10K) this molecule can be trapped in the dirty ice for a time longer than the age of the solar system. The comet nucleus interior therefore retains the same composition as that of its initial material. The recent discovery of the S_2 molecule in comet IRAS–Araki–Alcock 1983d favors the theory that comets have accreted out of interstellar dust particles.

Kometen.
See Abstr. 003.033.

Physics of comets.
See Abstr. 003.034.

Comets, meteors & asteroids – how they affect earth.
See Abstr. 003.099.

Kometer och kometskräck.
See Abstr. 003.107.

Der Komet.
See Abstr. 003.171.

Honneur aux comètes.
See Abstr. 003.186.

Kijk op kometen.
See Abstr. 003.200.

Comètes: histoire et histoires.
See Abstr. 004.002.

Astronomy in Tadjikistan.
See Abstr. 013.007.

Comets and the muse.
See Abstr. 015.033.

Astrophysical implications of amorphous ice – a microporous solid.
See Abstr. 022.004.

The production and sputtering of S_2 by keV ion bombardment.
See Abstr. 022.045.

On clathrate hydrates in comets.
See Abstr. 022.046.

Parents for the cometary CN radical.
See Abstr. 022.047.

Properties of filamentary sublimation residues from dispersions of clay in ice.
See Abstr. 022.063.

Comparative performances of an objective grating concept in UV spectroscopic observations of solar system objects. II. D/H ratio observations in cometary atmospheres.
See Abstr. 036.023.

A review of some digital image processing in cometary research.
See Abstr. 036.048.

Cómo observar cometas.
See Abstr. 036.097.

Lower–hybrid instabilities driven by an ion velocity ring.
See Abstr. 062.114.

Interaction of a solar wind magnetic cloud with comet 1975h.
See Abstr. 074.020.

Comets and three–dimensional solar wind structure.
See Abstr. 074.046.

Minimum interorbital distances of planets and long–period comets.
See Abstr. 091.024.

Some peculiarities in the Asteroidal Belt.
See Abstr. 098.006.

Spectroscopy of asteroids in unusual orbits.
See Abstr. 098.049.

Subtle gravitational effects in the motion of small bodies in the solar system.
See Abstr. 098.050.

Interconnection between small bodies of the solar system.
See Abstr. 098.066.

Planet X and the origins of the shower and steady state flux of short–period comets.
See Abstr. 101.028.

Carbon monoxide in cometary ice.
See Abstr. 103.009.

Probing the inner regions of the coma by UV observations of CS in comets.
See Abstr. 103.010.

The abundance of sulphur in comets.
See Abstr. 103.011.

Cometary dust: observational evidences and properties.
See Abstr. 103.016.

Volatiles in cometary grains.
See Abstr. 103.017.

Observations of cometary plasma.
See Abstr. 103.018.

Some remarks on cometary outbursts.
See Abstr. 103.020.

Simulation of ICE–Giacobini–Zinner and VEGA/GIOTTO–Halley encounters.
See Abstr. 103.030.

The intermediate comets and nongravitational effects.
See Abstr. 103.039.

Brightness asymmetry in comet Kopff.
See Abstr. 103.241.

On the anisotropic gas coma observed in comet IRAS–Araki–Alcock 1983d.
See Abstr. 103.261.

L'orbite de la comète de Halley.
See Abstr. 103.603.

Hydrodynamic expansion of cometary gas loaded with refractory and volatile grains.
See Abstr. 103.631.

Condensation and sublimation of ice particles in the circumnuclear region of a comet.
See Abstr. 103.632.

A model of cometary gas and dust production and nongravitational forces with application to P/Halley.
See Abstr. 103.656.

The AMPTE artificial comet experiments.
See Abstr. 106.017.

Optical observations of the AMPTE artificial comet from the Northern Hemisphere.
See Abstr. 106.018.

***In situ* magnetic field observations of the AMPTE artificial comet.**
See Abstr. 106.019.

UKS plasma measurements near the AMPTE artificial comet.
See Abstr. 106.020.

Plasma waves and wave–particle interactions seen at the UKS spacecraft during the AMPTE artificial comet experiment.
See Abstr. 106.021.

Dynamics of the AMPTE artificial comet.
See Abstr. 106.022.

Künstliche Kometen.
See Abstr. 106.035.

Interstellar material in the solar system. A working group report.
See Abstr. 106.059.

Upper limits to the masses of objects in the solar comet cloud.
See Abstr. 107.031.

Has Nemesis' orbit been detected?
See Abstr. 107.032.

Origin and formation environment of cometary nuclei.
See Abstr. 107.033.

Particle bombardment of minor bodies during eruptive phases of the early solar evolution.
See Abstr. 107.034.

Encounters between the solar system and giant molecular clouds.
See Abstr. 107.036.

On the number of comets around other single stars.
See Abstr. 112.040.

Diatoms on Earth, comets, Europa and in interstellar space.
See Abstr. 131.196.

103 Comets (Individual Objects)

103.001 **Comets in 1983.**
B. G. Marsden.
Q. J. R. Astron. Soc., Vol. 27, No. 1, p. 102 – 118 (1986).

103.002 **The nongravitational motion of comet P/Harrington and P/du Toit–Neujmin–Delporte.**
G. Forti.
Astron. Astrophys., Vol. 155, No. 1, p. 169 – 171 (1986).
The motion of the periodic comets Harrington and du Toit–Neujmin–Delporte has been studied. The effect of the nongravitational forces has been included. Since both comets show large observational gaps, the author tried to link the observations assuming that the nongravitational parameter A_2 was constant up to a fixed time and then changed in value and/or sign to another constant value.

103.003 **Polarimetry of comets: a review.**
O. V. Dobrovolsky (*O. V. Dobrovol'skij*),
N. N. Kiselev, G. P. Chernova.
Earth, Moon, Planets, Vol. 34, No. 2, p. 189 – 200 (1986).
Some attempts of polarimetric sounding of Comet Halley will be undertaken from the flyby probes. In order to facilitate the final planning and the future interpretation of these experiments the authors have done a thorough analysis of practically all available polarimetric observations. An emphasis is made on interpretation attempts and their discussion. A division into "gaseous" and "dusty" comets on polarimetric basis is revealed. The wavelength dependence of polarization is discussed. The numerous results of detailed polarimetry are compared to the negative results of attempts to detect the elliptical polarization.

103.004 **Occultations and appulses by Halley's comet.**
D. W. Dunham.
Occultation Newsl., Vol. 3, No. 15, p. 321 – 324 (1986).

103.005 **Observations visuelles de la comète Giacobini–Zinner et de la comète de Halley.**
J.–C. Merlin.
Astronomie, Vol. 100, p. 91 – 93 (1986).

103.006 **Désignations définitives des comètes de l'année 1983.**
A. C. Levasseur–Regourd.
Astronomie, Vol. 100, p. 94 – 95 (1986).

103.007 **Are cometary nuclei like asteroids?**
M. F. A'Hearn.
Asteroids, comets, meteors II, p. 187 – 190 (1986). – See Abstr. 012.010.
Simultaneous optical and infrared observations have been made to determine the radii, albedos, and spectral reflectivities of two cometary nuclei for comparison with asteroids. The cometary nuclei are dark, highly elongated, and larger than commonly assumed. The two comets are very different from each other. The reflection spectrum of the nucleus of comet Neujmin 1 is consistent only with that of an S asteroid but the albedo is much darker than for any S asteroid. The properties of the nucleus of comet Arend–Rigaux are consistent with the properties of a C or F asteroid.

103.008 **Presentation of an Atlas of Cometary Spectra.**
C. Arpigny, F. Dossin, A. Woszczyk, B. Donn,
J. Rahe, S. Wyckoff.
Asteroids, comets, meteors II, p. 259 – 262 (1986). – See Abstr. 012.010.
A working group set up by Commission 15 of the International Astronomical Union is preparing an Atlas of Cometary Spectra which will be a sequel to Swings and Haser's "Atlas of Representative Cometary Spectra" published in 1956. The main goal of the new atlas is to show the progress achieved in the meantime and to present a selection of the best of the existing material on cometary spectroscopy.

103.009 **Carbon monoxide in cometary ice.**
P. D. Feldman.
Asteroids, comets, meteors II, p. 263 – 267 (1986). – See Abstr. 012.010.
Carbon monoxide has been positively identified, by means of ultraviolet fluorescence in the Fourth Positive System, in two comets, West (1976 VI) and Bradfield (1979 X). In both cases, the spatial distribution of the emission is consistent with the CO being a "parent" molecule, present in the cometary ice, rather than a dissociation product of a polyatomic species such as CO_2 or HCOH. However, the abundance of CO relative to that of H_2O is more than an order of magnitude smaller for comet Bradfield than for comet West. Further data on CO are expected with the use of more sensitive ultraviolet instruments such as the ASTRO spectrographs which will observe P/Halley in March 1986.

103.010 **Probing the inner regions of the coma by UV observations of CS in comets.**
P. S. Butterworth, J. Russell, W. M. Jackson.
Asteroids, comets, meteors II, p. 269 – 272 (1986). – See Abstr. 012.010.
New vibrational bands of CS have been identified in several comets. The relative intensities of these bands are in agreement with a simple model, based upon most of the CS radicals being in the lowest vibrational level. The present observational accuracy is not sufficient to decide between this simple model and an earlier radiative equilibrium model. The observed intensity ratios appear to be similar in different comets, which means that they cannot depend strongly upon the total gas production rates.

103.011 **The abundance of sulphur in comets.**
G. Azoulay, M. C. Festou.
Asteroids, comets, meteors II, p. 273 – 277 (1986). – See Abstr. 012.010.
Atomic sulphur is detected in comet spectra by its triplet at 181.4 nm. The intensity of the components of this multiplet is shown to be mainly determined by optical depth effects. When the gas production rate of a comet is very high, collisions are likely to play an important role. Atomic sulphur is not produced by the sole dissociation of CS_2. The authors propose that OCS is the main source of S I atoms in comet atmospheres.

103.012 **Cometary observations at Onsala Space Observatory since 1980.**
L. Ekelund, C. Andersson, A. Ekelund, W. Irvine,
A. Winnberg.
Asteroids, comets, meteors II, p. 283 – 287 (1986). – See Abstr. 012.010.
At Onsala Space Observatory the authors have been involved in searches for a variety of molecules in several comets since 1980, using the 20 meter mm–wave telescope.

103.013 **Variations of photometric exponents of long–period comets at large heliocentric distances.**
J. Svoreň.
Asteroids, comets, meteors II, p. 323 – 326 (1986). – See Abstr. 012.010.
Photometric parameters of 67 long–period comets were determined on the basis of 2842 photometric observations at large heliocentric distances (beyond 2.5 AU), according to the exponential law.

103.014 **Light variations of comets.**
W. Z. Wisniewski, T. Fay, T. Gehrels.
Asteroids, comets, meteors II, p. 337 – 339 (1986). – See Abstr. 012.010.
Two periodic comets, P/Neujmin 1 and P/Arend–Rigaux, were monitored during 1984/1985. The obtained lightcurves are complex with an amplitude of 0.5 magnitude. The periodic variation is interpreted as a rotation of the nucleus and the observations are

best represented by periods of 1.053 days and 1.138 days, respectively.

103.015 **Expressions for the nongravitational force derived from thermal models of cometary nuclei.**
C. Froeschlé, H. Rickman.
Asteroids, comets, meteors II, p. 349 – 353 (1986). – See Abstr. 012.010.
Thermal models have been computed for cometary nuclei moving in the orbits of comets P/Kopff, P/Gehrels 3, P/Boethin, P/Crommelin, P/Halley, P/Encke and P/Schwassmann–Wachmann 1. Sublimation of pure H_2O ice is assumed to proceed from the surface at all times with no insulating dust coverage, and it is fed only by direct solar radiation. Variations of the resulting nongravitational forces with orbital position are presented, and some conclusions are drawn regarding the influence of physical and orbital parameters on these "force laws". The authors find no improvement to the standard force law; see Marsden et al. 1973. However, for cometary nuclei with very low thermal inertias and low equatorial obliquities the variations remain symmetric with respect to perihelion.

103.016 **Cometary dust: observational evidences and properties.**
P. L. Lamy.
Asteroids, comets, meteors II, p. 373 – 388 (1986). – See Abstr. 012.010.
Cometary dust in both the coma and the tail is observed by its scattered light and its thermal emission provided that they are carefully separated from the emission lines. Color, phase function, polarization and spectral signatures are the basic properties which help to characterize the dust grains. There do not appear systematic differences among various comets as the reddening, the strong forward scattering and the polarization behaviour appear to be wide–spread common features also shared by interplanetary dust grains. The interpretation of some of these features in terms of rough, slightly absorbing grains is discussed.

103.017 **Volatiles in cometary grains.**
W. A. Baum, T. J. Kreidl.
Asteroids, comets, meteors II, p. 397 – 402 (1986). – See Abstr. 012.010.
This paper is a status report on a continuing program concerned with the composition of cometary grains and their size distribution. Coma profiles of 15 comets have been observed with the Gunn–Westphal CCD camera system attached to the 1.8–meter Perkins reflector at Lowell Observatory. Results to date imply that most of the light seen in the continuum of the coma is reflected from the icy component of the grains, that the sublimation may shut off somewhere between 6 and 8 AU, and that the distribution of grain sizes is very broad.

103.018 **Observations of cometary plasma.**
K. Jockers.
Asteroids, comets, meteors II, p. 411 – 418 (1986). – See Abstr. 012.010.
The large–scale behaviour of cometary plasma is illustrated. Using a particularly well observed example of Comet Bradfield 1979 X, the formation of kinks, asymmetric tail rays and tail undulations is shown. The problem of large–scale tail disturbances (tail disconnections, tail substorms) is briefly addressed. The second part of the paper deals with observations of increased spectral resolution to separate the individual ions.

103.019 **Picture processing of weak ion–tail emission of H_2O in comets P/Crommelin and IRAS–Araki–Alcock.**
H. Spinrad, M. A. Strauss.
Asteroids, comets, meteors II, p. 443 – 448 (1986). – See Abstr. 012.010.
Two–dimensional CCD spectra of P/Crommelin, at $r \sim 0.8$ A.U., taken at Kitt Peak National Observatory (KPNO) and Lick Observatory, and the Earth–grazing little comet IRAS–Araki–Alcock at $r = 1.0$ A.U., taken at KPNO, reveal moderate to very weak bands of H_2O^+ [Wyckoff and Wehinger (1976)] in the red. These spectral features can best be displayed by monochromatic subtraction and ratioing over the local dust continua in small (17 Å) windows. The data are interpreted briefly in terms of extant cometary ionospheric models, and compared with data in the literature.

103.020 **Some remarks on cometary outbursts.**
S. Grudzińska.
Asteroids, comets, meteors II, p. 467 – 468 (1986). – See Abstr. 012.010.
Presumably some cometary outbursts are responsible for changes of the orbits of the comets. Most dramatic is the case of comet Kritzinger 1914 II, in which the cometary orbit changed from a nearly parabolic, elliptic one before the outburst into a slightly hyperbolic one afterwards.

103.021 **Coordinating work on P/Halley, the International Halley Watch.**
R. L. Newburn Jr., J. Rahe.
Asteroids, comets, meteors II, p. 473 – 476 (1986). – See Abstr. 012.010.
While retaining all its original goals, the IHW has matured significantly in the year since the trial run on P/Crommelin. The complete data handling system was tested in that run and is now fully operational, with the Crommelin Archive to appear in early 1986. Cooperation with the Halley flight projects is excellent, and the IHW now awaits data from the key organization members, the 900 astronomers in 47 countries who will make observations. Meanwhile Halley itself is the first periodic comet to be studied in the heliocentric distance range 11 – 5 AU. Results of this study are summarized.

103.022 **Comet Digest.**
J. E. Bortle.
Sky Telesc., Vol. 71, No. 1, p. 114 (1986).
Concerning comets: 1982i P/Halley, 1984e P/Giacobini–Zinner, 1985*l* Hartley–Good, 1985m Thiele.

103.023 **Comet Digest.**
J. E. Bortle.
Sky Telesc., Vol. 71, No. 2, p. 221 (1986).
Concerning periodic comet 1982i Halley.

103.024 **Comet Digest.**
J. E. Bortle.
Sky Telesc., Vol. 71, No. 3, p. 323 (1986).
Concerning comets: 1982i P/Halley, 1985*l* Hartley–Good.

103.025 **Comet Digest.**
J. E. Bortle.
Sky Telesc., Vol. 71, No. 4, p. 426 (1986).
Concerning comets: 1982i P/Halley, 1985n P/Boethin.

103.026 **Comet notes: VI.**
D. H. Levy.
Strolling Astron., Vol. 31, Nos. 7–8, p. 171 – 172 (1986).

103.027 **The intermediate period comets and nongravitational effects.**
D. K. Yeomans.
Bull. Am. Astron. Soc., Vol. 17, No. 4, p. 918 (1985). Abstract. – See Abstr. 010.063.

103.028 **Radio observations of the OH radical in comets P/Halley and P/Giacobini–Zinner.**
D. Bockelée–Morvan, G. Bourgois, J. Crovisier, E. Gérard.
Bull. Am. Astron. Soc., Vol. 17, No. 4, p. 920 (1985). Abstract. – See Abstr. 010.063.

103.029 **Radio observations of OH in P/Halley and P/Giacobini–Zinner.**
F. P. Schloerb, M. J. Claussen.
Bull. Am. Astron. Soc., Vol. 17, No. 4, p. 920 (1985). Abstract. – See Abstr. 010.063.

103.030 **Simulation of ICE–Giacobini–Zinner and VEGA/GIOTTO–Halley encounters.**
T. Roatsch, K. Sauer, K. Baumgärtel.
Earth, Moon, Planets, Vol. 35, No. 2, p. 119 – 123 (1986).
The formation of a weak shock at the interaction of the solar wind with cometary neutral gas is confirmed in a non–stationary gasdynamic model. No shocks will be detectable along the ICE path and if the VEGA/GIOTTO spacecraft enter the cometary atmosphere. A shock can only be expected at an exit from the Halley atmosphere.

103.031 **Désignation définitive des comètes de l'année 1984.**
A. C. Levasseur–Regourd.
Astronomie, Vol. 100, p. 316 (1986).

103.032 **Comet Digest.**
J. E. Bortle.
Sky Telesc., Vol. 71, No. 5, p. 527 (1986).
Concerning comets: 1982i P/Halley, 1985*l* Hartley–Good, 1985n P/Boethin, 1985q P/Wirtanen, 1986a P/Shoemaker 3.

103.033 **Recent news and research concerning comets.**
D. W. E. Green.
Int. Comet Q., Vol. 8, No. 1, p. 8, 39 – 40 (1986).
Concerning comets: 1982i P/Halley, 1984e P/Giacobini–Zinner, 1985e Machholz, 1985g P/Giclas, 1985m Thiele, 1985n P/Boethin, 1985o P/Kojima, 1985p P/Ciffréo, 1986a P/Shoemaker 3, SOLWIND 4, SOLWIND 5 (= 1984 XII).

103.034 **Tabulation of comet observations.**
Int. Comet Q., Vol. 8, No. 1, p. 9 – 38 (1986).
Concerning comets: 1982i P/Halley, 1984 XXIII Levy–Rudenko, 1984e P/Giacobini–Zinner, 1985a P/Ashbrook–Jackson, 1984f Shoemaker, 1985g P/Giclas, 1985*l* Hartley–Good, 1985m Thiele, 1985n P/Boethin, 1985p P/Ciffréo.

103.035 **Roman numeral designations of comets in 1984.**
Minor Planet Circ., No. 10330 (1986).

103.036 **Observations of comets.**
Minor Planet Circ., Nos. 10330 – 10360, 10457 – 10478, 10588 – 10592, 10669 – 10677, 10782 – 10795 (1986).
Observations made at the following stations are published: Abastuman, Alma Ata, Asiago, Baldone, Belgrade, Brescia, Bucharest, Bulgarian Natl. Obs., Burlington N.J., Burwash, Byurakan, Calar Alto, Cambridge, Cape, Carter Obs., Cavriana, Centro Astron. Yebes, Cerro el Roble, Chamberlin Obs., Chorzow, Colchester, Condor Brow, Córdoba, Cracow–Fort Skała, Crimean Astrophys. Obs., Dresden, Dushanbe, Eastfield, El Leoncito, Engelhardt Obs., Engelhardt Obs. Zelenchukskaya Stn., ESO, Fabra Obs., Figl Obs., Geisei, Gissar, Göttingen, Golosseevo–Kiev, Graz, Heidelberg–Königstuhl, Hoher List, Japal–Rangapur Obs., JCPM Sapporo Stn., Kambah (Canberra), Kharkov, Kiev, Kitab, Kitt Peak, Kleť, Kourovskaya, La Chaux de Fonds, Lahti, Lick Obs., Lowell Obs., Lowell Obs. Anderson Mesa Stn., McDonald Obs., Merate, Merida, Metzerlen Obs., Mt. Cuba Obs., Mt. John Obs., Mt. Palomar, Nice, Nikolaev, Oak Ridge Obs., Odessa, Odessa–Mayaki, Ordubad, Oxford, Perth Obs., Pino Torinese, Piszkéstetö, Poznan , Pulkovo, Purple Mountain Obs., Quonochontaug, Reintal, S. Vittore (Bologna), Sapporo Sci. Cent., Sedgefield, Sendai Obs., Shokin Majdanak, Siding Spring, Skalnaté Pleso, Skibotn, St. Andrews, St. Etienne, Stakenbridge, Sydney, Tarija, Tartu, Tashkent, Tautenburg, Torun–Piwnice, Tsingtao, Turku–Tuorla, Uccle, Uzhgorod, Victoria, Vienna, Zimmerwald, Zo–Sé, Zvenigorod.
1980 X P/Stephan–Oterma, 1980 XIII P/Tuttle, 1981 XXI SOLWIND 4, 1982i P/Halley, 1983 VII IRAS–Araki–Alcock, 1984 IV P/Crommelin, 1984 XI P/Faye, 1984 XII SOLWIND 5, 1984 XXIII Levy–Rudenko, 1984e P/Giacobini–Zinner, 1984f Shoemaker, 1984*l* P/Gehrels 3, 1984v Hartley, 1985a P/Ashbrook–Jackson, 1985g P/Giclas, 1985j P/Daniel, 1985k P/Maury, 1985*l* Hartley–Good, 1985m Thiele, 1985n P/Boethin, 1985o P/Kojima, 1985p P/Ciffréo, 1985q P/Wirtanen, 1986a P/Shoemaker 3, 1986b Shoemaker, 1986c P/Hartley 2, 1986d P/Singer Brewster, 1986e P/Machholz.

103.037 **Orbital elements of comets.**
Minor Planet Circ., Nos. 10329 – 10884 (1986).
The comets are listed according to their Roman numeral designation or preliminary designation. The names of the authors are given behind the respective M.P.C. numbers.
1973 VIII P/Wild 1, 1980 VI P/Forbes, 1980 VIII P/Reinmuth 1 10524 B. G. Marsden; 1985p P/Ciffréo 10817, 1986a P/Shoemaker 3 10817 D. W. E. Green; 1986c P/Hartley 2 10610, 10828, 1986d P/Singer Brewster 10759, 1986e P/Machholz 10759, 10829 B. G. Marsden.

103.038 **Ephemerides of comets.**
Minor Planet Circ., Nos. 10404 – 10405, 10551 – 10555, 10634 – 10637, 10774, 10850 (1986).
Concerning ephemerides of the following comets:
1978 XVII P/Comas Solá, 1979 IV P/Holmes, 1980 VI P/Forbes, 1981 IX P/Kohoutek, 1984 VI P/Encke, 1985a P/Ashbrook–Jackson, 1985h P/Whipple, 1985m Thiele, 1985p P/Ciffréo, 1986a P/Shoemaker 3, 1986b Shoemaker, 1986c P/Hartley 2, 1986d P/Singer Brewster, 1986e P/Machholz.

103.039 **The intermediate comets and nongravitational effects.**
D. K. Yeomans.
Astron. J., Vol. 91, No. 4, p. 971 – 973 (1986).
The motions of the intermediate–period comets Pons–Brooks, Olbers, Brorsen–Metcalf, and Westphal are investigated over their observed intervals.

103.040 **Observations of comet Shoemaker (1984f) and comet Schaumasse (1984m) at the Zelenchukskaya Astronomical Station.**
V. N. Kitkin.
Komet. Tsirk., No. 349 (1986). In Russian.

103.041 **Exact positions of comet Crommelin (1983n) and comet Levy–Rudenko (1984t) from observations at the Astronomical Observatory of the Ural University.**
T. I. Levitskaya.
Komet. Tsirk., No. 349 (1986). In Russian.

103.042 **Observations of comets at the Simeiz Astronomical Station of the Astronomical Council of the USSR Academy of Sciences, and at the Klet Observatory.**
Komet. Tsirk., No. 351 (1986). In Russian.
Concerning comets: Thiele (1985m), Hartley–Good (1985*l*), and P/Ciffréo (1985p).

103.043 **Recoveries of periodic comets.**
Br. Astron. Assoc. Circ., No. 655 (1986).
Concerning comets: P/Wirtanen 1985q and P/Kojima 1985o.

103.044 **Comets 1981 XXI (SOLWIND 4) and 1984 XII (SOLWIND 5).**
IAU Circ., No. 4230 (1986).

103.045 **Orbital elements of recent comets.**
Yamamoto Circ., No. 2051 (1986). In Japanese.
Concerning comets: P/Maury (1985k), Hartley (1984v), and Thiele (1985m).

103.046 **Comet 1981 XXI (SOLWIND 4) and comet 1984 XII (SOLWIND 5).**
Yamamoto Circ., No. 2063 (1986). In Japanese.

103.047 **The return comets in 1987 and observable comets in 1986.**
OAA Comput. Sect. Circ., NK 490, 4 pp. (1986).

Periodic comet Giacobini–Zinner

103.101 **The dynamics of charged dust in the tail of comet Giacobini–Zinner.**
M. Horanyi, D. A. Mendis.
J. Geophys. Res., Vol. 91, No. A1, p. 355 – 361 (1986).
The forthcoming NASA–ICE mission, which will fly through the tail of periodic comet Giacobini–Zinner on September 11, 1985, lends some urgency to the study of the dynamics of dust in its tail. In this paper the authors address this problem, while emphasizing the role of electromagnetic effects on the dust grains which must necessarily be electrically charged by virtue of being immersed in a radiative and plasma environment. The differences in the dust orbits and distributions, with and without electrical charging, are also discussed.

103.102 **MHD turbulence in the solar wind–comet interaction region.**
R. Z. Sagdeev, V. D. Shapiro, V. I. Shevchenko, K. Szego.
Geophys. Res. Lett., Vol. 13, No. 2, p. 85 – 88 (1986).
One of the most striking features discovered during the encounter of space probe ICE and comet Giacobini–Zinner was the presence of intensive MHD turbulence even at large distances from the comet. In this paper it is shown that the mass loading of the solar wind with cometary ions can cause such a phenomenon.

103.103 **Comet Giacobini–Zinner: plasma description.**
S. J. Bame, R. C. Anderson, J. R. Asbridge, D. N. Baker, W. C. Feldman, S. A. Fuselier, J. T. Gosling, D. J. McComas, M. F. Thomsen, D. T. Young, R. D. Zwickl.
Science, Vol. 232, No. 4748, p. 356 – 361 (1986).
A strong interaction between the solar wind and comet Giacobini–Zinner was observed on 11 September 1985 with the Los Alamos plasma electron experiment on the International Cometary Explorer (ICE) spacecraft. Periods of enhanced electron heat flux from the comet as well as almost continuous electron density fluctuations were measured. These effects are related to the strong electron heating observed in the cometary interaction region and to cometary ion pickup by the solar wind, respectively. No evidence for a conventional bow shock was found as ICE entered and excited the regions of strongest interaction of the solar wind with the cometary environment.

103.104 **Observations of energetic ions from comet Giacobini–Zinner.**
R. J. Hynds, S. W. H. Cowley, T. R. Sanderson, K.-P. Wenzel, J. J. Van Rooijen.
Science, Vol. 232, No. 4748, p. 361 – 365 (1986).
During the encounter with comet Giacobini–Zinner, the energetic particle anisotropy spectrometer on the International Cometary Explorer spacecraft observed large fluxes of energetic ions, believed to result principally from ionization of the cometary atmosphere followed by pickup and acceleration by the ambient flow of the solar wind. These heavy cometary ions were observed from approximately 1 day before closest approach to about 2 1/2 days afterward. Three regimes of differing ion characteristics have been identified.

103.105 **Comet Giacobini–Zinner: in situ observations of energetic heavy ions.**
F. M. Ipavich, A. B. Galvin, G. Gloeckler, D. Hovestadt, B. Klecker, M. Scholer.
Science, Vol. 232, No. 4748, p. 366 – 369 (1986).
Conclusive evidence is presented for the existence of energetic ($\sim$35,000 to 150,000 electron volts), heavy ($\gtrsim$12 atomic mass units), singly charged cometary ions within $\sim 1.5 \times 10^6$ kilometers of comet Giacobini–Zinner. The observations were made with the University of Maryland/Max–Planck–Institut ultralow–energy charge analyzer on the International Cometary Explorer spacecraft. The most direct evidence for establishing the mass of these ions was obtained from an analysis of the energy signals in one of the solid–state detectors; it is significant at the three–sigma level.

103.106 **Plasma diagnosis from thermal noise and limits on dust flux or mass in comet Giacobini–Zinner.**
N. Meyer–Vernet, P. Couturier, S. Hoang, C. Perche, J. L. Steinberg, J. Fainberg, C. Meetre.
Science, Vol. 232, No. 4748, p. 370 – 374 (1986).
Thermal noise spectroscopy was used to measure the density and temperature of the main (cold) electron plasma population during 2 hours (1.5×10^5 kilometers perpendicular to the tail axis) around the point of closest approach of the International Cometary Explorer (ICE) to comet Giacobini–Zinner. The time resolution was 18 seconds (370 kilometers) in the plasma tail and 54 seconds (1100 kilometers) elsewhere. Near the tail axis, the maximum plasma density was 670 per cubic centimeter and the temperature slightly above 1 electron volt. The tail was highly asymmetrical and showed much structure. On the other antenna, shot noise was recorded from the plasma particle impacts on the spacecraft body. No evidence was found of grain impacts on the antennas or spacecraft in the plasma tail. This yields an upper limit for the dust flux or particle mass, indicating either fluxes or masses in the tail smaller than implied by the models or an anomalous grain structure.

103.107 **Ion composition results during the International Cometary Explorer encounter with Giacobini–Zinner.**
K. W. Ogilvie, M. A. Coplan, P. Bochsler, J. Geiss.
Science, Vol. 232, No. 4748, p. 374 – 377 (1986).
The International Cometary Explorer spacecraft passed through the coma of comet Giacobini–Zinner about 7800 kilometers antisunward of the nucleus on 11 September 1985. Initial interpretation of the mesurements indicates the presence of H_2O^+, H_3O^+, probably CO^+ and HCO^+, and ions in the mass range 23 to 24; possible candidates are Na^+ and Mg^+. In addition to these heavy ions, He^{2+} of solar wind origin was measured. The results are in agreement with the picture of a comet primarily consisting of water ice, together with other material, that sublimes, streams away from the nucleus, becomes ionized, and interacts with the solar wind.

103.108 **Plasma wave observations at comet Giacobini–Zinner.**
F. L. Scarf, F. V. Coroniti, C. F. Kennel, D. A. Gurnett, W.–H. Ip, E. J. Smith.
Science, Vol. 232, No. 4748, p. 377 – 381 (1986).
The ICE measurements at Giacobini–Zinner indicate that many important cometary processes are influenced by microscopic plasma physics phenomena to an extent not anticipated in fluid dynamic or magnetohydrodynamic models.

103.109 **International Cometary Explorer encounter with Giacobini–Zinner: magnetic field observations.**
E. J. Smith, B. T. Tsurutani, J. A. Slavin, D. E. Jones, G. L. Siscoe, D. A. Mendis.
Science, Vol. 232, No. 4748, p. 382 – 385 (1986).
The vector helium magnetometer on the International Cometary Explorer observed the magnetic fields induced by the interaction of comet Giacobini–Zinner with the solar wind. A magnetic tail was penetrated $\sim$7800 kilometers downstream from the comet and was found to be 10^4kilometers wide. It consisted of two lobes, containing oppositely directed fields with strengths up to 60 nanoteslas, separated by a plasma sheet $\sim 10^3$kilometers thick containing a thin current sheet. The magnetotail was enclosed in an extended ionosheath characterized by intense hydromagnetic turbulence and interplanetary fields draped around the comet.

103.110 **Astrometria fotografica della cometa periodica Giacobini–Zinner (1984e).**
E. Colombini.
Astronomia, N. 1, p. 6 – 9 (1986).

103.111 **The encounter of comet Giacobini–Zinner (G–Z) by the International Cometary Explorer (ICE).**
J. C. Brandt.
Bull. Am. Astron. Soc., Vol. 17, No. 4, p. 873 (1985). Abstract. – See Abstr. 010.062.

103.112 **CCD filter photometry of P/Giacobini–Zinner.**
E. S. Barker, J. R. Green, C. B. Opal.
Bull. Am. Astron. Soc., Vol. 17, No. 4, p. 919 – 920 (1985). Abstract. – See Abstr. 010.063.

103.113 **Comet Giacobini–Zinner: a normal comet?**
A. L. Cochran.
Bull. Am. Astron. Soc., Vol. 17, No. 4, p. 920 (1985). Abstract. – See Abstr. 010.063.

103.114 **The bow wave of comet Giacobini–Zinner: ICE magnetic field observations.**
D. E. Jones, E. J. Smith, J. A. Slavin, B. T. Tsurutani, G. L. Siscoe, D. A. Mendis.
Geophys. Res. Lett., Vol. 13, No. 3, p. 243 – 246 (1986).
Fitting of a Mach 2 shock surface to the ICE magnetic field data obtained near comet Giacobini–Zinner has provided subsolar bow wave distances that infer neutral gas outflow rates comparable to previous measurements, and orientations of the bow wave symmetry axis that are consistent with the plasma measurements and motion of the comet relative to the solar wind.

103.115 **Heat flux observations and the location of the transition region boundary of Giacobini–Zinner.**
S. A. Fuselier, W. C. Feldman, S. J. Bame, E. J. Smith, F. L. Scarf.
Geophys. Res. Lett., Vol. 13, No. 3, p. 247 – 250 (1986).
Electron heat flux observations and associated plasma phenomena upstream from comet Giacobini–Zinner show many similarities to observations upstream from the Earth's magnetosheath. Two similarities are discussed. Heat flux events are used to compute the transition region boundary location. The computed subsolar standoff distance of 4×10^4km is commensurate with a comet gas production rate, G, of 3×10^{28}molecules/s, in accord with ground–based determinations of G.

103.116 **Cometary pick–up ions observed near Giacobini–Zinner.**
G. Gloeckler, D. Hovestadt, F. M. Ipavich, M. Scholer, B. Klecker, A. B. Galvin.
Geophys. Res. Lett., Vol. 13, No. 3, p. 251 – 254 (1986).
In–situ measurements of the rest–frame distribution functions of cometary pick–up water–group ions are used to derive the number and energy density of these ions as a function of distance from the Comet Giacobini–Zinner. The shapes of the distribution functions indicate that the interaction process of the solar wind with cometary ions is predominantly stochastic.

103.117 **Strong hydromagnetic turbulence associated with comet Giacobini–Zinner.**
B. T. Tsurutani, E. J. Smith.
Geophys. Res. Lett., Vol. 13, No. 3, p. 259 – 262 (1986).
It is the purpose of this paper to discuss the general nature and spatial distribution of the hydromagnetic turbulence surrounding comet Giacobini–Zinner. Although the detailed properties of this turbulence is under investigation at present (Tsurutani and Smith, 1986), the origin of the turbulence can be understood in a general way. It is one aspect of the intermingling of the cometary and solar wind plasmas or, expressed in other terms, a consequence of the assimilation of the cometary plasma into the solar wind.

103.118 **Hydromagnetic waves and instabilities associated with cometary ion pickup: ICE observations.**
B. T. Tsurutani, E. J. Smith.
Geophys. Res. Lett., Vol. 13, No. 3, p. 263 – 266 (1986).
The hydromagnetic turbulence surrounding comet Giacobini–Zinner is characterized by two dominant components: large amplitude, linearly polarized waves with periods near 10^2s and circularly polarized wave packets with periods near 3 s. The purpose of this paper is to present an analysis of the two dominant wave modes, to discuss their relationship and to explore possible generation mechanisms.

103.119 **Large amplitude, low frequency plasma fluctuations at comet Giacobini–Zinner.**
J. T. Gosling, J. R. Asbridge, S. J. Bame, M. F. Thomsen, R. D. Zwickl.
Geophys. Res. Lett., Vol. 13, No. 3, p. 267 – 270 (1986).
Very large amplitude fluctuations in electron density, temperature, and flow velocity were a prominent aspect of the solar wind interaction with comet Giacobini–Zinner during the ICE encounter in September 1985. These fluctuations, particularly fluctuations in electron density, are the subject of this paper.

103.120 **Plasma fluctuations and large–scale mixing near comet Giacobini–Zinner.**
D. N. Baker, W. C. Feldman, S. P. Gary, D. J. McComas, J. Middleditch.
Geophys. Res. Lett., Vol. 13, No. 3, p. 271 – 274 (1986).
Low–pass filtered data from the ICE electron spectrometer are used to examine relationships between plasma densities, flow speeds, flow directions, and temperatures at distances within $\sim 10^5$km of the comet Giacobini–Zinner nucleus.

103.121 **Ion distribution functions in the vicinity of comet Giacobini–Zinner.**
T. E. Cravens.
Geophys. Res. Lett., Vol. 13, No. 3, p. 275 – 278 (1986).
Photoionization of neutral molecules in the coma surrounding a comet produces heavy ions which contaminate and mass–load the solar wind. Cometary ion distribution functions in the vicinity of comet Giacobini–Zinner are calculated using a Monte Carlo method. The distribution function calculated behind the shock has both cold ($\approx$2 keV) and hot ($\approx$40 keV) components.

103.122 **Physical parameters for hot and cold electron populations in comet Giacobini–Zinner with the ICE radio experiment.**
N. Meyer–Vernet, P. Couturier, S. Hoang, C. Perche, J. L. Steinberg.
Geophys. Res. Lett., Vol. 13, No. 3, p. 279 – 282 (1986).
Thermal plasma noise spectroscopy using the ICE radio receiver yielded the parameters of the two Maxwellian populations approximating the comet plasma electron distribution. Within 150,000 km around closest approach, the authors observe three regions: (1) the inner one coincides with the low magnetic field region or plasma sheet. (2) From 9000 to 20,000 km from the nucleus, the density decreases as R^{-2}. (3) Further away, the density and temperature fluctuate widely around about solar wind values.

103.123 **Giacobini–Zinner magnetotail: ICE magnetic field observations.**
J. A. Slavin, E. J. Smith, B. T. Tsurutani, G. L. Siscoe, D. E. Jones, D. A. Mendis.
Geophys. Res. Lett., Vol. 13, No. 3, p. 283 – 286 (1986).
It is the conclusion of this investigation that the magnetotail observations made by the International Cometary Explorer (ICE) at Comet Giacobini–Zinner have confirmed the Alfvén field line draping theory of type I cometary tails.

103.124 **Statics and dynamics of Giacobini–Zinner magnetic tail.**
G. L. Siscoe, J. A. Slavin, E. J. Smith, B. T. Tsurutani, D. E. Jones, D. A. Mendis.
Geophys. Res. Lett., Vol. 13, No. 3, p. 287 – 290 (1986).
The authors analyze Giacobini–Zinner's magnetic tail in terms of a cometary analog of the more familiar planetary magnetotails.

103.125 **Dust particles detected near Giacobini–Zinner by the ICE plasma wave instrument.**
D. A. Gurnett, T. F. Averkamp, F. L. Scarf, E. Grün.
Geophys. Res. Lett., Vol. 13, No. 3, p. 291 – 294 (1986).
During the ICE flyby of Giacobini–Zinner the plasma wave instrument detected numerous impulsive signals caused by dust impacts on the spacecraft. Most of the impacts occurred within 30,000 km of the comet. The purpose of this paper is to present an initial analysis of these impacts.

103.126 **Groundbased infrared imaging of comet Giacobini–Zinner: the distribution of dust during the ICE flyby.**
H. Campins, C. M. Telesco, R. Decher, D. Mozurkewich, H. A. Thronson Jr.
Geophys. Res. Lett., Vol. 13, No. 3, p. 295 – 298 (1986).
A thermal–infrared image of comet Giacobini–Zinner has been used to determine the spatial distribution of dust grains in the comet's coma. The authors emphasize the results of this imaging which bear on the expected impact rate of dust grains on the International Cometary Explorer (ICE).

103.127 **Comet fly–by first results.**
L. J. Carter.
Spaceflight, Vol. 28, No. 3, p. 127 – 128 (1986).

103.128 **Detection of OH maser emission from comet Giacobini–Zinner.**
R. P. Norris, J. R. Forster, R. A. Duncan.
Proc. Astron. Soc. Aust., Vol. 6, No. 2, p. 180 – 181 (1985).
The authors have detected OH maser radiation from the comet Giacobini–Zinner using the Parkes radio telescope. The emission was detected in two transitions of the ground state of OH, at an intensity consistent with the predicted OH molecular production rate.

103.129 **ICE: mission to a comet.**
R. Berry.
Astronomy, Vol. 13, No. 12, p. 6 – 13 (1985). Abstr. in Phys. Abstr., Vol. 89, No. 1279, Entry 53808 (1986).

103.130 **Periodic comet Giacobini–Zinner (1984e).**
Komet. Tsirk., Nos. 348 – 350, 353 (1985/1986). In Russian.

103.131 **Periodic comet Giacobini–Zinner (1984e).**
IAU Circ., No. 4163 (1986).

Periodic comet Crommelin

103.151 **Observations of comet Crommelin.**
G. S. D. Babu, Shylaja, K. R. Sivaraman.
IHW Newsl., No. 8, p. 25 (1986).

103.152 **Detailed photometry of comet Crommelin (1983n).**
Yu. V. Sizonenko.
Kinematika Fiz. Nebesn. Tel, Tom 2, No. 2, p. 88 – 90 (1986). In Russian. English translation in Kinematics Phys. Celest. Bodies.
The procedure and computer–processed results of detailed photometry of comet Crommelin are described. The integral magnitude of the comet and that of the central condensation are determined. The analytical expression for dependences of the cometary surface brightness on the distance to the photometric centre along the chosen directions is obtained.

103.153 **Decametric radio emission from comets: an attempt at detection.**
T. E. Gergely, M. J. Mahoney.
Icarus, Vol. 66, No. 3, p. 575 – 578 (1986).
Radio continuum emission from comets may arise as thermal emission from ice or dust particles and free electrons, or as non-thermal emission due to plasma effects in the tail. Nonthermal emission is expected to become intense at the longer, possibly hectometer or even kilometer wavelengths. An attempt at the detection of a comet at the longest possible wavelength observable from the ground was made during the passage of P/Crommelin in March 1984 using the Teepee Tee array of the Clark Lake Radio Observatory. Although radio emission was not detected during the observations, the authors are able to place an upper limit on the thermal emission, and consequently on the electron density in the tail.

103.154 **Radio observations of comet P/Crommelin.**
E. Scalise Jr., Z. Abraham, J. L. Monteiro do Vale.
Rev. Mex. Astron. Astrofis., Vol. 12, Spec. Issue, p. 371 – 373 (1986). – See Abstr. 012.066.
The upper limits of the radio continuum and line measurements at ~ 22 GHz of comet P/Crommelin are presented briefly.

Comet 1985m Thiele

103.161 **Comet Thiele (1985m).**
IAU Circ., Nos. 4177, 4184 (1986).

Periodic comet Hartley 2

103.171 **Periodic comet Hartley 2 (1986c).**
IAU Circ., Nos. 4197, 4200 (1986).

103.172 **Periodic comet Hartley 2 (1986c).**
Yamamoto Circ., Nos. 2057, 2058 (1986). In Japanese.

Comet 1973 XII Kohoutek

103.181 **Absolute spectrophotometry of Comet Kohoutek 1973f during post–perihelion passage.**
P. S. Goraya, B. S. Rautela, B. B. Sanwal.
Earth, Moon, Planets, Vol. 34, No. 1, p. 77 – 91 (1986).
Absolute spectrophotometry of the coma of Comet Kohoutek 1973f during post–perihelion period has been presented for seven nights in January 1984. Moderately wide range of heliocentric distance (0.436 – 0.799 AU) covered during observations allowed to study the flux variation of emission bands with heliocentric distance. The emission features of CN, CH, C_2, C_3, and Na I have been identified in this comet. The abundances of CN and C_2 have been estimated and the production rates of CN, C_2 and C_3 have been derived.

Periodic comet Encke

103.191 **The spectrum of periodic comet Encke.**
P. S. Goraya, B. B. Sanwal, B. S. Rautela.
J. Br. Astron. Assoc., Vol. 96, No. 4, p. 210 – 211 (1986).
Periodic comet Encke was observed spectrophotometrically during its apparition in 1984 March, before perihelion. The authors observed emission features due to CN and C_2 molecules, but sodium emission was absent from the nucleus of the comet at $r = 0.38$ AU. The lack of sodium emission is probably due to a different composition for this comet and to its old age.

103.192 **Effects of the law for nongravitational forces on the precession model of Comet Encke.**
Z. Sekanina.
Astron. J., Vol. 91, No. 2, p. 422 – 431 (1986).
The precession model for P/Encke formulated by Whipple and Sekanina (1979) is re–examined in the light of the recent measurements of the OH and H production rates, which according to A'Hearn et al. (1985) indicate that the comet's visual light curve cannot be employed to approximate the variations with time in the nongravitational force that is being exerted on the nucleus by the comet's outgassing. The optimized precession solution derived here with the use of the A'Hearn et al. production law is shown to require unacceptably small nucleus dimensions and to offer an inferior fit to observed variations in the orientation of the comet's preperihelion fan–shaped coma. It is concluded that the Whipple–Sekanina model remains the best available precession solution for this comet.

103.193 **Periodic comet Encke 1971 II = 1974 V = 1977 XI = 1980 XI = 1984 VI.**
OAA Comput. Sect. Circ., NK 485, 4 pp. (1985).

Comet 1976 VI West

103.201 **Profile analysis of the (B–X) bands of CO^+ in comet West.**
K. S. Krishna Swamy.
Earth, Moon, Planets, Vol. 34, No. 3, p. 281 – 282 (1986).
The calculated profile of $(B^2\Sigma^+–X^2\Sigma^+)$ bands of CO^+ based on resonance fluorescence process is shown to be in agreement with the observed profile for comet West.

103.202 **Study of the spectrum of comet West (1976 VI). I. Spectrophotometry of the coma and nuclear fragments.**
V. K. Rozenbush.
Photometric and polarimetric investigations of celestial bodies, p. 14 – 21 (1985). In Russian. – See Abstr. 003.035.
Spectrograms of the coma and nuclear fragments A and D were obtained at the Special Astrophysical Observatory of the USSR Academy of Sciences by means of the 60–cm Zeiss telescope with an image intensifier. A comparative analysis of the spectra shows some differences which indicate spatial inhomogeneity of material in the parent single nucleus of the comet. The total energy fluxes were calculated from relative energy distributions for various emissions relative to the flux in the λ516.5 nm C_2 ($\Delta \upsilon = 0$) band. The time–dependent variations of abundances of various molecular components in the comet's atmosphere were found. The vibrational temperature (about 4000K) and transition probability (0.01 C^{-1}) for the C_2 molecule were determined. The column density of C_2, CN is estimated to be about 10^{13}mol/cm^2.

103.203 **Study of the spectrum of comet West (1976 VI). II. Spectrophotometry of the tail.**
V. K. Rozenbush.
Photometric and polarimetric investigations of celestial bodies, p. 21 – 26 (1985). In Russian. – See Abstr. 003.035.
Spectrograms of the tail of comet West 1976 VI were obtained by means of the 60–cm Zeiss telescope with an image intensifier at the Special Astrophysical Observatory of the USSR Academy of Sciences. The total energy fluxes in CO^+ and H_2O^+ emissions relative to the flux in the CN (0,0) band were calculated for three regions in the tail situated at 15′.8, 17′.4 and 19′ from the nucleus. Comparison of the observed relative intensities of CO^+ band with theoretical ones calculated by Krishna Swamy shows some differences in the CO^+ (1,0) band. The relative abundance of CO^+ ions for various transitions (υ', υ") is determined. The brightness profiles of CO^+ and H_2O^+ indicate the presence of ion condensation at a distance of 8.5×10^5km from the nucleus.

103.204 **Spectrophotometry of Comet West 1976 VI.**
V. K. Rosenbush (*V. K. Rozenbush*).
Icarus, Vol. 66, No. 2, p. 230 – 240 (1986).
On April 1, 2 and 7, 1976, spectra of the coma, nuclear fragments A and D, and the tail of Comet West 1976 VI were obtained. The main tasks of this paper are to reveal the possible differences between individual fragments of the nucleus and to study peculiarities in the tail of Comet West.

Periodic comet Clark

103.211 **Positions of comet Clark (1983w) in 1984.**
M. Ratag, B. Hidayat.
Acta Astron., Vol. 36, No. 1, p. 179 – 180 (1986).

Periodic comet Brooks 2

103.221 **The orbit of comet Brooks 2 in 1889 – 1933.**
Yu. D. Medvedev.
Kinematika Fiz. Nebesn. Tel, Tom 2, No. 2, p. 83 – 84 (1986). In Russian. English translation in Kinematics Phys. Celest. Bodies.
Seven apparitions of P/Brooks 2 covering 1889 –1933 are linked with the mean error of a normal place of 1″.57.

103.222 **Periodic comet Brooks 2 1960 VI = 1974 I = 1980 IX.**
OAA Comput. Sect. Circ., NK 483, 2 pp. (1985).

Periodic comet Schwassmann–Wachmann 1

103.231 **Periodic comet Schwassmann– Wachmann 1.**
IAU Circ., Nos. 4200, 4206 (1986).

103.232 **Periodic comet Schwassmann–Wachmann 1.**
Yamamoto Circ., Nos. 2058, 2059 (1986). In Japanese.

Periodic comet Kopff

103.241 **Brightness asymmetry in comet Kopff.**
A. L. Cochran.
Asteroids, comets, meteors II, p. 317 – 321 (1986). – See Abstr. 012.010.
As part of the Faint Comet Survey at McDonald Observatory, the author observed comet Kopff during its 1983 apparition. This comet is a possible target for NASA's CRAF mission. The observations proved to be quite interesting since Kopff showed a brightness asymmetry. Using both a non–equilibrium chemistry model and a simpler Haser (1957) model, the author modeled the data and shows that there is an excess of gas pre–perihelion and a depletion of gas post–perihelion.

103.242 **Comet Kopff: a case for mantle development.**
A. L. Cochran.
Astron. J., Vol. 91, No. 3, p. 646 – 649 (1986).
The author observed comet Kopff during its 1983 apparition. Intensified Dissector Scanner (IDS) spectra were obtained mapping the distribution of the gases in the coma. The comet appears to have a preperihelion brightness maximum in monochromatic light. The author shows this result to be model independent. These data are consistent with a partial mantle developing due to the high obliquity of the comet.

Periodic comet Singer Brewster

103.251 **Periodic comet Singer Brewster (1986d).**
IAU Circ., Nos. 4209, 4211, 4214, 4227 (1986).

103.252 **Comet Singer Brewster (1986d).**
Yamamoto Circ., No. 2059 (1986). In Japanese.

Comet 1983 VII IRAS–Araki–Alcock

103.261 **On the anisotropic gas coma observed in comet IRAS–Araki–Alcock 1983d.**
K. Jockers, H. L. F. Houpis, M. Geffert.
Asteroids, comets, meteors II, p. 331 – 335 (1986). – See Abstr. 012.010.
The anisotropic gas coma of comet IRAS–Araki–Alcock 1983d is analyzed in terms of the chemical differentiation model of the nucleus and a modified form of Haser's atmospheric model. The observations and theory correlate reasonably well.

103.262 **The distribution of dust in the inner coma of comet IRAS–Araki–Alcock (1983d).**
A. D. Storrs, A. T. Tokunaga, C. A. Christian, J. N. Heasley.
Icarus, Vol. 66, No. 1, p. 143 – 153 (1986).
High–resolution images of comet IRAS–Araki–Alcock taken on five separate nights through its closest approach to the Earth are presented. The images were taken in the red and near–infrared, and so consist primarily of continuum scattering from dust grains. The distribution of dust in the inner coma varied widely from day to day, although the dust was emitted primarily in the sunward direction on all days.

103.263 **The origin of CS in comet IRAS–Araki–Alcock 1983d.**
W. M. Jackson, P. S. Butterworth, D. Ballard.
Astrophys. J., Vol. 304, No. 1, p. 515 – 518 (1986).
It is shown that the IUE observations of CS in the atmosphere of comet IRAS–Araki–Alcock 1983d are consistent with its formation by the photodissociation of CS_2 released from the nucleus. The observed scale length can be matched by a calculated scale length, which is based on the photochemical lifetime of CS_2 at 1 AU and hydrodynamic models of the velocity field in the coma.

Periodic comet Machholz

103.271 **Periodic comet Machholz (1986e).**
IAU Circ., Nos. 4214, 4217, 4222, 4223, 4226 (1986).

103.272 **Periodic comet Machholz (1986e).**
Yamamoto Circ., Nos. 2060, 2061 (1986). In Japanese.

Periodic comet Tempel 1

103.281 **A preliminary analysis of IRAS AO's (*additional observations*) of Comet Tempel 1.**
J. C. Zarnecki, S. Chakaveh, D. Bowen, J. A. M. McDonnell.
Asteroids, comets, meteors II, p. 403 – 406 (1986). – See Abstr. 012.010.
A preliminary analysis of IRAS additional observations of Comet Tempel 1 is described. The overall spatial distribution of the dust component is obtained. The temperature of the grains determined from the integrated flux from the comet in the 12 and 25 μ bands yields a temperature some 10% above the equilibrium blackbody temperature. However a determination of the temperature distribution shows a variation over the range 220K to 270K.

Periodic comet Holmes

103.291 **Periodic comet Holmes (1986f)**
IAU Circ., No. 4225 (1986).

103.292 **Periodic comet Holmes (1986f).**
Yamamoto Circ., No. 2062 (1986). In Japanese.

Comet 1957 III Arend–Roland

103.301 **Defect in coincidence between dynamical and photometric centres for comet Arend–Roland (1956h) in May 1957.**
H. Debehogne.
Asteroids, comets, meteors II, p. 469 – 472 (1986). – See Abstr. 012.010.
The residuals of 107 observations obtained at Uccle between April 29 and June 1, 1957, by means of the Zeiss Double Astrograph (f = 2 m, d = 40 cm) of the comet Arend–Roland (1956h), during 16 nights, were computed. The elements of an hyperbolic orbit, by Hasegawa, were used. Before May 13, the residuals are oriented (directed) towards the sun. Later, they are 115° anticlockwise from the sun direction.

Comet 1983 XX SOLWIND 6

103.311 **Comet 1983 XX (SOLWIND 6).**
IAU Circ., No. 4229 (1986).

103.312 **Comet 1983 XX (SOLWIND) 6).**
Yamamoto Circ., No. 2063 (1986). In Japanese.

Comet 1982 VI Austin

103.321 **Photographic observations of comet Austin (1982g) in Tashkent.**
S. Kh. Azizov, Yu. M. Ivanov, E. Yu. Kolmakov.
Tsirk. Astron. Inst., Tashkent, No. 113, p. 37 – 39 (1985). In Russian.

103.322 **Position observations of comet Austin (1982q) in Tashkent.**
A. A. Latypov, M. R. Ehshmatov.
Komet. Tsirk., No. 349 (1986). In Russian.

Comet 1984 XIII Austin

103.331 **Spectroscopic observations of comet Austin 1984i.**
B. H. Foing, M. Festou, S. Char.
Rev. Mex. Astron. Astrofis., Vol. 12, Spec. Issue, p. 375 (1986). Abstract. – See Abstr. 012.066.

Periodic comet Churyumov–Gerasimenko

103.341 **Erratum: "The dust coma of periodic comet Churyumov–Gerasimenko (1982 VIII)" [Icarus, Vol. 64, No. 1, p. 11 – 19 (1981)].**
M. S. Hanner, E. Tedesco, A. T. Tokunaga, G. J. Veeder, D. F. Lester, F. C. Witteborn, J. D. Bregman, J. Gradie, L. Lebofsky.
Icarus, Vol. 65, No. 1, p. 158 (1986). See Abstr. 40.103.241.

Periodic comet Forbes

103.351 **Periodic comet Forbes (1986g).**
IAU Circ., No. 4229 (1986).

103.352 **Periodic comet Forbes (1986g).**
Yamamoto Circ., No. 2063 (1986). In Japanese.

103.353 **Periodic comet Forbes 1961 VI = 1974 IX = 1980 VI.**
OAA Comput. Sect. Circ., NK 478, 3 pp. (1985).

Comet 1970 II Bennett

103.361 **OAO–2 observations of comet Bennett (1970 II).**
C. F. Lillie.
Bull. Am. Astron. Soc., Vol. 17, No. 4, p. 874 (1985). Abstract. – See Abstr. 010.062.

103.362 **Scale–length's determination of the CN component of comet Bennett 1970 II.**
V. A. Dranevich, I. N. Matveev.
Komety Meteory, No. 37, p. 3 – 7 (1985). In Russian.

Comet 1986i Churyumov–Solodovnikov

103.371 **Comet Churyumov–Solodovnikov (1986i).**
Yamamoto Circ., No. 2064 (1986). In Japanese.

Periodic comet 1943 I Whipple–Fedtke–Tevzadze

103.381 **Photometric characteristics of comet 1943 I.**
M. Z. Markovich, R. S. Osherov.
Komety Meteory, No. 36, p. 38 – 41 (1984). In Russian.

Periodic comet Schwassmann–Wachmann 2

103.411 **Periodic comet Schwassmann–Wachmann 2 (1986h).**
Yamamoto Circ., No. 2064 (1986). In Japanese.

103.412 **Periodic comet Schwassmann–Wachmann 2 1961 VII = 1968 II = 1974 XIII = 1981 VI.**
OAA Comput. Sect. Circ., NK 482, 4 pp. (1985).

Periodic comet Ciffréo

103.421 **Periodic comet Ciffréo (1985p).**
Komet. Tsirk., Nos. 348, 349 (1985/1986). In Russian.

103.422 **Periodic comet Ciffréo (1985p).**
IAU Circ., Nos. 4158, 4163, 4195 (1986).

103.423 **Periodic comet Ciffréo (1985p).**
Yamamoto Circ., No. 2051 (1986). In Japanese.

Periodic comet Howell

103.431 **Periodic comet Howell 1981 X = 1981k.**
OAA Comput. Sect. Circ., NK 472, 2 pp. (1985).

Periodic comet Wirtanen

103.441 **Periodic comet Wirtanen (1985q).**
Komet. Tsirk., No. 349 (1986). In Russian.

103.442 **Periodic comet Wirtanen (1985q).**
IAU Circ., Nos. 4174, 4193 (1986).

103.443 **Periodic comet Wirtanen (1985q).**
Yamamoto Circ., No. 2057 (1986). In Japanese.

Periodic comet Jackson–Neujmin

103.451 **Periodic comet Jackson–Neujmin 1970 IX = 1978 XXVI.**
OAA Comput. Sect. Circ., NK 473, 1 p. (1985).

Periodic comet Stephan–Oterma

103.461 **Observations of comet Stephan–Oterma (1980g) in Alma–Ata.**
V. D. D'yakonova, A. A. Semenikin.
Komet. Tsirk., No. 349 (1986). In Russian.

Periodic comet Russell 2

103.471 **Periodic comet Russell 2 1980 III = 1980o.**
OAA Comput. Sect. Circ., NK 474, 1 p. (1985).

Periodic comet Shoemaker 3

103.481 **New short–period comet Shoemaker 3 (1986a).**
Komet. Tsirk., No. 350 (1986). In Russian.

103.482 **New periodic comet Shoemaker 1986a.**
Br. Astron. Assoc. Circ., No. 656 (1986).

103.483 **Periodic comet Shoemaker 3 (1986a).**
IAU Circ., Nos. 4165, 4166, 4167, 4171, 4180 (1986).

103.484 **Periodic comet Shoemaker 3 (1986a).**
Yamamoto Circ., Nos. 2052, 2053 (1986). In Japanese.

Periodic comet Klemola

103.491 **Periodic comet Klemola 1965 VI = 1976 X.**
OAA Comput. Sect. Circ., NK 475, 3 pp. (1985).

Periodic comet Ashbrook–Jackson

103.501 **Ephemeris of comet Ashbrook–Jackson (1948 IX).**
E. I. Kazimirchak–Polonskaya.
Komet. Tsirk., No. 351 (1986). In Russian.

Periodic comet West–Kohoutek–Ikemura

103.511 **Periodic comet West–Kohoutek–Ikemura 1975 IV = 1981 VIII.**
OAA Comput. Sect. Circ., NK 476, 2 pp. (1985).

Periodic comet Boethin

103.521 **Short–period comet Boethin (1985n).**
Komet. Tsirk., No. 351 (1986). In Russian.

103.522 **Periodic comet Boethin (1985n).**
IAU Circ., Nos. 4160, 4174, 4191, 4198 (1986).

103.523 **Periodic comet Boethin (1985n).**
Yamamoto Circ., Nos. 2052, 2057 (1986). In Japanese.

Periodic comet Kohoutek

103.531 **Periodic comet Kohoutek 1975 III = 1981 IX.**
OAA Comput. Sect. Circ., NK 477, 2 pp. (1985).

Comet 1985*l* Hartley–Good

103.541 **BVRI photometry of comet Hartley–Good (1985*l*).**
A. S. Miroshnichenko, N. Yu. Yutanov, K. I. Churyumov.
Komet. Tsirk., No. 351 (1986). In Russian.

103.542 **Comet Hartley–Good 1985*l*.**
Br. Astron. Assoc. Circ., No. 655 (1986).

103.543 **Comet Hartley–Good (1985*l*).**
IAU Circ., No. 4171 (1986).

Periodic comet du Toit–Hartley

103.551 **Periodic comet du Toit–Hartley 1982 II = 1982c.**
OAA Comput. Sect. Circ., NK 479, 2 pp. (1985).

Comet 1986b Shoemaker

103.561 **New comet Shoemaker (1986b).**
Komet. Tsirk., No. 353 (1986). In Russian.

103.562 **Comet Shoemaker (1986b).**
IAU Circ., Nos. 4191, 4193, 4201 (1986).

103.563 **Comet Shoemaker (1986b).**
Yamamoto Circ., Nos. 2055, 2056 (1986). In Japanese.

Periodic comet Grigg–Skjellerup

103.571 **Periodic comet Grigg–Skjellerup 1967 I = 1972 II = 1977 VI = 1982 IV.**
OAA Comput. Sect. Circ., NK 480, 3 pp. (1985).

Periodic comet Wolf

103.581 **The orbit of comet Wolf for the interval 1884 – 1925.**
Yu. D. Medvedev.
Komet. Tsirk., No. 353 (1986). In Russian.

Periodic comet Comas Solá

103.591 **Periodic comet Comas Solá 1969 VIII = 1978 XVII.**
OAA Comput. Sect. Circ., NK 481, 3 pp. (1985).

Periodic comet Halley

103.601 **Prospects for observing the occultation of pulsars by Halley's comet.**
R. G. Strom.
Astron. Astrophys., Vol. 154, No. 1/2, p. L33 – L34 (1986).
Pulsars, as compact, highly polarized low frequency radio emitters, are good potential probes of cometary atmospheres. Twelve catalogued pulsars are good occultation candidates during the current apparition of P/Halley, in most cases by its tail, but for PSR 1641–45 and PSR 1609–47 (part of) the coma will also be involved. Estimates of the dates and times of conjunction and tail occultation are provided.

103.602 **The theory of the motion of comet Halley.**
Eh. L. Akim, Ju. F. Koljuka (*Yu. F. Kolyuka*), V. V. Savtchenko (*V. V. Savchenko*), V. A. Stepanianz (*V. A. Stepan'yants*), N. M. Ivanov, V. F. Tikhonov.
Relativity in celestial mechanics and astrometry, p. 93 – 97 (1986). – See Abstr. 012.002 (IAU Symp. No. 114).
The theory of the motion of comet Halley is presented using ground optical positional observations of 1682, 1759, 1835, 1910 and 1982 – 84. The orbital elements of the comet and two coefficients characterizing non–gravitational perturbations were determined precisely as a result of the statistical treatment of the observations. The estimations of the accuracy of the theory and the data describing its agreement with the observations are presented. The theory has been applied to ballistic–navigational calculations for the Soviet Vega project.

103.603 **L'orbite de la comète de Halley.**
B. Morando.
C. R. Acad. Sci., Sér. Gén., Vie Sci., Tome 3, No. 1, p. 1 – 14 (1986).

103.604 **La comète de Halley: observation au sol et dans l'espace.**
T. Encrenaz.
C. R. Acad. Sci., Sér. Gén., Vie Sci., Tome 3, No. 1, p. 15 – 37 (1986).

103.605 **La cometa di Halley nel 1910.**
M. Di Martino.
Orione, Vol. 6, N. 1, p. 38 – 51 (1986).

103.606 **Update on P/Halley's light curve.**
C. S. Morris.
IHW Newsl., No. 8, p. 2 – 3 (1986).

103.607 **Dust tail orientation for P/Halley in 1985 – 86.**
T. Ozeki.
IHW Newsl., No. 8, p. 4 – 9 (1986).

103.608 **The apparent length of comet Halley's tail.**
S. J. Edberg.
IHW Newsl., No. 8, p. 10 (1986).

103.609 **Porträts des Kometen Halley. Die Geschichte vom kosmischen Vagabunden.**
R. Froböse.
Sterne Weltraum, 25. Jahrg., Nr. 2, p. 74 – 76 (1986).

103.610 **The first near–UV observation of P/Halley.**
M. C. Festou, C. Arpigny, J. L. Bertaux, W. Carey, T. Danks, R. Gilmozzi, D. W. Hughes, W. Ip, P. Patriarchi, G. P. Tozzi, M. K. Wallis, J. Zarnecki.
Astron. Astrophys., Vol. 155, No. 2, p. L17 – L18 (1986).
Comet Halley was observed on April 1, 1985 with the long wavelength camera of the IUE observatory. There is some indication that an OH emission was present at ~3085 Å. This emission would correspond to an OH production rate of about a few 10^{28} molecules per second.

103.611 **The historical records of Halley's comet in China.**
D. C. Wang, S. M. Gong.
Earth, Moon, Planets, Vol. 34, No. 1, p. 55 – 57 (1986).

103.612 **The comet Halley dust and gas environment.**
N. Divine, H. Fechtig, T. I. Gombosi, M. S. Hanner, H. U. Keller, S. M. Larson, D. A. Mendis, R. L. Newburn Jr., R. Reinhard, Z. Sekanina, D. K. Yeomans.
Space Sci. Rev., Vol. 43, Nos. 1/2, p. 1 – 104 (1986).
Quantitative descriptions of environments near the nucleus of comet *P*/Halley have been developed to support spacecraft and mission design for the flyby encounters in March, 1986. To summarize these models as they exist just before the encounters, the authors review the relevant data from prior Halley apparitions and from recent cometary research. Orbital elements, visual magnitudes, and parameter values and analysis for the nucleus, gas and dust are combined to predict Halley's position, production rates, gas and dust distributions, and electromagnetic radiation field for the current perihelion passage. The predicted numerical results have been useful for estimating likely spacecraft effects, such as impact damage and attitude perturbation. Sample applications are cited, including design of a dust shield for spacecraft structure, and threshold and dynamic range selection for flight experiments.

103.613 **The uncertainty of position predictions for comet Halley in March 1986.**
W. Landgraf.
Astron. Astrophys., Vol. 157, No. 2, p. 245 – 251 (1986).
With reference to the space missions to comet Halley, its position uncertainty for mid March 1986 is investigated. Without light biases in the observations, one can expect a position uncertainty of approximately 100 km, and in cases of light shifts which cannot be modelled, up to 700 km. The light shift between the center of light or the image center and the nucleus of the comet is the largest source of uncertainty, the nongravitational forces are of negligible importance compared to this. With increased observational effort during the first part of March 1986, the accuracy can improve significantly. A source of large systematic errors are the differences between the reference system for astrometric observations (FK4) and earth coordinates (DE 118) and the navigation frame of the space probes which should be corrected.

103.614 **The nature of comet Halley outbursts at large heliocentric distances.**
O. V. Dobrovol'skij.
Kinematika Fiz. Nebesn. Tel, Tom 2, No. 1, p. 66 – 69 (1986). In Russian. English translation in Kinematics Phys. Celest. Bodies.
Comet Halley outbursts at large heliocentric distances (≈8 AU) are interpreted as being a result of lifting and sinking of dust clouds, gravitationally coupled with the comet nucleus and influenced by reduced nonstationary thermal sublimation drag.

103.615 **Halley's comet as an outstanding target.**
R. Reinhard.
ESA Spec. Publ., ESA SP–1066, p. VII – XVI (1986). – See Abstr. 003.012.
Among more than 1000 known comets, Halley is the most outstanding target for a cometary mission as it is the only comet that has a well–known orbit and a high gas and dust production rate, comparable to that of new comets. A mission to Halley requires very little launch energy and the comet can also be observed from the Earth during the flybys. Because of its predictability and brightness Halley is also the most famous comet. For all of these reasons, Halley was selected as the target for the first cometary mission. Between 6 and 14 March 1986, five spacecraft from three space agencies will fly through the coma of Halley's comet. The encounters all take place in March 1986 because the launch energy required for these post–perihelion encounters is minimal and in March, four weeks after its perihelion passage,

Halley will be at its most active, displaying all known cometary phenomena.

103.616 **Halley's comet in history.**
D. K. Yeomans, J. Rahe, R. Freitag.
ESA Spec. Publ., ESA SP–1066, p. 1 – 24 (1986). – See Abstr. 003.012.

103.617 **The comet Halley ephemeris development effort.**
D. K. Yeomans.
ESA Spec. Publ., ESA SP–1066, p. 179 – 197 (1986). – See Abstr. 003.012.

103.618 **Vega spacecraft reveal more dust than expected.**
J. Maddox.
Nature, Vol. 320, No. 6058, p. 97 – 98 (1986).
A brief acount of the encounter of the two Soviet Vega space probes with comet Halley on March 6 and March 10, 1986 is presented.

103.619 **Telling the truth in real time.**
J. Maddox.
Nature, Vol. 320, No. 6058, p. 98 – 99 (1986).
This note presents a science reporter's experience of the activities at the ground control center in Moscow during the encounters of the Soviet Vega space probes with comet Halley.

103.620 **Japan's probe rocked by dust.**
A. Anderson.
Nature, Vol. 320, No. 6058, p. 99 (1986).
This note discusses some of the first preliminary results from the encounter of the Japanese space probe *Suisei* with comet Halley on March 9, 1986.

103.621 **Strong breathing of the hydrogen coma of comet Halley.**
E. Kaneda, K. Hirao, M. Takagi, O. Ashihara,
T. Itoh, M. Shimizu.
Nature, Vol. 320, No. 6058, p. 140 – 141 (1986).
Strong breathing of the hydrogen coma of comet Halley was observed by the Lyman–α imager of the Japanese interplanetary spacecraft, Suisei. Measurement of the peak–to–peak variation of central brightness yields a preliminary value of 2.2 ± 0.1 days for the period. Two outbursts from the active regions on the comet's surface could be the cause of this variation. A shell structure in the hydrogen coma was also detected. The outbursts appear to contribute overwhelmingly to the Lyman–α activity of this comet, at a pre–perihelion heliocentric distance of ~ 1.5 AU.

103.622 **Giotto confirms reports of dumbbell–shaped nucleus.**
P. Campbell.
Nature, Vol. 320, No. 6059, p. 202 – 203 (1986).
This note describes the events at the Darmstadt control center during the encounter of ESA's space probe Giotto with comet Halley on March 14, 1986 and comments on some immediately apparent scientific results from this mission.

103.623 **A year for comet studies.**
K. Bracher.
Phys. Today, Vol. 39, No. 1, p. S6 – S7 (1986). Short review paper. – See Abstr. 013.021.

103.624 **La comète de Halley est détectée pour la première fois en ondes radio avec le radiotélescope de Nançay.**
D. Bockelée–Morvan, G. Bourgois, P. Colom, J. Crovisier, E. Gérard.
Astronomie, Vol. 100, p. 151 – 152 (1986).

103.625 **Astrometry and photometry of comet Halley (1982i).**
S. I. Gerasimenko, N. N. Kiselev, G. P. Chernova.
Pis'ma Astron. Zh., Tom 12, No. 2, p. 149 – 155 (1986). In Russian. English translation in Sov. Astron. Lett., Vol. 12.
Results of astrometric and photometric observations of Halley's comet with an 1–m telescope are given. Photometric parameters M_0 and n are determined. It is obtained that the brightness of the comet increased slowly ($n \approx 1$) in the heliocentric distance range 11 – 8 a.u. and significantly more faster ($n \approx 6$) in the range 8 – 4 a.u. This behaviour at large distances may be explained by formation of dust clouds, so that not the nucleus itself but a pseudo–nucleus is observed the size of which diminishes as the gas sublimation rate rises. At small distances the brightness is controlled by the common gas–and–dust atmosphere.

103.626 **Motion of P/Halley in the interval 1759 – 1985.**
Yu. V. Batrakov, N. A. Belyaev, Yu. D. Medvedev,
Yu. A. Chernetenko.
Kinematika Fiz. Nebesn. Tel, Tom 2, No. 2, p. 92 – 94 (1986). In Russian. English translation in Kinematics Phys. Celest. Bodies.
The initial coordinates and velocity components of P/Halley are obtained along with non–gravitational parameters. Orbits are computed for two models of non–gravitational forces: the Marsden–Sekanina and impulse ones. Both orbits give reasonable residuals for normal places and observations and are close one to another at the initial moment. The prognosticated time of the perihelion passage in 1986 is different by 0.021 days.

103.627 **VEGA's 1 and 2 visit Halley.**
R. A. Kerr.
Science, Vol. 231, No. 4744, p. 1366 (1986).

103.628 **Review of positional observations of Halley's comet during its apparition in 1835.**
S. Szutowicz.
Postepy Astron., Tom 33, Zesz. 1–2, p. 51 – 57 (1985). In Polish.
The paper presents a review of positional observations of Halley's comet being made in 1835 – 1836. Problems connected with old observation sets are considered. Methods of observations made with a circular (or ring) micrometer and of observations made with equatorial and meridian instruments are described. The author discusses the sources of errors in observations: inaccurate positions of reference stars, failures in the reduction process, instrumental errors.

103.629 **Observations at La Silla of comet Halley after perihelion.**
R. M. West.
Messenger, No. 43, p. 5 – 6 (1986).

103.630 **Pourquoi la comète de Halley se montre–t–elle si discrète?**
R. Papy, A. Noels, I. Ottelet.
Ciel Terre, Vol. 102, No. 2, p. 45 – 47 (1986).

103.631 **Hydrodynamic expansion of cometary gas loaded with refractory and volatile grains.**
J. F. Crifo.
Asteroids, comets, meteors II, p. 389 – 392 (1986). – See Abstr. 012.010.
The expansion in free space of the mixture of gas and grains resulting from solar illumination of a comet nucleus is considered. It is assumed that the ice is not only vaporized, but also fragmented, so that volatile grains are carried along with refractory ones by the gas. Some results appropriate to comet Halley at the time of the flyby missions are presented. They illustrate the possible strong influence of icy grains on the thermodynamic state of the gas in the coma, and suggest molecular infrared band spectroscopy as a possible means of diagnostics for icy haloes.

103.632 **Condensation and sublimation of ice particles in the circumnuclear region of a comet.**
T. Yamamoto, O. Ashihara, Y. Kitamura.
Asteroids, comets, meteors II, p. 393 – 396 (1986). – See Abstr. 012.010.
It is found that H_2O molecules once sublimated from a cometary nucleus recondense into very fine pure ice particles in the immediate vicinity of the nucleus when the H_2O production rate is larger than $\sim 10^{29} s^{-1}$. This implies that a gas–free zone is achieved in the circumnuclear region of a comet. A discussion is given of the condensation and subsequent sublimation processes of these ice particles and their properties expected at the encounters of the spacecraft to the coming of Halley's comet.

103.633 **The accuracy of space ephemerides for Halley's comet.**
G. Sitarski.
Asteroids, comets, meteors II, p. 461 – 463 (1986). – See Abstr. 012.010.

Basing on the real moments of observation and the corresponding observational errors a set of fictitious observations was created for the apparitions of comet P/Halley in 1835 – 36, 1909 – 11 and 1982 – 86 according to an assumed mathematical model of the motion of the comet. Several improvements of the orbit of the comet were performed using these fictitious observations in different ways.

103.634 **The brightness variations of comet Halley at large heliocentric distances.**
K. R. Flammer, B. Jackson, H. L. F. Houpis, D. A. Mendis.
Asteroids, comets, meteors II, p. 477 – 480 (1986). – See Abstr. 012.010.

The irregular brightness variations of comet Halley observed between 11 and 8 AU is explained in terms of a model that involves electrostatic blow–off of fine dust from the surface. This blow–off is associated with the encounter of high–speed solar wind streams by the comet.

103.635 **Photometric and spectrophotometric activity of P/Halley during 1984 – 85.**
E. S. Barker, C. B. Opal.
Asteroids, comets, meteors II, p. 481 – 485 (1986). – See Abstr. 012.010.

Imaging and spectroscopic observations of P/Halley have been carried out at McDonald Observatory using an RCA CCD camera and an Image Dissector Scanner Spectrograph. The authors have obtained broadband CCD images on Sept. 28, Nov. 1, Dec. 22, 1984 and Feb. 18, 1985 and spectra on Dec. 23, 1984 and Feb. 15, 16, 17, 18 and Apr. 18, 1985. Halley activity was monitored photometrically on Nov. 1 and on Dec. 22, 1984; no periodic variations or consistent trends in brightness were observed. The coma was definitely diffuse during the Dec. 1984 and Feb. 1985 observations. P/Halley activity appears to begin around 6.23 au and varies as $r^{-5.3}$ for heliocentric distances between 5.8 and 4.2 au.

103.636 **Detection of CN emission in comet P/Halley (1982i).**
S. Wyckoff, P. A. Wehinger, M. J. S. Belton.
Asteroids, comets, meteors II, p. 487 – 490 (1986). – See Abstr. 012.010.

The authors report the first detection of CN emission in comet P/Halley (1982i), and describe the observing technique used to locate a faint moving object.

103.637 **Kitt Peak spectroscopy of P/Halley: October 1983 – March 1985.**
H. Spinrad, M. J. S. Belton, P. A. Wehinger, S. Wyckoff.
Asteroids, comets, meteors II, p. 491 – 496 (1986). – See Abstr. 012.010.

Four reasonably good low–resolution (15 Å) optical spectra (4600 – 7900 Å) of P/Halley were obtained at heliocentric distances between 8.78 and 4.83 A.U., using the Cryogenic Camera and R–C spectrograph on the 4–m Mayall reflector of Kitt Peak National Observatory. None of the early spectra show emission features of the usual cometary radicals; the reflection spectra probably indicate a progressive change in the color of the cometary reflection continuum. The authors discuss the substantial extent of the Halley dust coma, visible in the long slit spectra where cuts perpendicular to the dispersion show an asymmetric profile, on a spectrogram taken 30 Oct. 1984 U.T., at r = 5.84 A.U. The gaseous coma became spectroscopically active in February 1985. The authors derive an [O I]^{1}D production rate of about 3×10^{26}atoms s^{-1} (with a Haser model) on 18 Feb. 1985 U.T.

103.638 **Debris from comet Halley and the risk for space probes.**
A. Hajduk.
Asteroids, comets, meteors II, p. 497 – 500 (1986). – See Abstr. 012.010.

The mass distribution function of particles as determined from the radar and optical observations of meteor showers associated with comet Halley is studied. The particle distribution derived from the observed mass loss rate of the comet, the expansion rate of the dust envelope of ejected particles and the density variations in it suggest that the total influence of particles impacting comet Halley space probes is at least one order of magnitude higher than expected on the basis of earlier results. Hence the survival probability of these probes, especially with regard to maintaining the required stability in their orientation, appears to be considerably less than previously assumed.

103.639 **Observations of hydrogen cyanide in comet Halley.**
D. Despois, J. Crovisier, D. Bockelée–Morvan, J. Schraml, T. Forveille, E. Gérard.
Astron. Astrophys., Vol. 160, No. 2, p. L11 – L12 (1986).

The J 1→0 line of HCN at 88.6 GHz was detected in comet Halley during the Nov. 19 – Dec. 3, 1985 period. The observed line width corresponds to a coma expansion velocity of 0.85 km s^{-1}. The average HCN production rate was $5\times10^{25}s^{-1}$.

103.640 **Giotto finds a big black snowball at Halley.**
R. A. Kerr.
Science, Vol. 231, No. 4745, p. 1502 – 1503 (1986).

Halley's dust took its toll on Giotto but not before the spacecraft revealed that the comet's nucleus is larger, darker, and more encrusted with debris than expected.

103.641 **An inside look at Halley's comet.**
J. K. Beatty.
Sky Telesc., Vol. 71, No. 5, p. 438 – 443 (1986).

103.642 **Die Nacht der Begegnung.**
H. Fechtig.
Sterne Weltraum, 25. Jahrg., Nr. 5, p. 255 – 257 (1986).

103.643 **Comet Halley news.**
S. Wyckoff, P. Wehinger.
Mercury, Vol. 15, No. 1, p. 22 – 24, 31 (1986).

103.644 **Comet Halley news.**
S. Wyckoff, P. Wehinger.
Mercury, Vol. 15, No. 2, p. 51 – 52 (1986).

103.645 **Rotation of Halley.**
J. Watanabe.
Proceedings of the nineteenth symposium on celestial mechanics, p. 89 – 92 (1986). In Japanese. – See Abstr. 012.013.

103.646 **Physical observations of comet P/Halley using color filters.**
C. F. Capen.
Strolling Astron., Vol. 31, Nos. 5–6, p. 89 – 94 (1986).

The detection and study of physical properties of bright comets with the aid of high–transmission color filters is described. Orbital aspects and observing prospects for the P/Halley 1985/86 apparition are discussed.

103.647 **Pre–perihelion observations of comet P/Halley (1982i).**
S. Wyckoff, P. A. Wehinger, M. J. S. Belton, H. Spinrad.
Bull. Am. Astron. Soc., Vol. 17, No. 4, p. 873 – 874 (1985). Abstract. – See Abstr. 010.062.

103.648 **IUE observations of comet P/Halley in September 1985.**
P. D. Feldman, H. A. Weaver, M. F. A'Hearn, L. A. McFadden, M. C. Festou, C. Arpigny, J.–L. Bertaux, T. Encrenaz, W. Ip, H. U. Keller, P. Benvenuti, C. B. Cosmovici, W. Carey, J. Zarnecki, T. Danks, N. Evans, M. K. Wallis, K. S. Krishna–Swamy, D. Hughes, R. Gilmozzi, P. Patriarchi, G. P. Tozzi.
Bull. Am. Astron. Soc., Vol. 17, No. 4, p. 918 (1985). Abstract. – See Abstr. 010.063.

103.649 **Structures in the coma of comet Halley.**
E. Grün, U. Graser, L. Kohoutek, U. Thiele, L. Massonne, G. Schwehm.
Nature, Vol. 321, No. 6066, p. 144 – 147 (1986).

The authors report ground–based measurements of the optical brightness distribution of the coma of comet Halley, made in November 1985, when the comet was at a distance of 1.75 AU from the sun. From the total brightness the authors deduced a dust production rate of $1.5 \times 10^6 g\, s^{-1}$, but the results show that cometary dust and gas production, which seem to be correlated, are strongly anisotropic and time–variable. Structures in the coma can be revealed by comparison with a physical model of the coma, but astrometry of the comet nucleus is adversely affected by the anisotropic brightness distribution in the coma.

103.650 **"Portraits" of comet Halley.**
V. M. Linkin.
Zemlya Vselennaya, No. 3, p. 2 – 3 (1986). In Russian.

103.651 **Large–scale phenomena in the tail of comet Halley.**
K. I. Churyumov.
Zemlya Vselennaya, No. 3, p. 5 – 10 (1986). In Russian.

103.652 **Comet Halley before perihelion.**
N. S. Chernykh.
Zemlya Vselennaya, No. 3, p. 11 – 12 (1986). In Russian.

103.653 **Observations of Halley's comet in Tadzhikistan.**
O. V. Dobrovol'skij.
Zemlya Vselennaya, No. 3, p. 13 – 14 (1986). In Russian.

103.654 **Spectrophotometry of comet Halley.**
A. L. Gural'chuk, M. S. Dement'ev, V. M. Klimenko.
Kinematika Fiz. Nebesn. Tel, Tom 2, No. 3, p. 91 – 92 (1986). In Russian. English translation in Kinematics Phys. Celest. Bodies.

Preliminary results of a spectrophotometry of comet Halley from October 20 – November 13, 1985 are presented. The relative energy distribution is given in the spectral range 350 – 660 nm. The relative intensity of the C_2 emission bands was shown to be constant during the period of observations excepting the 5th and 6th of November.

103.655 **On outbursts of comet Halley at large distances from the sun.**
O. V. Dobrovol'skij.
Dokl. Akad. Nauk TadzhSSR, Tom 28, No. 6, p. 333 – 335 (1985). In Russian. Abstr. in Ref. Zh., 51. Astron., 4.51.214 (1986).

103.656 **A model of cometary gas and dust production and nongravitational forces with application to P/Halley.**
F. P. Fanale, J. R. Salvail.
Icarus, Vol. 66, No. 1, p. 154 – 164 (1986).

A program that computes gas and dust production rates and idealized nongravitational force components has been developed and applied to the case of Comet Halley. The authors use a modified form of their earlier comet model to which coma effects and a section on nongravitational forces have been added. The possibility of grain cohesion is also included. These models are used together with observations from 1910 and semiempirically derived data to investigate the effects of obliquity and thermal conductivity of the near surface nucleus on gas and dust production. A general method is presented for calculating the radii of cometary nuclei using theoretically derived and semiempirically derived nongravitational force components. This method is used to calculate possible radii for Comet Halley that depend on the model variation chosen.

103.657 **Max Wolfs Halley–Beobachtungen von 1910.**
G. Klare.
Sterne Weltraum, 25. Jahrg., Nr. 4, p. 203 – 205 (1986).

103.658 **Periodic comet Halley (1982i).**
Yamamoto Circ., Nos. 2051 – 2059, 2062 (1986). In Japanese.

103.659 **Halley's comet in 1910.**
H. J. P. Arnold.
Spaceflight, Vol. 28, No. 1, p. 30 – 34 (1986).

103.660 **P/Halley's light curve: some initial thoughts.**
C. S. Morris.
Int. Comet Q., Vol. 8, No. 1, p. 7 – 8 (1986).

103.661 **Vega spacecraft encounters with comet Halley.**
R. Z. Sagdeev, J. Blamont, A. A. Galeev, V. I. Moroz, V. D. Shapiro, V. I. Shevchenko, K. Szegö.
Nature, Vol. 321, No. 6067, p. 259 – 262 (1986).

The spacecraft Vega 1 and Vega 2 encountered comet Halley on 6 and 9 March 1986. Their scientific payload comprised 14 instruments, which collected data concerning the comet's optical characteristics, dust emission, and neutral gas, plasma and electromagnetic field environment. The main features of the encounters are described and highlights of the first scientific results are presented.

103.662 **Television observations of comet Halley from Vega spacecraft.**
R. Z. Sagdeev, F. Szabó, G. A. Avanesov, R. Cruvellier, L. Szabó, K. Szegö, A. Abergel, A. Balazs, I. V. Barinov, J.–L. Bertaux, J. Blamont, M. Detaille, E. Demarelis, G. N. Dul'nev, G. Endröczy, M. Gardos, M. Kanyo, V. I. Kostenko, V. A. Krasikov, T. Nguyen–Trong, Z. Nyitrai, I. Reny, P. Rusznyak, V. A. Shamis, B. Smith, K. G. Sukhanov, F. Szabó, S. Szalai, V. I. Tarnopolsky (*V. I. Tarnopolskij*), I. Toth, G. Tsukanova, B. I. Valníček, L. Varhalmi, Yu. K. Zaiko, S. I. Zatsepin, Ya. L. Ziman, M. Zsenei, B. S. Zhukov.
Nature, Vol. 321, No. 6067, p. 262 – 266 (1986).

The television system aboard the spacecraft Vega 1 and Vega 2 observed comet Halley from 4 to 11 March 1986, and transmitted $\sim$1,500 images to Earth. The best images of the cometary nucleus were obtained from distances of 8,000 – 9,000 km. The nucleus is an irregular potato–shaped body, $\sim$14x7.5x7.5 km in size, which rotates with a period of 53 ± 3 h. The surface is rather dark; the measured albedo is $\sim$0.04.

103.663 **Infrared sounding of comet Halley from Vega 1.**
M. Combes, V. I. Moroz, J. F. Crifo, J. M. Lamarre, J. Charra, N. F. San'ko, A. Soufflot, J. P. Bibring, S. Cazes, N. Coron, J. Crovisier, C. Emerich, T. Encrenaz, R. Gispert, A. V. Grigoryev (*A. V. Grigor'ev*), G. Guyot, V. A. Krasnopolsky (*V. A. Krasnopol'skij*), Yu. V. Nikolsky (*Yu. V. Nikol'skij*), F. Rocard.
Nature, Vol. 321, No. 6067, p. 266 – 268 (1986).

The IKS instrument carried by the Vega 1 spacecraft performed infrared sounding of the inner coma and nuclear region of comet Halley. An emissive centre, a few kilometers in size, was detected with a temperature significantly in excess of 300K. In the surrounding central coma, hydrocarbon species or other carbonaceous material seem to be present. The fluorescence signatures of H_2O and CO_2 were identified.

103.664 **Spectroscopic study of comet Halley by the Vega 2 three–channel spectrometer.**
V. A. Krasnopolsky (*V. A. Krasnopol'skij*), M. Gogoshev, G. Moreels, V. I. Moroz, A. A. Krys'ko, Ts. Gogosheva, K. Palazov, S. Sargoichev, J. Clairemidi, M. Vincent, J. L. Bertaux, J. E. Blamont, V. S. Troshin, B. Valníček.
Nature, Vol. 321, No. 6067, p. 269 – 271 (1986).

The three–channel spectrometer (TKS) aboard the Vega 2 spacecraft recorded infrared and visible spectra near the nucleus of comet Halley. Spectra in the range 0.95 – 1.9 µm at a distance of 300 km from the nucleus reveal the H_2O 1.38–µm band and OH bands, the latter excited by dissociation of a parent molecule. Their production rates are 4×10^{29} and 1.7×10^{30} molecules s^{-1}, respectively. The dust particle density is $0.35\, g\, cm^{-3}$.

103.665 **Near–ultraviolet and visible spectrophotometry of comet Halley from Vega 2.**
G. Moreels, M. Gogoshev, V. A. Krasnopolsky (*V. A. Krasnopol'skij*), J. Clairemidi, M. Vincent, J. P. Parisot, J. L. Bertaux, J. E. Blamont, M. C. Festou, Ts. Gogosheva, S. Sargoichev (*S. Sargojchev*), K. Palasov, V. I. Moroz, A. A. Krys'ko, V. Vanyšek.
Nature, Vol. 321, No. 6067, p. 271 – 273 (1986).
The authors report results obtained by the near–ultraviolet and visible channel (V) of the TKS instrument carried by Vega 2 during the hour preceding its closest approach to comet Halley. The brightness of the OH (0,0) band is determined at various distances from the nucleus. The OH gas production rate is estimated to have been $\sim 9x10^{29}$molecules s^{-1} on 9 March 1986.

103.666 **Neutral gas measurements of comet Halley from Vega 1.**
E. Keppler, V. V. Afonin, C. C. Curtis, A. V. D'yachkov, J. Erö Jr., C. Y. Fan, K. C. Hsieh, D. M. Hunten, W.–H. Ip, A. K. Richter, A. J. Somogyi, G. Umlauft.
Nature, Vol. 321, No. 6067, p. 273 – 274 (1986).
The neutral gas experiment (ING) aboard the Vega 1 spacecraft detected cometary gas within a distance of 60,000 km from the nucleus of comet Halley. A preliminary inspection of the data permits an analysis of the variation in neutral gas density. Fine structure was detected in the spatial distribution of lower–mass species.

103.667 **Dust coma structure of comet Halley from SP–1 detector measurements.**
O. L. Vaisberg (*O. L. Vajsberg*), V. N. Smirnov, L. S. Gorn, M. V. Iovlev, M. A. Balikchin, S. I. Klimov, S. P. Savin, V. D. Shapiro, V. I. Shevchenko.
Nature, Vol. 321, No. 6067, p. 274 – 276 (1986).
The SP–1 detectors aboard the Vega 1 and Vega 2 spacecraft measured the spectral and spatial distributions of dust grains, with the aim of investigating the dust release process of the comet nucleus. The dust particle mass spectra do not exhibit the expected low–mass cutoff at 10^{-14}g; instead, they continue to rise to $\sim 10^{-16}$g. The apex of the dust paraboloid resulting from scattering of sunlight is estimated to lie at a distance of 40 – $45x10^{3}$km from the nucleus.

103.668 **Comet Halley dust environment from SP–2 detector measurements.**
E. P. Mazets, R. L. Aptekar (*R. L. Aptekar'*), S. V. Golenetskii (*S. V. Golenetskij*), Yu. A. Gur'yan, A. V. D'yachkov, V. N. Ilyinskii (*V. N. Il'inskij*), V. N. Panov, G. G. Petrov, A. V. Savvin, R. Z. Sagdeev, I. A. Sokolov, N. G. Khavenson, V. D. Shapiro, V. I. Shevchenko.
Nature, Vol. 321, No. 6067, p. 276 – 278 (1986).
The SP–2 experiment carried by the Vega spacecraft during their encounter with comet Halley was designed to determine the spatial distribution of cometary dust particles and their mass distribution over the range $\sim 10^{-16} - 10^{-6}$g. The first dust particle impacts indicated that Vega 1 entered the cometary dust coma at a distance of $\sim$280,000 km from the nucleus. A number of remarkable features were revealed, such as the high density of sub–micrometre–sized particles and considerable variations in the particle mass distribution observed in different regions of the coma.

103.669 **Dust counter and mass analyser (DUCMA) measurements of comet Halley's coma from Vega spacecraft.**
J. A. Simpson, R. Z. Sagdeev, A. J. Tuzzolino, M. A. Perkins, L. V. Ksanfomality (*L. V. Ksanfomaliti*), D. Rabinowitz, G. A. Lentz, V. V. Afonin, J. Erö Jr., E. Keppler, J. Kosorokov, E. Petrova, L. Szabó, G. Umlauft.
Nature, Vol. 321, No. 6067, p. 278 – 280 (1986).
The Vega encounters with comet Halley allowed the first direct measurements of the spatial and temporal distributions of the masses and fluxes of dust particles originating from the comet nucleus. The dust coma whether quiescent (as seen by Vega 2) or containing a major jet structure (as seen by Vega 1), displays large, short–term variations throughout, which are at times quasi–periodic. The integral mass spectra increase in intensity to the lowest masses measured.

103.670 **Composition of comet Halley dust particles from Vega observations.**
J. Kissel, R. Z. Sagdeev, J. L. Bertaux, V. N. Angarov, J. Audouze, J. E. Blamont, K. Büchler, E. N. Evlanov, H. Fechtig, M. N. Fomenkova, H. von Hoerner, N. A. Inogamov, V. N. Khromov, W. Knabe, F. R. Krueger, Y. Langevin, V. B. Leonas, A. C. Levasseur–Regourd, G. G. Managadze, S. N. Podkolzin, V. D. Shapiro, S. R. Tabaldyev, B. V. Zubkov.
Nature, Vol. 321, No. 6067, p. 280 – 282 (1986).
The dust impact mass analyser (PUMA) carried by the spacecraft Vega 1 and Vega 2 has provided the first direct measurements of the physical and chemical properties of cometary dust. Particles of mass $< 10^{-14}$g are much more abundant than was predicted by models. Most of the particles are rich in light elements such as H, C, N and O, lending support to models that describe cometary material as consisting of radiation–processed ices.

103.671 **First *in situ* plasma and neutral gas measurements at comet Halley.**
K. I. Gringauz, T. I. Gombosi, A. P. Remizov, I. Apáthy, I. Szemerey, M. I. Verigin, L. I. Denchikova, A. V. D'yachkov, E. Keppler, I. N. Klimenko, A. K. Richter, A. J. Somogyi, K. Szegö, S. Szendrö, M. Tátrallyay, A. Varga, G. A. Vladimirova.
Nature, Vol. 321, No. 6067, p. 282 – 285 (1986).
The authors present the first *in situ* observations and a description of the large–scale behaviour of comet Halley's plasma environment. The PLASMAG–1 experiment, carried aboard the spacecraft Vega 1 and Vega 2, has observed a discontinuity (the 'cometopause') between the solar–wind–controlled cometosheath and heavy–ion mantle and the magnetized cometary plasma region. From the measured neutral gas density distribution the authors estimate a total gas production rate of $1.3x10^{30}$molecules s^{-1}.

103.672 **First observations of energetic particles near comet Halley.**
A. J. Somogyi, K. I. Gringauz, K. Szegö, L. Szabó, G. Kozma, A. P. Remizov, J. Erö Jr., I. N. Klimenko, I. T. Szücs, M. I. Verigin, J. Windberg, T. E. Cravens, A. V. D'yachkov, G. Erdõs, M. Faragó, T. I. Gombosi, K. Kecskeméty, E. Keppler, T. Kovács Jr., A. Kondor, Yu. I. Logachev, L. Lohonyai, R. Marsden, R. Redl, A. K. Richter, V. G. Stolpovskii (*V. G. Stolpovskij*), J. Szabó, I. Szentpétery, A. Szepesváry, M. Tátrallyay, A. Varga, G. A. Vladimirova, K. P. Wenzel, A. Zarándy.
Nature, Vol. 321, No. 6067, p. 285 – 288 (1986).
The TÜNDE–M energetic particle instrument aboard the Vega 1 spacecraft detected intense fluxes of energetic ($\geqslant$40 keV) ions in the vicinity of comet Halley, starting at a distance of 10^{7}km from closest approach. Three regions of differing ion characteristics have been identified. An outer region, several million kilometres in extent, contains pick–up ions in the solar wind. A second region, inside the bow shock (several hundred thousand kilometres in extent), contains the most intense fluxes, whereas the innermost region (several tens of thousands of kilometres) is characterized by lower intensities and sharp spikes near closest approach ($\sim$8,900 km from the nucleus).

103.673 **Magnetic field observations in comet Halley's coma.**
W. Riedler, K. Schwingenschuh, Y. G. Yeroshenko (*E. G. Eroshenko*), V. A. Styashkin (*V. A. Styazhkin*), C. T. Russell.
Nature, Vol. 321, No. 6067, p. 288 – 289 (1986).
During the encounter with comet Halley, the magnetometer (MISCHA) aboard the Vega 1 spacecraft observed an increased level of magnetic field turbulence, resulting from an upstream bow wave. Both Vega spacecraft measured a peak field strength of 70 – 80 nT and observed draping of magnetic field lines

around the cometary obstacle. An unexpected rotation of the magnetic field vector was observed.

103.674 **Observations of waves and plasma in the environment of comet Halley.**
R. Grard, A. Pedersen, J.–G. Trotignon, C. Beghin, M. Mogilevsky (*M. Mogilevskij*), Y. Mikhailov (*Yu. Mikhajlov*), O. Molchanov, V. Formisano.
Nature, Vol. 321, No. 6067, p. 290 – 291 (1986).
The high–frequency plasma–wave analyser (APV–V) carried by the spacecraft Vega 1 and Vega 2 measured electric fields at frequencies up to 300 kHz and characteristic parameters of the ambient plasma during the encounters with comet Halley. Electric field activity at frequencies of $\sim$100 Hz was already significant at a distance of $\sim$2.5x10^{6}km from the nucleus. The plasma density, measured with Langmuir probes, was $\sim$10^{4}cm^{-3} at closest approach.

103.675 **Extremely–low–frequency plasma waves in the environment of comet Halley.**
S. Klimov, S. Savin, Ya. Aleksevich, G. Avanesova, V. Balebanov, M. Balikchin, A. Galeev, B. Gribov, M. Nozdrachev, V. Smirnov, A. Sokolov, O. Vaisberg (*O. Vajsberg*), P. Oberc, Z. Krawczyk, S. Grzedzielski, J. Juchniewicz, K. Nowak, D. Orlowski, B. Parfianovich, D. Woźniak, Z. Zbyszynski, Y. Voita, P. Triska.
Nature, Vol. 321, No. 6067, p. 292 – 293 (1986).
The low–frequency plasma wave detector (APV–N) carried by the Vega 1 and Vega 2 spacecraft during their encounter with comet Halley registered an enhanced plasma–wave intensity identified with the inbound crossing of a quasi–perpendicular (Vega 1) or quasi–parallel (Vega 2) cometary bow shock. This letter also describes the characteristics of the plasma waves excited in the process of anomalously fast ionization of the cometary atmosphere in the inner coma.

103.676 **The Planet–A Halley encounters.**
K. Hirao, T. Itoh.
Nature, Vol. 321, No. 6067, p. 294 – 297 (1986).
Two spacecraft, Sakigake and Suisei, were launched in the Japanese Halley mission, the Planet–A project. Both spacecraft accomplished their task successfully. The plasma–wave probe aboard Sakigake, and the ultraviolet imager and the charged particle analyser aboard Suisei, carried out *in situ* measurements of the cometary environment. This article provides an overview of the project and the results obtained.

103.677 **Observation of comet Halley by the ultraviolet imager of Suisei.**
E. Kaneda, O. Ashihara, M. Shimizu, M. Takagi, K. Hirao.
Nature, Vol. 321, No. 6067, p. 297 – 299 (1986).
The ultraviolet imager aboard the Suisei spacecraft recorded the variation with time of the Lyman–α intensity of the coma of comet Halley. Strong breathing was still occurring at the hottest phase of the apparition. This observation confirms that the nuclear surface of Halley is largely covered with a hard substance. Two strong sources and at least four weak sources contributed to this periodic breathing.

103.678 **Plasma observation by Suisei of solar–wind interaction with comet Halley.**
T. Mukai, W. Miyake, T. Terasawa, M. Kitayama, K. Hirao.
Nature, Vol. 321, No. 6067, p. 299 – 303 (1986).
The Suisei probe made a closest approach of 151,000 km to comet Halley at 13:06 UT on 6 March 1986, and during the encounter Suisei passed through a strong interaction region, where the solar–wind flow was severely perturbed by picked–up ions of cometary origin. Plasma observations from Suisei clearly demonstrate the existence of the assimilation process of cometary ions in the solar–wind flow.

103.679 **Interaction between comet Halley and the interplanetary magnetic field observed by Sakigake.**
T. Saito, K. Yumoto, K. Hirao, T. Nakagawa, K. Saito.
Nature, Vol. 321, No. 6067, p. 303 – 307 (1986).
During 10 – 12 March 1986, the magnetometer carried by the spacecraft Sakigake detected clear multiple crossings of the nearly horizontal heliospheric neutral sheet. As there was no apparent disconnection event (DE) in comet Halley's ion tail during 11 – 14 March, an interplanetary magnetic field (IMF)–comet interaction model with a quasi–parallel neutral sheet is proposed to explain why a DE did not occur. Enhancements of hydromagnetic waves were observed near the time of closest approach to comet Halley.

103.680 **Discovery of cometary kilometric radiations and plasma waves at comet Halley.**
H. Oya, A. Morioka, W. Miyake, E. J. Smith, B. T. Tsurutani.
Nature, Vol. 321, No. 6067, p. 307 – 310 (1986).
The plasma–wave probe carried by the spacecraft Sakigake discovered discrete spectra of emissions from comet Halley in the frequency range 30 – 195 kHz. The observed cometary kilometric radiation appears to come from moving shocks in the coma region which are possibly associated with temporal variations of the solar wind. Waves due to plasma instabilities associated with the pick–up of cometary ions by the solar wind were observed within a region almost 10^{7}km from the comet nucleus.

103.681 **Was the solar wind decelerated by comet Halley?**
K.–i. Oyama, K. Hirao, T. Hirano, K. Yumoto, T. Saito.
Nature, Vol. 321, No. 6067, p. 310 – 313 (1986).
The velocity, ion density, flow direction and ion temperature of the solar–wind plasma were measured during Sakigake's approach to comet Halley from 10 to 13 March, with a distance of closest approach of 7x10^{6}km. Interplanetary magnetic field (IMF) observations show that Sakigake travelled through the solar plasma sheet and also through the lobes of the heliomagnetosphere. The data indicate that the solar wind was decelerated by the cometary atmosphere.

103.682 **The Giotto encounter with comet Halley.**
R. Reinhard.
Nature, Vol. 321, No. 6067, p. 313 – 318 (1986).
The Giotto spacecraft encountered comet Halley on 14 March 1986, approaching to within $\sim$600 km of the nucleus. Its scientific payload comprised ten hardware experiments: a narrow–angle camera, three mass spectrometers (for neutrals, ions and dust), various dust detectors a photopolarimeter and a set of plasma experiments. All experiments performed well and returned a wealth of new scientific results, of which perhaps the most important was the clear identification of the cometary nucleus.

103.683 **Pathfinder: accuracy improvement of comet Halley trajectory for Giotto navigation.**
R. E. Münch, R. Z. Sagdeev, J. F. Jordan.
Nature, Vol. 321, No. 6067, p. 318 – 320 (1986).
The 'Pathfinder' project represented a cooperative effort to fulfil the stringent requirement for Giotto to fly past comet Halley at a distance of $\sim$500 km on the sunward side. Observations obtained with the two Vega spacecraft were decisive for this purpose. These data combined with a very accurate determination of the Vega trajectories using VLBI, permitted a final fly–by distance of 600 km to be achieved with an uncertainty of only $\pm$40 km (1σ probability level).

103.684 **First Halley Multicolour Camera imaging results from Giotto.**
H. U. Keller, C. Arpigny, C. Barbieri, R. M. Bonnet, S. Cazes, M. Coradini, C. B. Cosmovici, W. A. Delamere, W. F. Huebner, D. W. Hughes, C. Jamar, D. Malaise, H. J. Reitsema, H. U. Schmidt, W. K. H. Schmidt, P. Seige, F. L. Whipple, K. Wilhelm.
Nature, Vol. 321, No. 6067, p. 320 – 326 (1986).
The first imaging results from the Halley Multicolour Camera during the Giotto fly–by of comet Halley provide images centred on the brightest part of the inner coma which show the silhouette of a large, solid and irregularly shaped cometary nucleus and jet–like dust activity visible in reflected sunlight. The nucleus is at

least 15 km long and ~10 km wide; its geometrical albedo is very low (< 4%). Only minor parts of the surface are active, most of it being covered by non–volatile material. Dust jets dominate the inner coma and are restricted to the sub–solar hemisphere. Preliminary interpretations of these observations are presented.

103.685 ***In situ* gas and ion measurements at comet Halley.**
D. Krankowsky, P. Lämmerzahl, I. Herrwerth, J. Woweries, P. Eberhardt, U. Dolder, U. Herrmann, W. Schulte, J. J. Berthelier, J. M. Illiano, R. R. Hodges, J. H. Hoffman.
Nature, Vol. 321, No. 6067, p. 326 – 329 (1986).

The neutral mass spectrometer experiment carried by the Giotto spacecraft was designed to determine the abundances and the chemical, elemental and isotopic composition of the gases and low–energy ions in the coma of comet Halley. Its first results show the predominance of water vapour with an H_2O density of $4.7x10^7$molecules cm^{-3} at 1,000 km. Limits on the abundances of CO_2, NH_3 and CH_4 relative to H_2O are given. The water–group ions H_3O^+, H_2O^+ and OH^+ have been unambiguously identified, along with the ions $^{12}C^+$, $^{12}CH^+$, $^{16}O^+$, Na^+, $^{12}C_2{}^+$, $^{32}S^+$, $^{34}S^+$ and $^{56}Fe^+$.

103.686 **Ion composition and dynamics at comet Halley.**
H. Balsiger, K. Altwegg, F. Bühler, J. Geiss, A. G. Ghielmetti, B. E. Goldstein, R. Goldstein, W. T. Huntress, W.–H. Ip, A. J. Lazarus, A. Meier, M. Neugebauer, U. Rettenmund, H. Rosenbauer, R. Schwenn, R. D. Sharp, E. G. Shelley, E. Ungstrup, D. T. Young.
Nature, Vol. 321, No. 6067, p. 330 – 334 (1986).

The ion mass spectrometer aboard the Giotto spacecraft measured the composition and velocity distributions of cometary ions at distances of ~$7.5x10^6$ to ~1,300 km from the nucleus of comet Halley. Well outside the bow shock, pick–up cometary H^+ ions were found. Heavier ions (C^+, H_2O^+–group, CO^+ and S^+) have been identified at $\leqslant 3x10^5$km. Solar–wind He^{2+} was found throughout the coma to as close as ~5,000 km; He^+ produced by charge exchange was seen inside ~$2x10^5$km. Deeper within the coma the main cometary hot–ion species identified were H^+, $H_2{}^+$, C^+, O^+, OH^+, H_2O^+, H_3O^+, CO^+ and S^+.

103.687 **Mass spectra of heavy ions near comet Halley.**
A. Korth, A. K. Richter, A. Loidl, K. A. Anderson, C. W. Carlson, D. W. Curtis, R. P. Lin, H. Rème, J. A. Sauvaud, C. d'Uston, F. Cotin, A. Cros, D. A. Mendis.
Nature, Vol. 321, No. 6067, p. 335 – 336 (1986).

The heavy–ion analyser aboard the Giotto spacecraft, detected the first cometary ions at a distance of ~$1.05x10^6$km from the nucleus of comet Halley. In the inner coma the major ions identified are associated with the H_2O, CO and CO_2 groups. Ions of larger atomic mass unit are also present, corresponding possibly to various hydrocarbons, heavy metals of the iron–group or to sulphur compounds.

103.688 **Composition of comet Halley dust particles from Giotto observations.**
J. Kissel, D. E. Brownlee, K. Büchler, B. C. Clark, H. Fechtig, E. Grün, K. Hornung, E. B. Igenbergs, E. K. Jessberger, F. R. Krueger, H. Kuczera, J. A. M. McDonnell, G. M. Morfill, J. Rahe, G. H. Schwehm, Z. Sekanina, N. G. Utterback, H. J. Völk, H. A. Zook.
Nature, Vol. 321, No. 6067, p. 336 – 337 (1986).

Mass spectra of cometary dust particles measured by the dust particle analyser aboard the Giotto spacecraft show some unexpected and striking features. First, small particles below 10^{-14}g are much more abundant than anticipated by models. Second, most of the particles are rich in light elements such as H, C, N and O, suggesting the validity of models that describe the cometary dust as including organic material. Third, the light elements specifically seem to have a low ratio of mass to volume.

103.689 **Dust density and mass distribution near comet Halley from Giotto observations.**
J. A. M. McDonnell, W. M. Alexander, W. M. Burton, E. Bussoletti, D. H. Clark, R. J. L. Grard, E. Grün, M. S. Hanner, D. W. Hughes, E. Igenbergs, H. Kuczera, B. A. Lindblad, J.–C. Mandeville, A. Minafra, G. H. Schwehm, Z. Sekanina, M. K. Wallis, J. C. Zarnecki, S. C. Chakaveh, G. C. Evans, S. T. Evans, J. G. Firth, A. N. Littler, L. Massonne, R. E. Olearczyk, G. S. Pankiewicz, T. J. Stevenson, R. F. Turner.
Nature, Vol. 321, No. 6067, p. 338 – 341 (1986).

The dust impact detection system carried by the Giotto spacecraft has measured the density and mass spectrum (in the range 10^{-20} – 10^{-8}kg) of the dust near comet Halley. Profiles of the dust spectrum obtained at a distance of 291,000 km from the Halley nucleus show a complex mass spectrum, depleted in intermediate and small masses. Near closest approach (~600 km from the nucleus), the dust activity increases dramatically and the spectrum is dominated by larger masses. The Halley dust surface emission rate is found to be typically $1x10^{-6}$kg $m^{-2}s^{-1}$ for masses $\leqslant 10^{-8}$kg, corresponding to a total dust production rate of $3.1x10^3$kg s^{-1}.

103.690 **Optical probing of comet Halley from the Giotto spacecraft.**
A. C. Levasseur–Regourd, J. L. Bertaux, R. Dumont, M. Festou, R. H. Giese, F. Giovane, P. Lamy, J. M. Le Blanc, A. Llebaria, J. L. Weinberg.
Nature, Vol. 321, No. 6067, p. 341 – 344 (1986).

The Halley optical probe experiment aboard the Giotto spacecraft has provided the first measurements of the optical properties of the dust and of some gaseous species (CN, C_2, CO^+ and OH) from inside the coma of comet Halley. The dust spatial distribution, inferred from sunlight scattering, obeys an r^{-2} law for distances from the nucleus $r \gtrsim 2,000$ km. The more rapid increase of dust density with distance observed in the innermost coma probably reflects the presence of a jet, or a surge of activity within a few hours before closest approach.

103.691 **Ion flow at comet Halley.**
A. Johnstone, A. Coates, S. Kellock, B. Wilken, K. Jockers, H. Rosenbauer, W. Studemann, W. Weiss, V. Formisano, E. Amata, R. Cerulli–Irelli, M. Dobrowolny, R. Terenzi, A. Egidi, H. Borg, B. Hultquist, J. Winningham, C. Gurgiolo, D. Bryant, T. Edwards, W. Feldman, M. Thomsen, M. K. Wallis, L. Biermann, H. Schmidt, R. Lüst, G. Haerendel, G. Paschmann.
Nature, Vol. 321, No. 6067, p. 344 – 347 (1986).

The three–dimensional positive ion analyser aboard the Giotto spacecraft has been used to study the interaction between protons and α–particles in the solar wind and positive ions from comet Halley. Three sharp transitions of relatively small amplitude can be identified on both the inbound and outbound legs of the trajectory. The outermost one, at ~10^6km from the nucleus, appears to be a multiple crossing of a weak bow shock. The innermost one, at 80,000 km, is the boundary where the flowing plasma becomes depleted.

103.692 **Energetic ions in the environment of comet Halley.**
S. McKenna–Lawlor, E. Kirsch, D. O'Sullivan, A. Thompson, K.–P. Wenzel.
Nature, Vol. 321, No. 6067, p. 347 – 349 (1986).

The energetic particle experiment carried aboard the Giotto spacecraft recorded *in situ* measurements of energetic charged particles on the sunward side of comet Halley, from a trajectory very close to the nucleus (closest approach 600 km). Preliminary studies suggest the presence on the inbound pass of an ion 'pick–up' region (~$6x10^6$km in extent) and a region inside the bow shock (< 10^6km in extent) where possible signatures of mass loading were evident.

103.693 **Comet Halley–solar wind interaction from electron measurements aboard Giotto.**
H. Rème, J. A. Sauvaud, C. d'Uston, F. Cotin, A. Cros, K. A. Anderson, C. W. Carlson, D. W. Curtis, R. P. Lin, D. A. Mendis, A. Korth, A. K. Richter.
Nature, Vol. 321, No. 6067, p. 349 – 352 (1986).

Measurements made by the three–dimensional electron spectrometer aboard the Giotto spacecraft describe the complex interaction between comet Halley and the solar wind. As Giotto approached the comet, it traversed several regions of differing plasma characteristics: a large ($\sim 10^7$km) upstream region, a foreshock, a bow wave with its apex at $\sim 4x10^5$km, a transition region, a region permanently and strongly contaminated by cometary ions, a collisionopause at a sub-solar distance of $\sim 5-6x10^4$km, and various inner regions up to the inner coma.

103.694 **First results from the Giotto magnetometer experiment at comet Halley.**
F. M. Neubauer, K. H. Glassmeier, M. Pohl, J. Raeder, M. H. Acuna, L. F. Burlaga, N. F. Ness, G. Musmann, F. Mariani, M. K. Wallis, E. Ungstrup, H. U. Schmidt.
Nature, Vol. 321, No. 6067, p. 352 – 355 (1986).
The Giotto magnetometer experiment at comet Halley has for the first time provided magnetic field measurements in all the important spatial regions characterizing the front-side interaction between the solar-wind magnetoplasma and a cometary atmosphere. A magnetic pile-up region has been identified inside $1.35x10^5$km, inbound, and $2.63x10^5$km, outbound, with fields up to 57 and 65 nT, respectively. A cavity region with essentially zero magnetic field has been discovered, with a width of 8,500 km along the trajectory around closest approach.

103.695 **First results from the Giotto Radio-Science Experiment.**
P. Edenhofer, M. K. Bird, J. P. Brenkle, H. Buschert, P. B. Esposito, H. Porsche, H. Volland.
Nature, Vol. 321, No. 6067, p. 355 – 357 (1986).
Doppler and ranging measurements, using the radio signals of the Giotto spacecraft and taken during the encounter with comet Halley, reveal a definite deceleration of the spacecraft due to drag in the cometary atmosphere. The total change in the radial velocity of the spacecraft was measured to be 16.7 cm s^{-1}. Using this velocity change, estimates of the total cometary mass striking the spacecraft range from 0.1 to 1 g.

103.696 **Dust jets in comet Halley observed by Giotto and from the ground.**
Z. Sekanina, S. M. Larson.
Nature, Vol. 321, No. 6067, p. 357 – 361 (1986).
A first map of discrete dust-emission sources on the nucleus of comet Halley was recently completed, based on a study of the morphology of 55 features (jets, spirals, envelopes and others) recognized and modelled on 34 digitally processed plates of the comet obtained during May and June of 1910. Here the authors assess the relevance of these results to the comet's current apparition, using data obtained by the Giotto spacecraft and from ground-based observations. The problems involved are possible variations in the spin vector over the 76-yr period, survival of the dust sources, and the positions of the prime meridian at the two returns to the Sun.

103.697 **IUE observations of comet Halley during the Vega and Giotto encounters.**
M. C. Festou, P. D. Feldman, M. F. A'Hearn, C. Arpigny, C. B. Cosmovici, A. C. Danks, L. A. McFadden, R. Gilmozzi, P. Patriarchi, G. P. Tozzi, M. K. Wallis, H. A. Weaver.
Nature, Vol. 321, No. 6067, p. 361 – 363 (1986).
The observations of comet Halley from the IUE satellite from 9 to 16 March 1986 allow a comprehensive study of gas and dust production by the comet in the week during which the fleet of six spacecraft from four space agencies encountered the comet. The gaseous output of the nucleus was found to vary over timescales of the order of one or two hours. The ultraviolet spectrum of comet Halley differs from that of other comets observed by the IUE satellite in that the continuum emission is very strong relative to the gaseous emissions, and the CO_2^+ violet bands, if present, are weak.

103.698 **Post-perihelion imaging of comet Halley at ESO.**
R. M. West, H. Pedersen, P. Monderen, R. Vio, P. Grosbøl.
Nature, Vol. 321, No. 6067, p. 363 – 365 (1986).
The near-nucleus and large-scale activity of comet Halley has been monitored at the ESO La Silla Observatory since mid-February 1986. During an 8-week period, images obtained with three telescopes documented the post-perihelion behaviour on a daily basis. The authors describe the instruments and procedures and provisionally summarize the major structures and events that were seen at the time of closest approach of the various spacecraft, between 5 March and 14 March 1986.

103.699 **First journey to a comet.**
J. Maddox.
Nature, Vol. 321, No. 6067, p. 366 (1986).
The several space probes passing by comet Halley in March this year have yielded a rich harvest of data in a short time. But many puzzles about this and other comets remain.

103.700 **Halley's comet.**
S. P. Maran.
J. Opt. Soc. Am. A, Vol. 2, No. 12, p. P33 – P34 (1985). Abstract.

103.701 **Investigation of comet Halley with the ultraviolet telescope of the astrophysical station "ASTRON" in December 1985.**
A. A. Boyarchuk, A. M. Zvereva, R. S. Kremnev, P. P. Petrov, K. G. Sukhanov, N. S. Chernykh, A. I. Shejkhet.
Kosm. Issled., Tom 24, Vyp. 3, p. 471 – 474 (1986). In Russian. English translation in Cosm. Res.

103.702 **Scientific experiment "Foton" for investigation of the micrometeoritic atmosphere of comet Halley and characteristics of its interaction with the space probe Vega.**
S. I. Anisimov, V. M. Kovtunenko, V. P. Karyagin, A. A. Leont'ev, A. L. Ni, Yu. A. Osip'yan, Yu. A. Ryzhov, R. Z. Sagdeev, S. B. Svirshchevskij, A. Z. Strukov, A. V. Terterashvili, V. E. Fortov, G. G. Shishkin.
Inst. vysok. temperatur Akad. Nauk SSSR. Prepr., No. 2/176, 28 pp. (1985). In Russian. Abstr. in Ref. Zh., 62. Issled. Kosm. Prostranstva, 5.62.110 (1986).

103.703 **A happening on the distribution of the Halley comet news.**
M. Sasaki.
Astron. Her., Vol. 79, No. 6, p. 150 – 153 (1986). In Japanese.

103.704 **Le retour de la comète de Halley.**
A. H. Delsemme.
Ciel, Vol. 48, p. 281 – 293 (1986).

103.705 **A kinematic analysis of the disconnection of the plasma tail of comet P/Halley, January 10, 1986.**
P. Guérin, S. Koutchmy.
C. R. Acad. Sci., Sér. II, Tome 303, No. 2, p. 147 – 152 (1986). In French.
Measurements show a sudden increase of the acceleration of the recession movement of the disconnected tail, affecting at the same time all the condensations visible along this tail, about 9 hrs. after the beginning of the separation. Later, a deceleration reduces the velocity in a few hours to the value it would have had if there had been no disconnection.

103.706 **The brightness distribution of OH around comet Halley.**
I. de Pater, P. Palmer, L. E. Snyder.
Astrophys. J., Lett. Ed., Vol. 304, No. 1, p. L33 – L36 (1986).
The first successful VLA observations of OH emission from any comet have been made for comet Halley. In 1985 November, 120″ and 60″ resolution images of the brightness distribution of the 1667 MHz line were obtained. The 120″ image shows that the peak of the OH emission is displaced 40″ to the west and 80″ to the south. The 60″ image shows that the emission is concentrated in "clumps", which surround a "hole" in the OH emission.

103.707 **Predicting that comet Halley is dark.**
J. M. Greenberg.
Nature, Vol. 321, No. 6068, p. 385 (1986).

103.708 **Space observations of comet Halley.**
J. C. Brandt.
Nature, Vol. 321, No. 6068, p. 391 – 393 (1986).
Following the encounters by Soviet, Japanese and European spacecraft, the International Cometary Explorer took measurements of comet Halley on 22 March. The comet has also been observed by the Solar Maximum Mission, the International Ultraviolet Explorer, the Pioneer Venus Orbiter, the Dynamics Explorer 1, and by sounding rockets.

103.709 **Observations of comet Halley at the Sarajevo Observatory.**
M. Muminović, M. Stupar.
Vasiona, Année 33, No. 5, p. 98 – 99 (1985). In Croatian.

103.710 **Some predictions on the nature of comet Halley.**
F. Hoyle, N. C. Wickramasinghe.
Astrophys. Relativ., Prepr. Ser., No. 122, A8, 9 pp. (1986).

103.711 **A preliminary study of the tail of comet Halley and the geomagnetic storm.**
L.–s. Yan, J.–x. Ye.
Kexue Tongbao, Vol. 30, No. 3, p. 356 – 360 (1985).
The authors study the return of comet Halley in 1910 with new concepts of natural science. The geomagnetic storm was of a gradual commencement type, its *K* index equals 5. The source of the storm probably did not come from solar activity but perhaps from the effect of the plasma tail of comet Halley. The authors try to analyse the structure of the plasma tail by means of the geomagnetic horizontal components from 17 – 19 May 1910.

103.712 **Half of the way passed through.**
M. N. Maksumov, O. V. Dobrovol'skij.
Priroda, No. 5, p. 106 – 107 (1986). In Russian.

103.713 **Detection of water vapor in Halley's comet.**
M. J. Mumma, H. A. Weaver, H. P. Larson, D. S. Davis, M. Williams.
Prepr. Steward Obs., No. 659, 27 pp. (1986). To appear in Science.

103.714 **The orbit of comet Halley, refined by current observations.**
Yu. F. Kolyuka, V. P. Tarasov, V. F. Tikhonov.
Sov. Astron. Lett., Vol. 11, No. 5, p. 328 – 331 (1985). English translation of 40.103.827.

103.715 **A Halley chronicle.**
J. E. Bortle.
Astronomy, Vol. 13, No. 10, p. 98 – 110 (1985). Abstr. in Phys. Abstr., Vol. 89, No. 1277, Entry 42207 (1986).

103.716 **Forecast of the physical characteristics of the nucleus of comet Halley.**
V. P. Karyagin, Yu. G. Shkuratov, V. P. Tishkovets.
Vestn. Khar'kov. Univ., No. 278, p. 56 – 65 (1985). In Russian.

103.717 **Near–infrared profiles of Comet Halley.**
J. P. Phillips, A. Mampaso, F. Garzon.
Astron. Astrophys., Vol. 161, No. 2, p. L17 – L19 (1986).
The authors present J, H, K, and L profiles of Comet Halley taken on January 15, 1986, and discuss the wavelength variability in apparent source structure, together with the spectral characteristics of the coma. The emission at J and H is found to be strongly dominated by grain scattering of the solar continuum, although grain thermal emission is strong at K and becomes increasingly dominant towards longer wavelengths. The mean temperature of the grains is found to be $\sim$730K for uniform emissivity, a value which is significantly in excess of equilibrium black–body temperatures. Similarly, the grain emission optical depth is determined to be $\tau(1.65\ \mu m) \cong 8.7 \times 10^{-5}$ (for albedo A = 0.3) within a 16 arcsecond aperture centred on the nucleus, declining to $\sim 2.2 \times 10^{-6}$ for an aperture 12.5 times greater. Both emission and scatter profiles are dominated by a central core.

103.718 **Discovery of the 3 µm ice band in comet Halley.**
J. D. Bregman, F. C. Witteborn, D. M. Rank, D. Wooden.
Bull. Am. Astron. Soc., Vol. 18, No. 1, p. 634 (1986). Abstract. – See Abstr. 010.066.

103.719 **Periodic comet Halley.**
Komet. Tsirk., Nos. 350 – 353 (1986). In Russian.

103.720 **The new look of Halley is black and lumpy.**
R. A. Kerr.
Science, Vol. 232, No. 4756, p. 1343 (1986).
The closeup view from the Giotto spacecraft that was recently released reveals a hill and craters spewing dust jets.

103.721 **Detection of water vapor in Halley's comet.**
M. J. Mumma, H. A. Weaver, H. P. Larson, D. S. Davis, M. Williams.
Science, Vol. 232, No. 4757, p. 1523 – 1528 (1986).
Gaseous, neutral H_2O was detected in the coma of comet Halley on 22.1 and 24.1 December 1985 Universal Time. Nine spectral lines of the ν_3 band (2.65 micrometers) were found by means of a Fourier transform spectrometer ($\lambda/\Delta\lambda \sim 10^5$) on the NASA–Kuiper Airborne Observatory. The numbers of spectral lines and their intensities are in accord with nonthermal–equilibrium cometary models. Rotational populations are derived from the observed spectral line intensities and excitation conditions are discussed.

103.722 **Halley's comet as on outstanding target.**
R. Reinhard.
ESA Bull., No. 46, p. 13 – 18 (1986). – See Abstr. 003.052.
Between 6 and 14 March 1986, five spacecraft from three space agencies flew through the coma of Halley's comet. The encounters all took place in March 1986, because a minimum of launch energy was required for these post–perihelion encounters. At that time, four weeks after its perihelion passage, Halley was also at its most active, exhibiting all of the well–known cometary phenomena.

103.723 **The Giotto encounter with Halley's comet – first scientific results.**
R. Reinhard.
ESA Bull., No. 46, p. 52 – 57 (1986). – See Abstr. 003.052.

103.724 **The history of Comet Halley.**
D. K. Yeomans, J. Rahe, R. S. Freitag.
J. R. Astron. Soc. Can., Vol. 80, No. 2, p. 62 – 86 (1986).
A summary is given of the dynamical history of Comet Halley (complete with ephemerides for each known apparition) and also of the physical observations made during the most recent apparitions.

103.725 **Infrared observations of comet Halley near perihelion.**
T. Le Bertre, P. Bouchet, A. Chalabaev, A. C. Danks, T. Encrenaz, N. Epchtein.
Messenger, No. 44, p. 9 – 11 (1986).

103.726 **Spatial distribution of constituents in the coma of comet Halley, an observing programme at the ESO 1–m telescope.**
K. Jockers, E. H. Geyer, A. Hänel.
Messenger, No. 44, p. 12 – 14 (1986).

103.727 **Optical spectroscopy of the coma of comet Halley at ESO.**
R. Falciani, L. A. Smaldone, G. P. Tozzi.
Messenger, No. 44, p. 15 – 16 (1986).

103.728 **Halley through the polaroids.**
R. Haefner, K. Metz.
Messenger, No. 44, p. 16 – 18 (1986).

103.729 **The recovery of comet Halley at Asiago Observatory.**
C. Barbieri, F. Bortoletto, C. Bonoli, M. D'Alessandro.
Mem. Soc. Astron. Ital., Vol. 56, No. 4, p. 777 – 779 (1985). – See Abstr. 010.821.

103.730 **On the first near UV observation of P/Halley.**
M. C. Festou, C. Arpigny, J. L. Bertaux, W. Carey, T. Danks, R. Gilmozzi, D. W. Hughes, W. Ip, P. Patriarchi, G. P. Tozzi, M. K. Wallis, J. Zarnecki.
Inst. Astrophys. Paris, Pré–Publ., No. 124, 6 pp. (1986). To appear in Astron. Astrophys.

103.731 **A kinematic analysis of the disconnection of the plasma tail of comet P/Halley, Jan. 10, 1986.**
P. Guerin, S. Koutchmy.
Inst. Astrophys. Paris, Pré–Publ., No. 144, 11 pp. (1986). In French. See Abstr. 103.705.

103.732 **Dynamical properties of a pre–perihelion comet tail. I. Dec. 6, 1985 observations of P/Halley 82i.**
P. Guerin, S. Koutchmy, J. C. Vial.
Inst. Astrophys. Paris, Pré–Publ., No. 148, 11 pp. (1986). To appear in Astron. Astrophys.

103.733 **Photometry and activity of the nucleus of P/Halley at heliocentric distances larger than 4.6 A.U., pre–perihelion.**
M. C. Festou, J. Lecacheux, J. L. Kohl, T. Encrenaz, J. Baudrand, M. Combes, R. Despiau, P. Laques, O. Le Fevre, J. P. Lemonnier, G. Lelievre, G. Mathez, M. Pierre, J. L. Vidal.
Inst. Astrophys. Paris, Pré–Publ., No. 150, 26 pp. (1986). To appear in Astron. Astrophys.

103.734 **Early photometry of comet P/Halley: development of the coma.**
K. J. Meech, D. Jewitt, G. R. Ricker.
Icarus, Vol. 66, No. 3, p. 561 – 574 (1986).

The authors present new photometry of P/Halley obtained when the comet was at $R = 5.9$ and 5.1 AU, during 1984 October and 1985 January, respectively. They compare and contrast the new observations with data obtained at larger heliocentric distances. The authors argue that sustained mass loss from the nucleus began at about $R = 5.9$ AU, at a rate which is consistent with production by sublimation from a predominantly water–ice nucleus. Observations at larger R are used to characterize the size and overall shape of the nucleus.

103.735 **Observations of appulses by Halley's comet.**
D. W. Dunham.
Occultation Newsl., Vol. 3, No. 16, p. 352 – 354 (1986).

103.736 **Mercury photofeature: the 1985 – 86 pass of comet Halley: a retrospective.**
S. Harrington.
Mercury, Vol. 15, No. 3, p. 80 – 85 (1986).

103.737 **Las condiciones observacionales del cometa Halley en 1985 – 86 y su brillo antes y despues del perihelio.**
I. Ferrin.
Rev. Mex. Astron. Astrofis., Vol. 12, Spec. Issue, p. 376 (1986). Abstract. – See Abstr. 012.066.

103.738 **Observations of comet Halley.**
S.–y. Gu, L.–s. Yan.
J. China Univ. Sci. Technol., Vol. 15, No. 3, p. 381 – 384 (1985). In Chinese.

103.739 **Beobachtungen des Kometen Halley in Bulgarien.**
B. Kovachev.
Mitt. Astron. Ges., No. 65, p. 178 – 179 (1986). Abstract. – See Abstr. 012.072.

103.740 **Periodic comet Halley (1982i).**
IAU Circ., Nos. 4159, 4161, 4162, 4167, 4169, 4173, 4175 – 4179, 4183, 4187 – 4189, 4194, 4196, 4198, 4203, 4205, 4209, 4212, 4226 (1986).

Periodic comet Reinmuth 2

103.761 **Periodic comet Reinmuth 2 1960 IX = 1967 XI = 1974 VI = 1981 III.**
OAA Comput. Sect. Circ., NK 484, 3 pp. (1985).

Periodic comet Borrelly

103.771 **Periodic comet Borrelly 1960 V = 1967 VIII = 1974 VII = 1981 IV.**
OAA Comput. Sect. Circ., NK 486, 2 pp. (1985).

Periodic comet Denning–Fujikawa

103.781 **Periodic comet Denning–Fujikawa 1978 XIX = 1978n.**
OAA Comput. Sect. Circ., NK 487, 2 pp. (1985).

Periodic comet Bus

103.791 **Periodic comet Bus 1981 XI = 1981b.**
OAA Comput. Sect. Circ., NK 488, 1 p. (1985).

Periodic comet Harrington

103.801 **Periodic comet Harrington 1953 VI = 1960 VII = 1980 XIV.**
OAA Comput. Sect. Circ., NK 489, 2 pp. (1985).

Catalogue of cometary orbits.
See Abstr. 002.024.

Long–term evolution of short–period comets.
See Abstr. 003.008.

Der Halleysche Komet im Jahre 1910. Eine Sammlung zeitgenössischer Berichte (1909 – 1911).
See Abstr. 003.031.

Kometen.
See Abstr. 003.033.

Comet Halley 1909 – 1911. Retrospect of observations.
See Abstr. 003.038.

Halley: the once–in–a–lifetime comet.
See Abstr. 003.054.

Guide de la comète de Halley.
See Abstr. 003.057.

Halley's comet.
See Abstr. 003.058.

Observing Halley's comet.
See Abstr. 003.085.

Halley's comet: a mysterious visitor from outer space.
See Abstr. 003.086.

Halley's comet for classrooms.
See Abstr. 003.100.

Halley's comet.
See Abstr. 003.109.

Halley, le roman des comètes.
See Abstr. 003.125.

A decade of comets.
See Abstr. 003.130.

Rendezvous with Halley's comet.
See Abstr. 003.134.

Halley's comet pop-up book.
See Abstr. 003.140.

Halleys komet.
See Abstr. 003.141.

Halley's comet in South Africa October 1985 – May 1986.
See Abstr. 003.155.

Capture a comet: an amateur photographer's guide.
See Abstr. 003.163.

Here's Halley.
See Abstr. 003.195.

Halley: Komeet op komst.
See Abstr. 003.199.

An encounter with Halley's comet 1985 – 1986.
See Abstr. 003.206.

Comet! The story behind Halley's comet.
See Abstr. 003.207.

Comet watch: the return of Halley's comet.
See Abstr. 003.212.

Comet Halley's first expected return: English public apprehensions, 1755 – 58.
See Abstr. 004.007.

Halley's comet: Canadian observations and reactions, 1835 – 36 and 1910.
See Abstr. 004.018.

Samuel Scott's comet Halley of 1759.
See Abstr. 004.058.

Nicholas T. Bobrovnikoff and the scientific study of Comet Halley 1910.
See Abstr. 005.015.

P/Halley activities in the German Democratic Republic.
See Abstr. 013.010.

The International Halley Watch (IHW) and its observing nets.
See Abstr. 013.015.

The Inter-Agency Consultative Group (IACG) and its associated working groups.
See Abstr. 013.016.

The Halley dust and gas environment (IACG WG-1).
See Abstr. 013.017.

Coordinated multispacecraft plasma science in the solar-wind and comet-plasma environments (IACG WG-2).
See Abstr. 013.018.

The large-scale phenomena network of the International Halley Watch.
See Abstr. 013.049.

The role of the Inter-Agency Consultative Group (IACG) and its associated working groups.
See Abstr. 013.051.

È difficile scoprire una cometa?
See Abstr. 013.052.

OPTOPUS observations of Halley's comet.
See Abstr. 034.029.

Shooting Halley.
See Abstr. 034.034.

A rocket experiment to make far-UV images of comet Halley.
See Abstr. 035.004.

The Giotto magnetic-field investigation.
See Abstr. 035.005.

The Giotto three-dimensional positive ion analyser.
See Abstr. 035.006.

The Giotto RPA-Copernic plasma experiment.
See Abstr. 035.007.

The Giotto Energetic Particle Experiment.
See Abstr. 035.008.

The Giotto Particulate Impact Analyser.
See Abstr. 035.009.

The Giotto Dust Impact Detection System.
See Abstr. 035.010.

The Giotto Neutral Mass Spectrometer.
See Abstr. 035.011.

The Giotto Ion Mass Spectrometer.
See Abstr. 035.012.

The Giotto Halley Multicolour Camera.
See Abstr. 035.013.

The Giotto Radio-Science Experiment.
See Abstr. 035.014.

The Giotto Optical Probe Experiment.
See Abstr. 035.015.

Rendezvous mit Giacobini-Zinner.
See Abstr. 051.003.

Heading for the Giotto encounter with comet Halley.
See Abstr. 051.008.

The Giotto project.
See Abstr. 051.011.

The Vega missions.
See Abstr. 051.012.

The Suisei/Sakigake (Planet-A/MS-T5) missions.
See Abstr. 051.013.

Impact-induced spacecraft charging and related effects during the Halley encounters.
See Abstr. 051.014.

The International Cometary Explorer (ICE) mission to comets Giacobini-Zinner and Halley.
See Abstr. 051.015.

The Astro–1 mission.
See Abstr. 051.016.

International Ultraviolet Explorer (IUE) observation of Halley's comet.
See Abstr. 051.017.

Observations of Halley's comet by the Solar Maximum Mission (SMM).
See Abstr. 051.018.

The Spartan mission.
See Abstr. 051.019.

NASA–supported smaller programmes to observe comet Halley.
See Abstr. 051.020.

The Pathfinder Concept.
See Abstr. 051.021.

Giotto mission background.
See Abstr. 051.022.

The Giotto project.
See Abstr. 051.023.

Alle ogen zijn gericht op komeet Halley.
See Abstr. 051.035.

The International Cometary Explorer mission to comets Giacobini–Zinner and Halley: an update.
See Abstr. 051.037.

Catching a volatile Halley before it's gone.
See Abstr. 051.038.

The International Cometary Explorer mission to comet Giacobini–Zinner.
See Abstr. 051.039.

The high road to Halley.
See Abstr. 051.040.

Giotto and Vega results on Halley's comet.
See Abstr. 051.041.

Comet rendezvous: the next stage in cometary exploration.
See Abstr. 051.063.

Ultraviolet spectroscopy of meteoric debris of comets.
See Abstr. 051.068.

Comet nucleus sample return missions with electrically propelled spacecraft.
See Abstr. 051.088.

Space mission "Vega": encounter with comet Halley.
See Abstr. 051.092.

Trajectories and orbital maneuvers for the ISEE–3/ICE comet mission.
See Abstr. 052.045.

Alternative gravity–assist sequences for the ISEE–3 escape trajectory.
See Abstr. 052.046.

Optimization of a multiple lunar–swingby trajectory sequence.
See Abstr. 052.047.

ISEE–3/ICE navigation analysis.
See Abstr. 052.049.

Interaction of a solar wind magnetic cloud with comet 1975h.
See Abstr. 074.020.

Planètes et petits corps du système solaire. Rapport thématique 1980 – 1984.
See Abstr. 091.049.

The detection of fast–moving asteroids and comets by IRAS.
See Abstr. 098.004.

Ephemerides of minor planets and comets.
See Abstr. 098.096.

Photographic positions of comets, minor planets and Pluto observed during 1975 – 1980.
See Abstr. 098.105.

Astrometria fotografica di pianetini e comete. Relazione annuale sull'attività svolta dagli astrofili dell'Osservatorio S. Vittore (Bologna) nel corso del 1984.
See Abstr. 098.126.

Thermal conduction in cometary nuclei and its effect on cometary activity.
See Abstr. 102.007.

Physical characteristics of comets 1980.
See Abstr. 102.015.

Elemental, isotopic and molecular abundances in comets.
See Abstr. 102.023.

The fluorescence spectrum of the CN radical in comets.
See Abstr. 102.025.

The derivation of OH gas production rates from visual magnitudes of comets.
See Abstr. 102.026.

Brightness curve and porosity of cometary nuclei.
See Abstr. 102.027.

Dust coma onset at large heliocentric distances.
See Abstr. 102.028.

Comet modeling.
See Abstr. 102.030.

Detection of submillimeter size particles in the inner coma of a comet through their forward scattering.
See Abstr. 102.058.

Nature of comets.
See Abstr. 102.059.

Comet–solar wind interaction: dynamical length scales and models.
See Abstr. 102.061.

The 18 cm OH lines in comets: preparation for Halley.
See Abstr. 102.066.

Metallic ions in cometary comae and plasma tails.
See Abstr. 102.068.

Cometary observations from space.
See Abstr. 102.078.

An improved thermal model for cometary nuclei.
See Abstr. 102.083.

S_2 formation in interstellar dust; a diagnostic of the maximum aggregation temperature for a comet.
See Abstr. 102.084.

On the meteor activity associated with the comet Sugano–Saigusa–Fujikawa 1983 V.
See Abstr. 104.005.

The relationship between comets and meteoroid streams.
See Abstr. 104.009.

Possibility of a meteor shower associated with comet Sugano–Saigusa–Fujikawa 1983e.
See Abstr. 104.014.

Asteroidal origin for the Geminid meteor stream.
See Abstr. 104.016.

The Draconid meteoroids.
See Abstr. 104.061.

Magnitudes of standard stars for IHW cometary photometry.
See Abstr. 113.008.

104 Meteors, Meteor Streams

104.001 **Radiants of meteor streams.**
H. Korpikiewicz.
Astron. Rep., Vol. 6, p. 33 – 36 (1981).

104.002 **Number of Perseid's in maximum of stream 12/13 Aug. 1981.**
T. Górski.
Astron. Rep., Vol. 6, p. 37 (1981).

104.003 **Giacobinid meteors 1985.**
B. A. Lindblad.
IHW Newsl., No. 8, p. 25 – 26 (1986).

104.004 **The origin of the sporadic meteoroid component.**
D. Olsson-Steel.
Mon. Not. R. Astron. Soc., Vol. 219, No. 1, p. 47 – 73 (1986).

The time–scales associated with close encounters with the planets (which can cause losses by planetary impact or ejection from the solar system, or a deflection on to a changed elliptical orbit), orbital collapse under the Poynting–Robertson effect, and catastrophic collisions with zodiacal dust particles are compared for 28 test orbits similar to well–known meteoroid streams. The results in this paper indicate that the flux of short–period comets must have been considerably higher $\sim 10^4$yr ago, as has been suggested by other researchers, in order to explain the supply of material to the sporadic complex and thence the zodiacal cloud.

104.005 **On the meteor activity associated with the comet Sugano–Saigusa–Fujikawa 1983 V.**
M. Šimek, P. Pecina.
Bull. Astron. Inst. Czech., Vol. 37, No. 2, p. 103 – 107 (1986).

The observations carried out in 1984 to prove the existence of a meteor shower associated with the comet Sugano–Saigusa–Fujikawa 1983 V are discussed. The presence of this shower is demonstrated on the increase of hourly rates in three duration groups in the expected time interval. This finding was also confirmed by the subsequent range–time analysis of meteoric activity observed within the time interval in question. The activity pattern observed cannot be produced by any known shower.

104.006 **Investigation of the structure of a meteor complex near the earth's orbit. II. Evaluation of spatially probable characteristics.**
Yu. I. Voloshchuk, B. L. Kashcheev, N. B. Nazarenko.
Astron. Vestn., Tom 20, No. 1, p. 50 – 60 (1986). In Russian. English translation in Sol. Syst. Res.

104.007 **Meteoroids inside the earth's orbit: the Eccentrids system.**
A. N. Simonenko, A. K. Terent'eva, I. V. Galibina.
Astron. Vestn., Tom 20, No. 1, p. 61 – 75 (1986). In Russian. English translation in Sol. Syst. Res.

There are data for a population of bodies on smallest orbits in the solar system of moderate and high eccentricity with aphelia near the earth's orbit. The population named Eccentrids includes three asteroids and 50 meteoroids registered by the photographic method. A deficiency of orbits penetrating into Mercury's orbit is noted. Some streams are revealed, members of which can be genetically related. An example is a stream on the orbit of the Aten 2340 Hathor. Arguments are found that the Pesyanoe meteorite is a fragment of the Aten 2100 Ra–Shalom.

104.008 **Hollow meteoroid streams.**
D. W. Hughes.
Nature, Vol. 319, No. 6056, p. 718 (1986).

104.009 **The relationship between comets and meteoroid streams.**
D. W. Hughes.
Asteroids, comets, meteors II, p. 503 – 519 (1986). – See Abstr. 012.010.

The decay of short period comets is, in the main, a gentle process which slowly feeds dust into a meteoroid stream. It is concluded that the cometary decay rate and the meteoroid stream formation rate are such that all active meteor showers should have streams with obvious, visible parent comets. Many don't, the prime examples being the Quadrantids. It is concluded that differential orbital perturbation is to blame. The orbital statistics of comets and meteoroid streams should be similar. Unfortunately both suffer from observational selection, this being especially severe in the case of the meteoroid streams. There is every indication that the removal of selection effects would improve the similarity.

104.010 **The effects of planetary perturbations on observations of meteor showers.**
K. Fox.
Asteroids, comets, meteors II, p. 521 – 525 (1986). – See Abstr. 012.010.

Many meteor streams are undergoing rapid orbital evolution due to planetary perturbations. In time periods as short as one hundred years some stream's orbits can be swept right across the Earth's orbit and will hence only produce a shower during this time. Others are undergoing much slower evolution and stay close to the Earth's orbit for thousands of years. Starting from the working list of meteor streams produced by A. F. Cook, 1973, the orbital evolution of every stream on a closed orbit has been calculated going both backwards and forwards from the present for one thousand years.

104.011 **Activity of the Lyrid meteor shower from radar observations.**
V. Porubčan.
Asteroids, comets, meteors II, p. 527 – 530 (1986). – See Abstr. 012.010.

Radar observations of the Lyrid meteor shower carried out by meteor radars at Springhill, Ondřejov and Budrio are presented. Mean cross–section curves of the shower derived for two different groups of overdense echoes are discussed. Maximum Lyrid activity is found to occur at solar longitude 31°7 (equinox 1950). A sudden outburst in Lyrid activity in 1982, observed by the Budrio meteor radar, peaked at solar longitude 31°369.

104.012 **Structure and activity of the Perseid meteor stream from visual observations 1953 – 81.**
B. A. Lindblad.
Asteroids, comets, meteors II, p. 531 – 535 (1986). – See Abstr. 012.010.
Visual observations of Perseid meteors 1953 – 81 indicate that the Perseid stream cross section consists of two components: a long duration rather flat component and a sharp spike–like feature which defines the Perseid maximum. Shower maximum for visual meteors occurs at solar longitude 139°4 (Equinox 1950). The Perseid shower shows large variations in activity from year to year, with more than a factor two variation in hourly rates at maximum.

104.013 **The activity curve of the Perseid meteor stream from Onsala radar observations 1953 – 78.**
B. A. Lindblad, M. Šimek.
Asteroids, comets, meteors II, p. 537 – 540 (1986). – See Abstr. 012.010.
The cross–section of the Perseid meteor stream is analyzed based on radar observations at the Onsala Space Observatory in 1953 – 78. The cross–section consists of two components: a flat long duration component and a sharp central core, which defines shower maximum. Maximum occurs at solar longitude 139°20 (Equinox 1950.0).

104.014 **Possibility of a meteor shower associated with comet Sugano–Saigusa–Fujikawa 1983e.**
M. Šimek, P. Pecina.
Asteroids, comets, meteors II, p. 541 – 544 (1986). – See Abstr. 012.010.
The comet 1983e, discovered in May, 1983 was suggested as a source of meteors with a theoretical radiant at $\alpha = 21°$, $\delta = +42°$. The closest approach of the Earth to the comet's orbit occurred on June 14.7, 1983, when the Earth passed 0.048 AU inside the comet's orbit, only 2.9 days after the comet has passed this point. Radio meteor observations were carried out using the 25 kW Ondřejov pulsed radar in 1983 and 1984. A small increase of the meteor rate was observed in the appropriate range interval. Results of the radar observation are described.

104.015 **Spurious maxima and minima in meteor stream profiles.**
B. A. Lindblad.
Asteroids, comets, meteors II, p. 545 – 548 (1986). – See Abstr. 012.010.
Meteor stream cross–sections derived from long–term observational records may exhibit spurious secondary maxima. These maxima, spaced at regular intervals in solar longitude, are related to the sampling frequency. Two cases based on radar recordings of the Perseid stream are presented.

104.016 **Asteroidal origin for the Geminid meteor stream.**
J. Hunt, K. Fox, I. P. Williams.
Asteroids, comets, meteors II, p. 549 – 553 (1986). – See Abstr. 012.010.
In previous papers, the authors have demonstrated that most of the characteristics of the Geminid meteor stream can be explained in terms of a model where the stream particles are ejected from a cometary nucleus. It now seems fairly certain that object 1983 TB is associated with the Geminid meteor stream but recent observations suggest that 1983 TB is not the nucleus of a comet. A model for the formation of the stream based on a collision between two rock like bodies is described. It is found that this model can explain two of the major observed characteristics but has more difficulty in explaining the observed distribution of aphelion distances.

104.017 **Numerical investigations on a possible gravitational breaking of the Quadrantid meteor stream.**
C. Froeschlé, H. Scholl.
Asteroids, comets, meteors II, p. 555 – 558 (1986). – See Abstr. 012.010.
The authors investigated numerically the dynamical evolution of a Quadrantid–like meteor stream orbiting the Sun in a 2/1 mean motion resonance with Jupiter. Only the main forces, the gravitational forces exerted by the Sun and by Jupiter were taken into account. Highly inclined streams located at resonance center may give rise to the formation of arcs. These arcs evolve separately due to the different possible modes for the motions of the stream particles nodal lines': regression or progression. These modes are due to resonant motion and are not caused by close approaches to Jupiter.

104.018 **Meteor storms.**
I. P. Williams, C. Johnson, K. Fox.
Asteroids, comets, meteors II, p. 559 – 563 (1986). – See Abstr. 012.010.
Meteor showers are generally seen as annual events, with the number of meteors seen per hour not varying dramatically from year to year. In contrast meteor storms, where the number of meteors seen per hour reaches values of the order of 10000, occur far less frequently. The authors develop a simple model for the rate of spreading of meteoroids around the mean orbit and apply this theory to the best known of the streams associated with a storm, namely the Leonids. Interesting constraints on the size of the parent comet are obtained.

104.019 **The distribution of sporadic meteor radiants and orbits.**
J. Štohl.
Asteroids, comets, meteors II, p. 565 – 574 (1986). – See Abstr. 012.010.
Sporadic meteors are outlined as the major part of the whole meteoric complex, with the ratio sporadic/shower meteors depending on the mass of meteoroids. Distributions of the apparent and true radiants with their seasonal variations are presented and their interrelation with the distribution of orbital parameters is analysed. Populations of sporadic meteoroids and their tentative diffuse streams and twin showers are discussed.

104.020 **Photographic fireball networks.**
Z. Ceplecha.
Asteroids, comets, meteors II, p. 575 – 582 (1986). – See Abstr. 012.010.
The initial interest and enthusiasm in large–scale meteoroid studies and in operation of photographic fireball networks has diminished during the last few years. However, more theoretical studies of the interaction of large meteoroids with the atmosphere, as well as more systematic observations of fireballs are needed to solve definitely the questions of meteorite relations to meteoroids before atmospheric entry. In total 3 meteorite falls have been photographed over the globe to date.

104.021 **Bi–telescopic, deep, simultaneous meteor observations.**
L. G. Taff.
Asteroids, comets, meteors II, p. 603 – 608 (1986). – See Abstr. 012.010.
The Experimental Test System of Lincoln Laboratory has been used to search for, detect, and discriminate meteors. The author presents a statistical summary of ten hours of observing sporadic meteors and two meteor showers.

104.022 **Meteor heights derived from radar–visual observations.**
B. A. Lindblad, J. Štohl.
Asteroids, comets, meteors II, p. 615 – 620 (1986). – See Abstr. 012.010.
Heights of 2748 meteors observed at the Onsala Space Observatory in 1953–75 both visually and by radar are studied. Mean mid–point heights of 102.2, 93.7 and 95.7 km are derived for Perseid, Delta Aquarid and sporadic meteors, respectively. The method of determining radar–visual meteor heights is discussed and observational errors are estimated.

104.023 **Radiative heat exchange of a meteor body in approximation of radiant thermal conductivity.**
N. N. Pilyugin, T. A. Chernova.
Kosm. Issled., Tom 24, Vyp. 1, p. 58 – 68 (1986). In Russian. English translation in Cosm. Res.

104.024 **The anisotropy of parameters of an ionized meteor trace associated with molecular processes in the atmosphere.**
I. A. Delov.
Geomagn. Aehron., Tom 26, No. 2, p. 265 – 269 (1986). In Russian. English translation in Geomagn. Aeron.

104.025 **Observing meteors: V – Draconid notes.**
D. H. Levy.
Strolling Astron., Vol. 31, Nos. 5–6, p. 95 (1986).

104.026 **Observing meteors: VI.**
D. H. Levy.
Strolling Astron., Vol. 31, Nos. 7–8, p. 172 (1986).

104.027 **Results of photographic observations of meteors in Kiev in the years 1967 – 1976. I. Kinematics.**
L. M. Sherbaum, V. V. Benyukh, A. M. Kazantsev, V. V. Kalenichenko, V. G. Kruchinenko, Yu. G. Taranukha, A. N. Shajdo, S. S. Tryashin.
Vestn. Kiev. Univ. Astron., No. 27, p. 73 – 77 (1985). In Russian.

104.028 **Diffusional dissipation of meteor trails containing various gaseous components.**
V. S. Tokhtas'ev.
Izv. Astron. Ehngel'gardt. Obs., No. 50, p. 200 – 213 (1985). In Russian.

104.029 **Using the regularization method for computation of the distribution of radiant density of sporadic meteors over the celestial sphere.**
T. K. Filimonova.
Izv. Astron. Ehngel'gardt. Obs., No. 51, p. 41 – 49 (1985). In Russian.

104.030 **The parameter of distribution of sporadic meteor bodies according to masses.**
P. B. Babadzhanov, R. Sh. Bibarsov.
Dokl. Akad. Nauk TadzhSSR, Tom 28, No. 6, p. 330 – 333 (1985). In Russian. Abstr. in Ref. Zh., 51. Astron., 3.51.304 (1986).

104.031 **Mean density of a stream of meteor bodies falling on the earth.**
P. B. Babadzhanov, R. Sh. Bibarsov, V. S. Getman, V. M. Kolmakov.
Dokl. Akad. Nauk SSSR. Ser. Mat. Fiz., Tom 284, No. 4, p. 824 – 826 (1985). In Russian. From Ref. Zh., 51. Astron., 3.51.306 (1986).

104.032 **Modelling of meteor matter distribution in the solar system.**
V. S. Zabolotnikov.
Astron. Vestn., Tom 20, No. 2, p. 134 – 141 (1986). In Russian. English translation in Sol. Syst. Res.
Using the phase function, determined in the space of Keplerian elements, the problem of modelling of meteor matter distribution in the solar system was defined and solved. On the basis of the McCrosky and Posen (1961) catalogue data for the ecliptic plane the dependence of the space density and the flux density upon heliocentric distance was obtained.

104.033 **On the spatial structure of the Geminid meteor shower.**
O. I. Bel'kovich.
Astron. Vestn., Tom 20, No. 2, p. 142 – 151 (1986). In Russian. English translation in Sol. Syst. Res.
A mathematical model of the three–dimensional distribution of masses and flux densities of Geminid meteor bodies in the plane normal to the mean velocity vector in the node is proposed. Elements of the orbit of the hypothetical parent body of the stream are found for epoch 1950.0. The full mass of the stream particles is 9×10^{14}g that corresponds to the radius of the spherical body of 1.2 km.

104.034 **Observations of the Orionids 1984.**
G. N. Sizonov.
Astron. Vestn., Tom 20, No. 2, p. 157 – 159 (1986). In Russian. English translation in Sol. Syst. Res.
By means of visual observations of the Orionid shower the influx of meteoric matter, the zenital hourly rates, the average distance between meteoric particles, the steep of the luminosity function, the mass–law parameter, the volume density were determined. The maximum of the shower has occurred on the night 22/23 October.

104.035 **Masses and densities of meteoroids of the Quadrantid shower 1979.**
R. Sh. Bibarsov, M. Narziev, R. P. Chebotarev.
Komety Meteory, No. 37, p. 20 – 23 (1985). In Russian.
The mean value of density of meteoroids of the Quadrantid shower 1979 is 1 g/cm^3 and their mean mass is 1.6×10^{-2}g.

104.036 **Some results of radar investigation of the structure of the Orionid meteor stream.**
Sh. O. Isamutdinov, R. P. Chebotarev.
Komety Meteory, No. 37, p. 24 – 27 (1985). In Russian.
The structure of the Orionid meteor stream was studied by radio observations at Dushanbe in 1974 – 1980. The Orionid meteor particles were shown to be distributed uneven in both directions along and across the orbit. Two maxima of stream activity at solar longitudes 207 – 208° and 210° can be distinguished in the mean activity data for the years of observations.

104.037 **Light and ionization curves of meteors according to combined TV and radar observations.**
P. B. Babadzhanov, I. F. Malyshev, M. Narziev, R. P. Chebotarev.
Komety Meteory, No. 37, p. 28 – 33 (1985). In Russian.
The apparatus parameters, the method of deriving light and ionization curves and the main parameters of 13 meteors from –3.3 to +2.5 absolute magnitude according to combined TV and radar observations are presented. It is noted that the shapes of light and ionization curves are in good accordance for individual meteors. For the Perseids the observed correlation of the absolute magnitude M and electron line density q is $M = 35–2.5 \lg q$.

104.038 **Generalizing Manning's method for measuring wind velocity components in the meteor zone.**
R. D. Gajnutdinova, K. A. Karimov.
Komety Meteory, No. 37, p. 40 – 48 (1985). In Russian.
Supplementary possibilities of the radiometer azimuthal method for measuring wind velocity (Manning's method) are examined. Calculating formulae are obtained for determination of the characteristics of the nonuniformity of the wind field in the meteor zone.

104.039 **Jaarverslag 1985.**
R. Veltman.
Radiant, Jaarg. 8, Nr. 1, p. 13, 10 (1986).
The paper contains a summary of the preliminary results of visual meteor observations in 1985 in the Netherlands. A table is given in which the totals of each campaign are listed including the totals over the whole year.

104.040 **Geminiden 1985.**
K. Jobse.
Radiant, Jaarg. 8, Nr. 1, p. 15 – 16 (1986).

104.041 **Simultaanmateriaal 1985. Deel 1. 15 sets baanelementen.**
H. Betlem, M. de Lignie, C. ter Kuile.
Radiant, Jaarg. 8, Nr. 2, p. 21 – 31 (1986).

104.042 **Orioniden 1985.**
R. Veltman.
Radiant, Jaarg. 8, Nr. 2, p. 34 – 38 (1986).

104.043 **Geminidenaktie 1985 te Puimichel.**
K. Jobse.
Werkgroepnieuws, Vol. 14, Nr. 1, p. 6 – 8 (1986).

104.044 **Geminids observed in France.**
P. Roggemans.
Werkgroepnieuws, Vol. 14, Nr. 1, p. 9 – 10 (1986).

104.045 **Photographic radiant positions obtained in August 1985.**
C. Steyaert.
Werkgroepnieuws, Vol. 14, Nr. 1, p. 11 – 13 (1986).

104.046 **Angular velocity of a meteor.**
R. Koschack.
Werkgroepnieuws, Vol. 14, Nr. 1, p. 13 – 16 (1986). In German.

104.047 **D.D.R.: Perseids and Orionids 1985.**
J. Rendtel.
Werkgroepnieuws, Vol. 14, Nr. 1, p. 16 – 18 (1986).

104.048 **Geminids 1985: Harderwijk – the Netherlands.**
K. Miskotte.
Werkgroepnieuws, Vol. 14, Nr. 1, p. 18 (1986).

104.049 **Australia: the Eta Aquariids in 1985.**
J. Wood.
Werkgroepnieuws, Vol. 14, Nr. 2, p. 39 – 40 (1986).

104.050 **Autumn 1985 observations of meteor streams from Finland.**
P. Parviainen.
Werkgroepnieuws, Vol. 14, Nr. 2, p. 40 – 42 (1986).

104.051 **The 1985 results from Norway.**
Werkgroepnieuws, Vol. 14, Nr. 2, p. 42 – 43 (1986).

104.052 **Recent observations from Florida.**
Werkgroepnieuws, Vol. 14, Nr. 2, p. 43 – 44 (1986).

104.053 **The photographic Lyrid radiant.**
C. Steyaert.
Werkgroepnieuws, Vol. 14, Nr. 2, p. 46 – 47 (1986).

104.054 **On the Geminid meteor stream in 1985.**
P. Roggemans.
Werkgroepnieuws, Vol. 14, Nr. 2, p. 48 – 63 (1986).

104.055 **The 1985 results of Norman McLeod (Florida, USA).**
N. McLeod.
Werkgroepnieuws, Vol. 14, Nr. 2, p. 63 – 65 (1986).

104.056 **The 1985 results of Robert Lunsford.**
R. Lunsford.
Werkgroepnieuws, Vol. 14, Nr. 2, p. 66 – 67 (1986).

104.057 **Radiowaarnemingen: Geminiden 1985.**
J. Van Wassenhove.
Werkgroepnieuws, Vol. 14, Nr. 3, p. 74 (1986).

104.058 **V.V.S. Meteor Section: annual report 1985.**
P. Roggemans.
Werkgroepnieuws, Vol. 14, Nr. 3, p. 79 – 89 (1986).

104.059 **The Netherlands: Quadrantids 1986.**
B. Rispens.
Werkgroepnieuws, Vol. 14, Nr. 3, p. 91 – 92 (1986).

104.060 **The Upsilon Pegasid meteor shower.**
H. Povenmire.
I.A.P.P.P. Commun., No. 23, p. 4 (1986). Abstract. – See Abstr. 010.521.

104.061 **The Draconid meteoroids.**
M. Beech.
Astron. J., Vol. 91, No. 1, p. 159 – 162 (1986).

The available observational data for the beginning and end heights of Draconid (Giacobinid) meteors are compared with the dustball meteor ablation theory of Hawkes and Jones, with which a good general agreement is found. Assuming that the fragile nature of the Draconid meteoroids indicates that the energy required to fragment unit mass of meteoroid material X is less than $\sim 10^5 \mathrm{Jkg}^{-1}$, the previously found minimum value, then the Draconid meteoroids have a bulk density greater than ~ 200 kg m^{-3}. The relation between comet P/Giacobini–Zinner and the Draconid meteoroids is briefly considered.

104.062 **Dissociation of oxygen molecules in meteor trains.**
R. Sh. Bibarsov.
Dokl. Akad. Nauk TadzhSSR, Tom 28, No. 8, p. 445 – 447 (1985). In Russian. Abstr. in Ref. Zh., 51. Astron., 5.51.386 (1986).

104.063 **Formation of ozone due to the interaction between a meteor body and the earth's atmosphere.**
R. Sh. Bibarsov.
Dokl. Akad. Nauk TadzhSSR, Tom 28, No. 9, p. 510 – 512 (1985). In Russian. Abstr. in Ref. Zh., 51. Astron., 5.51.387 (1986).

104.064 **On the structure of the Quadrantid meteor stream.**
G. V. Andreev, A. E. Epishova, L. N. Rubtsov, I. I. Tarasova.
Izv. Akad. Nauk TadzhSSR. Otd–nie fiz.–mat., khim. geol. nauk, No. 3, p. 32 – 41 (1985). In Russian. Abstr. in Ref. Zh., 51. Astron., 6.51.327 (1986).

104.065 **Perseid meteor stream: mean flux curve from radar observations.**
M. Šimek, B. A. McIntosh.
Bull. Astron. Inst. Czech., Vol. 37, No. 3, p. 146 – 155 (1986).

Sixteen years of observations of the Perseid meteor shower by the Ottawa (Canada) patrol radar have been used to determine the relative flux of the shower in particle size ranges producing radar meteor echoes having durations $\geqslant 1$ s and $\geqslant 8$ s.

104.066 **TV observation of 1985 Giacobinid meteor shower.**
K. Nagasawa, T. Kanda.
Tokyo Astron. Bull., Second Ser., No. 274, p. 3159 – 3162 (1986).

An extensive Giacobinid meteor shower was recognized in the evening of October 8th, 1985, all over Japan. The authors carried out a TV observation of the shower at Dodaira Station, Tokyo Astronomical Observatory. Although the observation was made in the decreasing stage of the shower, 8 meteors were recorded in total. From them, real radiant point was determined as $\alpha = 265\overset{\circ}{.}2$, $\delta = 55\overset{\circ}{.}3$ (J2000.0).

104.067 **Spectral–magnitude relations in Perseid meteors.**
J. A. Russell.
Astron. J., Vol. 91, No. 3, p. 640 – 645 (1986). With plate 60.

Perseid meteor counts, heights, magnitudes, and spectra are compared for the showers of 1977, 1978, 1980, 1981, and 1983. The counts were highest in 1980. Heights, magnitudes, and certain spectral characteristics appear correlated with the atmospheric density gradient along the path of the meteor, which is affected by modification of the vertical gradient but most strongly by the altitude of the meteor radiant. Inhomogeneity in the meteoroids may play an additional role in spectral–magnitude relations.

104.068 **Radar observations of the Lyrid meteor shower in 1980 – 1984.**
V. Porubčan.
Contrib. Astron. Obs. Skalnaté Pleso, Vol. 14, p. 105 – 109 (1986).

Radar observations of the Lyrid meteor shower carried out by the meteor radar at Ondřejov in 1980 – 1984 are analysed and discussed from the view-point of activity. The mean activity curves derived for two different sets of overdense echoes indicate no separation of particles according to their masses.

104.069 **Activity of the Geminid meteor shower in the years 1944 - 1974.**
J. Zvolánková.
Contrib. Astron. Obs. Skalnaté Pleso, Vol. 14, p. 111 - 119 (1986).
More than 8000 visual records of the Geminids from the Skalnaté Pleso Observatory, made over a period of 8 years, have been employed in specifying the activity of the Geminid shower in the years 1944 - 1974.

104.070 **Relation between bright nights and meteoritic showers.**
M. Herse, P. Rigaud.
Can. J. Phys., Vol. 63, No. 11, p. 1446 - 1448 (1985). Abstr. in Phys. Abstr., Vol. 89, No. 1278, Entry 47971 (1986).

104.071 **Analysis of meteor data on a microcomputer system.**
M. Koseki.
J. Br. Astron. Assoc., Vol. 96, No. 4, p. 232 - 240 (1986).
The compilation and use of a meteor data file held on a microcomputer system are discussed. The file contains details of 1894 meteor streams and 4431 photographic meteors. Identical streams have been identified, and investigation has been carried out into the relationships between meteor streams and comets or minor planets. The statistics of meteor orbits and of photographic streams have been examined and interesting results are presented for the Taurids, the Perseids and the Geminids.

104.072 **Simulatie meteoorwaarnemingen. De computer als visuele waarnemer.**
P. van de Veen.
Radiant, Jaarg. 8, Nr. 3, p. 41 - 45 (1986).

104.073 **Simultaanmateriaal 1985. Deel 2.**
H. Betlem, M. de Lignie, C. ter Kuile.
Radiant, Jaarg. 8, Nr. 3, p. 46 - 51 (1986).

104.074 **De grote vuurbol van 20-2-1986.**
P. Jenniskens.
Radiant, Jaarg. 8, Nr. 3, p. 52 - 54 (1986).

104.075 **De Geminiden van 1985.**
R. Veltman.
Radiant, Jaarg. 8, Nr. 3, p. 56 - 57 (1986).

104.076 **De struktuur van het Geminidenmaximum. De evolutie van een holle meteorenzwerm.**
P. Jenniskens.
Radiant, Jaarg. 8, Nr. 3, p. 58 - 59 (1986).

104.077 **Meteoorwaarnemingen bij totale maansverduistering.**
J. Wood.
Radiant, Jaarg. 8, Nr. 3, p. 60 - 61 (1986).

104.078 **On the scattering of radio waves from underdense meteor trains.**
G. G. Novikov, P. Pecina, A. V. Blokhin.
Bull. Astron. Inst. Czech., Vol. 37, No. 4, p. 189 - 195 (1986).
The paper deals with some problems of radio-wave scattering from underdense meteor trains. The diffusion equation is solved under the assumption that the ionization source terminates its action after the meteor is fully burnt out, which has not been taken into account so far. The finite train initial radius is also considered.

104.079 **Radiowaarnemingen: Quadrantiden 1986.**
J. Van Wassenhove.
Werkgroepnieuws, Vol. 14, Nr. 3, p. 74 - 75 (1986).

Comets, meteors & asteroids - how they affect earth.
See Abstr. 003.099.

Astronomy in Tadjikistan.
See Abstr. 013.007.

On the function inverse to the exponential integral function.
See Abstr. 021.002.

On the possibility of observing reflection phenomena in meteor trails.
See Abstr. 036.049.

Meteor photography limitations and the beginning heights of meteors.
See Abstr. 036.050.

Method of observation of meteor streams on adjustable antennae following the daily motion of the radiant.
See Abstr. 036.075.

On methods of radar measurements of the drift of meteor trains.
See Abstr. 036.076.

Alle waarnemers en hun lijnen. Een onderzoek naar individuele magnitudenschalen.
See Abstr. 036.099.

Limiting magnitude correction.
See Abstr. 036.100.

France, Puimichel = Perseids 1986 (I). Personal errors on visual magnitude estimates.
See Abstr. 036.101.

Gravitational splitting of Quadrantid like meteor streams in resonance with Jupiter.
See Abstr. 042.012.

Ultraviolet spectroscopy of meteoric debris of comets.
See Abstr. 051.068.

Debris from comet Halley and the risk for space probes.
See Abstr. 103.638.

Collisions in the Solar System - III. Meteoroid survival times.
See Abstr. 106.001.

On the problem of the origin of β meteoroids.
See Abstr. 106.070.

105 Meteorites, Meteorite Craters

105.001 **Unexpected Antarctic chemistry.**
G. W. Wetherill.
Nature, Vol. 319, No. 6052, p. 357 – 358 (1986).
This note comments on the recent finding by Dennison, Lingner and Lipschutz that the concentration of some trace elements in a sample of Antarctic meteorites is different from the concentrations in a similar-sized sample of non-Antarctic meteorites (see Abstr. 105.002).

105.002 **Antarctic and non-Antarctic meteorites form different populations.**
J. E. Dennison, D. W. Lingner, M. E. Lipschutz.
Nature, Vol. 319, No. 6052, p. 390 – 393 (1986).
The authors have studied the most frequently encountered Antarctic meteorites, H5 chondrites. They find that compositional differences between Antarctic and non-Antarctic specimens of this class, especially those involving thermally sensitive trace and ultra-trace elements, are so substantial that it seems extremely doubtful that the two sample populations are derived from the same parent population. This implies that the contemporary falls (less than 200 years old) and the substantially older (on average, 300,000 years for Victoria Land) Antarctic meteorites either differ in genetic history or sample different extraterrestrial parent populations, or both.

105.003 **Determination of cosmogenic ^{41}Ca in a meteorite with tandem accelerator mass spectrometry.**
P. W. Kubik, D. Elmore, N. J. Conard, K. Nishiizumi, J. R. Arnold.
Nature, Vol. 319, No. 6054, p. 568 – 570 (1986).
Calcium-41 with a half-life $T_{1/2} = 1.03 \times 10^5$yr has been proposed as a dating tool for determining the residence time of meteorites found on ablation areas of the Antarctic ice sheet. It occurs, however, in concentrations below the detection limit of conventional techniques. The authors report the first use of tandem accelerator mass spectrometry to measure ^{41}Ca in a natural sample – the Bogou iron meteorite. They find a $^{41}Ca/Ca$ ratio of $(3.8 \pm 0.6) \times 10^{-12}$ which corresponds to a ^{41}Ca activity of 6.9 ± 1.1 d.p.m per kg meteorite. This implies that ^{41}Ca can be successfully used to determine the irradiation history of meteorites.

105.004 **Oxidation state of iron in SNC meteorites as studied by Mössbauer spectroscopy.**
V. W. A. Vieira, T. V. V. Costa, H. G. Jensen, J. M. Knudsen, M. Olsen, L. Vistisen.
Phys. Scr., Vol. 33, No. 2, p. 180 – 186 (1986).
Mössbauer investigations of three meteorites of the type SNC-achondrites are reported. The results are compared with Mössbauer spectra of other meteorites, and selected terrestrial and lunar rocks. Through the oxidation state of iron in the pyroxenes and magnetic oxides studied, the work confirms that the mantle of the Earth is highly oxidized compared with the mantle of asteroidal bodies and lunar basalts.

105.005 **Fe-Mg-Mn relations of ureilite olivines and pyroxenes and the genesis of ureilites.**
D. W. Mittlefehldt.
Geochim. Cosmochim. Acta, Vol. 50, No. 1, p. 107 – 110 (1986).
It is the purpose of this article to show that the existing mineral composition data for ureilites show that a reducing agent, probably C, was present during the magmatic stage of ureilite genesis and that fractional crystallization played an important part in the genesis of some ureilites.

105.006 **Carbon isotope relationships in winonaites and forsterite chondrites.**
M. M. Grady, C. T. Pillinger.
Geochim. Cosmochim. Acta, Vol. 50, No. 2, p. 255 – 263 (1986).
Classification schemes for meteorites based upon physical, chemical and petrologic criteria alone have left many apparently "anomalous" meteorites which have resisted incorporation into precise groups. Here the authors discuss carbon isotope measurements made on several meteorites which have previously possessed long and confusing histories of classifications in both achondritic and chondritic groups, in an effort to evaluate further the usefulness of carbon studies.

105.007 **Chondrules in the Murray CM2 meteorite and compositional differences between CM-CO and ordinary chondrite chondrules.**
A. E. Rubin, J. T. Wasson.
Geochim. Cosmochim. Acta, Vol. 50, No. 2, p. 307 – 315 (1986).
Of the seven groups of carbonaceous and ordinary chondrites, all but the CI group contain chondrules. Chondrules have characteristic size ranges in each group: on average, CM and CO chondrules are smaller and CV chondrules larger than ordinary chondrite chondrules. The CM and CO chondrite groups appear to be closely related and probably formed at similar heliocentric distances.

105.008 **Survival of interstellar matter in meteorites: evidence from carbonaceous material.**
J. F. Kerridge, S. Chang.
Protostars & planets II, p. 738 – 754 (1985). – See Abstr. 012.008.
An interstellar origin has been attributed to certain organic and apparently elemental forms of C which occur in carbonaceous meteorites. Evidence for such an origin comes from anomalous isotopic composition either of the C itself or of elements, such as N, H, or the noble gasses, which are combined with or trapped within the carbonaceous material. Suggested sources for these anomalous compositions include atmospheres of red giant stars, novae and supernovae, and interstellar molecular clouds.

105.009 **Isotopic variations in solar system material: evaporation and condensation of silicates.**
R. N. Clayton, T. K. Mayeda, C. A. Molini-Velsko.
Protostars & planets II, p. 755 – 771 (1985). – See Abstr. 012.008.
Simultaneous determination of stable isotope abundances of oxygen and silicon in meteoritic chondrules and inclusions shows that these two abundant elements undergo fundamentally different cosmochemical processes. Silicon isotopic variations reflect processes of evaporation and condensation at high temperatures, probably during brief melting events. Oxygen isotopic variations reflect interaction between isotopically distinct reservoirs, probably involving a gaseous component of the solar nebula. There remains uncertainty as to the origin of the various reservoirs.

105.010 **Meteoritic evidence concerning conditions in the solar nebula.**
W. V. Boynton.
Protostars & planets II, p. 772 – 787 (1985). – See Abstr. 012.008.
Chondritic meteorites contain evidence of many fractionation processes, which separate groups of chemical elements with similar properties. These fractionations provide strong evidence that high temperatures (1400 – 1500K) occurred at least for short periods (hours) and that moderately high temperatures (1200K) occurred for longer times. These temperatures obtained as far from the Sun as the asteroid belt. Many episodes of condensation and isolation of grains from the gas are recorded and provide evidence of a chaotic environment. Nevertheless, the absence of Ca, Al-rich inclusions in ordinary chondrites suggests that mixing of solid materials over significant heliocentric distances did not occur.

105.011 **Application of the sphalerite cosmobarometer to group IAB iron meteorites.**
S. A. Kissin, H. P. Schwarcz, S. D. Scott.
Geochim. Cosmochim. Acta, Vol. 50, No. 3, p. 371 – 378 (1986).
The cosmobarometer shows that the group IA irons formed over a range of pressures from zero to 3.5 kbar, probably in a single parent object based on the chemical and structural proper-

ties of group IAB. The data also indicate that Ni content of the bulk metal phase increased with pressure, and, therefore, presumably with depth, in the parent object. This result must be considered in any theory that seeks to account for the origin of the group.

105.012 **A summary of extremes of isotopic variations in extraterrestrial materials.**
M. Shima.
Geochim. Cosmochim. Acta, Vol. 50, No. 4, p. 577 – 584 (1986).
In this comprehensive review of current research on isotopic variations of elements in extraterrestrial materials, the variations were classified in terms of the major process involved in the modification of the iostopic composition of the element concerned. Maximum isotopic variations of each element were retrieved from publications which were available in Tokyo up to December 1985, and are presented in tabular form.

105.013 **Rapid oxygen diffusion in melilite and its relevance to meteorites.**
T. Hayashi, K. Muehlenbachs.
Geochim. Cosmochim. Acta, Vol. 50, No. 4, p. 585 – 591 (1986).
The rapid exchange and diffusion of oxygen in melilite suggests that the isotopic composition of that mineral in the Ca–Al rich inclusions of carbonaceous chondrites records the isotopic composition of the last nebular gas to which they were exposed. Constraints on the earlier history of these inclusions may be deduced from the isotopic composition of the less exchangeable minerals.

105.014 **Noble gases in the diamond–free ureilite, ALHA 78019: the roles of shock and nebular processes.**
J. F. Wacker.
Geochim. Cosmochim. Acta, Vol. 50, No. 4, p. 633 – 642 (1986).
Primordial noble gases were measured in the unshocked ureilite ALHA 78019 and in the moderately shocked ureilite Kenna. Much of the gas (>25%) in 78019 is contained in fine–grained, amorphous carbon, where it is enriched 60–fold relative to that in the more abundant crystalline graphite. Isotopically, the gas in the gas–rich C resembles the planetary component ("phase Q") in chondrites, but it is not lost upon etching with HNO_3 and hence is not surface sited, unlike the chondritic gas. In most other respects, the noble gases are in the normal range for ureilites. These results suggest that the gases – and carbon – predate the shock event that produced the diamond, hence the gas–rich C in 78019 appears to be a good candidate for the hypothetical pre-shock, noble–gas–bearing C in diamondiferous ureilites. Furthermore, the results suggest that the shock event has no bearing on the origin of carbon in these meteorites.

105.015 **Relationship between dark rims, interchondrule matrix, and chondrules in U.O.C.s.**
C. Alexander, D. J. Barber, M. Davis, R. Hutchison.
Meteoritics, Vol. 20, No. 4, p. 600 (1985). Abstract. – See Abstr. 010.661.

105.016 **Chondrites and the solar nebula: a review.**
E. Anders.
Meteoritics, Vol. 20, No. 4, p. 601 – 602 (1985). Abstract. – See Abstr. 010.661.

105.017 **A subclassification of unequilibrated ordinary chondrites based on volatile element content.**
E. Anders, M. G. Zadnik.
Meteoritics, Vol. 20, No. 4, p. 602 (1985). Abstract. – See Abstr. 010.661.

105.018 **Ni–Pt–Ge–rich Fremdlinge: indicators of a turbulent early solar nebula.**
J. T. Armstrong, I. D. Hutcheon, G. J. Wasserburg.
Meteoritics, Vol. 20, No. 4, p. 603 – 604 (1985). Abstract. – See Abstr. 010.661.

105.019 **The unique ureilite, ALHA 82106–82130: evidence for progressive reduction during ureilite magmatic differentiation.**
J. L. Berkley, C. A. Goodrich, K. Keil.
Meteoritics, Vol. 20, No. 4, p. 607 – 608 (1985). Abstract. – See Abstr. 010.661.

105.020 **The Binningup (H6) chondrite: a new fall from Western Australia.**
A. W. R. Bevan, K. J. McNamara, J. C. Barton.
Meteoritics, Vol. 20, No. 4, p. 608 (1985). Abstract. – See Abstr. 010.661.

105.021 **Trace elements in Al–rich chondrules from Ybbsitz (H4).**
A. Bischoff, B. Spettel, H. Palme.
Meteoritics, Vol. 20, No. 4, p. 609 – 610 (1985). Abstract. – See Abstr. 010.661.

105.022 **Determining the chronology of breccia formation on meteorite parent bodies.**
D. Bogard.
Meteoritics, Vol. 20, No. 4, p. 611 (1985). Abstract. – See Abstr. 010.661.

105.023 **NO α–quartz–rich inclusions in ordinary chondrites and confirmation of fluid inclusions in chondritic phosphates.**
M. Bourot–Denise, J. Fabriès, C. Fiéni, D. C. Smith, P. Pellas, J. Touret, H. Boyer.
Meteoritics, Vol. 20, No. 4, p. 612 (1985). Abstract. – See Abstr. 010.661.

105.024 **Cosmic–ray–track studies of near–surface specimens of Jilin chondrite.**
M. Bourot–Denise, P. Pellas, C. Perron.
Meteoritics, Vol. 20, No. 4, p. 612 – 613 (1985). Abstract. – See Abstr. 010.661.

105.025 **Refractory rims: evidence for high temperature events in the post–formation history of Ca, Al–rich inclusions.**
W. V. Boynton, D. A. Wark.
Meteoritics, Vol. 20, No. 4, p. 613 – 614 (1985). Abstract. – See Abstr. 010.661.

105.026 **On the occurrence of silica in ordinary chondrites.**
F. Brandstätter, G. Kurat.
Meteoritics, Vol. 20, No. 4, p. 615 – 616 (1985). Abstract. – See Abstr. 010.661.

105.027 **Enstatite achondrites did not form from enstatite chondrites.**
R. Brett, K. Keil.
Meteoritics, Vol. 20, No. 4, p. 616 (1985). Abstract. – See Abstr. 010.661.

105.028 **Thermal migration: or how to transport iron sulfide in solid iron meteorites.**
V. F. Buchwald, T. Kjer, K. A. Thorsen.
Meteoritics, Vol. 20, No. 4, p. 617 – 618 (1985). Abstract. – See Abstr. 010.661.

105.029 **Sulphur isotopic studies in carbonaceous chondrites.**
R. Burgess, I. P. Wright, C. T. Pillinger.
Meteoritics, Vol. 20, No. 4, p. 618 – 619 (1985). Abstract. – See Abstr. 010.661.

105.030 **Structure of graphitic carbons: terrestrial vs. extraterrestrial origin.**
P. R. Buseck, H. Bo–Jun.
Meteoritics, Vol. 20, No. 4, p. 619 – 620 (1985). Abstract. – See Abstr. 010.661.

105.031 **Pre-compaction irradiation of Kapoeta: $^{40}Ar-^{39}Ar$ ages of individual grains.**
M. W. Caffee, C. M. Hohenberg, T. D. Swindle, J. N. Goswami.
Meteoritics, Vol. 20, No. 4, p. 620 – 622 (1985). Abstract. – See Abstr. 010.661.

105.032 **U–Th–Pb systematics in shergottites: young ages and low μ.**
J. H. Chen, G. J. Wasserburg.
Meteoritics, Vol. 20, No. 4, p. 623 – 624 (1985). Abstract. – See Abstr. 010.661.

105.033 **The contribution of preaccretionary alteration products of chondrules and inclusions to Allende matrix.**
M. Christophe–Michel–Lévy.
Meteoritics, Vol. 20, No. 4, p. 625 (1985). Abstract. – See Abstr. 010.661.

105.034 **Four ataxites from Antarctica.**
R. S. Clarke Jr., K. Yanai.
Meteoritics, Vol. 20, No. 4, p. 625 – 626 (1985). Abstract. – See Abstr. 010.661.

105.035 **Concentric objects in C2 and C3 carbonaceous chondrites.**
R. E. Cohen.
Meteoritics, Vol. 20, No. 4, p. 627 – 628 (1985). Abstract. – See Abstr. 010.661.

105.036 **Magnetic properties of Antarctic shergottites EETA 79001 and ALHA 77005.**
D. W. Collinson.
Meteoritics, Vol. 20, No. 4, p. 628 (1985). Abstract. – See Abstr. 010.661.

105.037 **Magnetic properties of the Olivenza chondrite – a cautionary tale.**
D. W. Collinson.
Meteoritics, Vol. 20, No. 4, p. 629 (1985). Abstract. – See Abstr. 010.661.

105.038 **Rare earth element concentrations of mesosiderite merrillite.**
G. Crozaz, E. Zinner, J. S. Delaney.
Meteoritics, Vol. 20, No. 4, p. 629 – 630 (1985). Abstract. – See Abstr. 010.661.

105.039 **Fe–Ni alloy phases in iron meteorites of high Ni content.**
J. Danon, R. B. Scorzelli.
Meteoritics, Vol. 20, No. 4, p. 631 – 632 (1985). Abstract. – See Abstr. 010.661.

105.040 **Microstructure of the Santa Catharina meteorite.**
J. Danon, R. B. Scorzelli, E. Galvão da Silva.
Meteoritics, Vol. 20, No. 4, p. 632 (1985). Abstract. – See Abstr. 010.661.

105.041 **Trace element abundances in OSCAR, a scandium–rich refractory inclusion from the Ornans meteorite.**
A. M. Davis, R. W. Hinton.
Meteoritics, Vol. 20, No. 4, p. 633 – 634 (1985). Abstract. – See Abstr. 010.661.

105.042 **The metamorphic history of ordinary chondrites according to cathodoluminescence.**
J. DeHart, D. W. G. Sears.
Meteoritics, Vol. 20, No. 4, p. 634 – 635 (1985). Abstract. – See Abstr. 010.661.

105.043 **Frontier Mountain: 34 new meteorites from antarctica.**
G. Delisle, H.–C. Höfle, R. Thierbach, O. Braun, L. Schultz, H. W. Weber, F. Wlotzka, U. Herpers, S. Vogt.
Meteoritics, Vol. 20, No. 4, p. 635 (1985). Abstract. – See Abstr. 010.661.

105.044 **Volatile/mobile trace elements in E3 chondrites.**
J. E. Dennison, P. W. Kaczaral, R. M. Verkouteren, M. E. Lipschutz.
Meteoritics, Vol. 20, No. 4, p. 636 – 637 (1985). Abstract. – See Abstr. 010.661.

105.045 **Monticello: a new eucrite from Kentucky.**
B. D. Dod, E. Olsen, P. Sipiera.
Meteoritics, Vol. 20, No. 4, p. 637 – 638 (1985). Abstract. – See Abstr. 010.661.

105.046 **The Qingzhen reaction: fingerprints of the EH planet?**
A. El Goresy.
Meteoritics, Vol. 20, No. 4, p. 639 (1985). Abstract. – See Abstr. 010.661.

105.047 **Araguainha impact structure, Mato Grosso, Brazil.**
W. von Engelhardt, J. Pohl, J. Walzebuck.
Meteoritics, Vol. 20, No. 4, p. 640 (1985). Abstract. – See Abstr. 010.661.

105.048 **REE and isotopic studies of a coarse grained CAI from Efremovka.**
A. Fahey, E. Zinner, G. Crozaz, A. S. Kornacki, A. A. Ulyanov.
Meteoritics, Vol. 20, No. 4, p. 643 – 644 (1985). Abstract. – See Abstr. 010.661.

105.049 **A nitrogen and carbon isotope study of the Bencubbin meteorite.**
I. A. Franchi, D. W. McGarvie, I. P. Wright, C. T. Pillinger.
Meteoritics, Vol. 20, No. 4, p. 645 – 646 (1985). Abstract. – See Abstr. 010.661.

105.050 **The making of chondrules.**
K. Fredriksson.
Meteoritics, Vol. 20, No. 4, p. 646 – 647 (1985). Abstract. – See Abstr. 010.661.

105.051 **A high resolution isotopic study of polymeric material in Murchison: comparisons with terrestrial organic matter.**
I. Gilmour, I. P. Wright, C. T. Pillinger.
Meteoritics, Vol. 20, No. 4, p. 647 (1985). Abstract. – See Abstr. 010.661.

105.052 **No evidence for multiple late Eocene tektite events.**
B. P. Glass.
Meteoritics, Vol. 20, No. 4, p. 648 (1985). Abstract. – See Abstr. 010.661.

105.053 **Mineralogy and origin of "white matrix" in Tieschitz (H3) chondrite: clues from bulk elemental compositions.**
J. L. Gooding.
Meteoritics, Vol. 20, No. 4, p. 648 – 650 (1985). Abstract. – See Abstr. 010.661.

105.054 **Roosevelt County 027: a low–shock ureilite with primary interstitial silicate liquid.**
C. A. Goodrich, K. Keil, J. L. Berkley, J. C. Laul, M. R. Smith, R. C. Clayton, T. K. Mayeda, J. F. Wacker.
Meteoritics, Vol. 20, No. 4, p. 650 – 651 (1985). Abstract. – See Abstr. 010.661.

105.055 **Concordant 3676 m.y. U–Pb formation age for the Kodaikanal iron meteorite.**
C. Göpel, G. Manhes, C. J. Allègre.
Meteoritics, Vol. 20, No. 4, p. 651 – 652 (1985). Abstract. – See Abstr. 010.661.

105.056 **Carbon isotopes in HF/HC1 residues of the unshocked ureilite ALHA 78019.**
M. M. Grady, C. T. Pillinger, J. F. Wacker.
Meteoritics, Vol. 20, No. 4, p. 652 – 653 (1985). Abstract. – See Abstr. 010.661.

105.057 **Light noble gas distributions in the L5 chondrite Knyahinya and an attempt to model their production.**
T. Graf, H. Baur, P. Signer, R. Wieler.
Meteoritics, Vol. 20, No. 4, p. 653 – 654 (1985). Abstract. – See Abstr. 010.661.

105.058 **Yamato 81020 and Yamato 82042; two new carbonaceous chondrites from Antarctica.**
A. L. Graham, K. Yanai, H. Kojima, S. Ikadai.
Meteoritics, Vol. 20, No. 4, p. 654 – 655 (1985). Abstract. – See Abstr. 010.661.

105.059 **Impact melt rocks of the Boltysh crater.**
R. A. F. Grieve, G. Reny, E. P. Grurov, V. A. Ryabenko.
Meteoritics, Vol. 20, No. 4, p. 655 – 656 (1985). Abstract. – See Abstr. 010.661.

105.060 **Semarkona: the least metamorphosed ordinary chondrite.**
J. N. Grossman.
Meteoritics, Vol. 20, No. 4, p. 656 – 657 (1985). Abstract. – See Abstr. 010.661.

105.061 **Carbon isotopes in bulk carbonaceous chondrites: Cold Bokkeveld.**
J. Halbout, T. K. Mayeda, R. N. Clayton.
Meteoritics, Vol. 20, No. 4, p. 657 – 658 (1985). Abstract. – See Abstr. 010.661.

105.062 **Relationship between petrography and silicon isotopes in Allende coarse–grained inclusions.**
A. Hashimoto, L. Grossman.
Meteoritics, Vol. 20, No. 4, p. 658 – 659 (1985). Abstract. – See Abstr. 010.661.

105.063 **Major, minor and trace element distribution in minerals of a coarse grained type B inclusion from Leoville.**
A. Havette, J. C. Lorin.
Meteoritics, Vol. 20, No. 4, p. 659 – 660 (1985). Abstract. – See Abstr. 010.661.

105.064 **Cosmogenic radionuclides in a near surface sample of the Jilin chondrite.**
G. Heusser.
Meteoritics, Vol. 20, No. 4, p. 661 – 662 (1985). Abstract. – See Abstr. 010.661.

105.065 **Xenon with atmospheric composition in Allende.**
D. Heymann, K. Liffman.
Meteoritics, Vol. 20, No. 4, p. 663 (1985). Abstract. – See Abstr. 010.661.

105.066 **A gas–rich matter in the Allende meteorite.**
D. Heymann, R. L. Palma.
Meteoritics, Vol. 20, No. 4, p. 663 – 664 (1985). Abstract. – See Abstr. 010.661.

105.067 **Induced alpha–track mapping of chondrites: a petrographic tool.**
M. D. Higgins, D. M. Shaw.
Meteoritics, Vol. 20, No. 4, p. 664 (1985). Abstract. – See Abstr. 010.661.

105.068 **A large negative titanium–50 anomaly in a refractory inclusion from the Murchison meteorite.**
R. W. Hinton, A. M. Davis, D. E. Scatena–Wachel.
Meteoritics, Vol. 20, No. 4, p. 664 – 665 (1985). Abstract. – See Abstr. 010.661.

105.069 **Further evidence for extensive metasommatism of the Allende parent body.**
R. M. Housley.
Meteoritics, Vol. 20, No. 4, p. 666 – 667 (1985). Abstract. – See Abstr. 010.661.

105.070 **The Cr isotopic composition of phosphates in IIIB iron meteorites: a search for ^{53}Mn.**
I. D. Hutcheon, J. Teshima, J. T. Armstrong, G. J. Wasserburg.
Meteoritics, Vol. 20, No. 4, p. 668 – 669 (1985). Abstract. – See Abstr. 010.661.

105.071 **Further evidence against a nebular origin for the ordinary chondrites.**
R. Hutchison, C. Alexander, D. J. Barber.
Meteoritics, Vol. 20, No. 4, p. 669 – 670 (1985). Abstract. – See Abstr. 010.661.

105.072 **Scanning electron microscopy of the Adelaide and Novo–Urei carbonaceous meteorites.**
M. Hyman, M. W. Rowe.
Meteoritics, Vol. 20, No. 4, p. 670 (1985). Abstract. – See Abstr. 010.661.

105.073 **Na–Ca zoning of chondrules in Allende and ALHA–77003 carbonaceous chondrites.**
Y. Ikeda, M. Kimura.
Meteoritics, Vol. 20, No. 4, p. 670 – 671 (1985). Abstract. – See Abstr. 010.661.

105.074 **Titanium isotopic "anomalies" in Murchison hibonites.**
T. R. Ireland, W. Compston, H. R. Heydegger.
Meteoritics, Vol. 20, No. 4, p. 671 (1985). Abstract. – See Abstr. 010.661. With a correction in Vol. 21, No. 1, p. 158 (1986).

105.075 **The composition of niningerite in EH chondrites: an indication of thermal history.**
A. V. Ivanov, A. I. Balabin.
Meteoritics, Vol. 20, No. 4, p. 672 (1985). Abstract. – See Abstr. 010.661.

105.076 **Nd and Sr isotope studies on Shergotty meteorite.**
E. Jagoutz, H. Wänke.
Meteoritics, Vol. 20, No. 4, p. 672 – 673 (1985). Abstract. – See Abstr. 010.661.

105.077 **A tale of two phosphates: REE reservoirs in the Shergotty meteorite.**
J. H. Jones, T. M. Benjamin, C. J. Duffy, P. S. Z. Rogers, M. Hollander, C. J. Maggiore.
Meteoritics, Vol. 20, No. 4, p. 674 – 675 (1985). Abstract. – See Abstr. 010.661.

105.078 **Meteorite Hg diffusion studies. II – Inferences regarding meteorite thermal events.**
S. Jovanovic, G. W. Reed Jr.
Meteoritics, Vol. 20, No. 4, p. 675 – 676 (1985). Abstract. – See Abstr. 010.661.

105.079 **Carbon–14 abundances in uranium ores and possible spontaneous exotic emission from U–series nuclides.**
A. J. T. Jull, D. Barker, D. J. Donahue.
Meteoritics, Vol. 20, No. 4, p. 676 (1985). Abstract. – See Abstr. 010.661.

105.080 **Ne–E in inclusions in apatite from Orgueil.**
M. H. A. Jungck, P. Eberhardt.
Meteoritics, Vol. 20, No. 4, p. 677 (1985). Abstract. – See Abstr. 010.661.

105.081 **Isotopic study on K from a Ca–Al–rich inclusion of Allende.**
T. Kaiser.
Meteoritics, Vol. 20, No. 4, p. 678 (1985). Abstract. – See Abstr. 010.661.

105.082 **Fission track ages of Viet–Nam tektites.**
L. L. Kashkarov, L. I. Genaeva, A. K. Lavrukhina, E. P. Izokh.
Meteoritics, Vol. 20, No. 4, p. 679 – 680 (1985). Abstract. – See Abstr. 010.661.

105.083 **Ti isotope abundances in whole meteorites.**
P. Kastenmayer, D. A. Papanastassiou, G. J. Wasserburg.
Meteoritics, Vol. 20, No. 4, p. 680 – 681 (1985). Abstract. – See Abstr. 010.661.

105.084 **An annealing study of Dhajala, an H3.8 chondrite, and application to the palaeothermometry of meteorites.**
B. D. Keck, R. K. Guimon, D. W. G. Sears.
Meteoritics, Vol. 20, No. 4, p. 681 – 682 (1985). Abstract. – See Abstr. 010.661.

105.085 **Iodine and chlorine in two fine–grained inclusions from the Allende meteorite.**
C. Kirschbaum.
Meteoritics, Vol. 20, No. 4, p. 683 – 684 (1985). Abstract. – See Abstr. 010.661.

105.086 **Track study of thermal effects in Kaidun meteorite minerals.**
N. N. Korotkova, G. G. Ruchman, A. K. Lavrukhina.
Meteoritics, Vol. 20, No. 4, p. 687 (1985). Abstract. – See Abstr. 010.661.

105.087 **Closure temperatures of subsolidus ionic exchange reactions for minerals from Yugoslavian chondrites.**
A. Kostić, L. Jovanovic, F. Brandstätter.
Meteoritics, Vol. 20, No. 4, p. 688 (1985). Abstract. – See Abstr. 010.661.

105.088 **Reduction and olivine–pyroxene fractionation during chondrule formation.**
A. Kracher.
Meteoritics, Vol. 20, No. 4, p. 688 – 689 (1985). Abstract. – See Abstr. 010.661.

105.089 **Metamorphic record in Shergottites.**
P. Lambert.
Meteoritics, Vol. 20, No. 4, p. 690 – 691 (1985). Abstract. – See Abstr. 010.661.

105.090 **Volatile behaviour during thermal metamorphism.**
J. W. Larimer.
Meteoritics, Vol. 20, No. 4, p. 691 – 692 (1985). Abstract. – See Abstr. 010.661.

105.091 **Extraterrestrial evidence regarding the frequency of collisional events in the solar system.**
B. Lavielle, S. Regnier, G. N. Simonoff, K. Marti, M. Woodard.
Meteoritics, Vol. 20, No. 4, p. 692 (1985). Abstract. – See Abstr. 010.661.

105.092 **Radioactivity of some stone meteorites.**
A. K. Lavrukhina, V. D. Gorin.
Meteoritics, Vol. 20, No. 4, p. 692 – 693 (1985). Abstract. – See Abstr. 010.661.

105.093 **Elemental composition of chondrules from Kainsaz CO carbonaceous chondrite.**
A. K. Lavrukhina, A. Yu. Ljul (*A. Yu. Lyul'*), G. M. Kolesov.
Meteoritics, Vol. 20, No. 4, p. 693 – 694 (1985). Abstract. – See Abstr. 010.661.

105.094 **Shock–induced behaviour Xe in experimentally shock–loaded matter of the Murchison meteorite.**
A. K. Lavrukhina, Z. K. Milnikova, S. V. Pershin, Dang Vu Minh, Yu. A. Shukolyukov.
Meteoritics, Vol. 20, No. 4, p. 694 – 695 (1985). Abstract. – See Abstr. 010.661.

105.095 **The meteoritic chromium isotopic composition and limits for radioactive Mn–53 in the early solar system.**
T. Lee, F. Tera.
Meteoritics, Vol. 20, No. 4, p. 695 – 696 (1985). Abstract. – See Abstr. 010.661.

105.096 **Revisitation of an XVIII century meteorite: Ploschkovitz.**
G. R. Levi–Donati, M. Prinz.
Meteoritics, Vol. 20, No. 4, p. 696 – 697 (1985). Abstract. – See Abstr. 010.661.

105.097 **Noble gases in the unique breccia, Bencubbin.**
R. S. Lewis.
Meteoritics, Vol. 20, No. 4, p. 698 – 699 (1985). Abstract. – See Abstr. 010.661.

105.098 **Algarrobo, a new iron meteorite from Chile.**
L. Lindner, V. F. Buchwald.
Meteoritics, Vol. 20, No. 4, p. 699 (1985). Abstract. – See Abstr. 010.661.

105.099 **Dynamic crystallization experiments on chondrule melts: the effects of kinetic growth factors on the chemistry of olivine and pyroxene.**
G. E. Lofgren.
Meteoritics, Vol. 20, No. 4, p. 699 – 700 (1985). Abstract. – See Abstr. 010.661.

105.100 **Isotopic heterogeneity in a coarse–grained Ca–Al–rich inclusion of Leoville.**
J. C. Lorin, A. Havette, G. Slodzian.
Meteoritics, Vol. 20, No. 4, p. 700 – 701 (1985). Abstract. – See Abstr. 010.661.

105.101 **Origin of silicate inclusions in the Landes meteorite.**
D. Luzius–Lange, H. Palme, H. J. Tobschall.
Meteoritics, Vol. 20, No. 4, p. 701 – 702 (1985). Abstract. – See Abstr. 010.661.

105.102 **Fine–grained phases in carbonaceous chondrites: Alais and Leoville.**
I. D. R. Mackinnon.
Meteoritics, Vol. 20, No. 4, p. 702 – 703 (1985). Abstract. – See Abstr. 010.661.

105.103 **Vigarano refractory inclusions: Allende unaltered, and a possible link with C2 inclusions.**
G. J. MacPherson.
Meteoritics, Vol. 20, No. 4, p. 703 – 704 (1985). Abstract. – See Abstr. 010.661.

105.104 **Early formation of the Ca–Al rich inclusions in chondrites: from the point of view of the U–Pb system.**
G. Manhes, C. Göpel, C. J. Allègre.
Meteoritics, Vol. 20, No. 4, p. 705 – 706 (1985). Abstract. – See Abstr. 010.661.

105.105 **Laser probe ^{40}Ar–^{39}Ar studies of the Peace River L6 chondrite.**
P. McConville, S. Kelley, G. Turner.
Meteoritics, Vol. 20, No. 4, p. 707 (1985). Abstract. – See Abstr. 010.661.

105.106 **Systematics in the release and isotopic composition of carbon in six CM2 meteorites between 800°C and 1300°C.**
D. W. McGarvie, M. M. Grady, C. T. Pillinger, I. P. Wright, E. K. Gibson Jr.
Meteoritics, Vol. 20, No. 4, p. 708 (1985). Abstract. – See Abstr. 010.661.

105.107 **Multiple meteoroid impacts: evidence in the geologic record.**
J. F. McHone, R. S. Dietz.
Meteoritics, Vol. 20, No. 4, p. 709 (1985). Abstract. – See Abstr. 010.661.

105.108 **Minor phases, Fe–rich pyroxene, and Shergotty chronology.**
G. McKay, S.–R. Yang, J. Wagstaff.
Meteoritics, Vol. 20, No. 4, p. 710 – 711 (1985). Abstract. – See Abstr. 010.661.

105.109 **Anomalous noble gases and presolar carbon in Murray (C2M) chondrite.**
T. Ming, R. S. Lewis, E. Anders.
Meteoritics, Vol. 20, No. 4, p. 712 – 713 (1985). Abstract. – See Abstr. 010.661.

105.110 **Another look at actinide chemistry in Ca–Al–rich inclusions.**
M. T. Murrell, D. S. Burnett.
Meteoritics, Vol. 20, No. 4, p. 713 – 714 (1985). Abstract. – See Abstr. 010.661.

105.111 **Nitrogen isotopic systematics in meteorites and in the solar wind.**
S. V. S. Murty, K. Marti.
Meteoritics, Vol. 20, No. 4, p. 714 – 715 (1985). Abstract. – See Abstr. 010.661.

105.112 **Experimental reduction of olivine and its application to the reduction in enstatite chondrites.**
H. Nagahara.
Meteoritics, Vol. 20, No. 4, p. 715 – 716 (1985). Abstract. – See Abstr. 010.661.

105.113 **Carbon isotope abundances in Murchison residue 2C10c.**
F. R. Niederer, P. Eberhardt, J. Geiss, R. S. Lewis.
Meteoritics, Vol. 20, No. 4, p. 716 – 718 (1985). Abstract. – See Abstr. 010.661.

105.114 **Systematics of Ti isotopes in carbonaceous chondrite bulk samples.**
S. Niemeyer.
Meteoritics, Vol. 20, No. 4, p. 718 – 719 (1985). Abstract. – See Abstr. 010.661.

105.115 **Cosmogenic ^{129}I in iron meteorites.**
K. Nishiizumi, J. R. Arnold, D. Elmore, P. W. Kubik.
Meteoritics, Vol. 20, No. 4, p. 719 – 720 (1985). Abstract. – See Abstr. 010.661.

105.116 **Comparative study of impact craters with layered targets.**
R. Ostertag, D. Stöffler, W. von Engelhardt.
Meteoritics, Vol. 20, No. 4, p. 720 – 721 (1985). Abstract. – See Abstr. 010.661.

105.117 **Martian meteorites: are they (all) from Mars: evidence from trapped noble gases.**
U. Ott, F. Begemann.
Meteoritics, Vol. 20, No. 4, p. 721 – 722 (1985). Abstract. – See Abstr. 010.661.

105.118 **s–process Xe and Kr and Ne–E in a ^{13}C rich Murchison sample; noble gas analysis by stepped combustion.**
U. Ott, J. Yang, S. Epstein.
Meteoritics, Vol. 20, No. 4, p. 722 – 723 (1985). Abstract. – See Abstr. 010.661.

105.119 **SCR Ne–21 in etched pyroxene separates of Kapoeta: a clue to ancient solar activity.**
J. T. Padia, M. N. Rao.
Meteoritics, Vol. 20, No. 4, p. 723 – 724 (1985). Abstract. – See Abstr. 010.661.

105.120 **Cr isotopic abundances.**
D. A. Papanastassiou, J. L. Zyskind, H. H. Ngo, G. J. Wasserburg.
Meteoritics, Vol. 20, No. 4, p. 724 – 725 (1985). Abstract. – See Abstr. 010.661.

105.121 **Refractory inclusions in the Efremovka meteorite and a comparison with other carbonaceous chondrites.**
J. M. Paque.
Meteoritics, Vol. 20, No. 4, p. 725 – 726 (1985). Abstract. – See Abstr. 010.661.

105.122 **Determination of Ca–41 production for meteorite studies by accelerator mass spectrometry.**
M. Paul, D. Fink, O. Meirav, S. Theis, P. Englert.
Meteoritics, Vol. 20, No. 4, p. 726 – 727 (1985). Abstract. – See Abstr. 010.661.

105.123 **Pu–244 fission–track cooling rates versus metallographic cooling rates: where is the truth?**
P. Pellas.
Meteoritics, Vol. 20, No. 4, p. 727 – 728 (1985). Abstract. – See Abstr. 010.661.

105.124 **Review of relative and absolute abundances of atmophilic gases in glass from the EETA 79001 shergottite, and comparison with the martian and terrestrial atmospheres.**
R. O. Pepin, R. H. Becker, R. C. Wiens.
Meteoritics, Vol. 20, No. 4, p. 728 – 729 (1985). Abstract. – See Abstr. 010.661.

105.125 **Chainpur (LL–3): fractionated siderophile elements in chondrules, fragments, and chondrite matrix.**
E. Pernicka, I. Herrwerth, G. Kurat, F. Brandstätter.
Meteoritics, Vol. 20, No. 4, p. 729 – 730 (1985). Abstract. – See Abstr. 010.661.

105.126 **Can tracks in meteorites be used to study ultraheavy cosmic rays?**
C. Perron, P. Pellas.
Meteoritics, Vol. 20, No. 4, p. 730 – 731 (1985). Abstract. – See Abstr. 010.661.

105.127 **ALHA 81189, a highly unequilibrated enstatite chondrite: evidence for a multistage history.**
M. Prinz, M. K. Weisberg, C. E. Nehru, J. S. Delaney.
Meteoritics, Vol. 20, No. 4, p. 731 – 732 (1985). Abstract. – See Abstr. 010.661.

105.128 **Layered chondrules: evidence for multistage histories during chondrule formation.**
M. Prinz, M. K. Weisberg, C. E. Nehru, J. S. Delaney.
Meteoritics, Vol. 20, No. 4, p. 732 – 733 (1985). Abstract. – See Abstr. 010.661.

105.129 **Are Ca and Ti isotopic anomalies correlated (or were they ever)?**
C. A. Prombo, G. W. Lugmair.
Meteoritics, Vol. 20, No. 4, p. 733 – 734 (1985). Abstract. – See Abstr. 010.661.

105.130 **Unusually high concentration of ^{10}Be in a cosmic spherule: possible evidence for irradiation outside the planetary solar system.**
G. M. Raisbeck, F. Yiou, D. Brownlee.
Meteoritics, Vol. 20, No. 4, p. 734 – 735 (1985). Abstract. – See Abstr. 010.661.

105.131 **Chemical, petrologic and isotopic studies of chondrules and inclusions from the Mokoia carbonaceous chondrite.**
R. S. Rajan, S. Hughes, R. A. Schmitt, I. D. Hutcheon.
Meteoritics, Vol. 20, No. 4, p. 735 – 736 (1985). Abstract. – See Abstr. 010.661.

105.132 **Origin of titanium anomalies.**
S. Ramadurai.
Meteoritics, Vol. 20, No. 4, p. 736 (1985). Abstract. – See Abstr. 010.661.

105.133 **On the thermal history of the mesosiderite parent body.**
K. L. Rasmussen, J. S. Delaney, M. Prinz.
Meteoritics, Vol. 20, No. 4, p. 738 – 739 (1985). Abstract. – See Abstr. 010.661.

105.134 **Cosmogenic nuclides and complex exposure histories of meteorites.**
R. C. Reedy.
Meteoritics, Vol. 20, No. 4, p. 739 – 740 (1985). Abstract. – See Abstr. 010.661.

105.135 **Post–impact modification is a deep (0,2 D) gravity compensation.**
J. Rondot.
Meteoritics, Vol. 20, No. 4, p. 747 – 748 (1985). Abstract. – See Abstr. 010.661.

105.136 **Studies of iron–nickel alloys in Ni–rich meteorites.**
H. Roy–Poulsen, L. Larsen, N. O. Roy–Poulsen, L. Vistisen, G. B. Jensen, J. M. Knudsen.
Meteoritics, Vol. 20, No. 4, p. 749 – 750 (1985). Abstract. – See Abstr. 010.661.

105.137 **Allende chondrules and their coarse–grained rims.**
A. E. Rubin, G. W. Kallemeyn.
Meteoritics, Vol. 20, No. 4, p. 750 – 751 (1985). Abstract. – See Abstr. 010.661.

105.138 **A search for extinct ^{53}Mn and ^{60}Fe in iron meteorites.**
D. E. Scatena–Wachel, R. W. Hinton, A. M. Davis.
Meteoritics, Vol. 20, No. 4, p. 751 – 752 (1985). Abstract. – See Abstr. 010.661.

105.139 **Trapped noble gases in some type 3 chondrites.**
N. Schelhaas, U. Ott, F. Begemann.
Meteoritics, Vol. 20, No. 4, p. 753 (1985). Abstract. – See Abstr. 010.661.

105.140 **Impact structures of Western Australia.**
E. M. Shoemaker, C. S. Shoemaker.
Meteoritics, Vol. 20, No. 4, p. 754 – 756 (1985). Abstract. – See Abstr. 010.661.

105.141 **The isotopic anomalous components of Xe in carbonaceous chondrites.**
Yu. A. Shukolyukov, A. K. Lavrukhina.
Meteoritics, Vol. 20, No. 4, p. 756 – 757 (1985). Abstract. – See Abstr. 010.661.

105.142 **Depth dependence of ^{10}Be in the iron meteorite Grant and an attempt to model ^{10}Be production.**
P. Signer, T. Graf, R. Wieler, R. Sarafin, U. Herpers, S. Vogt, G. Bonnani, W. Wölfli.
Meteoritics, Vol. 20, No. 4, p. 757 – 759 (1985). Abstract. – See Abstr. 010.661.

105.143 **Mineralogy and chemistry of the Spring Lake, Texas, meteorite.**
P. P. Sipiera, B. D. Dod, E. J. Olsen.
Meteoritics, Vol. 20, No. 4, p. 759 (1985). Abstract. – See Abstr. 010.661.

105.144 **Rare earth abundances in EETA 79001 phosphates.**
M. R. Smith, J. C. Laul.
Meteoritics, Vol. 20, No. 4, p. 760 – 761 (1985). Abstract. – See Abstr. 010.661.

105.145 **Spatial distribution of carbon in the Allende meteorite, and possible origin of its "elusiveness".**
C. C. A. H. van der Stap, R. D. Vis, H. Verheul, D. Heymann.
Meteoritics, Vol. 20, No. 4, p. 762 (1985). Abstract. – See Abstr. 010.661.

105.146 **Cathodoluminescence spectra of forsterite and variation with meteorite type.**
I. M. Steele.
Meteoritics, Vol. 20, No. 4, p. 762 – 764 (1985). Abstract. – See Abstr. 010.661.

105.147 **Shock history of the shergottites.**
D. Stöffler, R. Ostertag.
Meteoritics, Vol. 20, No. 4, p. 764 – 765 (1985). Abstract. – See Abstr. 010.661.

105.148 **The fission track age of high sodium/potassium Australites revisited.**
D. Storzer.
Meteoritics, Vol. 20, No. 4, p. 765 – 766 (1985). Abstract. – See Abstr. 010.661.

105.149 **Thermal history and possible degassing of meteorite parent bodies.**
N. Sugiura, J. Arkani–Hamed, D. W. Strangway, T. Matsui.
Meteoritics, Vol. 20, No. 4, p. 767 (1985). Abstract. – See Abstr. 010.661.

105.150 **I–Xe structures of chondrules, rims and matrix in Chainpur.**
T. D. Swindle, M. W. Caffee, C. M. Hohenberg.
Meteoritics, Vol. 20, No. 4, p. 767 – 768 (1985). Abstract. – See Abstr. 010.661.

105.151 **Rare gases in Japanese chondrites.**
N. Takaoka, M. Shima, F. Wakabayashi.
Meteoritics, Vol. 20, No. 4, p. 768 – 769 (1985). Abstract. – See Abstr. 010.661.

105.152 **A model for relating a pristine eucrite lava to ordinary eucrites by an impact event.**
H. Takeda, L. E. Nyquist.
Meteoritics, Vol. 20, No. 4, p. 769 (1985). Abstract. – See Abstr. 010.661.

105.153 **Concerning the impact events on the eucrites parent bodies.**
F. Tera, N. Z. Boctor.
Meteoritics, Vol. 20, No. 4, p. 770 – 771 (1985). Abstract. – See Abstr. 010.661.

105.154 **A petrographic and Mg isotopic study of CAI in Bali (C3V) and Coolidge (C4V).**
J. Teshima, I. D. Hutcheon, J. T. Armstrong, G. J. Wasserburg, G. Kurat.
Meteoritics, Vol. 20, No. 4, p. 771 – 772 (1985). Abstract. – See Abstr. 010.661.

105.155 **The track density depth profile of the Jilin core samples.**
K. Thiel, R.–y. Hu.
Meteoritics, Vol. 20, No. 4, p. 774 – 775 (1985). Abstract. – See Abstr. 010.661.

105.156 **Chondritic inclusions in Cumberland Falls and ALHA 78113 aubrites: different chips of the same old block.**
R. M. Verkouteren, M. E. Lipschutz, M. Prinz, M. K. Weisberg, C. E. Nehru.
Meteoritics, Vol. 20, No. 4, p. 776 – 777 (1985). Abstract. – See Abstr. 010.661.

105.157 **Noble gases in HF/HC1 residues of ALHA 78019: an unshocked ureilite.**
J. F. Wacker.
Meteoritics, Vol. 20, No. 4, p. 777 – 779 (1985). Abstract. – See Abstr. 010.661.

105.158 **Geochemistry of Lakangaon, the most Fe–rich eucrite.**
P. H. Warren, G. W. Kallemeyn, E. A. Jerde.
Meteoritics, Vol. 20, No. 4, p. 779 (1985). Abstract. – See Abstr. 010.661.

105.159 **Cooling rates of chondrules.**
J. T. Wasson.
Meteoritics, Vol. 20, No. 4, p. 780 (1985). Abstract. – See Abstr. 010.661.

105.160 **Mulga West, a C6 carbonaceous chondrite?**
G. Weckwerth, B. Spettel, F. Wlotzka.
Meteoritics, Vol. 20, No. 4, p. 781 (1985). Abstract. – See Abstr. 010.661.

105.161 **Siderophile elements and the relationship between type 3 and equilibrated ordinary chondrites.**
K. S. Weeks, D. W. G. Sears.
Meteoritics, Vol. 20, No. 4, p. 781 – 782 (1985). Abstract. – See Abstr. 010.661.

105.162 **Trace elements in two bediasite tektites.**
H. H. Weinke, C. Koeberl.
Meteoritics, Vol. 20, No. 4, p. 783 (1985). Abstract. – See Abstr. 010.661.

105.163 **Quantitative calculation of meteorite flux from various regions of the inner asteroid belt.**
G. W. Wetherill.
Meteoritics, Vol. 20, No. 4, p. 783 – 784 (1985). Abstract. – See Abstr. 010.661.

105.164 **Composition of large metal nodules in ordinary chondrites.**
E. Widom, A. E. Rubin, J. T. Wasson.
Meteoritics, Vol. 20, No. 4, p. 784 – 785 (1985). Abstract. – See Abstr. 010.661.

105.165 **Fine–grained chondrule rims.**
L. L. Wilkening, D. H. Hill.
Meteoritics, Vol. 20, No. 4, p. 785 – 786 (1985). Abstract. – See Abstr. 010.661.

105.166 **Terminal Cretaceous impact: search for bolide material and carbon from wildfires.**
W. S. Wolbach, R. S. Lewis, E. Anders.
Meteoritics, Vol. 20, No. 4, p. 786 – 787 (1985). Abstract. – See Abstr. 010.661.

105.167 **Identification of micron sized phases in meteorites by laser Raman microprobe spectroscopy.**
B. Wopenka, P. D. Swan.
Meteoritics, Vol. 20, No. 4, p. 788 – 789 (1985). Abstract. – See Abstr. 010.661.

105.168 **Cosmic–ray–produced–radioactive nuclides in freshly fallen chondrites, Aomori and Tomiya.**
S. Yabuki, M. Shima, S. Murayama.
Meteoritics, Vol. 20, No. 4, p. 789 – 790 (1985). Abstract. – See Abstr. 010.661.

105.169 **Yamato–82193 meteorite: the third lunar meteorite collected at the Yamato Mountains, Antarctica.**
K. Yanai, H. Kojima.
Meteoritics, Vol. 20, No. 4, p. 790 – 791 (1985). Abstract. – See Abstr. 010.661.

105.170 **A new type chondrite?: Yamato–75302 consists mostly of very high iron olivine.**
K. Yanai, H. Kojima, Y. Matsumoto.
Meteoritics, Vol. 20, No. 4, p. 791 (1985). Abstract. – See Abstr. 010.661.

105.171 **^{10}Be in iron type cosmic spherules: evidence for a differentiated parent body.**
F. Yiou, G. M. Raisbeck, D. Brownlee.
Meteoritics, Vol. 20, No. 4, p. 791 – 792 (1985). Abstract. – See Abstr. 010.661.

105.172 **The parent stratum and distribution of the Georgia tektites.**
H. Povenmire.
Meteoritics, Vol. 20, No. 4, p. 795 – 799 (1985).
Seventy–six field expeditions for Georgia tektites have been made to the vicinity of Dodge County, Georgia, in an effort to determine the parent formation and distribution pattern by locating new specimens. About 67 new specimens have been recovered; they represent about 9% of all known specimens. The primary area of distribution is a fan shaped area approximately 30 by 72 kilometers. It is likely that these tektites have been transported by water from the parent formation. No correlation could be found between size, shape, soil type, elevation and tektite location.

105.173 **News on the nature of the Tunguska Event.**
E. M. Kolesnikov.
Priroda, No. 1, p. 63 – 65 (1986). In Russian.

105.174 **Contamination of impactites of the Elgygytgyn astrobleme by meteoritic matter.**
I. G. Kapustkina, G. M. Kolesov, V. I. Fel'dman.
Dokl. Akad. Nauk SSSR, Tom 280, No. 3, p. 755 – 758 (1985). In Russian. Abstr. in Ref. Zh., 51. Astron., 1.51.260 (1986).

105.175 **Magnetic properties of the Olivenza meteorite – a cautionary tale.**
D. W. Collinson.
Geophys. J. R. Astron. Soc., Vol. 85, No. 1, p. 257 (1986). Abstract. See Abstr. 012.086.

105.176 **Hidden meteorite craters in Lithuania.**
A. Gaigals.
Zvaigžnota Debess, 1984. Gada Rudens, p. 15 – 18 (1984). In Latvian.

105.177 **Erratum: "SNC meteorites: clues to martian petrologic evolution?" [Rev. Geophys., Vol. 23, No. 4, p. 391 – 416 (1985)].**
H. Y. McSween Jr.
Rev. Geophys., Vol. 24, No. 1, p. 141 (1986). See Abstr. 40.105.094.

105.178 **Optical spectrometry of meteorites. IV. Comparative analysis of spectral properties of achondrites and ordinary chondrites.**
L. F. Golubeva, D. I. Shestopalov, A. Ya. Vokhmentsev, V. D. Kolomenskij, V. V. Titov.
Astron. Vestn., Tom 20, No. 2, p. 116 – 127 (1986). In Russian. English translation in Sol. Syst. Res.
Diffuse reflection spectra are taken as a basis of a comparative analysis of achondrites and ordinary chondrites spectral properties. It has been proposed a preliminary spectral classification of light stone meteorites by the presence or absence of the characteristic absorption bands Fe^{2+} (in olivine, pyroxene and feldspar) and Cr^{3+} (in pyroxene) in their spectra. The influence of reduced metal on the reflectance spectra and the limitations of optical

spectrometry for remote sensing of asteroids' surface composition is discussed too.

105.179 **Carbon in meteorites.**
A. A. Marakushev.
Vestn. MGU. Geol., No. 6, p. 3 – 13 (1985). In Russian. Abstr. in Ref. Zh., 51. Astron., 4.51.257 (1986).

105.180 **Search for traces of heavy and superheavy cosmic nuclei in olivines from meteorites.**
V. P. Perelygin, S. G. Stetsenko, G. N. Flerov.
Kratk. soobshch. OIYaI, Dubna, No. 7, p. 5 – 11 (1985). In Russian. Abstr. in Ref. Zh., 51. Astron., 4.51.273 (1986).

105.181 **Das Nördlinger Ries – ein Meteoritenkrater.**
W. Lüthi.
Orion, 44. Jahrg., Nr. 213, p. 67 – 70 (1986).

105.182 **Saturation magnetization measurements of carbonaceous chondrites.**
M. Hyman, M. W. Rowe.
Meteoritics, Vol. 21, No. 1, p. 1 – 22 (1986).
The saturation magnetization of some carbonaceous chondrites was studied using a Faraday balance. The authors present measurements on the magnetite contents of some CM2, CV3 and a CV5 chondrites. The method was also used to measure the content of metallic nickel–iron in Ornans, 3.4±0.3%. Of the CM2 chondrites examined, only Bells, Essebi and Haripura had magnetite contents over about 1% by weight. A number of CV chondrites have magnetite between 2.3 and 13%, with little or no metallic iron. Leoville and Vigarano contain both magnetite and metallic iron, complicating the saturation magnetization results. Arch and Allende have very little metallic iron or magnetite, probably <1% of either.

105.183 **Gallium–bearing sphalerite in a metal–sulfide nodule of the Qingzhen (EH3) chondrite.**
E. R. Rambaldi, R. S. Rajan, R. M. Housley, D. Wang.
Meteoritics, Vol. 21, No. 1, p. 23 – 31 (1986).
The Qingzhen (EH3) chondrite contains a population of spheroidal metal–sulfide nodules, which display textural evidence of reheating and melting. Evidence of metal sulfuration is also present, suggesting replacement of metal by sulfide during melting. This process has led to the nucleation of perryite along metal–sulfide interfaces. Gallium–bearing sphalerite and a Cu–sulfide of composition intermediate between chalcopyrite and cubanite as inclusions within the metal of some nodules. The Ga–bearing sphalerite may have formed by injection of molten sulfide droplets into the metal followed by subsolidus diffusion of Ga from the metal into the sulfide.

105.184 **The Maryville meteorite: a 1983 fall of an L6 chondrite.**
J. W. Shervais, L. A. Taylor, E.–H. Cirlin,
E. Jarosewich, J. C. Laul.
Meteoritics, Vol. 21, No. 1, p. 33 – 45 (1986).

105.185 **Jerslev, the first iron meteorite from Denmark.**
V. F. Buchwald.
Meteoritics, Vol. 21, No. 1, p. 47 – 58 (1986).

105.186 **Two new chondrite–falls in Japan.**
M. Shima, S. Murayama, F. Wakabayashi,
A. Okada, H. Yabuki.
Meteoritics, Vol. 21, No. 1, p. 59 – 78 (1986).
The authors present chemical compositions of major elements and some trace elements as well as the mineralogical and petrographical description of two new meteorites. They have also measured cosmic–ray–produced–radioactive nuclides and rare gases in those meteorites.

105.187 **Studies of kamacite, perryite and schreibersite in E–chondrites and aubrites.**
A. J. Easton.
Meteoritics, Vol. 21, No. 1, p. 79 – 93 (1986).
The bulk composition of metal (kamacite plus perryite) was determined in eleven E–chondrites and eight aubrites. The data are compatible with the subdivision of the E–chondrites into two groups (Yavnel, 1963; Anders, 1964), St Mark's and St Sauveur belonging to type I (Easton, 1985).

105.188 **Gas permeability of shocked chondrites.**
T. Matsui, N. Sugiura, N. S. Brar.
Meteoritics, Vol. 21, No. 1, p. 109 – 116 (1986).
The gas permeability of 11 ordinary chondrites was measured at various gas pressures (0.5 – 2.5 bars) under confining pressures up to 120 bars. The gas permeability ranges from less than a nanodarcy to a few millidarcies. There is a positive correlation between the permeability and the porosity. A linear relation between gas flow pressure dependence and confining pressure dependence of the gas permeability is observed, suggesting that on average, crack apertures are larger than pore spaces. Using the measured permeability data the authors estimated the size of a possible shocked–chondrite precursor body.

105.189 **Detection of high–pressure silica polymorphs in whole–rock samples from a Meteor Crater, Arizona, impact sample using solid–state silicon–29 nuclear magnetic resonance spectroscopy.**
W.–H. Yang, R. J. Kirkpatrick, N. Vergo, J. McHone,
T. I. Emilsson, E. Oldfield.
Meteoritics, Vol. 21, No. 1, p. 117 – 124 (1986).
The authors present the results of a high–resolution solid–state silicon–29 "magic–angle" sample–spinning (MASS) nuclear magnetic resonance (NMR) spectroscopic study of a sample of shocked Coconino sandstone from Meteor Crater, Arizona. The results indicate that MASS NMR can easily detect the presence of both coesite and stishovite in whole rock samples, and yields quantitative values for the coesite/stishovite ratio, an important parameter in examining the pressure and temperature history of shocked samples.

105.190 **Petrology and classification of the Garraf, Spain chondrite.**
K. Keil, G. H. Conrad, E. A. King, A. San Miguel.
Meteoritics, Vol. 21, No. 1, p. 125 – 129 (1986).
Microscopic and electron microprobe studies indicate that the Garraf meteorite is a highly–recrystallized chondrite of petrologic type 6. Olivine and low–Ca pyroxene compositions indicate that it belongs to the L–group. The authors conclude that Garraf is a highly recrystallized L6b chondrite that, after recrystallization, was cataclased and comminuted by shock.

105.191 **The classification of the Benares chondrite.**
B. Mason.
Meteoritics, Vol. 21, No. 1, p. 131 – 132 (1986).
The Benares meteorite is an LL4 chondrite, not LL6 as recorded in the literature. Some specimens labelled Benares are misidentified.

105.192 **The isotopic composition of tellurium in the Abee meteorite.**
C. L. Smith, J. R. De Laeter.
Meteoritics, Vol. 21, No. 1, p. 133 – 139 (1986).
The present work reports two sets of measurements of the tellurium isotopic composition of Abee, and compares the meteoritic data with a terrestrial tellurium standard. No isotopic anomalies can be distinguished within the error limits.

105.193 **Roedderite in the Qingzhen (EH3) chondrite.**
E. R. Rambaldi, R. S. Rajan, R. M. Housley.
Meteoritics, Vol. 21, No. 1, p. 141 – 149 (1986).
The authors report evidence for the occurrence of the rare mineral, roedderite in the Qingzhen enstatite chondrite. From its textural occurrence it appears that roedderite might represent the

precipitate of an alkali-rich fluid phase, possibly developed during planetary processes in the enstatite chondrite parent body.

105.194 **A nonmagmatic origin of group-IIE iron meteorites.**
J. T. Wasson, J. Wang.
Geochim. Cosmochim. Acta, Vol. 50, No. 5, p. 725 – 732 (1986).
The authors examine the questions of the genetic relationship among 14 irons having compositions in or near previously defined IIE fields on element Ni-diagrams, and the formational processes inferred for the most closely related irons.

105.195 **Primordial material in meteorites.**
J. F. Kerridge.
NASA Conf. Publ., NASA CP-2403, p. 71 – 107 (1986). – See Abstr. 012.030.

105.196 **Properties of the Guin ungrouped iron meteorite: the origin of Guin and of group-IIE irons.**
A. E. Rubin, E. A. Jerde, P. Zong, J. T. Wasson,
J. W. Westcott, T. K. Mayeda, R. N. Clayton.
Earth Planet. Sci. Lett., Vol. 76, No. 3/4, p. 209 – 226 (1985/86).

105.197 **Emission spectra of meteorites during thermoluminescence.**
J. A. Strain, P. D. Townsend, B. Jassemnejad,
S. W. S. McKeever.
Earth Planet. Sci. Lett., Vol. 77, No. 1, p. 14 – 19 (1986).
The emission spectra of four ordinary chondrite meteorites of petrological type 5 have been recorded during thermoluminescence. Spectra from all four samples are similar. A very broad emission structure peaked near 470 nm and extending from 300 to at least 700 nm is found for the glow peaks from ~150 to 250°C. The higher temperature peak, 300 – 500°C, exhibits a different character with a narrow emission band near 400 nm. A tentative model for a luminenscence site is offered.

105.198 **Evidence for successive episodes of condensation at high temperature in a part of the solar nebula.**
D. A. Wark.
Earth Planet. Sci. Lett., Vol. 77, No. 2, p. 129 – 148 (1986).

105.199 **The case for a martian origin of the shergottites, II. Trapped and indigenous gas components in EETA 79001 glass.**
R. C. Wiens, R. H. Becker, R. O. Pepin.
Earth Planet. Sci. Lett., Vol. 77, No. 2, p. 149 – 158 (1986).

105.200 **Magnetic properties of Antarctic shergottite meteorites EETA 79001 and ALHA 77005: possible relevance to a Martian magnetic field.**
D. W. Collinson.
Earth Planet. Sci. Lett., Vol. 77, No. 2, p. 159 – 164 (1986).
The possibility that the parent body of the SNC (*shergottite, nakhlite and chassignite*) meteorites is Mars implies that the magnetic properties of these meteorites may provide evidence concerning ancient Martian magnetic fields. EETA 79001 possesses a weak, very stable primary magnetization, the properties of which are consistent with its acquisition in an ambient magnetic field either during the meteorite's formation or during the severe shock event later in its history. The samples of ALHA 77005 studied possessed no measurable primary magnetization.

105.201 **Structural deformation of the Leoville chondrite.**
P. M. Cain, H. Y. McSween Jr., N. B. Woodward.
Earth Planet. Sci. Lett., Vol. 77, No. 2, p. 165 – 175 (1986).

105.202 **Chemical and physical studies of type 3 chondrites, VII. Annealing studies of the Dhajala H3.8 chondrite and the thermal history of chondrules and chondrites.**
B. D. Keck, R. K. Guimon, D. W. G. Sears.
Earth Planet. Sci. Lett., Vol. 77, No. 3/4, p. 419 – 427 (1986).

105.203 **The color of meteoritic hibonite: an indicator of oxygen fugacity.**
P. D. Ihinger, E. Stolper.
Earth Planet. Sci. Lett., Vol. 78, No. 1, p. 67 – 79 (1986).

105.204 **Impact glasses from Zhamanshin crater (U.S.S.R.): chemical composition and discussion of origin.**
C. Koeberl, K. Fredriksson.
Earth Planet. Sci. Lett., Vol. 78, No. 1, p. 80 – 88 (1986).

105.205 **Noble gas isotopic composition, cosmic ray exposure history, and terrestrial age of the meteorite Allan Hills A81005 from the Moon.**
O. Eugster, J. Geiss, U. Krähenbühl, S. Niedermann.
Earth Planet. Sci. Lett., Vol. 78, No. 2/3, p. 139 – 147 (1986).
The authors present a comprehensive study of the elemental and isotopic abundances of the noble gases He, Ne, Ar, Kr, and Xe in the meteorite Allan Hills A81005 from the Moon. In addition to a bulk sample five grain size fractions were analyzed. Chemical abundances relevant for the interpretation of the cosmic-ray-produced noble gases were determined and indicate that the grain size fractions are chemically uniform. The isotopic and elemental pattern of the trapped solar wind noble gases in A81005 are very similar to those observed in lunar soils and breccias.

105.206 **Possible transport of carbon in meteorite parent bodies.**
N. Sugiura, J. Arkani-Hamed, D. W. Strangway.
Earth Planet. Sci. Lett., Vol. 78, No. 2/3, p. 148 – 156 (1986).
Assuming that a meteorite parent body is initially made of type 3 ordinary chondritic material, the possibility of transportation of carbon as carbon monoxide gas during thermal metamorphism is considered. The thermal history of the parent body is constrained by the maximum metamorphic temperature and the duration of metamorphism.

105.207 **Formation of solid matter of the protoplanetary nebula and composition of chondrites.**
M. N. Izakov.
Inst. kosm. issled. Akad. Nauk SSSR, Prepr., No. 970, 23 pp. (1985). In Russian. Abstr. in Ref. Zh., 51. Astron., 6.51.337 (1986).

105.208 **Origin of the secondary changes in the crystalline foundation rocks within the Logoisk meteorite impact crater.**
N. N. Levykh, A. S. Makhnach.
Dokl. Akad. Nauk BSSR, Tom 29, No. 4, p. 355 – 358 (1985). In Russian. Abstr. in Ref. Zh., 51. Astron., 6.51.347 (1986).

105.209 **On the age of the Kamensk astrobleme and some peculiarities of its composition.**
E. V. Movshovich, A. E. Milyavskij.
Geol. Zh., Tom 45, No. 3, p. 67 – 75 (1985). In Russian. Abstr. in Ref. Zh., 51. Astron., 6.51.348 (1986).

105.210 **On the results of a study of the Tunguska meteorite.**
N. V. Vasil'ev.
Moskva, 50 let nauchn. deyatel'nosti VAGO, p. 115 – 128 (1985). In Russian. Abstr. in Ref. Zh., 51. Astron., 6.51.358 (1986).

105.211 **Investigation of the mechanical properties of the Sikhote-Alin meteorites with the method of holographic interferometry.**
V. P. Dzhapiashvili, G. V. Kurkhuli, G. L. Dalakishlivili,
A. M. Kvernadze.
Soobshch. Akad. Nauk GSSR, Tom 119, No. 3, p. 501 – 504 (1985). In Russian. Abstr. in Ref. Zh., 51. Astron., 6.51.359 (1986).

105.212 **Superparamagnetic component in the Orgueil meteorite and Mössbauer spectroscopy studies in applied magnetic fields.**
M. B. Madsen, S. Mørup, T. V. V. Costa, J. M. Knudsen, M. Olsen.
Nature, Vol. 321, No. 6069, p. 501 – 503 (1986).
The Orgueil meteorite is a carbonaceous chondrite of group CI. Wdowiak and Agresti have recently shown that very small magnetic particles, which exhibit superparamagnetic relaxation below 80K, are present in this meteorite. They suggested that these small particles consist of magnetite. Using Mössbauer spectroscopy in external magnetic fields the authors have found that the Orgueil meteorite does not contain superparamagnetic magnetite in any appreciable amount. The meteorite contains another, as yet unidentified iron (III) compound which exhibits superparamagnetic behaviour below 80K.

105.213 **Tektites and moldavites.**
I. Sondi.
Vasiona, Année 33, No. 5, p. 90 – 92 (1985). In Croatian.

105.214 **Ancient sun was highly active.**
J. T. Padia, M. N. Rao.
Curr. Sci., Vol. 54, No. 13, p. 625 – 627 (1985). Abstr. in Phys. Abstr., Vol. 89, No. 1273, Entry 18731 (1986).

105.215 **Determination of O, Si, Al, Mg, and Fe in some chondrites in China by 14 MeV neutron activation analysis.**
G.–h. Hu, H.–h. Zhong, H.–q. Xu, Y.–y. Ling, W.–x. Yi, D.–d. Wang.
Kexue Tongbao, Vol. 30, No. 3, p. 374 – 379 (1985). Abstr. in Phys. Abstr., Vol. 89, No. 1273, Entry 18732 (1986).

105.216 **Low temperature remanent magnetization of Fe–Ni alloy.**
K. Momose, H. Nagai.
J. Geomagn. Geoelectr., Vol. 37, No. 8, p. 817 – 821 (1985). Abstr. in Phys. Abstr., Vol. 89, No. 1274, Entry 23807 (1986).

105.217 **Manganiferous orthopyroxene and olivine in the Allende meteorite.**
A. E. Rubin.
Am. Mineral., Vol. 69, No. 9 – 10, p. 880 – 888 (1984). Abstr. in Phys. Abstr., Vol. 89, No. 1275, Entry 30199 (1986).

105.218 **Thermal history and trapped $^{40}Ar/^{36}Ar$ ratio on the Xinyang meteorite.**
S.–s. Wang, S.–l. Hu, H.–q. Sang, J. Qiu.
Chin. J. Space Sci., Vol. 5, No. 4, p. 308 – 313 (1985). In Chinese. Abstr. in Phys. Abstr., Vol. 89, No. 1277, Entry 42225 (1986).

105.219 **The Shergotty Consortium and SNC meteorites: an overview.**
J. C. Laul.
Geochim. Cosmochim. Acta, Vol. 50, No. 6, p. 875 – 887 (1986). – See Abstr. 012.053.
The Shergotty meteorite has a multi–phase (magmatic and shock) history. While the Shergotty picture is complex, consortium studies have advanced the knowledge and understanding of Shergotty and shergottites, nakhlites and Chassigny (SNC) meteorites. Martian origin for the SNC meteorites is strongly favored by several workers from the evidence of trapped noble gases and nitrogen compositions in glasses (lithology C) of the EETA 79001 meteorite, which compare well with the Martian atmosphere analysis made by the Viking Spacecraft. At least two scenarios for the ejection of SNC meteorites are possible: (1) ejection as a large body (>6 m size) by a single impact on Mars and then multiple breakup in the asteroidal belt at about 11 m.y. for Chassigny and nakhlites, at 2.5 m.y. for Shergotty, Zagami and ALHA 77005, and at 0.6 m.y. for EETA 79001; and (2) ejection of small objects (<0.5 m size) by multiple impacts on the Martian terrain at 11, 2.5 and 0.6 m.y. with no breakup in space.

105.220 **Shock metamorphism and petrography of the Shergotty achondrite.**
D. Stöffler, R. Ostertag, C. Jammes, G. Pfannschmidt, P. R. Sen Gupta, S. B. Simon, J. J. Papike, R. H. Beauchamp.
Geochim. Cosmochim. Acta, Vol. 50, No. 6, p. 889 – 903 (1986). – See Abstr. 012.053.
This paper has its main emphasis on a quantitative evaluation of the shock history of Shergotty but also includes the main petrographic and mineralogical characteristics of the new Shergotty subsamples.

105.221 **X–ray investigations related to the shock history of the Shergotty achondrite.**
F. Horz, R. Hanss, C. Serna.
Geochim. Cosmochim. Acta, Vol. 50, No. 6, p. 905 – 908 (1986). – See Abstr. 012.053.
X–ray diffraction techniques were employed to evaluate the shock stress suffered by naturally shocked materials from the Shergotty achondrite.

105.222 **Chemical systematics of the Shergotty meteorite and the composition of its parent body (Mars).**
J. C. Laul, M. R. Smith, H. Wänke, E. Jagoutz, G. Dreibus, H. Palme, B. Spettel, A. Burghele, M. E. Lipschutz, R. M. Verkouteren.
Geochim. Cosmochim. Acta, Vol. 50, No. 6, p. 909 – 926 (1986). – See Abstr. 012.053.
The authors report chemical data for 60 elements by INAA and RNAA in two bulk samples, for 30 elements in various mineral separates of Shergotty. Shergotty is homogeneous in major element composition but heterogeneous with respect to LIL trace elements (~20%). The heterogeneity is even greater for volatile and siderophile trace elements. The mineral data show that major minerals do not account for the rare earth elements in the bulk meteorite.

105.223 **Clinopyroxene REE distribution coefficients for shergottites: the REE content of the Shergotty melt.**
G. McKay, J. Wagstaff, S.–R. Yang.
Geochim. Cosmochim. Acta, Vol. 50, No. 6, p. 927 – 937 (1986). – See Abstr. 012.053.
REE abundances were computed for the Shergotty intercumulus melt using distribution coefficients measured for the natural phase compositions. These computed abundances display LREE depletion. Such LREE–depleted abundance patterns cannot be generated by simple partial melting of a mafic mineral assemblage having the LREE–enriched patterns required for a two–stage Sm–Nd evolution history. This result implies that either (1) the melting process was complex (*e.g.*, continuous melting), (2) the source region had a more complex history than that of the simple 2–stage model, or (3) Shergotty does not satisfy the closed–system assumption of the calculation, and has suffered post–magmatic LREE–enrichment.

105.224 **Sr and Nd isotopic systematics of Shergotty meteorite.**
E. Jagoutz, H. Wänke.
Geochim. Cosmochim. Acta, Vol. 50, No. 6, p. 939 – 953 (1986). – See Abstr. 012.053.
Strontium and neodymium isotopic measurements are reported on mineral separates, whole rock samples, and their leachates and residues of Shergotty meteorite. A very large variation of the Sm/Nd ratio in the phosphates of Shergotty is observed. Two sources of Nd are observed which are not mixed completely within Shergotty, resulting in two "ages": (1) an apparent age of 147 ± 20 m.y. from the whole rock leachate–residue Nd trend; and (2) 360 ± 16 m.y. from a well–defined Nd isochron of minerals and their leachates and residues.

105.225 **Formation ages and evolution of Shergotty and its parent planet from U–Th–Pb systematics.**
J. H. Chen, G. J. Wasserburg.
Geochim. Cosmochim. Acta, Vol. 50, No. 6, p. 955 – 968 (1986). – See Abstr. 012.053.
The authors report on a study of the isotopic composition of Pb and the concentrations of Pb, U and Th in samples of Shergot-

ty, Zagami and EETA 79001. This work was undertaken with the hope that the U–Th–Pb system might serve to elucidate further the origin and evolution of Shergotty and those other SNC meteorites which are related to it. The results clearly show that these shocked meteorites have been substantially metamorphosed at a young age (116 – 192 m.y.). However the precise ages of shock metamorphism, of planetary metamorphism and of primary crystallization from a melt have not been adequately defined.

105.226 **A discussion of isotopic systematics and mineral zoning in the shergottites: evidence for a 180 m.y. igneous crystallization age.**
J. H. Jones.
Geochim. Cosmochim. Acta, Vol. 50, No. 6, p. 969 – 977 (1986). – See Abstr. 012.053.

Proposed igneous and metamorphic ages of the shergottites, a suite of achondritic basaltic meteorites, are examined in terms of the shergottites' petrography and mineral chemistry. No metamorphic or hydrothermal event appears capable of resetting the refractory Sr and Nd isotopic systems of the shergottites since the time of their igneous crystallization. The approximately concordant Rb–Sr (180 ± 20 m.y.) and U–Th–Pb (190 ± 30 m.y.) internal "isochrons" are thus interpreted as representing the time of igneous crystallization. If the shergottites are indeed less than 200 m.y. old, they represent the youngest known extraterrestrial basalts. This conclusion is supportive of the hypothesis that the SNC meteorites are samples of the planet Mars.

105.227 **Oxygen isotopes in Shergotty.**
R. N. Clayton, T. K. Mayeda.
Geochim. Cosmochim. Acta, Vol. 50, No. 6, p. 979 – 982 (1986). – See Abstr. 012.053.

Oxygen isotopic fractionation among the major mineral phases of Shergotty corresponds to a temperature of last equilibration of about 1100°C. There are no large departures from equilibrium due to metamorphism, shock, or retrograde exchange. On the oxygen three–isotope plot, the shergottites, nakhlites, and chassignites form a unique group, consistent with their having originated in a common parent body. Too little is known about the spatial and temporal variation of oxygen isotope abundances in the early solar system to allow identification of the SNC meteorites with any particular parent body.

105.228 **Carbon abundance and isotopic studies of Shergotty and other shergottite meteorites.**
I. P. Wright, R. H. Carr, C. T. Pillinger.
Geochim. Cosmochim. Acta, Vol. 50, No. 6, p. 983 – 991 (1986). – See Abstr. 012.053.

Consortium samples of the Shergotty meteorite have been measured for C abundance and isotopic composition by stepped combustion and the results compared to different samples of the same meteorite and the other known shergottites. Clearly, the shergottite meteorites have a complex C chemistry and contain components of both low and high thermal stability. At high temperatures, two of the samples examined show evidence for a component of CO_2 trapped from the martian atmosphere, possibly indicating that shock–produced glass is heterogeneously distributed throughout the shergottite group. This interpretation is corroborated by N isotope measurements made on one of the specimens.

105.229 **Nitrogen and light noble gases in Shergotty.**
R. H. Becker, R. O. Pepin.
Geochim. Cosmochim. Acta, Vol. 50, No. 6, p. 993 – 1000 (1986). – See Abstr. 012.053.

Two samples of Shergotty and one of EETA 79001 lithology A have been analyzed for nitrogen and light noble gases. Nitrogen yields are of the order of one ppm, with isotopic ratios within a few per mil of the terrestrial atmospheric composition after correction for spallogenic nitrogen. The present study does not provide evidence for or against an origin of Shergotty on Mars.

105.230 **Xenon and other noble gases in shergottites.**
T. D. Swindle, M. W. Caffee, C. M. Hohenberg.
Geochim. Cosmochim. Acta, Vol. 50, No. 6, p. 1001 – 1015 (1986). – See Abstr. 012.053.

Shergotty and related meteorites have been scrutinized in recent years because of evidence that their common parent body may be Mars. While the authors will discuss the noble gases in shergottites within the context of the proposed Martian origin, they consider this only a working hypothesis. There is strong evidence for the Martian hypothesis, but, as the authors will point out, there are also observations that are difficult to understand if these meteorites are Martian.

105.231 **Exposure history of shergottites.**
K. Nishiizumi, J. Klein, R. Middleton, D. Elmore, P. W. Kubik, J. R. Arnold.
Geochim. Cosmochim. Acta, Vol. 50, No. 6, p. 1017 – 1021 (1986). – See Abstr. 012.053.

The cosmogenic nuclides ^{36}Cl ($t_{1/2}$ = 3.0x10^{5}years), ^{26}Al (7.05x10^{5}years), ^{10}Be (1.6x10^{6}years) and ^{53}Mn (3.7x10^{6}years) were measured in three shergottites, ALHA 77005, EETA 79001, and Shergotty by accelerator mass spectrometry and neutron activation. The cosmogenic nuclide data constrain the history and origin of shergottites. All cosmogenic nuclide data indicate that the three shergottites were never exposed to cosmic rays in their parent bodies, whether that parent body was Mars or an asteroid.

105.232 **Cosmogenic effects in shergottites.**
N. Bhandari, J. N. Goswami, R. Jha, D. Sengupta, P. N. Shukla.
Geochim. Cosmochim. Acta, Vol. 50, No. 6, p. 1023 – 1030 (1986). – See Abstr. 012.053.

Heavy nuclei tracks, induced radioactivity and thermoluminescence have been studied in Shergotty, ALHA 77005 and EETA 79001 achondrites with a view to understanding their thermal and cosmic ray exposure history. The track data indicate low ablation, on the average 1 to 3 cm, for all three, indicative of low geocentric velocities and small meteoroid masses (1 – 25 kg). The ^{26}Al activity of 91 dpm/kg in Shergotty, when compared with the expected production rates, suggest a small (effective radius ~ 12 cm) size of the meteoroid in space, consistent with the track data. Nuclear track, ^{26}Al and other cosmogenic effects in the three shergottites are consistent with their single stage exposure in space as small objects.

105.233 **Thermoluminescence and the shock and reheating history of meteorites – III. The shergottites.**
F. A. Hasan, M. Haq, D. W. G. Sears.
Geochim. Cosmochim. Acta, Vol. 50, No. 6, p. 1031 – 1038 (1986). – See Abstr. 012.053.

The shergottites are intensely shocked assemblages of pyroxene, maskelynite and sometimes olivine, with minor magnetite, pyrrhotite and whitlockite. Their young ages (1.3 Ga), igneous origin, Fe/Mn ratio and trapped gas composition has led many authors to conclude that the meteorites were ejected from Mars. In the present paper the authors report a TL study of the known shergottites, Shergotty, Allan Hills A 77005, Zagami and Elephant Moraine A 79001 (lithology A). Their hope was that such a study would shed further light on the shock history of this class.

105.234 **Nuclear tracks, Sm isotopes and neutron capture effects in the Elephant Moraine shergottite.**
R. S. Rajan, G. Lugmair, A. S. Tamhane, G. Poupeau.
Geochim. Cosmochim. Acta, Vol. 50, No. 6, p. 1039 – 1042 (1986). – See Abstr. 012.053.

Nuclear track studies, uranium concentration measurements and Sm–isotope studies have been performed on both lithologies A and B of the Elephant Moraine shergottite, EETA 79001. Track studies show that EETA 79001 was a rather small object in space with a preatmospheric radius of 12 ± 2 cm, corresponding to a preatmospheric mass of 28 ± 13 kg.

105.235 **Magnetic studies on Shergotty and other SNC meteorites.**
S. M. Cisowski.
Geochim. Cosmochim. Acta, Vol. 50; No. 6, p. 1043 – 1048 (1986). – See Abstr. 012.053.

This study involves both a description of the basic magnetic properties of various meteorites within the SNC group, and an attempt to estimate the strength of the magnetic field which produced the remanent magnetization of Shergotty meteorite. This latter effort has the potential to constrain the choices for the parent body of these objects, or at least to describe the magnetic environment in which they evolved.

105.236 **Martian volatiles in shergottite EETA 79001: new evidence from oxidized sulfur and sulfur–rich aluminosilicates.**
J. L. Gooding, D. W. Muenow.
Geochim. Cosmochim. Acta, Vol. 50, No. 6, p. 1049 – 1059 (1986). – See Abstr. 012.053.

This report summarizes initial results of a search for mineralogical and petrological clues to the origin of trapped gases in Lith–C. Evidence is presented that, although Antarctic weathering effects in EETA 79001 are more significant than has been previously recognized, it is possible to distinguish terrestrial from preterrestrial alteration effects through study of properly selected control samples. The major conclusion is that Lith–C formed in an atmosphere that was sufficiently oxidizing to form sulfate from sulfide and that at least a small portion of the sulfate probably exists as sulfur–rich, aluminosilicate mineraloids. The additional enrichment of chlorine in one variety of aluminosilicate denotes a possible petrogenetic link with the S– and Cl–rich, fine–grained materials that cover the Martian surface. Therefore, results reported here are consistent with the origin of EETA 79001 as a Martian rock.

105.237 **The parental magma of the Nakhla achondrite: ultrabasic volcanism on the shergottite parent body.**
A. H. Treiman.
Geochim. Cosmochim. Acta, Vol. 50, No. 6, p. 1061 – 1070 (1986). – See Abstr. 012.053.

The primary goal of this study is to determine the composition of Nakhla's *intercumulus magma*, the magma trapped among the cumulus minerals at the time of crystal settling. It is important to know whether the intercumulus magma was in equilibrium with the cumulus mineral, *i.e.* whether the intercumulus magma was *parental* to the Nakhla meteorite. Discussion of the intercumulus magma entails three topics: calculation of the magma composition, determination of equilibrium between magma and cumulus crystals, and evaluation of phase relations for the magma composition.

105.238 **Core formation in the Earth and shergottite parent body (SPB): chemical evidence from basalts.**
A. H. Treiman, M. J. Drake, M.–J. Janssens, R. Wolf, M. Ebihara.
Geochim. Cosmochim. Acta, Vol. 50, No. 6, p. 1071 – 1091 (1986). – See Abstr. 012.053.

Constraints on processes of core formation in terrestrial planets may be inferred from the abundances of siderophile and chalcophile elements in their mantles. Of particular interest is a comparison of processes of core formation in the Earth and in the shergottite parent body (SPB), a terrestrial planet tentatively inferred to be Mars. Element abundances in the mantles of the SPB and the Earth are estimated from element/element correlations observed among basalt samples. In basalts from the SPB (the SNC meteorites), four groups of covariant elements are observed. From correlations within these groups, the SPB mantle is found to be depleted in volatile elements, and strongly depleted in siderophile and chalcophile elements. For the Earth, the element/element correlation method gives a mantle composition similar to previous estimates.

105.239 **The solar–system abundances of Nb, Ta, and Y, and the relative abundances of refractory lithophile elements in differentiated planetary bodies.**
K. P. Jochum, H. M. Seufert, B. Spettel, H. Palme.
Geochim. Cosmochim. Acta, Vol. 50, No. 6, p. 1173 – 1183 (1986).

The purpose of the present contribution is to provide new abundance data in carbonaceous chondrites for Nb, Ta, Y, and Ho. By using additional literature data the authors shall establish a consistent set of chondritic abundances for Zr, Hf, Nb, Ta and Y, and define their ratios relative to REE. With these reference values it is possible to better characterize the type and extent of fractionation processes that separated these elements from each other during igneous processes on Earth, Moon and in parent bodies of differentiated meteorites.

105.240 **Experimental investigations of trace element fractionation in iron meteorites. III. Elemental partitioning in the system Fe–Ni–S–P.**
D. J. Malvin, J. H. Jones, M. J. Drake.
Geochim. Cosmochim. Acta, Vol. 50, No. 6, p. 1221 – 1231 (1986).

This contribution concentrates on establishing equilibrium values of solid metal/liquid metal partition coefficients and demonstrates that isothermal "static" experiments reliably yield equilibrium values. These experiments also permit reliable prediction of partition coefficients in the quaternary system Fe–Ni–S–P.

105.241 **Radar observation of the Giacobinid meteor shower 1985.**
M. Šimek.
Bull. Astron. Inst. Czech., Vol. 37, No. 4, p. 246 – 249 (1986).

105.242 **The wavelength dependence of phase coefficients.**
J. Gradie, J. Veverka.
Icarus, Vol. 66, No. 3, p. 455 – 467 (1986).

In this paper the authors use the results of laboratory measurements to investigate the wavelength dependence of disk–integrated phase coefficients of the following materials: a basalt, the ordinary chondrite Bruderheim, the carbonaceous chondrite Allende, and sulfur. These materials were chosen as approximate analogs to the surfaces of the Moon, S– and C–class asteroids, and of the Jovian satellite Io, respectively. The authors show that the observed wavelength dependence of phase coefficients can be predicted (at least qualitatively) once the spectrum of an object is known: the phase coefficients being relatively largest for wavelengths at which the object is darkest.

105.243 **The spectral and physical properties of metal in meteorite assemblages: implications for asteroid surface materials.**
M. J. Gaffey.
Icarus, Vol. 66, No. 3, p. 468 – 486 (1986).

The purposes of the present paper are (1) to define the spectral contribution of the nickel–iron metal component in meteoritic assemblages, (2) to elucidate the chemical, physical, and petrographic properties of the metal grains which affect their spectral signature in asteroid surface materials, and (3) to outline the most critical areas of additional investigation to better resolve the issues involved.

105.244 **Elemental abundances of Ningqiang chondrite.**
Z.–f. Chai, X.–y. Mao, S.–l. Ma, S.–c. Wang.
Kexue Tongbao, Vol. 31, No. 2, p. 111 – 114 (1986). Abstr. in Phys. Abstr., Vol. 89, No. 1282, Entry 73161 (1986).

105.245 **Cosmogenic radionuclides of Ningqiang chondrite from nondestructive gamma–ray spectrometry.**
Z.–f. Chai, Q. Su, X.–y. Mao, S.–l. Ma, S.–c. Wang, S.–m. Luo.
Kexue Tongbao, Vol. 31, No. 2, p. 115 – 118 (1986). Abstr. in Phys. Abstr., Vol. 89, No. 1282, Entry 73162 (1986).

105.246 **Anomalous fading of thermoluminescence in meteorites.**
F. A. Hasan, B. D. Keck, C. Hartmetz,
D. W. G. Sears.
J. Lumin., Vol. 34, No. 6, p. 327 – 335 (1986). Abstr. in Phys. Abstr., Vol. 89, Nr. 1284, Entry 83499 (1986).

105.247 **Petrology and chemistry of hyperferroan anorthosites and other clasts from lunar meteorite ALHA81005.**
C. A. Goodrich, G. J. Taylor, K. Keil, W. V. Boynton, D. H. Hill.
J. Geophys. Res., Vol. 89, Suppl., p. C87 – C94 (1984). – See Abstr. 012.065.
The purpose of this work was to characterize the population of lithic clasts in 81005 and their relation to known lunar rock types, to search for new lunar rock types, and to gain information about the 81005 source region.

105.248 **Oxygen isotopic compositions of enstatite chondrites and aubrites.**
R. N. Clayton, T. K. Mayeda, A. E. Rubin.
J. Geophys. Res., Vol. 89, Suppl., p. C245 – C249 (1984). – See Abstr. 012.065.
Oxygen isotopic compositions have been measured for ten EH chondrites, eight EL chondrites, and nine aubrites. The mean $\delta^{18}O$ and $\delta^{17}O$ values are: 5.31 and 2.74, 5.57 and 2.96, 5.24 and 2.70, respectively. Within the EH group, more equilibrated meteorites are fractionated toward higher heavy isotope abundances than the less equilibrated meteorites. Small departures of individual meteorites from a common mass fractionation occur for some of the EH samples. The aubrites may be genetically related to either the EH or EL group or to both. Almost all of the measured isotopic compositions are indistinguishable from the terrestrial oxygen isotope fractionation line, suggesting a genetic relationship among the earth, the moon, and the enstatite meteorites.

105.249 **The polymict eucrites.**
J. S. Delaney, M. Prinz, H. Takeda.
J. Geophys. Res., Vol. 89, Suppl., p. C251 – C288 (1984). – See Abstr. 012.065.
In this study, samples from all the polymict eucrite suites have been documented and comparisons made with other polymict eucrites and with the howardites, eucrites, cumulate eucrites, and diogenites.

105.250 **The case for a melt matrix in plagioclase–POIK mesosiderites.**
R. H. Hewins.
J. Geophys. Res., Vol. 89, Suppl., p. C289 – C297 (1984). – See Abstr. 012.065.
This paper offers a new interpretation of the plagioclase–POIK mesosiderites Bondoc, Budulan, and Mincy as poikilitic, thus substantially changing the percentage of mesosiderites in which the matrix appears to have been a melt. All mesosiderites, of course, contain clasts from preexisting igneous rocks.

105.251 **Petrographic studies of refractory inclusions from the Murchison meteorite.**
G. J. MacPherson, L. Grossman, A. Hashimoto, M. Bar–Matthews, T. Tanaka.
J. Geophys. Res., Vol. 89, Suppl., p. C299 – C312 (1984). – See Abstr. 012.065. With a correction in Vol. 90, Suppl., p. C843 (1985).
Mineralogical and petrographic data are presented for ten refractory inclusions recovered by freeze–thaw disaggregation and heavy liquid separation from the Murchison C2 chondrite.

105.252 **Crystal chemistry of meteoritic hibonites.**
R. G. Burns, V. M. Burns.
J. Geophys. Res., Vol. 89, Suppl., p. C313 – C321 (1984). – See Abstr. 012.065.
Structural features, influencing relative enrichments, cation stabilities, and colors of vanadium–, titanium–, and iron–bearing hibonite ($CaAl_{12}O_{19}$) in meteorites are examined.

105.253 **On the brittle–ductile behavior of iron meteorites: new experimental constraints.**
T. Matsui, P. H. Schultz.
J. Geophys. Res., Vol. 89, Suppl., p. C323 – C328 (1984). – See Abstr. 012.065.
High–velocity ($\sim$2 to $\sim$5 km/sec) impacts into octahedrite iron meteorites (Gibeon, El Sampal, and Arispe), steel, and iron targets were made at the NASA–Ames Vertical Gun Facility. The experiments clearly revealed the brittle behavior of meteorites at very low temperatures ($<$200K) as indicated by spallation of the crater rim and the development of deeply penetrating fractures in the meteorite sample. High–velocity ($\sim$5 km/sec) impacts resulted in a net mass loss regardless of the brittle–ductile behavior of the target or projectile, whereas low–velocity ($\sim$2 km/sec) impacts resulted in a net mass gain for ductile projectiles into brittle and ductile targets. Consequently, even if iron meteorites (octahedrites) were originally the core of a layered parent body, they might be subsequently destroyed in a brittle manner. Such results are consistent with the hypothesis by Matsui and Mizutani (1977) that the brittle behavior or iron–like planetesimals in the low–temperature asteroid zone prevents growth into full–size planets.

105.254 **Carbon–14 ages of Antarctic meteorites with accelerator and small–volume counting techniques.**
A. J. T. Jull, D. J. Donahue, T. H. Zabel, E. L. Fireman.
J. Geophys. Res., Vol. 89, Suppl., p. C329 – C335 (1984). – See Abstr. 012.065.
Carbon–14 terrestrial ages of six Antarctic meteorites have been measured by accelerator mass spectrometry. These six and three additional Antarctic meteorites were also measured by small counters. The results are in general agreement, with some differences that are related to the calibration samples used.

105.255 **Geochemistry of Muong–Nong type tektites. IV. Selected trace element correlations.**
C. Koeberl, F. Kluger, W. Kiesl.
J. Geophys. Res., Vol. 89, Suppl., p. C351 – C357 (1984). – See Abstr. 012.065.
The authors have analyzed nineteen Muong–Nong type tektites for 40 – 50 elements. They discuss fifteen nonvolatile elements.

105.256 **An apatite–rich, ferroan, mafic lithology from lunar meteorite ALHA81005.**
C. A. Goodrich, G. J. Taylor, K. Keil.
J. Geophys. Res., Vol. 90, Suppl., p. C405 – C414 (1985). – See Abstr. 012.065.
Lunar meteorite ALHA81005 contains a clast of ferroan anorthositic troctolite with (in vol %) 59% plagioclase, 27% olivine, 11% clinopyroxene, and 3% chlorapatite. This clast is probably pristine and is unique among lunar samples. Its texture is igneous, but is unusual in that pyroxene and apatite, which probably cocrystallized, have been brecciated and apparently resorbed, and olivine has precipitated around them. The authors discuss four possible explanations for this texture.

105.257 **Molybdenum in eucrites: evidence for a metal core in the eucrite parent body.**
H. E. Newsom.
J. Geophys. Res., Vol. 90, Suppl., p. C613 – C617 (1985). – See Abstr. 012.065.
Molybdenum concentrations from 15 to 70 ppb in eucrites have been determined with a new method of neutron activation. These values are very low compared to chondrites and are in agreement with the large depletions of other siderophile elements in eucrites.

105.258 **Basaltic volcanism on the eucrite parent body: petrology and chemistry of the polymict eucrite ALHA80102.**
A. H. Treiman, M. J. Drake.
J. Geophys. Res., Vol. 90, Suppl., p. C619 – C628 (1985). – See Abstr. 012.065.
This paper discusses the petrography, mineral compositions, and trace element compositions of clasts in ALHA80102 and

compares its lithologies with other eucrites to describe the diversity of eucritic magmatism.

105.259 **Chemical zoning and homogenization of Pasamonte-type pyroxene and their bearing on thermal metamorphism of a howardite parent body.**
M. Miyamoto, M. B. Duke, D. S. McKay.
J. Geophys. Res., Vol. 90, Suppl., p. C629 – C635 (1985). – See Abstr. 012.065.

The authors have examined the Mg–Fe zoning of pyroxenes in Pasamonte and Juvinas eucrites and have calculated the expected homogenization (1) to test the scale of Takeda et al. (1983), (2) to estimate the heating duration for homogenization, and (3) to better understand the conditions of the metamorphism in the surface layer of a eucrite/howardite parent body.

105.260 **The diogenite–eucrite links and the crystallization history of a crust of their parent body.**
H. Takeda, H. Mori.
J. Geophys. Res., Vol. 90, Suppl., p. C636 – C648 (1985). – See Abstr. 012.065.

To obtain a better understanding of the crystallization trends of differentiated planetary crusts, the authors investigated several Antarctic achondrites intermediate between diogenite and eucrite with electron microprobe, single crystal X ray diffraction, and analytical transmission electron microscope techniques. The crystallization trends have been compared with lunar and terrestrial trends. The results have been interpreted in terms of both the crystal fractionation model of Ikeda and Takeda (1984) and a partial melting model of Stolper (1977).

105.261 **A model for the origin of basaltic achondrites based on the Yamato 7308 howardite.**
Y. Ikeda, H. Takeda.
J. Geophys. Res., Vol. 90, Suppl., p. C649 – C663 (1985). – See Abstr. 012.065.

Four thin sections of Y–7308 have been examined petrographically. The authors first review briefly the lithological components relevant to their model and then propose their model of crystal fractionation of a single HED magma to account for crystallization of all known lithological components observed in the Y–7308 howardite. The proposed model permits better understanding of the evolution and origin of the HED parent body and can be tested using trace element data for HED meteorites.

105.262 **Carbon isotopes in three SNC meteorites.**
R. H. Carr, I. P. Wright, C. T. Pillinger.
J. Geophys. Res., Vol. 90, Suppl., p. C664 – C668 (1985). – See Abstr. 012.065.

Samples of three SNC meteorites – Shergotty, Chassigny, and Elephant Moraine (EETA 79001) – have been analysed for their contents and stable isotopic compositions of carbon using stepped combustion in an attempt to resolve indigenous components from terrestrial contaminants.

105.263 **Fluid inclusions in stony meteorites – a cautionary note.**
R. L. Rudnick, L. D. Ashwal, D. J. Henry,
E. K. Gibson Jr., E. Roedder, H. E. Belkin, M. T. Colucci.
J. Geophys. Res., Vol. 90, Suppl., p. C669 – C675 (1985). – See Abstr. 012.065.

Newly discovered fluid inclusions in thin sections of Bjurböle chondrules, shergottite EETA79001, lunar meteorite ALHA81005, and Apollo 16 glasses possess physical properties similar to those of fluid inclusions found in thin sections of five stony meteorites recently described by Warner et al. (1983). The distribution and physical properties of these new fluid inclusions indicate they may be artifacts of thin section preparation; the authors suggest that saw coolant was sucked into vacuum vesicles in glasses and minerals through submicroscopic features produced during sawing. The similarities between these fluid inclusions and fluid inclusions previously described by Warner et al. (1983) lead the authors to conclude that many of the fluid inclusions reported earlier may be artifacts.

105.264 **Analysis of segregation trends observed in iron meteorites using measured distribution coefficients.**
R. Sellamuthu, J. I. Goldstein.
J. Geophys. Res., Vol. 90, Suppl., p. C677 – C688 (1985). – See Abstr. 012.065.

The objectives of this study are (1) to explain the elemental distributions observed in various iron meteorite groups using the composition profiles measured in laboratory Fe–Ni alloys solidified by a plane front solidification technique, (2) to measure the distribution coefficients for Ni, P, Ir, Ge, and Cu as a function of the S and P contents in Fe–Ni alloys, and (3) to use these coefficients in the calculation of elemental distributions within a chemical group of iron meteorites.

105.265 **The evolution of partially differentiated planetesimals: evidence from iron meteorite groups IAB and IIICD.**
A. Kracher.
J. Geophys. Res., Vol. 90, Suppl., p. C689 – C698 (1985). – See Abstr. 012.065.

This paper summarizes some of the properties of meteorites thought to be derived from partially differentiated planetesimals. The bulk composition and thermal history of these bodies is constrained. Physical aspects that may have controlled parent body differentiation and affected the composition of the sulfide melt are discussed. The author deals with the chemical evolution of the parent body, in particular aspects not covered by Kracher (1982). Some observations that support the model presented in this paper are briefly discussed. Attempts to reinterpret barometric and chronometric data in light of the partial differentiation model are made, and finally, tentative conclusions and a discussion of questions left unanswered by the model are offered.

105.266 **Petrology of types 4 – 6 carbonaceous chondrites.**
E. R. D. Scott, G. J. Taylor.
J. Geophys. Res., Vol. 90, Suppl., p. C699 – C709 (1985). – See Abstr. 012.065.

The authors have studied the petrology of three type 4 carbonaceous chondrites, Coolidge, Karoonda, and Pecora Escarpment 82500, which was previously classified as an LL6 chondrite, and compared them to published descriptions of the other known metamorphosed C chondrites, Yamato 6903 and Mulga (west).

105.267 **Magnetite morphologies in the Essebi and Haripura CM chondrites.**
M. Hyman, E. B. Ledger, M. W. Rowe.
J. Geophys. Res., Vol. 90, Suppl., p. C710 – C714 (1985). – See Abstr. 012.065.

Many of the unusual magnetite morphologies previously observed only in the CI chondrites have been seen in the CM chondrites, Essebi and Haripura.

105.268 **Trapped Xe components in etched samples of the Murray (C2) and Murchison (C2) carbonaceous chondrites.**
C. M. Jones, G. R. Lumpkin, J. H. Reynolds.
J. Geophys. Res., Vol. 90, Suppl., p. C715 – C721 (1985). – See Abstr. 012.065.

Xe isotopic measurements are described. The samples analyzed were bulk chips of the Murray (C2) and Murchison (C2) carbonaceous chondrites that had been freeze–thaw disaggregated, etched with H_2O_2, and then split into colloidal and noncolloidal fractions using methanol. The Murray samples appeared to contain a binary mixture of U–Xe and H + L–Xe. The Murchison data concur with the Murray data for the heavy isotopes, but the structure at the light isotopes is obscured by spallation Xe. There is no evidence in these data for the separability of H–Xe and L–Xe. The fact that H_2O_2 etching removes Xe without prior demineralization supports the view that the carrier of a substantial portion of trapped Xe may reside on grain surfaces.

105.269 **A model for GCR–particle fluxes in stony meteorites and production rates of cosmogenic nuclides.**
R. C. Reedy.
J. Geophys. Res., Vol. 90, Suppl., p. C722 – C728 (1985). – See Abstr. 012.065.

A model is presented for the differential fluxes of galactic cosmic ray (GCR) particles with energies above 1 MeV inside any spherical stony meteorite as a function of the meteorite's preatmospheric radius and the sample's depth. This model is based on the Reedy–Arnold equations for the energy–dependent fluxes of GCR particles in the moon and is an extension of flux parameters that were derived for several meteorites of various sizes. This flux model is used to calculate the production rates of many cosmogenic nuclides as a function of radius and depth.

105.270 **NRM directions around a centimeter–sized dark inclusion in Allende.**
N. Sugiura, D. W. Strangway.
J. Geophys. Res., Vol. 90, Suppl., p. C729 – C738 (1985). – See Abstr. 012.065.

The natural remanent magnetization (NRM) carried by a centimeter–sized dark inclusion, the surrounding matrix, and individual chondrules in Allende was measured to determine the magnetic history of the meteorite.

Catalog of the Vatican Meteorite Collection.
See Abstr. 002.054.

The analysis of extraterrestrial materials.
See Abstr. 003.053.

Identification of implicit periodicity of historical recorded meteorite falls.
See Abstr. 004.033.

Isotope anomalies induced in laboratory distillation.
See Abstr. 022.003.

On the unification of thin target and thick target approaches to the interpretation of cosmogenic nuclides in meteorites.
See Abstr. 022.035.

A contribution to the low temperature Fe–Ni phase diagram.
See Abstr. 022.036.

Isotopic anomalies scattering processes and the Pauli exclusion principle.
See Abstr. 022.037.

Systematic studies of ^{10}Be and ^{26}Al production in meteorites: simulation by isotropically irradiated meteorite models.
See Abstr. 022.039.

Production of mass independently fractionated oxygen isotopic components by ultra–violet light.
See Abstr. 022.040.

Changes of chemical compositions of ejecta induced by hypervelocity impacts.
See Abstr. 022.129.

Radiation effects on grain materials.
See Abstr. 022.133.

The *s*–process branching at ^{85}Kr.
See Abstr. 061.013.

A possible unified interpretation of the solar ^{48}Ca/^{46}Ca abundance ratio and Ca – Ti isotopic anomalies in meteorites.
See Abstr. 061.134.

Planetary surface texture and albedo from parameter plots of optical polarization data.
See Abstr. 091.001.

Lunar sample 14425: not a lunar tektite.
See Abstr. 094.005.

Cosmic ray exposure age and terrestrial age of the lunar meteorite ALHA 81005.
See Abstr. 094.014.

Interconnection between small bodies of the solar system.
See Abstr. 098.066.

Photographic fireball networks.
See Abstr. 104.020.

An interplanetary dust particle analog to matrices of CO/CV carbonaceous chondrites and unmetamorphosed unequilibrated ordinary chondrites.
See Abstr. 106.014.

Interstellar material in the solar system. A working group report.
See Abstr. 106.059.

The meteoritic chromium isotopic composition and limits for radioactive ^{53}Mn in the early solar system.
See Abstr. 107.003.

Meteoritic constraints on processes in the solar nebula.
See Abstr. 107.012.

Short–lived nuclei in the early solar system.
See Abstr. 107.013.

Preplanetary nebula's solid matter formation and composition of chondrites.
See Abstr. 107.023.

"Solar flares" and chondrule formation in the primeval solar nebula.
See Abstr. 107.028.

Formation of chondrules by drag heating in the solar nebula: the latest word.
See Abstr. 107.029.

Interrelationships between interstellar and interplanetary grains. A working group report.
See Abstr. 131.229.

What predictions can be made on the nature of carbon and carbon–bearing compounds (hydrocarbons) in the interstellar medium based on studies of interplanetary dust particles?
See Abstr. 131.236.

106 Interplanetary Matter, Interplanetary Magnetic Field, Zodiacal Light

106.001 **Collisions in the Solar System – III. Meteoroid survival times.**
D. I. Steel, W. G. Elford.
Mon. Not. R. Astron. Soc., Vol. 218, No. 2, p. 185 – 199 (1986).
The lifetimes of meteoroids against catastrophic collisions with zodiacal dust particles is evaluated for specific orbits, including for the first time the effect of the orbital inclination. Although the results are for meteoroids of radius 1 mm, the collisional lifetimes for larger or smaller bodies can be found by the application of simple scaling factors. The figures can be used to estimate the collisional lifetime of any meteoroid in any particular orbit. The collisional lifetimes of 28 well–known streams are also calculated. Finally, an alternative method for the calculation of the spatial density of an object in an arbitrary orbit is presented.

106.002 **Three–dimensional properties of interplanetary disturbances in 1978 – 1981.**
T. Watanabe, T. Kakinuma.
Astrophys. Space Sci., Vol. 118, Nos. 1/2, p. 153 – 157 (1986). – See Abstr. 012.001.
This paper presents the average three–dimensional configuration of solar flare– or disappearing filament–associated interplanetary disturbances on the basis of IPS (interplanetary scintillation) and spacecraft observations in 1978 – 1981. The angular distribution of the propagation speed at 1 AU is largely isotropic over the range of 110° in solar longitude centered at the normal of the solar source. In the latitudinal direction, the characteristic angular extent is about 60°. Thus, the three–dimensional shape of an interplanetary disturbance can be approximated by a half of an ellipsoid having an axial ratio of about 1.8.

106.003 **Interplanetary shock collisions: forward with reverse shocks.**
Z. K. Smith, M. Dryer, S. M. Han.
Astrophys. Space Sci., Vol. 119, No. 2, p. 337 – 344 (1986).
When one interplanetary shock overtakes another, the structure that results depends upon the nature of the interacting shocks. The authors examine, numerically, the results of collisions of forward with reverse shocks, in two dimensions, and show that the results depend primarily upon shock strength. They also note that such interactions could explain why many energy outbursts on the Sun that would be expected to cause geomagnetic effects at Earth, do not.

106.004 **An interplanetary magnetic field ensemble at 1 AU.**
W. H. Matthaeus, M. L. Goldstein, J. H. King.
J. Geophys. Res., Vol. 91, No. A1, p. 59 – 69 (1986).
A method for calculating ensemble averages from magnetic field data is described. A data set comprising approximately sixteen months of nearly continuous ISEE–3 magnetic field data is used in this study. Individual subintervals of this data, ranging from 15 hours to 15.6 days comprise the ensemble. The sole condition for including each subinterval in the averages is the degree to which it represents a weakly time–stationary process. Averages obtained by this method are appropriate for a turbulence description of the interplanetary medium.

106.005 **Fractal structure of the interplanetary magnetic field.**
L. F. Burlaga, L. W. Klein.
J. Geophys. Res., Vol. 91, No. A1, p. 347 – 350 (1986).
It is shown that under some conditions the interplanetary magnetic field has the properties of a fractal curve, and calculation of its fractal dimension is an efficient and economical alternative to spectral analysis.

106.006 **Interplanetary scintillation and plasma density.**
S. J. Tappin.
Planet. Space Sci., Vol. 34, No. 1, p. 93 – 97 (1986).
Observations on a grid of 900 – 2500 radio sources have been combined with in situ spacecraft measurements to determine the relation between interplanetary scintillation and mean plasma–density along the line of sight. Studies carried out during a variety of disturbances, including shocks and corotating streams, indicate a relation $g \cong (N/9)^{0.5}$ where g is the scintillation enhancement factor and N (cm^{-3}) is the plasma density. The results show that scintillation can be used to provide a powerful and reliable method of monitoring the mean density along the line of sight for disturbances of many kinds, including post–shock flows.

106.007 **Interplanetary plasma turbulence and the Doppler detection of a gravitational wave background.**
M. Dobrowolny, L. Iess.
Astron. Astrophys., Vol. 157, No. 2, p. 346 – 352 (1986).
The authors discuss the role of the interplanetary plasma turbulence in the planned experiments for the detection of a gravitational wave background by means of a Doppler tracking of interplanetary spacecraft. They have in mind, in particular, the usual radio link configuration with a single frequency uplink and a double frequency downlink. They construct a new observable quantity (a linear combination of the S– and X–band Doppler shift) which strongly increases the signal to noise ratio at frequencies smaller than the reciprocal of an Earth–spacecraft light time. Next they employ a previously developed model of interplanetary turbulence to calculate the minimum energy spectral density of the background which can be detected against plasma noise. They point out the importance of the low frequency part of the spectrum, where the sensitivity of the experiment increases and the limits obtained for the gravitational background energy density should be more interesting.

106.008 **Random fluctuations versus Poynting–Robertson drag on interplanetary dust grains.**
M. K. Wallis.
Nature, Vol. 320, No. 6058, p. 146 – 148 (1986).
Stochastic forces on circumsolar grains arise from low–frequency fluctuations in surface charge and interplanetary fields. The author shows that stochastic heating can balance and indeed overwhelm the dissipative Poynting–Robertson drag for small grains ($\lesssim 1$ μm scale radius). Furthermore, the consequent drift to the outer Solar System may cancel the strong collisional source inferred for micrometre and sub–micrometre fragments, explaining how their present distribution is stable in time. More generally, the stochastic heating mechanism combined with collisional fragmentation sets limits on dust accretion onto cool stars.

106.009 **Microtomed thin–sections of individual interplanetary dust particles – new insight regarding textures and petrographic associations.**
J. P. Bradley, A. S. Teetsov, D. E. Brownlee.
Meteoritics, Vol. 20, No. 4, p. 614 – 615 (1985). Abstract. – See Abstr. 010.661.

106.010 **Microprobe analysis of sectioned mafic and sulfide micrometeorites.**
D. E. Brownlee, M. M. Wheelock.
Meteoritics, Vol. 20, No. 4, p. 617 (1985). Abstract. – See Abstr. 010.661.

106.011 **PIXE analysis of interplanetary dust particles.**
E. K. Jessberger, R. Wallenwein, H. Blank, K. Traxel.
Meteoritics, Vol. 20, No. 4, p. 673 (1985). Abstract. – See Abstr. 010.661.

106.012 **Composition of terrestrial and cosmic dust components in Greenland ice cores.**
G. W. Kallemeyn, K. L. Rasmussen.
Meteoritics, Vol. 20, No. 4, p. 678 – 679 (1985). Abstract. – See Abstr. 010.661.

106.013 **Micrometeorite material returned from space – a progress report on the analysis of Solar Max impact craters.**
D. S. McKay, L. S. Schramm, H. A. Zook.
Meteoritics, Vol. 20, No. 4, p. 709 – 710 (1985). Abstract. – See Abstr. 010.661.

106.014 **An interplanetary dust particle analog to matrices of CO/CV carbonaceous chondrites and unmetamorphosed unequilibrated ordinary chondrites.**
F. J. M. Rietmeijer, D. S. McKay.
Meteoritics, Vol. 20, No. 4, p. 743 – 744 (1985). Abstract. – See Abstr. 010.661.

106.015 **CAI's among the cosmic dust collection.**
M. E. Zolensky.
Meteoritics, Vol. 20, No. 4, p. 792 – 793 (1985). Abstract. – See Abstr. 010.661.

106.016 **Erratum: "Evidence of high cosmic dust concentrations in late Pleistocene polar ice (20.000 – 14.000 years BP)" [Meteoritics, Vol. 20, No. 3, p. 545 – 558 (1985)].**
P. A. LaViolette.
Meteoritics, Vol. 20, No. 4, p. 803 (1985). See Abstr. 40.106.092.

106.017 **The AMPTE artificial comet experiments.**
A. Valenzuela, G. Haerendel, H. Föppl, F. Melzner, H. Neuss, E. Rieger, J. Stöcker, O. Bauer, H. Höfner, J. Loidl.
Nature, Vol. 320, No. 6064, p. 700 – 703 (1986).

In July last year and in December 1984, barium clouds were injected into the solar wind from the AMPTE satellite. The clouds resembled those of natural comets in that a head and tail were clearly visible. Tail rays formed with speeds of a few tens of kilometres per second, and other structures separated at even higher speeds. The comet head, instead of moving slowly downstream, performed a lateral excursion of several hundred kilometres. The properties of the "artificial comet" are described.

106.018 **Optical observations of the AMPTE artificial comet from the Northern Hemisphere.**
D. Rees, T. J. Hallinan, H. C. Stenbaek–Nielsen, M. Mendillo, J. Baumgardner.
Nature, Vol. 320, No. 6064, p. 704 – 708 (1986).

The AMPTE artificial comet of 27 December 1984 developed a well–defined core (maximum size $\sim 160 \times 100$ km) of cold barium ions, surrounded by a diffuse coma, 500 km in diameter, consisting of ions extracted from the core. A plasma tail extended from the core, mainly directed away from the Sun. The plasma tail and diffuse coma consisted of energetic ions. Confounding predictions, the artificial comet survived for barely 5 minutes, and the core of the comet moved southward, rather than in the direction of the solar wind.

106.019 ***In situ* magnetic field observations of the AMPTE artificial comet.**
H. Lühr, D. J. Southwood, N. Klöcker, M. W. Dunlop, W. A. C. Mier–Jedrzejowicz, R. P. Rijnbeek, M. Six, B. Häusler, M. Acuña.
Nature, Vol. 320, No. 6064, p. 708 – 711 (1986).

Magnetometers aboard the IRM and UKS spacecraft monitored the magnetic field during the AMPTE artificial comet experiment of 27 December 1984. Rapid photoionization of the released barium vapour resulted in the formation of a magnetic cavity, shielded from the ambient magnetic field. The presence of this highly conductive obstacle caused draping and compression of the solar wind magnetic field.

106.020 **UKS plasma measurements near the AMPTE artificial comet.**
D. J. Rodgers, A. J. Coates, A. D. Johnstone, M. F. Smith, D. A. Bryant, D. S. Hall, C. P. Chaloner.
Nature, Vol. 320, No. 6064, p. 712 – 716 (1986).

UKS observations of the ionized barium and disturbed natural plasmas 170 km from the centre of the artificial comet release of 27 December 1984 lasted over 4 minutes. During this disturbance, solar–wind ions were retarded and deflected southwards and dawnwards, while barium was accelerated northwards. Substantial electron energization was seen throughout, varying with time and with the energy of the source electrons. Rapid electron and ion density changes were observed in later stages.

106.021 **Plasma waves and wave–particle interactions seen at the UKS spacecraft during the AMPTE artificial comet experiment.**
L. J. C. Woolliscroft, M. P. Gough, P. J. Christiansen, A. G. Darbyshire, H. G. F. Gough, D. S. Hall, D. Jones, S. R. Jones, A. J. Norris.
Nature, Vol. 320, No. 6064, p. 716 – 720 (1986).

Wave and particle correlator experiments on board the UKS revealed a wide range of plasma wave and wave–particle interactions associated with the 27 December 1984 artificial comet release just outside the main diamagnetic cavity. Shortly after the release strong bursts of plasma waves were recorded, and high–frequency electron oscillations in an enhanced–density plasma were observed later, along with evidence for interaction between a beam of electrons and the local plasma. At lower frequencies, extremely strong wave emissions were associated with what appeared to be the interface between the cloud and solar–wind plasmas.

106.022 **Dynamics of the AMPTE artificial comet.**
G. Haerendel, G. Paschmann, W. Baumjohann, C. W. Carlson.
Nature, Vol. 320, No. 6064, p. 720 – 723 (1986).

Sweeping the plasma before it, the interplanetary magnetic field quickly penetrated into an initially diamagnetic barium plasma cloud. The field was strongly compressed and extended in the flow direction. Ions accelerated by electric polarization fields formed a visible tail at the rear of the cloud; their recoil balancing the magnetic stresses. The dominant lateral motion of the head of the artificial comet is attributed to a recoil of ions extracted by the interplanetary electric field.

106.023 **Preliminary analysis of the IRAS solar system dust data.**
S. F. Dermott, P. D. Nicholson, B. Wolven.
Asteroids, comets, meteors II, p. 583 – 594 (1986). – See Abstr. 012.010.

Fourier analysis of the data in the IRAS Zodiacal History File has been used to separate the smooth, large–scale zodiacal background from the narrower dust bands. The geometry of the background zodiacal cloud appears to show features associated with both the forced inclinations and the forced eccentricities of the dust particle orbits. The authors have previously shown that the dust bands may be debris produced by the gradual comminution of the asteroids in the Hirayama asteroid families, in particular the Eos and Themis families, and they have predicted (1) that the ecliptic latitudes of the bands should vary with ecliptic longitude and (2) that the central dust band should be split. Evidence supporting these predictions is presented.

106.024 **Recent work on light scattering by small particles.**
R. Schiffer, K. O. Thielheim.
Asteroids, comets, meteors II, p. 595 – 598 (1986). – See Abstr. 012.010.

Analysing the light scattering properties of irregular surfaces is a rather intricate theoretical problem, and for a very general type of such surface, called "statistically rough", simple results are available only in two limiting cases: for large scale roughness, where the laws of geometrical optics are valid, and for slight roughness, where the height deviations from the smooth surface are much smaller than wavelength. The authors applied the results for both roughness types to the surface of large ($\gg \lambda$) convex particles with random orientation in order to compare the resulting scattering patterns with measurements of the zodiacal light – the object was the construction of a realistic surface model for the interplanetary dust particles.

106.025 **Atomic hydrogen and helium densities of the interstellar medium measured in the vicinity of the Sun.**
E. Chassefière, J. L. Bertaux, R. Lallement, V. G. Kurt.
Astron. Astrophys., Vol. 160, No. 2, p. 229 – 242 (1986).
Measurements of H I (121.6 nm) and He I (58.4 nm) resonance emission of the interplanetary gas have been obtained during the interplanetary cruise of Venera 11 and 12 spacecraft in 1978 – 1979 with an EUV spectrophotometer. Observations performed with the same instruments of H and He in the upper atmosphere of Venus during the flyby yielded the product of the calibration factor of the instrument by the exciting solar flux, allowing to measure with a minimal uncertainty the density of interstellar neutral hydrogen ($n_H = 0.065 \pm 0.010\ cm^{-3}$) and helium ($n_{He} = 0.0010 \pm 0.0045\ cm^{-3}$) far from the sun but inside the heliopause. These results are discussed in terms of local interstellar medium ionization state.

106.026 **A gas dynamic magnetosheath field model for unsteady interplanetary fields: application to the solar wind interaction with Venus.**
J. G. Luhmann, R. J. Warniers, C. T. Russell, J. R. Spreiter, S. S. Stahara.
J. Geophys. Res., Vol. 91, No. A3, p. 3001 – 3010 (1986).
The steady state gas dynamic model of magnetosheath magnetic fields previously developed by Spreiter and Stahara (1980) is generalized for the common situation of a temporally varying interplanetary field orientation. Examples for the particular case of Venus in the solar wind illustrate the application of the model to the passage of rotational discontinuities and MHD waves through planetary magnetosheaths. The results of this model illustrate how the field structure near a planetary magnetopause or ionopause can be affected by interplanetary field variations rather than by local processes because of the "pile up" of magnetosheath fields from a sequence of upstream fields. Changes in the spectrum of interplanetary waves on their transmission to the magnetopause are also indicated.

106.027 **On ion acceleration in interplanetary quasi–perpendicular shock waves.**
M. E. Pesses, R. B. Decker.
J. Geophys. Res., Vol. 91, No. A4, p. 4143 – 4148 (1986).
The kinematic and pitch angle scattering constraints for multiple shock encounters to occur are investigated. The results suggest that (1) large particle anisotropies and a large number of shock encounters are not mutually exclusive and (2) solar wind ions are not directly injected into the nearly perpendicular interplanetary shock acceleration process. Also, the average number of shock encounters required to accelerate an ion between two energies is calculated. The results are consistent with the shock geometry dependence observed in interplanetary shock spike and corotating interaction region associated energetic ion events.

106.028 **The inclination of the heliospheric neutral sheet and cosmic ray intensity at the earth.**
T. Saito, D. B. Swinson.
J. Geophys. Res., Vol. 91, No. A4, p. 4536 – 4542 (1986).
The authors uşe K–corona data to infer the inclination of the neutral sheet from 1971 to 1974 and show that these data are consistent with the IMF observed at the earth during this period. The inclination data are then compared with measurements of cosmic ray intensity at the earth from 1971 to 1975.

106.029 **On the multi–flux corotating structure of the interplanetary plasma.**
V. I. Vlasov.
Geomagn. Aehron., Tom 26, No. 1, p. 133 – 135 (1986). In Russian. English translation in Geomagn. Aeron.

106.030 **Cyclic variations of a connection between the interplanetary magnetic field sector structure, high–velocity streams and geomagnetic disturbances.**
V. N. Obridko, L. Yu. Konstantinova, M. M. Konakh, G. S. Mansurov.
Geomagn. Aehron., Tom 26, No. 2, p. 313 – 316 (1986). In Russian. English translation in Geomagn. Aeron.

106.031 **The effect of the sector structure of the interplanetary magnetic field on the intensity of auroral electrojets.**
A. L. Kotikov, O. A. Troshichev.
Geomagn. Aehron., Tom 26, No. 2, p. 338 – 340 (1986). In Russian. English translation in Geomagn. Aeron.

106.032 **Acid dissolution experiments: carbonates and the 6.8–micrometer bands in interplanetary dust particles.**
S. A. Sandford.
Science, Vol. 231, No. 4745, p. 1540 – 1541 (1986).
A chemical dissolution experiment on an interplanetary dust particle (IDP) showed that carbonates, not acid–insoluble organic compounds, were responsible for virtually all the absorption at 6.8 micrometers seen in the infrared spectra of this particle. The IDP examined had an infrared spectrum characteristic of layer–lattice silicates and belongs to a class of IDP's whose spectra resemble those of protostellar objects like W33 A, which also exhibit a band at 6.8 micrometers.

106.033 **A carbonate–rich, hydrated, interplanetary dust particle: possible residue from protostellar clouds.**
K. Tomeoka, P. R. Buseck.
Science, Vol. 231, No. 4745, p. 1544 – 1546 (1986).
Transmission electron microscopy of a hydrated interplanetary dust particle (IDP) indicates that it contains abundant magnesium–iron carbonates, primarily breunnerite and magnesian siderite. This IDP displays a strong absorption band at 6.8 micrometers in its infrared spectrum, similar to that in certain protostellar spectra. The carbonates probably account for the 6.8–micrometer band in the IDP spectrum, suggesting that carbonate also may occur in interstellar dust and be the source of the controversial 6.8–micrometer feature from the protostellar spectra.

106.034 **The formation and origin of the IRAS zodiacal dust bands as a consequence of single collisions between asteroids.**
M. V. Sykes, R. Greenberg.
Icarus, Vol. 65, No. 1, p. 51 – 69 (1986).
The zodiacal dust bands discovered by IRAS can be explained as products of single collisions between asteroids. Debris from such a collision is distributed about the plane of the ecliptic as particles experience differential precession of their ascending nodes due to dispersion of their semimajor axes. For each collision, two bands, one on each side of the ecliptic, are formed on time scales of 10^5 to 10^6years. The band pairs observed by IRAS are most likely the result of collisions between asteroids ~ 15 km in diameter that occurred within the last several million years.

106.035 **Künstliche Kometen.**
G. Haerendel.
Phys. Bl., 42. Jahrg., Heft 5, p. 134 – 137 (1986).

106.036 **Solar wind controlled pulsations: a review.**
T. J. Odera.
Rev. Geophys., Vol. 24, No. 1, p. 55 – 74 (1986).
Studies of the solar wind controlled Pc 3, 4 pulsations by early and recent researchers are highlighted. The review focuses on the recent observations, which cover the time during the International Magnetospheric Study (IMS). Results from early and recent observations agree on one point, that is, that the Pc 3, 4 pulsations are influenced by three main solar wind parameters, namely, the solar wind velocity, the IMF orientation, and magnitude. The results can be interpreted, preferably, in terms of an external origin for Pc 3, 4 pulsations.

106.037 **Three–dimensional configurations of interplanetary disturbances associated with coronal mass ejections.**
T. Watanabe, T. Kakinuma, M. Kojima.
The sun and the heliosphere in three dimensions, p. 123 – 128 (1986). – See Abstr. 012.016.
Three–dimensional propagation properties of interplanetary consequences of coronal mass ejections are studied on the basis of interplanetary scintillation and spacecraft observations in

1978 – 1981. Bubble–like interplanetary disturbances were associated with some white–light coronal mass ejections.

106.038 **The three–dimensional extent at 1 AU of interplanetary disturbances associated with disappearing solar filaments.**
C. S. Wright.
The sun and the heliosphere in three dimensions, p. 129 – 134 (1986). – See Abstr. 012.016.
This paper describes a technique that can provide additional information about the extent, averaged over many events, of interplanetary disturbances associated with the disappearance of solar filaments.

106.039 **Three–dimensional, time–dependent, MHD model of a solar flare–generated interplanetary shock wave.**
M. Dryer, S. T. Wu, S. M. Han.
The sun and the heliosphere in three dimensions, p. 135 – 140 (1986). – See Abstr. 012.016.
Three–dimensional model of the propagation of an interplanetary shock wave into a representative ambient three–dimensional heliospheric solar wind is demonstrated. The numerical MHD simulation is initialized by assuming a peak shock velocity of 1000 km sec^{-1} at the center of a right circular cone of 18° included angle at 18 solar radii. Examination of the shocked plasma and IMF parameters reveals several fundamental results that differ from the authors' earlier 2D and 2–1/2D simulations under similar input conditions that were confined to the ecliptic plane.

106.040 **Exploration of heliosphere by interplanetary scintillation.**
R. V. Bhonsle, S. K. Alurkar, S. S. Degaonkar, H. O. Vats, A. K. Sharma.
The sun and the heliosphere in three dimensions, p. 153 – 158 (1986). – See Abstr. 012.016.
The Physical Research Laboratory has been engaged in setting up a three station radio observatory operating at a frequency of 103 MHz in Western India for exploration of the heliosphere by the interplanetary scintillation (IPS) technique. IPS observations of different scintillating radio sources are made over a wide range of their solar elongations. Such IPS observations enable the determination of not only the solar wind velocity at different heliolatitudes and other travelling interplanetary disturbances but also the structure of heliosphere up to a distance of about 1.0 AU.

106.041 **3–D coronal and heliospheric structure from radio observations.**
J. L. Bougeret, S. Hoang, J. L. Steinberg.
The sun and the heliosphere in three dimensions, p. 213 – 228 (1986). – See Abstr. 012.016.
The authors review radio observations of the heliospheric 3–D structure from the mid–corona to the Earth orbit, made from the Earth and from space, using thermal emission, tracers in the form of radio bursts or interplanetary scintillation of radio sources.

106.042 **Latitude distribution of interplanetary magnetic field lines rooted in active regions.**
G. A. Dulk, J. L. Steinberg, S. Hoang, A. Lecacheux.
The sun and the heliosphere in three dimensions, p. 229 – 233 (1986). – See Abstr. 012.016.
Electrons, accelerated in solar active regions to $\geqslant 10$ keV, escape along open magnetic field lines and generate plasma radiation (Type III bursts) at frequencies ranging from ≈ 100 MHz near the Sun to ≈ 30 kHz near the Earth. With the radio astronomy receiver on the ISEE–3 spacecraft the authors have measured the positions of 120 Type III bursts in the range 0.03 to 1 AU (2 MHz to 30 kHz). They describe the latitude distribution of the field lines traced out by the Type III electrons.

106.043 **Heliospheric structure and multispacecraft observations of Type III radio bursts.**
J. L. Steinberg, S. Hoang, A. Lecacheux.
The sun and the heliosphere in three dimensions, p. 235 – 240 (1986). – See Abstr. 012.016.
The authors have tried to obtain 3–D positions of Type III radio burst sources from their direction measured from ISEE–3 and the difference in arrival times at ISEE–3 and the Voyager probes, 5 to 8 AU away from the Sun. They analyze one such event which is seen behind the limb on May 1, 1979; and suggest that most of the observed effects are due to scattering on the large scale density structures of the heliosphere.

106.044 **The relationship of the large–scale solar field to the interplanetary magnetic field: what will Ulysses find?**
J. T. Hoeksema.
The sun and the heliosphere in three dimensions, p. 241 – 254 (1986). – See Abstr. 012.016.
The author presents results of a model of the interplanetary magnetic field (IMF) for the present solar cycle and makes a qualitative prediction of the IMF structures Ulysses may encounter.

106.045 **The large–scale structure of the heliospheric magnetic field.**
A. Balogh.
The sun and the heliosphere in three dimensions, p. 255 – 266 (1986). – See Abstr. 012.016.
After a brief review of solar and coronal magnetic fields, the paper summarises the results obtained so far on spatial and temporal dependences of the interplanetary magnetic field. These are compared to Parker's model which proves to be an excellent first order approximation. Mention is made of recent results concerning the radial dependence of the field magnitude and of the potential consequences if the less–than–expected magnetic flux in the solar equatorial region is confirmed. Finally the aims of the magnetic field investigation on Ulysses are described briefly.

106.046 **The heliospheric current sheet: 3–dimensional structure and solar cycle changes.**
E. J. Smith, J. A. Slavin, B. T. Thomas.
The sun and the heliosphere in three dimensions, p. 267 – 274 (1986). – See Abstr. 012.016.
The reversal in polarity of the interplanetary current sheet/sector structure is investigated during the recent solar maximum. Multipoint observations by ISEE–3 and Pioneer 11 are presented. The current sheet appeared to be highly inclined during the ascending and descending phases of the solar cycle but was apparently too complex to describe as a simple inclined current sheet during solar maximum.

106.047 **Three–dimensional structure of the heliosphere as inferred from observations with a Japanese Halley spacecraft.**
T. Saito, K. Yumoto, K. Hirao, I. Aoyama, E. J. Smith.
The sun and the heliosphere in three dimensions, p. 281 – 286 (1986). – See Abstr. 012.016.
The present paper is an initial report of IMF observations with Sakigake, the first of two Japanese Comet Halley spacecraft. Evidence supporting the two–hemisphere model is presented using the Sakigake magnetometer data.

106.048 **Three–dimensional structure of the heliospheric current sheet.**
C. D. Fry, S.-I. Akasofu.
The sun and the heliosphere in three dimensions, p. 287 – 293 (1986). – See Abstr. 012.016.
Three–dimensional structure of the heliospheric current sheet out to 5 AU is constructed for actual observed source field configurations.

106.049 **Acceleration of energetic particles at solar wind shocks.**
M. A. Lee.
The sun and the heliosphere in three dimensions, p. 305 – 318 (1986). – See Abstr. 012.016.
A review of the acceleration of energetic ions at interplanetary shocks is presented with an emphasis on the theory of diffusive shock acceleration and interplanetary traveling shocks.

106.050 **Shock acceleration of nucleons at ⩾16° solar latitude associated with interplanetary corotating interaction regions.**
J. A. Simpson, E. J. Smith, B. Tsurutani.
The sun and the heliosphere in three dimensions, p. 319 – 324 (1986). – See Abstr. 012.016.

The analysis of magnetic field and nuclei measurements in the ~1 to ~10 MeV/nucleon range for protons and helium at ~16°N solar latitude near solar cycle minimum (1975 – 76) shows both: (1) that nucleons are accelerated locally in corotating shocks, and, (2) that the associated recurring corotating interaction regions (CIRs) at ~16° latitude have approximately the same temporal and spatial parameters found for CIRs in the equatorial plane at radial distances <5 AU. The evidence indicates that both the CIR structures and nucleon acceleration by shocks extend to intermediate solar latitudes and are likely to be observed during the Ulysses Mission out of the ecliptic plane.

106.051 **Latitude dependence of co–rotating shock acceleration in the outer heliosphere.**
R. E. Gold, L. J. Lanzerotti, C. G. Maclennan, S. M. Krimigis.
The sun and the heliosphere in three dimensions, p. 325 – 329 (1986). – See Abstr. 012.016.

Energetic particle observations ($E_i \gtrsim 30$ keV) have been made in the outer heliosphere ($\gtrsim$12 A.U.) by the LECP instruments on the Voyager 1 and Voyager 2 spacecraft. These observations show a definite latitude dependence of the number and intensities of particle enhancements produced by co–rotating interplanetary regions during a lengthy interval (approximately 200 days) in 1983 – 84 when no energetic solar particle events were observed. The particle enhancements are significantly fewer in number and tend to be less intense at higher (20° – 23°) heliolatitudes. The energy spectral shapes of the accelerated particles at the two spacecraft are similar, indicating that the acceleration process is the same at the two latitudes, but less intense at the higher latitudes.

106.052 **Neutral interstellar gases in the heliosphere: new aspects of the problem.**
H. J. Fahr.
The sun and the heliosphere in three dimensions, p. 421 – 434 (1986). – See Abstr. 012.016.

For more than a decade, it has been attempted to derive thermodynamical parameters of the local interstellar medium (LISM) from observations of the interplanetary glow patterns in helium and hydrogen resonance lines. However, reviewing the current status of these efforts, one has to admit that, besides perhaps the vector orientation of the interstellar wind and the unperturbed helium density, no other parameters of the LISM could be derived without ambiguity. In the following, the author gives theoretical and observational reasons for the present situation in this field.

106.053 **Interstellar gas parameters and solar wind anisotropies deduced from H and He observations in the solar system.**
J. L. Bertaux, R. Lallement, E. Chassefiere.
The sun and the heliosphere in three dimensions, p. 435 – 440 (1986). – See Abstr. 012.016.

The results of four space experiments looking at resonance scattering of H and He atoms in the solar system are reported. The characteristics of the Local Interstellar Medium are determined to be $T = 8000 \pm 500$K, and the wind velocity is 20 ± 1 km s^{-1}. Larger values are found for helium as a result of an artefact of the modeling. The degree of ionization is about 30% in the LISM. In addition, the solar wind mass flux was measured at all latitudes, showing a 50% decrease when going from the ecliptic to the pole in the 1976 – 1977 period, linked to the presence of a polar coronal hole. Predictions are made concerning some results which will be collected with Ulysses.

106.054 **The 3–dimensional structure of the interplanetary dust cloud.**
R. H. Giese, G. Kinateder.
The sun and the heliosphere in three dimensions, p. 441 – 454 (1986). – See Abstr. 012.016.

The three dimensional structure of the interplanetary dust cloud can be investigated by analysis of particle impacts and of zodiacal light. Optical models are presented and modifications by spatial variations of albedo are discussed. Relations between the spatial density and orbital inclinations of micrometeoroids provide the connection to the Ulysses dust experiment. In future missions optical probing at low elongations and at altitudes up to about 0.5 AU above the ecliptic plane are most promising for investigations of 3D–problems.

106.055 **The interaction of solid particle with the interplanetary medium.**
G. E. Morfill, E. Grün, C. Leinert.
The sun and the heliosphere in three dimensions, p. 455 – 474 (1986). – See Abstr. 012.016.

The structure of the interplanetary medium both electromagnetic and particulate are reviewed. The effects on the small particle population are discussed. Properties of the zodiacal dust cloud are described which may be explained by electromagnetic interactions. Stochastic perturbation theory is derived and conclusions on the large scale distribution of interplanetary particles are drawn.

106.056 **The relationship between the interplanetary magnetic field inhomogeneous structure and the distribution of large–scale magnetic fields in the photosphere (1969 – 1975).**
L. A. Plyusnina.
Sol. Phys., Vol. 102, Nos. 1/2, p. 191 – 201 (1985).

For the period 1969 – 1975, a study has been made of the dependence of the interplanetary magnetic field structure on the distribution and evolutionary properties of solar magnetic fields. By direct comparison of a sequence of synoptic charts of the photospheric magnetic field with the interplanetary magnetic field, and by applying the method of correlation analysis, it is shown that to areas with an unstable polarity of the interplanetary magnetic field there correspond regions with a complicated inverse polarity line that forms either narrow gulfs and islands against a background of the dominant polarity, or bipolar magnetic regions and their clusters. An asymmetry of the correlation between the interplanetary and photospheric magnetic field structures of different hemispheres is found.

106.057 **Superposed epoch analysis of corotating interaction regions at 0.3 and 1.0 AU: a comparative study.**
A. K. Richter, A. H. Luttrell.
J. Geophys. Res., Vol. 91, No. A5, p. 5873 – 5878 (1986).

Applying the superposed epoch analysis technique to 16 and to 31 well–defined, nonshock–associated stream–stream interaction regions observed by the Helios spacecraft in the distance ranges 0.3 to 0.4 AU and 0.9 to 1.0 AU, respectively, the authors obtain the average azimuthal variation in the solar wind density, velocity and temperature, in the magnetic field strength, and in the total proton plasma plus magnetic field pressure across CIRs at these two radial distances separately.

106.058 **Laboratory studies of interplanetary dust.**
R. M. Walker.
NASA Conf. Publ., NASA CP–2403, p. 55 – 69 (1986). – See Abstr. 012.030.

This brief review treats the analysis of dust particles collected in the stratosphere. These particles are the best available samples of interplanetary dust and have been studied using a variety of analytical techniques.

106.059 **Interstellar material in the solar system. A working group report.**
J. A. Wood.
NASA Conf. Publ., NASA CP–2403, p. WG 33 – WG 78 (1986). – See Abstr. 012.030.

106.060 **Anisotropy of particle fluxes accelerated in the interplanetary medium.**
M. F. Bakhareva.
Geomagn. Aehron., Tom 26, No. 3, p. 384 – 390 (1986). In Russian. English translation in Geomagn. Aeron.

106.061 **On turbulent heating of the interplanetary plasma at large heliocentric distances.**
I. V. Chashej.
Geomagn. Aehron., Tom 26, No. 3, p. 486 – 488 (1986). In Russian. English translation in Geomagn. Aeron.

106.062 **Mineralogy of interplanetary dust particles from the "olivine" infrared class.**
R. Christoffersen, P. R. Buseck.
Earth Planet. Sci. Lett., Vol. 78, No. 1, p. 53 – 66 (1986).

106.063 **Investigation of the fine structure of shock waves with the complex of plasma spectrometers BIFRAM.**
G. N. Zastenker, O. L. Vajsberg, Z. Nemeček, I. Šafránková, A. O. Fedorov, Yu. I. Ermolaev, A. V. Lejbov, M. Richter, V. Ganzal, N. L. Borodkova, A. A. Skal'skij, V. N. Smirnov, E. A. Gavrilova, N. V. Plyusnina, A. N. Omel'chenko.
Kosm. Issled., Tom 24, Vyp. 2, p. 151 – 165 (1986). In Russian. English translation in Cosm. Res.

106.064 **Investigation of plasma waves with the complex of combined wave diagnostic BUDVAR (Prognoz 10–Intercosmos).**
S. I. Klimov, M. N. Nozdrachev, P. Triska, Ya. Vojta, A. A. Galeev, Ya. N. Aleksevich, Yu. V. Afanas'ev, V. E. Baskakov, Yu. N. Bobkov, R. B Dunets, A. M. Zhdanov, V. E. Korepanov, S. A. Romanov, S. P. Savin, A. Yu. Sokolov, V. S. Shmelev.
Kosm. Issled., Tom 24, Vyp. 2, p. 177 – 184 (1986). In Russian. English translation in Cosm. Res.

106.065 **Investigation of energetic particles connected with shock waves in the framework of the INTERSHOCK project.**
M. Vandas, K. Kudela, V. N. Lutsenko, Z. Nemeček, M. Slivka, S. Fischer, I. Šafránková.
Kosm. Issled., Tom 24, Vyp. 2, p. 185 – 191 (1986). In Russian. English translation in Cosm. Res.

106.066 **Flare–associated shock wave propagation at large heliocentric distances.**
F.–s. Wei.
Kexue Tongbao, Vol. 30, No. 3, p. 361 – 364 (1985).
It is shown that the space at large heliocentric distances can be considered as the inertia–convection region for flare–associated interplanetary shock wave propagation, in which primary dynamical processes are the inertia–convection following the solar wind medium. It is possible that close relations exist between the three–dimensional propagation characteristics of the flare–associated shock wave and the magnetic configuration in the whole heliosphere from the sun to the interplanetary space.

106.067 **The time evolution effects in Fermi acceleration by shock waves. (II).**
D.–y. Wang, D.–y. Mao.
Kexue Tongbao, Vol. 30, No. 8, p. 1050 – 1055 (1985).
On the basis of two previous papers the present work investigates further the influence of various time distributions of source particles on the energy spectra of accelerated particles in the Fermi acceleration by shock waves.

106.068 **A method simulating random magnetic field in interplanetary space by an autoregressive method.**
M. Kato, T. Sakai.
J. Geomagn. Geoelectr., Vol. 37, No. 11, 1063 – 1070 (1985). Abstr. in Phys. Abstr., Vol. 89, No. 1281, Entry 67402 (1986).

106.069 **Interplanetary dust dynamics. I. Long–term gravitational effects of the inner planets on zodiacal dust.**
B. Å. S. Gustafson, N. Y. Misconi.
Icarus, Vol. 66, No. 2, p. 280 – 287 (1986).
The aim of this first paper in a series of papers on interplanetry dust dynamics is to establish the necessity of including effects of the inner planets in investigations of dust dynamics. In subsequent papers overall effects on the shape and other observable features of an interplanetary dust cloud are discussed.

106.070 **On the problem of the origin of β meteoroids.**
I. Kapišinský.
Contrib. Astron. Obs. Skalnaté Pleso, Vol. 14, p. 183 – 189 (1986).
This paper presents a brief outline of the possible origin of β particles by the co–action of destructive effects and mechanisms. This hypothesis essentially contains the idea of the integration of various non–gravitational effects of a destructive nature on meteoroids of ordinary size, which enables sufficient numbers of them to attain the phase of β meteoroids.

106.071 **IPS observations of transient interplanetary phenomena associated with solar filament activity in late August 1978.**
T. Watanabe, K. Marubashi.
J. Geomagn. Geoelectr., Vol. 37, No. 9, p. 869 – 882 (1985). Abstr. in Phys. Abstr., Vol. 89, No. 1278, Entry 47853 (1986).

106.072 **Ion bombardment of interplanetary dust.**
R. E. Johnson, L. J. Lanzerotti.
Icarus, Vol. 66, No. 3, p. 619 – 624 (1986).
The recent discovery of ion tracks in interplanetary dust and the increasing evidence for carbon and carburized materials in these objects are strongly suggestive that chemical processing by energetic charged–particle bombardment has occurred during the dust lifetimes. The track density gives a measure of the total ion fluence experienced by the grains.

106.073 **Sector structure of the interplanetary magnetic field and the ring current in the magnetosphere.**
A. A. Danilov, G. A. Makarov.
Physical processes in the circumterrestrial plasma, p. 28 – 40 (1984). In Russian. – See Abstr. 003.040.

106.074 **Collisionless shock waves in space.**
M. Scholer.
Proceedings of the 1984 International Conference on Plasma Physics, Vol. 2, p. 1019 – 1038 (1985). Abstr. in Phys. Abstr., Vol. 89, No. 1281, Entry 67306 (1986). – See Abstr. 012.056.

106.075 **The interplanetary magnetic field during solar cycle 21: ISEE–3/ICE observations.**
J. A. Slavin, G. Jungman, E. J. Smith.
Geophys. Res. Lett., Vol. 13, No. 6, p. 513 – 516 (1986).
Magnetic field observations from the JPL vector helium magnetometer on the ISEE–3/ICE spacecraft are used to investigate long term temporal variations in the interplanetary magnetic field during solar cycle 21.

106.076 **A comparison of the Poynting–Robertson drag with the charge drag in the Earth's plasma.**
G. B. Afonso.
Rev. Mex. Astron. Astrofis., Vol. 12, Spec. Issue, p. 444 (1986). Abstract. – See Abstr. 012.066.

106.077 **Radial and latitudinal gradients in the interplanetary magnetic field: evidence for meridional flux transport.**
B. T. Thomas, J. A. Slavin, E. J. Smith.
J. Geophys. Res., Vol. 91, No. A6, p. 6760 – 6764 (1986).
An analysis is presented of the magnetic field data obtained by the Pioneer 10 and 11 spacecraft in the outer heliosphere, with regard to radial and latitudinal gradients in the field components and magnitude.

106.078 **Observations of interplanetary scintillations of radio sources with flat spectrum at a frequency of 102.5 MHz.**
Yu. N. Vetukhnovskaya.
Fiz. inst. Akad. Nauk SSSR. Prepr., No. 213, 7 pp. (1985). In Russian. Abstr. in Ref. Zh., 51. Astron., 4.51.325 (1986).

106.079 **The zodiacal background in the IRAS data.**
M. G. Hauser, J. R. Houck.
Light on dark matter, p. 39 – 44 (1986). – See Abstr. 012.080.
The IRAS sky survey yielded a very extensive picture of bright diffuse thermal emission from interplanetary dust grains at wave-

lengths from 12 to 100 micrometers. These data provide insights into the character, spatial distribution, and perhaps origin of these particles which both confirm and expand upon those gained from previous zodiacal light studies. The zodiacal emission is a major component of the large scale infrared brightness of the sky at all IRAS wavelengths.

106.080 **Heliocentric dependences of zodiacal emission, temperature and albedo.**
R. Dumont, A. C. Levasseur-Regourd.
Light on dark matter, p. 45 – 46 (1986). – See Abstr. 012.080.
Values of the temperature, of the global emissivity, and (by comparison with zodiacal light optical observations) of the albedo of interplanetary dust are retrieved from IRAS data (Hauser et al., 1984) in two localized regions of the line–of–sight, near 0.98 and 1.40 a.u. heliocentric distance.

Collisionless shocks in the heliosphere. A tutorial review.
See Abstr. 003.185.

Collisionless shocks in the heliosphere. Review of current research.
See Abstr. 003.197.

Laboratory measurements of physical properties of submicron particles candidates for cosmic dust.
See Abstr. 022.111.

Radiation effects on grain materials.
See Abstr. 022.133.

Comparative performances of an objective grating concept in UV spectroscopic observations of solar system objects. III. Observations of diffuse emission lines.
See Abstr. 036.025.

The Ulysses mission.
See Abstr. 051.043.

On the use of a sunward libration–point–orbiting spacecraft as an interplanetary magnetic field monitor for magnetospheric studies.
See Abstr. 051.066.

Preparation, analysis, and release of simulated interplanetary grains into low earth orbit.
See Abstr. 051.067.

Optical properties of simulated astrophysical grains and their dynamics in the near–Earth environment.
See Abstr. 051.104.

Fast and optimal solution to the "Rankine–Hugoniot problem".
See Abstr. 062.010.

Solar cycle variation of heliospheric quantities.
See Abstr. 072.089.

The sudden decrease of the large flare occurrence and its manifestation in interplanetary space after the maximum of the solar cycle.
See Abstr. 073.024.

Radioastronomical observations of an interplanetary shock wave from the solar flare of August 18, 1979.
See Abstr. 073.044.

Reflection of the structure of the chromosphere and corona in the solar wind and interplanetary magnetic field.
See Abstr. 073.087.

Radial variation of interplanetary Alfvénic fluctuations: Pioneer 10 and 11 observations between 1 and 5 AU.
See Abstr. 074.006.

Development of a real–time algorithm for detection of solar wind discontinuities.
See Abstr. 074.012.

The heliospheric energy source.
See Abstr. 074.034.

HELIOS images of coronal mass ejections.
See Abstr. 074.040.

Relationship of coronal transients to interplanetary shocks: 3D aspects.
See Abstr. 074.041.

Interplanetary scintillation observations of the solar wind at high latitudes.
See Abstr. 074.042.

Evolution of turbulence and waves in the solar wind in radius and latitude.
See Abstr. 074.043.

Solar wind speed azimuthal variation along the heliospheric current sheet.
See Abstr. 074.049.

Plasma transport across the heliopause.
See Abstr. 074.073.

The mechanism for accelerating and heating the high speed solar wind between 0.3 AU and 1 AU.
See Abstr. 074.096.

Relationship of coronal transients to interplanetary shocks: 3D aspects.
See Abstr. 074.101.

On periodic components of the photospheric and interplanetary magnetic fields.
See Abstr. 075.014.

Interplanetary energetic particles, coronal flares, and hard X–ray microflares.
See Abstr. 076.013.

Interplanetary phenomena and solar radio bursts.
See Abstr. 077.052.

Three–dimensional gradients of solar particles inside 5 AU.
See Abstr. 078.007.

A spatially confined, long–lived stream of solar particles.
See Abstr. 078.008.

A sensitive radar search for small natural satellites of the earth.
See Abstr. 081.038.

A theoretical and empirical study of the response of the high latitude thermosphere to the sense of the "*Y*" component of the interplanetary magnetic field.
See Abstr. 082.003.

Earth's equatorial ionosphere and the heliospheric current sheet.
See Abstr. 083.012.

Relationship of the interplanetary electric field to the high–latitude ionospheric electric field and currents: observations and model simulation.
See Abstr. 083.033.

Geomagnetic disturbances and their connection with the parameters of the interplanetary magnetic field.
See Abstr. 084.028.

The relation of magnetospheric disturbances with the parameters of the interplanetary medium.
See Abstr. 084.035.

Modelling the geomagnetic effect of the ring current based on the hourly mean parameters of the interplanetary medium.
See Abstr. 084.037.

Ion release experiments in the solar wind.
See Abstr. 084.043.

Hook–shaped arcs in dayside polar cap and their relation to the IMF.
See Abstr. 084.065.

Micrometeoroid impact on planetary satellites as a magnetospheric mass source.
See Abstr. 091.044.

Cometary particles: thin sectioning and electron beam analysis.
See Abstr. 102.040.

Ionization of cometary atmospheres by the zodiacal dust cloud.
See Abstr. 102.047.

The origin of the sporadic meteoroid component.
See Abstr. 104.004.

Concerning the problem of collisional heating of the interstellar helium flow by solar wind protons.
See Abstr. 131.118.

Interrelationships between interstellar and interplanetary grains. A working group report.
See Abstr. 131.229.

What predictions can be made on the nature of carbon and carbon–bearing compounds (hydrocarbons) in the interstellar medium based on studies of interplanetary dust particles?
See Abstr. 131.236.

Numerical modeling of the 22–year variation of the cosmic ray intensity and anisotropy.
See Abstr. 144.006.

Two–dimensional numerical modeling of the cosmic ray storm.
See Abstr. 144.007.

Shock acceleration as explanation for steplike changes and hysteresis in cosmic ray modulation.
See Abstr. 144.008.

A theory of cosmic ray anisotropies of solar origin.
See Abstr. 144.009.

Interplanetary cosmic–ray radial and latitudinal gradients derived in 1984 using *IMP 8*, *Voyager*, and *Pioneer* data.
See Abstr. 144.028.

Differential measurement and model calculations of cosmic ray latitudinal gradient with respect to the heliosphere current sheet.
See Abstr. 144.032.

Variation of cosmic rays and solar wind properties with respect to the heliospheric current sheet. 2. Rigidity dependence of the latitudinal gradient of cosmic rays at 1 AU.
See Abstr. 144.033.

Cosmic rays near the heliospheric current sheet. 2. An ensemble approach to comparing theory and observation.
See Abstr. 144.034.

The effect of the heliospheric current sheet on cosmic ray intensities at solar maximum: two alternative hypotheses.
See Abstr. 144.035.

Cosmic ray variations and turbulent flow systems: 0.3 – 1.0 AU; 1977 – 1980.
See Abstr. 144.039.

Cosmic ray density gradients related to north–south asymmetry in activity on the sun.
See Abstr. 144.042.

Effects of three–dimensional heliospheric structures on cosmic–ray modulation.
See Abstr. 144.050.

The effect of the rotation of the sun in cosmic ray anisotropy during sporadic and recurrent electromagnetic perturbations in the interplanetary space.
See Abstr. 144.057.

IMF–sense–dependent cosmic ray anisotropy produced from diffusion–convection in heliosphere.
See Abstr. 144.058.

Dependence of cosmic ray solar daily variation (1st, 2nd, and 3rd) on heliomagnetic polarity reversals.
See Abstr. 144.059.

The effect of gradient and curvature drift on cosmic–ray Forbush decreases.
See Abstr. 144.064.

The role of the energy spectrum anisotropy in the variations of the cosmic–ray fluctuations power–spectrum before interplanetary medium disturbances.
See Abstr. 144.071.

On the law of variation of cosmic ray energy in the interplanetary space.
See Abstr. 144.100.

107 Cosmogony

107.001 **The effects of self-gravity on the solar nebula.**
C. Yuan, C.-k. Chou, T.-j. Lee.
Astrophys. Space Sci., Vol. 118, Nos. 1/2, p. 515 – 521 (1986). – See Abstr. 012.001.

The authors study the distribution and transport of angular momentum in a self-gravitating accretion disk formed during the collapse of a rotating gas cloud. Using the surface density for the low-viscosity models and minimum-mass models presented by Cassen and Summers, Poisson's equation is solved explicitly to determine the effects of self-gravitation of the protostellar disk. Analytic expressions for the angular momentum of the central star and other relevant quantities of interest during the formation stage are presented.

107.002 **Formation of the solar system.**
J. J. Rawal.
Astrophys. Space Sci., Vol. 119, No. 1, p. 159 – 166 (1986). – See Abstr. 012.001.

Prentice (1978), in his modern Laplacian theory of the origin of the solar system, has established a scenario in which he finds the ratio of the orbital radii of successively disposed gaseous rings to be a constant $\cong 1.69$. In an attempt to understand this law in an alternative way, Rawal (1984) assumes that during the collapse of the solar nebula the halts at various radii are brought about by the supersonic turbulent convection and arrives at the relation of the form $R_p = R_\odot a^p$, where $R_\odot$ is the radius of the present Sun and $a = 1.422$, is referred to, here, as the "Roche" constant. Kepler's third law assumes the form: $T_p = T_0(a^{3/2})^p$, T_0 being the rotational period of the Sun at the time it attained its present radius. R_p satisfy Laplace's resonance relation without any exception. The present paper investigates inter-relations among the concepts of supersonic turbulent convection, rotational instability, and Roche limit.

107.003 **The meteoritic chromium isotopic composition and limits for radioactive ^{53}Mn in the early solar system.**
T. Lee, F. Tera.
Geochim. Cosmochim. Acta, Vol. 50, No. 2, p. 199 – 206 (1986).

In this paper the authors report the result of a search for Cr isotopic anomalies in early solar system samples. The purpose of the study was to see if the radio-nuclide ^{53}Mn ($t_{1/2}$ = 3.7 million years) was present in the early solar system and whether there exist ^{54}Cr variations due to the incomplete mixing of nucleosynthetic components.

107.004 **Formation of the sun and its planets.**
W. M. Kaula.
Physics of the Sun, Vol. III, p. 1 – 32 (1986). – See Abstr. 003.001.

Contents: Star formation (galactic structure, stellar properties, planetary indications, interstellar clouds, observations of forming stars (T-Tauri variable stars, nonemission, nonvariable pre-main sequence stars, Herbig-Haro objects), conditions for cloud collapse, models for star formation, modeling of secondary features, dynamical evolution). Cosmochemistry (chronology, nuclide variations, chemical variations among chondritic meteorites, chemical variations among differentiated objects). Planet formation (*gd* gas-dust interaction, *CD* disk dynamics, *Cpp* planetesimal swarms, *CPD* planet-disk interactions, *CPP, CPPp* planetary systems, *CP(f,s)p* terrestrial planet formation, *P, Pg* gaseous protoplanet contraction, *WD, XD, BD* solar and external effects on the nebula). Implications for the formation of the sun and planets.

107.005 **Exogenic phenomena and the solar system.**
S. J. Poborski.
Astron. Rep., Vol. 7, p. 5 – 23 (1982).

107.006 **Late stages of planetary accretion.**
G. P. Horedt.
Icarus, Vol. 64, No. 3, p. 448 – 470 (1985).

The author derives first-order differential equations for the late stages of planetary accretion (planetesimal mass $>10^{13}$g). The effect of gravitational encounters, energy exchange, collisions, and gas drag has been included. Two simple models are discussed, namely (1) when all planetesimals have the same mass and (2) when there is one large planetesimal and numerous small planetesimals. Gravitational two-body encounters are modeled according to Chandrasekhar's classical theory from stellar dynamics. Characteristic time scales for the formation of the terrestrial planets are found for the most favorable models to be of order 10^8year.

107.007 **A calculation of the Rosseland mean opacity of dust grains in primordial solar system nebulae.**
J. B. Pollack, C. P. McKay, B. M. Christofferson.
Icarus, Vol. 64, No. 3, p. 471 – 492 (1985).

The Rosseland mean opacity was determined for an ensemble of dust grain species thought to have been present in the early solar nebula, as well as in the primordial nebulae that helped to form the outer planets. In performing these calculations, the authors have derived and used a more general equation for the Rosseland opacity that allows for anisotropic scattering. Calculations were performed for a very wide range of particle size distributions, including a nominal one based on that of interstellar dust grains. Results are presented for nebular temperatures varying from 10 to 2500K and for nebular gas densities varying from 10^{-14} to 1 g/cm^3.

107.008 **Further considerations on contracting solar nebula.**
J. J. Rawal.
Earth, Moon, Planets, Vol. 34, No. 1, p. 93 – 100 (1986).

It is shown that Kepler's third law of planetary system assumes the form: $T_p = T_0(a^{3/2})^p$, where T_p are the orbital periods at the radii R_p (radii of the solar nebula at various halts during the collapse), $T_0 \cong 0.1216$ d $\cong$ 3 h, and a the Roche constant. The author is inclined to interpret "T_0" to be the rotation period of the Sun at the time of its formation when it attained the present radius. It is also shown that the orbital periods T_p corresponding to the radii R_p submit themselves to the Laplace's resonance relation.

107.009 **The structure of Phaethon and detonation of its icy envelope.**
E. M. Drobyshevski (*Eh. M. Drobyshevskij*).
Earth, Moon, Planets, Vol. 34, No. 3, p. 213 – 222 (1986).

It is suggested that Phaethon – a hypothetical planet whose breakup gave rise to the asteroid belt – has a structure similar to that of Callisto, and thus consisted of a rocky core (40% in mass) and an ice envelope (60%). Total breakup of the planet becomes possible in an explosion of the electrolysis products accumulated in the ice in the form of a solid solution if the planet mass $M \lesssim 0.5\ M_{☾}$. Assuming $M = 0.5\ M_{☾}$ the author obtains 1750 km for the planet's radius with the envelope's thickness of 750 km.

107.010 **Formation of a "magma ocean" on the terrestrial planets due to the blanketing effect of an impact-induced atmosphere.**
T. Matsui, Y. Abe.
Earth, Moon, Planets, Vol. 34, No. 3, p. 223 – 230 (1986).

The authors show that the surface of a planet growing by planetesimal impact is heated over the melting temperature of the surface materials due to the blanketing effect of an impact induced H_2O atmosphere with the present H_2O abundance of the Earth even when the accretion time is as long as 10^8years. Hence a "magma ocean" covering the entire surface was formed on the Earth and Moon and other terrestrial planets during their formations.

107.011 **Dynamical and chemical evolution of the protoplanetary nebula.**
G. E. Morfill, W. Tscharnuter, H. J. Völk.
Protostars & planets II, p. 493 – 533 (1985). – See Abstr. 012.008.
The authors review the currently available information on the dynamical collapse of rotating protostellar clouds, the formation and properties of viscous accretion disks obtained from such numerical experiments and the resulting constraints and consequences for the protoplanetary environment. Using this information the authors develop a transport theory for dust and gas phases and present results obtained from this theory. Possible modifications by intermittent turbulence are discussed, chemical fractionation effects are analyzed and the heterogeneity on small scales as well as the homogeneity on large scales of the primitive bodies in our solar system is examined within the framework of this theory.

107.012 **Meteoritic constraints on processes in the solar nebula.**
J. A. Wood.
Protostars & planets II, p. 687 – 702 (1985). – See Abstr. 012.008.
Chondritic meteorites contain abundant components – chondrules, Ca, Al–rich inclusions – that clearly were formed as dispersed objects in the solar nebula by high–temperature processes. There is no consensus as to the nature of these processes. In spite of this, several broad conclusions can be drawn about chondrite–forming processes in the nebula.

107.013 **Short–lived nuclei in the early solar system.**
G. J. Wasserburg.
Protostars & planets II, p. 703 – 737 (1985). – See Abstr. 012.008.
The status of the ongoing search for presently extinct or exceedingly rare radioactive nuclei in the early solar system is reviewed.

107.014 **Formation of the giant planets and their satellite–ring systems: an overview.**
J. B. Pollack.
Protostars & planets II, p. 791 – 831 (1985). – See Abstr. 012.008.
This paper presents an overview of the nature of the outer giant planets and their satellite and ring systems, theories of their origins, and models of the planets' evolution.

107.015 **Observational constraints on models for giant planet formation.**
D. Gautier, T. Owen.
Protostars & planets II, p. 832 – 846 (1985). – See Abstr. 012.008.
The authors review current information about element abundances and isotope ratios in the atmospheres of Jupiter, Saturn, Uranus, and Neptune. The observed enhancement of C/H compared with the solar value favors models for the origin of these bodies that invoke the accretion and degassing of an ice–rock core followed by the accumulation of a solar composition envelope. Titan may represent an example of a core–forming planetesimal. Observations of D/H and other isotope ratios must be accommodated by these models in ways that are not yet completely clear. The authors suggest some additional tests.

107.016 **Evolution of the giant planets.**
P. Bodenheimer.
Protostars & planets II, p. 873 – 894 (1985). – See Abstr. 012.008.
The theory of the evolution of the giant planets is discussed with emphasis on detailed numerical calculations in the spherical approximation. Initial conditions are taken to be those provided by the two main hypotheses for the origin of the giant planets. Some observational constraints are described, and some problems with the two principal hypotheses are discussed.

107.017 **Rings and moons: clues to understanding the solar nebula.**
J. J. Lissauer, J. N. Cuzzi.
Protostars & planets II, p. 920 – 956 (1985). – See Abstr. 012.008.
Planetary satellites, ranging in size from ring particles to the Galilean moons of Jupiter, constitute a small but diverse component of the solar system. In some respects, the satellite systems of the giant planets can be thought of as miniature solar systems. The chemical and physical properties of these systems can yield clues to the conditions under which they may have formed, and thus can constrain models of the circumplanetary nebulae and the solar nebula. Planetary rings contain a great deal of structure, most strikingly displayed in Voyager spacecraft images. Understanding the causes of this structure can yield important insights into the dynamics of the protoplanetary disk. Taken together, rings and moons offer clues as to the nature of the processes which led to the formation of the planets themselves.

107.018 **Formation of the solar nebula and the planets.**
V. S. Safronov, T. V. Ruzmaikina
(*T. V. Ruzmajkina*).
Protostars & planets II, p. 959 – 980 (1985). – See Abstr. 012.008.
The authors turn their attention to questions which are important for the correct choice of the main mechanisms responsible for the formation of the solar system.

107.019 **On the dynamical origin of the solar system.**
D. N. C. Lin, J. Papaloizou.
Protostars & planets II, p. 981 – 1072 (1985). – See Abstr. 012.008.
In general, the basic physical properties of the primordial solar nebula are very similar to those of an accretion disk. The evolution of the nebula can be characterized by three stages: infall, diffusive, and clearing stages. During each stage, the structure, stability and evolution of the nebula are primarily determined by the efficiency of the angular momentum and heat transport processes. The dominant transport mechanism may be turbulence and convection. In this paper, the authors outline a quantitative model with which the distribution of temperature, density, pressure, thickness, and mass transfer rate of the nebula can be determined at any given epoch. Observational data, such as the spatial and mass distribution of the planets, as well as meteorite data are utilized to provide constraints on these models. The most favorable model is one in which the protogiant planets as well as the prototerrestrial planets formed through several stages of coagulation, accumulation, and accretion.

107.020 **Formation and evolution of the primitive solar nebula.**
A. G. W. Cameron.
Protostars & planets II, p. 1073 – 1099 (1985). – See Abstr. 012.008.
A disturbance mechanism is suggested to initiate the development of cores in dense molecular clouds and to bring into them the shorter–lived extinct radioactivities that were present in the early solar system. This mechanism involves a red–giant star of about 1 $M_\odot$ in its asymptotic giant branch which sheds its envelope and forms a planetary nebula. The collapse of a molecular cloud core to form the primitive solar nebula is discussed. Steady mass flow models of the primitive solar nebula are constructed having a range of masses and dissipation rates. With these as a guide, a history of the solar nebula is proposed into which a variety of phenomena are fitted.

107.021 **Formation of the solar system.**
C. Hayashi, K. Nakazawa, Y. Nakagawa.
Protostars & planets II, p. 1100 – 1153 (1985). – See Abstr. 012.008.
The authors review the over–all evolution of the solar system. Starting from a disk–like solar nebula with small mass composed of H_2 and He gases and dust grains, the authors persue multistep evolutionary processes through which the present planets, satellites, and asteroids have been formed. It is shown that interactions between solid bodies (i.e., small grains as well as large protoplanets) and the gas of the solar nebula are important in determining the various stages of the evolution. The basic philosophy of this study and remaining problems are briefly described.

107.022 **On the origin of small bodies of the solar system.**
B. A. Vorontsov–Vel'yaminov.
Astron. Zh., Tom 63, Vyp. 1, p. 181 – 183 (1986). In Russian. English translation in Sov. Astron., Vol. 30, No. 1.
Only an explosion of a planet some 3×10^9 years ago can produce the simultaneous formation of asteroids, comets and

crater–forming missiles. A planet Asteron between Mars and Jupiter must have been rather gaseous and of structure intermediate between the terrestrial and giant planets. Its extra mass dispersed into the solar system. The explosion might happen in the wrong course of evolution as it happens in stars and space rockets.

107.023 **Preplanetary nebula's solid matter formation and composition of chondrites.**
M. N. Izakov.
Astron. Vestn., Tom 20, No. 1, p. 35 – 49 (1986). In Russian. English translation in Sol. Syst. Res.
On the basis of the model of the preplanetary nebula formation as the accretion disk during the sun's formation the hypothesis was suggested that a considerable part of preplanetary nebula's solid matter was formed by means of successive condensation of components of the solar composition gas during its motion from a hot and dense region near the protosun to the nebula's periphery in the regions with less and less values of temperature and pressure. The hypothesis removes the contradiction between the presence of traces of high–temperature phenomena in chondritic matter and explains several properties of chondrites. It follows from the validity of this hypothesis that the nebula's mass and angular momentum were close to their minimal possible values and that the loss of the nebula gas began at the final stage of its formation.

107.024 **Isotopic anomalies in iron peak and the scenarios of the early solar system.**
C. J. Allègre, J. L. Birck.
Meteoritics, Vol. 20, No. 4, p. 601 (1985). Abstract. – See Abstr. 010.661.

107.025 **Chemical processing of the Earth's early atmosphere by large impacts.**
B. Fegley, H. Hartman, R. G. Prinn.
Meteoritics, Vol. 20, No. 4, p. 644 – 645 (1985). Abstract. – See Abstr. 010.661.

107.026 **Oxygen isotopes and pre–solar dust in the early solar system.**
G. R. Huss.
Meteoritics, Vol. 20, No. 4, p. 667 – 668 (1985). Abstract. – See Abstr. 010.661.

107.027 **The formation of planetary metal cores: the importance of metal silicate interaction.**
W. Klöck, H. Palme.
Meteoritics, Vol. 20, No. 4, p. 685 (1985). Abstract. – See Abstr. 010.661.

107.028 **"Solar flares" and chondrule formation in the primeval solar nebula.**
C. P. Sonett.
Meteoritics, Vol. 20, No. 4, p. 761 (1985). Abstract. – See Abstr. 010.661.

107.029 **Formation of chondrules by drag heating in the solar nebula: the latest word.**
J. A. Wood.
Meteoritics, Vol. 20, No. 4, p. 787 – 788 (1985). Abstract. – See Abstr. 010.661.

107.030 **On the capture origin of the satellites of the solar system.**
R. Dvorak.
Asteroids, comets, meteors II, p. 165 – 174 (1986). – See Abstr. 012.010.
The author concentrates on the purely gravitational capture problem. The question is: how can a small body of the solar system like a comet or an asteroid orbiting the Sun become a stable satellite of a planet? The work of the last twenty years on this topic is discussed critically.

107.031 **Upper limits to the masses of objects in the solar comet cloud.**
J. G. Hills.
Asteroids, comets, meteors II, p. 211 – 214 (1986). – See Abstr. 012.010.
The lack of a large steady stream of long–period comets with semimajor axes less than $2x10^4$AU rules out the Sun having a companion more massive than about 0.01 $M_\odot$ with a semimajor axis less than about $1x10^4$AU. Any companion with a semimajor axis between $1x10^4$AU and $5x10^4$AU has more than a 50% probability of having entered the planetary system during the lifetime of the Solar System. The lack of apparent damage to the planetary system rules out any companion more massive than about 0.02 $M_\odot$ with a semimajor axis less than about $5x10^4$AU.

107.032 **Has Nemesis' orbit been detected?**
A. H. Delsemme.
Asteroids, comets, meteors II, p. 215 – 218 (1986). – See Abstr. 012.010.
The orbital angular momenta of the system of young comets show a large anisotropy in a plane almost perpendicular to the ecliptic. Such an anisotropy would dissipate by orbital diffusion in a characteristic time of 10 to 30 million years. Therefore it cannot come from primordial or galactic effects and must be due to a recent impulsive event. The impulses due to fast–moving stars or molecular clouds have symmetry properties that do not introduce any anisotropy in the cometary orbits. Only a very slow massive body, with a velocity of the same order as that of the perturbed comets, can introduce a large anisotropy. This implies that it is likely to be bound to the solar system. Identification with the Nemesis proposed to explain the periodicity of the mass extinction of living species does not meet any inconsistency.

107.033 **Origin and formation environment of cometary nuclei.**
T. Yamamoto.
Asteroids, comets, meteors II, p. 219 (1986). Abstract. – See Abstr. 012.010.

107.034 **Particle bombardment of minor bodies during eruptive phases of the early solar evolution.**
V. Pirronello, C. D'Arrigo.
Asteroids, comets, meteors II, p. 235 – 239 (1986). – See Abstr. 012.010.
Some effects of eruptive phases in the early evolution of the Sun on icy solids (micron sized grains and cometary nuclei) present in the surrounding nebula are considered. Estimates and results are given.

107.035 **The role of kinetics of reactions of the CO and N_2 reduction in the chemical evolution of the protoplanetary nebula.**
R. A. Mendybaev, A. B. Makalkin, V. A. Dorofeeva, I. L. Khodakovskij, A. K. Lavrukhina.
Geokhimiya, No. 8, p. 1206 – 1217 (1985). In Russian. Abstr. in Ref. Zh., 51. Astron., 1.51.166 (1986).

107.036 **Encounters between the solar system and giant molecular clouds.**
M. J. Duncan.
Bull. Am. Astron. Soc., Vol. 17, No. 4, p. 838 (1985). Abstract. – See Abstr. 010.062.

107.037 **A new convective accretion disk model of the primordial solar nebula.**
W. Cabot, V. M. Canuto, O. Hubickyj, J. B. Pollack.
Bull. Am. Astron. Soc., Vol. 17, No. 4, p. 873 (1985). Abstract. – See Abstr. 010.062.

107.038 **Mechanism of the early differentiation of the earth.**
M. V. Gerasimov, L. M. Mukhin, Yu. P. Dikov, V. I. Rekharskij.
Vestn. Akad. Nauk SSSR., No. 9, p. 10 – 25 (1985). In Russian. Abstr. in Ref. Zh., 51. Astron., 3.51.240 (1986).

107.039 **Numerical realisation of a three–dimensional convective model of evolution of the relief of planets.**
A. V. Kobozev.
VUDal'nevost. nauchn. tsentra Akad. Nauk SSSR. Khabarovsk, 19 pp. (1985). In Russian. Abstr. in Ref. Zh., 51. Astron., 3.51.242 (1986).

107.040 **The supernova fragmentation model of solar system formation.**
W. K. Brown, L. A. Gritzo.
Astrophys. Space Sci., Vol. 123, No. 1, p. 161 – 181 (1986).
The authors have further developed Brown's model of solar system formation. In this model, each fragment of an ejected supernova shell evolves into a separate solar system. Specifically, the authors have formulated the reverse–flow hypothesis that may be responsible for the inner, earthlike planets. They have written a computer program with which it is possible to calculate mass distributions within a solar nebula. They have found mass distributions similar to our solar system over a wide range of the model parameters.

107.041 **Drag on a gravitating sphere moving through a gas.**
H. Takeda, T. Matsuda, K. Sawada, C. Hayashi.
Prog. Theor. Phys., Vol. 74, No. 2, p. 272 – 287 (1985). Abstr. in Phys. Abstr., Vol. 89, No. 1274, Entry 23769 (1986).

107.042 **A numerical simulation of the formation of the terrestrial planets.**
M. Lecar, S. J. Aarseth.
Astrophys. J., Vol. 305, No. 1, p. 564 – 579 (1986).
The authors have performed a numerical simulation of the accumulation of the terrestrial planets. They started with 200 lunar–size planetesimals distributed uniformly in a plane, in circular orbits around the Sun between 0.5 and 1.5 AU. It was assumed that when two bodies physically collided, they coalesced into one body with the same density. Initially, the eccentricities of the bodies were zero, but they were gradually built up by two–body gravitational encounters. After 50,000 yr, six bodies remained, the largest of which had $\sim 2/3\ M_{\delta}$. This simulation was confined to a plane. In three dimensions the time scale would increase to about 10^7yr.

107.043 **Fractionation of noble gases by thermal escape from accreting planetesimals.**
T. M. Donahue.
Icarus, Vol. 66, No. 2, p. 195 – 210 (1986).
A model for the selective loss of noble gases by thermal escape of the gases from planetesimals as they grow to form the terrestrial planets has been developed.

107.044 **Stability and evolution of primeval lunar satellite orbits.**
B. A. Conway.
Icarus, Vol. 66, No. 2, p. 324 – 329 (1986).
There are several independent sources of evidence which suggest that the multiring basins of the lunar surface were created by the impact of natural satellites of the moon, early in solar system history. If this hypothesis is correct the orbits of these primeval satellites would need to be stable for significant periods, to account for the known age differences of these basins. The stability of these primeval satellite orbits is considered. The author finds constraints on the satellite masses and initial orbits for long–term and short–term orbit stability. Dissipation due to lunar tidal friction may contribute significantly to the stability of close orbits.

107.045 **Protoearth mass shedding and the origin of the Moon.**
A. P. Boss.
Icarus, Vol. 66, No. 2, p. 330 – 340 (1986).
This paper has two purposes. The first purpose is to point out an even more restrictive argument against protoearth fission, which greatly lowers the limit on how inviscid the protoearth must have been for fission to have taken place. The second purpose is then to study the hydrodynamics of mass shedding from a rapidly rotating protoearth, in order to learn how much mass and angular momentum is ejected into protoearth orbit for a given instantaneous spin up of the protoearth. Mass shedding into the protoearth's equatorial plane provides another mechanism for supplying matter to a prelunar disk.

107.046 **Simulation of catastrophic fragmentation events: a review of the accelerating techniques and a resumé of the experiments performed until now.**
P. Cerroni.
Mem. Soc. Astron. Ital., Vol. 57, No. 1, p. 13 – 45 (1986). – See Abstr. 012.061.
Collisional processes are known to have played a fundamental role in the formation and the evolution of the Solar System. During the early accretionary stages direct collisions between planetesimals were crucial to determine their mass–distribution spectrum; collisional processes are also fundamental to the understanding of the collisional evolution of asteroids, the disruption of small satellites, the formation and evolution of planetary rings, and so on. The experimental approach is one of the most powerful tools to try to understand the possible outcomes of a catastrophic collisional event. The author deals with experiments where the projectile impacted against a finite–sized body, i.e. against a body of dimensions such that the reflections of the originary shock wave at the target free surface, and the interactions of these reflected waves, cannot be neglected but are determinant to the final outcome of the collisional event.

107.047 **Catastrophic disruption in the solar system: asteroid collisional history, origin of Hirayama families and disruption of small satellites.**
D. R. Davis.
Mem. Soc. Astron. Ital., Vol. 57, No. 1, p. 87 – 102 (1986). – See Abstr. 012.061.
The process of collisional catastrophic disruption has played a significantly role in structuring the solar system. Diverse populations of bodies such as the asteroid belt, small satellites of Jupiter and Saturn and perhaps even the rings of Saturn have been created or substantially changed by catastrophic disruption.

107.048 **Kosmogonie und Erdentstehung.**
K. Schmidt, H.–J. Treder.
Physik des Planeten Erde, p. 15 – 34 (1985). – See Abstr. 003.047.

107.049 **The formation of planetary systems – a simulation on the microcomputer.**
O. Kanamitsu, M. Hirokawa.
Bull. Fukuoka Univ. Educ., Vol. 35, Part III, p. 113 – 124 (1985).
Recent theories of the formation of planetary systems describe that protoplanets are formed by growth of planetesimals in a protonebula through mass aggregation mechanism. A test of this theory is carried out by a simulation on the microcomputer. Generated planetary systems possess principal features of the solar system as mass distribution of planets, spacing ratio of planets and distribution of Jupiter–type and Earth–type planets. Several discussions of the parameters are given compared with the solar system.

Die Kosmogonie Anton von Zachs.
See Abstr. 004.105.

Summary report on the International Workshop "Catastrophic disruption of asteroids and satellites" (Pisa, Italy, July 30 – August 2, 1985).
See Abstr. 011.034.

^{53}Mn in the early solar system.
See Abstr. 022.033.

The origin of the Moon and the single–impact hypothesis I.
See Abstr. 022.127.

Results obtained by laboratory simulations of catastrophic impact.
See Abstr. 022.128.

Monopole abundance in the Solar System and the intrinsic heat in the Jovian planets.
See Abstr. 061.074.

Accretion discs.
See Abstr. 062.133.

Protostellar angular momentum transport by spiral density waves.
See Abstr. 065.020.

Evolution of an impact–induced atmosphere and magma ocean on the accreting Earth.
See Abstr. 081.001.

Chemical effects of large impacts on the Earth's primitive atmosphere.
See Abstr. 081.002.

The structure of the terrestrial planets and tidal friction theory.
See Abstr. 081.032.

Structure of the primordial atmosphere surrounding the early–Earth.
See Abstr. 082.069.

Preferential perihelion and aphelion distances, and planetary formation.
See Abstr. 091.008.

What have we learned from modeling giant planet interiors?
See Abstr. 091.010.

Deuterium in the outer Solar System: evidence for two distinct reservoirs.
See Abstr. 091.014.

Structure and evolution of the terrestrial planets.
See Abstr. 091.030.

Tidal contribution of planets to removing angular momentum from the Sun.
See Abstr. 091.041.

Die chemische Zusammensetzung und Bildung der terrestrischen Planeten.
See Abstr. 091.055.

The origin of the moon.
See Abstr. 094.003.

Banded iron formations.
See Abstr. 094.022.

Lunar nodal tide and distance to the Moon during the Precambrian.
See Abstr. 094.023.

Reflectance spectrophotometry ($\sim 0.5 - 1.0$ μm) of outer–belt asteroids: implications for primitive, organic Solar System material.
See Abstr. 098.003.

Collisional evolution of asteroids constrained by their present rotational properties.
See Abstr. 098.039.

Asteroid strokes against the earth and their possible planetary geophysical consequences.
See Abstr. 098.051.

Scaling laws for the catastrophic collisions of asteroids.
See Abstr. 098.123.

Implications of the inferred compositions of asteroids for their collisional evolution.
See Abstr. 098.124.

Voyager Saturnian ring measurements and the early history of the solar system.
See Abstr. 100.006.

The composition of Saturn's atmosphere.
See Abstr. 100.031.

On the origin of Triton.
See Abstr. 101.012.

Formation of the Uranus system: pre–Voyager 2 predictions.
See Abstr. 101.054.

On the satellite system of Uranus.
See Abstr. 101.056.

The origin of comets: implications for planetary formation.
See Abstr. 102.008.

Isotopic variations in solar system material: evaporation and condensation of silicates.
See Abstr. 105.009.

Meteoritic evidence concerning conditions in the solar nebula.
See Abstr. 105.010.

Chondrites and the solar nebula: a review.
See Abstr. 105.016.

Ni–Pt–Ge–rich Fremdlinge: indicators of a turbulent early solar nebula.
See Abstr. 105.018.

The meteoritic chromium isotopic composition and limits for radioactive Mn–53 in the early solar system.
See Abstr. 105.095.

SCR Ne–21 in etched pyroxene separates of Kapoeta: a clue to ancient solar activity.
See Abstr. 105.119.

Evidence for successive episodes of condensation at high temperature in a part of the solar nebula.
See Abstr. 105.198.

Formation of solid matter of the protoplanetary nebula and composition of chondrites.
See Abstr. 105.207.

On the brittle–ductile behavior of iron meteorites: new experimental constraints.
See Abstr. 105.253.

The evolution of partially differentiated planetesimals: evidence from iron meteorite groups IAB and IIICD.
See Abstr. 105.265.

Interstellar material in the solar system. A working group report.
See Abstr. 106.059.

Supernova remnants.
See Abstr. 125.011.

Protostars and planets: overview from a planetary perspective.
See Abstr. 131.071.

Cosmic chemical memory: a new astronomy.
See Abstr. 131.163.

Interrelationships between interstellar and interplanetary grains. A working group report.
See Abstr. 131.229.

Comparison of a theory of sequential fragmentation with the initial mass function of stars.
See Abstr. 131.271.

Phase modulation of the Sun's z oscillations.
See Abstr. 155.034.

Stars

111 Parallaxes, Proper Motions, Radial Velocities, Space Motions, Distances

111.001 **Relativistic and perspective effects in proper motions and radial velocities of stars.**
P. Stumpff.
Relativity in celestial mechanics and astrometry, p. 193 – 198 (1986). – See Abstr. 012.002 (IAU Symp. No. 114).
The heliocentric motion of stars is investigated by taking into account all effects due to the geometry and due to the finite speed of light. Kapteyn's star serves as a numerical example. The ambiguous relativistic relation between Doppler shifts and radial velocities is discussed and demonstrated in a diagram. The ambiguity is solved with rigorous equations which allow to compute the inertial motion of a star from its proper motion, classical radial velocity, and distance.

111.002 **Statistical "Z" parallaxes.**
C. Jaschek.
Astron. Nachr., Vol. 307, No. 1, p. 53 – 54 (1986).
A new method for deriving statistical parallaxes is proposed. The method uses the symmetry in the distribution of the Z–component of the stellar space velocity vectors. It should be suited to derive the statistical parallaxes of stars of population II, for which the usual method fails.

111.003 **Photoelectric radial velocities. Paper XIII. 406 ninth–magnitude K0 stars in the Clube Selected Areas.**
R. F. Griffin.
Mon. Not. R. Astron. Soc., Vol. 219, No. 1, p. 95 – 109 (1986).
Radial velocities are given for 406 ninth–magnitude late–type stars. The stars are grouped in 10 small regions located on a systematic plan at Galactic latitudes of $\pm 35°$. Each star has been observed at least twice, and the mean velocities are accurate to $\lesssim 1$ km s^{-1}. Thirty spectroscopic binaries have been discovered; orbits have already been given for 13 of them, and observations of the others are continuing.

111.004 **Radial velocities of blue horizontal branch field stars in the outer Galactic halo.**
J. Sommer–Larson, P. R. Christensen.
Mon. Not. R. Astron. Soc., Vol. 219, No. 3, p. 537 – 544 (1986).
A sample of distant blue horizontal branch field (bhbf) stars has been identified. Thirty of the stars are located at Galactocentric distances between 15 and 40 kpc. Radial velocities of the bhbf stars have been obtained with good accuracy ($\sim$20 km s^{-1}). The radial velocity dispersion of a subsample of 17 stars with a mean Galactocentric distance of 19 kpc is 100 ± 18 km s^{-1} and the radial velocity dispersion of a subsample of 13 stars with a mean Galactocentric distance of 29 kpc is 73 ± 15 km s^{-1} (assuming $V_{\odot} = 225$ km s^{-1}).

111.005 **Distance determination for OH/IR stars: a geometrical method.**
J. Herman.
ESA J., Vol. 9, No. 4, p. 467 – 475 (1985).
Distances represent the most fundamental and also the most poorly resolved astronomical problems. The major task of the Hipparcos mission will be to increase the number of stars with well determined distances considerably, though still limited to a small volume around the Sun. Here, an accurate geometrical method is presented, which reaches much further out, but which is applicable for a specific class of objects, the OH/IR stars, only. This technique provides us with physical parameters for distant stars and also allows a good estimate of the size of our Galaxy to be made.

111.006 **De eigenbeweging van sterren.**
K. Velt.
Zenit, 13. Jaarg., Nr. 4, p. 152 – 155 (1986).

111.007 **Investigation of proper motions of stars relative to galaxies in two areas of the Pulkovo program.**
Yu. Baltabaev.
Tsirk. Astron. Inst., Tashkent, No. 115, 32 pp. (1985). In Russian.

111.008 **The binary frequency of the OB runaway stars.**
D. R. Gies, C. T. Bolton.
Bull. Am. Astron. Soc., Vol. 17, No. 4, p. 886 (1985). Abstract. – See Abstr. 010.062.

111.009 **Stellar motions in the solar neighborhood.**
A. R. Upgren.
Bull. Am. Astron. Soc., Vol. 17, No. 4, p. 898 (1985). Abstract. – See Abstr. 010.062.

111.010 **Radial velocities of standard stars.**
T. G. Barnes III, T. J. Moffett, M. H. Slovak.
Publ. Astron. Soc. Pac., Vol. 98, No. 600, p. 223 – 230 (1986).
Radial–velocity observations obtained over a five–year period with the McDonald Observatory photoelectric radial–velocity spectrometer are here reported for those stars not known to be velocity variable. There are 259 stars having 1650 observations included in the spectral type range F0 to M0 and brighter than $V = 6.5$ mag. For the best–observed 134 stars, the standard error of the mean velocity is typically better than ± 0.9 km s^{-1}. Sixteen stars are shown to be constant to a sufficient level to warrant standard–star status. Six possible spectroscopic binaries are found.

111.011 **The proper motion of stars.**
S. Bates.
Vasiona, Année 34, No. 1 – 2, p. 28 – 34 (1986). In Croatian.
A discussion of elementary ideas about stellar proper motion, together with examples concerning stars in Orion and Taurus is presented.

111.012 **A photometric study of the bright cloud B in Sagittarius: V. 185 new proper motion stars.**
A. Terzan, C. Turati, C. Ounnas.
ESO Sci. Prepr., No. 449, 28 pp. (1986). To appear in Astron. Astrophys.

111.013 **Carbon star radial velocities and dark matter in the universe.**
M. Jura.
Astron. J., Vol. 91, No. 3, p. 539 – 540 (1986).
Optical radial velocities of carbon stars in the Milky Way are compared to center–of–mass velocities derived from CO radio

emission produced in their circumstellar envelopes. It seems that there is an intrinsic velocity dispersion in the optically measured radial velocities. If the carbon stars in the dwarf spheroidals behave in a fashion similar to those in the Milky Way, then the use of their optical radial velocities to infer the mass-to-light ratio of dwarf spheroidal galaxies and the nature of the dark matter in the universe is suspect.

111.014 **Photometric parallaxes for selected stars of color class m from the NLTT catalog. II. The declination zone 0° to +20°.**
E. W. Weis.
Astron. J., Vol. 91, No. 3, p. 626 – 639 (1986).
VRI photometry and photometric parallaxes are presented for a sample of 702 stars of color class m in the NLTT catalog for which no trigonometric parallaxes have been measured. Additional VRI photometry of stars with trigonometric parallaxes is also presented. For the program stars, 97 have $\pi_{ph} \geqslant 0''.040$ but only four stars are found to have $\pi_{ph} \geqslant 0''.100$. These nearby stars have no significant effect on the local stellar luminosity function of Wielen.

111.015 **FK Comae – the kinematical determination of its age and evolutionary status.**
E. F. Guinan, C. R. Robinson.
Astron. J., Vol. 91, No. 4, p. 935 – 939 (1986).
In order to determine the evolutionary status of this star, the authors present an analysis of the space motion of FK Com. From its kinematics and location in the Galaxy, an approximate age is determined, thereby limiting possible evolutionary models.

111.016 **New subdwarfs. V. Radial velocities for 889 high–proper–motion stars measured with the Mount Wilson 100 inch reflector.**
G. Fouts, A. Sandage.
Astron. J., Vol. 91, No. 5, p. 1189 – 1208 (1986).
New radial velocities have been obtained from 2265 measurements of 889 high–proper–motion stars taken from the subdwarf candidate list of Paper IV (see Abstr. 113.053). Of the 878 stars with nonvariable velocities in the sample, 38 have radial velocities larger than $|v| = 200\ \mathrm{km\ s^{-1}}$, of which 22 are new. There is a correlation between radial velocity and reduced ultraviolet excess as expected from the previously known relations between space motion and chemical composition. The Strömberg asymmetric drift is directly visible in the radial velocity and the proper–motion data separately as each is displayed as a function of galactic longitude.

111.017 **Radial–velocity determinations of six LMC superluminous giant candidates.**
S. R. Baird, P. J. Flower.
Astron. J., Vol. 91, No. 6, p. 1336 – 1339 (1986).
Blue spectra at 121 Å $\mathrm{mm^{-1}}$ have been obtained for 12 candidate superluminous giant stars (SLGs) in ten clusters of the LMC. For ten of these stars, red spectra at 125 Å $\mathrm{mm^{-1}}$ were also obtained. Spectral classes are given for the 12 stars. Of the six measured for radial velocities, four are found to be foreground objects. The possibilities that all SLG stars are foreground objects or the combined light of more than one normal star are discussed.

111.018 **Trigonometric parallaxes determined with the Yerkes Observatory 40 inch refractor. VI. Measurements made with PDS microdensitometers.**
W. F. van Altena, L. H. Auer, C. L. Mora, E. U. Vilkki.
Astron. J., Vol. 91, No. 6, p. 1451 – 1455 (1986).
Thirteen Yerkes Observatory 40 in. parallax series have been measured on PDS microdensitometers and reduced using digital image–centering techniques and more rigorous reduction methods. An approach for determining a priori plate and star weights is discussed. New parallax–reduction programs have been written that solve for the plate constants and the parallaxes and proper motions iteratively, and realistic estimates have been derived for the error of the image positions. The application of these modifications to the above mentioned Yerkes parallax series yields an average accuracy for the parallaxes of $\pm 0''.0043$ (s.e) for a 24.5 plate single–exposure–per–plate series.

111.019 **Common proper motion stars in the AGK3.**
J. L. Halbwachs.
Bull. Inf. Cent. Données Stellaires, No. 30, p. 129 (1986).

111.020 **The binary frequency and origin of the OB runaway stars.**
D. R. Gies, C. T. Bolton.
Astrophys. J., Suppl. Ser., Vol. 61, No. 2, p. 419 – 454 (1986).
A radial velocity survey of 36 proposed runaway stars has been undertaken to determine their binary frequency. Twenty stars are probable high–velocity objects; two of these are binaries, and both are double–lined systems. Limits on velocity variability rule out the general occurrence of collapsed companions if their orbits are similar to those of the massive X–ray binaries. The results suggest that the runaways were ejected by close gravitational encounters with binary stars during the early dynamical evolution of associations.

111.021 **The proper motion of CR Gem.**
C. E. López.
Inf. Bull. Variable Stars, No. 2864, 2 pp. (1986).

111.022 **The proper motion of VX Sgr derived from library material.**
C. E. López.
Inf. Bull. Variable Stars, No. 2905, 1 p. (1986).

111.023 **On a correct algorithm for determining photographic relative proper motions of open clusters.**
X.–h. Zhou, J.–y. Jing.
Acta Astron. Sin., Vol. 27, No. 1, p. 9 – 15 (1986). In Chinese.

111.024 **Can a system be defined for the new edition of the Yale Parallax Catalogue?**
W. van Altena.
Astrometric techniques, p. 183 – 190 (1986). – See Abstr. 012.068 (IAU Symp. No. 109).
A preliminary analysis has been made of data in the forthcoming edition of the Yale Parallax Catalogue to study the possibility of defining a system of observatory zero–points and external accuracies. Based on this analysis and those of several other authors, the author concludes that it is not likely that a unique system of zero–points can be established.

111.025 **Proper motions with respect to galaxies.**
A. R. Klemola.
Astrometric techniques, p. 191 – 199 (1986). – See Abstr. 012.068 (IAU Symp. No. 109).
The Lick proper motion program with respect to galaxies is summarized. The subject of extragalactic astrometry is introduced with estimates of proper motions for selected members of the Local Group of galaxies and more distant systems. The problems of measuring proper motions for Magellanic Clouds is described in more detail.

111.026 **Astrometry of star clusters: problems, techniques, and opportunities.**
K. M. Cudworth.
Astrometric techniques, p. 201 – 208 (1986). – See Abstr. 012.068 (IAU Symp. No. 109).
The astrophysical problems that can be attacked via proper motions in star clusters are set forth and current techniques and opportunities are described. Cluster membership work continues to progress with recent work mixing plates from various telescopes, including large reflectors.

111.027 **Proper motion survey with the 48–inch Schmidt Telescope. LXVI. Proper motions for 7698 stars in the Hyades region.**
W. J. Luyten, S. Morris, G. Hill.
Separate Print, University of Minnesota, Minneapolis, Minn. 55455, USA, 131 pp. (1985).

111.028 **Proper motion survey with Schmidt telescopes. LXVII. Proper motions for 210 faint stars.**
W. J. Luyten, H. S. Hughes.
Separate Print, University of Minnesota, Minneapolis, Minn. 55455, USA, 5 pp. (1985).

111.029 **Proper motion survey with Schmidt telescopes. LXVIII. Reply to Eggen's misrepresentations.**
W. J. Luyten.
Separate Print, University of Minnesota, Minneapolis, Minn. 55455, USA, 3 pp. (1986).

Catalogue of proper motions of stars in the region of the galactic star cluster NGC 7243.
See Abstr. 002.026.

Catalogue of proper motions of PFKSZ–2 stars.
See Abstr. 002.048.

The Bibliographic Catalogue of Stellar Radial Velocities.
See Abstr. 002.067.

Parallaxes with large reflectors.
See Abstr. 032.049.

The distance determination of celestial objects by new technologies.
See Abstr. 036.085.

Statistical analysis of proper motion surveys.
See Abstr. 036.213.

Statistical problems encountered in using trigonometric parallaxes.
See Abstr. 036.214.

Expansion of the AGK3 system to the positions and proper motions of bright stars in the vicinity of galaxies.
See Abstr. 041.028.

Plans for the second epoch of the Southern Proper Motion Program.
See Abstr. 041.056.

Trigonometric parallaxes obtained with the UK Schmidt Telescope.
See Abstr. 041.057.

Variations in external parallax errors.
See Abstr. 041.059.

Astrometric desiderata for nearby stars.
See Abstr. 041.071.

The search for unwanted signals: systematic trends in proper motions of fundamental stars.
See Abstr. 043.018.

New subdwarfs. IV. UBV photometry of 1690 high–proper–motion stars.
See Abstr. 113.053.

Dwarf K and M stars of small proper motion found in a large spectroscopic survey.
See Abstr. 114.085.

Properties of stellar objects near the main sequence mass limit.
See Abstr. 115.020.

Supergiant γ Cyg pulsations with 8–d period of magnetic field variations.
See Abstr. 116.017.

Absolute magnitudes and kinematic properties of RR Lyrae stars.
See Abstr. 122.078.

The luminosity and kinematics of RR Lyrae stars in the solar neighbourhood – I. Statistical parallaxes of RR Lyraes.
See Abstr. 122.099.

Pseudocepheids. III. The low–mass stars.
See Abstr. 122.145.

ER 8: a very low luminosity degenerate star.
See Abstr. 126.062.

A search for ultraviolet–excess objects.
See Abstr. 142.036.

A proper–motion analysis of the cluster M35.
See Abstr. 153.033.

Proper motion of the open star cluster NGC 2301.
See Abstr. 153.045.

Proper motions in the field of the open star cluster NGC 2324.
See Abstr. 153.046.

High–precision radial velocities for faint giants: radial velocities of extreme halo systems and the mass of the Galaxy.
See Abstr. 154.019.

Dynamics of the globular cluster M2 (NGC 7089).
See Abstr. 154.051.

Kinematics of the Galaxy from a magnitude–limited proper–motion sample.
See Abstr. 155.107.

A kinematic study of K–type stars.
See Abstr. 155.113.

The stellar population of the Small Magellanic Cloud near NGC 121. I. The mean metallicity, metallicity spread, and radial velocity of SMC halo giants.
See Abstr. 156.023.

112 Stellar Environments (Chromospheres, Coronae, Stellar Winds, Shells, Masers, etc.)

112.001 **Far–infrared characteristics of K dwarfs with Hα emission.**
H. M. Johnson.
Astrophys. J., Vol. 300, No. 1, p. 401 – 405 (1986).

A sample of K dwarfs with Hα emission have counterparts in the *IRAS* Point Source Catalog. They show two types of infrared spectral behaviour, either net decline of emergent flux density from 0.55 μm to at least 12 μm, approximately like normal K dwarfs without Hα emission, or net rise in this spectral range, as though with radiation from cool circumstellar matter. Behaviour type is not well associated with Hα emission strength.

112.002 **First observations of circumstellar envelopes with the IRAM 30–m telescope.**
R. Lucas, A. Omont, S. Guilloteau, Nguyen–Q–Rieu.
Astron. Astrophys., Vol. 154, No. 1/2, p. L12 – L15 (1986).

Detection of SO_2 is reported for the first time in two circumstellar envelopes. The J = 1–0 line of the (0,2,0) vibrationally excited state of HCN is probably identified in IRC +10216, with an unusual profile which could be due to maser emission. $C^{33}S$ is detected in this source as well as 6 new unidentified lines. C_4H, SiC_2 and $H^{13}CCCN$ are detected in CRL 2688. In this source C_4H seems enhanced and silicon depleted with respect to IRC +10216.

112.003 **The far–infrared (IRAS) excess in HD 161796 and related stars.**
M. Parthasarathy, S. R. Pottasch.
Astron. Astrophys., Vol. 154, No. 1/2, p. L16 – L19 (1986). With a correction in Vol. 161, No. 2, p. 417 (1986).

The far infrared IRAS measurements of the high galactic latitude F–supergiant HD 161796 and related stars are found to show strong far infrared excesses, due to large amounts of dust around them. For HD 161796 the dust mass is found to be of the order of $10^{-2}M_\odot$ to $10^{-3}M_\odot$, and for HD 101584 it is of the order of $10^{-3}M_\odot$. These results suggest that HD 161796 and other similar high galactic latitude F–supergiants have suffered extensive mass loss in the past as a result of superwind phenomenon on their AGB stage of evolution.

112.004 **HR 4049: The hottest proto–planetary nebula star or a runaway hypergiant at high galactic latitude surrounded by a dust cloud?**
H. J. G. L. M. Lamers, L. B. F. M. Waters, C. D. Garmany, M. R. Perez, C. Waelkens.
Astron. Astrophys., Vol. 154, No. 1/2, p. L20 – L22 (1986).

The star HR 4049 has a visual spectrum which indicates that it is a B9.5 hypergiant at a surprisingly large distance above the galactic plane of Z = 1.3 kpc. The authors discovered that the star is surrounded by a dust cloud of T = 1250K, which produces a large IR excess and a severe UV deficiency. If the star is an extreme Pop. I hypergiant, it is the only known runaway hypergiant and the only one with such a large IR excess and UV deficiency. Alternatively, the star may be a less massive star with $M \lesssim 3\ M_\odot$, which is in transition between the Mira phase and a planetary nebula phase. The discovery of an emission feature at 11.3 μ, which is typical for planetary nebulae, supports this possibility.

112.005 **The outer atmosphere of Procyon (α CMi F5 IV–V): evidence of supergranulation or active regions.**
C. Jordan, A. Brown, F. M. Walter, J. L. Linsky.
Mon. Not. R. Astron. Soc., Vol. 218, No. 3, p. 465 – 476 (1986).

Observations made with the Einstein X–ray observatory and EXOSAT have shown Procyon (α CMi) to have a measurable X–ray flux. The authors discuss the interpretation of the X–ray data in the context of models made previously by Brown and Jordan (1981) based on spectra obtained with the IUE satellite. The X–ray data lead to a higher electron pressure than does the UV spectrum. These results support the conclusion by Brown and Jordan that the UV data cannot be reconciled with a hot corona in a spherically symmetric uniform model. Two types of models are explored; one in which the X–ray emission originates from an isothermal corona but the UV emission is restricted to limited regions of the disc, analogous to solar supergranulation structure; the other in which hot loop structures are embedded in a cool ($T_e \sim 3\times10^5$K) corona.

112.006 **Measurements of Balmer emission line profiles for southern Be stars. III. New data and radial velocities.**
J. Dachs, R. Hanuschik, D. Kaiser, D. Ballereau, P. Bouchet, R. Kiehling, J. Kozok, R. Rudolph, W. Schlosser.
Astron. Astrophys., Suppl. Ser., Vol. 63, No. 1, p. 87 – 141 (1986).

More than 500 Hα, Hβ, and Hγ line profiles were recorded between 1980 and 1983 for a total of 55 bright southern Be– and Be–shell stars, mostly of early type, by means of a photoelectric spectrum scanner providing nominal spectral resolutions of 1 or 2 Å. Equivalent widths and half–widths as obtained at different epochs are tabulated for each star, and an atlas is presented, containing more than 200 typical Balmer line profiles and demonstrating their variations. Results for individual stars are discussed and compared with those of earlier campaigns. A preliminary survey of results is used to determine average properties of Balmer line profiles and their general morphology.

112.007 **Stellar chromospheres, coronae, and winds.**
T. Simon.
Astrophys. Space Sci., Vol. 118, Nos. 1/2, p. 209 – 222 (1986). – See Abstr. 012.001.

Within the last few years there have been important advances in the understanding of the chromospheres, coronae, and winds of late–type stars. This progress has been the result of ultraviolet and X–ray observations obtained from space with the International Ultraviolet Explorer and Einstein satellites. In addition, ground–based astronomy has contributed much to synoptic studies of stellar activity. In this review the author summarizes some of the major recent developments in this field.

112.008 **High sensitivity SiO maser survey for Mira variables.**
S.–h. Cho, N. Kaifu, N. Ukita, M. Morimoto, M. Hayashi.
Astrophys. Space Sci., Vol. 118, Nos. 1/2, p. 237 – 239 (1986). – See Abstr. 012.001.

Simultaneous observations of the six transitions of SiO for 106 late–type stars were made. The SiO maser emission was detected in 83 stars. The υ = 3 maser emission was detected in eight stars, and the ^{29}SiO υ = 0 emission in six stars. The ^{29}SiO υ = 0 emission is stronger and narrower than that of ^{28}SiO, suggesting that the ^{29}SiO emission is masing.

112.009 **Active phenomena of the Be star EW Lac observed in 1978 – 1982.**
M. Suzuki, T. Kogure.
Astrophys. Space Sci., Vol. 119, No. 1, p. 69 – 72 (1986). – See Abstr. 012.001.

The long–term variations of EW Lac in the Hγ and Hβ line profiles are presented. The active phase observed in 1978 – 1982 is characterized by the strengthening of emission lines accompanied by variations of radial velocities and *V*/*R* asymmetries. A model of a rotating elliptic ring is proposed.

112.010 **Photometric behaviour of the Be star EW Lacertae.**
J. H. Jeong, C. W. Suh, I.–S. Nha.
Astrophys. Space Sci., Vol. 119, No. 1, p. 73 – 76 (1986). = Yonsei Univ. Obs. Contrib., No. 23. – See Abstr. 012.001.

The night–to–night and short–term variations in *UBV* light curves of EW Lac which were made during 1982 – 1984 at Yonsei

University Observatory are discussed. The long-term variability in the photometric behaviour of the star is examined with the authors' own data as well as those of Harmanec et al. (1980).

112.011 **Masers and circumstellar flows.**
R. J. Cohen.
Cosmical gas dynamics, p. 223 – 235 (1985). – See Abstr. 012.007.
Circumstellar flows can be studied with high precision by radio interferometer measurements of maser lines. The maser lines are produced in the dense envelopes of gas and dust which surround stars at the time of their birth, and also towards the end of their life when they undergo extensive mass loss. Recent radio observations are reviewed and compared with theoretical models of the circumstellar flows. Evidence is presented that stellar and interstellar magnetic fields affect the flow patterns.

112.012 **The nature of Be stars: periodic variables in the cluster NGC 3766.**
L. A. Balona, C. A. Engelbrecht.
Mon. Not. R. Astron. Soc., Vol. 219, No. 1, p. 131 – 144 (1986).
The authors present results of a continuous five-week photometric campaign on 15 stars in the galactic cluster NGC 3766. Of these stars, one is probably a β Cep variable. Of the seven Be stars observed, four show strictly periodic variations characteristic of the rotation period with indications of unequal maxima and minima. Two others are certainly variable but the amplitudes are too small to derive a period. A spotted star model is suggested as the most likely explanation for the variations.

112.013 **The spatial extent of heated dust around MWC 349.**
C. Leinert.
Astron. Astrophys., Vol. 155, No. 1, p. L6 – L7 (1986).
The infrared source MWC 349 was resolved by one-dimensional infrared speckle interferometry in east-west, and marginally in north-south direction. This is considered as direct evidence for a circumstellar dust distribution, possibly disk-shaped, postulated by earlier optical and radio investigations.

112.014 **The density structure of the wind of P Cygni (B1 Ia$^+$).**
L. B. F. M. Waters, P. R. Wesselius.
Astron. Astrophys., Vol. 155, No. 1, p. 104 – 112 (1986).
Near-IR and IRAS observations of the hypergiant P Cygni (B1 Ia$^+$) are used to derive the density structure of the wind of P Cygni. The authors find a linear velocity law that is identical to the one derived by Van Blerkom (1978) from an analysis of Balmer line profiles, and differs from the one derived by Barlow and Cohen (1977) based on their analysis of near-IR and radio data. Two models can explain the observed excess at 60 μ: a shell extending from 15 to 50 stellar radii with a density increase of a factor five relative to the quiescent wind, or cold dust with a typical temperature of 30K, situated far away from the star. The dust model can also explain the 100 μ flux if that flux can be attributed to the star and is not a part of the complicated background.

112.015 **Mass loss from Wolf-Rayet stars.**
M. Singh.
Astrophys. Space Sci., Vol. 120, No. 1, p. 115 – 119 (1986).
Mass loss rates for 9 LMC WR stars are determined using IUE, UV, and visible spectrophotometric observations. A good correlation of mass loss rate with effective temperature and luminosity is indicated by the data, in agreement with the theoretical predictions.

112.016 **Spectrophotometry of some Be stars.**
M. Singh.
Astrophys. Space Sci., Vol. 120, No. 1, p. 133 – 138 (1986).
The continuum energy distributions of six Be stars, namely 25 Cyg, 31 Peg, HR 8758, 14 Lac, 12 Vul, and β Psc, in the wavelength region λλ3200 – 7800 Å, are presented. Comparing the observed energy distributions with those of theoretical models given by Kurucz (1979), their effective temperatures are determined. A near-infrared excess emission at wavelengths above λ6000 Å is seen in most of the stars.

112.017 **On the properties of very young massive infrared sources.**
J. Gürtler, T. Henning, J. Dorschner, C. Friedemann.
Astron. Nachr., Vol. 306, No. 6, p. 311 – 327 (1985).
The data compiled for 34 very young massive compact infrared sources (Becklin-Neugebauer objects) in a separately published catalogue serve as the basis for an investigation of their general properties. The authors investigate relationships between the optical depths of the ice and silicate bands and correlations of these bands with such source parameters as shell temperature, luminosity and polarization in the infrared. Stellar wind models for the ionized region around the embedded star are favoured after the analysis of the intensities of the hydrogen infrared emission lines and the radio continuum. The BN objects are clearly more frequently associated with H_2O masers than OH masers. The linear separation of the infrared peaks from the masers are compatible with models where the H_2O masers are situated near the edges of the gas-dust shells of the BN objects.

112.018 **Condensation, growth and acceleration of grains in circumstellar shells: an observer's view.**
R. Papoular, B. Pégourié.
Astron. Astrophys., Vol. 156, No. 1/2, p. 199 – 206 (1986).
An attempt is made to apply the time-dependent homogeneous condensation theory to a number of well documented oxygen-rich red giants and supergiants. It does not lead to a satisfactory match of all the observed properties simultaneously. The purely dynamical part of the theory does not appear to be responsible for this failure. The problem lies in the tight relationship, imposed by the theory, between mass loss rate and grain size. If this is relaxed, the dynamical equations are able to account for the observations. The authors discuss possible improvements to the condensation and growth model, including grain coagulation in macro-turbulent atmospheres. Although these concepts are not easily made quantitative, it is possible to use them in a phenomenological way. When complemented with observational data, the model can then provide new quantitative information, such as star mass and mass loss rate. It also yields other quantities, e.g. grain velocity, whose observation could help in further assessing the validity of the theory.

112.019 **Observational evidence for disks around young stars.**
P. M. Harvey.
Protostars & planets II, p. 484 – 492 (1985). – See Abstr. 012.008.
The detection of energetic bipolar outflows from a variety of young stellar objects has prompted a wealth of new observational data on the geometry of circumstellar clouds. The techniques used to probe the circumstellar environment include both observations that have resolved a disk-like structure or its effects and observations in a single spatial resolution element for which modeling suggests the presence of a small circumstellar disk. The best evidence for disks around young stars comes from the high angular resolution optical, infrared, and radio images of bipolar nebulosity requiring the presence of an axially symmetric circumstellar cloud.

112.020 **Molecular ions in the circumstellar envelope of IRC +10216.**
A. E. Glassgold, R. Lucas, A. Omont.
Astron. Astrophys., Vol. 157, No. 1, p. 35 – 48 (1986).
The spatial distribution of the ionization in a C-rich circumstellar envelope is calculated on the basis of appropriate ion-molecule reactions. There are two main sources of ionization, cosmic ray ionization of molecular hydrogen, which yields H_3^+, and photoionization of acetylene and neutral carbon, which yield $C_2H_2^+$ and C^+. The most promising candidate for detection at millimeter wavelengths is HCO^+, and the authors expect that observations with higher sensitivity and with larger telescopes would detect HCO^+ in IRC +10216.

112.021 **A new free radical in IRC +10216.**
M. Guélin, J. Cernicharo, C. Kahane, J. Gomez–Gonzalez.
Astron. Astrophys., Vol. 157, No. 2, p. L17 – L20 (1986).

The authors report the discovery of a new "non terrestrial" molecule in the envelope of the carbon star IRC +10216. This molecule is linear or slightly asymmetric and must have an odd number of electrons; its electronic ground state is very probably split by spin–rotation interaction, i.e. must be $^2\Sigma$ if the molecule is linear. The rotation and spin–doubling constants are found to be 5966.8 MHz and ±15.1 MHz, respectively. The most likely candidates for such a radical are HSiCC (or its isomer HCCSi) and HSCC.

112.022 **Ultraviolet and visual variability of θ CrB during a normal B–phase following a shell phase (1980 – 1985).**
V. Doazan, J. M. Marlborough, C. Morossi, G. J. Peters, L. Rusconi, G. Sedmak, R. Stalio, R. N. Thomas, A. Willis.
Astron. Astrophys., Vol. 158, No. 1/2, p. 1 – 13 (1986).

The phenomena characterizing the end of a shell phase of θ CrB and the "normal" B phase which followed it are described, using coordinated far UV and visual observations made in 1980 – 1985. During this period, the line–spectrum is variable over the whole range of ionization observable in the IUE and visual spectra. The strength, the shape, and the velocity of the C IV, Si IV, and Al III resonance lines, and the Fe III lines of low excitation show associated variability patterns. Lines arising from the highest ionization levels exhibit the largest amplitude of variations.

112.023 **A long–term study of SiO (υ = 1, J = 2–1) maser emission from evolved stars and Orion A.**
L.–Å. Nyman, H. Olofsson.
Astron. Astrophys., Vol. 158, No. 1/2, p. 67 – 82 (1986).

The authors have observed the SiO ($\upsilon = 1$, $J = 2–1$) maser emission over a period of more than four years towards eight late–type stars and Orion A. This was done in order to study the general characteristics of the maser emission and in particular investigate any possible correlations with the optical variations of the central star. The implications of the results on the pump mechanism, the location of the masers, and the kinematics of the inner part of the circumstellar envelope are discussed.

112.024 **A–type shell stars and infrared sources.**
M. Jaschek, C. Jaschek, D. Egret.
Astron. Astrophys., Vol. 158, No. 1/2, p. 325 – 328 (1986).

Following the discovery of a dust cloud around an A–type shell star the authors examine the situation of all known Ae/A shell stars. They have found 19 stars of this group observed by IRAS, and of these 19, 8 are infrared anomalous emitters (IAE). The temperatures of the anomalous emitting sources are of the order of 50 – 400K, although three cases require two superposed cool sources. The proportion of IAE's among Ae/A shell stars is very high, when compared to a sample of non–shell A–type stars. This suggests a close link between the membership in the group and IAE.

112.025 **Periodicity of the radial–velocity and brightness variations of the hypergiant P Cygni (B1 Ia$^+$), and origin of the shell ejections.**
R. H. van Gent, H. J. G. L. M. Lamers.
Astron. Astrophys., Vol. 158, No. 1/2, p. 335 – 350 (1986).

A reanalysis of published radial–velocity measurements of the shell components of P Cyg shows that the variations are not due to periodic oscillations but to shell ejections with a characteristic timescale of about 50 to 100 days. A reanalysis of the photometric variations of P Cyg shows that most of the periodicities claimed in the literature (ranging from 0.5 days to 18 years) are not real. The photometric variations of P Cyg are not strictly periodic, but show a possible range of timescales from 25 to 60 days. The polarization measurements suggest non–periodic variations on timescales of 125 and possibly 12 days. The qualitative agreement between the pulsational timescale and the timescale for the shell ejections suggests that the shells are triggered by non–radial pulsations of the star. The comparison between the acceleration of the shells observed in the Balmer lines with those observed in the UV shows that the acceleration of shells is anticorrelated with their density, suggesting that the shells are accelerated by radiation pressure of optically thick lines.

112.026 **Circumstellar shells around luminous emission–line stars in the Large Magellanic Cloud.**
O. Stahl, B. Wolf.
Astron. Astrophys., Vol. 158, No. 1/2, p. 371 – 381 (1986).

High–resolution echelle spectrograms obtained with CASPEC of seven luminous emission–line stars in the LMC are discussed. The authors find narrow nebular emission lines in all stars of this sample. The emission lines are clearly resolved in all cases. The three S Dor variables R 71, S Dor and R 127 show velocity–doubled [N II] 6548, 6583 lines. The authors consider the presence of the nebular emission lines as evidence for the presence of supergiant planetary nebulae around the central stars. The density of the nebulae exceeds $10^5 cm^{-3}$ in several cases. The authors discuss the origin of the expanding shells. It seems most likely that they have been ejected during previous S Dor episodes of the central stars. This implies that a major fraction of the most luminous emission–line stars undergoes S Dor–like outbursts.

112.027 **Stellar winds and variability in hot luminous stars.**
P. S. Conti.
Publ. Astron. Soc. Pac., Vol. 98, No. 599, p. 41 – 43 (1986). – See Abstr. 012.009.

112.028 **Luminosity and colour variations of 88 Her through phase changes from the far UV to the visual region. I. A confrontation with traditional Be–star modelling.**
M. Barylak, V. Doazan.
Astron. Astrophys., Vol. 159, No. 1/2, p. 65 – 74 (1986).

The authors present and analyze the luminosity and colour variations of 88 Her in the wavelength range 1550 Å – 5500 Å during the period 1972 – 1983 where the star has exhibited two different phases: a quasi–normal B phase and a new Be–shell phase of increasing strength. They discuss the observed variations of the radiative flux in the framework of traditional Be–star modelling and conclude that model predictions are contradicted by the two observations: (1) Minimum luminosity is not observed at maximum development of the shell spectrum; on the contrary, it occurs when the shell spectrum is still very mild. (2) The development of the shell spectrum is not accompanied by a decrease in luminosity; on the contrary, the luminosity increases in all the investigated wavelengths.

112.029 **Luminosity and colour variations of 88 Her through phase changes from the far UV to the visual spectral regions. II. Empirical atmospheric modelling: an association between photospheric and envelope variations.**
V. Doazan, R. N. Thomas, M. Barylak.
Astron. Astrophys., Vol. 159, No. 1/2, p. 75 – 89 (1986).

The authors analyze and interpret the observations, presented in Paper I, of the luminosity variations, from the far–UV to the visual region, of 88 Her as it changes from the quasi–normal B phase to the Be–shell phase. The authors analyze the data without any preconceived assumptions on what are the atmospheric regions producing the luminosity changes; they admit that the observed luminosity variations may arise from changes of the structure and thermodynamic state of *both* the photosphere and the exophotospheric cool envelope. The authors present an iterative approach to such diagnostics of the comparative roles of the photosphere and envelope.

112.030 **The acceleration in the wind of the hypergiant P Cygni and the mass loss mechanism of P Cygni type stars.**
H. J. G. L. M. Lamers.
Astron. Astrophys., Vol. 159, No. 1/2, p. 90 – 100 (1986).

The acceleration of the wind of P Cygni between 1.1 R_* and 4 R_* is derived from the velocity law. The acceleration is due to an outward directed force, which barely exceeds the gravity and varies approximately as $r^{-1.8}$. The acceleration can be explained by radiation pressure due to a very large number ($\sim 10^3$) of weak lines in the Balmer continuum. These lines have been observed in the IUE spectra. The authors propose that the mass loss of P Cygni type stars and the luminosity upper limit in the HR-diagram is due to radiation pressure by weak metal lines in the Balmer continuum. This will produce an instability limit at $T_{eff} \lesssim 30,000K$ when $g_{eff} \lesssim 10^2 cm/s^2$.

112.031 **Time variation of the H_2O and SiO masers in late-type stars.**
A. M. Gómez Balboa, J. R. D. Lépine.
Astron. Astrophys., Vol. 159, No. 1/2, p. 166 – 174 (1986).
The authors present the results of up to six years monitoring of the 22 GHz H_2O maser, and less extensive monitoring of the 43 GHz SiO line, of ten late-type variables (o Cet, R Dor, L_2 Pup, VY CMa, R Crt, W Hya, U Her, AH Sco, VX Sgr, R Aql). The variability curves are compared with visual curves from AFOEV (Association Française des Observateurs d'Etoiles Variables). Aperiodic variability of the H_2O maser spanning several cycles of the stellar pulsation (R Crt), or superperiodic behaviour with periods equal to multiples of the fundamental period (R Aql, W Hya), have been noted. These results can be unterstood in terms of the models of Wood (1979) and Willson and Hill (1979) for shock wave propagation into the envelopes.

112.032 **Geometry of rotating envelopes around Be stars derived from comparative analysis of Hα emission line profiles.**
J. Dachs, R. Hanuschik, D. Kaiser, D. Rohe.
Astron. Astrophys., Vol. 159, No. 1/2, p. 276 – 290 (1986).
A large body of Balmer emission line profiles for a total of 55 bright southern and equatorial Be stars has been collected by Dachs et al. (1977, 1981, 1986) in order to study average properties of envelopes surrounding Be stars and their trends of variability. It is the purpose of the present publication to compare these profiles and to derive a coherent picture explaining the gross shapes of the profiles by means of a continuous sequence of disk-shaped model envelopes, with envelope diameter as a main parameter of the model.

112.033 **Long-wavelength emission lines from stellar winds.**
T. Nugis.
Tartu Astrofüüs. Obs. Teated, Nr. 81, 19 pp. (1985).
In the long-wavelength spectral region the mass outflow envelopes become partly opaque for continuous radiation. The radius of the opaque part of the envelope increases with λ. In this paper the author presents the necessary formulae for calculating the line profiles at long wavelengths. The analysis of some simple model cases showed that at long wavelengths, where the envelope already becomes partly opaque for continuous radiation the line profiles have some specific differences as compared with the optically thin envelope in the continuum. With the increase of wavelength greater and greater parts of the envelope become opaque for continuous radiation and only the lines corresponding to strong transitions with highly populated levels far from the star can be seen.

112.034 **Observation of the onset of an emission episode in the Be star Mu Centauri.**
G. J. Peters.
Astrophys. J., Lett. Ed., Vol. 301, No. 2, p. L61 – L65 (1986).
A series of recent CCD observations which shows the rapid development of Hα emission in the pulsating, transient Be star μ Cen in only 2 days and thus establishes a time scale for the physical changes in the photosphere that lead to the outburst, are presented. On 1985 March 25, Hα was in pure absorption in μ Cen but by 1985 March 27 a centrally reversed emission feature was observed. Postoutburst observations revealed that the Hα emission increased slightly in strength and then waned within a period 1 month. Complementary observations of He I 6678 as well as UV spectra from *IUE* are used to discuss the extent of photospheric activity associated with the outburst and line-of-sight mass flow into the circumstellar envelope.

112.035 **Interpretation of the OH-line profile 1667 MHz in the maser source M1-92 and the problem of rotation of circumstellar envelopes.**
V. P. Grinin.
Izv. Krymskoj Astrofiz. Obs., Tom 70, p. 139 – 143 (1985). In Russian. English translation in Bull. Crimean Astrophys. Obs., Vol. 70.
It is shown that the four-components profile of the OH line 1667 MHz in the maser source associated with the biconical nebula M1-92 may be interpreted in the frame of the model of an expanding and rotating disc-like envelope. The examples of other biconical nebulae which have direct or indirect evidences of rotation are given and its origin is discussed in short.

112.036 **Circumstellar, possibly protoplanetary disks.**
H. Shipman.
Phys. Today, Vol. 39, No. 1, p. S7 – S8 (1986). Short review paper. – See Abstr. 013.021.

112.037 **Narrow absorption components and variability in ultraviolet P Cygni profiles of early-type stars.**
R. K. Prinja, I. D. Howarth.
Astrophys. J., Suppl. Ser., Vol. 61, No. 2, p. 357 – 418 (1986).
An extensive set of velocity and column density measurements of both narrow absorption components and "underlying" P Cygni profiles is presented, based on 376 uniformly extracted high-dispersion IUE spectra of 24 OB stars. Mass-loss rates are given for 19 stars. Variability in $\dot{M}$ is usually at the 10% level, on time scales of a day or longer; changes of a factor of 2 or greater are not observed. Discrete absorption components are observed at least some of the time in *every* star with unsaturated P Cygni profiles. The authors investigate the observational characteristics of these features, and discuss models for their formation.

112.038 **Unusual burst of OH maser emission in the source VY CMa.**
N. A. Yudaeva.
Pis'ma Astron. Zh., Tom 12, No. 5, p. 361 – 365 (1986). In Russian. English translation in Sov. Astron. Lett., Vol. 12.
Results of observations of the source VY CMa, made from May 1983 to April 1985 with the RATAN-600 radio telescope in the main lines of the OH maser emission 1665 and 1667 MHz are described. In February – March 1985 the burst of OH maser emission occurred, which was accompanied by symmetrical satellite details and strong flux variation of the main component of the emission profile. The probable geometry of maser emission region is discussed.

112.039 **Rigel's circumstellar envelope structure.**
D. P. Hayes.
Astrophys. J., Vol. 302, No. 1, p. 403 – 409 (1986).
Intensive monitoring of the late B supergiant Rigel's optical continuum linear polarization over the 1984 – 1985 observing season reveals episodes of substantially enhanced aperiodic activity. Different activity episodes are centered in different sectors of the circumstellar envelope. These activity episodes are a manifestation of temporally and spatially variant mass loss. Localized disturbances originating at or below the stellar surface are suggested as the root cause of the polarization variations.

112.040 **On the number of comets around other single stars.**
C. Alcock, C. C. Fristrom, R. Siegelman.
Astrophys. J., Vol. 302, No. 1, p. 462 – 476 (1986).
The authors discuss an indirect method of inferring the presence or absence of Oort-type comet clouds in orbit around white dwarf stars. It is also shown that the detection of such clouds around white dwarfs can be directly linked to the presence of comets around the main-sequence progenitors of the white dwarfs. The method depends on the observation of the abundance of metals in the atmospheres of white dwarf stars and comparison of these abundances with those expected from model atmospheres, which predict that the composition will be strongly stratified by gravity-driven diffusion. As example, the authors suggest that the calcium reported in the white dwarf G 74-7 is due to recent comet accretion.

112.041 **New OH observations of the supergiant IRC +10420.**
B. M. Lewis, Y. Terzian, J. Eder.
Astrophys. J., Lett. Ed., Vol. 302, No. 1, p. L23 – L26 (1986).
The object IRC +10420 (V1302 Aql) was observed in 1985 with the Arecibo 305 m telescope in the 1612, 1665, and 1667 MHz lines of OH. The strongest feature in the 1612 MHz spectrum is 190 Jy, almost 3 times stronger than its intensity in 1982 April when last observed. The linear increase with time of the +55 km s^{-1} feature's intensity in both the 1665 and 1612 MHz lines has changed. The 1665 MHz feature is weakening while the 1612 MHz feature is growing more quickly than before. The observations are consistent with a linear growth of the mean flux from the two lines, which originate from the same excited level.

112.042 **Mass loss from cool stars.**
L. Goldberg.
From SU(3) to gravity, p. 279 – 288 (1985). – See Abstr. 003.014.
This review discusses (1) empirical determinations of mass loss rates for highly evolved cool stars, (2) some of the mechanisms that have been proposed to drive the flow, and (3) some aspects of stellar evolution, such as the formation of planetary nebulae.

112.043 **X-ray observations and the structure of stellar coronae.**
J. L. Culhane.
Observatory, Vol. 106, No. 1070, p. 5 – 6 (1986). Abstract. – See Abstr. 010.741.

112.044 **Non-thermal X-rays from the Wolf-Rayet star HD 193793.**
A. Pollock.
Observatory, Vol. 106, No. 1070, p. 6 – 7 (1986). Abstract. – See Abstr. 010.741.

112.045 **SiO and H_2O maser emission in OH/IR objects and late-type variable stars.**
L.-Å. Nyman, L. E. B. Johansson, R. S. Booth.
Astron. Astrophys., Vol. 160, No. 2, p. 352 – 366 (1986).
The authors have searched for 86 GHz SiO ($\upsilon = 1$, $J = 2$–1) and H_2O maser emission towards ~20 unidentified OH/IR objects and ~35 optically identified variable stars (mostly Mira variables) during a period of four years. Many of the sources were observed several times, giving some information on their temporal variations. 86 GHz SiO emission was clearly seen in only one OH/IR object, whereas eleven of the variable stars were detected, nine of them for the first time in this line. The authors have also searched for 43 GHz SiO ($\upsilon = 1$ and 2, $J = 1$–0) emission towards eleven OH/IR objects and five sources were detected in both lines. H_2O emission was detected in nine OH/IR objects, four of which are new detections.

112.046 **High-resolution maps of mass loss from carbon stars.**
R. Ball, L. G. Mundy.
Bull. Am. Astron. Soc., Vol. 17, No. 4, p. 841 (1985). Abstract. – See Abstr. 010.062.

112.047 **New lines of ethylene and a search for methylene in IRC+10216.**
D. M. Goldhaber, A. L. Betz.
Bull. Am. Astron. Soc., Vol. 17, No. 4, p. 841 (1985). Abstract. – See Abstr. 010.062.

112.048 **The effect of dust of the chromospheres of cool, luminous stars.**
K. G. Carpenter, R. E. Stencel, W. Hagen.
Bull. Am. Astron. Soc., Vol. 17, No. 4, p. 876 (1985). Abstract. – See Abstr. 010.062.

112.049 **Post-main sequence evolution of rotation and chromospheric activity.**
S. L. Baliunas, R. A. Donahue, J. H. Horne, D. K. Duncan, J. Frazer, H. Lanning, A. Misch, J. Mueller, A. H. Vaughan, L. Woodard.
Bull. Am. Astron. Soc., Vol. 17, No. 4, p. 877 (1985). Abstract. – See Abstr. 010.062.

112.050 **Hydrogen Lyman alpha fluxes for late-type dwarf stars.**
J. E. Neff, K. G. Carpenter, T. R. Ayres.
Bull. Am. Astron. Soc., Vol. 17, No. 4, p. 879 (1985). Abstract. – See Abstr. 010.062.

112.051 **A multi-component chromospheric model for AR Lacertae.**
D. M. Gibson, J. E. Neff, A. Brown, F. M. Walter.
Bull. Am. Astron. Soc., Vol. 17, No. 4, p. 880 (1985). Abstract. – See Abstr. 010.062.

112.052 **Kinematics of the Eta Carinae condensations – expanding shells and 10^3km s^{-1} He/N-rich ejected globules.**
R. J. Dufour, K. Davidson, N. R. Walborn.
Bull. Am. Astron. Soc., Vol. 17, No. 4, p. 884 – 885 (1985). Abstract. – See Abstr. 010.062.

112.053 **Chromospheric activity as an indicator of stellar age.**
D. C. Barry, R. H. Cromwell, E. K. Hege.
Bull. Am. Astron. Soc., Vol. 17, No. 4, p. 902 (1985). Abstract. – See Abstr. 010.062.

112.054 **Asteroids, comets, and other debris around other stars.**
C. Alcock, C. C. Fristrom.
Bull. Am. Astron. Soc., Vol. 17, No. 4, p. 919 (1985). Abstract. – See Abstr. 010.063.

112.055 **High-resolution infrared spectroscopy: a search for the 11.52-μm graphite feature.**
A. C. H. Glasse, W. A. Towlson, D. K. Aitken, P. F. Roche.
Mon. Not. R. Astron. Soc., Vol. 220, No. 1, p. 185 – 188 (1986).
Spectra are presented with a resolving power $\lambda/\Delta\lambda \approx 1100$ of the dusty Wolf-Rayet star Ve 2–45 and the southern H II region G333.6–0.2. An upper limit on the equivalent width of any features in the range 11.0 to 13.0 μm is determined to be 0.0022 μm towards Ve 2–45. This figure is compared with the predicted strength of the unobserved narrow feature at 11.52 μm which has been proposed to be associated with small graphite grains (Draine).

112.056 **The spatial extent and nature of the 3-μm emission features in HD 97048 and CPD -56°8032.**
P. F. Roche, D. A. Allen, J. A. Bailey.
Mon. Not. R. Astron. Soc., Vol. 220, No. 1, p. 7P – 11P (1986).
Speckle observations at the peaks of narrow dust emission features at 3.3 and 3.5 μm in the low-excitation planetary nebula CPD -56°8032 and the peculiar emission-line star HD 97048 are presented. In CPD -56°8032 the emission in the 3.28-μm band is extended on a scale of $\cong$1.3 arcsec, and this is probably the same as the extent of the ionized nebula. By contrast, the emission in the 3.53-μm band in HD 97048 is spatially unresolved and arises from a region <0.1 arcsec across. Dust grains within 0.05 arcsec of the star have temperatures $\geqslant$1000K, so that the proposed identification of the species emitting the strong 3.43- and 3.53-μm features with formaldehyde ice is untenable, and the materials responsible for the band emission remain unidentified.

112.057 **The X-ray spectrum of emission line stars (superluminous and symbiotic).**
G. B. Baratta, L. Piro, R. Viotti, A. Cassatella, A. Altamore, O. Ricciardi, M. Friedjung.
ESA Spec. Publ., ESA SP-239, p. 95 – 98 (1985). – See Abstr. 012.019.
The observation of the X-ray spectrum of peculiar emission line stars, whose atmospheric envelopes are characterized by extreme physical conditions, represents an important tool to investigate the processes of formation of high temperature plasmas. The authors discuss the case of the superluminous object η Car showing both high and low temperature regions and an asymmetric wind. They present new EXOSAT observations of symbiotic stars. AG Draconis observed at a minimum phase shows a low temperature X-ray spectrum. A weak flux from the Mira variable R Aqr observed near maximum has been also detected.

112.058 **The dust shells of NML Cygnus and IRC 10420: inner radius, temperature, and optical thickness.**
S. T. Ridgway, R. R. Joyce, D. Connors, J. L. Pipher, C. Dainty.
Astrophys. J., Vol. 302, No. 2, p. 662 – 674 (1986).
Infrared speckle interferometric observations of NML Cygnus and IRC 10420 at *K*, *L*, and *M* have been analyzed with a simple numerical model of the shells. The analysis yields the extinction function and the monochromatic optical depth of the shells for the wavelengths 2 – 5 μm, the temperature at the inner boundary, and the inner boundary radius. The inner boundary for the dust shell of NML Cyg is at a radius of $0''.045 \pm 0''.012$, and the dust there is at a temperature $\geqslant$1200K. The corresponding parameters for IRC 10420 are $0''.125 \pm 0''.015$ and 750 ±50K. Both stars apparently have a very extended (1″) shell which contributes ~10% of the flux at 2.2 μm.

112.059 **Two supergiants in the Large Magellanic Cloud with thick dust shells.**
J. H. Elias, J. A. Frogel, P. B. W. Schwering.
Astrophys. J., Vol. 302, No. 2, p. 675 – 679 (1986).
Ground–based observations from 0.6 to 20 μm have identified two luminous, evolved stars surrounded by thick dust shells among the Magellanic Cloud sources detected by IRAS. Their energy distribution resemble those of typical Galactic OH/IR stars, but they have bolometric magnitudes brighter than –9 and the small–amplitude variability of supergiants. One star, IRAS 04553–6825, has a spectral type of M7.5 and a dust shell which absorbs and reradiates roughly 75% of the star's luminosity. The second star, IRAS 05346–6949, has an even thicker dust shell, and the central star is not observable.

112.060 **Infrared photometry and high–resolution imaging of OH/IR stars.**
M. L. Cobb.
Diss. Abstr. Int., Sect. B, Vol. 46, No. 6, p. 1956 (1985). Thesis, The University of Iowa, 202 pp. (1985). Order No. DA8518813.

112.061 **O–B2 stars with large circumstellar absorption of light.**
R. M. Raznik.
Astron. Tsirk., No. 1401, p. 4 – 6 (1985). In Russian.

112.062 **Far infrared observations of Gamma Cassiopeiae from the *IRAS* satellite.**
M. J. Coe.
Publ. Astron. Soc. Pac., Vol. 98, No. 601, p. 334 – 337 (1986).
Detailed *IRAS* spectra are presented for the Be star γ Cas. The spectra are shown to be consistent with a Kurucz model atmosphere of 20,000K and agree well with previous, less–comprehensive observations. They confirm the existence of an infrared excess in this source, but do not agree in detail with the model of Poeckert and Marlborough based upon free–free emission from a circumstellar envelope.

112.063 **Hydrogen emission lines in the spectrum of γ Cas for 1979 and 1980.**
Z. K. Ivanova, T. S. Galkina.
Bull. Crimean Astrophys. Obs., Vol. 69, p. 30 – 35 (1986). English translation of 39.112.082.

112.064 **Patrol electrophotometric observations of Hα emission in the spectrum of γ Cas.**
N. I. Bondar, N. I. Shakhovskaya.
Bull. Crimean Astrophys. Obs., Vol. 69, p. 36 – 38 (1986). English translation of 39.112.083.

112.065 **Observational properties of nonradial oscillations in early–type stars and their possible effect on mass loss: the example of ζ Puppis (O4 If).**
D. Baade.
Seismology of the Sun and the distant stars, p. 465 – 466 (1986). Abstract. – See Abstr. 012.026.

112.066 **MERLIN observations of the circumstellar envelope of VX Sagittarius.**
J. M. Chapman, R. J. Cohen.
Mon. Not. R. Astron. Soc., Vol. 220, No. 2, p. 513 – 528 (1986).
MERLIN maps of the H_2O and OH maser emission from the circumstellar envelope of the supergiant star VX Sagittarius are presented. The authors discuss the spatial, velocity and magnetic field structure of the circumstellar envelope of VX Sgr.

112.067 **The infrared excess of Alpha Lyrae.**
B. G. Anandarao, D. B. Vaidya.
Astron. Astrophys., Vol. 161, No. 1, p. L9 – L11 (1986).
Recent measurements in far–infrared (12 – 200 μm) by IRAS (Aumann et al., 1984) and KAO (Harper et al., 1984) have shown evidence of infrared excess in the A0 star Alpha Lyrae which was interpreted as due to reradiation from circumstellar dust constituted by grains much larger ($\gtrsim$10 μm) than those of interstellar origin. The authors present a simple model to explain the far–infrared data in terms of multiple dust shells of different temperatures.

112.068 **On the origin of the cyanopolyynes in the envelope of IRC + 10216.**
I. K. Shmeld, V. S. Strel'nitskij, L. V. Gurvich.
Sov. Astron. Lett., Vol. 11, No. 4, p. 254 – 256 (1985). English translation of 40.112.005.

112.069 **Mass loss from carbon stars.**
M. Jura.
Astrophys. J., Vol. 303, No. 1, p. 327 – 332 (1986).
The author uses IRAS data to study the circumstellar envelopes around carbon stars. It is found that the grains are composed of amorphous carbon rather than graphite. The dust to gas ratio in carbon star outflows has an average value of $4.5 \times 10^{-3}(150/\chi_{60})$, where χ_{60} is the opacity ($cm^2 g^{-1}$) of the dust at 60 μm. Even carbon stars with relatively high outflow velocities appear to have mass loss that can be driven by radiation pressure on dust. Stars that are losing large amounts of mass are generally long–period variables ($P \geqslant 400$ days).

112.070 **Erratum: "Mass loss from evolved stars. III. Mass loss rates for fifty stars from CO J = 1–0 observations" [Astrophys. J., Vol. 292, No. 2, p. 640 – 669 (1985)].**
G. R. Knapp, M. Morris.
Astrophys. J., Vol. 303, No. 1, p. 521 (1986). See Abstr. 39.112.093.

112.071 **Ultraviolet observation of Be stars.**
D. N. Dawanas, R. Hirata, T. Kogure.
Galactic structure and variable stars, p. 95 – 101 (1985). – See Abstr. 012.029.

112.072 **Observational constraints on circumstellar dust.**
M. Jura.
NASA Conf. Publ., NASA CP–2403, p. 3 – 17 (1986). – See Abstr. 012.030.
There is an enormous range in the properties of stars that are losing mass. In this review the author concentrates on those red giants that are responsible for injecting roughly half or more of the material into the interstellar medium.

112.073 **Circumstellar grain formation.**
B. T. Draine.
NASA Conf. Publ., NASA CP–2403, p. 19 – 28 (1986). – See Abstr. 012.030.
The paper concentrates on dust formation around cool giants and supergiants.

112.074 **Circumstellar dust. A working group report.**
E. Dwek.
NASA Conf. Publ., NASA CP–2403, p. WG 1 – WG 17 (1986). – See Abstr. 012.030.

112.075 **Working group on circumstellar/interstellar relationships.**
A. E. Glassgold.
NASA Conf. Publ., NASA CP–2403, p. WG 79 – WG 84 (1986). – See Abstr. 012.030.

112.076 **Using thin circumstellar dust shells to study silicate grain opacities.**
W. Hagen.
NASA Conf. Publ., NASA CP–2403, p. A 7 (1986). – See Abstr. 012.030.

112.077 **IUE and IRAS observations of luminous M stars with varying gas–to–dust ratios.**
W. Hagen, K. G. Carpenter, R. E. Stencel.
NASA Conf. Publ., NASA CP–2403, p. A 8 (1986). – See Abstr. 012.030.

112.078 **Recent spectral variation of the Be star EW Lac.**
T. Kogure, Y. Asada, M. Suzuki.
Proceedings of the Second Asian–Pacific Regional Meeting on Astronomy, p. 504 – 511 (1984). – See Abstr. 012.031.
The inspection of spectral variation in the period of 1966 – 1981 is made with particular attention to the anomalous V/R variation which has appeared since 1975. An explanation of the anomalous V/R variation in terms of multiple elliptical ring model is suggested. Spectrophotometric analysis for the period 1977 – 1980 is also made. Some envelope parameters including the optical thickness in Hα, effective radius of envelope, mean electron density, and mass of envelope, are derived.

112.079 **Etude spectrophotométrique, paramètres physiques et géométriques de l'étoile Be Gamma Cassiopeiae.**
D. N. Dawanas.
Proceedings of the Second Asian–Pacific Regional Meeting on Astronomy, p. 560 (1984). Abstract. – See Abstr. 012.031.

112.080 **On the broad Balmer profiles from Car II and their possible origin as dust scattered lines from η Carinae.**
J. A. López, J. Meaburn.
Rev. Mex. Astron. Astrofis., Vol. 11, No. 2, p. 71 – 72 (1985).
It is suggested that the high velocity components present in the Hα and Hβ lines over the entire face of the Car II region, in the Carina nebula, could have their origin in the time varying emission from η Carinae scattered by dust in the Car II region.

112.081 **A search for radio continuum emission from evolved stars with asymmetric OH maser emission.**
L. F. Rodríguez, Y. Gómez, J. A. García–Barreto.
Rev. Mex. Astron. Astrofis., Vol. 11, No. 2, p. 139 – 142 (1985). = Contrib. Inst. Astron., Univ. Nac. Autónoma Mex., No. 182.
The coexistence of ionized gas and OH maser emission is possible during the transition from red supergiant star to planetary nebula. Under these circumstances the redshifted OH complex emission may be absorbed by the ionized core and will appear weaker than the blueshifted OH maser emission. Using the *VLA* at 6 and 2–cm the authors searched for radio continuum emission in a list of OH/IR stars with such an asymmetric OH maser emission. No detections were made, and the authors conclude that these stars are still in the superwind phase and that the OH asymmetry is caused by envelope asymmetries and not by the presence of an ionized core.

112.082 **Soft X–ray spectra of stellar–wind bubbles.**
N. G. Bochkarev.
Sov. Astron., Vol. 29, No. 5, p. 509 – 515 (1985). English translation of 40.112.062.

112.083 **Middle infrared spectropolarimetry of Eta Carinae.**
G. P. Briggs, D. K. Aitken.
Proc. Astron. Soc. Aust., Vol. 6, No. 2, p. 145 – 147 (1985).
Spectropolarimetric observations of the 10 arc second region surrounding Eta Carinae have been made in the 8 – 13 μm wavelength band. The observed polarization, of a few percent, is due to emission of radiation from aligned grains. The radial position angle of the polarization suggests that gas streaming is responsible for the grain alignment.

112.084 **Some regularities in the variation of the Balmer line profiles in the P Cyg spectrum.**
N. S. Markova.
Astrophys. Space Sci., Vol. 123, No. 1, p. 5 – 19 (1986).
In the coudé focus of the 2 m RCC telescope at the National Astronomical Observatory of the Bulgarian Academy of Science 47 spectrograms covering the range from 3500 to 4900 Å with a reciprocal dispersion of 9 Å mm^{-1} and a spectral resolution of 0.18 Å were obtained. The photometric and position behaviour of Balmer lines were investigated over a long time–scale, i.e. March 1981 to August 1982. It is found that the absorption consists of three and sometimes four components. Their position behaviour may be explained by the existence of two kinds of shells: pulsating and steadily expanding. The data suggest a periodic change in the radial velocities, intensity of each separate absorption component, total equivalent widths and profiles.

112.085 **A phenomenological interpretation of stellar chromospheres.**
G. A. Gurzadyan.
Astrophys. Space Sci., Vol. 123, No. 1, p. 67 – 101 (1986).
The following problems are examined: observed emission powers of 'magnesium' chromospheres on stars based on 2800 Mg II observations; dependence of chromosphere emission on spectral and luminosity classes; stellar chromospheres as an accidental event; chromospheres of components of binary systems; stars with the chromospheres of solar type and nonsolar type; distribution of stars by means of the type of their chromosphere on luminosity class; stars with superpower magnesium emission; emission measures for both the 'magnesium' and 'calcium' chromospheres; interrelation between chromosphere, transition zone and corona; chromospheric activity and rotation of stars; possibility of the existence of chromospheres on hot stars; phenomenological picture of stellar chromospheres; stars without the line 2800 Mg II; syndrome of red giant HD 4174. At the end, heating of stellar chromospheres is discussed.

112.086 **Carbon monoxide emission from stars in the IRAS and Revised AFGL catalogs. I. Mass loss driven by radiation pressure on dust grains.**
B. Zuckerman, H. M. Dyck.
Astrophys. J., Vol. 304, No. 1, p. 394 – 400 (1986).
The authors surveyed infrared bright stars in the IRAS and Revised AFGL catalogs in the $J = 2 \rightarrow 1$ rotational transition of carbon monoxide. Broad lines were detected in 39 objects not previously seen in CO emission, including the unusual supergiant VY CMa, the bright giant star W Hya, and some stars first discovered in the IRAS survey. A plot of momentum in the outflowing wind, deduced from a simple model for the CO emission, versus momentum in the stellar radiation field is presented.

112.087 **Carbon monoxide emission from stars in the IRAS and Revised AFGL catalogs. II. Massive carbon stars.**
B. Zuckerman, H. M. Dyck, M. J. Claussen.
Astrophys. J., Vol. 304, No. 1, p. 401 – 408 (1986).
The authors surveyed infrared bright stars in the IRAS and Revised AFGL catalogs in the $J = 1 \rightarrow 0$ and, to a lesser extent, $J = 2 \rightarrow 1$ rotational transitions of carbon monoxide. Broad lines were detected in 25 objects not previously seen in CO emission, including some stars first discovered in the IRAS survey. The authors discovered a number of carbon–rich stars at low Galactic latitude with large outflow velocity V_∞, which are very likely to be among the most massive and luminous carbon–rich objects in the Galaxy. The high–latitude prototypical carbon star with a large mass loss rate, IRC +10216, has a very modest V_∞ of ~ 15 km s^{-1}. It is suggested that IRC +10216 is considerably closer to Earth than is commonly believed, perhaps as close as 100 – 150 pc.

112.088 **Asymmetries in the CO lines of IRC +10216.**
P. J. Huggins, A. P. Healy.
Astrophys. J., Vol. 304, No. 1, p. 418 – 424 (1986).

The authors present high signal–to–noise spectra of the CO (1–0) and (2–1) lines in IRC +10216 to study the line shapes in detail. The profiles exhibit slight asymmetries which are consistent with those expected for optically thick lines formed in an expanding envelope in the presence of a small turbulent velocity in the gas. Analysis of the data yields values for the expansion and turbulent (FWHM) velocities of the envelope of 14.1 and 0.9 km s^{-1}.

112.089 **IRAS Low–Resolution Spectrograph observations of silicate and molecular SiO emission in Mira variables.**
M. S. Vardya, T. de Jong, F. J. Willems.
Astrophys. J., Lett. Ed., Vol. 304, No. 1, p. L29 – L32 (1986).

The authors have studied Low–Resolution Spectrograph spectra obtained by IRAS for a sample of about 20 bright M Mira variables with known mass–loss rates. It is found that the spectra of stars with asymmetric light curves show the well–known 9.7 μm and 20 μm silicate emission features while stars with more symmetric light curves show much weaker broad 12 μm and 20 μm emission features, presumably also due to some kind of silicate dust. The spectra of 80% of the stars in the sample further show a narrow 8 μm emission feature that is attributed to the $\upsilon = 1$–0 vibration–rotation band of SiO.

112.090 **The spatial structure of IRC +10216 and NGC 7027 in the far–infrared.**
D. F. Lester, P. M. Harvey, M. Joy.
Astrophys. J., Vol. 304, No. 2, p. 623 – 633 (1986).

High signal–to–noise ratio, far–infrared slit scans of the evolved mass outflow sources IRC +10216 (a carbon star) and NGC 7027 (a planetary nebula) are presented. Combined with well–known point source profiles, using nonlinear deconvolution techniques, these 50 and 100 μm scans give information on source structure that is on a scale approximately 10 times smaller than that previously achieved in the far infrared. The data lead to some significant changes in the picture of dust emission around IRC +10216, and have implications for the chemical equilibrium of molecular species in the extended envelope. For NGC 7027 the scans set limits on the temperature and spatial extent of the cold dust and show that the warmer dust grains responsible for the far–infrared emission are intimately associated with the ionized gas.

112.091 **SiO and H_2O maser emission in OH/IR objects and late type variable stars.**
L.–Å. Nyman, L. E. B. Johansson, R. S. Booth.
Onsala Space Obs., Prepr., No. 85:53, 36 pp. (1985). To appear in Astron. Astrophys.

112.092 **Circumstellar shells around luminous emission–line stars in the Large Magellanic Cloud.**
O. Stahl, B. Wolf.
ESO Sci. Prepr., No. 414, 27 pp. (1986). To appear in Astron. Astrophys.

112.093 **Las estrellas P Cygni.**
A. Cassatella.
Bol. Astron. Obs. Madr., Vol. 11, No. 3, p. 17 – 24 (1984).

112.094 **Máseres SiO asociados a estrellas variables de largo período.**
V. Bujarrabal.
Bol. Astron. Obs. Madr., Vol. 11, No. 3, p. 25 – 36 (1984).

112.095 **The outer atmosphere of the carbon star TX Piscium.**
K. Eriksson, B. Gustafsson, H. R. Johnson, F. Querci, M. Querci, J. H. Baumert, M. Carlsson, H. Olofsson.
Astron. Astrophys., Vol. 161, No. 2, p. 305 – 313 (1986).

The authors have obtained a high–resolution LWP IUE spectrum of the bright N–type carbon star TX Psc, which shows unusual profiles of the Mg II *h* and *k* emission lines and also a number of Fe II emission lines. They have detected the radio CO ($J = 1$–0) line from the circumstellar shell, and derived a column density of $\leqslant 10^{22} H_2$ molecules per cm^2 and a radial velocity close to that of the star; results that are consistent with those obtained from the ultraviolet data. The flux distribution in the far infrared indicates some dust emission from carbon grains. A significant fraction of all the carbon in the shell seems to be in this form. The mass–loss rate for the star, as judged from its envelope, is estimated to $\cong 10^{-6} - 10^{-8} M_\odot$ per year.

112.096 **Radio continuum emission from winds, chromospheres, and coronae of cool giants and supergiants.**
S. A. Drake, J. L. Linsky.
Astron. J., Vol. 91, No. 3, p. 602 – 620 (1986).

The authors present the results of a sensitive VLA radio continuum survey at 6 cm of 39 of the nearest, single cool giants and supergiants with spectral types in the range G0 – M5. They discuss their findings in the context of the various mechanisms that might be producing radio emission in these cool stars. They consider seven detected sources in detail, and present additional 2 cm data, which they have obtained in order to determine the radio spectral indices. The authors discuss the implications of the findings of their survey on the mass–loss rates and ionization fractions of stellar winds, as a function of spectral type and luminosity in the cool giant and supergiant portion of the H–R diagram.

112.097 **Spectrophotometry of the condensations around Eta Carinae.**
D. A. Kendall, R. J. Dufour.
Bull. Am. Astron. Soc., Vol. 18, No. 1, p. 637 (1986). Abstract. – See Abstr. 010.066.

112.098 **The optical counterpart of OH/IR 17.7–2.0.**
T. Le Bertre.
Messenger, No. 44, p. 6 – 9 (1986).

112.099 **The circumstellar gas cloud around Beta Pictoris. II.**
A. Vidal–Madjar, L. M. Hobbs, R. Ferlet, C. Gry, C. E. Albert.
Inst. Astrophys. Paris, Pré–Publ., No. 140, 34 pp. (1986). To appear in Astron. Astrophys.

112.100 **The Beta Pictoris circumstellar disk. IV. Redshifted UV lines.**
A. M. Lagrange, R. Ferlet, A. Vidal–Madjar.
Inst. Astrophys. Paris, Pré–Publ., No. 149, 13 pp. (1986). To appear in Astron. Astrophys.

112.101 **The Hα line in Ae and A shell stars.**
Y. Andrillat, C. Jaschek, M. Jaschek.
Bull. Inf. Etoiles Be, No. 13, p. 2 – 4 (1986). Abstract. – See Abstr. 010.563. To appear in Astron. Astrophys., Suppl. Ser.

112.102 **V/R variability in X–ray emitting Be star binaries.**
K. M. V. Apparao, S. P. Tarafdar.
Bull. Inf. Etoiles Be, No. 13, p. 4 – 5 (1986). Abstract. – See Abstr. 010.563. To appear in Astron. Astrophys.

112.103 **Mass loss and nonradial oscillations in Omega Orionis.**
D. Baade.
Bull. Inf. Etoiles Be, No. 13, p. 5 – 6 (1986). Abstract. – See Abstr. 010.563.

112.104 **Interferometric program on Be stars (II).**
P. Granes, C. Thom, F. Vakili.
Bull. Inf. Etoiles Be, No. 13, p. 7 – 9 (1986). Abstract. – See Abstr. 010.563.

112.105 **Be stars in open clusters.**
A. Slettebak.
Bull. Inf. Etoiles Be, No. 13, p. 10 (1986). Abstract. – See Abstr. 010.563.

112.106 **Geometry of rotating envelopes around Be stars as derived from comparative analysis of Hα emission line profiles.**
J. Dachs, R. Hanuschik, D. Kaiser, D. Rohe.
Bull. Inf. Etoiles Be, No. 13, p. 11 – 12 (1986). Abstract. – To appear in Astron. Astrophys.

112.107 **High–resolution emission–line spectroscopy of Be stars. 1. Evidence for a two–component structure of the Hα emitting envelope.**
R. W. Hanuschik.
Bull. Inf. Etoiles Be, No. 13, p. 14 (1986). Abstract.

112.108 **Influence of abundances on mass loss determination for WC stars.**
K. A. van der Hucht, J. P. Cassinelli, P. M. Williams.
Working Group Stars With Extended Atmospheres, Prepr., No. 77, 16+8 pp. (1986). To appear in Astron. Astrophys.

112.109 **Current microvariability for the Be star φ Persei.**
E. M. Halbedel.
Inf. Bull. Variable Stars, No. 2885, 2 pp. (1986).

112.110 **Recent mass loss episode of the Be star o And.**
K. K. Ghosh, K. Jayakumar, M. J. Rosario, M. Appakutty.
Inf. Bull. Variable Stars, No. 2897, 4 pp. (1986).

112.111 **Interstellar broadening of maser spots.**
M. Szymczak.
Acta Astron., Vol. 35, Nos. 3–4, p. 305 – 312 (1985).
A sample of OH maser spot sizes available from VLBI observations is analysed. Using the known parameters of pulsars near the maser sources, and the extended medium model, the angular scattering of the maser components is calculated. It is suggested, basing on a sample of eleven sources, that on the average, the apparent angular diameters of OH maser components are a factor of 9 ± 2 greater than the scattered diameters of point sources.

112.112 **The 1985 light curve of P Cygni.**
J. R. Percy, M. Richer, P. Harmanec, J. Horn, P. Koubsky, R. C. Reisenweber.
J. Am. Assoc. Variable Star Obs., Vol. 14, No. 2, p. 75 – 76 (1985). Abstract. – See Abstr. 010.021.

112.113 **Infrared recombination line profiles: a diagnostic probe of the velocity field in the S 106 stellar wind.**
R. P. Garden, T. R. Geballe.
Mon. Not. R. Astron. Soc., Vol. 220, No. 3, p. 611 – 617 (1986).
The authors have measured velocity profiles of the Brackett alpha and Pfund gamma hydrogen recombination lines towards the young exciting star of the S 106 bipolar H II region. Both profiles show the presence of velocities on the order of 200 km s^{-1} within the ionized circumstellar envelope. The authors infer that at the higher velocities both lines are at least partially optically thick. The asymmetry in the Pfγ profile may arise as a result of either obscuration by the stellar photosphere or the presence of a bipolar wind geometry close to the star. It appears that a decelerating velocity law is appropriate at distances from the star where the optically thick lines originate.

112.114 **Condensation of dust around a carbon star.**
I. K. Shmeld (*I. K. Šmelds*).
Sov. Astron., Vol. 29, No. 6, p. 716 – 717 (1985). English translation of 40.112.106.

112.115 **Near infrared photometry of stellar maser sources.**
K.–p. Zhou, Y.–x. Hao, P.–s. Chen, Y. Zhang, H. Gao.
Chin. Astron. Astrophys., Vol. 10, No. 1, p. 62 – 66 (1986). English translation of 40.112.105.

112.116 **Ultraviolet and visual wavelength spectroscopy of gas around Eta Carinae.**
K. Davidson, R. J. Dufour, N. R. Walborn, T. R. Gull.
Astrophys. J., Vol. 305, No. 2, p. 867 – 879 (1986).
The authors report observational results on η Carinae, especially spectroscopy of the outer "S condensation" supplemented by data on the homunculus and its core. The helium abundance at the surface of η Car appears to be roughly $Y \approx 0.4$, and most of the CNO is nitrogen. The often quoted temperature estimate $T \approx 30{,}000$K for the radiating surface is confirmed. The presently observed mass loss rate is probably less than $10^{-2.4} M_{\odot}\,yr^{-1}$ if the outflow is not strongly direction–dependent. Finally, the authors quote a largely forgotten but highly relevant historical conjecture concerning η Car.

112.117 **New observations of Alpha Orionis with a rotation shearing interferometer.**
F. Roddier, C. Roddier, R. Petrov, F. Martin, G. Ricort, C. Aime.
Astrophys. J., Lett. Ed., Vol. 305, No. 2, p. L77 – L80 (1986).
New two–dimensional maps of the visibility of fringes produced by Alpha Orionis have been obtained at several wavelengths using the McMath telescope at Kitt Peak in 1982 February. These maps show evidence for a time evolution of the dust envelope and indicate a possible stellar companion.

112.118 **Spectroscopic observations of Be stars.**
J. A. de Freitas Pacheco, S. J. C. Landaberry, D. F. Lopes.
Rev. Mex. Astron. Astrofis., Vol. 12, Spec. Issue, p. 185 – 187 (1986). – See Abstr. 012.066.
The authors report some preliminary results of an observational campaign for detecting Fe II lines in Be stars.

112.119 **High resolution observations of chromospheric lines in late–type dwarfs.**
B. H. Foing, J. Beckman, L. Crivellari, D. Galleguillos.
Rev. Mex. Astron. Astrofis., Vol. 12, Spec. Issue, p. 214 (1986). Abstract. – See Abstr. 012.066.

112.120 **The outer atmospheric layers of FK Comae stars.**
L. Bianchi, M. Grewing.
Mitt. Astron. Ges., No. 65, p. 196 – 198 (1986). – See Abstr. 012.072.

112.121 **η Carinae.**
IAU Circ., No. 4199 (1986).

112.122 **On the variations in the emission in Hα and Hβ lines in the spectra of some Be stars: γ Cas, ζ Tau, 66 Oph.**
T. S. Galkina.
Izv. Krymskoj Astrofiz. Obs., Tom 72, p. 72 – 84 (1985). In Russian. English translation in Bull. Crimean Astrophys. Obs., Vol. 72.
The behaviour of the emission in Hα and Hβ lines in the spectra of the Be stars γ Cas, ζ Tau, 66 Oph is studied. Short time–scale variations (30 – 60 minutes) in the relative intensities of V and R components of the Hα and Hβ emission lines are discovered. They are accompanied by changes in the radial velocities of the V component. It is noted that the activity of γ Cas increases when the relative intensities are being changed from $V/R > 1$ to $V/R < 1$. The structure of Hβ emission line profiles in the spectrum of γ Cas and the one of the Hα line in ζ Tau is multi–componental, what indicates tiny inhomogenieties of the envelope.

112.123 **Activity in the envelope of the G–type hypergiant HD 217476.**
J. Smoliński, J. L. Climenhaga, J. M. Fletcher.
Luminous stars and associations in galaxies, p. 89 – 90 (1986). – See Abstr. 012.079. (IAU Symp. No. 116).
The authors have observed HD 217476 with the same equipment during last fifteen years at the Dominion Astrophysical Observatory. Their present results are based upon several

hundreds of high dispersion spectra (6 and 10 Å/mm). A binary model is presented.

112.124 **IRAS observations of the cool galactic hypergiants.**
D. J. Stickland.
Luminous stars and associations in galaxies, p. 91 – 92 (1986). – See Abstr. 012.079. (IAU Symp. No. 116).
The apparent absence of very luminous red stars suggests either that stars never reach the right-hand side of the HR diagram during the course of their evolution at all, or that if they do, their sojourn there is very brief indeed. If the latter is the case, one might expect to see an increase in mass-loss activity with luminosity. An attempt has been made to search for one of the signatures of mass-loss – thermal reemission from ejected dust – using recently released data from the Infrared Astronomical Satellite.

112.125 **O-star winds: is rotation important?**
R. Prinja, I. Howarth.
Luminous stars and associations in galaxies, p. 93 – 94 (1986). – See Abstr. 012.079. (IAU Symp. No. 116).
The authors are working on an ultraviolet study of O star winds, and have preliminary mass-loss rates and terminal velocities for about 150 stars; here they examine the dependence of these data on rotation rate.

112.126 **The dependence of the stellar rate of mass loss on effective temperature and velocity.**
C. de Jager, H. Nieuwenhuijzen, K. A. van der Hucht.
Luminous stars and associations in galaxies, p. 109 – 110 (1986). – See Abstr. 012.079. (IAU Symp. No. 116).
From the existing literature data have been collected on the stellar rate of mass loss for 189 stars with known or derived values of the effective temperature T_{eff} and luminosity L. It appears that $\dot{M}$ depends only on T_{eff} and L for the O- through M-type stars brighter that about $3 \times 10^3 L_\odot$.

112.127 **P Cygni type stars: evolution and physical processes.**
H. J. G. L. M. Lamers.
Luminous stars and associations in galaxies, p. 157 – 182 (1986). – See Abstr. 012.079. (IAU Symp. No. 116).
The class of P Cygni type (PCT) stars is defined and the brightest members in our galaxy, LMC, M31 and M33 are identified. They are located near the upper luminosity limit in the HRD in the range of $8500 \leqslant T_{eff} \leqslant 27{,}000$K and $-11.0 \leqslant M_{bol} \leqslant -7.8$. The author suggests that all PCT stars are S Dor variables, and that the reverse may also be true. The basic parameters of the PCT stars are derived and compared with those of normal supergiants. The history and the physical processes in the star P Cygni are discussed.

112.128 **The massive stellar wind of the Hubble-Sandage variable S Doradus.**
C. Leitherer, I. Appenzeller, G. Klare, H. J. G. L. M. Lamers, O. Stahl, L. B. F. M. Waters, B. Wolf.
Luminous stars and associations in galaxies, p. 235 – 236 (1986). – See Abstr. 012.079. (IAU Symp. No. 116).
The authors present coordinated spectroscopic and photometric observations obtained during the present bright phase of the luminous variable star S Dor in the LMC. Their new observations support earlier suggestions that S Dor consists of a hot stellar core which during the maximum state is surrounded by a very extended optically thick gaseous envelope.

112.129 **MWC 300: a runaway hypergiant.**
B. Wolf, O. Stahl.
Luminous stars and associations in galaxies, p. 248 (1986). – See Abstr. 012.079. (IAU Symp. No. 116).
The authors have analysed highly resolved CASPEC spectra with a high S/N-ratio of the peculiar emission-line B-star MWC 300, which is surrounded by a circumstellar dust shell. These high quality spectra enabled to study the photospheric spectrum of MWC 300.

112.130 **New results on Eta Carinae. Evidence for an asymmetric, inhomogeneous wind.**
R. Viotti, L. Rossi, A. Altamore, C. Rossi, A. Cassatella.
Luminous stars and associations in galaxies, p. 249 – 252 (1986). – See Abstr. 012.079. (IAU Symp. No. 116).
The authors present new optical and UV observations of Eta Car and its small nebula with the aim of clearing the physical nature of its wind.

112.131 **Infrared spectroscopy of southern P Cygni stars.**
P. J. McGregor, A. R. Hyland, D. J. Hillier.
Luminous stars and associations in galaxies, p. 257 – 258 (1986). – See Abstr. 012.079. (IAU Symp. No. 116).
Investigations of the 1.0 – 2.5 μm spectra of high luminosity P Cygni stars reveal the presence of several new emission features.

112.132 **B[e]-stars of the Magellanic Clouds.**
F.-J. Zickgraf, B. Wolf, O. Stahl, C. Leitherer.
Luminous stars and associations in galaxies, p. 265 – 267 (1986). – See Abstr. 012.079. (IAU Symp. No. 116).
The authors carried out a spectroscopic and photometric study of the 8 presently known B[e]-stars of the Magellanic Clouds, i.e. R4 and R50 of the SMC, R66, R82, R126, Hen S12, Hen S22 and Hen S134 of the LMC.

112.133 **The mass loss rate of SK 80 (O7 Iaf) in the Small Magellanic Cloud.**
A. J. Willis, I. D. Howarth, K. Nandy, D. H. Morgan.
Luminous stars and associations in galaxies, p. 269 – 270 (1986). – See Abstr. 012.079. (IAU Symp. No. 116).
The authors have secured HIRES IUE and optical spectra of SK 80 and have attempted to derive the mass loss rate from these data.

112.134 **IRAS observations of cool excess around main sequence stars.**
F. C. Gillett.
Light on dark matter, p. 61 – 69 (1986). – See Abstr. 012.080.
IRAS observations of the main sequence stars α Lyr, α PsA, β Pic and ε Eri are reviewed and updated. All four stars exhibit cool excess emission that is attributed to thermal emission by orbiting circumstellar grains. A simple thermal model is used to determine surface areas and temperature ranges for the radiating material, and the state of this material is briefly discussed in terms of the formation and evolution of planetary systems.

112.135 **The infrared excess from stellar winds.**
H. J. G. L. M. Lamers.
Light on dark matter, p. 79 – 82 (1986). – See Abstr. 012.080.
The IR excess can be used to study the density or velocity structure in the lower parts of the winds of early type stars. The depth of the emitting region and the excess in magnitudes is given for typical O-stars, as a function of λ and $\dot{M}$. The first results of an ongoing study of the IRAS data are discussed briefly.

112.136 **The disc structure and mass loss rates of Be stars.**
L. B. F. M. Waters.
Light on dark matter, p. 83 – 86 (1986). – See Abstr. 012.080.
IRAS observations of four well-known Be stars, α Eri, φ Per, δ Cen and χ Oph are used to derive the density structure of the circumstellar discs and the mass loss rates. α Eri shows no sign of IR excess up to $\lambda = 60$ μm, but the other three have excesses in the order of a factor 10 to 70, caused by free-free radiation from the disc. The mass loss rates are of the order 7×10^{-9} to $2 \times 10^{-8} M_\odot$/yr.

112.137 **IRAS observations of Wolf-Rayet stars.**
K. A. van der Hucht, T. A. Jurriens, F. M. Olnon, P. S. Thé, P. R. Wesselius, P. M. Williams.
Light on dark matter, p. 87 – 89 (1986). – See Abstr. 012.080.
IRAS observations of Wolf-Rayet stars are used as diagnostics of hot circumstellar dust shells, cool dust in WR ring nebulae, the Ne/He abundance ratio, and the interstellar extinction. In two

cases the IR energy distributions indicate WR planetary nucleus status rather than population–I WR status.

112.138 **Dust formation in Wolf–Rayet stellar winds.**
K. A. van der Hucht, P. M. Williams, P. S. Thé.
Light on dark matter, p. 90 (1986). Abstract. – See Abstr. 012.080.

112.139 **Mass loss by cool stars.**
B. Zuckerman.
Light on dark matter, p. 93 – 100 (1986). – See Abstr. 012.080.
Mass loss by cool giant stars is very likely the most important mechanism for returning processed stellar material to the interstellar medium. This mass loss has profound effects on the evolution of individual intermediate mass stars and the chemical evolution of the Milky Way. The IRAS point source catalog is ideally suited for identifying those stars with large mass loss rates in dust grains. These stars, in turn, become prime targets for radio and infrared studies of outflowing gas. The total rate of return of processed material to the interstellar medium can then be obtained.

112.140 **Models of IRAS observations of circumstellar shells.**
M. Rowan–Robinson, A. Lock, D. W. Walker, S. Harris.
Light on dark matter, p. 101 – 102 (1986). – See Abstr. 012.080.
The authors have modelled the IRAS observations of the circumstellar dust shells around late–type stars previously studied by Rowan–Robinson and Harris (1982, 1983), using only high quality IRAS fluxes which do not appear extended in the raw data. Color–color diagrams for late M and carbon stars are shown, together with the predictions of simple model sequences. The spectra of a sample of the stars are presented.

112.141 **OH/IR catalogue and correlation with the IRAS data base.**
R. Breukers, W. E. C. J. van der Veen, P. te Lintel, M. Wiertz, H. Habing.
Light on dark matter, p. 105 – 106 (1986). Abstract.– See Abstr. 012.080.

112.142 **AGB stars with high mass loss rates in the bulge of our Galaxy.**
W. E. C. J. van der Veen.
Light on dark matter, p. 107 – 108 (1986). Abstract.– See Abstr. 012.080.

112.143 **The circumstellar envelope of VX Sagittarii.**
J. M. Chapman, R. J. Cohen.
Light on dark matter, p. 109 – 110 (1986). – See Abstr. 012.080.
For the luminous M–type supergiant star VX Sgr the authors have made MERLIN observations of the OH maser emission at 1612 MHz (LHC) and 1665/7 MHz (RHC) and the H_2O maser emission at 22 GHz (LHC). These have been combined with published data on the SiO masers (Lane 1984) to give a detailed description of the circumstellar envelope.

112.144 **The internal radius of CS shells around cool, oxygen–rich stars.**
R. Papoular, B. Pegourie.
Light on dark matter, p. 111 – 112 (1986). – See Abstr. 012.080.
Radiative models have been developed to describe the CS shells of cool giants and supergiants. Applying these models to the observed spectra of about 30 oxygen–rich stars, the authors have derived the common optical properties of the CS grains in the range 8 – 30 μm, as well as the optical thickness and internal radius of the shells.

112.145 **Ground–based and IRAS observations of proto–planetary nebulae.**
S. Kwok, B. J. Hrivnak, E. F. Milone, R. T. Boreiko.
Light on dark matter, p. 127 – 128 (1986). – See Abstr. 012.080.
Infrared photometric observations were made with the 3.6 m Canada–France–Hawaii Telescope in May and August of 1984. Sixteen IRAS sources with colour temperatures ranging from 150 to 400K were identified. Five sources show silicate features in emission or absorption and are likely to be oxygen–rich OH/IR stars. The others show no obvious spectral features and could be carbon stars. The authors also observed a number of compact ($\theta < 2''$) PN and found 8 to have definite far–infrared excesses. The colour temperatures of these nebulae are all <200K.

112.146 **IR CCD imaging of L 1551–IRS 5: direct observations of its circumstellar shell.**
A. Moneti, J. L. Pipher, W. J. Forrest, C. E. Woodward.
Light on dark matter, p. 315 – 318 (1986). – See Abstr. 012.080.
The authors present new, high spatial resolution images of L 1551–IRS 5 at 1.65, 2.2, and 3.8 μm. The observations are interpreted in terms of a flattened circumstellar shell that is viewed from about 20° above its equatorial plane. In this model, the central star is not seen directly, but only scattered light from one of the polar regions, where the shell is thinnest, is observed.

112.147 **A model for bipolar sources in molecular clouds.**
M. D. Smith.
Light on dark matter, p. 319 – 320 (1986). – See Abstr. 012.080.
A wind originating from a young star in a molecular cloud can be focussed into bipolar channels via shock deflection. The shocked material is here presumed to rapidly re–cool, flowing in a supersonic layer between the shock and the cloud. Stationary twin–jet flow patterns are possible even when the ambient pressure is isotropic about the central star. Single jet outflows can also occur when the stellar wind is interacting with the immediate protostellar environment.

A statistical analysis and overview of the IRAS point source catalog.
See Abstr. 002.092.

A statistical analysis of the LRS catalog.
See Abstr. 002.093.

Investigations of B and Be stars in the U.S.S.R. in 1985.
See Abstr. 013.054.

Report of activities concerning Be stars from Italy.
See Abstr. 013.055.

Photometric observing campaign for Be stars. Progress Report No. 9.
See Abstr. 013.056.

Stability of amorphous circumstellar silicate grains.
See Abstr. 022.023.

Proton threshold states in ^{26}Al and their role in astrophysics.
See Abstr. 022.097.

Low energy transmission grating spectrometer (LETGS) for AXAF.
See Abstr. 035.021.

Surface and circumstellar stellar imaging.
See Abstr. 036.246.

Energy balance in stellar coronae.
See Abstr. 051.045.

On pumping astronomical masers.
See Abstr. 062.046.

Astrophysical masers with magnetic fields and velocity gradients.
See Abstr. 063.043.

On the radiative–radiative pumping of circumstellar SiO masers.
See Abstr. 063.093.

Limits of interacting stellar winds model of planetary nebulae. A numerical study.
See Abstr. 064.013.

A simple explanation for the Linsky–Haisch boundary line for transition layers.
See Abstr. 064.015.

Atmospheric structure and mass loss from pulsation with speculative application to B and Be stars.
See Abstr. 064.016.

Summary comments on the state of stellar–wind theory as it relates to nonradial pulsations.
See Abstr. 064.017.

Models of transition regions in hybrid stars.
See Abstr. 064.021.

Konsistente sphärische NLTE–Modelle für Objekte mit Leuchtkräften entsprechend B–Sternen.
See Abstr. 064.028.

Stars in a gas–dust envelope.
See Abstr. 064.031.

Chromospheric dust formation, stellar masers and mass loss.
See Abstr. 064.034.

Shell phenomenon in Be stars.
See Abstr. 064.044.

Radiatively–driven stellar winds.
See Abstr. 064.051.

Two–dimensional models of stellar wind bubbles. V. The efficiency of momentum transfer between winds and dense clumps in free wind regions.
See Abstr. 064.055.

Alfvén wave driven wind in late type giants and supergiants. I. Critical solutions.
See Abstr. 064.057.

Mass loss from very young massive stars.
See Abstr. 064.059.

Coagulation and multiplication of dust particles in media with carbon excess.
See Abstr. 064.067.

Radiative transfer in circumstellar 1612 MHz OH masers.
See Abstr. 064.068.

Intrinsic parameters of hot blue stars.
See Abstr. 064.071.

Relation of chromospheric activity to convection, rotation, and pre–main–sequence evolution.
See Abstr. 065.002.

The Oort cloud in transition.
See Abstr. 102.019.

Random fluctuations versus Poynting–Robertson drag on interplanetary dust grains.
See Abstr. 106.008.

Distance determination for OH/IR stars: a geometrical method.
See Abstr. 111.005.

Carbon star radial velocities and dark matter in the universe.
See Abstr. 111.013.

The proper motion of VX Sgr derived from library material.
See Abstr. 111.022.

IRAS observations of Wolf–Rayet stars.
See Abstr. 113.045.

Infrared flux excesses of the warm carbon stars.
See Abstr. 113.052.

A search for infrared excesses in G–type stars.
See Abstr. 113.063.

IRAS observations of carbon stars.
See Abstr. 113.065.

Spectroscopic studies of Wolf–Rayet stars. III. The WC subclass.
See Abstr. 114.003.

Evolutionary effects on the surface abundances of an early–type supergiant.
See Abstr. 114.020.

Search for transition zone lines in early dwarf A stars.
See Abstr. 114.028.

Fe II in the UV spectrum of luminous emission line stars.
See Abstr. 114.037.

Ultraviolet observations of metal deficient field giants.
See Abstr. 114.056.

On the variation of Hα emission in the spectrum of the Be star 28 Cyg.
See Abstr. 114.104.

Comparison of spectral characteristics of three Be stars: φ Persei, θ Coronae Borealis and X Persei.
See Abstr. 114.105.

The profiles of Balmer lines in Ae/Be spectra.
See Abstr. 114.113.

A search for rotational modulation of the ultraviolet spectra of ψ^3Piscium and 31 Comae Berenices.
See Abstr. 114.117.

Spectral energy distributions of Be stars. I. Measurements of 26 southern and equatorial stars.
See Abstr. 114.126.

Steps to a new calibration of the spectral type – effective temperature relationship for early–type stars.
See Abstr. 114.140.

Optical and ultraviolet spectral morphology of luminous OB stars in the Galaxy and the Magellanic Clouds.
See Abstr. 114.143.

Parameters of Wolf–Rayet stars.
See Abstr. 115.031.

Luminosities of OH/IR stars.
See Abstr. 115.033.

Magnetic structure in cool stars. IX. Ultraviolet emission lines from chromospheres and transition regions.
See Abstr. 116.001.

EXOSAT detection of an X–ray flare from the solar type star π^1UMa.
See Abstr. 116.005.

A survey of chromospherically active stars.
See Abstr. 116.009.

Variability in H alpha, polarization, and magnetic fields in early–type stars.
See Abstr. 116.010.

The correlation between rotation and IR colour excess for B–type dwarfs.
See Abstr. 116.011.

Solar–like activity in warm stars.
See Abstr. 116.022.

An extensive survey of photospheric magnetic fields in G and K dwarfs.
See Abstr. 116.023.

Optical photometry and spectroscopy for five dwarf M stars.
See Abstr. 116.026.

Radio emission from galactic Wolf–Rayet stars and the structure of Wolf–Rayet winds.
See Abstr. 116.032.

An unbiased X–ray sampling of stars within 25 parsecs of the Sun.
See Abstr. 116.033.

Thermal radio emission from stars.
See Abstr. 116.034.

Evidence for a common magnetic driver for flares and quiescent coronae/chromospheres.
See Abstr. 116.035.

Activity on cool stars.
See Abstr. 116.036.

X–ray emission from normal stars.
See Abstr. 116.037.

X–ray coronae of late–type stars: theoretical implications.
See Abstr. 116.038.

The dependence of He I λ10830 absorption strength upon X–ray emission in late–type stars.
See Abstr. 116.047.

Comments on the hotspots model for the polarization of α Orionis.
See Abstr. 116.054.

Reply to "Comments on the hotspots model for the polarization of α Orionis".
See Abstr. 116.055.

X–ray, ultraviolet, optical and radio observations of cool stars.
See Abstr. 116.061.

On the rotation of 53 Be stars.
See Abstr. 116.065.

Multicolor polarimetry of selected Be stars: a long–term analysis.
See Abstr. 116.071.

Brackett–gamma–line profile of MWC 349.
See Abstr. 117.018.

The accretion of supergiant wind material onto the hot companion in Zeta Aurigae binaries.
See Abstr. 117.042.

X–rays and colliding winds in symbiotic novae.
See Abstr. 117.180.

Optical and UV spectroscopy of the peculiar RS CVn system RT Lacertae.
See Abstr. 117.237.

SW Lacertae, W Ursae Majoris, YY Eridani, and the prevalence of starspots in cool contact binaries.
See Abstr. 117.272.

IRAS observations of symbiotic objects.
See Abstr. 117.329.

A propos de la découverte d'une nouvelle planète.
See Abstr. 118.010.

An EXOSAT observation of quiescent and flare coronal X–ray emission from Algol.
See Abstr. 119.019.

HD 47755, a new eclipsing binary.
See Abstr. 119.060.

Circumstellar envelope around the binary system λ And.
See Abstr. 120.003.

Spectral similarities in the visible and UV spectrum of Herbig Ae/Be stars.
See Abstr. 121.001.

Brackett–gamma–line observations of G 45.13 + 0.14A.
See Abstr. 121.002.

A high resolution atlas of two T Tauri stars.
See Abstr. 121.006.

Studies of the Chamaeleon star–forming region. II. The pre–main–sequence stars HD 97048 and HD 97300.
See Abstr. 121.011.

Young stellar objects and their circumstellar dust: an overview.
See Abstr. 121.012.

Far–infrared photometry of low–mass pre–main–sequence stars with broad CO wings.
See Abstr. 121.019.

UBVR photoelectric observations of four Herbig Be–Ae stars.
See Abstr. 121.022.

Molecular outflows and mass loss in pre–main sequence stars.
See Abstr. 121.029.

The development of thermally driven winds around low mass protostars.
See Abstr. 121.030.

The IR line–to–luminosity relationship of young stellar outflows.
See Abstr. 121.031.

Infrared spectra of rotating protostars.
See Abstr. 121.034.

Multicolour photometry of Ae/Be Herbig stars.
See Abstr. 121.038.

Short–term spectral variability in AB Aurigae: clues for activity in Herbig Ae stars. I. The ultraviolet lines of Mg II and Fe II.
See Abstr. 121.042.

The circumstellar environment of LkHα 234.
See Abstr. 121.046.

Molecular outflows and mass loss in pre–main–sequence stars.
See Abstr. 121.047.

A detailed optical study of HH 32 and the highly collimated outflow from the T Tauri star AS 353A.
See Abstr. 121.052.

Photoelectric *UBV* photometry of the young star YY Orionis.
See Abstr. 121.065.

Polarimetric evidence for an evolving circumstellar cloud in L_2 Puppis.
See Abstr. 122.002.

New observational results of the LMC–S Dor variable R 127 during outburst.
See Abstr. 122.004.

Brightness variability investigation of V586 Orionis.
See Abstr. 122.013.

Episodic mass motions and surface phenomena.
See Abstr. 122.026.

Evidence for mass loss from IRAS observations of classical Cepheids.
See Abstr. 122.060.

Atmospheric parameters of RV Tauri variables.
See Abstr. 122.067.

Coordinated Exosat and spectroscopic observations of flare stars and coronal heating.
See Abstr. 122.114.

UBVR photometry and optical properties of the circumstellar shell of R CrB.
See Abstr. 122.134.

A brightening of the Be star HD 46380.
See Abstr. 122.158.

The hypergiant variable HR 8752 = HD 217476.
See Abstr. 122.193.

S Doradus type: Hubble–Sandage variables.
See Abstr. 122.196.

The evaporation of pre–existing grains around supernovae.
See Abstr. 125.013.

The atmosphere of subluminous B stars. II. Analysis of 10 helium poor subdwarfs and the birthrate of sdB stars.
See Abstr. 126.010.

Detection of interstellar vibrationally excited HCN.
See Abstr. 131.004.

The injection of energy into the interstellar medium by stars.
See Abstr. 131.019.

Hydrodynamical models of the Orion–KL Nebula.
See Abstr. 131.028.

The energetics of nebulae associated with Wolf–Rayet stars.
See Abstr. 131.050.

Protostars and planets: overview from an astronomical perspective.
See Abstr. 131.072.

Nebular complex in the region of Cas OB2 association; ring nebula Sh 157.
See Abstr. 131.112.

Magnetic braking in weakly ionized media.
See Abstr. 131.166.

X–ray spectroscopic study of interstellar medium.
See Abstr. 131.187.

Interrelationships between interstellar and interplanetary grains. A working group report.
See Abstr. 131.229.

"Clean" vs. "dirty" silicate grains and the state of carbon crystallization in interstellar and circumstellar dust.
See Abstr. 131.231.

HCN emission from bipolar reflection nebulae.
See Abstr. 131.255.

The kinematics of NGC 6164–5.
See Abstr. 131.272.

Star forming loops in the IRAS sky images.
See Abstr. 131.384.

Giant stellar–wind shell associated with the H II region M16.
See Abstr. 132.057.

A search for brown dwarf candidates among the IRAS serendipitous sources.
See Abstr. 133.009.

Photographic observation of AFGL sources in Cygnus.
See Abstr. 133.011.

The infrared spectrum of M8E: evidence for circumstellar CO.
See Abstr. 133.013.

The dust and molecular gas cloud surrounding a planetary nebula: NGC 7027.
See Abstr. 134.002.

On mass loss velocity of nuclei of planetary nebulae.
See Abstr. 134.031.

Ground–based and IRAS observations of compact planetary nebulae.
See Abstr. 134.034.

X–ray emission from the planetary nebula NGC 1360.
See Abstr. 134.038.

Wind velocity and mass loss rate for the nucleus of NGC 6543.
See Abstr. 134.063.

The three types of high–mass X–ray pulsator.
See Abstr. 142.098.

Ultraviolet observations of young stars in the Chamaeleon 1 association.
See Abstr. 152.003.

The Hyades: EXOSAT broad band spectroscopy.
See Abstr. 153.014.

The galactic distribution of stellar sources found by IRAS (with some emphasis on the bulge of the Galaxy).
See Abstr. 155.122.

A large sample of emission–line stars in the Large Magellanic Cloud: their location in the Hertzsprung–Russell Diagram.
See Abstr. 156.034.

113 Photometric Properties

113.001 ***VBLUW*** **photometry of two hypergiants HD 80077 (B2 Ia$^+$) and HD 74180 (F2 Ia$^+$) and of the open cluster Pismis 11.**
W. J. G. Steemers, A. M. van Genderen.
Astron. Astrophys., Vol. 154, No. 1/2, p. 308 – 312 (1986).
The authors present and discuss *VBLUW*–photometry of HD 80077 and HD 74180, two galactic hypergiants, and in addition also of the young open cluster Pismis 11, of which HD 80077 presumably is a member. Both hypergiants were observed for several months during two consecutive years, and appear to show small, irregular variations in brightness and colours. The results of the observations concerning distances and reddening do not contradict other studies of these stars.

113.002 **Five–colour photometry of early–type stars in the direction of galactic X–ray sources.**
J. van Paradijs, S. van Amerongen, E. Damen, H. van der Woerd.
Astron. Astrophys., Suppl. Ser., Vol. 63, No. 1, p. 71 – 85 (1986).
The authors present the results of five–colour (Walraven system) photometry of 551 O– and B–type stars located in 17 fields of a few square degrees around galactic X–ray sources. The photometric data are used to study the interstellar reddening in the direction of the X–ray sources.

113.003 ***uvbyβ*** **photometry of all stars earlier than G1 and brighter than *V* ~ 12 mag in Selected Area 132 containing Scorpius X–1.**
J. Knude.
Astron. Astrophys., Suppl. Ser., Vol. 63, No. 2, p. 313 – 321 (1986).
In a continuing photometric investigation of the spatial distribution of solid interstellar material all stars in SA 132 $(l, b) = (359.^\circ 0, +24.^\circ 5)$ brighter than $V = 12$ mag and earlier than G1 in the Potsdam Spektral Durchmusterung have been observed in the *uvbyβ* system with the Danish 1.5 m telescope at La Silla. The transformations to the standard *uvby* system are established from observations of primary standards through a gray filter. An extra transformation is thus introduced. The use of a neutral filter does, however, not reduce the accuracy of the programme stars. A preliminary indication of the character of the dust distribution in this direction may be obtained from the $(b–y)$ versus β diagram. The smallest reddening values seem to be caused by dust in some local, sheet–like structures.

113.004 **Photometric variability of some early–type stars.**
G. Mathys, J. Manfroid, P. Renson.
Astron. Astrophys., Suppl. Ser., Vol. 63, No. 3, p. 403 – 416 (1986).
Twelve stars of types B to early F, which had previously been used as comparison stars in studies of photometric variability of CP stars, proved to be variable. Photometric observations of these objects are presented, and the origin of their variations is discussed.

113.005 **Photometric variability of southern CP stars.**
J. Manfroid, G. Mathys.
Astron. Astrophys., Suppl. Ser., Vol. 64, No. 1, p. 9 – 19 (1986).
Photometric observations of eight southern CP stars are presented. Periods of variation and parameters of lightcurves are obtained by least–squares fitting.

113.006 **The broad λ–5200 continuum depression in the CP star HD 126515.**
H. Hensberge, H. M. Maitzen, F. A. Catalano, H. Schneider, K. Pavlovski, W. W. Weiss.
Astron. Astrophys., Vol. 155, No. 2, p. 314 – 318 (1986).
Δa photometry of HD 126515 shows the extraordinary variability of its continuum depression around 5200 Å. The Δa peculiarity is about $0.^m06$ over most of the rotation cycle, but drops to about half that value at phase of minimum surface field strength H_s and minimum line blocking. It is argued that the surface region which produces the high peculiarity values is associated with the regions of enhanced spectral line strength, rather than with local field strength.

113.007 **Metallicism among A and F giant stars.**
B. Hauck.
Astron. Astrophys., Vol. 155, No. 2, p. 371 – 379 (1986).
132 stars considered as A and F giants have been studied for their properties in the Geneva photometric system. Temperature, absolute magnitude and [Fe/H] values are derived for stars in this part of the HR diagram. 36% of the stars of this sample exhibit an enhanced value of Δm_2 that can be interpreted in terms of [Fe/H]. The red limit of stars having an enhanced [Fe/H] value is 0.225 in $B2–V1$ or 6500K in T_{eff}. This corresponds to the limit defined by Vauclair and Vauclair (1982) where the diffusion timescale is equal to the stellar lifetime and permits the assumption that the diffusion is the process responsible for the metallicism observed in the A and F giants.

113.008 **Magnitudes of standard stars for IHW cometary photometry.**
W. Pfau, B. Stecklum.
Astron. Nachr., Vol. 307, No. 1, p. 64 (1986). = Mitt. Univ.–Sternw. Jena, Nr. 173.
For stars to be used as flux standards and solar type analogues magnitudes in the system of the International Halley Watch (IHW) filters are given.

113.009 **An investigation of the micro–variations of highly luminous OBA type stars. V.**
A. M. van Genderen.
Astron. Astrophys., Vol. 157, No. 1, p. 163 – 170 (1986).
The optical micro variations of eight high luminous stars, five of type O, two of type B and one of type A are investigated. Temperatures, gravities and reddenings are determined with the aid of theoretical colours. At least seven stars are variable, but only for three of them a characteristic period or characteristic time scale ($\bar{P}$) could be found in the light curve.

113.010 **The purity parameter of photometric systems.**
C. Jaschek, S. Frankel.
Astron. Astrophys., Vol. 158, No. 1/2, p. 174 – 176 (1986).
To characterize the capability of a photometric system to single out stars sharing certain properties (spectral class, spectral peculiarities, luminosity) a new parameter, the purity, is introduced. The use of purity is illustrated for the case of UBV system.

113.011 **Multicolour photometry of carbon and zirconium stars.**
S. E. Nersisyan.
Nauchn. Inf., Vyp. 56, p. 109 – 112 (1984). In Russian.
The results of electrophotometric observations of 18 carbon and zirconium stars in the selected photometric system are given. The energy distribution curves of S, C and SC stars are presented for different values of the C/O index. A close correlation between the energy distribution curves and the C/O ratio has been found. The colours [λ] – [5167] are determined.

113.012 **Properties of the extreme carbon star CIT 6 = RW LMi.**
A. Alksnis, M. Eglite.
Nauchn. Inf., Vyp. 56, p. 125 – 128 (1984). In Russian.
Main properties of light variations of the extreme carbon star CIT 6 = RW LMi are presented based on the observations made during the last decade.

113.013 **The brightness distribution across the disk of red giants derived from lunar occultation observations.**
M. B. Bogdanov.
Nauchn. Inf., Vyp. 56, p. 134 – 138 (1984). In Russian.
The brightness distributions across the disk of the red giants λ Aqr, TX Psc and μ Gem are presented. The results are compared with the calculations of extended atmosphere models for late–type stars.

113.014 **Reddening of B stars leaving the main sequence.**
R. M. Raznik.
Nauchn. Inf., Vyp. 56, p. 142 – 145 (1984). In Russian.
It is shown that quantitative estimates of excess (above interstellar) reddening $\Delta E_{(B-V)}$ of B0–B3 spectral subtype stars of the main sequence are the same for cluster stars and for field stars. Excess reddening is caused by influence of intrinsic circumstellar and star rotation velocities.

113.015 **A low–dispersion, near–infrared survey for galactic M supergiants.**
D. J. MacConnell, R. F. Wing, E. Costa.
Luminous stars and associations in galaxies, p. 123 – 124 (1986). – See Abstr. 012.079. (IAU Symp. No. 116).
The authors have classified 27 new, red supergiants in an area of 443 sq deg along the southern galactic plane where 34 M supergiants were previously known.

113.016 ***UBVRI* photometry of red stars.**
L. Celis S.
Astrophys. J., Suppl. Ser., Vol. 60, No. 3, p. 879 – 891 (1986).
A total of 50 Mira– and SR–type variable stars were observed by means of the photoelectric photometry *UBVRI* Kron–Cousins system. In addition, 15 nearby red dwarf stars ($d < 7$ pc) having spectral subtypes similar to those of Mira stars at maximum were observed in order to show that the conversion of the natural system into the Landolt (1983) standard system can be made for stars as red as the Mira variables, in spite of the shortage of standard late M–type stars. The relationship function and spectral type–color index scale on the Johnson system was converted into the present system.

113.017 ***VBLUW* photometry of solar–type stars.**
A. Greve, A. M. van Genderen.
Astron. Astrophys., Vol. 160, No. 2, p. 392 – 397 (1986).
VBLUW photometric observations of 35 solar–type stars are presented and discussed in terms of abundance, T_{eff} and gravity (log *g*). Using the empirical transformation between the *VBLUW* and *UBV* systems, for four stars being very similar to the sun the authors derive $(B–V) = 0.656 \pm 0.011$.

113.018 **Low dispersion spectrophotometry of bright early–type stars.**
A. Gutiérrez–Moreno, H. Moreno, P. Loyola, G. Cortés.
Astron. Astrophys., Suppl. Ser., Vol. 64, No. 2, p. 205 – 224 (1986).
Intensity distributions for 79 stars of spectral types O9 to A3, plus an F3 star and 7 standard stars, are presented. The observations cover a wavelength range from 3300 to 6000 Å. The internal errors are discussed. Synthetic *uby* colors are obtained for all the stars observed, and are compared with observed published data.

113.019 **Seven–colour photoelectric photometry of bright stars in open clusters containing red giants.**
U. Dzērvitis, O. Paupers.
Issled. Solntsa Krasnykh Zvezd, No. 20, p. 46 – 84 (1984). In Russian.
Results of photoelectric photometry in the Vilnius seven–colour photometric system of bright stars in the open clusters NGC 129, NGC 6882/5, Cr 399, Tr 2, and the group of stars besides the carbon variable NQ Cas are given. The clusters are characterized by membership of giants and supergiants and in each of them 6 – 20 stars are measured. The results and existing photometric data in UBV are combined to determine spectral classes, the value of the individual reddening of the investigated stars and true distance moduli of the clusters. The determined values of reddening and data about proper motions and radial velocities are used to evaluate membership possibility for red giants and basing on the moduli of the clusters for the members the absolute magnitude is found.

113.020 **Seven–colour photoelectric photometry of bright stars in the open cluster NGC 6633.**
U. Dzērvitis, O. Paupers.
Issled. Solntsa Krasnykh Zvezd, No. 21, p. 29 – 67 (1984). In Russian.
Results of seven–colour photoelectric observations in the Vilnius photometric system of 35 bright stars in the galactic cluster NGC 6633 are given. The results are used to determine the reddening of stars in the cluster. Published photometric data and proper motions are used to select cluster stars, and by combining the main sequence of the cluster with the standard zero–age main sequence the true distance modulus $V_0–M_v = 7^m\!.75 \pm 0^m\!.05$ is found. From proper motions and values of the reddening, the membership of 6 G – K giants to the cluster is stated, and from the derived photometric modulus individual absolute magnitudes of the cluster giants are obtained.

113.021 **VRI photometry of IRC objects.**
J. Kižla.
Issled. Solntsa Krasnykh Zvezd, No. 21, p. 68 – 80 (1984). In Russian.
95 photoelectric VRI standards are presented for selected IRC objects.

113.022 **Light variations of some carbon–rich infrared objects.**
A. Alksnis, I. Daube, A. Rudzinskis.
Issled. Solntsa Krasnykh Zvezd, No. 22, p. 33 – 70 (1985). In Russian.
Based on photographic photometry, generally in R(0,63) and V passbands, new results of light variations for six infrared carbon stars are given. For 4 stars: AFGL 2232, AFGL 2699, AFGL 2881 and BC 56 cyclic light variations and their periods are found. For 3 of them – AFGL 2699, AFGL 2881 and BC 56 – secondary variations are studied. Magnitudes for comparison stars and finding charts are presented.

113.023 **Photoelectric photometry in the Vilnius system of bright stars in the open cluster NGC 6494.**
U. Dzērvitis, O. Paupers.
Issled. Solntsa Krasnykh Zvezd, No. 22, p. 71 – 84 (1985). In Russian.
Results of photoelectric observations in the Vilnius seven–colour photoelectric system of 19 brightest stars in the open cluster NGC 6494 are given. The results have been used to determine spectral classes and individual reddenings of the observed stars. By combining with already published observations of the cluster stars in the system UBV and measurements of proper motions the distance modulus $V_0–M_V = 8^m\!.9 \pm 0^m\!.1$ and the mean reddening $\bar{E}(B–V) = 0^m\!.38 \pm 0^m\!.03$ of the cluster have been found.

113.024 **New magnitudes and color indices of the Cygnus standard region stars in the Vilnius photometric system.**
K. Zdanavičius, E. Černienė.
Vilniaus Astron. Obs. Biul., Nr. 69, p. 3 – 8 (1985). In Russian.
New more exact values of magnitudes and color indices of 33 stars of the Cygnus standard region in the Vilnius photometric system are given. The stars were measured in 1982 – 1983 15 – 20 times each with the standard Vilnius system filters. The accuracy of the final mean values are of the order of $\pm 0.002 – 0.004$ mag. Variability of the star HD 186776 (M3 III) is confirmed.

113.025 **Photoelectric photometry of stars in the vicinity of the South Galactic Pole in the Vilnius system.**
S. Bartašiūtė.
Vilniaus Astron. Obs. Biul., Nr. 69, p. 24 – 32 (1985). In Russian.
Further photoelectric observations of stars with large proper motions and other peculiar properties in an area of 900□° in the wide vicinity of the South Galactic Pole have been carried out in the Vilnius seven–color system. The present catalogue of magnitudes and color indices includes 155 stars.

113.026 **Photometric properties of the superwide–band UJFN system.**
G. Kurilienė.
Vilniaus Astron. Obs. Biul., Nr. 70, p. 14 – 29 (1985). In Russian.
Photometric properties of the superwide–band UJFN system are studied. The band–width effects are found to be so great that they should be taken into account even in the work of photographic accuracy. The U–J, J–F diagram can be used to determine photometric redshifts of galaxies in the range of redshift z from 0.1 to 0.4 and the J–F, F–N diagram in the range of z from 0.4 to 1.0.

113.027 **On a system of stars of "primary" photometric standards.**
L. N. Knyazeva, V. M. Tereshchenko, A. V. Kharitonov.
Astron. Tsirk., No. 1406, p. 1 – 3 (1985). In Russian.

113.028 **Spectrophotometric standards for comet Halley. I. List of standard stars.**
V. M. Tereshchenko, A. V. Kharitonov.
Astron. Tsirk., No. 1406, p. 3 – 5 (1985). In Russian.

113.029 **Spectrophotometric standards for comet Halley. II. E(λ) distribution.**
V. M. Tereshchenko, E. A. Glushkova.
Astron. Tsirk., No. 1407, p. 3 – 6 (1985). In Russian.

113.030 **A comparison between Washington and DDO photometry of field red giants.**
G. H. Smith.
Publ. Astron. Soc. Pac., Vol. 98, No. 600, p. 231 – 237 (1986).
Washington CMT_1T_2 photometry has been obtained of a sample of field red giants for which DDO colors are also available from the literature. Washington line–blanketing indices $\Delta(M-T_1)$ and $\Delta(C-M)$ have been derived, and are found to correlate very well, suggesting that they are both useful indicators of overall metal abundance [A/H]. Both parameters also correlate quite well with the DDO metallicity index δc_m. Any [N/A] or [C/A] abundance variations which may exist among the present sample of old disk giants do not appear to significantly affect the $\Delta(C-M)$ index.

113.031 **The photometric variability of solar–type stars. V. The standard stars 10 and 11 Leonis Minoris.**
B. A. Skiff, G. W. Lockwood.
Publ. Astron. Soc. Pac., Vol. 98, No. 601, p. 338 – 341 (1986).
Light curves in Strömgren *b* and *y* for the 1985 season are presented for the *uvby* standard stars 10 LMi (HD 82635) and 11 LMi (HD 82885). The period of 10 LMi is 40.4 days, with an amplitude of 0.012 magnitude in *y*. The period of 11 LMi is 18.0 days, with an amplitude of 0.033 magnitude in *y*. The mechanism of the variability is interpreted as rotational modulation due to active regions on the stars. Observations of 11 LMi as a standard suggest the star was only weakly variable or constant during the interval from 1972 to 1984.

113.032 **The photometric variability of Lambda Eridani.**
J. R. Percy.
Publ. Astron. Soc. Pac., Vol. 98, No. 601, p. 342 – 345 (1986).
Photoelectric *uvby* and *UBV* observations of the Be star λ Eri have been obtained during several seasons. Although the observations are not well suited to period analysis, they are consistent with the spectroscopic period of $0^d.701538$. The amplitude ranges (in *b*) from $0^m.01$ to $0^m.08$, and on one occasion, was observed to change from the minimum value to the maximum value in only two days. The season–to–season variations are $0^m.05$ (in *b*) or less. An understanding of the relationship between the photometric variations and the other variations in this star will require systematic, long–term, coordinated observations.

113.033 **On the light variations of the hydrogen–deficient variable star BD -9°4395.**
A. U. Landolt, A. D. Grauer.
Publ. Astron. Soc. Pac., Vol. 98, No. 601, p. 346 – 353 (1986). = Contrib. Louisiana State Univ. Obs., No. 196.
Multicolor photometry of the hydrogen–deficient star BD -9°4395 has shown the star to be variable in light at optical wavelengths. Analysis indicates that any possible periods which exist must be longer than one hour.

113.034 **On the Johnson *U* passband.**
M. S. Bessell.
Publ. Astron. Soc. Pac., Vol. 98, No. 601, p. 354 – 359 (1986).
The Johnson *U* passband has been investigated by computing color indices for the Vilnius spectra using a variety of passbands, all filter defined on the red edge, and by direct observations of standard equatorial stars using the same filters. Comparison of these data enables the adoptions of a *U* passband which is filter defined on the red edge and tube and atmosphere defined on the blue edge, and which appears to match Johnson *U* for all spectral types.

113.035 **Electrophotometric observations of red stars in the vicinity of the open star clusters C 0027+5957, C 0448+4337, C 0646+4107, C 1940+4004, C 2010+2624 (NGC 129, 1664, 2281, 6819, 6882/5).**
P. E. Zakharova, V. P. Kozhevnikov, T. S. Polushina, T. P. Ermakova, N. V. Matkin, V. V. Kulagin, T. P. Gerasimenko.
Astronomical–geodetical investigations, p. 67 – 83 (1985). In Russian. Abstr. in Ref. Zh., 51. Astron., 4.51.525 (1986). – See Abstr. 003.024.

113.036 **On the interpretation of Strömgren photometry.**
J. B. Alexander.
Mon. Not. R. Astron. Soc., Vol. 220, No. 2, p. 473 – 483 (1986).
It is shown that at least four parameters are needed to explain the intrinsic colours of G dwarfs very accurately. This limits the precision with which [Fe/H] and surface gravity can be estimated from Strömgren photometry alone. Definite conclusions as to the origin of the Hyades anomaly must await more precise model atmospheres and a more accurate knowledge of the chemical composition of the stars concerned; however, it is likely that most of the anomaly can be attributed to differences in line blanketing.

113.037 **Intermediate and narrow band photometry of newly discovered CP 2 stars.**
H. Schneider.
Astron. Astrophys., Vol. 161, No. 1, p. 203 – 205 (1986).
From Bidelman's list (1983) of northern stars of astrophysical interest eleven chemically peculiar early type (CP 2) stars were photometrically investigated to check the reported CP 2 character by means of the Δ*a* photometry and to obtain Strömgren and Hβ values. The CP 2 character of all stars is verified, in one case the star belongs to a hotter subgroup than reported. One star (BD +24°3675) turns out to be variable and shows the largest Δ*a* value ever measured. Besides this, two stars from the Bright Star Catalogue were found to be CP 2 stars.

113.038 **Absolute *UBV* photometry at the Zacatecas Observatory.**
W. J. Schuster, L. Parrao, S. F. González Bedolla, M. Ríos Herrera, M. Ríos Berumen.
Rev. Mex. Astron. Astrofis., Vol. 11, No. 2, p. 55 – 59 (1985). = Contrib. Inst. Astron., Univ. Nac. Autónoma Mex., No. 132.
Atmospheric extinction data, color coefficients, errors of observation and a list of standard and comparison stars are given for *UBV* photometry at the Astronomical Observatory in Zacatecas. The observing and reduction procedures are discussed briefly, and the extinction coefficients and observing errors compared with previous values.

113.039 **Numerical convolution photometry: an example using the Strömgren four–color system.**
S. B. Howell.
Astron. J., Vol. 91, No. 1, p. 171 – 176 (1986).
The method of numerical convolution photometry is presented. This method is then applied to spectrophotometric observations of hot blue stars using the Strömgren four–color photometric system. New color relations are discussed as well as their

usefulness in determining some intrinsic properties of the program stars, in particular, the white dwarfs. A discussion of the errors involved and the accuracy of this method is also included.

113.040 **Is AS 431 a superluminous WR star?**
D. J. Stickland, C. Lloyd, A. J. Willis.
Luminous stars and associations in galaxies, p. 229 – 230 (1986). – See Abstr. 012.079. (IAU Symp. No. 116).

113.041 **Photometric investigation of stars in the field of NGC 7023.**
N. K. Krasnikov, O. I. Stal'bovskij.
Inst. kosm. issled. Akad. Nauk SSSR, Prepr. No. 1011, 23 pp. (1985). In Russian. Abstr. in Ref. Zh., 51. Astron., 5.51.834 (1986).

113.042 **The Wirtanen–Vysotskij photometric standards.**
M. S. Kazanasmas, L. A. Zavershneva, L. F. Tomak.
Odes. univ. Odessa, 13 pp. (1986). In Russian. Abstr. in Ref. Zh., 51. Astron., 6.51.467 (1986).

113.043 **Spectrophotometric standards of the 6.5 – 8.5 stellar magnitude.**
S. K. Panteleev.
Red. Zh., Kinematika Fiz. Nebesn. Tel, 12 pp. (1986). In Russian. Abstr. in Ref. Zh., 51. Astron., 6.51.513 (1986).

113.044 **A comparative study of barium peculiarity indices.**
M. Rego, M. Cornide, J. L. Fernández–Villacañas.
Astrophys. Space Sci., Vol. 123, No. 2, p. 313 – 324 (1986).
From DDO photometry an indicator of barium stars have been elaborated. This indicator is closely related to the barium line intensities derived by means of narrow–band photometry and spectroscopic observations. A better discrimination appears when the indicator is used instead of the qualitative Warner index.

113.045 **IRAS observations of Wolf–Rayet stars.**
K. A. van der Hucht, T. A. Jurriens, F. M. Olnon, P. S. Thé, P. R. Wesselius, P. M. Williams.
Luminous stars and associations in galaxies, p. 243 – 245 (1986). – See Abstr. 012.079. (IAU Symp. No. 116).
IRAS observations of Wolf–Rayet stars are used as diagnostics of hot circumstellar dust shells, cool dust in WR ring nebulae, the Ne/He abundance ratio, and the interstellar extinction. In two cases the IR energy distributions are indicative of a WR planetary nucleus status rather than a Population I WR status.

113.046 **Passbands and photometric systems.**
F. Rufener.
Space Telesc. Sci. Inst., Prepr. Ser., No. 100, p. 15 – 19 (1986). – See Abstr. 012.040.

113.047 **Observed stellar energy distributions for synthetic photometry.**
D. S. Hayes.
Space Telesc. Sci. Inst., Prepr. Ser., No. 100, p. 21 – 28 (1986). – See Abstr. 012.040.

113.048 **CCD photometry to V = 21 in a Puppis field.**
W. E. Harris, B. C. Reed, J. E. Hesser.
Publ. Astron. Soc. Pac., Vol. 98, No. 604, p. 599 – 604 (1986).
The authors present a photoelectric *UBV* sequence of 235 stars to $V = 21.25$ in the "Puppis Window" at $(l,b) = (245°,0°)$ obtained with the CCD camera on the CTIO 4–meter telescope.

113.049 **Fotometría uvby–β en el telescopio de 1.5 m del O.A.N.**
V. Reglero, A. de Castro, J. Fabregat.
Bol. Astron. Obs. Madr., Vol. 11, No. 4, p. 7 – 16 (1985).

113.050 **Photoelectric *UBV* photometry of the peculiar emission–line star Henize 2–467.**
V. P. Arkhipova, R. I. Noskova.
Sov. Astron. Lett., Vol. 11, No. 5, p. 297 – 300 (1985). English translation of 40.113.014.

113.051 **The helium–weak star 33 Geminorum.**
Yu. V. Glagolevskij, K. Panov, N. M. Chunakova.
Sov. Astron. Lett., Vol. 11, No. 5, p. 315 – 317 (1985). English translation of 40.113.025.

113.052 **Infrared flux excesses of the warm carbon stars.**
J. F. Dominy, D. L. Lambert, R. D. Gehrz, D. Mozurkewich.
Astron. J., Vol. 91, No. 4, p. 951 – 960 (1986).
The authors seek to examine the energy distributions of the brightest stars of type CH, Ba II, and the hydrogen–deficient R stars of high luminosity. Their goal is to map the infrared flux distribution to the longest possible wavelengths. They analyze the magnitudes through a short– versus long–baseline color–color diagram and by examination of the calibrated flux distributions. It is concluded that the distinctive energy distributions found in most of these stars are not caused by dust emission.

113.053 **New subdwarfs. IV. UBV photometry of 1690 high–proper–motion stars.**
A. Sandage, C. Kowal.
Astron. J., Vol. 91, No. 5, p. 1140 – 1188 (1986).
UBV photometry is given for 1690 high–proper–motion stars, providing a finding list for potential high–velocity stars of various metallicity values. The purpose of the survey is to obtain candidates for trigonometric programs on subdwarfs and to enlarge the sample with which to study the relation between stellar kinematics and metal abundance. Photometric parallaxes are obtained. A list of 113 stars with tangential space velocities $S_T \geqslant 300\ km\ s^{-1}$ is given. The highest tangential velocities relative to the Sun are $S_T \cong 630\ km\ s^{-1}$, but the local value of S_T(max) varies with galactic longitude. Using the variation of S_T(max) with longitude and adopting the galactic rotation at the solar circle to be $v_{rot} = 220\ km\ s^{-1}$, the authors estimate the rotation of the subdwarf system to be v_{rot}(halo) $\sim 0 \pm 50\ km\ s^{-1}$.

113.054 **Simultaneous determination of metallicity and O/C ratio for late–type giants in the Vilnius photometric system.**
V. Straizys, J. Sleivyte.
Bull. Etoiles Tardives Spectre Particulier, No. 3, p. 1 – 4 (1985).

113.055 **A note on V1286 Aquilae.**
D. P. Hube, M. Anderson.
Inf. Bull. Variable Stars, No. 2849, 4 pp. (1986).

113.056 **Variation of the CP2 star BD +24°3675.**
H. Schneider.
Inf. Bull. Variable Stars, No. 2870, 4 pp. (1986). With a correction in No. 2883.

113.057 **CD –30°5135.**
G. Welin, G. Westling.
Inf. Bull. Variable Stars, No. 2898, 1 p. (1986).

113.058 **Photometric investigation of the magnetic star 53 Camelopardalis.**
M. Muciek, P. North, F. Rufener, J. Gertner.
Acta Astron., Vol. 35, Nos. 3–4, p. 377 – 394 (1985).
In this paper the authors present new photometric results and combine them with earlier ones in order to present reappraisal of the photometric behaviour of this important star. The new photometry is presented and a model of brightness distribution over the star's surface is proposed. The pseudo–continuum and the λ5200 depression are discussed.

113.059 **An optimal photometric system for high resolubility of physical quantities.**
N.–K. Park, S.–W. Lee.
J. Korean Astron. Soc., Vol. 19, No. 1, p. 1 – 9 (1986).
The authors present a method for deriving most optimal filter system by which the accuracy of derived physical quantities can be maximized. Using Kurucz (1978)'s model atmospheres, an optimal four filter system for F and G dwarfs is suggested, for which mean wavelengths are located at 3400 Å, 3850 Å, 4190 Å, and 4600 Å with half–bandwidth of 200 Å. It is found that 35, 38

and 42 filters of the DDO system and the Strömgren u and v filters are close to those of the most optimal system.

113.060 **Narrow band photometry of bright stars.**
E. E. Mendoza V.
Rev. Mex. Astron. Astrofis., Vol. 12, Spec. Issue, p. 197 (1986). Abstract. – See Abstr. 012.066.

113.061 **Strömgren uvby photometry of E region stars.**
A. W. J. Cousins.
Mon. Notes Astron. Soc. S. Afr., Vol. 44, Nos. 7–8, p. 54 – 59 (1985).
Observations have been in progress at the S.A.A.O. in Cape Town since 1983 with the intention of providing secondary standards for Strömgren uvby photometry in the E regions because of the convenience of having sequences in that part of the sky and the need for fainter standards when using larger telescopes. The observations are being made with the 18–inch (0.47 m) reflector using 60″ and 40″ diaphragms in the photometer.

113.062 **The flux distribution of Vega for 10 µm < λ < 100 µm, and the calibration of IRAS at 12 µm and 25 µm.**
S. K. Leggett.
Light on dark matter, p. 71 – 74 (1986). – See Abstr. 012.080.
The flux distribution of Vega for 10 µm < λ < 100 µm is investigated using the IRAS data. Using the bands 1 and 2 fluxes, relative to Vega, of 23 stars known T_e, the Vega flux is adjusted to reproduce fluxes calculated by model atmospheres. These Vega fluxes agree with those given by the pointed IRAS observations and calibration, and show Vega in excess at all bands over modelled fluxes. A comparison with Sirius implies a similar excess, however a comparison with 33 A0–type survey stars suggests that Vega is normal at band 1.

113.063 **A search for infrared excesses in G–type stars.**
S. F. Odenwald.
Light on dark matter, p. 75 – 76 (1986). – See Abstr. 012.080.
The IRAS Point Source Catalogue (PSC) has been searched for infrared sources associated with stars of spectral class G. Of the 1069 IRAS sources identified, 15 have enhanced emission at λ > 25 µm. This group includes 4 supergiants, 8 variable stars, and 3 stars known to have dusty circumstellar medii (RU Cen, R Sge, HD 101584). The preliminary findings from this study are described.

113.064 **IRAS intrinsic colours of hot stars.**
J. Coté, L. B. F. M. Waters.
Light on dark matter, p. 77 – 78 (1986). – See Abstr. 012.080.
The O, B and A type stars observed with IRAS at 12 µm are used to derive an intrinsic $(B–V)_0$, $(V–[12])_0$ – relation for "normal" stars. The authors find that many of the O, B and A stars have excess fluxes at 12 µm. From a comparison of the relation to models they conclude that: (1) The extrapolation of the Kurucz (1979) models from the visual to 12 µm results in a 12 µm flux that is about 10 percent too low, and (2) the energy distribution for hot stars from 0.4 to 12 µm cannot be described with a one–temperature blackbody.

113.065 **IRAS observations of carbon stars.**
F. J. Willems.
Light on dark matter, p. 113 – 118 (1986). – See Abstr. 012.080.
The author reports on observations of cool carbon stars between 7 and 120 µm obtained by IRAS. He discusses the IRAS broad–band photometry and low resolution spectra of the carbon stars in his sample.

Bibliografia su metodologie di osservazione, riduzione dei dati e applicazioni della fotometria stellare multicolore UBV.
See Abstr. 002.008.

A spectrophotometric atlas of southern MK standards from 5800 Å to 10,200 Å.
See Abstr. 002.035.

Catalogue of magnitudes and colour indices of stars in the direction of NGC 7062.
See Abstr. 002.039.

A multicolor photometric catalog of galaxies and stars in the field of the rich cluster II Zw 1305.4+2941 at z = 0.24.
See Abstr. 002.052.

A multicolor photometric catalog of galaxies and stars in the field of the rich cluster II Zw 1305.4+2941 at z = 0.24.
See Abstr. 002.060.

The spectral energy distribution of early type stars. I. A catalogue of photometric data of 259 stars from 0.15 to 4.8 µm.
See Abstr. 002.079.

A statistical analysis and overview of the IRAS point source catalog.
See Abstr. 002.092.

Two–dimensional astronomical photometry techniques.
See Abstr. 034.153.

Synthetic photometry and the calibration of the Hubble Space Telescope.
See Abstr. 035.059.

Transformation equations and other aids for *VRI* photometry.
See Abstr. 036.021.

Principles and scope of synthetic photometry.
See Abstr. 036.129.

Principles and scope of synthetic photometry.
See Abstr. 036.130.

The possibilities of synthetic photometry.
See Abstr. 036.131.

On a possibility of realizing the UBVRI system band I using the photomultiplier FEhU–79.
See Abstr. 036.149.

Stellar photometry on glass copies of Schmidt photographic plates.
See Abstr. 036.171.

Semi–automated identification of IRAS point sources using UKST plates and the COSMOS measuring machine.
See Abstr. 036.258.

Red giant model atmospheres. IV. Colors of K and M giant models.
See Abstr. 064.022.

The absolute energy distributions of the Sun, of the 'solar analogs' 16 Cyg B, Hyades VB 64, 16 Cyg A, and of the standard stars α Lyr and 29 Psc.
See Abstr. 071.006.

Energy distribution in the spectrum of the sun as a star and UBV photometry.
See Abstr. 071.022.

A photometric study of the bright cloud B in Sagittarius: V. 185 new proper motion stars.
See Abstr. 111.012.

Photometric parallaxes for selected stars of color class m from the NLTT catalog. II. The declination zone 0° to +20°.
See Abstr. 111.014.

Etude spectrophotométrique, paramètres physiques et géométriques de l'étoile Be Gamma Cassiopeiae.
See Abstr. 112.079.

B[e]–stars of the Magellanic Clouds.
See Abstr. 112.132.

The infrared excess from stellar winds.
See Abstr. 112.135.

IRAS observations of Wolf–Rayet stars.
See Abstr. 112.137.

Models of IRAS observations of circumstellar shells.
See Abstr. 112.140.

Search for λ Bootis type candidates.
See Abstr. 114.005.

CH and barium metal–deficient giants in the Vilnius photometric system. I.
See Abstr. 114.063.

Spectral types from *uvbyβ* photometry.
See Abstr. 114.067.

Southern hemisphere spectrophotometric standards.
See Abstr. 114.089.

Spectrophotometric study of the suspected UU Herculis–star HR 4912.
See Abstr. 114.099.

On stellar spectrophotometric standards.
See Abstr. 114.109.

Absolute magnitudes of late–type stars.
See Abstr. 115.003.

The evolutionary state of the He–weak star HD 49333.
See Abstr. 115.006.

LHS 292 and the luminosity function of the nearby M dwarfs.
See Abstr. 115.023.

Optical photometry and spectroscopy for five dwarf M stars.
See Abstr. 116.026.

Multicolor polarimetry of selected Be stars: a long–term analysis.
See Abstr. 116.071.

Multicolor photometry of the binary central star LSS 2018.
See Abstr. 120.020.

Studies of the Chamaeleon star–forming region. II. The pre–main–sequence stars HD 97048 and HD 97300.
See Abstr. 121.011.

Some T Tauri stars and related objects. Results of UBVR photoelectric observations.
See Abstr. 121.021.

A spectroscopic survey of the Taurus–Auriga dark clouds for pre–main–sequence stars having Ca II H, K emission.
See Abstr. 121.051.

RR Lyrae, Delta Scuti, SX Phoenicis stars and Baade–Wesselink method. I. Photometric and radial velocity measurements of four field stars: RR Cet, DX Del, BS Aqr and DY Peg.
See Abstr. 122.010.

Variability of Magellanic Cloud supergiants.
See Abstr. 122.195.

Photometric variability of 27 Cygni.
See Abstr. 123.026.

The stars NGC 7538–IRS 5, 6, and 7, and a distance estimate to the nebula.
See Abstr. 131.206.

An Hα emission star survey in the field of the dark cloud complex Khavtassi 141.
See Abstr. 131.242.

Photographic observation of AFGL sources in Cygnus.
See Abstr. 133.011.

Distances to planetary nebulae. I. The reddening–distance method.
See Abstr. 134.006.

The chemical composition of the young open cluster NGC 6611.
See Abstr. 153.003.

The giant branch of the open cluster NGC 2539.
See Abstr. 153.017.

Fundamental parameters of the open cluster NGC 2567.
See Abstr. 153.027.

Photoelectric *UBVRI* sequences in the region of the galactic globular clusters NGC 2808, NGC 4372, and NGC 6397.
See Abstr. 154.029.

M13: main–sequence photometry and the mass function.
See Abstr. 154.032.

Near–infrared photometry of red giant and horizontal branch stars in M4.
See Abstr. 154.050.

Comparison between *UBV*– and *RGU*–photometrically determined space density functions in a starfield close to the galactic centre (SA 133).
See Abstr. 155.011.

Star counts and color indices in a south–galactic–pole field.
See Abstr. 155.066.

2000 Å UV imaging of a 6° diameter field around μ Aurigae.
See Abstr. 155.068.

Galactic structure examined from a faint object survey in a field in Aquarius.
See Abstr. 155.083.

The solar neighbourhood mass density.
See Abstr. 155.084.

Ground based observations of nuclear bulge stars.
See Abstr. 155.124.

A photometric determination of the metal content for F – G type supergiants in the Large Magellanic Cloud.
See Abstr. 156.006.

The properties of OB supergiants in the Large Magellanic Cloud.
See Abstr. 156.015.

The stellar population of the Small Magellanic Cloud near NGC 121. I. The mean metallicity, metallicity spread, and radial velocity of SMC halo giants.
See Abstr. 156.023.

A photometric study of the bright stars in M33.
See Abstr. 157.184.

Carbon and M stars in nearby galaxies: a preliminary survey using a photometric technique.
See Abstr. 157.202.

The brightest stars in nearby galaxies. VII. The Pegasus and Leo A Im V dwarfs.
See Abstr. 157.217.

The M supergiants in NGC 300.
See Abstr. 157.220.

The luminosities of M supergiants and the distances to M101, NGC 2403, and M81.
See Abstr. 157.226.

The distribution of young stars in nearby galaxies.
See Abstr. 157.309.

Red supergiants in NGC 300.
See Abstr. 157.311.

114 Spectra, Temperatures, Chemical Composition, etc.

114.001 **Abundances of carbon, nitrogen, and oxygen and their isotopes in the atmospheres of four SC stars.**
J. F. Dominy, G. Wallerstein, N. B. Suntzeff.
Astrophys. J., Vol. 300, No. 1, p. 325 – 338 (1986).

Fourier transform spectra in the 2.2 μm region have been analyzed to determine $^{12}C/^{13}C$, $^{16}O/^{17}O$, and CNO abundances in four SC stars: GP Ori, FU Mon, CY Cyg, and RZ Peg. The carbon isotopic ratio ranges from low (5.6, CY Cyg) to moderate (~30, GP Ori and FU Mon) values. The ratio $^{16}O/^{17}O$ ranges from 160 to about 400. The ratio $^{16}O/^{18}O$ is greater than 300. Carbon and oxygen abundances were derived from spectrum synthesis of CO and CN lines and the assumption C ≈ O. Suggestions concerning the evolutionary history of the SC stars are presented.

114.002 **Comments on using absolute spectrophotometry of Wolf–Rayet stars.**
A. B. Underhill.
Astrophys. J., Vol. 300, No. 1, p. 374 – 378 (1986).

It has been claimed by Garmany, Massey, and Conti (1984) on the basis of a study of spectrophotometric scans of Wolf–Rayet stars that the application of a standard reddening law to these scans gives spurious results in many cases for the effective temperatures. This would imply that the attempts by Underhill to determine the intrinsic continua and effective temperatures of Wolf–Rayet stars are in substantial error. The present paper argues that Garmany, Massey, and Conti have not convincingly demonstrated that the interstellar extinction law varies greatly from one Wolf–Rayet star to another. The effective temperatures given by Underhill, although not exact, probably do not contain errors greater than about ±10%.

114.003 **Spectroscopic studies of Wolf–Rayet stars. III. The WC subclass.**
A. V. Torres, P. S. Conti, P. Massey.
Astrophys. J., Vol. 300, No. 1, p. 379 – 395 (1986).

The authors present spectrophotometric data for the major optical emission lines of 64 Galactic and 18 Large Magellanic Cloud (LMC) WC stars. Using line ratios of O V λ5590, C III λ5696, and C IV λ5806 they quantify the subtype classification. A few Galactic stars are reclassified, and nearly all the LMC WC stars are found to be of type WC4. Thus there is even a greater discrepancy in the distribution of WC subtypes between the LMC and the Galaxy than previously assumed, since WC4 types in the Galaxy are rare. Terminal wind velocities are estimated from an excitation–line width relation. The terminal velocities range from 1000 km s^{-1} for the latest subtypes to 5000 km s^{-1} for the earliest types.

114.004 **The nitrogen abundance in Wolf–Rayet WC stars.**
H. Nussbaumer, W. Schmutz.
Astron. Astrophys., Vol. 154, No. 1/2, p. 100 – 102 (1986).

In order to clarify the important point of the presence of nitrogen in WC stars, the authors study the 1230 Å – 1250 Å region of the WC 8 star HD 192103. They conclude that there is no convincing evidence for N V $2s\ ^2S_{1/2}$–$2p\ ^2P^0_{3/2}$ λ1238.8 absorption in the spectrum of this star. Thus there is, up to now, no convincing evidence for the presence of N in WC stars.

114.005 **Search for λ Bootis type candidates.**
B. Hauck.
Astron. Astrophys., Vol. 154, No. 1/2, p. 349 – 351 (1986).

An extensive search for λ Bootis candidates has been made throughout the whole Geneva photometric catalogue and 17 stars have been found, matching the photometric conditions to be λ Bootis stars. 13 stars proposed by Abt (1984) as λ Bootis candidates, on the basis of the weakness of the λ4481 Mg II line, also match these conditions.

114.006 **Analysis of five high–latitude blue stars.**
D. Kilkenny, J. Lydon.
Mon. Not. R. Astron. Soc., Vol. 218, No. 2, p. 279 – 288 (1986).

A simple analysis of five high–latitude blue stars with sharp spectral lines is presented. Reticon spectra at 30 Å mm^{-1} and Strömgren (*uvby*) photometry are used to determine T_{eff} and $\log g$ by the "Kiel diagram" method and values of $V \sin i$ are estimated from the He I lines. The results are checked by fitting Kurucz (1979) models to the observed hydrogen line profiles. One star is probably an SMC supergiant, the other four appear to be population II stars, possibly related to the hot "*UV*–bright" stars detected in some globular clusters.

114.007 **Spectroscopic observations of three X–ray selected late–type stars.**
D. R. Silva, J. Liebert, J. T. Stocke, M. Aaronson.
Publ. Astron. Soc. Pac., Vol. 97, No. 597, p. 1096 – 1100 (1985).

The three X–ray selected mid–K stars with the highest values of (f_x/f_v) in the Einstein Medium Sensitivity Survey have been studied spectroscopically at moderate and high spectral resolution. One of these objects (1E1548.7+1125) is presented as a binary–star system with characteristics similar to the RS CVn class of binaries. A second of these stars (1E1751.0+7046) is a rapid rotator, while the third star (1E1339.9+6030) has not been shown to be either a binary system or a rapid rotator. The presence of both Ca II H and K emission core and high X–ray flux is attributed to chromospheric and coronal activity, but no single underlying cause is indicated.

114.008 **Relations between [Fe/H] and δm_1 for F and early–G stars.**
D. H. McNamara, J. M. Powell.
Publ. Astron. Soc. Pac., Vol. 97, No. 597, p. 1101 – 1103 (1985).

The authors have used the new catalog of [Fe/H] determinations of Cayrel de Strobel et al. (1985) to calibrate [Fe/H] in terms of δm_1. Three relations were found corresponding to F1 – F5 stars, F5 – F8 stars, and F8 – G2 stars. Quadratic equations were found necessary to represent the data of the two early groups. Serious errors in [Fe/H] determinations of early metal-poor F stars can result if the previous linear calibrations are used.

114.009 **Incidence of heavy element enhancements among carbon stars.**
J. F. Dominy.
Publ. Astron. Soc. Pac., Vol. 97, No. 597, p. 1104 – 1111 (1985).
The correlation of spectral line strength with temperature subclass has been examined in a survey of 40 carbon stars at a dispersion of 15 Å mm^{-1}. The two goals of this search were (1) to determine if additional cool N–type stars can be added to the list of N stars not having enhanced *s*–process elements, and (2) to confirm the anticorrelation of *s*–process enhancements with ^{13}C enhancement. Among the program stars one normal carbon star was found to have no, or only minor, enhancement of heavy elements associated with the *s*–process. The suspicion that increased surface ^{13}C is accompanied by weak line strengths of *s*–process elements is strongly confirmed.

114.010 **An atlas of high resolution IUE ultraviolet spectra of 14 Wolf–Rayet stars.**
A. J. Willis, K. A. van der Hucht, P. S. Conti, D. Garmany.
Astron. Astrophys., Suppl. Ser., Vol. 63, No. 3, p. 417 – 599 (1986).
The authors present an atlas of high resolution ($\Delta\lambda = 0.1 - 0.2$ Å) IUE ultraviolet spectra of 14 galactic Wolf–Rayet stars (9 WN and 5 WC). For each star the wavelength ranges covered are $\lambda\lambda 1150 - 2050$ and $\lambda\lambda 1850 - 3250$. The spectral data are presented in the form of line identification lists and spectral tracings. Where possible the authors have measured the central emission intensity of each feature in continuum units. In each sequence the relative line intensities within a chemical species, and from element to element, vary considerably with subclass and the ionization trends along both the WN and WC sequences are discussed. Generally at least three ionization stages of a given element are present within an individual spectrum.

114.011 **Turbulence, convection, and mixing in red giant stars: some empirical approaches based on high resolution spectroscopy.**
T. Tsuji.
Astrophys. Space Sci., Vol. 118, Nos. 1/2, p. 227 – 235 (1986). – See Abstr. 012.001.
The differential velocity field in cool stars can be measured effectively on photographic plates by the use of a PDS microdensitometer, from which it is shown that radial velocity gradients are larger for stars with larger turbulent velocities, which are determined from the high resolution echellograms. This indicates that the stellar turbulence may have something to do with the differential velocity field in stellar atmospheres. As an observational probe of the mixing processes in red giant stars, stellar abundances that are sensitive to mixing are determined on the basis of high resolution Fourier Transform spectroscopy.

114.012 **Effective temperatures, radii, and luminosities of O emission, Be and Ae stars.**
J.–O. Woo.
Astrophys. Space Sci., Vol. 119, No. 1, p. 61 – 63 (1986). – See Abstr. 012.001.
The author's de–reddened fluxes, together with the ultraviolet measurements of Thompson et al. (1978), have been compared with those of Kurucz's (1979) model atmospheres to derive effective temperatures of some O emission, Be and Ae stars. With the measured monochromatic fluxes the author determined their angular diameters and luminosities. It is found that the majority of the stars are cooler than the Zero–Age Main Sequence (ZAMS), suggesting that they are slightly more evolved than ZAMS stars.

114.013 **Der Kohlenstoffstern G77–61.**
H. Gass.
Sterne Weltraum, 25. Jahrg., Nr. 1, p. 23 – 25 (1986).

114.014 **Temperatures and gravities for horizontal–branch stars.**
D. S. Hayes.
News Lett. Astron. Soc. N.Y., Vol. 2, No. 9, p. 13 – 25 (1986). – See Abstr. 010.241.
Horizontal–branch stars are in the helium core–burning phase following the helium flash at the tip of the red giant branch. The author describes here a method for determining the effective temperature and surface gravity of these stars.

114.015 **Do UV–bright stars mimic normal (Population I) OB stars?**
R. F. Garrison, C. E. Albert.
Astrophys. J., Lett. Ed., Vol. 300, No. 2, p. L69 – L71 (1986). With plate L4.
To test this hypothesis, five globular cluster UV–bright stars were observed with CFHT at classification dispersion. Of the three which have well–exposed spectra, none can be confused with normal Population I stars.

114.016 **Optical region elemental abundance analyses of B and A stars. V. The normal stars Theta Leonis, Tau Herculis, 14 Cygni, and 5 Aquarii.**
S. J. Adelman.
Astron. Astrophys., Suppl. Ser., Vol. 64, No. 1, p. 173 – 187 (1986).
Abundance analyses using optical region data and fully line–blanketed model atmospheres have been performed for four sharp–lined normal B and A type stars. This work extends the results presented in the first and fourth papers of this series. The microturbulent velocities of all the stars studied in this series are found to follow the same relation with temperature. Some of the stars studied in this paper are found to have one or two anomalous abundances, such as the zirconium abundance of θ Leo.

114.017 **Low–resolution IUE observations of four unreddened, 11th magnitude B stars at high galactic latitudes.**
W. Tobin.
Astron. Astrophys., Vol. 155, No. 2, p. 326 – 332 (1986).
Low–resolution IUE spectra are presented for the four faint, unreddened, high–latitude, mid–B stars Feige 40 (≡FB 72), H.O. +23°B, H.O. +41°B (≡SS II 135) and PB 166. By comparison with stars in the Heck et al. atlas, the spectra are like those of normal disc stars of Population I metallicity. This suggests that the programme stars are massive, luminous objects, which, if true, means that they are situated several kiloparsecs outside the galactic plane. Effective temperatures and upper limits on the Population I ages are derived. Radial velocities are known for two of the stars: they are consistent with stellar birth in the galactic plane followed by ejection, at high speed, to the present locations within the evolutionary lifetimes. Predictions are made for the radial velocities of the other two stars.

114.018 **Observations of early–type stars at high galactic latitudes.**
F. P. Keenan, P. J. F. Brown, D. J. Lennon.
Astron. Astrophys., Vol. 155, No. 2, p. 333 – 338 (1986).
Using equivalent widths and line profiles measured from high resolution AAT spectra in conjunction with model atmosphere calculations, it is shown that a sample of eleven high galactic latitude early–type stars are normal, distant objects. Stellar velocities in the *z*–direction determined from the spectra, coupled with the galactic gravitational field $g(z)$, allow the times the stars would have taken to reach their present positions if they were ejected from the disc to be derived. A comparison of these times with the evolutionary ages indicate that at least four of the stars have been formed in situ, possibly from galactic fountain material.

114.019 **High–resolution spectra of "normal" A stars: evidence for compositional differences.**
H. Holweger, D. Gigas, M. Steffen.
Astron. Astrophys., Vol. 155, No. 1, p. 58 – 62 (1986).
High–resolution Reticon spectra of seven reputedly normal B9.5 – A2 V stars with $v \sin i < 50$ km s^{-1} were recorded using the ESO Coudé Echelle Spectrometer. A selection of these spectra

is presented here: two wavelength bands containing lines of He, C, Mg, the iron group, and Ba. Significant abundance variations from star to star are indicated for carbon and, apparently uncorrelated with C, for the heavier elements. One of the stars, HR 3383 (A1 V), has very sharp lines.

114.020 **Evolutionary effects on the surface abundances of an early–type supergiant.**
D. J. Lennon, P. L. Dufton.
Astron. Astrophys., Vol. 155, No. 1, p. 79 – 86 (1986).
An LTE abundance analysis has been performed using line–blanketed model atmospheres for two supergiants: HD 148379 and HD 157038. Atmospheric parameters (T_{eff}, log g) were deduced as follows: HD 148379 (20750K, 2.5) and HD 157038 (18500K, 2.3). Nitrogen appears to be overabundant in the atmosphere of HD 157038 compared with that of HD 148379 and of the sun by a factor of approximately 6. There is also evidence for a marginal helium overabundance in the same star. As the two supergiants have similar bolometric magnitudes ($M_{BOL} \sim -9.5$) and lie close together in the HR diagram, it is suggested that their evolutionary histories differ. HD 157038 may be the more evolved.

114.021 **The chemical composition of a G0 Ib star in the Small Magellanic Cloud.**
F. Thévenin, R. Foy.
Astron. Astrophys., Vol. 155, No. 1, p. 145 – 150 (1986).
The G0 supergiant AZV–121 in the Small Magellanic Cloud is analysed in detail, using moderately high dispersion spectrograms and model atmospheres. The authors find heavy elements are slightly underabundant (–0.3 dex) with respect to the Sun. This result strengthens the discrepancy between the abundance of heavy elements determined in stars and those of light elements determined in gaseous nebulae. Possible artefacts or physical reasons for this disagreement are discussed.

114.022 **High resolution spectroscopy of CO in the infrared spectra of cool stars: turbulent velocities and carbon abundance in M–giant stars.**
T. Tsuji.
Astron. Astrophys., Vol. 156, No. 1/2, p. 8 – 21 (1986).
A method of quantitative stellar spectroscopy is applied to the first overtone band of CO in high resolution and high quality infrared spectra of eighteen M–giant stars (plus two K–giants) obtained by Fourier transform spectroscopy. Helped by the favourable conditions associated with CO lines (e.g., wide intensity coverage, relatively well established f–value, applicability of LTE, ...), this attempt shows reasonable success for stars as late as M7 III. First, the micro–turbulent velocity is determined so that a consistent interpretation can be given to weak and medium–strong lines. Second, CO is a good indicator of the carbon abundance in cool oxygen–rich stars and the mean value of [C/H] for eighteen M–giant stars is -0.65 ± 0.22 (s.d.), which is about 0.3 dex smaller than that of G–K giant stars.

114.023 **The mid–ultraviolet spectrum of α^2CVn = HD 112413 = HR 4915.**
H. Hensberge, J. Van Santvoort, K. A. van der Hucht, T. H. Morgan.
Astron. Astrophys., Vol. 158, No. 1/2, p. 113 – 118 (1986).
Wavelength coincidence statistics (WCS) and traditional identification techniques have been used to search for atoms and ions absorbing in the mid–ultraviolet spectrum of α^2CVn, which has been obtained with the BUSS experiment near to phase of minimum rare–earth line strength. The authors identified at a 99.5% confidence level the ions Mg II, Si II, Cr II, Mn II, Fe II, Fe III, Co II, Ni II, Yb III and W II. The identification of doubly ionized ytterbium in a chemical peculiar (CP) star is unprecedented, but not unexpected. The occasion is used to discuss briefly the influence of correlations between laboratory lists on WCS.

114.024 **High resolution spectrum of the Li I 6708 region for a sample of CP2 and reference stars.**
R. Faraggiana, M. Gerbaldi, F. Castelli, M. Floquet.
Astron. Astrophys., Vol. 158, No. 1/2, p. 200 – 204 (1986).
The λ6708 region has been observed with high resolution in order to detect and analyse the abundance of Li in a sample of peculiar stars (mostly CP2 group). The observations were analysed by the synthetic spectrum method comparing the observed lines with the computed ones. The authors conclude that, with respect to the cosmic abundance (log $\varepsilon = -9$), Li may be over or underabundant or normal. At 6706.8 Å an intense feature has been detected in CP2 spectra; it remains unidentified.

114.025 **The temperatures of Alpha Centauri A and B.**
D. R. Soderblom.
Astron. Astrophys., Vol. 158, No. 1/2, p. 273 – 274 (1986).
High resolution and signal–to–noise observations of Hα in the Sun, α Cen A, and α Cen B are compared to models to derive temperatures: T_{eff}(α Cen A) = 5770 ± 20K and T_{eff}(α Cen B) = 5350 ± 50K. These are then used to calculate the radii of these stars: 1.23 and 0.80 $R_{\odot}$ for α Cen A and B, respectively.

114.026 **Statistical confirmation of the UV spectral classification system based on IUE low–dispersion stellar spectra.**
A. Heck, D. Egret, P. Nobelis, J. C. Turlot.
Astrophys. Space Sci., Vol. 120, No. 2, p. 223 – 237 (1986).
In the IUE Low–Dispersion Spectra Reference Atlas (Heck et al., 1984), a new spectral classification system specific to the UV had to be introduced because of the lack of one–to–one correspondence between the UV and visible ranges. It was elaborated from a classical morphological approach (Jaschek and Jaschek, 1984). This paper presents an independent confirmation of the correctness of this system.

114.027 **Variable absorption components in UV spectra of early–type stars.**
H. Henrichs.
Publ. Astron. Soc. Pac., Vol. 98, No. 599, p. 48 – 51 (1986). – See Abstr. 012.009.

114.028 **Search for transition zone lines in early dwarf A stars.**
R. Freire Ferrero.
Astron. Astrophys., Vol. 159, No. 1/2, p. 209 – 216 (1986).
The observation of a sample of early dwarf A stars with IUE allows the author to conclude that the features observed on regions of Si IV, C IV and He II lines may be explained by blends of metallic lines. In consequence, transition regions are probably not present or they have a very small thickness in dwarf A stars.

114.029 **Analysis of two field halo–dwarf stars, HD 99383 and HD 160617.**
M.–N. Perrin.
Astron. Astrophys., Vol. 159, No. 1/2, p. 239 – 245 (1986).
The spectroscopic fine analysis of two field halo–dwarf stars, HD 99383 and HD 160617 is presented. The main results are the following: (1) For HD 99383: $\theta_{eff} = 0.85$, log$g$ = 4.0, [Fe/H] = –1.60. (2) For HD 160617: $\theta_{eff} = 0.86$, log$g$ = 3.5, [Fe/H] = –1.60.

114.030 **High–resolution IUE spectrum of the WN6 star HD 192163: the Fe V and Fe VI dominated emission–line spectrum at $\lambda \lesssim 1720$ Å.**
T. Nugis, A. Sapar.
Tartu Astrofüüs. Obs. Teated, Nr. 80, 24 pp. (1985).
The results of a study of the high–resolution spectrum of the WN6 star HD 192163, observed by means of the International Ultraviolet Explorer satellite (IUE) in November 1978, are presented. Most of the previously unidentified spectral features in the region 1245 – 1720 Å turned out to be due to the overlapping of the pure or of the P Cygni type emission lines of Fe V and Fe VI. The estimated Fe/He ratio was found to be close to its solar value.

114.031 **Spectral observations of Be stars in the visual range. I. φ Per, ψ Per, 11 Cam, φ And.**
V. D. Krugov.
Kinematika Fiz. Nebesn. Tel, Tom 2, No. 1, p. 43 – 47 (1986). In Russian. English translation in Kinematics Phys. Celest. Bodies.
Spectral observations of four bright Be stars are analyzed. φ Per manifests considerable variations of emission in Hα and Hβ. Emission profiles of Hα and Hβ in ψ Per appear to vary with the period of several years. 11 Cam demonstrates complete disappearing of Hα emission. For φ And Hα emission does not show intensity variations during the observational period.

114.032 **The multi–component structure of spectral lines of P Cygni.**
N. Markova, I. Kolka.
Astrofizika, Tom 23, Vyp. 3, p. 539 – 550 (1985). In Russian. English translation in Astrophysics, Vol. 23, No. 3.
On the basis of a series of high–dispersion spectrograms of P Cygni the authors have concluded that there exists a relatively slowly varying background profile of the absorption lines (characteristic time 100^d – 120^d) and a structure of some rapidly varying small components (characteristic time some days). The fine structure is caused, most probably, by the inhomogeneity of matter outflow from P Cygni. The features of the background profile (e.g. the maximum of the whole absorption component) have a periodic trend from initial Doppler shift to the higher shift values with a rather sharp restoration of the initial state.

114.033 **On the periodicity of rapid variations of the Hα profile in the spectrum of the Be star 59 Cyg.**
A. E. Tarasov, A. G. Shcherbakov.
Izv. Krymskoj Astrofiz. Obs., Tom 70, p. 129 – 134 (1985). In Russian. English translation in Bull. Crimean Astrophys. Obs., Vol. 70.
128 spectrograms of the Be star 59 Cyg have been obtained during four nights in July – August 1981. During this period Hα had two emission components, V and R, with absorption reversal. The equivalent width of Hα emission was equal to about 5 Å. It has been shown that the equivalent width, as well as the intensities of line components are varying from night to night. Moreover, the intrinsic time of the observed rapid fluctuations of these parameters equals to dozens of minutes. It is found that the variation of the V/R ratio during the observational interval in 1981 occurred as a periodic process with the period 380 min being superimposed by a rapid fluctuation having also a quasiperiodic character in time intervals of 25 – 40 min.

114.034 **The formation of nitrogen and carbon lines in HD 50896 (WN5).**
D. J. Hillier.
Luminous stars and associations in galaxies, p. 261 – 264 (1986). – See Abstr. 012.079. (IAU Symp. No. 116).
The basic mechanisms for producing the nitrogen and carbon emission lines, which characterize a WR's spectrum, are unknown. To determine the mechanisms giving rise to these lines the authors are currently studying the formation of N and C emission lines in the WN5 star HD 50896.

114.035 **An investigation of the atmosphere of the supergiant δ CMa (F8 Ia) and determination of its evolutionary parameters.**
L. S. Lyubimkov, Z. A. Samedov.
Izv. Krymskoj Astrofiz. Obs., Tom 72, p. 99 – 106 (1985). In Russian. English translation in Bull. Crimean Astrophys. Obs., Vol. 72.
On the basis of the model atmospheres of Kurucz (1979) and using high dispersion spectroscopic data of Castley and Watson (1980) the authors have analysed the supergiant δ CMa (F8 Ia) and obtained the following values of the effective temperature and surface gravity: $T_{eff} = 6000 \pm 200$K, $\log g = 1.0 \pm 0.2$. It is shown that the microturbulent velocity in the upper layers of the atmosphere exceeds noticeably the sound velocity. By means of comparison with the evolution calculations the authors have found mass, radius, luminosity and the age of δ CMa.

114.036 **A model atmosphere analysis of α Lyr and γ Gem.**
I. S. Savanov, A. M. Khalilov.
Izv. Krymskoj Astrofiz. Obs., Tom 72, p. 106 – 119 (1985). In Russian. English translation in Bull. Crimean Astrophys. Obs., Vol. 72.
Using model atmospheres of Kurucz (1979) the authors have analysed high–dispersion spectrograms of α Lyr and γ Gem. They used for α Lyr the model with $T_{eff} = 9650$K and $\lg g = 3.9$. For γ Gem they obtained the following values: $T_{eff} = 9300 \pm 100$K, $\lg g = 3.40 \pm 0.15$. In both cases the microturbulence was found $\xi_t = 2.0 \pm 0.5$ km/s. The abundances of 14 and 20 elements in the atmospheres of α Lyr and γ Gem accordingly were obtained. Masses, radii, luminosities and ages of the stars were found using evolutionary calculations. They were compared with the data from direct measurements of angular diameters, parallaxes and absolute energy distributions.

114.037 **Fe II in the UV spectrum of luminous emission line stars.**
G. Muratorio, M. Friedjung, R. Viotti.
Luminous stars and associations in galaxies, p. 275 – 276 (1986). – See Abstr. 012.079. (IAU Symp. No. 116).

114.038 **A model atmosphere analysis of three A–type stars.**
I. S. Savanov.
Izv. Krymskoj Astrofiz. Obs., Tom 73, p. 92 – 101 (1985). In Russian. English translation in Bull. Crimean Astrophys. Obs., Vol. 73.
Using model atmospheres of Kurucz (1979) the analysis of high–dispersion spectrograms of η Oph, ε Ser and θ Leo was carried out. The obtained values of T_{eff}, $\lg g$ and ξ_t are presented in a table. All stars have chemical composition similar to that of metallic–line stars. Masses, radii, luminosities and ages of the stars were found using evolutionary calculations.

114.039 **On the abundance of technetium in the atmospheres of cool stars.**
M. Ya. Orlov, A. V. Shavrina.
Nauchn. Inf., Vyp. 56, p. 97 – 98 (1984). In Russian.
The abundance of technetium in the atmospheres of MS giants OP Her and R Hya has been determined by comparison of theoretical equivalent widths of resonance lines of neutral technetium with observed ones: $\lg N(\mathrm{Tc}) = (-2.0) - (-2.5)$ where $\lg N(\mathrm{H}) = 12.0$. An upper limit of the abundance of technetium in the atmospheres of some M giants $\lg N(\mathrm{Tc}) < -3.0$ is also derived.

114.040 **Energy distribution in the spectra of 9 variable carbon stars.**
R. Kh. Oganesyan, G. M. Tovmasyan.
Nauchn. Inf., Vyp. 56, p. 116 – 124 (1984). In Russian.
The observed energy distribution in the spectra of 9 variable carbon stars (spectral region 4000 – 7000 Å) has been determined. The final results of the measurements (erg/cm^2 sec Å) are given in the tables and on figures.

114.041 **Distribution of energy in the spectra of carbon stars.**
J.–I. Straume.
Nauchn. Inf., Vyp. 56, p. 129 – 133 (1984). In Russian.
The energy distribution in the spectra of carbon stars is calculated in the spectral range from 2000 Å to 20 microns. Continuous opacity by H, H^-, He, He^-, C^-, H_2^-, N_2^-, scattering on H, He, C, N, H_2, N_2, C_2 and molecular absorption of molecular CH, CN, C_2 is taken into account.

114.042 **Molecular absorption in the atmospheres of carbon stars. 2.**
J.–I. K. Straume.
Nauchn. Inf., Vyp. 57, p. 55 – 63 (1984). In Russian.
Column densities are calculated for selected molecules in atmospheres of carbon stars. Partial optical depths for various wavelengths of electronic–vibrational bands of diatomic molecules CN, C_2, CH, CS, SiS, SiH, SiN, MgH are obtained.

114.043 **The Wolf–Rayet star HD 16523 with strengthened oxygen lines.**
D. N. Rustamov, A. M. Cherepashchuk.
Pis'ma Astron. Zh., Tom 12, No. 5, p. 373 – 379 (1986). In Russian. English translation in Sov. Astron. Lett., Vol. 12.

New spectral observations of the peculiar WC5 star HD 16523 have confirmed the strong emission doublet O VI λλ3811, 3834. Rapid variability of these lines has been detected on ~1 hour intervals. The spectral type of HD 16523 is uncertain as estimated from different criteria. The dependence of emission line widths on the ionization potentials is characterized by considerable scattering. The spatial distribution of WR stars with strengthened oxygen lines and WR stars with probable relativistic companions are found to be similar.

114.044 **Carbon abundances in halo dwarfs.**
J. Tomkin, C. Sneden, D. L. Lambert.
Astrophys. J., Vol. 302, No. 1, p. 415 – 420 (1986).

The authors have made a spectrum synthesis analysis of Digicon observations of CH in 32 halo F and G dwarfs with iron abundances as low as [Fe/H] ≈ –2.5. The results show that carbon follows iron down to [Fe/H] ≈ –1.8; at lower metallicities there is evidence of a positive trend in [C/Fe]. It is suggested that the change in slope of the [C/Fe] versus [Fe/H] relationship at [Fe/H] ≈ –1.8 is due to the appearance of Type I supernovae. The sense of the change of slope is consistent with the prediction that Type I supernovae are inefficient producers of carbon.

114.045 **Atmospheric parameters and metallicity of θ Leo (A2 V).**
V. G. Klochkova, V. E. Panchuk, V. V. Tsymbal.
Bull. Spec. Astrophys. Obs. – North Caucasus, Vol. 19, p. 19 – 25 (1985). English translation of 39.114.062.

114.046 **Absolute spectrophotometry of carbon stars. I.**
I. Eglitis.
Issled. Solntsa Krasnykh Zvezd, No. 20, p. 26 – 45 (1984). In Russian.

Absolute energy distribution in the spectra of six carbon stars have been obtained. The spectral region is λ3225 – 8000 Å.

114.047 **Investigations of the spectral variability of carbon stars.**
I. Eglitis.
Issled. Solntsa Krasnykh Zvezd, No. 22, p. 100 – 126 (1985). In Russian.

195 lines in the spectra of semiregular carbon stars UX Dra, W Ori, Y Tau and irregular carbon stars ST Cam, DS Peg, TX Psc in the wave–length range λ6100 – 6800 Å are discussed. It is found that the intensity of C_2, CN bands and ionized rare–earth element lines vary with the light variations, and that the intensity of neutral metal lines does not vary with the phase of star brightness.

114.048 **Spectrophotometry of carbon stars. III. Molecular absorption bands: quantitative analysis.**
S. E. Nersisyan, R. Kh. Oganesyan.
Astrofizika, Tom 24, Vyp. 2, p. 245 – 255 (1986). In Russian. English translation in Astrophysics, Vol. 24, No. 2.

The indices of the C_2, CN, SiC_2 and CH molecular bands of 56 carbon stars are determined. A definite dependence between these indices was considered. The comparison of the *I*(5635) and *I*(4737) indices outlines a division of stars into two banches. On the $I(C^{12}N) - I(C^{13}N)$ diagram a separation of R and N stars is seen. A certain correlation is found between the Merrill–Sanford band and the sodium line indices. By the comparison of the C_2 molecule *I*(4737) and the CH molecule G–band indices a certain group of stars is revealed in the wavelength region 4480 – 4610 Å of the spectrum of which intensive CN isotopic bands are observed.

114.049 **Carbon stars in the Galaxy and in other stellar systems.**
Z. Alksne, A. Alksnis.
Zvaigžnota Debess, 1985. Gada Pavasaris, p. 12 – 17 (1985). In Latvian.

114.050 **A search for the tritium hyperfine line from nearby stars.**
F. Valdes, R. A. Freitas Jr.
Icarus, Vol. 65, No. 1, p. 152 – 157 (1986).

A search for the tritium hyperfine line at 1516 MHz from 108 assorted astronomical objects, with emphasis on 53 nearby stars, was conducted in June 1983. All stars within 20 light–years visible from the 26–m telescope at Hat Creek Radio Observatory were examined using 256 4883–Hz channels. Twelve stars were also examined using 1024 76–Hz channels. The wideband– and narrow–band–channel observations achieved sensitivities of $5 - 14 \times 10^{-24} W/m^2$/channel and $0.7 - 2 \times 10^{-24} W/m^2$/channel, respectively. No detections were made. The tritium frequency is highly attractive for SETI work.

114.051 **Do Kurucz's model fluxes fit IUE spectra of warm stars?**
R. A. Wade.
Bull. Am. Astron. Soc., Vol. 17, No. 4, p. 839 – 840 (1985). Abstract. – See Abstr. 010.062.

114.052 **The metallicity distribution of extreme halo stars.**
T. C. Beers, S. A. Shectman, G. W. Preston.
Bull. Am. Astron. Soc., Vol. 17, No. 4, p. 840 – 841 (1985). Abstract. – See Abstr. 010.062.

114.053 **Isotopic abundances of magnesium in metal poor stars.**
A. McWilliam, D. L. Lambert.
Bull. Am. Astron. Soc., Vol. 17, No. 4, p. 841 (1985). Abstract. – See Abstr. 010.062.

114.054 **Nitrogen and sulfur excesses in the post–AGB star HD 46703.**
H. E. Bond, R. E. Luck.
Bull. Am. Astron. Soc., Vol. 17, No. 4, p. 841 (1985). Abstract. – See Abstr. 010.062.

114.055 **On the properties of well–studied early type stars observed with the Einstein X–ray observatory.**
F. R. Harnden Jr., R. Rosner, M. Ramella, C. Morossi.
Bull. Am. Astron. Soc., Vol. 17, No. 4, p. 862 (1985). Abstract. – See Abstr. 010.062.

114.056 **Ultraviolet observations of metal deficient field giants.**
A. K. Dupree, L. Hartmann, G. Smith.
Bull. Am. Astron. Soc., Vol. 17, No. 4, p. 876 (1985). Abstract. – See Abstr. 010.062.

114.057 **A survey of Hα emission among metal–deficient giants.**
G. H. Smith, A. K. Dupree.
Bull. Am. Astron. Soc., Vol. 17, No. 4, p. 876 (1985). Abstract. – See Abstr. 010.062.

114.058 **Carbon isotope ratios in field population II giant stars.**
C. Sneden, C. A. Pilachowski, D. A. VandenBerg.
Bull. Am. Astron. Soc., Vol. 17, No. 4, p. 877 (1985). Abstract. – See Abstr. 010.062.

114.059 **Measurement of a velocity gradient in the first overtone band of CO observed in the red giant Chi Pegasi.**
D. Nadeau, J.–P. Maillard.
Bull. Am. Astron. Soc., Vol. 17, No. 4, p. 878 (1985). Abstract. – See Abstr. 010.062.

114.060 **Mass–line width relations from UV spectra of early type stars.**
K. Sekiguchi.
Bull. Am. Astron. Soc., Vol. 17, No. 4, p. 881 (1985). Abstract. – See Abstr. 010.062.

114.061 **A study of carbon, nitrogen and oxygen (CNO) abundances in chemically peculiar stars of the upper main–sequence.**
S. W. Roby.
Bull. Am. Astron. Soc., Vol. 17, No. 4, p. 902 (1985). Abstract. – See Abstr. 010.062.

114.062 **Synoptic observations of the Li I λ6707 line in a sample of solar–type stars.**
M. S. Giampapa.
Bull. Am. Astron. Soc., Vol. 17, No. 4, p. 902 (1985). Abstract. – See Abstr. 010.062.

114.063 **CH and barium metal–deficient giants in the Vilnius photometric system. I.**
J. Šleivytė.
Vilniaus Astron. Obs. Biul., Nr. 69, p. 9 – 23 (1985). In Russian.
The results of a photoelectric photometry of 27 CH and barium giants in the Vilnius photometric system are presented. For eleven of them color temperatures are determined using Baumert (1972) and Mendoza and Johnson (1965) infrared photometry corrected for interstellar reddening taken from Bartkevičius and Šleivytė (1983). In the dereddened $(U-P)_0$, $(V-S)_0$ diagram all these carbon–rich metal–deficient giants are situated above the sequence of solar–composition giants, contrary to oxygen–rich metal–deficient giants which lie below this sequence. Such difference in the behaviour of these types of stars probably is a consequence of their O/C abundance ratio.

114.064 **Metallicity determination of G–K giants from medium dispersion spectra.**
G. Tautvaišienė.
Vilniaus Astron. Obs. Biul., Nr. 70, p. 3 – 13 (1985). In Russian.
A method for metallicity determination of G–K giants using spectra of 50 Å/mm dispersion is proposed. It is based on theoretical interpretation of central intensities of strong Fe I lines using model stellar atmospheres. Temperatures and metallicities for 20 G–K metal–deficient giants and 7 normal giants are determined.

114.065 **Carbon, nitrogen, and oxygen in two lithium–rich supergiants.**
R. E. Luck, C. Sneden.
Publ. Astron. Soc. Pac., Vol. 98, No. 601, p. 310 – 318 (1986).
The authors present an analysis deriving CNO abundances for two lithium–rich supergiants (HR 7008 and HD 174104). They also rederive lithium abundances using high–resolution, high signal–to–noise ratio Reticon data. They find that these stars do not have CNO abundances typical of other supergiants, that is, processed through the CN cycle, nor do they have the CNO content which would be expected if they were unprocessed stars. In HR 7008, ΣCNO is enhanced relative to iron; whereas, in HD 174104 ΣCNO is deficient relative to iron. There is no known scenario in stellar evolution which can explain these objects.

114.066 **Atmosphere of the Am star 15 Vul.**
L. S. Lyubimkov, I. S. Savanov.
Bull. Crimean Astrophys. Obs., Vol. 69, p. 46 – 54 (1986). English translation of 39.114.044.

114.067 **Spectral types from *uvbyβ* photometry.**
T. Moon.
Astrophys. Space Sci., Vol. 122, No. 1, p. 173 – 176 (1986).
New diagrams are presented for the assimilation of MK spectral types with *uvbyβ* photometry, based upon reddening–free parameters calculated using revised reddening ratios and recently published values of mean *uvby* and β indices for a selection of MK spectral types.

114.068 **Determination of the Na content in the atmospheres of K giants.**
N. S. Komarov, T. V. Mishenina, V. D. Motrich.
Sov. Astron., Vol. 29, No. 4, p. 434 – 436 (1985). English translation of 40.114.027.

114.069 **The atmosphere of the supergiant α Cam. I. On the Hα profile.**
S. K. Zejnalov, F. A. Musaev, E. L. Chentsov.
Astrofiz. Issled. Izv. Spets. Astrofiz. Obs., Tom 21, p. 3 – 7 (1986). In Russian. English translation in Bull. Spec. Astrophys. Obs. – North Caucasus.
Time variations of the profiles of Hα and other lines of the red region of the α Cam spectrum are investigated from spectrograms with a dispersion of 14 Å/mm. The lower layers of the stellar wind are observed up to the level where their expansion velocity reaches 0.3 from the maximum. Periodicity of the radial velocity variations registered from the photospheric lines is not confirmed.

114.070 **The use of high members of hydrogen series in the investigation of A–star atmospheres.**
V. D. Bychkov, L. I. Snezhko, V. V. Sokolov.
Astrofiz. Issled. Izv. Spets. Astrofiz. Obs., Tom 21, p. 8 – 11 (1986). In Russian. English translation in Bull. Spec. Astrophys. Obs. – North Caucasus.
A calculation of the synthetic spectrum of H_{10}–H_{15} and Pa_{12}–Pa_{18} hydrogen lines has been fulfilled. It is shown that in the regions $3 \leqslant \lg g \leqslant 4$ and $8\times10^3 \leqslant T_{eff} \leqslant 13\times10^3$ the characteristics of the H_{11} lines referred to the observable continuous spectrum do not practically depend upon T_{eff} and lg *g*. In this region of atmosphere parameters the characteristics of the Pa_{14} line are not dependent upon T_{eff}, while the dependence upon lg *g* which is no less strong than for isolated lines, but of contrary sign, preserves.

114.071 **On the metallicity dispersion of K giants of the disk of the Galaxy.**
T. V. Mishenina, V. E. Panchuk.
Astrofiz. Issled. Izv. Spets. Astrofiz. Obs., Tom 21, p. 12 – 15 (1986). In Russian. English translation in Bull. Spec. Astrophys. Obs. – North Caucasus.
The homogeneity of a sample of K giants from the catalogue of spectroscopic [Fe/H] metallicity determinations is considered. From the catalogue data the metallicity dispersion may be reduced from 0.23 to 0.16 allowing for methodical errors. From the results of analysis of homogeneous spectroscopic observations the metallicity dispersion is equal to 0.11 with an error of Δ[Fe/H] = 0.10

114.072 **Non–LTE analysis of massive stars in the Magellanic Clouds.**
T. Gehren, D. Husfeld, R. P. Kudritzki, P. S. Conti, D. G. Hummer.
Luminous stars and associations in galaxies, p. 413 – 414 (1986). – See Abstr. 012.079. (IAU Symp. No. 116).
The massive stars of the Magellanic Clouds are of considerable current interest with regard to questions of initial mass function, star formation mechanisms, stellar evolution with mass loss and the chemical evolution of galaxies. The effective temperatures, surface gravities and helium abundances of 6 main sequence O–type stars, obtained by fitting non–LTE model atmospheres to high quality spectra, are presented.

114.073 **Statistical investigation of chemically peculiar stars. II. Stars with depression of the continuum spectrum at λ5200 Å.**
V. S. Lebedev.
Astrofiz. Issled. Izv. Spets. Astrofiz. Obs., Tom 21, p. 30 – 38 (1986). In Russian. English translation in Bull. Spec. Astrophys. Obs. – North Caucasus.
The data on the depression value in the band of 5200 Å for 209 chemically peculiar stars reduced to a united photometric system are presented. The value of depression Δ*a* correlates with the extremum of the magnetic field and silicon and europium content with the correlation coefficients 0.54, 0.33, 0.25 respectively.

114.074 **The Bp stars in the Orion OB1 association.**
V. G. Klochkova.
Sov. Astron. Lett., Vol. 11, No. 4, p. 209 – 213 (1985). English translation of 40.114.001.

114.075 **Determination of the physical parameters of the atmospheres of B and A stars in the spectral range near the Balmer discontinuity.**
S. A. Gulyaev, V. E. Panchuk, V. V. Pleshakov, S. G. Pyatkes.
Astrofiz. Issled. Izv. Spets. Astrofiz. Obs., Tom 22, p. 3 – 12

(1986). In Russian. English translation in Bull. Spec. Astrophys. Obs. – North Caucasus.

Basing on the analysis of the dependence of the kind of B– and A–stars spectra within λ3650 – 3900 Å on the effective temperature T_e and the gravity g methods have been developed for determination of these parameters. Spectra have been obtained and spectrophotometry has been made for 73 bright stars. The results of observations allow to connect the new system for T_e determination with the known effective temperature scale. The values obtained prove to be systematically by 1500 – 2000K overestimated. Possible causes for this discrepance are discussed.

114.076 **Chemical composition of K giants in the Hyades.**
T. V. Mishenina, V. E. Panchuk, N. S. Komarov.
Astrofiz. Issled. Izv. Spets. Astrofiz. Obs., Tom 22, p. 13 – 16 (1986). In Russian. English translation in Bull. Spec. Astrophys. Obs. – North Caucasus.

Spectra of four red giants, members of the Hyades open cluster, are analysed by the model atmosphere method. Comparison with the results of metallicity determinations of the K giants, members of other open clusters and dynamical groups, shows metallicity excess in the Hyades giants' atmospheres Δ[Fe/H] = 0.08.

114.077 **On the effective temperatures of chemically peculiar stars.**
Yu. V. Glagolevskij, N. M. Chunakova.
Astrofiz. Issled. Izv. Spets. Astrofiz. Obs., Tom 22, p. 39 – 53 (1986). In Russian. English translation in Bull. Spec. Astrophys. Obs. – North Caucasus.

A list of more than 300 chemically peculiar stars is given for which estimates of effective temperatures are made. The temperatures were derived on the basis of literature data about the energy distributions of these stars, using the index Q of *UBV* photometry and also the X and $(B2 - G)$ parameters of multicolour photometry. An analysis of the obtained temperatures has shown that these temperatures are sufficiently reliable for many statistical investigations.

114.078 **The system of standard wavelengths for white supergiants.**
V. M. Dobrichev, E. L. Chentsov, Z. U. Shkhagosheva.
Astrofiz. Issled. Izv. Spets. Astrofiz. Obs., Tom 22, p. 59 – 63 (1986). In Russian. English translation in Bull. Spec. Astrophys. Obs. – North Caucasus.

A list of spectral line wavelengths is made for determining the radial velocities from high–dispersion spectrograms of supergiants B9 – A3, including 153 values of wavelengths in the interval 3372 – 6563 Å, from which 93 values in the interval 3820 – 4635 Å are examined and specified according to the spectra of A stars of the main sequence.

114.079 **Lithium in late–type giants. IV. The subgiant CH stars.**
V. V. Smith, D. L. Lambert.
Astrophys. J., Vol. 303, No. 1, p. 226 – 232 (1986).

The authors report on lithium abundances in 10 subgiant CH stars and one giant CH star. In no star was the Li I resonance doublet detected, and upper limits to the lithium abundance in the range $\log \varepsilon(Li) \leqslant 0.7$ to $\log \varepsilon(Li) = 1.5$ [$\log \varepsilon(H) = 12.0$] were determined for this sample of stars. The consequences of this result for the evolutionary state of the CH stars are discussed. Since all these stars are binaries with low–luminosity companions, they may represent mass transfer from a former asymptotic giant branch star (now a white dwarf) onto a main–sequence companion (now a subgiant CH star).

114.080 **Interpretation of Δυ = 2 carbon monoxide lines in the spectrum of α Her.**
P. Höflich, R. P. Lowe, J. Moorhead, M. Scholz, W. Wehlau, R. Wehrse.
Mon. Not. R. Astron. Soc., Vol. 220, No. 2, p. 377 – 382 (1986).

CO Δυ = 2 line profiles of α Her obtained from a high–resolution Fourier–transform spectrum are analysed by means of spherically extended model atmospheres. It is found that (1) line scattering is very small in spite of the low density in the outer atmosphere, (2) the microturbulence is about 3 km s^{-1}, and (3) the star does not have an extended cool envelope. The carbon abundance appears to be solar.

114.081 **Faint early–type stars and emission–line stars in the Canis Majoris Complex.**
S. D. Wiramihardja, T. Kogure.
Galactic structure and variable stars, p. 44 – 52 (1985). – See Abstr. 012.029.

In order to study the distribution of OB stars in the Canis Majoris Complex, a survey and classification have been done for stars in the area. The study is based on spectral observations at the Kiso Schmidt Telescope.

114.082 **M stars: supergiants to dwarfs.**
J. B. Kaler.
Sky Telesc., Vol. 71, No. 5, p. 450 – 454 (1986).

114.083 **Non–membership of two superluminous giant candidates in the LMC.**
J. R. Sowell.
Astron. J., Vol. 91, No. 1, p. 79 – 80 (1986).

Two superluminous giant candidates in the LMC, identified from photometry, have been classified as G8 III and G5 V. The average radial velocities for both stars are approximately +15 km/s. It is evident that neither star is an SLG or is in the LMC.

114.084 **Spectral types for objects in the Kiso survey. II. Data for 61 stars.**
G. Wegner, R. K. McMahan.
Astron. J., Vol. 91, No. 1, p. 139 – 143 (1986).

Spectroscopic observations and spectral types for 61 blue objects, mostly from the Kiso Schmidt camera survey, are reported. The original spectroscopic data, obtained with the 1.3 m McGraw–Hill telescope, have 8 Å resolution and cover the wavelength range λλ4000 – 7200. Apparently normal main–sequence or horizontal–branch stars comprise the majority of the observed targets, but six subdwarfs, four new DA, one DC, and two DB white dwarfs, and five quasars and emission–line objects were found among the sample.

114.085 **Dwarf K and M stars of small proper motion found in a large spectroscopic survey.**
C. B. Stephenson.
Astron. J., Vol. 91, No. 1, p. 144 – 158 (1986).

More than 2000 candidates for K and M dwarf status, generally having proper motions that are either unknown or less than 0″.2 per annum, have been identified in an objective–prism survey covering the entire sky north of declination –25° and more than 10° from the galactic plane. More than 95% have no previous indication that they are dwarfs. Several dozen appear to be visual double stars.

114.086 **Abundances in field dwarf stars. III. Magnesium abundances.**
J. B. Laird.
Astrophys. J., Vol. 303, No. 2, p. 718 – 723 (1986).

Intermediate–dispersion spectra of 108 field dwarf stars, plus three Hyades dwarfs, have been used to derive magnesium abundances relative to iron. The program sample includes both disk and halo stars, spanning a range of +0.5 to –2.1 in [Fe/H]. Magnesium abundances have been determined from synthetic spectra of a region including the Mg I *b* lines. The results indicate that relatively more Mg was produced by stellar nucleosynthesis in the halo.

114.087 **Lithium in early F dwarfs.**
A. M. Boesgaard, M. J. Tripicco.
Astrophys. J., Vol. 303, No. 2, p. 724 – 739 (1986).

Abundances have been determined for Li in early F field dwarfs in order to determine the initial Li content for Population I stars. Reticon spectra were obtained at the coudé spectrograph of the CFH telescope of 75 F0 – F5 dwarfs with a spectral resolution of 0.11 Å. High S/N values allowed the observation of weak features ($W_\lambda \leqslant 2$ mÅ). Model atmospheres of Kurucz were used to calculate abundances from the Li I and Li I – Fe I blends, five Fe I lines, and one Ca I line. Lithium at the initial or "cosmic" abundance level was present in about one-third of the stars observed; the mean value is $Li/H = 1.02 \times 10^{-9}$. Lithium content and age are related with 62% of the stars younger than two billion years being Li-rich and 64% of the stars older than that being depleted by more than factors of 40. The relation of lithium abundance with metallicity, chromospheric activity and stellar rotation is discussed.

114.088 **Spectroscopic determination of the metallicity of F dwarfs in open clusters.**
V. G. Klochkova, V. E. Panchuk.
Pis'ma Astron. Zh., Tom 12, No. 6, p. 446 – 451 (1986). In Russian. English translation in Sov. Astron. Lett., Vol. 12.
The iron abundance for 18 F dwarfs, members of the open clusters Hyades, Pleiades and Coma Berenices is determined using spectroscopic observations with the 6-m telescope. It is shown that the «Hyades anomaly» for F dwarfs can be explained by strengthening of lines of middle intensity without hypotheses on a supermetallicity or on a low helium abundance for these stars.

114.089 **Southern hemisphere spectrophotometric standards.**
J. H. Spencer Jones.
Mon. Notes Astron. Soc. S. Africa, Vol. 44, Nos. 3 – 4, p. 33 – 35 (1985).
Spectral energy distributions are given for three stars, suitable for use as southern hemisphere spectrophotometric standards. The wavelength range covered is 3350 – 6800 Å, and the stars have magnitudes in the interval V ≈ 10 – 13.

114.090 **The absolute fluxes of six hot stars in the ultraviolet (912 – 1600 Ångstroms).**
T. N. Woods.
Diss. Abstr. Int., Sect. B, Vol. 46, No. 6, p. 1959 – 1960 (1985). Thesis, The Johns Hopkins University, 119 pp. (1985). Order No. DA8518544.

114.091 **The neutral sodium D_2 line in late-type stars.**
D. C. Boice.
Diss. Abstr. Int., Sect. B, Vol. 46, No. 7, p. 2345 (1986). Thesis, New Mexico State University, 229 pp. (1985). Order No. DA8519967.

114.092 **On the source of the violet depression in carbon stars (a study of the Balmer decrement in carbon long-period variable stars).**
M.-A. Orlati.
Diss. Abstr. Int., Sect. B, Vol. 46, No. 12, p. 4283 (1986). Thesis, Case Western Reserve University, 71 pp. (1985). Order No. DA8601957.

114.093 **The red giant metallicity function.**
V. A. Marsakov, A. A. Suchkov.
Sov. Astron., Vol. 29, No. 5, p. 495 – 498 (1985). English translation of 40.114.059.

114.094 **Chemical composition of the B star HD 204754 suspected of deficiency of iron.**
D. A. Ptitsyn, T. A. Ryabchikova.
Astron. Zh., Tom 63, Vyp. 3, p. 527 – 533 (1986). In Russian. English translation in Sov. Astron., Vol. 30, No. 3.
A chemical analysis of the B star HD 204754 has been carried out. According to Cowley et al. (1982) this star is suspected to have decreased abundance of iron. Parameters of the atmosphere $T_e = 13000$K, $\lg g = 3.5$, and $v_t = 4$ km/s are obtained. Abundances of 9 elements from C to Ni are calculated by means of the model atmosphere technique. No deficit of iron exceeding 0.3 dex has been found. Abundances of all other elements practically do not differ from both solar abundances and those in the normal B star ν Cap of spectral type similar to that of the star under study.

114.095 **The central depths of the lines Hγ and Hδ in low-dispersion spectrograms – additional criterion in recognizing Ap stars.**
Ts. S. Radoslavova.
Dokl. Bolg. Akad. Nauk, Vol. 38, No. 10, p. 1277 – 1278 (1985). In Russian. Abstr. in Ref. Zh., 51. Astron., 6.51.484 (1986).

114.096 **Equivalent widths of the lines in the spectra of 10 giants.**
T. V. Mishenina.
Odes. univ. Odessa, 14 pp. (1985). In Russian. Abstr. in Ref. Zh., 51. Astron., 6.51.506 (1986).

114.097 **Mercury-manganese stars observed in the ultraviolet region.**
K. Sadakane.
Astron. Her., Vol. 79, No. 6, p. 140 – 144 (1986). In Japanese.

114.098 **Infrared excess in Wolf-Rayet stars.**
M. Singh.
Astrophys. Space Sci., Vol. 123, No. 1, p. 155 – 159 (1986).
On the basis of continuum energy distributions in the wavelength region λλ3200 – 11000 Å, effective temperatures for 14 Wolf-Rayet stars are estimated by comparing with Kurucz (1979) model atmospheres. The continuum energy distribution curve shows strong infrared excess emissions above of λ5000 Å in every star.

114.099 **Spectrophotometric study of the suspected UU Herculis-star HR 4912.**
D. D. Sasselov, D. Z. Kolev.
Astrophys. Space Sci., Vol. 123, No. 2, p. 363 – 374 (1986).
The authors have obtained high-dispersion blue spectrograms and BV photometry of the high-latitude F-supergiant HR 4912 = HD 112374. As a result of a curve-of-growth analysis they conclude that it appears to have the parameters of a F-supergiant with slight metal deficiency ([Fe/H] = -0.7). The latter depends much upon the derived temperature of HR 4912. The blue spectra and new BV photometry made it possible to obtain four independent estimates of the temperature. HR 4912 seems to stand close to the stars of the UU Herculis type.

114.100 **The abundance of gallium in B-type chemically peculiar stars.**
M. Takada-Hidai, K. Sadakane, J. Jugaku.
Astrophys. J., Vol. 304, No. 1, p. 425 – 435 (1986).
Quantitative analyses of the Ga abundances in 27 Hg-Mn stars, 11 Si (magnetic) stars, eight He-weak stars, and seven normal stars are made with the resonance lines of Ga II at 1414 Å and Ga III at 1495 Å in IUE spectra. The Ga overabundances are confirmed as a genuine anomaly in many peculiar stars. The Ga abundances in normal stars are all upper limits which roughly correspond to the solar value.

114.101 **Isotopic abundances of magnesium in the metal-poor subgiant ν Indi.**
D. L. Lambert, A. McWilliam.
Astrophys. J., Vol. 304, No. 1, p. 436 – 442 (1986).
An isotopic abundance analysis for Mg in the metal-poor subgiant ν Indi is reported. A high-resolution, high signal-to-noise spectrum centered near 5131 Å in combination with synthetic spectra gives the ratios $^{25}Mg/^{24}Mg$ and $^{26}Mg/^{24}Mg \leqslant 3\%$. The ^{25}Mg and ^{26}Mg isotopes are obviously underabundant relative to solar system material, where $^{25}Mg/^{24}Mg = 13\%$ and $^{26}Mg/^{24}Mg = 14\%$. An elemental abundance analysis is presented for the elements: Li, C, O, Mg, Si, Ca, V, Mn, Ti, Fe, Co, Ni, Y, Ba, La, Ce, and Eu.

114.102 **An O3 star in the Small Magellanic Cloud H II region NGC 346.**
N. R. Walborn, J. C. Blades.
Astrophys. J., Lett. Ed., Vol. 304, No. 1, p. L17 – L19 (1986). With plate L1.
Two new O stars have been found in NGC 346, the largest H II region of the Small Magellanic Cloud. One of these has a spectrum of type O3 III(f*), identical to that of HDE 269810 in the

Large Magellanic Cloud; it is a major source of ionization in NGC 346.

114.103 **On the synthetic spectra computation for stars and the sun.**
L. S. Lyubimkov.
Izv. Krymskoj Astrofiz. Obs., Tom 74, p. 3 – 14 (1986). In Russian. English translation in Bull. Crimean Astrophys. Obs., Vol. 74.
A technique for synthetic spectra calculation is described, which has been realized in the computer program SYNTHEL used at the Crimean Astrophysical Observatory for stellar and solar spectra analysis. The problems concerning the account of the star rotation and the instrumental profile are considered. As an example a sample of synthetic spectra in the region 2580 – 2590 Å for an A0 V star were calculated corresponding to different rotational velocities, different half–widths of the instrumental profile, and the combined effect of stellar rotation and instrumental resolution. The ASTRON spectrum of the star η Leo (A0 Ib) is compared with the synthetic one computed in the range of 2550 – 2590 Å.

114.104 **On the variation of Hα emission in the spectrum of the Be star 28 Cyg.**
A. E. Tarasov, A. G. Shcherbakov.
Izv. Krymskoj Astrofiz. Obs., Tom 74, p. 19 – 22 (1986). In Russian. English translation in Bull. Crimean Astrophys. Obs., Vol. 74.
Eighty–nine spectrograms of the Be star 28 Cyg were obtained during three nights in August 1981. During this period of time the equivalent width of Hα emission has been varying from 12 to 17 Å, whereas the V/R ratio remained constant. The period of brightness variations was detected being equal to $0\overset{d}{.}35$ (or $0\overset{d}{.}70$). The model of the magnetic oblique rotator is considered as a possible one to interpret the observational data.

114.105 **Comparison of spectral characteristics of three Be stars: φ Persei, θ Coronae Borealis and X Persei.**
T. S. Galkina.
Izv. Krymskoj Astrofiz. Obs., Tom 74, p. 23 – 34 (1986). In Russian. English translation in Bull. Crimean Astrophys. Obs., Vol. 74.
The results of spectroscopic observations of the Be stars φ Per, θ CrB and X Per are reported. The profiles of hydrogen and helium lines were investigated and their variations with time have been found. Some parameters of the emission hydrogen lines in the spectra of φ Per and X Per and those of the absorption lines in the spectrum of θ CrB are estimated.

114.106 **Spectrophotometric study of the star ζ Tau from observations in 1983 – 1984. I. On the variability of the spectrum.**
T. M. Rachkovskaya, Ch. M. Nasibova.
Izv. Krymskoj Astrofiz. Obs., Tom 74, p. 34 – 41 (1986). In Russian. English translation in Bull. Crimean Astrophys. Obs., Vol. 74.
The spectrum of ζ Tau has been studied from observations during the period March 1983 – September 1984. The general configuration of the H_α and H_β line profiles has been conserved during the observation period and there was satisfied the relation $I_V/I_R > 1$. It was observed both the stationary and the variable fine structure of the violet and red H_α emissions. Equivalent widths of the total H_α emission varied irregularly. The rotational velocity $v \cdot \sin i = 116$ km/s measured on the He I lines in the red region is lower than $v \cdot \sin i = 240$ km/s measured on the He I lines in the blue region of the spectrum.

114.107 **An O3 star in the Small Magellanic Cloud H II region NGC 346.**
N. R. Walborn, J. C. Blades.
Space Telesc. Sci. Inst., Prepr. Ser., No. 94, 12 pp. (1986). To appear in Astrophys. J., Lett. Ed.

114.108 **On the relationship of the variable stars AG Car and HDE 269852 to the Ofpe/WN 9 objects.**
O. Stahl.
ESO Sci. Prepr., No. 428, 21 pp. (1986). To appear in Astron. Astrophys.

114.109 **On stellar spectrophotometric standards.**
Eh. I. Terez.
Photometric and polarimetric investigations of celestial bodies, p. 52 – 55 (1985). In Russian. – See Abstr. 003.035.
A list of 17 stationary stars of early spectral classes is proposed. The stars are recommended as primary spectrophotometric standards.

114.110 **Absolute calibration of energy distribution in the Vega spectrum. I. Spectral region 310 – 750 nm.**
Eh. I. Terez.
Photometric and polarimetric investigations of celestial bodies, p. 55 – 64 (1985). In Russian. – See Abstr. 003.035.
Results of experimental investigations of the absolute energy distribution in the Vega spectrum in the region of 310 – 750 nm are described. Absolute calibration is based on the new method consisting in comparison of stellar radiation with the brightness of a diffusely scattering screen illuminated by a laboratory standard light source. The atmospheric extinction was determined by the Bouger equations surplus system solution. The obtained values of the extra–atmospheric energy illumination from Vega are tied to the Soviet State Standard of spectral density of energy brightness.

114.111 **Absolute calibration of energy distribution in the Vega spectrum. II. Infrared region 750 – 1117.5 nm.**
A. A. Arkharov, Eh. I. Terez.
Photometric and polarimetric investigations of celestial bodies, p. 65 – 69 (1985). In Russian. – See Abstr. 003.035.
The results of experimental investigations of the absolute energy distribution of Vega in the near infrared region of 750 – 1117.5 nm are given. The absolute calibration was based on a new method consisting in comparison of star radiation with the brightness of a diffusely scattering screen illuminated by a laboratory standard light source. The method of correction for the atmospheric extinction in the water vapour band ϱ (870 – 1000 nm) was used to obtain a detailed spectrum of Vega in this region. The obtained values of the extra–atmospheric energy illumination from Vega are related to the USSR State Standard of the spectral density of energy brightness.

114.112 **Energy distribution in the emission spectra of G, K, and M giants.**
A. V. Dragunova, N. S. Komarov, T. V. Mishenina.
Photometric and polarimetric investigations of celestial bodies, p. 69 – 78 (1985). In Russian. – See Abstr. 003.035.
The energy distribution in radiation of 18 G–, K–, M–type giants in the wavelength region from 300 to 900 nm is given. These stars are proposed as standards for investigating physical–chemical parameters of the atmospheres of cool stars.

114.113 **The profiles of Balmer lines in Ae/Be spectra.**
M. A. Pogodin.
Astrofizika, Tom 24, Vyp. 3, p. 491 – 506 (1986). In Russian. English translation in Astrophysics, Vol. 24, No. 3.
A procedure of the construction of theoretical profiles of emission lines fit for any kinematical model of the envelopes is described. It has been shown that the model of expanding and simultaneously rotating envelopes can be successfully used for the interpretation of the observed profiles of Balmer lines in the spectra of Ae/Be stars.

114.114 **Analysis of the ultraviolet spectrum (2000 Å to 3000 Å) of Vega.**
K. Sadakane, M. Nishimura, R. Hirata.
Publ. Astron. Soc. Jpn., Vol. 38, No. 2, p. 215 – 224 (1986).
A high–dispersion spectrum of Vega between 2000 Å and 3000 Å is analyzed. The continuum level is determined by searching for line–free windows, comparing the observed spectrum with

a synthesized one. The abundances of five iron peak elements (Cr through Ni) obtained from the UV lines are found to agree with those obtained in the visual region. The deficiency of iron in the atmosphere of Vega is reconfirmed.

114.115 **A detailed kinematic and abundance analysis of old disk giants.**
P. L. Cottrell, C. Sneden.
Astron. Astrophys., Vol. 161, No. 2, p. 314 – 326 (1986).
Prompted by the work of various investigators (Lambert and Ries, 1981; Kjaergaard et al., 1982) the authors have undertaken a semiautomated detailed spectroscopic analysis of a relatively large (34) sample of evolved high-velocity giant stars (metal-poor members of the disk population). CNO and overall metal abundances, as well as the $^{12}C/^{13}C$ ratio, were derived. The conclusions of both the above papers were confirmed and extended, namely that for old disk giants the C/N ratios have approximately the solar ratio and the O/Fe ratio is enhanced relative to the sun. The evolutionary consequences of these more detailed results in terms of the kinematic, mass and convective mixing characteristics of these objects are discussed.

114.116 **The spectrum of the Am star HR 178: line identifications and equivalent widths in the range 3800 to 6700 Å.**
M. F. Coupry, C. Van't Veer-Menneret, C. Burkhart.
Astron. Astrophys., Suppl. Ser., Vol. 64, No. 3, p. 477 – 486 (1986).
The authors present line identifications and equivalent width measurements used for the detailed abundance analysis of the giant Am star HR 178.

114.117 **A search for rotational modulation of the ultraviolet spectra of ψ^3Piscium and 31 Comae Berenices.**
T. Simon.
Astron. J., Vol. 91, No. 5, p. 1233 – 1237 (1986).
IUE observations were made daily over a 7 day interval to search for rotational variability in the ultraviolet emission lines of two rapidly rotating, early G giants, ψ^3Psc and 31 Com. No evidence was found to suggest that the chromospheric and transition-region lines of these stars, or their ultraviolet continua near 1725 Å, are modulated by rotation. The uniformity of the ultraviolet emission of these giants can be accounted for if their outer atmospheres are heated by acoustic waves. If the heating is a manifestation of dynamo-generated magnetic fields as on the Sun, then regions of magnetic activity must be rather evenly distributed in longitude over the stellar surface.

114.118 **On Wolf-Rayet stars with unshifted absorption lines.**
A. F. J. Moffat, R. Lamontagne, M. M. Shara, H. A. McAlister.
Astron. J., Vol. 91, No. 6, p. 1392 – 1399 (1986).
The authors examine a sample of 16 galactic Wolf-Rayet stars that exhibit unshifted absorption lines, but for which orbital motion involving the WR star and an O-type companion is either unknown or thought to be absent. If all of these are binaries, this would increase the estimate of the binary frequency for WR stars by about 10%. Speckle interferometry of three bright non-WNL+abs stars in Cygnus fails to reveal any astrometric companions down to a separation of 0.05″. Thus, either the absorption lines are intrinsic to the WR star in these three cases or, more likely, they arise in a bound companion. The search for wide spectroscopic orbits for WR 137 and the WN3+abs star WR 3 is re-examined.

114.119 **μ Cep as an indicator of the telluric H_2O lines in the photographic infrared spectral region.**
Y. Fujita.
Proc. Jpn. Acad., Ser. B, Vol. 61, No. 6, p. 237 – 241 (1985). Abstr. in Phys. Abstr., Vol. 89, No. 1279, Entry 53760 (1986).

114.120 **Hydrogen deficiency and AGB M stars.**
G. C. Augason, F. C. Witteborn, K. O'Brien, H. R. Johnson.
Bull. Am. Astron. Soc., Vol. 18, No. 1, p. 635 (1986). Abstract. – See Abstr. 010.066.

114.121 **Four eleventh magnitude B stars at high galactic latitudes that appear to belong to Population I.**
W. Tobin.
Bull. Am. Astron. Soc., Vol. 18, No. 1, p. 638 (1986). Abstract. – See Abstr. 010.066.

114.122 **Three dwarf stars having exceptionally strong calcium hydride bands.**
C. B. Stephenson.
Publ. Astron. Soc. Pac., Vol. 98, No. 602, p. 467 (1986).
Three stars of the author's recent *General Catalog of Galactic S Stars* are in fact M dwarfs, apparently having exceptionally strong red CaH bands which on low-resolution objective-prism plates were mistaken for ZrO bands.

114.123 **The remarkable spectral line asymmetries of F and G Ib supergiant stars.**
D. F. Gray, C. G. Toner.
Publ. Astron. Soc. Pac., Vol. 98, No. 603, p. 499 – 503 (1986).
The authors observed a group of supergiants spanning the F5 Ib to K2 Ib interval. The line asymmetries of the F supergiants are found to be large and opposite in direction from the normal case seen in cooler and less-luminous stars, and in the Sun. There is a continuous and monotonic change in the bisector behavior with spectral type, with the transition from the anomalous to the normal asymmetry occurring near G1 Ib. The authors reproduce the observations with numerical simulations based on a simple two-stream model, and from this deduce velocities of rise of the hot material to be ~ -25 km s^{-1} for the F Ib supergiants.

114.124 **Oxygen abundances in horizontal branch stars.**
R. Nesci.
Messenger, No. 44, p. 4 – 6 (1986).

114.125 **Some recent results on high dispersion spectroscopy.**
L. Gratton, R. G. Gratton.
Mem. Soc. Astron. Ital., Vol. 56, No. 4, p. 599 – 612 (1985). – See Abstr. 012.054.
Some recent work on high dispersion spectroscopy is discussed. Problems in the analysis of spectra of late type stars are briefly reviewed. The connections between high dispersion spectroscopy and stellar and galactic evolution are explored by means of some explicative cases. The need for high quality data is underlined.

114.126 **Spectral energy distributions of Be stars. I. Measurements of 26 southern and equatorial stars.**
D. Kaiser.
Bull. Inf. Etoiles Be, No. 13, p. 13 (1986). Abstract. – To appear in Astron. Astrophys., Suppl. Ser.

114.127 **Objective-prism discoveries in a region of the Milky Way.**
T. Radoslavova.
Inf. Bull. Variable Stars, No. 2845, 2 pp. (1986).

114.128 **The nature of Balmer line variability in chemically peculiar stars.**
B. Musielok.
Acta Astron., Vol. 36, No. 1, p. 131 – 146 (1986).
Results of photometric measurements of the β-index during the rotational period for six Ap stars (56 Ari, 41 Tau, φ Dra, HD 188041, HD 215441, HD 221568) and one He-rich star (HD 184927) are presented. All these stars show variations of the β-index within ranges from 0.01 to 0.05 mag. For stars with well known periods the minimum of the β-index occurs at the phase of maximum light in the wavelength region longward of λ_0 = 5450Å. Such phase coincidence can be explained by the blanketing mechanism.

114.129 **On the surface gravities of Am stars.**
M. M. Dworetsky, T. T. Moon.
Mon. Not. R. Astron. Soc., Vol. 220, No. 3, p. 787 – 792 (1986).
Using uvbyβ photometry for 34 Am stars and 48 normal stars in five open clusters, the authors show that the method of Moon & Dworetsky, for obtaining T_{eff} and log g, is relatively insensitive to δm_0. They find the surface gravities of Am stars in open clusters to be the same as those of normal stars of similar effective temperatures. A small correction for metallicity should be applied to surface gravities of star with $T_{eff} \leqslant 8500$K.

114.130 **Flares on dMe stars: IUE and optical observations of AT Mic, and comparison of far–ultraviolet stellar and solar flares.**
G. E. Bromage, K. J. H. Phillips, P. L. Dufton, A. E. Kingston.
Mon. Not. R. Astron. Soc., Vol. 220, No. 4, p. 1021 – 1046 (1986).
IUE observations of AT Mic in quiescent state indicate surface fluxes in transition–region lines that are 10 – 40 times quiet–Sun values, and only a few times for lower temperature lines, though the correlation with temperature is less clear than in previous data. The He II 1640–Å line is much stronger than in the Sun, and the optical 4686–Å line may be present; their intensity ratio agrees with the theoretical recombination value.

114.131 **Spectrophotometry of standard stars: physical parameters of the atmospheres and the scale of effective temperatures.**
I. N. Glushneva.
Sov. Astron., Vol. 29, No. 6, p. 659 – 663 (1985). English translation of 40.114.143.

114.132 **High–resolution Hα observations of M dwarf stars: implications for stellar dynamo models and stellar kinematic properties at faint magnitudes.**
M. S. Giampapa, J. Liebert.
Astrophys. J., Vol. 305, No. 2, p. 784 – 794 (1986).
The authors present high–dispersion observations of the Hα spectral region for a sample of very late M dwarf stars. Radial velocities and total space motions are derived. The Hα emission line properties of their sample are thereby examined according to membership in stellar kinematic populations. There are comparable numbers showing and not showing Hα emission for $M_v > +15$ in this sample of large proper motion stars. However dMe stars are the dominant class of objects for absolute magnitudes $M_v \geqslant +17$. Kinematically, the dMe stars tend to belong to the young disk population, while nearly all the M dwarfs lacking Hα emission are members of either the old disk or halo populations. The implications for stellar dynamo theory and for the nature of the stellar luminosity function at faint magnitudes are discussed.

114.133 **An observational study of Si–λ4200 stars.**
P. Didelon.
Rev. Mex. Astron. Astrofis., Vol. 12, Spec. Issue, p. 163 – 172 (1986). – See Abstr. 012.066.
Intensity measurements and line identifications are presented of the blue spectral region (3700 – 4900 Å) for 10 Si stars and for several standard stars. Combining these observations with available photometric data, the characteristics of the group are studied and linked with the groups of magnetic chemically peculiar stars.

114.134 **Nitrogen–to–carbon ratio in 70 dwarf halo stars.**
D. F. Carbon, R. P. Kraft, B. Barbuy, E. Friel, N. B. Suntzeff.
Rev. Mex. Astron. Astrofis., Vol. 12, Spec. Issue, p. 173 – 176 (1986). – See Abstr. 012.066.

114.135 **Yellow supergiants with hot companions.**
A. Arellano Ferro, B. F. Madore.
Rev. Mex. Astron. Astrofis., Vol. 12, Spec. Issue, p. 189 (1986). Abstract. – See Abstr. 012.066.

114.136 **Morphology of low–resolution spectra of late–type stars in the near–infrared.**
F. J. Fuenmayor, P. H. Bulka.
Rev. Mex. Astron. Astrofis., Vol. 12, Spec. Issue, p. 225 – 227 (1986). – See Abstr. 012.066.
Preliminary results on a procedure intended to characterize the low–dispersion spectra of M and carbon stars using image–processing techniques are presented.

114.137 **Spectroscopic study of visual spectrum of ζ Capricorni.**
M. Hirai, T. Murayama.
Bull. Fukuoka Univ. Educ., Vol. 35, Part III, p. 105 – 112 (1985).
The visual spectrum of ζ Capricorni is observed. The spectrogram covered the wavelength region from 5500 Å to 5950 Å with the mean dispersion of 5.11 Å/mm. Using the empirical curve of growth method the chemical composition in ζ Cap for nineteen elements is derived on basis of Tech's high dispersion spectral analysis of this star (1971). The results are consistent with Tech's in the view of atmospheric parameters and overabundances of rare earth elements. Several discussions are given for the barium stars including ζ Cap.

114.138 **High–dispersion spectroscopy of the brightest emission–line stars of the Magellanic Clouds.**
O. Stahl, B. Wolf, M. de Groot, C. Leitherer.
Luminous stars and associations in galaxies, p. 247 – 248 (1986). – See Abstr. 012.079. (IAU Symp. No. 116).
The authors present an atlas of high dispersion spectra of 24 of the brightest peculiar emission–line stars of the Magellanic Clouds.

114.139 **The absorption feature at 1920 Å in the spectra of early–type supergiants in the Small Magellanic Cloud.**
D. H. Morgan, K. Nandy, G. I. Thompson.
Luminous stars and associations in galaxies, p. 95 – 96 (1986). – See Abstr. 012.079. (IAU Symp. No. 116).
The 1600 – 2500 Å spectra of many early–type supergiants are dominated by a broad depression of the continuum near 1920 Å. The 1920 Å feature is also prominent in LMC early–type supergiants. Early–type supergiants in the SMC have metal absorption lines generally weaker in strength than those of their LMC counterparts. The authors study whether the 1920 Å feature, attributed to Fe III lines, is also weak in the SMC stars.

114.140 **Steps to a new calibration of the spectral type – effective temperature relationship for early–type stars.**
B. Bohannan, D. C. Abbott, S. A. Voels, D. G. Hummer.
Luminous stars and associations in galaxies, p. 111 – 112 (1986). – See Abstr. 012.079. (IAU Symp. No. 116).
The spectral type – effective temperature calibration is a cornerstone of the understanding of massive stars and their environment. The authors have begun a new determination of the basic stellar parameters of massive stars with an analysis of ζ Pup (O4f).

114.141 **Photospheric and "chromospheric" C IV lines in B–main–sequence stars.**
F. Castelli, C. Morossi, R. Stalio.
Luminous stars and associations in galaxies, p. 113 – 116 (1986). – See Abstr. 012.079. (IAU Symp. No. 116).
No photospheric model, constructed under standard equations, can predict the existence of C IV ions over the temperature range indicated by the observations. Thus the question is: at what spectral type in a main sequence star photospheric C IV ceases to contribute to the observed profile? In order to answer this question the authors compared IUE spectra of a set of B–main–sequence stars with synthetic photospheric spectra. The spectral range investigated was from 1541 to 1553 Å. It is concluded that the boundary at which photospheric C IV disappears is between spectral types B3 and B6 in main–sequence stars.

114.142 **Voyager observations of hot luminous stars.**
R. S. Polidan, R. Stalio.
Luminous stars and associations in galaxies, p. 117 – 119 (1986). – See Abstr. 012.079. (IAU Symp. No. 116).

This is a progress report on a study of the far–UV spectra of hot stars observed with the Voyager ultraviolet spectrometers (UVS); the authors discuss UV flux distribution and variability of a set of early type high luminosity stars.

114.143 **Optical and ultraviolet spectral morphology of luminous OB stars in the Galaxy and the Magellanic Clouds.**
N. R. Walborn.
Luminous stars and associations in galaxies, p. 185 – 198 (1986). – See Abstr. 012.079. (IAU Symp. No. 116).

Three areas are discussed. (1) New spectroscopic observations of the 30 Doradus central cluster confirm the presence of numerous very early O–type members, including several of type O3. In combination with new direct imagery of the luminous central object R 136 these results have evident implications for understanding the ionization of the supergiant H II region, as well as for the interpretation of R 136 itself and of the apparently similar regions seen in more distant galaxies. (2) A further member of the Ofpe/WN9 category in the LMC has been identified. (3) An extensive survey of IUE high–resolution data has revealed a strong correlation between the ultraviolet stellar wind features and the optical spectral classifications for the majority of normal O stars.

Third catalogue of Am stars with known spectral types.
See Abstr. 002.012.

Catalogue of 900 faint–star ultraviolet spectra based on observed data of the Space Observatory Orion–2.
See Abstr. 002.027.

Documentation for the machine–readable version of *The Henry Draper Catalogue*. (Edition 1985).
See Abstr. 002.028.

New carbon stars in the IRAS Point Source Catalogue.
See Abstr. 002.033.

Continuing results from the Michigan Catalogue of MK types for the HD stars (–90° to –12°) and future plans.
See Abstr. 002.034.

A spectrophotometric atlas of southern MK standards from 5800 Å to 10,200 Å.
See Abstr. 002.035.

Catalogue of ultraviolet spectra of 900 faint stars. Observational material of the space station Orion 2.
See Abstr. 002.047.

The Palomar–Green Catalog of Ultraviolet–Excess Stellar Objects.
See Abstr. 002.074.

The spectral energy distribution of early type stars. I. A catalogue of photometric data of 259 stars from 0.15 to 4.8 μm.
See Abstr. 002.079.

Origins of the spectral sequence.
See Abstr. 004.046.

Microturbulence and stellar spectral lines.
See Abstr. 004.122.

On the structure of the energy spectrum of Tc I.
See Abstr. 022.041.

Measurement of the oscillator strengths of very weak 1–eV Fe I lines.
See Abstr. 022.078.

Oscillator strength measurements in the vacuum–ultraviolet. I. The strong 1243, 1493 and 1743 Å multiplets of neutral nitrogen.
See Abstr. 022.080.

Rotational analysis of the red electronic emission spectrum of molybdenum nitride (MoN).
See Abstr. 022.110.

Quantitative two–dimensional spectral classification for the prism spectrograph of the Shemakha Astrophysical Observatory.
See Abstr. 034.104.

Intermediate mass stars undergoing a very hot phase: can we measure their temperatures?
See Abstr. 036.014.

Coments on techniques for spectral deconvolution.
See Abstr. 036.083.

Automatische Klassifikation von Spektren geringer Dispersion.
See Abstr. 036.232.

High resolution spectroscopy with ST.
See Abstr. 051.095.

The puzzle of lithium in evolved stars.
See Abstr. 061.122.

Magnesium isotopes and galactic evolution.
See Abstr. 061.129.

Observational evidence of element formation processes in stars.
See Abstr. 061.146.

The statistical equilibrium of hydrogen and helium in a radiation field, with an application to interpreting Wolf–Rayet spectra.
See Abstr. 063.004.

Influence of ion–atom collisions on the absorption of radiation.
See Abstr. 063.016.

Intrinsic parameters of hot blue stars.
See Abstr. 064.071.

Stellar technetium and niobium abundances as a measure of the lifetime of AGB stars in the third dredge–up phase.
See Abstr. 065.049.

The low–luminosity J–type carbon stars.
See Abstr. 065.121.

Radial–velocity determinations of six LMC superluminous giant candidates.
See Abstr. 111.017.

Measurements of Balmer emission line profiles for southern Be stars. III. New data and radial velocities.
See Abstr. 112.006.

Mass loss from Wolf–Rayet stars.
See Abstr. 112.015.

Spectrophotometry of some Be stars.
See Abstr. 112.016.

Ultraviolet and visual variability of θ CrB during a normal B–phase following a shell phase (1980 – 1985).
See Abstr. 112.022.

Circumstellar shells around luminous emission–line stars in the Large Magellanic Cloud.
See Abstr. 112.026.

Narrow absorption components and variability in ultraviolet P Cygni profiles of early–type stars.
See Abstr. 112.037.

New lines of ethylene and a search for methylene in IRC + 10216.
See Abstr. 112.047.

Hydrogen Lyman alpha fluxes for late–type dwarf stars.
See Abstr. 112.050.

Recent spectral variation of the Be star EW Lac.
See Abstr. 112.078.

Etude spectrophotométrique, paramètres physiques et géométriques de l'étoile Be Gamma Cassiopeiae.
See Abstr. 112.079.

Some regularities in the variation of the Balmer line profiles in the P Cyg spectrum.
See Abstr. 112.084.

High–resolution emission–line spectroscopy of Be stars. 1. Evidence for a two–component structure of the Hα emitting envelope.
See Abstr. 112.107.

The dependence of the stellar rate of mass loss on effective temperature and velocity.
See Abstr. 112.126.

MWC 300: a runaway hypergiant.
See Abstr. 112.129.

New results on Eta Carinae. Evidence for an asymmetric, inhomogeneous wind.
See Abstr. 112.130.

Infrared spectroscopy of southern P Cygni stars.
See Abstr. 112.131.

B[e]–stars of the Magellanic Clouds.
See Abstr. 112.132.

The mass loss rate of SK 80 (O7 Iaf) in the Small Magellanic Cloud.
See Abstr. 112.133.

IRAS observations of Wolf–Rayet stars.
See Abstr. 112.137.

Models of IRAS observations of circumstellar shells.
See Abstr. 112.140.

A low–dispersion, near–infrared survey for galactic M supergiants.
See Abstr. 113.015.

Low dispersion spectrophotometry of bright early–type stars.
See Abstr. 113.018.

Numerical convolution photometry: an example using the Strömgren four–color system.
See Abstr. 113.039.

A comparative study of barium peculiarity indices.
See Abstr. 113.044.

Simultaneous determination of metallicity and O/C ratio for late–type giants in the Vilnius photometric system.
See Abstr. 113.054.

IRAS observations of carbon stars.
See Abstr. 113.065.

Absolute magnitudes of late–type stars.
See Abstr. 115.003.

The effective temperatures, diameters and luminosities of 22 bright stars by application of the infrared flux method.
See Abstr. 115.007.

On the luminosity, effective temperature and radius of P Cygni.
See Abstr. 115.009.

The three prototype FHB stars: standards for Pop II.
See Abstr. 115.015.

LHS 292 and the luminosity function of the nearby M dwarfs.
See Abstr. 115.023.

Parameters of Wolf–Rayet stars.
See Abstr. 115.031.

Metallicity effect on absolute magnitude determination.
See Abstr. 115.032.

A survey of chromospherically active stars.
See Abstr. 116.009.

Optical photometry and spectroscopy for five dwarf M stars.
See Abstr. 116.026.

The dependence of He I λ10830 absorption strength upon X–ray emission in late–type stars.
See Abstr. 116.047.

Spectrum of the symbiotic star CH Cygni in 1982.
See Abstr. 117.097.

Spectral classification of the close binary V453 Cyg.
See Abstr. 117.138.

Optical and UV spectroscopy of the peculiar RS CVn system RT Lacertae.
See Abstr. 117.237.

Rapid ultraviolet spectral variations in HD 50896 (WN5 + ?).
See Abstr. 117.327.

The spectroscopic study of μ Sgr.
See Abstr. 119.029.

On the peculiarities in the spectrum of the eclipsing binary AR Aur.
See Abstr. 119.030.

Ultraviolet survey for hot companions among nonvariable yellow supergiants.
See Abstr. 120.011.

A spectroscopic survey of the Taurus–Auriga dark clouds for pre–main–sequence stars having Ca II H, K emission.
See Abstr. 121.051.

Metal abundances and radial velocities of short–period variable stars in the direction of the Large Magellanic Cloud.
See Abstr. 122.005.

Spectral and luminosity variation of long–period red variable stars.
See Abstr. 122.109.

Energy distribution in the spectrum of Mira (o Ceti).
See Abstr. 122.112.

An analysis of the spectrum of Zeta Geminorum.
See Abstr. 122.113.

The chemical composition of Polaris.
See Abstr. 122.126.

Carbon and oxygen abundances of field RR Lyrae stars. II. Oxygen abundances.
See Abstr. 122.144.

The chemical composition of Polaris.
See Abstr. 122.150.

Metal abundances in helium–rich white dwarf atmospheres.
See Abstr. 126.009.

Time–averaged ultraviolet observations of pulsating DA white dwarfs from the IUE observatory.
See Abstr. 126.072.

Interstellar molecular carbon in distant supergiants.
See Abstr. 131.011.

Interstellar absorption lines between 2000 and 3000 Å in nearby stars observed with BUSS.
See Abstr. 131.092.

Surveys for emission–line objects with Schmidt telescopes.
See Abstr. 131.223.

Distribution of Mg II in the local interstellar medium towards eight cool giants.
See Abstr. 131.316.

Evidence for a 12 micron water–ice absorption band in the IRAS LRS spectra of protostars and late type stars.
See Abstr. 131.371.

CO observations of sources from IRAS Circular No. 9.
See Abstr. 133.010.

Photographic observation of AFGL sources in Cygnus.
See Abstr. 133.011.

Distances to planetary nebulae. I. The reddening–distance method.
See Abstr. 134.006.

Breaking through the wall.
See Abstr. 134.019.

Detection of the very hot central star in NGC 2440.
See Abstr. 134.020.

New colour and Zanstra temperatures for 15 central stars of planetary nebulae.
See Abstr. 134.029.

Optical study of the Geminga candidate field.
See Abstr. 143.021.

No optical pulses from the Geminga candidate.
See Abstr. 143.022.

Some statistical characteristics of Ap and Am stars in regions occupied by OB associations.
See Abstr. 152.009.

The chemical composition of the young open cluster NGC 6611.
See Abstr. 153.003.

Carbon isotope ratios in giant stars of the old galactic cluster NGC 7789.
See Abstr. 153.006.

Bright giant stars in intermediate–age galactic clusters.
See Abstr. 153.010.

The lithium abundance gap in the Hyades F stars: the signature of diffusion.
See Abstr. 153.016.

The giant branch of the open cluster NGC 2539.
See Abstr. 153.017.

Lithium in the Hyades Cluster.
See Abstr. 153.018.

Old stellar populations. III. The metallicities of M5, M71, and M67.
See Abstr. 153.034.

TiO continuum placement effects in 47 Tucanae giants.
See Abstr. 154.043.

High–resolution CCD spectra of stars in globular clusters. I. Oxygen in M13.
See Abstr. 154.054.

The C, N, and O abundances of giant stars in ω Centauri.
See Abstr. 154.065.

Late M stars found in a hemispheric survey.
See Abstr. 155.040.

Chemical evolution of the Galaxy: a comparison of the abundances of light metals in disk and halo dwarfs.
See Abstr. 155.047.

Abundances and kinematics of K giants in the galactic nuclear bulge.
See Abstr. 155.059.

2000 Å UV imaging of a 6° diameter field around μ Aurigae.
See Abstr. 155.068.

The chemical evolution of the solar neighborhood: age–metallicity relation of F–stars.
See Abstr. 155.111.

Ground based observations of nuclear bulge stars.
See Abstr. 155.124.

A photometric determination of the metal content for F – G type supergiants in the Large Magellanic Cloud.
See Abstr. 156.006.

The stellar population of the Small Magellanic Cloud near NGC 121. I. The mean metallicity, metallicity spread, and radial velocity of SMC halo giants.
See Abstr. 156.023.

A new survey for Hα–emission–line stars in the Small Magellanic Cloud.
See Abstr. 156.033.

Bright stars in the SMC clusters.
See Abstr. 156.037.

Bright stars in five LMC star clusters.
See Abstr. 156.040.

New carbon stars in spheroidal galaxies: II. Draco, Ursa Minor and new data on Sculptor, Carina and Leo I.
See Abstr. 157.196.

An objective–prism survey of emission–line objects in M33 and IC 1613.
See Abstr. 157.198.

The M supergiants in NGC 300.
See Abstr. 157.220.

Wolf–Rayet stars in nearby galaxies.
See Abstr. 157.312.

115 Luminosities, Masses, Diameters, HR and other Diagrams

115.001 **On the nature of the dwarf carbon star G77–61.**
D. S. P. Dearborn, J. Liebert, M. Aaronson, C. C. Dahn, R. Harrington, J. Mould, J. L. Greenstein.
Astrophys. J., Vol. 300, No. 1, p. 314 – 324 (1986).
A variety of new astrometric, photometric, and spectrophotometric data are presented for the low–luminosity carbon star G77–61. The improved parallax yields $M_v = +10.08 \pm 0.43$, while the revised temperature estimate from infrared, optical, and ultraviolet fluxes in 4100 (+200, –300)K, yielding $\log L/L_\odot = -1.75 \pm 0.16$. Most importantly, radial velocity variations have been detected, with the binary period 245 days. The unseen companion probably is a cool white dwarf of much higher mass than the visible object. Both the spectrum and the H–R diagram position below the main sequence require G77–61 to be either extremely metal poor, or helium enriched and at least moderately metal poor.

115.002 **A survey of emission–line stars in the CMa star formation region.**
S. D. Wiramihardja, T. Kogure, M. Nakano, S. Yoshida, K. Tatematsu.
Astrophys. Space Sci., Vol. 118, Nos. 1/2, p. 311 – 315 (1986). – See Abstr. 012.001.
A survey of emission–line stars was carried out in the CMa star formation region, based on spectral observations with the Kiso Schmidt telescope. In total, 128 emission–line stars were detected, ranging from $V = 6$ to 15, and the majority is found with V between 11 and 14. The two–colour ($U–B$, $B–V$) diagram is also shown for the detected stars.

115.003 **Absolute magnitudes of late–type stars.**
T. Mikami.
Astrophys. Space Sci., Vol. 119, No. 1, p. 65 – 68 (1986). – See Abstr. 012.001.
The improved calibration of absolute magnitude versus spectral type and luminosity class is presented for carbon stars and for K– to M–type giants. It is based on the recent results estimated by applying the maximum–likelihood method and other statistical methods to the homogeneous kinematical data including the proper motions in the AGK3 system.

115.004 **The case of α Centauri: mass, age and *p*–mode oscillation spectrum.**
P. Demarque, D. B. Guenther, W. F. van Altena.
Astrophys. J., Vol. 300, No. 2, p. 773 – 778 (1986).
The authors have investigated the astrometric and physical properties of the α Centauri system, a nearby visual binary which shares a common space motion with the nearest star Proxima Cen. First, the parallax of the binary system is estimated to be $0''.7506 \pm 0''.0046$ (s.e.). Using this absolute parallax, the masses of α Cen A and B are respectively 1.09 ± 0.01 and 0.90 ± 0.01 $M_\odot$. Second, the evolutionary status of α Cen A and B is determined with the help of theoretical stellar models. Evolutionary tracks yield the age of the α Cen system as a function of metallicity. The discussion points to a consistent model with an age in the vicinity of $4 - 4.5 \times 10^9$yr. Third, the nonradial oscillations of α Cen A are considered. The *p*–mode spectrum of the models of α Cen A is compared to that of the Sun and to the observations of Fossat et al. It is concluded that Fossat et al.'s observations are inconsistent with *p*–mode spectra constructed from standard theoretical models.

115.005 **The position of ET And in the Hertzsprung–Russell diagram.**
G. Scholz.
Astron. Nachr., Vol. 307, No. 1, p. 21 – 25 (1986).
The interstellar lines of Ca II (K) and Na I (D) are used to estimate the distance of the Ap star ET And in order to derive its position in the HRD, because the classical methods for the determination of M are affected by some restrictions. By these means the star lies distinctly above the main sequence. The post–main sequence evolutionary stage suggests to assume that pulsations are the reason for the observed short–time spectroscopic and photometric variations. Possibly, the 1.61–day period is also caused by pulsations.

115.006 **The evolutionary state of the He–weak star HD 49333.**
H. M. Maitzen.
Astron. Astrophys., Vol. 158, No. 1/2, p. 205 – 206 (1986).
It is pointed out that existing photometry in the Geneva and the uvby–system does not support HD 49333 to be a giant. It is a foreground star at roughly half the distance to NGC 2287. Therefore, the anomalously large radius derived by Lanz (1985) reduces to the proper value for mid to late–B type stars of luminosity class V.

115.007 **The effective temperatures, diameters and luminosities of 22 bright stars by application of the infrared flux method.**
S. K. Leggett, C. M. Mountain, M. J. Selby, D. E. Blackwell, A. J. Booth, D. J. Haddock, A. D. Petford.
Astron. Astrophys., Vol. 159, No. 1/2, p. 217 – 222 (1986).
The infrared flux method is used to determine the effective temperatures and angular diameters of 22 bright stars of various spectral types. Their parallaxes are combined with the angular diameters to give linear radii, and with the observed integrated fluxes to give luminosities. The ratios of stellar integrated flux to infrared monochromatic flux required by the method were computed using published Kurucz model atmospheres and STARLINK generated MARCS models. These are compared to the observed values of the ratios, derived using ultraviolet and infrared data taken from the literature, and visible data obtained by the authors in Tenerife in 1982. A new absolute infrared calibration of Vega is used.

115.008 **Of stars in young open clusters.**
A. Feinstein, R. A. Vázquez, O. G. Benvenuto.
Astron. Astrophys., Vol. 159, No. 1/2, p. 223 – 230 (1986).
The authors obtained from the literature photometric and spectroscopic data for 54 Of–type stars that belong to 21 young open clusters. Their location in the HR diagram shows that all of them are high luminosity objects slightly evolved from the ZAMS. The Of objects are more common, but more difficult to detect than WR stars in open clusters. The open clusters that have several Of stars allowed the authors to check that their positions in the M_{bol}–$\log T_{eff}$ diagram are in a quite good agreement with theoretical evolutionary tracks; accordingly, the Of stars might provide a quick way to derive the ages of very young open clusters.

115.009 **On the luminosity, effective temperature and radius of P Cygni.**
L. Luud.
Proc. Acad. Sci. Estonian SSR, Phys., Math., Vol. 34, No. 2, p. 198 – 207 (1985). = Tartu Astrofüüs. Obs. Teated, Nr. 77.
For the further modelling of P Cygni's envelope and to specify our knowledge of P Cygni's physics and evolutionary stage we must know the star's fundamental physical parameters. The author discusses the luminosity, effective temperature and radius of the star, taking into account the observational data about the far ultraviolet, optical, infrared and radio fluxes.

115.010 **Absolute magnitudes and evolutionary status of Wolf–Rayet stars.**
J. Breysacher.
Astron. Astrophys., Vol. 160, No. 1, p. 185 – 194 (1986).
From a combination of both galactic and LMC data, an improved visual absolute–magnitude calibration for W–R stars is proposed. The occurrence of W–R stars in the OB associations and H II regions of the LMC is also investigated as well as the frequency of W–R binaries and the relative number of WN to WC type stars. The systematic tendency for WN8 stars to deviate

from the general behaviour of late WN–type stars is underlined. The peculiarities of the W–R star distribution in the LMC are interpreted as a consequence of the metal deficiency of this galaxy. No new clues about the influence of the initial mass function (IMF) on the W–R star population of a galaxy could be derived, however.

115.011 **The influence of star luminosity and temperature calibration on the determination of the initial mass function.**
V. I. Myakutin.
Nauchn. Inf., Vyp. 56, p. 52 – 58 (1984). In Russian.
An estimate is made of the influence of small systematic differences in the definition of luminosity and effective temperature in the 4–colour Strömgren photometric system and MK spectral classification on the inclination of the initial mass function. It is shown that the exponent α defined by $uvby\beta$ photometry is greater by 0.2 – 0.3 than obtained by spectral classification. Spectral classification for the B star of the Orion OB1 association gives the value of $\alpha = 2.5$.

115.012 **The gaps of the luminosity function for B and A stars.**
S. V. Vereshchagin, A. Eh. Piskunov.
Nauchn. Inf., Vyp. 57, p. 76 – 79 (1984). In Russian.
A statistical investigation of the field stars luminosity function was carried out on the basis of some photometric and spectral machine–readable catalogues. The investigation shows the existence of the significant deficiency of B stars (with $M_V \cong -3^m$) and late A stars ($M_V \cong 2^m\!.0$). Similar features were found for luminosity functions of open cluster members and mass distributions of binary stars by other investigators.

115.013 **R 136 – a supermassive star?**
V. G. Surdin.
Priroda, No. 4, p. 103 (1986). In Russian.

115.014 **Unscrambling the mess at the left hand side of the HR diagram.**
H. L. Shipman.
Bull. Am. Astron. Soc., Vol. 17, No. 4, p. 853 (1985). Abstract. – See Abstr. 010.062.

115.015 **The three prototype FHB stars: standards for Pop II.**
A. G. D. Philip.
Bull. Am. Astron. Soc., Vol. 17, No. 4, p. 878 (1985). Abstract. – See Abstr. 010.062.

115.016 **How well do we know stellar parameters?**
S. Frandsen.
Seismology of the Sun and the distant stars, p. 473 – 476 (1986). – See Abstr. 012.026.
A small table is presented containing estimates of the standard errors in the best determinations of fundamental parameters for stars. Only for a small sample of late type main sequence stars do we have values for the mass, the radius and the luminosity or the combination of these parameters, the surface gravity g and the effective temperature T_e that are known with a precision getting near 1 percent. It should be investigated in detail, how much asteroseismology (the extension of helioseismology to stars) can improve the situation, once the sensitivity of the instruments have improved so much, that regular observations of stellar oscillations are possible.

115.017 **"Chemical structure" in the red giant Hertzsprung–Russell diagram, and the age of the galactic disk.**
V. A. Marsakov, A. A. Suchkov.
Sov. Astron., Vol. 29, No. 4, p. 403 – 409 (1985). English translation of 40.115.002.

115.018 **The formation and evolution of carbon stars.**
J. Mould, M. Aaronson.
Astrophys. J., Vol. 303, No. 1, p. 10 – 18 (1986).
In the Magellanic Clouds carbon stars form at the top of the asymptotic giant branch in star clusters younger than ~8 billion yr. The mechanism, devised 10 yr ago by Iben, is termed "third dredge up", and raises carbon produced in helium shell flashes to the stellar surface. Clusters younger than ~0.8 billion yr, however, have giant branches populated by M stars. It seems that thermal pulses are not able to establish themselves in stars of 5 $M_\odot$ or more before the stellar envelope is completely eroded away by rapid mass loss. Spectroscopy and photometry are presented of carbon stars in NGC 6822, IC 1613, M31, WLM, M33, Andromeda II, and NGC 205. This will permit comparison of evolutionary processes on the asymptotic giant branch in a number of different environments.

115.019 **Absolute magnitudes and evolutionary status of Wolf–Rayet stars.**
J. Breysacher.
ESO Sci. Prepr., No. 415, 41 pp. (1986). To appear in Astron. Astrophys.

115.020 **Properties of stellar objects near the main sequence mass limit.**
J. W. Liebert, C. C. Dahn.
Prepr. Steward Obs., No. 616, 7 pp. (1986). Paper presented at "Astrophysics of brown dwarfs", a workshop held at George Mason University, Virginia, October 14 – 15, 1985.

115.021 **Diameter and limb–darkening measures for α Ori.**
A. Y. S. Cheng, E. K. Hege, E. N. Hubbard, L. Goldberg, P. A. Strittmatter, W. J. Cocke.
Prepr. Steward Obs., No. 654, 30 pp. (1986). To appear in Astrophys. J.

115.022 **Two dimensional images of Alpha Orionis.**
J. C. Hebden, J. C. Christou, A. Y. S. Cheng, E. K. Hege, P. A. Strittmatter, J. M. Beckers, H. P. Murphy.
Prepr. Steward Obs., No. 655, 35 pp. (1986). To appear in Astrophys. J.

115.023 **LHS 292 and the luminosity function of the nearby M dwarfs.**
C. C. Dahn, J. Liebert, R. S. Harrington.
Astron. J., Vol. 91, No. 3, p. 621 – 625 (1986).
Preliminary astrometry, broadband BVI photometry, and low–resolution (10 Å) spectrophotometry covering λλ4800 – 7800 Å are presented that demonstrate that LHS 292 is the lowest–luminosity M dwarf known within 5.2 pc of the Sun. The results show $\pi_{abs} = 0\farcs217 \pm 0\farcs005$ (m.e.), $V = 15.60 \pm 0.01$ (m.e.), $M_V = 17.28 \pm 0.05$ (m.e.), and the spectral type is M6.5–7 V. Since the spectrum shows strong emission lines and the star has a relatively low space motion, it should be considered a flare star candidate.

115.024 **Angular diameters by the lunar occultation technique. VII.**
P. C. Schmidtke, J. L. Africano, G. H. Jacoby, R. R. Joyce, S. T. Ridgway.
Astron. J., Vol. 91, No. 4, p. 961 – 970 (1986).
Lunar occultation observations of 18 events for 12 late–type stars have yielded 36 new angular diameters. With the addition of a three–channel optical photometer to the instrumental setup, measurements can be simultaneously obtained at six wavelengths in the range 0.45 – 2.28 μm. The diameters of M0.5 to M4 stars are systematically smaller at 1.6 μm than at 2.2 μm, confirming initial observations in Paper V for a smaller sample of stars. The diameter of Y Tau (C6,4) and FY Lib (gM5) may vary with time, but the size of SW Vir (M7 III:) appears constant.

115.025 **Stellar radius determination from IRAS 12 μm fluxes.**
M.–N. Perrin, H. Karoji.
Inst. Astrophys. Paris, Pré–Publ., No. 151, 23 pp. (1986). To appear in Astron. Astrophys.

115.026 **HD 161796 – estimate of parameters based on pulsation calculations.**
J. Zalewski.
Acta Astron., Vol. 36, No. 1, p. 63 – 68 (1986).
Pulsation computations have been performed for HD 161796 in order to determine its probable mass and luminosity. An inter-

pretation of HD 161796 as an evolved, low mass star is suggested. The observed periods are identified with a first and second overtone of radial pulsations.

115.027 **On the gravity of the subgiant 31 Aquilae.**
L. da Silva.
Rev. Mex. Astron. Astrofis., Vol. 12, Spec. Issue, p. 230 (1986). Abstract. – See Abstr. 012.066.

115.028 **Normal supergiants and their imitations. I. Sodium as a mass indicator?**
D. D. Sasselov.
Publ. Astron. Soc. Pac., Vol. 98, No. 604, p. 561 – 571 (1986).
The author has discussed in some detail the possible ways for a low–mass evolved object to attain high luminosity for a sufficiently long time and mimic spectroscopically and photometrically a normal supergiant (massive star at the core helium burning stage). He has briefly reviewed the observational evidence for that from a couple of such known imitations. In an attempt to find a reliable method to distinguish them from the normal supergiants, the author has investigated one case based on the changes occurring in the surface abundances of some isotopes resulting from nuclear burning at different conditions, depending on the initial stellar mass.

115.029 **Absolute magnitudes of Ap stars from low–dispersion spectrograms.**
Ts. S. Radoslavova.
Dokl. Bolg. Akad. Nauk, Vol. 38, No. 9, p. 1109 – 1111 (1985). In Russian. Abstr. in Ref. Zh., 51. Astron., 5.51.646 (1986).

115.030 **Brightest stars in galaxies as distance indicators.**
A. Sandage.
Luminous stars and associations in galaxies, p. 31 – 41 (1986). – See Abstr. 012.079. (IAU Symp. No. 116).
Four areas of application of the brightest star data are discussed as (1) determining M_B of the Eddington limit to be ~ -10 for blue supergiants, (2) calibration of $M_B(\max) = -20.0 \pm 0.4$ for type I supernovae, leading to a Hubble constant of $H_0 = 43 \pm 10$ km s^{-1}Mpc^{-1}, (3) detection of the deceleration of the cosmological expansion by the Local Group leading to a Local Group mass of $4 \times 10^{11} M_\odot$ and a mass–to–blue light ratio of 3, and (4) use of the brightest stars to map the Virgo cluster velocity perturbation of the Hubble flow.

115.031 **Parameters of Wolf–Rayet stars.**
P. S. Conti.
Luminous stars and associations in galaxies, p. 199 – 214 (1986). – See Abstr. 012.079. (IAU Symp. No. 116).
Parameters of stars describe such intrinsic characteristics as stellar luminosities, effective temperatures, chemical composition, radii, masses, and for hot stars such as the mass loss rates. In this paper the author briefly touches on these topics, in particular those in which some recent new results are pertinent to current work. It is commonly held that W–R stars are the decendants of massive O–type stars in which the initial convective cores have become visible due to mass loss and mixing in the progenitor object. After discussing the distribution of W–R stars in our galaxy, the author indicates the arguments for believing that W–R stars are a natural consequence of massive star evolution.

115.032 **Metallicity effect on absolute magnitude determination.**
P. Dubois.
Luminous stars and associations in galaxies, p. 415 – 416 (1986). – See Abstr. 012.079. (IAU Symp. No. 116).

115.033 **Luminosities of OH/IR stars.**
J. H. Burger, J. Herman.
Light on dark matter, p. 103 – 104 (1986). Abstract.– See Abstr. 012.080.

Photons, galaxies and stars. Selected papers of R. Hanbury Brown, Raman Professor, 1974.
See Abstr. 003.103.

Shapley's calibration of the P–L curve.
See Abstr. 004.052.

How no(w) brown dwarfs?
See Abstr. 011.008.

Results and limitations of ground–based optical arrays.
See Abstr. 032.033.

Statistical problems encountered in using trigonometric parallaxes.
See Abstr. 036.214.

Advancements in the stellar evolution theory: the role of convective overshooting all across the HR diagram.
See Abstr. 065.050.

Evolution of massive spinning stars.
See Abstr. 065.059.

Stellar evolution in the upper Hertzsprung–Russell diagram.
See Abstr. 065.076.

Lower bounds on the masses of rapidly rotating white dwarfs.
See Abstr. 065.115.

Integrated colours and synthetic HR diagrams for LMC clusters.
See Abstr. 065.154.

The cooler supergiants (A to M): crucial signposts in the lifecycles of massive stars.
See Abstr. 065.160.

Synthetic HR diagrams for luminous stars: a test of stellar evolution theory.
See Abstr. 065.162.

Effects of convective overshooting, mass loss (and chemical composition) across the HR diagram.
See Abstr. 065.167.

The absolute energy distributions of the Sun, of the 'solar analogs' 16 Cyg B, Hyades VB 64, 16 Cyg A, and of the standard stars α Lyr and 29 Psc.
See Abstr. 071.006.

Mass loss from Wolf–Rayet stars.
See Abstr. 112.015.

Interferometric program on Be stars (II).
See Abstr. 112.104.

Ultraviolet and visual wavelength spectroscopy of gas around Eta Carinae.
See Abstr. 112.116.

A comparative study of barium peculiarity indices.
See Abstr. 113.044.

Effective temperatures, radii, and luminosities of O emission, Be and Ae stars.
See Abstr. 114.012.

The temperatures of Alpha Centauri A and B.
See Abstr. 114.025.

On the properties of well–studied early type stars observed with the Einstein X–ray observatory.
See Abstr. 114.055.

Spectral types from *uvbyβ* photometry.
See Abstr. 114.067.

Spectrophotometry of standard stars: physical parameters of the atmospheres and the scale of effective temperatures.
See Abstr. 114.131.

High-resolution Hα observations of M dwarf stars: implications for stellar dynamo models and stellar kinematic properties at faint magnitudes.
See Abstr. 114.132.

X-ray emission from normal stars.
See Abstr. 116.037.

The relationship between X-ray luminosity and Rossby number for a sample of late-type stars.
See Abstr. 116.041.

Masses of the components of the strongly interacting binary system RX Cas.
See Abstr. 117.093.

The ages and dimensions of Trapezium systems.
See Abstr. 118.027.

RR Lyrae, Delta Scuti, SX Phoenicis stars and Baade-Wesselink method. I. Photometric and radial velocity measurements of four field stars: RR Cet, DX Del, BS Aqr and DY Peg.
See Abstr. 122.010.

ER 8: a very low luminosity degenerate star.
See Abstr. 126.062.

Quelques nouvelles brèves du "survey" IR du plan galactique austral.
See Abstr. 133.015.

X-ray selected M dwarfs and the diffuse X-ray background.
See Abstr. 142.029.

Mass and age distributions of stars in young open clusters.
See Abstr. 153.025.

A new approach to zero-age main sequence fitting.
See Abstr. 153.029.

Investigation of colour-magnitude diagrams of globular clusters with the method of main components.
See Abstr. 154.020.

Age-metallicity relation for F-stars.
See Abstr. 155.012.

Age and mass of M giants in the galactic bulge.
See Abstr. 155.050.

Statistical properties of barium stars.
See Abstr. 155.052.

The solar neighbourhood mass density.
See Abstr. 155.084.

Luminous blue stars: distribution and numbers.
See Abstr. 155.119.

The WR/OB number ratio within 2.5 kpc from the sun.
See Abstr. 155.120.

A large sample of emission-line stars in the Large Magellanic Cloud: their location in the Hertzsprung-Russell Diagram.
See Abstr. 156.034.

Two faint WC stars near 30 Doradus.
See Abstr. 156.036.

The AGB population of nearby galaxies.
See Abstr. 157.089.

The luminosities of M supergiants and the distances to M101, NGC 2403, and M81.
See Abstr. 157.226.

Carbon and M stars in nearby galaxies: a preliminary survey using a photometric technique.
See Abstr. 157.257.

116 Rotation, Magnetic Fields, Activity, Polarization, Radio Radiation

116.001 **Magnetic structure in cool stars. IX. Ultraviolet emission lines from chromospheres and transition regions.**
B. J. Oranje.
Astron. Astrophys., Vol. 154, No. 1/2, p. 185 – 196 (1986).
Ultraviolet fluxes measured by the International Ultraviolet Explorer in various chromospheric and transition-region lines are compiled for 132 stars. The author plots the fluxes against each other in various combinations, and derives power-law fits. Special attention is paid to emission-line Me dwarfs, short-period binaries. W UMa contact binaries and FK Comae stars. The results are discussed in detail.

116.002 **Faint optical flares from the dwarf M star Gliese 812 (Ross 193).**
J. G. Doyle, P. B. Byrne.
Astron. Astrophys., Vol. 154, No. 1/2, p. 370 – 372 (1986).
The authors report on a search for flaring on the Balmer emission line star, Gliese 812. Three flares were detected in a total monitoring time of 11.66 h. The time-averaged flare energy in the Johnson *U*-band is substantially smaller than in other stars of similar absolute *U* magnitude, such as AD Leo, EQ Peg and EV Lac.

116.003 **Magnetodynamic phenomena in the solar and stellar outer atmospheres.**
Y. Uchida.
Astrophys. Space Sci., Vol. 118, Nos. 1/2, p. 127 – 148 (1986). – See Abstr. 012.001.
Magnetic effects causing anomalous heating and drastic flarings in the atmospheres of various types of stars are discussed. The best studied examples of these magnetic effects with spatially resolved observations are those in the case of the Sun. The author examines possible effects of the magnetic field in the respective situations of several types of stars which show X-ray and radio emissions indicating the presence of such activities, and it is concluded that the magnetic field may play important roles in producing anomalous heating and drastic flarings also in those stars having very different physical conditions, in ways seemingly very different from, but intrinsically related to, those in the case of the Sun.

116.004 **Solar and stellar magnetic activity.**
R. W. Noyes.
Physics of the Sun, Vol. III, p. 125 – 153 (1986). – See Abstr. 003.001.

Contents: Solar and stellar magnetic activity: a phenomenological comparison (surface magnetic fields and their effects on stellar radiative flux, direct detection of magnetic fields on stars like the Sun, Ca II H and K emission as indicators of stellar magnetic fields, coronal active regions, magnetic activity cycles). The rotation/activity/age connection (the aging of magnetic activity and rotation, rotation as the fundamental determinant of magnetic activity, the influence of convection zone properties, the Vaughan–Preston gap, the evolution of rotation and magnetic activity on the Sun). Avenues for future research (observational studies of solar magnetic activity, observational studies of stellar magnetic activity, theoretical studies).

116.005 **EXOSAT detection of an X–ray flare from the solar type star π^1UMa.**
M. Landini, B. C. Monsignori Fossi, R. Pallavicini, L. Piro.
Astron. Astrophys., Vol. 157, No. 2, p. 217 – 222 (1986).

An X–ray flare has been detected on the G0 dwarf star π^1UMa using the European satellite EXOSAT. With the exception of the Sun, this is the first time that an X–ray flare has been observed in a normal solar–type star, which is not a member of a binary system and is not a classical flare star. As a member of the Ursa Major Cluster, π^1UMa is a young, rapidly rotating star with a high level of chromospheric and coronal activity. The energy emitted in X–rays by the observed flare is at least one order of magnitude higher than the total energy released in large solar flares. The authors interpret the event as occurring in one or more magnetic loops which occupy only a small fraction of the star surface. The flare is found to cool both by radiation and conduction. The authors derive physical parameters for the emitting region and compare them with transient events occurring on the Sun, on dMe flare stars and on RS CVn and Algol–type binaries.

116.006 **Rotational braking of late–type main sequence stars.**
J. Bohigas, L. Carrasco, C. A. O. Torres, G. R. Quast.
Astron. Astrophys., Vol. 157, No. 2, p. 278 – 292 (1986).

The authors present a model for the angular momentum loss of late type main sequence stars. This loss is thought to be due to a magnetic torque operating on the stellar wind. Assuming that only the convective envelope is braked, an excellent agreement is obtained between the model and the data for the secular decrease of the rotational velocity of G type stars. The observed behaviour of the rotational period between spectral types F2 to G2 is also adequately reproduced. Furthermore, using the observed relationship between chromospheric or coronal activity, and the rotational velocity, the behaviour of both types of activity can be reproduced as a function of stellar mass, amongst relatively young stars. On the other hand, the authors find that braking is underestimated for the oldest and least massive stars. This might indicate that the area covered by magnetically active regions is larger in less massive stars.

116.007 **The stellar seismology of ε Eridani.**
D. B. Guenther, P. Demarque.
Astrophys. J., Vol. 301, No. 1, p. 207 – 212 (1986).

This calculated *p*–mode oscillation spectrum of several stellar models of ε Eridani are compared with the observational data of Noyes et al. Specifically, models with different masses, metallicities, and mixing length parameters are constructed, and for each model the low–degree *p*–mode oscillation spectrum is calculated and its characteristic frequency spacing is compared with the observed value of 172 μHz. Of all the models constructed, the model with a mass of 0.80 $M_\odot$, $Z = 0.02$, mixing length parameter $\alpha = 1.00$, an age of 11.5 Gyr, and a radius of 0.80 $R_\odot$ best fits the observed luminosity, effective temperature, and oscillation data. The characteristic frequency spacing of the oscillation spectrum of this model is 170 μHz. It is concluded that the structure of this model must be very close to the actual physical structure of ε Eri.

116.008 **Variable polarization and activity in Arcturus.**
J. C. Kemp, G. D. Henson, D. J. Kraus, I. S. Beardsley, L. C. Carroll, D. K. Duncan.
Astrophys. J., Lett. Ed., Vol. 301, No. 1, p. L35 – L38 (1986).

High–sensitivity measurements of the blue light (*B*) polarization of this bright, nearby star on 65 nights have led to the discovery of intrinsic variable polarization, with amplitude about 0.005%. There may be a period or semiperiod of around 45 days, and there is also more rapid variation. Concurrent Ca II H and K line emission data clearly show variability, indicating activity, during the time interval represented.

116.009 **A survey of chromospherically active stars.**
F. C. Fekel, T. J. Moffett, G. W. Henry.
Astrophys. J., Suppl. Ser., Vol. 60, No. 2, p. 551 – 576 (1986).

Photometric and/or spectroscopic observations have been obtained of 52 late–type stars which are suspected or known to be chromospherically active. Although not all types of observations were obtained for each star, these observations include all–sky *BVRI* Johnson photometry, ultraviolet spectrograms, low–dispersion blue–wavelength spectrograms, and high–dispersion red–wavelength spectrograms. From the spectroscopic observations the authors have determined *v* sin *i*'s, radial velocities, and the appearance of the Ca II H and K emission lines as well as the Hα line. The photometric observations indicate that chromospherically active stars have *V–R* and *V–I* color excesses.

116.010 **Variability in H alpha, polarization, and magnetic fields in early–type stars.**
P. K. Barker.
Publ. Astron. Soc. Pac., Vol. 98, No. 599, p. 44 – 47 (1986). – See Abstr. 012.009.

116.011 **The correlation between rotation and IR colour excess for B–type dwarfs.**
L. B. F. M. Waters.
Astron. Astrophys., Vol. 159, No. 1/2, p. L1 – L4 (1986).

The author presents the correlation between IR colour excess and spectral type, and between IR colour excess and v sin i for a large number of B0 to B9.5 type stars of luminosity class IV and V, including emission line stars and peculiar stars. He shows that a clear correlation exists between v sin i and 12 μ colour excess (derived from IRAS observations), indicating that rotation facilitates mass loss for B–type dwarfs, but that one or more mechanisms are necessary to trigger mass loss from a rapidly rotating star.

116.012 **Magnetic structure in cool stars. X. The Ca II H+K flux density and rotation rate for main–sequence stars.**
R. G. M. Rutten.
Astron. Astrophys., Vol. 159, No. 1/2, p. 291 – 298 (1986).

This paper studies the relation between the Ca II H+K line–core flux density and the rotation rate for main–sequence stars. Accurate rotation periods have been determined from rotation modulation in the Ca II H+K line–core flux, from orbital periods of synchronized spectroscopic binaries, and from variations of the visual broad–band fluxes from BY Draconis–type stars. For a larger sample of stars rotation periods are estimated from the projected equatorial velocities *v* sin *i* derived from spectral–line broadening. The sample of rotation periods presented in this paper enables the choice of the parameters characterizing Ca II emission and stellar rotation which yield the simplest possible relation.

116.013 **The photometric and spectroscopic period of the Ap star HD 4778.**
H. Hensberge.
Astron. Astrophys., Vol. 159, No. 1/2, p. 331 – 332 (1986).

The analysis of published photometric and spectroscopic data shows that the variations are best represented with a period of 2.5481 days, although a period of 1.64417 days cannot be ruled out.

116.014 **Polarization of star radiation in the Mon R1 association.**
L. A. Pavlova, F. K. Rspaev.
Astrofizika, Tom 23, Vyp. 3, p. 521 – 525 (1985). In Russian. English translation in Astrophysics, Vol. 23, No. 3.

New polarization data for stars in the Mon R1 association are given. 17 stars were observed in the *B* photometric system and 9 of them are the cores of reflection nebulae. Large polarization ~10% is obtained for VY Mon. The connection of the reflection nebulae structure and of the polarization of stars with the magnetic field of the cloud in which they are placed is discussed.

116.015 **Active regions on HR 1099 from a high resolution spectroscopic study of photospheric (Fe I) and chromospheric (Ca II and Hα) lines.**
P. Gondoin.
Astron. Astrophys., Vol. 160, No. 1, p. 73 – 94 (1986).
The Doppler imaging technique applied to photospheric (Fe I) absorption lines and chromospheric (Ca II IR and Hα) emission lines of the HR 1099 (V711 Tau) RS CVn system shows that, on the primary star, photospheric spots 1000K cooler than the surrounding photosphere cover about 20% of the stellar disk and are overlapped with bright solar–like chromospheric plages. The location and extent of these active centers suggest the existence of important magnetic flux at relatively high latitudes (~60°) of the K1 IV primary star. A speculative model in which material is driven from one active region to the other along magnetic field lines is suggested by the geometry of the system and can especially account for the Hα profile structure and the degree of polarization at radio frequencies observed in HR 1099.

116.016 **Astroseismology.**
H. M. Van Horn.
Phys. Today, Vol. 39, No. 1, p. S8 (1986). Short review paper. – See Abstr. 013.021.

116.017 **Supergiant γ Cyg pulsations with 8–d period of magnetic field variations.**
S. I. Plachinda, E. L. Chentsov.
Izv. Krymskoj Astrofiz. Obs., Tom 72, p. 120 – 124 (1985). In Russian. English translation in Bull. Crimean Astrophys. Obs., Vol. 72.
Periodic variations of the radial velocity with 8–d magnetic field variations for γ Cyg were registered in the core of H_α and Ca I λ6499.644 Å lines. The amplitude of the radial velocity curve variations according to H_α is about 1.5 km/s and in the Ca I–line constitutes nearly 0.5 km/s. The gas responsible for the H_α–core formation is accreted onto the star. The authors conclude that the radial velocity variations are caused by pulsations.

116.018 **The magnetic field of Canopus.**
W. W. Weiss.
Astron. Astrophys., Vol. 160, No. 2, p. 243 – 250 (1986).
The detection of magnetic fields in evolved earlier type stars are rare and usually controversial. Additional Zeeman spectrograms for Canopus (F0 Ib – II) corroborate evidence for a variable weak magnetic field. A new period of 6.90 days for the magnetic field variation of Canopus is proposed. The relation of the MSHIFT technique (Weiss et al., 1978) with the standard cross correlation method which is used to determine systematic shifts of left– and righthand circularly polarized spectrum components is discussed.

116.019 **Investigation of the mean surface magnetic fields of Ap stars.**
Yu. V. Glagolevskii (*Yu. V. Glagolevskij*), V. D. Bychkov, I. I. Romanyuk, N. M. Chunakova.
Bull. Spec. Astrophys. Obs. – North Caucasus, Vol. 19, p. 26 – 35 (1985). English translation of 39.116.024.

116.020 **A search for magnetic fields in stars with anomalous helium lines.**
Yu. V. Glagolevskii (*Yu. V. Glagolevskij*), N. M. Chunakova.
Bull. Spec. Astrophys. Obs. – North Caucasus, Vol. 19, p. 36 – 39 (1985). English translation of 39.116.025.

116.021 **A study of the periodic variable radio star LSI + 61°303 and its environs.**
D. A. Frail, E. R. Seaquist, A. R. Taylor.
Bull. Am. Astron. Soc., Vol. 17, No. 4, p. 855 (1985). Abstract. – See Abstr. 010.062.

116.022 **Solar–like activity in warm stars.**
F. M. Walter, J. L. Linsky.
Bull. Am. Astron. Soc., Vol. 17, No. 4, p. 879 (1985). Abstract. – See Abstr. 010.062.

116.023 **An extensive survey of photospheric magnetic fields in G and K dwarfs.**
S. H. Saar, J. L. Linsky.
Bull. Am. Astron. Soc., Vol. 17, No. 4, p. 879 (1985). Abstract. – See Abstr. 010.062.

116.024 **Stellar activity in F stars: the age dependence of Ca II flux and rotation.**
J. R. Varsik.
Bull. Am. Astron. Soc., Vol. 17, No. 4, p. 902 (1985). Abstract. – See Abstr. 010.062.

116.025 **The effects of lines on the ultraviolet surface polarization of cool stars: normal A stars.**
L. R. Doherty, B. K. Armstrong.
Bull. Am. Astron. Soc., Vol. 17, No. 4, p. 902 – 903 (1985). Abstract. – See Abstr. 010.062.

116.026 **Optical photometry and spectroscopy for five dwarf M stars.**
J. G. Doyle, P. B. Byrne, J. W. Menzies.
Mon. Not. R. Astron. Soc., Vol. 220, No. 1, p. 223 – 231 (1986).
The authors report on a search for flaring and BY Dra (spotted) variations on five M–dwarf stars, Gliese 1, 461, 825, 899 and 908. The results for Gl 461 and 825 are compared with predictions of activity levels based on the measured quiescent X–ray flux. The chromospheric radiative loss rate in the Ca H and K lines are determined for Gl 461.

116.027 **Stellar activity: the X–ray picture.**
S. Serio.
ESA Spec. Publ., ESA SP–239, p. 59 – 67 (1985). – See Abstr. 012.019.
This paper reviews the observational status of stellar X–ray astronomy, focussing on the perspectives offered by future X–ray observations.

116.028 **The rotation effect – a mechanism for measuring granulation velocities in stars.**
D. F. Gray.
Publ. Astron. Soc. Pac., Vol. 98, No. 601, p. 319 – 324 (1986).
The Doppler effects of stellar granulation, when combined with those of the star's rotation, have been predicted to enhance the asymmetries in spectral line profiles, a process referred to as the rotation effect. A dozen F stars were observed for the rotation effect. The effect is seen in some, but not all, of them. The asymmetries, as expressed in the line bisectors, are compared to numerical simulations from which it is possible to obtain the average velocity of rise of the hot granules on an absolute velocity scale. Values ~ –1.5 to –3.0 km s^{-1} are deduced.

116.029 **γ Cyg – a periodic magnetic variable?**
S. I. Plachinda, E. S. Dmitrienko, A. B. Severnyi (*A. B. Severnyj*).
Bull. Crimean Astrophys. Obs., Vol. 69, p. 39 – 45 (1986). English translation of 39.116.022.

116.030 **Observations of certain stars at millimeter wavelengths.**
N. S. Nesterov.
Bull. Crimean Astrophys. Obs., Vol. 69, p. 67 – 73 (1986). English translation of 39.116.040.

116.031 **On the possible fine structure of the magnetic field of the chemically peculiar stars α^2CVn and β CrB.**
I. I. Romanyuk.
Astrofiz. Issled. Izv. Spets. Astrofiz. Obs., Tom 22, p. 25 – 38 (1986). In Russian. English translation in Bull. Spec. Astrophys. Obs. – North Caucasus.
Magnetic fields of the chemically peculiar stars α^2CVn and β CrB have been investigated in the spectral range 3300 – 4000 Å.

The phase shift by 0.15 period found earlier at the magnetic field measurements in the spectral ranges with wavelength larger and smaller than the wavelength of the Balmer jump 3647 Å is confirmed for β CrB. For α^2CVn a decrease of the field value and variation amplitude on the spectral interval only at the short-wave side from the jump is found; that can be an indication for existence of a magnetic field radial gradient. The measurements made from the lines of some chemical elements show that the magnetic field defined from the iron and chromium lines decreases at $\lambda < 3647$ Å, and for titanium lines no field decrease is noticed.

116.032 **Radio emission from galactic Wolf–Rayet stars and the structure of Wolf–Rayet winds.**
D. C. Abbott, J. H. Bieging, E. Churchwell, A. V. Torres.
Astrophys. J., Vol. 303, No. 1, p. 239 – 261 (1986).

The authors have observed 23 Wolf–Rayet stars with the VLA at 6 cm and reobserved four stars at both 2 and 6 cm. When combined with previous data, these observations form an essentially complete sample of 40 W–R stars within a distance of 3 kpc and north of declination –47°. Available measurements of optical emission lines are used to estimate the volume emission measure, ionization balance, and terminal velocity V_∞ of the stellar wind for the sample stars. In the sample 33 stars (77%) are probable or definite thermal wind sources and five stars (12%) are probable or definite nonthermal sources. Stellar mass loss rates for the thermal sources lie in the narrow range $0.8 \times 10^{-5} \lesssim \dot{M} \lesssim 8.0 \times 10^{-5} M_\odot \mathrm{yr}^{-1}$, with a mean value of $2 \times 10^{-5} M_\odot \mathrm{yr}^{-1}$. It is speculated that the narrow range in observed $\dot{M}$ is caused by a narrow range in intrinsic mass for the sample stars.

116.033 **An unbiased X–ray sampling of stars within 25 parsecs of the Sun.**
H. M. Johnson.
Astrophys. J., Vol. 303, No. 1, p. 470 – 479 (1986).

The paper reports a search of all the Einstein Observatory IPC and HRI fields for untargeted stars in the Woolley et al. catalog of the nearby stars. Optical data and IPC coordinates, flux density F_x, and luminosity L_x, or upper limits, are tabulated for 126 single or blended systems, and HRI results for a few of them. IPC luminosity functions are derived for the systems, for 193 individual stars in the system (with L_x shared equally among blended components), and for 63 individual M dwarfs.

116.034 **Thermal radio emission from stars.**
A. E. Wright.
Proceedings of the Second Asian–Pacific Regional Meeting on Astronomy, p. 61 – 73 (1984). – See Abstr. 012.031.

At the present time (mid–1981) over 50 thermally emitting objects are known and the first results of high resolution VLA observations of these objects are available. Such observations, at a variety of radio frequencies, when correlated with optical and infrared variability data, are likely to provide very useful data about mass–loss and thus a hitherto poorly understood phase of stellar evolution.

116.035 **Evidence for a common magnetic driver for flares and quiescent coronae/chromospheres.**
A. Skumanich.
Aust. J. Phys., Vol. 38, No. 6, p. 971 – 974 (1985). Paper presented at the R. G. Giovanelli Commemorative Colloquium, Part II, Tucson, Arizona, 17 – 18 January 1985.

The mean power and rate of optical flares in dMe stars are found to be correlated with their quiet coronal X–ray luminosity. The mean flare luminosity in the photometric U–band appears to scale linearly with the X–ray luminosity with a slope of 0.04. This coincidence suggests that quiet and flare activity in coronae/chromospheres may be related by one and the same flaring mechanism. The author proposes that 'quiet' activity is due to microflares – a low yield but high frequency continuation of flares.

116.036 **Activity on cool stars.**
R. D. Robinson, O. B. Slee.
Aust. J. Astron., Vol. 1, No. 3, p. 105 – 128 (1986).

The authors examine the characteristics of solar–like activity, which is a term applied to a wide range of magnetically related phenomena occurring in the atmospheres of cool stars. The authors describe in general terms four basic types of activity: (1) active chromospheres, (2) activity in transition regions and coronae, (3) starspots, and (4) flares. Each of these phenomena are observable over nearly the entire electromagnetic spectrum, and, when possible, the authors attempt to show how data in various wavelength regions relate to one another. They then examine the stellar parameters which seem to influence the creation of activity and how the form of this activity changes as the star evolves.

116.037 **X–ray emission from normal stars.**
S. Serio.
X–ray astronomy '84, p. 1 – 15 (1985). – See Abstr. 012.033.

The distribution of coronal emission from main sequence stars is reviewed, together with the attempts for a theoretical understanding of its properties based on models. The fact that X–ray emission in late–type stars shows a dependence on rotational velocity, while it is directly related to bolometric luminosity in early–type stars, suggests to consider separately the mechanisms responsible for coronae in the two classes of stars.

116.038 **X–ray coronae of late–type stars: theoretical implications.**
J. H. M. M. Schmitt.
X–ray astronomy '84, p. 17 – 22 (1985). – See Abstr. 012.033.

The Einstein Observatory provided evidence for the ubiquity of coronae in late type stars. The hypothesis that loop–like structures, as seen in the solar corona, are also present in the coronae of other stars, is consistent with the available X–ray spectra.

116.039 **EXOSAT observations of late–type stars: preliminary results.**
M. Landini, B. C. Monsignori–Fossi, R. Pallavicini.
X–ray astronomy '84, p. 31 – 34 (1985). – See Abstr. 012.033.

The authors present preliminary results of a program of observations of late–type stars using EXOSAT. They derive temperatures and emission measures and compare their results with previous Einstein observations of the same targets.

116.040 **X–ray emission from solar–type stars: X–ray luminosity function of late F and G stars.**
A. Maggio, J. Bookbinder, F. R. Harnden Jr., L. Golub, P. Majer, R. Rosner, G. S. Vaiana.
X–ray astronomy '84, p. 39 – 42 (1985). – See Abstr. 012.033.

The authors present preliminary results from a volume–limited survey of X–ray emission from late F and G dwarf stars. They have obtained count rates or upper limits at the locations of catalogued stars within 25 parsecs for which an observation of the Einstein Imaging Proportional Counter was available. They have constructed X–ray luminosity functions for the whole sample and for selected subsamples of binary and non–binary sources.

116.041 **The relationship between X–ray luminosity and Rossby number for a sample of late–type stars.**
G. Micela, S. Sciortino, S. Serio.
X–ray astronomy '84, p. 43 – 46 (1985). – See Abstr. 012.033.

The authors investigate the relation between the X–ray luminosity and the ratio of rotational period to convective turn–over time (i.e. Rossby number) for a sample of stars later than F7 belonging to three groups with different ages: nearby stars, Hyades and Pleiades.

116.042 **Velocities of rotation of chemically peculiar stars.**
V. G. Klochkova, I. M. Kopylov.
Sov. Astron., Vol. 29, No. 5, p. 549 – 554 (1985). English translation of 40.116.035.

116.043 **Rotational velocities of metallic stars.**
V. Dobrichev.
Astrofiz. issled., Tom 4, p. 40 – 46 (1985). In Russian. Abstr. in Ref. Zh., 51. Astron., 6.51.486 (1986).

116.044 **A study of the rapidly rotating variable star HD 36705 (AB Doradus).**
J. L. Innis, R. D. Robinson, D. W. Coates, K. Thompson.
Proc. Astron. Soc. Aust., Vol. 6, No. 2, p. 156 – 160 (1985).
Photoelectric B and V light curves and high dispersion spectroscopic observations were obtained in 1985 February for the rapidly rotating spotted star HD 36705. There is evidence for a broadband flare of ~0.05 magnitude in V and 0.07 magnitude in B. This occurred near maximum light. Several interpretations of the nature of HD 36705 are discussed; however at present none is completely satisfactory and further observations are required.

116.045 **Anomalous radio-emission from the direction of a stellar object.**
J. Heidmann.
C. R. Acad. Sci., Sér. II, Tome 303, No. 1, p. 47 – 49 (1986). In French.
The author discovered a 5 mJy radio continuum emission at 21 cm wavelength coming from the direction of a stellar object with magnitude V = 14.5. This object is red and appears to have a solar type spectrum. If it is a star its distance is 100 pc and its radio emission is a million times stronger than the one of the quiet sun. It had the same level in 1980 and 1981 and its visible magnitude was the same in 1953 and 1979 – 1982. It could be a pure chance perspective effect, a new type of powerful radio star or an emission of artificial origin.

116.046 **X-ray-selected M dwarfs and the diffuse X-ray background.**
J.-P. Caillault, D. J. Helfand, J. A. Nousek, L. O. Takalo.
Astrophys. J., Vol. 304, No. 1, p. 318 – 325 (1986).
The authors have carried out a systematic optical identification program of serendipitous X-ray sources discovered by the Einstein Observatory. A total of 104 were detected above the threshold flux of 3×10^{-13}ergs cm^{-2}s^{-1} in the 114 deg^2 survey region; 11 of these 104 sources are dM stars. Including X-ray selected M dwarfs from the LMC survey, the Medium Sensitivity Survey, and the Einstein Deep Survey, the authors have constructed an M dwarf X-ray luminosity function and conclude that the integrated contribution to the diffuse soft (0.28 – 1.0 keV) X-ray background from such stars is less than 10%.

116.047 **The dependence of He I λ10830 absorption strength upon X-ray emission in late-type stars.**
D. M. Zarro, H. Zirin.
Astrophys. J., Vol. 304, No. 1, p. 365 – 370 (1986).
The authors present measurements of He I λ10830 absorption equivalent widths and fractional X-ray luminosities L_x/L_{bol} for a large sample of late-type stars. A correlation of increasing λ10830 strength with increasing L_x/L_{bol} is evident in dwarfs, giants, and supergiants later than and including spectral type F7. No equivalent correlation is apparent in main-sequence stars earlier than dF7. The data suggest that coronal X-ray irradiation constitutes an important source of He I excitation in late-type stars.

116.048 **Measurements of the magnetic field of β CrB.**
S. I. Plachinda.
Izv. Krymskoj Astrofiz. Obs., Tom 74, p. 47 – 49 (1986). In Russian. English translation in Bull. Crimean Astrophys. Obs., Vol. 74.
The results of measurements of the magnetic field of β CrB that have been obtained using the spectral line λ4254.33 Cr I are presented.

116.049 **Additional photometric data for the X-ray source KR Aurigae during 1971 – 1980.**
V. N. Popov, Z. T. Kraicheva, M. D. Popova.
Eruptive phenomena in stars, p. 373 – 374 (1986). – See Abstr. 012.039.

116.050 **Recent photometric data for the X-ray source KR Aurigae.**
A. P. Antov, V. N. Popov, M. D. Popova.
Eruptive phenomena in stars, p. 375 – 376 (1986). – See Abstr. 012.039.

116.051 **On the last cycle of optical variability of X-ray source KR Aurigae.**
Z. T. Kraicheva, V. N. Popov, M. D. Popova, A. P. Antov.
Eruptive phenomena in stars, p. 377 – 378 (1986). – See Abstr. 012.039.

116.052 **Modeling of coronal X-ray emission from active cool stars. I. Hyades cluster.**
R. A. Stern, S. K. Antiochos, F. R. Harnden Jr.
Astrophys. J., Vol. 305, No. 1, p. 417 – 432 (1986).
X-ray pulse height spectra of the most active cool stars in the Hyades cluster obtained with the *Einstein* IPC cannot be satisfactorily fitted using isothermal thin plasma emission models. A more physically meaningful set of coronal parameters is provided by models which consist of an ensemble of loops with a single maximum temperature, but with the temperature distribution within the loop determined by loop physics. Such models have been successfully fitted to the IPC pulse height spectra. A consistent physical description is an ensemble of small ($<10^{10}$cm), high-pressure (>400 dynes cm^{-2}) loops of similar maximum temperature ($\sim 10 - 15\text{x}10^6$K).

116.053 **Brightness variations of a disc with two precessing spots on the surface at an arbitrary inclination of the precession axis slope.**
D. P. Kyurkchieva, V. G. Shkodrov.
Dokl. Bolg. Akad. Nauk, Tome 38, No. 10, p. 1273 – 1276 (1985).

116.054 **Comments on the hotspots model for the polarization of α Orionis.**
J. A. Marcondes-Machado.
Astron. Astrophys., Vol. 161, No. 2, p. L15 – L16 (1986).
The model presented by Schwarz and Clarke (1984) and Clarke and Schwarz (1984) for the polarization of α Ori is discussed. It is shown that, if the scattering region is outside the stellar atmosphere, the use of limb darkening in the calculation of the polarization is not correct, and that a proper use will lead to increases (and not decreases) in the polarization across TiO bands.

116.055 **Reply to "Comments on the hotspots model for the polarization of α Orionis".**
H. E. Schwarz.
Astron. Astrophys., Vol. 161, No. 2, p. L16 (1986). See Abstr. 116.054 for the original paper.

116.056 **Polarization measurements of β Cep stars.**
D. Clarke.
Astron. Astrophys., Vol. 161, No. 2, p. 412 – 416 (1986).
Polarimetric measurements are reported for three β Cephei stars, β Cru, α Vir and σ Sco. Although the data analysis suggests that periodicities close to fundamental photometric variations are present at a statistically significant level, these are shown to be spurious resulting from the way the measurements have been assembled, covering a few hours observation per night on a small number of nights. There is, however, an indication that α Vir is displaying a variable polarization possibly connected either with its instability or its binary nature.

116.057 **New photometry for the rapidly rotating Pleiades K dwarf HII 1883.**
J. R. Stauffer, J. D. Dorren, J. L. Africano.
Astron. J., Vol. 91, No. 6, p. 1443 – 1450 (1986).
The authors have obtained a number of optical light curves for the rapidly rotating spotted Pleiades K dwarf HII 1883 during the period January 1983 through November 1985. BVRIJHK photometry for HII 1883 allows an estimate of the size and temperature of the spot group(s) responsible for the photometric variability during 1983 and 1984. With a simple two-spot model,

a spot temperature 900K cooler than the mean photospheric temperature, and a spot area about 8% of the total surface area of the star are derived.

116.058 **The rich oscillation spectrum of the F0p star HD 60435.**
J. M. Matthews, D. W. Kurtz, W. H. Wehlau.
Bull. Am. Astron. Soc., Vol. 18, No. 1, p. 637 (1986). Abstract. – See Abstr. 010.066.

116.059 **Chromospheric activity in the Hyades triple system BD +23°635.**
B. W. Bopp, J. L. Africano, B. D. Goodrich.
Publ. Astron. Soc. Pac., Vol. 98, No. 602, p. 457 – 460 (1986).
The Hyades K–dwarf BD +23°635 is a triple system, consisting of a single–line spectroscopic binary with a 2.4–day orbital period and a third star at much larger distance. Observations of the red region of BD +23°635 that the authors obtained in 1981 – 82 show Hα to be a weak emission feature, exhibiting large velocity shifts with respect to photospheric lines. They identify the spectroscopic primary as the source of the Hα emission, implying a high level of chromospheric activity, presumably induced by rapid rotation in the synchronously rotating short–period pair. BD +23°635 appears to show small changes in brightness that are characteristic of BY Draconis stars.

116.060 **Stellar activity in the evolutionary framework: the Space Telescope perspective.**
S. Serio.
Mem. Soc. Astron. Ital., Vol. 56, No. 4, p. 673 – 679 (1985). – See Abstr. 012.054.

116.061 **X–ray, ultraviolet, optical and radio observations of cool stars.**
R. Pallavicini.
Mem. Soc. Astron. Ital., Vol. 56, No. 4, p. 813 – 822 (1985). – See Abstr. 010.821.
This paper describes the Arcetri program to study the physical properties of cool stars using a variety of observation techniques.

116.062 **Light variations in HD 36705 in October 1982.**
M. Kubiak.
Acta Astron., Vol. 35, Nos. 3–4, p. 369 – 375 (1985).
Photoelectric observations made on five nights in the uvby system confirm rapid (on time–scale of days) variations of the light–curve. Moments of minima observed till now can be tied together with a period 0.51339 d which probably is also a period of rotation.

116.063 **Polarimetry of VV Puppis.**
M. Cropper, B. Warner.
Mon. Not. R. Astron. Soc., Vol. 220, No. 3, p. 633 – 645 (1986).
Broadband simultaneous circular and linear polarization observations of VV Puppis during the years 1983 – 85 with time resolutions down to 20 s are presented. From the observed variations around orbit, an inclination ~78° and a magnetic colatitude of the principal accreting pole ~155° are derived. The question of whether VV Pup is an eclipsing system is discussed but the situation remains unresolved. It is shown that the ratio of linear to circular polarization is a rough indicator of magnetic field strength for broadband observations. Strong flaring activity on a time–scale ~25 s is shown to be well correlated with flares observed simultaneously with EXOSAT. The fractional circular polarization during the flares remains constant.

116.064 **Measurement of spin in twenty binary systems.**
H.–s. Tan.
Chin. Astron. Astrophys., Vol. 10, No. 1, p. 54 – 61 (1986). English translation of 40.116.071.

116.065 **On the rotation of 53 Be stars.**
W.–s. Gao, H.–l. Cao.
Acta Astrophys. Sin., Vol. 6, No. 2, p. 143 – 153 (1986). In Chinese.
The spectra of 53 Be stars from a *list of early–type shell stars* which is edited by Bidelman (1976) have been observed since 1982. The authors obtained a correlation between the spectral types and the quantum number of the highest Balmer line observed in Be stars. The ultraviolet flux of stars in the region longward of the Lyman jump increases with the effective temperature. This region may be important to the formation of stellar winds. The distribution of mean rotational velocities in spectral type intervals is also shown. The incidence of the Be phenomenon has a maximum in the B2 – B4.9 interval, which includes the spectral type, B2, with the maximum relative fraction of Be/B.

116.066 **Rotational measurement and study of RS CVn stars.**
H.–s. Tan, X.–f. Liu.
Acta Astron. Sin., Vol. 27, No. 2, p. 130 – 136 (1986). In Chinese.
The rotational velocities for 33 primaries and 5 secondaries of 33 RS CVn binary systems are determined. The method to measure the rotation is that the standard star spectrum was rotationally broadened with a number of different $V \sin i$ and then compared with the chosen line profiles of program stars.

116.067 **Rotational braking of late type main sequence stars.**
J. Bohigas, L. Carrasco, C. A. O. Torres,
G. R. Quast.
Rev. Mex. Astron. Astrofis., Vol. 12, Spec. Issue, p. 208 (1986). Abstract. – See Abstr. 012.066.

116.068 **Rotation in lower main–sequence stars.**
G. R. Quast, C. A. Torres.
Rev. Mex. Astron. Astrofis., Vol. 12, Spec. Issue, p. 209 – 212 (1986). – See Abstr. 012.066.
The authors calculated rotational periods for F and later stars. The known periods of BY Dra stars allowed to calibrate their relation for redder stars. Samples of stars with different kinematical ages are well separated, showing that older stars are rotating slower than younger ones in a way that depends on spectral type. If age is the main parameter governing the braking of rotation in single stars, then BY Dra stars must be very young.

116.069 **Coordinated IUE and ground–based observations of stellar flares: YZ CMi, Proxima Cen and AD Leo.**
B. H. Foing, M. Rodono, G. Cutispoto, S. Catalano,
J. L. Linsky, D. M. Gibson, A. Brown, B. M. Haisch,
C. J. Butler, P. B. Byrne, A. D. Andrews, J. G. Doyle,
D. E. Gary, G. W. Henry, G. Russo, A. Vittone, F. Scaltriti.
Rev. Mex. Astron. Astrofis., Vol. 12, Spec. Issue, p. 213 (1986). Abstract. – See Abstr. 012.066.

116.070 **Polarimetry of Wolf–Rayet binaries.**
M. T. Meliani, A. M. Magalhães.
Rev. Mex. Astron. Astrofis., Vol. 12, Spec. Issue, p. 407 (1986). Abstract. – See Abstr. 012.066.

116.071 **Multicolor polarimetry of selected Be stars: a long–term analysis.**
D. McDavid.
Publ. Astron. Soc. Pac., Vol. 98, No. 604, p. 572 – 576 (1986).
New observations of linear polarization in the *UBVRI* system for six bright northern Be stars are combined with existing measurements from two previous epochs in an analysis of long–term variation over nearly two decades. No indication of large short–term variability is found within the individual data sets, but ζ Tau and π Aqr have shown steady change on the longer time scale.

116.072 **Some aspects of the physics of stellar X–ray emission.**
R. Rosner.
Mitt. Astron. Ges., No. 65, p. 79 – 93 (1986). – See Abstr. 012.072.

116.073 **ω Orionis.**
IAU Circ., No. 4195 (1986).

116.074 **HD 32918.**
IAU Circ., No. 4226 (1986).

116.075 **Statistical investigation of chemically peculiar stars. I. Stars with known periods.**
V. S. Lebedev.
Astrofiz. Issled. Izv. Spets. Astrofiz. Obs., Tom 21, p. 21 – 29 (1986). In Russian. English translation in Bull. Spec. Astrophys. Obs. – North Caucasus.

The data on the rotation periods of 220 chemically peculiar stars of the upper part of the main sequence have been collected from the literature. Stellar radii, equatorial rotation velocities and inclination angles of the rotational axis to the line of sight are estimated from the effective temperatures obtained from photometric and spectral data. It is found that the rotation velocity distribution is exponential with scale of 80 km/s, and the orientation of the rotation axes does not contradict the assumption about the same probability of their distribution.

116.076 **Some results of a narrow-band photometry of the magnetic star β CrB.**
V. I. Burnashev, V. P. Malanushenko, N. S. Polosukhina.
Izv. Krymskoj Astrofiz. Obs., Tom 74, p. 42 – 47 (1986). In Russian. English translation in Bull. Crimean Astrophys. Obs., Vol. 74.

Observations of the emission flux from the magnetic star β CrB in the depression range $\lambda_0 = 5215$ Å were carried out. The analysis of the observational data showed that, first, short-period variations of flux do exist with the periods $196^m_.9$, $158^m_.3$ and $58^m_.57$ with amplitude of variations equal to $0^m_.005$, $0^m_.004$ and $0^m_.004$, respectively. These variations can be attributed to the model of an oblique magnetic pulsator. Secondly, sharp variations of flux ($\Delta m = 0.02 - 0.03$) with the phase of star rotation are revealed that can be described in terms of a conventional model of an oblique rotator.

A new Fabry–Perot radial velocity spectrometer.
See Abstr. 034.063.

Interferometric stellar oscillation spectrometry.
See Abstr. 034.064.

Statistical methods of stellar polarimetry.
See Abstr. 036.074.

Absolute astronomical accelerometry.
See Abstr. 036.084.

Optimized variability analysis in non periodic sources: application to X–ray emission from normal stars.
See Abstr. 036.116.

Plasma and radiation processes.
See Abstr. 063.007.

Magnetic fields in starspots on late–type giants.
See Abstr. 064.056.

Stellar magnetic activity and long term luminosity variations of late type stars. I. On generalization of Öpik's convection theory to a rotating medium.
See Abstr. 064.060.

Relation of chromospheric activity to convection, rotation, and pre–main–sequence evolution.
See Abstr. 065.002.

Rapidly rotating core–collapse models.
See Abstr. 065.084.

Radio method of mapping large–scale coronal magnetic fields on active stars.
See Abstr. 075.010.

FK Comae – the kinematical determination of its age and evolutionary status.
See Abstr. 111.015.

The outer atmosphere of Procyon (α CMi F5 IV–V): evidence of supergranulation or active regions.
See Abstr. 112.005.

Periodicity of the radial–velocity and brightness variations of the hypergiant P Cygni (B1 Ia$^+$), and origin of the shell ejections.
See Abstr. 112.025.

Post–main sequence evolution of rotation and chromospheric activity.
See Abstr. 112.049.

Chromospheric activity as an indicator of stellar age.
See Abstr. 112.053.

Observational properties of nonradial oscillations in early–type stars and their possible effect on mass loss: the example of ζ Puppis (O4 If).
See Abstr. 112.065.

A phenomenological interpretation of stellar chromospheres.
See Abstr. 112.085.

Radio continuum emission from winds, chromospheres, and coronae of cool giants and supergiants.
See Abstr. 112.096.

The outer atmospheric layers of FK Comae stars.
See Abstr. 112.120.

O–star winds: is rotation important?
See Abstr. 112.125.

Photometric investigation of the magnetic star 53 Camelopardalis.
See Abstr. 113.058.

On the properties of well–studied early type stars observed with the Einstein X–ray observatory.
See Abstr. 114.055.

On the variation of Hα emission in the spectrum of the Be star 28 Cyg.
See Abstr. 114.104.

A search for rotational modulation of the ultraviolet spectra of ψ^3Piscium and 31 Comae Berenices.
See Abstr. 114.117.

High–resolution Hα observations of M dwarf stars: implications for stellar dynamo models and stellar kinematic properties at faint magnitudes.
See Abstr. 114.132.

The case of α Centauri: mass, age and *p*–mode oscillation spectrum.
See Abstr. 115.004.

How well do we know stellar parameters?
See Abstr. 115.016.

X–ray spectra and the rotation–activity connection of RS Canum Venaticorum binaries.
See Abstr. 117.003.

Detection of optical variability of LSI + 61°303.
See Abstr. 117.004.

Simultaneous optical and X–ray observations of a flare on BY Draconis.
See Abstr. 117.040.

Differential rotation and activity cycles in RS CVn binaries. III. UV Piscium and UX Arietis.
See Abstr. 117.041.

Spot cycles in the RS CVn variable AY Cet.
See Abstr. 117.051.

Evidence for extended radio emission surrounding RX Puppis.
See Abstr. 117.087.

On the variability of Sigma Geminorum – an interesting candidate for X–ray studies.
See Abstr. 117.131.

X–ray spectra and the rotation–activity connection of RS CVn binaries.
See Abstr. 117.132.

X–ray observations of the AM Her star CW 1103 + 254.
See Abstr. 117.174.

The soft X–ray superoutburst of VW Hydri: 14 second periodicity.
See Abstr. 117.175.

EXOSAT observations of AM Her objects – preliminary results.
See Abstr. 117.176.

EXOSAT observations of intermediate polars: preliminary results.
See Abstr. 117.177.

X–rays and colliding winds in symbiotic novae.
See Abstr. 117.180.

Ultraviolet, optical, infrared, and microwave observations of HR 5110.
See Abstr. 117.186.

Cyclotron radiation from magnetic cataclysmic variables.
See Abstr. 117.195.

A 6 centimeter radio survey of short–period active binary stars.
See Abstr. 117.241.

X–ray emission from the symbiotic system CH Cyg.
See Abstr. 117.246.

The magnetic fields of accreting white dwarfs.
See Abstr. 117.279.

Stellar activity in a model of RS CVn systems observed with the IUE satellite.
See Abstr. 117.282.

Evidence for p–modes on Alpha Centauri and Procyon.
See Abstr. 118.016.

An EXOSAT observation of quiescent and flare coronal X–ray emission from Algol.
See Abstr. 119.019.

EXOSAT hard X–ray observations of EX Hydrae.
See Abstr. 119.054.

Orbital inclinations of late B–type spectroscopic binaries.
See Abstr. 120.014.

Measurements of rotation for 18 binaries.
See Abstr. 120.029.

Short–term spectral variability in AB Aurigae: clues for activity in Herbig Ae stars. I. The ultraviolet lines of Mg II and Fe II.
See Abstr. 121.042.

Rotation in T Tauri stars. I. Observations and immediate analysis.
See Abstr. 121.054.

Study of quiescent state X–ray emission from flare stars.
See Abstr. 122.017.

Polarimetric observations of the RV Tauri star AC Herculis.
See Abstr. 122.024.

Narrow–band, slowly varying decimetric radiation from the dwarf M flare star YZ Canis Minoris.
See Abstr. 122.059.

Simultaneous optical, infrared and microwave observations of the flare star AT Mic.
See Abstr. 122.077.

The magnetic field of the BY Draconis flare star EQ Virginis.
See Abstr. 122.079.

A review of the rapidly oscillating Ap stars.
See Abstr. 122.096.

Frequency analysis of the 5.65–min oscillations in the rapidly oscillating Ap star HD 134214.
See Abstr. 122.098.

Microwave emission from dMe flare stars.
See Abstr. 122.105.

Quiescent state X–ray emission from flare stars.
See Abstr. 122.110.

Coordinated Exosat and spectroscopic observations of flare stars and coronal heating.
See Abstr. 122.114.

Millisecond radio spikes from the dwarf M flare star AD Leonis.
See Abstr. 122.143.

A test of the dipole model for the rotating magnetic white dwarf Feige 7.
See Abstr. 126.013.

Determination of the white dwarf's magnetic axis from X–ray orbital light curves.
See Abstr. 126.057.

The new magnetic white dwarf PG 1031 + 234: polarization and field structure at more than 500 million Gauss.
See Abstr. 126.066.

Interstellar polarization from observations of A and F stars in high and intermediate galactic latitudes, and from stars in the Mathewson and Ford polarization catalogue.
See Abstr. 131.319.

The wavelength of maximum polarization in the Chamaeleon dark cloud.
See Abstr. 131.365.

Long–term X–ray variability in the Hyades cluster.
See Abstr. 153.015.

Positions of 127 Hyads and 6–cm observations of 320 Hyads.
See Abstr. 153.023.

Chromospheric emission and rotation in open clusters.
See Abstr. 153.037.

Durchmusterung offener Sternhaufen nach CP2–Sternen.
See Abstr. 153.047.

Infrared polarimetry of the stars in the inner Galaxy.
See Abstr. 155.020.

117 Close Binaries (Observations, Theory)

117.001 **Hard X–ray observations of Cygnus X–1 with the MISO telescope.**
F. Perotti, A. Della Ventura, G. Villa, L. Bassani, R. C. Butler, G. Di Cocco, R. E. Baker, A. J. Dean, C. G. Hanson.
Astrophys. J., Vol. 300, No. 1, p. 297 – 303 (1986).
The authors present the results of hard X–ray, soft gamma–ray observations of the galactic black hole candidate, Cyg X–1, taken with the MISO telescope in 1979 October and in 1980 May. They confirm the "superlow" state measured during 1979 September – October by the HEAO 3 satellite. The 1980 observation coincides with a low– to high–state transition. No gamma–ray counting–rate excess above 200 keV was recorded in either observation.

117.002 **A light synthesis program for binary stars. II. Light curve and color curve effects in a contact system.**
A. P. Linnell.
Astrophys. J., Vol. 300, No. 1, p. 304 – 313 (1986).
This paper studies various models which attempt to explain light curves and color curves for eclipsing W Ursae Majoris stars of type W. Observed color curves for VW Cephei are in disagreement with the Mullan starspot model to explain W–type light curves. An alternative starspot model, with starspots located on the averted hemisphere of the larger star, represents the light curves and color curves for 1980 August 21 with good accuracy. The observed light curves and color curves of VW Cephei agree with theoretical curves for a Rucinski hot secondary model. A single spot, added to the underlying hot secondary model, provides a reasonably accurate representation of *UBVRI* photometric data.

117.003 **X–ray spectra and the rotation–activity connection of RS Canum Venaticorum binaries.**
P. Majer, J. H. M. M. Schmitt, L. Golub, F. R. Harnden Jr., R. Rosner.
Astrophys. J., Vol. 300, No. 1, p. 360 – 373 (1986).
Results are presented from a survey of RS CVn binaries which were observed with IPC on board the Einstein Observatory. Spectral analyses of the IPC pulse height spectra show that the coronae of RS CVn binaries always contain hot gas with temperatures in excess of 10^7K, and that at least two temperature components are necessary to account for the higher quality IPC spectra. It is argued that these bimodal temperature distributions found by the IPC are indicative of true distributions of emission measure versus temperature that are continuous. It is also shown that none of the derivable X–ray characteristics of RS CVn binaries depend on rotation period.

117.004 **Detection of optical variability of LSI + 61°303.**
J. M. Paredes, F. Figueras.
Astron. Astrophys., Vol. 154, No. 1/2, p. L30 – L32 (1986).
The authors report the detection of optical variability of the star LSI + 61°303 obtained from UBVRI photometric observations. The optical light varies by about 0.1 magnitude over a 10–day time interval. An analysis of the spectral type and luminosity class of the star is presented and discussed. Some results of radio observations are also presented.

117.005 **Optical light curves of massive X–ray binaries.**
S. A. Tjemkes, E. J. Zuiderwijk, J. van Paradijs.
Astron. Astrophys., Vol. 154, No. 1/2, p. 77 – 91 (1986).
The authors use a simple geometric model to analyse the optical light curves of massive X–ray binaries. This model is based on a description of the tidal and rotational distortion of the primary star in terms of an equipotential surface, and gravity darkening using Von Zeipel's theorem. In addition the model includes the effects of X–ray heating of the primary, the presence of an accretion disk, and eccentricity of the orbit. The authors discuss a number of the simplifying assumptions underlying their model, and the possible limitations of its applicability. The model is applied to three X–ray binary pulsars, whose orbital parameters are well known (Vela X–1, SMC X–1 and Cen X–3), and for which accurate optical light curves are available.

117.006 **Spectrophotometry of the eclipsing dwarf nova V2051 Ophiuchi.**
D. J. Watts, J. Bailey, P. W. Hill, J. G. Greenhill, C. McCowage, T. Carty.
Astron. Astrophys., Vol. 154, No. 1/2, p. 197 – 213 (1986).
A detailed spectrophotometric study of the eclipsing dwarf nova V2051 Oph is presented. Using a new data acquisition technique for the spectrograph, all detected events were recorded individually allowing simultaneous high resolution temporal and spectral studies to be made. The first UV observations using IUE are also presented. The optical results and the UV data are discussed in terms of an accretion disk model. This evidence for a disc asymmetry is presented and the solution for the system dynamics is given. The data are compared with a model Keplerian disc. The authors discuss the UV to optical continuum and its compatibility with disc models.

117.007 **The ultraviolet variability of the symbiotic star HBV 475. II. Study of line flux variations.**
B. E. A. Mueller, H. Nussbaumer, W. Schmutz.
Astron. Astrophys., Vol. 154, No. 1/2, p. 313 – 320 (1986).
The symbiotic star HBV 475 (= V1329 Cyg) is one of the few symbiotics where periodic variations in the ultraviolet line fluxes have been observed. The authors collect the IUE observations between June 1979 and December 1984. Extending the work of Nussbaumer and Schmutz (1983), who presented a detailed study of this object, the authors establish the periodicity in the ultraviolet line and continuum fluxes and in velocity shifts. There can thus be no doubt about the binary nature of this object. The line flux variations result in a period of 975 days.

117.008 **Radio emission of cataclysmic variable stars.**
E. Fürst, A. Benz, W. Hirth, A. Kiplinger, M. Geffert.
Astron. Astrophys., Vol. 154, No. 1/2, p. 377 – 378 (1986).
Eight cataclysmic variable stars were observed at 6 cm wavelength using the Very Large Array (VLA). The objects were: CN Ori, SS Aur, YZ Cnc, SU UMa, Z Cam, V603 Aql, EM Cyg, and RZ Sge. Most of these objects were in optical high stage, but none were detected beyond flux limits between 0.1 and 0.3 mJy.

117.009 **A spectroscopic study of the X–ray binary V1727 Cygni.**
K. Horne, F. Verbunt, D. P. Schneider.
Mon. Not. R. Astron. Soc., Vol. 218, No. 1, p. 63 – 74 (1986).
Time–resolved high–resolution spectroscopy is used to study the binary X–ray source V1727 Cyg (= 4U2129 + 47). The He II 4686 emission line has a double–peaked line profile, a strong *S*–wave distortion, and distinctive eclipse behaviour–features suggesting that the line is formed in an accretion disc around the compact object. Hβ absorption and a sharp emission component at N III 4640 arise on the heated face of companion star. The radial velocities of the lines, the duration of the X–ray eclipse, and the separation of the He II emission peaks constrain the masses at $0.6 \pm 0.2\ M_\odot$ for the compact star and $0.4 \pm 0.2\ M_\odot$ for the companion. If the compact component is a neutron star, as seems probable from the X–ray luminosity, its derived mass is significantly lower than previously determined neutron star masses.

117.010 **A study of the continuum and iron *K* line emission from low–mass X–ray binaries.**
N. E. White, A. Peacock, G. Hasinger, K. O. Mason, G. Manzo, B. G. Taylor, G. Branduardi–Raymont.
Mon. Not. R. Astron. Soc., Vol. 218, No. 1, p. 129 – 138 (1986).
The spectral properties of a sample of six low–mass X–ray binaries are studied using data obtained with the Gas Scintillation Proportional Counter on EXOSAT. The various continuum models proposed for these sources are investigated. An iron

K emission line centred between 6.4 and 6.8 keV was detected in five out of the six sources with equivalent widths of 70 – 130 eV and a full–width half–maximum of 1 keV. These results are similar to those of Sco X–1 obtained by the same instrument.

117.011 **Contact and near–contact binary systems – III. The semi–detached system YY Ceti.**
T. M. McFarlane, D. J. King, R. W. Hilditch.
Mon. Not. R. Astron. Soc., Vol. 218, No. 1, p. 159 – 169 (1986).
Reticon spectroscopy and $BVRI_c$ photometry are presented for the eclipsing binary system YY Ceti. It is shown that the system, which has an orbital period of 0.79 day, is composed of a normal main–sequence star of spectral type A8 and an evolved late G–type secondary which completely fills its Roche lobe. The system is expected to evolve into contact within $\sim 10^9$yr.

117.012 **E1405–451: three seasons of polarimetry and photometry.**
M. Cropper, J. W. Menzies, S. Tapia.
Mon. Not. R. Astron. Soc., Vol. 218, No. 2, p. 201 – 222 (1986).
The authors report three seasons of observations of the AM Her–type variable star E1405–451. Orbital and long–term intensity and polarization variations are investigated. Evidence is found for irregular variations in the position of the accretion column on the surface of the primary. The authors use the derived orientations and observed ratios of linear to circular polarization to estimate a magnetic field of approximately 30 MG in the cyclotron emitting region. Various cyclotron emission calculations are compared, and the shock front calculations of Wickramasinghe and Meggitt are found to provide excellent quantitative fits to the observations. The authors confirm the existence of the 1 – 3 s quasi–periodic oscillations found by Mason et al.

117.013 **Tidal spin–up and magnetic braking in cataclysmic variables.**
J. van Paradijs.
Mon. Not. R. Astron. Soc., Vol. 218, No. 2, p. 31P – 35P (1986).
Based on proposed models for the tidal spin–up and magnetic braking of stars with a convective outer envelope the author suggests that the rotation of secondaries in cataclysmic variables is not necessarily synchronized with the orbital revolution. This may provide an explanation for the observed large range in the mass transfer rate (at the same orbital period) of cataclysmic variables above the period gap.

117.014 **Accretion flows in the non–synchronous magnetic cataclysmic variables.**
J.–M. Hameury, A. R. King, J.–P. Lasota.
Mon. Not. R. Astron. Soc., Vol. 218, No. 4, p. 695 – 710 (1986).
The authors investigate the accretion flows in the non–synchronous magnetic cataclysmic variables (intermediate polars). They show that for orbital periods <5 hr, the white dwarf magnetic field is likely to prevent the formation of an accretion disc, and channels the infalling matter on to large fractions of the white dwarf surface. The authors show that spin data for these systems suggest magnetic moments $\mu \sim 10^{33} - 10^{34}$G cm^{-3} for the white dwarfs. The picture provides natural explanations for the orbital effects found in hard X–ray observations of EX Hya and AO Psc (=H2252–035).

117.015 **A radio outburst and jet from the symbiotic star CH Cyg.**
A. R. Taylor, E. R. Seaquist, J. A. Mattei.
Nature, Vol. 319, No. 6048, p. 38 – 41 (1986).
As part of a monitoring programme on symbiotic stars, observations with the VLA have been made on CH Cyg over the past year at intervals of a few months. During the period April 1984 to May 1985 this monitoring revealed a strong radio outburst coincident with the production of a multi–component jet. The jet is expanding at a rate of 1.1 arc s yr^{-1}. The onset of the radio outburst coincides in time with a remarkable decline in the visual brightness of the star. The luminosity of the star during the optically active state and the energetics of the mass loss suggest that the outburst is the result of accretion at, or near, the Eddington limit onto the surface of a white dwarf.

117.016 **A new spectroscopic orbit of the W Ursae Majoris system U Pegasi.**
W. Lu.
Publ. Astron. Soc. Pac., Vol. 97, No. 597, p. 1086 – 1091 (1985).
High–dispersion spectra have been obtained for the W UMa system U Peg. The orbital elements for both components of the system have been determined to be: $T_0 = 2445956.983$, $V_0 = -28.5$ km s^{-1}, $K_1 = 77.6$ km s^{-1}, $K_2 = 246.5$ km s^{-1}. A circular orbit is adopted. From the semiamplitudes of the two components, an accurate mass ratio $m_1/m_2 = 3.18$ is deduced. Spectrophotometric measurements were also carried out from which the luminosity ratio (determined from line–strength ratios) is deduced to be 1.47, corresponding to a difference of $0^m\!.42$ in the blue region of the spectrum. Both stars are found to be on the main sequence; their spectral types are each close to G2 with the secondary a little earlier.

117.017 **Optical studies of the X–ray transient EXO 0748–676.**
R. A. Wade, H. Quintana, K. Horne, T. R. Marsh.
Publ. Astron. Soc. Pac., Vol. 97, No. 597, p. 1092 – 1095 (1985).
The authors provide optical identification, positional information, photometry, and spectroscopy for the recently discovered X–ray transient EXO 0747–676.

117.018 **Brackett–gamma–line profile of MWC 349.**
M. Tanaka, T. Yamashita, S. Sato, M. Nishida, N. Ukita, H. Okuda.
Publ. Astron. Soc. Pac., Vol. 97, No. 597, p. 1115 – 1118 (1985).
The Bγ line emission of MWC 349 was observed in May 1984 with a velocity resolution of 25 km s^{-1}. The spectrum shows two distinct emission components. The main component has a FWHM of 115 km s^{-1}at $V_{LSR} = 5$ km s^{-1}. The observed line profile is well reproduced with a model in which the Bγ emission arises from a stellar wind with a constant velocity of 55 km s^{-1}. However, the profile shows a slight asymmetry. Comparison of the observed profile with theoretical ones suggests a blue–shifted absorption dip which might be produced in an outer decelerated region or in an expanding disk.

117.019 **Classical Algol systems.**
E. Budding.
Astrophys. Space Sci., Vol. 118, Nos. 1/2, p. 241 – 255 (1986). – See Abstr. 012.001.
Progress in the study of Algol–type binaries is looked at both in the light of more general efforts at compilation and categorization, and also some specific new and important observational results. Results can be referred to the well known underlying model of the semi–detached configuration and Roche–lobe overflow mechanism. Successes and limitations of this model are considered. Lines of future research are suggested.

117.020 **Massive X–ray binaries: their physics and evolution.**
W. Sutantyo.
Astrophys. Space Sci., Vol. 118, Nos. 1/2, p. 257 – 270 (1986). – See Abstr. 012.001.
The author reviews various aspects of the evolutionary history of massive X–ray binaries. It is expected that moderately massive close binaries evolve to Be X–ray binaries, while very massive systems evolve to standard X–ray binaries. The compact objects are formed through supernova explosions. A way to represent the rotational and magnetic evolution of neutron stars in close binary systems is presented. The evolutionary relationship between massive X–ray binaries, binary pulsars, and millisecond pulsars is also discussed.

117.021 **Rapid oscillations in cataclysmic variables.**
B. Warner.
Astrophys. Space Sci., Vol. 118, Nos. 1/2, p. 271 – 277 (1986). – See Abstr. 012.001.
Rapid coherent and quasi–periodic oscillations in the optical and X–ray fluxes of cataclysmic variable stars are described.

117.022 **Short–period noncontact binaries.**
A. Yamasaki, A. Okazaki, M. Kitamura.
Astrophys. Space Sci., Vol. 118, Nos. 1/2, p. 279 – 282 (1986). – See Abstr. 012.001.

Five short–period noncontact binaries (GR Tau, UU Lyn, BV Eri, RU Eri, and BW Eri) are studied. It is found from photometric and spectroscopic observations that (1) both components are noncontact to each other but are very close to the respective Roche lobes; (2) period change does not appear or it is very small; and (3) component stars of UU Lyn and RU Eri are normal main sequence stars while those of GR Tau and BV Eri are smaller and less massive components indicating in an advanced evolutionary state having experienced mass reversal.

117.023 **The electron temperature of AM Herculis type systems.**
D. T. Wickramasinghe, S. Meggitt.
Astrophys. Space Sci., Vol. 118, Nos. 1/2, p. 287 – 289 (1986). – See Abstr. 012.001.

The authors argue that the polarised optical emission from AM Herculis–type systems arises from a compact high temperature ($T \gtrsim 10$ keV) region closely associated with the accretion shock.

117.024 **Photometry, polarimetry, and spectroscopy of AM Herculis variables.**
I. R. Tuohy, N. Visvanathan, D. T. Wickramasinghe.
Astrophys. Space Sci., Vol. 118, Nos. 1/2, p. 291 – 294 (1986). – See Abstr. 012.001.

The authors report AAT observations of two southern AM Herculis variables, H0139–68 and E1405–451. H0139–68 was found to be in an unusually faint state ($V \gtrsim 17$), and this has permitted the first measurement of the magnetic field of the white dwarf, $B_p \sim 3 \times 10^7$G. Linear polarimetry of E1405–451 has revealed a substantial position angle variation with phase, corresponding to an orbital inclination of $\sim 60°$ and a colatitude of $\sim 20°$ for the magnetic pole of the white dwarf.

117.025 **Line profiles in symbiotic stars.**
B. L. Webster, L. H. Taaffe, A. J. Kalnajs.
Astrophys. Space Sci., Vol. 118, Nos. 1/2, p. 295 – 298 (1986). – See Abstr. 012.001.

High resolution profiles of permitted and forbidden lines in symbiotic stars are described, with particular reference to the broad wings. An interpretation of the data in terms of an optically thin accretion disc around a hot subdwarf is presented.

117.026 **Light curves of V711 Tauri (HR 1099).**
I.–S. Nha, J. Y. Oh.
Astrophys. Space Sci., Vol. 118, Nos. 1/2, p. 299 – 303 (1986). – See Abstr. 012.001.

V and *B* observations of V711 Tauri (HR 1099) for the year 1982 – 1983 are presented. The light curves made with the data of this season are compared with those of the previous observations of the 1981 – 1982 season. The single–peaked maximum light had shifted slightly and the peak amplitude had decreased to 0.04 mag in *V*. No clear sign of colour, *B–V*, variation in comparison with the low temperature dark–spot is detected.

117.027 **Tenma observation of the X–ray pulsar 4U1626–67.**
T. Kii, S. Hayakawa, F. Nagase.
Astrophys. Space Sci., Vol. 118, Nos. 1/2, p. 375 – 377 (1986). – See Abstr. 012.001.

Energy–dependent pulse profiles from the X–ray pulsar 4U1626–67, observed with Tenma in May 1983, were studied using an anisotropic radiation transfer model in an accretion column on a strongly magnetized neutron star. The result indicates that the strength of the magnetic field is about 8×10^{12}G at the surface of the neutron star.

117.028 **X–ray emissions from Vela X–1 during its eclipsing period.**
N. Sato, S. Hayakawa, F. Nagase.
Astrophys. Space Sci., Vol. 119, No. 1, p. 81 – 84 (1986). – See Abstr. 012.001.

The X–ray pulsar Vela X–1 was observed with TENMA throughout the entire period of eclipse in March 1984. The energy spectra at several phase angles showed the iron emission line at about 6.4 keV throughout the eclipse of Vela X–1 and its intensity at the mid–eclipse to be about 10% of that in the non–eclipsing phase, whereas the continuum intensity was about 1%. The result provides means of studying the contribution of the stellar wind to the fluorescence line and the structure of the stellar atmosphere.

117.029 **Electrodynamical synchronization of AM Herculis–type stars.**
O. Kaburaki.
Astrophys. Space Sci., Vol. 119, No. 1, p. 85 – 88 (1986). – See Abstr. 012.001.

A new mechanism is proposed for the explanation of synchronous rotations in AM Herculis–type binaries. In contrast to the resistive mechanism of Joss et al. (1979), the author's model predicts that an asynchronous rotation, if it is present, damps exponentially and the damping time is inversely proportional to the electrical conductivity. It is shown by using the parameters which fit for AM Herculis that a typical damping time is very short compared with the lifetime of the system.

117.030 **The system of DO Cassiopeiae: photoelectric observations and light curves analysis.**
H. Rovithis–Livaniou, P. Rovithis.
Astrophys. Space Sci., Vol. 119, No. 2, p. 381 – 393 (1986).

The new *B* and *V* photoelectric observations of DO Cas, obtained during 1979 and 1980, are presented, analysed, and discussed. The observations were made at the National Observatory of Athens, Greece, and their analysis was based on the frequency domain techniques. New geometric, photometric, and absolute elements for the system are given.

117.031 **R Aquarii, une variable unique: Mira, symbiotique, algolide et nova récurrente.**
E. Schweitzer.
Bull. Assoc. Fr. Obs. Etoiles Variables, No. 35, p. 2 – 8 (1986).

117.032 **An elliptical binary orbit model of GX 1+4.**
E. P. Cutler, B. R. Dennis, J. F. Dolan.
Astrophys. J., Vol. 300, No. 2, p. 551 – 556 (1986).

The X–ray source GX 1+4 (4U 1728–24) was observed by the high–energy X–ray spectrometer on OSO 8 for 5 days from 1978 September 20 to 24. X–rays in the energy range 16 – 72 keV were analyzed. The pulsar rotation period at 1978 September 21.0 (JD 2,443,772.5) is measured to be 228.508 ± 0.011 s, with a value of $\dot{P}/P = -0.035 \pm 0.007$ yr^{-1}. When these results are combined with the measurements of other observers, evidence is obtained for the rotation period of GX 1+4 being approximately 220 s rather than 110 s, and for an orbital period of GX 1+4 about a companion star of approximately 304 days.

117.033 **The spectrum of SS Cygni during a dwarf nova eruption.**
F. V. Hessman.
Astrophys. J., Vol. 300, No. 2, p. 794 – 803 (1986).

Analyses of spectroscopic observations made during the declining phase of a short, symmetric eruption of SS Cygni in 1981 are presented. The profiles of the broad Hβ and Hγ absorption lines from the optically thick disk remained constant throughout much of the decline. This behavior is difficult to reconcile with any of the currently accepted theoretical models for dwarf novae eruptions. The equivalent width of the He II λ4686 emission is roughly constant during the decline. A significant fraction of the increased λ4686 emission comes from a disk "hot spot", implying that the outburst is accompanied by an increased mass flux from the companion star.

117.034 **The optical continua of magnetic variables.**
G. D. Schmidt, H. S. Stockman, S. A. Grandi.
Astrophys. J., Vol. 300, No. 2, p. 804 – 818 (1986).

Optical circular spectropolarimetry and spectroscopy with full phase coverage are presented for four well–known AM Herculis magnetic variables. The results yield new estimates of surface field strength on the primaries of two systems: 36 ± 6 MG on PG 1550+191 and ~ 35 MG on AN UMa (MG = 10^6G). In addition, a polar field of 32 MG on VV Pup is confirmed through the detection of cyclotron harmonics in polarization from the "backside" magnetic pole. The authors propose a refinement to the "standard" accretion model for magnetic variables. The new picture suggests an accretion profile which is not perfectly collimated, where the optical/IR continuum can be dominated by a low-density halo around the compact hard X-ray shock. The role of stellar winds from the secondaries in these systems is also discussed.

117.035 **EXOSAT observations of the 35 day cycle of Hercules X-1: evidence for neutron star precession.**
J. Trümper, P. Kahabka, H. Ögelman, W. Pietsch, W. Voges.
Astrophys. J., Lett. Ed., Vol. 300, No. 2, p. L63 – L67 (1986).
The authors have observed Her X-1 with EXOSAT throughout a 35 day cycle at about 4 day sampling intervals. These measurements include for the first time an accurate determination of 1.24 s pulse profiles during the short-on state at the middle of the 35 day cycle. The data show a large change in the pulse profiles between the main and short-on states, suggesting a precession of the magnetic dipole axis of the rotating neutron star. It is proposed that the clock mechanism resides in the free precession of the neutron star.

117.036 **Periodic optical line intensity variations in X Persei.**
K. M. V. Apparao, S. P. Tarafdar.
Astron. Astrophys., Vol. 155, No. 2, p. 423 – 424 (1986).
A model in which a neutron star orbits around the Be star, immersed in a ring of material, is used to calculate the optical line intensities emitted by the ionized region produced by the absorption of X-rays from the neutron star. About 2×10^{31} erg s^{-1} of H_α emission is found to be produced by this process and varies with the X-ray period as is observed. Most of the He II λ4686 line is produced by the ionized region and varies with the X-ray period, but constitutes about 10^{-4} of the continuum in this region.

117.037 **Spiral shocks on a Roche lobe overflow in a semi-detached binary system.**
K. Sawada, T. Matsuda, I. Hachisu.
Mon. Not. R. Astron. Soc., Vol. 219, No. 1, p. 75 – 88 (1986).
Two-dimensional hydrodynamic calculations of a Roche lobe overflow in a semi-detached binary system with a mass ratio of unity are carried out. The region of the computation covers both a mass-losing star filling its critical Roche lobe and a mass-accreting compact star. Gas ejected from the mass-losing star with specified energy flows through the L1 point to form an elephant trunk and an accretion ring. It is found that spiral-shaped shocks are formed on the accretion ring. It is suggested that the gas in the accretion ring loses angular momentum at the shocks and spirals in towards the compact star even without viscosity.

117.038 **CH Cygni – one year after the drop in brightness.**
M. Mikołajewski, T. Tomov.
Mon. Not. R. Astron. Soc., Vol. 219, No. 1, p. 13P – 18P (1986).
The authors present moderately high-resolution spectroscopic observations in the blue spectral region of CH Cygni, following the sudden drop in brightness in 1984 July. On the basis of these data they have distinguished five periods of the spectral evolution. The star has finally reached a nebular phase and forbidden lines of [O III], [Ne III], [S II] and [Fe II] are the strongest in the spectrum.

117.039 **The minimum orbital period for ultra-compact binaries with helium burning secondaries.**
G. J. Savonije, M. de Kool, E. P. J. Van Den Heuvel.
Astron. Astrophys., Vol. 155, No. 1, p. 51 – 57 (1986).
The authors present evolutionary calculations of an ultracompact binary system undergoing mass transfer, consisting initially of a 0.6 $M_\odot$ nondegenerate helium-star accompanied by a 1.3 $M_\odot$ compact star in a tight orbit of 37 minutes. The binary parameters were chosen to model the ultra-compact binary 1E 2259+586 which is currently thought to have an orbital period of 38 min.

117.040 **Simultaneous optical and X-ray observations of a flare on BY Draconis.**
C. de Jager, J. Heise, S. Avgoloupis, G. Cutispoto, K. Kjeboom, R. B. Herr, M. Landini, A. F. Langerwerff, L. N. Mavridis, A. S. Melkonian, R. Molenaar, B. C. Monsignori-Fossi, H. L. Nations, R. Pallavicini, V. Piirola, M. Rodonò, M. A. Seeds, G. H. J. van den Oord, O. Vilhu, C. Waelkens.
Astron. Astrophys., Vol. 156, No. 1/2, p. 95 – 100 (1986).
The authors present a first report on a campaign for simultaneous observations in the visual, radio and X-ray ranges of stellar flares on the binary BY Dra. During two nights of observations they observed one significant flare, simultaneously in soft X-rays and visible wavelengths, and one or two smaller bursts. The main flare impulsively heated an area of $\lesssim 2 \times 10^7 km^2$ of the stellar photosphere to a brightness temperature $\gtrsim$25000K, during about 5 min, and generated hot plasma, emitting soft X-rays. This emission reached maximum after the impulsive burst and lasted for about an hour; it covered a much larger area on the star than the optical emission. There was no significant hard X-ray component. The authors ascribe the optical continuous emission to the flare's impulsive phase, and the soft X-ray emission to the gradual phase.

117.041 **Differential rotation and activity cycles in RS CVn binaries. III. UV Piscium and UX Arietis.**
M. Busso, F. Scaltriti, A. Cellino.
Astron. Astrophys., Vol. 156, No. 1/2, p. 106 – 110 (1986).
New photoelectric observations of the spotted binaries UX Arietis and UV Piscium are presented and studied on the basis of a previously established technique for analyzing unevenly sampled light curves. The new data, when combined with already published observations, allows to reconstruct the recent evolution of the spot activity in the system studied by means of the method described in Busso et al. (1985). The results confirm the correlations between the differential rotation, the orbital period, and the duration of spot cycles.

117.042 **The accretion of supergiant wind material onto the hot companion in Zeta Aurigae binaries.**
A. Che-Bohnenstengel, D. Reimers.
Astron. Astrophys., Vol. 156, No. 1/2, p. 172 – 180 (1986).
This paper is an attempt to interpret the behaviour of the high excitation C IV and Si IV resonance lines in the course of a period of the Zeta Aurigae binary systems. Three distinct phenomena have been observed: shock front emission, accretion wake absorption and accretion disc emission. At the C IV and Si IV emission regions the electron temperature lies between 50000 and 80000K. The shock front electron density is of the order of 10^8 to $10^9 cm^{-3}$, while N_e is probably of the order of $10^{10} cm^{-3}$ at the rotating disc where part of the wind material has been captured by the accreting star. The authors deal with an accretion shock where the shock cone angle is large ($\beta \approx 25 - 45°$) even when the Mach number is very large, because the accreting material is heated by its passage through the shock front. The gas inside the cone is hot enough for the existence of C^{3+} and Si^{3+} ions. The density in the accretion wake is 10^7 to $10^8 cm^{-3}$ – more than an order of magnitude denser than in the surrounding stellar wind.

117.043 **Period variations of binary systems as a possible source of information about motions in the stellar core.**
F. van't Veer.
Astron. Astrophys., Vol. 156, No. 1/2, p. 181 – 185 (1986).
The author discusses the size of the period variations currently observed for the W UMa and RS CVn type eclipsing binaries. From a theoretical discussion, taking into account the different modes of mass loss and angular momentum loss, he concludes that it is impossible to believe these mechanisms responsible for all the period variations observed. On the basis of recent results of solar research he proposes an alternative mechanism using matter flows in the radiative core of these synchronously rotating binary components.

117.044 **The symbiotic star CH Cygni: end of active phase or eclipse phenomenon?**
M. Mikołajewski, R. Biernikowicz.
Astron. Astrophys., Vol. 156, No. 1/2, p. 186 – 190 (1986).
The spectroscopic behaviour of CH Cygni during the 1977 – 84 outburst is described. Hβ profile variations in the period 1982 – 84 are presented and discussed. Sudden changes in the radial velocities and intensities of Hβ profile components as well as a drop of the *UBV* brightness in August – December 1984 may be interpreted as the eclipse ingress of the Balmer emission region by the M6 giant. This event coincides with the epoch of eclipse predicted from the spectroscopic ephemeris given by Yamashita and Maehara (1979).

117.045 **The soft X–ray superoutburst of VW Hydri.**
H. van der Woerd, J. Heise, F. Bateson.
Astron. Astrophys., Vol. 156, No. 1/2, p. 252 – 260 (1986).
The dwarf nova VW Hyi has been monitored with EXOSAT, during and shortly after the optical superoutburst of November 1983. VW Hyi appears to be a very soft X–ray emitter, and is marginally detected in the hard (1 to 6 keV) X–ray range. The soft X–ray emission increased by two orders of magnitude during outburst, with the dominant emission at wavelengths longer than 170 Å. VW Hyi is now the third dwarf–nova system, in which a strong soft X–ray component is detected during outburst. An increase of the soft X–ray flux was observed, with a delay of 2.5 days with respect to the onset of the optical outburst. This delay is larger than the typical delay of ≈1 day between the UV/soft X–ray outburst and optical outburst, observed at other dwarf–nova outbursts. The soft X–ray outburst ended 1.5 to 2.5 days before the optical had returned to its quiescent level. This is consistent with other simultaneous observations.

117.046 **Multiple optical orbital sidebands in intermediate polars.**
B. Warner.
Mon. Not. R. Astron. Soc., Vol. 219, No. 2, p. 347 – 356 (1986).
Amplitudes and phases of the Fourier components of the light curve of the standard model of an intermediate polar are evaluated. For a white dwarf angular rotation frequency ω and orbital frequency Ω it is shown that as well as the well–known components with frequencies ω and ω–Ω there can exist sidebands at ω+Ω and ω–2Ω. The intermittent presence of the ω–2Ω component in V1223 Sgr and FO Aqr (H2215–035) provides an explanation of results for these stars which have previously been dismissed as spurious. Analyses of the stars V1223 Sgr, FO Aqr and AO Psc show that the observed orbital modulations of their brightnesses arise predominantly from the same source as in other cataclysmic variables, viz, a hotspot on the accretion disc, and not from reflection effect from the secondary star.

117.047 **The attainment of synchronism in the presence of accretion in the AM Herculis binaries.**
C. G. Campbell.
Mon. Not. R. Astron. Soc., Vol. 219, No. 3, p. 589 – 596 (1986).
Conditions are derived for which the white dwarf component in an AM Herculis binary can attain synchronism with the orbit in the presence of accretion. A stable locking mechanism must exist and, for the case of an initially over–synchronous white dwarf, there is a maximum value of the magnetic diffusivity of the secondary, η_0, above which exact corotation cannot be attained. In the absence of a locking mechanism the primary will reach a slightly asynchronous state with a typical synodic period of a few tens of years. If η is significantly less than η_0 the instantaneous angular velocity of the primary in this state will vary about a central value.

117.048 **The evolution of polar caps in magnetic cataclysmic variables.**
J. Frank, G. Chanmugam.
Mon. Not. R. Astron. Soc., Vol. 219, No. 3, p. 597 – 602 (1986).
A simple analysis of the evolution of the size of the magnetic polar cap in accreting white dwarfs is made on the basis of current theories of the secular evolution of magnetic cataclysmic variables. For white dwarfs with dipolar fields it is shown that the size of the polar cap in DQ Her binaries is larger than in AM Her binaries. The size of the former is, however, smaller than deduced from interpretation of their X–ray light curves, while that of the latter is in rough agreement. If the white dwarf contains an aligned magnetic quadrupole the size of the polar caps of the DQ Her binaries is significantly increased. Magnetic field decay of the quadrupole moment in the older AM Her binaries implies that their fields are predominantly dipolar.

117.049 **High–speed photometry of the dwarf nova Z Cha in quiescence.**
J. Wood, K. Horne, G. Berriman, R. Wade, D. O'Donoghue, B. Warner.
Mon. Not. R. Astron. Soc., Vol. 219, No. 3, p. 629 – 655 (1986).
High–speed white light photometry of the cataclysmic variable star Z Cha is used to study the structure of the system and to derive the properties of its components.

117.050 **Semidetached systems: evolutionary viewpoints and observational constraints.**
J. P. De Greve.
Space Sci. Rev., Vol. 43, Nos. 1/2, p. 139 – 210 (1986).
The present knowledge on the evolution of semidetached systems is reviewed. Characteristics of observed systems are discussed and general properties tested by the behaviour of theoretical models. New models of mass accreting companion stars are computed. The results are compared with observed systems with masses of the gainers located within the theoretical range. Up to now no computations exist for the evolution of medium mass close binaries including overshooting of the convective core. However some of the influences of extended convective mixing on the interaction of close binaries are investigated. A larger probability for the occurrence of case A of mass exchange and a larger remnant mass at the end of the process are the most important results. Finally the investigation into the origin of individual systems (in mass, mass ratio and period) is discussed.

117.051 **Spot cycles in the RS CVn variable AY Cet.**
E. Poretti, L. Mantegazza, D. S. Hall, M. B. K. Sarma, B. D. Ausekar, B. V. N. S. Prakasa Rao, L. J. Boyd, G. W. Henry, H. J. Landis, R. E. Fried, J. L. Hopkins, P. Nielsen, H. Louth, T. R. Renner, P. A. Brooks, R. Miles, H. J. Stelzer, L. Pazzi.
Astron. Astrophys., Vol. 157, No. 1, p. 1 – 5 (1986).
New UBV photometry of 39 Cet ≡ AY Cet from 1981 – 1982, 1982 – 83, and 1983 – 84 is presented and analyzed along with previously published data from 1971, 1972, 1980 – 81, and 1981 – 82. The V data between 1980 and 1984 can be represented as a combination of three sine curves with the following semi–amplitudes and periods: 0.06 mag at P_1 = 77.22 d, 0.03 mag at P_2 = 79.36 d, and 0.07 mag at P_3 = 1820 d. The complex photometric behaviour is discussed in terms of starspots, chromospheric activity, and a possible activity cycle. The determination of P_1, which the authors interpret as the rotation period of the G5 III spectroscopy primary, confirms that AY Cet is a second example of clearly asynchronous rotation in an RS CVn binary.

117.052 **Optical and X–ray observations of 4U2129+47/V1727 Cyg in an extended quiescent state.**
W. Pietsch, H. Steinle, M. Gottwald, U. Graser.
Astron. Astrophys., Vol. 157, No. 1, p. 23 – 28 (1986).
Observations of the $5^h\!.2$ h X–ray binary 4U2129+47/V1727 Cyg for more than one orbital cycle on September 29 and October 4, 1983 with the ESA X–ray satellite EXOSAT failed to detect the source in the low and medium X–ray energy band. Quasi–simultaneous UBV photometry using the 2.2 m telescope at the German–Spanish Astronomical Center, Calar Alto, did not show the large amplitude light curve reported in earlier observations but showed the optical companion in a low intensity state. The large amplitude light curve has been interpreted as due to X–ray heating of the optical star by the X–ray source. The optical observations indicate that the heating X–ray source has been shut off.

117.053 **The activity of CH Cygni in the years 1981 – 82.**
A. Skopal.
Bull. Astron. Inst. Czech., Vol. 37, No. 1, p. 18 – 26 (1986).
Eight high–dispersion spectrograms (8.5 Å/mm) from 1981 and 1982 were used to measure the radial velocities and for qualitative analysis of spectral line profiles. The development of the absorption components of the spectral lines was observed during the whole observation period and, in the summer of 1981, also rapid changes of the red–wing profiles of the absorption components, which lead to the separation of metals into two components. In 1982, an increase in the intensity of emission lines, of the blue continuum and of the velocity gradient of the absorption components of Balmer's series was observed. For the purpose of explaining the phenomena observed, CH Cygni was assumed to be of binary substance.

117.054 **A self–consistent treatment of the mechanical structure of contact binaries.**
H. Kähler.
Astron. Astrophys., Vol. 157, No. 2, p. 329 – 334 (1986).
A self–consistent treatment of the mechanical structure of contact binaries is achieved within the framework of two assumptions: (1) The system is in uniform rotation. (2) The components are spherical stars. The method is shown to be determined uniquely by these assumptions. It is also shown to be virtually equivalent to non–spherical techniques. Centrifugal forces are allowed for by introducing effective gravities and potentials for the two components. They are given as simple analytical expressions.

117.055 **Generalized mass–radius relations and mass–entropy relations for components of contact binaries.**
B. Matraka.
Astron. Astrophys., Vol. 157, No. 2, p. 341 – 345 (1986).
The relations between mass and radius and between mass and entropy in the outer convective zone are presented for homogeneous stars which are subject to an additional luminosity transfer (positive or negative). This luminosity transfer is supposed to take place in the adiabatic part of the outer convective zone. Such models can simulate the components of contact binaries. For a wide range of coupling conditions between the components, contact binary models in thermal equilibrium can be constructed on the basis of these results and the influence of the various parameters can be discussed. This will be demonstrated by examples which concern in particular the dependence of the angular momentum on the mass ratio and on the coefficients in the transport equation.

117.056 **Detection of 10^{15}eV gamma rays from Cygnus X–3.**
T. Kifune, K. Nishijima, T. Hara, Y. Hatano, N. Hayashida, M. Honda, K. Kamata, Y. Matsubara, M. Mori, M. Nagano, G. Tanahashi, M. Teshima.
Astrophys. J., Vol. 301, No. 1, p. 230 – 234 (1986).
Muon–poor air showers received at Akeno have been analyzed to search for γ–rays from Cyg X–3, based on its 4.8 hr binary period. A concentration of events is seen near phase 0.6, the time of the source X–ray maximum. The probability that a uniform background can create the observed distribution is 0.2%. The time–averaged integral γ–ray flux is estimated as $(1.1 \pm 0.4) \times 10^{-14}$ $cm^{-2}s^{-1}$ for $E_0 \geqslant 10^{15}$eV and $(8.8 \pm 5.0) \times 10^{-14}$ $cm^{-2}s^{-1}$ for $E_0 \geqslant 6 \times 10^{14}$eV.

117.057 **High–energy neutrinos from Cygnus X–3.**
V. S. Berezinsky (*V. S. Berezinskij*), C. Castagnoli, P. Galeotti.
Astrophys. J., Vol. 301, No. 1, p. 235 – 239 (1986).
The high–energy neutrino flux from Cygnus X–3 is calculated on the basis of the binary model of the source. The possibility of observing such a flux through an excess of muons in small underground detectors is discussed. Constraints on the cosmic–ray luminosity of Cygnus X–3 are derived by comparing the expected neutrino flux with the gamma–ray observations of this source.

117.058 **Discovery of a new short–period, eclipsing cataclysmic variable.**
R. A. Downes, M. Mateo, P. Szkody, D. C. Jenner, B. Margon.
Astrophys. J., Vol. 301, No. 1, p. 240 – 251 (1986).
The authors report photometry and spectroscopy of a newly recognized 14th mag eclipsing cataclysmic variable, KPD 1911+1212 (=SVS 8130, V1315 Aql). The system exhibits deep (1.7 mag) eclipses with period 0.1397 day. The spectrum is that of a high–excitation old nova and shows dramatic variability of the emission line strengths through the eclipse. There is no direct evidence for the secondary. The authors present a model for the system consistent with current data, which implies a mass for the primary and secondary stars of 0.9 and 0.4 $M_\odot$ respectively and an inclination of $i = 78°$.

117.059 **WZ Sagittae: time–resolved spectroscopy during quiescence.**
R. L. Gilliland, E. Kemper, N. Suntzeff.
Astrophys. J., Vol. 301, No. 1, p. 252 – 261 (1986).
High time resolution spectroscopy has been used to obtain an orbital velocity amplitude for the primary component of the dwarf nova WZ Sagittae, an 82 minute binary, of $K_1 = 49 \pm 5$ km s^{-1}. The higher orbital velocity yields standard solutions for the system masses of $M_1 \approx 1.1\ M_\odot$, $M_2 \approx 0.1\ M_\odot$. Implications of the new K_1 value and observed line profile variations for the physical model of WZ Sge are discussed. Relative phasing of the velocity and photometric eclipse does not agree with the canonical hot spot model.

117.060 **Z Camelopardalis: outburst P Cygni profiles and quiescent continuum.**
P. Szkody, M. Mateo.
Astrophys. J., Vol. 301, No. 1, p. 286 – 296 (1986).
The authors present the first orbital study of the C IV P Cygni profile in the dwarf nova Z Cam, a system of intermediate inclination. A modulation of the absorption equivalent width is apparent, with phases suggesting an effect from the mass–transfer stream. Over the outburst cycle, the strength of the P Cygni absorption and emission components is greater after outburst than during the standstill configuration, while the terminal velocities and central absorption wavelength are similar. The continuum flux distributions during standstill and decline are similar, consistent with an $\alpha = -2.0$ power law, while the quiescent–state UV flux distribution can be separated into two components. The best fits are a 14,000K blackbody and an $\alpha = -3.6$ power law, conceivably representing contributions from the hot spot and the inner disk/boundary layer.

117.061 ***EXOSAT* results on V426 Ophiuchi: indications of a DQ Herculis system.**
P. Szkody.
Astrophys. J., Lett. Ed., Vol. 301, No. 1, p. L29 – L33 (1986).
EXOSAT data reveal a sinusoidal 1 hr modulation during the 6 hr binary orbit of the cataclysmic variable V426 Oph. The amplitude and shape of this modulation, the X–ray fluxes and spectrum, as well as the time scales involved argue for a DQ Her interpretation. In this case, the white dwarf in V426 Oph has the longest rotation period among the known DQ Her systems with orbital periods greater than 3 hr.

117.062 **Calculation of linear polarization of radiation of the tidally distorted component in the X–ray binary system Cyg X–1 ≡ V1357 Cyg.**
N. G. Bochkarev, E. A. Karitskaya, V. M. Loskutov, V. V. Sokolov.
Astron. Zh., Tom 63, Vyp. 1, p. 71 – 77 (1986). In Russian. English translation in Sov. Astron., Vol. 30, No. 1.
For the system Cyg X–1 ≡ V1357 Cyg the linear polarization variation caused by tidal distortion and possible shallow partial eclipse of the optical star has been calculated. The distributions have been taken into account of the sources of true radiation and of the single–scattering albedo with the optical depth for the model atmosphere with $T_{eff} = 32900$K, $\lg g = 3.1$. For a number of wavelengths the angular distributions of intensity and polarization of radiation from unit area of the star surface have

been obtained. The change of the expected variations of polarization with wavelength is investigated.

117.063 **Variability of absolute intensities of emission lines in the spectrum of SS433.**
S. S. Asadullaev, A. M. Cherepashchuk.
Astron. Zh., Tom 63, Vyp. 1, p. 94 – 112 (1986). In Russian. English translation in Sov. Astron., Vol. 30, No. 1.
Using the spectroscopic data of Vittone et al. (1983) and Margon et al. (1984) the absolute intensities of the stationary and moving emission lines in the spectrum of SS433 are determined in different phases of orbital ($P = 13^{d}.086$) and precessional ($P = 164^{d}$) periods, taking into account the strong interstellar reddening $A_v = 7^{m}.85$ and the variability of the continuum of this close binary system. It is found that the intensities of the stationary H_α, H_β, He I emission lines vary regularly with phases of the orbital and precessional periods. It proves directly that the main part of intensities of the stationary emission lines is formed in the supercritical "precessing" accretion disk surrounding the relativistic object.

117.064 **Four–colour photometry of HZ Herculis in 1982 – 1984. Optical variability of the system during the period of anomalous X–ray behaviour.**
A. V. Mironov, V. G. Moshkalev, E. M. Trunkovskij, A. M. Cherepashchuk.
Astron. Zh., Tom 63, Vyp. 1, p. 113 – 122 (1986). In Russian. English translation in Sov. Astron., Vol. 30, No. 1.
Photoelectric *WBVR* observations of HZ Her carried out in February – April 1982, September – October 1983 and February – March 1984 have shown that there has been no significant change of the amplitude of the "reflection effect" in the system, despite of the anomalous prolonged "turn–off" state of the X–ray source in 1983 and the following "turn–on" of the source in 1984.

117.065 **Photometric investigation of the eclipsing cataclysmic variable AC Cancri.**
A. V. Bajdak, S. Yu. Shugarov.
Astron. Zh., Tom 63, Vyp. 1, p. 123 – 127 (1986). In Russian. English translation in Sov. Astron., Vol. 30, No. 1.
Photoelectric *UBVR* observations and photographic *BVR* observations of the variable AC Cnc have been carried out. The analysis of the observations showed that the system AC Cnc consists of a cold K4 V star and a white dwarf surrounded by an accretion disc. The distance between the components was estimated as 0.01 a.u.; the size, brightness and temperature of the disc were estimated to be 0.5 $R_\odot$, 1.3×10^{33}erg/s and 10^4K, respectively. A possible connection with the X–ray object H 0850+13 is discussed.

117.066 **Photometry of V1329 Cyg in 1980 – 1983.**
O. G. Taranova, B. F. Yudin.
Astron. Zh., Tom 63, Vyp. 1, p. 151 – 154 (1986). In Russian. English translation in Sov. Astron., Vol. 30, No. 1.
Results of *UBVRJHK* photometry of V1329 Cyg are discussed. The variations of the optical brightness agree with the period of 950.07 days. The infrared flux showed fluctuations of $\sim 0^{m}.7$ in the *J* band. The cool component in the V1329 Cyg system is an M6 giant. The 1.25–μm band flux density implies a distance for V1329 Cyg of ~9.7 kpc.

117.067 **The mass spectrum of the white dwarfs in cataclysmic binaries.**
H. Ritter, A. Burkert.
Astron. Astrophys., Vol. 158, No. 1/2, p. 161 – 173 (1986).
Using a simple model for the secular evolution of cataclysmic binaries the authors investigate observational selection in favour of massive white dwarfs. For this they consider a *V*–magnitude–limited sample of cataclysmic binaries with bright accretion disks, i.e. nova–like systems and dwarf novae in outburst. They find that the selection effect is strong enough to account fully for the high mean white dwarf mass observed and that therefore the intrinsic mass spectrum of the white dwarfs in cataclysmic binaries need not be significantly different from that of single white dwarfs. Furthermore they find the following trends: the selection in favour of massive white dwarfs becomes stronger with increasing orbital period. The main reason for the strong selection is found to be the steep increase of the white dwarfs' gravitational potential with mass. In addition, the authors find that only a small fraction of cataclysmic binaries is actually observed. If the mass spectrum of the white dwarfs is not too different from that of single white dwarfs, the intrinsic space density must be of order $1-2\times10^{-4}pc^{-3}$.

117.068 **The eclipsing dwarf nova OY Carinae. IV. Photometry of advanced stages of a superoutburst.**
R. Schoembs.
Astron. Astrophys., Vol. 158, No. 1/2, p. 233 – 246 (1986).
Photometric observations of a superoutburst of the SU UMa star OY Carinae are presented. Starting 6 days after maximum light, the observations continued until almost the faint quiescent state was reached. The secular, orbital and short time photometric behaviour is described. OY Car shows stronger cyclic variations of its mean brightness than low inclination SU UMa stars. The data suggest a location for the superhump source within the disk primary configuration. No secondary eclipse was detected. Several observational facts indicate the existence of irradiation induced variable mass transfer rate and an eccentric precessing disk. Dwarf nova oscillations were present during the rapid transition to quiescent brightness level. The oscillations are eclipsed. The eclipses of the oscillation are delayed with respect to the primary eclipse and suggest a reprocessing effect operating near the position of the quiescent bright spot.

117.069 **More about active and inactive states of HZ Her = Her X–1.**
R. Hudec, W. Wenzel.
Astron. Astrophys., Vol. 158, No. 1/2, p. 396 – 398 (1986).
Measurements on 1132 plates yield new details about the active and inactive states of Her X–1/HZ Her in the past. The transitions between *A* and *I* states proceed within several days. The orbital light curve for the inactive state is presented and discussed. Altogether 4*A* and 4*I* states were found between 1928 and 1979.

117.070 **Flare features of AR Lacertae.**
R. K. Srivastava.
Astrophys. Space Sci., Vol. 120, No. 2, p. 239 – 252 (1986).
BV optical flares of AR Lacertae, found in 1979 observations, are presented and their features are discussed. The effect of wave migration and the significance levels of flares originating around both eclipses have been estimated. Flare features suggest that both the components of AR Lac have spotted regions and show activity. The varying amplitudes of flares indicate varying degrees of activity. These results are well supported by observations in other wavelength regions.

117.071 **A photoelectric study of the W UMa–system U Pegasi.**
S. J. Lafta, J. F. Grainger.
Astrophys. Space Sci., Vol. 121, No. 1, p. 61 – 81 (1986).
The W UMa–system U Pegasi was observed photoelectrically in two wavebands (*B* and *V*) in 1984. The light curves of this system are analysed using Kopal's frequency–domain technique and the optimisation method. New geometrical and physical elements have been determined and the period–variations discussed.

117.072 **Further evidence for the binary nature of HR 3084.**
R. Haefner, H. Drechsel.
Astrophys. Space Sci., Vol. 121, No. 2, p. 205 – 212 (1986).
New photometric and spectroscopic observations yield further evidence for the binary nature of the early–type star HR 3084. Shape and small amplitude of the visual and IR light curves indicate that the light variations are essentially due to ellipsoidal effects rather than to β Cep variability.

117.073 **Nature of the symbiotic binary V1329 Cyg.**
D. Chochol, A. Vittone.
Astrophys. Space Sci., Vol. 121, No. 2, p. 225 – 240 (1986).
The published photometric and spectroscopic data of the symbiotic binary V1329 Cyg are interpreted. It is shown, that V1329 Cyg is an eclipsing binary with an elliptical orbit ($e = 0.28$). The cooler component fills up the Roche–lobe at periastron. A model of moving gaseous structures in the system is proposed and their influence on the radial velocity curve is shown. The following characteristics of the system are derived: the cooler component is an M6 giant with mass 7.9 $M_\odot$, radius 339 $R_\odot$ and luminosity $M_{bol} = -5.42$, the hot component is a white dwarf surrounded by an accretion disk. The mean distance between the components is 842 $R_\odot$ and in periastron it decreases to 605 $R_\odot$.

117.074 ***UBV* photometry of FK Comae.**
A. Cellino, F. Scaltriti, M. Busso.
Astrophys. Space Sci., Vol. 121, No. 2, p. 265 – 271 (1986).
UBV photometry of the peculiar star FK Comae, obtained during the period January – June 1983, is presented. Colour curves indicate that the star is redder at minimum light; this fact is consistent with both the models presented to explain the properties of FK Comae. A Fourier analysis of the light curves available in the literature has been performed; the phases of minimum and maximum light turn out to be noticeably stable in time. This fact is difficult to be reconciled with the hypothesis of dark spots that are observed to migrate in other active stars. Rather, it could be explained by the model of Walter and Basri (1982), implying an extremely evolved Algol–type system accreting mass on its primary component.

117.075 **Wave minimum of AR Lacertae.**
R. K. Srivastava.
Astrophys. Space Sci., Vol. 121, No. 2, p. 397 – 401 (1986).
U,B, and *V* observations of AR Lacertae, obtained in 1981, have been presented alongwith the colour indices. A distortion wave minimum is found to lie at $0^P\!.21$. The amplitude of the wave minimum in *B* filter is too weak, while it is stronger in *U* and *V* filters, the strongest being in *V* filter. The period of the migration wave turns out to be 2.53 yr.

117.076 **Choice of the model of Z Andromedae.**
B. F. Yudin.
Astron. Zh., Tom 63, Vyp. 1, p. 137 – 150 (1986). In Russian. English translation in Sov. Astron., Vol. 30, No. 1.
The analysis of the observations of the symbiotic star Z And has shown that at present these do not allow to make a definite choice for its model. If the hot component is an accreting white dwarf with a hydrogen–burning shell source, then the gas envelope must have $T_e \approx 2.6 \times 10^4$K. If the hot component is a main–sequence star with an accretion disk around it, then it is classified as a red dwarf.

117.077 **H–alpha variability in X–ray transient source X0331 + 53.**
M. Iye, K. Kodaira.
Publ. Astron. Soc. Pac., Vol. 97, No. 598, p. 1186 – 1188 (1985).
Spectroscopic observations of X0331 + 53 between October 1984 and March 1985 revealed that the equivalent width of Hα emission decreased by a factor of about four relative to that observed one year ago. This finding, in association with the decline of X–ray activity of the source reported by the EXOSAT group, is interpreted to support the binary model of an OB star and a pulsar in which the Hα emission originates in a direct vicinity of the massive primary star.

117.078 **The very low state of V794 Aquilae.**
R. K. Honeycutt, E. M. Schlegel.
Publ. Astron. Soc. Pac., Vol. 97, No. 598, p. 1189 – 1191 (1985).
The cataclysmic variable V794 Aql was found to have dropped from its normal magnitude range of B ≈ 14 – 17 to B ≈ 20. In this very low state the accretion–disk emission lines were replaced by very narrow Balmer emission. This behavior is very similar to that of MV Lyr and a few other systems in which the accretion apparently switches off for brief periods of time.

117.079 **Pulse–phased spectroscopy of AO Piscium (H2252–035) and GK Persei.**
J. B. Hutchings, T. J. Cote.
Publ. Astron. Soc. Pac., Vol. 98, No. 599, p. 104 – 109 (1986).
Spectroscopic observations are presented of these two slow X–ray pulsars. Spectra are accumulated in 3 or 4 phase bins per pulsation, to study spectroscopic variations. Emission–line variations are found with pulse phase in both systems, with flux amplitudes of ⩽50%, and velocity amplitudes of 50 – 100 km s^{-1}. Pulse phasing of the variations is similar in both systems. These variations are discussed in terms of X–ray reprocessing. Possible orbital parameters, including a refined period, are derived for AO Psc.

117.080 **Recent optical spectra of the symbiotic object R Aquarii.**
G. Wallerstein.
Publ. Astron. Soc. Pac., Vol. 98, No. 599, p. 118 – 122 (1986).
Spectroscopic observations of R Aqr and its nearby "jet" between 1980 and 1984 are presented. The significant difference between the velocity of the jet and the inner nebula associated with the star is confirmed. Spectra at high resolution show that many emission lines are double with one component near that of the star and one near that of the jet; however, forbidden lines favored by high density do not show the latter component. Profiles of the single [Ne III] lines from the inner nebula indicate an expansion velocity no greater than 12.5 km s^{-1} in 1984. An assembly of absorption–line velocities for the long–period variable obtained during the past 40 years is consistent with a 44–year period, and a velocity amplitude of 7 km s^{-1}. A highly eccentric orbit, which might explain the outbursts by a mass–transfer event at periastron, is not likely to be consistent with the absorption–line data.

117.081 **The symbiotic star BX Monocerotis.**
R. Viotti, A. Altamore, M. Ferrari–Toniolo,
M. Friedjung, P. Persi, C. Rossi, L. Rossi.
Astron. Astrophys., Vol. 159, No. 1/2, p. 16 – 21 (1986).
BX Mon is a peculiar star characterized by a composite spectrum, and very long period photometric variability. To investigate its nature, the authors have included BX Mon in an extensive program of ESO observations of emission line stars. In this paper they discuss new results based on coordinated optical (IDS) and infrared observations made in February 1983, on high resolution Hα spectrophotometry made with CAT–CES in February 1984, and on infrared photometry of March 1985.

117.082 **CH Cygni: the outburst of 1977 – 85.**
M. Hack, L. Rusconi, G. Sedmak, C. Aydin,
S. Engin, N. Yilmaz.
Astron. Astrophys., Vol. 159, No. 1/2, p. 117 – 128 (1986).
The visual and UV spectroscopic observations of CH Cyg in Nov. 84, Jan. and May 85 indicate that the outburst that started in 1977 has almost ended. The authors describe the characteristics of the spectrum and its variations during these eight years of activity. The RV measurements strongly suggest that CH Cyg is a binary system, with *P* of about 13 – 16 yr, formed of the M6 III primary and of a collapsed companion of about the same mass as the primary, surrounded by an accretion disk where the outbursts take place. Evidence of mass loss is given by the resonance lines of Ca II and Mg II, by the recently observed IR excess and by the radio sources associated with CH Cyg.

117.083 **YY Eri revisited.**
R. Nesci, C. Maceroni, L. Milano, G. Russo.
Astron. Astrophys., Vol. 159, No. 1/2, p. 142 – 146 (1986).
The RV Cass spectrograph at ESO has been used to observe the W UMa star YY Eri, in order to obtain the spectroscopic mass ratio of its components. A new technique has been successfully applied for the measure of radial velocities from intermediate dispersion spectra. The fit of the radial velocities has been done using a Roche lobe model of stellar surfaces which takes into account the distortion of the radial velocity curves due

to tidal effects. As a by-product of the procedure a new photometric solution of YY Eri has been obtained through a simultaneous treatment of B and V bands data.

117.084 **Contact binaries. I. A computational scheme, thermal stability, and linear series.**
H. Kähler, B. Matraka, A. Weigert.
Astron. Astrophys., Vol. 159, No. 1/2, p. 317 – 327 (1986).
An efficient scheme for calculating the structure, evolution and thermal stability of contact binaries is described. By means of a generalized Henyey method the system is calculated as a whole, including both components and their interaction. As representative applications, evolutionary calculations are presented for thermal relaxation oscillations and for a chemical evolution. The formulation of the thermal stability problem of contact binaries shows that the eigenvalues can also be calculated by means of this scheme making use of a contour-integral method. General properties of linear series of equilibrium systems are discussed, and the connection with the stability problem is shown.

117.085 **Spectrophotometric observations of V1016 Cyg in 1980 – 1985.**
A. P. Ipatov, B. F. Yudin.
Astrofizika, Tom 23, Vyp. 3, p. 527 – 537 (1985). In Russian. English translation in Astrophysics, Vol. 23, No. 3.
Spectrophotometric data were obtained for V1016 Cyg in the range of 0.33 to 0.75 µm during the period from 1980 – 1985. On the background of continuous increase of excitation of a gaseous shell emission spectrum in 1984 a compact, $r \gtrsim 10^{14}$cm, gaseous and dust formation was detected with high self-absorption in the Balmer lines. This new formation effectively screened hard L_c-radiation of the hot component and dissipated in general in the middle of 1985. The authors have also presented some physical parameters of V1016 Cyg estimated from observational data.

117.086 **Dwarf nova burst asymmetry and the physics of accretion disks.**
J. K. Cannizzo, J. C. Wheeler, R. S. Polidan.
Astrophys. J., Vol. 301, No. 2, p. 634 – 640 (1986).
Voyager far-ultraviolet and optical data are presented of an outburst of SS Cygni which show clearly for the first time in a classical U Gem type dwarf nova that for an asymmetric burst, with faster rise than decay, the far-UV flux increases ~0.5 day later than that of the optical. Theoretical burst profiles based on global thermal instabilities in the disk show that bursts which are initiated at large radii produce asymmetric light curves for which the far-UV flux lags the optical (V band) flux by ~0.5 day, as observed. It is concluded that to reproduce the common asymmetric outbursts observed in dwarf novae, the thermal instability model must provide a natural explanation for the initiation of outbursts near the outer edge of the disk.

117.087 **Evidence for extended radio emission surrounding RX Puppis.**
J. M. Hollis, R. J. Oliversen, M. Kafatos, A. G. Michalitsianos.
Astrophys. J., Vol. 301, No. 2, p. 877 – 880 (1986).
The authors report evidence for ~1″ extended structure in 6 cm continuum emission emanating from the symbiotic star system RX Puppis. They did not detect hourly continuum flux changes as suggested in previous radio experiments by others. The observations indicate that the predominant nature of the radio emission is thermal and consistent with an optically thick stellar wind emanating from the symbiotic star system.

117.088 **Circular polarization in the DQ Herculis binary 3A 0729+103.**
W. R. Penning, G. D. Schmidt, J. Liebert.
Astrophys. J., Vol. 301, No. 2, p. 881 – 885 (1986).
The authors present optical circular polarimetry of the 13th magnitude long-period DQ Her star 3A 0729+103. For the first time, circular polarization has been detected from a member of this class of objects. Observations in the λλ6400 – 8600 spectral region in January and March of 1985 produced results of −0.347±0.067% and −0.236±0.057%, respectively, with an overall result of −0.239±0.030%. No modulation with the rotational period of the white dwarf is seen. Possible polarization mechanisms in the magnetic field of the white dwarf primary star are discussed.

117.089 **Long-term observations of Cygnus X-3 with *Vela 5B*.**
W. Priedhorsky, J. Terrell.
Astrophys. J., Vol. 301, No. 2, p. 886 – 893 (1986).
The authors present long-term (1969 – 1979) observations of Cygnus X-3, obtained by the *Vela 5B* satellite at 3 – 12 keV, with 10^d time resolution. The object has a 4.8 hr period which is thought to be orbital. Long-term periodicities of 17^d, 20^d, and 33^d – 34^d have been reported by several authors. The authors of this paper do not observe the 17^d and 33^d – 34^d variations, and set upper limits significantly lower than the reported amplitude of the 33^d – 34^d variation. The 4.8 hr variation is evident in the data, appearing with the doubly aliased period of $20^d.1$. The nonperiodic longer fluctuations in the X-ray intensity of Cyg X-3 are suggestive of randomly occurring ~100^d outbursts (shot noise).

117.090 **Detection of an X-ray flare in the RS CVn binary Sigma Coronae Borealis.**
P. C. Agrawal, A. R. Rao, G. R. Riegler.
Mon. Not. R. Astron. Soc., Vol. 219, No. 4, p. 777 – 784 (1986).
The detection of an X-ray flare in the RS CVn binary Sigma Coronae Borealis with the Monitor Proportional Counter on the Einstein Observatory is described. During the 513 min of observation, an X-ray flare of 208 min duration was detected at a significance level of 26σ in the 1.19 – 10.16 keV band. The rise time of the flare is between 25 and 70 min and the decay time is ⩾ 34 min. The X-ray luminosity at the flare maximum is found to be 6×10^{30}erg s^{-1} and the total energy radiated in X-rays during the flare is 2×10^{34}erg. The energy spectrum in the flaring state is found to be harder (temperature $T \sim 2.5\times10^7$K) compared to the one observed in the quiescent state ($T \sim 6\times10^6$K). Applying the coronal loop model, the loop parameters are calculated and compared for the X-ray flares observed in the various RS CVn binaries and the Sun. The significance of the differences in the observed and derived parameters of the X-ray flares is briefly discussed.

117.091 **A dynamical model for the dwarf nova AH Herculis.**
K. Horne, R. A. Wade, P. Szkody.
Mon. Not. R. Astron. Soc., Vol. 219, No. 4, p. 791 – 808 (1986).
The authors develop a model for the AH Her binary based on spectroscopic observations. They find emission line radial velocity variations that confirm the photometric period and allow them to determine an accurate spectroscopic ephemeris. By folding data on the spectroscopic period, the authors examine fine structure in the emission line profiles and derive a radial velocity curve from the emission line wings that are thought to reflect the orbital motion of the white dwarf primary. They detect the weak absorption spectrum of the companion star (AH Her B) and use a cross-correlation analysis to measure its rotation and orbital motion. Lastly, the authors examine a dynamical model for the AH Her system.

117.092 **On short- and long-period variations in radial velocities of SS433 relativistic lines.**
T. A. Somova.
Pis'ma Astron. Zh., Tom 12, No. 3, p. 205 – 211 (1986). In Russian. Englisch translation in Sov. Astron. Lett., Vol. 12.
Using the results of spectral observations of the object SS433 by the 6-meter telescope in combination with published data the nutation and precession periods of this system in 1980 – 1982 are specified by means of Fourier analysis. They appeared to be $P_n = 6^d.284\pm0^d.001$ and $P_p = 163^d.34\pm0^d.05$, respectively. During this time in the nutation variations of the disk the components with periods of $6^d.384\pm0^d.005$, $5^d.84\pm0^d.01$ and others were also present. It is suspected that the variation of the nutation period at time intervals of several months is discrete. The analysis of the results on determination of the precession period at shorter intervals of time (of the order of one period) has shown that precession instability is nonmonotonous. Secular variations of the precession period are absent.

117.093 **Masses of the components of the strongly interacting binary system RX Cas.**
V. Ya. Alduseva.
Pis'ma Astron. Zh., Tom 12, No. 3, p. 212 – 218 (1986). In Russian. Englisch translation in Sov. Astron. Lett., Vol. 12.

Results of 1979 – 1982 spectroscopic observations of RX Cas are presented. It is shown that the absorption A2 – A5 spectrum primary component is a spectrum of an extended accretion disk, while the secondary component has a K0(II–III) spectrum. The radial velocity curves derived from the lines of primary and secondary components are symmetrical with $\gamma = -25$ km/s and semi-amplitudes $K_1 = 14$ km/s and $K_2 = 87$ km/s. The masses of the components are estimated to be $M_1 = 3.00\, M_\odot$ and $M_2 = 0.48\, M_\odot$. From the mass ratio $q = M_2/M_1 = 0.16$ it is evident that RX Cas is near the final stage of the case B of mass exchange.

117.094 **UBV photometry of dwarf novae at minimum light. II. The change of accretion disk dimensions in SS Cygni.**
I. B. Voloshina.
Pis'ma Astron. Zh., Tom 12, No. 3, p. 219 – 228 (1986). In Russian. Englisch translation in Sov. Astron. Lett., Vol. 12.

New observations of the dwarf nova SS Cyg at minimum light carried out in 1983 and 1984 confirm the early conclusion about the existence of eclipses in this system at conjunctions. The dependence of the shape of orbital light curves on the outburst cycle phase allows to distinguish between two quiescent states: (1) steady state (15 – 20 days before an outburst), (2) unsteady state (a few days before an outburst). The difference in light curves at these states can be explained by changes in disk dimensions.

117.095 **Observations of the IR emission peak of the HZ Her/Her X–1 system near the phase of an X–ray eclipse.**
Yu. N. Gnedin, V. V. Kir'yan, A. V. Krat, M. A. Pogodin, A. E. Tarasov.
Pis'ma Astron. Zh., Tom 12, No. 4, p. 300 – 303 (1986). In Russian. English translation in Sov. Astron. Lett., Vol. 12.

IR observations of the HZ Her/Her X–1 system showed an essential excess of emission in the range 1 – 2 µm near the phase of an X–ray emission eclipse ($\varphi = 0.91$). An evaluation of the lower limit of its power attains 10^{36}erg/s. The nature of the excess cannot be explained in terms of the existing models of the system in question.

117.096 **On spectral duplicity of the runaway star α Camelopardalis.**
S. K. Zejnalov, F. A. Musaev.
Pis'ma Astron. Zh., Tom 12, No. 4, p. 304 – 310 (1986). In Russian. English translation in Sov. Astron. Lett., Vol. 12.

Duplicity of the runaway star α Cam is revealed on the basis of radial velocity measurements of 31 high dispersion (5 – 9 Å/mm) spectrograms. The radial velocity varies with the period $3\overset{d}{.}6784$ and semiamplitude 9 km/s, the orbital eccentricity being 0.45.

117.097 **Spectrum of the symbiotic star CH Cygni in 1982.**
T. S. Galkina.
Izv. Krymskoj Astrofiz. Obs., Tom 73, p. 101 – 111 (1985). In Russian. English translation in Bull. Crimean Astrophys. Obs., Vol. 73.

The results of spectroscopic observations of the symbiotic star CH Cygni obtained from September to December 1982 are discussed. In the spectrum of CH Cygni in 1982 many sharp absorption lines of Ti II, Ca I, Fe I, Ti I, Sc II and others are present. The identification and the measurements of the equivalent widths of all absorption lines as well as of the emission lines were made. The spectrum CH Cygni 1982 shows similarities to the recurrent nova RS Ophiuchi between outburst.

117.098 **On the problem of the gap of the orbital periods in cataclysmic variables.**
A. V. Fedorova, L. R. Yungel'son.
Nauchn. Inf., Vyp. 57, p. 64 – 75 (1984). In Russian.

The evolution of a white dwarf – main sequence low–mass star binary under the influence of angular momentum loss via magnetic stellar wind and gravitational radiation has been investigated numerically. Magnetic stellar wind with rate exceeding $(10^{-10} - 10^{-9})\, M_\odot$/year is able to drive the evolution of such systems with mass exchange. If the time scale of switch–off of magnetic wind, when the secondary becomes fully convective is not greater than $\sim 10^6$years, mass exchange interrupts due to contraction of the secondary and the system becomes temporary unobservable. Mass exchange starts again when still continuing momentum loss via gravitational radiation brings components close enough for repeated filling of the Roche lobe by the secondary. Location and width of the thus arising gap in orbital period distribution satisfactorily agree with those observed.

117.099 **Evolutionary status of W UMa–type binaries.**
J. Kałużny.
Acta Astron., Vol. 35, Nos. 3–4, p. 313 – 326 (1985).

Absolute parameters of 18 W Ursae Majoris systems have been collected. The evolutionary status of these systems is discussed on the basis of the mass–radius and mass–luminosity diagrams. It is concluded that the majority of binaries from analysed sample lie close to ZAMS line on both diagrams. Four systems seem to be evolved and to have unexpectedly large temperatures as for their masses.

117.100 **The influence of the accretion column's inclination onto the light curve of polars. Effects of geometry.**
I. L. Andronov.
Astron. Zh., Tom 63, Vyp. 2, p. 274 – 278 (1986). In Russian. English translation in Sov. Astron., Vol. 30, No. 2.

It is shown that the asymmetry of orbital light curves in the range from X–ray to infrared, which is observed in AM Herculis–type objects, may be qualitatively explained if the accretion column is inclined to the white dwarf's surface. In this model the asymmetry may be observed, if the column's axis is not crossing the rotation axis of the white dwarf and its base is eclipsed by the compact star. The asymmetry of the radial–velocity and polarization curves also may be explained by the inclination of the column. The results are compared with the observations of polars.

117.101 **Sporadic formation of a dust envelope in the system V1016 Cygni in summer 1983.**
O. G. Taranova, B. F. Yudin.
Astron. Zh., Tom 63, Vyp. 2, p. 317 – 325 (1986). In Russian. English translation in Sov. Astron., Vol. 30, No. 2.

The authors analyse photometric observations of the symbiotic star V1016 Cyg made in 1978 – 1984 and conclude therefrom that in summer 1983 around the cool component of the system a dust envelope formed, with the optical depth, the inner–boundary temperature, and the internal radius being respectively $\geqslant 2.5$, 1200K, and 1×10^{14}cm. The mean velocity of expansion of the envelope is ~ 60 km/s, the mean dust mass in it is $\sim 4 \times 10^{-7} M_\odot$. The analysis of optical properties of dust particles in the "stationary" envelope surrounding the cool component in the system V1016 Cyg showed that this envelope probably consists of purely silicate particles. The V1016 Cyg distance, estimated by the J magnitude of the cool component in the middle phase of its variability, is ~ 2.8 kpc. The nature of the hot component of V1016 Cyg is discussed.

117.102 **Evolution of low–mass close binaries with orbital momentum loss.**
A. V. Tutukov, A. V. Fedorova, E. V. Ergma, L. R. Yungel'son.
Astrofizika, Tom 24, Vyp. 1, p. 85 – 96 (1986). In Russian. English translation in Astrophysics, Vol. 24, No. 1.

The evolution of the secondary components of close binaries with compact primaries has been computed taking into account angular momentum losses due to gravitational radiation and magnetic stellar wind. Stars with chemical composition corresponding to disk and bulge populations have been considered. The influence of induced stellar wind on the evolution has been investigated. The results of computations allow to explain the distribution of cataclysmic binaries over periods and mass exchange rates observed in cataclysmic binaries and low-mass X-ray systems.

117.103 **Evidence for a bipolar wind in the cataclysmic variable PG 1012–029.**
R. K. Honeycutt, E. M. Schlegel, R. H. Kaitchuck.
Astrophys. J., Vol. 302, No. 1, p. 388 – 402 (1986).
PG 1012–029 is a nova-like variable which undergoes deep eclipses by the secondary star. Whole-orbit, time-resolved spectrophotometry has been used to study the size, shape, and kinematics of the accretion disk. Analysis of line and continuum eclipse light curves are used to determine the radius and the elevation above the plane of the disk for three characteristic emitting regions: (1) the high-excitation lines, (2) the low-excitation lines, and (3) the continuum. It is argued that a nonrotating component dominates the line profiles and eclipse behavior in this system, and the authors attribute this component to a bipolar wind from the accretion disk. Such a model can explain the apparent non-Keplerian behavior of the emission lines as they are eclipsed.

117.104 **H 0538+608: a bright AM Herculis-type X-ray source.**
R. A. Remillard, H. V. Bradt, J. E. McClintock, J. Patterson, W. Roberts, D. A. Schwartz, S. Tapia.
Astrophys. J., Lett. Ed., Vol. 302, No. 1, p. L11 – L15 (1986). With plate L1.
The authors report the discovery of an AM Herculis-type magnetic variable, H 0538+608. The object was located during a program to optically identify the X-ray sources detected with HEAO 1. The X-ray flux (2 – 10 keV) of H 0538+608 is exceeded by only two of the 10 previously observed AM Herculis-type systems. The optical spectrum shows very strong and broad emission lines of H, He I and He II. Photometric observations show modulations by a factor of 2, with *B* magnitudes between 15 and 16, and the source exhibits rapid flickering at a time scale of 10 s. Circular polarization measurements indicate variable optical polarization between 0 and 10%. The modulations in the polarization and the photometric magnitude suggest an orbital period of 3.1 ± 0.2 hr.

117.105 **The most reliable black hole yet.**
M. G. Watson.
Nature, Vol. 321, No. 6065, p. 16 (1986).
This note reviews current observational evidence for the presence of a black hole in the X-ray binary A 0620–00 identified with Nova Monocerotis 1975.

117.106 **Periodic dips in X-ray binaries.**
A. N. Parmar.
Observatory, Vol. 106, No. 1070, p. 7 – 8 (1986). Abstract. – See Abstr. 010.741.

117.107 **Her X-1: the 35-day cycle.**
J. Trümper.
Observatory, Vol. 106, No. 1070, p. 8 – 9 (1986). Abstract. – See Abstr. 010.741.

117.108 **T CrB, a "Seyfert 1 nucleus" at 1200 ± 100 parsecs?**
P. L. Selvelli, J. Clavel, A. Cassatella, M. Hack.
Structure and evolution of active galactic nuclei, p. 709 – 712 (1986). – See Abstr. 012.011.
Several symbiotic stars are characterized by UV spectral signatures which are similar, in several aspects, to those observed in the broad-line region of Seyfert galaxies. The authors suggest to consider the symbiotic stars as a convenient laboratory for the study of the spectroscopic mechanisms which are at work in Seyfert 1 nuclei.

117.109 **Brightness variations of a spherical star with two hot processing spots and arbitrary inclination of the precession axis.**
D. P. Kyurkchieva, V. G. Shkodrov.
Dokl. Bolg. Akad. Nauk, Vol. 38, No. 5, p. 529 – 532 (1985). Abstr. in Ref. Zh., 51. Astron., 1.51.363 (1986).

117.110 ***UBVRI* photometry of the dwarf nova SS Aur.**
Yu. S. Efimov, G. G. Tovmasyan, N. M. Shakhovskoj.
Astrofizika, Tom 24, Vyp. 2, p. 227 – 237 (1986). In Russian. English translation in Astrophysics, Vol. 24, No. 2.
The results of simultaneous *UBVRI* observations of the dwarf nova SS Aur are presented. The decrease of brightness to $V = 15^{m}.5$ in comparison with the usual quiescence magnitude $V = 14^{m}.7$ is noted. In the low brightness stage, the distribution of energy is best fitted by the sum of two blackbody curves of 25000K and 3000K temperatures which correspond to the radiation of the double system components consisting of white and red dwarfs. The additional radiation which is responsible for the brightening up of the system to the usual value $V = 14^{m}.7$ is in good agreement with steady-state accretion-disk radiation. Besides, the system undergoes rapid light variations in the time scale of 20 – 40 minutes with amplitudes increasing to the shorter wavelengths and attaining $0^{m}.7$ in the *U* band. The intensity of these variations increases with general brightening of the system.

117.111 **Characteristics of the cool components of symbiotic stars.**
L. Luud, L. Leedjärv.
Astrofizika, Tom 24, Vyp. 2, p. 265 – 278 (1986). In Russian. English translation in Astrophysics, Vol. 24, No. 2.
Using the Blackwell-Shallis method the luminosities, temperatures and radii for cool components of symbiotic stars and for a sample of field red giants have been determined by means of infrared photometric observations. It turned out that the cool components of symbiotic stars do not differ from the normal red giants of the asymptotic branch. The masses of cool components of symbiotic stars have been found to be close to 3 $M_{\odot}$. The cool components of symbiotic stars do not fill their Roche lobes. About 10 times more carbon stars than the normal value in the vicinity of the sun have been found among the cool components of symbiotic stars.

117.112 **Evolution of close binaries under the assumption of momentum loss by magnetic stellar wind.**
Z. T. Krajcheva, A. V. Tutukov, L. R. Yungel'son.
Astrofizika, Tom 24, Vyp. 2, p. 287 – 300 (1986). In Russian. English translation in Astrophysics, Vol. 24, No. 2.
The authors suggest a simple method to compute the evolution of semi-detached close binaries having secondaries with degenerate helium cores and losing momentum via magnetic stellar wind. They employ the results of computations to estimate the initial parameters of some low-mass ($M_1 + M_2 \lesssim 5\ M_{\odot}$) Algols under assumptions of conservative and non-conservative mass exchange. It has been shown that the assumption of conservative mass exchange leads to contradicting initial parameters, while assumption of momentum loss via magnetic stellar wind allows to resolve these contradictions.

117.113 **The strange star SS433.**
A. Černins (*A. Chernin*).
Zvaigžnota Debess, 1984. Gada Rudens, p. 2 – 6 (1984). In Latvian.

117.114 **UV study of U Gem throughout an entire quiescent interval.**
P. Szkody, A. Kiplinger.
Bull. Am. Astron. Soc., Vol. 17, No. 4, p. 839 (1985). Abstract. – See Abstr. 010.062.

117.115 **Hot components in binary systems.**
J. C. Raymond.
Bull. Am. Astron. Soc., Vol. 17, No. 4, p. 853 (1985). Abstract. – See Abstr. 010.062.

117.116 **The black-hole candidate A0620–00.**
J. E. McClintock, R. A. Remillard.
Bull. Am. Astron. Soc., Vol. 17, No. 4, p. 854 (1985). Abstract. – See Abstr. 010.062.

117.117 **Sco X–1: an evolving double radio source.**
B. J. Geldzahler, E. B. Fomalont.
Bull. Am. Astron. Soc., Vol. 17, No. 4, p. 854 (1985). Abstract. – See Abstr. 010.062.

117.118 **Inverse Compton models of SS 433 and radio quiet AGN.**
D. L. Band, J. E. Grindlay.
Bull. Am. Astron. Soc., Vol. 17, No. 4, p. 867 (1985). Abstract. – See Abstr. 010.062.

117.119 **Effects of the X–ray scattering halo on the observational properties of Cygnus X–3.**
C. W. Mauche, L. A. Molnar.
Bull. Am. Astron. Soc., Vol. 17, No. 4, p. 867 (1985). Abstract. – See Abstr. 010.062.

117.120 **New radio and X–ray results on Cyg X–3.**
L. A. Molnar.
Bull. Am. Astron. Soc., Vol. 17, No. 4, p. 867 (1985). Abstract. – See Abstr. 010.062.

117.121 **Contact binaries in X–rays.**
O. Vilhu, J. Heise.
Bull. Am. Astron. Soc., Vol. 17, No. 4, p. 868 (1985). Abstract. – See Abstr. 010.062.

117.122 **The optical counterpart of the X–ray source in the core of M15.**
P. A. Charles, D. C. Jones, T. Naylor.
Bull. Am. Astron. Soc., Vol. 17, No. 4, p. 868 (1985). Abstract. – See Abstr. 010.062.

117.123 **Echelle spectroscopy of the RS CVn system II Pegasi.**
L. W. Ramsey, D. P. Huenemoerder.
Bull. Am. Astron. Soc., Vol. 17, No. 4, p. 880 (1985). Abstract. – See Abstr. 010.062.

117.124 **Far–UV observations of the symbiotic star AG Peg.**
R. S. Polidan, C. D. Keyes.
Bull. Am. Astron. Soc., Vol. 17, No. 4, p. 886 (1985). Abstract. – See Abstr. 010.062.

117.125 **The exotic helium variable PG 1346+082.**
M. A. Wood, D. E. Winget, R. E. Nather, J. Liebert, F. V. Hessman.
Bull. Am. Astron. Soc., Vol. 17, No. 4, p. 886 (1985). Abstract. – See Abstr. 010.062.

117.126 **Observations of the accretion disk in the dwarf nova HT Cassiopeia.**
E. L. Robinson, E.–H. Zhang.
Bull. Am. Astron. Soc., Vol. 17, No. 4, p. 886 (1985). Abstract. – See Abstr. 010.062.

117.127 **Ejection of bipolar winds from cataclysmic variable systems.**
M. V. Torbett.
Bull. Am. Astron. Soc., Vol. 17, No. 4, p. 887 (1985). Abstract. – See Abstr. 010.062.

117.128 **Cyclic variation of RS CVn binaries, AR Lac and V711 Tau.**
Y. W. Kang, R. E. Wilson.
Bull. Am. Astron. Soc., Vol. 17, No. 4, p. 887 (1985). Abstract. – See Abstr. 010.062.

117.129 **Solution of the asymmetric light curve of TY Mensae.**
T. A. Nagy.
Bull. Am. Astron. Soc., Vol. 17, No. 4, p. 887 (1985). Abstract. – See Abstr. 010.062.

117.130 **Optical, UV and X–ray observations of X–ray binaries.**
L. Bianchi, M. Pakull.
ESA Spec. Publ., ESA SP–239, p. 99 – 106 (1985). – See Abstr. 012.019.

The authors have observed a number of X–ray binary candidates in the Galaxy and in the Magellanic Clouds in the optical, far ultraviolet and X–ray bands. They present preliminary results on two sources for which the photometric and spectroscopic data suggest the presence of an accretion disk. For another source (LMC X–1) the authors show several observational evidences of interaction of the strong X–ray source with the surrounding H II region and the discovery of an X–ray ionized He III region.

117.131 **On the variability of Sigma Geminorum – an interesting candidate for X–ray studies.**
Ø. Elgarøy, O. Engvold, P. Jorås.
ESA Spec. Publ., ESA SP–239, p. 117 – 120 (1985). – See Abstr. 012.019.

New observations of σ Gem in UV show that high excitation line fluxes increase with the emergence of a dark region from behind the approaching limb. A moderate increase in X–ray intensity occurs together with the coming into view of another, weaker spot region.

117.132 **X–ray spectra and the rotation–activity connection of RS CVn binaries.**
P. Majer, J. H. M. M. Schmitt, L. Golub, F. R. Harnden Jr., R. Rosner.
ESA Spec. Publ., ESA SP–239, p. 141 – 147 (1985). – See Abstr. 012.019.

Results are presented from a survey of RS CVn binaries which were observed with the Imaging Proportional Counter (IPC) on board the Einstein Observatory. Spectral analyses of the IPC pulse height spectra show that the coronae of RS CVn binaries always contain hot gas with temperatures in excess of 10^7K, similar to active late–type main sequence stars, and that at least two temperature components are necessary to account for the higher quality IPC spectra.

117.133 **Einstein OGS observations of Cyg X–1.**
M. Bałucińska, T. Chlebowski.
ESA Spec. Publ., ESA SP–239, p. 201 – 203 (1985). – See Abstr. 012.019.

117.134 **Study of the nonstationary state of detached eclipsing binaries with main–sequence components. I.**
R. A. Botsula.
Izv. Astron. Ehngel'gardt. Obs., No. 48, p. 3 – 14 (1984). In Russian.

The peculiarities of the photoelectric light curves of 35 systems without large tidal distortion were studied. Their modifications with time and colour, spectral characteristics and the behaviour of periods of revolution were examined. Three hypotheses are considered to interpret the results: the variability of one of the components, spotting of the components and existence of gaseous matter in the system.

117.135 **Nonstationary state in detached binary systems with main–sequence solar–type components. ER Vul.**
R. A. Botsula.
Izv. Astron. Ehngel'gardt. Obs., No. 49, p. 57 – 61 (1985). In Russian.

The nonstationary state of the system ER Vul is studied, starting with light curves and spectral characteristics. Comparing the nonstationary states of UV Leo, BH Vir, ER Vul and FL Lyr, the author concludes that 1) at least one component is active in all these systems; 2) gaseous matter exists in all these systems; 3) not component proximity, but physical characteristics of the components play the principal role in the appearance of the nonstationary state.

117.136 **Study of the nonstationary state of detached eclipsing binaries with main–sequence components. II.**
R. A. Botsula.
Izv. Astron. Ehngel'gardt. Obs., No. 49, p. 62 – 75 (1985). In Russian.
Among 35 systems with good photoelectric light curves and without large tidal distortion 17 are nonstationary systems. Three from them possibly have nonstationary components (Ap, δ Sct), two systems probably have spots on the surface, 15 systems indicate some symptoms of gaseous matter in the systems. Two types are found mostly among their light curves.

117.137 **Quantitative analysis of the effect of light refraction in the atmospheres of close binaries.**
I. Kudzej.
Problems of astronomy, p. 26 – 31 (1985). In Russian. Abstr. in Ref. Zh., 51. Astron., 3.51.548 (1986). – See Abstr. 003.023.

117.138 **Spectral classification of the close binary V453 Cyg.**
L. V. Glazunova.
Problems of astronomy, p. 19 – 25 (1985). In Russian. Abstr. in Ref. Zh., 51. Astron., 3.51.723 (1986). – See Abstr. 003.023.

117.139 **On the time of motion of plasma in magnetic close binary systems.**
I. L. Andronov.
Problems of astronomy, p. 2 – 13 (1985). In Russian. Abstr. in Ref. Zh., 51. Astron., 3.51.727 (1986). – See Abstr. 003.023.

117.140 **SS433: new results, new problems.**
A. M. Cherepashchuk.
Zemlya Vselennaya, No. 1, p. 21 – 29 (1986). In Russian.

117.141 **Binaries and their role in present–day astronomy.**
D. Ya. Martynov.
Zemlya Vselennaya, No. 2, p. 25 – 31 (1986). In Russian.

117.142 **The character of rapid brightness variations of CH Cyg.**
V. Vansevičius.
Vilniaus Astron. Obs. Biul., Nr. 70, p. 30 – 34 (1985). In Russian.
Non–synchronous photoelectric measurements of rapid variation of CH Cyg in the U, P, V, and S bands of the Vilnius photometric system are presented. The irregular character of rapid variation of 1 to 15 min duration is confirmed. No difference is observed between the mean rates of increase and decrease of the magnitude of the star.

117.143 **On the change of the orbital period of AM Herculis.**
V. P. Smykov, L. I. Shakun.
Astron. Tsirk., No. 1384, p. 3 – 5 (1985). In Russian.

117.144 **Photographic observations of HZ Her in 1983 – 1984.**
V. P. Smykov, L. I. Shakun.
Astron. Tsirk., No. 1387, p. 6 – 7 (1985). In Russian.

117.145 **Strong growth of excitation in the spectrum of the gaseous shell of HM Sge in 1984.**
A. P. Ipatov, B. F. Yudin.
Astron. Tsirk., No. 1392, p. 1 (1985). In Russian.

117.146 **Elements of the spectroscopic orbit of V448 Cygni.**
L. V. Glazunova.
Kinematika Fiz. Nebesn. Tel, Tom 2, No. 3, p. 94 – 96 (1986). In Russian. English translation in Kinematics Phys. Celest. Bodies.
Elements of the orbit of the close binary star V448 Cygni have been reconsidered. It is shown that the anomaly of the radial–velocity curve of the more massive star of the system is due to a complicated structure of line profiles. New values of masses of the components of the system differ greatly from those determined earlier.

117.147 **CH Cygni in 1983 – 84: transition to a real symbiotic star?**
G. Wallerstein, M. Bolte, P. Whitehill–Bates, M. Mateo.
Publ. Astron. Soc. Pac., Vol. 98, No. 601, p. 330 – 333 (1986).
The authors report the end of the shell spectrum phase of CH Cyg and its transformation into a symbiotic star. A Reticon spectrum of moderate dispersion showing forbidden and permitted emission lines as well as strong absorption by Ca I and Al I is displayed. The absence of Sr II absorption in the shell during prior years may be explained by ionization from the first excited state by Lα photons. Sr II absorption gradually reappeared in 1983 – 84.

117.148 **Spectral and electrophotometric observations of the peculiar star CH Cygni during 1980 – 1981.**
T. S. Galkina, N. I. Shakhovskaya, N. I. Bondar.
Bull. Crimean Astrophys. Obs., Vol. 69, p. 15 – 25 (1986). English translation of 39.117.101.

117.149 **Symbiotic eclipsing variables CI Cyg and V1329 Cyg. Comparison of light curves.**
V. P. Arkhipova, T. S. Belyakina.
Bull. Crimean Astrophys. Obs., Vol. 69, p. 26 – 29 (1986). English translation of 39.117.102.

117.150 **On the character of variability of the spectrum of SS433 with 6–day period (June – August 1981).**
I. M. Kopylov, R. N. Kumajgorodskaya, N. N. Somov, T. A. Somova, S. N. Fabrika.
Spets. astrofiz. obs. Akad. Nauk SSSR. Prepr., No. 24 L, 36 pp. (1985). In Russian. Abstr. in Ref. Zh., 51. Astron., 4.51.576 (1986).

117.151 **On short– and long–period fluctuations in the radial velocities of the relativistic lines of SS433.**
T. A. Somova.
Spets. astrofiz. obs. Akad. Nauk SSSR. Prepr., No. 25 L, 14 pp. (1985). In Russian. Abstr. in Ref. Zh., 51. Astron., 4.51.577 (1986).

117.152 **Photometric and geometric elements of early–type close binaries.**
I. I. Bondarenko.
Astronomical–geodetical investigations, p. 113 – 127 (1985). In Russian. Abstr. in Ref. Zh., 51. Astron., 4.51.591 (1986). – See Abstr. 003.024.

117.153 **The ellipsoidal variables. II. The main–sequence distribution.**
M. Beech.
Astrophys. Space Sci., Vol. 122, No. 1, p. 57 – 61 (1986).
The author performs a Monte Carlo analysis of the eclipse probabilities of short–period binary systems in an attempt to estimate the distribution of ellipsoidal variable systems with main–sequence components. The results suggest that possibly as many as one in four of the non–eclipsing, spectroscopic binaries with $P(d) \leqslant 10$, and main–sequence components with $M_V \leqslant 10$ could be ellipsoidal variable systems. This result, while an upper limit, would seem to hold irrespective of primary spectral type (B5 to G5).

117.154 **BV Centauri: dwarf or classical nova?**
J. W. Menzies, D. O'Donoghue, B. Warner.
Astrophys. Space Sci., Vol. 122, No. 1, p. 73 – 80 (1986).
The similarity in properties of GK Per and BV Cen suggests that the latter may be a classical nova remnant rather than a dwarf nova. The authors' Hα and V CCD photometry show no sign of a nova shell but disclose the presence of a close and probably physical companion. $BV(RI)_c$ CCD photometry of the companion, assumed to be a dwarf, gives $M_v = 3.1 \pm 0.3$ for BV Cen at quiescence. Even though half of the light in BV Cen comes from the secondary this is much brighter than expected for a dwarf nova and supports the conclusion that BV Cen was an unrecorded nova.

117.155 **RS CVn binaries.**
J. B. Srivastava.
Binary and multiple systems, p. 5 – 19 (1984). – See Abstr. 012.025.
The author discusses the observational properties of thirty–five RS CVn binaries which are known to eclipse.

117.156 **Models for RS CVn binaries.**
P. Vivekananda Rao, M. B. K. Sarma.
Binary and multiple systems, p. 21 – 26 (1984). – See Abstr. 012.025.
The authors review and discuss observational properties and models of RS CVn binaries.

117.157 **Reflection effect in close binaries.**
A. Peraiah.
Binary and multiple systems, p. 27 – 46 (1984). – See Abstr. 012.025.
The author describes an initial attempt of how the radiation field is calculated from the irradiated surface of the component in a binary system.

117.158 **Cygnus X–1: a search for eclipse and precession effects.**
V. M. Lyutyi (*V. M. Lyutyj*).
Sov. Astron., Vol. 29, No. 4, p. 429 – 433 (1985). English translation of 40.117.036.

117.159 **Triple star evolution and the formation of short–period, low–mass X–ray binaries.**
P. P. Eggleton, F. Verbunt.
Mon. Not. R. Astron. Soc., Vol. 220, No. 2, p. 13P – 18P (1986).
To explain the presence of a relatively massive black hole in 2A0620–00 the authors propose a new evolutionary scenario for the formation of short–period, low–mass X–ray binaries. The progenitor in this scenario is a triple star in which a massive close binary is accompanied at large distance by a late dwarf. After the evolution of the close binary into an ordinary X–ray binary, the compact object is engulfed by its expanding massive companion, and spirals in to settle at its centre. The resulting Thorne–Żytkow supergiant gradually expands until it attains the size of the late–dwarf orbit. Then a second spiral–in phase ensues, leading to the formation of a low–mass close binary. Depending on the initial parameters of the massive binary, this close low–mass binary may contain either a black hole or a neutron star.

117.160 **The radius of the accretion disc in Z Cha between outbursts.**
D. O'Donoghue.
Mon. Not. R. Astron. Soc., Vol. 220, No. 2, p. 23P – 26P (1986).
Published measurements of the radius of the accretion disc in the dwarf nova Z Cha and new observations made at significant times during the outburst cycle, exhibit a decline in disc radius between outbursts, similar to that found for U Gem. The relationship has least scatter when considered as a function of fractional time between outbursts.

117.161 **Non–conservative evolution of close binaries.**
R. Q. Huang, X. Xie.
Astron. Astrophys., Vol. 161, No. 1, p. 142 – 144 (1986).
A way to treat non–conservative evolution of close binaries is suggested. Mass and angular momentum loss are introduced by means of one parameter. Detailed evolutionary computations are described for a binary of $8M_\odot + 2M_\odot$ in case B.

117.162 **On the apsidal motion in binary stars.**
N. I. Shakura.
Sov. Astron. Lett., Vol. 11, No. 4, p. 224 – 226 (1985). English translation of 40.117.006.

117.163 **The spectrum of the novalike binary UX Ursae Majoris.**
N. F. Vojkhanskaya.
Sov. Astron. Lett., Vol. 11, No. 4, p. 259 – 261 (1985). English translation of 40.117.022.

117.164 **Polarimetry of UV Aurigae.**
T. N. Khudyakova.
Sov. Astron. Lett., Vol. 11, No. 4, p. 262 – 263 (1985). English translation of 40.117.023.

117.165 **Polarization in massive X–ray binaries. I. A low–inclination model for Cygnus X–1.**
D. B. Friend, J. P. Cassinelli.
Astrophys. J., Vol. 303, No. 1, p. 292 – 299 (1986).
The authors investigate the possibility that variable linear polarization in massive X–ray binaries is produced by electron scattering in an asymmetric stellar wind. The stellar wind is asymmetric because of the gravitational field of the secondary (X–ray source). The degree of asymmetry and the magnitude of the linear polarization are controlled by the degree to which the primary star fills its Roche lobe. For the well–observed X–ray binary Cyg X–1, the model can produce the correct magnitude for the polarization. If the inclination of the system is less than $\sim 20°$, the model should also predict the correct phase dependence of the polarization.

117.166 **Photoelectric observations of RU Pegasi during the 1984 August outburst.**
B. F. Marino, W. S. G. Walker.
South. Stars, Vol. 31, No. 4, p. 193 – 201 (1985).
Photoelectric observations of RU Pegasi during a 12–day outburst are presented. These are part of an international programme to determine whether superhumps are confined to the short period SU UMa subclass of dwarf novae. No evidence of superhump modulation of the light curves is found in the observations.

117.167 **X–ray binaries – formation and evolution.**
E. P. J. van den Heuvel, S. Rappaport.
Proceedings of the Second Asian–Pacific Regional Meeting on Astronomy, p. 32 – 54 (1984). – See Abstr. 012.031.
The observed characteristics of the type I (massive) and type II (low–mass) binary X–ray sources are reviewed and their evolutionary status and possible formation mechanisms are discussed.

117.168 **On the masses of Wolf–Rayet components in close binary system.**
W. Sutantyo, L. Dermawan.
Proceedings of the Second Asian–Pacific Regional Meeting on Astronomy, p. 475 – 478 (1984). – See Abstr. 012.031.
Massey (1981) indicates that the average mass of the *WR* components in double line *WR* binaries is about 20 solar masses. The authors argue that if such large mass *WR* stars eventually explode as supernova, the systems will generally be accelerated to velocities considerably in excess of 50 – 65 km/s. Such large velocities are incompatible with the moderate galactic z values of massive X–ray binaries and single line *WR* stars/binaries which are believed to be massive binaries in the post supernova stage. It is concluded that if the estimate of Massey for the masses of *WR* stars is correct, a considerable amount of mass (about 10 solar masses) should have been lost from the star before the explosion.

117.169 **UU Lyncis: a non–contact binary system with a short period.**
A. Okazaki, A. Yamasaki, M. Kitamura.
Proceedings of the Second Asian–Pacific Regional Meeting on Astronomy, p. 479 – 481 (1984). – See Abstr. 012.031.
The physical properties of the short–period binary UU Lyn (P = 0.468 days) are investigated on the basis of photo–electric and spectroscopic observations. It is found that UU Lyn is a non–contact system.

117.170 **Binary nature of HD 62001.**
N. Kameswara Rao, D. P. Gilra.
Proceedings of the Second Asian–Pacific Regional Meeting on Astronomy, p. 524 (1984). Abstract. – See Abstr. 012.031.

117.171 **Close binaries: physical models and intuition.**
R. E. Wilson.
I.A.P.P.P. Commun., No. 23, p. 5 (1986). Abstract. – See Abstr. 010.521.

117.172 **Period and cone–angle changes in SS 433.**
G. H. Newsom, G. W. Collins II.
Astron. J., Vol. 91, No. 1, p. 118 – 124 (1986).
Using Doppler–shift data from the moving–line spectra of SS 433 collected over 6 yr, the authors find that the secular changes in the precessional period and cone angle can be approximated with a simple analytic form. This significantly reduces the errors in the fit to the data, allowing a more accurate ephemeris to be calculated and improved limits on the orbital–period changes and light–time effects to be determined.

117.173 **A light–curve distortion–wave analysis of eight RS Canum Venaticorum systems.**
D. B. Caton.
Astron. J., Vol. 91, No. 1, p. 132 – 138 (1986). = Contrib. Dep. Astron., Univ. Florida, No. 87.
A program of differential *U,B,V* photometry of 14 RS CVn systems carried out at Rosemary Hill Observatory in 1978 – 1981 is described, and an analysis of the light curves for the characteristic distortion wave for eight of the systems is presented. The *V* light curves of the systems are shown. Significant waves were observed in RS CVn, RZ Eri, and RW UMa. No significant waves were found in UX Com, GK Hya, AR Lac, LX Per, or TY Pyx. Unusual light curve distortions were observed in UX Com and AR Lac.

117.174 **X–ray observations of the AM Her star CW 1103 + 254.**
K. Beuermann, L. Stella, J. Krautter.
X–ray astronomy '84, p. 27 – 30 (1985). – See Abstr. 012.033.
The authors report the detection of soft and hard X–ray emission from the AM Her star CW 1103 + 254. Spectral analysis reveals a hard bremsstrahlung component with $kT > 12$ keV and a soft (blackbody) component with $18 \text{ eV} < kT < 40$ eV. The soft component is consistent with being reprocessed X–ray and cyclotron radiation from the polar cap of the white–dwarf atmosphere.

117.175 **The soft X–ray superoutburst of VW Hydri: 14 second periodicity.**
H. van der Woerd, J. Heise, F. Paerels.
X–ray astronomy '84, p. 55 – 58 (1985). – See Abstr. 012.033.
EXOSAT observations of the dwarf nova VW Hydri reveal a strong soft X–ray flux during optical superoutburst. The onset of the X–ray outburst was delayed by 2.5 days compared to the optical outburst. A modulation of the extreme soft X–ray flux was detected, consistent with a coherent ($Q > 10^7$) pulsation with a period of 14.07 seconds, probably reflecting the rotation period of the white dwarf. If this is indeed the case, VW Hyi is the fastest white dwarf detected so far.

117.176 **EXOSAT observations of AM Her objects – preliminary results.**
J. Osborne, L. Maraschi, K. Beuermann, J. M. Bonnet–Bidaud, P. A. Charles, L. Chiappetti, C. Motch, M. Mouchet, E. G. Tanzi, A. Treves, K. O. Mason.
X–ray astronomy '84, p. 59 – 62 (1985). – See Abstr. 012.033.
The authors present preliminary EXOSAT soft X–ray light curves of the AM Her objects VV Puppis, E2003 + 225, AN UMa and E1405–451. These are the first X–ray light curves with continuous and complete orbital coverage. For the last three objects these light curves show remarkable similarities each having two short eclipses per orbit. Only E1405–451 was detected in the medium energy experiment, a folded light curve is presented.

117.177 **EXOSAT observations of intermediate polars: preliminary results.**
J. Osborne, K. O. Mason, J. M. Bonnet–Bidaud, K. Beuermann, S. Rosen.
X–ray astronomy '84, p. 63 – 66 (1985). – See Abstr. 012.033.
EXOSAT medium energy data on the intermediate polars V1223 Sgr and H2215–086 are presented.

117.178 **Simultaneous EXOSAT and optical observations of the pulsing X–ray binary H2252–035/AO Psc.**
W. Pietsch, M. Pakull, S. Tjemkes, W. Voges, E. Kendziorra, J. van Paradijs.
X–ray astronomy '84, p. 67 – 72 (1985). – See Abstr. 012.033.

117.179 **X–ray emission from Be star binaries.**
K. M. V. Apparao.
X–ray astronomy '84, p. 73 – 76 (1985). – See Abstr. 012.033.
A model in which a neutron star in an eccentric inclined orbit around a Be star passes through a ring of matter around the Be star, has been used to explain the recurrent flares of X–ray emission from these systems. The model is applied to the X–ray source A0538–66.

117.180 **X–rays and colliding winds in symbiotic novae.**
D. A. Leahy, S. Kwok.
X–ray astronomy '84, p. 81 – 84 (1985). – See Abstr. 012.033.
The X–ray luminosities and spectra of 3 symbiotic novae were measured from observations with the Imaging Proportional Counter on the Einstein Observatory. The observed X–ray emissions indicate a hot ($\sim 10^7$K) gas surrounding the M giant–white dwarf binary that comprises a symbiotic nova. A model of the slow stellar wind from the M giant and the fast wind from the white dwarf colliding in a shock predicts such a hot circumstellar gas. The measured decrease in X–ray luminosity from youngest to oldest symbiotic novae is also in agreement with the model.

117.181 **High and low luminosity globular cluster X–ray sources.**
P. Hertz.
X–ray astronomy '84, p. 85 – 99 (1985). – See Abstr. 012.033.
There are (at least) two classes of globular cluster X–ray sources: high luminosity ($L_x > 10^{36}$erg/s) and low luminosity ($L_x < 10^{34}$erg/s). A coherent picture of the nature and origin of these sources is presented. Both classes of sources are close binaries containing a compact object accreting matter from a low mass companion star. Both classes are formed via tidal capture of the compact object by the companion star in the cores of globular clusters. The high luminosity sources contain a neutron star as the primary and the low luminosity sources contain a white dwarf.

117.182 **X–ray sources of population II: moderate mass black holes accreting gas from extended low density envelopes.**
A. Finzi.
X–ray astronomy '84, p. 101 – 102 (1985). – See Abstr. 012.033.
The author interprets X–ray sources of galactic population II as black holes accreting gas from low density envelopes of outer radius $\sim 10^{16.5}$cm. At intermediate radii $\sim 10^{13}$cm the envelope is supported mainly by trapped line radiation, primarily Lα; at still smaller radii the envelope is fully ionized and collapsing onto the black hole.

117.183 **Instabilities in irradiated accretion disks – an alternative model for the 35^d cycle in HZ Her/Her X–1.**
F. Meyer, E. Meyer–Hofmeister.
X–ray astronomy '84, p. 243 – 246 (1985). – See Abstr. 012.033.
Self–excited non–linear mass flow oscillations in an irradiated accretion disk provide a physical cause for the 35^d cycle.

117.184 **Photometrical observations indicative of a black hole in SS433.**
E. M. Leibowitz.
X–ray astronomy '84, p. 247 – 250 (1985). – See Abstr. 012.033.
The brightness of the star SS433 has been found to vary periodically in the two major cycles of the system. The structure of the light curve in each period is modulated in the other cycle. This feature of the system, along with other photometric and spectroscopic characteristics can be explained on the basis of an eclipsing binary model for the system. When combined with the mass function that has been determined from spectroscopy it implies that the compact object in the SS433 system is a black hole of 10 – 100 $M_\odot$.

117.185 **A common origin for Type I supernovae and low–mass binary X–ray sources.**
R. Canal, J. Isern, J. Labay.
X–ray astronomy '84, p. 293 – 296 (1985). – See Abstr. 012.033.
It is widely accepted that both Type I supernovae and low–mass binary X–ray sources originate from mass–accreting white dwarfs in close binary systems.

117.186 **Ultraviolet, optical, infrared, and microwave observations of HR 5110.**
I. R. Little–Marenin, T. Simon, T. R. Ayres, N. L. Cohen, P. A. Feldman, J. L. Linsky, S. J. Little, R. Lyons.
Astrophys. J., Vol. 303, No. 2, p. 780 – 790 (1986).
HR 5110 is a close binary system ($P = 2^{d}.6$) with the characteristics of an Algol system. Because the primary star is relatively cool (F2 IV) and there is no apparent emission from an accretion disk, the authors were able to detect in *IUE* spectra the emission of an active chromosphere and transition region of the cooler (K0 IV) secondary. The surface fluxes of the UV emission lines of the K star are similar to those of active RS CVn binaries. The line fluxes appear to vary with orbital phase and are interpreted as emission from an active region on the K star. Two large radio flares were detected. VLBI observations during one of these flares indicated that half the emission came from a region more than 4 times the binary separation.

117.187 **The eclipse and mass of Cen X–3.**
G. W. Clark, J. R. Minato.
Japan – U.S. seminar on galactic and extragalactic compact X–ray sources, p. 41 – 44 (1985). – See Abstr. 012.034.
Variations of the X–ray intensity of Cen X–3 during numerous eclipse transitions observed with the SAS–3 X–ray observatory can be fitted well by a formula derived from an exponential model of the atmosphere of the primary star. From the fits in the energy range from 10 to 18 keV the authors derive an eclipse half–angle of $32.5° \pm 1.6°$, which is smaller than previous determinations. When this angle is used with the best available values of the other observables of Cen X–3 in a calculation of the mass range of the neutron star, the result is $1.21 \pm 0.60\ M_{\odot}$, which is larger than previous estimates.

117.188 **X–ray emission and the winds of cataclysmic variables.**
F. A. Córdova.
Japan – U.S. seminar on galactic and extragalactic compact X–ray sources, p. 119 – 125 (1985). – See Abstr. 012.034.
The author summarizes the observed changes in the X–ray and UV continuum and spectral features through the outbursts of the dwarf novae. Then he discusses how the modeling of these data has refined our ideas about the location and nature of the emissions, and the source of the outburst. He then compares X–ray and UV properties of cataclysmic variables with similar phenomena in other astronomical systems.

117.189 **Low–mass X–ray binaries.**
K. Makishima, K. Mitsuda.
Japan – U.S. seminar on galactic and extragalactic compact X–ray sources, p. 127 – 136 (1985). – See Abstr. 012.034.
This paper encompasses a class of neutron–star X–ray binaries in which the magnetic field of the neutron star has negligible effects upon the mass accretion flow. Such sources preferrentially have low–mass primaries. The X–ray emission mechanism and accretion–disk structure in these low–mass X–ray binaries are discussed.

117.190 **Accretion and precession in SS 433.**
B. Margon.
Japan – U.S. seminar on galactic and extragalactic compact X–ray sources, p. 137 – 142 (1985). – See Abstr. 012.034.
There are now more than six years of intensive observations of SS 433 on hand. The author summarizes some of the current inferences available from these data regarding the accretion and precession mechanisms in the system, and how these conclusions may be relevant to other compact galactic X–ray sources. Despite its highly unusual properties, SS 433 can be valuable for exploring these issues because of the ease of following temporal changes in the system via ground based spectroscopy and photometry with small telescopes.

117.191 **Accretion disk theory.**
R. Hoshi.
Japan – U.S. seminar on galactic and extragalactic compact X–ray sources, p. 143 – 151 (1985). – See Abstr. 012.034.
Recent Tenma observations of four low mass X–ray binaries provide us with an important clue to the structure of an inner accretion disk surrounding a weakly magnetized neutron star. Motived from suggested emitting regions, thermal instabilities in the inner disk are examined with a generalized formula for viscous stress. The outer boundary of the unstable region is found to depend critically on the formula for viscous stress. Observational constraints to model parameters and a possible alternative model to account for the observed soft component are also discussed.

117.192 **Optical outburst of HDE 245770–A0535+26 near the expected X–ray maximum.**
K. L. Maslennikov.
Pis'ma Astron. Zh., Tom 12, No. 6, p. 458 – 460 (1986). In Russian. English translation in Sov. Astron. Lett., Vol. 12.
UBV photometry of the optical component of the X–ray binary HDE 245770–A0535+26 was carried out in April 12 – 18, 1985. The brightness increase (by $0^{m}.25$ in the *U* band) was observed four days before an X–ray maximum of A0535+26 predicted from the 111–day period.

117.193 **Spectral investigation of the AM Herculis–type binary system E2003+225.**
N. F. Vojkhanskaya.
Pis'ma Astron. Zh., Tom 12, No. 6, p. 468 – 475 (1986). In Russian. English translation in Sov. Astron. Lett., Vol. 12.
Results of spectral observations in 1984 of the new AM Her–type binary E2003+225 are presented. Changes of equivalent widths of various lines during the orbital period are different. The radial velocity curves obtained from the line peaks are distorted by matter–escaping from the system. The radial velocity curves obtained from the lines bases are shifted with respect to the peak curves obtained from peaks by 0.1*P*. It is shown that the mass ratio of the components is very small in the E2003+225 system. The matter ejected from the system absorbs the X–ray emission of the main component, thus leading to a low X–ray luminosity and an extremely soft X–ray spectrum.

117.194 **Interacting stellar wind models of eruptive symbiotic systems.**
T. M. Girard.
Diss. Abstr. Int., Sect. B, Vol. 46, No. 5, p. 1589 (1985). Thesis, Iowa State University, 90 pp. (1985), Order No. DA8514403.

117.195 **Cyclotron radiation from magnetic cataclysmic variables.**
P. E. Barrett.
Diss. Abstr. Int., Sect. B, Vol. 46, No. 6, p. 1955 – 1956 (1985). Thesis, The Louisiana State University and Agricultural and Mechanical Col., 188 pp. (1985). Order No. DA8517723.

117.196 **The binary frequency of the OB runaway stars.**
D. R. Gies.
Diss. Abstr. Int., Sect. B, Vol. 46, No. 9, p. 3090 – 3091 (1986). Thesis, University of Toronto (1985).

117.197 **Starpot cycles of RS Canum Venaticorum type stars.**
Y. W. Kang.
Diss. Abstr. Int., Sect. B, Vol. 46, No. 9, p. 3091 (1986). Thesis, The University of Florida, 154 pp. (1985). Order No. DA8523845.

117.198 **B and V photoelectric photometry and light curve solution of the interacting binary systems HI Puppis, BL Eridani and SY Horologii.**
J. R. Kern.
Diss. Abstr. Int., Sect. B, Vol. 46, No. 10, p. 3481 (1986). Thesis, Clemson University, 265 pp. (1985). Order No. DA8527702.

117.199 **Spectrophotometric observations of symbiotic stars.**
A. P. Ipatov, B. F. Yudin.
Sov. Astron., Vol. 29, No. 5, p. 538 – 543 (1985). English translation of 40.117.149.

117.200 **A spectral investigation of XZ Cephei – a close binary star of an early spectral class.**
L. V. Glazunova, V. G. Karetnikov.
Sov. Astron., Vol. 29, No. 5, p. 544 – 548 (1985). English translation of 40.117.150.

117.201 **The parameters of the X–ray binary system V779 Cen = 4U 1119–60 (Cen X–3).**
T. S. Khruzina, A. M. Cherepashchuk.
Astron. Zh., Tom 63, Vyp. 3, p. 494 – 507 (1986). In Russian. English translation in Sov. Astron., Vol. 30, No. 3.
The analysis of optical light curves of the system Cen X–3 in the «standard» Roche model with an accretion disk showed that the amplitude of optical variations can be explained if the presence of an accretion disk around the neutron star is taken into account with radius of $(3.5\pm1.2)R_{\odot}$ and the optical luminosity of $(1-3)\times10^{35}$erg/s, whose major part is probably connected with reradiation of the O7 supergiant's ultraviolet and optical light. Masses of the optical and X–ray components are equal to $(17.7\pm0.2)M_{\odot}$ and $(1.1\pm0.2)M_{\odot}$ respectively, the inclination of the orbital plane of the system is $i = 82° - 85°$; the filling by the optical star of its Roche lobe is close to unity, $\mu = 0.995\pm0.005$. The reason of the long–period–variation of the light–curve amplitude of Cen X–3 analysed in this model is a change of the radius of the accretion disk from 3.3 $R_{\odot}$ to 4.4 $R_{\odot}$.

117.202 **Spectral investigations of the polar AN Ursae Majoris.**
N. F. Vojkhanskaya.
Astron. Zh., Tom 63, Vyp. 3, p. 516 – 526 (1986). In Russian. English translation in Sov. Astron., Vol. 30, No. 3.
The results of spectral observations carried out in 1981 and 1983 are presented. Variations of the equivalent widths, profiles, intensities and radial velocities of emission lines are studied during the orbital period. These proved to be different in different years.

117.203 **An optical and radio investigation of the active RS CVn star HD 127535.**
J. L. Innis, D. W. Coates, K. Thompson, G. J. Nelson, O. B. Slee, A. E. Wright.
Proc. Astron. Soc. Aust., Vol. 6, No. 2, p. 160 – 164 (1985).
The authors present some preliminary results of an optical and radio study of the very active RS CVn binary HD 127535. Photometric measurements show the presence of a large amplitude wave which exhibits marked changes in shape and range on time scales as short as a few months. This photometric variation is almost certainly due to large cool starspots on the cooler, more luminous component. Photometry obtained two cycles before the 8.4 GHz observations suggest a possible correlation between the radio emission and the photometric wave, i.e. spot visibility.

117.204 **Models and theoretical spectra of accretion discs in dwarf novae.**
S. Kříž, I. Hubený.
Bull. Astron. Inst. Czech., Vol. 37, No. 3, p. 129 – 142 (1986).
The authors present a new method of modelling accretion discs in dwarf novae devised to determine self–consistently the structure of a disc together with the radiation field. The basic idea is to divide a (cylindrically symmetric) disc into a set of concentric rings, each of them behaving like an independent planeparallel radiating slab.

117.205 **New technique to compute the geometrical elements of the eclipsing of W UMa–type.**
S. R. Jabbar, H. A. Fleyeh, J. M. Al Razzaz.
Astrophys. Space Sci., Vol. 123, No. 2, p. 325 – 333 (1986).
The aim of this paper is to present a general method for computation of the geometrical elements of totally eclipsing W UMa (close, or in actual contact) binary systems in the frequency–domain. The results are compared with the results calculated by different methods, that have used the same observation references. The improved photometric perturbation terms have been discussed and its effects are computed for both (primary and secondary) minima.

117.206 **Orbital elements of the binary X–ray pulsar GX 301–2.**
N. Sato, F. Nagase, N. Kawai, R. L. Kelley, S. Rappaport, N. E. White.
Astrophys. J., Vol. 304, No. 1, p. 241 – 248 (1986).
X–ray outbursts from GX 301–2 were observed with Hakucho on two occasions during 1982 April and May and with Tenma in 1984 April. Pulse arrival times measured with Hakucho were combined with those previously measured with Ariel 5 and SAS 3. The joint timing analysis yields an orbital period of 41.5 days; this is consistent with the period derived from the X–ray flaring events of the source. The orbital elements of the binary system are presented.

117.207 **The optical spectrum of HDE 226868 = Cygnus X–1. II. Spectrophotometry and mass estimates.**
D. R. Gies, C. T. Bolton.
Astrophys. J., Vol. 304, No. 1, p. 371 – 388 (1986).
The authors have carried out spectrophotometric measures on blue spectra of HDE 226868 obtained by orbital phase averaging of 55 low signal–to–noise–spectra into 10 phase bins to search for phase–dependent variations. There is no evidence for statistically significant variations of the spectral type about the mean value of O9.7 Iab. The authors show how the rotational broadening of the photospheric absorption lines can be used to place further constraints on the component masses. The minimum masses of the visible star and its unseen companion are 20 and 7 $M_{\odot}$, with probable values of 33 and 16 $M_{\odot}$, respectively.

117.208 **The optical spectrum of HDE 226868 = Cygnus X–1. III. A focused stellar wind model for He II λ4686 emission.**
D. R. Gies, C. T. Bolton.
Astrophys. J., Vol. 304, No. 1, p. 389 – 393 (1986).
The authors have computed theoretical He II λ4686 emission line profiles for Friend and Castor's focused stellar wind model for the HDE 226868 = Cyg X–1 system. These were added to theoretical photospheric profiles and compared to the observed profiles. The composite profiles reproduce most of the significant features of the variations of the observed profiles if the visible star is close to filling its Roche lobe.

117.209 **Ultraviolet observations of V471 Tauri: evidence for a nova–like outbursts?**
F. C. Bruhweiler, E. M. Sion.
Astrophys. J., Lett. Ed., Vol. 304, No. 1, p. L21 – L24 (1986).
High–resolution ultraviolet spectroscopy obtained with the IUE reveal what appears to be evidence for high–velocity expanding circumbinary gas around the Hyades red dwarf–white dwarf binary system V471 Tauri. In particular, a sharp absorption feature for C II λ1334 and Si II λ1260 appears at -590 km s^{-1} in the line of sight to this system. The data indicate that this absorption occurs at a considerable distance from the binary. Since there is some historical evidence that V471 Tauri may have undergone a nova–like eruption, the authors speculate that the origin of these features might be linked to such an event.

117.210 **Solar oscillations and rotation of close binary systems in the Galaxy: 160 minute period.**
V. A. Kotov.
Izv. Krymskoj Astrofiz. Obs., Tom 74, p. 69 – 84 (1986). In Russian. English translation in Bull. Crimean Astrophys. Obs., Vol. 74.

The previous conclusion that the 160^m period of the global oscillations of the sun might be related to the external 160^m gravitational wave from an unknown source is checked by statistical analysis of the period distribution of close binary stars. It is found that the observed distribution of about 4000 binaries shows a significant (3.5 σ confidence) lack of periods $\sim 2 \cdot z \cdot P_0$ and an excess (2.5 σ confidence) of periods $\sim (2 \cdot z + 1) \cdot P_0$, where z is integer and $P_0 = 160^m$. The overall significance of this 160^m effect is about 4.1 σ. A precise determination of the characteristic period leads to the value $160^m0 \pm 0^m1$ which is in fair agreement with the solar oscillation period $160^m0101 \pm 0^m0007$. A similar analysis of the distribution of 97 catalysmic binaries shows that the same frequency, P_0^{-1}, is characteristic for these objects. It is conjectured that the frequency $P_0^{-1} = 104.160$ μHz in the distribution of orbital rates of the galactic binaries is a consequence of the 160^m gravitational radiation which might be cosmological (relict) by nature.

117.211 **The nature of dwarf novae.**
A. Schwarzenberg–Czerny.
Eruptive phenomena in stars, p. 339 – 349 (1986). – See Abstr. 012.039.

117.212 **Analysis of light curves of DQ Herculis based on five-colour photometry in 1982 – 1985.**
E. S. Dmitrienko.
Eruptive phenomena in stars, p. 357 – 358 (1986). – See Abstr. 012.039.

117.213 **The amplitude – cycle length relation of long–cyclic cataclysmic binaries.**
G. A. Richter.
Eruptive phenomena in stars, p. 361 – 362 (1986). – See Abstr. 012.039.

117.214 **Mass transfer bursts in the symbiotic binary system CH Cyg during the maximum of its activity in the year 1982.**
D. Chochol, A. Skopal, T. S. Galkina.
Eruptive phenomena in stars, p. 363 – 364 (1986). – See Abstr. 012.039.

117.215 **On the period of the symbiotic star AG Pegasi.**
R. Luthardt.
Eruptive phenomena in stars, p. 365 – 366 (1986). – See Abstr. 012.039.

117.216 **Outburst activity in cataclysmic binaries: parallel evolution or activity cycles?**
Z. Urban.
Eruptive phenomena in stars, p. 367 – 368 (1986). – See Abstr. 012.039.

117.217 **Influence of the accretion column's asymmetry on the orbital variability of polars.**
I. L. Andronov.
Eruptive phenomena in stars, p. 369 – 370 (1986). – See Abstr. 012.039.

117.218 **Influence of the magnetic field on accretion in close binary systems.**
I. L. Andronov.
Eruptive phenomena in stars, p. 371 – 372 (1986). – See Abstr. 012.039.

117.219 **Starspot problems.**
K. Oláh.
Eruptive phenomena in stars, p. 393 – 407 (1986). – See Abstr. 012.039.

117.220 **On the dust jet model of SS 433.**
H. L. Helfer, M. P. Savedoff.
Astrophys. J., Vol. 304, No. 2, p. 581 – 584 (1986).

The dust jet model for production of narrow γ–ray lines from SS 433 is examined and shown to be implausable. Sputtering rates at disk temperatures are high, and self–absorption rates in the disk and jet are important; both imply that dust particles would easily be destroyed. The energy deposited in the dust jet particles by collisions with the ambient protons should be radiated in the optical and infrared in amounts greatly exceeding that observed. At present, there is no viable model for the steady state production of the γ–ray line radiation.

117.221 **Phase of periodic variations in SS 433.**
L.–z. Fang, S.–x. Li.
Kexue Tongbao, Vol. 30, No. 4, p. 468 – 470 (1985). Abstr. in Phys. Abstr., Vol. 89, No. 1273, Entry 18799 (1986).

117.222 **Accretion disk instabilities.**
F. Meyer.
MPA Rep., No. 227, 19 pp. (1986). Invited paper presented at IAU Colloquium No. 89 "Radiation hydrodynamics in stars and compact objects", Copenhagen, June 11 – 20, 1985.

117.223 **Numerical studies of wind accretion.**
U. Anzer, G. Börner, J. J. Monaghan.
MPA Rep., No. 230, 33 pp. (1986). Submitted to Astron. Astrophys.

117.224 **The discovery of 3.9 hour periodic dips in the X–ray intensity of XB 1254–690.**
T. J.–L. Courvoisier, A. N. Parmar, A. Peacock, M. Pakull.
ESO Sci. Prepr., No. 438, 40 pp. (1986). To appear in Astrophys. J.

117.225 **Detection of an extremely active state of AM Canum Venaticorum.**
K. R. Narayanan Kutty, T. M. K. Marar, V. N. Padmini, S. Seetha, K. Kasturirangan, U. R. Rao, J. C. Bhattacharyya, S. Mohin, K. Jayakumar.
Indian Inst. Astrophys., Prepr., No. 199, 7 pp. (1985). To appear in IAU Colloq. No. 87.

117.226 **Observations and light curve of NSV 7530.**
J. Pensado.
Bol. Astron. Obs. Madr., Vol. 11, No. 4, p. 17 – 23 (1985).

Blue and yellow photoelectric observations of the eclipsing binary system NSV 7530 were made on fifteen nights during the 1984 – 85 seasons. A listing of all the observations and the corresponding light curve and ephemeris is given.

117.227 **Dwarf novae in outburst: modelling the observations.**
J. E. Pringle, F. Verbunt, R. A. Wade.
Prepr. Steward Obs., No. 634, 56 pp. (1986). To appear in Mon. Not. R. Astron. Soc.

117.228 **A spectroscopic study of the cataclysmic binary star SU Ursae Majoris.**
J. T. Thorstensen, R. A. Wade, J. B. Oke.
Prepr. Steward Obs., No. 658, 34 pp. (1986).

117.229 **Postthermonuclear runaway angular momentum loss in cataclysmic binaries.**
J. MacDonald.
Astrophys. J., Vol. 305, No. 1, p. 251 – 260 (1986).

During the common envelope phases that follow hydrogen shell flashes in the white dwarf component of cataclysmic binaries, angular momentum and energy are transferred from the orbit to the common envelope by friction processes. The common envelope is ejected due to the frictional heating, carrying away angular momentum from the binary system. An estimate of the rate of this frictional angular momentum loss shows that it can be the dominant angular momentum loss mechanism for cataclysmic binaries with orbital period $P_{orb} \lesssim 6$ hr.

117.230 **Mass transfer in cataclysmic variables: clues from the dwarf nova period distribution.**
A. W. Shafter, J. C. Wheeler, J. K. Cannizzo.
Astrophys. J., Vol. 305, No. 1, p. 261 – 266 (1986).
Evidence is presented in support of the hypothesis that the mean mass–transfer rate at a given orbital period, $\langle \dot{M}(P) \rangle$, is not continuous across the 2 – 3 hr gap in the orbital period distribution for cataclysmic variables. The authors point out that although dwarf novae comprise nearly half (48%) of all disk systems with orbital periods less than 10 hr, only three systems out of the 22 with periods between 3 and 4 hr appear to be dwarf novae. It is argued that mass–transfer rates must be generally higher for systems with orbital periods greater than 3 hr relative to systems with periods less than 2 hr.

117.231 **A phenomenological triple star scenario for SS 433.**
A. C. Fabian, P. P. Eggleton, P. Hut, J. E. Pringle.
Astrophys. J., Vol. 305, No. 1, p. 333 – 335 (1986).
The authors suggest that SS 433 is a triple star, containing a short–period ($\sim 1\overset{d}{.}5$) binary like Sco X–1, orbiting a massive OB star in 13^{d}. Polar jets produced by the Sco X–1–like close binary can precess, with the period observed, if this binary is misaligned with the larger binary. The possible evolutionary history of such a system is discussed.

117.232 **Structure of the accretion disk in the dwarf nova IP Pegasi.**
V. P. Goranskij, V. M. Lyutyj, S. Yu. Shugarov.
Sov. Astron. Lett., Vol. 11, No. 5, p. 293 – 297 (1985). English translation of 40.117.074.

117.233 **Synthetic light–curve method applied to the W UMa systems SY Hologii and VY Ceti.**
E. Lapasset, J. J. Clariá.
Astron. Astrophys., Vol. 161, No. 2, p. 264 – 270 (1986).
Photoelectric observations of SY Hor and VY Cet have yielded complete light curves in the UBV system. These new observations are used to derive up–dated ephemeris and to provide photometric solutions by the Wilson and Devinney's technique. Both systems are found to be W–type contact binaries with a shallow percentage of overcontact. Concerning the A and W classification for W UMa stars, a new photometric criterion – based on the asymmetry in the light curves at maximum light – is presented and applied to both binaries. This asymmetry is explained as being due to the limb darkening effect when applied to the Roche model configuration.

117.234 ***V*/*R* variability in X–ray emitting Be star binaries.**
K. M. V. Apparao, S. P. Tarafdar.
Astron. Astrophys., Vol. 161, No. 2, p. 271 – 274 (1986).
In Be star binaries with a compact object, the X–ray emission from the compact object forms an ionized sphere around itself. The line emission from this ionized sphere moves across the Be star emission line profile during the orbital motion of the compact object, leading to a variation of the ratio of intensities of the "Violet" and "Red" components of the line profile.

117.235 **The symbiotic star RR Tel.**
M. A. Hayes, H. Nussbaumer.
Astron. Astrophys., Vol. 161, No. 2, p. 287 – 295 (1986).
The authors investigate the time variability of RR Tel in the stronger ultraviolet emission lines. They find that there is a decrease in the UV line fluxes over the period 1978 – 1984, but they find no evidence for correlation with the infrared fluxes which have a 387 day periodicity. The authors obtain estimates of the electron temperature and density from emission line intensity ratios and try to model the region where the emission lines originate.

117.236 **Contact binaries. II. Metastable systems and limit cycles.**
H. Kähler, B. Matraka, A. Weigert.
Astron. Astrophys., Vol. 161, No. 2, p. 296 – 302 (1986).
A variety of linear and non–linear thermal phenomena of contact binary models is analyzed in the neighbourhood of the border between thermally stable and unstable systems. The transition between stability and instability proceeds via an intermediate region of metastability. When sufficiently perturbed, metastable and unstable systems evolve into periodic relaxation oscillations. The metastability arises from a subcritical bifurcation in the sequence of steady solutions. It occurs where, in the linear stability analysis, the real part of a conjugate complex pair of eigenvalues vanishes. The imaginary part of these eigenvalues is shown to give the period of the relaxation oscillations.

117.237 **Optical and UV spectroscopy of the peculiar RS CVn system RT Lacertae.**
D. P. Huenemoerder, S. C. Barden.
Astron. J., Vol. 91, No. 3, p. 583 – 589 (1986).
The authors have obtained spectra in the Hα and Hβ regions of the peculiar double–lined RS CVn binary RT Lacertae in the fall of 1984. Limited IUE long–wavelength low– and high–resolution spectra were obtained concurrently. The consistency of the data with both theory and other observations strongly suggest that RT Lac is a mass–transfer system, and that this is the primary cause of its activity. The phase behavior of the low–resolution data indicates "traditional" chromospheric activiy as well.

117.238 **The single–line Wolf–Rayet star EZ CMa revisited: additional evidence for a 3.8 day binary period.**
R. Lamontagne, A. F. J. Moffat, A. Lamarre.
Astron. J., Vol. 91, No. 4, p. 925 – 934 (1986).
Precision continuum photometry shows that the variable WN 5 star EZ CMa has exhibited a complex but stably modulated light curve over the past five years, with one and only one period $P = 3.766 \pm 0.001$ days. Together with new spectroscopic data, as well as previously published spectroscopic, photometric, and polarimetric observations, these photometric variations support the hypothesis that EZ CMa is a second–stage Wolf–Rayet binary probably containing a neutron star.

117.239 **Orbital studies of cataclysmic binaries. II. Three objects from the Palomar–Green sample.**
J. R. Thorstensen.
Astron. J., Vol. 91, No. 4, p. 940 – 950 (1986).
The authors obtained radial–velocity spectra of PG 0834+488, PG 1003+678 = CH UMa, PG 1711+336 = V795 Her and determined their orbital periods.

117.240 **FO Virginis: a pre–contact binary.**
S. W. Mochnacki, J. D. Fernie, R. Lyons, F. H. Schmidt, R. O. Gray.
Astron. J., Vol. 91, No. 5, p. 1221 – 1228 (1986).
The star FO Vir, an eclipsing binary with a period of 0.776 days, has been observed photometrically and spectroscopically. Light–curve and line–profile synthesis shows that FO Vir is semidetached with a mass ratio of 0.15 ± 0.02. The A7 V primary component very nearly fills its Roche lobe. Radial velocities from the observed and synthesized profiles have been measured by cross correlation. Masses and the mean densities of the components have been obtained. The period and mass ratio are typical of the more extreme "A–type" W UMa systems, which suggests that FO Vir will evolve into a contact binary of this type.

117.241 **A 6 centimeter radio survey of short–period active binary stars.**
S. A. Drake, T. Simon, J. L. Linsky.
Astron. J., Vol. 91, No. 5, p. 1229 – 1232 (1986).
The authors observed 13 binaries with periods in the range of 0.2 – 2.0 days at 6 cm wavelength with the VLA. Eight out of these 13 systems were detected, of which seven are RS Canum Venaticorum systems and one is an Algol system, with observed fluxes in the range of 0.3 – 5.0 mJy. The authors briefly discuss the individual characteristics of the detected sources. As a group, relative to active binaries of longer orbital periods, the short–period active binaries have a slightly lower mean radio luminosity. There is also a clear correlation of high radio luminosity with high X–ray luminosity evident in these short–period systems.

117.242 **Spectroscopic orbits of symbiotic stars: preliminary results.**
M. R. Garcia.
Astron. J., Vol. 91, No. 6, p. 1400 – 1415 (1986).
Preliminary results of a systematic search for radial–velocity variations due to orbital motion in symbiotic stars are reported. Seventeen symbiotics were observed and radial–velocity variations in the M–giant primary were observed in nine. In eight of the stars, the photospheric absorption lines of the M giant could not be seen clearly and radial velocities could not be measured. The observed radial–velocity curves are commensurate with those expected from a group of binaries having random sin i and $\sim 20\ \mathrm{km\ s^{-1}}$ orbital velocities. The curves suggest orbital periods for seven stars. Knowledge of the orbits allows identification of those symbiotics that are close to filling their Roche lobes.

117.243 **Variations in mass transfer during the activity maximum of the binary symbiotic system CH Cygni.**
D. Chochol, A. Skopal, T. S. Galkina.
Contrib. Astron. Obs. Skalnaté Pleso, Vol. 14, p. 87 – 95 (1986).
Based on the study of the variations in the radial velocities of absorption spectral lines of ionized metals Ti II and Sc II of the symbiotic star CH Cyg, it is demonstrated that variations in mass transfer between the components of the system occurred on a time scale of 20 days during the activity maximum of this object in 1982.

117.244 **An IUE search for white dwarf companions to field halo blue stragglers.**
B. W. Carney, H. E. Bond.
Bull. Am. Astron. Soc., Vol. 18, No. 1, p. 635 (1986). Abstract. – See Abstr. 010.066.

117.245 **Orbital periodicity in the low–mass X–ray binary 4U 1957 + 11.**
J. R. Thorstensen.
Bull. Am. Astron. Soc., Vol. 18, No. 1, p. 636 (1986). Abstract. – See Abstr. 010.066.

117.246 **X–ray emission from the symbiotic system CH Cyg.**
D. A. Leahy, A. R. Taylor.
Bull. Am. Astron. Soc., Vol. 18, No. 1, p. 636 (1986). Abstract. – See Abstr. 010.066.

117.247 **TT Arietis: high–state spectroscopy and the phase–magnitude relationship.**
J. B. Hutchings, B. Thomas, R. Link.
Publ. Astron. Soc. Pac., Vol. 98, No. 603, p. 507 – 510 (1986).
Measurements have been made on ~50 spectra of TT Ari taken during high states in 1979 and 1981. The phasing of the velocity variation appears to be the same for all high–state spectra, and varies smoothly with mean system magnitude in lower states. The emission–line equivalent widths and their orbital phase dependence do not appear to be related only to system state. There is normally an orbital modulation of line equivalent widths when the system is bright, but its phasing is variable. The velocity amplitude in high state appears to be variable by at least a factor two.

117.248 **SU Ursae Majoris, 1970 – 79.**
J. E. Isles.
J. Br. Astron. Assoc., Vol. 96, No. 4, p. 217 – 222 (1986).
Outbursts of the dwarf nova SU UMa in 1970 – 79 are listed, and the star's behaviour since 1926 is discussed. The range, $10\overset{m}{.}4 - 14\overset{m}{.}6$ is greater than is given in the catalogue, and the mean cycle length, $12\overset{d}{.}7$, is much shorter. Since the early 1960s, short outbursts have become fainter and more frequent, but supermaxima have become less frequent. Supermaxima show oscillations with an apparent period of 3 days.

117.249 **Symbiotische Sterne.**
R. Luthardt.
Astron. Raumfahrt, 24. Jahrg., Heft 3, p. 66 – 72 (1986).

117.250 **The use of ST in research about evolutionary connections among various types of close binary systems.**
D. Chochol, S. Mancuso, L. Milano, M. Rigutti, G. Russo, A. Vittone.
Mem. Soc. Astron. Ital., Vol. 56, No. 4, p. 653 – 663 (1985). – See Abstr. 012.054.
Some particular open problems on the origin and evolution of close binary systems are presented. Suggestions are made to use the ST facilities to contribute to their solution.

117.251 **Stream models and period variations for contact binaries.**
F. van't Veer.
Inst. Astrophys. Paris, Pré–Publ., No. 123, 8 pp. (1986). Paper presented at the Beijing meeting on close binary systems, 7 – 13 November 1985.

117.252 **Evolution of the spectra of the symbiotic star Z And.**
R. Gravina.
Inf. Bull. Variable Stars, No. 2844, 2 pp. (1986).

117.253 **The variability of HD 17433.**
J. A. Eaton, C. H. Poe.
Inf. Bull. Variable Stars, No. 2846, 3 pp. (1986).

117.254 **HD 174853 – a new eclipsing binary.**
D. P. Hube, M. Anderson.
Inf. Bull. Variable Stars, No. 2848, 4 pp. (1986).

117.255 **Photoelectric times of minimum light of BL Eridani.**
J. R. Kern, B. B. Bookmyer.
Inf. Bull. Variable Stars, No. 2850, 2 pp. (1986).

117.256 **Optical behaviour of the polar AM Her in 1985.**
W. Götz.
Inf. Bull. Variable Stars, No. 2851, 2 pp. (1986).

117.257 **Improved ephemeris for the short period, eclipsing cataclysmic variable V1315 Aquilae.**
J. T. Annis.
Inf. Bull. Variable Stars, No. 2858, 4 pp. (1986).

117.258 **A photoelectric light curve and period study for CZ Aqr.**
J. R. Bruton.
Inf. Bull. Variable Stars, No. 2865, 4 pp. (1986).

117.259 **CH Cygni in eclipse? An appeal for observations.**
T. Tomov.
Inf. Bull. Variable Stars, No. 2866, 4 pp. (1986).

117.260 **Photoelectric light curves of RR Leporis.**
B. B. Bookmyer, D. R. Faulkner, R. G. Samec.
Inf. Bull. Variable Stars, No. 2873, 2 pp. (1986).

117.261 **The Mg II light curve of AR Lac.**
E. Derman, Ü. Kiziloglu, F. Tokdemir.
Inf. Bull. Variable Stars, No. 2881, 3 pp. (1986).

117.262 **Observations of W Serpentis stars.**
N. L. Markworth, R. L. Gann, J. B. Rafert, E. Smith, R. E. Wilson.
Inf. Bull. Variable Stars, No. 2887, 2 pp. (1986).

117.263 **Behaviour of the X–ray binary V1727 Cygni = 4U 2129 + 47 in 1985.**
W. Götz.
Inf. Bull. Variable Stars, No. 2895, 2 pp. (1986).

117.264 **A secondary photometric period of TT Arietis.**
I. Semeniuk, A. Schwarzenberg–Czerny, H. Duerbeck, M. Hoffmann, J. Tremko.
Inf. Bull. Variable Stars, No. 2899, 3 pp. (1986).

117.265 **1985/86 multi–color photoelectric observations of the starspot activity on V711 Tauri (HR 1099).**
S. W. Wacker, E. F. Guinan.
Inf. Bull. Variable Stars, No. 2903, 4 pp. (1986).

117.266 **Photometric observations of visual double star ADS 9537 (BV Dra, BW Dra) by area scanning technique.**
S. Yu. Gorda.
Inf. Bull. Variable Stars, No. 2906, 4 pp. (1986).

117.267 **Variations of the light curve of VW Cep.**
F. Santiago, J. A. Quesada, F. Palacios,
J. A. Sanchez, M. Cobo.
Inf. Bull. Variable Stars, No. 2907, 4 pp. (1986).

117.268 **Photometric solutions for seven close binaries of intermediate spectral types: BL And, ST Aqr, DO Cas, TT Her, TZ Lyr, SW Lyn and RU UMi.**
J. Kałużny.
Acta Astron., Vol. 35, Nos. 3–4, p. 327 – 350 (1985).
Light curves of seven close binaries of intermediate spectral types were analysed with the Wilson–Devinney method. The problem of uniqueness of solutions obtained with the aid of the light curve synthesis technique is discussed. It is demonstrated that the obtained parameters and configurations are sensitive to the choice of the gravity darkening coefficient.

117.269 **On the S–wave components of the emission lines in the spectra of cataclysmic variables.**
J. Smak.
Acta Astron., Vol. 35, Nos. 3–4, p. 351 – 356 (1985).
Emission lines in the spectra of cataclysmic variables consist of several components. One of them is the so–called S–wave component. The author uses geometric and kinematic evidences to determine more precisely the location of the S–wave emitting region and then uses the available stream and disk models to explain the dichotomy.

117.270 **Statistical analysis of outbursts and superoutbursts of VW Hyi.**
J. Smak.
Acta Astron., Vol. 35, Nos. 3–4, p. 357 – 367 (1985).
The amplitude and width of normal outbursts of VW Hyi are correlated with the length of the preceding cycle, but not with the length of the following cycle. Two types of supercycles are identified: Type L, in which the average length of the last two cycles is longer than 30 days and Type S, in which it is shorter than 23 days. No obvious relation of the two types to other parameters of supercycles or superoutbursts seems to exist.

117.271 **A comparison between the Roche model and the model of triaxial ellipsoids in close binary systems.**
P. G. Niarchos.
Acta Astron., Vol. 35, Nos. 3–4, p. 433 – 440 (1985).
The Roche model for close binary systems is compared with the model of homogeneous triaxial ellipsoids as far as the position of the inner Lagrangian point L_1 is concerned. The result of such a comparison is that the position of L_1 is approximately the same in both models.

117.272 **SW Lacertae, W Ursae Majoris, YY Eridani, and the prevalence of starspots in cool contact binaries.**
J. A. Eaton.
Acta Astron., Vol. 36, No. 1, p. 79 – 103 (1986).
Detailed light curve analyses have been carried out for three cool contact binaries to determine (1) what makes the larger component appear cooler and (2) how a period change affects the structure of the common envelope.

117.273 **Contact binaries with components in poor thermal contact. II. WZ Cephei.**
J. Kałużny.
Acta Astron., Vol. 36, No. 1, p. 105 – 111 (1986).
B and *V* photoelectric light curves of the short period eclipsing binary WZ Cep have been analysed with the Wilson–Devinney method. The author's solutions indicate that most probably WZ Cep is a contact binary with the components having significantly different effective temperatures.

117.274 **Contact binaries with components in poor thermal contact. III. AU Serpentis and FT Lupi.**
J. Kałużny.
Acta Astron., Vol. 36, No. 1, p. 113 – 120 (1986).
Photoelectric light curves of AU Ser and FT Lup are analysed by means of the Wilson–Devinney code. Both systems are found to be contact binaries. The secondary components of both systems have disturbed brightness distributions. The author suggests that the pronounced asymmetry of the light curves of AU Ser is due to a "hot spot" localized at the neck between the stars.

117.275 **Contact binaries with components in poor thermal contact. IV. AG Virginis.**
J. Kałużny.
Acta Astron., Vol. 36, No. 1, p. 121 – 129 (1986).
New photometric observations of the short period eclipsing binary AG Vir were obtained in 1984. The light curve still exhibits its characteristic asymmetry which has been observed by several observers in the past. The orbital period of AG Vir has not varied significantly since 1960. The Binnendijk's (1969) light curves are analysed by means of the Wilson–Devinney code. The system is found to be in contact.

117.276 **FT Lupi: UBV photometry and synthetic solution.**
S. L. Lipari, R. F. Sisteró.
Mon. Not. R. Astron. Soc., Vol. 220, No. 4, p. 883 – 894 (1986).
The authors present 1423 UBV observations of the eclipsing binary FT Lupi observed over 6 yr at Bosque Alegre Station of Córdoba Observatory. A period study based on 32 photoelectric minima and on 181 previous photographic and visual minima covering a time base of 80 yr gives a period change of $\dot{P} = -0.016\ s\ yr^{-1}$. The Wilson–Devinney analysis leads to a solution of a semi–detached configuration. The primary component is a 'normal' F2 V star filling its Roche lobe. The secondary one is a K5 – K7 object overluminous and oversized as compared with a main–sequence star, but fractionally smaller than its lobe. A hot area on the following hemisphere of the secondary component could account for the observed asymmetries in the light curve. The conservative mass transfer as inferred from the observed $\dot{P}/P$ is $\dot{M} = -1.3x10^{-7} M_{\odot} yr^{-1}$.

117.277 **The flicker spectrum of YZ Cancri.**
Y. Elsworth, J. F. James.
Mon. Not. R. Astron. Soc., Vol. 220, No. 4, p. 895 – 899 (1986).
High–speed Fourier Transform photometry has been used to observe the light curve of YZ Cancri. The flicker spectrum exhibits an inverse relation between amplitude and frequency with a variable power law, with evidence for a high–frequency cut–off point which is also variable. The authors regard this as evidence that the source of the flickering lies within an optically dense region of variable size, and that the size changes when the star is in outburst.

117.278 **Neutron star evolution in eccentric–orbit binaries.**
V. S. Gnusareva, V. M. Lipunov.
Sov. Astron., Vol. 29, No. 6, p. 645 – 649 (1985). English translation of 40.117.301.

117.279 **The magnetic fields of accreting white dwarfs.**
K. A. Postnov.
Sov. Astron., Vol. 29, No. 6, p. 650 – 654 (1985). English translation of 40.117.302.

117.280 **Black holes in massive close binaries: observational data and evolutionary status.**
A. V. Tutukov, A. M. Cherepashchuk.
Sov. Astron., Vol. 29, No. 6, p. 654 – 658 (1985). English translation of 40.117.303.

117.281 **The determination of the radial velocity and the physical parameters of the envelope of the binary system β Lyrae.**
G. Rodríguez Caderot, M. J. Fernandez–Figueroa.
An. Fis., Ser. A, Vol. 81, No. 3, p. 168 – 179 (1985). In Spanish. Abstr. in Phys. Abstr., Vol. 89, No. 1281, Entry 67523 (1986).

117.282 **Stellar activity in a model of RS CVn systems observed with the IUE satellite.**
E. de Castro, M. J. Fernandez–Figueroa, A. Gimenez.
An. Fis., Ser. A, Vol. 81, No. 3, p. 190 – 199 (1985). In Spanish. Abstr. in Phys. Abstr., Vol. 89, No. 1281, Entry 67524 (1986).

117.283 **Photometric investigation of the semidetached binary system TW Dra.**
J. Papoušek, J. Tremko, M. Vetesnik.
Folia Fac. Sci. Nat. Univ. Purkynianae Brun., Phys., Vol. 25, No. 4, p. 5 – 64 (1984). Abstr. in Phys. Abstr., Vol. 89, No. 1283, Entry 78437 (1986).

117.284 **X–ray binaries.**
C. de Loore.
Europhys. News, Vol. 17, No. 2, p. 21 – 24 (1986). Abstr. in Phys. Abstr., Vol. 89, Nr. 1284, Entry 83613 (1986).

117.285 **The period variation of YZ Bootis.**
S.–y. Jiang.
Chin. Astron. Astrophys., Vol. 10, No. 2, p. 115 – 118 (1986). English translation of 40.117.299.

117.286 **Observation and study for Hα emission of 33 RS CVn stars.**
X.–f. Liu, H.–s. Tan.
Acta Astrophys. Sin., Vol. 6, No. 2, p. 132 – 142 (1986). In Chinese.

117.287 **The ejective process of particles due to a sudden increase of initial velocities in close binary systems.**
H.–n. Zhou, T. Zhu.
Acta Astron. Sin., Vol. 27, No. 2, p. 121 – 129 (1986). In Chinese.
Some mechanisms of mass transfer in detached binaries are discussed. It is assumed that there is a disk of particles around the primary. The particles defining the outer rim of the disk are ejected due to a sudden increase of initial velocities. The increase of initial velocities is divided into radial and transverse velocities. The path of particles under different initial conditions have been obtained by means of numerical integration subroutines.

117.288 **What do we know about Cyg X–1.**
P. C. Agrawal.
Cosmic pathways, p. 207 – 228 (1986). – See Abstr. 003.045.
The current state of observational knowledge of the peculiar X–ray binary Cyg X–1 is reviewed and the evidence indicating the presence of a black hole in this system is critically examined.

117.289 **On the discordant measurements of the radial velocity curve of the late–type star in the dwarf nova SS Cygni.**
E. L. Robinson, E.–H. Zhang, R. J. Stover.
Astrophys. J., Vol. 305, No. 2, p. 732 – 739 (1986).
The purpose of this paper is to investigate and eliminate the discrepancies among the various radial velocity measurements of the K5 V star in the dwarf nova SS Cyg. It is shown that the discrepancies have been caused (1) by erroneous measurements; and (2) by the heating of the K5 V star by the white dwarf and its accretion disk in SS Cyg. After correcting for the effects of residual heating and geometric distortions, the K velocity of the center of mass of the K5 V star is found to be 158 ± 3 km s^{-1}.

117.290 **The eclipses of cataclysmic variables. I. HT Cassiopeiae.**
E.–H. Zhang, E. L. Robinson, R. E. Nather.
Astrophys. J., Vol. 305, No. 2, p. 740 – 758 (1986).
HT Cas is an eclipsing dwarf nova with an orbital period of 106 minutes. The authors have obtained high–speed, multicolor photometry of HT Cas on 15 nights between 1982 October and 1984 October and on another two nights in 1985 January. HT Cas did not erupt from 1982 to 1984. In 1985 January an eruption was observed showing all characteristics of a supermaximum. To analyze the photometry, the authors have written a program to synthesize the light curves of eclipsing cataclysmic variables. The program includes the effects of the accretion disks. Physical parameters of the accretion disk in HT Cas are derived from this analysis.

117.291 **SS 433: a 6 year photometric record.**
J. C. Kemp, G. D. Henson, D. J. Kraus, L. C. Carroll, I. S. Beardsley, K. Takagishi, J. Jugaku, M. Matsuoka, E. M. Leibowitz, T. Mazeh, H. Mendelson.
Astrophys. J., Vol. 305, No. 2, p. 805 – 813 (1986).
A homogeneous record of optical photoelectric photometric measurements of this special object is presented, including over 800 observations on ~700 dates during 1979 – 1985, in the V band. A complete data table is furnished. The authors present: (1) a power spectrum identifying the dominant periodicities, $13^{d}.081$ and $162^{d}.5$; (2) updated ephemerides for these periods; and (3) a mean, binned light curve on the 13 day period; (4) a new synthesis of the average 162 day "precession" light curves broken down into four phase intervals on the 13 day cycle.

117.292 **Outburst of SW UMa.**
Br. Astron. Assoc. Circ., No. 656 (1986).

117.293 **The light curve of the dwarf nova, VW Hydri.**
F. M. Bateson, R. McIntosh.
Publ. Variable Star Sect., R. Astron. Soc. N.Z., No. 13 (C85), p. 1 – 46 (1986).
A light curve for VW Hyi covering the interval J.D. 2,442,600 to 2,444,502 is presented. Seventy–two observed outbursts are tabulated and individual light curves for most outbursts are reproduced on enlarged scales. The mean interval between successive outbursts, irrespective of their types, is $26^{d}.59$. The mean interval between successive super–outbursts is $174^{d}.90$.

117.294 **Polarized radiation from AM Herculis binaries.**
J. B. G. Canalle, R. Opher.
Rev. Mex. Astron. Astrofis., Vol. 12, Spec. Issue, p. 331 – 338 (1986). – See Abstr. 012.066.
Is the accretion column of AM Herculis binaries a circular or a flattened cylinder? The authors show that polarization data can answer this question. They evaluate the polarized radiation emitted by a circular cylinder accretion column and compare it with a flattened cylinder.

117.295 **Monochromatic reflection effect and temperature distributions on the distorted surfaces of close binary systems.**
A. Yamasaki.
Publ. Astron. Soc. Jpn., Vol. 38, No. 3, p. 449 – 461 (1986).
In the present work the author calculates the irradiated atmospheres in LTE and radiative equilibrium over the surface of the distorted secondary component in a model close binary system. The temperature distribution and the monochromatic albedos are computed. Also, the irradiation effect is evaluated at the substellar point in the primary component.

117.296 **High–resolution spectrum of the peculiar optical counterpart of the X–ray binary pulsar 4U 1907+09.**
M. Iye.
Publ. Astron. Soc. Jpn., Vol. 38, No. 3, p. 463 – 473 (1986).
A 0.5–Å resolution spectrum of the optical counterpart of the X–ray binary pulsar 4U 1907+09 was obtained. Stellar absorption lines of He I 6678 Å and probably of He I 5876 Å are identified. The distinct Hα emission of equivalent width 6.3 ± 0.3 Å shows a broad symmetric line profile. These lines indicate a primary star of spectral type B2 III–Ve. A geometrical model of the binary system 4U 1907+09 is constructed considering the fact that there are two X–ray flares, 0.45 orbital phase apart, in each orbital period.

117.297 **IRAS observations of binaries with compact objects.**
B. E. Schaefer.
Publ. Astron. Soc. Pac., Vol. 98, No. 604, p. 556 – 560 (1986).
The author reports on the infrared emission of over 200 binary systems which contain compact objects (either white dwarfs, neutron stars, or possibly black holes). The sample contains all the classic galactic X–ray binaries, many of the known cataclysmic variables, and a wide selection of other interesting objects. The infrared data were obtained by the *Infrared Astronomy Satellite* (*IRAS*) in four broad–band filters centered on 12, 25, 60, and 100 microns. The author finds evidence for a static dust cloud with a radius of roughly one astronomical unit which has formed around the classical nova RR Pic since its 1925 eruption. He presents the first far infrared light curve of a classical nova following eruption.

117.298 **The spectroscopic orbit of the W Ursae Majoris system V508 Ophiuchi.**
W.–x. Lu.
Publ. Astron. Soc. Pac., Vol. 98, No. 604, p. 577 – 580 (1986).
Intensified Reticon spectrograms of the W UMa system V508 Oph have been obtained at high dispersion and time resolution. They were measured by cross–correlation techniques. The spectroscopic orbital elements, determined for the first time, are $T_0 = 2446254.8384$, $K_1 = 119.1$ km s^{-1}, and $K_2 = 229.2$ km s^{-1}. A circular orbit is adopted.

117.299 **Accretion disks in mass–exchange binaries.**
J. E. McClintock.
Plasma penetration into magnetospheres, p. 70 – 84 (1986). – See Abstr. 012.071.
The phenomenology of cataclysmic variables is discussed; a recent photometric study of Z Cha is used to illustrate the main points. Accretion disks in low–mass X–ray binaries are discussed and photometric and spectroscopic data are presented. Observational data on accretion disk coronae and on accretion disks in X–ray novae, X–ray pulsars and burst sources are discussed briefly.

117.300 **Accretion disks in binary stars.**
E. Meyer–Hofmeister.
Plasma penetration into magnetospheres, p. 85 – 100 (1986). – See Abstr. 012.071.
The author shortly describes the condition for the formation of an accretion disk and shows where accretion disks can exist during the evolution of binary stars. The physics of accretion disks is discussed, and in particular the equations for the vertical structure of quasi–stationary disks are derived. Results of disk structure computations show the dependence on the parameters. Finally the influence of irradiation and a magnetic field are investigated.

117.301 **Accretion onto magnetic white dwarfs.**
G. Chanmugam.
Plasma penetration into magnetospheres, p. 153 – 164 (1986). – See Abstr. 012.071.
Accretion processes in magnetized cataclysmic variables, with emphasis on models for cyclotron emission and time dependent accretion, are discussed.

117.302 **Mass ejection in R Aquarii.**
A. G. Michalitsianos, M. Kafatos, J. M. Hollis.
Plasma penetration into magnetospheres, p. 215 – 222 (1986). – See Abstr. 012.071.
The symbiotic star R Aquarii embodies a number of distinguishing properties, the most significant being a collimated jet that extends approximately out to 6–arcsec NE from the star. The authors have obtained low resolution ultraviolet spectra of the R Aquarii jet with the International Ultraviolet Explorer (IUE). Their most recent IUE observations indicate the ionization state of the jet is increasing. Sub–arcsecond VLA observations of the R Aquarii have resolved the radio continuum structure into discrete parcels of emission, which are extended and highly collinear. R Aquarii provides evidence that indicates stellar jet activity is not unique to objects associated with high energy emission processes alone.

117.303 **On the origin of ultra high γ–rays from Cygnus X–3 and related sources.**
D. Kazanas, D. C. Ellison.
Plasma penetration into magnetospheres, p. 223 – 227 (1986). – See Abstr. 012.071.
The authors consider diffusive shock acceleration of ions as the mechanism responsible for the ultra high energy γ–ray emission from Cygnus X–3 and the several other binary X–ray sources that have been observed at γ–ray energies $\gtrsim 10^{15}$eV.

117.304 **Origin and evolution of X–ray binaries in globular clusters.**
J. E. Grindlay.
The evolution of galactic X–ray binaries, p. 25 – 41 (1986). – See Abstr. 012.076.
The relative numbers of low luminosity to high luminosity X–ray sources in globular clusters may be understood if the neutron stars in the latter were formed from high mass transfer onto the white dwarfs in a fraction of the former. This paper describes the possible evolutionary history of white dwarf and neutron star binaries in globular clusters and the final fate of these binaries as the cluster itself evolves and eventually disrupts. The author discusses new results, both observational and theoretical, for the bright bulge source GX 17+2 which suggest that the evolution of cluster white dwarf binaries into galactic bulge X–ray bursters does in fact occur.

117.305 **A note on the nature of the secondaries in ultra–compact binaries.**
G. J. Savonije.
The evolution of galactic X–ray binaries, p. 43 – 50 (1986). – See Abstr. 012.076.
This paper discusses the evolutionary state of the secondary stars in close binaries with orbital periods smaller than 50 – 80 min.

117.306 **The three types of high mass X–ray pulsator.**
R H. D. Corbet.
The evolution of galactic X–ray binaries, p. 63 – 74 (1986). – See Abstr. 012.076.
High mass X–ray pulsators are shown to be divisible into three distinct groups on the basis of their observational properties: (1) the Be star systems, (2) non–Be star systems with long pulse periods ($\sim 10^2$ – 10^3s), and (3) non–Be star systems with short pulse periods ($\sim 10^0$ – 10^1s). Differences in mass transfer processes between the groups are invoked to account for the different properties. The separation of high mass pulsators into three clearly defined groups enables some of the properties of a pulsator to be predicted if only a few of its parameters are known.

117.307 **Orbital period changes in massive X–ray binaries.**
R. L. Kelley.
The evolution of galactic X–ray binaries, p. 75 – 86 (1986). – See Abstr. 012.076.
This paper discusses possible causes for the decay of the orbital period observed in Cen X–3. Specifically, it studies effects of an orbiting third body on the period, tidal interactions, and effects of mass loss/mass transfer. Prospects for measuring changes in the orbital periods of several other X–ray binaries are also considered.

117.308 **Some remarks on classifications of low–mass X–ray binaries.**
J. van Paradijs, W. H. G. Lewin.
The evolution of galactic X–ray binaries, p. 187 – 193 (1986). – See Abstr. 012.076.
According to current evolutionary scenarios for low–mass X–ray binaries one would expect that most of these objects fall into either one of two groups: (1) very luminous ($\sim 10^{38}$erg s^{-1}) systems, consisting of an evolved companion and a relatively young (age up to a few times 10^7yr), relatively strongly magnetized ($>10^9$G) neutron star; (2) less luminous systems with an un-

evolved companion and a very old (age up to a few times 10^9yr), very weakly magnetized neutron star.

117.309 **The evolution of low mass binary systems.**
R. E. Taam.
The evolution of galactic X–ray binaries, p. 207 – 220 (1986). – See Abstr. 012.076.

This review discusses the evolution of low mass X–ray binary systems in which the mass losing star is unevolved. It briefly studies the stability of the mass transfer process, and summarizes the various angular momentum loss mechanisms. The results of model calculations are presented. These results are compared with low mass binary evolutionary scenarios for the millisecond radio pulsar.

117.310 **Eclipse of a bright spot in the disk of Her X–1.**
H.–C. Thomas, H. U. Schmidt, R. Schoembs.
The evolution of galactic X–ray binaries, p. 221 – 225 (1986). – See Abstr. 012.076.

Photometric observations of HZ Herculis obtained in June 1984 show a well defined eclipse ingress about two hours before primary minimum. The data are compared with observations obtained earlier. It is concluded that this phenomenon is produced by an X–ray beam which illuminates only certain parts of this disk.

117.311 **The orbital periods of the low mass X–ray binaries.**
N. E. White.
The evolution of galactic X–ray binaries, p. 227 – 237 (1986). – See Abstr. 012.076.

The orbital period distribution of the low mass X–ray binaries (LMXRB) is reviewed with particular emphasis on the impact of recent EXOSAT observations. A comparison with the period distribution of the cataclysmic variables indicates the two to be similar, except that there are no LMXRB with periods between 1 and 2 hr on the short period side of the 2 – 3 hr gap found in the CV period distribution. The various sub–classes of LMXRB are considered. The orbital periods of the X–ray burst sources predominantly lie in the range of 3 – 7 hr.

117.312 **EXOSAT observations of the 35–day cycle of Her X–1; evidence for neutron star precession.**
J. Truemper, P. Kahabka, H. Oegelman, W. Pietsch, W. Voges.
The evolution of galactic X–ray binaries, p. 239 – 251 (1986). – See Abstr. 012.076.

The authors have observed Her X–1 with the EXOSAT satellite throughout a 35–day cycle at about four day sampling intervals. These measurements include for the first time an accurate measurement of the 1.24 sec pulse profiles seen during the short–on state at the middle of the 35–day cycle. The data show a large change in the pulse profiles between the main and short–on states, suggesting a precession of the magnetic dipole axis of the rotating neutron star.

117.313 **On the evolution of Her X–1.**
W. Sutantyo, T. J. van der Linden,
E. P. J. van den Heuvel.
The evolution of galactic X–ray binaries, p. 261 – 262 (1986). – See Abstr. 012.076.

117.314 **Secular evolution of cataclysmic binaries.**
H. Ritter.
The evolution of galactic X–ray binaries, p. 271 – 293 (1986). – See Abstr. 012.076.

Recent developments and ideas concerning the secular evolution of cataclysmic binaries are summarized. In particular the following topics are addressed: basic observational properties, stability against mass transfer, mechanisms for driving the mass transfer, the reaction of the secondary to mass loss, the period gap model with its implications, the minimum orbital period, the secular evolution of the white dwarf primary, the mass spectrum of the white dwarfs in cataclysmic binaries, precataclysmic binaries, and finally the evolution of systems containing an evolved secondary.

117.315 **The evolution of magnetic cataclysmic variables.**
J. Frank.
The evolution of galactic X–ray binaries, p. 295 – 300 (1986). – See Abstr. 012.076.

This paper reviews the evolution of magnetic cataclysmic variables. It discusses the sizes of polar caps in these systems, how they vary as a result of the orbital evolution of the binary and how this could explain the main characteristics of the X–ray light curves of these systems.

117.316 **CH Cygni.**
IAU Circ., No. 4157 (1986).

117.317 **R Aquarii.**
IAU Circ., Nos. 4157, 4169 (1986).

117.318 **AG Draconis.**
IAU Circ., No. 4165 (1986).

117.319 **TT Arietis.**
IAU Circ., No. 4165 (1986).

117.320 **IP Pegasi.**
IAU Circ., No. 4165 (1986).

117.321 **SW Ursae Majoris.**
IAU Circ., Nos. 4188, 4193, 4198, 4206, 4227 (1986).

117.322 **V439 Cygni.**
IAU Circ., No. 4208 (1986).

117.323 **VY Aquarii.**
IAU Circ., Nos. 4209, 4210, 4222 (1986).

117.324 **AM Canum Venaticorum.**
IAU Circ., No. 4210 (1986).

117.325 **Dwarf nova SW UMa.**
Yamamoto Circ., Nos. 2055, 2059 (1986). In Japanese.

117.326 **Bolometric corrections and magnitudes of WR + O stars.**
P. Hellings.
Luminous stars and associations in galaxies, p. 221 – 222 (1986). – See Abstr. 012.079. (IAU Symp. No. 116).

In this study bolometric corrections and M_v are computed for a number of WR + O binaries.

117.327 **Rapid ultraviolet spectral variations in HD 50896 (WN5 + ?).**
A. J. Willis, I. D. Howarth, P. S. Conti, C. D. Garmany.
Luminous stars and associations in galaxies, p. 259 – 260 (1986). – See Abstr. 012.079. (IAU Symp. No. 116).

HD 50896 has become one of the most promising candidate WR + collapsar systems (Moffat 1982). UV spectra of analogous OB X–ray binaries show P Cygni profile changes as a function of binary phase which are well understood in terms of anisotropic wind ionisation caused by X–ray emission from the accreting neutron star – the Hatchett & McCray (1976) effect. The authors have obtained IUE spectra of HD 50896 in sequential NASA/VILSPA shifts over 6 consecutive days in 1983, together with complementary IUE data secured at various epochs during 1978–82, to search for the Hatchett & McCray effect in this star.

117.328 **The evolution of massive close binaries with mass loss and overshooting. An application to V729 Cyg (= BD + 40°4220).**
C. H. B. Sybesma.
Luminous stars and associations in galaxies, p. 399 – 400 (1986). – See Abstr. 012.079. (IAU Symp. No. 116).

Calculations for massive systems undergoing case A of mass transfer have been performed for two different primary masses: 80 and 40 $M_\odot$, with an initial mass ratio of 0.5.

117.329 **IRAS observations of symbiotic objects.**
P. A. Whitelock.
Light on dark matter, p. 323 – 324 (1986). – See Abstr. 012.080.
The three groups of symbiotics, defined according to their near IR colours (S: stellar, D: hot dust, D': cool dust), are shown to have distinctly different IRAS colours. The D–type symbiotics divide into two groups, those with and without 10 μm silicate features.

Observational study of close binary stars.
See Abstr. 013.039.

Lunar occultation techniques in investigations of binary stars.
See Abstr. 036.104.

Motions in the field of two rotating magnetic dipoles. I. Equilibrium points.
See Abstr. 042.041.

Areas of equatorial motion in the magnetic–binary problem.
See Abstr. 042.042.

On the dynamical evolution of binary systems.
See Abstr. 042.069.

Proposed test of a strange matter pulsar by observation of the neutrino flux from Cygnus X–3.
See Abstr. 061.117.

Can neutrinos from Cygnus X–3 be seen by proton decay detectors?
See Abstr. 061.142.

Accretion discs.
See Abstr. 062.133.

Plasma and radiation processes.
See Abstr. 063.007.

Two–photon emission in X–ray pulsars. I. Basic formulas.
See Abstr. 063.018.

Two–photon emission in X–ray pulsars. II. Radiation transfer.
See Abstr. 063.019.

The radiative acceleration of astrophysical jets: line locking in SS 433.
See Abstr. 063.025.

The formation of the continuum spectrum in X–ray pulsars.
See Abstr. 063.027.

Theoretical X–ray spectra of accreting magnetized neutron stars and quasars.
See Abstr. 063.066.

Magnetically controlled stellar winds in close binary systems.
See Abstr. 064.003.

Emission line formation in accretion discs.
See Abstr. 064.005.

Bombardment solutions to the "soft X–ray puzzle" in radial white dwarf accretion.
See Abstr. 064.019.

On the dimensions of accretion disks in cataclysmic variables.
See Abstr. 064.039.

Structure of the accretion disks in cataclysmic variables.
See Abstr. 064.058.

The structure and stability of the accretion disc boundary layer.
See Abstr. 064.062.

The effect of back pressure on the contact evolution of a close binary system.
See Abstr. 065.008.

Rotational and tidal perturbations of linear isentropic oscillations of a gaseous star.
See Abstr. 065.016.

Observational implications of the explosion of white dwarfs.
See Abstr. 065.019.

Fission of rapidly rotating fluid systems.
See Abstr. 065.025.

The evolution of massive close binary systems with mass loss and overshooting. II. Case A of mass transfer in systems with high mass ratios.
See Abstr. 065.034.

Models for Type I supernova. I. Detonations in white dwarfs.
See Abstr. 065.038.

The evolution of WR + O binaries into massive X–ray binaries.
See Abstr. 065.072.

On the evolutionary state of cataclysmic variable white dwarfs and their recurrence times.
See Abstr. 065.075.

Hydrodynamic models for novae with ejecta rich in oxygen, neon, and magnesium.
See Abstr. 065.077.

On the evolution of binary components which first fill their Roche lobes after the exhaustion of central helium.
See Abstr. 065.091.

Relaxation oscillations of a helium–burning star: helium shell flashes not triggered by the accumulation of a critical mass.
See Abstr. 065.092.

The evolution of ultrashort period binary systems.
See Abstr. 065.093.

On the behavior of double degenerate binaries associated with Type I supernovae.
See Abstr. 065.107.

Common envelope evolution.
See Abstr. 065.155.

Wind fed X–ray binaries.
See Abstr. 065.156.

From wide to close binaries?
See Abstr. 065.157.

Evolution of close binaries and the formation of neutron stars.
See Abstr. 065.158.

The duration of the detached phase in low–mass X–ray binaries in which the neutron star was formed by accretion–induced collapse of a white dwarf.
See Abstr. 065.159.

On the occurrence of WR + O binaries.
See Abstr. 065.164.

Binary evolution in the upper HRD.
See Abstr. 065.168.

The average X–ray lifetime of massive X–ray binaries.
See Abstr. 065.169.

Gravitational light deflection effects in eclipsing binary pulsars.
See Abstr. 066.107.

Relativistic Kepler's third law.
See Abstr. 066.235.

Origin of ultra–high–energy γ–rays from Cygnus X–3 and related sources.
See Abstr. 067.005.

Quasi–periodic oscillations from accretion disk coronas.
See Abstr. 067.006.

A qualitative model for GX5–1.
See Abstr. 067.008.

The lower mass limit of the compact object in LMC X–3.
See Abstr. 067.019.

Can comet clouds around neutron stars explain gamma–ray bursts?
See Abstr. 067.020.

Production of ultra–high–energy γ–rays by accreting neutron stars.
See Abstr. 067.024.

Accreting neutron stars in close binary systems.
See Abstr. 067.059.

The evolution of neutron stars.
See Abstr. 067.070.

Pulse frequency variations in pulsars.
See Abstr. 067.076.

How do X–ray pulsars make their X–rays?
See Abstr. 067.077.

Accretion flow.
See Abstr. 067.079.

Neutron star formation in close binaries: collapse of accreting white dwarfs.
See Abstr. 067.080.

Cosmic X–ray jets.
See Abstr. 067.083.

Magnetic field decay and the origin of neutron star binaries.
See Abstr. 067.112.

On the stream–disk interaction in accreting compact objects.
See Abstr. 067.114.

Maser instability theory of emission lines from Her X–1.
See Abstr. 067.117.

Cyclotron emission near stellar mass black holes.
See Abstr. 067.145.

Optical reprocessing of gamma–ray bursts.
See Abstr. 067.148.

Quasi–periodic oscillations as a probe of galactic–bulge X–ray sources.
See Abstr. 067.159.

On the nature of the quasiperiodic oscillations of low mass X–ray binaries.
See Abstr. 067.160.

Do neutron star magnetic fields decay?
See Abstr. 067.161.

The binary frequency of the OB runaway stars.
See Abstr. 111.008.

FK Comae – the kinematical determination of its age and evolutionary status.
See Abstr. 111.015.

The X–ray spectrum of emission line stars (superluminous and symbiotic).
See Abstr. 112.057.

Far infrared observations of Gamma Cassiopeiae from the *IRAS* satellite.
See Abstr. 112.062.

A phenomenological interpretation of stellar chromospheres.
See Abstr. 112.085.

Activity in the envelope of the G–type hypergiant HD 217476.
See Abstr. 112.123.

Photoelectric *UBV* photometry of the peculiar emission–line star Henize 2–467.
See Abstr. 113.050.

Spectroscopic observations of three X–ray selected late–type stars.
See Abstr. 114.007.

Lithium in late–type giants. IV. The subgiant CH stars.
See Abstr. 114.079.

On Wolf–Rayet stars with unshifted absorption lines.
See Abstr. 114.118.

Magnetic structure in cool stars. IX. Ultraviolet emission lines from chromospheres and transition regions.
See Abstr. 116.001.

Magnetodynamic phenomena in the solar and stellar outer atmospheres.
See Abstr. 116.003.

Active regions on HR 1099 from a high resolution spectroscopic study of photospheric (Fe I) and chromospheric (Ca II and Hα) lines.
See Abstr. 116.015.

X–ray emission from solar–type stars: X–ray luminosity function of late F and G stars.
See Abstr. 116.040.

Polarization measurements of β Cep stars.
See Abstr. 116.056.

Polarimetry of VV Puppis.
See Abstr. 116.063.

Polarimetry of Wolf–Rayet binaries.
See Abstr. 116.070.

The spectroscopic orbits of BV and BW Draconis.
See Abstr. 118.007.

A photometric study of the Algol system Y Hydri.
See Abstr. 119.003.

The eclipsing system ER Orionis.
See Abstr. 119.008.

Light curve variations in V471 Tauri.
See Abstr. 119.011.

Photoelectric study of AW UMa.
See Abstr. 119.012.

Photoelectric study of the low–mass semidetached system XX Cephei.
See Abstr. 119.014.

Equivalent widths of the spectral lines of the β Lyrae system and their changes.
See Abstr. 119.016.

HD 47755, a new eclipsing binary.
See Abstr. 119.060.

Photometry of long–period Algol binaries. I. Five–color light curves of the interacting eclipsing binary RS Cephei.
See Abstr. 119.061.

Photometry of long–period Algol binaries. II. Anomalies in five–color photometric solutions of RS Cephei.
See Abstr. 119.062.

Study of the eclipsing system V388 Cyg.
See Abstr. 119.065.

Spectroscopic and polarimetric parameters of the runaway WN7 binary system HD 197406: is the secondary an X–ray–quiet black hole?
See Abstr. 120.018.

Study of quiescent state X–ray emission from flare stars.
See Abstr. 122.017.

Investigation of variable stars WY Virginis and PG 1341–079.
See Abstr. 122.047.

Simultaneous optical, infrared and microwave observations of the flare star AT Mic.
See Abstr. 122.077.

Quiescent state X–ray emission from flare stars.
See Abstr. 122.110.

The historical optical variation of DO Dra (= 3A 1148 + 719).
See Abstr. 122.159.

Sigma Scorpii, a bright close binary.
See Abstr. 122.174.

The hypergiant variable HR 8752 = HD 217476.
See Abstr. 122.193.

The outburst of RS Ophiuchi in 1985.
See Abstr. 124.001.

On the nature of the recurrent nova T Pyxidis.
See Abstr. 124.008.

Do novae hibernate during most of the millenia between eruptions? Links between dwarf and classical novae, and implications for the space densities and evolution of cataclysmic binaries.
See Abstr. 124.009.

The orbit of GK Persei.
See Abstr. 124.161.

Spectroscopic observations of PU Vulpeculae.
See Abstr. 124.245.

Period derivative and orbital eccentricity of binary pulsar 1953 + 29.
See Abstr. 126.002.

Feasibility of a low–mass binary source progenitor for PSR1937 + 214.
See Abstr. 126.003.

The atmosphere of subluminous B stars. II. Analysis of 10 helium poor subdwarfs and the birthrate of sdB stars.
See Abstr. 126.010.

On mass distribution of white dwarfs in cataclysmic variables.
See Abstr. 126.039.

Are many pulsars processed in binaries?
See Abstr. 126.047.

Determination of the white dwarf's magnetic axis from X–ray orbital light curves.
See Abstr. 126.057.

The outcome of mass transfer in a carbon–oxygen white dwarf binary system.
See Abstr. 126.070.

Binary pulsars 0655 + 64 and 0820 + 02.
See Abstr. 126.078.

New millisecond binary pulsar.
See Abstr. 126.079.

Measuring interstellar dust grains from the haloes of binary X–ray sources.
See Abstr. 131.018.

Kinematic structure of NGC 3132: the planetary nebula with a binary nucleus.
See Abstr. 134.048.

EXOSAT measurements of the quiescent X–ray spectrum of the burster 2S 1636–536: evidence for Comptonization in a hot corona.
See Abstr. 142.001.

Quasi–periodic oscillations in the X–ray flux of Cyg X–2.
See Abstr. 142.002.

Iron K–emission line from Cygnus X–2.
See Abstr. 142.004.

Pulsating X–ray sources.
See Abstr. 142.006.

The X–ray source in M15: a binary composed of a neutron star plus a post main–sequence star?
See Abstr. 142.010.

Rapid, intensity–dependent quasi–periodic oscillations in GX 5–1.
See Abstr. 142.014.

The discovery of X–ray bursts from Cir X–1.
See Abstr. 142.015.

Observations of quasi–periodic oscillations from GX 5–1 with the Einstein (HEAO–2) Observatory.
See Abstr. 142.021.

5 Hz quasi–periodic oscillations in Cygnus X–2.
See Abstr. 142.022.

EXOSAT and Einstein observations of the X–ray pulsar 2S 1145–61.
See Abstr. 142.026.

Quasi–periodic oscillations in galactic X–ray sources.
See Abstr. 142.027.

Energy spectra of non–pulsating low mass X–ray binaries.
See Abstr. 142.037.

Spectral studies of low–mass X–ray binaries observed by EXOSAT.
See Abstr. 142.038.

Discovery of regularities in the cycle–to–cycle variability of Cygnus X–3.
See Abstr. 142.039.

Detection of iron K–emission lines from low–mass binary X–ray sources.
See Abstr. 142.040.

EXOSAT observations of 4U 1705–44.
See Abstr. 142.041.

X–ray bursts.
See Abstr. 142.042.

Quasi–periodicity of the Rapid Burster in 1983.
See Abstr. 142.043.

Optical evidence for the binary nature of the X–ray burster X 1254–69.
See Abstr. 142.045.

X–ray bursts – some remaining key issues.
See Abstr. 142.047.

X–ray pulsars and black–hole candidates.
See Abstr. 142.048.

The transient X–ray source A 0535 + 26: coordinated observations in X–ray, UV, optical and IR regions.
See Abstr. 142.053.

TENMA observations of X–ray pulsars.
See Abstr. 142.056.

Properties of the iron line from Vela X–1.
See Abstr. 142.058.

EXOSAT observations of the X–ray transient V 0332 + 53.
See Abstr. 142.059.

Energy spectra of X–ray pulsars.
See Abstr. 142.067.

Type I X–ray bursts.
See Abstr. 142.068.

Recent progress in the theory of double peaked bursts.
See Abstr. 142.069.

Theoretical studies of X–ray burst sources.
See Abstr. 142.070.

X–ray burst sources, some remaining problems in 1985.
See Abstr. 142.072.

Iron line from compact X–ray sources.
See Abstr. 142.075.

The spectra of Sco X–1 and similar bulge sources.
See Abstr. 142.076.

Iron emission line from low–mass binary X–ray sources.
See Abstr. 142.077.

How common are long–term periodicities in X–ray binaries?
See Abstr. 142.078.

Soft and hard X–ray flares of Cir X–1 observed by Tenma. Preliminary results.
See Abstr. 142.079.

Cygnus X–3: the dependence of the iron–line parameters on orbital phase.
See Abstr. 142.080.

The spectral variation and the iron line in Cygnus X–3.
See Abstr. 142.081.

Compact X–ray binaries in and out of globular clusters.
See Abstr. 142.083.

Optical counterparts of HEAO–1 X–ray sources.
See Abstr. 142.087.

The evolution of the 1984 outburst of the transient X–ray source 4U 1630–47.
See Abstr. 142.088.

2S 0918–549: optical identification and study of a new distant low–mass X–ray binary.
See Abstr. 142.092.

Observation of an X–ray flare from Sco X–1.
See Abstr. 142.100.

A 6 second periodic X–ray source in Carina.
See Abstr. 142.101.

Quasi–periodic oscillations in bright galactic X–ray sources: preliminary results from recent EXOSAT observations.
See Abstr. 142.103.

Quasi–periodic oscillations in GX 5–1 and Sco X–1.
See Abstr. 142.105.

Quasi–periodic oscillations in the X–ray flux of Cyg X–2 and their relation to the source spectra.
See Abstr. 142.106.

Two spectral components and their luminosity–related changes of low–mass binary X–ray sources.
See Abstr. 142.107.

Timing analysis of the galactic bulge X–ray source GX17 + 2.
See Abstr. 142.108.

Synchrotron emission models for gamma–ray burst sources, the determination of the orientations of AM Herculis binaries, and non–adiabatic shocks on magnetic white dwarfs.
See Abstr. 143.024.

A fast pulsar in Cygnus X–3.
See Abstr. 143.027.

Gamma ray astronomy experiments at energies greater than 1 TeV.
See Abstr. 143.030.

Pulsed flux of 5×10^{14}eV gamma–rays from Cygnus X–3.
See Abstr. 143.031.

Accreting binaries, diffuse radio emission and galactic cosmic rays.
See Abstr. 144.010.

Is Cygnus X–3 a quark star?
See Abstr. 144.025.

Signals from Cygnus X–3.
See Abstr. 144.026.

Underground muon energy spectrum and its relevance for muon astronomy.
See Abstr. 144.083.

Can a light gluino solve the Cyg X–3 puzzle?
See Abstr. 144.084.

Nature of the high–energy particles from Cygnus X–3.
See Abstr. 144.086.

The high-energy muon flux in air showers from Cygnus X-3.
See Abstr. 144.087.

Could Cyg X-3 muons indicate a light supersymmetric particle?
See Abstr. 144.089.

Cygnus X-3 events and very-high-energy photonuclear cross sections.
See Abstr. 144.106.

The acceleration of cosmic rays.
See Abstr. 144.121.

The role of binaries in globular cluster evolution.
See Abstr. 151.039.

The fate of wide binaries in the solar neighborhood: encounters with stars and molecular clouds.
See Abstr. 151.050.

Tidal binary formation in globular clusters.
See Abstr. 151.057.

2000 Å UV imaging of a 6° diameter field around μ Aurigae.
See Abstr. 155.068.

Daily observations of compact radio sources in 1983.
See Abstr. 158.204.

118 Visual Binaries, Multiple Stars, Astrometric Binaries

118.001 **Erratum: "Alpha Tauri CD – a well-known Hyades binary" [Publ. Astron. Soc. Pac., Vol. 97, No. 595, p. 858 – 859 (1985)].**
R. F. Griffin.
Publ. Astron. Soc. Pac., Vol. 97, No. 597, p. 1127 (1985). See Abstr. 40.118.029.

118.002 **Trapezium-type multiple systems.**
G. N. Salukvadze.
Astrophys. Space Sci., Vol. 118, Nos. 1/2, p. 321 – 323 (1986). – See Abstr. 012.001.
Trapezium-type multiple systems are one of the youngest systems in the galaxy. Ambartsumian (1954) regards many of the Trapezium-type multiple systems as having positive total energy. The examination of a table, compiled from the Abastumani Catalogue of Trapezia (Salukvadze, 1978; Poveda et al., 1977) and the Index Catalogue of Visual Binary Stars, shows that real trapezia are met among multiple stars with their primaries of O and B spectral types.

118.003 **Extra solar planetary systems II: Habitable planets in the Galaxy.**
M. J. Fogg.
J. Br. Interplanet. Soc., Vol. 39, No. 3, p. 99 – 109 (1986).
A lengthy analysis of the output of the "Silicon Creation" model is presented. Seven thousand five hundred microcomputer simulations of stars in the mass range 0.6 – 1.3 $M_\odot$ suggest that habitable planets are to be expected to occur around stars of between 0.8 – 1.2 $M_\odot$, and most commonly around stars between 0.95 – 1.05 $M_\odot$. An extended simulation of an evolving volume of Galactic disc space, 270 ly in diameter, containing 100,000 star systems of maximum age 10^{10} years was performed, assuming constant stellar birthrate and rising "metallicity" with time. 86 habitable planets were produced giving a value for the ratio of habitable planets to stars in the disk $N_{HP}/N_{*disc} = 8.6 \times 10^{-4}$. The average separation between habitable planets is therefore ~50 ly and the number of habitable planets in the Galaxy is approximately 90 million.

118.004 **Etoiles doubles nouvelles.**
G. M. Popovic, P. Muller, P. Couteau.
Circ. Inf., Nos. 98 – 99 (1986).

118.005 **Orbits of 20 visual binaries.**
W. D. Heintz.
Astron. Astrophys., Suppl. Ser., Vol. 64, No. 1, p. 1 – 7 (1986).
Elements have been determined for twenty pairs from observations in 1984 – 85.

118.006 **Visual measurements of southern double stars. II.**
G. Torres.
Astron. Astrophys., Suppl. Ser., Vol. 64, No. 1, p. 105 – 110 (1986).
A group of 248 micrometric observations of 104 double stars are presented, forming the second series of a program carried out at Córdoba with a 30 cm refractor. Usually neglected IDS pairs south of –60° are selected for observation, most of them having been measured only one or two times previously. The measurements are compared with all the earlier observations to derive estimates for the personal equation in P.A. and separation, and the nature of many pairs is discussed by making use of catalogue proper motions.

118.007 **The spectroscopic orbits of BV and BW Draconis.**
A. H. Batten, W.-x. Lu.
Publ. Astron. Soc. Pac., Vol. 98, No. 599, p. 92 – 98 (1986).
Reticon spectra of high dispersion and time resolution have been obtained of BV an BW Draconis, which together form a visual binary and each of which is a W UMa system. The spectra were measured by cross-correlation. The orbital elements of BV Dra have been revised to the values T_0 = 2446127.6888, K_1 = 93.9 km s^{-1} and K_2 = 233.6 km s^{-1}, while those of BW Dra, determined for the first time, are T_0 = 2446135.8067, K_1 = 70.5 km s^{-1} and K_2 = 251.5 km s^{-1}. A circular orbit is adopted for each system. Different values of V_0 are obtained from each component of each system, but V_0 for each pair is close to –60 km s^{-1}. Radial velocities confirm the proper motions in identifying the two systems as a genuine binary (ADS 9537) whose space velocity with respect to the Sun is over 90 km s^{-1}. From spectrophotometry, the components of BV Dra are classified as F9 (more massive star) and F8, with a luminosity ratio (in blue light) of 1.78 (Δm = 0.63), while the components of BW Dra are G3 (more massive star) and G0, with a luminosity ratio (also in blue light) of 1.52 (Δm = 0.45). All four stars are of luminosity class V.

118.008 **VB 8B, the first brown dwarf outside our solar system.**
F. J. Low.
Phys. Today, Vol. 39, No. 1, p. S6 (1986). Short review paper. – See Abstr. 013.021.

118.009 **Où en est l'astronomie des étoiles doubles?**
P. Baize.
Astronomie, Vol. 100, p. 17 – 22 (1986).

118.010 **A propos de la découverte d'une nouvelle planète.**
A. Chelli.
Astronomie, Vol. 100, p. 61 – 70 (1986).

118.011 **Search for a planet.**
A. A. Tokovinin.
Priroda, No. 3, p. 39 – 44 (1986). In Russian.

118.012 **Search for other planetary systems.**
Z. Alksne.
Zvaigžnota Debess, 1984. Gada Rudens, p. 7 – 12 (1984). In Latvian.

118.013 **A study of visual double stars with early–type primaries. V. Post–T Tauri secondaries.**
K. P. Lindroos.
Astron. Astrophys., Vol. 156, No. 1/2, p. 223 – 233 (1986).

In the previous articles of this series about 100 secondaries were found to be physical companions to O and B type primaries. The majority of the secondaries are of spectral type F, G or K and in the present work their properties are investigated in relation to their evolutionary status. The ages of the systems have been determined from *uvbyβ* photometry of the primaries and it is found that all are younger than 150 million years and that half of them are less than 30 million years old. More than 50% of the F, G and K secondaries exhibit spectroscopic features (Ca II H, K and Hα emission and strong lithium absorption) typical of young stars and reminiscent of T Tauri stars although not as conspicuous. Also, in contrast to the T Tauri stars they do not possess any significant infrared or ultraviolet excess emission and they only show a small degree of variability. These properties and the fact that they are older and lie closer to the ZAMS make it appropriate to classify them as post–T Tauri stars.

118.014 **Stars to search for planetary systems.**
V. A. Zakhozhaj, T. V. Ruzmajkina.
Astron. Vestn., Tom 20, No. 2, p. 128 – 133 (1986). In Russian. English translation in Sol. Syst. Res.

A catalogue of stars worth to be studied in order to search for planetary systems is presented. The stars were selected from the solar vicinity of radius 10 parsec. Stars of next types are chosen: single stars of the spectral classes F5 – G9; stars which are suspected to have satellites of low masses; stars at distances shorter than 10 light years; wide binaries, with distances between components greater than 30 AU; stars with IR excesses.

118.015 **Multiplicity among eclipsing binaries.**
K. D. Abhyankar.
Binary and multiple systems, p. 1 – 4 (1984). – See Abstr. 012.025.

Light time effect can be used for detecting additional components in eclipsing binaries. One way to test their existence would be to consider the stability of such hierarchical systems. It is found that the triple systems are stable, but higher order multiple systems are not. Hence the three body model of AK Her proposed by Barker and Herczeg (1979) is more plausible than the four body model suggested by Abhyankar and Panchatsaram (1982).

118.016 **Evidence for p–modes on Alpha Centauri and Procyon.**
E. Fossat, B. Gelly, G. Grec.
Seismology of the Sun and the distant stars, p. 405 – 416 (1986). – See Abstr. 012.026.

Helioseismology has proved to be a powerful tool to probe the internal structure of the Sun. With a new adapted optical resonance spectrophotometer, an extension has been attempted to Alpha Centauri A and Procyon. In both cases, solar–like pressure oscillations have been detected. The complete analysis is presented, which leads to determine essentially five parameters: the mean equidistance Δv, the frequency range, the amplitudes, the departure around equidistance (curvature of the Echelle diagram) and the fine frequency spacing between modes of degrees 0 and 2.

118.017 **Wide binaries in the direction of Andromeda.**
J. N. Bahcall, B. F. Jones, K. U. Ratnatunga.
Astrophys. J., Suppl. Ser., Vol. 60, No. 4, p. 939 – 943 (1986).

A statistically well–defined sample of candidate binary stars with separations that are expected to be mostly in the range 0.01 – 0.1 pc is presented. The 36 candidate pairs are all brighter than apparent visual magnitude 12; about half of the projected pairs are expected to be physically associated. Spectroscopic and photometric observations are required to establish which pairs are physical and what their properties are.

118.018 **First orbits for the visual double stars ADS 328, ADS 9043, and ADS 13048.**
J. A. Docobo, J. M. Costa.
Astrophys. J., Suppl. Ser., Vol. 60, No. 4, p. 945 – 948 (1986).

The orbital elements of the visual double stars ADS 328, ADS 9043, and ADS 13048 have been calculated for the first time, together with their dynamical parallaxes and masses.

118.019 **Geometrische Bestimmung der Bahnelemente von Doppelsternen aus der scheinbaren Bahn (2).**
H. Blatter.
Orion, 44. Jahrg., Nr. 212, p. 30 – 31 (1986).

118.020 **Sur deux étoiles doubles à courte période 51 Tauri et Nu Bootis.**
P. Baize.
Astronomie, Vol. 100, p. 291 – 295 (1986).

118.021 **The radial velocities of the visual binary systems.**
G. Ishida.
Proceedings of the Second Asian–Pacific Regional Meeting on Astronomy, p. 59 – 60 (1984). – See Abstr. 012.031.

118.022 **On the determination of the orbit and the mass of visual binary COU 292 = SAO 099125.**
S. Siregar.
Proceedings of the Second Asian–Pacific Regional Meeting on Astronomy, p. 561 (1984). Abstract. – See Abstr. 012.031.

118.023 **Orbit, mass ratio, and parallax of 99 Herculis.**
K. W. Kamper, W. R. Beardsley.
Astron. J., Vol. 91, No. 2, p. 419 – 421 (1986).

From measures of 131 Allegheny parallax plates, combined with the visual measurements of the system, the authors derive a new orbit, parallax, and mass ratio for 99 Her. They obtain a relative parallax of 0.0660 ± 0.0049, which supersedes the two previous Allegheny determinations. Combining this result with determinations from other observatories, a mean absolute parallax of 0.064 is obtained. With the values of $a = 1.077$, $P = 55.54$, and $B = 0.355$, the authors obtain individual masses of 0.95 and 0.52 $M_{\odot}$.

118.024 **The orbits of ten visual and interferometric binaries.**
A. A. Tokovinin.
Pis'ma Astron. Zh., Tom 12, No. 6, p. 480 – 485 (1986). In Russian. English translation in Sov. Astron. Lett., Vol. 12.

The orbits of Cou 2031, ADS 4392, HR 2425, ADS 6185 and HR 7536 have been computed for the first time. The previously known orbits of ADS 1183, Cou 79 have been considerably modified and corrections have been made to the orbits of ADS 3210, Rst 5225 and ADS 16098 to improve their agreement with recent interferometric observations. For main–sequence stars dynamical parallaxes and masses are derived.

118.025 **Speckle interferometric measurements of binaries with the BTA telescope.**
V. N. Dudinov, S. G. Kuz'menkov, V. V. Konichek, V. S. Tsvetkova, V. S. Rylov, V. N. Erokhin.
Astron. Zh., Tom 63, Vyp. 3, p. 606 – 615 (1986). In Russian. English translation in Sov. Astron., Vol. 30, No. 3.

Results of 42 angular–separation and position–angle measurements for 28 binaries are presented obtained with speckle interferometric technique at the BTA 6–meter telescope. The system ADS 11311 (φ Dra) earlier has not been measured with speckle interferometric technique. The system ADS 11149 has been identified as being triple for the first time in speckle interferometric measurements. The procedure of recording initial images and their further processing is presented. Comparison of the results presented in this work with ephemerides and results of other observers indicates the absence of serious systematic errors.

118.026 **Two visual observations of relevance to the search for optical counterparts of gamma–ray sources.**
B. Warner.
Astrophys. Space Sci., Vol. 122, No. 2, p. 397 – 400 (1986).
The authors draw attention to a visual observation of a brief flash from ζ Lyrae, observed by Heis in 1850, which resembles the optical burst detected electronically by Wdowiak and Clifton (1985) from β Cam in 1969. Visual observation by the author of a second magnitude flash of very short duration is shown to originate from planar reflection from a very distant satellite. Such flashes will contribute to the "noise" in all–sky searches for optical counterparts of γ–ray bursters.

118.027 **The ages and dimensions of Trapezium systems.**
H. A. Abt.
Astrophys. J., Vol. 304, No. 2, p. 688 – 694 (1986).
The author has obtained MK spectral types and hence memberships of 120 stars in 31 systems thought to be Trapezium systems, i.e., multiple–star systems with several similar separations. After the removal of optical components and hierarchical systems there remain 11 Trapezium systems. All but one of these systems are the same age or younger than the Hyades. Hence the dynamically–unstable Trapezium systems are less than $10^{8.9}$yr old. The maximum radii of Trapezium plus hierarchical systems show a pronounced decrease with primary spectral type or age T, ranging from roughly 50,000 AU for OB primaries to 1000 AU for G dwarfs. The systems evolve and lose outer components such that the maximum separations decrease as $T^{-0.3}$.

118.028 **The search for substellar companions to nearby stars: infrared imaging from the ground and from space.**
D. W. McCarthy Jr.
Prepr. Steward Obs., No. 619, 13 pp. (1986). To appear in "The astrophysics of brown dwarfs", 1986, Cambridge University Press.

118.029 **Coordonnées équatoriales précises de composantes connues ou nouvelles d'une quatrième série de quelques 200 systèmes doubles et multiples.**
G. Soulié.
Bull. Inf. Cent. Données Stellaires, No. 30, p. 91 – 101 (1986).
Accurate equatorial coordinates of known or new components of a fourth series of some 200 double and multiple systems are given.

118.030 **Coordonnées équatoriales précises de composantes connues ou nouvelles d'une cinquième série de quelque 200 systèmes doubles et multiples.**
G. Soulié.
Bull. Inf. Cent. Données Stellaires, No. 30, p. 103 – 111 (1986).
Accurate equatorial coordinates of known or new components of a fifth series of some 200 double and multiple systems are given.

118.031 **Una stella doppia per ogni ora di ascensione retta: I.**
S. Baroni.
Coelum, Vol. 54, N. 3, p. 127 – 135 (1986).

118.032 **Search of extra–solar planets.**
C. Barbieri, A. Nota.
Mem. Soc. Astron. Ital., Vol. 56, No. 4, p. 583 – 585 (1985). – See Abstr. 012.054.
The authors present a brief account of the possibility to search for extra–solar planets with the Space Telescope.

118.033 **Galaxian trapezia.**
P. L. Bernacca.
Mem. Soc. Astron. Ital., Vol. 56, No. 4, p. 687 – 690 (1985). – See Abstr. 012.054.
Properties of Trapezium systems in the Galaxy are briefly reviewed and prospects for observations of these objects with the Hubble Space Telescope are examined.

118.034 **Micrometer measurements of visual double stars (3rd List).**
J. A. Docobo.
Acta Astron., Vol. 36, No. 1, p. 175 – 178 (1986).

118.035 **Eléments orbitaux corrigés de l'étoile double visuelle ADS 9626 BC – STF 1938.**
M. Scardia.
Astron. Nachr., Vol. 307, No. 2, p. 149 – 154 (1986).
Orbital elements of the visual binary star ADS 9626 BC are given. The dynamical parallax and total mass of the system have been calculated.

118.036 **Orbites nouvelles.**
Circ. Inf., Nos. 98 – 99 (1986).

General catalogue of ephemerides and apparent orbits of 736 visual binary stars.
See Abstr. 002.050.

The new index catalog of visual double stars – WDS.
See Abstr. 002.085.

Dark companions of stars. Astrometric commentary on the lower end of the main sequence.
See Abstr. 003.026.

A space telescope for infrared spectroscopy of earthlike planets.
See Abstr. 035.068.

Chromatic position difference: a technique for studying double stars.
See Abstr. 036.094.

Lunar occultation techniques in investigations of binary stars.
See Abstr. 036.104.

The potential precision of speckle interferometric measurements. Binary stars.
See Abstr. 036.121.

Application de la méthode par déconvolution intégrée au cas des occultations d'étoiles doubles.
See Abstr. 036.215.

Infrared speckle interferometry: a sensitive technique for physical measurements of unseen companions to nearby stars.
See Abstr. 036.222.

The formation of planetary systems – a simulation on the microcomputer.
See Abstr. 107.049.

New observations of Alpha Orionis with a rotation shearing interferometer.
See Abstr. 112.117.

The case of α Centauri: mass, age and *p*–mode oscillation spectrum.
See Abstr. 115.004.

Photometric observations of visual double star ADS 9537 (BV Dra, BW Dra) by area scanning technique.
See Abstr. 117.266.

Spectroscopic orbital elements of $\varkappa^2$Bootis B.
See Abstr. 120.022.

The T Tauri radio source. II. The winds of T Tauri.
See Abstr. 121.041.

Binary survey among high velocity stars using GSU speckle camera.
See Abstr. 155.061.

CCD observations of young stellar associations and multiple systems in the Magellanic Clouds.
See Abstr. 156.035.

119 Eclipsing Binaries

119.001 **Epsilon Aurigae: polarization, light curves, and geometry of the 1982 – 1984 eclipse.**
J. C. Kemp, G. D. Henson, D. J. Kraus, I. S. Beardsley,
L. C. Carroll, T. B. Ake, T. Simon, G. W. Collins.
Astrophys. J., Lett. Ed., Vol. 300, No. 1, p. L11 – L14 (1986).
Polarization and light curves of the recent eclipse of ε Aur are presented and modeled, using a tilted, opaque secondary disk and a model atmosphere for the F Ia primary star. That model qualitatively explains the striking asymmetries in the observed curves. But out-of-eclipse polarimetry shows the primary to be a nonradial pulsator, and this will have to be included in a truly quantitative picture of the eclipse and the system.

119.002 **Optical observations of the radio star β Persei at the Paris Observatory astrolabe from 1980 to 1983.**
S. Débarbat, S. K. Lam, P. Texier, M. Tomas.
Astron. Astrophys., Suppl. Ser., Vol. 63, No. 2, p. 305 – 306 (1986). In French.
β Persei is one of the radio stars to be used for linking of "radio" and "optical" reference system. Δα and Δδ (with the meaning "Astrolabe-FK4") together with the mean square error for each campaign are given.

119.003 **A photometric study of the Algol system Y Hydri.**
H. W. Duerbeck, A. Yamasaki.
Astrophys. Space Sci., Vol. 119, No. 2, p. 345 – 352 (1986).
297 *UBV* observations of the Algol type eclipsing binary Y Hyi are presented and analyzed by a synthetic light curve technique. The system is semi-detached and shows complete eclipses.

119.004 **The nature of the eclipses of VZ Hydrae.**
T. D. Padalia.
Astrophys. Space Sci., Vol. 119, No. 2, p. 395 – 397 (1986).
Further study of the observed *U*, *B*, *V* light curves of VZ Hya reveals that the primary eclipse is an occultation.

119.005 **A visual light curve and minima of EG Cephei.**
H. Grzelczyk.
Astron. Rep., Vol. 6, p. 25 – 26 (1981).

119.006 **Two minima of UW Bootis.**
H. Grzelczyk.
Astron. Rep., Vol. 6, p. 27 (1981).

119.007 **V471 Tauri: a three-body orbit.**
W. I. Beavers, T. J. Herczeg, A. Lui.
Astrophys. J., Vol. 300, No. 2, p. 785 – 787 (1986).
A model is proposed for V471 Tauri to explain the apparent period changes currently assigned to this K2 V+ white dwarf eclipsing system. It is shown that the light-time effects of a model in which the eclipsing pair orbits with an as yet undetected third body can fit the observed eclipse timings. The model features a constant period $P_1 = 0.52118334$ days for the eclipsing pair, with the third-body orbit of period $P_0 = 24.6$ yr, eccentricity $e = 0.36$, and probably of low inclination. Present observations imply an upper limit to the third-body mass of about 0.8 $M_\odot$.

119.008 **The eclipsing system ER Orionis.**
P. Rovithis, H. Rovithis-Livaniou.
Astron. Astrophys., Vol. 155, No. 1, p. 46 – 50 (1986).
In this paper the light curves of the eclipsing binary system ER Orionis are analysed. The analysis is made using Frequency Domain techniques and is based on the authors' *B* and *V* photoelectric observations made in Greece during 1982 and 1983. A new ephemeris formula is proposed and new photometric and absolute elements for the system are given.

119.009 **A low-mass model of Epsilon Aurigae.**
M. Takeuti.
Astrophys. Space Sci., Vol. 120, No. 1, p. 1 – 7 (1986).
The eclipsing binary system Epsilon Aurigae, containing an unseen component, is discussed on the basis of its orbital elements. The masses of the components can be approximately 4 $M_\odot$ and 1 $M_\odot$, respectively. It is shown that there is no reason to reject a low-mass model. An opaque elongated eclipsing body is briefly discussed as one of the components which is an accretion disc surrounding a central main-sequence star.

119.010 **Apsidal motion in the eclipsing binary system of BW Aqr.**
Kh. F. Khaliullin, V. S. Kozureva (*V. S. Kozyreva*).
Astrophys. Space Sci., Vol. 120, No. 1, p. 9 – 16 (1986).
The first photoelectric light curve of the eclipsing binary system BW Aqr discovered by Miss Leavitt at the beginning of the century, was obtained. The photometric elements were determined. The components of this system are considerably evolved stars: the age of the system is about 2×10^9yr. The zero-age spectral classes of components were F2 V and F1 V. The system has an elliptical orbit with the eccentricity $e = 0.18$. The angular rate of the apsidal motion ($\omega_{obs} = 0.070$ deg yr^{-1}) and the corresponding value of the apsidal parameter $k_2 = 0.0090$ (the relativistic term included) were found.

119.011 **Light curve variations in V471 Tauri.**
S. Evren, C. Ibanoğlu, Z. Tunca, O. Tümer.
Astrophys. Space Sci., Vol. 120, No. 1, p. 97 – 106 (1986).
The light curves of V471 Tau obtained in four successive years from 1981 to 1984 are presented. The light variations due to the tidally distorted red dwarf have been computed and subtracted out from the light curves obtained so far. A pure shape of the wave-like distortion has therefore been derived. The period of the retrograde migration has been found to be 182.17 days. On the other hand, the mean brightness of the system seems to vary with a period of about 18 years. The amplitude of the variation is greater at longer wavelength. The periodic variation of the mean brightness resembles the solar cycle of about 11.2 years, which depends upon the number of spots or spot groups.

119.012 **Photoelectric study of AW UMa.**
R. K. Srivastava, T. D. Padalia.
Astrophys. Space Sci., Vol. 120, No. 1, p. 121 – 131 (1986).
UBV photoelectric photometry of the eclipsing binary system AW UMa has been presented. A slightly improved period of 0.^{d}4387304 has been given. The colour of the system has been discussed. Eccentricity is present in the system. Light and colour curves show that intrinsic light variations may be present in the system. The presence of mass transfer is possible.

119.013 **Light curves analysis of AB Andromedae.**
H. Rovithis-Livaniou, P. Rovithis.
Astron. Nachr., Vol. 307, No. 1, p. 17 – 19 (1986).
The new *B* and *V* light curves of the eclipsing variable AB And obtained during 1979 are analysed. The frequency domain technique is used and new geometric and photometric elements are given.

119.014 **Photoelectric study of the low-mass semidetached system XX Cephei.**
A. Bonifazi, A. Guarnieri.
Astron. Astrophys., Vol. 156, No. 1/2, p. 38 – 50 (1986).
A *V* and *B* photoelectric light curve for the eclipsing binary XX Cep is presented and the photometric solution deduced and discussed also from the point of view of the previously assumed semidetached-detached configuration in an eccentric orbit. The system, which is a low-mass one, appears to be normal semidetached with an underluminous subgiant secondary in a circular orbit.

119.015 **The chemical composition of Algol systems – III. Beta Lyrae-nucleosynthesis revealed.**
S. Balachandran, D. L. Lambert, J. Tomkin, M. Parthasarathy.
Mon. Not. R. Astron. Soc., Vol. 219, No. 3, p. 479 – 494 (1986).
This paper presents a model atmosphere analysis of neutral and ionized lines of C, N and O in the spectra of β Lyr and a

standard star of similar spectral type, and concludes with a discussion of the present composition and its relationship to the predicted nucleosynthesis in a main sequence star.

119.016 **Equivalent widths of the spectral lines of the β Lyrae system and their changes.**
V. Bahýl.
Bull. Astron. Inst. Czech., Vol. 37, No. 1, p. 42 – 63 (1986).
The equivalent widths of 82 absorption spectral lines were measured on spectrograms well spread over the entire light curve. The dependence of the equivalent widths on phase and the dependence of the central intensities on phase were studied for selected spectral lines. The constructed phase diagrams were qualitatively interpreted in accordance with the generally accepted model of the β Lyrae system.

119.017 **1H. Cas = AR Cas.**
W. B. Somerville.
Observatory, Vol. 106, No. 1071, p. 40 – 42 (1986).

119.018 **Four–colour photometry of eclipsing binaries. XXIA. Photometric analysis and apsidal motion study of V346 Centauri.**
A. Giménez, J. V. Clausen, J. Andersen.
Astron. Astrophys., Vol. 160, No. 2, p. 310 – 320 (1986).
New accurate *uvby* light curves of the detached early B–type eclipsing binary V346 Cen (P = 6.32 days) are analyzed and photometric elements determined by means of the Wood (1972) model. The results provide accurate relative dimensions of the system, the ratio of the radii being close to 0.5 and the orbital inclination near 84°. The secondary eclipse is total, allowing an unambiguous determination of the luminosity ratio. The system has a high orbital eccentricity (0.288), and its apsidal motion is discussed from a study of the available times of minimum. Supplementary spectroscopic observations provide improved estimates of the spectroscopic orbital elements, the spectral types and rotational velocities of the components. Finally, the relation between V346 Cen and the open cluster Stock 14 is discussed in light of all available photometric and spectroscopic informations.

119.019 **An EXOSAT observation of quiescent and flare coronal X–ray emission from Algol.**
N. E. White, J. L. Culhane, A. N. Parmar, B. J. Kellett, S. Kahn, G. H. J. van den Oord, J. Kuijpers.
Astrophys. J., Vol. 301, No. 1, p. 262 – 274 (1986).
X–ray emission from the Algol system is believed to originate in a corona associated with the K star. The authors have used the EXOSAT Observatory to make a 35 hr continuous observation centered on the occultation of the K star by the B star primary. The spectrum of the quiescent emission in the 1 – 10 keV band gives a temperature of 2.5×10^7K. This spectrum, extrapolated to lower energies, can account for more than 80% of the observed count rate. No obvious X–ray eclipse was seen. An X–ray flare was detected with a rise time of ~1700 s and an exponential decay of ~7000 s. The 0.1 – 10 keV peak luminosity was 1.4×10^{31}ergs s^{-1}. The peak temperature was 6×10^7K, with an iron K line confirming the thermal character of the emission.

119.020 **IUE observations of accretion columns in two Zeta Aurigae–type systems.**
I. A. Ahmad.
Astrophys. J., Vol. 301, No. 1, p. 275 – 285 (1986).
This paper reports the existence of an accretion column in 32 Cyg. It also confirms the presence of the accretion column in ζ Aur and reports the detection of secular variations. Mg II observations support the parameters derived by Ahmad, Chapman, and Kondo for ζ Aur but suggest that the column may be curved. A C IV and Si IV emission component is seen in ζ Aur but not in 32 Cyg. Resonance line equivalent widths have been calculated and the implications for flow geometry, secular variability, and plasma density are discussed.

119.021 **The 1983 eclipse of the high–latitude eclipsing system BL Tel.**
A. M. van Genderen.
Astron. Astrophys., Vol. 158, No. 1/2, p. 361 – 366 (1986).
Strömgren (uvby)– and Walraven (VBLUW) photometry of the 1983 eclipse of BL Tel is presented and discussed. Significant differences in the depth of the minimum and the colour variations during the eclipse compared with those of the previous recorded minima, are found and discussed. The metallicity of the bright F–type companion is presumably nearly solar and its gravity amounts to log g ~ 1 – 2.

119.022 **Light–time orbit and apsidal motion of the eclipsing binary U Ophiuchi.**
B.–C. Kämper.
Astrophys. Space Sci., Vol. 120, No. 2, p. 167 – 189 (1986).
A comprehensive period study of the times of minima observed from 1881 to 1985 has been performed. Previous interpretations of the O–C diagram based on light–time effect are confirmed. The light–time orbit of U Oph has been revised using a differential correction procedure. The author gets an eccentric orbit with $e' = 0.22\pm0.06$, $P' = 38.7\pm0.2$ yr, and a mass function $f(M) = 0.0044\pm0.0005\ M_\odot$. In addition, the analysis has revealed short–period apsidal motion in a slightly eccentric close orbit, allowing a reliable determination of the density concentration coefficient. A comparison with stellar evolutionary models calculated by Jeffery (1984) yields the helium content $Y = 0.28\pm0.05$ and an age of 3×10^7yr for the components of U Oph.

119.023 **An accretion disc surrounding a component of Epsilon Aurigae.**
M. Takeuti.
Astrophys. Space Sci., Vol. 121, No. 1, p. 127 – 135 (1986).
The low–mass model of Epsilon Aurigae consisting of an F supergiant star and a main–sequence B star which is surrounded by a cool accretion disc has been investigated. The disc requires a high accretion rate if one assumes the disc to be completely heated up by friction alone. One the other hand, discs with low accretion rates are possible when the disc are heated up from the radiation of both stars. A swollen crescent portion produced by the radiation of the F star may form a bulky body sufficient to obscure the radiation of the F star.

119.024 ***BV* observations of the eclipsing binary WX Eridani.**
J. B. Srivastava, C. D. Kandpal.
Astrophys. Space Sci., Vol. 121, No. 2, p. 301 – 308 (1986).
Photoelectric observations of WX Eridani in B and V filters have been discussed. The maximum magnitudes of the system are $B = 9^m.70$ and $V = 9^m.46$. The $B–V$ colours of the components are $0^m.16$ and $0^m.80$ for the primary and the secondary, respectively, while their spectral types are A5 + K0. The period has been found to be constant and its revised value is $0^d.82327076$.

119.025 **Period variations in BZ Eridani.**
R. K. Srivastava.
Astrophys. Space Sci., Vol. 121, No. 2, p. 317 – 320 (1986).
The O–C diagram of BZ Eri has been presented for the first time, and the period variations present in the system have been analysed. In all, eight period decreases and eight period increases are noticed. Of these, four period decreases and seven period increases are appreciable. The strongest period changes are noticed in the interval 1960 to 1962. The total period change in different portions of the O–C diagram ranges from 1.17×10^{-3}d to 3.96×10^{-6}d. The trend of the period variation appears to have reversed around the year 1980.

119.026 **Emission in the hydrogen–Brackett lines of Epsilon Aurigae during eclipse.**
D. E. Backman, T. Simon, K. H. Hinkle.
Publ. Astron. Soc. Pac., Vol. 97, No. 598, p. 1163 – 1171 (1985).
The authors report high–resolution infrared spectroscopy of the Bα and Bγ lines of H I performed before, during and after the recent eclipse of ε Aur. In both lines, emission components are detected that are not eclipsed by the secondary. The absorption–

and emission–line velocities and the emission–line strengths and widths are determined. The Bα emission has an integrated flux of 9×10^{-15}W m^{-2}, a FWHM of 60 km s^{-1}, and a radial velocity that follows more closely the predicted velocity curve of the primary star than that of the secondary. However, the emission cannot be definitely identified with either component. The mean Bα/Bγ flux ratio was approximately unity on the three dates both lines were observed, indicating optically thick emission. The authors conclude that the emission originates in a flow rather than from material in Keplerian revolution; the Brackett–line emission probably has the same source as the previously observed Hα emission.

119.027 **A survey for transient accretion disks in short–period Algol systems. II.**
R. H. Kaitchuck, R. K. Honeycutt, E. M. Schlegel.
Publ. Astron. Soc. Pac., Vol. 97, No. 598, p. 1178 – 1185 (1985).
Time–resolved spectroscopy during primary eclipse was utilized in order to search for emission lines from accretion disks surrounding the primary component in short–period Algol systems. The authors present the results of observations of 46 systems surveyed during 80 eclipses. When combined with an earlier survey this makes a total of 52 systems and 104 eclipses. A distinction is made between the disks in long–period systems and those in short–period systems. In the former case the gas stream from the secondary star moves around the trailing hemisphere of the primary and forms a permanent disk. However, in the latter case the gas stream strikes the primary star. Disks formed by this latter process are found to be transient and unstable. The authors report the discovery of two new transient disk systems: RW Mon and RR Dra. When the relative radius of the primary star is plotted versus mass ratio, the systems with permanent disks, transient disks, and without disks segregate into three distinct regions of the plot. This implies that the differences between systems with transient disks and those without disks may be simply a matter of the location of the stream impact on the primary star.

119.028 **Four–colour photometry of eclipsing binaries. XXIV. Apsidal motion of QX Carinae, ξ Phoenicis and NO Puppis.**
A. Giménez, J. V. Clausen, K. S. Jensen.
Astron. Astrophys., Vol. 159, No. 1/2, p. 157 – 165 (1986).
New accurate photoelectric times of minimum light have been obtained for the detached southern–hemisphere eclipsing binaries QX Car, ξ Phe and NO Pup, previously analyzed within this series. The apsidal motion rates have been derived and the corresponding internal structure constants are discussed in the light of the available theoretical models for stellar interiors.

119.029 **The spectroscopic study of μ Sgr.**
L. Leedjärv.
Proc. Acad. Sci. Estonian SSR, Phys., Math., Vol. 34, No. 2, p. 208 – 218 (1985). = Tartu Astrofüüs. Obs. Teated, Nr. 78. In Russian.
The results of the study of the binary star μ Sgr are presented. Twenty–one high–dispersion spectrograms have been used. The microturbulent velocities and the chemical composition of the atmosphere of the primary component have been determined by the curve–of–growth method. The radial velocities and orbital elements are in accordance with the data obtained by Kohl (1932). The distance of the star is about 1.36 kpc, colour–excess $E(B–V) = 0\overset{m}{.}32$ and $M_v = -7\overset{m}{.}8$. It has been found that $R = 105\ R_\odot$ and $T_{eff} = 13000$K for the primary component. Therefore, its spectral class ought to be B6. The masses of the components have been found to be approximately 25 $M_\odot$ for the primary star and 17 $M_\odot$ for the secondary star.

119.030 **On the peculiarities in the spectrum of the eclipsing binary AR Aur.**
T. M. Rachkovskaya.
Izv. Krymskoj Astrofiz. Obs., Tom 70, p. 134 – 139 (1985). In Russian. English translation in Bull. Crimean Astrophys. Obs., Vol. 70.
The spectrum of the eclipsing binary AR Aur has been studied and radial velocities were measured using the spectrograms with dispersions 15 and 36 Å/mm (the 1968 observations) and 12 Å/mm (the 1981 – 1982 observations). On the basis of the relation $W(\lambda4246$ Sc II$)/W(\lambda4215$ Sr II$) < 1$ and the fact that the most intensive helium lines are not observed, but metallic lines are present in the spectrum of AR Aur, it has been concluded that AR Aur is a Am star. The parameters of the AR Aur system have been determined: the velocity of the system $\gamma = +27$ km/s, the semi–amplitudes of radial velocity $K_1 = K_2 = 114$ km/s and the mass–ratio of the components $q = 1.0$.

119.031 **Period study of the RS CVn eclipsing binary SV Camelopardalis.**
C. V. S. R. Sarma, M. B. K. Sarma, K. D. Abhyankar.
Bull. Astron. Soc. India, Vol. 13, No. 4, p. 346 – 358 (1985).
Using the available photographic and photoelectric times of minima, the authors have revised the elements of the light–time orbit of SV Camelopardalis, an RS CVn type eclipsing binary. Taking a mass function of 0.118 $M_\odot$ and the inclination of $89\overset{\circ}{.}84$ for the eclipsing pair, the authors obtain the third body mass of 0.18, 0.21 and 0.38 $M_\odot$ for the assumed inclination of the third body orbit of 90°, 60° and 30° respectively.

119.032 **Verbesserte Elemente für vier Bedeckungsveränderliche.**
W. Braune, D. Lichtenknecker.
BAV Rundbrief, 35. Jahrg., Nr. 1, p. 1 – 7 (1986).
Within a discussion about the period behaviour of different eclipsing binary stars of the BAV programs new elements are published for QQ Cas, DK Cyg, AW Lac and CF Tau.

119.033 **Photometry and spectroscopy of V Sagittae through eclipse.**
G. A. Williams, A. R. King, A. K. Uomoto, W. A. Hiltner.
Mon. Not. R. Astron. Soc., Vol. 219, No. 4, p. 809 – 814 (1986).
The authors present simultaneous spectroscopy and UBV photometry obtained for the binary system V Sagittae through several eclipses. The spectroscopy clearly reveals a rotational disturbance similar to those seen in eclipsing cataclysmic variables, but with a duration longer than can be explained as a disc occulted by a secondary star. An uneclipsed component to the line suggests a high–velocity wind is being expelled from the system. If the rotational disturbance represents disc material being occulted, its velocity implies that the primary must be a compact object, and not a main–sequence star or helium star. The difficulties in modelling this system are discussed.

119.034 **Research of two small–amplitude massive eclipsing binaries V729 Cygni and CX Cephei.**
N. E. Kurochkin.
Perem. Zvezdy, Tom 22, No. 2, p. 219 – 224 (1985). In Russian.
The stability of periods of two eclipsing binaries is studied on the basis of photographic observations. The periode of V729 Cygni decreased by $0\overset{d}{.}00014$ for 70 years. The upper limit of period variations is obtained for CX Cep: $\Delta P/P < 4 \times 10^{-5}$.

119.035 **New variable star in Cassiopeia SVS 2511.**
L. N. Berdnikov.
Perem. Zvezdy, Tom 22, No. 2, p. 248 (1985). In Russian.
The comparison star "*b*" by Parenago and Kukarkin, 1940 of CG Cas was found to be an eclipsing variable with the period of $4\overset{d}{.}94208$. The results of a photographic study of the star based on Moscow collection of plates are given.

119.036 **RU Monocerotis: recent results.**
D. Ya. Martynov, A. I. Khaliullina.
Astron. Zh., Tom 63, Vyp. 2, p. 288 – 297 (1986). In Russian. English translation in Sov. Astron., Vol. 30, No. 2.
In the eclipsing binary system RU Monocerotis since its discovery at the beginning of this century the apsides line has rotated by 90°, and photometric as well as spectroscopic observations have allowed to investigate anew its physical parameters. At present, this is one of the best studied systems with apsidal motion. The authors determined the new value of the parameter $k_2 = 0.0068$ reflecting the internal constitution of the star.

119.037 **Flare activity of the eclipsing binary V502 Cyg.**
M. M. Zakirov.
Tsirk. Astron. Inst., Tashkent, No. 114, p. 32 – 46 (1985). In Russian.

119.038 **The eclipsing binaries BV and GR Tauri.**
T. J. Brelstaff, J. E. Isles.
J. Br. Astron. Assoc., Vol. 96, No. 2, p. 98 – 101 (1986).

119.039 **The 1982 – 84 eclipse of Epsilon Aurigae.**
J. E. Isles.
J. Br. Astron. Assoc., Vol. 96, No. 3, p. 148 – 151 (1986).
Visual observations during the period 1979 – 84 are in reasonable agreement with published photoelectric data. Possible flare activity on 1983 July 30 – August 2 is reported; observers are invited to re-examine any photographs or photoelectric records for this period.

119.040 **IUE spectra of the VV Cephei star BD + 63°0003.**
J. S. Shaw, E. F. Guinan.
Bull. Am. Astron. Soc., Vol. 17, No. 4, p. 875 – 876 (1985). Abstract. – See Abstr. 010.062.

119.041 **Observation and analysis of the Hyades binary HD 27130.**
S. J. Schiller, B. J. Hrivnak, E. F. Milone.
Bull. Am. Astron. Soc., Vol. 17, No. 4, p. 886 (1985). Abstract. – See Abstr. 010.062.

119.042 **Photoelectric observations and photometric orbit of BS Dra.**
R. A. Botsula.
Izv. Astron. Ehngel'gardt. Obs., No. 48, p. 15 – 26 (1984). In Russian.
Two-colour photoelectric observations (501 blue, 408 yellow) have been made in the period 1971 – 1978. Photometric elements were determined by Lavrov's method. The system has small orbital eccentricity. The primary star's limb darkening coefficient (0.4) is less than the theoretical one. The components have nearly identical luminosity. Absolute elements were determined with Popper's spectral orbit.

119.043 **Investigation of circumstellar envelopes of eclipsing binaries with emissions in the spectrum.**
V. G. Karetnikov, L. V. Glazunova, S. V. Kutsenko, E. V. Menchenkova.
Problems of astronomy, p. 14 – 17 (1985). In Russian. Abstr. in Ref. Zh., 51. Astron., 3.51.568 (1986). – See Abstr. 003.023.

119.044 **Once again on the apsidal motion in the HS Herculis system.**
D. Ya. Martynov.
Astron. Tsirk., No. 1401, p. 1 – 4 (1985). In Russian.

119.045 **Photoelectric observations of CC Cas.**
T. S. Polushina.
Astronomical-geodetical investigations, p. 97 – 112 (1985). In Russian. Abstr. in Ref. Zh., 51. Astron., 4.51.597 (1986). – See Abstr. 003.024.

119.046 **Investigation of the eclipsing variable BW Draconis. I. Photoelectric observations.**
A. V. Mironov, E. I. Staritsyn.
Astronomical-geodetical investigations, p. 84 – 91 (1985). In Russian. Abstr. in Ref. Zh., 51. Astron., 4.51.605 (1986). – See Abstr. 003.024.

119.047 **Investigation of the eclipsing variable BW Draconis. II. Relative and absolute elements.**
L. F. Istomin, E. I. Staritsyn.
Astronomical-geodetical investigations, p. 92 – 96 (1985). In Russian. Abstr. in Ref. Zh., 51. Astron., 4.51.606 (1986). – See Abstr. 003.024.

119.048 **Spectral investigation of the eclipsing binary star V367 Cygni.**
V. G. Karetnikov, E. V. Menchenkova.
Sov. Astron., Vol. 29, No. 4, p. 436 – 439 (1985). English translation of 40.119.012.

119.049 **Spectroscopic investigations of the Wolf-Rayet eclipsing binary CQ Cep. Spectroscopic orbit and spectral classification of the satellite.**
T. A. Kartasheva, L. I. Snezhko.
Sov. Astron., Vol. 29, No. 4, p. 440 – 444 (1985). English translation of 40.119.013.

119.050 **Photoelectric photometry of the eclipsing binary BM Cassiopeiae.**
E. A. Smith.
I.A.P.P.P. Commun., No. 23, p. 2 (1986). Abstract. – See Abstr. 010.521.

119.051 **The spectrographic orbit of the eclipsing binary HH Carinae.**
C. H. Mandrini, R. H. Méndez, O. E. Ferrer, V. S. Niemela.
Rev. Mex. Astron. Astrofis., Vol. 11, No. 2, p. 99 – 102 (1985).
The authors present a radial velocity study of the eclipsing binary system HH Carinae, and determine for the first time its spectrographic orbital elements. Using the results of a previous photometric study by Söderhjelm, the authors also determine the values of the masses and dimensions of the binary components.

119.052 **Properties of main-sequence eclipsing binaries: into the G stars with HS Aurigae, FL Lyrae, and EW Orionis.**
D. M. Popper, C. H. Lacy, M. L. Frueh, A. E. Turner.
Astron. J., Vol. 91, No. 2, p. 383 – 404 (1986).
Radial velocities and *V*, *R* photometry of HS Aur (G8 + K0), FL Lyr (F8 + G8), and EW Ori (G0 + G5) are analyzed to obtain the fundamental properties of the components of these detached binaries.

119.053 **Archival period study of the eclipsing binary KT Hydrae.**
O. C. St. Cyr, S. B. St. Cyr.
Astron. J., Vol. 91, No. 2, p. 416 – 418 (1986).
The authors have undertaken an archival survey of the photographic-plate collections of Harvard College Observatory to search for times of minima of the eclipsing binary KT Hydrae. They report here on the result of that survey, which includes an improved determination of the system's period. A new finding is that the period of KT Hydrae has been variable over the past century. The authors have also obtained a spectral classification for the variable (mid to late A).

119.054 **EXOSAT hard X-ray observations of EX Hydrae.**
K. Beuermann, J. Osborne.
X-ray astronomy '84, p. 23 – 26 (1985). – See Abstr. 012.033.
The authors present light curves and spectra of EX Hydrae in the 1 – 10 keV energy range obtained with the medium energy experiment of EXOSAT. The data support an intermediate-polar model of EX Hydrae.

119.055 **An analysis of the rotation effect in the radial velocity curve of the eclipsing binary Delta Librae.**
T. F. Worek.
Diss. Abstr. Int., Sect. B, Vol. 46, No. 7, p. 2347 (1986). Thesis, University of Pittsburgh, 251 pp. (1985). Order No. DA8519464.

119.056 **Possible shrinking of the primary component of Epsilon Aurigae.**
M. Saitō, M. Kitamura.
Astrophys. Space Sci., Vol. 122, No. 2, p. 387 – 393 (1986).
Radial velocities of the primary star of ε Aur are presented for phases outside the eclipse between 1975 and 1985. A violent collapse of the atmosphere has occurred with a velocity of about 40 km s^{-1} on October 1977, indicating a decrease of 0.07 AU in the radius for three days. A shrinking of the primary star is suggested from the characteristics of eclipse light curves that the duration of totality has become longer and the duration of eclipse

shorter from eclipse to eclipse. The decrement of radius is estimated to be about 16%, 0.2 AU, in recent 27 years.

119.057 **Epsilon Aurigae in eclipse: the light and colour variations.**
M. Parthasarathy, M. L. Frueh.
Astrophys. Space Sci., Vol. 123, No. 1, p. 31 – 48 (1986).

The *uvby* and *VRI* observations of ε Aur during the ingress and early totality phases obtained at the McDonald Observatory are analysed. The light and colour variations of ε Aur during the 1982 – 1984 eclipse are compared with those found during the 1955 – 1957 eclipse.

119.058 **Determinacy of the eclipse elements from a part of the light curves of eclipsing binaries.**
O. Demircan.
Astrophys. Space Sci., Vol. 123, No. 2, p. 305 – 312 (1986).

A part of the light curves of two detached (CD Tau and V909 Cyg) and one semi–detached (Algol) binaries have been analysed with the emphasis on the determinacy of the parameters of different systems from a part of their eclipse curves.

119.059 **Four–colour photometry of eclipsing binaries. XXIIA. Photometric elements and apsidal motion of GL Carinae.**
A. Giménez, J. V. Clausen.
Astron. Astrophys., Vol. 161, No. 2, p. 275 – 286 (1986).

The authors have obtained an accurate set of photometric elements of the B–type eclipsing binary GL Car from a detailed analysis of new uvby light curves. The high quality determination of the apsidal motion rate led to a value of the internal structure constant in good agreement with theoretical predictions based on the estimated absolute dimensions. GL Car is found to be a very young and relatively massive eclipsing binary which seems to be associated to the Carina OB complex surrounding the open cluster NGC 3572. The small degree of evolution for the estimated masses of the components of the system makes GL Car a good candidate for the discrimination between internal structure models based on different opacities.

119.060 **HD 47755, a new eclipsing binary.**
R. H. Koch, D. H. Bradstreet, B. J. Hrivnak, R. J. Pfeiffer, P. M. Perry.
Astron. J., Vol. 91, No. 3, p. 590 – 597 (1986).

IUE spectra have been studied for the close binary HD 47755. The line spectra are described and equivalent widths tabulated. IUE and *UBV* fluxes, when dereddened for $E(B-V) = 0.09$, yield $T_e = 16500$K. The ultraviolet and blue line spectra permit approximate evaluations of the temperatures of each star when used with the monochromatic fluxes from model atmospheres. It is suggested that the wind (discovered on the IUE images) from at least one of the stars causes the light curve complexity. In terms of the orbital radius the stars are not small and the system is not widely detached or a ZAMS one. The weight of the nonastrometric evidence is that HD 47755 is a cluster member of NGC 2264.

119.061 **Photometry of long–period Algol binaries. I. Five–color light curves of the interacting eclipsing binary RS Cephei.**
E. C. Olson, C. A. Stoehr.
Astron. J., Vol. 91, No. 6, p. 1418 – 1420 (1986).

The authors present differential uvby I photometry of RS Cephei obtained between September 1980 and August 1985 at the Prairie and Mount Laguna Observatories and compare these new observations with those of Hall et al. (1984), obtained between November 1965 and February 1972.

119.062 **Photometry of long–period Algol binaries. II. Anomalies in five–color photometric solutions of RS Cephei.**
E. C. Olson.
Astron. J., Vol. 91, No. 6, p. 1421 – 1427 (1986).

New photometric solutions, essentially consistent at five wavelengths, are derived for the long–period totally eclipsing binary RS Cephei. The relatively small perturbations to the continuum light from the accretion disk surrounding the hot gainer are isolated and discussed, and a disk temperature of roughly 6000K is found. Light perturbations suggest that the disk is asymmetrical. Occasional brightenings of the system at all wavelengths are identified as accretion episodes. Large color anomalies, variable from eclipse to eclipse, were detected across the disk of the gainer during partial phases of primary eclipse.

119.063 **Rediscussion of eclipsing binaries. XV. Alpha Coronae Borealis, a main–sequence system with components of types A and G.**
J. Tomkin, D. M. Popper.
Astron. J., Vol. 91, No. 6, p. 1428 – 1437 (1986).

The presented analysis of α CrB is based on the Ebbighausen (1976) spectroscopic observations of the primary component, on the McDonald Reticon observations of the secondary, and on the Kron–Gordon (1953) photometric observations at 0.72 μm. The primary component, with mass 2.6 $M_\odot$ and spectral type A0, lies about 0.4 mag above the zero–age main sequence, while the secondary, with mass 0.92 $M_\odot$, has a radiative flux corresponding to a type near G5, and must lie close to the ZAMS. The system is unique among analyzed main–sequence systems with respect to its small mass ratio and large magnitude difference between the components.

119.064 **A study of unitary light curves of AR Lacertae.**
E.–H. Lee, K.–Y. Chen, I.–S. Nha.
Astron. J., Vol. 91, No. 6, p. 1438 – 1442 (1986). = Contrib. Dep. Astron. Univ. Florida, No. 89. = Yonsei Univ. Obs. Contrib., No. 40.

During the 1982 – 1983 season, the eclipsing binary AR Lac was observed photoelectrically with yellow and blue filters at Yonsei University Observatory. As has often happened, the light curves show changes in the depth of primary minimum and apparently extraneous distortion outside eclipses. Solutions on the Russell model were obtained; the geometric and photometric elements agree well with those of Chambliss (1976) and Park (1984).

119.065 **Study of the eclipsing system V388 Cyg.**
J. M. Kreiner, J. Tremko.
Contrib. Astron. Obs. Skalnaté Pleso, Vol. 14, p. 145 – 162 (1986).

Photoelectric observations of the eclipsing system V388 Cyg were made at the Astronomical Observatory of Skalnaté Pleso in the blue region. Published epochs of minima and these new observations were used to study the changes in period of the system. Linear elements with a sudden change in period as well as non–linear elements were considered. The continuous decrease of the period agrees with the theory of mass transfer between the components of binaries. New light elements were calculated.

119.066 **Das langperiodische Bedeckungssystem V1068 Cygni.**
S. Rößiger.
Mitt. Veränderliche Sterne, Band 10, Heft 7, p. 150 – 152 (1985).

Recent photoelectric observations show that V1068 Cygni is an EB/GS type eclipsing binary. Approximate system constants from the V–light curve are presented.

119.067 **Beobachtung des Bedeckungsminimums von Zeta Aurigae 1985.**
D. Böhme.
Mitt. Veränderliche Sterne, Band 10, Heft 7, p. 160 (1985).

119.068 **Epsilon Aurigae in eclipse II. Optical absorption lines from the secondary.**
D. L. Lambert, S. R. Sawyer.
Publ. Astron. Soc. Pac., Vol. 98, No. 602, p. 389 – 402 (1986).

Optical high–resolution spectra of ε Aur obtained during the 1982 – 85 eclipse are discussed. The K I λλ7664 and 7699 lines strengthen greatly during the eclipse as gas in orbit around the secondary passes in front of the primary. Anlaysis of the radial velocities shows that the mass of the secondary is probably in the range $3 \lesssim M_2/M_\odot \lesssim 6$ and, then, the primary mass is $M_1 \lesssim 3\ M_\odot$. Many other lines confirm the velocity pattern provided by the K I lines. It is suggested that the primary is a low–

mass post–AGB star which, in less than 10^4yrs, will become a white dwarf. Earlier, this star lost mass through a wind and by Roche–lobe transfer to the secondary. This scenario is shown to be rather more consistent with available data.

119.069 **The light curve of the eclipsing binary AR Bootis.**
J. C. Houck, J. T. Pollock.
Publ. Astron. Soc. Pac., Vol. 98, No. 602, p. 461 – 463 (1986).
A photographic *B* light curve of the eclipsing binary AR Boo is presented. The previous period of 0.416718 day by Kurochkin (1960) is incorrect. The newly determined period is $0^d.34470 \pm 0^d.00001$ and the eclipse depths are found to be unequal.

119.070 **Revidierte Lichtwechselelemente für AH Aur und AP Aur.**
D. Lichtenknecker.
BAV Rundbrief, 35. Jahrg., Nr. 2, p. 41 – 46 (1986).
For the eclipsing binary AH Aur new elements were calculated. The elements given in the GCVS 1985 are wrong. New instantaneous elements were derived for AP Aur. A rapid change in the period is supposed.

119.071 **The O'Connell effect systems CX Canis Majoris, TU Crucis, AQ Monocerotis, and DQ Velorum.**
E. F. Milone.
Astrophys. J., Suppl. Ser., Vol. 61, No. 2, p. 455 – 464 (1986). = Publ. Rothney Astrophys. Obs., No. 36.
UBV photometry and classification spectroscopy have been carried out for four suspected O'Connell effect systems in the southern hemisphere. Although the light curves are incomplete, they represent the first photoelectric data on these systems and reveal the degree to which the O'Connell effect is present. Phases of minima were computed for all four systems, and for previously published light curves for AQ Mon. Improved elements were derived for the latter system. Fourier representations for the four systems were also determined.

119.072 **On the period of YY Geminorum.**
E. Zsoldos.
Inf. Bull. Variable Stars, No. 2860, 3 pp. (1986).

119.073 **Revised elements for three eclipsing binaries.**
T. Berthold.
Inf. Bull. Variable Stars, No. 2861, 2 pp. (1986).

119.074 **BP Muscae – UBV light curve and ephemeris.**
Z. Kviz.
Inf. Bull. Variable Stars, No. 2863, 4 pp. (1986).

119.075 **The eclipsing binary V651 Cassiopeiae.**
W.–x. Lu.
Inf. Bull. Variable Stars, No. 2868, 4 pp. (1986).

119.076 **U Cephei: an unusual development.**
J. B. Rafert, E. Smith, N. L. Markworth.
Inf. Bull. Variable Stars, No. 2872, 4 pp. (1986).

119.077 **Photoelectric UBV observations of AQ Tau and the problem of its period.**
X.–f. Liu, H.–s. Tan.
Inf. Bull. Variable Stars, No. 2875, 2 pp. (1986).

119.078 **AH Cephei: a new photometric triple star.**
P. Mayer, M. Wolf.
Inf. Bull. Variable Stars, No. 2886, 4 pp. (1986).

119.079 **HαN and HαW observations of the eclipsing binary system VV Orionis (HR 1868).**
B. M. Davan.
Inf. Bull. Variable Stars, No. 2892, 4 pp. (1986).

119.080 **Evolutionary status of V356 Sgr.**
J. Ziółkowski.
Acta Astron., Vol. 35, Nos. 3–4, p. 199 – 212 (1985).
Different possible evolutionary models of V356 Sgr are investigated. All models with contact components evolving on nuclear time scale are ruled out. The evolutionary calculations lead to two conclusions: (1) contact component is burning hydrogen in a shell and (2) the evolution cannot be conservative. There are also two paradoxes: (1) the established evolutionary status of the system is at conflict with the observed *O–C* diagram and (2) the theoretical models require non–conservative evolution with unrealistically small loss of angular momentum from the system.

119.081 **112th list of minima of eclipsing binaries.**
BBSAG Bull., No. 79, p. 1 – 4 (1986).

119.082 **Eclipsing variable stars in neighbouring galaxies. II. Discussion of the light curve of V 55 in NGC 2403.**
A. Gaspani.
GEOS Circ., EB 14, 8 pp. (1986).
The light curve of the eclipsing variable V 55 in the spiral galaxy NGC 2403 is discussed by means of an analysis in the frequency domain in order to obtain photometric orbital elements. The solution of the orbit shows that V 55 is a detached system whose components should belong to the earlier spectral classes. The binary system appears to be the most remote system so far discovered.

119.083 **Eclipsing variable stars in neighbouring galaxies. III. Discussion of the light curve of V 31 in IC 1613.**
A. Gaspani.
GEOS Circ., EB 15, 6 pp. (1986).
The light curve of the only eclipsing variable so far discovered in IC 1613 is discussed by applying a method of analysis in the frequency domain in order to obtain the photometric orbital elements. The solution shows that V 31 is an Algol–type system in which the larger component is probably more evolved than its companion.

119.084 **Corrections and improvements in the effectiveness and extension of the applicability range of the WINK model of eclipsing binaries.**
L. P. R. Vaz.
Rev. Mex. Astron. Astrofis., Vol. 12, Spec. Issue, p. 177 – 183 (1986). – See Abstr. 012.066.
Errors and inconsistencies still present in the program WINK have been detected and corrected. Modifications of the program have been implemented in order to improve the effectiveness of the model during the solution of the inverse problem of finding the parameters of a system from the light curves. Other modifications are introduced in order to extend the validity range of the model to closer and more deformed systems. After the introduction of the modifications, the WINK solutions for such systems become much closer to the solutions produced by models which are more realistic geometrically (eg. the Wilson–Devinney model) than the solutions produced by the original WINK model.

119.085 **Is the companion disk of ε Aur becoming flat?**
T. Ôki, Y. Sakai.
Sci. Rep. Fukushima Univ., No. 36, p. 35 – 39 (1985).
The authors observed the recent eclipse of ε Aur by U, B and V wideband photometry. They estimate the durations of totality and the depths of eclipses. The results show that the durations are getting longer, and the depths are getting shallower. If the companion is an obscuring disk as accepted by many authors, it is suggested that the disk's thickness is becoming thin and its diameter is becoming large.

119.086 **Photographic photometry of the eclipsing binary U Cephei with small equipment.**
F. Sato, F. Ito.
Hyogo University of Teacher Education Journal, Ser. 3 (Natural Sciences and Practical Life Studies), Vol. 4, p. 85 – 100 (1984).
Photographic photometry of U Cephei with a 10–cm refractor and a 35–mm camera has manifested characteristics of its light

variation and determined times of its primary minima. It has been found that the period of U Cep in 1983 is longer than that in 1979 – 80. Photographic photometries of bright eclipsing binaries with such small equipment can easily be made in extracurricular activities at lower and upper secondary schools.

An Apple Pascal simulation of eclipsing binary star systems.
See Abstr. 014.034.

Elemente–Ableitungsmöglichkeiten am Beispiel EZ Lyr.
See Abstr. 036.028.

Chromatic position difference: a technique for studying double stars.
See Abstr. 036.094.

A program for observations of the nearest eclipsing binaries from the HIPPARCOS satellite.
See Abstr. 051.091.

Multilevel line formation in the expanding envelope of ζ Aur/ VV Cep systems.
See Abstr. 064.001.

The reflection effect in model stellar atmospheres. I. Grey atmospheres with convection.
See Abstr. 064.069.

Boundary layer theory and its application to Algol systems: a model for the pseudo–photosphere stars.
See Abstr. 065.005.

A multi–component chromospheric model for AR Lacertae.
See Abstr. 112.051.

Photometric variability of some early–type stars.
See Abstr. 113.004.

A light synthesis program for binary stars. II. Light curve and color curve effects in a contact system.
See Abstr. 117.002.

Spectrophotometry of the eclipsing dwarf nova V2051 Ophiuchi.
See Abstr. 117.006.

Contact and near–contact binary systems – III. The semi–detached system YY Ceti.
See Abstr. 117.011.

A new spectroscopic orbit of the W Ursae Majoris system U Pegasi.
See Abstr. 117.016.

Classical Algol systems.
See Abstr. 117.019.

R Aquarii, une variable unique: Mira, symbiotique, algolide et nova récurrente.
See Abstr. 117.031.

Period variations of binary systems as a possible source of information about motions in the stellar core.
See Abstr. 117.043.

Flare features of AR Lacertae.
See Abstr. 117.070.

Nature of the symbiotic binary V1329 Cyg.
See Abstr. 117.073.

***UBV* photometry of FK Comae.**
See Abstr. 117.074.

The very low state of V794 Aquilae.
See Abstr. 117.078.

Observations of the accretion disk in the dwarf nova HT Cassiopeia.
See Abstr. 117.126.

Binaries and their role in present–day astronomy.
See Abstr. 117.141.

RS CVn binaries.
See Abstr. 117.155.

UU Lyncis: a non–contact binary system with a short period.
See Abstr. 117.169.

Binary nature of HD 62001.
See Abstr. 117.170.

Photometrical observations indicative of a black hole in SS433.
See Abstr. 117.184.

The eclipse and mass of Cen X–3.
See Abstr. 117.187.

New technique to compute the geometrical elements of the eclipsing of W UMa–type.
See Abstr. 117.205.

Synthetic light–curve method applied to the W UMa systems SY Horologii and VY Ceti.
See Abstr. 117.233.

FO Virginis: a pre–contact binary.
See Abstr. 117.240.

HD 174853 – a new eclipsing binary.
See Abstr. 117.254.

Contact binaries with components in poor thermal contact. II. WZ Cephei.
See Abstr. 117.273.

Contact binaries with components in poor thermal contact. III. AU Serpentis and FT Lupi.
See Abstr. 117.274.

Contact binaries with components in poor thermal contact. IV. AG Virginis.
See Abstr. 117.275.

The eclipses of cataclysmic variables. I. HT Cassiopeiae.
See Abstr. 117.290.

Multiplicity among eclipsing binaries.
See Abstr. 118.015.

Frequency of Be and Ae stars in spectroscopic binaries.
See Abstr. 120.004.

Discovery of an atmospheric eclipse of τ Per.
See Abstr. 120.026.

A β Cephei variable in an eclipsing binary in the cluster NGC 3293.
See Abstr. 122.018.

New data on the variable stars in the region of M56.
See Abstr. 122.041.

Study of 11 stars in Lyra.
See Abstr. 122.042.

Hα profile variations in variable luminous yellow supergiants.
See Abstr. 122.108.

Where is AK Canis Minoris?
See Abstr. 123.024.

120 Spectroscopic Binaries

120.001 **Speckle interferometry of spectroscopic and Hyades binary stars.**
J. Ebersberger, G. Weigelt, R. B. Orellana.
Astron. Astrophys., Suppl. Ser., Vol. 64, No. 1, p. 131 – 133 (1986).
The authors report on 39 high–resolution speckle observations of 18 spectroscopic, 9 Hyades and 7 other binary stars. Measurements were performed with the Danish 1.5–m and the ESO 3.6–m telescope at La Silla, Chile.

120.002 **Photometric and spectroscopic observations of the variable star ET And. II.**
G. Scholz, E. Gerth, K. P. Panov.
Astron. Nachr., Vol. 306, No. 6, p. 329 – 337 (1985).
New spectroscopic and photometric observations of the variable Ap star ET And carried out 1982 and 1983 confirm once again the existence of short–time variations. For the radial velocity variations a period of 0.19771 days and for the light variations a period of 0.0989 days was found. Some spectrograms show a velocity progression of the Balmer lines. The gradient is of such a kind that the radial velocity values increase from the line H_γ to H_{15}. Different models for the interpretion of the velocity progression are discussed.

120.003 **Circumstellar envelope around the binary system λ And.**
R. Glebocki, J. Sikorski, E. Bielicz, M. Krogulec.
Astron. Astrophys., Vol. 158, No. 1/2, p. 392 – 394 (1986).
Asymmetries in emission peaks and relative shifts of h3 and k3 absorptions in Mg II resonant lines are analysed in 31 spectra of λ And (long period RS CVn type star) obtained in the years 1978 – 1983. An evident correlation is found between these values and the orbital phase. The only reasonable explanation of this correlation is the existence of a circumstellar envelope around this binary system.

120.004 **Frequency of Be and Ae stars in spectroscopic binaries.**
T. Kogure, M. Suzuki.
Astrophys. Space Sci., Vol. 120, No. 2, p. 253 – 262 (1986).
The frequency of Be and Ae stars in spectroscopic binaries and Algol binaries is examined based on the available catalogue data, and compared with that of stars in the Bright Star Catalogue. It is shown that in binary systems Be and Ae stars reveal different statistical behaviours. The frequency of Be stars in spectroscopic binaries shows a notable peak in its orbital–period distribution in the range of 100 – 300 days, suggesting a statistical group separated from Algol systems. In contrast, most of Ae stars in spectroscopic binaries belong to the Algol systems.

120.005 **Spectroscopic observations of 112 Herculis (HD 174933).**
T. S. Galkina, N. Kh. Minikulov.
Izv. Krymskoj Astrofiz. Obs., Tom 73, p. 111 – 121 (1985). In Russian. English translation in Bull. Crimean Astrophys. Obs., Vol. 73.
Spectroscopic observations of the spectroscopic binary 112 Her were analysed on the basis of 44 spectrograms in the region λ3650 – 4950 Å obtained from June to September 1975. The radial velocity measurements from Balmer absorption lines and Mg II λ4481 show differences in the amplitude of the variation of v_{r-} and γ_0–velocity, respectively. The variations of the equivalent widths of H I, Mg II, Fe II, He I lines with phase as well as the variations of the spectral type of the primary component were found. The latter varies from B8.5 to A0.5.

120.006 **A spectroscopic orbit for the unresolved subsystem of the Hyades binary HD 30869.**
D. G. Turner, R. W. Lyons, C. T. Bolton.
Observatory, Vol. 106, No. 1070, p. 13 – 16 (1986).

120.007 **Spectroscopic binary orbits from photoelectric radial velocities. Paper 66: HR 2259.**
R. F. Griffin.
Observatory, Vol. 106, No. 1070, p. 16 – 18 (1986).

120.008 **Spectroscopic binary orbits from photoelectric radial velocities. Paper 67: HR 5053.**
R. F. Griffin.
Observatory, Vol. 106, No. 1071, p. 35 – 37 (1986).

120.009 **The bright, newly discovered BY Draconis system HD 80715.**
S. C. Barden, H. L. Nations.
Bull. Am. Astron. Soc., Vol. 17, No. 4, p. 879 – 880 (1985). Abstract. – See Abstr. 010.062.

120.010 **The massive binary system DH Cephei.**
A. C. Theokas, E. F. Guinan, H. C. Lines, R. D. Lines.
Bull. Am. Astron. Soc., Vol. 17, No. 4, p. 880 – 881 (1985). Abstract. – See Abstr. 010.062.

120.011 **Ultraviolet survey for hot companions among nonvariable yellow supergiants.**
A. Arellano Ferro, B. F. Madore.
Astrophys. J., Vol. 302, No. 2, p. 767 – 776 (1986).
Twenty nonvariable yellow supergiants have been observed with the IUE satellite, with the aim of discovering hot companions. Two newly discovered systems are announced: HD 74395 (G2 I + B9.5 IV–V) and HD 193469 (later than G0 + B8 IV–V). Some stars independently known to have a hot companion were included in order to study the properties of the companions. Atmospheric models were used to fit the overall energy distribution of the binary system, allowing an estimate of T_e, log g, and the spectral type of the companion. Absolute magnitudes M_v were estimated for the supergiants in the binary systems, and their positions in the H–R diagram were compared with theoretical evolutionary models.

120.012 **The orbits of six spectroscopic Hg–Mn binary stars.**
V. S. Lebedev.
Astron. Tsirk., No. 1393, p. 7 – 8 (1985). In Russian.

120.013 **Revised spectroscopic orbits for the metallic–line binaires: ADS 8710 A and 41 Sextantis.**
T. F. Worek, E. R. Zizka, M. W. King, W. R. Beardsley.
Publ. Astron. Soc. Pac., Vol. 98, No. 600, p. 238 – 242 (1986).
The authors present improved spectroscopic orbital elements for the metallic, double–line binary ADS 8710 A. They argue that it is not very likely that ADS 8710 A is an eclipsing binary as some researchers have suggested but they do in turn propose that it might possibly form a triple system with its closest optical companion, ADS 8710 B. They also verify the spectroscopic orbital solutions published by Worek, Beardsley, and King (1978) and Abt and Levy (1985) for the metallic, single–line binary 41 Sextantis.

120.014 **Orbital inclinations of late B–type spectroscopic binaries.**
B. N. G. Guthrie.
Mon. Not. R. Astron. Soc., Vol. 220, No. 2, p. 559 – 569 (1986).
Information on the orbital inclinations j of late B–type spectroscopic binaries with periods between 3 and 50 day is obtained from the masses M_1 of their primary components, derived from $uvby\beta$ photometry, and the values of $M_1 \sin^3 j$.

120.015 **The WN4 + O8V spectroscopic binary HD 94546.**
V. S. Niemela, C. H. Mandrini, R. H. Méndez.
Rev. Mex. Astron. Astrofis., Vol. 11, No. 2, p. 143 – 147 (1985).
The authors report new spectrographic observations of HD 94546 and present an improved orbital solution.

120.016 **The identity of the primary component of Capella.**
R. Griffin, R. Griffin.
J. Astrophys. Astron., Vol. 7, No. 1, p. 45 – 51 (1986).
Ever since the discovery in 1899 that Capella is a double–lined spectroscopic binary system the sharp–lined component, which has the later spectral type, has always been regarded as the primary (more luminous) star. However, traces obtained with radial–velocity spectrometers show that the broad–lined, earlier–type component is actually the primary.

120.017 **Spectroscopic binary orbits from photoelectric radial velocities. Paper 68: HD 182593.**
R. F. Griffin.
Observatory, Vol. 106, No. 1072, p. 67 – 69 (1986).

120.018 **Spectroscopic and polarimetric parameters of the runaway WN7 binary system HD 197406: is the secondary an X–ray–quiet black hole?**
L. Drissen, R. Lamontagne, A. F. J. Moffat, P. Bastien, M. Seguin.
Astrophys. J., Vol. 304, No. 1, p. 188 – 200 (1986).
New spectroscopic and polarimetric data for the single–line spectroscopic binary HD 197406 are presented. New orbital elements and the mass function $f(m) = 0.28\,M_{\odot}$ yield a mass of $12.4\,M_{\odot}$ for the unseen companion if one supposes $M_{WN7} = 60\,M_{\odot}$ for the Wolf–Rayet primary. This system is shown to be a runaway, and in many respects resembles the OB supergiant + black hole (BH) binary Cyg X–1, except that HD 197406 is not an X–ray source. If the unseen companion of HD 197406 is a BH, the recoil from a preceding supernova explosion may be the cause of its present runaway status.

120.019 **BD –3°2525: a spectroscopic binary subdwarf.**
J. L. Greenstein, A. Saha.
Astrophys. J., Vol. 304, No. 2, p. 721 – 727 (1986).
As a result of a Palomar radial velocity program of high–velocity stars, the sdG star BD –3°2525 was found to be a single–lined spectroscopic binary. The period is 113 days; the semiamplitude is small, 14.4 km s^{-1}, and the eccentricity is large, $e = 0.5$. The authors place limits on the brightness, nature, and mass of the invisible secondary from multichannel spectrophotometry and from the orbit. A brief discussion of the frequency of binaries in population II suggests that close binaries are rare. The pair in BD –3°2525 is relatively wide and noninteracting.

120.020 **Multicolor photometry of the binary central star LSS 2018.**
A. U. Landolt, J. S. Drilling.
Astron. J., Vol. 91, No. 6, p. 1372 – 1375 (1986). = Contrib. Louisiana State Univ. Obs., No. 197.
Multicolor photometry on the Kron–Cousins UBVRI photometric system is presented for the binary central star of the planetary nebula LSS 2018.

120.021 **On the spectroscopic orbit of the bright WC binary γ²Velorum, with implications for the evolutionary scenario of Wolf–Rayet stars.**
A. F. J. Moffat, N. Vogt, G. Paquin, R. Lamontagne, L. H. Barrera.
Astron. J., Vol. 91, No. 6, p. 1386 – 1391 (1986).
From a series of over 50 spectroscopic plates distributed over three orbital cycles, the authors confirm the radial–velocity orbit for this 78.5 day, WC 8 + O9 I binary, as determined recently by Niemela and Sahade (1980). No evidence is found for a third, low–mass body in the system, as proposed elsewhere in the literature. Comparison with other WC + O binaries reveals a trend of mass ratio with spectral subclass, such that the cooler, late–type, WCL stars are more massive relative to their O companions than the hotter, early–type, WCE stars. Therefore, the finite lifetime of the WC phase combined with the observed high mass–loss rate for W–R stars implies that WCL stars may evolve into WCE stars.

120.022 **Spectroscopic orbital elements of ϰ²Bootis B.**
G. A. Bakos.
Astron. J., Vol. 91, No. 6, p. 1416 – 1417 (1986).
Radial–velocity measurements extending over a period of 70 yr were used to derive the orbital elements of this spectroscopic binary. The period of the system is 1790 days and the semiamplitude of the velocity curve is 7.2 km s^{-1}.

120.023 **On the spectroscopic orbit of the bright WC binary γ²Velorum, with implications for the evolutionary scenario of Wolf–Rayet stars.**
A. F. J. Moffat, N. Vogt, G. Paquin, R. Lamontagne, L. H. Barrera.
Astrophys. Prepr. Ser., No. 14, 23 pp. (1986). To appear in Astron. J.

120.024 **The spectroscopic orbit of the double–lined binary 29 Arietis.**
W.–x. Lu.
Publ. Astron. Soc. Pac., Vol. 98, No. 602, p. 468 – 470 (1986).
The star 29 Ari, listed as a suspected single–lined spectroscopic binary in the *Bright Star Catalogue*, turns out to be double lined. The orbital elements are determined to be: $P = 19.3787$ days, $V_0 = +8.84$ km s^{-1}, $K_1 = 24.04$ km s^{-1}, $K_2 = 31.18$ km s^{-1}, $e = 0.391$, $\omega = 184.2$ degrees, and $T =$ JD2446335.76.

120.025 **The spectroscopic orbits of HD 44780 and 65 Geminorum.**
R. F. Griffin.
J. R. Astron. Soc. Can., Vol. 80, No. 2, p. 91 – 108 (1986).
Binary star systems in which both components are normal late–type giants are rare. HD 44780 and 65 Gem are shown to be such systems; they have periods of 577 and 1601 days respectively, and their spectra are conspicuously double–lined for considerable parts of their orbits. Their mass ratios are very close to unity, although their luminosity ratios are not. Attention is drawn to the imminent series of lunar occultations of 65 Gem.

120.026 **Discovery of an atmospheric eclipse of τ Per.**
T. B. Ake, W. S. Barksdale, R. E. Fried, J. L. Hopkins, F. C. Fekel Jr., D. S. Hall, H. J. Landis, H. Louth.
Inf. Bull. Variable Stars, No. 2847, 4 pp. (1986).

120.027 **Spectroscopic observations of ν Centauri.**
B. N. Ashoka, R. Surendiranath, N. Kameswara Rao.
Acta Astron., Vol. 35, Nos. 3–4, p. 395 – 399 (1985).
Presented are radial velocity observations of the single line spectroscopic binary and β Cephei–type variable ν Cen obtained in the period 1983 – 84. New orbital elements have been determined by combining these observations with observations obtained eighty years earlier.

120.028 **Structure of the outer atmosphere of Capella.**
M. M. Katsova.
Sov. Astron., Vol. 29, No. 6, p. 664 – 668 (1985). English translation of 40.120.027.

120.029 **Measurements of rotation for 18 binaries.**
H.–s. Tan.
Acta Astrophys. Sin., Vol. 6, No. 2, p. 154 – 159 (1986). In Chinese.
Rotational V·sin *i* of 18 binaries with spectral type B have been determined by means of the He I 5876 Å line in accordance with the method of Tomkin and Tan (1985).

120.030 **The spectrum of HD 72754.**
J. A. de Freitas Pacheco, S. J. C. Landaberry.
Rev. Mex. Astron. Astrofis., Vol. 12, Spec. Issue, p. 184 (1986). Abstract. – See Abstr. 012.066.

120.031 **A radial velocity study of two early O stars.**
G. R. Solivella, V. S. Niemela.
Rev. Mex. Astron. Astrofis., Vol. 12, Spec. Issue, p. 188 (1986). Abstract. – See Abstr. 012.066.

120.032 **Spectrophotometric study of two stars HD 400 and HD 140283.**
S. Abulazm, M. Hamdy, M. Marie, M. K. Aly, A. Aiad.
Bull. Fac. Sci., Cairo Univ., Vol. 54, p. 761 – 768 (1986).

A spectrophotometric study of the two peculiar stars HD 400 and HD 140283 is represented. The spectrum of these stars was found to be composite and their peculiarity arises from the fact that they are physical binaries and could be classified as (B3 V + F5 V) and (B5 V + F5 V) respectively.

On the errata in the "14ème Catalogue Complémentaire de Binaires Spectroscopiques".
See Abstr. 002.066.

Machine-readable version of the "14ème Catalogue Complémentaire de l'Observatoire de Toulouse" by A. Pedoussaut, N. Ginestet and J. M. Carquillat.
See Abstr. 002.070.

Application of the Fourier transform for solving and improving spectroscopic binary star orbits.
See Abstr. 036.069.

Photoelectric radial velocities. Paper XIII. 406 ninth-magnitude K0 stars in the Clube Selected Areas.
See Abstr. 111.003.

Radial velocities of standard stars.
See Abstr. 111.010.

The binary frequency and origin of the OB runaway stars.
See Abstr. 111.020.

On Wolf-Rayet stars with unshifted absorption lines.
See Abstr. 114.118.

On the nature of the dwarf carbon star G77-61.
See Abstr. 115.001.

Chromospheric activity in the Hyades triple system BD +23°635.
See Abstr. 116.059.

Further evidence for the binary nature of HR 3084.
See Abstr. 117.072.

The ellipsoidal variables. II. The main-sequence distribution.
See Abstr. 117.153.

HD 174853 – a new eclipsing binary.
See Abstr. 117.254.

Où en est l'astronomie des étoiles doubles?
See Abstr. 118.009.

The spectrographic orbit of the eclipsing binary HH Carinae.
See Abstr. 119.051.

EXOSAT hard X-ray observations of EX Hydrae.
See Abstr. 119.054.

The Cepheid mass problem and Cepheid binaries.
See Abstr. 122.102.

Spectroscopic binaries in the LMC.
See Abstr. 156.031.

121 Early-stage Stars (T Tauri Stars, Herbig-Haro Objects, etc.)

121.001 **Spectral similarities in the visible and UV spectrum of Herbig Ae/Be stars.**
C. Catala, J. Czarny, P. Felenbok, F. Praderie.
Astron. Astrophys., Vol. 154, No. 1/2, p. 103 – 114 (1986).

The pre-main sequence Herbig Ae/Be stars can be divided into three subclasses, according to their Hα profile. The authors address the question of whether the envelopes of the stars belonging to a same subclass are similar. Therefore, they have carried out high resolution spectroscopic observations of four members of the "P Cygni" subclass. In addition to Hα, the authors have found that Mg II resonance lines, the Ca II K line, the Ca II infrared triplet lines, and the Na I D lines have similar shapes in the spectra of these stars. Since these lines are formed in distinct regions in the envelopes of the stars, these similarities show that the envelopes of the stars of the "P Cygni" subclass have basically the same structure.

121.002 **Brackett-gamma-line observations of G 45.13+0.14A.**
M. Tanaka, T. Yamashita, S. Sato, M. Nishida.
Publ. Astron. Soc. Pac., Vol. 97, No. 597, p. 1112 – 1114 (1985).

Velocity-resolved observations of G 45.13+0.14A revealed that the Bγ(2.17 μm) emission line shows a peak velocity of 100 ± 10 km s^{-1}(LSR) and a deconvolved width of 100 ± 15 km s^{-1}(FWHM). Such a broad width was first detected for a highly luminous young stellar object. This width suggests that dense ionized gas flows outward from the central object.

121.003 **Erratum: "Spectroscopic observations of 33 T Tau stars in the blue" [Astron. Astrophys., Suppl. Ser., Vol. 62, No. 2, p. 309 – 316 (1985)].**
Y.-l. Sun, M. Jaschek, Y. Andrillat, C. Jaschek.
Astron. Astrophys., Suppl. Ser., Vol. 63, No. 3, p. 602 (1986). See Abstr. 40.121.015.

121.004 **Observations of bipolar flow sources in ammonia lines.**
T. Takano.
Astron. Her., Vol. 79, No. 1, p. 4 – 9 (1986). In Japanese.

121.005 **Pre-main-sequence stars in the Serpens molecular cloud.**
E. Churchwell, J. Koornneef.
Astrophys. J., Vol. 300, No. 2, p. 729 – 736 (1986). With plate 6.

The authors have mapped the central 10 $arcmin^2$ region of the Serpens molecular cloud at 2.2 μm with a spatial resolution of $\sim 7\farcs5$. More than 13 stars with $K \lesssim 13.4$ mag were detected in this area, only three of which have so far been identified with optical counterparts. Additional infrared photometry as well as spectrophotometry around 3.1 μm were obtained for the four brightest stars. The flux distributions of all four stars peak at approximately 3 μm, and the 3.07 μm "ice" feature was detected in absorption toward all four stars. The primary result of this study is the detection of at least three pre-main-sequence stars. The range of possible luminosities, radii, and temperatures for these stars as derived from the authors' near-infrared observations are consistent with those expected from models of low mass, pre-main-sequence stars. The near-infrared radiation is predicted to originate from circumstellar dust at a radius of up to 100 $R_{\odot}$ with an effective temperature of $\lesssim$2100K.

121.006 **A high resolution atlas of two T Tauri stars.**
I. Appenzeller, I. Jankovics, R. Jetter.
Astron. Astrophys., Suppl. Ser., Vol. 64, No. 1, p. 65 – 104 (1986).
Using an echelle spectrograph and a linear detector the authors obtained high resolution ($R \approx 2 \times 10^4$) spectrograms of the two strong–emission line T Tauri stars S CrA and VV CrA. From these spectrograms they derive the basic properties of the different types of emission lines and of the photospheric absorption spectra of these two examples of, respectively, the «YY Ori» and the «P Cyg» subgroup of the strong–emission–line T Tauri stars. In spite of some distinct spectral differences it appears possible to explain the emission line spectra of both types of T Tauri stars with the same basic physical model seen under different inclination angles.

121.007 **The jets and discs associated with R and T CrA.**
R. D. Wolstencroft.
Cosmical gas dynamics, p. 251 – 257 (1985). – See Abstr. 012.007.
Optical jets are associated with the Herbig Ae/Be stars R and T CrA. Polarization mapping of the reflection nebula NGC 6729, in which R and T CrA are embedded, reveals parallel bands of polarization vectors across the stars which can be explained by dust discs in which the grains are aligned by toroidal magnetic fields. The author suggests that the optical jets are collimated by a small inner circumstellar disc, which has decoupled from the magnetic field in the outer regions because of ambipolar diffusion, and whose orientation is governed principally by the angular momentum. The large outer interstellar disc, in which the grains are aligned by a toroidal magnetic field, is inclined obliquely to the inner disc and collimates the CO bipolar outflow.

121.008 **Herbig–Haro objects.**
J. Cantó.
Cosmical gas dynamics, p. 267 – 280 (1985). – See Abstr. 012.007.
A brief overview of the most relevant aspects of the Herbig–Haro objects is given. In particular, the most recent observations of these objects are reviewed. Emphasis is given to their possible relationship with the high–velocity molecular flows often found in the same region as the optical objects. The models advanced to explain these objects are also discussed. A model in which the optical objects represent either bright spots on the walls of cavities created by action of a stellar wind or the tips of these cavities is favored. This model is also likely to account for the high–velocity molecular flows and the filamentary (jet–like) emission features often associated with the HH–objects.

121.009 **Radio continuum emission from T Tauri stars and associated structures.**
A. Brown.
Cosmical gas dynamics, p. 287 – 292 (1985). – See Abstr. 012.007.
The current status of radio continuum observations of low mass, pre–main sequence stars (T Tauri stars) is discussed. The majority of T Tauri radio sources have been detected by the VLA telescope and a guide to both these and earlier observations is presented. Both nonthermal and thermal radio continuum emission has been detected from T Tauri stars. The likely emission processes and specific examples of their occurrence are outlined.

121.010 **The high velocity outflow in Herbig–Haro 7–11.**
J. F. Lightfoot, W. M. Glencross.
Cosmical gas dynamics, p. 293 – 298 (1985). – See Abstr. 012.007.
A map of the Herbig–Haro 7–11 outflow made in shock excited lines of H_2 is presented. The outflow is much bigger than previously suspected and seems to be well collimated on a large scale. The authors suggest that HH 7–11 are shocks inside the flow and that they also produce most of the H_2 emission. The CO "outflow" appears to be a secondary feature, outside the main flow.

121.011 **Studies of the Chamaeleon star–forming region. II. The pre–main–sequence stars HD 97048 and HD 97300.**
P. S. Thé, P. R. Wesselius, H. R. E. Tjin A Djie, H. Steenman.
Astron. Astrophys., Vol. 155, No. 2, p. 347 – 355 (1986).
Photometric observations in the visual, near infrared, and far infrared (IRAS), and UV(IUE) and IDS spectra are presented of the pre–main–sequence stars HD 97048 and HD 97300. The spectral energy distribution between 0.1 and 100 μm is obtained, and corrected for foreground and circumstellar extinction. Distances are derived and the infrared excess fluxes are interpreted as thermal re–radiation of circumstellar dust. For both stars the integrated infrared excess flux is less than half the flux absorbed in the ultraviolet and visual by circumstellar dust. This discrepancy can be explained in terms of either a flattened circumstellar dust shell, or a significant contribution of the Chamaeleon dust cloud to the total extinction towards the stars.

121.012 **Young stellar objects and their circumstellar dust: an overview.**
A. E. Rydgren, M. Cohen.
Protostars & planets II, p. 371 – 385 (1985). – See Abstr. 012.008.
The basic observational classes of young stellar objects are briefly reviewed and some of the theoretical and observational problems in pre–main sequence stellar evolution are discussed. The authors then summarize the evidence which indicates that the infrared excesses in recently formed stars are primarily due to thermal emission from circumstellar dust. Various indirect lines of evidence on the spatial distribution of the circumstellar dust are examined, and the authors conclude that the concentration of the dust in a large disk structure is the most viable model at this time.

121.013 **The ambient radiation field of young solar systems: ultraviolet and X–ray emission from T Tauri stars.**
M. S. Giampapa, C. L. Imhoff.
Protostars & planets II, p. 386 – 404 (1985). – See Abstr. 012.008.
The authors review the principal results that have emerged from ultraviolet and X–ray observations of T Tauri stars. They emphasize that, as seen in the ultraviolet, T Tauri stars are characterized by strongly enhanced emission. The X–ray emission and its variability, when detected, is also enhanced and indicate of the occurrence of violent flare activity near the stellar surface. The authors discuss possible effects of these enhanced emissions on conditions in early solar systems and the chemical evolution of early planetary atmospheres.

121.014 **The nature and origin of Herbig–Haro objects.**
R. D. Schwartz.
Protostars & planets II, p. 405 – 413 (1985). – See Abstr. 012.008.
A brief description of the nature of Herbig–Haro nebulae is given, and the shock–wave origin of the nebulae is discussed. Kinematical evidence suggests that Herbig–Haro objects are ejected in bipolar flows from young stars. Evidence from infrared observations of the stars that excite Herbig–Haro objects is summarized; these stars appear to be T Tauri stars. The origin of these nebulae is discussed emphasizing energy required to power them, and a number of questions are posed pertaining to outflow mechanisms associated with the exciting stars.

121.015 **Highly collimated mass outflows from young stars.**
R. Mundt.
Protostars & planets II, p. 414 – 433, 1186 (1985). – See Abstr. 012.008.
The observational evidence for highly collimated (bipolar) flows from young stars of low to medium luminosity ($0.1\ L_{\odot} \lesssim L \lesssim 100\ L_{\odot}$) is reviewed, with focus on optical studies of Herbig Haro objects and jets. These flows have typical lengths of 0.01 to 0.2 pc, opening angles of about 3 to 10 deg, and velocities of several hundred km s^{-1}. The author discusses the possibility that these outflows are continuous and concludes that for several sources quasi–continuous outflow activity is indicated over time scales of up to several 10^4yr. He considers that many HH objects may be powered by collimated flows. For some HH objects available kinematical data strongly suggest that they

are powered by a jet emanating from a nearby T Tauri star or infrared source.

121.016 **The double radio source associated with L1551 IRS 5: binary system or ionized circumstellar torus?**
L. F. Rodríguez, J. Cantó, J. M. Torrelles, P. T. P. Ho.
Astrophys. J., Lett. Ed., Vol. 301, No. 1, p. L25 – L28 (1986).
Using the VLA at 2 cm, the authors mapped the core of L1551 with an angular resolution of ~0.″15. It is found that the two unresolved sources detected by Bieging and Cohen and reported in 1985 are centered in a region of extended (~2″ × 1″) emission. This suggests that the two compact sources may be the ionized inner part of a confining torus with radius of ~25 AU seen in projection, and not two stars as was previously surmised. The authors propose that the energy source of the system is a star located between the two radio peaks and as yet undetected in the radio. The observed ionization can be explained if a small fraction of a predominantly neutral stellar wind from the central object shocks against the confining torus. The density of the gaseous torus is estimated to be $\sim 10^{10} cm^{-3}$.

121.017 **Molecular hydrogen emission in HH complexes. I. AS 353A/HH 32.**
W. J Zealey, P. M. Williams, K. N. R. Taylor, J. W. V. Storey, G. Sandell.
Astron. Astrophys., Vol. 158, No. 1/2, p. L9 – L11 (1986).
High resolution mapping (5 arc sec) of the bipolar outflow region AS 353A and HH 32A,B,C in the υ = 1–0 S(1) line of H_2 shows that the shock excited H_2 emission is closely associated with the visible Herbig–Haro objects. The authors find strong H_2 emission centred on HH 32A and B and some evidence for emission toward HH 32C. They also find a faint bridge of emission extending from AS 353A to HH 32A,B, which has a visible counterpart on high contrast CCD images.

121.018 **DN Tauri: a spotted T Tauri star.**
J. Bouvier, C. Bertout, P. Bouchet.
Astron. Astrophys., Vol. 158, No. 1/2, p. 149 – 157 (1986).
The authors present simultaneous optical and infrared photometric, as well as spectroscopic, measurements of the T Tauri star DN Tauri, which displayed a periodic optical light curve during the observing run. They interpret the variability of DN Tauri as rotational modulation by a large spot on the stellar surface, and present a quantitative analysis of the spot properties.

121.019 **Far–infrared photometry of low–mass pre–main–sequence stars with broad CO wings.**
N. J. Evans II, R. M. Levreault, P. M. Harvey.
Astrophys. J., Vol. 301, No. 2, p. 894 – 900 (1986).
A sample of 14 low–mass pre–main–sequence stars has been studied in the far–infrared. There is evidence for outflowing material around 13 of these stars; in most cases, the evidence includes broad wings on the profiles of the CO $J = 2 \rightarrow 1$ line. Eleven of the objects were detected ($> 5\sigma$) at 100 μm, and seven were mapped at 50 and 100 μm. The far–infrared emission makes a significant contribution to the total luminosity of these objects (25% – 100%). Even including this contribution, the force available from radiation pressure is not sufficient to drive the outflows.

121.020 **A search for magnetic fields in the T Tauri stars GW Ori, CoD –34°7151, and RU Lup.**
R. M. Johnstone, M. V. Penston.
Mon. Not. R. Astron. Soc., Vol. 219, No. 4, p. 927 – 941 (1986).
The authors present new high resolution circular spectropolarimetry, taken with the RGO Spectrograph and Pockel's Cell Polarimeter at the AAT, of the T Tauri stars GW Ori, CoD -34°7151, and RU Lupi. These spectra have been subjected to a statistical Zeeman analysis and a 2.3 sigma result for an effective longitudinal magnetic field of –548 Gauss is found in RU Lupi. This field strength is close to that predicted from a model of the wind driving mechanism for this star developed by Lago in which momentum is imparted to the wind by Alfvén waves propagating through a large–scale magnetic field. Three sigma upper limits of 1110 G and 2022 G are set for the strength of the effective longitudinal magnetic field in the stars GW Ori and CoD -34°7151 respectively.

121.021 **Some T Tauri stars and related objects. Results of UBVR photoelectric observations.**
V. I. Kardopolov, G. K. Filip'ev.
Perem. Zvezdy, Tom 22, No. 2, p. 103 – 121 (1985). In Russian.
Photoelectric observations of variable stars BM And, RW Aur, UY Aur, GM Aur, VX Cas, YZ Cep, EH Cep, DI Cep, R CrA, S CrA, V425 Cyg, BH Lac, RU Lup, SY Sgr, VV Ser, T Tau, RY Tau, LkHα 127, LkHα 373, LkHα 374, CoD -33°10685, CoD -35°10525 carried out at 48–cm and 60–cm reflectors of Majdanak Observatory are presented.

121.022 **UBVR photoelectric observations of four Herbig Be–Ae stars.**
A. F. Shajmieva, N. A. Shutemova.
Perem. Zvezdy, Tom 22, No. 2, p. 167 – 171 (1985). In Russian.
Photoelectric observations of AB Aur, T Ori, BF Ori and RR Tau are presented.

121.023 **Photoelectric UBVR photometry of V1057 Cygni.**
A. F. Shajmieva, N. A. Shutemova.
Perem. Zvezdy, Tom 22, No. 2, p. 172 – 175 (1985). In Russian.
Results of the photoelectric UBVR photometry of the FU Ori–type star V1057 Cyg are given. The presence of microvariability within $0^m.1$ in V is confirmed. The brightness shows wave–like changes in the time scale of about 10 – 13 days. During 1976 – 1981 a general slow decrease in brightness for V1057 Cyg is seen.

121.024 **Two years photoelectric UBVR observations of DF and DG Tauri.**
A. F. Shajmieva, N. A. Shutemova.
Perem. Zvezdy, Tom 22, No. 2, p. 176 – 179 (1985). In Russian.

121.025 **The relation between brightness and *B–V*, *U–B* colour indices for RY Tauri.**
G. V. Zajtseva.
Pis'ma Astron. Zh., Tom 12, No. 5, p. 380 – 382 (1986). In Russian. English translation in Sov. Astron. Lett., Vol. 12.
It is shown that the dependence of colour indices *B–V* and *U–B* on the *V*–brightness has a heterogeneous character. The *V*–brightness decreases from its maximum ($9^m.4$) to some intermediate value ($10^m.6$) is accompanied by the increase of colour indices with their following decrease when the brightness drops up to $V \approx 11^m.7$. Such a dependence, found earlier for several irregular variable stars, indicates to a relation of these two types of variables.

121.026 **Objects associated with low–mass star formation in the Gum nebula.**
J. A. Graham.
Astrophys. J., Vol. 302, No. 1, p. 352 – 362 (1986).
The Gum nebula contains a number of small, dense molecular clouds, some of which are still producing stars of low mass. Observations are presented of three small nebulae and one peculiar emission line star which are located in or near the dust clouds Sandqvist 109 and 111. The three nebulae are each coincident with point sources in the *IRAS* catalog. Two nebulae, Re 4 and Re 5, were discovered in a survey by Reipurth. The third is the Herbig–Haro object HH 46. The spectrum of Re 4 shows emission lines characteristic of HH objects superposed on a very red continuum. Re 5 shows only a red continuum with a few absorption lines. Broad emission lines in HH 46 indicate a complex velocity pattern near the associated *IRAS* source. The principal peculiarity of the M–type emission line dwarf is a small, jet–like appendage, with a purely emission line spectrum.

121.027 **Extended far–IR emission associated with young star molecular outflows.**
S. Strom, R. Snell, T. Jarrett, S. Edwards, K. Strom, C. Beichman.
Bull. Am. Astron. Soc., Vol. 17, No. 4, p. 837 (1985). Abstract. – See Abstr. 010.062.

121.028 **The relationship between disks, mass outflows and global cloud structures from young stellar objects.**
K. Strom, S. Strom, M. Wenz, F. Vrba, S. Wolff, T. Simon.
Bull. Am. Astron. Soc., Vol. 17, No. 4, p. 837 – 838 (1985). Abstract. – See Abstr. 010.062.

121.029 **Molecular outflows and mass loss in pre-main sequence stars.**
R. M. Levreault.
Bull. Am. Astron. Soc., Vol. 17, No. 4, p. 861 – 862 (1985). Abstract. – See Abstr. 010.062.

121.030 **The development of thermally driven winds around low mass protostars.**
S. Terebey, K. B. MacGregor.
Bull. Am. Astron. Soc., Vol. 17, No. 4, p. 862 (1985). Abstract. – See Abstr. 010.062.

121.031 **The IR line-to-luminosity relationship of young stellar outflows.**
H. A. Smith.
Bull. Am. Astron. Soc., Vol. 17, No. 4, p. 862 (1985). Abstract. – See Abstr. 010.062.

121.032 **Bowshock models of Herbig-Haro objects.**
P. Hartigan, J. Raymond, L. Hartmann.
Bull. Am. Astron. Soc., Vol. 17, No. 4, p. 862 (1985). Abstract. – See Abstr. 010.062.

121.033 **Kinematics and hydrodynamics of the HH 32 complex.**
K.-H. Böhm, J. Solf, A. Raga.
Bull. Am. Astron. Soc., Vol. 17, No. 4, p. 863 (1985). Abstract. – See Abstr. 010.062.

121.034 **Infrared spectra of rotating protostars.**
F. C. Adams, F. H. Shu.
Bull. Am. Astron. Soc., Vol. 17, No. 4, p. 863 (1985). Abstract. – See Abstr. 010.062.

121.035 **The exciting star of HH 57.**
M. Cohen, M. A. Dopita, R. Schwartz.
Astrophys. J., Lett. Ed., Vol. 302, No. 2, p. L55 – L58 (1986).
The authors provide the first spectral classification for the exciting star of HH 57. The star is an F8 III and suffers about 5 mag of visual extinction, far greater than that toward the HH 57A knot, only a few arc seconds away. This difference supports the previous identification of a flattened geometry for the circumstellar dust around this star. Using *IRAS* photometry the authors determine a distance of 940 pc to HH 57 by bolometrically matching the observed luminosity of the star to those of FU Ori and V1057 Cyg.

121.036 **Fine jets observed in some T Tauri-type stars and cometary nebulae.**
A. L. Gyul'budagyan.
Dokl. Akad. Nauk SSSR. Ser. Mat. Fiz., Tom 284, No. 6, p. 1331 – 1332 (1985). In Russian. Abstr. in Ref. Zh., 51. Astron., 3.51.751 (1986).

121.037 **Observations of the molecular bands in the spectrum of DI Cephei.**
Z. A. Ismailov, N. Z. Ismailov.
Astron. Tsirk., No. 1395, p. 1 – 3 (1985). In Russian.

121.038 **Multicolour photometry of Ae/Be Herbig stars.**
Yu. K. Bergner, A. S. Miroshnichenko, R. V. Yudin, N. Yu. Yutanov, K. S. Kuratov, D. B. Mukanov.
Astron. Tsirk., No. 1396, p. 1 – 3 (1985). In Russian.

121.039 **Th 28: a new bipolar Herbig-Haro jet.**
J. Krautter.
Astron. Astrophys., Vol. 161, No. 1, p. 195 – 200 (1986).
A bipolar Herbig-Haro jet system associated with the emission-line object Th 28 has been discovered from direct CCD-imaging, long-slit spectroscopy, and broadband infrared observations. The observations strongly suggest that Th 28 is a new case of a bipolar, well-collimated high-velocity outflow from a stellar object in a star-forming region.

121.040 **Gas-dust plasma in the envelopes of T Tauri stars.**
Yu. N. Gnedin, N. P. Red'kina.
Sov. Astron. Lett., Vol. 11, No. 4, p. 215 – 218 (1985). English translation of 40.121.001.

121.041 **The T Tauri radio source. II. The winds of T Tauri.**
P. R. Schwartz, T. Simon, R. Campbell.
Astrophys. J., Vol. 303, No. 1, p. 233 – 238 (1986).
High-resolution radio observations show that T Tau is a multiple star consisting of at least two interacting pre-main-sequence objects. Both stars have powerful ionized stellar winds. T Tau (S), the invisible infrared companion, is the dominant object insofar as the system's interaction with its immediate environment is concerned. It probably drives the associated molecular flow and excites the nearby Herbig-Haro objects.

121.042 **Short-term spectral variability in AB Aurigae: clues for activity in Herbig Ae stars. I. The ultraviolet lines of Mg II and Fe II.**
F. Praderie, T. Simon, C. Catala, A. M. Boesgaard.
Astrophys. J., Vol. 303, No. 1, p. 311 – 326 (1986).
The Herbig Ae star AB Aur was monitored with the IUE satellite in the long-wavelength, high-resolution mode, first during 40 consecutive hours in 1982 and second intermittently over a span of 150 hr in 1984. The Mg II λ2795 wind profile and a number of Fe II ultraviolet lines were studied for variability during the two epochs. Temporal variations in the Mg II blue wing velocity by 30% – 50% are interpreted as rotational modulation. This modulation is the first direct evidence of a nonaxisymmetric wind in an A-type pre-main-sequence object. A qualitative model for the expanding envelope is proposed, with recurrent corotating fast and slow streams, a model analogous to the solar wind in the inner heliosphere.

121.043 **Staubscheiben und Molekülflüsse bei jungen Sternen.**
T. Bührke, K.-W. Hodapp.
Sterne Weltraum, 25. Jahrg., Nr. 4, p. 206 – 210 (1986).

121.044 **Optical images of the region around HH1 and HH2 taken with the MEPSICRON system.**
J. Bohigas, J. M. Torrelles, J. Echevarría, J. Cantó, R. Enríquez, C. Firmani, L. Gutiérrez, E. Ruiz, L. Salas.
Rev. Mex. Astron. Astrofis., Vol. 11, No. 2, p. 149 – 157 (1985). With 3 plates. = Contrib. Inst. Astron., Univ. Nac. Autónoma Mex., No. 185.
Using the MEPSICRON detection system, the authors obtained images of the region around Herbig-Haro objects 1 and 2. They found two conical nebulosities uniting the central radio continuum source with HH1 and HH2. The emission lines (Hα and [S II]) are produced in situ, being probably excited by a shock wave created by the stellar wind emerging from the central source. Continuum emission is likely reflected light from the same source.

121.045 **Coordinated X-ray, ultraviolet and optical observations of T Tauri stars.**
F. Giovannelli, G. S. Bisnovatyj-Kogan, I. M. Golynskaya, V. G. Kurt, S. A. Lamzin, T. A. Mizyakina, E. Yu. Shafer, V. M. Shamolin, E. K. Sheffer, A. S. Smirnoff (*A. S. Smirnov*), A. A. Vittone.
X-ray astronomy '84, p. 77 – 80 (1985). – See Abstr. 012.033.
Within the framework of a large campaign of observations of RU Lupi in X-ray, UV and optical, the authors present some partial results of their observations. Some theoretical considerations on T Tauri stars are given.

121.046 **The circumstellar environment of LkHα 234.**
B. A. Wilking, L. G. Mundy, R. D. Schwartz.
Astrophys. J., Lett. Ed., Vol. 303, No. 2, p. L61 – L65 (1986).
The authors present high-resolution radio continuum and molecular-line observations of the circumstellar environment of

the emission–line star LkHα 234. These observations reveal a compact continuum source at 15, 23, and 98 GHz coincident with the star and a 10,000 by 17,000 A.U. ridge of CS $J = 2 \rightarrow 1$ emission which peaks $\sim 4''$ east of the star. The radio continuum emission may arise from a partially ionized stellar wind. The inferred mass loss rate from a spherically symmetric wind is $\geqslant 2x10^{-6} M_{\odot} yr^{-1}$.

121.047 **Molecular outflows and mass loss in pre–main–sequence stars.**
R. M. Levreault.
Diss. Abstr. Int., Sect. B, Vol. 46, No. 10, p. 3482 (1986). Thesis, The University of Texas at Austin, 359 pp. (1985). Order No. DA8529878.

121.048 **Peculiarities of the circumstellar envelopes of some Herbig Ae/Be stars.**
M. A. Pogodin.
Sov. Astron., Vol. 29, No. 5, p. 531 – 538 (1985). English translation of 40.121.018.

121.049 **Interpretation of the photometric and polarimetric observations of T Tauri stars.**
N. P. Red'kina, K. V. Tarasov, N. N. Kiselev, G. P. Chernova.
Eruptive phenomena in stars, p. 441 – 442 (1986). – See Abstr. 012.039.

121.050 **Les étoiles Ae, Be de Herbig.**
V. Doazan.
Bol. Astron. Obs. Madr., Vol. 11, No. 3, p. 1 – 4 (1984).

121.051 **A spectroscopic survey of the Taurus–Auriga dark clouds for pre–main–sequence stars having Ca II H, K emission.**
G. H. Herbig, F. J. Vrba, A. E. Rydgren.
Astron. J., Vol. 91, No. 3, p. 575 – 582 (1986). With plate 59. = Lick Obs. Bull., No. 1031.

A large area of the Tau–Aur dark clouds was surveyed to about $B = 15.0$ in a search for stars having the Ca II H, K lines in emission. Essentially all the known T Tauri stars (TTS) in the area were rediscovered (51 in all), as well as three of the four known X–ray sources; H, K emission was detected in about 20 other faint, red stars. Slit spectrograms and multicolor photometry have been obtained for most of the latter, which turn out to be low–luminosity K and M stars with weak Hα emission. A majority are believed to be pre–main–sequence association members. The location of the H, K emission stars in an H–R diagram shows them to be mixed among the conventional TTS. They appear to be TTS that have Hα emission too weak to have been detected in the conventional surveys.

121.052 **A detailed optical study of HH 32 and the highly collimated outflow from the T Tauri star AS 353A.**
P. Hartigan, R. Mundt, J. Stocke.
Astron. J., Vol. 91, No. 6, p. 1357 – 1371 (1986). With plates 98 – 100.

The authors present new deep emission–line CCD frames of the AS 353A/HH32 region and high–resolution spectra of AS 353A and its associated HH objects and optical jet. They discuss how these new data constrain existing models of HH objects and mass outflow.

121.053 **The Herbig stars.**
J. K. Davies.
Astronomy, Vol. 13, No. 12, p. 90 – 95 (1985). Abstr. in Phys. Abstr., Vol. 89, No. 1279, Entry 53864 (1986).

121.054 **Rotation in T Tauri stars. I. Observations and immediate analysis.**
J. Bouvier, C. Bertout, W. Benz, M. Mayor.
Inst. Astrophys. Paris, Pré–Publ., No. 135, 33 pp. (1986). To appear in Astron. Astrophys.

121.055 **Accretion disks and outflows around young stellar objects.**
C. Bertout.
Inst. Astrophys. Paris, Pré–Publ., No. 136, 47 pp. (1986). Paper presented at the CNRS workshop on "Anisotropic mass–loss and accretion phenomena in stellar systems".

121.056 **Photometric behavior of DR Tauri in the season 1985/86.**
W. Götz.
Inf. Bull. Variable Stars, No. 2894, 2 pp. (1986).

121.057 **The interesting young star LkHα 324: the 1983 – 1985 observations.**
E. A. Kolotilov.
Pis'ma Astron. Zh., Tom 12, No. 6, p. 475 – 479 (1986). In Russian. English translation in Sov. Astron. Lett., Vol. 12.

Spectral (covering Hα and Hβ regions) and *UBV* photometric observations of the young star LkHα 324 carried out in 1983 – 1985 are presented. The spectrum of the star shows strong Hα emission, whereas the contribution of Hβ emission completely «blurs» the absorption line profile. The brightness variation of LkHα 324 did not exceed $0^m.3$, the mean values of $\bar{V} \approx 12^m.75$, $\overline{B-V} \approx +1^m.13$ and $\overline{U-B} \approx +0^m.44$. The comparison of the *UBV* data with the bulk of previous photometric results allows to conclude that within the period from 1975 to 1980 the star has experienced a remarkable brightening with amplitude $\Delta V > 1^m$. Significant changes in its spectrum are probably connected with this phenomenon.

121.058 **Kinematical and hydrodynamical study of the HH 32 complex.**
J. Solf, K.-H. Böhm, A. C. Raga.
Astrophys. J., Vol. 305, No. 2, p. 795 – 804 (1986). With plate 11.

The authors have obtained coudé spectra of the HH 32 complex with high spectral and spatial resolution, using a 117″ long slit and eight different position angles. The observations cover the four condensations: A, B, C, and D. The authors have studied the (rather complex) spatially resolved line profiles ("position–velocity diagrams") of all four condensations. It is shown that the qualitative properties of the observed position–velocity diagrams agree to a large extent with the theoretical predictions for a bow shock whose axis forms an angle of 20° with the line of sight.

121.059 **On the nature of the young Orion population star LkHα198.**
C. Chavarría-K., E. de Lara.
Rev. Mex. Astron. Astrofis., Vol. 12, Spec. Issue, p. 198 (1986). Abstract. – See Abstr. 012.066.

121.060 **The emergence of stellar activity: T Tauri stars.**
N. Calvet.
Rev. Mex. Astron. Astrofis., Vol. 12, Spec. Issue, p. 217 – 223 (1986). – See Abstr. 012.066.

In this review, the problem of stellar activity is examined into an evolutionary context. The activity observed in T Tauri stars is considered, and it is shown that it can be explained with a model which is consistent with that appropriate for more evolved stars

121.061 **Emission–line stars in the Canis Major star–formation region.**
S. D. Wiramihardja, T. Kogure, M. Nakano, S. Yoshida.
Publ. Astron. Soc. Jpn., Vol. 38, No. 3, p. 395 – 447 (1986).

In this paper the authors present their survey observations for Hα–emission stars in the CMa star–formation region and its adjacent reference area. For the detected Hα–emission stars they have carried out photographic photometry in the UBV system and derived some color properties. The purposes of the study are first to detect Hα–emission stars which may be composed of reddened OB stars, Herbig Be/Ae stars, and/or a brighter part of T Tauri stars, and second to investigate the relationship between Hα–emission stars and the star–formation region.

121.062 **Bipolar jets in Herbig–Haro No. 24.**
J. Solf.
Mitt. Astron. Ges., No. 65, p. 186 – 187 (1986). – See Abstr. 012.072.

121.063 **HH 34: Die Kopfwelle eines Jets?**
T. Bührke, R. Mundt.
Mitt. Astron. Ges., No. 65, p. 188 – 190 (1986). – See Abstr. 012.072.

121.064 **Herbig–Haro 46.**
IAU Circ., No. 4213 (1986).

121.065 **Photoelectric *UBV* photometry of the young star YY Orionis.**
E. A. Kolotilov.
Astron. Zh., Tom 63, Vyp. 2, p. 298 – 309 (1986). In Russian. English translation in Sov. Astron., Vol. 30, No. 2.

The pre–main–sequence variable star YY Ori was observed in 1979 – 1984 photoelectrically in the *UBV* system. The dependences of *U*, *B* magnitudes and *(U–B)* and *(B–V)* colors upon *V* brightness supplemented by the data of other authors are analysed; the following correlations are found: $\Delta U = (1.08 \pm 0.09)\Delta V$ and $\Delta B = (1.06 \pm 0.04)\Delta V$. A permanent existence of UV and blue excess radiation follows from the location of YY Ori on the *(U–B)*–*(B–V)* diagram. The distribution of the photographic brightness of the star based on observations over 233 nights shows a significant asymmetry. It is suggested that the optical variability of YY Ori is caused by flares associated with the accretion from the circumstellar gas envelope and by the existence of a dust envelope ($R \approx 15$), which also contains a number of large massive condensations.

121.066 **Observations of young (Orion–type) stars with IRAS.**
H. J. Walker, P. L. Marsden.
Light on dark matter, p. 91 – 92 (1986). – See Abstr. 012.080.

Forty–one young objects from Cohen's (1973, 1974) lists were found to have fluxes in more than one of the IRAS bands. Some of the sources are plotted.

Amorphous bronzite – a silicate of astronomical importance.
See Abstr. 022.002.

Bipolar hydromagnetic winds from disks around protostellar objects.
See Abstr. 064.020.

Stellar winds from young stars.
See Abstr. 064.047.

Continuum spectra of collimated, ionized stellar winds.
See Abstr. 064.048.

A pulsed jet as the source of momentum in bipolar molecular outflows.
See Abstr. 064.052.

Two–dimensional models of stellar wind bubbles. V. The efficiency of momentum transfer between winds and dense clumps in free wind regions.
See Abstr. 064.055.

Relation of chromospheric activity to convection, rotation, and pre–main–sequence evolution.
See Abstr. 065.002.

Stellar chromospheres, coronae, and winds.
See Abstr. 112.007.

Observational evidence for disks around young stars.
See Abstr. 112.019.

IR CCD imaging of L 1551–IRS 5: direct observations of its circumstellar shell.
See Abstr. 112.146.

A model for bipolar sources in molecular clouds.
See Abstr. 112.147.

Magnetodynamic phenomena in the solar and stellar outer atmospheres.
See Abstr. 116.003.

A study of visual double stars with early–type primaries. V. Post T Tauri secondaries.
See Abstr. 118.013.

The structure of the R Monocerotis, NGC 2261, and HH39 nebular complex.
See Abstr. 131.025.

Sandqvist 187: a dense molecular cloud in Norma.
See Abstr. 131.043.

Protostars and planets: overview from an astronomical perspective.
See Abstr. 131.072.

Star formation: an overview.
See Abstr. 131.079.

Infrared spectroscopy of protostars and young stellar objects.
See Abstr. 131.080.

Protostellar disks and star formation.
See Abstr. 131.086.

Infrared studies of the Herbig–Haro object 1–2 region.
See Abstr. 131.101.

First light from a young star?
See Abstr. 131.143.

A high latitude cloud in Draco as a matrix of protostars.
See Abstr. 131.178.

The dust and gas surrounding LkHα 101.
See Abstr. 131.211.

Temperature determination of surrounding molecular clouds in bipolar flow sources.
See Abstr. 131.212.

High–resolution mapping of molecular outflows in NGC 2071, W49, and NGC 7538.
See Abstr. 131.217.

Surveys for emission–line objects with Schmidt telescopes.
See Abstr. 131.223.

An Hα emission star survey in the field of the dark cloud complex Khavtassi 141.
See Abstr. 131.242.

IRAS 04238+5336: a young reflection nebula surrounding a double star.
See Abstr. 131.274.

The geometry of anisotropic CO outflows.
See Abstr. 131.275.

Are bipolar jets precessing?
See Abstr. 131.279.

Hydrodynamical model of the Orion KL Nebula.
See Abstr. 131.294.

Point sources in the Orion complex.
See Abstr. 131.379.

Young stars and high density condensations in the Horsehead region.
See Abstr. 131.381.

Mapping of the Coronae Austrinae star forming region.
See Abstr. 131.382.

A. Analysis of point sources in the Ophiuchus and Perseus clouds. B. CPC observations of NGC 1333.
See Abstr. 131.383.

Young stars and dense cores in nearby dark clouds.
See Abstr. 131.385.

Compact radio sources associated with molecular outflows.
See Abstr. 132.039.

The discovery of high–speed ionized knots in the H II regions M16 (NGC 6611) and the Rosette nebula (NGC 2237–2246).
See Abstr. 132.047.

Comparison of CO and IR emission of IRAS unidentified sources.
See Abstr. 133.019.

Ultraviolet observations of young stars in the Chamaeleon 1 association.
See Abstr. 152.003.

122 Intrinsic Variables (Pulsating Variables, Spectrum Variables, etc.)

122.001 **Frequency analysis of the rapidly oscillating peculiar A star HD 60435.**
J. M. Matthews, D. W. Kurtz, W. H. Wehlau.
Astrophys. J., Vol. 300, No. 1, p. 348 – 359 (1986).
Rapid *B* photometry of the cool Ap star HD 60435 was carried out over 18 nights in early 1984. Frequency analysis of these data reveals oscillations (some transient) with periods ranging from 4 to 15 minutes. These represent the shortest and longest periods yet observed in the class of rapidly oscillating Ap stars. Use of the oblique pulsator model to interpret fine–scale frequency splittings and comparison of observed frequency spacings with those predicted for models of A stars, suggest that the star is undergoing nonradial pulsations of both odd and even degree. The analysis also points to a rotation period of approximately 8 days for HD 60435.

122.002 **Polarimetric evidence for an evolving circumstellar cloud in L_2 Puppis.**
A. M. Magalhães, G. V. Coyne, S. J. Codina–Landaberry, C. Gneiding.
Astron. Astrophys., Vol. 154, No. 1/2, p. 1 – 7 (1986).
New polarimetric observations of the red semiregular variable star, L_2 Puppis, over a period of three years, 1980 – 1982, mostly in intermediate and narrow bands between 0.36 μm and 0.80 μm, are presented. Some wide–band observations from the period 1975 – 1979 are also given. In combination with previously published observations, systematic changes in both the percentage and the direction of polarization with both wavelength and time over a four year period are seen. These variations are especially noteworthy in the atomic line of Ca I 4226 Å and in various TiO bands. The observations are best explained by a combination of photospheric effects, which include a non–uniform distribution of Ca across the stellar disk, and of scattering from grains in an asymmetric cloud about the star, where there is a systematic time variation in grain size with distance from the star.

122.003 **The flare star EV Lac. I. The activity cycle.**
L. N. Mavridis, S. Avgoloupis.
Astron. Astrophys., Vol. 154, No. 1/2, p. 171 – 175 (1986).
A discussion is made of the long–term variability of the flare activity and the quiet–state luminosity of EV Lac. The existence of an activity cycle with a duration of about 5 yr is demonstrated, with the help of data concerning both the quiet–state luminosity and the integrated flare activity of the star. A discussion is made of the behaviour of the various types of flares (I–IV) separately during the activity cycle.

122.004 **New observational results of the LMC–S Dor variable R 127 during outburst.**
O. Stahl, B. Wolf.
Astron. Astrophys., Vol. 154, No. 1/2, p. 243 – 248 (1986).
The LMC–S Dor variable R 127 is photometrically monitored by the group for "Long–term photometry of variables" at ESO. It turned out that R 127 had an extremely bright phase in Aug. 1984. CASPEC and IUE high–dispersion spectra were taken just during this phase and are compared in this paper with coudé and IUE spectra during phases of intermediate brightness in the time span 1982 to 1985. It is found that during this brightest phase R 127 resembles particularly closely S Dor during maximum. The complex shell phenomenon reported earlier is further investigated.

122.005 **Metal abundances and radial velocities of short–period variable stars in the direction of the Large Magellanic Cloud.**
H. A. Smith.
Publ. Astron. Soc. Pac., Vol. 97, No. 597, p. 1053 – 1057 (1985).
Spectra have been obtained for five short–period ($P < 1^d$) variable stars in the direction of the Large Magellanic Cloud. Radial velocities and ΔS metallicity indices have been determined for the variables. Most of the variables appear to be foreground RR Lyrae stars of the Galaxy and not members of the LMC. The stars range in [Fe/H] from –0.6 to –1.7. The radial velocities of these variables and those of two similar variables observed by Connolly are discussed. The line–of–sight velocity dispersion for these stars is consistent with both spherical and cylindrical descriptions of the halo velocity dispersion.

122.006 **Classical Cepheid light curves revisited.**
N. R. Simon, T. J. Moffett.
Publ. Astron. Soc. Pac., Vol. 97, No. 597, p. 1078 – 1085 (1985).
Fourier decompositions are made of the *B, V, R*, and *I* classical Cepheid observations of Moffett and Barnes. The Fourier diagrams for *V* agree very well with the corresponding plots of Simon and Lee. The *B, R*, and *I* diagrams resemble the plots for *V*, but with small, systematic shifts in the phases ϕ_{21} and ϕ_{31}. The 2.5–day star, DT Cyg, is confirmed as a likely overtone pulsator by its anomalous position in the Fourier plots. Finally, the authors study the Fourier phase quantity ϕ_{41}, and introduce phase–phase diagrams to look at the Hertzsprung progression. Discontinuities in these diagrams support the idea that the long–period and short–period Cepheids may reach their limit cycles in different ways.

122.007 **Oscillations and pulsations in the sun and stars.**
Y. Osaki, H. Shibahashi.
Astrophys. Space Sci., Vol. 118, Nos. 1/2, p. 195 – 207 (1986). – See Abstr. 012.001.

In this review, the general properties of stellar eigenmode oscillations are first discussed. Recent observational and theoretical developments in the solar seismology and on pulsations and oscillations of newly discovered variable stars are then reviewed.

122.008 **Unusual variable stars in the globular cluster M4.**
B.-a. Yao.
Astrophys. Space Sci., Vol. 119, No. 1, p. 41 – 44 (1986). – See Abstr. 012.001.

For the first time new unusual variable stars have been found in the central part of the red giant branch of the globular cluster M4.

122.009 **Visual observations of four classical cepheids.**
J. Speil.
Astron. Rep., Vol. 7, p. 25 – 28 (1982).

Results of visual observations of SU Cas, δ Cep, η Aql, and X Cyg are presented.

122.010 **RR Lyrae, Delta Scuti, SX Phoenicis stars and Baade–Wesselink method. I. Photometric and radial velocity measurements of four field stars: RR Cet, DX Del, BS Aqr and DY Peg.**
G. Meylan, G. Burki, F. Rufener, M. Mayor, M. Burnet, E. Ischi.
Astron. Astrophys., Suppl. Ser., Vol. 64, No. 1, p. 25 – 51 (1986).

Simultaneous measurements in the Geneva seven–colour photometry and in radial velocities with the spectrophotometer CORAVEL for two RR Lyrae, one Delta Scuti and one SX Phoenicis field stars were obtained in order to apply the Baade–Wesselink method to these kinds of variable stars. As a first step, data are presented regarding RR Cet, DX Del, BS Aqr and DY Peg. The target of this study will consist in determining the physical parameters (temperature, gravity, metal content, mass luminosity) and distances of these stars.

122.011 **The RCB star RY Sagittarii as a pulsating star.**
T. Lloyd Evans.
Mon. Not. R. Astron. Soc., Vol. 219, No. 1, p. 191 – 196 (1986).

Measurements of CN and C_2 bandstrengths in the spectrum of RY Sagittarii in 1969 and 1970 show that the bands vary in strength in the 38.6–day pulsation period. The variations follow the phasing of the *B–V* and *U–B* colour curves rather than the *V* light curve, and must be determined largely by the photospheric temperature as in the case of the carbon–rich Cepheid V553 Centauri. This is supported by a comparison of the ranges in colour and in bandstrength, though the bands are stronger at a given colour than in typical class Ib supergiants.

122.012 **HD 47147: a small–amplitude extreme metal–poor RRab pulsating variable.**
M. Grenon, C. Waelkens.
Astron. Astrophys., Vol. 155, No. 1, p. 24 – 32 (1986).

The ninth magnitude star HD 47147 is an RR Lyrae pulsating variable with a period of 0.79732 days and an amplitude A_v of 0.17 mag. A rough analysis of the spectrum, an application of the metallicity calibration of the Geneva system for F–G–type stars, and the dependence on metallicity of the period–amplitude relation for RRab pulsators all consistently point to a very low metal content for HD 47147. A reasonable average value of all abundance determinations is [Fe/H] = –2.3. The extreme Population II character of HD 47147 is confirmed by its kinematical properties. The evolutionary stage of this star appears to be rather exceptional.

122.013 **Brightness variability investigation of V586 Orionis.**
A. F. Pugach, G. U. Kovalchuck (*G. U. Koval'chuk*).
Astron. Nachr., Vol. 307, No. 1, p. 13 – 15 (1986).

Photoelectric *UBVR* observations of the young irregular variable star V586 Ori are presented and discussed. The authors conclude that variability of V586 Ori as well as of other antiflare stars is caused by the sporadic appearance of circumstellar dust. The dust veils the photospheric radiation of the star resulting in a brightness decrease.

122.014 **RR Lyrae, Delta Scuti, SX Phoenicis stars and Baade–Wesselink method. II. Physical parameters and distance of RR Cet.**
G. Burki, G. Meylan.
Astron. Astrophys., Vol. 156, No. 1/2, p. 131 – 136 (1986).

The observations of the RR Lyrae star RR Cet in Geneva photometry and radial velocity with CORAVEL are analysed. Three different versions of the Baade–Wesselink method are applied to derive the mean radius. Validity of the method is discussed. It is shown that the interval of phase corresponding to the rising light (minimum radius) must be excluded because at this phase, the colour index [*B–V*] is not uniquely related to the effective temperature. The physical parameters of RR Cet are determined, in particular $R_0 = 6.7 \pm 0.2\, R_\odot$ and $\bar{M}_v = 0.28 \pm 0.14$. The distance from the sun is 760 ± 40 pc. The total uncertainties on these quantities are roughly twice as large as the values given here which result only from the mathematical analysis.

122.015 **Flare activity and BY Draconis–type variability on the late–type dMe star Gliese 867B.**
J. G. Doyle, P. B. Byrne, C. J. Butler.
Astron. Astrophys., Vol. 156, No. 1/2, p. 283 – 288 (1986).

The authors report on a search for flaring and BY Draconis type variability on the M–dwarf Balmer emission line star, Gliese 867B. The time–averaged flare energy in the Johnson *U*–band is similar to that measured in 1977 and 1978. A typical wave–like variation of 0.15 magnitudes, seemingly due to rotational modulation of large spotted areas, was observed in the *V*–band with a period 1.95 days.

122.016 **Studies of Cepheid–type variability. III. Fourier description of light curves of Type II Cepheids of periods 1 – 3 days.**
J. O. Petersen, R. Diethelm.
Astron. Astrophys., Vol. 156, No. 1/2, p. 337 – 346 (1986).

Fourier decomposition parameters for Type II Cepheids with periods $1^d - 3^d$ are derived from *UBVRI* data and photographic mean light curves collected from the literature. Although data of high quality exist for only 4 – 5 stars, it is shown that available light curves for about 35 Type II variables (BL Her and RRd stars) allow sufficiently accurate determinations of Fourier parameters for an overview of their pulsation properties. The basic pulsation properties of the Type II Cepheids seem to be very similar to those of the classical Cepheids. However, small but significant differences are shown to be present. The RRd group has larger amplitude ratios than other Cepheid variables, and the bump progression sequence, expected to correspond to the conspicuous Hertzsprung sequence in the classical Cepheids, is much less pronounced in the Type II variables. Possible reasons for these differences are briefly discussed.

122.017 **Study of quiescent state X–ray emission from flare stars.**
P. C. Agrawal, A. R. Rao, B. V. Sreekantan.
Mon. Not. R. Astron. Soc., Vol. 219, No. 2, p. 225 – 239 (1986).

Results on the X–ray emission from flare stars observed with the imaging proportional counter (IPC) on the Einstein Observatory are described. The authors present in list form a summary of the X–ray properties of all the detected flare stars as well as other dMe and dM stars. They compare the X–ray characteristics of the flare stars with those of the non–flaring dMe and dM stars and point out similarities in X–ray emission characteristics of the flare stars with those of the RS CVn binaries. The authors discuss the correlation between X–ray luminosity L_x and bolometric luminosity L_{Bol} as well as between L_x and luminosities in chromospheric and transition region lines. They also discuss the dependence of L_x on the rotation periods of the flare stars. Finally, the authors briefly discuss similarities between the spectra of the flare stars and those of the RS CVn binaries.

122.018 **A β Cephei variable in an eclipsing binary in the cluster NGC 3293.**
C. A. Engelbrecht, L. A. Balona.
Mon. Not. R. Astron. Soc., Vol. 219, No. 3, p. 449 – 456 (1986).
The authors present photometric observations of HD 92024 in NGC 3293 which is found to be both a β Cep variable with oscillation periods of 0.177 and 0.130 day and a member of an eclipsing binary with a period of 8.323 day. The solution of the eclipsing light curve indicates that the secondary is probably a main–sequence star of spectral type B8. The oscillation frequencies have a ratio close to what is expected for first overtone and second overtone radial pulsation, but a 1 cycle day^{-1} ambiguity in one of the frequencies also allows the identification of the two modes as quadrupole travelling waves.

122.019 **The pulsation mode and period–luminosity relationship of cool variables in globular clusters.**
P. A. Whitelock.
Mon. Not. R. Astron. Soc., Vol. 219, No. 3, p. 525 – 536 (1986).
The period–luminosity–temperature relationship for globular cluster red and yellow variables is examined. The results suggest that the higher temperature, more metal–deficient cluster variables pulsate in the fundamental mode, while the lower temperature more metal–rich variables pulsate in the first overtone. On the assumption that this is correct, a relationship between fundamental period and bolometric magnitude is derived for cluster variables with observed periods of between 1 and 300 days.

122.020 **Cepheids in the open cluster NGC 6067.**
T. J. Moffett, T. G. Barnes III.
Mon. Not. R. Astron. Soc., Vol. 219, No. 3, p. 45P – 48P (1986).
Fourier decomposition of the light curve indicates that the Cepheid HD 144972 is an overtone pulsator. Using the visual surface brightness technique, the authors find a distance to HD 144972 which places it in the open cluster NGC 6067. The evidence is strong that NGC 6067 contains two Cepheids – HD 144972 and CPD –53°7400p. The nearby Cepheid GU Nor appears not to be a cluster member.

122.021 **HR 7671: another UU Herculis star?**
J. D. Fernie.
Astrophys. J., Vol. 301, No. 1, p. 302 – 307 (1986).
HR 7671 is an F giant or supergiant at a comparatively high galactic latitude of –21°.5. During the second half of 1984, $uvby\beta(RI)_{KC}$ photometry of HR 7671 was obtained. Thirty–two observations show a range in y magnitude of ~0.3 mag, confirming the variability reported earlier but it has not been possible to derive a satisfactory period or periods. The color indices, corrected for $E(B–V)$ = 0.10, show HR 7671 to have a low surface gravity ($\log g < 1$), like UU Her and HD 161796, but also to have [Fe/H] ≈ –2, unlike other UU Her stars. A new MK spectral classification yields F2 p (shell). This conflicts with the intrinsic colors of the star, which correspond to F7, a conflict almost identical to the case of the UU Her star HD 161796.

122.022 **A possible second period for HD 200925 ≡ V1719 Cyg.**
L. Mantegazza, E. Poretti.
Astron. Astrophys., Vol. 158, No. 1/2, p. 389 – 391 (1986).
All the available photometry of HD 200925 ≡ V1719 Cyg is discussed here and a possible second period P_2 = 0.2138d is derived. In this case the star could be a double mode pulsator in the first and second radial overtones.

122.023 **Photometry and period determination of the Delta Scuti star YZ Bootis.**
R. Peniche, S. F. González, J. H. Peña.
Publ. Astron. Soc. Pac., Vol. 97, No. 598, p. 1172 – 1177 (1985).
Differential photoelectric photometry in the V filter of the δ Sct star YZ Boo was carried out in 1982 and 1984. With a total time span of 16,398 days and 77 times of maximum light new ephemerides were found.

122.024 **Polarimetric observations of the RV Tauri star AC Herculis.**
G. D. Henson, J. C. Kemp, D. J. Kraus.
Publ. Astron. Soc. Pac., Vol. 97, No. 598, p. 1192 – 1195 (1985).
Polarimetric observations in the B band were obtained over one cycle of the 75.46–day period and on four nights in the V band. A correlation between the light and polarization changes is clearly evident. The variations in the polarization exhibit first and third harmonic behavior with an amplitude of approximately 1.0%. The maximum polarization occurs just after the primary light minimum. The path of the polarization on the equatorial Q, U plane is elongated along an axis which makes an angle 2θ = 145°±3° with the $^{+}$Q axis. This suggests either nonradial pulsations in the star or radial pulsations propagating into an asymmetrical circumstellar medium.

122.025 **Nonradial pulsations in massive stars; observations and oddities.**
M. A. Smith.
Publ. Astron. Soc. Pac., Vol. 98, No. 599, p. 33 – 34 (1986). – See Abstr. 012.009.
This summary of the observational status of nonradial pulsation in OB stars, as interpreted from spectral line profile variations, is in large part a result of ongoing work reported by D. Penrod, D. Baade, and the UBC group.

122.026 **Episodic mass motions and surface phenomena.**
G. D. Penrod.
Publ. Astron. Soc. Pac., Vol. 98, No. 599, p. 35 – 36 (1986). – See Abstr. 012.009.

122.027 **RR Lyrae, δ Scuti, SX Phoenicis stars and Baade–Wesselink method. III. Physical parameters and distance of DX Del, and the problem of the mean absolute magnitude of RR Lyrae stars.**
G. Burki, G. Meylan.
Astron. Astrophys., Vol. 159, No. 1/2, p. 255 – 260 (1986).
The observations of the RR Lyrae star DX Del in Geneva photometry and radial velocity with CORAVEL are analysed. Three different versions of the Baade–Wesselink method are applied to derive the mean radius. The physical parameters are determined, in particular $R_0 = 5.5 \pm 0.2\,R_\odot$ and $\bar{M}_v = 0.49 \pm 0.16$. The distance to the sun is 750±45 pc. The various determinations of the mean absolute magnitude of the RR Lyrae stars are compared.

122.028 **RR Lyrae, δ Scuti, SX Phoenicis stars and Baade–Wesselink method. IV. Physical parameters and distances of BS Aqr and DY Peg, and the period–radius relations for pulsating stars.**
G. Burki, G. Meylan.
Astron. Astrophys., Vol. 159, No. 1/2, p. 261 – 268 (1986).
The observations of the δ Scuti star BS Aqr and of the SX Phoenicis star DY Peg in Geneva photometry and radial velocity with CORAVEL are analysed. Three different versions of the Baade–Wesselink method are applied and the following values of the physical parameters and distance are derived: (1) BS Aqr: $R_0 = 3.2\,R_\odot$, $\bar{T}_{eff}$ = 6960K, $\bar{M}_{v_0} = 1^m.4$, [Fe/H] = –0.4, mass = 1.8 $M_\odot$, distance = 410 pc. (2) DY Peg: $R_0 = 1.4\,R_\odot$, $\bar{T}_{eff}$ = 7240K, $\bar{M}_{v_0} = 3^m.2$, [Fe/H] = –0.8, distance = 250 pc. A general $\log R$–$\log P$ diagram is constructed, collecting the Baade–Wesselink radii for seventy stars in six classes of pulsating variables.This observational diagram is discussed on the basis of theoretical predictions from models of pulsating stars.

122.029 **The two periods of the Cepheid CO Aurigae.**
E. Antonello, L. Mantegazza, E. Poretti.
Astron. Astrophys., Vol. 159, No. 1/2, p. 269 – 275 (1986).
A comparison of different techniques of analysis applied to the observational data of CO Aur emphasizes the importance of a careful choice of the statistical techniques for the analysis of time series. A new set of observations performed in 1985 gives an independent confirmation of the previous results: CO Aur is a double–mode Cepheid with $P1 = 1^d.78304$ and $P2 = 1^d.42776$.

The ratio between the two periods (0.8007) renders this star unique among the variable stars of this class.

122.030 **An observational study of 16 EN Lacertae.**
N. Sato, T. Hayasaka.
Publ. Astron. Soc. Jpn., Vol. 38, No. 1, p. 47 – 61 (1986).
Photometry of the β Cephei–type variable 16 EN Lacertae was carried out at Akita and radial velocities were derived from the coudé spectrograms obtained at Okayama in 1978. An analysis of the data has been performed and compared with observations previously reported. The three known periods are confirmed and it is concluded that an abrupt increase in the light and velocity amplitudes of the tertiary period occurred during the term from late 1977 to early 1978. The light amplitude of the tertiary period is now dominant, but the velocity amplitude of the primary period is still the largest among the three components in 1978. The secular change in the primary period is also discussed.

122.031 **On the variability of Hα emission at the brightness maximum of WW Vul.**
V. G. Zajtseva, A. E. Tarasov, A. G. Shcherbakov.
Izv. Krymskoj Astrofiz. Obs., Tom 72, p. 85 – 89 (1985). In Russian. English translation in Bull. Crimean Astrophys. Obs., Vol. 72.
Photometric and spectroscopic observations of the rapid variable Ae star WW Vul were carried out in April – October,1983. Within the observational period the equivalent widths of Hα emission were varying 15 times, whereas the brightness of the star remained constant. The radial velocity of absorption of the Hα two–component profile has been varying from 0 to –140 km/s.

122.032 **Atmospheres of Delta Scuti pulsating variables. II. The analysis of 44 Tau and V644 Her spectra.**
L. S. Lyubimkov, T. M. Rachkovskaya.
Izv. Krymskoj Astrofiz. Obs., Tom 72, p. 89 – 99 (1985). In Russian. English translation in Bull. Crimean Astrophys. Obs., Vol. 72.
Using spectrograms with dispersion 8 and 12 Å/mm a model atmosphere analysis of the pulsating stars 44 Tau and V644 Her has been carried out. The values of the effective temperature, surface gravity and microturbulent velocity obtained for these stars appeared to be very close: T_{eff} = 7100K, $\log g = 3.8$ and ξ_t = 5.2 km/s for 44 Tau; T_{eff} = 7050K, $\log g = 3.8$ and ξ_t = 5.4 km/s for V644 Her. The atmospheric abundances were determined. The chemical composition of both stars differs from the solar one. The estimated values of T_{eff} and $\log g$ compared with the results of evolution calculations permitted to find masses, radii, luminosities and the age of 44 Tau and V644 Her; these parameters for both stars are the same.

122.033 **Distribution of flares in UV Cet–type stars in open clusters and in the solar vicinity from optical radiation energies.**
S. A. Korotin, V. I. Krasnobabtsev.
Izv. Krymskoj Astrofiz. Obs., Tom 73, p. 131 – 143 (1985). In Russian. English translation in Bull. Crimean Astrophys. Obs., Vol. 73.
A statistical analysis of flares in UV Cet–type stars in four open clusters is presented. The comparison of the flare energy spectra of stars in clusters and in the solar vicinity confirms their similarity. Clear evidence of the evolution effect in flare activity is found: the spectral index of the flare energy spectra decreases with age, and it means that a relative portion of weak flares decreases in total flare energy.

122.034 **Four years of photoelectric UBVR observations of WW Vulpeculae.**
V. I. Kardopolov, G. K. Filip'ev.
Perem. Zvezdy, Tom 22, No. 2, p. 122 – 125 (1985). In Russian.

122.035 **Some regularities of the activity of Is variable stars. I. V517 Cygni is a photometric analogue of VV Serpentis.**
V. I. Kardopolov, G. K. Filip'ev.
Perem. Zvezdy, Tom 22, No. 2, p. 126 – 136 (1985). In Russian.
The results of BVR observations of the irregular variable star V517 Cyg = LkHα 371 in 1981 are given. The V517 Cyg photoelectric data are compared with those for VV Ser in 1979 – 80. The V517 Cyg and VV Ser light curves display some regularities and remarkable similarity. This likeness confirms that these two objects have identical ages. Intrinsic stellar perturbations, the perturbations combined with the dynamics of the circumstellar envelope, the effects in semi–detached eclipsing systems are discussed to explain the photometric features of V517 Cyg and VV Ser.

122.036 **Some regularities of the activity of Is variable stars. II. Recurrent phenomena in the light curve of SV Cephei.**
V. I. Kardopolov, L. V. Timoshenko, G. K. Filip'ev.
Perem. Zvezdy, Tom 22, No. 2, p. 137 – 152 (1985). In Russian.
The results of photoelectric UBVR measurements of the Isa variable star SV Cep in 1981 and 1982 are given. On the basis of the photometric data collected during 3 years it was found that the star exhibits some semiregular wavelike light variations. There exist stellar brightness variations with cycles of 700^d or more, $40 - 60^d$ and near 10^d. The equivalent widths of some lines of metals in the spectral range λ3650 – 5000 Å have been measured for spectral classification. Spectroscopic features suggest that the spectral type of SV Cep is A0 III near the star's light maximum. Both the spectral and photoelectric observations indicate in SV Cep the same characteristics as those of Be stars.

122.037 **Brightness of VX Cassiopeiae, SV Cephei, EH Cephei and VV Serpentis in 1980.**
V. I. Kardopolov, G. K. Filip'ev.
Perem. Zvezdy, Tom 22, No. 2, p. 153 – 157 (1985). In Russian.

122.038 **On the light variability of RS Sagittae.**
V. I. Kardopolov, G. K. Filip'ev.
Perem. Zvezdy, Tom 22, No. 2, p. 158 – 160 (1985). In Russian.
The results of photoelectric BVR measurements of the RV Tau–type variable star RS Sge in 1981 are given. The authors confirm its being an RV Tau–type variable. The complete cycle of the star brightness variations is confined to 90^d. Algol–type dips of stellar brightness were found for two nights. It seems that RS Sge may be also classified as an eclipsing variable star.

122.039 **On the constancy of brightness of eight variable stars in the Large Orion nebula.**
V. I. Kardopolov, N. A. Shutemova.
Perem. Zvezdy, Tom 22, No. 2, p. 161 – 165 (1985). In Russian.
Photoelectric UBV observations of eight variable stars in the Large Orion nebula: KS, LP, MX, NU, NV, V359, V361 and V372 Ori were carried out in 1977 – 1979. The results of the observations are summarized in a table. Using analogous data of other authors the conclusion was obtained that the brightness of the investigated stars was practically constant during 23 – 28 years.

122.040 **Photometric features of RZ Piscium.**
G. V. Zajtseva.
Perem. Zvezdy, Tom 22, No. 2, p. 181 – 190 (1985). In Russian.
Results are presented of photoelectric UBV photometry of the irregular variable RZ Piscium carried out in 1974 – 1983. 13 minima were observed, for 8 of them continuous photometry was obtained. A heterogeneous relation between the V brightness and (B–V), (U–B) colours is found which indicates the presence of several sources of the variability. Minima fainter than 13^m obey the period of $12\overset{d}{.}67$.

122.041 **New data on the variable stars in the region of M56.**
N. E. Kurochkin.
Perem. Zvezdy, Tom 22, No. 2, p. 201 – 218 (1985). In Russian.
Variable stars in the region of M56 discovered or investigated earlier by the author are re–examined. New photographic data

(JD 2433031–44192, about 240 plates) and electronic computer programs were used. For the majority of stars new light elements and light curves have been obtained.

122.042 **Study of 11 stars in Lyra.**
M. P. Galkina, S. Yu. Shugarov.
Perem. Zvezdy, Tom 22, No. 2, p. 225 – 235 (1985). In Russian.

122.043 **On the Blazhko effect of V421 Herculis.**
V. P. Tsesevich, O. E. Mandel'.
Perem. Zvezdy, Tom 22, No. 2, p. 237 – 244 (1985). In Russian.
From photographic and visual observations the elements of this RR Lyrae–type star have been specified: Max hel = JD 2437172.4281 + $0^{d}.5567545$·E. Mean light curves have been constructed. The star shows Blazhko effect especially noticeable near maximum of the light curve. The ratio of the secondary period to the primary one is equal to 0.79.

122.044 **RR Lyrae–type variables V817 Oph and V1071 Oph.**
D. A. Mitrofanov.
Perem. Zvezdy, Tom 22, No. 2, p. 245 – 247 (1985). In Russian.

122.045 **Investigation of the red variable DI Sagittae and RR Lyrae–type variables DH and ER Sagittae.**
A. Yu. Pogosyants.
Perem. Zvezdy, Tom 22, No. 2, p. 249 – 256 (1985). In Russian.
Light curves of three variable stars were derived on the basis of estimates of about 350 plates in the interval of 1960 – 1981. A more precise period of DI Sge is $201^{d}.1$. Period change was found in the case of DH Sge. Spasmodic variations of the period of ER Sge occurred near JD 2438400 and JD 2441800. The Blazhko effect in the last two variables is not confirmed being lower than the accuracy of the observations.

122.046 **FV Serpentis is a Mira Ceti–type variable star.**
A. Yu. Pogosyants.
Perem. Zvezdy, Tom 22, No. 2, p. 257 – 259 (1985). In Russian.
The examination of about hundred plates (1974 – 1982) showed absence of R Coronae Borealis type variability. The star is a Mira Ceti–type variable with the period of 365^{d}. Photographic (B) observations and a finding chart with comparison stars magnitudes are given.

122.047 **Investigation of variable stars WY Virginis and PG 1341–079.**
E. Yu. Osminkin.
Perem. Zvezdy, Tom 22, No. 2, p. 261 – 263 (1985). In Russian.

122.048 **Investigation of three new RR Lyrae–type variable stars in Virgo.**
I. V. Khatuntsev.
Perem. Zvezdy, Tom 22, No. 2, p. 265 – 267 (1985). In Russian.

122.049 **The Mira–type variable star RZ Serpentis.**
A. M. Chernenko.
Perem. Zvezdy, Tom 22, No. 2, p. 269 – 270 (1985). In Russian.

122.050 **Co–operative observations of long–period variable stars in the H_2O radio line and in the infrared range.**
I. I. Berulis, A. S. Gladyshev, E. E. Lekht, M. I. Pashchenko, G. M. Rudnitskij, R. L. Sorochenko, G. V. Khozov.
Nauchn. Inf., Vyp. 56, p. 92 – 96 (1984). In Russian.
Preliminary results of observations of the long–period variable stars R Aql and U Ori in the water–vapour emission radio line at the wavelength of 1.35 cm as well as in the *RIHK* spectral bands made during 1980 – 1982 are presented. From the infrared data the new value for the period of infrared light variations found for R Aql is 280^{d}. A comparison of the flux variations in the H_2O line with the infrared variability of both stars is carried out.

122.051 **On the efficiency of grains sweep out by the radiation from variable red giants.**
N. V. Voshchinnikov, V. B. Il'in.
Nauchn. Inf., Vyp. 56, p. 99 – 103 (1984). In Russian.
The sweep out of dust grains by radiation from Mira Ceti (oxygen giant) and V CrB (carbon giant) has been considered for maximum and minimum of the light curve. The sweep efficiency β (the ratio of radiation pressure force to the gravitational one) has been computed for cylindrical and spherical dust grains of various chemical composition.

122.052 **Variations of infrared radiation of W Cyg and T Cep.**
T. A. Movsesyan.
Nauchn. Inf., Vyp. 56, p. 113 – 115 (1984). In Russian.
For W Cyg and T Cep results of the observations in the 1 – 24 μm region are presented. The variability of H_2O absorption bands near 1.4 and 1.9 μm and of the continuum near 2.2 μm is discovered.

122.053 **Photometric observations of UV Aur.**
T. N. Khudyakova.
Nauchn. Inf., Vyp. 56, p. 139 – 141 (1984). In Russian.
This paper presents the results of observation of the carbon variable UV Aur in the *B, V, R, I* photometric system. According to the photometric characteristics this star belongs to long–period variables. A new epoch of maximum is reported.

122.054 **Models of the active stage of R Coronae Borealis–type stars.**
V. A. Okorokov.
Nauchn. Inf., Vyp. 57, p. 3 – 16 (1984). In Russian.
A number of theoretical light curves for models of R Coronae Borealis phenomena was constructed in the spherical symmetry approximation. The dynamics of the ejected envelope was investigated for various luminosities, gas to dust ratios and graphite particles sizes. The main features of observational light curves are well reproduced in the frame of the model accepted.

122.055 **RV Tauri stars.**
A. B. Fokin.
Nauchn. Inf., Vyp. 57, p. 17 – 29 (1984). In Russian.
Main observational properties and their theoretical interpretation of RV Tau variables are reviewed. The mass problem of these stars and their relations to population II cepheids and long–period variables are discussed. For double–period semiregular variables a relation between the main and the secondary periods is found, similar to double–period RV Tau stars.

122.056 **A study of flare stars in the Taurus region.**
A. S. Khodzhaev.
Astrofizika, Tom 24, Vyp. 1, p. 65 – 78 (1986). In Russian. English translation in Astrophysics, Vol. 24, No. 1.
The results of the search for flare stars and their photometric, H_α spectroscopic and statistical study in the Taurus region are presented. By means of photographic observations carried out during 1980 – 1984 92 new flare stars were discovered. The spatial distribution of these stars was considered and the problem of their membership was discussed. Comparative analysis of the data of flare stars in the Taurus region with that of other systems has been carried out. The Hertzsprung–Russell and two–colour (*U–B, B–V*) diagrams for the Taurus flare stars are similar to the diagrams of stellar clusters and associations (Pleiades, Orion etc.). The estimated total number of flare stars in this region is larger than 500.

122.057 **Brightness decrease of R CrB stars: model calculations.**
Yu. K. Melik–Alaverdyan.
Astrofizika, Tom 24, Vyp. 1, p. 79 – 84 (1986). In Russian. English translation in Astrophysics, Vol. 24, No. 1.
It has been shown that dust formation in the atmospheres of R CrB stars can be explained under the assumption of the presence of some flare phenomena in these atmospheres.

122.058 **A study of resonances in Cepheid II models.**
A. B. Fokin.
Astrofizika, Tom 24, Vyp. 1, p. 109 – 122 (1986). In Russian. English translation in Astrophysics, Vol. 24, No. 1.
The Fourier spectra of nonlinear pulsations of Cepheids II models are investigated. It has been found that the resonance

between the fundamental mode and some overtone leads to the mechanical energy of this overtone to be increased by a factor of 2 – 3. It has been shown that a pulse is generated on each cycle in the outermost part of the envelope. A secondary bump on the velocity and light curves exists until the fundamental mode and the second overtone are in resonance. Yet, the bump rise coincides with the return of the pulse reflected from the core to the surface. The origin of the bump seems to be related with both these mechanisms.

122.059 **Narrow–band, slowly varying decimetric radiation from the dwarf M flare star YZ Canis Minoris.**
K. R. Lang, R. F. Willson.
Astrophys. J., Lett. Ed., Vol. 302, No. 1, p. L17 – L21 (1986).
Narrow–band, slowly varying microwave radiation has been detected from the dwarf M star YZ Canis Minoris at frequencies near 1465 MHz. This quiescent, or nonflaring, emission cannot be attributed to gyroresonant radiation from coronal loops or to nonthermal gyrosynchrotron radiation. It may be explained by coherent burst mechanisms like electron–cyclotron masers or coherent plasma radiation. The slow variation of the narrow–band emission might be explained by the stochastic nature of the continued low–level, coherent burst activity.

122.060 **Evidence for mass loss from IRAS observations of classical Cepheids.**
H. Deasy, C. J. Butler.
Nature, Vol. 320, No. 6064, p. 726 – 728 (1986).
Evidence of anomalous Cepheid mass loss was sought by making a comparison between the infrared emission of Cepheids and that of nonvariable supergiants in the same luminosity and effective temperature range. A search of the IRAS catalogue found a number of Cepheids and stable supergiants which showed emission in at least one of the IRAS wavelength bands. Some long–period Cepheids showed infrared excesses with respect to their non–pulsating counterparts, while emission from Cepheids with periods of less than 10 days was comparable to the levels seen in the stable supergiants. Mass loss rates of up to $7x10^{-7}M_{\odot}yr^{-1}$ were derived from the infrared excesses, which is sufficiently high to have a major effect on the evolution of these stars.

122.061 **RR Lyrae, Delta Scuti, SX Phoenicis stars and the Baade–Wesselink method.**
G. Meylan, G. Burki.
Messenger, No. 43, p. 16 – 23 (1986).

122.062 **Physical parameters of pulsating variables with periods between one and three days. I. Photometry and metallicities.**
R. Diethelm.
Astron. Astrophys., Suppl. Ser., Vol. 64, No. 2, p. 261 – 274 (1986).
Based on a new set of observations in the Walraven VBLUW system $[Fe/H]_{VBLU}$ values of 57 pulsating variables, mostly with periods between one and three days, are determined. The distribution of the metal abundances is shown to be essentially bi–modal. All the variables with $[Fe/H] < -0.4$ were classified by Diethelm (1983) as RRd stars. They are identified as the galactic field counterpart of the Cepheids in the same period range found in globular clusters.

122.063 **Variations séculaires d'amplitude dans les étoiles β CMa.**
E. Chapellier.
Astron. Astrophys., Suppl. Ser., Vol. 64, No. 2, p. 275 – 285 (1986).
Secular variations of the pulsational amplitudes of 13 β CMa stars have been examined. After discussion of different observational limitations on amplitude determinations in radial velocity and in light variations, the author confirms the long term amplitude variations in BW Vul, δ Cet, α Vir, β CMa and 16 Lac and points out their existence in four other stars: β Cep, θ Oph, 15 CMa, ξ_1 CMa. The remaining four stars do not exhibit this phenomenon. A general character of the observed variations seems to be their cyclic behaviour with time scales ranging from 30 years to several centuries. No interpretation of this phenomenon could be found.

122.064 **On the variability of the SC–type Mira variable LX Cyg.**
Z. Alksne, A. Alksnis.
Issled. Solntsa Krasnykh Zvezd, No. 22, p. 85 – 99 (1985). In Russian.
The light variability of LX Cyg was studied based on three–colour photographic observations in red, visual and blue obtained during the last 15 years. The observed data of the maximum light including earlier results published by other investigators indicated continuous increasing of the cycle length of the star.

122.065 **On the V351 Ori quasi–periodic microvariability.**
G. U. Koval'chuk.
Astrofizika, Tom 24, Vyp. 2, p. 213 – 225 (1986). In Russian. English translation in Astrophysics, Vol. 24, No. 2.
In the variable star V351 Ori light variations a quasi–periodic component with $P = 326^s$ and variable amplitude has been found. Its amplitude is $0^m.06$ (U) in normal light and increased to $0^m.15$ (U) in minimum light. The quasi–periodic component amplitude is possibly highest in the ultraviolet in comparison with the blue region of the spectrum.

122.066 **On the nature of variability of TZ Ori.**
Eh. S. Parsamyan, G. A. Pogosyan.
Astrofizika, Tom 24, Vyp. 2, p. 239 – 244 (1986). In Russian. English translation in Astrophysics, Vol. 24, No. 2.
Among the irregular brightness variations of TZ Ori resembling "rapid" and "slow" flares are distinguished. A statistical analysis has allowed to detect a five–day period of brightness variation. Colour indices $U–B$, $B–V$ changed between $-0^m.2$ – $+0^m.5$, $0^m.3$ – $1^m.5$ respectively.

122.067 **Atmospheric parameters of RV Tauri variables.**
G. M. Wahlgren.
Bull. Am. Astron. Soc., Vol. 17, No. 4, p. 875 (1985). Abstract. – See Abstr. 010.062.

122.068 **The Mg II 2800 Å spectral feature in RV Tauri stars.**
D. W. Dawson, D. L. DuPuy.
Bull. Am. Astron. Soc., Vol. 17, No. 4, p. 875 (1985). Abstract. – See Abstr. 010.062.

122.069 **Photometric and spectroscopic observations of R Scuti during a strong–TiO minimum and subsequent recovery.**
R. F. Wing, G. M. Wahlgren, G. Fontaine, F. Wesemael.
Bull. Am. Astron. Soc., Vol. 17, No. 4, p. 875 (1985). Abstract. – See Abstr. 010.062.

122.070 **Period determinations for 11 southern hemisphere red variables.**
D. J. Bord, S. J. Shawl.
Bull. Am. Astron. Soc., Vol. 17, No. 4, p. 876 (1985). Abstract. – See Abstr. 010.062.

122.071 **Feinheit functions and cepheid distance determination.**
T. G. Barnes, T. J. Moffett.
Bull. Am. Astron. Soc., Vol. 17, No. 4, p. 877 (1985). Abstract. – See Abstr. 010.062.

122.072 **An IUE, optical and EXOSAT rotational modulation study of the active flare star EV Lac.**
C. W. Ambruster, B. R. Pettersen.
Bull. Am. Astron. Soc., Vol. 17, No. 4, p. 880 (1985). Abstract. – See Abstr. 010.062.

122.073 **SU Cas: an overtone pulsator.**
N. R. Simon, T. Aikawa.
Bull. Am. Astron. Soc., Vol. 17, No. 4, p. 894 – 895 (1985). Abstract. – See Abstr. 010.062.

122.074 **Pioneer 10 observations of the β Cephei stars γ Pegasi and δ Ceti.**
G. J. Peters, H. Ogawa, K. S. Judge, D. L. Judge.
Bull. Am. Astron. Soc., Vol. 17, No. 4, p. 903 (1985). Abstract. – See Abstr. 010.062.

122.075 **The modal characteristics of δ Scuti.**
S. Barch, B. J. McNamara.
Bull. Am. Astron. Soc., Vol. 17, No. 4, p. 903 (1985). Abstract. – See Abstr. 010.062.

122.076 **CCD observations of some RR Lyrae stars and a standard sequence in Baade's window.**
A. R. Walker, P. Mack.
Mon. Not. R. Astron. Soc., Vol. 220, No. 1, p. 69 – 89 (1986).
CCD measurements of 11 RR Lyrae variables and a standard sequence near NGC 6522 are used to recalibrate photographic photometry of the RR Lyrae variables in Baade's window and then to determine the distance to the galactic centre, which is found to be $R_0 = 8.1 \pm 0.4$ kpc if $<M_V(\mathrm{RR})> = 0.6$. The data tend to deny the existence of an absolute magnitude–metallicity relation for RR Lyrae stars. The colours of the RR Lyrae variables, together with other reddening indicators, imply that $A_V = 1.78 \pm 0.10$, rather larger than usually assumed. A colour–magnitude diagram for the giant branch of NGC 6522 is consistent with the cluster being moderately metal poor and at approximately the same distance as the galactic centre. At least some of the RR Lyrae variables adjacent to the cluster are likely to be members.

122.077 **Simultaneous optical, infrared and microwave observations of the flare star AT Mic.**
G. J. Nelson, R. D. Robinson, O. B. Slee, M. C. B. Ashley, A. R. Hyland, I. R. Tuohy, I. Nikoloff, A. E. Vaughan.
Mon. Not. R. Astron. Soc., Vol. 220, No. 1, p. 91 – 106 (1986).
The authors report the first extensive observations of the binary dMe flare star AT Mic made simultaneously at optical, infrared and microwave wavelengths. It is concluded that the data can be best accounted for in terms of a hot ($\sim 10^5$K) plasma flare model which produces a relatively flat energy–spectrum over a range of wavelengths determined by the temperature, density and depth of the source region. This range may therefore vary from flare to flare and from star to star. Microwave observations during a large number of optical flares indicate a variation of at least several orders of magnitude in the ratio of optical to microwave flare luminosities. This is taken to imply that different plasmas are responsible for the two emissions.

122.078 **Absolute magnitudes and kinematic properties of RR Lyrae stars.**
S. L. Hawley, W. H. Jefferys, T. G. Barnes III, L. Wan.
Astrophys. J., Vol. 302, No. 2, p. 626 – 631 (1986).
A maximum–likelihood statistical analysis of several subgroups of the field RR Lyrae stars was performed to determine the relative solar motion, velocity ellipsoid parameters, and mean absolute visual magnitude for each group. The full sample of 159 stars was taken from a recent Chinese proper–motion survey, and new mean radial velocities were used for 46 of the stars. The best estimate for the mean absolute visual magnitude is $<M_v> = 0.76 \pm 0.14$ mag for the entire RR *ab*–type sample.

122.079 **The magnetic field of the BY Draconis flare star EQ Virginis.**
S. H. Saar, J. L. Linsky, J. M. Beckers.
Astrophys. J., Vol. 302, No. 2, p. 777 – 784 (1986).
The authors apply a new Zeeman analysis procedure, which includes radiative transfer effects and compensation for blends, to high–resolution, high–signal–to–noise line profiles of the BY Draconis–type flare star EQ Vir obtained with the Multiple Mirror Telescope. Using a number of lines with effective Landé *g* factors ranging from 0.5 to 2.5, and two different analysis methods, the authors find a mean field of 2500 ± 300 G covering $80\% \pm 15\%$ of EQ Vir.

122.080 **On a mechanism of distortion of the forms of observed radial velocity curves of pulsating stars.**
S. M. Andrievskij.
Problems of astronomy, p. 39 – 44 (1985). In Russian. Abstr. in Ref. Zh., 51. Astron., 3.51.549 (1986). – See Abstr. 003.023.

122.081 **Radial velocities and variation of the radius of RR Lyrae.**
S. N. Udovichenko.
Problems of astronomy, p. 45 – 50 (1985). In Russian. Abstr. in Ref. Zh., 51. Astron., 3.51.580 (1986). – See Abstr. 003.023.

122.082 **Photometric investigation of the RR Lyrae–type star V363 Cassiopeiae.**
V. P. Bezdenezhnyj.
Problems of astronomy, p. 51 – 55 (1985). In Russian. Abstr. in Ref. Zh., 51. Astron., 3.51.581 (1986). – See Abstr. 003.023.

122.083 **On some peculiarities of yellow semiregular stars.**
E. N. Makarenko, S. M. Andrievskij.
Problems of astronomy, p. 32 – 38 (1985). In Russian. Abstr. in Ref. Zh., 51. Astron., 3.51.583 (1986). – See Abstr. 003.023.

122.084 **Results of photoelectric observations of four variable stars.**
L. N. Berdnikov.
Astron. Tsirk., No. 1385, p. 8 (1985). In Russian.

122.085 **On brightness variability of BD +12°4134 in Aquila.**
V. V. Golovatyj, Yu. F. Mal'kov.
Astron. Tsirk., No. 1386, p. 6 – 7 (1985). In Russian.

122.086 **BH Herculis – an RR Lyr–type variable.**
M. V. Pavlov, S. Yu. Shugarov.
Astron. Tsirk., No. 1386, p. 8 (1985). In Russian.

122.087 **EV Aql.**
L. N. Berdnikov.
Astron. Tsirk., No. 1387, p. 8 (1985). In Russian.

122.088 **The luminosities of cepheids.**
L. N. Berdnikov, Yu. N. Efremov.
Astron. Tsirk., No. 1388, p. 1 – 4 (1985). In Russian.

122.089 **On the variable S Vul.**
G. R. Ivanov.
Astron. Tsirk., No. 1389, p. 7 – 8 (1985). In Russian.

122.090 **On the radial velocity of R Cas.**
T. E. Derviz.
Astron. Tsirk., No. 1394, p. 6 – 8 (1985). In Russian.

122.091 **On the Blazhko effect in TT Cancri.**
B. N. Firmanyuk, I. L. Andronov, M. A. Levina.
Astron. Tsirk., No. 1395, p. 6 – 7 (1985). In Russian.

122.092 **MV Herculis.**
V. V. Pravdikova.
Astron. Tsirk., No. 1395, p. 7 – 8 (1985). In Russian.

122.093 **New variable stars in Ursa Major.**
V. P. Goranskij.
Astron. Tsirk., No. 1407, p. 7 – 8 (1985). In Russian.

122.094 **RR Lyrae stars as distance indicators.**
R. F. Jameson.
Vistas Astron., Vol. 29, Part 1, p. 17 – 26 (1986).
This paper attempts to review the most significant recent developments concerned with understanding the absolute magnitude of the RR Lyrae stars. Coupled with this are some new results by the author and his collaborators on Baade–Wesselink analysis of four field RR Lyrae stars using both optical and infrared observations.

122.095 **Observations of the Hβ and He II λ4686 lines in the spectra of UV Cet–type flare stars.**
P. P. Petrov, P. F. Chugainov (*P. F. Chugajnov*), A. G. Shcherbakov.
Bull. Crimean Astrophys. Obs., Vol. 69, p. 1 – 14 (1986). English translation of 39.122.100.

122.096 **A review of the rapidly oscillating Ap stars.**
D. W. Kurtz.
Seismology of the Sun and the distant stars, p. 417 – 440 (1986). – See Abstr. 012.026.
The rapidly oscillating Ap stars are cool magnetic Ap stars which oscillate with periods between 4 and 15 min with peak–to–peak light variations of B $\leqslant$ 0.016 mag. There are eleven of these stars known. They are listed along with some of their observational characteristics and references to literature on them.

122.097 **Nonradial modes in line–profile variable stars.**
Y. Osaki.
Seismology of the Sun and the distant stars, p. 453 – 463 (1986). – See Abstr. 012.026.
Nonradial oscillation modes in line–profile variable stars are discussed. The line–profile modeling based on nonradial oscillations is first reviewed and then general observational properties of line–profile variable stars are presented. A question is raised about the mode identification of nonradial oscillations based on the conventional line–profile modeling. It is suggested that wave phenomena observed in line–profile variable stars could be superficial phenomena confined more or less in the atmospheric layers of these stars.

122.098 **Frequency analysis of the 5.65–min oscillations in the rapidly oscillating Ap star HD 134214.**
T. J. Kreidl, D. W. Kurtz.
Mon. Not. R. Astron. Soc., Vol. 220, No. 2, p. 313 – 320 (1986).
High–speed photometric observations of HD 134214 obtained during 35 hr of observation in 1985 are presented. A frequency analysis of these data indicates the presence of only one frequency of oscillation in this star at $f_1 = 2.94960 \pm 0.00004$ mHz ($P = 5.650 \pm 0.001$ min). This is the highest frequency which is demonstrably not a harmonic of a lower frequency yet discovered in a rapidly oscillating Ap star. This frequency is above the critical frequency calculated for A star models by Shibahashi & Saio. The phase shift $\Delta\phi(V,B-V) = 19 \pm 6°$ has been calculated for HD 134214 from simultaneous B and V observations. The interpretation of this phase shift is problematical.

122.099 **The luminosity and kinematics of RR Lyrae stars in the solar neighbourhood – I. Statistical parallaxes of RR Lyraes.**
P. Strugnell, N. Reid, C. A. Murray.
Mon. Not. R. Astron. Soc., Vol. 220, No. 2, p. 413 – 427 (1986).
The authors have used an improved maximum–likelihood statistic to determine the absolute magnitude and kinematics of local RR Lyrae variables from a statistical and secular parallax analysis of proper motions and radial velocities. The analysis indicates a "best–buy" absolute magnitude (at mean intensity) of $<M_V> = +0.75 \pm 0.2$. The authors find no statistically significant evidence for any correlation between absolute magnitude and metallicity. However, there is a correlation between systemic rotation and both metallicity and period.

122.100 **Titanium oxide radial velocities of long–period variables.**
T. E. Derviz, S. V. Milyutikova.
Sov. Astron. Lett., Vol. 11, No. 4, p. 218 – 220 (1985). English translation of 40.122.029.

122.101 **The spectroscopic investigation of AC Herculis. 1. The parameters of the atmosphere and chemical composition.**
Yu. V. Borisov, V. E. Panchuk.
Astrofiz. Issled. Izv. Spets. Astrofiz. Obs., Tom 22, p. 17 – 24 (1986). In Russian. English translation in Bull. Spec. Astrophys. Obs. – North Caucasus.
Two spectrograms of the variable star AC Herculis for the two phases of the light curve (φ = 0\.P4 and 0\.P0) have been carried out in summer 1984. The parameters of the AC Herculis atmosphere $T_e = 5800$K, lg $g = 1.2$, $\xi_t = 6.3$ km/s and metallicity relative to the sun [Fe/H] = –1.25 are determined by the model atmosphere method analysing the Fe I and Fe II line groups of middle and weak intensities. The chemical abundances of seven elements are determined.

122.102 **The Cepheid mass problem and Cepheid binaries.**
E. Böhm–Vitense.
Astrophys. J., Vol. 303, No. 1, p. 262 – 272 (1986).
The author examines existing mass determination for Cepheids with different periods. Wesselink masses are independent of the adopted distance scale. For short periods (< 6 days) they follow the sequence of evolutionary masses. For periods longer than 10 days they are lower by up to a factor of 2. The lower mass branch joins up with the bump masses. The new pulsational masses agree with the Wesselink masses for periods longer than 6 days. Dynamical masses determined for S Mus and V636 Sco also agree with the low Wesselink and giant companion masses.

122.103 **A photometric and radial velocity study of six southern Cepheids. II. Binarity, radii, luminosities, and masses.**
I. M. Coulson, J. A. R. Caldwell, W. P. Gieren.
Astrophys. J., Vol. 303, No. 1, p. 273 – 279 (1986).
The data now available for the galactic Cepheids AQ Car, XX Cen, XY Car, TT Aql, XX Car, and XZ Car are analyzed for binarity, radius, luminosity, and mass. There is weak evidence only for binarity of XX Cen, XX Car, and XZ Car, and it is concluded that all six Cepheids may be considered to be single. The radii are derived by an improved version of the Baade–Wesselink technique and are found to depend upon the color index used to define the surface brightness. The *V–I* radii, for instance, are 20% – 25% larger than the *B–V* radii.

122.104 **Essai de détermination d'éphémérides d'étoiles de type Mira peu suivies.**
J. Gunther.
Bull. Assoc. Fr. Obs. Etoiles Variables, No. 36, p. 7 – 8 (1986).

122.105 **Microwave emission from dMe flare stars.**
O. B. Slee, W. H. Allen, D. W. Coates, A. A. Page, P. J. Quinn.
Proceedings of the Second Asian–Pacific Regional Meeting on Astronomy, p. 74 – 81 (1984). – See Abstr. 012.031.
The present paper attempts to summarize the Australasian microwave and optical observations of flare stars up to October 1980.

122.106 **A new search for RR Lyrae variables in Baade's window.**
B. Blanco, V. M. Blanco.
Proceedings of the Second Asian–Pacific Regional Meeting on Astronomy, p. 110 – 116 (1984). – See Abstr. 012.031.

122.107 **Abundances of Cepheids in the Galaxy and the Magellanic Clouds.**
J. W. Pel.
Proceedings of the Second Asian–Pacific Regional Meeting on Astronomy, p. 411 – 422 (1984). – See Abstr. 012.031.
Over the last decade there has been increasing emphasis on the importance of abundance determinations for Cepheids. Information on the chemical composition of these stars appears to be important both for the practical use of Cepheids as distance indicators, and for the theoretical understanding of Cepheid pulsation and evolution. In the context of chemical evolution of galaxies, Cepheids are powerful probes, because they can be studied over large distances. The knowledge of Cepheid abundances in the Galaxy and the Magellanic Clouds has very recently been improved substantially. This paper discusses these new results, and some of their implications.

122.108 **Hα profile variations in variable luminous yellow supergiants.**
A. Arellano Ferro.
Rev. Mex. Astron. Astrofis., Vol. 11, No. 2, p. 113 – 120 (1985). = Contrib. Inst. Astron., Univ. Nac. Autónoma Mex., No. 183.
Hα profiles of the variable luminous yellow supergiants HD 161796, 89 Her, ε Aur, V509 Cas and ϱ Cas, are presented. Profile variations in time scales of months and weeks are detected. The nature and variability of the profiles is discussed. The profiles of HD 161796 are most peculiar due to incipient emission centered on the line. Spectacular variations are found in the blue wing of Hα in the eclipsing binary ε Aur, even before the first contact of the 1982 – 1984 eclipse, these variations may be connected with the pulsation of the primary star.

122.109 **Spectral and luminosity variation of long–period red variable stars.**
L. Celis S.
Astron. J., Vol. 91, No. 2, p. 405 – 415 (1986).
An analysis of the visual (V) apparent–magnitude observations with simultaneous determination of the spectral subtype (S) of a long–period type Me stars, mainly Mira variables, is presented. For each star, there is a linear correlation between V and S ($dM_V = GdS$, $\sigma_G = 0.25$). For a total of 16 stars, the slopes G were determined in the range of 0.30 to 2.11 mag per spectral subtype. For any moment of observation, the absolute magnitude obeys a linear relation $M_V = M_{Va} + G(S-S_a)$, where M_{Va} and S_a are the absolute magnitudes and spectrum at maximum light, respectively. Thus, it is possible to conclude that the instantaneous M_V of any given Mira star depends only on the physical characteristics that involve the instantaneous spectral subtype S. Moreover, a best estimation of M_{Va} is given by the function $M_V(P, S_a)$, where P is the period.

122.110 **Quiescent state X–ray emission from flare stars.**
P. C. Agrawal, A. R. Rao, B. V. Sreekantan.
X–ray astronomy '84, p. 51 – 54 (1985). – See Abstr. 012.033.
Observations of quiescent state X–ray emission from flare stars are discussed. From analysis of all the X–ray observed flare stars it is found that distribution of L_x/L_{Bol} for flare stars is similar to that of the RS CVn binaries but dissimilar to that of non–flaring dM stars. A correlation between the X–ray and bolometric luminosities of flare stars is found. It is suggested that coronal activities of the flare stars and RS CVn binaries are of similar nature.

122.111 **On the amplitudes of bump Cepheids.**
J. D. Fernie, S.–k. J. Chan.
Astrophys. J., Vol. 303, No. 2, p. 766 – 768 (1986).
Over 100 Galactic classical Cepheids in the period range 5 – 15 days having well–determined photoelectric V mag light curves have been examined in terms of the amplitude–period diagram. The authors verify the long–known dip at log $P \approx 0.95$ in the upper envelope of the amplitude distribution but find in addition a coincident rise in the lower envelope, so that all Cepheids with periods between ~7.2 and 11 days have very similar amplitudes. It is tentatively concluded that all classical Galactic Cepheids in this period range show bumps or shoulders in their lightcurves.

122.112 **Energy distribution in the spectrum of Mira (o Ceti).**
E. A. Depenchuk, V. F. Karamysh, R. I. Chuprina.
Odes. univ. Odessa, 14 pp. (1985). In Russian. Abstr. in Ref. Zh., 51. Astron., 6.51.508 (1986).

122.113 **An analysis of the spectrum of Zeta Geminorum.**
B. B. Sanwal, B. S. Rautela, G. C. Joshi.
Astrophys. Space Sci., Vol. 123, No. 1, p. 183 – 192 (1986).
Four spectrograms of the cepheid ζ Geminorum at different phases have been analysed for the determination of the abundances of various elements. The analysis shows that the atmosphere of ζ Gem has an essentially solar composition.

122.114 **Coordinated Exosat and spectroscopic observations of flare stars and coronal heating.**
C. J. Butler, M. Rodono, B. H. Foing, B. M. Haisch.
Nature, Vol. 321, No. 6071, p. 679 – 682 (1986).
The X–ray flux of dMe stars is thought to arise from two distinct mechanisms, one involving a continuous "quiescent" emission from a high–temperature plasma and the other involving dramatic flare events. The authors present here some results of simultaneous monitoring of the two flare stars, UV Ceti and EQ Peg, with Exosat and ground–based optical spectroscopy. They observe short–timescale variability in the 0.1 – 2 keV emission from both these objects and, in the case of UV Ceti, find a strong correlation between the soft X–ray and Hγ fluctuations. The implication is that much of the low–level X–ray flux previously considered "quiescent" probably originates from small flare events.

122.115 **Atmospheres of Delta Scuti pulsating variables. IV. Thorium abundance in the stars 20 CVn, 28 And, V644 Her and δ Del.**
L. S. Lyubimkov, T. M. Rachkovskaya.
Izv. Krymskoj Astrofiz. Obs., Tom 74, p. 14 – 19 (1986). In Russian. English translation in Bull. Crimean Astrophys. Obs., Vol. 74.
For four δ Sct variables the region of 4017 – 4021 Å containing the Th II λ4019.12 line was analysed using the method of spectral synthesis. While calculating the synthetic spectra the stellar rotation and the instrumental profile were taken into account. The thorium abundance has been determined by means of comparison of the computed spectra with the observed ones.

122.116 **Visual observations of the variable S Sagittae.**
M. Fonović.
Vasiona, Année 33, No. 5, p. 99 – 101 (1985). In Croatian.
The author presents the light curve of S Sge based on observations made in 1983.

122.117 **Flare stars – physics and evolution.**
L. V. Mirzoyan.
Eruptive phenomena in stars, p. 409 – 421 (1986). – See Abstr. 012.039.

122.118 **Photoelectric observations of EV Lac in 1984: fast flare activity?**
M. K. Tsvetkov, A. P. Antov, A. G. Tsvetkova.
Eruptive phenomena in stars, p. 423 – 424 (1986). – See Abstr. 012.039.

122.119 **On the flare activity variations of HII 2411.**
G. Szécsényi–Nagy.
Eruptive phenomena in stars, p. 425 – 426 (1986). – See Abstr. 012.039.

122.120 **Flare stars' count in Orion.**
R. Sh. Natsvlishvili.
Eruptive phenomena in stars, p. 427 – 428 (1986). – See Abstr. 012.039.

122.121 **Photographic photometry of new flare stars in the Orion nebula.**
A. Tsvetkova, S. Tsvetkov.
Eruptive phenomena in stars, p. 431 – 432 (1986). – See Abstr. 012.039.

122.122 **Photographic photometry of flare stars in Pleiades.**
J. Kelemen.
Eruptive phenomena in stars, p. 433 – 434 (1986). – See Abstr. 012.039.

122.123 **Space distribution of the Hα and flare stars in the Orion OB1d association.**
K. G. Gasparian (*K. G. Gasparyan*).
Eruptive phenomena in stars, p. 439 – 440 (1986). – See Abstr. 012.039.

122.124 **Flare activity of antiflare stars.**
G. U. Koval'chuk.
Eruptive phenomena in stars, p. 443 – 444 (1986). – See Abstr. 012.039.

122.125 **Analysis of R Coronae Borealis variability from photoelectric observations.**
R. I. Goncharova.
Eruptive phenomena in stars, p. 445 – 446 (1986). – See Abstr. 012.039.

122.126 **The chemical composition of Polaris.**
R. E. Luck, H. E. Bond.
Space Telesc. Sci. Inst., Prepr. Ser., No. 96, 19 pp. (1986). Submitted to Publ. Astron. Soc. Pac.

122.127 **Flares on dMe stars: IUE and optical observations of AT Mic, and comparison of far–ultraviolet stellar and solar flares.**
G. E. Bromage, K. J. H. Phillips, P. L. Dufton, A. E. Kingston.
Rutherford Appleton Lab., Rep., RAL–86–009, 63 pp. (1986).
The authors describe observations and analysis of a large flare event on a dMe star, detected by the International Ultraviolet Explorer (IUE). They compare its far–ultraviolet spectrum with those of other stellar flares and also a large solar flare recorded by the Skylab mission in 1973. The authors also discuss the quiescent–phase optical and ultraviolet spectrum of the same dMe flare star.

122.128 **Observational studies of Cepheids. IV. The Feinheit functions.**
T. J. Moffett, T. G. Barnes III.
Astrophys. J., Vol. 304, No. 2, p. 607 – 616 (1986).
The authors examine the properties of the Feinheit functions and their use in Cepheid distance determination. Three Feinheit–type functions have been proposed: the *W* function of van den Bergh, the Wesenheit function of Madore, and the random–phase *F* function of Madore. The *W* and Wesenheit functions are shown to have the properties envisioned by their creators, i.e., a color–independent distance parameter and a reddening–free index, respectively. On the other hand, it is concluded that the fundamental assumption of the Feinheit Method is false and that application of the Feinheit Method may lead to distance moduli in error by as much as several tenths of a magnitude.

122.129 **Pulsational mode typing in line–profile variables. VII. Commensurable modes and nonuniform period behavior in Delta Scorpii.**
M. Smith.
Astrophys. J., Vol. 304, No. 2, p. 728 – 738 (1986).
On three occasions the author has observed the Si III $\lambda\lambda$4552 – 4574 line profiles in the typical B0.3 IV star δ Scorpii to vary over short time scales. The analysis of these variations suggests the presence of a total of six nonradial modes; no more than two are usually active at one time. These modes have short periods, large amplitudes, and are characterized by nonradial indices $-m = l$. Four of these modes seem to be members of a "commensurability" family. The physical pulsational amplitudes appear to be unequal from cycle to cycle. Some of the periods are "modulated" around a mean value. There is also a modulation pattern of periods which repeats during every superperiod (360° circuit of a wave around the star).

122.130 **A frequent optical flasher.**
B. Katz.
Astronomer, Vol. 22, No. 258, p. 95 – 96 (1985). Abstr. in Phys. Abstr., Vol. 89, No. 1274, Entry 23839 (1986).

122.131 **Nonradial and radial oscillations observed in non–emission line OB dwarfs and giants.**
D. Baade.
ESO Sci. Prepr., No. 418, 11 pp. (1986). To appear in 'Highlights of Astronomy', Vol. 7.

122.132 **Spectroscopy of RR Lyrae stars in Baade's window and ω Centauri.**
R. G. Gratton, A. Tornambè, S. Ortolani.
ESO Sci. Prepr., No. 426, 49 pp. (1986). To appear in Astron. Astrophys.

122.133 **Spectral observations of WW Vul at different brightness levels.**
L. V. Timoshenko.
Photometric and polarimetric investigations of celestial bodies, p. 86 – 96 (1985). In Russian. – See Abstr. 003.035.
Spectral observations of the rapid irregular variable WW Vul have been carried out at different brightness levels: at the minimum ($V \approx 11^m\!.5$), at the increased light level ($V \approx 10^m\!.8$) and at the maximum ($V \approx 10^m\!.5$). The spectra were obtained in September – October of 1981. The two–dimensional quantitative spectral classification of this star permits determining variations of spectral characteristics at different brightness levels. Different criteria based on metal lines give considerable differences in the spectral type of WW Vul. Hydrogen and *K* Ca II lines cannot serve as criteria of spectral classification since different hydrogen lines give quite different spectral types for the same spectrum. The author concludes that these lines are of non–photospheric origin. The spectral type WW Vul at maximum brightness is A3 III. At the minimum brightness it ranges from A3 to F4 along the luminosity line. It is concluded that dust is not the main cause of the observed minimum of WW Vul, and the variability is explained in terms of physical inhomogeneity of the stellar surface.

122.134 **UBVR photometry and optical properties of the circumstellar shell of R CrB.**
A. F. Pugach, R. I. Goncharova, G. U. Koval'chuk.
Photometric and polarimetric investigations of celestial bodies, p. 96 – 102 (1985). In Russian. – See Abstr. 003.035.
The results of four–colour UBVR observations of R CrB obtained for the period 1977 – 1982 are given. Analysis of colour variations in ascending branches of 1977 and 1982 minimum permitted to determine the dependence of the extinction coefficient of the dust cloud on the wavelength.

122.135 **Colour variations of the light of the star R CrB at brightness maximum.**
R. I. Goncharova.
Photometric and polarimetric investigations of celestial bodies, p. 102 – 104 (1985). In Russian. – See Abstr. 003.035.
Peculiarities of colour indexes *U–B*, *B–V*, *V–R* variations with respect to the light curve of R CrB during *JD* 2444(104 – 154) were investigated. $\Delta m_V = 0^m\!.36$ variations of amplitude are found. According to photometric properties pulsation is similar to classical cepheids and differs from pulsations of RY Sgr. It is shown that variations of light are caused by variation of temperature.

122.136 **Spectrophotometric observations of R CrB during 1972, 74 minima.**
N. Kameswara Rao, R. Vasundhara, B. N. Ashoka.
Indian Inst. Astrophys., Prepr., No. 194, 4 pp. (1985). To appear in IAU Colloq. No. 87.

122.137 **The Large Magellanic Cloud R CrB star – HV12842.**
D. H. Morgan, K. Nandy, N. Kameswara Rao.
Indian Inst. Astrophys., Prepr., No. 197, 4 pp. (1985). To appear in IAU Colloq. No. 87.

122.138 **3.0 to 3.5 micron spectrum of V348 Sgr and R CrB.**
K. Nandy, N. Kameswara Rao, D. H. Morgan.
Indian Inst. Astrophys., Prepr., No. 198, 3 pp. (1985). To appear in IAU Colloq. No. 87.

122.139 **Photometric and radial velocity variations of R CrB near maximum light.**
A. V. Raveendran, B. N. Ashoka, N. Kameswara Rao.
Indian Inst. Astrophys., Prepr., No. 201, 5 pp. (1985). To appear in IAU Colloq. No. 87.

122.140 **Preliminary analysis of the broad He I emission lines in R CrB.**
R. Surendiranath, K. E. Rangarajan, N. Kameswara Rao.
Indian Inst. Astrophys., Prepr., No. 214, 4 pp. (1986). To appear in IAU Colloq. No. 87.

122.141 **Aspectos observacionales de las estrellas variables "Mira".**
M. López Arroyo.
Bol. Astron. Obs. Madr., Vol. 11, No. 3, p. 5 – 17 (1984).

122.142 **Detection of cepheid variables by the Infrared Astronomical Satellite (IRAS).**
C. W. McAlary, D. L. Welch.
Prepr. Steward Obs., No. 635, 44 pp. (1986).

122.143 **Millisecond radio spikes from the dwarf M flare star AD Leonis.**
K. R. Lang, R. F. Willson.
Astrophys. J., Vol. 305, No. 1, p. 363 – 368 (1986).
This paper presents Arecibo observations of highly lefthand circularly polarized (up to 100%) spikes from AD Leonis at 1415 MHz with rise times $\tau_R \leqslant 5$ ms. These rise times provide an upper limit of $L \leqslant 1.5\text{x}10^8$cm, and a lower limit to $T_B \geqslant 10^{16}$K for a symmetric emitter. Some of the spikes were part of a quasi–periodic train of spikes with a mean periodicity of $\tau_P = 32 \pm 5$ ms and a total duration of $\tau_D = 150$ ms. This spike train was itself one pulse in a quasi–periodic sequence of pulsations with a mean periodicity of $\tau_P = 3.2 \pm 0.3$ s and total duration of $\tau_D = 25$ s; the pulsations were superposed upon a longer lasting (50 s) burst. The high brightness temperatures and high circular polarization of the spikes are interpreted in terms of coherent maser emission processes. One possible explanation of the quasi–periodic pulsations and spikes is magnetoacoustic oscillations in a coronal loop that modulate the maser action.

122.144 **Carbon and oxygen abundances of field RR Lyrae stars. II. Oxygen abundances.**
D. Butler, J. B. Laird, K. Eriksson, A. Manduca.
Astron. J., Vol. 91, No. 3, p. 570 – 574 (1986).
From non–LTE analysis of the neutral oxygen triplet (7771 – 7775 Å) absorption–line strengths on KPNO 4 m echelle plates, the authors have determined oxygen abundances and oxygen–to–iron ratios for a large number of field RR Lyrae stars covering the range in iron abundance $+0.1 \geqslant$ [Fe/H] $\geqslant -2.2$. They have found that (1) RR Lyrae stars at the metal–poor end of the range have [O/Fe] values near zero. (2) RR Lyrae stars at the metal–rich end of the range have [O/Fe] values between +0.25 and +0.95. (3) The abundance sum C+N+O has been determined. For most stars the ratio (C+N+O)/Fe is high relative to the solar value. Possible explanations for these results are discussed.

122.145 **Pseudocepheids. III. The low–mass stars.**
O. J. Eggen.
Astron. J., Vol. 91, No. 4, p. 890 – 918 (1986).
Light and color curves in four–color, Hβ, and (*RI*) photometric systems are presented for 20 low–mass pseudocepheids. They have catalog types of "W Vir" (five stars), "semiregular" (nine), "irregular" (one), "R CrB" (three), "eclipsing" (one), and "classical Cepheid" (one). All proper motions are given on the FK4 system with precessional corrections, and space–motion vectors are directed away from the galactic center, in the direction of galactic rotation and toward the North Galactic Pole, respectively.

122.146 **Time–dependent spectropolarimetry of the red variable V CVn.**
A. M. Magalhães, G. V. Coyne, E. K. Benedetti.
Astron. J., Vol. 91, No. 4, p. 919 – 924 (1986).
Systematic spectropolarimetric observations over the cycle of the semiregular red variable V CVn were obtained. The observed variations across the 4955 Å TiO band give further credence to the hypothesis of a photospheric origin for most of the optical polarization in this object and, in particular, they support the existence of an intermediate scattering layer in the stellar photosphere.

122.147 **Neue Klassifikation von SU Columbae.**
H. Geßner.
Mitt. Veränderliche Sterne, Band 10, Heft 7, p. 155 (1985).

122.148 **Beobachtungen des Mira–Sternes BV Cygni auf Platten der Sonneberger Himmelsüberwachung.**
E. Zische.
Mitt. Veränderliche Sterne, Band 10, Heft 7, p. 159 – 160 (1985).

122.149 **Discovery of an RR Lyrae star in the halo of M31.**
S. van den Bergh, C. J. Pritchet.
Bull. Am. Astron. Soc., Vol. 18, No. 1, p. 635 (1986). Abstract. – See Abstr. 010.066.

122.150 **The chemical composition of Polaris.**
R. E. Luck, H. E. Bond.
Publ. Astron. Soc. Pac., Vol. 98, No. 602, p. 442 – 447 (1986).
From high–signal–to–noise coudé Reticon spectra the authors have determined the chemical composition of the bright Population I Cepheid Polaris (α UMi). The chemical abundances in Polaris are found to be completely typical of normal F–G supergiants. The iron abundance is solar and other metals are present in essentially solar ratios. Carbon is deficient ([C/H] = –0.4) and nitrogen overabundant ([N/H] = +0.4) relative to solar abundances indicating the presence of CN–cycle products at the surface. Oxygen shows an underabundance ([O/H] = –0.2) that is also typical of F–G supergiants, but is not as easily understood.

122.151 **Photometric observations of the dwarf Cepheid XX Cygni.**
A. C. Sadun, M. E. Ressler.
Publ. Astron. Soc. Pac., Vol. 98, No. 603, p. 504 – 506 (1986).
Photometric observations were made during the summer of 1984 of the short–period dwarf Cepheid XX Cyg. The light curves of XX Cyg taken with *B* and *V* filters are presented, as well as current estimates of various parameters inferred from them. Among these parameters are an accurate determination of the period and the change in visual light output. In addition, the light curves suggest a small migrating bump that slowly changes in shape and position over many periods, and may even disappear altogether.

122.152 **S 10804 Boo – RR Lyrae star now confirmed.**
S. Rössiger.
Inf. Bull. Variable Stars, No. 2855, 2 pp. (1986).

122.153 **On the type of variability of S Vul.**
L. N. Berdnikov, G. R. Ivanov.
Inf. Bull. Variable Stars, No. 2856, 4 pp. (1986).

122.154 **LQ And (HD 224559): the 0.62–day (or the 0.31–day) period confirmed.**
P. Harmanec.
Inf. Bull. Variable Stars, No. 2859, 7 pp. (1986).

122.155 **AX Ursae Majoris.**
G. Romano.
Inf. Bull. Variable Stars, No. 2862, 3 pp. (1986).

122.156 **On the light curve of AI Scl.**
H. Schneider, W. Kollatschny.
Inf. Bull. Variable Stars, No. 2869, 2 pp. (1986).

122.157 **Photometric behavior of HR 8752 = V509 Cas. II.**
E. M. Halbedel.
Inf. Bull. Variable Stars, No. 2876, 2 pp. (1986).

122.158 **A brightening of the Be star HD 46380.**
E. M. Halbedel.
Inf. Bull. Variable Stars, No. 2877, 2 pp. (1986).

122.159 **The historical optical variation of DO Dra (= 3A 1148 + 719).**
M. L. Hazen.
Inf. Bull. Variable Stars, No. 2880, 2 pp. (1986).

122.160 **Correct period of the Mira star BV Cygni.**
E. Splittgerber.
Inf. Bull. Variable Stars, No. 2883, 1 p. (1986).

122.161 **Photoelectric observations of BY Dra in 1984 – 85.**
K. Oláh, M. Ivanova, L. Iliev.
Inf. Bull. Variable Stars, No. 2889, 3 pp. (1986).

122.162 **The light curve of KR Aurigae in the season 1985/86.**
W. Götz.
Inf. Bull. Variable Stars, No. 2896, 2 pp. (1986).

122.163 **High dispersion spectroscopy on TY Vir.**
D. Proust.
Inf. Bull. Variable Stars, No. 2900, 4 pp. (1986).

122.164 **Spectral estimation of luminosity and other characteristics of the Delta Scuti stars AI CVn (= 4 CVn) and δ Ser.**
Ts. G. Tsvetkov.
Inf. Bull. Variable Stars, No. 2902, 3 pp. (1986).

122.165 **Strömgren uvby–H_β photometry of the solar type star HD 206860.**
V. Reglero, J. Fabregat, A. De Castro.
Inf. Bull. Variable Stars, No. 2904, 4 pp. (1986).

122.166 **Photoelectric observations of R CrB during the recent light minimum.**
B. N. Ashoka, S. Pukalenthi.
Inf. Bull. Variable Stars, No. 2908, 2 pp. (1986).

122.167 **Infrared excess of cepheid binaries.**
L. Szabados.
Inf. Bull. Variable Stars, No. 2910, 3 pp. (1986).

122.168 **On the correlation between abrupt period change and period.**
M. Jerzykiewicz.
Acta Astron., Vol. 36, No. 1, p. 147 – 152 (1986).
The correlation between $\log|\Delta P/P|$ and $\log P$, where ΔP is an abrupt period change and P is the period, found by Hoffleit (1976) to be obeyed by various kinds of variable stars, is shown to have arisen as a result of observational selection.

122.169 **On the estimate of cycle length for long–cyclic U Geminorum stars.**
W. Wenzel, G. A. Richter.
Astron. Nachr., Vol. 307, No. 3, p. 209 – 212 (1986). In German.
Several methods for estimating very long mean cycle lengths of U Geminorum type variables and the influence of observational gaps are discussed. The well–observed star SW UMa serves as an example for comparing statistical findings with more direct evaluations.

122.170 **On the periodicity of UZ Comae Berenices.**
D. Svetlov.
J. Am. Assoc. Variable Star Obs., Vol. 14, No. 2, p. 55 – 58 (1985).
A study of the periodicity of UZ Comae Berenices employing data gathered over the past 75 years has revealed a constantly increasing period from 1957 to 1985. Discrimination between a changing and constant period is ambiguous from the data in the interval 1910 to 1957.

122.171 **Large period changes in V1176 Sagittarii.**
K. A. Gloria.
J. Am. Assoc. Variable Star Obs., Vol. 14, No. 2, p. 59 – 60 (1985).
The previously published period for the variable star V1176 Sagittarii, an RR Lyrae star of type RRab, has been confirmed for the years 1942 through 1966. Since 1966, V1176 Sgr has undergone at least two period changes. New periods and epochs have been determined for 1967 – 76 and 1977 – 80.

122.172 **Period adjustment of TT Scuti.**
S. E. Cederbloom.
J. Am. Assoc. Variable Star Obs., Vol. 14, No. 2, p. 61 – 62 (1985).
The RR Lyrae variable TT Scuti was studied using photographic photometry. The O–C data show a possible downwards curve, which would indicate a decreasing period, but the results are not conclusive. New linear elements are derived.

122.173 **Period test for AA Serpentis.**
N. Eisenman.
J. Am. Assoc. Variable Star Obs., Vol. 14, No. 2, p. 63 (1985).
The Cepheid variable star AA Serpentis was studied on 598 Maria Mitchell Observatory plates. The new data were checked for deviations from the previously assigned period of 17.1412 days. The period remains the same, 17.14120 days ± 0.00004, but a new epoch is offered.

122.174 **Sigma Scorpii, a bright close binary.**
D. W. Dunham.
Occultation Newsl., Vol. 3, No. 16, p. 346 – 350 (1986).

122.175 **Velocities and radii of five Magellanic Cloud cepheids.**
J. A. R. Caldwell, I. M. Coulson, J. H. Spencer Jones, C. A. Black, M. W. Feast.
Mon. Not. R. Astron. Soc., Vol. 220, No. 3, p. 671 – 678 (1986).
New radial velocities and photometry are presented for five Magellanic Cloud cepheids. Baade–Wesselink radii are determined which agree with the galactic period–radius relation.

122.176 **Flares and preflare phenomena in EV Lacertae.**
G. Sh. Rojzman, G. I. Kabichev.
Sov. Astron., Vol. 29, No. 6, p. 639 – 644 (1985). English translation of 40.122.147.

122.177 **UBV photometry of a new RR Lyrae star in Leo.**
H.–s. Tan.
Chin. Astron. Astrophys., Vol. 10, No. 2, p. 119 – 120 (1986). English translation of 40.122.145.

122.178 **Analysis of 1961 – 1982 maxima of VZ Draconis, an RRc with a variable period.**
T. Berthold, P. Ralincourt.
GEOS Circ., RR 9, 7 pp. (1985).
Figer (1982) gave a first ephemeris of VZ Dra: Max. = Hel.J.D. 2,443,534.7546 + 0.3210250 E. The observations discussed are composed of two series. The first series consists of brightness estimates made from plates exposed at Sonneberg Observatory between 1961 and 1979. The second series is a batch of over 5000 visual observations made between 1980 and 1982. Both series confirm Figer's ephemeris independently, though the O–C's are fairly important and alternately positive or negative. It follows therefore that the period of VZ Dra varies fairly rapidly.

122.179 **Determinations of maxima for V526 Aql, BD Cas and V514 Cyg; important O–C values observed for V526 Aql and BD Cas.**
J. Busquets.
GEOS Circ., Cep 1, 7 pp. (1986).
The paper gives the results of the visual observations made by GEOS on V526 Aql, BD Cas and V514 Cyg. The following maxima were determined: V526 Aql: Max = 2,445,178.43 $\pm$ 0.45, BD Cas: Max = 2,445,089.52 $\pm$ 0.19, V514 Cyg: Max = 2,444,532.36 $\pm$ 0.09. These determinations establish im-

portant O–C values for V526 Aql and BD Cas. As for V526 Aql, the O–C is so important that the number E of elapsed cycles cannot be known with certainty. The O–C found for V514 Cyg is quite small and might not be significant at all.

122.180 **A light curve for R Reticuli, a Mira variable.**
B. J. Poppleton.
Publ. Variable Star Sect., R. Astron. Soc. N.Z., No. 13 (C85), p. 47 – 57 (1986).
A light curve based on ten day means from 50 years of visual observations of R Reticuli has been plotted with the aid of a micro–computer. A table of observed maxima and minima is presented. The determined period is $278\overset{d}{.}0$ with root mean square deviation of $8\overset{d}{.}7$. A plot of O–C residuals suggests there is a second order effect operating and the period is steadily increasing.

122.181 **Light curve and period of AA Tucanae.**
A. W. Dodson.
Publ. Variable Star Sect., R. Astron. Soc. N.Z., No. 13 (C85), p. 58 – 66 (1986).
A light curve, from three decades of visual observations, is presented for the Mira variable, AA Tuc. A period from the observed maxima is given and it is shown that most minima were fainter than the threshold of most instruments used.

122.182 **Flare stars and solar bursts: high resolution in time and frequency.**
K. R. Lang.
Sol. Phys., Vol. 104, No. 1, p. 227 – 233 (1986). – See Abstr. 012.067.
Coronal loops on the sun and nearby stars are investigated using observations at 20 cm wavelength with high resolution in time and frequency. Observations of the dwarf M star AD Leonis with high time resolution using the Arecibo Observatory have resulted in the discovery of a quasi–periodic train of circularly polarized spikes with a mean periodicity of 32 ± 5 ms and a total duration of 150 ms. Observations at closely spaced wavelengths, or high frequency resolution, using the Very Large Array have revealed narrow–band structure ($\Delta\nu/\nu \leqslant 0.01$) in solar bursts and in the slowly–varying radiation of the dwarf M star YZ Canis Minoris. Observations with high resolution in time and frequency suggest coherent processes in the coronae of the sun and dwarf M stars. The scientific potential of these discoveries may be best fulfilled by the construction of a solar–stellar synthesis radiotelescope.

122.183 **Revised photometric properties for the multiperiodic variable HD 200925.**
S. B. Johnson, M. D. Joner.
Publ. Astron. Soc. Pac., Vol. 98, No. 604, p. 581 – 598 (1986).
The authors have presented photometric evidence that HD 200295 is a very interesting star that warrants further study. The light curve of this star has subtle variations from cycle to cycle indicating that this star is multiperiodic. Both photometric and spectroscopic evidence show that this star has a heavily blanketed spectrum and may be metal rich. The problem lies in its m_1 values which vary with respect to temperature change. This problem is not found in other dwarf Cepheids. This change is also opposite to the change predicted by models for the spectral range of HD 200295.

122.184 **R Coronae Borealis in 1985.**
J. D. Fernie, J. R. Percy, M. G. Richer.
Publ. Astron. Soc. Pac., Vol. 98, No. 604, p. 605 – 608 (1986).
Photoelectric photometry of R CrB during 1985 is presented and discussed briefly. The low–amplitude pulsation is well defined, and the beginnings of a deep minimum were recorded in detail. The $(U–B)/(B–V)$ behavior of the star during this early decline was consistent with similar data for the star during the emergence from an earlier deep minimum and is not much different from the law of interstellar reddening.

122.185 **R Leonis.**
IAU Circ., No. 4164 (1986).

122.186 **U Aquarii.**
IAU Circ., No. 4165 (1986).

122.187 **Novalike object in Centaurus.**
IAU Circ., No. 4180 (1986).

122.188 **AR Puppis.**
IAU Circ., No. 4188 (1986).

122.189 **R Coronae Borealis.**
IAU Circ., Nos. 4192, 4199, 4214 (1986).

122.190 **Kuwano's object in Aquila = CGSS 1175.**
IAU Circ., Nos. 4221, 4224, 4225, 4227 (1986).

122.191 **Kuwano's object in Aquila = CGSS 1175.**
Yamamoto Circ., No. 2061 (1986). In Japanese.

122.192 **Atmospheres of Delta Scuti–type pulsating variables. III. An analysis of the 28 And spectrum.**
L. S. Lyubimkov, T. M. Rachkovskaya.
Izv. Krymskoj Astrofiz. Obs., Tom 73, p. 82 – 92 (1985). In Russian. English translation in Bull. Crimean Astrophys. Obs., Vol. 73.
Using spectrograms a model atmosphere analysis of the pulsating star 28 And has been carried out. The following values of effective temperature, surface gravity and microturbulent velocity were obtained: $T_{eff} = 7900 \pm 200$K, $\log g = 4.1 \pm 0.2$ and $\xi_t = 5.5 \pm 0.5$ km/s. The chemical composition of 28 And differs from the solar one. Having compared the estimated values of T_{eff} and log g with the results of evolutionary calculations the authors found the mass, the radius, the luminosity and the age of 28 And.

122.193 **The hypergiant variable HR 8752 = HD 217476.**
E. Zsoldos.
Luminous stars and associations in galaxies, p. 87 – 88 (1986). – See Abstr. 012.079. (IAU Symp. No. 116).
The observed and derived parameters of HR 8752 are as follows: Semi–period 400 days, $\log(L/L_\odot) = 5.62$, $\log T_{eff} = 3.70$, Mass $= 30\ M/M_\odot$, $\log Q_\odot = -1.060$, $\log R/R_\odot = 2.93$. These quantities are based upon the assumption of radial pulsation. The values are in good agreement with those of Arellano Ferro (1985).

122.194 **The light variations of the red supergiant Mu Cephei.**
L. Mantegazza.
Luminous stars and associations in galaxies, p. 97 – 98 (1986). – See Abstr. 012.079. (IAU Symp. No. 116).

122.195 **Variability of Magellanic Cloud supergiants.**
G. R. Grieve, B. F. Madore.
Luminous stars and associations in galaxies, p. 99 (1986). – See Abstr. 012.079. (IAU Symp. No. 116).
Statistics on the variability of 127 high–luminosity, intermediate spectral–type supergiants in the Magellanic Clouds are presented.

122.196 **S Doradus type: Hubble–Sandage variables.**
B. Wolf.
Luminous stars and associations in galaxies, p. 151 – 156 (1986). – See Abstr. 012.079. (IAU Symp. No. 116).
The luminous S Dor variables or Hubble–Sandage variables have gained considerable interest during the past few years for their exceptional stellar wind properties. They seem to represent a short intermezzo in the evolution of the very massive stars prior to becoming WR stars. More recent results on S Dor variables (mainly of the LMC) are briefly summarized.

Mean light and colour (B–V and U–B) curves of 210 field RR Lyrae type stars.
See Abstr. 002.081.

A new Fabry–Perot radial velocity spectrometer.
See Abstr. 034.063.

Atmospheric structure and mass loss from pulsation with speculative application to B and Be stars.
See Abstr. 064.016.

Mode identification from line profile variations.
See Abstr. 065.017.

A non–local theory of time–dependent convection for numerical application to the radial pulsation of stars.
See Abstr. 065.026.

Nonradial pulsation theory of massive stars.
See Abstr. 065.033.

Amplitude instability in pulsating stars: a one–zone model.
See Abstr. 065.035.

A simple model for coupled convection/pulsation.
See Abstr. 065.056.

Ionization driving mechanisms in pulsating stars.
See Abstr. 065.057.

Pulsational mass loss and cepheid evolution.
See Abstr. 065.058.

Evolutionary period changes in variable helium–rich white dwarfs.
See Abstr. 065.063.

Asteroseismology for certain Ap stars.
See Abstr. 065.069.

The structure of the thermal modes in pulsating stars.
See Abstr. 065.080.

The effects of a 2:1 resonance in nonlinear radial stellar pulsations.
See Abstr. 065.081.

The analysis of modal coupling in a hydrodynamic model of Cepheids by the maximum entropy method.
See Abstr. 065.087.

Hydrodynamic models of RR*ab* variables: comparison of observed and theoretical light curves.
See Abstr. 065.094.

Modulation of amplitudes in oscillating stars due to resonant mode coupling.
See Abstr. 065.118.

Stellar oscillations in the presence of a magnetic field.
See Abstr. 065.120.

Instability in massive stars: an overview.
See Abstr. 065.163.

From Miras to planetary nebulae: a model of mass loss.
See Abstr. 065.175.

High sensitivity SiO maser survey for Mira variables.
See Abstr. 112.008.

The nature of Be stars: periodic variables in the cluster NGC 3766.
See Abstr. 112.012.

Stellar winds and variability in hot luminous stars.
See Abstr. 112.027.

Time variation of the H_2O and SiO masers in late–type stars.
See Abstr. 112.031.

Observation of the onset of an emission episode in the Be star Mu Centauri.
See Abstr. 112.034.

The X–ray spectrum of emission line stars (superluminous and symbiotic).
See Abstr. 112.057.

Infrared photometry and high–resolution imaging of OH/IR stars.
See Abstr. 112.060.

Mass loss from carbon stars.
See Abstr. 112.069.

IRAS Low–Resolution Spectrograph observations of silicate and molecular SiO emission in Mira variables.
See Abstr. 112.089.

Máseres SiO asociados a estrellas variables de largo período.
See Abstr. 112.094.

Near infrared photometry of stellar maser sources.
See Abstr. 112.115.

Activity in the envelope of the G–type hypergiant HD 217476.
See Abstr. 112.123.

P Cygni type stars: evolution and physical processes.
See Abstr. 112.127.

The massive stellar wind of the Hubble–Sandage variable S Doradus.
See Abstr. 112.128.

An investigation of the micro–variations of highly luminous OBA type stars. V.
See Abstr. 113.009.

***UBVRI* photometry of red stars.**
See Abstr. 113.016.

CD –30°5135.
See Abstr. 113.057.

A search for infrared excesses in G–type stars.
See Abstr. 113.063.

Investigations of the spectral variability of carbon stars.
See Abstr. 114.047.

Yellow supergiants with hot companions.
See Abstr. 114.135.

The position of ET And in the Hertzsprung–Russell diagram.
See Abstr. 115.005.

Brightest stars in galaxies as distance indicators.
See Abstr. 115.030.

Evidence for a common magnetic driver for flares and quiescent coronae/chromospheres.
See Abstr. 116.035.

Polarization measurements of β Cep stars.
See Abstr. 116.056.

X–ray, ultraviolet, optical and radio observations of cool stars.
See Abstr. 116.061.

R Aquarii, une variable unique: Mira, symbiotique, algolide et nova récurrente.
See Abstr. 117.031.

Spectral and electrophotometric observations of the peculiar star CH Cygni during 1980 – 1981.
See Abstr. 117.148.

IRAS observations of symbiotic objects.
See Abstr. 117.329.

Photometric and spectroscopic observations of the variable star ET And. II.
See Abstr. 120.002.

Spectroscopic observations of ν Centauri.
See Abstr. 120.027.

Search for flare stars with the ESO GPO astrograph in La Silla.
See Abstr. 123.005.

Detection of Cepheid variables by the Infrared Astronomical Satellite.
See Abstr. 123.007.

Spectrophotometric and fast photometric studies of stars near the PG 1159–035 instability strip.
See Abstr. 126.031.

IUE observations of pulsating DA white dwarfs and the effective temperature – period correlation of ZZ Ceti stars.
See Abstr. 126.034.

Southern ZZ Ceti stars.
See Abstr. 126.044.

On mode changes in ZZ Ceti stars.
See Abstr. 126.045.

Time–averaged ultraviolet observations of pulsating DA white dwarfs from the IUE observatory.
See Abstr. 126.072.

K1–16 revisited: mode structure and identification as an O VI central star.
See Abstr. 134.023.

Fast transients: a search in X–rays for short flares, bursts, and related phenomena.
See Abstr. 142.066.

On the distribution of the frequency of stellar flares in stellar aggregates.
See Abstr. 152.007.

A new approach to zero–age main sequence fitting.
See Abstr. 153.029.

RR Lyrae stars in globular clusters: better distances from infrared measurements?
See Abstr. 154.018.

Period changes for variables in M28.
See Abstr. 154.056.

Spectroscopy of c–type RR Lyrae stars in the globular cluster NGC 6638.
See Abstr. 154.058.

Cepheids and the spiral structure of the Galaxy.
See Abstr. 155.062.

Red variables as distance indicators and the distance to the galactic centre.
See Abstr. 155.076.

Classical cepheids and the local chemical inhomogeneities in the galactic disc.
See Abstr. 155.117.

Variables, the galactic bulge and IRAS.
See Abstr. 155.123.

The geometry and distance of the Magellanic Clouds from Cepheid variables.
See Abstr. 156.002.

Discovery of Cepheids in M101.
See Abstr. 157.054.

A search for cepheids in M101.
See Abstr. 157.119.

Two cepheid variables in the Fornax dwarf galaxy.
See Abstr. 157.120.

The distance to M31 from infrared photometry of its Cepheids.
See Abstr. 157.253.

123 Variable Stars (Surveys, Lists of Observations, Charts, etc.)

123.001 **Tableaux des observations reçues à l'AFOEV pendant les mois d'octobre, novembre et décembre 1985.**
Bull. Assoc. Fr. Obs. Etoiles Variables, No. 35, p. 10 – 43 (1986).

123.002 **Table for the conversion of the Argelander designations of variables into the ordinal numeration.**
BAV Rundbrief, 35. Jahrg., Nr. 1, p. 13 (1986).

123.003 **Tableaux des observations reçues à l'AFOEV pendant les mois de janvier, février et mars 1986.**
Bull. Assoc. Fr. Obs. Etoiles Variables, No. 36, p. 10 – 28 (1986).

123.004 **Suspected variables for southern observers.**
C. Bembrick.
Aust. J. Astron., Vol. 1, No. 3, p. 129 – 135 (1986).
The eight suspected variables detailed here are ideally suited for photoelectric photometry from the southern hemisphere. All are suspected on spectral and other evidence of being chromospherically active stars, possibly of the RS CVn type. They are all bright enough ($V > 9.0$) to be observable with modest telescopes and would constitute a useful observing programme for amateurs and others interested in photoelectric photometry.

123.005 **Search for flare stars with the ESO GPO astrograph in La Silla.**
M. K. Tsvetkov, W. C. Seitter, H. Duerbeck.
Eruptive phenomena in stars, p. 429 – 430 (1986). – See Abstr. 012.039.

123.006 **On the variability of NSV 5501.**
J. Pensado.
Bol. Astron. Obs. Madr., Vol. 11, No. 4, p. 1 – 5 (1985).
The observations of the suspected variable NSV 5501 show that if this star is variable it does not belong to the types RR Lyrae or W Ursae Majoris, as referenced in the literature.

123.007 **Detection of Cepheid variables by the Infrared Astronomical Satellite.**
C. W. McAlary, D. L. Welch.
Astron. J., Vol. 91, No. 5, p. 1209 – 1220 (1986).
Of the more than 700 Cepheid variables listed in the third edition of the General Catalogue of Variable Stars, 68 confirmed detections have been made in the Infrared Astronomical Satellite point–source catalog.

123.008 **Zur Variabilität einiger heißer Sterne im Orionnebel.**
S. Rößiger.
Mitt. Veränderliche Sterne, Band 10, Heft 7, p. 145 – 149 (1985).
Recent photoelectric UBV observations obtained by the author during the years 1981 to 1985 confirm the result of Kardopolov that the early–type stars KX, LP, NU, V359, V361 and V372 Orionis in the Orion nebula show brightness constancy within the observational errors. The same holds for the stars KS and LZ Orionis.

123.009 **Bearbeitung von 40 Veränderlichen im Feld β Aurigae.**
E. Splittgerber.
Mitt. Veränderliche Sterne, Band 10, Heft 7, p. 153 (1985).

123.010 **Bearbeitung von 22 Veränderlichen (Feld γ Aquilae, Teil III).**
H. Geßner.
Mitt. Veränderliche Sterne, Band 10, Heft 7, p. 154 (1985).

123.011 **Neuer Veränderlicher S 10901 Lacertae.**
H. Geßner.
Mitt. Veränderliche Sterne, Band 10, Heft 7, p. 154 (1985).

123.012 **Visuelle und photovisuelle Beobachtungen von drei Veränderlichen.**
D. Böhme, E. Zische.
Mitt. Veränderliche Sterne, Band 10, Heft 7, p. 161 – 163 (1985).

123.013 **Beobachtungsergebnisse des Arbeitskreises "Veränderliche Sterne" im Kulturbund der DDR (Teil XII).**
Mitt. Veränderliche Sterne, Band 10, Heft 7, p. 163 – 168 (1985).

123.014 **Photométrie photoélectrique d'étoiles variables brillantes.**
E. Schweitzer.
Bull. Inf. Cent. Données Stellaires, No. 30, p. 79 – 84 (1986).
A statistical study of variable stars brighter than m = 10 at maximum has been undertaken. It is based upon the Fourth Edition of the General Catalogue of Variable Stars (Moscow, 1985), first part, which contains the variables located in the constellations Andromeda, Antlia, etc. to Orion. For a total of 18151 stars, 2207 (12%) are brighter than m = 10 at maximum. 1240 out of this number (i.e. more than half) have been observed photoelectrically.

123.015 **Search for supernovae with the Schmidt telescope of the Radioastrophysical Observatory.**
L. A. Začs.
Nauchn. Inf., Vyp. 57, p. 101 – 105 (1984). In Russian.
Monitoring for supernovae revealed two new variable stars (No. 1 and No. 2). The equatorial coordinates and results of photographic photometry are given. The observed amplitudes of light variations in the *B* passband reached the values $1^m.7$ and $0^m.7$ for variables No. 1 and No. 2, respectively. It seems that variable No. 1 is nova–like, but No. 2 a nova–like or U Gem–type variable.

123.016 **IAU archives of unpublished photoelectric observations of variable stars (1985/6).**
M. Breger.
Inf. Bull. Variable Stars, No. 2853, 3 pp. (1986).

123.017 **A possible variable star near the gamma ray burst source GBS 0526–66.**
J. Flohrer, J. Greiner.
Inf. Bull. Variable Stars, No. 2854, 2 pp. (1986).

123.018 **Photometry of comparison stars.**
S. F. McLaughlin.
Inf. Bull. Variable Stars, No. 2857, 4 pp. (1986).

123.019 **A new blue variable star.**
M. Hamuy, J. Maza.
Inf. Bull. Variable Stars, No. 2867, 5 pp. (1986).

123.020 **Variable stars in Lynds 810.**
D. G. Turner.
Inf. Bull. Variable Stars, No. 2874, 4 pp. (1986).

123.021 **New variable stars in Cassiopeia.**
L. Dahlmark.
Inf. Bull. Variable Stars, No. 2878, 5 pp. (1986).

123.022 **Seven new variable stars in the field of 2 Lacertae.**
H. Gessner.
Inf. Bull. Variable Stars, No. 2882, 1 p. (1986).

123.023 **The variability of BD +60°562.**
E. M. Halbedel.
Inf. Bull. Variable Stars, No. 2884, 2 pp. (1986).

123.024 **Where is AK Canis Minoris?**
R. M. Robb, T. B. Lowe, J. D. Moffatt, D. D. Balam.
Inf. Bull. Variable Stars, No. 2888, 3 pp. (1986).

123.025 **Nonvariability of SVS 1740.**
P. Broglia, P. Conconi.
Inf. Bull. Variable Stars, No. 2890, 2 pp. (1986).

123.026 **Photometric variability of 27 Cygni.**
J. R. Percy, M. G. Richer, H. Božić, P. Harmanec, J. Horn, P. Koubský, S. Kříž, L. A. Kelsey, D. B. Hoff, H. J. Landis, R. C. Reisenweber, R. Wasson.
Inf. Bull. Variable Stars, No. 2891, 4 pp. (1986). With a correction in No. 2908.

123.027 **Photoelectric observations of γ Cas, X Per and BU Tau.**
D. Böhme.
Inf. Bull. Variable Stars, No. 2893, 2 pp. (1986).

123.028 **Once again: on the identity of AK Canis Minoris.**
W. Wenzel.
Inf. Bull. Variable Stars, No. 2909, 1 p. (1986).

123.029 **Tau Cassiopeiae: not a variable star.**
J. R. Percy.
J. Am. Assoc. Variable Star Obs., Vol. 14, No. 2, p. 52 – 54 (1985).

123.030 **Was Harlow Shapley right about V Cephei?**
D. Hoffleit.
J. Am. Assoc. Variable Star Obs., Vol. 14, No. 2, p. 64 – 66 (1985).
V Cephei needs extensive UBV data before it can be decided whether it is really a cst or an Ia type of star.

123.031 **Photoelectric observations of seventeen variable stars reported by Archer.**
H.–s. Tan, Z.–s. Zhang, Y.–l. Zhang.
Chin. Astron. Astrophys., Vol. 10, No. 2, p. 143 – 147 (1986). English translation of Acta Astrophys. Sin., Vol. 6, No. 1, p. 12 – 19 (1986).
Seventeen of the 24 variables in Coma, announced by Archer (1959), were observed in Feb., and Mar. 1984. No variation larger than $0^{m}.07$ in V magnitude was found for any of these stars, and the authors conclude that No. 3, 4, 7, 14, 16, 17, 18, 19, 21, 22, and 23 are not variables.

123.032 **Identifications for some neglected southern variables.**
M. Morel, R. H. McNaught.
Publ. Variable Star Sect., R. Astron. Soc. N.Z., No. 13 (C85), p. 68 – 74 (1986).
Improved identification charts are presented for 27 southern variables, mainly of the Mira type. The results of research into the plate records is also presented in the form of notes.

123.033 **Photoelectric observations of variable & suspected variable stars.**
L. J. Williamson.
Publ. Variable Star Sect., R. Astron. Soc. N.Z., No. 13 (C85), p. 75 – 79 (1986).
Presented are photoelectric observations of a number of variables and suspected variables made with the 76 cm Reynolds reflector at the Mount Stromlo Observatory.

123.034 **Photoelectric sequences in variable star fields.**
L. J. Williamson.
Publ. Variable Star Sect., R. Astron. Soc. N.Z., No. 13 (C85), p. 82 – 84 (1986).
Photoelectric magnitudes are listed for some of the sequence stars in ten southern variable star fields.

123.035 **Suspected photovisual variables in the field of NGC 2324.**
A. Aiad.
Bull. Fac. Sci., Cairo Univ., Vol. 54, p. 745 – 750 (1986).
The comparison of two visual photographic plates taken with the Alger astrograph for the field of NGC 2324 indicated 29 suspected variable stars.

123.036 **Very blue variable in Orion.**
IAU Circ., Nos. 4172, 4190 (1986).

L'Association Française des Observateurs d'Etoiles Variables.
See Abstr. 010.102.

Photometric variability of some early–type stars.
See Abstr. 113.004.

112th list of minima of eclipsing binaries.
See Abstr. 119.081.

Revised photometric properties for the multiperiodic variable HD 200925.
See Abstr. 122.183.

124 Novae

124.001 **The outburst of RS Ophiuchi in 1985.**
M. F. Bode, F. D. Kahn.
Cosmical gas dynamics, p. 111 – 117 (1985). – See Abstr. 012.007.
The most recent outburst of the recurrent nova RS Ophiuchi took place on January 26. The subsequent development of the nova remnant was followed at a variety of wavelengths, from radio to X–ray; previously such observations had been confined to optical wavelengths. The new data help to describe a model for the remnant, having many analogies with supernova remnants.

124.002 **Search for novae towards the galactic center.**
R. Lukas.
BAV Rundbrief, 35. Jahrg., Nr. 1, p. 8 – 12 (1986).

124.003 **Identification of ancient novae.**
K. P. Hertzog.
Observatory, Vol. 106, No. 1071, p. 38 – 40 (1986).

124.004 **Gamma ray lines from nova envelopes.**
M. D. Leising.
Bull. Am. Astron. Soc., Vol. 17, No. 4, p. 867 (1985). Abstract. – See Abstr. 010.062.

124.005 **The unusual radio source produced by the 1985 outburst of the recurrent nova RS Ophiuchi.**
R. M. Hjellming, J. van Gorkom, R. J. Davis, M. F. Bode.
Bull. Am. Astron. Soc., Vol. 17, No. 4, p. 886 – 887 (1985). Abstract. – See Abstr. 010.062.

124.006 **Les deux novæ de 1984 dans le Petit Renard.**
E. Schweitzer.
Astronomie, Vol. 100, p. 247 – 255 (1986).

124.007 **Spectroscopic orbits for symbiotic stars. I. The recurrent nova T Coronae Borealis.**
S. J. Kenyon, M. R. Garcia.
Astron. J., Vol. 91, No. 1, p. 125 – 131 (1986).
The authors report new radial–velocity measurements of the giant component in the recurrent nova T Coronae Borealis. These data have been combined with existing observations to derive a new orbital solution and confirm previous estimates for the component masses. The orbital data and recent optical and ultraviolet spectra provide new support for a model in which a lobe–filling gM3 star loses material to a main–sequence companion star at a rate of $\sim 10^{-6} M_{\odot} yr^{-1}$. T CrB appears to be at the onset of runaway case C mass exchange, and the authors expect an increase in the luminosity of the hot object over the next few decades.

124.008 **On the nature of the recurrent nova T Pyxidis.**
Z. Urban.
Eruptive phenomena in stars, p. 359 – 360 (1986). – See Abstr. 012.039.

124.009 **Do novae hibernate during most of the millenia between eruptions? Links between dwarf and classical novae, and implications for the space densities and evolution of cataclysmic binaries.**
M. M. Shara, M. Livio, A. F. J. Moffat, M. Orio.
Space Telesc. Sci. Inst., Prepr. Ser., No. 112, 36 pp. (1986). To appear in Astrophys. J.

124.010 **Rapid accretion and hibernation in the pre–outburst history of classical novae.**
D. Prialnik, M. M. Shara.
Space Telesc. Sci. Inst., Prepr. Ser., No. 113, 43 pp. (1986). To appear in Astrophys. J.

124.011 **The differences between RS Oph and classical novae.**
M. Friedjung.
Inst. Astrophys. Paris, Pré–Publ., No. 128, 7 pp. (1986). To appear in "RS Oph (1985) and the recurrent nova phenomenon".

124.012 **The spectrum of RS Ophiuchi 2–1/2 years prior to its recent outburst.**
V. V. Smith, G. Wallerstein.
Inf. Bull. Variable Stars, No. 2879, 1 p. (1986).

124.013 **Radio observations of the 1985 outburst of RS Ophiuchi.**
R. M. Hjellming, J. H. van Gorkom, A. R. Taylor, E. R. Seaquist, S. Padin, R. J. Davis, M. F. Bode.
Astrophys. J., Lett. Ed., Vol. 305, No. 2, p. L71 – L75 (1986).
The authors report observations of the first radio source observed in association with a recurrent nova event, the 1985 January 26 outburst of RS Ophiuchi. They describe VLA measurements of the rise, peak and decline of a two–component, four–frequency radio spectrum. At 15 GHz the source could be resolved and the angular size and the rate of expansion are determined. A distance of 1.6 kpc is derived from the H I 21 cm absorption line.

124.014 **The interaction of nova shells with the interstellar medium.**
H. Duerbeck.
Mitt. Astron. Ges., No. 65, p. 207 (1986). – See Abstr. 012.072.

124.015 **The recurrent nova T Pyxidis.**
W. C. Seitter.
Mitt. Astron. Ges., No. 65, p. 209 – 210 (1986). – See Abstr. 012.072.

124.016 **Novae.**
IAU Circ., No. 4164 (1986).
Concerning: nova Vulpeculae 1984 No. 1 = PW Vulpeculae and nova Vulpeculae 1984 No. 2.

124.017 **Novae.**
Yamamoto Circ., No. 2053 (1986). In Japanese.
Concerning: nova Vulpeculae 1984 No. 1 = PW Vul and nova Vulpeculae 1984 No. 2.

124.018 **Novae detected in the IRAS Point Source Catalog.**
H. L. Dinerstein, E. L. Robinson.
Light on dark matter, p. 145 – 148 (1986). – See Abstr. 012.080.
Five classical novae are found to have counterparts in the IRAS Point Source Catalog. The IRAS measurements confirm the trend that slow novae have larger amounts of dust than fast novae and support the picture that dust grains condense in expanding nova shells.

124.019 **IRAS observations of classical novae.**
C. M. Callus, J. S. Albinson, A. Evans, M. F. Bode.
Light on dark matter, p. 149 – 150 (1986). – See Abstr. 012.080.
IRAS observations of classical novae are presented, with additional ground based IR photometry. Infrared emission from dust appears to arise from a wide range of conditions in the circumnova medium.

124.020 **Collisional heating of dust in the 1985 outburst of RS Ophiuchi.**
J. S. Albinson, C. M. Callus, A. Evans.
Light on dark matter, p. 151 – 152 (1986). Abstract.– See Abstr. 012.080.

Nova Cygni 1975 = V1500 Cygni

124.101 **Spectroscopic observations of Nova Cygni 1975: the coronal line region revisited.**
G. J. Ferland, D. L. Lambert, J. H. Woodman.
Astrophys. J., Suppl. Ser., Vol. 60, No. 1, p. 375 – 392 (1986).
A synopsis of the McDonald observations of V1500 Cyg is presented. It is shown that (1) the remnant maintained a luminosity at or above the Eddington limit for at least a year, (2) free–free emission from the coronal line region made a significant contribution to the optical continuum, (3) the coronal line region was a significant source of ionizing radiation, and (4) the energetics are dominated by the lift–off energy from the white dwarf and radiation from the coronal line region.

Nova Serpentis 1983

124.121 **Nova Serpentis 1983.**
E. M. Schlegel, R. K. Honeycutt, R. H. Kaitchuck.
Publ. Astron. Soc. Pac., Vol. 97, No. 597, p. 1075 – 1077 (1985).
The authors present a spectrum of Nova Ser 1983 in outburst, and a light curve summarizing the reported magnitude estimates of this nova.

Nova Aquilae 1982

124.141 **Spectroscopic and photometric observations of Nova Aquilae 1982.**
A. Okazaki, A. Yamasaki.
Astrophys. Space Sci., Vol. 119, No. 1, p. 89 – 92 (1986). – See Abstr. 012.001.
Nova Aquilae 1982 was observed spectroscopically and photometrically in January, February, and April 1982. The results of these observations, including those obtained a few days after the discovery, are presented.

Nova Persei 1901 = GK Persei

124.161 **The orbit of GK Persei.**
D. Crampton, A. P. Cowley, W. A. Fisher.
Astrophys. J., Vol. 300, No. 2, p. 788 – 793 (1986).

New spectroscopic observations of the old nova GK Per reveal the orbital period to be 1.996803 ± 0.000007 days. The orbit is circular and the late-type component has a semiamplitude of 124 ± 2 km s^{-1}. The hydrogen emission lines show a velocity variation in antiphase with the absorption lines with a semiamplitude of 34 ± 5 km s^{-1}. The most probable masses are 0.25 $M_\odot$ for the K star and 0.9 $M_\odot$ for the white dwarf. The K star has obviously lost a lot of mass and is continuing to do so.

124.162 **Phenomenology of the optical outbursts of the classical old-nova GK Persei (1901).**
A. Bianchini, F. Sabbadin, G. C. Favero, I. Dalmeri.
Astron. Astrophys., Vol. 160, No. 2, p. 367 – 373 (1986).

The phenomenology of the outburst activity of the old-nova GK Per is reviewed and discussed in the light of current theoretical models of dwarf nova explosions. The presence of a high energy precursor indicates that the outbursts of GK Per start at the inner edge of the accretion disc where an unstable region can be formed if $\dot{M} \cong 10^{16}$g s^{-1}, as suggested in a previous paper. The observed long recurrence times ($\sim n \cdot 400$ days) can be explained by assuming that such an unstable region is occasionally formed by a modulation of the mass transfer rate from the secondary. An analogy between GK Per, Population II X-ray binaries and the superoutburst activity of SU UMa stars is also suggested.

124.163 **The accretion disk limit cycle mechanism in GK Per.**
J. K. Cannizzo.
Bull. Am. Astron. Soc., Vol. 17, No. 4, p. 887 (1985). Abstract. – See Abstr. 010.062.

Nova Vulpeculae 1968 No. 1 = LV Vulpeculae

124.181 **The spectral evolution of nova Vulpeculae No. 1, 1968 (LV Vul) after maximum brightness.**
Y. Andrillat, L. I. Antipova, M. B. Babaev.
Astron. Zh., Tom 63, Vyp. 1, p. 128 – 136 (1986). In Russian. English translation in Sov. Astron., Vol. 30, No. 1.

An analysis of the spectral evolution of LV Vulpeculae 1968 after the maximum brightness was carried out. The principal conclusions are: 1) The pre-maximum spectrum was replaced by the principal one in less than 24^h. 2) The diffuse enhanced spectrum and the Orion one existed already when the nova brightness has decreased only by $0^m.4$ and $0^m.5$, respectively. 3) The radial velocities of the Orion spectrum coincided with those of the diffuse enhanced one during the total observational period. 4) The Orion spectrum consists of the lines of He I, N II, O II and may be H I. 5) The appearance of two additional components is probably due to splitting of the principal and diffuse enhanced spectrum.

124.182 **The evolution of the Nova LV Vulpeculae spectrum.**
D. Raikova, V. Dobrichev.
Astrofiz. issled., Tom 4, p. 33 – 39 (1985). Abstr. in Ref. Zh., 51. Astron., 6.51.615 (1986).

Nova Serpentis 1970 = FH Serpentis

124.201 **Researches on novae and related objects. I. The spectral evolution of Nova FH Ser (1970).**
L. Rosino, F. Ciatti, M. Della Valle.
Astron. Astrophys., Vol. 158, No. 1/2, p. 34 – 44 (1986).

Beginning a series on novae and related objects, the authors give a short account of the spectral evolution of Nova FH Ser (1970) based on observations carried out regularly at Asiago from the first appearance of the nova to its minimum attained near 1978. The optical light curve of the nova with a deep minimum in the transition phase, is akin to that of Nova XX Tau (1927).

Nova Pictoris 1925 = RR Pictoris

124.221 **Light curves of RR Pic (Nova Pictoris 1925) 1972 – 84.**
B. Warner.
Mon. Not. R. Astron. Soc., Vol. 219, No. 4, p. 751 – 758 (1986).

Light curves of RR Pic obtained over the period 1972 – 84 show that whereas in the 1970s a strong orbital modulation of brightness was present this has been replaced in the 1980s by an irregular, shallow eclipse superimposed on a flickering background. The disappearance of the orbital modulation coincided with decline in mean brightness of the system. The cause of these structural changes, taking place 50 yr after the nova eruption, cannot be determined until a better model for the mass transfer and accretion disc structure is developed.

Nova Vulpeculae 1979 = PU Vulpeculae

124.241 **The study of the Kuwano-Honda peculiar object (PU Vul).**
T. S. Belyakina, N. I. Bondar', R. E. Gershberg, Yu. S. Efimov, V. I. Krasnobabtsev, P. P. Petrov, V. Piirola, I. S. Savanov, K. K. Chuvaev, N. I. Shakhovskaya, N. M. Shakhovskoj, V. I. Shenavrin.
Izv. Krymskoj Astrofiz. Obs., Tom 72, p. 3 – 72 (1985). In Russian. English translation in Bull. Crimean Astrophys. Obs., Vol. 72.

Results of photometric, spectrophotometric, spectral and polarimetric observations of the Kuwano-Honda peculiar object (PU Vul) that have been carried out during the phases of an ascent from the deep minimum in 1980 – 81 and of the 4 following brightness maxima till the end of 1982 are given.

124.242 **A dynamical model for the dusty envelope around the symbiotic nova PU Vulpeculae.**
A. B. Men'shchikov, A. V. Tutukov, B. M. Shustov, E. V. Ergma.
Sov. Astron. Lett., Vol. 11, No. 4, p. 221 – 224 (1985). English translation of 40.124.241.

124.243 **Kuwano-Honda's peculiar object (PU Vul) and some problems of extremely slow novae.**
R. E. Gershberg.
Eruptive phenomena in stars, p. 351 – 354 (1986). – See Abstr. 012.039.

124.244 **Peculiar slow nova-like object PU Vul – facts and interpretation.**
C. Chochol, J. Grygar.
Eruptive phenomena in stars, p. 355 – 356 (1986). – See Abstr. 012.039.

124.245 **Spectroscopic observations of PU Vulpeculae.**
S. J. Kenyon.
Astron. J., Vol. 91, No. 3, p. 563 – 569 (1986).

Optical and infrared spectrophotometric observations demonstrate that the nova-like variable PU Vulpeculae is a symbiotic binary composed of an M4-5 giant and a hot companion whose expanding atmosphere mimics an A supergiant. Fits to the optical/infrared data suggest the system is reddened by $E_{B-V} = 0.4 - 0.5$. The 1979 outburst appears to be the result of a thermonuclear runaway on the surface of a white dwarf. Dust condensation in the expanding envelope ejected in the eruption, or an eclipse of the hot component by the M giant companion, can explain many features of the 1980 minimum, but neither mechanism provides a satisfactory explanation for all the available data.

Nova Vulpeculae 1983

124.261 **Nova Vulpeculae 1983.**
A. Sh. Khatisashvili, R. Ya. Inasaridze.
Astron. Tsirk., No. 1397, p. 3 – 4 (1985). In Russian.

Nova Cygni 1978 = V1668 Cygni

124.281 **Extended light curves for Nova Cygni 1978.**
J. B. Kaler.
Publ. Astron. Soc. Pac., Vol. 98, No. 600, p. 243 – 245 (1986).

Differential magnitudes of Nova Cygni 1978 observed with a wide Hβ (*w*) filter are presented over a period of 829 days following the event, along with terminal values of Strömgren *y*. The *y* (continuum) and *w* (Hβ and [O III] emission + continuum) light curves rapidly separate, with *w* over four magnitudes brighter than *y* by day 311, by which time the star appears to have returned to minimum. The decay in the nebular radiation in magnitude vs. time is linear from at least day 432 onward, with an exponential decay time of 220 days. The data provide a constraint on theoretical models of the development of the star with its expanding nebula.

Nova Vulpeculae 1984 No. 1 = PW Vulpeculae

124.301 **The light curve of nova Vulpeculae 1984a.**
R. I. Noskova, G. V. Zajtseva, E. A. Kolotilov.
Sov. Astron. Lett., Vol. 11, No. 4, p. 257 – 258 (1985). English translation of 40.124.123.

124.302 **Beobachtungen der Nova Vulpeculae 1984/1 durch den AKV.**
Mitt. Veränderliche Sterne, Band 10, Heft 7, p. 156 – 159 (1985).

Nova Vulpeculae 1984 No. 2

124.311 **Nova Vulpeculae 1984 No. 2.**
IAU Circ., Nos. 4184, 4221, 4227 (1986).

Nova Scorpii 1985

124.321 **Nova Scorpii 1985.**
R. Lukas.
Inf. Bull. Variable Stars, No. 2852, 2 pp. (1986).

124.322 **Spectroscopy of Nova Scorpii 1985.**
T. Richtler, W. Liller.
Inf. Bull. Variable Stars, No. 2871, 2 pp. (1986).

124.323 **Nova Scorpii 1985.**
IAU Circ., No. 4184 (1986).

An atlas of nova shells.
See Abstr. 002.089.

Effects of a supercritical wind on nova observations.
See Abstr. 064.054.

Explosive hydrogen burning in novae.
See Abstr. 065.040.

On the evolutionary state of cataclysmic variable white dwarfs and their recurrence times.
See Abstr. 065.075.

Hydrodynamic models for novae with ejecta rich in oxygen, neon, and magnesium.
See Abstr. 065.077.

Nucleosynthesis accompanying classical nova outbursts.
See Abstr. 065.128.

Reaction rates in the *rp*–process and nucleosynthesis in novae.
See Abstr. 065.129.

R Aquarii, une variable unique: Mira, symbiotique, algolide et nova récurrente.
See Abstr. 117.031.

The most reliable black hole yet.
See Abstr. 117.105.

BV Centauri: dwarf or classical nova?
See Abstr. 117.154.

A search for ultraviolet–excess objects.
See Abstr. 142.036.

First optical candidate for a recovered classical nova in a globuar cluster: Nova 1938 in M14.
See Abstr. 154.042.

Novae as distance indicators.
See Abstr. 161.057.

125 Supernovae, Supernova Remnants

125.001 **The supernova remnant W50 at 5 GHz.**
A. J. B. Downes, T. Pauls, C. J. Salter.
Mon. Not. R. Astron. Soc., Vol. 218, No. 3, p. 393 – 407 (1986).
Observations of the supernova remnant W50, which contains SS 433, have been made at 5 GHz with the Effelsberg 100–m telescope. These are combined with previous results at 1.7 and 2.7 GHz to investigate the intrinsic linear polarization and the variation of spectral index across the source. The spectrum of W50 is typical for old shell remnants and shows no evidence for the injection of flat–spectrum fast particles from SS 433. The northern boundary of the remnant is bright and has a flat spectrum ($\alpha \approx 0.4$). The polarization there is strong (up to 50 per cent of the total intensity). The magnetic field direction, where it can be deduced, is tangential to the shell. The bright northern edge appears to be a typical example of compression of a large–scale interstellar magnetic field. The shell of W50 has bright regions which coincide with the position angles of the precession cone of the jets from SS 433, seen in projection.

125.002 **Observations of 3C 58 at 151 and 2695 MHz.**
D. A. Green.
Mon. Not. R. Astron. Soc., Vol. 218, No. 3, p. 533 – 540 (1986).
New radio observations of 3C 58 at 151 and 2695 MHz are discussed with reference to current theories of 'filled–centre' remnants, in particular whether or not shells of fast ejecta surround them. The low–resolution 151–MHz observations and the Einstein IPC image set a low upper limit on the luminosity of any shell around 3C 58 – the X–ray limit is only one–hundredth of the luminosity of the shell of the SN of AD 1006, the faintest 'historical' shell remnant known. This result, together with the well–defined boundary of the remnant revealed by the high–resolution 2.7–GHz observations, suggests that the parent supernova of 3C 58 (that of AD 1181) produced no fast shell of ejecta. These results, the spectrum of 3C 58 and the nature of any compact source within the remnant are discussed.

125.003 **Observations of post–shock neutral hydrogen in four evolved supernova remnants.**
R. Braun, R. G. Strom.
Astron. Astrophys., Suppl. Ser., Vol. 63, No. 2, p. 345 – 401 (1986).
Observations in the λ21 cm line of neutral hydrogen of four evolved supernova remnants: IC 443, VRO 42.05.01, CTB 109 and G 78.2+2.1 are presented and briefly discussed. Many H I features display a remarkably high degree of morphological correspondence with radio continuum and optical filaments, proving that the neutral clouds observed are intimately connected to the remnants. The authors find evidence for accelerated, recombined gas associated with each object. In some cases a large fraction of the gas has received only modest acceleration implying high pre–shock densities. The morphologies of neutral filaments and clouds are classified and discussed in terms of an imbedded magnetic field model. Systematic structural changes with velocity are observed for some features which seem to imply a radially decreasing density profile within the preshock clouds.

125.004 **X–ray detection of the Monoceros supernova remnant.**
D. A. Leahy, S. Naranan, K. P. Singh.
Astrophys. Space Sci., Vol. 119, No. 1, p. 249 – 252 (1986). – See Abstr. 012.001.
The Monoceros nebula is seen in the optical and in radio as a 3°.5 degree diameter ring. It is believed to be an old supernova remnant. Here is reported the detection of X–rays from the Monoceros nebula, confirming its supernova remnant nature. Einstein imaging proportional counter observations of 0.2 to 5 keV X–rays were analyzed to produce a surface brightness map. Preliminary modeling of the Monoceros supernova remnant yields an age of 50000 years. A large age is expected for such a large remnant. However, the remnant is found to be still in the adiabatic blast wave stage of evolution.

125.005 **The Cygnus Loop: a detailed comparison of X–ray and optical emission.**
J. J. Hester, D. P. Cox.
Astrophys. J., Vol. 300, No. 2, p. 675 – 697 (1986).
A comparison is presented between X–ray and optical emission for a field on the southeast edge of the Cygnus Loop supernova remnant (SNR) using Einstein high–resolution imager and narrow–band image–tube data. The two brightest regions of X–ray emission in the field are associated with optical emission. The edge of the brightest X–ray feature is coincident with the leading edge of the optical emission over a length of several minutes of arc. Several possible explanations for the presence of bright X–ray emission in the vicinity of optical emission are presented. These mechanisms involve evaporation, gradual variations and gradients in the density of the preshock intercloud medium, and additional compression of material which has already been heated to X–ray temperatures by the adiabatic blast wave.

125.006 **The X–ray spectrum of SN 1006.**
A. J. S. Hamilton, C. L. Sarazin, A. E. Szymkowiak.
Astrophys. J., Vol. 300, No. 2, p. 698 – 712 (1986).
The X–ray spectrum of SN 1006 is reanalyzed on the assumption that the origin of the spectrum is thermal. It is argued that the power–law continuum and absence of line emission are consistent with the spectrum of a reverse shock into SN ejecta with a roughly uniform density, and consisting of heavy elements stratified in layers. Detailed nonequilibrium model spectra of a blast wave plus reverse shock are computed and fitted to published X–ray spectral data. A satisfactory fit to the spectrum is achieved with a three–layer model of ejecta consisting of an outer layer of carbon, a middle layer of processed heavy elements, and an inner layer of iron. If the outer layer of ejecta is mainly carbon, then the results suggest a total ejecta mass of 1.4 $M_\odot$. The model predicts strong line emission from carbon Lyα.

125.007 **The X–ray spectrum of Tycho.**
A. J. S. Hamilton, C. L. Sarazin, A. E. Szymkowiak.
Astrophys. J., Vol. 300, No. 2, p. 713 – 721 (1986).
The X–ray spectrum of Tycho's supernova remnant is reanalyzed in the context of a self–consistent model in which the emission arises both from a blast wave into a uniform medium, and from a reverse shock into uniform–density supernova ejecta consisting of pure heavy elements stratified in layers. Detailed nonequilibrium model spectra are computed and fitted to published X–ray spectral data. The results are consistent with a total ejecta mass of 1.4 $M_\odot$. A satisfactory fit to the spectrum is achieved with a three–layer model of ejecta consisting of an outer layer of carbon or carbon–oxygen, a mixed middle layer containing O, Mg, Al, Si, S, Ar, Ca, Fe, and Ni, and an inner layer of Fe–Ni. The marked difference between the X–ray spectra of Tycho and SN 1006 is attributed mainly to a higher density of the ambient medium around Tycho, which causes the gas in Tycho to be more ionized than in SN 1006.

125.008 **On the relative frequencies of the kinds of Type I supernovae.**
D. Branch.
Astrophys. J., Lett. Ed., Vol. 300, No. 2, p. L51 – L54 (1986).
Examination of the optical spectra of 58 Type I supernovae has revealed that 31 had spectra of the "ordinary" or "classical" variety (Type Ia). No further members of the subtype exemplified by SN 1983n in M83 and SN 1984*l* in NGC 991 (Type Ib) have been identified. Five Type I supernovae (Type I pec) showed individual peculiarities, and the subtypes of the remaining 18 cannot be established from the spectra available. The observed fractions of Type I supernovae of subtypes, Ia, Ib, and I pec are 0.78 ± 0.14, 0.10 ± 0.05, and 0.12 ± 0.06, respectively, and the mean absolute magnitudes are -18.2 ± 0.2, -16.6 ± 0.3, and -16.7 ± 0.1.

125.009 **X–ray spectral observations of the Crab–like supernova remnants G21.5–0.9 and 3C 58.**
J. Davelaar, A. Smith, R. H. Becker.
Astrophys. J., Lett. Ed., Vol. 300, No. 2, p. L59 – L62 (1986).
Broad–band (2 – 10 keV) spectral data of the Crab–like supernova remnants 3C 58 and G21.5–0.9 have been obtained from the medium–energy instrument on EXOSAT. Both objects show a simple power–law spectrum with a spectral index of 0.72 ± 0.12 for G21.5–0.9 and 1.30 ± 0.26 for 3C 58. VLA 21 cm absorption measurements for G21.5–0.9 indicate a distance of 5.5 – 9.3 kpc.

125.010 **G291.0–0.1, a centrally bright supernova remnant, in Carina.**
R. S. Roger, D. K. Milne, J. L. Caswell, A. G. Little.
Mon. Not. R. Astron. Soc., Vol. 219, No. 4, p. 815 – 822 (1986).
The authors describe new high–resolution observations at 843 MHz, 5 and 8.4 GHz of the galactic supernova remnant G291.0–0.1. The emission from the remnant is dominated by an elongated centrally bright component surrounded by weaker extended emission and subsidiary peaks. Polarization of the main component implies an orderly magnetic field aligned along its major axis. The radio emission has a relatively flat spectrum ($S \propto \nu^{-0.29}$), in common with other filled–centre remnants. The peak of the X–ray emission lies on the major axis of the elongated radio component.

125.011 **Supernova remnants.**
T. W. Hartquist.
Cosmical gas dynamics, p. 77 – 92 (1985). – See Abstr. 012.007.
Observations of the thermal emission and the processes establishing the ionization structure and producing the emission in remnants are reviewed briefly as are hydrodynamic studies, made prior to April, 1985, of remnant evolution; the need for solutions of the coupled time dependent ionization and hydrodynamic equations for studies of radiative remnants is stressed. The nature of the event that produced the hot gas around the Sun is discussed. The question of whether the event that most recently heated the local gas could have produced the Al^{26} seen in gamma–ray emission and the relevance of that emission for the hypothesis that solar system formation was triggered by a supernova are addressed. First order Fermi acceleration of cosmic rays in shocks and remnants is discussed. Evidence for second order Fermi acceleration in an old remnant is also discussed.

125.012 **Simplified models for supernova remnant evolution using energetic particle hydrodynamics.**
H. J. Völk, L. O'C. Drury, E. Dorfi.
Cosmical gas dynamics, p. 101 – 109 (1985). – See Abstr. 012.007.
The role of cosmic rays as a high energy component of the interstellar medium is discussed with regard to the evolution of supernova remnants. Emphasis is given to the basic characteristics of the interaction between thermal gas and the energetic particles. The hydrodynamic approximation for the dynamical equations is applied in a simple 2–shell model which allows the discussion of diffusive particle acceleration at the outer shock, of cloud evaporation in the interior, external pressure effects, as well as radiative cooling. It is argued that particle acceleration goes on throughout all of the different phases of a supernova remnant until it dissolves.

125.013 **The evaporation of pre–existing grains around supernovae.**
G. Pearce, A. J. Mayes.
Astron. Astrophys., Vol. 155, No. 2, p. 291 – 296 (1986).
The cavity formed by an erupting supernova due to the evaporation of grains existing around a progenitor is investigated. A variety of grain types are considered for both type I and type II supernovae. The variation with time of the cavity radius due to grain evaporation, and forces acting on the grains are investigated.

125.014 **Secular changes within Cassiopeia A at 5 GHz – I.**
R. J. Tuffs.
Mon. Not. R. Astron. Soc., Vol. 219, No. 1, p. 13 – 38 (1986).
A detailed comparison of radio maps of Cassiopeia A at 5 GHz, obtained with the Cambridge 5–km radio telescope in 1974 and 1978, is described. Accurate proper motions and brightness changes are presented for 342 distinct radio peaks, ranging in angular size from 1 arcmin for fragments of the intense ring of emission to less than the beam size of 2.0 arcsec × 2.3 arcsec for compact structure.

125.015 **Detection of X–ray emission from the supernova remnant PKS 0646+06.**
D. A. Leahy.
Astron. Astrophys., Vol. 156, No. 1/2, p. 191 – 193 (1986).
Einstein IPC imaging observations east of the Monoceros supernova remnant reveal a region of enhanced X–ray emission. This is identified with the supernova remnant PKS 0646+06. A remnant distance of ~3 – 11 kpc and age of ~60000 yr are derived from the observed flux and temperature using a Sedov model, with the lower distance preferred on the basis of optical data. The derived low density of $\sim 0.04\ cm^{-3}$ implies the remnant is still in the adiabatic blast wave phase.

125.016 **Neutral hydrogen towards 3C 10, the remnant of Tycho's supernova.**
J. S. Albinson, R. J. Tuffs, E. Swinbank, S. F. Gull.
Mon. Not. R. Astron. Soc., Vol. 219, No. 2, p. 427 – 439 (1986).
Aperture synthesis observations of neutral hydrogen towards the remnant of Tycho's supernova (AD 1572), 3C 10, have been made with the Cambridge Half–Mile Telescope. Details of the absorption features indicate that the distance to 3C 10 is in the range 1.7 – 3.7 kpc. The neutral hydrogen in a cloud in the local arm has a spin temperature of ~80K and an optical depth of order unity. Numerous arc and filamentary structures are seen in emission; some have been analysed in terms of expanding shells. Stellar winds are proposed as energy sources for two of the arcs because three open clusters are near the centre of curvature of one, and a bright star near the centre of the other. An open cluster, three 9th mag stars and two CO clouds are seen to coincide with various minima in the H I emission. Two other CO clouds coincide with bright H I emission features, as does the small H II region S175.

125.017 **The energy–diameter relation of supernova remnants in the Large Magellanic Cloud.**
A. Preite–Martinez, R. Fusco–Femiano.
Astron. Astrophys., Vol. 157, No. 1, p. 6 – 10 (1986).
An apparent relation between the initial explosion energy E_0 and observed diameter D has been found for the supernova remnants in the Large Magellanic Cloud, as well as in other galaxies. Once the size of the remnant and the electron density in its cooling filaments are known, the determination of the initial explosion energy depends on a number of assumptions that are discussed in this paper. The authors also present the E_0–D relation expected from a population of adiabatic remnants, expanding in media of different densities and with a spread in initial energies derived from optical and X–ray data. It is shown that such a population of adiabatic remnants can account for the observed E_0–D relation, as for the Σ–D and N–D relations discussed in a previous paper.

125.018 **Spectroscopy of supernova–remnants in the 0.6 – 1.05 μ spectral range.**
M. Dennefeld.
Astron. Astrophys., Vol. 157, No. 2, p. 267 – 277 (1986).
Spectra of two galactic and three LMC SNR are presented in the 0.6 – 1.05 μ range. They reveal interesting lines of carbon, sulfur, iron, nickel, calcium, chlorine and perhaps chromium. The reddening determined from the red [S II] lines is consistent with the one derived from the Balmer decrement. In the absence of specific model predictions, empirical comparisons of line ratios are used to derive abundances. The value derived for carbon is consistent with previous determinations showing an underabundance of a factor of about 5 in LMC with respect to the

Galaxy. The detection of the IR calcium triplet implies a minimum optical thickness in the emitting medium. The required column–density implies a calcium abundance only slightly below the cosmic value, therefore showing that grain destruction must have occurred. Finally, a straight application of simple atomic physics reveals a Ni/Fe ratio about 10 times above cosmic. A critical discussion of this ratio is made.

125.019 **The condensation of grains in the ejecta of supernovae.**
G. Pearce.
Astron. Astrophys., Vol. 157, No. 2, p. 335 – 337 (1986).
The condensation of grains around supernovae is considered. The mechanism by which grains could condense from supernovae ejecta to form a circumstellar dust shell is investigated. It is discovered that the depletion factor of grain forming material from the surrounding gas is important.

125.020 **A theoretical interpretation of the Σ–D relation for supernova remnants.**
N. Duric, E. R. Seaquist.
Astrophys. J., Vol. 301, No. 1, p. 308 – 311 (1986).
A theoretical interpretation of the Σ–D relation is presented for supernova remnants (SNRs) in the adiabatic expansion phase. By incorporating the Sedov blast wave solution for SNR expansion, the generation and evolution of the magnetic field as outlined by Gull, and the acceleration of relativistic electrons by shocks, as formulated by Bell, a model is synthesized for the evolution of the radio emission of an SNR. A theoretical Σ–D relation is constructed and shown to agree with the observed relation. The implications and limitations of the model are discussed.

125.021 **Comments on expansion parallaxes of supernovae.**
K. Hempe.
Astron. Astrophys., Vol. 158, No. 1/2, p. 329 – 334 (1986).
Supernovae can be used to determine expansion parallaxes. The parallaxes depend on temperature and angular diameters of the SN. Errors of the angular diameter caused by uncertainties of the temperature have been analyzed using a spherical gray extended atmosphere. The velocity at the photosphere has been determined from line profiles. Hydrogen Balmer line profiles have been used for an analysis of SN 1983 in NGC 1448. The results of this study are discussed under the aspect of distance determination. It has been found that expansion parallaxes are only useful if the temperature estimates and the determination of the photospheric radius are very accurate.

125.022 **A new approach to the possibilities of Σ(D)–dependence and distances of the supernova remnants.**
A. O. Allakhverdiyev, O. H. Guseinov (*O. Kh. Gusejnov*), F. K. Kasumov, I. M. Yusifov.
Astrophys. Space Sci., Vol. 121, No. 1, p. 21 – 44 (1986).
A new way of choosing the calibrators for constructing $\Sigma(D)$–dependence of supernova remnants is suggested. It is shown that neither more accurate determination of distance to calibrators, nor an increase in their number, nor an increase in the precision of the remnant angular size and radio flux determination can lead to a considerable improvement of a simple $\Sigma(D)$–dependence suggested in the given work. For remnants with small surface brightnesses there is no $\Sigma(D)$–dependence at all. The most accurate remnant distances currently known, are given.

125.023 **High–resolution X–ray spectroscopy of the Crab Nebula and the oxygen abundance of the interstellar medium.**
M. L. Schattenburg, C. R. Canizares.
Astrophys. J., Vol. 301, No. 2, p. 759 – 771 (1986).
The authors have measured a high–resolution spectrum of the Crab Nebula in the 0.5 – 3 keV band using the Einstein Observatory. The oxygen K–edge is prominent in the measured spectrum. The detailed spectrum in the vicinity of this edge gives marginal evidence for a resonant $1s-2p$ absorption line and hints of a K–edge from singly ionized oxygen. Model fitting to the overall spectrum yields a hydrogen column density of $3.45 \pm 0.42 \times 10^{21} cm^{-2}$ and an oxygen column density of $2.78 \pm 0.55 \times 10^{18} cm^{-2}$, corresponding to an oxygen abundance of 1.1 ± 0.3 times solar. The absence of any evident oxygen depletion effects suggests that most interstellar grains are reasonably transparent to soft X–rays, implying sizes <0.4 μm.

125.024 **Radio supernovae.**
K. W. Weiler, R. A. Sramek, N. Panagia, J. M. van der Hulst, M. Salvati.
Astrophys. J., Vol. 301, No. 2, p. 790 – 812 (1986).
The authors present extensive new results on radio supernovae (RSNs) and discuss in detail their observation, calibration, and analysis. They collect, for completeness, most other available results and discuss their accuracy and applicability. With this extensive data set the authors investigate the class properties of the RSNs and derive descriptive parameters for the gross features of their "light curves". These are then applied to the several models available for the phenomenon to investigate their relative merits and their relations to the various types and subtypes of supernovae. In addition to investigation of the observations and models, possible radio supernova–supernova remnant (RSN–SNR) relations are briefly discussed and new lines of study are suggested.

125.025 **Supernova observed in its brightening stage.**
K. Bracher.
Phys. Today, Vol. 39, No. 1, p. S8 – S9 (1986). Short review paper. – See Abstr. 013.021.

125.026 **Type II supernova outburst as a result of two successive ejections of matter.**
E. K. Grasberg, D. K. Nadezhin.
Pis'ma Astron. Zh., Tom 12, No. 2, p. 168 – 175 (1986). In Russian. English translation in Sov. Astron. Lett., Vol. 12.
A model of type II supernova outburst is proposed that is based on the hypothesis of two successive explosions, the second explosion being much stronger than the first one. The model can explain the behaviour of subsystems of relatively narrow lines observed in spectra of some type II supernovae. The model is applied to the supernova SN 1983k in NGC 4696, for which the time interval between the explosions is estimated to be 1 – 2 months. The first explosion may be produced by a relatively weak shock wave appearing due to perturbation of stellar structure in the course of gravitational collapse of the central core, whereas the second explosion may be a result of the transport of energy from the new–born neutron star to the presupernova envelope.

125.027 **Is there any radio shell around the Crab Nebula?**
S. A. Trushkin.
Pis'ma Astron. Zh., Tom 12, No. 3, p. 198 – 204 (1986). In Russian. Englisch translation in Sov. Astron. Lett., Vol. 12.
Based on RATAN–600 observations at wavelengths 1.38 and 2.08 cm and a VLA–map at 20 cm by Velusamy (1985) it is shown that the spectral index along drift scans remains constant along the whole Crab Nebula. The results obtained deny the existence of any radio shell around the Crab Nebula.

125.028 **Radio image of the Crab nebula at 750 MHz.**
M. I. Agafonov, A. M. Aslanyan, A. G. Gulyan, V. P. Ivanov, R. M. Martirosyan, O. A. Podvojskaya, K. S. Stankevich.
Pis'ma Astron. Zh., Tom 12, No. 4, p. 275 – 280 (1986). In Russian. English translation in Sov. Astron. Lett., Vol. 12.
Using the Bracewell algorithm and a method of standard cleaning a Crab nebula radio map is synthesized with the resolution $20'' \times 35''$ from strip distributions obtained during lunar occultations in 1982 – 1983. Dimensions of the extended elliptical structure amount to $300'' \times 400''$ at the 4% level of the peak intensity. The brightness distribution has a maximum at $10''$ to the south–east from the pulsar and a number of local brightenings. At the distance of $\sim 20''$ to the north–west from the center a gap in the intensity distribution is observed in the form of a strip with the width $\lesssim 15''$ and PA $\approx 12°$. The radio emission at the level $\gtrsim 0.5$ comes from a region inside $\sim 70''$ from the center, except the north–west direction in which it extends up to $120''$.

125.029 **The relative abundance of nitrogen and oxygen in type II supernovae.**
N. N. Chugaj.
Pis'ma Astron. Zh., Tom 12, No. 4, p. 293 – 299 (1986). In Russian. English translation in Sov. Astron. Lett., Vol. 12.

Two estimates of the upper limit of the N/O ratio are obtained by two methods from optical spectra of the type II supernovae SN 1970g and SN 1969*l*. In both cases $N/O < 1$ unlike the higher ratio $N/O > 2$ found earlier for SN 1979c. This difference is suggested to be due to the smaller mass of the envelope of the SN 1979c as compared with the former two supernovae.

125.030 **Observations of 11 supernovae.**
D. Yu. Tsvetkov.
Perem. Zvezdy, Tom 22, No. 2, p. 191 – 199 (1985). In Russian.

Photographic observations of 11 supernovae obtained at the Sternberg Astronomical Institute from 1963 to 1976 are reported. Observations of 6 supernovae were not elaborated earlier, for other supernovae new magnitudes were obtained based on new determination of magnitudes of comparison stars. Parameters of light curves and absolute magnitudes are derived.

125.031 **Meridional motions in supernovae.**
G. Ruben.
Nauchn. Inf., Vyp. 56, p. 49 – 51 (1984). In Russian.

Qualitative results of calculations of the collapse of a rotating core (two–dimensional problem) are presented. After the collapse a reflecting shock wave and ejection of an envelope are noted. Because of the star's rotation the envelope is asymmetric, but corresponding matter meridional motions in the envelope in direction to the rotating axis later on decreases this asymmetry.

125.032 **New selection effect of statistical investigations of supernova remnants.**
A. O. Allakhverdiev, O. Kh. Gusejnov, F. K. Kasumov.
Astrofizika, Tom 24, Vyp. 1, p. 97 – 107 (1986). In Russian. English translation in Astrophysics, Vol. 24, No. 1.

The influence of H II regions on the parameters of supernova remnants (SNR) is investigated. It has been shown that the projection of such regions on the SNRs leads to: a) local changes of morphological structure of young shell–type SNRs and b) considerable distortions of integral parameters of evolved shell–type SNRs (with $D > 10$ pc) and plerions, up to their complete undetectability on the background of classical and gigantic H II regions. A new selection effect, in fact, arises from these factors connected with additional limitations made by the real structure of the interstellar medium on the statistical investigations of SNRs. The influence of this effect on the statistical completeness of objects has been estimated.

125.033 **Supernovae.**
IAU Circ., Nos. 4219, 4227 (1986).

Concerning supernovae: 1986I in NGC 4254, 1986H in NGC 5645, and 1986E in NGC 4302.

125.034 **Resolution of Crab nebula filaments into chains of quasistellar knots.**
S. van den Bergh, C. J. Pritchet.
Nature, Vol. 321, No. 6065, p. 46 – 47 (1986).

The authors have clearly resolved a number of the Hα emitting filaments in the Crab nebula into chains of knots with diameters $\lesssim 1.0$ arc s. Individual filaments resemble beads on a string which shows that these objects are rope–like structures. In the brightest filament close to the Crab pulsar the quasistellar knots are slightly displaced from the spine of the low–density filament on which they are superposed.

125.035 **Radio studies of extragalactic supernovae.**
K. W. Weiler, R. A. Sramek, N. Panagia.
Science, Vol. 231, No. 4743, p. 1251 – 1254 (1986).

The authors summarize the observational data and theoretical knowledge presently available on radio supernovae, with primary concentration on the newest examples for which the vast majority of information is available.

125.036 **EXOSAT observations of supernova remnants.**
A. Smith.
Observatory, Vol. 106, No. 1070, p. 10 (1986). Abstract. – See Abstr. 010.741.

125.037 **Photoionization model of emission and mean abundance of chemical elements in the filaments of the Crab nebula.**
V. V. Golovatyj, B. S. Novosyadlyj.
Vestn. L'vov. univ. Ser. astron., No. 59, p. 20 – 23 (1985). In Russian. Abstr. in Ref. Zh., 51. Astron., 2.51.855 (1986).

125.038 **Neutron stars in supernova remnants.**
F. D. Seward.
Sky Telesc., Vol. 71, No. 1, p. 6 – 10 (1986).

125.039 **Is there a Σ(D)–dependency for supernova remnants?**
A. O. Allakhverdiev, O. Kh. Gusejnov, F. K. Kasumov.
Astrofizika, Tom 24, Vyp. 2, p. 397 – 410 (1986). In Russian. English translation in Astrophysics, Vol. 24, No. 2.

For 25 years the dependence of the surface brightness Σ on the diameter D of the remnants is used to determine the distances to SNRs. In this paper it is shown that only shell–type remnants evolved in a comparably dense medium may be the calibrators for the construction of a $\Sigma(D)$–dependency. Using the $\Sigma(D)$–dependency one can estimate distances and see the evolution of the overwhelming majority of SNRs.

125.040 **Expansion degradation of ultraviolet photons in supernova ejecta.**
S. A. Colgate, A. G. Petschek.
Bull. Am. Astron. Soc., Vol. 17, No. 4, p. 847 – 848 (1985). Abstract. – See Abstr. 010.062.

125.041 **Probing the dark matter with supernovae.**
R. V. Wagoner, E. Linder, G. Efstathiou.
Bull. Am. Astron. Soc., Vol. 17, No. 4, p. 871 (1985). Abstract. – See Abstr. 010.062.

125.042 **Variations in X–ray spectra across Puppis A.**
A. E. Szymkowiak.
Bull. Am. Astron. Soc., Vol. 17, No. 4, p. 883 (1985). Abstract. – See Abstr. 010.062.

125.043 **The X–ray structure and spectrum of IC 443.**
R. Petre, A. E. Szymkowiak, F. D. Seward, C. R. Canizares, R. Willingale, D. Rolf, N. Woods.
Bull. Am. Astron. Soc., Vol. 17, No. 4, p. 883 (1985). Abstract. – See Abstr. 010.062.

125.044 **The electron temperature and velocity structure of the central region of CTB 80.**
W. P. Blair, R. A. Fesen, R. H. Becker.
Bull. Am. Astron. Soc., Vol. 17, No. 4, p. 883 – 884 (1985). Abstract. – See Abstr. 010.062.

125.045 **Kinematic detection of supernova remnants in giant H II regions.**
R. C. Kennicutt, Y.–H. Chu.
Bull. Am. Astron. Soc., Vol. 17, No. 4, p. 884 (1985). Abstract. – See Abstr. 010.062.

125.046 **The velocity signature of sheets in the Cygnus Loop SNR.**
J. J. Hester, J. C. Raymond, B. E. Woodgate.
Bull. Am. Astron. Soc., Vol. 17, No. 4, p. 884 (1985). Abstract. – See Abstr. 010.062.

125.047 **New measurements of the expansion of the Cygnus Loop.**
P. Shull Jr., H. H. Hippelein.
Bull. Am. Astron. Soc., Vol. 17, No. 4, p. 884 (1985). Abstract. – See Abstr. 010.062.

125.048 **Spectroscopy of the recent supernovae in NGC 2748 and NGC 4045A.**
R. K. McMahan, G. Wegner.
Bull. Am. Astron. Soc., Vol. 17, No. 4, p. 885 (1985). Abstract. – See Abstr. 010.062.

125.049 **Detection of additional QSF's and extended diffuse emission outside the supernova remnant Cas A.**
R. A. Fesen, R. H. Becker, W. P. Blair.
Bull. Am. Astron. Soc., Vol. 17, No. 4, p. 885 (1985). Abstract. – See Abstr. 010.062.

125.050 **On the mechanism for Type II supernova explosions.**
W. D. Arnett.
Bull. Am. Astron. Soc., Vol. 17, No. 4, p. 908 (1985). Abstract. – See Abstr. 010.062.

125.051 **IRAS sources associated with shocked–gas regions in SNRs.**
Y.–L. Huang, R. L. Dickman, R. L. Snell.
Bull. Am. Astron. Soc., Vol. 17, No. 4, p. 909 (1985). Abstract. – See Abstr. 010.062.

125.052 **X–ray spectroscopy of supernova remnants.**
R. G. Strom.
ESA Spec. Publ., ESA SP–239, p. 43 – 47 (1985). – See Abstr. 012.019.
Properties of the X–ray spectra of some twenty galactic supernova remnants and a further half–dozen in the Large Magellanic Cloud are reviewed, and some of the interpretations briefly discussed. Consideration is given to the possible uses of such data, including deciding whether the mission is predominantly thermal or nonthermal, determining excitation temperatures, and studying SNR kinematics. Attention is paid to the instrumental requirements deemed important for X–ray spectroscopy of SNR. The requirements which such considerations place on the XMM are briefly discussed.

125.053 **Non ionization equilibrium effects in the SNR RCW 86.**
J. J. Claas, A. Peacock.
ESA Spec. Publ., ESA SP–239, p. 107 – 110 (1985). – See Abstr. 012.019.
By using non ionization models, the analysis of EXOSAT low energy X–ray data (0.4 – 2 keV) of the supernova remnant RCW 86 reveals the following properties: a shell–like structure, consisting of two separate parts, producing relatively hard and soft X–rays respectively. The soft X rays come from C–K and O–K emission lines which implies a low $N_H \lesssim 10^{20}$atoms/cm^2. The harder X–rays probably result from the Fe–L complex and Ne–K line. A low N_H value may also imply the distance to be 1.6 kpc or slightly more.

125.054 **Interpretation of the oxygen VII soft X–ray spectrum from the Puppis–A supernova remnant.**
A. H. Gabriel, L. W. Acton, F. Bely–Dubau, P. Faucher.
ESA Spec. Publ., ESA SP–239, p. 137 – 140 (1985). – See Abstr. 012.019.
The soft X–ray line spectrum from Puppis–A recorded with the Einstein observatory shows line ratios different from those anticipated. Previous interpretations were based upon high temperatures $> 5 \times 10^6$K and departures from ionization equilibrium. It is shown that an alternative model can fit the observations, in which 1% of fast electrons are present in an otherwise thermal plasma at 1 to 2×10^6K. This plasma can be in steady–state ionization balance.

125.055 ***EXOSAT* and *Einstein* X–ray observations of the SNR 1E 1149.4–6209 in Crux: a unified picture?**
G. F. Bignami, P. A. Caraveo, A. Goldwurm, S. Mereghetti, G. G. C. Palumbo.
Astrophys. J., Vol. 302, No. 2, p. 606 – 610 (1986). With plates 18 – 20.
New *EXOSAT* and *Einstein Observatory* X–ray data from a region in Crux ($\alpha \approx 11^h45^m$, $\delta \approx 62°$) are reported. Morphological as well as rough spectral analysis of a large (~1°) diffuse feature is performed. The diffuse emission is identified with a SNR, a conclusion supported by comparison with existing radio and optical data. The earlier data (1981) of Markert *et al.*, who discovered the feature, are also discussed.

125.056 **X–ray observations of supernova remnants resembling the Crab Nebula.**
A. S. Wilson.
Astrophys. J., Vol. 302, No. 2, p. 718 – 726 (1986).
The three supernova remnants G 291.0–0.1 (MSH 11–62), G 308.7+0.0, and G 328.4+0.2 (MSH 15–57), which resemble the Crab Nebula in many of their radio properties, have been observed with the IPC on the *Einstein* X–ray observatory. Only G 291.0–0.1 was detected. This source shows very extended (> 20′ diameter) X–ray emission, with strong central brightening. The spectral data for the bright central region are consistent with a power law of photon index $\alpha \approx 1.8$. The X–ray spectrum is considerably softer ($\alpha \approx 3.6$) in the outer parts of the object. Given the large distances and hydrogen column densities to G 308.7+0.0 and G 328.4+0.2, their X–ray limits are consistent with membership in the Crab Nebula–like group. The evolutions of radio luminosity L_R, X–ray luminosity, L_X, and their ratio L_X/L_R with remnant diameter are reassessed for this class of supernova remnant.

125.057 ***IRAS* sources associated with shocked gas regions in IC 443.**
Y.–L. Huang, R. L. Dickman, R. L. Snell.
Astrophys. J., Lett. Ed., Vol. 302, No. 2, p. L63 – L66 (1986).
Although many supernova remnants are expected to interact with surrounding clouds, only IC 443 exhibits convincing evidence of such interactions, characterized by highly perturbed gas with very large ($\geqslant$20 km s^{-1}) velocity dispersions. Large–scale, high–resolution CO observations with the 14 m telescope of the FCRAO have revealed seven such regions, five newly discovered, in IC 443. A comparison with the *IRAS Point Source Catalog* shows that three bright, unidentified far–infrared pointlike sources display good positional coincidence with clumps of highly perturbed gas. Chance superposition of the clumps and the *IRAS* sources is shown to be highly unlikely.

125.058 **Radio emission of supernova remnants on the radiative phase of expansion.**
K. V. Bychkov.
Astrofiz. Issled. Izv. Spets. Astrofiz. Obs., Tom 21, p. 58 – 64 (1986). In Russian. English translation in Bull. Spec. Astrophys. Obs. – North Caucasus.
Thermal radio emission is connected with the swept up material, it consists of 21–cm neutral hydrogen emission (brightness temperature of some Kelvin degrees) and the bremsstrahlung 10 – 100 mJy of ionized plasma, both are being located in concentric rings on the supernova remnant periphery.

125.059 **An X–ray study of the Monoceros supernova remnant.**
D. A. Leahy, S. Naranan, K. P. Singh.
Mon. Not. R. Astron. Soc., Vol. 220, No. 2, p. 501 – 511 (1986).
Twelve Einstein IPC exposures within the region of the Monoceros supernova remnant are combined into a 0.2 – 4 keV X–ray map with 80 arcsec resolution. The X–ray bright regions correlate well with the bright optical filaments. From the IPC pulse height spectra, the bright regions are found to have a 2 – 6 keV high–temperature component, while all regions show a fainter 0.2 – 0.3 keV temperature component. A model with the supernova shock wave expanding into a three–component interstellar medium with a low–density (0.003 cm^{-3}) hot component can fit the observed properties of Monoceros. Preferred values for the distance and age are 1500 pc and 3×10^4yr. The overall observed expansion of 50 km s^{-1} can be explained by outward moving clouds interior to the shock front.

125.060 **Peculiar infrared and optical behaviour in Type I supernovae and the origin of the 1.2 μm feature.**
J. R. Graham.
Mon. Not. R. Astron. Soc., Vol. 220, No. 2, p. 27P – 31P (1986).
A small number of Type I supernovae exhibit well–defined peculiarities. In particular some Type I supernovae do not have the characteristic Si II 6150 Å feature, and some do not have the 1.2 μm absorption. It is noted that all supernovae which lack the infrared absorption also lack the 6150 Å feature. Therefore it is proposed that these supernovae constitute a single sub–class of Type I supernovae and that the absence of Si is responsible for the lack of both the 6150 Å and 1.2 μm absorptions. Si is an excellent candidate as the cause of the infrared absorption because Si I has a dense array of strong lines near 1.2 μm.

125.061 **A deep Hα image of faint Balmer–line filaments in the northeast Cygnus Loop supernova remnant.**
J. J. Hester, J. C. Raymond, G. E. Danielson.
Astrophys. J., Lett. Ed., Vol. 303, No. 1, p. L17 – L19 (1986). With plates L2 – L4.
A deep Hα image of a field outside the optically bright northeast limb of the Cygnus Loop supernova remnant is presented. Two filaments which are just visible on the red POSS print form part of two long and unbroken lines of filaments which generally bound a region of diffuse emission and fainter filaments. The brightest filaments are resolved into multiple components with widths < 2″. Comparison with the POSS print leads to a rough estimate of the proper motion of 6″ per century. The morphology of the emission strongly suggests that it arises in a single large sheet.

125.062 **1410 MHz continuum and H I line observations towards the SNR G296.5 + 10.0 and nearby sources. Evidences of two SNRs tunneling through the interstellar medium.**
G. M. Dubner, F. R. Colomb, E. B. Giacani.
Astron. J., Vol. 91, No. 2, p. 343 – 353 (1986).
1410 MHz radio continuum and H I 21 cm line observations have been performed on an extended area that contains the SNR G296.5 + 10.0. Two additional radio–continuum sources have been identified in the 1410 MHz map, namely (1) G300.1 + 9.4, which is classified as a faint SNR, located at a distance comparable to that of G296.5 + 10.0, and (2) G300.8 + 11.7, with a broken spectrum and a possible double structure unresolvable with the present resolution. The observations reproduce the theoretically expected outcome of two interacting supernova remnants. In addition, some small H I clouds accelerated to relatively high velocities are observed to be positionally coincident with the SNRs.

125.063 **Supernovae and supernova remnants.**
C. R. Canizares.
X–ray astronomy '84, p. 275 – 283 (1985). – See Abstr. 012.033.
Supernova remnants are the most photogenic and spectrogenic of X–ray objects. The author gives a sampling of the highlights, concentrating almost entirely on the results from the Einstein observatory and limiting himself to consideration of younger SNRs that are still dominated by their ejecta.

125.064 **X–ray spectra of supernova remnants Cas A and Tycho.**
K. Koyama, H. Tsunemi.
X–ray astronomy '84, p. 285 – 287 (1985). – See Abstr. 012.033.
X–ray spectra of the supernova remnants Cas A and Tycho have been obtained by Tenma. The energy of K–emission lines of silicon, sulphur, argon, calcium and iron indicate clearly the ionization non equilibrium for Tycho.

125.065 **EXOSAT observations of the supernova remnant Cas A.**
F. A. Jansen, S. P. McKechnie, P. A. J. de Korte, J. A. M. Bleeker, E. Gronenschild, A. Peacock, G. Manzo, G. Branduardi–Raymont, B. Kellett.
X–ray astronomy '84, p. 289 – 292 (1985). – See Abstr. 012.033.
The young supernova remnant Cas A has been observed with the imaging proportional counter of the EXOSAT observatory. High quality spatially resolved, spectral data allow for the first time the determination of the temperature structure of the remnant. Preliminary results on the distribution of temperature and emission measure over the remnant are presented.

125.066 **The X–ray structure of the Crab Nebula.**
B. Aschenbach, W. Brinkmann, A. Langmeier, G. Hasinger, T. Bork.
X–ray astronomy '84, p. 301 – 304 (1985). – See Abstr. 012.033.

125.067 **EXOSAT observations of SN 1006.**
L. R. Jones, J. P. Pye, J. L. Culhane.
X–ray astronomy '84, p. 305 – 308 (1985). – See Abstr. 012.033.
Comparison of EXOSAT images of the whole of the remnant, with greater spatial resolution (~30 arcsec) than obtained previously, with an Einstein Observatory image and a radio map, do not support a synchrotron origin for the soft X–ray emission.

125.068 **Einstein observation of supernova remnant HB 3.**
D. A. Leahy, D. Venkatesan, S. Naranan, K. S. Long.
X–ray astronomy '84, p. 309 – 312 (1985). – See Abstr. 012.033.
A contour map of the X–ray brightness of the remnant and spectra of various regions across it have been obtained. Supernova remnant models have been applied to HB 3 and the spatial structure compared with the radio morphology.

125.069 **Hard X–ray observation of the Crab Nebula.**
Y. Ogawara, K. Makishima, M. Oda, S. Miyamoto, M. R. Pelling, W. S. Paciesas, L. E. Peterson.
X–ray astronomy '84, p. 313 – 318 (1985). – See Abstr. 012.033.
The two–dimensional distribution of the hard X–ray emission from the Crab Nebula was observed with a balloon–borne modulation collimator. The authors present a model for the energy transport and the high energy processes in the nebula, in relation to the structures observed at various wavelengths.

125.070 **Light curves of Type I supernovae.**
R. López, J. Isern, R. Canal, J. Labay.
X–ray astronomy '84, p. 329 – 332 (1985). – See Abstr. 012.033.
The authors discuss models of Type I supernovae.

125.071 **Photoionization model of luminescence and dispersion of chemical element abundances in the Crab nebula filaments.**
V. V. Golovatyj, B. S. Novosyadlyj.
Pis'ma Astron. Zh., Tom 12, No. 6, p. 440 – 445 (1986). In Russian. English translation in Sov. Astron. Lett., Vol. 12.
The observed filaments spectrum has been separated into three groups. The photoionization models for each group of filaments is constructed. In each case the agreement of calculated and observed line intensity is good. The results of calculation show that the dispersion of the abundances $He/H = 0.15 - 1.20$, $N/H = (0.28 - 0.92) \times 10^{-4}$, $S/H = (0.10 - 0.64) \times 10^{-4}$ and, possibly, $O/H = (1.35 - 2.3) \times 10^{-4}$ is real.

125.072 **What makes the peculiar type I supernovae differ from the normal ones?**
N. N. Chugaj.
Pis'ma Astron. Zh., Tom 12, No. 6, p. 461 – 467 (1986). In Russian. English translation in Sov. Astron. Lett., Vol. 12.
The bolometric light curve of the peculiar type I supernova (SN I) SN 1983n and of the normal SN I SN 1972e have been calculated. The optical depth of the envelope for gamma–quanta in the case of SN 1983n is shown to be approximately 2.6 times larger than in the case of SN 1972e. The restrictions on the values of the mass and expansion velocity of the envelope of a peculiar SN I relative to the values for a normal SN I are obtained. In particular the expansion velocity on the edge of the incineration zone in a peculiar SN I is 2 – 3 times smaller than in a normal SN I. The author predicts the appearance of the strong emission lines from the unburnt matter of a supernova in the spectra of a peculiar SN I on the late time $t > 40^d$. These lines must contain about 40% of the overall optical luminosity of the supernova.

125.073 **The Cygnus Loop supernova remnant: new observations and a framework for understanding its structure and evolution.**
J. J. Hester.
Diss. Abstr. Int., Sect. B, Vol. 46, No. 5, p. 1957 (1985). Thesis, Rice University, 275 pp. (1985). Order No. DA8517205.

125.074 **Supernova remnants and molecular clouds.**
Y.-l. Huang.
Diss. Abstr. Int., Sect. B, Vol. 46, No. 8, p. 2693 (1986). Thesis, Columbia University, 197 pp. (1985). Order No. DA8523181.

125.075 **On explosions of extended stars as type I supernovae.**
W. T. G. Glen.
Diss. Abstr. Int., Sect. B, Vol. 46, No. 12, p. 4282 (1986). Thesis, McMaster University, Canada (1985).

125.076 **A comparative study of type I and type II supernovae.**
O. S. Bartunov, D. Yu. Tsvetkov.
Astrophys. Space Sci., Vol. 122, No. 2, p. 343 – 354 (1986).
Revised photometric data are used to compare the light and colour curves of type I and type II supernovae (SNe I, SNe II); their statistical properties are also compared. No significant difference between SNe I and SNe II has been found in their radial distribution and frequency of outbursts in spiral galaxies. The comparison of light and colour curves reveals several features common to both types and the possibility of transition between types.

125.077 **Mechanism of regular acceleration and evolution of supernova remnants.**
A. O. Allakhverdiyev, A. I. Asvarov, O. H. Guseinov (*O. Kh. Gusejnov*), F. K. Kasumov.
Astrophys. Space Sci., Vol. 123, No. 2, p. 237 – 258 (1986).
The possibility of realization of a mechanism of regular acceleration of charged particles at the shock front in shell-type supernova remnants is demonstrated.

125.078 **Dispersion of the chemical abundances in Crab nebula gas filaments.**
V. V. Golovaty (*V. V. Golovatyj*), V. I. Pronik.
Eruptive phenomena in stars, p. 381 – 382 (1986). – See Abstr. 012.039.

125.079 **Observations of supernovae.**
N. Panagia.
Space Telesc. Sci. Inst., Prepr. Ser., No. 83, 18 pp. (1985). Invited review presented at the NATO Advanced Study Institute on "High energy phenomena around collapsed stars", Cargèse, Corsica (France), 2 – 13 September 1985.
The observations of supernovae are reviewed and their main implications are shortly discussed.

125.080 **The terrain of evolution of isotropic adiabatic supernova remnants.**
D. P. Cox.
Astrophys. J., Vol. 304, No. 2, p. 771 – 779 (1986).
Models of the evolution of supernova remnants in isotropic media are reviewed and generalized. It is shown that properties of remnants are almost history-independent functions of their instantaneous average density and postshock temperature. These properties, presented as a contour diagram in the density-temperature plane, thus define the territory through which remnant properties evolve. In different models, remnants merely take different paths across the available terrain.

125.081 **Recent developments on the Crab Nebula.**
L. Woltjer.
ESO Sci. Prepr., No. 411, 13 pp. (1986). To appear in the Proceedings of the Advanced Study Institute on 'High energy phenomena around collapsed stars', Cargèse, 2 – 13 September 1985.

125.082 **Type I SNe from double degenerate CO dwarfs and their rate in the solar neighbourhood.**
A. Tornambè, F. Matteucci.
ESO Sci. Prepr., No. 446, 22 pp. (1986). To appear in Mon. Not. R. Astron. Soc.

125.083 **A catalog of the [Hα + N II] emission features in the Kepler SNR.**
S. D'Odorico, R. Bandiera, J. Danziger, P. Focardi.
Astron. J., Vol. 91, No. 6, p. 1382 – 1385 (1986). With plates 101 – 104.
A deep image of the Kepler SNR has been obtained in the light of Hα + [N II] with the faint object spectrograph and camera (EFOSC) and a CCD detector at the ESO 3.6 m telescope. Features as faint as 2% of the night-sky brightness have been identified and a half-shell of emission is clearly revealed. The total Hα luminosity of the ionized gas is estimated to be 8.3×10^{45} photons/s at a distance of 5 kpc implying a mass of ionized hydrogen of $\cong 0.02\ M_{\odot}$. A catalog with accurate celestial positions and approximate fluxes for 64 emission-line knots is given.

125.084 **CCD observations of supernova remnants.**
S. van den Bergh, C. J. Pritchet.
Publ. Astron. Soc. Pac., Vol. 98, No. 602, p. 448 – 452 (1986).
The high quantum efficiency of CCD detectors makes it possible to study faint features in supernova remnants that are only barely detectable on the longest photographic exposures. The advantage of the CCD over the photographic plate is particularly striking in the near infrared. The authors present CCD observations of the supernova remnants Cas A, the Crab nebula, and CTB 80. Observations of the Crab nebula in sub-arc-second seeing show that the Hα filaments in this object consist of chains of stellar or quasi-stellar knots with diameters $\lesssim 1.0$ arc sec.

125.085 **Precise positions for two supernovae.**
A. R. Klemola.
Publ. Astron. Soc. Pac., Vol. 98, No. 602, p. 464 – 466 (1986).
Precise positions are measured for SN 1961v in NGC 1058 and SN 1976k in NGC 3226. Positional coincidence is confirmed for SN 1961v with the radio source observed by Branch and Cowan and with the presupernova object visible on a plate taken in 1937.

125.086 **The optical structure of Kepler supernova remnant.**
R. Bandiera, I. J. Danziger, S. D'Odorico, P. Focardi.
Mem. Soc. Astron. Ital., Vol. 56, No. 4, p. 773 – 776 (1985). – See Abstr. 010.821.
Deep optical images, obtained at the ESO 3.6 m telescope, of Kepler's supernova remnant are presented and compared with radio and X-ray data.

125.087 **The hydrodynamics of clouds overtaken by supernova remnants.**
G. Tenorio-Tagle, M. Rozyczka.
Rev. Mex. Astron. Astrofis., Vol. 12, Spec. Issue, p. 293 (1986). Abstract. – See Abstr. 012.066.

125.088 **Distribution and kinematics of neutral hydrogen around the SNR G296.5 + 10.0.**
G. Dubner, F. Colomb, E. Giacani.
Rev. Mex. Astron. Astrofis. Vol. 12, Spec. Issue, p. 302 (1986). Abstract. – See Abstr. 012.066.

125.089 **Type I supernovae.**
J. C. Wheeler.
Nucleosynthesis and its implications on nuclear and particle physics, p. 113 – 120 (1986). – See Abstr. 012.069.
Several projects involving Type I supernovae and their impact on nucleosynthesis are described: (1) A major effort is underway to compute the spectrum at epochs near maximum light. (2) Dynamical calculations have been done exploring the range of possible outcomes of the evolution of binary white dwarfs. (3) Consideration is given to the need to eject iron from the Galaxy to avoid overproduction by Type I events. (4) The nature of the class of peculiar Type I supernovae is explored.

125.090 **Optical supernova remnants.**
I. J. Danziger.
Nucleosynthesis and its implications on nuclear and particle physics, p. 233 – 242 (1986). – See Abstr. 012.069.
A review of some unsolved problems in supernova remnant research is presented.

125.091 **Ist Cas A Überrest eines 100 $M_\odot$–Sternes?**
M. F. El Eid, N. Langer.
Mitt. Astron. Ges., No. 65, p. 211 – 212 (1986). – See Abstr. 012.072.

125.092 **Supernovae.**
Yamamoto Circ., No. 2054 (1986). In Japanese.
Concerning supernovae: 1986A in NGC 3367 and 1986B in NGC 5101.

125.093 **Supernovae.**
Yamamoto Circ., No. 2062 (1986). In Japanese.
Concerning supernovae: 1986H in NGC 5645 and 1986I in NGC 4254.

125.094 **Infrared observations of Tycho using IRAS.**
P. L. Marsden.
Light on dark matter, p. 153 – 154 (1986). – See Abstr. 012.080.

125.095 **Shock–heated dust in young supernova remnants.**
R. Braun, R. G. Strom, H. van der Laan, H. Greidanus.
Light on dark matter, p. 155 – 158 (1986). – See Abstr. 012.080.
Infrared emission in young supernova remnants is interpreted as coming from shock–heated dust. Using models and data from other wavelength regimes, many physical parameters of the remnants can accurately be derived.

125.096 **Collisionally heated dust in LMC supernova remnants.**
J. R. Graham, W. P. S. Meikle, A. Evans, M. F. Bode, J. S. Albinson.
Light on dark matter, p. 397 – 398 (1986). – See Abstr. 012.080.
Preliminary IRAS additional observations are presented, which show that luminous ($\sim 10^5 L_\odot$) infrared sources are associated with supernova remnants. Comparison of the infrared and X–ray data show that the infrared emission of these sources is due to collisionally heated dust.

Supernova in NGC 5236 (1983n)

125.201 **Discovery of a large mass of iron in a type I supernova.**
J. R. Graham, W. P. S. Meikle, D. A. Allen, A. J. Longmore, P. M. Williams.
Mon. Not. R. Astron. Soc., Vol. 218, No. 1, p. 93 – 102 (1986).
The authors have discovered $\sim 0.3\ M_\odot$ of iron in the type I supernova 1983n in the galaxy M83 about one year after explosion. This is the first time that a massive quantity of iron has been directly and unambiguously detected in a supernova. It was accomplished by the observation of strong [Fe II] line emission in the infrared at 1.644 μm from the optically thin ejecta. This result was predicted by the ^{56}Ni–^{56}Co–^{56}Fe radioactive decay scenario, and provides strong support for its credibility.

Supernova in NGC 991 (1984*l*)

125.221 **Subluminous, radio emitting Type I supernovae.**
N. Panagia, R. A. Sramek, K. W. Weiler.
Astrophys. J., Lett. Ed., Vol. 300, No. 2, p. L55 – L58 (1986).
The supernova SN 1984*l* in NGC 991 has been established as a subluminous Type I supernova similar to SN 1983n. Evidence that these two supernovae strongly resemble one another and represent a distinguishable subclass of Type I supernovae is strengthened by the detection of radio emission from this new object. Comparison of the radio properties of SN 1984*l* with those found earlier for SN 1983n shows that the spectral indices, initial thermal absorption, and luminosities can be considered to be identical, albeit to within relatively large errors.

Supernova in NGC 224 (1885a)

125.241 **Erratum: "S Andromedae 1885: a centennial review" [Astrophys. J., Vol. 295, No. 2, p. 287 – 304 (1985)].**
G. de Vaucouleurs, H. G. Corwin Jr.
Astrophys. J., Vol. 302, No. 1, p. 517 (1986). See Abstr. 40.125.225.

Supernova in NGC 6946 (1980k)

125.261 **Supernova 1980k in NGC 6946.**
G. M. Hurst, M. D. Taylor.
J. Br. Astron. Assoc., Vol. 96, No. 2, p. 102 – 105 (1986).
The results of visual and photographic observations of the supernova 1980k in NGC 6946 are presented. Photographic results obtained by M. Swan for the UK Nova/Supernova Patrol prior to the discovery announcement show faint images of the supernova during its rise to maximum. Photographic magnitude estimates suggest maximum occurred on 1980 November 3 at photovisual magnitude 11.1.

Supernova in NGC 4618 (1985F)

125.281 **SN 1985F: death of a Wolf–Rayet star.**
M. C. Begelman, C. L. Sarazin.
Astrophys. J., Lett. Ed., Vol. 302, No. 2, p. L59 – L62 (1986).
From an analysis of the optical spectrum of SN 1985F, the authors show that the supernova ejecta contain $\gtrsim 5\ M_\odot$ of oxygen and very little hydrogen. It is suggested that the explosion resulted from the pair instability supernova of a $\sim 50\ M_\odot$ WO Wolf–Rayet star. The optical luminosity of the supernova is powered by the radioactive decay of ^{56}Co synthesized in the explosion. From the rate of decay of the optical emission it is estimated that the explosion occurred ~ 350 days before it was discovered in 1985 February. The authors suggest that some of the oxygen–rich (Cas A type) supernova remnants may also have been produced by exploding WO stars.

125.282 **The unique supernova (1985f) in NGC 4618.**
A. V. Filippenko, W. L. W. Sargent.
Astron. J., Vol. 91, No. 4, p. 691 – 696 (1986). With plates 61, 62.
The complete optical spectrum (3200 – 10000 Å) of supernova 1985f in NGC 4618, obtained one to two months after discovery, is presented for the first time. The spectrum of SN 1985f does not resemble any previously published spectra of supernovae. The authors postulate that the progenitor of this peculiar object was a massive Wolf–Rayet star that expelled its outer atmosphere of H (and probably He) prior to the supernova explosion. Several photographs of SN 1985f illustrate its location in very bright H II regions near the nucleus of NGC 4618.

Supernova in NGC 5128 (1986G)

125.301 **ESO observations of bright supernova in Centaurus A.**
Messenger, No. 44, p. 1 – 2 (1986).

125.302 **CCD observations of supernova 1986G in Cen A.**
G. Galletta.
Messenger, No. 44, p. 2 (1986).

125.303 **Low resolution spectroscopy of the supernova 1986G near maximum brightness.**
S. di Serego Alighieri.
Messenger, No. 44, p. 3 (1986).

125.304 **Supernova 1986G in NGC 5128.**
IAU Circ., Nos. 4208, 4210, 4215, 4216, 4224, 4227, 4230 (1986).

125.305 **Supernova in NGC 5128, 1986G.**
Yamamoto Circ., Nos. 2059, 2064 (1986). In Japanese.

Supernova in NGC 3367 (1986A)

125.321 **Supernova 1986A in NGC 3367.**
IAU Circ., Nos. 4173, 4175, 4177, 4179, 4195, 4201 (1986).

Supernova in NGC 5101 (1986B)

125.341 **Supernova 1986B in NGC 5101.**
IAU Circ., Nos. 4177, 4185, 4190, 4191, 4201 (1986).

125.342 **Supernova 1986B in NGC 5101.**
Yamamoto Circ., Nos. 2056, 2058 (1986). In Japanese.

Supernova in NGC 1433 (1985P)

125.361 **Supernova 1985P in NGC 1433.**
IAU Circ., No. 4181 (1986).

Supernova in UGC 6697 (1986C)

125.381 **Supernova 1986C in UGC 6697.**
IAU Circ., No. 4191 (1986).

Supernova in NGC 3034 (1986D)

125.401 **Supernova 1986D in M82.**
IAU Circ., Nos. 4197, 4202 (1986).

125.402 **Supernova in M82, 1986D.**
Yamamoto Circ., No. 2057 (1986). In Japanese.

Supernova in NGC 4302 (1986E)

125.421 **Supernova 1986E in NGC 4302.**
IAU Circ., Nos. 4202, 4206 (1986).

125.422 **Supernova 1986E in NGC 4302.**
Yamamoto Circ., No. 2058 (1986). In Japanese.

Supernova in NGC 5645 (1986H)

125.441 **Supernova 1986H in NGC 5645.**
IAU Circ., No. 4213 (1986).

Supernova in NGC 4254 (1986I)

125.461 **Supernova 1986I in NGC 4254.**
IAU Circ., No. 4225 (1986).

The exploding suns: the secrets of the supernovas.
See Abstr. 003.056.

Experimental observations with the radio telescope RATAN–600 using the method «estafette with zoning».
See Abstr. 036.089.

Spectral decomposition of IRAS maps.
See Abstr. 036.259.

Non–steady radiative shocks.
See Abstr. 062.013.

Spectral flux from low–density photospheres: numerical results.
See Abstr. 064.014.

Spectral flux from low–density photospheres: approximate results.
See Abstr. 064.038.

Carbon deflagration models for type–I supernovae and theoretical optical spectra.
See Abstr. 065.009.

The phase transitions of superdense matter and supernova explosions.
See Abstr. 065.011.

Observational implications of the explosion of white dwarfs.
See Abstr. 065.019.

Explosive nucleosynthesis in carbon deflagration models of Type I supernovae.
See Abstr. 065.030.

Models for Type I supernova. I. Detonations in white dwarfs.
See Abstr. 065.038.

Supernova nucleosynthesis in low–metallicity populations.
See Abstr. 065.039.

The evolution of WR + O binaries into massive X–ray binaries.
See Abstr. 065.072.

Precursor heating and acceleration by Langmuir waves: the type II supernova case.
See Abstr. 065.089.

On the evolution of binary components which first fill their Roche lobes after the exhaustion of central helium.
See Abstr. 065.091.

How stars explode.
See Abstr. 065.101.

Nucleosynthesis in He–flashes on accreting white dwarfs.
See Abstr. 065.104.

On the behavior of double degenerate binaries associated with Type I supernovae.
See Abstr. 065.107.

Supernovae and nuclear structure: electron capture and the nuclear incompressibility.
See Abstr. 065.109.

Supernova matter: a semiclassical approach.
See Abstr. 065.110.

On the evolution of very luminous stars. I. Presupernova evolution.
See Abstr. 065.116.

On the evolution of very luminous stars. II. Pair creation supernova in massive Wolf–Rayet stars.
See Abstr. 065.117.

Peculiar supernova of type I as a result of explosive helium burning in an accreting white dwarf.
See Abstr. 065.124.

Nucleosynthesis and Type I supernovae.
See Abstr. 065.130.

Explosive nucleosynthesis in carbon deflagration models of Type I supernovae.
See Abstr. 065.131.

Theoretical models for Type I and Type II supernovae.
See Abstr. 065.133.

Pair creation supernovae with rotation.
See Abstr. 065.134.

Light curves of exploding Wolf–Rayet stars.
See Abstr. 065.141.

Type I SNe from binary systems: consequences on galactic chemical evolution.
See Abstr. 065.145.

Supernova theory.
See Abstr. 065.147.

Das Schicksal eines 100 $M_\odot$–Sternes.
See Abstr. 065.152.

Wind fed X–ray binaries.
See Abstr. 065.156.

Luminous content of galaxies: inferences from supernovae.
See Abstr. 065.173.

The equation of state and general relativity in supernovae.
See Abstr. 066.114.

Confronting X–ray observations of young supernova remnants with neutron star cooling models.
See Abstr. 067.115.

Brightest stars in galaxies as distance indicators.
See Abstr. 115.030.

A common origin for Type I supernovae and low–mass binary X–ray sources.
See Abstr. 117.185.

Search for supernovae with the Schmidt telescope of the Radioastrophysical Observatory.
See Abstr. 123.015.

The injection of energy into the interstellar medium by stars.
See Abstr. 131.019.

The hydrodynamics of clouds overtaken by supernova remnants. I. Cloud crushing phenomena.
See Abstr. 131.063.

Interstellar scattering of compact radio sources near supernova remnants.
See Abstr. 131.097.

Aperture synthesis observations of the 21 centimeter Zeeman effect toward Cassiopeia A.
See Abstr. 131.098.

The dynamical evolution of H I shells and the initial mass function of OB associations.
See Abstr. 131.122.

An empirical study of ring–shaped nebulae in the Galaxy and in the LMC.
See Abstr. 131.145.

Supernova remnants associated with an H I "supershell" in the Perseus spiral arm.
See Abstr. 131.219.

On the carbon enrichment of the interstellar medium.
See Abstr. 131.244.

Cold H I clouds near the supernova remnant W44.
See Abstr. 131.250.

Lupus Loop: a possible case of star formation induced by a supernova event?
See Abstr. 131.344.

G70.68 + 1.20: a thermal source, not a supernova remnant.
See Abstr. 132.015.

Decameter wavelength observations of the Rosette Nebula and the Monoceros Loop supernova remnant.
See Abstr. 132.019.

G 179.0 + 2.7, an unusual galactic radio source?
See Abstr. 141.001.

Multifrequency radio continuum observations of extended galactic objects. I. Nine objects from the 2695 MHz Effelsberg galactic plane survey.
See Abstr. 141.004.

WSRT observations of radio sources in the galactic plane near $l = 54°$.
See Abstr. 141.016.

Iron K X–ray emission lines from cosmic X–ray sources observed with Tenma.
See Abstr. 142.034.

The creation of compact objects in the Local Group.
See Abstr. 142.082.

A supernova for a neighbour?
See Abstr. 144.003.

Supernova origin of cosmic rays.
See Abstr. 144.021.

Shocks and cosmic ray acceleration.
See Abstr. 144.022.

Time–dependent cosmic ray/gas interactions.
See Abstr. 144.023.

Enhancement of high–energy cosmic–ray spectrum by type–II supernovae.
See Abstr. 144.075.

The dynamical evolution of expanding H I shells and initial mass function of OB associations.
See Abstr. 152.011.

Relative rôles of type I and II supernovae in the chemical enrichment of the interstellar gas.
See Abstr. 155.006.

Low–latitude radio continuum structure in the fourth galactic quadrant.
See Abstr. 155.028.

Distribution of stars in the Orion Barnard Loop.
See Abstr. 155.079.

Diffuse X–rays from the galactic plane.
See Abstr. 155.089.

Three–dimensional simulations of supernovae–dominated interstellar media in disk galaxies.
See Abstr. 157.179.

Limits on supernova rates and interstellar densities from X–ray observations of M101.
See Abstr. 157.193.

Redshifts of parent galaxies of supernovae. II.
See Abstr. 157.216.

Metal–enhanced galactic winds. I.
See Abstr. 157.258.

Far–infrared properties of multiple nucleus galaxies.
See Abstr. 158.325.

Young supernovae in the starburst galaxy M82.
See Abstr. 158.326.

126 Degenerate Stars, White Dwarfs, Pulsars

126.001 **Spectral signatures in the soft γ–ray emission from PSR 0833–45.**
M. Salvati.
Astron. Astrophys., Vol. 154, No. 1/2, p. 379 – 380 (1986).
Tümer et al. (1984) have detected soft γ–rays in the range $0.3 < E_\gamma < 30$ MeV from the Vela pulsar, PSR 0833–45. Their measurements are consistent with a differential number spectrum of slope 1.5. This value was predicted on theoretical grounds, and is expected to occur between two characteristic photon energies, related univocally to the location of the active region within the pulsar's magnetosphere. The author stresses the importance of such spectral signatures, and points out ways of exploiting to the fullest extent their information content.

126.002 **Period derivative and orbital eccentricity of binary pulsar 1953+29.**
L. A. Rawley, J. H. Taylor, M. M. Davis.
Nature, Vol. 319, No. 6052, p. 383 – 384 (1986).
PSR 1953+29, the second fastest of more than 400 known pulsars, has a period of 6.133 ms and is a member of a binary system with an orbital period of 117.35 days. The authors have determined two important parameters of the system, the pulsar period derivative $\dot{P}$ and the orbital eccentricity e. They find a small but non–zero eccentricity, $e = (330.3 \pm 0.4) \times 10^{-6}$, and a spindown rate of $\dot{P} = (1 \pm 6) \times 10^{-20}\mathrm{s\,s^{-1}}$, smaller than that measured for any other pulsar. These values are consistent with an evolutionary history involving spin–up during a mass accretion phase.

126.003 **Feasibility of a low–mass binary source progenitor for PSR1937+214.**
L. C. Jeffrey.
Nature, Vol. 319, No. 6052, p. 384 – 386 (1986).
One suggested progenitor system for the 1.5–ms pulsar PSR1937+214 is a close binary containing, in the late stages of its evolution, a neutron star and a very–low–mass ($\lesssim 0.1\ M_\odot$) degenerate dwarf helium star. The two main questions that affect the plausibility of this scenario are: (1) whether the system's evolution would in fact leave the neutron star at the observed high spin rate; and (2) whether the secondary could be disrupted, leaving a single pulsar. The authors have analysed such a system and find that the equilibrium rotation period predicted for the neutron star immediately before the disruption of the secondary is so large that matter accreted from a disk after disruption would be unlikely to spin the pulsar up to the observed period. It is thus unlikely that PSR1937+214 originated in a binary system in which the companion had an extremely low mass at disruption.

126.004 **The colors and luminosities of white dwarfs.**
O. J. Eggen.
Publ. Astron. Soc. Pac., Vol. 97, No. 597, p. 1029 – 1049 (1985).
Photoelectrically determined color indices of white dwarfs are reviewed with particular emphasis on the possibility that a large percentage of these objects in the solar neighborhood are members of the Hyades and Sirius superclusters. The astrometric, photometric, and spectroscopic data now available for the best–established cases of such membership are discussed. The color–luminosity array obtained from supercluster members is tested against that obtained with trigonometric parallaxes determined at the U.S. Naval Observatory.

126.005 **The pulsar velocities and their binary origins.**
V. Radhakrishnan, C. S. Shukre.
Astrophys. Space Sci., Vol. 118, Nos. 1/2, p. 329 – 336 (1986). – See Abstr. 012.001.
The observed proper–motions are shown to indicate that most pulsars come from binary systems. Due to a remarkable relationship between the speed of a pulsar and that of its progenitor in orbit, it is concluded that these binaries went through a common–envelope phase. This provides independent support to similar conclusions drawn from stellar evolution considerations.

126.006 **Low–luminosity stars.**
V. Trimble.
Q. J. R. Astron. Soc., Vol. 27, No. 1, p. 38 – 59 (1986).
The stars near the bottom of and below the main sequence, while individually inconspicuous, are the commonest self–luminous objects in our Galaxy and will eventually be the only ones shining. A number of issues concerning them are reviewed here. These issues include (1) existence, numbers and masses of stars just at or below the limit for hydrogen burning, (2) magnitudes and effects of white dwarf magnetic fields, (3) pulsating white dwarfs as probes of interior structure and composition, (4) determination and explanation of white dwarf surface compositions, and (5) white dwarf statistics.

126.007 **Null transition times, quantized drift modes, and no memory across nulls for PSR 1944+17.**
W. T. S. Deich, J. M. Cordes, T. H. Hankins, J. M. Rankin.
Astrophys. J., Vol. 300, No. 2, p. 540 – 550 (1986).
The authors have studied the behavior of the drifting subpulses and nulling in PSR 1944+17 ($P = 0.441$ s), an object which allows a large number of null transitions to be examined, since 60% of the pulses are null pulses. No evidence for phase memory is found, but there are differences between the last pulse before a null and the other pulses in a burst. Two time scales are observed for the onset of a null. There is a decline of the pulse intensity for several pulse periods, followed by a rapid drop to the null state on a time scale 2 ms $< \tau <$ 30 ms. The implications of the measurements are briefly discussed in terms of polar cap models.

126.008 **X–rays from the magnetic white dwarf PG 1658+441.**
S. H. Pravdo, F. E. Marshall, N. E. White, P. Giommi.
Astrophys. J., Vol. 300, No. 2, p. 819 – 823 (1986).
X–ray emission has been detected from PG 1658+441, a hot, isolated, magnetic white dwarf. The source was first detected in an Einstein IPC observation and was later identified in an EXOSAT channel multiplier array observation. The broad–band spectrum of this DA white dwarf can be explained as emission from a homogeneous, high–gravity, pure hydrogen atmosphere with a temperature near 28,000K. The observations of PG 1658+441 support the suggestion that a correlation exists between temperature and helium abundance in white dwarf atmospheres.

126.009 **Metal abundances in helium–rich white dwarf atmospheres.**
E.–M. Zeidler–K. T., V. Weidemann, D. Koester.
Astron. Astrophys., Vol. 155, No. 2, p. 356 – 370 (1986).
IUE observations of L745–46A, L119–34 and Ross 640 reveal the presence of Si, Fe, C, and very strong Mg II lines. These spectra, together with optical observations of white dwarfs with Ca II lines (type DZ) are analyzed within a network of model atmospheres with varying H/He, C/He and metals/He ratios. The influence of non–negligible H content (electron donor) is discussed. The final abundances are presented. Carbon is present in amounts comparable, perhaps somewhat smaller, than in white dwarfs of spectral type DQ (C_2 and UV C I lines). Its presence supports the carbon dredge–up hypothesis.

126.010 **The atmosphere of subluminous B stars. II. Analysis of 10 helium poor subdwarfs and the birthrate of sdB stars.**
U. Heber.
Astron. Astrophys., Vol. 155, No. 1, p. 33 – 45 (1986).
Ultraviolet and visual spectra of 10 helium poor hot subdwarfs (sdB's and sdOB's) are analyzed for effective temperatures, gravities and helium abundances. Effective temperatures range from 25000K to about 75000K. Helium is strongly depleted, the number fraction varying from less than 0.07% to 6%. Three of the program stars are binaries. It is concluded from the ($g - T_{\mathrm{eff}}$)–diagram that the subluminous B– and OB–stars form a single evolutionary group of extended horizontal branch stars which behave like helium main sequence stars with masses

$M \approx 0.5\ M_{\odot}$. The author derives statistical properties of the sdB and sdOB stars from a complete and homogeneous survey in a 840 deg^2 field at the south galactic pole. SdB and sdOB stars evolve directly into white dwarfs.

126.011 **CD –24°9052: a new bright and helium–rich subdwarf O star.**
D. Kilkenny, U. Heber, K. Hunger.
Astron. Astrophys., Vol. 155, No. 1, p. 175 – 177 (1986).
CD –24°9052 has been discovered to be a bright subdwarf O star ($m_v \approx 9.6$). A spectral analysis based on SAAO Reticon spectra and on a grid of NLTE model atmospheres yields the following parameters: $T_{eff} = 42500$K, $\log g = 5.0$, and the helium to hydrogen ratio of 0.5 by numbers. Accordingly, CD –24°9052 belongs to the subclass of helium rich sdO's.

126.012 **The radio luminosity and magnetic field of pulsars.**
S. Pineault.
Astrophys. J., Vol. 301, No. 1, p. 145 – 154 (1986).
Recent developments in pulsar theory provide the framework for a reanalysis of the relationship between radio luminosity and magnetic field. The statistical significance of a gap in the observed luminosity distribution near $\log L = 27.5$ is first discussed. A tentative explanation, involving both standard and disk pulsars, is found to be inconclusive. Within the framework of the isolated neutron star model, the luminosity of pulsars shows a good and nearly linear correlation with the observed magnetic field if one assumes that the braking index evolves from $n = 3$ to $n = 0$. In such a hybrid model, pulsars initially lose energy by the emission of magnetic dipole radiation, but, at longer periods, a new mechanism such as magnetic dipole radiation by superfluid neutron pairs (Huang et al. 1982) becomes dominant.

126.013 **A test of the dipole model for the rotating magnetic white dwarf Feige 7.**
B. Martin, D. T. Wickramasinghe.
Astrophys. J., Vol. 301, No. 1, p. 177 – 184 (1986).
Liebert et al. have presented observations of flux and circular polarization for Feige 7. They interpreted the star as a hot, rotating magnetic white dwarf and inferred values of the magnetic field strength, field geometry, and rotation angle. In this paper the authors use a detailed mathematical model of Feige 7 to test Liebert et al.'s conclusions. They generally find these inferences to be sound, with several important qualifications and alterations. In particular, it seems impossible to model the observed periodic variation in the positions of the absorption lines using a dipole field.

126.014 **Calculation of the spectra and integrated profiles of the radio emission of pulsars.**
A. D. Kuz'min, A. G. Solov'ev.
Astron. Zh., Tom 63, Vyp. 1, p. 62 – 70 (1986). In Russian. English translation in Sov. Astron., Vol. 30, No. 1.
A quantitative analysis of pulsars' radio emission in the framework of the hollow–cone and curvature–radiation model was made. The authors indicate a possibility of fitting this model to the observed spectra, including decreasing steepness at lower frequencies followed by low–frequency cutoff, and to the observed integrated profiles including their frequency evolution. They have determined the frequency dependence of the radius of the emitting region, the energy spectrum of emitting particles and its radial distribution which provided such fitting. They propose to use the scheme considered for mathematical modeling of the characteristics of pulsar radio emission.

126.015 **A new determination of the luminosity function of pulsars.**
O. Kh. Gusejnov, I. M. Yusifov.
Astron. Zh., Tom 63, Vyp. 1, p. 78 – 83 (1986). In Russian. English translation in Sov. Astron., Vol. 30, No. 1.
Pulsars' luminosities are calculated on the basis of new distances to them. Possible values of intrinsic (true) widths of radiation diagrams, but not visible widths of pulses, are used. The luminosity function of pulsars, taking into account the change of the beaming factor with increase of the rotation period is found.

126.016 **Investigation of the correlation of micropulse emission at close frequencies.**
T. V. Smirnova, V. A. Soglasnov, M. V. Popov, A. Yu. Novikov.
Astron. Zh., Tom 63, Vyp. 1, p. 84 – 93 (1986). In Russian. English translation in Sov. Astron., Vol. 30, No. 1.
The pulsars PSR 0809+74 and PSR 1133+16 were observed at the frequencies 102.5 and 101 MHz simultaneously with 10 μs time resolution. It is shown from a cross–correlation analysis that the microstructure with timescales exceeding 300 μs is highly correlated at both frequencies, but such a correlation is absent for microstructure with small timescale (<300 μs). This difference between short timescale and long timescale microstructure can be explained in the polar cap model, due to the complicated structure of the pulsar's emitting region. For PSR 0809+74 a new value of the dispersion measure is obtained from time delay between micropulses at the two frequencies, $DM = 5.751 \pm 0.003\ cm^{-3}$pc.

126.017 **Velocities and birthplaces of pulsars.**
P. R. Amnuel (*P. R. Amnuehl'), O. H. Guseinov (O. Kh. Gusejnov*), Yu. S. Rustamov.
Astrophys. Space Sci., Vol. 121, No. 1, p. 1 – 20 (1986).
The values of the proper motions of 36 pulsars are analysed. It is shown that the birthplaces of pulsars are located within the OB–associations and/or in the galactic arms. Eleven cases of pulsar formation in associations, and fourteen cases of formation in the galactic arms, have been established. More precise parameters of pulsars – such as: distances, ages, spatial velocity components –, and the spatial velocity distribution of pulsars are constructed.

126.018 **KPD 2146+4117: the coolest DBA white dwarf.**
R. A. Downes, H. L. Shipman.
Publ. Astron. Soc. Pac., Vol. 98, No. 599, p. 99 – 101 (1986).
The authors report the discovery and analysis of a new member of the small class of DBA stars – white dwarfs with He–dominated atmospheres that show visible Balmer lines in their spectra. This object, KPD 2146+4117, has $T_{eff} = 12{,}500\text{K} \pm 500\text{K}$ and $\text{H/He} = (1.2 \pm 0.4) \times 10^{-4}$, and is the coolest DBA discovered to date.

126.019 **High–radio–frequency survey for young and millisecond pulsars.**
T. R. Clifton, A. G. Lyne.
Nature, Vol. 320, No. 6057, p. 43 – 45 (1986).
A high–radio–frequency survey for distant pulsars is being carried out at Jodrell Bank. An area of 200 square degrees along the galactic plane has been observed with high sensitivity for objects having both short periods ($P \geqslant 0.004$ s) and large dispersion measures ($\leqslant 35{,}000\ P\ cm^{-3}$pc). The authors have confirmed 32 new pulsars, including one of period 85 ms and two objects of longer period which both have characteristic ages of the order of 20,000 yr. Only three known pulsars are younger than this. Most of the new pulsars have very high dispersion measure ($\geqslant 400\ cm^{-3}$pc), and many display significant multi–path scattering.

126.020 **Toward an empirical theory of pulsar emission. III. Mode changing, drifting subpulses, and pulse nulling.**
J. M. Rankin.
Astrophys. J., Vol. 301, No. 2, p. 901 – 922 (1986).
Three well–known pulse–sequence phenomena – mode changing, drifting subpulses, and pulse nulling – are considered in this paper with the intent of examining their implications for the empirical model developed in the first paper of this study. Although this original analysis was confined to pulsar average profiles, the resulting model should be pertinent to the study of individual–pulse sequences as well. It it shown that, indeed, the species–specific characteristics of the three effects provide strong, independent confirmation of many of the premises in the first paper.

126.021 **The direct detection of the slow pace of stellar evolution.**
H. Shipman.
Phys. Today, Vol. 39, No. 1, p. S11 – S12 (1986). Short review paper. – See Abstr. 013.021.

126.022 **The geometry of emission from the Vela and the Crab pulsars.**
F. G. Smith.
Mon. Not. R. Astron. Soc., Vol. 219, No. 4, p. 729 – 736 (1986).
It is well known that the Crab Pulsar emits narrow beams of radiation with very similar characteristics over a very wide range of the electromagnetic spectrum, but the geometric conditions which determine the beaming have so far not been properly understood. The Vela Pulsar emits beams which are completely different in the radio and the gamma–ray spectral regions, and several authors have suggested that entirely different emission regions are involved. This note presents a unified geometrical scheme for the location of the emission regions, which applies to both pulsars. The radio emission is located close to one magnetic pole, while the gamma–ray emission is located in the field lines diverging from the other magnetic pole, at a radial distance of 0.9 of the velocity–of–light radius.

126.023 **Newly observed degenerate stars and related objects from the Praesepe, Case, and Kiso surveys.**
R. M. Wagner, E. M. Sion, J. Liebert, S. G. Starrfield, N. Zotov.
Publ. Astron. Soc. Pac., Vol. 98, No. 604, p. 552 – 555 (1986).
The authors report the discovery of 15 white dwarfs from among blue candidates in the Low Dispersion Northern Sky Case Survey and the Kiso Schmidt camera survey. The observations were made at the 2.3–m Steward telescope, the KPNO 2.1–m telescope, and the Multiple Mirror Telescope. Spectroscopy of Praesepe white–dwarf candidates revealed a DA star, and a possible DC star, which if confirmed would be only the second known non–DA white dwarf in an open cluster. One Praesepe white–dwarf candidate is a QSO with $z = 2.1$.

126.024 **Space distribution of pulsars in the Galaxy and the nature of gamma–sources.**
B. M. Vladimirskij.
Izv. Krymskoj Astrofiz. Obs., Tom 72, p. 125 – 131 (1985). In Russian. English translation in Bull. Crimean Astrophys. Obs., Vol. 72.
Pulsars probably connected with γ–sources are concentrated in spiral arms of the Galaxy. This fact was used to estimate the distance to the objects in the 2 COS–B–catalogue. It was found that the mean distance to 18 objects from this catalogue is about 2.9 – 3.3 kpc. The mean distance to the galactic plane is equal to 50 – 70 pc, their averaged gamma–luminosity $L_\gamma(E > 100\ \mathrm{MeV}) = (5-9)\times 10^{34}\mathrm{erg\ s^{-1}}$. The gamma–luminosity function may be presented as $N(>L_\gamma) \propto L_\gamma^{0.2\pm0.1}$.

126.025 **On the magnetic field of pulsars.**
Yu. I. Neshpor.
Izv. Krymskoj Astrofiz. Obs., Tom 72, p. 131 – 134 (1985). In Russian. English translation in Bull. Crimean Astrophys. Obs., Vol. 72.
On the basis of an analysis of the published data the author concludes that the full strength of the magnetic field of a pulsar is not age dependent. The magnetic field component $H_\perp$ decreases with the distance to the galactic plane. It might be explained by the variation of the angle between the magnetic field axis and the rotation axis of the pulsar which decreases with its age.

126.026 **Evolution of pulsars.**
O. Kh. Gusejnov, I. M. Yusifov.
Astron. Zh., Tom 63, Vyp. 2, p. 265 – 273 (1986). In Russian. English translation in Sov. Astron., Vol. 30, No. 2.
It is shown that most pulsars lose their rotation energy due to their magnetic–dipole radiation. The characteristic timescale of the decrease of the magnetic field component perpendicular to the rotation axis is $\sim 10^7$ years. The evolutionary tracks of pulsars on the P–$\dot{P}$ diagram are considered. The time of evolution of most known pulsars is $\sim(5-10)\times 10^6$ years. This time for concrete pulsars may vary from 10^6 to 10^8 years.

126.027 **Mode–switching, nulling and drifting subpulses in PSR 1112+50.**
G. A. E. Wright, W. Sieber, A. Wolszczan.
Astron. Astrophys., Vol. 160, No. 2, p. 402 – 405 (1986).
The authors present a single–pulse analysis of the little–known pulsar PSR 1112+50. When not in the null state the pulsar adopts one of three modes of emission. In one of these modes the subpulses in the trailing component drift to later times with a periodicity (P_3) of about six times the rotation period. The double–peaked integrated profile and the asymmetric mode behaviour are reminiscent of the younger pulsar PSR 1822–09.

126.028 **Longitudinal plasma waves generated in the atmosphere of pulsars.**
M. A. Mamedov.
Izv. Akad. Nauk AzSSR. Ser. fiz.–tekh. i mat. nauk, Tom 6, No. 2, p. 84 – 93 (1985). In Russian. Abstr. in Ref. Zh., 51. Astron., 2.51.686 (1986).

126.029 **Voyager observations of PG 1159 objects.**
J. Basile, J. B. Holberg.
Bull. Am. Astron. Soc., Vol. 17, No. 4, p. 838 (1985). Abstract. – See Abstr. 010.062.

126.030 **EXOSAT soft X–ray observations of PG 1159–035.**
J. B. Holberg, M. A. Barstow.
Bull. Am. Astron. Soc., Vol. 17, No. 4, p. 838 (1985). Abstract. – See Abstr. 010.062.

126.031 **Spectrophotometric and fast photometric studies of stars near the PG 1159–035 instability strip.**
J. Liebert, A. D. Grauer, R. F. Green.
Bull. Am. Astron. Soc., Vol. 17, No. 4, p. 838 – 839 (1985). Abstract. – See Abstr. 010.062.

126.032 **Detection and analysis of photospheric metals in the peculiar DOZ1 degenerate, KPD 0005+5106.**
E. M. Sion, R. Downes, J. W. Liebert.
Bull. Am. Astron. Soc., Vol. 17, No. 4, p. 839 (1985). Abstract. – See Abstr. 010.062.

126.033 **Upcoming EUV/FUV observations of hot white dwarfs with the Hopkins Ultraviolet Telescope.**
R. A. Kimble, K. S. Long, A. F. Davidsen.
Bull. Am. Astron. Soc., Vol. 17, No. 4, p. 839 (1985). Abstract. – See Abstr. 010.062.

126.034 **IUE observations of pulsating DA white dwarfs and the effective temperature – period correlation of ZZ Ceti stars.**
F. Wesemael, R. Lamontagne, G. Fontaine.
Bull. Am. Astron. Soc., Vol. 17, No. 4, p. 839 (1985). Abstract. – See Abstr. 010.062.

126.035 **A review of recent progress in the study of hot, pulsating white dwarf stars.**
D. E. Winget.
Bull. Am. Astron. Soc., Vol. 17, No. 4, p. 853 (1985). Abstract. – See Abstr. 010.062.

126.036 **Model atmospheres of DA white dwarfs. II. Catalogue.**
A. Galdikas.
Vilniaus Astron. Obs. Biul., Nr. 72, 91 pp. (1985). In Russian.
A catalogue of 60 model atmospheres of DA white dwarfs with $T_e = 6000-18000\mathrm{K}$ and $\log g = 7-9$ is given. Other tables contain data on the emergent energy fluxes of the models in the wavelength interval from 912 Å to 300 μm in units $\mathrm{erg\ cm^{-2}sec^{-1}Å^{-1}}$ and their color indices in the UBV and uvby photometric systems.

126.037 **Model atmospheres of DQ white dwarfs. I. Details of computation.**
A. Galdikas.
Vilniaus Astron. Obs. Biul., Nr. 73, p. 3 – 23 (1985).
A method used to calculate model atmospheres of cool helium–rich white dwarfs with admixture of carbon is described. Details of computation of carbon line profiles and Swan bands of the C_2 molecule are given. A method for determination temperature corrections in the convection zone is presented.

126.038 **Model atmospheres of DQ white dwarfs. II. Catalogue.**
A. Galdikas.
Vilniaus Astron. Obs. Biul., Nr. 73, p. 24 – 65 (1985).
A catalogue of 24 model atmospheres of helium–rich white dwarfs with admixture of carbon is given. The model atmospheres are computed for T_e = 7000 – 12000K, log g = 8, and log (C/He) = –7.0, –6.0, –5.0, and –4.0. Other tables contain data on the emergent energy fluxes of the models and their color indices in the UBV and uvby photometric systems.

126.039 **On mass distribution of white dwarfs in cataclysmic variables.**
Kh. I. Novruzova.
Astron. Tsirk., No. 1399, p. 6 – 7 (1985). In Russian.

126.040 **Observations of pulsars at high frequencies.**
A. D. Kuz'min, V. M. Malofeev, V. A. Izvekova, V. Ziver, R. Wielebinski.
Fiz. inst. Akad. Nauk SSSR. Prepr., No. 210, 52 pp. (1985). In Russian. Abstr. in Ref. Zh., 51. Astron., 4.51.555 (1986).

126.041 **Investigation of the microstructure of the pulsars PSR 0809+74, 0950+08, and 1133+16 in the 67 – 102 MHz range.**
M. V. Popov, T. V. Smirnova, V. A. Soglasnov.
Fiz. inst. Akad. Nauk SSSR. Prepr., No. 288, 35 pp. (1985). In Russian. Abstr. in Ref. Zh., 51. Astron., 4.51.556 (1986).

126.042 **Investigation of the effect of "mode switch" on the polarization and temporal properties of pulses of the pulsar PSR 0943+10 at 103 MHz frequency.**
S. A. Sulejmanova, T. V. Tabanova, V. D. Pugachev.
Fiz. inst. Akad. Nauk SSSR. Prepr., No. 262, 35 pp. (1985). In Russian. Abstr. in Ref. Zh., 51. Astron., 4.51.558 (1986).

126.043 **The fields of pulsars and magnetic white dwarf stars.**
C. C. Leiby Jr.
Astrophys. Space Sci., Vol. 122, No. 1, p. 151 – 156 (1986).
A vector theory of electromagnetism and gravitation has indicated a possible equivalence of gravitational energy and electric charge. If true, gravitating masses should exhibit external electric (E) and magnetic (B) fields proportional to $M\varrho$ and $M\varrho v$, respectively. (M, mass; ϱ, density; and v, rotational velocity.) Accordingly, pulsars should exhibit surface fields $3\times10^{21} \gtrsim E \gtrsim 3\times10^{19}$V m^{-1} (consistent with inferred possible particle energies $\sim10^{25}$eV) and $6\times10^{14} \gtrsim B \gtrsim 10^{10}$G (consistent with $B\sim5\times10^{12}$G, inferred from magnetic dipole radiation effects). For magnetic white dwarfs, predicted fields are $E \lesssim 2\times10^{16}$V m^{-1} (consistent with the featureless spectra of DC stars) and $B \lesssim 1.2\times10^{8}$G (consistent with observations).

126.044 **Southern ZZ Ceti stars.**
D. O'Donoghue.
Seismology of the Sun and the distant stars, p. 467 – 472 (1986). – See Abstr. 012.026.
The author presents preliminary results of recent observations of the low amplitude southern ZZ Ceti, MY Aps (= L 19–2) which confirm the inequality in the separation of the frequency triplet comprising the 192 s oscillation seen in this star. This inequality is in reasonable agreement with the calculation by Chlebowski and 2nd order effects in the removal of pulsation mode degeneracy by rotation. The author also presents an initial analysis of the frequency structure of a large amplitude southern ZZ Ceti, VY Hor (= BPM 31594).

126.045 **On mode changes in ZZ Ceti stars.**
D. O'Donoghue.
Mon. Not. R. Astron. Soc., Vol. 220, No. 2, p. 19P – 22P (1986).
The author discusses the observations of three ZZ Ceti stars which most strongly suggest that oscillation mode changes occur in these stars on a very short time–scale. He demonstrates (1) that the reported mode change in VY Hor (= BPM 31594) is spurious; (2) that the evidence for a mode change in GD 385 is at best tentative and (3) that an alternative, plausible explanation in terms of beating is viable for BG CVn (= GD 154).

126.046 **A comparison of high–frequency and low–frequency characteristics of pulsars.**
A. D. Kuz'min, V. M. Malofeev, V. A. Izvekova, W. Sieber, R. Wielebinski.
Astron. Astrophys., Vol. 161, No. 1, p. 183 – 194 (1986).
The authors have measured integrated pulse profiles of 27 pulsars at 4.6 GHz and the same number at 10.7 GHz, determining their integrated flux values. During these observations 16 pulsars were observed quasi–simultaneously at 102.7 MHz so that phase shifts in the arrival times of the integrated pulse profiles could be determined. The authors discuss the validity of the current dispersion measure determinations which rely exclusively on integrated pulse profiles. The experimental results are discussed and interpreted in the framework of a two–component magnetic field model. It is suggested that the dipole fields dominate at large distances from the neutron star, while at small distances quadrupole components of the magnetic field prevail.

126.047 **Are many pulsars processed in binaries?**
V. Radhakrishnan, G. Srinivasan.
Proceedings of the Second Asian–Pacific Regional Meeting on Astronomy, p. 423 – 431 (1984). – See Abstr. 012.031.
Although many pulsars must be born in binary systems, only those that have been "processed" in close binaries are likely to have distinguishable characteristics. These result from the field decay during the time between the two explosions, and the spin–up history during the X–ray phase. It is shown that in addition to the binary pulsar 1913+16, there are two others which must also have had a close–binary history. It is predicted that they should show high proper motions. It is concluded that these three together account fairly well for the observed number of recycled pulsars expected from the known statistics of binary systems.

126.048 **Flux densities, spectra and variability of pulsars at metre wavelengths.**
O. B. Slee, S. K. Alurkar, A. D. Bobra.
Aust. J. Phys., Vol. 39, No. 1, p. 103 – 114 (1986).
The authors present the results of two–frequency flux density measurements of 74 pulsars with the Culgoora circular array. They show that the spectral index of a typical pulsar steepens markedly from 80 to 1400 MHz, but they found no significant relationship between the metre–wave spectral index and the published pulsar parameters.

126.049 **Discovery of a 50 msec pulsar in the Large Magellanic Cloud.**
F. D. Seward, F. R. Harnden Jr., D. J. Helfand.
X–ray astronomy '84, p. 297 – 300 (1985). – See Abstr. 012.033.
The authors report here the discovery of a new pulsed X–ray source in the Large Magellanic Cloud supernova remnant 0540–693 that, in many ways, is remarkably similar to the Crab Nebula.

126.050 **Discovery of a Crab–like pulsar in the LMC.**
F. R. Harnden Jr.
X–ray astronomy '84, p. 319 (1985). Abstract. – See Abstr. 012.033.

126.051 **Recent results on the Crab–pulsar X–ray light curve.**
G. Hasinger.
X–ray astronomy '84, p. 321 – 324 (1985). – See Abstr. 012.033.
Recent X–ray data of NP 0532 from the HEXE balloon experiment and the EXOSAT observatory are presented. The measurements are compared with a pulsar model, in which the two pulses

are generated by synchrotron radiation near the last closed field lines above a single pole of a magnetized rotating neutron star.

126.052 **Fast rotating pulsars.**
J. H. Taylor.
Japan – U.S. seminar on galactic and extragalactic compact X-ray sources, p. 1 – 5 (1985). – See Abstr. 012.034.
The author reviews the observational facts concerning the galactic population of radio pulsars, with special emphasis on the newly discovered category of unusually fast rotating pulsars.

126.053 **X-rays from non-accreting pulsars.**
F. R. Harnden Jr.
Japan – U.S. seminar on galactic and extragalactic compact X-ray sources, p. 239 – 243 (1985). – See Abstr. 012.034.
Most neutron stars are found in such rich environments that their surface temperatures are undoubtedly determined by external factors rather than by the physics of their interiors. Isolated pulsars found in young supernova remnants, however, may provide constraints on theories of nuclear matter.

126.054 **A search for short period pulsars and its implication for the period distribution of pulsars.**
G. H. Stokes.
Diss. Abstr. Int., Sect. B, Vol. 46, No. 9, p. 3092 (1986). Thesis, Princeton University, 94 pp. (1985). Order No. DA8523995.

126.055 **Pulse timing and neutron star structure.**
R. Nandkumar.
Diss. Abstr. Int., Sect. B, Vol. 46, No. 11, p. 3882 (1986). Thesis, University of Illinois at Urbana-Champaign, 134 pp. (1985). Order No. DA8600274.

126.056 **The ultraviolet spectra of DA white dwarfs.**
E. P. Nelan.
Diss. Abstr. Int., Sect. B, Vol. 46, No. 11, p. 3882 (1986). Thesis, Dartmouth College, 213 pp. (1985). Order No. DA8601400.

126.057 **Determination of the white dwarf's magnetic axis from X-ray orbital light curves.**
I. L. Andronov.
Astron. Zh., Tom 63, Vyp. 3, p. 508 – 515 (1986). In Russian. English translation in Sov. Astron., Vol. 30, No. 3.
The beamshape of soft X-ray emission produced in the «polar cap» on a white dwarf's surface is derived, taking into account the absorption in the axially symmetric accretion column, homogeneous along its height. An algorithm for the determination of orientation of the magnetic axis of a compact star from orbital curves of soft X-ray flux is suggested. The values of the orbital inclination ($51° \leqslant i \leqslant 64°$) and of the angle between the rotational and magnetic axes ($30° \leqslant \delta \leqslant 34°$) were calculated for the polar AM Herculis for different values of model parameters.

126.058 **Are pulsars born as slow rotators?**
R. A. Chevalier, R. T. Emmering.
Astrophys. J., Vol. 304, No. 1, p. 140 – 153 (1986).
The authors have modeled the observed properties of pulsars on the assumption that they evolve by emitting magnetic dipole radiation and have exponentially decaying magnetic fields. The observed properties are affected by the dependence of radio luminosity (L) on period (P) and period derivative ($\dot{P}$). If one adopts the recently determined relation $L \propto P^{-1}\dot{P}^{1/3}$, then the observed period distribution is well fitted by a model which assumes flat initial distributions with initial periods ranging from 0.09 to 0.25 s. The authors have also estimated the distribution of pulsar z-velocities from observations and used it with their model to calculate the variation of pulsar number, period, and characteristic age with height z.

126.059 **Frequency dependence of the main pulse to interpulse separation for seven pulsars.**
T. H. Hankins, L. A. Fowler.
Astrophys. J., Vol. 304, No. 1, p. 256 – 264 (1986).
In the course of a study of pulsar interpulses the authors searched 43 bright pulsars for interpulse emission. They found a previously undetected interpulse in PSR 1944+17 and set improved upper limits on the others. The authors measured the separation between the main pulse (MP) and interpulse (IP) and the IP to MP amplitude ratio S_{IP}/S_{MP} for seven of the pulsars known to have interpulses. The ratio S_{IP}/S_{MP} decreases with frequency for all seven objects, although for the Crab and millisecond pulsar the IP is distinctly different from the others in that the ratio is much larger and less strongly frequency dependent.

126.060 **The frequency of hydrogen white dwarfs as observed at high signal to noise ratio.**
J. L. Greenstein.
Astrophys. J., Vol. 304, No. 1, p. 334 – 355 (1986).
Spectra of white dwarfs with the double CCD spectrograph at Palomar were studied at unusually high signal-to-noise ratio, ~20,000 detected photons per pixel. Of 140 stars, 121 were classified to give composition type and a refined temperature scale. The number of line-free (DC) degenerates was drastically reduced. Many had weak C_2 bands or He I lines. Detection of weak and sharp Hα lead to a discussion of the relative frequency of hydrogen and helium-dominated atmospheres in these old stars. Predominance of non-DA stars raises questions as to the fate of the accreted and residual hydrogen in postasymptotic giant branch evolution.

126.061 **An X-ray survey of hot white dwarf stars: evidence for a n(He)/n(H) versus T_{eff} correlation.**
R. Petre, H. L. Shipman, C. R. Canizares.
Astrophys. J., Vol. 304, No. 1, p. 356 – 364 (1986).
The authors report on observations of 13 white dwarf and subdwarf stars using the Einstein Observatory. Included are stars of classes DA, DB, DAV, sDO, and sDB, with optically determined effective temperatures in the range $1 - 6 \times 10^4$K. X-ray emission was detected from two of the 13: the very hot DA1 star WD 2309+105 and the relatively cool DA3 star WD 1052–273. The effective temperatures determined from ultraviolet and optical observations are used to place limits on the He content of the white dwarf photospheres. When one combines the results with those available from the literature there is evidence for a correlation between T_{eff} and n(He)/n(H), in which HZ 43 is a conspicuous exception to the general trend.

126.062 **ER 8: a very low luminosity degenerate star.**
M. T. Ruiz, J. Maza, M. Wischnjewsky,
L. E. González.
Astrophys. J., Lett. Ed., Vol. 304, No. 1, p. L25 – L28 (1986).
In a search for faint, high proper motion stars, a very cold ($T_{BB} = 3500 \pm 100$K) degenerate star with a featureless spectra has been found. Its proper motion is $\mu = 2.128 \pm 0.008$ arcsec yr^{-1}, and its apparent visual magnitude is $m_v = 17.2 \pm 0.1$. Available data suggest that ER 8 is a possible member of the Hyades Supercluster implying a distance $d \leqslant 4.5$ pc, a luminosity $L \leqslant 5.9 \times 10^{-6} L_\odot$, and an age less than 10^9yr.

126.063 **On the angles between the axis of the magnetic dipole and the axis of rotation of pulsars.**
I. F. Malov.
Astrofizika, Tom 24, Vyp. 3, p. 507 – 522 (1986). In Russian. English translation in Astrophysics, Vol. 24, No. 3.
A critical survey of known methods for the determination of the angle β between the magnetic field and the rotation axis of pulsars is given. More accurate estimates of this angle are obtained on the basis of observed profile widths for 282 pulsars and observed position angles for 63 pulsars. Polarization data and structure of observed profiles gave the possibility to estimate the angle β for pulsars with interpulses: 5 pulsars have small β (<10°), for 3 pulsars β ~ 90°. In addition 4 pulsars with small β and two pulsars with β ≳ 80° may have interpulses or interpulse radiation.

126.064 **The chemical composition of the cool white dwarf LP701–29.**
S. Kapranidis, J. Liebert.
Prepr. Steward Obs., No. 617, 17 pp. (1986). To appear in Astrophys. J.

126.065 **The luminosity function of DA white dwarfs.**
T. A. Fleming, J. Liebert, R. F. Green.
Prepr. Steward Obs., No. 637, 45 pp. (1986).

126.066 **The new magnetic white dwarf PG 1031+234: polarization and field structure at more than 500 million Gauss.**
G. D. Schmidt, S. C. West, J. Liebert, R. F. Green, H. S. Stockman.
Prepr. Steward Obs., No. 639, 52 pp. (1986).

126.067 **Temperatures for hot and pulsating DB white dwarfs obtained with the IUE Observatory.**
J. Liebert, F. Wesemael, C. J. Hansen, G. Fontaine, H. L. Shipman, E. M. Sion, D. E. Winget, R. F. Green.
Prepr. Steward Obs., No. 650, 47 pp. (1986). To appear in Astrophys. J.

126.068 **An analysis of pulsar timing irregularities: is it induced by a source of gravitational waves in the Galaxy?**
M. Cerdonio, S. Vitale.
Nuovo Cimento C, Vol. 8C, Ser. 1, No. 4, p. 410 – 414 (1985). Abstr. in Phys. Abstr., Vol. 89, No. 1276, Entry 36203 (1986).

126.069 **Pulsar radiation.**
E. Asseo.
Recherche, Vol. 16, No. 171, p. 1314 – 1323 (1985). In French. Abstr. in Phys. Abstr., Vol. 89, No. 1276, Entry 36204 (1986).

126.070 **The outcome of mass transfer in a carbon–oxygen white dwarf binary system.**
A. M. Khokhlov.
Sov. Astron. Lett., Vol. 11, No. 5, p. 318 – 321 (1985). English translation of 40.126.026.

126.071 **Pulsar PSR 0656+14: period and spindown.**
D. Domingue, J. M. Rankin, J. M. Weisberg, P. R. Backus.
Astron. Astrophys., Vol. 161, No. 2, p. 303 – 304 (1986).
On July 7, 1984 pulsar 0656+14 was found to exhibit a larger period than expected. Using observations from a number of sources, the authors have attempted to explain this discrepancy. No indication of orbital motion was found. The available evidence indicates that the star has a period derivative of $54x10^{-15}ss^{-1}$ rather than $1.6x10^{-15}ss^{-1}$ as previously reported.

126.072 **Time–averaged ultraviolet observations of pulsating DA white dwarfs from the IUE observatory.**
F. Wesemael, R. Lamontagne, G. Fontaine.
Astron. J., Vol. 91, No. 6, p. 1376 – 1381 (1986).
IUE observations in the range 1200 – 1900 Å have been obtained for four ZZ Ceti stars. The spectra reveal the 1400 and 1600 Å absorption features, common to all DA stars in that temperature range, but also previously unreported structures near 1520 Å. These observations, together with available archival data for three additional ZZ Ceti stars, are analyzed with model atmosphere fluxes incorporating recently identified quasimolecular opacities. The instability strip is found to cover the approximate range $11400K \lesssim T_e \lesssim 13000K$. The results also confirm the existence of a period–effective temperature correlation for ZZ Ceti stars, as predicted by pulsation theory.

126.073 **Geometrical investigation of the linear polarization position angle of the pulsar's radiation.**
X.-j. Wu, G.-j. Qiao, X. y. Xia.
Kexue Tongbao, Vol. 30, No. 12, p. 1597 – 1600 (1985).
Polarization is one of the striking characteristics of pulsar's emission. An important problem is whether the observed polarization data and the apparent beam width are in quantitative agreement with the geometrical relationship of the polar–cap model. The statistical result of the observed parameters can be fitted by a theoretical relation. It shows that the polar–cap model is supported by the observational data.

126.074 **The chemical composition of the cool white dwarf LP 701–29.**
S. Kapranidis, J. Liebert.
Astrophys. J., Vol. 305, No. 2, p. 863 – 866 (1986).
The authors present new observations and a new synthetic spectrum of the cool white dwarf LP 701–29. The analysis suggests that the atmosphere of this star is helium dominated, with enhanced metal abundances compared to the usual abundances of white dwarfs. The best fit is achieved with an effective temperature of 4500K and an abundance of heavy elements of the order 10^{-2} of the solar ratio of metals to helium.

126.075 **Nonlinear temporal modulation of pulsar radio emission.**
A. C.-L. Chian.
Rev. Mex. Astron. Astrofis., Vol. 12, Spec. Issue, p. 330 (1986). Abstract. – See Abstr. 012.066.
The whole paper is published in Astrophys. Space Sci., Vol. 97, p. 9 – 18 (1983).– See Abstr. 34.141.117.

126.076 **A measurement of secular evolution in the pre–white–dwarf star PG 1159–035.**
D. E. Winget, S. O. Kepler, E. L. Robinson, R. E. Nather, D. O'Donoghue.
Rev. Mex. Astron. Astrofis., Vol. 12, Spec. Issue, p. 342 (1986). Abstract. – See Abstr. 012.066.

126.077 **Light and line profile variations due to r–mode pulsations. With an application to the ZZ Ceti star G117–B15A.**
S. O. Kepler.
Rev. Mex. Astron. Astrofis., Vol. 12, Spec. Issue, p. 343 (1986). Abstract. – See Abstr. 012.066.

126.078 **Binary pulsars 0655+64 and 0820+02.**
IAU Circ., No. 4160 (1986).

126.079 **New millisecond binary pulsar.**
IAU Circ., No. 4162 (1986).

The Palomar–Green Catalog of Ultraviolet–Excess Stellar Objects.
See Abstr. 002.074.

Black holes, white dwarfs and neutron stars. The physics of compact objects.
See Abstr. 003.178.

Measurements of pulsar parallaxes by VLBI.
See Abstr. 041.031.

Pulsar time scale–PT.
See Abstr. 044.078.

Studies of isolated neutron stars, pulsars and pulsar–driven nebulae with the AXAF.
See Abstr. 051.047.

Strahlungstransport in den Außenschichten weißer Zwerge mit Linienabsorption unter Einwirkung starker Magnetfelder.
See Abstr. 063.106.

Bombardment solutions to the "soft X–ray puzzle" in radial white dwarf accretion.
See Abstr. 064.019.

Evolution of white dwarfs.
See Abstr. 065.003.

Observational implications of the explosion of white dwarfs.
See Abstr. 065.019.

The effects of diffusion due to gravity and due to composition gradients on the rate of hydrogen burning in a cooling degenerate dwarf. II. Dependence on initial metallicity and on buffer mass.
See Abstr. 065.027.

Models for Type I supernova. I. Detonations in white dwarfs.
See Abstr. 065.038.

Evolutionary period changes in variable helium–rich white dwarfs.
See Abstr. 065.063.

On the evolution of binary components which first fill their Roche lobes after the exhaustion of central helium.
See Abstr. 065.091.

Relaxation oscillations of a helium–burning star: helium shell flashes not triggered by the accumulation of a critical mass.
See Abstr. 065.092.

Diffusion coefficients for stellar plasmas.
See Abstr. 065.096.

Diffusion in white dwarfs: new results and comparative study.
See Abstr. 065.097.

Nucleosynthesis in He–flashes on accreting white dwarfs.
See Abstr. 065.104.

The origin and evolution of helium–rich white dwarfs.
See Abstr. 065.106.

Carbon pollution in helium–rich white dwarfs: time–dependent calculations.
See Abstr. 065.114.

Lower bounds on the masses of rapidly rotating white dwarfs.
See Abstr. 065.115.

Peculiar supernova of type I as a result of explosive helium burning in an accreting white dwarf.
See Abstr. 065.124.

General relativity confronts experiment.
See Abstr. 066.030.

Timing observations of the pulsar signal propagations in the galactic gravitational field as precision tests of the Einstein Equivalence Principle.
See Abstr. 066.103.

Electron–positron pair production by ultrarelativistic electrons in a soft photon field.
See Abstr. 066.115.

Estimation of the gravitational radiation (GR) from the pulsar PSR 1937 + 214.
See Abstr. 066.121.

Properties of condensed matter in very strong magnetic fields.
See Abstr. 067.003.

Estimation of some parameters of pulsars and their applications.
See Abstr. 067.009.

Pulsar magnetosphere with conspicuous trans–field flow.
See Abstr. 067.010.

Energetic radiation from rapidly spinning pulsars. I. Outer magnetosphere gaps.
See Abstr. 067.013.

Energetic radiation from rapidly spinning pulsars. II. Vela and Crab.
See Abstr. 067.014.

Models of magnetic white dwarfs.
See Abstr. 067.026.

Cosmic superfluidity: the evidence for superfluidity in neutron stars.
See Abstr. 067.038.

About observational manifestation of pion stars.
See Abstr. 067.049.

The origin of high velocity pulsars.
See Abstr. 067.071.

A pulsar magnetosphere model with streaming.
See Abstr. 067.073.

Some comments on neutron star formation.
See Abstr. 067.081.

Discussion about pulsar's braking index n.
See Abstr. 067.107.

Magnetic field decay and the origin of neutron star binaries.
See Abstr. 067.112.

Confronting X–ray observations of young supernova remnants with neutron star cooling models.
See Abstr. 067.115.

Oscillation spectra of neutron stars with strong magnetic fields.
See Abstr. 067.147.

Photoionization of positronium in strong magnetic fields.
See Abstr. 067.158.

Prospects for observing the occultation of pulsars by Halley's comet.
See Abstr. 103.601.

New subdwarfs. V. Radial velocities for 889 high–proper–motion stars measured with the Mount Wilson 100 inch reflector.
See Abstr. 111.016.

On the number of comets around other single stars.
See Abstr. 112.040.

Interstellar broadening of maser spots.
See Abstr. 112.111.

New subdwarfs. IV. UBV photometry of 1690 high–proper–motion stars.
See Abstr. 113.053.

Spectral types for objects in the Kiso survey. II. Data for 61 stars.
See Abstr. 114.084.

On the nature of the dwarf carbon star G77–61.
See Abstr. 115.001.

Unscrambling the mess at the left hand side of the HR diagram.
See Abstr. 115.014.

Massive X–ray binaries: their physics and evolution.
See Abstr. 117.020.

Photometry, polarimetry, and spectroscopy of AM Herculis variables.
See Abstr. 117.024.

The optical continua of magnetic variables.
See Abstr. 117.034.

The mass spectrum of the white dwarfs in cataclysmic binaries.
See Abstr. 117.067.

X–ray observations of the AM Her star CW 1103+254.
See Abstr. 117.174.

The soft X–ray superoutburst of VW Hydri: 14 second periodicity.
See Abstr. 117.175.

Detection of an extremely active state of AM Canum Venaticorum.
See Abstr. 117.225.

An IUE search for white dwarf companions to field halo blue stragglers.
See Abstr. 117.244.

The magnetic fields of accreting white dwarfs.
See Abstr. 117.279.

Secular evolution of cataclysmic binaries.
See Abstr. 117.314.

The evolution of magnetic cataclysmic variables.
See Abstr. 117.315.

Oscillations and pulsations in the sun and stars.
See Abstr. 122.007.

Spectroscopic observations of PU Vulpeculae.
See Abstr. 124.245.

Neutron stars in supernova remnants.
See Abstr. 125.038.

Evidence for sustained multiple imaging of pulsars from the interstellar medium.
See Abstr. 131.160.

Refractive effects in pulsar scintillation.
See Abstr. 131.182.

Evidence for a high–density electron layer from the spatial distribution of pulsars.
See Abstr. 131.183.

Breaking through the wall.
See Abstr. 134.019.

Detection of the very hot central star in NGC 2440.
See Abstr. 134.020.

K1–16 revisited: mode structure and identification as an O VI central star.
See Abstr. 134.023.

A search of steep–spectrum radio sources for millisecond pulsars.
See Abstr. 141.017.

A search for ultraviolet–excess objects.
See Abstr. 142.036.

A very high energy γ–ray burst from the Crab pulsar.
See Abstr. 143.001.

Analysis of imaging data from a balloon–borne coded aperture gamma–ray telescope.
See Abstr. 143.013.

Synchrotron emission models for gamma–ray burst sources, the determination of the orientations of AM Herculis binaries, and non–adiabatic shocks on magnetic white dwarfs.
See Abstr. 143.024.

How promising is the search for gamma–ray pulsars?
See Abstr. 143.025.

A fast pulsar in Cygnus X–3.
See Abstr. 143.027.

Gamma ray astronomy experiments at energies greater than 1 TeV.
See Abstr. 143.030.

Ultraviolet–excess objects in the field of NGC 6121 (M4).
See Abstr. 154.010.

On the nature of objects detected as faint long–term variables.
See Abstr. 159.030.

Interstellar Matter, Nebulae

131 Interstellar Matter (Molecular Clouds, Reflection Nebulae, etc.), Star Formation

131.001 **Far–infrared observations of Sagittarius B2: reconsideration of source structure.**
H. A. Thronson Jr., D. A. Harper.
Astrophys. J., Vol. 300, No. 1, p. 396 – 400 (1986).
New moderate angular resolution, far–infrared observations of the Sagittarius B2 star–forming region are presented, discussed, and compared with recent radio molecular and continuum observations of this source. In contrast to previous analyses, the authors interpret its far–infrared spectrum as the result of a massive frigid cloud overlying a more–or–less normal infrared source, a natural explanation for the object's previously noted peculiarities. The characteristics derived for the obscuring cloud are similar to those found for the W51 MAIN object.

131.002 **An infrared reflection nebula surrounding SGS 1 in the NGC 1333 dark cloud.**
M. W. Castelaz, J. A. Hackwell, G. L. Grasdalen, R. D. Gehrz.
Astrophys. J., Vol. 300, No. 1, p. 406 – 414 (1986).
Surface brightness maps at *JHKL* and polarimetric images at *K* of the object SGS 1 in the NGC 1333 dark cloud were made with the 2.3 m Wyoming infrared telescope. The images show that SGS 1 is a single–lobed asymmetric reflection nebula detectable at *H*, *K*, and *L*. At *K*, the nebula extends 34″ along the major axis, and the illuminating star has a magnitude of 10.95. The data suggest the presence of optically thick obscuring matter along the line of sight to the illuminating star, but optically thin obscuring matter between the star and the nebulosity and along the line of sight to the nebulosity. A model is presented in which a tilted accretion disk surrounds a star and obscures one polar reflection lobe.

131.003 **Chemistry of chlorine in dense interstellar clouds.**
G. A. Blake, V. G. Anicich, W. T. Huntress Jr.
Astrophys. J., Vol. 300, No. 1, p. 415 – 419 (1986).
Laboratory experiments and theoretical modeling show that the chemistry of chlorine is fairly simple in dense interstellar clouds, with Cl and HCl as the only species whose fractional abundances are significant. The estimated fraction of gas–phase chlorine present as HCl lies between 25 – 65%, in good agreement with the recent observations of the ground state HCl transition by Blake, Keene, and Phillips. These results, combined with the observational limits on HCl, indicate that chlorine is not severely depleted in dense interstellar clouds.

131.004 **Detection of interstellar vibrationally excited HCN.**
L. M. Ziurys, B. E. Turner.
Astrophys. J., Lett. Ed., Vol. 300, No. 1, p. L19 – L23 (1986).
Vibrationally excited HCN has been observed for the first time in the interstellar medium. The J = 3–2 rotational transitions of the *l*–doubled $(0,1^{1d,1c},0)$ bending mode of HCN have been detected toward Orion–KL and IRC +10216. In Orion, the overall column density in the (0,1,0) mode, which exclusivley samples the "hot core", is $1.7 \times 10^{16} cm^{-2}$ and can be understood in terms of the "doughnut" model for Orion. Radiative excitation by 14 μm flux from IRc2 accounts for the (0,1,0) population, provided the hot core is $\sim 6-7 \times 10^{16}$cm distant from IRc2. Toward IRC +10216 the authors have detected J = 3–2 transitions of both $(0,1^{1c,1d},0)$ and $(0,2^0,0)$ excited states.

131.005 **Ultraviolet extinction toward five open clusters, and the accuracy of IUE extinction measurements.**
D. Massa, E. L. Fitzpatrick.
Astrophys. J., Suppl. Ser., Vol. 60, No. 1, p. 305 – 321 (1986).
The authors analyze UV extinction curves for 33 main–sequence B stars in five open clusters. An analytic fitting function is used to codify the structure of the 2175 Å extinction bump. The wavelength dependence of the extinction toward four clusters is nonvariable. The data for these clusters determine the observational uncertainties affecting extinction parameters for which a priori error estimates cannot be made. All of the parameters which describe the shapes of UV extinction curves differ significantly from cluster to cluster except for the central position of the 2175 Å bump.

131.006 **The rotational emission–line spectrum of Orion A between 247 and 263 GHz.**
G. A. Blake, E. C. Sutton, C. R. Masson, T. G. Phillips.
Astrophys. J., Suppl. Ser., Vol. 60, No. 1, p. 357 – 374 (1986).
Results are presented from a molecular line survey of the core of the Orion molecular cloud between 247 and 263 GHz. The spectrum contains a total of 243 resolvable lines from 23 different chemical species. When combined with the earlier survey of Orion from 215 to 247 GHz by Sutton et al. (1985), the complete data set includes over 780 emission features from 29 distinct molecules. Only 33 of the more than 780 lines remain unidentified, of which 16 occur in the upper frequency band.

131.007 **IRAS observations of the L 1551 bipolar outflow.**
F. O. Clark, R. J. Laureijs.
Astron. Astrophys., Vol. 154, No. 1/2, p. L26 – L29 (1986).
IRAS "Sky Flux" survey images have been used to study infrared emission in the region of the bipolar outflow associated with L 1551. 60 μm sky flux data show an emission feature of 50′ extent which is well correlated with high velocity CO emission from the bipolar flow. From images at 60 μm and 100 μm are derived a dust color temperature of 24K and a total mass of 3 $M_\odot$. The color temperature suggests that the dust emission arises in the interface region, between the bipolar flow and the ambient material. Using the known interface geometry, a gas density of 2000 cm^{-3} for the interface region is derived from the IRAS data.

131.008 **Galaxy and star counts in Taurus and Perseus.**
F. Lebrun, I. Grenier.
Astron. Astrophys., Vol. 154, No. 1/2, p. 181 – 184 (1986).
The star counts of MacCuskey (1938) in Taurus and Perseus are compared with the Lick galaxy counts of Shane and Wirtanen (1967) for $b < -10°$. It is found that the absorption derived from star counts suffers from serious zero level errors. This is due to the fact that none of the reference fields used by MacCuskey is really free of interstellar extinction. Furthermore, this comparison allows a more reliable calibration of the absorption derived from galaxy counts. Finally, a corrected map of the interstellar extinction in Taurus and Perseus is presented in galactic coordinates.

131.009 **Lynds 379: a new source of bipolar molecular flow.**
J. Hilton, G. J. White, N. J. Cronin, R. Rainey.
Astron. Astrophys., Vol. 154, No. 1/2, p. 274 – 278 (1986).
A study has been made incorporating molecular line CO $J = 2\text{–}1$ and optical star counting techniques of the sharp–edged Lynds dark cloud 379. It is found to lie at a distance of 200 pc, and to have an average extinction, A_B, of <2 mag. Molecular mapping has revealed the presence of a new bipolar molecular flow source coincident with a strong IRAS far infrared peak. This indicates the presence of active star formation at the edge of the cloud – possibly as a consequence of the mechanism (such as an interstellar shock) which has led to compression of the edge of L 379.

131.010 **Warm dust in the R CrA molecular cloud.**
A. Leene.
Astron. Astrophys., Vol. 154, No. 1/2, p. 295 – 302 (1986).
Infrared maps made by IRAS of a part of the R CrA molecular cloud are presented. The data show extended warm dust of 400K around the star HR 7129. There are only two likely heating sources: the star HR 7129 and the interstellar radiation field. Both heating sources are unable to explain this high colour temperature by normal equilibrium processes. A possible explanation of the warm 12 and 25 micron radiation can be found in the emission of very small grains, which are transiently heated to very high temperatures. This model would imply that a large amount of mass is hidden in small graphite grains.

131.011 **Interstellar molecular carbon in distant supergiants.**
R. Gredel, G. Münch.
Astron. Astrophys., Vol. 154, No. 1/2, p. 336 – 342 (1986).
The results of a search for interstellar C_2 lines in the spectrum of southern reddened supergiants are reported. The C_2 excitation parameters along the lines of sight to HD 169454 and BD –14°5037, both in the Sct OB3 association, have been determined. Radiative excitation effects of high rotational levels were found in HD 169454. The C_2 lines in BD –14°5037 are clearly double. A spectrum of HD 92693, in Car OB1, shows C_2 lines weak in proportion to its color excess E_{B-V}, but otherwise the column densities $N(C_2)$ for all stars so far observed appear linearly correlated to E_{B-V}. Through the known relation of E_{B-V} with molecular hydrogen column density $N(H_2)$, a mean columnar abundance $N(C_2)/N(H_2) = 3.0 \times 10^{-8}$ has been estimated, which agrees with theoretical predictions for the C_2 abundance in dense molecular clouds.

131.012 **A bipolar nebula in Orion identified with the IRAS source 05329–0505.**
R. D. Wolstencroft, S. M. Scarrott, R. F. Warren–Smith, H. J. Walker, B. Reipurth, A. Savage.
Mon. Not. R. Astron. Soc., Vol. 218, No. 1, p. 1P – 6P (1986).
The IRAS source 05329–0505 is found to be the central illuminating star of a bipolar nebula. The relationship between this source and the structure and illumination of the nebula is discussed.

131.013 **Characteristics of the H_2CO maser emission in Sgr B2.**
F. F. Gardner, J. B. Whiteoak, J. R. Forster, V. Pankonin.
Mon. Not. R. Astron. Soc., Vol. 218, No. 3, p. 385 – 391 (1986).
VLA observations of the 5–GHz H_2CO maser emission towards Sgr B2, made with high spectral and angular resolution (0.76 km s^{-1}, 1.25 arcsec respectively), show that most of the emission consists of multiple narrow components with velocity halfwidths between 0.6 and 3.0 km s^{-1}. The minimum maser gains range from 0.2 to 60. All five masers are located near OH or H_2O masers, suggesting a common mechanism for maser production.

131.014 **Dissipation of magnetic fields in very dense interstellar clouds – I. Formulation and conditions for efficient dissipation.**
T. Nakano, T. Umebayashi.
Mon. Not. R. Astron. Soc., Vol. 218, No. 4, p. 663 – 684 (1986).
The authors investigate the dissipation processes of magnetic fields in dense interstellar clouds, and derive the drift velocity of magnetic fields relative to the neutrals which contains the effects of plasma drift and Joule dissipation. Comparing the drift velocity with the free–fall velocity, the authors investigate the dissipation of magnetic fields in the clouds of various configurations with density up to 10^{15}cm^{-3} and clarify at what stage of contraction the magnetic field decouples from the gas. Some related problems are discussed.

131.015 **Theoretical studies of interstellar molecular shocks – II. Molecular hydrogen cooling and rotational level populations.**
D. R. Flower, G. Pineau–des–Forets, T. W. Hartquist.
Mon. Not. R. Astron. Soc., Vol. 218, No. 4, p. 729 – 741 (1986).
The authors have compared the results of calculations of magnetohydrodynamic shocks in which the rotational level populations of H_2 were (1) in a stationary state and (2) in "statistical equilibrium" (the static limit). In the models studied, the gas consisted mainly of molecular hydrogen, which dominates the radiative cooling of the shocked region. This comparison shows that significant deviations from the static limit are to be anticipated only in J–type shocks, in which the neutral gas passes through a discontinuity.

131.016 **Ionic origins of carbenes in space.**
D. K. Bohme.
Nature, Vol. 319, No. 6053, p. 473 – 474 (1986).
Among the many molecules now known to enrich interstellar space are carbenes, molecules which have two electrons available on a carbon atom for chemical bonding. The origin of these molecules may involve ionizing radiation which can drive reactions between ions and molecules towards the formation of ionic carbene precursors. The author reports experiments which suggest that a much higher proportion of the ion chemistry in partially ionized regions in space is directed towards the formation of carbene molecules than has previously been recognized. He identifies plausible reaction paths which lead to both known and previously unknown carbenes, such as the cyclopropenylidene ring and long carbon–chain molecules. The cyclopropenylidene ring has recently been reported to be ubiquitous in the Galaxy.

131.017 **Is IRAS cirrus cloud emission largely fine–structure radiation?**
M. Harwit, J. R. Houck, G. J. Stacey.
Nature, Vol. 319, No. 6055, p. 646 – 647 (1986).
One of the more interesting observations made by the infrared astronomical satellite (IRAS) mission is the discovery of a diffuse 'cirrus' component seen at 60 and 100 μm. It straddles the galactic plane and extends, in patches, to latitudes of $\pm 80°$. Comparison of the observed brightness from these clouds with [O I] 63–μm and [O III] 88–μm fine–structure line emission, calculated on the basis of observed [C II] emission from the galactic plane, shows the flux levels observed by IRAS to be of the same order as the expected fine–structure line emission. The authors propose that fine–structure line emission, from neutral gas at 63 μm and from diffuse ionized clouds at 88 μm, is a significant component and in some clouds perhaps the prime source of the cirrus radiation.

131.018 **Measuring interstellar dust grains from the haloes of binary X–ray sources.**
Y. Xu, R. McCray, R. Kelley.
Nature, Vol. 319, No. 6055, p. 652 – 653 (1986).
Overbeck first predicted that the images of cosmic X–ray sources might have observable haloes due to coherent forward scattering by interstellar dust grains. Observations with the Einstein Observatory have provided evidence of faint haloes extending to a few arc minutes beyond the X–ray images of supernova remnants and several compact galactic sources. In this letter it is shown how observations of the time–dependence of the halo X–ray emission may be analysed to determine the location, size distribution, and composition of interstellar grains. The authors suggest that such haloes probably account for the soft X–rays

that have been observed during eclipses or dips in several binary systems.

131.019 **The injection of energy into the interstellar medium by stars.**
C. F. McKee.
Astrophys. Space Sci., Vol. 118, Nos. 1/2, p. 383 – 393 (1986). – See Abstr. 012.001.
Stars inject energy into the interstellar medium (ISM) by radiation, stellar winds, and supernova explosions. This energy injection causes the ISM to be inhomogeneous, which in turn alters the manner in which the energy is transferred through the ISM. A significant fraction of the energy is injected by massive stars, which form H II regions in the ISM. The structure and evolution of H II regions in a cloudy medium differs significantly from that in a homogeneous one. The strong stellar winds produced by massive stars form bubbles in the ISM, and the structure of these bubbles is often dominated by the structure of the H II region in which they are embedded. Finally, when the star explodes as a supernova, the evolution and appearance of the resulting remnant is determined by the structure of the bubble and H II region formed by the star during its lifetime.

131.020 **High–resolution observations of the Orion bright bar.**
T. Omodaka, M. Hayashi, S. Suzuki, T. Hasegawa, R. Miyawaki.
Astrophys. Space Sci., Vol. 118, Nos. 1/2, p. 401 – 404 (1986). – See Abstr. 012.001.
High resolution strip maps of CS (J = 1–0) and H51α line emission across the Orion bright bar are presented. They reveal the existence of a high density molecular layer (*molecular sheet*) plane parallel to the ionization front. This molecular sheet is redshifted relative to the ambient molecular cloud by about 2 km s^{-1}. The rapid decrease of the CS emission at about 50 arc sec from the bar suggests that a shock front exists here and the sheet is a post shock layer.

131.021 **Excitation of interstellar molecules in the Ori–KL source.**
M. Ohishi, N. Kaifu, H. Suzuki, M. Morimoto.
Astrophys. Space Sci., Vol. 118, Nos. 1/2, p. 405 – 407 (1986). – See Abstr. 012.001.
A molecular spectral line survey of the title source detected 166 molecular lines from 18 interstellar molecules in the frequency ranges of 34.25 – 50.00, 83.50 – 84.50, and 86.00 – 91.50 GHz. For each molecule, gaussian decomposition of the velocity components in the transition profiles gave consistent radial velocity and linewidth. Rotation diagrams were drawn for each velocity component.

131.022 **Large–scale structures of high galactic latitude dark clouds.**
Y. Tomita.
Astrophys. Space Sci., Vol. 118, Nos. 1/2, p. 409 – 410 (1986). – See Abstr. 012.001.
The obscuring material distribution of an area of about 2000 square degrees was derived by analyzing about 130 plates, using the starcounts method.

131.023 **A quenched carbonaceous composite (QCC) grain model for the interstellar 220 nm extinction hump.**
T. Onaka, Y. Nakada, T. Tanabe, A. Sakata, S. Wada.
Astrophys. Space Sci., Vol. 118, Nos. 1/2, p. 411 – 413 (1986). – See Abstr. 012.001.
The dependence of the wavelength of peak absorption of dust grains on the grain size is investigated analytically by using an oscillator model for the absorption band.

131.024 **Star formation associated with high–velocity mass outflows.**
T. Hasegawa.
Astrophys. Space Sci., Vol. 118, Nos. 1/2, p. 421 – 434 (1986). – See Abstr. 012.001.
Energetic mass outflows have been detected in molecular line observations towards young stellar objects. In this review the author takes the Orion–KL as an example to discuss the overall structure of a high–velocity outflow and its environment. The kinematics of the high–velocity molecular emission show clear evidence of a bipolar jet which originates in the vicinity of IRc2, a massive protostar.

131.025 **The structure of the R Monocerotis, NGC 2261, and HH39 nebular complex.**
J. R. Walsh.
Astrophys. Space Sci., Vol. 118, Nos. 1/2, p. 439 – 441 (1986). – See Abstr. 012.001.
NGC 2261 is a reflection nebula that shows variations in brightness and structural changes of up to 1/4 arc sec day^{-1}. UB observation of R Mon and NGC 2261 revealed a strongly blue–shifted line profile variation of the reflected lines with distance from R Mon.

131.026 **CS, $C^{34}S$, and CH_3OH observations of the molecular cloud associated with NGC 7538.**
O. Kameya, T. Hasegawa, N. Hirano, M. Seki, M. Tosa, Y. Taniguchi, K. Takakubo.
Astrophys. Space Sci., Vol. 118, Nos. 1/2, p. 449 – 451 (1986). – See Abstr. 012.001.
CS(J = 1–0), $C^{34}S$(J = 1–0), and $CH_3OH(1_0–0_0$ A and E) emission lines in the core region of the molecular cloud associated with NGC 7538 have been surveyed at angular resolution of 33″. Distribution of CS column density shows two prominent peaks. Both blue and red wings in CS line are most prominent at 30″ northwest of IRS 11.

131.027 **N–body simulation of giant molecular clouds in a galaxy.**
M. Tosa, M. Fukunaga.
Astrophys. Space Sci., Vol. 118, Nos. 1/2, p. 463 – 465 (1986). – See Abstr. 012.001.
The authors made N–body simulation of giant molecular clouds in a model galaxy and confirmed that their random motion is generated and accelerated by purely gravitational interaction among them in some 10^{7-8}yr. The results of the present simulation agree well with those of the previous theoretical calculation by Fukunaga (1984).

131.028 **Hydrodynamical models of the Orion–KL Nebula.**
T. Okuda, S. Ikeuchi.
Astrophys. Space Sci., Vol. 119, No. 1, p. 115 – 121 (1986). – See Abstr. 012.001.
The interaction between an isotropic protostellar wind and an ambient molecular cloud is investigated by two–dimensional hydrodynamic computations. The wind–cloud interaction model is promising for explanation of the observed properties of the Orion–KL Nebula.

131.029 **The Scutum ring of H II regions.**
T. Handa, Y. Sofue.
Astrophys. Space Sci., Vol. 119, No. 1, p. 127 – 130 (1986). – See Abstr. 012.001.
A ring of compact radio continuum sources was found at l = 24°.6, b = 0°.0, which the authors call the Scutum ring. Radio continuum, H I line, and CO line observations suggest that it is a star–forming region triggered by an expanding diffuse H II region.

131.030 **On the radial distributions of molecular clouds in galaxies.**
M. Fukunaga.
Astrophys. Space Sci., Vol. 119, No. 1, p. 143 – 146 (1986). – See Abstr. 012.001.
The dynamics of the interstellar gas (the IS gas) driven by the viscous torque of a system of giant molecular clouds (the MC gas) is considered with the infinitesimally thin disk layer approximation. The flow explains the radial distributions of molecules observed in galaxies.

131.031 **Hydrostatic models of Bok globules.**
T. Hasegawa.
Astrophys. Space Sci., Vol. 119, No. 1, p. 151 – 154 (1986). – See Abstr. 012.001.

Hydrostatic models of small, isolated dark clouds are presented. A dynamical, thermal, and chemical equilibrium is considered in these models.

131.032 **Cep A: a possible proto–cluster.**
S. S. Hayashi, N. Kaifu, T. Hasegawa.
Astrophys. Space Sci., Vol. 119, No. 1, p. 155 – 157 (1986). – See Abstr. 012.001.

The authors present high resolution CS and CO maps of Cep A region made with the 45 m telescope at Nobeyama. The CS map shows that a dense cloud surrounding the proto–star cluster extends in the North–South direction and is probably rotating. The bipolar molecular flow apparent in the CO maps is well–collimated along East–West direction within 0.2 pc from the proto–stars. The dense cloud is gravitationally unstable and appears to be in a contracting phase to form a cluster of massive stars.

131.033 **On the nature of the interstellar grains.**
F. Hoyle, N. C. Wickramasinghe.
Q. J. R. Astron. Soc., Vol. 27, No. 1, p. 21 – 37 (1986).

This article reviews the arguments for considering interstellar grains to have a biological connotation. The astronomical observations on which the discussion is based cover the waveband from about 2000 Å in the ultraviolet to 100 μm in the far infrared.

131.034 **Interstellar chemistry: gas and surface phase processes; time dependence.**
D. A. Williams.
Q. J. R. Astron. Soc., Vol. 27, No. 1, p. 64 – 70 (1986).

The astrophysical information content in data on interstellar molecules may be tapped by studying chemical models of interstellar clouds. The considerations to be made in setting up such a model are briefly described. Such models do not lead to unique descriptions, and a procedure useful in minimizing the ambiguities in some diffuse cloud models is given. Dense clouds also require careful treatment. However, the situation in them is complicated by many time–dependent factors, including especially the problem of accretion of gas phase molecules onto grain surfaces.

131.035 **Chemistry in interstellar flows.**
T. W. Hartquist.
Q. J. R. Astron. Soc., Vol. 27, No. 1, p. 71 – 82 (1986).

Dissipation in flows often heats molecular gas to temperatures at which many reactions that are endothermic or that have activation barriers can proceed. H_2 observations have been used to diagnose the conditions in flows in molecular clouds, but theoretical studies of the H_2 level population problem are very incomplete. Much of the H_2 emission and many of the chemical species formed arise in the precursor regions of MHD shocks. The abundance of CH^+ in MHD shocks in diffuse molecular clouds should exceed that in ordinary hydrodynamic shocks but shock formation models for the observed CH^+ abundances face several difficulties. In dense clouds shocks in T–Tauri wind blown bubbles may play important rôles in establishing the observed chemical composition in "apparently quiescent" regions.

131.036 **Is interstellar chemistry useful?**
A. Dalgarno.
Q. J. R. Astron. Soc., Vol. 27, No. 1, p. 83 – 89 (1986).

A brief review is presented of the present status of interstellar chemistry with particular reference to its value as a source of astronomically useful information.

131.037 **Unbiased survey of star formation regions.**
Y. Fukui.
Astron. Her., Vol. 79, No. 2, p. 33 – 35 (1986). In Japanese.

131.038 **Molecular gas interacting with the Orion nebula.**
K. Sugitani.
Astron. Her., Vol. 79, No. 2, p. 37 – 40 (1986). In Japanese.

131.039 **Filamentary structure of dark clouds.**
A. Mizuno.
Astron. Her., Vol. 79, No. 2, p. 41 – 43 (1986). In Japanese.

131.040 **Organic molecules in interstellar space.**
Å. Hjalmarson.
J. Br. Interplanet. Soc., Vol. 39, No. 3, p. 137 (1986). Abstract. – See Abstr. 012.003.

131.041 **Star formation in dark globules.**
C. M. Leung.
News Lett. Astron. Soc. N.Y., Vol. 2, No. 9, p. 5 – 6 (1986). Abstract. – See Abstr. 010.241.

131.042 **Modelling of infrared emission from dark clouds.**
R. Dubisch, C. M. Leung.
News Lett. Astron. Soc. N.Y., Vol. 2, No. 9, p. 7 – 8 (1986). Abstract. – See Abstr. 010.241.

131.043 **Sandqvist 187: a dense molecular cloud in Norma.**
H. Alvarez, L. Bronfman, R. Cohen, G. Garay, J. Graham, P. Thaddeus.
Astrophys. J., Vol. 300, No. 2, p. 756 – 765 (1986).

The authors present and discuss observations of Sandqvist 187, an elongated dust cloud in the southern constellation Norma. The cloud contains two Herbig–Haro objects, HH 56 and HH 57. HH 57 currently displays on its NE edge a 17th mag variable star of the FU Ori type. The authors have mapped the region in the 115 GHz CO line and found an extended CO cloud which envelops and is elongated along the optical dust cloud. The position of maximum CO emission coincides with HH 56 and HH 57. Assuming a distance of 0.7 kpc, the total mass of the cloud is found to be close to 500 $M_\odot$. Photographs and CCD images obtained mostly with the CTIO 4 m telescope show the detailed optical structure of the dark cloud's core region.

131.044 **Interstellar molecular line searches at 1.5 centimeters.**
H. E. Matthews, T. J. Sears.
Astrophys. J., Vol. 300, No. 2, p. 766 – 772 (1986).

Observations at the frequencies of low–J transitions of a number of important molecules are reported. The authors confirm an earlier result for CH_3CN, and have made a detection of the $1_{01}\rightarrow0_{00}$ line of ketene (CH_2CO) in two sources, Sgr B2 and TMC–1. The latter is the first detection of ketene in a cold dust cloud. Upper limits were obtained for $CH_3{}^{13}CN$ ($1_0\rightarrow0_0$), CH_3NC ($1_0\rightarrow0_0$), NH_2CN ($1_{01}\rightarrow0_{00}$), C_2H_5CN ($2_{11}\rightarrow1_{10}$), and $CH_2(CN)_2$ ($1_{10}\rightarrow1_{01}$). Column densities are derived for TMC–1 and the implications for interstellar chemistry in cold dust clouds are briefly discussed.

131.045 **New interstellar masers in nonmetastable ammonia.**
S. C. Madden, W. M. Irvine, H. E. Matthews, R. D. Brown, P. D. Godfrey.
Astrophys. J., Lett. Ed., Vol. 300, No. 2, p. L79 – L84 (1986).

The authors report the first astronomical detections of several ammonia inversion transitions involving nonmetastable levels with energies as high as 1090K above the ground state. The (J,K) = (9,6) inversion transition shows maser–like emission in the directions of W51, NGC 7538, W49, and DR 21(OH). The NH_3 (6,3) line exhibits similar characteristics in W51 but is seen in absorption in NGC 7538. These are the first definite detections of ammonia masers in space. The masers are found in regions containing compact H II regions and strong infrared sources with associated H_2O and OH masers; i.e., in regions of active star formation.

131.046 **Enhancement of CS in the high–velocity wing emission in NGC 2071.**
T. Takano.
Astrophys. J., Lett. Ed., Vol. 300, No. 2, p. L85 – L87 (1986).
High–velocity wing emission in CS and NH_3 spectra towards the bipolar flow source NGC 2071 are compared. Column densities of CS and NH_3 in the high–velocity wing emission are comparable, which means that CS is overabundant relative to NH_3 by one to two orders of magnitude. The large enhancement of CS may be caused by an interstellar shock wave generated by the high–velocity outflow from the central infrared source(s).

131.047 **Detection of clump and interclump gas in the Rosette Molecular Cloud Complex.**
L. Blitz, A. A. Stark.
Astrophys. J., Lett. Ed., Vol. 300, No. 2, p. L89 – L93 (1986).
High–resolution, high–sensitivity CO and ^{13}CO observations have been made of the Rosette Molecular Cloud Complex. The clumps are shown to be easily identified entities with densities ~10 times the volume–averaged density of the entire complex. The clumps are shown to be embedded in pervasive molecular "interclump" gas of low average volume density ($\langle n \rangle \approx 2\ cm^{-3}$) which has not previously been observed. The velocity dispersion of the interclump gas is about 3 times greater than that of the clumps, implying that the gas is not gravitationally bounded to the complex.

131.048 **Structure of cosmic ray modified shocks.**
L. O'C. Drury.
Cosmical gas dynamics, p. 131 – 136 (1985). – See Abstr. 012.007.

131.049 **Chemical and physical conditions that favour continuous rather than jump MHD shocks.**
D. R. Flower, G. Pineau des Forets, T. W. Hartquist.
Cosmical gas dynamics, p. 163 – 171 (1985). – See Abstr. 012.007.
Efficient cooling and the removal of atomic ions to form molecular ions, which recombine rapidly, lead MHD shocks propagating into diffuse clouds over a large range of speeds to be of C type. The nature of MHD shocks in dense clouds may be affected by how much inertia grains provide to the ionized component. General differences should exist between the chemistries in C type and J type shocks.

131.050 **The energetics of nebulae associated with Wolf–Rayet stars.**
J. E. Dyson, L. J. Smith.
Cosmical gas dynamics, p. 173 – 188 (1985). – See Abstr. 012.007.
The expansion of nebulae around Wolf–Rayet stars is driven by the interaction of a fast stellar wind with its surroundings. However, the kinetic energy content of the moving gas is always far less than anticipated from theoretical models of the interaction process. The authors consider four WR nebulae in detail and conclude that there may be four distinct reasons for this apparent deficit.

131.051 **Echelle observations of high speed flows in the interstellar medium.**
J. Meaburn, J. R. Walsh, C. A. Clayton.
Cosmical gas dynamics, p. 189 – 203 (1985). – See Abstr. 012.007.
Initial results from the use of a very efficient echelle spectrometer on the Anglo Australian Telescope are summarised. Examples of supersonic motions in Herbig–Haro like objects in the Rosette nebula, in the lobes of the bi–polar nebula MZ–3 and in the H II complexes of M17 and 30 Doradus are reported.

131.052 **The Cone Nebula and the infrared source GL989.**
N. Brosch, J. M. Greenberg.
Cosmical gas dynamics, p. 213 – 221 (1985). – See Abstr. 012.007.
The authors show that published and recently acquired data indicate that the Cone Nebula may be generated through the interaction of a channelled wind from GL989 and a Bok globule. This is in accord with the data on the molecular cloud dynamics as well.

131.053 **Optical polarimetry of nebulae with emission line jets.**
S. M. Scarrott, R. F. Warren–Smith, P. W. Draper, T. M. Gledhill.
Cosmical gas dynamics, p. 281 – 286 (1985). – See Abstr. 012.007.
The authors present optical linear polarisation maps of two nebulae with emission line jets. They interpret the data in terms of the identification of illuminating sources and propose geometries for the nebular systems.

131.054 **Interstellar collimation in the bipolar outflow source G35.2N.**
W. R. F. Dent.
Cosmical gas dynamics, p. 299 – 305 (1985). – See Abstr. 012.007.
Results of a study of the massive star formation region G35.2N in molecular lines, near infrared and radio continuum are presented. Assuming normal CO abundance, they show at least 80 $M_\odot$ of high velocity molecular gas moving out from a compact near–infrared source. Both the red and blue shifted material have elongated morphologies with semi–major axes of 0.6 pc and axial ratios of 4. Perpendicular to the outflow axis is a dense, flat rotating disc observed nearly edge–on. It is believed that the dense material of the disc is responsible for the collimation of a high velocity stellar wind. Transfer of momentum from this wind to the denser ambient cloud will result in the observed CO outflow.

131.055 **High resolution measurements of cyanoacetylene in dark clouds.**
R. L. Sorochenko, A. M. Tolmachev, G. Winnewisser.
Astron. Astrophys., Vol. 155, No. 2, p. 237 – 241 (1986).
Emission from the $J = 4 \to 3$ transition of cyanoacetylene has been observed towards seven dark clouds with a spectral resolution of $0.062\ km\ s^{-1}$. These observations were made with the 22 m–radiotelescope of the Lebedev Physcial Institute at Puschino. The optical depth and excitation temperature of the $J = 4$ level of HCCCN were obtained for each cloud by measuring the intensity ratio of the hyperfine components of the line. Densities and HCCCN abundances were derived.

131.056 **Principal heating sources of dust in the galactic disk.**
P. Cox, E. Krügel, P. G. Mezger.
Astron. Astrophys., Vol. 155, No. 2, p. 380 – 396 (1986).
The authors present a model of the IR/sub mm emission from the galactic disk, which explains quantitatively all relevant observations available to date. The emission comes from dust which is heated by absorption of star light. The spectrum of the dust emission from the inner part of the Galaxy ($R \lesssim 8$ kpc), expressed in units of λI_λ, attains two maxima at ~100 µm and between 4 and 20 µm. This spectrum can be explained by contributions from different dust components, which are characterized by their modified Planck or color temperatures. The principal contribution to the heating of these dust components comes from the general interstellar radiation field and from the more massive stars of the galatic disk population in specific evolutionary stages.

131.057 **Deuterium and the local interstellar cloud(s?).**
A. Vidal–Madjar, R. Ferlet, C. Gry, R. Lallement.
Astron. Astrophys., Vol. 155, No. 2, p. 407 – 412 (1986).
The authors underline the present situation of deuterium abundances evaluation in interstellar space, which seems to vary from one direction to another. Studying in more detail the λ Sco line of sight and having observed two Na I and Ca II interstellar components toward that star, they can show that the D/H evaluation made toward λ Sco is in fact related to the local interstellar medium (less than 15 pc from the sun). Because this evaluation is also $\leqslant 10^{-5}$, it is in striking contrast with that made toward α Aur (D/H $\geqslant 1.8 \times 10^{-5}$) confirming the fact that the deuterium abundance in the local interstellar medium varies by at least a factor of two over a few parsecs. Strong evidences are also given for the presence of two clouds or for a complicated interstellar morphology in the immediate vicinity of the solar system.

131.058 **MERLIN observations of OH outflow in W75N.**
E. E. Baart, R. J. Cohen, R. D. Davies, R. P. Norris, P. R. Rowland.
Mon. Not. R. Astron. Soc., Vol. 219, No. 1, p. 145 – 155 (1986).
MERLIN observations are presented of the 1665–MHz OH maser emission from W75N. The measurements reveal many masers not seen in previous maps of W75N. The new data are not consistent with the rotating disc model which has previously been proposed for this source. The 1665–MHz masers are concentrated around the western and southern edges of the associated compact H II region, and their velocities indicate that outflow rather than rotation is the dominant motion. It is most likely that the masers lie in a compressed region behind an expanding shock front, as suggested by Elitzur & de Jong (1978). A magnetic field of ~ 3 mG is detected in the maser regions, which may affect both the gas dynamics and the shock chemistry. Clustering of the masers is also observed.

131.059 **The structure and kinematics of the DR21 region.**
K. J. Richardson, G. J. White, J. P. Phillips, L. W. Avery.
Mon. Not. R. Astron. Soc., Vol. 219, No. 1, p. 167 – 190 (1986).
Observations of the DR21/DR21(OH) region have been made in the molecular lines CO $J = 2$–1, CO $J = 3$–2, HCN $J = 4$–3, HCO^+ $J = 4$–3, $H^{13}CO^+$ $J = 4$–3 and CS $J = 7$–6, supplemented by continuum observations of DR21 and DR21(OH) at 350 μm and a 20 μm map of DR21. The CO observations show a high–velocity wing region around DR21, which is blue shifted with respect to the -3 km s^{-1} central velocity of the source, and may be due to an outflow. The CS, HCN, HCO^+ and $H^{13}CO^+$ intensities, spatial extents and line profiles, in addition to the self–absorptions seen in some CO lines, can be understood on the basis of a simple 2–component model similar to that of Norman & Silk and consisting of dense ($10^5 - 10^6 cm^{-3}$) clumps surrounded by a less dense ($\sim 10^3 cm^{-3}$) interclump gas of higher temperature.

131.060 **The mass distribution of dark clouds in external galaxies.**
H. C. Bhatt, I. P. Williams.
Mon. Not. R. Astron. Soc., Vol. 219, No. 1, p. 217 – 221 (1986).
By making use of published observational data on dark clouds in other galaxies, the distribution function for the cross–section of such clouds has been obtained. Models are discussed to convert to a mass distribution. If $N(M)dM$ denotes the number of clouds with mass in the range M to $M + dM$, the authors have plotted log $N(M)$ against log M for the clouds in each of the galaxies M31 (Andromeda), the Large Magellanic Cloud and the Small Magellanic Cloud. For the LMC and M31 the distribution indicates a mass function of the form $N(M) \propto M^{-s}$, with the index s having a value of around 1.5 for most models. Such a value is consistent with the value for the corresponding index found for clouds in our own Galaxy.

131.061 **Erratum: "Submillimetre line and continuum observations of the S255 molecular cloud" [Mon. Not. R. Astron. Soc., Vol. 216, No. 3, p. 713 – 733 (1985)].**
K. J. Richardson, G. J. White, G. Gee, M. J. Griffin, C. T. Cunningham, P. A. R. Ade, L. W. Avery.
Mon. Not. R. Astron. Soc., Vol. 219, No. 1, p. 223 (1986). See Abstr. 40.131.095.

131.062 **ON2: a complex star formation region.**
N. Matthews, M. Andersson, G. H. Macdonald.
Astron. Astrophys., Vol. 155, No. 1, p. 99 – 103 (1986).
The region centred on ON2 has been extensively mapped in the $J = 1 \to 0$ ^{12}CO and ^{13}CO transitions, with a spatial resolution of $\sim 30''$. The emission encloses three main regions of activity with cloud masses of $> 0.3 - 1 \times 10^4 M_\odot$. High velocity outflow was not unambiguously detected in the complex line profiles of ^{12}CO, but evidence is found for possible interactions between the H II region G75.77+0.34 and the surrounding cloud medium.

131.063 **The hydrodynamics of clouds overtaken by supernova remnants. I. Cloud crushing phenomena.**
G. Tenorio–Tagle, M. Rozyczka.
Astron. Astrophys., Vol. 155, No. 1, p. 120 – 128 (1986).
The hydrodynamical events resulting from the interaction of a supernova remnant with a high density condensation – a cloudlet, are investigated by means of high resolution two dimensional hydrodynamical calculations. Spherical and cylindrical cloudlets initially immersed in a constant density medium are considered. The evolution and final fate of the cloudlets is shown to depend not only on their original shape, but also on the distance to the explosion site, i.e., on whether the remnant, prior to the interaction, evolves along its Sedov or its radiative phase. However, none of the performed calculation leads to density enhancements which could clearly be related to the filamentary structure, typical of supernova remnants.

131.064 **Cyanoacetylene in the Sgr A molecular clouds.**
C. M. Walmsley, R. Güsten, P. Angerhofer, E. Churchwell, L. Mundy.
Astron. Astrophys., Vol. 155, No. 1, p. 129 – 136 (1986).
The authors have used the Effelsberg 100–m, Kitt Peak 11–m, and Texas 4.9–m telescopes to study cyanoacetylene in the galactic center (Sgr A) molecular clouds. The cyanoacetylene density determination suggests that the Sgr A molecular clouds (if stable against tidal disruption) are at least 75 pc from the galactic nucleus and do not directly interact with the region surrounding Sgr A West.

131.065 **A possible cause of the noncircular pattern of the linear polarization in some (bipolar) reflection nebulae.**
P. Notni.
Astron. Nachr., Vol. 306, No. 5, p. 265 – 272 (1985).
The scattering of linearly polarized light in reflection nebulae is investigated. The distortion of the centrosymmetric pattern of the polarization vectors is calculated; it is found that linear polarization of the illuminating source in an optically thin nebula leads to an "elliptical" pattern. If foreward scattering angles are strongly favoured, arrays of nearly parallel polarizations may arise. In addition, circular polarization will always occur. Both pattern types have been observed in bipolar nebulae.

131.066 **Effects of overshooting on the chemical enrichment of the interstellar medium.**
L. Greggio, M. Tosi.
Astron. Astrophys., Vol. 156, No. 1/2, p. L1 – L4 (1986).
Numerical models for the chemical evolution of the Galaxy have been computed in order to check the effect of overshooting from stellar cores on the interstellar medium enrichment. It is found that the observed CNO abundance distributions, rather than by standard models, are better reproduced by models which take overshooting into account. In this case, the estimated upper mass cut off for intermediate mass stars is $M_{up} \cong 6 M_\odot$. The best agreement with the observations is obtained for an adopted mixing length parameter $\alpha \cong 1.5$.

131.067 **The wavelength dependence of the interstellar polarization of o Sco, σ Sco and η Aql.**
D. Clarke.
Astron. Astrophys., Vol. 156, No. 1/2, p. 213 – 216 (1986).
Accurate spectropolarimetry has been performed on the three stars o Sco, σ Sco and η Aql with a view to determining the peakiness, K, of the $p(\lambda)$ curve. A high value of K appears to be associated with o Sco relative to the values for σ Sco and η Aql. A newly discovered dispersion of position angle, $\theta(\lambda)$ is reported for η Aql and this star can now be considered only as a secondary standard for polarimetric calibration purposes. The determined values of K have been considered in terms of the two–cloud model but no firm conclusions can be drawn.

131.068 **A study of bipolar and compact nebulae at radio, submillimetre and infrared wavelengths.**
G. J. White, G. Gee.
Astron. Astrophys., Vol. 156, No. 1/2, p. 301 – 309 (1986).
Radio observations have been made of a sample of optically selected bipolar nebulae, to examine their spatial structures and to determine the nature of the sources which excite them. IRAS point source survey data, sub–millimetre wavelength molecular line and continuum measurements have been combined with the radio data to constrain the estimates of the nature of the excitation sources. Two thirds of the sample are found to be associated with luminous ($L > 1000\ L_{\odot}$) objects which are characteristed as ZAMS B0–B2 stars. Several cases show characteristics of ionised outflowing gas, and may be pre–main sequence objects. The remaining third of the sample have ZAMS exciting stars between B3 and A5.

131.069 **Electric charge on ionic grains.**
A. P. Jones, D. A. Williams.
Mon. Not. R. Astron. Soc., Vol. 219, No. 2, p. 441 – 447 (1986).
The charge and potential of small interstellar oxide grains in a variety of locations are calculated. The nature of the charge on such grains is shown to be localized at defect centres of both positive and negative character. The electrostatic potential in the vicinity of various sites is explored, and it is shown that positive ions are channelled by the electrostatic field to impact with high energy on sites of overall negative charge, even though the grains may themselves be overall positive. Thus ion–grain collisions are not suppressed in diffuse clouds where grains are normally positive. The consequences of these energetic collisions for surface reactions are briefly described.

131.070 **An infrared image of the Hourglass region of M8.**
D. A. Allen.
Mon. Not. R. Astron. Soc., Vol. 219, No. 2, p. 35P – 38P (1986).
An infrared image of the Hourglass region of M8 shows several features of interest. The young O star Herschel 36 has a companion which is probably pre–main sequence. The nebulosity which links Herschel 36 to the Hourglass is more extensive than previously realized. A central source has been found in the Hourglass, and is probably responsible for its illumination. Finally, many stars are found to the west of Herschel 36, with a projected density comparable to that in the densest part of the Orion Nebula.

131.071 **Protostars and planets: overview from a planetary perspective.**
E. H. Levy.
Protostars & planets II, p. 3 – 16 (1985). – See Abstr. 012.008.
This paper summarizes and examines some of the implications of ideas about the formation of stars and planetary systems.

131.072 **Protostars and planets: overview from an astronomical perspective.**
S. E. Strom.
Protostars & planets II, p. 17 – 29 (1985). – See Abstr. 012.008.
Current problems in understanding the formation and early evolution of stars are reviewed. First discussed are observational studies of external galaxies which may provide insight into the factors controlling star–forming efficiency and the initial mass function. Advances in the classification of molecular cloud properties in the Milky Way are summarized. Considerable emphasis is given to a discussion of recent observations of mass outflows from young stellar objects and the possibility that such outflows may provide a guide to the role played by magnetic fields during the prestellar collapse phases. Recent observations of possible disk structures surrounding young stellar objects are presented along with a picture of one such disk, that surrounding the T Tauri star, HL Tau.

131.073 **Molecular clouds and star formation: an overview.**
B. G. Elmegreen.
Protostars & planets II, p. 33 – 58 (1985). – See Abstr. 012.008.
Theoretical work on star formation has progressed rapidly in the last decade because molecular–line observations have provided detailed information about star–forming clouds. This overview summarizes the most general aspects of these clouds, such as why they are molecular, what observations of the molecules have revealed about star formation, what observations of cloud masses and the distribution of clouds in the Galaxy suggest about star formation on a large scale and about the evolution and lifetime of molecular clouds, and why future observations of extremely high densities and clumping in cloud cores are essential to the understanding of the star formation mechanism.

131.074 **Star formation in a galactic context: the location and properties of molecular clouds.**
P. M. Solomon, D. B. Sanders.
Protostars & planets II, p. 59 – 80, 1179 – 1183 (1985). – See Abstr. 012.008.
Giant molecular clouds, the site of star formation, are the largest and most massive objects in the Galaxy. Their physical properties and distribution are best traced by millimeter–wave observations of CO. The authors review the overall galactic distribution of molecular clouds and compare it with other galaxies.

131.075 **Molecular cloud cores.**
P. C. Myers.
Protostars & planets II, p. 81 – 103 (1985). – See Abstr. 012.008.
The author describes recent progress in observations and interpretations of molecular cloud cores, i.e. the dense ($n \gtrsim 10^4 cm^{-3}$) parts of molecular clouds.

131.076 **Physical conditions in isolated dark globules.**
C. M. Leung.
Protostars & planets II, p. 104 – 136 (1985). – See Abstr. 012.008.
Isolated dark globules, also known as Bok globules or Barnard objects, are nearby dense interstellar clouds of gas and dust with no significant internal heat source. The current status of observational and theoretical research on the physical conditions in these objects is reviewed. In particular, the physics of the gas and dust components is discussed. Observations of both line and continuum radiation at optical, infrared and millimeter wavelengths are shown to provide important constraints on their physical structure, dynamical state and evolutionary status. While most globules appear to be gravitationally bound and in virial equilibrium, recent infrared and radio observations suggest that low–mass star formation may be taking place in some globules, a well–studied example being B335.

131.077 **Rotation in dark clouds.**
P. F. Goldsmith, R. Arquilla.
Protostars & planets II, p. 137 – 149 (1985). – See Abstr. 012.008.
Recent observations of dark interstellar clouds are analyzed with regard to the role of rotation in determining the structure and energetics of these objects. Observational techniques are discussed. A sample of sixteen objects or condensations which show evidence for rotation have been obtained.

131.078 **Turbulence in molecular clouds.**
R. L. Dickman.
Protostars & planets II, p. 150 – 174 (1985). – See Abstr. 012.008.
The basic aim of this paper is to offer a primer of basic concepts and methods of analysis for observationally–oriented individuals who wish to work in the rapidly developing area of molecular cloud turbulence. First the difficulties which beset early attempts to determine the nature of gas motions within molecular clouds are reviewed. Some aspects of turbulence as a hydrodynamic phenomenon are considered next along with an introduction to the statistical vocabulary of the subject which is required to understand the methods for analyzing observational data. A simple and useful approximation for estimating the velocity correlation length of a molecular cloud is also described. The paper concludes with a final perspective, which considers the extent to which size–velocity dispersion correlations can serve as a probe of chaotic velocity fields in molecular clouds.

131.079 **Star formation: an overview.**
N. J. Evans II.
Protostars & planets II, p. 175 – 187 (1985). – See Abstr. 012.008.
More detailed study of star formation regions is now beginning, and traditional interpretations are being questioned. Some compact sources are now thought to be density enhancements which are not self-luminous. Infrared excesses around young stars may not always be caused by circumstellar dust; speckle measurements have shown that at least some of the excess toward T Tauri is caused by an infrared companion. Consequently, interpretation of infrared excesses in terms of disks should be viewed with caution. Spectroscopic studies of the dense, star-forming cores and of the compact objects themselves have uncovered a wealth of new phenomena, including the widespread occurrence of energetic outflows. New discoveries with the Infrared Astronomical Satellite and with other planned infrared telescopes will continue to advance this field.

131.080 **Infrared spectroscopy of protostars and young stellar objects.**
N. Z. Scoville.
Protostars & planets II, p. 188 – 200 (1985). – See Abstr. 012.008.
Near-infrared spectroscopy offers a unique opportunity to observe young stellar objects (possibly protostars) and their circumstellar nebula. The atmospheric windows at $\lambda = 2$ to 5 μm enable the study of molecular rotation-vibration bands of CO and H_2 in addition to recombination lines of H I. Recent observations giving kinematic resolution of 6 to 20 km s^{-1} are reviewed for the Becklin-Neugebauer (BN) object in Orion. The observations suggest lingering accretion in a neutral shell ($R \approx 25$ AU) surrounding the star and its outflowing ionized wind.

131.081 **Fragmentation and hierarchical structure in the interstellar medium.**
J. M. Scalo.
Protostars & planets II, p. 201 – 296 (1985). – See Abstr. 012.008.
Observational and theoretical work related to the fragmentation of interstellar structures is reviewed. An attempt is made to clarify the essential features of cool interstellar structures related to star formation. One way is shown in which observations can be used to quantify the structure. The numerous proposed physical processes which may lead to fragmentation are reviewed. Theories for the origin of the largest-scale structures are discussed.

131.082 **The formation of bound stellar clusters.**
B. A. Wilking, C. J. Lada.
Protostars & planets II, p. 297 – 319, 1184 – 1185 (1985). – See Abstr. 012.008.
The authors review the properties of visible associations and open clusters and discuss what these properties infer about the processes by which these stellar aggregates are formed. To assist in the identification of young embedded clusters, the authors discuss the required efficiency of conversion of gas into stars and the molecular cloud environment necessary to produce bound clusters. Quantitative estimates for the final star formation efficiency in bound systems are presented from both analytic and numerical studies. Observational techniques which are used to identify regions forming groups of young stellar objects are reviewed. Existing observations of young stellar groups are presented and compared to the proposed scenarios for association and cluster formation. A preliminary study of the luminosity of the young dust-embedded cluster in the ϱ Ophiuchi dark cloud is presented.

131.083 **Magnetic fields.**
L. Mestel.
Protostars & planets II, p. 320 – 339 (1985). – See Abstr. 012.008.
The principal topics discussed are: (1) the dynamical and observational consequences of the anisotropic collapse of cool gas clouds permeated by the local galactic magnetic field; (2) magneto-gravitational equilibria of such clouds with subcritical mass-flux ratios, especially in the thin disk approximations; (3) magnetic braking of both subcritical and supercritical masses; and (4) the consequences of flux leakage during the molecular phase, including the effect on field topology.

131.084 **Processes and problems in secondary star formation.**
R. I. Klein, R. W. Whitaker, M. T. Sandford II.
Protostars & planets II, p. 340 – 367 (1985). – See Abstr. 012.008.
Recent developments relating the conditions in molecular clouds to star formation triggered by a prior stellar generation are reviewed. Primary processes are those that lead to the formation of a first stellar generation. The secondary processes that produce stars in response to effects caused by existing stars are compared and evaluated in terms of observational data presently available. The triggering processes of supernovae, stellar winds, and H II regions are discussed with emphasis on the consequences for radiation-driven implosion as a promising secondary star formation mechanism. Detailed two-dimensional, radiation-hydrodynamic calculations of radiation-driven implosion are discussed.

131.085 **High mass versus low mass star formation.**
R. Thompson.
Protostars & planets II, p. 434 – 447 (1985). – See Abstr. 012.008.
Advances in observational techniques, especially in the radio and infrared, have led to observations of high- and intermediate-mass young stellar objects while they are still embedded in their natal dust clouds. At the same time, these techniques, along with new optical observations, have been applied to the study of low-mass stars which are no longer completely obscured by dust. Some of these new observations, especially for higher-mass stars, are examined in this paper both to afford a better view of the stellar formation process and to illustrate the differences between the birth of high-mass and low-mass stars.

131.086 **Protostellar disks and star formation.**
P. Cassen, F. H. Shu, S. Terebey.
Protostars & planets II, p. 448 – 483 (1985). – See Abstr. 012.008.
The configuration of the solar system, as well as recent observations of young stellar objects, suggests that disks are a natural byproduct of the formation of stars from rotating interstellar clouds. The authors review current theories of angular momentum and mass transport in protostellar disks, present a model for their formation from the gravitational collapse of molecular cloud cores, and discuss some different evolutionary pathways possible for the combined protostar-disk systems that result.

131.087 **The chemical state of dense interstellar clouds: an overview.**
W. M. Irvine, F. P. Schloerb, Å. Hjalmarson, E. Herbst.
Protostars & planets II, p. 579 – 620 (1985). – See Abstr. 012.008.
The authors list the currently known interstellar molecules and isotopes, discuss procedures for determining relative chemical abundances in molecular clouds, and present current best estimates for such abundances in regions of differing physical properties. Among the results are a general chemical similarity across a range of density and temperature for quiescent clouds, and some striking differences among regions which are not easily related to such physical parameters and may instead reflect cloud history and evolution. The possibility of constraining chemical models via measurements of relative abundances for the isomeric pairs HNC/HCN, CH_3NC/CH_3CN, and HOC^+/HCO^+ is discussed in detail.

131.088 **Grain evolution in dark clouds.**
B. T. Draine.
Protostars & planets II, p. 621 – 640 (1985). – See Abstr. 012.008.
Observations pertaining to the sizes and composition of grains in diffuse clouds and dark clouds are briefly reviewed. The physical processes acting to change the composition and size distribution of grains in dark clouds are examined.

131.089 **Time–dependent interstellar chemistry.**
A. E. Glassgold.
Protostars & planets II, p. 641 – 649 (1985). – See Abstr. 012.008.
The author has attempted to deal with some current problems in interstellar chemistry in the context of time–dependent calculations.

131.090 **Electron abundances and ionization rates in interstellar clouds.**
W. D. Langer.
Protostars & planets II, p. 650 – 667 (1985). – See Abstr. 012.008.
Observations of the isotopes of H_2 and HCO^+ make it possible to estimate bounds on the electron abundance and cosmic ray and X–ray ionization rates in interstellar clouds. The theory behind these methods is reviewed and the current state of the observations summarized.

131.091 **Gas–phase carbon chemistry in dense clouds.**
E. Herbst.
Protostars & planets II, p. 668 – 684 (1985). – See Abstr. 012.008.
The literature concerning the chemistry of complex molecules in dense interstellar clouds is reviewed and recent gas–phase studies are emphasized. Detailed ion–molecule pathways leading to large interstellar molecules are presented. These pathways include radiative association reactions; therefore these reactions are discussed in some detail. The few existing models of dense clouds that include complex molecules are compared. Finally, the chemistry of collapsing protostellar regions is briefly explored.

131.092 **Interstellar absorption lines between 2000 and 3000 Å in nearby stars observed with BUSS.**
K. S. de Boer, H. Lenhart, K. A. van der Hucht, T. M. Kamperman, Y. Kondo, F. C. Bruhweiler.
Astron. Astrophys., Vol. 157, No. 1, p. 119 – 128 (1986).
Spectra obtained between 2000 and 3000 Å with the BUSS payload have been searched for interstellar absorption lines. In bright stars, with spectral types between O9 V and F5 V, such lines have been measured of Mg I, Mg II, Cr II, Mn II, Fe II and Zn II. Column densities are derived and interstellar abundances are determined for the observed species.

131.093 **Fragmented molecular clouds: statistical analysis of the ^{13}CO (J = 1–0) emission distribution.**
M. Pérault, E. Falgarone, J. L. Puget.
Astron. Astrophys., Vol. 157, No. 1, p. 139 – 147 (1986).
A quantitative analysis of the structure of fragmented molecular clouds is attempted through the computation of the 3–dimensional (l,b,v) autocorrelation function of the ^{13}CO emission measured in a few nearby complexes.

131.094 **Ammonia observations and temperatures in the S 140/L 1204 molecular cloud.**
H. Ungerechts, C. M. Walmsley, G. Winnewisser.
Astron. Astrophys., Vol. 157, No. 2, p. 207 – 216 (1986).
The authors have mapped the NH_3(1,1) and (2,2) transitions in the molecular cloud L 1204 associated with the H II region S 140, the number density is close to $5 \times 10^5 cm^{-3}$. The derived column density, and the rotational temperature T_{12} peak at the position of the embedded infrared sources S 140–IRS. The kinetic temperature $T_{kin}(NH_3)$ is in close agreement with earlier measurements of T_{kin}(CO) and the dust color temperature and decreases smoothly from a value of 40K at S 140–IRS to 20K within the neighbourhood ($r < 1.4 \times 10^{18}$cm) of the infrared source.

131.095 **The centimeter transitions of *E*–type methanol.**
K. M. Menten, C. M. Walmsley, C. Henkel, T. L. Wilson.
Astron. Astrophys., Vol. 157, No. 2, p. 318 – 328 (1986).
The authors have used the 100–m telescope at a frequency of 25 GHz to observe the $2_2 \rightarrow 2_1$, $3_2 \rightarrow 3_1$, $4_2 \rightarrow 4_1$, $5_2 \rightarrow 5_1$, $6_2 \rightarrow 6_1$, and $9_2 \rightarrow 9_1$ transitions of *E*–type methanol towards a selection of galactic molecular–line sources. They have found several sources of strong narrow line emission in these transitions which are believed to be masers similar to those known in the Orion–KL region. The authors also observe absorption in these lines towards W3(OH) and NGC 7538 showing that the transitions are not necessarily inverted under interstellar conditions. Finally, most strong molecular line sources show emission in these transitions with a profile similar to that known from observations of other molecules. The rotation temperatures estimated for such emission features range between 20 and 100K. The authors conclude that they are observing hot (100K) compact (10^{17}cm) clumps immersed in the general molecular cloud material, which is preferentially sampled, for example, in surveys of the $J = 2 \rightarrow 1$ methanol transitions.

131.096 **Temperature distribution of interstellar dust.**
F. Pajot, P. Boissé, R. Gispert, J. M. Lamarre, J. L. Puget, G. Serra.
Astron. Astrophys., Vol. 157, No. 2, p. 393 – 399 (1986).
Using the infrared spectrum (2 μm to 1 mm) of three luminous galactic regions and the absorption properties of interstellar dust, the authors derive its temperature distribution. They also compute the energy output of interstellar dust as a function of temperature. These distributions are bimodal and do not agree with predictions based on standard models, the amount of high temperature dust being much too large.

131.097 **Interstellar scattering of compact radio sources near supernova remnants.**
S. R. Spangler, R. L. Mutel, J. M. Benson, J. M. Cordes.
Astrophys. J., Vol. 301, No. 1, p. 312 – 319 (1986).
The authors report a multifrequency, VLBI search for interstellar scattering of extragalactic radio sources near supernova remnants. VLBI observations at 610, 1663, and 4991 MHz were made of compact sources near the supernova remnants CTA 1, G33.6+0.1, G74.9+1.2, and HB 21, and 610 MHz observations made of a source near HB 9. These observations were motivated by the possibility of enhanced cosmic–ray–induced turbulence in front of supernova remnants as expected in "diffusive" theories of shock wave acceleration. Angular broadening is definitely seen in the case of the source 2013+370, which lies within 4′ of the supernova remnant G74.9+1.2. The source 1849+005 appears to be highly scattered since no fringes were detected.

131.098 **Aperture synthesis observations of the 21 centimeter Zeeman effect toward Cassiopeia A.**
U. J. Schwarz, T. H. Troland, J. S. Albinson, J. D. Bregman, W. M. Goss, C. Heiles.
Astrophys. J., Vol. 301, No. 1, p. 320 – 330 (1986).
The authors have used the Westerbork synthesis radio telescope to perform an improved sensitivity study of the 21 cm H I Zeeman effect toward Cas A. They present synthesis maps of the line–of–sight magnetic field strength for velocities corresponding to the Perseus arm absorption features. Considerable spatial structure exists in these maps, with measured field strengths exceeding 40 μG at some positions in front of the source. No significant reversals in field direction are apparent. Peaks in magnetic field strength often coincide in position with clumps of molecular gas. The magnetic field peaks arise in clumps of higher density H I gas that are associated with the molecular clumps.

131.099 **Zeeman splitting of 18 centimeter OH lines toward Cassiopeia A and other sources.**
C. Heiles, M. Stevens.
Astrophys. J., Vol. 301, No. 1, p. 331 – 338 (1986).
The authors have detected Zeeman splitting of the OH absorption lines toward Ori B, W22, and Cas A. For W22, the derived magnetic field strength is –11 μG. For Ori B, the inferred magnetic field strength is consistent in magnitude, but not in direction, with the 38 μG derived previously by Crutcher and Kazes. For Cas A, the average field in four velocity components is +9 μG, in the same direction as, but weaker than, the field strengths measured in single–dish H I data. They are considerably weaker than the field strengths in less dense H I clumps that are associated with the OH clumps. This circumstance raises serious interpretive difficulties.

131.100 **Interstellar magnetic field strengths and gas densities: observational and theoretical perspectives.**
T. H. Troland, C. Heiles.
Astrophys. J., Vol. 301, No. 1, p. 339 – 345 (1986).
The authors present an updated compilation of observational data concerning the relationship between the interstellar magnetic field strength and the gas density. Pulsar and Zeeman–effect data provide the only reliable information about the (B, n) relationship, and they now span nearly six orders of magnitude in gas density. Field strengths show no evidence of increase over the density range 0.1 – $\sim 100\ cm^{-3}$. At higher densities, a modest increase in field strength is observed in some regions, in line with theoretical expectations for selfgravitating clouds. Despite the consistency between observations and theory, questions still exist about how the magnetic field strength remains constant for densities up to $\sim 100\ cm^{-3}$.

131.101 **Infrared studies of the Herbig–Haro object 1–2 region.**
P. M. Harvey, M. Joy, D. F. Lester, B. A. Wilking.
Astrophys. J., Vol. 301, No. 1, p. 346 – 354 (1986).
Far–infrared (50 – 100 μm) photometric mapping and near–infrared spectroscopic mapping of the Herbig–Haro (HH) 1–2 region are presented. Both lobes of the double–lobed far–infrared source in this region are found to be nearly coincident with recently discovered radio continuum sources. Near–infrared spectroscopic maps made at 2.12 μm ($\Delta\lambda$ = 0.06 μm) and 2.2 μm ($\Delta\lambda$ = 0.5 μm) reveal widespread molecular hydrogen emission [$\upsilon = 1 \rightarrow 0$, $S(1)$] associated with the optical nebulae HH 1, HH 2, and in a region between the two HH sources. In addition, there is probably some scattered continuum visible near and southeast of the Cohen–Schwartz star. The data are consistent with a picture in which each far–infrared lobe contains a highly obscured young star surrounded by a highly nonspherically symmetric dust cloud. The direction of the temperature gradients, however, are such that some heating by the Cohen–Schwartz star and/or the HH nebulae themselves is not ruled out.

131.102 **Abundances of interstellar atoms from ultraviolet absorption lines.**
E. B. Jenkins, B. D. Savage, L. Spitzer Jr.
Astrophys. J., Vol. 301, No. 1, p. 355 – 379 (1986).
The equivalent widths of interstellar absorption lines of Mg II, P II, Cl I and II, Mn II, Fe II, Cu II, and Ni II, obtained in a Copernicus survey, have been analyzed to yield column densities along the lines of sight. The measured depletions are clearly correlated with $\langle n_H \rangle$, the mean hydrogen column density along the line of sight. Depletions also seem to be weakly correlated with various ratios of hydrogen to extinction by dust grains and also variations of extinction with wavelength. The authors interpret the apparent coupling of depletion to the mean density in terms of an idealized model (Spitzer) where each element has one value of depletion in the low–density, warm, neutral gas and an enhanced, different value in cold clouds. The difference of apparent depletions for Mg, P, Cl, Mn, and Fe between warm and cold clouds averages 0.44 ± 0.12 (rms) dex.

131.103 **The extent of the local H I halo.**
F. J. Lockman, L. M. Hobbs, J. M. Shull.
Astrophys. J., Vol. 301, No. 1, p. 380 – 394 (1986).
Forty–five high–latitude, OB stars have been observed in the Lyα and 21 cm lines of H I in an effort to map out the vertical distribution and extent of the local H I halo. The 25 stars for which a reliable H I column density can be obtained from Lyα lie between 60 and 3100 pc of the plane. The principal result is that the total column density of H I at $|z| > 1$ kpc is, on the average, $5 \pm 3 \times 10^{19} cm^{-2}$, or 15% of the total N_{HI}. A model with half of the H I mass in clouds having radii of a few parsecs and a Gaussian vertical distribution with σ_z = 135 pc, and half of the mass in an exponential component with a scale height of 500 pc, gives a satisfactory fit to the data.

131.104 **High–latitude molecular clouds: completeness of the local sample and implications for molecular surveys.**
L. Magnani, E. A. Lada, L. Blitz.
Astrophys. J., Vol. 301, No. 1, p. 395 – 397 (1986).
The results of an unbiased survey of CO emission at high galactic latitudes ($|b| \geqslant 25°$) are presented. In this paper, the value for ε, the fractional completeness of the Blitz, Magnani, and Mundy high–latitude cloud survey, is obtained and found to be 0.55 ± 0.15. Using this value, the following properties of the molecular component of the local interstellar medium are determined: the number of individual clouds within 100 pc of the Sun is ~ 120; the surface filling fraction is $\sim 4.5 \times 10^{-3}$; the total mass is of order $5 \times 10^3 M_\odot$; and the mass surface density is $0.2\ M_\odot pc^{-2}$.

131.105 **Molecular clouds and star formation in the inner Galaxy: a comparison of CO, H II, and far–infrared surveys.**
P. C. Myers, T. M. Dame, P. Thaddeus, R. S. Cohen,
R. F. Silverberg, E. Dwek, M. G. Hauser.
Astrophys. J., Vol. 301, No. 1, p. 398 – 422 (1986).
The authors have compared surveys of the galactic plane over $-1° \leqslant b \leqslant 1°$, $12° \leqslant l \leqslant 60°$, in the CO line at 2.6 mm, in the far–infrared (FIR) continuum at 150 μm and 250 μm, and in the radio continuum and H 110α recombination line at 6 cm. They identify 54 molecular cloud complexes, with mean mass $\sim 10^6 M_\odot$. Most FIR sources are coincident with H II regions, and nearly all H II regions in turn are associated with molecular clouds. The stellar content of a cloud is estimated by assuming that the associated FIR and radio continuum emission result from clusters whose mass distribution is the same as the initial mass function. The star formation efficiency (SFE) for the entire sample lies near 0.02, but some massive clouds are extremely star–poor, with SFE less than 10^{-3}. The stellar mass in a cloud appears correlated with the cloud mass, as $M_* \propto M_{cloud}^{0.5 \pm 0.2}$.

131.106 **The photoionization of the diffuse Galactic gas.**
J. S. Mathis.
Astrophys. J., Vol. 301, No. 1, p. 423 – 429 (1986).
Observations recently published by Reynolds show that there is diffuse ionized gas (DIG) both in the Galactic plane and at high Galactic latitudes. Relative to normal H II regions, this gas shows strong [S II]/Hα, normal [N II]/Hα, and very weak [O III]/Hα. Models are presented for this gas, assuming that it is photoionized by O stars in the Galactic plane, that the ionized zone is in a steady state as regards both ionization and thermal balance, and that the DIG is also responsible for the pulsar dispersion measures. It is possible to obtain an excellent fit to the observations with a plausible mixture of spectral types for the ionizing stars. It is also shown that shocks are unlikely to be producing the diffuse emission lines.

131.107 **Errata: "GSS 30: an infrared reflection nebula in the Ophiuchus dark cloud" [Astrophys. J., Vol. 290, No. 1, p. 261 – 272 (1985)].**
M. W. Castelaz, J. A. Hackwell, G. L. Grasdalen, R. D. Gehrz, C. Gullixson.
Astrophys. J., Vol. 301, No. 1, p. 502 (1986). See Abstr. 39.131.122.

131.108 **$^{12}C^{18}O$ in OMC–1: kinematics, molecular column density, and kinetic temperature distribution.**
T. L. Wilson, E. Serabyn, C. Henkel, C. M. Walmsley.
Astron. Astrophys., Vol. 158, No. 1/2, p. L1 – L4 (1986).
A fully sampled map of size ~1'x3' (R.A.xDec.), centered on BN–KL, has been made in the J = 1–0 line of $^{12}C^{18}O$ with 21" angular resolution. The $^{12}C^{18}O$ emission is concentrated in a ⩽40" wide continuous strip running S to NE.

131.109 **Compact ammonia cloud in the bipolar flow source CRL 2591.**
T. Takano, J. Stutzki, Y. Fukui, G. Winnewisser.
Astron. Astrophys., Vol. 158, No. 1/2, p. 14 – 16 (1986).
The authors have observed the bipolar flow source CRL 2591 in the NH_3 (J,K) = (1,1) and (2,2) lines with the Effelsberg 100 m telescope. A compact ammonia cloud of ~0.6 pc diameter around the central infrared source CRL 2591 was found and its mass was estimated to be $\sim 40\ M_\odot$. A velocity gradient in the direction orthogonal to the CO bipolar flow suggests that this

compact ammonia cloud is rotating around the symmetry axis of the CO flow.

131.110 **Time-dependent chemistry in dense molecular clouds. II. Ultraviolet photoprocessing and infrared spectroscopy of grain mantles.**
L. B. d'Hendecourt, L. J. Allamandola, R. J. A. Grim, J. M. Greenberg.
Astron. Astrophys., Vol. 158, No. 1/2, p. 119 – 134 (1986).
The authors have performed a series of experiments in which mixtures of gases condensed on a substrate at 10K are irradiated with ultraviolet light in order to probe the photochemical processing of interstellar ices. The evolution of these simulated grain mantles is followed by infrared spectroscopy (2.5 – 20 μm). Upon photolysis, the appearance of new bands in the spectra is observed and their possible identification is discussed. Rough quantitative estimates of the photo-production rate of new molecules are given and some infrared cross sections of a few of these bands are discussed.

131.111 **Excitation in the ionized diffuse interstellar medium.**
J.-P. Sivan, G. Stasińska, J. Lequeux.
Astron. Astrophys., Vol. 158, No. 1/2, p. 279 – 287 (1986).
The authors present new observations of optical line intensity ratios in large, diffuse, ionized regions of the Sagittarius–Carina arm of the Galaxy and in the Orion–Eridanus complex. They collect some observational information on a few classical H II regions, the diffuse ionized medium and some supernova remnants. They examine the pertinency of model calculations for explaining observations and present a series of model H II regions computed in order to interpret their data. The excitation of the diffuse medium is discussed.

131.112 **Nebular complex in the region of Cas OB2 association; ring nebula Sh 157.**
T. A. Lozinskaya, T. G. Sitnik, A. I. Lomovskii (*A. I. Lomovskij*).
Astrophys. Space Sci., Vol. 121, No. 2, p. 357 – 385 (1986).
This paper presents the results of monochromatic [O III], [N II], and [S II] observations of ring nebula Sh 157 around the star HD 219460 belonging to the Ba 3 cluster. The observations suggest that the Sh 157 ring structure may arise as a result of the HD 219460 stellar wind blowing the surrounding H II region, and the bubble age is found to be $t \cong (2-5) \times 10^5$yr. Three outer envelopes have been distinguished: a weaker extended emission shell apparently blown out by the wind from B-stars of the Ba 3 cluster, and two dust shells which are likely to be associated with NGC 7510 and Cas OB2. The paper is also concerned with the discussion of young star aggregates Ba 3, NGC 7510, Cas OB1, OB2, OB4, OB5, OB7, and Cep OB1 and the associated H II regions, shells and supershells of gas and dust, molecular clouds, and supernova remnants which may be probable members of a single giant stellar complex where the star formation process is in progress.

131.113 **An investigation of diffuse cloud chemistry. III. Towards a logical approach in chemical modelling.**
A. P. C. Mann, D. A. Williams.
Astrophys. Space Sci., Vol. 121, No. 2, p. 387 – 395 (1986).
Chemical modelling of diffuse interstellar clouds is capable of providing a complete description of their physical condition. However, in the present state of astronomical and chemical data, ambiguities arise and unique descriptions cannot be obtained. This paper argues that a logical approach to chemical modelling is required, and offers a simple procedure which minimizes the ambiguities arising in diffuse cloud models. The case of interstellar CN is illustrated.

131.114 **Analytical, time-dependent mass function and fragmentation rate due to gravitational instability.**
A. Di Fazio.
Astron. Astrophys., Vol. 159, No. 1/2, p. 49 – 64 (1986).
An analytical, theoretical, time-dependent initial mass function is derived for the objects created in the fragmentation of a gravitationally unstable gas protocloud. The mass spectrum depends on the chemical-dynamical-radiative evolution of the protocloud and it peaks at a mass slightly greater than the minimum Jeans mass attained throughout the evolution. A fragmentation rate mass spectrum is also analytically derived. By using an evolutionary model presented in this paper, it is shown that the fragmentation process implies the existence of several generations of fragments, with different ranges of mass. A protogalactic evolution model, using the authors' new theory, predicts that the first stars are created in the third generation of fragments. These can represent the "zero-metal" population. The theoretical mass functions predicted by the evolutionary models and the fragmentation rate are quantitatively compared with the observations, yielding a remarkably good agreement.

131.115 **CO J = 4-3 observations of molecular clouds.**
G. J. White, J. P. Phillips, K. J. Richardson, R. H. Harten.
Astron. Astrophys., Vol. 159, No. 1/2, p. 309 – 316 (1986).
A survey of the CO J = 4-3 transition has been made towards a sample of seven galactic molecular clouds. These data have been combined with spectra for the lower CO transitions to examine how they vary in line shape with different transition. Such variations are examined using LVG modelling techniques to obtain estimates of the physical state of the emitting gas. The authors discuss the implications and importance that observations of these submillimetre wavelength transitions have in refining knowledge of the gas in molecular clouds.

131.116 **Nature of very small grains: PAH molecules or silicates?**
F. X. Désert, F. Boulanger, A. Léger, J. L. Puget, K. Sellgren.
Astron. Astrophys., Vol. 159, No. 1/2, p. 328 – 330 (1986).
The predictions of the model of Puget et al. (1985) for the emission from very small grains (VSGs) including both graphitic (PAHs) and silicate components are compared with published 8 – 13 μm observations of astronomical sources. The VSGs are found to be mainly graphitic and an upper limit is placed on the relative mass of silicates based on lack of the 9.7 μm silicate emission feature of M82 and NGC 2023. This dissymmetry in the composition of VSGs supports the suggestion that they are formed in grain-grain collisions where the behaviors of graphite and silicate grains are expected to be quite different.

131.117 **LiH molecules in dense interstellar clouds.**
B. V. Vajner, Yu. A. Shchekinov, I. A. Yudilevich.
Kinematika Fiz. Nebesn. Tel, Tom 2, No. 1, p. 36 – 42 (1986). In Russian. English translation in Kinematics Phys. Celest. Bodies.
The abundances of LiH molecules in interstellar clouds are estimated. The excitation temperatures of rotational levels and the expected radiation fluxes in rotational transitions are calculated. The possibility of observation of these lines is discussed.

131.118 **Concerning the problem of collisional heating of the interstellar helium flow by solar wind protons.**
M. A. Gruntman.
Planet. Space Sci., Vol. 34, No. 4, p. 387 – 389 (1986).
In the past heating by elastic collisions with the solar wind protons of the interstellar helium flow into the heliosphere has been a subject of consideration by several authors. The approaches used in this context are critically analysed in this paper and it is shown that the concept of a continuous momentum transfer is not likely to be applicable. It is argued in contrast that large angle deflections though being rare events give the predominant contributions to the momentum transfer to the neutral helium atoms. This may have consequences for a progressed modelling of the interstellar helium distribution in the heliosphere.

131.119 **Flicker of extragalactic radio sources and refractive interstellar scintillation.**
R. Blandford, R. Narayan, R. W. Romani.
Astrophys. J., Lett. Ed., Vol. 301, No. 2, p. L53 – L56 (1986).
Recent work has identified variability of flat-spectrum extragalactic radio sources at $\lambda \approx 10$ cm with rms amplitude of $\sim 2\% - 3\%$ and time scale of days. The authors show that this "flicker" is consistent with intensity fluctuations caused by re-

fractive scintillation in an extended interstellar medium in our Galaxy. Further observation of flicker may allow the structure of suitable sources to be partially resolved on angular scales smaller than those probed by VLBI.

131.120 **Radio observations of CH in three dark nebulae and the correlation of CH with optical extinction and molecular hydrogen.**
K. Mattila.
Astron. Astrophys., Vol. 160, No. 1, p. 157 – 170 (1986).
Observations of the 3335 MHz (9 cm) CH line have been carried out in the dark nebulae L134, L1780, and L1590 using the 43–m radio telescope at Green Bank. CH column densities have been derived and are compared with optical extinctions derived from star counts and convolved with a Gaussian response function corresponding to the beam of the radio telescope. The application of the star count technique to clouds at high latitudes is discussed and errors introduced by usage of the van Rhijn (1929) lg$N(m)$–curves are pointed out. A good correlation is found between the CH column density and optical extinction in L1590 and L1780. A comparison of lg$N$(CH) with lg$N(H_2)$ is presented for diffuse, dark, and molecular cloud data. Finally, the structure of the three nebulae is briefly discussed in the light of the present CH and extinction data.

131.121 **New HC_7N cloudlets in Taurus.**
J. Cernicharo, R. Bachiller, G. Duvert.
Astron. Astrophys., Vol. 160, No. 1, p. 181 – 184 (1986).
The authors report the detection of the $J = 17–16$ line of HC_7N towards six cloudlets of the Taurus dark clouds complex. The HC_7N column densities towards these globules range between $0.8x10^{12}$ and $4x10^{12}cm^{-2}$, i.e., 4 – 20 times lower than towards TMC1. However, the average column density ratio $N(HC_5N)/N(HC_7N)$ is 3.6, i.e., similar to the TMC1 one. Little variations of this abundance ratio are found from cloud to cloud whatever the total cyanopolyyne column density is. One of these cloudlets, L1495, is a new cyanopolyyne source and physical conditions towards it are derived from NH_3, HC_3N, and HC_5N observations.

131.122 **The dynamical evolution of H I shells and the initial mass function of OB associations.**
S. A. Silich.
Prepr. Inst. Teor. Fiz. Akad. Nauk Ukr. SSR, ITP–85–48E, 9 pp. (1985).
The evolution of expanding supershells produced by supernovae explosions in OB associations is considered. It is shown that the narrow range of the observed velocities is naturally explained in a model. The supershells expanding velocities and star mass spectral index obtained from the condition of velocity constancy agree well with observational data.

131.123 **A revised theory of the ^{26}Mg/Al correlation.**
D. D. Clayton.
Meteoritics, Vol. 20, No. 4, p. 626 – 627 (1985). Abstract. – See Abstr. 010.661.

131.124 **On the nature of interstellar grains and the interpretation of the IRAS background radiation.**
M. Rowan–Robinson.
Mon. Not. R. Astron. Soc., Vol. 219, No. 4, p. 737 – 749 (1986).
The interstellar extinction curve has been accurately fitted with a simple composite model for interstellar grains consisting of small ($\leqslant$0.01 and 0.03 μm radius) grains of graphite and silicate, and larger ($\sim$0.1 μm) grains of amorphous carbon and amorphous silicate. This model is also consistent with observations of circumstellar dust shells around late–type stars and accounts naturally for the far–infrared spectra of hot–centred clouds associated with newly formed massive stars. The interstellar "cirrus" found by IRAS can be identified as due to the small grains, primarily the carbon. An upper limit of 0.8 MJy/sr can be set to the contribution of interstellar grains to the 100 μm background intensity, outside obvious regions of cirrus and neutral hydrogen. This leaves a residual background of 5 – 6 MJy/sr unaccounted for by interplanetary or interstellar dust.

131.125 **Submillimetre observations of a disc around the embedded source GL 490.**
W. K. Gear, G. Gee, E. I. Robson, P. A. R. Ade, W. D. Duncan.
Mon. Not. R. Astron. Soc., Vol. 219, No. 4, p. 835 – 840 (1986).
The authors present the results of mapping the 350 μm emission around the bipolar outflow source GL 490. They also report IRAS observations of this source. The 350 μm data show the source to be extended north–south. The authors interpret these results in terms of a disc of dust and gas surrounding the central embedded young star but show that it is incapable of channelling the bipolar molecular outflow. The outflow must therefore be collimated on scales smaller than 10^{17}cm.

131.126 **PAH molecules and carbon dust in interstellar clouds.**
W. W. Duley, D. A. Williams.
Mon. Not. R. Astron. Soc., Vol. 219, No. 4, p. 859 – 864 (1986).
An analysis of possible chemical reactions between PAH molecules and atoms and ions in the interstellar gas shows that a variety of destruction channels exist for these molecules. PAH molecules will have lifetimes of $\lesssim 10^5$yr under diffuse cloud conditions and therefore are unlikely to be a ubiquitous component of interstellar dust clouds. The destruction of PAH molecules is enhanced by temperature fluctuations, although PAH molecules may be observable as transient species in shocked regions. The relationship between PAH and carbon dust is also discussed.

131.127 **The variation of interstellar extinction in the ultraviolet.**
D. J. Carnochan.
Mon. Not. R. Astron. Soc., Vol. 219, No. 4, p. 903 – 926 (1986).
One hundred and fifty–four reddened stars that are apparently normal in the visible were selected from the S2/68 Ultraviolet Sky Survey. The ultraviolet data for 92 of these cannot be explained in terms of a fixed interstellar extinction law. Between 1400 and 2740 Å the extinction curve for each star can be well represented by two parts, a straight–line scattering component and a Lorentzian 2200 Å absorption feature. Independent variations are found in both parts and these cannot be explained by photometric or spectral classification errors. Both parts vary smoothly implying that there is no fixed extinction law, and one star in three is found to depart from the mean law by more than 1 mag at either 1500 of 2200 Å. The two variations allow not only all 154 stars to be explained but also anomalous stars reported by other authors.

131.128 **Distribution of the interstellar absorbing matter in the direction $l \sim 207°$, $b \sim -3°$.**
N. G. Guseva, M. D. Metreveli.
Kinematika Fiz. Nebesn. Tel, Tom 2, No. 2, p. 3 – 12 (1986). In Russian. English translation in Kinematics Phys. Celest. Bodies.
The distribution of the interstellar absorption matter in the direction of the star formation region 2 Mon ($l = 207° \pm 2.5°$; $b = -3° \pm 2.5°$) is investigated. The mean curves of interstellar absorption for seven regions of the area investigated are derived using two methods of averaging. A selection effect in the observations is taken into account. The area is characterized by low mean absorption ($1.2^m - 1.4^m$)kpc^{-1} within the local spiral arm. The regions near the galactic plane contain dense interstellar clouds.

131.129 **Ionization of the interstellar medium in the vicinity of non–stationary ionizing radiation sources.**
A. A. Suchkov.
Kinematika Fiz. Nebesn. Tel, Tom 2, No. 2, p. 21 – 26 (1986). In Russian. English translation in Kinematics Phys. Celest. Bodies.
The process of setting and relaxation of the ionic composition in the interstellar gas with $T = 10^4$K in the vicinity of a non–stationary ionizing radiation source is considered. Results of computation are presented for time–dependent fractional abundances of the highly charged ions C IV, N V, O VI, Si III, Si IV, S IV, and S VI in a non–stationary H II zone.

131.130 **Investigation of the NGC 2264 star formation region.**
N. B. Kalandadze, V. I. Kuznetsov, V. I. Voroshilov.
Kinematika Fiz. Nebesn. Tel, Tom 2, No. 2, p. 27 – 34 (1986). In Russian. English translation in Kinematics Phys. Celest. Bodies.
The distribution of the absorbing matter in NGC 2264 and its relation to the star formation indicators and molecular clouds has been investigated on the basis of *B, V* magnitudes and spectral classes of 1670 O, B, A stars and 2700 F, G, K, M stars. Three dark clouds have been detected: the north–western, northern and central ones. A space–time picture of star formation in the complex is given.

131.131 **Exciting stars and distances of diffuse nebulae.**
V. S. Avedisova, G. I. Kondratenko.
Nauchn. Inf., Vyp. 56, p. 59 – 91 (1984). In Russian.
The early–type stars exciting 220 diffuse nebulae were found from extensive literature survey, as well as from extensive study of the magnetic tape catalogues of the Soviet astronomical data center. The photometric distances of the stars and nebulae were determined from *UBV* photometry and spectral classification of the stars. The transformation of the spectral classification into absolute visual magnitudes were made by means of the averaged calibration constructed by the authors. The distribution of the diffuse nebulae in the galactic plane shows the internal, local and two external arms of the Galaxy.

131.132 **Ortho–H_2O and the energy equilibrium in molecular clouds.**
C.–y. Zhang, Y. Tang, D.–l. Xiang.
Chin. Astron. Astrophys., Vol. 10, No. 1, p. 45 – 53 (1986). English translation of 40.131.272.

131.133 **Two cometary nebulae.**
S. A. Akopyan.
Astrofizika, Tom 24, Vyp. 1, p. 203 – 205 (1986). In Russian. English translation in Astrophysics, Vol. 24, No. 1.
This paper presents two probable cometary nebulae discovered on POSS charts.

131.134 **Temperature fluctuations and infrared emission from dust particles in a hot gas.**
E. Dwek.
Astrophys. J., Vol. 302, No. 1, p. 363 – 370 (1986).
The author calculates the effect of temperature fluctuations in very small dust particles caused by electronic collisions with an ambient hot gas. The dust–temperature distribution differs strongly from that derived on the basis of equilibrium heating models, peaking at lower temperatures in the stochastically heated case. The resulting infrared spectrum is broadened at short wavelengths, giving much more emission than expected from equilibrium calculations, and at long wavelengths exhibits a lower color temperature. Dust–temperature fluctuations are thus expected to play an important role in determining the infrared spectrum of dusty, X–ray–emitting plasmas.

131.135 **Measurements of X–ray scattering from interstellar grains.**
C. W. Mauche, P. Gorenstein.
Astrophys. J., Vol. 302, No. 1, p. 371 – 387 (1986).
The authors report on the results of an *Einstein Observatory* investigation of the halos produced by the scattering of X–rays from interstellar grains of four compact Galactic (low–latitude) and two extragalactic (high–latitude) X–ray sources. The intensity of these halos correlates well with (1) the amount of visual extinction and (2) the distance through the Galaxy's dust layer. A number of grain parameters is derived in the context of two alternate grain size distributions: the Mathis–Rumpl–Nordsieck (MRN) and the Oort–van de Hulst distributions, either of which is capable of producing the observed halos. Though no single grain size is capable of producing the shapes observed for these halos, a mean size of $\sim 0.1\ \mu m$ and a number density of $\sim 10^{-12}$grains cm^{-3} produce the correct halo intensities.

131.136 **Colors of reflection nebulae. III. Ultraviolet scattering by small particles in reflection nebulae associated with 17, 20, and 23 Tauri.**
A. N. Witt, R. C. Bohlin, T. P. Stecher.
Astrophys. J., Vol. 302, No. 1, p. 421 – 431 (1986).
New surface brightness measurements in the reflection nebulae associated with the Pleiades stars 17, 20, and 23 Tau have been carried out with the *IUE* satellite. Each nebula was observed at two angular offsets south of the respective illuminating star. At the larger offsets each nebula exhibits a flux distribution which rises much more steeply with decreasing wavelengths than the corresponding stellar spectrum. At the smaller offsets the nebular spectra are more similar to those of the stars and fail to show the steep rise at $\lambda < 2200$ Å. It is suggested that the reflection nebulosity near the Pleiades stars is caused by a thin sheet of interstellar matter in front of the stars, at an approximate distance of a few $x10^{-2}$pc. The data indicate that scattering in the UV by particles in the 10^{-6}cm size range appears more important than is generally recognized by current models.

131.137 **The structure of galactic H I in directions of low total column density.**
F. J. Lockman, K. Jahoda, D. McCammon.
Astrophys. J., Vol. 302, No. 1, p. 432 – 449 (1986).
The authors have made a detailed 21 cm study of areas that have the smallest known amount of H I in the northern sky. These observations have an estimated uncertainty in N_{HI} of $\leqslant 5x10^{18} cm^{-2}$. The region of main interest, around $\alpha = 10^h45^m$, $\delta = 57°20'$, has a minimum N_{HI} of $4.5x10^{19} cm^{-2}$. Spectra taken at 21′ resolution over a field 4°x3° in this direction show up to four H I line components. Two, near 0 and –50 km s^{-1}, are ubiquitous. There is also a narrow component at –10 km s^{-1} attributable to a diffuse cloud covering half of the field, and scattered patches of H I at $v < -100$ km s^{-1}. Eight other low column density directions were also observed. The data indicate that it is unlikely that the H I column density ever drops significantly below $4.5x10^{19} cm^{-2}$ anywhere in the sky.

131.138 **A theoretical steady state model for the distribution of gas and magnetic field in M31 and in our Galaxy.**
E. Battaner, M. L. Sánchez–Saavedra.
Astrophys. J., Vol. 302, No. 1, p. 450 – 458 (1986).
A time–independent model for the radial distributions of gas and magnetic field in M31 and the Galaxy has been developed. Total gas pressure profiles taken from observations were used as input in the MHD equations. From the model the magnetic field responsible for (and maintained by) the gas distribution has been obtained. The field has an azimuthal component only, and can be used to predict the synchrotron radio emission, in good agreement with observations. A nucleus of unseen gas is required at the center of both galaxies.

131.139 **Visual absorption lines in dark clouds: highly reddened stars near the center of the Ophiuchus molecular cloud.**
J. A. Cardelli, G. Wallerstein.
Astrophys. J., Vol. 302, No. 1, p. 492 – 503 (1986).
New high–resolution spectroscopic observations of four stars deeply embedded in the ρ Ophiuchi dark cloud complex are presented. The stars are located near the center of the molecular cloud and include HD 147889, the most heavily obscured star observed in the Ophiuchus cloud. The data include molecular lines of CN, CH, and CH^+, as well as atomic lines of Fe I, Ca I, and Ca II. Radial velocities and column densities were derived from the line measurements. The dependence of CN and CH densities on that of H_2 is discussed. Electron densities and depletion factors were determined from the atomic line measurements.

131.140 **$HCNH^+$: a new interstellar molecular ion.**
L. M. Ziurys, B. E. Turner.
Astrophys. J., Lett. Ed., Vol. 302, No. 1, p. L31 – L36 (1986).
A new molecular ion, $HCNH^+$, or protonated HCN, has been detected in the interstellar medium. The J = 1–0, 2–1, and 3–2 rotational transitions of this species have been observed toward Sgr B2. Using a large velocity gradient model calculation, the column density of $HCNH^+$ is found to be $\sim 4 \times 10^{14} cm^{-2}$. Such

a column density implies a fractional abundance for $HCNH^+$ in Sgr B2 of $\sim 3 \times 10^{-10}$, one to several orders of magnitude greater than the values predicted by ion–molecule models of interstellar chemistry.

131.141 **The Columbia CO survey of molecular clouds in the first Galactic quadrant.**
R. S. Cohen, T. M. Dame, P. Thaddeus.
Astrophys. J., Suppl. Ser., Vol. 60, No. 3, p. 695 – 818 (1986).
Using the Columbia millimeter–wave telescope, the authors have made a survey of 2.6 mm CO emission from the Galactic disk from $l = 12°$ to $l = 60°$ and $|b| \leqslant 1°$. Spectra were taken every 0°125, approximately every beamwidth, for $|b| \leqslant$ 0°5, and every 0°25 elsewhere. The calibration of the instrument was compared with that of several other telescopes and found to agree well. This paper presents the entire collection of spectra as well as spatial and longitude–velocity maps.

131.142 **A survey of the millimeter–wave spectrum of Sagittarius B2.**
S. E. Cummins, R. A. Linke, P. Thaddeus.
Astrophys. J., Suppl. Ser., Vol. 60, No. 3, p. 819 – 878 (1986).
More than 60% of the spectrum of Sgr B2 (OH) between 70 and 150 GHz was surveyed with the AT&T Bell Laboratories 7 m telescope. The rms noise in antenna temperature typically lies between 0.015 and 0.06K. Of the 457 lines tabulated, 31 remain unidentified. Column densities and rotational temperatures are estimated for 21 molecules. Most rotational temperatures are subkinetic, but the detectability of high–energy SO_2 and CH_3OH as well as vibrationally excited HC_3N and CH_3CN offers evidence for hot spots in the source.

131.143 **First light from a young star?**
B. Reipurth, J. Bally.
Nature, Vol. 320, No. 6060, p. 336 – 338 (1986).
The authors report the sudden appearance of a highly variable, conical nebula 1.5 arc min south of a 250 $L_\odot$ infrared source in the southern part of the L1641 cloud in Orion. The nebula coincides with the edge of the approaching lobe of a 10^5–yr–old bipolar molecular outflow. Variability of the infrared source generates light pulses observed to propagate through the surrounding ambient cloud, producing a fluctuating illumination pattern. This emission may represent the first optical radiation to have emerged from this newborn star.

131.144 **The discovery of an interstellar $^{15}NH_3$ maser.**
R. Mauersberger, T. L. Wilson, C. Henkel.
Astron. Astrophys., Vol. 160, No. 2, p. L13 – L16 (1986).
From a comparison between the (J,K) = (1,1), (2,2) absorption lines, the (3,3) emission line and limits for the (4,4) line, the authors conclude that maser emission is present in the (J,K) = (3,3) inversion transition of $^{15}NH_3$ toward the compact H II region NGC 7538–IRS1. Model computations show that the maser emission may originate from the hot molecular condensation described by Wilson et al. (1983). The probable maser pump is collisional, involving a transfer of population from the K = 0 to the K = 3 ladder of $^{15}NH_3$. Such a process places constraints on the parity selection rules in collision rates. The $^{14}NH_3/^{15}NH_3$ abundance ratio is found to be $\sim$500.

131.145 **An empirical study of ring–shaped nebulae in the Galaxy and in the LMC.**
M. Rosado.
Astron. Astrophys., Vol. 160, No. 2, p. 211 – 228 (1986).
Samples of optical supernova remnants, nebulae around WR and Of stars and nebulae driven by the action of the stellar winds of OB stars, in our Galaxy and in the LMC, have been selected on the basis of the knowledge of some of their properties derived from previous observations. The confrontation of those empirical data with the predictions of theoretical evolutionary models allows to discriminate between them. The comparison of the properties of galactic and LMC nebulae turns out to be an important tool in the study of these galaxies.

131.146 **Grain temperature fluctuations: a key to infrared spectra.**
F. X. Désert, F. Boulanger, S. N. Shore.
Astron. Astrophys., Vol. 160, No. 2, p. 295 – 300 (1986).
Interstellar grains of radius less than 0.03 μm are known to have a fluctuating temperature because their internal energy is comparable to the energy of UV and visible photons. The authors give the exact resultant grain temperature distribution by studying the multiphoton absorption processes. This distribution satisfies an homogeneous integral equation which can be solved by an iteration technique. These small grains generally emit at shorter wavelengths than big cold grains and therefore could help in the interpretation of the infrared spectra (1 – 1000 μm) of various astrophysical objects.

131.147 **Chemical equilibrium from diffuse to dense clouds. II. Interstellar clouds in the Magellanic Clouds.**
N. Bel, Y. P. Viala, I. Guidi.
Astron. Astrophys., Vol. 160, No. 2, p. 301 – 309 (1986).
The authors present the column densities of some selected molecules for interstellar clouds in the Magellanic Clouds (CO, OH, CH, NH_3, H_2CO, and HCO^+). Chemical steady state equilibrium is supposed to be achieved; it is obtained with the chemical scheme as developed by Viala (1985). The sketch of the physical conditions in the Magellanic Clouds, namely the element abundances, some grain properties, and the energy density of the radiation field are obtained from the most recent observational data. The general trend is a deficiency of the molecular content of the Magellanic Clouds compared with our Galaxy. This deficiency is discussed in detail.

131.148 **λ1300 μm dust emission from giant molecular clouds close to the galactic center.**
P. G. Mezger, R. Chini, E. Kreysa, H.–P. Gemünd.
Astron. Astrophys., Vol. 160, No. 2, p. 324 – 330 (1986).
The authors present maps of the dust and free–free emission at λ1300 μm of the giant molecular clouds M –0.13–0.08, M –0.02–0.07 and Sgr A West, made with an angular resolution of $\theta_A = 90''$. They derive reliable dust temperatures and estimates of hydrogen column densities and cloud masses, which they compare with results obtained by molecular spectroscopy. The authors estimate molecular abundances relative to hydrogen column densities.

131.149 **Interstellar molecular clouds.**
J. Bally.
Science, Vol. 232, No. 4747, p. 185 – 193 (1986).
The interstellar medium in our galaxy contains matter in a variety of states ranging from hot plasma to cold and dusty molecular gas. The molecular phase consists of giant clouds, which are the largest gravitationally bound objects in the galaxy, the primary reservoir of material for the ongoing birth of new stars, and the medium regulating the evolution of galactic disks.

131.150 **Cosmic masers.**
R. Norris.
Sky Telesc., Vol. 71, No. 3, p. 248 – 252 (1986).

131.151 **Light extinction and polarization by dust grains in the interstellar medium: the interstellar extinction curves.**
N. V. Voshchinnikov, A. E. Il'in, V. B. Il'in.
Astrofizika, Tom 24, Vyp. 2, p. 307 – 320 (1986). In Russian. English translation in Astrophysics, Vol. 24, No. 2.
The interstellar extinction curves in visual and near infrared regions and the total–to–selective extinction ratio R_V are computed for core–mantle (astronomical silicate–dirty ice) cylindrical interstellar grains with perfect or imperfect Davies–Greenstein alignment. The dependence of R_V on the degree and the direction of alignment and the grain size are discussed.

131.152 **Physical processes in star formation.**
J. Silk.
Spectral evolution of galaxies, p. 31 – 46 (1986). – See Abstr. 012.015.

In this brief review of star formation processes, the author emphasizes aspects that are most relevant for constructing models of galaxy evolution. In particular, he discusses constraints from observations of molecular clouds, and the initial mass function, large–scale star formation processes, and cloud support and fragmentation.

131.153 **Angular momentum in a turbulent interstellar medium.**
R. C. Fleck.
Bull. Am. Astron. Soc., Vol. 17, No. 4, p. 835 (1985). Abstract. – See Abstr. 010.062.

131.154 **Hierarchical structure and the interstellar medium.**
P. Houlahan, J. M. Scalo.
Bull. Am. Astron. Soc., Vol. 17, No. 4, p. 835 (1985). Abstract. – See Abstr. 010.062.

131.155 **IRAS surface brightness image of reflection nebulae in the Pleiades.**
M. W. Castelaz, M. W. Werner, K. Sellgren.
Bull. Am. Astron. Soc., Vol. 17, No. 4, p. 835 (1985). Abstract. – See Abstr. 010.062.

131.156 **Clumpy molecular cloud cores.**
H. M. Martin, E. A. Lada.
Bull. Am. Astron. Soc., Vol. 17, No. 4, p. 836 (1985). Abstract. – See Abstr. 010.062.

131.157 **H I, OH, and H_2CO emission from high latitude molecular clouds.**
L. Magnani, L. Blitz, J. Wouterloot.
Bull. Am. Astron. Soc., Vol. 17, No. 4, p. 836 (1985). Abstract. – See Abstr. 010.062.

131.158 **A molecular survey of L134N.**
D. A. Swade, F. P. Schloerb, W. M. Irvine, R. L. Snell.
Bull. Am. Astron. Soc., Vol. 17, No. 4, p. 836 (1985). Abstract. – See Abstr. 010.062.

131.159 **VLBI studies of interstellar turbulence.**
S. R. Spangler, R. L. Mutel, J. M. Benson, J. M. Cordes.
Bull. Am. Astron. Soc., Vol. 17, No. 4, p. 836 (1985). Abstract. – See Abstr. 010.062.

131.160 **Evidence for sustained multiple imaging of pulsars from the interstellar medium.**
A. Wolszczan, J. M. Cordes.
Bull. Am. Astron. Soc., Vol. 17, No. 4, p. 836 – 837 (1985). Abstract. – See Abstr. 010.062.

131.161 **Predictions of ultraviolet polarization and the constraints imposed on dust parameters.**
M. Balcells, J. S. Mathis.
Bull. Am. Astron. Soc., Vol. 17, No. 4, p. 837 (1985). Abstract. – See Abstr. 010.062.

131.162 **HEAO 3 search for interstellar ^{44}Ti.**
W. A. Mahoney, J. C. Higdon, J. C. Ling, W. A. Wheaton, A. S. Jacobson.
Bull. Am. Astron. Soc., Vol. 17, No. 4, p. 849 (1985). Abstract. – See Abstr. 010.062.

131.163 **Cosmic chemical memory: a new astronomy.**
D. D. Clayton.
Bull. Am. Astron. Soc., Vol. 17, No. 4, p. 854 (1985). Abstract. – See Abstr. 010.062.

131.164 **High resolution maps of bipolar molecular outflow regions: results from the OVRO millimeter wave array.**
M. Barsony, N. Z. Scoville.
Bull. Am. Astron. Soc., Vol. 17, No. 4, p. 862 (1985). Abstract. – See Abstr. 010.062.

131.165 **High angular resolution far–infrared scans of the S 106 dust disk.**
P. M. Harvey, D. F. Lester, M. Joy.
Bull. Am. Astron. Soc., Vol. 17, No. 4, p. 863 (1985). Abstract. – See Abstr. 010.062.

131.166 **Magnetic braking in weakly ionized media.**
A. Königl.
Bull. Am. Astron. Soc., Vol. 17, No. 4, p. 863 (1985). Abstract. – See Abstr. 010.062.

131.167 **Molecular clouds and star formation in the outer Galaxy.**
K. N. Mead.
Bull. Am. Astron. Soc., Vol. 17, No. 4, p. 870 (1985). Abstract. – See Abstr. 010.062.

131.168 **A giant molecular cloud associated with the η Carina nebula.**
D. A. Grabelsky, R. S. Cohen, P. Thaddeus.
Bull. Am. Astron. Soc., Vol. 17, No. 4, p. 870 (1985). Abstract. – See Abstr. 010.062.

131.169 **The physical and kinematic structures of molecular clouds.**
J. Kwan, D. B. Sanders.
Bull. Am. Astron. Soc., Vol. 17, No. 4, p. 870 (1985). Abstract. – See Abstr. 010.062.

131.170 **Observations and interpretations of CO isotope ratios in interstellar molecular clouds.**
C. M. Leung, M. L. Kutner, N. J. Evans II, B. A. Rice.
Bull. Am. Astron. Soc., Vol. 17, No. 4, p. 870 (1985). Abstract. – See Abstr. 010.062.

131.171 **The 809 GHz fine structure line of neutral atomic carbon in dense molecular clouds.**
J. Zmuidzinas, A. L. Betz, D. M. Goldhaber.
Bull. Am. Astron. Soc., Vol. 17, No. 4, p. 871 (1985). Abstract. – See Abstr. 010.062.

131.172 **Chemistry in dynamically evolving interstellar clouds: an extended model.**
S. S. Prasad, S. P. Tarafdar, W. T. Huntress Jr., K. R. Villere, D. C. Black.
Bull. Am. Astron. Soc., Vol. 17, No. 4, p. 871 (1985). Abstract. – See Abstr. 010.062.

131.173 **Detection of interstellar $HCNH^+$.**
L. M. Ziurys, B. E. Turner.
Bull. Am. Astron. Soc., Vol. 17, No. 4, p. 871 (1985). Abstract. – See Abstr. 010.062.

131.174 **Detection of interstellar CH in the far–infrared.**
G. J. Stacey, J. B. Lugten, R. Genzel.
Bull. Am. Astron. Soc., Vol. 17, No. 4, p. 871 (1985). Abstract. – See Abstr. 010.062.

131.175 **Spectra and composition of NGC 6164–65 – an oxygen–poor stellar ejected nebula?**
K. G. Henize, R. A. R. Parker, R. J. Dufour.
Bull. Am. Astron. Soc., Vol. 17, No. 4, p. 885 (1985). Abstract. – See Abstr. 010.062.

131.176 **Emission – absorption observations of the interstellar medium.**
S. W. J. Colgan, E. E. Salpeter, Y. Terzian.
Bull. Am. Astron. Soc., Vol. 17, No. 4, p. 890 – 891 (1985). Abstract. – See Abstr. 010.062.

131.177 **Distribution and ionization of the warm gas clouds at high galactic latitude.**
Z. Wang, L. L. Cowie.
Bull. Am. Astron. Soc., Vol. 17, No. 4, p. 891 (1985). Abstract. – See Abstr. 010.062.

131.178 **A high-latitude cloud in Draco as a matrix of protostars.**
H. M. Johnson.
Bull. Am. Astron. Soc., Vol. 17, No. 4, p. 891 (1985). Abstract. – See Abstr. 010.062.

131.179 **Detection of the OH Zeeman effect toward Orion A.**
T. H. Troland, R. M. Crutcher, I. Kazès.
Bull. Am. Astron. Soc., Vol. 17, No. 4, p. 892 (1985). Abstract. – See Abstr. 010.062.

131.180 **Optical images of sources in the NGC 7538 molecular cloud complex.**
B. Campbell, S. E. Persson.
Bull. Am. Astron. Soc., Vol. 17, No. 4, p. 892 (1985). Abstract. – See Abstr. 010.062.

131.181 **A study of the ground state hydroxyl maser emissions associated with eleven regions of star formation.**
R. A. Gaume, R. L. Mutel.
Bull. Am. Astron. Soc., Vol. 17, No. 4, p. 892 (1985). Abstract. – See Abstr. 010.062.

131.182 **Refractive effects in pulsar scintillation.**
R. W. Romani, R. Narayan, R. Blandford.
Mon. Not. R. Astron. Soc., Vol. 220, No. 1, p. 19 – 49 (1986).
Recent studies have focused attention on the refractive effects of long-wavelength ($\lesssim 10^{14}$cm) electron density fluctuations in the interstellar medium upon radio observations of pulsars and compact extragalactic radio sources. In earlier work, a simple scattering model was introduced by the authors. In this work, they extend the analysis to include refraction-induced fluctuations in intrinsically diffractive quantities such as the scintillation time-scale, t_s, and the decorrelation bandwidth, ν_{dc}. They then use the theory to study the drifting bands in dynamic scintillation spectra caused by the dispersive steering of the diffraction pattern. They also estimate the fluctuations in the position of the image on the sky, rates of variation of intensity and position, and the root mean square elongation of the scatter-broadened image.

131.183 **Evidence for a high-density electron layer from the spatial distribution of pulsars.**
S. Krishnamohan.
Mon. Not. R. Astron. Soc., Vol. 220, No. 1, p. 119 – 124 (1986).
Pulsars are asymmetrically distributed perpendicular to the galactic plane. While their number decreases smoothly with the increasing distance above the plane, it varies irregularly below. There is a deficiency of pulsars with $-0.2 \leqslant z(\mathrm{kpc}) < 0$ and an excess with $-0.6 \leqslant z < -0.3$ compared to the numbers expected on the basis of the distribution above the plane. A high-density electron layer placed about 50 pc below the galactic plane can explain the irregularity. The product of the excess density and the thickness of the layer needed to explain the irregularity is about 6 cm^{-3}pc.

131.184 **An extremely luminous bipolar H_2 flow in the DR21 star-forming region.**
R. Garden, T. R. Geballe, I. Gatley, D. Nadeau.
Mon. Not. R. Astron. Soc., Vol. 220, No. 1, p. 203 – 221 (1986).
The authors have mapped the spatial distribution of molecular hydrogen (H_2) emission in the DR21 molecular cloud in the $\upsilon = 1$–0 $S(1)$ line at 2.12 μm. The emission delineates a highly collimated, bipolar structure extending over a projected distance of $\sim$5 pc, centred on the DR21 H II region/molecular cloud core. This is probably the most luminous and largest galactic H_2 source presently known. The extended bipolar morphology and considerable luminosity in shocked H_2 line emission most probably derive from the dynamical interaction of an energetic protostellar outflow with the surrounding ambient cloud medium. The neighbouring star-forming region W75S was similarly mapped in H_2 line emission and also appears to be double-structured; it is both smaller and less luminous than the DR21 source. A supersonic mass-loss flow associated with the heavily reddened young stellar object W75 (IRS1) is proposed as the driving agent for the observed shocks in this region.

131.185 **Rapid variability of H_2O masers in Cepheus A.**
P. R. Rowland, R. J. Cohen.
Mon. Not. R. Astron. Soc., Vol. 220, No. 1, p. 233 – 251 (1986).
The 22-GHz H_2O maser emission from Cepheus A has been monitored in three separate experiments in 1982 – 83 using single telescopes at Chilbolton and Darnhall, and the MERLIN array. Spectacular variations in the H_2O maser lines are reported. Time-scales for spectral features to change in intensity by a factor of two were as short as 2.4 day, making these some of the fastest variations in H_2O masers yet reported. The observed variability puts constraints on possible pumping schemes for the masers, and provides strong support for the hot dust-cold gas scheme.

131.186 **Interstellar zinc revisited: possible depletion in high-density sight-lines.**
A. W. Harris, J. M. Mas Hesse.
Mon. Not. R. Astron. Soc., Vol. 220, No. 1, p. 271 – 278 (1986).
The question of the non-depletion of interstellar zinc has been re-examined in the light of recent results which show that the depletions of many elements show better correlations with the mean volume density of hydrogen, $\bar{n}(\mathrm{H})$, than with hydrogen column density. Zinc abundances from an earlier study have been combined with newly derived data for high mean density sight-lines [$\bar{n}(\mathrm{H}) > 1\ \mathrm{cm}^{-3}$]. Although the data quality is insufficient to rule out the possibility that zinc is universally undepleted, the results are consistent with a trend of decreasing gas-phase abundance with increasing $\bar{n}(\mathrm{H})$. A possible interpretation in terms of a recently published model of interstellar gas distribution, and consequences for the use of zinc as a tracer of metallicity, are discussed.

131.187 **X-ray spectroscopic study of interstellar medium.**
S. Hayakawa.
ESA Spec. Publ., ESA SP-239, p. 69 – 76 (1985). – See Abstr. 012.019.
Recent results of low resolution X-ray spectroscopy are described to show how the properties of interstellar and circumstellar media are obtained. Further information to be obtained with higher spectral resolutions is discussed by reference to the observations of emission lines, absorption lines and absorption edges.

131.188 **Comparison of 2.1 and 3.8 micron line profiles of shocked H_2 in the Orion molecular cloud.**
T. R. Geballe, S. E. Persson, T. Simon, C. J. Lonsdale, P. J. McGregor.
Astrophys. J., Vol. 302, No. 2, p. 693 – 700 (1986).
Velocity profiles of the 2.12 μm 1–0 $S(1)$ and 3.81 μm 1–0 $O(7)$ lines of H_2 have been measured at a number of locations in the shocked gas region of the Orion molecular cloud. The data imply that although some extinction through the outflow is present, the asymmetries in the H_2 lines are due in part to actual line-of-sight distributions of velocities within the shocked gas. These distributions appear consistent with the sense of the bipolar outflow observed in millimeter CO lines between the lobes of H_2 line emission.

131.189 **Submillimeter wavelength molecular spectroscopy of the Orion molecular cloud.**
G. J. White, L. W. Avery, K. J. Richardson, J. C. G. Lesurf.
Astrophys. J., Vol. 302, No. 2, p. 701 – 710 (1986).
A submillimeter wavelength spectroscopic study of the Orion molecular cloud has been made in the $J = 4$–3 HCN, $H^{13}CN$, HCO^+, $H^{13}CO^+$, and $J = 7$–6 CS molecular transitions. Densities of up to a few times $10^6\mathrm{cm}^{-3}$ are found coupled with high inferred brightness temperatures, indicating kinetic temperatures of 120K. Evidence for lower densities in the surrounding ambient molecular cloud is presented along with maps of HCN and HCO^+ emission.

131.190 The infrared emission bands. I. Correlation studies and the dependence on C/O ratio.
M. Cohen, L. Allamandola, A. G. G. M. Tielens, J. Bregman, J. P. Simpson, F. C. Witteborn, D. Wooden, D. Rank.
Astrophys. J., Vol. 302, No. 2, p. 737 – 749 (1986).
The authors present new airborne measurements of the unidentified (UIR) emission bands between 5 and 8 μm in eight planetaries, five reflection nebulae, and four H II regions (including the Orion Bar). These data are compared with preexisting and new ground–based observations of the 3.3, 8.7, and 11.3 μm bands. Good correlations are found betweeen the strengths of all pairs of bands, including two newly established weak bands at 5.6 and 6.9 μm. The fraction of total, far–infrared luminosity radiated by a planetary in the strongest UIR feature at 7.7 μm is strongly correlated with the nebular C/O ratio. This strongly suggests that hydrocarbons are the carriers of these features.

131.191 Molecular clouds in Orion and Monoceros.
R. J. Maddalena.
NASA Tech. Memo., NASA TM–87786, 6+203 pp. (1986). = Thesis, Columbia University, New York.
Results of a new CO survey sampling an 850 deg^2 field in Orion and Monoceros with the Columbia University radio telescope are reported. This survey outlines the full extent of the molecular material associated with the Orion A and Orion B molecular clouds. The network of molecular features discovered in the present survey includes two long, thin (10° × 1/2°) molecular filaments and an expanding ring of molecular clouds surrounding the H II region assoaciated with λ Ori. A giant molecular cloud associated with Mon R2 may be physically related to the Orion complex. At a distance of 3 kpc a large (250 × 100 pc) and massive ($\sim 10^6 M_\odot$) cloud was discovered with an unusual combination of low temperatures ($T_R < 2.7$K) and wide spectral lines (~ 7 km s^{-1}). The 15 large CO clouds found in the outer Galaxy ($l \sim 206 - 220°$) probably lie in two spiral arms. A comparison of the properties of these clouds with those of local clouds in the Orion arm is presented.

131.192 Detection of weak emission in the Hβ line from three high–velocity clouds.
A. S. Kutyrev.
Astron. Tsirk., No. 1396, p. 3 – 5 (1985). In Russian.

131.193 HCN, CO and continuum observations of some galactic molecular clouds at 2.6 mm and 3.4 mm.
A. B. Burov, V. N. Voronov, I. I. Zinchenko, A. G. Kislyakov, A. A. Krasil'nikov, Eh. P. Kukina, A. V. Lapinov, L. E. Pirogov, V. F. Vdovin, V. M. Demkin.
Astron. Tsirk., No. 1404, p. 1 – 4 (1985). In Russian.
A number of molecular clouds have been observed in the J = 1–0 HCN line. Preliminary results of these observations are reported. The observing program included an HCN survey of Sharpless clouds (40 objects), observations of dark clouds (15 objects), bipolar sources, some H II regions, and carbon stars. The HCN line has been detected in 11 Sharpless clouds, in some other H II regions, in 4 dark clouds, in all bipolar sources, and in the carbon star CIT6. Some sources have been mapped (S 106, G 35.2–0.74, and others). The 3.4 mm continuum emission has been detected in S 106 and in L 1551. The frequency dependence of the dust grain absorption efficiency has been estimated for L 1551.

131.194 Star formation regions as galactic dissipative structures.
G. Bodifée.
Astrophys. Space Sci., Vol. 122, No. 1, p. 41 – 56 (1986).
The theory of dissipative structures, applied to star formation systems, provides a conceptual framework for the study of the behaviour and evolution of these systems. As shown by an analysis of a model star formation process system, prolonged stationary star formation in localized areas and repetitive bursting star formation events can be understood as different behavioural modes of galactic dissipative structures. Young stellar associations with their H II regions and molecular clouds are manifestations of the ordered distribution of matter participating in the star formation processes. A self–organization with the appearance of ordered structures is, in general, to be expected in nonequilibrium systems in which nonlinear processes occur.

131.195 On the nature of the particles causing the 2200 Å peak in the extinction of starlight.
F. Hoyle, N. C. Wickramasinghe.
Astrophys. Space Sci., Vol. 122, No. 1, p. 181 – 184 (1986).
The recently reported discovery of the essentially spherical molecule C_{60} provides a possible explanation of the 2180 Å peak in the extinction of starlight, an explanation that appears to be free of objections to the earlier graphite hypothesis.

131.196 Diatoms on Earth, comets, Europa and in interstellar space.
R. B. Hoover, F. Hoyle, N. C. Wickramasinghe, M. J. Hoover, S. Al–Mufti.
Earth, Moon, Planets, Vol. 35, No. 1, p. 19 – 45 (1986).
There exists a close correspondence between the measured infrared properties of diatoms and the infrared spectrum of interstellar dust as observed in the Trapezium nebula and toward the galactic center source GC–IRS 7. Diatoms and bacteria also exhibit an absorbance peak near 2200 Å, which is found to agree with the observed ultraviolet absorbance properties of interstellar grains. The authors review the observational data and consider the known properties of diatoms and bacteria. It is suggested that these characteristics are consistent with the concept of a cosmic microbiological system in which these or similar microorganisms might exist on comets, Europa and in interstellar space.

131.197 On the non–uniformity of the star formation rate.
V. I. Korchagin, A. D. Ryabtsev.
Kinematika Fiz. Nebesn. Tel, Tom 2, No. 3, p. 3 – 7 (1986). In Russian. English translation in Kinematics Phys. Celest. Bodies.
The star formation process in a system consisting of stars and interstellar clouds is considered. Two mechanisms of star formation are taken into account: cloud collision and interaction of stars and clouds. In the framework of the model the non–uniformity of the star formation rate occurs when interactions of stars and clouds predominate over cloud collisions. The conclusions obtained are in qualitative agreement with the results of numerical simulation of the star formation process.

131.198 A study of the 2 Mon star formation region: selection of interstellar dust clouds.
N. G. Guseva.
Kinematika Fiz. Nebesn. Tel, Tom 2, No. 3, p. 22 – 29 (1986). In Russian. English translation in Kinematics Phys. Celest. Bodies.
The interstelllar dust clouds of various density have been selected in the direction of the Rosette nebula. The obtained parameters of the clouds (absorption, sizes, masses, spatial densities) show that they are the known types of diffuse and dark clouds. The absorption in dark clouds situated within supernova remnants exceeds significantly the absorption in the rest of clouds. That may be the result of clouds compression by a shock wave from the supernova explosion.

131.199 Non–LTE process in the interstellar medium.
K. R. Anantharamaiah.
Binary and multiple systems, p. 119 – 125 (1984). – See Abstr. 012.025.
The author first recalls few basic concepts regarding the thermodynamic equilibrium. He then discusses LTE and departures from it with a few examples of their occurrence in the interstellar medium. Finally he discusses the importance of non–LTE processes in formation of low frequency radio recombination lines such as those observed recently using the Ooty radio telescope.

131.200 Implications on X–ray and UV radiation on interstellar medium.
K. P. Singh.
Binary and multiple systems, p. 127 – 141 (1984). – See Abstr. 012.025.
Four principal components or 'phases' of the interstellar medium have been identified in our galaxy. These components are: (1) Hot ionised gas with temperature (T) $\sim 10^5 - 10^6$K known as

the hot interstellar medium. (2) Warm and intercloud gas with $T \sim 10^3 - 10^4$K. (3) Cool neutral atomic H I with $T \sim 100$K which is confined to the disk having scale height of $\sim$150 pc. (4) Cold molecular gas H_2 with $T \sim 10$K confined to the disk having scale height of $\sim$60 pc near the sun.

131.201 **Dynamics of local dust clumps in a protostellar cloud.**
V. S. Kessel'man, A. B. Frolov.
Sov. Astron., Vol. 29, No. 4, p. 417 – 421 (1985). English translation of 40.131.058.

131.202 **Radiation of interstellar shock waves. IV. The limits of using the isothermic approximation.**
I. S. Balinskaya, K. V. Bychkov, O. V. Fedorova.
Astrofiz. Issled. Izv. Spets. Astrofiz. Obs., Tom 21, p. 39 – 48 (1986). In Russian. English translation in Bull. Spec. Astrophys. Obs. – North Caucasus.

A very common situation in the interstellar medium is considered when the flow velocities of atoms, ions and electrons may be considered as equal, but it is necessary to distinguish two temperatures: atom–ionic and electronic ones. By solving the model system of non–stationary equations the use of «isothermic» approximation, i. e. an assumption on the rapid equalization of temperatures is investigated.

131.203 **Nonequilibrium excitation of rotational levels of interstellar cyanoacetylene.**
I. I. Zinchenko, V. K. Khersonskij.
Astrofiz. Issled. Izv. Spets. Astrofiz. Obs., Tom 21, p. 49 – 57 (1986). In Russian. English translation in Bull. Spec. Astrophys. Obs. – North Caucasus.

Results of calculations of the rotational levels populations for the interstellar cyanoacetylene are presented. The populations of 30 rotational levels are investigated. The calculations were made for the gas concentration interval $N \approx 10^2 - 10^7 \mathrm{cm}^{-3}$ and the kinetic temperature range $T_k \approx 10 - 60$K. The case of small optical depth in the rotational lines is considered. The authors also discuss the dependence of the excitation temperatures and absorption coefficients for the transitions under consideration upon N and T_k and separate the regions of the thermalization and inverse population of the rotational levels.

131.204 **Non–equilibrium ionization in evaporating clouds. Constraints on the interstellar medium.**
J. Ballet, M. Arnaud, R. Rothenflug.
Astron. Astrophys., Vol. 161, No. 1, p. 12 – 22 (1986).

The authors have modelled non–equilibrium ionization processes in the evaporative flows from clouds surrounded by hot gas. Equilibrium is not reached in most realistic interstellar situations. That model was specifically applied to the ions O VI, N V, C IV and Si IV observed by UV experiments. The predicted densities are way above the observations. That raises doubts about evaporation theories. Alternatively, for small clouds and low pressure, the elements reach the Li–like state so far from the clouds that the high densities predicted might be real but hidden because of turbulence in the ambient gas.

131.205 **The effect of heating and cooling on the velocity fluctuations in the ISM induced by the system of stars.**
B. M. Deiss, W. H. Kegel.
Astron. Astrophys., Vol. 161, No. 1, p. 23 – 30 (1986).

The gravitational interaction with stars leads to the excitation of velocity fluctuations in the interstellar gas. Expressions for these fluctuations and the stability criterion of the coupled system have been derived by Kegel and Völk (1983) for a polytropic gas. The present work shows that when dissipative thermal effects are taken into account the corresponding expressions are modified considerably. The results are relevant to the interpretation of interstellar lines, the large scale flow of the interstellar matter, and the collapse of interstellar clouds.

131.206 **The stars NGC 7538–IRS 5, 6, and 7, and a distance estimate to the nebula.**
M. A. Moreno, C. Chavarría–K.
Astron. Astrophys., Vol. 161, No. 1, p. 130 – 138 (1986).

The authors present optical (9 filters in the 13 colour system) and near infrared (*JHKL*) photometry of NGC 7538–IRS 5, 6, and 7. [2.23 μm] mapping of the central part of the nebula yielded new infrared sources and the extended nature of NGC 7538–IRS 5. This star has a significant excess in the [3.8 μm] passband, probably caused by circumstellar dust of $T_{\mathrm{dust}} \leqslant 800$K, and is less luminous than NGC 7538–IRS 6, being most probably of spectral type O9. Photometry of stars NGC 7538–IRS 5 and 6 gives a Cygnus type interstellar extinction law, with $A_v/E(B-V) = 3.6$, and a photometric distance estimate to the nebula of 2.2 kpc. The authors' kinematic and photographic study of NGC 7538 confirms previous results by Deharveng et al. (1979). Intrinsic colours of hot and intermediate hot stars in the (optical) 13 colour and infrared (*JHKL'M'*) systems are also discussed.

131.207 **A Burgers model of interstellar dynamics.**
J. I. Katz, M. L. Green.
Astron. Astrophys., Vol. 161, No. 1, p. 139 – 141 (1986).

The authors consider a very simplified model of interstellar dynamics based on the Burgers equation. The gas is assumed isobaric, and randomly occurring supernova explosions are assumed to give impulses to their surroundings. Voids appear and expand until a single void and two dense clouds fill the entire grid. This may qualitatively describe the development of the observed structure of cloud and hot coronal gas.

131.208 **Correlations between NH_3 (1,1) and far–IR emission.**
L. G. Stenholm, B. Baud, R. Mauersberger.
Astron. Astrophys., Vol. 161, No. 1, p. 150 – 154 (1986).

The regions around six cool IRAS sources, embedded in molecular clouds, have been mapped in the NH_3 (1,1) and (2,2) lines. A strong correlation between the 100 μm flux and the NH_3 line intensity is present. The star–cloud–turbulence theory suggested by Henriksen and Turner (1984) is supported by the data. A simple way to identify very young embedded objects is suggested. The gas around the object has higher kinetic energy than gas in more distant cloud regions. The protostellar deposition of kinetic energy in the gas is strongly time–dependent.

131.209 **The ionization balance in H I regions.**
D. Péquignot, S. M. V. Aldrovandi.
Astron. Astrophys., Vol. 161, No. 1, p. 169 – 176 (1986).

Some of the atomic data relevant to the ionization balance of atoms in H I regions are considered. Photoionization rates are given for several interstellar radiation fields. An updated table of radiative recombination coefficients is given for a number of atoms in the temperature range 10 – 1000K. Attention is drawn to the probable importance of charge exchange reactions and a tentative table of rate coefficients is proposed. These data are used to comment briefly on some observations.

131.210 **A search for interstellar glycine.**
I. I. Berulis, G. Winnewisser, V. V. Krasnov, R. L. Sorochenko.
Sov. Astron. Lett., Vol. 11, No. 4, p. 251 – 253 (1985). English translation of 40.131.029.

131.211 **The dust and gas surrounding LkHα 101.**
R. O. Redman, T. B. H. Kuiper, J. J. Lorre, J. E. Gunn.
Astrophys. J., Vol. 303, No. 1, p. 300 – 310 (1986). With plates 3 – 4.

The linear polarization of the reflection nebula NGC 1579 and the CO (1→0) emission from the associated molecular gas have been mapped for several minutes of arc around the exciting star LkHα 101. These maps show conclusively that LkHα 101 is the sole significant source of illumination in the region. The dust in the reflection nebula appears to be uniform over the illuminated region and is uniformly illuminated by LkHα 101. The star may have formed out of a placental cloud whose remnants now include four molecular cloud fragments, two in front of and two behind the reflection nebula.

131.212 **Temperature determination of surrounding molecular clouds in bipolar flow sources.**
T. Takano.
Astrophys. J., Vol. 303, No. 1, p. 349 – 355 (1986).
The kinetic temperature of surrounding molecular clouds in typical bipolar flow sources has been determined using NH_3 $(J,K) = (1,1)$ and (2,2) line observations. The kinetic temperature shows a strong correlation with the luminosity of the exciting infrared sources. The surrounding molecular clouds may be heated by the central infrared sources via hot dust grains, and the contribution of collisional heating by outflowing material seems to be negligible. The data suggest that the molecular gas is thermally coupled with the dust grains in sources of $\lesssim 10^4 L_\odot$.

131.213 **A detailed examination of the kinematics of rotating dark clouds.**
R. Arquilla, P. F. Goldsmith.
Astrophys. J., Vol. 303, No. 1, p. 356 – 374 (1986). = Five College Astron. Dep., Contrib., No. 602.
The velocity fields of eight molecular clouds, six of which are described in the literature as rotating, are analyzed using two–dimensional maps of high–quality carbon monoxide spectra. Using a simple analytical technique, the directions are identified along which significant velocity gradients occur across the clouds. Kinematic models of the sources are developed from an examination of the relevant spatial–velocity diagrams. The data strongly support the presence of rotation in only three of the six clouds originally believed to possess it. If rotation is present, however, it plays a major role in the energetics of a cloud.

131.214 **The large system of molecular clouds in Orion and Monoceros.**
R. J. Maddalena, M. Morris, J. Moscowitz, P. Thaddeus.
Astrophys. J., Vol. 303, No. 1, p. 375 – 391 (1986). With plates 5 – 6.
A region centered on Orion and Monoceros that covers 850 deg^2 of the celestial sphere was surveyed in the $J = 1 \rightarrow 0$ line of CO, about 1/8 of that region showing emission, most of which arises from the giant molecular clouds associated with Orion A, Orion B, and Mon R2. Other objects found and studied include two filamentary features more than 10° in length, a series of clouds distributed symmetrically around the H II region S264 apparently expanding away from λ Ori, and clouds associated with Barnard's loop and CMa OB1.

131.215 **A reanalysis of the HCO^+/HOC^+ abundance ratio in dense interstellar clouds.**
M. F. Jarrold, M. T. Bowers, D. J. DeFrees, A. D. McLean, E. Herbst.
Astrophys. J., Vol. 303, No. 1, p. 392 – 400 (1986).
New theoretical and experimental results have prompted a reinvestigation of the HCO^+/HOC^+ abundance ratio in dense interstellar clouds. These results pertain principally but not exclusively to the reaction between HOC^+ and H_2, which was previously calculated by DeFrees, McLean, and Herbst to possess a large activation energy barrier. New calculations, reported here, indicate that this activation energy barrier is quiete small and may well be zero. In addition, experimental results at higher energy and temperature indicate strongly that the reaction proceeds efficiently at interstellar temperatures.

131.216 **On the CH^+ ion in diffuse interstellar clouds.**
D. L. Lambert, A. C. Danks.
Astrophys. J., Vol. 303, No. 1, p. 401 – 415 (1986).
A survey of 65 lines of sight in the southern Milky Way for the interstellar line of CH^+ at 4232 Å yielded 30 detections. The CH^+ column density $N(CH^+)$ is shown to be well correlated with the rotationally excited H_2 detected by surveys with the Copernicus satellite. This correlation is strong evidence that CH^+ is formed in warm gas where the reaction $C^+ + H_2 \rightarrow CH^+ + H$ is effective. Shocks probably provide the necessary heating.

131.217 **High–resolution mapping of molecular outflows in NGC 2071, W49, and NGC 7538.**
N. Z. Scoville, A. I. Sargent, D. B. Sanders, M. J. Claussen, C. R. Masson, K. Y. Lo, T. G. Phillips.
Astrophys. J., Vol. 303, No. 1, p. 416 – 432 (1986).
The Owens Valley millimeter–wave interferometer has been used to map at high resolution (7″) the molecular gas and continuum emission in the cores of NGC 2071, W49, and NGC 7538. These data and single–dish maps obtained on the FCRAO telescope are used to study the high–velocity molecular gas flows. In each of the three regions the interferometer CO maps reveal a high–velocity, bipolar outflow within 10″ of the central infrared sources with orientations agreeing well with those in larger scale, single–telescope maps. The masses of high–velocity gas detected by the interferometer are 0.2, 138, and 12 $M_\odot$ in NGC 2071, W49, and NGC 7538 respectively. In NGC 7538 the data suggest the existence of a rotating disk of mass 100 – 500 $M_\odot$ collimating the flow into the polar directions. Continuum emission at 2.7 mm was detected in all the sources; this emission coincides with the origins of the molecular gas outflows.

131.218 **Observations of neutral atoms in the Rho Ophiuchi cloud.**
T. P. Snow, C. L. Joseph, D. M. Meyer.
Astrophys. J., Vol. 303, No. 1, p. 433 – 438 (1986).
New observations of lines from neutral atoms in the lines of sight to three stars in the ϱ Oph cloud are combined with literature data to look for evidence of density–enhanced regions with higher elemental depletions than those characterizing the cloud as a whole. The depletions inferred from neutral species, however, are comparable to those measured directly from observations of ions, suggesting that no strong density enhancement exist in the observed lines of sight. This conclusion is supported by electron density data.

131.219 **Supernova remnants associated with an H I "supershell" in the Perseus spiral arm.**
M. Fich.
Astrophys. J., Vol. 303, No. 1, p. 465 – 469 (1986).
A large (280 pc diameter) H I shell has been found in the Perseus spiral arm of our Galaxy. Three old supernova remnants are associated with the shell. These are the first supernova remnants directly linked to large H I shells. All three of these supernova remnants are located within the H I shell, at least two of them are near its edge. This suggests that these supernovae are actively engaged in enlarging the H I shell, which was presumably "blown" to its present dimensions by several previous supernova remnants and/or the sustained action of interior stars with substantial stellar winds.

131.220 **Erratum: "Extinction to ionized gas at the galactic center" [Astrophys. J., Vol. 265, No. 2, p. 760 – 765 (1983)].**
S. P. Willner, J. L. Pipher.
Astrophys. J., Vol. 303, No. 1, p. 521 (1986). See Abstr. 33.131.039.

131.221 **Molecular outflows, gas density distribution, and the effects of star formation in the dark cloud Barnard 5.**
P. F. Goldsmith, W. D. Langer, R. W. Wilson.
Astrophys. J., Lett. Ed., Vol. 303, No. 1, p. L11 – L15 (1986).
The authors have detected high–velocity molecular outflows associated with three and possibly four of the IRAS sources in B5. $C^{18}O$ mapping indicates that the cloud contains at least five dense fragments, but only IRS 1 appears to be located at a molecular column density maximum. From the time scales of the outflows, the authors conclude that star formation has been taking place over an extended period of approximately a few $\times 10^5$yr. The consequences of the mechanical energy input to the cloud from the outflows are discussed.

131.222 **Moln mellan stjärnorna.**
L. Spitzer.
Astron. Tidsskr., Årg. 19, Nr. 2, p. 57 – 69 (1986).

131.223 **Surveys for emission–line objects with Schmidt telescopes.**
K. Ogura, H. Maehara.
Galactic structure and variable stars, p. 35 – 43 (1985). – See Abstr. 012.029.

131.224 **The Southern Coalsack – a dark cloud complex with inactive star formation.**
H. Ohtani.
Galactic structure and variable stars, p. 53 – 58 (1985). – See Abstr. 012.029.
The literature relating to the problem on star formation in the Southern Coalsack is briefly reviewed. From a recently determined total mass of dust contained in the object it is shown that the cloud is gravitationally unstable if the normal gas–to–dust ratio, 100, is adopted. This implies that active star formation may take place in the cloud contrary to the negative result of search for young stellar population. This contradiction would be removed by assuming a small gas–to–dust ratio in the Coalsack as recently suggested.

131.225 **Observations and theories of interstellar dust.**
J. S. Mathis.
NASA Conf. Publ., NASA CP–2403, p. 29 – 35 (1986). – See Abstr. 012.030.
The author summarizes the observational properties of dust, as based on (1) the extinction over a factor of 100 in wavelength (0.1 μm – 10 μm), (2) on polarization, both linear and circular, (3) on rather narrow emission and absorption features in the spectrum, and (4) on reflection nebulae. He then discusses theories.

131.226 **Shock processing of interstellar grains.**
C. G. Seab, J. M. Shull.
NASA Conf. Publ., NASA CP–2403, p. 37 – 53 (1986). – See Abstr. 012.030.
There is strong evidence for the destruction of dust grains in fast shock waves in the interstellar medium. This paper discusses some of this evidence and reviews the mechanisms by which the grains are destroyed in shocks. Grain growth in shock waves is a more controversial subject, although it is potentially as important as the destruction processes. A few comments on growth mechanisms are also made.

131.227 **Experimental investigations relating to the properties and formation of cosmic grains.**
B. Donn.
NASA Conf. Publ., NASA CP–2403, p. 109 – 134 (1986). – See Abstr. 012.030.
The interpretation of observations or theoretical analyses of interstellar processes requires a sound knowledge of relevant data. In many instances this can only be obtained by experiments carried out under appropriate conditions. This report is a general survey of the availability of such data.

131.228 **Interstellar grains. A working group report.**
T. P. Snow.
NASA Conf. Publ., NASA CP–2403, p. WG 19 – WG 32 (1986). – See Abstr. 012.030.

131.229 **Interrelationships between interstellar and interplanetary grains. A working group report.**
D. D. Clayton.
NASA Conf. Publ., NASA CP–2403, p. WG 85 – WG 120 (1986). – See Abstr. 012.030.

131.230 **Ionized polycyclic aromatic hydrocarbons in space.**
L. Allamandola, J. Barker, M. Crawford, X. Tielens, G. van der Zwet.
NASA Conf. Publ., NASA CP–2403, p. A 1 – A 2 (1986). – See Abstr. 012.030.

131.231 **"Clean" vs. "dirty" silicate grains and the state of carbon crystallization in interstellar and circumstellar dust.**
P. R. Buseck.
NASA Conf. Publ., NASA CP–2403, p. A 3 – A 4 (1986). – See Abstr. 012.030.

131.232 **Excess depletion of Al, Ca, Ti from interstellar gas.**
D. D. Clayton.
NASA Conf. Publ., NASA CP–2403, p. A 5 (1986). – See Abstr. 012.030.

131.233 **Are aromatic hydrocarbons the carriers of the diffuse interstellar bands in the visible?**
A. Léger, L. d'Hendecourt.
NASA Conf. Publ., NASA CP–2403, p. A 16 (1986). – See Abstr. 012.030.

131.234 **Grain alignment by ferromagnetic impurities.**
J. S. Mathis.
NASA Conf. Publ., NASA CP–2403, p. A 17 (1986). – See Abstr. 012.030.

131.235 **The effects of charged dust grains on IR molecular excitation.**
R. Puetter, I. Krinsky.
NASA Conf. Publ., NASA CP–2403, p. A 19 – A 22 (1986). – See Abstr. 012.030.

131.236 **What predictions can be made on the nature of carbon and carbon–bearing compounds (hydrocarbons) in the interstellar medium based on studies of interplanetary dust particles?**
F. J. M. Rietmeijer.
NASA Conf. Publ., NASA CP–2403, p. A 23 – A 27 (1986). – See Abstr. 012.030.

131.237 **Why do interstellar grains exist?**
C. G. Seab, D. J. Hollenbach, C. F. McKee, A. G. G. M. Tielens.
NASA Conf. Publ., NASA CP–2403, p. A 28 (1986). – See Abstr. 012.030.

131.238 **Interstellar grain mantles.**
A. G. G. M. Tielens, L. J. Allamandola, J. Bregman, F. C. Witteborn.
NASA Conf. Publ., NASA CP–2403, p. A 40 (1986). – See Abstr. 012.030.

131.239 **Excimer emission of polyaromatic hydrocarbons as the mechanism of broad–band visible fluorescence of interstellar dust.**
T. J. Wdowiak.
NASA Conf. Publ., NASA CP–2403, p. A 41 – A 44 (1986). – See Abstr. 012.030.

131.240 **Molecular clouds and star formation.**
J. Silk.
Proceedings of the Second Asian–Pacific Regional Meeting on Astronomy, p. 126 – 138 (1984). – See Abstr. 012.031.
Two principal issues are addressed: the fragmentation of molecular clouds into units of stellar mass and the impact of star formation on molecular clouds. The observational evidence for fragmentation is summarized, and the gravitational instability described of a uniform spherical cloud collapsing from rest. The role of magnetic fields is discussed. Interactions between fragments are described.

131.241 **On the statistics of galactic H I clouds.**
K. R. Anantharamaiah, V. Radhakrishnan, P. A. Shaver.
Proceedings of the Second Asian–Pacific Regional Meeting on Astronomy, p. 139 – 156 (1984). – See Abstr. 012.031.
The number density and random motions of interstellar H I clouds have been studied using an entirely novel method involving the comparison of the terminal velocities of H I absorption

spectra in the direction of H II regions with their recombination line velocities. The authors confirm the conclusions of an earlier study attributing these velocity differences mainly to the random motions of the clouds.

131.242 **An Hα emission star survey in the field of the dark cloud complex Khavtassi 141.**
Y. Tomita, H. Ohtani.
Proceedings of the Second Asian–Pacific Regional Meeting on Astronomy, p. 177 – 181 (1984). – See Abstr. 012.031.
Hα emission stars were surveyed by taking objective prism spectra for a region toward the dark cloud complex Khavtassi 141 and an unobscured region neighbouring to the cloud. As result, 88 stars were detected in the above field of about 70 square degree in total. It has been found that the number ratio of the emission line stars to general stars is greater by an order of magnitude in the dark cloud region than in the unobscured region. In the cloud region, the emission line stars show a tendency of clustering. Two of the emisson line stars seen in the cloud field are known as T Tauri stars.

131.243 **The extinction and the H I content of the dark cloud complex Khavtassi 141.**
Y. Tomita, T. Saito, H. Ohtani.
Proceedings of the Second Asian–Pacific Regional Meeting on Astronomy, p. 490 – 491 (1984). – See Abstr. 012.031.

131.244 **On the carbon enrichment of the interstellar medium.**
A. Sarmiento, M. Peimbert.
Rev. Mex. Astron. Astrofis., Vol. 11, No. 2, p. 73 – 81 (1985). = Contrib. Inst. Astron., Univ. Nac. Autónoma Mex., No. 165.
The interstellar and the solar C/H and $^{12}C/^{13}C$ ratios provide important constraints for the study of the chemical evolution of galaxies. It has been suggested in the literature that supernovae of Type II (massive stars), planetary nebulae (intermediate mass stars), and novae contribute to the interstellar enrichment of ^{12}C and ^{13}C. It is the purpose of this paper to explore the relative importance that these three types of objects have on the enrichment of C in the interstellar medium.

131.245 **Multifrequency observations of the region associated with the cometary nebula GM24.**
M. Tapia, M. Roth, L. F. Rodríguez, J. Cantó, P. Persi, M. Ferrari–Toniolo, J. A. López.
Rev. Mex. Astron. Astrofis., Vol. 11, No. 2, p. 83 – 89 (1985). With 1 plate. = Contrib. Inst. Astron., Univ. Nac. Autónoma Mex., No. 181.
Near and mid–infrared (1 – 20 μm) observations, new radio continuum and line–emission searches and measurements as well as high resolution spectroscopy in the Hα line spectral region of GM24 are presented. The observations are described and the results presented. A discussion of the properties of the region and the individual sources is given.

131.246 **Near infrared photometry of violent star formation regions.**
J. Melnick, R. Terlevich, M. Moles.
Rev. Mex. Astron. Astrofis., Vol. 11, No. 2, p. 91 – 97 (1985).
Near infrared broad band photometry and CO indices for a significant number of violent star formation regions are presented. The existence of a narrow correlation between W(Hβ) and IR colour is confirmed. The interpretation of this relation as an age sequence implies a correlation between CO index and W(Hβ) which is not found. It is argued however that this failure is most likely a consequence of using narrow band filters to determine CO indices in objects with strong emission–line spectra.

131.247 **Formaldehyde observations in the direction of extragalactic sources.**
S. W. J. Colgan, E. E. Salpeter, Y. Terzian.
Astron. J., Vol. 91, No. 1, p. 107 – 112 (1986).
Absorption by galactic H_2CO has been observed with the λ6 cm system at Arecibo towards the extragalactic sources 3C 123 and 3C 154. Towards 3C 123, which lies behind part of the Taurus cloud complex, there is both a broad and a remarkably narrow spectral component. The narrow component is at a velocity of 4.40 km s^{-1}. It has a fitted width of only 0.12 km s^{-1}. There is no evidence for H_2CO at a velocity of 5.5 km s^{-1}, where other molecules have been observed. A VLA synthesis map of the H_2CO absorption has been obtained towards 3C 123. This map shows that the optical depth across 3C 123 varies less than 15% on angular scales of 3″ – 20″. At the estimated distance of the molecular cloud, these correspond to 0.003 – 0.02 pc. The velocity peak and width are constant to 0.15 km s^{-1} over the extent of 3C 123. Upper limits are given for H_2CO absorption towards 3C 75, 3C 138, 3C 348, 3C 390, 3C 410, and 3C 454.3.

131.248 **Polarimetry of the infrared reflection nebulae in Cepheus A and S140.**
R. R. Joyce, T. Simon.
Astron. J., Vol. 91, No. 1, p. 113 – 117 (1986).
New 2.2 μm polarization maps of the Cepheus A and S140 molecular clouds show significant levels of polarization, reaching maximum values of 78% in Cepheus A and 43% in S140. The centrosymmetric pattern of the polarization vectors in each region is characteristic of reflection nebulae illuminated by a central source. The center of the polarization pattern in Cepheus A lies within a few arcseconds of the ultracompact H II regions and molecular masers believed to be the source of the bipolar outflow observed in CO and H_2. The linear arrangement of the H II regions suggests that they are shock–heated jets that have been ejected from a young star located at the position of the polarization centroid, deeply embedded in the molecular cloud. In S140 the polarization is axisymmetric about the position of IRS 1 and reaches a maximum value at the position of IRS 2. This symmetry suggests that the outflow of high–velocity gas in S140 may be viewed nearly pole–on.

131.249 **On optical studies of high–velocity clouds.**
D. G. York, G. S. Burks, T. B. Gibney.
Astron. J., Vol. 91, No. 2, p. 354 – 377 (1986).
Lists of distant objects that can be used to study physical conditions in, and distances of, 21 cm (Oort) high–velocity clouds are presented. Recent published observations are used to compile positions, velocities, and distances of the clouds.

131.250 **Cold H I clouds near the supernova remnant W44.**
F. Sato.
Astron. J., Vol. 91, No. 2, p. 378 – 382 (1986).
The cold H I clouds near the supernova remnant W44 are investigated by the use of the Maryland – Green Bank Survey (Westerhout 1973). Several clouds with a mean diameter of ~20 pc are distributed in the region. They do not seem to make a shell around W44, contrary to the suggestion by Knapp and Kerr (1974). Some of them form a chain, ~100 pc in length, extending approximately along the galactic equator. The major constituent of the clouds is probably the hydrogen molecule, and the total mass of the entire complex amounts to 2.5×10^4 to $8.1 \times 10^4 M_{\odot}$. It is concluded that the cold H I cloud complex near W44 is a giant molecular cloud at an early evolutionary stage.

131.251 **Rotation measures of extragalactic radio sources viewed through the molecular clouds Cepheus A and L1551.**
J. H. Simonetti, J. M. Cordes.
Astrophys. J., Vol. 303, No. 2, p. 659 – 666 (1986).
Faraday rotation measures of background extragalactic radio sources behind high–velocity molecular flows were determined to help constrain magnetic fields in and near the molecular flows. Rotation measures of seven sources seen through Cep A fall in two groups separated by ~150 rad m^{-2}. It is not clear whether the variation of rotation measure is determined by the high–velocity material or by larger scale structure in the Cep A region. For L1551, a ΔRM ≈ 30 rad m^{-2} is observed between a line of sight directly through the NE lobe of the high–velocity CO and another 7′ away. Based upon a discussion of various models for the molecular flow, it seems likely that the magnetic field strength in the flow region may be as great as ~0.1 mG.

131.252 **^{13}CO in Orion.**
K. Sugitani, Y. Fukui, H. Ogawa, K. Kawabata.
Astrophys. J., Vol. 303, No. 2, p. 667 – 682 (1986).
New ^{13}CO (J = 1–0) measurements of the Orion molecular cloud are reported. Data were taken with the new 4 m millimeter-wave telescope of Nagoya University having a beamwidth of 2'.7. The velocity resolution of 0.1 km s^{-1} has revealed significant details of the ^{13}CO emission toward the H II region. It is found that the ^{13}CO emission consists of 14 separate features and the main N–S ridges. The results indicate that an appreciable portion of the molecular gas toward the Orion nebula is interacting with the H II region.

131.253 **Small-scale structure of interstellar H I clouds.**
E. W. Greisen, H. S. Liszt.
Astrophys. J., Vol. 303, No. 2, p. 702 – 717 (1986).
The authors have used the VLA to synthesize the spatial distribution of λ21 cm H I absorption from some 15 galactic clouds occulting 3C 111, 3C 161, and 3C 348. The angular resolution of the observations is between 3".5 and 4".4. On scales of 1' substantial but smooth variations in channel and line properties are the norm, rather than the exception. The observations suggest that the internal structure of H I clouds is ordered, in the sense that any underlying chaotic behavior occurs on scales much smaller than the resolution.

131.254 **IUE observations of interstellar hydrogen and deuterium toward Alpha Centauri B.**
W. B. Landsman, J. Murthy, R. C. Henry, H. W. Moos, J. L. Linsky, J. L. Russell.
Astrophys. J., Vol. 303, No. 2, p. 791 – 796 (1986).
The authors present a profile of the Lyα emission line of α Cen B (K1 V, d = 1.3 pc), obtained by addition of two IUE small-aperture, high-dispersion images. The square-well-shaped interstellar H I absorption profile shows excellent agreement with that previously observed in the Lyα emission of α Cen A. Modeling of the data has yielded improved limits on the heliocentric bulk velocity ($v_{HI} = -15.5 \pm 1.9$ km s^{-1}), the average number density ($n_{HI} < 0.21$ cm^{-3}), and the velocity dispersion ($b_{HI} > 13.8$ km s^{-1}) of the neutral gas along the α Cen sightline.

131.255 **HCN emission from bipolar reflection nebulae.**
S. Deguchi, M. J. Claussen, P. F. Goldsmith.
Astrophys. J., Vol. 303, No. 2, p. 810 – 815 (1986).
The authors have detected the HCN J = 1–0 emission in bipolar reflection nebulae and OH/IR sources. The HCN line profiles of bipolar nebulae are found to be composed of broad and narrow velocity features, indicating the presence of HCN in the bipolar flow as well as in the disk surrounding the infrared source. HCN emission has also been detected in the planetary nebula NGC 7027, the S-type star χ Cyg, and the carbon star CRL 2135. The abundance of HCN is of the order of 10^{-7} to 10^{-8} per H_2 molecule in the circumstellar envelopes. The authors also report the detection of SiO J = 2–1, v = 0, thermal emission in several OH/IR stars.

131.256 **Interstellar HCO: detection of the missing 3 millimeter quartet.**
M. S. Schenewerk, L. E. Snyder, Å. Hjalmarson.
Astrophys. J., Lett. Ed., Vol. 303, No. 2, p. L71 – L74 (1986).
All four of the strongest 1_{01}–0_{00} hyperfine lines of the formyl radical, HCO, have been detected for the first time in any interstellar source. Comparisons between these hyperfine lines and the J = 1–0 transition of $H^{13}CO^+$, which was also within the band, suggest that the column densities of HCO in the two typical molecular clouds which were sampled, DR21 and NGC 2024, are comparable to those of $H^{13}CO^+$ and in excellent agreement with current theoretical chemical cloud models.

131.257 **Origin and evolution of ionized hydrogen regions in the process of formation of massive stars.**
S. G. Kravchuk.
Inst. teor. fiz. Akad. Nauk USSR, Prepr. No. 121 R, 27 pp. (1985). In Russian. Abstr. in Ref. Zh., 51. Astron., 5.51.788 (1986).

131.258 **Correlation analysis of the Taurus Molecular Cloud Complex.**
S. C. Kleiner.
Diss. Abstr. Int., Sect. B, Vol. 46, No. 5, p. 1957 – 1958 (1985). Thesis, University of Massachusetts, 261 pp. (1985). Order No. DA8517120.

131.259 **Polarization and infrared imaging of regions of star formation.**
A. Moneti.
Diss. Abstr. Int., Sect. B, Vol. 46, No. 6, p. 1958 (1985). Thesis, The University of Rochester, 236 pp. (1985). Order No. DA8511908.

131.260 **Observational constraints on interstellar depletion mechanisms in lines of sight exhibiting peculiar extinction curves.**
C. L. Joseph.
Diss. Abstr. Int., Sect. B, Vol. 46, No. 9, p. 3091 (1986). Thesis, University of Colorado at Boulder, 141 pp. (1985). Order No. DA8522673.

131.261 **A polarimetric study of the interstellar medium in Taurus Dark Clouds.**
J.-c. Hsu.
Diss. Abstr. Int., Sect. B, Vol. 46, No. 10, p. 3481 (1986). Thesis, The University of Texas at Austin, 180 pp. (1985). Order No. DA8527580.

131.262 **Gas and dust in regions of recent star formation.**
J. A. Cardelli.
Diss. Abstr. Int., Sect. B, Vol. 46, No. 12, p. 4281 – 4282 (1986). Thesis, University of Washington, 305 pp. (1985). Order No. DA8529890.

131.263 **CO radio-line profiles in cold interstellar clouds.**
I. I. Zinchenko, A. V. Lapinov.
Sov. Astron., Vol. 29, No. 5, p. 502 – 507 (1985). English translation of 40.131.150.

131.264 **Influence of cosmic rays on the stability and large-scale dynamics of the interstellar medium.**
V. D. Kuznetsov.
Astron. Zh., Tom 63, Vyp. 3, p. 446 – 457 (1986). In Russian. English translation in Sov. Astron., Vol. 30, No. 3.
In the framework of the diffusion-convection description of galactic cosmic rays their influence on the stability and dynamics of the interstellar medium retained in equilibrium by the gravitational field of stars is considered. The dependence of the stability boundary and growth rate of MHD instability on the cosmic rays energy density and on their diffusion coefficient is investigated.

131.265 **Molecular ions in dense interstellar clouds.**
M. G. Opendak, D. A. Varshalovich.
Astron. Zh., Tom 63, Vyp. 3, p. 458 – 464 (1986). In Russian. English translation in Sov. Astron., Vol. 30, No. 3.
Ion kinetics in dense molecular clouds are considered. The molecular ion abundances are calculated using observational data on neutral molecules. H_3^+, CH_3^+, HCO^+, $C_2H_2^+$, H_3O^+, HS^+, N_2H^+ are shown to be the most abundant. As far as ion-dust interactions are concerned, the only essential process is sticking of ions on the grain surface.

131.266 **Ion composition of cooling interstellar gas with variable density.**
A. A. Suchkov, Yu. A. Shchekinov.
Astron. Zh., Tom 63, Vyp. 3, p. 470 – 475 (1986). In Russian. English translation in Sov. Astron., Vol. 30, No. 3.
The results of calculations are given of the ion composition of low-density interstellar gas cooling from $T \cong 5 \times 10^5 - 10^6$K to $T = 10^4$K in adiabatic and isobaric regimes. It is shown that the ratio of concentrations of ions N(C IV)/N(O VI) can serve to reveal the character of non-stationarity in the gas with temperature $T > 10^4$K.

131.267 **The size distribution function of solid particles in protostellar clouds.**
V. S. Kessel'man, A. B. Frolov.
Astron. Zh., Tom 63, Vyp. 3, p. 476 – 482 (1986). In Russian. English translation in Sov. Astron., Vol. 30, No. 3.
A numerical solution, taking into account a real sticking process of dust particles, for the size distribution function of solid charged particles in protostellar clouds is found.

131.268 **Interstellar Na I absorption towards the stellar association Ara OB1.**
J. B. Whiteoak, F. F. Gardner.
Proc. Astron. Soc. Aust., Vol. 6, No. 2, p. 164 – 171 (1985).
Observations have been made of the interstellar D–line absorption of Na I towards a loose grouping of OB stars (containing the association Ara OB1 and probably the Hα emission regions RCW 107 and 108). The individual absorption spectra contain several velocity components which show well–defined groupings at –37, –18 and 0 km s^{-1}. The corresponding median optical depths are 2, 7 and 6. The evidence indicates that both the –37 and –18 km s^{-1} Na I clouds are located in the Carina–Sagittarius spiral arm. Several H II regions are located at similar longitudes, but most are close to zero latitude and are more distant than the arm.

131.269 **Ammonia in the hot galactic centre cloud G 1.6–0.025.**
F. F. Gardner, J. B. Whiteoak, J. R. Forster, W. L. Peters, T. B. H. Kuiper.
Proc. Astron. Soc. Aust., Vol. 6, No. 2, p. 176 – 180 (1985).
As part of an extensive southern survey of interstellar NH_3 with the Parkes 64–m radio telescope (with a beamwidth of 81 arcsec), the (1,1), (2,2), and (3,3) transitions have been observed towards the galactic centre molecular cloud G 1.6–0.025. The cloud has an overall size of 10 arcmin, and contains several concentrations with differing velocities. It has several features also observed in other galactic centre clouds, e.g. high optical depths and kinetic temperatures above 50K.

131.270 **Fragmentation in collapsing magnetic gas clouds – non–uniform initial fields.**
G. J. Phillips.
Proc. Astron. Soc. Aust., Vol. 6, No. 2, p. 205 – 207 (1985).
The results found in this work are little different from those already discovered when the initial field was uniform. The effect of the magnetic field is to resist the growth of density enhancements whilst allowing the material to flow down the field lines. As a result of the latter the collapse eventually becomes dominated by gravity as the cloud collapses to a thin disc. In addition to the magnetic force, the pressure also acts to dissipate density gradients which develop.

131.271 **Comparison of a theory of sequential fragmentation with the initial mass function of stars.**
W. K. Brown.
Astrophys. Space Sci., Vol. 122, No. 2, p. 287 – 298 (1986).
The author has formulated a theory of sequential fragmentation based on his theory of solar system formation. The theory is validated using the data from an explosive aerosolization experiment. Comparison of the fragmentation theory with the initial mass function of stars shows excellent agreement. This agreement does not validate the author's solar system formation theory, but it does add support to it and perhaps more importantly, does not falsify it.

131.272 **The kinematics of NGC 6164–5.**
G. J. Carranza, E. L. Agüero.
Astrophys. Space Sci., Vol. 123, No. 1, p. 59 – 65 (1986).
Accurate interferometric kinematic data suggest the interpretation of the brightest regions of NGC 6164–5 as a two–branched helical structure related to HC 148937 and having a systematic radial velocity nearly coincident with that of the tenuous surrounding medium; the gas in the helix could be the result of non–equatorial ejections from the central star. Morphologic, spectroscopic, and kinematic data would also indicate that the distance of 1.4 kpc proposed by Westerlund (1960) is substantially correct.

131.273 **A modified model for interstellar extinction.**
N. L. Jabir, S. R. Jabbar, S. A. H. Salih, Q. S. Majeed.
Astrophys. Space Sci., Vol. 123, No. 2, p. 351 – 362 (1986).
It is shown that available optical and ultraviolet data relating to the interstellar dust are elegantly explained on the basis of a modified microbal grain model. A model comprised of two biologically derived components, modified under interstellar conditions. Data on interstellar extinction, albedo and polarization may be accounted for by this model.

131.274 **IRAS 04238 + 5336: a young reflection nebula surrounding a double star.**
C. G. Wynn–Williams, J. N. Heasley, D. L. DePoy, G. J. Hill, E. E. Becklin.
Astrophys. J., Vol. 304, No. 1, p. 409 – 417 (1986).
IRAS 04238 + 5336P03 is identified with a 17th magnitude compact nebulous object which contains a 1″.6 double star. About 99% of the 450 $L_{\odot}$ luminosity of the source is radiated longward of 1 μm. The source displays an A–type visible spectrum, strong molecular hydrogen emission lines at 2.4 μm, and a dust emission feature at 3.3 μm. It lies within a small molecular cloud. The object is one of the first clear examples of a binary system young enough to be still interacting with the interstellar cloud out of which it formed.

131.275 **The geometry of anisotropic CO outflows.**
R. Liseau, G. Sandell.
Astrophys. J., Vol. 304, No. 1, p. 459 – 465 (1986).
A simple geometrical model for the space motions of the bipolar high–velocity CO outflows in regions of recent, active star formation is proposed. From observations of the spatial distribution and from the characteristics of the line shape of the high–velocity molecular gas emission the geometry of the line–emitting regions can be inferred, i.e., the diretion in space and the collimating angle of the flow. The model is applied to regions where Herbig–Haro objects are associated with the CO flows. The results provide convincingly strong evidence for the physical association of CO outflows and Herbig–Haro objects.

131.276 **CO observations of southern high–latitude clouds.**
E. R. Keto, P. C. Myers.
Astrophys. J., Vol. 304, No. 1, p. 466 – 480 (1986). With plates 5 – 7.
The authors report on CO observations of 15 high–latitude clouds, $|b| > 20°$, and three small clouds in the Chamaeleon complex. The clouds have an estimated distance of 100 pc with a median mass, radius, and velocity dispersion of 2 $M_{\odot}$, 0.7 pc, and 0.5 km s^{-1}. The properties of clouds found at high galactic latitude, away from star–forming regions, are similar to the properties of diffuse clouds found in the Chamaeleon complex, a region of active star formation. The clouds do not appear to be gravitationally bound, but they appear to be in quasi–static equilibrium with the intercloud medium.

131.277 **H_2 spectroscopy as an agent for extinction determinations: the near–infrared curve for the Orion molecular cloud.**
D. S. Davis, H. P. Larson, R. Hofmann.
Astrophys. J., Vol. 304, No. 1, p. 481 – 489 (1986).
A near–infrared (1.8 μm $\leqslant \lambda \leqslant$ 3.5 μm) extinction curve for the Orion molecular cloud is presented. The curve is derived from high–resolution spectra of the Orion H_2 source recorded from the Kuiper Airborne Observatory. The data reveal that the Orion extinction law is indistinguishable from a λ^{-1} form in the near–infrared, except for strongly enhanced extinction near $\lambda \sim 3$ μm. The implications of these results, in the context of current interstellar grain models, are discussed.

131.278 **Molecular clouds associated with compact H II regions. III. Spin-up and collapse in the core of G 10.6–0.4.**
P. T. P. Ho, A. D. Haschick.
Astrophys. J., Vol. 304, No. 1, p. 501 – 514 (1986).
The authors report aperture synthesis observations of the $(J, K) = (1,1)$ NH_3 emission in the core of G 10.6–0.4. There is strong evidence for a collapsing molecular cloud core that is forming a central cluster of young massive stars. With 9″ resolution, the data show a rapidly rotating core with a velocity gradient of ~ 10 km $s^{-1}pc^{-1}$. This core region is embedded within a slowly rotating envelope with a gradient of ~ 1 km $s^{-1}pc^{-1}$ in the same sense as the core. There is an absorption feature redshifted by ~ 5 km s^{-1} from the velocity centroid of the rotating core. The combination of a rapidly rotating core, a slowly rotating envelope, evidence of collapse, and a central cluster of young stars suggests that one is seeing the spin-up of collapsing material while conserving angular momentum.

131.279 **Are bipolar jets precessing?**
J. Fukue, T. Yokoo.
Nature, Vol. 321, No. 6073, p. 841 – 842 (1986).
Whether bipolar jets from the star-forming regions precess or not is indicative of their model of origin and formation. If the recent observation of L723 represents precession, then this is probably the orbital precession of a torus (disk) around a central oblate star, or the forced precession of a torus around a central star by its companion. The free precession of a central star as well as the precession of a torus may be responsible in the case of L1551 IRS5.

131.280 **The Massachusetts–Stony Brook galactic plane CO survey: statistical clustering properties of molecular cloud cores.**
A. R. Rivolo, P. M. Solomon, D. B. Sanders.
Space Telesc. Sci. Inst., Prepr. Ser., No. 80, 15 pp. (1985). To appear in Astrophys. J., Lett. Ed.

131.281 **High resolution optical and ultraviolet absorption line studies of interstellar gas.**
L. L. Cowie, A. Songaila.
Space Telesc. Sci. Inst., Prepr. Ser., No. 84, 57 pp. (1985). To appear in Annu. Rev. Astron. Astrophys., Vol. 24.

131.282 **A survey of interstellar sulphur abundance and implications for P II oscillator strengths.**
A. W. Harris, J. M. Mas Hesse.
Rutherford Appleton Lab., Rep., RAL-86-017, 20 pp. (1986). To be published in Astrophys. J.

131.283 **Submillimeter observations of CS in M17.**
R. L. Snell, N. R. Erickson, P. F. Goldsmith, B. L. Ulich, C. J. Lada, R. N. Martin, A. Schulz.
Astrophys. J., Vol. 304, No. 2, p. 780 – 786 (1986). = Five Coll. Astron. Dep., Contrib. No. 616.
Observations of M17 in the emission of the $J = 7 \rightarrow 6$ transition of CS at a wavelength of 874 μm are presented. The MMT was used as a phase-coherent array to map the CS emission from the dense core in M17 with an angular resolution of 26″. Observations of the $J = 4 \rightarrow 3$ transition were made at selected positions using the FCRAO 14 m telescope with comparable angular resolution. The comparison of the emission from the two CS transitions suggests that the density within the core is more than $10^6 cm^{-3}$, and the present data resolve the core into several condensations, each with a size ~ 0.7 pc and mass ~ 1000 $M_\odot$.

131.284 **High-resolution IRAS observations of the Rho Ophiuchi cloud core.**
E. T. Young, C. J. Lada, B. A. Wilking.
Astrophys. J., Lett. Ed., Vol. 304, No. 2, p. L45 – L49 (1986). With plate L2.
The authors report high-resolution IRAS pointed observations of a significant portion of the dense core of the Rho Ophiuchi dark cloud at 12 μm, 25 μm, 60 μm, and 100 μm. From the 12 μm data 18 individual embedded sources with a surface brightness of greater than 1 Jy $arcmin^{-2}$ were identified in the field. The far-infrared luminosities range between 0.2 and 21 $L_\odot$, confirming the low-luminosity nature of the cluster. Using ground-based infrared observations, the authors have identified with some confidence the near-infrared counterparts for 13 of the IRAS sources.

131.285 **Protostellar condensations within the Orion ridge.**
L. G. Mundy, N. Z. Scoville, L. B. Bååth, C. R. Masson, D. P. Woody.
Astrophys. J., Lett. Ed., Vol. 304, No. 2, p. L51 – L55 (1986). With plate L3.
Aperture synthesis observations of the CS $J = 2 \rightarrow 1$ line and 95 GHz continuum emission from a 2′ × 3′ region around IRc2 in the OMC-1 molecular cloud have been obtained at 7″.5 resolution with the OVRO millimeter interferometer. These observations reveal four distinct condensations within the OMC-1 molecular ridge. The condensation lie along the edge of the bright central part of the H II region. Their masses and densities, $M \approx 50$ $M_\odot$ and $n \approx 10^6 - 10^7 cm^{-3}$, suggest that the condensations will undergo further massive star formation.

131.286 **Detection of the OH Zeeman effect toward Orion A.**
T. H. Troland, R. M. Crutcher, I. Kazès.
Astrophys. J., Lett. Ed., Vol. 304, No. 2, p. L57 – L60 (1986).
The authors have detected the Zeeman effect in the 18 cm OH absorption spectrum of Orion A. They derive a line-of-sight magnetic field strength of -125 ± 20 μG. At the same position, an H I Zeeman effect indicates a magnetic field of -49 ± 4 μG. Thus, the magnetic field in the molecular gas toward Orion A is significantly stronger than that in the atomic gas.

131.287 **Spectroscopy in radio astronomy and the study of the interstellar medium.**
J. Crovisier.
Ciel Espace, No. 205, p. 12 – 17 (1985). In French. Abstr. in Phys. Abstr., Vol. 89, No. 1273, Entry 18884 (1986).

131.288 **On the collision of high-velocity clouds with a galactic disk.**
G. Tenorio-Tagle, P. Bodenheimer, M. Różyczka, J. Franco.
MPA Rep., No. 235, 22 pp. (1986). Submitted to Astron. Astrophys.

131.289 **Recent Onsala observations related to star formation.**
Å. Hjalmarson.
Onsala Space Obs., Prepr., No. 86:5, 31 pp. (1986).

131.290 **CO observations of dark clouds in Lupus.**
D. C. Murphy, R. Cohen, J. May.
Onsala Space Obs., Prepr., No. 86:24, 26 pp. (1986). To appear in Astron. Astrophys.

131.291 **Star formation in Bok globules and low-mass clouds. III. Barnard 62.**
B. Reipurth, G. Gee.
ESO Sci. Prepr., No. 445, 26 pp. (1986). To appear in Astron. Astrophys.

131.292 **Radiation polarization of galactic nebulae.**
N. V. Voshchinnikov.
Photometric and polarimetric investigations of celestial bodies, p. 111 – 118 (1985). In Russian. – See Abstr. 003.035.
A summary of polarimetric observations of diffuse, reflection, cometary, and bipolar nebulae obtained in 1965 – 1983 is presented. Characteristics of the observed polarization of the nebulae, theoretical models, and their interpretation are discussed.

131.293 **Light extinction and polarization by dust grains in the interstellar medium: interstellar linear polarization.**
N. V. Voshchinnikov, A. E. Il'in, V. B. Il'in.
Astrofizika, Tom 24, Vyp. 3, p. 523 – 534 (1986). In Russian. English translation in Astrophysics, Vol. 24, No. 3.
The wavelength dependence of interstellar linear polarization is computed for core-mantle (astronomical silicate – dirty ice) cylindrical grains with Davies-Greenstein alignment. The effects

of variation of the grain size, the degree and the direction of grain alignment are discussed. It has been found that the wavelength of maximum polarization λ_{max} is determined mainly by the grain size. Dependence of the ratio $P_{max}/E(B-V)$ on the degree and the direction of grain alignment is strong. The width of the normalized polarization curve is only slightly sensitive to the parameters.

131.294 **Hydrodynamical model of the Orion KL Nebula.**
T. Okuda, S. Ikeuchi.
Publ. Astron. Soc. Jpn., Vol. 38, No. 2, p. 199 – 214 (1986).
As a possible model of the Orion KL Nebula, the authors investigate the interaction between isotropic stellar wind from an embedded protostar and an ambient molecular cloud by two-dimensional hydrodynamical computations. The wind–cloud interaction model is promising for the explanation of the bipolar flow of the Orion KL Nebula.

131.295 **The rotational excitation of OH in the Orion shocked region.**
J. H. Black, E. F. van Dishoeck.
Prepr. Steward Obs., No. 629, 11 pp. (1986). To appear in Masers, molecules, and molecular outflows in star forming regions, Proceedings of a conference at Haystack Observatory, May 1985.

131.296 **Detection of the (3,0) Phillips band of interstellar C_2 toward ζ Oph.**
E. F. van Dishoeck, J. H. Black.
Prepr. Steward Obs., No. 630, 19 pp. (1986). To appear in Astrophys. J.

131.297 **Theoretical studies of diffuse cloud chemistry.**
J. H. Black.
Prepr. Steward Obs., No. 631, p. 1 – 9 (1986). To appear in IAU Symp. No. 120.

131.298 **Fluorescent line emission of molecular hydrogen.**
J. H. Black, E. F. van Dishoeck.
Prepr. Steward Obs., No. 631, p. 10 – 12 (1986). To appear in IAU Symp. No. 120.

131.299 **The rotational excitation of OH in Orion.**
E. F. van Dishoeck, G. J. Melnick, J. H. Black.
Prepr. Steward Obs., No. 631, p. 13 – 15 (1986). To appear in IAU Symp. No. 120.

131.300 **Comprehensive models of diffuse interstellar clouds: physical conditions and molecular abundances.**
E. F. van Dishoeck, J. H. Black.
Prepr. Steward Obs., No. 641, 87 pp. (1986). To appear in Astrophys. J., Suppl. Ser.

131.301 **Star formation: from OB associations to protostars.**
C. J. Lada.
Prepr. Steward Obs., No. 647, 17 pp. (1986). To appear in IAU Symp. No. 115.

131.302 **Refractive interstellar scintillation in 1741–038.**
R. M. Hjellming, R. Narayan.
Prepr. Steward Obs., No. 661, 20 pp. (1986). To appear in Astrophys. J.

131.303 **The reddening at the North Galactic Pole.**
J. Knude.
Mat.–Fys. Medd. Dan. Vidensk. Selsk., Vol. 41, No. 1 – 10, p. 71 – 101 (1985). Abstr. in Phys. Abstr., Vol. 89, No. 1275, Entry 30430 (1986).

131.304 **Star formation in colliding gas flows.**
J. H. Hunter Jr., M. T. Sandford II, R. W. Whitaker, R. I. Klein.
Astrophys. J., Vol. 305, No. 1, p. 309 – 332 (1986).
Gas flows colliding within molecular clouds produce regions of compressed gas and dust that are hydrodynamically unstable. Fragmentation of the compressed region produces clumps in the mass range $0.01 \leqslant m \leqslant 5.0\ M_\odot$, in time $t_b \approx 5 \times 10^{12}$s. Gravitational instability of the fragment distribution occurs on the time scale $t_c \approx 10^{13}$s. The gravitational attraction of the clumps and the disk material can form protostars in the mass range $1.0 \leqslant m_f \leqslant 65.0\ M_\odot$. The conditions for dynamical fragmentation to occur do not require that the overall molecular cloud be Jeans unstable, or even that it be collapsing; there must only be a source for the kinetic energy of the gas flows that interact. These results support the conclusion that the star formation process in molecular clouds can be triggered by high–velocity gas–flow interactions within molecular clouds.

131.305 **Images of star–forming regions. I. Optical and radio morphology of the bipolar outflow source GL 490.**
B. Campbell, S. E. Persson, P. J. McGregor.
Astrophys. J., Vol. 305, No. 1, p. 336 – 352 (1986). With plates 4 – 7.
The authors present new, deep optical CCD images of the young stellar object GL 490. A large–scale diffuse cone of emission extends out to 0.8 pc from the source itself and generally corresponds to the blueshifted part of the bipolar CO outflow. The core image of GL 490 is easily resolved, with a core size of 700 – 1000 AU. This image is elongated in a direction orthogonal to the CO outflow. Also presented is a 2 cm radio continuum map of the source obtained at the VLA, which indicates the presence of a very dense core ($n_e \geqslant 10^6 cm^{-3}$) and thin extended emission covering some 3″. A scattering mechanism is proposed to account for the observed optical image of GL 490.

131.306 **The structure of the W49 A molecular cloud complex: burst of star formation in the $10^5 M_\odot$ core.**
R. Miyawaki, M. Hayashi, T. Hasegawa.
Astrophys. J., Vol. 305, No. 1, p. 353 – 362 (1986).
The authors have observed the CS (J = 1–0), $C^{34}S$ (J = 1–0), and H51α emission toward the W49 A molecular cloud complex in an area of 3′x2′ with a resolution of 33″. The CS emission is strong toward the compact continuum sources, and its distribution is similar to that of the 1 mm continuum. The data show that there are two dense molecular clouds toward W49 A, one of these associated with the infrared/maser source complex W49 A IRS. The estimated core mass of the W49 A molecular cloud is $(0.5 - 2.5) \times 10^5 M_\odot$. This mass is closely packed in a small region 3.4 pc in diameter. Implications of the data for the star formation history of the W49 A complex are discussed.

131.307 **Ultraviolet–pumped infrared fluorescent molecular hydrogen emission in reflection nebulae.**
K. Sellgren.
Astrophys. J., Vol. 305, No. 1, p. 399 – 404 (1986).
Strong molecular hydrogen emission at 2.41 µm has been observed in three out of six reflection nebulae surveyed. A spectrum of one nebula, Parsamyan 18, shows several H_2 lines whose intensity ratios have values agreeing with those predicted if the excitation is due to UV–pumped fluorescence and disagreeing with those predicted for shock and X–ray excitations.

131.308 **The excitation, abundance, and distribution of HNCO in Sagittarius B2.**
E. Churchwell, D. Wood, P. C. Myers, R. V. Myers.
Astrophys. J., Vol. 305, No. 1, p. 405 – 416 (1986).
The authors report observations of the $1_{01} - 0_{00}$, $2_{02} - 1_{01}$, $7_{07} - 6_{06}$, and $10_{010} - 9_{09}$ transitions of HNCO toward Sgr B2. Partial maps were obtained in the $7_{07} - 6_{06}$ and $10_{010} - 9_{09}$ transitions. The authors discuss the velocity structure and distribution of HNCO in Sgr B2, and consider the mechanism of HNCO excitation. A total of 14 observed lines, involving excitation energies up to 192K above ground, are used in the analysis. The main implication of the derived population distribution is that far–infrared radiation from warm dust is responsible for the excitation of HNCO in Sgr B2.

131.309 **Shapes and correlations as observational discriminants for the origin of diffuse bands.**
G. Chlewicki, G. P. van der Zwet, L. J. van Ijzendoorn, J. M. Greenberg, P. P. Alvarez.
Astrophys. J., Vol. 305, No. 1, p. 455 – 466 (1986).
The authors have measured the intensities of several diffuse interstellar bands in intermediate resolution spectra (0.35 – 0.65 Å) of heavily reddened early–type stars in three associations: Cas OB6, Be 87, and Cyg OB2. The extinction is attributed mostly to material at intermediate densities between molecular clouds and the diffuse medium. The degree of correlation with E_{B-V} separates the lines into at least two groups. The analysis of line profiles reveals a wide variety of shapes even among strongly correlated lines, ranging from a nearly symmetric Lorentzian for λ6284 to a highly complex multiple structure of λ6203 + λ6205. These observational facts appear to preclude the origin of the lines in bulk solid particles and favor their explanation as vibronic transitions in large molecules.

131.310 **Detection of scattering in the 2175 Å interstellar band.**
A. N. Witt, R. C. Bohlin, T. P. Stecher.
Astrophys. J., Lett. Ed., Vol. 305, No. 1, p. L23 – L27 (1986).
Spectrophotometric observations of two reflection nebulae and their illuminating stars obtained with the IUE Satellite have provided the first evidence for the presence of a scattering contribution to the 2175 Å interstellar extinction band. Lower than normal far–UV extinction for the stars embedded in the nebulae indicates that the nebulae have a dust particle size distribution that is dominated by larger particles. The strength of the 2175 Å band is larger than normal in both cases. The scattering is found to dominate the long–wavelength wing of the band, without shifting the central wavelength of the band by more than 20 Å toward longer wavelengths.

131.311 **Charged–particle acceleration by turbulent neutral gas in galactic molecular clouds.**
A. V. Gurevich, V. A. Dogel', K. P. Zybin, Ya. N. Istomin.
Sov. Astron. Lett., Vol. 11, No. 5, p. 284 – 288 (1985). English translation of 40.131.071.

131.312 **The slow interstellar scintillation of extragalactic radio sources.**
N. Ya. Shapirovskaya.
Sov. Astron. Lett., Vol. 11, No. 5, p. 289 – 291 (1985). English translation of 40.131.207.

131.313 **Comparative analyses of ^{13}CO with ^{12}CO and H I emission in the galactic plane.**
D.–l. Xiang.
Publ. Purple Mt. Obs., Vol. 4, No. 2, p. 7 – 30 (1985). In Chinese.
The longitudinal distributions of integrated intensities of ^{13}CO and H I, and the galactic radial distributions of column abundance of ^{13}CO, ^{12}CO and H I as well as the column abundance–galactic longitude diagrams are derived from ^{13}CO observations and data sets of ^{12}CO. Statistics indicates that the most peaks of ^{13}CO integrated emissions and the corresponding H I intensities present anticorrelations. The radial distributions of column abundance of ^{13}CO shows three definite peaks. But it does not seem plausible to consider these peaks evidences for spiral arms.

131.314 **A search for rotationally–excited CH in galactic sources.**
H. E. Matthews, M. B. Bell, T. J. Sears, B. E. Turner, L. J. Rickard.
Astron. Astrophys., Vol. 161, No. 2, p. 329 – 333 (1986).
The authors report the results of sensitive searches for rotationally–excited transitions of CH in galactic sources, most of which are known to contain CH in its lowest rotational level. No detections were made, although the sensitivity achieved in many instances is approaching that for which thermal emisson might be detectable. The implications of these results are discussed in terms of the expected excitation and in the light of recent observations of the lowest excited state of CH by Ziurys and Turner (1985).

131.315 **Ammonia in the neutral ring around the galactic center.**
E. Serabyn, R. Güsten.
Astron. Astrophys., Vol. 161, No. 2, p. 334 – 338 (1986).
Several metastable NH_3 lines have been detected from the molecular gas which orbits about the center of our galaxy at radii $\gtrsim$2 pc. A rotational temperature of ~95K is estimated for this gas, implying a kinetic temperature of 240±100K. Several spectral features arising from gas along the line of sight to the center have also been detected. Gas at 40 km s^{-1} is seen in absorption in the higher lying inversion transitions. From a comparison of the excitation temperatures of the levels with the background continuum temperature, the density of this gas is estimated to be ~1 – 3x10^{4}cm^{-3}. Also seen are NH_3 absorption features at –135, –51, –30 and –3 km s^{-1}, associated with the "expanding molecular ring at 300 pc" and with the spiral structure of our galaxy. The three lower velocity features have $T_{rot} \lesssim 14 - 19$K.

131.316 **Distribution of Mg II in the local interstellar medium towards eight cool giants.**
P. Molaro, G. Vladilo, J. E. Beckman.
Astron. Astrophys., Vol. 161, No. 2, p. 339 – 346 (1986).
IUE spectra of eight cool (G4 – K0) giants are analyzed in a search for Mg II interstellar components. The analysis shows that interstellar Mg II absorption features can affect severely the chromospheric emission profiles for cool stars even if they are very close to the Sun, if they are located behind the high density region. This last result has repercussions on the use of Mg II profiles for chromospheric modelling.

131.317 **Chemical equilibrium from diffuse to dense interstellar clouds. I. Galactic molecular clouds.**
Y. P. Viala.
Astron. Astrophys., Suppl. Ser., Vol. 64, No. 3, p. 391 – 437 (1986).
A steady state model of the chemical composition of interstellar clouds is presented. The chemical reaction scheme involves 1074 reactions among 80 species formed from H, C, N and O. The heavy elements He, Mg, Si, Fe and S are also included because they play an important role in the ionization equilibrium. The model is restricted to the simplest radicals and molecules observed or thought to be present in interstellar clouds. The chemical reaction scheme is relevant to diffuse clouds as well as to dense dark clouds.

131.318 **Time dependent chemistry in dense molecular clouds. III. Infrared band cross sections of molecules in the solid state at 10K.**
L. B. d'Hendecourt, L. J. Allamandola.
Astron. Astrophys., Suppl. Ser., Vol. 64, No. 3, p. 453 – 467 (1986).
Thin film transmission infrared spectra and a list of integrated absorbance values (band strengths) of bands in the 2.5 to 20 micron range of various molecular solids deposited on a CsI window cooled to 10K are presented. These solids include molecules known or suspected to be present on interstellar grains such as H_2O, CO, NH_3 and CH_4 and mixtures of these molecules as well as various hydrocarbons. The method is described by which the absorbance values derived can be used to estimate column densities of species absorbed on interstellar grains and other solids associated with celestial IR sources.

131.319 **Interstellar polarization from observations of A and F stars in high and intermediate galactic latitudes, and from stars in the Mathewson and Ford polarization catalogue.**
T. Korhonen, A. Reiz.
Astron. Astrophys., Suppl. Ser., Vol. 64, No. 3, p. 487 – 494 (1986).
Tables of linear polarization measured in the *B* band of the *UBV* system are given for 358 A and F stars in high and intermediate galactic latitudes, and for 118 stars in the Mathewson and Ford polarization catalogue (1970). For most of the A and F stars, reliable *ubvy* and Hβ photometry is available and, therefore, distances are known with good accuracy. The programme aims at studying the interstellar magnetic field in selected areas as

well as correlations between polarization and distance and colour excess, respectively. Instrumental errors are believed to be smaller than 10^{-4}. Comparison between the Mathewson and Ford polarization values and the present data shows that the external error in the authors' catalogue does not exceed ± 0.03 per cent.

131.320 **The Orion star–forming region: far–infrared and radio molecular observations.**
H. A. Thronson Jr., D. A. Harper, J. Bally, M. Dragovan, D. Mozurkewich, M. A. Greenhouse, P. R. Schwartz, H. A. Smith, J. H. Bieging, R. F. Loewenstein, C. J. Lada.
Astron. J., Vol. 91, No. 6, p. 1350 – 1356 (1986). With plates 96, 97.

Three new maps of the Orion star–forming region are presented and discussed: a 60 μm 34″ resolution map of the core of the Orion Molecular Cloud 1 (Ori MC 1), a 60 μm large–area image of the region from IRAS data, and a $J = 1 \rightarrow 0$ CO antenna–temperature map. Most of the luminosity of the region, and therefore most of the heating of the molecular ridge, arises ultimately from the Trapezium stars. The authors estimate a total far–infrared luminosity for Ori MC 1 of $2.5 \times 10^5 L_\odot$, with an uncertainty of $\pm 30\%$. Over the same region, there is at least 16 $M_\odot$ of dust in emission. The authors estimate a luminosity of $4 - 5 \times 10^4 L_\odot$ for the most energetic member of the BN–KL embedded cluster IRc 2. The observations also include the M43 region, which is explained as a cavity on the side of the dense molecular ridge.

131.321 **A high resolution $C^{18}O$ map of OMC 1.**
E. Serabyn, T. L. Wilson, C. M. Walmsley, C. Henkel.
Bull. Am. Astron. Soc., Vol. 18, No. 1, p. 634 – 635 (1986). Abstract. – See Abstr. 010.066.

131.322 **The distribution of atomic carbon in the Rho Ophiuchi dark cloud.**
J. Keene, G. A. Blake, T. G. Phillips.
Bull. Am. Astron. Soc., Vol. 18, No. 1, p. 637 (1986). Abstract. – See Abstr. 010.066.

131.323 **Structure in dark clouds as revealed by deep CCD images and IRAS maps.**
L. G. Mundy, E. J. Shaya, A. I. Sargent, N. Z. Scoville, D. Zaritsky.
Bull. Am. Astron. Soc., Vol. 18, No. 1, p. 637 – 638 (1986). Abstract. – See Abstr. 010.066.

131.324 **Infrared imaging polarimetry of AFGL 2591.**
M. S. Burns, P. Johnson, H. A. Thronson, T. Hayward.
Bull. Am. Astron. Soc., Vol. 18, No. 1, p. 638 (1986). Abstract. – See Abstr. 010.066.

131.325 **UV observations of galactic halo gas kinematics.**
L. Danly.
Bull. Am. Astron. Soc., Vol. 18, No. 1, p. 638 (1986). Abstract. – See Abstr. 010.066.

131.326 **The local interstellar medium over 5 pc in the direction of α Aquilae.**
R. Ferlet, R. Lallement, A. Vidal–Madjar.
Inst. Astrophys. Paris, Pré–Publ., No. 132, 15 pp. (1986). To appear in Astron. Astrophys.

131.327 **A model of molecular emission in bipolar flows and its physical implications.**
S. Cabrit.
Inst. Astrophys. Paris, Pré–Publ., No. 137, 15 pp. (1986). Communication presented at the CNRS workshop on "Anisotropic mass–loss and accretion phenomena in stellar systems".

131.328 **Multi–component velocity structure of the local interstellar medium.**
R. Lallement, A. Vidal–Madjar, R. Ferlet.
Inst. Astrophys. Paris, Pré–Publ., No. 142, 25 pp. (1986). To appear in Astron. Astrophys.

131.329 **Amorphous ice in astrophysics.**
J. Klinger.
J. Phys. Colloq., Vol. 46, No. C–8, p. 657 – 660 (1985). Abstr. in Phys. Abstr., Vol. 89, No. 1280, Entry 61021 (1986).

131.330 **Density structure and stability of the subcomponents in giant molecular cloud complexes.**
C. W. Yoo, S. S. Hong.
J. Korean Astron. Soc., Vol. 19, No. 1, p. 33 – 49 (1986).

Radial distribution of internal density has been determined for thirteen subclouds in the three giant molecular cloud complexes accompanying Mon OB1, Mon OB2 and CMa OB1 associations. The authors modeled their radial density structures with the density distribution of isothermal gas spheres. Most of the subclouds, nine out of the thirteen, are well described by isothermal spheres of single component, while the rest four require an additional component. Total mass and potential energy of each subcloud are also derived from the radial density structure.

131.331 **Interstellar absorption lines in the directions of extragalactic objects – II. Analysis of 25 sight lines.**
D. C. Morton, J. C. Blades.
Mon. Not. R. Astron. Soc., Vol. 220, No. 4, p. 927 – 948 (1986).

Observations of galactic Ca II, and in some cases Na I, interstellar absorption with resolutions as fine as 17 km s^{-1} FWHM are reported for the sources Fairall 9, Akn 120, NGC 4593, IC 4329A, 2155–304, 2251–178, 2302–089, the 1981 supernovae in NGC 1316 and 4536 and the star K996 in the globular cluster NGC 7078. All objects have a Ca II K line with $|v_{LSR}| < 17$ km s^{-1} and an equivalent width in the range 80 – 400 mÅ, which may be attributed to local disc gas. Comparison of the sight lines to the extragalactic objects with stars at various distances and galactic latitudes shows that the Ca II is concentrated in a thick disc with a scale height of about 1(+1,–0.5) kpc.

131.332 **The kinetic temperature gradient and the structure of a thin molecular disk in Cepheus A.**
J. M. Torrelles, P. T. P. Ho, L. F. Rodríguez, J. Cantó.
Astrophys. J., Vol. 305, No. 2, p. 721 – 731 (1986).

With the VLA the authors observed the (2,2) inversion transition line of NH_3 toward Cepheus A. They detected five small clumps within the Cep A–1 and Cep A–3 ammonia condensations reported previously. Cep A–1 and Cep A–3 may be physically connected, thus constituting a very elongated structure with a length–to–width ratio of 13 to 1. The stellar activity center appears embedded within this structure and seems to heat the molecular gas close ($\lesssim 3''$) to it. This result supports the interpretation that the knots seen in continuum are bright rims excited by one or two central stars. The geometry of the molecular disk and its relation to the high–velocity CO outflow are discussed.

131.333 **A physical model for the 2175 Å interstellar extinction feature.**
J. H. Hecht.
Astrophys. J., Vol. 305, No. 2, p. 817 – 822 (1986).

Recent *IUE* observations have shown that the 2175 Å interstellar extinction feature is constant in wavelength but varies in width. A model has been constructed to explain these results. It is proposed that the 2175 Å feature will only be seen when there is extinction due to carbon grains which have lost their hydrogen. In particular, the feature is caused by a separate population of small (< 50 Å radius), hydrogen–free carbon grains. The variations in width would be due to differences in either their temperature, size distribution, or impurity content.

131.334 **Effects of interstellar dust on the photometric properties of faint disk galaxies.**
G. Bruzual A., G. Magris C.
Rev. Mex. Astron. Astrofis., Vol. 12, Spec. Issue, p. 159 (1986). Abstract. – See Abstr. 012.066.

131.335 **The effects of interstellar dust on the spectrum of disk galaxies.**
G. Magris C., G. Bruzual A.
Rev. Mex. Astron. Astrofis., Vol. 12, Spec. Issue, p. 160 (1986). Abstract. – See Abstr. 012.066.

131.336 **Magnetic fields and the orientation of spin angular momentum in protostellar clouds.**
J. Franco.
Rev. Mex. Astron. Astrofis., Vol. 12, Spec. Issue, p. 224 (1986). Abstract. – See Abstr. 012.066.

131.337 **Linear polarization of the water maser in Orion.**
Z. Abraham, J. W. S. V. Boas.
Rev. Mex. Astron. Astrofis., Vol. 12, Spec. Issue, p. 228 (1986). Abstract. – See Abstr. 012.066.

131.338 **High velocity OH near young stellar objects: L1551.**
I. F. Mirabel, L. F. Rodríguez, J. Cantó, E. M. Arnal.
Rev. Mex. Astron. Astrofis., Vol. 12, Spec. Issue, p. 229 (1986). Abstract. – See Abstr. 012.066.

131.339 **The distribution of interstellar dust.**
A. Clocchiatti, H. G. Marraco.
Rev. Mex. Astron. Astrofis., Vol. 12, Spec. Issue, p. 264 – 268 (1986). – See Abstr. 012.066.
A statistical method to analyse the interstellar reddenings to obtain information about the interstellar dust distribution is presented. The properties of the interstellar matter structural functions and some ideal dust distribution models are described.

131.340 **Photoelectric photometry of CED 128 – a reflection nebula.**
M. de Oliveira, W. J. Maciel.
Rev. Mex. Astron. Astrofis., Vol. 12, Spec. Issue, p. 273 – 282 (1986). – See Abstr. 012.066.
BV photometric measurements are presented of the reflection nebula CED 128 along four radial directions relative to the illuminating star, ν Sco.

131.341 **Interaction between high–velocity clouds and the Galaxy: the case of the Orion molecular cloud complex.**
J. Franco.
Rev. Mex. Astron. Astrofis., Vol. 12, Spec. Issue, p. 287 – 291 (1986). – See Abstr. 012.066.
The location of the Orion molecular complex, about 150 pc away from the galactic plane, and some of the features of the region are discussed. Two possible origins are analyzed: (1) expanding shells from OB associations, and (2) a collision between the disk and a high–velocity cloud.

131.342 **A possible existence criterion for molecular clouds.**
D. P. Cox, J. Franco.
Rev. Mex. Astron. Astrofis., Vol. 12, Spec. Issue, p. 295 (1986). Abstract. – See Abstr. 012.066.

131.343 **Turbulent motions in molecular clouds.**
G. A. P. Franco, R. D. Tarsia, R. J. Quiroga.
Rev. Mex. Astron. Astrofis., Vol. 12, Spec. Issue, p. 297 (1986). Abstract. – See Abstr. 012.066.

131.344 **Lupus Loop: a possible case of star formation induced by a supernova event?**
M. Arnal, G. Dubner.
Rev. Mex. Astron. Astrofis., Vol. 12, Spec. Issue, p. 298 (1986). Abstract. – See Abstr. 012.066.

131.345 **Electrodynamic forces in stellar formation.**
L. C. Jafelice, R. Opher.
Rev. Mex. Astron. Astrofis., Vol. 12, Spec. Issue, p. 311 – 318 (1986). – See Abstr. 012.066.
Due to dynamo action of a rotating collapsing protostellar cloud in an ambient interstellar magnetic field, electrodynamic forces are created. The authors study the electrodynamic forces for a range of protostellar parameters, in particular, as a function of the ambient magnetic field and the radius, mass, angular velocity and temperature of the protostellar cloud.

131.346 **Dust in high velocity clouds.**
B. P. Wakker.
Light on dark matter, p. 379 – 380 (1986). – See Abstr. 012.080.
The facilities available in Groningen to produce IRAS maps were used to make comparatively high–quality maps of two fields where a high–velocity cloud (HVC) is observed in the 21–cm line. The HVC's are not discernable in these maps although one would expect to see them if the relation between 100 micron emission and H I column density found by Boulanger, Baud & van Albada (1985) is used. The implications of this non–detection are discussed in this contribution.

131.347 **High energy particles in dark molecular clouds.**
H. Reeves.
Nucleosynthesis and its implications on nuclear and particle physics, p. 23 – 26 (1986). – See Abstr. 012.069.
The role of galactic cosmic rays in generating the light elements lithium, beryllium and boron by bombardment of interstellar matter is examined. It is shown that the local irradiation in OB associations cannot explain the reported variations of the $^{6}Li/^{7}Li$ isotopic ratio in interstellar matter.

131.348 **Cyanoacetylene observations of B335.**
T. I. Hasegawa, N. Hirano, O. Kameya, Y. Taniguchi, M. Tosa, K. Takakubo, M. Seki.
Publ. Astron. Soc. Jpn., Vol. 38, No. 3, p. 379 – 394 (1986).
The purpose of the observations is to investigate the detailed structure of the high–density central region of B335 with HC_3N (J = 5–4 and 4–3) lines at higher angular resolutions. The authors also intended to get information on the cyanopolyyne abundance in B335 to study the variation of the abundance among various dark clouds by comparing it with previous observations of other dark clouds.

131.349 **Ammonia toward H_2O maser regions.**
R. Mauersberger, C. Henkel, T. L. Wilson, C. M. Walmsley.
Mitt. Astron. Ges., No. 65, p. 190 (1986). Abstract. – See Abstr. 012.072.

131.350 **The first detection of a $^{15}NH_3$ maser toward NGC 7538–IRS1.**
T. L. Wilson, R. Mauersberger, C. Henkel.
Mitt. Astron. Ges., No. 65, p. 191 (1986). Abstract. – See Abstr. 012.072.

131.351 **On the IMF of violent star forming regions.**
J. Melnick.
Star–forming dwarf galaxies and related objects, p. 171 – 180 (1986). – See Abstr. 012.074.
The IMF of very massive, young star clusters is investigated by comparing the luminosity function of the ionizing cluster of the 30 Doradus nebula to that of young galactic clusters.

131.352 **Modelling violent star forming regions.**
R. J. Terlevich.
Star–forming dwarf galaxies and related objects, p. 395 – 402 (1986). – See Abstr. 012.074.
Violent star forming regions are ideal laboratories to study formation and evolution of massive stars. The observed regularities between electron temperature, equivalent width of Hβ and abundance can be explained by a simple change in the slope of the IMF with metal abundance. The scatter in the observed regular-

ities seems to be mainly due to an age effect. A method is described to remove the age effect.

131.353 **Il ruolo della formazione stellare nell'evoluzione galattica.**
P. Mazzei.
G. Astron., Vol. 11, N. 4, p. 441 – 448 (1985).

131.354 **Physical processes in massive star formation.**
J. Silk.
Luminous stars and associations in galaxies, p. 301 – 315 (1986). – See Abstr. 012.079. (IAU Symp. No. 116).
The gravitational fragmentation theory of star formation is reviewed. Theoretical arguments are presented which suggest that the lower stellar mass cut–off to the IMF in giant H II regions may be as high as 10 $M_{\odot}$. Mechanisms for bimodal star formation are described in the context of a coagulation model for formation of the giant molecular clouds, and application is made to starbursting galaxies.

131.355 **Oscillating star formation.**
G. Bodifée.
Luminous stars and associations in galaxies, p. 397 – 398 (1986). – See Abstr. 012.079. (IAU Symp. No. 116).
As a consequence of positive feedback effects in interstellar chemical and star forming processes, a star formation region may undergo nonlinear oscillations.

131.356 **Propagating star formation induced by superbubbles.**
M. Kafatos, R. McCray.
Luminous stars and associations in galaxies, p. 411 – 412 (1986). – See Abstr. 012.079. (IAU Symp. No. 116).

131.357 **Luminous stellar content of the Galaxy: inferences from radio and infrared data.**
P. G. Mezger.
Luminous stars and associations in galaxies, p. 479 – 495 (1986). – See Abstr. 012.079. (IAU Symp. No. 116).
Lyman continuum (Lyc) photon production rates can be estimated from radio free–free emission and used to estimate the star formation rate (SFR) of O stars. If this SFR is linked to the total SFR through a constant IMF ($m \gtrsim 0.1\ m_{\odot}$) one derives for our Galaxy a present–day SFR of $\sim 10\ m_{\odot}\ yr^{-1}$, which is close to the average SFR over the age of the galactic disk. Observations of radio recombination of H and He show that the Lyc photon radiation field gets softer with decreasing galactocentric distance. This effect is linked to an increase of the metal abundance towards the galactic center. This can be attributed to the combined effect of UV line blanketing, increase of stellar radius and decrease of the upper mass limit of the IMF.

131.358 **Observations of infrared cirrus.**
T. N. Gautier.
Light on dark matter, p. 49 – 54 (1986). – See Abstr. 012.080.
This paper reviews the observations and the current understanding of infrared cirrus, describing the morphology and observed spatial distribution of the cirrus and compares the IRAS data with observations at other wavelengths. The measured properties of the cirrus material are related to models of interstellar dust. Finally some of the unique observational opportunities presented by the cirrus are mentioned along with a review of some recent observational results.

131.359 **Dust at the North Galactic Pole.**
J. Knude.
Light on dark matter, p. 55 – 58 (1986). – See Abstr. 012.080.
Results on the most reddened part of the NGP are reported. Large fractions of the pole zone $b > +70°$ are apparently reddened by more than $E(b{-}y) = 0.05$ mag. The reddened areas may be organized in parallel string–like features. The IRAS clouds B and C are located in the NGP zone. The gas/dust ratio varies across the polar area.

131.360 **Grains, what do we know?**
H. C. van de Hulst.
Light on dark matter, p. 161 – 170 (1986). – See Abstr. 012.080.
This paper is an attempt to summarize the questions on which the answers are believed to be known and the questions on which we should like to have the answers, all in the area of interstellar grains.

131.361 **IRAS cirrus observations and the nature of dust.**
J. S. Mathis.
Light on dark matter, p. 171 – 176 (1986). – See Abstr. 012.080.
Two theories regarding the nature of interstellar dust are considered. One, by Mathis, Rumpl, and Nordsieck (1977), has bare silicate and carbon particles. The other, by Greenberg and his associates (1983), has silicates with refractory mantles providing the extinction in the visible region of the spectrum. Some relevant observations are reviewed.

131.362 **Dust in diffuse clouds: one stage in a cycle.**
J. M. Greenberg.
Light on dark matter, p. 177 – 188 (1986). – See Abstr. 012.080.
An evolutionary point of view together with application of basic observational constraints suggest that the major solid particle constituents of the diffuse cloud medium are: silicate core – organic refractory mantle particles (~ 0.1 µm); small disordered carbon particles ($\lesssim 0.01$ µm); small silicates ($\lesssim 0.01$ µm). Implications of likely distributions of such particles in the near infrared extinction and the far infrared emission are considered. An extension of interstellar dust to comet dust optical properties provides a basis for searching for a hitherto unexpected source of infrared emission within the solar system.

131.363 **Infrared extinction in molecular clouds: the form of the curve in Orion.**
R. Hofmann, D. S. Davis, H. P. Larson.
Light on dark matter, p. 189 – 190 (1986). – See Abstr. 012.080.
The authors have determined an empirical extinction curve for the Orion molecular cloud in the near–IR spectral region using shock–excited H_2 emission lines as the "background" source. Comparison of the observed line intensities with theoretical predictions yielded the wavelength–dependent extinction measurements.

131.364 **Ultraviolet extinction as a key to grain optical properties in the infrared and ultraviolet.**
G. Chlewicki, J. M. Greenberg.
Light on dark matter, p. 191 – 196 (1986). – See Abstr. 012.080.
The paper summarizes the results of a large study of UV extinction based on IUE spectra. The analysis of variations in UV extinction between individual lines of sight is used to separate the contributions of individual grain populations and to constrain their properties. The temperatures of particles whose possible presence in the interstellar medium is indicated by grain models are compared with recent IRAS results.

131.365 **The wavelength of maximum polarization in the Chamaeleon dark cloud.**
D. C. B. Whittet, J. H. Hough, J. A. Bailey, M. F. Rouse, T. M. Kirrane.
Light on dark matter, p. 197 – 200 (1986). – See Abstr. 012.080.
The authors conclude from this study that either (1) five of the stars in their sample of seven have circumstellar infrared excess emission, or (2) the normal correlation between the ratio of total to selective extinction and the wavelength of maximum polarization breaks down in the Chamaeleon cloud. The former explanation seems unlikely in view of the fact that none of these stars have emission lines in their optical spectra, precluding a hot gaseous shell. HD 97300 has a cool dust shell detected by IRAS but this produces negligible contamination at 2.2 µm. If the second explanation is correct, this could indicate an unusual grain chemistry in the Chamaeleon region.

131.366 **Three principal heating sources of dust in the galactic disk.**
P. Cox, E. Krügel, P. G. Mezger.
Light on dark matter, p. 201 – 202 (1986). – See Abstr. 012.080.
The authors present the results of a model of the galactic infrared emission where in all presently available observations can be accounted for. The model assumes a cylindrical symmetry around the galactic center and adopts the MRN (Mathis et al., 1977) dust mixture of silicate and graphite grains, including the recent modifications as described in Draine and Lee (1984). The galactic infrared emission is a contribution from three main dust components with different temperatures and heating sources. The authors review the relevant characteristics of these three components and refer the readers to Cox et al. (1985) for a more detailed discussion.

131.367 **Mid–IR emission of the interstellar medium.**
F. Boulanger, M. Pérault, J. L. Puget.
Light on dark matter, p. 203 – 208 (1986). – See Abstr. 012.080.
IRAS observations of the local interstellar medium and the galactic plane are presented; for all components, the radiation observed at 12 μm represents a large fraction of the far–IR emission. The 12/25 μm color of the diffuse IR emission of the galactic plane increases from the inner to the outer parts while the 60/100 μm ratio decreases, a color–color anticorrelation observed also for giant H II regions. The implications of these observations are discussed.

131.368 **Optical luminescence from reflection nebulae?**
G. Olofsson.
Light on dark matter, p. 209 – 212 (1986). – See Abstr. 012.080.

131.369 **Infrared spectra and dust temperature fluctuations.**
F. X. Désert.
Light on dark matter, p. 213 – 216 (1986). – See Abstr. 012.080.
A model of dust emission is built which contains graphite, silicate grains and the recently discovered PAH (polycyclic aromatic hydrocarbons) molecules. The resulting infrared spectra are compared with IRAS observations. It seems likely that there is a gap in the size distribution between graphite grains and PAH molecules. Furthermore, this model can account for the anticorrelation seen in our Galaxy in the color–color diagram F(12 μ)/F(25 μ) vs F(60 μ)/F(100 μ).

131.370 **Non–equilibrium emission from small particles.**
K. Sellgren.
Light on dark matter, p. 217 – 220 (1986). – See Abstr. 012.080.
The idea of non–equilibrium thermal emission from very small particles, whether grains or molecules, has a growing interest for infrared astronomers as more and more examples of its potential applicability are found in the IRAS data set and other observations. This paper reviews the observational development of these ideas, as well as discusses recent observations and suggests possible future directions for theory and observations of this phenomenon.

131.371 **Evidence for a 12 micron water–ice absorption band in the IRAS LRS spectra of protostars and late type stars.**
M. de Muizon, L. B. d'Hendecourt, C. Perrier.
Light on dark matter, p. 221 – 224 (1986). – See Abstr. 012.080.
The authors present evidence for a 12 micron absorption band due to water ice in some IRAS LRS spectra. This identification is based on the 3 micron spectrophotometry of the same objects showing the presence of the 3 micron water ice absorption band.

131.372 **Formation of organic molecules on interstellar dust particles.**
W. Schutte, J. M. Greenberg.
Light on dark matter, p. 229 – 232 (1986). – See Abstr. 012.080.
The authors show how an organic grain mantle is produced in a molecular cloud. It is estimated that half of the volatile molecules containing O, C, N are converted to organics in the lifetime of an average cloud. The spectrum of an organic material made by laboratory simulation of this process closely resembles the N.I.R. spectrum of the galactic center source IRS7. The authors estimate the extinction towards the galactic center, finding a value which agrees well with other determinations.

131.373 **Polycyclic aromatic hydrocarbons and the diffuse interstellar bands.**
G. P. van der Zwet, L. J. Allamandola.
Light on dark matter, p. 233 – 236 (1986). – See Abstr. 012.080.
The profiles, widths and positions of the Diffuse Interstellar Bands (DIB's) are discussed. It is shown that the carriers of the DIB's are molecular rather than impurities in grains. Polycyclic Aromatic Hydrocarbons (PAH's) are attractive candidates for the DIB's, in view of their likely abundance, their extreme stability, and the fact that these molecules, and their corresponding radicals, show visible transitions.

131.374 **Identification of polycyclic aromatic hydrocarbons.**
A. Léger, L. d'Hendecourt.
Light on dark matter, p. 237 – 240 (1986). – See Abstr. 012.080.
The nature of the Very Small Grains evidenced by K. Sellgren (1985) is discussed. Their stability suggests that they are graphitic material and specifically Polycyclic Aromatic Hydrocarbons (PAHs). The expected IR emission of a typical PAH, coronene, gives an impressive spectroscopic agreement with the five observed "Unidentified IR Emission Features", leading to an unambiguous identification. Those PAHs are the most abundant organic molecules detected to this date ($f \sim 10^{-5}$). Several astrophysical implications are reviewed.

131.375 **Silicate absorption strength; polarized emission and absorption by aligned grains.**
D. K. Aitken.
Light on dark matter, p. 241 – 244 (1986). – See Abstr. 012.080.

131.376 **Reflection nebulae, non–equilibrium thermal emission, and IRAS.**
K. Sellgren, L. J. Allamandola, J. D. Bregman, M. W. Werner, D. H. Wooden.
Light on dark matter, p. 261 – 262 (1986). – See Abstr. 012.080.
The 1 – 13 micron spectra of visual reflection nebulae are presented, and their implications for the IRAS 12 micron band discussed.

131.377 **Theories of star formation confronted by IRAS data.**
B. G. Elmegreen.
Light on dark matter, p. 265 – 276 (1986). – See Abstr. 012.080.
The importance to star formation theories of IRAS observations of the dust distribution around young stars, the luminosities of protostars and embedded clusters, and the size distribution of dust particles is discussed. The important role of the smallest grains in determining the ionization fraction and magnetic diffusion rate in dense molecular clouds is illustrated by new solutions to the equations of charge equilibrium.

131.378 **IMF in starburst regions.**
H. Zinnecker.
Light on dark matter, p. 277 – 278 (1986). – See Abstr. 012.080.

131.379 **Point sources in the Orion complex.**
C. A. Beichman.
Light on dark matter, p. 279 – 292 (1986). – See Abstr. 012.080.
A 667 sq.deg. region in Orion contains thousands of IRAS sources, including stars, embedded sources, galaxies, as well as clumps in more extended emission (cirrus). Two particular types of sources, visible stars with infrared excesses characteristic of T Tauri stars and invisible, cool, embedded objects ($L \sim 25\ L_{\odot}$) are probably related to the formation of low mass stars in Orion.

131.380 **The IR emission of the Orion–Monoceros molecular clouds.**
F. Boulanger, R. J. Maddalena, P. Thaddeus.
Light on dark matter, p. 293 – 294 (1986). – See Abstr. 012.080.
The present investigation is based on both IRAS and Columbia CO observations of a region ($200° \leqslant l \leqslant 225°$ and $-25° \leqslant b \leqslant -7.°5$) which includes all molecular clouds observed by Maddalena et al. (1985), except the one around λ Ori.

131.381 **Young stars and high density condensations in the Horsehead region.**
G. Sandell, B. Reipurth, C. Menten, M. Walmsley, H. Ungerechts.
Light on dark matter, p. 295 – 296 (1986). – See Abstr. 012.080.
The authors have mapped the Horsehead in ammonia and made partial maps in the CO and ^{13}CO lines, both with a resolution of 30″. They find strong molecular emission close to the ionization front, and several NH_3 condensations both in the head and down at the base of the Horsehead. Within the ammonia map there are four IRAS sources. Two are deeply embedded, one of them close to a NH_3 condensation. The other two are identifiable with visible stars, one with the nebulous star B33–1 in front of the Horsehead, the other with V615 Ori, which the authors have found to be a luminous G–type emission line star associated with several Herbig–Haro objects.

131.382 **Mapping of the Coronae Austrinae star forming region.**
A. Evans, J. S. Albinson, M. F. Bode, D. C. B. Whittet.
Light on dark matter, p. 297 – 298 (1986). – See Abstr. 012.080.
The authors describe observations of the Coronae Austrinae (CrA) star forming region. With the exception of R CrA, TY CrA and HH100 no infrared sources were detected in the region surveyed.

131.383 **A. Analysis of point sources in the Ophiuchus and Perseus clouds. B. CPC observations of NGC 1333.**
R. E. Jennings, W. Cudlip, C. J. Hirst, D. H. M. Cameron.
Light on dark matter, p. 299 – 300 (1986). – See Abstr. 012.080.
Sources have been chosen from the IRAS Point Source Catalogue which are associated with particular molecular clouds, in this case the Ophiuchus and Perseus clouds. The region lying south of the reflection nebula NGC 1333 in the Perseus molecular cloud has been scanned at 50 and 100 μm with the Chopped Photometric Channel on IRAS. The area scanned has a number of Herbig–Haro objects and embedded infrared sources.

131.384 **Star forming loops in the IRAS sky images.**
P. R. Schwartz.
Light on dark matter, p. 301 – 302 (1986). – See Abstr. 012.080.
Loops of heated dust and embedded young stars are seen in IRAS images of the galaxy. Some of these loops are undoubtedly OB association "bubbles": a conclusion which revives the possibility of sequential star formation.

131.385 **Young stars and dense cores in nearby dark clouds.**
P. C. Myers.
Light on dark matter, p. 307 – 312 (1986). – See Abstr. 012.080.
The author reviews recent progress in the study of young stars and dense cores in dark clouds, as revealed by gas emission line observations from NH_3 (1.3 cm), CO (2.6 mm); by dust and stellar continuum observations with the IRAS satellite (12 – 100 μm), and with ground–based telescopes (1 – 20 μm).

131.386 **Water masers coincident with IRAS sources.**
J. G. A. Wouterloot, C. M. Walmsley.
Light on dark matter, p. 313 – 314 (1986). – See Abstr. 012.080.
The authors have started a survey of H_2O masers towards IRAS sources apparently associated with two molecular cloud complexes in Orion and Cepheus. They selected infrared sources towards the clouds L 1630 and L 1641 in Orion as well as the concentration of clouds and H II regions in the Perseus arm at $l = 110°$, which also contains the Cep OB3 complex. The authors have detected 3 new masers in the Orion clouds, 2 in the Cep OB3 cloud and 7 masers situated in the Perseus arm as well as confirming several known masers.

131.387 **A large scale OH survey in Orion and Monoceros.**
P. M. M. Jenniskens, H. J. Habing, J. G. A. Wouterloot, P. te Lintel–Hekkert, A. Blaauw.
Light on dark matter, p. 325 – 326 (1986). – See Abstr. 012.080.
Large scale mapping in both the 1665 and 1667 MHz molecular lines of OH shows overlap of two clouds near $l = 211°$ and $b = -19°$ in Orion South. Clouds associated with Monoceros R2 show a shell like structure. VB80 is associated with a large molecular cloud.

131.388 **The galactic morphology of the interstellar dust detected by IRAS.**
W. B. Burton, E. R. Deul, H. J. Walker, A. A. W. Jongeneelen.
Light on dark matter, p. 357 – 372 (1986). – See Abstr. 012.080.
The authors address following questions: Is the IRAS–detected dust confined, like the molecules, to dark regions of high density and low volume–filling factor, or is it much more pervasively distributed, like the atomic gas? Is the dust confined to a very thin layer, like the molecules and the most recently formed stars? Does it show the same sort of central deficiency as the gas? How far does the dust distribution extend in radius?

131.389 **The association of clouds in the Carina arm with IRAS spline maps.**
H. J. Walker, E. R. Deul, H. M. Butner, W. B. Burton.
Light on dark matter, p. 373 – 374 (1986). – See Abstr. 012.080.
The position of the emission peaks at 100 μm were compared with molecular clouds, identified using CO and H II regions.

131.390 **Dark cloud statistics.**
J. V. Feitzinger, J. A. Stüwe.
Light on dark matter, p. 375 – 376 (1986). – See Abstr. 012.080.
The dark cloud catalogues of Lynds (1962) and Feitzinger and Stüwe (1984) are combined. The authors deduce for the Northern (N), Anticenter (A) and Southern (S) sectors the obscuration percentages and the cloud distribution in galactic latitude.

131.391 **High latitude molecular clouds: completeness of the survey and implications for molecular surveys.**
L. Magnani, L. Blitz, E. Lada.
Light on dark matter, p. 377 – 378 (1986). – See Abstr. 012.080.
The fractional completeness of the Blitz, Magnani, and Mundy (1984) high latitude molecular cloud catalog is obtained and used to determine the contribution of the clouds to the local interstellar medium and to the Galaxy as a whole.

A catalogue of southern dark clouds.
See Abstr. 002.002.

A statistical analysis and overview of the IRAS point source catalog.
See Abstr. 002.092.

Die Entstehung der Sterne. Interstellare Materie.
See Abstr. 003.017.

Astrophysical implications of amorphous ice – a microporous solid.
See Abstr. 022.004.

Einstein *A*–values for rotational transitions in the H_2CO molecule.
See Abstr. 022.005.

Production and loss of HC_3N in interstellar clouds: some relevant laboratory measurements.
See Abstr. 022.015.

Stability of amorphous circumstellar silicate grains.
See Abstr. 022.023.

Millimeter wave spectrum of tricarbon oxide sulfide, OCCCS.
See Abstr. 022.024.

Millimeter wave spectrum of ^{13}C substitutions of cyclopropenylidene C_3H_2.
See Abstr. 022.027.

Molecular electrostatic potentials as a first approach to protonated interstellar molecules.
See Abstr. 022.029.

2.5 – 300 µm laboratory observations of submicron SiC particles as cosmic dust candidates.
See Abstr. 022.042.

Theoretical study of CS–containing molecules I. *Ab initio* Hartree–Fock predictions for C_2S_2.
See Abstr. 022.044.

The laboratory millimeter–wave spectrum of methyl formate in its ground torsional *E* state.
See Abstr. 022.049.

Statistics of equivalent width data and new oscillator strengths for Si II, Fe II and Mn II.
See Abstr. 022.053.

Fine–structure population ratios for the ^2P ground state of C II.
See Abstr. 022.079.

The reaction of C^+ with HCN: isomeric identity of the C_2N^+ product.
See Abstr. 022.081.

The rotational spectrum of the C_3H radical.
See Abstr. 022.082.

An experimental study of plasma grain interactions.
See Abstr. 022.084.

Simulation of cosmic dust spectra.
See Abstr. 022.085.

Laboratory studies of the infrared small–particle extinction of amorphous silicates.
See Abstr. 022.086.

The electrostatics of a dusty plasma.
See Abstr. 022.087.

On a similarity between clusters obtained by SIMS and molecules observed in astrophysics.
See Abstr. 022.104.

Collisional excitation of molecules in dense interstellar clouds.
See Abstr. 022.112.

The microwave and far–infrared spectra of the ^{18}OH radical.
See Abstr. 022.113.

Calculated forbidden rotational spectra of $H_3{}^+$.
See Abstr. 022.114.

C–N bound formation in the gas phase reaction of C^+ with NH_3.
See Abstr. 022.116.

Cyanoallene, $CH_2=C=CH–CN$: astrophysical predictions.
See Abstr. 022.118.

Ultraviolet photoprocessing and infrared spectroscopy of laboratory simulated grain mantles.
See Abstr. 022.132.

Radiation effects on grain materials.
See Abstr. 022.133.

Direct measurement of the fine–structure interval and g_J factors of singly ionized atomic carbon by laser magnetic resonance.
See Abstr. 022.135.

Preparation, analysis, and release of simulated interplanetary grains into low earth orbit.
See Abstr. 051.067.

Formation of astrophysical jets by a contracting magnetic accretion disk.
See Abstr. 062.004.

Effect of magnetic field on a shock–induced thermal instability.
See Abstr. 062.005.

Stratification of emission from interstellar shock waves.
See Abstr. 062.031.

Multicomponent reacting MHD flows – a reply.
See Abstr. 062.044.

On pumping astronomical masers.
See Abstr. 062.046.

On the structure and stability of radiative shock waves.
See Abstr. 062.083.

Magnetized thermal conduction fronts.
See Abstr. 062.086.

Theoretical studies of interstellar molecular shocks – III. The formation of CH^+ in diffuse clouds.
See Abstr. 062.107.

Thermochemical instabilities in reacting gases. II. Plasma of hydrogen in presence of a magnetic field.
See Abstr. 062.128.

Circular polarization of astrophysical masers due to overlap of Zeeman components.
See Abstr. 063.003.

On pumping astronomical masers.
See Abstr. 063.014.

Astrophysical water masers in the presence of magnetic fields.
See Abstr. 063.055.

The analysis of ensembles of moderately saturated interstellar lines.
See Abstr. 063.075.

New results in the optics of grains with resonant impurities.
See Abstr. 063.089.

A cloud collision model for water maser excitation.
See Abstr. 063.090.

A method of analysis of the pumping mechanisms of cosmic masers.
See Abstr. 063.094.

Bipolar hydromagnetic winds from disks around protostellar objects.
See Abstr. 064.020.

Stellar winds from young stars.
See Abstr. 064.047.

A pulsed jet as the source of momentum in bipolar molecular outflows.
See Abstr. 064.052.

Mass loss from very young massive stars.
See Abstr. 064.059.

An exact solution for a rotating selfgravitating gas mass with oscillatory density.
See Abstr. 065.004.

Protostellar angular momentum transport by spiral density waves.
See Abstr. 065.020.

Fission of rapidly rotating fluid systems.
See Abstr. 065.025.

Further studies of a fragmentation problem.
See Abstr. 065.031.

Initial mass function for zero–metal stars.
See Abstr. 065.037.

Primordial stellar evolution: the protostar phase.
See Abstr. 065.064.

Supermassive star formation – the early phases.
See Abstr. 065.082.

Dynamic fission instabilities in rapidly rotating $n = 3/2$ polytropes: a comparison of results from finite–difference and smoothed particle hydrodynamics codes.
See Abstr. 065.108.

Plasma transport across the heliopause.
See Abstr. 074.073.

Uranian H Ly–α emission: the interstellar wind source.
See Abstr. 101.015.

Fluffy comets.
See Abstr. 102.021.

Interstellar S_2 in comets.
See Abstr. 102.022.

S_2 formation in interstellar dust; a diagnostic of the maximum aggregation temperature for a comet.
See Abstr. 102.084.

Survival of interstellar matter in meteorites: evidence from carbonaceous material.
See Abstr. 105.008.

Atomic hydrogen and helium densities of the interstellar medium measured in the vicinity of the Sun.
See Abstr. 106.025.

A carbonate–rich, hydrated, interplanetary dust particle: possible residue from protostellar clouds.
See Abstr. 106.033.

Neutral interstellar gases in the heliosphere: new aspects of the problem.
See Abstr. 106.052.

Interstellar gas parameters and solar wind anisotropies deduced from H and He observations in the solar system.
See Abstr. 106.053.

Interstellar material in the solar system. A working group report.
See Abstr. 106.059.

The effects of self–gravity on the solar nebula.
See Abstr. 107.001.

Formation of the sun and its planets.
See Abstr. 107.004.

Dynamical and chemical evolution of the protoplanetary nebula.
See Abstr. 107.011.

Formation and evolution of the primitive solar nebula.
See Abstr. 107.020.

Encounters between the solar system and giant molecular clouds.
See Abstr. 107.036.

On the properties of very young massive infrared sources.
See Abstr. 112.017.

Observational evidence for disks around young stars.
See Abstr. 112.019.

Working group on circumstellar/interstellar relationships.
See Abstr. 112.075.

On the broad Balmer profiles from Car II and their possible origin as dust scattered lines from η Carinae.
See Abstr. 112.080.

Interstellar broadening of maser spots.
See Abstr. 112.111.

Near infrared photometry of stellar maser sources.
See Abstr. 112.115.

IRAS observations of Wolf–Rayet stars.
See Abstr. 112.137.

Mass loss by cool stars.
See Abstr. 112.139.

IR CCD imaging of L 1551–IRS 5: direct observations of its circumstellar shell.
See Abstr. 112.146.

A model for bipolar sources in molecular clouds.
See Abstr. 112.147.

Five–colour photometry of early–type stars in the direction of galactic X–ray sources.
See Abstr. 113.002.

***uvbyβ* photometry of all stars earlier than G1 and brighter than $V \sim 12$ mag in Selected Area 132 containing Scorpius X–1.**
See Abstr. 113.003.

Is AS 431 a superluminous WR star?
See Abstr. 113.040.

IRAS observations of Wolf–Rayet stars.
See Abstr. 113.045.

Comments on using absolute spectrophotometry of Wolf–Rayet stars.
See Abstr. 114.002.

A survey of emission–line stars in the CMa star formation region.
See Abstr. 115.002.

Polarization of star radiation in the Mon R1 association.
See Abstr. 116.014.

A propos de la découverte d'une nouvelle planète.
See Abstr. 118.010.

Observations of bipolar flow sources in ammonia lines.
See Abstr. 121.004.

Pre–main–sequence stars in the Serpens molecular cloud.
See Abstr. 121.005.

The jets and discs associated with R and T CrA.
See Abstr. 121.007.

The high velocity outflow in Herbig–Haro 7–11.
See Abstr. 121.010.

Studies of the Chamaeleon star–forming region. II. The pre–main–sequence stars HD 97048 and HD 97300.
See Abstr. 121.011.

The nature and origin of Herbig–Haro objects.
See Abstr. 121.014.

The double radio source associated with L1551 IRS 5: binary system or ionized circumstellar torus?
See Abstr. 121.016.

Objects associated with low–mass star formation in the Gum nebula.
See Abstr. 121.026.

Extended far–IR emission associated with young star molecular outflows.
See Abstr. 121.027.

The relationship between disks, mass outflows and global cloud structures from young stellar objects.
See Abstr. 121.028.

The development of thermally driven winds around low mass protostars.
See Abstr. 121.030.

Fine jets observed in some T Tauri–type stars and cometary nebulae.
See Abstr. 121.036.

The circumstellar environment of LkHα 234.
See Abstr. 121.046.

A spectroscopic survey of the Taurus–Auriga dark clouds for pre–main–sequence stars having Ca II H, K emission.
See Abstr. 121.051.

Kinematical and hydrodynamical study of the HH 32 complex.
See Abstr. 121.058.

Emission–line stars in the Canis Major star–formation region.
See Abstr. 121.061.

Variable stars in Lynds 810.
See Abstr. 123.020.

The interaction of nova shells with the interstellar medium.
See Abstr. 124.014.

High–resolution X–ray spectroscopy of the Crab Nebula and the oxygen abundance of the interstellar medium.
See Abstr. 125.023.

Kinematic detection of supernova remnants in giant H II regions.
See Abstr. 125.045.

***IRAS* sources associated with shocked gas regions in IC 443.**
See Abstr. 125.057.

1410 MHz continuum and H I line observations towards the SNR G296.5+10.0 and nearby sources. Evidences of two SNRs tunneling through the interstellar medium.
See Abstr. 125.062.

Supernova remnants and molecular clouds.
See Abstr. 125.074.

The terrain of evolution of isotropic adiabatic supernova remnants.
See Abstr. 125.080.

Distribution and kinematics of neutral hydrogen around the SNR G296.5+10.0.
See Abstr. 125.088.

1.3 mm continuum observations of compact H II regions detected by IRAS I.
See Abstr. 132.001.

Surface photometry of simple H II regions.
See Abstr. 132.005.

Observations of the Scutum ring of H II regions.
See Abstr. 132.006.

High angular resolution infrared mapping of the compact H II regions W51 and DR 21/W75.
See Abstr. 132.008.

The M17 complex in the light of the near infrared line [S III] λ9531.
See Abstr. 132.011.

1.3 mm continuum observations of compact H II regions detected by IRAS. II.
See Abstr. 132.016.

Detection of [Si II] (34.8 micron) emission in Orion–KL: a measurement of the silicon abundance in dense interstellar gas.
See Abstr. 132.020.

The Rosette nebula. III. Interstellar dust extinction and a model of the molecular cloud complex.
See Abstr. 132.026.

On the connection of H II regions with molecular clouds of the Galaxy.
See Abstr. 132.028.

Near IR observations of the Sharpless regions S 269, S 271, S 307, and S 311.
See Abstr. 132.032.

Multiband far–infrared observations of the NGC 6334 complex.
See Abstr. 132.037.

Observations of the 157.7 micron [C II] emission from the galactic H II regions W3 and W51.
See Abstr. 132.038.

Compact radio sources associated with molecular outflows.
See Abstr. 132.039.

Observations of far–infrared line profiles in the Orion–KL region.
See Abstr. 132.040.

Star formation, giant H II regions and spiral structure.
See Abstr. 132.043.

The discovery of high–speed ionized knots in the H II regions M16 (NGC 6611) and the Rosette nebula (NGC 2237–2246).
See Abstr. 132.047.

Formation of OB clusters: radiation–driven implosion?
See Abstr. 132.049.

Effect of the dust on relative [O II] and [S II] intensities in H II regions.
See Abstr. 132.054.

Giant stellar–wind shell associated with the H II region M16.
See Abstr. 132.057.

The Scutum ring of H II regions, G24.6+0.0, associated with CO and H I shells.
See Abstr. 132.058.

The exciting stars of giant extragalactic H II regions.
See Abstr. 132.060.

Sub–mm continuum observations of sources from the IRAS point–source catalogue.
See Abstr. 132.061.

Three micron spectroscopy of the galactic centre source IRS 7.
See Abstr. 133.001.

Infrared reflection nebulae.
See Abstr. 133.002.

Infrared studies of southern AFGL sources. II. GL4176 and GL4182: two sources in star forming regions.
See Abstr. 133.004.

An unusual high velocity molecular outflow in the Rho Ophiuchi cloud.
See Abstr. 133.007.

CO observations of sources from IRAS Circular No. 9.
See Abstr. 133.010.

Comparison of CO and IR emission of IRAS unidentified sources.
See Abstr. 133.019.

The dust and molecular gas cloud surrounding a planetary nebula: NGC 7027.
See Abstr. 134.002.

Distances to planetary nebulae. II. H I absorption observations.
See Abstr. 134.007.

X–ray emission from bursts of star formation.
See Abstr. 142.104.

^{26}Al experimental results.
See Abstr. 143.032.

Cosmic rays and the interstellar medium.
See Abstr. 144.020.

A mathematical model of the initial stage in the formation of a disk galaxy.
See Abstr. 151.011.

Gas in galaxies.
See Abstr. 151.015.

Predicting star formation in interacting and colliding galaxies; the special case of ring galaxies.
See Abstr. 151.016.

Simulations of close encounter between galaxies: behaviour of interstellar gas clouds and enhancement of star formation rate.
See Abstr. 151.023.

Velocity dispersion of stars and giant molecular clouds.
See Abstr. 151.036.

Formation of giant molecular clouds by coagulation of small clouds and spiral structure.
See Abstr. 151.037.

***N*–body simulation of giant molecular clouds in a galaxy. I. Generation and acceleration of the random velocity of giant molecular clouds.**
See Abstr. 151.118.

Radial systems of dark globules in the association Cep OB2.
See Abstr. 152.001.

Systems of stellar associations in galaxies.
See Abstr. 152.010.

The dynamical evolution of expanding H I shells and initial mass function of OB associations.
See Abstr. 152.011.

The potential for supernova–induced chemical enrichment of protoglobular cluster clouds.
See Abstr. 154.038.

Massachusetts–Stony Brook Galactic Plane CO Survey. I. (b, V) maps of the first Galactic quadrant.
See Abstr. 155.001.

Massachusetts–Stony Brook Galactic Plane CO Survey. II. (l, V) maps of the first Galactic quadrant.
See Abstr. 155.002.

The radial distribution of galactic gamma rays. III. The distribution of cosmic rays in the Galaxy and the CO–H_2 calibration.
See Abstr. 155.003.

Relative rôles of type I and II supernovae in the chemical enrichment of the interstellar gas.
See Abstr. 155.006.

Infrared spectropolarimetry of the galactic centre: magnetic alignment in the discrete sources.
See Abstr. 155.007.

Star formation and some aspects of the chemical evolution of the Galaxy.
See Abstr. 155.008.

An estimation of the radial distribution of molecular hydrogen in the Galaxy from the star formation rates.
See Abstr. 155.009.

Distance to the anti–center shell.
See Abstr. 155.018.

CO observations, geometry, and galactic structure.
See Abstr. 155.019.

The mass of gas within 500 pc of the galactic centre.
See Abstr. 155.025.

Gamma ray estimates of galactic gas contents.
See Abstr. 155.026.

The dynamics of the interstellar gas and the kinetics of the galactic magnetic field.
See Abstr. 155.027.

The Massachusetts–Stony Brook galactic plane CO survey: statistical clustering properties of molecular cloud cores.
See Abstr. 155.035.

The distance to the center of the Galaxy.
See Abstr. 155.058.

2000 Å UV imaging of a 6° diameter field around μ Aurigae.
See Abstr. 155.068.

Abundance gradient in interstellar medium.
See Abstr. 155.070.

On warping of the old stellar disk in our Galaxy.
See Abstr. 155.080.

Galactic structure examined from a faint object survey in a field in Aquarius.
See Abstr. 155.083.

On the stability of the galactic magnetic field in the presence of a magnetic monopole halo.
See Abstr. 155.096.

The distribution of atomic hydrogen and diffuse gas in the Galaxy.
See Abstr. 155.103.

The projected dark cloud distribution of the Milky Way.
See Abstr. 155.105.

The largest molecular cloud complexes in the first galactic quadrant.
See Abstr. 155.114.

Classical cepheids and the local chemical inhomogeneities in the galactic disc.
See Abstr. 155.117.

Polarisierte Radiostrahlung der galaktischen Ebene bei 11 cm Wellenlänge.
See Abstr. 155.118.

Star formation in the Wing of the Small Magellanic Cloud.
See Abstr. 156.001.

Star formation in the Magellanic Cloud. III. IR observations of giant H II regions.
See Abstr. 156.009.

Age and processes of star formation in very young LMC associations.
See Abstr. 156.014.

Carbon monoxide in the Magellanic Clouds.
See Abstr. 156.019.

Gas in and toward the Magellanic Clouds.
See Abstr. 156.020.

Star formation of star clusters in the SMC and their adjoining fields.
See Abstr. 156.032.

Far–infrared emission from clumpy irregular galaxies.
See Abstr. 157.007.

Bimodal star formation and remnant–dominated galactic models.
See Abstr. 157.010.

Large–scale dissociation of molecular gas in galaxies by newly formed stars.
See Abstr. 157.011.

Evolution of disk galaxies regulated by supernova remnants.
See Abstr. 157.019.

Wind–driven shells in E/S0 galaxies: a research note.
See Abstr. 157.025.

Gas flow in the barred spiral galaxy NGC 5383.
See Abstr. 157.026.

Submillimetre continuum observations of NGC 253.
See Abstr. 157.028.

NGC 4038/39: interacting spiral galaxies with enhanced extended radio emission.
See Abstr. 157.029.

A physical mechanism for bursts of star formation.
See Abstr. 157.041.

Stellar populations and star formation in irregular galaxies.
See Abstr. 157.050.

Mapping dark nebulae in elliptical galaxies.
See Abstr. 157.052.

Galaxy evolution: some issues and questions.
See Abstr. 157.077.

Towards a theory of galaxy formation.
See Abstr. 157.078.

Theoretical and observational evidence for the existence of a single fragmentation law, and galactic evolution.
See Abstr. 157.079.

Star formation tracers in galaxies.
See Abstr. 157.080.

On the relationship between neutral hydrogen mass, luminosity and color in irregular galaxies.
See Abstr. 157.081.

I Zw 18 and the search for very metal poor galaxies.
See Abstr. 157.082.

Bolometric luminosity evolution.
See Abstr. 157.083.

Regions of massive star formation in two nearby spiral galaxies.
See Abstr. 157.122.

Distributions of ^{13}CO emission in the disks of late–type spiral galaxies.
See Abstr. 157.131.

A feature of the spiral pattern and star formation in the galaxy NGC 1566.
See Abstr. 157.133.

Three–dimensional simulations of supernovae–dominated interstellar media in disk galaxies.
See Abstr. 157.179.

CO observations of infrared bright galaxies: the efficiency of star formation.
See Abstr. 157.187.

IRAS observations of the nuclear bulge of M31.
See Abstr. 157.192.

Limits on supernova rates and interstellar densities from X–ray observations of M101.
See Abstr. 157.193.

A high resolution hydrogen line survey of Messier 31. III. H I holes in the interstellar medium.
See Abstr. 157.197.

Two–color CCD mapping of the luminous type I irregular galaxy NGC 4449.
See Abstr. 157.218.

Molecules in galaxies. I. CO observations in a spiral arm of M31.
See Abstr. 157.259.

Formaldehyde absorption and maser emission in galaxies.
See Abstr. 157.260.

Surveys for H II galaxies and related star forming systems.
See Abstr. 157.274.

Blue compact galaxies: a mixing bag.
See Abstr. 157.279.

Bursts of star formation in ring galaxies. Some results from CCD images.
See Abstr. 157.280.

Observational properties of the gas and the young stars in BCG.
See Abstr. 157.287.

H I – H II relationship in dwarf star–forming galaxies.
See Abstr. 157.296.

A search for Wolf–Rayet stars in giant extragalactic bursts of star formation.
See Abstr. 157.319.

Giant H II regions in M81.
See Abstr. 157.320.

The relationship between blue and FIR luminosities of spiral galaxies.
See Abstr. 157.326.

Starbursts in non–interacting galaxies.
See Abstr. 157.328.

Gas in galactic centres.
See Abstr. 158.031.

Star–formation rates in the nuclei of violently interacting galaxies.
See Abstr. 158.205.

Detection of the hydrocarbon ring molecule C_3H_2 in the radio galaxy Centaurus A (= NGC 5128).
See Abstr. 158.220.

A search for the CO line emission toward two clumpy irregular galaxies, Markarian 8 and 297.
See Abstr. 158.259.

Star formation in NGC 5128.
See Abstr. 158.322.

Young supernovae in the starburst galaxy M82.
See Abstr. 158.326.

Cosmic background radiation temperature from CN absorption.
See Abstr. 161.218.

132 H II Regions, Emission Nebulae

132.001 **1.3 mm continuum observations of compact H II regions detected by IRAS I.**
R. Chini, E. Kreysa, P. G. Mezger, H.–P. Gemünd.
Astron. Astrophys., Vol. 154, No. 1/2, p. L8 – L11 (1986).
The authors report on new 1.3 mm observations of 57 strong FIR sources, recently observed by IRAS. All sources can be identified with heavily obscured compact H II regions. The authors derive IR luminosities of the embedded stars and temperatures of the surrounding dust clouds.

132.002 **Spectrophotometry of H II regions in irregular galaxies.**
G. Stasińska, G. Comte, L. Vigroux.
Astron. Astrophys., Vol. 154, No. 1/2, p. 352 – 356 (1986).
The authors present spectrophotometric observations of 15 H II regions in 10 irregular galaxies. They are analyzed in a standard way, but using up–to–date atomic data, in order to derive the chemical abundances of O, N, Ne and He in these objects.

132.003 **H II regions in M33. III. Physical properties.**
F. Viallefond, W. M. Goss.
Astron. Astrophys., Vol. 154, No. 1/2, p. 357 – 369 (1986).
The new WSRT survey of M33 in the radio continuum (Viallefond et al., 1985) is used to study the physical properties of the sources which have been identified in the optical Hα Balmer line. The luminosity and diameter functions for these sources are also determined. A comparison with an Hα survey is used to determine from a simple model the abundance of dust; this is correlated with the excitation as measured by the [O III]/Hβ line ratio. The authors briefly discuss the properties of H I – H II associations in M33. These results are compared with those obtained for very bright nebulae as those observed in M101 and very faint galactic H II regions.

132.004 **Radio recombination lines of hydrogen atoms associated with stimulated emissions in ionized regions.**
M. Shinohara.
Astrophys. Space Sci., Vol. 119, No. 1, p. 123 – 125 (1986). – See Abstr. 012.001.
In this paper, the maximum intensity ratio of stimulated radio recombination lines to the continuum in ionized regions at electron temperatures ranging from 10^4K to 8×10^4K are shown.

132.005 **Surface photometry of simple H II regions.**
S. Yoshida, M. Nakano, T. Kogure, T. Sasaki, S. D. Wiramihardja, S. Mizuno, K. Sakka.
Astrophys. Space Sci., Vol. 119, No. 1, p. 131 – 133 (1986). – See Abstr. 012.001.
The authors have carried out surface photometry of six simple H II regions which are characterized by the existence of an exciting star earlier than B0, and a nearly round shape. On the basis of calibrated image data and adopting a spherical model, they derive the distributions of gas and dust densities in the H II regions.

132.006 **Observations of the Scutum ring of H II regions.**
T. Handa.
Astron. Her., Vol. 79, No. 2, p. 44 – 47 (1986). In Japanese.

132.007 **Hα velocity widths of giant extragalactic H II regions.**
J.–R. Roy, R. Arsenault, G. Joncas.
Astrophys. J., Vol. 300, No. 2, p. 624 – 638 (1986).
Integrated Hα line profiles of 47 giant extragalactic H II regions have been obtained with a Fabry–Perot spectrometer. The observed H II regions belong to 16 nearby late–type spirals and Magellanic irregulars. A relationship of the form $D \propto W^n$ is found between the linear diameters, D, of the H II regions and the velocity width, W, of the profile in excess of the thermal broadening. The power index n is found to be smaller than 1.6, which is inconsistent with virialized motions in a gravitationally bound system. The mean velocity width $\langle W \rangle$ of the three largest H II regions is strongly correlated with the absolute magnitude of the parent galaxy. Hydrodynamic turbulence is suggested as the most viable mechanism to explain the observed relationships between velocity dispersion, H II region diameter, and galaxian absolute magnitude.

132.008 **High angular resolution infrared mapping of the compact H II regions W51 and DR 21/W75.**
P. M. Harvey, M. Joy, D. F. Lester, B. A. Wilking.
Astrophys. J., Vol. 300, No. 2, p. 737 – 744 (1986).
Far–infrared maps of W51/G49.5–0.4, W51/G49.4–0.3, and the DR 21/W75 S regions are presented. The simultaneous 50 and 100 μm maps enable a determination of the dust temperature and density structure in these areas. The data show that there are large temperature and density gradients in the dust clouds associated with these H II regions. In addition, several new far–infrared

sources have been found for which photometry and luminosity estimates are presented.

132.009 **"Strömgren region" analysis of cometary H II regions.**
A. C. Raga.
Astrophys. J., Vol. 300, No. 2, p. 745 – 748 (1986).
The author has modeled an H II region in which the ionizing star is in motion with respect to the surrounding medium. He presents a simple method of comparing these models with observations and shows a comparison with observations of a "cometary" H II region by Reid and Ho.

132.010 **Observations of a high–excitation transition of SO in galactic H II regions.**
G. D. Watt, T. J. Millar, G. J. White, R. H. Harten.
Astron. Astrophys., Vol. 155, No. 2, p. 339 – 343 (1986).
High–excitation observations of the 5_6–4_5 transition of SO at 251 GHz have been made. The central four arcminutes of the Orion–KL region have been mapped and a survey of 6 other galactic sources has been performed. Detailed chemical kinetic models have been utilised to investigate the chemistry of sulphur in view of recent estimates of sulphur depletion and the possibility of a gas phase carbon to oxygen ratio greater than 1. The authors find a link between the SO/SO_2 and C/O ratios and that their data are consistent with the high C I abundance detected in the Orion ridge component. In the plateau source the enhanced abundances of SO and SO_2 may be caused by a molecular outflow from an oxygen–rich star. In addition a previously undetected methanol line and an unidentified line appear in the Orion data.

132.011 **The M17 complex in the light of the near infrared line [S III] λ9531.**
H. H. Hippelein, C. Goudis.
Astron. Astrophys., Vol. 155, No. 1, p. 6 – 10 (1986).
Measurements of profiles and intensities of the [S III] λ9531 line in M17 are presented. An extinction map is derived. Line splitting up to 30 km s^{-1} is observed in the central region of M17, caused by ionized gas sheets streaming away from the ionization fronts of the cone shaped H II region. Close to the interface region between ionized and molecular clouds, where A_v reaches 10 mag, internal extinction in the H II gas of ~1 mag is found. In the eastern arc–like filament a velocity component at -25 km s^{-1} with respect to the M17 complex is detected.

132.012 **Physical conditions in NGC 6164–5.**
E. L. Agüero, G. J. Carranza.
Astrophys. Space Sci., Vol. 120, No. 1, p. 17 – 25 (1986).
From spectroscopic observations of the emission lines Hβ, λ4959 and λ5007 [O III], Hα, λ6584 [N II], λ6717 and λ6731 [S II] and some interferometric data in Hα and [N II] it was established that NGC 6164 and NGC 6165 have characteristics of normal H II regions for Ne and Te. They are embedded in a very tenuous medium nearly coincident with the Strömgren sphere of the central star HD 148937, corresponding to the outer peripheral structure described by Westerlund. Variatons of the relative abundance of N, O, and S are discussed.

132.013 **H II region age indicators.**
M. V. F. Copetti, M. G. Pastoriza, H. A. Dottori.
Astron. Astrophys., Vol. 156, No. 1/2, p. 111 – 120 (1986).
The evolution of some properties of H II regions such as the relative volume R of the He^+ and H^+ zones, the equivalent width $W_{H\beta}$ of the H_β emission line and the ratio [O III]/H_β of the lines [O II] λλ4959, 5007 to H_β, was studied through models which take into account: a single burst for the formation of the ionizing association with different initial mass function ($1 \leqslant \chi \leqslant 3$) and upper stellar mass limits ($30 \leqslant M_u/M_\odot \leqslant 120$); different chemical composition models of stellar evolution with and without mass loss. It was found that $W_{H\beta}$, R and [O III]/H_β decrease monotonically as a function of time and consequently they are good H II region age indicators.

132.014 **Lyman line pumping and the modelling of H II regions.**
D. Hoang–Binh.
Astron. Astrophys., Vol. 156, No. 1/2, p. 393 – 394 (1986).
The author shows that Lyman line pumping (Hoang–Binh, 1983) reduces the maser action in radio recombination lines emitted by H II regions. This effect is equivalent to an increase of the electron density N_e without increasing the electron impact broadening. In the modelling of H II regions by means of radio line data, the variation of the parameter N_e could be significantly influenced by this pump mechanism.

132.015 **G70.68+1.20: a thermal source, not a supernova remnant.**
D. A. Green.
Mon. Not. R. Astron. Soc., Vol. 219, No. 3, p. 39P – 43P (1986).
G70.68+1.20 is a small shell source, ~20 arcsec in extent, which has recently been proposed as a very young galactic supernova remnant. 5–GHz observations with the 5–km telescope at a resolution of ~2×4 arcsec² (RA×Dec) are presented, and discussed together with other radio, optical and IR data. These favour the interpretation that the source is thermal, rather than a young supernova remnant.

132.016 **1.3 mm continuum observations of compact H II regions detected by IRAS. II.**
R. Chini, E. Kreysa, P. G. Mezger, H.–P. Gemünd.
Astron. Astrophys., Vol. 157, No. 1, p. L1 – L2 (1986).
The authors report on new 1.3 mm observations of 30 heavily obscured compact H II regions, which have also been detected by IRAS. From the observed energy distributions they derive IR luminosities, which imply spectral types between O4 and B1 for the embedded stars. The surrounding dust clouds have temperatures between 21 and 32K; their optical depths at 1.3 mm range from $(2-100)\times10^{-4}$. The observations provide a lower limit of 1.7 for the wavelength dependence of dust opacity in the submm range.

132.017 **A study of the dust distribution and extinction law in Mon R2.**
A. Natta, S. Beckwith, N. J. Evans II, S. C. Beck,
A. F. M. Moorwood, E. Oliva.
Astron. Astrophys., Vol. 158, No. 1/2, p. 143 – 148 (1986).
The authors have observed infrared hydrogen recombination lines with a number of different beam sizes toward the H II region Mon R2 IRS 1. Two distinctly different models with variations in the extinction are proposed.

132.018 **The chemical composition of NGC 2363.**
M. Peimbert, M. Peña, S. Torres–Peimbert.
Astron. Astrophys., Vol. 158, No. 1/2, p. 266 – 272 (1986).
Photoelectric spectrophotometry in the 3400 – 7400 Å is presented for the metal–poor extragalactic H II region NGC 2363. Based on these data, on IUE observations and on ionization structure models the authors have obtained the chemical abundances. The C/O ratio is a factor of two smaller than in the solar vicinity and under reasonable assumptions implies that most of the C in the solar vicinity is produced by intermediate mass stars. The N/C ratio is the same in NGC 2363 and in the Orion nebula, but the N/H ratio is 17 times smaller than in the Orion nebula. From line ratios involving five He emission lines it is found that $N(He)/N(H) = 0.077\pm0.006(3\sigma)$, which corresponds to $Y = 0.235\pm0.013(3\sigma)$.

132.019 **Decameter wavelength observations of the Rosette Nebula and the Monoceros Loop supernova remnant.**
N. Odegard.
Astrophys. J., Vol. 301, No. 2, p. 813 – 824 (1986).
Observations are presented at 20.6, 25.6, and 30.9 MHz of a field containing the Rosette Nebula and the Monoceros Loop supernova remnant. From the maps presented here, it appears that radiation from the supernova remnant (SNR) is being absorbed by the Rosette Nebula, suggesting that the SNR is located behind the H II region but still within the Mon OB2 association, at a distance of ~1.6 kpc. An electron temperature on the order of 8000K is obtained for the diffuse ionized gas in the Monoceros

Loop, and it is suggested that this gas is photoionized by members of the Mon OB2 association. A mean electron temperature of 8700±1300K is derived for the Rosette Nebula.

132.020 **Detection of [Si II] (34.8 micron) emission in Orion–KL: a measurement of the silicon abundance in dense interstellar gas.**
M. R. Haas, D. J. Hollenbach, E. F. Erickson.
Astrophys. J., Lett. Ed., Vol. 301, No. 2, p. L57 – L60 (1986).
The authors report the first detection of the ground–state fine–structure transition of Si^+ at a rest wavelength determined to be 34.815±0.004 μm. A 6'NW–SE strip scan across the Orion–KL region shows [Si II] emission from both the extended photodissocation region surrounding θ^1Ori C and from the shocked gas NW of BN–KL. The inferred gas–phase silicon elemental abundance relative to hydrogen in the dense ($\sim 10^5 cm^{-3}$) primarily neutral photodissociation region is approximately $2.6x10^{-6}$, a factor of 0.075 times the solar value and 3.4 times greater than in the moderate density ($\lesssim 10^3 cm^{-3}$) cloud toward ζ Oph. The silicon abundance in the shocked gas is approximately solar.

132.021 **Hollow H II regions. I. Effects of stellar evolution and wind energy dissipation.**
H. Dorland, T. Montmerle, C. Doom.
Astron. Astrophys., Vol. 160, No. 1, p. 1 – 17 (1986).
Extended H II regions often exhibit a hollow structure: these bright nebulae consist of thick (e.g., Rosette nebula, Carina nebula), or thin (e.g. NGC 6888) ionized shells with an inner cavity and surrounded by neutral material. The exciting stars belong to OB associations; they are massive (up to several 10 $M_\odot$), and characterized by strong stellar winds. However, current models, in particular the classical "hot interstellar bubble", fail to account for the structure of these nebulae. The observations in fact suggest that the shock at the edge of the wind cavity is not adiabatic, like in these models, but must on the contrary be strongly dissipative. To solve this problem, the authors present in this work models based on two new developments. They apply their model to a number of definite objects: hollow H II regions like the Carina and Rosette nebulae, as well as thin "W–nebulae" around WR stars. The results show a good agreement with a variety of widely differing and independent observational constraints.

132.022 **Observations of the giant H II region complex W51 at decameter wavelengths.**
A. A. Deshpande, C. V. Sastry.
Astron. Astrophys., Vol. 160, No. 1, p. 129 – 131 (1986).
Radio continuum absorption observations at 34.5 MHz with a spatial resolution of 34'x48' have been obtained for the giant H II region complex W51. A mean electron temperature of 9500±600K and a background temperature of 28,000±2600K have been derived. The proportion of non–thermal emission originating on the near side of W51 corresponds to a mean emissivity of 2K pc^{-1}. The present observations also indicate that the ring towards the east of the nebula (W51C) is non–thermal.

132.023 **The velocity structure of the core of the giant H II region NGC 3603.**
C. A. Clayton.
Mon. Not. R. Astron. Soc., Vol. 219, No. 4, p. 895 – 902 (1986).
Hα and [N II] line profiles have been obtained along two lines across the core of the giant H II region NGC 3603. The main features detected and their likely origins are as follows: (1) Several continuous components are visible across the entire region observed, with peak–to–peak velocity differences of up to 150 km s^{-1}. These large–scale motions were most probably set up before the formation of the central cluster, possibly by the action of earlier supernovae. (2) Two smaller shell–like structures have also been detected and one of these is visible on photographs of NGC 3603. These shells are believed to be wind–driven bubbles of order 10^4yr old. The larger bubble may be driven by a Wolf–Rayet star.

132.024 **Observations of the ring nebula NGC 7635 around the star BD +60°2522.**
A. I. Lomovskij, A. Yu. Klement'eva.
Astron. Zh., Tom 63, Vyp. 2, p. 246 – 254 (1986). In Russian. English translation in Sov. Astron., Vol. 30, No. 2.
The distribution of the brightness of the monochromatic emission of the nebula NGC 7635 in the [N II], [S II] and [O III] lines is studied. The monochromatic photographs of the nebula and the isophote systems in the [N II] and [O III] lines in absolute units of intensity, in the [S II] line in relative units of intensity, are presented. The nature of the nebula is discussed.

132.025 **Les gaz ionisés dans notre galaxie (Revue des régions H II, He II, C II, S II interstellaires).**
J. P. Vallée.
J. R. Astron. Soc. Can., Vol. 80, No. 1, p. 16 – 50 (1986).
Nature has created a large variety of types of ionised regions (H II, He II, C II, S II) in the interstellar medium: cocoon and small shell, jet and sphere, champagne and blister, shell and wind, bubble and half–bubble, complexes and diffuse interstellar medium. For each type of region, this review examines the observations at radio wavelengths concerning: form and distribution, dynamics and recombination, energy source and environment, and physical modelling and theories.

132.026 **The Rosette nebula. III. Interstellar dust extinction and a model of the molecular cloud complex.**
W. E. Celnik.
Astron. Astrophys., Vol. 160, No. 2, p. 287 – 294 (1986).
Hα emission line and radio continuum intensities of the Rosette nebula have been compared and the distribution of interstellar dust in front of the H II region is obtained. From the distribution of dust grains and of CO molecules, a detailed model of the Rosette nebula molecular cloud complex can be established in three dimensions, including a reasonable velocity field. The mass of the complex is also estimated.

132.027 **Turbulent gas motions in giant H II regions. II. The luminosity–velocity dispersion relation.**
H. H. Hippelein.
Astron. Astrophys., Vol. 160, No. 2, p. 374 – 384 (1986).
The velocity dispersion for 43 giant extragalactic H II regions has been measured in Hα and in [O III] λ5007 in order to reinvestigate the relation between luminosity $L(H\alpha)$ and velocity dispersion σ. A correlation $L(H\alpha) \propto \sigma(H\alpha)^{6.6}$ is indicated. The velocity dispersions derived from the [O III] line obey a similar relation but are somewhat smaller. The slope of both relations is steeper than expected for gravitationally bound systems. Several contributors to the kinematical state of giant H II regions have to be considered. Stellar winds seem to play the dominant role for the objects with $10^{38} < L(H\alpha) < 10^{40}$erg s^{-1}.

132.028 **On the connection of H II regions with molecular clouds of the Galaxy.**
L. V. Mirzoyan, V. V. Ambaryan.
Astrofizika, Tom 24, Vyp. 2, p. 257 – 264 (1986). In Russian. English translation in Astrophysics, Vol. 24, No. 2.
The results of radio observations of optical H II regions in the CO line ($J = 1 - 0$) are discussed. Arguments are given which testify the genetic connection of H II regions with molecular clouds. It has been shown that the absence of the associated CO emission in some H II regions can be explained as a consequence of either observational selection (low surface brightness) or the fact that the corresponding H II regions are excited by OB stars which have already left their "placental" molecular clouds. The genetic connection of optical H II regions with molecular clouds can be considered as an evidence in favour of the idea that the molecular clouds are components of stellar associations.

132.029 **The M8 Hourglass core: a region of recent star formation?**
C. E. Woodward, J. L. Pipher, H. L. Helfer, W. J. Forrest.
Bull. Am. Astron. Soc., Vol. 17, No. 4, p. 837 (1985). Abstract. – See Abstr. 010.062.

132.030 **The fine velocity structure at the center of the Orion Nebula.**
H. O. Castañeda.
Bull. Am. Astron. Soc., Vol. 17, No. 4, p. 891 (1985). Abstract. – See Abstr. 010.062.

132.031 **Turbulent motion in galactic H II regions.**
C. R. O'Dell.
Bull. Am. Astron. Soc., Vol. 17, No. 4, p. 891 (1985). Abstract. – See Abstr. 010.062.

132.032 **Near IR observations of the Sharpless regions S 269, S 271, S 307, and S 311.**
K. Shivanandan, P. Persi, M. Ferrari–Toniolo, L. Springolio.
Bull. Am. Astron. Soc., Vol. 17, No. 4, p. 892 (1985). Abstract. – See Abstr. 010.062.

132.033 **Equidensitometry of the Trifid nebula (M20).**
I. A. Dubyago, L. A. Urasin.
Izv. Astron. Ehngel'gardt. Obs., No. 51, p. 58 – 59 (1985). In Russian.

132.034 **Pregraphitic molecules and the Red–Rectangle emission.**
L. d'Hendecourt, A. Léger.
NASA Conf. Publ., NASA CP–2403, p. A 6 (1986). – See Abstr. 012.030.

132.035 **Surface photometry of some small H II regions.**
S. Mizuno, K. Sakka, T. Sasaki, T. Kogure, M. Nakano, S. D. Wiramihardja.
Proceedings of the Second Asian–Pacific Regional Meeting on Astronomy, p. 496 – 503 (1984). – See Abstr. 012.031.

The surface photometry of H II regions S 237 and S 254 – S 257 has been carried out by means of the image processing technique. The calibrated maps in absolute brightness are given in both Hα emission and V–continuum. Central concentration of both lights is noticed in S 237, while not in S 254 – S 257. Some physical parameters are derived with the aid of spectral data.

132.036 **Nebulosas de gas ionizado: las regiones H II.**
P. Planesas Bigas.
Anuario del Observatorio Astronómico de Madrid para 1986, p. 357 – 416 (1985). – See Abstr. 046.012.

132.037 **Multiband far–infrared observations of the NGC 6334 complex.**
L. Loughran, B. McBreen, G. G. Fazio, T. N. Rengarajan, C. W. Maxson, S. Serio, S. Sciortino, T. P. Ray.
Astrophys. J., Vol. 303, No. 2, p. 629 – 637 (1986).

The authors present multiband far–infrared (20 – 155 μm) observations of the large H II region/molecular cloud complex NGC 6334. The observations were performed simultaneously in four bands with effective wavelengths 21, 42, 71, and 134 μm. Eight far–infrared sources were detected which cover the range from sources with little or no continuum radio emission to well–developed H II regions. Maps of dust temperature and far–infrared optical depth of the complex are presented.

132.038 **Observations of the 157.7 micron [C II] emission from the galactic H II regions W3 and W51.**
G. Melnick, G. J. Stacey, P. J. Viscuso, C. E. Fuller.
Astrophys. J., Vol. 303, No. 2, p. 638 – 648 (1986).

The authors have detected the 157.7 μm ground–state fine–structure transition of singly ionized carbon in the galactic H II regions W3 and W51. These observations show that the 157.7 μm [C II] emission is extended beyond both the compact H II regions and the carbon recombination region in each source. These observations establish a lower limit to the luminosity in the 157.7 μm [C II] line of 360 $L_\odot$ from W3, assuming a distance of 2 kpc, and 410 $L_\odot$ from W51, assuming a distance of 7 kpc. It is concluded that much of the observed emission originates in the thin carbon recombination line–emitting region bordering on the H II regions and the warm gas at the surface of the adjacent molecular clouds.

132.039 **Compact radio sources associated with molecular outflows.**
R. L. Snell, J. Bally.
Astrophys. J., Vol. 303, No. 2, p. 683 – 701 (1986).

The authors report the detection of compact radio continuum sources from 12 regions containing molecular outflows. All infrared sources associated with these outflows having a bolometric luminosity in excess of 100 $L_\odot$ were detected; in addition, three infrared sources with bolometric luminosities under 100 $L_\odot$ were also seen. If the radio emission is formed by a steady, fully ionized wind, the implied mass loss parameters fail by one or two orders of magnitude to explain the energetics of the associated molecular outflows. Detailed, high–resolution radio maps were obtained toward two clusters of luminous ($L_{bol} > 10^3 L_\odot$) infrared sources near NGC 2071 and S255. The authors also present 1.4 GHz maps of the extended H II regions S254, S255, S257, and S187, which are located near molecular outflows.

132.040 **Observations of far–infrared line profiles in the Orion–KL region.**
M. K. Crawford, J. B. Lugten, W. Fitelson, R. Genzel, G. Melnick.
Astrophys. J., Lett. Ed., Vol. 303, No. 2, p. L57 – L60 (1986).

The authors report measurements of several far–infrared emission–line profiles in the Orion–KL region. The data were taken with a new high–resolution Fabry–Perot spectrometer ($\lambda/\Delta\lambda \approx 3x10^4$) on the Kuiper Airborne Observatory. Toward the BN–KL and H_2 peak 1 positions, the observed CO, OH, and [O I] emission lines (CO $J = 16 \rightarrow 15$ at 163 μm; OH $^2\Pi_{1/2}$, $J = 3/2 \rightarrow 1/2$ at 163 μm; and [O I] $^3P_1 \rightarrow {}^3P_2$ at 63 μm) are well resolved with intrinsic line widths of 30 – 50 km s^{-1}. The emission probably comes from dense, hot molecular and atomic gas in the Orion–KL shock. The authors have also spectrally resolved the [O I] 63 μm and [C II] $^2P_{3/2} \rightarrow {}^2P_{1/2}$ 158 μm emission from UV photodissociated gas near Orion–Trapezium.

132.041 **Turbulent motion in galactic H II regions.**
C. R. O'Dell.
Astrophys. J., Vol. 304, No. 2, p. 767 – 770 (1986).

An investigation of radial velocities across the face of NGC 1499, NGC 7000, and S252 reveals that the fine–scale (turbulent) motions are well ordered from the limit of spatial resolution to the full size of the nebulae. As in higher spatial resolution studies of S142 and M17, analysis of the dispersion relation reveals that there is agreement with Kolmogorov theory. However, the structure function increases with separation much more slowly than theoretically expected, calling into question the application of Kolmogorov theory.

132.042 **Detection and study of two H II blobs in the LMC giant H II region N160 and investigation of their nebular and stellar environment.**
M. Heydari–Malayeri, G. Testor.
ESO Sci. Prepr., No. 413, 52 pp. (1986). To appear in Astron. Astrophys.

132.043 **Star formation, giant H II regions and spiral structure.**
M. Rosa.
ESO Sci. Prepr., No. 427, 10 pp. (1986). To appear in 'Highlights of astronomy', Vol. 7.

132.044 **Modelling for O and N infrared line observations of H II blisters.**
P. A. Aannestad, R. J. Emery.
Astron. Astrophys., Vol. 161, No. 2, p. 347 – 356 (1986).

The use of far–infrared line intensities for determining ionic abundances of O and N and determining the effective temperature of the ionizing star has been investigated by model calculations for H II blisters with an exponential density gradient. The ionizing radiation corresponds to line–blanketed model atmospheres (Kurucz, 1979) with effective temperatures of 35,000K or 50,000K. The effects of charge exchange and of absorption by O and N on the ionizing radiation, as well as the effect of nebular dust are included.

132.045 **Optical, radio, and infrared observations of compact H II regions. V. The Hourglass in M8.**
C. E. Woodward, J. L. Pipher, H. L. Helfer, S. Sharpless, A. Moneti, D. Kozikowski, M. Oliveri, S. P. Willner, M. G. Lacasse, T. Herter.
Astron. J., Vol. 91, No. 4, p. 870 – 889 (1986). With plates 74, 75.
Multiwavelength observations of the inner core of the M8 Hourglass region are presented, including VLA interferometric maps, 2 – 4 μm and 8 – 13 μm spectroscopy, photometric mapping in the K (2.2 μm) and L (3.45 μm) bands and in the 3.28 μm dust–emission feature, optical CCD imaging, and optical and infrared polarimetry. The observations favor models in which feature emission comes from small refractory grains rather than from fluorescence or thermal emission from volatile mantles.

132.046 **The association of neutral hydrogen with H II regions and radio sources.**
Y. K. Minn.
J. Korean Astron. Soc., Vol. 18, No. 2, p. 100 – 110 (1985).
The H I features associated with H II regions and radio sources in the galactic plane are searched in the Maryland–Green Bank Galactic 21–cm Line Survey. Among the twenty–eight such objects, twenty–five show H I depression features. The depth of the H I depression is proportional to the radio continuum brightness temperature. The angular dimensions of the H I feature and radio source are comparable.

132.047 **The discovery of high–speed ionized knots in the H II regions M16 (NGC 6611) and the Rosette nebula (NGC 2237–2246).**
J. Meaburn, J. R. Walsh.
Mon. Not. R. Astron. Soc., Vol. 220, No. 3, p. 745 – 757 (1986).
Filter photographs have revealed bright ionized knots in the cores of the galactic H II regions M16 and the Rosette nebula. Both low– and very high–dispersion spectra of these have now shown that several have high–speed components which are bright in the emission lines from lowly ionized species. In the Rosette nebula it is proposed that they could be Herbig–Haro like objects driven by newly formed stars embedded in the approaching side of an expanding shell. Alternatively, they could be bow shocks around globules overrun by this shell and exposed to the powerful wind from the central OB association or collimated flows in the channels between adjacent globules but still driven by the central wind.

132.048 **Decameter observations of H II regions in the galactic disk, 147° < l < 153°.**
E. A. Abramenkov.
Sov. Astron., Vol. 29, No. 6, p. 616 – 620 (1985). English translation of 40.132.045.

132.049 **Formation of OB clusters: radiation–driven implosion?**
P. T. P. Ho, R. I. Klein, A. D. Haschick.
Astrophys. J., Vol. 305, No. 2, p. 714 – 720 (1986).
New high dynamic range (1000 – 2000) radio continuum maps were obtained for two regions of vigorous star formation, G 10.6–0.4 and W33. Ionized gas due to an earlier epoch of star formation appears to flow around dense neutral condensations. Bright rims are found facing toward the central continuum core where the most recent episode of star formation is taking place. This supports a model where radiation–driven ionization shock fronts from first–generation OB stars lead to the efficient implosion of nearby neutral condensations and the formation of second generation stars.

132.050 **Grain heating in H II regions.**
S. de Oliveira, W. J. Maciel.
Rev. Mex. Astron. Astrofis., Vol. 12, Spec. Issue, p. 257 – 263 (1986). – See Abstr. 012.066.
The heating of H II regions by electrons ejected from grains after absorption of stellar and diffuse photons is considered. This mechanism is compared with photoionization heating, taking into account central stars in different positions on the HR diagram.

132.051 **On the nature of the unusual nebula PL1547.3–5612.**
M. T. Ruiz.
Rev. Mex. Astron. Astrofis., Vol. 12, Spec. Issue, p. 269 – 272 (1986). – See Abstr. 012.066.
In an attempt to understand the nature of PL1547.3–5612, a comparison is made between the observed characteristics of this nebula and the several different hypotheses for its origin.

132.052 **Age determination of galactic H II regions.**
M. V. F. Copetti, H. A. Dottori, M. G. Pastoriza, J. Melnick.
Rev. Mex. Astron. Astrofis., Vol. 12, Spec. Issue, p. 283 – 284 (1986). – See Abstr. 012.066.

132.053 **Radio and infrared observations of the faint nebula GM24.**
L. F. Rodríguez, M. Roth, M. Tapia, J. Cantó, P. Persi, M. Ferrari–Toniolo.
Rev. Mex. Astron. Astrofis. Vol. 12, Spec. Issue, p. 285 – 286 (1986). Abstract. – See Abstr. 012.066.

132.054 **Effect of the dust on relative [O II] and [S II] intensities in H II regions.**
M. E. Mendes, J. R. D. Lépine.
Rev. Mex. Astron. Astrofis., Vol. 12, Spec. Issue, p. 296 (1986). Abstract. – See Abstr. 012.066.

132.055 **Determination of temperature in gaseous nebulae.**
S. M. V. Aldrovandi, R. B. Gruenwald.
Rev. Mex. Astron. Astrofis., Vol. 12, Spec. Issue, p. 300 (1986). Abstract. – See Abstr. 012.066.

132.056 **Abundance determination from H II regions in NGC 2403.**
J. Fierro, S. Torres–Peimbert, M. Peimbert.
Rev. Mex. Astron. Astrofis., Vol. 12, Spec. Issue, p. 308 (1986). Abstract. – See Abstr. 012.066.

132.057 **Giant stellar–wind shell associated with the H II region M16.**
Y. Sofue, T. Handa, E. Fürst, W. Reich, P. Reich.
Publ. Astron. Soc. Jpn., Vol. 38, No. 3, p. 347 – 360 (1986).
Radio continuum observations at 2.7 and 10 GHz of the star forming complex M16 revealed a large loop of diameter 1°2 (60 pc) with a flat (thermal) spectrum. The loop is likely to be a shell of H II gas which expands from M16 toward the south. Optical (Hα) filaments are found associated with this loop. An H I shell of diameter 70 pc is found surrounding continuum shell, and is expanding at a velocity of 4 km s^{-1}. A giant molecular cloud is located in contact with the northern edge of the shell. A stellar–wind–bubble model is proposed for the formation of the shell.

132.058 **The Scutum ring of H II regions, G24.6 + 0.0, associated with CO and H I shells.**
T. Handa, Y. Sofue, W. Reich, E. Fürst, I. Suwa, Y. Fukui.
Publ. Astron. Soc. Jpn., Vol. 38, No. 3, p. 361 – 378 (1986).
Radio observations of the ring complex G24.6+0.0 or the Scutum ring were made at 10, 5, and 2.7 GHz in continuum and at 115.27 GHz in the CO (J = 1–0) line emission. The complex is composed of several compact H II regions associated with a diffuse extended radio emission. The spectrum of the diffuse component is flat, indicating its thermal origin. CO observations have revealed association of an arc–shaped molecular cloud of mass $3 \times 10^5 M_\odot$ aligned concentric to the H II ring having a sharp inner edge, most likely being shock–compressed from inside. This molecular arc is further surrounded by a larger arc of H I gas of radius 70 pc and a mass of $2 \times 10^4 M_\odot$. The positional relationship of the H II, H_2 (CO), and H I gases suggests a typical sequential star formation on a shock–compressed CO shell.

132.059 **The luminous stellar content of 30 Doradus and NGC 3603 – the nearest visible giant H II regions.**
A. F. J. Moffat, M. M. Shara, W. Seggewiss.
Luminous stars and associations in galaxies, p. 233 – 234 (1986). – See Abstr. 012.079. (IAU Symp. No. 116).
30 Dor in the LMC and NGC 3603 in the Galaxy are the nearest visible examples of similar giant H II regions. Wolf–Rayet stars were detected there using CCD image subtraction. The authors confirm that the number ratio of WR to O stars in 30 Dor and NGC 3603 is normal ($\sim$0.05) compared to other less exotic regions.

132.060 **The exciting stars of giant extragalactic H II regions.**
M. Rosa, S. D'Odorico.
Luminous stars and associations in galaxies, p. 355 – 367 (1986). – See Abstr. 012.079. (IAU Symp. No. 116).
Recent investigations of the stellar content of giant H II regions and star formation bursts in nearby galaxies are summarized. The preliminary results of a spectroscopic survey of 78 H II regions in 15 galaxies are presented.

132.061 **Sub–mm continuum observations of sources from the IRAS point–source catalogue.**
R. Chini, E. Kreysa, E. Krügel, P. G. Mezger, H.–P. Gemünd.
Light on dark matter, p. 29 – 30 (1986). – See Abstr. 012.080.
More than 120 galactic and extragalactic objects, taken from the IRAS point–source catalogue have been observed at 350 and 1300 μm. Supplementary near IR data and extensive model fits over this large spectral range allow the determination of the total luminosity, dust temperatures, the wavelength dependence of dust opacity, the total amount of extinction towards embedded stellar objects and the density distribution of dust around central heating sources within H II regions and dust clouds.

132.062 **Models for IRAS observations of galactic H II regions.**
J. Crawford, M. Rowan–Robinson.
Light on dark matter, p. 303 – 304 (1986). – See Abstr. 012.080.
The authors have investigated the transfer of radiation through a spherically symmetric distribution of dust surrounding a centrally located hot star. Comparison with observation is made via IRAS colour–colour diagrams for compact H II regions.

132.063 **Groundbased infrared observations of Magellanic Cloud H II regions.**
J. Koornneef, F. P. Israel.
Light on dark matter, p. 399 – 400 (1986). – See Abstr. 012.080.
The authors have obtained near–infrared JHKL'(M) photometry of some fifteen small ($\leqslant$10″) H II regions and H II regions knots in both the LMC and the SMC. The IRAS observations indicate the presence of large amounts of additional cool dust.

A statistical analysis and overview of the IRAS point source catalog.
See Abstr. 002.092.

A statistical analysis of the LRS catalog.
See Abstr. 002.093.

Atomic data for the S IV 10.5–μm fine–structure transition.
See Abstr. 022.054.

A fast and reliable method for computing free–bound emission coefficients for hydrogenic ions.
See Abstr. 022.088.

Atomic structure calculations and nebular diagnostics.
See Abstr. 022.100.

Far–infrared (100 – 200 μm) photometry of H II regions with a 1 m balloon borne telescope.
See Abstr. 035.100.

On photoproduction of ultraviolet emission lines.
See Abstr. 063.104.

Konsistente sphärische NLTE–Modelle für Objekte mit Leuchtkräften entsprechend B–Sternen.
See Abstr. 064.028.

How to form a 200 $M_\odot$ star.
See Abstr. 065.165.

Masers and circumstellar flows.
See Abstr. 112.011.

High–resolution infrared spectroscopy: a search for the 11.52–μm graphite feature.
See Abstr. 112.055.

Infrared recombination line profiles: a diagnostic probe of the velocity field in the S 106 stellar wind.
See Abstr. 112.113.

Ultraviolet and visual wavelength spectroscopy of gas around Eta Carinae.
See Abstr. 112.116.

An O3 star in the Small Magellanic Cloud H II region NGC 346.
See Abstr. 114.102.

Optical and ultraviolet spectral morphology of luminous OB stars in the Galaxy and the Magellanic Clouds.
See Abstr. 114.143.

Optical, UV and X–ray observations of X–ray binaries.
See Abstr. 117.130.

The double radio source associated with L1551 IRS 5: binary system or ionized circumstellar torus?
See Abstr. 121.016.

Kinematic detection of supernova remnants in giant H II regions.
See Abstr. 125.045.

The unique supernova (1985f) in NGC 4618.
See Abstr. 125.282.

Far–infrared observations of Sagittarius B2: reconsideration of source structure.
See Abstr. 131.001.

The rotational emission–line spectrum of Orion A between 247 and 263 GHz.
See Abstr. 131.006.

The injection of energy into the interstellar medium by stars.
See Abstr. 131.019.

High–resolution observations of the Orion bright bar.
See Abstr. 131.020.

The Scutum ring of H II regions.
See Abstr. 131.029.

Echelle observations of high speed flows in the interstellar medium.
See Abstr. 131.051.

Principal heating sources of dust in the galactic disk.
See Abstr. 131.056.

MERLIN observations of OH outflow in W75N.
See Abstr. 131.058.

ON2: a complex star formation region.
See Abstr. 131.062.

A study of bipolar and compact nebulae at radio, submillimetre and infrared wavelengths.
See Abstr. 131.068.

Ammonia observations and temperatures in the S 140/L 1204 molecular cloud.
See Abstr. 131.094.

Molecular clouds and star formation in the inner Galaxy: a comparison of CO, H II, and far–infrared surveys.
See Abstr. 131.105.

Excitation in the ionized diffuse interstellar medium.
See Abstr. 131.111.

Nebular complex in the region of Cas OB2 association; ring nebula Sh 157.
See Abstr. 131.112.

Ionization of the interstellar medium in the vicinity of non–stationary ionizing radiation sources.
See Abstr. 131.129.

High resolution maps of bipolar molecular outflow regions: results from the OVRO millimeter wave array.
See Abstr. 131.164.

A giant molecular cloud associated with the η Carina nebula.
See Abstr. 131.168.

A study of the ground state hydroxyl maser emissions associated with eleven regions of star formation.
See Abstr. 131.181.

An extremely luminous bipolar H_2 flow in the DR21 star–forming region.
See Abstr. 131.184.

Comparison of 2.1 and 3.8 micron line profiles of shocked H_2 in the Orion molecular cloud.
See Abstr. 131.188.

Submillimeter wavelength molecular spectroscopy of the Orion molecular cloud.
See Abstr. 131.189.

The infrared emission bands. I. Correlation studies and the dependence on C/O ratio.
See Abstr. 131.190.

Star formation regions as galactic dissipative structures.
See Abstr. 131.194.

Non–LTE process in the interstellar medium.
See Abstr. 131.199.

The stars NGC 7538–IRS 5, 6, and 7, and a distance estimate to the nebula.
See Abstr. 131.206.

The dust and gas surrounding LkHα 101.
See Abstr. 131.211.

On the statistics of galactic H I clouds.
See Abstr. 131.241.

Multifrequency observations of the region associated with the cometary nebula GM24.
See Abstr. 131.245.

^{13}CO in Orion.
See Abstr. 131.252.

H_2 spectroscopy as an agent for extinction determinations: the near–infrared curve for the Orion molecular cloud.
See Abstr. 131.277.

Molecular clouds associated with compact H II regions. III. Spin–up and collapse in the core of G 10.6–0.4.
See Abstr. 131.278.

Submillimeter observations of CS in M17.
See Abstr. 131.283.

Protostellar condensations within the Orion ridge.
See Abstr. 131.285.

The structure of the W49 A molecular cloud complex: burst of star formation in the $10^5 M_\odot$ core.
See Abstr. 131.306.

On the IMF of violent star forming regions.
See Abstr. 131.351.

Mid–IR emission of the interstellar medium.
See Abstr. 131.367.

The IR emission of the Orion–Monoceros molecular clouds.
See Abstr. 131.380.

Water masers coincident with IRAS sources.
See Abstr. 131.386.

The association of clouds in the Carina arm with IRAS spline maps.
See Abstr. 131.389.

Infrared studies of southern AFGL sources. II. GL4176 and GL4182: two sources in star forming regions.
See Abstr. 133.004.

The 4 μm spectra of compact infrared sources.
See Abstr. 133.005.

Chemical abundances of planetary nebulae in M31.
See Abstr. 134.044.

Planetary nebulae and the galactic temperature gradient.
See Abstr. 134.060.

Multifrequency radio continuum observations of extended galactic objects. I. Nine objects from the 2695 MHz Effelsberg galactic plane survey.
See Abstr. 141.004.

WSRT observations of radio sources in the galactic plane near $l = 54°$.
See Abstr. 141.016.

22 GHz map of the Carina region.
See Abstr. 141.022.

NGC 2070 and NGC 3603.
See Abstr. 152.012.

Dynamics of the ionized gas within 2 pc of the galactic center. I. First results of a radio recombination line survey.
See Abstr. 155.033.

Abundance gradient in interstellar medium.
See Abstr. 155.070.

On the large scale structure in the Cep–Cas region.
See Abstr. 155.073.

Extinction and reddening of H II regions in the Large Magellanic Cloud.
See Abstr. 156.005.

Star formation in the Magellanic Cloud. III. IR observations of giant H II regions.
See Abstr. 156.009.

Age determination of Magellanic Clouds and isolated extragalactic H II regions.
See Abstr. 156.030.

Gas in early-type galaxies.
See Abstr. 157.024.

H II regions in M33. II. Radio continuum survey.
See Abstr. 157.073.

The Hα velocity widths of giant H II regions as distance indicators.
See Abstr. 157.128.

A feature of the spiral pattern and star formation in the galaxy NGC 1566.
See Abstr. 157.133.

Kinematics of the ionized hydrogen in the irregular galaxy NGC 4449.
See Abstr. 157.233.

Photometric measurements of WH_β and [O III]/H_β of H II regions in M83.
See Abstr. 157.267.

Surveys for H II galaxies and related star forming systems.
See Abstr. 157.274.

A search for Wolf–Rayet stars in giant extragalactic bursts of star formation.
See Abstr. 157.319.

Giant H II regions in M81.
See Abstr. 157.320.

Mid infrared spectroscopy of IRAS bright galaxies.
See Abstr. 157.325.

H II region spectroscopy and two dimensional stellar photometry of the barred Seyfert NGC 1566.
See Abstr. 158.048.

Star formation in NGC 5128.
See Abstr. 158.322.

Giant H II regions as distance indicators I: relations between global parameters for the local calibrators.
See Abstr. 161.217.

133 Infrared Sources

133.001 **Three micron spectroscopy of the galactic centre source IRS 7.**
I. Butchart, A. D. McFadzean, D. C. B. Whittet, T. R. Geballe, J. M. Greenberg.
Astron. Astrophys., Vol. 154, No. 1/2, p. L5 – L7 (1986).

A new 3 micron band spectrum of the infrared source IRS 7 in Sagittarius A has been obtained. The profiles of the two broad absorption features, centred at 3.0 and 3.4 microns, are well characterized in this spectrum. Of the narrow absorption features previously reported, only that at 3.48 microns is detected. No other narrow absorption features with equivalent widths greater than 0.001 microns are resolved. The spectrum provides a firm basis for models of interstellar grains based on extinction curves in this spectral region.

133.002 **Infrared reflection nebulae.**
T. Yamashita.
Astron. Her., Vol. 79, No. 3, p. 66 – 70 (1986). In Japanese.

133.003 **IRAS 18333–2357: an unusual source in M22.**
F. C. Gillett, G. Neugebauer, J. P. Emerson, W. L. Rice.
Astrophys. J., Vol. 300, No. 2, p. 722 – 728 (1986).

IRAS 18333–2357 is a pointlike source located about 1′ from the center of the globular cluster M22 (NGC 6566). Its infrared colors are cooler than virtually all OH/IR stars and hotter than almost all galaxies. If the source is physically associated with M22, its infrared luminosity is 700 $L_\odot$ and the dust mass in the radiating region is in the range 6×10^{-4} to $1\times10^{-2} M_\odot$. It is speculated that IRAS 18333–2357 may be in a very early phase of planetary nebulae evolution. One possible model for the source is that of a rapidly evolving bipolar nebula, similar to OH 0739–14, with carbon abundance enriched by more than 20 times compared to abundances in M22 as a whole.

133.004 **Infrared studies of southern AFGL sources. II. GL4176 and GL4182: two sources in star forming regions.**
P. Persi, M. Ferrari–Toniolo, L. Spinoglio.
Astron. Astrophys., Vol. 157, No. 1, p. 29 – 34 (1986).

An infrared study of two southern AFGL sources, GL4176 and GL4182, including new photometry from 1.25 to 20 μm and CVF spectra between 2 – 4 μm and 8 – 13 μm, is presented. Several IR sources are found in the region around GL4182. Irs 1 is the reddest and brightest source in the field. The data are consistent with an earlier suggestion that the object is a compact H II region.

133.005 **The 4 μm spectra of compact infrared sources.**
R. Hofmann, H. P. Larson, U. Fink.
Astron. Astrophys., Vol. 160, No. 1, p. 18 – 24 (1986).

The authors present high resolution spectra in the 4 μm region of three very luminous ($L > 10^6 L_\odot$) compact near infrared sources: K3–50, W51–IRS 2 East, and G333.6–0.2. These data represent about an order of magnitude increase in resolving power over previous observations of these sources in this wavelength region. The authors use the observed intensity ratios of the Brα, Pfβ, and Huε lines in connection with infrared and radio maps to discriminate between different line excitation mechanisms.

133.006 **Infrared polarimetry of highly polarized multiple sources in W 33, Cep A, and S 255.**
P. A. Heckert.
Bull. Am. Astron. Soc., Vol. 17, No. 4, p. 835 (1985). Abstract. – See Abstr. 010.062.

133.007 **An unusual high velocity molecular outflow in the Rho Ophiuchi cloud.**
C. K. Walker, C. J. Lada, E. T. Young, M. Margulis, B. A. Wilking.
Bull. Am. Astron. Soc., Vol. 17, No. 4, p. 835 (1985). Abstract. – See Abstr. 010.062.

133.008 **The identification of IRAS sources.**
W. P. Bidelman.
Bull. Am. Astron. Soc., Vol. 17, No. 4, p. 841 – 842 (1985). Abstract. – See Abstr. 010.062.

133.009 **A search for brown dwarf candidates among the IRAS serendipitous sources.**
R. M. Cutri, F. J. Low, E. T. Young, S. G. Kleinmann, F. C. Gillett.
Bull. Am. Astron. Soc., Vol. 17, No. 4, p. 878 – 879 (1985). Abstract. – See Abstr. 010.062.

133.010 **CO observations of sources from IRAS Circular No. 9.**
R. Arquilla, D. A. Leahy, S. Kwok.
Mon. Not. R. Astron. Soc., Vol. 220, No. 1, p. 125 – 131 (1986).
The authors report observations of 14 sources from the IRAS Circular No. 9 in the $J = 1 \rightarrow 0$ line of ^{12}CO. They detect wide lines in two sources and identify these objects as previously unknown carbon stars. Strong ^{12}CO lines are also observed towards two additional sources near which H_2O masers have previously been detected. These sources mark sites where star formation has recently occurred.

133.011 **Photographic observation of AFGL sources in Cygnus.**
S. D. Wiramihardja, H. Ohtani, K. Takayanagi.
Proceedings of the Second Asian–Pacific Regional Meeting on Astronomy, p. 563 – 566 (1984). – See Abstr. 012.031.
For AFGL souces located in Cygnus the authors made the optical identification, the photometric measurement, and the spectral classification. The area were photographed at V, R, and I. Most of the sources are of M types or highly reddened stars. Some show much infrared excess, probably caused by circumstellar dust.

133.012 **Contrapartidas ópticas de tres fuentes del catálogo AFGL.**
J. F. Lahulla.
Bol. Astron. Obs. Madr., Vol. 11, No. 3, p. 37 – 53 (1984).
Results are given of a search for optical counterparts of three unidentified sources of the AFGL infrared rocket survey: AFGL 1026, AFGL 1058, and AFGL 2053. UBVRI photoelectric photometry has been carried out for these counterparts.

133.013 **The infrared spectrum of M8E: evidence for circumstellar CO.**
H. P. Larson, R. Hofmann, U. Fink.
Prepr. Steward Obs., No. 644, 19 pp. (1986). To appear in Astrophys. J.

133.014 **The 4 μm spectra of compact infrared sources.**
R. Hofmann, H. P. Larson, U. Fink.
Prepr. Steward Obs., No. 645, 22 pp. (1986). To appear in Astron. Astrophys.

133.015 **Quelques nouvelles brèves du "survey" IR du plan galactique austral.**
N. Epchtein.
Bull. Etoiles Tardives Spectre Particulier, No. 3, p. 5 – 6 (1985).

133.016 **The kinematics of OH/IR maser sources.**
R.–s. Peng, J. Sun.
Acta Astrophys. Sin., Vol. 6, No. 2, p. 124 – 131 (1986). In Chinese.
This paper analysed the kinematics of 127 OH/IR sources. The authors found out deviations from the Galaxy circular motion. They also derived a model for the velocity distribution of OH/IR sources from the true spatial density distribution. For the OH maser sources associated with recently identified infrared sources (not IRC sources), there is a radial expanding velocity of 50 km/s at $R \sim 4.2$ kpc from the galactic center. Finally, they discussed the difference and relationship between identified and unidentified OH/IR maser sources.

133.017 **Progress of the survey at 2.2 μm on southern infrared sources near the galactic plane.**
N. Epchtein, T. Le Bertre, O. T. Matsuura, J. R. D. Lépine, E. Picazzio, P. Marques dos Santos, P. Boscolo.
Rev. Mex. Astron. Astrofis., Vol. 12, Spec. Issue, p. 215 (1986). Abstract. – See Abstr. 012.066.

133.018 **Spectra of some IRAS sources.**
J. W. Menzies, P. A. Whitelock, I. M. Coulson.
Light on dark matter, p. 129 – 130 (1986). – See Abstr. 012.080.
Three objects with peculiar spectra have been found in a small sample of IRAS sources identified from JHKL observations.

133.019 **Comparison of CO and IR emission of IRAS unidentified sources.**
F. Boulanger, F. Casoli, F. Combes, C. Dupraz, M. Gerin.
Light on dark matter, p. 321 – 322 (1986). – See Abstr. 012.080.
The authors study the correlation between IR and CO emission for a sample of 27 IRAS unidentified sources in the Galaxy. These sources seem to be associated with protostars or young stars still embedded in molecular clouds; in five cases, the authors detect high velocity flows of molecular gas associated with a forming star. There are fairly good correlations between the ^{12}CO and ^{13}CO luminosities and the total IR luminosity. Therefore, the luminosity of the central object seems to be related to the mass of the parent molecular clump.

Semi–automated identification of IRAS point sources using UKST plates and the COSMOS measuring machine.
See Abstr. 036.258.

IRAS und der neue Infrarothimmel. Teil 1: Die technische Meisterleistung.
See Abstr. 051.069.

The spatial extent of heated dust around MWC 349.
See Abstr. 112.013.

On the properties of very young massive infrared sources.
See Abstr. 112.017.

A–type shell stars and infrared sources.
See Abstr. 112.024.

Two supergiants in the Large Magellanic Cloud with thick dust shells.
See Abstr. 112.059.

Carbon monoxide emission from stars in the IRAS and Revised AFGL catalogs. I. Mass loss driven by radiation pressure on dust grains.
See Abstr. 112.086.

Carbon monoxide emission from stars in the IRAS and Revised AFGL catalogs. II. Massive carbon stars.
See Abstr. 112.087.

IR CCD imaging of L 1551–IRS 5: direct observations of its circumstellar shell.
See Abstr. 112.146.

Objects associated with low–mass star formation in the Gum nebula.
See Abstr. 121.026.

Extended far–IR emission associated with young star molecular outflows.
See Abstr. 121.027.

IRAS sources associated with shocked–gas regions in SNRs.
See Abstr. 125.051.

***IRAS* sources associated with shocked gas regions in IC 443.**
See Abstr. 125.057.

A bipolar nebula in Orion identified with the IRAS source 05329–0505.
See Abstr. 131.012.

Is IRAS cirrus cloud emission largely fine–structure radiation?
See Abstr. 131.017.

Hydrodynamical models of the Orion–KL Nebula.
See Abstr. 131.028.

The Cone Nebula and the infrared source GL989.
See Abstr. 131.052.

First light from a young star?
See Abstr. 131.143.

An extremely luminous bipolar H_2 flow in the DR21 star–forming region.
See Abstr. 131.184.

The stars NGC 7538–IRS 5, 6, and 7, and a distance estimate to the nebula.
See Abstr. 131.206.

Temperature determination of surrounding molecular clouds in bipolar flow sources.
See Abstr. 131.212.

High–resolution mapping of molecular outflows in NGC 2071, W49, and NGC 7538.
See Abstr. 131.217.

Molecular outflows, gas density distribution, and the effects of star formation in the dark cloud Barnard 5.
See Abstr. 131.221.

IRAS 04238 + 5336: a young reflection nebula surrounding a double star.
See Abstr. 131.274.

High–resolution IRAS observations of the Rho Ophiuchi cloud core.
See Abstr. 131.284.

Star formation in Bok globules and low–mass clouds. III. Barnard 62.
See Abstr. 131.291.

Images of star–forming regions. I. Optical and radio morphology of the bipolar outflow source GL 490.
See Abstr. 131.305.

Infrared imaging polarimetry of AFGL 2591.
See Abstr. 131.324.

Evidence for a 12 micron water–ice absorption band in the IRAS LRS spectra of protostars and late type stars.
See Abstr. 131.371.

Point sources in the Orion complex.
See Abstr. 131.379.

Young stars and high density condensations in the Horsehead region.
See Abstr. 131.381.

A. Analysis of point sources in the Ophiuchus and Perseus clouds. B. CPC observations of NGC 1333.
See Abstr. 131.383.

Water masers coincident with IRAS sources.
See Abstr. 131.386.

Multiband far–infrared observations of the NGC 6334 complex.
See Abstr. 132.037.

Compact radio sources associated with molecular outflows.
See Abstr. 132.039.

A study of the far infrared counterparts of new candidates for planetary nebulae.
See Abstr. 134.043.

Infrared spectropolarimetry of the galactic centre: magnetic alignment in the discrete sources.
See Abstr. 155.007.

Sagittarius A* and the positions of infrared sources in the galactic center.
See Abstr. 155.041.

Stellar contents contributing to the near–infrared radiation of the Galaxy.
See Abstr. 155.078.

The galactic distribution of stellar sources found by IRAS (with some emphasis on the bulge of the Galaxy).
See Abstr. 155.122.

Studies of *IRAS* sources at high galactic latitudes – I. Source counts at $|b| > 60°$ and evidence for a north–south anisotropy of cosmological significance.
See Abstr. 157.033.

Studies of IRAS sources at high galactic latitudes. II. Results from a redshift survey at $b > 60°$: distribution in depth, luminosity function, and physical nature of IRAS galaxies.
See Abstr. 157.038.

The luminosity function and space density of the most luminous galaxies in the *IRAS* survey.
See Abstr. 157.173.

The identification of IRAS galaxies.
See Abstr. 157.324.

The smoothness of the 2.2 micron background: constraints on models of primeval galaxies.
See Abstr. 161.046.

134 Planetary Nebulae

134.001 **Seven new planetary nebulae in the direction of Baade's Window.**
R. A. Shaw, A. Wirth.
Publ. Astron. Soc. Pac., Vol. 97, No. 597, p. 1071 – 1074 (1985).
The authors present seven new planetary nebula candidates that were discovered with interference–filter photographs, and that lie in or near Baade's Window ($l = 1°0$, $b = -3°9$). Two of them are angularly resolved, and are probably relatively nearby disk objects. All are faint and suffer from high interstellar reddening.

134.002 **The dust and molecular gas cloud surrounding a planetary nebula: NGC 7027.**
H. A. Thronson Jr., J. Bally.
Astrophys. J., Vol. 300, No. 2, p. 749 – 755 (1986).
The authors have searched for millimeter–wave emission from nine molecules in the planetary nebula NGC 7027: CN, CS, C_2H, HC_3N, SiS, $C^{18}O$, SiCC, $^{13}CH_3OH$, and H_2CO. A search for the $J,K = (1,1)$ line of NH_3 and the $J = 21 \rightarrow 20$ transition of HC_7N was undertaken at 1.3 cm. Only the $N = 1 \rightarrow 0$ line of CN was positively detected. Limits on integrated line strengths, column densities, and relative abundances are calculated for most of the molecules. Abundances calculated for the planetary nebula are compared with those found for the evolved carbon star, IRC +10°216. An analysis of the extinction in the circumstellar cloud shows that the abundance ratio, [dust]/[CO], has nearly the same value in NGC 7027, a number of highly evolved stars, and diffuse interstellar molecular clouds.

134.003 **Abundances in the planetary nebula NGC 6153.**
S. R. Pottasch, M. Dennefeld, J.–e. Mo.
Astron. Astrophys., Vol. 155, No. 2, p. 397 – 401 (1986).
Spectra in the visual, infrared and ultraviolet, of the planetary nebula NGC 6153 are presented. They are analysed to obtain the abundance and physical conditions in the nebula. It is found that NGC 6153 is a very unusual nebula, its abundance of many elements being higher than any other nebula.

134.004 **Late stages of the expansion of planetary nebulae.**
F. Gieseking, H. Hippelein, R. Weinberger.
Astron. Astrophys., Vol. 156, No. 1/2, p. 101 – 105 (1986).
The authors present expansion velocities and radial space velocities of 15 very large planetary nebulae with linear radii between 0.4 and 0.8 pc, deduced from Fabry–Pérot–spectrometry. The results are discussed in terms of the interaction of the envelopes with the surrounding interstellar material during the late stages of nebular expansion.

134.005 **Outer haloes of planetary nebulae as probes of a fast luminosity decline in their nuclei.**
R. Tylenda.
Astron. Astrophys., Vol. 156, No. 1/2, p. 217 – 222 (1986).
Theoretical models of planetary nebula nuclei burning hydrogen quiescently predict a fast luminosity decline when the star leaves the horizontal part of its evolutionary track and descends towards the white dwarf sequence. Time–dependent photoionization models of nebulae show that this event should be followed by the appearance of a double–envelope structure in the surrounding nebula: an inner, bright, high–excitation (optically thick to the ionizing radiation) portion of the nebula is surrounded by a faint, low–excitation, recombining halo fading with time. An observational search for recombining haloes in planetary nebulae would, therefore, allow to confirm whether the central stars really experience a rapid luminosity drop during their evolution. This would, in turn, help to answer the question if the planetary nebula formation process takes place during quiescent H–burning or whether it is triggered by a helium shell flash.

134.006 **Distances to planetary nebulae. I. The reddening–distance method.**
R. Gathier, S. R. Pottasch, J. W. Pel.
Astron. Astrophys., Vol. 157, No. 1, p. 171 – 190 (1986).
Distances to 12 planetary nebulae are derived from the increase of interstellar reddening with distance along the lines of sight of the nebulae. The reddening–distance relations are determined using Walraven *VBLUW* photometry of stars in small fields around the planetary nebulae. With a theoretical calibration of the *VBLUW* colours the effective temperature, suface gravity and reddening are determined for early type stars. The reddenings of the planetary nebulae are determined using a number of different methods. The amount of internal reddening is estimated. The planetary nebula distances derived from the reddening–distance relations have accuracies of 10 to 40%. As a byproduct, the distance to the young open cluster NGC 2453 is determined.

134.007 **Distances to planetary nebulae. II. H I absorption observations.**
R. Gathier, S. R. Pottasch, W. M. Goss.
Astron. Astrophys., Vol. 157, No. 1, p. 191 – 203 (1986).
H I absorption measurements at 21 cm of 24 planetary nebulae are obtained using the Westerbork Synthesis Radio Telescope. The observations are used to determine kinematic distances to the nebulae. These distances are combined with distance information from other sources. For 12 planetary nebulae the final distance derived has an uncertainty of less than $\sim 50\%$; this is a significant improvement with respect to previous distance determinations. H I is found associated with the planetary nebula NGC 6790. An estimate of the total H I mass is $\geqslant 0.01\ M_\odot$. Flux densities at 21 and 49 cm are presented for the planetary nebulae and for nearby background sources.

134.008 **Study of IRAS observations of faint planetary nebulae.**
K. V. K. Iyengar.
Astron. Astrophys., Vol. 158, No. 1/2, p. 89 – 96 (1986).
Far Infrared (FIR) photometric data at 12, 25, 60, and 100 μm from the Infrared Astronomical Satellite (IRAS) on 46 optically faint planetary nebulae are used with related data on radio fluxes, angular sizes and distances to determine their nebular parameters such as dust temperature, total infrared flux, infrared excesses (IRE), IR luminosity, mass of dust and optical depth at 25 μm. The temperature of dust grains emitting FIR in these nebulae ranges from 60 – 240K as in bright nebulae (Pottasch et al., 1984). The nebulae are all optically thin at 25 μm. The optical depth at 25 μm, τ_{25}, decreases when the size of the nebula increases, and is seen to be well correlated with it. Also the infrared luminosity of these planetary nebulae is correlated with the radio luminosity (at 5 GHz). The temperature of the dust in these nebulae and the infrared luminosity appear to decrease as the nebular size increases.

134.009 **Abundance gradients from type II planetary nebulae.**
M. Faúndez–Abans, W. J. Maciel.
Astron. Astrophys., Vol. 158, No. 1/2, p. 228 – 232 (1986).
The existence of galactic abundance gradients is analysed taking into account a sample of type II planetary nebulae. Helium, oxygen, nitrogen, sulphur, and carbon gradients are considered, on the basis of recently published abundances and distances. Measurable radial gradients of He/H, O/H, N/H, S/H, S/O, and C/H are detected. The results are compared with the electron temperature gradient recently determined for type II planetary nebulae. From an analysis of the main cooling agents in the ionized gas, it is shown that the abundance gradients alone are able to explain the observed temperature gradient.

134.010 **A survey of the central–star characteristics of mostly faint planetary nebulae.**
L. H. Aller, C. D. Keyes.
Publ. Astron. Soc. Pac., Vol. 97, No. 598, p. 1142 – 1148 (1985).
From a survey of 70 planetary nebulae whose spectra have been studied with the image–tube scanner (ITS) at the Shane 3–m

telescope of Lick Observatory, the authors assess the characteristics of the central stars. About two dozen appear to have central stars with continuous spectra; a slightly large number is classified as weak–emission–line objects with N III, C III, and C IV lines; seven are of the Wolf–Rayet type. Most of the observed objects lie in the direction of the galactic bulge; the others were scattered in galactic longitude to the anticenter. There appears to be no evidence for a systematic concentration of the types with respect to galactic longitude.

134.011 **Origin of planetary nebulae: morphology, carbon–to–oxygen abundance ratios, and central star multiplicity.**
B. Zuckerman, L. H. Aller.
Astrophys. J., Vol. 301, No. 2, p. 772 – 789 (1986).
The authors have examined data pertaining to the morphology, carbon–to–oxygen abundance ratio, and central star multiplicity of 139 planetary nebulae (PNs). Among 108 nonstellar PNs, $\sim 50\%$ display a bipolar symmetry, and an additional $\sim 30\%$ display elliptical symmetry. Comparison of the sizes of extreme red giant stars and separations of main–sequence, solar–type binary stars suggests that the gravitational field of a companion star is unlikely to be responsible for the mass ejection in all or even a majority of these cases. However, it is perhaps conceivable that a secondary star is responsible for most of the above–mentioned symmetries. Based on optical, ultraviolet, infrared, and radio data, the authors find 42 C–rich and 26 O–rich PNs in the set. These numbers agree with the relative numbers of C– and O–rich red giant stars with large mass loss rates and are consistent with the idea that the progenitors of PNs are this class of red giant stars.

134.012 **Observations of the Bowen fluorescent mechanism in planetary nebulae.**
L. Likkel, L. H. Aller.
Astrophys. J., Vol. 301, No. 2, p. 825 – 833 (1986).
The authors present measurements of Bowen fluorescent lines in 17 planetary nebulae and compare the relative line strengths with theory. There is generally good agreement with the line intensities expected by Saraph and Seaton. The largest deviation from theory is in the O III $I(\lambda 3429)/I(\lambda 3444)$ ratio, which is consistently lower than predicted. Certain lines are enhanced by charge transfer, and the predicted contribution is derived. The Bowen efficiency is calculated for each of the planetary nebulae considered; the range of values indicates differences between the nebulae.

134.013 **The central star of NGC 2346: further photometric observations.**
G. Jasniewicz, A. Acker.
Astron. Astrophys., Vol. 160, No. 1, p. L1 – L3 (1986).
New photometric U, B, V, observations of the central star of the bipolar planetary nebula NGC 2346 have been done in September 1985. The observations confirm the progressive dephasing of the light–curve and show a "looping" evolution of the colours from 1982 to 1985. The model proposed by Mendez et al. (1982) of a clumpy dust cloud passing in front of the stellar system is confirmed. The authors suggest: (1) a part of the dusty material was ejected in early 1982 by the sdO star near the apastre in the direction of the A–star's orbit; (2) the obscuring cloud will probably leave the central star system in 1986.

134.014 **Statistical evolutionary properties of planetary nebulae: thicknesses of the nebulae and temperatures of the stars.**
F. Sabbadin.
Astron. Astrophys., Vol. 160, No. 1, p. 31 – 38 (1986).
The evolutionary behaviours of the Zanstra temperatures of the central stars of planetary nebulae depend on the nebular mass and/or on the mass of the exciting star and confirm the validity of the division scheme suggested by Greig (1971, 1972) for this class of objects. Planetary nebuale of the subclass B of Greig tend to have higher stellar masses and/or higher nebular masses than planetary nebulae of the subclass C.

134.015 **Evidence for intrinsic polarization in the optical radiation of planetary nebulae.**
J. L. Leroy, J. F. Le Borgne, J. Arnaud.
Astron. Astrophys., Vol. 160, No. 1, p. 171 – 180 (1986).
The authors present new polarimetric measurements in several regions of five planetary nebulae which indicate that there is a small intrinsic linear polarization in nebular radiation.

134.016 **Monochromatic and spectral observations of the thin–filament nebula YM 29.**
T. A. Lozinskaya, T. G. Sitnik, M. S. Toropova.
Astron. Zh., Tom 63, Vyp. 2, p. 255 – 264 (1986). In Russian. English translation in Sov. Astron., Vol. 30, No. 2.
Spectral and monochromatic observations have been carried out of YM 29, an old planetary nebula of type PN I according to Peimbert's classification. Monochromatic images and isophotes in the lines [O III] 5007 Å, [N II] 6584 Å, and [S II] 6717 + 6732 Å have been obtained. Morphology differences have been revealed: diffuse emission in the [O III] line and thin filaments in the [N II] and [S II] lines. Stratification of emission has been found: the regions bright in the [O III] line are located by 10″ to 40″ (0.01 – 0.04 pc at the distance of 250 pc) nearer to the nucleus than those bright in the [N II] and [S II] lines.

134.017 **The influence of stellar duplicity on the form of planetary nebulae.**
I. G. Kolesnik, L. S. Pilyugin.
Astron. Zh., Tom 63, Vyp. 2, p. 279 – 287 (1986). In Russian. English translation in Sov. Astron., Vol. 30, No. 2.
Formation of planetary nebulae's spatial structures is considered. A simple expression for the angular distribution of density in planetary nebulae is obtained. Bipolar structures are formed effectively in binary systems in which the velocity of the expanding shell around the main star is smaller than the orbital velocity of the satellite. Masses of satellites lie in the range 0.1 – 0.4 $M_\odot$. The theoretical isophotal contour map for the model of the planetary nebula NGC 3587 is consistent with observational data. It is shown that central stars of planetary nebulae are usually binary systems.

134.018 **Angular expansion measurement with the VLA: the distance to NGC 7027.**
C. R. Masson.
Astrophys. J., Lett. Ed., Vol. 302, No. 1, p. L27 – L30 (1986).
The author describes a new method of using the VLA for measuring small changes in the structures of bright radio sources. As a test of the method, he has measured the angular expansion of NGC 7027, with a signal–to–noise ratio of 30, in a period of only 2.8 yr. Based on a simple model for the dynamics of the nebula, a distance of 940 ± 200 pc is derived. The ability to detect velocities as low as 20 km s^{-1} at a distance of 1 kpc opens up many exciting areas of research.

134.019 **Breaking through the wall.**
J. B. Kaler.
Nature, Vol. 320, No. 6061, p. 394 – 395 (1986).
This note dicusses the significance of the detection of the extremely hot central star in the planetary nebula NGC 2440 by Atherton, Reay and Pottasch (see Abstr. 134.020).

134.020 **Detection of the very hot central star in NGC 2440.**
P. D. Atherton, N. K. Reay, S. R. Pottasch.
Nature, Vol. 320, No. 6061, p. 423 – 425 (1986).
The authors report the detection of the central star of the planetary nebula NGC 2440 in a narrow–band continuum image using a CCD camera at the prime focus of the Anglo–Australian telescope. Its visual magnitude has been measured as 18.9 ± 0.2. Its Zanstra temperature is about 350,000K, so that it is one of the hottest stars ever observed. Its radius corresponds to that of a nearly degenerate star. Theoretical calculations can explain the observed temperature and luminosity if it has a mass of at least 1.0 $M_\odot$ and is in its cooling stage, but the predicted age of the nebula is considerably less than that required by the calculations. Futhermore, the present estimates of the progenitor mass of a 1 $M_\odot$ white dwarf are brought into question.

134.021 **An accurate position and radial velocity for the planetary nebula in the Fornax dwarf galaxy.**
J. C. McDowell, P. J. Godwin.
Observatory, Vol. 106, No. 1070, p. 19 (1986).
The authors have observed the Fornax planetary nebula with a long-slit spectrograph. Its heliocentric radial velocity is 50 ± 4 km s^{-1}. This result resolves earlier uncertainty in the radial velocity of the Fornax system. The position of the planetary nebula has been measured from UK Schmidt plates.

134.022 **Influence of inhomogeneities of gas on the ionization of atoms and the linear emission spectrum of a planetary nebula.**
V. V. Golovatyj, O. S. Yatsyk.
Vestn. L'vov. univ. Ser. astron., No. 59, p. 15 – 19 (1985). In Russian. Abstr. in Ref. Zh., 51. Astron., 2.51.776 (1986).

134.023 **K1–16 revisited: mode structure and identification as an O VI central star.**
A. D. Grauer, H. E. Bond, J. W. Liebert, R. F. Green, T. Fleming.
Bull. Am. Astron. Soc., Vol. 17, No. 4, p. 839 (1985). Abstract. – See Abstr. 010.062.

134.024 **Balmer line evidence for weak shocks in planetary nebulae.**
R. W. Olson.
Bull. Am. Astron. Soc., Vol. 17, No. 4, p. 885 (1985). Abstract. – See Abstr. 010.062.

134.025 **Multiple shell planetary nebulae: morphologies and frequency of occurrence.**
Y.–H. Chu, R. Arendt, G. H. Jacoby.
Bull. Am. Astron. Soc., Vol. 17, No. 4, p. 885 (1985). Abstract. – See Abstr. 010.062.

134.026 **Detection of [O I] 63 μm line emission from planetary nebulae.**
H. L. Dinerstein, H. B. Ellis Jr., M. R. Haas, M. W. Werner.
Bull. Am. Astron. Soc., Vol. 17, No. 4, p. 908 (1985). Abstract. – See Abstr. 010.062.

134.027 **Hydrogenated amorphous carbon (α–C:H) in the planetary nebula NGC 7027.**
J. H. Goebel.
Bull. Am. Astron. Soc., Vol. 17, No. 4, p. 908 – 909 (1985). Abstract. – See Abstr. 010.062.

134.028 **Observations of the extremely highly ionized planetary nebula NGC 2242.**
D. R. Garnett, H. L. Dinerstein.
Bull. Am. Astron. Soc., Vol. 17, No. 4, p. 909 (1985). Abstract. – See Abstr. 010.062.

134.029 **New colour and Zanstra temperatures for 15 central stars of planetary nebulae.**
J. A. de Freitas Pacheco, S. J. Codina, L. Viadana.
Mon. Not. R. Astron. Soc., Vol. 220, No. 1, p. 107 – 117 (1986).
Spectrophotometric observations of the continuum emission from 15 nuclei of planetary nebulae are presented. From these data, colour and Zanstra temperatures are derived. The characteristics of each object are briefly discussed and the results are compared with previous studies.

134.030 **Halos around planetary nebulae.**
D. C. Jewitt, G. E. Danielson, P. N. Kupferman.
Astrophys. J., Vol. 302, No. 2, p. 727 – 736 (1986). With plates 22 – 27.
The authors report preliminary results of a CCD survey designed to detect and investigate faint halos around planetary nebulae. A TI 800x800 pixel CCD was used to take deep exposures of 44 planetary nebulae. The exposures were obtained through an Hα filter at the Cassegrain focus of the Palomar 1.5 m telescope. The images reveal numerous planetary nebula halos. About 2/3 of the studied nebulae possess extensive outer halos, here defined as any extended emission beyond the 10% isophote. Ionized sulphur electron density measurements show that in some nebulae, the mass in the halo is comparable to the mass contained in the primary H II region. The origin of these halos is discussed.

134.031 **On mass loss velocity of nuclei of planetary nebulae.**
V. A. Krol'.
Inst. teor. fiz. Akad. Nauk USSR. Prepr., No. 118, 12 pp. (1985). In Russian. Abstr. in Ref. Zh., 51. Astron., 4.51.640 (1986).

134.032 **Small grains in nebulae: the case of IC 418.**
J. P. Phillips, A. Mampaso, J. M. Vilchez, P. Gomez.
Astrophys. Space Sci., Vol. 122, No. 1, p. 81 – 96 (1986).
The authors present near infrared maps of the planetary nebula IC 418 taken at 5 arc sec spatial resolution. The possibility that the near infrared continuum derives from small grains with low thermal capacity is discussed, and synthetic spectra are derived. The relevant grains are required to be refractory and, given the large C/O ratio in this nebula, probably consist of graphite.

134.033 **The puzzling object Henize 2–442.**
V. P. Arkhipova, V. F. Esipov, B. F. Yudin.
Sov. Astron. Lett., Vol. 11, No. 4, p. 213 – 215 (1985). English translation of 40.134.001.

134.034 **Ground–based and IRAS observations of compact planetary nebulae.**
S. Kwok, B. J. Hrivnak, E. F. Milone.
Astrophys. J., Vol. 303, No. 1, p. 451 – 464 (1986). = Publ. Rothney Astrophys. Obs., No. 35.
A group of compact planetary nebulae suspected of being young objects from their radio properties were observed in the infrared using the CFH telescope. Combining the results with the data from the IRAS satellite, it is found that most have far–infrared excesses with color temperatures of ~200K. The free–free emission components are clearly distinguished from the dust components. The total bolometric fluxes of the nebulae have been measured, and the results suggest that the dust in these compact nebulae are heated by direct starlight rather than by nebular Lyman–α photons. The infrared spectra of these young nebulae support the suggestion that the far–infrared dust emission of planetary nebulae originates from the remnants of the circumstellar envelopes of their red giant progenitors.

134.035 **Spectral characteristics of dust in planetary nebulae.**
H. Moseley, R. F. Silverberg.
NASA Conf. Publ., NASA CP–2403, p. A 18 (1986). – See Abstr. 012.030.

134.036 **Neutral hydrogen and continuum *VLA* observations of the planetary nebulae NGC 6302 and NGC 2440.**
L. F. Rodríguez, J. A. García–Barreto, Y. Gómez.
Rev. Mex. Astron. Astrofis., Vol. 11, No. 2, p. 109 – 112 (1985). = Contrib. Inst. Astron., Univ. Nac. Autónoma Mex., No. 169.
A high angular resolution map of the –40 km s^{-1} H I absorption feature in NGC 6302 confirms previous results in the sense that the absorbing gas is associated with the planetary nebula. The continuum map of NGC 6302 shows structures that may be associated with the extended (~1′) optical lobes. In the case of NGC 2440 both H I absorption features at 0 and 25 km s^{-1} produce uniform attenuation across the face of the planetary nebula, suggesting that they are being caused by line–of–sight interstellar clouds unrelated to NGC 2440. The authors' continuum map of this nebula is similar to the Hα isophotal map of Phillips, Reay, and Worswick (1980), supporting the notion that the extinction is uniform over the extent of the nebula.

134.037 **Erratum: "The progressive occultation of the binary central star of NGC 2346 by a dense dust cloud" [Rev. Mex. Astron. Astrofis., Vol. 10, Spec. Issue, p. 187 – 197 (1985)].**
R. H. Méndez, B. F. Marino, J. J. Clariá, W. van Driel.
Rev. Mex. Astron. Astrofis., Vol. 11, No. 2, p. 159 (1985). See Abstr. 40.134.012.

134.038 **X–ray emission from the planetary nebula NGC 1360.**
P. A. J. de Korte, J. J. Claas, F. A. Jansen, S. P. McKechnie.
X–ray astronomy '84, p. 35 – 38 (1985). – See Abstr. 012.033.

The EXOSAT observatory has detected the nucleus of NGC 1360 in four photometric energy bands. The data rules out that the emission is from blackbody origin. Initial fits made with LTE model atmosphere spectra require the presence of highly ionized oxygen and neon in the stellar atmosphere.

134.039 **The morphology of the planetary nebula NGC 3242.**
R. W. Olson.
Diss. Abstr. Int., Sect. B, Vol. 46, No. 7, p. 2346 (1986). Thesis, New Mexico State University, 109 pp. (1985). Order No. DA8519961.

134.040 **The evolution of planetary nebula nuclei.**
R. A. Shaw.
Diss. Abstr. Int., Sect. B, Vol. 46, No. 11, p. 3882 – 3883 (1986). Thesis, University of Illinois at Urbana–Champaign, 154 pp. (1985). Order No. DA8600311.

134.041 **Infrared emission lines from planetary nebulae.**
M. A. Shure.
Diss. Abstr. Int., Sect. B, Vol. 46, No. 12, p. 4283 (1986). Thesis, Cornell University, 156 pp. (1985). Order No. DA8525805.

134.042 **The spatial structure of planetary nebulae with two nuclei.**
L. S. Pilyugin.
Inst. teor. fiz. Akad. Nauk. USSR, Prepr., No. 7R, 17 pp. (1986). In Russian. Abstr. in Ref. Zh., 51. Astron., 6.51.697 (1986).

134.043 **A study of the far infrared counterparts of new candidates for planetary nebulae.**
K. V. K. Iyengar.
Astrophys. Space Sci., Vol. 122, No. 2, p. 257 – 262 (1986).

A search (using the Infrared Astronomical Satellite "IRAS" point source catalogue) for infrared counterparts of the fourteen new candidates for planetary nebulae of low surface brightness detected by Hartl and Tritton (1985), resulted in only five identifications. The infrared sources of four of these candidate nebulae are found within 5 arc sec of their optical position and the fifth one within 1 arc min.

134.044 **Chemical abundances of planetary nebulae in M31.**
G. H. Jacoby, H. C. Ford.
Astrophys. J., Vol. 304, No. 1, p. 490 – 500 (1986). With plates 8 – 9.

The authors have measured chemical abundances in M31's disk and halo by observing three planetary nebulae and an H II region at projected distances of 3.5, 18 and 33 kpc. The measurements of O/H in a disk planetary nebula and H II region show that abundances are high in M31's interstellar medium in comparison with the Galaxy. In spite of the fact that the present star formation rate in M31 is lower than in the Galaxy, the abundances in M31 at $R = 18$ kpc are similar to those in the Orion nebula. These data show that M31 experienced considerably more enrichment of its interstellar medium during its early life than did the Galaxy. It is also shown that, in contrast to the Galaxy, there is little evidence for a relationship between kinematics and metallicity in the halo of M31.

134.045 **Planetary nebulae.**
G. Stasinska.
Recherche, Vol. 16, No. 169, p. 1018 – 1025 (1985). In French. Abstr. in Phys. Abstr., Vol. 89, No. 1273, Entry 18895 (1986).

134.046 **The planetary nebula NGC 3918.**
R. E. S. Clegg, J. P. Harrington, M. J. Barlow, J. R. Walsh.
Anglo–Aust. Obs., Prepr., No. 213, 84 pp. (1986). Submitted to Astrophys. J.

134.047 **CO in the Helix nebula.**
P. J. Huggins, A. P. Healy.
Astrophys. J., Lett. Ed., Vol. 305, No. 1, p. L29 – L32 (1986).

The authors report the detection of CO (2–1) emission from the Helix planetary nebula. The spatial distribution and kinematics of the CO emission indicate that the molecular gas is closely associated with the prominent ring structures seen in optical photographs. The CO velocity varies systematically with position in the nebula and is consistent with the simple kinematics of two expanding toroids. Mass estimates indicate that at least one–tenth of the nebular material has not yet been ionized by the central star.

134.048 **Kinematic structure of NGC 3132: the planetary nebula with a binary nucleus.**
K. C. Sahu, J. N. Desai.
Astron. Astrophys., Vol. 161, No. 2, p. 357 – 362 (1986).

The authors have carried out a detailed study of the velocity structure of the southern planetary nebula NGC 3132 by obtaining [O III] λ5007 Å line profiles at 5 positions of the nebula using a Fabry–Perot spectrometer. The main results of the study are the following: (1) The observations are consistent with an ellipsoidal model with a velocity asymmetry. (2) The velocity structure favours a binary star hypothesis for the nucleus of the nebula. As a consequence, HD 87892 and the nebula could be physically associated.

134.049 **IRAS spectra of planetary nebulae. III.**
S. R. Pottasch, A. Preite–Martinez, F. M. Olnon, J.–e. Mo, S. Kingma.
Astron. Astrophys., Vol. 161, No. 2, p. 363 – 375 (1986).

Measurements of the IRAS spectra of planetary nebulae are presented. The spectra cover the wavelength 7.5 μm to 23 μm. The lines observed are mainly from ions of neon and sulfur. The abundance of these two elements is determined and compared to results found in other astronomical objects. It is also found that the Ne^{4+} ion, when present, is formed in region whose density is an order of magnitude higher than the region where most of the line emission originates. IUE measurements of the Ne^{3+} line of many of the nebulae are presented.

134.050 **Planetary nebulae at known distance.**
F. Sabbadin.
Astron. Astrophys., Suppl. Ser., Vol. 64, No. 3, p. 579 – 589 (1986).

The author has compared statistical distances of planetary nebulae given by different authors (assuming a constant ionized nebular mass or an empirical evolutionary behaviour of the ionized mass) with distances obtained independently on assumptions on the ionized mass. Statistical distance scales which consider a fully ionized plasma (independently on the evolutionary phase of the object) overestimate distances of planetary nebulae in the first evolution (which are optically thick). Central stars of planetary nebulae of the sub–class B of Greig (1971, 1972) are statistically more massive than those of nebulae belonging to the sub–class C. The author's data point out the existence of a temporal discrepancy between the observed early evolution of the central stars and that predicted by theory.

134.051 **Studies of southern planetary nebulae. II. Electron temperatures and densities.**
A. Gutiérrez–Moreno, H. Moreno, G. Cortés.
Publ. Astron. Soc. Pac., Vol. 98, No. 603, p. 488 – 498 (1986).

A set of spectrophotometric observations of 14 bright planetary nebulae was presented in Paper I. In this paper these results are used to derive electron temperatures and densities by comparing observed intensity ratios with their theoretical values.

134.052 **Significance of ST for studies of planetary nebulae.**
M. Perinotto.
Mem. Soc. Astron. Ital., Vol. 56, No. 4, p. 645 – 651 (1985). – See Abstr. 012.054.

134.053 **Kinematic structure of NGC 3132: the planetary nebula with a binary nucleus.**
K. C. Sahu, J. N. Desai.
Inst. Astrophys. Paris, Pré–Publ., No. 130, 23 pp. (1986). To appear in Astron. Astrophys.

134.054 **CO in the planetary nebulae NGC 2346 and 6720.**
P. J. Huggins, A. P. Healy.
Mon. Not. R. Astron. Soc., Vol. 220, No. 3, p. 33P – 37P (1986).
The authors report detection of the CO (2–1) line in the planetary nebulae NGC 2346 and 6720. Estimates of the amounts of molecular material associated with the CO emission yield ratios of molecular/ionized masses of $\gtrsim 0.13$ and $\gtrsim 0.07$ for NGC 2346 and 6720, respectively. Thus a significant fraction of each nebula has not yet been ionized by the central star.

134.055 **Hβ fluxes from planetary nebulae.**
L. Viadana, J. A. De Freitas Pacheco.
Rev. Bras. Fis., Vol. 15, No. 1, p. 70 – 80 (1985). Abstr. in Phys. Abstr., Vol. 89, No. 1281, Entry 67581 (1986).

134.056 **Detection of atomic hydrogen in the young planetary nebula IC 4997.**
D. R. Altschuler, S. E. Schneider, C. Giovanardi, P. R. Silverglate.
Astrophys. J., Lett. Ed., Vol. 305, No. 2, p. L85 – L88 (1986).
The authors report the detection of 21 cm neutral atomic hydrogen in absorption toward the young planetary nebula IC 4997 (PK 58–10.1) at the velocity of the approaching edge of the nebular shell. The optical depth of the line is 0.13 ± 0.02, and the gas is well distinguished from any galactic H I. A nominal mass in neutral hydrogen of $\sim 10^{-3} M_\odot$ is derived.

134.057 **Photoelectric scanner observations of central stars of planetary nebulae.**
S. J. C. Landaberry, J. A. de Freitas Pacheco, L. Viadana, B. Bazzanella.
Rev. Mex. Astron. Astrofis., Vol. 12, Spec. Issue, p. 191 – 192 (1986). – See Abstr. 012.066.
The authors have derived color temperatures and visual magnitudes for a sample of 12 central stars of planetary nebulae, combining data for the optical continuum recently obtained with the UV fluxes measured by the ANS satellite. Zanstra temperatures are also calculated and compared with the results.

134.058 **Extinction distances to the planetary nebulae NGC 6565 and NGC 5979.**
W. J. Maciel, M. Faúndez–Abans, M. de Oliveira.
Rev. Mex. Astron. Astrofis., Vol. 12, Spec. Issue, p. 233 – 239 (1986). – See Abstr. 012.066.
Colour excesses and distances of stars of known spectral types close to the planetary nebulae NGC 6565 and NGC 5979 in the plane of the sky are obtained from UBV photoelectric photometry. The colour excess x distance relations are used to determine the nebular distances, which are compared to other estimates available in the literature.

134.059 **Galactic abundance gradients from planetary nebulae.**
M. Faúndez–Abans, W. J. Maciel.
Rev. Mex. Astron. Astrofis., Vol. 12, Spec. Issue, p. 240 – 248 (1986). – See Abstr. 012.066.
Radial and vertical gradients of He/H, O/H, N/H, N/O, S/H and C/H are studied, considering a sample of 40 type II galactic planetary nebulae, with known abundances and distances. Radial gradients have been detected for He/H and O/H, for which reliable abundance determinations are available.

134.060 **Planetary nebulae and the galactic temperature gradient.**
W. J. Maciel, M. Faúndez–Abans.
Rev. Mex. Astron. Astrofis., Vol. 12, Spec. Issue, p. 249 – 256 (1986). – See Abstr. 012.066.
Galactic electron temperature gradients derived both from H II regions and planetary nebulae are studied. The effects of (1) systematic variations of the central star temperature, (2) dust grains, and (3) heavy–element abundance gradients on the electron temperature are considered. It is concluded that the heavy elements are the main responsible for the temperature gradient, and an estimate is made of the galactic O/H gradient.

134.061 **On the planetary nebula NGC 3918.**
M. Peña, S. Torres–Peimbert.
Rev. Mex. Astron. Astrofis., Vol. 12, Spec. Issue, p. 292 (1986). Abstract. – See Abstr. 012.066.

134.062 **Ultraviolet spectrum and probable chemical composition of the high–excitation planetary nebula M1–1.**
L. H. Aller, C. D. Keyes, W. A. Feibelman.
Proc. Natl. Acad. Sci. U.S.A., Vol. 83, p. 2777 – 2781 (1986).
One of the highest excitation planetary nebulae known, M1–1, was studied with the image–tube scanner on the Shane 3–m telescope at Lick Observatory and with the International Ultraviolet Explorer. Large fractions of abundant elements such as C, N, O, S, and Ar exist in unobservable stages of ionization. The logarithmic abundance values of various elements are compared with those of the sun. In contrast to NGC 6537, the composition of M1–1 does not appear to differ markedly from that of the sun. N may be enhanced but there is no enhancement of He or C. In spite of its high excitation and its presumed origin from a relatively massive star, M1–1 shows no evidence for pronounced nuclear processing.

134.063 **Wind velocity and mass loss rate for the nucleus of NGC 6543.**
L. Bianchi, M. Grewing, S. Cerrato.
Mitt. Astron. Ges., No. 65, p. 200 – 202 (1986). – See Abstr. 012.072.

134.064 **NGC 40: a low excitation nebula around a hot central star.**
M. Grewing, G. Baessgen, M. Baessgen, L. Bianchi, M. Gutekunst.
Mitt. Astron. Ges., No. 65, p. 203 – 204 (1986). – See Abstr. 012.072.

134.065 **IRAS measurements of planetary nebulae.**
S. R. Pottasch.
Light on dark matter, p. 131 – 142 (1986). – See Abstr. 012.080.
The low resolution IRAS spectra are described and a discussion of their importance is given. The survey measurements are then discussed. The interpretation of the continuum emission as dust radiation is considered. The energy balance is used as a probe for the source of dust heating. An evolution of the dust 'temperature' is shown to exist and simple interpretations are given. The gas to dust mass ratio is presented and its evolution is discussed. Finally the possibility of using the IRAS data to substantially extend the known number of planetary nebulae is examined.

134.066 **IR observations of an extended planetary nebula: NGC 7293 – the Helix Nebula.**
A. Leene, S. R. Pottasch.
Light on dark matter, p. 143 – 144 (1986). – See Abstr. 012.080.
In this paper it is shown that the broadband fluxes as measured by IRAS in the nebula NGC 7293 are mainly due to atomic line emission; this is probably even true for the long wavelength bands. All the emission in these bands can be explained by line emission, but it is likely that some emission is also due to dust.

A statistical analysis of the LRS catalog.
See Abstr. 002.093.

Atomic structure calculations and nebular diagnostics.
See Abstr. 022.100.

Limits of interacting stellar winds model of planetary nebulae. A numerical study.
See Abstr. 064.013.

From Miras to planetary nebulae: a model of mass loss.
See Abstr. 065.175.

HR 4049: The hottest proto–planetary nebula star or a runaway hypergiant at high galactic latitude surrounded by a dust cloud?
See Abstr. 112.004.

The spatial extent and nature of the 3–μm emission features in HD 97048 and CPD –56°8032.
See Abstr. 112.056.

The spatial structure of IRC + 10216 and NGC 7027 in the far–infrared.
See Abstr. 112.090.

IRAS observations of Wolf–Rayet stars.
See Abstr. 112.137.

Ground–based and IRAS observations of proto–planetary nebulae.
See Abstr. 112.145.

Unscrambling the mess at the left hand side of the HR diagram.
See Abstr. 115.014.

Binary nature of HD 62001.
See Abstr. 117.170.

Multicolor photometry of the binary central star LSS 2018.
See Abstr. 120.020.

The infrared emission bands. I. Correlation studies and the dependence on C/O ratio.
See Abstr. 131.190.

On the carbon enrichment of the interstellar medium.
See Abstr. 131.244.

HCN emission from bipolar reflection nebulae.
See Abstr. 131.255.

Determination of temperature in gaseous nebulae.
See Abstr. 132.055.

IRAS 18333–2357: an unusual source in M22.
See Abstr. 133.003.

Spectra of some IRAS sources.
See Abstr. 133.018.

Dynamics of the outer disk and halo of M31.
See Abstr. 157.205.

Radio Sources, X-ray Sources, Cosmic Rays

141 Radio Sources (Surveys, etc.)

141.001 **G 179.0 + 2.7, an unusual galactic radio source?**
E. Fürst, W. Reich.
Astron. Astrophys., Vol. 154, No. 1/2, p. 303 – 307 (1986).
Radio continuum observations of a new nonthermal large diameter source in the galactic anticentre direction are reported. It consists of a nearly circular shell–like structure with a diameter of ≈ 70' and radial magnetic field, possibly a supernova remnant. At the centre a polarized, nonthermal triple source is seen. It might be extragalactic, but an association with the extended emission cannot be ruled out.

141.002 **A radio continuum survey of the northern sky at 1420 MHz – Part II.**
P. Reich, W. Reich.
Astron. Astrophys., Suppl. Ser., Vol. 63, No. 2, p. 205 – 292 (1986).
The second part of the Stockert 21 cm radio continuum survey of the whole northern sky is presented (see 03.141.085). These observations cover the area $0^h \leqslant \alpha_{50} \leqslant 24^h$ and $-19° \leqslant \delta_{50} \leqslant 24°$. The results are given in the form of contour maps. The effective resolution of the survey is about 35', while the effective sensitivity is about 50 mK T_B (full beam).

141.003 **Erratum: "Precise optical positions for radio/optical astrometric sources south of $\delta \sim +12°$" [Astron. Astrophys., Suppl. Ser., Vol. 58, No. 2, p. 193 – 196 (1984)].**
C. Torres, H. Wroblewski, E. Costa.
Astron. Astrophys., Suppl. Ser., Vol. 63, No. 3, p. 603 (1986). See Abstr. 38.141.016.

141.004 **Multifrequency radio continuum observations of extended galactic objects. I. Nine objects from the 2695 MHz Effelsberg galactic plane survey.**
W. Reich, E. Fürst, P. Reich, Y. Sofue, T. Handa.
Astron. Astrophys., Vol. 155, No. 1, p. 185 – 192 (1986).
The recently published 2696 MHz galactic plane survey with the Effelsberg 100–m telescope has been inspected for extended, unidentified objects. Nine of these objects have subsequently been observed at 1420, 4750 MHz and 10 GHz. Based on the available continuum radio data, the authors propose the following most likely identifications: two supernova remnants, five H II–regions, one radio galaxy, and one object resolved into an H II–region seen in front of a supernova remnant.

141.005 **A radio survey in the Origem Loop region.**
V. V. Vitkovskij, N. A. Nizhel'skij, S. A. Trushkin.
Soobshch. Spets. Astrofiz. Obs., Vyp. 48, p. 61 – 79 (1985).
The regions near the possible supernova remnant PKS 0607 + 17 have been investigated. This extended source may be a part of the Origem Loop – giant ring–like feature in the radio continuum emission. Observations have been carried out in the region from 5^h50^m to 6^h30^m in R.A. and from 16° to 18° in Dec. About 50 sources have been detected in this survey. Most of the compact radio sources were identified with the sources from the Texas survey at 365 MHz. The distribution of spectral indices of the sample is normal with the mean value of 0.7. Properties of the sources and their spatial distribution relative to the galactic plane were discussed. No excess of sources in the ring of the Origem Loop was detected.

141.006 **Estimation of the angular pattern of radiation of double radio sources at the frequency 1400 MHz.**
V. R. Amirkhanyan.
Pis'ma Astron. Zh., Tom 12, No. 3, p. 194 – 197 (1986). In Russian. Englisch translation in Sov. Astron. Lett., Vol. 12.
The analysis of double radio sources components flux density ratios has shown that the angular radiation pattern of these objects have, beside the spherical background, a lobe of about 30° wide directed along the major axis of the source.

141.007 **Anisotropy of radio source distribution from a deep sky survey.**
V. A. Korolev, L. A. Yakubtsev.
Pis'ma Astron. Zh., Tom 12, No. 4, p. 266 – 269 (1986). In Russian. English translation in Sov. Astron. Lett., Vol. 12.
From a deep survey of radio sources at 21 cm with the limit flux density of 0.6 mJy the anisotropy of radio source distribution is found. The existence of a supercluster with $z \sim 0.54$ is confirmed.

141.008 **Calibration of radio sources in the millimeter wave range.**
I. G. Moiseev, N. S. Nesterov.
Izv. Krymskoj Astrofiz. Obs., Tom 73, p. 154 – 165 (1985). In Russian. English translation in Bull. Crimean Astrophys. Obs., Vol. 73.
The results of millimeter wave calibration of radio sources observations made in 1973 – 1983 at the frequencies 22.2 and 36.7 GHz are presented. It is shown that no variations exceeding several percents of the mean flux density have been in the emission of Jupiter, Saturn and 3C 274. The published flux measurements of DR 21, 3C 274, NGC 7027, Jupiter, Saturn and Mars are reconsidered and analytical approximations of their spectra are given. These approximations determine the spectra of the sources mainly in the frequency range 20 – 150 GHz with precision about 3%.

141.009 **Faint galactic radio sources and the spiral structure in the solar neighbourhood.**
T. B. Pyatunina.
Pis'ma Astron. Zh., Tom 12, No. 5, p. 353 – 360 (1986). In Russian. English translation in Sov. Astron. Lett., Vol. 12.
For selected fields of the Milky Way in I and III quadrants of galactic longitudes the differential statistics of radio sources detected at 7.6 cm with RATAN–600 are compared. Nearly one half of the previously detected population of weak galactic radio sources consists of radio sources likely connected with star formation regions within (1 – 2) kpc from the sun. Probable luminosity of the sources is (30 – 50) mJy kpc^2.

141.010 **Erratum: "Optical counterpart candidates of radio sources 1320–446, 1622–297, 1908–202, and 2106–413" [Astron. Astrophys., Vol. 141, No. 1, p. 271 – 273 (1984)].**
C. Torres, H. Wroblewski.
Astron. Astrophys., Vol. 160, No. 2, p. 406 (1986). See Abstr. 38.141.017.

141.011 **Determination of the flux densities of radio sources on the set of broadband continuous–spectrum radiometers for the RATAN–600 radio telescope.**
K. D. Aliakberov, M. G. Mingaliev, M. N. Naugol'naya, S. A. Trushkin, L. M. Sharipova, S. N. Yusupova.
Bull. Spec. Astrophys. Obs. – North Caucasus, Vol. 19, p. 59 – 65 (1985). English translation of 39.141.005.

141.012 **Can superconducting cosmic strings explain anomalous radio sources?**
D. N. Spergel, A. Vilenkin, G. B. Field, E. M. Chudnofsky (*E. M. Chudnovskij*).
Bull. Am. Astron. Soc., Vol. 17, No. 4, p. 856 (1985). Abstract. – See Abstr. 010.062.

141.013 **Radio source surveys of the Milky Way and M33.**
N. Duric, P. C. Gregory.
Bull. Am. Astron. Soc., Vol. 17, No. 4, p. 858 – 859 (1985). Abstract. – See Abstr. 010.062.

141.014 **Radio patrol of the northern Milky Way.**
P. C. Gregory.
Bull. Am. Astron. Soc., Vol. 17, No. 4, p. 904 (1985). Abstract. – See Abstr. 010.062.

141.015 **The Zelenchuk sky survey of the declination zone 5°50′ – 9° at the 8.7 GHz frequency.**
V. R. Amirkhanyan, A. G. Gorshkov, A. A. Kapustkin, V. K. Konnikova, A. N. Lazutkin, M. G. Larionov, A. S. Nikanorov, V. N. Sidorenkov, L. S. Ugol'kova.
Astron. Tsirk., No. 1403, p. 1 – 3 (1985). In Russian.

141.016 **WSRT observations of radio sources in the galactic plane near $l = 54°$.**
T. Velusamy, W. M. Goss, E. M. Arnal.
J. Astrophys. Astron., Vol. 7, No. 2, p. 105 – 112 (1986).
A radio continuum map of a $1\overset{\circ}{.}5 \times 1\overset{\circ}{.}5$ region in the galactic plane near $l = 54°$ is presented at 49 cm with a resolution of 100 arcsec × 200 arcsec. The shell source G 54.4–0.3 has the characteristics of a supernova remnant while the second large ring structure G 53.9+0.3 is an H II ring consisting of W52 and several small–diameter thermal sources.

141.017 **A search of steep–spectrum radio sources for millisecond pulsars.**
N. D'Amico, R. N. Manchester, J. M. Durdin, W. C. Erickson.
Proc. Astron. Soc. Aust., Vol. 6, No. 2, p. 174 – 176 (1985).
A total of 18 radio sources selected on the basis of steep low–frequency radio spectra have been searched for the presence of millisecond pulsars using the Molonglo Observatory synthesis telescope. The search covered pulsar periods down to 2 ms with a limiting sensitivity of approximately 10 mJy. No pulsars were detected.

141.018 **The MIT–Green Bank (MG) 5 GHz survey.**
C. L. Bennett, C. R. Lawrence, B. F. Burke, J. N. Hewitt, J. Mahoney.
Astrophys. J., Suppl. Ser., Vol. 61, No. 1, p. 1 – 104 (1986).
The MIT–Green Bank (MG) 5 GHz survey covers 1.87 sr of sky with $-00°30'13'' < \delta(B1950.0) < 19°29'47''$ and $|b| > 10°$. The MG catalog contains 5974 sources with a signal–to–noise ratio greater than 5. A supplemental catalog contains 3836 possible detections with a signal–to–noise ratio less than 5. The MG catalog is essentially complete to its 5 σ limit, which varies with declination from 53 to 106 mJy.

141.019 **5 GHz radio structure and optical identifications of sources from the MG survey. II. Maps and finding charts.**
C. R. Lawrence, C. L. Bennett, J. N. Hewitt, G. I. Langston, S. E. Klotz, B. F. Burke, K. C. Turner.
Astrophys. J., Suppl. Ser., Vol. 61, No. 1, p. 105 – 157 (1986). With plates 1–13.
Nearly 1000 sources from the 5 GHz MIT–Green Bank (MG) survey have been observed with the VLA. The authors give radio maps for 460 resolved sources, finding charts for 390 optical identifications, and spectral indices between 1.4 and 5 GHz for 632 sources.

141.020 **A 1400 MHz sky survey. II. Confusion–limited maps covering $19^h30^m < \alpha < 7^h30^m$, $-5° < \delta < +82°$.**
J. J. Condon, J. J. Broderick.
Astron. J., Vol. 91, No. 5, p. 1051 – 1057 (1986).
The NRAO 91 m transit telescope and rebuilt four–feed receiver were used to make a 1400 MHz continuum survey with 12.7 arcmin x 11.1 arcmin resolution. New maps covering $19^h30^m < \alpha < 7^h30^m$ supplement those in the right ascension range $7^h30^m < \alpha < 19^h30^m$ (Condon and Broderick 1985) to complete the ≈ 6.8 sr declination band $-5° < \delta < +82°$. Both sets of maps are confusion limited and contain $\approx 3 \times 10^3$ sources per sr stronger than 0.15 Jy.

141.021 **Optical identifications of radiosources.**
G. Grueff.
Mem. Soc. Astron. Ital., Vol. 56, No. 4, p. 755 – 758 (1985). – See Abstr. 012.054.
This paper reports on a project to obtain optical identifications for two new deep radio source catalogues: (1) the Bologna B2 catalogue and (2) the Cambridge 5C12 catalogue; both performed at 408 MHz.

141.022 **22 GHz map of the Carina region.**
C. Tateyama, Z. Abraham.
Rev. Mex. Astron. Astrofis., Vol. 12, Spec. Issue, p. 216 (1986). Abstract. – See Abstr. 012.066.

The designation of radio sources observed with the DRAO synthesis telescope.
See Abstr. 002.019.

A review of problems in the fundamental radio–source catalog.
See Abstr. 002.041.

The IAU numbering system of radio sources.
See Abstr. 002.086.

Galactic astronomy at millimeter wavelength in Japan.
See Abstr. 013.042.

Automation of investigations of extended radio sources with the radio telescope of the Institute of Physics of the Academy of Sciences.
See Abstr. 036.080.

Electron–cyclotron maser emission from the Sun and stars: variations with plasma temperature and density.
See Abstr. 077.053.

Observations of interplanetary scintillations of radio sources with flat spectrum at a frequency of 102.5 MHz.
See Abstr. 106.078.

The unusual radio source produced by the 1985 outburst of the recurrent nova RS Ophiuchi.
See Abstr. 124.005.

Flicker of extragalactic radio sources and refractive interstellar scintillation.
See Abstr. 131.119.

VLBI studies of interstellar turbulence.
See Abstr. 131.159.

Refractive interstellar scintillation in 1741–038.
See Abstr. 131.302.

The association of neutral hydrogen with H II regions and radio sources.
See Abstr. 132.046.

The galactic center: the nearest Seyfert nucleus.
See Abstr. 155.065.

Study of non–thermal emission from the galactic center.
See Abstr. 155.116.

Extragalactic nature of G227.1 + 1.0.
See Abstr. 158.064.

Neutral hydrogen absorption towards G227.1 + 1.0.
See Abstr. 158.065.

X–ray and radio observations of steep spectrum radio sources: I. 4 fields from the Clark Lake Observatory 26 MHz survey.
See Abstr. 158.153.

The extragalactic nature of G227.1 + 1.0.
See Abstr. 158.173.

142 UV Sources, X-ray Sources, X-ray Background

142.001 **EXOSAT measurements of the quiescent X–ray spectrum of the burster 2S 1636–536: evidence for Comptonization in a hot corona.**
L. M. Breedon, M. J. L. Turner, A. R. King, T. J.–L. Courvoisier.
Mon. Not. R. Astron. Soc., Vol. 218, No. 3, p. 487 – 496 (1986).
Measurement of the X–ray spectrum of the quiescent flux from the burster 2S 1636–536 with EXOSAT on 1983 July 17 – 18 gave as a best–fitting model a power law with a high–energy cut–off. The flux (2 – 10 keV) was seen to increase from 1.6×10^{-9} erg cm^{-2}s^{-1} to 2.5×10^{-9} erg cm^{-2}s^{-1} during the observation and this increase was accompanied by a steepening power–law index (1.7 to 2.1 and a decreasing high–energy cut–off). The observed power–law spectrum is consistent with a model where soft photons emitted from the surface of the neutron star are Comptonized in a hot plasma surrounding the star (neutron star corona). The increasing power–law index observed as the intensity rises can be interpreted as Compton cooling of the corona when the flux of photons from the neutron star rises in response to increasing accretion rate.

142.002 **Quasi–periodic oscillations in the X–ray flux of Cyg X–2.**
G. Hasinger, A. Langmeier, M. Sztajno, J. Trümper, W. H. G. Lewin, N. E. White.
Nature, Vol. 319, No. 6053, p. 469 – 471 (1986).
The authors have discovered quasi–periodic X–ray oscillations (QPOs) (30 – 45 Hz) and low–frequency noise (below 15 Hz) in Cyg X–2. The centroid frequency of the oscillations is correlated with the source intensity: the frequency increases when the source intensity goes up. The amplitude of the QPOs decreases while that of the low–frequency noise increases with increasing source intensity. The authors describe the spectrum of the persistent emission as the sum of a variable 1.2 keV black–body component, a more or less constant 7 keV thermal bremsstrahlung component and an iron emission line.

142.003 **X–ray spectra of various bursts from the rapid burster.**
Y. Tawara, S. Hayakawa, H. Kunieda, D.–Y. Wang.
Astrophys. Space Sci., Vol. 118, Nos. 1/2, p. 317 – 320 (1986). – See Abstr. 012.001.
About 20 type–I bursts and 3000 type–II bursts were observed with Tenma and Hakucho in 1983. Spectral evolutions were analyzed for both types of bursts and results are presented.

142.004 **Iron K–emission line from Cygnus X–2.**
T. Hirano, S. Hayakawa, F. Nagase, Y. Tawara.
Astrophys. Space Sci., Vol. 119, No. 1, p. 77 – 80 (1986). – See Abstr. 012.001.
The iron K–emission line of Cyg X–2 was detected from the TENMA observation. The centre energy, the equivalent width and the intrinsic line width are 6.7 keV, ~30 eV, and less than 0.6 keV, respectively. These properties indicate that the observed line is mostly due to helium–like iron and the electron scattering depth of the line emitting region is smaller than three.

142.005 **Structure of the X–ray background on arc–minute scales.**
T. T. Hamilton, D. J. Helfand.
News Lett. Astron. Soc. N.Y., Vol. 2, No. 9, p. 49 – 50 (1986). Abstract. – See Abstr. 010.241.

142.006 **Pulsating X–ray sources.**
J. Trümper.
Cosmic radiation in contemporary astrophysics, p. 217 – 225 (1986). – See Abstr. 012.006.
The author briefly summarizes some of the major results and discusses some very recent results obtained with EXOSAT.

142.007 **X–ray burst sources.**
W. H. G. Lewin.
Cosmic radiation in contemporary astrophysics, p. 227 – 240 (1986). – See Abstr. 012.006.
Contents: Low–mass X–ray binaries. The Rapid Burster (= MXB 1730–335). Spectral features, absorption lines. Formation of low–mass X–ray binaries outside globular clusters. Highly irregular type I burst intervals. 5 10 minute type I burst intervals. The super Eddington limit problems. Type II bursts from the Rapid Burster.

142.008 **X–ray sky surveys and the ROSAT mission.**
J. Trümper.
Cosmic radiation in contemporary astrophysics, p. 241 – 247 (1986). – See Abstr. 012.006.

142.009 **The interpretation of blackbody spectra and radii during X–ray bursts.**
J. van Paradijs, W. H. G. Lewin.
Astron. Astrophys., Vol. 157, No. 2, p. L10 – L12 (1986).
If the neutron star in an X–ray burst source is sufficiently hot to give a significant (blackbody or blackbody–like) contribution to the total persistent X–ray flux, blackbody radii obtained from standard analyses of X–ray bursts contain systematic errors which can be very large near the end of a burst when the burst flux reaches low values. The blackbody temperature calculated from the burst flux spectra (i.e., from the radiation in excess of the pre– and post–burst persistent emission) is then directly related to the quiescent temperature of the neutron star, and is independent of the excess burst flux. As a consequence, the calculated apparent blackbody radii go to zero as the burst flux goes to zero. Therefore, a decrease in the apparent blackbody radius near the end of a burst (this has been observed) does not necessarily reflect an actual decrease in the area of a burst emission region. Systematic errors can, in principle, also be introduced

when persistent spectra of one source observed at different times are subtracted, and blackbody fits are made to the difference spectrum.

142.010 **The X-ray source in M15: a binary composed of a neutron star plus a post main-sequence star?**
M. Aurière, A. Maucherat, J.-P. Cordoni, B. Fort, J. P. Picat.
Astron. Astrophys., Vol. 158, No. 1/2, p. 158 – 160 (1986).
The authors present new observations of AC 211, the variable "UV" object included in the Einstein satellite error circle for the M15 globular cluster X-ray source. *U*, *B*, and *V* images were obtained in excellent seeing using either a CCD camera or a micro-channel-plate photographic camera at the *F*/25 Cassegrain focus of the 2 m telescope of Pic du Midi Observatory. The authors observed brightness variations greater than 1 magnitude in *U*, and of about 0.5 magnitude in *B* and *V*. They thus confirm an earlier report of the photometric characteristics of AC 211. They conclude that it is very likely that AC 211 is the optical counterpart of the X-ray source and show that their data are consistent with this object being a binary composed of a neutron star plus a post main-sequence companion.

142.011 **Angular distribution and polarization of X-ray burster radiation.**
I. I. Lapidus, R. A. Syunyaev, L. G. Titarchuk.
Astrofizika, Tom 23, Vyp. 3, p. 503 – 520 (1985). In Russian. English translation in Astrophysics, Vol. 23, No. 3.
The angular distribution and polarization of X-ray burster radiation have been calculated. It has been shown that the accretion disk around the neutron star intercepts and reradiates 23% of the burster luminosity during a flash. The result is a strong dependence of radiation flux from the burster during the flash on the inclination angle of the system. The degree of linear polarization of burster radiation during the flash reaches the maximum value $\approx 3.7\%$ in the direction of $\approx 72\overset{\circ}{.}4$ from the normal. In periods between flashes the stationary X-radiation is provided mainly by the boundary layer. The disk reradiates up to 50% of the boundary layer radiation. The degree to which the radiation from the system is polarized depends rather weakly on the inclination angle of the system and is about 6%. Analytical formulae are obtained for the calculation of linear polarization of fluorescent K-line radiation of heavy elements coming from outer regions of accretion disks.

142.012 **The discovery of X-ray bursts from Cir X-1.**
A. F. Tennant, A. C. Fabian, R. A. Shafer.
Mon. Not. R. Astron. Soc., Vol. 219, No. 4, p. 871 – 881 (1986).
A sequence of eight X-ray flares was detected during an EXOSAT observation of the low-state Circinus X-1 (= BR Cir). The authors identify these flares with X-ray bursts from Cir X-1. The recurrence time, ratio of persistent flux to the time-averaged flux and indications of cooling during the bursts are all consistent with these bursts being type I, although the authors cannot rule out the possibility that the bursts are type II. Type I bursts would be strong evidence that Cir X-1 is a neutron star and not a black hole.

142.013 **Co-ordinated X-ray and optical observations of 4U 1735-44.**
A. P. Smale.
Observatory, Vol. 106, No. 1070, p. 7 (1986). Abstract. – See Abstr. 010.741.

142.014 **Rapid, intensity-dependent quasi-periodic oscillations in GX 5-1.**
M. van der Klis.
Observatory, Vol. 106, No. 1070, p. 9 (1986). Abstract. – See Abstr. 010.741.

142.015 **The discovery of X-ray bursts from Cir X-1.**
A. F. Tennant.
Observatory, Vol. 106, No. 1070, p. 9 – 10 (1986). Abstract. – See Abstr. 010.741.

142.016 **Serendipitous EXOSAT sources in the field of the Coma cluster.**
G. Branduardi-Raymont.
Observatory, Vol. 106, No. 1070, p. 10 – 11 (1986). Abstract. – See Abstr. 010.741.

142.017 **Optical observations of 1E 0630+178, and a possible optical counterpart of Geminga.**
G. F. Bignami, P. A. Caraveo, L. Salotti, S. Kulkarni, S. Djorgovski, J. A. Paul, L. Vigroux.
Bull. Am. Astron. Soc., Vol. 17, No. 4, p. 848 (1985). Abstract. – See Abstr. 010.062.

142.018 **The 1974 recurrence of X-ray transient 4U 0115+63.**
L. Whitlock, D. Roussel-Dupre, W. C. Priedhorsky.
Bull. Am. Astron. Soc., Vol. 17, No. 4, p. 849 (1985). Abstract. – See Abstr. 010.062.

142.019 **Spectral evolution of Type II X-ray bursts.**
N. Kawai.
Bull. Am. Astron. Soc., Vol. 17, No. 4, p. 849 (1985). Abstract. – See Abstr. 010.062.

142.020 **A dependence between burst properties and the strength of the persistent emission in the X-ray burster EXO 0748-676.**
M. Gottwald, F. Haberl, A. N. Parmar, N. E. White.
Bull. Am. Astron. Soc., Vol. 17, No. 4, p. 849 (1985). Abstract. – See Abstr. 010.062.

142.021 **Observations of quasi-periodic oscillations from GX 5-1 with the Einstein (HEAO-2) Observatory.**
R. F. Elsner, M. C. Weisskopf, W. Darbro, S. Miyaji, B. Ramsey, A. C. Williams, P. G. Sutherland, J. E. Grindlay.
Bull. Am. Astron. Soc., Vol. 17, No. 4, p. 854 (1985). Abstract. – See Abstr. 010.062.

142.022 **5 Hz quasi-periodic oscillations in Cygnus X-2.**
J. P. Norris, K. S. Wood.
Bull. Am. Astron. Soc., Vol. 17, No. 4, p. 855 (1985). Abstract. – See Abstr. 010.062.

142.023 **Cyg X-3 X-ray fluctuations, 1969 – 1979.**
J. Terrell, W. C. Priedhorsky.
Bull. Am. Astron. Soc., Vol. 17, No. 4, p. 855 (1985). Abstract. – See Abstr. 010.062.

142.024 **The Cyg X-3 radio outburst of 1985.**
K. J. Johnston, J. H. Spencer, E. B. Waltman, G. G. Pooley, R. E. Spencer, P. E. Angerhofer, D. R. Florkowski, F. J. Josties, D. D. McCarthy, D. N. Matsakis, G. Herford, D. Gary, R. M. Hjellming, C. Schalinski, L. Molnar, M. Reid.
Bull. Am. Astron. Soc., Vol. 17, No. 4, p. 855 (1985). Abstract. – See Abstr. 010.062.

142.025 **Structure of the X-ray background on arc-minute scales.**
T. T. Hamilton, D. J. Helfand.
Bull. Am. Astron. Soc., Vol. 17, No. 4, p. 857 (1985). Abstract. – See Abstr. 010.062.

142.026 **EXOSAT and Einstein observations of the X-ray pulsar 2S 1145-61.**
S. Mereghetti, G. F. Bignami, P. A. Caraveo, A. Goldwurm.
Bull. Am. Astron. Soc., Vol. 17, No. 4, p. 867 – 868 (1985). Abstract. – See Abstr. 010.062.

142.027 **Quasi-periodic oscillations in galactic X-ray sources.**
F. K. Lamb, N. Shibazaki, M. A. Alpar, J. Shaham.
Bull. Am. Astron. Soc., Vol. 17, No. 4, p. 868 (1985). Abstract. – See Abstr. 010.062.

142.028 **EXOSAT 1985 observations of the ~60 sec periodicity of 1E 0630+178.**
P. A. Caraveo, G. F. Bignami, A. M. T. Pollock.
Bull. Am. Astron. Soc., Vol. 17, No. 4, p. 868 (1985). Abstract. – See Abstr. 010.062.

142.029 **X-ray selected M dwarfs and the diffuse X-ray background.**
J.-P. Caillault, D. J. Helfand, J. A. Nousek, L. O. Takalo.
Bull. Am. Astron. Soc., Vol. 17, No. 4, p. 879 (1985). Abstract. – See Abstr. 010.062.

142.030 **HEAO-1 LED maps of the Eridanus soft X-ray enhancement.**
D. N. Burrows, J. A. Nousek, R. J. Truax, G. P. Garmire, K. P. Singh.
Bull. Am. Astron. Soc., Vol. 17, No. 4, p. 883 (1985). Abstract. – See Abstr. 010.062.

142.031 **The optical identification and spectral analysis of HEAO-1 X-ray sources.**
R. A. Remillard.
Bull. Am. Astron. Soc., Vol. 17, No. 4, p. 889 (1985). Abstract. – See Abstr. 010.062.

142.032 **The X-ray background.**
P. Guilbert.
ESA Spec. Publ., ESA SP-239, p. 15 – 18 (1985). – See Abstr. 012.019.

The spectrum of the X-ray background has now been fairly accurately measured, particularly in the 3 – 100 keV energy range but its origin is still uncertain. A major task of XMM will be to measure the spectra of faint extragalactic sources at energies above 3 keV in order to identify the kinds of sources which make up the background.

142.033 **EXOSAT CMA observations of X0512-401 = NGC 1851.**
E. N. Ercan, U. Kiziloglu, A. Esendemir, F. Tokdemir, G. Branduardi-Raymont.
ESA Spec. Publ., ESA SP-239, p. 121 – 125 (1985). – See Abstr. 012.019.

EXOSAT CMA observations of X0512-401 = NGC 1851 are reported. The observations covered $\sim 2\times10^4$seconds and were carried out with Lexan thin, Aluminum/Parylene and Boron filters. The authors give preliminary results of "3-colour" spectroscopy making use of these CMA + filters data. In addition they present the first long (~6 hours), uninterrupted, simultaneous soft and medium energy X-ray light curves of X0512-401.

142.034 **Iron K X-ray emission lines from cosmic X-ray sources observed with Tenma.**
M. Matsuoka.
ESA Spec. Publ., ESA SP-239, p. 149 – 154 (1985). – See Abstr. 012.019.

Gas scintillation proportional counters on board Tenma have detected iron K X-rays at an energy of 6.4 keV from some binary X-ray pulsars. This is satisfactorily explained by a picture in which continuum X-rays originating from a compact region excite fluorescent K X-rays in the surrounding cold matter. Intense K X-rays at 6.4 keV have also been detected from the Seyfert galaxy NGC 4151. The emission line at an energy of 6.7 keV is observed from non-pulsating low mass X-ray binaries and Cyg X-3. This is also probably explained by fluorescent mechanism in a hot gas region. On the other hand intense 6.7 keV lines are detected from supernova remnants, clusters of galaxies and the galactic ridge on the plane. Various emission lines as well as the 6.7 keV lines are produced from hot plasma region in these sources.

142.035 **Eddington luminosities and photospheric radius expansion during X-ray bursts from 4U/MXB 1820-30.**
W. D. Vacca, W. H. G. Lewin, J. van Paradijs.
Mon. Not. R. Astron. Soc., Vol. 220, No. 2, p. 339 – 349 (1986).

The authors present the analysis of six X-ray bursts from 4U/MXB 1820-30. In all cases they observed photospheric contraction during burst decay. The contraction manifests itself as a distinct hardening of the burst spectra, a phenomenon which was previously noted but misinterpreted. The photospheric expansion which must have preceded the contraction takes place in less than one second. The authors also find that the bolometric burst peak luminosities are all near the Eddington limit. This burst source has the best-known distance of all burst sources.

142.036 **A search for ultraviolet-excess objects.**
T. Noguchi, H. Maehara, M. Kondo.
Proceedings of the Second Asian-Pacific Regional Meeting on Astronomy, p. 352 – 357 (1984). – See Abstr. 012.031.

A search for ultraviolet-excess objects has been carried out with the 105-cm Schmidt telescope at the Kiso Observatory. Proper motions of about 80 objects are determined in two pilot fields. The spectroscopic observation of about 50 bright constituents is made with the Cassegrain I.I. spectrograph attached to the 188-cm telescope at the Okayama Astrophysical Observatory. A tentative classification indicates that more than half of them are possibly identified with white dwarfs, while some others are with O, B subdwarfs and blue horizontal-branch stars.

142.037 **Energy spectra of non-pulsating low mass X-ray binaries.**
M. Matsuoka, K. Mitsuda.
X-ray astronomy '84, p. 103 – 110 (1985). – See Abstr. 012.033.

Detailed analysis of the spectra has been done for four low mass X-ray binaries; Sco X-1, 4U 1608-52, GX 5-1 and GX 349+2. The authors present new results. For the full paper see Abstr. 38.142.074.

142.038 **Spectral studies of low-mass X-ray binaries observed by EXOSAT.**
M. Sztajno, J. Trümper, A. Langmeier.
X-ray astronomy '84, p. 111 – 114 (1985). – See Abstr. 012.033.

The authors have studied the X-ray emission of five low-mass X-ray binaries: two bursters (MXB 1702-43, MXB 1728-34), one flaring source (GX 17+2) and two bright bulge sources (4U 1705-44 and 4U 1755-338) which show evidence for orbital periods.

142.039 **Discovery of regularities in the cycle-to-cycle variability of Cygnus X-3.**
M. van der Klis, F. Jansen.
X-ray astronomy '84, p. 115 – 118 (1985). – See Abstr. 012.033.

The authors have found quasi-periodic variations in the X-ray flux of Cyg X-3 during recent observations of this source with EXOSAT. Amplitudes and periods of the variations range from 5% to 20% of the total flux and from 50 to 1500 s, respectively. In this paper the authors show examples of the phenomenon and give a preliminary discussion of the possible origin of the quasi-oscillations.

142.040 **Detection of iron K-emission lines from low-mass binary X-ray sources.**
H. Inoue.
X-ray astronomy '84, p. 119 (1985). Abstract. – See Abstr. 012.033. For the full paper see Abstr. 38.142.075.

142.041 **EXOSAT observations of 4U 1705-44.**
A. Langmeier, M. Sztajno, J. Trümper.
X-ray astronomy '84, p. 121 – 124 (1985). – See Abstr. 012.033.

During a 7-hour observation of 4U 1705-44, the authors detected 6 intensity dips, which show a quasi-periodicity of 1.3 hours. In additon, there are intensity fluctuations in the range of 100 – 500 seconds. These timescales may be interpreted in terms of an orbital period and oscillations of an accretion disc, respectively.

142.042 **X–ray bursts.**
Y. Tanaka.
X–ray astronomy '84, p. 125 – 139 (1985). – See Abstr. 012.033.
Investigation of X–ray burst may be divided into two objectives: (1) What are X–ray bursts? (2) Investigating neutron stars and their environment by means of X–ray bursts.

142.043 **Quasi–periodicity of the Rapid Burster in 1983.**
H. Kunieda.
X–ray astronomy '84, p. 141 – 144 (1985). – See Abstr. 012.033.
The Japanese X–ray astronomy satellites Hakucho and Tenma observed the activity of the rapid burster MXB 1730–335 in 1983.

142.044 **EXOSAT observations of 1636–536.**
L. M. Breedon, M. J. L. Turner.
X–ray astronomy '84, p. 145 – 148 (1985). – See Abstr. 012.033.
An EXOSAT observation of MXB 1636–536 in July 1983 detected 7 bursts with untypical rise and decay (e–folding) times, and found persistent flux steady over the 1st 12,000 seconds but gradually increasing over the remainder of the observation. Preliminary analysis reveals that the best fitting model to the persistent flux is a power law with a high energy cut–off, consistent with comptonization of soft photons by a hot plasma.

142.045 **Optical evidence for the binary nature of the X–ray burster X 1254–69.**
C. Motch, H. Pedersen, S. A. Ilovaisky, C. Chevalier, M. W. Pakull.
X–ray astronomy '84, p. 149 – 152 (1985). – See Abstr. 012.033.
The authors present V light curves of the optical counterpart of X 1254–69 obtained from a large number of CCD images with 5 minute average integration time. The optical behaviour of X 1254–69 resembles much that of MXB 1636–53 and the orbital period is within the range of the periods proposed for other X–ray bursters. Details of the light curve and possible mechanisms for the optical modulation are discussed.

142.046 **Recent results of EXOSAT observations of 2S 1254–690.**
T. J.–L. Courvoisier, A. Parmar, A. Peacock.
X–ray astronomy '84, p. 153 – 156 (1985). – See Abstr. 012.033.
EXOSAT observed 2S 1254–690 on February 5 and May 15, 1984. The first Type 1 X–ray burst from this source has been seen during the observation. Also observed were a set of deep dips.

142.047 **X–ray bursts – some remaining key issues.**
W. H. G. Lewin.
X–ray astronomy '84, p. 157 – 164 (1985). – See Abstr. 012.033.
The global picture of X–ray bursts and the low–mass binary systems in which they occur seems clear. Several remaining issues which are interesting and puzzling are reviewed.

142.048 **X–ray pulsars and black–hole candidates.**
K. Makishima.
X–ray astronomy '84, p. 165 – 180 (1985). – See Abstr. 012.033.
Several recent topics concerning binary X–ray pulsars are presented. Also X–ray observations of galactic black–hole candidates are compared with those of X–ray pulsars.

142.049 **Variation of the pulse profile of Her X–1.**
T. Ohashi.
X–ray astronomy '84, p. 187 – 192 (1985). – See Abstr. 012.033.
Her X–1 was observed during the on–state of the 35 day cycle in May, 1983. The pulse profile, the energy spectrum, and the iron emission line were studied.

142.050 **Pulse phase spectroscopy of Her X–1 with EXOSAT.**
P. Kahabka, W. Pietsch, J. Trümper, W. Voges, E. Kendziorra, R. Staubert.
X–ray astronomy '84, p. 193 – 196 (1985). – See Abstr. 012.033.
During March/April 1984 Her X–1 has been observed in a series of 10 pointings spread approximately evenly through a 35 day cycle. The authors discuss the results of an analysis of pulse–profiles and source spectra. They propose that the interpulse of the 1.24 second variation is caused by a reflection of the primary beam from the inner edge of the accretion disk or a disk corona.

142.051 **EXOSAT observations of the 35^d intensity variations of Her X–1.**
H. Ögelman, P. Kahabka, W. Pietsch, J. Trümper, W. Voges, E. Kendziorra, R. Staubert.
X–ray astronomy '84, p. 197 – 200 (1985). – See Abstr. 012.033.
The low energy telescope data in the .04 to 2 keV range represents the first 35^d cycle sampling of the soft X–rays. The observed light curve is similar to the medium energy light curve.

142.052 **Transient dips of Cygnus X–1 observed from Tenma.**
Y. Tanaka.
X–ray astronomy '84, p. 201 – 203 (1985). – See Abstr. 012.033.

142.053 **The transient X–ray source A 0535+26: coordinated observations in X–ray, UV, optical and IR regions.**
F. Giovannelli, M. Ferrari Toniolo, P. Persi, I. M. Golinskaya (*I. M. Golynskaya*), V. G. Kurt, T. A. Mizyakina, E. Yu. Shafer, V. M. Shamolin, A. S. Smirnoff (*A. S. Smirnov*), E. K. Sheffer, G. V. Zaytzeva (*G. V. Zajtseva*), E. L. van Dessel, H. van Diest, H. Hensberge, M. Burger, C. de Loore, C. Bartolini, A. Guarnieri, M. Lolli, A. Piccioni, Yu. N. Gnedin, G. V. Khozov, V. M. Larionov, N. I. Shakovskaya (*N. I. Shakhovskaya*).
X–ray astronomy '84, p. 205 – 208 (1985). – See Abstr. 012.033.
Multispectral coordinated observations of the Be/X–ray binary HDE 245770/A 0535+26 are briefly presented and discussed. A strong X–ray event was revealed and the period of the X–ray pulsar measured during a quiescent phase.

142.054 **Further evidence on the increasing ~1 minute period in the X–ray data of Geminga (1E 0630+178).**
G. F. Bignami, P. A. Caraveo, L. Salotti.
X–ray astronomy '84, p. 209 – 212 (1985). – See Abstr. 012.033.

142.055 **EXOSAT observation of "Geminga".**
P. A. Caraveo, G. F. Bignami, S. Mereghetti.
X–ray astronomy '84, p. 213 – 216 (1985). – See Abstr. 012.033.
1E 0630+178, the unique X–ray source discovered in the error–box of the bright gamma–ray source Geminga (2CG 195+04) has been observed by EXOSAT for about 10 hours in Sept. 1983.

142.056 **TENMA observations of X–ray pulsars.**
F. Nagase, T. Kii, N. Sato.
X–ray astronomy '84, p. 217 – 223 (1985). – See Abstr. 012.033.
Characteristic features of energy spectra were summarized for the X–ray pulsars observed by TENMA. Fluorescent emission line of iron was found for most X–ray pulsars at the energy 6.4 keV. The 6.4 keV emission line was also observed during the eclipse of Vela X–1. The pulse profile of 4U1626–67 observed by TENMA was discussed and the magnetic field at the surface of the neutron star is estimated to be $\sim 8 \times 10^{12}$gauss.

142.057 **EXOSAT observations of 3A 1954+319.**
M. C. Cook, R. S. Warwick, M. G. Watson.
X–ray astronomy '84, p. 225 – 228 (1985). – See Abstr. 012.033.
An EXOSAT observation of 3A 1954+319 [4U 1954+31] revealed a source with a hard X–ray spectrum and a light curve dominated by flaring behaviour on timescales of 1 – 10 minutes. Spectral variations were also observed on these timescales. No evidence was found for the existence of coherent pulsations. The X–ray properties indicate a high mass system accreting from a highly inhomogeneous medium.

142.058 **Properties of the iron line from Vela X–1.**
T. Ohashi.
X–ray astronomy '84, p. 229 – 233 (1985). – See Abstr. 012.033.
Properties of the iron emission line from Vela X–1 are studied with the gas scintillation proportional counters onboard Tenma. The energy and the width of the iron line are determined to be 6.43 ± 0.05 keV and less than 0.5 keV (FWHM), respectively. The iron line intensity during eclipse suggests the size of the line

emitting region to be less than 5×10^{11}cm from the neutron star. A remarkable finding is that the iron line intensity is constant over the 283 sec pulse phase of Vela X–1. This implies a special geometry of the matter near the neutron star.

142.059 **EXOSAT observations of the X–ray transient V 0332+53.**
J. Davelaar, N. E. White, L. Stella, A. Parmar, R. J. Blissett, M. van der Klis, H. Schrijver.
X–ray astronomy '84, p. 235 – 238 (1985). – See Abstr. 012.033.
EXOSAT observed three outbursts from V 0332+53, which revealed coherent 4.4 s X–ray pulsations in addition to the reported rapid time variability (Tanaka et al., 1983). The Doppler variations during the recurrent outbursts indicate a 34.25 day orbit with moderate eccentricity of 0.31. The X–ray spectra and pulse profiles are similar to other X–ray pulsars.

142.060 **Observations of LMC X–3 with EXOSAT.**
A. Treves, J. M. Bonnet–Bidaud, L. Chiappetti, L. Maraschi, L. Stella, E. G. Tanzi, M. van der Klis.
X–ray astronomy '84, p. 259 – 262 (1985). – See Abstr. 012.033.
The authors report on observations performed with EXOSAT on 1983 Dec. 11.

142.061 **Search for new X–ray pulsars in the galactic plane by the HEAO–1 Large Area Sky Survey experiment.**
J. P. Norris, K. S. Wood.
X–ray astronomy '84, p. 267 – 270 (1985). – See Abstr. 012.033.
A search for X–ray pulsars in the HEAO A–1 pointed observations is being performed using Fast Fourier Transform techniques. The survey will ultimately include much of the region within 15° of the galactic plane as well as the Magellanic Clouds. A new pulsar, H 0850–42, with a period of 1.78 seconds has already been discovered in the survey.

142.062 **From the Universe at large to the stars near us: an X–ray survey at faint fluxes.**
I. M. Gioia.
X–ray astronomy '84, p. 349 – 354 (1985). – See Abstr. 012.033.
The author has used the current MSS (*Medium Sensitivity Survey*) sample to investigate in detail the behavior of the different classes of X–ray selected objects. Moreover she has discovered several unusual X–ray emitters of high individual interest, thus the MSS has become also one of the most effective means to discover the unexpected in the faint X–ray sky. The author briefly summarizes the major results of the survey and considers those obtainable with a much enlarged sample of about 1000 sources.

142.063 **Identifications of X–ray point sources in the LMC.**
A. P. Cowley, D. Crampton, J. B. Hutchings, D. J. Helfand, T. T. Hamilton, J. R. Thorstensen, I. R. Thompson.
X–ray astronomy '84, p. 399 – 402 (1985). – See Abstr. 012.033.
X–ray positions from Einstein Observatory data are presented for 48 point sources in the direction of the Large Magellanic Cloud. The authors present data on the optical counterparts and characteristics of these X–ray sources.

142.064 **Optical counterparts of M31 X–ray sources.**
D. Crampton, J. B. Hutchings, D. J. Schade, A. P. Cowley, L. P. van Speybroeck.
X–ray astronomy '84, p. 403 – 406 (1985). – See Abstr. 012.033.
Einstein Observatory X–ray positions for 117 sources in the direction of M31 have been examined on 1000 Å mm^{-1} slitless spectra plates obtained with the CFHT. Optical counterparts for about a third of the sources are suggested.

142.065 **New identifications of bright X–ray sources with the HEAO–1 scanning modulation collimator.**
D. A. Schwartz, H. Bradt, D. Buckley, J. Patterson, R. Remillard, W. Roberts, I. Tuohy.
X–ray astronomy '84, p. 455 – 458 (1985). – See Abstr. 012.033.
The authors present 3 candidate Be star systems, and two of the BL Lac objects they found.

142.066 **Fast transients: a search in X–rays for short flares, bursts, and related phenomena.**
A. Connors, P. J. Serlemitsos, J. H. Swank.
Astrophys. J., Vol. 303, No. 2, p. 769 – 779 (1986).
A survey of HEAO 1 A–2 source–free scanning data has uncovered five new short transients above a threshold (6 σ) of 4 millicrabs, with durations between 60 and 2000 s. Their peak fluxes ranged from $\sim 10^{-10}$ to 10^{-9} erg $cm^{-2}s^{-1}$ (2 – 20 keV). Together with three transients found previously, this small number constrains the total number of such events per year to be between 10^4 and 2×10^5 over the whole sky above the present threshold. Spectral data for these transients are also presented. It is suggested that most of the new events were hard flares from dMe–dKe stars.

142.067 **Energy spectra of X–ray pulsars.**
F. Nagase.
Japan – U.S. seminar on galactic and extragalactic compact X–ray sources, p. 7 – 17 (1985). – See Abstr. 012.034.
Properties of an iron emission line from X–ray pulsars are investigated using the gas scintillation proportional counters aboard the Tenma satellite. The ionization state of circumstellar matter surrounding the neutron star is discussed based on the observed energies of the emission line and the absorption edge of iron. Soft–excess features in the energy spectra observed from Vela X–1 can be accounted for in terms of the inhomogeneous absorption by clumpy matter. Iron line emissions at 6.4 keV measured during the eclipse period of Vela X–1 suggest a significant contribution of the extended stellar wind to the fluorescent emissions of iron line.

142.068 **Type I X–ray bursts.**
M. Matsuoka.
Japan – U.S. seminar on galactic and extragalactic compact X–ray sources, p. 45 – 59 (1985). – See Abstr. 012.034.
Recent problems on type I X–ray bursts are presented, mainly based on Hakucho and Tenma data. A remarkable result of absorption lines in X–ray burst spectra, if a gravitational redshift is determined from the result, is quite important to give the mass to radius ratio of neutron stars.

142.069 **Recent progress in the theory of double peaked bursts.**
T. Hanawa.
Japan – U.S. seminar on galactic and extragalactic compact X–ray sources, p. 61 – 67 (1985). – See Abstr. 012.034.
Apparent expansion during a double peaked burst is discussed according to the model that the outer envelope of an X–ray bursting neutron star is pushed upward by radiation between the two peaks. From observations of color temperature, one can derive some physical quantities such as the mass of the hydrogen–rich envelope, the mass loss rate by the stellar wind, and so on.

142.070 **Theoretical studies of X–ray burst sources.**
R. E. Taam.
Japan – U.S. seminar on galactic and extragalactic compact X–ray sources, p. 69 – 78 (1985). – See Abstr. 012.034.
Recent progress in the theoretical understanding of type I and type II X–ray bursts is presented. Results of model atmosphere calculations for X–ray bursting neutron stars are reported in which the effects of Comptonization, as well as non gray opacities are fully taken into account. The spectra are briefly discussed in the context of the super Eddington flux problem. In addition, a study of the non linear development of accretion disk instabilities based upon the standard α disk model is described which may be relevant to the behavior exhibited by the rapid burster MXB 1730–335.

142.071 **Spectra of X–ray bursts.**
J. van Paradijs.
Japan – U.S. seminar on galactic and extragalactic compact X–ray sources, p. 79 – 88 (1985). – See Abstr. 012.034.
In this paper a brief review is given of various models which have been used to describe the spectral energy distributions observed in type I X–ray bursts. These models are discussed in order of increasing complexity, from the very simplest (i.e. blackbody)

models to the numerical model atmospheres which are now becoming available. The latter provide a reasonable first–order explanation of the deviations from blackbodies, evidence for which has been found from recent Tenma and Exosat observations of X–ray burst spectra of MXB 1636–53.

142.072 **X–ray burst sources, some remaining problems in 1985.**
W. H. G. Lewin.
Japan – U.S. seminar on galactic and extragalactic compact X–ray sources, p. 89 – 100 (1985). – See Abstr. 012.034.

Contents: Spectral features – absorption lines. Formation of low–mass X–ray binaries outside globular clusters. Highly irregular type I burst intervals. 5 – 10 minute type I burst intervals. Burst peak luminosity ≫ Eddington limit? Type II bursts from the Rapid Burster. Blackbody radii derived from the persistent X–ray emission.

142.073 **Type–I X–ray burst from the Rapid Burster.**
Y. Tawara, D. Y. Wang.
Japan – U.S. seminar on galactic and extragalactic compact X–ray sources, p. 109 – 112 (1985). – See Abstr. 012.034.

Twenty–eight type–I X–ray bursts from the Rapid Burster were observed with Tenma in August 1983. Burst properties were found to be similar to those of bursts from other burst sources. A remarkable change in burst properties was found before and after August 20.2 UT. Comparison with other bursters suggest that it is common for many bursters to have two groups of individual burst properties.

142.074 **Time–scale invariant profile of the type II bursts from the Rapid Burster.**
N. Kawai, H. Inoue, Y. Ogawara, Y. Tanaka, Y. Tawara, H. Kunieda.
Japan – U.S. seminar on galactic and extragalactic compact X–ray sources, p. 113 – 117 (1985). – See Abstr. 012.034.

The Tenma satellite observed the Rapid Burster in August 1983 and July 1984. The analysis shows that the time profile in the decay portion of most type II bursts is essentially similar to each other except for the absolute time scale. This profile has a characteristic feature with three to four humps following the broad maximum, which is reproduced to a fine detail for the bursts with the durations of 2 to 50 s. The time scale of the structure is scaled proportionally to the duration of individual burst.

142.075 **Iron line from compact X–ray sources.**
K. Koyama.
Japan – U.S. seminar on galactic and extragalactic compact X–ray sources, p. 153 – 164 (1985). – See Abstr. 012.034.

Properties of iron line emission is discussed based on observations by gas scintillation proportional counters on board Tenma. Iron lines from most binary X–ray sources and active galactic nuclei are due to fluorescent X–rays. Dense matter near the Alfvén shell is most likely the source of fluorescent lines for X–ray pulsars. The puzzle of Cyg X–3 and Cir X–1 is discussed. Iron lines from thin thermal plasma can be a good tracer for star formation regions. The possibility of a new class of X–ray sources is suggested by the observation of an intense iron line in diffuse emission on the galactic plane.

142.076 **The spectra of Sco X–1 and similar bulge sources.**
J. H. Swank, P. J. Serlemitsos.
Japan – U.S. seminar on galactic and extragalactic compact X–ray sources, p. 175 – 179 (1985). – See Abstr. 012.034.

OSO–8 data (2 – 60 keV) and HEAO 2 SSS (0.5 – 4.0 keV) data are compared for Sco X–1 and a model is developed taking into account a combination of a black body and a thermal bremsstrahlung.

142.077 **Iron emission line from low–mass binary X–ray sources.**
K. Mitsuda.
Japan – U.S. seminar on galactic and extragalactic compact X–ray sources, p. 181 (1985). – See Abstr. 012.034.

Iron K–emission line was observed from two low–mass binary X–ray sources, Sco X–1 and 4U 1608–52, with the GSPC's on board Tenma.

142.078 **How common are long–term periodicities in X–ray binaries?**
J. A. Petterson.
Japan – U.S. seminar on galactic and extragalactic compact X–ray sources, p. 181 – 186 (1985). – See Abstr. 012.034.

Periodicities on a time scale of at least 10 times the binary period have now been detected and confirmed in 4 X–ray binary systems: Her X–1, LMC X–4, SS 433, and Cyg X–1. It remains uncertain whether the long period is caused by the same mechanism in all 4 cases, but if it is, the favorite mechanism at present is accretion disk precession.

142.079 **Soft and hard X–ray flares of Cir X–1 observed by Tenma. Preliminary results.**
S. Miyamoto, S. Kitamoto.
Japan – U.S. seminar on galactic and extragalactic compact X–ray sources, p. 187 – 193 (1985). – See Abstr. 012.034.

142.080 **Cygnus X–3: the dependence of the iron–line parameters on orbital phase.**
M. van der Klis.
Japan – U.S. seminar on galactic and extragalactic compact X–ray sources, p. 195 – 199 (1985). – See Abstr. 012.034.

Cygnus X–3 was observed with the GSPC on board EXOSAT on several occasions, one observation lasting for 7 orbital cycles. The width W and centroid energy E of the iron emission feature near 6.7 keV show a smooth, correlated, sinusoidal–type modulation, the iron line being widest and E being lowest just before X–ray maximum. The line flux is modulated similarly to the total flux. Two possible explanations for the phase variations of the line parameters are briefly discussed.

142.081 **The spectral variation and the iron line in Cygnus X–3.**
S. Kitamoto.
Japan – U.S. seminar on galactic and extragalactic compact X–ray sources, p. 201 – 205 (1985). – See Abstr. 012.034.

Cygnus X–3 was observed with the Scintillation Proportional Counters on board Tenma in Sept. 1983 and in Sept. 1984 in the energy range of 1.5 to 20 keV. Spectral variation depending on the 4.8–hour modulation and the phase independent flux of the iron line gives constraint to models for Cygnus X–3.

142.082 **The creation of compact objects in the Local Group.**
D. J. Helfand.
Japan – U.S. seminar on galactic and extragalactic compact X–ray sources, p. 207 – 214 (1985). – See Abstr. 012.034.

The author reviews the observational evidence, derived mainly from X–ray observations, that supernovae create compact objects – neutron stars and black holes. There are now ten cases in which a point source of radiation is more or less definitely associated with a diffuse supernova remnant. However, the large majority ($\gtrsim 75\%$) of supernova remnants exhibit no evidence of a central compact object. The author summarizes the status of X–ray surveys of Local Group galaxies.

142.083 **Compact X–ray binaries in and out of globular clusters.**
J. E. Grindlay.
Japan – U.S. seminar on galactic and extragalactic compact X–ray sources, p. 215 – 225 (1985). – See Abstr. 012.034.

The current status of observations of compact X–ray sources in globular clusters is discussed. Newly derived optical structure parameters for the globular clusters containing high luminosity sources are presented. The proposed optical candidate for the source in the cluster M15 has been observed. The suggestion that compact X–ray binaries in globular clusters can tidally capture distant giant companions and survive the eventual disruption and evaporation of the cluster to become the observed "G–star" bursters in the galactic bulge is discussed.

142.084 **New insight on the evolution of binary X–ray sources from the X–ray observations of nearby spiral galaxies.**
G. Fabbiano.
Japan – U.S. seminar on galactic and extragalactic compact X–ray sources, p. 233 – 237 (1985). – See Abstr. 012.034.

The study of spiral galaxies is a very powerful tool for resolving some of the fundamental outstanding questions of classical galactic X–ray astronomy. Future X–ray missions with increased sensitivity and angular resolutions (like AXAF and next generation telescopes) will be essential to extend this work.

142.085 **The soft X–ray background as a blast wave viewed from inside.**
R. J. Edgar.
Diss. Abstr. Int., Sect. B, Vol. 46, No. 10, p. 3480 – 3481 (1986). Thesis, The University of Wisconsin – Madison, 146 pp. (1985). Order No. DA8519761.

142.086 **Shock waves in the problem of generation of cosmic X–ray emission.**
V. A. Krol'.
Probl. nelinejn. i turbulent. protsessov v fiz. Tr. 2 Mezhdunar. rab. gruppy, Kiev, 1983. Chast' 2. Kiev, p. 307 – 310 (1985). In Russian. Abstr. in Ref. Zh., 51. Astron., 6.51.633 (1986).

142.087 **Optical counterparts of HEAO–1 X–ray sources.**
D. A. H. Buckley, I. R. Tuohy, R. A. Remillard.
Proc. Astron. Soc. Aust., Vol. 6, No. 2, p. 147 – 151 (1985).
The authors present preliminary results of a search for the optical counterparts of faint but energetic (1 – 13 keV) X–ray sources observed by the HEAO–1 satellite. The objects they have identified include active galactic nuclei, cataclysmic variables, Be binaries, and RS CVn systems. A description is also given of the identification techniques and the X–ray database, which represents the most recent flux–limited all–sky survey of hard X–ray sources.

142.088 **The evolution of the 1984 outburst of the transient X–ray source 4U 1630–47.**
A. N. Parmar, L. Stella, N. E. White.
Astrophys. J., Vol. 304, No. 2, p. 664 – 670 (1986).
The authors present the results of a series of EXOSAT observations of the X–ray transient 4U 1630 47 during an outburst in 1984. The source decayed over an interval of ~100 days from a maximum 1 – 50 keV luminosity of $\sim 2.8 \times 10^{38}$ to $\sim 4 \times 10^{36}$ergs s^{-1} (for an assumed distance of 10 kpc). The 1 – 50 keV spectrum was best represented by a soft Wien–like spectrum with a temperature ~1 keV and a high–energy power–law tail. At maximum, short time scale 7% intensity variations (1 – 7 keV) were seen with a characteristic time scale of ~20 s.

142.089 **High resolution spectrum of the peculiar optical counterpart of the X–ray binary pulsar 4U1907 + 09.**
M. Iye.
ESO Sci. Prepr., No. 430, 25 pp. (1986). To appear in Publ. Astron. Soc. Jpn.

142.090 ***Einstein* observations of the soft X–ray background close to the Galactic plane.**
S. M. Kahn, J.–P. Caillault.
Astrophys. J., Vol. 305, No. 1, p. 526 – 533 (1986). = Columbia Astrophys. Lab. Contrib., No. 300.
A map of the soft X–ray emission along a Galactic meridian within $\pm 3°$ of the Galactic plane is presented. An excess of diffuse *M*–band emission is found at low latitudes which exhibits a resolved, double–peaked profile that is roughly symmetric about $b = 0°$ with a half–width ~1° – 2°. The integrated intensity derived for that band for this component is 2×10^{-8}ergs $cm^{-2}s^{-1}sr^{-1}$. Assuming that this excess is characteristic of the whole Galactic plane, the authors perform model fits to derive constraints on the scale height and volume emissivity of the X–ray emitting material.

142.091 **GX 339–4: EXOSAT observations in the off and soft states.**
S. A. Ilovaisky, C. Chevalier, C. Motch, L. Chiappetti.
Obs. Haute Provence, Pré–Publ., No. 7, 19 pp. (1986). Submitted to Astron. Astrophys.

142.092 **2S 0918–549: optical identification and study of a new distant low–mass X–ray binary.**
C. Chevalier, S. A. Ilovaisky.
Obs. Haute Provence, Pré–Publ., No. 9, 14 pp. (1986). Submitted to Astron. Astrophys.

142.093 **A progress report on the full HEAO 1 A–2 fast X–ray transient survey.**
A. Connors, P. J. Serlemitsos, J. H. Swank.
Bull. Am. Astron. Soc., Vol. 18, No. 1, p. 636 (1986). Abstract. – See Abstr. 010.066.

142.094 **An intermediate–sensitivity survey at high galactic latitude with HEAO A–1.**
C. R. Shrader, K. S. Wood, T. Matilsky.
Astrophys. J., Suppl. Ser., Vol. 61, No. 2, p. 353 – 356 (1986).
A 314 square degree region of sky at high galactic latitude was mapped in 0.5 – 20 keV X–rays with data from the HEAO A–1 experiment. Twenty–one sources were detected above a 5σ threshold. The limiting flux is ~0.2 UFU. A table of source positions, intensities, corresponding uncertainties and proposed identifications is presented. A map of the survey region is displayed. The authors have constructed a number–flux curve representing all high galactic latitude X–ray sources detected by HEAO A–1.

142.095 **Optical photometry of A 0538–66 during quiescence.**
S. van Amerongen, F. Tesselaar, J. van Paradijs.
Inf. Bull. Variable Stars, No. 2901, 4 pp. (1986).

142.096 **Rotational status of X–ray pulsars.**
J. Ziółkowski.
Acta Astron., Vol. 35, Nos. 3–4, p. 185 – 198 (1985).
Arguments relevant for testing the hypothesis that most X–ray pulsars are rotating at equilibrium periods are presented. These arguments come from the interpretation of the observed time scales of spin–up and spin–down of X–ray pulsars and from direct observations of pulse period behaviour of some X–ray pulsars.

142.097 **Variability of emitting area during bursts from Serpens X–1.**
M. Bałucińska, M. Czerny.
Acta Astron., Vol. 35, Nos. 3–4, p. 291 – 303 (1985).
The authors have analyzed about 30 bursts from the source MXB 1837+05/Ser X–1. The data are grouped in one–second bins and they present the colour temperature and the flux measured near the earth in the energy range 1 – 20 keV (corrected for interstellar absorption) and the colour temperature. The authors try to verify the following hypothesis: (1) the colour temperature obtained from the fitting of the black–body spectrum to the counts of photons in different energy channels is equal to the effective temperature; (2) the radiating area after burst maximum is constant.

142.098 **The three types of high–mass X–ray pulsator.**
R. H. D. Corbet.
Mon. Not. R. Astron. Soc., Vol. 220, No. 4, p. 1047 – 1056 (1986).
The author considers all high–mass X–ray pulsators to show that they can be categorized into three groups which lie in different regions of the orbital versus spin period diagrams. These groups also differ in their X–ray luminosities, spectral type of the primary star, mean eccentricity and spin period variability. The existence of these groups is explained by differences in mass accretion processes between the groups.

142.099 **HEAO results on the cosmic X–ray background.**
E. Boldt.
Cosmic pathways, p. 229 – 232 (1986). – See Abstr. 003.045.
HEAO–1 results on the cosmic X–ray background are examined within the context of the dipole anisotropy measured for the 2.7K cosmic microwave background.

142.100 **Observation of an X-ray flare from Sco X-1.**
W. H. G. Lewin, G. W. Clark, W. B. Smith.
Cosmic pathways, p. 233 – 242 (1986). – See Abstr. 003.045.

The authors have observed a sudden increase in the X-ray intensity of Sco X-1 in the course of an X-ray survey of the southern sky carried out in a balloon flight on October 15, 1967. They suggest a model of the X-ray flare in which a relatively condensed mass of plasma was suddenly heated to a temperature near 30×10^6K so that its initial intense "flare" radiation was that of a gray body with its maximum emission near 7 keV. The plasma mass then exploded, decreasing rapidly in its X-ray emissivity as it expanded into a corona to which it contributed a major portion of its initial energy.

142.101 **A 6 second periodic X-ray source in Carina.**
F. D. Seward, P. A. Charles, A. P. Smale.
Astrophys. J., Vol. 305, No. 2, p. 814 – 816 (1986). With plate 12.

A serendipitous source, 1E 1048.1–5937, was discovered during Einstein imaging observations of the Carina nebula. On 1979 July 13, this source had an intensity of 0.14 IPC counts s^{-1}, and the signal was 65% pulsed with a period of 6.44 s. An earlier observation failed to detect any source with strength greater than 1/10 the above signal. The source is therefore highly variable, perhaps transient. An *EXOSAT* observation of this source on 1985 June 20 confirmed the pulse period and refined the source position to an accuracy of 10″. On the basis of the position, the authors tentatively identify the source with a $V \approx 19$ optical counterpart. The X-ray characteristics are consistent with an accretion-powered Be star binary.

142.102 **GX 9+9.**
IAU Circ., No. 4221 (1986).

142.103 **Quasi-periodic oscillations in bright galactic X-ray sources: preliminary results from recent EXOSAT observations.**
L. Stella.
Plasma penetration into magnetospheres, p. 199 – 214 (1986). – See Abstr. 012.071.

The author presents preliminary results of a series of EXOSAT observations of bright galactic X-ray sources carried out between 1985 August 28 and September 23 which led to the discovery of fast (2 – 30 Hz) quasi-periodic oscillations (QPO) in GX 17+2 and 4U1820–30, confirmed the presence of QPO in GX 349+2 and revealed a series of very interesting new features in the QPO from Rapid Burster (MXB 1730–33), when the source was in its flat-topped type II bursting mode. These new results are compared with the properties of QPO in the other three galactic X-ray sources for which this type of fast activity has been found (GX 5–1, Sco X–1 and Cyg X–2) and a classification scheme is proposed.

142.104 **X-ray emission from bursts of star formation.**
A. C. Fabian.
Star-forming dwarf galaxies and related objects, p. 383 – 391 (1986). – See Abstr. 012.074.

The X-ray emission from normal galaxies and regions of star formation is reviewed. Some extragalactic H II regions are X-ray luminous. The possible role and influence of binary stars on the evolution of a cluster of massive stars is discussed.

142.105 **Quasi-periodic oscillations in GX 5–1 and Sco X–1.**
M. van der Klis, F. Jansen.
The evolution of galactic X-ray binaries, p. 129 – 137 (1986). – See Abstr. 012.076.

This note briefly describes the frequency range, the amplitudes and correlations with source flux of quasi-periodic oscillations in the X-ray fluxes of GX 5–1 and Sco X–1.

142.106 **Quasi-periodic oscillations in the X-ray flux of Cyg X-2 and their relation to the source spectra.**
G. Hasinger.
The evolution of galactic X-ray binaries, p. 139 – 149 (1986). – See Abstr. 012.076.

The author reports the discovery of quasi-periodic oscillations in the X-ray flux of Cyg X–2 in the frequency range 30 – 45 Hz. The centroid frequency of a broad peak in the power spectrum shifts towards higher frequencies as the source intensity increases. In addition, low-frequency noise is present in the range below 15 Hz.

142.107 **Two spectral components and their luminosity-related changes of low-mass binary X-ray sources.**
K. Mitsuda, Y. Tanaka.
The evolution of galactic X-ray binaries, p. 195 – 205 (1986). – See Abstr. 012.076.

This paper reports results from an analysis of the spectral changes in X-ray binaries. These changes are correlated with intensity variations of the sources.

142.108 **Timing analysis of the galactic bulge X-ray source GX17+2.**
A. Langmeier, M. Sztajno, W. D. Vacca, J. Truemper, W. Pietsch.
The evolution of galactic X-ray binaries, p. 253 – 259 (1986). – See Abstr. 012.076.

A timing analysis of the low mass X-ray binary system GX17+2 (4U 1813–14) is presented. The analysis of 3 individual data sets, each lasting 7 hr, reveals a 1.5% modulation of the source intensity with a period of about 1.4 ± 0.1 hr. This is probably the orbital period of the system. Some evolutionary implications of such a short orbital period are discussed.

142.109 **Variability of apparent radii in the decay part of type I X-ray bursts.**
M. Matsuoka.
The evolution of galactic X-ray binaries, p. 301 – 309 (1986). – See Abstr. 012.076.

It is shown that apparent radii in the decay part of type I X-ray bursts from X 1608–52 differ from one burst to another, whereas the color temperature remains constant. Furthermore, it is shown that the apparent sizes of the emission regions differ systematically for different high-luminosity bursters. Implications of these results for neutron star models of the bursters are briefly considered.

142.110 **GX 3+1.**
IAU Circ., No. 4170 (1986).

142.111 **EXO 020528+1454.8.**
IAU Circ., Nos. 4172, 4207 (1986).

142.112 **Terzan 2.**
IAU Circ., No. 4174 (1986).

The Palomar-Green Catalog of Ultraviolet-Excess Stellar Objects.
See Abstr. 002.074.

New results from Exosat.
See Abstr. 011.001.

Old soldier's tale – how the "HAKUCHO" satellite went into orbit.
See Abstr. 013.059.

1962 – 1972 (up through UHURU).
See Abstr. 013.060.

Light weight aluminum optics.
See Abstr. 031.009.

Instrumentation for X-ray spectroscopy.
See Abstr. 035.034.

XMM imaging Bragg spectrometer with a doubly curved crystal.
See Abstr. 035.038.

Dispersive X–ray spectroscopy with transmission gratings.
See Abstr. 036.067.

X–ray imaging techniques.
See Abstr. 036.162.

Space astronomy in Japan.
See Abstr. 051.002.

Capabilities of spectroscopy in the context of X–ray astronomy in the 1990's.
See Abstr. 051.057.

The Pinhole/Occulter as a tool for X–ray astronomy.
See Abstr. 051.060.

Two–photon emission in X–ray pulsars. I. Basic formulas.
See Abstr. 063.018.

Two–photon emission in X–ray pulsars. II. Radiation transfer.
See Abstr. 063.019.

The formation of the continuum spectrum in X–ray pulsars.
See Abstr. 063.027.

Theoretical X–ray spectra of accreting magnetized neutron stars and quasars.
See Abstr. 063.066.

Radiative transfer in the accretion column of X–ray pulsars: effects from the hot spot.
See Abstr. 063.069.

About explanations for X–ray bursts with double peaks and precursors.
See Abstr. 063.077.

Wind fed X–ray binaries.
See Abstr. 065.156.

The duration of the detached phase in low–mass X–ray binaries in which the neutron star was formed by accretion–induced collapse of a white dwarf.
See Abstr. 065.159.

Binary evolution in the upper HRD.
See Abstr. 065.168.

The average X–ray lifetime of massive X–ray binaries.
See Abstr. 065.169.

Quasi–periodic oscillations from accretion disk coronas.
See Abstr. 067.006.

A qualitative model for GX5–1.
See Abstr. 067.008.

The lower mass limit of the compact object in LMC X–3.
See Abstr. 067.019.

X–ray bursts from soft X–ray transients.
See Abstr. 067.029.

Steady mass loss from X–ray bursting neutron stars with H–rich envelope.
See Abstr. 067.030.

Nonradial g–mode oscillations in X–ray bursting neutron stars.
See Abstr. 067.051.

Accreting neutron stars in close binary systems.
See Abstr. 067.059.

X–ray spectra of bursting neutron stars: effects of Comptonization and bound–free opacity.
See Abstr. 067.060.

Pulse frequency variations in pulsars.
See Abstr. 067.076.

How do X–ray pulsars make their X–rays?
See Abstr. 067.077.

X–ray spectra of bursting neutron stars and color temperature – luminosity diagram.
See Abstr. 067.078.

Cosmic X–ray jets.
See Abstr. 067.083.

On the interpretation of multi–peak structure observed in some X–ray bursts from 4U/MXB 1636–53.
See Abstr. 067.101.

The mass and radius of a neutron star in X–ray burster MXB 1636–536.
See Abstr. 067.113.

Classical and relativistic models of X–ray bursters.
See Abstr. 067.129.

Accretion disk instabilities in the Rapid Burster.
See Abstr. 067.144.

Cyclotron emission near stellar mass black holes.
See Abstr. 067.145.

X–ray spectra of bursting neutron stars and the color temperature versus luminosity diagram.
See Abstr. 067.149.

Centrifugally driven winds from X–ray pulsars.
See Abstr. 067.152.

Hydrodynamics of accretion columns.
See Abstr. 067.156.

Quasi–periodic oscillations as a probe of galactic–bulge X–ray sources.
See Abstr. 067.159.

On the nature of the quasiperiodic oscillations of low mass X–ray binaries.
See Abstr. 067.160.

Non–thermal X–rays from the Wolf–Rayet star HD 193793.
See Abstr. 112.044.

Spectroscopic observations of three X–ray selected late–type stars.
See Abstr. 114.007.

EXOSAT detection of an X–ray flare from the solar type star π^1UMa.
See Abstr. 116.005.

Stellar activity: the X–ray picture.
See Abstr. 116.027.

An unbiased X–ray sampling of stars within 25 parsecs of the Sun.
See Abstr. 116.033.

X–ray–selected M dwarfs and the diffuse X–ray background.
See Abstr. 116.046.

The dependence of He I λ10830 absorption strength upon X–ray emission in late–type stars.
See Abstr. 116.047.

Light variations in HD 36705 in October 1982.
See Abstr. 116.062.

Hard X–ray observations of Cygnus X–1 with MISO telescope.
See Abstr. 117.001.

X–ray spectra and the rotation–activity connection of RS Canum Venaticorum binaries.
See Abstr. 117.003.

A spectroscopic study of the X–ray binary V1727 Cygni.
See Abstr. 117.009.

A study of the continuum and iron *K* line emission from low–mass X–ray binaries.
See Abstr. 117.010.

Massive X–ray binaries: their physics and evolution.
See Abstr. 117.020.

Tenma observation of the X–ray pulsar 4U1626–67.
See Abstr. 117.027.

X–ray emissions from Vela X–1 during its eclipsing period.
See Abstr. 117.028.

An elliptical binary orbit model of GX 1 + 4.
See Abstr. 117.032.

EXOSAT observations of the 35 day cycle of Hercules X–1: evidence for neutron star precession.
See Abstr. 117.035.

Periodic optical line intensity variations in X Persei.
See Abstr. 117.036.

Simultaneous optical and X–ray observations of a flare on BY Draconis.
See Abstr. 117.040.

The soft X–ray superoutburst of VW Hydri.
See Abstr. 117.045.

Optical and X–ray observations of 4U2129 + 47/V1727 Cyg in an extended quiescent state.
See Abstr. 117.052.

***EXOSAT* results on V426 Ophiuchi: indications of a DQ Herculis system.**
See Abstr. 117.061.

More about active and inactive states of HZ Her = Her X–1.
See Abstr. 117.069.

H–alpha variability in X–ray transient source X0331 + 53.
See Abstr. 117.077.

Long–term observations of Cygnus X–3 with *Vela 5B*.
See Abstr. 117.089.

Detection of an X–ray flare in the RS CVn binary Sigma Coronae Borealis.
See Abstr. 117.090.

H 0538 + 608: a bright AM Herculis–type X–ray source.
See Abstr. 117.104.

The most reliable black hole yet.
See Abstr. 117.105.

Periodic dips in X–ray binaries.
See Abstr. 117.106.

Her X–1: the 35–day cycle.
See Abstr. 117.107.

The optical counterpart of the X–ray source in the core of M15.
See Abstr. 117.122.

Einstein OGS observations of Cyg X–1.
See Abstr. 117.133.

X–ray binaries – formation and evolution.
See Abstr. 117.167.

Simultaneous EXOSAT and optical observations of the pulsing X–ray binary H2252–035/AO Psc.
See Abstr. 117.178.

X–ray emission from Be star binaries.
See Abstr. 117.179.

High and low luminosity globular cluster X–ray sources.
See Abstr. 117.181.

X–ray sources of population II: moderate mass black holes accreting gas from extended low density envelopes.
See Abstr. 117.182.

X–ray emission and the winds of cataclysmic variables.
See Abstr. 117.188.

Low–mass X–ray binaries.
See Abstr. 117.189.

Accretion and precession in SS 433.
See Abstr. 117.190.

Orbital elements of the binary X–ray pulsar GX 301–2.
See Abstr. 117.206.

The optical spectrum of HDE 226868 = Cygnus X–1. II. Spectrophotometry and mass estimates.
See Abstr. 117.207.

The optical spectrum of HDE 226868 = Cygnus X–1. III. A focused stellar wind model for He II λ4686 emission.
See Abstr. 117.208.

Accretion disk instabilities.
See Abstr. 117.222.

The discovery of 3.9 hour periodic dips in the X–ray intensity of XB 1254–690.
See Abstr. 117.224.

***V*/*R* variability in X–ray emitting Be star binaries.**
See Abstr. 117.234.

Behaviour of the X–ray binary V1727 Cygni = 4U 2129 + 47 in 1985.
See Abstr. 117.263.

A secondary photometric period of TT Arietis.
See Abstr. 117.264.

X–ray binaries.
See Abstr. 117.284.

What do we know about Cyg X–1.
See Abstr. 117.288.

High–resolution spectrum of the peculiar optical counterpart of the X–ray binary pulsar 4U 1907 + 09.
See Abstr. 117.296.

Origin and evolution of X–ray binaries in globular clusters.
See Abstr. 117.304.

The three types of high mass X–ray pulsator.
See Abstr. 117.306.

Orbital period changes in massive X–ray binaries.
See Abstr. 117.307.

Some remarks on classifications of low–mass X–ray binaries.
See Abstr. 117.308.

Eclipse of a bright spot in the disk of Her X–1.
See Abstr. 117.310.

The orbital periods of the low mass X–ray binaries.
See Abstr. 117.311.

EXOSAT observations of the 35–day cycle of Her X–1; evidence for neutron star precession.
See Abstr. 117.312.

On the evolution of Her X–1.
See Abstr. 117.313.

V439 Cygni.
See Abstr. 117.322.

Study of quiescent state X–ray emission from flare stars.
See Abstr. 122.017.

X–ray detection of the Monoceros supernova remnant.
See Abstr. 125.004.

X–ray spectral observations of the Crab–like supernova remnants G21.5–0.9 and 3C 58.
See Abstr. 125.009.

High–resolution X–ray spectroscopy of the Crab Nebula and the oxygen abundance of the interstellar medium.
See Abstr. 125.023.

X–ray spectroscopy of supernova remnants.
See Abstr. 125.052.

***EXOSAT* and *Einstein* X–ray observations of the SNR 1E 1149.4–6209 in Crux: a unified picture?**
See Abstr. 125.055.

X–ray observations of supernova remnants resembling the Crab Nebula.
See Abstr. 125.056.

X–rays from non–accreting pulsars.
See Abstr. 126.053.

An X–ray survey of hot white dwarf stars: evidence for a n(He)/n(H) versus T_{eff} correlation.
See Abstr. 126.061.

Measuring interstellar dust grains from the haloes of binary X–ray sources.
See Abstr. 131.018.

Measurements of X–ray scattering from interstellar grains.
See Abstr. 131.135.

X–ray spectroscopic study of interstellar medium.
See Abstr. 131.187.

Hot super–low state gamma–ray spectrum of Cygnus X–1.
See Abstr. 143.014.

Optical study of the Geminga candidate field.
See Abstr. 143.021.

No optical pulses from the Geminga candidate.
See Abstr. 143.022.

Search for high–energy muons from Cygnus X–3.
See Abstr. 144.070.

The Einstein soft X–ray survey of the Pleiades.
See Abstr. 153.002.

Thermal X–ray emission with intense 6.7–keV iron line from the galactic ridge.
See Abstr. 155.042.

Soft X–ray maps of the Large Magellanic Cloud.
See Abstr. 156.018.

Cooling flows and the X–ray emission of elliptical galaxies.
See Abstr. 157.125.

The X–ray properties of normal galaxies. What do we know and what can we learn with the next generation of X–ray telescopes?
See Abstr. 157.126.

High–resolution X–ray and radio observations of the giant elliptical galaxies NGC 4636 and 4649.
See Abstr. 157.141.

The X–ray properties of normal galaxies.
See Abstr. 157.268.

X–ray observation of AGN's from TENMA.
See Abstr. 158.022.

VELA observations of Centaurus A X–ray activity, 1969 – 1979.
See Abstr. 158.028.

The X–ray spectrum and variability of Markarian 421.
See Abstr. 158.033.

The discovery of eight bright active galactic nuclei observed with the *HEAO 1* scanning modulation collimator.
See Abstr. 158.068.

Fast X–ray variability in NGC 4051.
See Abstr. 158.088.

The X–ray spectra of active galaxies.
See Abstr. 158.089.

X–ray spectroscopy of active galactic nuclei.
See Abstr. 158.175.

1E 1415.6 + 2557: an X–ray–selected BL Lacertae object in a luminous galaxy.
See Abstr. 158.176.

Cosmological evolution of AGNs in the X–ray band.
See Abstr. 158.190.

Rapid X–ray and optical variability in the X–ray–selected BL Lacertae object 1E 1402.3 + 0416.
See Abstr. 158.216.

Broad band X–ray spectra and time variability of selected active galactic nuclei observed with EXOSAT.
See Abstr. 158.221.

Fluorescent iron lines from active galactic nuclei.
See Abstr. 158.222.

Workshop on active galactic nuclei.
See Abstr. 158.223.

Active galactic nuclei and electron–positron pairs.
See Abstr. 158.224.

Some comments on iron emission line from NGC 4151: beam model.
See Abstr. 158.225.

X–ray spectroscopy of five BL Lacertae objects.
See Abstr. 158.267.

Detection of an intense iron line at 6.4 keV in the X–ray spectrum of NGC 4151.
See Abstr. 158.312.

***EINSTEIN* IPC spectra of quasars: first results.**
See Abstr. 159.108.

The X–ray properties of high redshift, optically selected QSOs.
See Abstr. 159.112.

Some properties of X–ray selected quasars.
See Abstr. 159.136.

X–ray observations of the Bright Quasar Survey.
See Abstr. 159.137.

Counts, evolution, and background contribution of X–ray quasars and other extragalactic X–ray sources.
See Abstr. 159.138.

X–ray properties of optically selected QSOs.
See Abstr. 159.139.

The X–ray structure of the 3C 295 cluster: a cooling flow at a redshift of 0.5.
See Abstr. 160.035.

High resolution X–ray spectroscopy of clusters of galaxies.
See Abstr. 160.059.

A hot intergalactic medium?
See Abstr. 160.065.

HEAO 2 X–ray observations of clusters of galaxies.
See Abstr. 160.067.

X–ray studies of galactic and intergalactic gas in the Pegasus I cluster.
See Abstr. 160.089.

143 Gamma-ray Sources, Gamma-ray Background

143.001 **A very high energy γ–ray burst from the Crab pulsar.**
P. N. Bhat, P. V. Ramanamurthy, B. V. Sreekantan, P. R. Vishwanath.
Nature, Vol. 319, No. 6049, p. 127 – 128 (1986).

The Crab pulsar, a source of pulsed photons over a wide range of energies, has been observed in the TeV energy range by several groups using the atmospheric Cerenkov technique. The authors (Ooty group) have identified a 15–min interval on 23 January 1985 during which pulsed TeV γ–rays were detected at a 5.1 sigma level at the same phase as the radio main pulse. The inference of transient pulsed emission is strengthened by the absence of a signal in a nearby control telescope tilted 8° away from the pulsar and evidence for signal in another telescope 11 km away. The authors present an analysis of the data from the Ooty 1984 – 1985 observations.

143.002 **The relativistic effects in localization of gamma–burst sources.**
G. A. Mersov.
Relativity in celestial mechanics and astrometry, p. 215 – 222 (1986). – See Abstr. 012.002 (IAU Symp. No. 114).

The paper discusses influence of relativistic effects in gamma–burst propagation and the time sync. It is shown that the phenomenon of gamma–source aberration, while using the localization algorithm based on the estimates of gamma–burst arrival time instants should be regarded as relativistic effect. Criterions are derived which determine whether it is necessary to account for the relativistic effects, depending on the localization accuracy.

143.003 **Interdisciplinary effort towards understanding the galactic gamma–ray source population.**
G. F. Bignami.
Cosmic radiation in contemporary astrophysics, p. 175 – 191 (1986). – See Abstr. 012.006.

Gamma–ray sources have recently, after the COS–B mission, become an astronomical reality and tend so far to remain as puzzling as they are challenging to the astrophysical interpretation. In order to best understand them, it is useful review the concept of a GRS and its time evolution since the beginning of GRA, and then move on to the GR aspect of already well–known objects. Finally, the full story is presented for the so far best case for the identification at other wavelengths of an object discovered through its gamma–ray emission: 2CG 195+04 "Geminga".

143.004 **Ultra high energy gamma–ray astronomy.**
J. Wdowczyk.
Cosmic radiation in contemporary astrophysics, p. 193 – 206 (1986). – See Abstr. 012.006.

The experimental data on ultra high energy γ–rays are reviewed and a comparison of the properties of photon and proton initiated shower is made. The consequences of the existence of the strong ultra high energy γ–ray sources for other observations is analysed and possible mechanisms for the production of ultra high energy γ–rays in the sources are discussed. It is demonstrated that if the γ–rays are produced via cosmic ray interactions the sources have to produce very high fluxes of cosmic ray particles. In fact it is possible that a small number of such sources can supply the whole Galactic cosmic ray flux.

143.005 **Secondary production of neutral pi–mesons and the diffuse galactic gamma radiation.**
C. D. Dermer.
Astron. Astrophys., Vol. 157, No. 2, p. 223 – 229 (1986).

Accelerator data is used to compare isobaric and scaling model predictions of the secondary spectra of neutral pi–mesons produced in proton–proton collisions at energies between threshold and a few GeV. The isobaric model is found superior, and is used in conjunction with a scaling model representation at high energies to recalculate the contribution of the π^0 γ–radiation to the diffuse galactic γ background. Cosmic ray–induced π^0 production of photons with energies greater than 100 MeV proceeds at a rate of 1.53×10^{-25} photons/(s – H atom), with an uncertainty of less than 15% associated with the strong interaction data. Comparison of the results is made with previous calculations of this process and with COS–B observations of the diffuse galactic γ radiation.

143.006 **Spectral evolution of pulse structures in gamma-ray bursts.**
J. P. Norris, G. H. Share, D. C. Messina, B. R. Dennis, U. D. Desai, T. L. Cline, S. M. Matz, E. L. Chupp.
Astrophys. J., Vol. 301, No. 1, p. 213 – 219 (1986).

The Hard X-Ray Burst Spectrometer (HXRBS) and the Gamma-Ray Spectrometer (GRS) on the SMM satellite have independently monitored cosmic gamma-ray bursts since launch in 1980 February. Ten relatively intense events detected by the GRS (four detected by HXRBS) have been studied in an attempt to reveal a basic pattern of spectral variability in gamma-ray bursts. In each of these bursts there appear pulses $\gtrsim 1$ s in duration which exhibit a hard-to-soft spectral evolution. There is no evidence for soft-to-hard evolution in any of the pulses. It is demonstrated that this hard-to-soft evolution is not due to time-varying absorption features at energies below 100 keV. Details of the spectral evolution of pulses are presented for two events, 1982 March 1 and November 4.

143.007 **Selection effects on the size and frequency distribution of cosmic gamma-ray bursts.**
T. Yamagami, J. Nishimura.
Astrophys. Space Sci., Vol. 121, No. 2, p. 241 – 253 (1986).

The authors investigate selection effects on the size and frequency relation (log$N$-log$S$ curve) of cosmic gamma-ray bursts. They find an indication that an effect is caused by the different time profiles of the bursts. The effect is important for small bursts. In order to avoid this selection effect, the authors mention that it is essential to use the log$N$-log$P$ relation instead of the log$N$-log$S$ relation, as already suggested by other authors. Here, P is the peak flux of the burst. After an analysis of the published data, the authors find a distribution near $P^{-3/2}$ in the range above a peak flux of 40 counts/0.25 s. They also show that the relation between the log$N$-log$P$ curve and the observed celestial distribution for the bursts can easily be explained by general arguments on a disk-like spatial distribution of burst sources in the Galaxy.

143.008 **Implications of the observation of PeV γ-rays. II. The production of high latitude Compton photons.**
C. L. Bhat, T. Kifune, A. W. Wolfendale.
Astron. Astrophys., Vol. 159, No. 1/2, p. 299 – 304 (1986).

A variety of data are used in order to determine the number of Cygnus X-3-like γ-ray sources in the Galaxy, both at the present time and in the past. The firmest number relates to contemporary sources and the number here is of order twenty. A method of studying the numbers which have existed in the past – by way of high latitude Inverse Compton photons – is enunciated in some detail; application to the currently available data suggests that there may have been the equivalent of the order of 200 sources (presumably a much smaller number of much more powerful sources) over the period $10^4 - 10^5$yr ago.

143.009 **Constraints on neutron star models of gamma-burst sources from the Einstein Observatory.**
G. Pizzichini, M. Gottardi, J.-L. Atteia, C. Barat, K. Hurley, M. Niel, G. Vedrenne, J. G. Laros, W. D. Evans, E. E. Fenimore, R. W. Klebesadel, T. L. Cline, U. D. Desai, V. G. Kurt, A. V. Kuznetsov, V. M. Zenchenko.
Astrophys. J., Vol. 301, No. 2, p. 641 – 649 (1986).

Six Einstein observations of five gamma-ray burst sources are presented and discussed. With one possible exception, no point source was detected in any of the observations. The data are interpreted in the framework of neutron star models for gamma bursters. Upper limits are derived for the surface temperatures of the neutron stars assumed to be responsible for the bursts. It is shown that the lack of soft X-ray emission may impose stringent constraints on accretion rates onto neutron stars.

143.010 **Angular sizes of EAS Cerenkov flashes with energies $>10^{13}$eV.**
B. M. Vladimirskij, Yu. L. Zyskin, Yu. I. Neshpor, A. A. Stepanyan, V. P. Fomin, V. G. Shitov.
Izv. Krymskoj Astrofiz. Obs., Tom 70, p. 152 – 158 (1985). In Russian. English translation in Bull. Crimean Astrophys. Obs., Vol. 70.

Angular sizes of EAS Cerenkov flashes were measured using the elements of the gamma-telescope GT-48. The authors present a short description of the installation and the method of measurements. The analysis of about 3000 events observed during September – October 1982 shows that for vertical showers with energies of about $2\text{x}10^{13}$eV, 75% of light from the flash is concentrated on the cone with full angle $1^\circ\!.10 \pm 0^\circ\!.20$.

143.011 **Radio radiation of cosmic gamma-ray bursts.**
L. P. Gorbachev, A. Yu. Matronchik.
Izv. Vyssh. Uchebn. Zaved., Radiofiz., Tom 28, No. 8, p. 947 – 952 (1985). In Russian. Abstr. in Ref. Zh., 51. Astron., 2.51.725 (1986).

143.012 **Les sursauts gamma cosmiques.**
J.-L. Atteia, K. Hurley.
C. R. Acad. Sci., Sér. Gén., Vie Sci., Tome 3, No. 2, p. 81 – 104 (1986).

143.013 **Analysis of imaging data from a balloon-borne coded aperture gamma-ray telescope.**
M. L. McConnell, P. P. Dunphy, D. J. Forrest, E. L. Chupp.
Bull. Am. Astron. Soc., Vol. 17, No. 4, p. 849 – 850 (1985). Abstract. – See Abstr. 010.062.

143.014 **Hot super-low state gamma-ray spectrum of Cygnus X-1.**
J. C. Ling, W. A. Mahoney, W. A. Wheaton, A. S. Jacobson.
Bull. Am. Astron. Soc., Vol. 17, No. 4, p. 850 (1985). Abstract. – See Abstr. 010.062.

143.015 **The spatial distribution of gamma-ray bursts and spectral selection effects.**
J. C. Higdon.
Bull. Am. Astron. Soc., Vol. 17, No. 4, p. 850 (1985). Abstract. – See Abstr. 010.062.

143.016 **Spectral analysis of the long-duration burst GB840304.**
M. Itoh, E. E. Fenimore, M. Fikani, R. W. Klebesadel, J. G. Laros, S. R. Kane.
Bull. Am. Astron. Soc., Vol. 17, No. 4, p. 850 (1985). Abstract. – See Abstr. 010.062.

143.017 **The gamma-ray background from a superposition of quasar jet sources.**
A. C. Sadun.
Bull. Am. Astron. Soc., Vol. 17, No. 4, p. 856 (1985). Abstract. – See Abstr. 010.062.

143.018 **Investigations of gamma-radiation and quasi-periodic pulsations of charged particles made with the low-energy gamma-telescope Nataliya 1.**
A. S. Glyanenko, A. F. Iyudin, V. A. Kaplin, V. G. Kirillov-Ugryumov, A. P. Klarin, Yu. D. Kotov, L. V. Kurnosova, V. T. Samojlenko, Yu. V. Smirnov, V. G. Tyshkevich, M. I. Fradkin, E. P. Chichkova, V. N. Yurov, G. S. Gokhale, S. V. Damle, P. K. Kunte, B. V. Sreekantan.
Tr. fiz. inst. Akad. Nauk SSSR, Tom 162, p. 6 – 54 (1985). In Russian. Abstr. in Ref. Zh., 62. Issled. Kosm. Prostranstva, 3.62.345 (1986).

143.019 **Variability of the high-energy γ-ray flux from Cyg X-3.**
Yu. I. Neshpor, Yu. L. Zyskin, B. M. Vladimirskii (*B. M. Vladimirskij*), A. A. Stepanyan, V. P. Fomin.
Bull. Crimean Astrophys. Obs., Vol. 69, p. 55 – 59 (1986). English translation of 39.143.027.

143.020 **Periodicity of gamma rays from the source 2CG 195+04.**
Yu. L. Zyskin, D. B. Mukanov.
Bull. Crimean Astrophys. Obs., Vol. 69, p. 63 – 66 (1986). English translation of 39.143.028.

143.021 **Optical study of the Geminga candidate field.**
S. Djorgovski, S. R. Kulkarni.
Astron. J., Vol. 91, No. 1, p. 90 – 97 (1986).
The authors have observed the field of the X-ray source 1E 0630+178, proposed as a probable counterpart of the γ-ray source 2CG 195+04 (Geminga). Deep CCD images of the field show two very faint optical sources within the X-ray error circle, as well as the 21 mag star (hereafter referred to as the G candidate) proposed earlier as the possible optical counterpart. The G candidate does not have a proper motion greater than 0.15 arcsec yr^{-1}. The spectrum of this object shows no strong emission or absorption features. The authors conclude that the G candidate is most probably a distant G dwarf or a nearby, cool white dwarf. They believe that one of the fainter objects within the X-ray circle is more likely to be the true counterpart of the X-ray source, and possibly of the γ-ray source as well. This result is consistent with the interpretation of Geminga as a nearby neutron star, if all the γ rays are generated by some nonthermal mechanism.

143.022 **No optical pulses from the Geminga candidate.**
S. R. Kulkarni, S. Djorgovski.
Astron. J., Vol. 91, No. 1, p. 98 – 106 (1986).
The authors have conducted several optical studies of a 21 mag star suggested to be the optical counterpart of the pulsed X-ray source 1E 0630+178 and the γ-ray source 2CG 195+04 (Geminga). They report the absence of significant optical pulsed emission over the entire period range claimed from the analysis of high-energy data. The authors find no significant variability on time scales from tens of minutes to a few months. They suggest that the true optical counterpart of the X-ray source is either one of the two fainter stars in the X-ray source error box or an as yet undetected fainter star. In either case, 1E 0630+178 is a unique X-ray source because its ratio of $L_x(0.2-2.0\ \mathrm{keV})/L_v(3000-7000\ \text{Å})$ is at least 1100, which is characteristic of galactic-bulge X-ray sources. However, its absolute magnitude is at least about $M_v \sim 20$ and not $M_v \sim 3$ mag, typical of the bulge sources.

143.023 **On long-period oscillations of the gamma-ray source Cyg X-3.**
Yu. I. Neshpor, Yu. L. Zyskin.
Pis'ma Astron. Zh., Tom 12, No. 6, p. 452 – 457 (1986). In Russian. English translation in Sov. Astron. Lett., Vol. 12.
The analysis of gamma-ray emission $E \gtrsim 2\mathrm{x}10^{12}$eV of Cyg X-3 observations showed the existence of gamma-quanta flux variations with the period $T = 19.58 \pm 0.02$ days. Indications of other possible variations with the period $T = 328$ days are also found.

143.024 **Synchrotron emission models for gamma-ray burst sources, the determination of the orientations of AM Herculis binaries, and non-adiabatic shocks on magnetic white dwarfs.**
J. J. Brainerd.
Diss. Abstr. Int., Sect. B, Vol. 46, No. 7, p. 2346 (1986). Thesis, Harvard University, 261 pp. (1985). Order No. DA8520165.

143.025 **How promising is the search for gamma-ray pulsars?**
V. Schönfelder.
Naturwissenschaften, 72. Jahrg., Heft 3, p. 133 – 138 (1985). In German.
Radiopulsars are rotating neutron stars. At present more than 330 of these objects are known. From two of them (Crab and Vela) pulsed emission has been observed at γ-ray energies, too. In spite of intensive searches in the γ-ray range no further pulsars have been found yet. Indeed, theoretical estimates on the γ-ray luminosity indicate that only the next generation of γ-ray telescopes will be sufficiently sensitive to see more of them.

143.026 **Polarimetry of cosmic sources of hard gamma-ray emission.**
A. A. Kozlenkov, I. G. Mitrofanov.
Sov. Astron., Vol. 29, No. 5, p. 591 – 593 (1985). English translation of 40.143.010.

143.027 **A fast pulsar in Cygnus X-3.**
V. A. Dogel'.
Priroda, No. 6, p. 107 (1986). In Russian.

143.028 **Chasing the Ogre (*Optical Gamma Ray Emitter*).**
N. Katz.
Astronomy, Vol. 14, No. 1, p. 99 – 101 (1986). Abstr. in Phys. Abstr., Vol. 89, No. 1279, Entry 53865 (1986).

143.029 **Observation of a gamma ray burst from Spacelab 2.**
C. A. Meegan, G. J. Fishman, R. B. Wilson, W. S. Paciesas.
Bull. Am. Astron. Soc., Vol. 18, No. 1, p. 639 (1986). Abstract. – See Abstr. 010.066.

143.030 **Gamma ray astronomy experiments at energies greater than 1 TeV.**
P. V. Ramana Murthy.
Cosmic pathways, p. 255 – 271 (1986). – See Abstr. 003.045.
Ultra-high-energy gamma rays were detected in a number of experiments from pulsars (Crab and Vela), X-ray binaries (Cyg X-3, Her X-1, Vela X-1) and extragalactic sources (Cen A, M31). Most of these objects emit TeV energy gamma rays at intensity levels of $10^{-11\pm1}\gamma\ \mathrm{cm}^{-2}\mathrm{s}^{-1}$. This review paper presents a summary of currently available observational results.

143.031 **Pulsed flux of 5×10^{14}eV gamma-rays from Cygnus X-3.**
C. L. Bhat, M. L. Sapru, H. Razdan.
Cosmic pathways, p. 272 – 286 (1986). – See Abstr. 003.045.
Atmospheric Čerenkov observations, using a wide-angle photomultiplier system, have revealed the presence of a pulsed signal from Cygnus X-3 with the characteristics modulation period of 4.8 h, duty-cycle ~5% and a detected flux of $(2.6\pm0.4)\times10^{12}\gamma\ \mathrm{cm}^{-2}\mathrm{s}^{-1}$ at $\geqslant 5\times10^{14}$eV. This is a factor of ~7.5 higher than that based on the extrapolation of the flux observed at 2×10^{15}eV by the Kiel group and suggests either a source distance of ~20 kpc or a distance of 11.6 kpc and long-term reduction in Cygnus X-3 luminosity at the rate of a factor of 2 per year.

143.032 **^{26}Al experimental results.**
P. Durouchoux.
Nucleosynthesis and its implications on nuclear and particle physics, p. 331 – 338 (1986). – See Abstr. 012.069.
Observations of galactic gamma-ray line sources are reviewed. These lines, due to the deexcitation of ^{24}Mg* in SS 433 and the beta decay reaction of ^{26}Al in the interstellar medium provide useful information on nuclear reactions in astrophysical sites and on nucleosynthesis in supernovae and on stellar winds of O and WR stars. A new high resolution gamma-ray spectrometer is briefly described.

143.033 **Cosmic gamma ray bursts.**
K. Hurley.
Plasma penetration into magnetospheres, p. 165 – 179 (1986). – See Abstr. 012.071.
The observational properties of gamma ray bursts – their time histories, energy spectra, statistical properties, and counterpart searches – are reviewed. The observational evidence for a neutron star origin is discussed. The characteristics of some of the time histories and energy spectra argue strongly in favor of such an origin, but firm evidence, say in the form of an identifiable counterpart, is still lacking.

143.034 **Gamma ray bursts.**
J. Heyvaerts.
Plasma penetration into magnetospheres, p. 180 – 198 (1986). – See Abstr. 012.071.

Gamma-ray astronomy.
See Abstr. 003.030.

Very high-energy gamma-ray sources.
See Abstr. 014.038.

Cos–B – a mission fully accomplished.
See Abstr. 051.009.

Cosmic quarkonium: a probe of dark matter.
See Abstr. 061.034.

Can neutrinos from Cygnus X–3 be seen by proton decay detectors?
See Abstr. 061.142.

Multiple Compton scattering in strongly magnetized plasma.
See Abstr. 062.116.

^{26}Al produced by Wolf–Rayet stars and the 1.8 MeV line emission of the Galaxy.
See Abstr. 065.146.

Origin of ultra–high–energy γ–rays from Cygnus X–3 and related sources.
See Abstr. 067.005.

The neutron–star oscillations model for gamma–ray bursts.
See Abstr. 067.017.

Can comet clouds around neutron stars explain gamma–ray bursts?
See Abstr. 067.020.

Production of ultra–high–energy γ–rays by accreting neutron stars.
See Abstr. 067.024.

A flare–induced cascade model of γ–ray bursts.
See Abstr. 067.039.

Gamma–ray burst annihilation lines and neutron star structure.
See Abstr. 067.096.

Gamma ray bursts.
See Abstr. 067.100.

Optical reprocessing of gamma–ray bursts.
See Abstr. 067.148.

Detection of 10^{15}eV gamma rays from Cygnus X–3.
See Abstr. 117.056.

High–energy neutrinos from Cygnus X–3.
See Abstr. 117.057.

On the origin of ultra high γ–rays from Cygnus X–3 and related sources.
See Abstr. 117.303.

V439 Cygni.
See Abstr. 117.322.

Two visual observations of relevance to the search for optical counterparts of gamma–ray sources.
See Abstr. 118.026.

A frequent optical flasher.
See Abstr. 122.130.

A possible variable star near the gamma ray burst source GBS 0526–66.
See Abstr. 123.017.

Spectral signatures in the soft γ–ray emission from PSR 0833–45.
See Abstr. 126.001.

Space distribution of pulsars in the Galaxy and the nature of gamma–sources.
See Abstr. 126.024.

Optical observations of 1E 0630 + 178, and a possible optical counterpart of Geminga.
See Abstr. 142.017.

The acceleration of cosmic rays.
See Abstr. 144.121.

The radial distribution of galactic gamma rays. III. The distribution of cosmic rays in the Galaxy and the CO–H_2 calibration.
See Abstr. 155.003.

Beobachtung von Centaurus A im MeV–Gammabereich.
See Abstr. 158.248.

The bremsstrahlung and π^0 decay as gamma–ray source in 3C 273 and Geminga.
See Abstr. 159.102.

Gravitino decay and the cosmic gamma–ray background.
See Abstr. 161.146.

144 Cosmic Rays

144.001 **The cosmic–ray source composition.**
S. H. Margolis.
Astrophys. J., Vol. 300, No. 1, p. 20 – 31 (1986).

The cosmic–ray source composition is shown to be related to the observed cosmic–ray abundances by a set of linear equations which are valid for either ad hoc path length distributions or quantitative solutions of Galactic propagation models. In the limit of no ionization energy loss, this method is exact for all path length distributions; it provides a simple way to calculate directly the source composition implied by a particular set of cross sections, observed abundances, and propagation model. This method is used to determine the source abundances of the Galactic cosmic rays in a model–independent manner.

144.002 **Propagation of injected cosmic rays under distributed reacceleration.**
M. Simon, W. Heinrich, K. D. Mathis.
Astrophys. J., Vol. 300, No. 1, p. 32 – 40 (1986).

In the stochastic cosmic–ray reacceleration model it is assumed that the particles obtain their energy not from discrete cosmic–ray sources but from subsequent reaccelerations in the Galaxy by encountering blast waves from supernova remnants. The authors present a calculation in which they assume cosmic–ray pre–acceleration by the sources, but also allow for reacceleration of propagating particles. It is shown that such a model can be accommodated by the data, but it requires a modification of the path–length distribution by predicting a weaker energy dependence of the mean traversed matter.

144.003 **A supernova for a neighbour?**
A. Wolfendale.
Nature, Vol. 319, No. 6049, p. 99 (1986).

Recently Clayton, Cox and Curtis have suggested that about 10^5a ago a supernova exploded at a distance of 15 kpc from the solar system. This note considers the implications of the hypothetical event for the origin of cosmic rays.

144.004 **Where do antiprotons come from?**
A. Buffington.
Nature, Vol. 319, No. 6050, p. 178 – 179 (1986).
This note comments on new measurements of the cosmic ray antiproton anomaly and considers a possible explanation of this anomaly proposed by C. D. Dermer and R. Ramaty (see Abstr. 144.005).

144.005 **Secondary antiproton production in relativistic plasmas.**
C. D. Dermer, R. Ramaty.
Nature, Vol. 319, No. 6050, p. 205 – 206 (1986).
Recent measurements of the low–energy flux of antiprotons ($\bar{p}$) in cosmic rays give ratios of $\bar{p}/p$ which are greater by a factor of 3 – 5 than predicted by the simple "leaky box" model of cosmic–ray propagation. The authors report on an investigation of the possibility that the low–energy $\bar{p}$ excess reflects an origin in p–p collisions in relativistic plasmas. Because of both target and projectile motion in such plasmas, the antiproton production threshold in the frame of the plasma is much lower than the threshold of antiproton production in cosmic ray interactions with ambient matter. The spectrum of the resultant antiprotons therefore extends to much lower energy than in the cosmic ray case. Sites at which matter is accreted by compact objects could be sources of the low energy antiprotons.

144.006 **Numerical modeling of the 22–year variation of the cosmic ray intensity and anisotropy.**
A. Kadokura, A. Nishida.
J. Geophys. Res., Vol. 91, No. A1, p. 1 – 11 (1986).
The authors have solved the two–dimensional time–dependent diffusion–convection equation numerically to obtain the distribution and anisotropy of cosmic rays in the heliosphere. They have assumed that the parallel and perpendicular mean free paths are proportional to the particle Larmor radius, and they have treated each proportionality constant (a,b) as a parameter. The authors have found that the set $(a,b) = (4,2)$ gives the steady state solution compatible with observations on the intensity and the solar diurnal anisotropy of cosmic rays in 0.5– to 10–GeV range as obtained at the earth. They discuss the nature of the solution in order to understand the effect of the density gradient drift motion on the cosmic ray distribution.

144.007 **Two–dimensional numerical modeling of the cosmic ray storm.**
A. Kadokura, A. Nishida.
J. Geophys. Res., Vol. 91, No. A1, p. 13 – 29 (1986).
A numerical model of the cosmic ray storm in the two–dimensional heliosphere is constructed incorporating the drift effect. The authors estimate the effect of a flare–associated interplanetary shock and the disturbed region behind it (characterized by enhancement in velocity and magnetic field, and decrease in mean free path) on the density and anisotropy of cosmic rays in the heliosphere.

144.008 **Shock acceleration as explanation for steplike changes and hysteresis in cosmic ray modulation.**
P. H. Stoker, H. Moraal.
J. Geophys. Res., Vol. 91, No. A2, p. 1355 – 1360 (1986).
When the counting rates of neutron monitors at two different cutoff rigidities are plotted against each other, steplike changes in the rigidity dependence of the modulation are observed, leading to the well–known hysteresis effect. These steplike changes are generally related to the modulation effects of solar flare–induced shocks propagating through the heliosphere. Shock acceleration as a mechanism for the steplike changes has however not yet been considered explicitly. Therefore, in this paper it is shown that the acceleration effects of multiple shocks in the outer heliosphere may indeed explain these steps.

144.009 **A theory of cosmic ray anisotropies of solar origin.**
K. Munakata, K. Nagashima.
Planet. Space Sci., Vol. 34, No. 1, p. 99 – 116 (1986).
The authors theoretically derive the first three–order anisotropies in space, based on the simulation of the diffusive propagation of cosmic rays through the model heliosphere, and show some numerical solutions to the first three harmonics of solar daily variation. They present also some dependences of the solution to the modulation parameters of the model heliosphere. In the discussion the results are compared with former studies of the higher order cosmic ray anisotropies.

144.010 **Accreting binaries, diffuse radio emission and galactic cosmic rays.**
D. J. Helfand.
News Lett. Astron. Soc. N.Y., Vol. 2, No. 9, p. 47 – 48 (1986).
Abstract. – See Abstr. 010.241.

144.011 **High–energy and ultra–high–energy cosmic rays.**
C. E. Fichtel, J. Linsley.
Astrophys. J., Vol. 300, No. 2, p. 474 – 487 (1986).
An analysis of the cosmic–ray energy spectrum, the anisotropy, and particularly the composition, taken together with our current knowledge of the Galaxy, seem to show clearly that Galactic cosmic rays must come from two different source types. Somewhere, between 10^6 and 10^7GeV the more familiar source type with its mixture of protons, about 10% He nuclei, and about 1% heavy nuclei no longer dominates, and a different source type consisting predominantly of singly charged particles prevails up to an energy of 3×10^9 or 10^{10}GeV.

144.012 **Cosmic ray abundances, elemental and isotopic.**
N. Lund.
Cosmic radiation in contemporary astrophysics, p. 1 – 26 (1986). – See Abstr. 012.006.
The relative abundances of the elements and isotopes in the cosmic radiation provides us with essential information about the origin and propagation history of this radiation. It is shown that it is very difficult to formulate a unified picture of the origin of GeV–energy cosmic rays which encompasses both the heavy nuclei and the proton–helium component.

144.013 **Stochastic particle acceleration in cosmic objects.**
R. Schlickeiser.
Cosmic radiation in contemporary astrophysics, p. 27 – 55 (1986). – See Abstr. 012.006.
A simple model for the evolution of energetic particle momentum spectra in cosmic objects is developed. Particles are accelerated by stochastic resonant scattering off hydromagnetic waves and by shock waves, but lose momentum in ionization, bremsstrahlung, radiation and spallation interactions with ambient matter, radiation and electromagnetic fields. A fundamental transport equation for the equilibrium momentum distribution function is derived which is a generalized heat–conduction equation with convection and diffusion terms both in momentum and position space. Exact analytical solutions of the latter equations in the steady–state and time–dependent case are given and compared with observations.

144.014 **Energetics of cosmic ray origin.**
A. W. Wolfendale.
Cosmic radiation in contemporary astrophysics, p. 87 – 97 (1986). – See Abstr. 012.006.
The question of the origin of cosmic rays is still an open one and further work is needed. The present article deals with the question of energetics, both by way of Galactic and extragalactic sources and processes. Within the Galaxy, attention is directed to acceleration in pulsars, supernova remnants, collapsing giant molecular clouds, novae and flare stars. Explosive events at the Galactic Centre are also considered and the possibility of origin in Galactic collapse. The relevance of the recent discovery of sources of ultra high energy γ–rays is also examined. Turning to extragalactic origin, quasars, Seyfert galaxies and a variety of cosmological processes are discussed.

144.015 **The composition, propagation and origin of high energy ($E \geqslant 10^{12}$eV/u) cosmic rays.**
R. Silberberg.
Cosmic radiation in contemporary astrophysics, p. 99 – 111 (1986). – See Abstr. 012.006.

The origin and propagation of high energy cosmic rays presents crucial problems: The shock wave acceleration theory in interstellar space and supernova remnants breaks down near 10^{14}eV. Are the high–energy cosmic rays then accelerated at pulsars or accreting neutron stars? The statistically limited JACEE data and the somewhat indirect measurements of the Maryland group of Yodh indicate an increase of nuclei with $Z \geqslant 6$ at energies $\geqslant 10^{14}$eV. Various propagation models can be discriminated above 10^{12}eV/u, due to mutually divergent predictions. Further experiments – those under design as well as those not yet planned – which should provide answers to the above problems are outlined. At 10^{15} to 10^{16}eV, the composition appears normal, while above 10^{16}eV, protons appear to dominate strongly.

144.016 **Nuclear cross sections, cosmic ray propagation and source composition.**
R. Silberberg, C. H. Tsao, J. R. Letaw.
Cosmic radiation in contemporary astrophysics, p. 113 – 133 (1986). – See Abstr. 012.006.

Most cosmic rays with atomic number $Z \geqslant 6$ suffer nuclear collisions in the interstellar gas, with transformation of nuclear composition. The isotopic and elemental source composition has to be inferred from the observed composition near the earth. The uncertainty in the inferred source composition is largely due to uncertainties in cross sections – especially for nuclides that have large secondary components. These uncertainties will be explored by comparing the calculated and observed abundances of the secondary components. Various models of cosmic ray propagation are critically explored.

144.017 **Synthesis of ideas on cosmic ray origin and propagation.**
A. W. Wolfendale.
Cosmic radiation in contemporary astrophysics, p. 135 – 147 (1986). – See Abstr. 012.006.

An attempt is made to synthesise ideas which have been put forward on cosmic ray origin and propagation. The conclusions drawn are as follows. The bulk of cosmic rays detected at earth appear to be of Galactic origin, many probably having come from supernova remnants, at least at the lowest energies. Only above 10^{19}eV does an extragalactic origin appear likely and here the Virgo cluster at the centre of our supercluster is a likely source. Concerning the controversy about continuous or "quick" particle acceleration, it appears necessary to separate origin and acceleration.

144.018 **Anisotropies of high energy cosmic rays.**
J. Wdowczyk.
Cosmic radiation in contemporary astrophysics, p. 149 – 159 (1986). – See Abstr. 012.006.

Experimental data on cosmic ray anisotropies are reviewed. It is shown that when the data are combined a significant anisotropy exists at all energies considered (above 10^{14}eV). It is interesting to note that in some energy intervals the first harmonic dominates and in others the second harmonic is more important. The existence of a significant anisotropy alone points towards a Galactic origin of cosmic rays in the energy interval $10^{14} - 10^{19}$eV. It is suggested that the Galactic cosmic rays should be described by some sort of two component model.

144.019 **New evidence on the galactic cosmic ray distribution from COS–B gamma–ray data.**
G. F. Bignami.
Cosmic radiation in contemporary astrophysics, p. 161 – 174 (1986). – See Abstr. 012.006.

The end of the COS–B mission and the near completion of the first analysis stage of its data base prompts a critical review of the work done on the topic of the diffuse gamma–ray emission from the IS galactic gas. The local (<1 kpc) ISM features are mapped for the first time in gamma–ray light, requiring a CR intensity similar to that observed at the Sun. On a wider galactic scale, however, and especially in the anticenter direction it now seems that the COS–B gamma–ray data do not reveal the expected radial decrease of the proton CR component, as expected from the standard galactic origin theory.

144.020 **Cosmic rays and the interstellar medium.**
A. W. Wolfendale.
Cosmic radiation in contemporary astrophysics, p. 207 – 216 (1986). – See Abstr. 012.006.

It is inevitable that there is a close connection between cosmic rays and the ISM insofar as the propagation of cosmic rays is conditioned by the magnetic field in the ISM and the cosmic rays interact with the gas (and photon fluxes) in this medium. This paper deals with both topics. Propagation effects manifest themselves as an anisotropy in arrival directions and a review is given of anisotropy measurements and their interpretation. The status of studies of cosmic ray interactions is examined with particular reference to the information about the ISM itself which comes from observations of the flux of secondary γ–rays produced by cosmic ray interactions with gas, the situation regarding molecular gas in the Inner Galaxy being of particular concern.

144.021 **Supernova origin of cosmic rays.**
C. L. Bhat, B. P. Houston.
Cosmical gas dynamics, p. 93 – 100 (1985). – See Abstr. 012.007.

The authors have recently presented evidence from γ–ray observations for the acceleration of cosmic rays in the Loop I supernova remnant. The cosmic ray intensity thus inferred is in agreement with the predictions of the shock acceleration model of Blandford and Cowie. The authors examine this model further, specifically by comparing its predictions with the presently available information on the cosmic–ray pre–history as well as with the cosmic ray anisotropy measurements in the energy range $10^{9} - 10^{15}$eV.

144.022 **Shocks and cosmic ray acceleration.**
W. I. Axford.
Cosmical gas dynamics, p. 119 – 130 (1985). – See Abstr. 012.007.

For more than 20 years it has been known that collisionless shocks in the interplanetary medium can modulate and accelerate particles to energies very much higher than the characteristic thermal energy to be expected from a shock. It has since been recognized that this process is capable of producing the correct source spectrum and sufficiently efficient energy conversion to account for galactic cosmic rays, provided the shock waves are those associated with supernova remnants. As might be expected, various problems have emerged and the whole question of how energetic particles interact with shock waves in the interstellar medium has been found to be very complex. It is impossible to review all aspects of shock acceleration in these pages and so the discussion is restricted to the "linear" situation where the effects of cosmic ray pressure and streaming on the plasma motion and turbulence are not considered.

144.023 **Time–dependent cosmic ray/gas interactions.**
E. A. Dorfi.
Cosmical gas dynamics, p. 137 – 142 (1985). – See Abstr. 012.007.

Time–dependent effects of cosmic ray acceleration are discussed for plane shocks and for a supernova–remnant evolution. The non–linear reactions on the background plasma modify the shock structure significantly. For a simple model of a supernova–remnant one gets up to 12% of E_{SN} in high energy particles at $t = 1.9 \times 10^{14}$sec having used the parameters, supernova energy $E_{SN} = 10^{51}$erg and mean diffusion coefficient for the cosmic rays $\varkappa = 10^{27}\text{cm}^2\text{sec}^{-1}$.

144.024 **Sporadic acceleration of cosmic rays.**
R. Cowsik.
Astron. Astrophys., Vol. 155, No. 2, p. 344 – 346 (1986).

The spectra of primary and secondary cosmic rays generated by the Fermi–acceleration process is derived analytically, exact to all orders. The observed *L*/*M* ratios in galactic cosmic rays constrain the frequency of encounters and the amount of acceleration by this process so severely as to imply that cosmic rays at high energies suffer negligible reacceleration in the interstellar medium.

144.025 **Is Cygnus X–3 a quark star?**
M. M. Waldrop.
Science, Vol. 231, No. 4736, p. 336 – 338 (1986).
From a distance of 37,000 light years, the most luminous X–ray source in the Galaxy seems to be showering the earth with a new kind of particle; could it be quark matter?

144.026 **Signals from Cygnus X–3.**
T. K. Gaisser.
Phys. Today, Vol. 39, No. 1, p. S10 – S11 (1986). Short review paper. – See Abstr. 013.021.

144.027 **Cosmic ray generation in the interplanetary space.**
B. A. Tverskoj.
Priroda, No. 1, p. 11 – 19 (1986). In Russian.

144.028 **Interplanetary cosmic–ray radial and latitudinal gradients derived in 1984 using *IMP 8*, *Voyager*, and *Pioneer* data.**
W. R. Webber, J. A. Lockwood.
Astrophys. J., Vol. 302, No. 1, p. 511 – 516 (1986).
The authors have examined the interplanetary cosmic–ray radial and latitudinal gradients during a period of relatively stable cosmic–ray intensity in 1984 when the intensity was recovering from the minimum in solar cycle 21. Data from cosmic–ray telescopes on *IMP 8*, *Voyager 2*, and *Pioneer 10*, all essentially in the ecliptic plane, and *Voyager 1* at an average heliographic latitude ~24°, are used. It is concluded that: (1) the integral radial gradient was significantly less than it was throughout the period 1977 – 1982; (2) in 1984 this gradient was a function of radius, decreasing from 2.23% AU^{-1} at 14.5 AU to 1.77% AU^{-1} at 34.0 AU; (3) data from *Voyager 1* at 20.2 AU and 24° north of the ecliptic plane indicate a small latitude gradient of ~–0.04% deg^{-1}, not inconsistent with zero.

144.029 **Cosmic–ray shock acceleration in oblique MHD shocks.**
G. M. Webb, L. O'C. Drury, H. J. Völk.
Astron. Astrophys., Vol. 160, No. 2, p. 335 – 346 (1986).
A one–dimensional, steady–state hydrodynamical model of cosmic–ray acceleration at oblique MHD shocks is presented. Upstream of the shock the incoming thermal plasma is subject to the adverse pressure gradient of the accelerated particles, the **JxB** force, as well as the thermal gas pressure gradient. The efficiency of the acceleration of cosmic–rays at the shock as a function of the upstream magnetic field obliquity and upstream plasma beta is investigated. Astrophysical applications of the results are briefly discussed.

144.030 **Characteristic recovery times of Forbush–type decreases in the cosmic radiation. 1. Observations at earth at different energies.**
J. A. Lockwood, W. R. Webber, J. R. Jokipii.
J. Geophys. Res., Vol. 91, No. A3, p. 2851 – 2857 (1986). Paper presented at the Workshop on Solar Cycle Modulation of Cosmic Rays.
The authors report on the recovery characteristics of about 30 Forbush–type decreases which occurred from 1972 to 1984, using data from a cosmic ray telescope on IMP near earth and neutron monitors, located on earth. The relative amplitudes of these decreases, their characteristic recovery times, and the energy dependence of these quantities are examined.

144.031 **The dependence of solar modulation on the sign of the cosmic ray particle charge.**
M. Garcia–Munoz, P. Meyer, K. R. Pyle, J. A. Simpson, P. Evenson.
J. Geophys. Res., Vol. 91, No. A3, p. 2858 – 2866 (1986). Paper presented at the Workshop on Solar Cycle Modulation of Cosmic Rays.
The authors compare the solar modulation of galactic cosmic ray helium and electrons at 1 AU, within the 600– to 1000–MV magnetic rigidity interval, during a period covering 20 years, from the solar minimum of 1965 through 1984. Comparison of the time–intensity variations during the two solar maxima around 1970 and 1981 shows that after 1970 the intensity of helium recovers earlier than that of the electrons, whereas after 1981 the intensity of electrons recovers earlier than that of helium. Within the measured magnetic rigidity interval the flux of helium to electrons (He/*e*) undergoes a major increase during 1969 – 1971 and a major decrease during 1979 – 1983, periods that include the times of reversal of the polarity of the solar polar magnetic fields. A conventional model of solar modulation can describe the modulation of protons and helium from 1965 through 1984, but its predictions depart significantly from the observed modulation of electrons around and after the 1981 solar maximum.

144.032 **Differential measurement and model calculations of cosmic ray latitudinal gradient with respect to the heliosphere current sheet.**
S. P. Christon, A. C. Cummings, E. C. Stone, K. W. Behannon, L. F. Burlaga, J. R. Jokipii, J. Kota.
J. Geophys. Res., Vol. 91, No. A3, p. 2867 – 2877 (1986). Paper presented at the Workshop on Solar Cycle Modulation of Cosmic Rays.
Simultaneous magnetic field and charged particle measurements from the Voyager spacecraft with heliographic latitude separations of >10° are used to investigate the distribution of ~1–GeV galactic cosmic ray protons with respect to the heliospheric current sheet in the outer solar system. By comparing the ratio of cosmic ray flux at Voyager 1 to that at Voyager 2 during periods of relatively quiet interplanetary conditions when the spacecraft are either both north or both south of the heliospheric current sheet, the authors derive an average latitude component of the gradient of the cosmic ray flux on opposite sides of the current sheet under restricted interplanetary conditions of –0.22±0.03%/deg, equivalent to a decrease of ~1%/AU away from the current sheet at ~12 AU. The results for these limited periods are in qualitative agreement with propagation models incorporating particle drifts.

144.033 **Variation of cosmic rays and solar wind properties with respect to the heliospheric current sheet. 2. Rigidity dependence of the latitudinal gradient of cosmic rays at 1 AU.**
G. Newkirk Jr., J. Asbridge, J. A. Lockwood, M. Garcia–Muñoz, J. A. Simpson.
J. Geophys. Res., Vol. 91, No. A3, p. 2879 – 2884 (1986). Paper presented at the Workshop on Solar Cycle Modulation of Cosmic Rays.
The authors discuss the role which empirical determinations of the latitudinal variation of cosmic rays with respect to the current sheet may have in illuminating the importance of the cross–field drift of particles in the large–scale heliospheric magnetic field. Using *K* coronameter observations and measured solar wind speeds, they have determined the latitudinal gradients with respect to the current sheet for cosmic rays in four rigidity ranges. Gradients vary between approximately –2 and –50% per astronomical unit. The rigidity dependence of the decrease of cosmic ray flux with distance from the current sheet lies between $P^{-0.72}$ and $P^{-0.86}$, with the exact dependence being determined by the definition used for the median rigidity of each monitor.

144.034 **Cosmic rays near the heliospheric current sheet. 2. An ensemble approach to comparing theory and observation.**
J. R. Jokipii, J. Kota.
J. Geophys. Res., Vol. 91, No. A3, p. 2885 – 2888 (1986). Paper presented at the Workshop on Solar Cycle Modulation of Cosmic Rays.
The authors report on a quantitative comparison between theoretical predictions and observations of the intensity of galactic cosmic rays near the interplanetary current sheet. In the present work, comparison of their model calculations is made with a statistical analysis of observations of galactic cosmic rays at the earth and the simultaneous position of the current sheet. The authors find agreement between theory and observation at energies of the order of a few GeV.

144.035 **The effect of the heliospheric current sheet on cosmic ray intensities at solar maximum: two alternative hypotheses.**
B. T. Thomas, B. E. Goldstein, E. J. Smith.
J. Geophys. Res., Vol. 91, No. A3, p. 2889 – 2895 (1986). Paper presented at the Workshop on Solar Cycle Modulation of Cosmic Rays.
There is now a growing awareness that solar cycle related changes in the large–scale structure of the heliospheric current sheet may play an important role in the modulation of galactic cosmic rays. To date, attention has been focused on the configuration of the current sheet at times near solar minimum when the current sheet structure is relatively simple. This paper attempts to extend the previous analyses into the period near solar maximum. Two alternative hypotheses are explored: (1) that the tilt of the current sheet continues to increase as solar maximum approaches, finally becoming vertical and overturning, and (2) that the single sheet structure breaks down near solar maximum and the sun at this time sheds the poloidal flux of the previous cycle and develops a new field structure of the opposite polarity. It is found that both hypotheses lead to variations in cosmic ray intensity comparable to those actually observed over the solar cycle.

144.036 **Changes in the energy spectrum of anomalous oxygen during 1977 – 1985.**
A. C. Cummings, E. C. Stone.
J. Geophys. Res., Vol. 91, No. A3, p. 2896 – 2902 (1986). Paper presented at the Workshop on Solar Cycle Modulation of Cosmic Rays.
The authors have used data from the cosmic ray experiments on the Voyager and Pioneer 10 spacecraft during the period 1977 – 1985 to measure the energy spectrum of oxygen with 4 – 125 MeV/nucleon, with particular emphasis on the energy interval from ~4 to 30 MeV/nucleon, where the "anomalous" component of cosmic rays is dominant. They find that the spectrum of anomalous oxygen changes dramatically after the middle of 1980, with the peak or plateau region of the differential spectrum shifting to a higher energy. This change does not appear to be due to the increasing radial distance of the spacecraft in this time period, nor does it appear to be simply correlated with the level of modulation. The energy shift also appears to be significantly greater in the ecliptic plane than at 24° north latitude.

144.037 **A perturbation approach to cosmic ray transients in interplanetary space.**
P. P. Chih, M. A. Lee.
J. Geophys. Res., Vol. 91, No. A3, p. 2903 – 2913 (1986). Paper presented at the Workshop on Solar Cycle Modulation of Cosmic Rays.
A general linear response analysis is presented for investigating cosmic ray transients in interplanetary space. The diffusion equation for energetic particle transport in the solar wind is first linearized about a stationary configuration. The perturbation in the particle omnidirectional distribution function resulting from perturbations to the solar wind velocity, drift velocity, or spatial diffusion tensor corresponding to transient solar wind conditions may then be found by solving an inhomogenous boundary value problem. As illustrative examples of the general analysis, transient perturbations in the cosmic ray number density due to variations in the spatial diffusion coefficient are computed in a one–dimensional solar wind based on the convection–diffusion equation, which disregards cosmic ray energy changes in the solar wind.

144.038 **Phase propagation of the solar modulation of galactic cosmic rays.**
M. A. Forman, F. C. Jones, J. S. Perko.
J. Geophys. Res., Vol. 91, No. A3, p. 2914 – 2916 (1986). Paper presented at the Workshop on Solar Cycle Modulation of Cosmic Rays.
The phase of the 11–year galactic cosmic ray variation, due to a varying rate of emission of long–lived, propagating regions of enhanced scattering, travels faster than the scattering regions themselves. The radial speed of the 11–year phase in the quasi–steady force field approximation is exactly twice the speed of the individual episodic decreases. A time–dependent numerical solution for 1–GeV protons at 1 and 30 AU gives a phase speed which is 1.85 times the propagation speed of the individual decreases.

144.039 **Cosmic ray variations and turbulent flow systems: 0.3 – 1.0 AU; 1977 – 1980.**
L. F. Burlaga, M. L. Goldstein, F. B. McDonald, A. J. Lazarus, F. Mariani, F. M. Neubauer, R. Schwenn.
J. Geophys. Res., Vol. 91, No. A3, p. 2917 – 2927 (1986). Paper presented at the Workshop on Solar Cycle Modulation of Cosmic Rays.
Relations among the relative counting rates of cosmic ray protons and the magnetic fields and flows observed by Helios 1 between 0.3 AU and 1 AU in the period January 1977 to August 1980 are discussed.

144.040 **Particle acceleration at a termination shock. 1. Application to the solar wind and the anomalous component.**
J. R. Jokipii.
J. Geophys. Res., Vol. 91, No. A3, p. 2929 – 2932 (1986). Paper presented at the Workshop on Solar Cycle Modulation of Cosmic Rays.
The results of a numerical study of the diffusive acceleration of charged particles at the termination shock of the solar wind are reported. In the model a realistic magnetic field structure is employed which is similar to that observed in the solar wind. In addition to causing spatial variation of the diffusion tensor, a major effect of the magnetic field is to cause guiding–center drifts of the accelerated particles, both in the solar wind and at the shock. It is demonstrated that the inclusion of the drifts has a large effect on the acceleration. It is concluded, furthermore, that acceleration at the termination shock, in conjunction with drifts, can explain several observed features of the anomalous component.

144.041 **Latitudinal extent of the heliospheric current sheet and modulation of galactic cosmic rays.**
E. J. Smith, B. T. Thomas.
J. Geophys. Res., Vol. 91, No. A3, p. 2933 – 2942 (1986). Paper presented at the Workshop on Solar Cycle Modulation of Cosmic Rays.
This paper reports the results of an analysis intented to investigate the possible relation between observed changes in the current sheet inclination and cosmic ray intensities observed at the earth and in space by spacecraft during the recent solar maximum.

144.042 **Cosmic ray density gradients related to north–south asymmetry in activity on the sun.**
D. B. Swinson, M. A. Shea, J. E. Humble.
J. Geophys. Res., Vol. 91, No. A3, p. 2943 – 2949 (1986).
Data are presented from underground cosmic ray telescopes in both the northern and southern hemispheres from 1965 to 1983, analyzed as a function of interplanetary magnetic field sense, to examine the long–term properties of the cosmic ray density gradient perpendicular to the ecliptic plane.

144.043 **Characteristics of large Forbush–type decreases in the cosmic radiation. 2. Observations at different heliocentric radial distances.**
W. R. Webber, J. A. Lockwood, J. R. Jokipii.
J. Geophys. Res., Vol. 91, No. A4, p. 4103 – 4110 (1986).
Cosmic ray data from IMP 8, Voyager 1 and 2, Pioneer 10 are used to investigate the heliocentric radial dependence of the characteristics of about 20 Forbush–type transient decreases which occurred from 1978 to 1984. These characteristics include (1) the recovery time, (2) the amplitude, and (3) the time to decrease to minimum. A model of the Forbush–type decrease is proposed to explain the observed radial dependence of the recovery time and time to minimum of the decrease. The implications of these results for understanding the relationship between Forbush–type decreases and the 11–year variation are discussed.

144.044 **The enigma of cosmic ray zonal modulation.**
T. N. Charakhch'yan.
Geomagn. Aehron., Tom 26, No. 1, p. 10 – 17 (1986). In Russian. English translation in Geomagn. Aeron.

144.045 **Dynamics of the size of the 11–year modulation region of galactic cosmic rays.**
T. N. Charakhch'yan.
Geomagn. Aehron., Tom 26, No. 2, p. 191 – 196 (1986). In Russian. English translation in Geomagn. Aeron.

144.046 **Possible interpretation of changes in the shape of the cosmic ray fluctuation power spectrum before interplanetary medium disturbances.**
L. I. Dorman, I. Ya. Libin, K. F. Yudakhin.
Geomagn. Aehron., Tom 26, No. 2, p. 204 – 208 (1986). In Russian. English translation in Geomagn. Aeron.

144.047 **Height dependence of secondary cosmic ray variations.**
A. V. Belov, L. I. Dorman, I. V. Sirotina.
Geomagn. Aehron., Tom 26, No. 2, p. 209 – 213 (1986). In Russian. English translation in Geomagn. Aeron.

144.048 **The maximum entropy principle in cosmic ray transport theory.**
P. Hick, G. Stevens, J. van Rooijen.
The sun and the heliosphere in three dimensions, p. 355 – 358 (1986). – See Abstr. 012.016.

The authors describe a procedure to obtain an approximate solution to the cosmic ray transport equation, which, contrary to the familiar diffusion approximation, is valid also for large anisotropies.

144.049 **Modulation of galactic cosmic rays in the heliosphere.**
R. B. McKibben.
The sun and the heliosphere in three dimensions, p. 361 – 374 (1986). – See Abstr. 012.016.

Observations of the intensity of galactic cosmic rays and anomalous components near earth and in the outer heliosphere are providing important tests for theories of solar modulation. The most recent observations show effects that seem to require that models for modulation include gradient and curvature drifts as well as the conventional processes of convection, adiabatic deceleration, and diffusion. New observations are required to define the interplay of these processes in the three–dimensional heliosphere.

144.050 **Effects of three–dimensional heliospheric structures on cosmic–ray modulation.**
J. R. Jokipii.
The sun and the heliosphere in three dimensions, p. 375 – 387 (1986). – See Abstr. 012.016.

Extrapolation of present understanding of the regions near the heliospheric equator to high heliographic latitudes leads to the conclusion that particle drift in the large–scale magnetic field plays an important role in cosmic–ray modulation. The large–scale, three–dimensional structure of the interplanetary magnetic field is therefore very important in understanding cosmic rays. Several key observed modulation effects are summarized, each of which is a natural consequence of drift, but which requires special assumptions if drift plays no role. It is concluded that particle drifts play an important and possibly dominant role in transport in the heliosphere.

144.051 **Measurement of radial and latitudinal gradients of cosmic ray intensity during the decreasing phase of sunspot cycle 21.**
D. Venkatesan, R. B. Decker, S. M. Krimigis.
The sun and the heliosphere in three dimensions, p. 389 – 394 (1986). – See Abstr. 012.016.

This study over a 4–year period (1981 – 84) during the decreasing phase of sunspot cycle 21 utilizes data from spacecraft Voyagers 1 and 2 and the earth–orbiting satellite IMP–8 to estimate the radial and latitudinal cosmic ray intensity gradients.

144.052 **North/south asymmetry in solar activity and its effects on the high energy cosmic ray diurnal variation.**
M. A. Shea, D. F. Smart, D. B. Swinson, J. E. Humble.
The sun and the heliosphere in three dimensions, p. 395 – 400 (1986). – See Abstr. 012.016.

Using a data base extending from 1955 through 1980 the authors find that the northern hemisphere of the sun was decidedly more active than the southern hemisphere for the period 1959 through 1970. From 1971 through 1980 there was no systematic asymmetry. Examination of the diurnal variation amplitude present in the extremely high energy (approximately 100 GeV) cosmic radiation measurements indicates a consistent north/south asymmetry until about 1971 which they suggest is the result of the asymmetric solar activity.

144.053 **The anomalous component, its variation with latitude and related aspects of modulation.**
L. A. Fisk.
The sun and the heliosphere in three dimensions, p. 401 – 411 (1986). – See Abstr. 012.016.

The theory for the origin of the anomalous cosmic ray component and several possible mechanisms for its acceleration are reviewed. The predictions of these mechanisms for the latitude variations of the anomalous component that should be seen by Ulysses are discussed.

144.054 **Pick–up ions in the solar wind as a source of suprathermal particles.**
D. Hovestadt, E. Möbius, B. Klecker, G. Gloeckler, F. M. Ipavich, M. Scholer.
The sun and the heliosphere in three dimensions, p. 413 – 418 (1986). – See Abstr. 012.016.

This paper presents first direct observations of the distribution function of freshly ionized helium in the solar wind which likely has its origin in the neutral interstellar wind and probably represent the source of the anomalous helium component in cosmic rays.

144.055 **The detection of trans TeV electrons from nearby sources using the Particle Astrophysics Superconducting Magnet Facility.**
J. F. Ormes.
Bull. Am. Astron. Soc., Vol. 17, No. 4, p. 856 (1985). Abstract. – See Abstr. 010.062.

144.056 **Cosmic ray intensity variations with periods close to the 27–day periods in the 19th and 20th solar cycles.**
G. A. Bazilevskaya, E. S. Vernova, M. I. Tyasto.
Probl. zem. magn., Moskva, p. 132 – 138 (1985). In Russian. Abstr. in Ref. Zh., 51. Astron., 4.51.348 (1986).

144.057 **The effect of the rotation of the sun in cosmic ray anisotropy during sporadic and recurrent electromagnetic perturbations in the interplanetary space.**
B. D. Naskidashvili, G. L. Tsereteli, L. Kh. Shatashvili.
Tr. inst. geofiz. Akad. Nauk GSSR, Tom 53, p. 78 – 83 (1985). In Russian. Abstr. in Ref. Zh., 51. Astron., 4.51.349 (1986).

144.058 **IMF–sense–dependent cosmic ray anisotropy produced from diffusion–convection in heliosphere.**
K. Nagashima, K. Munakata, R. Tatsuoka.
Planet. Space Sci., Vol. 34, No. 5, p. 453 – 467 (1986).

An IMF–sense–dependent first order (or unidirectional) anisotropy of cosmic rays, which is produced perpendicularly to the ecliptic from the radial density gradient in the solar system, has been confirmed by Swinson. In the present paper, the authors point out the existence of IMF–sense–dependent higher order anisotropies, based on the simulation of cosmic ray diffusion–convection in the heliomagnetosphere. In order to confirm their existence, the authors demonstrate some examples of the observed cosmic ray daily variation which is supposed to be due to these anisotropies.

144.059 **Dependence of cosmic ray solar daily variation (1st, 2nd, and 3rd) on heliomagnetic polarity reversals.**
K. Nagashima, R. Tatsuoka, K. Munakata.
Planet. Space Sci., Vol. 34, No. 5, p. 469 – 482 (1986).
Using 696 station–years of neutron monitor data during the period of 1964 – 1983 and those from muon telescopes at surface and underground stations, a long term variation of cosmic ray solar daily variations (1st, 2nd, and 3rd) has been analysed in order to study its dependence on the heliomagnetic polarity reversals.

144.060 **Globular cosmic–ray diffusion in the Galaxy.**
E. G. Berezhko.
Sov. Astron. Lett., Vol. 11, No. 4, p. 250 – 251 (1985). English translation of 40.144.009.

144.061 **Arrival directions of cosmic rays of $E > 0.4$ EeV.**
R. M. Baltrusaitis, G. L. Cassiday, R. Cooper, J. W. Elbert, P. R. Gerhardy, S. Ko, E. C. Loh, Y. Mizumoto, P. Sokolsky, D. Steck.
Astrophys. J., Vol. 303, No. 1, p. 333 – 335 (1986).
The arrival directions of cosmic rays observed by the Utah Fly's Eye detector have been studied in an effort to find the sources of this high–energy radiation. No statistically significant departure from isotropy has been observed for energies greater than 0.4 EeV (1 EeV = 10^{18}eV).

144.062 **Measurements of hydrogen and helium isotopes in galactic cosmic rays from 1978 through 1984.**
R. Kroeger.
Astrophys. J., Vol. 303, No. 2, p. 816 – 828 (1986).
The author has measured the flux of the hydrogen and helium isotopes in galactic cosmic rays using the University of Chicago Instrument on the ISEE 3 spacecraft. The data were obtained during periods of solar quiet time from 1978 August through 1984 December and cover the energy range from 26 to 138 MeV per nucleon for both ^{1}H and ^{4}He, from 24 to 89 MeV per nucleon for ^{2}H, and from 43 to 146 MeV per nucleon for ^{3}He. During the observations, solar activity modulated the observed flux by about an order of magnitude. However, the abundance ratios ^{2}H/^{4}He and ^{3}He/^{4}He remained constant within the experimental uncertainties.

144.063 **Solar cycle variation of cosmic–ray north–south anisotropy and radial gradient.**
J. W. Bieber, M. A. Pomerantz.
Astrophys. J., Vol. 303, No. 2, p. 843 – 848 (1986).
Measurements from neutron monitors in Thule, Greenland, and McMurdo, Antarctica, have been used to determine yearly values of the cosmic–ray north–south anisotropy over the period 1961 – 1983. The results support earlier suggestions that the steady state north–south anisotropy has its origin in the cosmic–ray radial gradient. The long–term variation of the anisotropy may be characterized as an ~10 yr solar cycle variation of amplitude 0.028% superposed upon a mean anisotropy of 0.053%. This in turn implies that the cosmic–ray radial gradient at 10 GV rigidity oscillates between ~0.8% and ~2.6% per AU.

144.064 **The effect of gradient and curvature drift on cosmic–ray Forbush decreases.**
M. S. Mulder, H. Moraal.
Astrophys. J., Lett. Ed., Vol. 303, No. 2, p. L75 – L78 (1986).
When Forbush decreases are analyzed separately in the periods 1959 – 1969 and 1971 – 1980, it is found that during the former period there typically is a slower reset than in the latter. This is in agreement with what one would expect from drift motions of charged particles in the interplanetary magnetic field during these two periods. The effect is quite small, however, and it is barely statistically significant.

144.065 **Cosmic ray dynamics in arbitrary magnetic fields.**
V. S. Ptuskin.
Probl. nelinejn. i turbulent. protsessov v fiz. Tr. 2 Mezhdunar. rab. gruppy. Kiev, 1983, Chast' 2. Kiev, p. 322 – 324 (1985). In Russian. Abstr. in Ref. Zh., 62. Issled. Kosm. Prostranstva, 5.62.443 (1986).

144.066 **The distribution function of cosmic rays before a Forbush decrease.**
L. I. Dorman, V. Kh. Shogenov.
Geomagn. Aehron., Tom 26, No. 3, p. 496 – 497 (1986). In Russian. English translation in Geomagn. Aeron.

144.067 **Abundances and energy spectra of high energy heavy cosmic–ray nuclei.**
S. D. Barthelmy.
Diss. Abstr. Int., Sect. B, Vol. 46, No. 10, p. 3480 (1986). Thesis, Washington University, 188 pp. (1985). Order No. DA8529963.

144.068 **An investigation of cosmic ray scintillations in muons near sea level.**
R. H. Benson.
Diss. Abstr. Int., Sect. B, Vol. 46, No. 10, p. 3480 (1986). Thesis, Texas A&M University, 115 pp. (1985). Order No. DA8528302.

144.069 **Abundances of heavy cosmic rays at energies up to 1000 GeV/amu.**
M. D. Jones.
Diss. Abstr. Int., Sect. B, Vol. 46, No. 10, p. 3481 (1986). Thesis, Washington University, 172 pp. (1985). Order No. DA8529973.

144.070 **Search for high–energy muons from Cygnus X–3.**
Y. Oyama, K. Arisaka, T. Kajita, M. Koshiba, M. Nakahata, A. Suzuki, M. Takita, Y. Totsuka, T. Kifune, T. Suda, N. Sato, K. Takahashi, K. Miyano.
Phys. Rev. Lett., Vol. 56, No. 9, p. 991 – 994 (1986).
A total of 1.9×10^6 good–quality high–energy muon events were analyzed from the data of the underground Kamioka nucleon–decay experiment. No definite sign of excess flux was observed in the direction of Cygnus X–3 and in its phase interval of 0.7 to 0.8.

144.071 **The role of the energy spectrum anisotropy in the variations of the cosmic–ray fluctuations power–spectrum before interplanetary medium disturbances.**
L. I. Dorman, I. Ya. Libin, K. F. Yudakhin.
Astrophys. Space Sci., Vol. 123, No. 1, p. 53 – 58 (1986).
The variations in the form of the cosmic–ray fluctuation power spectrum as an interplanetary shock wave approaches the earth have been calculated for different values of cosmic ray anisotropy. The relevant experimental estimates of the power spectra are inferred from the data of cosmic ray detection with the ground–based neutron monitors at cosmic–ray stations. A comparison between the theoretical and experimental estimates has demonstrated an important role of the cosmic ray anisotropy spectrum in the generation of the power spectrum as the latter is rearranged before the interplanetary medium disturbances.

144.072 **The origin of cosmic rays. I. Observations of the spiral galaxy NGC 3310.**
N. Duric, E. R. Seaquist, P. C. Crane, L. E. Davis.
Astrophys. J., Vol. 304, No. 1, p. 82 – 95 (1986).
An observational technique is presented for addressing the problem of the origin of cosmic rays in galaxies by determining and comparing the distributions of cosmic rays, stars, and thermal gas. It is argued that optical continuum, emission line, and radio continuum imaging can be used to determine the distributions of the major stellar populations, the ionized gas, and the cosmic rays in galactic disks. An application of the technique is demonstrated by presenting and discussing observations of the spiral galaxy, NGC 3310.

144.073 **The origin of cosmic rays. II. The cosmic–ray distribution and the spiral structure of NGC 3310.**
N. Duric.
Astrophys. J., Vol. 304, No. 1, p. 96 – 110 (1986).
An analysis of the spiral structure of NGC 3310 points to the presence of a strong density wave and leads to a number of constraints regarding the origin and distribution of relativistic

electrons. The role of supernova remnants in the production of relativistic particles (cosmic rays) is investigated and is found to be inconsistent with the constraints. It is shown that global spiral shocks can generate cosmic rays through a Fermi–type shock acceleration process. Classical density wave theory is used to model this effect and to predict the distribution of synchrotron radiation in NGC 3310.

144.074 **The origin of cosmic rays. III. Particle acceleration by global spiral shocks.**
N. Duric.
Astrophys. J., Vol. 304, No. 1, p. 111 – 114 (1986).

Cosmic–ray acceleration by global spiral shocks in galactic disks is investigated using existing theory. The diffusive shock acceleration (DSA) mechanism is discussed in the low–velocity regime represented by density wave shocks. It is found that a medium of $T \approx 10^4$K partially to fully ionized, supports the DSA mechanism. It is concluded that existing theory permits particle acceleration by global, slow moving shocks as suggested by observational work on the spiral galaxy NGC 3310.

144.075 **Enhancement of high–energy cosmic–ray spectrum by type–II supernovae.**
Y. Takahashi, S. Miyaji, T. A. Parnell, M. C. Weisskopf, T. Hayashi, K. Nomoto.
Nature, Vol. 321, No. 6073, p. 839 – 841 (1986).

The cosmic–ray spectrum has an intensity enhancement in the energy range 10^{14}– 10^{16}eV per nucleus. Recent observations of heavy cosmic rays in this energy range indicate that the Ca/Fe ratio may be as large as 10 times the solar value. The authors suggest that pulsars in type–II supernova remnants are the origin of this component of the cosmic–ray spectrum.

144.076 **The primary nucleon spectrum in the energy range $1 < E < 100$ TeV.**
O. C. Allkofer, D. P. Bhattacharyya, J. N. Capdeville.
Lett. Nuovo Cimento, Vol. 44, Ser. 2, No. 4, p. 227 – 230 (1985). Abstr. in Phys. Abstr., Vol. 89, No. 1273, Entry 18609 (1986).

144.077 **Feynman scaling and hadron–muon density in air showers.**
J. Sidhanta, R. Roychoudhury.
Can. J. Phys., Vol. 63, No. 7, p. 973 – 975 (1985). Abstr. in Phys. Abstr., Vol. 89, No. 1273, Entry 18612 (1986).

144.078 **Study of high–energy hadrons in EAS.**
H. Sakuyama, N. Suzuki, K. Watanabe.
Lett. Nuovo Cimento, Vol. 44, Ser. 2, No. 4, p. 253 – 256 (1985). Abstr. in Phys. Abstr., Vol. 89, No. 1273, Entry 18613 (1986).

144.079 **Search for cosmic nuclearites at sea level.**
K. Nakamura, H. Horie, T. Takahashi, T. Tanimori.
Phys. Lett. B, Vol. 161B, No. 4 – 6, p. 417 – 421 (1985). Abstr. in Phys. Abstr., Vol. 89, No. 1274, Entry 23730 (1986).

144.080 **Measurement of heavy particle mass distribution in cosmic rays at sea–level.**
D.–r. Wang, C.–j. Wang, X.–l. Wang, D.–w. Wang, N.–z. Liu, Y.–s. Li, Z.–p. Li, C.–z. Li, Z.–g. Li, J.–f. Lin, Y.–l. Xu, Y.–b. Chen, Z.–p. Chen, R.–f. Xu, J.–w. Xi, Z.–y. Yao, H.–q. Zhao, Y.–y. Jiang, J.–s. Lou, Y.–g. Xie, J.–q. Zhang.
Phys. Energ. Fortis Phys. Nucl., Vol. 9, No. 5, p. 513 – 519 (1985). In Chinese. Abstr. in Phys. Abstr., Vol. 89, No. 1275, Entry 30127 (1986).

144.081 **The vertical differential momentum spectrum of the muon at 3200 m above sea–level.**
Y.–g. Li, H.–n. Li, S.–p. Wu, R.–t. Zheng, Y.–k. Yuan, K.–y. Qin, A.–x. Huo.
Phys. Energ. Fortis Phys. Nucl., Vol. 9, No. 5, p. 531 – 536 (1985). In Chinese. Abstr. in Phys. Abstr., Vol. 89, No. 1275, Entry 30129 (1986).

144.082 **The origin of galactic protons and the spectrum of primary protons.**
C.–x. Xu.
Chin. J. Space Sci., Vol. 5, No. 2, p. 123 – 128 (1985). In Chinese. Abstr. in Phys. Abstr., Vol. 89, No. 1276, Entry 35905 (1986).

144.083 **Underground muon energy spectrum and its relevance for muon astronomy.**
A. Castellina, B. D'Ettorre Piazzoli, G. Mannocchi, P. Picchi, S. Vernetto, H. Bilokon, E. Meroni.
Lett. Nuovo Cimento, Vol. 44, Ser. 2, No. 6, p. 401 – 404 (1985). Abstr. in Phys. Abstr., Vol. 89, No. 1276, Entry 35906 (1986).

144.084 **Can a light gluino solve the Cyg X–3 puzzle?**
G. Auriemma, L. Maiani, S. Petrarca.
Phys. Lett. B, Vol. 164B, No. 1 – 3, p. 179 – 183 (1985). Abstr. in Phys. Abstr., Vol. 89, No. 1276, Entry 36236 (1986).

144.085 **Near–Earth charge composition of galactic cosmic rays.**
M.–y. Chen, G.–x. Ren, Y.–z. Zhou, R.–q.–Huang.
Chin. J. Space Sci., Vol. 5, No. 4, p. 314 – 316 (1985). In Chinese. Abstr. in Phys. Abstr., Vol. 89, No. 1277, Entry 42110 (1986).

144.086 **Nature of the high–energy particles from Cygnus X–3.**
A. Dar, J. J. Lord, R. J. Wilkes.
Phys. Rev. D, Vol. 33, No. 1, p. 303 – 306 (1986). Abstr. in Phys. Abstr., Vol. 89, No. 1277, Entry 42309 (1986).

144.087 **The high–energy muon flux in air showers from Cygnus X–3.**
A. Castellina, B. D'Ettorre Piazzoli, G. Mannocchi, P. Picchi, S. Vernetto, H. Bilokon, E. Meroni.
Lett. Nuovo Cimento, Vol. 44, Ser. 2, No. 7, p. 491 – 502 (1985). Abstr. in Phys. Abstr., Vol. 89, No. 1277, Entry 42358 (1986).

144.088 **Stochastic space–time and the primary energy spectrum of cosmic rays at $E_0 \geqslant 10^{19}$eV.**
M. Sinha, S. Roy.
Nuovo Cimento C, Vol. 8C, Ser. 1, No. 5, p. 531 – 537 (1985). Abstr. in Phys. Abstr., Vol. 89, No. 1278, Entry 47850 (1986).

144.089 **Could Cyg X–3 muons indicate a light supersymmetric particle?**
R. N. Mohapatra, S. Nussinov, J. W. F. Valle.
Phys. Lett. B, Vol. 165B, No. 4 – 6, p. 417 – 421 (1985). Abstr. in Phys. Abstr., Vol. 89, No. 1278, Entry 48050 (1986).

144.090 **An upper limit on the muon flux at energies above 100 TeV determined from horizontal air showers observed at Akeno.**
M. Nagano, H. Yoshii, T. Hara, N. Hayashida, M. Honda, K. Kamata, S. Kawaguchi, T. Kifune, Y. Matsubara, G. Tanahashi, M. Teshima.
J. Phys. G, Vol. 12, No. 1, p. 69 – 84 (1986). Abstr. in Phys. Abstr., Vol. 89, No. 1279, Entry 53701 (1986).

144.091 **Energy spectrum of cosmic–ray iron nucleus oberved with emulsion chambers.**
Y. Sato, E. Shimada, I. Ohta, S. Tasaka, S. Tanaka, H. Sugimoto, K. Taira, N. Tateyama.
J. Phys. Soc. Jpn., Vol. 54, No. 12, p. 45 – 52 (1985). Abstr. in Phys. Abstr., Vol. 89, No. 1279, Entry 53702 (1986).

144.092 **Thickness of the particle swarm in cosmic–ray air showers.**
J. Linsley.
J. Phys. G, Vol. 12, No. 1, p. 51 – 57 (1986). Abstr. in Phys. Abstr., Vol. 89, No. 1279, Entry 53703 (1986).

144.093 **Analysis of cosmic ray multi–μ phenomena in the energy region 10^{14} – 10^{16}eV.**
C.–l. Jing, G.–r. Jing, L.–k. Ding, Q.–q. Zhu.
Chin. Phys., Vol. 5, No. 4, p. 910 – 917 (1985). English translation of 40.144.016. Abstr. in Phys. Abstr., Vol. 89, No. 1279, Entry 53705 (1986).

144.094 **A study of the contribution of prompt cosmic–ray muons to total muon fluxes.**
H. Inazawa, K. Kobayakawa, T. Kitamura.
J. Phys. G, Vol. 12, No. 1, p. 59 – 67 (1986). Abstr. in Phys. Abstr., Vol. 89, No. 1279, Entry 53706 (1986).

144.095 **High energy cosmic ray events of ultra–relativistic nucleus–nucleus collisions.**
T. H. Burnett, S. Dake, M. Fuki, J. C. Gregory, T. Hayashi, R. Holynski, J. Iwai, W. V. Jones, A. Jurak, J. J. Lord, O. Miyamura, H. Oda, T. Ogata, T. A. Parnell, T. Saito, S. Strausz, M. Szarska, T. Tabuki, Y. Takahashi, T. Tominaga, J. W. Watts, J. P. Wefel, B. Wilczynska, R. J. Wilkes, W. Wolter, B. Wosiek.
Nucl. Phys. A, Vol. A447, p. 189 – 195 (1986). Abstr. in Phys. Abstr., Vol. 89, No. 1280, Entry 54856 (1986). – See Abstr. 012.057.

144.096 **Excessive production of electron pairs in low multiplicity interactions.**
T. H. Burnett, S. Dake, M. Fuki, J. C. Gregory, T. Hayashi, R. Holynski, J. Iwai, W. V. Jones, A. Jurak, J. J. Lord, O. Miyamura, T. Ogata, T. A. Parnell, T. Saito, S. Strausz, T. Tabuki, Y. Takahashi, T. Tominaga, J. Watts, B. Wilczynska, R. J. Wilkes, W. Wolter, B. Wosiek.
Nucl. Phys. A, Vol. A447, p. 197 – 202 (1986). Abstr. in Phys. Abstr., Vol. 89, No. 1280, Entry 54857 (1986). – See Abstr. 012.057.

144.097 **The cosmic–ray anisotropy and the relative fluxes of heavy nuclei and protons in high–energy cosmic radiation.**
T. Kifune, J. Wdowczyk, A. W. Wolfendale.
J. Phys. G, Vol. 12, No. 2, p. 143 – 150 (1986). Abstr. in Phys. Abstr., Vol. 89, No. 1280, Entry 60569 (1986).

144.098 **Anisotropy of the arrival direction of extensive air showers observed at Akeno.**
T. Kifune, T. Hara, Y. Hatano, N. Hayashida, M. Honda, K. Kamata, M. Nagano, K. Nishijima, G. Tanahashi, M. Teshima.
J. Phys. G, Vol. 12, No. 2, p. 129 – 142 (1986). Abstr. in Phys. Abstr., Vol. 89, No. 1280, Entry 60570 (1986).

144.099 **Longitudinal development of air–shower muons studied from the arrival time distributions measured at 900 m above sea level.**
F. Kakimoto, T. Tamura, Y. Matsumoto, T. Enoki, I. Tsuchimoto, K. Nishi, K. Suga.
J. Phys. G, Vol. 12, No. 2, p. 151 – 160 (1986). Abstr. in Phys. Abstr., Vol. 89, No. 1280, Entry 60573 (1986).

144.100 **On the law of variation of cosmic ray energy in the interplanetary space.**
E. G. Berezhko.
Physical processes in the circumterrestrial plasma, p. 3 – 8 (1984). In Russian. – See Abstr. 003.040.

144.101 **Modulation of the anisotropy of pressure of galactic cosmic rays by the solar wind.**
S. M. Kamoldinov.
Physical processes in the circumterrestrial plasma, p. 9 – 15 (1984). In Russian. – See Abstr. 003.040.

144.102 **Propagation and secondary production of low–energy antiprotons in the atmosphere.**
T. Bowen, A. Moats.
Phys. Rev. D, Vol. 33, No. 3, p. 651 – 656 (1986). Abstr. in Phys. Abstr., Vol. 89, No. 1281, Entry 61948 (1986).

144.103 **Ultra high energy cosmic rays.**
A. A. Watson.
Sci. Prog., Vol. 70, No. 277, Part 1, p. 1 – 15 (1986). Abstr. in Phys. Abstr., Vol. 89, No. 1281, Entry 67286 (1986).

144.104 **Cosmic ray density gradient and its dependence on the north–south asymmetry in solar activity.**
Badruddin, R. S. Yadav, N. R. Yadav.
Indian J. Radio Space Phys., Vol. 14, No. 6, p. 151 – 156 (1985). Abstr. in Phys. Abstr., Vol. 89, No. 1281, Entry 67289 (1986).

144.105 **Estimation of prompt cosmic–ray muon flux at high energies.**
Y. Minorikawa, K. Mitsui.
Lett. Nuovo Cimento, Vol. 44, Ser. 2, No. 8A, p. 651 – 657 (1985). Abstr. in Phys. Abstr., Vol. 89, No. 1281, Entry 67292 (1986).

144.106 **Cygnus X–3 events and very–high–energy photonuclear cross sections.**
W. Ochs, L. Stodolsky.
Phys. Rev. D, Vol. 33, No. 5, p. 1247 – 1251 (1986). Abstr. in Phys. Abstr., Vol. 89, No. 1281, Entry 67544 (1986).

144.107 **Cosmic ray acceleration by collisionless shocks.**
W. I. Axford.
Proceedings of the 1984 International Conference on Plasma Physics, Vol. 2, p. 1067 – 1075 (1985). Abstr. in Phys. Abstr., Vol. 89, No. 1281, Entry 67287 (1986). – See Abstr. 012.056.

144.108 **Research on the possibility of detecting cosmic ray particles by acoustic techniques.**
Y.–l. Jiang, Y.–k. Yuan, Y.–g. Li, D.–b. Chen, R.–t. Zheng, J.–n. Song.
Phys. Energ. Fortis Phys. Nucl., Vol. 10, No. 1, p. 9 – 14 (1986). In Chinese. Abstr. in Phys. Abstr., Vol. 89, No. 1283, Entry 78243 (1986).

144.109 **The longitudinal asymmetry of the interplanetary perturbation producing Forbush decreases.**
N. Iucci, S. Pinter, M. Parisi, M. Storini, G. Villoresi.
Nuovo Cimento C, Vol. 9C, Ser. 1, No. 1, p. 39 – 50 (1986). Abstr. in Phys. Abstr., Vol. 89, Nr. 1284, Entry 83393 (1986).

144.110 **High–energy physics with cosmic rays.**
S. C. Tonwar.
Cosmic pathways, p. 67 – 88 (1986). – See Abstr. 003.045.
This review discusses recent results from air shower observations of very high–energy cosmic ray primary particles. The derivation of particle interaction cross sections and comparisons with QCD predictions are described.

144.111 **Super high–energy cosmic ray events and properties of hadronic interactions.**
G. B. Yodh.
Cosmic pathways, p. 89 – 109 (1986). – See Abstr. 003.045.
This review summarizes the results from large–scale emulsion chamber experiments carried out to study super high energy interactions using the cosmic ray beam. These events have a total visible energy greater than 100 TeV. Over 1000 events have been observed at mountain laboratories around the world. The results strongly suggest a dominance of nuclei with atomic mass between $20 \leqslant A \leqslant 56$ in the primary cosmic ray beam at $10^6 - 10^7$GeV per nucleus, in the region of the "bend" of the all–particle cosmic ray spectrum.

144.112 **Cosmic ray cascades in the atmosphere.**
T. K. Gaisser.
Cosmic pathways, p. 110 – 117 (1986). – See Abstr. 003.045.
This paper reports the results of cosmic ray cascade calculations for: (1) the flux of atmospheric neutrinos below 5 GeV; and (2) the properties of extensive air showers around 10^9GeV.

144.113 **New facets in the study of neutral cosmic ray particles.**
A. Subramanian, S. D. Verma.
Cosmic pathways, p. 126 – 138 (1986). – See Abstr. 003.045.
This paper discusses results from cosmic ray experiments which seem to suggest an unknown neutral hadron component in the cosmic ray primary particle flux.

144.114 **Cosmic rays: there are only 27 years left before the centenary.**
V. L. Ginzburg.
Cosmic pathways, p. 141 – 147 (1986). – See Abstr. 003.045.

144.115 **Cosmic ray sources: another swing of the pendulum?**
A. M. Hillas.
Cosmic pathways, p. 148 – 161 (1986). – See Abstr. 003.045.
This paper discusses possible cosmic acceleration mechanisms for very high energy ($\gtrsim 10^{16}$ eV) cosmic rays. The impact of Cyg X–3 on studies in this field is examined.

144.116 **Synthesis of ideas on cosmic ray origin and propagation.**
A. W. Wolfendale.
Cosmic pathways, p. 162 – 172 (1986). – See Abstr. 003.045.
The bulk of cosmic rays detected at the surface of the earth appear to be of galactic origin, many probably having come from supernova remnants, at least at the lowest energies. Only above 10^{19}eV does an extra–galactic origin appear likely and here the Virgo cluster at the centre of our Supercluster is a likely source. Although extragalactic cosmic rays are not present to a large extent their energy density could well be significant and the case is made for it being about 10^{-4}eV cm^{-3}.

144.117 **Primary cosmic ray composition at very high energy – an old idea revisited.**
M. Koshiba.
Cosmic pathways, p. 173 – 174 (1986). – See Abstr. 003.045.

144.118 **Composition of primaries of extensive air showers.**
M. V. S. Rao.
Cosmic pathways, p. 175 – 200 (1986). – See Abstr. 003.045.
Results obtained on the size spectrum and the muon component of extensive air showers from experiments carried out at Ootacamund and Kolar Gold Fields by the Tata Institute group are summarized. To understand these results, in particular the size spectrum and the variation of high–energy muons with shower size, a model of primary composition is proposed, in which galactic nuclei are removed at around 7.5×10^4GeV/nucleon due to photo–disintegration and protons continue up to 6.7×10^6GeV where their spectrum steepens due to leakage from the galaxy and continues up to at least 10^{10}GeV. In this model, the mean mass number remains nearly constant in the energy range $1 \times 10^5 - 5 \times 10^6$GeV even though the abundance of protons as well as iron nuclei steadily increases with energy due to continuous change of composition.

144.119 **Primaries of extensive air showers.**
B. Peters.
Cosmic pathways, p. 201 – 204 (1986). – See Abstr. 003.045.

144.120 **Erratum: "Searches for narrow–angle anisotropies in the primary energy range 0.1 – 10 TeV" [Astrophys. J., Vol. 291, No. 2, p. 468 – 470 (1980)].**
O. C. Allkofer, W. D. Dau, H. Jokisch, G. Klemke, R. C. Uhr, G. Bella, Y. Oren.
Astrophys. J., Vol. 305, No. 2, p. 954 (1986). See Abstr. 39.144.052.

144.121 **The acceleration of cosmic rays.**
J. J. Quenby.
Mitt. Astron. Ges., No. 65, p. 133 – 148 (1986). – See Abstr. 012.072.
The theory of pure shock and diffusive shock acceleration is described with emphasis on ways of enhancing the acceleration rate. Relevant observations within the solar system and using γ–rays are discussed. Test particle and non–linear shock models for the energetic particle spectra are reviewed in relation to measurements of spectral slope. Direct accelerators are briefly discussed.

Experimental particle physics in the age of innocence. (The decay of mesotrons: 1939 – 1943).
See Abstr. 004.101.

Gravitation, cosmology and cosmic–ray physics.
See Abstr. 013.023.

Fast timing of air shower fronts.
See Abstr. 034.080.

A Cerenkov–ΔE–Cerenkov detector for high energy cosmic ray isotopes and an accelerator study of ^{40}Ar&^{56}Fe fragmentation.
See Abstr. 034.093.

The Utah Fly's Eye detector.
See Abstr. 034.100.

Observatory for ultra high–energy processes: the Fly's Eye.
See Abstr. 034.177.

Detector system for the measurement of the muon array density spectrum in cosmic ray showers.
See Abstr. 034.179.

High energy gamma ray observatories for the study of cosmic ray electrons above 10^{14}eV.
See Abstr. 035.064.

Response of high resolution drift tubes to relativistic heavy ions.
See Abstr. 035.065.

Cosmic radiation and the earth's rotation.
See Abstr. 044.006.

Particle acceleration by magnetic reconnection.
See Abstr. 061.012.

Searching for supersymmetry in high–energy cosmic–ray interactions.
See Abstr. 061.108.

Transmission and damping of hydromagnetic waves behind a strong shock front: implications for cosmic ray acceleration.
See Abstr. 062.003.

Nucleosynthesis in massive stars: winds from WR stars and isotopic anomalies in cosmic rays.
See Abstr. 065.138.

Cosmic ray acceleration by binary neutron stars.
See Abstr. 067.015.

Particle acceleration by pulsars.
See Abstr. 067.016.

Spectral analysis of variations of the solar activity and cosmic ray intensity at the decay phase of the 19th solar cycle.
See Abstr. 072.059.

Cosmogenic helium in a terrestrial igneous rock.
See Abstr. 081.016.

Annual variations of cosmic rays in the lower atmosphere.
See Abstr. 082.021.

Temporal dependences of the short–range cosmic ray component in the stratosphere.
See Abstr. 082.022.

Correlation between upper atmospheric density and intensity of galactic cosmic radiation.
See Abstr. 082.054.

Cosmic ray exposure histories and ^{235}U–^{136}Xe dating of Apollo 11, Apollo 12, and Apollo 17 mare basalts.
See Abstr. 094.054.

Origin of Jovian cosmic rays.
See Abstr. 099.035.

Can tracks in meteorites be used to study ultraheavy cosmic rays?
See Abstr. 105.126.

A model for GCR–particle fluxes in stony meteorites and production rates of cosmogenic nuclides.
See Abstr. 105.269.

The inclination of the heliospheric neutral sheet and cosmic ray intensity at the earth.
See Abstr. 106.028.

Acceleration of energetic particles at solar wind shocks.
See Abstr. 106.049.

The time evolution effects in Fermi acceleration by shock waves. (II).
See Abstr. 106.067.

High–energy neutrinos from Cygnus X–3.
See Abstr. 117.057.

Supernova remnants.
See Abstr. 125.011.

Simplified models for supernova remnant evolution using energetic particle hydrodynamics.
See Abstr. 125.012.

Structure of cosmic ray modified shocks.
See Abstr. 131.048.

Influence of cosmic rays on the stability and large–scale dynamics of the interstellar medium.
See Abstr. 131.264.

Ultra high energy gamma–ray astronomy.
See Abstr. 143.004.

Secondary production of neutral pi–mesons and the diffuse galactic gamma radiation.
See Abstr. 143.005.

The radial distribution of galactic gamma rays. III. The distribution of cosmic rays in the Galaxy and the CO–H_2 calibration.
See Abstr. 155.003.

The distribution and origin of cosmic rays in spiral galxies: a test study of the galaxy NGC 3310.
See Abstr. 158.226.

Stellar Systems, Galaxy, Extragalactic Objects, Cosmology

151 Stellar Systems (Kinematics, Dynamics)

151.001 **Orbital decay of satellite galaxies in spherical systems.**
M. D. Weinberg.
Astrophys. J., Vol. 300, No. 1, p. 93 – 111 (1986).
The rate of orbital decay of a dwarf satellite moving in a spherical parent galaxy is considered using a theory of dynamical friction applicable to spherical systems. In this theory, the orbits are detemined exactly in a given spherical potential $U(r)$. In contrast, in Chandrasekhar's theory of dynamical friction, which is frequently applied in this case, the background medium is infinite and homogeneous and the trajectories are straight lines. The effective logarithmic factor appearing in Chandrasekhar's theory is derived from the spherical theory. The response density of the sphere to the satellite (i.e., the wake of the satellite) is computed and compared with the wake in the infinite homogeneous case.

151.002 **Dynamical instabilities in spherical stellar systems.**
J. Barnes, J. Goodman, P. Hut.
Astrophys. J., Vol. 300, No. 1, p. 112 – 131 (1986).
The first numerical examples of spherical stellar systems in equilibrium which are unstable on a dynamical time scale were found by Hénon using *N*–body code with enforced spherical symmetry. The authors have reexamined Hénon's models using a code which includes nonradial forces to quadrupole order. They confirm the radial instability reported by Hénon; in addition, two nonradial instabilities occur. In the first kind, seen in models with predominantly radial orbits, the system permanently loses spherical symmetry and settles into a strongly triaxial ellipsoid. In the second kind, which appears in models with nearly circular orbits, the mass distribution exhibits quadrupole–mode oscillations.

151.003 **Formation of elliptical galaxies and massive halos.**
R. G. Carlberg, G. Lake, C. A. Norman.
Astrophys. J., Lett. Ed., Vol. 300, No. 1, p. L1 – L4 (1986).
The authors report *N*–body models for the formation of spheroids of galaxies. Models with and without dissipation are considered. The dissipative models are better fits to the $r^{1/4}$ profiles of elliptical galaxies. The densities match the $r^{1/4}$ function for up to two decades in radius. The models are extended to include a dissipationless, massive, "dark halo" with a mass 10 times that of the gas. As the amount of dissipation is increased the inner region of the dark matter develops a power–law density distribution with a slope near –2 and nearly isotropic velocity dispersions.

151.004 **Finite length–scale anti–gravity and observations of mass discrepancies in galaxies.**
R. H. Sanders.
Astron. Astrophys., Vol. 154, No. 1/2, p. 135 – 144 (1986).
The modification of Newtonian attraction suggested by Sanders (1984) contains a repulsive Yukawa component which is characterised by two physical parameters: a coupling constant, α, and a length scale, r_0. Although this form of the gravitational potential can result in flat rotation curves for a galaxy (or a point mass) it is not obvious that any modification of gravity associated with a definite length scale can reproduce the observed rotation curves of galaxies covering a wide range of mass and size. Here it is shown that the rotation curves of galaxies ranging in size from 5 to 40 kpc can be reproduced by this modified potential. Moreover, the implied mass–to–light ratios for a larger sample of galaxies are reasonable (one to three) and show no systematic trend with the size of the galaxy. The observed infrared Tully–Fisher law is shown to be consistent with the prediction of this revised gravity. The modified potential permits the X–ray emitting halos observed around elliptical galaxies to be bound without the addition of dark matter.

151.005 **The spheroids of galaxies before and after disc formation.**
J. Binney, A. May.
Mon. Not. R. Astron. Soc., Vol. 218, No. 4, p. 743 – 760 (1986).
Disc galaxies probably started out as undressed spheroids, the discs forming gradually over several billion years. The authors show that most spheroid stars are probably on non–resonant orbits, and that the actions characterizing these orbits are reasonably invariant when a massive disc is slowly inserted into the spheroid.

151.006 **Dynamics of disk galaxies.**
K. C. Freeman.
Astrophys. Space Sci., Vol. 118, Nos. 1/2, p. 337 – 342 (1986). – See Abstr. 012.001.
In order to detect the gravitational effect of the dark corona component of disk galaxies, it is necessary to have surface photometry and rotation data that extend well beyond three disk–scale lengths. Such data exists now for several pure disk galaxies. There is very strong evidence for dark coronas in these galaxies. It has been possible to estimate the scale length and density for their dark components, which can be well represented as isothermal spheres. The pure disk galaxies that have been studied so far have absolute magnitudes in the range $-17 > M_B > -19.5$. It would be very interesting to know whether the fainter dwarf irregular galaxies (which are also disk systems) have dark coronas: are these coronas necessary for disk galaxies to form?

151.007 **Post–collapse evolution of small–*N* clusters.**
S. Inagaki.
Astrophys. Space Sci., Vol. 118, Nos. 1/2, p. 367 – 370 (1986). – See Abstr. 012.001.
N–body simulations were made to study post–collapse evolution of small–*N* clusters. The qualitative behaviour of the dynamical evolution is the same as in the Fokker–Planck approximation. The flatter the mass spectrum is the more violent is the activity of binaries and so are the expansion and contraction of the cluster core.

151.008 **The shape of spiral arms.**
S.–A. Sørensen.
Astrophys. Space Sci., Vol. 118, Nos. 1/2, p. 395 – 399 (1986). – See Abstr. 012.001.
The response by a gaseous disc to perturbations is investigated. Spectral analysis of the resulting pattern indicates that information about the nature of the perturber is maintained only if the perturbation is strong.

151.009 **Elliptical galaxies under perturbation.**
G.–x. Song.
Astrophys. Space Sci., Vol. 118, Nos. 1/2, p. 457 – 458 (1986). – See Abstr. 012.001.
Three experiments on the initially slightly flattened spheroids ($\varepsilon = 0.45$) have been made. The first one is run without perturbation and the others with perturbation. The results have shown that the spheroid without perturbation will evolve to nearly sphere and the other two will almost keep the originally projected shape after perturbation and relaxation.

151.010 **Global instability of thin stellar discs.**
T. Fujiwara, S. Hozumi.
Astrophys. Space Sci., Vol. 119, No. 1, p. 199 – 200 (1986). – See Abstr. 012.001.
The authors have determined the dominant global modes of stellar discs by integrating the linearized collisionless Boltzmann equation. The models examined are the Kuzmin discs with two types of the distribution function. It is found that the growth rate correlates well with the central value of Q.

151.011 **A mathematical model of the initial stage in the formation of a disk galaxy.**
B. Basu.
Astrophys. Space Sci., Vol. 119, No. 1, p. 201 – 206 (1986). – See Abstr. 012.001.
The collapse of a homogeneous, initially spheroidal halo under self gravitation, has been considered. A thick stellar disk is formed within a time–scale $< 3 \times 10^9$yr. Brisk star formation takes place while the collapse of the gaseous disk is still in progress. This gives rise to the halo stars with low metal content and high Z–motion. A bulge is formed at the centre simultaneously. This is the first phase of formation of a disk galaxy. The thin disk is formed at a later stage as the remaining primordial gas and the gas released by the evolution of stars in the thick disk gradually settles on to it. The presented model is rather a crude one.

151.012 **Simulation of compact groups of galaxies.**
T. Ishizawa.
Astrophys. Space Sci., Vol. 119, No. 1, p. 221 – 225 (1986). – See Abstr. 012.001.
Self–consistent simulations of seven groups of galaxies with halos have been performed to find a constraint upon the size of missing halos around spiral galaxies.

151.013 **Evolution of galaxies: one–zone model with a birth–rate stellar function depending on gas density and temperature.**
R. Caimmi, L. Secco.
Astrophys. Space Sci., Vol. 119, No. 2, p. 315 – 336 (1986).
In this paper closed one–zone models with a birth–rate stellar function depending on gas density and temperature are built up, in order to make a zeroth order approximation of thermo–chemical evolution of galaxies. To get a more realistic picture, the models are allowed first to expand with the universe and then to contract according to three successive time–scales, in order to simulate violent relaxation, rapid contraction, and slow contraction processes.

151.014 **A new approach to a relativistic model of galaxies.**
M. R. Roberts.
Astrophys. Space Sci., Vol. 119, No. 2, p. 405 – 407 (1986).
The observation that stars in galaxies have constant orbital period for a large region from the centre is used to place a constraint on the form of the line element in that region.

151.015 **Gas in galaxies.**
W. W. Roberts Jr.
Cosmical gas dynamics, p. 1 – 12 (1985). – See Abstr. 012.007.
This review discusses the dynamics of galactic gas flows and the role played by the gas in the structure and dynamics of galaxies. Computational simulation utilizing continuum gas dynamics and cloud/particle gas dynamics is playing a major role in the progress currently being achieved toward a deeper understanding of the interstellar medium and galaxies. Normal and barred spiral galaxies are a major focus because of the wealth of improved and detailed observational data that has become available in recent years, now making possible new definitive tests and comparisons between theory and observations.

151.016 **Predicting star formation in interacting and colliding galaxies; the special case of ring galaxies.**
P. N. Appleton, C. Struck–Marcell, P. A. Foster.
Cosmical gas dynamics, p. 33 – 38 (1985). – See Abstr. 012.007.
Models of a rotating disc galaxy are presented which show its response to the gravitational perturbation caused by a massive companion passing through the rotational centre of the system. The results suggest that strong transient but periodic bursts of star formation would occur initially at the centre, but later in an expanding ring of material if certain physical conditions in the interstellar medium are met.

151.017 **Ring formation in gaseous disks.**
P. A. Foster.
Cosmical gas dynamics, p. 39 – 41 (1985). – See Abstr. 012.007.

151.018 **Can galactic bars end well before corotation?**
M. Petrou, T. Papayannopoulos.
Mon. Not. R. Astron. Soc., Vol. 219, No. 1, p. 157 – 166 (1986).
It is shown that a bar which possesses a 1:1 resonance is unlikely to extend close to corotation and suggested that at least some bars in galaxies end because of this resonance. In this case the spiral arms are confined within the corotation radius. This could explain why dust lanes along the spiral arms often appear on the trailing side of the spiral.

151.019 **Stellar dynamics of spiral galaxies: nonlinear effects at the 4/1 resonance.**
G. Contopoulos, P. Grosbøl.
Astron. Astrophys., Vol. 155, No. 1, p. 11 – 23 (1986).
The authors find that the main families of orbits in a realistic model of a spiral galaxy enhance the spiral arms up to the resonance $\varkappa/(\Omega-\Omega_s) = 4/1$. Beyond this resonance the response density is out of phase, if the density excess is strong enough. Therefore strong spirals seem to terminate near the 4/1 resonance. If the amplitude is very small the response is in phase up to corotation; thus the authors recover the results of the linear theory. The same results are derived by considering the motions of gas clouds that undergo collisions at each spiral arm.

151.020 **Periodic orbits in elliptical galaxies. II. Rotation about the axis of symmetry.**
E. Davoust.
Astron. Astrophys., Vol. 156, No. 1/2, p. 152 – 156 (1986).
The investigation of periodic orbits in rotating oblate and prolate elliptical galaxies reveals three families of orbits at resonances; the trajectories are circular epicycles or circles in the equatorial plane coupled with vertical oscillations. They are continuations of two families of periodic orbits found in the non rotating case. The epicyclic orbits bifurcate with direct circular ones at, and do not exist beyond, a maximum angular velocity of rotation. The three families also exist outside resonances. An additional integral of motion is found.

151.021 **Computation of the quasi–steady gas flow in a spiral galaxy by means of a multigrid method.**
W. A. Mulder.
Astron. Astrophys., Vol. 156, No. 1/2, p. 354 – 380 (1986).
An efficient numerical method for the integration of the Euler equations for gas dynamics is presented. The key assumption is the existence of a stationary solution. The several elements of the method, most of which have been described elsewhere, are reviewed in some detail to provide a reference in the astronomical literature and to promote its use in astrophysics. The method is applied to the well–studied problem of the quasi–steady isothermal gas flow in the gravitational potential of a weakly barred galaxy. Convergence on a 64×64 grid typically takes 100 to 300 multigrid cycles for a second–order–accurate solution. Comparison with first–order–accurate solutions shows that second–order schemes are to be preferred, despite their slower convergence.

There are two fundamentally different types of quasi–steady solutions, depending on the strength of the bar and the size of the sound speed. Type I has only shocks inside co–rotation. Type II solutions have shocks inside and outside co–rotation. The solutions display strong non–circular motions. It is suggested that the tilted–ring model for the observed neutral–hydrogen velocity fied of disk galaxies may just as well be interpreted as a rotated–ellipse model. A set of edge–on rotation curves is presented that can be compared with those based on neutral–hydrogen observations.

151.022 **H–functions and mixing in violent relaxation.**
S. Tremaine, M. Hénon, D. Lynden–Bell.
Mon. Not. R. Astron. Soc., Vol. 219, No. 2, p. 285 – 297 (1986).

An H–function is any function of the phase space distribution function $F(x,v)$ which is non–decreasing with time. In collisionless systems Boltzmann's H–function $-\int F \log F \, dx \, dv$ is only one of a variety of H–functions of the form $-\int C(F) \, dx \, dv$, where C is any convex function. Every equilibrium stellar system in which the distribution function is a decreasing function of the energy alone is a stationary point of some H–function of this form. During violent relaxation, all such H–functions must increase, and the distribution function is said to become "more mixed". The authors give a simple criterion for determining whether a given distribution function is more mixed than another; this criterion is used to show that a violently relaxed galaxy resembles observed elliptical galaxies only if the initial state is cold or clumpy.

151.023 **Simulations of close encounter between galaxies: behaviour of interstellar gas clouds and enhancement of star formation rate.**
M. Noguchi, S. Ishibashi.
Mon. Not. R. Astron. Soc., Vol. 219, No. 2, p. 305 – 331 (1986).

Numerical simulations are performed of close encounters between a parent galaxy with a disc system and a perturbing galaxy. Cloud particles are used in constructing the disc of the parent galaxy, in order to investigate the behaviour of the interstellar gas clouds and the influence of the encounter on the star formation rate.

151.024 **Are polar rings self–gravitating?**
L. S. Sparke.
Mon. Not. R. Astron. Soc., Vol. 219, No. 3, p. 657 – 670 (1986).

In a number of S0 galaxies, rings of gas, dust and stars are observed to lie roughly perpendicular to the galactic disc. These polar rings appear nearly flat, and undisturbed by differential precession in the gravitational field of the galaxy. It is shown that a self–gravitating polar ring may precess uniformly as a solid body about an axisymmetric galaxy, if the ring is sufficiently massive and its tilt varies suitably with radius. Both stable and unstable equilibria exist and are discussed in the paper.

151.025 **Theoretical considerations on the rotation curves of galactic halos.**
A. Malagoli, R. Ruffini.
Astron. Astrophys., Vol. 157, No. 2, p. 293 – 300 (1986).

The influence of the visible matter of the galaxy on the galactic halo is examined in the two limiting configurations of a classical isothermal sphere and a fully degenerate configuration. The authors infer the masses of the particles composing the galactic halos. The possible role of dwarf spheroidal galaxies in probing the structure of the galactic halos at large galactic distance is considered.

151.026 **Angular momentum distributions and energies of rotating selfgravitating discs.**
F. Schmitz.
Astron. Astrophys., Vol. 157, No. 2, p. 365 – 371 (1986).

Angular momentum distributions and energies for a variety of rotating infinitely thin discs are calculated. The infinitely thin discs are mainly considered to be zero order approximations of flat disk–like selfgravitating gases with differential rotation. The angular momentum distributions and the energies of different discs are compared under the assumption of the same total mass and total angular momentum. Discs with various rotation laws show nearly identical angular momentum distributions. The total energies of these discs are also nearly equal. The influence of a finite disc thickness on the angular momentum distribution is briefly discussed. For comparison, angular momentum distributions of rotating cylinders with an infinite uniform extent in the direction of the rotation axis are calculated using the rotation laws of the discs. The influence of a massive nonrotating sphere surrounding a disc is discussed. In particular an exponential disc embedded in a halo with a density power law, $\varrho \sim r^{-2}$ is considered. Conclusions are drawn regarding the relevance of angular momentum conservation for the hydrodynamics of gaseous discs.

151.027 **Specific entropy and the evolution of unconfined stellar systems.**
H. E. Kandrup.
Astron. Astrophys., Vol. 157, No. 2, p. 387 – 392 (1986).

This paper examines the time rate of change in number, energy, and entropy appropriate for a cluster described initially by a King model, and, in particular, obtains analytic expressions for the changes in number density n, energy density ε, and entropy density ζ valid in the limit that the local escape velocity is much greater or much less than the rms velocity of the cluster as a whole, i.e., near the center or near the edge of the system. Not that surprisingly, one finds that n and ζ both decrease as time goes by. What is more interesting, however, is the behaviour of the specific entropy $\sigma \equiv \zeta/n$. Near the edge of the system, σ is constant to leading order. Alternatively, in the limit of large escape velocities, σ decreases as time goes by: losses due to evaporation exceed gains resulting from a lack of thermalization.

151.028 **Massive central galaxies, globular cluster swapping, and tidal accretion.**
J. C. Muzzio.
Astrophys. J., Vol. 301, No. 1, p. 23 – 26 (1986).

The results of numerical simulations of clusters of galaxies which have globular clusters orbiting around them are used to investigate the role of massive central galaxies in globular cluster swapping. Those galaxies tend to *lose* globular clusters rather than gain them: on the average, after 10^{10}yr, they lose more than 70% of their globulars, but those that they capture help to lower the total loss to less than 20%; in other words, in addition to mergers, which are not included in the present models, *tidal accretion* helps massive galaxies to balance losses due to tidal stripping.

151.029 **Dynamical effects of stellar evolution in globular clusters.**
J. H. Applegate.
Astrophys. J., Vol. 301, No. 1, p. 132 – 144 (1986).

The dynamical role of stellar evolution in globular clusters is explored using simple models. The cluster initial mass function is divided into mass groups spanning the range 4.0 – 0.35 $M_\odot$. The evolution of the cluster is calculated incorporating the modification of the mass function by stellar evolution, the weakening of the cluster potential due to the escape of shed stellar mass from the cluster, the collisional transfer of energy between the various mass groups, and evaporation. It is found that the dynamical evolution of clusters born rich in massive stars or with long initial relaxation times is dominated by stellar mass loss. Massive clusters of this type will have long pre–core collapse lives. Clusters born poor in massive stars or with short initial relaxation times are dominated by collisional evolution; the role of stellar evolution in these systems is to delay core collapse. These clusters suffer rapid core collapse, typically in 1 – 5 Gyr.

151.030 **The formation of faint ring structures.**
T. K. Chatterjee.
Astrophys. Space Sci., Vol. 121, No. 2, p. 213 – 224 (1986).

In a previous paper (Chatterjee, 1984) the author studied the formation of ring galaxies due to head–on collisions between disk and spherical galaxies. The rings so formed are quiet prominent and well defined. However, in the domain of rings, there are many faint rings which are ill defined and are so faint that they lie on the borderline of detectability. In this paper the formation of such rings is investigated.

151.031 **A special case of two-component polytropes with rigid rotation.**
R. Caimmi.
Astron. Astrophys., Vol. 159, No. 1/2, p. 147 – 156 (1986).
The author first defines ECN polytropes as Emden–Chandrasekhar concentric, copolar, axisymmetric, solid–body rotating, N–component polytropes and then builds up the basic theory of EC2 polytropes following the same line of thought expressed in a previous paper for EC1 polytropes. Sequences of equilibrium configurations characterized by constant mass and angular momentum of each component are treated. Exact solutions for the equilibrium equations relative to the special case of uniform density in both subsystems ($n_i = n_j = 0$) are outlined and simple approximations are used to express all physical parameters in analytical form. An application of these last results to galaxies – modelled as EC2 polytropes with $n_i = n_j = 0$ – is reported and discussed.

151.032 **A tidal model for the interacting pair of galaxies VV 117.**
K. S. V. S. Narasimhan, S. M. Alladin.
Bull. Astron. Soc. India, Vol. 13, No. 4, p. 359 – 362 (1985).
Limber's (1965) tidal model for the interacting pair of galaxies VV 117 (= Arp 143 = NGC 2444–45) is improved upon in the light of subsequent H I observations. It is proposed that the spiral Sc component is being disrupted by the tidal effects of the elliptical (which is about twice as massive) in a close collision in a bound orbit.

151.033 **On the formation of tails and bridges in interacting galaxies.**
P. M. S. Namboodiri, R. K. Kochhar.
Bull. Astron. Soc. India, Vol. 13, No. 4, p. 363 – 371 (1985).
Interacting galaxies produce peculiar features like tails, bridges, and counterarms of various sizes and shapes. Numerical experiments have been performed to investigate as to how the formation of tails and bridges depends on mass ratio of the two interacting galaxies. Results indicate that tail formation is favoured when the mass ratio is close to unity. As a result of the encounter, a sharp decrease in the density of the main galaxy at the tidal radius is numerically observed.

151.034 **Evolution and propagation of density waves on galactic disk. (II) An approximation of the thin transition layer.**
J. j. Xu.
Sci. Sin., Ser. A, Vol. 29, No. 1, p. 67 – 77 (1986).

151.035 **Neutral hydrogen in M31. VI. The tidal interaction between M31 and NGC 205.**
N. R. Sato, T. Sawa.
Publ. Astron. Soc. Jpn., Vol. 38, No. 1, p. 63 – 75 (1986).
To explain the origin of the observed warping in the outer H I disk of M31, the tidal interaction between M31 and its companion galaxy NGC 205 is simulated by assuming the two–body problem of M31 and NGC 205. Spherical–symmetric gravitational potential fields are assumed for both galaxies. An orbit of NGC 205 is found by which the observed warping in the outer H I disk of M31 is well reproduced by tidal interaction. The simulation can reproduce not only the shape of the warped disk and the asymmetric distribution of the outer H I gas but also the velocity deviation of the outer H I gas from the circular rotation.

151.036 **Velocity dispersion of stars and giant molecular clouds.**
H. Kamahori, M. Fujimoto.
Publ. Astron. Soc. Jpn., Vol. 38, No. 1, p. 77 – 94 (1986).
An attempt is made to reproduce the age dependence of the velocity dispersion of stars based on the stochastic gravitational interaction between stars and the giant molecular clouds. The authors employ the theory of random walk in the field of harmonic oscillation and formulate the velocity dispersion under the dynamical friction of stars in their orbital motion.

151.037 **Formation of giant molecular clouds by coagulation of small clouds and spiral structure.**
K. Tomisaka.
Publ. Astron. Soc. Jpn., Vol. 38, No. 1, p. 95 – 109 (1986).
The formation of giant molecular clouds (GMCs) in a spiral gravitational potential is investigated based on a coagulation theory of molecular clouds. In this model, small clouds collide with each other and grow to massive ones. Finally GMCs are formed, which have a finite lifetime. The author simulates this ensemble of clouds by a collisional *N*–body method, which includes destruction of GMCs.

151.038 **Effects of viscous drag on a slowly rotating supermassive star due to an embedded stellar system.**
F. Hagio.
Publ. Astron. Soc. Jpn., Vol. 38, No. 1, p. 111 – 120 (1986).
The author studies the effects of stars in a dense stellar system on the rotation of a slowly rotating supermassive star (SMS). The hydrodynamic interaction between the stars and the gas of the SMS produces a torque removing angular momentum from the SMS. If the plunging rate of stars into the SMS is calculated in an $n\sigma v$ approximation, the SMS spins down during a quasi–pile stage and becomes general relativistically unstable within a time scale of 10^{4-5}yr from the formation of the SMS. But, in a loss–cone approximation the SMS suffers little from its influence and spins up and finally becomes rotationally unstable and leads to mass loss.

151.039 **The role of binaries in globular cluster evolution.**
P. Hut.
The evolution of galactic X–ray binaries, p. 1 – 11 (1986). – See Abstr. 012.076.
In dense cores of globular clusters binaries will be formed dynamically, either by two–body tidal capture or in an encounter of three independent stars. These binaries provide a heat source in globular clusters. Globular cluster evolution calculations require gravitational reaction rates, for encounters between binaries and single stars. These encounters extract binding energy from the binaries which is used to heat the stellar population as a whole. A review is given of recent efforts to determine these gravitational reaction rates, and their applications to late stages of globular cluster evolution, past core collapse.

151.040 **Model calculations of the evolution of "clumpy" gas in elliptical galaxies.**
R. Kunze, H. W. Yorke, H.–H. Loose.
Mitt. Astron. Ges., No. 65, p. 232 (1986). Abstract. – See Abstr. 012.072.

151.041 **Strong nonlinear waves in gaseous disks. I. Stationary periodic waves.**
M. G. Abramyan, E. A. Mikhajlova, A. G. Morozov.
Astrofizika, Tom 24, Vyp. 1, p. 167 – 177 (1986). In Russian. English translation in Astrophysics, Vol. 24, No. 1.
The exact nonlinear equations of short–scale perturbations of a rotating gaseous disk are numerically integrated. Nonlinear stationary periodic waves restricted from the above amplitude have been obtained. The limiting value of amplitude increases with the velocity of wave propagation. The parameters of the 3–kpc arm, the 135 km/s feature of the Galaxy and of the fine structure of Saturn rings have been estimated.

151.042 **Remarks on stellar clusters.**
E. Teller.
From SU(3) to gravity, p. 321 – 324 (1985). – See Abstr. 003.014.

151.043 **Collective relaxation of stellar systems.**
V. G. Gurzadyan, G. K. Savvidy (*G. K. Savvidi*).
Astron. Astrophys., Vol. 160, No. 2, p. 203 – 210 (1986).
The problem of relaxation of stellar systems is investigated from the point of view of ergodic theory. It is shown that an exponential instability peculiar to Kolmogorov K–systems and leading to equilibrium exists in general stellar systems. The relaxation time for real stellar systems is calculated. That time is

substantially smaller than the binary relaxation time. The physical difference between the two relaxation times is discussed.

151.044 **A method for integrating the equations that govern linear isentropic oscillations of an incompressible Maclaurin spheroid embedded in a spheroidal halo.**
P. Smeyers.
Astron. Astrophys., Vol. 160, No. 2, p. 385 – 391 (1986).
A method is presented for the solution of the eigenvalue problem that governs linear isentropic oscillations of an incompressible Maclaurin spheroid embedded in a spheroidal halo by direct and analytical integration of a finite system of differential equations. The validity of the method is verified by determining the time–dependent second–harmonic modes and by comparing the results with those derived by Chandrasekhar (1969) and Durisen (1978) with the tensor virial equation method. Furthermore, the connection is made with the modes found in the limiting case of the non–rotating, self–gravitating equilibrium sphere with the same uniform density in the absence of a halo.

151.045 **Equilibrium magnetized disks.**
G. S. Bisnovatyj–Kogan, Z. F. Seidov.
Inst. kosm. issled Akad. Nauk SSSR. Prepr., No. 994, 31 pp. (1985). In Russian. Abstr. in Ref. Zh., 51. Astron., 1.51.624 (1986).

151.046 **On the dynamics of close pairs of galaxies. I. Effects of contactless interaction.**
D. A. Verner, A. D. Chernin.
Fiz.–tekh. inst. Akad. Nauk SSSR. Prepr., No. 957, 20 pp. (1985). In Russian. Abstr. in Ref. Zh., 51. Astron., 1.51.670 (1986).

151.047 **On the stability of the spherical collapse of collisionless gravitating systems.**
E. A. Malkov.
Astrofizika, Tom 24, Vyp. 2, p. 377 – 385 (1986). In Russian. English translation in Astrophysics, Vol. 24, No. 2.
The stability of the family of nonstationary models that have identical total energy but nonidentical energy of choatic motion has been investigated. This family includes the stationary distribution function of the homogeneous sphere by Camm and the dusty distribution function corresponding to the Friedmann cosmological solution with negative total energy as limit cases. All models with the exception of the stationary one are instable relative to radial perturbation while these models are instable relative to the quadrupole mode when the ratio of the chaotic energy to the total energy is less than 0.62.

151.048 **On evaluating the orbital eccentricity.**
S. Ninkovich.
Astrofizika, Tom 24, Vyp. 2, p. 411 – 416 (1986). In Russian. English translation in Astrophysics, Vol. 24, No. 2.
A criterion for the orbital eccentricity evaluated for objects moving in a spherically symmetric force field with mass distribution $M(r) \propto r^{\beta}$, based on the radial component rate in total kinetic energy is proposed. The latter quantity preserves a nearly constant value over a sufficiently large fraction of the orbit, admitting thus its instantaneous value to be a measure of the eccentricity.

151.049 **Stability of a pulsating rotating model of a stellar system.**
E. A. Malkov.
Astrofizika, Tom 24, Vyp. 2, p. 416 – 420 (1986). In Russian. English translation in Astrophysics, Vol. 24, No. 2.
The stability of a nonstationary two–parametric model of the Freeman sphere with respect to a perturbation leading the sphere to an ellipsoidal form is investigated.

151.050 **The fate of wide binaries in the solar neighborhood: encounters with stars and molecular clouds.**
M. D. Weinberg, S. L. Shapiro, I. Wasserman.
Icarus, Vol. 65, No. 1, p. 27 – 36 (1986).
The time evolution of wide binaries by stochastic encounters with field stars and molecular clouds in the solar neighbourhood is computed with a hybrid Monte Carlo scheme. Specifically, a Fokker–Planck analysis is used for distant encounters and the effects of close encounters are treated encounter by encounter. As an example, the authors consider the lifetime and stability of the proposed solar companion.

151.051 **Three–dimensional solution of Poisson equation in stellar systems.**
R. Nagai.
Proceedings of the nineteenth symposium on celestial mechanics, p. 99 – 101 (1986). In Japanese. – See Abstr. 012.013.

151.052 **An extention of Newton's equation of motion (X): Rotation curves of disk galaxies.**
K. J. Cheon.
Proceedings of the nineteenth symposium on celestial mechanics, p. 102 – 106 (1986). In Japanese. – See Abstr. 012.013.

151.053 **Toroidal dynamic instability of rotating polytropes: n = 0.8 – 1.8.**
H. A. Williams, J. E. Tohline.
Bull. Am. Astron. Soc., Vol. 17, No. 4, p. 834 (1985). Abstract. – See Abstr. 010.062.

151.054 **Colliding galaxies – modeling bursts of star formation in ring galaxies.**
P. N. Appleton, C. Struck–Marcell.
Bull. Am. Astron. Soc., Vol. 17, No. 4, p. 845 (1985). Abstract. – See Abstr. 010.062.

151.055 **Triaxial models of highly flattened elliptical galaxies.**
H. F. Levison, D. O. Richstone.
Bull. Am. Astron. Soc., Vol. 17, No. 4, p. 860 (1985). Abstract. – See Abstr. 010.062.

151.056 **Ring galaxies as a laboratory for studying star formation and gas dynamics in galaxies.**
C. Struck–Marcell, P. N. Appleton.
Bull. Am. Astron. Soc., Vol. 17, No. 4, p. 861 (1985). Abstract. – See Abstr. 010.062.

151.057 **Tidal binary formation in globular clusters.**
S. L. W. McMillan, P. N. McDermott, R. E. Taam.
Bull. Am. Astron. Soc., Vol. 17, No. 4, p. 881 – 882 (1985). Abstract. – See Abstr. 010.062.

151.058 **What warped disks can tell us about galaxy halos and gravity.**
D. M. Christodoulou, J. E. Tohline.
Bull. Am. Astron. Soc., Vol. 17, No. 4, p. 893 (1985). Abstract. – See Abstr. 010.062.

151.059 **Global spiral modes in stellar disks with gas.**
G. Bertin, A. B. Romeo.
Bull. Am. Astron. Soc., Vol. 17, No. 4, p. 893 (1985). Abstract. – See Abstr. 010.062.

151.060 **Dynamical friction and orbit circularization.**
S. Casertano, E. S. Phinney, J. V. Villumsen.
Bull. Am. Astron. Soc., Vol. 17, No. 4, p. 904 (1985). Abstract. – See Abstr. 010.062.

151.061 **Recent investigations on disk galaxies in massive halos.**
R. H. Miller.
Celest. Mech., Vol. 37, No. 3, p. 307 – 324 (1985). – See Abstr. 012.017.
Numerical experiments undertaken to investigate the longevity and behavior of dark–lane elliptical galaxies are described. This is dynamically the same problem as a disk galaxy in a massive halo. Spiral galaxies are disks from a dynamical point of view. A disk of particles embedded in a self–consistent galaxy provides the basic model used for the experiments. This model is applicable to ordinary disk galaxies if the disk is interpreted as the visible

galaxy and the "galaxy" is interpreted as the massive halo thought to be present around disk galaxies. Fully three-dimensional fully self-consistent n-body computer programs that can handle 100,000 particles are used for the experiments.

151.062 **Global simulations of gas flow in disc galaxies – I. Response to a spiral density wave.**
T. C. Johns, A. H. Nelson.
Mon. Not. R. Astron. Soc., Vol. 220, No. 1, p. 165 – 183 (1986).
The gas flow in a simple model of a disc galaxy with a mild logarithmic spiral perturbation in the force field due to a postulated low-amplitude density wave is investigated numerically. Simple models with various pattern speeds, pitch angles and perturbation strengths are calculated using two codes; the first is of a standard two-dimensional finite-difference type, the second makes use of fluid particles moving within a three-dimensional mesh with interparticle forces providing the pressure effects.

151.063 **Collisionless Boltzmann simulation of stellar disks having a small bulge.**
M. T. Nishida.
Astrophys. J., Vol. 302, No. 2, p. 611 – 616 (1986).
The dynamical behavior of an infinitely thin stellar disk is investigated for a bisymmetric perturbation by a collisionless Boltzmann simulation. The disk considered has a small-mass (about 1/20 of that of the disk) bulge component with a small size and has two inner Lindblad resonances (ILRs). The disk does not show strong bar instability if stellar motion in the ILR region is treated with enough accuracy in the simulation.

151.064 **Solutions for the modified Newtonian dynamics field equation.**
M. Milgrom.
Astrophys. J., Vol. 302, No. 2, p. 617 – 625 (1986).
The author describes general properties of the solutions for the modified Newtonian dynamics field equation suggested by Bekenstein and Milgrom. He also describes a numerical scheme for solving this equation for axisymmetric mass configurations and presents numerical results. Examples of numerical solutions are presented for: (1) The force law between two point masses (of obvious use in analyzing binary-galaxy dynamics). (2) Rotation curves of various model disk galaxies. (3) The field of a point mass in a constant external acceleration field.

151.065 **Is violent relaxation possible in collisionless stellar systems?**
L. P. Osipkov.
Astron. Tsirk., No. 1399, p. 1 – 2 (1985). In Russian.

151.066 **How could a protogalaxy contract?**
L. P. Osipkov.
Astron. Tsirk., No. 1399, p. 3 – 4 (1985). In Russian.

151.067 **On the disruption of star clusters by interstellar clouds.**
T. S. Kozhanov, L. P. Osipkov.
Astron. Tsirk., No. 1401, p. 7 – 8 (1985). In Russian.

151.068 **On the dynamics of spherical stellar systems.**
V. L. Polyachenko.
Astron. Tsirk., No. 1405, p. 1 – 4 (1985). In Russian.

151.069 **Instability of collapsing collisionless systems.**
V. L. Polyachenko.
Astron. Tsirk., No. 1405, p. 4 – 7 (1985). In Russian.
N-body simulations of collapsing collisionless systems show growing ellipsoidal deformations. In connection with this instability the well-known Lynden-Bell's violent relaxation should be revised.

151.070 **Development of spiral density waves and evaluation of their parameters: a computer experiment.**
E. M. Grivnev.
Sov. Astron., Vol. 29, No. 4, p. 400 – 403 (1985). English translation of 40.151.015.

151.071 **The equilibrium and stability of inhomogeneous Riemann ellipsoids with anisotropic pressure.**
F. Pacheco, G. Pucacco, R. Ruffini.
Astron. Astrophys., Vol. 161, No. 1, p. 39 – 46 (1986).
Extending the tensor virial equations analysis the authors obtain new equilibrium solutions for inhomogeneous Riemann ellipsoids. It is shown that, taking into account anisotropic velocity dispersion and a confocal ellipsoidal density distribution, it is possible to find stable ellipsoidal figures of equilibrium, for which the existence of a massive galactic halo as a means of stabilizing the structure against bar-like modes of oscillation is not required. At the same time the amount of velocity dispersion necessary in the model is not far from that actually observed in our galaxy. These results, obtained for specific values of the angular velocity, the vorticity and the axial ratios, naturally lead to triaxial galaxy configurations.

151.072 **Anisotropy in stellar dynamics.**
E. Bettwieser, R. Spurzem.
Astron. Astrophys., Vol. 161, No. 1, p. 102 – 112 (1986).
Generation of stellar velocity anisotropy in spherical star clusters is examined by linear perturbation theory applied to hydrostatic equilibrium and by numerical evolution simulations. The tensor of inverse specific heat for a singular isothermal sphere is computed under admission of perturbations in anisotropy. Relaxation is treated in the limit of very many small angle encounters, which cause the redistribution of energy. In the scheme of a gaseous model numerical evolution calculations are done for a massive globular cluster of $10^6 M_\odot$. Anisotropy generation, its time evolution and spatial variation are examined.

151.073 **Simple model barred spirals with a triaxial bulge inside the bar.**
M. G. Abramyan.
Sov. Astron. Lett., Vol. 11, No. 4, p. 245 – 247 (1985). English translation of 40.151.082.

151.074 **Mouvements stellaires et évolution dynamique des galaxies.**
L. Martinet.
Orion, 44. Jahrg., Nr. 213, p. 63 – 66 (1986).

151.075 **Dynamical evolution of spherical stellar systems.**
S. Inagaki, S. Kato, P. Wiyanto.
Galactic structure and variable stars, p. 67 – 77 (1985). – See Abstr. 012.029.
Dynamical evolution of two-component star clusters and of single-component King's clusters is studied by numerical integration of the isotropized orbit-averaged Fokker-Planck equation. Equipartition in the core of two-component clusters is achieved if the fraction of the more massive stars is sufficiently small. Early phase evolution of King's clusters with small ratio of tidal radius, r_t, to core radius, r_c, takes place along the sequence of King's model clusters. In the late phase the evolution deviates from King's sequence and gravothermal instability sets in. King's clusters with large r_t/r_c cause gravothermal instability from the beginning.

151.076 **Vlasov simulations of stellar disks.**
M. T. Nishida, Y. Watanabe, S. Inagaki, S. Kato.
Proceedings of the Second Asian-Pacific Regional Meeting on Astronomy, p. 189 – 196 (1984). – See Abstr. 012.031.
This paper reviews some results of Vlasov simulations of disk galaxies recently made by the authors. As in the case of N-body simulations, locally stable stellar disks are nearly stable to global axisymmetric perturbations, but they are not stable to global bar mode perturbations. The bar mode instability in the Vlasov simulations is, however, much milder than that in N-body simulations. The difference between the results of N-body simulations and those of Vlasov simulations are ascribed to the largeness of artificial fluctuations in N-body simulations.

151.077 **Dynamical evolution of two–component stellar systems: gravitational focusing effect.**
S. Inagaki, P. Wiyanto.
Proceedings of the Second Asian–Pacific Regional Meeting on Astronomy, p. 492 – 495 (1984). – See Abstr. 012.031.
To examine the effect of gravitational focusing, the dynamical evolution of two–component stellar systems if followed. This is done by solving the Fokker–Planck equation numerically. The results show that the gravitational focusing is not so effective as to inhibit the mass segregation instability.

151.078 **Hierarchy of equations of the dynamics of gravitating systems.**
L. P. Osipkov.
Astronomical–geodetical investigations, p. 9 – 18 (1985). In Russian. Abstr. in Ref. Zh., 51. Astron., 5.51.809 (1986). – See Abstr. 003.024.

151.079 **On modelling spherical instationary collisionless stellar systems.**
V. M. Danilov, A. P. Ryazanov.
Astronomical–geodetical investigations, p. 19 – 47 (1985). In Russian. Abstr. in Ref. Zh., 51. Astron., 5.51.812 (1986). – See Abstr. 003.024.

151.080 **Comparison of rapid and collision relaxations in instationary star clusters.**
V. M. Danilov, G. V. Beshenov.
Astronomical–geodetical investigations, p. 48 – 55 (1985). In Russian. Abstr. in Ref. Zh., 51. Astron., 5.51.819 (1986). – See Abstr. 003.024.

151.081 **A study on the formation and dynamics of galaxies.**
J. A. Fillmore.
Thesis, California Institute of Technology, 160 pp. (1985). Available from University Microfilms International, Ann Arbor, Mich., USA, Order No. DA8522612.
The compilation of two projects is contained within this thesis. The first part is a study on the growth of scale–free perturbations in an Einstein–de Sitter Universe. Chapter One considers the gravitational collapse of overdense perturbations (see Abstr. 37.161.240), and Chapter Two follows the development of spherical voids (see Abstr. 37.161.241). Chapter Three presents detailed fits of two component axisymmetric models to long–slit observations of early–type spirals. Chapter Four entails the author's efforts to find self–consistent models to oblate spheroids, apply them to a sample of bulges and ellipticals, and consider which observations can yield information about the kinematic structure of these systems.

151.082 **Gas dynamical stripping of galaxies.**
T. J. Gaetz.
Diss. Abstr. Int., Sect. B, Vol. 46, No. 9, p. 3090 (1986). Thesis, Cornell University, 198 pp. (1985). Order No. DA8525692.

151.083 **Stability of spiral–wave packets in a gaseous disk.**
M. G. Abramyan, S. V. Arutyunyan.
Sov. Astron., Vol. 29, No. 5, p. 507 – 509 (1985). English translation of 40.151.055.

151.084 **Computer modelling of the evolution of galaxies.**
J. Palous.
Budushchee nauk. Mezhdunar ezhegod., Vyp. 17. Moskva, p. 82 – 91 (1984). In Russian. From Ref. Zh., 51. Astron., 6.51.29 (1986).

151.085 **On the relationship between lenses and inner rings in spiral galaxies.**
M. P. Schwarz.
Proc. Astron. Soc. Aust., Vol. 6, No. 2, p. 202 – 205 (1985).
In simulations of gas flow in the gravitational field of model barred galaxies structures resembling inner rings have formed. The model rings encircle the bar as observed in real galaxies, but are more elongated than the average inner ring. In this paper the author shows that the addition of a lens–like component to the background field results in much rounder rings.

151.086 **Simple three–dimensional periodic orbits in a galactic–type potential.**
G. Contopoulos, P. Magnenat.
Celest. Mech., Vol. 37, No. 4, p. 387 – 414 (1985). In French.
The authors study the families of simple periodic orbits in a three–dimensional system that represents the inner parts of a perturbed triaxial galaxy.

151.087 **Bifurcations in systems of three degrees of freedom.**
G. Contopoulos.
Celest. Mech., Vol. 38, No. 1, p. 1 – 22 (1986).
The author studies the bifurcations of families of double and quadruple period orbits in a simple Hamiltonian system of three degrees of freedom.

151.088 **Violent relaxation and mixing in non–uniform one–dimensional gravitational systems.**
G. Severne, M. Luwel.
Astrophys. Space Sci., Vol. 122, No. 2, p. 299 – 325 (1986).
A series of numerical experiments on non–uniform one–dimensional gravitational systems have been performed with the aim of testing the conflicting models of violent relaxation, developed by Lynden–Bell and by Shu.

151.089 **Star–gas density waves in spiral galaxies.**
S. H. Lubow, S. A. Balbus, L. L. Cowie.
Space Telesc. Sci. Inst., Prepr. Ser., No. 104, 70 pp. (1986). To appear in Astrophys. J.

151.090 **Cloud collisional damping of spiral density waves in galaxies.**
S. H. Lubow.
Space Telesc. Sci. Inst., Prepr. Ser., No. 110, 16 pp. (1986). To appear in Astrophys. J., Lett. Ed.

151.091 **Theoretical considerations on the rotation curves of galactic halos.**
A. Malagoli, R. Ruffini.
Galaxies, quasars and cosmology, p. 11 – 24 (1985). – See Abstr. 012.042.
The observations that the rotation curves of some spiral galaxies are flat in regions extending from $R \sim 10$ kpc all the way up to $\sim 50 - 100$ kpc is the single most important evidence either for the existence of massive galactic halos of dark matter. The authors infer from the constancy of the rotation curves properties of the "dark" matter component of the galactic halos.

151.092 **Rotation and vorticity in galactic models.**
F. Pacheco, G. Pucacco, R. Ruffini.
Galaxies, quasars and cosmology, p. 25 – 28 (1985). – See Abstr. 012.042.

151.093 **Heating of large stellar systems by binaries.**
M. Giersz.
Postepy Astron., Tom 32, Zesz. 3/4, p. 249 – 265 (1984). In Polish.
The following topics are discussed: the formation of binaries in two–body and three–body interactions, cooling and heating of stellar systems connected with the formation of binaries and with their subsequent interactions with field stars.

151.094 **Simple three–dimensional periodic orbits in a galactic–type potential.**
G. Contopoulos, P. Magnenat.
ESO Sci. Prepr., No. 433, 50 pp. (1986). To appear in Celest. Mech.

151.095 **Qualitative changes in 3–dimensional dynamical systems.**
G. Contopoulos.
ESO Sci. Prepr., No. 436, 56 pp. (1986). To appear in Astron. Astrophys.

151.096 **What value must have the gradient of dispersion of radial velocities of stars in disks of galaxies?**
A. G. Morozov, A. V. Khoperskov.
Astrofizika, Tom 24, Vyp. 3, p. 467 – 476 (1986). In Russian. English translation in Astrophysics, Vol. 24, No. 3.
The dispersion of radial velocities of stars c_* (which is a minimum necessity for gravitational stability of the stellar disk) has been calculated as a function of radial inhomogeneity scales of surface density L_σ, dispersion of radial velocities of stars L_c and other disk parameters. It has been shown tat the quantity c_* reaches its minimum value with $\eta = L_\sigma/L_c \cong 1$. This result presents a satisfactory agreement with the observational data.

151.097 **The structural–parameter tables of King's star–cluster models.**
S.–i. Ichikawa.
Ann. Tokyo Astron. Obs., Second Ser., Vol. 21, No. 1, p. 77 – 84 (1986).
The structural parameters of King's star–cluster models are calculated and presented in the form of tables. These tables contain spatial density distributions, surface (projected) density distributions, and integrated profiles as functions of the normalized radius for 9 values of the concentration index. The method is described by which concentration index, core radius, and total apparent magnitude observed for an object are transformed into the photometric parameters for use in a quantitative classification system.

151.098 **Bisymmetric spiral configuration of magnetic fields in spiral galaxies. I. Local theory.**
T. Sawa, M. Fujimoto.
Publ. Astron. Soc. Jpn., Vol. 38, No. 2, p. 133 – 150 (1986).
Parker (1970, 1971) introduced the dynamo action in the galactic gaseous disk to maintain the magnetic field against its turbulent diffusion. The field configuration that Parker (1971) assumed is, however, only axisymmetric. In the present paper the authors extend this dynamo theory so that it can be applied locally for the bisymmetric spiral field prevailing in the gas of finite conductivity in the differential rotation.

151.099 **Tidal triggering of Seyfert galaxies and quasars – perturbed galaxy disk models versus observations.**
G. G. Byrd, V. J. Valtonen, B. Sundelius, L. Valtaoja.
Rep. Ser., Dep. Phys. Sci., Univ. Turku, No. FTL–R92, 35 pp. (1986) = Turku Univ. Obs., Informo No. 100. Submitted to Astron. Astrophys.

151.100 **Models of axisymmetric galaxies.**
J. L. Bishop.
Astrophys. J., Vol. 305, No. 1, p. 14 – 27 (1986).
A method for the construction of models of axisymmetric galaxies is developed. In this formulation a mass density and the corresponding gravitational potential are given, and a numerical solution for the distribution function is determined. The method applies to potentials in which the stellar motions admit of three isolating integrals. The method is illustrated with the construction of the "perfect" oblate–spheroidal galaxy. The solution for the distribution function is not unique, and various solutions are determined. Rotation curves and velocity ellipsoids are presented.

151.101 **Bar formation in disk galaxies, and the normalized angular momentum of the bulge.**
A. V. Zasov.
Sov. Astron. Lett., Vol. 11, No. 5, p. 277 – 279 (1985). English translation of 40.151.026.

151.102 **Orbital angular momenta of binary galaxies.**
D. A. Verner.
Sov. Astron. Lett., Vol. 11, No. 5, p. 280 – 282 (1985). English translation of 40.151.027.

151.103 **Statistical equilibrium of collisionless systems with Coulomb and Newton interactions.**
V. M. Pergamenshchik.
Phys. Lett. A, Vol. 113A, No. 4, p. 225 – 230 (1985). Abstr. in Phys. Abstr., Vol. 89, No. 1277, Entry 42316 (1986).

151.104 **Qualitative changes in 3–dimensional dynamical systems.**
G. Contopoulos.
Astron. Astrophys., Vol. 161, No. 2, p. 244 – 256 (1986).
The author studies the main qualitative changes that appear in the types of orbits that are encountered in 3–dimensional dynamical systems. As a prototype he considers the potential $V = 1/2\,(Ax^2+By^2+Cz^2)–\varepsilon xz^2–\eta yz^2$, that is assumed to represent the inner parts of distorted 3–axial elliptical galaxies. He studies the changes in the stability and bifurcations of some of the main types of periodic orbits as some of the parameters vary.

151.105 **A method for obtaining Hamiltonians with simple periodic solutions.**
N. Caranicolas, A. Diplas.
Astron. J., Vol. 91, No. 3, p. 660 – 664 (1986).
The authors present a method for obtaining Hamiltonian systems with accurate periodic solutions. The general solution of a degenerate integrable Hamiltonian is expressed in a polynomial form when the ratio of the frequencies $q = \omega_1/\omega_2$ is an integer, or as a power series when q is a rational number. Adding this expression as a perturbation in the above Hamiltonian, a new nontrivial Hamiltonian with one exact periodic solution is obtained. Some possible applications of these findings are also discussed.

151.106 **Observable properties of oblate spheroids.**
J. A. Fillmore.
Astron. J., Vol. 91, No. 5, p. 1096 – 1107 (1986).
Self–consistent solutions of the stellar hydrodynamic equations are presented for systems with isodensity surfaces which are concentric oblate spheroids and have constant mass–to–light ratios. Various models are investigated in order to determine whether observable parameters can distinguish the different configurations. The simulations indicate that observations can uniquely characterize those galaxies that lie at certain kinematic extremes, particularly those with excessive velocity dispersions in one component (e.g., tangential or radial). In restricted cases this diagnostic can put an upper limit on the true flattening of the spheroid.

151.107 **Possible evidence for gravitational Bohr orbits in double galaxies.**
M. DerSarkissian.
Lett. Nuovo Cimento, Vol. 44, Ser. 2, No. 8, p. 629 – 636 (1985). Abstr. in Phys. Abstr., Vol. 89, No. 1279, Entry 53735 (1986).

151.108 **The criterion for equipartition in clusters with a mass spectrum.**
W. R. Stoeger.
Mem. Soc. Astron. Ital., Vol. 56, No. 4, p. 759 – 762 (1985). – See Abstr. 012.054.

151.109 **Two–body capture in large N body systems. II.**
M. Giersz.
Acta Astron., Vol. 35, Nos. 3–4, p. 401 – 420 (1985).
The rates of formation of binaries in the following processes are considered: (1) tidal two–body interactions. (2) self–hardening of soft binaries formed in three–body interactions, (3) three–body interactions, (4) two–body interactions in which the emission of gravity waves takes place.

151.110 **Three–dimensional waves in disks.**
Z. Loska.
Acta Astron., Vol. 36, No. 1, p. 43 – 61 (1986).
Propagation of three–dimensional waves in a thin Keplerian polytropic disk is described. The resulting partial differential equation describing linear adiabatic nonaxisymmetric perturbations of the disk is separated into the vertical and radial parts by

the use of the WKBJ method. The vertical part forms a Sturm–Liouville–type equation for vertical perturbations. The solutions depend on the radial wavenumbers which are eigenvalues of the problem. The character of resulting waves depends on their location relative to the Lindblad resonances.

151.111 **Dynamical friction on a satellite of a disc galaxy.**
G. G. Byrd, S. Saarinen, M. J. Valtonen.
Mon. Not. R. Astron. Soc., Vol. 220, No. 3, p. 619 – 631 (1986).
Gravitational drag on satellite galaxies of a disc galaxy has been studied. A variety of computer models were used to simulate the galaxy: a "cold" Mestel disc with no velocity dispersion, and a more realistic "hot" disc with a velocity dispersion inside an inert spherical halo were studied. About 60000 particles were used in the disc and satellites of 0.05 and 0.1 disc mass were studied. The authors find for hot and cold discs with or without haloes that the energy change due to tidal interaction is always negative, i.e. the satellite becomes more strongly bound.

151.112 **Stellar dynamics and the description of stellar systems.**
H. Dejonghe.
Phys. Rep., Vol. 133, No. 3 – 4, p. 217 – 313 (1986). Abstr. in Phys. Abstr., Vol. 89, No. 1281, Entry 67547 (1986).

151.113 **A further study of the solution of Poisson's equation for three–dimensional spiral galaxies.**
H.–g. Bi, Q.–h. Peng, Z.–y. Li.
Chin. Astron. Astrophys., Vol. 10, No. 2, p. 162 – 167 (1986). English translation of 40.151.170.

151.114 **Hydrodynamical models of elliptical galaxies.**
V. G. Ortega, J. A. de Freitas Pacheco.
Rev. Mex. Astron. Astrofis., Vol. 12, Spec. Issue, p. 97 – 100 (1986). – See Abstr. 012.066.
Stellar hydrodynamical (Jeans) equations and the de Vaucouleurs' $r^{1/4}$ law for the intensity profile are used to obtain expressions for the projected rotational velocity and dispersion profile along the major axis of elliptical galaxies. The numerical results are applied to the elliptical galaxy NGC 3379.

151.115 **Evolution of velocity dispersions in some collapsing spheroidal distributions of mass.**
E. Moreno, P. Pismis.
Rev. Mex. Astron. Astrofis., Vol. 12, Spec. Issue, p. 109 – 113 (1986). – See Abstr. 012.066.
Numerical integration of a sample of orbits performed in the gravitational field of some collapsing spheroidal distributions of mass leads to the following main results: (1) neither the time of collapse nor the mass distribution in the system seem to play a dominant role in the dynamical evolution of the calculated sample of orbits. (2) the form of the velocity distribution, initially "spherical", is approximately conserved in time.

151.116 **On the tidal origin of angular momentum in galaxies.**
M. B. Mosconi, D. García Lambas, J. L. Sérsic.
Rev. Mex. Astron. Astrofis., Vol. 12, Spec. Issue, p. 131 (1986). Abstract. – See Abstr. 012.066.

151.117 **A global model for violent relaxation.**
J. L. Sérsic, D. García Lambas, M. B. Mosconi.
Rev. Mex. Astron. Astrofis., Vol. 12, Spec. Issue, p. 132 (1986). Abstract. – See Abstr. 012.066.

151.118 ***N*–body simulation of giant molecular clouds in a galaxy. I. Generation and acceleration of the random velocity of giant molecular clouds.**
M. Tosa, M. Fukunaga.
Publ. Astron. Soc. Jpn., Vol. 38, No. 3, p. 335 – 346 (1986).
In the present paper the authors make an *N*–body simulation of a giant molecular cloud (GMC) system in a galaxy to see how the random motion of GMCs is generated and accelerated through gravitational interaction and compare the results with those of previous calculations based on binary encounters among GMCs by Fukunaga (1984).

151.119 **Anisotropy and self–similar evolution of star clusters.**
R. Spurzem, L. Lange, U. Fritze, E. Bettwieser.
Mitt. Astron. Ges., No. 65, p. 223 (1986). Abstract. – See Abstr. 012.072.

151.120 **Heating and cooling in dense stellar systems.**
U. Fritze, K. J. Fricke.
Mitt. Astron. Ges., No. 65, p. 223 (1986). Abstract. – See Abstr. 012.072.

Computer recreations: how close encounters with star clusters are achieved with a computer telescope.
See Abstr. 014.001.

Local dynamic thermal instability.
See Abstr. 062.066.

Theory of galactic disks and its relevance to the solar wind magnetic configuration.
See Abstr. 062.108.

Simulations of the brightest stars in galaxies as distance indicators.
See Abstr. 065.161.

How to form a 200 $M_\odot$ star.
See Abstr. 065.165.

The birth of AGNs and quasars via the collapse of dense star clusters to supermassive black holes.
See Abstr. 067.042.

Effect of dynamical friction on the escape of a supermassive black hole ejected from the center of a galaxy.
See Abstr. 067.044.

N–body simulation of giant molecular clouds in a galaxy.
See Abstr. 131.027.

On the radial distributions of molecular clouds in galaxies.
See Abstr. 131.030.

On the collision of high–velocity clouds with a galactic disk.
See Abstr. 131.288.

The Lindblad diagram for open clusters.
See Abstr. 153.028.

Hydrodynamic simulation of a white dwarf–main sequence star collision in a globular cluster core.
See Abstr. 154.040.

Dynamical evolution of globular clusters. II. Binaries – method.
See Abstr. 154.061.

Construction of a global gas–dynamical model for our Galaxy.
See Abstr. 155.031.

On stellar streaming motions in triaxial objects: the case of the galactic bulge.
See Abstr. 155.093.

Structure and dynamics of the galactic system. Report 1981 – 1984.
See Abstr. 155.100.

Colliding and merging galaxies.
See Abstr. 157.020.

The dynamical influence of the nucleus on the bulge of M31.
See Abstr. 157.034.

Contraction of dark matter galactic halos due to baryonic infall.
See Abstr. 157.040.

Gas flow in spiral galaxies: effects of rotation, self–gravitation and thermal processes.
See Abstr. 157.046.

The disks of spiral galaxies.
See Abstr. 157.138.

Binary and multiple galaxies.
See Abstr. 157.140.

A case against baryons in galactic halos.
See Abstr. 157.150.

Elliptical galaxies with dust lanes.
See Abstr. 157.156.

Stellar kinematics and the stability of disks in spiral galaxies.
See Abstr. 157.172.

Towards a physical model for galactic rotation curves.
See Abstr. 157.183.

Angular momentum–limited degenerate and semidegenerate neutrino halos.
See Abstr. 157.256.

Kinematics of spiral galaxies. An overview.
See Abstr. 157.263.

Dwarf galaxies and galaxy formation.
See Abstr. 157.306.

On the structure and evolution of dwarf galaxies.
See Abstr. 157.307.

Stellar dynamics of Cen A.
See Abstr. 158.005.

Tidal triggering of Seyfert galaxies and quasars–models.
See Abstr. 158.145.

On the effects of compression of a gaseous disc by thermal and dynamical pressures of intergalactic gas.
See Abstr. 160.008.

Compact configurations within small evolving groups of galaxies.
See Abstr. 160.082.

The dynamical evolution of clusters of galaxies.
See Abstr. 160.121.

Thermal–chemical instability in a pre–galactic gas cloud.
See Abstr. 161.019.

How relaxed is the distribution of galaxies?
See Abstr. 161.044.

Hierarchical clustering: angular momentum versus density.
See Abstr. 161.047.

Dynamical effects of dark matter in systems of galaxies.
See Abstr. 161.172.

152 Stellar Associations

152.001 **Radial systems of dark globules in the association Cep OB2.**
A. L. Gyul'budagyan, L. F. Rodriguez, J. Cantó.
Astrofizika, Tom 24, Vyp. 1, p. 201 – 203 (1986).
Results of ^{12}CO and ^{13}CO observations of dark globules in the Cep OB2 association are presented.

152.002 **VBLUW photometry of OB associations: SPECTER at La Silla.**
E. de Geus, J. Lub, T. de Zeeuw.
Messenger, No. 43, p. 7 – 10 (1986).

152.003 **Ultraviolet observations of young stars in the Chamaeleon 1 association.**
A. Brown, J. L. Linsky, F. Walter, C. Jordan, P. Judge, G. Gahm, E. D. Feigelson.
Bull. Am. Astron. Soc., Vol. 17, No. 4, p. 837 (1985). Abstract. – See Abstr. 010.062.

152.004 **The OB 112 association in the M33 galaxy.**
P. Kynchev, G. Ivanov.
Astron. Tsirk., No. 1386, p. 1 – 4 (1985). In Russian.

152.005 **Proper motions of some associations in the Perseus spiral arm.**
T. P. Gerasimenko.
Astron. Tsirk., No. 1392, p. 4 – 7 (1985). In Russian.

152.006 **Ara OB1: an H I study of a very young southern complex.**
E. M. Arnal, J. C. Cersosimo.
Rev. Mex. Astron. Astrofis., Vol. 12, Spec. Issue, p. 299 (1986). Abstract. – See Abstr. 012.066.

152.007 **On the distribution of the frequency of stellar flares in stellar aggregates.**
M. A. Mnatsakanyan.
Astrofizika, Tom 24, Vyp. 3, p. 621 – 623 (1986). In Russian. English translation in Astrophysics, Vol. 24, No. 3.
The analytic time–representation of the multiplicitiy of flares in stellar aggregates based on observational data at present is given under the condition of independency of flares from each other. They are exactly approximated by two Poisson groups with stars in each of them having the same frequency: $N_1 = 670$, $\nu_1 = 1.1$ flares at all observational time, $N_2 = 60$, $\nu_2 = 9$ for the Pleiades, and by one Poisson process with $N = 1250$, $\nu = 0.49$ for Orion, $N = 330$, $\nu = 0.37$ for the dark nebulae of Taurus. The total number of flare stars in the Pleiades is nearly equal or less than 750.

152.008 **Stellar associations in M31 arm S4.**
G. R. Ivanov, P. Z. Kynchev.
Sov. Astron. Lett., Vol. 11, No. 5, p. 310 – 312 (1985). English translation of 40.152.005.

152.009 **Some statistical characteristics of Ap and Am stars in regions occupied by OB associations.**
Ts. S. Radoslavova.
Dokl. Bolg. Akad. Nauk, Tome 38, No. 8, p. 961 – 962 (1985).

152.010 **Systems of stellar associations in galaxies.**
P. Hodge.
Luminous stars and associations in galaxies, p. 369 – 379 (1986). – See Abstr. 012.079. (IAU Symp. No. 116).
This paper begins with an attempt to examine the problem of identifying stellar associations in galaxies in a consistent way, so that meaningful physical comparisons can be made for the popu-

lation of stellar associations of different galaxies. A compilation of the existing data on associations in other galaxies is given and their properties compared. Questions relating to star formation in stellar associations are discussed, and then the issue of the initial mass function of core clusters, especially those located in giant H II regions, is briefly examined.

152.011 **The dynamical evolution of expanding H I shells and initial mass function of OB associations.**
S. A. Silich.
Luminous stars and associations in galaxies, p. 393 – 394 (1986). – See Abstr. 012.079. (IAU Symp. No. 116).
The evolution of expanding supershells produced by SN explosion in OB associations is considered. The expansion velocities of the shells and OB associations IMF spectral index are obtained. The results agree well with observational data.

152.012 **NGC 2070 and NGC 3603.**
Y.–H. Chu, P. Seitzer, N. Walborn.
Luminous stars and associations in galaxies, p. 515 – 516 (1986). – See Abstr. 012.079. (IAU Symp. No. 116).
The authors have obtained CCD images in UBVRI for NGC 2070, and BVRI for NGC 3603 to determine their color–magnitude diagrams and luminosity functions. They have also obtained 4 m prime focus plates with interference filters centered on the He II 4686 line ($\Delta\lambda$ = 65 Å) and the adjacent blue continuum at 4765 Å ($\Delta\lambda$ = 70 Å) to identify the Wolf–Rayet (WR) stars in their cores. The authors discuss in these proceedings only the WR stars resolved in the core of NGC 3603.

The binary frequency and origin of the OB runaway stars.
See Abstr. 111.020.

The Bp stars in the Orion OB1 association.
See Abstr. 114.074.

Polarization of star radiation in the Mon R1 association.
See Abstr. 116.014.

Space distribution of the Hα and flare stars in the Orion OB1d association.
See Abstr. 122.123.

Velocities and birthplaces of pulsars.
See Abstr. 126.017.

The formation of bound stellar clusters.
See Abstr. 131.082.

Nebular complex in the region of Cas OB2 association; ring nebula Sh 157.
See Abstr. 131.112.

The dynamical evolution of H I shells and the initial mass function of OB associations.
See Abstr. 131.122.

Star formation regions as galactic dissipative structures.
See Abstr. 131.194.

Interstellar Na I absorption towards the stellar association Ara OB1.
See Abstr. 131.268.

Star formation: from OB associations to protostars.
See Abstr. 131.301.

Star forming loops in the IRAS sky images.
See Abstr. 131.384.

Hollow H II regions. I. Effects of stellar evolution and wind energy dissipation.
See Abstr. 132.021.

Formation of OB clusters: radiation–driven implosion?
See Abstr. 132.049.

Problems in the evolution of young stars in open clusters and associations. Perspectives of future work.
See Abstr. 153.036.

The chemical composition of six southern clusters and associations.
See Abstr. 153.039.

Age and processes of star formation in very young LMC associations.
See Abstr. 156.014.

CCD observations of young stellar associations and multiple systems in the Magellanic Clouds.
See Abstr. 156.035.

Stellar associations in the galaxy NGC 2403.
See Abstr. 157.013.

Two–color CCD mapping of the luminous type I irregular galaxy NGC 4449.
See Abstr. 157.218.

A criterion determining the association limits in the Andromeda galaxy.
See Abstr. 157.252.

Wolf–Rayet stars in M31's OB associations.
See Abstr. 157.314.

New stellar associations in M31.
See Abstr. 157.315.

The associations OB 110 and OB 112 in M33 galaxy.
See Abstr. 157.316.

On the arm pattern of M33.
See Abstr. 157.317.

153 Open Clusters

153.001 **The abundance of lithium in old galactic clusters. I. NGC 7789.**
C. Pilachowski.
Astrophys. J., Vol. 300, No. 1, p. 289 – 296 (1986).
A comprehensive study of the abundance of lithium in the moderately old galactic–cluster NGC 7789 is reported. The abundance of lithium in stars on the main–sequence turnoff is cosmic [log ε(Li) = 3.0]. Stars just beyond the turnoff show some evidence of lithium depletion or dilution. Rotational velocities are consistent with observed velocities for early F stars. The decline of lithium abundance with effective temperature in the cluster giants can probably be attributed to destruction of lithium at the base of the convective envelope. The abundance of lithium in the giant star Küstner 301 is much higher than in a normal giant, and this star is tentatively identified as a weak G–band star.

153.002 **The Einstein soft X–ray survey of the Pleiades.**
J.–P. Caillault, D. J. Helfand.
News Lett. Astron. Soc. N.Y., Vol. 2, No. 9, p. 29 – 46 (1986). – See Abstr. 010.241.
The authors report on a soft X–ray survey of the Pleiades undertaken with the imaging instruments of the Einstein Observatory. They discuss the results of this survey in the context of recent work on the correlation of X–ray emission with age and rotation and on coronal variability.

153.003 **The chemical composition of the young open cluster NGC 6611.**
P. J. F. Brown, P. L. Dufton, D. J. Lennon, F. P. Keenan, D. Kilkenny.
Astron. Astrophys., Vol. 155, No. 1, p. 113 – 119 (1986).
Four colour photometric and spectroscopic observations for eight early–type main sequence stars in the cluster NGC 6611 were obtained at the South African and Anglo–Australian observatories. Equivalent widths and line profiles were measured for lines of hydrogen, helium and seven other elements present in the spectra of five stars. Effective temperatures were derived from ionization equilibria while surface gravities were estimated from profiles fitted to Hδ and Hε absorption lines. Abundances, deduced from comparison with LTE and non–LTE model atmosphere calculations are found to be consistent with normal early–type stellar abundances. The present study implies also the possibility of abundance variations over relatively small distance scales.

153.004 **Photometry of open cluster NGC 1931.**
A. K. Pandey, H. S. Mahra.
Astrophys. Space Sci., Vol. 120, No. 1, p. 107 – 113 (1986).
Photoelectric *UBV* magnitudes and colours have been determined for stars in the field of NGC 1931. The reddening across the cluster varies from $0^m.33$ to $1^m.20$. A distance of 2.16 kiloparsecs has been estimated for the cluster. It is concluded that the age of the cluster lies between the ages of NGC 6231 and NGC 2362 groups.

153.005 **Evolution of massive stars: comparison of cluster sequences and models with mass loss.**
J.–C. Mermilliod, A. Maeder.
Astron. Astrophys., Vol. 158, No. 1/2, p. 45 – 59 (1986).
A detailed comparison between theoretical isochrones of models with mass loss and observed sequences of 25 young clusters is undertaken. These clusters have been carefully discussed for reddening, distances, membership, binarity and stellar peculiarities. The number counts are compared with the theoretically expected star numbers in various parts of the isochrones. The effects of differences in mass loss rates and differences in stellar opacities are also considered. Great caution is to be given to the effects of unresolved binaries and of rapidly rotating stars. These comparisons indicate differences which can not be attributed only to mass loss rates. On the basis of various recent models, the authors would rather propose that some extra–mixing, overshooting or turbulent diffusion is at the origin of the observed differences. If one attributes the whole difference to the overshooting process (which is by no means certain), a ratio $d_{over}/H_p \cong 0.3$ of the overshooting distance to the pressure scale height at the edge of the classical core could be tentatively estimated on the basis of massive stars in the range of 9 to 15 $M_\odot$.

153.006 **Carbon isotope ratios in giant stars of the old galactic cluster NGC 7789.**
C. Sneden, C. A. Pilachowski.
Astrophys. J., Vol. 301, No. 2, p. 860 – 867 (1986).
Carbon isotope ratios have been derived for seven giant stars in the old open cluster NGC 7789. High–resolution, high signal–to–noise spectra of the CN red system bands in these stars were obtained with the KPNO 4 m Cassegrain echelle spectrograph and a CCD camera. ^{13}CN features were detected easily in all cluster stars, and isotope ratios in the range $^{12}C/^{13}C = 10 - 30$ were derived. The isotope ratios of most stars are normal for K giants possessing convective envelopes and show no correlations with effective temperatures and with lithium abundances.

153.007 **Faint second photometric standards in the R colour system in the cluster NGC 188.**
V. G. Medvedev, E. P. Pavlenko, V. V. Prokof'eva.
Izv. Krymskoj Astrofiz. Obs., Tom 73, p. 122 – 130 (1985). In Russian. English translation in Bull. Crimean Astrophys. Obs., Vol. 73.
The 149 stellar magnitudes of faint stars ($13^m.2 < R < 15^m.6$) were measured in the central zone of the cluster NGC 188. The stellar brightnesses were defined by measuring television stellar images obtained during seven sets of observation in the autumn of 1981 and spring of 1982. The internal accuracy of the catalogue stellar magnitude in the instrumental system is equal to $\pm 0^m.05$ and the external accuracy may be up to $\pm 0^m.2$.

153.008 **On the application of the *P–Q* diagram method for investigation of star clusters.**
V. I. Kuznetsov.
Kinematika Fiz. Nebesn. Tel, Tom 2, No. 2, p. 45 – 50 (1986). In Russian. English translation in Kinematics Phys. Celest. Bodies.
Intrinsic colour indices obtained from photometric data and from spectral classification of stars up to $V = 14^m.0$ in the region around NGC 2244 are compared. The classification is carried out by the method developed for unwidened stellar spectra. The *P–Q* photometric diagram method is shown to be inapplicable for the determination of intrinsic colour indices, for isolation of probable members of clusters, as well as for investigation of physical parameters of star clusters.

153.009 **Carbon and nitrogen abundances along the evolutionary sequence of M67.**
J. A. Brown.
Bull. Am. Astron. Soc., Vol. 17, No. 4, p. 840 (1985). Abstract. – See Abstr. 010.062.

153.010 **Bright giant stars in intermediate–age galactic clusters.**
J. Sowell.
Bull. Am. Astron. Soc., Vol. 17, No. 4, p. 878 (1985). Abstract. – See Abstr. 010.062.

153.011 **MK spectral classification for the lower main sequence of Praesepe.**
C. J. Corbally, R. F. Garrison.
Bull. Am. Astron. Soc., Vol. 17, No. 4, p. 880 (1985). Abstract. – See Abstr. 010.062.

153.012 **The capture of field stars by open clusters: can it explain the blue stragglers?**
P. J. T. Leonard.
Bull. Am. Astron. Soc., Vol. 17, No. 4, p. 882 – 883 (1985). Abstract. – See Abstr. 010.062.

153.013 **The structure and internal motions of the open cluster M35.**
B. J. McNamara, K. Sekiguchi.
Bull. Am. Astron. Soc., Vol. 17, No. 4, p. 903 (1985). Abstract. – See Abstr. 010.062.

153.014 **The Hyades: EXOSAT broad band spectroscopy.**
N. J. Westergaard, H. U. Nørgaard–Nielsen,
L. Hansen, H. E. Jørgensen, I. L. Rasmussen,
H. W. Schnopper.
ESA Spec. Publ., ESA SP–239, p. 159 – 162 (1985). – See Abstr. 012.019.

Two EXOSAT observation fields of the Hyades star cluster have been combined with Einstein IPC data for 3 band filter spectroscopy. Although temporal variations disturb the picture the coronal temperatures in 9 stars are confined to the few million degree range. The column density is probably 1 or $2 \times 10^{19} cm^{-2}$.

153.015 **Long–term X–ray variability in the Hyades cluster.**
M.–C. Zolcinski, R. A. Stern.
ESA Spec. Publ., ESA SP–239, p. 163 – 169 (1985). – See Abstr. 012.019.

Twenty Hyades stars have been observed over a period of one and a half years. Reliable results are obtained for seven of them. Only one star, the K0 III giant θ^1 Tau does not exhibit X–ray variability. This result casts some doubt about the activity cycle being the explanation for the wide range of X–ray luminosities seen among the Hyades giants. The six other dwarfs exhibit variability on a time scale of few days to one and a half years. Assuming that the X–ray emission is of coronal origin, implications on the stellar coverage variations are drawn.

153.016 **The lithium abundance gap in the Hyades F stars: the signature of diffusion.**
G. Michaud.
Astrophys. J., Vol. 302, No. 2, p. 650 – 655 (1986).

Boesgaard and Tripicco have recently obtained high–precision spectra of stars of the Hyades showing the existence of an abundance gap of lithium in the F stars of that cluster. The observed gap is restricted to a narrow effective temperature range ($6400 < T_{eff} < 7000$K). It is shown here that in this effective temperature range the depth of the convection zone increases rapidly as the effective temperature is reduced. Diffusion then leads to a normal lithium abundance in stars of $\sim$7000K, while it leads to underabundances in stars between 6800 and 6400K. The Li abundance is not modified by diffusion in stars cooler than 6400K. This narrow abundance gap is used to put constraints on mass loss in the Hyades.

153.017 **The giant branch of the open cluster NGC 2539.**
J. J. Clariá, E. Lapasset.
Astrophys. J., Vol. 302, No. 2, p. 656 – 661 (1986).

Observations in the DDO and CMT_1T_2 photometric systems have been obtained for 13 G and K stars in the field of NGC 2539. A reddening $E(B–V) = 0.08 \pm 0.02$ mag, a distance modulus of 9.8 ± 0.5 mag, and an age of $(6.4 \pm 0.8) \times 10^8$yr are derived for the cluster. There is no firm evidence for any intrinsic differences in CN strengths among the cluster giants. It appears that the cluster giants have enhanced CNO abundances relative to "normal" stars. The masses of the red cluster giants were found to be systematically lower than those predicted by the theoretical models.

153.018 **Lithium in the Hyades Cluster.**
A. M. Boesgaard, M. J. Tripicco.
Astrophys. J., Lett. Ed., Vol. 302, No. 2, p. L49 – L53 (1986).

Twenty F dwarfs in the Hyades Cluster have been observed for Li content with the CFH telescope and the UH 2.2 m telescope at high spectral resolution and high signal–to–noise ratios. Abundances have been determined for Li/H and Fe/H from a standard model atmosphere abundance routine. There is a remarkable variation of Li/H with stellar surface temperature: stars with T_{eff} near 6600K show Li depletions by factors >100 relative to stars 300K hotter and cooler. The hottest stars ($T > 7000$K) show Li/H values of 10^{-9}, the maximum observed in F field stars, meteorites, T Tauri stars, etc.

153.019 **On an estimate of the dynamical parameters of multiple open star clusters.**
V. M. Danilov, G. V. Beshenov.
Astron. Tsirk., No. 1397, p. 1 – 3 (1985). In Russian.

153.020 **The intermediate–age open cluster Mel 71.**
M. W. Pound, K. A. Janes.
Publ. Astron. Soc. Pac., Vol. 98, No. 600, p. 210 – 217 (1986).

The authors have obtained a color–magnitude diagram of the open cluster Mel 71 using plates from the Mauna Kea 2.2–meter telescope of the University of Hawaii. From a Hyades ZAMS fit the authors deduce that Mel 71 has a reddening of $E(B–V) = 0.10$, a distance modulus of $(m–M)_0 = 11.88$, and a corrected main–sequence turnoff color index of $(B–V)_0 = 0.18$. They also find its age to be about 9.3×10^8years, somewhat older than previous estimates. Thus Mel 71 appears to be a typical outer–disk cluster.

153.021 **Bright yellow giants in the open cluster Berkeley 82.**
D. Forbes.
Publ. Astron. Soc. Pac., Vol. 98, No. 600, p. 218 – 222 (1986).

UBV photometry is presented for stars in the previously unstudied open cluster Berkeley 82. Fourteen members of this poor cluster have been identified photometrically within the cluster radius of 5 arcmin (1.4 pc) derived from star counts. The cluster suffers a mean reddening $E(B–V) = 1.01 \pm 0.01$ (m.e.) at a distance of 980 ± 70 pc derived from ZAMS fitting. The two brightest likely cluster members have spectral types of G2 II/I*b* and K2 II with absolute magnitudes of –2.8 and –2.4, respectively, assuming cluster membership. Berkeley 82 is found to have an evolutionary age of $(6 - 8) \times 10^7$yr, and is similar to other poor clusters of comparable age, some of which have Cepheid members.

153.022 **MK classification of the brighter Praesepe stars.**
H. A. Abt.
Publ. Astron. Soc. Pac., Vol. 98, No. 601, p. 307 – 309 (1986).

The author provides the first, in most cases, two–dimensional spectral classifications for the 42 brightest stars in Praesepe. The types agree very well with the photoelectric colors. The author confirms Bidelman's discovery of nine Am stars. The frequency (26%) of those among the late A stars is consistent with results for field stars. But there is only one Ap star.

153.023 **Positions of 127 Hyads and 6–cm observations of 320 Hyads.**
J. N. Crain, W. L. Sanders, E. B. Fomalont, R. A. Sramek.
Publ. Astron. Soc. Pac., Vol. 98, No. 601, p. 325 – 329 (1986).

Accurate positions are given for 127 Hyades cluster stars which previously had only approximate positions known. The fields of 320 Hyads, including 55 X–ray stars, were observed with the VLA to a 6–cm detection level of approximately 0.5 mJy. There were no detections.

153.024 **Decay dynamics of nonstationary open clusters.**
V. M. Danilov.
Sov. Astron., Vol. 29, No. 4, p. 413 – 417 (1985). English translation of 40.153.011.

153.025 **Mass and age distributions of stars in young open clusters.**
R. Sagar, A. E. Piskunov (*A. Eh. Piskunov*), V. I. Myakutin, U. C. Joshi.
Mon. Not. R. Astron. Soc., Vol. 220, No. 2, p. 383 – 403 (1986).

Mass and age distributions of stars, based on homogeneous photoelectric *UBV* data and reliable cluster membership, are studied in 11 young open clusters. Stellar masses and ages are determined by fitting the theoretical stellar models to each star's position in the HR diagram by means of an interpolation procedure.

153.026 **Positions and magnitudes in the area of the open cluster NGC 752.**
J. Stock.
Rev. Mex. Astron. Astrofis., Vol. 11, No. 2, p. 103 – 108 (1985).
Positions and approximate visual magnitudes are derived for 458 stars in the area of the open cluster NGC 752.

153.027 **Fundamental parameters of the open cluster NGC 2567.**
J. J. Clariá, E. Lapasset.
Astron. J., Vol. 91, No. 2, p. 326 – 337 (1986). With plates 39, 40.
New photoelectric *UBV* photometry of 164 stars brighter than $V = 14.6$ as well as DDO and CMT_1T_2 photometry of five red stars in the open cluster NGC 2567 are presented and discussed. By using both color–magnitude and color–color diagrams, 81 stars are found to be cluster members and 15 stars to be probable members. A foreground reddening $E(B–V) = 0.13$ mag, a distance modulus $V_0–M_v = 11.05$ mag ($d = 1620$ pc), and an age of 2.9×10^8yr have been derived. NGC 2567 appears to have a nearly solar composition.

153.028 **The Lindblad diagram for open clusters.**
S. A. Kutuzov.
Astronomical–geodetical investigations, p. 56 – 66 (1985). In Russian. Abstr. in Ref. Zh., 51. Astron., 5.51.830 (1986). – See Abstr. 003.024.

153.029 **A new approach to zero–age main sequence fitting.**
M. H. Pedreros.
Diss. Abstr. Int., Sect. B, Vol. 46, No. 5, p. 1589 – 1590 (1985). Thesis, University of Toronto (1984).

153.030 **Analysis of color–magnitude diagrams from three Large Magellanic Cloud clusters.**
J. H. Jones.
Diss. Abstr. Int., Sect. B, Vol. 46, No. 10, p. 3481 (1986). Thesis, Clemson University, 230 pp. (1985). Order No. DA8527701.

153.031 **The open cluster NGC 1960.**
K. A. Barkhatova, P. E. Zakharova, L. P. Shashkina, L. K. Orekhova.
Sov. Astron., Vol. 29, No. 5, p. 499 – 501 (1985). English translation of 40.153.022.

153.032 **On the metallicity of the Hyades.**
V. G. Klochkova, V. E. Panchuk.
Sov. Astron. Lett., Vol. 11, No. 5, p. 291 – 293 (1985). English translation of 40.153.013.

153.033 **A proper–motion analysis of the cluster M35.**
B. McNamara, K. Sekiguchi.
Astron. J., Vol. 91, No. 3, p. 557 – 562 (1986). With plate 58.
Relative proper motions and membership assignments are presented for 526 stars located within a radius of 14′ of the center of M35 and extending to $V \sim 14.7$. Ninety–nine of these stars were not included in Cudworth's (1971) study of this cluster. A vastly greater plate collection and longer epoch difference results in a three–fold increase in proper–motion precision over the earlier analysis. Three post–main–sequence stars are tentatively identified as belonging to the cluster. New photographic photometry casts doubt upon the claimed abundance anomalies within M35 reported by Vidal (1973).

153.034 **Old stellar populations. III. The metallicities of M5, M71, and M67.**
D. Burstein, S. M. Faber, J. J. Gonzalez.
Astron. J., Vol. 91, No. 5, p. 1130 – 1139 (1986). = Lick Obs. Bull., No. 1036.
A moderate–resolution iron index ‹ΔFe› has been measured for individual stars in the clusters M5, M71, and M67. Iron abundances based on ‹ΔFe› for the three clusters are presented. This paper is one of a series based on old stellar population data taken with the Lick image–dissector scanner. It is an important feature of this program that all spectra were taken with the same equipment and the same observing and reduction procedures.

153.035 **Hyades and Sirius supercluster members brighter than magnitude (V) 7.1. II. Right ascension six to twelve hours.**
O. J. Eggen.
Publ. Astron. Soc. Pac., Vol. 98, No. 602, p. 423 – 441 (1986).
This paper lists the probable members of the Hyades and Sirius superclusters among stars brighter than $V = 7.1$ mag. The star sample is contained in the Bright Star Catalogue and its Supplement, augmented with a further supplement of 788 stars found during various observing programs over the past 40 years. Accurate, four–color and Hβ, or (*RI*), photometry is available for most of the supercluster members. The criteria for membership are the comparisons of the proper motion, radial velocity, and luminosity obtained from the supercluster parameters with the observed motions and the luminosity derived from the photometric parameters. New proper motions, based on all available catalogs, have been derived for the additional 788 stars discussed, as well as for all supercluster members.

153.036 **Problems in the evolution of young stars in open clusters and associations. Perspectives of future work.**
G. Bertelli, A. G. Bressan, C. Chiosi, E. Nasi.
Mem. Soc. Astron. Ital., Vol. 56, No. 4, p. 619 – 635 (1985). – See Abstr. 012.054.
The authors discuss several observational problems relative to young stars, which cannot find an easy solution in the context of classical stellar models. The study of the effects of convective overshooting and opacity enhancement in the region of CNO ionization performed by Bertelli et al. (1984) in the domain of massive stars is extended to the range of intermediate mass stars. Several preliminary results are presented and compared with observations.

153.037 **Chromospheric emission and rotation in open clusters.**
S. Catalano, E. Marilli.
Mem. Soc. Astron. Ital., Vol. 56, No. 4, p. 665 – 672 (1985). – See Abstr. 012.054.
The problem of rotation and stellar angular momentum evolution of late main sequence stars is discussed. It is shown that the rotation of late type stars can be studied as a function of the mass and the age in open clusters as old as 5 billion years and as far as 1 kpc by observing the Mg II and C IV ultraviolet emission lines by means of the FOS of the Space Telescope.

153.038 ***uvbyβ* photometry of southern clusters – VII. NGC 2547.**
R. R. Shobbrook.
Mon. Not. R. Astron. Soc., Vol. 220, No. 4, p. 825 – 831 (1986).
Strömgren four–colour and β photometry has been obtained for the 13 brightest members of the young galactic cluster NGC 2547. The data are generally in good agreement with other photometry for these stars and the mean distance modulus obtained from several methods, including the new Balona & Shobbrook $\beta/c_0/M_V$ calibration for B stars, is 8.1.

153.039 **The chemical composition of six southern clusters and associations.**
P. J. F. Brown, P. L. Dufton, D. J. Lennon, F. P. Keenan.
Mon. Not. R. Astron. Soc., Vol. 220, No. 4, p. 1003 – 1020 (1986).
The observational data are presented for 71 stars in five clusters, namely NGC 4755, 3293, 6531, IC 2944, NGC 2362 and a loose association previously investigated by Dufton (1972, 1979). Atmospheric parameters and element abundances are deduced for each star from model atmosphere analyses, and finally the results are discussed.

153.040 **Statistical boundaries for open clusters.**
V. M. Danilov, N. V. Matkin, O. P. Pyl'skaya.
Sov. Astron., Vol. 29, No. 6, p. 621 – 625 (1985). English translation of 40.153.043.

153.041 **The Hyades moving group: new members, kinematics, age.**
J. Bubeníček, J. Palouš, A. Eh. Piskunov.
Sov. Astron., Vol. 29, No. 6, p. 625 – 627 (1985). English translation of 40.153.044.

153.042 **Erratum: "The open cluster NGC 2158" [Astrophys. J., Vol. 299, No. 2, p. 683 – 694 (1985)].**
C. A. Christian, J. N. Heasley, K. A. Janes.
Astrophys. J., Vol. 305, No. 2, p. 954 (1986). See Abstr. 40.153.050.

153.043 **Hα, O I–photometry of the Hyades and Coma Berenices clusters.**
E. E. Mendoza V.
Rev. Mex. Astron. Astrofis., Vol. 12, Spec. Issue, p. 193 – 196 (1986). – See Abstr. 012.066.
This paper presents α(16)Λ(9)–photometry of 30 cluster members of the Hyades and 29 Coma Berenices stars.

153.044 **Polarimetry of reddened stars in M16 and NGC 6334/6357.**
A. M. Magalhães, W. F. Velloso Jr., E. Krugel, R. Chini.
Rev. Mex. Astron. Astrofis., Vol. 12, Spec. Issue, p. 408 (1986). Abstract. – See Abstr. 012.066.

153.045 **Proper motion of the open star cluster NGC 2301.**
A. Aiad.
Bull. Fac. Sci., Cairo Univ., Vol. 54, p. 729 – 744 (1986).
Proper motions of 190 stars in the field of NGC 2301 were determined from three combinations on two photographic plates taken by the Alger astrograph in the years 1898 and 1973 with an epoch difference of 75.72 years. The absolute central proper motions and position angle were found to be $\mu_\alpha \cos\delta = (-0.07 \pm 0.02)$ and $\mu_\delta = (-0.37 \pm 0.02)''$/century and 190°.7, respectively. 16 member– and 23 nonmember stars were confirmed from which a diameter of 15′ was estimated for the cluster. Proper motion members found in the outer regions implies a new cluster diameter of 21′.

153.046 **Proper motions in the field of the open star cluster NGC 2324.**
A. Aiad.
Bull. Fac. Sci., Cairo Univ., Vol. 54, p. 753 – 760 (1986).
Proper motions of 98 stars in the field of the open star cluster NGC 2324 were determined using three combinations on two photographic plates taken by the Alger astrograph in the years 1905 and 1973 with an epoch difference of 68.75278 years. The author discusses the membership of the stars in the cluster. The member stars revealed a proper motion center of $\mu_\alpha \cos\delta = (0.0006 \pm 0.04)''$/century,
$\mu_\delta = (0.0159 \pm 0.02)''$/century, and a position angle of 2°.2.

153.047 **Durchmusterung offener Sternhaufen nach CP2–Sternen.**
H. M. Maitzen.
Mitt. Astron. Ges., No. 65, p. 222 (1986). – See Abstr. 012.072.

IDAM: a catalogue of DM, HD(E) and other cross–identifications for stars in open clusters (magnetic tape).
See Abstr. 002.005.

Integrated spectral energy distribution and colors of young stellar populations.
See Abstr. 065.153.

Integrated colours and synthetic HR diagrams for LMC clusters.
See Abstr. 065.154.

On a correct algorithm for determining photographic relative proper motions of open clusters.
See Abstr. 111.023.

Astrometry of star clusters: problems, techniques, and opportunities.
See Abstr. 111.026.

The nature of Be stars: periodic variables in the cluster NGC 3766.
See Abstr. 112.012.

Be stars in open clusters.
See Abstr. 112.105.

***VBLUW* photometry of two hypergiants HD 80077 (B2 Ia$^+$) and HD 74180 (F2 Ia$^+$) and of the open cluster Pismis 11.**
See Abstr. 113.001.

Seven–colour photoelectric photometry of bright stars in open clusters containing red giants.
See Abstr. 113.019.

Seven–colour photoelectric photometry of bright stars in the open cluster NGC 6633.
See Abstr. 113.020.

Photoelectric photometry in the Vilnius system of bright stars in the open cluster NGC 6494.
See Abstr. 113.023.

On the interpretation of Strömgren photometry.
See Abstr. 113.036.

Chemical composition of K giants in the Hyades.
See Abstr. 114.076.

Spectroscopic determination of the metallicity of F dwarfs in open clusters.
See Abstr. 114.088.

Of stars in young open clusters.
See Abstr. 115.008.

"Chemical structure" in the red giant Hertzsprung–Russell diagram, and the age of the galactic disk.
See Abstr. 115.017.

The relationship between X–ray luminosity and Rossby number for a sample of late–type stars.
See Abstr. 116.041.

Modeling of coronal X–ray emission from active cool stars. I. Hyades cluster.
See Abstr. 116.052.

New photometry for the rapidly rotating Pleiades K dwarf HII 1883.
See Abstr. 116.057.

Four–colour photometry of eclipsing binaries. XXIA. Photometric analysis and apsidal motion study of V346 Centauri.
See Abstr. 119.018.

Observation and analysis of the Hyades binary HD 27130.
See Abstr. 119.041.

HD 47755, a new eclipsing binary.
See Abstr. 119.060.

Speckle interferometry of spectroscopic and Hyades binary stars.
See Abstr. 120.001.

A β Cephei variable in an eclipsing binary in the cluster NGC 3293.
See Abstr. 122.018.

Cepheids in the open cluster NGC 6067.
See Abstr. 122.020.

Distribution of flares in UV Cet–type stars in open clusters and in the solar vicinity from optical radiation energies.
See Abstr. 122.033.

The luminosities of cepheids.
See Abstr. 122.088.

Photographic photometry of flare stars in Pleiades.
See Abstr. 122.122.

Suspected photovisual variables in the field of NGC 2324.
See Abstr. 123.035.

Ultraviolet extinction toward five open clusters, and the accuracy of IUE extinction measurements.
See Abstr. 131.005.

The formation of bound stellar clusters.
See Abstr. 131.082.

Nebular complex in the region of Cas OB2 association; ring nebula Sh 157.
See Abstr. 131.112.

Investigation of the NGC 2264 star formation region.
See Abstr. 131.130.

High–resolution IRAS observations of the Rho Ophiuchi cloud core.
See Abstr. 131.284.

On the IMF of violent star forming regions.
See Abstr. 131.351.

Distances to planetary nebulae. I. The reddening–distance method.
See Abstr. 134.006.

Specific entropy and the evolution of unconfined stellar systems.
See Abstr. 151.027.

Deep(est) colour–magnitude diagrams of clusters.
See Abstr. 154.003.

2000 Å UV imaging of a 6° diameter field around μ Aurigae.
See Abstr. 155.068.

The initial mass function of massive stars in galaxies: empirical evidence.
See Abstr. 155.121.

Distribution of spectral types of stars in the SMC clusters.
See Abstr. 156.011.

Ages of two LMC clusters and an LMC distance determination using features of an LMC field CMD.
See Abstr. 156.016.

CCD photometry of Large Magellanic Cloud clusters. II. The intermediate–age cluster H4.
See Abstr. 156.021.

Evolutionary population synthesis and the Magellanic Cloud clusters.
See Abstr. 156.026.

Bright stars in the SMC clusters.
See Abstr. 156.037.

The distance modulus of LMC.
See Abstr. 156.041.

154 Globular Clusters

154.001 **Metallicity determinations for globular clusters through spectrophotometry of their integrated light.**
J. P. Brodie, D. A. Hanes.
Astrophys. J., Vol. 300, No. 1, p. 258 – 278 (1986).
Using a combination of 16 indices of absorption line strength measured in low–dispersion spectra of the integrated light of globular clusters, the authors determine metallicities [Fe/H] for thirty–six clusters in the Galaxy. The results confirm the suggestion that Zinn's (1980) scale suffers a systematic error in the region of intermediate metallicity and support an explanation in which his metallicity–indicative Q_{39} index has been diluted by excess ultraviolet light in clusters with anomalously rich blue horizontal branches. The methods presented here produce estimates of metallicity which are insensitive to this problem. A similar method with a modified calibration is applied to determine metallicities for the nuclei of six galaxies.

154.002 **The metallicity of globular clusters associated with the giant elliptical galaxy NGC 4486 (M87).**
D. A. Hanes, J. P. Brodie.
Astrophys. J., Vol. 300, No. 1, p. 279 – 288 (1986).
The authors have derived cluster metallicities from measurements of absorption features in low–dispersion spectra acquired at the AAT for five globular clusters associated with the giant elliptical galaxy M87, a member of the Virgo Cluster. They have augmented the sample to six through the recalibration and inclusion of the results presented by Racine, Oke, and Searle (1978) for three clusters, two of which are common to both studies. The mean metallicity in the total sample is [Fe/H] = -0.5 ± 0.4, considerably more metal–rich than the average (~ -1.2) for clusters in the Galaxy. Moreover, it appears that a significant fraction of the globular clusters in M87 are more metal–rich than any known in the Galaxy. There is no correlation of cluster metallicity with projected galactocentric distance.

154.003 **Deep(est) colour–magnitude diagrams of clusters.**
C. A. Christian, J. N. Heasley.
Astrophys. Space Sci., Vol. 118, Nos. 1/2, p. 453 – 455 (1986). – See Abstr. 012.001.
Color–magnitude diagrams are presented for the open cluster NGC 2158 and globular clusters Pal 4 and NGC 1904.

154.004 **Colour gradients in globular clusters.**
M. G. Pastoriza, E. L. D. Bica, M. V. F. Copetti, H. A. Dottori.
Astrophys. Space Sci., Vol. 119, No. 2, p. 279 – 285 (1986).
Radial colour gradients in the globular clusters NGC 2808, NGC 1851, NGC 5139, NGC 6388, NGC 6441, NGC 6541, NGC 6723, and NGC 7099 were searched by *B–V* and DDO photoelectric photometry, using a set of circular diaphragms.

154.005 **A photoelectric *BVRI* sequence in the field of the globular cluster NGC 6205 (M13).**
D. Forbes, P. C. Dawson.
Publ. Astron. Soc. Pac., Vol. 98, No. 599, p. 102 – 103 (1986).
Photoelectric *BVRI* observations of 20 stars in the field of the galactic globular cluster NGC 6205 (M13) are presented. These

stars have been observed in the *UBV* system by Arp and Johnson (1955) and should be useful in the calibration of photometry obtained by other means.

154.006 **CCD photometry in globular clusters. I. The age of NGC 6752.**
R. Buonanno, C. E. Corsi, G. Iannicola, F. Fusi Pecci.
Astron. Astrophys., Vol. 159, No. 1/2, p. 189 – 196 (1986).
A *B,V* CCD–photometry of the main sequence of the globular cluster NGC 6752 is presented, and the problem of the determination of cluster ages is critically analysed. The use of different methods converge for NGC 6752 to an absolute age of 16 ± 2 Gyr with a metallicity of [Fe/H] = -1.5 ± 0.2, a primordial helium abundance of $Y = 0.23 \pm 0.02$, and a distance modulus of 13.3 ± 0.2. The resulting absolute magnitude of the horizontal branch at the RR Lyrae gap turns out to be $M_v = +0.40 \pm 0.20$.

154.007 **Deep photometry of globular clusters. III. The globular cluster in Eridanus GlC 0423–21.**
S. Ortolani, R. G. Gratton.
Astron. Astrophys., Vol. 160, No. 1, p. 25 – 30 (1986).
B and *V* photometry and star counts have been derived from CCD images and a couple of direct plates of the globular cluster in Eridanus GlC 0423–21. The authors present a deep ($V_{\rm lim} \sim 23.5$) colour–magnitude diagram and a new derivation of the structural parameters. They discuss the main features of the *c–m* diagram, including the red horizontal branch and the steep giant branch.

154.008 **Simulated populations of the asymptotic giant branch of globular clusters in the Magellanic Clouds.**
Yu. L. Frantsman.
Pis'ma Astron. Zh., Tom 12, No. 3, p. 229 – 237 (1986). In Russian. Englisch translation in Sov. Astron. Lett., Vol. 12.
Simulated populations of the asymptotic giant branch (AGB) in clusters of different age are obtained for different laws of mass loss on the AGB stage and for different initial chemical compositions. The results are compared with the observations of two clusters in the Magellanic Clouds. There are two classes of stars on the AGB: early–AGB stars and thermally pulsating stars. Position of stars of these two classes on the HR diagram differ by luminosity and effective temperatures. Theoretical numbers of M and C stars in the clusters are presented for different ages, mass loss laws and initial chemical compositions.

154.009 **The ages of globular clusters in the Magellanic Clouds.**
Yu. L. Frantsman.
Pis'ma Astron. Zh., Tom 12, No. 4, p. 281 – 285 (1986). In Russian. English translation in Sov. Astron. Lett., Vol. 12.
By comparing the observed luminosities of the asymptotic giant branch stars in globular clusters of the Magellanic Clouds with theoretical ones the ages of 12 clusters in the SMC and 22 clusters in the LMC are obtained. The use of the theoretical calculations of stellar evolution with large mass loss on the AGB stage leads to good agreement with ages determined from the colour–luminosity diagram. The results obtained evidence that the origin of clusters is different in LMC and SMC.

154.010 **Ultraviolet–excess objects in the field of NGC 6121 (M4).**
E. Chan, H. B. Richer.
Astrophys. J., Vol. 302, No. 1, p. 257 – 271 (1986).
The authors have attempted to locate a hot white dwarf component in the globular cluster M4 by performing a blink survey designed to select ultraviolet–excess objects for a 50′ diameter field centered on the cluster. The search area between 3'9 and 23'7 from the cluster center was found to contain 1096 such objects brighter than $U = 22.3$ mag. These objects exhibited a radial concentration toward the center of the cluster. Subtraction of the estimated background unveiled a radial gradient similar to that of King, Hedemann, and Hodge's star counts in the same region. It is estimated that ~240 white dwarfs are present in the area searched brighter than $M_V \approx 11$. *UBV* photometry of 74 of the objects is also presented, showing that at least six objects have colors and photometric parallaxes consistent with those expected for very hot cluster white dwarfs.

154.011 **Catalogue of stars of the central region of the M13 globular cluster.**
Z. I. Kadla, A. N. Gerashchenko, T. M. Natsvlishvili, N. V. Yablokova.
Glav. astron. obs. Akad. Nauk SSSR. Leningrad, 41 pp. (1985). In Russian. Abstr. in Ref. Zh., 51. Astron., 1.51.579 (1986).

154.012 **On the carbon abundance of subgiant stars in the globular cluster M92.**
G. E. Langer, R. P. Kraft, D. F. Carbon, E. Friel, J. B. Oke.
Bull. Am. Astron. Soc., Vol. 17, No. 4, p. 881 (1985). Abstract. – See Abstr. 010.062.

154.013 **TiO continuum placement effects in 47 Tuc giants.**
D. P. Geisler.
Bull. Am. Astron. Soc., Vol. 17, No. 4, p. 881 (1985). Abstract. – See Abstr. 010.062.

154.014 **Metal abundances of globular clusters.**
H. A. Smith.
Bull. Am. Astron. Soc., Vol. 17, No. 4, p. 882 (1985). Abstract. – See Abstr. 010.062.

154.015 **Velocity dispersions and M/L's for seven globular clusters.**
C. P. Pryor, R. D. McClure, J. E. Hesser.
Bull. Am. Astron. Soc., Vol. 17, No. 4, p. 882 (1985). Abstract. – See Abstr. 010.062.

154.016 **Kinematical and dynamical properties of ω Cen and 47 Tuc.**
G. Meylan, M. Mayor.
Bull. Am. Astron. Soc., Vol. 17, No. 4, p. 882 (1985). Abstract. – See Abstr. 010.062.

154.017 **Radial velocities of 91 globular star clusters.**
J. E. Hesser, S. J. Shawl, J. E. Meyer.
Bull. Am. Astron. Soc., Vol. 17, No. 4, p. 882 (1985). Abstract. – See Abstr. 010.062.

154.018 **RR Lyrae stars in globular clusters: better distances from infrared measurements?**
A. J. Longmore, J. A. Fernley, R. F. Jameson.
Mon. Not. R. Astron. Soc., Vol. 220, No. 1, p. 279 – 287 (1986).
An investigation into the suitability of near–infrared photometry for the study of distances to globular clusters has been carried out. Single–phase observations at 2.2 μm of RR Lyrae stars in three clusters have been made. A clear period – [2.2 μm (*K*) luminosity] relation is observed in all three clusters (M3, M5 and M107). Using distance moduli for the three clusters determined on the precepts of recent studies by Sandage, the absolute *K* magnitude, normalized to a "standard" period, was calculated for the RR Lyrae stars in each cluster. Based on a value of $M_V = 0\overset{m}{.}80$ for the RR Lyrae stars in M3 the absolute *K* magnitudes at $\log P = -0.3$ for the stars in M3, 5 and 107 were $-0\overset{m}{.}25$, $-0\overset{m}{.}26$ and $-0\overset{m}{.}24$ respectively. This remarkable small scatter suggests that infrared photometry of RR Lyrae stars could prove a very powerful tool in the study of distances within the Galaxy.

154.019 **High–precision radial velocities for faint giants: radial velocities of extreme halo systems and the mass of the Galaxy.**
E. W. Olszewski, R. C. Peterson, M. Aaronson.
Astrophys. J., Lett. Ed., Vol. 302, No. 2, p. L45 – L48 (1986).
The authors present radial velocities of stars in the distant halo globular clusters Palomar 4, 14, and 15. These MMT echellette data are combined with earlier results for a compendium of velocities for 16 remote systems ($R > 50$ kpc). From these, the authors reassess the mass of the Galaxy, and obtain $M_G = 5 \pm 2 \times 10^{11} M_\odot$ if *velocities are distributed isotropically*. The quoted uncertainty in this estimate merely reflects the few remaining ambiguous velocities.

154.020 **Investigation of colour–magnitude diagrams of globular clusters with the method of main components.**
A. M. Ehjgenson, O. S. Yatsyk.
Vestn. L'vov. univ. Ser. astron., No. 59, p. 30 – 33 (1985). In Russian. Abstr. in Ref. Zh., 51. Astron., 3.51.863 (1986).

154.021 **On the formation of bound open clusters in the Galaxy.**
V. M. Danilov.
Astron. Tsirk., No. 1389, p. 1 – 4 (1985). In Russian.

154.022 **Blue and visual concentric aperture photometry of the globular clusters.**
C. J. Peterson.
Publ. Astron. Soc. Pac., Vol. 98, No. 600, p. 192 – 209 (1986).
Visual and blue photoelectric measurements obtained at Cerro Tololo Inter-American Observatory of 101 globular clusters in the Galaxy and one in the Large Magellanic Cloud are presented. This extensive multiaperture photometry has been analyzed with all other available (*B–V*) color-index data to address the question of color gradients in globular clusters.

154.023 **Globular cluster color excesses and the cosecant law.**
A. V. Zasov, I. A. Trifalenkov, A. A. Bryukhanov.
Sov. Astron., Vol. 29, No. 4, p. 409 – 413 (1985). English translation of 40.154.008.

154.024 ***BVRI* CCD photometry of the globular cluster NGC 2298.**
G. Alcaino, W. Liller.
Astron. Astrophys., Vol. 161, No. 1, p. 61 – 69 (1986).
From 48 *BVRI* CCD frames the authors have constructed five color magnitude diagrams. From an inspection of the five CMD's, they find that the main sequence turnoffs are consistently located at $V_{TO} = 19.50 \pm 0.1$ and the colour turnoffs at: $B-V = 0.60 \pm 0.02$, $V-R = 0.27 \pm 0.03$, $R-I = 0.39 \pm 0.03$, $V-I = 0.71 \pm 0.03$ and $B-I = 1.30 \pm 0.03$. The magnitude difference between the turnoff and the horizontal branch for all diagrams is 3.4 ± 0.2. The *V* vs. *B–V* diagram has been matched to the recently improved isochrones of VandenBerg and Bell (1985); the best fit is produced using $Y = 0.2$, [Fe/H] $= -1.27$, $\alpha = 1.65$, a distance modulus of $(m-M)_V = 15.5$ and $E(B-V) = 0.21$, from which the authors deduce an age of $17 \pm 1.5 \times 10^9$yr.

154.025 **Photoelectric photometry of star clusters in M31. XI.**
A. S. Sharov, V. M. Lyutyj.
Sov. Astron. Lett., Vol. 11, No. 4, p. 248 – 249 (1985). English translation of 40.154.005.

154.026 **Color–magnitude diagram of Palomar 4: CCD photometry.**
C. A. Christian, J. N. Heasley.
Astrophys. J., Vol. 303, No. 1, p. 216 – 225 (1986). With plates 1 – 2.
Photometry of the globular cluster Pal 4 was obtained with the RCA CCD camera on the 3.6 m CFH Telescope on Mauna Kea. The color–magnitude diagram of the cluster has been sampled to the main-sequence turnoff region ($V = 25$). A comparison with theoretical models leads to the cluster parameters $(m-M)_0 = 20.1 \pm 0.1$ mag, $E(B-V) = 0.02 \pm 0.02$, [Fe/H] $= -1.7 \pm 0.1$, and an age of 15 ± 1 Gyr.

154.027 **Distribution of BHB (*Blue Horizontal Branch*) stars in the cluster ω Centauri and its effect on the radial change in colour and ellipticity of the cluster.**
K. K. Scaria, M. K. V. Bappu.
Proceedings of the Second Asian-Pacific Regional Meeting on Astronomy, p. 449 – 459 (1984). – See Abstr. 012.031.
The authors have studied the radial change in the distribution of giant branch stars and the horizontal branch stars in ω Centauri. They show how the observed change in the distribution is related to the radial change in the colour of the cluster and also the radial change in its ellipticity.

154.028 **CCD photometry of the distant halo cluster Palomar 4.**
B. C. Reed, W. E. Harris.
Astron. J., Vol. 91, No. 1, p. 81 – 86 (1986). With plate 8.
The authors report on *BV* photometry of the distant globular cluster Palomar 4, obtained with the CCD system at the Cassegrain focus of the CFHT. The resulting color–magnitude diagram, to *V*(lim) ~ 24.8, suggests that the main-sequence turnoff of Pal 4 appears at $V \sim 23.8$, $(B-V) \sim 0.45$. Approximate isochrone fits indicate that the cluster is of intermediate metallicity, and is probably only slightly younger ($\lesssim 2 - 4$ Gyr) than the more familiar, inner-halo globular clusters. In agreement with another recent CCD study by Christian and Heasley (1985), the authors estimate the distance of Pal 4 to be 100 kpc.

154.029 **Photoelectric *UBVRI* sequences in the region of the galactic globular clusters NGC 2808, NGC 4372, and NGC 6397.**
G. Alcaino, W. Liller.
Astron. J., Vol. 91, No. 1, p. 87 – 89 (1986). With plates 9 – 11.
The authors present *UBVRI* photoelectric sequences in the galactic globular clusters NGC 2808, NGC 4372, and NGC 6397. Previously, no *RI* magnitudes were available for these clusters, but for NGC 2808, most of the stars were previously observed in *UBV* by Harris (1978), for NGC 4372 by Hartwick and Hesser (1973), and for NGC 6397 by Alcaino (1977).

154.030 ***BVRI* CCD photometry of the globular cluster NGC 6362.**
G. Alcaino, W. Liller.
Astron. J., Vol. 91, No. 2, p. 303 – 311 (1986). With plates 36, 37.
The authors have obtained 78 *BVRI* CCD frames with the 1.54 m Danish telescope at ESO, La Silla, and have constructed *V* vs *B–V*, *V* vs *V–R*, *V* vs *R–I*, *V* vs *V–I*, and *V* vs *B–I* color–magnitude diagrams in a 4′ × 2.′5 field of the globular cluster NGC 6362. They find that the VandenBerg and Bell isochrones (1985) yield a consistent age for NGC 6362 in all color indexes of $16 \pm 1.5 \times 10^9$yr. The solar distance to the cluster is 7.7 kpc and the galactic distance is 5.6 kpc assuming $R_0 = 9$ kpc.

154.031 **Accurate optical positions for the centers of galactic globular clusters.**
S. J. Shawl, R. E. White.
Astron. J., Vol. 91, No. 2, p. 312 – 316 (1986).
The authors present optical-center positions internally precise to ~1 arcsec for 109 galactic globular clusters.

154.032 **M13: main-sequence photometry and the mass function.**
R. H. Lupton, J. E. Gunn.
Astron. J., Vol. 91, No. 2, p. 317 – 325 (1986). With plate 38.
The authors discuss the path used to go from Thuan–Gunn magnitudes to masses, and tests for completeness of the luminosity function. Next comes a section on the color–magnitude and two-color diagrams for M13, in which the authors present three-color photometry for selected stars on the main sequence. This section is followed by a discussion of the luminosity function and the derived mass function.

154.033 **Theoretical stellar luminosity functions and globular cluster ages and compositions.**
S. J. Ratcliff.
Diss. Abstr. Int., Sect. B, Vol. 46, No. 10, p. 3482 (1986). Thesis, Princeton University, 287 pp. (1985). Order No. DA8529046.

154.034 **The spatial structure of the globular cluster M56.**
R. M. Rusev, Z. I. Pejkov.
Astron. Zh., Tom 63, Vyp. 3, p. 483 – 493 (1986). In Russian. English translation in Sov. Astron., Vol. 30, No. 3.
The spatial structure of the globular cluster M56 in dependence on the limit star magnitude up to which star counts are made is investigated. The analysis of the apparent *F(r)* and the spatial *f(r)* star-density functions up to limit magnitudes $m_{pg} = 15$, 15.5, 16.ᵐ04 and $B = 17$, 19, 19.5, 21, 21.ᵐ5 showed that the radii of the spatial zones of the cluster, described by Kholopov, increase systematically with B_{lim}, following the same law below the

main–sequence turn–off. Experimental relationships $R_i = R_i(B)$ and $f = f(R_i, B)$ (R_i being the radius of the ith zone) are obtained. It is shown that the relative zone radii $r_i = R_i(B)/R_4(B)$ are independent of B_{lim}. A tendency for distribution of the brighter stars nearer to the cluster centre is observed.

154.035 **Evolution of globular clusters.**
N. Spasova, P. Dimitrov.
Astrofiz. issled., Tom 4, p. 80 – 85 (1985). In Russian. Abstr. in Ref. Zh., 51. Astron., 6.51.779 (1986).

154.036 **The age of the Large Magellanic Cloud globular cluster NGC 1651.**
J. R. Mould, G. S. Da Costa, M. D. Crawford.
Astrophys. J., Vol. 304, No. 1, p. 265 – 272 (1986). With plates 1 – 3.
The age of the LMC red globular cluster NGC 1651 is estimated as $2.0 \pm 0.8 \times 10^9$yr by photometry of the main–sequence turnoff. The accuracy of the estimate is limited chiefly by uncertainty in the distance modulus of the cluster. Here the uncertainties stem from a metallicity estimate of one–third solar abundance and from uncertainties relating to the effects of convective overshooting on stellar evolution.

154.037 **Deep CCD photometry in globular clusters. IV. M13.**
H. B. Richer, G. G. Fahlman.
Astrophys. J., Vol. 304, No. 1, p. 273 – 282 (1986). With plate 4.
CCD photometry in *U*, *B*, and *V* is presented for a single field measuring 2.'0 × 3.'5 in the globular cluster M13. The photometry reaches fainter than $V = 25$. It is estimated that the metallicity of the cluster is [Fe/H] = –1.4. The distance to M13 was determined by fitting a main sequence defined by subdwarfs to the M13 main–sequence fiducial. This results in $(m–M)_V = 14.50 \pm 0.26$. An age of 16 ± 2 Gyr is derived for M13 from an overlay of theoretical isochrones. The main–sequence luminosity function is found to be still steeply rising at the limit of the photometry ($M_V = 10.5$).

154.038 **The potential for supernova–induced chemical enrichment of protoglobular cluster clouds.**
M. A. Dopita, G. H. Smith.
Astrophys. J., Vol. 304, No. 1, p. 283 – 294 (1986).
The authors seek to explain the large internal abundance variations that are seen in the globular cluster ω Cen in terms of supernova–induced chemical enrichment that occurred when the cluster was still largely in a gaseous phase and star formation was continuing. Using a simple power–law density model of this protoglobular gas cloud, it is shown that supernova–induced secondary star formation can be efficient only if the total cloud mass is at least $10^7\ M_\odot$.

154.039 **Erratum: "The globular cluster system of the Galaxy. IV. The halo and disk subsystems" [Astrophys. J., Vol. 293, No. 2, p. 424 – 444 (1985)].**
R. Zinn.
Astrophys. J., Vol. 304, No. 1, p. 579 (1986). See Abstr. 39.154.072.

154.040 **Hydrodynamic simulation of a white dwarf–main sequence star collision in a globular cluster core.**
M. M. Shara, O. Regev.
Space Telesc. Sci. Inst., Prepr. Ser., No. 88, 29 pp. (1986). To appear in Astrophys. J.

154.041 **The distribution of globular clusters associated with M87 = NGC 4486.**
C. Grillmair, C. Pritchet, S. van den Bergh.
Space Telesc. Sci. Inst., Prepr. Ser., No. 98, 38 pp. (1986). To appear in Astron. J.

154.042 **First optical candidate for a recovered classical nova in a globuar cluster: Nova 1938 in M14.**
M. M. Shara, A. F. J. Moffat, M. Potter, H. Sawyer Hogg, A. Wehlau.
Space Telesc. Sci. Inst., Prepr. Ser., No. 109, 21 pp. (1986). To appear in Astrophys. J.

154.043 **TiO continuum placement effects in 47 Tucanae giants.**
D. Geisler.
Astrophys. J., Lett. Ed., Vol. 304, No. 2, p. L41 – L44 (1986).
High–resolution echelle SIT Vidicon spectra have been obtained of giants in the metal–rich globular cluster 47 Tuc. The new spectra show that all of these giants have TiO absorption, which was not recognized in earlier echelle studies. The presence of TiO absorption indicates that the abundances derived by these previous investigations are underestimates because of the underestimation of the continuum. The underestimates are not large, however, amounting to 0.2 dex or less. An abundance for 47 Tuc of [Fe/H] ≈ –0.9 is indicated.

154.044 **Galactic globular clusters as star systems and satellites of the Galaxy.**
P. Giannone.
Galaxies, quasars and cosmology, p. 29 – 60 (1985). – See Abstr. 012.042.
Galactic globular clusters (and dwarf spheroidal galaxies) can be studied as isolated star systems, or as star systems orbiting the Galaxy. In both cases they give information on a large number of topics, including age, original chemical composition, galactic structure and Galaxy evolution. Stellar evolution provides data on age and chemical composition; a stability analysis of globular clusters gives indications of the galactic structure.

154.045 **A spectroscopic study of HB (*horizontal–branch*) stars in the galactic globular cluster NGC 6752.**
U. Heber, R. P. Kudritzki, V. Caloi, V. Castellani, J. Danziger, R. Gilmozzi.
ESO Sci. Prepr., No. 416, 30 pp. (1986). To appear in Astron. Astrophys.

154.046 **High–precision radial velocities for faint giants. Radial velocities of extreme halo systems and the mass of the Galaxy.**
E. W. Olszewski, R. C. Peterson, M. Aaronson.
Prepr. Steward Obs., No. 621, 18 pp. (1986). To appear in Astrophys. J., Lett. Ed.

154.047 **Accurate radial velocities for three Fornax globular clusters.**
M. Aaronson, E. W. Olszewski.
Prepr. Steward Obs., No. 660, 23 pp. (1986). To appear in Astron. J.

154.048 **The age of the Small Magellanic Cloud globular cluster NGC 411.**
G. S. Da Costa, J. R. Mould.
Astrophys. J., Vol. 305, No. 1, p. 214 – 227 (1986). With plate 3.
The age of the SMC red globular cluster NGC 411 is estimated as 1.5 ± 0.5 Gyr from a C–M diagram that reaches below the cluster main–sequence turnoff. The accuracy of this estimate is limited primarily by uncertainty in the distance modulus of the cluster: at $(m–M)_0 = 18.8$, the age is 1.8 ± 0.3 Gyr, while at $(m–M)_0 = 19.3$ it is 1.2 ± 0.3 Gyr. For the SMC field in the vicinity of the cluster, zero–age main–sequence fits strongly support the adoption of the smaller (18.8) modulus.

154.049 **The rotation of M3.**
Z. I. Kadla, A. A. Strugatskaya.
Sov. Astron. Lett., Vol. 11, No. 5, p. 282 – 284 (1985). English translation of 40.154.010.

154.050 **Near-infrared photometry of red giant and horizontal branch stars in M4.**
J. P. Phillips, C. Martinez Roger, C. Sanchez Magro, C. Lazaro Vilchez.
Astron. Astrophys., Vol. 161, No. 2, p. 257 – 263 (1986).
The authors have presented J, H, and K photometry for the red-giant and horizontal branches in M4. The red giant branch confirms earlier work in implying an intermediate to high cluster metallicity. The horizontal branch intersects the RGB at $(V-K)_0 \cong 2.0$, and continues to increase a further two or more magnitudes in K towards the BHB. Fitting these results with the models of Kurucz (1979) indicates a mean HB luminosity $\log(L_{HB}/L_\odot) \sim 1.65$ for $E(B-V) \cong 0.46$, an extinction some 0.1 magnitudes greater than implied by previous visual estimates.

154.051 **Dynamics of the globular cluster M2 (NGC 7089).**
C. Pryor, R. D. McClure, J. M. Fletcher, F. D. A. Hartwick, J. Kormendy.
Astron. J., Vol. 91, No. 3, p. 546 – 556 (1986).
The authors have obtained radial velocities for 69 stars in the globular cluster M2. M2's rotation axis is, within sizeable uncertainties, perpendicular to the major axis determined by the flattening. The ratio of rotational to random kinetic energy agrees with that predicted from the ellipticity assuming an oblate figure and an isotropic velocity-dispersion tensor. The authors have fitted King-Michie models to determine M2's mass, the exponent of an assumed power-law mass function, and the anisotropy radius. The models favor mass functions similar to or shallower than the Salpeter initial-mass function and a moderate amount of velocity anisotropy.

154.052 **CCD photometry of the globular cluster Palomar 5.**
G. H. Smith, R. D. McClure, P. B. Stetson, J. E. Hesser, R. A. Bell.
Astron. J., Vol. 91, No. 4, p. 842 – 854 (1986).
Deep V and B photometry for the outer-halo cluster Palomar 5 has been obtained with the prime-focus CCD camera at the CTIO 4 m telescope. The color-magnitude diagram reaches several magnitudes below the main-sequence turnoff. Although this is a sparse cluster, no evidence for a binary sequence was detected. The luminosity function on the main sequence shows a turnover like that found for E3 by McClure et al. This implies a mass function with a very low power-law index, which indicates either that there has been considerable dynamical evolution in the cluster or that the initial mass function had a very peculiar form. The matching of the observed color-magnitude diagram with theoretical isochrones is discussed.

154.053 **CCD and photographic photometry of NGC 1904.**
J. N. Heasley, K. A. Janes, C. A. Christian.
Astron. J., Vol. 91, No. 5, p. 1108 – 1116 (1986). With plates 81 – 84.
New CCD and photographic BV photometry is presented for the halo globular cluster NGC 1904 = M79 (C0522-245). The photometry extends well below the main-sequence turnoff. The CCD observations of selected fields in the cluster are used to provide a large number of high-precision standard stars for calibrating the photographic photometry. The results largely confirm the suggestion originally made by Stetson and Harris that NGC 1904 is similar to M13. The cluster appears to be a "normal" globular located 20 kpc from the galactic center.

154.054 **High-resolution CCD spectra of stars in globular clusters. I. Oxygen in M13.**
E. M. Leep, G. Wallerstein, J. B. Oke.
Astron. J., Vol. 91, No. 5, p. 1117 – 1120 (1986).
High-resolution (0.3 Å) CCD spectra obtained at the 200 in. coudé spectrograph have been analyzed for the abundances of O, Sc, Fe, and La in four stars in the globular cluster M13. The authors find [Fe/H] = –1.6, as found by many other observers of this cluster. For three stars they find [O/Fe] = $+0.3 \pm 0.1$, which is similar to O/Fe ratios in other globular clusters and metal-poor field stars. For star II-67 no oxygen line is visible at λ6300 and it is found [O/Fe] $\leqslant -0.4$ for a high carbon content and $\leqslant -0.7$ for a low carbon content. The latter is more likely to be correct. The authors briefly discuss possible explanations of the oxygen deficiency in II-67.

154.055 **Implications of a mixing origin for globular cluster bimodal CN variations.**
G. H. Smith, R. A. Bell.
Astron. J., Vol. 91, No. 5, p. 1121 – 1129 (1986).
An analytic formalism for calculating the variation of CN opacity with carbon abundance is discussed and is compared with the results of spectrum-synthesis calculations. Several aspects of the behavior of CN within the framework of the mixing hypothesis are then discussed. Before presenting this discussion, the results of new spectroscopic observations of red giants in the globular cluster NGC 6934 are reported. These serve to illustrate the basic observational facts considered subsequently.

154.056 **Period changes for variables in M28.**
A. Wehlau, H. Sawyer Hogg, R. Moorhead, P. Rice.
Astron. J., Vol. 91, No. 6, p. 1340 – 1349 (1986).
Magnitudes for variables have been measured on 240 plates of M28 taken in the northern hemisphere from 1939 through 1975. These magnitudes have been combined with measures from 69 plates taken in Chile from 1974 through 1985 and used to determine period changes.

154.057 **Systematic reinvestigation of the radial velocities of the galactic globular clusters: image-tube results.**
J. E. Hesser, S. J. Shawl, J. E. Meyer.
Publ. Astron. Soc. Pac., Vol. 98, No. 602, p. 403 – 422 (1986).
Radial velocities measured from ~260 image-tube spectrograms at 120 Å mm^{-1} of 90 galactic globular clusters and NGC 121 in the Small Magellanic Cloud are presented and compared with previous major surveys. From the authors' data the solar apex is found to lie at $\alpha = 21^h10^m \pm 52^m$ and $\delta = 44.^\circ9 \pm 9.^\circ8$ or $l = 88.^\circ1$ and $b = -2.^\circ2$. Evidence for the G-type clusters rotating more rapidly than the F-type ones persists. The availability of new spectral classifications reveals more distinct differences in velocity dispersions from the solar motion solutions for F- and G-type clusters, ~128 km s^{-1} and ~86 km s^{-1}, respectively, than heretofore recognized. The velocity dispersion for the cluster system as a whole is ~120 km s^{-1}.

154.058 **Spectroscopy of c-type RR Lyrae stars in the globular cluster NGC 6638.**
H. A. Smith, L. L. Stryker.
Publ. Astron. Soc. Pac., Vol. 98, No. 602, p. 453 – 456 (1986).
The authors have obtained low-resolution spectra for six of the supposed c-type RR Lyrae stars. In all six cases the spectra are consistent with the c-type designation. The authors have determined radial velocities from the spectra, measured metal abundances by the ΔS method, and discuss membership criteria. The ΔS measures confirm that NGC 6638 is only moderately metal deficient: [Fe/H] = -0.82 ± 0.2. The large ratio of c-types to ab-type variables in NGC 6638 can be attributed to a gap in the color distribution of horizontal-branch stars.

154.059 **On the carbon abundance of subgiant stars in the globular cluster M92.**
G. E. Langer, R. P. Kraft, D. F. Carbon, E. Friel, J. B. Oke.
Publ. Astron. Soc. Pac., Vol. 98, No. 603, p. 473 – 485 (1986). = Lick Obs. Bull., No. 1039.
The authors find that ‹[C/Fe]› ≈ 0.0 in stars at the base of the subgiant branch and approaches +0.1 in stars near the main-sequence turnoff in the metal-poor globular cluster M92. The carbon abundance falls on the average by about 0.5 dex by the time stars reach $M_V{}^0 \sim +1.5$. These carbon depletions occur early enough and are approximately the right size to account for the large average nitrogen enhancement ([N/Fe] ~ +0.65) observed in M92 subgiants near $M_V{}^0 \sim +1.5$. Although the present observations do not absolutely prove that C→N conversion and mixing alone control the C- and N-abundances seen in M92 giants, neither is there any longer a compelling reason to believe there were large primordial nitrogen abundance variations in M92. The mechanism responsible for the large (Δ[C/Fe] ≈ 1.0)

carbon depletions observed as stars ascend the M92 giant branch to the red–giant tip remains unknown.

154.060 **ST to the discovery of the galactic halo.**
V. Castellani.
Mem. Soc. Astron. Ital., Vol. 56, No. 4, p. 613 – 617 (1985). – See Abstr. 012.054.

This note briefly outlines the prospective impact of the Space Telescope on the investigation of galactic globular clusters and field halo stars.

154.061 **Dynamical evolution of globular clusters. II. Binaries – method.**
J. S. Stodółkiewicz.
Acta Astron., Vol. 36, No. 1, p. 19 – 41 (1986).

A numerical method of investigation of evolution of globular clusters containing binaries is described. The clusters can be composed of stars with a mass distribution. Due to the stellar evolution two kinds of stars (evolved ones, which can be treated as mass points, and unevolved ones with finite radii) are present in the system. The method is an extension of Stodółkiewicz's version of Hénon's Monte–Carlo scheme. It allows to consider the binaries in the cluster during their whole life. Two kinds of effects of binaries on the cluster dynamics are considered: (1) an exchange of energy and angular momentum of binary mass centers and field stars (the relaxation process), and (2) a heating of the cluster by the recoil energy released as a result of the interactions of binaries with single stars.

154.062 **The evolution of galactic globular clusters. II. Helium abundances and ages.**
S.–W. Lee.
J. Korean Astron. Soc., Vol. 18, No. 2, p. 41 – 69 (1985).

For well observed 16 globular clusters with known metal abundance (Z), the helium abundance (Y) and ages are determined by various methods, and the relations between Y, Z and the age are examined.

154.063 **CCD photometry of the globular cluster NGC 6752.**
A. J. Penny, R. J. Dickens.
Mon. Not. R. Astron. Soc., Vol. 220, No. 4, p. 845 – 867 (1986).

A V, $B–V$ colour–magnitude diagram for the globular cluster NGC 6752 is presented. The use of a CCD has resulted in a colour–magnitude diagram with a very well–defined sequence going fainter than previously published work, and showing small scatter and little contamination. The distance to the cluster has been accurately tied to the trigonometrical parallaxes of local subdwarfs, giving a true distance modulus of 13.17 ± 0.17 mag. An age of 16 ± 1 Gyr and a metallicity of $[Fe/H] = -1.5 \pm 0.2$ are derived, and the age spread of stars in the cluster must be less than 0.5 Gyr.

154.064 **Stellar velocity dispersions for four low–concentration globular clusters.**
R. C. Peterson, D. W. Latham.
Astrophys. J., Vol. 305, No. 2, p. 645 – 650 (1986).

Mean velocities and velocity dispersions are presented and discussed for four clusters of low central concentration: NGC 5466, NGC 6121 (M4), NGC 6656 (M22), an NGC 6838 (M71). The measured dispersions range from 1.7 ± 0.6 km s^{-1} for NGC 5466 to 8.5 ± 1.9 km s^{-1} for M22. These data, in conjunction with those of several other recent investigations of globular clusters, show that the clusters' mass–to–light ratios fall in the range $0.5 \leqslant M/L \leqslant 4$, whatever their total mass or concentration.

154.065 **The C, N, and O abundances of giant stars in ω Centauri.**
J. G. Cohen, R. A. Bell.
Astrophys. J., Vol. 305, No. 2, p. 698 – 713 (1986).

Indices measuring the strength of CH, C_2, CN, and NH bands and Ca II and Ca I lines have been determined for 72 stars in ω Cen. These data, and narrow–band photometry of the 2.4 μm CO bands, are calibrated by using spectral synthesis techniques. In addition to the four C stars, there are two groups of giants in ω Cen. Both groups show a wide range of abundances of the heavy elements, Z. The N–enhanced stars have N/Z enhanced by a factor of 5 to 15, while both C/Z and O/Z are depleted by a factor which may be as much as 10. The N–normal stars have N/Z and C/Z within a factor of 2 of solar.

154.066 **A preliminary survey of collapsed cores in globular clusters.**
S. Djorgovski, I. R. King.
Astrophys. J., Lett. Ed., Vol. 305, No. 2, p. L61 – L65 (1986).

This is a preliminary report on surface photometry of the major fraction of known globular clusters, to see which of them show the signs of a collapsed core. The authors find more than 20 examples, or one–fifth of the total. Core classifications are given for all clusters examined. The fraction of collapsed–core clusters may be an index of how long a cluster takes to reexpand after collapse.

154.067 **The age of galactic globular clusters.**
M. L. Quarta.
Rev. Mex. Astron. Astrofis., Vol. 12, Spec. Issue, p. 199 – 207 (1986). – See Abstr. 012.066.

The problem of the galactic globular cluster ages is revisited, discussing the possible sources of errors and indeterminations. The role of RR Lyrae variables is discussed in the frame of the evolutionary scenario. Application of a general fitting procedure to the galactic globular clusters M15 and M4 gives respectively $t = (13 \pm 3) \times 10^9$yr and $t = (12.2 \pm 0.2) \times 10^9$yr. A similar age for the globular cluster M3 is also suggested.

154.068 **Integrated photometry and light distribution of 21 southern–hemisphere galactic globular clusters.**
G. E. Kron, K. C. Gordon.
Publ. Astron. Soc. Pac., Vol. 98, No. 604, p. 544 – 548 (1986).

Multiaperture observations of 21 galactic globular clusters, made in 1960 – 61, are presented and analyzed to determine, from these data alone, total magnitudes and brightness distributions. ($B–V$) and ($V–I$) measures are presented.

Gli ammassi globulari. III.
See Abstr. 002.009.

Performance tests of DAOPHOT/INVENTORY photometry programmes in dense stellar fields.
See Abstr. 036.045.

Convective overshooting: new integrated colours vs age relations for star clusters.
See Abstr. 065.174.

Radial–velocity determinations of six LMC superluminous giant candidates.
See Abstr. 111.017.

Astrometry of star clusters: problems, techniques, and opportunities.
See Abstr. 111.026.

Temperatures and gravities for horizontal–branch stars.
See Abstr. 114.014.

Do UV–bright stars mimic normal (Population I) OB stars?
See Abstr. 114.015.

High and low luminosity globular cluster X–ray sources.
See Abstr. 117.181.

Origin and evolution of X–ray binaries in globular clusters.
See Abstr. 117.304.

Unusual variable stars in the globular cluster M4.
See Abstr. 122.008.

The pulsation mode and period–luminosity relationship of cool variables in globular clusters.
See Abstr. 122.019.

CCD observations of some RR Lyrae stars and a standard sequence in Baade's window.
See Abstr. 122.076.

RR Lyrae stars as distance indicators.
See Abstr. 122.094.

Spectroscopy of RR Lyrae stars in Baade's window and ω Centauri.
See Abstr. 122.132.

IRAS 18333–2357: an unusual source in M22.
See Abstr. 133.003.

EXOSAT CMA observations of X0512–401 = NGC 1851.
See Abstr. 142.033.

Eddington luminosities and photospheric radius expansion during X-ray bursts from 4U/MXB 1820–30.
See Abstr. 142.035.

Compact X-ray binaries in and out of globular clusters.
See Abstr. 142.083.

Terzan 2.
See Abstr. 142.112.

Dynamical instabilities in spherical stellar systems.
See Abstr. 151.002.

Specific entropy and the evolution of unconfined stellar systems.
See Abstr. 151.027.

Massive central galaxies, globular cluster swapping, and tidal accretion.
See Abstr. 151.028.

Dynamical effects of stellar evolution in globular clusters.
See Abstr. 151.029.

The role of binaries in globular cluster evolution.
See Abstr. 151.039.

Remarks on stellar clusters.
See Abstr. 151.042.

Tidal binary formation in globular clusters.
See Abstr. 151.057.

Anisotropy in stellar dynamics.
See Abstr. 151.072.

Dynamical evolution of spherical stellar systems.
See Abstr. 151.075.

The criterion for equipartition in clusters with a mass spectrum.
See Abstr. 151.108.

Old stellar populations. III. The metallicities of M5, M71, and M67.
See Abstr. 153.034.

A galactic window at $l = 355°$, $b = -1°$ and its implications for further observations.
See Abstr. 155.085.

Colour–magnitude diagrams of star clusters in the Magellanic Clouds from wide-field electronography. II. NGC 2210.
See Abstr. 156.004.

Ages and metallicities of LMC and SMC red clusters through Hβ and G band photometry.
See Abstr. 156.007.

Distribution of spectral types of stars in the SMC clusters.
See Abstr. 156.011.

Identification of new carbon star candidates in SMC globular cluster NGC 419.
See Abstr. 156.013.

CCD photometry of Large Magellanic Cloud clusters. II. The intermediate-age cluster H4.
See Abstr. 156.021.

Star formation of star clusters in the SMC and their adjoining fields.
See Abstr. 156.032.

Bright stars in the SMC clusters.
See Abstr. 156.037.

Tidal radii and masses of the clusters in the LMC.
See Abstr. 156.038.

The distribution of bright stars in the SMC clusters.
See Abstr. 156.039.

Bright stars in five LMC star clusters.
See Abstr. 156.040.

Search for the globular clusters in nearby galaxies. I. NGC 253.
See Abstr. 157.006.

Globular clusters and galactic evolution.
See Abstr. 157.015.

Globular clusters in the Andromeda galaxy M31.
See Abstr. 157.154.

Globular clusters in galaxies beyond the Local Group. VII. The S0 galaxy NGC 3115.
See Abstr. 157.190.

The abundances and velocities of the Leo dwarf spheroidals.
See Abstr. 157.204.

Two-color CCD mapping of the luminous type I irregular galaxy NGC 4449.
See Abstr. 157.218.

Globular clusters in galaxies beyond the Local Group. V. The giant ellipticals reconsidered.
See Abstr. 157.227.

The abundances and velocities of the Leo dwarf spheroidals.
See Abstr. 157.231.

The core of the M87 globular cluster system.
See Abstr. 158.195.

Evidence for unusual dynamical effects in the halo of NGC 5128 (Centaurus A).
See Abstr. 158.219.

The distribution of globular clusters associated with M87 = NGC 4486.
See Abstr. 158.280.

Possible mechanism for generating almost isothermal baryon-density perturbations.
See Abstr. 161.141.

The age of the universe, primordial galaxies, and Space Telescope.
See Abstr. 161.306.

155 Galaxy

155.001 **Massachusetts–Stony Brook Galactic Plane CO Survey. I. (b, V) maps of the first Galactic quadrant.**
D. B. Sanders, D. P. Clemens, N. Z. Scoville, P. M. Solomon.
Astrophys. J., Suppl. Ser., Vol. 60, No. 1, p. 1 – 296 (1986).
. The FCRAO 14 m telescope has been used for a new high–resolution, CO $J = 1\text{–}0$ survey of the first Galactic quadrant. The survey consists of 40,551 spectra obtained in the longitude range $l = 8° - 90°$ from $b = -1.°05$ to $b = +1.°05$. Data were obtained on a 3′ grid between $l = 18°$ and $l = 55°$ and on a 6′ grid outside this longitude range. Spectral coverage is 300 km s^{-1} at 1 km s^{-1} resolution, sufficient to include all Galactic emission out to $R = 2\ R_{\odot}$. The reduced data are presented in the form of latitude–velocity (b, V) contour maps at each longitude. The longitude and latitude coverage of the maps is sufficient to include more than 90% of the emission in the first Galactic quadrant (excluding the Galactic center, $R < 0.15\ R_{\odot}$). The small grid spacing enables detection of all clouds larger than ~15 pc diameter in the $R = 0.4 - 0.8\ R_{\odot}$ molecular cloud ring.

155.002 **Massachusetts–Stony Brook Galactic Plane CO Survey. II. (l, V) maps of the first Galactic quadrant.**
D. P. Clemens, D. B. Sanders, N. Z. Scoville, P. M. Solomon.
Astrophys. J., Suppl. Ser., Vol. 60, No. 1, p. 297 – 303 (1986). With plates 1 – 27.
The Massachusetts–Stony Brook Galactic Plane CO Survey data are presented in (l, V) format as seven gray scale maps and in (l, b) format as 17 contour and gray scale maps of integrated CO intensity. Also included are one (l, V) contour map for the $b = 0°$ data subset and plots of integrated CO intensity versus longitude for seven latitude ranges. These presentations are most useful for comparison with other first–quadrant surveys (e.g., IRAS).

155.003 **The radial distribution of galactic gamma rays. III. The distribution of cosmic rays in the Galaxy and the CO–H_2 calibration.**
J. B. G. M. Bloemen, A. W. Strong, L. Blitz, R. S. Cohen, T. M. Dame, D. A. Grabelsky, W. Hermsen, F. Lebrun, H. A. Mayer–Hasselwander, P. Thaddeus.
Astron. Astrophys., Vol. 154, No. 1/2, p. 25 – 41 (1986).
High–energy (>70 MeV) gamma–ray observations are compared to H I and CO surveys over more than half of the Milky Way. The kinematics of both H I and CO are used as a distance indicator to determine, in combination with COS–B gamma–ray data, the galacto–centric distribution of the gamma–ray emissivity (the production rate per H atom) for three gamma–ray energy intervals. The ratio between H_2 column density and integrated CO line intensity is calibrated independently of excitation and abundance effects. The galacto–centric distributions of cosmic–ray electrons and nuclei are derived separately from the gamma–ray emissivity distributions.

155.004 **Submillimetric photometry of the integrated galactic emission.**
F. Pajot, R. Gispert, J. M. Lamarre, R. Peyturaux, J. L. Puget, G. Serra, N. Coron, G. Dambier, J. Leblanc, J. P. Moalic, J. C. Renault, R. Vitry.
Astron. Astrophys., Vol. 154, No. 1/2, p. 55 – 60 (1986).
The authors present results obtained with the experiment EMILIE which was designed for ground–based observations with a large field of view (0.5°) at submillimetric wavelengths (0.7 – 2.0 mm). The authors describe the experimental set–up, the absolute internal calibration and the signal processing. Then they present new results on the emission of the galactic plane in the range of galactic longitude between 0° and 30°.

155.005 **22 GHz continuum and spectral index maps of the galactic center region.**
C. E. Tateyama, F. M. Strauss, Z. Abraham.
Astron. Astrophys., Vol. 154, No. 1/2, p. 176 – 180 (1986).
The authors present a map of the galactic center region obtained with the Itapetinga radio telescope in the 22 GHz continuum with 4.′1 resolution. They compare these data to a 5 GHz map obtained with similar angular resolution (Whiteoak and Gardner, 1973), and a magnitude equivalent to the spectral index is derived. These relative spectral indices were obtained for the area mapped, and reveal the existence of sources exhibiting pronounced flux reduction at 22 GHz compared to the 5 GHz fluxes, which are suggested to be associated with non–thermal sources.

155.006 **Relative rôles of type I and II supernovae in the chemical enrichment of the interstellar gas.**
F. Matteucci, L. Greggio.
Astron. Astrophys., Vol. 154, No. 1/2, p. 279 – 287 (1986).
The authors have computed the evolution of several chemical elements (^{12}C, ^{16}O, ^{24}Mg, ^{28}Si, ^{56}Fe) in the solar neighbourhood, by means of a chemical evolution model in which the nucleosynthesis products coming from type I and II SNe are taken into account.

155.007 **Infrared spectropolarimetry of the galactic centre: magnetic alignment in the discrete sources.**
D. K. Aitken, P. F. Roche, J. A. Bailey, G. P. Briggs, J. H. Hough, J. A. Thomas.
Mon. Not. R. Astron. Soc., Vol. 218, No. 2, p. 363 – 384 (1986).
The authors present the results of 8 – 13 μm spectropolarimetric observations of galactic centre sources. All the sources show interstellar polarization due to absorption by aligned siliate grains and these grains appear to be spatially separated along the line–of–sight from those which produce the polarization at shorter wavelengths. The line of sources IRS 1, 10, 5 and 8 comprising the northern arc all show strong intrinsic polarization due to thermal emission from aligned grains of amorphous silicate–like material. This polarization is very uniform among these sources with position angle closely normal to the line of the arc. It is shown that the data favour alignment by a strong magnetic field >10 mG directed along and linking the northern arc of sources. A field of this magnitude will have significant influence on the structure and evolution of sources in the galactic centre, and some possible implications are discussed.

155.008 **Star formation and some aspects of the chemical evolution of the Galaxy.**
N. C. Rana, D. A. Wilkinson.
Mon. Not. R. Astron. Soc., Vol. 218, No. 3, p. 497 – 532 (1986).
This study aims to develop a consistent model of chemical evolution of the Galaxy, particularly the solar neighbourhood, with a view to formulating the star formation rate. The star formation rate is shown to be correlated with the surface density of molecular hydrogen, and to have remained practically constant over the lifetime of the disc. This is explained through a correlation between the molecular hydrogen fraction and the average metallicity of any given region of the disc, which suggests negligible infall and finite initial metallicity. A consistent model is proposed to solve the G–dwarf problem, the metallicity gradient and the stellar age–metallicity relation, leading to the prediction of some initial amount of pre–disc processing of gas into visible and dark matter.

155.009 **An estimation of the radial distribution of molecular hydrogen in the Galaxy from the star formation rates.**
N. C. Rana, D. A. Wilkinson.
Mon. Not. R. Astron. Soc., Vol. 218, No. 4, p. 721 – 728 (1986).
The authors' previous work (Rana and Wilkinson, 1985) shows that the star formation rate ψ_s in the Galaxy could be empirically related to the surface density of molecular hydrogen Σ_{H_2} by a relationship $\psi_s \propto \Sigma^k_{H_2}$, where $k(=1.12 \pm 0.03)$ is obtained from the observed metallicity gradient in the Galaxy. The authors have used here various available radial distributions of the star formation rate in the above relation in order to derive a radial distribution of molecular hydrogen, independently of the CO surveys. A comparison has also been made with the present controversial estimates of Σ_{H_2} based on various CO surveys.

155.010 **Is the galactic centre black hole a dwarf?**
D. A. Allen, R. H. Sanders.
Nature, Vol. 319, No. 6050, p. 191 – 194 (1986).
A 2.2–μm image of IRS 16, the relatively blue infrared source near the galactic centre, is presented at the highest spatial resolution yet reported. It comprises several discrete, extended sources. By intercomparison of radio and infrared images of the region, the non–thermal radio source Sgr A* can be located relative to IRS 16 without the uncertainty of intermediate astrometric frames. No component of IRS 16 is coincident with Sgr A*. These data preclude the presence of a massive ($>10^6 M_\odot$) black hole at the galactic centre. Any black hole present has a mass nearer to 100 $M_\odot$.

155.011 **Comparison between *UBV*– and *RGU*–photometrically determined space density functions in a starfield close to the galactic centre (SA 133).**
R. Fenkart, L. Topaktas, S. Boydağ, G. Kandemir.
Astron. Astrophys., Suppl. Ser., Vol. 63, No. 1, p. 49 – 57 (1986).
Three high–latitude fields of the Basle Halo Program have been treated with the Basle three–colour photometric methods independently in *RGU* and *UBV*, so far (Fenkart and Esin–Yilmaz 1983, 1984, 1985), in order to obtain space density functions of the photometrically identifiable and separable star types. The satisfactory results in all three fields encouraged the present investigation of another test–field of the named program, SA 133. The separation of three photometrically identifiable star types: "disk stars", "halo–stars" and "giants" and the determination of their space densities are possible in both systems, and the comparison of the system–independent overall density functions again confirms, as in the previous investigations, about equally reliable performance of the involved methods in both systems.

155.012 **Age–metallicity relation for F–stars.**
H. B. Ann, Y. H. Kang.
Astrophys. Space Sci., Vol. 118, Nos. 1/2, p. 325 – 327 (1986). – See Abstr. 012.001.
The authors derived the age–metallicity relation for the solar neighbourhood from the uvby, Hβ photometry of F–stars using the evolutionary models of Hejlesen (1980). It is shown that disk formation is divided into two phases according to the fast and slow collapse which accompany the initial rapid enrichment and slow increase of the metallicity of the galactic disk, respectively.

155.013 **The galactic centre.**
A. R. Hyland.
Astrophys. Space Sci., Vol. 118, Nos. 1/2, p. 343 – 364 (1986). – See Abstr. 012.001.
Recent observations of the galactic centre at radio and infrared wavelengths are reviewed. Particular emphasis is placed on the new large scale radio structures – the galactic centre lobe, and the filamentary ribbon structures in the radio arc – which have just been reported. New observations of the two parsec ring of molecular hydrogen, and the relationship of the ring to the interior distribution of ionised gas are discussed. The kinematics and distribution of neutral material, shocked molecular hydrogen and plasma in the inner few parsecs of the nucleus are considered in terms of a central luminous mass–losing object, together with rotation and infall of the neutral material.

155.014 **Expected and observed late giant star counts in the Milky Way bulge.**
V. M. Blanco, B. M. Blanco.
Astrophys. Space Sci., Vol. 118, Nos. 1/2, p. 365 – 366 (1986). – See Abstr. 012.001.
In this paper, late M giants star counts in the Milky Way central bulge are compared with those expected from the Hubble and de Vaucouleurs surface brightness distributions.

155.015 **The galactic center radio lobe – a cosmic jet in our Galaxy?**
Y. Sofue, T. Handa, I. Suwa, Y. Fukui.
Astrophys. Space Sci., Vol. 118, Nos. 1/2, p. 371 – 374 (1986). – See Abstr. 012.001.
The galactic center radio lobe found in the radio continuum is interpreted as due to a channeled exhaust of material perpendicular to the galactic plane driven by an activity in the nuclear region. A common characteristic of the lobe to extragalactic more energetic jets is pointed out.

155.016 **A model of the galactic centre with magnetic monopoles.**
D.–y. Wang, Q.–h. Peng, T.–y. Chen.
Astrophys. Space Sci., Vol. 118, Nos. 1/2, p. 379 – 382 (1986). – See Abstr. 012.001.
A model of the galactic centre with magnetic monopoles has been presented here. The positron can be produced continuously through magnetic monopoles to induce baryon decay (Rubakov catalytic reaction) and a lot of energy can be released as well. The calculation results show that even if the galactic centre contains only a few magnetic monopoles ($\zeta = N_M/N_B \approx 10^{-24}$), this massive object can not collapse into a black hole. This model can explain the observed intensities of the annihilation line and higher energy photons of $E_\gamma > 511$ keV from the galactic centre.

155.017 **Star count of the North Galactic Pole region in the UBV colour bands.**
R. S. Stobie, K. Ishida, Y. Yoshii, H. T. MacGillivray.
Astrophys. Space Sci., Vol. 118, Nos. 1/2, p. 419 – 420 (1986). – See Abstr. 012.001.
Star count data in the magnitude range $12 \leqslant V \leqslant 18$ covering 24 square degrees of the North Galactic Pole region have been obtained in the UBV colour bands. The photographic plates were taken with the Kiso Schmidt telescope, were subsequently scanned with the COSMOS machine, and were reduced at the Royal Observatory, Edinburgh.

155.018 **Distance to the anti–center shell.**
S. R. Kulkarni, R. Mathieu.
Astrophys. Space Sci., Vol. 118, Nos. 1/2, p. 531 – 533 (1986). – See Abstr. 012.001.
Heiles (1984) has identified a large angular diameter (~30°), anomalous velocity H I structure in the anti–center region of the Galaxy. The distance to this structure is the key to understanding of the energetics and, hence, the origin. The authors report the absence of any anomalous velocity Ca K absorption features in high sensitivity echelle spectra towards 17 stars located within the boundaries of this structure. They conclude that the structure is definitely beyond 500 pc and barring usually low Ca II abundance is more distant than 2.5 kpc. This result supports the hypothesis that the anti–center structure is the result of collision of high–velocity clouds with the galactic H I disk.

155.019 **CO observations, geometry, and galactic structure.**
B. J. Robinson, R. N. Manchester, W. H. McCutcheon.
Astrophys. Space Sci., Vol. 119, No. 1, p. 111 – 114 (1986). – See Abstr. 012.001.
2.6 mm carbon monoxide and 21 cm hydrogen line observations of the Galaxy define six directions where gas is seen tangentially along a spiral feature and four directions where extended structures cross the $R = R_0$ circle. These directions provide a geometrical framework for any spiral structure pattern. Also simple equations are derived to define the normalized radius R/R_0 of gas with a particular radial velocity in terms of the observed tangential velocity of gas with the same $R/R_0 = \sin l_T$, without specific knowledge of $\omega(R)$, ω_0 or R_0.

155.020 **Infrared polarimetry of the stars in the inner Galaxy.**
Y. Kobayashi, J. Jugaku, H. Okuda, S. Sato, T. Nagata.
Astrophys. Space Sci., Vol. 119, No. 1, p. 135 – 139 (1986). – See Abstr. 012.001.
Near–infrared photometric and polarimetric observations were made for stars embedded in the galactic plane. The sources were picked up from two selected regions in the galactic plane ($l = 20°$ and 30°). In total, 59 sources were observed, 35 of which were observed with polarimetry in the K–band (2.2 μm) and 44 were observed with photometry by the near–infrared broad band

system. The observations were made by using the 2.2 m telescope of the University of Hawaii on Mauna Kea.

155.021 **A survey of population types at the south galactic pole.**
A. G. D. Philip.
News Lett. Astron. Soc. N.Y., Vol. 2, No. 9, p. 26 – 28 (1986). – See Abstr. 010.241.

155.022 **Black holes and local dark matter.**
D. J. Hegyi, E. W. Kolb, K. A. Olive.
Astrophys. J., Vol. 300, No. 2, p. 492 – 495 (1986).
Two independent constraints are placed on the amount of dark matter in black holes contained in the galactic disk. First, gas accretion by black holes leads to X–ray emission which cannot exceed the observed soft X–ray background. Second, metals produced in stellar processes that lead to black hole formation cannot exceed the observed disk metal abundance. Based on these constraints, it appears unlikely that the missing disk mass could be contained in black holes. A consequence of this conclusion is that at least two different types of dark matter are needed to solve the various missing mass problems.

155.023 **A low–energy jet emanating from the Galactic nucleus?**
F. Yusef–Zadeh, M. Morris, O. B. Slee, G. J. Nelson.
Astrophys. J., Lett. Ed., Vol. 300, No. 2, p. L47 – L50 (1986). With plate L3.
A narrow ridge of low–frequency radio emission is found to lie along the rotation axis of the Galaxy and to extend ~30 pc from the nucleus toward negative latitudes. No positive latitude counterpart is evident. This radio feature is observed at 160 MHz using the Culgoora Circular Array, and the constraints on its spectral index indicate that it is nonthermal. The authors argue that it results from the continuous, or frequent, ejection of mildly relativistic particles from the Galactic nucleus.

155.024 **Stellar populations in the Milky Way: comparisons of a synthetic model with star counts in nine fields.**
A. Robin, M. Crézé.
Astron. Astrophys., Suppl. Ser., Vol. 64, No. 1, p. 53 – 64 (1986).
The model of population synthesis described in Robin and Crézé (1985) is shown to predict correctly star counts in magnitudes and colours in nine fields, at quite different latitudes and longitudes, where observations are available. It can be used to derive reliable predictions in any other galactic direction but the centre. The authors also compare the ability of different Galaxy models (Bahcall et al., 1984, 1985; Gilmore et al., 1984, 1985) and theirs to fit existing star counts.

155.025 **The mass of gas within 500 pc of the galactic centre.**
C. L. Bhat, C. J. Mayer, A. W. Wolfendale.
Cosmical gas dynamics, p. 13 – 19 (1985). – See Abstr. 012.007.
The problem of the mass of gas within 500 pc of the galactic centre is addressed from a variety of standpoints. It is argued that all methods give roughly the same mass: $3 \times 10^7 M_\odot$, a value that is much less than the value quoted by many CO observers. Arguments supporting this conclusion are made on the basis of observations of other galaxies.

155.026 **Gamma ray estimates of galactic gas contents.**
C. J. Mayer.
Cosmical gas dynamics, p. 21 – 26 (1985). – See Abstr. 012.007.
Estimates of the mass of molecular hydrogen (H_2) in the inner Galaxy based on observations of CO differ by factors of 3 to 4. Estimates based on the fluxes of γ–rays favour the lower range of H_2 masses, i.e. comparable or less than the mass of atomic hydrogen (H I) $\sim 10^9 M_\odot$. A metallicity correction is invoked to reconcile the γ–ray and CO estimates for the Galaxy and some consequences of extrapolating such a correction to external galaxies are explored.

155.027 **The dynamics of the interstellar gas and the kinetics of the galactic magnetic field.**
P. A. H. Seymour.
Cosmical gas dynamics, p. 43 – 47 (1985). – See Abstr. 012.007.
This paper considers the coupling between the dynamics of the interstellar gas and the galactic magnetic field. After a brief historical introduction to the various attempts to relate field structure to the large and small scale gas dynamics, the author considers the use of spherical harmonics to unravel local irregularities from large scale structures. The paper ends with a proposal for the use of spherical harmonics to elucidate further the kinematics of the galactic magnetic field.

155.028 **Low–latitude radio continuum structure in the fourth galactic quadrant.**
J. L. Jonas.
Mon. Not. R. Astron. Soc., Vol. 219, No. 1, p. 1 – 12 (1986).
A number of new galactic continuum emission features have been found at low latitudes between longitudes 310° and 5°. These sources are in the form of loops, spurs and arcs which have angular dimensions between 2 and 6°. An attempt is made to classify each of the new sources, and in some cases determine their distance from the Sun.

155.029 **Near infrared stellar counts and a model of galactic distribution.**
F. Garzon, C. Sanchez–Magro, M. Prieto.
Astron. Astrophys., Vol. 155, No. 1, p. 63 – 71 (1986).
A model of the galactic stellar distribution has been developed and the surface stellar densities in the near infrared are compared with observations along the galactic plane between $l = 0°$ and $l = 10°$. The model, based on data taken from the literature, is in reasonable agreement with the observational results. However, the mass distribution obtained from the star counts is much lower than derived by other methods. As a result, the authors have also obtained the expected integrated brightness at 2.2 μm close to the galactic centre. The results are in very good agreement with the experimental data.

155.030 **Stellar populations in the Milky Way: a synthetic model.**
A. Robin, M. Crézé.
Astron. Astrophys., Vol. 157, No. 1, p. 71 – 90 (1986).
The authors propose a detailed model of the stellar content of our Galaxy. It uses a scheme of galactic evolution for deriving the age distribution of stars in the solar neighbourhood. Then it links metallicities and scale heights to ages through kinematics. In the course of model elaboration, the authors give or confirm a number of quantitative results; the most outstanding ones concern the densities and space distributions of Halo and Intermediate Population II stars.

155.031 **Construction of a global gas–dynamical model for our Galaxy.**
W. A. Mulder, B. T. Liem.
Astron. Astrophys., Vol. 157, No. 1, p. 148 – 158 (1986).
This paper describes an initial effort to construct a global gas–dynamical model for our Galaxy. The model is a two–dimensional quasi–steady solution of the gas–dynamical equations in the gravitational potential of a weakly barred galaxy. From this solution the authors construct longitude–velocity diagrams that are compared with observations of neutral hydrogen in our Galaxy.

155.032 **Surface photometry of edge–on spiral galaxies. V. The distribution of luminosity in the disk of the Galaxy derived from the Pioneer 10 background experiment.**
P. C. van der Kruit.
Astron. Astrophys., Vol. 157, No. 2, p. 230 – 244 (1986).
In this paper the surface brightness of the Galactic background in the Pioneer 10 background starlight experiment is analysed in terms of recent models for the distribution of stars in the Galaxy. Only areas of low Galactic extinction are used and the data are corrected for contributions of diffuse Galactic light. This restricts the analysis to those areas of sky with $|b| \gtrsim 20°$. The procedure provides important, independent information on the photometric

parameters of the old disk population. Comparison is made of the scalelength and photometric dimension of the Galaxy with external galaxies for which detailed surface photometry is available. Comparison of the model with star counts obtained in the plan of Selected Areas early this century shows that the relative counts across the sky are in excellent agreement with the model.

155.033 **Dynamics of the ionized gas within 2 pc of the galactic center. I. First results of a radio recombination line survey.**
P. G. Mezger, J. E. Wink.
Astron. Astrophys., Vol. 157, No. 2, p. 252 – 266 (1986).
The gas within a radius of 2 pc (=41″) of the galactic center appears to be almost completely ionized. It forms the H II region Sgr A West, which consists of a compact central H II region embedded in an extended H II region of lower surface brightness. The authors have mapped the H 66α radio recombination line emission at 1.3 cm wavelength using the MPIfR 100–m radio telescope with an angular resolution of 42″. The observations show a rotation of the ionized gas in the plane $\Delta b = 0$ in the sense of general galactic rotation but around a dynamical center, which is displaced by $\Delta l \sim -5''$ to $-10''$ from the center position. The observed profiles are compared with model line profiles computed for two different mass distributions. The positions of center of gravity of the central stellar cluster and point mass, respectively, do not appear to coincide. If this point mass is in fact a supermassive object, a supermassive star can be ruled out on the basis of the low bolometric and Lyc–photon luminosities derived from observations. An accreting black hole may be a better candidate for this central point mass. A very preliminary survey of the ionized gas perpendicular to the plane $\Delta b = 0$ has also been made.

155.034 **Phase modulation of the Sun's *z* oscillations.**
W. L. H. Shuter, C. Klatt.
Astrophys. J., Vol. 301, No. 1, p. 471 – 477 (1986).
The Sun's oscillations perpendicular to the Galactic disk have been invoked to explain apparent periodicities in the dates of terrestrial mass extinctions, the ages of impact craters, and the times of reversals of the Earth's magnetic field. In this work it is shown that these oscillations are expected to be phase modulated, as a result of the epicyclic motion of the Sun about the Galactic center. Consequences of the phase modulation are that the period of the z oscillations averaged over the past 250 Myr (the time scale of the geological data) is $\sim 8\%$ greater than the present value, and that it is possible to devise a phase correction to enhance the periodicities in the geological data. The analysis suggests a possible connection between the dates of mass extinctions and the Sun's z oscillations. It also confirms the existence of a highly significant periodicity in the geomagnetic data, which may be related to the Sun's Galactic rotation period or the epicyclic period.

155.035 **The Massachusetts–Stony Brook galactic plane CO survey: statistical clustering properties of molecular cloud cores.**
A. R. Rivolo, P. M. Solomon, D. B. Sanders.
Astrophys. J., Lett. Ed., Vol. 301, No. 1, p. L19 – L23 (1986).
The authors employ the formalism of two–point covariance functions to analyze the statistical clustering properties of ~ 2000 molecular cloud cores identified in the Massachusetts–Stony Brook CO survey of the galactic plane between galactic longitudes 20° and 50°. Evidence is presented that the warmest cores are strongly clustered into groups with a characteristic size of $\sim 50 - 150$ pc, which are in turn confined into constant galactic radius bands separated by ~ 2.5 kpc. A characteristic core–core velocity dispersion of ~ 12 km s^{-1}(three–dimensional) is identified for the warmest cores and is associated with the clustering scale leading to a virial mass estimate of $\sim 2 - 6 \times 10^6 M_\odot$ for these clusters.

155.036 **The rotation curve of the Galaxy.**
K. Rohlfs, R. Chini, J. E. Wink, R. Böhme.
Astron. Astrophys., Vol. 158, No. 1/2, p. 181 – 190 (1986).
The increase of the number of galactic H II regions with well determined photometric distances and radial velocities made a new determination of the galactic rotation curve possible. From suitably selected local H II regions new values for the local rotation constants were derived resulting in $r_0 = 7.9 \pm 0.7$ kpc, $A_0 = 16.0 \pm 1.3$ km s^{-1}kpc^{-1}. The resulting rotation curve for $r < r_0$ relies mainly on radio data, and in the innermost 4 kpc non circular gas motion was taken into account. For $r > r_0$ the curve is based on the H II region data. The resulting rotation curve is remarkably flat on a large scale although it is decreasing from 200 km s^{-1} to 170 km s^{-1} between $r = 6$ kpc and $r = 10$ kpc and again rises back to $\theta = 200$ km s^{-1} at $r = 16$ kpc. It thus bears great similarity to the rotation curves observed for other large Sb and Sc galaxies.

155.037 **The results of a surface brightness determination of the Milky Way in the *V* system.**
Yu. M. Zavarzin.
Astrofizika, Tom 23, Vyp. 3, p. 473 – 477 (1985). In Russian. English translation in Astrophysics, Vol. 23, No. 3.
The results of the surface photometry of the northern Milky Way in the V system are presented. The surface brightness has been determined for the galactic latitudes –20°, –15°, –10°, 10°, 15° and 20° at the galactic longitudes 30° – 150°.

155.038 **Formation of the superhigh energy electron spectrum in the Galaxy.**
F. A. Agaronyan, A. S. Ambartsumyan.
Astrofizika, Tom 23, Vyp. 3, p. 479 – 485 (1985). In Russian. English translation in Astrophysics, Vol. 23, No. 3.
The formation of the superhigh energy electron spectrum in the disk of the Galaxy and the halo is considered. A different behaviour of the electron spectrum within the framework of capture models in disk or halo in the energy region $E \gtrsim 10^5$GeV is revealed due to the account of relativistic corrections in the energy loss of electrons during inverse Compton scattering. A comparison with the existing experimental data is carried out.

155.039 **The infrared response to electron–positron pair annihilation at the Galactic center.**
A. Sternberg.
Astrophys. J., Vol. 301, No. 2, p. 923 – 926 (1986).
It is argued that the annihilation of electrons and positrons at the Galactic center may result in observable infrared line emission by atomic hydrogen. A correlation between the intensity variations of the 511 keV line and the hydrogen infrared lines emitted by the annihilation region is predicted. The observed infrared fluxes from compact infrared sources at the Galactic center may be used to constrain theories of pair production there.

155.040 **Late M stars found in a hemispheric survey.**
C. B. Stephenson.
Astrophys. J., Vol. 301, No. 2, p. 927 – 937 (1986).
An objective–prism survey of somewhat more than half the sky, being the portion more than 10° from the Galactic plane and north of declination –25°, has been completed, using the visual–red spectral region. From this survey this paper lists 583 late M stars (mostly type M6 or later) not contained in the third edition of the variable star catalog or its supplements. The great majority of the stars are likely to be giants. For an assumed visual absolute magnitude of –0.9, the distance of the stars from the Galactic plane is well represented by an exponential distribution with scale height 1800 pc; or 900 pc for an assumed absolute magnitude of zero. The sample may well be related, therefore, to the recently discussed "thick disk" component of Galactic structure.

155.041 **Sagittarius A* and the positions of infrared sources in the galactic center.**
W. J. Forrest, J. L. Pipher, W. A. Stein.
Astrophys. J., Lett. Ed., Vol. 301, No. 2, p. L49 – L52 (1986). With plates L8 – L9.
Observations of the central region of the Galaxy have been made with the University of Rochester 32x32 element InSb array camera in the 1 – 5 μm wavelength range. The 2.2 μm images have been used to determine positions of infrared features in comparison with that of the galactic center compact radio source Sgr A*. No infrared source lies precisely at the Sgr A* position in these images. However, the infrared source closest to the Sgr A* position is a previously little studied source directly south of IRS 7 that the authors have called IRS 16NW. IRS 16NW is coincident with one of the very red objects seen in various 1 μm images of this region.

155.042 **Thermal X–ray emission with intense 6.7–keV iron line from the galactic ridge.**
K. Koyama, K. Makishima, Y. Tanaka, H. Tsunemi.
Publ. Astron. Soc. Jpn., Vol. 38, No. 1, p. 121 – 131 (1986).
An intense emission feature attributed to iron has been discovered in the unresolved excess emission along the galactic ridge. The mean line energy is 6.71 ± 0.04 keV, which indicates emission from iron of helium–like ions. The observed spectra are in good agreement with that predicted from a model of thermal bremsstrahlung from an optically thin plasma. The temperature is found to vary from region to region in the range of $kT = 5 - 10$ keV. The observed intensity distribution is consistent with a disk–shaped emission region with a scale height of 100 – 300 pc and a radius of 10 kpc. This excess emission is considered to be of discrete origin. Within the constraints imposed by the observational results, possible classes of sources are discussed, including white dwarf binaries, RS CVn's, emission nebulae as well as unidentified supernova remnants.

155.043 **The stability of a hypothetical binary in the galactic center.**
A. S. Baranov.
Astron. Zh., Tom 63, Vyp. 2, p. 220 – 232 (1986). In Russian. English translation in Sov. Astron., Vol. 30, No. 2.
The paper deals with the stability of a hypothetical binary system of black holes in the galactic nucleus (Kardashev's model). On the assumption that the rotation of the pair is in steady state synchronized with that of the surrounding stellar system, the two cases have been studied: 1) the pair is within a homogeneous triaxial ellipsoid; 2) each of the mass points is surrounded by its own sphere which are not overlapping. It is shown that in the first case the binary is dynamically stable if both components are on the principal axis of the ellipsoid and their centers of gravity coincide. In the second case which describes a certain general tendency in the variation of the stability nature, when considering heterogeneous models. The binary is constantly dynamically stable. It is actually always stable in the secular aspect as well.

155.044 **On the properties of the galactic halo gas.**
A. A. Suchkov, Yu. A. Shchekinov.
Astrofizika, Tom 24, Vyp. 1, p. 123 – 130 (1986). In Russian. English translation in Astrophysics, Vol. 24, No. 1.
The possible mechanisms of creating ions C IV and Si IV, absorption lines of which are seen in the spectra of stars in the galactic halo at $z \sim 1 - 3$ kpc, have been considered. It has been shown that the observed quantities of these ions can be explained in the model of the absorbing region in the halo with gas temperature $T = 10^4$K, number density of particles $n \lesssim 0.01$ cm^{-3}, and ionization of elements by ultraviolet background radiation.

155.045 **Current status of the galactic center positron–annihilation source.**
M. Leventhal, C. J. MacCallum, A. F. Huters, P. D. Stang.
Astrophys. J., Vol. 302, No. 1, p. 459 – 461 (1986).
A balloon–borne germanium gamma–ray telescope was again flown over Alice Springs, Australia, to search for 511 keV positron–annihilation radiation from the galactic center direction. The result for the line flux was $(0.6 \pm 4.4) \times 10^{-4}$ photons cm^{-2}s^{-1}, indicating that the source was in an "off", or "low", state in the fall of 1984.

155.046 **The explanation of the positron–electron annihilation line at our Galactic Center under the model of galactic nuclei with monopoles.**
D.–y. Wang, Q.–h. Peng, Z.–y. Li.
Structure and evolution of active galactic nuclei, p. 737 – 739 (1986). – See Abstr. 012.011.
The observational data on the intense positron–electron annihilation line at our Galactic Center, detected by balloons and satellites are summarised. The authors explain these data, using the model, based on the induced nucleon decay by monopole, $pM \rightarrow \pi^0 e^+ M$ (85%) or $\mu^+\mu^-e^+\mu$ (15%), in their last paper.

155.047 **Chemical evolution of the Galaxy: a comparison of the abundances of light metals in disk and halo dwarfs.**
P. François.
Astron. Astrophys., Vol. 160, No. 2, p. 264 – 276 (1986).
High resolution and high signal–to–noise spectra of 36 field dwarf stars have been obtained in the red and used to derive light metal (Na, Al, Mg, Si) abundances relative to iron. Values of iron abundance vary from [Fe/H] = –2.60 to [Fe/H] = +0.30. The results show that in halo stars the relative abundances of light metals, such as [Mg/Fe] and [Si/Fe], are constant, i.e., independent of [Fe/H]. The relative abundances [Al/Mg] and [Na/Mg] are probably constant also: their values show that the odd–even effect is enhanced in the halo. In the dwarfs of the disk, the relative abundances of metals change progressively from halo–type ratios to disk (near solar) ratios. A preliminary interpretation of these results is given.

155.048 **Galactic center update.**
A. Killian.
Sky Telesc., Vol. 71, No. 3, p. 255 (1986).

155.049 **M giants in the galactic nuclear bulge.**
J. A. Frogel.
Spectral evolution of galaxies, p. 143 – 156 (1986). – See Abstr. 012.015.
The author argues that the giants in the galactic nuclear bulge are our best chance for learning, from the ground, about the cool stellar component of spheroidal systems and describes what has been learned to date about these stars.

155.050 **Age and mass of M giants in the galactic bulge.**
A. E. Whitford.
Spectral evolution of galaxies, p. 157 – 170 (1986). – See Abstr. 012.015.
This paper compares the modeling of the cool–star component in certain classical population syntheses with data on M giants in the galactic bulge. It also examines the central question "Are the late M giants in the bulge young or old?" in the light of recent observations.

155.051 **The chemical evolution of galaxies.**
R. Güsten.
Spectral evolution of galaxies, p. 449 – 467 (1986). – See Abstr. 012.015.
After a brief introduction into the evolution of spiral galaxies, the observational evidence for temporal and spatial abundance variations across the galactic disk is discussed. The theoretical framework of chemical evolution models is summarised, followed by a critical discussion of available constraints. Solutions applying to the special cases of the solar vicinity and the galactic abundance gradients are presented. The author summarizes evidence for a bimodal nature of star formation, and emphasizes that if in regions and periods of high star formation activity the formation of low–mass stars is suppressed, a self–consistent model of star formation and chemical evolution of the Milky Way disk can be presented.

155.052 **Statistical properties of barium stars.**
J. Hakkila.
Bull. Am. Astron. Soc., Vol. 17, No. 4, p. 840 (1985). Abstract. – See Abstr. 010.062.

155.053 **Search for ^{26}Al line radiation from the galactic plane.**
C. MacCallum, M. Leventhal.
Bull. Am. Astron. Soc., Vol. 17, No. 4, p. 851 (1985). Abstract. – See Abstr. 010.062.

155.054 **Secondary metallicity in analytic models of chemical evolution.**
I. Pantelaki, D. D. Clayton.
Bull. Am. Astron. Soc., Vol. 17, No. 4, p. 859 – 860 (1985). Abstract. – See Abstr. 010.062.

155.055 **Brackett–α imaging of the galactic center using a cooled Fabry–Perot array spectrometer.**
J. Fischer, E. Albert, J. Arens, J. Lacy, D. Mozurkewich, R. Russell, H. A. Smith, W. Waltman.
Bull. Am. Astron. Soc., Vol. 17, No. 4, p. 860 (1985). Abstract. – See Abstr. 010.062.

155.056 **Color–magnitude diagrams of the galactic bulge.**
D. Terndrup.
Bull. Am. Astron. Soc., Vol. 17, No. 4, p. 860 (1985). Abstract. – See Abstr. 010.062.

155.057 **The inner Milky Way: ring–like or spiral–like?**
D. P. Clemens, D. B. Sanders, N. Z. Scoville.
Bull. Am. Astron. Soc., Vol. 17, No. 4, p. 860 (1985). Abstract. – See Abstr. 010.062.

155.058 **The distance to the center of the Galaxy.**
C. R. Gwinn, M. J. Reid, M. Schneps, J. M. Moran, R. Genzel, D. Downes.
Bull. Am. Astron. Soc., Vol. 17, No. 4, p. 903 (1985). Abstract. – See Abstr. 010.062.

155.059 **Abundances and kinematics of K giants in the galactic nuclear bulge.**
R. M. Rich.
Bull. Am. Astron. Soc., Vol. 17, No. 4, p. 903 – 904 (1985). Abstract. – See Abstr. 010.062.

155.060 **Kinematics and chemical properties of the old disk.**
J. R. Lewis, K. C. Freeman.
Bull. Am. Astron. Soc., Vol. 17, No. 4, p. 904 (1985). Abstract. – See Abstr. 010.062.

155.061 **Binary survey among high velocity stars using GSU speckle camera.**
P. K. Lu, P. Demarque, W. van Altena, W. Hartkopf, H. McAlister.
Bull. Am. Astron. Soc., Vol. 17, No. 4, p. 904 (1985). Abstract. – See Abstr. 010.062.

155.062 **Cepheids and the spiral structure of the Galaxy.**
L. N. Berdnikov.
Astron. Tsirk., No. 1388, p. 4 – 5 (1985). In Russian.

155.063 **On the thermal equilibrium in the galactic halo gas.**
A. A. Suchkov, I. G. Kovalenko, Yu. A. Shchekinov.
Astron. Tsirk., No. 1390, p. 1 – 3 (1985). In Russian.

155.064 **Programme of the main meridional section of the Galaxy.**
J. E. Einasto, V. D. Malyuto, N. V. Kharchenko.
Astron. Tsirk., No. 1394, p. 1 – 6 (1985). In Russian.

155.065 **The galactic center: the nearest Seyfert nucleus.**
K.–Y. Lo.
Publ. Astron. Soc. Pac., Vol. 98, No. 600, p. 179 – 184 (1986). Review paper. – See Abstr. 012.022.

Recent observations of the central 10 pc of the Galaxy are reviewed. The compact nonthermal radio source has a diameter <20 AU and an elongated source structure, which has been observed for the first time at $\lambda = 3.6$ cm. It is a unique radio source in the Galaxy, but it is similar to the nuclear radio sources in external galaxies. Observations of the ionized gas in Sgr A West at 0\".3 resolution support an infall hypothesis for its origin in the central 3 pc. Similarities of the galactic center to Seyfert nuclei are pointed out.

155.066 **Star counts and color indices in a south–galactic–pole field.**
L. Infante.
Publ. Astron. Soc. Pac., Vol. 98, No. 601, p. 360 – 363 (1986).

Faint star counts and color indices have been obtained in a south–galactic–pole field. The counts are presented over an area of 0.55 deg^2 and span a magnitude range, $19.25 \leqslant J \leqslant 22.25$. The color distribution was obtained for a 0.33 deg^2 subarea for the interval $20 \leqslant J \leqslant 22$. The data are in very good agreement with the *standard model* suggested by Bahcall and Soneira and by Pritchet. The following main results are highlighted. (1) The color of the blue peak suggests an intermediate metallicity (similar to that of M13) for the halo component. (2) The data are not consistent with the Gilmore and Reid normalization of the thick disk component, although it does not rule out the existence of this component. (3) The author notes that the number of observed objects at $(J–r) \leqslant 0.5$ is consistent with the density of QSOs suggested by Koo and Kron. (4) The good agreement with the northern star counts suggests that the observed field–to–field variations in the galaxy number counts are real.

155.067 **Observations of the galactic disk at ultrahigh energies at the Tyan Shan High –Altitude Station.**
D. B. Mukanov, V. P. Fomin.
Bull. Crimean Astrophys. Obs., Vol. 69, p. 60 – 62 (1986). English translation of 39.155.080.

155.068 **2000 Å UV imaging of a 6° diameter field around μ Aurigae.**
M. Golay, X. Rabattu, N. Cramer, D. Huguenin, A. Blecha.
Astrophys. Space Sci., Vol. 122, No. 1, p. 109 – 149 (1986).

UV imagery of a 6° diameter field centered on μ Aurigae in the galactic plane was obtained with the SCAP 2000 experiment dedicated to an UV sky survey. Photometric UV magnitudes at 2000 Å are given here for more than 500 stars detected in the field. A star–counting method is applied to give a fine description of the interstellar matter distribution up to 1000 pc. The mean space density for B8 – A3 stars is derived. Stars with an UV excess are detected and spectral reclassification of some single or binary stars is suggested.

155.069 **Additional evidence for the accuracy of galactic reddenings from ScI galaxies and quasars.**
S. Haarala, P. Teerikorpi.
Astrophys. Space Sci., Vol. 122, No. 1, p. 163 – 171 (1986).

The relevance of the galactic reddenings *E(B–V)* derived by Teerikorpi (1978, 1981) from ScI galaxies and quasars, is further confirmed by considerations of close pairs in the sky, the Lick counts, and the counts/NH I ratio. The compact dust clouds suggested by Teerikorpi (1981) could not be detected directly by the counts. However, the counts/NH I ratio showed a behaviour expected if the reddenings really describe the division of the dust in the Local Spiral into compact clouds and a more diffuse medium, in the way suggested in the previous study.

155.070 **Abundance gradient in interstellar medium.**
S. Giridhar.
Binary and multiple systems, p. 143 – 155 (1984). – See Abstr. 012.025.

The author discusses in this review the implications of observed abundance gradient in the interstellar medium on the chemical evolution of the Galaxy.

155.071 **Red giant survey toward the galactic central direction.**
T. Ichikawa, K. Hamajima, K. Ishida, M. Raharto, B. Hidayat.
Galactic structure and variable stars, p. 25 – 34 (1985). – See Abstr. 012.029.

The main purposes of the authors' observation are (1) to study the galactic structure in the direction of $l = 355°$, $b = -1°$ where the region of 2° × 1° area shows the most anomalous enhancement of the infrared radiation at 2.4–μm wavelength, (2) to search for galactic windows, or fields with extraordinarily low interstellar extinction in the direction of the galactic center, (3) to study the space distribution of dark clouds, and (4) to find candidates of M supergiants.

155.072 **H I and CO studies of galactic structure.**
F. J. Kerr.
Proceedings of the Second Asian–Pacific Regional Meeting on Astronomy, p. 82 – 87 (1984). – See Abstr. 012.031.

Radio spectral line studies provide the best method for investigating the spiral structure of the whole Galaxy. For the most part, these studies have been carried out with the 21–cm hyperfine transition of neutral hydrogen. More recently, the 2.6–mm rotational transition of carbon monoxide has provided an alternative way of delineating the spiral structure. It requires more elaborate receiving equipment and a long integration time for each observed point, but it has the distinct advantage that it is *not* detectable at every point and every velocity, and therefore it is easier to separate individual structural features.

155.073 **On the large scale structure in the Cep–Cas region.**
M. Nakano, T. Kogure, S. D. Wiramihardja.
Proceedings of the Second Asian–Pacific Regional Meeting on Astronomy, p. 88 – 97 (1984). – See Abstr. 012.031.

The authors redetermined the smooth rotation curve of the Galaxy in the range of the galactocentric distance R = 10 – 17 kpc, by making use of available optical data of H II regions. They then applied their rotation curve to H I survey data to deduce the space distribution of H I gas outside the solar circle. In deriving the space density of H I gas n_H as a function of heliocentric distance, the effect of noncircular motion of the Perseus arm has been taken into account. The authors examined the bending of the galactic disk in this direction and derived less steep inclination than previously derived ones.

155.074 **Surface brightness and colors of the galactic disk.**
K. Ishida, T. Mikami.
Proceedings of the Second Asian–Pacific Regional Meeting on Astronomy, p. 98 – 100 (1984). – See Abstr. 012.031.

The space number density in the galactic plane and density distribution perpendicular to the plane, of stars of various spectral and luminosity classes, were examined and used to generate surface brightness and colors of the galactic disk, to see if they reproduce those of external galaxies.

155.075 **Infrared emission from the galactic center.**
C. G. Wynn–Williams.
Proceedings of the Second Asian–Pacific Regional Meeting on Astronomy, p. 101 – 103 (1984). – See Abstr. 012.031.

155.076 **Red variables as distance indicators and the distance to the galactic centre.**
M. W. Feast.
Proceedings of the Second Asian–Pacific Regional Meeting on Astronomy, p. 104 – 109 (1984). – See Abstr. 012.031.

The author concentrates on the Mira variables and in particular on the determination of the distance to the galactic centre and he concludes with a summary of the current position regarding red supergiant variables.

155.077 **Survey of red giants near the galactic center.**
T. Ichikawa, K. Hamajima, K. Ishida, B. Hidayat, M. Raharto.
Proceedings of the Second Asian–Pacific Regional Meeting on Astronomy, p. 117 – 121 (1984). – See Abstr. 012.031.

A survey of red giants in the field of $330° \leqslant l \leqslant 30°$ through $l = 0°$, between $b = -2°$ and $+2°$ has been undertaken since 1979 at the Bosscha Observatory on the basis of objective–prism spectra in the near–infrared. The main purposes of the survey are (1) to study the galactic structure in the direction of $l = 355°$, $b = -1°$, (2) to search for galactic windows, fields with low interstellar extinction, near the galactic center, (3) to study the global structure of the disk of late–type stars, and (4) to find candidates of M supergiants.

155.078 **Stellar contents contributing to the near–infrared radiation of the Galaxy.**
T. Mikami, K. Kawara.
Proceedings of the Second Asian–Pacific Regional Meeting on Astronomy, p. 122 – 125 (1984). – See Abstr. 012.031.

The purpose of the present analyses is to examine the contents of near–infrared sources using the observational data at the K band.

155.079 **Distribution of stars in the Orion Barnard Loop.**
S. Isobe, Y. Nishino.
Proceedings of the Second Asian–Pacific Regional Meeting on Astronomy, p. 157 – 161 (1984). – See Abstr. 012.031.

From the photographic U, B, and V observations of the stars in the Orion Barnard Loop, it is found that the number of blue stars is larger on the Loop than that outside the Loop. This result suggests that stars on the Loop are younger than those outside the Loop and that the stars were formed during some explosive events which made the Loop.

155.080 **On warping of the old stellar disk in our Galaxy.**
T. Ichikawa, T. Sasaki.
Proceedings of the Second Asian–Pacific Regional Meeting on Astronomy, p. 182 – 188 (1984). – See Abstr. 012.031.

A total number of about 1500 late M giants, classified on the Case system, are found in a field of 130 square degrees near $l = 116°$, between $b = -11°$ and $+10°$; in this direction the H I–gas disk viewed from the sun is apparently most warped. The surface distribution of these stars as a function of the galactic latitude shows no evident warp of the old stellar disk, represented by late M giants with an age of several times 10^9years, in the direction of Cassiopeia; this phenomenon is different from that of the H I–gas disk in the outer part of our Galaxy.

155.081 **The galactic distribution of Wolf–Rayet stars.**
B. Hidayat, K. Supelli, K. A. van der Hucht.
Proceedings of the Second Asian–Pacific Regional Meeting on Astronomy, p. 197 – 212 (1984). – See Abstr. 012.031.

On the basis of the most recent compilation of narrow–band photometry and absolute visual magnitudes of Wolf–Rayet stars, and adopting a normal interstellar extinction law in all directions, the galactic distribution of 132 of the 159 known galactic WR stars is presented and discussed.

155.082 **Star counts and galactic structure.**
D. C. Morton.
Proceedings of the Second Asian–Pacific Regional Meeting on Astronomy, p. 248 – 263 (1984). – See Abstr. 012.031.

This review describes how recent star counts to faint magnitudes can provide new information on galactic structure.

155.083 **Galactic structure examined from a faint object survey in a field in Aquarius.**
K. P. Tritton, D. C. Morton.
Proceedings of the Second Asian–Pacific Regional Meeting on Astronomy, p. 364 – 367 (1984). – See Abstr. 012.031.

An area in the constellation of Aquarius, one third of a square degree, has been selected for study because of its low interstellar extinction and the availability of good photometric calibration. An attempt has been made to identify and classify all objects in the area to a magnitude B = 20.0. For the results presented in this report UK Schmidt and Anglo–Australian telescope photographs were used to separate stellar and non–stellar objects and provide UBVRI photometry of the stars. With absolute magnitudes derived from the colours, the stars in this area have been compared in detail with the Galactic model of Bahcall and

Soneira (1980). Considerable discrepancies are revealed which imply defects in the adopted model of the spheroidal component.

155.084 **The solar neighbourhood mass density.**
G. Gilmore, N. Reid.
Proceedings of the Second Asian–Pacific Regional Meeting on Astronomy, p. 368 – 374 (1984). – See Abstr. 012.031.
The authors have completed a BVRI photometric study towards the South Galactic Pole, which is complete for all stars brighter than I = 17.0 in 18.4 square degrees. Absolute magnitudes have been derived from an M_V/V–I relation. The authors have derived the solar neighbourhood main sequence luminosity function to $M_V = +19$, some 5 magnitudes fainter than any previous photometric determination. They find no evidence for any excess of M dwarfs relative to the kinematically derived function. The authors have also derived the stellar density gradient towards the South Galactic Pole, as a function of absolute magnitude.

155.085 **A galactic window at $l = 355°$, $b = -1°$ and its implications for further observations.**
K. Hamajima, K. Ishida, T. Ichikawa, B. Hidayat, M. Raharto.
Proceedings of the Second Asian–Pacific Regional Meeting on Astronomy, p. 532 – 535 (1984). – See Abstr. 012.031.
It is most probable that a galactic window exists in the direction of $l = 355°$ and $b = -1°$ according to the authors' recent work. The window is considered to penetrate over 4 kpc from the sun judging from the surface density of M giants. The window will be useful for the study of galactic structure. Here the authors discuss two kinds of distant objects: globular clusters and M supergiants.

155.086 **Erratum: "Cinemática galáctica local: un modelo isotérmico" [Rev. Mex. Astron. Astrofis., Vol. 8, No. 2, p. 91 – 108 (1983)].**
J. Núñez.
Rev. Mex. Astron. Astrofis., Vol. 11, No. 2, p. 159 (1985). See Abstr. 37.155.045.

155.087 **Giants of type M1 or later in Baade's Window.**
V. M. Blanco.
Astron. J., Vol. 91, No. 2, p. 290 – 292 (1986). With plate 34.
Spectral types, approximate *V* magnitudes, and a finding chart are presented for 174 M giants in a 42 square arcminutes area within Baade's Window. The stars of type M1 or later and those presented here supplement a previously published list of very late–type giants. Their most significant feature is their degree of completeness to a distance past that of the galactic center.

155.088 **A photometric survey of two high–galactic–latitude fields and comparison to galaxy models.**
E. D. Friel, K. M. Cudworth.
Astron. J., Vol. 91, No. 2, p. 293 – 302 (1986). With plate 35. = Lick Obs, Bull., No. 1028.
The authors have obtained stellar photographic photometry over a 4 square degree area in each of two high–galactic–latitude fields, in SA 94 ($l = 175°$, $b = -49°$) and in Serpens ($l = 37°$, $b = +51°$). Observed magnitude and color distributions are compared to the star–count predictions from the standard Bahcall–Soneira Galaxy model. Agreement is good in general although the model predicts far more blue stars in the Serpens region than are found. The color distributions are reproduced best by galaxy models that include a spheroid color–magnitude diagram from an intermediate–metallicity globular cluster like M3. There is some discrepancy between the authors' conclusions and those based on deeper surveys in these same fields.

155.089 **Diffuse X–rays from the galactic plane.**
K. Koyama.
X–ray astronomy '84, p. 325 – 328 (1985). – See Abstr. 012.033.
Diffuse X–ray spectra on the galactic plane are well described by thin thermal plasmas of several keV. They show pronounced iron K–emissions at about 6.7 keV. The temperature is different in each place on the galactic plane. The possible origin of these diffuse X–rays is supernova remnants in tenuous medium.

155.090 **The Massachusetts–Stony Brook galactic plane CO survey: observations, galactic structure, and cloud identification.**
D. P. Clemens.
Diss. Abstr. Int., Sect. B, Vol. 46, No. 6, p. 1956 (1985). Thesis, University of Massachusetts, 235 pp. (1985). Order No. DA8517090.

155.091 **Far–infrared line emission from the Galaxy.**
G. J. Stacey.
Diss. Abstr. Int., Sect. B, Vol. 46, No. 6, p. 1959 (1985). Thesis, Cornell University, 228 pp. (1985). Order No. DA8504504.

155.092 **A new determination of the luminosity function of the galactic halo.**
P. C. Dawson.
Diss. Abstr. Int., Sect. B, Vol. 46, No. 8, p. 2693 (1986). Thesis, Victoria University, Toronto, Canada (1984).

155.093 **On stellar streaming motions in triaxial objects: the case of the galactic bulge.**
M. Vietri.
Diss. Abstr. Int., Sect. B, Vol. 46, No. 10, p. 3483 (1986). Thesis, Princeton University, 69 pp. (1985). Order No. DA8529049.

155.094 **On the large–scale distribution of neutral hydrogen in the Galaxy.**
I. V. Petrovskaya.
Vestn. Leningr. Univ., Mat. Mekh. Astron., No. 1, p. 126 – 130 (1986). In Russian. Abstr. in Ref. Zh., 51. Astron., 6.51.786 (1986).

155.095 **Distribution of stars of spectral types of F7 and earlier in a field around NGC 7686.**
L. G. Balázs, M. Paparó, I. Tóth.
Commun. Konkoly Obs., No. 85, p. 291 – 329 (1985).
A statistical study was performed on 993 stars of spectral type F7 and earlier in 19.5 sq. deg. field centred on NGC 7686 down to a limiting magnitude of 12.5. The observational material was obtained with Konkoly Observatory's 60/90/180 cm Schmidt telescope. The stars were classified utilizing small scale spectra (580 Å/mm at Hγ) and photographic UBV colours were measured. The interstellar absorption in this area was determined to be $A_V = 1.2$ mag. and the space densities in the spectral groups of B8–A1, A2–A3, A4–A7, A8–F2 and F3–F7 were obtained using the matrix method of Dolan. These empirical space distributions were compared with an exponential model, with Camm's models, with Bahcall's model and with that of Woolley and Stewart. The best fit was obtained by an isothermal model and this fitting enabled the authors to estimate the total mass density near the Sun resulting 0.104 $M_\odot/pc^3$ using the B8–A1 group and 0.159 $M_\odot/pc^3$ using the A type stars.

155.096 **On the stability of the galactic magnetic field in the presence of a magnetic monopole halo.**
D. F. Chernoff, S. L. Shapiro, I. Wasserman.
Astrophys. J., Vol. 304, No. 2, p. 799 – 820 (1986).
The self–consistent problem of gravitationally interacting monopoles in the presence of a large–scale magnetic field is studied numerically and analytically. For a toroidal magnetic field geometry with a fixed gravitational potential, the authors study the characteristic oscillatory frequency and damping rates of the magnetic field. It is shown that even in the presence of monopole density gradients, there exists, for fixed magnetic field geometry, a large–scale mode of the system which oscillates, to zeroth order, at a suitably averaged monopole plasma frequency. It is demonstrated with simple dynamo models that it is not possible, in general, to use the survival of galactic magnetic fields to establish bounds on the magnetic monopole flux.

155.097 **The positron–electron annihilation line at the galactic center and a model of massive objects with magnetic monopoles.**
D.–y. Wang, Q.–h. Peng, Z.–y. Li.
Kexue Tongbao, Vol. 30, No. 2, p. 210 – 214 (1985). From Phys. Abstr., Vol. 89, No. 1273, Entry 18959 (1986).

155.098 **The effect of the new $^{12}C(\alpha,\gamma)^{16}O$ rate on the chemical evolution of the solar neighbourhood.**
F. Matteucci.
ESO Sci. Prepr., No. 434, 16 pp. (1986). To appear in Astrophys. J., Lett. Ed.

155.099 **A search for million degree gas in the galactic halo and the LMC through [Fe X] λ6375 absorption.**
M. Pettini, S. D'Odorico.
ESO Sci. Prepr., No. 441, 36 pp. (1986). To appear in Astrophys. J.

155.100 **Structure and dynamics of the galactic system. Report 1981 – 1984.**
R. Wielen.
Astron. Rechen–Inst. Heidelb., Mitt., Ser. A, Nr. 176, 7 + 127 pp. (1986).
Contents: Basic data and calibrations. Local galactic structure. Overall galactic structure. Kinematics. Dynamics. Galactic environment. Report of the Working Group on galactic constants. Resolution on galactic constants adopted by IAU Commission 33 on 21 November 1985.

155.101 **Additional evidence for the accuracy of galactic reddenings from Sc I galaxies and quasars.**
S. Haarala, P. Teerikorpi.
Rep. Ser., Dep. Phys. Sci., Univ. Turku, No. FTL–R86, 10 pp. (1985). = Turku Univ. Obs., Informo No. 97. Submitted to Astrophys. Space Sci.

155.102 **Rotation curve of the outer parts of our Galaxy from neutral hydrogen 21 cm line profiles.**
I. V. Petrovskaya, P. Teerikorpi.
Rep. Ser., Dep. Phys. Sci., Univ. Turku, No. FTL–R94, 13 pp. (1986). = Turku Univ. Obs., Informo No. 101. Submitted to Astron. Astrophys.

155.103 **The distribution of atomic hydrogen and diffuse gas in the Galaxy.**
C. Heiles.
Nuovo Cimento C, Vol. 8C, Ser. 1, No. 4, p. 329 – 344 (1985). Abstr. in Phys. Abstr., Vol. 89, No. 1276, Entry 36340 (1986).

155.104 **Effect of binarity on the local stellar mass density.**
M. Mezzetti, G. Giuricin, F. Mardirossian.
Nuovo Cimento C, Vol. 8C, Ser. 1, No. 4, p. 415 – 417 (1985). Abstr. in Phys. Abstr., Vol. 89, No. 1276, Entry 36343 (1986).

155.105 **The projected dark cloud distribution of the Milky Way.**
J. V. Feitzinger, J. A. Stüwe.
Astrophys. J., Vol. 305, No. 1, p. 534 – 545 (1986).
The dark cloud catalogs of Lynds for the northern and Feitzinger and Stüwe for the southern hemisphere are combined. By calibrating the opacity classes, the authors demonstrate that the above data base is complete and uniform. They deduce for the total Milky Way band and for the northern, anticenter, and southern sectors the obscuration percentages of the sky, average cloud areas, cloud distributions in Galactic longitude and latitude, and mean spatial distances. Using the stellar–statistical integral equation, the authors calculate the intrinsic dark cloud area distribution and absolute number densities, and compare the results with the size distribution of molecular clouds.

155.106 **Erratum: "The gamma–ray spectrum of the galactic center region" [Astrophys. J., Lett. Ed., Vol. 294, No. 1, p. L13 – L15 (1985)].**
G. R. Riegler, J. C. Ling, W. A. Mahoney, W. A. Wheaton, A. S. Jacobson.
Astrophys. J., Lett. Ed., Vol. 305, No. 1, p. L33 (1986). See Abstr. 40.155.005.

155.107 **Kinematics of the Galaxy from a magnitude–limited proper–motion sample.**
R. F. G. Wyse, G. Gilmore.
Astron. J., Vol. 91, No. 4, p. 855 – 869 (1986).
The authors have analyzed components of the *UVW* Galactic space motions resulting from the accurate proper–motion surveys of Chiu (1980), which contain several hundred stars complete to $V = 20.5$ in each of three fields. Two of these fields lie close to the Galactic center – Sun meridional plane, so that derived space motions are nearly independent of the unknown radial velocities. These data show a rotating, intermediate velocity dispersion component of the spheroid, in addition to the well–studied thin (old) disk and the extreme spheroid.

155.108 **High resolution imaging of the galactic center with an IRCCD.**
E. Tollestrup, R. W. Capps, E. E. Becklin.
Bull. Am. Astron. Soc., Vol. 18, No. 1, p. 640 (1986). Abstract. – See Abstr. 010.066.

155.109 **Observational constraints to the chemical evolution of the galactic halo from the Space Telescope.**
M. Busso, R. Gallino.
Mem. Soc. Astron. Ital., Vol. 56, No. 4, p. 699 – 707 (1985). – See Abstr. 012.054.
Results from numerical computations of the chemical enrichment of the galactic halo are shown and compared with presently available observations. A number of problems involved in the models of both stellar and galactic evolution are reviewed, and key observations, which will be possible with the various instruments of the Space Telescope, are discussed.

155.110 **Local mass density of halo stars.**
S.–G. Lee.
J. Korean Astron. Soc., Vol. 18, No. 2, p. 70 – 78 (1985).
From a kinematically unbiased sample of halo stars, the local mass density of halo dwarfs is estimated as $6.0 - 6.3 \times 10^{-4} m_{\odot}/pc^3$ by adopting a color–magnitude relation and a mass–luminosity relation. The derived halo mass density is not much different from the results of previous studies, which were derived from the kinematically biased sample of halo stars. Therefore it is confirmed that the local mass density of halo stars is far less than that required by Ostriker–Peebles to stabilize the galactic disk against barlike instabilities.

155.111 **The chemical evolution of the solar neighborhood: age–metallicity relation of F–stars.**
H. B. Ann, Y. H. Kang.
J. Korean Astron. Soc., Vol. 18, No. 2, p. 79 – 85 (1985).
From the uvby, Hβ photometry of intermediate population II F–stars in the catalogue of Olsen (1983), the authors derived age–metallicity relations, using Hejlesen's (1980) isochrone. The derived age–metallicity relations well coincide with the theoretical predictions by the unclosed two–zone model of Lee and Ann (1981).

155.112 **Very old stars of the Kapteyn group.**
R. B. Shatsova.
Nauchn. Inf., Vyp. 56, p. 104 – 108 (1984). In Russian.
The traces of the torus–like structure of the Kapteyn group are discovered. The files of stars with similar characteristics such as RR Lyrae–type variables with periods $0^{d}\!.5 - 0^{d}\!.65$, or F, G subdwarfs with almost equally small [Fe/H] are their elements. This group contains a great number of high–velocity clouds, judging by their location on the sky and by their radial velocity. It permits evolutionary interpretation.

155.113 **A kinematic study of K–type stars.**
K.–p. Tian, J.–l. Zhao.
Chin. Astron. Astrophys., Vol. 10, No. 2, p. 98 – 102 (1986). English translation of Acta Astrophys. Sin., Vol. 6, No. 1, p. 6 – 11 (1986).

Recently available radial velocity data of K–stars are used for an analysis of their kinematics in the solar neighbourhood. Solution for the mean solar motion is $u_0 = -9.1 \pm 2.2$, $v_0 = -20.4 \pm 2.6$, $w_0 = -5.2 \pm 2.5$ km/s, in fair agreement with previous determinations. Another solution using proper motions from the SAO Catalogue shows no significant differences. The mean solar motion found for these nearby K–stars appears to be consistent with the kinematics of the Gould's Belt.

155.114 **The largest molecular cloud complexes in the first galactic quadrant.**
T. M. Dame, B. G. Elmegreen, R. S. Cohen, P. Thaddeus.
Astrophys. J., Vol. 305, No. 2, p. 892 – 908 (1986). With plate 13.

The Columbia CO survey of the first galactic quadrant was used to determine the locations and physical properties of the largest molecular complexes in the inner Galaxy. Within the range of the survey (l = 12° – 60°), 26 complexes were detected with masses greater than $5 \times 10^5 M_\odot$, and roughly several hundred such complexes are deduced to exist throughout the Galaxy within the solar circle. These complexes are the parent objects of much of the Population I in the Galaxy. Distances to most of the complexes were determined kinematically. The largest complexes are good tracers of spiral structure, the Sagittarius arm in particular being delineated with unprecedented clarity.

155.115 **Velocity ellipses in the galactic plane: regional variations and spectral distributions.**
S. M. Freitas.
Rev. Mex. Astron. Astrofis., Vol. 12, Spec. Issue, p. 134 (1986). Abstract. – See Abstr. 012.066.

155.116 **Study of non–thermal emission from the galactic center.**
C. Tateyama, Z. Abraham.
Rev. Mex. Astron. Astrofis., Vol. 12, Spec. Issue, p. 301 (1986). Abstract. – See Abstr. 012.066.

155.117 **Classical cepheids and the local chemical inhomogeneities in the galactic disc.**
S. Giridhar.
J. Astrophys. Astron., Vol. 7, No. 2, p. 83 – 98 (1986).

Places of formation have been derived for a sample of 23 cepheids with well–determined atmospheric abundances in an attempt to study the chemical inhomogeneities in the local interstellar medium. The abundance data available for the sample cepheids is compiled. The author finds that the most conspicuous irregularity in the metallicity is exhibited by stars that are born in the local arm or in the interarm region. He proposes a scenario to explain these local variations in terms of supernova–induced star formation in interstellar gas enriched by massive stars formed in the density wave.

155.118 **Polarisierte Radiostrahlung der galaktischen Ebene bei 11 cm Wellenlänge.**
N. Junkes.
Mitt. Astron. Ges., No. 65, p. 224 – 225 (1986). – See Abstr. 012.072.

155.119 **Luminous blue stars: distribution and numbers.**
C. D. Garmany.
Luminous stars and associations in galaxies, p. 19 – 30 (1986). – See Abstr. 012.079. (IAU Symp. No. 116).

This talk deals with massive stars within 3 kpc of the Sun and in the LMC and SMC. Observations required to resolve these stars and the calibrations used to construct an H–R diagram are discussed.

155.120 **The WR/OB number ratio within 2.5 kpc from the sun.**
D. Vanbeveren.
Luminous stars and associations in galaxies, p. 237 – 238 (1986). – See Abstr. 012.079. (IAU Symp. No. 116).

Comparing the observed and theoretical HR diagram for massive stars, it is concluded that at least 34% of the stars within 2.5 kpc from the sun with initial ZAMS mass larger than 35 $M_\odot$ have not yet been detected. It is argued that the number of core helium burning pop I WR stars within 2.5 kpc from the sun may be as low as 26. If the progenitors of WR stars have initial ZAMS masses larger than 35 $M_\odot$, it is then concluded that the observed number ratio (core helium burning pop I WR stars)/(WR progenitors) within 2.5 kpc from the sun ranges between 0.05 and 0.09.

155.121 **The initial mass function of massive stars in galaxies: empirical evidence.**
J. M. Scalo.
Luminous stars and associations in galaxies, p. 451 – 466 (1986). – See Abstr. 012.079. (IAU Symp. No. 116).

Observational constraints on the form of the high–mass stellar IMF are reviewed. The evidence includes star counts in the solar neighborhood, individual and composite star clusters, and nearby galaxies, and arguments based on integrated light and chemical evolution modeling. There is no convincing evidence for any systematic variations of the shape of the high–mass IMF. However, the various determinations are very uncertain, and do not allow any firm estimate of the logarithmic slope of the upper IMF; the appropriate value is somewhere between –1.3 and –2.3, with region–to–region variations smaller than about ± 0.5.

155.122 **The galactic distribution of stellar sources found by IRAS (with some emphasis on the bulge of the Galaxy).**
H. J. Habing.
Light on dark matter, p. 329 – 338 (1986). – See Abstr. 012.080.

Most of the point sources detected by IRAS at its shortest wavelength, 12 μm, are circumstellar shells around late type stars. Maps of the galactic distribution of these sources for different flux density intervals and different 25 to 12 μm colours show that sources are being found several kiloparsecs deep into the interior of the Galaxy. The faintest and reddest sources show the bulge of the Galaxy. A large fraction of the sources are variable.

155.123 **Variables, the galactic bulge and IRAS.**
M. W. Feast.
Light on dark matter, p. 339 – 348 (1986). – See Abstr. 012.080.

Evidence from the Baade windows suggests that the majority of IRAS sources which delineate the bulge are Mira variables (including probably their long period extension, the OH/IR variables). No observational differences, at a given period, have yet been firmly established between bulge Miras and local or globular cluster Miras.

155.124 **Ground based observations of nuclear bulge stars.**
J. A. Frogel.
Light on dark matter, p. 349 – 356 (1986). – See Abstr. 012.080.

Infrared photometry and CCD spectroscopy have been obtained for about 400 M giants at several latitudes in the Galactic nuclear bulge. Their colors and luminosities differ significantly from M giants in the solar neighborhood which are used in stellar synthesis models for ellipticals and the bulges of spirals. Initial examination of the IRAS "disk and bulge" stars indicates that rather than representing a new class of objects, the disk stars, at least, are quite similar to and extend to redder colors the latest and most luminous group of spectroscopically identified M giants.

New Hα emission stars found above 10° galactic latitude.
See Abstr. 002.011.

How no(w) brown dwarfs?
See Abstr. 011.008.

Statistical analysis of proper motion surveys.
See Abstr. 036.213.

The search for unwanted signals: systematic trends in proper motions of fundamental stars.
See Abstr. 043.018.

Effects of metal–dependent stellar models on the yield of nitrogen.
See Abstr. 061.027.

Some considerations on the origin of nitrogen.
See Abstr. 061.067.

Constraints to the radiative lifetime of a light neutral fermion in the galactic halo from IUE.
See Abstr. 061.079.

Detection rates for "invisible"–axion searches.
See Abstr. 061.094.

Can scalar neutrinos or masive Dirac neutrinos be the missing mass?
See Abstr. 061.109.

Direct–annihilation fraction of positrons in a neutral low–density galactic environment.
See Abstr. 061.119.

The effect of the new $^{12}C(\alpha,\gamma)^{16}O$ rate on the chemical evolution of the solar neighborhood.
See Abstr. 061.121.

Magnesium isotopes and galactic evolution.
See Abstr. 061.129.

Certainties and uncertainties in long–lived chronometers.
See Abstr. 061.131.

More about nucleocosmochronology: the reliability of the long–lived chronometers, and the production of extinct radioactivities.
See Abstr. 061.132.

Chronometer studies with initial galactic enrichment.
See Abstr. 061.133.

Observational evidence of element formation processes in stars.
See Abstr. 061.146.

Intrinsic parameters of hot blue stars.
See Abstr. 064.071.

Type I SNe from binary systems: consequences on galactic chemical evolution.
See Abstr. 065.145.

^{26}Al produced by Wolf–Rayet stars and the 1.8 MeV line emission of the Galaxy.
See Abstr. 065.146.

Gravitational microlensing by the galactic halo.
See Abstr. 066.136.

Episodic accretion onto massive black holes.
See Abstr. 067.022.

Supermassive stars in galactic nuclei.
See Abstr. 067.023.

An upper limit on the mean mass of black holes in the galactic halo.
See Abstr. 067.106.

Tidal gravitational forces: the infall of "new" comets and comet showers.
See Abstr. 102.041.

The influence of the Galactic tidal field on the Oort comet cloud.
See Abstr. 102.042.

Radial velocities of blue horizontal branch field stars in the outer Galactic halo.
See Abstr. 111.004.

Stellar motions in the solar neighborhood.
See Abstr. 111.009.

A photometric study of the bright cloud B in Sagittarius: V. 185 new proper motion stars.
See Abstr. 111.012.

Photometric parallaxes for selected stars of color class m from the NLTT catalog. II. The declination zone 0° to +20°.
See Abstr. 111.014.

New subdwarfs. V. Radial velocities for 889 high–proper–motion stars measured with the Mount Wilson 100 inch reflector.
See Abstr. 111.016.

AGB stars with high mass loss rates in the bulge of our Galaxy.
See Abstr. 112.142.

CCD photometry to $V = 21$ in a Puppis field.
See Abstr. 113.048.

New subdwarfs. IV. UBV photometry of 1690 high–proper–motion stars.
See Abstr. 113.053.

Low–resolution IUE observations of four unreddened, 11th magnitude B stars at high galactic latitudes.
See Abstr. 114.017.

Observations of early–type stars at high galactic latitudes.
See Abstr. 114.018.

A search for the tritium hyperfine line from nearby stars.
See Abstr. 114.050.

The red giant metallicity function.
See Abstr. 114.093.

A detailed kinematic and abundance analysis of old disk giants.
See Abstr. 114.115.

"Chemical structure" in the red giant Hertzsprung–Russell diagram, and the age of the galactic disk.
See Abstr. 115.017.

LHS 292 and the luminosity function of the nearby M dwarfs.
See Abstr. 115.023.

Parameters of Wolf–Rayet stars.
See Abstr. 115.031.

An unbiased X–ray sampling of stars within 25 parsecs of the Sun.
See Abstr. 116.033.

X–ray sources of population II: moderate mass black holes accreting gas from extended low density envelopes.
See Abstr. 117.182.

CCD observations of some RR Lyrae stars and a standard sequence in Baade's window.
See Abstr. 122.076.

Absolute magnitudes and kinematic properties of RR Lyrae stars.
See Abstr. 122.078.

A new search for RR Lyrae variables in Baade's window.
See Abstr. 122.106.

Space distribution of pulsars in the Galaxy and the nature of gamma–sources.
See Abstr. 126.024.

Fast rotating pulsars.
See Abstr. 126.052.

An analysis of pulsar timing irregularities: is it induced by a source of gravitational waves in the Galaxy?
See Abstr. 126.068.

Galaxy and star counts in Taurus and Perseus.
See Abstr. 131.008.

Is IRAS cirrus cloud emission largely fine–structure radiation?
See Abstr. 131.017.

Principal heating sources of dust in the galactic disk.
See Abstr. 131.056.

Cyanoacetylene in the Sgr A molecular clouds.
See Abstr. 131.064.

Star formation in a galactic context: the location and properties of molecular clouds.
See Abstr. 131.074.

The extent of the local H I halo.
See Abstr. 131.103.

High–latitude molecular clouds: completeness of the local sample and implications for molecular surveys.
See Abstr. 131.104.

Molecular clouds and star formation in the inner Galaxy: a comparison of CO, H II, and far–infrared surveys.
See Abstr. 131.105.

The photoionization of the diffuse Galactic gas.
See Abstr. 131.106.

The structure of galactic H I in directions of low total column density.
See Abstr. 131.137.

A theoretical steady state model for the distribution of gas and magnetic field in M31 and in our Galaxy.
See Abstr. 131.138.

The Columbia CO survey of molecular clouds in the first Galactic quadrant.
See Abstr. 131.141.

An empirical study of ring–shaped nebulae in the Galaxy and in the LMC.
See Abstr. 131.145.

λ1300 μm dust emission from giant molecular clouds close to the galactic center.
See Abstr. 131.148.

Interstellar molecular clouds.
See Abstr. 131.149.

HEAO 3 search for interstellar ^{44}Ti.
See Abstr. 131.162.

Evidence for a high–density electron layer from the spatial distribution of pulsars.
See Abstr. 131.183.

Molecular clouds in Orion and Monoceros.
See Abstr. 131.191.

Implications on X–ray and UV radiation on interstellar medium.
See Abstr. 131.200.

Supernova remnants associated with an H I "supershell" in the Perseus spiral arm.
See Abstr. 131.219.

The reddening at the North Galactic Pole.
See Abstr. 131.303.

Comparative analyses of ^{13}CO with ^{12}CO and H I emission in the galactic plane.
See Abstr. 131.313.

Ammonia in the neutral ring around the galactic center.
See Abstr. 131.315.

Distribution of Mg II in the local interstellar medium towards eight cool giants.
See Abstr. 131.316.

UV observations of galactic halo gas kinematics.
See Abstr. 131.325.

The local interstellar medium over 5 pc in the direction of α Aquilae.
See Abstr. 131.326.

Interstellar absorption lines in the directions of extragalactic objects – II. Analysis of 25 sight lines.
See Abstr. 131.331.

Luminous stellar content of the Galaxy: inferences from radio and infrared data.
See Abstr. 131.357.

Three principal heating sources of dust in the galactic disk.
See Abstr. 131.366.

Mid–IR emission of the interstellar medium.
See Abstr. 131.367.

The galactic morphology of the interstellar dust detected by IRAS.
See Abstr. 131.388.

Dark cloud statistics.
See Abstr. 131.390.

High latitude molecular clouds: completeness of the survey and implications for molecular surveys.
See Abstr. 131.391.

Three micron spectroscopy of the galactic centre source IRS 7.
See Abstr. 133.001.

The kinematics of OH/IR maser sources.
See Abstr. 133.016.

Progress of the survey at 2.2 μm on southern infrared sources near the galactic plane.
See Abstr. 133.017.

Distances to planetary nebulae. II. H I absorption observations.
See Abstr. 134.007.

Faint galactic radio sources and the spiral structure in the solar neighbourhood.
See Abstr. 141.009.

Radio source surveys of the Milky Way and M33.
See Abstr. 141.013.

Radio patrol of the northern Milky Way.
See Abstr. 141.014.

Search for new X–ray pulsars in the galactic plane by the HEAO–1 Large Area Sky Survey experiment.
See Abstr. 142.061.

Iron line from compact X–ray sources.
See Abstr. 142.075.

***Einstein* observations of the soft X–ray background close to the Galactic plane.**
See Abstr. 142.090.

Propagation of injected cosmic rays under distributed reacceleration.
See Abstr. 144.002.

High–energy and ultra–high–energy cosmic rays.
See Abstr. 144.011.

New evidence on the galactic cosmic ray distribution from COS–B gamma–ray data.
See Abstr. 144.019.

Evolution of galaxies: one–zone model with a birth–rate stellar function depending on gas density and temperature.
See Abstr. 151.013.

Velocity dispersion of stars and giant molecular clouds.
See Abstr. 151.036.

Formation of giant molecular clouds by coagulation of small clouds and spiral structure.
See Abstr. 151.037.

Strong nonlinear waves in gaseous disks. I. Stationary periodic waves.
See Abstr. 151.041.

The fate of wide binaries in the solar neighborhood: encounters with stars and molecular clouds.
See Abstr. 151.050.

High–precision radial velocities for faint giants: radial velocities of extreme halo systems and the mass of the Galaxy.
See Abstr. 154.019.

On the formation of bound open clusters in the Galaxy.
See Abstr. 154.021.

Galactic globular clusters as star systems and satellites of the Galaxy.
See Abstr. 154.044.

High–precision radial velocities for faint giants. Radial velocities of extreme halo systems and the mass of the Galaxy.
See Abstr. 154.046.

Systematic reinvestigation of the radial velocities of the galactic globular clusters: image–tube results.
See Abstr. 154.057.

ST to the discovery of the galactic halo.
See Abstr. 154.060.

Bimodal star formation and remnant–dominated galactic models.
See Abstr. 157.010.

Large–scale configuration of the magnetic field in spiral galaxies.
See Abstr. 157.017.

The origin of nitrogen and the chemical evolution of spiral galaxies.
See Abstr. 157.047.

Warped galactic disks and optical projection effects.
See Abstr. 157.048.

The stellar populations of galaxies in the Local Group.
See Abstr. 157.088.

Nitrogen and oxygen evolution in nearby spiral galaxies.
See Abstr. 157.106.

A possible relationship between ring magnetism and H I gas in spiral galaxies.
See Abstr. 157.222.

Investigation of the existence of large–scale anisotropy in the metagalactic space.
See Abstr. 161.084.

Cosmic ray antimatter from supersymmetric dark matter.
See Abstr. 161.300.

156 Magellanic Clouds

156.001 **Star formation in the Wing of the Small Magellanic Cloud.**
M. Pierre, M. Viton, J. P. Sivan, G. Courtès.
Astron. Astrophys., Vol. 154, No. 1/2, p. 249 – 254 (1986).

The very–wide–field photographic survey performed during the first Spacelab mission has yielded images at 1680 and 1950 Å of the Wing of the Small Magellanic Cloud, a region which has been known for a long time to contain many blue stars. On the basis of *UBV* photometry, the authors determine the HR diagram of the region and derive the corresponding luminosity function and initial mass function. Theoretical ultraviolet fluxes calculated from this stellar population agree well with the observed fluxes at 1680 and 1950 Å. A good agreement is also found between predicted and observed Hα fluxes. The authors find a relatively flat initial mass function ($x = 1$) and the value of 30 $M_\odot$ for the upper mass limit.

156.002 **The geometry and distance of the Magellanic Clouds from Cepheid variables.**
J. A. R. Caldwell, I. M. Coulson.
Mon. Not. R. Astron. Soc., Vol. 218, No. 2, p. 223 – 246 (1986).

The present work uses the photoelectric results assembled for 73 LMC and 63 SMC Cepheids to obtain solutions for the PL, PC and PLC relations in the Clouds, and the Clouds' geometry and distance.

156.003 **Dynamics of the Magellanic system and the Galaxy – present status of theoretical understanding.**
T. Murai, M. Fujimoto.
Astrophys. Space Sci., Vol. 119, No. 1, p. 169 – 171 (1986). – See Abstr. 012.001.

In an attempt to interpret the Magellanic Stream (Mathewson et al., 1974), a new tidal simulation for the triple system, the Galaxy, the Large Magellanic Cloud (LMC), and the Small Magellanic Cloud (SMC), has been performed to represent the past 10^{10}yr, on relaxing a simplifying assumption of the bound state of the SMC with respect to the LMC for the whole 10^{10}yr. The parameters for the SMC orbit at the present epoch are varied around those of the best fit orbit of a previous simulation (Murai and Fujimoto, 1980).

156.004 **Colour–magnitude diagrams of star clusters in the Magellanic Clouds from wide–field electronography. II. NGC 2210.**
J. Andersen, A. Blecha, M. F. Walker.
Astron. Astrophys., Suppl. Ser., Vol. 64, No. 1, p. 189 – 202 (1986). = Lick Obs. Bull., No. 1026.
The authors report photometric observations in B and V to $V = 23.1$ for 235 stars in an annular field $39'' \leqslant R \leqslant 60''$ centered on the LMC cluster NGC 2210. Corrected for the effects of non–member stars, the colour–magnitude diagram (CMD) of NGC 2210 is found to closely resemble the diagram for the old, very metal–poor galactic globular cluster M92. The detailed CMD for the cluster then confirms the assignment of NGC 2210 by Searle et al. (1980) to the oldest, most metal–poor population in the LMC (SWB class VII), based on its integrated colours. From the fit of the CMD to those of M92 and M3, the authors find for NGC 2210 a distance modulus of $(m\text{–}M)_0 = 17.9 - 18.4$ under various assumptions concerning the reddening and metal abundance of the cluster and the absolute magnitudes of its HB stars.

156.005 **Extinction and reddening of H II regions in the Large Magellanic Cloud.**
J. Caplan, L. Deharveng.
Astron. Astrophys., Vol. 155, No. 2, p. 297 – 313 (1986).
Absolute Hα and Hβ fluxes of H II regions in the Large Magellanic Cloud are used in conjunction with radio continuum, 21–cm line, and stellar *UBV* data to study the dust associated with these regions. The authors examine various hypotheses concerning the optical properties and the spatial distribution of the dust, and then confront predicted characteristics with observations. Uniform interstellar extinction cannot explain the data. They suggest the following: some of the extinction is due to clumped dust well outside of the emission zone, and the rest is caused by dust located closer to the ionized gas where scattering effects are important. This latter dust probably lies mostly just outside the boundary of the H II region, but some of the dust may be internal.

156.006 **A photometric determination of the metal content for F – G type supergiants in the Large Magellanic Cloud.**
A. M. van Genderen, W. van Driel, H. Greidanus.
Astron. Astrophys., Vol. 155, No. 1, p. 72 – 78 (1986).
VBLUW photometry of roughly 100 F– and G–type supergiants in the Galaxy and the LMC is used to derive the metal content of the LMC stars. For this purpose the empirical Cepheid locus of Pel (1978) in the *V–B/B–L* diagram has been used in conjunction with the relative dependence of theoretical colours on the metal abundance. The result is that the LMC stars show a metal deficiency of ~1.5 compared to the stars in the solar neighbourhood. The reddenings for most of the galactic supergiants could be derived.

156.007 **Ages and metallicities of LMC and SMC red clusters through Hβ and G band photometry.**
E. Bica, H. Dottori, M. Pastoriza.
Astron. Astrophys., Vol. 156, No. 1/2, p. 261 – 267 (1986).
The authors present narrow band integrated photometry of the Hβ and G band absorption features for 41 LMC and 10 SMC red star clusters. An age–metallicity calibration is provided for the color–color diagram. SWB types are derived between IV and VII for 23 unclassified clusters and their distribution in the ages vs metallicity plane is discussed. The authors study the chemical evolution of the Magellanic Clouds: the LMC presents a steeper chemical enrichment slope. An intrinsic metallicity dispersion is found in the LMC chemical evolution, indicating that the gas has been inhomogeneous at any time, prevailing a local enrichment over a global one. One zone models describe the evolution of both Clouds, being the efficiency of star cluster formation larger in the LMC. The LMC presents a burst of star cluster formation at $t = 4.5 \times 10^9$yr. The authors also present new *B–V* data for fainter SMC clusters, providing an essentially complete color histogram for clusters with globular cluster appearance.

156.008 **A distance modulus of the LMC from *uvby* β photometry of B supergiants.**
R. R. Shobbrook.
Mon. Not. R. Astron. Soc., Vol. 219, No. 3, p. 495 – 504 (1986).
Strömgren four–colour and Hβ photometry has been obtained for 44 B supergiants in the Large Magellanic Cloud. As expected, nearly one–half of them have β indices which suggest emission at Hβ. From 24 stars which appear to be free from this emission, the $\beta/c_0/M_V$ calibration of Balona & Shobbrook (1984), applied with a correction to the β index for the mean radial velocity of the LMC stars, indicates a distance modulus of 18.8 ± 0.3.

156.009 **Star formation in the Magellanic Cloud. III. IR observations of giant H II regions.**
T. J. Jones, A. R. Hyland, S. Straw, P. M. Harvey,
B. A. Wilking, M. Joy, I. Gatley, J. A. Thomas.
Mon. Not. R. Astron. Soc., Vol. 219, No. 3, p. 603 – 627 (1986).
The authors present the results of far–infrared and near–infrared mapping of several H II regions in the LMC and SMC. They also present the results of a continuing search for protostars using NIR techniques. The FIR map of N 159 shows strong 100 and 50 μm emission over the entire 2×4 arcmin^2 area seen in Hα, with two major peaks closely coinciding with the 843 MHz radio map. A second new IR point source showing no evidence of ionizing radiation (protostar) has been found in N 159. N 160A is nearly as luminous as N 159, but is far more compact. All four LMC H II regions studied in the FIR (N 159, N 160A, N 59A, N 158) produce more ionizing radiation than expected for their luminosity. The LMC H II regions do not have the high surface brightness cores seen in galactic H II regions such as W49, but on the other hand are more compact than very diffuse galactic H II regions such as DR23.

156.010 **A new radio continuum survey of the Magellanic Clouds at 1.4 GHz. I. Observations and data analysis.**
R. F. Haynes, U. Klein, R. Wielebinski, J. D. Murray.
Astron. Astrophys., Vol. 159, No. 1/2, p. 22 – 32 (1986).
New radio continuum surveys of the Large and Small Magellanic Cloud at 1.4 GHz have been made with the Parkes 64–m radio telescope with a beamwidth of 15:0 and an rms noise of ~30 mJy per beam area in the final maps. This paper describes the observations and presents the results. A brief discussion of their radio continuum morphology as well as lists of sources contained in both fields are also given.

156.011 **Distribution of spectral types of stars in the SMC clusters.**
E. Kontizas, M. Kontizas, A. Dapergolas.
Astron. Astrophys., Vol. 159, No. 1/2, p. 305 – 308 (1986).
Spectral classification of stars brighter than $B \sim 18.5$ mag in 15 SMC clusters was carried out using U.K. Schmidt Telescope objective prism plates. The studied clusters represent all ages of clusters (disk, intermediate and halo) and are located at various areas of the parent galaxy. All classified stars are situated within the tidal radius of each cluster. For comparison, stars in adjoining fields of each cluster were also studied.

156.012 **The structure of the Small Magellanic Cloud.**
D. S. Mathewson, V. L. Ford, N. Visvanathan.
Astrophys. J., Vol. 301, No. 2, p. 664 – 674 (1986).
B, *V*, and near–infrared photometry was done for 161 Cepheids in the SMC to derive their relative distances to an accuracy of ±4%. The Cepheids extend from ~43 to 75 kpc with a maximum concentration at 59 kpc. The line–of–sight distribution of the younger Cepheids with periods greater than 10 days (for which there is almost a complete sample) splits into two components, each of depth of about 6 kpc and with centers 12 kpc apart. Recent radial velocity mesurements of stars and their interstellar Ca II absorption lines show convincingly that the near and far components should be identified with the low– and high–velocity entities in the SMC H I distribution respectively. The results are in good agreement with the "tidal" model of Murai and Fujimoto, which has the SMC colliding with the LMC 2×10^8yr ago.

156.013 **Identification of new carbon star candidates in SMC globular cluster NGC 419.**
M. Azzopardi, B. Dumoulin, J. Quebatte, E. Rebeirot.
Messenger, No. 43, p. 12 – 15 (1986).

156.014 **Age and processes of star formation in very young LMC associations.**
I. Tarrab.
Spectral evolution of galaxies, p. 75 – 80 (1986). – See Abstr. 012.015.

Theoretical time evolution models of the UV–flux, the number of Lyman continuum photons, the effective temperature of the exciting stars and of the equivalent width of Hβ have been computed for three star formation rates (burst, constant and decreasing rate), considering that they should lead to differences in the observed global parameters. This method was applied to some OB–associations in the Large Magellanic Cloud, 30 Dor and the neighboring regions. Coherent results were obtained for a burst of star formation and for a decreasing rate. A constant star formation is not consistent with observational data.

156.015 **The properties of OB supergiants in the Large Magellanic Cloud.**
E. L. Fitzpatrick.
Bull. Am. Astron. Soc., Vol. 17, No. 4, p. 881 (1985). Abstract. – See Abstr. 010.062.

156.016 **Ages of two LMC clusters and an LMC distance determination using features of an LMC field CMD.**
J. H. Jones.
Bull. Am. Astron. Soc., Vol. 17, No. 4, p. 883 (1985). Abstract. – See Abstr. 010.062.

156.017 **An X–ray image of the Large Magellanic Cloud.**
D. J. Helfand, T. T. Hamilton.
Bull. Am. Astron. Soc., Vol. 17, No. 4, p. 893 (1985). Abstract. – See Abstr. 010.062.

156.018 **Soft X–ray maps of the Large Magellanic Cloud.**
K. P. Singh, J. A. Nousek, D. N. Burrows, G. P. Garmire.
ESA Spec. Publ., ESA SP–239, p. 131 – 136 (1985). – See Abstr. 012.019.

The authors present soft (0.18 – 2.8 keV) X–ray maps of the Large Magellanic Cloud (LMC) obtained from nine months of scanning observations with the HEAO–1 low energy detectors. The 1/4 keV X–ray map, with a total exposure of 1.25×10^5s, is the most sensitive survey of the LMC yet reported. A detailed comparison of the 1/4 keV X–ray observations with the neutral hydrogen column densities in the LMC obtained from a new 21 cm line survey is performed.

156.019 **Carbon monoxide in the Magellanic Clouds.**
F. P. Israel, T. de Graauw, H. van de Stadt, C. P. de Vries.
Astrophys. J., Vol. 303, No. 1, p. 186 – 197 (1986).

The authors have detected ^{12}CO (2–1) emission at 11 of 22 positions in the LMC and, for the first time, at 6 of 16 positions in the SMC. All but one of the presently known Magellanic Cloud masers are associated with CO clouds; CO emission was also detected in the direction of several dark clouds and bright H II regions. CO emission from Magellanic Cloud objects tends to be weaker than that from Galactic objects; in addition, it is less widespread in the SMC than in either the Galaxy or the LMC. The authors suggest that the combination of relatively low C and O abundances, relatively low gas–to–dust ratios, and relatively strong mean UV radiation fields in the Magellanic Clouds is the principal cause of these differences.

156.020 **Gas in and toward the Magellanic Clouds.**
A. Songaila, J. C. Blades, E. M. Hu, L. L. Cowie.
Astrophys. J., Vol. 303, No. 1, p. 198 – 215 (1986).

The authors present Ca II K absorption spectra, at resolutions of 8 – 17 km s^{-1}, of 48 early–type supergiants in the Large and Small Magellanic Clouds. These data are used to study the possible fragmentary nature of the Clouds (particularly the SMC); and the distance to the gas responsible for the absorption lines at 60 and 120 km s^{-1} (LSR) seen in UV spectra of Magellanic Cloud stars. In both Clouds, velocity components in the neutral hydrogen distribution lie at different distances along the line of sight. There is also lower column density material, similar to the 60 and 120 km s^{-1} features, which can be demonstrated to lie within the Clouds. Good agreement is found between the spread in radial velocity shown in absorbing material and the radial velocity spread predicted by Murai and Fujimoto's dynamical model of the formation of the Magellanic Stream by tidal stripping.

156.021 **CCD photometry of Large Magellanic Cloud clusters. II. The intermediate–age cluster H4.**
M. Mateo, P. Hodge.
Astrophys. J., Suppl. Ser., Vol. 60, No. 4, p. 893 – 937 (1986). With plates 57 – 60.

UBVR CCD frames of the LMC star cluster H4 have resulted in V, $(B–V)$, and/or $(V–R)$ photometry of 3096 stars and $(U–B)$ colors of 310 stars. It is concluded that (1) $(m–M)_0 = 18.1 \pm 0.3$ on the basis of comparisons with NGC 2420 *and* the theoretical isochrones of VandenBerg; (2) $E(B–V) = 0.07 \pm 0.03$; (3) [Fe/H] = -0.7 ± 0.3; (4) the age of H4 is $(2.5 \pm 0.4)\times 10^9$yr using the present modulus, or $(1.5 \pm 0.3)\times 10^9$yr for $(m–M)_{0,LMC} = 18.7$; (5) the cluster seems to be dynamically relaxed; and (6) the field stars around H4 range in age from $\lesssim 0.6\times 10^9$ to $\gtrsim 5\times 10^9$yr.

156.022 **The cool–giants populations in the Magellanic Clouds.**
V. M. Blanco, M. F. McCarthy.
Proceedings of the Second Asian–Pacific Regional Meeting on Astronomy, p. 402 – 410 (1984). – See Abstr. 012.031.

The authors discuss the observational techniques used to survey cool giants in the Magellanic Clouds and how these techniques have given data for complete sample of cool giants at reliably known distances. The authors review the findings and summarize their relevance in the evolutionary theory of cool giants. For simplicity they limit their discussion to the more common and easily detected cool giants.

156.023 **The stellar population of the Small Magellanic Cloud near NGC 121. I. The mean metallicity, metallicity spread, and radial velocity of SMC halo giants.**
N. B. Suntzeff, E. Friel, A. Klemola, R. P. Kraft, J. A. Graham.
Astron. J., Vol. 91, No. 2, p. 275 – 289 (1986). With plate 33. = Lick Obs. Bull., No. 1029.

The authors examine the giant stars of the SMC halo in the vicinity of the SMC cluster NGC 121, and report on the chemical composition and radial velocity of these objects.

156.024 **A drag dominated model of the Magellanic stream.**
G. R. Meurer, G. V. Bicknell, R. A. Gingold.
Proc. Astron. Soc. Aust., Vol. 6, No. 2, p. 195 – 198 (1985).

The authors present here the best of a series of models of the Magellanic stream. The dominant force in these models is gas drag. Gaseous cloudlets are torn from the bridge between the Large and Small Magellanic Clouds as the Magellanic system passes through a hot gaseous halo about our galaxy. The cloudlets are then stretched apart from each other by tidal and drag forces to form the Magellanic stream. The best model closely reproduces the position of the stream on the sky and the run of radial velocities along the Magellanic stream. The agreement is almost as good as the best purely tidal model.

156.025 **Gas in and towards the Magellanic Clouds.**
A. Songaila, J. C. Blades, E. M. Hu, L. L. Cowie.
Space Telesc. Sci. Inst., Prepr. Ser., No. 81, 44 pp. (1985). To appear in Astrophys. J.

156.026 **Evolutionary population synthesis and the Magellanic Cloud clusters.**
P. Battinelli, R. Capuzzo–Dolcetta.
Mem. Soc. Astron. Ital., Vol. 56, No. 4, p. 789 – 790 (1985). – See Abstr. 010.821.
The authors have computed synthetic models in a large interval of ages (10^7a $\lesssim t \lesssim 10^{10}$a) and for two values of the metallicity ($Z = 10^{-3}$, $Z = 0.02$), with the aim of reproducing the photometric properties of the stellar clusters of the Magellanic Clouds.

156.027 **Estudio interferometrico de la Nube Menor de Magallanes.**
G. Torres, G. J. Carranza.
Rev. Mex. Astron. Astrofis., Vol. 12, Spec. Issue, p. 114 – 118 (1986). – See Abstr. 012.066.
Hα interferometric observations of the general radial velocity field in the Small Magellanic Cloud are being carried out. The authors present preliminary results in reasonable agreement with H I measurements.

156.028 **The 2.6 mm CO survey of the Magellanic Clouds.**
M. Rubio, R. Cohen, J. Montani.
Rev. Mex. Astron. Astrofis., Vol. 12, Spec. Issue, p. 303 – 307 (1986). – See Abstr. 012.066.
CO emission in the LMC appears associated with other population I tracers. Preliminary analysis of four CO complexes in the LMC indicates that these complexes are correlated in velocity and position with the highest density of neutral hydrogen. In the SMC, CO has been detected only in the southwest part of the bar and correlates with the lower velocity component of the H I gas. In this area the highest density of neutral hydrogen is observed. Estimations of the molecular mass indicate that the ratio of molecular hydrogen to atomic hydrogen is less than 1 in the Magellanic Clouds.

156.029 **Analyse der Autokorrelations– und Strukturfunktion in der Großen Magellanschen Wolke.**
J. Spicker, J. V. Feitzinger.
Mitt. Astron. Ges., No. 65, p. 226 (1986). Abstract. – See Abstr. 012.072.

156.030 **Age determination of Magellanic Clouds and isolated extragalactic H II regions.**
M. V. Copetti, H. Dottori, M. Pastoriza.
Star–forming dwarf galaxies and related objects, p. 243 – 249 (1986). – See Abstr. 012.074.
H II region evolution models were used to obtain information on star formation in the Magellanic Clouds and in isolated extragalactic H II regions. The main results are: (1) IMFs with x = 3.0 and 2.5 are inconsistent with a large number of H II regions; (2) IMFs with x = 2 give a more uniform age distribution for the isolated extragalactic H II regions; (3) age estimation indicates a burst of star formation about $5.5 \pm 1.0 \times 10^6$ years ago in the LMC and about $2.3 \pm 0.9 \times 10^6$ years ago for the SMC.

156.031 **Spectroscopic binaries in the LMC.**
V. S. Niemela.
Luminous stars and associations in galaxies, p. 85 – 86 (1986). – See Abstr. 012.079. (IAU Symp. No. 116).
Preliminary results are presented of an observing programme aimed to obtain estimates of the stellar masses from studies of spectroscopic binary systems in the Large Magellanic Cloud. These are the first steps with the final purpose to determine an empirical mass–luminosity relation in a galaxy other than our own.

156.032 **Star formation of star clusters in the SMC and their adjoining fields.**
M. Kontizas, E. Kontizas.
Luminous stars and associations in galaxies, p. 101 – 102 (1986). – See Abstr. 012.079. (IAU Symp. No. 116).
Photometric and recent spectroscopic studies of the SMC have shown that the differences observed in the SMC clusters and those of our Galaxy could be attributed to differences in metallicity, star formation rate and/or the Initial Mass Function (Humphries, 1983). The studied clusters NGC 152 and KRON 3 are located at the west side of the bar of the SMC and their adjoining fields represent the halo population of this galaxy.

156.033 **A new survey for Hα–emission–line stars in the Small Magellanic Cloud.**
M. Azzopardi, N. Meyssonnier.
Luminous stars and associations in galaxies, p. 225 – 226 (1986). – See Abstr. 012.079. (IAU Symp. No. 116).
The authors' SMC survey – which is in an early stage – has resulted in the visual identification of about 3000 Hα–emission–line objects, quintupling the number of these objects found during the previous most complete survey (593 objects) by Lindsay (1961). Work is in progress.

156.034 **A large sample of emission–line stars in the Large Magellanic Cloud: their location in the Hertzsprung–Russell Diagram.**
B. Bohannan.
Luminous stars and associations in galaxies, p. 227 – 228 (1986). – See Abstr. 012.079. (IAU Symp. No. 116).
In this investigation, mass–losing stars in a galaxy are studied through a sample of 59 H alpha emission–line stars in the Large Magellanic Cloud. The most luminous, most massive stars in the LMC, like their counterparts in the Galaxy, are losing mass at a rate that significantly alters their spectroscopic appearance and that affects their evolution.

156.035 **CCD observations of young stellar associations and multiple systems in the Magellanic Clouds.**
M. C. Lortet, M. Heydari–Malayeri, G. Testor.
Luminous stars and associations in galaxies, p. 401 – 402 (1986). – See Abstr. 012.079. (IAU Symp. No. 116).
The nebular and stellar observations were carried out at the Cassegrain focus of the Danish 1.50 m telescope at the ESO La Silla Observatory in August and September 1984, using a RCA CCD (512 × 320 pixels) with a pixel size of 0".47. The frames were reduced using the VAX/MIDAS software at ESO Garching and Meudon Observatory (INVENTORY automatic program).

156.036 **Two faint WC stars near 30 Doradus.**
D. H. Morgan, A. R. Good.
Luminous stars and associations in galaxies, p. 403 – 404 (1986). – See Abstr. 012.079. (IAU Symp. No. 116).
The faintest WC star in Breysacher's (1981) catalogue is Br74 which has $m_v = 15.6$. The authors have now detected two faint WC stars at $m_v \sim 18.6$, three magnitudes fainter than Br74.

156.037 **Bright stars in the SMC clusters.**
E. Kontizas, A. Dapergolas, M. Kontizas.
Luminous stars and associations in galaxies, p. 405 – 406 (1986). – See Abstr. 012.079. (IAU Symp. No. 116).
The bright stars for 15 SMC clusters were classified in oder to derive the distribution of various spectral types. The studied clusters represent all evolutionary ages (disk, intermediate and halo) and are located at various places of the parent galaxy.

156.038 **Tidal radii and masses of the clusters in the LMC.**
M. Kontizas, M. Chrysovergis, E. Kontizas, D. Hadjidimitriou.
Luminous stars and associations in galaxies, p. 407 – 408 (1986). – See Abstr. 012.079. (IAU Symp. No. 116).
Masses and tidal radii of star clusters in the LMC have been derived by means of star counts from U.K. Schmidt plates. Two groups of clusters according to their distance from the rotation centre of the LMC were measured. The tidal radii of the central clusters vary from 58 to 85 pc and those of the most distant clusters from 33 to 86 pc whereas masses were found to vary from 10^5 to $4 \times 10^5 M_\odot$ and from 10^4 to $2 \times 10^5 M_\odot$ respectively.

156.039 **The distribution of bright stars in the SMC clusters.**
E. Kontizas, E. Theodosiou, M. Kontizas.
Luminous stars and associations in galaxies, p. 409 – 410 (1986). – See Abstr. 012.079. (IAU Symp. No. 116).

Twenty five disk, three intermediate and eighteen halo clusters of the SMC were measured for studying the segregation effect by means of star counts. Short and long exposure red plates (IIIaF + RG630) were used, taken with the 1.2 m U.K. Schmidt telescope.

156.040 **Bright stars in five LMC star clusters.**
E. Kontizas, E. Xiradaki, M. Kontizas.
Luminous stars and associations in galaxies, p. 417 – 418 (1986). – See Abstr. 012.079. (IAU Symp. No. 116).

The bright stars of five LMC clusters were classified for deriving the distribution of various spectral types. The studied clusters are very young (NGC 2098), young (NGC 1818, NGC 2157), intermediate (NGC 1831), and old (NGC 1806) (Van den Bergh, 1981).

156.041 **The distance modulus of LMC.**
C. Chiosi, L. Pigatto.
Luminous stars and associations in galaxies, p. 513 – 514 (1986). – See Abstr. 012.079. (IAU Symp. No. 116).

Deep CCD photometry of the star clusters NGC 2162 and NGC 2190 in LMC presented by Schommer et al. (1984) is used together with new evolutionary models computed by Bertelli et al. (1985) which take into account overshooting from convective cores, to derive the clusters ages and the distance modulus of LMC.

156.042 **IRAS observations of the Magellanic Clouds.**
F. P. Israel, P. B. Schwering.
Light on dark matter, p. 383 – 396 (1986). – See Abstr. 012.080.

First results of the IRAS coverage of the Magellanic Clouds are briefly discussed in terms of the overall infrared properties of the Clouds.

Nomenclature for objects in the Magellanic Clouds.
See Abstr. 002.020.

A new camera and a CCD detector for the Coudé Echelle Spectrograph.
See Abstr. 034.031.

Integrated spectral energy distribution and colors of young stellar populations.
See Abstr. 065.153.

Integrated colours and synthetic HR diagrams for LMC clusters.
See Abstr. 065.154.

Convective overshooting: new integrated colours vs age relations for star clusters.
See Abstr. 065.174.

Gravitational microlensing by the galactic halo.
See Abstr. 066.136.

Radial–velocity determinations of six LMC superluminous giant candidates.
See Abstr. 111.017.

Circumstellar shells around luminous emission–line stars in the Large Magellanic Cloud.
See Abstr. 112.026.

Two supergiants in the Large Magellanic Cloud with thick dust shells.
See Abstr. 112.059.

Circumstellar shells around luminous emission–line stars in the Large Magellanic Cloud.
See Abstr. 112.092.

The massive stellar wind of the Hubble–Sandage variable S Doradus.
See Abstr. 112.128.

B[e]–stars of the Magellanic Clouds.
See Abstr. 112.132.

The mass loss rate of SK 80 (O7 Iaf) in the Small Magellanic Cloud.
See Abstr. 112.133.

Spectroscopic studies of Wolf–Rayet stars. III. The WC subclass.
See Abstr. 114.003.

The chemical composition of a G0 Ib star in the Small Magellanic Cloud.
See Abstr. 114.021.

Fe II in the UV spectrum of luminous emission line stars.
See Abstr. 114.037.

Non–LTE analysis of massive stars in the Magellanic Clouds.
See Abstr. 114.072.

Non–membership of two superluminous giant candidates in the LMC.
See Abstr. 114.083.

An O3 star in the Small Magellanic Cloud H II region NGC 346.
See Abstr. 114.102.

An O3 star in the Small Magellanic Cloud H II region NGC 346.
See Abstr. 114.107.

High–dispersion spectroscopy of the brightest emission–line stars of the Magellanic Clouds.
See Abstr. 114.138.

The absorption feature at 1920 Å in the spectra of early–type supergiants in the Small Magellanic Cloud.
See Abstr. 114.139.

Optical and ultraviolet spectral morphology of luminous OB stars in the Galaxy and the Magellanic Clouds.
See Abstr. 114.143.

Absolute magnitudes and evolutionary status of Wolf–Rayet stars.
See Abstr. 115.010.

The formation and evolution of carbon stars.
See Abstr. 115.018.

Absolute magnitudes and evolutionary status of Wolf–Rayet stars.
See Abstr. 115.019.

Metallicity effect on absolute magnitude determination.
See Abstr. 115.032.

Optical light curves of massive X–ray binaries.
See Abstr. 117.005.

Optical, UV and X–ray observations of X–ray binaries.
See Abstr. 117.130.

New observational results of the LMC–S Dor variable R 127 during outburst.
See Abstr. 122.004.

Metal abundances and radial velocities of short–period variable stars in the direction of the Large Magellanic Cloud.
See Abstr. 122.005.

Abundances of Cepheids in the Galaxy and the Magellanic Clouds.
See Abstr. 122.107.

The Large Magellanic Cloud R CrB star – HV12842.
See Abstr. 122.137.

Velocities and radii of five Magellanic Cloud cepheids.
See Abstr. 122.175.

Variability of Magellanic Cloud supergiants.
See Abstr. 122.195.

S Doradus type: Hubble–Sandage variables.
See Abstr. 122.196.

The energy–diameter relation of supernova remnants in the Large Magellanic Cloud.
See Abstr. 125.017.

Collisionally heated dust in LMC supernova remnants.
See Abstr. 125.096.

Discovery of a 50 msec pulsar in the Large Magellanic Cloud.
See Abstr. 126.049.

Discovery of a Crab–like pulsar in the LMC.
See Abstr. 126.050.

The mass distribution of dark clouds in external galaxies.
See Abstr. 131.060.

An empirical study of ring–shaped nebulae in the Galaxy and in the LMC.
See Abstr. 131.145.

Chemical equilibrium from diffuse to dense clouds. II. Interstellar clouds in the Magellanic Clouds.
See Abstr. 131.147.

On the IMF of violent star forming regions.
See Abstr. 131.351.

Detection and study of two H II blobs in the LMC giant H II region N160 and investigation of their nebular and stellar environment.
See Abstr. 132.042.

The luminous stellar content of 30 Doradus and NGC 3603 – the nearest visible giant H II regions.
See Abstr. 132.059.

Groundbased infrared observations of Magellanic Cloud H II regions.
See Abstr. 132.063.

Search for new X–ray pulsars in the galactic plane by the HEAO–1 Large Area Sky Survey experiment.
See Abstr. 142.061.

Identifications of X–ray point sources in the LMC.
See Abstr. 142.063.

Systems of stellar associations in galaxies.
See Abstr. 152.010.

Analysis of color–magnitude diagrams from three Large Magellanic Cloud clusters.
See Abstr. 153.030.

Simulated populations of the asymptotic giant branch of globular clusters in the Magellanic Clouds.
See Abstr. 154.008.

The ages of globular clusters in the Magellanic Clouds.
See Abstr. 154.009.

The age of the Large Magellanic Cloud globular cluster NGC 1651.
See Abstr. 154.036.

The age of the Small Magellanic Cloud globular cluster NGC 411.
See Abstr. 154.048.

Systematic reinvestigation of the radial velocities of the galactic globular clusters: image–tube results.
See Abstr. 154.057.

Red variables as distance indicators and the distance to the galactic centre.
See Abstr. 155.076.

A search for million degree gas in the galactic halo and the LMC through [Fe X] λ6375 absorption.
See Abstr. 155.099.

Luminous blue stars: distribution and numbers.
See Abstr. 155.119.

The nature of the emission–line object S103 near the LMC.
See Abstr. 157.058.

The stellar populations of galaxies in the Local Group.
See Abstr. 157.088.

Global properties of stellar populations and the spectral evolution of galaxies.
See Abstr. 157.091.

Nonthermal radio continuum emission from dwarf galaxies.
See Abstr. 157.301.

Wolf–Rayet stars in nearby galaxies.
See Abstr. 157.312.

157 Normal Galaxies (Structure, Evolution, Pairs, etc.)

157.001 **Stellar spiral arms in smooth–armed galaxies.**
R. C. Kennicutt Jr., B. K. Edgar.
Astrophys. J., Vol. 300, No. 1, p. 132 – 141 (1986).

The authors have obtained deep multicolor CCD surface photometry for three extreme examples of smooth–armed spirals, galaxies with a global spiral pattern but with little or no interstellar gas, dust, or recent star formation. Two of the galaxies, NGC 7743 and NGC 495, possess moderate–amplitude, red arms which are identical in color to the underlying disk. This implies that at least 70% of the observed arm amplitudes in these galaxies arise from a surface density perturbation in the disk, and that the visible light effectively traces the structure of the density wave. In NGC 718 one of the smooth arms is noticeably bluer than the disk, but the observed arm colors and amplitudes indicate that roughly half of the observed amplitude is due to a spiral density perturbation. These galaxies strongly resemble recently published N–body models of bar–driven stellar disks.

157.002 **New evidence for galaxy luminosity evolution.**
M. J. Lebofsky, P. R. M. Eisenhardt.
Astrophys. J., Vol. 300, No. 1, p. 151 – 158 (1986).

Near–infrared photometry for 61 elliptical galaxies is used with previously published data to demonstrate that little or no color evolution is observed at infrared wavelengths, while galaxies near $z = 1$ are brighter than those at low redshift. Optically selected galaxies are compared with radio galaxies, and no differences in infrared properties are found. Evolution of a stellar population with an initial mass function similar to that observed in the solar neighborhood coupled with a value for q_0 near 0.5 provides the best fit to the color and luminosity data.

157.003 **The molecular morphology of the Sab galaxy NGC 4736.**
L. E. Garman, J. S. Young.
Astron. Astrophys., Vol. 154, No. 1/2, p. 8 – 14 (1986).

CO emission has been mapped along the major axis of the Sab galaxy NGC 4736. These millimeter line data reveal the presence of a molecular ring near the center of this galaxy. The authors discuss the relation of the molecular morphology of NGC 4736 to the other peculiar features of this galaxy, as well as to other galaxies which exhibit central holes in their CO distributions.

157.004 **Search for ultraviolet emission lines from a hot gaseous halo in the edge–on galaxy NGC 4244.**
J.–M. Deharveng, J. Bixler, M. Joubert, S. Bowyer, R. Malina.
Astron. Astrophys., Vol. 154, No. 1/2, p. 119 – 124 (1986).

The halo of the edge–on galaxy NGC 4244 has been observed with the short and long wavelength cameras of the International Ultraviolet Explorer (IUE) in a search for emission lines from a hot galactic corona. The spectra exhibit four features. Two of these correspond in wavelength to prominent emission features found in hot ($\geqslant 10^5$K) astrophysical plasmas. Due to the nature of the IUE instrumental background and faintness of the features, a positive detection cannot be claimed. The upper limit on line emission constrains any coronal halo gas at $T < 10^5$K to be either highly clumped or to have a pressure lower than that found in the plane of the Galaxy. If the two spectral features are real emission lines, then the NGC 4244 halo gas observed is at $T \cong 10^{5.2}$K. This is hotter than the Galactic halo gas temperature derived from IUE measurements assuming steady state collisional ionization. This gas would be consistent with the gas parameters derived from rocket observations of diffuse high latitude ultraviolet emission.

157.005 **Kinematics of elliptical galaxies with minor axis dust lanes.**
C. Möllenhoff, G. Marenbach.
Astron. Astrophys., Vol. 154, No. 1/2, p. 219 – 226 (1986).

The elliptical galaxies NGC 5266, 0609–331, 0632–629, and 0641–412 show dust lanes along their minor axis. Spectroscopic kinematical findings are presented. All objects rotate about their short axis, orthogonal to the dust–lane rotation. The combination of kinematic data of the stars and of the dust lane allows conclusions to the intrinsic shape of the galaxies. Arguments for a triaxial, near prolate shape of NGC 5266 and 0609–331 are presented. These arguments are especially strong for 0609–331 which shows significant rotation about major axis and minor axis and minor axis as well.

157.006 **Search for the globular clusters in nearby galaxies. I. NGC 253.**
A. Blecha.
Astron. Astrophys., Vol. 154, No. 1/2, p. 321 – 335 (1986).

25 reliable candidates for globular clusters were detected in a radius of 13′ around the galaxy NGC 253 in the Sculptor group. 3 candidates are blue clusters and 4 are redder than the globular clusters in our galaxy. The spatial distribution of these clusters follows the $D^{1/4}$ law. A study of the appearance of the globular clusters at large (over 1 Mpc) distances is carried out using the new cluster size classification. A revised distance of 2.88 Mpc for NGC 253 is proposed. As a by–product the field stars luminosity function and the B–V distribution are presented and compared with the model predictions. Similar results are presented for field galaxies. The total number of clusters is evaluated at 80 ± 42. A comparison with other known systems in spiral galaxies gives a simple law for the frequency coefficient in spirals: $S = 4.6$ (K_t–0.1). The number of clusters per unit luminosity of the halo component appears to be constant for spiral galaxies of all types ranging from Sa to Sc.

157.007 **Far–infrared emission from clumpy irregular galaxies.**
U. Klein, J. Heidmann, R. Wielebinski, E. Wunderlich.
Astron. Astrophys., Vol. 154, No. 1/2, p. 373 – 375 (1986).

The four clumpy irregular galaxies Mkn 8, Mkn 296, Mkn 297 and Mkn 325 have been observed by IRAS. The ratios of $S_{100\mu}/S_{60\mu}$ are comparable to those of H II regions or violently star forming galaxies. The average star formation rate in clumpy irregular galaxies is of the order of a few $M_\odot \mathrm{yr}^{-1}$. The continuum spectra of the galaxies over a range of 7 decades in frequency are presented and compared to those of the starbust galaxy M82 and the giant H II region NGC 604 in M33.

157.008 **NGC 4418; a very extinguished galaxy.**
P. F. Roche, D. K. Aitken, C. H. Smith, S. D. James.
Mon. Not. R. Astron. Soc., Vol. 218, No. 1, p. 19P – 23P (1986).

The authors present observations of the nucleus of NGC 4418 which show it to be perhaps the first true infrared galaxy. NGC 4418 was found to be very bright in the infrared by IRAS, and ground–based observations have been employed to investigate the nature of the infrared emission. Optical spectra show little evidence of the luminous infrared core of the galaxy, revealing only stellar absorption features and very weak emission lines. The infrared spectra have a very deep minimum at 9.7 μm, implying a visual extinction of $\gg 50$ mag and indicating that the source producing the infrared luminosity is completely obscured in the visible. From the relatively warm IRAS colours, and the non–detection of dust or ionic line emission structure in the 8 – 13 μm spectrum, it is likely that NGC 4418 harbours a very heavily obscured active nucleus.

157.009 **CCD photometry of apparent dwarf galaxies in Fornax.**
S. Phillipps, P. L. Grimley, M. J. Disney, M. G. M. Cawson, E. J. Kibblewhite.
Mon. Not. R. Astron. Soc., Vol. 218, No. 2, p. 289 – 295 (1986).

The authors present blue and red CCD surface photometry of two apparent dwarf galaxies in the Fornax cluster region. Luminosity profiles are derived and their form discussed. The fainter galaxy resembles an archetypal diffuse dwarf elliptical but the brighter of the pair is either an unusual red dwarf or a background galaxy in chance juxtaposition.

157.010 **Bimodal star formation and remnant–dominated galactic models.**
R. B. Larson.
Mon. Not. R. Astron. Soc., Vol. 218, No. 3, p. 409 – 428 (1986).
It is proposed that the initial mass function (IMF) for star formation in galaxies is not a monotonic function of stellar mass but is double–peaked or bimodal. In models with a bimodal IMF the star formation rate can be strongly decreasing function of time, and stellar remnants can constitute an important or dominant fraction of the total mass. A bimodal model is constructed that accounts for all of the unseen mass in the solar neighbourhood as remnants, and it is shown that this model is consistent with all of the available constraints on the evolution and stellar content of the solar neighbourhood. The properties of the inner discs of our Galaxy and M83 can be accounted for by similar models in which the high–mass mode of star formation is more dominant than in the solar neighbourhood, and remnants constitute a large fraction of the mass.

157.011 **Large–scale dissociation of molecular gas in galaxies by newly formed stars.**
R. J. Allen, P. D. Atherton, R. P. J. Tilanus.
Nature, Vol. 319, No. 6051, p. 296 – 298 (1986).
Molecular hydrogen is thought to form from atomic hydrogen on the surface of dust grains in the interstellar medium, and to be dissociated primarily by ultraviolet radiation from hot stars. Observational evidence that molecular hydrogen is indeed dissociated in shells around young OB stars in the Galaxy has been presented recently. The authors present new high–resolution observations of the spiral galaxy M83 both in H I and in the Hβ emission line. These data suggest that the same dissociation process occurs on the kpc scale in active star–forming regions of some galaxies, and that this dissociation may strongly affect the observed morphology of atomic hydrogen in spiral arms.

157.012 **Implications of the dwarf spheroidal galaxy mass–metallicity relation.**
G. H. Smith.
Publ. Astron. Soc. Pac., Vol. 97, No. 597, p. 1058 – 1064 (1985).
The presence of a mass, metallicity relation among the dwarf spheroidal galaxies is taken to mean that these systems were able to support a degree of chemical enrichment. The properties of this relation are discussed in terms of a model which assumes that internal chemical evolution of the dwarf spheroidals was promoted by supernova activity, which at the same time, however, caused a gradual loss of mass from these galaxies. It is suggested that dwarf spheroidals and ellipticals had different formation histories, in accord with Kormendy's observation that these two types of systems exhibit different trends between surface brightness and luminosity.

157.013 **Stellar associations in the galaxy NGC 2403.**
P. Hodge.
Publ. Astron. Soc. Pac., Vol. 97, No. 597, p. 1065 – 1070 (1985).
Stellar associations in the Sc galaxy NGC 2403 are identified and compared to similar samples in Local Group galaxies. Their properties, including the total number delineated, their sizes, the average number of supergiant stars they contain, and their radial distribution are tabulated and are shown to be similar to those of M33 and the LMC. The stellar associations have an exponential radial–density distribution, the scale length of which is 2.0 ± 0.2 kpc, which is the same as that of the H II regions, larger than that of the blue background galaxian light, but considerably smaller than that of the H I distribution.

157.014 **H I–observations of galaxies in the Kraan–Korteweg – Tammann catalogue of nearby galaxies. I. The data.**
W. K. Huchtmeier, O. G. Richter.
Astron. Astrophys., Suppl. Ser., Vol. 63, No. 2, p. 323 – 343 (1986).
H I–data are presented for almost all non–elliptical galaxies listed in the Kraan–Korteweg – Tammann (1980) catalogue of nearby galaxies and some additional galaxies satisfying the selection criterion $v \leqslant 500\ \mathrm{km\ s^{-1}}$, i.e. the most complete sample of nearby galaxies within $D \leqslant 10$ Mpc. Data for a few galaxies without previously determined radial velocities are included. Due to their morphological appearance these were suspected to be very nearby.

157.015 **Globular clusters and galactic evolution.**
S. van den Bergh.
Astrophys. Space Sci., Vol. 118, Nos. 1/2, p. 435 – 438 (1986). – See Abstr. 012.001.
Observations of the frequency with which globular clusters surround galaxies are reviewed. With a single exception the data do not favour the view that elliptical galaxies are formed from merging disc systems. Attention is also drawn to the unusually high frequency of globular clusters in central galaxies of rich clusters.

157.016 **Emission line velocity survey of spiral galaxies with bright nuclei.**
A. J. Turtle, W. D. Pence.
Astrophys. Space Sci., Vol. 119, No. 1, p. 173 – 175 (1986). – See Abstr. 012.001.
High resolution spectrographic observations of 220 nearby galaxies with bright nuclei have been made mainly at Hβ and [O III] λ5007 to search for signs of nuclear activity. In many cases evidence for weak activity has been found.

157.017 **Large–scale configuration of the magnetic field in spiral galaxies.**
Y. Sofue, U. Klein, R. Beck, R. Wielebinski.
Astrophys. Space Sci., Vol. 119, No. 1, p. 191 – 194 (1986). – See Abstr. 012.001.
Global configuration of magnetic field in several spiral galaxies were determined by analysing characteristic variation of Faraday rotation within the galaxy disks. The majority has an open spiral, bisymmetric field configuration, while some (10 – 20%) have a ring field.

157.018 **Correlation between the physical parameters and morphological type of spiral galaxies.**
S.–W. Kim, M.–S. Chun.
Astrophys. Space Sci., Vol. 119, No. 1, p. 195 – 197 (1986). – See Abstr. 012.001.
From PDS scans of the late–type spiral galaxies NGC 1313, 1365, 6946, and 7793, the authors determine the surface brightness distributions. These distributions enable them to calculate physical parameters such as $R_e{}^*$, μ_e, α^{-1}, μ_0, D/B, C_{21},and M/L. The values of some of these parameters were also compiled from other studies, and all were compared with morphological type T. One of the meaningful results shows that there is a certain correlation between D/B and T, which is constant until $T \leqslant 1$ or 2, and then increases sharply for $T \geqslant 1$ or 2. This may support the validity of the intrinsic formation theory.

157.019 **Evolution of disk galaxies regulated by supernova remnants.**
Y. D. Tanaka, S. Ikeuchi, A. Habe.
Astrophys. Space Sci., Vol. 119, No. 1, p. 207 – 210 (1986). – See Abstr. 012.001.
Assuming that a disk galaxy is composed of an ambient pervasive gas, small clouds, molecular clouds and stars, its evolution is studied through examining the interchange processes among them. Main results obtained are: (1) The star formation rate is directed by the formation process of molecular clouds. (2) Depending upon the parameters there may be three or four types of evolution of disk galaxies: the no star formation case, the active in the past and inactive at present star formation case, the burst–like star formation case and the very active star formation case.

157.020 **Colliding and merging galaxies.**
F. Schweizer.
Science, Vol. 231, No. 4735, p. 227 – 234 (1986).
Aided by advances in computer technology and observations from space, astronomers have begun to unravel the mysteries of galaxy formation and evolution. Galaxies evolve by interacting with their environment and especially with each other. During

brief but often fierce galactic encounters, gravitational forces generate strong tides that survive as telltale signatures for billions of years. Because these so-called collisions dissipate orbital energy, galaxies on bound orbits may eventually merge. Collisions and mergers are responsible for a great variety of phenomena, including the triggering of widespread star formation in galaxies and the fueling of nuclear activity in quasars. Evidence is accumulating that not all galaxies formed shortly after the Big Bang. A sizable fraction of them may have formed later, and many are still experiencing significant dynamical evolution.

157.021 **I Zw 18 and the existence of very metal poor blue compact dwarf galaxies.**
D. Kunth, W. L. W. Sargent.
Astrophys. J., Vol. 300, No. 2, p. 496 – 499 (1986).
No dwarf emission-line galaxies are known that have heavy-element abundances less than that of I Zw 18 about 1/30 of the solar value. The authors show that this fact is not a result of observational selection. It is a consequence of using H II regions for deriving the compositions of gas-rich systems. It is pointed out that, provided the bursts of star formation in the blue compact galaxies last much longer than the evolutionary time scale of a typical massive star, a minimum metal abundance close to that observed in I Zw 18 must be found in any H II region. The reason is that the same massive stars produce both heavy elements and ionizing photons, while on short time scales mixing of newly formed heavy elements is confined to the H II region.

157.022 **DDO 13: a low surface brightness galaxy with a well-formed H I disk and two tiny companions.**
F. H. Briggs.
Astrophys. J., Vol. 300, No. 2, p. 613 – 623 (1986). With plate 3.
The Irregular V galaxy DDO 13 has a well-developed H I disk. There is a hint of a warp and evidence for nonplanar motions in the H I at the outer edge of the disk. DDO 13 appears to form a bound system with UGC 1171 and another small feature, Dw 0137+1541, which has been identified in the H I data and which may be entirely hydrogen. The system demonstrates the usual excess of binding mass above both the luminous mass and the dynamical masses derived for the individual galaxies from their rotation. This system may in turn be bound to the nearly ScI NGC 628.

157.023 **Erratum: "Hot coronae around early-type galaxies" [Astrophys. J., Vol. 293, No. 1, p. 102 – 119 (1985)].**
W. Forman, C. Jones, W. Tucker.
Astrophys. J., Vol. 300, No. 2, p. 836 (1986). See Abstr. 40.157.202.

157.024 **Gas in early-type galaxies.**
E. M. Sadler.
Cosmical gas dynamics, p. 27 – 32 (1985). – See Abstr. 012.007.
Despite their reputation as gas-poor systems, recent observational work makes it clear that most elliptical and S0 galaxies have a substantial gaseous component. Studies of gas dynamics in these galaxies are important because they can provide an effective probe of the mass distribution and true shape of ellipticals. Observational data obtained for a large galaxy sample show that bright elliptical galaxies are characterised by weak emission from a rotating, kiloparsec-scale gaseous disc containing 10^3– $10^5 M_\odot$ of ionised gas, often in association with dust. Low-luminosity elliptical galaxies often show emission from H II regions, sometimes clumped in a gaseous ring centred on the nucleus.

157.025 **Wind-driven shells in E/S0 galaxies: a research note.**
A. Wilkinson, M. E. Bailey.
Cosmical gas dynamics, p. 63 – 64 (1985). – See Abstr. 012.007.
The possibility that many shell galaxies possess X-ray coronae has led the authors to reconsider the theories of how the shells are formed.

157.026 **Gas flow in the barred spiral galaxy NGC 5383.**
O. Bienaymé, A. Bosma, E. Athanassoula.
Cosmical gas dynamics, p. 205 – 211 (1985). – See Abstr. 012.007.
New H I-observations obtained at Westerbork with an angular resolution of 13 × 19 arcsec have motivated the authors to reanalyse the gas flow in the barred spiral galaxy NGC 5383. They use and extend a method described first by Tremaine and Weinberg (1984).

157.027 **A color dependence in the distance moduli derived from the *B* and *H* band Tully–Fisher relations.**
E. H. Giraud.
Astron. Astrophys., Vol. 155, No. 2, p. 283 – 290 (1986).
The distance moduli derived from the *B*-band version of the Tully–Fisher relations, $\mu_0^W(V_M)$, are compared with those derived from the *H*-band version $\mu_0(W)$. It has been shown that: (1) the difference $\mu_0^W(V_M)$–$\mu_0(W)$ in the Bottinelli et al., catalogue depends on the hybrid color B_T^0–$H^c_{-0.5}$. (2) For the sample the distance moduli would not be color dependent if the difference in slope between the *B* and *H* Tully–Fisher relation is larger than 5.4, which is the mean slope of the [B_T^0–$H^c_{-0.5}$, log $W^c(0)$] color relation. (3) The differences in slope between the *B* and *H* Tully–Fisher relation (first-sided, second-sided, and mean regressions) for that sample do not support such a condition. (4) The distance moduli derived in the *V*(5500 Å) and *IV*(10500 Å) bands from Visvanathan are also color-dependent.

157.028 **Submillimetre continuum observations of NGC 253.**
W. K. Gear, G. Gee, E. I. Robson, P. A. R. Ade, W. D. Duncan.
Mon. Not. R. Astron. Soc., Vol. 219, No. 1, p. 19P – 22P (1986).
The authors have mapped the 350-μm emission along the major axis of the dusty galaxy NGC 253 and find this to have an extent of about 95 arcsec. The total mass of interstellar material in the central kpc is $\sim 3 \times 10^8 M_\odot$. Comparison of the results with CO observations suggests that the interstellar material in NGC 253 is in the form of massive clouds, similar to those in our Galaxy.

157.029 **NGC 4038/39: interacting spiral galaxies with enhanced extended radio emission.**
E. Hummel, J. M. van der Hulst.
Astron. Astrophys., Vol. 155, No. 1, p. 151 – 160 (1986).
The radio continuum emission from the interacting galaxies NGC 4038/39 (The "Antennae") was observed with the Very Large Array at 1.465 GHz and 4.885 GHz with a resolution of 6″. These observations suggest that an enhanced star formation rate and maybe a compression of the magnetic field in between the galaxies are responsible for the enhanced radio emission.

157.030 **H_2O and OH maser emission from bright IRAS galaxies.**
C. Henkel, J. G. A. Wouterloot, J. Bally.
Astron. Astrophys., Vol. 155, No. 1, p. 193 – 199 (1986).
H_2O maser emission with an isotropic luminosity of 1 $L_\odot$ has been detected in IC 10. The maser is the most powerful yet detected in a compact irregular galaxy. The authors first discuss the properties of the H_2O maser in IC 10. Then they analyse a sample of 43 bright IRAS galaxies.

157.031 **Search for dwarf galaxies in the IC 342 complex of galaxies.**
F. Börngen, V. E. Karachentseva.
Astron. Nachr., Vol. 306, No. 5, p. 301 – 304 (1985).
A survey of dwarf galaxies was carried out in the inner part of the IC 342 complex of galaxies, which is a subgroup of the M81 group, in an area of about 42 sq. degrees by means of Tautenburg 52-in. Schmidt plates. Fifteen new dwarf candidates and four extended "cloudy" objects of very low surface brightness not yet classified were discovered. Coordinates, linear dimensions as well as remarks on the nature of the cloudy objects are given. A deeper analysis with 6-meter plates is planned to study a part of the new objects more exactly.

157.032 **Galaxy photometry in the Virgo cluster. II.**
K. Fritze.
Astron. Nachr., Vol. 307, No. 1, p. 27 – 33 (1986).
For two fields of the Virgo cluster of galaxies a photometry in *B*, *V*, and *R* has been performed. A total of 72 galaxies has been identified and their magnitudes and diameters have been determined.

157.033 **Studies of *IRAS* sources at high galactic latitudes – I. Source counts at $|b| > 60°$ and evidence for a north–south anisotropy of cosmological significance.**
M. Rowan–Robinson, D. Walker, T. Chester, T. Soifer, J. Fairclough.
Mon. Not. R. Astron. Soc., Vol. 219, No. 2, p. 273 – 283 (1986).
Source counts and identifications of *IRAS* sources at $|b| > 60°$ are discussed. It is shown that emission from interstellar dust on the *IRAS* point–source angular scale at 100 μm is localized to a few small areas of the galactic polar caps. At 12 and 25 μm the sky is dominated by stars, and at 60 and 100 μm by galaxies, at these latitudes. After exclusion of the Virgo cluster and of regions affected by emission from interstellar dust, comparison of the 60– and 100–μm source counts at $b > 60°$ and $b < -60°$ shows that the source density is 20 per cent higher in the north than in the south. The authors argue that this is not due to any known instrumental effects and that it represents a cosmologically significant anisotropy in the galaxy distribution.

157.034 **The dynamical influence of the nucleus on the bulge of M31.**
O. E. Gerhard.
Mon. Not. R. Astron. Soc., Vol. 219, No. 2, p. 373 – 386 (1986).
Massive compact structures in the centres of triaxial stellar systems may subject stars on box orbits to large–angle deflections and thereby disrupt the triaxiality of their parent galaxies. In this paper the author investigates whether the nucleus of M31 can affect the dynamics of the bulge in this way.

157.035 **The extremely low dust content of blue compact galaxies: results of *IRAS* observations.**
P. M. Gondhalekar, D. H. Morgan, M. Dopita, R. S. Ellis.
Mon. Not. R. Astron. Soc., Vol. 219, No. 3, p. 505 – 510 (1986).
Far–infrared observations of three blue compact galaxies, Tololo 1924–416, Mkn 36 and IZw 18, are described. These galaxies, undergoing intense bursts of star formation, have low infrared luminosities. If the infrared luminosities are due to dust re–radiation, then the dust content of these galaxies is exceptionally low, perhaps because the galaxies are young and have not evolved sufficiently to build up the dust mass seen in normal galaxies.

157.036 **Radio observations of the S0 galaxy NGC 1218 (3C 78).**
D. J. Saikia, C. R. Subrahmanya, A. R. Patnaik, S. W. Unger, T. J. Cornwell, D. A. Graham, T. P. Prabhu.
Mon. Not. R. Astron. Soc., Vol. 219, No. 3, p. 545 – 553 (1986).
The authors present VLA observations of the S0 galaxy NGC 1218 at λ20 and 6 cm with both the A and D arrays, at λ20 cm with the B array and at λ2 and 1.3 cm with the A array. They also present VLBI observations at λ6 cm with the European Network. The high–resolution radio observations reveal a flat–spectrum core and a one–sided jet whose width increases with distance from the nucleus. Assuming Faraday rotation to be small at λ6 cm, the magnetic field lines in the outer region of the jet are inclined to its axis and appear to diverge. The large–scale structure does not exhibit well–defined lobes but forms an extended halo with a largely circumferential magnetic field. The authors suggest that the observed one–sidedness of the jet could perhaps be due to a difference in the collimation of the jets on opposite sides of the nucleus, in addition to any possible effects of relativistic beaming.

157.037 **Radio sources in giant E and S0 galaxies.**
R. A. Cordey.
Mon. Not. R. Astron. Soc., Vol. 219, No. 3, p. 575 – 588 (1986).
The author presents an investigation of the pattern of radio activity in giant E and S0 galaxies. His aims are to set limits on the duration of periods of source production, to determine the fraction of its life that a giant galaxy spends as a radio source, and to investigate factors that influence activity in such galaxies.

157.038 **Studies of IRAS sources at high galactic latitudes. II. Results from a redshift survey at $b > 60°$: distribution in depth, luminosity function, and physical nature of IRAS galaxies.**
A. Lawrence, D. Walker, M. Rowan–Robinson, K. J. Leech, M. V. Penston.
Mon. Not. R. Astron. Soc., Vol. 219, No. 3, p. 687 – 701 (1986).
The authors have identified 530 IRAS sources from a complete sample, flux limited at S(60 μm) = 0.5 Jy, in an area of sky covering 844 square degrees at the North Galactic Pole. After exclusion of 34 bright stars, 99 per cent of the remainder coincide with galaxies, mostly in the magnitude range 14 – 17, but with a tail of faint objects. Spectroscopic observations have been made of candidates for 183 IRAS sources. Together with published data, the authors have redshifts for two overlapping complete sub–samples, totalling 303 galaxies. They derive several important results.

157.039 **Interpretation of the UV luminosity of elliptical galaxies.**
R. Mochkovitch.
Astron. Astrophys., Vol. 157, No. 2, p. 311 – 317 (1986).
The energy distribution curves of elliptical galaxies, in the visible and infrared ranges, can be accounted for by an old stellar population of red giants and small mass dwarfs. In this simple model, however, the upturn observed in the far–UV ($\lambda < 2500$ Å) is unexpected and must be produced by another class of star with high effective temperature. Several possibilities have been proposed for this hot component, involving either young or evolved blue stars. The author discusses a model where the UV luminosity is a sum of three contributions: (1) young stars, formed in short bursts separated by long inactive periods, (2) nuclei of planetary nebulae and, (3) accreting white dwarfs. He shows that good fits of the energy curves from the UV to the infrared are possible if the progenitors of type I supernovae represent only a few percent of all accreting white dwarfs.

157.040 **Contraction of dark matter galactic halos due to baryonic infall.**
G. R. Blumenthal, S. M. Faber, R. Flores, J. R. Primack.
Astrophys. J., Vol. 301, No. 1, p. 27 – 34 (1986). = Lick Obs. Bull., No. 1017.
Varied evidence suggests that galaxies consist of roughly 10% baryonic matter by mass and that baryons sink dissipatively by about a factor of 10 in radius during galaxy formation. The authors show that such infall strongly perturbs the underlying dark matter distribution, pulling it inward and creating cores that are considerably smaller and denser than would have evolved without dissipation. Any discontinuity between the baryonic and dark matter mass distributions is smoothed out by the coupled motions of the two components. If dark halos have large core radii in the absence of dissipation, the above infall scenario yields rotation curves that are flat over large distances, in agreement with observations of spiral galaxies.

157.041 **A physical mechanism for bursts of star formation.**
J. M. Scalo, C. Struck–Marcell.
Astrophys. J., Vol. 301, No. 1, p. 77 – 82 (1986).
Star formation in galaxies is often associated with large–scale disturbances such as density waves and direct and tidal galaxy encounters. The authors suggest an explanation based on the nonlinear behaviour of the Oort model for interstellar cloud evolution and star formation when subjected to such perturbations. The behavior depends critically on only one parameter: the ratio of the time delay between the formation of massive star–forming clouds and the disruption of these clouds by internal stellar winds and explosions to the equilibrium collision time scale. The authors discuss the possible relevance of this model to a variety of galactic phenomena, especially star formation bursts induced by tidal encounters between galaxies.

157.042 **Magnesium gradients in elliptical galaxies. I. Method and initial results.**
W. A. Baum, B. Thomsen, B. L. Morgan.
Astrophys. J., Vol. 301, No. 1, p. 83 – 91 (1986).
This paper reports on an effort to determine stellar population gradients in elliptical galaxies by using the magnesium index as population indicator. Using a state–of–the–art CCD camera, medium bandwidth filters, selected observing conditions, unusual calibration procedures, and special image processing steps, the authors obtain a photometric accuracy equivalent to 0.2% of the sky brightness. Observations have been obtained in the Coma Cluster and the authors present results for NGC 4874, one of the two giant ellipticals dominating, the center of the cluster. The magnesium index gradient in NGC 4874 is -0.04 ± 0.01 mag arcsec$^{-1/3}$, suggesting that the metallicity drops a factor of 3–4 between the nuclear region and the outer halo.

157.043 **UGC 12591: the most rapidly rotating disk galaxy.**
R. Giovanelli, M. P. Haynes, V. C. Rubin, W. K. Ford Jr.
Astrophys. J., Lett. Ed., Vol. 301, No. 1, p. L7 – L11 (1986). With plate L5.
Observations at 21 cm and optical observations of the S0/Sa galaxy UGC 12591 show that this galaxy possesses the largest rotational velocity of any disk system known so far. At 500 km s^{-1}, it exceeds by $\sim$30% the largest values previously measured. It is not yet certain whether UGC 12591 is a unique object at the extremum of the distribution, or one of a class of very massive, early–type disk galaxies, yet unrecognized because of instrumental limitations.

157.044 **Photometry of resolved galaxies. V. NGC 6822.**
J. G. Hoessel, N. Anderson.
Astrophys. J., Suppl. Ser., Vol. 60, No. 2, p. 507 – 542 (1986). With plates 40 – 49.
Three–color *GRI* CCD photometry for 3475 resolved stars in the Local Group irregular galaxy NGC 6822 is discussed. General agreement is found in a comparison between these data and earlier photographic results, but zero–point offsets of $\sim$0.1 mag are present. Color–magnitude diagrams are constructed, and the stellar luminosity function is compared with those of 10 other nearby irregular galaxies. The bright–end stellar mass function slope is similar to that in other Local Group members, but NGC 6822 is presently forming stars at a slower rate for its mass than Sextans A and the Magellanic Clouds.

157.045 **The rotation curves and masses of 5 double interacting galaxies (VV 41, 50, 254, 330, 810).**
V. P. Arkhipova, V. A. Mineva, R. I. Noskova.
Astron. Zh., Tom 63, Vyp. 1, p. 16 – 24 (1986). In Russian. English translation in Sov. Astron., Vol. 30, No. 1.
The redshifts, rotation curves and masses of components of double interacting galaxies VV 41, 50, 254, 330, 810 were derived. The galaxies studied have mainly flat rotation curves and a relatively low velocity gradient in the solid–rotation part. Non–circular motions were observed only in the nuclear region of the bright component of VV 50a. The mean mass–luminosity ratio was found to be $f = 4.2$, and the mass ratio of the components to lie in the range between 3 and 18.

157.046 **Gas flow in spiral galaxies: effects of rotation, self–gravitation and thermal processes.**
V. G. Berman, L. S. Marochnik, Yu. N. Mishurov, A. A. Suchkov, P. N. Timonin.
Astron. Zh., Tom 63, Vyp. 1, p. 31 – 37 (1986). In Russian. English translation in Sov. Astron., Vol. 30, No. 1.
The motion of the interstellar gas in the gravitational field of a spiral density wave is investigated. By numerical integration of non–linear gas dynamic and thermal processes equations it is shown that the rotation of the galactic disk is a necessary condition for a phase transition and formation of a cloud structure in the gas, flowing in the field of a spiral arm by thermal instability. In a thin, dense gas disk, the combined effect of the gas self–gravitation and thermal processes is found to lead to formation of structural features characterized by rapidly growing density and falling temperature; however, no indication on gravitational fragmentation within these dense, cool regions is found.

157.047 **The origin of nitrogen and the chemical evolution of spiral galaxies.**
A. I. Diaz, M. Tosi.
Astron. Astrophys., Vol. 158, No. 1/2, p. 60 – 66 (1986).
Numerical models of chemical evolution reproducing the observed oxygen abundances through the discs of the Galaxy and eight nearby spirals have been computed to predict their corresponding nitrogen abundances under different assumptions on the stellar nucleosynthesis involved in the computation of the yields. It is found that a primary fraction of nitrogen between 30 and 60% (corresponding to a mixing length $1.0 \leqslant \alpha_{RV} \leqslant 1.5$) is required in order to reproduce simultaneously the observed oxygen and nitrogen abundances. The best agreement between model results and observational data is attained for α_{RV} closer to 1.5 and an upper limit for intermediate mass stars of $M_{up} \cong 5 - 6\, M_{\odot}$ instead of the standard 8 $M_{\odot}$. This combination of stellar evolution parameters is also required to reproduce the $^{12}C/^{13}C$ isotopic ratio observed in galactic molecular clouds and the N/C vs. C/H distribution derived from stellar observations in the solar neighbourhood.

157.048 **Warped galactic disks and optical projection effects.**
R. L. Oldershaw.
Astrophys. Space Sci., Vol. 121, No. 2, p. 273 – 281 (1986).
Direct and indirect evidence for global warping in galactic disks is reviewed. The existing data suggest that inversion symmetric warping is common in spiral galaxies, that it can occur throughout the inner and outer disk regions, and that it can involve stellar, dust, and gas components. The amplitude of the warping tends to be proportional to the radial distance from the galactic center and varies from instances wherein the disk is essentially flat to cases wherein the *z*–axis displacement is at least 30% of the galactocentric distance. Projection effects that are inherent features of warped optical disks are discussed.

157.049 **The ionized gas in NGC 4945.**
E. L. Agüero, G. J. Carranza.
Astrophys. Space Sci., Vol. 121, No. 2, p. 403 – 411 (1986).
Physical conditions and relative abundance ratios of the ionized medium of NGC 4945 are discussed. The spectral characteristics indicate for $T_e = 10^4$K a relatively low electron density ($N_e \leqslant 1250$ cm^{-3}) decreasing outwards. N(N)/N(H) and N(S)/(H) diminish with the distance from the nucleus while N(N)/N(S) is nearly constant. These abundance gradients, the lack of kinematical evidences of a bar and the observational difficulties to grasp its morphology, make it improbable that NGC 4945 could be classified as a barred galaxy.

157.050 **Stellar populations and star formation in irregular galaxies.**
D. A. Hunter, J. S. Gallagher III.
Publ. Astron. Soc. Pac., Vol. 98, No. 599, p. 5 – 28 (1986). Invited review article.
The authors review and intercompare data on various morphologically–chosen groups of irregular galaxies in order to understand the underlying physical mechanisms that differentiate these systems. They particularly discuss the observational clues to the stellar content, star–formation processes, and star–formation histories of these galaxies, with an emphasis on the uncertainties and the many unanswered questions.

157.051 **Near–infrared surface photometry of barred spiral galaxies.**
C. W. Baumgart, C. J. Peterson.
Publ. Astron. Soc. Pac., Vol. 98, No. 599, p. 56 – 69 (1986).
Near–infrared photographic plates of eleven barred spiral galaxies have been studied to determine the parameters which characterize the luminosity structures of the stellar bars of these systems. Although the galaxies under study are all SB forms, they show relatively few strong similarities in their radial luminosity profiles. Eight of the eleven exhibit some isophotal twisting in their central regions which suggests that the bulges of these sys-

tems are triaxial in nature. A similar fraction of the galaxies exhibits isophotal twisting within the bars. About one–half of the sample possesses radial luminosity profiles in which the disk component, if extrapolated into the central regions of the galaxy, would contribute more luminosity than is actually observed. The data have been compared to the theoretical stellar bar model of the three–dimensional *N*–body simulation of Miller and Smith (1979). The observed radial luminosity profiles are exponential in qualitative agreement with the model, but the behavior of the ellipticity differs significantly both in terms of the average value and in its change with increasing radius.

157.052 **Mapping dark nebulae in elliptical galaxies.**
J. S. Gallagher III.
Publ. Astron. Soc. Pac., Vol. 98, No. 599, p. 81 – 91 (1986).

The Kitt Peak National Observatory video camera has been used on the 2.1–m telescope to provide imaging data for a survey of dark nebulae in 15 classical elliptical galaxies. Dark nebulae in ellipticals trace cool, dense interstellar matter and are potential sites for star formation. Spatial filtering techniques allow previously known dark nebulae in NGC 404, NGC 708, NGC 1052, NGC 4374, NGC 5363, and NGC 7052 to be readily mapped from the digital images. An extended dark nebular component is also found in NGC 5363. The majority of classical ellipticals in the sample, however, do not contain large dark nebulae. These observations are discussed in terms of processes which might control the amount and distribution of cool gas in elliptical galaxies. In particular, the presence of dust suggests that some of the gas has not been heated to the high temperatures characteristic of the X–ray emitting atmospheres that are found in many elliptical galaxies.

157.053 **The inclination dependence of far–infrared emission from Sc galaxies.**
D. Burstein, M. J. Lebofsky.
Astrophys. J., Vol. 301, No. 2, p. 683 – 688 (1986).

A sample of 2000 Sc galaxies chosen from the UGC is used to show that the detection of Sc galaxies by IRAS is apparently a function of the inclination of a galaxy to the line of sight. Sc galaxies with axial ratios $\log(a/b) < 0.1$ (face–on) are detected at a rate twice that of Sc galaxies with axial ratios $\log(a/b) > 0.7$ (edge–on). The ratio of IRAS 100 μm flux to optical size, expressed as a hybrid "surface brightness" F_{100}/a^2, is a function of inclination, consistent with the difference in detection rates. In an attempt to interpret this effect, the authors conclude that spiral galaxies become optically thick at 100 μm when they are viewed at high angles of inclination. A simple model is proposed for the 60 and 100 μm emission from spiral galaxies that explains the observed infrared properties of galaxies. In its present form this model has far–reaching consequences for the interpretation of already published IRAS results, as well as serious consequences for future choices in the analysis of IRAS data on galaxies.

157.054 **Discovery of Cepheids in M101.**
K. H. Cook, M. Aaronson, G. Illingworth.
Astrophys. J., Lett. Ed., Vol. 301, No. 2, p. L45 – L48 (1986). With plate L7.

The authors report the first definite detection of Cepheid variables in M101, a galaxy of fundamental importance to the extragalactic distance scale. The Cepheids are the most distant ever found, and their discovery illustrates the ready feasibility of sampling through modern digital means these most primary of standard candles over a wide volume of the universe.

157.055 **The radio continuum – far infrared correlation and magnetic fields in Sbc galaxies.**
E. Hummel.
Astron. Astrophys., Vol. 160, No. 1, p. L4 – L6 (1986).

The radio continuum power at 1.465 GHz of Sbc galaxies correlates tightly with the 100 μm power. The author notices a coincidence of the frequency distribution of the deviations from the radio power – 100 μm power correlation and of the frequency distribution of $B^{1.9}$ where B, the magnetic field strength, is calculated by assuming minimum energy conditions. The coincidence implies that the minimum energy condition is fulfilled in almost all the galaxies. This shows that the spread in the radio – infrared correlation is due to the magnetic field. It would further suggest that the star formation histories of most of these galaxies have been similar during the past $\sim 10^8$yr.

157.056 **Optical and near–infrared observations of IRAS galaxies.**
A. F. M. Moorwood, M.–P. Véron–Cetty, I. S. Glass.
Astron. Astrophys., Vol. 160, No. 1, p. 39 – 50 (1986).

Long slit spectra, CCD images and near infrared photometry have been obtained for 22 out of a complete sample of 86 optically identified galaxies detected at 60 μm during the IRAS minisurvey (Soifer et al., 1984). Redshifts are found to be in the range 0.005 to 0.074 leading to 40 – 120 μm luminosities of $6x10^9 - 3x10^{11} L_{\odot}$. The spectra, obtained at 2.5 Å resolution around Hα, are typical of H II regions except for three cases where additional weak Seyfert activity is suspected in the nucleus. Apart from one ring galaxy and one possible starbursting irregular or interacting system, all the galaxies appear to be spirals exhibiting no particular morphological peculiarities.

157.057 **The brightest stars in galaxies: a theoretical simulation.**
L. Greggio.
Astron. Astrophys., Vol. 160, No. 1, p. 111 – 115 (1986).

The brightest stars in late type galaxies have been frequently used as distance indicators, although statistical effects are likely to be important, due to the relatively low number of objects populating the upper part of the HR diagram. This paper presents the results of theoretical simulations of the probability distribution of the absolute magnitude, in several photometric bands, of the three brightest objects in a population of massive stars. Two UV bands, centered at 1400 and 1750 Å, and the *B* and *V* bands are considered. The experiments are performed for different choices of the initial mass function and for a constant star formation rate. The use of the brightest red supergiants as distance indicators is also briefly discussed.

157.058 **The nature of the emission–line object S103 near the LMC.**
M. Dennefeld.
Astron. Astrophys., Vol. 160, No. 1, p. 199 – 202 (1986).

The emission–line object S103, first classified as a planetary nebula belonging to the Large Magellanic Cloud, is in fact an H II–region–type dwarf galaxy with velocity 10,400 km s^{-1}. The abundances of oxygen and neon are quite similar to the SMC ones, but nitrogen is at least twice as abundant as in SMC, while helium is probably more abundant too. This object has the largest N/O ratio of the published sample of metal–poor galaxies, thus suggesting it had a quite different evolutionary history of intermediate–mass stars.

157.059 **Optical and high–resolution H I observations of the massive and unusual lenticular galaxy NGC 5084.**
S. T. Gottesman, T. G. Hawarden.
Mon. Not. R. Astron. Soc., Vol. 219, No. 4, p. 759 – 775 (1986).

The edge–on, $S0^+$galaxy NGC 5084 has been observed with the VLA telescope in the 21–cm line of atomic hydrogen. The gas is found to be in a flat annulus containing some structure. Closely associated with the H I is a very faint optical disc containing some clumpy structure in its outer parts and a weak dust lane in the inner one–third of its extent. The galaxy is very massive, exceeding $10^{12} M_{\odot}$. It has an exceptionally large mass–to–light ratio, $M/L_B \geqslant 65\, M_{\odot}/L_{B\odot}$, and, for its type, an unusually high ratio of H I mass to blue luminosity $M_H/L_B \cong 0.35$ solar units. The galaxy seems to be an extreme member of a class of massive, H I–rich, early–type disc galaxies. The symmetry axes of the inner lens and the H I disc coincide accurately. It is therefore most unlikely that the atomic hydrogen has been acquired by accretion.

157.060 **Discovery of the optical variability of the nucleus of M33.**
V. M. Lyutyj, A. S. Sharov.
Pis'ma Astron. Zh., Tom 12, No. 3, p. 187 – 193 (1986). In Russian. Englisch translation in Sov. Astron. Lett., Vol. 12.

Results of three-year photoelectric UBV observations of the nucleus of the Local Group galaxy M33 are presented. Optical variability with amplitude up to $\Delta U \sim 1^m$ is found. Characteristics of the optical variability of the nucleus of M33 resemble those of the well-known Seyfert galaxy NGC 4051. According to the published data and these measurements the total amplitude of the variability of the nucleus of M33 during ~20 years may by about $\Delta V \sim 1^m$.

157.061 **Most distant galaxy.**
S. F. Shandarin.
Priroda, No. 3, p. 101 – 102 (1986). In Russian.

157.062 **Spectrophotometric investigation of NGC 4922.**
N. K. Andreasyan, Eh. E. Khachikyan.
Astrofizika, Tom 24, Vyp. 1, p. 17 – 24 (1986). In Russian. English translation in Astrophysics, Vol. 24, No. 1.

The results of a spectrophotometric investigation of the binary galaxy NGC 4922 are presented. One of the components appears to be an elliptical galaxy with a spectrum corresponding to the spectral classes G0 – G3, the other has spectral properties of low ionization Seyfert nuclei. Physical conditions and chemical abundances in this component resemble those of Sy 2 galaxies. The mass of the Seyfert-like component within the radius of 1 kpc is nearly $0.4 \times 10^8 M_\odot$. The mass of the whole system is $\sim 2.2 \times 10^{11} M_\odot$.

157.063 **On the radio emission of nuclei of barred galaxies.**
A. T. Kalloglyan, R. A. Kandalyan.
Astrofizika, Tom 24, Vyp. 1, p. 47 – 55 (1986). In Russian. English translation in Astrophysics, Vol. 24, No. 1.

It has been shown that the high percentage of radio emission in galaxies of Byurakan classes 2, 2s, 4 and 5 in comparison with classes 1 and 3 is reliably observed only in barred spirals due to galaxies with both central and disk radio components. The radio emission of barred spirals more often than in other types of spirals is connected with the central regions of those galaxies.

157.064 **Gas equilibrium in the gravitational field of cD galaxies.**
E. V. Volkov.
Astrofizika, Tom 24, Vyp. 1, p. 57 – 64 (1986). In Russian. English translation in Astrophysics, Vol. 24, No. 1.

The possibility to apply the solutions of hydrostatic and energy-equilibrium equations to the description of the gas distribution in the gravity field of cD galaxies has been investigated. It has been shown that the energy-equilibrium condition without the energy fluxes are most probably unfulfilled. The radiation losses in the center of cD galaxies are then balanced by the electron thermal conductivity and by the energy fluxes from the hotter outer regions of cD galaxies.

157.065 **Internal kinematics of spiral galaxies: gas and stellar rotation curves and dispersion profiles.**
J. A. Fillmore, T. A. Boroson, A. Dressler.
Astrophys. J., Vol. 302, No. 1, p. 208 – 233 (1986).

The authors present emission- and absorption-line observations of six moderately inclined spiral galaxies. The mean rotation and velocity dispersion for the gas and the stars have been measured. Two component mass models are fitted to surface brightness measurements and are used to fit kinematic models to the velocity data. Models with constant mass-to-light ratios and isotropic velocity dispersions roughly fit the inner regions of spiral galaxies.

157.066 **Anomalous redshift companion galaxies: NGC 7603.**
N. A. Sharp.
Astrophys. J., Vol. 302, No. 1, p. 245 – 256 (1986). With plates 3 – 7.

The bright galaxy NGC 7603 has an extended spiral arm which appears to end exactly on a small, high surface brightness companion (referred to here as NGC 7603B). However, the companion has a much larger redshift, placing it conventionally almost twice as far away. Deep plates presented here show that NGC 7603 actually extends for some considerable distance beyond the companion, and that there is evidence for past interactions which can explain the disturbed structure. Narrow-band Hα images show no sign of strong disturbances in the feature apparently leading to NGC 7603B, which would be expected if tidal interaction is currently taking place. It is concluded that there are no glaring inconsistencies in the usual interpretation of the system as a background projection effect.

157.067 **Extent of ionized calcium in the outer parts of galaxies.**
D. C. Morton, D. G. York, E. B. Jenkins.
Astrophys. J., Vol. 302, No. 1, p. 272 – 279 (1986).

The authors have searched for Ca II H and K absorption in the spectra of seven QSOs shining through the outer parts of 12 galaxies at projected distances of 9 – 200 $(100/H_0)$ kpc from their centers. No Ca II was found with 2 σ equivalent width limits ranging from $W_\lambda(\mathrm{K}) = 57$ to 240 mÅ. The authors have analyzed these results along with previously reported observations of Ca II in the spectra of QSOs, supernovae, and other extragalactic sources. In terms of the diameter $D(0)_{25}$ at $B = 25$ mag arcsec^{-2}, it appears that Ca II is detectable out to an angular distance of $\sim 0.6 - 1.4\ D(0)_{25}$ from the center of a galaxy.

157.068 **Far-infrared structure in the nuclear region of M51.**
D. F. Lester, P. M. Harvey, M. Joy.
Astrophys. J., Vol. 302, No. 1, p. 280 – 286 (1986).

Far-infrared (100 μm) scans across the nucleus of M51 are presented. These scans show that much of the far-infrared luminosity of this galaxy comes from a small, sharply bounded region within a 30″ radius (700 pc) of the nucleus. This radius corresponds to the innermost spiral structure. A nonlinear image restoration analysis of the data yields information on the spatial structure of star formation near the nucleus that exceeds the diffraction limit of the telescope at this wavelength. This "superresolution" treatment shows that star formation is enhanced at the outer edge of this region compared to the nucleus, which appears to be relatively devoid of young stars.

157.069 **De Andromedanevel. Schoolvoorbeeld van een spiraalstelsel.**
E. Brinks.
Zenit, 13. Jaarg., Nr. 2, p. 64 – 67 (1986).

157.070 **De Andromedanevel: een gatenkaas?**
E. Brinks.
Zenit, 13. Jaarg., Nr. 3, p. 105 – 109 (1986).

157.071 **The infrared (*H*) surface brightness of galaxies as a function of the morphological type.**
I. Manousoyannaki, G. Chincarini.
Astron. Astrophys., Vol. 160, No. 2, p. 331 – 334 (1986).

Using $H_{0.5}$ magnitudes and UGC diameters the authors compute the total infrared surface brightness SH of galaxies of different morphological type. They show that the infrared surface brightness so determined is strongly correlated with morphological type.

157.072 ***VRI* aperture photometry of southern E and S0 galaxies.**
P. Poulain.
Astron. Astrophys., Suppl. Ser., Vol. 64, No. 2, p. 225 – 236 (1986).

VRI aperture photometry of 115 E and S0 southern galaxies is presented. Out of this number, 86 objects had never been observed through the *I* filter. The results are in good agreement with other observations.

157.073 **H II regions in M33. II. Radio continuum survey.**
F. Viallefond, W. M. Goss, J. M. van der Hulst, P. C. Crane.
Astron. Astrophys., Suppl. Ser., Vol. 64, No. 2, p. 237 – 246 (1986).

The authors present a new WSRT survey of M33 in the continuum at 1.4 GHz. They have identified 112 radio sources with Hα nebulosities. From a comparison of this survey data and 5 GHz observations obtained with the VLA and WSRT, the authors have determined the spectral indices of 17 sources. Only three of these sources have non–thermal spectra. A compact source is discovered 1 arcmin south of the nucleus of M33. This source has a flat radio spectrum and no Hα counterpart. This source is either a heavily obscured compact H II region or, more likely, a source similar to the Crab nebula.

157.074 **Accurate positions of Zwicky galaxies. I.**
G. Vettolani, G. Palumbo, N. Santagata.
Astron. Astrophys., Suppl. Ser., Vol. 64, No. 2, p. 247 – 260 (1986).

Accurate optical positions are given for 1007 galaxies in the Zwicky Catalogue fields 502–505, 519–523, 536–539.

157.075 **Surface brightnesses of isolated and double galaxies.**
I. D. Karachentsev, V. E. Karachentseva, A. L. Shcherbanovskii (*A. L. Shcherbanovskij*).
Bull. Spec. Astrophys. Obs. – North Caucasus, Vol. 19, p. 1 – 11 (1985). English translation of 39.157.081.

157.076 **Spectrophotometric investigation of NGC 3353.**
A. N. Burenkov, Eh. E. Khachikyan.
Astrofizika, Tom 24, Vyp. 2, p. 349 – 361 (1986). In Russian. English translation in Astrophysics, Vol. 24, No. 2.

Results of detailed spectrophotometric investigations of NGC 3353 are presented. Four separate condensations, the brightest of which is Mark 35, are studied. In the spectrum of Mark 35 the emission lines from Hα to H_{12}, He I λλ7065, 6678, 5876, 4922, 4472, 3820 and forbidden lines [O I] λλ6300/64, [O II] λ3727, [O III] λλ5007, 4959, 4363, [Ne III] λ3869, [N II] λλ6584/48, [S II] λλ6717/31, [S III] λ6310, [Ar III] λ7136 are detected. In the second central condensation, called "nucleus", emission lines are weaker and beginning with Hβ the absorption components appear which become stronger in the late members of Balmer lines. The forbidden lines in the nucleus are strong: [Ne III], [O III], [O II], [S II], [N II]. The relative intensities and equivalent widths of emission lines as well as the chemical composition of Mark 35 and "nucleus" are estimated. Both condensations according to their physical properties look like super-associations. It has been concluded that the source of excitation are young stars. NGC 3353 is probably a net of super-associations.

157.077 **Galaxy evolution: some issues and questions.**
J. Silk.
Spectral evolution of galaxies, p. 3 – 11 (1986). – See Abstr. 012.015.

157.078 **Towards a theory of galaxy formation.**
J. Silk.
Spectral evolution of galaxies, p. 15 – 29 (1986). – See Abstr. 012.015.

This paper describes some aspects of how star formation processes are relevant to galaxy formation. The author commences by discussing the formation of the first stars in the collapsing protogalactic gas cloud. He then examines the characteristic parameters of the old spheroidal components of galaxies, and shows how important a role star formation processes must have played in their development and evolution.

157.079 **Theoretical and observational evidence for the existence of a single fragmentation law, and galactic evolution.**
A. Di Fazio, R. Capuzzo Dolcetta.
Spectral evolution of galaxies, p. 47 – 55 (1986). – See Abstr. 012.015.

Mass functions for various objects are elaborated from observational data. Their shapes are compared among themselves quantitatively. A striking similarity suggests the hypothesis that all the different kinds of astrophysical objects considered were formed by essentially the same process. Theoretical mass functions, based on the hypothesis of fragmentation due to gravitational instability (Di Fazio, 1985), are compared with the observations, yielding a remarkable good agreement.

157.080 **Star formation tracers in galaxies.**
J. Lequeux.
Spectral evolution of galaxies, p. 57 – 73 (1986). – See Abstr. 012.015.

This paper deals in a critical way with various methods of determining the present rate of star formation and the history of star formation in galaxies, with emphasis on quantitative aspects and on irregular galaxies.

157.081 **On the relationship between neutral hydrogen mass, luminosity and color in irregular galaxies.**
G. Comte.
Spectral evolution of galaxies, p. 81 – 86 (1986). – See Abstr. 012.015.

The author presents evidence for a dependence on the luminosity of the neutral hydrogen mass to luminosity ratio in a large sample of irregular galaxies of moderate or low surface brightness. This is not valid for active bright irregulars of the Markaryan class. Active star–forming galaxies are segregated from quiescent ones in an (U–V) versus (log M_H/L) plot. This diagram can be interpreted as an evolutionary diagram if one supposes that the formation in these galaxies proceeds by recurrent bursts.

157.082 **I Zw 18 and the search for very metal poor galaxies.**
D. Kunth.
Spectral evolution of galaxies, p. 87 – 90 (1986). – See Abstr. 012.015.

157.083 **Bolometric luminosity evolution.**
C. J. Lonsdale.
Spectral evolution of galaxies, p. 91 – 96 (1986). – See Abstr. 012.015.

The observations made by IRAS are capable of detecting evolution in the OB star formation rates of galaxies at cosmologically interesting redshifts if that evolution takes the form of star formation rates ten or more times higher than seen in nearby galaxies. Spectroscopic follow–up observations of IRAS–selected galaxies, and IRAS source counts, indicate that there may indeed exist a rare population of galaxies with such high star formation rates. The most luminous found so far approach quasars in luminosity.

157.084 **Galaxy encounters and the Holmberg effect.**
B. F. Madore.
Spectral evolution of galaxies, p. 97 – 102 (1986). – See Abstr. 012.015.

It is pointed out that interacting double galaxies are the most common form of peculiar galaxy found in recent compilations, indicating that interactions which substantially alter a galaxy's morphology are relatively common. Continuing effects of the encounters on the star formation rate may be revealing themselves in the Holmberg Effect, which is a strong color correlation and coupling of apparent Hubble types found among binary galaxies.

157.085 **The effects of galaxy–galaxy interactions on nuclear activity.**
J. M. van der Hulst, E. Hummel, W. C. Keel, R. C. Kennicutt Jr.
Spectral evolution of galaxies, p. 103 – 110 (1986). – See Abstr. 012.015.

The authors have undertaken a systematic optical and radio study of a large sample of interacting spiral galaxies. They chose spiral galaxies because they have significant amounts of gas which enable them to trace the activity whether in the form of enhanced star formation or non–thermal nuclear activity. The purpose of this contribution is to merely summarize the main results the authors have obtained so far.

157.086 **Galaxies in the infrared.**
T. de Jong.
Spectral evolution of galaxies, p. 111 – 126 (1986). – See Abstr. 012.015.

The author reviews IRAS studies of galaxies and attempts to systematically present the new insights that have resulted from these studies.

157.087 **IRAS far–infrared observations of interacting galaxies.**
K. Brink.
Spectral evolution of galaxies, p. 127 – 130 (1986). – See Abstr. 012.015.

Using the large IRAS database, the author investigates the effect of interaction between galaxies on their far–infrared properties by comparing the infrared–to–blue luminosity ratios of optically selected samples of interacting and field galaxies, and by briefly discussing the results of additional mapping with IRAS of a few interacting galaxy pairs.

157.088 **The stellar populations of galaxies in the Local Group.**
J. Mould.
Spectral evolution of galaxies, p. 133 – 141 (1986). – See Abstr. 012.015.

157.089 **The AGB population of nearby galaxies.**
M. Aaronson, K. H. Cook, J. Norris.
Spectral evolution of galaxies, p. 171 – 181 (1986). – See Abstr. 012.015.

The authors report the results of a preliminary survey for carbon and M stars in Local Group galaxies using a newly developed photometric technique. The surveyed objects include M31, M33, NGC 6822, IC 1613, and WLM. The C/M ratio is seen to be strongly dependent on parent galaxy absolute magnitude, a finding that has important bearing on our understanding of galaxian star formation history.

157.090 **M33: radial distributions and a comparison of its global luminosity function with other nearby galaxies.**
W. L. Freedman.
Spectral evolution of galaxies, p. 183 – 192 (1986). – See Abstr. 012.015.

The upper end of the luminosity functions for luminous blue stars is presented for a sample of 10 nearby galaxies. The slope of this luminosity function does not show significant variation from galaxy to galaxy, nor as a function of radius in M33. The blue–to–red supergiant and Wolf–Rayet to red supergiant ratios are also consistent with showing no radial trend. The radial surface distributions of neutral hydrogen, H II regions, OB stars, 6 cm radio continuum, Wolf–Rayet stars and blue surface brightness are compared. With the exception of the relatively flat neutral hydrogen distribution, all of these other components exhibit very similar scale lengths.

157.091 **Global properties of stellar populations and the spectral evolution of galaxies.**
A. Renzini, A. Buzzoni.
Spectral evolution of galaxies, p. 195 – 235 (1986). – See Abstr. 012.015.

Several relevant properties of evolving stellar populations are discussed from a theoretical viewpoint. The authors introduce the concept of phase transition in the stellar content of a population, an event which is produced by the first appearence of stars with either C–O or He degenerate cores. The spectral energy distribution of a population is then predicted to suffer a (major) change at each of these phase transitions. Some uncertainties in both current stellar and population evolutionary models, and in population synthesis techniques are then briefly discussed, and the need is emphasized for appropriate checks and calibrations using template stellar populations.

157.092 **Spectral evolution of galaxies.**
G. Bruzual A.
Spectral evolution of galaxies, p. 263 – 282 (1986). – See Abstr. 012.015.

A summary of the current situation in the subject of spectral evolution of galaxies is presented. First the existing uncertainties in the interpretation of the stellar populations in nearby galaxies are analyzed and the model dependency of the results is discussed. Then a few samples of faint galaxies with measured redshift and photometric properties are used to derive the degree of evolution observable in these samples with respect to nearby galaxies. Finally, recent results that claim that 3CR radio galaxies show evidence of dramatic evolution are discussed. It is concluded that for the vast majority of galaxies spectral evolution is a slow process which may escape detection. In a given volume element galaxies of very different populations coexist, producing a wide spread in the observable properties at any redshift.

157.093 **Spectrophotometric models of galaxies.**
G. Barbaro, F. M. Olivi.
Spectral evolution of galaxies, p. 283 – 307 (1986). – See Abstr. 012.015.

This paper analyses the basic assumptions for the computation, by the evolutionary synthesis, of galaxy spectral energy distributions. The input from the theory of stellar atmospheres is analysed. As intermediate step in the process of galaxy synthesis, models are presented for single generations of stars of different ages; for old generations two chemical compositions are used: $Z = 10^{-4}$ and $Z = 10^{-2}$. Such models are compared with the observed spectra of clusters in the LMC and of globular clusters of our galaxy. Galaxy models are then constructed by changing the values of some parameters and allowing the presence of two populations.

157.094 **Photometric evolution of elliptical galaxies in the color–magnitude diagram.**
N. Arimoto, Y. Yoshii.
Spectral evolution of galaxies, p. 309 – 318 (1986). – See Abstr. 012.015.

Photometric evolutions of elliptical (E) galaxies are investigated by introducing galactic winds into the evolutionary population synthesis models. It is shown that E galaxies are one parameter family of their initial masses. E galaxies are composite stellar systems composed of stars with a wide variety of metallicity, and have the same age and the same initial mass function. The color–magnitude relation for E galaxies comes from their different average metallicities. Both color and luminosity evolutions of E galaxies depend strongly on their average metallicities and, hence, on their initial masses.

157.095 **Analysis of stellar populations at large lookbacks.**
R. W. O'Connell.
Spectral evolution of galaxies, p. 321 – 343 (1986). – See Abstr. 012.015.

The capabilities of optimizing spectral synthesis for interpretation of stellar populations are reviewed with emphasis on the growing evidence that the light of nearby E/S0 galaxies and Sb bulges is dominated by a population significantly younger than expected from the study of globular cluster color–magnitude diagrams. The implications of these results for the study of high redshift galaxies are considered. The observational requirements and analytical tools necessary for proper interpretation of populations in galaxies viewed at large lookbacks are discussed in the context of the capabilities of the Hubble Space Telescope and the 10–m telescope.

157.096 **Population synthesis and epochs of star formation in NGC 1316 (Fornax A).**
A. Pickles.
Spectral evolution of galaxies, p. 345 – 356 (1986). – See Abstr. 012.015.

Population synthesis models are presented for the galaxy NGC 1316 with the main emphasis being placed on deriving epochs of star formation in this disturbed and possibly merged system. IRAS and radio results indicate that the nucleus of this dusty galaxy is optically obscured, and that the observed optical light is emitted along a line of sight which becomes opaque about 10 arcsec in front of the nucleus. The star formation history qualitatively indicates extended periods or recurrent bursts of star formation, with strong current star formation as well. The detailed synthesis results depend on the internal reddening, and results are tabulated for a range of reddenings encompassing the best estimate of E(B–V) = 0.13 mag.

157.097 **Far–UV stellar populations of S0 galaxies.**
B. Rocca–Volmerange, B. Guiderdoni.
Spectral evolution of galaxies, p. 357 – 362 (1986). – See Abstr. 012.015.

IUE spectra of S0 galaxies are analyzed in correlation with their gas content. For two respectively gas–rich (NGC 5102) and gas–poor (NGC 3115) S0 galaxies, a preliminary stellar population synthesis with IUE Atlas shows that the old population (A or G spectral type) is surprisingly uniform and of one single type. It contributes mostly to the 2000 Å – 3000 Å flux and also partly to the 1500 Å – 2000 Å range. For the two galaxies, far–UV excesses are low and strongly increasing with shorter wavelengths.

157.098 **The UV energy distribution of elliptical galaxies.**
F. Bertola.
Spectral evolution of galaxies, p. 363 – 367 (1986). – See Abstr. 012.015.

157.099 **UV spectra of normal ellipticals.**
R. Nesci.
Spectral evolution of galaxies, p. 369 – 371 (1986). – See Abstr. 012.015.

157.100 **Studies of cluster galaxies at large lookback times.**
A. Dressler.
Spectral evolution of galaxies, p. 375 – 389 (1986). – See Abstr. 012.015.

157.101 **Observational tests for galaxy evolution.**
D. Hamilton.
Spectral evolution of galaxies, p. 391 – 401 (1986). – See Abstr. 012.015.

A discussion of recent optical observational work using the look–back time method is presented. This discussion includes brief outlines of the techniques of and problems associated with broad–band counts of galaxies, broad–band photometric comparisons, and spectrophotometric comparisons. The importance of the spectrophotometric comparisons is emphasized as well as the most recent work in this field. The evidence for evolution has not been solid.

157.102 **Colors of 3CR and first–ranked high redshift galaxies.**
P. R. M. Eisenhardt.
Spectral evolution of galaxies, p. 403 – 410 (1986). – See Abstr. 012.015.

Forty radio selected and thirty nine optically selected giant elliptical galaxies of known redshifts ranging from 0.019 to 1.2 have been observed simultaneously in four colors from 0.4 to 2.3 microns. No evidence for differences between the colors of radio and optically selected galaxies is found. The infrared H–K colors are best fit by a passively evolving model with little residual star formation. The optical–IR colors are contained by a red envelope well fit by a passively evolving model, but some galaxies show strong blue deviations for z > 0.4. This behavior is most easily explained by episodes of star formation involving small fractions of the total number of stars.

157.103 **Evolution of disk galaxies in high–redshift clusters.**
B. Guiderdoni, B. Rocca–Volmerange.
Spectral evolution of galaxies, p. 411 – 418 (1986). – See Abstr. 012.015.

The authors propose an evolutionary scenario for spiral galaxies in rich, compact clusters. Interactions with the environment, probably dominated by the processes in the hot gaseous intergalactic medium, remove the gas from the disks of blue objects infalling into cluster cores from outer layers. Predictions are made for UBV colors of cluster spirals versus the redshift z, accounting for cosmological and evolutionary effects, by means of a model of spectrophotometric evolution.

157.104 **Quests for primeval galaxies: a review of optical surveys.**
D. C. Koo.
Spectral evolution of galaxies, p. 419 – 438 (1986). – See Abstr. 012.015.

Optical surveys explicitly designed to detect continuum or line emission from individual primeval galaxies are reviewed. In addition to such searches, a variety of recent redshift surveys of very faint quasars, radio sources, and galaxies are also useful in constraining the epoch, number, and brightness of galaxies during their birth. Presently the strongest constraints come from long–slit spectroscopy of blank sky to search for Lyman α emission. The lack of positive detections suggests that galaxy formation occurred at redshifts z > 5, a result consistent with the lack of primeval galaxies among samples of quasars fainter than B = 20 and with the mild evolution predicted by models that fit faint galaxy counts and colors.

157.105 **New observations of galaxy number counts.**
P. R. F. Stevenson, T. Shanks, R. Fong.
Spectral evolution of galaxies, p. 439 – 446 (1986). – See Abstr. 012.015.

157.106 **Nitrogen and oxygen evolution in nearby spiral galaxies.**
M. Tosi, A. I. Diaz.
Spectral evolution of galaxies, p. 469 – 472 (1986). – See Abstr. 012.015.

157.107 **Surface brightness distributions in two edge–on spiral galaxies.**
M. A. Shaw, G. F. Gilmore.
Spectral evolution of galaxies, p. 477 – 484 (1986). – See Abstr. 012.015.

157.108 **Morphology of cD, D, and gE galaxies.**
J. M. Schombert.
Bull. Am. Astron. Soc., Vol. 17, No. 4, p. 858 (1985). Abstract. – See Abstr. 010.062.

157.109 **Results of an H I survey of a sample of IRAS galaxies.**
J. T. Armstrong, H. A. Wootten.
Bull. Am. Astron. Soc., Vol. 17, No. 4, p. 858 (1985). Abstract. – See Abstr. 010.062.

157.110 **Two normal spiral nuclei: NGC 2903 and NGC 4321.**
J. W. Goad, J. S. Gallagher, M. R. Laycox, A. M. Albert.
Bull. Am. Astron. Soc., Vol. 17, No. 4, p. 859 (1985). Abstract. – See Abstr. 010.062.

157.111 **Computer processed color images of galaxies.**
P. E. Seiden, D. M. Elmegreen, B. G. Elmegreen, A. Mobarak.
Bull. Am. Astron. Soc., Vol. 17, No. 4, p. 859 (1985). Abstract. – See Abstr. 010.062.

157.112 **An I vs B–I color–magnitude relation for galaxies.**
M. J. Pierce.
Bull. Am. Astron. Soc., Vol. 17, No. 4, p. 859 (1985). Abstract. – See Abstr. 010.062.

157.113 **Abundance gradients in barred spiral galaxies.**
D. L. Talent.
Bull. Am. Astron. Soc., Vol. 17, No. 4, p. 859 (1985). Abstract. – See Abstr. 010.062.

157.114 **Model UBVR colours of disk galaxies.**
J. M. Yamanaka.
Bull. Am. Astron. Soc., Vol. 17, No. 4, p. 859 (1985). Abstract. – See Abstr. 010.062.

157.115 **IUE observations of luminous blue star associations in irregular galaxies.**
S. A. Lamb, D. A. Hunter, J. S. Gallagher.
Bull. Am. Astron. Soc., Vol. 17, No. 4, p. 860 – 861 (1985). Abstract. – See Abstr. 010.062.

157.116 **The spatial distribution of star forming regions in interacting spiral galaxies.**
H. A. Bushouse.
Bull. Am. Astron. Soc., Vol. 17, No. 4, p. 861 (1985). Abstract. – See Abstr. 010.062.

157.117 **Optical and infrared surface photometry of NGC 3077.**
J. S. Price, C. A. Gullixson.
Bull. Am. Astron. Soc., Vol. 17, No. 4, p. 861 (1985). Abstract. – See Abstr. 010.062.

157.118 **Two color CCD mapping of the luminous type I irregular galaxy NGC 4449.**
G. D. Bothun.
Bull. Am. Astron. Soc., Vol. 17, No. 4, p. 861 (1985). Abstract. – See Abstr. 010.062.

157.119 **A search for cepheids in M101.**
K. H. Cook, M. Aaronson, G. Illingworth.
Bull. Am. Astron. Soc., Vol. 17, No. 4, p. 872 (1985). Abstract. – See Abstr. 010.062.

157.120 **Two cepheid variables in the Fornax dwarf galaxy.**
R. M. Light, T. E. Armandroff, R. J. Zinn.
Bull. Am. Astron. Soc., Vol. 17, No. 4, p. 883 (1985). Abstract. – See Abstr. 010.062.

157.121 **Radio and X-ray properties of elliptical and S0 galaxies.**
G. Fabbiano, U. Klein, G. Trinchieri, R. Wielebinski.
Bull. Am. Astron. Soc., Vol. 17, No. 4, p. 889 (1985). Abstract. – See Abstr. 010.062.

157.122 **Regions of massive star formation in two nearby spiral galaxies.**
S. D. Lord.
Bull. Am. Astron. Soc., Vol. 17, No. 4, p. 892 – 893 (1985). Abstract. – See Abstr. 010.062.

157.123 **The halo and disk population of M31.**
A. P. S. Crotts.
Bull. Am. Astron. Soc., Vol. 17, No. 4, p. 893 (1985). Abstract. – See Abstr. 010.062.

157.124 **Radio continuum observations of the central region of the barred galaxy NGC 4314.**
J. A. García-Barreto, P. Pismis.
Bull. Am. Astron. Soc., Vol. 17, No. 4, p. 893 (1985). Abstract. – See Abstr. 010.062.

157.125 **Cooling flows and the X-ray emission of elliptical galaxies.**
C. L. Sarazin.
Bull. Am. Astron. Soc., Vol. 17, No. 4, p. 907 (1985). Abstract. – See Abstr. 010.062.

157.126 **The X-ray properties of normal galaxies. What do we know and what can we learn with the next generation of X-ray telescopes?**
G. Fabbiano.
ESA Spec. Publ., ESA SP-239, p. 33 – 41 (1985). – See Abstr. 012.019.

157.127 **Systematics of bulge-to-disk ratios.**
F. Simien, G. de Vaucouleurs.
Astrophys. J., Vol. 302, No. 2, p. 564 – 578 (1986).
Decompositions of the blue-band luminosity profiles of 98 galaxies into spheroidal ($r^{1/4}$) and disk (exponential) components on a homogeneous system are used to study the systematics of bulge-to-disk ratios and related parameters. The mean dependences on morphological type of the fractional luminosity of the spheroid, of the effective radii and specific intensities, and of the mean absolute magnitudes of each component are established. The correlations between effective radius and specific intensity for ellipticals and the spheroidal components of lenticulars and spirals are compared. The data confirm that the Hubble classification sequence is primarily one of decreasing bulge-to-disk ratios.

157.128 **The Hα velocity widths of giant H II regions as distance indicators.**
J.-R. Roy, R. Arsenault.
Astrophys. J., Vol. 302, No. 2, p. 579 – 584 (1986).
The mean velocity widths $<W>$ of the Hα line profiles of the largest giant H II regions in 10 or more galaxies are compared with the mean diameters $<D>$ of the same H II regions to evaluate the relative merits of $<D>$ and $<W>$ as distance indicators. It is shown that the mean velocity widths deduced from integrated Hα line profiles correlate more strongly with the absolute magnitudes of galaxies than the mean diameters of the three largest H II regions; this also holds when isophotal diameters of H II regions are used.

157.129 **Very extended infrared emission at the center of M51: the role of star formation.**
C. M. Telesco, R. Decher, I. Gatley.
Astrophys. J., Vol. 302, No. 2, p. 632 – 639 (1986).
The authors have detected low surface brightness 10 μm emission in M51 extending more than 1′, or 3 kpc, in RA. The emission has roughly uniform brightness and originates from dust in a region that is prominent visually and centered on the nucleus. Complementary near-infrared scans, which are dominated by the light from red stars, also show notable structure from hot dust. The dust is heated predominantly by young stars. Simple starburst models imply that the star formation rate per square parsec in the extended infrared disk is at least two to three orders of magnitude higher than in the solar neighborhood.

157.130 **Radio observations of starburst galaxies.**
R. A. Sramek, D. W. Weedman.
Astrophys. J., Vol. 302, No. 2, p. 640 – 649 (1986).
Scaled array observations were made with the VLA at 1465 MHz and 4885 MHz of starburst regions having various morphologies. These are blue compact galaxies Mrk 59 and II Zw 40; starburst nuclei Mrk 52, 357, 603, 1089, 1259, and 1344; and giant H II regions NGC 595, 604, 5455, 5461, and 5471 in M33 and M101. Radio spectra are deconvolved into thermal and nonthermal components, which relate to the ionizing radiation from the starburst and the number of supernova remnants from the starburst. Starburst models based on various forms of the initial mass function are compared to the radio observations and to *IRAS* data.

157.131 **Distributions of ^{13}CO emission in the disks of late-type spiral galaxies.**
J. S. Young, D. B. Sanders.
Astrophys. J., Vol. 302, No. 2, p. 680 – 692 (1986).
The authors have observed the radial distributions of ^{13}CO and ^{12}CO emission in six late-type spiral galaxies using the 14 m millimeter telescope of the FCRAO (HPBW = 45″). Emission from the ^{13}CO molecule was detected at a total of 27 out of 32 positions observed in these galaxies, out to radii of 4 – 5 kpc in IC 342, NGC 6946, and M83, and out to 5 – 7 kpc in NGC 1068, NGC 4321, and M51. Each of the galaxies shows the highest ^{13}CO integrated intensity in the center. The mean ^{12}CO/^{13}CO integrated intensity ratio, R, in the six galactic centers is 11.0. The global mean value of this ratio for each galaxy ranges between 8.3 ± 2.0 and 14.7 ± 2.0, with the higher value found in NGC 1068 and the lower values in NGC 4321 and IC 342. There is strong evidence that R is higher in galaxies with higher dust temperatures. The values of R in the external galaxies closer than 10 Mpc show greater than 3 σ variations at adjacent positions, corresponding to size scales of 1 – 2 kpc; these variations are similar to those seen in the disk of the Milky Way.

157.132 **Physical conditions in nuclei of galaxies with emission lines. Rotation, mass and parameters of the nucleus of the galaxy NGC 1084.**
G. T. Petrov, V. A. Mineva, G. A. Kyazoumov (*G. A. Kyazumov*).
Dokl. Bolg. Akad. Nauk, Vol. 38, No. 6, p. 699 – 702 (1985). Abstr. in Ref. Zh., 51. Astron., 3.51.909 (1986).

157.133 **A feature of the spiral pattern and star formation in the galaxy NGC 1566.**
I. I. Pasha.
Astron. Tsirk., No. 1387, p. 4 – 6 (1985). In Russian.
The NGC 1566 spiral pattern is modelled upon data on H II regions with account of their specific arrangement in the arms. The corotation radius is found to be $66'' \pm 10''$, the pattern speed $\Omega_p = 24 \pm 3$ km/s/kpc. It is suggested that the centrifugal instability in the internal H II ring (R = 35″) may excite the bright arms.

157.134 **On the galaxy NGC 6207.**
A. N. Deutsch.
Astron. Tsirk., No. 1396, p. 6 – 8 (1985). In Russian.

157.135 **Discovery of the optical variability of the nucleus of M33.**
V. M. Lyutyj, A. S. Sharov.
Astron. Tsirk., No. 1408, p. 1 – 2 (1985). In Russian.

157.136 **Optical emission–line gas in the nuclei of normal galaxies: implications for nuclear activity.**
T. M. Heckman.
Publ. Astron. Soc. Pac., Vol. 98, No. 600, p. 159 – 162 (1986). Review paper. – See Abstr. 012.022.
This review focuses on the properties of the optical emission–line gas in normal galactic nuclei and how these properties relate to the more extreme forms of nuclear activity seen in Seyfert galaxies, radio galaxies, QSOs, etc.

157.137 **Nuclei of normal galaxies.**
R. W. O'Connell.
Publ. Astron. Soc. Pac., Vol. 98, No. 600, p. 163 – 170 (1986). Review paper. – See Abstr. 012.022.
The author reviews recent research on the environment in which active nuclei reside, namely the dense stellar populations found in the centers of elliptical galaxies and spiral bulges. He discusses new developments concerning the structure and dynamics of nuclei, recent results on the age and chemical composition of nuclear populations, the possibility that important changes in the initial mass function occur in nuclei, limits which can be placed on recent star formation in old populations, and the detection of "submerged" active nuclei in outwardly normal galaxies. Emphasis is placed on problems for which space astronomy experiments are well suited.

157.138 **The disks of spiral galaxies.**
D. J. Adams.
Vistas Astron., Vol. 29, Part 1, p. 1 – 15 (1986).
This article aims to give an overview of the present state of knowledge about the disks of spiral galaxies obtained from a very large number of observations at various wavelengths. Theoretical ideas which are essential for an understanding of much of the observational material are sketched and observations described. An attempt has been made to give a picture at a uniform "low resolution", avoiding an in–depth treatment of selected topics. References given are biassed towards recent, rather than the original, mention of a topic, and it is hoped that the reader will so find his way into the literature on a chosen topic.

157.139 **Static properties of galaxies in the Shapley–Centaurus cluster. I. The catalogue.**
R. J. Dodd, F. P. Andrews, H. T. MacGillivray.
Astrophys. Space Sci., Vol. 122, No. 1, p. 63 – 72 (1986).
A catalogue of the positions, sizes, orientations, ellipticities, and morphological descriptions of 127 galaxies measured in the Southern Sky Survey Field 444 is given. The measurement errors, both internal and external, are discussed.

157.140 **Binary and multiple galaxies.**
G. M. Ballabh.
Binary and multiple systems, p. 47 – 53 (1984). – See Abstr. 012.025.
The author discusses the statistical studies which are important in knowing the masses, mass–to–luminosity ratios, morphological type relation etc. of binary and multiple galaxies. Then he looks at the dynamical aspects which are of current interest as more and more phenomena associated with the galaxies are being attributed to their mutual interactions.

157.141 **High–resolution X–ray and radio observations of the giant elliptical galaxies NGC 4636 and 4649.**
V. J. Stanger, R. S. Warwick.
Mon. Not. R. Astron. Soc., Vol. 220, No. 2, p. 363 – 376 (1986).
Einstein HRI observations of the giant elliptical galaxies NGC 4636 and 4649 reveal that both are extended soft X–ray sources with luminosities $\sim 1.5 \times 10^{41}$ erg s^{-1}. Most of the emission probably arises from hot gas with $kT \sim 1$ keV. Deprojection of the X–ray data yields gas parameters which are consistent with the presence of cooling accretion flows throughout both galaxies. VLA observations reveal extended radio sources with luminosities $\sim 10^{38}$ erg s^{-1} in both galaxies. The thermal pressure of the ambient hot gas can strongly confine these radio sources and through pressure gradients may give rise to the twisted morphology of the radio source in NGC 4636. The in–falling cooling gas may also fuel the radio sources by accretion on to a massive compact object.

157.142 **The properties of the nuclei of the two hot spot galaxies NGC 1097 and 2997.**
J. R. Walsh, K. Nandy, G. I. Thompson, J. Meaburn.
Mon. Not. R. Astron. Soc., Vol. 220, No. 2, p. 453 – 471 (1986).
NGC 1097 and 2997 are barred–spiral galaxies with hot spots in annuli around their central red nuclei. *IUE* low–dispersion spectra of two hot spots in NGC 1097 and of the ensemble of knots in the annulus around the central nucleus of NGC 2997 are presented. In addition, low–dispersion visible spectra of the hot spots and nuclei of both galaxies are analysed, and near–infrared photometry has been obtained. The nature of these hot spots and their possible relation to the inner nuclei is determined.

157.143 **Optical polarization in the spiral galaxy NGC 7331.**
D. J. King.
Mon. Not. R. Astron. Soc., Vol. 220, No. 2, p. 485 – 491 (1986).
An optical linear polarization map is presented for the highly inclined spiral galaxy NGC 7331. Along the major axis of the galaxy the polarization appears to arise from the scattering of light from the bright central nucleus, but elsewhere the dominant polarizing mechanism appears to be the transmission of light through grains aligned by the galactic magnetic field and the Davis–Greenstein mechanism, in a similar way to the polarization of starlight in our own Galaxy.

157.144 **The galactic wind in the circum–nuclear zone of NGC 6946.**
J. E. Beckman, C. Mũnoz Tuñon, E. Battaner, M. Prieto, M. L. Sanchez Saavedra.
Astron. Astrophys., Vol. 161, No. 1, p. 55 – 60 (1986).
The authors have obtained high resolution ($\Delta\lambda \cong 0.35$ Å) long slit spectra of the central zone of the galaxy NGC 6946. Velocity curves were derivable from the strong Hα emission which comes from the circum–nuclear H II regions. These curves show the striking feature that the radial velocity gradient, and the limiting radial velocity along the minor axis are much greater than those along the major axis, which is in direct contrast to the prediction of the distribution due to a rotating galaxy. Using CO spectra from the literature the authors conclude that the central zone of NGC 6946, within 1 kpc of the centre, is in a state of expansion. The mechanisms, which drive this expansion, are discussed in detail.

157.145 **Near–infrared mapping of spiral galaxies. II. *J, H, K* profiles of M31.**
E. Battaner, J. E. Beckman, E. Mediavilla, M. Prieto,
C. Sanchez Magro, C. Muñoz Tuñon, M. L. Sanchez Saavedra.
Astron. Astrophys., Vol. 161, No. 1, p. 70 – 74 (1986).

The authors present four radial profiles of M31 in the near infrared. Three of these, in the *J, H, K* bands extend 20 kpc eastward from the galactic centre, while they also have a *K* band profile from the centre to 20 kpc westward. The authors have combined these data with published visual observations. Intrinsic magnitudes and colours were obtained after removing the effects of the extinction in our own galaxy and intrinsic extinction within M31. At its maximum the latter is close to 1 mag in the visible. A bluer zone of stellar population coincident with the known position of the gaseous ring in the disc of M31 is clearly shown. The photometric properties of the M31 blue zone are comparable to those found by Prieto et al. (1985), for the ring in NGC 2841.

157.146 **Star formation rates, galaxy morphology, and the Hubble sequence.**
A. Sandage.
Astron. Astrophys., Vol. 161, No. 1, p. 89 – 101 (1986).

The fraction of the mass ever made into stars that is forever locked–up in those stars is calculated. The shape of the integrated birth rate (IBR) and the integrated lock–up rate (ILUR) functions as the termination mass at the faint end of the initial mass function (IMF) is varied is discussed. The observed IBR and the derived ILUR for the LMC, SMC, and Sextans A galaxies are listed. The time to consume the remaining gas in LMC, SMC, and Sextans A, and the factor by which this time can be increased if bimodal star formation exists for various faint termination masses, M_T, of the IMF is given. An empirical calibration of the Scoville and Young (1983) method of obtaining the IBR (and from it, the ILUR) using the total luminosity of the young stars is given. Application of these calculations to the Virgo cluster Sm/Im dwarfs is presented. The gas consumption times for bimodal star formation with M^i(limit) $= 2\ M_{\odot}$ are set out. Finally, the evolution of the Hubble sequence with time for various Hubble types is discussed in terms of the time variation of the star formation rate.

157.147 **Study of a complete sample of galaxies. III. Discussion and results.**
P. Véron, M.–P. Véron–Cetty.
Astron. Astrophys., Vol. 161, No. 1, p. 145 – 149 (1986).

The authors have undertaken a photoelectric and spectroscopic study of the nuclear region of a complete sample of nearby galaxies. They discuss these two sets of data with the aim of clarifying which conditions are necessary in the nucleus of a galaxy to give rise either to a Seyfert–like nebulosity or to an H II region.

157.148 **Multi–frequency radio continuum observations of dwarf irregular galaxies.**
U. Klein, R. Gräve.
Astron. Astrophys., Vol. 161, No. 1, p. 155 – 168 (1986).

The authors report radio continuum observations of nearby dwarf irregular galaxies carried out with the Effelsberg 100–m telescope. All sample galaxies have been mapped at 4.75 GHz (λ6.3 cm) to obtain accurate integrated flux densities. The three radio–brightest galaxies in the sample, IC 10, NGC 4449, and NGC 1569, have in addition been mapped at 10.7 GHz (λ2.8 cm) and 24.5 GHz (λ1.2 cm) to enable a more detailed study of their high–frequency radio continuum morphologies and to study their spectral indices. The sample represents a wide range of current star–forming activity of dwarf irregular galaxies.

157.149 **Anisotropies in the orientation of the Uppsala and ESO/Uppsala galaxies.**
A. V. Mandzhos, V. V. Tel'nyuk–Adamchuk, A. Ya. Gregul'.
Sov. Astron. Lett., Vol. 11, No. 4, p. 206 – 209 (1985). English translation of 40.157.002.

157.150 **A case against baryons in galactic halos.**
D. J. Hegyi, K. A. Olive.
Astrophys. J., Vol. 303, No. 1, p. 56 – 65 (1986).

The possibility that galactic halos are composed of baryonic matter is discussed in detail. Specifically, halos composed of snowballs, dust and rocks, planets, stars, dead stellar remnants, and hot and cold gas are considered. The serious problems that would arise for each of these types of matter lead to the conclusion that halos cannot plausibly contain substantial amounts of such matter. This conclusion is considered in relation to big bang nucleosynthesis. Various nonbaryonic candidates for halos are also discussed.

157.151 **Evidence for nonaxisymmetric nuclear bulges in spiral galaxies.**
D. Zaritsky, K. Y. Lo.
Astrophys. J., Vol. 303, No. 1, p. 66 – 75 (1986).

The $\lambda = 0.8$ μm surface brightness distribution of the nuclear region of 11 galaxies was analyzed to look for evidence of deviation from axisymmetry. Nuclear bulges are clearly present in all the galaxies, including the Sc galaxies. Variation of the position angle and ellipticity with semimajor axis length of the elliptical isophotes of the nuclear region was found in all the galaxies. This suggests that nonaxisymmetric nuclear bulges could be very common in ordinary spiral galaxies.

157.152 **Photometric redshifts of galaxies.**
E. D. Loh, E. J. Spillar.
Astrophys. J., Vol. 303, No. 1, p. 154 – 161 (1986).

The authors describe a photometric method for measuring thousands of redshifts of 22 mag galaxies with a charge–coupled detector. All objects in a field are photometered between 400 and 950 nm with six filters. The data for each object are matched with colors of a set of fiducial objects, which are several types of galaxies at many redshifts and several types of stars; the object is identified as the galaxy or the star represented by the fiducial object that matches most closely. The photometric redshifts of the galaxies in the field of the cluster 0024+1654 are compared with the spectroscopic redshifts of Dressler, Gunn, and Schneider. Agreement is found for 30 out of 34 galaxies.

157.153 **IRAS observations of a small sample of blue irregular galaxies.**
D. A. Hunter, F. C. Gillett, J. S. Gallagher III, W. L. Rice,
F. J. Low.
Astrophys. J., Vol. 303, No. 1, p. 171 – 185 (1986).

The authors use IRAS infrared observations to explore the properties of the dust in a small sample of irregular galaxies chosen to cover a wide range in levels of star formation activity. The irregulars are found to have relatively normal infrared properties. The ratios of 25 μm to 12 μm flux densities are similar among this sample of irregulars in spite of the differences in levels of star formation activity. The integrated far–IR (60, 100 μm) dust color–temperatures range from 26 to 53K and cover the range of values for normal spiral galaxies. The ratio L_{IR}/L_B varies widely among the sample. The $L_{IR}/L_{H\alpha}$ ratios are roughly consistent with luminosities measuring the integrated OB stellar luminosity output.

157.154 **Globular clusters in the Andromeda galaxy M31.**
M. Iye, J. Watanabe.
Astron. Her., Vol. 79, No. 5, p. 112 – 116 (1986). In Japanese.

157.155 **Ultraviolet–excess galaxies.**
B. Takase, H. Maehara, T. Noguchi.
Galactic structure and variable stars, p. 59 – 66 (1985). – See Abstr. 012.029.

General remarks on the observation and research of ultraviolet–excess galaxies which have recently been carried out in Japan are presented. Preliminary results from follow–up observations are given and the nature of the ultraviolet–excess phenomena is discussed.

157.156 **Elliptical galaxies with dust lanes.**
D. Carter, R. M. Sharples, T. G. Hawarden, A. J. Longmore.
Proceedings of the Second Asian–Pacific Regional Meeting on Astronomy, p. 304 – 307 (1984). – See Abstr. 012.031.
The orientation and dynamics of dust lanes in elliptical galaxies does indeed give a clue as to the three dimensional potential of the stellar component. The authors emerge with a picture of NGC 5363 as a prolate bar, lying more or less in the plane of the sky and tumbling through it. The dynamics of the stars and gas are consistent with this general picture.

157.157 **The fossil shell of NGC 1344.**
D. A. Allen, D. Carter, D. F. Malin.
Proceedings of the Second Asian–Pacific Regional Meeting on Astronomy, p. 523 (1984). Abstract. – See Abstr. 012.031.

157.158 **Surface photometry of edge–on galaxies.**
M. Hamabe, S. Okamura.
Proceedings of the Second Asian–Pacific Regional Meeting on Astronomy, p. 551 – 555 (1984). – See Abstr. 012.031.
The authors have observed nine edge–on or nearly edge–on galaxies, NGC 3115 ($S0^-$), 4350 (S0), 4762 ($S0^0$), 4710 ($S0^+$), 5866 ($S0^+$), 4594 (Sa), 7814 (Sab), 4565 (Sb), 4244 (Scd), and decomposed their luminosity profiles into the spheroid and the disk components by means of the semi–empirical three–dimensional model and investigated the global characteristics of these two components.

157.159 **A study for the quantitative classification of galaxies.**
S. Okamura, M. Watanabe, K. Kodaira.
Proceedings of the Second Asian–Pacific Regional Meeting on Astronomy, p. 556 – 559 (1984). – See Abstr. 012.031.
A new method of reducing the photometric data of galaxies obtained with the Kiso Schmidt telescope is developed for a purpose of analyzing galaxy properties. Twenty galaxies in the Virgo cluster are analyzed in this new system.

157.160 **Population synthesis in the nuclei of Fornax, E, S0 galaxies.**
A. J. Pickles.
Proceedings of the Second Asian–Pacific Regional Meeting on Astronomy, p. 562 (1984). Abstract. – See Abstr. 012.031.

157.161 **The statistical distribution of the neutral–hydrogen content of S0 galaxies.**
M. Wardle, G. R. Knapp.
Astron. J., Vol. 91, No. 1, p. 23 – 48 (1986).
The distribution of relative global H I content M_{HI}/L_B has been derived for galaxies of types S0 and S0/a using a data set derived from recent H I observations in the literature. The relative H I content of these galaxies is found to show transitional properties between elliptical and spiral galaxies. The distribution of M_{HI}/L_B for S0/a's resembles that for spirals, and these galaxies may represent "fossil" spirals, i.e., galaxies whose gas has been severely depleted by star formation. The distribution for S0's, however, resembles that for ellipticals. The form of this distribution strongly suggests an external origin for most of the H I in S0 galaxies.

157.162 **Further radio observations of *IRAS* extreme infrared galaxies.**
R. R. J. Antonucci, E. W. Olszewski.
Astron. J., Vol. 91, No. 1, p. 56 – 57 (1986).
Aaronson and Olszewski (1984) have identified five *IRAS* infrared sources, previously considered to be blank fields, with faint galaxies. The authors reported previously the results of their VLA D–array observations at 6 cm (Antonucci and Olszewski, 1985), which resulted in detections of all objects at the mJy level. The sources were unresolved by the ~16 arcsec beam. The present paper reports on B–array observations at 6 and 20 cm, made in order to determine or limit the source angular sizes, and to measure the spectral indices. The source 0358 + 223 has an angular size of ~3 arcsec at 20 cm, but no redshift is available for this object; also 0404 + 101 is marginally resolved at 20 cm. The other sources are unresolved by the ~1.1 arcsec resolution deep 6 cm maps, implying linear sizes <2 kpc. The spectra are steep, indicating that the radiation mechanism is optically thin synchrotron emission.

157.163 **A search for box– and peanut–shaped bulges.**
B. J. Jarvis.
Astron. J., Vol. 91, No. 1, p. 65 – 69 (1986). With plates 2 – 6.
A search for disk galaxies with box– or peanut–shaped bulges from the ESO/SERC *J* sky survey south of declination –18° has revealed 30 such galaxies. Adding to these 11 other galaxies already known to have a similar morphology yields a bimodal frequency distribution with Hubble type: there is a strong tendency for box– and peanut–shaped bulges to occur in S0 and Sab to Sb type galaxies. Furthermore, the disk–to–bulge luminosity ratios of the S0 galaxies are similar to the spiral galaxies that have box– and peanut–shaped bulges, i.e., Sab und Sb Hubble types. Of the total sample of 41 galaxies, 54% are possible members of clusters and 63% have peanut–shaped bulges, compared to box–shaped bulges. Galaxy counts show that about 1.2% of all disk galaxies have box– or peanut–shaped bulges.

157.164 **IC 3475: a stripped dwarf galaxy in the Virgo cluster.**
L. Vigroux, T. X. Thuan, J. P. Vader, M. Lachieze–Rey.
Astron. J., Vol. 91, No. 1, p. 70 – 75 (1986). With plate 7.
The authors have obtained *B* and *R* CCD and H I observations of the Virgo dwarf galaxy IC 3475. The galaxy is remarkable for its very large diameter (~10 kpc for a Virgo distance modulus of 31) and is comparable in size to the large dwarfs discussed by Sandage and Binggeli. Its light profile is best fitted by an exponential law, characteristic of a dwarf Magellanic irregular galaxy. Despite possessing the photometric structure of a dwarf Magellanic irregular galaxy, IC 3475 contains less than $5.3 \times 10^6 M_\odot$ of neutral hydrogen. Its hydrogen mass to blue light ratio is less than 0.01, ~60 times less than the mean value observed for dwarf Magellanic irregulars. It is most likely that IC 3475, which is located near the core of the Virgo cluster, is a stripped dwarf galaxy. The very large size of the galaxy (its diameter is ~1.8 times larger than that of "normal" dwarfs) appears to rule out evolution of IC 3475 from a normal dwarf irregular or to a normal dwarf elliptical.

157.165 **The absolute magnitude–effective radius relation for luminous elliptical galaxies.**
W. Romanishin.
Astron. J., Vol. 91, No. 1, p. 76 – 78 (1986).
Using published surface photometry from three separate sources for ~70 luminous elliptical galaxies, the author derives the following relation between absolute blue magnitude (M_B) and effective radii (r_e) for luminous ellipticals ($M_B \lesssim -20.5$ for $H_0 = 50$ km $s^{-1}Mpc^{-1}$): $M_B = 2.06 \log r_e$–19.77. The slope is similar to that previously derived by Kormendy. This relation should be useful for modeling the host galaxies of quasars and BL Lac objects. The derived slope shows that the M_B–$\log r_e$ relation can be used to map nonuniformities in the Hubble flow.

157.166 **Galaxy counts and colors in a South Galactic Pole field.**
L. Infante, C. Pritchet, H. Quintana.
Astron. J., Vol. 91, No. 2, p. 217 – 225 (1986). With plate 31.
Counts and colors of faint galaxies have been obtained for a 0.5 deg^2 area near the South Galactic Pole. Both counts and colors of galaxies are consistent with models requiring some galaxy evolution. Observations of a faint, rich cluster in the field show a break in the integral luminosity function consistent with a redshift $z \cong 0.3$, and a significant population of blue galaxies, in agreement with the recent suggestion by Butcher and Oemler that some galaxy evolution has occurred in galaxy clusters at recent epochs.

157.167 **Nucleated spheroidal galaxies.**
S. van den Bergh.
Astron. J., Vol. 91, No. 2, p. 271 – 274 (1986). With plate 32.
Early–type galaxies belong to two form families: ellipticals and spheroidals. On average, nucleated spheroids are found to be

rounder than non–nucleated ones. This suggests that nuclei might have been formed by gas sinking to the centers of the most slowly rotating spheroidals. The frequency distributions of apparent flattening values for ellipticals and for spheroidals do not differ significantly. This result is, perhaps, unexpected if spheroidals are swept–disk systems. Finally it is shown that nucleated and non–nucleated spheroidal galaxies in the Virgo cluster have differing spatial distributions. A major clumping of such nucleated objects is centered on the giant elliptical M86.

157.168 **X–ray properties of normal galaxies.**
G. Fabbiano.
X–ray astronomy '84, p. 333 – 348 (1985). – See Abstr. 012.033.
The following topics are discussed: (1) Spiral galaxies, including the properties of the "normal" galactic component, and new unexpected results. (2) Peculiar starburst galaxies, to explore the effect of large scale star formation on the X–ray emission. (3) Elliptical galaxies.

157.169 **EXOSAT observations of the Andromeda nebula.**
S. P. McKechnie, F. A. Jansen, P. A. J. de Korte, F. W. H. Hulscher, M. van der Klis, J. A. M. Bleeker, K. O. Mason.
X–ray astronomy '84, p. 373 – 376 (1985). – See Abstr. 012.033.
EXOSAT has observed the central 1.5° region of the Andromeda nebula (M31) for two periods separated by 25 days, and totaling 70 hours observing time. The authors present some results from the analysis of high spatial resolution images in which several tens of point–like sources are seen, many of which can be identified with globular clusters.

157.170 **High resolution X–ray and radio observations of NGC 4636 and NGC 4649.**
V. J. Stanger, R. S. Warwick, J. Schwarz.
X–ray astronomy '84, p. 377 – 380 (1985). – See Abstr. 012.033.
NGC 4636 and NGC 4649 are giant elliptical galaxies located in the outskirts of the Virgo Cluster. Einstein IPC observations indicate that $\sim 10^{41}$erg s^{-1} of extended X–ray emission is associated with each galaxy.

157.171 **Details and structure of M83 in X–rays.**
G. Trinchieri, G. Fabbiano, G. G. C. Palumbo.
X–ray astronomy '84, p. 451 – 454 (1985). – See Abstr. 012.033.
Two major conclusions are derived from the X–ray observations of M83. (1) The X–ray emitting population in the plane of the galaxy is closely connected with the distribution of the smooth disk population. The X–ray sources are most likely the result of the evolution of binary systems belonging to a population with ages $\lesssim 10^9$yr. (2) A much shorter age, $\tau \lesssim 10^7$yr, is derived for the burst of star formation in the nuclear region. The detection of thermal emission from hot gas ejected from the nucleus and seen projected on it is also suggested.

157.172 **Stellar kinematics and the stability of disks in spiral galaxies.**
P. C. van der Kruit, K. C. Freeman.
Astrophys. J., Vol. 303, No. 2, p. 556 – 572 (1986).
The authors have observed the velocity dispersion of the old disk population in two spiral galaxies. In the face–on Sc spiral NGC 5247, the radial decline of the vertical velocity dispersion $\langle V_z{}^2\rangle^{1/2}$ from the center to about two disk scale lengths is consistent with that predicted for a disk of constant thickness and M/L ratio. Combined with earlier results, the $(M/L)_B$ value for the old disk is 6 ± 2 solar units. For the edge–on Sb spiral NGC 7184, the authors have measured the asymmetric drift and the absorption line profile widths, which show that the radial velocity dispersion $\langle V_R{}^2\rangle^{1/2}$ also decreases with radius. In the inner regions, the velocity dispersion contributes significantly to the global disk stability. The observational results are compared with various theoretical stability parameters and numerical simulations of model galaxies.

157.173 **The luminosity function and space density of the most luminous galaxies in the *IRAS* survey.**
B. T. Soifer, D. B. Sanders, G. Neugebauer, G. E. Danielson, C. J. Lonsdale, B. F. Madore, S. E. Persson.
Astrophys. J., Lett. Ed., Vol. 303, No. 2, p. L41 – L44 (1986).
The local space density of galaxies with luminosity νL_ν (60 μm) $\gtrsim 10^{10} L_\odot$ is derived from a sample of bright galaxies detected in the *IRAS* survey. The sample is complete to 5 Jy at 60 μm and covers one–quarter of the sky. With a maximum redshift in the sample of 0.081, this sample represents a survey of the infrared characteristics for infrared bright galaxies in the local universe. The space density of infrared bright galaxies appears comparable to or larger than other classes of extragalactic objects for total luminosities greater than $10^{11} L_\odot$. It is suggested that a substantial fraction of all galaxies with $L > 10^{10} L_\odot$ undergo a period of extreme infrared activity.

157.174 **NGC 4261: a prolate elliptical galaxy.**
R. L. Davies, M. Birkinshaw.
Astrophys. J., Lett. Ed., Vol. 303, No. 2, p. L45 – L49 (1986). With plate L5.
Long–slit spectra for the E2 galaxy NGC 4261 show that its projected rotation axis lies only $6° \pm 4°$ from its projected major axis. This observation suggests that the intrinsic figure of the galaxy is prolate: if the galaxy is oblate, then its rotation axis is almost perpendicular to its symmetry axis. The angular momentum vector of the stars may be parallel to either the shortest or the longest axis of the galaxy, so that NGC 4261 may be either a bar or a spindle.

157.175 **On some peculiarities of the morphology of spiral galaxies.**
K. F. Ogorodnikov.
Astronomical–geodetical investigations, p. 3 – 8 (1985). In Russian. Abstr. in Ref. Zh., 51. Astron., 5.51.902 (1986). – See Abstr. 003.024.

157.176 **The resolution of star–like objects in the Arp ring around M81.**
Yu. N. Efremov, I. D. Karachentsev, V. E. Karachentseva.
Pis'ma Astron. Zh., Tom 12, No. 6, p. 434 – 439 (1986). In Russian. English translation in Sov. Astron. Lett., Vol. 12.
Blue diffuse and star–like objects are found in the Arp ring around M81 on plates obtained with the 6–m telescope. The light of the Arp ring is stellar but not synchrotron radiation. The parallel straight absorption bands across the spiral pattern in the northern part of M81 are suggested to be the projection of the dust matter of the ring onto the background of the M81 disk. The density of detected objects is proportional to the density of H I. They are mainly large young clusters connected with H II regions. The ongoing star formation in the Arp ring and its very existence may be related in some way with interaction between M81 and M82 and even with the galactic wind from the latter.

157.177 **OH megamaser in III Zwicky 35.**
IAU Circ., Nos. 4180, 4185 (1986).

157.178 **The young stellar content of nearby resolved galaxies.**
W. L. Freedman.
Diss. Abstr. Int., Sect. B, Vol. 46, No. 5, p. 1589 (1985). Thesis, University of Toronto (1984).

157.179 **Three–dimensional simulations of supernovae–dominated interstellar media in disk galaxies.**
D. F. Cioffi.
Diss. Abstr. Int., Sect. B, Vol. 46, No. 8, p. 2692 (1986). Thesis, University of Colorado at Boulder, 166 pp. (1985). Order No. DA8522652.

157.180 **FST observations of NGC 4945 and the Circinus galaxy.**
J. B. Whiteoak, J. D. Bunton.
Proc. Astron. Soc. Aust., Vol. 6, No. 2, p. 171 – 174 (1985).
The Fleurs synthesis telescope, which provides 20 arcsec resolution at 1.4 GHz, was used to map the continuum emission in NGC 4945 and the Circinus galaxy. Both objects have prominent

small–diameter radio nuclei, containing 50% to 75% of the total intensity, superimposed on extended emission associated with the outer regions of the galaxies. The scale of the nuclei, together with the large velocity widths of the associated spectral–line profiles, are not unlike those encountered in the central region of the Galaxy.

157.181 **A statistical study on the difference between normal and starburst galaxies.**
Y. Taniguchi.
Sci. Rep. Tôhoku Univ., Eighth Ser., Vol. 6, No. 3, p. 181 – 205 (1986).
Observational properties are compared between normal galaxies (NGs) and starburst ones (SBGs) in detail in order to investigate the nature of starburst activity in galaxies. Photometric, spectroscopic, and neutral hydrogen data are compiled from available literature for 64 NGs and 41 SBGs. The comparisons are made for each morphological type of galaxies. The main conclusions are: (1) The difference of H I properties between the SBGs and the NGs are more significant for early type (E, S0, and Sa) galaxies than for late type ones. (2) The charactertistics of the starburst activity may be controlled by the local properties of the pre–starburst region rather than by the global properties of the underlying galaxies. (3) The new measure of degree of UV excess in U–B, Δ(U–B), which is a function of the underlying color of the galaxy, provides a quantitative measure to evaluate the strength of the starburst activity.

157.182 **Einzelobjekte im Andromedanebel, Teil 1.**
H. Eckstein, H. Mandel.
Sterne Weltraum, 25. Jahrg., Nr. 6, p. 336 – 339 (1986).

157.183 **Towards a physical model for galactic rotation curves.**
R. Cowsik, P. Ghosh.
J. Astrophys. Astron., Vol. 7, No. 1, p. 17 – 28 (1986).
Extensive and meticulous observations of the rotation curves of galaxies show that they are either flat or gently going up, but rarely decreasing, at large galactocentric distances. Here the authors show that the gravitational potential which would lead to such rotation curves arises naturally when the visible matter modelled as a collisionless Maxwellian gas is embedded in a dark halo of collisionless particles with a much higher dispersion in velocities.

157.184 **A photometric study of the bright stars in M33.**
P. Z. Kunchev, G. R. Ivanov.
Astrophys. Space Sci., Vol. 122, No. 2, p. 235 – 245 (1986).
The *UBV* photometry of 105 bright stars and the *U* magnitudes of 36 bright stars in M33 taken by the 2 m RCC telescope of the Rozhen Observatory were used to obtain colour–magnitude and colour–colour diagrams. A criterion for *V*, *B–V*, and *U–B* is suggested which facilitates the separation of the member stars from the foreground stars. The vast majority of the bright blue stars with strong negative *U–B* are M33 members. The existence of objects with an unusual combination of colours ($B-V > 0.5$ and strong negative *U–B*) has been confirmed.

157.185 ***BV* surface photometry of southern galaxies. I. NGC 4507.**
T. S. Bergmann, M. G. Pastoriza.
Astrophys. J., Vol. 304, No. 1, p. 305 – 311 (1986).
From *B* and *V* photographic surface photometry the authors have obtained photometric parameters, mean–luminosity profiles, and *B–V* color indices for the barred spiral galaxy NGC 4507. The decomposition of the mean–luminosity *B* profile shows the existence of an $r^{1/4}$ bulge, an exponential disk, a bar, and possibly a lens. The dimensions and colors of these structures are derived.

157.186 **The luminosity function for field galaxies in the infrared.**
G. H. Rieke, M. J. Lebofsky.
Astrophys. J., Vol. 304, No. 1, p. 326 – 333 (1986).
The infrared properties of an unbiased, volume–limited sample of galaxies are estimated from the IRAS point source catalog. The ratio of infrared to optical luminosities and 60 to 100 μm fluxes both increase with increasing luminosity. Seyfert galaxies have relatively hot colors from 60 to 25 μm, but they nonetheless appear to have substantial excesses above the spectra expected from nonthermal processes. The infrared luminosity function is derived; it has a much stronger weighting toward very energetic galaxies than does the optical luminosity function.

157.187 **CO observations of infrared bright galaxies: the efficiency of star formation.**
J. S. Young, F. P. Schloerb, J. D. Kenney, S. D. Lord.
Astrophys. J., Vol. 304, No. 1, p. 443 – 458 (1986).
The authors have detected CO emission in each of 14 of the infrared bright galaxies listed on IRAS Circular 15 using the 14 m millimeter telescope of the FCRAO. They have compared the total CO luminosity for each galaxy with its IR luminosity and found a correlation of the two which also depends on the temperature of the dust. The authors present a simple model for the CO and IR emission, and deduce that most of the IR flux in these galaxies originates in molecular clouds and is related to star formation. It is found that the efficiency of star formation may vary by almost two orders of magnitude from galaxy to galaxy. Galaxies which have been previously identified as mergers are among those with the highest observed star formation efficiencies and dust temperatures.

157.188 **UV observations of the nucleus of M31 with the ESA Photon Counting Detector (PCD).**
J.–L. Nieto, F. D. Macchetto, M. A. C. Perryman, S. di Serego Alighieri, G. Lelièvre.
Space Telesc. Sci. Inst., Prepr. Ser., No. 90, 35 pp. (1986). To be published in Astron. Astrophys.

157.189 **Spectrophotometry of shell galaxies.**
W. D. Pence.
Space Telesc. Sci. Inst., Prepr. Ser., No. 106, 28 pp. (1986). Submitted to Astrophys. J.

157.190 **Globular clusters in galaxies beyond the Local Group. VII. The S0 galaxy NGC 3115.**
D. A. Hanes, W. E. Harris.
Astrophys. J., Vol. 304, No. 2, p. 599 – 606 (1986). With plates 16 – 17.
The authors describe the total population and spatial distribution of globular clusters associated with the edge–on S0 galaxy NGC 3115, derived from star counts carried out upon CFHT and AAT prime focus plates. The cluster system is of specific frequency $S \approx 2$, and exhibits circular symmetry within the observational errors. The authors derive a preliminary integrated luminosity function and deduce a distance in the range of 8 – 12 Mpc. A nucleated dwarf companion of NGC 3115, which itself has a globular cluster system, is also identified.

157.191 **VLA mapping of the region around NGC 4319.**
J. W. Sulentic.
Astrophys. J., Vol. 304, No. 2, p. 617 – 622 (1986). With plates 18 – 20.
The region around NGC 4319 has been mapped at 6 and 20 cm with the VLA. The optically luminous "connection" between the galacxy NGC 4319 and the quasar Markarian 205 was not detected. The nucleus of NGC 4319 was detected at a level that implied a radio luminosity ($H_0 = 75$) of $\log L_{20} = 19.1$. Three resolved radio sources overlap partially or entirely the optical image of the galaxy. The most surprising result is that two of these sources are closely aligned across the nucleus of NGC 4319. This implies that NGC 4319 is an active spiral galaxy exhibiting scaled–down radio galaxy behavior.

157.192 **IRAS observations of the nuclear bulge of M31.**
B. T. Soifer, W. L. Rice, J. R. Mould, F. C. Gillett, M. Rowan–Robinson, H. J. Habing.
Astrophys. J., Vol. 304, No. 2, p. 651 – 656 (1986).
IRAS observations of the nuclear bulge of M31 are reported. The 12 μm and 25 μm emission is attributed to circumstellar dust emission from late–type stars, while the 60 and 100 μm emission is attributed to interstellar dust emission. The total input rate of

circumstellar gas and dust into the interstellar medium is estimated to be 0.015 $M_{\odot}yr^{-1}$. The mass of dust in the interstellar medium estimated from the far-infrared flux is $\sim 1500\ M_{\odot}$. The color temperature of the far-infrared-emitting dust is 45K.

157.193 **Limits on supernova rates and interstellar densities from X-ray observations of M101.**
D. P. Cox, D. McCammon.
Astrophys. J., Vol. 304, No. 2, p. 657 – 663 (1986).
Measurements of the X-ray surface brightness of a face-on-disk galaxy M101 have previously been used to place upper limits on the power radiated by a hot corona. These X-ray measurements are here used to constrain the properties of the population of supernova remnants evolving in the disk. The strength of this limit depends strongly on the density history of the remnants (e.g., evaporative vs. nonevaporative evolution) and on the assumed supernova rate. For standard Sedov evolutions, the density limit marginally admits 0.2 cm^{-3}.

157.194 **The intrinsic shape of galaxies.**
F. Bertola.
Galaxies, quasars and cosmology, p. 1 – 9 (1985). – See Abstr. 012.042.
Recent developments in the study of the three-dimensional shapes of elliptical and disk galaxies are reviewed.

157.195 **Imaging spectroscopy of a spiral arm of the galaxy NGC 1566.**
J.-R. Roy, J. R. Walsh.
Anglo-Aust. Obs., Prepr., No. 212, 54 pp. (1986). Submitted to Mon. Not. R. Astron. Soc.

157.196 **New carbon stars in spheroidal galaxies: II. Draco, Ursa Minor and new data on Sculptor, Carina and Leo I.**
M. Azzopardi, J. Lequeux, B. E. Westerlund.
ESO Sci. Prepr., No. 425, 17 pp. (1986). To appear in Astron. Astrophys.

157.197 **A high resolution hydrogen line survey of Messier 31. III. H I holes in the interstellar medium.**
E. Brinks, E. Bajaja.
ESO Sci. Prepr., No. 429, 57 pp. (1986). To appear in Astron. Astrophys.

157.198 **An objective-prism survey of emission-line objects in M33 and IC 1613.**
J. Lequeux, N. Meyssonnier, M. Azzopardi.
ESO Sci. Prepr., No. 439, 34 pp. (1986). To appear in Astron. Astrophys., Suppl. Ser.

157.199 **The dependence of the Tully-Fisher relation on morphological type.**
E. Giraud.
ESO Sci. Prepr., No. 442, 42 pp. (1986). To appear in Astrophys. J.

157.200 **Low-resolution spectra of Kiso ultraviolet-excess galaxies.**
H. Maehara, T. Noguchi, B. Takase.
Ann. Tokyo Astron. Obs., Second Ser., Vol. 21, No. 1, p. 85 – 106 (1986).
The authors present low-resolution spectra of 57 ultraviolet-excess galaxies. It is shown that about 85 percent of this sample exhibits conspicuous emission lines similar to galactic H II regions or emission nebulae.

157.201 **Kiso survey for ultraviolet-excess galaxies. IV.**
B. Takase, N. Miyauchi-Isobe.
Ann. Tokyo Astron. Obs., Second Ser., Vol. 21, No. 1, p. 127 – 180 (1986).
Presented are the fourth list and identification charts of the ultraviolet-excess galaxies which have been detected on the multi-color plates taken with the Kiso Schmidt telescope for 10 survey fields. In the sky area of some 300 square degrees 752 objects are catalogued down to the photographic magnitude of about 18.

157.202 **Carbon and M stars in nearby galaxies: a preliminary survey using a photometric technique.**
K. H. Cook, M. Aaronson, J. Norris.
Prepr. Steward Obs., No. 613, 43 pp. (1986).

157.203 **The distance to M31 from infrared photometry of its cepheids.**
D. L. Welch, C. W. McAlary, R. A. McLaren, B. F. Madore.
Prepr. Steward Obs., No. 615, 33 pp. (1986).

157.204 **The abundances and velocities of the Leo dwarf spheroidals.**
N. B. Suntzeff, M. Aaronson, E. W. Olszewski, K. H. Cook.
Prepr. Steward Obs., No. 633, 25 pp. (1986). To appear in Astron. J.

157.205 **Dynamics of the outer disk and halo of M31.**
R. Nolthenius, H. C. Ford.
Prepr. Steward Obs., No. 636, 94 pp. (1986).

157.206 **Proper motions and bright-star photometry in the Ursa Minor dwarf galaxy.**
K. M. Cudworth, E. W. Olszewski, R. A. Schommer.
Prepr. Steward Obs., No. 665, 40 pp. (1986). To appear in Astron. J.

157.207 **Dynamics of the elliptical galaxy NGC 7097 measured from gas and stars.**
N. Caldwell, R. P. Kirshner, D. O. Richstone.
Astrophys. J., Vol. 305, No. 1, p. 136 – 147 (1986).
The E4 galaxy NGC 7097 appears to be an ordinary elliptical galaxy. Spectra taken at the CTIO 4 m show that NGC 7097 has extended [O II] and [O III] emission which provides an unusual opportunity to use the motion of the gas to probe the radial mass distribution of an elliptical galaxy. If the gas velocities are circular, they can be used to determine $M(r)$ for this galaxy. An analysis of this mass distribution leads to the hypothesis that NGC 7097 may possess a dark matter halo.

157.208 **Is the distribution of mass within spiral galaxies a function of galaxy environment?**
D. Burstein, V. C. Rubin, W. K. Ford Jr., B. C. Whitmore.
Astrophys. J., Lett. Ed., Vol. 305, No. 1, p. L11 – L14 (1986).
Rotation curves have been derived for 20 spiral galaxies that are known to be members of large clusters. The forms of mass distributions for these galaxies are compared to those previously determined for 60 relatively isolated spiral galaxies. Hubble types Sa, Sb, and Sc are equally represented in field and cluster samples. Yet, cluster galaxies have statistically different forms of rotation curves, and hence different forms of mass distributions, than field galaxies. The evidence suggests that the form of mass distribution for a spiral galaxy is more a function of environment than of Hubble type.

157.209 **On the dependence of far-infrared and radio continuum luminosities on Hubble type in spiral galaxies.**
G. Gavazzi, A. Cocito, G. Vettolani.
Astrophys. J., Lett. Ed., Vol. 305, No. 1, p. L15 – L18 (1986).
The analysis of a complete sample of ~ 2000 galaxies surveyed in the radio continuum and in the far-infrared (FIR) shows that nonthermal radio and thermal FIR luminosities of spiral galaxies are closely correlated suggesting that they have a common origin. There is direct evidence that the mean radio and FIR luminosities of spiral galaxies are a function of the Hubble type. Star formation is probably the process that provides the energy sources for both types of radiation.

157.210 **Galaxy halo/disk mass ratios estimated from a local disk stability criterion.**
A. V. Zasov.
Sov. Astron. Lett., Vol. 11, No. 5, p. 307 – 310 (1985). English translation of 40.157.078.

157.211 **The gas/dust ratio in M33.**
A. S. Sharov.
Sov. Astron. Lett., Vol. 11, No. 5, p. 313 – 314 (1985). English translation of 40.157.079.

157.212 **On the arm pattern of M33.**
G. R. Ivanov.
Dokl. Bolg. Akad. Nauk, Tome 39, No. 2, p. 9 – 11 (1986).

157.213 **New carbon stars in spheroidal galaxies. II. Draco, Ursa Minor and new data on Sculptor, Carina, and Leo I.**
M. Azzopardi, J. Lequeux, B. E. Westerlund.
Astron. Astrophys., Vol. 161, No. 2, p. 232 – 236 (1986).
The authors report on the continuation of their objective grating survey of dwarf spheroidal galaxies; observations with the CFH telescope yielded one new candidate star in Draco, found no new carbon star in Ursa Minor and added two others in Sculptor. Most of the candidate carbon stars in Sculptor, Carina and Leo I have been confirmed by spectroscopy observations at the ESO 3.6 m telescope. Some interesting objects were found serendipitously in Carina.

157.214 **Near–infrared mapping of M31.**
C. Martinez Roger, J. P. Phillips, C. Sanchez Magro.
Astron. Astrophys., Vol. 161, No. 2, p. 237 – 243 (1986).
The authors have mapped M31 at 5″ and 20″ resolution, and at wavelengths 1.25, 1.65, and 2.2 μm. The surface brightness variation away from the nucleus is adequately represented by a de Vaucouleurs or Hubble type law. The variation of surface brightness is steeper in the infrared than it is at visual wavelengths, a trend which can be interpreted in terms of a radially varying bulge metallicity. The metallicity throughout much of the bulge is found to be high if various simple models of stellar evolution are adopted, and this would also be consistent with a range of independent optical data. The position angle of the bulge major axis appears to be largely invariant with wavelength, although ellipticity is certainly larger in the NIR.

157.215 **Photometric and spectroscopic observations of 56 southern compact galaxies.**
N. Bergvall, K. Olofsson.
Astron. Astrophys., Suppl. Ser., Vol. 64, No. 3, p. 469 – 476 (1986).
The authors present Cousins UBVRI and Johnson UBVJHK multiaperture photometry of 56 southern compact galaxies, along with basic properties for 35 of these, derived from spectroscopic and photometric observations.

157.216 **Redshifts of parent galaxies of supernovae. II.**
R. Barbon, M. Capaccioli.
Astron. Astrophys., Suppl. Ser., Vol. 64, No. 3, p. 503 – 506 (1986).
Redshifts of thirty–six parent galaxies of supernovae have been measured on IDS spectra taken with the ESO 1.5 m telescope. An accurate position is provided for each object together with a detailed description of the main features detected in each spectrum.

157.217 **The brightest stars in nearby galaxies. VII. The Pegasus and Leo A Im V dwarfs.**
A. Sandage.
Astron. J., Vol. 91, No. 3, p. 496 – 506 (1986). With plates 42 – 45.
Photometry of the brightest resolved stars in the nearby dwarf Im V galaxies of Pegasus and Leo A is given to magnitude limits near $V = 21.5$, $B = 23$. The color–magnitude diagrams for both systems are similar to those of other dwarf galaxies such as Ho IX, Sextans A, Sextans B, and WLM. The data suggest the uncertain distance modulus values of $m–M = 27 \pm 1$ for Pegasus and $m–M = 26 \pm 1$ for Leo A, giving the absolute blue magnitudes of $M_{B_T} = -14.4$ and $M_{B_T} = -12.9$, respectively.

157.218 **Two–color CCD mapping of the luminous type I irregular galaxy NGC 4449.**
G. D. Bothun.
Astron. J., Vol. 91, No. 3, p. 507 – 516 (1986). With plates 46 – 53.
The author presents two color (*B–R*) difference maps of NGC 4449. The dispersion in (*B–R*) is indicative of a mixture of stellar populations of differing ages. Moreover, examination of the structure of NGC 4449 as a function of (*B–R*) color reveals that the locus of present–day star formation is occurring in a highly flattened and connected configuration. Embedded within the main body of NGC 4449 are 150 – 200 blue, resolved images, which the author identifies as OB associations. Distributed outside of the main body is a small population of globular clusters. Assuming a cluster luminosity function similar to our galaxy's yields a distance modulus of 27.3 ± 0.1 to NGC 4449.

157.219 **Molecules in galaxies. II. The disk of NGC 4565.**
M. W. Richmond, G. R. Knapp.
Astron. J., Vol. 91, No. 3, p. 517 – 521 (1986).
Emission in the $J = 1–0$ line of CO has been mapped along the major axis of the large edge–on Sb galaxy NGC 4565. The shape of the CO distribution is very similar to that of the radio continuum emission, and of the disk light. The CO emission is much more centrally peaked than is the H I, but the total (H I + H_2) gas surface density may have a smooth radial distribution. The total molecular hydrogen mass is similar to that of the H I. The asymmetry seen in the H I emission in this galaxy is also seen in the CO emission.

157.220 **The M supergiants in NGC 300.**
R. M. Humphreys, J. A. Graham.
Astron. J., Vol. 91, No. 3, p. 522 – 529 (1986). With plates 54 – 57.
The authors have surveyed the nearby Sc spiral NGC 300 for its brightest red stars and have confirmed six of the brightest candidates to be definite M supergiants. The authors have used far–red spectroscopy and the strength of the Ca II triplet ($\approx\lambda 8500$) as a luminosity discriminant between supergiants and foreground red dwarfs, plus near–infrared (JHK) photometry to derive accurate extinction and bolometric luminosities. Adopting Graham's (1984) true distance modulus of 26.1 mag for NGC 300 it is found that its brightest M supergiants have M_v near –8 mag and $M_{Bol} \leqslant -9.5$ mag. A comparison with the brightest red supergiants in the morphologically similar galaxies M33 and NGC 2403 suggests that these galaxies have very similar populations of evolved massive stars.

157.221 **Blue and visual photometry of the Rubin et al. (1976) sample of faint galaxies.**
C. J. Peterson, C. W. Baumgart.
Astron. J., Vol. 91, No. 3, p. 530 – 538 (1986).
Blue and visual multiaperture photoelectric photometry of 112 faint spiral galaxies from the list of Rubin et al. (1976) is presented. The authors briefly compare total galaxy magnitudes with the photographic magnitudes of the Zwicky et al. (1961 – 1968) *Catalogue of Galaxies and Clusters of Galaxies*. From a reanalysis for the motion of the Local Group, it is found that use of the new galaxy magnitudes increases the difference between the motion vector derived from Rubin et al. and the motions derived from other samples of galaxies and that implied by the anisotropy of the cosmic background radiation.

157.222 **A possible relationship between ring magnetism and H I gas in spiral galaxies.**
J. P. Vallée.
Astron. J., Vol. 91, No. 3, p. 541 – 545 (1986).
Polarization observations of spiral galaxies have revealed so far only two types of large–scale distribution for their magnetic fields, either mostly ring shaped or mostly spiral shaped. A newly found relationship is reported here, namely: the observation of ring magnetism in the presence of a "large" quantity of H I gas mass, and the observation of spiral magnetism in the presence of a "small" quantity of H I gas mass. A previously known relationship is strengthened here, namely: the observation of spiral mag-

netism in the presence of a large tidal interaction with a companion galaxy, and the observation of ring magnetism in the presence of a small tidal interaction.

157.223 **A study of the large-scale structure in the distribution of galaxies in a region centered about the Cancer cluster. I. Initial observational results.**
M. D. Bicay, R. Giovanelli.
Astron. J., Vol. 91, No. 4, p. 705 – 731 (1986).
The initial observational results are presented from a study of the large-scale structure in the distribution of galaxies in an ~4000 deg^2 region nominally centered about the Cancer cluster. A sample of 1555 galaxies is defined and the observational methods used in the study are summarized. The spectra of 250 galaxies detected in H I, and the associated measured parameters and computed properties, are given. An appendix contains the results from an investigation of the errors associated with the parameters measured from galaxian spectral-line profiles as a function of the signal-to-noise ratio.

157.224 **A study of the large-scale structure in the distribution of galaxies in a region centered about the Cancer cluster. II. Further observational results.**
M. D. Bicay, R. Giovanelli.
Astron. J., Vol. 91, No. 4, p. 732 – 750 (1986).
Further observational results are presented from a study of the large-scale structure in the distribution of galaxies in an ~4000 deg^2 region nominally centered about the Cancer cluster. The spectra of 260 sample galaxies detected in H I, and their measured parameters and computed properties, are given.

157.225 **A photometric study of M104.**
M. S. Burkhead.
Astron. J., Vol. 91, No. 4, p. 777 – 790 (1986). With plates 70, 71.
The data reported in this study consist of *UBVRI* photoelectric drift scans, *BVRI* CCD observations, and deep *R* Schmidt plates. These data provide a magnitude profile of the galaxy extending from 0.25 to 1700 arcsec and from $\mathbf{B} = 16$ to 30 $mag/arcsec^2$. The acquisition and reduction of these data are discussed. The major- and minor-axis magnitude profiles, the determination of the ellipticity of the galaxy, and the determination of the ellipsoidal and disk parameters are given. An attempt to correct the nucleus for the effects of seeing and the presentation of the nuclear photometric data follow. A brief discussion of the observational effects of the dust is presented and the integrated magnitude of the galaxy is given.

157.226 **The luminosities of M supergiants and the distances to M101, NGC 2403, and M81.**
R. M. Humphreys, M. Aaronson, M. Lebofsky, C. W. McAlary, S. E. Strom, R. W. Capps.
Astron. J., Vol. 91, No. 4, p. 808 – 821 (1986).
The authors have obtained near-infrared photometry plus some far-red spectroscopy of the red stars now identified in M101, NGC 2403, and M81. Their observations are described and the extinction, colors, and bolometric corrections are derived. The authors discuss the M supergiant luminosities and the implications for the distances to these three spiral galaxies.

157.227 **Globular clusters in galaxies beyond the Local Group. V. The giant ellipticals reconsidered.**
W. E. Harris.
Astron. J., Vol. 91, No. 4, p. 822 – 841 (1986).
The large-scale spatial structures of the globular cluster halo systems around the giant ellipticals M87 and M49 in Virgo, and NGC 3311 in Hydra I, are analyzed. One of the primary motivations for the present work was to develop better methodology to deduce the structural profile of an arbitrary cluster system starting from the raw observations. These methods are described, revised star-count data for M87, M49, and NGC 3311 are presented, and new derivations of their halo system profiles are made. Finally, these results are combined with those from other galaxies into a broader picture of the relation between globular clusters and galactic halos.

157.228 **The mass of the isolated elliptical NGC 720 as determined from the dynamics of its companions.**
A. Dressler, P. L. Schechter, J. A. Rose.
Astron. J., Vol. 91, No. 5, p. 1058 – 1061 (1986). With plate 78.
Redshifts have been measured for 14 galaxies brighter than $V = 16$ in a 1.5 degree square field centered on the relatively isolated elliptical galaxy NGC 720. The authors consider the possibility that five of the six, with redshifts significantly less than that of NGC 720, form a foreground group. With a half-light radius for NGC 720 of 38 arcsec, and an internal velocity dispersion of 214 km s^{-1}, the mass computed for the galaxy alone is smaller than the mass computed from the dynamics of its companions by a factor of 44 (albeit with considerable uncertainty).

157.229 **Ionized gas in elliptical and S0 galaxies. I. A survey for Hα and [N II] emission.**
M. M. Phillips, C. R. Jenkins, M. A. Dopita, E. M. Sadler, L. Binette.
Astron. J., Vol. 91, No. 5, p. 1062 – 1085 (1986).
The authors report the results of a spectroscopic survey of 203 southern E and S0 galaxies for ionized gas. The data complement a new and more sensitive survey for [O II] λ3727 emission in elliptical galaxies carried out concurrently by Caldwell (1984), but go beyond the latter in providing valuable information on linewidths, fluxes, and relative intensities for the [N II] and Hα lines (as well as the [O I] λ6300 and [S II] λλ6716, 6731 lines in a few cases).

157.230 **Optical rotation velocities and images of the spiral galaxy NGC 3198.**
D. A. Hunter, V. C. Rubin, J. S. Gallagher III.
Astron. J., Vol. 91, No. 5, p. 1086 – 1090 (1986). With plates 79, 80.
Hα rotation velocities obtained from an echelle spectrogram to a radius of 2′ for the spiral galaxy NGC 3198 agree well with the 21 cm rotation velocities of van Albada et al. (1985). The rotation curve is typical of Sc type galaxies in spite of NGC 3198's extended H I disk. From optical *R* and Hα-emission images, used to search for star-forming regions in the outer parts of this galaxy, several Hα knots are identified as far out as the Holmberg radius. With increasing radial distance out to 2′, the *R*-band surface brightness falls, the Hα is approximately constant, and the H I surface brightness rises.

157.231 **The abundances and velocities of the Leo dwarf spheroidals.**
N. B. Suntzeff, M. Aaronson, E. W. Olszewski, K. H. Cook.
Astron. J., Vol. 91, No. 5, p. 1091 – 1095 (1986).
Moderate (6 Å) resolution spectra obtained with the Multiple Mirror Telescope are reported for two red giants in the Leo I dwarf spheroidal galaxy and three giants in the Leo II dwarf. The calcium and magnesium line strengths imply mean metallicities of -1.9 ± 0.2 dex for Leo II and -1.5 ± 0.25 dex for Leo I, values that support the existence of an abundance/absolute-magnitude relation that extends smoothly through the lowest-luminosity early-type galaxies. The radial velocities derived from the spectra are in good agreement with earlier measurements based on carbon stars. The authors also report the first measured velocity for the Galactic globular NGC 5053.

157.232 **Is NGC 4569 located near the core of the Virgo cluster?**
J. R. Stauffer, J. D. Kenney, J. S. Young.
Astron. J., Vol. 91, No. 6, p. 1286 – 1294 (1986). With plate 92.
The authors note that at least two of the arguments for placing NGC 4569 significantly foreground to the rest of the Virgo cluster – the Tully-Fisher relation using the H I linewidth, and the bulge luminosity-line index relation postulated by Cowley, Crampton, and McClure – are flawed due to peculiarities in the nucleus and disk of that galaxy. Revised Tully-Fisher distance estimates, based on new optical rotation-curve data, place NGC 4569 within 2 σ of the mean Virgo cluster distance. An important consequence of this result is that the widely used H I linewidth Tully-Fisher relation may be inapplicable to some environmentally modified cluster spirals.

157.233 **Kinematics of the ionized hydrogen in the irregular galaxy NGC 4449.**
E. M. Malumuth, T. B. Williams, R. A. Schommer.
Astron. J., Vol. 91, No. 6, p. 1295 – 1300 (1986). With plates 93, 94.
The authors have used a Fabry–Perot imaging interferometer to determine the velocity and velocity–dispersion fields of the ionized hydrogen gas in the LMC–type irregular galaxy NGC 4449. They find, in addition to the previously known H II regions and filaments, an extensive system of diffuse, low–intensity Hα emission throughout the galaxy. The velocities of the bright H II regions in the bar–like region of the galaxy are in good agreement with published velocities. The diffuse gas in the bar as well as the H II regions indicate a rotation of about ± 15 km s^{-1}. The velocities and velocity dispersions are in qualitative agreement with a spherical patchy inflow of gas from a large H I cloud that surrounds the galaxy.

157.234 **Dark matter in spiral galaxies. I. Galaxies with optical rotation curves.**
S. M. Kent.
Astron. J., Vol. 91, No. 6, p. 1301 – 1327 (1986).
Luminosity profiles measured with a CCD camera are presented for 37 Sb and Sc galaxies with existing optical rotation curves. A new method is developed for decomposing luminosity profiles into bulge and disk components. The mass distribution in these galaxies is modeled as the sum of bulge and disk components and a dark halo. The rotation curves are used to determine the mass/light ratios of the bulge and disk components and two halo parameters. Three methods of constraining the halo or bulge/disk parameters are presented.

157.235 **Galaxy distance determinations: combining two distant independent observables.**
E. J. Shaya.
Bull. Am. Astron. Soc., Vol. 18, No. 1, p. 633 – 634 (1986). Abstract. – See Abstr. 010.066.

157.236 **The absolute magnitude–effective radius relation for E galaxies and the redshifts of BL Lac objects.**
W. Romanishin.
Bull. Am. Astron. Soc., Vol. 18, No. 1, p. 640 (1986). Abstract. – See Abstr. 010.066.

157.237 **The evolution of galaxies in dense clusters: spectral synthesis models for post–burst galaxies.**
M. V. Newberry.
Bull. Am. Astron. Soc., Vol. 18, No. 1, p. 640 (1986). Abstract. – See Abstr. 010.066.

157.238 ***BV* photometry of southern elliptical galaxies.**
C. J. Peterson.
Publ. Astron. Soc. Pac., Vol. 98, No. 603, p. 486 – 487 (1986).
Multiaperture photoelectric photometry (*V* and (*B–V*) magnitudes) for 78 southern elliptical galaxies is presented.

157.239 **The UV energy distribution of elliptical galaxies.**
F. Bertola.
Mem. Soc. Astron. Ital., Vol. 56, No. 4, p. 711 – 715 (1985). – See Abstr. 012.054.
A summary of IUE observations of ~40 elliptical galaxies is presented.

157.240 **Starbursts in non–interacting galaxies.**
T. G. Hawarden, R. D. Joseph, S. K. Leggett, C. M. Mountain, R. Wade.
Rutherford Appleton Lab., Rep., RAL 85–086, p. 13 – 20 (1985). – See Abstr. 012.060.
In many barred spirals between S0/a and Sc the bar contains shock fronts, visible in the optical as almost–radial dust lanes which are known to be very efficient at driving an inward flow of ISM. Small complexes of bright H II regions are frequently observed surrounding the nucleus. In at least one galaxy – NGC 1097 – this ring–like complex is a very powerful infrared starburst. Examination of IRAS results for a complete sample of nearby galaxies suggest that, in the absence of strong interactions, IR–luminous starbursts occur *only* in barred galaxies.

157.241 **Near infrared imaging of galaxies.**
J. M. Stewart.
Rutherford Appleton Lab., Rep., RAL 85–086, p. 21 – 31 (1985). – See Abstr. 012.060.
Images are presented here of spiral galaxies in the near infrared (*JHK*). These wavelengths trace the evolved stellar component of galaxies, which tends to represent the mass distribution, rather than regions of recent star formation. The discussion here is primarily concerned with the colour distribution. This allows the properties of interstellar dust in these galaxies to be studied and the distribution of the interstellar medium to be determined.

157.242 **The infrared photometric properties of the average galaxy.**
J. H. Fairclough.
Rutherford Appleton Lab., Rep., RAL 85–086, p. 32 – 51 (1985). – See Abstr. 012.060.
The Revised Shapley Ames (RSA) catalogue is compared with the IRAS Point Source Catalogue, and the typical infrared behaviour of galaxies is assessed. 95% of the spirals in the RSA catalogue occur in the IRAS catalogue. Spiral galaxies have infrared luminosities that are 50% of their optical (*B*) luminosity. 38% of the lenticular galaxies in the RSA catalogue occur in the IRAS catalogue and only 5% of the ellipticals. These galaxies are, on the whole, much less infrared luminous than spirals but show more variety in their behaviour. Irregular galaxies are well represented and are infrared luminous.

157.243 **Near infrared spectroscopy of galaxies.**
G. S. Wright, R. D. Joseph, R. Wade, J. R. Graham, I. Gatley.
Rutherford Appleton Lab., Rep., RAL 85–086, p. 52 – 59 (1985). – See Abstr. 012.060.
The authors have obtained spectra in the *H* and *K* windows for about 12 galaxies, and have measured several lines for most of these galaxies. This paper presents a broad overview of the results and their implications. The detection of massive quantities of shocked H_2 gas in the galaxy NGC 6240 is described in some detail.

157.244 **A near infrared study of Virgo dwarf ellipticals.**
H. Zinnecker, R. D. Cannon.
Rutherford Appleton Lab., Rep., RAL 85–086, p. 60 – 67 (1985). – See Abstr. 012.060.
JHK photometry of a sample of 11 dwarf elliptical galaxies (dE's) in the Virgo Cluster is reported, together with 5 giant ellipticals (gE's). The total *B*–magnitudes of the dE–objects span the range 14 – 17. *B* surface photometry is available for most of the dE galaxies so that an optical–infrared colour can be constructed. The *J–H* vs *H–K* two colour diagram suggests that the metallicity of the dwarf ellipticals is intermediate between that of the central regions of giant ellipticals and that of several metal poor globular clusters. Implications for age estimates of the dwarf ellipticals are discussed.

157.245 **Infrared observations of blue compact galaxies.**
P. M. Gondhalekar.
Rutherford Appleton Lab., Rep., RAL 85–086, p. 68 – 86 (1985). – See Abstr. 012.060.
Near and far infrared observations of blue compact galaxies are described. The near infrared colours and the infrared–to–radio energy distributions of these galaxies suggest that in some blue compact galaxies the population of low mass evolved stars is small and these galaxies may be young objects. There is good correlation between far infrared luminosity and colours, neutral hydrogen mass and radio emission of these galaxies. The dust mass of these galaxies is extremely low.

157.246 **Infrared emission and extragalactic starbursts.**
C. M. Telesco.
Rutherford Appleton Lab., Rep., RAL 85–086, p. 87 – 102 (1985). – See Abstr. 012.060.

This paper reviews the observational basis for the belief that recent star formation plays a significant role in determining many of the infrared properties of galaxies. It also considers the most salient features of recently developed models which describe the evolution of starbursts.

157.247 **Infrared maps of M81: an example of the IRAS Close Galaxy Programme.**
J. A. Abolins.
Rutherford Appleton Lab., Rep., RAL 85–086, p. 103 – 113 (1985). – See Abstr. 012.060.

The infrared maps from the IRAS pointed–mode Close Galaxy Programme provide an important new data base for the study of nearby galaxies. The aim and scope of the programme are described and, as an example of what potential users of the data can expect, maps of the M81 region are presented and briefly discussed.

157.248 **Models for IRAS galaxies.**
M. Rowan–Robinson, J. Crawford.
Rutherford Appleton Lab., Rep., RAL 85–086, p. 171 – 173 (1985). – See Abstr. 012.060.

The authors have modelled IRAS 12 – 100 μm observations of galaxies as a mixture of 3 components: a cool "disc" component, a warmer "starburst" component and a "Seyfert" component peaking at 25 μm.

157.249 **Mass–to–light ratios in elliptical galaxies.**
K. A. Arnaud, G. Gilmore.
Mon. Not. R. Astron. Soc., Vol. 220, No. 3, p. 759 – 780 (1986).

The authors have obtained medium–resolution infrared spectra of the 2.3–μm CO absorption feature sensitive to stellar luminosity class in a sample of elliptical galaxies and standard stars. They introduce a synthesis technique based on data on Baade's window and investigate the limits on numbers of low–mass stars. Resulting mass–to–light ratios are $2.5 \leqslant M/L_{\odot\mathrm{bol}} \leqslant 5.0$.

157.250 **Three–dimensional structure in field 349 of the southern sky survey – I. Redshifts for a magnitude–limited sample of galaxies from slit spectra.**
Q. A. Parker, H. T. MacGillivray, P. W. Hill, R. J. Dodd.
Mon. Not. R. Astron. Soc., Vol. 220, No. 4, p. 901 – 925 (1986).

Redshifts are presented for a magnitude–limited sample ($b_J \leqslant 16.5$) of 107 galaxies detected from scans with the COSMOS machine in the central 29 square degrees of the ESO/SERC southern sky survey field 349 (centred at 00^h00^m, $-35°00'$, 1950 coordinates). The observed sample represents a completeness of ~80 per cent for all the galaxies detected by COSMOS down to $b_J = 16.5$ and provides a significant data base for analysis of the brighter end of the galaxy distribution in this field.

157.251 **A cooling flow in the giant, elliptical galaxy NGC 4472.**
P. A. Thomas.
Mon. Not. R. Astron. Soc., Vol. 220, No. 4, p. 949 – 969 (1986).

X–ray observations of elliptical galaxies show that a large fraction of them contain substantial quantities of hot gas. This gas is part of a cooling flow which is almost hydrostatic in the outer regions. In the case of NGC 4472 the gas cannot be contained by the visible matter and so the author postulates a dark halo of non–luminous matter to provide a suitable potential well.

157.252 **A criterion determining the association limits in the Andromeda galaxy.**
S. S. Zlatev, G. R. Ivanov.
Dokl. Bolg. Akad. Nauk, Vol. 38, No. 7, p. 823 – 824 (1985). Abstr. in Ref. Zh., 51. Astron., 3.51.904 (1986).

157.253 **The distance to M31 from infrared photometry of its Cepheids.**
D. L. Welch, C. W. McAlary, R. A. McLaren, B. F. Madore.
Astrophys. J., Vol. 305, No. 2, p. 583 – 590 (1986).

The distance to M31 has been determined by using H (1.65 μm) photometry of seven classical Cepheids in fields III and IV. Using the galactic calibration of Welch et al. and mean H magnitudes of 38 LMC Cepheids, a weighted least–squares solution for the H P–L relation results in an apparent distance modulus of $(m-M)_H = 24.31 \pm 0.08$ mag. Using an estimate of $A_H = 0.05 \pm 0.02$ mag, the authors obtain a true distance modulus of $(m-M)_0 = 24.26 \pm 0.08$ mag. Astrometry for 82 Cepheids in fields I, II, III, an IV is also presented.

157.254 **The stellar population in the halos of M31 and M33.**
J. Mould, J. Kristian.
Astrophys. J., Vol. 305, No. 2, p. 591 – 599 (1986). With plates 8 – 9.

The authors present deep two–color (V and I) CCD photometry of stars in the halos of M31 and M33. The color–magnitude diagrams show the brightest 2 mag of the giant branches in the two galaxies. The M31 halo giant branch is very broad, indicating a large dispersion in metallicty, and the M33 halo giant branch is much narrower and less metal–rich. Comparison of the observed brightness of the tips of the giant branches with the calibration of Frogel, Cohen, and Persson give distance moduli of 24.4 ± 0.25 for M31 and 24.8 ± 0.3 for M33.

157.255 **The mass and halo dispersion profile of M32.**
R. Nolthenius, H. Ford.
Astrophys. J., Vol. 305, No. 2, p. 600 – 608 (1986).

The authors present radial velocities of 15 planetary nebulae in the halo of M32 and use them to put constraints on the form of the stellar orbits and on the mass. These velocities are compared with trial velocity dispersion profiles derived from the models of Jaffe and Merritt with varying amounts of anisotropy. Assuming constant M/L with radius, the data favor an isotropic dispersion tensor in the halo. The resulting mass is $7.0 \times 10^8 M_\odot$ for a Jaffe model galaxy and $9.4 \times 10^8 M_\odot$ for a de Vaucouleurs model. These correspond to M/L_B values of 2.8 and 3.7, respectively.

157.256 **Angular momentum–limited degenerate and semidegenerate neutrino halos.**
T. A. Chubb.
Astrophys. J., Vol. 305, No. 2, p. 609 – 622 (1986).

An equation of state is derived for degenerate neutrino matter in a spherical spatial configuration in which there is zero occupancy of high angular momentum states. Hydrostatically supported galaxy halos using this equation of state are characterized by galaxy rotation curves having nearly constant circular rotation velocities over an extended range of R. Nonvirialized degenerate halos appear to be consistent with a low–entropy very early universe in which an entropy–increasing global phase change occurred just prior to nucleogenesis. An equation of state is also derived for semidegenerate neutrino matter corresponding to the "relic" distribution used by Schramm and Steigman.

157.257 **Carbon and M stars in nearby galaxies: a preliminary survey using a photometric technique.**
K. H. Cook, M. Aaronson, J. Norris.
Astrophys. J., Vol. 305, No. 2, p. 634 – 644 (1986).

A new photometric method for identifying very distant carbon and late–M stars has been developed which utilizes two intermediate–band filters centered on CN and TiO absorption bands. The authors have successfully tested the technique on galactic field stars and in eight regions in the disks of five nearby galaxies: M31, M33, NGC 6822, IC 1613, and WLM. The resulting carbon to M star ratios are directly related to parent galaxy absolute magnitude, presumably reflecting a decreasing abundance in the lower mass objects. The potential use of the carbon star luminosity function as a distance indicator is discussed.

157.258 **Metal–enhanced galactic winds. I.**
J. P. Vader.
Astrophys. J., Vol. 305, No. 2, p. 669 – 678 (1986).
Supernova–driven gas loss during the early evolution of elliptical galaxies is considered as a possible explanation for the correlations among the observed chemical and structural properties of these systems. Mass loss from systems with a chemically homogeneous interstellar medium does not work. It is shown that supernova–driven winds are in fact metal–enhanced with respect to the star–forming gas because the metal production of any supernova that drives the wind is directly flushed out of the galaxy. The fraction of the total metal production lost in the wind is thus at least as large as the fraction of supernovae that power the wind, independently of the total mass loss.

157.259 **Molecules in galaxies. I. CO observations in a spiral arm of M31.**
B. S. Ryden, A. A. Stark.
Astrophys. J., Vol. 305, No. 2, p. 823 – 829 (1986).
The authors have mapped a 4′ × 20′ spiral arm region of M31 in ^{12}CO ($J = 1 \rightarrow 0$). The emission is confined to an arm 1 – 2 kpc wide. The surface density of H_2, averaged over the mapped region, is 2.2 $M_{\odot}pc^{-2}$, projected onto the plane of the galaxy. On the outer edge of the arm, the CO shows a systematic velocity deviation of up to 30 km s^{-1}. The molecular gas is accelerated as it crosses the spiral arm. Many spectra are double–peaked, with one peak at the velocity expected from circular rotation, and the other peak shifted to greater velocity.

157.260 **Formaldehyde absorption and maser emission in galaxies.**
W. A. Baan, R. Güsten, A. D. Haschick.
Astrophys. J., Vol. 305, No. 2, p. 830 – 836 (1986).
A sensitive search for the $1_{10} - 1_{11}$ (6.2 cm) H_2CO transition toward eight galaxies with prominent H I absorption and OH absorption or emission has resulted in the detection of three weak absorption lines (NGC 660, NGC 3079, and NGC 3628) and, in addition, of H_2CO emission toward the OH megamaser galaxy IC 4553. This feature is the first extragalactic H_2CO maser, and is at least 7 orders of magnitudes more powerful than the two known galactic maser sources. Possible excitation mechanisms for this maser are discussed.

157.261 **On the ratio of [N II]/Hα in the nucleus of M33 and in the nuclei of other galaxies.**
V. C. Rubin, W. K. Ford Jr.
Astrophys. J., Lett. Ed., Vol. 305, No. 2, p. L35 – L37 (1986). With plate L1.
Spectra of the inner 1.″77 (radial distance <4 pc) of M33 are unique in showing Hα in deep sharp absorption and [N II] λ6548 and λ6583 in emission. Just beyond the nucleus, strong disk Hα and [N II] emission are present. Hence if M33 were located at a distance several times as great, a spectrum through the 1.″77 aperture would show Hα and [N II] in emission, as the contamination from the disk emission overwhelms the nuclear Hα absorption. For galaxies at greater distances, the measured [N II]/Hα ratio in the central region is a complex function of nuclear and disk light which enters the aperture.

157.262 **Molecular gas in high–luminosity IRAS galaxies.**
D. B. Sanders, N. Z. Scoville, J. S. Young,
B. T. Soifer, F. P. Schloerb, W. L. Rice, G. E. Danielson.
Astrophys. J., Lett. Ed., Vol. 305, No. 2, p. L45 – L49 (1986). With plate L2.
CO($J = 1 \rightarrow 0$) emission has been detected from 15 high–luminosity IRAS galaxies selected from a complete list of bright IRAS galaxies in the northern sky, with $L(40 - 400\ \mu m) = 6 \times 10^{10} - 6 \times 10^{11} L_{\odot}$. All of the observed galaxies are rich in molecular gas with H_2 masses in the range $(4 \times 10^9) - (4 \times 10^{10}) M_{\odot}$. Their primary luminosity source appears to be star formation in molecular clouds. The majority, if not all, of the most luminous IRAS galaxies ($L_{FIR} > 10^{11} L_{\odot}$) appear to be strongly interacting systems.

157.263 **Kinematics of spiral galaxies. An overview.**
P. Pismis.
Rev. Mex. Astron. Astrofis., Vol. 12, Spec. Issue, p. 79 – 88 (1986). – See Abstr. 012.066.
The kinematics of spiral galaxies are reviewed with special emphasis on rotation curves, the flat maxima and undulations of these curves in late type spirals. The two dimensional velocity fields from the H I 21–cm line as well as from optical lines support the finding that the maxima and minima of the rotation correspond reasonably well to arm and interarm regions respectively.

157.264 **The dynamics of galactic bulges: NGC 7814 and NGC 4594.**
B. Jarvis, K. C. Freeman.
Rev. Mex. Astron. Astrofis., Vol. 12, Spec. Issue, p. 89 – 96 (1986). – See Abstr. 012.066.
Stationary, self–consistent axisymmetric rotating models with energy and angular momentum integrals have been constructed for the bulges of the near edge–on disk galaxies NGC 7814 and NGC 4594. The models are capable of accurately fitting the observed kinematics with an assumption of constant M/L. M/L values are derived for NGC 7814 and NGC 4594 and found to be consistent with recently observed values for elliptical galaxies.

157.265 **Structure of S0's in clusters.**
R. R. de Carvalho, L. N. da Costa, P. S. Pellegrini.
Rev. Mex. Astron. Astrofis., Vol. 12, Spec. Issue, p. 101 – 108 (1986). – See Abstr. 012.066.
The authors briefly describe some properties of lenticular galaxies in the central regions of clusters. The discussion is based on the results of an ongoing surface photometry program of S0 galaxies.

157.266 **BV surface photometry of the southern galaxies NGC 6769, 6770 and 6771.**
T. S. Bergmann, M. F. S. Schröder, M. G. Pastoriza.
Rev. Mex. Astron. Astrofis., Vol. 12, Spec. Issue, p. 133 (1986). Abstract. – See Abstr. 012.066.

157.267 **Photometric measurements of WH_{β} and [O III]/H_{β} of H II regions in M83.**
M. V. F. Copetti, H. A. Dottori, M. G. Pastoriza.
Rev. Mex. Astron. Astrofis., Vol. 12, Spec. Issue, p. 137 – 138 (1986). – See Abstr. 012.066.
WH_{β} and [O III]/H_{β} were measured photoelectrically in 12 H II regions of M83. It was not found any systematic variation of WH_{β} or [O III]/H_{β} across M83. The very low excitation of the emission nebulae in M83 may be a consequence of the time evolution of the H II regions.

157.268 **The X–ray properties of normal galaxies.**
G. Fabbiano.
Publ. Astron. Soc. Pac., Vol. 98, No. 604, p. 525 – 543 (1986).
In this paper the author tries to give an up–to–date review of the present state of knowledge of the X–ray properties of "normal" galaxies, including spiral systems, starburst nuclei and blue peculiar galaxies, and elliptical and S0 galaxies. He summarizes the unresolved questions and discusses their possible resolution.

157.269 **Das Magnetfeld in IC 342.**
R. Gräve, R. Beck.
Mitt. Astron. Ges., No. 65, p. 227 (1986). – See Abstr. 012.072.

157.270 **Das Magnetfeld in M33.**
U. R. Buczilowski.
Mitt. Astron. Ges., No. 65, p. 228 – 229 (1986). – See Abstr. 012.072.

157.271 **Morphology and structure of blue compact dwarf galaxies.**
H.–H. Loose, T. X. Thuan.
Mitt. Astron. Ges., No. 65, p. 231 (1986). Abstract. – See Abstr. 012.072.

157.272 **Surface photometry of elliptical galaxies.**
L. Colina, K. J. Fricke, H.-H. Loose, I. Perez-Fournon, K. M. Thieme.
Mitt. Astron. Ges., No. 65, p. 231 – 232 (1986). Abstract. – See Abstr. 012.072.

157.273 **Dust in early-type galaxies – an IRAS perspective.**
U. C. Peppel, S. K. Ghosh.
Mitt. Astron. Ges., No. 65, p. 233 – 236 (1986). – See Abstr. 012.072.

157.274 **Surveys for H II galaxies and related star forming systems.**
C. Hazard.
Star-forming dwarf galaxies and related objects, p. 9 – 29 (1986). – See Abstr. 012.074.
An outline is given of the properties of star forming galaxies and of the various survey techniques used to find them. A more detailed account is then presented of surveys based on low dispersion UK Schmidt Telescope IIIaJ and IIIaF objective prism plates particularly as applied to searches for compact H II galaxies. Attention is drawn to some particularly interesting objects which have been found in such surveys.

157.275 **The classification and evolutionary sequence of galaxies.**
A. Sandage.
Star-forming dwarf galaxies and related objects, p. 31 – 40 (1986). – See Abstr. 012.074.
The modern generalized version of the Hubble classification sequence for galaxies is claimed to be Baconian in its inductive power. Evolution along the sequence is calculated to be slow – the forms we see now are nearly the forms that were laid down soon after the initial dissipative collapse for spirals and the almost dissipationless collapse of E0 ellipticals. The features of the whole of the Hubble sequence can be understood by the variation of the star-formation rate with time, driven by some fundamental parameter (such as density and/or angular momentum) yet to be identified.

157.276 **Remarks on the luminosity function of faint galaxies.**
G. A. Tammann.
Star-forming dwarf galaxies and related objects, p. 41 – 51 (1986). – See Abstr. 012.074.
The luminosity functions of individual types of galaxies are intercompared for a Virgo cluster sample and a sample comprising field and group galaxies. While the relative type frequencies vary strongly with the galaxy density of the surrounding field, the shape of the luminosity functions shows little density effect. There is marginal evidence that the mean luminosity of cluster spirals is brighter than that of field/group spirals.

157.277 **On the relationship between dwarf irregular and dwarf elliptical galaxies.**
B. Binggeli.
Star-forming dwarf galaxies and related objects, p. 53 – 67 (1986). – See Abstr. 012.074.
Data from the recently compiled Virgo Cluster Catalog are used to compare the basic photometric properties of dwarf irregular and dwarf elliptical galaxies.

157.278 **The morphology and structure of blue compact dwarf galaxies from CCD observations.**
H.-H. Loose, T. X. Thuan.
Star-forming dwarf galaxies and related objects, p. 73 – 88 (1986). – See Abstr. 012.074.
The authors have obtained CCD observations of ~50 blue compact dwarf galaxies (BCD's) from the list of Thuan and Martin. They use the CCD material to propose a classification scheme for BCD's based on the regularity of the isophotal shapes of both the high-surface-brightness star-forming regions and the low-surface-brightness outer regions. They can divide BCD's into 4 types. The most common type has several star-forming regions near the center embedded in an elliptical or circular halo of light. Details of the surface photometry of two BCD's of different types (Haro 2 and Mrk 86) are discussed.

157.279 **Blue compact galaxies: a mixing bag.**
D. Kunth, J. M. Martin, S. Maurogordato, L. Vigroux.
Star-forming dwarf galaxies and related objects, p. 89 – 95 (1986). – See Abstr. 012.074.
Multicolor surface photometry performed on a selected sample of nearby (V ~ 1000 km/s) blue compact galaxies shows that this galaxy class contains very different objects: interacting galaxies, irregular, or elliptical-like galaxies. Only some of them have small sizes (<2 kpc) and can be suspected to be genuine young galaxies. In the largest ones, in addition to the previously known star-formation regions, the authors find other knots, geometrically ordered from the redder to the bluer. This is a good evidence for a propagating star formation process.

157.280 **Bursts of star formation in ring galaxies. Some results from CCD images.**
C. Bonoli.
Star-forming dwarf galaxies and related objects, p. 97 – 101 (1986). – See Abstr. 012.074.
First results from a CCD photometry project on a sample of ring galaxies are presented. The galaxies appear to be the consequence of a collision with another object; that seems to have triggered a strong stellar formation. The observations support the theory of a burst of star formation undergone some 10^8 years ago.

157.281 **The young stellar populations in blue compact dwarf galaxies: spectral synthesis of ultraviolet spectra.**
T. X. Thuan.
Star-forming dwarf galaxies and related objects, p. 105 – 123 (1986). – See Abstr. 012.074.
The young stellar populations in blue compact dwarf galaxies can be probed directly by ultraviolet observations. A brief review is given of the general properties of ultraviolet spectra of blue compact galaxies. Evolutionary and optimizing synthesis techniques applied to ultraviolet spectra are discussed. Ultraviolet synthesis, when combined with optical and infrared synthesis, shows the promise of yielding information on the Initial Mass Function and the star formation history in these galaxies.

157.282 **The older stellar population of dwarf galaxies.**
M. Aaronson.
Star-forming dwarf galaxies and related objects, p. 125 – 143 (1986). – See Abstr. 012.074.
Study of the older stars is key for understanding the origin and evolution of dwarf systems. In this review, after briefly discussing composite light observations, the author turns to the wealth of information accumulated in the last few years on nearby resolvable dwarfs. Even the simplest of these galaxies appears to have a far more fascinating and complicated star formation history than was once thought. In particular, evidence suggesting the low surface brightness dwarf ellipticals and spheroidals share a heredity closer to dwarf irregulars than to brighter ellipticals continues to mount.

157.283 **Ultraviolet and infrared properties of blue compact galaxies.**
P. M. Gondhalekar.
Star-forming dwarf galaxies and related objects, p. 145 – 153 (1986). – See Abstr. 012.074.
Ultraviolet and infrared observations of a small sample of blue compact galaxies have been described. The ultraviolet spectra of these galaxies suggest that there has been a burst of star formation in these galaxies at a recent epoch. The age of the burst and the initial mass function of the burst can vary from one galaxy to next. The far infrared data of these galaxies indicate that the dust mass in these galaxies is very low and dust extinction should not be a major constraint in modelling the ultraviolet spectra of these galaxies. The low dust content of these galaxies suggests that the integrated rate of star formation (assuming dust is formed in stellar atmosphere) has been very low in these galaxies.

157.284 Infrared evidence on the metallicity and the history of star formation of Virgo dwarf elliptical galaxies.
H. Zinnecker, R. D. Cannon.
Star-forming dwarf galaxies and related objects, p. 155 – 169 (1986). – See Abstr. 012.074.

JHK photometry of a sample of 11 dwarf elliptical galaxies (dE's) in the Virgo Cluster is reported, together with 5 giant ellipticals (gE's). B surface photometry is available for most of the dE galaxies so that an optical–infrared colour can be constructed. The J–H vs H–K two colour diagram suggests that the metallicity of the dwarf ellipticals is intermediate between that of the central regions of giant ellipticals and that of several metal poor globular clusters. Judging from the B–H vs J–K two colour diagram, the age of some dwarf ellipticals could be as young as 2 – 5 Gyr. Alternatively, it is possible that star formation started in dE's at the same time as in gE's ($\sim$10 Gyr ago) but the star formation rate declined much more slowly in dE's than in gE's. In the Appendix it is shown that NGC 205 (a dE companion of M31) would appear nucleated, if displaced from the Local Group to the Virgo Cluster. It is conjectured that the nuclei of Virgo dE's, like the nucleus of NGC 205, are active star forming regions.

157.285 Abundances in blue compact galaxies.
D. Kunth.
Star-forming dwarf galaxies and related objects, p. 183 – 195 (1986). – See Abstr. 012.074.

About 100 blue compact and irregular galaxies have now been observed for abundance determinations. Oxygen is the most common and better determined species. The abundance distribution shows that the galaxies are underabundant and exhibit little or no metallicity gradient.

157.286 Helium in dwarf emission–line galaxies.
G. A. Shields.
Star-forming dwarf galaxies and related objects, p. 197 – 205 (1986). – See Abstr. 012.074.

Dwarf irregular galaxies provide a potential opportunity to measure the primordial abundance of helium, Υp, because of their strong emission lines and low abundances of heavy elements. Recent work suggests values ranging from $\Upsilon p = 0.22$ to 0.24. Because values $\Upsilon p < 0.24$ are difficult to reconcile with the standard Big Bang cosmological model, there is great interest in narrowing the observational margin of error. Uncertainties to be addressed include the absorption–line equivalent widths, the ionization correction for neutral helium, and the degree of stellar enrichment. Observations of objects with very low abundances of heavy elements and relatively litte O^{+} may hold the greatest promise.

157.287 Observational properties of the gas and the young stars in BCG.
F. Viallefond.
Star-forming dwarf galaxies and related objects, p. 207 – 215 (1986). – See Abstr. 012.074.

It is shown that extragalactic H II regions in some spirals and irregulars have properties very similar to those in BCGs; they form the so–called H I – H II associations. By contrast DIGs have a very different morphology in their distributions of H I. It is suggested that the SFR depends primarily on the gas content in these associations. BCGs have a very large star formation fractional involvement as compared to irregulars such as the LMC or NGC 2366. From the excitation of these regions it is suggested that the IMF must be enriched in very massive stars for metal poor H II nebulae. This might be related to the dust content. The infrared data also suggest variations of the IMF for the very massive stars relative to the non ionizing less massive stars.

157.288 A spectroscopic analysis of the bright knots in Mk 170 and Mk 178.
J. Zamorano, M. Rego, R. González–Riestra.
Star-forming dwarf galaxies and related objects, p. 217 – 225 (1986). – See Abstr. 012.074.

The authors present long–slit spectroscopic observations of the blue compact galaxies Mk 170 and Mk 178. The images show the existence of several emission knots in both galaxies. The authors have computed the distribution of the emission–line and the continuum intensities along the slit. The preliminary results show that their behaviour is very different, implying that the regions of maximum star concentration and those with the most massive stars do not coincide.

157.289 Analysis of new Hα emission line galaxies.
A. Prieto, M. Rego, J. Zamorano, R. González–Riestra.
Star–forming dwarf galaxies and related objects, p. 227 – 235 (1986). – See Abstr. 012.074.

Low resolution spectrometric observations of five new emission line galaxies are presented.

157.290 New abundance values in a BCG sample.
J. Zamorano, M. Rego.
Star–forming dwarf galaxies and related objects, p. 237 – 241 (1986). – See Abstr. 012.074.

From a five level atomic model and new atomic data collected from the literature, abundance values have been obtained for a BCG sample. The authors found lower values for the O^{+}, O^{++}, N^{+} and S^{+} and higher values for the Ne^{++}.

157.291 The H I structures of extreme dwarf galaxies.
W. L. W. Sargent, K. Y. Lo.
Star–forming dwarf galaxies and related objects, p. 253 – 262 (1986). – See Abstr. 012.074.

For the most extreme dwarf, gas–rich systems, such as M81 dwA, the kinematics of the gas are dominated by chaotic motions rather than by rotation. The authors have obtained H I maps of 9 extreme dwarfs ($M_{pg} > -14$) using the C and D configurations of the VLA. These show that the annular gas distribution inferred for M81 dwA is not a characteristic feature of these galaxies. Instead, the gas seems to be chaotically distributed relative to the stars and there is no common overall H I structure. Although there is a trend that the H I contents of dwarf irregular galaxies rise with decreasing optical luminosity, the gas only rarely dominates the total mass. All the extreme dwarfs have high mass–to–light ratios – $M/L = 10 - 30$ – implying that even dwarf galaxies contain 'dark matter'. Attention is drawn to the peculiar properties of the galaxy LGS 3. Some form of self–regulated star–formation, probably occurring in bursts, is implied by the observations.

157.292 H I observations of Sextans A.
E. D. Skillman, R. Terlevich, H. van Woerden.
Star forming dwarf galaxies and related objects, p. 263 – 269 (1986). – See Abstr. 012.074.

Interferometric H I observations of the dwarf irregular galaxy Sextans A are presented. The H I distribution shows a strong correspondence with the optical structure, and the velocity field is dominated by solid body rotation. However, the velocity field is not aligned with the optical axis of symmetry, and this renders the mass distribution analysis untrustworthy. The H I is clumpy, and the clumps are associated with regions of recent star formation. It is possible that the star formation process will proceed for quite some time as the inner part of the galaxy is replenished with neutral gas from a low density halo. Measurements of the surface density support the postulate of Gallagher and Hunter that OB star formation requires an H I surface density of about $5x10^{20}$atoms/cm^2.

157.293 VLA observations of dwarf galaxies.
G. Comte, J. Lequeux, F. Viallefond.
Star–forming dwarf galaxies and related objects, p. 273 – 279 (1986). – See Abstr. 012.074.

High–resolution neutral hydrogen observations of the dwarf irregular galaxies DDO 43, DDO 47 and Sextans B (DDO 70) made with the Very Large Array synthesis telescope have revealed gas distributions of regular shapes and large scale axisymmetric ordered motions. The maximum rotation velocity is attained near the detected edge of the H I disk (DDO 47) or beyond it (Sextans B), this implies the presence of massive dark halos around the galaxies.

157.294 **VLA continuum and H I observations of blue compact dwarf galaxies.**
E. Brinks, U. Klein.
Star–forming dwarf galaxies and related objects, p. 281 – 290 (1986). – See Abstr. 012.074.

Recently performed VLA continuum observations of 6 BCDGs, Mkn 314, Mkn 527, Mkn 600, Haro 20, Haro 21, and II Zw 40, at 20 cm and 6 cm wavelength provide further evidence that their radio continuum properties are markedly different from those of normal spiral galaxies. Four BCDGs were selected to be observed in H I with the VLA. Mkn 527 proved to be too weak. The others, Mkn 314, Haro 21, and II Zw 40, could be mapped in H I. In each galaxy the H I is resolved in several clumps which seem to avoid the brightest optical features.

157.295 **Quiescent and active irregular galaxies in a color versus M_H/L diagram and starburst evolution.**
G. Comte.
Star–forming dwarf galaxies and related objects, p. 291 – 297 (1986). – See Abstr. 012.074.

Active star–forming and quiescent galaxies are well separated in an (U–V) color versus (M_H/L) ratio diagram. Evolutionary tracks drawn over such a diagram for generalized bursts of stellar formation explain quite well this separation. The corresponding distribution of galaxies in the (U–B) versus (B–V) color–color diagram is briefly examined, and color isochrones are indicated.

157.296 **H I – H II relationship in dwarf star–forming galaxies.**
J. Lequeux.
Star–forming dwarf galaxies and related objects, p. 299 – 306 (1986). – See Abstr. 012.074.

The morphological and kinematical relationship between H I clouds and H II regions is discussed for blue compact dwarf galaxies. Then a discussion of the rate of formation of massive stars in irregular galaxies is given. It is shown that apart from strong and probably rare bursts of star formation, this rate is roughly proportional to the mass of gas contained in the galaxy.

157.297 **Constraints on the molecular content.**
F. Combes.
Star–forming dwarf galaxies and related objects, p. 307 – 315 (1986). – See Abstr. 012.074.

A summary is presented of all CO observations towards dwarfs and irregulars, either detected or upper limits. The rate of detection of this type of galaxies is 1 out of 7. They present very low CO emission with respect to spiral galaxies of the same blue luminosity, but this could be explained by a low metallicity and different excitation conditions. Some perspectives are given with future telescopes.

157.298 **IRAS observations of dwarf galaxies with low surface brightness.**
G. Helou.
Star–forming dwarf galaxies and related objects, p. 319 – 329 (1986). – See Abstr. 012.074.

An overview is given of IRAS survey data on the DDO low surface brightness galaxies and blue compact dwarf galaxies comparing them to bright spirals in Virgo.

157.299 **IRAS observations of blue compact emission line galaxies.**
D. Kunth, F. Sèvre.
Star–forming dwarf galaxies and related objects, p. 331 – 349 (1986). – See Abstr. 012.074.

A preliminary statistical study of the most studied blue compact dwarf galaxies using the IRAS point source catalogue is presented.

157.300 **Star formation in IRAS galaxies: from blue dwarfs to red super–star bursters.**
M. Dennefeld, H. Karoji, P. Belfort.
Star–forming dwarf galaxies and related objects, p. 351 – 367 (1986). – See Abstr. 012.074.

The authors present preliminary results of a ground–based follow–up of faint uncatalogued galaxies, counterparts of IRAS point sources. Most of them reveal low–ionization emission lines, a low excitation and are classified as H II region type objects. The striking fact is the large reddening derived from the Balmer decrement which, if applied to the entire blue magnitude, brings the IR/B luminosity ratios back to more "classical" values. Instead, the dominant characteristic is the high IR luminosity (up to a few $10^{12}L_\odot$) derived for a few objects. A strong correlation is found between IR and Hα luminosities and holds over more than four decades, including comparison samples of "normal" and blue compact galaxies. This is interpreted as evidence for a common powering mechanism which is massive star formation occurring in the central regions.

157.301 **Nonthermal radio continuum emission from dwarf galaxies.**
U. Klein.
Star–forming dwarf galaxies and related objects, p. 371 – 382 (1986). – See Abstr. 012.074.

The properties of the nonthermal radio continuum emission from dwarf galaxies is reviewed. A short summary of the radio continuum emission from the Magellanic Clouds is given. Radio observations of Local Group and nearby dwarf galaxies show that they lack the thick nonthermal disks which are usually found in normal spiral galaxies. Measurements of a larger sample of I(B)m/S(B)m dwarf galaxies suggest that they are scaled–down versions of normal spirals as far as their integrated radio emission is concerned. Blue compact dwarf galaxies generally show a significant excess of radio emission in comparison to normal spirals. Their radio spectra can probably be explained in terms of nonthermal emission from young SNRs plus a considerable fraction of free–free emission from optically thin gas.

157.302 **Chemical evolution of galaxies: H, He, C, N and O.**
M. Peimbert.
Star–forming dwarf galaxies and related objects, p. 403 – 415 (1986). – See Abstr. 012.074.

Four problems related to the chemical evolution of dwarf irregular galaxies are briefly discussed: the variation of p(SM), the "observed" heavy element yield, the $\Delta Y/\Delta Z$ ratio, the C/O, O/H diagram and the N enrichment of the interstellar medium.

157.303 **The origin of nitrogen and the evolution of dwarf irregular galaxies.**
M. Tosi, F. Matteucci.
Star–forming dwarf galaxies and related objects, p. 417 – 424 (1986). – See Abstr. 012.074.

The chemical evolution of dwarf irregular galaxies has been computed mainly to investigate the origin of nitrogen. In their numerical models the authors assume that star formation proceeds in bursts and that galactic winds powered by supernova explosions are present in these galaxies. They conclude that, in order to explain the observed features of dwarf irregular galaxies, (1) nitrogen should be both a primary and secondary element, (2) various amounts of primary nitrogen should be synthesized in different galaxies, (3) galactic wind rates ranging from 0 to 30 times the star formation rate should be required.

157.304 **Some inferences on chemical evolution from a study of irregular and blue compact galaxies.**
L. Vigroux, G. Stasińska, G. Comte.
Star–forming dwarf galaxies and related objects, p. 425 – 429 (1986). – See Abstr. 012.074.

157.305 **Lyman α in emission line galaxies.**
J. M. Deharveng, M. Joubert, D. Kunth.
Star–forming dwarf galaxies and related objects, p. 431 – 437 (1986). – See Abstr. 012.074.

Three new IUE observations are reported which supplement the published sample of Lyman α observations in blue galaxies. The Lyα/Hβ ratios are much lower than the recombination theory predictions and reveal no trend as a function of the Balmer decrement. Explanations may be traced back to large Lyman α optical depths and dust absorption enhancement. The data are used as empirical inputs for an evaluation of the space density of the galaxies that a long slit spectroscopy might detect at high redshift by their Lyman α emission.

157.306 **Dwarf galaxies and galaxy formation.**
J. Silk.
Star–forming dwarf galaxies and related objects, p. 465 – 476 (1986). – See Abstr. 012.074.
The role of dwarf galaxies in galaxy formation is described. Dwarfs are likely to be the left–over fragments remaining from a model wherein a luminous galaxy formed by the interactions of many gas clouds. Comparison with observational data on galaxy surface brightnesses and stellar velocity dispersions suggests that both dissipational and dynamical relaxation processes play important roles. The energy input from massive stars is crucial in determining the binding energy of the final galaxy, and drives winds from dwarf galaxies. This enables the chemical and dynamical evolution of dwarfs to be understood provided that their dynamics is presently dominated by dark matter. Some cosmological implications of this scenario are mentioned.

157.307 **On the structure and evolution of dwarf galaxies.**
C. A. Norman.
Star–forming dwarf galaxies and related objects, p. 477 – 486 (1986). – See Abstr. 012.074.
The dynamical evolution of dwarf galaxies, the effects of star formation, and interaction with their environment including the role of collisions and mergers are discussed. A prediction is made that dwarf elliptical galaxies with intense nuclei should be rounder as a class than those with no such central mass concentration.

157.308 **NGC 3226.**
IAU Circ., No. 4212 (1986).

157.309 **The distribution of young stars in nearby galaxies.**
W. L. Freedman.
Luminous stars and associations in galaxies, p. 61 – 80 (1986). – See Abstr. 012.079. (IAU Symp. No. 116).
This review concentrates on studies of photometry in galaxies near enough that stars can be resolved in them. Studies of luminosity functions, spatial distributions of young stars, and comparison with the distribution of gas, and radial distributions of stars in external galaxies are discussed.

157.310 **A stellar content study of NGC 2403 with the automated plate scanner.**
T. R. Manley.
Luminous stars and associations in galaxies, p. 83 – 84 (1986). – See Abstr. 012.079. (IAU Symp. No. 116).
The author illustrates how the Minnesota Automated Plate Scanner (APS) is being used to study the luminous stellar content of NGC 2403. Presented are a brief description of the APS, examples of photometric calibration, separation of stellar and non-stellar images, and a preliminary color–magnitude diagram.

157.311 **Red supergiants in NGC 300.**
J. A. Graham, R. M. Humphreys.
Luminous stars and associations in galaxies, p. 103 (1986). – See Abstr. 012.079. (IAU Symp. No. 116).
Red supergiant stars that are members of the nearby Sc galaxy NGC 300 have been isolated by spectrophotometry and infrared photometry from a blink comparator survey of red stars in the general field of this highly resolved stellar system. Photometry of 4 M supergiants indicates absorption corrections in V averaging $0\overset{m}{.}7$. Adopting a true distance modulus of $26\overset{m}{.}1$ for NGC 300, a mean absolute visual magnitude of $-7\overset{m}{.}9$ is found for these stars. The IR photometry leads to a mean absolute magnitude at K of $-11\overset{m}{.}1$ and a mean bolometric absolute magnitude of $-8\overset{m}{.}5$.

157.312 **Wolf–Rayet stars in nearby galaxies.**
P. Massey.
Luminous stars and associations in galaxies, p. 215 – 218 (1986). – See Abstr. 012.079. (IAU Symp. No. 116).

157.313 **Wolf–Rayet stars in M31.**
M. M. Shara, A. F. J. Moffat.
Luminous stars and associations in galaxies, p. 231 – 232 (1986). – See Abstr. 012.079. (IAU Symp. No. 116).
The authors have completed a survey of the closest giant spiral galaxy beyond our own for Wolf–Rayet stars, using narrow band imaging techniques.

157.314 **Wolf–Rayet stars in M31's OB associations.**
T. E. Armandroff, P. Massey, P. S. Conti.
Luminous stars and associations in galaxies, p. 239 – 241 (1986). – See Abstr. 012.079. (IAU Symp. No. 116).
The primary goal of this study is to use the observed WR content to deduce global measures of the massive star population in M31.

157.315 **New stellar associations in M31.**
Yu. N. Efremov, G. R. Ivanov, N. S. Nikolov.
Luminous stars and associations in galaxies, p. 389 – 390 (1986). – See Abstr. 012.079. (IAU Symp. No. 116).

157.316 **The associations OB 110 and OB 112 in M33 galaxy.**
P. Z. Kunchev, N. S. Nikolov.
Luminous stars and associations in galaxies, p. 391 – 392 (1986). – See Abstr. 012.079. (IAU Symp. No. 116).
The observations of OB 110 and OB 112 were obtained in the period 1983 – 1984 by 2 m Ritchey–Chrétien telescope in the Rhodopes Mountains (Bulgaria).

157.317 **On the arm pattern of M33.**
G. R. Ivanov, P. Z. Kunchev.
Luminous stars and associations in galaxies, p. 419 – 420 (1986). – See Abstr. 012.079. (IAU Symp. No. 116).
By comparison of U and B plates of M33 the authors have extended the numerations of Humphreys and Sandage (1980) adding 54 new OB associations.

157.318 **Luminosity functions in dwarf irregulars.**
J. G. Hoessel.
Luminous stars and associations in galaxies, p. 439 – 449 (1986). – See Abstr. 012.079. (IAU Symp. No. 116).
Observing luminosity functions and color–magnitude diagrams is a first step towards understanding the luminous stellar populations in galaxies. The galaxies discussed in this paper in no way constitute a proper statistical sample. Data are taken from the literature where they exist and include unpublished results available to the author. There are certainly strong selections applied, including an emphasis on galaxies which have very low surface brightness.

157.319 **A search for Wolf–Rayet stars in giant extragalactic bursts of star formation.**
A. W. Campbell, L. J. Smith.
Luminous stars and associations in galaxies, p. 499 – 501 (1986). – See Abstr. 012.079. (IAU Symp. No. 116).
The authors have begun a detailed optical/infrared investigation of a sample of "H II galaxies" (galaxies experiencing a relatively very luminous burst of star formation) with the objective of studying the occurrence and evolution of WR stars as functions of the age, mass and abundance of the ionizing cluster.

157.320 **Giant H II regions in M81.**
M. Kaufman, R. C. Kennicutt, F. N. Bash.
Luminous stars and associations in galaxies, p. 503 – 504 (1986). – See Abstr. 012.079. (IAU Symp. No. 116).

157.321 **Dust in M31 – observations in the IR with IRAS.**
R. A. M. Walterbos, P. B. W. Schwering.
Light on dark matter, p. 401 – 404 (1986). – See Abstr. 012.080.
The authors have analysed IRAS observations of M31 at 12, 25, 60, and 100 micron.

157.322 **Statistical properties of IRAS galaxies.**
G. Helou.
Light on dark matter, p. 405 – 414 (1986). – See Abstr. 012.080.
Statistical approaches to the study of galaxies seen by IRAS are reviewed, stressing the contrast between infrared complete and optically complete samples. A statistical overview is given of the Cataloged Galaxies and Quasars Observed in the IRAS Survey.

157.323 **Infrared radiation from normal galaxies.**
E. E. Becklin.
Light on dark matter, p. 415 – 420 (1986). – See Abstr. 012.080.
The author reviews the possible sources of thermal infrared radiation in a complete sample of optically bright normal galaxies. The sample comes primarily from the Virgo cluster but is augmented by field galaxies at the same distance. Ground based 10 μm measurements of the nuclear regions and the IRAS survey are both reviewed.

157.324 **The identification of IRAS galaxies.**
R. D. Wolstencroft, R. G. Clowes, M. Kalafi, S. K. Leggett, H. T. MacGillivray, A. Savage.
Light on dark matter, p. 425 – 434 (1986). – See Abstr. 012.080.
The authors report on the first results of a large–scale program to identify sources in the IRAS Point Source (IRPS) Catalogue. The positions and magnitudes of candidate identifications are obtained from UK Schmidt Telescope J plates, which are measured by the automatic high–speed measuring machine COSMOS.

157.325 **Mid infrared spectroscopy of IRAS bright galaxies.**
D. K. Aitken, P. F. Roche, C. Smith.
Light on dark matter, p. 435 – 436 (1986). – See Abstr. 012.080.
The authors have made spectroscopic observations between 8 – 13 μm of a small sample (6) of galaxies found to be bright in the IRAS bands but without apparent sign of nuclear activity. On closer inspection three of these show the presence of nuclear H II regions and the remaining three show very deep silicate absorption.

157.326 **The relationship between blue and FIR luminosities of spiral galaxies.**
K. V. K. Iyengar, T. N. Rengarajan, R. P. Verma.
Light on dark matter, p. 437 – 438 (1986). – See Abstr. 012.080.
The authors have studied optical and infrared properties of 86 galaxies from IRAS circulars, identified in RC–2 and UGC. They find a correlation between L_{FIR} and L_B. The observed correlation implies a relationship between the mass of a galaxy and its young star formation activity.

157.327 **Preliminary results of an H I survey of a sample of IRAS galaxies.**
J. T. Armstrong, A. Wootten.
Light on dark matter, p. 439 – 440 (1986). – See Abstr. 012.080.
The authors report H I detections in 46 of a sample of 110 galaxies selected from the IRAS circulars and from the point source catalog.

157.328 **Starbursts in non–interacting galaxies.**
T. G. Hawarden, J. H. Fairclough, R. D. Joseph, S. K. Leggett, C. M. Mountain.
Light on dark matter, p. 455 – 462 (1986). – See Abstr. 012.080.
Examination of IRAS results for a complete sample of nearby galaxies with morphological class T between 0 and 6 (inclusive) suggests that episodes of star formation vigorous enough to dominate the IR colours of a system occur only in barred galaxies. This is understandable if the location of these "starbursts" are the small complexes of H II regions which commonly surround the nuclei of barred galaxies of these types and which probably mark circumnuclear stagnation zones in a bar–driven inward flow of gas. The luminosity of these complexes will be regulated by the supply of gas to the bar and this model appears capable of explaining most of the mid–IR properties of bright spiral galaxies.

Nomenclature for objects in the galaxy M33.
See Abstr. 002.021.

A multicolor photometric catalog of galaxies and stars in the field of the rich cluster II Zw 1305.4+2941 at z = 0.24.
See Abstr. 002.052.

A multicolor photometric catalog of galaxies and stars in the field of the rich cluster II Zw 1305.4+2941 at z = 0.24.
See Abstr. 002.060.

Catalogue of isolated galaxies.
See Abstr. 002.069.

The Observatorio Nacional – Center for Astrophysics Redshift Survey. A status report.
See Abstr. 002.084.

A statistical analysis and overview of the IRAS point source catalog.
See Abstr. 002.092.

A statistical analysis of the LRS catalog.
See Abstr. 002.093.

Galaxies.
See Abstr. 003.050.

Galaxy distances and deviations from universal expansion.
See Abstr. 011.009.

Exercises for distance estimates of the Local Group galaxies using Cepheid variables.
See Abstr. 014.047.

Machine measurements of the objective–prism spectra of faint galaxies – I. Plate measurements and initial data reduction.
See Abstr. 036.017.

Narrow–band photometry and the evolution of galaxies.
See Abstr. 036.024.

Image restoration and superresolution in the far infrared.
See Abstr. 036.062.

Design of an expert system for object classification through an application to the classification of galaxies.
See Abstr. 036.135.

A statistical analysis of errors in the 21–cm line redshifts of galaxies.
See Abstr. 036.185.

Semi–automated identification of IRAS point sources using UKST plates and the COSMOS measuring machine.
See Abstr. 036.258.

A monopole model for galactic nuclei.
See Abstr. 061.077.

Initial mass function for zero–metal stars.
See Abstr. 065.037.

Massive stars in galaxies.
See Abstr. 065.053.

Integrated spectral energy distribution and colors of young stellar populations.
See Abstr. 065.153.

The cooler supergiants (A to M): crucial signposts in the lifecycles of massive stars.
See Abstr. 065.160.

Gravitational microlensing at large optical depth.
See Abstr. 066.068.

Proper motions with respect to galaxies.
See Abstr. 111.025.

The formation and evolution of carbon stars.
See Abstr. 115.018.

Brightest stars in galaxies as distance indicators.
See Abstr. 115.030.

Eclipsing variable stars in neighbouring galaxies. II. Discussion of the light curve of V 55 in NGC 2403.
See Abstr. 119.082.

Eclipsing variable stars in neighbouring galaxies. III. Discussion of the light curve of V 31 in IC 1613.
See Abstr. 119.083.

Discovery of an RR Lyrae star in the halo of M31.
See Abstr. 122.149.

On the relative frequencies of the kinds of Type I supernovae.
See Abstr. 125.008.

Discovery of a large mass of iron in a type I supernova.
See Abstr. 125.201.

On the radial distributions of molecular clouds in galaxies.
See Abstr. 131.030.

The mass distribution of dark clouds in external galaxies.
See Abstr. 131.060.

Star formation in a galactic context: the location and properties of molecular clouds.
See Abstr. 131.074.

A theoretical steady state model for the distribution of gas and magnetic field in M31 and in our Galaxy.
See Abstr. 131.138.

Physical processes in star formation.
See Abstr. 131.152.

Effects of interstellar dust on the photometric properties of faint disk galaxies.
See Abstr. 131.334.

The effects of interstellar dust on the spectrum of disk galaxies.
See Abstr. 131.335.

Il ruolo della formazione stellare nell'evoluzione galattica.
See Abstr. 131.353.

Spectrophotometry of H II regions in irregular galaxies.
See Abstr. 132.002.

H II regions in M33. III. Physical properties.
See Abstr. 132.003.

Hα velocity widths of giant extragalactic H II regions.
See Abstr. 132.007.

Turbulent gas motions in giant H II regions. II. The luminosity–velocity dispersion relation.
See Abstr. 132.027.

Star formation, giant H II regions and spiral structure.
See Abstr. 132.043.

Abundance determination from H II regions in NGC 2403.
See Abstr. 132.056.

The exciting stars of giant extragalactic H II regions.
See Abstr. 132.060.

Sub–mm continuum observations of sources from the IRAS point–source catalogue.
See Abstr. 132.061.

An accurate position and radial velocity for the planetary nebula in the Fornax dwarf galaxy.
See Abstr. 134.021.

Chemical abundances of planetary nebulae in M31.
See Abstr. 134.044.

Optical counterparts of M31 X–ray sources.
See Abstr. 142.064.

The creation of compact objects in the Local Group.
See Abstr. 142.082.

New insight on the evolution of binary X–ray sources from the X–ray observations of nearby spiral galaxies.
See Abstr. 142.084.

X–ray emission from bursts of star formation.
See Abstr. 142.104.

The origin of cosmic rays. I. Observations of the spiral galaxy NGC 3310.
See Abstr. 144.072.

The origin of cosmic rays. II. The cosmic–ray distribution and the spiral structure of NGC 3310.
See Abstr. 144.073.

The origin of cosmic rays. III. Particle acceleration by global spiral shocks.
See Abstr. 144.074.

Orbital decay of satellite galaxies in spherical systems.
See Abstr. 151.001.

Formation of elliptical galaxies and massive halos.
See Abstr. 151.003.

Finite length–scale anti–gravity and observations of mass discrepancies in galaxies.
See Abstr. 151.004.

Evolution of galaxies: one–zone model with a birth–rate stellar function depending on gas density and temperature.
See Abstr. 151.013.

Periodic orbits in elliptical galaxies. II. Rotation about the axis of symmetry.
See Abstr. 151.020.

Are polar rings self–gravitating?
See Abstr. 151.024.

Theoretical considerations on the rotation curves of galactic halos.
See Abstr. 151.025.

Massive central galaxies, globular cluster swapping, and tidal accretion.
See Abstr. 151.028.

A tidal model for the interacting pair of galaxies VV 117.
See Abstr. 151.032.

Neutral hydrogen in M31. VI. The tidal interaction between M31 and NGC 205.
See Abstr. 151.035.

Triaxial models of highly flattened elliptical galaxies.
See Abstr. 151.055.

Ring galaxies as a laboratory for studying star formation and gas dynamics in galaxies.
See Abstr. 151.056.

Collisionless Boltzmann simulation of stellar disks having a small bulge.
See Abstr. 151.063.

Theoretical considerations on the rotation curves of galactic halos.
See Abstr. 151.091.

Rotation and vorticity in galactic models.
See Abstr. 151.092.

Bisymmetric spiral configuration of magnetic fields in spiral galaxies. I. Local theory.
See Abstr. 151.098.

Models of axisymmetric galaxies.
See Abstr. 151.100.

Observable properties of oblate spheroids.
See Abstr. 151.106.

Dynamical friction on a satellite of a disc galaxy.
See Abstr. 151.111.

The OB 112 association in the M33 galaxy.
See Abstr. 152.004.

Stellar associations in M31 arm S4.
See Abstr. 152.008.

Systems of stellar associations in galaxies.
See Abstr. 152.010.

The metallicity of globular clusters associated with the giant elliptical galaxy NGC 4486 (M87).
See Abstr. 154.002.

Photoelectric photometry of star clusters in M31. XI.
See Abstr. 154.025.

The distribution of globular clusters associated with M87 = NGC 4486.
See Abstr. 154.041.

Accurate radial velocities for three Fornax globular clusters.
See Abstr. 154.047.

The chemical evolution of galaxies.
See Abstr. 155.051.

Additional evidence for the accuracy of galactic reddenings from ScI galaxies and quasars.
See Abstr. 155.069.

The initial mass function of massive stars in galaxies: empirical evidence.
See Abstr. 155.121.

A survey of neutral hydrogen absorption in the nuclei of active spiral galaxies.
See Abstr. 158.001.

NGC 4410a/b: a strong radio source associated with interacting galaxies.
See Abstr. 158.038.

A comparative study of the stellar content in normal and active nuclei of spiral galaxies.
See Abstr. 158.058.

A deficiency of normal elliptical galaxies among Markarian galaxies.
See Abstr. 158.077.

MAGN; mildly active galactic nuclei.
See Abstr. 158.133.

Interaction and emission morphology.
See Abstr. 158.140.

A search for OH maser emission from spiral galaxies.
See Abstr. 158.174.

The distribution and some examples of UV–bright galaxies.
See Abstr. 158.199.

The strong UV source in the active E galaxy NGC 4552.
See Abstr. 158.218.

Surface brightness distribution in Seyfert galaxies. II. Multiaperture photometry.
See Abstr. 158.256.

Ultraviolet spectrophotometry of three LINERs.
See Abstr. 158.262.

The nuclear velocity fields of NGC 1068 and M51: spatially resolved [N II] emission associated with radio jets and bubbles.
See Abstr. 158.286.

On the origin of extragalactic infrared radiation.
See Abstr. 158.293.

Models for IRAS galaxies.
See Abstr. 158.324.

The Mg II absorption system in the QSO PKS 2128–12: a galaxy disc/halo with a radius of 65 kpc.
See Abstr. 159.027.

CCD imaging of the QSO–galaxy pair PKS 1327–206, ESO 1327–2041.
See Abstr. 159.093.

Damped Lyman–alpha absorption by disk galaxies with large redshifts. I. The Lick survey.
See Abstr. 159.155.

Morphological segregation in the Pisces–Perseus supercluster.
See Abstr. 160.001.

New polar ring galaxies in rich clusters of galaxies.
See Abstr. 160.007.

Alignment of spiral galaxies in the Local Supercluster.
See Abstr. 160.009.

High–resolution imaging from Mauna Kea: surface photometry of galaxies in CL 0024+16.
See Abstr. 160.011.

H I observations of galaxies in the Virgo cluster of galaxies. I. The data.
See Abstr. 160.013.

Investigation of clusters of galaxies. I. Explanation of photometric scale errors by light diffusion in a photographic emulsion.
See Abstr. 160.016.

Machine measurements of the objectve–prism spectra of faint galaxies – II. Interactive redshift measurements.
See Abstr. 160.019.

The blue Tully–Fisher relation in a variety of environments.
See Abstr. 160.024.

The morphological types and orbits of H I–deficient spirals in clusters of galaxies.
See Abstr. 160.025.

Blue disk galaxies in the Coma Cluster: analogs to $z = 0.5$ cluster members?
See Abstr. 160.026.

The two–point galaxy correlation function of the Local Supercluster.
See Abstr. 160.027.

CO in H I – deficient Virgo cluster spiral galaxies.
See Abstr. 160.028.

Surface photometry of dwarf elliptical galaxies in the Virgo cluster.
See Abstr. 160.029.

The structure of brightest cluster members. I. Surface photometry.
See Abstr. 160.042.

Redshifts of galaxies in the south–west region of the Perseus supercluster.
See Abstr. 160.046.

CO in H I–deficient Virgo cluster spiral galaxies.
See Abstr. 160.050.

Cooling flows in clusters and galaxies.
See Abstr. 160.056.

Static properties of galaxies in the Shapley–Centaurus cluster. II. The analysis.
See Abstr. 160.088.

X–ray studies of galactic and intergalactic gas in the Pegasus I cluster.
See Abstr. 160.089.

Spectroscopy of three Butcher–Oemler clusters of galaxies at a redshift of 0.2.
See Abstr. 160.090.

The detection of distant cooling flows and the formation of dark matter.
See Abstr. 160.103.

The large–scale distribution of galaxies in the Lynx–Gemini region.
See Abstr. 160.106.

Detection of large–scale alignment of Lick counts around Abell clusters.
See Abstr. 160.107.

The nearby Abell clusters. I. Abell 168.
See Abstr. 160.108.

The orbits of H I–deficient spiral galaxies in the Virgo cluster.
See Abstr. 160.115.

The Centaurus cluster of galaxies – I. The data.
See Abstr. 160.118.

The mean properties of Virgo cluster dwarf galaxies as a function of distance from the center of the cluster.
See Abstr. 160.126.

H I observations of blue compact dwarf galaxies in the Virgo cluster.
See Abstr. 160.127.

Isolated galaxies.
See Abstr. 161.008.

Pushing back the redshift limit.
See Abstr. 161.015.

Galaxy and cluster formation in a universe dominated by cold dark matter.
See Abstr. 161.031.

Pancakes of galaxies and the nature of dark matter.
See Abstr. 161.032.

Where do galaxies form?
See Abstr. 161.033.

The Malmquist bias and the value of H_0 from the Tully–Fisher relation.
See Abstr. 161.041.

How relaxed is the distribution of galaxies?
See Abstr. 161.044.

The distance moduli from the *B*–band Tully–Fisher relation and the velocity field.
See Abstr. 161.045.

The smoothness of the 2.2 micron background: constraints on models of primeval galaxies.
See Abstr. 161.046.

The origin of large scale cosmic structure.
See Abstr. 161.049.

New perspectives on galaxy formation.
See Abstr. 161.052.

Large–scale anisotropy in the Hubble flow.
See Abstr. 161.075.

An estimate of the total space density of dwarf galaxies.
See Abstr. 161.085.

The masses of galaxies and clusters of galaxies.
See Abstr. 161.090.

A distance scale from the infrared magnitude/H I velocity–width relation. V. Distance moduli to 10 galaxy clusters, and positive detection of bulk supercluster motion toward the microwave anisotropy.
See Abstr. 161.096.

The distance scale: present status and future prospects.
See Abstr. 161.115.

The extragalactic distance scale derived from "sosie" galaxies. II. Solar motion and Hubble constant.
See Abstr. 161.116.

The origin of dwarf galaxies, cold dark matter, and biased galaxy formation.
See Abstr. 161.118.

The neutrino hypothesis for dark matter in dwarf spheroidals.
See Abstr. 161.160.

Dynamical effects of an extended cloud of dark matter on dwarf spheroidals.
See Abstr. 161.161.

Dark matter – systematics of its distribution.
See Abstr. 161.162.

On the nature of dark matter in dwarf galaxies.
See Abstr. 161.163.

Rotational curves of galaxies and neutrino halos.
See Abstr. 161.215.

On large–scale quantization.
See Abstr. 161.252.

Dark matter in the universe.
See Abstr. 161.307.

A new test for the cosmological evolution of extragalactic sources.
See Abstr. 161.308.

The primordial helium abundance determination from metal–poor galaxies.
See Abstr. 161.315.

An infrared search for primeval galaxies.
See Abstr. 161.339.

The galaxy luminosity function and the redshift–distance controversy. (A review).
See Abstr. 161.399.

Zur dreidimensionalen Verteilung entfernter Galaxien.
See Abstr. 161.409.

158 Active Galaxies (Seyfert Galaxies, BL Lacertae Objects, Radio Galaxies)

158.001 **A survey of neutral hydrogen absorption in the nuclei of active spiral galaxies.**
J. M. Dickey.
Astrophys. J., Vol. 300, No. 1, p. 190 – 202 (1986).

Nineteen spiral and irregular galaxies with bright continuum nuclei have been observed with the VLA to search for absorption at $\lambda = 21$ cm by cool atomic hydrogen in or in front of the nuclei. Most of these galaxies show activity at other wavelengths, such as Seyfert nuclei, intense far–IR emission, and X–ray emission. Fifteen absorption lines were detected, and six tentatively detected, in 15 of these galaxies. The line centers are offset from the systemic velocities by as much as 500 km s^{-1}. Infall is more common than outflow at these high velocities, although the more common lines with offsets of less than 100 km s^{-1} are evenly distributed between infall and outflow.

158.002 **A double–lobed radio galaxy with spiral structure?**
S. M. Simkin, A. Michel.
Astrophys. J., Lett. Ed., Vol. 300, No. 1, p. L5 – L9 (1986). With plate L1.

The authors report evidence for the presence of distinct, four–armed, spiral structure in the inner optical region (14 – 18 kpc) of the classical double radio galaxy 3C 33. On deep images, this spiral can be traced to a distance of 37 kpc from the nucleus. In addition, there is evidence for an asymmetric offset between the spiral pattern and the dominant, underlying DE galaxy, as well as an outer, asymmetric extension of the galaxy. One explanation for these asymmetries is that this galaxy has experienced strong tidal distortion, supporting the currently popular "merger" hypothesis of radio galaxies.

158.003 **The Parkes selected regions: powerful radio galaxies and quasars at high redshifts.**
A. J. B. Downes, J. A. Peacock, A. Savage, D. R. Carrie.
Mon. Not. R. Astron. Soc., Vol. 218, No. 1, p. 31 – 62 (1986).

A complete sample of extragalactic radio sources with flux density $S \geqslant 0.1$ Jy at 2.7 GHz has been compiled to probe the radio luminosity function to redshifts $z \sim 3 - 4$. VLA maps are presented for the 188 sources in this flux density range; 178 sources remain in the complete sample after examination for confusion. The new radio data define the structures and positions sufficiently well for a re–examination of existing plate material to have yielded an identification content of 56 per cent of the sample down to $V \sim 21$. The results from this programme will strongly constrain the behaviour of the radio population at high redshifts. The present incomplete data already provide some indications that a redshift cut–off is required for both steep– and flat–spectrum sources.

158.004 **Extra–nulcear molecular hydrogen in NGC 1068.**
N. K. Reay, P. D. Atherton, N. A. Walton.
Mon. Not. R. Astron. Soc., Vol. 218, No. 1, p. 13P – 17P (1986).

The authors report observations of the $\upsilon = 1 - 0$ S(1) line of molecular hydrogen in the nucleus and inner spiral arms of the Seyfert galaxy NGC 1068. Spectra of the nucleus, for which the $\upsilon = 2 - 1$ S(1) line was also observed, confirm that the lines are collisionally excited. Significant detections were made at two of the seven positions at which spectra were recorded in the inner spiral arms. The inner region of NGC 1068 is known to be turbulent. The authors have suggested in a previous publication that this turbulence may be the result of nuclear outflow interacting with the galactic disc. They suggest here that the molecular hydrogen is excited by a shock driven by this outflow.

158.005 **Stellar dynamics of Cen A.**
A. Wilkinson, R. M. Sharples, R. A. E. Fosbury, P. T. Wallace.
Mon. Not. R. Astron. Soc., Vol. 218, No. 2, p. 297 – 329 (1986). With a correction in Vol. 220, No. 3, p. 799 (1986).

The authors have made the first complete map of the stellar velocity field within 100 arcsec of the nucleus of the bright elliptical galaxy NGC 5128 (Cen A) and obtained complementary long–slit spectra out to 400 arcsec along the optical major axis. These data show that the ellipsoidal stellar component is rotating slowly (maximum line–of–sight velocity 40 km s^{-1}) approximately perpendicular to the dust lane about a projected axis lying in pa $\sim 135°$ and in the opposite sense to that expected if the warp were due to the dust lying in stable orbits round a triaxial potential. The velocity field may be explained by two alternative models presented in the paper. It is shown that the dust lane can significantly affect the steepness of the rotation curve, the skew of the velocity field and the magnitude of the dispersion.

158.006 **Infrared photometry of the nuclei of early–type radio galaxies.**
W. B. Sparks, J. H. Hough, D. J. Axon, J. Bailey.
Mon. Not. R. Astron. Soc., Vol. 218, No. 3, p. 429 – 444 (1986).

The authors have obtained near– and mid–infrared photometry of the nuclei of both strong and weak radio galaxies. The implication of these data are that a relatively small percentage, $\cong 15$ per cent, of radio ellipticals have strong infrared emission. Although as a class they do have a 10–μm flux which is statistically greater than zero, especially the optically brightest, a large fraction of this excess can be attributed to starlight from the galaxy. Those galaxies which are detected at 10 μm tend to be bright at longer infrared wavelengths, and have a spectrum which is flatter than that typical of dust–lane ellipticals. Unlike Seyfert galaxies, there is no correlation between 10–μm flux and radio flux; the infrared fluxes for the compact active radio galaxies require a spectrum which turns over between 5 GHz and 10 μm. Only the galaxies with strongest 10–μm radiation have near infrared colours which are significantly redder than the norm.

158.007 **Some new bright southern Seyfert galaxies.**
A. P. Fairall.
Mon. Not. R. Astron. Soc., Vol. 218, No. 3, p. 453 – 455 (1986).
Four new Seyfert galaxies have been found in the Hydra–Centaurus supercluster. One is ESO 323–G77 with a 13 mag Seyfert 1 nucleus having strong Fe II emission. The others, IC 2560 and NGC 4785 and 4968, are Seyfert 2 galaxies – the last having also been recently identified by de Grijp et al.

158.008 **Some implications of excess soft X–ray emission from Seyfert 1 galaxies.**
A. C. Fabian, P. W. Guilbert, K. A. Arnaud, R. A. Shafer, A. F. Tennant, M. J. Ward.
Mon. Not. R. Astron. Soc., Vol. 218, No. 3, p. 457 – 464 (1986).
The X–ray spectrum of Seyfert 1 galaxies is characterized by a hard power–law spectrum. It is often postulated that this maintains a Compton–heated two–phase Broad–Line Region (BLR) around the central source. The authors show that the strong excess soft X–ray emission observed in MKN 841 and other Seyfert galaxies invalidates this model if the BLR is spherically symmetric. Alternatives are proposed. Weak excess soft X–ray emission from the Compton–hot intercloud medium may be observed when the central source is obscured by a broad–line cloud, such as occurs in NGC 4151. Clouds cooling from the hot phase can increase this soft X–ray flux and in future may also be detected by ultraviolet and X–ray absorption studies. Strong excess soft X–ray emission could be common in many galaxies.

158.009 **IUE observations of the highly luminous Seyfert galaxy ESO 113–IG45 (Fairall 9).**
J. Masegosa, M. Moles, M. V. Penston.
Mon. Not. R. Astron. Soc., Vol. 218, No. 3, p. 541 – 550 (1986).
Ultraviolet variations of the continuum in ESO 113–IG45 with a doubling time of 50 days indicate a maximum size of 1×10^{17}cm to the emitting region. Power law and blackbody fits to the continuum are both possible. The power–law fit is preferred because of the observed correlation between flux and spectral index which is similar to that found in other Seyferts. The rapid variability is hard to explain with an accretion disc since the variation time–scale is faster than the orbit time in disc models.

158.010 **High–resolution mapping of the nucleus of NGC 5128 (Cen A) at *J*, *H* and *K*.**
A. B. Giles.
Mon. Not. R. Astron. Soc., Vol. 218, No. 4, p. 615 – 627 (1986).
High–resolution near–IR maps of the nucleus of Cen A show a pronounced asymmetry for wavelengths of 1.2 and 1.6 μm. The axis of this extension lies perpendicular to the dust lane along the X–ray jet axis. Comparisons with CCD images show the optical nucleus to lie 5.8 ± 0.3 arcsec to the NE of the 2.2–μm nucleus at PA $\sim 45°$, so it may possibly be identified as source N on the VLA radio maps. The 2.2–μm source is tentatively identified with a faint object on an Hα photograph and must be considered to be the true or principal nucleus of Cen A. The radial intensity of the 2.2–μm source is well fitted by de Vaucouleurs' law to within 1 – 2 arcsec of the nucleus and radial $J–H/H–K$ variations show the existence of a $\sim$24–arcsec diameter core within the optical dust lane.

158.011 **Narrow line profiles and interacting gas flows in active galaxies.**
J. Mardaljevic, D. J. Raine, M. D. Smith.
Mon. Not. R. Astron. Soc., Vol. 218, No. 4, p. 651 – 662 (1986).
The "catapult model" of the narrow line region involves both infalling and outflowing clouds interacting with an outflowing wind. The authors derive the emission line profiles in this model and compare the results with observation. Supersonic and subsonic winds in the interaction zone produce distinct line profiles, both of which are observed. Flat–topped profiles, the presence of broad wings, asymmetry and line shifts are discussed. A relationship between profile shape and X–ray properties is predicted.

158.012 **The soft X–ray spectrum of NGC 4151.**
K. A. Pounds, R. S. Warwick, J. L. Culhane, P. A. J. de Korte.
Mon. Not. R. Astron. Soc., Vol. 218, No. 4, p. 685 – 694 (1986).
EXOSAT observations of the Seyfert galaxy NGC 4151 over the period 1983 July – 1984 April have revealed a steeply decreasing flux in the 2 – 10 keV band. Other observations have shown that this coincided with a fall in the whole infrared–to–ultraviolet continuum and a marked weakening of the broad emission lines. The X–ray spectral shape, however, remained unchanged, although – in accord with previous measurements – a power–law spectrum attenuated with a simple column of cold gas does not provide a satisfactory spectral fit below about 3 keV, where a lower relative opacity is required. Inclusion of additional low energy data from the EXOSAT telescopes allows the modelling of the absorbing column to be improved and reveals a second, separate, soft X–ray spectral component. It is suggested that this soft component may be thermal emission from a hot intercloud medium in the narrow emission line region.

158.013 **The radio source in the very luminous gamma–ray Seyfert galaxy MCG 8–11–11.**
J. S. Ulvestad, A. S. Wilson.
Mon. Not. R. Astron. Soc., Vol. 218, No. 4, p. 711 – 719 (1986).
The Seyfert 1.5 galaxy MCG 8–11–11 has been observed at 2 and 6 cm with the highest resolution configuration of the VLA. The centre of the galaxy contains a triple radio source having a size of 400 pc; the central component of that source is less than 30 pc in diameter and presumably can be identified with the core of the active galaxy. The triple source is embedded in more diffuse emission $\sim$2 kpc in size. This diffuse emission has a major–axis position angle that is similar to that of the bar in the optical galaxy but differs from the position angle of the triple source by 50°. The overall morphology of the source is S–shaped, but the curvature is quiete abrupt at the ends of the triple source. The 2–kpc–scale emission must be largely non–thermal. It appears that the core radio emission in Seyfert 1 and 1.5 galaxies may be weakly correlated with their X–ray emission.

158.014 **The parsec–scale structure of the radio nucleus of NGC 4151.**
B. Harrison, A. Pedlar, S. W. Unger, P. Burgess, D. A. Graham, E. Preuss.
Mon. Not. R. Astron. Soc., Vol. 218, No. 4, p. 775 – 784 (1986).
The authors present MK III European VLBI Network observations of the nucleus of NGC 4151 with an angular resolution of 20 milliarcsec (1.3 pc). They have also monitored the flux density of the nucleus at a wavelength of 18 cm using the MERLIN network, and find no evidence for variability greater than 10 mJy. The structure is a 450–milliarsec collinear triple source along PA $\cong 83°$. The outer components are resolved and have sizes of $\cong$30 and $\cong$60 milliarsec (2 and 4 pc). The stronger of these components is elongated in PA $\cong 57°$, which is close to the PA of the arcminute–scale optical structure, but is misaligned with the arcsecond–scale radio structure. This suggests that the UV radiation responsible for ionizing the extended gas is collimated on a scale of <1 pc, whilst the final collimation of the radio jet occurs on a scale of >1 pc.

158.015 **Optical synchrotron emission in the southern lobe of 3C33.**
K. Meisenheimer, H.-J. Röser.
Nature, Vol. 319, No. 6053, p. 459 – 463 (1986).
Charge coupled device polarimetry at a wavelength of 660 nm has detected a highly–polarized optical source, which is coincident with the radio hotspot at the leading edge of the southern lobe of 3C33. The optical polarization data perfectly match the high–frequency Very Large Array observations, thus providing the first direct evidence for optical synchrotron radiation from a classic double radio source. The extended structure of the emission region requires a huge particle accelerator boosting electrons

to highly relativistic energies (up to 100 GeV) over a region several kiloparsecs in extent.

158.016 **How old is Cygnus A?**
M. J. Spinks, W. G. Rees, P. J. Duffett-Smith.
Nature, Vol. 319, No. 6053, p. 471 – 473 (1986).

The authors have mapped the extragalactic radio source Cygnus A at 81.5 MHz using a two-element interferometer with variable orientation and spacing (maximum 45 km). This is the lowest frequency at which the radio source has been mapped with sufficient resolution (16 arc s) to resolve the bridge of emission joining the two outer lobes. By comparing the map with one made at 2.7 GHz, the authors have measured the two-point spectral index as a function of distance along the source axis. They estimate the age of the radio source by using a model in which two oppositely directed beams ram their way out through the surrounding medium. The analysis suggests that Cyg A is ~ 3 Myr old.

158.017 **Optical variability of Seyfert nuclei. I. *VBLUW* photometry.**
H. R. de Ruiter, J. Lub.
Astron. Astrophys., Suppl. Ser., Vol. 63, No. 1, p. 59 – 69 (1986).

The authors present the data on *VBLUW* photometry of 15 Seyfert galaxies obtained between December 1979 and December 1982 and discuss their quality. The variability of Seyfert nuclei on timescales of ~ 1 day to ~ 1 year is investigated. They find in none of the Seyfert galaxies variations on timescales of ~ 1 day. Seyfert nuclei of type 2 do not show significant long term (~ 1 year) variability in the period between 1979 and 1982, whereas 10 out of 11 type 1 Seyfert nuclei that were observed during these three years have varied significantly in V, with amplitudes ranging from 0.1 to 0.6 magnitude. The number of objects with variable colour indices decreases monotonically going from $(V-B)$ to the far ultraviolet colour index $(U-W)$; in fact, only one object, Mkn 609, has possibly varied in $(U-W)$. The authors present various two-colour diagrams and show that type 1 Seyfert nuclei have a well-defined locus in these diagrams with a remarkably small scatter (~ 0.06 to 0.10 mag).

158.018 **An absorption line study of galaxies at high redshift.**
H. S. Murdoch, R. W. Hunstead, J. C. Blades, M. Pettini.
Astrophys. Space Sci., Vol. 118, Nos. 1/2, p. 501 – 503 (1986). – See Abstr. 012.001.

Complex velocity structure is revealed in several absorption systems in the spectrum of 0215+015 ($z = 1.715$) at resolutions of 20 – 30 km s^{-1} FWHM. Striking differences are found in the relative strengths of low to high ions among the components, reinforcing the interpretation of these systems as intervening galaxies.

158.019 **Echelle spectroscopy of narrow line regions of Seyfert galaxies.**
M. Iye, M.-H. Ulrich.
Astrophys. Space Sci., Vol. 118, Nos. 1/2, p. 523 – 528 (1986). – See Abstr. 012.001.

Preliminary results of spectroscopic studies of selected Seyfert galaxies employing a new Cassegrain echelle spectrograph completed at ESO are presented. Subjects reported are the asymmetric profiles of narrow emission lines, spatial variation of narrow emission line profiles, and the interstellar absorption lines.

158.020 **Optical light variation of the Seyfert galaxy NGC 4151.**
H. Ohtani, J. Meaburn, C. Goudis, A. El-Bassuny, M. Soliman.
Astrophys. Space Sci., Vol. 119, No. 1, p. 177 – 179 (1986). – See Abstr. 012.001.

The optical light variation of the nucleus of the galaxy NGC 4151 was monitored mainly using the 74 inch telescope of the Helwan Observatory, Egypt.

158.021 **Emission line velocity field in the central region of M82.**
M. Sasaki, M. Saitō.
Astrophys. Space Sci., Vol. 119, No. 1, p. 181 – 184 (1986). – See Abstr. 012.001.

Velocity field in the central region of M82 has drawn in detail mainly from Hα. This velocity field is consistent with that of Ne II 12.8 μm except for the region near a huge dust lane, and it cannot be explained by pure rotation around the galactic center. Origin of peculiarities of the velocity field is briefly discussed.

158.022 **X-ray observation of AGN's from TENMA.**
S. Miyoshi, S. Hayakawa, H. Kunieda, F. Nagase, Y. Tawara.
Astrophys. Space Sci., Vol. 119, No. 1, p. 185 – 190 (1986). – See Abstr. 012.001.

In order to investigate the structure and mechanism of active galactic nuclei through X-ray observation, detailed energy spectra are examined with the gas scintillation proportional counters on board TENMA. For three selected objects, PKS 2155–304, Centaurus A and IC 4329 A, the observational results in the intensity variation, the spectral shapes and the iron line features are presented.

158.023 **Knots, wiggles, and flaring in ballistic and stochastic jets.**
D. A. Roberts.
Astrophys. J., Vol. 300, No. 2, p. 568 – 576 (1986).

This paper gives a variety of arguments to support two contentions: (1) the morphology of an extragalactic jet, and in particular the appearance of flaring, knots, and wiggles, does not necessarily imply that it is confined; and (2) stochastic effects may be fundamental to understanding jets, whether or not they are free. These issues are demonstrated to be related through a simple computer simulation that shows the seemingly significant structures can result from random effects. Some specific sources, in particular the jets in NGC 6251 and M87, are examined in the light of these ideas.

158.024 **On wind-type flows in astrophysical jets. II. Propagation outside the nucleus and the case of M87.**
A. Ferrari, E. Trussoni, R. Rosner, K. Tsinganos.
Astrophys. J., Vol. 300, No. 2, p. 577 – 590 (1986).

This paper presents solutions of the basic polytropic flow equations appropriate to the physical conditions of extragalactic jets well outside the acceleration region within their parent active galactic nuclei. The solutions are based on observationally determined galactic mass distributions and quasi-periodic brightness enhancements along the jets. The authors demonstrate that such a model is able to describe many features of the observed morphology of jets.

158.025 **A model for the surface brightness of a turbulent, low Mach number jet. II. The global energy budget and radiative losses.**
G. V. Bicknell.
Astrophys. J., Vol. 300, No. 2, p. 591 – 604 (1986).

Detailed consideration is given to the energy balance in extragalactic radio sources, in particular the Fanaroff-Riley class I radio sources whose morphology is likely to be due to jets being stopped by entrainment of the surrounding medium. Two important features of the energy balance are the demonstration that, given the fractional composition of the plasma in relativistic electrons, the ratio of lobe luminosity to jet energy flux, $\varkappa$, can be estimated from observational parameters and that in low Mach number sources the jet energy flux is dominated by the enthalpy flux. The parameter $\varkappa$ depends upon the ratio E_e/E_L of relativistic electron energy to total lobe energy and it is likely that this is determined by the initial composition of plasma ejected from the core and the amount of entrained nonrelativistic material. Spectral index variations in jets of arbitrary velocity are also considered since this provides a means of determining whether a jet is dissipative over a given region or not. The theoretical results are applied to analyze further the class I source 3C 31.

158.026 **Beam models for radio sources. V. Collimation in more realistic galactic potentials.**
P. J. Wiita, M. J. Siah.
Astrophys. J., Vol. 300, No. 2, p. 605 – 612 (1986).
The evolution of relativistic fluid jets is examined using an extension of a previously developed boundary–following code. A constant source of relativistic plasma is injected into an isothermal cloud whose distribution is determined by the gravitational potential of the galaxy. The confining gas isobars are modeled as modified power laws. Several different effective angular momentum distributions are considered by allowing for different degrees of depression of the isobars of the confining medium along the rotation axis. It is found that, in general, more powerful sources yield jets directly, while weaker ones yield bubbles.

158.027 **IUE spectra and a resulting model of Seyfert 2 galaxies.**
G. J. Ferland, D. E. Osterbrock.
Astrophys. J., Vol. 300, No. 2, p. 658 – 668 (1986). = Lick Obs. Bull., No. 1016.
Nearly simultaneous optical and ultraviolet observations of several Seyfert 2 galaxies are presented. These, and archival data, are used to derive mean emission–line spectra and ultraviolet to X–ray spectral indices for several narrow–line objects. The composite emission–line spectrum is compared with a simple photoionization model, using the observed, nearly power–law continuum, and filamentary gas clouds, a geometry which may apply to some narrow–line objects. The results suggest that an obscuring layer does not hide an underlying broad–line region, and that narrow–line objects such as Seyfert 2 and narrow–line radio galaxies are physically different from broad–line objects, at least in not having an inner broad–line region.

158.028 **VELA observations of Centaurus A X–ray activity, 1969 – 1979.**
J. Terrell.
Astrophys. J., Vol. 300, No. 2, p. 669 – 674 (1986).
The X–ray luminosity of Cen A (NGC 5128) was monitored by the VELA 5B satellite for 10 yr, beginning in 1969 May. The resultant data greatly augment the knowledge of the time history of the X–ray strength of Cen A. The 1973 – 1976 period of exceptional activity is characterized by numerous rapid changes in source intensity, indicating that the nucleus of Cen A is small, massive, and at times an intense source of X–ray emission. The power spectrum for the most active period is similar to that for shot noise produced by 50 day outbursts. A number of outbursts lasting ~ 50 days are, in fact, quite evident in the new data.

158.029 **Limb brightening and filamentation in the inner radio jet of Centaurus A.**
D. A. Clarke, J. O. Burns, E. D. Feigelson.
Astrophys. J., Lett. Ed., Vol. 300, No. 2, p. L41 – L46 (1986). With plate L2.
New high–resolution total intensity and polarization VLA maps at 2 and 6 cm of the inner jet in Centaurus A are presented. The combination of linear resolution (7.5 pc) and dynamic range (11,000:1) in these maps is unprecedented in extended radio jet observations. A distinct side–to–side edge brightening of the jet is observed. It is proposed that this structure arises from internal oblique shocks and/or external shocks produced by the interaction between the jet and the interstellar medium. The 6 cm data also reveal a filamentary structure in the jet where knots in the jet seem to be connected by what the authors call "streamers".

158.030 **VLA observations of low luminosity radio galaxies. I. Sources with angular size smaller than two arcminutes.**
P. Parma, H. R. de Ruiter, C. Fanti, R. Fanti.
Astron. Astrophys., Suppl. Ser., Vol. 64, No. 1, p. 135 – 171 (1986).
Fifty seven radio sources with angular size smaller than about two arcmin, selected from two complete samples of low luminosity B2 radio galaxies, have been observed at 20 cm with the VLA in the B configuration, with an angular resolution of about 3.5 arcsec. The reduction procedure is described and the quality of the radio maps is discussed. The dynamic range in the final CLEANed maps is usually around 100:1 or better. Observational parameters and intrinsic properties derived for each source are given in separate tables. Contour maps are presented for resolved sources. Information on individual sources is provided, particularly in case of interesting structures.

158.031 **Gas in galactic centres.**
M. E. Bailey.
Cosmical gas dynamics, p. 49 – 62 (1985). – See Abstr. 012.007.
The origin and evolution of the gas in galactic centres is briefly reviewed, emphasising the increasing evidence for infall leading ultimately to nuclear activity. A particular model is described in which inflowing stellar mass loss from the surrounding galaxy as a whole accumulates as a dense cold nuclear disc until it eventually flows into the centre to form a supermassive star or spinar.

158.032 **The effect of expanding radio components in Seyfert 2 nuclei.**
A. Pedlar, S. W. Unger, J. E. Dyson, J. Meaburn.
Cosmical gas dynamics, p. 65 – 75 (1985). – See Abstr. 012.007.
The authors describe a model in which the expansion of the radio emitting components in Seyfert 2 nuclei compresses and accelerates the ambient medium by means of shock waves. When the expansion velocity of the component has decreased to approximately 400 km/s the compressed gas cools and radiates, producing the forbidden line region. This readily accounts for the intimate relationship between the radio continuum emission and the properties of the forbidden line region. The authors discuss the model specifically in relation to NGC 1068.

158.033 **The X–ray spectrum and variability of Markarian 421.**
R. S. Warwick, I. M. George, I. McHardy, K. A. Pounds.
Mon. Not. R. Astron. Soc., Vol. 219, No. 1, p. 39 – 45 (1986).
The X–ray bright BL Lac object Markarian 421 (Mrk 421) was observed by EXOSAT four times over a period of 6 days in 1984 February. The source exhibited a very soft power–law X–ray spectrum with an extremely low intrinsic column density. Significant X–ray variability (~40 per cent) was apparent on a time–scale of the order of a day with an indication of a softening of the X–ray spectrum as the source flux decreased. The relatively rapid X–ray variability and overall (radio to X–ray) spectrum of Mrk 421 are discussed in terms of the synchrotron self–Compton model. A high degree of relativistic beam is not a requirement for this source.

158.034 **New ultra–luminous galaxies.**
J. H. Fairclough.
Mon. Not. R. Astron. Soc., Vol. 219, No. 1, p. 1P – 4P (1986).
The IRAS survey has revealed seven more galaxies with infrared luminosities of at least $10^{12} L_{\odot}$. One galaxy, Mrk 1014, equals Mrk 231 in infrared luminosity. All of the galaxies are in optical, ultraviolet and radio catalogues, and all but one show signs of activity or structural peculiarity.

158.035 **Variations of the Seyfert galaxy Fairall 9.**
I. S. Glass.
Mon. Not. R. Astron. Soc., Vol. 219, No. 1, p. 5P – 11P (1986).
JHKL observations of F9 made between 1976 and 1985 are presented. Until about 1980 the infrared output increases slowly, but during the following five years it dropped by ~1.3 mag at K(2.2 μm). The simultaneous decline at UV wavelengths was much more severe. In recent months a recovery has been noted in J, H and K but not L. The overall spectral distribution of F9 from infrared to UV has changed considerably during this period, resembling at various times the spectra of several other Seyfert 1 galaxies. The nuclear flux excess of F9 shows two components; a highly variable UV–optical continuum and a less variable infrared one. The latter may have its origin in dust heated by the former.

158.036 **PKS 2005–489: a very bright BL Lac object in a nearby galaxy.**
J. V. Wall, I. J. Danziger, M. Pettini, R. S. Warwick, W. Wamsteker.
Mon. Not. R. Astron. Soc., Vol. 219, No. 1, p. 23P – 29P (1986).
The galaxy identified with the flat–spectrum radio source PKS 2005–489 has a bright and variable stellar nucleus with a brightest observed V magnitude of ~ 13. The authors report radio, optical, UV and X–ray observations which indicate variability and power–law continua in each of these wavebands, leading to the conclusion that PKS 2005–489 is one of the brightest BL Lac objects known. The ultraviolet data show unusually strong absorption lines formed in the interstellar medium of our Galaxy, but no obvious features due to the host galaxy. The X–ray output in the range 0.04 – 6 keV is far in excess of a simple power–law extrapolation from lower frequencies.

158.037 **VLA observations of three high IR–luminosity IRAS galaxies.**
H. Karoji, M. Dennefeld, N. Ukita.
Astron. Astrophys., Vol. 155, No. 1, p. L3 – L5 (1986).
VLA observations of 3 IRAS galaxies are presented. Two of them (09061–1248, 09111–1007) have large luminosities at 20 cm, reaching values of about $20 - 30 \times 10^{21} \mathrm{WHz^{-1}sr^{-1}}$. This, together with their far–infrared luminosities of $2 \times 10^{12} L_{\odot}$, places them in the class of Arp 220 and NGC 6240, i.e., among the most luminous and active galaxies discovered by IRAS. All the available data suggest that these galaxies have a high rate of star–formation at present. The ratio of $L_{bol}(W)$ to $L_{20cm}(\mathrm{WHz^{-1}sr^{-1}})$ ranges from 2.5×10^{16} to 4.6×10^{16}, following rather well the correlation found in previous work for the galaxies with star burst nuclei.

158.038 **NGC 4410a/b: a strong radio source associated with interacting galaxies.**
E. Hummel, C. G. Kotanyi, J. H. van Gorkom.
Astron. Astrophys., Vol. 155, No. 1, p. 161 – 168 (1986).
The tidally interacting pair of galaxies NGC 4410a/b is in a group of galaxies. Two members of the group form with NGC 4410 a chain linked by bridges of diffuse, but knotty, optical emission. The radio source associated with NGC 4410 is studied in detail with the aid of observations done with the Very Large Array. The source has a radio luminosity ($L \cong 7 \times 10^{40} \mathrm{erg\,s^{-1}}$) comparable to low luminosity radio galaxies and is very asymmetric with respect to NGC 4410. Its size is much larger than the optical size of NGC 4410. Some features, if not all, of the radio source are probably related to the interaction.

158.039 **The relative luminosities of the disk and the invisible nucleus of M82: a rediscussion of the polarization pattern.**
P. Notni.
Astron. Nachr., Vol. 306, No. 5, p. 273 – 281 (1985).
The origin of the noncircular pattern of the polarization vectors in the halo of scattered light around M82 is discussed anew. This pattern had been interpreted by Solinger and Markert (1975) to indicate a high contribution of the disk to the illumination of the halo, whereas the blue colour of the halo demands a significant contribution from a very blue illuminating source, the nucleus. The discussion of Solinger and Markert has been modified in two respects: (1) an extinction in the disk of M82, distributed similar to the stars, is assumed; this strengthens the disk's ability to turn the polarization vectors parallel to itself, and (2) an original polarization of the nuclear light parallel to the galaxy's plane is assumed; this also causes a corresponding turning of the polarization vectors.

158.040 **BL Lac objects as an evolving population.**
A. Cavaliere, E. Giallongo, F. Vagnetti.
Astron. Astrophys., Vol. 156, No. 1/2, p. 33 – 37 (1986).
The authors discuss the apparently anomalous behaviour of the number counts of BL Lacertae objects, currently estimated to be much flatter than the counts of the other active galactic nuclei. They show how that peculiarity can be understood and quantitatively accounted for, if these objects stem from a population of conventionally evolving active galactic nuclei, but are singled out by a fraction of their power channelled into a randomly oriented, modestly relativistic jet of emitting particles. The authors present explicit predictions to be compared with future, statistically significant optical counts at $B \cong 16 - 17$, and X–ray counts at fluxes $\lesssim 10^{-12} \mathrm{erg\,cm^{-2}s^{-1}}$ at a few keV, that can elicit the underlying intrinsic evolutionary behaviour.

158.041 **The narrow line active galaxies NGC 3081 and NGC 4507: from the infrared to the UV.**
F. Durret, J. Bergeron.
Astron. Astrophys., Vol. 156, No. 1/2, p. 51 – 63 (1986).
The authors present optical data on NGC 3081 and NGC 4507, acquired as part of a long slit spectroscopic search for extended ionized line emission around active galactic nuclei. For both AGNs an extended ionized region surrounding the nucleus was detected together with ionized gas in the ring of NGC 3081 and in the spiral arms of NGC 4507. The authors determine the relative contribution of the stellar population, the dust and the non–thermal source to the continuum. From the emission line intensities and their spatial gradients, they derive the characteristics of the ionized regions. They discuss the link between the inner diffuse ionized regions as well as the outer emissive regions associated to the spiral arms or to the ring, and the central non thermal source. They finally try to outline the possible physical conditions responsible for the presence of extended ionized nebulosities.

158.042 **Centrifugally driven MHD–winds in active galactic nuclei.**
M. Camenzind.
Astron. Astrophys., Vol. 156, No. 1/2, p. 137 – 151 (1986).
When the prime mover in quasars is a supermassive magnetized and rapidly rotating object, the centrifugal instability can drive strong MHD–winds away from the equatorial region and extract angular momentum and rotational energy from the central object. The author shows that the necessary magnetic fields are produced, when this central object has been formed from the central part of a galactic disc. The position of the light cylinder for these objects requires a relativistic description for the corresponding MHD–winds. The author discusses the relevant equations for any stationary and axisymmetric spacetime and derives explicitly the position of the Alfvén point in the flow. He finds that centrifugally driven winds from supermassive objects carry a Poynting flux comparable with the kinetic energy flux. In addition, the magnetic field structure in the open wind zone requires the existence of a global current topology, which might explain the necessary magnetic collimation for escaping jet material. As a result, centrifugally driven winds from rapidly rotating supermassive objects carry the energy necessary to power the non–thermal emission of BLR AGNs and the material required to fuel the broad emission line clouds and the thermal jets.

158.043 **A multi–frequency polarization study of 3C 66B.**
J. P. Leahy, W. J. Jägers, G. G. Pooley.
Astron. Astrophys., Vol. 156, No. 1/2, p. 234 – 251 (1986).
The nearby twin–jet radio source 3C 66B has been studied in total intensity and polarization at 5 frequencies and at resolutions ranging from 30″ to 0.″35. The authors derive the distribution of spectral index, projected magnetic field and Faraday rotation across the source. For the first time a correlation is detected between depolarization and rotation measure, implying that some of the rotation must be occurring within the radio–emitting components. There is also good evidence for rotation in a foreground screen localised near the centre of the host galaxy. The minimum densities derived for the extended components are similar to those in the X–ray halo surrounding the radio source, strongly suggesting that matter enters the radio lobes via entrainment from the confining medium. The brighter jet gives textbook examples of the characteristic features known in such low–luminosity sources. Large–scale magnetic–field rings occur where the jets flare into brighter lobes, reinforcing the idea of an abrupt change in the jet flow.

158.044 **The asymmetry of extragalactic double radio sources and their foreshortening.**
A. D. Fokker.
Astron. Astrophys., Vol. 156, No. 1/2, p. 315 – 322 (1986).
A common and quiet reasonable interpretation of the asymmetry presented by several extragalactic double radio sources is, to ascribe it to the effect that is bound to occur when the two sources move away from the central object with a velocity component in the direction of the line of sight which is a sizable fraction of the velocity of light. If this interpretation is correct, the double sources with an outspoken asymmetry are seen foreshortened and so, as a group, they should present physical sizes that are statistically smaller than the sizes of the more symmetrical sources. It is shown that such a difference, if existent, can be ascertained in spite of the large spread of intrinsic extents of the objects. In this paper the author looks for the presence of a difference between the (apparent) size distributions of symmetrical and asymmetrical double radio sources. However, no effect of any significance is found.

158.045 **1011+496 and 1217+348: two new candidate BL Lacertae objects.**
W. Z. Wisniewski, M. L. Sitko, A. K. Sitko.
Mon. Not. R. Astron. Soc., Vol. 219, No. 2, p. 299 – 304 (1986).
This paper presents recent optical and infrared observations of the extragalactic radio sources 1011+496 and 1217+348. These observations indicate that the two sources are BL Lacertae objects.

158.046 **Compact radio components in Seyfert galaxies.**
S. W. Unger, A. Pedlar, R. V. Booler, B. A. Harrison.
Mon. Not. R. Astron. Soc., Vol. 219, No. 2, p. 387 – 402 (1986).
The authors present new MERLIN observations of the Seyfert galaxies NGC 5506, Mkn 6, Mkn 463, NGC 7674 (Mkn 533) and MCG 8–11–11 at 408 and 1666 MHz. Most of the objects show linear or double radio structure consistent with collimated ejection but the radio emission from NGC 5506 is diffuse, suggestive of a starburst. Compact off–nuclear radio components (<0.25 arcsec) are seen, with internal pressures substantially larger than that of the surrounding thermal gas. In order to study the overall properties of Seyfert radio components, the authors have used both new and published observations to obtain a sample of Seyfert galaxies with well–resolved radio sources.

158.047 **An ultraviolet and optical study of the broad–line radio galaxy 3C 382.**
C. N. Tadhunter, E. Pérez, R. A. E. Fosbury.
Mon. Not. R. Astron. Soc., Vol. 219, No. 3, p. 555 – 574 (1986).
Ultraviolet spectra covering 13 separate epochs are presented for the extreme broad–line radio galaxy 3C 382. These show marked variability in both lines and continuum on a time–scale $\lesssim$1 yr. The C IV λ(1550) and L$\alpha\lambda$(1216) lines follow the changes in the ultraviolet continuum but appear to have different profiles, the Lα(1216) having a relatively stronger core. Lα(1216) becomes weaker relative to the Balmer line as the ultraviolet continuum diminishes – an observation that places important constraints on ionization models for the BLR. A long–slit spectrum obtained with the newly commissioned INT telescope shows extended emission lines at up to 17 arcsec from the nucleus and demonstrates a large velocity gradient in the narrow–line gas across the nucleus.

158.048 **H II region spectroscopy and two dimensional stellar photometry of the barred Seyfert NGC 1566.**
J. E. Beckman, S. G. Bransgrove, J. P. Phillips.
Astron. Astrophys., Vol. 157, No. 1, p. 49 – 58 (1986).
Radial velocities of 14 H II regions were obtained using reticon spectroscopy at Hα for the barred spiral NGC 1566. The same spectra yield N II/H and S II/H ratios, and give values of the electron densities in the range 10 cm^{-3}. Kinematic and dynamic arguments are used to show that non–radial gas velocities are present, especially outflow at the ends of the bar. Two–dimensional isophotes at high angular resolution ($\cong$ 1 arcsec) are presented in V, R and I. Comparison of M/L ratios based on surface photometry, and based on the inferred rotational dynamics shows that no missing mass is required in NGC 1566 out to 13.5 kpc from the centre. The photometry also shows that the Seyfert nucleus is bluer by over a magnitude than the rest of the central bulge.

158.049 **The eruptive BL Lac object 0846+51W1: a gravitationally lensed QSO?**
L. Nottale.
Astron. Astrophys., Vol. 157, No. 2, p. 383 – 386 (1986).
The eruptive BL Lac object 0846+51W1 was observed by Arp et al. (1979) to vary by 4 mag in less than a month and to have its optical spectrum turning to that of a QSO of redshift 1.86 when faintest. Moreover, it lies only 12″ south of a spiral galaxy of redshift 0.072. The author shows in this paper that most of the observed characteristics of 0846+51W1 may be explained by the effects of gravitational lensing by a compact object lying in the intervening galaxy, passing in front of the QSO. After having developed the theory of extended source amplification, the author shows that a fit of the light curve based on this model is consistent with a lensing object of mass $\sim 10^{-2} M_\odot$ acting on a QSO with core dimension $\sim 2\times10^{-4}$pc for the central region emitting the continuum.

158.050 **Morphology and asymmetry of active galaxies from cavity stability.**
A. J. Allen.
Astrophys. J., Vol. 301, No. 1, p. 44 – 49 (1986).
It is shown that hot cavities in active galactic nuclei must break out into twin exhausts at subparsec distances and that they could have considerable influence in producing radio source morphology not readily explained by intrinsic collimation near the central hole itself. Thermal instabilities in accreting gas and Rayleigh–Taylor instability of the cavity boundary together allow ingestion of denser baryonic material into the outflow, and portions of the flow may naturally be related to narrow–line and broad–line regions. Low–luminosity sources are constrained to be classical symmetric twin exhausts on breakout, while at higher luminosities it is suggested that asymmetric double structure can be expected.

158.051 **Application of a relativistic jet model to the study of BL Lacertae objects.**
D. J. Hutter, S. L. Mufson.
Astrophys. J., Vol. 301, No. 1, p. 50 – 56 (1986).
The authors describe the development of a relativistic jet model and its use in fitting coordinated observations of the BL Lacertae objects Mrk 180, Mrk 421, and Mrk 501. This model is a modified version of one by Königl. The model fits show that the properties of the jets in all three objects are similar. The variation in the angle between the jet axis and the observer's line of sight accounts for the differing degrees of variability of the sources. The variation of particle density and magnetic field strength along the jets is similar in all three sources, and suggests that the energy flux is dominated by the magnetic field in the inner regions and by the particle energy at large distances from the jet apex.

158.052 **NGC 1320: a feeble, high–ionization Seyfert 2 galaxy.**
M. M. De Robertis, D. E. Osterbrock.
Astrophys. J., Vol. 301, No. 1, p. 98 – 104 (1986). = Lick Obs. Bull., No. 1020.
Emission–line strengths and widths are reported for NGC 1320, a Seyfert 2 galaxy that is unusual in having both a quite high ionization spectrum, as well as rather narrow lines and a relatively weak featureless continuum. The narrow–line region is evidently quite small in this galaxy, and the observed gas has not been accelerated to velocities as high as the velocities observed in most other Seyfert galaxies.

158.053 **Observations of upper–level hydrogen lines in Seyfert 1 galaxies.**
C. W. McAlary, G. H. Rieke, M. J. Lebofsky, J. T. Stocke.
Astrophys. J., Vol. 301, No. 1, p. 105 – 115 (1986).
The authors have observed the Paα (1.875 µm) and Brγ (2.166 µm) lines of hydrogen in a number of Seyfert 1 galaxies. In agreement with the findings published by Lacy et al. in 1982, it is

found that the Paα line can usually be used as a reddening diagnostic. However, particularly for objects with steep Balmer decrements, the available hydrogen recombination line strengths occasionally deviate sharply from those expected from case B radiative recombination. Further observations of the Paβ line (1.28 μm) may be very useful in studying the effects of extreme physical conditions versus those of dust

158.054 **On the relationship between optically selected and X–ray–selected active galactic nuclei.**
A. Franceschini, I. M. Gioia, T. Maccacaro.
Astrophys. J., Vol. 301, No. 1, p. 124 – 131 (1986).

The authors have used the available information on optically selected active galactic nuclei (AGNs), together with the knowledge of their X–ray to optical luminosity ratio, to predict the overall properties of X–ray–selected active galactic nuclei (number counts, evolution, luminosity and redshift distributions, etc.). These predicted properties are then compared with those observed. Similarly, predictions for the optical properties of the X–ray–selected AGN parent population are derived. Several significant inconsistencies between predictions and observations are noted and possible explanations are outlined.

158.055 **Erratum: "The nuclear activity of interacting galaxies" [Astrophys. J., Suppl. Ser., Vol. 57, No. 4, p. 643 – 664 (1985)].**
O. Dahari.
Astrophys. J., Suppl. Ser., Vol. 60, No. 2, p. 601 (1986). See Abstr. 39.158.181.

158.056 **On the interaction of extragalactic jets with interstellar and intergalactic gas.**
V. N. Fedorenko, A. S. Zentsova.
Astron. Zh., Tom 63, Vyp. 1, p. 38 – 44 (1986). In Russian. English translation in Sov. Astron., Vol. 30, No. 1.

The authors consider the interaction of a supersonic conical gaseous beam with a quiet medium gas. They derive an equation for the beam braking distance and also find an equation describing the motion of the gaseous cloud inflated by the beam. Comparison of the theory with observational data enables to put some restrictions on the values of the parameters of extragalactic jets.

158.057 **Observations of the lunar occultations of the sources 4C 24.11 and 4C 17.43 in the decameter radio wave range.**
V. P. Bovkun, I. N. Zhuk.
Astron. Zh., Tom 63, Vyp. 1, p. 45 – 49 (1986). In Russian. English translation in Sov. Astron., Vol. 30, No. 1.

The results of observations of the lunar occultations of the sources 4C 24.11 and 4C 17.43 at the frequencies 20 and 25 MHz are presented. The coordinates, flux densities and upper limits of angular size are determined. For the source 4C 24.11 $\alpha(1950.0) = 06^h01^m51^s.8 \pm 0^s.3$, $\delta(1950.0) = 24°30'12'' \pm 10''$ the angular size is smaller than 20″, the flux density is 27 ± 4 and 23 ± 4 Jy at the frequencies 20 and 25 MHz. For 4C 17.43 $\alpha(1950.0) = 08^h12^m43^s.1 \pm 0^s.8$, $\delta(1950.0) = 17°51'23'' \pm 16''$ the size is smaller than 30″, the flux density is 27 ± 5 and 22 ± 4 Jy at 20 and 25 MHz respectively.

158.058 **A comparative study of the stellar content in normal and active nuclei of spiral galaxies.**
H. A. Dottori, M. G. Pastoriza.
Astrophys. Space Sci., Vol. 121, No. 1, p. 147 – 160 (1986).

The authors present spectrophotometry and *UBV* photometry of the central region of spiral galaxies. The sample consists of the Seyfert galaxies NGC 1566, 3783 and IC 4329A; NGC 1097, 2997, and 5236 containing peculiar nuclei and NGC 5530, 5643 and 6699 with normal nuclei. The stellar composition of the nuclei was derived by treating the equivalent widths of a sample of absorption lines with the method of constrained nonlinear optimization. The age of the last cycle of star formation, the internal absorption and the theoretical relations M/L and H/K Ca II present differences among the three types of nuclei.

158.059 **Optical spectra of IRAS "warm" galaxies.**
D. E. Osterbrock, M. M. de Robertis.
Publ. Astron. Soc. Pac., Vol. 97, No. 598, p. 1129 – 1141 (1985). = Lick Obs. Bull., No. 1024.

Redshifts, relative emission–line–intensity measurements, and optical spectral classifications are presented for 30 galaxies selected by the IRAS observers as Seyfert candidates on the basis of their infrared properties. The authors confirm earlier results that a large number of these objects are indeed previously unknown Seyfert galaxies, most of them Seyfert 2s. These IRAS Seyfert 2 galaxies as a group are at relatively small redshift, and thus represent a low–luminosity tail of the Seyfert galaxy luminosity function. There are in addition a significant number of Liners among them, as well as H II region galaxies. Many of the IRAS Seyfert 2 and H II region galaxies have line spectra that are significantly more heavily reddened than samples of previously known objects of these types.

158.060 **Continuum and line fluxes of OJ 287 at minimum light.**
M. L. Sitko, V. T. Junkkarinen.
Publ. Astron. Soc. Pac., Vol. 97, No. 598, p. 1158 – 1162 (1985).

The authors report optical–infrared continuum observations of OJ 287 during its lowest recorded flux level. The spectrum was steeper than previously seen, but does not unambiguously show the presence of any underlying galaxy. It is suggested that this is primarily due to the object's still being too bright to make such a detection in this manner. The authors also report spectroscopic observations of OJ 287 showing the presence of Hα, Hβ, and [O III] emission lines. These lines confirm a redshift $z = 0.306$. The data are also used to compare the emitted luminosities in these lines to those of other active extragalactic objects. These lines are intrinsically underluminous compared to quasars of similar ionizing luminosities. The data are consistent with OJ 287 being deficient in gas compared to normal quasars or with the observed continuum flux having a Doppler factor of $\delta \sim 4$.

158.061 **Echelle observations of the extended ionized gas around the nucleus of the Seyfert galaxy NGC 1068.**
J. Meaburn, A. Pedlar.
Astron. Astrophys., Vol. 159, No. 1/2, p. 336 – 341 (1986).

Observations of [O III] profiles with an echelle spectrograph have revealed particularly well the 150 km s^{-1} expansion of the outer region of the Seyfert nucleus of NGC 1068 as well as the velocity structure in the inner broad–line region. These results complement several observations made by previous authors. A schematic model is devised to relate the radio and optical features of this nucleus.

158.062 **A study of radio objects with continuous optical spectra. II. Daily brightness and spectrum variability from data of a four–color photometry.**
S. A. Pustil'nik, L. A. Pustil'nik, S. I. Neizvestnyj, V. M. Lyutyj.
Soobshch. Spets. Astrofiz. Obs., Vyp. 48, p. 27 – 60 (1985). In Russian.

The results are given of UBVR electrophotometry for three radio objects with continuous optical spectra (OI 090.4, OJ 049, OQ 530) and two BL Lac–type objects (OJ 287, W Com). The measurements have been made in December 1982. Brightness variations of four objects have been detected. Variability times were in the range from one to seven days and amplitudes were $0^m.2 - 0^m.8$. Rapid decrease of brightness in all bands was registered for OJ 049 at $0^m.4 - 0^m.8$ for only three hours. The forms of the most four–point observed spectra ($\cong 80\%$) were well approximated by a power law. Conclusions are drawn on the possible mechanisms generating observable optical emission of the objects OJ 049 and OJ 287 by using the data on spectral variations during the process of brightness variations.

158.063 **Two–dimensional photometry of the circumnuclear region of the Seyfert galaxy NGC 1275.**
L. P. Metik, I. I. Pronik.
Astrofizika, Tom 23, Vyp. 3, p. 451 – 457 (1985). In Russian. English translation in Astrophysics, Vol. 23, No. 3.

The results of two–dimensional photometric investigations of the NGC 1275 galaxy circumnucleus region are presented. Maps of the brightness distribution in the circumnuclear region of NGC 1275 have been obtained. The elongation of the NGC 1275 nucleus in north–east was confirmed. It can be caused by the interaction of the nucleus with the star–like detail which is located at 7″ in north–east. The nucleus of NGC 1275 may possibly consist of two or more components. The parts of the NGC 1275 nucleus stellar and nonstellar radiation were estimated.

158.064 **Extragalactic nature of G227.1 + 1.0.**
G. A. Chanan, D. J. Helfand, H. Spinrad, K. Ebneter.
Nature, Vol. 320, No. 6057, p. 41 – 42 (1986).
Green and Gull have presented evidence that G227.1 + 1.0, one of the small–diameter radio sources in the Molonglo survey of Clark and Crawford, is a Crab–like supernova remnant. The authors report the discovery of a bright ($V = 17.4$), diffuse optical object within a few arc seconds of the centre of the radio source, but they present spectroscopic evidence that it is a luminous external galaxy at $z = 0.073$. The authors argue, furthermore, that the original radio emission represents not a supernova remnant, but a radio halo surrounding this galaxy, similar to that around M87.

158.065 **Neutral hydrogen absorption towards G227.1 + 1.0.**
D. A. Green, S. F. Gull.
Nature, Vol. 320, No. 6057, p. 42 – 43 (1986).
The radio source G227.1 + 1.0 was observed in a search for young galactic supernova remnants, and its peculiar structure indicated that it might be a very young remnant. Chanan et al., however, have recently found a bright optical object with $z = 0.073$ close to the radio centre, which represents strong circumstantial evidence that the source is in fact a radio galaxy. The authors present a neutral hydrogen absorption profile observed with the VLA which confirms the extragalactic nature of this object.

158.066 **Very Large Array observations of rapid non–periodic variations in OJ 287.**
J. W. Dreher, D. H. Roberts, J. Lehár.
Nature, Vol. 320, No. 6059, p. 239 – 242 (1986).
The authors observed the BL Lacertae object OJ 287 with the VLA at 5, 15 and 22 GHz several times in early 1983. The 5–GHz data show OJ 287 to have decreased in flux by about 1.8% in an irregular way over a span of about 7 h, and to have fluctuated by about 0.5% on a timescale of about 15 min. Brightness temperatures derived from causality constraints range between 10^{16} and 10^{20}K. Fourier spectra of the time series were analysed using a novel one–dimensional complex CLEAN algorithm to remove the effects of the data sampling. At 5 GHz there was no harmonic component to the intensity of OJ 287 which exceeded 0.1% of the mean for any period between 80 and 5,000 s. The limits at 15 and 22 GHz were 2% and 0.8%, respectively.

158.067 **An analysis of the narrow line profiles in Seyfert 2 galaxies.**
M. M. De Robertis, D. E. Osterbrock.
Astrophys. J., Vol. 301, No. 2, p. 727 – 741 (1986). = Lick Obs. Bull., No. 1021.
An analysis of the line profiles in 18 high–ionization Seyfert 2 galaxies is presented. Two–thirds of the sample show a correlation between emission–line width and critical density. Few objects show a correlation between line width and ionization potential. The data argue strongly for a model in which optically thick emission–line clouds occupy regions of highest density and velocity dispersion. A correlation was found between the Hβ emission–line equivalent width and the line widths for all emission lines. According to the relation, all narrow–line widths tend to 320 ± 25 km s^{-1} as the Hβ emission–line equivalent width approaches zero. A comparison between Seyfert 2 line profiles and the corresponding Seyfert 1 narrow–line profiles is included. The significant profile differences between Seyfert 1s and 2s argue in a statistical sense against evolution between types on short time scales, and that Seyfert 2s are not highly reddened Seyfert 1s.

158.068 **The discovery of eight bright active galactic nuclei observed with the *HEAO 1* scanning modulation collimator.**
R. A. Remillard, H. V. Bradt, D. A. H. Buckley, W. Roberts, D. A. Schwartz, I. R. Tuohy, K. Wood.
Astrophys. J., Vol. 301, No. 2, p. 742 – 752 (1986). With plate 2.
Eight active galactic nuclei (AGNs; seven type 1 Seyferts and one QSO) are reported. They are 14th and 15th mag objects with broad hydrogen lines and redshifts in the range $0.026 < z < 0.138$. The AGNs were discovered in a systematic program to identify the fainter X–ray sources of the *HEAO 1* survey. The X–ray and optical luminosities and many of the optical emission features of these AGNs resemble other bright and nearby Seyfert 1 galaxies. However, three of the galaxies exhibit very strong Fe II emission, and two additionally show very weak forbidden line spectra similar to I Zw 1. One of the I Zw 1 types, the previously unidentified radio source PKS 0558–504, is a QSO with unusually narrow hydrogen lines. Photometric magnitudes and the intensities of selected emission lines are given for each of the AGNs.

158.069 **Detection of [Ni II] λ7378 in six Seyfert galaxies.**
J. P. Halpern, J. B. Oke.
Astrophys. J., Vol. 301, No. 2, p. 753 – 758 (1986). = Columbia Astrophys. Lab., Contrib., No. 295.
The authors detected a line due to [Ni II] λ7378 in the Seyfert galaxies NGC 1068, NGC 2110, NGC 3227, NGC 4151, NGC 5506, and Arp 102 B. The average Ni abundance is about 2 times solar, which is 5 times less than in the filaments of the Crab Nebula. This argues for nucleosynthetic processing in the latter. The [Ni II] line is spatially resolved in NGC 1068, and shows at least a factor of 4 enhancement in the Ni abundance away from the nucleus. The off–nuclear abundance of Ni in NGC 1068 approaches that of the Crab, which strongly suggests that Type II supernovae enriched the off–nuclear gas clouds.

158.070 **VLBI and VLA observations of the BL Lacertae–type object Markarian 501: evidence for grossly misaligned small–scale and large–scale radio structure.**
W. van Breugel, R. Schilizzi.
Astrophys. J., Vol. 301, No. 2, p. 834 – 840 (1986).
The authors present VLBI and VLA observations of the core of the BL Lac–type object Mrk 501. The VLBI observations show a small (10 pc) one–sided jet structure which is grossly misaligned with the previously known extended (40 kpc) structure of this source. The VLA observations show that the change in alignment must occur within 40 pc of the nucleus. The 10 pc jet does align with the preferred optical polarization angle measured by Angle and Stockman. There is no evidence for radio variability in the core on scales as small as a few parsec. Constraints for relativistic jet models of this object are derived from the data.

158.071 **Multifrequency VLA observations of the prototypical narrow–angle tail radio source NGC 1265.**
C. P. O'Dea, F. N. Owen.
Astrophys. J., Vol. 301, No. 2, p. 841 – 859 (1986). With plates 3 – 8.
VLA observations of NGC 1265 (3C 83.1B) at 21, 6, and 2 cm are reported. Maps of the jets and tails with a wide range of resolutions (0″.45 – 12″) are presented, and the total intensity, spectral index, and polarization structure of this prototypical narrow–angle tail source are discussed.

158.072 **The spectra of radio emission of several extragalactic radio sources in the range 11.6 – 36.8 GHz.**
E. Valtaoja, M. Valtonen, V. A. Efanov, H. Lehto, I. G. Moiseev, N. S. Nesterov, E. Salonen, M. Tiuri, S. Urpo, S. Haarala.
Izv. Krymskoj Astrofiz. Obs., Tom 70, p. 144 – 152 (1985). In Russian. English translation in Bull. Crimean Astrophys. Obs., Vol. 70.
The results of coordinated observations of 20 extragalactic radio sources in the frequency range 11.6 – 36.8 GHz are presented. The radio emission spectra are given for 12 sources. The magnetic field strength is estimated for sources 0235 + 16,

OH 471, OJ 287, and BL Lac. That is in the range of $10^{-3}-10^{-2}$G.

158.073 **Study of the cosmological evolution of extragalactic radio sources observed in decametric range.**
K. P. Sokolov.
Pis'ma Astron. Zh., Tom 12, No. 4, p. 259 – 265 (1986). In Russian. English translation in Sov. Astron. Lett., Vol. 12.
The differential relation "sources number – flux density $n_{25}(S)$" obtained from observations at 25 MHz is analysed. The classes of objects determining the strong evolution observed within various surface flux density intervals in the decametric wavelength range have been established.

158.074 **The [O III] line profile and the central gaseous cloud in the nucleus of NGC 1068.**
V. I. Pronik.
Izv. Krymskoj Astrofiz. Obs., Tom 72, p. 135 – 142 (1985). In Russian. English translation in Bull. Crimean Astrophys. Obs., Vol. 72.
Arguments confirming the presence of a gaseous formation with dimension of 10 parsec in the center of the NGC 1068 nucleus are presented. The author tried to find among numerous details of the [O III] line profile the main peak corresponding to the central cloud. It is shown that the maximum is actually a rather weak detail of a line profile with a radial velocity of about 50 km/s with respect to the galaxy.

158.075 **A brightness variation analysis of the object OJ 287.**
V. T. Doroshenko, Yu. S. Efimov, V. Yu. Terebizh, N. M. Shakhovskoj.
Izv. Krymskoj Astrofiz. Obs., Tom 73, p. 143 – 153 (1985). In Russian. English translation in Bull. Crimean Astrophys. Obs., Vol. 73.
UBVRI photoelectric photometry of the object OJ 287 was performed during the interval December 1982 – April 1983. The data on color indices variations with time are presented. The observations of 24 hours duration were used in search for short–time periodic variations of the optical flux. No evidence for presence of such variarions in the 5^m to 60^m period range was found at about 0.1% confidence level.

158.076 **Investigation of the galaxy Markarian 306.**
A. R. Petrosyan, M. Turatto, Eh. E. Khachikyan.
Astrofizika, Tom 24, Vyp. 1, p. 5 – 16 (1986). In Russian. English translation in Astrophysics, Vol. 24, No. 1.
The results of spectroscopic and spectrophotometric observations of Markarian 306 are presented. By means of intensity ratios of observed emission line T_e, N_e the abundances of O, N, S, the masses and volumes of the emitting gas and the numbers of the O9 type stars for both nuclei and for the condensation in the north spiral arm of the galaxy are estimated. All three formations are H II region type objects. The possibility to consider the north spiral arm of Markarian 306 as a separate galaxy, the nucleus of which is its condensation, has been discussed. The reality of double–nucleus structure of the galaxy has also been discussed.

158.077 **A deficiency of normal elliptical galaxies among Markarian galaxies.**
M. M. Jõeveer.
Astrofizika, Tom 24, Vyp. 1, p. 25 – 32 (1986). In Russian. English translation in Astrophysics, Vol. 24, No. 1.
A study of morphological types of galaxies in the UGCG catalogue reveals a strong deficiency of normal elliptical galaxies among the Markarian galaxies. In the Markarian sample only $\lesssim 1\%$ of the objects are normal elliptical galaxies against 10% among other galaxies. There is an excess of barred spirals in the Markarian sample, approximately one–half of Markarian Sa–Scd spirals are barred, among non–Markarian spirals $\sim 1/3$ are barred.

158.078 **The luminosity function of Seyfert galaxies.**
V. P. Reshetnikov.
Astrofizika, Tom 24, Vyp. 1, p. 33 – 46 (1986). In Russian. English translation in Astrophysics, Vol. 24, No. 1.
The luminosity function of Seyfert galaxies in the range $-18.5 \geqslant M_p \geqslant -24.5$ has been determined from a sample of 219 galaxies, and the total space density ($\sim 2 \times 10^{-4}$Mpc^{-3}) has been found.

158.079 **The interaction of interstellar clouds with a "quasar wind" and its relation to the narrow emission line regions of Seyfert galaxies.**
A. V. R. Schiano.
Astrophys. J., Vol. 302, No. 1, p. 81 – 94 (1986).
This paper investigates the interaction between a "quasar wind" and interstellar clouds in a Seyfert galaxy. The propagation of the wind into the low–density gas component of the interstellar medium (ISM) was studied in an earlier paper, and the wind velocity field determined there is used to determine the physical structure and motion of interstellar clouds. The cloud component of the ISM of a Seyfert galaxy is modeled by assuming that these galaxies are similar to the Milky Way and other nearby "nonactive" galaxies. The results of these calculations suggest that the narrow emission line regions of Seyfert galaxies are the result of the interaction of active galactic nuclei ionizing photons *and* a thermal wind on dense, massive interstellar clouds.

158.080 **Narrow emission line profiles in a "quasar wind".**
A. V. R. Schiano.
Astrophys. J., Vol. 302, No. 1, p. 95 – 107 (1986).
Emission–line profiles are calculated on the basis of the recently proposed narrow emission line region model of Schiano (1986). A Monte Carlo–type calculation is used to determine emission–line profiles of an ensemble of photo–ionized interstellar clouds in a supersonic, thermal wind. These theoretical line profiles are shown to agree favorably with observations.

158.081 **The dynamics of the active galaxy NGC 1052.**
R. L. Davies, G. D. Illingworth.
Astrophys. J., Vol. 302, No. 1, p. 234 – 244 (1986). With plates 1 – 2.
NGC 1052 is an elliptical galaxy with an active nucleus. The authors present velocity maps for both the stars and ionized gas, the stellar velocity dispersion profile, and surface photometry of this galaxy. The stellar component rotates about an axis close to the projected minor axis, whereas the ionized gas rotates about an axis 69° away from that. There is a small stellar rotation on the photometric minor axis. The rotation axis of this gas disk is the same as that reported by van Gorkom *et al.* for the H I. This rotation axis remains constant over a factor of 10 in radius. The authors suggest that the galaxy has a triaxial figure and that the gas could be on stable orbits rotating about the long axis of the triaxial ellipsoid. The large misalignment of the gas and stellar rotation axes suggests that the interstellar medium in NGC 1052 originated elsewhere, perhaps from the nearby spiral NGC 1042.

158.082 **The jet of PKS 0521–36: an aging counterpart of M87?**
W. C. Keel.
Astrophys. J., Vol. 302, No. 1, p. 296 – 305 (1986). With plates 8 – 9.
Imaging photometry and VLA mapping at 2 and 6 cm have been used to study the structure and spectral energy distribution of the jet in PKS 0521–36. The jet resembles that of M87 in morphology. Its optical spectrum is notably steeper, probably implying a lower energy cutoff for the radiating electrons. The jet's optical–to–radio spectrum steepens by $\Delta\alpha \approx 0.07$ going outward along the jet, in an apparently continuous manner. This stands in contrast to the constant value reported in M87. The data suggest that this difference is mostly due to a lower average magnetic field in PKS 0521–36. In both objects, a significant flaring of the jet occurs at a projected distance of ~ 2 kpc from the nucleus; this might represent interaction between the jet and the interstellar medium of each galaxy.

158.083 **The radio galaxy IC 4296 (PKS 1333–33). I. Multifrequency Very Large Array observations.**
N. E. B. Killeen, G. V. Bicknell, R. D. Ekers.
Astrophys. J., Vol. 302, No. 1, p. 306 – 336 (1986). With plates 10 – 14.
The authors present VLA observations of the E1 radio galaxy IC 4296 (PKS 1333–33) at 1.3, 2, 6, and 20 cm, with a best resolution of 3."2. The source consists of an unresolved core (coincident with the nucleus of IC 4296) from which issue two symmetric, curved jets. Both jets terminate in slightly edge–brightened lobes. The total extent of the source is approximately 35′ (360 kpc, for $H_0 = 100$ km s^{-1} Mpc^{-1}). Physical characteristics of the jets are derived from the total intensity maps, from spectral index variations and from radio polarization maps. Equipartition parameters are then calculated from these data.

158.084 **H0323 + 022: a new BL Lacertae object with extremely rapid variability.**
E. D. Feigelson, H. Bradt, J. McClintock, R. Remillard, C. M. Urry, S. Tapia, B. Geldzahler, K. Johnston, W. Romanishin, P. A. Wehinger, S. Wyckoff, G. Madejski, D. A. Schwartz, J. Thorstensen, B. E. Schaefer.
Astrophys. J., Vol. 302, No. 1, p. 337 – 351 (1986).
H0323 + 022 is a highly variable X–ray source discovered in the *HEAO 1* all–sky survey coincident with a 16th mag stellar object. Optical nebulosity around the object strongly suggests that it is a BL Lac object. This paper presents the results of a multiband, multiepoch series of observations that demonstrate beyond reasonable doubt that H0323 + 022 is a remarkably variable BL Lac object. The optical nebulosity around the stellar object is confirmed, the stellar optical radiation is polarized, and the radio characteristics and the overall electromagnetic spectrum resemble those of other BL Lac objects. Variability is seen in all bands, including radio and broad–band optical variations in days, optical polarization changes on time scales of tens of minutes, and X–ray changes on time scales of ~ 30 s. H0323 + 022 is thus an active galactic nucleus with much of the observed emission emerging from a region $\lesssim 10^{12} - 10^{14}$cm in size.

158.085 **Remarkable emission–line profiles in the Seyfert galaxy ESO 263–G13 (= F 427).**
A. S. Wilson, J. A. Baldwin.
Astrophys. J., Lett. Ed., Vol. 302, No. 1, p. L7 – L10 (1986).
It has recently been claimed that the profiles of the nuclear [O III] λλ4959, 5007 emission lines in the type 2 Seyfert galaxy ESO 263–G13 = F 427 are variable over intervals as short as 30 minutes. The authors report here long–slit spectroscopic observations at and around the nucleus of the galaxy and find these lines to be spatially extended over at least 8″(N–S), with velocity profiles which vary rapidly with position. The profiles exhibit strongly enhanced red wing asymmetries on the north side and blue wing asymmetries on the south side of the nucleus. It is argued that the claimed time variability is not real but reflects small differences in slit centering and seeing from exposure to exposure.

158.086 **A hidden quasar revealed.**
C. M. Gaskell.
Nature, Vol. 320, No. 6061, p. 398 – 399 (1986).
This note discusses the implications of recent spectropolarimetric observations of NGC 1068 by Antonucci and Miller (see Abstr. 40.158.154) for the nature of Seyfert 2 nuclei.

158.087 **Limits of X–ray variability in active galactic nuclei.**
P. Barr, R. F. Mushotzky.
Nature, Vol. 320, No. 6061, p. 421 – 423 (1986).
The authors show that there is a significant correlation between the X–ray luminosity and timescale of X–ray variability for Seyfert galaxies and quasars. They interpret this as evidence that the emitting plasma is near the limit of being dominated by electron–positron pairs. BL Lac objects do not follow this pattern; this may be due either to relativistic beaming or to the differing importance of the pair production process.

158.088 **Fast X–ray variability in NGC 4051.**
A. Lawrence.
Observatory, Vol. 106, No. 1070, p. 11 (1986). Abstract. – See Abstr. 010.741.

158.089 **The X–ray spectra of active galaxies.**
R. S. Warwick.
Observatory, Vol. 106, No. 1070, p. 12 (1986). Abstract. – See Abstr. 010.741.

158.090 **Explosieve stervorming in Messier 82.**
P. Biermann.
Zenit, 13. Jaarg., Nr. 1, p. 14 – 17 (1986).

158.091 **IRAS observations of active galaxies.**
G. Neugebauer, B. T. Soifer, M. Rowan–Robinson.
Structure and evolution of active galactic nuclei, p. 11 – 19 (1986). – See Abstr. 012.011.
The IRAS survey gives an unbiased view of the infrared properties of the active galaxies. Seyfert galaxies occupy much the same area in color–color plots as do normal infrared bright galaxies, but extend the range towards flatter 60 – 25 μm slopes. Statistically the Seyfert 1 galaxies can be distinguished from the Seyfert 2 galaxies, lying predominantly closer to the area with constant slopes between 25 and 100 μm. The infrared measurements of the Seyfert galaxies cannot distinguish between the emission mechanisms in these objects although they agree with the currently popular ideas; they do provide a measure of the total luminosity of the Seyferts. The quasar's position in the color–color diagrams continue the trend of the Seyferts.

158.092 **The properties of "dwarf" Seyfert nuclei in nearby galaxies.**
A. V. Filippenko, W. L. W. Sargent.
Structure and evolution of active galactic nuclei, p. 21 – 45 (1986). – See Abstr. 012.011.
The authors describe preliminary results from a large survey of bright, nearby galaxies which is designed to quantify the faint end of the luminosity function. Many high–quality spectra are shown to emphasize the great variety of physical conditions in the nuclei and to illustrate the methods used in a detailed analysis of the data. The characteristics of Seyfert galaxies are detected in a surprisingly large number of objects. A very wide range of densities is found in the narrow–line regions of certain low–ionization nuclei, just as in many QSOs and Seyfert 1 galaxies. These results suggest that intrinsically faint ("dwarf"), but nevertheless active, nuclei are much more common than previously believed, and that they may have evolved from luminous quasars.

158.093 **Spectral properties of Blazars observed at ultraviolet and X–ray frequencies.**
A. Treves, G. Ghisellini, L. Maraschi, E. G. Tanzi.
Structure and evolution of active galactic nuclei, p. 63 – 80 (1986). – See Abstr. 012.011.
The large majority of all Blazars observed in the far ultraviolet (26 objects) are considered. Broad band spectral data are collected choosing measurements close in time and corrected for contamination from the surrounding galaxy. It is found that the overall spectrum of this class of objects steepens substantially between radio and infrared and between infrared and ultraviolet frequencies but not between ultraviolet and X–ray frequencies. The spectral slopes in different bands are correlated from X–ray to IR frequencies.

158.094 **Internal dynamics and formation of emission clouds in active galactic nuclei.**
I. Shlosman, P. A. Vitello, G. Shaviv.
Structure and evolution of active galactic nuclei, p. 95 – 111 (1986). – See Abstr. 012.011.
The authors show that cool UV accretion disks, if present in active galactic nuclei, can be sources of line driven winds. Continuum subcritical disks are shown to possess optically thin continuum winds which are accelerated by radiation pressure in lines. Results of numerical simulation of outflows from disks around supermassive black holes are presented. The winds are character-

ized by steep velocity gradients and high asymptotic velocities v/c ~ 1/30, as observed in broad emission–line regions of QSO's and Seyfert galaxies, and are thermally unstable as they heat up. This results in a two–phase equilibrium: cold clouds embedded in a hot, Compton–heated background. The hot wind provides the pressure confinement for the dense and massive clouds.

158.095 **The possible mechanism of the formation of the hard spectrum of active galactic nuclei.**
I. D. Novikov, B. E. Stern.
Structure and evolution of active galactic nuclei, p. 149 – 171 (1986). – See Abstr. 012.011.
The mechanism of the formation of the hard spectrum of active galactic nuclei due to the conversion of the energy of charged particles accelerated up to ultrarelativistic velocity is considered.

158.096 **Seyfert galaxies in the Universe.**
D. E. Osterbrock.
Structure and evolution of active galactic nuclei, p. 193 – 213 (1986). – See Abstr. 012.011.
Recent optical spectroscopy of Seyfert galaxies, particularly at Lick Observatory, is discussed. The frame of reference is the luminosity function of Seyfert galaxies, the incompleteness of available surveys, and how complete samples can be found by combining several different types of surveys. The importance of spectral classification, or differentiating between various classes of active galactic nuclei, is stressed. All Seyfert galaxies and AGNs cannot be fitted into a single one–parameter sequence. Starburst galaxies and Liners are briefly discussed. At the end some recent results on the frequencies of Seyfert galaxies with companions and in interacting systems are briefly summarized.

158.097 **Evolutionary connection of Seyfert galaxies and quasars.**
D. W. Weedman.
Structure and evolution of active galactic nuclei, p. 215 – 229 (1986). – See Abstr. 012.011.
Various recent data are summarized to yield a local luminosity function for Seyfert 1 galaxies. This luminosity function is evolved in different ways to find fits to other observations of faint, high redshift quasars. It is found that a simple form of pure luminosity evolution, exponential in look–back time, is consistent with all existing data. Extrapolations to redshift 4 and magnitude 26 are given for future tests. It is pointed out that such tests from ground or space must continue to rely on the unresolved quasar. An angular diameter–redshift diagram implies that there is no realistic possibility of detecting galactic envelopes of quasars at high redshift.

158.098 **Evolutionary continuity of the optical AGNs.**
A. Cavaliere, E. Giallongo, F. Vagnetti.
Structure and evolution of active galactic nuclei, p. 231 – 252 (1986). – See Abstr. 012.011.
With the optical data base now rapidly expanding, it is time to begin bridging the gap between astrophysical issues and phenomenological descriptions concerning the evolutionary properties of the AGNs. The authors review problems, achievements and perspectives of such intermediate models that are parameterized in terms of quantities of direct astrophysical meaning. Specific issues here discussed are: the continuity between Seyfert 1 nuclei and QSOs; a first parameterization of the interactions as triggers of activity; the rise of the QSOs; the evolutionary peculiarity of BL Lac Objects.

158.099 **Surveys of local AGN's.**
P. Véron.
Structure and evolution of active galactic nuclei, p. 253 – 273 (1986). – See Abstr. 012.011.
The author reviews the optical, X–ray and infrared surveys of local AGN's, discussing the limitations and biases of each of them. He concludes that no single survey may lead to a complete sample but that all of them must be used simultaneously to build a meaningful luminosity function.

158.100 **Discovery of narrow and variable lines in the ultraviolet spectrum of the Seyfert galaxy NGC 4151, and an outline of our previous results.**
M. H. Ulrich, A. Altamore, A. Boksenberg, G. E. Bromage, J. Clavel, A. Elvius, M. V. Penston, G. C. Perola, M. A. J. Snijders.
Structure and evolution of active galactic nuclei, p. 275 – 296 (1986). – See Abstr. 012.011.
Extensive observations of NGC 4151 when the nucleus is at a minimum have led to the discovery of two emission lines which have a full width at half maximum of less than 7 and 16 Å respectively and which vary in intensity by a factor of 3 in less than 10 days. These lines appear on either side of the C IV λ1550 line at λ_{rest} = 1518.5, 1594.4 Å. These lines are too narrow to be emitted by the whole broad line region. They must come instead from two localized regions which have a special excitation mechanism. No identifications for these lines have been found. An attractive – but speculative – possibility is that the regions emitting them are associated with a two–sided jet reminiscent of SS 433.

158.101 **Large scale ionized gas in radio galaxies and quasars.**
R. A. E. Fosbury.
Structure and evolution of active galactic nuclei, p. 297 – 308 (1986). – See Abstr. 012.011.
Many early–type active galaxies and quasars are surrounded by extensive regions of low density, highly ionized gas which can be considered the analogy of the much smaller Narrow Line Region in the spiral Seyferts. Observations of this gas carry information about the galaxy and its active nucleus and can now be interpreted in the context of X–ray observations of hot gas and 21 cm observations of cold gas in the same systems.

158.102 **Host galaxies of quasars.**
J. W. Fried.
Structure and evolution of active galactic nuclei, p. 309 – 315 (1986). – See Abstr. 012.011.
Properties of host galaxies of low redshift quasars are reviewed. There is a good evidence that radio–loud quasars sit in bright elliptical galaxies, whereas optical and X–ray quasars reside in spiral galaxies. The host galaxies of radio–loud quasars are brighter by ~2 mag than those of radio–quiet ones; the brightest ones have luminosities comparable to those of brightest cluster galaxies. The neighborhood of quasars has higher than average galaxy density; tidal interactions with the host galaxies are frequently observed. Very deep surfacephotometry of 2 radio–loud low–redshift quasars shows that in one case (1635+119, z = 0.146) the host galaxy can very well be fitted by an elliptical galaxy over 3 orders in intensity; the host galaxy of 4C 14.83 (z = 0.237) is strongly asymmetric and can not be well fitted by either elliptical or spiral galaxy.

158.103 **Spectral evolution in young active galactic nuclei.**
E. Boldt, D. Leiter.
Structure and evolution of active galactic nuclei, p. 383 – 396 (1986). – See Abstr. 012.011.
The spectral evolution of AGNs is discussed within the context of a scenario where the cosmic X–ray background (CXB) is dominated by these sources. Attention is drawn to the fact that individually observed AGN X–ray spectra are significantly steeper than that of the CXB. The remarkably flat spectrum thereby required for the "as–yet" unresolved sources of the residual CXB is interpreted as an observational constraint on an earlier stage of AGN evolution. Assuming black hole disk accretion, a picture emerges where young AGNs are compact Eddington limited thermal X–ray sources and where canonical AGNs represent later stages in which they have become appreciably less compact, exhibiting the importance of non–thermal disk–dynamo processes.

158.104 **Radio galaxy populations: a progress report.**
H. van der Laan, P. Katgert, M. J. A. Oort.
Structure and evolution of active galactic nuclei, p. 437 – 445 (1986). – See Abstr. 012.011.
This is a progress report on the project on faint radio galaxy populations that is being carried out at Leiden Observatory. The

aim of the project is to chart the active (radio) galaxy population in space and radio luminosity. The final goal is to use these data in a study of the evolutionary behaviour of individual active galaxies, population characteristics like birth rate functions etc. In this paper the authors report on the current status of the radio and optical work; the present information on the nature of the radio– and optically–faint radio galaxies; and on the population evolution as described by the redshift dependence of the radio luminosity function.

158.105 **Some theoretical aspects of AGNs.**
M. J. Rees.
Structure and evolution of active galactic nuclei, p. 447 – 457 (1986). – See Abstr. 012.011.
Some comments are made on the primary power source in QSOs and radio galaxies. The implications of high–z QSOs for theories of galaxy formation are briefly discussed.

158.106 **Megamasers in nuclei of galaxies.**
W. A. Baan.
Structure and evolution of active galactic nuclei, p. 461 – 463 (1986). – See Abstr. 012.011.
Powerful maser emission has been detected in a small number of galaxies. Most of these have Seyfert or starburst nuclei and exhibit other signs of galactic activity. The galaxies with maser emission are usually prominent infrared emitters. Some have perturbed disks with dust lanes and are showing evidence of bursts of star formation; others are part of a strongly interacting system.

158.107 **Liners: the low luminosity end of AGN.**
L. Binette.
Structure and evolution of active galactic nuclei, p. 475 – 478 (1986). – See Abstr. 012.011.
Discussion of the evidence for photoionisation as the excitation mechanism in Liners and its relationship to nuclear activity.

158.108 **Observations of active nuclei at Calar Alto Observatory.**
K. Birkle, P. Rafanelli, U. Thiele.
Structure and evolution of active galactic nuclei, p. 479 – 482 (1986). – See Abstr. 012.011.
CCD images and photographic spectra of Seyfert galaxies have been recently obtained by the authors at the Cassegrain focus of the 2.2 m, F/8 telescope of the Calar Alto Observatory. The main purposes of the observing programme were essentially three: (1) To obtain medium resolution spectra and high quality images of interacting active nuclei in order to study in detail their physical and morphological features and to verify whether nuclear activity can be related to some external disturbances. (2) To monitor some previously observed Seyfert–1 galaxies in order to detect changes in their emission line spectra. (3) To verify the spectral features of some objects remarkable within the Seyfert class for their peculiar line profiles and/or intensity line ratios.

158.109 **Frequency dependent polarization in blazars.**
C.–I. Björnsson.
Structure and evolution of active galactic nuclei, p. 483 – 486 (1986). – See Abstr. 012.011.
It is argued that the observed properties of the intrinsically frequency–dependent polarization (FDP) in blazars suggest a one – rather than a multi – component source model. Assuming a synchrotron origin of the emitted radiation, a well ordered magnetic field and a sharp break or cutoff in the electron distribution are both obligatory in such models; non–uniform pitch angle distribution or synchrotron losses enhance the resulting FDP. It is emphasized that the existence of such conditions in blazars finds observational support. Furthermore, a specific model is presented in order to illustrate the salient features of one–component models.

158.110 **CCD photometry of a sample of Seyfert galaxies: a progress report.**
C. Bonoli, F. Bortoletto, F. Bònoli, F. Delpino, V. Zitelli.
Structure and evolution of active galactic nuclei, p. 487 – 489 (1986). – See Abstr. 012.011.
A project of surface photometry of a sample of Seyfert galaxies was undertaken with the CCD camera of Asiago Observatory, for the study of the properties both of the nuclei and of the underlying galaxies. Here the authors present some results from the first observing runs.

158.111 **The relative frequency of compact extragalactic objects.**
P. Brosche.
Structure and evolution of active galactic nuclei, p. 499 – 502 (1986). – See Abstr. 012.011.
Central regions of galaxies are seen as the product of a random agglomeration of clouds. The resulting distribution of angular momentum is a Maxwellian one. Small angular momentum is a necessary condition for compact cores. For a given mass in virial equilibrium, a small angular momentum means a fast rotation velocity. The author assumes the broad line width to be caused mainly by rotation. A relation between the relative frequency and the angular momentum contrast of the central region is given (relative with respect to all galaxies of the same luminosity). The data for Seyfert 1 galaxies are not in conflict with this relation. The application to quasars seems possible.

158.112 **Ram pressure confinement of the emission line gas and the associated radio emission in AGN's.**
T. J. Carroll.
Structure and evolution of active galactic nuclei, p. 509 – 511 (1986). – See Abstr. 012.011.
The inferred high temperature and small size of the broad–line emitting clouds in quasars and Seyfert galaxies requires some confining mechanism to prevent rapid dissipation of those clouds in a time much shorter than their dynamical time scale. Ram pressure confinement by a supersonic wind provides such confinement while avoiding the mass budget and dynamical drag problems of a static confining medium. Reasonable limits on the mass outflow in the wind restricts the size of the high density BLR (broad–line region) to the canonical 1 pc. The limited range in observed ionization parameter also follows given that the mass accretion rate is similar to the mass outflow rate. Flat spectrum radio emission from the BLR is also expected to be produced in the bow shocks around the confined clouds.

158.113 **Ultraviolet spectra of 3 Seyfert 1 galaxies with optical Fe II emission lines.**
F.–Z. Cheng, C. A. Grady, P. L. Selvelli.
Structure and evolution of active galactic nuclei, p. 513 – 516 (1986). – See Abstr. 012.011.
UV spectra of 3 Seyfert 1 galaxies, VII Zw 118, I Zw 1, and III Zw 2, which are known to have strong to moderate optical Fe II emissions are presented.

158.114 **The behaviour of the C IV emission line strength in variable Seyfert I galaxies.**
L. Colina, W. Wamsteker.
Structure and evolution of active galactic nuclei, p. 525 – 529 (1986). – See Abstr. 012.011.
The variation of the C IV 1550 Å line strength in variable Seyfert I galaxies based on low resolution IUE data is discussed. The results suggest that the Baldwin–relation might not be a reliable distance indicator for QSO's. Seyfert I galaxies vary in a way indicating that the emission line gas is matter bounded when in a high state. For F–9 the transition from photon–bounded to matter bounded occurs at $\log\{L(1350)\} \approx 29.7$ erg/sec/Hz.

158.115 **The small wide angle tail NGC 4874 in the center of Coma cluster.**
L. Feretti, G. Giovannini.
Structure and evolution of active galactic nuclei, p. 541 – 542 (1986). – See Abstr. 012.011.

158.116 **Collimated winds from active galactic nuclei.**
A. Ferrari, E. Trussoni, R. Rosner, K. Tsinganos.
Structure and evolution of active galactic nuclei, p. 543 – 546 (1986). – See Abstr. 012.011.
The authors present a wind–type model for the modulation of jet brightness (knots) on kpc scale. It is shown that the mass

distribution of the parent galaxy and the development of unstable modes on the surface of the beam can lead to the periodic formation of internal shocks that can be associated with enhanced particle acceleration. The results are compared specifically with the data of the jet of M87.

158.117 **Comparison between optically selected and X–ray selected active galactic nuclei.**
A. Franceschini, I. M. Gioia, T. Maccacaro.
Structure and evolution of active galactic nuclei, p. 547 – 550 (1986). – See Abstr. 012.011.
The authors compare statistical properties on optically selected and X–ray selected active galactic nuclei by using the available data on the corresponding distributions of the X–ray to optical luminosity ratio. They show that these computations critically depend on the width of these distributions.

158.118 **First results from a catalog of data concerning AGN – FWHM and FWZI versus L_0 for C IV and Hβ.**
M. Joly.
Structure and evolution of active galactic nuclei, p. 559 – 563 (1986). – See Abstr. 012.011.
From a statistical analysis on a large sample of Seyfert 1 nuclei and quasars, the author puts into evidence the existence of a homogeneous class of objects defined by a FWHM(Hβ) larger than 1700 km/s. For these AGN, there is a tight correlation between FWZI(Hβ) and the optical luminosity. From this relation one can deduce that the radiation process in low luminosity object occurs far below the Eddington limit with an efficiency factor independent of the mass while at high luminosity the masses involved are very high. It is shown also that in low luminosity objects the C IV line arises in a region with higher velocity dispersion than Hβ.

158.119 **Cosmological evolution in the extent of double radio galaxies.**
V. K. Kapahi.
Structure and evolution of active galactic nuclei, p. 565 – 568 (1986). – See Abstr. 012.011.
A comparison of the sizes of radio galaxies of similar radio luminosity at different redshifts is shown to provide strong evidence for size evolution with epoch independent of any luminosity–size relationship.

158.120 **The spectral index – flux density relation and the cosmological evolution of radio sources.**
V. K. Kapahi, V. K. Kulkarni.
Structure and evolution of active galactic nuclei, p. 569 – 572 (1986). – See Abstr. 012.011.
A careful redetermination of the α–S relation for radio sources selected from metre–wavelengths surveys does not show any significant flattening of α_{median} below $S_{408} \sim 1$ Jy contrary to earlier claims. Constraints on evolutionary models of the luminosity function are briefly discussed.

158.121 **Classification of optical jets in galaxies.**
W. C. Keel.
Structure and evolution of active galactic nuclei, p. 579 – 582 (1986). – See Abstr. 012.011.
New, deep images of galaxies reported in various catalogs to have optical jets show that most are not jets, but tidal features, polar rings, or superpositions. From those objects that do seem to represent nuclear ejecta, criteria of brightness relative to the parent galaxy, width, and location have been derived, which should produce much purer samples of optical jets. Jet–like features in ESO 0610–23, UGC 3995, and NGC 1598 are described. Finally, a new search of the SRC IIIa–J survey is described, which was intended to incorporate the new criteria and provide a large sample of optically–selected jets for detailed study.

158.122 **An interpretation of the lightcurve of BL Lac.**
M. R. Kidger, J. E. Beckman.
Structure and evolution of active galactic nuclei, p. 587 – 590 (1986). – See Abstr. 012.011.
Extensive monitoring is now available for a good sample of the brighter quasars and BL Lacs. In some cases, lightcurves are well enough sampled to permit stochastic modelling. The authors present here, a simple model of the lightcurve of the prototype BL Lacertae object, BL Lac in terms of a series of random events, each with rapid rise and exponential decline. Whilst unable to explain very rapid "optically violent" activity, it seems to provide a satisfactory fit to the long term trends in the lightcurve.

158.123 **Double nucleus galaxies.**
W. Kollatschny, K. J. Fricke, J. Hellwig.
Structure and evolution of active galactic nuclei, p. 605 – 609 (1986). – See Abstr. 012.011.
Double nucleus galaxies may represent late stages of galaxy mergers and may therefore be key objects for understanding galaxy evolution and activity. Here the authors present spectroscopic data on these objects and their individual nuclei.

158.124 **The properties of X–ray selected BL Lac objects.**
T. Maccacaro, I. M. Gioia, D. Maccagni, J. T. Stocke.
Structure and evolution of active galactic nuclei, p. 615 – 618 (1986). – See Abstr. 012.011.
The optical and radio properties of the X–ray selected BL Lacs in the Einstein medium sensitivity survey are presented and compared with those of the HEAO 1 A–2 sample and with those of radio selected BL Lacs. As a class, X–ray selected BL Lacs have the largest ratio of X–ray to optical flux of any AGN yet discovered. The authors have also found that the number–flux relation for BL Lacs shows a drastic flattening below fluxes of the order of 10^{-12}ergs $cm^{-2}s^{-1}$ and they have interpreted this as evidence that BL Lacs do not share the evolutionary properties of quasars.

158.125 **Kiso surveys of ultraviolet–excess objects.**
H. Maehara, T. Noguchi, M. Kondo, N. Miyauchi–Isobe, B. Takase.
Structure and evolution of active galactic nuclei, p. 619 – 622 (1986). – See Abstr. 012.011.
Kiso surveys of ultraviolet–excess objects have been carried out with the use of the 105 cm Schmidt telescope, in which about 1,200 stellar (KUVs) and 1,100 galaxy (KUGs) images are catalogued from the (U,G,R) triple image plates. Follow–up observations in the optical, infrared, and radio wavelengths reveal the results below in relation to active galaxies: (1) a few percent of KUVs belong to quasars, (2) about 80 percent of KUGs possess conspicuous emission lines, (3) there are some Seyfert–like objects among KUGs.

158.126 **Steep spectrum radio sources showing low frequency variability.**
F. Mantovani, I. Browne, R. Fanti, A. Ficarra, T. Muxlow, L. Padrielli, J. Romney.
Structure and evolution of active galactic nuclei, p. 623 – 626 (1986). – See Abstr. 012.011.
The Bologna group has been carrying on a monitoring program at 408 MHz of more than a hundred sources since 1975. Among sources found to be variable or possibly variable (Fanti et al., 1983) a small sample (10) has a steep spectrum ($\alpha \sim 0.6$, $S \propto \nu^{-\alpha}$) in the frequency range 0.4 – 15 GHz. The authors present here flux density measurements, spectral index behaviour, arcsecond and milliarcsecond observations of some of these sources.

158.127 **Near infrared surface photometry of NGC 1068.**
E. Mediavilla Gradolph, C. Sanchez Magro.
Structure and evolution of active galactic nuclei, p. 633 – 636 (1986). – See Abstr. 012.011.
NGC 1068 has been observed at the near infrared bands J, H, K, L. Near infrared scans of NGC 1068 with 20 and 5 arcsec apertures have been used to isolate the nuclear emission. The nuclear near infrared emission can be explained by a combination of dust clouds at different positions and temperatures. The same scans have been used for a tentative detection of the ring in the L band.

158.128 **Active galaxies in high density environments.**
T. K. Menon, P. Hickson.
Structure and evolution of active galactic nuclei, p. 637 – 640 (1986). – See Abstr. 012.011.
A radio survey of 401 galaxies in small groups with local galaxy space densities ranging from 10^2 to $10^6 Mpc^{-3}$ has detected 41 sources above a flux limit of 1.5 mJy at 18 cm. These radio sources fall into two categories according to the optical morphology of the host galaxies. Radio sources in elliptical galaxies occur only in the first–ranked (in optical luminosity) galaxies. They are compact nuclear sources with no detectable extended emission. Those in spirals are more extended and show no preference for optical rank. The authors' interpretation is that the emission in spiral galaxies is a result of star formation activity and that the sources in elliptical galaxies result from inflow onto a central compact object.

158.129 **Towards the luminosity function of Seyfert galaxy nuclei.**
E. J. A. Meurs.
Structure and evolution of active galactic nuclei, p. 641 – 644 (1986). – See Abstr. 012.011.
The optical and radio luminosity functions of Seyfert galaxies for total emission of the systems are briefly reviewed. A project of CCD–photometry aimed at determining a luminosity function for Seyfert nuclei is described. The merits of integrated and nuclear magnitudes with respect to Seyfert luminosity functions are discussed. Finally, a few remarks on systematic survey incompletenesses are made.

158.130 **The optical luminosity of the invisible nucleus of M82.**
P. Notni.
Structure and evolution of active galactic nuclei, p. 649 – 652 (1986). – See Abstr. 012.011.
A new estimate of the relative luminosity of the hidden nucleus of M82 compared to the disk is given. This estimate is a factor of 10 higher than a previous one; the very blue colour of the scattered halo light can now be explained. An approximate conversion to absolute luminosity shows that this high nuclear luminosity is also compatible with some of the models of Rieke et al (1980).

158.131 **An evolutionary link between Seyfert I and II galaxies?**
E. Pérez, M. V. Penston.
Structure and evolution of active galactic nuclei, p. 669 – 674 (1986). – See Abstr. 012.011.
First spectra of NGC 4151 and 3C 390.3 from the newly sited Isaac Newton Telescope on the island of La Palma, when compared with old spectra of the same objects taken with the INT at Herstmonceux leads to the conclusion that these two Seyfert galaxies have faded over the years and are adopting Seyfert II–like properties.

158.132 **The possible determination of two new BL Lac redshifts.**
M. Persic, P. Salucci.
Structure and evolution of active galactic nuclei, p. 675 – 677 (1986). – See Abstr. 012.011.
The authors present tentative identifications of very faint emission lines in the spectra of the BL Lacertae objects 0754 + 100 and 1514 + 197 as given in the atlas of spectra of QSOs and related objects by Wilkes et al. (1983). The proposed identifications, quite common in BL Lac spectra, are discussed.

158.133 **MAGN; mildly active galactic nuclei.**
P. Pismis.
Structure and evolution of active galactic nuclei, p. 679 – 681 (1986). – See Abstr. 012.011.
The author proposes the designation MAGN (mildly active galactic nuclei) for all galactic nuclei the degree of activity of which falls, on the average, below that of Seyfert 2's.

158.134 **On the duplicity of Seyfert galaxy NGC 1275 nucleus.**
I. Pronik, L. Metik.
Structure and evolution of active galactic nuclei, p. 683 – 687 (1986). – See Abstr. 012.011.
Recently the authors have investigated the nuclear region of NGC 1275 by a two–colour photometry method. Their data testify in favour of two hypotheses. They are as follows: (1) two interacting galaxies, (2) the gas accretion from the intergalactic space.

158.135 **PCD spectroscopy of Mkn 463, a double nucleus Seyfert–2 galaxy.**
P. Rafanelli, S. di Serego Alighieri.
Structure and evolution of active galactic nuclei, p. 689 – 692 (1986). – See Abstr. 012.011.
PCD spectroscopy of Mkn 463, a double nucleus Seyfert–2 galaxy, has been performed at the 2.2 m telescope of the ESO Observatory in order to verify the spatial resolution and the sensitivity of the detecting system. A preliminary analysis of the spectra shows that the asymmetric profiles of the emission lines arising from the Seyfert 2 nucleus are more likely due to dynamical effects rather than to dust extinction.

158.136 **Narrow band photometry and the reddening of Cyg A.**
K. D. Rakos, N. Fiala.
Structure and evolution of active galactic nuclei, p. 693 – 696 (1986). – See Abstr. 012.011.
The authors' photometric system vz, bz, yz is very similar to the Stroemgren v, b, y passbands. It has many advantages in connection with a CCD camera compared to broad band UBV photometry. The authors have supposed that the light of Cyg A comes from an average elliptical galaxy with average solar chemical composition without strong nonthermal components, and that the standard reddening law can be applied to the dark matter within elliptical galaxies. They found E(bz–yz) = 0.37 for Cyg A, which corresponds to AV = 1.53 or AB = 2.03. The dereddened color of Cyg A is the same to that of 3C 293.

158.137 **Evolution of powerful AGN and the collapse of rich clusters.**
N. Roos.
Structure and evolution of active galactic nuclei, p. 697 – 700 (1986). – See Abstr. 012.011.
The cosmological evolution of active galactic nuclei (AGN) may be described by a luminosity function which depends on redshift and on the (stellar) luminosity or mass of the galaxy. The author's mathematical expression for the luminosity function of AGN is based on the "loss–cone filling" model for the fuelling of central black holes in AGN during galaxy mergers. He assumes that galactic nuclei containing massive central black holes were formed at $z \cong 3$ and that the subsequent cosmological evolution of AGN is mainly due to a decrease in the merging rate among galaxies during the subsequent epoch of cluster formation. The rapid evolution of powerful AGN is interpreted as an evolution of the merging rate of the brightest galaxies during the collapse and virialization of rich clusters.

158.138 **Low–luminosity active nuclei in nearby elliptical galaxies.**
E. M. Sadler.
Structure and evolution of active galactic nuclei, p. 701 – 704 (1986). – See Abstr. 012.011.
The ability of an elliptical galaxy to produce a central radio source appears to be determined by both its luminosity and the gas supply in the inner ~1 kpc of the galaxy. At least half of all nearby ellipticals contain $10^3 - 10^5 M_\odot$ of ionised gas in their central regions (often in association with dust) and show weakly active nuclei characterised by a Liner–like spectrum and radio emission at a level of $10^{20} - 10^{24}$W/Hz ($10^{37} - 10^{41}$erg/s).

158.139 **Kinematics in the narrow line region of NGC 4151 – evidence for a merger?**
H. Schulz.
Structure and evolution of active galactic nuclei, p. 705 – 708 (1986). – See Abstr. 012.011.
The velocity structure in the narrow line region of the Seyfert galaxy NGC 4151 is studied by means of various long–slit coudé spectra. Spectral line profiles in the NE and SW quadrants display a 2–component structure while those in the NW and SE are

narrower and smoother. One component fits to the general rotation pattern of the disk of NGC 4151. The other component may refer to a bipolar radial flow or to gas spiraling inwards within a plane which is perpendicular to the main galactic disk. It is suggested that merging gaseous disks are a common feature of narrow line regions in Seyferts.

158.140 **Interaction and emission morphology.**
N. A. Sharp.
Structure and evolution of active galactic nuclei, p. 713 – 716 (1986). – See Abstr. 012.011.

A simple comparison between broad V–band images and narrow band images tuned to the redshifted wavelength of Hα enables to make a quick survey of the distribution of emission in galaxies. The appearance of interacting systems can then be contrasted with that of isolated objects. Those systems for which data are presently available seem to split into three classes: (1) nothing happening, despite apparent connections, (2) emission distributed either smoothly or lumpily throughout, (3) a central peak of emission in an otherwise smooth profile, for one or more of the galaxies.

158.141 **Emission lines indicating a universal L/M ratio for the central engine in AGN?**
A. Wandel.
Structure and evolution of active galactic nuclei, p. 727 – 735 (1986). – See Abstr. 012.011.

Assuming the velocity dispersion of the emission–line clouds is induced by gravity, the authors calculate the dynamic mass of the central object in a sample of ~50 Seyfert 1 nuclei and ~40 nearby quasars. Typical values are $10^8 M_\odot$ for Seyferts and $10^9 M_\odot$ for quasars. The dynamic mass is found to be proportional to the continuum luminosity (at the ~20σ confidence level), which suggests a universal value of L/L_{Edd}. For the optical band the authors find $\log < L_c/L_{Edd} > = -2.0 \pm 0.3$. Applying the same technique to the O III line, one can estimate the dynamic mass within the radius of the narrow–line region (~100 pc). For some 50 objects of the sample above, this mass is found to be well correlated with the continuum luminosity and with the dynamic mass estimated from the Hβ line.

158.142 **Influence of the cluster environment on Seyfert galaxies.**
R. Nesci.
Astron. Astrophys., Vol. 160, No. 2, p. 259 – 263 (1986).

The frequency of Seyfert galaxies in Zwicky clusters has been explored on the basis of a fairly complete sample of 78 objects. No differences have been found between the distributions of Seyferts of type 1 and type 2. The distribution of Seyferts members of a cluster as a function of the distance from the cluster center is found consistent with a uniform space density of these objects within a cluster, at variance with the density of normal galaxies. The general Seyfert frequency in clusters of different compactness and in the field is found to be extremely similar to that of late type spirals, and markedly different from those of ellipticals and lenticulars. The suggestion is made that the hosts of Seyferts were originally normal late type spirals, morphologically perturbed by the onset of the nuclear activity.

158.143 **Surface brightness distribution in Seyfert galaxies. I. The sample. Results of scanning.**
V. L. Afanas'ev, V. T. Doroshenko, V. Yu. Terebizh.
Astrofizika, Tom 24, Vyp. 2, p. 333 – 348 (1986). In Russian. English translation in Astrophysics, Vol. 24, No. 2.

A sample of 45 spiral galaxies including 19 Seyfert objects is formed within distances less than 100 Mpc ($H_0 = 75$ km s^{-1}Mpc^{-1}). Small aperture measurements in B with a scanning photoelectric photometer were made to obtain the surface brightness distributions along the major axes in 14 galaxies. The characteristic parameters such as the surface brightness and its gradient at the distance of 1 kpc from the center of galaxies are calculated.

158.144 **High resolution spectra of narrow line profiles in Seyfert nuclei.**
J. F. Kielkopf.
Bull. Am. Astron. Soc., Vol. 17, No. 4, p. 829 (1985). Abstract. – See Abstr. 010.062.

158.145 **Tidal triggering of Seyfert galaxies and quasars–models.**
G. Byrd, M. Valtonen, B. Sundelius, L. Valtaoja.
Bull. Am. Astron. Soc., Vol. 17, No. 4, p. 829 (1985). Abstract. – See Abstr. 010.062.

158.146 **A radio loop near the nucleus of Seyfert galaxy NGC 5506.**
A. E. Wehrle, M. Morris.
Bull. Am. Astron. Soc., Vol. 17, No. 4, p. 830 (1985). Abstract. – See Abstr. 010.062.

158.147 **NGC 4319 – a spiral galaxy with radio lobes?**
J. W. Sulentic.
Bull. Am. Astron. Soc., Vol. 17, No. 4, p. 830 (1985). Abstract. – See Abstr. 010.062.

158.148 **Optical polarization and brightness variability in BL Lacertae objects and OVV (*optically violent variable*) quasars.**
T. J. Balonek, P. S. Smith, R. Elston, P. A. Heckert.
Bull. Am. Astron. Soc., Vol. 17, No. 4, p. 830 (1985). Abstract. – See Abstr. 010.062.

158.149 **Phased–VLA polarimetry of 3C 371 (1807+698).**
J. M. Wrobel, K. R. Lind.
Bull. Am. Astron. Soc., Vol. 17, No. 4, p. 830 – 831 (1985). Abstract. – See Abstr. 010.062.

158.150 **Evidence for intrinsically asymmetric radio structure in the nucleus of NGC 6251.**
D. L. Jones.
Bull. Am. Astron. Soc., Vol. 17, No. 4, p. 831 (1985). Abstract. – See Abstr. 010.062.

158.151 **The jet in 3C 120: a continuous structure from parsec to 100 kpc scales.**
R. C. Walker, J. M. Benson, S. C. Unwin.
Bull. Am. Astron. Soc., Vol. 17, No. 4, p. 831 (1985). Abstract. – See Abstr. 010.062.

158.152 **Morphology of the jet in 1759+211.**
P. J. Wiita, D. J. Saikia.
Bull. Am. Astron. Soc., Vol. 17, No. 4, p. 831 (1985). Abstract. – See Abstr. 010.062.

158.153 **X–ray and radio observations of steep spectrum radio sources: I. 4 fields from the Clark Lake Observatory 26 MHz survey.**
D. E. Harris, P. E. Dewdney, C. H. Costain, I. McHardy, A. G. Willis.
Bull. Am. Astron. Soc., Vol. 17, No. 4, p. 831 (1985). Abstract. – See Abstr. 010.062.

158.154 **Spatially resolved 10–micron images of the nucleus of Seyfert galaxy NGC 1068.**
R. Tresch–Fienberg, G. G. Fazio, D. Y. Gezari, G. M. Lamb, P. Shu, W. F. Hoffmann, C. R. McCreight.
Bull. Am. Astron. Soc., Vol. 17, No. 4, p. 845 (1985). Abstract. – See Abstr. 010.062.

158.155 **The far–infrared and optical structure of Arp 220.**
M. Joy, D. F. Lester, P. M. Harvey, M. Frueh.
Bull. Am. Astron. Soc., Vol. 17, No. 4, p. 845 (1985). Abstract. – See Abstr. 010.062.

158.156 **Star–burst nuclei in "S0" galaxies.**
L. L. Dressel.
Bull. Am. Astron. Soc., Vol. 17, No. 4, p. 845 – 846 (1985). Abstract. – See Abstr. 010.062.

158.157 **Infrared luminosity function of galaxies.**
D. W. Weedman.
Bull. Am. Astron. Soc., Vol. 17, No. 4, p. 846 (1985). Abstract. – See Abstr. 010.062.

158.158 **Ultraviolet and optical observations of the Seyfert 1.8 galaxy Markarian 609.**
R. D. Cohen, R. J. Rudy, T. Ake.
Bull. Am. Astron. Soc., Vol. 17, No. 4, p. 846 (1985). Abstract. – See Abstr. 010.062.

158.159 **Beam size dependence of the ratio of [Ne II] λ12.8 and the 11.3 micron feature in the starburst galaxy M82.**
P. D. LeVan, S. D. Price.
Bull. Am. Astron. Soc., Vol. 17, No. 4, p. 846 (1985). Abstract. – See Abstr. 010.062.

158.160 **Detection of nickel in six Seyfert galaxies.**
J. P. Halpern, J. B. Oke.
Bull. Am. Astron. Soc., Vol. 17, No. 4, p. 846 (1985). Abstract. – See Abstr. 010.062.

158.161 **He I 10830 measurements of Seyfert II galaxies.**
R. J. Rudy, G. S. Rossano, R. D. Cohen, R. C. Puetter.
Bull. Am. Astron. Soc., Vol. 17, No. 4, p. 846 (1985). Abstract. – See Abstr. 010.062.

158.162 **Comparison of nearly simultaneous optical and ultraviolet spectra of Seyfert 1 galaxies.**
R. L. Ptak, R. E. Stoner.
Bull. Am. Astron. Soc., Vol. 17, No. 4, p. 846 – 847 (1985). Abstract. – See Abstr. 010.062.

158.163 **The spatial distribution of active galactic nuclei I. The density of Seyfert galaxies.**
R. Burg, J. Huchra.
Bull. Am. Astron. Soc., Vol. 17, No. 4, p. 865 (1985). Abstract. – See Abstr. 010.062.

158.164 **Optical and radio observations of the cooling–flow galaxy NGC 708.**
P. J. McCarthy, W. J. M. van Breugel, T. Heckman, G. K. Miley, A. V. Filippenko.
Bull. Am. Astron. Soc., Vol. 17, No. 4, p. 865 (1985). Abstract. – See Abstr. 010.062.

158.165 **Velocity dispersion profiles of the cD galaxies A2052 and A2589.**
E. M. Malumuth, G. A. Kriss, K. D. Borne.
Bull. Am. Astron. Soc., Vol. 17, No. 4, p. 866 (1985). Abstract. – See Abstr. 010.062.

158.166 **Seyfert galaxies, Liners, and H II region galaxies among the IRAS warm galaxies.**
D. E. Osterbrock, M. M. De Robertis.
Bull. Am. Astron. Soc., Vol. 17, No. 4, p. 868 – 869 (1985). Abstract. – See Abstr. 010.062.

158.167 **The 0.3 to 100 micron continua of type 1 Seyferts.**
M. Ward, M. S. Elvis, G. Fabbiano, N. Carleton, S. Willner, A. Lawrence.
Bull. Am. Astron. Soc., Vol. 17, No. 4, p. 869 (1985). Abstract. – See Abstr. 010.062.

158.168 **Sizes of continuum and line emitting regions in Seyfert galaxies.**
C. M. Gaskell.
Bull. Am. Astron. Soc., Vol. 17, No. 4, p. 869 (1985). Abstract. – See Abstr. 010.062.

158.169 **Interpretation of the Einstein X–ray spectra of BL Lac objects.**
G. M. Madejski.
Bull. Am. Astron. Soc., Vol. 17, No. 4, p. 889 (1985). Abstract. – See Abstr. 010.062.

158.170 **A minimum rate of change analysis of X–ray variability in active galactic nuclei.**
D. A. Schwartz.
Bull. Am. Astron. Soc., Vol. 17, No. 4, p. 889 (1985). Abstract. – See Abstr. 010.062.

158.171 **The shocking events in 3C 279.**
P. A. Hughes, H. D. Aller, M. F. Aller.
Bull. Am. Astron. Soc., Vol. 17, No. 4, p. 889 – 890 (1985). Abstract. – See Abstr. 010.062.

158.172 **Wavelength–dependent polarization in BL Lacertae objects and OVV quasars.**
P. S. Smith.
Bull. Am. Astron. Soc., Vol. 17, No. 4, p. 890 (1985). Abstract. – See Abstr. 010.062.

158.173 **The extragalactic nature of G227.1 + 1.0.**
G. A. Chanan, D. J. Helfand, H. Spinrad, K. Ebneter.
Bull. Am. Astron. Soc., Vol. 17, No. 4, p. 909 (1985). Abstract. – See Abstr. 010.062.

158.174 **A search for OH maser emission from spiral galaxies.**
S. W. Unger, J. M. Chapman, R. J. Cohen, T. G. Hawarden, C. M. Mountain.
Mon. Not. R. Astron. Soc., Vol. 220, No. 1, p. 1P – 6P (1986).

A sample of 40 nearby spiral galaxies with infrared colours dominated by star formation activity has been searched for OH mainline maser emission. No new detections were obtained, showing that a high level of such activity is not a sufficient condition for megamaser emission. Galaxies in which maser emission has been detected have significantly different infrared colours from the sample galaxies, showing strong excesses at both 25 and 60 μm. This is likely to be related to far–infrared pumping of the OH masers.

158.175 **X–ray spectroscopy of active galactic nuclei.**
R. Mushotzky.
ESA Spec. Publ., ESA SP–239, p. 19 – 23 (1985). – See Abstr. 012.019.

X–ray spectroscopy of active galactic nuclei (AGN) has the potential for determining many of the fundamental properties of AGN. The author discusses the scientific goals of such observations with respect to the underlying energy source, the regions responsible for the optical emission lines, the different "types" of active galaxies and cosmology. The requirements for an X–ray mission of broad band width, large collecting area, modest spatial resolution and good spectral resolution are outlined. It is concluded that the proposed XMM mission meets these requirements.

158.176 **1E 1415.6 + 2557: an X–ray–selected BL Lacertae object in a luminous galaxy.**
J. P. Halpern, C. D. Impey, G. D. Bothun, S. Tapia, E. D. Skillman, A. S. Wilson, E. J. A. Meurs.
Astrophys. J., Vol. 302, No. 2, p. 711 – 717 (1986). = Columbia Astrophys. Lab. Contrib., No. 298. With plate 21.

A bright *Einstein* X–ray source is identified with a new BL Lacertae object in a highly luminous galaxy. The resolved component is well fitted by an exponential disk with scale length 18 kpc and absolute *r* magnitude $M_r \approx -24.2$ ($H_0 = 50$). The redshift of 0.237 is derived from stellar absorption features. Decomposition of the optical spectrum into a standard galaxy plus a power law yields a spectral index of $\alpha_0 = 0.5 \pm 0.5$. Linear polarization of the nonstellar component is ~6% in the wavelength range 4500 – 7000 Å. The X–ray luminosity in the 0.3 – 3.5 keV band is 3.5×10^{45} ergs s^{-1}.

158.177 **The optical variation and outburst of OJ 287.**
K.–h. Li, G.–z. Xie, M.–x. Bao, P.–j. Hau.
Acta Astron. Sin., Vol. 27, No. 2, p. 137 – 143 (1986). In Chinese.
UBV magnitudes and color indices have been obtained during a continuous 4–year monitoring program. The main results are as follows: (1) A proposed variation on a time scale of about 8 days with an amplitude of $B = 0.37$ was confirmed. (2) During the period of 1980 – 1982, OJ 287 was stable. The authors recorded an outburst of $B = 14^m25$ in March 1983. (3) An optical light curve for the period 1968 – 1984, constructed from all published photographic data shows evidence for the existence of an optical variation with larger amplitude ($\Delta B = 1^m25 - 4^m00$) and a long time scale (2 – 4 years) except the variability on a time scale of about 8 days with an amplitude of $\Delta B = 0^m37$.

158.178 **Spectra of galaxies obtained with the "Rojen" 2–m telescope of the Bulgarian Academy of Sciences.**
Kh. S. Markov, S. A. Zhekov, G. T. Petrov, Z. I. Tsvetanov.
Astron. Tsirk., No. 1387, p. 1 – 3 (1985). In Russian.

158.179 **Theory of active galactic nuclei.**
G. A. Shields.
Publ. Astron. Soc. Pac., Vol. 98, No. 600, p. 130 – 133 (1986).
Review paper. – See Abstr. 012.022.

158.180 **Gas in active galactic nuclei.**
G. M. MacAlpine.
Publ. Astron. Soc. Pac., Vol. 98, No. 600, p. 134 – 139 (1986).
Review paper. – See Abstr. 012.022.

158.181 **The optical continuum emission of active galactic nuclei.**
S. L. O'Dell.
Publ. Astron. Soc. Pac., Vol. 98, No. 600, p. 140 – 147 (1986).
Review paper. – See Abstr. 012.022.
This article reviews general properties of active galactic nuclei (AGNs) and distinctions among the various classes of AGNs.

158.182 **Current issues in the X–ray properties of active galactic nuclei.**
M. Elvis.
Publ. Astron. Soc. Pac., Vol. 98, No. 600, p. 148 – 151 (1986).
Review paper. – See Abstr. 012.022.
Some issues raised by new X–ray spectra and timing observations of active galaxies and quasars are presented.

158.183 **Radio characteristics of galactic nuclei.**
J. J. Condon.
Publ. Astron. Soc. Pac., Vol. 98, No. 600, p. 152 – 158 (1986).
Review paper. – See Abstr. 012.022.
The observed characteristics of compact radio sources in galactic nuclei are complicated by orientation–dependent relativistic beaming and by refractive focusing in the interstellar medium of our galaxy. Incoherent synchrotron radiation is still the preferred emission mechanism, so the "central engine" is probably forever hidden from direct radio observation by synchrotron self–absorption in compact sources. Nonetheless, constraints set by synchrotron theory and radio observations of both compact and extended radio sources favor a single massive object not supported by radiation pressure – either a spinar or a black hole – as the energy source in radio–galaxy nuclei.

158.184 **Gravitational processes in galactic nuclei.**
W. C. Saslaw.
Publ. Astron. Soc. Pac., Vol. 98, No. 600, p. 171 – 178 (1986).
Review paper. – See Abstr. 012.022.
Some of the important gravitational physical processes in galactic nuclei are sketched: violent relaxation, gentle relaxation, and the redistribution of stars through gravithermal instability, evaporation, mass segregation and orbit segregation.

158.185 **Evidence for a low–density component in the broad–line region of Seyfert 1 galaxies.**
D. M. Crenshaw, B. M. Peterson.
Publ. Astron. Soc. Pac., Vol. 98, No. 600, p. 185 – 191 (1986).
On the basis of attempts to deconvolve the complex blend of broad lines in the vicinity of Hβ in Seyfert 1 galaxy spectra, the authors conclude that broad [O III] λλ4959, 5007 is present. This argues for the existence of clouds with electron density $\sim 10^8 cm^{-3}$ in the broad–line region.

158.186 **Faint galaxies and cosmology.**
H. Spinrad.
Publ. Astron. Soc. Pac., Vol. 98, No. 601, p. 269 – 284 (1986).
This paper reviews progress in two areas of optical research on faint galaxies. First the author discusses the philosophy and techniques needed to employ radio galaxies as evolutionary and cosmological probes out to redshifts now approaching 2.0. The distant radio galaxies seem to be systematically more active than their nearby counterparts. Radio galaxies can also be studied as galaxies of evolving stars; the present interpretations of visual and IR galaxy colors and magnitudes suggest modest lingering star formation over the past 10^{10}years in many luminous radio galaxies. The author concludes with a discussion on the location and utilization of faint radio–quiet red galaxies; these are giant E systems at $z \geqslant 0.5$.

158.187 **Observations of extragalactic radio sources at the wavelengths of 8.2 and 13.5 mm.**
V. A. Efanov, I. G. Moiseev, N. S. Nesterov.
Bull. Crimean Astrophys. Obs., Vol. 69, p. 74 – 83 (1986). English translation of 39.158.073.

158.188 **Investigation of the structure of extragalactic radio sources at 3.5 m wavelength.**
Yu. V. Volodin, A. G. Rubanov, R. D. Dagkesamanskij.
Fiz. inst. Akad. Nauk SSSR. Prepr., No. 184, 44 pp. (1985). In Russian. Abstr. in Ref. Zh., 51. Astron., 4.51.815 (1986).

158.189 **Multifrequency observations of the wide–angle tail radio source 1313+073.**
A. R. Patnaik, M. A. Malkan, C. J. Salter.
Mon. Not. R. Astron. Soc., Vol. 220, No. 2, p. 351 – 362 (1986).
The authors present further VLA observations of the wide–angle tail (WAT) source 1313+073 in the C array at 20 and 6 cm and the redshift measurement of the galaxy. These observations when combined with the D array observations reported earlier, give detailed information on the distributions of spectral index, rotation measure and depolarization across the source. Some recent models for the formation of bent tails in WATs are discussed in the context of this source. The authors also present optical spectroscopic data suggesting that 1313+073 lies in a cluster of galaxies at a redshift $z = 0.0507$, located in the background of the Zwicky cluster 1313.7+0721 at $z = 0.0212$.

158.190 **Cosmological evolution of AGNs in the X–ray band.**
L. Danese, G. De Zotti, G. Fasano, A. Franceschini.
Astron. Astrophys., Vol. 161, No. 1, p. 1 – 6 (1986).
The authors present a new analysis of the cosmological evolution of X–ray selected active galactic nuclei (AGNs), taking into account both the soft and the hard X–ray data. They show that an increase of the published soft (0.5 – 4.5 keV) X–ray fluxes by a factor $\cong 1.5$ is required to bring them into the hard X–ray (2 – 10 keV) flux scale. While the available samples of X–ray selected AGNs are found to provide rather loose limits on evolutionary models, strong constraints follow from the observed properties (spectrum and intensity) of the X–ray background.

158.191 **Gamma–ray observations and relativistic jets in active galactic nuclei.**
L. Bassani, A. J. Dean.
Astron. Astrophys., Vol. 161, No. 1, p. 85 – 88 (1986).
Independent lines of evidence for the presence of relativistic jets in active galactic nuclei (AGN) include high resolution radio maps, superluminal expansion measurements, brightness temperature limits, synchrotron self Compton predictions of the X–ray flux density. Here, the authors investigate the consequences of relativistic bulk motion on the absorption of γ–rays by pair producing photon–photon collisions with X–rays. They show that γ–ray spectral measurements of AGN can yield independent estimates of the Doppler factor of the jet. Furthermore, use of the γ–ray data in conjunction with X–ray and radio measurements

enable a unique set of parameters to be obtained which describe the geometry and the kinematics of the jet motion.

158.192 **The space distribution of emission–line galaxies.**
D. A. Stepanyan.
Sov. Astron. Lett., Vol. 11, No. 4, p. 242 – 244 (1985). English translation of 40.158.015.

158.193 **Active galactic nuclei. I. Central source as a cluster of accreting black holes.**
A. G. Pacholczyk, W. R. Stoeger.
Astrophys. J., Vol. 303, No. 1, p. 76 – 86 (1986).
A generalization of an accreting supermassive black hole model for the central nuclear engine of QSOs and active galactic nuclei (AGNs) is suggested in terms of a dense cluster of many black holes and compact objects. The cluster possesses a mass spectrum, with each hole surrounded by an accretion disk. The fitting of the thermal radiation spectrum of such a cluster to the observed optical – X–ray power–law spectrum of QSOs and AGNs determines the total mass, number of black holes, and mass spectrum index for the cluster. Their dynamics are also discussed, and the consumption and evaporation rates of member holes are estimated.

158.194 **X–ray absorption in active galaxies: constraints on the broad–line emitting regions.**
G. A. Reichert, R. F. Mushotzky, S. S. Holt.
Astrophys. J., Vol. 303, No. 1, p. 87 – 93 (1986).
X–ray observations of intrinsic absorption in active galactic nuclei (AGN) provide direct information concerning the amount and physical properties of cool (un–ionized) material along the line of sight. The X–ray absorption is assumed to occur in the broad–line emitting gas. The authors use the results of Einstein Observatory observations to derive detailed physical constraints concerning the geometry of the broad–line region. Physical implications of these constraints are also discussed.

158.195 **The core of the M87 globular cluster system.**
T. R. Lauer, J. Kormendy.
Astrophys. J., Lett. Ed., Vol. 303, No. 1, p. L1 – L4 (1986). With plate L1.
High–resolution CCD observations have been used to identify 124 globular clusters with $m_B < 23.6$ within 60″ of the center of M87. The luminosity function of these clusters is indistinguishable from that measured further out by other investigators. The core radius r_c of the cluster system is an order of magnitude larger than that of the underlying galaxy surface brightness distribution. The present data combined with those of other investigators imply $r_c = 88'' \pm 5''$ and a central surface density of 72 ± 4 clusters arcmin^{-2} with $m_B < 23.6$.

158.196 **Studies of active nuclei in early type galaxies.**
G. F. O. Schnur.
Proceedings of the Second Asian–Pacific Regional Meeting on Astronomy, p. 280 (1984). – See Abstr. 012.031.
In an extensive program of spectroscopic, morphological, IR and radio studies of early type galaxies the author has observed a large number of S0–galaxies in order to study the physical nature of the nuclear regions. Large scale photographs have been obtained with the 3.6 m and 1.54 m telescopes at ESO–La Silla as well as spectroscopic information.

158.197 **Recent observations and evolution of extended extragalactic sources.**
G. Swarup.
Proceedings of the Second Asian–Pacific Regional Meeting on Astronomy, p. 297 – 303 (1984). – See Abstr. 012.031.
In this review the author discusses some of the intrinsic and extrinsic factors that affect the evolution of the extended extragalactic sources, particularly those based on recent observations of the hot spots and the jets seen in these sources.

158.198 **Infrared polarimetry and photometry of BL Lac objects.**
C. D. Impey, P. W. J. L. Brand, R. D. Wolstencroft, P. M. Williams.
Proceedings of the Second Asian–Pacific Regional Meeting on Astronomy, p. 308 – 334 (1984). – See Abstr. 012.031.
Infrared polarimetry and photometry have been obtained for a sample of 18 BL Lac objects. Large and variable degrees of polarization are a common property of the sample. Two BL Lac objects show wavelength dependent polarization and two objects show evidence for position angle rotations over a five–day period. The BL Lac objects cover an enormous range of infrared luminosity; the three most luminous having $L_{IR} > 10^{46}$ergs s^{-1} and the other end of the range having infrared luminosities similar to normal elliptical galaxies.

158.199 **The distribution and some examples of UV–bright galaxies.**
B. Takase, H. Maehara.
Proceedings of the Second Asian–Pacific Regional Meeting on Astronomy, p. 347 – 351 (1984). – See Abstr. 012.031.
Some 1,100 UV–bright galaxies, which the authors call KUG's (Kiso UV–bright galaxies), were detected in 20 fields of about 650 sq. degrees on the plates taken with the Kiso 105 cm Schmidt telescope by means of the UGR three image photograph. Detectable magnitude is estimated to be 17 – 18 mag, varying from plate to plate. The ratio of the number of KUG's to that of all galaxies found on each plate is 0.25 on average. The follow–up observations for a dozen of selected KUG's with larger scale reflector reveals that most of them exhibit some peculiarities in their appearance and two thirds have emission–line spectra.

158.200 **Radio galaxies.**
A. Downes.
Mercury, Vol. 15, No. 2, p. 34 – 45, 57 (1986).

158.201 **Luminosity dependence of binary radio sources on the distance between the components.**
V. R. Amirkhanyan.
Astrofiz. Issled. Izv. Spets. Astrofiz. Obs., Tom 22, p. 54 – 58 (1986). In Russian. English translation in Bull. Spec. Astrophys. Obs. – North Caucasus.
The analysis of sources of the 3CR catalogue allowed to draw the conclusion that the luminosity of binary radio sources decreases weakly with increasing distance between the components. This effect can be explained by the existence of a nonspheric component of the emission diagram of such objects.

158.202 **The redshift of the highly variable BL Lac object H0323 + 022.**
A. V. Filippenko, S. Djorgovski, H. Spinrad, W. L. W. Sargent.
Astron. J., Vol. 91, No. 1, p. 49 – 55 (1986).
New optical spectra are used to derive the redshift of H0323 + 022, a bright, highly variable X–ray source whose nature has been subject of controversy. Most of the spectra are dominated by a featureless nonstellar continuum, as found in BL Lac objects. Photometric variations of $\gtrsim 2$ mag have been seen over a time interval of only 2 days. The nonstellar light is sometimes faint enough to allow the detection of absorption lines at a redshift $z = 0.1471 \pm 0.0005$. These lines are clearly due to stars in the host galaxy, which is easily visible in deep images of the object. The galaxy is probably a relatively normal elliptical with $M_V \approx -21$ to -22, except that a noticeable extension is evident in one quadrant. There is now no doubt that H0323 + 022 is a BL Lac object, rather than a peculiar Galactic star.

158.203 **The optical–radio structure of the southwest hotspot of 3C 33: astrometry.**
J. W. Dreher, S. M. Simkin.
Astron. J., Vol. 91, No. 1, p. 58 – 64 (1986). With plate 1.
The authors report astrometric measurements for the radio image of the SW radio hotspot of 3C 33 and compare them with new astrometric measurements for both red (6650 Å) and blue (4400 – 4950 Å) optical images. The brightest peaks in the radio and red optical images are coincident to within the measurement errors. However, the bluer optical features appear to be aligned

but not congruent with the ridge of the radio hotspot emission. This suggests a similar origin for the radio and red optical radiation but a dissimilar one for the blue optical radiation.

158.204 **Daily observations of compact radio sources in 1983.**
E. B. Waltman, B. J. Geldzahler, K. J. Johnston, J. H. Spencer, P. E. Angerhofer, D. R. Florkowski, F. J. Josties, D. D. McCarthy, D. N. Matsakis.
Astron. J., Vol. 91, No. 2, p. 231 – 254 (1986).
The authors report flux densities at 3.7 and 11.1 cm for 36 compact variable radio sources that were monitored on a daily basis during 1983 with the Green Bank interferometer. The majority of the sources, with the exception of SS 433, 0235+164, BL Lac, 3C 446, and 3C 454.3, did not vary by more than 15%. Previous results for BL Lac are also included to show its daily variability over a period of almost 5 yr. Possible implications relevant to models of the outbursts in BL Lac are discussed.

158.205 **Star–formation rates in the nuclei of violently interacting galaxies.**
H. A. Bushouse.
Astron. J., Vol. 91, No. 2, p. 255 – 270 (1986).
This paper presents results of a spectrophotometric survey of the nuclear regions of a large sample of violently interacting galaxies. Other components of this project include emission–line imaging, global H I properties, and radio continuum mapping.

158.206 **Erratum: "Companions of Seyfert galaxies. II" [Astron. J., Vol. 90, No. 9, p. 1772 – 1782 (1985)].**
O. Dahari.
Astron. J., Vol. 91, No. 2, p. 438 (1986). See Abstr. 40.158.024.

158.207 **Multifrequency observations of MCG 8–11–11.**
L. Maraschi, L. Chiappetti, J. Danziger, H. Fink, R. Falomo, E. G. Tanzi, A. Treves, U. Zimmerman.
X–ray astronomy '84, p. 407 (1985). Abstract – See Abstr. 012.033.

158.208 **EXOSAT observations of active galactic nuclei.**
K. A. Pounds, I. M. McHardy, G. Stewart, R. S. Warwick.
X–ray astronomy '84, p. 409 – 417 (1985). – See Abstr. 012.033.
Preliminary results are reported on a continuing X–ray spectral survey of AGNs with EXOSAT.

158.209 **Evolution and astrophysics of active galactic nuclei.**
A. Cavaliere, E. Giallongo, F. Vagnetti.
X–ray astronomy '84, p. 435 – 449 (1985). – See Abstr. 012.033.
A review of the ongoing developments in the description and the understanding of the AGN evolution, which are induced by recent statistical data from optical and X–ray observations of local Seyfert–type nuclei and of distant quasars, and by the settling of the conceptions about the physics of the primary energy source and the radiative processes in these objects.

158.210 **EXOSAT observations of three bright BL Lac objects.**
R. S. Warwick, I. M. McHardy, K. A. Pounds.
X–ray astronomy '84, p. 467 – 470 (1985). – See Abstr. 012.033.
The BL Lac objects Mkn 421, Mkn 501 and 2A 1219+305 were observed by EXOSAT during February and March 1984. All three sources exhibited soft power law X–ray spectra characteristic of BL Lac objects. The low energy absorption of each source was found to be consistent with estimates of the line of sight column density through our Galaxy. Variability on a timescale $\lesssim$days was apparent in Mkn 421 and 2A 1219+305 and on a somewhat longer timescale for Mkn 501.

158.211 **Internal reddening and the ratio of X–ray to optical luminosity for Seyfert 1 nuclei.**
F. Z. Cheng, L. Danese, G. De Zotti, F. Lucchin.
X–ray astronomy '84, p. 471 – 474 (1985). – See Abstr. 012.033.
The authors point out that there are several independent indications of significant obscuration of Seyfert 1 nuclei by a flattened dusty region in the equatoral plane of the surrounding galaxy, while high luminosity quasars show little, if any, evidence of dust absorption.

158.212 **IUE–EXOSAT observations of NGC 4151.**
G. C. Perola, A. Altamore, A. Boksenberg, G. E. Bromage, J. Clavel, A. Elvius, M. V. Penston, M. Pettini, L. Piro, M. A. J. Snijders, M. Tarenghi, M. H. Ulrich.
X–ray astronomy '84, p. 475 – 478 (1985). – See Abstr. 012.033.
Preliminary results obtained with IUE and the ME and LE EXOSAT instruments indicate the existence of a long term correlation between the UV and the X–ray continuum intensities, and confirm the presence of the soft X–ray excess discovered by Holt et al. (1980).

158.213 **The soft X–ray emission from ON 235 and MK 766.**
D. Maccagni, B. Garilli, A. Rampini, L. Chiappetti, P. Giommi.
X–ray astronomy '84, p. 479 – 482 (1985). – See Abstr. 012.033.
In the framework of a program of extensive monitoring of BL Lac objects in soft X rays, the authors obtained a 460 min exposure of ON 235 with the EXOSAT Observatory. The same CMA field contains also the X–ray emitting Seyfert 1 galaxy MK 766.

158.214 **Centrifugally driven MHD–winds and the non–thermal emission regions in AGNs.**
M. Camenzind, T. J.–L. Courvoisier.
X–ray astronomy '84, p. 487 – 490 (1985). – See Abstr. 012.033.

158.215 **Observations of three BL Lacertae objects: constraints on inhomogeneous relativistically beamed jet models.**
D. M. Worrall, J. M. Rodríguez–Espinosa, W. Z. Wiśniewski, H. R. Miller, F. C. Bruhweiler, M. F. Aller, H. D. Aller.
Astrophys. J., Vol. 303, No. 2, p. 589 – 595 (1986).
The authors present contemporaneous multifrequency observations of three BL Lac objects, including the first ultraviolet measurements of 0912+297 and 1219+285. An 8 day optical flare was measured from 0422+004 in which the visual flux density doubled, and the other sources exhibited smaller daily fluctuations. The 5 – 15 GHz radio flux from these three objects remained roughly constant for 2 months. The authors discuss the constraints provided by these and other BL Lac data for inhomogeneous relativistically beamed jet models. It is found that a tapered jet model provides a satisfactory fit to the data for each source.

158.216 **Rapid X–ray and optical variability in the X–ray–selected BL Lacertae object 1E 1402.3+0416.**
P. Giommi, P. Barr, I. M. Gioia, T. Maccacaro, R. Schild, B. Garilli, D. Maccagni.
Astrophys. J., Vol. 303, No. 2, p. 596 – 600 (1986).
Results from X–ray and optical observations of the X–ray–discovered BL Lac object 1E 1402.3+0416 are reported. The X–ray measurements, carried out with EXOSAT, revealed a factor of 2 decrease in intensity on a time scale of a few hours. At maximum flux, the source was several times brighter than at the time of the Einstein discovery observation. The 2 – 6 keV X–ray spectrum has been determined. A comparison with Einstein data shows that the source has varied by a factor of $\sim$5 over a time scale of years. If the redshift of this object is greater than $\sim$0.2, super–Eddington luminosity or anisotropic emission mechanisms must be invoked.

158.217 **The stellar population of NGC 6166.**
F. Bertola, M. D. Gregg, J. E. Gunn, A. Oemler Jr.
Astrophys. J., Vol. 303, No. 2, p. 624 – 628 (1986).
The authors have obtained new optical and ultraviolet observations of NGC 6166, a cD galaxy and strong radio source in which both a large cooling flow and ongoing star formation have been detected. The spectrophotometry has been analyzed by a technique of differential spectral synthesis. Both the optical and ultraviolet spectra are best fitted by a combination of metal–rich elliptical plus $\sim$4% young population, the latter enriched in early– to mid–O stars by a factor of 20 over that observed in the solar neighborhood. Such a young population corresponds to a

star–formation rate of 0.1 $h^{-2}M_{\odot}yr^{-1}$ within 3.2 h^{-1}kpc of the center.

158.218 **The strong UV source in the active E galaxy NGC 4552.**
R. W. O'Connell, T. X. Thuan, J. J. Puschell.
Astrophys. J., Lett. Ed., Vol. 303, No. 2, p. L37 – L40 (1986).
The authors have obtained 1200 – 3200 Å *IUE* spectra of the nucleus of NGC 4552 (M89) in order to investigate the nature of the strong 10 µm source in this galaxy. There is a strong, extended UV source in NGC 4552 which has a spatial distribution nearly identical with that at optical wavelengths and is undoubtedly stellar in origin. There is no evidence for a nonthermal point source in the UV. It appears unlikely that the 10 µm emission is from heated dust grains. Instead, the 10 µm radiation is probably nonthermal in origin.

158.219 **Evidence for unusual dynamical effects in the halo of NGC 5128 (Centaurus A).**
J. E. Hesser, H. C. Harris, G. L. H. Harris.
Astrophys. J., Lett. Ed., Vol. 303, No. 2, p. L51 – L55 (1986). With plate L6. = Publ. Univ. Waterloo Obs., No. 125.
New observations bring to 35 the number of spectroscopically confirmed globular clusters in NGC 5128. At radii less than ~11′ the ensemble–average cluster velocity appears to exceed by ~4σ the systemic velocity of the galaxy, while at larger radii the two values agree. It is shown that this peculiar result is not likely to have arisen from a statistical fluctuation. It is suggested that whatever phenomenon produced the warped dust lane and/or the faint optical "shells" around NGC 5128 may also be related to the unexpected cluster velocities.

158.220 **Detection of the hydrocarbon ring molecule C_3H_2 in the radio galaxy Centaurus A (= NGC 5128).**
E. R. Seaquist, M. B. Bell.
Astrophys. J., Lett. Ed., Vol. 303, No. 2, p. L67 – L70 (1986).
The authors report on a search, mainly at 18.3 GHz, for the hydrocarbon ring molecule cyclopropenylidene (C_3H_2) in a number of active galaxies. A strong absorption line at the 1_{10}–1_{01} transition at 18.343 GHz was found against the nuclear continuum of the nearby radio galaxy NGC 5128 (= Centaurus A). An analysis of this feature and comparison with similar measurements of neutral hydrogen and formaldehyde provide evidence that the C_3H_2 feature is produced in the disk responsible for the band of obscuration seen in NGC 5128.

158.221 **Broad band X–ray spectra and time variability of selected active galactic nuclei observed with EXOSAT.**
K. A. Pounds.
Japan – U.S. seminar on galactic and extragalactic compact X–ray sources, p. 261 – 281 (1985). – See Abstr. 012.034.
Results are reported from observations of selected AGN made during the first year of the EXOSAT programme. Among several AGN of particular interest, regular monitoring of the Seyfert galaxy NGC 4151 has established a physically separate soft X–ray spectral component, for the first time in an AGN. The low luminosity Seyfert, NGC 4051, was found to be remarkably variable during an extended 8–hour EXOSAT observation, with an apparent quasi–period of ~4000 seconds. Early results from the multi–wavelength monitoring of several bright BL Lac objects and OVV – QSOs are also reviewed.

158.222 **Fluorescent iron lines from active galactic nuclei.**
H. Inoue.
Japan – U.S. seminar on galactic and extragalactic compact X–ray sources, p. 283 – 289 (1985). – See Abstr. 012.034.
Properties of iron emission lines from two active galaxies: NGC 4151 and Cen A are presented. Although the iron emission lines are probably produced through the fluorescence of continuum X–rays by ambient cool matter, the line from NGC 4151 is too intense to be interpreted by a simple model that isotropic continuum X–rays illuminate a uniformly surrounding matter with solar abundance of iron. The reason why NGC 4151 is peculiar in the intense iron line emission is discussed.

158.223 **Workshop on active galactic nuclei.**
R. Mushotzky.
Japan – U.S. seminar on galactic and extragalactic compact X–ray sources, p. 311 – 313 (1985). – See Abstr. 012.034.
In this brief introduction to the workshop on active galactic nuclei the author comments on the major theoretical and observational problems concerning AGN that can be attacked with the help of X–ray observations.

158.224 **Active galactic nuclei and electron–positron pairs.**
F. Takahara.
Japan – U.S. seminar on galactic and extragalactic compact X–ray sources, p. 315 – 319 (1985). – See Abstr. 012.034.
A view is given on the emission mechanisms and the structure of active galactic nuclei. The central problem, i.e., how the released gravitational energy is converted to the emergent radiation and mass outflow remains to be solved. As one of the most important processes to attack the problem, the author investigates the electron–positron pair production in an accretion plasma.

158.225 **Some comments on iron emission line from NGC 4151: beam model.**
S. Tsuruta.
Japan – U.S. seminar on galactic and extragalactic compact X–ray sources, p. 321 – 325 (1985). – See Abstr. 012.034.
It has been suggested that unobserved strong X–ray beams can be naturally produced in a certain model involving a rotating black hole and that these beams can produce the unusually strong iron emission line detected from NGC 4151 by Tenma and other satellites.

158.226 **The distribution and origin of cosmic rays in spiral galaxies: a test study of the galaxy NGC 3310.**
N. Duric.
Diss. Abstr. Int., Sect. B, Vol. 46, No. 5, p. 1589 (1985). Thesis, University of Toronto (1985).

158.227 **An analysis of the broad emission line profiles of Seyfert 1 galaxies.**
D. M. Crenshaw.
Diss. Abstr. Int., Sect. B, Vol. 46, No. 6, p. 1956 – 1957 (1985). Thesis, The Ohio State University, 129 pp. (1985). Order No. DA8518929.

158.228 **A surface photometric study of NGC 5128.**
R. L. Pennington.
Diss. Abstr. Int., Sect. B, Vol. 46, No. 6, p. 1959 (1985). Thesis, Rice University, 130 pp. (1985). Order No. DA8517230.

158.229 **The influence of galaxy interactions on nuclear activity.**
R. M. Cutri.
Diss. Abstr. Int., Sect. B, Vol. 46, No. 9, p. 3090 (1986). Thesis, The University of Arizona, 187 pp. (1985). Order No. DA8525594.

158.230 **Infrared studies of Seyfert galaxies and of the irregular galaxy M82.**
J. M. Rodriguez Espinosa.
Diss. Abstr. Int., Sect. B, Vol. 46, No. 9, p. 3092 (1986). Thesis, University of California, San Diego, 213 pp. (1985). Order No. DA8524578.

158.231 **The impact of the local environment on the properties of radio galaxies: a study of galaxies identified from the Texas Survey.**
E. P. Bozyan.
Diss. Abstr. Int., Sect. B, Vol. 46, No. 10, p. 3480 (1986). Thesis, The University of Texas at Austin, 514 pp. (1985). Order No. DA8527527.

158.232 **The distribution and kinematics of neutral hydrogen gas and the radio continuum emission in barred spiral galaxies.**
M. P. Ondrechen.
Diss. Abstr. Int., Sect. B, Vol. 46, No. 10, p. 3482 (1986). Thesis, University of Minnesota, 222 pp. (1985). Order No. DA8526496.

158.233 **Collimation of radio galaxy jets.**
A. H. Krautter.
Diss. Abstr. Int., Sect. B, Vol. 46, No. 12, p. 4282 (1986). Thesis, Queen's University at Kingston, Canada (1986).

158.234 **Determination of the parameters of the spatial distribution of extragalactic radio sources observed in the decameter wavelength range. Direct count of radio sources at 25 MHz.**
K. P. Sokolov.
Astron. Zh., Tom 63, Vyp. 3, p. 426 – 433 (1986). In Russian. English translation in Sov. Astron., Vol. 30, No. 3.
Samples of radio sources from a decameter–wave band survey are used to determine the parameters of the differential number–flux–density relation $n(S)dS = kS^{-\gamma}dS$ at the frequency 25 MHz by the direct count method. For flux densities $80 \gtrsim S \gtrsim 20$ Jy, the parameter γ is $\gamma_1 = 2.84 \pm 0.08$. This value is in good agreement with similar estimates obtained for higher frequencies. The number of sources with flux densities $S > 80$ Jy observed at 25 MHz decreases more rapidly ($\gamma_2 > 3.7$) than at higher frequencies which is evidence for a strong evolution of the extended radio sources with steep spectra. It is shown that for sources with fluxes below the direct count limit ($S < 20$ Jy) the parameters of the $n(S)$ relation should differ from those presented.

158.235 **Model of a subsonic turbulent jet for FR1–type radio galaxies.**
S. S. Komissarov.
Fiz. inst. Akad. Nauk SSSR, Prepr., No. 263, 22 pp. (1985). In Russian. Abstr. in Ref. Zh., 51. Astron., 6.51.838 (1986).

158.236 **The relationship of the Fanaroff–Riley classification of extragalactic radio sources to jet physics.**
G. V. Bicknell.
Proc. Astron. Soc. Aust., Vol. 6, No. 2, p. 130 – 137 (1985). Paper presented at the Alec Little Memorial Symposium.
The Class I/Class II division of extragalactic radio sources by Fanaroff–Riley is a manifestation of important physical differences existing in radio sources. It is proposed that the division essentially arises from the differing Mach numbers in Class I and Class II jets. An analysis of the energy balance in both types of source reveals jet velocities of the order of 5 – 10,000 km s^{-1} for Class I jets and mildly relativistic velocities for Class II jets. The important role of optical and X–ray observations in determining the gravitational field of pressure distribution in radio galaxies is discussed with examples given of NGC 1399 and IC 4296.

158.237 **A test for models of radio source evolution.**
S. E. Okoye, L. I. Onuora.
Astrophys. Space Sci., Vol. 122, No. 2, p. 267 – 279 (1986).
A relationship between observable parameters for non–evolving radio sources has been worked out and used as a standard for investigating models of radio source evolution of a given radio source sample. Using the 3CR sample, it was found that for radio galaxies both luminosity–linear size evolution and linear size–redshift evolution are probably occurring. On the other hand for quasars the best results were obtained when it was assumed that there was no cosmological evolution of linear size. Furthermore, if a correlation between mean luminosity and redshift was included for quasars, the luminosity was found to be independent of linear size.

158.238 **Multifrequency observations of blazars. II. The variability of the 1 μm to 2 mm continuum.**
W. K. Gear, L. M. J. Brown, E. I. Robson, P. A. R. Ade, M. J. Griffin, M. G. Smith, I. G. Nolt, J. V. Radostitz, G. Veeder, L. Lebofsky.
Astrophys. J., Vol. 304, No. 1, p. 295 – 304 (1986).
The authors present the results of monitoring, over a period of two years, a sample of 12 blazars, in 11 wave bands between 1 μm and 2 mm. All sources exhibit some variability, both in flux density and spectral shape. The infrared variability is consistent with repeated injections or reaccelerations of electrons, which subsequently suffer radiative losses. Variations in some objects are inconsistent with most models for compact extragalactic sources, but are naturally explained by shock waves travelling in an adiabatically expanding relativistic jet (Marscher and Gear).

158.239 **PKS 1345 + 125: a Seyfert nucleus merging with a powerful radio galaxy.**
G. Gilmore, M. A. Shaw.
Nature, Vol. 321, No. 6072, p. 750 – 753 (1986).
The bright compact radio source PKS 1345+125 (4C 12.50) has been optically identified with a magnitude 17 galaxy with a redshift $z = 0.122$ and an unusual spectrum. The authors show it to be an elliptical galaxy with a compact radio source coincident with its optically undistinguished nucleus. An unresolved radio–quiet Seyfert nucleus with the same redshift as the elliptical galaxy is offset by 1.8 arc s from the radio source. The Seyfert nucleus is remarkable for its extreme far–infrared flux. The authors present a model of 1345+125 involving the dynamical merger of an elliptical galaxy with a spiral.

158.240 **A compact radio source in the nucleus of M87.**
R. E. Spencer, W. Junor.
Nature, Vol. 321, No. 6072, p. 753 – 755 (1986).
The authors report high–resolution, very–long–baseline interferometric (VLBI) observations of the nucleus of M87 at 1.35 cm wavelength, which reveal the existence of a compact core ($\lesssim$0.2 marc s) and an associated extended feature (0.3 marc s), pointing in the same direction as the large–scale jet. These observations show that the radio jet is formed on a size scale of $\lesssim 10^{-2}$pc, thus strengthening the evidence for the presence of a massive black hole in the nucleus of M87.

158.241 **Deep radio maps of BL Lacertae and 3C 446.**
R. R. J. Antonucci.
Astrophys. J., Vol. 304, No. 2, p. 634 – 636 (1986).
Deep 20 cm VLA maps have been made of the blazars BL Lacertae and 3C 446. Dynamic ranges of 5000 – 6300 (peak/3σ) have been achieved. The diffuse emission associated with BL Lacertae has a flux of ~29 mJy and an extent of ~40″ = 48 kpc. The data are consistent with the hypothesis of BL Lacertae is a normal double radio galaxy seen from the jet direction. For 3C 446 any diffuse emission well sparated from the core must be weaker than 1.3 mJy per 4.4″ (25 kpc) beam.

158.242 **Variability of the BL Lacertae objects PKS 2155–304 and OJ 287 in the far–ultraviolet.**
L. Maraschi, G. Tagliaferri, E. G. Tanzi, A. Treves.
Astrophys. J., Vol. 304, No. 2, p. 637 – 645 (1986).
All the ultraviolet spectra of the two bright BL Lacertae objects PKS 2155–304 and OJ 287 taken with the IUE in the period 1978 – 1984 are examined. For each spectrum the best–fitting power law is determined and a correlation between spectral slope and intensity is searched for. The correlation, if present, is weak. This is discussed in terms of models of the continuum emission of active galactic nuclei.

158.243 **Simultaneous submillimeter and infrared observations of flat–spectrum radio sources.**
T. L. Roellig, E. E. Becklin, C. D. Impey, M. W. Werner.
Astrophys. J., Vol. 304, No. 2, p. 646 – 650 (1986).
The authors present observations of 12 extragalactic non-thermal radio emission sources made at eight wavelengths between 1.25 and 1000 μm, including the important submillimeter wavelength of 350 μm. Nine of the objects were detected at 350 μm. The results, combined with other studies, show that these objects generally have flat energy distributions at wavelengths longer than 1 mm that break in most cases between 350 μm and 1 mm. At wavelengths shorter than 1 mm, the energy distributions fall with a rough power–law behavior with a spectral index of ~ –1.

158.244 **The highly obscured nucleus of 3C 219.**
G. Fabbiano, S. P. Willner, N. P. Carleton, M. Elvis.
Astrophys. J., Lett. Ed., Vol. 304, No. 2, p. L37 – L40 (1986).
The authors report the detection of a strong, and possibly broad, Paschen–α line from the narrow–line radio galaxy 3C 219. The detected flux is larger than predicted from the Hα line and the case B recombination. This implies the presence of a highly reddened line–emitting region in the nucleus.

158.245 **The optical luminosity of type I Seyfert galaxy nuclei.**
Y.–y. Zhou, Z.–g. Deng.
Kexue Tongbao, Vol. 30, No. 4, p. 448 – 451 (1985).
Based on Sandage's composite model and the behaviour of type I Seyfert galaxies and normal galaxies on the colour–colour diagram, the authors have calculated the nuclear luminosity of type I Seyfert galaxies. The result is coincident with the observed one.

158.246 **Active galactic nuclei: a review of optical properties.**
D. R. Altschuler.
Rev. Colomb. Fis., Vol. 15, No. 3, p. 357 – 412 (1981). Abstr. in Phys. Abstr., Vol. 89, No. 1274, Entry 23905 (1986).

158.247 **Infrared polarimetry of the nucleus of Centaurus A: the nearest blazar?**
J. Bailey, W. B. Sparks, J. H. Hough, D. J. Axon.
Anglo–Aust. Obs., Prepr., No. 211, 16 pp. (1986). Submitted to Nature.

158.248 **Beobachtung von Centaurus A im MeV–Gammabereich.**
P. von Ballmoos.
MPE Rep., No. 193, 114 pp. (1985).

158.249 **Relation of the jet in M87 to nearby galaxies in the Virgo cluster.**
H. Arp.
MPA Rep., No. 224, p. 1 – 12 (1986).
The massive and active galaxies in the core of the Virgo cluster were shown in 1968 to be aligned on either side of M87, along the direction of the jet and counter jet. Detailed X–ray maps of M86 show that this galaxy, which is part of the alignment, is probably blown by a wind from M87 due to its close alignment with the jet. But the large radio E galaxy, M84, which is 1°.4 away from M87, and exactly along the line of the jet, is shown by the compression of its X–ray isophotes to be actually moving out along the line of the jet. This evidence furnishes rather spectacular support for the earlier conclusion that M84 had originated as a proto–galaxy within M87 and had been ejected out along the line of the jet.

158.250 **Superluminal radio sources.**
K. I. Kellermann.
Comments Astrophys., Vol. 11, No. 2, p. 69 – 81 (1985). = Natl. Radio Astron. Obs., Repr. Ser. B, No. 541.

158.251 **Recurrent outbursts in the broad line region of NGC 1566.**
D. Alloin, D. Pelat, M. M. Phillips, R. A. E. Fosbury, K. Freeman.
ESO Sci. Prepr., No. 423, 45 pp. (1986). To appear in Astrophys. J.

158.252 **Extended non–nuclear infrared emission from Seyfert galaxies.**
J. M. Rodriguez Espinosa, R. J. Rudy, B. Jones.
ESO Sci. Prepr., No. 435, 16 pp. (1986). To appear in Astrophys. J.

158.253 **3.28 μm feature and continuum emission in galaxy nuclei.**
A. F. M. Moorwood.
ESO Sci. Prepr., No. 444, 28 pp. (1986). To appear in Astron. Astrophys.

158.254 **Simultaneous multiband observations of rapid variability in NGC 4051.**
G.–z. Xie, M.–x. Bao, R.–w. Lu, P.–j. Hao, F.–z. Cheng, M. S. Elvis.
Publ. Yunnan Obs., No. 1, p. 88 – 93 (1986). In Chinese.
The main results are: (1) The soft X–ray flux had decreased by a factor of thirteen, and hard X–rays were not detected. (2) The U band results show that NGC 4051 has a very short characteristic time scale of approximately 1200 s for optical variability with an amplitude of $\Delta U \approx 0^m.11$.

158.255 **Polarization of radiation in extragalactic objects in the optical spectral range.**
V. A. Hagen–Thorn.
Photometric and polarimetric investigations of celestial bodies, p. 105 – 111 (1985). In Russian. – See Abstr. 003.035.
Recent results of polarimetric investigations of extragalactic objects in the optical spectral range are described.

158.256 **Surface brightness distribution in Seyfert galaxies. II. Multiaperture photometry.**
V. L. Afanas'ev, V. T. Doroshenko, V. Yu. Terebizh.
Astrofizika, Tom 24, Vyp. 3, p. 425 – 435 (1986). In Russian. English translation in Astrophysics, Vol. 24, No. 3.
Photoelectric surface brightness distributions are calculated using multiaperture photometry in B in 10 normal and Seyfert galaxies of the sample discussed in the first paper. A comparison of surface brightness profiles from photoelectric scans along the major axis with those of multiaperture photometry shows that the latter are accurate within about $0^m.2$/sq. arc sec for galaxies with low or moderate inclination. No statistical significant differences were found between the color indexes of the central regions of normal and Seyfert galaxies.

158.257 **A study of the profiles of bright emission lines in the nucleus of the Seyfert galaxy Markarian 533.**
V. A. Lipovetskij, A. I. Shapovalova, O. A. Dobrodij.
Astrofizika, Tom 24, Vyp. 3, p. 437 – 444 (1986). In Russian. English translation in Astrophysics, Vol. 24, No. 3.
The profiles of bright emission lines in the nucleus of the Seyfert galaxy Markarian 533 have been studied by the gauss–analysis method from the spectra obtained at BTA in 1977 and 1981. Blue components are found in the lines [O III], He II, Hβ, Hα, shifted by –800 km/s and –1500 km/s relative to the central components. It has been shown that the high–velocity gas is situated nearer to the ionization source than the emission region of the central components. Continuum variability of the Markarian 533 nucleus is noted and the variability of the blue components Hβ and Hα is suspected.

158.258 **Condition of the existence of a cooling flow in galaxies.**
E. V. Volkov.
Astrofizika, Tom 24, Vyp. 3, p. 477 – 489 (1986). In Russian. English translation in Astrophysics, Vol. 24, No. 3.
A criterion for the existence of subsonic spherical symmetrical flow of cooling gas in galaxies has been found. Some equations are given describing the behaviour of gas in the gravitational field of a galaxy in the framework of a stationary accretion model. The results of numerical calculations of a nonstationary accretion of gas on a cD galaxy are presented. The gas is initially in a hydrostatic equilibrium.

158.259 **A search for the CO line emission toward two clumpy irregular galaxies, Markarian 8 and 297.**
Y. Sofue, T. Hasegawa, N. Nakai, T. Handa, M. Hayashi, J. Heidmann.
Publ. Astron. Soc. Jpn., Vol. 38, No. 2, p. 161 – 165 (1986).
A deep search for the ^{12}CO line emission at 2.6 mm ($J = 1\text{–}0$) was made of two clumpy irregular galaxies using the 45–m telescope at Nobeyama. Upper limits to the column density of CO molecules of $1.3 \times 10^{17} cm^{-2}$ and $1.1 \times 10^{17} cm^{-2}$ were obtained toward the most active regions in Markarian 8 and 297, respectively. The ratio of CO mass to H I mass in these galaxies is about a factor of 5 smaller than that for our Galaxy. The quantities are compared with those for the star–bursting galaxy M82.

158.260 **Optical monitoring of quasars and BL Lac objects.**
A. Sillanpää, S. Haarala, T. Korhonen.
Rep. Ser., Dep. Phys. Sci., Univ. Turku, No. FTL–R89, 25 pp. (1986). = Turku Univ. Obs., Informo No. 98. Submitted to Astron. Astrophys.

158.261 **A model for the surface brightness of a turbulent low Mach number jet. III. Adiabatic jets of arbitrary density ratio: application to NGC 315.**
G. V. Bicknell.
Astrophys. J., Vol. 305, No. 1, p. 109 – 130 (1986).
The approach of using jet spreading rate data to infer the velocity variation of extragalactic jets is generalized to jets of arbitrary density ratio. The dissipative heating of the jet plasma by turbulence is neglected. In this adiabatic model the surface brightness variation of a Fanaroff–Riley class I jet is due to the compression and expansion of plasma and magnetic field as the jet decelerates and accelerates under the opposing influences of entrainment and buoyancy. The model is applied to the main jet in NGC 315, and it is shown that good fits to the surface brightness of this jet can be obtained from low Mach number, low density ratio models with the temperature of the background medium approximately equal to the virial temperature.

158.262 **Ultraviolet spectrophotometry of three LINERs.**
R. W. Goodrich, W. C. Keel.
Astrophys. J., Vol. 305, No. 1, p. 148 – 156 (1986). = Lick. Obs. Bull., No. 1030.
Three galaxies known to be LINERs were observed spectroscopically in the ultraviolet in an attempt to detect the presumed nonthermal continuum source thought to be the source of photoionization in the nuclei. NGC 4501 was too faint for study with the *IUE*, while NGC 5005 had an extended ultraviolet light profile, which is probably due to a component of hot stars in the nucleus. In NGC 4579 the authors find a point source contributing most of the ultraviolet excess, much different than the optical light distribution. The emission–line spectrum of NGC 4579 is consistent with a photoionization model based on this UV continuum source.

158.263 **Molecular gas in the type 1 Seyfert galaxy NGC 7469: implications for nuclear activity.**
T. M. Heckman, S. Beckwith, L. Blitz, M. Skrutskie, A. S. Wilson.
Astrophys. J., Vol. 305, No. 1, p. 157 – 166 (1986).
The authors report the results of a search for 2.12 μm H_2 1–0 $S(1)$ line emission from the nuclear ($\leqslant$ 1 kpc) region in 11 active galaxies. The galaxies observed included type 1 and type 2 Seyferts, LINERs, and starburst nucleus galaxies. H_2 emission was detected in one galaxy, the classical type 1 Seyfert NGC 7469. The authors also report the detection of very luminous ^{12}CO 2.6 mm emission from NGC 7469, and present optical narrow–band images and long–slit spectroscopy data concerning the ionized gas in this galaxy. These data imply that NGC 7469 not only contains a highly active compact nucleus but also is undergoing a luminous circumnuclear starburst. Most of the other galaxies with known H_2 emission also seem to be composite "monster"/starburst systems and/or interacting galaxies.

158.264 **Far–ultraviolet and optical spectrophotometry of X–ray selected Seyfert galaxies.**
J. T. Clarke, S. Bowyer, M. Grewing.
Astrophys. J., Vol. 305, No. 1, p. 167 – 174 (1986).
The authors have performed near–simultaneous far–ultraviolet and optical spectrophotometry of five X–ray selected Seyfert galaxies (previously unobserved by the *IUE*) to further test the models for excitation of the emission lines by X–ray and ultraviolet continuum photoionization. The X–ray to optical continuum spectra and most emission–line properties of the nuclei of these Seyfert I galaxies appear very similar to those of previous samples, but the observed ratio Lyα/Hβ in the present sample averages 22, which is significantly higher than that observed in previous samples.

158.265 **Line variations in quasars and Seyfert galaxies.**
C. M. Gaskell, L. S. Sparke.
Astrophys. J., Vol. 305, No. 1, p. 175 – 186 (1986).
The authors present quantitative estimates of the sizes of broad–line regions (BLRs) in three well–studied Seyfert I galaxies and discuss how these differ from previously reported size estimates for NGC 4151. It is shown that the high–ionization lines in NGC 4151, Akn 120 and Mrk 509 arise from regions about 5 to 10 light days from the central sources. These results imply that the densities in broad–line regions are a couple of orders of magnitude higher than in current standard models.

158.266 **The structure of the Cygnus A radio galaxy.**
M. J. Pierce, A. Stockton.
Astrophys. J., Vol. 305, No. 1, p. 204 – 213 (1986). With plates 1 – 2.
The authors present interference–filter CCD images of Cygnus A that show clearly the very different morphological structures seen in emission–line radiation on the one hand and continuum radiation on the other. They also discuss variations in the emission–line and continuum spectra with position in the galaxy, as derived from long–slit image–intensifier spectroscopy obtained under good seeing conditions. The authors suggest an alternative possibility to the Cen A analog for Cyg A: that the double structure present in the continuum image is largely the result of scattered radiation escaping from a bright nucleus along the poles of a dense, flattened dust cloud that hides the nucleus from direct view.

158.267 **X–ray spectroscopy of five BL Lacertae objects.**
C. M. Urry, R. F. Mushotzky, S. S. Holt.
Astrophys. J., Vol. 305, No. 1, p. 369 – 398 (1986).
Five X–ray bright BL Lacertae objects were observed in the energy range 0.6 – 4.5 keV with the solid state spectrometer of the *Einstein Observatory*. These data do not reveal the presence of large amounts of thermal gas, with upper limits for the equivalent width of individual spectral features typically $\lesssim$100 eV. The spectra indicate low–energy absorption from rather hot, highly ionized material, probably intrinsic to the sources. The X–ray continua suggest that two–component X–ray spectra, steep at low energies and flat at high energies, are a common feature of BL Lac objects. Dramatic intensity and spectral variability is evident from the data.

158.268 **Linear polarization observations in selected celestial zones. The Abell 2319 area.**
N. W. Broten, J. P. Vallée, J. M. MacLeod.
Astron. Astrophys., Suppl. Ser., Vol. 64, No. 3, p. 499 – 501 (1986).
The integrated linear polarization has been measured for several extragalactic radio sources in a restricted celestial angular zone encompassing the cluster of galaxies Abell 2319. These observations were made at wavelengths of 6.2, 18.2, and 20.1 cm using the Very Large Array. These polarization measurements complement and extend a continuing program to detect localized effects from specific zones in the sky.

158.269 **New redshifts of strong radio sources.**
C. R. Lawrence, T. J. Pearson, A. C. S. Readhead, S. C. Unwin.
Astron. J., Vol. 91, No. 3, p. 494 – 495 (1986).
New redshifts are reported for eight sources in the Pearson–Readhead 6 cm VLBI survey. Tentative redshifts are given for two objects with almost featureless spectra, two previously reported redshifts are corrected, and one tentative redshift confirmed.

158.270 **Erratum: "Photoelectric comparison sequences in the fields of five Seyfert galaxies" [Astron. J., Vol. 86, No. 1, p. 87 – 89 (1981)].**
H. R. Miller.
Astron. J., Vol. 91, No. 3, p. 665 (1986). See Abstr. 29.158.047.

158.271 **Observations of the low–luminosity broad–line radio galaxy 1717+49.**
J. J. Puschell, R. Moore, R. D. Cohen, F. N. Owen, A. C. Phillips.
Astron. J., Vol. 91, No. 4, p. 751 – 754 (1986).
The elliptical galaxy 1717+49 (= A1718+49A = Arp 102B = VV 10 = K508) contains an active nucleus characterized by a compact radio source, nonstellar visual–infrared emission, variable optical polarization, and strong broad emission lines. When the remarkable nature of this object became clear the authors decided to look at it in more detail in order to gain a better understanding of how the activity in this galaxy is related to other active galactic nuclei.

158.272 **Morphology of the nuclear region of M82 at 2.2 μ.**
R. D. Dietz, J. Smith, J. A. Hackwell, R. D. Gehrz, G. L. Grasdalen.
Astron. J., Vol. 91, No. 4, p. 758 – 760 (1986).
High–resolution 2.2 μ images of the central arcminute of M82 (NGC 3034) reveal a complex morphology. The authors report coordinates and surface brightnesses for six features, one of which is near the position of the brightest centimeter source, 41.9+58. The 2.2 μ center of M82 coincides with the radio kinematic center to within the errors of measurement.

158.273 **The distribution and kinematics of H I in the active elliptical galaxy NGC 1052.**
J. H. van Gorkom, G. R. Knapp, E. Raimond, S. M. Faber, J. S. Gallagher.
Astron. J., Vol. 91, No. 4, p. 791 – 807 (1986). With plates 72, 73.
The H I distribution in the active elliptical galaxy NGC 1052 has been mapped at a resolution of 1 arcmin with the VLA. The H I structure is about three times the size of the optical galaxy and is roughly perpendicular to its major axis. The H I distribution is unusually irregular. The presence of dust associated with the H I and the distribution and kinematics of the H I are consistent with capture of gas from a gas–rich dwarf or spiral. The redshifted absorption seen in this galaxy, and seen towards the nuclei of a few other early–type radio galaxies, suggests that the accretion of gas into the center might be a general phenomenon in radio galaxies.

158.274 **The third image, the redshift of the lens, and new components of the gravitational lens 2016+112.**
D. P. Schneider, J. E. Gunn, E. L. Turner, C. R. Lawrence, J. N. Hewitt, M. Schmidt, B. F. Burke.
Astron. J., Vol. 91, No. 5, p. 991 – 997 (1986). With plate 76.
A third lensed image has been found in the gravitational lens system 2016+112. It is located near 2016+112 C, the strongest radio source in the system, and was detected by observations with the Palomar 5 m telescope in a narrowband filter centered on the Lyman α emission of the two previously known images. Resolved, low–surface–brightness companions of the two brighter images (A and B) were also discovered in the narrowband image. Simple model–independent calculations indicate $M \geqslant 1.2x10^{12}h^{-1}M_\odot$ and $M/L_B \geqslant 21\ h(M_\odot/L_\odot)$ for the lensing object. Furthermore, this mass must be concentrated within a circle of radius 8.3 h^{-1}kpc, or the total lens mass must greatly exceed this lower bound.

158.275 **Weak nuclei of powerful radio sources – spectra and polarizations.**
L. Rudnick, T. W. Jones, R. Fiedler.
Astron. J., Vol. 91, No. 5, p. 1011 – 1018 (1986).
Observations are presented of the spectra and polarizations of the weak cores of extended extragalactic radio sources. The authors comment briefly on the relation between these weak cores and other classes of compact sources, and the implications for "unified" models based on relativistic beaming. They conclude that there is a distinct class of compact objects, different from relativistic jets, which appear in the form of separate weak or dominant cores, or as the base component in superluminal jets.

158.276 **NGC 6212: an X–ray–selected active elliptical galaxy.**
J. P. Halpern, A. V. Filippenko.
Astron. J., Vol. 91, No. 5, p. 1019 – 1025 (1986).
Optical observations of the elliptical galaxy NGC 6212 show it to be a type 1 Seyfert galaxy at $z = 0.030$, not a BL Lac object as suggested by Biermann, Strom, and Bartel (1985). The X–ray luminosity in the 0.3 – 3.5 keV band varies between $2x10^{42}$ and $8x10^{42}$erg s^{-1} over a period of months. One purpose of this paper is to emphasize that the X–ray variability and relatively weak radio emission of NGC 6212 are not unusual for Seyfert galaxies. In the absence of optical spectroscopy, the a priori probability that an object with the X–ray and radio properties of NGC 6212 will turn out to be a BL Lac object is much smaller than the probability that it will be a Seyfert galaxy.

158.277 **The extended narrow emission–line region of NGC 7469 revisited.**
M. M. De Robertis, R. W. Pogge.
Astron. J., Vol. 91, No. 5, p. 1026 – 1037 (1986). With plate 77. = Lick Obs. Bull., No. 1035.
The authors have undertaken a re–examination of the central region of NGC 7469 using high–resolution spectrophotometry and direct imaging with more sensitive modern detectors. In addition to re–examining the gas kinematics, they have also set out to obtain a more quantitative idea of the physical nature of this gas and to investigate its possible relationship to the gas of the "classic" narrowline region.

158.278 **VLBI observations of 416 extragalactic radio sources.**
D. D. Morabito, A. E. Niell, R. A. Preston, R. P. Linfield, A. E. Wehrle, J. Faulkner.
Astron. J., Vol. 91, No. 5, p. 1038 – 1050 (1986).
Very long baseline interferometry (VLBI) observations of 416 radio sources with declinations north of –45 deg have been conducted at frequencies of 2.3 and 8.4 GHz. At 2.3 GHz, 323 of 391 radio sources observed were detected with a fringe spacing of 3 milliarcseconds and a detection limit of ~0.1 Jy. At 8.4 GHz, 278 of 416 radio sources were detected with a fringe spacing of 1 milliarcsec and a detection limit of ~0.1 Jy. This survey was conducted primarily to determine the strength of compact components at 8.4 GHz for radio sources previously detected with VLBI at 2.3 GHz.

158.279 **15 GHz VLA observations of the radio galaxies 3C 166 and 3C 411.**
A. L. Fey, S. R. Spangler, S. T. Myers.
Astron. J., Vol. 91, No. 6, p. 1279 – 1285 (1986).
The authors present 15 GHz Very Large Array observations of the luminous double radio galaxies 3C 166 and 3C 411. The data confirm the spectral steepening in the lobes of both sources indicated by previous lower–frequency observations. The spectral steepening is interpreted as being due to synchrotron–radiation losses (synchrotron aging) and is used to infer a radiative age for electrons in both lobes of 3C 411 and the southern lobe of 3C 166. The results are presented as a relative speed of separation between the hot spot and a parcel of lobe material. The observations also suggest that for both sources geometric depolarization is responsible for the reduction in fractional linear polarization at frequencies of 5 GHz and above.

158.280 **The distribution of globular clusters associated with M87 = NGC 4486.**
C. Grillmair, C. Pritchet, S. van den Bergh.
Astron. J., Vol. 91, No. 6, p. 1328 – 1335 (1986). With plate 95.
Using five hours of CCD observations of the core region of M87, the authors have examined the distribution of globular clusters in M87 down to $B = 24.4$ as a function of both magnitude and projected radius. They find that there is no strong evidence for a systematic radial trend in the luminosity function. The data also indicate that the run of globular cluster surface density with radius is considerably shallower than the galaxy luminosity profile in the same region. Simple model calculations indicate that the core region of the cluster distribution is probably observed, rather than the effects of such processes as dynamical friction and tidal disruption.

158.281 **Polarimetry of the peculiar elliptical galaxy, NGC 5128.**
D. S. Berry.
Thesis, Univ. Durham, England (1985). Abstr. in Phys. Abstr., Vol. 89, No. 1279, Entry 53964 (1986).

158.282 **Spectral energy distributions of active galactic nuclei between 0.1 and 100 μm.**
R. A. Edelson, M. A. Malkan.
Bull. Am. Astron. Soc., Vol. 18, No. 1, p. 633 (1986). Abstract. – See Abstr. 010.066.

158.283 **Far infrared spectroscopy of M82: electron densities and N^{++}/O^{++} ratios.**
P. B. Duffy, E. F. Erickson, M. R. Haas, J. R. Houck.
Bull. Am. Astron. Soc., Vol. 18, No. 1, p. 639 (1986). Abstract. – See Abstr. 010.066.

158.284 **Extended ionized gas and radio emission in active galactic nuclei.**
O. Dahari, H. C. Ford, R. Ciardullo, G. Jacoby, P. C. Crane.
Bull. Am. Astron. Soc., Vol. 18, No. 1, p. 640 (1986). Abstract. – See Abstr. 010.066.

158.285 **A comparison of the environments of Seyfert galaxies and non–Seyfert Markarian galaxies.**
J. W. MacKenty.
Bull. Am. Astron. Soc., Vol. 18, No. 1, p. 640 (1986). Abstract. – See Abstr. 010.066.

158.286 **The nuclear velocity fields of NGC 1068 and M51: spatially resolved [N II] emission associated with radio jets and bubbles.**
G. Cecil, R. B. Tully, J. Bland.
Bull. Am. Astron. Soc., Vol. 18, No. 1, p. 640 – 641 (1986). Abstract. – See Abstr. 010.066.

158.287 **High resolution spectroscopy and photometry of galactic nuclei associated with radio jets.**
A. Ferrari, L. Zaninetti.
Mem. Soc. Astron. Ital., Vol. 56, No. 4, p. 725 – 736 (1985). – See Abstr. 012.054.

The authors review briefly the problems of radio jets in connection with their parent nuclei and indicate some of the new insights which may be gained when the Space Telescope become operational.

158.288 **Emission lines in active galactic nuclei.**
G. C. Perola.
Mem. Soc. Astron. Ital., Vol. 56, No. 4, p. 737 (1985). Abstract. – See Abstr. 012.054.

158.289 **Multifrequency observations of BL Lac objects.**
L. Maraschi, E. G. Tanzi, A. Treves.
Mem. Soc. Astron. Ital., Vol. 56, No. 4, p. 739 – 746 (1985). – See Abstr. 012.054.

Preliminary results of a program to monitor the energy distributions of BL Lac objects at X–ray, UV, optical and infrared wavelengths are reported.

158.290 **Seyfert galaxies: a project of CCD photometry.**
C. Bonoli, F. Bortoletto, F. Bonoli, F. Delpino, V. Zitelli.
Mem. Soc. Astron. Ital., Vol. 56, No. 4, p. 785 – 788 (1985). – See Abstr. 010.821.

This paper presents some results from a project of CCD photometry of Seyfert galaxies. This surface photometry will provide accurate values of axial ratios, of nuclear and galactic luminosities and will show whether the underlying galaxies actually behave as "normal" galaxies.

158.291 **Ionization state of the absorption systems in the BL Lac object 0215+015 and properties of low excitation absorbers.**
J. Bergeron, S. D'Odorico.
Inst. Astrophys. Paris, Pré–Publ., No. 126, 22 pp. (1986). To appear in Mon. Not. R. Astron. Soc.

158.292 **On the star–burst model for Markarian 8.**
Q.–f. Yin.
Kexue Tongbao, Vol. 30, No. 10, p. 1365 – 1367 (1985).

It is indicated that the VLA observations of radio emissions from the envelope and the clumps support the star–burst model for Markarian 8.

158.293 **On the origin of extragalactic infrared radiation.**
H. W. Yorke, W. Kollatschny.
Rutherford Appleton Lab., Rep., RAL 85–086, p. 1 – 12 (1985). – See Abstr. 012.060.

Many aspects of activity in galactic nuclei and their infrared manifestations can be understood in terms of star formation and the corresponding stellar evolution. In particular, starbust nuclei, the Seyfert 2 and Liner phenomena could represent different evolutionary phases during and after a very strong outburst of massive star formation. Many stellar evolution details are at present uncertain, however. Models involving stellar evolution alone cannot be applied to the Seyfert 1–, QSO–, BL Lac–phenomena where clearly a very strong non–thermal central component is present.

158.294 **Infrared emission from active nuclei.**
M. Ward.
Rutherford Appleton Lab., Rep., RAL 85–086, p. 114 – 129 (1985). – See Abstr. 012.060.

This short review summarises some recent ground–based infrared observations from 1 – 20 microns of two major classes of active nuclei: Liner's and Seyfert type 1 galaxies. In addition the author presents some IRAS data on an X–ray selected sample of Seyfert 1 galaxies. He discusses the origin of their near IR (1 – 3.5 micron) and mid–IR (10 – 20 micron) emission, and the characteristic shape of their 0.3 – 100 micron continua.

158.295 **Ionization state of the absorption systems in the BL Lac object 0215+015 and properties of low–excitation absorbers.**
J. Bergeron, S. D'Odorico.
Mon. Not. R. Astron. Soc., Vol. 220, No. 4, p. 833 – 843 (1986).

Spectra of the BL Lac object 0215+015 have been obtained with the ESO 3.6–m telescope at a resolution of 18 and 29 km s^{-1} from 5650 to 6850 Å. Absorption lines of Mg I, Mg II, Mn II and Fe II are detected in the system at $z = 1.345$ known from previous observations. These lines break up into four components spread over 180 km s^{-1} with observed equivalent widths of isolated features ranging from 35 to 2800 mÅ. Accurate estimates of the velocity structure and column densities of the singly ionized and neutral elements mentioned above are derived for the multiple system at $z = 1.345$ using a line–profile fitting procedure.

158.296 **Interpretation of the variability of the X–ray emission of nuclei of type 1 Seyfert galaxies and quasars.**
A. S. Zentsova.
Sov. Astron., Vol. 29, No. 6, p. 714 – 715 (1985). English translation of 40.158.242.

158.297 **Rapid flux variations in compact radio sources.**
S.–j. Qian, R.–y. Zhao, N. Yu.
Chin. Astron. Astrophys., Vol. 10, No. 1, p. 38 – 44 (1986). English translation of 40.158.241.

158.298 **An objective prism survey for emission–line galaxies.**
J.–y. Liu, Y.–w. Huang, X.–c. Feng.
Acta Astrophys. Sin., Vol. 6, No. 2, p. 111 – 118 (1986). In Chinese.

The paper contains the first list of emission–line objects detected in an objective–prism survey. A 60/90 cm Schmidt telescope with a 5°.3 objective prism and III_aJ emulsion have been used. On the plates, covering about 220 square degrees of sky, 50 emission-line objects are detected. 47 of them are galaxies and 3 are planetary nebulae. Finding charts are given.

158.299 **Large beam radio observations toward M87.**
G.-z. Xie, C. Ceccarelli, F. Melchiorri, L. Pietranera, N. Mandolesi, G. Morigi.
Acta Astrophys. Sin., Vol. 6, No. 2, p. 119 – 123 (1986). In Chinese.
The authors show how combined large-beam infrared and radio observations towards M87 may be explained by assuming the existence of a large (10^5K) dark matter halo surrounding the giant galaxy M87. The total mass of the more extended halo is $10^{13}M_{\odot}$. Such a halo would be able to close the cluster gravitationally.

158.300 **A model for rapid flux variations in extragalactic radio sources.**
S.-j. Qian, R.-y. Zhao.
Acta Astron. Sin., Vol. 27, No. 2, p. 158 – 163 (1986). In Chinese.
The authors discuss some emission properties of a relativistically expanding plasmon, which is ejected obliquely with relativistic speed. Numerical results are presented. It is pointed out that this type of dual-relativistic effect provides more possibilities to account for the apparent superluminal flux variations in extragalactic radio sources. Moreover, with suitable selection of the parameters, the apparent superluminal separation and flux variations may occur simultaneously even when the viewing angles are considerably large.

158.301 **High dynamic range VLBI observations of NGC 6251.**
D. L. Jones, S. C. Unwin, A. C. S. Readhead, W. L. W. Sargent, G. A. Seielstad, R. S. Simon, R. C. Walker, J. M. Benson, R. A. Perley, A. H. Bridle, I. I. K. Pauliny-Toth, J. Romney, A. Witzel, P. N. Wilkinson, L. B. Bååth, R. S. Booth, D. N. Fort, J. A. Galt, R. L. Mutel, R. P. Linfield.
Astrophys. J., Vol. 305, No. 2, p. 684 – 697 (1986).
The large radio galaxy NGC 6251 has been observed with an 11 telescope VLBI array, including the full VLA, at 18 cm. The authors present high dynamic range maps of both the VLBI- and VLA-scale radio structure from these observations, as well as hybrid maps from earlier VLBI experiments at 6 and 13 cm. The 18 cm VLBI hybrid map has an unusually high dynamic range of ~300 to 1. A lower limit of 80 to 1 for the ratio of jet/counterjet brightness at parsec scales can be set. The authors discuss the physical conditions in the inner few parsecs of the NGC 6251 jet and their relation to the conditions in the extraordinarily long and well collimated kiloparsec-scale jets.

158.302 **An optically-thin, strong magnetic field model for the energy distribution and time-variability of blazars.**
J. R. D. Lépine, M. A. Braż, N. Epchtein.
Rev. Mex. Astron. Astrofis., Vol. 12, Spec. Issue, p. 53 – 58 (1986). – See Abstr. 012.066.
The energy distributions and time-scales of variability of "blazars" are interpreted in terms of an optically thin synchrotron radiation model, in which the electrons are injected into the source in the form of high-energy bursts, and flow down along the energy spectrum due to radiation loss. The model requires magnetic fields of the order of 10 – 100 gauss.

158.303 **Photopolarimetric observations of the BL Lac object PKS 2155-304.**
H. G. Luna.
Rev. Mex. Astron. Astrofis., Vol. 12, Spec. Issue, p. 73 (1986). Abstract. – See Abstr. 012.066.

158.304 **Spectrophotometry of the nucleus of NGC 7552.**
H. Dottori, M. Pastoriza, S. J. C. Landaberry, J. A. de Freitas Pacheco.
Rev. Mex. Astron. Astrofis., Vol. 12, Spec. Issue, p. 119 – 123 (1986). – See Abstr. 012.066.
The authors present new spectrophotometric measurement of the Sérsic-Pastoriza nucleus of NGC 7552, giving absolute fluxes of the continuum between 3650 Å and 6900 Å and intensities of the lines H_2, Hβ, [O III] 4959 + 5007 Å and Hα + N II. The synthesis of the absorption spectrum indicates 3 or 4 bursts of star formation.

158.305 **Narrow emission line region of active galactic nuclei.**
S. M. V. Aldrovandi.
Rev. Mex. Astron. Astrofis., Vol. 12, Spec. Issue, p. 124 – 130 (1986). – See Abstr. 012.066.
An overview of theoretical models for the narrow emission line region of the active galactic nuclei is presented and their possibility to account for the observations is discussed.

158.306 **M82 (NGC 3034): the nuclear region.**
L. Carrasco, E. Recillas-Cruz, I. Cruz-González, J. Melnick.
Rev. Mex. Astron. Astrofis., Vol. 12, Spec. Issue, p. 135 – 136 (1986). Abstract. – See Abstr. 012.066.

158.307 **On the effect of relativistic electrons in the narrow line regions of active galactic nuclei.**
S. M. V. Aldrovandi, M. L. César, R. B. Gruenwald.
Rev. Mex. Astron. Astrofis., Vol. 12, Spec. Issue, p. 139 – 142 (1986). – See Abstr. 012.066.
The effect of relativistic electrons on a photoionized emission-line region is investigated, considering optically thick models. The main effect of these electrons is to heat the gas while their influence on the ionization structure is not very important. The calculated emission-line fluxes are compared to the available observational data.

158.308 **An analysis of the infrared and optical continua of Seyfert 2 galaxies.**
R. O. Lagua, S. M. V. Aldrovandi, P. Benevides-Soares.
Rev. Mex. Astron. Astrofis., Vol. 12, Spec. Issue, p. 153 – 156 (1986). – See Abstr. 012.066.
The infrared-optical continua of nine Seyfert 2 nuclei are analysed. The observed spectra are fitted by three components: a power law, a low temperature (T ≈ 200K) and an intermediate temperature (T ≈ 3800K) black-body.

158.309 **High resolution spectroscopy of southern Seyfert galaxies.**
I. C. Busko, J. E. Steiner.
Rev. Mex. Astron. Astrofis., Vol. 12, Spec. Issue, p. 157 – 158 (1986). Abstract. – See Abstr. 012.066.

158.310 **Maps of Cen A at 22 GHz and 44 GHz.**
C. Tateyama, Z. Abraham.
Rev. Mex. Astron. Astrofis., Vol. 12, Spec. Issue, p. 294 (1986). Abstract. – See Abstr. 012.066.

158.311 **Seyfert galaxies in the IRAS survey and JHKL photometry.**
I. S. Glass.
Mon. Notes Astron. Soc. S. Afr., Vol. 44, Nos. 7-8, p. 60 – 65 (1985).
The near- to far-infrared properties of an optically selected sample of ordinary spiral galaxies, H2 (starburst) galaxies and Seyferts types 1 and 2 are summarized. The spectral indices derived from photometry at 1 and 100 μm are shown to be almost independent of galaxy type. New JHKL photometry for a sample of Seyferts found from IRAS data by de Grijp et al. (1985) is given.

158.312 **Detection of an intense iron line at 6.4 keV in the X-ray spectrum of NGC 4151.**
M. Matsuoka, T. Ikegami, H. Inoue, K. Koyama.
Publ. Astron. Soc. Jpn., Vol. 38, No. 3, p. 285 – 294 (1986).
The authors present the X-ray spectrum of NGC 4151 observed with the gas scintillation proportional counters (GSPC) on board Tenma. The energy resolution of the GSPC is twice that of the proportional counters used in the past observations, which enables one to obtain detailed properties of the iron emission line from NGC 4151. The GSPC detected a clear emission line at 6.4 keV from NGC 4151 in spite of the low X-ray intensity level

of the source. The authors examine the intense iron emission line. To this end they evaluate the iron column density from the depth of the iron K–edge absorption and compare this to the hydrogen column density derived from the low energy cutoff. The authors then discuss the abundance of iron and the geometrical configuration of the matter responsible for the iron line in relation to the derived column densities.

158.313 **Erratum: "A third update of the status of the 3CR sources: further new redshifts and new identifications of distant galaxies" [Publ. Astron. Soc. Pac., Vol. 97, No. 596, p. 932 – 961 (1985)].**
H. Spinrad, S. Djorgovski, J. Marr, L. Aguilar.
Publ. Astron. Soc. Pac., Vol. 98, No. 604, p. 628 (1986). See Abstr. 40.158.246.

158.314 **Relation of the jet in M87 to nearby galaxies in the Virgo Cluster.**
H. Arp.
J. Astrophys. Astron., Vol. 7, No. 2, p. 71 – 75 (1986).
The massive and active galaxies in the core of the Virgo Cluster were shown in 1968 to be aligned on either side of M87, along the direction of the jet and counter–jet. Detailed X–ray maps of M86 show that this galaxy, which is part of the alignment, is probably blown by a wind from M87 due to its close alignment with the jet. But the large radio E galaxy, M84, which is 1.4 degrees away from M87, and exactly along the line of the jet, is shown by the compression of its X–ray isophotes to be actually moving out along the line of the jet. This evidence furnishes rather spectacular support for the earlier conclusion that M84 had originated as a proto–galaxy within M87 and had been ejected out along the line of the jet.

158.315 **Compact radio structure of active galactic nuclei.**
R. W. Porcas.
Mitt. Astron. Ges., No. 65, p. 95 – 107 (1986). – See Abstr. 012.072.
This review contains a survey of recent radio observations relating to the compact radio structure of active galaxies and quasars. It is mainly based on published and unpublished material resulting from the technique of VLBI.

158.316 **The starburst galaxy Markarian 788.**
K. J. Fricke, W. Kollatschny, H. Netzer, W. K. Huchtmeier.
Mitt. Astron. Ges., No. 65, p. 230 (1986). – See Abstr. 012.072.

158.317 **Kurzzeit–Variabilität extragalaktischer Radioquellen.**
A. Witzel, D. S. Heeschen, C. Schalinski, T. Krichbaum.
Mitt. Astron. Ges., No. 65, p. 239 – 241 (1986). – See Abstr. 012.072.

158.318 **Ein vollständiger CCD Survey von 44 radio–hellen BL Lac Objekten.**
H. Kühr.
Mitt. Astron. Ges., No. 65, p. 248 – 250 (1986). – See Abstr. 012.072.

158.319 **Markarian 507.**
IAU Circ., No. 4202 (1986).

158.320 **OJ 287.**
IAU Circ., No. 4208 (1986).

158.321 **VLBI observations of the compact components in M82.**
N. Bartel, M. I. Ratner, A. E. E. Rogers, I. I. Shapiro, R. J. Bonometti, N. L. Cohen, M. V. Gorenstein, J. M. Marcaide.
Cent. Astrophys., Prepr. Ser., No. 2262, 5 pp. (1985). To appear in Highlights of Astronomy (1986).

158.322 **Star formation in NGC 5128.**
R. L. Pennington.
Luminous stars and associations in galaxies, p. 395 – 396 (1986). – See Abstr. 012.079. (IAU Symp. No. 116).
Star formation in the dust lane of NGC 5128 (Cen A) has been studied using digitized CTIO 4 m plates. The digital images were calibrated to the standard U, B and V passbands using photoelectric photometry (van den Bergh 1976). Ages for the dominant stellar component were derived by de–reddening each pixel along an R = 3.3 reddening law to its intercept with a theoretical cluster evolution track in the U–B, B–V plane (Davis 1979). The H II regions of Hodge and Kennicutt (1983) were used as tracers for the most recent star formation.

158.323 **Warmers: massive stars in the nuclei of galaxies.**
J. Melnick, R. Terlevich, M. Moles.
Luminous stars and associations in galaxies, p. 505 – 509 (1986). – See Abstr. 012.079. (IAU Symp. No. 116).
The authors have computed evolutionary models for the emission line spectra of constant density, spherically symmetric nebulae photoionized by coeval clusters of massive stars with solar–neighborhood IMF. These models naturally explain the different types of active nuclei as an evolutionary sequence determined by the total mass of the ionizing cluster (or starburst) and the mass–loss rates of the ionizing stars.

158.324 **Models for IRAS galaxies.**
M. Rowan–Robinson, J. Crawford.
Light on dark matter, p. 421 – 424 (1986). – See Abstr. 012.080.
The authors have modelled IRAS 12 – 100 μ observations of galaxies as a mixture of 3 components: a cool "disc" component, a warmer "starburst" component and a "Seyfert" component peaking at 25 μ.

158.325 **Far–infrared properties of multiple nucleus galaxies.**
K. J. Fricke, W. Kollatschny.
Light on dark matter, p. 441 – 442 (1986). – See Abstr. 012.080.
The double nucleus Markarian galaxies have properties as expected from advanced stages of galaxy mergers. From their far–infrared, radio–continuum, and H I fluxes the authors conclude that these systems are sites of violent starburst activity due to massive star formation and supernova explosions.

158.326 **Young supernovae in the starburst galaxy M82.**
S. W. Unger, A. Pedlar, D. J. Axon, P. N. Wilkinson, P. N. Appleton.
Light on dark matter, p. 443 – 444 (1986). – See Abstr. 012.080.
MERLIN observations at 1666 MHz detect 18 compact (< 0.25 arcsec) radio sources within the central few hundred parsecs of M82. M82 is well known as a "starburst" galaxy, and it is probable that these objects are very young supernova remnants. The distribution of radio sources appears asymmetric with respect to the centre of the galaxy, and the authors suggest that this is due to the current region of star formation lying within a molecular ring.

158.327 **The amazing tail of NGC 2146.**
H. C. M. Caspers, W. W. Shane.
Light on dark matter, p. 445 – 446 (1986). – See Abstr. 012.080.
NGC 2146 is a disturbed looking SbII pec galaxy, distance 14.5 Mpc. Optically there are seen a nucleus, two spiral arms and a prominent dust lane almost crossing the nucleus. IRAS data now show that it is a very strong infrared source, suggesting a recent burst of star formation. The galaxy was observed in 1981 with the WSRT in the 21–cm line of H I. Well beyond the optical image the authors find two long extensions to the south–east and south, without known optical counterparts. Most remarkable is the gently curved southern tail.

158.328 **Starbursts in the nuclei of interacting and merging galaxies.**
R. D. Joseph.
Light on dark matter, p. 447 – 454 (1986). – See Abstr. 012.080.
The author reviews the evidence for IR activity in interacting and merging galaxies. Interactions appear to induce bursts of star

formation of exceptional IR luminosity with very high efficiency. Mergers produce "super–starbursts" of even greater luminosity and spatial extent. As a result, interacting galaxies are exceptionally prominent among IRAS galaxies. Finally, the author indicates some of the larger implications of interaction–induced starbursts for extra–galactic astronomy and observational cosmology.

158.329 **New ground–based studies of two active IRAS galaxies.**
C. G. Wynn–Williams, E. E. Becklin, D. L. DePoy, J. N. Heasley, G. J. Hill, J. W. MacKenty, C. A. Beichman.
Light on dark matter, p. 463 – 466 (1986). – See Abstr. 012.080.
New data are presented for IRAS 0421+040P06 and NGC 6240. These two galaxies have properties that differ from both Seyfert and starburst galaxies.

158.330 **Spectroscopy of active and starburst galaxies between 8 – 13 μm.**
P. F. Roche.
Light on dark matter, p. 467 – 470 (1986). – See Abstr. 012.080.
Spectra at 8 – 13 μm of more than 50 galaxy nuclei have been secured and many have been published. The division of the galaxies into two classes, active or starburst, is reflected in their 8 – 13 μm spectra. The starburst galaxies generally have spectra dominated by the well known, but as yet unidentified, narrow dust emission features at 11.25, 8.65 and 7.7 μm seen in a variety of Galactic sources together with strong ionic line emission from the 12.8 μm [Ne II] fine–structure line. By contrast, these features are seen only rarely in the active nuclei which instead have 8 – 13 μm spectra well represented by a power law; in some cases there is evidence for silicate absorption.

158.331 **IRAS observations of active galaxies – a review.**
G. Miley, R. de Grijp.
Light on dark matter, p. 471 – 486 (1986). – See Abstr. 012.080.
IRAS observations of AGNs are reviewed. The authors consider both the IRAS results on known AGNs and ground–based followup work on candidate–AGNs from the IRAS Catalogue.

158.332 **Seyfert galaxies in the IRAS survey and JHKL photometry.**
I. S. Glass.
Light on dark matter, p. 487 – 488 (1986). – See Abstr. 012.080.
The near– to far–infrared properties of an optically selected sample of ordinary spiral galaxies, H2 (starburst) galaxies and Seyferts types 1 and 2 are summarized. The spectral indices between 1 and 100 μm are shown to be almost independent of galaxy type. New JHKL photometry for a sample of Seyferts found from IRAS data by de Grijp et al. (1985) is given. No essential differences are found in the near–infrared behaviour of Seyferts selected by the two differential methods. The de Grijp et al. criteria are shown to exclude many Seyfert galaxies known from optical work.

158.333 **The most luminous galaxies.**
R. P. Norris, P. F. Roche, D. A. Allen.
Light on dark matter, p. 489 – 490 (1986). – See Abstr. 012.080.

158.334 **Submillimetre to infrared observations of active galaxies.**
L. M. J. Brown.
Light on dark matter, p. 491 – 492 (1986). – See Abstr. 012.080.
The author has investigated the submillimetre to infrared spectra and variability of a sample of active galaxies comprising 13 blazars, the quasar 3C 273 and the peculiar Seyfert 2 galaxy NGC 1275.

158.335 **Active galactic nuclei in the IRAS PSC.**
M. Dennefeld, M. P. Véron–Cetty.
Light on dark matter, p. 493 – 496 (1986). – See Abstr. 012.080.
The authors have investigated the far–infrared properties of previously known active galactic nuclei detected by IRAS.

The second Byurakan spectral sky survey. IV. The field $\alpha = 12^h22^m$, $\delta = +55°00'$.
See Abstr. 002.014.

A new catalogue of quasi–stellar objects.
See Abstr. 002.036.

A catalogue of Seyfert galaxies and related objects.
See Abstr. 002.042.

Low energy transmission grating spectrometer (LETGS) for AXAF.
See Abstr. 035.021.

X–ray spectroscopy of AGN with the AXAF "Microcalorimeter".
See Abstr. 035.024.

A proposed method for the determination of the bivariate luminosity function utilizing an incomplete sample with an application to Seyfert galaxies.
See Abstr. 036.020.

Experimental observations with the radio telescope RATAN–600 using the method «estafette with zoning».
See Abstr. 036.089.

Optical astrometry of extragalactic radio sources.
See Abstr. 041.009.

Celestial reference coordinate systems: submilliarcsecond precision demonstrated with VLBI observations.
See Abstr. 043.014.

Quasars and galaxies observable with the Space Telescope.
See Abstr. 051.034.

A monopole model for galactic nuclei.
See Abstr. 061.024.

Formation of collimated beams.
See Abstr. 062.006.

The reconfinement of jets.
See Abstr. 062.015.

The relativistic short wavelength instabilities in astrophysical jets. II. The case of magnetofluid–dynamics.
See Abstr. 062.017.

The relativistic maximum instability in astrophysical jets.
See Abstr. 062.026.

Numerical modeling of magnetized jets.
See Abstr. 062.035.

Evolution of magnetized jets in power–law potentials.
See Abstr. 062.036.

Instabilities of thick accretion tori and redistribution of angular momentum.
See Abstr. 062.042.

Helical twisting on an adiabatically expanding jet. II. Spatial growth.
See Abstr. 062.058.

On the hydrodynamical equations of astrophysical jets.
See Abstr. 062.063.

Optically thin, adiabatic electron–positron pair wind.
See Abstr. 062.092.

Reaction instability in electron–positron pair equilibrium plasmas.
See Abstr. 062.093.

Magnetohydrodynamic solitons and radio knots in jets.
See Abstr. 062.096.

Beams and jets in astrophysical plasmas.
See Abstr. 062.100.

Is magnetic energy the source of synchrotron radiation in jets?
See Abstr. 062.120.

Three–dimensional structure and dynamics of a supersonic jet.
See Abstr. 062.124.

Magnetohydrodynamic effects in radio jets.
See Abstr. 062.125.

Magnetization effects in an incoherent synchrotron source.
See Abstr. 063.015.

Relativistic particle streams in AGN.
See Abstr. 063.045.

Monte Carlo spectra from pair–equilibrium, weakly magnetized thermal plasmas.
See Abstr. 063.062.

Radiation pressure and emission clouds around active galactic nuclei.
See Abstr. 063.087.

On the origin of the infrared and X–ray continua of active galactic nuclei.
See Abstr. 063.088.

On the correlation between X–ray flux and hydrogen line intensities in quasars and active galactic nuclei.
See Abstr. 063.102.

A two–dimensional model for relativistic gas jets.
See Abstr. 066.113.

A gravitational lens origin of AGN–variability? Consequences of micro–lensing.
See Abstr. 066.173.

Morphological aberration in gravitational–lens images.
See Abstr. 066.197.

Pair production instabilities as a source of X–ray flares from accreting black holes.
See Abstr. 067.007.

Electron–positron pairs in a mildly relativistic plasma in active galactic nuclei.
See Abstr. 067.011.

Episodic accretion onto massive black holes.
See Abstr. 067.022.

Supermassive stars in galactic nuclei.
See Abstr. 067.023.

The birth of AGNs and quasars via the collapse of dense star clusters to supermassive black holes.
See Abstr. 067.042.

An idealized model for dust accretion in active galactic nuclei.
See Abstr. 067.043.

Floppy discs: a recipe for the "observers' dream" model.
See Abstr. 067.045.

Rotating wind solution and the acceleration of the cosmic radio jets.
See Abstr. 067.075.

The physics of black holes.
See Abstr. 067.082.

Cosmic X–ray jets.
See Abstr. 067.083.

The central engine of quasars and active galactic nuclei: hadronic interactions of shock–accelerated relativistic protons.
See Abstr. 067.093.

Black hole models for quasar–like objects: some selected topics.
See Abstr. 067.097.

Central engine or locomotive?
See Abstr. 067.105.

Accretion disks and periodic outbursts of active galactic nuclei.
See Abstr. 067.111.

Accretion disks and periodic outbursts of AGN.
See Abstr. 067.127.

Dynamics of an affine star model in a black hole tidal field.
See Abstr. 067.128.

T CrB, a "Seyfert 1 nucleus" at 1200 ± 100 parsecs?
See Abstr. 117.108.

Inverse Compton models of SS 433 and radio quiet AGN.
See Abstr. 117.118.

Nature of very small grains: PAH molecules or silicates?
See Abstr. 131.116.

Flicker of extragalactic radio sources and refractive interstellar scintillation.
See Abstr. 131.119.

Formaldehyde observations in the direction of extragalactic sources.
See Abstr. 131.247.

Small–scale structure of interstellar H I clouds.
See Abstr. 131.253.

The slow interstellar scintillation of extragalactic radio sources.
See Abstr. 131.312.

IMF in starburst regions.
See Abstr. 131.378.

Determination of temperature in gaseous nebulae.
See Abstr. 132.055.

The exciting stars of giant extragalactic H II regions.
See Abstr. 132.060.

Multifrequency radio continuum observations of extended galactic objects. I. Nine objects from the 2695 MHz Effelsberg galactic plane survey.
See Abstr. 141.004.

Radio source surveys of the Milky Way and M33.
See Abstr. 141.013.

The MIT–Green Bank (MG) 5 GHz survey.
See Abstr. 141.018.

5 GHz radio structure and optical identifications of sources from the MG survey. II. Maps and finding charts.
See Abstr. 141.019.

Optical identifications of radiosources.
See Abstr. 141.021.

Iron K X–ray emission lines from cosmic X–ray sources observed with Tenma.
See Abstr. 142.034.

New identifications of bright X–ray sources with the HEAO–1 scanning modulation collimator.
See Abstr. 142.065.

Iron line from compact X–ray sources.
See Abstr. 142.075.

Optical counterparts of HEAO–1 X–ray sources.
See Abstr. 142.087.

Predicting star formation in interacting and colliding galaxies; the special case of ring galaxies.
See Abstr. 151.016.

Model calculations of the evolution of "clumpy" gas in elliptical galaxies.
See Abstr. 151.040.

Colliding galaxies – modeling bursts of star formation in ring galaxies.
See Abstr. 151.054.

Tidal triggering of Seyfert galaxies and quasars – perturbed galaxy disk models versus observations.
See Abstr. 151.099.

The criterion for equipartition in clusters with a mass spectrum.
See Abstr. 151.108.

The initial mass function of massive stars in galaxies: empirical evidence.
See Abstr. 155.121.

New evidence for galaxy luminosity evolution.
See Abstr. 157.002.

NGC 4418; a very extinguished galaxy.
See Abstr. 157.008.

Globular clusters and galactic evolution.
See Abstr. 157.015.

Emission line velocity survey of spiral galaxies with bright nuclei.
See Abstr. 157.016.

NGC 4038/39: interacting spiral galaxies with enhanced extended radio emission.
See Abstr. 157.029.

Radio sources in giant E and S0 galaxies.
See Abstr. 157.037.

The inclination dependence of far–infrared emission from Sc galaxies.
See Abstr. 157.053.

Spectrophotometric investigation of NGC 4922.
See Abstr. 157.062.

Spectrophotometric investigation of NGC 3353.
See Abstr. 157.076.

The effects of galaxy–galaxy interactions on nuclear activity.
See Abstr. 157.085.

Spectral evolution of galaxies.
See Abstr. 157.092.

Population synthesis and epochs of star formation in NGC 1316 (Fornax A).
See Abstr. 157.096.

Colors of 3CR and first–ranked high redshift galaxies.
See Abstr. 157.102.

Radio observations of starburst galaxies.
See Abstr. 157.130.

Optical emission–line gas in the nuclei of normal galaxies: implications for nuclear activity.
See Abstr. 157.136.

Nuclei of normal galaxies.
See Abstr. 157.137.

The properties of the nuclei of the two hot spot galaxies NGC 1097 and 2997.
See Abstr. 157.142.

Study of a complete sample of galaxies. III. Discussion and results.
See Abstr. 157.147.

Ultraviolet–excess galaxies.
See Abstr. 157.155.

Further radio observations of *IRAS* extreme infrared galaxies.
See Abstr. 157.162.

Nucleated spheroidal galaxies.
See Abstr. 157.167.

The luminosity function and space density of the most luminous galaxies in the *IRAS* survey.
See Abstr. 157.173.

A statistical study on the difference between normal and starburst galaxies.
See Abstr. 157.181.

The luminosity function for field galaxies in the infrared.
See Abstr. 157.186.

VLA mapping of the region around NGC 4319.
See Abstr. 157.191.

Globular clusters in galaxies beyond the Local Group. V. The giant ellipticals reconsidered.
See Abstr. 157.227.

The absolute magnitude–effective radius relation for E galaxies and the redshifts of BL Lac objects.
See Abstr. 157.236.

The evolution of galaxies in dense clusters: spectral synthesis models for post–burst galaxies.
See Abstr. 157.237.

Near infrared spectroscopy of galaxies.
See Abstr. 157.243.

Infrared emission and extragalactic starbursts.
See Abstr. 157.246.

Models for IRAS galaxies.
See Abstr. 157.248.

Molecular gas in high–luminosity IRAS galaxies.
See Abstr. 157.262.

The morphology and structure of blue compact dwarf galaxies from CCD observations.
See Abstr. 157.278.

Blue compact galaxies: a mixing bag.
See Abstr. 157.279.

The young stellar populations in blue compact dwarf galaxies: spectral synthesis of ultraviolet spectra.
See Abstr. 157.281.

Ultraviolet and infrared properties of blue compact galaxies.
See Abstr. 157.283.

Abundances in blue compact galaxies.
See Abstr. 157.285.

Helium in dwarf emission–line galaxies.
See Abstr. 157.286.

Observational properties of the gas and the young stars in BCG.
See Abstr. 157.287.

A spectroscopic analysis of the bright knots in Mk 170 and Mk 178.
See Abstr. 157.288.

Analysis of new Hα emission line galaxies.
See Abstr. 157.289.

New abundance values in a BCG sample.
See Abstr. 157.290.

VLA continuum and H I observations of blue compact dwarf galaxies.
See Abstr. 157.294.

Quiescent and active irregular galaxies in a color versus M_H/L diagram and starburst evolution.
See Abstr. 157.295.

H I – H II relationship in dwarf star–forming galaxies.
See Abstr. 157.296.

IRAS observations of blue compact emission line galaxies.
See Abstr. 157.299.

Star formation in IRAS galaxies: from blue dwarfs to red super–star bursters.
See Abstr. 157.300.

Lyman α in emission line galaxies.
See Abstr. 157.305.

Statistical properties of IRAS galaxies.
See Abstr. 157.322.

The relationship between blue and FIR luminosities of spiral galaxies.
See Abstr. 157.326.

Starbursts in non–interacting galaxies.
See Abstr. 157.328.

Optical identifications of radio sources with accurate positions using the United Kingdom Schmidt Telescope (UKST) IIIA–J plates.
See Abstr. 159.023.

log *N*–log *S* of radio sources at 10 GHz with flat spectra.
See Abstr. 159.024.

Quasars: cosmological evolution and X–ray background contribution.
See Abstr. 159.026.

Quasars near NGC 520.
See Abstr. 159.036.

Radio observations of extended sources suspected of having steep–spectrum cores.
See Abstr. 159.044.

Evolution of the luminosity function of quasars.
See Abstr. 159.051.

Continuity in the observed properties of QSOs, high–redshift emission–line radio galaxies, BL Lac objects, N systems and Seyfert galaxies, and possible interpretations.
See Abstr. 159.052.

Optical variability in quasars.
See Abstr. 159.053.

The low excitation lines in quasars and AGN in the frame work of photoionization models.
See Abstr. 159.061.

The I.A.C./Q.M.C. catalogue of quasar multiband spectra.
See Abstr. 159.063.

The luminosity function of quasars and low luminosity active galactic nuclei.
See Abstr. 159.066.

Flux variations and structural changes in extragalactic radio sources.
See Abstr. 159.067.

Origin of the thermal continuum in AGN's and QSO's.
See Abstr. 159.074.

Q/G 0903 + 175: the closest projected quasar/galaxy pair yet seen.
See Abstr. 159.075.

Ly α galaxy companions of quasars at $z > 2.5$.
See Abstr. 159.077.

QSOs and galaxies.
See Abstr. 159.078.

Redshifts and observed colors of QSOs and active galaxies.
See Abstr. 159.097.

***EINSTEIN* IPC spectra of quasars: first results.**
See Abstr. 159.108.

A study of the optical and X–ray properties of quasars.
See Abstr. 159.109.

A relativistic blast wave model for outbursts in quasars and active galactic nuclei.
See Abstr. 159.110.

The effect of a quasi–stellar object on its host galaxy: dynamical and physical processes in the interstellar medium around a QSO.
See Abstr. 159.111.

Counts, evolution, and background contribution of X–ray quasars and other extragalactic X–ray sources.
See Abstr. 159.138.

A search for quasars and emission–line galaxies in the Sculptor region.
See Abstr. 159.147.

Evidence for evolution of quasar environments.
See Abstr. 159.151.

Blazars, quasars and starbursts.
See Abstr. 159.158.

Positions of extragalactic objects in the vicinity of M3 and M31 (5C3 radio field) and in the 5C1 radio field.
See Abstr. 159.159.

Extragalactic sources with asymmetric radio structure. II. Further observations of the quasar B2 1320+299.
See Abstr. 159.173.

The inner structure of quasars and active galactic nuclei.
See Abstr. 159.174.

Stand der VLBI–Untersuchungen extragalaktischer Quellen nördlich 70° Deklination.
See Abstr. 159.175.

Gruppen um aktive Galaxien.
See Abstr. 159.177.

The X–ray structure of the 3C 295 cluster: a cooling flow at a redshift of 0.5.
See Abstr. 160.035.

Groups around bright Seyfert nuclei.
See Abstr. 160.043.

Old radio galaxies and cluster radio halos.
See Abstr. 160.064.

Radio continuum survey of the Coma/A1367 supercluster. I. 610 MHz observations of CGCG galaxies in four groups.
See Abstr. 160.072.

Radio continuum survey of the Coma/A1367 supercluster. II. 1.5 GHz observations of 396 CGCG galaxies.
See Abstr. 160.073.

Hot gas in clusters and early–type galaxies.
See Abstr. 160.074.

An optical and X–ray study of the radio source 3C 75.
See Abstr. 160.086.

Large–beam radio and infra–red observations toward the Virgo Cluster.
See Abstr. 160.102.

The protogalactic connection.
See Abstr. 161.079.

159 Quasi-stellar Objects

159.001 **The statistics of gravitational lensing. III. Astrophysical consequences of quasar lensing.**
J. P. Ostriker, M. Vietri.
Astrophys. J., Vol. 300, No. 1, p. 68 – 76 (1986).
The authors combine the method of Schmidt and Green for calculating the luminosity function of quasars with the theory of gravitational lensing developed in previous papers of this series to compute expected properties of lensed systems. Multiple quasar images produced by galaxies will be of order 10^{-3} of the observed quasars, with the numbers over the whole sky calculated to be (0.86, 120, 1600) to limiting B magnitudes of (16, 19, 22). The overall luminosity function derived from observations has not been greatly modified by lensing; the amount of "false evolution" is small except for an interesting subset of apparently bright, large–redshift objects for which minilensing by starlike objects may be important.

159.002 **CCD camera brightness monitoring of Q0957+561 A, B.**
R. E. Schild, B. Cholfin.
Astrophys. J., Vol. 300, No. 1, p. 209 – 215 (1986).
The authors present results of continued monitoring of the twin QSO 0957+561 with data obtained with a CCD camera from 1983 February through 1984 March. The 0.2 mag increase previously reported to have occurred from 1982 January through 1983 January in the A component was seen to occur during 1983 January through 1984 January in the B component. From an autocorrelation of the CCD data, the time of arrival difference is determined to be $\Delta t = 1.03 \pm 0.1$ yr; the A image precedes the B image.

159.003 **The ultraviolet excess of quasars. III. The highly polarized quasars PKS 0736+017 and PKS 1510–089.**
M. A. Malkan, R. L. Moore.
Astrophys. J., Vol. 300, No. 1, p. 216 – 223 (1986).
The authors analyze ultraviolet, optical, and infrared spectrophotometry of the highly polarized quasars PKS 0736+017 and PKS 1510–089. A "blazar" continuum component like that in BL Lac objects contributes about half the visual light of 1510–089, and at least three–quarters of that in 0736+017. The remaining light has the same spectrum as normal (low–polarization) quasars, including an "ultraviolet excess" or "blue bump". The line fluxes do vary, but not as much as the continuum.

159.004 **Comparative statistical study of the chronometric and Friedmann cosmologies based on the Palomar Bright Quasar and other complete quasar surveys.**
I. E. Segal, J. F. Nicoll.
Astrophys. J., Vol. 300, No. 1, p. 224 – 241 (1986).
The Palomar Bright Quasar Survey and other complete samples derived by Schmidt and Green from quasar surveys are used for comparative cosmological studies and for predictions of quasar counts. Systematic self–tests and mutual cross–tests of the chronometric and Friedmann cosmologies are made. Nonparametric estimates of the normalized luminosity function are made that involve no assumption as to the homogeneity of the spatial distribution of quasars. The homogeneity question is studied independently by V/V_m tests. The chronometric cosmology is found to fit the observations quite satisfactorily without either luminosity or density evolution, and without the use of any adjustable parameters such as q_0 or Λ.

159.005 **CCD photo–polarimetry of the jet of 3C 273.**
H.–J. Röser, K. Meisenheimer.
Astron. Astrophys., Vol. 154, No. 1/2, p. 15 – 24 (1986).
Spatially resolved CCD photo–polarimetry in three band-passes centered at 420, 610, and 860 nm revealed three different emission regions along the jet of 3C 273: (1) The radio hot spot at 21″.2 from the quasar which has a sharp continuum fall off above 7×10^{13}Hz and is therefore clearly detected in the I, but hardly visible in the R band. (2) The optical knot at 19″.5 from the core terminating the visible jet image. The knot is 13% polarized at position angle 140° in the R band, degree and angle beeing identical with 5 GHz data. This and a radio–to–optical powerlaw $f_\nu \sim \nu^{-0.96}$ are strong indications for a synchrotron origin of the knot emission. (3) The inner 11″ to 17″ of the jet, where no polarization stronger than 8% is detectable. This is in contrast to VLA observations at 5 GHz and is attributed to an additional unpolarized component observed at short wavelengths.

159.006 **The optical variability of 3C 345.**
M. R. Kidger, J. E. Beckman.
Astron. Astrophys., Vol. 154, No. 1/2, p. 288 – 294 (1986).
B–magnitudes have been compiled from the literature for 3C 345, covering a period from 1965 to 1980, 905 values in all. Since the last major study of the light–curve of this object (Barbieri et al., 1977), new and also revised data have been presented, which extend the light–curve and improve the temporal resolution. A compilation of the observations and a detailed Fourier analysis are presented in the paper and compared with previous studies.

159.007 **Automated quasar detection in the SGP field: a clustering study.**
R. G. Clowes.
Mon. Not. R. Astron. Soc., Vol. 218, No. 1, p. 139 – 157 (1986).
The Automated Quasar Detection (AQD) process has been applied to a field (SGP) containing the south galactic pole, which is one of the two undisputed 'selected areas' in the southern hemisphere. A comparison with confirmed quasars shows that AQD re–discovered all 30 quasars with non–overlapped objective–prism spectra that were found by the visual search of Clowes & Savage, and re–discovered 12 of the 13 non–overlapped quasars that were missed by their search but found independently by others. A relatively small, high–probability subset, for which the present absence of follow–up spectroscopy is of minimum importance, shows no definite evidence of clustering. The paper shows that redshifts are essential for clustering studies, and that very much more progress could be made if they were obtained from low–resolution spectroscopy.

159.008 **Lyman absorption lines in quasar spectra: evidence for gravitationally–confined gas in dark minihaloes.**
M. J. Rees.
Mon. Not. R. Astron. Soc., Vol. 218, No. 1, p. 25P – 30P (1986).
Primordial gas photoionized by a UV background can be gravitationally confined, in stable equilibrium, by minihaloes of dark matter with virial velocities ~ 30 km s^{-1}, which are an inevitable concomitant of the cold dark matter cosmogony. The calculated properties of these clouds suggest that they could account for the forest of Lyman absorption lines found in the spectra of high–z quasars.

159.009 **The statistics of radio emission from quasars.**
J. A. Peacock, L. Miller, M. S. Longair.
Mon. Not. R. Astron. Soc., Vol. 218, No. 2, p. 265 – 278 (1986).
The radio properties of quasars have traditionally been discussed in terms of the radio–to–optical flux–density ratio R, implying a correlation between emission in these wavebands. The authors show that, for bright quasars, this apparent correlation is largely due to an abrupt change in the radio properties of the quasar population near absolute magnitude $M_B = -24$. They suggest that this change is due to the existence of two classes of quasar with different host galaxies. Furthermore, the fraction of quasars detected above a given radio luminosity threshold decreases at higher redshifts (at constant M_B): this may also be explained by two separately evolving populations of quasars. Within this framework, the authors demonstrate that existing data cannot establish whether or not radio and optical luminosities are correlated, as the optical luminosity functions for the two classes are in general different.

159.010 **Studies of broad emission line profiles in QSOs – II. Properties of a large, predominantly radio selected sample.**
B. J. Wilkes.
Mon. Not. R. Astron. Soc., Vol. 218, No. 2, p. 331 – 361 (1986).
Emission line parameters for 214 QSOs are reported. The majority were originally selected according to the flatness of their radio spectra from the Parkes 2700, 5000 MHz surveys. The optical spectra of these objects were presented graphically in an earlier paper. Preliminary studies of the properties of this sample, including line velocity width and equivalent width distributions, have been made. Specific emission lines have been investigated in more detail to confirm or refute earlier results.

159.011 **On the completeness of a sample of bright quasars selected by colour excess in the direction of the North Galactic Pole.**
A. Warnock III, P. D. Usher, K. J. Mitchell, S. B. Howell.
Mon. Not. R. Astron. Soc., Vol. 218, No. 3, p. 445 – 451 (1986).
The Medium Bright Quasar Survey (MBQS) shows evidence for a dearth of bright quasars in a Palomar Schmidt field centred on Selected Area (SA) 57 near the North Galactic Pole, compared to similar fields centred on SA 28, 29, 55, and 94. The SA 57 field has been searched again for bright quasar candidates with the help of a second survey plate exposed according to a slightly modified Haro–Luyten three–colour (Tonantzintla) prescription. Candidates so selected have both a blue and ultraviolet excess (B–UVX). The main result of the paper is that there appear to be no B–UVX quasars in the SA 57 field that are brighter than $B = 17.25$ mag. The significance of this apparent anomaly is briefly discussed.

159.012 **Alignments of clustered quasars.**
K. Subramanian, J. D. Barrow.
Mon. Not. R. Astron. Soc., Vol. 218, No. 3, p. 587 – 592 (1986).
The authors derive an analytic expression for the expected number of triplet alignments in a clustered field in terms of the two– and three–point correlation functions of the field. The existence of non–random clustering increases the number of alignments expected to arise by chance. The authors apply this general result to the quasar fields examined previously for alignments and conclude that, although the possibility that a small fraction of quasars do exist in physically associated triplets cannot be entirely excluded by statistical arguments, there is no significant evidence for the existence of more well–aligned triplets of quasars than would arise purely by chance if their redshifts were all cosmological.

159.013 **A quasar with $z = 3.71$ and limits on the number of more distant objects.**
J. S. Dunlop, A. J. B. Downes, J. A. Peacock, A. Savage, S. J. Lilly, F. G. Watson, M. S. Longair.
Nature, Vol. 319, No. 6054, p. 564 – 567 (1986).
Spectroscopy of objects identified with radio sources from the Parkes Selected Regions has shown the compact, flat–spectrum source 1351–018 to be a quasar with redshift 3.71, the second most distant discovered to date. Compared with known quasars at such high redshifts, the object is optically very faint ($B = 20.9$), red in colour ($B–R = 1.6$) and is distinguishable photometrically from faint galaxies only by its relatively blue optical–infrared colour ($R–K = 2.1$). In common with other high–redshift, radio–loud quasars, 1351–018 has weak emission lines when compared with objects selected optically. Because it is a member of a complete sample, 1351–018, in fact, confirms the recent suggestions that the quasar radio luminosity function declines for redshifts $z > 2$.

159.014 **Quasar candidates in the field of SA 94 (2^h53^m, $+0°20'$).**
C. Barbieri, S. Cristiani.
Astron. Astrophys., Suppl. Ser., Vol. 63, No. 1, p. 1 – 26 (1986).
A survey of quasars has been carried out in the field of SA 94 on UK Schmidt objective–prism plates, selecting 208 candidates in 39 sq. deg. Magnitudes for those located in the central 22.9 sq. deg. of the field and slit–spectroscopy information for 44 of them are provided. The characteristics of the survey are analyzed in comparison with other previous samples.

159.015 **Erratum: "*UBV* photometry of quasars and quasar candidates II" [Astron. Astrophys., Suppl. Ser., Vol. 61, No. 2, p. 225 – 235 (1985)].**
G. Adam.
Astron. Astrophys., Suppl. Ser., Vol. 63, No. 3, p. 601 (1986). See Abstr. 40.159.003.

159.016 **QSO absorption lines.**
J.–s. Chen.
Astrophys. Space Sci., Vol. 118, Nos. 1/2, p. 473 – 486 (1986). – See Abstr. 012.001.
Some techniques and selection effects in establishing uniform samples of absorption line systems are presented. Summaries of

the basic properties of various absorbers are given. Finally, attention is paid to some special, interesting problems which are now widely discussed.

159.017 **The environment of QSOs.**
A. Stockton.
Astrophys. Space Sci., Vol. 118, Nos. 1/2, p. 487 – 497 (1986). – See Abstr. 012.001.
The evidence for the association of QSOs with groups of galaxies and the available data on the global properties of such groups are reviewed. At least for redshifts less than 0.5, QSOs tend to occur in richer–than–average environments. Observed correlations between radio properties, optical spectra, and environment are mentioned briefly. The present evidence tends to support suggestions that there are two classes of QSOs, one being the extension of the Seyfert population to higher luminosities and having spiral host galaxies, the other being closely related to the broad–line radio galaxies.

159.018 **Heavy element abundances in absorption line systems towards Q2206–199N.**
J. G. Robertson, R. F. Carswell, P. A. Shaver.
Astrophys. Space Sci., Vol. 118, Nos. 1/2, p. 499 – 500 (1986). – See Abstr. 012.001.
Spectroscopy of the $z = 2.0763$ absorption line system in Q2206–199N shows that it has a very large hydrogen column density ($5 \times 10^{20} cm^{-2}$) but extremely weak absorption due to heavy elements. The authors discuss the possibility that the gas cloud causing this aborption system may be exceptionally metal–poor.

159.019 **QSOs as probes of the early Universe.**
R. W. Hunstead, H. S. Murdoch, M. Pettini, J. C. Blades.
Astrophys. Space Sci., Vol. 118, Nos. 1/2, p. 505 – 507 (1986). – See Abstr. 012.001.
The spectrum of the highest redshift QSO 2000–330 ($z = 3.78$) contains four heavy–element absorption systems with $z_{abs} > 3.0$. Interesting features include velocity structure at $z_{abs} = 3.552$ which suggests a cluster origin and a purely low ionization system at $z_{abs} = 3.1881$ typical of a galactic disk sightline.

159.020 **Motion of quasar images by gravitational lens galaxies.**
M. Fujishita.
Astrophys. Space Sci., Vol. 119, No. 1, p. 227 – 230 (1986). – See Abstr. 012.001.
Rotational motions of twin quasar images due to the motion of a gravitational lens galaxy are studied. A weak gravity field and a point mass lens are assumed. Conclusions: (1) Quasar images move a few hundred milli–arc–seconds per year or more, if the lens galaxy passes within a thousandth parsec from the line connecting the observer with the quasar, (2) the distance between the galactic center and the quasar images is about ten kilo–parsec at the most. As an example, the velocities of 0957 + 561 A, B1 and B2 are calculated.

159.021 **Optical identification of QSO's with flat radio spectrum.**
Y.–q. Chu, X.–f. Zhu, H. Butcher.
Astrophys. Space Sci., Vol. 119, No. 1, p. 231 – 232 (1986). – See Abstr. 012.001.
Preliminary results of a program to identify optically a sample of flat spectrum radio sources are described. The identifications are based only on positional coincidences, and have yielded at least one object with a very high redshift.

159.022 **United Kingdom Schmidt Telescope (UKST) objective prism search for quasars.**
A. Savage.
Astrophys. Space Sci., Vol. 119, No. 1, p. 233 – 237 (1986). – See Abstr. 012.001.
Several optical techniques are being used to find quasars using UKST direct and objective prism plates. Six already much–studied fields are proposed as "standard fields" for the comparison of techniques. The completeness of objective prism samples, and their applicability to studies of three–dimensional quasar clustering, are discussed.

159.023 **Optical identifications of radio sources with accurate positions using the United Kingdom Schmidt Telescope (UKST) IIIA–J plates.**
A. Savage.
Astrophys. Space Sci., Vol. 119, No. 1, p. 239 – 242 (1986). – See Abstr. 012.001.
Several programmes are making use of UKST Sky Survey plates to identify southern radio sources. The fine–grain modern plates and accurate radio positions give a much improved identification rate. It seems that it will very soon be possible to determine whether or not there is a quasar redshift cut–off at $z \sim 4$. There is an urgent need for more accurate fundamental reference star positions in the South.

159.024 **log *N*–log *S* of radio sources at 10 GHz with flat spectra.**
K. Aizu, H. Tabara, T. Kato, M. Inoue.
Astrophys. Space Sci., Vol. 119, No. 1, p. 257 – 261 (1986). – See Abstr. 012.001.
A log *N*–log *S* relation at 10 GHz is constructed for sources with the flat spectra $\alpha \leqslant 0.5$ (flux density $S \propto \nu^{-\alpha}$) from observations at NRO, MPIfR and others. Based on the source distribution on log(Luminosity) $\sim$ volume plane the authors obtain an epoch–depending luminosity function, which explains the above relation.

159.025 **1642 + 690: a superluminal quasar.**
T. J. Pearson, P. D. Barthel, C. R. Lawrence, A. C. S. Readhead.
Astrophys. J., Lett. Ed., Vol. 300, No. 2, p. L25 – L29 (1986).
The radio source 1642 + 690 (4C 69.21) has previously been found to contain a dominant, compact core with a jet extending 5″ to the south. This letter presents two 5 GHz VLBI maps of the core, made in 1980 and 1983. A milliarcsecond jet is closely aligned with the arcsecond jet, and a "knot" in the jet is moving outwards along the jet at 0.34 mas yr^{-1}. The authors have found that the associated optical object is a quasar with redshift $z = 0.75$. At this redshift, the displacement of the knot implies an apparent transverse velocity of $9.3\,h^{-1}c$ ($H_0 = 100\,h$ km $s^{-1}Mpc^{-1}$, $q_0 = 0.05$).

159.026 **Quasars: cosmological evolution and X–ray background contribution.**
M. Schmidt, R. F. Green.
Cosmic radiation in contemporary astrophysics, p. 249 – 259 (1986). – See Abstr. 012.006.
The luminosity function of quasars varies with redshift or cosmic epoch. The authors discuss how the luminosity function and its evolution can be determined from complete samples of quasars. They first concentrate on optical survey of quasars. For quasars of medium luminosity, the co–moving space density rises very steeply with redshift. Evidence for a cutoff of quasar redshift and for a possible dependence of the cutoff on luminosity is discussed. The authors evaluate X–ray counts of quasars and show the need for negative X–ray luminosity evolution in order to explain the counts and the low average redshifts of X–ray quasars. They discuss other extragalactic contributors to the X–ray background and conclude that they, together with the quasars, contribute about 60 percent of the observed background. About half of this is contributed by active galactic nuclei with optical luminosities below those of quasars.

159.027 **The Mg II absorption system in the QSO PKS 2128–12: a galaxy disc/halo with a radius of 65 kpc.**
J. Bergeron.
Astron. Astrophys., Vol. 155, No. 1, p. L8 – L11 (1986).
The absorption line systems observed in QSO spectra, at least those with detected Mg II doublets, are assumed to arise in intervening galaxies. To verify this assumption the author started an imaging survey of fields around QSOs with low redshift narrow Mg II absorption line systems. He found a galaxy 8.6 arcsec away on the sky from the QSO PKS 2128–12 with a magnitude m(V) = 21.5. It is the only resolved object between radii of 1.7

and 38 arcsec around the QSO. The absorber has a redshift, z = 0.4299, smaller than the QSO emission redshift, z = 0.5003, and it should belong to the class of intervening systems.

159.028 **Radiatively–driven shock waves in quasar envelopes.**
L. Mestel, D. W. Moore.
Astron. Astrophys., Vol. 156, No. 1/2, p. 121 – 130 (1986).
The authors study the response of a quasar envelope to periodic disturbances occurring below the surface, using a simple form for the density–dependent radiative acceleration. Quasi–steady winds flow, with strong shock waves propagating outwards as the gas absorbs momentum from the quasar radiation field. The compressed gas behind a shock has high speed in the quasar frame and so can yield broad emission lines. A short wave approximation method yields self–consistent solutions in which strong shocks forming near the surface decelerate outwards and steadily weaken, but does not describe rigorously solutions with shock–trains that simultaneously accelerate and strengthen. Micro–physical limitations on the parameters of the macro–solutions are discussed. Difficulties emerge in attempting to account for the broadest observed emission lines, suggesting that radiation driving plays an auxiliary rather than a dominant role in the dynamics of quasar envelopes.

159.029 **Spectroscopy of the QSOs Q0118–031 A,B,C.**
J. G. Robertson, P. A. Shaver, J. Surdej, J. P. Swings.
Mon. Not. R. Astron. Soc., Vol. 219, No. 2, p. 403 – 416 (1986).
This paper presents observations of three QSOs which form a triplet in the plane of the sky, Q0118–031 A,B,C. Low dispersion (emission line) spectroscopy is given for all three objects, and intermediate dispersion spectroscopy for Q0118–031 B and C. A detailed discussion is given of the rich absorption spectrum of Q0118–031 B.

159.030 **On the nature of objects detected as faint long–term variables.**
M. R. S. Hawkins.
Mon. Not. R. Astron. Soc., Vol. 219, No. 2, p. 417 – 426 (1986).
Two complete samples of objects in the range $17 < B < 18.5$ are selected on the basis of optical variability and ultraviolet excess. Multicolour photometry, objective prism and slit spectra are used to ascertain the composition of the two samples. The sample of variables is found to be composed almost entirely of quasars. The sample of UVX objects appears to contain about 50 per cent white dwarfs or blue subdwarfs, and 50 per cent quasars. About 30 per cent of the quasars are included in the sample of variables and most of the remainder show significant variability.

159.031 **The spectral shape and variability of the blazar 3C 446.**
L. M. J. Brown, E. I. Robson, W. K. Gear,
R. P. Crosthwaite, I. M. McHardy, C. G. Hanson,
B. J. Geldzahler, J. R. Webb.
Mon. Not. R. Astron. Soc., Vol. 219, No. 3, p. 671 – 686 (1986).
The authors discuss the radio to X–ray spectrum and variability of 3C 446 and show that the entire infrared–ultraviolet wavelength region exhibits strongly correlated variations, with the infrared and optical fluxes varying together on a time–scale of weeks or less. The variations in spectral shape which occur in the submillimetre to optical wavelength region are discussed in detail.

159.032 **On the abundance of metals and the ionization state in absorbing clouds toward QSOs.**
F. H. Chaffee Jr., C. B. Foltz, J. Bechtold, R. J. Weymann.
Astrophys. J., Vol. 301, No. 1, p. 116 – 123 (1986).
The authors present photoionization models for clouds of material illuminated by the integrated radiation of QSOs over a range of epochs and predict the column densities of the most abundant ions for clouds with $N(\mathrm{H\ I}) = 5\times10^{16}\mathrm{cm}^{-2}$. For a wide density range, C III is predicted to have the highest column density of all metallic ions, and it is searched for in the $z = 3.321$ absorption system toward S5 0014+81. The observations set an upper limit of the metal–to–hydrogen ratio of $10^{-3.5}$ compared to the solar value. The authors discuss the likely ionization mechanism and ionization state of such clouds in light of a recent unified picture of the Lyman–α and metal line QSO absorption systems proposed by Tytler.

159.033 **A probable absorption feature in the radio spectrum of the quasar PHL 61: some interpretations of this absorption.**
S. A. Levshakov, V. K. Khersonskij, D. A. Varshalovich.
Astron. Zh., Tom 63, Vyp. 1, p. 25 – 30 (1986). In Russian. English translation in Sov. Astron., Vol. 30, No. 1.
An analysis of the radio spectrum of the quasar PHL 61 obtained by Takahara et al. (1983) permits to reveal the presence of a probable absorption feature at the frequency ν = 39.317 GHz. Its width is 5.2 MHz and the corresponding optical depth is 0.15. The authors suggest two possible identifications of the feature as the rotational line $J = 0\rightarrow1$ of the $^{12}C^{16}O$ molecule at the redshift $z_a(^{12}C^{16}O) = 1.9318$ or the $^{12}C^{17}O$ molecule at $z_a(^{12}C^{17}O) = 1.8578$. The possibility of the identification of this absorption feature with hyperfine lines in $K = 0\rightarrow1$ transition of CN at $z_a(\mathrm{CN}) = 1.8784$ is also discussed. Some tests for the proposed identifications are considered.

159.034 **Variability of the gravitational mirage PG 1115+080.**
C. Vanderriest, G. Wlérick, G. Lelièvre, J. Schneider,
H. Sol, D. Horville, L. Renard, B. Servan.
Astron. Astrophys., Vol. 158, No. 1/2, p. L5 – L8 (1986). In French.
Photometric data on the "triple quasar" PG 1115+080 were recorded, mostly from electronographic plates, between 1981 and 1985. They show that: (1) The source is variable by more than 0.3 magnitude. (2) The time delays Δt(A,B) and Δt(A,C) are probably less than 2.5 months. (3) The time delay $\Delta t(A_1,A_2)$ is probably more than 4 days. The authors conclude that the detected variability of PG 1115+080 strongly supports the gravitational lens hypothesis and allows to put constraints on the current models.

159.035 **Apparent superluminal expansion in a quasar neutrino halo.**
J.–l. Zhang, S.–p. Xiang.
Astrophys. Space Sci., Vol. 120, No. 2, p. 159 – 166 (1986).
The authors deduce the apparent velocity formula of general relativity for a source moving in a spherical neutrino halo and show that, considering gravitational effects on the motion of photons, the neutrino halo around a quasar may cause apparent superluminal phenomena. Sample calculations illustrating this result are shown.

159.036 **Quasars near NGC 520.**
H. Arp, O. Duhalde.
Publ. Astron. Soc. Pac., Vol. 97, No. 598, p. 1149 – 1157 (1985).
Quasars are reported in an area 2.1 square degrees around the bright, disturbed galaxy NGC 520. For quasars brighter than $m = 19.0$ mag, areas to the southwest of NGC 520, in the direction of previously identified radio ejection, have a density of 2.9 per square degree or about 1.5 times the expected background density. Near NGC 520 the density increases until it reaches 15 times the expected background density for the four nearest quasars to the galaxy. Six of the seven nearest quasars to NGC 520 are aligned so accurately across the galaxy as to demonstrate physical association.

159.037 **Optical manifestation of superluminal motion in the quasar 3C 345.**
M. K. Babadzhanyants, E. T. Belokon'.
Astrofizika, Tom 23, Vyp. 3, p. 459 – 471 (1985). In Russian. English translation in Astrophysics, Vol. 23, No. 3.
All the 1971 – 1982 data (5 – 15 GHz) on superluminal motions in the VLBI structure of 3C 345 are compared with its optical variability in the B band. The extrapolated moments of ejection of the compact jet components well coincide with the initial phases of the "slow" optical flares (with characteristic times of 1 – 2 ys) observed since 1966. The beginning of the last optical s–flare with maximum brightness in 1982 – 1983 is consistent with the time of ejection of the newly detected jet compo-

nent. The s–flares correlate with the total flux density variations at 10.7 GHz with time delays of about 400 – 1000 days. The authors conclude that part of the optical variability of extragalactic objects must be connected with jet events in their nuclei.

159.038 **VLBI limits on the proper motion of the 'core' of the superluminal quasar 3C345.**
N. Bartel, T. A. Herring, M. I. Ratner, I. I. Shapiro, B. E. Corey.
Nature, Vol. 319, No. 6056, p. 733 – 738 (1986).
VLBI observations between 1971 and 1983 have been used to determine the positions of the 'core' of the quasar 3C345 relative to the more distant compact quasar NRAO512 with a fractional uncertainty as small as 2 parts in 10^8. The core of 3C345 appears stationary in right ascension to within 20 arc μs yr^{-1}, a subluminal bound corresponding to $0.7c$. The apparent velocities of the jets are superluminal, up to $14c$ in magnitude.

159.039 **Variability of Lyman–alpha and the ultraviolet continuum of 3C 446.**
J. N. Bregman, A. E. Glassgold, P. J. Huggins, A. L. Kinney.
Astrophys. J., Vol. 301, No. 2, p. 698 – 702 (1986).
IUE observations were made of the violently variable quasar 3C 446 at intervals of 1.2, 2.2, 0.5, and 0.4 yr, starting in 1980 June. The continuum slope in the range 1830 – 3175 Å has a slope of -3.0 ± 0.3. Strong absorption of the continuum was found below 1830 Å and probably corresponds to a Lyman edge at $z = 1.00 \pm 0.01$. The Lyα emission line was detected in five spectra and the weaker Lyβ + O VI λ1035 blend was detected in the summed spectrum. Relative to the number of ionizing photons, the line strengths are the same as in normal quasars. The variability of the line fluxes provides strong constraints on the geometry of the broad emission line region.

159.040 **On the X–ray and ultraviolet spectrum of the high–redshift quasar 3C 446.**
B. Garilli, G. Tagliaferri.
Astrophys. J., Vol. 301, No. 2, p. 703 – 707 (1986).
Existing IUE and Einstein observations of the high–redshift ($z = 1.414$) quasar 3C 446 have been analyzed. In the UV the continuum is characterized by a steep slope. The lack of signal shortward of 2300 Å is probably due to the Lyman continuum absorption at the source. The X–ray spectrum shows the "universal" slope of active galactic nuclei coupled with a hydrogen columnar density in excess of the Galactic value, suggesting that the absorption occurs in or around the source.

159.041 **Multifrequency observations of the superluminal quasar 3C 345.**
J. N. Bregman, A. E. Glassgold, P. J. Huggins, G. Neugebauer, B. T. Soifer, K. Matthews, J. Elias, J. Webb, J. T. Pollock, A. J. Pica, R. J. Leacock, A. G. Smith, H. D. Aller, M. F. Aller, P. E. Hodge, W. A. Dent, T. J. Balonek, R. E. Barvainis, T. P. L. Roellig, W. Z. Wiśniewski, G. H. Rieke, M. J. Lebofsky, B. J. Wills, D. Wills, W. H.–M. Ku, J. D. Bregman, F. C. Witteborn, D. F. Lester, C. D. Impey, J. A. Hackwell.
Astrophys. J., Vol. 301, No. 2, p. 708 – 726 (1986).
The authors have investigated the continuum properties of the superluminal quasar 3C 345 with monitoring studies at radio, optical, infrared, and X–ray frequencies. The data presented here include optical, infrared, and radio measurements for a period of 20 yr, far–infrared fluxes obtained with IRAS, and X–ray fluxes taken with the Einstein Observatory. The second aspect of this work is the presentation and discussion of four simultaneous multifrequency spectra. These spectra, which extend from radio to ultraviolet frequencies, are analyzed to determine spectral shapes, total power, the connection between the spectral regions, and the details of spectral variation during outbursts. Simple theoretical considerations are applied to these data to estimate conditions in the emitting region. The ultraviolet and optical emission lines are compared to those of normal quasars and found to be an important discrimination for the issue of relativistic beaming.

159.042 **Distribution of QSO emission line redshifts and cosmological redshift.**
Z.–g. Deng, Y.–y. Zhou, H.–j. Dai.
Sci. Sin., Ser. A, Vol. 29, No. 3, p. 312 – 325 (1986).
The authors calculate the redshift distribution on the assumption of the cosmological redshift, using the standard cosmological model. The results are in good agreement with observations. Especially, for the QSOs obtained with the slitless technique, the distribution obtained from calculations are almost the same as that from observations. It does not only show that the peaks and dips appeared in the redshift distribution of QSOs are consistent with the cosmological principle, but also may estimate the cosmological parameter q_0 from the observed redshift distribution of QSOs.

159.043 **Les mirages gravitationnels: un outil de choix pour sonder l'univers.**
C. Vanderriest.
Astronomie, Vol. 100, p. 3 – 16 (1986).

159.044 **Radio observations of extended sources suspected of having steep–spectrum cores.**
D. J. Saikia, V. K. Kulkarni, R. W. Porcas.
Mon. Not. R. Astron. Soc., Vol. 219, No. 4, p. 719 – 728 (1986).
This paper presents VLA and VLBI observations of a sample of extended radio sources suspected of having steep–spectrum cores (SSCs). For these sources, the authors find that either previous estimates of the core spectral index are too high or the cores have not been identified correctly. This work suggests that true SSCs associated with the nuclei of extended radio sources are very rare.

159.045 **Curvature effects on tests for quasar alignment.**
M. G. Edmunds, G. H. George, S. Phillipps.
Mon. Not. R. Astron. Soc., Vol. 219, No. 4, p. 865 – 869 (1986).
Effects are discussed which affect searches for triplet (or higher order) alignments of objects in the sky, particularly relevant for quasar alignments. An effect which can readily be corrected for is the incautious use of a Cartesian projection of a spherical polar coordiante system when the alignments in fact lie on great circles. A much more significant effect is the random fluctuation in positions induced by non–homogeneous clumping of matter in the Universe. It appears that this effect would seriously dilute any statistical test for alignments in which the central quasar is at a moderate cosmological distance. From isotropy arguments the linearity of alignments cannot be affected by the overall curvature of the Universe.

159.046 **A sample of powerful radio sources for VLBI studies.**
T. V. Cawthorne, P. A. G. Scheuer, I. Morison, T. W. B. Muxlow.
Mon. Not. R. Astron. Soc., Vol. 219, No. 4, p. 883 – 893 (1986).
High–resolution observations are presented of several powerful extragalactic radio sources. They are members of a sample of sources, whose flux density (excluding any contribution from compact central components) exceeds 10 Jy at 178 MHz and whose luminosity exceeds 10^{28}W $Hz^{-1}sr^{-1}$ at 178 MHz (emitted frequency), which were poorly resolved on previously published maps. Such a sample might be used as part of the systematic investigation of proper motion of components in VLBI jets, since it is free from any selection effects that might prejudice the outcome. The constraints that VLBI observations might place on simple models for the "superluminal velocities" are briefly discussed.

159.047 **Study of the structure of the quasar 3C 345 at decimeter wavelengths.**
L. I. Matveenko, I. I. K. Pauliny–Toth, B. Sherwood, L. Bååth, A. Kus.
Pis'ma Astron. Zh., Tom 12, No. 2, p. 156 – 167 (1986). In Russian. English translation in Sov. Astron. Lett., Vol. 12.
The structure of the quasar 3C 345 has been studied at 18 and 49 cm wavelengths by the VLBI method. The structure has a nucleus and a jet, which changes direction with distance from the nucleus. The continuum emission of the nucleus and the jet re-

gions correspond to a source with large optical depth, the cut frequence goes down with increasing distance from the nucleus.

159.048 **High-resolution radio observations of the most luminous quasar.**
H. Kühr, J. T. Stocke, P. A. Strittmatter, N. Bartel, A. Eckart, C. Schalinski, A. Witzel, P. Biermann.
Astrophys. J., Vol. 302, No. 1, p. 52 – 55 (1986).
The extremely high energy output of S5 0014+81 ($5x10^{48}$ergs s^{-1}) and the presence of a high column density metal absorption system at $z = 1.11$ in the optical spectrum suggest that a gravitational lens might be present. Indeed, a weak second component was found with the VLA at 6 cm, 0".6 south of the unresolved main component. Further VLA and VLBI observations reveal clearly only the one bright source, but with an elongation in the direction of the second component. Therefore, the radio structure is probably a jetlike phenomenon, and it does not provide evidence for or against a gravitational lens.

159.049 **Relativistic beaming and quasar emission lines.**
B. J. Wills, I. W. A. Browne.
Astrophys. J., Vol. 302, No. 1, p. 56 – 63 (1986).
A highly significant correlation has been found for quasars, between the line widths (FWHM) of broad Hβ lines and R, the ratio of radio core flux density to the extended radio lobe flux density. The authors discuss how this might be accounted for by a simple relativistic beaming model in which the motion of the broad–line gas is confined to a disk whose axis is perpendicular to the radio axis. The authors comment on how this relation may explain why Fe II lines are apparently weak in many objects with strong extended radio lobes. They also show how the very asymmetric distribution of FWHM may be accounted for in this model. Finally the dependence of the intensity of the [O III] lines on relative core strength is discussed.

159.050 **The emission–line regions in broad absorption line quasars.**
G. F. Hartig, J. A. Baldwin.
Astrophys. J., Vol. 302, No. 1, p. 64 – 80 (1986).
The properties of the emission–line regions in 16 broad absorption line (BAL) quasars are compared to those of non–BAL quasars. It is found that about half the BAL quasars, those having detached BALs, are significantly different from the authors' comparison samples of normal quasars. They have large Al III/C III] intensity ratios, strong Fe II emission, and unusually weak C IV emission lines whose profiles do not agree well with those of the lower ionization emission lines.

159.051 **Evolution of the luminosity function of quasars.**
M. Schmidt.
Structure and evolution of active galactic nuclei, p. 3 – 10 (1986). – See Abstr. 012.011.
The luminosity function of quasars shows a strong dependence on redshift. The author discusses alternate descriptions of this evolution. Density evolution is assumption–free if the density variation with redshift is also allowed to be a function of absolute magnitude. Luminosity evolution and, in particular, pure luminosity evolution make a priori assumptions about the shape of the luminosity function. The author illustrates the derivation of density evolution and its strong dependence on optical luminosity. He discusses work by Schmidt and Green (1985) on X–ray counts of quasars and the need for negative X–ray luminosity evolution to explain the counts and the low average redshift of X–ray quasars. As a consequence the quasar contribution to the 2 keV X–ray background is only around 8 – 13%. The evolution of active galactic nuclei of lower optical luminosity is tightly constrained by the observed X–ray background.

159.052 **Continuity in the observed properties of QSOs, high–redshift emission–line radio galaxies, BL Lac objects, N systems and Seyfert galaxies, and possible interpretations.**
G. Burbidge.
Structure and evolution of active galactic nuclei, p. 47 – 62 (1986). – See Abstr. 012.011.
The evidence for continuity between the observed properties of QSOs and the active nuclei of galaxies of various types is reviewed. It is concluded that the continuity argument is compatible with many of the observations, provided (1) that the idea of directed bulk relativistic motions in many objects which show apparent superlight velocities or very large rapid flux variations is accepted, and (2) that all of the evidence for the physical associations of QSOs and galaxies, and pairs of QSOs, with very different redshifts, is ignored.

159.053 **Optical variability in quasars.**
S. Cristiani.
Structure and evolution of active galactic nuclei, p. 81 – 93 (1986). – See Abstr. 012.011.
Analysis of QSO light curves shows variations often on time–scales of one year and, in some cases, of a few days, implying dimensions <1 pc for the central source. Very simple computations indicate that, in this situation, the underlying energy source must be gravitational in nature (Cavaliere 1981) as modelled by Zeldovich and Novikov and by Salpeter as early as 1964. But the wealth of information for theoreticians coming from studies of variability in quasars is not limited to these simple results. The author recalls here the phenomenon of superluminal velocities (see Cohen and Unwin 1982).

159.054 **Observations of faint quasars: shape and evolution of the luminosity function.**
D. C. Koo.
Structure and evolution of active galactic nuclei, p. 317 – 337 (1986). – See Abstr. 012.011.
The author highlights results from a study of faint quasars in SA 57 using 4–m photographic multicolor photometry, astrometry, variability, and CCD spectroscopy. Quasar candidates with $B < 22.6$ were chosen as stellar–like objects with colors unlike normal stars. Together, these techniques allow additional corrections for white dwarfs, hot subdwarfs, and narrow–emission–line galaxies in the sample and yield a quasar density of about $150/deg^2$ to $B = 22.6$. A major surprise was that no quasars were found with redshifts above 2.54 among the 8 bonafide quasars fainter than $B = 21.5$. The author discusses the importance of their faint counts and redshifts in defining the shape and evolution of the luminosity function of quasars.

159.055 **A new complete sample of faint, optically selected quasars: comparison with previous results.**
B. Marano, G. Zamorani, V. Zitelli.
Structure and evolution of active galactic nuclei, p. 339 – 352 (1986). – See Abstr. 012.011.
The authors present the results on a new sample of faint, optically selected quasar candidates. Two different search techniques (multicolour and grism) have been applied to the same field of 0.69 square degrees. The cross–check of the two lists of candidates allows to estimate the level of completeness of each technique. The authors find about 50 quasars per square degree with $J < 21.0$. In the magnitude range $J = 21 - 22$ the results are in very good agreement with those of other existing surveys.

159.056 **Quasar evolution and luminosity function: an X–ray perspective.**
Y. Avni.
Structure and evolution of active galactic nuclei, p. 397 – 420 (1986). – See Abstr. 012.011.
X–ray properties of optically–selected QSOs are reviewed, and their implications for the QSO evolution and luminosity function are summarized. Analyses of the explicit dependence of $\alpha_{0,x}$, the X–ray to optical luminosity ratio parameter, on redshift z and optical luminosity L_0, are described. Previous results obtained from the study of heterogeneous samples are confirmed. It is shown that the large majority, probably all, of optically–selected QSOs are X–ray loud.

159.057 **Properties of the sharp metal–rich absorption lines observed in QSO spectra.**
J. Bergeron.
Structure and evolution of active galactic nuclei, p. 421 – 435 (1986). – See Abstr. 012.011.
The author considers here the sharp metal–rich absorption systems. She outlines their properties, with emphasis on their degree of ionization, and compares them to those of halos or extended disks around nearby galaxies. She presents some evidence for cosmological evolution of the density per unit redshift z of these systems and tries to find whether the differences between the high z absorbers and galaxy halos are a consequence of cosmological evolution in the ionization state or whether there are two populations or two phases among the metal–rich systems.

159.058 **Quasar candidates in the field of SA 94 ($2^h53^m + 0°20'$).**
C. Barbieri, S. Cristiani.
Structure and evolution of active galactic nuclei, p. 465 – 470 (1986). – See Abstr. 012.011.
The authors present the results obtained from the visual inspection of two UK–Schmidt plates covering a field centered on the SA 94. The search produced 208 candidates for 40 of which slit spectra were taken subsequently with the ESO telescopes at La Silla (4 were already known in the literature to be quasars). An effort was also made to put the magnitude scale on a firm basis.

159.059 **Improvement of the Hubble diagram for QSO's leading towards q_0.**
J. E. Beckman, M. Kidger.
Structure and evolution of active galactic nuclei, p. 471 – 474 (1986). – See Abstr. 012.011.
Using the predictions of the Lynden Bell/Hills model of the energy source of QSO's in which discrete packets of matter (e.g. stars) are converted into radiation near a massive black hole the authors show how the variability observations of QSO's over long periods may be used to upgrade the Hubble diagram for these objects.

159.060 **A new faint QSO survey.**
B. J. Boyle, R. Fong, T. Shanks, B. A. Peterson.
Structure and evolution of active galactic nuclei, p. 491 – 498 (1986). – See Abstr. 012.011.
The authors are compiling a large catalogue of QSOs with complete spectroscopic identification. Candidate QSOs are selected from COSMOS machine measurements of UK Schmidt photographic plates using the ultra–violet excess (UVX) technique. These candidates are subsequently observed spectroscopically using the new 40 arc minute fibre optic coupler at the Anglo Australian Telescope which allows a maximum of 50 objects to be observed simultaneously. To date the authors have obtained spectra for 400 (B < 21) UVX objects of which approximately 170 are emission line objects.

159.061 **The low excitation lines in quasars and AGN in the frame work of photoionization models.**
S. Collin–Souffrin.
Structure and evolution of active galactic nuclei, p. 531 – 535 (1986). – See Abstr. 012.011.
Up to now, the photoionization models, which account well for the intensities of high excitation lines (C IV, C III], L_α...) in quasars and AGN, are less successful in explaining the low excitation lines (Fe II, Mg II, Balmer lines). In this paper a work is described in which a large number of situations, especially suited for low excitation line emission, are explored, in the framework of photoionization models.

159.062 **Cosmological evolution of optically selected QSOs.**
L. Danese, G. De Zotti, A. Franceschini.
Structure and evolution of active galactic nuclei, p. 537 – 540 (1986). – See Abstr. 012.011.
The authors report on an updated analysis of the cosmological evolution of optically selected QSOs. Both luminosity dependent density evolution (DE) models and luminosity evolution (LE) models have been tested. The log of the local luminosity function has been expressed as a third order polynomial whose four coefficients have been treated as free parameters. LE models require only one additional free parameter, while four of them have been used for DE models.

159.063 **The I.A.C./Q.M.C. catalogue of quasar multiband spectra.**
M. R. Kidger, J. E. Beckman.
Structure and evolution of active galactic nuclei, p. 591 – 600 (1986). – See Abstr. 012.011.
One of the major problems using quasars to investigate the Hubble diagram is an uncertainty as to the size of the bolometric correction. Some estimates have suggested that it may be as high as 6 magnitudes. To this end, the authors have compiled data from a wide range of sources, to investigate the behaviour of quasar spectra over the range from radio to X–rays, and hence to be able to calculate this important parameter. Here they present a sample of the compiled spectra, whilst a complete catalogue of around two hundred spectra is in preparation.

159.064 **The optical variability of 3C 345.**
M. R. Kidger, J. E. Beckman.
Structure and evolution of active galactic nuclei, p. 601 – 604 (1986). – See Abstr. 012.011.
The publication by Lloyd (1984) of new and revised data for a sample of active nuclei brought the number of published observations of 3C 345 up to over 900, between 1965 and 1980. Since the last major study, by Barbieri et al. (1977), sufficient new data has been produced to merit a complet revision of previous analysis. A compilation of the lightcurve and Fourier Transform analysis are presented here and compared with previous studies. The authors find no strong evidence for long lasting major periodicities, in contradiction to some other published conclusions.

159.065 **A test for the gravitational lens hypothesis in the most luminous quasar: S5 0014+81.**
H. Kühr.
Structure and evolution of active galactic nuclei, p. 611 – 614 (1986). – See Abstr. 012.011.
The most luminous quasar known, S5 0014+81 (z = 3.41), was extensively studied with direct CCD imaging, and with radio interferometric techniques using the VLA as well as MK II and MK III VLBI systems. Optical spectroscopic observations revealed a high column density metal absorption system at z = 1.11, and VLA maps displayed a weak second radio component. Thus evidence for a possible existence of a gravitational lens is given, which may have boosted the observed flux of the quasar towards an energy output of about 5×10^{48}erg/s. However, results from subsequent VLA and VLBI observations are not compatible with such interpretation, instead the weak radio component is probably associated with a jet–like phenomenon.

159.066 **The luminosity function of quasars and low luminosity active galactic nuclei.**
H. L. Marshall.
Structure and evolution of active galactic nuclei, p. 627 – 632 (1986). – See Abstr. 012.011.
The quasar luminosity function is extended to include the nuclear luminosities of Seyfert 1 galaxies. The model is fitted to the (total) luminosity data for the active galaxies from a complete flux–limited survey. A good fit is obtained with $dN/dL \propto L^{-1.2}$, matching onto the steeper quasar luminosity function near $M_B = -22$. The feature thus produced conforms to the luminosity functions of Koo and Kron using a pure luminosity evolution model derived from brighter quasar data.

159.067 **Flux variations and structural changes in extragalactic radio sources.**
L. Padrielli, M. F. Aller, H. D. Aller, N. Bartel, C. Fanti, R. Fanti, A. Ficarra, L. Gregorini, F. Mantovani, L. Matveenko, G. D. Nicolson, J. D. Romney, K. W. Weiler.
Structure and evolution of active galactic nuclei, p. 653 – 657 (1986). – See Abstr. 012.011.
The authors have established the existence of at least 3 classes of the low frequency variability (LFV). On a preliminary basis,

they might interpret the observed different behaviours as the manifestation of two distinct phenomena. Together with the relativistic bulk motion responsible for the high frequency activity sometimes extending to the meter wavelengths, another mechanism, probably extrinsic to the source, must coexist to account for the variability in the low frequency range of the sources with uncorrelated LF activity.

159.068 **Two newly discovered OVV objects.**
Y. Zhan, J. S. Chen.
Astron. Astrophys., Vol. 160, No. 2, p. 321 – 323 (1986).
Two newly discovered OVV (optically violent variable) objects showing emission line features on low disperson objective prism spectra during the bright phases of their outbursts are reported together with the failure detection in deep CCD observations, suggesting they have very large variations.

159.069 **Search for "antipodal" images of quasars.**
V. S. Lebedev, I. A. Lebedeva.
Bull. Spec. Astrophys. Obs. – North Caucasus, Vol. 19, p. 12 – 13 (1985). English translation of 39.159.038.

159.070 **Search for periodicity in the spatial distribution of quasars.**
V. S. Lebedev, I. A. Lebedeva.
Bull. Spec. Astrophys. Obs. – North Caucasus, Vol. 19, p. 14 – 18 (1985). English translation of 39.159.039.

159.071 **Possible role of dwarf galaxies and globular clusters in the formation of L_α and 21–cm absorption lines in the spectra of distant QSOs.**
B. V. Komberg.
Astrofizika, Tom 24, Vyp. 2, p. 321 – 331 (1986). In Russian. English translation in Astrophysics, Vol. 24, No. 2.
This paper considers the main observational characteristics of absorption lines of classes I and II which are formed in the spectra of distant quasars when the line of sight crosses the gaseous complexes in the intervening objects. Emphasis has been put on the discussion of the possible nature of objects responsible for numerous class–II absorption lines (the so–called "L_α–forest" lines). The paper considers the possible nature of clouds where 21–cm absorption lines are formed that are sometimes observed in radio spectra of quasars. The luminosity, size and mass of the objects responsible for the formation of 21–cm lines are estimated. The objects are similar to globular clusters in terms of these characteristics, though the former have relatively larger amounts of gas ($\sim 50\ M_\odot$).

159.072 **Latest advances in quasar investigations.**
A. Balklavs.
Zvaigžnota Debess, 1985. Gada Pavasaris, p. 17 – 26 (1985). In Latvian.

159.073 **Are there optically–selected X–ray–quiet QSOs?**
Y. Avni, H. Tananbaum.
Bull. Am. Astron. Soc., Vol. 17, No. 4, p. 829 (1985). Abstract. – See Abstr. 010.062.

159.074 **Origin of the thermal continuum in AGN's and QSO's.**
R. E. Stoner, R. L. Ptak.
Bull. Am. Astron. Soc., Vol. 17, No. 4, p. 829 (1985). Abstract. – See Abstr. 010.062.

159.075 **Q/G 0903+175: the closest projected quasar/galaxy pair yet seen.**
S. Djorgovski, P. McCarthy.
Bull. Am. Astron. Soc., Vol. 17, No. 4, p. 830 (1985). Abstract. – See Abstr. 010.062.

159.076 **X–ray emission from the jet in 3C 273.**
C. P. Stern, D. E. Harris.
Bull. Am. Astron. Soc., Vol. 17, No. 4, p. 831 – 832 (1985). Abstract. – See Abstr. 010.062.

159.077 **Ly α galaxy companions of quasars at z > 2.5.**
H. Spinrad, S. Djorgovski, M. Strauss, P. J. McCarthy.
Bull. Am. Astron. Soc., Vol. 17, No. 4, p. 865 (1985). Abstract. – See Abstr. 010.062.

159.078 **QSOs and galaxies.**
A. Stockton.
Bull. Am. Astron. Soc., Vol. 17, No. 4, p. 868 (1985). Abstract. – See Abstr. 010.062.

159.079 **Variability of hydrogen emission lines of some QSOs.**
W. Zheng.
Bull. Am. Astron. Soc., Vol. 17, No. 4, p. 869 (1985). Abstract. – See Abstr. 010.062.

159.080 **EXOSAT X–ray and VLBI radio observations of the quasar NRAO 140.**
A. P. Marscher.
Bull. Am. Astron. Soc., Vol. 17, No. 4, p. 890 (1985). Abstract. – See Abstr. 010.062.

159.081 **The multiple source 0023+171: a dark gravitational lens?**
J. N. Hewitt, E. L. Turner, C. R. Lawrence, D. P. Schneider, J. E. Gunn, M. Schmidt, J. H. Mahoney, G. I. Langston, B. F. Burke.
Bull. Am. Astron. Soc., Vol. 17, No. 4, p. 907 (1985). Abstract. – See Abstr. 010.062.

159.082 **Absorption lines and the radio structure of quasars.**
G. Swarup, D. J. Saikia, M. Beltrametti, R. P. Sinha, C. J. Salter.
Mon. Not. R. Astron. Soc., Vol. 220, No. 1, p. 1 – 18 (1986).
The authors present high angular resolution radio observations at λ6 cm of 20 quasars. They summarize the radio structures for the entire sample of 66 sources of Weymann et al. and investigate whether the radio structure might provide clues on understanding the origin of the absorption lines in the range $z_{em} - z_{ab} \sim 3000 - 18000$ km s^{-1}. The frequency of occurrence of absorption lines appears to be similar for both radio and optically selected quasars. The authors find a marginal trend for the radio–selected sources in the above range of $z_{em} - z_{ab}$ to have compact radio structures. An investigation of a larger sample of quasars from the compilation by Hewitt & Burbidge is suggestive of a similar trend.

159.083 **X–ray and radio core emission in radio quasars.**
A. Kembhavi, E. D. Feigelson, K. P. Singh.
Mon. Not. R. Astron. Soc., Vol. 220, No. 1, p. 51 – 67 (1986).
In order to investigate the physical relationship between X–ray and radio core emission in radio–selected quasars, the authors have observed with the Very Large Array, 35 radio quasars at 6 and 20 cm. The sample was chosen from a list of radio quasars with known X–ray luminosity but poorly known radio properties. The authors have sought statistical association between radio core luminosity and X–ray luminosity, and find that there is a strong correlation.

159.084 **High–resolution imaging from Mauna Kea: the triple quasar in 0.3–arc s seeing.**
J. P. Henry, J. N. Heasley.
Nature, Vol. 321, No. 6066, p. 139 – 142 (1986).
The "triple" quasar, PG 1115+08 (which actually contains at least four objects: A_1, A_2, B and C) was the second gravitational lens object to be found. So far, the lensing object(s) have not been identified. The authors report imaging observations made on a night of exceptional seeing, as well as spectroscopy of two galaxies near the quasar. They find that there is a galaxy centred approximately midway between the two A components. The properties of of this galaxy are consistent with it being the brightest member of a small group at a group redshift of 0.305. Although detailed modelling is required, it is likely that the galaxy and its group are the lenses.

159.085 **An apparent gravitational lens with an image separation of 2.6 arc min.**
E. L. Turner, D. P. Schneider, B. F. Burke, J. N. Hewitt, G. I. Langston, J. E. Gunn, C. R. Lawrence, M. Schmidt.
Nature, Vol. 321, No. 6066, p. 142 – 144 (1986).
The authors report observations which confirm Paczynski's speculation that the previously known pair of 19th magnitude quasars 1146+111B,C are actually two images of a single object produced by a gravitational lens. The image splitting is 157 arc s, thus indicating an exceptionally massive lensing object. The data supporting the lens hypothesis are remarkably similar, high signal–to–noise, moderate–resolution spectra of the two components. Both spectra show strong Mg II λ2798 emission at $z = 1.012 \pm 0.001$ with indistinguishable redshifts, widths and detailed profile shapes. If the foreground galaxy clustering apparent in a deep *R* band CCD image proves insufficient to explain the large image splitting, other possibilities such as massive dark objects (for example, a $\sim 10^{15} M_{\odot}$ black hole) or a cosmic string may be indicated.

159.086 **Quasars to *B* > 22.5 in Selected Area 57: a catalog of multicolor photometry, variability, and astrometry.**
D. C. Koo, R. G. Kron, K. M. Cudworth.
Publ. Astron. Soc. Pac., Vol. 98, No. 601, p. 285 – 306 (1986).
The authors present a catalog of 77 quasar candidates at the north galactic pole (Selected Area 57) to a blue magnitude limit of 22.5 over an area of 0.3 square degree. The catalog is based on data from 15 deep 4–m plates and includes photographic *UBVI* photometry, an index of variability, and proper motions for all but two of the objects. Spectroscopy of over half the sample yields 27 redshifts for galaxies and quasars. Candidates are selected photometrically. The lack of large numbers of non-ultraviolet–excess candidates at faint magnitudes suggests that very faint quasars with redshifts greater than 2.5 are rare. The counts similarly limit the existence of a large population of primeval galaxies that otherwise resemble quasars in color and image size. The authors demonstrate that variability and proper–motion data are potentially powerful and complementary techniques to separate true quasars from peculiar subdwarfs and white dwarfs.

159.087 **Gravitational lenses and multiple quasars.**
S. M. Chitre.
Binary and multiple systems, p. 63 – 68 (1984). – See Abstr. 012.025.
The VLBI features observed in the quasar Q0957+561 A, B are incorporated in a gravitational lens model. The effective time delay between components A and B is shown to be of the order of one year and the observed intensity variation in B is demonstrated to arise from the action of minilensing by a low–mass star.

159.088 **Jets in quasars.**
D. J. Saikia.
Binary and multiple systems, p. 69 – 84 (1984). – See Abstr. 012.025.
Although jets have provided the most direct evidence in favour of the 'beam models' of radio sources, the detailed understanding of these features is still at the initial stages. More observations aimed at obtaining high dynamic range maps and also detailed rotation measure and spectral index maps for a large sample of these sources would be welcome.

159.089 **Radio and X–ray emission in radio–loud quasars.**
A. Kembhavi.
Binary and multiple systems, p. 85 – 94 (1984). – See Abstr. 012.025.

159.090 **Correlated radiation parameters of quasars and Seyfert nuclei, and the synchro–Compton reactor.**
A. P. Mosin, Eh. A. Dibaj, V. M. Charugin.
Sov. Astron., Vol. 29, No. 4, p. 389 – 393 (1985). English translation of 40.159.023.

159.091 **Photographic photometry of compact extragalactic objects: optical variability of the quasar 3C 345.**
M. K. Babadzhanyants, E. T. Belokon', N. S. Denisenko, E. V. Semenova.
Sov. Astron., Vol. 29, No. 4, p. 394 – 399 (1985). English translation of 40.159.024.

159.092 **The remarkable absorption–line systems in the quasar Tololo 1037–27.**
M. H. Ulrich, M. A. C. Perryman.
Mon. Not. R. Astron. Soc., Vol. 220, No. 2, p. 429 – 438 (1986).
The authors present the first spectroscopic observations and analysis of the absorption–line systems in the quasar Tololo 1037–27. The observations show that this quasar has a rich and remarkable absorption spectrum. Six absorption–line systems are found between $z_{abs} = 1.971$ and $z_{abs} = 2.138$. The two strongest of these systems are highly ionized and have Lyα and C IV linewidth between 600 and 1100 km s^{-1} wide. These lines could be the superposition of the absorption lines produced in the haloes of galaxies in two separate clusters, where the absorbing material is ionized by the quasar. Alternatively these absorption systems could be intrinsic to the quasars and represent two outflows of gas.

159.093 **CCD imaging of the QSO–galaxy pair PKS 1327–206, ESO 1327–2041.**
E. Giraud.
Astron. Astrophys., Vol. 161, No. 1, p. 206 – 208 (1986).
CCD images of the QSO–galaxy pair, PKS 1327–206, ESO 1327–2041 show that the galaxy is very disturbed. A strong plume or jet emerges from the galaxy as well as luminous extensions. A filament or a ring structure, asymmetric and possibly related to a polar ring may be present. The ring is superimposed on the image of the quasar. It is possible that the structure at this point could be a gas rich region with recent star formation.

159.094 **Results of a preliminary search for quasars in the Virgo cluster region.**
X. T. He, R. D. Cannon, M. G. Smith.
Proceedings of the Second Asian–Pacific Regional Meeting on Astronomy, p. 375 – 385 (1984). – See Abstr. 012.031.
Results of a preliminary survey for optically selected quasars in a central 5°x5° region of the Virgo cluster are presented. 53 emission–line quasar candidates and 18 blue stellar objects (possible low–redshift quasars) have so far been found within the 25 square degree region. Illustrations of some of the objective–prism spectra are given.

159.095 **Quasar research with Schmidt telescopes.**
A. Savage.
Proceedings of the Second Asian–Pacific Regional Meeting on Astronomy, p. 386 – 397 (1984). – See Abstr. 012.031.
This paper presents a very general review of the way Schmidt telescope plate material can be used in quasar research, with a bias towards research with the United Kingdom 1.2 m Schmidt telescope.

159.096 **Absolute spectrophotometry of quasars: TON 469 (3C 232) and 3C 249.1.**
D. Dultzin–Hacyan.
Rev. Mex. Astron. Astrofis., Vol. 11, No. 2, p. 121 – 126 (1985). = Contrib. Inst. Astron., Univ. Nac. Autónoma Mex., No. 178.
The author analyzes in some detail the spectrophotometric properties of quasars TON 469 (3C 232) and 3C 249.1 and compares the results with previous observations of these objects.

159.097 **Redshifts and observed colors of QSOs and active galaxies.**
D. Basu.
Astron. J., Vol. 91, No. 2, p. 226 – 230 (1986).
The present work studies the relationship between colors and redshifts of the two classes of objects, QSOs and active galaxies, individually and combined. A statistically significant mathematical relation between color index and redshift has been determined for active galaxies. The present analysis confirms the nonlinear

variation of color indices with redshifts of QSOs to be certainly due to the effect of emission lines. Interpretation of the results also suggests that active galaxies and QSOs are possibly manifestations of the same type of object observed at different redshifts.

159.098 **The nature of the emission and the cosmological evolution of quasars: what are the X–ray observations telling us?**
G. Zamorani.
X–ray astronomy '84, p. 419 – 433 (1985). – See Abstr. 012.033.
The author tries to show how the X–ray observations of quasars obtained in the past few years can be used to gain some insight into the problem of the dominant emission mechanisms in these objects. Then he discusses the present state of the art of the X–ray cosmological evolution of quasars and the related problem of their contribution to the diffuse soft X–ray background.

159.099 **X–ray and optical evolution of quasars.**
H. L. Marshall.
X–ray astronomy '84, p. 463 – 465 (1985). – See Abstr. 012.033.
Quasar surveys with complete redshift information are analyzed by a likelihood method which treats the luminosity function in a non–parametric but time–independent manner. Confidence regions are obtained for the evoutionary parameters and tests are performed to check for changes in the evolution rate as a function of luminosity and/or redshift. The consistency of the results from the two spectral bands is examined.

159.100 **The high energy X–ray spectrum of 3C 273.**
R. Staubert, M. Bezler, E. Kendziorra, G. Hasinger, W. Pietsch, C. Reppin, J. Trümper, W. Voges.
X–ray astronomy '84, p. 483 – 486 (1985). – See Abstr. 012.033.
The quasar 3C 273 was observed in the hard X–ray range (18 – 240 keV) with a balloon–borne detector on 1981 September 28. The source was found to be quite strong and to emit a very hard spectrum. Compared to other observations in 1978/79 the luminosity increased by a factor of three, representing the first observation of time variability of 3C 273 in this energy range. The authors briefly discuss this behaviour with respect to synchrotron–self–Compton and thermal Comptonization models.

159.101 **The core region in quasars.**
M. Camenzind, T. J.–L. Courvoisier.
X–ray astronomy '84, p. 495 – 498 (1985). – See Abstr. 012.033.
The authors present a particular two–component model which includes (1) a stationary source consisting of the central thermal UV–emission, synchrotron and X–ray emission from an extended shell, as well as (2) time–dependent beamed emission from moving shock fronts in a stationary jet.

159.102 **The bremsstrahlung and π^0 decay as gamma–ray source in 3C 273 and Geminga.**
W. Tkaczyk, S. Karakula, F. Giovannelli.
X–ray astronomy '84, p. 499 – 502 (1985). – See Abstr. 012.033.
The experimental data from COS B sources can be fitted with the expected gamma–ray spectrum from π^0 decay and bremsstrahlung for the thermal and non–thermal models in the case of matter spherically accreting onto collapsed objects. The authors compare spectra from thermal and non–thermal models with experimental measurements in the gamma and X–ray regions for 2CG 289 + 64 (3C 273) and 2CG 195 + 04 (Geminga) gamma–ray sources. The accreting process can explain the observed spectrum of 3C 273 but not that of Geminga. The identification of Geminga with a local source is more probable.

159.103 **Detection of the lensing galaxy in PG 1115 + 08.**
S. B. Shaklan, E. K. Hege.
Astrophys. J., Vol. 303, No. 2, p. 605 – 613 (1986).
The authors have fitted models to good–seeing (0".6) post-guided observations of the triple quasar PG 1115 + 08 in blue and red light. The four components (A, A', B, and C) all have similar point spread functions in the blue fit. The fits to components A, A', and C in the red image are consistent, but that of the B component suggests that there is an excess red flux from an area between all four components. The residuals show that this flux has a peak magnitude of 19.8 ± 0.3 mag in the *R* band and a *V–R* color of at least 2.7 ± 0.5 mag. The position of this flux excess is consistent with that of the lensing galaxy in models of gravitational imaging.

159.104 **On the completeness of the Medium Sensitivity Survey quasar sample.**
T. Maccacaro, I. M. Gioia.
Astrophys. J., Vol. 303, No. 2, p. 614 – 616 (1986).
The Medium Sensitivity Survey (MSS) quasar sample is being used more and more by authors to study quasar properties in general and to analyze the relationship between X–ray–selected and optically selected quasars in particular. The authors have undertaken a detailed examination of possible causes of incompleteness of the MSS quasar sample, paying particular attention to the problem of interstellar absorption within the Galaxy. It is found that there is no evidence of a loss of sources due to the effects considered, and that the MSS quasar sample is statistically complete.

159.105 **Distorted structure in the small high–redshift radio source 4C 29.50.**
C. J. Lonsdale, P. D. Barthel.
Astrophys. J., Vol. 303, No. 2, p. 617 – 623 (1986).
The powerful radio source 4C 29.50, which is identified with a quasar at redshift 1.927, has been mapped with a variety of instruments, frequencies, and resolutions. The highest resolution is 30 mas, corresponding to about 150 pc at this redshift. There are apparent sudden deflections of a jet emerging from the quasar core, and it is proposed that these are due to interaction with a dense, clumpy interstellar environment. The great power of the source and the small scales on which the deflections occur yield constraints on the properties of this environment.

159.106 **The quasar pair Tololo 1037–27 and 1038–27: evidence for correlated absorption on megaparsec scales.**
P. Jakobsen, M. A. C. Perryman, M. H. Ulrich, F. Macchetto, S. di Serego Alighieri.
Astrophys. J., Lett. Ed., Vol. 303, No. 2, p. L27 – L30 (1986).
The authors present medium–resolution observations of the absorption–line spectrum of the quasar Tol 1038–27 ($z = 2.32$) and point out its remarkable similarity to that of the nearby quasar Tol 1037–27 ($z = 2.18$). Both objects display a broad C IV system centered near $z = 2.08$ in addition to three neighboring C IV systems that coincide in redshift in the two quasars to within 2000 km s^{-1}. At an angular separation of 17'.9, the projected distance between the lines of sight to the two objects is of order ~4 Mpc. The absorption therefore may be caused by a massive intervening supercluster.

159.107 **The unusual ultraviolet variability of the QSO 3C 232.**
F. C. Bruhweiler, M. Kafatos, U. J. Sofia.
Astrophys. J., Lett. Ed., Vol. 303, No. 2, p. L31 – L35 (1986).
After examining ultraviolet *IUE* spectra of 18 QSOs with $z > 0.35$, the authors find that one object, 3C 232 ($z = 0.533$), stands out as exhibiting highly unusual spectral and continuum variability. The H I–Lyα shows at least 50% variability on a time scale of roughly a month, while the ultraviolet continuum exhibited approximately a factor of 2 decrease in a day. Several spectral features at wavelengths shortward of H I–Lyα in the rest frame are identified.

159.108 ***EINSTEIN* IPC spectra of quasars: first results.**
M. Elvis.
Japan – U.S. seminar on galactic and extragalactic compact X–ray sources, p. 291 – 300 (1985). – See Abstr. 012.034.
The diversity of the *Einstein* Imaging Proportional Counter spectra of quasars is presented and discussed in the context of the complete infrared–to–X–ray continuum. Several different emission mechanisms are seen to be necessary to explain the results.

159.109 **A study of the optical and X-ray properties of quasars.**
S. E. Bradley.
Diss. Abstr. Int., Sect. B, Vol. 46, No. 6, p. 1956 (1985). Thesis, University of California, San Diego, 143 pp. (1985). Order No. DA8517893.

159.110 **A relativistic blast wave model for outbursts in quasars and active galactic nuclei.**
E. B. Hessel.
Diss. Abstr. Int., Sect. B, Vol. 46, No. 6, p. 1957 (1985). Thesis, Brandeis University, 170 pp. (1985). Order No. DA8517037.

159.111 **The effect of a quasi-stellar object on its host galaxy: dynamical and physical processes in the interstellar medium around a QSO.**
C.-A. Chang.
Diss. Abstr. Int., Sect. B, Vol. 46, No. 11, p. 3881 – 3882 (1986). Thesis, University of Pittsburgh, 144 pp. (1985). Order No. DA8600621.

159.112 **The X-ray properties of high redshift, optically selected QSOs.**
S. F. Anderson.
Diss. Abstr. Int., Sect. B, Vol. 46, No. 12, p. 4281 (1986). Thesis, University of Washington, 173 pp. (1985). Order No. DA8603097.

159.113 **High resolution spectroscopy of quasar absorption lines.**
J. E. Bechtold.
Diss. Abstr. Int., Sect. B, Vol. 46, No. 12, p. 4281 (1986). Thesis, The University of Arizona, 342 pp. (1985). Order No. DA8603142.

159.114 **QSO 1146+111 B,C.**
IAU Circ., No. 4227 (1986).

159.115 **Possible evidence for a massive black hole in a quasar with double redshift systems.**
Y.-z. Liu, Z.-g. Deng, Y.-y. Zhou, J.-h. Huang.
Astrophys. Space Sci., Vol. 123, No. 1, p. 137 – 144 (1986).
With the aid of a simple black hole model of quasars the authors have found that the majority of the distinguishable emission lines in the spectrum of the quasar 1604+179 can be assigned to two redshift systems, $z_r = 3.712$ and $z_b = 2.701$. The appearance of double emission redshifts means that this quasar might be a massive black hole (of mass $10^8 M_\odot \lesssim M \lesssim 10^{11} M_\odot$) with a ring-like emission line region (of radius 1 light-day $\lesssim r_0 \lesssim$ 1 light-year) in its accretion disk.

159.116 **The linear polarization of 3C 345 at milliarcsecond resolution.**
J. F. C. Wardle, D. H. Roberts, R. I. Potash, A. E. E. Rogers.
Astrophys. J., Lett. Ed., Vol. 304, No. 1, p. L1 – L4 (1986).
A polarization-sensitive 6 cm map of the superluminal quasar 3C 345 with angular resolution 2.5 × 4.7 mas was made using four stations of the US VLBI network. The optically thick core is essentially unpolarized (fractional linear polarization $m < 1\%$), while both of the knots in the jet are highly polarized. The inner knot C3 has $m = 11\%$, and the outer knot C2 has $m = 6\%$.

159.117 **Disappointment over double quasar.**
Nature, Vol. 321, No. 6068, p. 377 (1986).
This note comments on new observations of the quasar QSO 1146+111 A, B presenting evidence that the two objects separated by more than two arc min may not be gravitationally lensed images of a single quasar (see Abstr. 159.085).

159.118 **Is there a black hole in the sky?**
B. Paczyński.
Nature, Vol. 321, No. 6068, p. 419 – 420 (1986).
The recent discovery that the two quasars, Hazard 1146+111 B, C, separated by 157 arc s, are in fact two images of the same quasar has led the author to the speculation that the lensing object may be a supermassive black hole. Constraints on the properties of the black hole and the geometry of the gravitational lensing are derived from the observational data.

159.119 **Is QSO 1146+111 B, C due to lensing by a cosmic string?**
J. R. Gott III.
Nature, Vol. 321, No. 6068, p. 420 – 421 (1986).
A predicted consequence of cosmic strings is that they should produce equal-brightness double-lens images of distant quasars with separations of up to several minutes of arc. The author discusses the newly discovered lens candidate QSO 1146+111 B, C, which appears to consist of two images of equal brightness of a quasar at redshift $z = 1.01$, which are separated by 2.6 arc min. If this is produced by a cosmic string, its mass per unit length is $\mu \gtrsim 4.0 \times 10^{23}$ g cm^{-1}. Some implications of the existence of such a massive cosmic string are outlined.

159.120 **Q1146+111B, C quasar pair: illusion or delusion?**
E. S. Phinney, R. D. Blandford.
Nature, Vol. 321, No. 6070, p. 569 – 570 (1986).
The quasar pair Q1146+111B, C has been re-observed by Turner et al. who argue that it comprises two gravitationally lensed images of a single source. In this letter, the authors show that one is probably observing two distinct though neighbouring quasars.

159.121 **Test of the gravitational lens hypothesis for the quasar pair 1146+111B, C.**
P. A. Shaver, S. Cristiani.
Nature, Vol. 321, No. 6070, p. 585 – 586 (1986).
It has recently been suggested that the quasars 1146+111B, C, separated on the sky by 2.6 arc min, are actually two images of a single object produced by an extremely massive gravitational lens. The basis for this suggestion is the similarity in redshifts (within 100 – 200 km s^{-1}) and spectral features. Here the authors report differences between the two spectra in a hitherto unobserved range, suggesting that the two images are in fact distinct quasars located close together in space.

159.122 **Absorption spectrum of the z = 3.78 QSO 2000–330. I. The Lyman α forest region at 1.5 Å resolution.**
R. W. Hunstead, H. S. Murdoch, B. A. Peterson, J. C. Blades, D. L. Jauncey, A. E. Wright, M. Pettini, A. Savage.
Space Telesc. Sci. Inst., Prepr. Ser., No. 86, 39 pp. (1986). To appear in Astrophys. J.

159.123 **Broad absorption line QSOs.**
D. A. Turnshek.
Space Telesc. Sci. Inst., Prepr. Ser., No. 97, 22 pp. (1986). To appear in IAU Symp. 119.

159.124 **Absorption spectrum of the z = 3.78 QSO 2000–330: II. The redshift and equivalent width distributions of primordial hydrogen clouds.**
H. S. Murdoch, R. W. Hunstead, M. Pettini, J. C. Blades.
Space Telesc. Sci. Inst., Prepr. Ser., No. 103, 48 pp. (1986). To appear in Astrophys. J.

159.125 **Quasars measured by the Infrared Astronomical Satellite.**
G. Neugebauer, G. K. Miley, B. T. Soifer, P. E. Clegg.
Space Telesc. Sci. Inst., Prepr. Ser., No. 108, 49 pp. (1986). To appear in Astrophys. J.

159.126 **Ultraviolet spectra of quasars in the redshift interval 0.3 to 2.0.**
P. M. Gondhalekar, P. O'Brien, R. Wilson.
Rutherford Appleton Lab., Rep., RAL-86-034, 61 pp. (1986). To appear in Mon. Not. R. Astron. Soc.

159.127 **High redshift QSO's selected from IIIaJ objective prism plates of the UK Schmidt Telescope.**
C. Hazard, D. C. Morton, R. G. McMahon, W. L. W. Sargent, R. Terlevich.
Anglo–Aust. Obs., Prepr., No. 210, 56 pp. (1986). Submitted to Mon. Not. R. Astron. Soc.

159.128 **Quasars in the central region of the Virgo cluster.**
H. Arp.
MPA Rep., No. 224, p. 13 – 23 (1986).
Quasars found from objective prism searches have been reported to show no associaton with galaxies in the Virgo cluster. A simple analysis shows significant associations of the brightest of these quasars with core galaxies in the Virgo cluster.

159.129 **Further investigation of the pair of quasars Q 0107–025 A and B.**
J. Surdej, H. Arp, E. Gosset, A. Kruszewski, J. G. Robertson, P. A. Shaver, J. P. Swings.
ESO Sci. Prepr., No. 417, 29 pp. (1986). To appear in Astron. Astrophys.

159.130 **The iron spectra of PG 1700+518 and PG 2302+029.**
E. J. Wampler.
ESO Sci. Prepr., No. 421, 28 pp. (1986). To appear in Astron. Astrophys.

159.131 **Statistics of quasar pairs.**
P. A. Shaver.
ESO Sci. Prepr., No. 422, 7 pp. (1986). To appear in Nature.

159.132 **The remarkable absorption–line systems in the quasar Tololo 1037–27.**
M. H. Ulrich, M. A. C. Perryman.
ESO Sci. Prepr., No. 431, 21 pp. (1986). To appear in Mon. Not. R. Astron. Soc.

159.133 **Radio asymmetry vs. size for quasars with different asymmetry types.**
P. Teerikorpi.
Rep. Ser., Dep. Phys. Sci., Univ. Turku, No. FTL–R95, 9 pp. (1986). = Turku Univ. Obs., Informo No. 102. Submitted to Astron. Astrophys.

159.134 **C IV absorption systems in QSO spectra: is the character of systems with $z_{abs} \cong z_{em}$ different from those with $z_{abs} \ll z_{em}$?**
C. B. Foltz, R. J. Weymann, B. M. Peterson, L. Sun, M. A. Malkan, F. H. Chaffee Jr.
Prepr. Steward Obs., No. 614, 66 pp. (1986).

159.135 **The absorption–line spectrum of GC 1556+335: ejected or intervening material?**
S. L. Morris, R. J. Weymann, C. B. Foltz, D. A. Turnshek, S. Shectman, C. Price, T. A. Boroson.
Prepr. Steward Obs., No. 626, 45 pp. (1986).

159.136 **Some properties of X–ray selected quasars.**
R.–l. Liu, H.–s. Yang.
Chin. J. Space Sci., Vol. 5, No. 2, p. 84 – 91 (1985). In Chinese. Abstr. in Phys. Abstr., Vol. 89, No. 1276, Entry 36363 (1986).

159.137 **X–ray observations of the Bright Quasar Survey.**
H. Tananbaum, Y. Avni, R. F. Green, M. Schmidt, G. Zamorani.
Astrophys. J., Vol. 305, No. 1, p. 57 – 67 (1986).
Using the *Einstein Observatory*, the authors have carried out X–ray observations of ~80% of the quasars in the Bright Quasar Survey located outside the declination range defined by $30° < \delta < 60°$. X–ray emission has been detected above the 3σ level for 57 out of 66 quasars observed. This paper describes the observations and data reduction.

159.138 **Counts, evolution, and background contribution of X–ray quasars and other extragalactic X–ray sources.**
M. Schmidt, R. F. Green.
Astrophys. J., Vol. 305, No. 1, p. 68 – 82 (1986).
The authors are presenting a study of the statistics of extragalactic X–ray sources, in particular quasars and active galactic nuclei of lower luminosity. First, they discuss the X–ray observations of optically selected quasars in the Bright Quasar Survey sample and in the faint Braccesi sample and review available samples of extragalactic X–ray sources for which optical identifications are essentially complete. The authors use evolution models for optically selected quasars to derive the evolution of X–ray quasars, as well as their source counts and background contribution. They employ essentially the same method to discuss AGNs of low luminosity, BL Lac objects, clusters of galaxies, and galaxies. Finally, the composition of the X–ray background and the spectral index of the unexplained component are discussed.

159.139 **X–ray properties of optically selected QSOs.**
Y. Avni, H. Tananbaum.
Astrophys. J., Vol. 305, No. 1, p. 83 – 99 (1986).
The dependence of the X–ray to optical luminosity ratio on optical luminosity and redshift for optically selected QSOs is studied, largely on the basis of two, complete, magnitude–limited samples (Bright Quasar Survey and Braccesi Faint) which were observed with the *Einstein Observatory*. The previous results of Avni and Tananbaum for $\alpha_{0,x}(z, L_{opt})$ are confirmed and strengthened, and the numerical sensitivity to changes in the values of q_0 and of the optical spectral index is evaluated. It is shown that the large majority, probably all, of optically selected QSOs are X–ray loud; no more than a few percent can be X–ray quiet.

159.140 **Structure and stability of quasar clouds.**
W. G. Mathews.
Astrophys. J., Vol. 305, No. 1, p. 187 – 203 (1986). = Lick. Obs. Bull., No. 1027.
The structure of broad emission–line clouds is discussed, including radiative forces and dynamical pressure. A simplified model is developed for the radial variation of pressure and density and the associated radial equation of motion of the cloud. The discussion is focused on the possible vulnerability of rim regions of clouds to selective acceleration by radiative and dynamical forces that act throughout the broad–line region. Optically thin quasar clouds accelerated primarily by radiative momentum may be significantly elongated in the radial direction due to the weak dependence of the radiative force on the column density. Optically thick clouds cannot be accelerated to $\sim 10^4 km\ s^{-1}$ in a few sound crossing times by radiative forces alone. This suggests that optically thick clouds may be radially accelerated primarily by dynamical pressure.

159.141 **The optical and near–infrared polarization properties of the OVV quasar 3C 345.**
P. S. Smith, T. J. Balonek, P. A. Heckert, R. Elston.
Astrophys. J., Vol. 305, No. 1, p. 484 – 495 (1986).
Detailed multiwavelength, optical and near–infrared photometric, and polarimetric monitoring of 3C 345, from early 1983 to the middle of 1984, reveals several properties of this OVV quasar not observed in other highly polarized extragalactic objects. The polarization of the optical emission was strongly correlated with brightness during this period. When 3C 345 was close to its peak brightness during the intense 1982 – 1983 outburst, the fractional polarization was at the highest recorded level for this quasar (35%). Strong wavelength–dependent fractional polarization was also observed throughout this period. The degree of optical linear polarization increased with wavelength. The polarization wavelength dependence is consistent with dilution of the synchrotron light by unpolarized emission from thermal gas. The authors interpret the wavelength–dependent near–infrared polarization as evidence for a second synchrotron component with different polarization characteristics than the component dominating at shorter wavelengths.

159.142 **Absorption spectrum of the z = 3.78 QSO 2000–330. I. The Lyman–alpha forest region at 1.5 Å resolution.**
R. W. Hunstead, H. S. Murdoch, B. A. Peterson, J. C. Blades, D. L. Jauncey, A. E. Wright, M. Pettini, A. Savage.
Astrophys. J., Vol. 305, No. 1, p. 496 – 512 (1986).

Spectra of the $z = 3.78$ QSO 2000–330 have been obtained at the AAT with a resolution of $\sim$1.5 Å in the wavelength region 4100 – 5900 Å. 280 absorption lines are listed, and identifications are made in 45 Lyman–series systems and four heavy–element systems, the latter systems ranging in redshift from $z_{abs} = 3.1881$ to 3.5519. The $z_{abs} = 3.1881$ system is unusual in being purely low ionization, with no detectable Si IV or C IV. For this system the authors estimate [O/H] $\geqslant -1.1$ dex relative to solar, indicating an oxygen abundance at this early epoch that is surprisingly close to values typical of neutral interstellar clouds in the Galaxy.

159.143 **Erratum: "Two–point correlation function of QSO metal absorption systems from an inhomogeneous catalog of observations" [Astrophys. J., Vol. 298, No. 2, p. 732 – 742 (1985)].**
A. P. S. Crotts.
Astrophys. J., Vol. 305, No. 1, p. 581 (1986). See Abstr. 40.159.093.

159.144 **Further investigation of the pair of quasars Q 0107–025 A and B.**
J. Surdej, H. Arp, E. Gosset, A. Kruszewski, J. G. Robertson, P. A. Shaver, J. P. Swings.
Astron. Astrophys., Vol. 161, No. 2, p. 209 – 216 (1986).

Low and intermediate dispersion spectroscopy of Q 0107–025 A ($z = 0.956$, $g \cong 17.9$ mag) and Q 0107–025 B ($z = 0.952$, $g \cong 17.3$ mag) obtained at ESO, the AAT and the MMT enables one to definitely reject the possibility that the pair of objects are gravitational lens images of a single quasar. CCD photometry reveals the possible presence of a highly redshifted ($z \gtrsim 0.5$) cluster of galaxies in the field of Q 0107–025 A and B. Spectroscopy of the brightest galaxies is required in order to investigate the possible association between the galaxy and quasar clusters.

159.145 **The iron spectra of PG 1700+518 and PG 2302+029.**
E. J. Wampler.
Astron. Astrophys., Vol. 161, No. 2, p. 223 – 231 (1986).

Ground–based spectra of the broad absorption line quasar PG 1700+518 are combined with published IUE and infrared observations to disentangle the various spectral components. In the wavelength interval $1500 < \lambda < 5500$ Å the two major components are a power law continuum with a spectral index of –1.2 and strong Fe II emission bands. Except that PG 2302+029 shows no broad Mg II λ2798 absorption its spectrum is nearly identical to that of PG 1700+518. For PG 2302+029 the total flux emitted in the Fe II lines is estimated to be about $6\text{x}10^{46}$ erg s^{-1}. The distribution of Fe II emission with wavelength agrees with published models. Several emission peaks in the spectrum of PG 2302+029 can be identified with Fe III multiplets.

159.146 **VLA observations of B2 quasars. I. Extended sources.**
A. Rogora, L. Padrielli, H. R. de Ruiter.
Astron. Astrophys., Suppl. Ser., Vol. 64, No. 3, p. 557 – 577 (1986).

VLA observations at 6 cm of radio sources selected from a complete sample of B2 quasars are presented. The angular resolution of the radio maps is about 3.5 arcsec. The authors describe the selection of the sources and the reduction procedure. The dynamic range in the final CLEANed maps is usually better than 100 to 1. Contour maps of the resolved sources are presented.

159.147 **A search for quasars and emission–line galaxies in the Sculptor region.**
D. Kunth, W. L. W. Sargent.
Astron. J., Vol. 91, No. 4, p. 761 – 776 (1986). With plates 63 – 69.

A spectroscopic survey of emission–line candidates found in three objective–prism plates has been carried out resulting in 76 quasars, 28 emission–line galaxies, and three carbon stars. The new sample of quasars doubles the number of known quasars in this region. 80% have redshifts larger than 1.6 and were found from their Lyα line emission. A preliminary study shows that quasar clustering is absent on any scale from 10 to 3500 Mpc.

159.148 **The GB/GB2 sample of intermediate–strength quasars.**
J. Machalski, J. J. Condon.
Astron. J., Vol. 91, No. 5, p. 998 – 1010 (1986).

This paper is concerned with the optical and radio properties of quasars identified with intermediate–strength radio sources from the 1400 MHz GB and GB2 catalogs (Machalski and Condon 1985).

159.149 **Deep CCD images of 2345+007: lensing by dark matter.**
J. A. Tyson, P. Seitzer, R. J. Weymann, C. Foltz.
Astron. J., Vol. 91, No. 6, p. 1274 – 1278 (1986). With plates 87 – 91.

Deep CCD images of the lensed QSO 2345+007 taken in the blue, red, and near infrared show no galaxy between the QSO images to $J = 25.5$, $R = 26$, and $I = 24.5$ mag. There are a few 23.8 I mag galaxies within 10 arcsec of the QSO. Even if one of these field galaxies was near the line of sight to the QSO, there must be other objects of very high M/L assisting the lensing. Two absorption–line systems at $z = 1.5$, separated by 1000 km/s, are seen in the QSO spectra. The authors would have detected images of galaxies at $z = 1.5$, in the QSO vicinity. This is the strongest case, among the seven confirmed lensed QSOs, for lensing by dark compact objects. A nonluminous mass of $10^{13} M_{\odot}$ is required.

159.150 **The cosmic evolution of quasars at high redshifts.**
P. Veron.
Obs. Haute Provence, Pré–Publ., No. 8, 14 pp. (1986). Submitted to Astron. Astrophys.

159.151 **Evidence for evolution of quasar environments.**
H. K. C. Yee, R. F. Green.
Bull. Am. Astron. Soc., Vol. 18, No. 1, p. 639 (1986). Abstract. – See Abstr. 010.066.

159.152 **A cosmic string in Leo?**
M. M. Waldrop
Science, Vol. 232, No. 4753, p. 936 (1986).

A new gravitational lens system may be evidence for an ultramassive cosmic string winding from one side of the universe to the other.

159.153 **Snarls from the cosmic string in Leo.**
M. M. Waldrop.
Science, Vol. 232, No. 4758, p. 1601 (1986).

The case for an ultramassive gravitational lens is suddenly much weaker than it was; astronomers may have to settle for a more mundane explanation.

159.154 **The nature of the X–ray emission in quasars.**
G. Zamorani.
Mem. Soc. Astron. Ital., Vol. 56, No. 4, p. 747 – 753 (1985). – See Abstr. 012.054.

The correlations between X–ray luminosity and other intrinsic quantities, the spectral shape of the X–ray continuum, the presence or absence of intrinsic low energy absorption, the frequency and timescales of observed X–ray variability are different tools at our disposal in the study of the physics of X–ray emission in quasars. This paper briefly summarizes the present knowledge, from an observational point of view, of some of these aspects. Some order of magnitude estimates of the size of the X–ray emitting region as a function of the X–ray luminosity are presented.

159.155 **Damped Lyman–alpha absorption by disk galaxies with large redshifts. I. The Lick survey.**
A. M. Wolfe, D. A. Turnshek, H. E. Smith, R. D. Cohen.
Astrophys. J., Suppl. Ser., Vol. 61, No. 2, p. 249 – 304 (1986).
Low–resolution ($\Delta\lambda$ = 10 Å) spectra of 68 high–z QSOs are used to locate the large column density of neutral hydrogen in the disks of foreground galaxies through the detection of Lyα absorption lines broadened by radiation damping. The authors present a list of absorption features that are candidates for "Lyα disks". This list comprises 47 absorption features with rest–frame equivalent widths W > 5 Å. Investigations of 26 of these show that 15 are in fact damped Lyα and that 10 are blends of multiple narrow Doppler components.

159.156 **A search for quasars and emission line galaxies in the Sculptor region.**
D. Kunth, W. L. W. Sargent.
Inst. Astrophys. Paris, Pré–Publ., No. 125, 55 pp. (1986). To appear in Astron. J.

159.157 **The energy puzzle of the broad line region in quasars.**
S. Collin–Souffrin.
Inst. Astrophys. Paris, Pré–Publ., No. 145, 13 pp. (1986). To appear in Astron. Astrophys.

159.158 **Blazars, quasars and starbursts.**
E. I. Robson.
Rutherford Appleton Lab., Rep., RAL 85–086, p. 130 – 147 (1985). – See Abstr. 012.060.
This paper reviews recent submillimetre observations of quasars, BL Lacertae objects, Seyfert nuclei and galaxies where massive starburst activity is thought to occur.

159.159 **Positions of extragalactic objects in the vicinity of M3 and M31 (5C3 radio field) and in the 5C1 radio field.**
W. R. Dick, H. Lorenz, P. Notni, E. Schilbach, R.–D. Scholz, G. A. Richter.
Astron. Nachr., Vol. 307, No. 2, p. 85 – 88 (1986).
The authors have determined the positions of 65 compact extragalactic objects with magnitudes between 16^m and $21^m.7$. The mean error of one position is about 0".3 in $\alpha \cos \delta$ and in δ. Some hitherto unpublished redshifts are also listed.

159.160 **Far–infrared observations of the quasar 0241 + 62.**
M. J. Coe, A. J. Dean, S. Sembay, M. Ferrari–Toniolo, P. Persi, L. Spinoglio, L. Bassani, B. Elsmore.
Mon. Not. R. Astron. Soc., Vol. 220, No. 3, p. 781 – 785 (1986).
Far–infrared observations of the QSO 0241 + 62 taken around 1983 September are reported. They include detailed IRAS observations and maps of the region. The exact spectral turnover point in the optical–IR region is determined. Simple model–fitting using the established synchrotron self–Compton mechanism convincingly links the X–ray and the IR emission. The infrared colours, however, differ from published IRAS results for other quasars.

159.161 **Comparison of the Hubble diagrams among X–ray–, optically– and radio–selected quasars.**
F.–h. Cheng, Y.–y. Zhou, S.–z. Bao.
Kexue Tongbao, Vol. 31, No. 2, p. 81 – 84 (1986). Abstr. in Phys. Abstr., Vol. 89, No. 1282, Entry 73422 (1986).

159.162 **On the alignment of quasars.**
X.–f. Zhu.
Chin. Astron. Astrophys., Vol. 10, No. 2, p. 94 – 97 (1986). English translation of Acta Astrophys. Sin., Vol. 6, No. 1, p. 1 – 5 (1986).
Taking the two Savage–Bolton 5° × 5° regions of optical quasar patrol as samples, the authors made a systematic analysis of the number of aligned quasars and compared with the random data generated by Monte Carlo method. The statistical result is that, at least for these two samples, there is no clear evidence for alignment.

159.163 **Preliminary results of 6 cm VLBI observation of 3C 454.3 with 9 radio telescopes.**
S.–y. Wu, I. Pauliny–Toth, R. Porcas, A. Zensus.
Acta Astrophys. Sin., Vol. 6, No. 2, p. 89 – 100 (1986). In Chinese.
3C 454.3 is one of the most active variable quasars. It has a core–jet structure on the m.a.s. scale. The core showed superluminal brightening after a large flux–density outburst in 1981. VLBI observations at 6 cm with 9 telescopes have confirmed the core–jet structure and a superluminal expansion of the core between 1981.6 and 1983.9. They also show a systematic curvature of the structure on scales from m.a.s. to arcsec.

159.164 **Correlation between Lyman alpha absorption line density and the emission line redshift of a QSO.**
Y.–l. Bian, J.–s. Chen, Z.–l. Zou.
Acta Astrophys. Sin., Vol. 6, No. 2, p. 101 – 110 (1986). In Chinese.
This paper presents some results of the statistic study of the Lyman alpha forest in the absorption line spectra of high redshift QSOs. The samples used for this work consist of the spectra of 16 objects which have been widely studied. The main results can be summarized as follows: (1) The mean line density in the observed pure Lyman alpha wavelength region increases with increasing z_{em} of the QSO. (2) The mean line density in the same wavelength interval increases with increasing z_{em} of the QSO, too. (3) The analysis shows that the Lyman alpha absorption line density depends significantly on z_{em}.

159.165 **Long–term optical behaviour of the radio–quiet quasar PHL 938.**
H.–s. Yang.
Acta Astron. Sin., Vol. 27, No. 1, p. 61 – 68 (1986). In Chinese.
Photometric data for the radio–quiet quasar PHL 938 have been obtained over a period of more than 20 years. The data and light curve are presented. The results show that the brightness of this radio–quiet and absorption–line quasar has a long–term trend of variation of small amplitude with short–term deviations. The size of the emitting region of PHL 938 is 0.32 pc.

159.166 **Clustering of quasar distribution.**
X.–t. He, Y.–q. Chu, L.–z. Fan, M. G. Smith.
Acta Astron. Sin., Vol. 27, No. 2, p. 144 – 150 (1986). In Chinese.
Three objective–prism plates and one direct plate centred at $01^h44^m$and –40°00′ were taken. 1092 quasar candidates were picked up in the 40 square degree field. The two–dimensional power spectrum analysis method is adopted to test the clustering of quasar candidates in the field. The intervals of reciprocal wavelength are $\Delta(1/\lambda) = 0.2$. The clustering scale is about 0.5 degree. If one takes the average redshift Z = 2, the diameter of the supercluster is about 35 Mpc.

159.167 **Optical spectra and redshifts for 15 radio sources with flat spectra.**
Y.–q. Chu, X.–f. Zhu, H. Butcher.
Acta Astron. Sin., Vol. 27, No. 2, p. 151 – 157 (1986). In Chinese.
Spectra are presented for 15 QSO candidates which were selected from optical identifications of flat–spectrum radio sources. 6 of them have obvious emission–line features and confirmed to be quasars. The largest redshift is 3.45 for PKS 0335–122. The selection effects in search for quasars with redshifts greater than 3.5 are discussed.

159.168 **Spectroscopically confirmed QSOs in a 1.5 deg² area near the south galactic pole.**
L. E. Campusano.
Rev. Mex. Astron. Astrofis., Vol. 12, Spec. Issue, p. 49 – 52 (1986). – See Abstr. 012.066.
The author determines the presently known number of spectroscopically confirmed QSOs in a 1.5 deg² region centered at the already mentioned coordinates. This region is inside a larger one that have been surveyed independently by spectroscopic and color criteria, and the author has also used a number of QSO candidates coming from a grism search the author has conducted

in the 1.5 deg^2 area. The spectroscopic confirmation of the candidates was taken from the literature when available.

159.169 **Relativistic motion in quasars.**
M. H. Cohen.
Rev. Mex. Astron. Astrofis., Vol. 12, Spec. Issue, p. 75 (1986). Abstract. – See Abstr. 012.066.

159.170 **The Cerenkov radiation in atom–atomic ion clouds and line–spectra of QSOs.**
Z.–y. Feng, M.–q. Wang.
Sci. Sin., Ser. A, Vol. 29, No. 6, p. 631 – 638 (1986).
The line–spectrum of the Cerenkov radiation in atom–atomic ion gas–clouds is discussed. The effect of the N_V (1240 Å) line on the $L\alpha$ (1216 Å) line is calculated. The $N_V/L\alpha$ intensity ratio of the spectra of QSOs is explained. The profile of the $L\alpha$–N_V line in the spectrum of quasar UM663 is analyzed, and the redshift of the $L\alpha$ line of this quasar is estimated.

159.171 **Quasars in the central region of the Virgo Cluster.**
H. Arp.
J. Astrophys. Astron., Vol. 7, No. 2, p. 77 – 81 (1986).
Quasars found from objective prism searches have been reported to show no association with galaxies in the Virgo Cluster. A simple analysis shows significant association of the brightest of these quasars with core galaxies in the Virgo Cluster.

159.172 **The luminosity–volume test and the local hypothesis of quasars.**
F. Hoyle, J. V. Narlikar.
J. Astrophys. Astron., Vol. 7, No. 2, p. 99 – 103 (1986).
It is shown that the luminosity–volume test for optically selected objects has an in–built bias towards increasing the average value of V/V_m above the Euclidean value of 1/2. A more satisfactory bias–free statistic is suggested in the form of $\ln(V/V_m)$. The result of applying the test to a sample from the Bright Quasar Survey shows that the local hypothesis of quasars is consistent with the data.

159.173 **Extragalactic sources with asymmetric radio structure. II. Further observations of the quasar B2 1320 + 299.**
T. J. Cornwell, D. J. Saikia, P. Shastri, L. Feretti, G. Giovannini, P. Parma, C. J. Salter.
J. Astrophys. Astron., Vol. 7, No. 2, p. 119 – 129 (1986).
The authors present VLA A–array observations at λ20, 6 and 2 cm and B–array observations at λ20 and 6 cm of the quasar B2 1320 + 299, which has a very unusual radio structure. In addition to a component, A, coincident with the quasar, there are two lobes of radio emission, B and C, on the same side of A. These are located at distances of ≅25 and 50 arcsec respectively from A. The present observations show that A has a flat–spectrum component coincident with the quasar and a weak outer component at a distance of ≅4 arcsec along PA ≅ 100°. The morphology of B resembles a head–tail type of structure with its tail towards the north–east. Component C exhibits some extension towards the north–west. The authors discuss the possible nature of B2 1320 + 299 and suggest that while A appears to be an independent source, the relation between B and C, if they are associated at all, is unclear.

159.174 **The inner structure of quasars and active galactic nuclei.**
H. Netzer.
Mitt. Astron. Ges., No. 65, p. 109 – 121 (1986). – See Abstr. 012.072.
Recent observational development in the area of quasars and AGN enable one to probe their innermost structure in a more quantitative way. In this review the author has pointed out the progress in observing the continuum over a large frequency range, the new results on line and continuum variability and the possible correlation of cloud velocity with the continuum luminosity.

159.175 **Stand der VLBI–Untersuchungen extragalaktischer Quellen nördlich 70° Deklination.**
A. Witzel, A. Eckart, P. Biermann, C. Schalinski, K. Johnston.
Mitt. Astron. Ges., No. 65, p. 242 – 243 (1986). – See Abstr. 012.072.

159.176 **Der Quasar 1928 + 738.**
C. Schalinski, A. Witzel, P. Biermann, A. Eckart, K. J. Johnston.
Mitt. Astron. Ges., No. 65, p. 244 – 245 (1986). – See Abstr. 012.072.

159.177 **Gruppen um aktive Galaxien.**
W. Kollatschny, K. J. Fricke, U. Fritze–v. Alvensleben, J. Hellwig.
Mitt. Astron. Ges., No. 65, p. 246 – 247 (1986). – See Abstr. 012.072.

159.178 **PG 1115 + 080.**
IAU Circ., Nos. 4182, 4185 (1986).

159.179 **QSO 0716 + 332.**
IAU Circ., No. 4195 (1986).

The second Byurakan spectral sky survey. IV. The field $\alpha = 12^h22^m$, $\delta = +55°00'$.
See Abstr. 002.014.

A new catalogue of quasi–stellar objects.
See Abstr. 002.036.

Quasars to B > 22.5 in Selected Area 57: a catalog of multicolor photometry, variability, and astrometry.
See Abstr. 002.053.

The Palomar–Green Catalog of Ultraviolet–Excess Stellar Objects.
See Abstr. 002.074.

Was sind Quasare?
See Abstr. 003.082.

Quasar astronomy.
See Abstr. 003.208.

Our place in the Universe. II. To the boundaries of the Universe with radioastronomy.
See Abstr. 033.067.

Evolution of the luminosity function of extragalactic objects.
See Abstr. 036.051.

The role of the automated plate measuring (APM) machine in developing a program to discover QSO's automatically.
See Abstr. 036.242.

Measurements of pulsar parallaxes by VLBI.
See Abstr. 041.031.

QSO's in the post–fit residual of the JPL VLBI data.
See Abstr. 044.028.

Quasars and galaxies observable with the Space Telescope.
See Abstr. 051.034.

A monopole model for galactic nuclei.
See Abstr. 061.024.

The relativistic maximum instability in astrophysical jets.
See Abstr. 062.026.

A numerical investigation of approximation procedures for optically thick resonance lines.
See Abstr. 063.026.

Cerenkov line radiation – a new interpretation of the emission lines of quasars.
See Abstr. 063.046.

Theoretical X–ray spectra of accreting magnetized neutron stars and quasars.
See Abstr. 063.066.

Radiation pressure and emission clouds around active galactic nuclei.
See Abstr. 063.087.

On the origin of the infrared and X–ray continua of active galactic nuclei.
See Abstr. 063.088.

On the correlation between X–ray flux and hydrogen line intensities in quasars and active galactic nuclei.
See Abstr. 063.102.

On observations in Schwarzschild background.
See Abstr. 066.027.

Gravitational lenses: model fitting, time delay and Hubble parameter.
See Abstr. 066.043.

Gravitational microlensing at large optical depth.
See Abstr. 066.068.

'Missing image' in gravitational lens systems?
See Abstr. 066.083.

Mikro–Gravitationslinsen und ihre Bedeutung für die Astrophysik.
See Abstr. 066.106.

Superluminal separation caused by gravitational lens' effect of neu trino objects.
See Abstr. 066.131.

Statistical gravitational lensing: influence of compact objects on the observed quasar luminosity function.
See Abstr. 066.167.

Gravitational lensing – models.
See Abstr. 066.179.

The birth of AGNs and quasars via the collapse of dense star clusters to supermassive black holes.
See Abstr. 067.042.

Buoyancy limits on magnetic viscosity stress law scalings in quasi–stellar object accretion disk models.
See Abstr. 067.087.

The central engine of quasars and active galactic nuclei: hadronic interactions of shock–accelerated relativistic protons.
See Abstr. 067.093.

Black hole models for quasar–like objects: some selected topics.
See Abstr. 067.097.

Central engine or locomotive?
See Abstr. 067.105.

Accretion disks and periodic outbursts of active galactic nuclei.
See Abstr. 067.111.

Spectral types for objects in the Kiso survey. II. Data for 61 stars.
See Abstr. 114.084.

The MIT–Green Bank (MG) 5 GHz survey.
See Abstr. 141.018.

5 GHz radio structure and optical identifications of sources from the MG survey. II. Maps and finding charts.
See Abstr. 141.019.

Optical identifications of radiosources.
See Abstr. 141.021.

A search for ultraviolet–excess objects.
See Abstr. 142.036.

The gamma–ray background from a superposition of quasar jet sources.
See Abstr. 143.017.

Tidal triggering of Seyfert galaxies and quasars – perturbed galaxy disk models versus observations.
See Abstr. 151.099.

Additional evidence for the accuracy of galactic reddenings from ScI galaxies and quasars.
See Abstr. 155.069.

Extent of ionized calcium in the outer parts of galaxies.
See Abstr. 157.067.

Quests for primeval galaxies: a review of optical surveys.
See Abstr. 157.104.

VLA mapping of the region around NGC 4319.
See Abstr. 157.191.

Statistical properties of IRAS galaxies.
See Abstr. 157.322.

The Parkes selected regions: powerful radio galaxies and quasars at high redshifts.
See Abstr. 158.003.

Knots, wiggles, and flaring in ballistic and stochastic jets.
See Abstr. 158.023.

On wind–type flows in astrophysical jets. II. Propagation outside the nucleus and the case of M87.
See Abstr. 158.024.

The eruptive BL Lac object 0846+51W1: a gravitationally lensed QSO?
See Abstr. 158.049.

On the relationship between optically selected and X–ray–selected active galactic nuclei.
See Abstr. 158.054.

Continuum and line fluxes of OJ 287 at minimum light.
See Abstr. 158.060.

The discovery of eight bright active galactic nuclei observed with the *HEAO 1* scanning modulation collimator.
See Abstr. 158.068.

The interaction of interstellar clouds with a "quasar wind" and its relation to the narrow emission line regions of Seyfert galaxies.
See Abstr. 158.079.

Narrow emission line profiles in a "quasar wind".
See Abstr. 158.080.

A hidden quasar revealed.
See Abstr. 158.086.

Limits of X–ray variability in active galactic nuclei.
See Abstr. 158.087.

IRAS observations of active galaxies.
See Abstr. 158.091.

The properties of "dwarf" Seyfert nuclei in nearby galaxies.
See Abstr. 158.092.

Spectral properties of Blazars observed at ultraviolet and X–ray frequencies.
See Abstr. 158.093.

Internal dynamics and formation of emission clouds in active galactic nuclei.
See Abstr. 158.094.

Evolutionary connection of Seyfert galaxies and quasars.
See Abstr. 158.097.

Evolutionary continuity of the optical AGNs.
See Abstr. 158.098.

Large scale ionized gas in radio galaxies and quasars.
See Abstr. 158.101.

Host galaxies of quasars.
See Abstr. 158.102.

Some theoretical aspects of AGNs.
See Abstr. 158.105.

Frequency dependent polarization in blazars.
See Abstr. 158.109.

The relative frequency of compact extragalactic objects.
See Abstr. 158.111.

Ram pressure confinement of the emission line gas and the associated radio emission in AGN's.
See Abstr. 158.112.

The behaviour of the C IV emission line strength in variable Seyfert I galaxies.
See Abstr. 158.114.

First results from a catalog of data concerning AGN – FWHM and FWZI versus L_0 for C IV and Hβ.
See Abstr. 158.118.

The spectral index – flux density relation and the cosmological evolution of radio sources.
See Abstr. 158.120.

Double nucleus galaxies.
See Abstr. 158.123.

Kiso surveys of ultraviolet–excess objects.
See Abstr. 158.125.

Steep spectrum radio sources showing low frequency variability.
See Abstr. 158.126.

MAGN; mildly active galactic nuclei.
See Abstr. 158.133.

Emission lines indicating a universal L/M ratio for the central engine in AGN?
See Abstr. 158.141.

Tidal triggering of Seyfert galaxies and quasars–models.
See Abstr. 158.145.

NGC 4319 – a spiral galaxy with radio lobes?
See Abstr. 158.147.

Optical polarization and brightness variability in BL Lacertae objects and OVV (*optically violent variable*) quasars.
See Abstr. 158.148.

Wavelength–dependent polarization in BL Lacertae objects and OVV quasars.
See Abstr. 158.172.

Gas in active galactic nuclei.
See Abstr. 158.180.

The optical continuum emission of active galactic nuclei.
See Abstr. 158.181.

Current issues in the X–ray properties of active galactic nuclei.
See Abstr. 158.182.

Active galactic nuclei. I. Central source as a cluster of accreting black holes.
See Abstr. 158.193.

X–ray absorption in active galaxies: constraints on the broad–line emitting regions.
See Abstr. 158.194.

Recent observations and evolution of extended extragalactic sources.
See Abstr. 158.197.

Evolution and astrophysics of active galactic nuclei.
See Abstr. 158.209.

The impact of the local environment on the properties of radio galaxies: a study of galaxies identified from the Texas Survey.
See Abstr. 158.231.

A test for models of radio source evolution.
See Abstr. 158.237.

Multifrequency observations of blazars. II. The variability of the 1 μm to 2 mm continuum.
See Abstr. 158.238.

Simultaneous submillimeter and infrared observations of flat–spectrum radio sources.
See Abstr. 158.243.

Active galactic nuclei: a review of optical properties.
See Abstr. 158.246.

Optical monitoring of quasars and BL Lac objects.
See Abstr. 158.260.

Line variations in quasars and Seyfert galaxies.
See Abstr. 158.265.

Weak nuclei of powerful radio sources – spectra and polarizations.
See Abstr. 158.275.

On the origin of extragalactic infrared radiation.
See Abstr. 158.293.

Interpretation of the variability of the X–ray emission of nuclei of type 1 Seyfert galaxies and quasars.
See Abstr. 158.296.

Compact radio structure of active galactic nuclei.
See Abstr. 158.315.

IRAS observations of active galaxies – a review.
See Abstr. 158.331.

Submillimetre to infrared observations of active galaxies.
See Abstr. 158.334.

Pushing back the redshift limit.
See Abstr. 161.015.

The baryon clump within an extended dark matter region.
See Abstr. 161.018.

Dark matter and the formation of large-scale structure in the universe.
See Abstr. 161.021.

Statistical gravitational lensing and quasar-galaxy associations.
See Abstr. 161.038.

The protogalactic connection.
See Abstr. 161.079.

Distribution of quasars and the formation of large scale structure in the Universe.
See Abstr. 161.080.

An upper limit to the density inhomogeneity in the Universe.
See Abstr. 161.164.

160 Galaxy Groups, Clusters of Galaxies, Superclusters, Intergalactic Matter

160.001 **Morphological segregation in the Pisces-Perseus supercluster.**
R. Giovanelli, M. P. Haynes, G. L. Chincarini.
Astrophys. J., Vol. 300, No. 1, p. 77 – 92 (1986).

Evidence for the continuous morphological segregation among galaxies along the entire Hubble sequence is sought for the wide range of galaxian densities represented by the Pisces-Perseus supercluster. The two-point autocorrelation function is computed for various subsets of the galaxy sample. Early-type galaxies in the Pisces-Perseus region tend to cluster on smaller angular scales than do later types. Furthermore, the correlation function for Sa and Sab galaxies is steeper than that of Sb or Sc galaxies. As is seen in rich clusters and groups, the population fractions of elliptical and S0 galaxies grow with increasing density throughout the supercluster, while the spiral fraction correspondingly decreases. The variation in the population fractions can be traced smoothly to regions of very low galaxian density so that a gradient in the population fraction is evident at nearly all density regimes.

160.002 **A search for the Sunyaev-Zel'dovich effect at $\lambda = 3$ millimeters.**
S. J. E. Radford, P. E. Boynton, B. L. Ulich, R. B. Partridge, R. A. Schommer, A. A. Stark, R. W. Wilson, S. S. Murray.
Astrophys. J., Vol. 300, No. 1, p. 159 – 166 (1986).

The authors have searched for the Sunyaev-Zel'dovich effect in seven clusters of galaxies at a wavelength of 3 mm using the NRAO 12 m telescope on Kitt Peak and the Bell Laboratories 7 m telescope at Crawford Hill. This wavelength is advantageous because flux from sources in the clusters is negligible and the Sunyaev-Zel'dovich signal is near maximum amplitude. These measurements set upper limits on the magnitude of the Sunyaev-Zel'dovich effect that are consistent with all previous, longer wavelength, searches. The consistency of existing observations and the prospects for future measurements are discussed.

160.003 **Electronographic BV photometry of three distant clusters of galaxies. I. Observations and reduction techniques.**
O. Lefèvre, A. Bijaoui, G. Mathez, J. P. Picat, G. Lelièvre.
Astron. Astrophys., Vol. 154, No. 1/2, p. 92 – 99 (1986).

The authors present the reduction techniques and a complete BV photometry of the galaxies in three distant clusters: A 992 ($z = 0.247$), 3C 299 ($z = 0.367$) and 1305+295 ($z = 0.28$ or 0.95). A mainly automatic procedure is described to detect the objects, to fit their isophotes to elliptical contours and to integrate the intensity within elliptical contours and to integrate the intensity within elliptical annuli. An original procedure has been used to discriminate stars from galaxies. The catalogue with nine parameters including V_{25}, $B_{26.5}$ and B–V color indices will be published in Astron. Astrophys., Suppl. Ser.

160.004 **A prominent string of galaxies in Bootes: evidence for a Lagrangian singularity?**
E. Tago, J. Einasto, E. Saar.
Mon. Not. R. Astron. Soc., Vol. 218, No. 2, p. 177 – 184 (1986).

An outstanding string of galaxies and of small groups in Bootes has been found during a study of bridges between the Coma and the Local superclusters. It lies on the boundary of a large void between the Hercules and the Local superclusters, has a length of 50 Mpc, is only a few megaparsecs wide and consists of non-elliptical galaxies. A possible explanation by a specific Langrangian singularity at the formation, the imprint of which has survived in the present galaxy distribution, is proposed.

160.005 **Space distribution of nearby superclusters of galaxies.**
M. Kalinkov, I. Kuneva.
Mon. Not. R. Astron. Soc., Vol. 218, No. 3, p. 49P – 55P (1986).

Two lists of rich and poor nearby superclusters of galaxies have been compiled and it is shown that the superclusters are almost uniformly distributed. The spatial supercluster-supercluster correlation function, which has nothing to do with the correlation functions of galaxies or clusters of galaxies, is calculated. In general, the correlation function of superclusters of galaxies is small, $|\xi| < 0.5$. There is also an indication of the existence of multiple rich superclusters. An anticorrelation between rich and poor superclusters has been found.

160.006 **A photometric study of poor clusters of galaxies.**
T. Yamagata, H. Maehara.
Astrophys. Space Sci., Vol. 118, Nos. 1/2, p. 459 – 462 (1986). – See Abstr. 012.001.

Sixteen poor clusters of galaxies with suspected cD galaxies are studied using V-band photographs taken with the 105 cm Kiso Schmidt telescope. Galaxian images were automatically detected using an image detection software developed for the present study. It is found from the luminosity functions that the richness of the clusters are well correlated with the optical properties of central galaxies and the X-ray luminosities of the clusters.

160.007 **New polar ring galaxies in rich clusters of galaxies.**
Y. Taniguchi, K. Shibata, K.-I. Wakamatsu.
Astrophys. Space Sci., Vol. 118, Nos. 1/2, p. 529 – 530 (1986). – See Abstr. 012.001.

Four polar ring galaxies discovered in rich clusters of galaxies are presented. Brief comments on their structural properties are given.

160.008 **On the effects of compression of a gaseous disc by thermal and dynamical pressures of intergalactic gas.**
Y. Kumai, M. Tosa.
Astrophys. Space Sci., Vol. 119, No. 1, p. 211 – 212 (1986). – See Abstr. 012.001.

The authors consider a galaxy moving in a cluster of galaxies and study the effects of compression of the gas disc by both thermal and dynamical pressures of intergalactic gas. As the result of compression, massive gas clouds are formed. They are not stripped by dynamical pressure of the intergalactic gas but stay in the galaxy until they are disrupted by formation of massive stars.

160.009 **Alignment of spiral galaxies in the Local Supercluster.**
H. T. MacGillivray, R. J. Dodd.
Astrophys. Space Sci., Vol. 119, No. 1, p. 253 – 256 (1986). – See Abstr. 012.001.

Evidence is presented indicating the non–random alignment of spiral galaxies in the Local Supercluster. The form of this effect is such that the spin angular momentum vectors of intermediate–type spirals are coherently aligned in space. The results suggest the formation of galaxies in the Local Supercluster according to the fragmentation hypothesis.

160.010 **Density cusps in clusters of galaxies.**
T. C. Beers, J. L. Tonry.
Astrophys. J., Vol. 300, No. 2, p. 557 – 567 (1986).

The authors find that the density of galaxies in rich clusters usually has a central cusp. The derived density profiles depend crucially on the choice of cluster centers, which cannot be obtained unambiguously from galaxy positions alone. Centers may be specified that are independent of the galaxy distribution by using X–ray maxima or the locations of D and cD galaxies. It is shown that galaxy density profiles constructed about either of these external centers are power laws with slope –1. These power laws extend from the cluster center out to about $1\,h^{-1}$Mpc ($H_0 = 100{\cdot}h$ km s^{-1}Mpc^{-1}).

160.011 **High–resolution imaging from Mauna Kea: surface photometry of galaxies in CL 0024+16.**
L. A. Thompson.
Astrophys. J., Vol. 300, No. 2, p. 639 – 657 (1986). With plates 4 – 5.

A deep CCD image with resolution 0″.6 FWHM has been obtained of the rich and distant galaxy cluster CL 0024+16 at $z = 0.394$. Even though most of the galaxies have major diameters less than 5″, the excellent image quality of the CCD frame and careful model fitting make it possible to determine the bulge and disk photometric characteristics for 26 galaxies. The nine blue "Butcher–Oemler" galaxies in the field of view show a wide diversity of a structure: one is an elliptical, one a small bulge spiral, and the other seven have moderate to large bulge–to–disk ratios ($0.5 \leqslant B/D \leqslant 2.2$).

160.012 **Are superclusters correlated on a very large scale?**
N. A. Bahcall, W. S. Burgett.
Astrophys. J., Lett. Ed., Vol. 300, No. 2, p. L35 – L39 (1986).

A positive spatial correlation is observed among a complete sample of superclusters on a scale of $\sim 100\,h^{-1}$Mpc. The correlation is significant at about the 95% – 99% confidence level, suggesting the possible existence of the largest structures yet detected. The correlation appears to be stronger than that of galaxies and galaxy clusters, indicating a continuation in the increasing strength of clustering with richness.

160.013 **H I observations of galaxies in the Virgo cluster of galaxies. I. The data.**
W. K. Huchtmeier, O.–G. Richter.
Astron. Astrophys., Suppl. Ser., Vol. 64, No. 1, p. 111 – 130 (1986).

New H I–data for a large number of bright galaxies inside the 10° radius area of the Virgo cluster of galaxies have been obtained with the 100 m radiotelescope at Effelsberg. 234 galaxies were observed for the first time. Among them 53 have been detected providing new accurate radial velocities. Data from the literature have been compiled. Together with the new data they form a (nearly homogeneous) set of H I observations for more than 450 galaxies.

160.014 **VV 172: an effect of gravitational amplification by a massive halo?**
F. Hammer, L. Nottale.
Astron. Astrophys., Vol. 155, No. 2, p. 420 – 422 (1986).

The authors show that the well known quintet of galaxies VV 172, which contains a discrepant redshift member, may be fully understood when accounting for the effects of gravitational amplification by the halo of the foreground quartet on the more distant galaxy. Gravitational lensing increases the luminosity, diameter and chance probability of association of background galaxies in such a way that the presently known number of similar associations containing a discrepant redshift is not any more found to be improbable.

160.015 **New results on large–scale structures.**
M. Kalinkov, K. Stavrev, I. Kuneva.
Astron. Nachr., Vol. 306, No. 5, p. 283 – 300 (1985).

A new version of the magnetic–tape catalog of Abell clusters of galaxies is used to obtain redshift estimators and to generate two samples of clusters. A procedure for searching for superclusters of galaxies is applied and the results are given in tabular and graphic form. For a limited homogeneous sample (distance 60 – 275 Mpc, galactic latitude $B > 35°$), 12 multiplets, having member clusters with known redshifts, are found. It is shown that the spatial covariance function for rich clusters has the form $\xi = (r_0/r)^\gamma$ with $r_0 = 22.4 \pm 1.8$ Mpc and $\gamma = 1.90 \pm 0.25$ for 3 Mpc $\lesssim r \lesssim$ 80 Mpc.

160.016 **Investigation of clusters of galaxies. I. Explanation of photometric scale errors by light diffusion in a photographic emulsion.**
F. W. Baier, R. Ziener.
Astron. Nachr., Vol. 307, No. 1, p. 35 – 40 (1986).

A photographic photometry in B and V was performed for 326 galaxies in the central field of the Coma cluster. The authors have found an explanation (diffusion of light in the emulsion according to an earlier investigation of Ziener) for the systematic deviation between the results from Rood and Baum and from Godwin and Peach in the V pass band. They conclude that the photometry of Rood and Baum is true. Diffusion also can explain the results of Fong, Godwin, and Metcalfe.

160.017 **A corrected velocity for the local standard of rest by fitting to the mean redshift of Local Group galaxies.**
H. Arp.
Astron. Astrophys., Vol. 156, No. 1/2, p. 207 – 212 (1986).

The amplitude of the solar velocity correction to the centroid of the Local Group which is currently in use is 300 km s^{-1}. If those galaxies which are considered certain members of the Local Group are fitted with a solution which minimizes their mean deviation, they give a correction of only 251 km s^{-1}. This leads to a motion of the local standard of rest in the direction of rotation of about $v_\theta = 239$ km s^{-1}, in much better agreement with the measured value of galactic rotation of $\theta_0 = 220$ km s^{-1}. The one remaining anomally which has always troubled this subject is the large differential blueshift between our own Galaxy and M31. This effect is now explained empirically by the systematic negative redshifts which dominant galaxies have been recently shown to have with respect to their companions. The periodicity in redshifts of 72 km s^{-1} (Tifft, 1976) now emerges for the first time from the M31 and the M81 groups of galaxies as: period = 72 ± 2 km s^{-1} (p.e.).

160.018 **A large–scale magnetic feature in the galaxy cluster A 2319.**
J. P. Vallée, J. M. MacLeod, N. W. Broten.
Astron. Astrophys., Vol. 156, No. 1/2, p. 386 – 390 (1986).

The authors report the discovery of a large–scale magnetic feature in the cluster of galaxies Abell 2319, which has been observed from a programme to determine accurate Faraday rotation measures (RM) from a large sample of extragalactic radio sources. In the case of A 2319, the radio sources chosen are located far behind the galaxy cluster, with their lines of sight through or adjacent to the galaxy cluster. An average excess Faraday RM $\approx$ 120 rad m^{-2} is observed for these radio sources behind the galaxy cluster but shining through the cluster, as opposed to the sources adjacent to the galaxy cluster. The authors propose an optical identification for the composite radio source system 13W 315 – 13W 318. A faint optical object is seen at the location of a newly discovered weak radio source at $\approx 19^h21^m30\overset{s}{.}7 + 43°43'08''$, peaking near the center of an elongated radio source system.

160.019 **Machine measurements of the objectve–prism spectra of faint galaxies – II. Interactive redshift measurements.**
S. M. Beard, J. A. Cooke, D. Emerson, H. T. MacGillivray, B. D. Kelly.
Mon. Not. R. Astron. Soc., Vol. 219, No. 2, p. 251 – 271 (1986).
See also 036.017.

The redshift survey in this paper covers only a small area of the sky containing a small corner of the Indus supercluster. None the less, interesting results have been obtained concerning the structure of the south–western corner and the "tail". Evidence has been found that a structure does exist at redshifts around $Z \sim 0.08$ and that this extends at least 4° from the rich clusters in the north–east corner of the field studied here. No evidence for a filamentary "tail" to this supercluster is found, and there is no evidence for any connection between the supercluster and the nearer cluster 2131–6215 in the south. A bridge of galaxies is detected joining the rich clusters 2151–5805 and 2143–5732 which is estimated to be around 6.8 h^{-1}Mpc long.

160.020 **Structure of superclusters and supercluster formation – V. Spatial correlation and voids.**
J. Einasto, A. A. Klypin, E. Saar.
Mon. Not. R. Astron. Soc., Vol. 219, No. 3, p. 457 – 478 (1986).

In this paper the authors study the spatial correlation function of galaxies in detail. They calculate this function for catalogues of different absolute magnitude limit and different volumes or depth. A more recent version of the Center for Astrophysics compilation of galaxy redshifts is used. To compare results with theoretical calculations, the spatial correlation function was also found for computer–generated catalogues, obtained on the basis of N–body dynamical evolution simulations. Both the adiabatic and isothermal scenarios were used. In fair samples of the Universe covering several superclusters and voids the correlation length for galaxies is at least twice the conventional value and approaches the correlation length for clusters of galaxies.

160.021 **Taxonomical analysis of superclusters – II. The A1367/Coma supercluster.**
J. Perea, M. Moles, A. del Olmo.
Mon. Not. R. Astron. Soc., Vol. 219, No. 3, p. 511 – 524 (1986).

The non–hierarchical taxonomical method has been applied to a sample of 185 galaxies, complete to $m \leqslant 15$, in the A1367/Coma region. Once the fore– and background galaxies have been identified, the analysis of the A1367 cluster shows it composed by two gravitationally bound clumps at essentially the same redshift. Central Coma is traced by three different groups whose gravitational binding does not appear clearly. However, accepting the unique dynamical entity of Central Coma, some other neighbouring groups would be part of the cluster. In that case, the average radial velocity and dispersion of the Coma cluster would be 7013 and 423 km s^{-1}, respectively. The M/L ratios for the groups evidenced by the taxonomy range from 13 to 170. The relationship found in other systems between the redshift, the morphological type and the radioemissivity is also present in the sample.

160.022 **Groups of galaxies and the Local Supercluster.**
G. Giuricin, F. Mardirossian, M. Mezzetti, A. Pisani, M. Ramella.
Astron. Astrophys., Vol. 157, No. 1, p. 129 – 132 (1986).

Loose groups of galaxies have been considered as tracers of the Local Supercluster structure. There is observational evidence that, within 20 Mpc from the Virgo cluster, the nearer the groups are, the shorter their crossing times are. Thus some interaction between the Local Supercluster and the dynamical status of these loose groups of galaxies is suggested.

160.023 **Complex structure in A496?**
A. Mazure, D. Gerbal, D. Proust, H. V. Capelato.
Astron. Astrophys., Vol. 157, No. 1, p. 159 – 162 (1986).

The authors present evidence that the A496 cluster has complex structure. In particular it is composed of a main spherical cluster, a group or a subcluster to which belongs the cD galaxy, and a satellite group.

160.024 **The blue Tully–Fisher relation in a variety of environments.**
G. Giuricin, F. Mardirossian, M. Mezzetti.
Astron. Astrophys., Vol. 157, No. 2, p. 400 – 402 (1986).

The authors have examined the blue absolute luminosities and the corrected 21 cm line widths of $\sim$460 disk galaxies belonging to galaxy groups and to several clusters (Virgo, Pegasus I, A 262, A 1367, A 1656, Zwicky 74–23), in order to investigate the possible environmental dependences of the form of the blue Tully–Fisher relation. Within the limits of the present data, the authors have found no firm evidence of appreciable environmental effects. Some implications of this result are briefly discussed.

160.025 **The morphological types and orbits of H I–deficient spirals in clusters of galaxies.**
A. Dressler.
Astrophys. J., Vol. 301, No. 1, p. 35 – 43 (1986).

Further analysis of the Giovanelli and Haynes catalog of the H I content of cluster spirals reveals that H I deficiency is strongly correlated with morphological type. Those spirals deficient in H I are predominantly early–type galaxies; spirals of type Sbc and later are rarely found to be gas deficient, even though they often reside in the same dense environments. Comparison of the velocity dispersion as a function of radius for gas–poor and gas–rich spirals indicates that the gas–poor spirals are on radial orbits which carry them into the dense environment of the cluster core, while the orbits of gas–rich spirals are more likely to be isotropic or even circular. The simplest interpretation of these data is that ram pressure stripping of spiral galaxies on radial orbits is responsible for H I deficiency.

160.026 **Blue disk galaxies in the Coma Cluster: analogs to z = 0.5 cluster members?**
G. D. Bothun, A. Dressler.
Astrophys. J., Vol. 301, No. 1, p. 57 – 64 (1986). With plate 1.

Long–slit spectroscopy of blue disk galaxies that are located in the central portion of the Coma Cluster reveals the presence of Balmer absorption lines with equivalent widths comparable to those observed for some blue galaxies in the 3C 295 cluster. These anomalously strong Balmer absorption lines indicate the presence of a large population of main sequence A–F stars that has presumably arisen from a relatively recent star formation event. We may be witnessing in Coma the same phenomenon that occurred several billion years ago in the 3C 295 cluster. The authors offer a qualitative model of how enhanced star formation, caused by some interaction of the galaxy with the cluster environment, may be responsible for the prevalence of blue cluster galaxies at high redshift.

160.027 **The two–point galaxy correlation function of the Local Supercluster.**
A. R. Rivolo.
Astrophys. J., Vol. 301, No. 1, p. 70 – 76 (1986).

A direct three–dimensional determination of the local two–point galaxy correlation function $\zeta(r)$ in the interval $200 \leqslant H_0 r \leqslant 8000$ km s^{-1} using the data of the Revised Shapley–Ames catalog is presented. The sample space spans a volume centered on the Galaxy and has a radius of $H_0 r \approx 4000$ km s^{-1}. The result of these analyses yields a correlation function which is in good agreement with global determinations, thus implying that this volume of space is a fair sample of the universe at large.

160.028 **CO in H I – deficient Virgo cluster spiral galaxies.**
J. D. Kenney, J. S. Young.
Astrophys. J., Lett. Ed., Vol. 301, No. 1, p. L13 – L17 (1986). With plate L6.

The authors have mapped CO emission along the major axis of 23 Virgo cluster spiral galaxies, including many of the H I – deficient spirals. These galaxies appear to have roughly normal CO luminosities and extents, relative to optical properties. The existence of galaxies with unusually large ratios of molecular gas to atomic gas indicate that some process has removed the low–density atomic gas and left the denser molecular gas relatively unscathed. This conclusion is supported by the observation that the ratio of CO to H I diameters decreases away from M87. These results are consistent with a process which selectively strips low–density material, such as ram pressure stripping, thermal evaporation, or turbulent viscous stripping.

160.029 **Surface photometry of dwarf elliptical galaxies in the Virgo cluster.**
S.–I. Ichikawa, K.–I. Wakamatsu, S. Okamura.
Astrophys. J., Suppl. Ser., Vol. 60, No. 2, p. 475 – 506 (1986). With plates 28 – 39.

Photographic surface photometry is carried out for 69 dwarf ellipticals in the central region of the Virgo cluster. It is confirmed that the population of dwarf ellipticals is not dominated by flat disk systems like spiral galaxies, in spite of their exponential profiles. Two diagrams, of concentration index versus absolute magnitude and of diameter versus surface brightness, are presented which indicate a structural discontinuity in the sequence of spheroidal stellar systems consisting of giant ellipticals, dwarf ellipticals, and globular clusters.

160.030 **Radio emission of Abell clusters of galaxies with redshifts from 0.02 to 0.075 at 102.5 MHz. III. The local radio luminosity function.**
A. G. Gubanov.
Astron. Zh., Tom 63, Vyp. 1, p. 1 – 15 (1986). In Russian. English translation in Sov. Astron., Vol. 30, No. 1.

The local radio luminosity function of Abell clusters of galaxies is constructed at 102.5 MHz from $\sim 10^{24}$ to $\sim 10^{27}$W Hz^{-1} and for total radio luminosities from $\sim 10^{40}$ to $\sim 10^{43}$erg s^{-1}. The behaviour of the clusters' radio luminosity function agrees well with the known properties of the radio luminosity function of galaxies.

160.031 **The compared specific energies of galaxies and ICM in the Coma cluster.**
D. Gerbal, G. Mathez, A. Mazure, E. Salvador–Sole.
Astron. Astrophys., Vol. 158, No. 1/2, p. 177 – 180 (1986).

An attempt is made to reconcile the observed low ratio of galaxy to ICM specific energies in clusters of galaxies with present ideas on their dynamical evolution. In view of non–isothermal multi mass models, the "scale height problem" appears to be resolved for the Coma cluster.

160.032 **Is the surface brightness spectrum of the Coma cluster radio halo inverted?**
B. Dennison.
Astron. Astrophys., Vol. 159, No. 1/2, p. 251 – 254 (1986).

Comparison of published maps of the Coma cluster radio halo at 430 and 1400 MHz (Hanisch, 1980) reveals that the spectrum of the surface brightness is inverted or flat over most of the extent of the 1400 MHz diffuse emission. This is very unusual for a highly extended source thought to emit via the synchrotron mechanism. Various explanations are considered, including a relatively flat electron–energy spectrum, free–free emission, emission by numerous compact sources, as well as instrumental effects.

160.033 **A search for the Syunyaev–Zel'dovich effect in X–ray clusters of galaxies.**
K. D. Aliakberov, V. I. Dokuchaev, Yu. N. Parijskij.
Soobshch. Spets. Astrofiz. Obs., Vyp. 48, p. 81 – 90 (1985).

The results of deep λ7.6 cm observations, searching for the Syunyaev–Zel'dovich effect towards some X–ray emitting Abell clusters of galaxies are presented. The authors find no evidence for the effect in the present observations.

160.034 **Optical and infrared studies of galaxy clusters with cooling accretion flows.**
W. Romanishin.
Astrophys. J., Vol. 301, No. 2, p. 675 – 682 (1986).

What is the final fate of the gas that X–ray observations indicate is accreting onto the brightest cluster members in some galaxy clusters? To search for star formation in the accretion flow gas, new *V–K* colors are presented for eight brightest cluster members in X–ray clusters with a range of accretion flow rates, and published optical data for other clusters is analyzed. Except for A426 (Perseus), color data do not provide evidence for star formation in any accreting cluster. Two other consequences of star formation in accretion flows, changes in galaxy absolute magnitude and changes in M/L ratio, are discussed. Again, only in A426 evidence for the "final" fate of the accreting gas could be found.

160.035 **The X–ray structure of the 3C 295 cluster: a cooling flow at a redshift of 0.5.**
J. P. Henry, M. J. Henriksen.
Astrophys. J., Vol. 301, No. 2, p. 689 – 697 (1986).

The authors describe the X–ray morphology of the 3C 295 cluster. All X–ray and optical data are consistent with an isothermal hydrostatic equilibrium model with parameters similar to nearby clusters. There is evidence for a cooling flow of approximately 145 $M_{\odot}$yr^{-1} onto 3C 295 itself. From a comparison of the cooling times of low–redshift clusters with the 3C 295 cluster, the authors determine that the Hubble constant is $H_0 \leqslant 20\,(+27,-9)$ km s^{-1}Mpc^{-1}. The on–orbit performance characteristics of the high–resolution imager of the Einstein Observatory is also presented.

160.036 **A sparse–sampling strategy for the estimation of large–scale clustering from redshift surveys.**
N. Kaiser.
Mon. Not. R. Astron. Soc., Vol. 219, No. 4, p. 785 – 790 (1986).

The form of the power spectrum of galaxy clustering at large wavelengths provides a powerful test of theories for galaxy formation. The author shows that estimating the two–point correlation function from a complete, magnitude limited redshift survey provides a rather inefficient way to apply this test. By sampling only a randomly chosen fraction of the galaxies, but to a fainter magnitude limit, one can significantly reduce the uncertainty in the two–point function, ξ, for a given investment of telescope time.

160.037 **Gas in a cosmic void: properties and origin.**
L. M. Ozernoj, V. V. Chernomordik.
Pis'ma Astron. Zh., Tom 12, No. 4, p. 325 – 336 (1986). In Russian. English translation in Sov. Astron. Lett., Vol. 12.

The analysis of absorption lines with redshift of the cosmic void in Bootes, which were discovered by Brosch and Gondhalekar (1984), enables to propose the following alternative as to where the gas producing the lines is located: (1) in numerous dwarf galaxies provided their concentration in the void be greater than 3×10^2Mpc^{-3}, (2) in a substantial part of the overall volume of the void. The choice between these possibilities depends on which of the mechanisms, photoionization or collisional ionization, is responsible for the ionization state of Si and C. Some tests to prove the expected shell form for both galaxies and the processed gas distribution are proposed; discovery of such a bubble would be the decisive argument for the explosional origin of the void.

160.038 **Structure of the observed field of objects. Determination of multiplicity of clusters of galaxies.**
T. A. Agekyan, S. P. Yakimov.
Astron. Zh., Tom 63, Vyp. 2, p. 214 – 219 (1986). In Russian. English translation in Sov. Astron., Vol. 30, No. 2.

The method of specially selected characteristics of a field of objects, being stochastic quantities, is proposed to determine the presence of some structure in the field and to estimate the parameters of the structure, if it exists. The method is applied to the Abell catalogue of clusters of galaxies. The presence of tendency to group in multiclusters is confirmed. An estimate of grouping parameters is obtained.

160.039 **Revised radial velocities for 12 near Zwicky clusters.**
M. Turatto, A. R. Petrosyan.
Astrofizika, Tom 24, Vyp. 1, p. 205 – 207 (1986). In Russian. English translation in Astrophysics, Vol. 24, No. 1.
On the basis of new radial velocities of Markarian galaxies, the members of 12 near Zwicky clusters and their radial velocities are recalculated.

160.040 **Expected source counts for microwave radiation sources associated with rich clusters of galaxies.**
V. A. Korolev, R. A. Syunyaev, L. A. Yakubtsev.
Pis'ma Astron. Zh., Tom 12, No. 5, p. 339 – 352 (1986). In Russian. English translation in Sov. Astron. Lett., Vol. 12.
It is shown that distant clusters of galaxies may give dominant contribution to source counts in the microwave spectral band. The expected source counts drastically differ from the results of source counts in the centimeter and decimeter bands.

160.041 **The X–ray spectrum of the Coma Cluster of galaxies.**
M. J. Henriksen, R. F. Mushotzky.
Astrophys. J., Vol. 302, No. 1, p. 287 – 295 (1986).
The authors present the *HEAO 1* A–2 and *OSO 8* observations of the Coma Cluster of galaxies in the 2 – 60 keV energy band. The spectrum is not well fitted by an isothermal bremsstrahlung model with an Fe line. The authors have performed fits of polytropic models and find that the gas is well fitted by an equation of state with polytropic index ranging between 1.38 and 1.6. This analysis reduces the total cluster mass by a factor of 3 – 5 below the optically derived virial mass. The gas is found to be a significant fraction (25%) of the total cluster mass. The Fe Kα line blend is strongly detected, indicating a 0.18 – 0.36 solar abundance in the X–ray emitting gas.

160.042 **The structure of brightest cluster members. I. Surface photometry.**
J. M. Schombert.
Astrophys. J., Suppl. Ser., Vol. 60, No. 3, p. 603 – 693 (1986).
Faint surface photometry of 261 bright ellipticals in the cores of 63 clusters and the field is presented. Brightest cluster members are found to have enlarged characteristic radii and shallow profile slopes, which distinguishes them structurally from normal ellipticals as a class of objects. Two non–first–ranked cD galaxies were found in A262 and A2151; however, none were found in the field, indicating that a cluster environment is a prerequisite for extended envelope formation.

160.043 **Groups around bright Seyfert nuclei.**
K. J. Fricke, W. Kollatschny, H. H. Loose.
Structure and evolution of active galactic nuclei, p. 551 – 554 (1986). – See Abstr. 012.011.
The authors have determined the nuclear galaxy spectra of the members of wide groups ($\sim$1 Mpc across) around bright Seyfert galaxies. They find a tight relation between the "activity classes" of the nuclei as defined from their emission line spectra and the distance from the central Seyfert galaxy. This relation is characterized by a critical radius.

160.044 **Statistics of bright galaxies and cluster morphology.**
G. Giuricin, F. Mardirossian, M. Mezzetti.
Structure and evolution of active galactic nuclei, p. 555 – 558 (1986). – See Abstr. 012.011.
The authors have investigated the luminosity distributions of the first–ranked, second–ranked, and third–ranked cluster galaxies for different cluster morphological types.

160.045 **The Leo intergalactic neutral hydrogen cloud.**
Y. Terzian, S. E. Schneider, E. E. Salpeter.
Structure and evolution of active galactic nuclei, p. 723 – 725 (1986). – See Abstr. 012.011.
The current status of the neutral hydrogen λ21 cm radio observations of the intergalactic cloud in Leo are described. These include both H I observations at Arecibo with an angular resolution of $\sim$3 arcmin, and VLA D–array observations with an effective resolution of 45 arcsec.

160.046 **Redshifts of galaxies in the south–west region of the Perseus supercluster.**
R. Merighi, P. Focardi, B. Marano, G. Vettolani.
Astron. Astrophys., Vol. 160, No. 2, p. 398 – 401 (1986).
Redshifts for 33 galaxies have been measured allowing to study the spatial distribution of a complete sample of 125 galaxies brighter than 14.5 mag in a large area west of the Perseus supercluster. The authors show that the Perseus supercluster connects to another large scale structure: the A 2634 – A 2666 supercluster.

160.047 **Multiaperture spectroscopy of clusters of galaxies at a redshift of 0.2.**
R. J. Lavery, J. P. Henry.
Bull. Am. Astron. Soc., Vol. 17, No. 4, p. 858 (1985). Abstract. – See Abstr. 010.062.

160.048 **The nearby Abell clusters.**
W. R. Oegerle, J. G. Hoessel, M. S. Jewison.
Bull. Am. Astron. Soc., Vol. 17, No. 4, p. 858 (1985). Abstract. – See Abstr. 010.062.

160.049 **The Corona Borealis supercluster.**
M. Postman, M. J. Geller, J. P. Huchra.
Bull. Am. Astron. Soc., Vol. 17, No. 4, p. 858 (1985). Abstract. – See Abstr. 010.062.

160.050 **CO in H I–deficient Virgo cluster spiral galaxies.**
J. D. Kenney, J. S. Young.
Bull. Am. Astron. Soc., Vol. 17, No. 4, p. 865 (1985). Abstract. – See Abstr. 010.062.

160.051 **Spectral analysis of Spartan–1 X–ray observations of the Perseus cluster.**
M. P. Ulmer, R. G. Cruddace, G. G. Fritz, W. A. Snyder, E. E. Fenimore.
Bull. Am. Astron. Soc., Vol. 17, No. 4, p. 866 (1985). Abstract. – See Abstr. 010.062.

160.052 **The 1 – 12 keV X–ray morphology of the Perseus cluster of galaxies, as determined from Spartan–1 observations.**
W. A. Snyder, R. G. Cruddace, G. G. Fritz, M. P. Ulmer, J. Middleditch, E. E. Fenimore.
Bull. Am. Astron. Soc., Vol. 17, No. 4, p. 866 (1985). Abstract. – See Abstr. 010.062.

160.053 **Large–scale clustering of galaxies with massive dark halos.**
A. E. Evrard.
Bull. Am. Astron. Soc., Vol. 17, No. 4, p. 866 (1985). Abstract. – See Abstr. 010.062.

160.054 **The difference between the galaxy and cluster correlation functions: a manifestation of tails of galaxy clusters.**
N. A. Bahcall.
Bull. Am. Astron. Soc., Vol. 17, No. 4, p. 866 (1985). Abstract. – See Abstr. 010.062.

160.055 **Systematic properties of clusters of galaxies: cosmological scenario versus observations.**
M. J. West, A. Dekel, A. Oemler.
Bull. Am. Astron. Soc., Vol. 17, No. 4, p. 866 – 867 (1985). Abstract. – See Abstr. 010.062.

160.056 **Cooling flows in clusters and galaxies.**
A. C. Fabian.
Bull. Am. Astron. Soc., Vol. 17, No. 4, p. 888 (1985). Abstract. – See Abstr. 010.062.

160.057 **The X-ray emitting gas in clusters of galaxies.**
M. J. Henriksen.
Bull. Am. Astron. Soc., Vol. 17, No. 4, p. 906 (1985). Abstract. – See Abstr. 010.062.

160.058 **Cosmological H II regions and the photoionization of the IGM.**
P. R. Shapiro, M. L. Giroux, I. Wasserman.
Bull. Am. Astron. Soc., Vol. 17, No. 4, p. 906 (1985). Abstract. – See Abstr. 010.062.

160.059 **High resolution X-ray spectroscopy of clusters of galaxies.**
H. U. Nørgaard-Nielsen.
ESA Spec. Publ., ESA SP-239, p. 25 – 31 (1985). – See Abstr. 012.019.
The prospects for studying the X-ray emission from clusters of galaxies with XMM are discussed.

160.060 **EXOSAT observations of the poor cluster of galaxies 2A 0335+096.**
H. U. Nørgaard-Nielsen, N. J. Westergaard, L. Hansen, H. E. Jørgensen, H. W. Schnopper.
ESA Spec. Publ., ESA SP-239, p. 155 – 157 (1985). – See Abstr. 012.019.
The authors have observed 2A 0335+096 for 10^h with EXOSAT. The LE image shows an irregular structure, and the X-ray emission is not centered on the dominant galaxy. Optical emission lines are seen from a large region, and they indicate normal composition, whereas the ME spectrum suggests a very low metal content, implying that the ionized gas seen optically is not a product of the cooling flow inferred from previous X-ray observations.

160.061 **The difference between the galaxy and cluster correlation functions: a manifestation of tails of galaxy clusters.**
N. A. Bahcall.
Astrophys. J., Lett. Ed., Vol. 302, No. 2, p. L41 – L44 (1986).
A phenomenological model is presented that explains the observed difference between the galaxy and rich cluster correlation functions, based on the fraction of galaxies that participate in the clustering. It is shown that if $\sim 25\%$ of all galaxies are associated with rich (richness $\geqslant 1$) clusters, and the other galaxies are only weakly correlated, the galaxy correlation function will be, as observed, ~ 18 times weaker than the cluster correlations. The model suggests that the fraction of galaxies associated with rich clusters is larger than previously expected; most of these galaxies are distributed in the outer tails of the clusters which may extend to at least $\sim 30\ h^{-1}$Mpc.

160.062 **On a probable mechanism of formation of breaks in the luminosity function of galaxy clusters.**
L. I. Vinokurov, A. V. Kats, V. M. Kontorovich, A. P. Miroshnichenko.
Astron. Tsirk., No. 1390, p. 5 – 8 (1985). In Russian.
The breaks in the luminosity functions which correspond to the breaks in the mass functions for clusters of galaxies can be due to the relaxation mechanism replacement or the interaction replacement. The mass function indices and the break locations are in satisfactory agreement with the observed data for clusters of galaxies.

160.063 **Programme "Northern Cone". III. The results of additional redshift measurements of rich clusters of galaxies inside the cone with the 6-m telescope.**
A. I. Kopylov, V. A. Lipovetskij, T. S. Fetisova, V. F. Shvartsman.
Astron. Tsirk., No. 1393, p. 1 – 3 (1985). In Russian.

160.064 **Old radio galaxies and cluster radio halos.**
S. S. Komissarov.
Sov. Astron., Vol. 29, No. 4, p. 382 – 388 (1985). English translation of 40.160.022.

160.065 **A hot intergalactic medium?**
P. W. Guilbert, A. C. Fabian.
Mon. Not. R. Astron. Soc., Vol. 220, No. 2, p. 439 – 451 (1986).
The authors have reinvestigated the possibility that a hot intergalactic medium produces the 5 – 200 keV X-ray background, including relativistic corrections for the gas emissivity and thermodynamics and Compton scattering on the microwave background.

160.066 **Two suspected rings of nearby galaxies.**
G. H. I. Johansson.
Astron. Astrophys., Vol. 161, No. 1, p. 113 – 118 (1986).
It is investigated if the two groups of galaxies containing NGC 55 and NGC 2403, respectively, form rings or not. It is shown that both rings are probably due to chance projections. Statistical methods are used for the NGC 2403-group. The area around the south galactic pole has been searched (in part semiautomatically) for nearby dwarf galaxies physically related to NGC 55 and 9 new candidates have been found. These galaxies are listed and described. Semiautomated survey methods versus manual ones are discussed.

160.067 **HEAO 2 X-ray observations of clusters of galaxies.**
M. P. Ulmer, M. P. Kowalski, R. G. Cruddace.
Astrophys. J., Vol. 303, No. 1, p. 162 – 170 (1986).
The authors provide a summary of results of their Einstein satellite observations of clusters of galaxies. They report X-ray luminosities or upper limits of 27 clusters. Newly reported clusters with interesting morphologies are presented, and a brief discussion of the data in relation to theories of cluster formation and evolution is given.

160.068 **The local supercluster.**
R. B. Tully.
Proceedings of the Second Asian-Pacific Regional Meeting on Astronomy, p. 295 – 296 (1984). – See Abstr. 012.031.
An attempt is made to illustrate the three-dimensional distribution of nearby galaxies. There is an evident overdensity of galaxies in the north galactic hemisphere that has been called the local supercluster. It is argued that this system is comprised of two distinct components: a disk component with 60% of the luminous, and a halo component with 40% of the luminous galaxies.

160.069 **Luminosity functions of clusters of galaxies.**
T. Yamagata, H. Maehara, S. Okamura, B. Takase.
Proceedings of the Second Asian-Pacific Regional Meeting on Astronomy, p. 525 – 531 (1984). – See Abstr. 012.031.
Direct photographs of a rich cluster A 1553 and poor clusters MKW 1 and MKW 8 were taken with the 105 cm Schmidt telescope at the Kiso Observatory. The luminosity function is obtained from the number of images included in the cluster region after applying the background/foreground correction. A 1553 is more extended (r = 11') than the result of Bucknell et al. (1979). The general shape of luminosity function of the two MKW clusters is similar to that of rich clusters, though numbers and radii of these clusters are smaller than those of A 1553.

160.070 **Redshift observations in the Centaurus-Hydra supercluster region. I.**
L. N. da Costa, M. A. Nunes, P. S. Pellegrini, C. Willmer, G. Chincarini, J. J. Cowan.
Astron. J., Vol. 91, No. 1, p. 6 – 12 (1986).
The authors present 111 new redshifts in the region of the Centaurus-Hydra supercluster.

160.071 **High-resolution observations of the intergalactic H I cloud in the M96 group.**
S. E. Schneider, E. E. Salpeter, Y. Terzian.
Astron. J., Vol. 91, No. 1, p. 13 – 22 (1986).
The large cloud complex of intergalactic H I in the M96 group has been observed with the D array of the Very Large Array. The main cloud has been resolved into a number of clumps which account for most of the H I previously detected at Arecibo. Additionally, high-resolution spectra have been made at Arecibo that

show evidence of narrower subcomponents within the $\sim 30-50$ km s^{-1} wide H I profiles previously found at Arecibo. In general, the kinematics of the gas follow the overall pattern found in the original observations, but there are some indications that the clumps may be dynamically independent.

160.072 **Radio continuum survey of the Coma/A1367 supercluster. I. 610 MHz observations of CGCG galaxies in four groups.**
W. Jaffe, G. Gavazzi, E. Valentijn.
Astron. J., Vol. 91, No. 2, p. 199 – 203 (1986). With plates 12 – 18.
Radio continuum observations obtained with the Westerbork Radio Synthesis Telescope at 0.6 GHz of four groups of galaxies in the Coma/A1367 supercluster area are presented. Ninety–nine CGCG galaxies were surveyed, yielding the detection of 21 objects. A wide–angle–tail radio galaxy, NGC 4061, is found in the NGC 4065 group. Analysis of this source suggests a relatively low value ($n_e T \sim 1000$ cm^{-3}K) for the intracluster gas pressure in this group.

160.073 **Radio continuum survey of the Coma/A1367 supercluster. II. 1.5 GHz observations of 396 CGCG galaxies.**
W. Jaffe, G. Gavazzi.
Astron. J., Vol. 91, No. 2, p. 204 – 216 (1986). With plates 19 – 30.
1.5 GHz VLA radio continuum observations of 396 relatively isolated CGCG galaxies in the Coma/A1367 supercluster yielded the detection of 95 objects. These observations, added to the ones presented in previous papers, form a complete sample of optically selected objects with $m_p \leqslant 15.3$. Two wide–angle–tailed sources have been found in smaller groups within the supercluster.

160.074 **Hot gas in clusters and early–type galaxies.**
C. Jones, W. Forman.
X–ray astronomy '84, p. 355 – 371 (1985). – See Abstr. 012.033.
The authors summarize the properties of the hot, X–ray emitting gas observed in clusters and early–type galaxies. In particular they review the determination of the total mass in the halo of M87, the detection of microwave decrements in distant Coma–type clusters, the observations of radiative cooling in cluster cores, and the use of galactic coronae to measure the total mass of early–type galaxies.

160.075 **X–ray measurements of the aspherical mass distribution in the cluster of galaxies A 2256.**
D. Fabricant, G. Rybicki, P. Gorenstein.
X–ray astronomy '84, p. 381 – 386 (1985). – See Abstr. 012.033.
The authors describe a technique for deriving the underlying mass distribution from X–ray surface brightness maps of rich clusters of galaxies with a smooth X–ray morphology and approximate bilateral symmetry. They apply this technique to A 2256 and find significant departures from spherical symmetry.

160.076 **X–ray studies of superclusters.**
M. P. Ulmer.
X–ray astronomy '84, p. 387 – 390 (1985). – See Abstr. 012.033.
The author has used X–ray observations to make the following types of analyses: (1) nearest neighbor distances; (2) angular correlation analysis; (3) the luminosity function of clusters that are in physically associated multiple clusters; and (4) X–ray pictures of possible superclusters.

160.077 **Some clues for the iron origin in galaxy clusters.**
R. Rothenflug, M. Arnaud, O. Boulade, L. Vigroux.
X–ray astronomy '84, p. 391 – 394 (1985). – See Abstr. 012.033.
In galaxy clusters, the mass in iron can be considered as directly proportional to the mass in gas. On the other side, the gas mass seems proportional to the virial mass. The authors propose that this virial mass is made of stars and/or star remnants which produced the iron in the past. The implications of this hypothesis are examined in terms of chemical evolution model of a mass of gas.

160.078 **Energy content and temperature behaviour of intracluster medium.**
D. Gerbal, G. Mathez, E. Salvador–Solé.
X–ray astronomy '84, p. 395 – 398 (1985). – See Abstr. 012.033.
It is shown that the intracluster medium (ICM) energy content is overestimated by the usual energy indicators provided that the ICM temperature is typically decreasing outwards. One of them can be used as a tracer of such a temperature gradient.

160.079 **Redshift asymmetries in systems of galaxies and the missing mass.**
M. J. Valtonen, G. G. Byrd.
Astrophys. J., Vol. 303, No. 2, p. 523 – 534 (1986).
The authors use redshift asymmetries to study the dynamical state of binary galaxies and groups and clusters of galaxies. They confirm their earlier suggestion that groups are composed largely of an expanding population. It is found that a massive central binary ($\sim 10^{14} M_\odot$) dominates the Coma Cluster, making it dynamically unstable. There is evidence that other rich clusters have similar massive cores and are also largely unbound. The authors examine these results and other information regarding the missing mass problem. Excluding a possible uniform component of missing mass, they estimate the cosmological density parameter $\Omega \approx 0.02$. Missing mass in amounts large enough to close the universe is apparently not concentrated in galaxies or clusters of galaxies.

160.080 **Luminosity functions for nine Abell clusters.**
P. M. Lugger.
Astrophys. J., Vol. 303, No. 2, p. 535 – 555 (1986).
A study of the luminosity functions of nine Abell clusters of galaxies has been carried out using photographic plates obtained with the Palomar 1.2 m Schmidt telescope. Luminosity functions were determined for the nine clusters from square regions 4.6 Mpc on a side. For this sample of nine clusters, measures of characteristic galaxy magnitude (e.g., the Schechter function parameter M^*) show no evidence for a correlation with cluster morphology, central density, or magnitude of the first–ranked galaxy.

160.081 **The structure of brightest cluster galaxies.**
J. M. Schombert.
Diss. Abstr. Int., Sect. B, Vol. 46, No. 5, p. 1590 (1985). Thesis, Yale University, 208 pp. (1984). Order No. DA8514685.

160.082 **Compact configurations within small evolving groups of galaxies.**
G. A. Mamon.
Thesis, Princeton University, 8 + 287 pp. (1985). Available from University Microfilms International, Ann Arbor, Mich., USA, Order No. DA8518639.
The evolution of small groups of galaxies from virialized initial conditions has been simulated with a gravitational *N*–body code, specifically written for this purpose. Dense groups of galaxies with massive dark haloes coalesce into a single galaxy and lose their compact group appearance in approximately 3 group half–mass crossing times, while dense groups of galaxies without massive haloes survive the merger instability for 15 half–mass crossing times. These timescales correspond to roughly $H_0^{-1}/8$ and $H_0^{-1}/4$, respectively, and despite strong stochasticity, almost never reach $H_0^{-1}/2$. Galaxies without haloes first decay into the background before merging, while galaxies with massive haloes merge first. The instability times are relatively insensitive to uncertainties in the collisional stripping and merger cross–sections, and to the presence of mean–field limitation, but can be larger if mean–field limitation is evaluated with the usual and incorrect assumption that the galaxies travel along nearly circular orbits. Almost all of the loose groups retain their general aspect over H_0^{-1}. Compact subsystems occur in roughly 10% of the projected loose groups, and the large majority are chance superpositions of galaxies, rather than physical cores. The author suggests that most observed compact groups are either isolated dense systems of galaxies devoid of massive dark haloes and younger than 1 billion years, or, what seems more likely, chance superpositions within larger loose groups.

160.083 **Limits to the extragalactic distance scale from integrated properties of Local Group galaxies.**
D. J. Westpfahl Jr.
Diss. Abstr. Int., Sect. B, Vol. 46, No. 7, p. 2347 (1986). Thesis, Montana State University, 173 pp. (1985). Order No. DA8519362.

160.084 **Probing the large–scale structure of the universe: an analysis of 55 bright southern clusters of galaxies.**
R. P. Olowin.
Diss. Abstr. Int., Sect. B, Vol. 46, No. 12, p. 4283 (1986). Thesis, The University of Oklahoma, 562 pp. (1985). Order No. DA8603517.

160.085 **Two–point correlation for rich clusters of galaxies.**
S. Otto, H. D. Politzer, M. B. Wise.
Phys. Rev. Lett., Vol. 56, No. 17, p. 1878 – 1881 (1986). With a correction in No. 25, p. 2772.
The rich–cluster two–point correlation function is examined in a model where rich clusters of galaxies form at high peaks of primordial Gaussian mass–density fluctuations (averaged over a suitable volume). Particular attention is paid to the case when the primordial fluctuations have a Zel'dovich power spectrum.

160.086 **An optical and X–ray study of the radio source 3C 75.**
S. Caganoff, G. V. Bicknell, D. Carter.
Proc. Astron. Soc. Aust., Vol. 6, No. 2, p. 151 – 156 (1985).
3C 75 consists of a pair of twin–jetted radio galaxies near the centre of the cluster Abell 400 which is a strong extended X–ray source. Fibre spectroscopy from the AAT, combined with number counts of the galaxies in Abell 400 give the gravitational potential of the cluster which is used to model the X–ray source. The inferred pressure of the X–ray gas suggests that the lobes of the radio source are probably confined.

160.087 **Modelling clusters of galaxies.**
P. F. Teague, D. Carter.
Proc. Astron. Soc. Aust., Vol. 6, No. 2, p. 198 – 202 (1985).
Thorough investigation of clusters of galaxies involves the complete modelling of their dynamics and structure. Presented here is a description of such a substantial project utilizing a sample of several rich southern clusters of galaxies.

160.088 **Static properties of galaxies in the Shapley–Centaurus cluster. II. The analysis.**
R. J. Dodd, F. P. Andrews, H. T. MacGillivray.
Astrophys. Space Sci., Vol. 123, No. 1, p. 145 – 154 (1986).
The data of paper I are examined in a search for non–random effects. A global trend of bimodal alignment of galaxies in field 444 of the Southern Sky Survey is found. This effect is present even for galaxies of large apparent ellipticity ($e \geqslant 0.5$) and is interpreted as revealing a real physical phenomenon on the large scale in this field. Analysis of nearest neighbouring pairs of galaxies reveals evidence also for small–scale interactions. The difference in apparent ellipticities between close pairs is larger than would be expected if the galaxies were randomly distributed. Nearest neighbouring galaxies also show a strong preference to be aligned with their major axes parallel. The results are discussed with regard to theories for galaxy formation.

160.089 **X–ray studies of galactic and intergalactic gas in the Pegasus I cluster.**
C. R. Canizares, M. E. Donahue, G. Trinchieri, G. C. Stewart, T. A. McGlynn.
Astrophys. J., Vol. 304, No. 1, p. 312 – 317 (1986).
Observations of Pegasus I with the Einstein Observatory IPC reveal hot gas in a low–density intergalactic medium and in a galactic medium within each of the dominant elliptical galaxies NGC 7619 and NGC 7626. The galaxies have X–ray luminosities of 8 and 3×10^{41}ergs s^{-1} respectively, temperatures of $\sim 10^7$K, mean gas densities of $\sim 10^{-3}$cm^{-3} and gas masses of $\sim 10^{10} M_\odot$. The intergalactic gas has a luminosity of 1×10^{42}ergs s^{-1}, a density of $\sim 2 \times 10^{-4}$cm^{-3}, a mass of $\sim 4 \times 10^{11} M_\odot$, and a temperature of $\sim 10^7$K.

160.090 **Spectroscopy of three Butcher–Oemler clusters of galaxies at a redshift of 0.2.**
R. J. Lavery, J. P. Henry.
Astrophys. J., Lett. Ed., Vol. 304, No. 1, p. L5 – L9 (1986).
The authors present the results of spectroscopic observations of three concentrated Abell clusters of galaxies at a redshift of 0.2: A223, A963, and A2111. The data confirm the blue galaxy fraction for these clusters determined photometrically by Butcher and Oemler. The blue galaxy population is dominated by galaxies whose spectra show features due to young and intermediate–age stars. The authors present evidence that these galaxies have undergone extensive bursts of star formation. They conclude that the "Butcher–Oemler effect" is due primarily to bursts of star formation occurring in galaxies in concentrated clusters at redshifts out to 0.5.

160.091 **The nearby Abell clusters. I. Abell 168.**
W. R. Oegerle, J. G. Hoessel, R. M. Ernst.
Space Telesc. Sci. Inst., Prepr. Ser., No. 87, 31 pp. (1986). To appear in Astron. J.

160.092 **The difference between the galaxy and cluster correlation functions: a manifestation of tails of galaxy clusters.**
N. A. Bahcall.
Space Telesc. Sci. Inst., Prepr. Ser., No. 92, 12 pp. (1986). To appear in Astrophys. J., Lett. Ed.

160.093 **Radio continuum survey of the Coma/A1376 supercluster. III. Radio properties of galaxies in different density environments.**
G. Gavazzi, W. Jaffe.
Space Telesc. Sci. Inst., Prepr. Ser., No. 102, 70 pp. (1986). Submitted to Astrophys. J.

160.094 **Kinematic evidence of satellite galaxy populations in the potential wells of first ranked cluster galaxies.**
L. L. Cowie, E. M. Hu.
Space Telesc. Sci. Inst., Prepr. Ser., No. 107, 18 pp. (1986). To appear in Astrophys. J., Lett. Ed.

160.095 **The Shane–Wirtanen counts: systematics and two–point correlation function.**
V. de Lapparent, M. J. Kurtz, M. J. Geller.
Astrophys. J., Vol. 304, No. 2, p. 585 – 598 (1986). With plates 10 – 15.
Both the break in the correlation function and the filamentary appearance of the Shane–Wirtanen map are suspect because of residual systematic errors in the Seldner et al. (SSGP) map. This paper analyzes the systematic errors and shows them to be caused by the applied counting pattern and plate vignetting. These errors cause a residual rms fluctuation of 11% in the SSGP–corrected counts. The authors examine techniques for reconstruction of the correlation function in catalogs which are affected by systematic biases related to the plates. They discuss limits which must be met by new surveys designed to measure the large–scale behavior of the correlation function directly.

160.096 **The large scale distribution of galaxies in the Lynx–Gemini region.**
P. Focardi, B. Marano, G. Vettolani.
ESO Sci. Prepr., No. 420, 19 pp. (1986). To appear in Astron. Astrophys.

160.097 **Statistical investigation of the structure of observed multiple systems. I. Compact groups of galaxies.**
S. A. Malykh, V. V. Orlov.
Astrofizika, Tom 24, Vyp. 3, p. 445 – 451 (1986). In Russian. English translation in Astrophysics, Vol. 24, No. 3.
A method for the statistical study of multiple–system configurations is proposed permitting to judge their static and dynamic properties: oblation or prolation, hierarchy, ring–like structure, rotation, etc. The method has been applied to the investigation of a sample of 100 isolated compact galaxy groups from the Hickson list. It has been shown that the groups have a prolate ten-

dency. Their mean apparent sphericity is equal to approximately 0.4. Hierarchy of the structure has not been discovered.

160.098 **A paradox of the distribution of galaxy clusters.**
B. I. Fesenko.
Astrofizika, Tom 24, Vyp. 3, p. 453 – 465 (1986). In Russian. English translation in Astrophysics, Vol. 24, No. 3.
Owing to the supersensibility effect considered in the first part of this paper, nearly all clusters included in the catalogues were discovered in interstellar–medium windows. For that reason the latitude dependence of colour excesses of the brightest members is weak and the number density of distant clusters in the vicinity of the nearer ones increases. Apparent and space distributions of 702 rich clusters with $z < 0.2$ and $b \geqslant 40°$ are investigated. Application of the covariance function method leads to a clear indication on the existence of multiple clusters. But a careful analysis does not confirm this result. It has been shown for the first time how irregular extinction provides the observed dependence of a covariance function from the distance to the observer.

160.099 **Photometry of 18 clusters of galaxies.**
T. Yamagata.
Ann. Tokyo Astron. Obs., Second Ser., Vol. 21, No. 1, p. 31 – 76 (1986).
Poor clusters of galaxies identified by Morgan, Kayser, and White (1975), and Albert, White, and Morgan and 2 rich clusters are observed. Measured area of each cluster is 1 square degree centered on the first ranked galaxy on plate of the *V* color band taken by the Kiso Schmidt telescope. Positions, apparent magnitudes, ellipticities, and position angles of major axis for 6,300 galaxies of down to 18.0 magnitude in 16 poor cluster and 2 rich cluster fields are presented.

160.100 **The mass of the Coma cluster.**
L. S. The, S. D. M. White.
Prepr. Steward Obs., No. 664, 20 pp. (1986). To appear in Astron. J.

160.101 **Intergalactic obscuring matter and holes between cluster structures of galaxies.**
K. Rudnicki, T. Grabinska, M. Zabierowski.
Nuovo Cimento C, Vol. 8C, Ser. 1, No. 4, p. 368 – 372 (1985). Abstr. in Phys. Abstr., Vol. 89, No. 1276, Entry 36341 (1986).

160.102 **Large–beam radio and infra–red observations toward the Virgo Cluster.**
C. Ceccarelli, P. De Bernardis, S. Masi, F. Melchiorri, G. Moreno, L. Pietranera, G. Dall'Oglio, B. Melchiorri, N. Mandolesi, G. Morigi.
Nuovo Cimento C, Vol. 8C, Ser. 1, No. 4, p. 374 – 378 (1985). Abstr. in Phys. Abstr., Vol. 89, No. 1276, Entry 36342 (1986).

160.103 **The detection of distant cooling flows and the formation of dark matter.**
A. C. Fabian, K. A. Arnaud, P. E. J. Nulsen, R. F. Mushotzky.
Astrophys. J., Vol. 305, No. 1, p. 9 – 13 (1986).
Cooling flows involving substantial mass inflow rates appear to be common in many nearby rich and poor clusters and in isolated galaxies. The extensive optical and ultraviolet filaments produced by the thermal instability of large flows are detectable out to redshifts greater than 1. The authors propose that this may explain the extended optical line emission reported in, and around, many distant radio galaxies, narrow–line quasars, and even nearby normal and active galaxies. Cooling flows are characterized by the presence of extensive regions at high thermal pressure.

160.104 **Erratum: "The A2197 and A2199 galaxy clusters" [Astrophys. J., Vol. 286, No. 2, p. 422 – 436 (1984)].**
S. A. Gregory, L. A. Thompson.
Astrophys. J., Vol. 305, No. 1, p. 580 (1986). See Abstr. 38.160.141.

160.105 ***N*–point correlations for biased galaxy formation.**
L. G. Jensen, A. S. Szalay.
Astrophys. J., Lett. Ed., Vol. 305, No. 1, p. L5 – L9 (1986).
A general series expansion of arbitrary accuracy is given for the correlation function of "biased" regions in terms of the mass autocorrelations. The first term of the expansion agrees with the limit reported by Kaiser in 1984. It is shown that the "biased" correlation function stays zero, whenever the mass correlation is zero. A simple formula to calculate the *N*–point correlations to arbitrary accuracy is given as well.

160.106 **The large–scale distribution of galaxies in the Lynx–Gemini region.**
P. Focardi, B. Marano, G. Vettolani.
Astron. Astrophys., Vol. 161, No. 2, p. 217 – 222 (1986).
The authors have studied the redshift distribution of a sample of 283 galaxies, corresponding to 93% of the objects brighter than $m_z = 14.5$ in a 1800 square degrees area in the Lynx–Gemini region. New redshifts have been obtained for 59 galaxies. Evidence for the existence of several large scale structures has been found.

160.107 **Detection of large–scale alignment of Lick counts around Abell clusters.**
P. C. Argyres, E. J. Groth, P. J. E. Peebles, M. F. Struble.
Astron. J., Vol. 91, No. 3, p. 471 – 477 (1986).
The authors find that galaxy counts are systematically high along the line defined by the long axis of a cluster or of its dominant member, the effect extending to at least $\sim 15\,h^{-1}$Mpc from the cluster ($H = 100\,h$ km s^{-1}Mpc^{-1}).

160.108 **The nearby Abell clusters. I. Abell 168.**
W. R. Oegerle, J. G. Hoessel, R. M. Ernst.
Astron. J., Vol. 91, No. 4, p. 697 – 704 (1986).
As part of a large program to study the spatial and luminosity distributions of galaxies in the nearby Abell clusters, the results of a photometric study of Abell 168 are presented and compared to the previous results of Dressler (1978).

160.109 **The cluster–cluster correlation function.**
M. Postman, M. J. Geller, J. P. Huchra.
Astron. J., Vol. 91, No. 6, p. 1267 – 1273 (1986). With plates 85, 86.
The clustering properties of the Abell and Zwicky cluster catalogs are studied using the two–point angular and spatial correlation functions.

160.110 **Erratum: "A new application of Binggeli's test for large–scale alignment of clusters of galaxies" [Astron. J., Vol. 90, No. 4, p. 582 – 589 (1985)].**
M. F. Struble, P. J. E. Peebles.
Astron. J., Vol. 91, No. 6, p. 1474 (1986). See Abstr. 39.160.038.

160.111 **The Centaurus cluster of galaxies. III. Its structure and the distribution of the different galaxy types.**
J. R. Lucey, M. J. Currie, R. J. Dickens.
Rutherford Appleton Lab., Rep., RAL–86–047, 41 pp. (1986).
The authors have previously shown that the Centaurus cluster is composed of two velocity components, Cen 30 (mean velocity 3000 km s^{-1}) and Cen 45 (mean velocity 4500 km s^{-1}), which very probably lie within one cluster. In this paper the authors describe the internal structure of the cluster and discuss the spatial and velocity distributions of the different galaxy types within the cluster.

160.112 **Clustering and voids.**
G. Vettolani, R. E. de Souza, B. Marano, G. Chincarini.
Mem. Soc. Astron. Ital., Vol. 56, No. 4, p. 717 – 724 (1985). – See Abstr. 012.054.
The authors describe an algorithm used to detect the regions void of galaxies. The catalogue of voids, derived from a catalogue of galaxies with $m \leqslant 14.5$ is then compared with a catalogue of voids compiled from non–dynamical hierarchical models generated by following the prescription by Soneira and Peebles. It is

shown that the hierarchical models reproduce fairly well the observed distribution function of voids.

160.113 **A two-color Tully–Fisher relation.**
E. Giraud.
Inst. Astrophys. Paris, Pré–Publ., No. 129, 39 pp. (1986). To appear in Astron. Astrophys.

160.114 **Perturbation of the nearby extragalactic velocity field by the Local Group.**
E. Giraud.
Inst. Astrophys. Paris, Pré–Publ., No. 138, 34 pp. (1986). To appear in Astron. Astrophys.

160.115 **The orbits of H I–deficient spiral galaxies in the Virgo cluster.**
E. Giraud.
Inst. Astrophys. Paris, Pré–Publ., No. 139, 36 pp. (1986). To appear in Astron. Astrophys.

160.116 **Compilation of clusters of galaxies with published redshifts.**
K.–H. Schmidt.
Astron. Nachr., Vol. 307, No. 2, p. 69 – 83 (1986).
The spectroscopic measured redshifts of 863 clusters of galaxies have been compiled from literature.

160.117 **Empirical properties of nearby groups of galaxies.**
J. Vennik.
Astron. Nachr., Vol. 307, No. 3, p. 157 – 169 (1986).
Virial parameters of individual groups and of synthesized groups of galaxies are determined on the basis of a new list of nearby groups of galaxies ($V_0 < 3200$ km/s). Small values of the crossing time allow to consider the groups of being gravitationally bound systems. The cumulative mass of the synthesized groups increases up to ~1 Mpc from the center, giving the total mass–to–luminosity ratio $f = 97 f_{\odot}$ ($H_0 = 75$ km/s/Mpc). Statistically significant correlations between virial parameters are found $f \propto \sigma_v^{1.5}$ and $f \propto R_{VT}^{0.7}$; σ_v and R_{VT} are uncorrelated.

160.118 **The Centaurus cluster of galaxies – I. The data.**
R. J. Dickens, M. J. Currie, J. R. Lucey.
Mon. Not. R. Astron. Soc., Vol. 220, No. 3, p. 679 – 711 (1986).
The observations obtained from an extensive study of the Centaurus cluster of galaxies are presented and described. Redshifts for 259 galaxies in the region were measured. Two–band photographic photometry (*G*: IIIa–J + GG 455, *F*: IIIa–F + RG 630) for 319 galaxies brighter than $G_{26.5} = 17.5$ mag were measured. Within 1° of the cluster centre the redshift sample is essentially complete for galaxies brighter than $G_{26.5} = 16.5$ mag. The authors present an extensive catalogue of galaxies in the region giving positions, magnitudes, morphological types, redshifts and other parameters.

160.119 **Heating and cooling in clusters of galaxies.**
L. Miller.
Mon. Not. R. Astron. Soc., Vol. 220, No. 3, p. 713 – 722 (1986).
The role of galaxy motions in the heating of intracluster gas is discussed. It is shown that if galaxies in the cores of clusters have mass–to–light ratios M/L_B greater than about 30 then heating by gravitational drag can replace the heat lost by radiation, even in clusters where the cooling time is short compared with the Hubble time. The optically emitting filaments observed in clusters of galaxies may be due to thermal instability of the intracluster gas, but a high rate of mass accretion is not required to explain their luminosities if they are photoionized by surrounding warm gas.

160.120 **X–ray emission from clusters of galaxies.**
C. L. Sarazin.
Rev. Mod. Phys., Vol. 58, No. 1, p. 1 – 115 (1986).
In this paper the X–ray observations of clusters of galaxies are reviewed. Related optical and radio measurements of clusters are also summarized. Theories for the physical state, distribution, origin, and evolution of the intracluster medium are extensively discussed. Finally, the author comments on the prospects for future X–ray observations.

160.121 **The dynamical evolution of clusters of galaxies.**
A. Cavaliere, P. Santangelo, G. Tarquini, N. Vittorio.
Astrophys. J., Vol. 305, No. 2, p. 651 – 667 (1986). With plate 10.
The authors have simulated the dynamical evolution of clusters of galaxies, using a direct *N*–body integrator, from a large set of statistically equivalent initial conditions. They have also mapped the corresponding X–ray emission from the intracluster plasma at various evolutionary stages in a full three–dimensional description. The simulations are compared with recent observations both in the optical and in the X–ray band, with the purpose of testing the predictive ability for morphologies of the hierarchical clustering scenario. The results describe closely the observed variety of cluster morphologies and predict that only < 30% of the clusters as massive as Coma should be in a relaxed phase, the rest being still in a slowly evolving, clumpy or multicomponent and unrelaxed stage.

160.122 **Kinematic evidence of satellite galaxy populations in the potential wells of first–ranked cluster galaxies.**
L. L. Cowie, E. M. Hu.
Astrophys. J., Lett. Ed., Vol. 305, No. 2, p. L39 – L43 (1986).
The authors have measured the velocities of 38 centrally positioned galaxies ($r \ll 100$ kpc) relative to the velocity of the first–ranked galaxy in 14 rich clusters. Analysis of the velocity distribution function of this sample and of previous data shows that the population cannot be fit by a single Gaussian. An adequate fit is obtained if 60% of the objects lie in a Gaussian with $\sigma = 250$ km s^{-1} and the remainder in a population with $\sigma = 1400$ km s^{-1}. This suggests that there is a bound population of galaxies in the potential well of the central galaxy in addition to the normal population of the cluster core.

160.123 **Luminosity segregation and local density of galaxies in clusters.**
L. Sodré, H. V. Capelato, J. C. Penereiro, J. E. Steiner.
Rev. Mex. Astron. Astrofis., Vol. 12, Spec. Issue, p. 59 – 62 (1986). – See Abstr. 012.066.
The luminosity segregation of galaxies in clusters is studied and it is shown that the D and cD galaxies are the only morphological species that show a segregated distribution.

160.124 **The relative motion of the Local Group of galaxies towards the Virgo cluster.**
J. A. de Freitas Pacheco.
Rev. Mex. Astron. Astrofis., Vol. 12, Spec. Issue, p. 74 (1986). – See Abstr. 012.066.
The author derives a peculiar Virgocentric velocity $V_P = 200 \pm 50$ km s^{-1} from a sample of 67 E–galaxies members of groups and clusters, with radial velocities in the range 1000 – 7000 km s^{-1}.

160.125 **Cooling flows in clusters of galaxies.**
A. C. Fabian.
Mitt. Astron. Ges., No. 65, p. 123 – 132 (1986). – See Abstr. 012.072.
Cooling flows of diffuse gas can occur in the cores of clusters of galaxies by the effects of radiative cooling due to X–ray emission. Observations of this phenomenon show that it is reasonably common and may cause substantial masses to be deposited in the cluster core. The evidence for these flows and the implications for galaxy formation are discussed.

160.126 **The mean properties of Virgo cluster dwarf galaxies as a function of distance from the center of the cluster.**
G. Reaves.
Star–forming dwarf galaxies and related objects, p. 69 – 71 (1986). – See Abstr. 012.074.
The properties of individual Virgo cluster dwarf galaxies – dwarf ellipticals and IC 3475–types – appear to be independent of the location of these galaxies within the cluster.

160.127 **H I observations of blue compact dwarf galaxies in the Virgo cluster.**
G. L. Hoffman, G. Helou, E. E. Salpeter.
Star-forming dwarf galaxies and related objects, p. 271 – 272 (1986). Abstract. – See Abstr. 012.074.

Electronographic *BV* photometry of three distant clusters of galaxies. II. The catalogue.
See Abstr. 002.006.

A multicolor photometric catalog of galaxies and stars in the field of the rich cluster II Zw 1305.4 + 2941 at z = 0.24.
See Abstr. 002.052.

A multicolor photometric catalog of galaxies and stars in the field of the rich cluster II Zw 1305.4 + 2941 at z = 0.24.
See Abstr. 002.060.

Poisson clusters and Poisson voids.
See Abstr. 021.015.

The AXAF High Resolution Camera (HRC) and its use for observations of distant clusters of galaxies.
See Abstr. 035.023.

Narrow-band photometry and the evolution of galaxies.
See Abstr. 036.024.

Automatic identification of faint images on the ESO/SRC Sky Atlas.
See Abstr. 036.209.

A statistical search for intergalactic dust.
See Abstr. 036.233.

Distance of the far galaxies and scattering of the photons in the space.
See Abstr. 063.047.

Brightest stars in galaxies as distance indicators.
See Abstr. 115.030.

Anisotropy of radio source distribution from a deep sky survey.
See Abstr. 141.007.

Serendipitous EXOSAT sources in the field of the Coma cluster.
See Abstr. 142.016.

Iron K X-ray emission lines from cosmic X-ray sources observed with Tenma.
See Abstr. 142.034.

Orbital decay of satellite galaxies in spherical systems.
See Abstr. 151.001.

Simulation of compact groups of galaxies.
See Abstr. 151.012.

Massive central galaxies, globular cluster swapping, and tidal accretion.
See Abstr. 151.028.

Gas dynamical stripping of galaxies.
See Abstr. 151.082.

Dynamics of the Magellanic system and the Galaxy – present status of theoretical understanding.
See Abstr. 156.003.

CCD photometry of apparent dwarf galaxies in Fornax.
See Abstr. 157.009.

DDO 13: a low surface brightness galaxy with a well-formed H I disk and two tiny companions.
See Abstr. 157.022.

Search for dwarf galaxies in the IC 342 complex of galaxies.
See Abstr. 157.031.

Galaxy photometry in the Virgo cluster. II.
See Abstr. 157.032.

Studies of cluster galaxies at large lookback times.
See Abstr. 157.100.

Evolution of disk galaxies in high-redshift clusters.
See Abstr. 157.103.

Static properties of galaxies in the Shapley-Centaurus cluster. I. The catalogue.
See Abstr. 157.139.

Binary and multiple galaxies.
See Abstr. 157.140.

Photometric redshifts of galaxies.
See Abstr. 157.152.

A study for the quantitative classification of galaxies.
See Abstr. 157.159.

Population synthesis in the nuclei of Fornax, E, S0 galaxies.
See Abstr. 157.160.

IC 3475: a stripped dwarf galaxy in the Virgo cluster.
See Abstr. 157.164.

Galaxy counts and colors in a South Galactic Pole field.
See Abstr. 157.166.

Nucleated spheroidal galaxies.
See Abstr. 157.167.

Is the distribution of mass within spiral galaxies a function of galaxy environment?
See Abstr. 157.208.

A study of the large-scale structure in the distribution of galaxies in a region centered about the Cancer cluster. I. Initial observational results.
See Abstr. 157.223.

A study of the large-scale structure in the distribution of galaxies in a region centered about the Cancer cluster. II. Further observational results.
See Abstr. 157.224.

Globular clusters in galaxies beyond the Local Group. V. The giant ellipticals reconsidered.
See Abstr. 157.227.

The mass of the isolated elliptical NGC 720 as determined from the dynamics of its companions.
See Abstr. 157.228.

Is NGC 4569 located near the core of the Virgo cluster?
See Abstr. 157.232.

The evolution of galaxies in dense clusters: spectral synthesis models for post-burst galaxies.
See Abstr. 157.237.

A near infrared study of Virgo dwarf ellipticals.
See Abstr. 157.244.

Structure of S0's in clusters.
See Abstr. 157.265.

On the relationship between dwarf irregular and dwarf elliptical galaxies.
See Abstr. 157.277.

Infrared evidence on the metallicity and the history of star formation of Virgo dwarf elliptical galaxies.
See Abstr. 157.284.

Some new bright southern Seyfert galaxies.
See Abstr. 158.007.

Seyfert galaxies in the Universe.
See Abstr. 158.096.

The small wide angle tail NGC 4874 in the center of Coma cluster.
See Abstr. 158.115.

Active galaxies in high density environments.
See Abstr. 158.128.

Influence of the cluster environment on Seyfert galaxies.
See Abstr. 158.142.

Velocity dispersion profiles of the cD galaxies A2052 and A2589.
See Abstr. 158.165.

Multifrequency observations of the wide–angle tail radio source 1313 + 073.
See Abstr. 158.189.

The stellar population of NGC 6166.
See Abstr. 158.217.

Relation of the jet in M87 to nearby galaxies in the Virgo cluster.
See Abstr. 158.249.

Linear polarization observations in selected celestial zones. The Abell 2319 area.
See Abstr. 158.268.

The distribution and kinematics of H I in the active elliptical galaxy NGC 1052.
See Abstr. 158.273.

Relation of the jet in M87 to nearby galaxies in the Virgo Cluster.
See Abstr. 158.314.

QSO absorption lines.
See Abstr. 159.016.

The environment of QSOs.
See Abstr. 159.017.

Results of a preliminary search for quasars in the Virgo cluster region.
See Abstr. 159.094.

The quasar pair Tololo 1037–27 and 1038–27: evidence for correlated absorption on megaparsec scales.
See Abstr. 159.106.

Quasars in the central region of the Virgo cluster.
See Abstr. 159.128.

Further investigation of the pair of quasars Q 0107–025 A and B.
See Abstr. 159.144.

Quasars in the central region of the Virgo Cluster.
See Abstr. 159.171.

The effect of clustering of galaxies on the statistics of gravitational lenses.
See Abstr. 161.002.

Galactic voids may be statistical.
See Abstr. 161.017.

Galaxy and cluster formation in a universe dominated by cold dark matter.
See Abstr. 161.031.

Pancakes of galaxies and the nature of dark matter.
See Abstr. 161.032.

Where do galaxies form?
See Abstr. 161.033.

How relaxed is the distribution of galaxies?
See Abstr. 161.044.

Hierarchical clustering: angular momentum versus density.
See Abstr. 161.047.

The origin of large scale cosmic structure.
See Abstr. 161.049.

New perspectives on galaxy formation.
See Abstr. 161.052.

Discovering a bubbly Universe.
See Abstr. 161.060.

Measuring the Hubble constant.
See Abstr. 161.066.

A slice of the Universe.
See Abstr. 161.074.

Abell clusters as tracers of large–scale structure.
See Abstr. 161.088.

The masses of galaxies and clusters of galaxies.
See Abstr. 161.090.

The gravitational collapse of triaxial protoclusters.
See Abstr. 161.094.

A distance scale from the infrared magnitude/H I velocity–width relation. V. Distance moduli to 10 galaxy clusters, and positive detection of bulk supercluster motion toward the microwave anisotropy.
See Abstr. 161.096.

Voids in the Center for Astrophysics catalog.
See Abstr. 161.098.

Galaxy clustering.
See Abstr. 161.107.

Alignment of clusters and galaxies on scales up to 0.1 *c*.
See Abstr. 161.117.

The origin of dwarf galaxies, cold dark matter, and biased galaxy formation.
See Abstr. 161.118.

The angular correlation function of faint galaxies.
See Abstr. 161.124.

Anisotropy of the galaxies detected by *IRAS*.
See Abstr. 161.125.

The Hubble diagram for brightest cluster galaxies and the cosmological constant.
See Abstr. 161.149.

Infrared line emission by protogalaxies and protocluster pancakes.
See Abstr. 161.152.

A new estimate of the Hubble constant using the Virgo cluster distance.
See Abstr. 161.158.

Dark matter – systematics of its distribution.
See Abstr. 161.162.

Galaxy clustering and thermodynamics: quasi–static evolution of *b*(*t*).
See Abstr. 161.175.

The significance of voids.
See Abstr. 161.177.

Development of voids in the thin–wall approximation. II. Radiation–filled voids in a flat background.
See Abstr. 161.178.

On the application of a spherical infall model to the Local Supercluster.
See Abstr. 161.179.

The decaying particles scenario and the Virgocentric infall.
See Abstr. 161.259.

Topics in observational cosmology.
See Abstr. 161.316.

The Sunyaev–Zeldovich effect at 1.15 mm wavelength.
See Abstr. 161.340.

The galaxy luminosity function and the redshift–distance controversy. (A review).
See Abstr. 161.399.

On what scale can the universe be considered homogeneous.
See Abstr. 161.407.

Zur dreidimensionalen Verteilung entfernter Galaxien.
See Abstr. 161.409.

161 Universe, Cosmology, Background Radiation

161.001 **The universe between $z = 10$ and $z = 1000$: spectral constraints on reheating.**
A. Stebbins, J. Silk.
Astrophys. J., Vol. 300, No. 1, p. 1 – 19 (1986).
The possible thermal histories of the universe during the epoch $z = 10 - 1000$ are studied. Expressions for the fractional ionization and electron temperature are given in the case of homogeneous heating as a function of the parameters of arbitrary ionizing sources. It is shown that present and future limits on spectral distortions to the microwave background radiation do not provide very restrictive constraints on possible thermal histories of the universe. If $H_0 = 50$ km s^{-1}Mpc^{-1}, the universe could have been completely ionized throughout its history with no observable spectral effects.

161.002 **The effect of clustering of galaxies on the statistics of gravitational lenses.**
N. Anderson, C. Alcock.
Astrophys. J., Vol. 300, No. 1, p. 56 – 67 (1986).
The authors examine the question: Can clustering of galaxies significantly alter the statistical properties of gravitational lenses? They consider only models of clustering that resemble the observed distribution of galaxies in the properties of the two–point correlation function. Monte–Carlo simulations of the imaging process are described. It is found that the effect of clustering is too small to be significant, unless the mass of the deflectors is so large that gravitational lenses become common occurrences.

161.003 **Erratum: "Shell crossings and the Tolman model" [Astrophys. J., Vol 290, No. 2, p. 381 – 387 (1985)].**
C. Hellaby, K. Lake.
Astrophys. J., Vol. 300, No. 1, p. 461 (1986). See Abstr. 39.161.141.

161.004 **Superstring gravity and the early universe.**
M. J. Bowick, L. C. R. Wijewardhana.
Gen. Relativ. Gravitation, Vol. 18, No. 1, p. 59 – 66 (1986).
Ten–dimensional superstring theories have been proposed as candidates for a unified description of all the forces of nature. These theories reduce to Einstein gravity coupled to Yang–Mills interactions at energy scales small compared to the string tension. The phenomenologically promising superstring theory, the heterotic string, is investigated at the high temperatures and short distances relevant in the early universe.

161.005 **Bianchi type–V perfect fluid models.**
B. K. Nayak, G. B. Bhuyan.
Gen. Relativ. Gravitation, Vol. 18, No. 1, p. 79 – 91 (1986).
A study of the nontilted diagonal Bianchi type–V perfect fluid models is undertaken in the nonlocally rotationally symmetric case. It is shown that in such a model, the singularity is necessarily velocity dominated and the upper bound for the ratio of the shear to the expansion is $1/\sqrt{3}$. A class of such exact models is also presented; the structure of the singularity and the effect of the anisotropy on the scale are discussed.

161.006 **Dimensional reduction in an oscillatory Kaluza–Klein cosmology.**
A. Tomimatsu, H. Ishihara.
Gen. Relativ. Gravitation, Vol. 18, No. 2, p. 161 – 171 (1986).
The authors study the dynamical evolution of a seven–dimensional, homogeneous, cosmological model. Anisotropy is prevailed all over the spatial dimensions with the topology $S^3 \times S^3$, and leads to repetitious (nonchaotic) oscillations of the scale factors. The universe evolves from an initial Kasner phase to a final one, from which the dimensional reduction results.

161.007 **A physical characterization of the adiabatic vacuum in Robertson–Walker universe.**
M. Castagnino, L. Chimento.
Gen. Relativ. Gravitation, Vol. 18, No. 2, p. 193 – 205 (1986).
As the red shift shows no sign of oscillatory behavior in an expanding universe the authors postulate that the energy Ω_k of a scalar field does not oscillate in such a universe. In the massive case, expanding the differential equation for Ω_k in mass powers, the nonoscillatory solution coincides with the adiabatic solution up to m^6 order. It is demonstrated that this solution is unique for all the evolution of the universe with cosmological interest. Using the nonoscillatory solution as the one that defines the good vacuum state, the authors compute the particle and energy creation for a simple model of the expanding universe.

161.008 **Isolated galaxies.**
G. Vettolani, R. de Souza, G. Chincarini.
Astron. Astrophys., Vol. 154, No. 1/2, p. 343 – 348 (1986).
Using a catalogue of galaxies complete to $m_z \leqslant 14.5$, the authors investigate the possibility of extracting a sample of nonclustered galaxies and identifying it with a homogeneous background of objects. They conclude, in agreement with previous findings, that such a background does not exist observationally and stress the fact that the sample of isolated galaxies listed in this

paper represents galaxies whose properties do not depend on an environmental related evolution.

161.009 **Dust voids?**
G. Chincarini.
Astron. Astrophys., Vol. 154, No. 1/2, p. 376 (1986).
The author shows that the two large voids of galaxies detected by Einasto et al. are in the direction of the Orion–Cygnus spiral arm and therefore they might be due to galactic absorption.

161.010 **Isocurvature cold dark matter fluctuations.**
G. Efstathiou, J. R. Bond.
Mon. Not. R. Astron. Soc., Vol. 218, No. 1, p. 103 – 121 (1986).
Recently it has become fashionable to assume that galaxies and clusters formed from scale–invariant adiabatic perturbations. This idea is supported by Grand Unified Theories, which predict that primordial irregularities in the ratio of baryons to photons would not survive the epoch of baryosynthesis, and by inflationary models for the origin of perturbations. However, if axions (or some other pseudo–Goldstone bosons with similar properties) dominate the mass density of the Universe, scale–invariant isocurvature perturbations could have played an important role in galaxy formation. The authors investigate this possibility in detail, with special emphasis on the constraints set by observations of the microwave background radiation.

161.011 **A radio search for primordial pancakes.**
D. H. O. Bebbington.
Mon. Not. R. Astron. Soc., Vol. 218, No. 3, p. 577 – 585 (1986).
According to the pancake theory of formation of large–scale structure in the Universe, gaseous pancakes are expected to form in the range of redshifts 5 – 10. One of the ways in which they should manifest their presence is through emission in the 21–cm line by their neutral hydrogen component. An aperture synthesis radio telescope has been used at 151 MHz to map the sky north of declination 82°. Data sampling was arranged to permit a differential mapping technique to be used, which enabled the continuum emission in the field to be almost entirely cancelled, to leave a residual rms level of 5 mJy. At the 2σ level, no sources were found in a 1 per cent redshift range around $z = 8.4$, the nominal survey redshift. The observations are found to provide new limits on the pancake population under certain Friedmann cosmologies.

161.012 **Observational homogeneity of the Universe.**
W. B. Bonnor, G. F. R. Ellis.
Mon. Not. R. Astron. Soc., Vol. 218, No. 4, p. 605 – 614 (1986).
A new approach to observational homogeneity is presented. The observation that stars and galaxies in distant regions appear similar to those nearby may be taken to imply that matter has had a similar thermodynamic history in widely separated parts of the Universe (the Postulate of Uniform Thermal Histories, or PUTH). If one supposes that similar thermodynamic histories imply similar dynamical histories, then the distant apparent similarity is evidence for spatial homogeneity of the Universe. Using General Relativity to test this idea the authors take a perfect fluid model and implement PUTH by the condition that the density and entropy per baryon shall be the same function of the proper time along all galaxy world–lines. The standard models of cosmology, the Bianchi models, and the Kantowski–Sachs models all satisfy this condition, and are all spatially homogeneous. The authors now ask whether this condition ensures spatial homogeneity in general. Somewhat to their surprise, they find examples which show that it does not.

161.013 **Gravitational radiation from cosmological phase transitions.**
C. J. Hogan.
Mon. Not. R. Astron. Soc., Vol. 218, No. 4, p. 629 – 636 (1986).
Cavitation during a first–order phase transition, which may have occurred in the early Universe as a consequence of QCD or electroweak interactions, would have produced gravitational radiation in two ways: by generating acoustic noise in relativistic plasma, and by perturbing the expansion law on large scales. The author estimates the spectrum of the resulting stochastic background, its dependence on the parameters governing the phase transition, and the possibility of observing it above instrumental noise and other gravitational wave backgrounds of local and cosmological origin.

161.014 **Cosmology with unstable dark matter.**
Y. Suto, H. Kodama, K. Sato.
Mon. Not. R. Astron. Soc., Vol. 218, No. 4, p. 637 – 649 (1986).
Cosmology with unstable dark matter is investigated particularly from the viewpoint of formation of structures. For galactic scales, it is shown that the amplitude of isothermal density perturbations of dark matter should be smaller than 3×10^{-4} at the time of recombination. As a result, isothermal density perturbations cannot account for the non–linear structures in the universe. On the other hand, for adiabatic density perturbations, although their amplitude is also restricted to be smaller than 10^{-2}, non–linear effect enables the formation of galactic size objects before $Z = 4$. However, the formation of supercluster scale structures seems to be difficult.

161.015 **Pushing back the redshift limit.**
C. M. Gaskell.
Nature, Vol. 319, No. 6054, p. 539 (1986).
This note comments on the importance of the recent discoveries of a galaxy with redshift $z = 3.218$ by S. Djorgovski et al. (see Abstr. 40.158.270) and of the quasar PKS 1351–018 with redshift $z = 3.71$ (see Abstr. 159.013).

161.016 **Will cosmic strings be discovered using the Space Telescope?**
B. Paczýnski.
Nature, Vol. 319, No. 6054, p. 567 – 568 (1986).
Cosmic strings are topologically stable defects in the vacuum which may be produced at a phase transition in the early Universe. Some may form closed loops, and some may be open, extending over the entire observable Universe. This letter suggests that observations of very distant galaxies may be useful as a means for their discovery. A galaxy may appear as cut by a sharp edge if there is a cosmic string between the galaxy and the observer. It is argued that if there is only one string out to a redshift $z \approx 1$, then the probability that it crosses a random image obtained with the Wide Field Camera of the Space Telescope is $\sim 10^{-4}$. Therefore, there is a fair chance of discovering a string, after taking 10^4 images.

161.017 **Galactic voids may be statistical.**
J. Maddox.
Nature, Vol. 319, No. 6053, p. 445 (1986).
Excitement in the past few years about the apparent clumpiness of galactic clusters, leaving empty spaces in between, may be misplaced.

161.018 **The baryon clump within an extended dark matter region.**
S. Ikeuchi.
Astrophys. Space Sci., Vol. 118, Nos. 1/2, p. 509 – 514 (1986). – See Abstr. 012.001.
As a model of the Lyman α absorbing systems of quasar lights, the gravitational equilibrium of a baryon clump within an extended dark matter composed of collisonless particles is examined.

161.019 **Thermal–chemical instability in a pre–galactic gas cloud.**
Y. Sabano, M. Tosa.
Astrophys. Space Sci., Vol. 119, No. 1, p. 167 – 168 (1986). – See Abstr. 012.001.
The growth of thermal–chemical instability in a pre–galactic medium is followed by numerical simulation of gas dynamics. Results show that a primordial gas cloud breaks into self–gravitating subcondensations with the mass of normal stars.

161.020 **Formation of a void and galaxies in a neutrino–dominated universe.**
M. Umemura, S. Ikeuchi.
Astrophys. Space Sci., Vol. 119, No. 1, p. 243 – 245 (1986). – See Abstr. 012.001.
The recent discovery of the large "honeycomb" structure of the universe has triggered many models of the universe dominated by dark matter. The neutrino–dominated universe is a favorable model for explaining the size of the large–scale structure and the dark matter of the larger scale than the galactic one. The authors' calculations on the evolution of density perturbations in a two–component universe composed of neutrinos and dissipative gas on a spherically–symmetric model have shown that the galactic scale does correlate the scale of a void of galaxies: if a neutrino has the mass of some tens eV, galaxies of the typical size from surrounding a typical void.

161.021 **Dark matter and the formation of large–scale structure in the universe.**
L.–z. Fang, S.–p. Xiang, S.–x. Li, Y.–q. Chu, X.–f. Zhu.
Astrophys. Space Sci., Vol. 119, No. 1, p. 247 – 248 (1986). – See Abstr. 012.001.
A new scenario of clustering in a two component dark matter universe is discussed, which yields the difference beween the distribution of quasars and galaxies on the scale of 10 – 100 Mpc and the difference between the distributions of quasars with $Z > 2$ and $Z < 2$. Several analyses on quasars distribution are in good agreement with these predictions.

161.022 **Dynamical equation for the degree of ionization of the cosmic substratum before the formation of the Galaxy.**
E. Nowotny, J. Klapp.
Astrophys. Space Sci., Vol. 119, No. 2, p. 287 – 294 (1986).
The authors discuss the variable degree of ionization x of a hydrogen–helium plasma in the early Universe for a quasi–static expansion in thermal equilibrium. The final equation for the degree of ionization can be reduced to a polynomial of fourth order in x with known coefficients depending on the temperature. Restricting to a pure hydrogen–plasma applied to the recombination era, where the main ionization effects are due to photo–electric and collisional processes, the authors study the dynamical evolution of the degree of ionization for nonstatic and nonequilibrium situations. The calculation can be reduced to a pure quadrature.

161.023 **Comments on the BDT–FRW perfect fluid solutions.**
D. Lorenz–Petzold.
Astrophys. Space Sci., Vol. 119, No. 2, p. 399 – 400 (1986).
It is shown that the BDT–FRW perfect fluid solutions given some years ago by Chauvet and Obregón (1979) are identical to the solutions first given by Jordan (1955) and later by Brill (1962).

161.024 **Comments on the BDT–FRW solutions with a cosmological constant.**
D. Lorenz–Petzold.
Astrophys. Space Sci., Vol. 119, No. 2, p. 401 – 403 (1986).
Some critical comments are made on the power–law solutions recently given by Pimentel (1985) on the basis of the Brans–Dicke field equations for Friedmann–Robertson–Walker space–times.

161.025 **Die Suche nach dem Urzeitpfeil.**
H. D. Zeh.
Phys. Bl., 42. Jahrg., Heft 3, p. 80 – 83 (1986).

161.026 **On the Sivaram and Sinha model of the universe and hadrons.**
M. Zabierowski.
Astron. Rep., Vol. 6, p. 29 – 32 (1981).

161.027 **Remnants from compactification.**
E. W. Kolb.
Particles and gravity, p. 129 – 138 (1984). – See Abstr. 012.004.
Possible observable effects of the existence of compact dimensions are discussed. The effects are remnants from the early stages of the Big Bang when the Universe was at a temperature comparable to the inverse size of the extra dimensions.

161.028 **Towards successful phase transitions in the early universe.**
J. S. Kim.
Particles and gravity, p. 139 – 146 (1984). – See Abstr. 012.004.
The authors suggests a way out of the phase transition problem observed by Breit, Gupta and Zaks and Moss in the new inflationary universe theory. As a by–product, a severe constraint is derived for the choice of scalar representations in grand unification theories that employ the Coleman–Weinberg symmetry breaking mechanism or the Higgs mechanism with cubic terms included.

161.029 **Supersymmetric inflationary cosmology.**
P. Ramond.
Particles and gravity, p. 147 – 155 (1984). – See Abstr. 012.004.
This paper demonstrates that, with a definite cosmological model for inflation, the existence of baryogenesis and the absence of proton decay pose severe constraints on any supersymmetric quantum interacting dynamics.

161.030 **Generation and evolution of energy density fluctuations in inflationary universe models.**
R. H. Brandenberger.
Particles and gravity, p. 157 – 174 (1984). – See Abstr. 012.004.
Spatial correlations in the vacuum state wave functional of a scalar quantum field generate classical matter perturbations in the de Sitter phase of inflationary universe models. The primordial energy density fluctuation spectrum at initial Hubble radius crossing is scale invariant. This paper analyzes the evolution of perturbations outside the Hubble radius, demonstrates that the spectrum remains independent of scale when measured at the time the scales reenter the present horizon, and gives a general formula to determine the amplitude of the fluctuation spectrum for a wide class of particle physics models.

161.031 **Galaxy and cluster formation in a universe dominated by cold dark matter.**
J. R. Primack.
Particles and gravity, p. 175 – 190 (1984). – See Abstr. 012.004.
A universe with ~ 10 times as much cold dark (DM) matter as baryonic matter provides a remarkably good fit to the observed universe. This model predicts roughly the observed mass range of galaxies, the dissipational nature of galaxy collapse, and the observed Faber–Jackson and Tully–Fisher relations. It also gives dissipationless galactic halos and clusters. In addition, it may also provide natural explanations for galaxy–environment correlations and for the differences in angular momenta between ellipticals and spiral galaxies. Finally, the cold DM picture seems reasonably consistent with the observed large–scale clustering, including superclusters and voids.

161.032 **Pancakes of galaxies and the nature of dark matter.**
A. Dekel.
Particles and gravity, p. 191 – 217 (1984). – See Abstr. 012.004.
The distribution of galaxies on large scales is used to constrain the linear density perturbations at the onset of the gravitational instability era, and hence the nature of the dominant mass component and its density. The major theoretical scenarios, represented by N–body models, are confronted with the observations through a series of tests including the flattening of individual superclusters, the galaxy correlation function, the alignment of clusters, the null alignment of galaxies, and the cluster correlation function. Evidence is found for $\sim 30\,h^{-1}$Mpc "pancakes", elongated or flattened, which have formed only recently (at a redshift $z \sim 1$), slightly before rich clusters. A baryonic open universe, with $\Omega \sim 0.1$, can explain the required features in the perturbations spectrum. A comparable contribution ($\Omega \sim 0.1$) from "cold" particles may help forming galaxies and pancakes.

161.033 **Where do galaxies form?**
A. S. Szalay.
Particles and gravity, p. 219 – 228 (1984). – See Abstr. 012.004.
The present theories of galaxy formation are reviewed. The relation between peculiar velocities and the correlation function of galaxies points to the possibility that galaxies do not form uniformly everywhere. None of the present theories of galaxy formation can account for this fact in a natural way. Theoretical implications of several observations as Lyman–α clouds, correlations of faint galaxies are discussed.

161.034 **Problems of dimensional reduction and inflationary cosmology.**
D. Appel, M. Dresden.
Particles and gravity, p. 228 – 257 (1984). – See Abstr. 012.004.
The authors discuss various aspects of the relationship between Kaluza–Klein theories and cosmology, with particular emphasis on the process of dimensional reduction. Some features of higher–dimensional Friedmann–Robertson–Walker–type universes are discussed, such as the compatibility of the Einstein equations, their connection with inflationary scenarios, and their consequences for the critical density of the observable four–dimensional universe. The role of the Casimir effect as a possible mechanism for dimensional compactification is explored.

161.035 **Supersymmetry and cosmology.**
S. Raby.
Supersymmetry, supergravity and related topics, p. 226 – 270 (1985). – See Abstr. 012.005.
Contents: (1) Review of minimal low energy supergravity (MLES) spectrum. (2) R–parity and discuss candidates for lightest supersymmetric partner (LSP). (3) Standard "Big Bang" cosmology and the relativistic Boltzmann equation. (4) Cosmological bounds for the mass of the LSP. (5) Cosmological bounds for the gravitino.

161.036 **Supersymmetric inflationary cosmology.**
P. Ramond.
Supersymmetry, supergravity and related topics, p. 335 – 345 (1985). – See Abstr. 012.005.
The author reviews the general requirements imposed on cosmology by particle physics. In particular, he discusses a model which includes inflation, supersymmetry and grand unification. It is shown how this model can be used to put an upper bound on the proton lifetime.

161.037 **2.7K radiation and the isotropy of the universe.**
S. Bajtlik, R. Juszkiewicz, M. Prószyński, P. Amsterdamski.
Astrophys. J., Vol. 300, No. 2, p. 463 – 473 (1986).
The authors study the temperature distribution of the microwave background radiation expected in flat and open cosmological models with shear. In the flat (Bianchi type I) model, the anisotropy is quadrupole–like. In the open (type V) model, the anisotropy is focused into a small region of the sky: the "hot spot". The measurements of the temperature distribution of the microwave background radiation constrain the present value of the shear to be less than $10^{-6} \times H_0$ for $0.1 \leqslant \Omega_0 \leqslant 1$, whatever the ionization history of the primordial hydrogen may be.

161.038 **Statistical gravitational lensing and quasar–galaxy associations.**
P. Schneider.
Astrophys. J., Lett. Ed., Vol. 300, No. 2, p. L31 – L34 (1986).
The influence of gravitational light deflection by compact objects within a foreground galaxy on the apparent surface number density of distant sources around it is considered. The effects of finite source extent, the galactic gravitational field at the location of the mini–lenses, and the simultaneous influence of several stars on the light beam are taken into account. The resulting number density enhancement depends sensitively on the source size and the shape of the intrinsic luminosity function of the sources. Although the number density of quasars around galaxies can be increased significantly, a large sample of galaxies has to be explored to isolate this effect from statistical noise.

161.039 **Friedmann–Robertson–Walker stiff matter solutions in the scale covariant theory.**
A. Beesham.
Astrophys. Space Sci., Vol. 120, No. 1, p. 151 – 154 (1986).
The author derives Friedmann–Robertson–Walker stiff matter solutions in the scale covariant theory and discusses some properties of the solutions.

161.040 **Spherical simulations of large–scale structure of the Universe.**
J. P. Mücket.
Astron. Nachr., Vol. 306, No. 6, p. 305 – 309 (1985).
The author considers a spherical symmetric perturbation in an initially homogeneous dust filled universe within the framework of the Bondi–Tolman solution. The decoupling of the perturbed region from the background evolution naturally leads to shell–crossing in a certain outer environment, which describes energy relaxation. The behaviour of the outgoing front shell and different parameters of the shell–crossing region can be analytically determined in the approximation of Newtonian gravity. On the basis of the obtained results the author gives a possible explanation of the "large voids" in the observed galaxy distribution.

161.041 **The Malmquist bias and the value of H_0 from the Tully–Fisher relation.**
L. Bottinelli, L. Gouguenheim, G. Paturel, P. Teerikorpi.
Astron. Astrophys., Vol. 156, No. 1/2, p. 157 – 171 (1986).
A large sample (n = 395) of spiral (Sab to Sd type) galaxies having corrected apparent magnitudes $B_T{}^0$ and 21–cm line data (H I line widths and radial velocities) is used to investigate in a new way the influence of the Malmquist bias on the determination of the extragalactic distance scale and the Hubble constant derived from application of the B–band Tully–Fisher relation. This effect is clearly identified by using relative kinematic distances derived from a classical local velocity field model and the concept of normalized relative kinematic distance. It results in an unbiased estimate of the Hubble constant H_0 which appears quite insensitive to the parameters (mean velocity of Virgo and infall velocity of the Local Group towards Virgo) adopted for the local velocity field model. A similar effect is found from a sample of galaxies (n = 72) which are "sosies" of 14 primary galaxies. According to the local distance scale adopted – due to de Vaucouleurs (deV), or Sandage and Tammann (ST) – the results are (in km s^{-1}Mpc^{-1}): $H_0 = 72 \pm 3$ (deV scale), $H_0 = 56 \pm 3$ (old ST scale), $H_0 = 63 \pm 3$ (new ST scale). It is suggested that the presently derived H_0 represents the global value of the Hubble constant.

161.042 **Are microwave background small–scale fluctuations substantially overestimated?**
S. A. Bonometto, X. Shouping.
Astron. Astrophys., Vol. 157, No. 2, p. L7 – L9 (1986).
The variation of the ionization rate, due to the passage of a fluctuation through the recombination layer, may alter the last scattering band, depressing the temperature anisotropies of the MW background due to those very fluctuations by quite a large factor. The authors present here preliminary computations which seem to support this idea and debate the problems arising if a full proper computation is to be performed.

161.043 **Primordial fluctuations in a universe dominated by quark nuggets.**
R. Schaeffer.
Astron. Astrophys., Vol. 157, No. 2, p. L13 – L16 (1986).
It has been recently suggested by E. Witten that quark matter can exist under the form of droplets of finite size, the "quark nuggets" and be more stable than nuclear matter, or at least be metastable. This provides another possible explanation for the missing mass in the universe. The author investigates here the form of the primordial fluctuation spectrum within this hypothesis.

161.044 **How relaxed is the distribution of galaxies?**
P. Crane, W. C. Saslaw.
Astrophys. J., Vol. 301, No. 1, p. 1 – 6 (1986).
The authors compare the distribution functions $f(N)$ predicted for a quasi–relaxed gravitating system with the observed distributions of galaxies in the Zwicky catalog. There is good agreement between them. The value of the ratio of gravitational correlation energy to thermal energy which best fits the observations is $b = 0.70 \pm 0.05$. This agrees with the values found previously in N–body computer experiments. It suggests that the observed distribution of galaxies is mostly the relaxed result of gravitational clustering.

161.045 **The distance moduli from the *B*–band Tully–Fisher relation and the velocity field.**
E. Giraud.
Astrophys. J., Vol. 301, No. 1, p. 7 – 16 (1986).
A sample of 255 Sab to Scd spiral galaxies have precise $B_T{}^0$ magnitudes and 21 cm line widths is used to investigate the apparent increase of the Hubble constant with respect to the kinematic distances, based on the application of the *B*–band Tully–Fisher relation with slope of 5 and an infall model, reported in an earlier paper. The conclusions are: (1) If the dispersion of the *B*–band Tully–Fisher relation is only $\sigma = 0.55$ mag, the contribution of the Malmquist bias is about 0.45 mag and can explain one–third of the effect. (2) It seems a priori possible to remove the increase of H_0 by using a slope of 7 based on the calibration of Sandage and Tammann, but contradictory results are obtained. (3) Most of the objects with small log H_0 at small velocity are located in the Canes Venatici or Leo Minor clouds; an anomaly of the velocity field at the position of these clouds can be suggested. (4) The problem of the increase of the Hubble ratios with respect to the Hubble moduli is also raised with data on more distant clusters.

161.046 **The smoothness of the 2.2 micron background: constraints on models of primeval galaxies.**
S. P. Boughn, P. R. Saulson, J. M. Uson.
Astrophys. J., Vol. 301, No. 1, p. 17 – 22 (1986).
The authors have searched for fluctuations in the near–infrared background in the K band ($\lambda = 2.2$ μm). The data show no statistically significant signal and imply 90% confidence level upper limits on brightness fluctuations of 7×10^{-20}ergs $s^{-1}cm^{-2}Hz^{-1}sr^{-1}$ on scales from 10″ to 30″ and 3×10^{-20}ergs $s^{-1}cm^{-2}Hz^{-1}sr^{-1}$ on scales from 60″ to 300″. The implications of these measurements on models of primeval galaxies are discussed.

161.047 **Hierarchical clustering: angular momentum versus density.**
Y. Hoffman.
Astrophys. J., Vol. 301, No. 1, p. 65 – 69 (1986).
The growth of angular momentum in a general clustering scenario is analyzed and its correlation with the primordial density perturbation δ_0 is calculated. High–density peaks are found to evolve into low–angular momentum systems. The basic mechanism behind this anticorrelation is that anisotropic structure, which leads to angular momentum gain, is statistically uncorrelated with the primordial perturbation. The collapse time, on the other hand, is correlated with it. This time scale is the one that determines the angular momentum of a bound object.

161.048 **The dipole anisotropies of the *IRAS* galaxies and the microwave background radiation.**
A. Yahil, D. Walker, M. Rowan–Robinson.
Astrophys. J., Lett. Ed., Vol. 301, No. 1, p. L1 – L5 (1986). With plates L1 – L4.
IRAS 60 μm sources are used to map the local ($\lesssim 200\ h^{-1}$Mpc) gravitational field, and to determine its dipole component, on the assumption that the infrared radiation traces the matter. The dipole moment is found to point in the direction $l = 248° \pm 9°$ and $b = 40° \pm 8°$, or 26° away from the direction of the velocity of the Local Group relative to the microwave background radiation. Comparison of the two anisotropies, using the luminosity function of the *IRAS* galaxies and linear perturbation theory, yields an estimate of the cosmological density parameter, $\Omega_0 = 0.85 \pm 0.16$, with nonlinear effects increasing Ω_0 by $\sim$0.15%.

161.049 **The origin of large scale cosmic structure.**
B. J. T. Jones, P. L. Palmer.
Galaxies, axisymmetric systems and relativity, p. 3 – 22 (1985). – See Abstr. 003.007.
Contents: (1) Introduction. (2) The evolution of density perturbations. (3) The nonlinear regime – Zel'dovich's solution and others. (4) The Gott and Rees clustering hierarchy. (5) The statistics of random density fluctuations. (6) The spectrum of condensations. (7) Biased galaxy formation. (8) Where do we go from here?

161.050 **The problem of origin of the primordial perturbations and the modern cosmology.**
V. N. Lukash, I. D. Novikov.
Galaxies, axisymmetric systems and relativity, p. 23 – 53 (1985). – See Abstr. 003.007.
The theory of the parametric amplification of potential perturbations in the isotropic universe is reviewed. The problem of production of the primordial cosmological perturbations due to the parametric effect is considered. The related fluctuations of the relic background radiation are analysed. The spottiness structure of the large–scale background anisotropy is discussed.

161.051 **The automorphism group and field equations for Bianchi universes.**
W. L. Roque, G. F. R. Ellis.
Galaxies, axisymmetric systems and relativity, p. 54 – 73 (1985). – See Abstr. 003.007.
In the case of vacuum Bianchi universes, simplification of the metric, and hence the field equations, is possible through use of the automorphism group of the symmetry group. This is because some of the time–variation of the metric components is transferred to the basis chosen, without changing the canonical structure constants. The resulting diagonalisation of the final metric tensor components particularly simplifies the finding of the inverse metric components; while use of the "gauge variables" simplifies the field equations, effectively reducing second–order to first order and algebraic equations. In the case of matter–filled universes, much of the simplification is maintained when the fluid flow is orthogonal to the surfaces of homogeneity.

161.052 **New perspectives on galaxy formation.**
J. Silk.
Galaxies, axisymmetric systems and relativity, p. 74 – 96 (1985). – See Abstr. 003.007.
This paper briefly reviews the linearized theory of fluctuation growth in the early universe, and describes the constraints imposed by the microwave background isotropy on the various possible primordial fluctuation spectra. The author discusses recent ideas about the non–linear theory, involving gravitational collapse, dynamical relaxation, and hydrodynamical dissipation processes en route to forming galaxies.

161.053 **Physical constraints on the bipartition model of the Universe.**
F. X. Désert, E. Schatzman.
Astron. Astrophys., Vol. 158, No. 1/2, p. 135 – 142 (1986).
The presence of a void zone surrounding us, as found by Fliche, Souriau and Triay (1982) (FST) suggests a model of a spherical, closed and bipartitioned Universe, one half matter, the other half anti–matter. Taking for granted the matter–antimatter bipartition, the authors try to check the physical consistency of the model. There is no evidence that the model of FST is physically inconsistent, but the general difficulties of any physical model of the void zone are underlined. Further observational data are needed.

161.054 **Production of massive spin–1/2 particles in Robertson–Walker universes with external electromagnetic fields.**
K. H. Lotze.
Astrophys. Space Sci., Vol. 120, No. 2, p. 191 – 204 (1986).
The author investigates the combined influence of both cosmological and electromagnetic particle creation mechanisms upon massive particles with spin–1/2 on the basis of general covariant Dirac theory. Curved space–time, a radiation–dominated Friedmann universe, is treated as an unquantized gravitational field and the low–frequency part of the 2.7K background radiation is approximated by homogeneous, constant, and parallel external electric and magnetic fields. The author calculates the number density of spin–1/2 particles with mass m which are created under the influence of both these external fields. He finds that the electric field and the magnetic field both amplify the genuine, purely gravitational particle production.

161.055 **cD galaxies of apparent supergiant sizes due to the curvature of space.**
J. Souček.
Astrophys. Space Sci., Vol. 121, No. 1, p. 161 – 191 (1986).
By combining two two–dimensional subspaces, closed into themselves due to curvature, it is possible to create a model of three–dimensional space of the same properties. If the Universe is a space of this type, its effect is that of a monstrous lens. Close objects are observed to diminish according to the normal law of perspective; however, the remote galaxies are seen to be very highly magnified.

161.056 **Universal fragmentation.**
W. K. Brown.
Astrophys. Space Sci., Vol. 121, No. 2, p. 351 – 355 (1986).
The hypothesis is tested that the Universe underwent hydrodynamic fragmentation at an early stage after the Big Bang. Two emipirical functions used to describe such fragmentation on the laboratory scale are compared with the present distribution of total galactic luminosities (with M/L = constant). The comparison supports the hypothesis.

161.057 **Novae as distance indicators.**
S. van den Bergh, C. J. Pritchet.
Publ. Astron. Soc. Pac., Vol. 98, No. 599, p. 110 – 115 (1986).
From published photographic observations the dispersion in the maximum magnitude versus rate of decline relation for novae in M31 is $\sigma = 0.28$ mag, from which the intrinsic dispersion is found to be $\sigma_0 < 0.14$ mag. This is far smaller than the 1.5 mag range in distance moduli that corresponds to $50 < H < 100\ \mathrm{km\,s^{-1}Mpc^{-1}}$. These results suggest that the m_{pg}(max) versus rate of decline relationship for novae is potentially a powerful tool for the calibration of the extra–galactic distance scale. Observations of the nova T Sco in the metal–poor globular cluster M80 show that the maximum magnitude versus rate of decline relationship for novae is not very sensitive to metallicity. Two new techniques are suggested for using novae as distance indicators. Both of these involve the assumption that the relative frequency of novae with differing rates of decline is constant from galaxy to galaxy.

161.058 **The evolution of the hot Universe with massive unstable leptons.**
B. V. Vajner, Yu. A. Shchekinov, M. B. Ehntehl'.
Astrofizika, Tom 23, Vyp. 3, p. 621 – 636 (1985). In Russian. English translation in Astrophysics, Vol. 23, No. 3.
The various aspects of massive unstable neutrino presence in the early Universe are analysed. The heat balance and cosmological nucleosynthesis are calculated. The limits on the mean baryon density, for the mass and the decay constant of massive leptons are obtained. Models with $m_H = 5 - 100$ MeV and $\tau_H = 10^3 - 10^4$s are in good accordance with modern observations if $\Omega_b = 0.03 - 0.06$ and the Hubble constant $H_0 = 50$ km/s Mpc.

161.059 **Large–scale homogeneity of the Universe measured by the microwave background.**
N. Mandolesi, P. Calzolari, S. Cortiglioni, F. Delpino, G. Sironi, P. Inzani, G. De Amici, J.–E. Solheim, L. Berger, R. B. Partridge, P. L. Martenis, C. H. Sangree, R. C. Harvey.
Nature, Vol. 319, No. 6056, p. 751 – 753 (1986).
The authors report measurements of the isotropy of the cosmic–microwave background at a wavelength of 3 cm. At the 95% confidence level, the results restrict departures from isotropy to $< 5 - 7 \times 10^{-4}$ on angular scales of $\sim 2 - 10°$. The angular scale of these observations is of particular cosmological significance because regions of size $\sim 1°$ and above were not causally connected at the epoch of last scattering, at least in conventional big–bang models.

161.060 **Discovering a bubbly Universe.**
J. Silk.
Nature, Vol. 320, No. 6057, p. 12 – 13 (1986).
This note comments on theoretical attempts to explain the origin of the large galaxy voids recently observed in surveys of the large–scale galaxy distribution.

161.061 **Symmetry changes in Friedmann cosmologies.**
S. S. Bayin.
Astrophys. J., Vol. 301, No. 2, p. 517 – 521 (1986).
The author introduces a suggestive model wherein certain quantities in Friedmann models are treated like their thermodynamic counterparts: temperature, entropy, Gibbs energy, and so on. Within this model, changes in the Bianchi symmetry of the universe are interpreted as first– or second–order phase transitions. This model also offers a new way of determining the symmetry of the universe. The results obtained are independent of the details of the local equation of state.

161.062 **Evolution of the intergalactic medium: what happened during the epoch $z = 3 - 10$?**
S. Ikeuchi, J. P. Ostriker.
Astrophys. J., Vol. 301, No. 2, p. 522 – 543 (1986).
The authors attempt to model in a consistent way the thermal and dynamical history of the intergalactic medium from the era of reheating ($z = 10 - 5$) to the present, and to provide a unified explanation for the origin of ordinary galaxies, blue compact objects and Lyα clouds. First, the thermal history of the intergalactic medium is explored, assuming it to be uniform in temperature and density, considering radiative and Compton cooling, adiabatic cooling, and the heating produced by the diffuse UV flux and shock heating. Second, the paper investigates the formation and the evolution of intergalactic clouds. It is suggested that massive shells fragment gravitationally and produce clouds with masses $10^{11.3} \gtrsim M/M_\odot \gtrsim 10^{9.5}$ which collapse to galaxies; those with $10^{9.5} - 10^{8.5} M_\odot$ have collapsed recently and are observed as young blue compact galaxies. Still lower mass shells would have formed clouds in the mass range $10^{8.5} - 10^{6.0} M_\odot$ which are now expanding and detected as the quasar Lyα absorbers.

161.063 **Is the universe really expanding?**
P. A. LaViolette.
Astrophys. J., Vol. 301, No. 2, p. 544 – 553 (1986).
The no–evolution, tired–light model and the no–evolution, $q_0 = 0$, expanding universe cosmology are compared against observational data on four kinds of cosmological tests. On all four tests the tired–light model is found to make the better fit to the data without requiring the ad hoc introduction of assumptions about rapid galaxy evolution. The data may be interpreted in the simplest fashion if space is assumed to be Euclidean, galaxies cosmologically static, evolutionary effects relatively insignificant, and photon energy nonconserved, with photons losing about 5% – 7% of their energy for every 10^9 light years of distance traveled through intergalactic space.

161.064 **Evidence for a nonvanishing energy density of the vacuum (or cosmological constant).**
H. V. Klapdor, K. Grotz.
Astrophys. J., Lett. Ed., Vol. 301, No. 2, p. L39 – L43 (1986).
In recent years the better understanding of the synthesis of heavy elements by the *r*–process and the investigation of globular clusters led to a larger age of the universe of $\gtrsim 17\text{x}10^9$yr. It is discussed in this letter that with the assumption of inflationary expansion at the beginning of the evolution of the universe such a number leads to a nonvanishing cosmological constant $\Lambda = (4.7 - 19)\text{x}10^{-57}\text{cm}^{-2}$ in the Friedmann equation corresponding to an energy density of the vacuum of $\varrho_v = (2.5 - 10)\text{x}10^{-30}\text{g cm}^{-3}$. Such a model leads to stringent limits on the amount of dark matter and especially on the neutrino mass.

161.065 **Bianchi type–II cosmological model with viscous fluid.**
A. Banerjee, S. B. Duttachoudhury, A. K. Sanyal.
Gen. Relativ. Gravitation, Vol. 18, No. 5, p. 461 – 477 (1986).
A spatially homogeneous and locally rotationally symmetric Bianchi type–II cosmological model under the influence of both shear and bulk viscosity is studied. Exact solutions are obtained with a barotropic equation of state and considering the linear relationships between the fluid density, the expansion, and the shear scalars, respectively. Special cases with vanishing bulk viscosity coefficients and with the perfect fluid in the absence of viscosity are also studied.

161.066 **Measuring the Hubble constant.**
B. Schechter.
Phys. Today, Vol. 39, No. 1, p. 20 – 21 (1986).

161.067 **Interpreting the wave function of the Universe.**
F. J. Tipler.
Phys. Rep., Vol. 137, No. 4, p. 231 – 275 (1986).
The Many–Worlds Interpretation of quantum mechanics is used to determine the meaning of the universal wave function of quantum cosmology. More precisely, the Many–Worlds Interpretation is used to distinguish those quantities in quantum cosmology which are measureable, and hence physically meaningful, from those which are not. A number of rather surprising conclusions are drawn from the analysis. All conclusions are illustrated with a closed Friedmann universe quantized in conformal time. The author's quantization procedure allows only one solution to Schrödinger's equation, and this solution solves the Flatness Problem. He shows that the ADM quantization method plus the Hartle–Hawking initial foundary condition gives the same result.

161.068 **Survival of the structures in the universe dominated by unstable dark matter.**
Y. Suto, M. Noguchi.
Mon. Not. R. Astron. Soc., Vol. 219, No. 4, p. 841 – 857 (1986).
Dynamical evolution of non–linear structures in the universe dominated by unstable particles is studied by the numerical integration of the Vlasov equation. Their dynamical behaviour is determined by two parameters; one is the ratio of energy densities of baryons and unstable particles β and the other is the ratio of the lifetime of unstable particles τ_X to the initial free–fall time–scale $t_{ff,i}$. Non–linear structures satisfying the survival condition – $\tau_X/t_{ff,i} > 1/(2\beta^2)$ – remains as bound systems irrespective of the decay of unstable particles. As long as one considers the cosmologically restricted range of τ_X and β, galaxies satisfy the survival condition and their evolution is well understood in terms of adiabatic behaviour. On the other hand, clusters of galaxies may be disrupted in future due to decay of unstable particles. A possible restriction to the formation of structures and the consequence of this scenario to the observed angular size of galaxies are discussed quantitatively.

161.069 **Dynamics of small perturbations in a multi–component cosmological medium.**
I. S. Nurgaliev.
Pis'ma Astron. Zh., Tom 12, No. 3, p. 179 – 186 (1986). In Russian. Englisch translation in Sov. Astron. Lett., Vol. 12.
Analytical solutions of the equation of small perturbations in a three–component cosmological medium are obtained. The addition of dust and polytropic–type components do not change the analytical expressions for corresconding modes. It is concluded that the dynamics of small perturbations is described in the case of a medium consisting of $\gamma = 4/3$ components by power laws, in the case of identical $\gamma \neq 4/3$ components by Bessel functions, and in the case of dust and $\gamma = 4/3$ components with only one $\gamma \neq 4/3$ component by Meier functions.

161.070 **The isotropization in spatially homogeneous cosmological models with a two–component medium.**
G. G. Ivanov.
Astron. Zh., Tom 63, Vyp. 2, p. 209 – 213 (1986). In Russian. English translation in Sov. Astron., Vol. 30, No. 2.
The exact solutions of the collisionless Vlasov equation are obtained for anisotropic cosmological models of the I and IX Bianchi types, compatible with Einstein's equations with the energy–momentum tensor of a collisionless gas. On the basis of these solutions the evolution of small perturbations for the space–flat and closed Friedmann models is investigated. It is shown that at the nonrelativistic stage of the expansion of the mixture of ultrarelativistic collisionless gas and a nonrelativistic perfect fluid the interaction of the tensor perturbations with the collisionless component leads to the residual anisotropy of the metric and of the energy–momentum tensor of the collisionless gas.

161.071 **The dependence of baryon asymmetry of the Universe on the equation of state.**
Yu. A. Beletskij, A. I. Bugrij, A. A. Trushevskij.
Astrofizika, Tom 24, Vyp. 1, p. 179 – 189 (1986). In Russian. English translation in Astrophysics, Vol. 24, No. 1.
The baryon asymmetry of the Universe, originated at the beginning of the hadron era, might change its value during evolution. The variation depends on the character of the equation of state and permits to reconcile the observed baryon asymmetry with predictions of GUT. The evolution of baryon asymmetry for the case of phase transitions in matter of the early Universe is studied.

161.072 **Heating of primordial gas by background radiation if molecular hydrogen is available.**
V. K. Khersonskij.
Astrofizika, Tom 24, Vyp. 1, p. 191 – 199 (1986). In Russian. English translation in Astrophysics, Vol. 24, No. 1.
A small impurity of molecular hydrogen can exist in primordial gas in the cosmological epochs $z \cong 1000$. The effect of heating of the gas by background radiation in the epochs $z \lesssim 150$ is considered. The processing of radiation energy to thermal energy of the gas takes place at transitions between low rotational levels of H_2 molecules. This process leads to the increase of gas temperature to about some scores of Kelvins. The possibility of an observational test for this effect is discussed.

161.073 **Causality constraints on cosmological perturbations.**
L. F. Abbott, J. Traschen.
Astrophys. J., Vol. 302, No. 1, p. 39 – 42 (1986).
Using recently derived integral constraints, the authors give a general–relativistic proof of the k^4 power spectrum for perturbations generated by local, causal processes in a Robertson–Walker universe. If galactic clusters formed from such fluctuations and if light traces mass, then in order to reconcile the integral constraints with observed cluster–cluster correlations, these perturbations must have been created when the age of the universe was greater than a few thousand years. Inflation provides a way out of this bound.

161.074 **A slice of the Universe.**
V. de Lapparent, M. J. Geller, J. P. Huchra.
Astrophys. J., Lett. Ed., Vol. 302, No. 1, p. L1 – L5 (1986).
The authors describe recent results obtained as part of the extension of the Center for Astrophysics redshift survey to $m_B = 15.5$. The new sample contains 1100 galaxies (584 new redshifts were measured) in a $6° \times 117°$ strip going through the Coma cluster. Several features of the data are striking. The galaxies appear to be on the surfaces of bubble–like structures. The

bubbles have a typical diameter of $\sim 25\,h^{-1}$Mpc. The largest bubble in the survey has a diameter of $\sim 50\,h^{-1}$Mpc, comparable with the most recent estimates of the diameter of the void in Bootes. The galaxy density in the region of the largest void contained in the survey is only 0.20 of the mean. The edge of the largest void in the survey is remarkably sharp.

161.075 **Large–scale anisotropy in the Hubble flow.**
C. A. Collins, R. D. Joseph, N. A. Robertson.
Nature, Vol. 320, No. 6062, p. 506 – 508 (1986).

Detection of anisotropy in the Hubble flow is one of the outstanding problems of observational cosmology. The authors have obtained infrared photometry for an all–sky sample of ScI–II galaxies identified by Rubin et al. at a mean redshift of 5,100 km s^{-1}. At this distance these galaxies provide information on the Hubble flow well outside the Local Supercluster. From observations of half the sample, well distributed across the sky, the authors find a systematic streaming velocity of these galaxies of $\sim 1{,}000 \pm 300$ km s^{-1}. In the context of current speculation regarding non–baryonic particle species which may dominate the mass of the Universe, such large velocities on these spatial scales appear to exclude cold dark matter models.

161.076 **Galaxy superclusters and cosmic strings.**
C. Hogan.
Nature, Vol. 320, No. 6063, p. 572 (1986).

161.077 **The mean mass density of the Universe.**
P. J. E. Peebles.
Nature, Vol. 321, No. 6065, p. 27 – 32 (1986).

Observations of galaxy clustering indicate that the mean mass density is only about one–third of that predicted by the popular Einstein – de Sitter cosmological model. Theory and observation can be reconciled if there has been large–scale segregation of galaxies from mass, but the gravitational instability of the expanding universe would tend to destroy such a segregation. Thus either conventional ideas about galaxy formation are wrong, or the Einstein – de Sitter model is not a useful approximation.

161.078 **Le principe de Weyl et l'invention d'une cosmologie non–statique.**
P. Kerszberg.
Arch. Hist. Exact Sci., Vol. 35, No. 1, p. 1 – 89 (1986).

161.079 **The protogalactic connection.**
J. Silk, R. F. G. Wyse.
Structure and evolution of active galactic nuclei, p. 173 – 192 (1986). – See Abstr. 012.011.

The authors argue that the physics underlying active galactic nuclei and the quasar phenomenon will not be understood without first having knowledge of the processes governing the gas to star transformation that occurs during galaxy formation. Galaxies actually in the process of forming are extremely difficult to observe, or perhaps even recognise, and the authors take a more indirect approach, utilising the observed scaling relations between various parameters of present day galaxies to infer information about the physical conditions at the epoch of galaxy formation.

161.080 **Distribution of quasars and the formation of large scale structure in the Universe.**
Y.–q. Chu, L.–z. Fang.
Structure and evolution of active galactic nuclei, p. 517 – 519 (1986). – See Abstr. 012.011.

Statistical evidence show that the distribution of quasars with $Z < 2$ is different from that of $Z > 2$, the former has large scale clustering while the later has not. All the results are in very good agreement with the developed new scenario of formation of large scale structure in a universe with several components of dark matter.

161.081 **Erratum: "Early Brans–Dicke axisymmetric universe with magnetic field" [Astron. Astrophys., Vol. 121, No. 1, p. 1 – 4 (1983)].**
S. P. Chakravarti, U. K. De.
Astron. Astrophys., Vol. 160, No. 2, p. 407 – 408 (1986). See Abstr. 33.161.092.

161.082 **The effect of spontaneous disturbance of gauge symmetry in isotropic cosmological models with torsion.**
A. V. Minkevich, Nguen Hong Tyong, F. I. Fedorov.
Dokl. Akad. Nauk BSSR, Tom 29, No. 8, p. 697 – 699 (1985). In Russian. Abstr. in Ref. Zh., 51. Astron., 2.51.1063 (1986).

161.083 **Redshift surveys of galaxies find a bubbly universe.**
B. Schwarzschild.
Phys. Today, Vol. 39, No. 5, p. 17 – 20 (1986).

161.084 **Investigation of the existence of large–scale anisotropy in the metagalactic space.**
R. R. Andreasyan.
Astrofizika, Tom 24, Vyp. 2, p. 363 – 375 (1986). In Russian. English translation in Astrophysics, Vol. 24, No. 2.

The rotation measures of 247 extragalactic radio sources with known redshifts are used for simultaneous investigation of magnetic fields of the Galaxy and the Metagalaxy. It has been shown that the direction of the metagalactic magnetic field $(l_0, b_0) = (100°, 25°)$ nearly coincides with the direction of the galactic magnetic field in the southern hemisphere and is opposite to it in the northern hemisphere. The strength of the metagalactic magnetic field is estimated to be of the order of 10^{-9}Gauss. It has been proposed that the presence of a certainly distinguished direction may be connected with large–scale rotation of the Metagalaxy. The angle between the rotation axes of the Metagalaxy and the metagalactic magnetic field is very large.

161.085 **An estimate of the total space density of dwarf galaxies.**
N. D. Tyson, J. M. Scalo.
Bull. Am. Astron. Soc., Vol. 17, No. 4, p. 857 (1985). Abstract. – See Abstr. 010.062.

161.086 **The possible protogalactic role of eddies in a turbulent early universe.**
W. K. Brown.
Bull. Am. Astron. Soc., Vol. 17, No. 4, p. 857 (1985). Abstract. – See Abstr. 010.062.

161.087 **A slice of the universe.**
J. P. Huchra, V. de Lapparent, M. J. Geller, M. J. Kurtz, E. Horine, J. Peters, S. Tokarz.
Bull. Am. Astron. Soc., Vol. 17, No. 4, p. 857 – 858 (1985). Abstract. – See Abstr. 010.062.

161.088 **Abell clusters as tracers of large–scale structure.**
D. J. Batuski.
Bull. Am. Astron. Soc., Vol. 17, No. 4, p. 871 – 872 (1985). Abstract. – See Abstr. 010.062.

161.089 **Simulation and analysis of galaxy distributions in model universes.**
A. Chokshi.
Bull. Am. Astron. Soc., Vol. 17, No. 4, p. 872 (1985). Abstract. – See Abstr. 010.062.

161.090 **The masses of galaxies and clusters of galaxies.**
R. Mushotzky.
Bull. Am. Astron. Soc., Vol. 17, No. 4, p. 888 (1985). Abstract. – See Abstr. 010.062.

161.091 **Fragmentation of expanding isothermal sheets.**
R. Carlberg.
Bull. Am. Astron. Soc., Vol. 17, No. 4, p. 906 (1985). Abstract. – See Abstr. 010.062.

161.092 **Hydrogen molecules and the radiative cooling of pregalactic shocks.**
H. Kang, P. R. Shapiro.
Bull. Am. Astron. Soc., Vol. 17, No. 4, p. 906 (1985). Abstract. – See Abstr. 010.062.

161.093 **The Hubble plot of quasars with redshift z > 3.**
J. M. Barnothy.
Bull. Am. Astron. Soc., Vol. 17, No. 4, p. 907 (1985). Abstract. – See Abstr. 010.062.

161.094 **The gravitational collapse of triaxial protoclusters.**
J. G. More, A. F. Heavens, J. A. Peacock.
Mon. Not. R. Astron. Soc., Vol. 220, No. 1, p. 189 – 202 (1986).
The gravitational collapse of non–rotating, homogeneous, triaxial ellipsoids is followed through and beyond the formation of a two–dimensional caustic surface ("pancake"). In previous work on this problem, "filaments" (one–dimensional structures) correspond only to the degenerate case of collapsing prolate spheroids; all other shapes, except spheres, collapse to pancake. In this work, the authors find that, assuming dissipation or relaxation processes keep the surface relatively thin after formation, pancakes undergo further anisotropic collapse in their planes, forming either filamentary structures or more isotropic configurations which the authors call clusters.

161.095 **Cooling of Population III objects in a pressure supported collapse.**
O. Lahav.
Mon. Not. R. Astron. Soc., Vol. 220, No. 1, p. 259 – 269 (1986).
The cooling and collapse of Population III objects is investigated in the framework of "flat" standard cosmological model. The author has assumed that the significant component of fluctuations was isothermal and that the cooling of the collapsing cloud was due to the H_2 molecule. He has taken into account the pressure inside the cloud. The only parameter in his models is the cloud's mass.

161.096 **A distance scale from the infrared magnitude/H I velocity–width relation. V. Distance moduli to 10 galaxy clusters, and positive detection of bulk supercluster motion toward the microwave anisotropy.**
M. Aaronson, G. Bothun, J. Mould, J. Huchra, R. A. Schommer, M. E. Cornell.
Astrophys. J., Vol. 302, No. 2, p. 536 – 563 (1986).
The authors use the relation between infrared magnitudes and 21 cm velocity widths to derive relative distance moduli to 10 nearby galaxy clusters. The sample ranges in redshift from Pegasus at $V \sim 4000$ km s^{-1} up to Hercules at $V \sim 11{,}000$ km s^{-1}. Several selection biases in the data set are identified and discussed. The final cluster Hubble ratios exhibit considerable scatter, signifying the presence of an unaccounted for Local Group streaming motion. Formal solution leads to a velocity in close agreement with the 3K dipole anisotropy, which can now be understood to arise from two principal components – Local Group motion within the Supercluster, and bulk motion of the Supercluster itself. Both components are ~ 300 km s^{-1} in size. The expansion rate is discussed and it is stated that the best current evidence favors a large value for the Hubble constant, $H_0 \sim 90$ km s^{-1}Mpc^{-1}.

161.097 **Molecular processes and gravitational collapse in intergalactic shocks.**
M.–M. Mac Low, J. M. Shull.
Astrophys. J., Vol. 302, No. 2, p. 585 – 589 (1986).
A major problem in galaxy formation theory is how the primordial intergalactic medium can cool and collapse with zero metallicity. Cooling from trace molecules such as H_2, HD, and LiH may solve this problem. The authors examine molecule formation behind 20 – 200 km s^{-1} intergalactic shocks produced by explosions at large redshift or by cloud collisions during infall into gravitational potential wells. It is shown that shocks can produce a substantial molecular fraction, which cools the postshock gas below 300K. The sensitivity of molecule formation to the preshock density, shock velocity, and ambient quasar radiation field is discussed.

161.098 **Voids in the Center for Astrophysics catalog.**
F. R. Bouchet, M. Lachièze–Rey.
Astrophys. J., Lett. Ed., Vol. 302, No. 2, p. L37 – L40 (1986).
The authors present a general method to estimate the count probabilities of finding *m* galaxies in a given volume, as a function of the volume size, for various *m* simultaneously. The method is applied to the two–dimensional CfA catalog, with *m* ranging from 0 to 5. The measured void probability ($m = 0$) is compared with the theoretical predictions of the model published by Schaeffer in 1984. As a result, the free parameter of this model is determined to be unity, which leads to a definite prediction for the amplitude of the correlation functions at all orders.

161.099 **The contemporary cosmology and the doctrine of the creation.**
A. Tursunov.
Zemlya Vselennaya, No. 3, p. 51 – 54 (1986). In Russian.

161.100 **On the possibility of detecting molecular hydrogen radiation from protogalaxies.**
Yu. I. Izotov.
Kinematika Fiz. Nebesn. Tel, Tom 2, No. 3, p. 34 – 40 (1986). In Russian. English translation in Kinematics Phys. Celest. Bodies.
The emission fluxes in the molecular hydrogen lines from large–scale density fluctuations ("pancakes") are calculated. The flux densities in the most intensive rotational and vibrational lines from the "pancakes" having redshifts $z = 3 - 10$ are shown to be equal to $10^{-16} - 10^{-17}$Wm^{-2}. The intensity fluctuations of the infrared background radiation due to H_2 emission may be of $10^{-1} - 10^{-2}$ for the spectral resolution $\Delta\nu \sim 10^9$Hz.

161.101 **A test for the explosion origin of cosmic voids.**
L. M. Ozernoj, V. V. Chernomordik.
Astron. Tsirk., No. 1407, p. 1 – 3 (1985). In Russian.
Recently the Lyα, C IV and Si IV absorption lines at the redshifts of the voids in Perseus–Pisces and Bootes have been detected in the UV spectra of background quasars. A detailed analysis of the data for the void in Bootes undertaken by the authors 1985 has shown that the gas could be contained either in large numbers of dwarf galaxies in the void or be distributed within the void. Both possibilities suggest an explosional origin of the cosmic void. Here the authors propose a test to check the expected shell–like structure of the gas: the column density on the edge of the void should be much larger than near its centre.

161.102 **Unsaturated character of the Lyα absorption line observed from cosmic voids.**
L. M. Ozernoj, V. V. Chernomordik.
Astron. Tsirk., No. 1408, p. 3 – 5 (1985). In Russian.
The authors show that all the observed lines, including Lyα, should be unsaturated otherwise the total column density of hydrogen which is determined mainly by the ionized component becomes tremendous. The physical reason for the Lyα line to be unsaturated lies in its large equivalent width that might be due to the Hubble expansion of the void.

161.103 **Quantity limitation of antimatter in the early universe from data of interaction between antiprotons and ^{4}He.**
Yu. A. Batusov, S. A. Bunyatov, F. Nikitiu, D. B. Pontekorvo, M. G. Sapozhnikov, I. V. Falomkin, F. Balestra, M. P. Bussa, L. Busso, L. Ferrero, D. Dzhervino, D. Panzieri, G. Piradzhino, F. Tosello, D. Bendisholi, V. Filippini, E. Lodi Rizzini, A. Rotondi, A. Zenone, K. Guaraldo, A. Madzhiora, M. Vaskon, D. Zanella, M. Yu. Khlopov.
Kratk. soobshch. OIYaI, Dubna, No. 6, p. 11 – 16 (1985). In Russian. Abstr. in Ref. Zh., 51. Astron., 4.51.858 (1986).

161.104 **Vacuum and a nonsingular inflationary Universe.**
A. V. Minkevich.
Dokl. Akad. Nauk BSSR, Tom 29, No. 11, p. 998 – 1001 (1985). In Russian. Abstr. in Ref. Zh., 51. Astron., 4.51.908 (1986).

161.105 **The inflationary Universe from the point of view of relativity theory.**
I. G. Dymnikova.
Fiz.-tekh. inst. Akad. Nauk SSSR. Prepr., No. 953, 17 pp. (1985). In Russian. Abstr. in Ref. Zh., 51. Astron., 4.51.909 (1986).

161.106 **Comment on stiff matter solutions in cosmological models based on Lyra's manifold.**
A. Beesham.
Astrophys. Space Sci., Vol. 122, No. 1, p. 177 – 179 (1986).
It is pointed out that the anisotropic cosmological model in Lyra's manifold given recently by Reddy and Innaiah (1985) is, apart from a minor difference in energy density and pressure, identical to the stiff matter anisotropic model of general relativity without a cosmological constant. Some comments are made on stiff matter solutions based on Lyra's manifold.

161.107 **Galaxy clustering.**
W. C. Saslaw.
Binary and multiple systems, p. 55 – 61 (1984). – See Abstr. 012.025.

161.108 **Gravitational instability of a two-component medium in an expanding universe.**
L. V. Solov'eva, A. A. Starobinskij.
Sov. Astron., Vol. 29, No. 4, p. 367 – 371 (1985). English translation of 40.161.079.

161.109 **Relativistic kinetics of baryon production in the big bang.**
Yu. G. Ignat'ev.
Sov. Astron., Vol. 29, No. 4, p. 371 – 374 (1985). English translation of 40.161.080.

161.110 **Evolution of adiabatic irregularities during the recombination era, and the missing mass in the universe.**
N. A. Zabotin, P. D. Nasel'skij.
Sov. Astron., Vol. 29, No. 4, p. 375 – 381 (1985). English translation of 40.161.081.

161.111 **Development of structure in a model universe containing unstable particles.**
A. G. Doroshkevich, A. A. Klypin, M. Yu. Khlopov.
Sov. Astron. Lett., Vol. 11, No. 4, p. 201 – 204 (1985). English translation of 40.161.002.

161.112 **Polarization of the unresolved-source microwave background.**
M. V. Sazhin, V. A. Korolev.
Sov. Astron. Lett., Vol. 11, No. 4, p. 204 – 206 (1985). English translation of 40.161.003.

161.113 **Fluctuations of the cosmic background temperature in unstable-particle cosmologies.**
A. G. Doroshkevich, M. Yu. Khlopov.
Sov. Astron. Lett., Vol. 11, No. 4, p. 236 – 238 (1985). English translation of 40.161.039.

161.114 **Radio brightness distribution within cosmic background fluctuations.**
M. V. Sazhin.
Sov. Astron. Lett., Vol. 11, No. 4, p. 239 – 241 (1985). English translation of 40.161.040.

161.115 **The distance scale: present status and future prospects.**
M. Aaronson, J. Mould.
Astrophys. J., Vol. 303, No. 1, p. 1 – 9 (1986).
The Hubble constant is arguably the single most important quantity in observational cosmology. Unfortunately, after more than half a century of intensive study, the value of H_0 obtained via calibration of the extragalactic distance scale remains a source of tremendous controversy. Currently the best global distance indicator appears to be the relation between galaxian infrared luminosity and rotation speed as measured by the velocity width of the 21 cm profile. Results obtained to date from exploitation of the IR/H I relation are summarized. Correction for all velocity deviation leads to a high value ($H_0 \approx 90$ km s^{-1}Mpc^{-1}) for the expansion rate. The principal uncertainty in this result (or any other current estimate for the Hubble constant) remains calibration of the local distance scale.

161.116 **The extragalactic distance scale derived from "sosie" galaxies. II. Solar motion and Hubble constant.**
G. de Vaucouleurs, W. L. Peters.
Astrophys. J., Vol. 303, No. 1, p. 19 – 24 (1986).
The redshifts of 140 galaxies having radial velocities $V < 4000$ km s^{-1} and distances $2 \leqslant \Delta \leqslant 28$ Mpc, derived in Paper I by the method of "sosies", are analyzed to evaluate the solar motion and the Hubble constant. Solutions are made by Jefferys's generalized least-squares technique which allows for errors in both variables. The data are divided into distance intervals, with mean distances $\langle\Delta\rangle = 8.1$ and 18.8 Mpc, to test for the possible presence of a Malmquist bias in the solutions. None is indicated. The mean Hubble ratio is $\langle H^* \rangle = 115 \pm 5$ (internal m.e.) in the first distance interval, $\langle H^* \rangle = 105 \pm 4$ in the second, and $\langle H^* \rangle = 109 \pm 4$ for the total sample. Including the mean error of the zero point of the distance modulus scale defined by 14 nearby galaxies, the external mean error is ~10%.

161.117 **Alignment of clusters and galaxies on scales up to 0.1 *c*.**
R. B. Tully.
Astrophys. J., Vol. 303, No. 1, p. 25 – 38 (1986).
Three remarkable properties of the large-scale distribution of galaxies appear to have escaped attention: (1) The structure in the vicinity of the Virgo Cluster that has traditionally been called the Local Supercluster seems to be appended to a very large agglomeration that includes the Coma/A1367, Hydra-Centaurus, Perseus-Pisces, and Pisces-Cetus Superclusters. The whole entity includes 48 known Abell-class clusters. (2) This very large structure is flattened, with axial ratios 4:2:1; and the pole, defined by the orientation of the short axis, is coincident with the pole of the plane of the traditional Local Supercluster to within 10°. (3) Within 40 h_{75}^{-1}Mpc of our location, where we have reasonable completion, there is a suggestion that the distribution of galaxies is stratified into four layers. Within a radius of 0.1 *c* there are two other very large superclusters and many voids.

161.118 **The origin of dwarf galaxies, cold dark matter, and biased galaxy formation.**
A. Dekel, J. Silk.
Astrophys. J., Vol. 303, No. 1, p. 39 – 55 (1986).
The formation of dwarf, diffuse, metal-poor galaxies as a result of supernova-driven winds is reexamined in view of the accumulating data on the systematic properties of dwarfs in the Local Group and in the Virgo Cluster. The observed luminosity-radius-metallicity relations are found to be produced naturally inside dominant halos, with a mass-radius relation that resembles the predictions of the "cold" dark matter cosmological scenario. The critical condition for global gas loss as a result of the first burst of star formation is that the virial velocity be below a critical value on the order of 100 km s^{-1}. In any hierarchial scenario for galaxy formation, this condition leads to two distinct classes of galaxies as observed: (1) the diffuse dwarfs which mostly originate from typical density perturbations; and (2) the normal, brighter galaxies which can originate only from the highest density peaks. This provides a statistical biasing mechanism for the preferential formation of bright galaxies in denser regions (clusters and superclusters).

161.119 **Anisotropic fluids and cosmology.**
S. S. Bayin.
Astrophys. J., Vol. 303, No. 1, p. 101 – 110 (1986).
The author considers applications of anisotropic fluids to cosmology, where the anisotropic pressure is generated by the presence of two noninteracting perfect fluids, which are in relative motion with respect to each other. The author discusses the extent of anisotropic pressure that can occur with various equations of state. He also presents a perturbative analysis of the amount of anisotropy induced in the pressure distribution by a small

(spherically symmetric) deviation from the Friedmann metric. It is shown that such anisotropies could grow faster than the expansion of the universe.

161.120 **Die Welt als Friedmann-Kosmos.**
H. Meusinger.
Astron. Raumfahrt, 24. Jahrg., Heft 2, p. 34 – 42 (1986).

161.121 **The age and size of the Universe.**
S. van den Bergh.
Proceedings of the Second Asian-Pacific Regional Meeting on Astronomy, p. 264 (1984). – See Abstr. 012.031.

161.122 **Aspects of the extragalactic distance scale.**
D. A. Hanes.
Proceedings of the Second Asian-Pacific Regional Meeting on Astronomy, p. 265 – 279 (1984). – See Abstr. 012.031.
The author gives some general comments and then a few specific remarks about three distance determination methods in particular: the Tully-Fisher relation, globular clusters, the traditional chain. He points out some critical problems.

161.123 **Some aspects of modern cosmology.**
J. V. Narlikar.
Proceedings of the Second Asian-Pacific Regional Meeting on Astronomy, p. 281 – 294 (1984). – See Abstr. 012.031.
The author divides his review of cosmology into four areas: (1) geometrical cosmology, (2) physical cosmology, (3) non-standard cosmology, (4) observational cosmology.

161.124 **The angular correlation function of faint galaxies.**
C. Pritchet, L. Infante.
Astron. J., Vol. 91, No. 1, p. 1 – 5 (1986).
The angular correlation function of faint galaxies has been measured down to a limiting magnitude of $J = 23.5$ in a 0.55 deg^2 area near the South Galactic Pole. This function follows a power law with exponent 0.7 – 0.8 for $J < 22.5$; at fainter magnitudes there is evidence for a flattening in the power-law slope, in agreement with the observations of Koo and Szalay. The authors' correlation amplitudes are in reasonable agreement with those of Koo and Szalay at $J < 22.5$, but are a factor of ~ 2 larger at $J = 23.5$. If real, this result favors models with a lower level of galaxy evolution; however, this conclusion appears to be weakened by the observation that differences in A_ω among different groups are correlated with systematic differences in galaxy number counts.

161.125 **Anisotropy of the galaxies detected by *IRAS*.**
A. Meiksin, M. Davis.
Astron. J., Vol. 91, No. 2, p. 191 – 198 (1986).
The *IRAS* point-source catalog provides a unique opportunity to generate a galaxy catalog uniformly selected over most of the sky (>9.5 sr). The authors have generated such a catalog of 6730 sources and find a small (4% – 7%) but robust dipole anisotropy in the galaxy distribution that points towards $l = 235°$, $b = 45°$, within 30° of the microwave dipole anisotropy. It is suggested that the *IRAS* galaxies roughly trace the large-scale mass distribution, that the microwave dipole velocity is mostly induced by the local supercluster ($cz < 3000$ km s^{-1}), and that the indicated cosmological density is high ($\Omega \approx 0.5$).

161.126 **Cosmological evolution in the X-ray, optical and radio bands.**
L. Danese, G. De Zotti, A. Franceschini.
X-ray astronomy '84, p. 459 – 462 (1985). – See Abstr. 012.033.
A comparative study of the cosmological evolution in the X-ray, optical and radio bands has revealed strong similarities, but also important quantitative differences, in the evolutionary patterns. In particular, the authors find that, for luminosity evolution models, the timescales are impressively close to each other, with the only exception of the X-ray band, for which a lower evolution rate seems to be indicated. The authors discuss possible reasons for this difference.

161.127 **Cosmic strings and cold matter.**
A. Stebbins.
Astrophys. J., Lett. Ed., Vol. 303, No. 2, p. L21 – L25 (1986).
The author presents a model of the formation of bound objects induced by loops of cosmic string in a universe dominated by cold matter. It is shown that accretion before Z_{eq} is important. Large objects the size of the Virgo supercluster will have collapsed by the present epoch. Collapse will continue to occur for $z < \Omega_0^{-1}$. Galaxies are formed predominantly by secondary infall. The velocity dispersion of collapsed objects as a function of mass seems tenable.

161.128 **Effects of quantum gravity in particle physics and cosmology.**
V. A. Kuz'min.
Kvarki 84. Mater. Semin., Tbilisi, 15 – 17 maya, 1984, Tom 1. Moskva, p. 132 – 144 (1985). In Russian. Abstr. in Ref. Zh., 51. Astron., 5.51.970 (1986).

161.129 **Formation of a new vacuum and evolution of bubbles in the early Universe.**
V. A. Berezin, V. A. Kuz'min, I. I. Tkachev.
Kvarki 84. Mater. Semin., Tbilisi, 15 – 17 maya, 1984, Tom 1. Moskva, p. 150 – 169 (1985). In Russian. Abstr. in Ref. Zh., 51. Astron., 5.51.971 (1986).

161.130 **Evolution of fluctuations in the early Universe with phase transitions.**
Yu. A. Beletskij.
Probl. nelinejn. i turbulent. protsessov v fiz. Tr. 2 Mezhdunar. rab. gruppy, Kiev, 1983. Chast' 2. Kiev, p. 264 – 266 (1985). In Russian. Abstr. in Ref. Zh., 51. Astron., 5.51.972 (1986).

161.131 **Nonlinear and turbulent processes in galaxy formation.**
A. G. Doroshkevich.
Probl. nelinejn. i turbulent. protsessov v fiz. Tr. 2 Mezhdunar. rab. gruppy, Kiev, 1983. Chast' 2. Kiev, p. 295 – 297 (1985). In Russian. Abstr. in Ref. Zh., 51. Astron., 5.51.973 (1986).

161.132 **Rich clusters and superclusters of galaxies – observable appearances of large-scale structures in the Universe.**
S. F. Shandarin.
Probl. nelinejn. i turbulent. protsessov v fiz. Tr. 2 Mezhdunar. rab. gruppy, Kiev, 1983. Chast' 2. Kiev, p. 350 – 351 (1985). In Russian. Abstr. in Ref. Zh., 51. Astron., 5.51.974 (1986).

161.133 **On stability of the Minkowski space-time and on the dynamical chaos in cosmology.**
O. I. Batsula.
Probl. nelinejn. i turbulent. protsessov v fiz. Tr. 2 Mezhdunar. rab. gruppy, Kiev, 1983. Chast' 2. Kiev, p. 261 – 263 (1985). In Russian. Abstr. in Ref. Zh., 51. Astron., 5.51.985 (1986).

161.134 **Large-scale anisotropy of the relic background radiation in cosmological models with unstable particles.**
L. A. Kofman, D. Yu. Pogosyan, A. A. Starobinskij.
Pis'ma Astron. Zh., Tom 12, No. 6, p. 419 – 428 (1986). In Russian. English translation in Sov. Astron. Lett., Vol. 12.
The large-scale anisotropy of the temperature of the relic background electromagnetic radiation caused by adiabatic perturbations with flat initial spectrum in the flat Friedmann model with unstable massive particles is calculated. The multipole structure of $\Delta T/T$ differs significantly from those obtained in the standard neutrino model or the cold particle model. The upper limit on the moment of decay of unstable neutrinos or gravitinos $Z_d < 3 - 4$ is obtained using observational upper bounds on $\Delta T/T$ over the range 6 – 180°.

161.135 **A closed world and the gravitational field.**
L. P. Grishchuk, A. N. Petrov.
Pis'ma Astron. Zh., Tom 12, No. 6, p. 429 – 433 (1986). In Russian. English translation in Sov. Astron. Lett., Vol. 12.
It is explicitely shown that the geometry of a closed world which satisfies the classical Einstein equations can be presented as a field configuration on a topologically trivial and geometrical-

ly flat «space–time» background. The total energy, momentum and angular momentum of the configuration are equal to zero. The possible connection of this approach with the problem of the quantum «birth» of the Universe is discussed.

161.136 **The large scale anisotropy of the cosmic microwave background radiation.**
R. K. Schaefer.
Diss. Abstr. Int., Sect. B, Vol. 46, No. 6, p. 1959 (1985). Thesis, Brandeis University, 108 pp. (1985). Order No. DA8517051.

161.137 **Topics in inflationary cosmology.**
R. N. Kahn.
Diss. Abstr. Int., Sect. B, Vol. 46, No. 7, p. 2346 (1986). Thesis, Harvard University, 351 pp. (1985). Order No. DA8520223.

161.138 **The spectrum of the cosmic background radiation from 2.3 cm^{-1} to 11.0 cm^{-1}.**
J. B. Peterson.
Diss. Abstr. Int., Sect. B, Vol. 46, No. 9, p. 3092 (1986). Thesis, University of California, Berkeley, 123 pp. (1985). Order No. DA8525087.

161.139 **A measurement of the cosmic background radiation temperature at 3.3 mm wavelength.**
C. Witebsky.
Diss. Abstr. Int., Sect. B, Vol. 46, No. 9, p. 3093 (1986). Thesis, University of California, Berkeley, 165 pp. (1985). Order No. DA8525158.

161.140 **Properties and consequences of topological structures in the early universe.**
T. Vachaspati.
Diss. Abstr. Int., Sect. B, Vol. 46, No. 10, p. 3483 (1986). Thesis, Tufts University, 132 pp. (1985). Order No. DA8526853.

161.141 **Possible mechanism for generating almost isothermal baryon–density perturbations.**
M. Fukugita, V. A. Rubakov.
Phys. Rev. Lett., Vol. 56, No. 9, p. 988 – 990 (1986).
There appears a possibility that the correct cosmological baryon asymmetry is generated in the late baryogenesis scenario combined with electroweak baryon–number nonconservation at finite temperatures. In this scenario the authors expect almost isothermal baryon–density perturbations, which can be large enough to make the nonlinearities needed for the formation of globular clusters.

161.142 **Improved bounds on the dimension of space–time.**
B. Müller, A. Schäfer.
Phys. Rev. Lett., Vol. 56, No. 12, p. 1215 – 1218 (1986).
The authors treat the perihelion shift of the planetary motion and the Lamb shift in hydrogen in an arbitrary number of space dimensions. Comparison with experimental data shows that the deviation from dimensionality four of space–time less than 10^{-9} and 3.6×10^{-11}, respectively, on the length scales associated with these phenomena.

161.143 **Quantum instability of de Sitter space.**
I. Antoniadis, J. Iliopoulos, T. N. Tomaras.
Phys. Rev. Lett., Vol. 56, No. 13, p. 1319 – 1322 (1986).
The graviton propagator in a de Sitter background is found to be divergent. The authors show that as a consequence of this divergence, de Sitter space is not a solution of the equations of motion of the complete theory. If one starts from de Sitter space as a classical ground state, quantum corrections change it into flat Minkowski space.

161.144 **Infrared behavior and finite–size effects in inflationary cosmology.**
B. L. Hu, D. J. O'Connor.
Phys. Rev. Lett., Vol. 56, No. 15, p. 1613 – 1616 (1986).

161.145 **Comment on "Nonlinear gravitational clustering in cosmology".**
A. L. Melott.
Phys. Rev. Lett., Vol. 56, No. 18, p. 1992 (1986).
For the original paper by Hausel et al. see Abstract 40.161.187.

161.146 **Gravitino decay and the cosmic gamma–ray background.**
F. W. Stecker.
Phys. Rev. Lett., Vol. 56, No. 23, p. 2551 (1986). With a response by Olive and Silk on p. 2552.
This paper comments on a recent publication by Olive and Silk – see Abstract 40.161.189.

161.147 **Spin–dominated inflation in the Einstein–Cartan theory.**
M. Gasperini.
Phys. Rev. Lett., Vol. 56, No. 26, p. 2873 – 2876 (1986).
In the framework of the simplest Einstein–Cartan generalization of the standard cosmological model, it is shown that an inflationary phase may occur at a sufficiently early epoch, during which the dominant contributions to the effective energy–momentum tensor are represented by the spin density of the matter sources. The general features of this inflationary scenario are briefly discussed, and in particular the possibility of obtaining an expansion rate different from that of the standard exponential inflation is stressed.

161.148 **Particle creation in early Universe.**
S. N. Pandey.
Sir Arthur Eddington Centenary Symposium, Vol. 3: Gravitational radiation and relativity, p. 139 – 150 (1986). – See Abstr. 012.036.
An attempt is made to outline a complete scenario of evolution of the Universe with an emphasis on particle creation. This is attempted both by considering the quantum effects in an isotropic (Friedmann) Universe and also classically by considering the gravitational wave equations which are not conformally invariant.

161.149 **The Hubble diagram for brightest cluster galaxies and the cosmological constant.**
V. E. Yakimov.
Sir Arthur Eddington Centenary Symposium, Vol. 3: Gravitational radiation and relativity, p. 283 – 291 (1986). – See Abstr. 012.036.
The least–squares best–fit values of observational parameters to Friedmann models have been computed for a sample of 99 first–ranked cluster galaxies. The effects of the evolutionary correction, photon scattering by free electrons and extinction by intergalactic dust on the redshift–magnitude relation have been taken into account. An upper limit on the value of the cosmological constant is found.

161.150 **A nonsingular vacuum model for pre–inflationary universe.**
K. A. Bronnikov, V. N. Mel'nikov.
Sir Arthur Eddington Centenary Symposium, Vol. 3: Gravitational radiation and relativity, p. 386 – 394 (1986). – See Abstr. 012.036.
The authors outline a cosmological scenario based on a semiclassical nonsingular open model taking into account such vacuum effects as spontaneous symmetry breaking for a conformal Higgs field and vacuum polarization for any set of massless matter fields, including an effective Λ term which provides inflation.

161.151 **Candidate missing–mass carriers in an inflationary universe.**
P. D. Nasel'skij, A. G. Polnarev.
Sov. Astron., Vol. 29, No. 5, p. 487 – 491 (1985). English translation of 40.161.124.

161.152 **Infrared line emission by protogalaxies and protocluster pancakes.**
Yu. A. Shchekinov, M. B. Éntél' (*M. B. Ehntehl'*).
Sov. Astron., Vol. 29, No. 5, p. 491 – 495 (1985). English translation of 40.161.125.

161.153 **Evolution of perturbations in cosmological models with unstable neutrinos.**
A. G. Doroshkevich, A. A. Klypin, Eh. V. Kotok.
Astron. Zh., Tom 63, Vyp. 3, p. 417 – 425 (1986). In Russian. English translation in Sov. Astron., Vol. 30, No. 3.
A cosmological model with heavy unstable neutrinos (rest mass 30 – 400 eV, decay time 10^{15}– 10^{16}s) is analysed. The equations describing the growth of perturbations in this model are obtained. The evolution of perturbations at the nonlinear stage is studied numerically in the one–dimensional approximation.

161.154 **Propagation of a spherical detonation wave in the expanding universe with account of gravitation.**
Ya. M. Kazhdan.
Astron. Zh., Tom 63, Vyp. 3, p. 434 – 445 (1986). In Russian. English translation in Sov. Astron., Vol. 30, No. 3.
The self–similar solution of the equations describing a gas flow behind a divergent detonation wave in matter expanding according to the Hubble law is carried out. For a narrow range of the adiabatic exponent $\varkappa$: $1 < \varkappa < \varkappa_1$ $(1.2 < \varkappa_1 < 1.21)$ there is a continuous solution in the whole region from the center to the front of the detonation wave with a weak discontinuity along the characteristic surface. If $\varkappa > 4/3$ a shock is formed in the flow behind the detonation wave front. If $\varkappa_1 < \varkappa \leqslant 4/3$, there is no self–similar solution to satisfy the physical conditions of the problem.

161.155 **Automated data retrieval in observational cosmology.**
M. Kalinkov, K. Stavrev, I. Kuneva.
Sbornik tr. Mezhdunar. konf. na temu: Avtomatiz. nauchn. issled., Plovdiv, 15 – 20 okt., 1984, Tom 2, Sofiya, p. 550 – 557). In Russian. Abstr. in Ref. Zh., 51. Astron., 6.51.888 (1986).

161.156 **Fluctuations of angular sizes of sources in an inhomogeneous universe.**
V. A. Korolev, M. V. Sazhin.
Inst. kosm. issled. Akad. Nauk SSSR., Prepr., No. 1060, 23 pp. (1986). In Russian. Abstr. in Ref. Zh., 51. Astron., 6.51.891 (1986).

161.157 **Self–consistent mega–cosmology.**
R. G. Zaycoff.
Astrofiz. issled., Tom 4, p. 111 – 142 (1985). Abstr. in Ref. Zh., 51. Astron., 6.51.904 (1986).

161.158 **A new estimate of the Hubble constant using the Virgo cluster distance.**
N. Visvanathan.
Proc. Astron. Soc. Aust., Vol. 6, No. 2, p. 140 – 144 (1985).
A new distance modulus (31.23) to the Virgo cluster is derived using the distances to nearby galaxies given by Sandage and Tammann, de Vaucouleurs, the author and DDO observers. This when combined with the undisturbed mean Virgo cluster velocity 1182 km s^{-1}, gives a value for the global Hubble constant as 67 ± 4 km $s^{-1}Mpc^{-1}$.

161.159 **Aufbau und Entwicklung des Weltalls. VIII. Strukturbildung im Kosmos.**
H.–E. Fröhlich.
Sterne, 62. Band, Heft 2, p. 97 – 104 (1986).

161.160 **The neutrino hypothesis for dark matter in dwarf spheroidals.**
R. Cowsik.
J. Astrophys. Astron., Vol. 7, No. 1, p. 1 – 6 (1986).
The theoretical expectation of the high mass of $\gtrsim 400$ eV for the particles constituting the dark matter in dwarf–spheroidals as an artifact of the implicit assumption that the density of particles vanishes at the visible edge. On the contrary if our galaxy and the dwarf–spheroidals are embedded in a neutrino condensation of the dimensions of the cluster then $m_\nu \cong 10$ eV can accommodate all the observations.

161.161 **Dynamical effects of an extended cloud of dark matter on dwarf spheroidals.**
R. Cowsik, P. Ghosh.
J. Astrophys. Astron., Vol. 7, No. 1, p. 7 – 15 (1986).
If the dwarf spheroidals are embedded in an extended cloud of dark matter then their density profiles can be reproduced by assuming a Maxwellian distribution of velocities for the constituent stars. The observed luminosity profiles of dwarf spheroidals imply densities for the dark matter in the range 10^{-26} to 10^{-25}g cm^{-3}, and mass–to–luminosity ratios which are typically an order of magnitude greater than those of globular clusters. Neutrinos of mass ~ 10 eV and $\langle v \rangle \sim 1000$ kms^{-1} can provide this requisite density for the background.

161.162 **Dark matter – systematics of its distribution.**
R. Cowsik, M. Vasanthi.
J. Astrophys. Astron., Vol. 7, No. 1, p. 29 – 38 (1986).
The authors conclude that neutrino condensates with $M \sim 10^{16} M_\odot$ were the first objects to be formed in the universe at a redshift of $\sim 10^4$. Subsequent to formation they expanded much slower than the rest of the universe and fluctuations in the density of baryonic matter grew effectively with the formation of the galaxies. The dynamics of these galaxies and that of the stars within galaxies are sensitive to the background gas. Assuming that neutrinos have a rest mass of ~ 10 eV/c^2 details of the dynamical motions of galaxies and stars can be understood quantitatively.

161.163 **On the nature of dark matter in dwarf galaxies.**
J. Melnick, R. Terlevich.
Observatory, Vol. 106, No. 1072, p. 69 – 71 (1986).

161.164 **An upper limit to the density inhomogeneity in the Universe.**
L.–z. Fang, Y.–q. Chu, X.–f. Zhu.
Astrophys. Space Sci., Vol. 122, No. 2, p. 213 – 218 (1986).
An upper limit to the amplitude of the overall density fluctuation has been found by means of the gravitational lensing effect of the density inhomogeneity on the luminosities of quasars with larger redshifts. The observed differences of luminosities of quasars located at different directions are partially given by the lensing effect, therefore, a useful upper limit to the inhomogeneity can be derived if the luminosity distribution of quasars is uniform enough. The authors obtain that, in the case of the density parameter of the Universe $\Omega = 1$, the overall matter should be less clustered than the luminous matter by a factor of at least 3. This result may not favour the biased dark matter scenario for the formation of large–scale structure in the Universe.

161.165 **On the causal geometry of spacetime: some applications to cosmology.**
A. Kartsaklis.
Astrophys. Space Sci., Vol. 122, No. 2, p. 219 – 229 (1986).
Zeeman's fundamental theorem is reformulated and proved by means of simpler concepts. Also a generalization of this theorem, an application to de Sitter's universes, and an extension to the general relativity theory are given.

161.166 **Bianchi type–I Lyttleton–Bondi universe.**
D. R. K. Reddy, P. Innaiah.
Astrophys. Space Sci., Vol. 122, No. 2, p. 231 – 234 (1986).
The exterior field of the Bianchi type–I metric in the Lyttleton–Bondi universe is considered and an exact solution, which is a simple and elegant generalisation of the Kasner model, is presented. Some physical properties of the model are discussed.

161.167 **Robertson–Walker type Lyttleton–Bondi universe.**
D. R. K. Reddy.
Astrophys. Space Sci., Vol. 122, No. 2, p. 263 – 265 (1986).
The exterior field of the Robertson–Walker type metric in the Lyttleton–Bondi universe is considered and an exact solution is obtained. Some physical properties of the solution are discussed.

161.168 **An analysis of the derivation of Friedmann's equation.**
P. Voráček.
Astrophys. Space Sci., Vol. 122, No. 2, p. 327 – 341 (1986).
A deeper specification is made of the physical meaning of the terms in the Friedmann Equation (FE). It is deduced that the FE for the radiation–dominated phase of the evolution of the Universe (which is considered to be closed) is valid even for the matter–dominated era. On the other hand, such a form of the FE is shown to be irrelevant when used for deducing how the frequency of the 3K–photons depends on the radius of the Universe. The deduction of the critical density–value is revised, and the new result shortly discussed. It is concluded that there is only one possible value for the maximum radius of our Universe. A modified "soapbubble" analogy with the first law of thermodynamics is used to describe "the dynamics" of the cosmological expansion.

161.169 **Thermodynamics of the vacuum.**
L. Diósi, B. Lukács, K. Martinás, G. Paál.
Astrophys. Space Sci., Vol. 122, No. 2, p. 371 – 386 (1986).
The fact that the energy–momentum tensor has nonzero vacuum expectation value is some space–times (the so–called reaction) indicates a nontrivial thermodynamics of such vacua. A consequent thermodynamic analysis of the problem in Robertson–Walker space–times shows that, in the generic case, the number of the independent extensives is 2, in spite of the fact that the energy density is completely determined by the time–evolution of the geometry since the preparation of the vacuum state. The appearance of a second independent extensive seems to be in direct connection with the extra entropy term in the Generalized Second Law of Bekenstein and Hawking.

161.170 **A plane symmetric cosmological model in Lyra manifold.**
D. R. K. Reddy, P. Innaiah.
Astrophys. Space Sci., Vol. 123, No. 1, p. 49 – 52 (1986).
A non–static plane–symmetric cosmological model in Lyra's manifold is obtained when the source of the gravitational field is a perfect fluid with pressure equal to energy density. Some properties of the model have also been discussed.

161.171 **The production of light elements in unconventional cosmologies with VMOs.**
Yu. A. Shchekinov, B. V. Vainer (*B. V. Vajner*).
Astrophys. Space Sci., Vol. 123, No. 1, p. 103 – 115 (1986).
The abundances of light elements (deuterium, helium and lithium) are calculated in unconventional cosmologies in which these elements are produced by interaction of fast α–particles ejected by massive pregalactic stars with a purely hydrogeneous cosmological substrate. The number density and energetics of such stars needed for suitable production of light elements are estimated. It is shown that the models always lead to overproduction of lithium. The intensity of ultraviolet, X–ray, and γ–ray emission caused by explosions of pregalactic massive stars is calculated. The distortions of cosmic background radiation by Compton losses of electrons heated by explosive stars are also calculated.

161.172 **Dynamical effects of dark matter in systems of galaxies.**
J. F. Navarro, D. Garcia Lambas, J. L. Sérsic.
Astrophys. Space Sci., Vol. 123, No. 1, p. 117 – 124 (1986).
Several *N*–body experiments were performed in order to simulate the dynamical behaviour of systems of galaxies gravitationally dominated by a massive dark background.

161.173 **The Bianchi type–V vacuum cosmological model in the scale covariant theory.**
A. Beesham.
Astrophys. Space Sci., Vol. 123, No. 2, p. 389 – 391 (1986).
A solution of the field equations in the scale–covariant theory for a Bianchi type–V spacetime is obtained. It is shown that this solution represents a generalization of the general relativistic one.

161.174 **Dynamical friction and formation of structures: some cosmological and cosmogonical aspects.**
V. M. Lipunov, S. G. Simakov.
Astrophys. Space Sci., Vol. 123, No. 2, p. 393 – 402 (1986).
The influence of dynamical friction on peculiar velocities in a two–component medium is considered. The solutions for the stationary as well as for the cosmologically expanding medium are obtained. These solutions illustrate the possibility for clusters of galaxies to have super–low velocities. It is emphasized that the presence of two components leads to the formation of individual strings and general string structure in one of the component. A string–like structure of the Universe becomes inevitable in the case when an initial perturbation spectrum has a specific scale, and falls down abruptly for larger and smaller distances. Such a picture of the formation of large–scale structure of the Universe is similar to the formation of particle tracks in the Wilson chamber.

161.175 **Galaxy clustering and thermodynamics: quasi–static evolution of *b*(*t*).**
W. C. Saslaw.
Astrophys. J., Vol. 304, No. 1, p. 11 – 14 (1986).
Combining the cosmic energy equation with the BBGKY clustering hierarchy gives a simple evolution equation for $b(t)$, the ratio of gravitational correlation energy to kinetic energy in the expanding universe. It applies most accurately to the case of an initially unclustered distribution. The asymptotic value of this ratio for nonlinear clustering is $(2\alpha+1)/(2\alpha+2)$, where the cosmic scale length $R(t) \propto t^{\alpha}$. This agrees with the observed value of $b = 0.7 \pm 0.05$ for the distribution of galaxies in our universe for α near the Einstein–de Sitter value of 2/3.

161.176 **The statistics of peaks of Gaussian random fields.**
J. M. Bardeen, J. R. Bond, N. Kaiser, A. S. Szalay.
Astrophys. J., Vol. 304, No. 1, p. 15 – 61 (1986).
Cosmological density fluctuations are often assumed to be Gaussian random fields. The local maxima of such fields are obvious sites for the formation of nonlinear structures. The statistical properties of the peaks can be used to predict the abundances and clustering properties of objects of various types. In this paper, the authors derive (1) the number density of peaks of various heights; (2) the factor by which the peak density is enhanced in large–scale overdense regions; (3) the n–point peak–peak correlation function in the limit that the peaks are well separated; and (4) the density profiles centered on peaks. This semianalytical approach is applied to galaxy formation and galaxy clustering in the early universe.

161.177 **The significance of voids.**
S. Otto, H. D. Politzer, J. Preskill, M. B. Wise.
Astrophys. J., Vol. 304, No. 1, p. 62 – 74 (1986).
The statistical significance of voids in the spatial distribution of galaxies or clusters of galaxies is studied. The probability per unit volume of finding a large void is expressed in terms of correlation functions. Numerical simulations are used to estimate the likelihood of observing a large void in the distribution of rich clusters of galaxies, assuming that rich clusters arose wherever suitably averaged primordial density fluctuations were unusually large.

161.178 **Development of voids in the thin–wall approximation. II. Radiation–filled voids in a flat background.**
R. Pim, K. Lake.
Astrophys. J., Vol. 304, No. 1, p. 75 – 81 (1986).
The evolution of spherical radiation–filled voids in a spatially flat Robertson–Walker background is studied within the context of the general relativistic thin–wall approximation. This work extends previous discussion of vacuum voids. It is found that the inclusion of radiation within the void has significant quantitative and important qualitative effects on the evolution of the void.

161.179 **On the application of a spherical infall model to the Local Supercluster.**
H. Lee, Y. Hoffman, C. Ftaclas.
Astrophys. J., Lett. Ed., Vol. 304, No. 1, p. L11 – L15 (1986).
The application of a spherical, nonlinear infall model to the Local Supercluster (LSC), and in particular its reliability as a cosmological probe of Ω_0, are studied by means of *N*–body simulations. An *N*–body model of a flat universe with an initial Poisson distribution has been analyzed by identifying objects similar to the LSC, and calculating the apparent values of Ω_0, Ω_{st} by the standard (nonlinear spherical infall) model. The distribution of the values of Ω_{st} ranges from 0.65 to 1.05 with a mean of 0.92, compared to the $\Omega_0 = 1.0$ of the simulation. An implication of the present study is that the standard cosmological scenario of a flat universe is incompatible with the observed LSC.

161.180 **Galaxy formation by mock gravity.**
C. J. Hogan, S. D. M. White.
Nature, Vol. 321, No. 6070, p. 575 – 578 (1986).
Absorbing particles in an isotropic radiation field experience an attractive force due to their mutual shadowing. If the pre–galactic universe contained suitable sources of radiation and absorbing material, an instability caused by this effect may have led to the formation of galaxies and galaxy clusters. Other consequences include the segregation of galaxies from dark matter and a strong anisotropic far–infrared background.

161.181 **Entropy in an oscillating universe in the assumption of universe "splitting" into numerous smaller "daughter" universes.**
M. A. Markov.
Proceedings of the Third Seminar on Quantum Gravity, p. 3 – 15 (1985). – See Abstr. 012.038.
The author considers the possibility that in the collapse of a closed universe "daughter" universes (with smaller mass and entropy) may form in the regions of the mother universe where the gravitational mass defect due to fluctuations reaches certain values. The model describes transformation of the classical Friedmann universe into the classical de Sitter universe under the assumption that the gravitational constant is a decreasing function of the matter density.

161.182 **Quantum fluctuations as the cause of inhomogeneity in the universe.**
J. Halliwell, S. Hawking.
Proceedings of the Third Seminar on Quantum Gravity, p. 509 – 565 (1985). – See Abstr. 012.038.
This paper derives the canonical quantization of an inflationary cosmological model without any assumptions of homogeneity or isotropy. It is shown that initial quantum fluctuations give rise to a scale free spectrum of density fluctuations that could explain the origin of galaxies and other structures in the universe. The anisotropy in the microwave background would be within the upper limit set by observation if the mass of the scalar field that drives the inflation is less than 10^{14}Gev.

161.183 **Inflationary stages in cosmological models with a scalar field.**
V. A. Belinsky (*V. A. Belinskij*), L. P. Grishchuk,
Ya. B. Zel'dovich, I. M. Khalatnikov.
Proceedings of the Third Seminar on Quantum Gravity, p. 566 – 590 (1985). – See Abstr. 012.038.

161.184 **The present status of the inflationary universe scenario.**
A. D. Linde.
Proceedings of the Third Seminar on Quantum Gravity, p. 591 – 604 (1985). – See Abstr. 012.038.
Different versions of the inflationary universe scenario are considered and it is argued that the most natural realization of the idea of inflation can be achieved in the context of the chaotic inflation scenario. The possibility of generation of isothermal density perturbations and a new mechanism of baryosynthesis in the inflationary universe are suggested.

161.185 **Dynamics of inflating bubbles in the early universe.**
V. A. Berezin, V. A. Kuz'min, I. I. Tkachev.
Proceedings of the Third Seminar on Quantum Gravity, p. 605 – 622 (1985). – See Abstr. 012.038.
The authors show that the inflating fluctuation in the early universe can be equivalent only to a wormhole (such a possibility can take place both in the open universe and in the closed one) or the field fluctuation occupies more than a half of a closed universe. Other possibilities are highly exotic.

161.186 **Inflation?**
W. G. Unruh.
Proceedings of the Third Seminar on Quantum Gravity, p. 623 – 627 (1985). – See Abstr. 012.038.
Questions are raised about the use of the effective potential in determining whether or not inflation occurs in the early universe due to a supercooled phase transition.

161.187 **Kaluza–Klein cosmology and the actual state of the universe.**
U. Bleyer, D.–E. Liebscher.
Proceedings of the Third Seminar on Quantum Gravity, p. 662 – 674 (1985). – See Abstr. 012.038.
The authors consider generalized Friedmann–Robertson–Walker cosmologies in $4+D$ dimensions. By use of a canonical decomposition of matter distributions they give a systematic analysis of the possibilities to compensate the large curvature of the internal space necessary today to make the additional dimensions undetectable. This leads to general statements about the existence of solutions with static internal space.

161.188 **Anisotropy of the relic radiation as a test of the early universe theories.**
V. N. Lukash, P. D. Nasel'skij, I. D. Novikov.
Proceedings of the Third Seminar on Quantum Gravity, p. 675 – 689 (1985). – See Abstr. 012.038.
The correlation and multipole characteristics of anisotropy in the intensity distribution $\Delta T/T$ of the microwave background are calculated for a number of cosmological models with $\Omega \leqslant 1$. Different spectra of primordial perturbations and different assumptions as to the nature of the missing mass are considered. The authors emphasize the importance of measurements of the sky $\Delta T/T$ – pattern on large angular scales ($\alpha > 5°$) and compare the results of calculations with the available observational data.

161.189 **Primordial black holes and observational restrictions on quantum gravity.**
M. Yu. Khlopov, P. D. Nasel'skij, A. G. Polnarev.
Proceedings of the Third Seminar on Quantum Gravity, p. 690 – 705 (1985). – See Abstr. 012.038.
Astrophysical upper limits on the primordial black hole (PBH) spectrum restrict the parameters of quantum gravity, cf supergravity. A possible contribution of PBHs to the dark matter of the present universe is discussed.

161.190 **Particle physics and the standard cosmology.**
S. Sarkar.
Rutherford Appleton Lab., Rep., RAL–85–115, 29 pp. (1985). To appear in "Superstrings, supergravity and unified theories", Proc. ICTP Workshop on Theoretical Physics, Trieste, June – July 1985.
The author reviews the constraints imposed by the standard cosmological model on physics beyond the standard $SU(3)_C \times SU(2)_L \times U(1)_Y$ model, with particular attention to supersymmetry and supergravity.

161.191 **The infrared background radiation: new observational approaches. I. The noise of microwave background.**
P. de Bernardis, S. Masi.
Galaxies, quasars and cosmology, p. 95 – 101 (1985). – See Abstr. 012.042.
The authors describe a new method of investigating the features of the cosmic background radiation, searching for its temporal fluctuations by means of an infrared correlator. Prelimi-

nary results of a balloon flight devoted to this measurement are presented.

161.192 **The infrared background radiation: new observational approaches. II. Search for extragalactic backgrounds.**
S. Masi, P. de Bernardis.
Galaxies, quasars and cosmology, p. 103 – 116 (1985). – See Abstr. 012.042.
The authors describe the theoretical motivations for a new, balloon borne experiment, devoted to the search for extragalactic background's anisotropies, together with the experimental problems connected with it. Preliminary results of a balloon experiment are discussed.

161.193 **Production of inos in the big bang.**
E. W. Kolb.
Galaxies, quasars and cosmology, p. 117 – 146 (1985). – See Abstr. 012.042.
The production of elementary particles in the big bang is discussed. In particular, the possibility that some elementary particle ("ino") is produced in the big bang and is present today in sufficient numbers to contribute to the overall mass density of the universe in the form of dark matter is illustrated with several examples. The production of neutrinos, axions and baryons are covered in detail.

161.194 **Hawking's theory of quantum cosmology.**
L. Z. Fang, Z. C. Wu.
Galaxies, quasars and cosmology, p. 147 – 179 (1985). – See Abstr. 012.042.
Recently Hartle and Hawking put forward a ground state proposal for the quantum state of the universe which incorporates the idea that the universe must come from nothing. Many models have been discussed in quantum cosmology with this boundary condition. It has been shown that every model is a step towards to a realistic universe, i.e. a 4–dimensional isotropic universe with a long inflationary stage.

161.195 **Inflationary cosmology from a classical gauge field.**
A. Aurilia, G. Denardo, F. Legovini, E. Spallucci.
Galaxies, quasars and cosmology, p. 181 – 200 (1985). – See Abstr. 012.042.
The role of the cosmological constant and its relation to the vacuum energy in the inflationary phase of the expanding universe is considered. An interesting way of generating the cosmological constant in Einstein's field equations is coupling gravity with a fully antisymmetric third rank abelian gauge field. Such a field in four dimensions has zero degrees of freedom and has some similarity with the "vacuum" as it does not represent any propagating field.

161.196 **The inflationary universe.**
J. Ellis.
Europhys. News, Vol. 16, No. 9, p. 6 – 8 (1985). Abstr. in Phys. Abstr., Vol. 89, No. 1273, Entry 18966 (1986).

161.197 **Bianchi type–VI_0 solutions in modified Brans–Dicke cosmology.**
S. Ram.
J. Math. Phys., Vol. 26, No. 11, p. 2915 – 2917 (1985). Abstr. in Phys. Abstr., Vol. 89, No. 1273, Entry 18968 (1986).

161.198 **Bianchi type I cosmological model with a viscous fluid.**
A. Banerjee, S. B. Duttachoudhury, A. K. Sanyal.
J. Math. Phys., Vol. 26, No. 11, p. 3010 – 3015 (1985). Abstr. in Phys. Abstr., Vol. 89, No. 1273, Entry 18969 (1986).

161.199 **Correlation of orientations and two scenarios for clustering of matter in the universe.**
L.–z. Fang, Y.–q. Chu, X.–f. Zhu.
Kexue Tongbao, Vol. 30, No. 2, p. 172 – 175 (1985). Abstr. in Phys. Abstr., Vol. 89, No. 1273, Entry 18970 (1986).

161.200 **Inflationary cosmology and 4–index tensor fields.**
R. G. Moorhouse, J. Nixon.
Nucl. Phys. B, Part. Phys., Vol. B261, No. 1, p. 172 – 184 (1985). Abstr. in Phys. Abstr., Vol. 89, No. 1273, Entry 18973 (1986).

161.201 **Semiclassical wave function of the universe at small three–geometries.**
K. Schleich.
Phys. Rev. D, Vol. 32, No. 8, p. 1889 – 1898 (1985). Abstr. in Phys. Abstr., Vol. 89, No. 1273, Entry 18974 (1986).

161.202 **Quantum mechanics of the scalar field in the new inflationary universe.**
A. H. Guth, S.–y. Pi.
Phys. Rev. D, Vol. 32, No. 8, p. 1899 – 1920 (1985). Abstr. in Phys. Abstr., Vol. 89, No. 1273, Entry 18975 (1986).

161.203 **Stability of self–consistent higher–dimensional cosmological solutions.**
E. J. Copeland, D. J. Toms.
Phys. Rev. D, Vol. 32, No. 8, p. 1921 – 1927 (1985). Abstr. in Phys. Abstr., Vol. 89, No. 1273, Entry 18976 (1986).

161.204 **Creation–field cosmology: a possible solution to singularity, horizon, and flatness problems.**
J. V. Narlikar, T. Padmanabhan.
Phys. Rev. D, Vol. 32, No. 8, p. 1928 – 1934 (1985). Abstr. in Phys. Abstr., Vol. 89, No. 1273, Entry 18977 (1986).

161.205 **Entropy production due to gravitational–wave viscosity in a Kaluza–Klein inflationary universe.**
K. Tomita, H. Ishihara.
Phys. Rev. D, Vol. 32, No. 8, p. 1935 – 1941 (1985). Abstr. in Phys. Abstr., Vol. 89, No. 1273, Entry 18978 (1986).

161.206 **Quantum states and the Hadamard form. III. Constraints in cosmological space–times.**
A. H. Najmi, A. C. Ottewill.
Phys. Rev. D, Vol. 32, No. 8, p. 1942 – 1948 (1985). Abstr. in Phys. Abstr., Vol. 89, No. 1273, Entry 18979 (1986).

161.207 **Baryogenesis in the inflationary universe.**
N. Sato, H. Kodama, K. Sato.
Prog. Theor. Phys., Vol. 74, No. 2, p. 405 – 408 (1985). Abstr. in Phys. Abstr., Vol. 89, No. 1274, Entry 19452 (1986).

161.208 **The Universe on a large scale. The large structures and their formation.**
A. Mazure.
Ann. Phys. (Paris), Vol. 10, No. 5, p. 415 – 473 (1985). Abstr. in Phys. Abstr., Vol. 89, No. 1274, Entry 23915 (1986).

161.209 **Quark–hadron phase transition and thermodynamics of anisotropic cosmology.**
E. Giannetto, P. Salucci.
Phys. Lett. A, Vol. 112A, No. 8, p. 374 – 376 (1985). Abstr. in Phys. Abstr., Vol. 89, No. 1274, Entry 23919 (1986).

161.210 **Tunneling transitions with gravitation: breakdown of the quasiclassical approximation.**
G. V. Lavrelashvili, V. A. Rubakov, P. G. Tinyakov.
Phys. Lett. B, Vol. 161B, No. 4 – 6, p. 280 – 284 (1985). Abstr. in Phys. Abstr., Vol. 89, No. 1274, Entry 23920 (1986).

161.211 **Nucleosynthesis problems for string models of galaxy formation.**
R. L. Davis.
Phys. Lett. B, Vol. 161B, No. 4 – 6, p. 285 – 288 (1985). Abstr. in Phys. Abstr., Vol. 89, No. 1274, Entry 23921 (1986).

161.212 **Nonsingular cosmological model with torsion induced by vacuum quantum effects.**
I. L. Buchbinder, S. D. Odintsov, I. L. Shapiro.
Phys. Lett. B, Vol. 162B, No. 1 – 3, p. 92 – 96 (1985). Abstr. in Phys. Abstr., Vol. 89, No. 1274, Entry 23922 (1986).

161.213 **Evolution of the Universe.**
G. Swarup.
Proc. Indian Natl. Sci. Acad., Part A, Vol. 51, No. 2, p. 295 – 315 (1985). Abstr. in Phys. Abstr., Vol. 89, No. 1274, Entry 23923 (1986).

161.214 **Towards vacuum structure.**
J. Lopez de Lerma.
Speculations Sci. Technol., Vol. 8, No. 4, p. 241 – 255 (1985). Abstr. in Phys. Abstr., Vol. 89, No. 1274, Entry 23924 (1986).

161.215 **Rotational curves of galaxies and neutrino halos.**
R. Paganini, N. Straumann, D. Wyler.
MPA Rep., No. 231, 21 pp. (1986).
Assuming that massive neutrinos dominate the dark matter in galactic halos, the authors calculate the mass radius relations and rotational curves for an isothermal neutrino distribution as well as for a distribution with a spatial cutoff (King model). For $m_\nu \gtrsim 40$ eV, galaxies between 10 and 100 kpc radius can be reproduced with satisfactory (flat) rotational curves in the isothermal model.

161.216 **The Hubble relation: differences between galaxy types Sb and Sc.**
H. Arp.
MPA Rep., No. 236, 51 pp. (1986). Submitted to Astron. J.

161.217 **Giant H II regions as distance indicators I: relations between global parameters for the local calibrators.**
J. Melnick, M. Moles, R. Terlevich, J.–M. Garcia–Pelayo.
ESO Sci. Prepr., No. 440, 51 pp. (1986). To appear in Mon. Not. R. Astron. Soc.

161.218 **Cosmic background radiation temperature from CN absorption.**
P. Crane, D. J. Hegyi, N. Mandolesi, A. C. Danks.
ESO Sci. Prepr., No. 447, 23 pp. (1986). To appear in Astrophys. J.

161.219 **Uncertainty principle limits on the cosmological constant.**
C. Sivaram.
Indian Inst. Astrophys., Prepr., No. 211, 8 pp. (1986). Submitted to Int. J. Theor. Phys.

161.220 **"Relicts" of the early universe and secondary ionization of pregalactic matter.**
E. I. Dorosheva, P. D. Nasel'skij.
Astrofizika, Tom 24, Vyp. 3, p. 561 – 577 (1986). In Russian. English translation in Astrophysics, Vol. 24, No. 3.
Dynamics of secondary ionization in models with a background of relict gravitating particles is investigated. High–energy quanta created by decays of massive unstable particles or primary black holes are assumed to be the source of hydrogen ionization. It has been shown that distortion of recombination dynamics occurs only if the lifetime of the particles is greater than 3×10^{13}s. It is stated that the rate of change of the Thomson optical plasma depth depends on the ionizing power and its lifetime. It is concluded that the nonequilibrium hydrogen ionization at redshifts $z < 10^3$ is an effective mechanism of suppression of the small–scale anisotropy of the relict radiation.

161.221 **Infrared background radiation generated by population III stars.**
Yu. A. Shchekinov.
Astrofizika, Tom 24, Vyp. 3, p. 579 – 590 (1986). In Russian. English translation in Astrophysics, Vol. 24, No. 3.
The spectrum of infrared background radiation produced by primordial stars is calculated. It has been shown that the IR flux is of the order of 50 Jy/sr at wavelengths $\lambda = 15 - 30$ µm and 200 Jy/sr at $\lambda = 150 - 300$ µm, provided these stars have enriched prepopulation II with metals $Z \lesssim 10^{-4}$. The flux of the nearby IR background ($\lambda = 2 - 5$ µm) generated by small massive stars ($M = 0.1 - 1\ M_\odot$) and massive stars ($M = 60\ M_\odot$) is estimated. It has been shown that this flux is too small compared with the observable one.

161.222 **Galaxy formation by mock gravity.**
C. J. Hogan, S. D. M. White.
Prepr. Steward Obs., No. 620, 21 pp. (1986). Submitted to Nature.

161.223 **Runaway cosmic strings.**
C. J. Hogan.
Prepr. Steward Obs., No. 662, 9 pp. (1986). Submitted to Phys. Rev. Lett.

161.224 **Clusters, filaments and voids in a universe dominated by cold dark matter.**
S. D. M. White, C. S. Frenk, M. Davis, G. Efstathiou.
Prepr. Steward Obs., No. 666, 31 pp. (1986). To appear in Astrophys. J.

161.225 **Homogeneous viscous universes with magnetic field. I. Basic equations.**
W. A. van Leeuwen, G. A. Q. Salvati.
Ann. Phys. (N.Y.), Vol. 165, No. 1, p. 214 – 236 (1985). Abstr. in Phys. Abstr., Vol. 89, No. 1275, Entry 24190 (1986).

161.226 **Non–oscillatory behaviour in vacuum Kaluza–Klein cosmologies.**
J. Demaret, M. Henneaux, P. Spindel.
Phys. Lett. B, Vol. 164B, No. 1 – 3, p. 27 – 30 (1985). Abstr. in Phys. Abstr., Vol. 89, No. 1275, Entry 24200 (1986).

161.227 **Path–integral quantum cosmology. I. Vacuum Bianchi type I.**
B. K. Berger, C. N. Vogeli.
Phys. Rev. D, Vol. 32, No. 10, p. 2477 – 2484 (1985). Abstr. in Phys. Abstr., Vol. 89, No. 1275, Entry 24201 (1986).

161.228 **Path–integral quantum cosmology. II. Bianchi type I with volume–dependent source.**
B. K. Berger.
Phys. Rev. D, Vol. 32, No. 10, p. 2485 – 2488 (1985). Abstr. in Phys. Abstr., Vol. 89, No. 1275, Entry 24202 (1986).

161.229 **Expansion isotropization during the inflationary era.**
O. Gron.
Phys. Rev. D, Vol. 32, No. 10, p. 2522 – 2527 (1985). Abstr. in Phys. Abstr., Vol. 89, No. 1275, Entry 24203 (1986).

161.230 **Quantum effect on Kaluza–Klein cosmologies: high–temperature case.**
K. Maeda.
Phys. Rev. D, Vol. 32, No. 10, p. 2528 – 2538 (1985). Abstr. in Phys. Abstr., Vol. 89, No. 1275, Entry 24204 (1986).

161.231 **Wave packet in quantum cosmology.**
Y. Kazama, R. Nakayama.
Phys. Rev. D, Vol. 32, No. 10, p. 2500 – 2510 (1985). Abstr. in Phys. Abstr., Vol. 89, No. 1275, Entry 24217 (1986).

161.232 **Classical and quantum cosmology of the Starobinsky inflationary model.**
A. Vilenkin.
Phys. Rev. D, Vol. 32, No. 10, p. 2511 – 2521 (1985). Abstr. in Phys. Abstr., Vol. 89, No. 1275, Entry 24218 (1986).

161.233 **Arrow of time in cosmology.**
S. W. Hawking.
Phys. Rev. D, Vol. 32, No. 10, p. 2489 – 2495 (1985). Abstr. in Phys. Abstr., Vol. 89, No. 1275, Entry 24588 (1986).

161.234 **Will entropy decrease if the Universe recollapses?**
D. N. Page.
Phys. Rev. D, Vol. 32, No. 10, p. 2496 – 2499 (1985). Abstr. in Phys. Abstr., Vol. 89, No. 1275, Entry 24589 (1986).

161.235 **Hydrodynamic fluctuations in the matter dominated era before hydrogen recombination.**
W. Zimdahl.
Exp. Tech. Phys., Vol. 33, No. 5, p. 397 – 402 (1985). Abstr. in Phys. Abstr., Vol. 89, No. 1275, Entry 30487 (1986).

161.236 **Expanding Universe solutions for the Yang–Mills–Higgs system coupled to the Poincaré gauge theory of gravitation.**
H.–H. Chen, R.–R. Hsu, W. B. Yeung.
Classical Quantum Gravity, Vol. 2, No. 6, p. 927 – 934 (1985). Abstr. in Phys. Abstr., Vol. 89, No. 1275, Entry 30494 (1986).

161.237 **Inhomogeneous cosmologies: the cosmic peeling–off property of gravity.**
M. Carmeli, A. Feinstein.
Int. J. Theor. Phys., Vol. 24, No. 10, p. 1009 – 1017 (1985). Abstr. in Phys. Abstr., Vol. 89, No. 1275, Entry 30495 (1986).

161.238 **On Komar integrals in asymptotically anti–de Sitter space–times.**
A. Magnon.
J. Math. Phys., Vol. 26, No. 12, p. 3112 – 3117 (1985). Abstr. in Phys. Abstr., Vol. 89, No. 1275, Entry 30498 (1986).

161.239 **Anisotropic cosmological model in Nordtvedt's scalar–tensor theory of gravitation.**
A. Banerjee, N. Banerjee, N. O. Santos.
J. Math. Phys., Vol. 26, No. 12, p. 3125 – 3130 (1985). Abstr. in Phys. Abstr., Vol. 89, No. 1275, Entry 30500 (1986).

161.240 **A statistical mechanical approach to mass density fluctuations in the early universe.**
S. Sen.
Lett. Nuovo Cimento, Vol. 44, Ser. 2, No. 5, p. 270 – 274 (1985). Abstr. in Phys. Abstr., Vol. 89, No. 1275, Entry 30501 (1986).

161.241 **Clustering in an Einstein–Straus Universe. A Newtonian approach.**
S. Filippi.
Nuovo Cimento B, Vol. 90B, Ser. 11, No. 1, p. 1 – 14 (1985). Abstr. in Phys. Abstr., Vol. 89, No. 1275, Entry 30502 (1986).

161.242 **Five–dimensional cosmological models with massless scalar fields.**
J. D. Gegenberg, A. Das.
Phys. Lett. A, Vol. 112A, No. 9, p. 427 – 430 (1985). Abstr. in Phys. Abstr., Vol. 89, No. 1275, Entry 30506 (1986).

161.243 **Initial conditions for inflation.**
A. D. Linde.
Phys. Lett. B, Vol. 162B, No. 4 – 6, p. 281 – 286 (1985). Abstr. in Phys. Abstr., Vol. 89, No. 1275, Entry 30507 (1986).

161.244 **Inflation and broken Lorentz symmetry in the very early Universe.**
M. Gasperini.
Phys. Lett. B, Vol. 163B, No. 1 – 4, p. 84 – 86 (1985). Abstr. in Phys. Abstr., Vol. 89, No. 1275, Entry 30508 (1986).

161.245 **Avoiding the secondary magnetic monopole problem in the inflation theories: the 75 of SU(5).**
C. W. Kim, J. E. Kim, J. S. Kim.
Phys. Lett. B, Vol. 163B, No. 1 – 4, p. 87 – 90 (1985). Abstr. in Phys. Abstr., Vol. 89, No. 1275, Entry 30509 (1986).

161.246 **Topologically massive planar universe.**
I. Vuorio.
Phys. Lett. B, Vol. 163B, No. 1 – 4, p. 91 – 95 (1985). Abstr. in Phys. Abstr., Vol. 89, No. 1275, Entry 30510 (1986).

161.247 **The distortions of the microwave background radiation, heavy neutral decaying leptons and decaying gravitinos.**
P. Salati.
Phys. Lett. B, Vol. 163B, No. 1 – 4, p. 236 – 242 (1985). Abstr. in Phys. Abstr., Vol. 89, No. 1275, Entry 30511 (1986).

161.248 **Inflation in universes with a massive scalar field.**
T. Piran, R. M. Williams.
Phys. Lett. B, Vol. 163B, No. 5 – 6, p. 331 – 335 (1985). Abstr. in Phys. Abstr., Vol. 89, No. 1275, Entry 30512 (1986).

161.249 **Comments on rotational perturbations of Friedmann models.**
S. S. Bayin.
Phys. Rev. D, Vol. 32, No. 8, p. 2241 – 2242 (1985). Abstr. in Phys. Abstr., Vol. 89, No. 1275, Entry 30513 (1986).

161.250 **Technocosmology – could a technibaryon excess provide a 'natural' missing mass candidate?**
S. Nussinov.
Phys. Lett. B, Vol. 165B, No. 1 – 3, p. 55 – 58 (1985). Abstr. in Phys. Abstr., Vol. 89, No. 1276, Entry 31090 (1986).

161.251 **Bounds on the neutrino mass from the dark matter in the Universe.**
C. Castagnoli, P. Galeotti.
Nuovo Cimento C, Vol. 8C, Ser. 1, No. 4, p. 450 – 460 (1985). Abstr. in Phys. Abstr., Vol. 89, No. 1276, Entry 35966 (1986).

161.252 **On large–scale quantization.**
J. C. Carvalho.
Lett. Nuovo Cimento, Vol. 44, Ser. 2, No. 6, p. 337 – 342 (1985). Abstr. in Phys. Abstr., Vol. 89, No. 1276, Entry 36327 (1986).

161.253 **Diffuse galactic and extragalactic radiation in the far infra–red.**
C. Ceccarelli, G. Dall'Oglio, P. De Bernardis, S. Masi, B. Melchiorri, F. Melchiorri, G. Moreno, L. Pietranera.
Nuovo Cimento C, Vol. 8C, Ser. 1, No. 4, p. 353 – 367 (1985). Abstr. in Phys. Abstr., Vol. 89, No. 1276, Entry 36372 (1986).

161.254 **On detecting diffuse ultraviolet radiation of cosmological origin.**
F. Paresce.
Nuovo Cimento C, Vol. 8C, Ser. 1, No. 4, p. 379 – 396 (1985). Abstr. in Phys. Abstr., Vol. 89, No. 1276, Entry 36373 (1986).

161.255 **Quantum cosmology in 2 + 1 and 3 + 1 dimensions.**
T. Banks, W. Fischler, L. Susskind.
Nucl. Phys. B, Part. Phys., Vol. B262, No. 1, p. 159 – 186 (1985). Abstr. in Phys. Abstr., Vol. 89, No. 1276, Entry 36385 (1986).

161.256 **Entropy production in an expanding universe.**
M. Morikawa, M. Sasaki.
Phys. Lett. B, Vol. 165B, No. 1 – 3, p. 59 – 62 (1985). Abstr. in Phys. Abstr., Vol. 89, No. 1276, Entry 36386 (1986).

161.257 **Towards higher–dimensional homogeneous cosmologies.**
H. Ishihara.
Prog. Theor. Phys., Vol. 74, No. 3, p. 490 – 500 (1985). Abstr. in Phys. Abstr., Vol. 89, No. 1276, Entry 36387 (1986).

161.258 **An independent test of biased galaxy formation with cold particles: the three–point function.**
A. L. Melott, J. N. Fry.
Astrophys. J., Vol. 305, No. 1, p. 1 – 8 (1986).
Identification of only high peaks of mass density with visible galaxies pushes up the amplitude of the two–point correlation of critical density cold–particle universe simulations into better agreement with the observed distribution of galaxies. The three–point function, or bispectrum in transform space, permits a fully independent test of this hypothesis. In a formulation of biasing which can treat a connected density maximum as more than a single galaxy, the authors find that sufficient amplification of the underlying density field is achieved with a 1 σ bias, and the value

of the three–point correlation is $Q \approx 1$, over length scales $1(\Omega h^2)^{-1}$Mpc $< r < 25(\Omega h^2)^{-1}$Mpc.

161.259 **The decaying particles scenario and the Virgocentric infall.**
Y. Hoffman.
Astrophys. J., Lett. Ed., Vol. 305, No. 1, p. L1 – L4 (1986).
The dynamics of Local Supercluster–like objects is studied for the case of a flat universe whose mass–energy density is dominated by relativistic remnants of some massive, unstable, and weakly interacting particles. Recent analytical, quasi–linear calculations have been employed to derive an apparent value of Ω_0. The decaying particles scenario is found to be consistent with current observations of the Local Supercluster.

161.260 **Effect of the cosmological constant on large–scale anisotropies in the microwave background.**
L. A. Kofman, A. A. Starobinskij.
Sov. Astron. Lett., Vol. 11, No. 5, p. 271 – 274 (1985). English translation of 40.161.099.

161.261 **Small–scale microwave–background anisotropies in a universe dominated by nonrelativistic particles.**
A. G. Doroshkevich.
Sov. Astron. Lett., Vol. 11, No. 5, p. 304 – 307 (1985). English translation of 40.161.133.

161.262 **Repulsive gravitation and inflationary universe models.**
O. Gron.
Am. J. Phys., Vol. 54, No. 1, p. 46 – 52 (1986). Abstr. in Phys. Abstr., Vol. 89, No. 1277, Entry 36439 (1986).

161.263 **A class of algebraically general inhomogeneous cosmologies.**
D. Papadopoulos, J. L. Sanz.
Nuovo Cimento B, Vol. 90B, Ser. 11, No. 2, p. 191 – 203 (1985). Abstr. in Phys. Abstr., Vol. 89, No. 1277, Entry 36602 (1986).

161.264 **Cosmological solutions with Calabi–Yau compactification.**
K. Maeda.
Phys. Lett. B, Vol. 166B, No. 1, p. 59 – 64 (1986). Abstr. in Phys. Abstr., Vol. 89, No. 1277, Entry 36643 (1986).

161.265 **Goldstone bosons in string models of galaxy formation.**
R. L. Davis.
Phys. Rev. D, Vol. 32, No. 12, p. 3172 – 3177 (1985). Abstr. in Phys. Abstr., Vol. 89, No. 1277, Entry 42343 (1986).

161.266 **A (4+1)–dimensional cosmological model.**
J.–m. Zhu, S.–c. Zhu, H.–y. Guo.
Acta Phys. Sin., Vol. 34, No. 10, p. 1342 – 1347 (1985). In Chinese. Abstr. in Phys. Abstr., Vol. 89, No. 1277, Entry 42359 (1986).

161.267 **Cosmological models with both radiation and matter.**
A. Coley.
Can. J. Phys., Vol. 63, No. 9, p. 1199 – 1200 (1985). Abstr. in Phys. Abstr., Vol. 89, No. 1277, Entry 42363 (1986).

161.268 **The big bang and quantum cosmology.**
J. V. Narlikar.
Curr. Sci., Vol. 54, No. 17, p. 825 – 831 (1985). Abstr. in Phys. Abstr., Vol. 89, No. 1277, Entry 42364 (1986).

161.269 **An introduction to subquantum kinetics. III. The cosmology of subquantum kinetics.**
P. A. Laviolette.
Int. J. Gen. Syst., Vol. 11, No. 4, p. 329 – 345 (1985). Abstr. in Phys. Abstr., Vol. 89, No. 1277, Entry 42365 (1986).

161.270 **The cosmological red–shift can only be due to the universe expansion and the galaxies recession velocities are given by the nonrelativistic Doppler formula. 1.**
G. Cavalleri, S. Filippi.
Lett. Nuovo Cimento, Vol. 44, Ser. 2, No. 7, p. 519 – 522 (1985). Abstr. in Phys. Abstr., Vol. 89, No. 1277, Entry 42366 (1986).

161.271 **Cosmology and the large–mass problem of the five–dimensional Kaluza–Klein theory.**
B. Lukacs, T. Pacher.
Phys. Lett. A, Vol. 113A, No. 4, p. 200 – 202 (1985). Abstr. in Phys. Abstr., Vol. 89, No. 1277, Entry 42367 (1986).

161.272 **Kinematical properties of generalized inflation.**
F. Lucchin, S. Matarrese.
Phys. Lett. B, Vol. 164B, No. 4 – 6, p. 282 – 286 (1985). Abstr. in Phys. Abstr., Vol. 89, No. 1277, Entry 42368 (1986).

161.273 **Termination of the GUT phase transition in the new inflationary universe.**
M. Sakagami, A. Hosoya.
Phys. Lett. B, Vol. 165B, No. 4 – 6, p. 315 – 320 (1985). Abstr. in Phys. Abstr., Vol. 89, No. 1277, Entry 42370 (1986).

161.274 **Large–scale anisotropies of the cosmic background radiation in generalized inflationary cosmologies.**
R. Fabbri, F. Lucchin, S. Matarrese.
Phys. Lett. B, Vol. 166B, No. 1, p. 49 – 53 (1986). Abstr. in Phys. Abstr., Vol. 89, No. 1277, Entry 42371 (1986).

161.275 **Bianchi type–I model with cosmological constant in a generalized scalar–tensor theory of gravitation.**
A. Banerjee, S. B. Duttachoudhury, N. Banerjee.
Phys. Rev. D, Vol. 32, No. 12, p. 3096 – 3100 (1985). Abstr. in Phys. Abstr., Vol. 89, No. 1277, Entry 42372 (1986).

161.276 **Three–dimensional closed universes without collapse in a five–dimensional Kaluza–Klein theory.**
R. A. Matzner, A. Mezzacappa.
Phys. Rev. D, Vol. 32, No. 12, p. 3114 – 3117 (1985). Abstr. in Phys. Abstr., Vol. 89, No. 1277, Entry 42373 (1986).

161.277 **Quantum particle production in an expanding universe and Kaluza–Klein theories.**
P. Gribosky.
Phys. Rev. D, Vol. 32, No. 12, p. 3118 – 3123 (1985). Abstr. in Phys. Abstr., Vol. 89, No. 1277, Entry 42374 (1986).

161.278 **Gödel cosmology in Riemann–Cartan spacetime with spin density.**
L. L. Smalley.
Phys. Rev. D, Vol. 32, No. 12, p. 3124 – 3127 (1985). Abstr. in Phys. Abstr., Vol. 89, No. 1277, Entry 42375 (1986).

161.279 **Vacuum states in de Sitter space.**
B. Allen.
Phys. Rev. D, Vol. 32, No. 12, p. 3136 – 3149 (1985). Abstr. in Phys. Abstr., Vol. 89, No. 1277, Entry 42376 (1986).

161.280 **"Isothermal" density perturbations in an axion–dominated inflationary universe.**
D. Seckel, M. S. Turner.
Phys. Rev. D, Vol. 32, No. 12, p. 3178 – 3183 (1985). Abstr. in Phys. Abstr., Vol. 89, No. 1277, Entry 42377 (1986).

161.281 **Cosmological heavy–neutrino problem.**
J. Bernstein, L. S. Brown, G. Feinberg.
Phys. Rev. D, Vol. 32, No. 12, p. 3261 – 3267 (1985). Abstr. in Phys. Abstr., Vol. 89, No. 1277, Entry 42378 (1986).

161.282 **Minimum amount of inflation to solve the flatness and horizon problems.**
C. W. Kim, P. Murphy.
Phys. Rev. D, Vol. 32, No. 12, p. 3303 – 3306 (1985). Abstr. in Phys. Abstr., Vol. 89, No. 1277, Entry 42379 (1986).

161.283 **The inflationary Universe.**
E. J. Weinberg.
Recent developments in quantum field theory, p. 169 – 184 (1985). Abstr. in Phys. Abstr., Vol. 89, No. 1277, Entry 42383 (1986). – See Abstr. 012.047.

161.284 **The extra–galactic background light: a modern version of Olber's paradox. (Or: Why the space between galaxies is dark).**
P. S. Wesson.
Space Sci. Rev., Vol. 44, Nos. 1/2, p. 169 – 176 (1986).
A non–technical discussion is given of the energy density E of the extra–galactic background light. The fact that E is small means that the space between galaxies is dark, which is a modern version of a classical problem in astronomy known as Olbers' paradox. It is seen that the order of magnitude of E is fixed by the order of magnitude of the lifetime of the galaxies, as pointed out by Harrison; but that the expansion of the universe can affect E by a smaller factor, typically about 2. These comments should help to end persistent confusion about the effects of the lifetime of the galaxies and the expansion of the universe on the darkness of the night sky. It is hoped that Olbers' so–called paradox can now rest in peace.

161.285 **On the uniqueness of the Einstein–de Sitter cosmology in spinor–tetrad theory.**
J. T. Lynch.
Classical Quantum Gravity, Vol. 3, No. 1, p. 103 – 116 (1986). Abstr. in Phys. Abstr., Vol. 89, No. 1278, Entry 42582 (1986).

161.286 **Thermodynamic stability of cosmological models.**
G. Horwitz, D. Weil.
Phys. Rev. D, Vol. 33, No. 2, p. 567 – 569 (1986). Abstr. in Phys. Abstr., Vol. 89, No. 1278, Entry 42587 (1986).

161.287 **String–dominated universe.**
T. W. B. Kibble.
Phys. Rev. D, Vol. 33, No. 2, p. 328 – 332 (1986). Abstr. in Phys. Abstr., Vol. 89, No. 1278, Entry 42604 (1986).

161.288 **A mechanism for baryogenesis in supersymmetric inflationary cosmologies.**
S. Mahajan.
Phys. Rev. D, Vol. 33, No. 2, p. 338 – 345 (1986). Abstr. in Phys. Abstr., Vol. 89, No. 1278, Entry 42606 (1986).

161.289 **Gauge theories, geometry, and cosmology.**
C. A. Lucey.
Phys. Rev. D, Vol. 33, No. 2, p. 346 – 353 (1986). Abstr. in Phys. Abstr., Vol. 89, No. 1278, Entry 42607 (1986).

161.290 **Behaviour of homogeneous five–dimensional space–times.**
P. Halpern.
Phys. Rev. D, Vol. 33, No. 2, p. 354 – 362 (1986). Abstr. in Phys. Abstr., Vol. 89, No. 1278, Entry 42608 (1986).

161.291 **Time variation of coupling constants in Kaluza–Klein cosmologies reexamined.**
M. Gleiser, J. G. Taylor.
Phys. Rev. D, Vol. 33, No. 2, p. 570 – 571 (1986). Abstr. in Phys. Abstr., Vol. 89, No. 1278, Entry 42612 (1986).

161.292 **Gravitational solitons and spatial flatness.**
A. Feinstein, C. Charach.
Classical Quantum Gravity, Vol. 3, No. 1, p. L5 – L8 (1986). Abstr. in Phys. Abstr., Vol. 89, No. 1278, Entry 48105 (1986).

161.293 **The application of the homogeneous Hilbert problem of Hauser and Ernst to cosmological models with spatial axes of symmetry.**
D. W. Kitchingham.
Classical Quantum Gravity, Vol. 3, No. 1, p. 133 – 153 (1986). Abstr. in Phys. Abstr., Vol. 89, No. 1278, Entry 48106 (1986).

161.294 **Some anisotropic universes and relativistic cosmologies.**
M. Bray.
Ann. Fac. Sci. Toulouse Math., Vol. 7, Ser. 5, No. 1, p. 75 – 85 (1985). In French. Abstr. in Phys. Abstr., Vol. 89, No. 1279, Entry 48270 (1986).

161.295 **Two–fluid cosmological models.**
A. A. Coley, B. O. J. Tupper.
J. Math. Phys., Vol. 27, No. 1, p. 406 – 416 (1986). Abstr. in Phys. Abstr., Vol. 89, No. 1279, Entry 48277 (1986).

161.296 **Does thermodynamics require a new expansion after the "big crunch" of our cosmos?**
E. Recami, V. Tonin–Zanchin.
Report 343, Univ. Estadual de Campinas, Brazil, 13 pp. (1985). Abstr. in Phys. Abstr., Vol. 89, No. 1279, Entry 53743 (1986).

161.297 **Maximal and minimal acceleration in quantum mechanics and in cosmic quantum mechanics.**
C. Massa.
Lett. Nuovo Cimento, Vol. 44, Ser. 2, No. 8, p. 609 – 611 (1985). Abstr. in Phys. Abstr., Vol. 89, No. 1279, Entry 53987 (1986).

161.298 **Bulk viscosity and the inflationary Universe.**
G. Papini, M. Weiss.
Lett. Nuovo Cimento, Vol. 44, Ser. 2, No. 8, p. 612 – 616 (1985). Abstr. in Phys. Abstr., Vol. 89, No. 1279, Entry 53988 (1986).

161.299 **A model of the Universe. I. The theory of chaotons.**
L. H. Reis.
Lett. Nuovo Cimento, Vol. 44, Ser. 2, No. 8, p. 637 – 640 (1985). Abstr. in Phys. Abstr., Vol. 89, No. 1279, Entry 53989 (1986).

161.300 **Cosmic ray antimatter from supersymmetric dark matter.**
J. S. Hagelin, G. L. Kane.
Nucl. Phys. B, Part. Phys., Vol. B263, No. 2, p. 399 – 412 (1986). Abstr. in Phys. Abstr., Vol. 89, No. 1279, Entry 53995 (1986).

161.301 **Models for inflation with a low supersymmetry–breaking scale.**
P. Bienetruy, S. Mahajan.
Nucl. Phys. B, Part. Phys., Vol. B263, No. 2, p. 413 – 432 (1986). Abstr. in Phys. Abstr., Vol. 89, No. 1279, Entry 53996 (1986).

161.302 **Self–consistent Gödel cosmology with spin–density in Riemann–Cartan spacetime.**
L. L. Smalley.
Phys. Lett. A, Vol. 113A, No. 9, p. 463 – 466 (1986). Abstr. in Phys. Abstr., Vol. 89, No. 1279, Entry 53997 (1986).

161.303 **The strong *CP* problem in a compact Robertson–Walker universe.**
K. Cahill.
Z. Phys., C, Vol. 29, No. 4, p. 537 – 539 (1985). Abstr. in Phys. Abstr., Vol. 89, No. 1279, Entry 54000 (1986).

161.304 **Anisotropy of the cosmic blackbody radiation.**
D. T. Wilkinson.
Science, Vol. 232, No. 4757, p. 1517 – 1522 (1986).
The universe is filled with thermal radiation having a current temperature of 2.75K. Originating in the very early universe, this radiation furnishes strong evidence that the Big Bang cosmology best describes our expanding universe. The model can be used to extrapolate our physics backward in time to predict events whose effects might be observable in the 2.75K radiation today. The spectrum and isotropy are being studied with sophisticated microwave radiometers on the ground, in balloons, and in satellites. The results are as predicted by the simple theory. However, a problem is emerging. Primordial fluctuations in the mass density, which later became the great clusters of galaxies that we see today, should have left an imprint on the 2.75K radiation–bumpiness on the sky at angular scales of about 10 arc minutes. They have not yet been seen.

161.305 **Time, vacuum and cosmos.**
W. McCrea.
Q. J. R. Astron. Soc., Vol. 27, No. 2, p. 137 – 152 (1986).
The author considers quantum transitions and their significance; special relativity in this context; the Casimir effect as a fulfilled prediction from these ideas. Then he extends such ideas to particles other than photons, and thence to the physics of time–keeping. General relativity must be mentioned as a prototype field theory. Then he looks at photon waves and electron waves in the vacuum and at the significance of all this for physics in the large – for the cosmos.

161.306 **The age of the universe, primordial galaxies, and Space Telescope.**
A. Renzini, A. Buzzoni, F. Fusi Pecci.
Mem. Soc. Astron. Ital., Vol. 56, No. 4, p. 637 – 644 (1985). – See Abstr. 012.054.
Theoretical aspects concerning (1) the age of galactic globular clusters and (2) the evolution of stellar populations in high redshift galaxies are examined with a view on possible observations with the Space Telescope.

161.307 **Dark matter in the universe.**
R. Sancisi.
Mem. Soc. Astron. Ital., Vol. 56, No. 4, p. 709 – 710 (1985). – See Abstr. 012.054.
This note briefly summarizes the evidence for dark matter galactic halos derived from rotation curves of spiral galaxies.

161.308 **A new test for the cosmological evolution of extragalactic sources.**
L. Danese, G. de Zotti, A. Franceschini.
Mem. Soc. Astron. Ital., Vol. 56, No. 4, p. 763 – 768 (1985). – See Abstr. 012.054.
The authors describe a new test for the cosmological evolution of extragalactic sources, which appears to be particularly well suited for probing the existence of a redshift cutoff.

161.309 **Supersymmetry and cosmology.**
J. Ellis.
Relativity, supersymmetry and cosmology, p. 81 – 129 (1985). – See Abstr. 012.055.
This paper reviews cosmological aspects of supersymmetric theories. It discusses the inflationary phase of the universe, baryosynthesis, the magnetic monopole problem, and various relic particles from the supersymmetric phase, that may account for the dark matter in the universe.

161.310 **Self–consistent cosmogenesis.**
E. Gunzig.
Relativity, supersymmetry and cosmology, p. 181 – 212 (1985). – See Abstr. 012.055.
The cosmological history owes its origin to a phase transition mechanism due to the puzzling nature of the semi–classical gravity. It appears indeed that the existence of our universe is unavoidable: it represents a necessary response of the quantum behaviour of matter fields in the presence of classical gravitational interaction. The latter induces a profound alteration of the traditional Minkowski ground state therefore provoking a spontaneously broken symmetry breakdown mechanism which forces a phase transition between the traditional empty flat Minkowski space and the true ground state: the inflationary universe.

161.311 **Toy cosmology: a laboratory to study cosmological puzzles.**
R. Ferraro, M. Castagnino.
Relativity, supersymmetry and cosmology, p. 238 – 242 (1985). – See Abstr. 012.055.
The authors briefly describe their "toy cosmology" which is based on a new definition of a vacuum depending on the applied reference system.

161.312 **De Sitter self consistent cosmology with adiabatic regularization.**
J. P. Paz, D. D. Harari, M. A. Castagnino.
Relativity, supersymmetry and cosmology, p. 243 – 247 (1985). – See Abstr. 012.055.
The relation between the scalar curvature and the mass of the quantum scalar free field which acts as the source of Einstein eqs. is analyzed for the de Sitter space, which is a solution of the back–reaction problem.

161.313 **On the meaning of quantum cosmology.**
M. P. Ryan Jr.
Relativity, supersymmetry and cosmology, p. 296 – 307 (1985). – See Abstr. 012.055.

161.314 **Weyl (integrable) spacetimes as a model for our cosmos.**
L. A. R. Oliveira.
Relativity, supersymmetry and cosmology, p. 340 – 345 (1985). – See Abstr. 012.055.
The author presents a discussion of Weyl (integrable) spacetimes as a viable model for our cosmos, and of the physical implications for systems non–minimally coupled to gravity.

161.315 **The primordial helium abundance determination from metal–poor galaxies.**
D. Kunth.
Inst. Astrophys. Paris, Pré–Publ., No. 143, 13 pp. (1986). To appear in Publ. Astron Soc. Pac.

161.316 **Topics in observational cosmology.**
M. J. Kurtz.
Bull. Inf. Cent. Données Stellaires, No. 30, p. 171 – 173 (1986). = Newsl. Working Group Mod. Astron. Methodol., No. 2.

161.317 **Cosmic Lorentz transformation.**
M. Carmeli.
Int. J. Theor. Phys., Vol. 25, No. 2, p. 195 – 198 (1986). Abstr. in Phys. Abstr., Vol. 89, No. 1280, Entry 54170 (1986).

161.318 **On the origin of the universe in the context of string models.**
Y. Aharonov, A. Casher.
Phys. Lett. B, Vol. 166B, No. 3, p. 289 – 291 (1986). Abstr. in Phys. Abstr., Vol. 89, No. 1280, Entry 54272 (1986).

161.319 **Primordial shear and the questions of inflation.**
E. Martinez–Gonzalez, B. J. T. Jones.
Phys. Lett. B, Vol. 167B, No. 1, p. 37 – 42 (1986). Abstr. in Phys. Abstr., Vol. 89, No. 1280, Entry 54275 (1986).

161.320 **Pyrgon annihilation in a five–dimensional model universe.**
H. Ishihara, A. Tomimatsu, M. Den.
Prog. Theor. Phys., Vol. 74, No. 5, p. 1045 – 1052 (1985). Abstr. in Phys. Abstr., Vol. 89, No. 1280, Entry 54277 (1986).

161.321 **Operator ordering and the flatness of the Universe.**
S. W. Hawking, D. N. Page.
Nucl. Phys. B, Part. Phys., Vol. B264, No. 1, p. 185 – 196 (1986). Abstr. in Phys. Abstr., Vol. 89, No. 1280, Entry 54284 (1986).

161.322 **Inflation with superstrings.**
P. Oh.
Phys. Lett. B, Vol. 166B, No. 3, p. 292 – 294 (1986). Abstr. in Phys. Abstr., Vol. 89, No. 1280, Entry 54290 (1986).

161.323 **Cosmological constraints on some supergravity–induced low–energy electroweak models.**
G. Girardi, P. Salati.
Nucl. Phys. B, Part. Phys., Vol. B264, No. 4, p. 621 – 639 (1986). Abstr. in Phys. Abstr., Vol. 89, No. 1280, Entry 54696 (1986).

161.324 **Is the Universe made of strings?**
R. Walgate.
Recherche, Vol. 17, No. 173, p. 116 – 118 (1986). In French. Abstr. in Phys. Abstr., Vol. 89, No. 1280, Entry 54711 (1986).

161.325 **Cosmic time and chaos.**
M. Heller, M. Szydlowski, A. Wosczyna.
Acta Phys. Pol., Ser. B, Vol. B17, No. 1, p. 11 – 18 (1986). Abstr. in Phys. Abstr., Vol. 89, No. 1280, Entry 61130 (1986).

161.326 **Stochastic time scale for the Universe.**
M. Szydlowski, M. Heller, Z. Golda.
Acta Phys. Pol., Ser. B, Vol. B17, No. 1, p. 19 – 24 (1986). Abstr. in Phys. Abstr., Vol. 89, No. 1280, Entry 61131 (1986).

161.327 **Exact Bianchi type–I vacuum cosmological model in a scalar–tensor theory of gravitation.**
D. R. K. Reddy, P. Innaiah.
Indian J. Pure Appl. Phys., Vol. 23, No. 12, p. 624 – 625 (1985). Abstr. in Phys. Abstr., Vol. 89, No. 1280, Entry 61138 (1986).

161.328 **The hadronization transition in the early universe.**
G. Baym.
Nucl. Phys. A, Vol. A447, p. 463 – 473 (1986). Abstr. in Phys. Abstr., Vol. 89, No. 1280, Entry 61140 (1986). – See Abstr. 012.057.

161.329 **Evolution of the Kaluza–Klein universe.**
Y. Okada.
Nucl. Phys. B, Part. Phys., Vol. B264, No. 1, p. 197 – 220 (1986). Abstr. in Phys. Abstr., Vol. 89, No. 1280, Entry 61141 (1986).

161.330 **Graviton production in inflationary cosmology.**
L. F. Abbott, D. D. Harari.
Nucl. Phys. B, Part. Phys., Vol. B264, No. 2 – 3, p. 487 – 492 (1986). Abstr. in Phys. Abstr., Vol. 89, No. 1280, Entry 61142 (1986).

161.331 **Baryon excess in a preon model.**
P. K. Mohapatra, J. C. Pati, H. Stremnitzer.
Phys. Lett. B, Vol. 166B, No. 4, p. 407 – 412 (1986). Abstr. in Phys. Abstr., Vol. 89, No. 1280, Entry 61144 (1986).

161.332 **Comments on upper bounds on the inflationary Hubble constant.**
C. W. Kim, P. Murphy.
Phys. Lett. B, Vol. 167B, No. 1, p. 43 – 45 (1986). Abstr. in Phys. Abstr., Vol. 89, No. 1280, Entry 61145 (1986).

161.333 **Higher–dimensional Bianchi cosmologies.**
D. Lorenz–Petzold.
Phys. Lett. B, Vol. 167B, No. 2, p. 157 – 162 (1986). Abstr. in Phys. Abstr., Vol. 89, No. 1280, Entry 61146 (1986).

161.334 **Inflation with a non–minimally coupled scalar field.**
F. Lucchin, S. Matarrese, M. D. Pollock.
Phys. Lett. B, Vol. 167B, No. 2, p. 163 – 168 (1986). Abstr. in Phys. Abstr., Vol. 89, No. 1280, Entry 61147 (1986).

161.335 **Axion cosmology in automatic $E_6 \times U(1)$ models.**
R. Holman, T. W. Kephart.
Phys. Lett. B, Vol. 167B, No. 2, p. 169 – 172 (1986). Abstr. in Phys. Abstr., Vol. 89, No. 1280, Entry 61148 (1986).

161.336 **The stability of some Kaluza–Klein cosmological models.**
J. D. Barrow, J. Stein–Schabes.
Phys. Lett. B, Vol. 167B, No. 2, p. 173 – 177 (1986). Abstr. in Phys. Abstr., Vol. 89, No. 1280, Entry 61149 (1986).

161.337 **Slow rotation of compressible viscous fluid in cosmological universe.**
K. M. Singh, K. S. Bhamra.
Pramāna, Vol. 26, No. 2, p. 117 – 127 (1986). Abstr. in Phys. Abstr., Vol. 89, No. 1280, Entry 61150 (1986).

161.338 **Is the Universe really expanding? II. Red–shift without expansion.**
R. A. Waldron.
Speculations Sci. Technol., Vol. 8, No. 5, p. 315 – 327 (1985). Abstr. in Phys. Abstr., Vol. 89, No. 1280, Entry 61152 (1986).

161.339 **An infrared search for primeval galaxies.**
C. A. Collins, R. D. Joseph.
Rutherford Appleton Lab., Rep., RAL 85–086, p. 148 – 158 (1985). – See Abstr. 012.060.
This paper presents preliminary results from a near IR (*JHK*) search for galaxies at early epochs of the universe ($z \sim 2 - 8$).

161.340 **The Sunyaev–Zeldovich effect at 1.15 mm wavelength.**
S. T. Chase, R. D. Joseph, N. A. Robertson, P. A. R. Ade.
Rutherford Appleton Lab., Rep., RAL 85–086, p. 159 – 170 (1985). – See Abstr. 012.060.
The Sunyaev–Zeldovich (S–Z) effect is a spectral distortion of the cosmic background radiation induced by inverse Compton scattering of the microwave photons by energetic ($\approx$10 keV) electrons in hot astrophysical plasmas. The S–Z effect has considerable potential as a diagnostic tool for observational cosmology and also for the study of the intra–cluster medium in distant clusters of galaxies. The authors report the first ever attempt to detect the S–Z effect at a wavelength (1150 μm) where a flux increment is predicted.

161.341 **Cosmological results from IRAS.**
M. Rowan–Robinson.
Rutherford Appleton Lab., Rep., RAL 85–086, p. 174 – 181 (1985). – See Abstr. 012.060.
IRAS is shown to be a powerful cosmological tool. Results from galactic polar cap source counts at 60 μm and 100 μm are presented and it is shown that the IRAS survey has a characteristic depth of 200 Mpc. The local gravitational field has been mapped using IRAS 60 μm sources and analyzed in spherical harmonics. The dipole component agrees well in direction with that of the cosmic microwave background anisotropy. The 100 μm background radiation has been deconvolved into contributions from interplanetary dust, interstellar dust and a residual, relatively isotropic component, which may be of cosmological origin.

161.342 **Extended thermodynamics in the early Universe.**
B. Lukács, K. Martinás, T. Pacher.
Astron. Nachr., Vol. 307, No. 3, p. 171 – 181 (1986).
It is a general belief that in some early stages of the evolution of the Universe nonequilibrium situations played an important role. In order to incorporate some deviations from thermal equilibrium into the description of the evolution, an extension of the thermodynamic formalism is given, where new terms are introduced into the Gibbs–Duhem relation for representing the deviation. Two situations are investigated in simplified models: the primordial thermalisation in the symmetric phase of GUT, and the development of a nonthermal distribution for massive neutrinos.

161.343 **Polarization and anisotropy induced in the microwave background by cosmological gravitational waves.**
A. G. Polnarev.
Sov. Astron., Vol. 29, No. 6, p. 607 – 613 (1985). English translation of 40.161.307.

161.344 **Statistical evidence from fine–scale anisotropies in the cosmic background radiation.**
N. A. Zabotin, P. D. Nasel'skij.
Sov. Astron., Vol. 29, No. 6, p. 614 – 615 (1985). English translation of 40.161.308.

161.345 **Remarks on Dersarkissian's cosmic quantum mechanics.**
C. Massa.
Lett. Nuovo Cimento, Vol. 44, Ser. 2, No. 8A, p. 671 – 677 (1985). Abstr. in Phys. Abstr., Vol. 89, No. 1281, Entry 61355 (1986).

161.346 **Comment on horizons in Dirac cosmology.**
A. Beesham.
Classical Quantum Gravity, Vol. 3, No. 2, p. L23 – L25 (1986). Abstr. in Phys. Abstr., Vol. 89, No. 1281, Entry 61378 (1986).

161.347 **Stability and attractor in a higher-dimensional cosmology. I.**
K. Maeda.
Classical Quantum Gravity, Vol. 3, No. 2, p. 233 – 247 (1986). Abstr. in Phys. Abstr., Vol. 89, No. 1281, Entry 61403 (1986).

161.348 **The quantum cosmology of Einstein–Maxwell theory in six dimensions.**
J. J. Halliwell.
Nucl. Phys. B, Part. Phys., Vol. B266, No. 1, p. 228 – 244 (1986). Abstr. in Phys. Abstr., Vol. 89, No. 1281, Entry 61406 (1986).

161.349 **The effective early universe in the presence of extra dimensions.**
A. Davidson, J. Sonnenshein, A. H. Vozmediano.
Phys. Lett. B, Vol. 168B, No. 3, p. 183 – 186 (1986). Abstr. in Phys. Abstr., Vol. 89, No. 1281, Entry 61414 (1986).

161.350 **Supersymmetry cosmology with a gauge singlet.**
B. R. Greene, P. J. Miron.
Phys. Lett. B, Vol. 168B, No. 3, p. 226 – 232 (1986). Abstr. in Phys. Abstr., Vol. 89, No. 1281, Entry 61772 (1986).

161.351 **Superunification, phase transitions and cosmology.**
C. E. Vayonakis.
Surv. High Energy Phys., Vol. 5, No. 2, p. 87 – 197 (1986). Abstr. in Phys. Abstr., Vol. 89, No. 1281, Entry 61806 (1986).

161.352 **Analytic solutions of Friedmann equation for spatially opened universes with cosmological constant and radiation pressure.**
M. Dabrowski, J. Stelmach.
Ann. Phys. (N.Y.), Vol. 166, No. 2, p. 422 – 442 (1986). Abstr. in Phys. Abstr., Vol. 89, No. 1281, Entry 67633 (1986).

161.353 **Cosmological models with coherent scalar field oscillations in fourth-order gravity.**
S. Gottlöber, V. Müller.
Classical Quantum Gravity, Vol. 3, No. 2, p. 183 – 192 (1986). Abstr. in Phys. Abstr., Vol. 89, No. 1281, Entry 67636 (1986).

161.354 **Galaxy formation, nonbaryonic matter and cosmic strings.**
D. N. Schramm, N. Vittorio.
Comments Nucl. Part. Phys., Vol. 15, No. 1, p. 1 – 8 (1985). Abstr. in Phys. Abstr., Vol. 89, No. 1281, Entry 67637 (1986).

161.355 **The cosmology/particle physics interface.**
K. A. Olive, D. N. Schramm.
Comments Nucl. Part. Phys., Vol. 15, No. 2, p. 69 – 97 (1985). Abstr. in Phys. Abstr., Vol. 89, No. 1281, Entry 67638 (1986).

161.356 **Penrose diagrams for the Einstein, Eddington–Lemaitre, Eddington–Lemaitre–Bondi, and anti-de-Sitter universes.**
F. J. Tipler.
J. Math. Phys., Vol. 27, No. 2, p. 559 – 561 (1986). Abstr. in Phys. Abstr., Vol. 89, No. 1281, Entry 67639 (1986).

161.357 **Perfect fluid models of Bianchi type-VI_0 in modified Brans–Dicke cosmology.**
S. Ram.
J. Math. Phys., Vol. 27, No. 2, p. 650 – 652 (1986). Abstr. in Phys. Abstr., Vol. 89, No. 1281, Entry 67640 (1986).

161.358 **Maximal acceleration hypothesis and cosmological constraints on particle masses.**
C. Massa.
Lett. Nuovo Cimento, Vol. 44, Ser. 2, No. 8A, p. 694 – 696 (1985). Abstr. in Phys. Abstr., Vol. 89, No. 1281, Entry 67641 (1986).

161.359 **Cosmic nucleosynthesis and nonlinear inhomogeneities.**
J. Centrella, R. A. Matzner, T. Rothman, J. R. Wilson.
Nucl. Phys. B, Part. Phys., Vol. B266, No. 1, p. 171 – 227 (1986). Abstr. in Phys. Abstr., Vol. 89, No. 1281, Entry 67643 (1986).

161.360 **On the creation of an inflationary universe from 'nothing' in a higher-dimensional theory of gravity with higher-derivative terms.**
M. D. Pollock.
Phys. Lett. B, Vol. 167B, No. 3, p. 301 – 306 (1986). Abstr. in Phys. Abstr., Vol. 89, No. 1281, Entry 67644 (1986).

161.361 **Primordial nucleosynthesis, additional neutrinos and neutral currents from the superstring.**
J. Ellis, K. Enqvist, D. V. Nanopoulos, S. Sarkar.
Phys. Lett. B, Vol. 167B, No. 4, p. 457 – 463 (1986). Abstr. in Phys. Abstr., Vol. 89, No. 1281, Entry 67645 (1986).

161.362 **Time variation of fundamental constants, primordial nucleosynthesis, and the size of extra dimensions.**
E. W. Kolb, M. J. Perry, T. P. Walker.
Phys. Rev. D, Vol. 33, No. 4, p. 869 – 871 (1986). Abstr. in Phys. Abstr., Vol. 89, No. 1281, Entry 67646 (1986).

161.363 **Evolution of cosmic strings.**
D. P. Bennett.
Phys. Rev. D, Vol. 33, No. 4, p. 872 – 888 (1986). Abstr. in Phys. Abstr., Vol. 89, No. 1281, Entry 67647 (1986).

161.364 **Cosmic and local mass density of 'invisible' axions.**
M. S. Turner.
Phys. Rev. D, Vol. 33, No. 4, p. 889 – 896 (1986). Abstr. in Phys. Abstr., Vol. 89, No. 1281, Entry 67648 (1986).

161.365 **Invisible axions and the QCD phase transition in the early Universe.**
T. DeGrand, T. W. Kephart, T. J. Weiler.
Phys. Rev. D, Vol. 33, No. 4, p. 910 – 914 (1986). Abstr. in Phys. Abstr., Vol. 89, No. 1281, Entry 67650 (1986).

161.366 **Lee–Weinberg bound reexamined.**
E. W. Kolb, K. A. Olive.
Phys. Rev. D, Vol. 33, No. 4, p. 1202 – 1203 (1986). Abstr. in Phys. Abstr., Vol. 89, No. 1281, Entry 67651 (1986).

161.367 **Transition of a rotating Bianchi type-IX cosmological model into an inflationary era.**
O. Gron.
Phys. Rev. D, Vol. 33, No. 4, p. 1204 – 1205 (1986). Abstr. in Phys. Abstr., Vol. 89, No. 1281, Entry 67652 (1986).

161.368 **The formation of the Universe.**
T. X. Thuan.
Recherche, Vol. 17, No. 174, p. 172 – 181 (1986). In French. Abstr. in Phys. Abstr., Vol. 89, No. 1281, Entry 67653 (1986).

161.369 **Exact Kantowski–Sachs and Bianchi types I and III cosmological models with a conformally invariant scalar field.**
A. J. Accioly.
Rev. Bras. Fis., Vol. 15, No. 2, p. 167 – 173 (1985). Abstr. in Phys. Abstr., Vol. 89, No. 1281, Entry 67654 (1986).

161.370 **Bulk-viscosity-driven inflationary model.**
I. Waga, R. C. Falcao, R. Chanda.
Phys. Rev. D, Vol. 33, No. 6, p. 1839 – 1841 (1986). Abstr. in Phys. Abstr., Vol. 89, No. 1282, Entry 67890 (1986).

161.371 **On the relic, cosmic abundance of stable, weakly interacting massive particles.**
R. J. Scherrer, M. S. Turner.
Phys. Rev. D, Vol. 33, No. 6, p. 1585 – 1589 (1986). Abstr. in Phys. Abstr., Vol. 89, No. 1282, Entry 73035 (1986).

161.372 **Cosmology and galaxy formation.**
J. Silk.
Proceedings of the 1985 International Symposium on Lepton and Photon Interactions at High Energies, p. 590 – 610 (1986). Abstr. in Phys. Abstr., Vol. 89, No. 1282, Entry 73040 (1986). – See Abstr. 012.063.

161.373 **Distortion of cosmic microwave background spectrum by dust heated by population III stars.**
G.–z. Xie.
Kexue Tongbao, Vol. 31, No. 3, p. 192 – 196 (1986). Abstr. in Phys. Abstr., Vol. 89, No. 1282, Entry 73431 (1986).

161.374 **In search of the missing mass.**
J. Gribbin.
New Sci., Vol. 109, No. 1490, p. 37 – 40 (1986). Abstr. in Phys. Abstr., Vol. 89, No. 1282, Entry 73437 (1986).

161.375 **Viscous phenomena in cosmology. I. Lepton era.**
F. Hoogeveen, W. A. van Leeuwen, G. A. Q. Salvati, E. E. Schelling.
Physica A, Vol. 134A, No. 2, p. 458 – 473 (1986). Abstr. in Phys. Abstr., Vol. 89, No. 1282, Entry 73439 (1986).

161.376 **Quantum creation of multidimensional universes.**
E. Alvarez.
Z. Phys., C, Vol. 30, No. 1, p. 157 – 159 (1986). Abstr. in Phys. Abstr., Vol. 89, No. 1282, Entry 73440 (1986).

161.377 **Cosmology and particle physics.**
M. S. Turner.
Techniques and concepts of high–energy physics III, p. 153 – 233 (1985). Abstr. in Phys. Abstr., Vol. 89, No. 1282, Entry 73442 (1986). – See Abstr. 012.062.

161.378 **A cosmological model of matter in the framework of special relativity.**
N. Ionescu–pallas.
Rev. Roum. Phys., Vol. 31, No. 2, p. 101 – 105 (1986). Abstr. in Phys. Abstr., Vol. 89, No. 1283, Entry 73606 (1986).

161.379 **Professor Wheeler and the crack of doom: closed cosmologies in the 5–d Kaluza–Klein theory.**
R. A. Matzner, A. Mezzacappa.
Found. Phys., Vol. 16, No. 3, p. 227 – 248 (1986). Abstr. in Phys. Abstr., Vol. 89, No. 1283, Entry 73613 (1986).

161.380 **The fissioning universe: topological inflation and Kaluza–Klein cosmologies.**
M. Kaku, J. Lykken.
Nucl. Phys. B, Part. Phys., Vol. B268, No. 2, p. 489 – 508 (1986). Abstr. in Phys. Abstr., Vol. 89, No. 1283, Entry 73623 (1986).

161.381 **Cosmological difficulties for intermediate scales in superstring models.**
K. Enqvist, D. V. Nanopoulos, M. Quiros.
Phys. Lett. B, Vol. 169B, No. 4, p. 343 – 346 (1986). Abstr. in Phys. Abstr., Vol. 89, No. 1283, Entry 73625 (1986).

161.382 **The effect of radiative decay of massive particles on the spectrum of the microwave background radiation.**
M. Kawasaki, K. Sato.
Phys. Lett. B, Vol. 169B, No. 2 – 3, p. 280 – 284 (1986). Abstr. in Phys. Abstr., Vol. 89, No. 1283, Entry 78536 (1986).

161.383 **Does Newton's world model revive?**
A. Meszaros.
Acta Phys. Hung., Vol. 58, No. 3 – 4, p. 187 – 190 (1985). Abstr. in Phys. Abstr., Vol. 89, No. 1283, Entry 78537 (1986).

161.384 **Obvious, incredible... (*role of neutrinos in cosmology*).**
V. N. Lukash, I. D. Novikov.
Found. Phys., Vol. 16, No. 3, p. 209 – 225 (1986). Abstr. in Phys. Abstr., Vol. 89, No. 1283, Entry 78542 (1986).

161.385 **Cosmology with time–varying *G*.**
A. Alfonso–Faus.
Int. J. Theor. Phys., Vol. 25, No. 3, p. 293 – 317 (1986). Abstr. in Phys. Abstr., Vol. 89, No. 1283, Entry 78543 (1986).

161.386 **Cosmological quark hadron phase transition and formation of isothermal density fluctuations.**
K. Iso, H. Kodama, K. Sato.
Phys. Lett. B, Vol. 169B, No. 4, p. 337 – 342 (1986). Abstr. in Phys. Abstr., Vol. 89, No. 1283, Entry 78547 (1986).

161.387 **Inflation of the Universe.**
A. Bouquet.
Recherche, Vol. 17, No. 176, p. 448 – 456 (1986). In French. Abstr. in Phys. Abstr., Vol. 89, No. 1283, Entry 78548 (1986).

161.388 **Dust universes with spatial spherical symmetry and euclidean initial hypersurfaces.**
E. Laserra.
Meccanica, Vol. 20, No. 4, p. 267 – 271 (1985). Abstr. in Phys. Abstr., Vol. 89, Nr. 1284, Entry 83690 (1986).

161.389 **Observational and dynamical properties of small universes.**
G. F. R. Ellis, G. Schreiber.
Phys. Lett. A, Vol. 115, No. 3, p. 97 – 107 (1986). Abstr. in Phys. Abstr., Vol. 89, Nr. 1284, Entry 83693 (1986).

161.390 **Formation of black holes in quantum cosmology.**
L. Z. Fang, M. Li.
Phys. Lett. B, Vol. 169B, No. 1, p. 28 – 31 (1986). Abstr. in Phys. Abstr., Vol. 89, Nr. 1284, Entry 83694 (1986).

161.391 **Quantum fields can provide enough energy density to close the universe.**
P. R. Anderson.
Phys. Lett. B, Vol. 169B, No. 1, p. 31 – 35 (1986). Abstr. in Phys. Abstr., Vol. 89, Nr. 1284, Entry 83695 (1986).

161.392 **A Bianchi–V universe with radiation.**
J.–m. Zhu, S.–c. Zhu.
Chin. Astron. Astrophys., Vol. 10, No. 1, p. 12 – 16 (1986). English translation of 40.161.306.

161.393 **The invisible universe.**
R. Cowsik.
Cosmic pathways, p. 289 – 310 (1986). – See Abstr. 003.045.
This paper shows that neutrinos were generated copiously in the "big bang" that originated the universe and that these neutrinos, even if they should have a tiny mass of 10 eV/c^2 would dominate the gravitational dynamics of the universe and ensure its closure, explain the formation of large clusters of galaxies and their virial discrepancy, generate the observed luminosity profiles of galaxies as well as the so–called flat rotation curves of spiral galaxies, and finally impose a Maxwellian distribution of velocities on the stars in ellipticals.

161.394 **Baryon non–conservation and cosmology.**
J. V. Narlikar.
Cosmic pathways, p. 311 – 317 (1986). – See Abstr. 003.045.

161.395 **Cosmology and particle physics.**
G. Steigman.
Rev. Mex. Astron. Astrofis., Vol. 12, Spec. Issue, p. 21 – 30 (1986). – See Abstr. 012.066.

161.396 **Formation of structures in the very early Universe.**
O. B. Neto, G. C. Marques, I. Ventura.
Rev. Mex. Astron. Astrofis., Vol. 12, Spec. Issue, p. 31 – 37 (1986). – See Abstr. 012.066.

The authors sketch an alternative picture of cosmological phase transition and study its implications on the formation of structures in the very early Universe. They show that the condensation of walls at high temperatures might lead to fluctuations which are in accordance to all necessary conditions to the formation of structures in the Universe.

161.397 **Growth of isothermal perturbations during and after the recombination era.**
J. C. N. de Araujo, R. Opher.
Rev. Mex. Astron. Astrofis., Vol. 12, Spec. Issue, p. 38 – 45 (1986). – See Abstr. 012.066.

A study of the evolution of isothermal perturbations is given during and after the recombination era. Effects due to recombination, nonlinear growth and the decoupling of matter and radiation are taken into account. A result of particular interest is that isothermal fluctuations of mass very much smaller than the Jeans mass at the beginning of the recombination era (i.e. $M \ll M_J \cong 10^5 M_\odot$) are not damped to extinction. In particular, isothermal fluctuations of mass 25 – 200 $M_\odot$ survive. The results thus indicate that massive Population III stars in the mass range 25 – 200 $M_\odot$ need not have formed by fragmentation from massive clouds, but could have formed from perturbations which survived directly from the recombination era.

161.398 **A cosmological model with cylindrical symmetry.**
F. M. Gomide, M. S. Berman, R. L. Garcia.
Rev. Mex. Astron. Astrofis., Vol. 12, Spec. Issue, p. 46 – 48 (1986). – See Abstr. 012.066.

The authors exhibit the assumptions and preliminary results as to the possibility of Mach's principle being valid in a universe with expansion and rotation.

161.399 **The galaxy luminosity function and the redshift–distance controversy. (A review).**
E. E. Salpeter, G. L. Hoffman Jr.
Proc. Natl. Acad. Sci. U.S.A., Vol. 83, p. 3056 – 3063 (1986).

The mean relation between distance and redshift for galaxies is reviewed as an observational question. The luminosity function for galaxies is an important ingredient and is given explicitly. The authors discuss various observational selection effects that are important for comparison of the linear and quadratic distance–redshift laws. Several lines of evidence are reviewed, including the distribution of galaxy luminosities in various redshift ranges, the luminosities of brightest galaxies in groups and clusters at various redshifts, and the Tully–Fisher correlation between neutral hydrogen velocity widths and luminosity. All of these strongly favor the linear law over the quadratic.

161.400 **Primordial nucleosynthesis in 1985.**
H. Reeves.
Nucleosynthesis and its implications on nuclear and particle physics, p. 3 – 12 (1986). – See Abstr. 012.069.

This paper reviews recent developments in theories of primordial nucleosynthesis. Emphasis is on the use of ^{7}Li as a cosmological probe and on big bang yields as a function of the present cosmic baryonic density.

161.401 **Standard big bang nucleosynthesis and chemical evolution of D and ^{3}He.**
P. Delbourgo–Salvador, G. Malinie, J. Audouze.
Nucleosynthesis and its implications on nuclear and particle physics, p. 27 – 35 (1986). – See Abstr. 012.069.

Effects of stellar and galactic evolution on the abundances of D and ^{3}He are examined in an attempt to relate the observations of these two light elements with their primordial abundances. It is shown that in order to reconcile the baryonic densities deduced respectively from ^{4}He and D in the frame of standard big bang nucleosynthesis models, the ratio between the primordial and the present abundance of D should be as large as 7 – 10.

161.402 **Early–photoproduction of D and ^{3}He and pregalactic nucleosynthesis of the light elements.**
J. Audouze, D. Lindley, J. Silk.
Nucleosynthesis and its implications on nuclear and particle physics, p. 57 – 63 (1986). – See Abstr. 012.069.

Two unconventional scenarios concerning the early nucleosynthesis of the lightest elements (D, ^{3}He, ^{4}He, ^{7}Li) which attempt to alleviate the strong predictions of the standard big bang nucleosynthesis are briefly reviewed here. The first one consists in a "big bang photosynthesis" of D and ^{3}He induced on ^{4}He and ^{7}Li by energetic photons coming from the decay of hypothetical massive neutrinos and gravitinos. The second one consists in a pregalactic formation of D, ^{3}He and ^{7}Li by the interaction of energetic ^{4}He nuclei produced by a first generation of massive stars with the primordial gas made only of pure hydrogen.

161.403 **Influence of quark nuggets on primordial nucleosynthesis.**
R. Schaeffer, P. Delbourgo–Salvador, J. Audouze.
Nucleosynthesis and its implications on nuclear and particle physics, p. 65 – 70 (1986). – See Abstr. 012.069.

The hypothesis that the universe is populated by stable or metastable quark nuggets is consistent with the observed abundances of light elements, despite the strong interactions of baryons and quark nuggets, provided the average baryon number of these nuggets is quite large ($A \sim 2 \times 10^{15}$). It leads to a universe with 98% of the matter in the form of nuggets and 2% in the form of baryons.

161.404 **Primordial nucleosynthesis with generic particles.**
T. P. Walker, E. W. Kolb, M. S. Turner.
Nucleosynthesis and its implications on nuclear and particle physics, p. 71 – 78 (1986). – See Abstr. 012.069.

The authors discuss a revision of the standard model for big bang nucleosynthesis which allows for the presence of generic particle species. The primordial production of ^{4}He and D + ^{3}He is calculated as a function of the mass, spin degrees of freedom, and spin statistics of the generic particle for masses in the range $10^{-2} \lesssim m/m_e \lesssim 10^2$. The particular case of the Gelmini and Roncadelli majoron model for massive neutrinos is discussed.

161.405 **Dark matter and cosmological nucleosynthesis.**
D. N. Schramm.
Nucleosynthesis and its implications on nuclear and particle physics, p. 79 – 85 (1986). – See Abstr. 012.069.

It is shown that big bang nucleosynthesis requires at least some of the baryons in the universe must be dark. All current solutions seem to require at least two ad hoc assumptions. Of these more complex but complete solutions, the ability to produce large scale free structures where these dark baryons lie may be the best way to resolve the different proposed solutions to the several different dark matter problems.

161.406 **Nucleosynthetic consequences of Population III stars.**
B. J. Carr, W. Glatzel.
Nucleosynthesis and its implications on nuclear and particle physics, p. 87 – 94 (1986). – See Abstr. 012.069.

It is argued that the Population III stars with the most explanatory power are very massive objects in the mass range $10^2 < M/M_\odot < 10^5$. Their evolution is reviewed, with particular emphasis on their nucleosynthetic consequences, and this leads to constraints on the mass spectrum of Population III stars. The possible relevance of Population III stars for some elemental anomalies and the cosmic helium abundance is discussed.

161.407 **On what scale can the universe be considered homogeneous.**
A. P. Fairall.
Mon. Notes Astron. Soc. S. Afr., Vol. 44, Nos. 11–12, p. 114 – 119 (1985).

Whilst superclusters and voids reflect a texture scale of approximately 75 Mpc (H = 75 km/sec per Mpc), dominant superclusters are more widely scattered and underdense inhomogeneities on a scale > 200 Mpc are suspected. Unique structures of this nature are found within a sample "volume" of dimension

500 Mpc. Homogeneity, which requires repetition of structure, may therefore only occur on a scale of at least several hundred Megaparsecs (i.e. $z \approx 0.15$).

161.408 **Constraints on unstable heavy neutrinos from cosmology.**
M. M. Vasanthi, T. Padmanabhan.
J. Astrophys. Astron., Vol. 7, No. 2, p. 59 – 70 (1986).
Cosmological scenarios with massive unstable neutrinos are discussed. Restrictions on the mass and the lifetime of the unstable neutrino are derived from (a) age and mass density of the universe and (b) the growth of primordial fluctuations. It will not be possible to accommodate unstable neutrinos with masses above ~1 keV in standard cosmology unless they have exceedingly small lifetime.

161.409 **Zur dreidimensionalen Verteilung entfernter Galaxien.**
P. Schücker.
Mitt. Astron. Ges., No. 65, p. 251 – 253 (1986). – See Abstr. 012.072.

161.410 **Space and time in the microstructure of the matter and basic questions of the universe in the theory of Ruder Bošković.**
B. Jovanović.
Publ. Astron. Soc. "Rudjer Bošković", No. 4, p. 15 – 17 (1985). – See Abstr. 012.078. In Croatian.

161.411 **Cosmological results from IRAS.**
M. Rowan–Robinson.
Light on dark matter, p. 499 – 506 (1986). – See Abstr. 012.080.
IRAS is shown to be a powerful cosmological tool. At high galactic latitudes the problem of emission from interstellar dust is greatly reduced and virtually all 60 μm sources are identifiable with galaxies. Source–counts at 60 and 100 μm show that there is a 20% higher source–density at $b > 60°$ than at $b < -60°$. The local gravitational field has been mapped using IRAS 60 μm sources and analyzed in spherical harmonics. The dipole component agrees well in direction with that of the cosmic microwave background anisotropy. The 100 μm background radiation has been deconvolved into contributions from interplanetary dust, interstellar dust and a residual, relatively isotropic component of intensity 5.7 ± 2.5 MJy/sr.

161.412 **Evidence from IRAS data for large–scale anisotropy in the Hubble flow.**
C. A. Collins, R. D. Joseph, N. A. Robertson.
Light on dark matter, p. 507 – 508 (1986). – See Abstr. 012.080.
The authors have been engaged in an investigation of the isotropy of the Hubble flow using near–IR photometry for an all–sky sample of spiral galaxies. Herein they describe the result of using IRAS 100 μm photometry for this same galaxy sample, and compare the solutions obtained with those using near–IR and optical photometry.

Methods in the qualitative theory of dynamical systems in astrophysics and gas dynamics.
See Abstr. 003.028.

The universe.
See Abstr. 003.078.

A random walk in relativity and cosmology.
See Abstr. 003.079.

Kosmologie. Geschichte und Systematik in philosophischer Perspektive.
See Abstr. 003.114.

Space, time, infinity.
See Abstr. 003.194.

Newton and the infinite universe.
See Abstr. 004.012.

Galaxy distances and deviations from universal expansion.
See Abstr. 011.009.

Supersymmetry, supergravity and related topics. Proceedings of the XVth GIFT International Seminar on Theoretical Physics.
See Abstr. 012.084.

Gravitation, cosmology and cosmic–ray physics.
See Abstr. 013.023.

Where does time asymmetry come from?
See Abstr. 015.017.

Physical cosmology and the anthropic principle.
See Abstr. 015.018.

Speculation in cosmology.
See Abstr. 015.022.

Alcune considerazioni sui numeri cosmici.
See Abstr. 015.040.

Poisson clusters and Poisson voids.
See Abstr. 021.015.

Possible experiments for direct detection of particle candidates for the galactic dark matter.
See Abstr. 034.099.

Expected AXAF mirror characteristics and their implications for measurements of the Hubble constant using the Sunyaev Zel'dovich effect.
See Abstr. 035.022.

Evolution of the luminosity function of extragalactic objects.
See Abstr. 036.051.

A look at the primeval explosion.
See Abstr. 051.096.

Towards unification of elementary particle physics and cosmology in 10–dimensions.
See Abstr. 061.007.

Limits on mass of missing mass.
See Abstr. 061.014.

Inflation and shadow matter.
See Abstr. 061.015.

York's cosmic time versus proper time as relevant to changes in the dimensionless "constants", *K*–meson decay, and the unity of black hole and big crunch.
See Abstr. 061.020.

Quantum tunnelling towards an exploding universe?
See Abstr. 061.022.

How does the SU(2) ⊗ U(1) symmetry break in the early universe?
See Abstr. 061.032.

Eddington's numbers and quantum gravity.
See Abstr. 061.039.

Is nature supersymmetric?
See Abstr. 061.045.

Numerical nucleosynthesis results for Bianchi type–V cosmological models.
See Abstr. 061.048.

On the many–worlds interpretation of quantum theory.
See Abstr. 061.051.

The axion problem.
See Abstr. 061.062.

Saving the axions in superstring models.
See Abstr. 061.064.

The fate of relic neutrinos.
See Abstr. 061.066.

Bianchi–Kantowski–Sachs solutions of $N = 1$ supergravity in $d = 11$ dimensions for a torus space.
See Abstr. 061.070.

Cosmological instability in ten–dimensional supergravity.
See Abstr. 061.071.

Geometrical aspects of magnetic monopoles.
See Abstr. 061.072.

Cosmological constraints on heavy unstable particles.
See Abstr. 061.075.

A monopole model for galactic nuclei.
See Abstr. 061.077.

And if neutrinos do have mass!
See Abstr. 061.080.

Massive–neutrino decays and the cosmological nucleosynthesis.
See Abstr. 061.081.

Fermion masses and cosmology in multidimensional unified theories.
See Abstr. 061.082.

Supergravity limit of superstring theory and Friedmann–Robertson–Walker cosmology.
See Abstr. 061.083.

Primordial two–component maximally symmetric inflation.
See Abstr. 061.084.

Non–minimal supergravity at finite temperatures and cosmological applications.
See Abstr. 061.085.

Constraints on Hamiltonian lattice formulations of field theories in an expanding universe.
See Abstr. 061.086.

Search for possible superheavy particles in sodium nuclei.
See Abstr. 061.088.

Tuning the cosmological constant in $N = 1$ supergravity.
See Abstr. 061.090.

Production of massive spin–1/2 particles in anisotropic spacetimes.
See Abstr. 061.092.

Auger–like formation of monopolium.
See Abstr. 061.093.

Quantum field theory of mutually interacting particles in expanding universes.
See Abstr. 061.096.

Scalar particle production in anisotropic metrics.
See Abstr. 061.098.

World–sheet instantons and the Peccei–Quinn symmetry.
See Abstr. 061.101.

Quantum instabilities and the cosmological constant.
See Abstr. 061.102.

Axion constraints from white dwarf cooling times.
See Abstr. 061.104.

Implications of a non–zero cosmological constant on self–gravitating fermionic systems.
See Abstr. 061.105.

$N = 1$ supergravity grand unified theories with noncanonical kinetic energy terms.
See Abstr. 061.106.

The electron neutrino mass double beta decay and cosmology.
See Abstr. 061.107.

Cosmology and the causal interpretation of quantum mechanics.
See Abstr. 061.111.

The invisible axion in a SUSY GUT SO(10) model.
See Abstr. 061.112.

Tunneling without barriers.
See Abstr. 061.113.

Grand unification of fermion generations: neutrino masses and cosmological constraints.
See Abstr. 061.114.

Strange matter.
See Abstr. 061.115.

On the use of SSC as a detector of certain dark matter candidates.
See Abstr. 061.118.

The survival of helium–3 in stars.
See Abstr. 061.123.

How degenerate can we be?
See Abstr. 061.124.

Large–basis shell–model technology in nucleosynthesis and cosmology.
See Abstr. 061.137.

Nuclear matter under extreme conditions and the early universe.
See Abstr. 061.144.

Field of an accelerating celestial body with a large number of magnetic monopoles.
See Abstr. 061.147.

Local dynamic thermal instability.
See Abstr. 062.066.

Destruction of ^{3}He in stars.
See Abstr. 065.048.

Primordial stellar evolution: the protostar phase.
See Abstr. 065.064.

The behavior of the gravitational field near the initial singularity.
See Abstr. 066.014.

Friedmann–Robertson–Walker vacuum solutions in the scale covariant theory.
See Abstr. 066.019.

Local measurements and cosmological background.
See Abstr. 066.026.

Gauge theory in curved space.
See Abstr. 066.037.

Gravitational lenses: model fitting, time delay and Hubble parameter.
See Abstr. 066.043.

On the distance measure for gravitational lenses.
See Abstr. 066.076.

Central charges in the canonical realization of asymptotic symmetries: an example from three dimensional gravity.
See Abstr. 066.084.

Note on equivalence of cosmological space–times.
See Abstr. 066.095.

Linking Maxwell and Einstein through the Large Numbers Hypothesis.
See Abstr. 066.104.

Model systems of fields in the theory of gravitation and the problems of cosmology.
See Abstr. 066.125.

An attempt for unification of microcosmos and mega(macro)-cosmos on the base of the order (causality).
See Abstr. 066.129.

Comment on the GRT–LRS Bianchi type V solutions.
See Abstr. 066.130.

Comment on the big–bang singularity in the scale–covariant theory.
See Abstr. 066.135.

Relativistic theory of gravitation.
See Abstr. 066.137.

Quantum gravity effects in particle physics and cosmology.
See Abstr. 066.148.

On the vacuum definition in curved space–time.
See Abstr. 066.154.

Tunneling transitions with gravitation: breaking of the quasiclassical approximation.
See Abstr. 066.155.

The gravitational arrow of time.
See Abstr. 066.160.

Symmetry breaking, phase transition and gravity.
See Abstr. 066.161.

Relation between the gravitational and the cosmological constant.
See Abstr. 066.163.

Consistency of the Kaluza–Klein ansatz for cosmological solutions.
See Abstr. 066.166.

Statistical gravitational lensing: influence of compact objects on the observed quasar luminosity function.
See Abstr. 066.167.

Gravitational radiation near spatial and null infinity.
See Abstr. 066.172.

Constraints on the constant in Rosen's bimetric general relativity theory.
See Abstr. 066.177.

Twistors, minimal surfaces and strings.
See Abstr. 066.180.

Spacetimes with a transitive similarity group.
See Abstr. 066.181.

Induction of correct centrifugal force in a rotating mass shell.
See Abstr. 066.182.

Spontaneous compactification with quadratic and cubic curvature terms.
See Abstr. 066.185.

Some exact inhomogeneous solutions of Einstein's equations with symmetries on the hypersurfaces t = const.
See Abstr. 066.186.

Static planar domain wall in general relativity with a cosmological constant.
See Abstr. 066.190.

A class of inhomogeneous Gödel–type models.
See Abstr. 066.191.

Regular reduction of relativistic theories of gravitation with a quadratic Lagrangian.
See Abstr. 066.192.

Isometries of homogeneous Gödel–type spacetimes.
See Abstr. 066.193.

Canonical quantum Kaluza–Klein theory.
See Abstr. 066.194.

Kaluza–Klein solutions with non–compact internal spaces.
See Abstr. 066.195.

Spontaneous compactification in quantum Kaluza–Klein theories.
See Abstr. 066.199.

A simple coupling with cosmological implications.
See Abstr. 066.201.

Scalar fields and the cosmological constant.
See Abstr. 066.207.

Spherically symmetric perfect fluid solutions in relativistic astrophysics.
See Abstr. 066.212.

The quantization of matter and gravitational fields – the cosmological aspect.
See Abstr. 066.215.

Relativistic neutral superfluids. Exact solutions.
See Abstr. 066.218.

On analytical continuation of the Kerr–Newman metric with cosmological constant.
See Abstr. 066.221.

Dimensional reduction caused by a cosmological constant.
See Abstr. 066.226.

Gravitational radiation in electromagnetic universes.
See Abstr. 066.227.

On the vanishing of the cosmological constant.
See Abstr. 066.228.

Dimensionality of space and the pulsating Universe.
See Abstr. 066.229.

Thermodynamic instability of de Sitter space.
See Abstr. 066.230.

Birkhoff–type theorem in self–creation cosmology.
See Abstr. 066.232.

Symmetric solutions to the Gauss–Bonnet extended Einstein equations.
See Abstr. 066.233.

Quantum state for a black hole in a de Sitter universe.
See Abstr. 067.110.

Exterior solutions for electric and magnetic stars in 2+1 dimensions.
See Abstr. 067.124.

The Hawking radiation of charged Dirac particles in Kerr–Newman–de Sitter spacetime.
See Abstr. 067.140.

Inversion symmetric initial data for *N* charged black holes.
See Abstr. 067.162.

Does dark matter affect the solar neutrino flux?
See Abstr. 080.025.

Interplanetary plasma turbulence and the Doppler detection of a gravitational wave background.
See Abstr. 106.007.

Carbon star radial velocities and dark matter in the universe.
See Abstr. 111.013.

Brightest stars in galaxies as distance indicators.
See Abstr. 115.030.

Probing the dark matter with supernovae.
See Abstr. 125.041.

Can superconducting cosmic strings explain anomalous radio sources?
See Abstr. 141.012.

HEAO results on the cosmic X–ray background.
See Abstr. 142.099.

Formation of elliptical galaxies and massive halos.
See Abstr. 151.003.

A mathematical model of the initial stage in the formation of a disk galaxy.
See Abstr. 151.011.

Evolution of galaxies: one–zone model with a birth–rate stellar function depending on gas density and temperature.
See Abstr. 151.013.

A study on the formation and dynamics of galaxies.
See Abstr. 151.081.

The potential for supernova–induced chemical enrichment of protoglobular cluster clouds.
See Abstr. 154.038.

Studies of *IRAS* sources at high galactic latitudes – I. Source counts at $|b| > 60°$ and evidence for a north–south anisotropy of cosmological significance.
See Abstr. 157.033.

Galaxy evolution: some issues and questions.
See Abstr. 157.077.

Quests for primeval galaxies: a review of optical surveys.
See Abstr. 157.104.

A case against baryons in galactic halos.
See Abstr. 157.150.

Galaxy counts and colors in a South Galactic Pole field.
See Abstr. 157.166.

A study of the large–scale structure in the distribution of galaxies in a region centered about the Cancer cluster. I. Initial observational results.
See Abstr. 157.223.

A study of the large–scale structure in the distribution of galaxies in a region centered about the Cancer cluster. II. Further observational results.
See Abstr. 157.224.

Angular momentum–limited degenerate and semidegenerate neutrino halos.
See Abstr. 157.256.

Dwarf galaxies and galaxy formation.
See Abstr. 157.306.

Study of the cosmological evolution of extragalactic radio sources observed in decametric range.
See Abstr. 158.073.

Spectral evolution in young active galactic nuclei.
See Abstr. 158.103.

Some theoretical aspects of AGNs.
See Abstr. 158.105.

Cosmological evolution in the extent of double radio galaxies.
See Abstr. 158.119.

The spectral index – flux density relation and the cosmological evolution of radio sources.
See Abstr. 158.120.

Evolution of powerful AGN and the collapse of rich clusters.
See Abstr. 158.137.

Faint galaxies and cosmology.
See Abstr. 158.186.

Cosmological evolution of AGNs in the X–ray band.
See Abstr. 158.190.

A test for models of radio source evolution.
See Abstr. 158.237.

The statistics of gravitational lensing. III. Astrophysical consequences of quasar lensing.
See Abstr. 159.001.

Comparative statistical study of the chronometric and Friedmann cosmologies based on the Palomar Bright Quasar and other complete quasar surveys.
See Abstr. 159.004.

Lyman absorption lines in quasar spectra: evidence for gravitationally–confined gas in dark minihaloes.
See Abstr. 159.008.

Alignments of clustered quasars.
See Abstr. 159.012.

A quasar with $z = 3.71$ and limits on the number of more distant objects.
See Abstr. 159.013.

QSOs as probes of the early Universe.
See Abstr. 159.019.

Quasars: cosmological evolution and X–ray background contribution.
See Abstr. 159.026.

Distribution of QSO emission line redshifts and cosmological redshift.
See Abstr. 159.042.

Cosmological evolution of optically selected QSOs.
See Abstr. 159.062.

Ly α galaxy companions of quasars at z > 2.5.
See Abstr. 159.077.

The multiple source 0023+171: a dark gravitational lens?
See Abstr. 159.081.

Is QSO 1146+111 B, C due to lensing by a cosmic string?
See Abstr. 159.119.

Counts, evolution, and background contribution of X–ray quasars and other extragalactic X–ray sources.
See Abstr. 159.138.

The cosmic evolution of quasars at high redshifts.
See Abstr. 159.150.

Comparison of the Hubble diagrams among X–ray–, optically– and radio–selected quasars.
See Abstr. 159.161.

Morphological segregation in the Pisces–Perseus supercluster.
See Abstr. 160.001.

A search for the Sunyaev–Zel'dovich effect at λ = 3 millimeters.
See Abstr. 160.002.

A prominent string of galaxies in Bootes: evidence for a Lagrangian singularity?
See Abstr. 160.004.

Space distribution of nearby superclusters of galaxies.
See Abstr. 160.005.

Are superclusters correlated on a very large scale?
See Abstr. 160.012.

New results on large–scale structures.
See Abstr. 160.015.

Structure of superclusters and supercluster formation – V. Spatial correlation and voids.
See Abstr. 160.020.

The blue Tully–Fisher relation in a variety of environments.
See Abstr. 160.024.

The two–point galaxy correlation function of the Local Supercluster.
See Abstr. 160.027.

A sparse–sampling strategy for the estimation of large–scale clustering from redshift surveys.
See Abstr. 160.036.

Gas in a cosmic void: properties and origin.
See Abstr. 160.037.

Systematic properties of clusters of galaxies: cosmological scenario versus observations.
See Abstr. 160.055.

The X–ray emitting gas in clusters of galaxies.
See Abstr. 160.057.

Cosmological H II regions and the photoionization of the IGM.
See Abstr. 160.058.

The difference between the galaxy and cluster correlation functions: a manifestation of tails of galaxy clusters.
See Abstr. 160.061.

Redshift asymmetries in systems of galaxies and the missing mass.
See Abstr. 160.079.

Limits to the extragalactic distance scale from integrated properties of Local Group galaxies.
See Abstr. 160.083.

Probing the large–scale structure of the universe: an analysis of 55 bright southern clusters of galaxies.
See Abstr. 160.084.

Two–point correlation for rich clusters of galaxies.
See Abstr. 160.085.

***N*–point correlations for biased galaxy formation.**
See Abstr. 160.105.

The large–scale distribution of galaxies in the Lynx–Gemini region.
See Abstr. 160.106.

The cluster–cluster correlation function.
See Abstr. 160.109.

Clustering and voids.
See Abstr. 160.112.

Author Index

The authors are listed in alphabetical order according to the initial letter following the first names. Author names which are not transliterated according to our scheme are indicated by †.

Zhao

Subject Index

The Subject Index provides an alphabetical subject key to the papers abstracted in *Astronomy and Astrophysics Abstracts*. In order to explain the structure of this Index some general remarks on the construction principles of our key words will follow.

Whenever possible a key word was formed in such a way as to combine two different descriptors characterizing an abstracted article. Such a combination represents the contents of a paper more precisely than both components regarded as single entries. As an example we take the two terms Minor Planets and Orbits and the combination

Minor Planets
  Orbits

Efforts were made to choose combinations which can be inverted in order to increase the usefulness of the Index. That means the two components represent terms of equal status which both serve as primary entries. In the given example we find the two entries

Minor Planets
  Orbits

and

Orbits
  Minor Planets

But there are also two-term key words which cannot be reversed. The conditions for non-inversibility are: the secondary descriptor is either a very specific term (e.g. LTE Analyses) or a very general one (e.g. Models). Thus we distinguish two kinds of descriptors, the primary ones, serving as direct entries and the secondary ones as auxiliary entries to the Index. Our intention is to choose comprehensive as well as significant formulations.

Some further aids to the use of this Subject Index are given by the following concordances:

Asteroids	see Minor Planets
Earth: Magnetic Field	see Geomagnetic Field
Galactic...	see also Galaxies, Galaxy
Lunar...	see also Moon
Planetary...	see also Planets
Solar...	see also Sun
Stellar...	see also Stars

Efforts were made to standardize the key words. This idea led to the development of our vocabulary of astronomical and astrophysical terms. From time to time we enlarged our vocabulary and at present version 10.86 consists of 2,305 items.

The list of key words was examined in order to standardize the key words and to confine some of the synonymous entries. For example, according to our vocabulary the key words Catalogues, Positions, Star Catalogues, and Star Positions were admissible terms, likewise the combinations

Catalogues
  Star Positions

Star Catalogues
  Positions

Star Catalogues
  Star Positions

The user is – in any case – kindly requested to look for synonymous entries, because further references to the topic looked for might exist elsewhere in the Index under another current astronomical term.

Some key words concerning comets, novae, and solar eclipses are sorted by a special routine. Comets are listed according to their Roman numeral designation or preliminary designation, novae according to their stellar constellation and solar eclipses according to the date of the event.

Due to the introduction of a special object index containing a list of single objects, special object designations (e.g. SS433, BL Lacertae, Algol, etc) have been removed from the subject index and transferred to the object index (see SS433, BL Lac, β Per).

Exceptions have been made where major planets and their moons, comets, novae, and supernovae are concerned, which are given in the subject index only. Some objects (name or catalog designation) are contained in both indexes

Andromeda Galaxy	M31, NGC 224
Magellanic Clouds	LMC, SMC
Pleiades	M45
Hyades	Melotte 25
Crab Nebula	M1

In order to obtain the most comprehensive information possible the use of both indexes is recommended.

Starting with Volume 30 of *Astronomy and Astrophysics Abstracts,* this Index is given in a four-column arrangement. The overall size further has been optimized by suppression of a repetition of the first key word in a two-term combination. Thus, the primary term is printed only once (in semi-boldface) for all the following secondary key words which belong to this header.

The following List of Key Words might serve as a further guidance to the use of this Subject Index. This list, a subset of our vocabulary 10.86, consists of 1,212 entries and comprises all one-term key words and all primary terms of the two-term ones actually used in this Subject Index.

List of Key Words

A Dwarfs
A Stars
Aberration
Absolute Magnitudes
Absorption
Absorption Coefficient
Absorption Lines
Absorption Spectra
Accretion
Accretion Disk
Accretion Disks
Achondrites
Acoustic Waves
Active Galactic Nuclei
Active Galaxies
Adaptive Optics
Ae Stars
Aerosols
Air Showers
Airglow
Albedo
Alfven Waves
Algol Systems
AM Herculis Stars
Am Stars
Andromeda Galaxy
Angular Diameters
Angular Momentum
Antennas
Antimatter
Ap Stars
Aperture Synthesis
Apochromats
Apodization
Apollo Objects
Apsidal Motion
Aquarids
Arakelian Galaxies
Archaeoastronomy
Artificial Satellites
Asteroidal Belt
Astrodynamics
Astrographic Catalogue
Astrographs
Astrolabes
Astrometric Binaries
Astrometric Reflectors
Astrometry
Astronomical Constants
Astronomical Geodesy
Astronomical Optics
Aten-Type Objects
Atlases
Atmospheres
Atomic Clocks
Atomic Processes
Atomic Spectra
Atomic Time
Aurorae
Auroral Arcs
Auroral Jets
Auxiliary Instrumentation
Azimuth Determination

B Dwarfs
B Stars
B Supergiants
Background Radiation
Balmer Decrement
Balmer Lines
Barium Stars
Barred Spirals
Baryons
Be Stars
Beta Cephei Stars
Beta CMa Stars
Beta Lyrae Stars
Binaries
Binary Pulsars
Bipolar Nebulae
Bipolar Outflows
BL Lacertae Objects
Black Holes
Black-Body Radiation
Blazhko Effect
Blue Galaxies
Blue Objects
Blue Stars
Blue Stragglers
Blue Variables
Bolometers
Bolometric Corrections
Bp Stars
Bremsstrahlung
Bright Galaxies
Bright Giants
Bright Stars
Brightness Temperature
Broad-Band Photometry
BV Photometry
BY Draconis Stars

C-M Diagrams
Calendars
Cameras
Carbon Stars
Cataclysmic Binaries
Cataclysmic Variables
Catalogues
CCD Cameras
CCD Detectors
CCD Observations
CCD Photometry
Celestial Bodies
Celestial Mechanics
Central Stars
Cepheids
Cerenkov Radiation
CH Stars
Chandler Wobble
Chondrites
Chondrules
Circumstellar Clouds
Circumstellar Disks
Circumstellar Dust
Circumstellar Envelopes
Circumstellar Grains
Circumstellar Matter
Circumstellar Shells
Clocks
Close Binaries
Clusters of Galaxies
Collapse
Collapsed Stars
Collapsing Stars
Color Excesses
Color Indices
Colors
Cometary Atmospheres
Cometary Comae
Cometary Dust
Cometary Envelopes
Cometary Ionospheres
Cometary Nebulae
Cometary Nuclei
Cometary Plasma
Cometary Tails
Comets
Compact Galaxies
Compact Objects
Compton Effect
Compton Scattering
Contact Binaries
Continental Drift
Convection
Convective Zones
Cool Giants
Cool Stars
Coronographs
Cosmic Dust
Cosmic Rays
Cosmochronology
Cosmogony
Cosmological Constant
Cosmology
Coude Spectra
Coude Spectrographs
Crab Nebula
Curves-of-Growth
Cyclotron Radiation
Cygnus Loop

Dark Clouds
Dark Matter
Dark Nebulae
Data Bases
Data Centers
Data Processing
DDO Photometry
Degenerate Dwarfs
Degenerate Matter
Degenerate Stars
Delta Scuti Stars
Dense Clouds
Dense Matter
Densities
Densitometers
Density
Density Waves
Detached Systems
Detector Arrays
Detectors
Diameter
Diameters
Differential Rotation
Diffuse Nebulae
Diode Arrays
Disk Galaxies
Disk Population
Disk Stars
Distance
Distance Indicators
Distance Moduli
Distance Scale
Distances
Doppler Velocities
Draconids
Dust
Dust Clouds
Dust Grains
Dwarf Cepheids
Dwarf Galaxies
Dwarf Novae
Dwarfs
Dynamical Parallaxes
Dynamo Theory

Early Universe
Early-Type Galaxies
Early-Type Stars
Early-Type Supergiants
Earth
Earth Atmosphere
Earth Core
Earth Interior
Earth Ionosphere
Earth Magnetosphere
Earth Mantle
Earth Mesosphere
Earth Plasmasphere
Earth Radiation Belts
Earth Rotation
Earth Stratosphere
Earth Thermosphere
Earth Troposphere
Earth-Moon System
Echelle Spectra
Echelle Spectrographs
Eclipses
Eclipsing Binaries
Effective Temperatures
Electron Densities
Electron Temperatures
Electronographic Cameras
Electrophotometers
Elementary Particles
Ellipsoidal Binaries
Elliptical Galaxies
Emission Lines
Emission Nebulae
Emission Spectra
Emission-Line Galaxies
Emission-Line Objects
Emission-Line Stars
Ephemerides
Equidensitometry
Equilibrium Figures
Eruptive Stars
Europa
Europa Surface
Extinction
Extragalactic Objects
Extraterrestrial Intelligence
Extraterrestrial Life
Extreme UV
Extreme UV Astronomy
Extreme UV Spectra

F Dwarfs
F Giants
F Stars
F Supergiants
Fabry-Perot Interferometers
Fabry-Perot Interferometry
Fabry-Perot Spectrometers
Faint Galaxies
Faint Stars
Fiber Optics
Field Theories
Filamentary Nebulae
Filters
Fireballs
FK4
FK4 System
FK5
FK5 System
Flare Stars
Fluid Dynamics
Fluid Spheres
Flux Densities
Forbidden Lines
Forbush Effect
Four-Body Problem
Four-Color Photometry
Fraunhofer Lines
Frequency Standards
FU Orionis Stars
Fundamental Astrometry
Fundamental Catalogues
Fundamental Stars
Fundamental Systems

G Dwarfs
G Giants
G Stars
G Supergiants
Galactic Anticenter
Galactic Bars
Galactic Center
Galactic Centers
Galactic Coronae
Galactic Disk
Galactic Disks
Galactic Halo
Galactic Halos
Galactic Magnetic Field
Galactic Nuclei
Galactic Nucleus
Galactic Plane
Galactic Poles

Object Index

The usual bibliographic information accompanying the papers in the fields of astronomy and astrophysics certainly allows for their exact identification but the selection process of information retrieval is facilitated by supplementary information. Phenomena of all kinds, methods of observation and reduction as well as properties of classes of objects are referred to in the Subject Index.

Beginning with Vol. 39 *Astronomy and Astrophysics Abstracts* will present an additional index of astronomical objects dealt with in papers compiled. Basically all individual objects appearing in the title and the author's abstract are listed, but objects treated in the paper proper have also been included into the index. The number of single objects listed usually has been limited to 20. Papers discussing more than 20 individual objects have not been evaluated in the Object Index.

It is recommended to use the Object Index together with the key words of the Subject Index. Individual comets, novae, supernovae, major planets and their satellites are exclusively referred to in the Subject Index. A number of familiar astronomical objects will continue to appear in the Subject Index with their common designation. These objects are listed in the Object Index with their standard acronyms:

M31, NGC 224	Andromeda Galaxy
LMC, SMC	Magellanic Clouds
M45	Pleiades
Melotte 25	Hyades
M1	Crab Nebula

Novae identified with variables are referred to in the Object Index with their variable name:

GK Per	Nova Persei 1901

Astronomy and Astrophysics Abstracts supports the attempts of the IAU in standardizing the nomenclature of astronomical objects. Acronyms used follow closely the recommendations of "The First Dictionary of the Nomenclature of Celestial Objects" (A. Fernandez, M.-C. Lortet, F. Spite; Astron. Astrophys., Suppl. Ser., Vol. 52, No. 4 (1983)).

In sorting astronomical objects deviations from the lexicographic sequence are unavoidable in many cases. Sorting of the Object Index is based on the IBM standard routine SORT, which sorts blanks followed by special characters, Latin small and capital letters, and finally numbers. The three-letter mnemonic code of the 88 stellar constellations adopted by the IAU defines the highest rank within the sorting hierarchy. This implies that all stellar and non-stellar objects, whose designations explicitely refer to a constellation, are listed under the constellation's abbreviation. Stars within a constellation are sorted according to their Greek letter, followed by the standard variable nomenclature, and finally the Flamsteed numbers as well as other catalogue designations:

Cyg A
Cyg OB2
Cyg X-3
α Cyg
χ Cyg
BF Cyg
NML Cyg
Z Cyg
V367 Cyg
V1727 Cyg
6 Cyg

Designations without reference to stellar constellations are treated – as a rule – by sorting letters before numbers. An exception has been made for the minor planets. These objects have been retained in a particular section following after the letter 'Z'. The internal sequence is (1) permanently numbered planets, (2) objects with provisional designations, and (3) Palomar-Leiden objects:

(9) Metis
(29) Amphitrite
1979 VA
1984 HA_1
6344 P-L

Efforts were made to sort objects from individual catalogues according to the internal order of these catalogues. In Durchmusterungen the stars are arranged from the equator to the pole, taking the northward direction first. Object designations which contain coordinates are sorted first according to the usual catalogue abbreviation (B2, HMS, PKS, ZwCl, etc.) or object class designator (GX, PSR, SNR, etc.) and then – if present – according first to equatorial and then to galactic coordinates. Objects with catalogue designations beginning with a number (1E, 3C, 4U, etc.) are listed under the respective number. If only a position is given, the object is referred to by the number sequence, subdivided in equatorial and galactic coordinates, at the end of the Index. Note that Roman numbers are treated as Latin capital letters. For instance, objects from the II Zw and V Zw lists have to be searched under the letter I respectively V. A consequence of the applied sorting routine is, e.g., that Messier objects will be found at the end of the list of objects beginning with the letter 'M'.

Attempts have been made to avoid ambiguous acronyms. For instance, in the astronomical literature the acronym VB is used to designate (1) planetary nebula listed by Van den Bergh, (2) Bologna variables, (3) Van Bueren's list of Hyades stars, (4) Van Biesbroeck's list of faint stars. This ambiguity is partially resolved by using 'Van Bueren ...' for objects in (3), and 'Van Biesbroeck ...' for those in (4). With respect to the problem of multiple designations for a single object, a concordance relation is beyond the scope of the present Object Index. The user is therefore advised to search for possible alternative designations of a particular object. For example, the supergiant α Ori may also appear under BD +7°1055 or HD 39801, or the radio source Virgo A may be referred under 1218+1240, 3C 274, NGC 4486 or M87.

ASTRONOMY AND ASTROPHYSICS ABSTRACTS

A Publication of the Astronomisches Rechen-Institut Heidelberg

Editors: S. Böhme, U. Esser, H. Hefele, I. Heinrich, W. Hofmann, D. Krahn, V. R. Matas, L. D. Schmadel, G. Zech

Volume 1	Literature 1969, Part 1	X+ 435 pp. (1969)
Volume 2	Literature 1969, Part 2	X+ 516 pp. (1970)
Volume 3	Literature 1970, Part 1	X+ 490 pp. (1970)
Volume 4	Literature 1970, Part 2	X+ 562 pp. (1971)
Volume 5	Literature 1971, Part 1	X+ 505 pp. (1971)
Volume 6	Literature 1971, Part 2	X+ 560 pp. (1972)
Volume 7	Literature 1972, Part 1	X+ 526 pp. (1972)
Volume 8	Literature 1972, Part 2	X+ 594 pp. (1973)
Volume 9	Literature 1973, Part 1	X+ 610 pp. (1973)
Volume 10	Literature 1973, Part 2	X+ 661 pp. (1974)
Volume 11	Literature 1974, Part 1	X+ 579 pp. (1974)
Volume 12	Literature 1974, Part 2	X+ 699 pp. (1975)
Volume 13	Literature 1975, Part 1	X+ 632 pp. (1975)
Volume 14	Literature 1975, Part 2	X+ 747 pp. (1976)
Volume 15/16	Author and Subject Indexes to Volumes 1–10 Literature 1969–1973	 VII+ 655 pp. (1976)
Volume 17	Literature 1976, Part 1	XII+ 645 pp. (1976)
Volume 18	Literature 1976, Part 2	X+ 859 pp. (1977)
Volume 19	Literature 1977, Part 1	X+ 732 pp. (1977)
Volume 20	Literature 1977, Part 2	X+ 786 pp. (1978)
Volume 21	Literature 1978, Part 1	X+ 834 pp. (1978)
Volume 22	Literature 1978, Part 2	X+ 849 pp. (1979)
Volume 23/24	Author and Subject Indexes to Volumes 11–14 and 17–22 Literature 1974–1978	 VIII+1127 pp. (1979)
Volume 25	Literature 1979, Part 1	X+ 872 pp. (1979)
Volume 26	Literature 1979, Part 2	X+ 794 pp. (1980)
Volume 27	Literature 1980, Part 1	X+ 939 pp. (1980)
Volume 28	Literature 1980, Part 2	X+ 841 pp. (1981)
Volume 29	Literature 1981, Part 1	X+ 853 pp. (1981)
Volume 30	Literature 1981, Part 2	X+ 792 pp. (1982)
Volume 31	Literature 1982, Part 1	X+ 776 pp. (1982)
Volume 32	Literature 1982, Part 2	X+ 848 pp. (1983)
Volume 33	Literature 1983, Part 1	X+ 815 pp. (1983)
Volume 34	Literature 1983, Part 2	X+ 921 pp. (1984)
Volume 35/36	Author and Subject Indexes to Volumes 25–34 Literature 1979–1983	 XIII+ 891 pp. (1984)
Volume 37	Literature 1984, Part 1	X+ 937 pp. (1984)
Volume 38	Literature 1984, Part 2	X+ 919 pp. (1985)
Volume 39	Literature 1985, Part 1	X+1149 pp. (1985)
Volume 40	Literature 1985, Part 2	XIV+1096 pp. (1986)
Volume 41	Literature 1986, Part 1	X+1114 pp. (1986)

Published for Astronomisches Rechen-Institut by
Springer-Verlag Berlin Heidelberg New York

GPSR Compliance

The European Union's (EU) General Product Safety Regulation (GPSR) is a set of rules that requires consumer products to be safe and our obligations to ensure this.

If you have any concerns about our products, you can contact us on ProductSafety@springernature.com

In case Publisher is established outside the EU, the EU authorized representative is:

Springer Nature Customer Service Center GmbH
Europaplatz 3
69115 Heidelberg, Germany

Batch number: 10838053

Printed by Printforce, the Netherlands

Springer Graduate Texts in Philosophy

Volume 1

The Springer Graduate Texts in Philosophy offers a series of self-contained textbooks aimed towards the graduate level that covers all areas of philosophy ranging from classical philosophy to contemporary topics in the field. The texts will, in general, include teaching aids (such as exercises and summaries), and covers the range from graduate level introductions to advanced topics in the field. The publications in this series offer volumes with a broad overview of theory in core topics in field and volumes with comprehensive approaches to a single key topic in the field. Thus, the series offers publications for both general introductory courses as well as courses focused on a sub-discipline within philosophy.

The series publishes:

- All of the philosophical traditions
- Includes sourcebooks, lectures notes for advanced level courses, as well as textbooks covering specialized topics
- Interdisciplinary introductions – where philosophy overlaps with other scientific or practical areas

We aim to make a first decision within 1 month of submission. In case of a positive first decision the work will be provisionally contracted: the final decision about publication will depend upon the result of the anonymous peer review of the complete manuscript. We aim to have the complete work peer-reviewed within 3 months of submission. Proposals should include:

- A short synopsis of the work or the introduction chapter
- The proposed Table of Contents
- CV of the lead author(s)
- List of courses for possible course adoption.

The series discourages the submission of manuscripts that are below 65,000 words in length.

For inquiries and submissions of proposals, authors can contact Ties.Nijssen@Springer.com

More information about this series at http://www.springer.com/series/13799

Horacio Arló-Costa • Vincent F. Hendricks
Johan van Benthem
Editors

Readings in Formal Epistemology

Sourcebook

Assistant Editors
Henrik Boensvang
Rasmus K. Rendsvig

Springer

Editors
Horacio Arló-Costa (deceased)

Johan van Benthem
University of Amsterdam
Amsterdam, The Netherlands

Stanford University
Stanford, United States

Vincent F. Hendricks
Center for Information and Bubble Studies
University of Copenhagen
Copenhagen, Denmark

Assistant Editors
Henrik Boensvang
University of Copenhagen
Copenhagen, Denmark

Rasmus K. Rendsvig
Lund University
Lund, Sweden

Center for Information and Bubble Studies
University of Copenhagen
Copenhagen, Denmark

Springer Graduate Texts in Philosophy
ISBN 978-3-319-20450-5 ISBN 978-3-319-20451-2 (eBook)
DOI 10.1007/978-3-319-20451-2

Library of Congress Control Number: 2016935072

Springer Cham Heidelberg New York Dordrecht London

Printed on acid-free paper

Springer International Publishing AG Switzerland is part of Springer Science+Business Media (www.springer.com)

On June 1, 2011, we received an email from Horacio, expressing his pleasure in our successful cooperation that was drawing to a close and his eager anticipation of the published book. This would turn out to be the last correspondence we received from our friend. Horacio died on July 14, 2011. This anthology would not have seen the light of day without Horacio's encyclopedic knowledge and fine scholarship, his original angle of approach to philosophy, and his constant ambition to push the formal epistemology agenda forward. Horacio will never see our joint work in print, but we will continue his quest and honor his legacy by dedicating this book to him.

Vincent F. Hendricks
Johan van Benthem

Preface

"Formal epistemology" is a term coined in the late 1990s for a new constellation of interests in philosophy, merging traditional epistemological concerns with new influences from surrounding disciplines like linguistics, game theory, and computer science. Of course, this movement did not spring to life just then. Formal epistemological studies may be found in the classic works of Carnap, Hintikka, Levi, Lewis, Kripke, Putnam, Quine, and many others.

Formal epistemology addresses a growing agenda of problems concerning knowledge, belief, certainty, rationality, deliberation, decision, strategy, action, and agent interaction – and it does so using methods from logic, probability theory, computability theory, decision theory, game theory, and elsewhere. The use of these formal tools is to rigorously formulate, analyze, and sometimes solve important issues of interest to philosophers but also to researchers in other disciplines, from the natural sciences and humanities to the social and cognitive sciences and sometimes even the realm of technology. This makes formal epistemology an interdisciplinary endeavor practiced by philosophers, logicians, mathematicians, computer scientists, theoretical economists, social scientists, cognitive psychologists, etc.

Although a relative newcomer, formal epistemology is already establishing itself in research environments and university curricula. There are conferences, workshops, centers, and jobs in formal epistemology, and several institutions offer courses or seminars in the field.

Yet no volume is in existence comprising canonical texts that define the field by exemplars. Lecturers and students are forced to collect influential classics and seminal contemporary papers from uneven sources, some of them hard to obtain even for university libraries. There are excellent anthologies in mainstream epistemology, but these are not tuned to new fruitful interactions between the mainstream and a wider spectrum of formal approaches.

Readings in Formal Epistemology is intended to remedy this situation by presenting some three dozen key texts, divided into five subsections: Bayesian Epistemology, Belief Change, Decision Theory, Logics of Knowledge and Belief, and Interactive Epistemology. The selection made is by no means complete but hopefully representative enough for an accurate picture of the landscape. This collection will hopefully serve as a study and research companion while also helping shape and stimulate a flourishing new field in philosophy and its broader intellectual environment.

Pittsburgh, PA, USA	Horacio Arló-Costa
Copenhagen, Denmark	Vincent F. Hendricks
Amsterdam, The Netherlands	Johan van Benthem
Copenhagen, Denmark	Henrik Boensvang
Lund, Sweden	Rasmus K. Rendsvig

Acknowledgments

On the way to compiling this volume, we have been assisted by many people. Jeffrey Helzner and Gregory Wheeler gave us valuable suggestions for texts to include. We are grateful for their assistance in the selection process. Many authors represented in this volume provided us with essential copies of their papers while also giving important input on the organization of this collection. We thank them for their kind help. We would have liked to have included even more seminal papers, but due to limitations of space, and the fact that some copyrights were either impossible trace or too expensive to obtain, we ended up with the current selection. We are furthermore indebted to Springer Science and Business Media for taking on this project, especially Ties Nijssen, Christi Lue, and Werner Hermens. The editors also acknowledge the generous funding provided by the Elite Research Prize from the Danish Ministry of Science, Technology, and Innovation awarded to Vincent F. Hendricks in 2008.

Finally, this volume would not have seen the light of day without the constant efforts of Henrik Boensvang and Rasmus K. Rendsvig in communicating with relevant parties, collecting the required permissions, and compiling all the papers patiently and efficiently while paying painstaking attention to detail. In the process, they have more than earned the right to the title of assistant editors of *Readings in Formal Epistemology*.

Copyright Acknowledgments

We would like to thank authors, editors, publishers, copyright holders, permissions officers, and other parties who allowed us to reprint the papers found in this volume.

Helzner, J. and Hendricks, V.F. (2010). "Agency and Interaction: What we are and what we do in formal epistemology," *Journal for the Indian Council of Philosophical Research*, vol. XXVII:2, 2010: 44–71, special issue on *Logic and Philosophy Today*, guest edited by Amitabha Gupta and Johan van Benthem.

Part I. Bayesian Epistemology

Ramsey, F.P. (1926) "Truth and Probability," in Ramsey, F.P. (1931), *The Foundations of Mathematics and other Logical Essays*, Ch. VII, p.156–198, edited by R.B. Braithwaite, London: Kegan, Paul, Trench, Trubner & Co., New York: Harcourt, Brace and Company.

Jeffrey, R.C. (1968), "Probable Knowledge," in *The Problem of Inductive Logic*, ed. I. Lakatos, 166–180, Amsterdam: North-Holland. Courtesy of Edith Jeffrey.

Van Fraassen, B.C. (1995) "Fine-Grained Opinion, Probability, and the Logic of Full Belief." *Journal of Philosophical Logic* 24 (4).

Gaifman, H. (1986) "A theory of higher order probabilities," in *Theoretical Aspects of Reasoning About Knowledge: Proceedings of the 1986 conference on Theoretical aspects of reasoning about knowledge,* pp. 275—292, Morgan Kaufmann Publishers Inc. (Monterey, California).

Levi, I. (1974), "*On Indeterminate Probabilities*," *The Journal of Philosophy,* 71, 391–418.

Glymour, C. (1981) "Why I'm not a Bayesian," excerpt from Glymour, C. (1981) *Theory and Evidence,* Chicago University Press, 63–93.

Skyrms, B. (1993) "A Mistake in Dynamic Coherence Arguments – Discussion." *Philosophy of Science*, 60(2):320–328.

Arntzenius, F. (2003), "Some problems for conditionalization and reflection," *The Journal of Philosophy*, Vol. C, No. 7, 356–371.

M. J. Schervish, T. Seidenfeld and J. B. Kadane (2004) "Stopping to Reflect," *The Journal of Philosophy*, Vol. 101, No. 6, 315–322.

Part II. Belief Change

Alchourron, C.E., Gardenfors, P., and Makinson, D. (1985) "On the Logic of Theory Change: Partial Meet Contraction and Revision Functions." *Journal of Symbolic Logic,* 50(2): 510–530. Reprinted with the permission of the copyright holders, the Association of Symbolic Logic.

Hansson, S.O. (1993) "Theory Contraction and Base Contraction Unified." *Journal of Symbolic Logic,* 58(2): 602–625. Reprinted with the permission of the copyright holders, the Association of Symbolic Logic.
Levi, I. How Infallible but Corrigible Full Belief is Possible, hitherto unpublished.
Rott, H. (1993) "Belief Contraction in the Context of the General Theory of Rational Choice." *Journal of Symbolic Logic*, 58(4): 1426–1450. Reprinted with the permission of the copyright holders, the Association of Symbolic Logic.
Spohn, W. (2009) "A Survey of Ranking Theory." In Franz Huber and Christoph Schmidt-Petri (eds.). *Degrees of Belief.* Dordrecht: Springer.

Part III. Decision Theory

Savage, L. (1972) "Allais's Paradox" *The Foundations of Statistics*, Dover Publications, Inc., New York, 101–103.
Seidenfeld, T. (1988), "Decision Theory without 'Independence' or without 'Ordering': What is the Difference," *Economics and Philosophy*, 4: 267–290.
Gilboa, I. and M. Marinacci (forthcoming) "Ambiguity and the Bayesian Paradigm," *Advances in Economics and Econometrics: Theory and Applications, Tenth World Congress of the Econometric Society*.
Schervish, M.J., Seidenfeld, T. and Kadane, J.B. (1990) "State-Dependent Utilities," *Journal of the American Statistical Association*, Vol. 85, No. 411, 840–847.
Gibbard, A. and Joyce, J.M. (1998) "Causal Decision Theory." In Salvador Barberà, Peter J. Hammond, and Christian Seidl, eds., *Handbook of Utility Theory*, Vol. 1: Principles, pp. 701–740. Dordrecht & Boston: Kluwer.
Tversky, A. and Kahneman, D. (1992). "Advances in prospect theory: Cumulative representation of uncertainty." *Journal of Risk and Uncertainty* 5: 297–323.

Part IV. Logics of Knowledge and Belief

Hintikka, J. (2007) "Epistemology without Knowledge and without Belief" in *Socratic Epistemology: Explorations of Knowledge-Seeking by Questioning*, Cambridge University Press.
Dretske, F.I. (1970) "Epistemic Operators." *The Journal of Philosophy*, Vol. 67, No. 24, 1007–1023.
Lewis, D. (1996) "Elusive Knowledge," *Australasian Journal of Philosophy*, Vol. 74(4), 549–567. Courtesy of Stephanie Lewis.
Nozick, R. (1981) "Knowledge and Skepticism" In *Philosophical Explanations,* Harvard University Press, 167–169, 172–179, 197–211, 679–690. Reprinted by permission of the publisher from "Knowledge and Skepticism," in PHILOSOPHICAL EXPLANATIONS by Robert Nozick, pp. 167–169, 172–179, 197–211, 679–690, Cambridge, Mass.: The Belknap Press of Harvard University Press, Copyright © 1981 by Robert Nozick.
Stalnaker, R. (2006) "On Logics of Knowledge and Belief," *Philosophical Studies* 120, 169–199.
Parikh, R. (2008) "Sentences, Belief and Logical Omniscience, or What Does Deduction Tell Us?." *The Review of Symbolic Logic*, 1(4). Reprinted with the permission of the copyright holders, the Association of Symbolic Logic.
Artemov, S.N. (2008) "The logic of justification." *The Review of Symbolic Logic*, 1(4):477–513. Reprinted with the permission of the copyright holders, the Association of Symbolic Logic.
Kelly, K. (2004) "Learning Theory and Epistemology" in *Handbook of Epistemology*, I. Niiniluoto, M. Sintonen, and J. Smolenski (eds.), Dordrecht: Kluwer.
Williamson, T. (2004) "Some Computational Constraints in Epistemic Logic," in *Logic, Epistemology and the Unity of Science*, S. Rahman et al (eds). Dordrecht: Kluwer Academic Publishers: 437–456.

Part V. Interactive Epistemology

Lewis, D. (1969) *Convention: A Philosophical Study*, Harvard University Press, 24–42 (excerpt). Courtesy of Stephanie Lewis.

Barwise, J. (1988) "Three Views of Common Knowledge." In *Proc. TARK'88*: 365–379, Morgan Kaufmann Publishers.

Baltag, A. and Smets, S. (2008) "A Qualitative Theory of Dynamic Interactive Belief Revision," in G. Bonanno, W. van der Hoek, M. Wooldridge (eds.), *Logic and the Foundations of Game and Decision Theory*, Texts in Logic and Games, Vol 3, 9–58, Amsterdam University Press.

Aumann, R. (1976) "Agreeing to Disagree," *Annals of Statistics* 4, 1236–1239.

Aumann, R. and Brandenburger, A. (1995) "Epistemic Conditions for Nash Equilibrium," *Econometrica,* Vol. 63, No. 5,1161–1180.

Stalnaker, R. (1996), "*Knowledge, belief and counterfactual reasoning in games*," *Economics and Philosophy* 12: 133–163

Halpern, J.Y., (2001) "*Substantive Rationality and Backward Induction*," *Games and Economic Behavior*, Elsevier, vol. 37(2), 425–435.

Contents

Part II Belief Change

Part III Decision Theory

Part IV Logics of Knowledge and Belief

Part V Interactive Epistemology

Contributors

Carlos E. Alchourrón (deceased) Universidad de Buenos Aires, Buenos Aires, Argentina

Horacio Arló-Costa (deceased) Carnegie Mellon University, Pittsburgh, PA, USA

Frank Arntzenius University of Oxford, Oxford, UK

Sergei Artemov Graduate Center CUNY, New York, NY, USA

Robert J. Aumann Federmann Center for the Study of Rationality, The Hebrew University of Jerusalem, Jerusalem, Israel

Alexandru Baltag ILLC, University of Amsterdam, The Netherlands

Jon Barwise (deceased) Stanford University, Stanford, CA, USA

Henrik Boensvang University of Copenhagen, Copenhagen, Denmark

Adam Brandenburger Stern School of Business, Tandon School of Engineering, NYU Shanghai, New York University, New York, NY, USA

Fred I. Dretske (deceased) University of Wisconsin, Madison, WI, USA

Haim Gaifman Columbia University, New York, NY, USA

Peter Gärdenfors Lund University, Lund, Sweden

Allan Gibbard University of Michigan, Ann Arbor, MI, USA

Itzhak Gilboa HEC, Paris, France

Tel-Aviv University, Tel Aviv, Israel

Clark Glymour Carnegie Mellon University, Pittsburgh, PA, USA

Joseph Y. Halpern Computer Science Department, Cornell University, Ithaca, NY, USA

Sven Ove Hansson Division of Philosophy, KTH, Stockholm, Sweden

Jeffrey Helzner AIG, New York, NY, USA

Vincent F. Hendricks Center for Information and Bubble Studies, University of Copenhagen, Copenhagen, Denmark

Jaakko Hintikka (deceased) Boston University, Helsinki, Finland

Richard C. Jeffrey (deceased) Princeton University, Boston, MA, USA

James M. Joyce University of Michigan, Ann Arbor, MI, USA

Joseph B. Kadane Departments of Statistics and Social and Decision Sciences, Carnegie Mellon University, Pittsburgh, PA, USA

Daniel Kahneman Princeton University, Princeton, NJ, USA

Kevin T. Kelly Carnegie Mellon University, Pittsburgh, PA, USA

Isaac Levi Columbia University, New York, NY, USA

David Lewis (deceased) Princeton University, Princeton, NJ, USA

David Makinson London School of Economics, London, UK

Massimo Marinacci Università Bocconi, Milano, Italy

Lawrence S. Moss Mathematics Department, Indiana University, Bloomington, IN, USA

Robert Nozick (deceased) Harvard University, Boston, MA, USA

Rohit Parikh City University of New York, New York, NY, USA

Frank P. Ramsey (deceased) University of Cambridge, Cambridge, UK

Rasmus K. Rendsvig Lund University, Lund, Sweden

Center for Information and Bubble Studies, University of Copenhagen, Copenhagen, Denmark

Hans Rott Department of Philosophy, University of Regensburg, Regensburg, Germany

Leonard Savage (deceased) Princeton University, New York, NY, USA

Mark J. Schervish Department of Statistics, Carnegie Mellon University, Pittsburgh, PA, USA

Teddy Seidenfeld Departments of Philosophy and Statistics, Carnegie Mellon University, Pittsburgh, PA, USA

Brian Skyrms Department of Philosophy, University of California, Irvine, CA, USA

Sonja Smets ILLC, University of Amsterdam, The Netherlands

Sławomir Solecki Mathematics Department, Indiana University, Bloomington, IN, USA

Wolfgang Spohn Fachbereich Philosophie, Universität Konstanz, Konstanz, Germany

Robert Stalnaker Department of Linguistics and Philosophy, MIT, Cambridge, MA, USA

Johan van Benthem University of Amsterdam, Amsterdam, The Netherlands

Stanford University, Stanford, United States

Bas C. van Fraassen San Francisco State University, San Francisco, CA, USA

Amos Tversky (deceased) Stanford University, Stanford, CA, USA

Timothy Williamson University of Oxford, Oxford, UK

About the Editors

The late Horacio Arló-Costa was Professor of Philosophy at Carnegie Mellon University, Pennsylvania. Arló-Costa served as editor for the *Review of Symbolic Logic*, as area editor in epistemology for *Synthese*, and as member of the editorial board for the *Journal of Philosophical Logic*.

Vincent F. Hendricks is Professor of Formal Philosophy at the University of Copenhagen and Director of the Center for Information and Bubble Studies (CIBS). His recent publications include *Handbook of Formal Philosophy (2015), Infostorms* (2014), *Mainstream and Formal Epistemology* (2007), and *The Convergence of Scientific Knowledge* (2001). He served as editor-in-chief of *Synthese* from 2005 to 2015.

Johan van Benthem is University Professor of Logic at Amsterdam University, Henry Waldgrave Stuart Professor of Philosophy at Stanford University, and Distinguished Foreign Expert at Tsinghua University, Beijing. His recent publications include *Logical Dynamics of Information and Interaction* (2011), *Modal Logic for Open Minds* (2010), *Exploring Logical Dynamics* (1996), and *Language in Action* (1995). Van Benthem is coeditor, with Alice ter Meulen, of the *Handbook of Logic and Language* (1997).

Chapter 1
Agency and Interaction What We Are and What We Do in Formal Epistemology

Jeffrey Helzner and Vincent F. Hendricks

Introduction

Formal epistemology is a recent field of study in formal philosophy dating back only a couple of decades or so (Helzner and Hendricks 2011; Hendricks 2006). The point of departure of this essay is rooted in two philosophically fundamental and interrelated notions central to formal epistemology;

- *agency* – what agents are, and
- *interaction* – what agents do.

Agents may be individuals, or they may be groups of individuals working together. In each of the sections that follow, assumptions are made concerning the relevant features of the agents at issue. For example, such relevant features may include the agent's beliefs about its environment, its desires concerning various possibilities, the methods it employs in learning about its environment, and the strategies it adopts in its interactions with other agents in its environment. Fixing these features serves to bound investigations concerning interactions between the agent and its environment. The agent's beliefs and desires are assumed to inform its decisions. Methods employed by the agent for the purposes of learning are assumed to track or approximate or converge upon the facts of the agent's environment. Strategies adopted by the agent are assumed to be effective in some sense.

J. Helzner (✉)
AIG, New York, NY, USA
e-mail: jeffreyhelzner@yahoo.com

V.F. Hendricks
Center for Information and Bubble Studies, University of Copenhagen, Copenhagen, Denmark
e-mail: vincent@hum.ku.dk

H. Arló-Costa et al. (eds.), *Readings in Formal Epistemology*, Springer Graduate Texts in Philosophy 1, DOI 10.1007/978-3-319-20451-2_1

In what follows is an attempt to locate predominant paradigms in formal epistemology – e.g., probability theory, belief revision theory, decision theory, logics of knowledge and belief and finally interactive epistemology – within the framework of agency and interaction.

Probability

Probabilities are very useful in formal epistemology. They are used to measure key epistemic components like belief and degrees thereof, the strength of expectations and predictions, may be used to describe actual occurrent frequencies in nature or in the agent's environment and of course probabilities play a paramount in accounting for notions of (Bayesian) confirmation (Earman 1992). Current cognitive models apply probabilities to represent aggregated experience in tasks involving language acquisition, problem solving and inductive learning, conditionalization and updating beliefs and scientific hypotheses.

What sorts of internal states are essential to the agent's representation of its environment? Various doxastic notions e.g., according to which the agent simply believes or is certain of propositions, in contrast to believing the proposition to some degree, are a traditional interest within mainstream epistemology. Some philosophers, e.g. Jeffrey (1992), have argued in favor of a doctrine known as *radical probabilism*. A central tenet of this doctrine is that various propositional attitudes of epistemic interest, especially full belief, are reducible to credal judgments. There are several ways that one might attempt such a reduction. Perhaps the most obvious is to identify full belief with maximal partial belief. For example, if it is assumed that the agent's credal state can be represented by a probability measure, then such a reduction would identify those propositions that are fully believed by the agent with those propositions that have maximal probability according to this representing measure. Following this proposal, it would seem that a proposition counts as a serious possibility for the agent just in case its negation is not assigned maximal probability according to the probability measure representing the agent's credal judgments. Hence, by the probability axioms, a proposition counts as seriously possible for the agent just in case it has nonzero probability under the representing measure. This leads to certain difficulties. For example, if the agent is concerned to estimate the height of an object that is sufficiently distant, then the agent might regard a continuum of values as possible – e.g., the height of the object is judged to be between three and four feet. According to the suggested reduction, such a continuum of possible values for the height of the object could not serve as a set of serious possibilities, since it is a mathematical fact that no probability measure can distribute positive probability to each element of such a continuum. The interested reader is urged to consult van Fraassen (1995) and Arlo-Costa (2001) for more sophisticated versions of probabilism.

Following Levi (1980), one may assume that the agent is, at each point in time, committed to full belief in some set of propositions concerning its environment.

Where the agent is not committed to full belief in a given proposition, the negation of that proposition is a serious possibility for the agent. The agent may judge some serious possibilities to be more probable than others. What can be said about these judgments? The received view, following a tradition that goes back to the work of Ramsey (1931), maintains that such credal judgments ought to be representable by a probability measure. This view has been criticized as being too weak and as being too strong. As for being too weak, the simple requirement that such judgments be representable by a probability measure says little about the extent to which these subjective probabilities should approximate objective probabilities, e.g., limiting frequencies in the sense of von Mises (1957) or perhaps even propensities in the sense of Popper (1959). Various principles have been offered in order to require that the subjective probabilities of a rational agent are informed by that agent's knowledge of objective probabilities (Kyburg 1974; Levi 1978; Lewis 1980). As for being too strong, requiring credal judgments to be representable by a probability measure implies, among other things, that such credal judgments are complete. A consequence of such a requirement is that, for any given pair of serious possibilities, the agent either judges one of the possibilities to be more probable than the other or the agent regards the possibilities as being equally probable. Thus, the requirement bars situations in which the agent, because of a lack of information, is unable to supply such a judgment. Such considerations, which to some extent echo earlier, related concerns of Keynes (1921) and Knight (1921), have motivated some – e.g., Kyburg (1968), Levi (1974) and Walley (1990) – to consider indeterminate probabilities, either in the form of interval-valued measures or sets of traditional measures, in representing rational credences.

Belief Change

As already hinted, some probability theorists tend to think that belief, as opposed to knowledge, may be good enough for action, deliberation and decision. Thus beliefs may suffice as they can serve important epistemic purposes while holding the information, expectations and conjectures that agents act on. Beliefs may be used for making creative leaps beyond what is logically implied by available information, evidence or knowledge and are crucial in agent interaction models representing what agents think about moves, strategies, payoffs and beliefs of other agents and what agent rationality amounts to Finally, beliefs and belief revision are prime vehicles for understanding the mechanism of learning by trial-and-error, one of the main motors of scientific inquiry in general.

Initially, an agent has beliefs about the environment with which it interacts. Sometimes these interactions are such that the agent, on pain of irrationality, must revise its beliefs. The classic example is that of a scientific agent who has beliefs about the world that might need to be revised in light of new data. The study of this sort of example has a long history in the philosophy of science, where it is often discussed at a relatively informal level in connection with topics such as underdetermination. In the context of formal epistemology, the study of belief revision has been generalized to include various sorts of epistemic agents. Questions such as the following suggest the range of theoretical options that are available in connection with such investigations:

How are the potential belief states to be interpreted? One might take the belief states to represent partial beliefs; e.g., the agent has a certain degree of belief in proposition *P*. Alternatively, one might be interested in states of full belief, expectation or plain belief; e.g., the agent fully believes *P*, expects *P*, etc. Further refinements have been considered. For example, one might consider those full beliefs with respect to which the agent manifests some level of awareness; e.g., I am aware of my belief that I am presently writing the words of this sentence. In contrast to a focus on conscious beliefs, one might consider those propositions that the agent is committed to fully believing; e.g., all of those propositions that are deducible from my conscious beliefs.

How are the potential belief states to be represented? The answers to this question depend, at least to some extent, on how the previous question is answered. For example, if partial beliefs are the issue, then probability distributions might be taken as the basis for the representation so that a potential belief state is represented as a probability measure over the possible states of nature. On the other hand, if the problem is the representation of commitments to full belief (expectation, plain belief), then one might specify a suitably formalized language and represent each potential belief state as a theory formulated over the given language so that membership in the theory indicates full belief.

How are revisions interpreted? If credal states are the concern, then modifications of the credal state might be understood in terms of something like conditionalization. The interested reader is urged to consult (Halpern 2003) for a survey of various proposals concerning the representation and modification of credal states. What about revising or contracting states of full belief? When an instance of belief contraction concerning full beliefs is the result of the agent selecting from a set of (full) belief states that the agent recognizes as potential alternatives, then such an instance may be regarded as the resolution of a decision problem. Isaac Levi has developed a decision-theoretic approach to belief change; important discussions of Levi's approach include (1980), which considers belief change in the context Levi's general approach to epistemology, and Arlo-Costa and Levi (2006), Arlo-Costa and Liu (2011) which gives greater emphasis to the formal details concerning Levi's approach. Different connections between choice and belief revision are emphasized in Rott (1993). Rott demonstrates an important correspondence between the "AGM" account of belief revision offered in Alchourron et al. (1985) and the economists' study of rational choice functions. Finally, it is worth noting that where both partial and full beliefs are considered, there may be significant dependencies between the modification of these two sorts of belief states. For example, if the credal judgments of rational agents are a function of their judgments of full belief, as some philosophers assume, then changes to the latter may result in changes to the former.

There are other alternative interpretations of doxastic change. Spohn have considered change from the point of view of entire epistemic states rather than mere beliefs in terms of ranking functions and plausibility representations (Spohn 2009) while Hansson have considered change from the point of belief bases as finite sets of sentences that likewise are possible axiomatic bases for a given theory (Hansson 1993).

Decision Theory

An agent interacts with its environment through the choices it makes. Choice presupposes alternatives, and a theory of rational choice should, at least, distinguish some of the available alternatives as admissible. As an example, consider those accounts of rational choice that are built on the concept of preference. One such account assumes that the agent has complete and transitive preferences over the set of available alternatives. Those alternatives that are optimal with respect to the given preference ranking are taken as admissible. This abstract preference-based account says nothing about the way in which preferences are informed by the agent's beliefs about its environment. Subjective expected utility theory [SEU], which is at the center of modern-day decision theory, provides significantly more detail than the abstract theory of preference optimization. SEU assumes that alternatives are acts, which, following Savage's classic formulation of SEU in Savage (1972), are functions from states to consequences. Drawing upon the earlier work of Ramsey (1931) on subjective probability and the work of von Neumann and Morgenstern (1947) on utility, Savage provides conditions on the agent's preferences over acts that guarantee the existence of a probability measure p and a utility function u such that the agent's preferences can be regarded as if they were the result of maximizing utility u with respect to probability p. According to the intended interpretation, the probability measure p represents the agent's degrees of belief concerning the possible states and the utility function u represents the extent to which the agent values the possible consequences.

The assumptions of SEU may be questioned in various ways. We focus on two ways that have generated significant interest among philosophers. First, why should it be that the rational agent's degrees of belief can be represented by a probability distribution p? As already noted, it is not clear why such an assumption should obtain in cases where the agent has very little information concerning the possible states. Second, in SEU it is assumed that the agent's subjective probability concerning the states is independent of the act that is chosen. Some question this assumption and offer examples in which a modification of SEU that provides for such dependencies, through the use of conditional probabilities, is supposed to give an irrational recommendation. The first line of questioning has led some – e.g., Ellsberg (1961), Levi (1974, 1977), and Gardenfors and Sahlin (1982) – to use indeterminate probabilities in their normative accounts of decision making under uncertainty. The second line of questioning has led some – e.g., Gibbard and Harper (1978), Lewis (1981), and Joyce (1999) – to investigate causal decision theory.

Logics of Knowledge and Belief

What is now known as epistemic logic started with the study of proper axiomatizations for knowledge, belief, certainty and other epistemic attitudes. Hintikka inaugurated the field with his seminal book Hintikka (1962) which focuses on axiomatizing knowledge and belief in mainly mono-agent systems. Agents are syntactically represented as indices on epistemic operators in a formal logical language. From the semantic perspective, to be an agent is to be an index on an accessibility relation between possible worlds representing the epistemic alternatives over which the agent has to succeed in order to know some proposition (interesting alternative semantics to Kripke semantics have been developed by Arlo-Costa and Pacuit 2006, Baltag and Moss 2004 and others). Like many other philosophical logics in their infancy, interesting axiomatizations governing the logics of knowledge and belief took center stage in the beginning together with nailing down important logical properties for these new logics. The field was living a somewhat isolated life remote from the general concerns of mainstream epistemology. Hintikka himself (and a few others like Lenzen 1978) was a notable exception and insisted on telling a better story, not about what agents are in the logical language, but about what they do and the meaning of epistemic axioms for epistemology (Stalnaker 2006). Accordingly, Hintikka took axioms of epistemic logic to describe a certain kind of strong rationality much in sync with the autoepistemological tradition of G.E. Moore and especially Norman Malcolm. Axioms of epistemic logic are really prescriptions of rationality in mono-agent systems. Epistemic logic has since been used address a number of important philosophical problems including for instance the Fitch Paradox (Brogaard and Salerno 2009), the problem of logical omniscience (Duc 1997; Parikh 2005), and various conceptual characterizations of knowledge and other epistemic attitudes (Kraus and Lehmann 1988).[1]

But rationality considerations are not only central to the singular agent acting in some environment, call it nature, but likewise, and perhaps especially, central to agents when in presence of other agents and interacting with these. Thus mono-agent systems had to be extended to multi-modal systems in order to get both agency and interaction off the epistemological ground for real. A sea-change took place in epistemic logic in the late 1980s and the beginning of the 1990s especially due to the work of Joseph Halpern and his collaborators (Fagin et al. 1995) and others (Meyer and Hoek 1995). Multiple agents were introduced into the logical language which, along with multiple epistemic accessibility relations on the semantic level, gave rise to a precise and adequate representation of the flow of information through an agent system, together with the nature of various protocols governing such systems. In this setting, possible worlds are to be understood as the states of the system taken as a whole, or sometimes the possible histories or consecutive runs of the system as a

[1]For solid overviews refer to De Bruin (2008) and Gochet and Gribomont (2006).

whole, that are compatible with the state transition directives which rule the system. Stalnaker has recently summarized the consequences of this sea-change precisely:

> The general lesson I drew from this work was that it was useful for epistemology to think of communities of knowers, exchanging information and interacting with the world, as (analogous to) distributed computer systems. (Hendricks and Roy 2010: 78)

Agent systems can now be thought of as encompassing everything from a group of robots on an assembly line to a group of poker players in Texas Hold 'Em. In turn, there is much more to what agents are nowadays, but also much more to what they do dynamically (as opposed to statically in terms of, say, (van Ditmarsch et al. 2008) epistemic axioms describing the rationality of single agents). Dynamic epistemic logic is a rich blend of studies ranging multi-agent axiomatizations of knowledge, belief, common knowledge and belief (Barwise 1988) certainty, uncertainty, doubt, ignorance and a host of other epistemic attitudes; models of the interplay between knowledge and games (Benthem 2001, 2007), knowledge and justification in mainstream epistemology (Artemov and Nogina 2005), social software (Parikh 2002), knowledge and public announcement of information (Baltag et al. 2002), knowledge intertwined with preferences, actions and decisions (Liu 2011); knowledge acquisition in light of formal learning theory, logical reliability, methods of scientific inquiry and computability studies (Gierasimczuk 2009; Hendricks 2001; Kelly 1996, 2004), belief revision (Baltag and Smets 2008), models of agent interaction in multi-agent systems; combined multi-agent and multi-modal systems in which for instance the development of knowledge over time may be scrutinized (Kraus and Lehmann 1988), relations between knowledge and deontic commitments investigated, divisions of cognitive labor modeled and so forth (for epistemic logic paired up with mainstream epistemological concerns, refer to Williamson (2006), Hendricks (2006) and Hendricks and Pritchard (2007)).

Interactive Epistemology

Theoretical economics is to a significant extent about understanding, anticipating and modeling phenomena like trading, stock speculation, real-estate dealing, hostile company take-overs, shareholding, convention and so forth. Obviously, agency and interaction play a paramount role here and seen from this perspective economics is about multi-individual and collective action balancing information and utility.

Independently, but informed by the developments in epistemic logic, economists have used game theory to scrutinize an extensive spread of the mentioned phenomena. By way of example, in 1976 the later Nobel Prize Laureate Robert Aumann published his famous Agreement Theorem in "Agreeing to Disagree" in which he describes conditions under which two "like minded" agents or players cannot "agree to disagree" in the sense that if the two players' posteriors of some event are common knowledge then they must coincide. In other words, in order to make trade possible, agents have to agree to disagree (Aumann 1976). That is agency in terms of players, interaction in terms of games.

On the way to this result Aumann made a host of assumptions about the nature knowledge much in tune with what is to be found in epistemic logic like the axiomatic strength of knowledge in order to infer the backwards induction equilibrium and assumptions about what is common knowledge among the players (Halpern 2001). In 1999, Aumann coined a term for these kinds of study in theoretical economics: "Interactive epistemology" (Aumann 1999). It denotes an epistemic program studying shared knowledge and belief given more than one agent or player in an environment and has, as already suggested, strong ties to game theoretic reasoning and questions of common knowledge and belief, backward induction, various forms of game equilibria and strategies in games, (im)perfect information games, (bounded) rationality etc (Aumann and Brandenburger 1995; Stalnaker 1996, 2006).

Given its inauguration with Aumann, the program was in the beginning dominated by scholars drawn from theoretical economics and computer science rather than philosophy and logic, but recently philosophers and logicians have begun to pay close attention to what is going on in this striving program of formal epistemology. And for good reason too; social epistemology focuses on knowledge acquisition and justification in groups or institutions (Goldman 1999) and the extent to which exactly institutions may be viewed as genuine agents (List and Pettit 2011) while the interactive epistemological approach to agency and interaction also have close shaves with the major new focal points in dynamic epistemic logic (Benthem 2011) and much of the technical machinery is a common toolbox for both paradigms (Brandenburger 2007).

Formal Epistemology

Formal epistemology is the study of crucial concepts in mainstream epistemology including knowledge, belief (-change), certainty, rationality, reasoning, decision, justification, learning, agent interaction and information processing using formal tools from three streams; probability, logic and computability. In particular, the tools may come from tool boxes like modal logic, probability calculus, game theory, decision theory, formal learning theory, computability theory and distributed computing. Practitioners of formal epistemology include philosophers, computer scientists, social scientists, cognitive psychologists, theoretical economists, mathematicians, and theoretical linguists but also scholars from the empirical sciences like cognitive science, engineering and biology are onboard. This mixed bag of practitioners is surely a witness to the thoroughly interdisciplinary nature of formal epistemology and its wide range of applications in natural science, social science, humanities and the technical sciences.

Formal epistemology is right in the middle; between mainstream epistemology's fairly abstract theories on the one hand and the more concrete cognitive sciences devoted to the empirical reality of agency and interaction on the other. In formal epistemology we are walking the fine line between theory and reality. This is as it should be: The hallmark of a progressive research program.

This is an edited and reorganized version of the paper "Agency and Interaction: What We Are and What We Do in Formal Epistemology", *Journal for the Indian Council of Philosophical Research*, nr. 2, vol. XXVII, 2010: 44–71, special issue on *Logic and Philosophy Today*, guest edited by Amitabha Gupta and Johan van Benthem.

References

Alchourron, C., Gardenfors, P., & Makinson, D. (1985). On the logic of theory change: Partial meet functions for contraction and revision. *Journal of Symbolic Logic, 50*, 510–530.

Arlo-Costa, H. (2001). Bayesian epistemology and epistemic conditionals: On the status of the export-import laws. *The Journal of Philosophy, 98*(11), 555–593.

Arlo-Costa, H., & Levi, I. (2006). Contraction: On the decision-theoretical origins of minimal change and entrenchment. *Synthese, 152*(1), 129–154.

Arlo-Costa, H., & Liu, H. (2011). Value-based contraction: A representation result. In *Proceedings of TARK'11*. New York: ACM Digital Library (ACM DL).

Arlo-Costa, H., & Pacuit, E. (2006). First order classical modal logic. *Studia Logica, 84*(2), 171–210.

Artemov, S., & Nogina, E. (2005). Introducing justification into epistemic logic. *Journal of Logic and Computation, 15*, 1059–1073.

Aumann, R. (1976). Agreeing to disagree. *The Annals of Statistics, 4*(6), 1236–1239.

Aumann, R. (1999). Interactive epistemology I. *International Journal of Game Theory, 28*(3), 263–300.

Aumann, R., & Brandenburger, A. (1995). Epistemic conditions for nash equilibrium. *Econometrica, 63*(5), 1161–1180.

Baltag, A., & Moss, L. (2004). Logics for epistemic programs. *Synthese/Knowledge, Rationality and Action, 139*, 165–224.

Baltag, A., & Smets, S. (2008). A qualitative theory of dynamic interactive belief revision. In *Logic and the foundation of game and decision theory* (LOFT7 2008) (Vol. 3, pp. 13–60). Amsterdam: Amsterdam University Press.

Baltag, A., Moss, L. S., & Solecki, S. (2002). The logic of public annoucements, common knowledge, and private suspicion. In *Proceedings of TARK 1998* (pp. 43–56). Los Altos: Morgan Kaufmann Publishers.

Barwise, J. (1988). Three theories of common knowledge. In *Proceedings of the 2nd conference on theoretical aspects of reasoning about knowledge* (pp. 365–379). Pacific Grove: California.

Benthem, J. v. (2001). Games in dynamic epistemic logic. *Bulletin of Economic Research, 53*(4), 219–224.

Benthem, J. v. (2007). Logic games, from tools to models of interaction. In A. Gupta, R. Parikh, & J. van Benthem (Eds.), *Logic at the crossroads* (pp. 283–317). Mumbai: Allied Publishers.

Benthem, J. v. (2011). *Logical dynamics of information and interaction*. Cambridge: Cambridge University Press.

Brandenburger, A. (2007). The power of paradox: Some recent developments in interactive epistemology. *International Journal of Game Theory, 35*(4), 465–492.

Brogaard, B., & Salerno, J. (2009). Fitch's paradox of knowability. In E. Zalta (Ed.), *The stanford encyclopedia of philosophy*. Palo Alto: Stanford University.

De Bruin, B. (2008). Epistemic logic and epistemology. In V. F. Hendricks & D. Prichard (Eds.), *New waves in epistemology*. London: Palgrave Macmillan.

Duc, H. N. (1997). Reasoning about rational, but not logically omniscient agents. *Journal of Logic and Computation, 7*, 633–648.
Earman, J. (1992). *Bayes or bust?* Cambridge: MIT.
Ellsberg, D. (1961). Risk, ambiguity, and the savage axioms. *The Quarterly Journal of Economics, 75*, 643–669.
Fagin, R., Halpern, J. Y., Moses, Y., & Vardi, M. Y. (1995). *Reasoning about knowledge*. Cambridge: MIT.
Gardenfors, P., & Sahlin, N. E. (1982). Unreliable probabilities, risk taking, and decision making. *Synthese, 53*, 361–386.
Gibbard, A., & Harper, W. (1978). Counterfactuals and two kinds of expected utility". In C. A. Hooker, J. J. Leach, & E. F. McClennen (Eds.), *Foundations and applications of decision theory* (pp. 125–162). Dordrecht: Reidel.
Gierasimczuk, N. (2009). Learning by erasing in dynamic epistemic logic. In A. H. Dediu, A. M. Ionescu, & C. Martín-Vide *Language and automata theory and applications* (Lecture notes in computer science, Vol. 5457, pp. 362–373). Berlin/Heidelberg: Springer.
Gochet, P., & Gribomont, P. (2006). Epistemic logic. In *Logic and the modalities in the twentieth century* (Handbook of the history of logic, pp. 99–185). Amsterdam: Elsevier.
Goldman, A. (1999). *Knowledge in a social world*. New York: Oxford University Press.
Halpern, J. Y. (2001). Substantive rationality and backwards induction. *Games and Economic Behavior, 37*(2), 425–435.
Halpern, J. Y. (2003). *Reasoning about uncertainty*. Cambridge: MIT.
Hansson, S. O. (1993). Theory contraction and base contraction unified. *Journal of Symbolic Logic, 58*(2), 602–625.
Helzner, J., & Hendricks, V. F. (2011, forthcoming). Formal epistemology. In *Oxford bibliography online*. http://www.oxfordbibliographies.com/view/document/obo-9780195396577/obo-9780195396577-0140.xml?rskey=a6uyIx&result=81
Hendricks, V. F. (2001). *The convergence of scientific knowledge – A view from the limit*. Dordrecht: Springer.
Hendricks, V. F. (2006). *Mainstream and formal epistemology*. New York: Cambridge University Press.
Hendricks, V. F., & Pritchard, D. (Eds.) (2007). *Epistemology: 5 questions*. New York: Automatic Press/London: VIP.
Hendricks, V. F., & Roy, O. (Eds.) (2010). *Epistemic logic: 5 questions*. New York: Automatic Press/London: VIP.
Hintikka, J. (1962). *Knowledge and belief: An introduction to the logic of the two notions*. Ithaca: Cornell University Press.
Jeffrey, R. (1992). *Probability and the art of judgment*. New York: Cambridge University Press.
Joyce, J. (1999). *The foundations of causal decision theory*. Cambridge: Cambridge University Press.
Kelly, K. T. (1996). *The logic of reliable inquiry*. New York: Oxford University Press.
Kelly, K. T. (2004). Learning theory and epistemology. In I. Niiniluoto, M. Sintonen, & J. Smolenski (Eds.), *Handbook of epistemology*. Dordrecht: Kluwer.
Keynes, J. M. (1921). *A treatise on probability*. London: MacMillan.
Knight, F. H. (1921). *Risk, uncertainty and profit*. Boston/New York: Houghton-Mifflin.
Kraus, S., & Lehmann, D. (1988). Knowledge, belief and time. *Theoretical Computer Science, 58*, 155–174.
Kyburg, H. E. (1968). Bets and beliefs. *American Philosophical Quarterly, 5*(1), 54–63.
Kyburg, H. E. (1974). *The logical foundations of statistical inference*. Dordrecht: Reidel.
Lenzen, W. (1978). *Recent work in epistemic logic* (Acta philosophica fennica, Vol. 30, pp. 1–219). Amsterdam: North-Holland
Levi, I. (1974). On indeterminate probabilities. *The Journal of Philosophy, 71*, 391–418.
Levi, I. (1977). Direct inference. *The Journal of Philosophy, 74*, 5–29.
Levi, I. (1978). In A. Hooker, J. J. Leach, & E. F. McClennen (Eds.), *Foundations and applications of decision theory* (pp. 263–273). Dordrecht/Boston: D. Reidel.

Levi, I. (1980). *The enterprise of knowledge*. Cambridge: MIT.
Lewis, D. (1980). A subjectivist's guide to objective chance. In R. Jeffrey (Ed.), *Studies in inductive logic* (Vol. II). Berkeley/Los Angeles: University of California Press.
Lewis, D. (1981). Causal decision theory. *Australasian Journal of Philosophy, 59*, 5–30.
List, C., & Pettit, P. (2011). *Group agency*. New York: Oxford University Press.
Liu, F. (2011). *Reasoning about preference dynamics*. Dordrecht: Springer.
Meyer, J. -J. Ch., & Hoek, W. van der (1995). *Epistemic logic for AI and computer science* (Cambridge tracts in theoretical computer science, Vol. 41). Cambridge: Cambridge University Press.
Parikh, R. (2002). Social software. *Synthese, 132*, 187–211.
Parikh, R. (2005). Logical omniscience and common knowledge: WHAT do we know and what do WE know? In *Proceedings of TARK05* (pp. 62–77). Singapore: National University of Singapore.
Popper, K. R. (1959). The propensity interpretation of probability. *British Journal for the Philosophy of Science, 10*, 26–42.
Ramsey, F. P. (1931). Truth and probability. In R. B. Braithwaite (Ed.) *The foundations of mathematics and other logical essays*. London: Routledge and Kegan Paul.
Rott, H. (1993). Belief contraction in the context of the general theory of rational choice. *The Journal of Symbolic Logic, 58*(4), 1426–1450.
Savage, L. J. (1972) *The foundations of statistics*. New York: Dover. The Dover edition is a republication of the 1954 work.
Spohn, W. (2009). A survey of ranking theory. In F. Huber & C. Schmidt-Petri (Eds.), *Degrees of belief* (pp. 185–228). Dordrecht: Springer.
Stalnaker, R. (1996). Knowledge, belief and counterfactual reasoning in games. *Economics and Philosophy, 12*, 133–163.
Stalnaker, R. (2006). On logics of knowledge and belief. *Philosophical Studies, 128*(1), 169–199.
van Ditmarsch, H., Hoek, W. v. d., & Kooi, B. (2008). *Dynamic epistemic logic*. Dordrecht: Springer.
van Fraassen, B. (1995). Fine-grained opinion, probability, and the logic of full belief. *Journal of Philosophical Logic, 24*(4), 349–377.
von Mises, R. (1957). *Probability, statistics and truth* (Revised English ed.). New York: Macmillan.
von Neumann, J., & Morgenstern, O. (1947). *Theory of games and economic behavior*. Princeton: Princeton University Press.
Walley, P. (1990). *Statistical reasoning with imprecise probabilities*. New York: Chapman and Hall.
Williamson, T. (2006). *Knowledge and its limits*. Oxford: Oxford University Press.

Part I
Bayesian Epistemology

Chapter 2
Introduction

Horacio Arló-Costa, Vincent F. Hendricks, and Johan van Benthem

There are various possible ways of articulating what Bayesian epistemology is and how it relates to other branches of formal and mainstream epistemology. Following the steps of Ramsey, Richard Jeffrey outlines in his article "Probable Knowledge" a possible way of constructing an epistemology grounded on Bayesian theory. While knowledge is a central notion in traditional epistemology (and in various branches of formal epistemology) Jeffrey suggests an epistemology where knowledge does not have the importance generally attributed to it. The idea is "[...] to try to make the concept of belief do the work that philosophers have generally assigned to the grander concept" (knowledge). Moreover the notion of belief is pragmatically analyzed along the lines proposed by Ramsey: "the kind of measurement of belief with which probability is concerned is .... a measurement of belief qua basis of action". The result of this move is to conceive the logic of partial belief as a branch of decision theory. So, the first two essays in this section are also quite relevant for the section of decision theory presented below (Ramsey's essay contains the first axiomatic presentation of decision theory). Both Jeffrey and Ramsey present the foundations of an epistemology which is deeply intertwined with a theory of action.

Horacio Arló-Costa was deceased at the time of publication.

H. Arló-Costa (deceased)
Carnegie Mellon University, Pittsburgh, PA, USA

V.F. Hendricks (✉)
Center for Information and Bubble Studies, University of Copenhagen, Copenhagen, Denmark
e-mail: vincent@hum.ku.dk

J. van Benthem
University of Amsterdam, Science Park 904, 1098 XH Amsterdam, The Netherlands

Stanford University, Stanford, United States
e-mail: johan@science.uva.nl

H. Arló-Costa et al. (eds.), *Readings in Formal Epistemology*, Springer Graduate Texts in Philosophy 1, DOI 10.1007/978-3-319-20451-2_2

This move has a behaviorist pedigree but perhaps the behavioral inspiration is not an essential ingredient of an interpretation of the formal theory that thus arises.

The notion of certainty or full belief does not play a central role in Jeffrey's epistemology either. According to him we are only certain of mathematical and logical truths and the truths related immediately to experience. The rest is the domain of probable knowledge. To be coherent with this view Jeffrey has to propose a modification of the classical notion of learning by conditioning on data (which occupies a central role in various versions of Bayesian Epistemology as used in the social sciences like economics or psychology). In fact, according to conditioning when one learns a new piece of evidence this information acquires measure one in spite of being based on perfectly fallible evidence. A modification of conditioning that permits to update on uncertain evidence is presented in "Probable Knowledge". The version of Jeffrey's article reprinted here contains as well comments by L. Hurwicz and P. Suppes and responses by Jeffrey. Some of Suppes' comments point in the direction of constructing a theory of rationality that is sensible to our cognitive limitations. The possibility of constructing a "bounded" theory of rationality only started with the seminal work of Herb Simon (Rational choice and the structure of the environment, *Psychological Review*, Vol. 63 No. 2, 129–138. 1956) and is today an active area of investigation in economics, psychology and philosophy.

The uses of Bayesianism in epistemology are usually dismissed by realist philosophers for delivering a subjective picture of rationality that is not sufficiently sensible to the way in which the behavior of rational agents is connected with the structure of the environment. Simon's work was certainly sensible nevertheless to ecological considerations. And Ramsey's essay ends with programmatic ideas differentiating what he called "the logic of consistency" from the "logic of truth". Even Bruno de Finetti who is usually presented as a precursor of Jeffrey's radical probabilism, had philosophical ideas about certainty that clashed with this view (he thought that certainty has to be assumed as a primitive alongside probability, and that we can be certain of more than mere tautologies). Moreover his mature philosophical work veered towards a more objective point of view. For example he dismissed the use of Dutch Book arguments and embraced the use of foundational arguments in terms of scoring rules, a methodological move favored today by many "objective" Bayesians (a presentation of de Finetti's mature views appear in: *Philosophical Lectures on Probability*: collected, edited, and annotated by Alberto Mura, Synthese Library, Springer, 2008).

Van Fraassen introduces in his essay a version of radical probabilism (the term was coined by Jeffrey) where the only epistemological primitive is a notion of conditional probability. Van Fraassen sees this notion as encoding a notion of supposition from which he derives a non-trivial notion of full belief. According to this view it is perfectly possible to be sure of the contingent propositions of science and everyday knowledge. One can see van Fraassen's theory as introducing paradox-free acceptance rules that link probability and belief (some of the usual acceptance rules of this type like high probability rules are known to be the victim of various forms of paradoxes, like the paradox of the lottery first proposed by Henry Kyburg (*Probability and the Logic of Rational Belief*, Wesleyan University

Press, 1961)). Jeffrey renounced to the use of any form of acceptance rules of this type and therefore proposed to eliminate any notion of qualitative belief without a probabilistic origin. Van Fraassen has exactly the opposite intention: namely to tend bridges between traditional and Bayesian epistemology via the use of novel acceptance rules.

Most of the models of probability update considered above deal with synchronic or suppositional change. Is it possible to extend these models to cover cases of genuine changes of probabilistic belief? David Lewis, Bas van Fraassen and Paul Teller provided in the 1970s various dynamic coherence arguments showing that one should update diachronically via conditioning on pain of incoherence (see references in the section on "Further reading" below). If we denote by $P_t(B|A)$ the conditional probability of B given A at time t and by $P_t(.)$ the monadic probability P at time t, we can denote by $P_{t'}$ the monadic probability at time t' where the total evidence gathered between these two times is A. Then the idea that one should update diachronically via conditioning can be expressed formally by: $P_{t'}(B) = P_t(B|A)$. These arguments in favor of this diachronic principle are dynamic versions of the static Dutch Book arguments first proposed in Ramsey's essay. Unfortunately these arguments are considerably more controversial than the well know static Dutch Book argument. The article by Brian Skyrms summarizes almost 30 years of arguments pro and con dynamic book arguments and offers a temperate and more modest version of these arguments that he thinks is valid. This debate is nevertheless still open.

Levi's essay is embedded on his own version of Bayesian epistemology where the notion of full belief is taken as a primitive alongside probability. But the central goal of the article reprinted here is to study the case where probabilities are indeterminate, imprecise or vague. Levi also thinks that a theory of partial belief (precise or imprecise) should be conceived as a branch of decision theory and therefore proposes rules for deciding in conditions of uncertainty. Currently there is a fair amount of work in this area not only in philosophy but also in statistics, computer science and economics.

Gaifman's article focuses on characterizing the structure of higher order probability. In particular he investigates a form of the so-called Miller's principle by which a rational agent adjusts his or her probabilities in accordance to the probabilities of an expert. So, we have a principle of this form:

(Miller's Principle) $P_{you}(A \mid P_{expert}(A) = r) = r$[1]

van Fraassen proposed a diachronic version of this principle for a single agent:

(Reflection) $P_{now}(A \mid P_{later}(A) = r) = r$

[1]Actually Gaifman's formulation of the principle is formally cleaner. He defines $PR(A, r) = \{x \text{ in } W \mid p_x(A) = r\}$, where p_x is an expert function at world x, i.e. the distribution chosen by the expert at that world. Then he formulates the principle as follows: $P(A \mid PR(A, r)) = r$.

Arntzenius' article presents five puzzles showing that rational people can update their degrees of belief in manners that violate Bayesian conditioning and Reflection. But the article by M.J. Schervish, T. Seidenfeld and J. Kadane disputes that Arntzenius' examples impose any new restrictions or challenges to conditioning or Reflection beyond what is already familiar about these principles.

Suggested Further Reading

- An excellent introduction to Ramsey's philosophy in general and to the essay reprinted here in particular can be found in the corresponding chapters of: *The Philosophy of F.P. Ramsey*, by Nils-Eric Sahlin, Cambridge University Press, 2008. The classical introduction to Richard Jeffrey's decision theory is his: *The Logic of Decision*, University Of Chicago Press: 2nd edition (July 15, 1990). A detailed articulation of radical probabilism can be found in Probability and the Art of Judgment, Cambridge Studies in Probability, Induction and Decision Theory (Mar. 27, 1992). The theory of probability cores presented in van Fraassen's article has been slightly modified and extended in a paper by Horacio Arlo-Costa and Rohit Parikh: "Conditional Probability and Defeasible Inference," *Journal of Philosophical Logic* 34, 97-119, 2005. The best axiomatic presentation of primitive conditional probability is given by Lester E. Dubins in his article Finitely Additive Conditional Probabilities, Conglomerability and Disintegrations, *The Annals of Probability*, 3(1):89–99, 1975. Teddy Seidenfeld wrote an accessible note presenting recent results in this area in: Remarks on the theory of conditional probability: Some issues of finite versus countable additivity, *Probability Theory*, V.F. Hendricks et al. (eds.) 2001, pp. 167-178. Alan Hájek articulated a philosophical defense of the use of primitive conditional probability in: What Conditional Probability Could Not Be, *Synthese*, Vol. 137, No. 3, Dec., 2003. Finally there is an interesting article by David Makinson linking conditional probability and central issues in belief change: Conditional probability in the light of qualitative belief change, to appear in a 2011 issue of the *Journal of Philosophical Logic* marking 25 years of AGM. References to other classical articles in this area by Karl Popper, Alfred Renyi and Bruno de Finetti appear in the aforementioned articles.
- Brian Skyrms has also contributed to the theory of higher order probability. One accessible article is: "Higher Order Degrees of Belief," in D. H. Mellor (ed.), *Prospects for Pragmatism.* Cambridge: Cambridge University Press, 109–13. Isaac Levi has articulated his theory of indeterminate probabilities in various books and articles. One of the classical sources is: *The Enterprise of Knowledge*, MIT Press, Cambridge, 1983. More information about Levi's version of decision theory under uncertainty appears in section 7 on Decision Theory below.
- There are two classical sources for the formulation of dynamic Dutch books. One is: Teller, P. (1973), "Conditionalization and Observation", *Synthese* 26: 218-258. The other is: van Fraassen, Bas (1984), "Belief and the Will," *Journal of Philosophy* 81: 235–256. The second piece introduces also a theory of second order probability that complements the writings of Skyrms and Gaifman. Van Fraassen introduces there the Reflection Principle. The original formulation of some of the puzzles discussed by Arntzenius and Seidenfeld is a brief piece by Adam Elga: "Self-Locating Belief and the Sleeping Beauty problem, *Analysis*, 60(2): 143-147, 2000. More detailed reference to the work by Carnap on induction and confirmation can be found in the bibliography of Maher's paper. The so-called Raven's Paradox appeared for the first time in a seminal article by Carl Hempel: "Studies in the Logic of Confirmation (I.)," *Mind,* New Series, Vol. 54, No. 213 (Jan., 1945), pp. 1-26. Branden Fitelson and James Hawthorne offer an alternative and interesting Bayesian account of the paradox in: "How Bayesian Confirmation Theory Handles the Paradox of the Ravens," in E. Eels and J. Fetzer (eds.), *The Place of Probability in Science*, Chicago: Open Court. Further information about confirmation theory can be found in a classical book by John Earman: *Bayes or Bust? A Critical*

Examination of Bayesian Confirmation Theory, MIT Press, 1992. Another classical source is: *Scientific Reasoning: The Bayesian Approach*, by Colin Howson and Peter Urbach, Open Court; 3rd edition, 2005. A interesting book touching a cluster of issues recently discussed in this area like coherence and the use of Bayesian networks in epistemology is: *Bayesian Epistemology* by Luc Bovens and Stephan Hartmann, Oxford University Press,. 2004.

- Another important formal epistemological issue is investigated by Timothy Williamson in his paper, "Conditionalizing on Knowledge", *British Journal for the Philosophy of Science* 49 (1), 1998: 89-121, which intends to integrate the theory of probability and probability kinematics, with other epistemological notions like the notion of knowledge. The theory of *evidential probability* that thus arises is based on two central ideas: (1) the evidential probability of a proposition is its probability conditional on the total evidence (or conditional on evidence propositions); (2) one's total evidence is one's total knowledge. The tools of epistemic logic are used in order to represent the relevant notion of knowledge.
- Jeffrey does not adopt (1) but according to his modified notion of updating once a proposition has evidential probability 1, it keeps it thereafter (monotony). This is a feature shared by Jeffrey's updating and the classical notion of updating. Williamson does embrace (1) but develops a model of updating that abandons monotony. This seems a very promising strategy given the limited applicability of a cumulative model of growth of knowledge. Similarly motivated models (that are nevertheless formally quite different) have been proposed by Isaac Levi, Peter Gärdenfors. Gärdenfors' model appears in his book *Knowledge in Flux* (see the corresponding reference in the bibliographical references of chapter 6). Levi presents his account in *The Enterprise of Knowledge* (the reference appears in the bibliographical section below). Both models appeal directly not only to qualitative belief but also to models of belief change (*contraction* and *revision* - see chapter 6).
- Philosophers of science have traditionally appealed to Bayesian theory in order to provide a Carnapian explication of the notoriously vague, elusive and paradox-prone notion of *confirmation* or *partial justification* in science. Patrick Maher revives in his article, "Probability Captures the Logic of Scientific Confirmation," in *Contemporary Debates in the Philosophy of Science*, ed. Christopher Hitchcock, Blackwell, 69–93, the Carnapian program of inductive inference in order to provide one of these explications. In contrast Clark Glymour and Kevin Kelly argue in their article, "Why Probability Does Not Capture the Logic of Scientific Justification", in Christopher Hitchcock, ed., *Contemporary Debates in the Philosophy of Science*, London: Blackwell, 2004, that Bayesian confirmation cannot deliver the right kind of account of the logic of scientific confirmation. One of the reasons for this skepticism is that they think that scientific justification should reflect how intrinsically difficult is to find the truth and how efficient one's methods are at finding it. So, their skepticism arises because they think that Bayesian confirmation captures neither aspect of scientific justification. While deploying their arguments the two articles discuss the well-known paradox of confirmation first proposed by Hempel, Carnap's research program on the philosophy of probability and induction and the possible application of learning theory in order to offer a non-Bayesian account of scientific justification. The article by Glymour and Kelly continues Glymour's earlier critique of the applications of Bayesianism in philosophy of science (also reprinted here). This earlier piece contains the original versions of some influential and much-discussed conundra engendered by Bayesian confirmation (like the problem of Old Evidence).

Chapter 3
Truth and Probability

Frank P. Ramsey

To say of what is that it is not, or of what is not that it is, is false, while to say of what is that it is and of what is not that it is not is true.
– *Aristotle.*

When several hypotheses are presented to our mind which we believe to be mutually exclusive and exhaustive, but about which we know nothing further, we distribute our belief equally among them .... This being admitted as an account of the way in which we actually do distribute our belief in simple cases, the whole of the subsequent theory follows as a deduction of the way in which we must distribute it in complex cases if we *would be consistent.*
– *W. F. Donkits.*

The object of reasoning is to find out, from the consideration of what we already know, something else which we do not know. Consequently, reasoning is good if it be such as to give a true conclusion from true premises, and not otherwise.
– *C. S. Peirce.*

Truth can never be told so as to be understood, and not be believed.
– *W. Blake.*

Foreword

In this essay the Theory of Probability is taken as a branch of logic, the logic of partial belief and inconclusive argument; but there is no intention of implying that this is the only or even the most important aspect of the subject. Probability is of fundamental importance not only in logic but also in statistical and physical science, and we cannot be sure beforehand that the most useful interpretation of it

---

Frank P. Ramsey was deceased at the time of publication.

F.P. Ramsey (deceased)
University of Cambridge, Cambridge, UK

H. Arló-Costa et al. (eds.), *Readings in Formal Epistemology*, Springer Graduate Texts in Philosophy 1, DOI 10.1007/978-3-319-20451-2_3

in logic will be appropriate in physics also. Indeed the general difference of opinion between statisticians who for the most part adopt the frequency theory of probability and logicians who mostly reject it renders it likely that the two schools are really discussing different things, and that the word 'probability' is used by logicians in one sense and by statisticians in another. The conclusions we shall come to as to the meaning of probability in logic must not, therefore, be taken as prejudging its meaning in physics.

The Frequency Theory

In the hope of avoiding some purely verbal controversies, I propose to begin by making some admissions in favour of the frequency theory. In the first place this theory must be conceded to have a firm basis in ordinary language, which often uses 'probability' practically as a synonym for proportion; for example, if we say that the probability of recovery from smallpox is three-quarters, we mean, I think, simply that that is the proportion of smallpox cases which recover. Secondly, if we start with what is called the calculus of probabilities, regarding it first as a branch of pure mathematics, and then looking round for some interpretation of the formulae which shall show that our axioms are consistent and our subject not entirely useless, then much the simplest and least controversial interpretation of the calculus is one in terms of frequencies. This is true not only of the ordinary mathematics of probability, but also of the symbolic calculus developed by Mr. Keynes; for if in his $\frac{a}{h}$, a and h are taken to be not propositions but propositional functions or class-concepts which define finite classes, and $\frac{a}{h}$ is taken to mean the proportion of members of h which are also members of a, then all his propositions become arithmetical truisms.

Besides these two inevitable admissions, there is a third and more important one, which I am prepared to make temporarily although it does not express my real opinion. It is this. Suppose we start with the mathematical calculus, and ask, not as before what interpretation of it is most convenient to the pure mathematicism, but what interpretation gives results of greatest value to science in general, then it may be that the answer is again an interpretation in terms of frequency; that probability as it is used in statistical theories, especially in statistical mechanics – the kind of probability whose logarithm is the entropy – is really a ratio between the numbers, of two classes, or the limit of such a ratio. I do not myself believe this, but I am willing for the present to concede to the frequency theory that probability as used in modern science is really the same as frequency.

But, supposing all this admitted, it still remains the case that we have the authority both of ordinary language and of many great thinkers for discussing under the heading of probability what appears to be quite a different subject, the logic of partial belief. It may be that, as some supporters of the frequency theory have maintained, the logic of partial belief will be found in the end to be merely the study of frequencies, either because partial belief is definable as, or by reference to, some sort of frequency, or because it can only be the subject of logical treatment when it is grounded on experienced frequencies. Whether these contentions are valid can, however, only be decided as a result of our investigation into partial belief, so that I propose to ignore the frequency theory for the present and begin an inquiry into the logic of partial belief. In this, I think, it will be most convenient if, instead of straight away developing my own theory, I begin by examining the views of Mr Keynes, which are so well known and in essentials so widely accepted that readers probably feel that there is no ground for re-opening the subject *de novo* until they have been disposed of.

Mr. Keynes' Theory

Mr Keynes[1] starts from the supposition that we make probable inferences for which we claim objective validity; we proceed from full belief in one proposition to partial belief in another, and we claim that this procedure is objectively right, so that if another man in similar circumstances entertained a different degree of belief, he would be wrong in doing so. Mr Keynes accounts for this by supposing that between any two propositions, taken as premiss and conclusion, there holds one and only one relation of a certain sort called probability relations; and that if, in any given case, the relation is that of degree α, from full belief in the premiss, we should, if we were rational, proceed to a belief of degree α in the conclusion.

Before criticising this view, I may perhaps be allowed to point out an obvious and easily corrected defect in the statement of it. When it is said that the degree of the probability relation is the same as the degree of belief which it justifies, it seems to be presupposed that both probability relations, on the one hand, and degrees of belief on the other can be naturally expressed in terms of numbers, and then that the number expressing or measuring the probability relation is the same as that expressing the appropriate degree of belief. But if, as Mr. Keynes holds, these things are not always expressible by numbers, then we cannot give his statement that the degree of the one is the same as the degree of the other such a simple interpretation, but must suppose him to mean only that there is a one-one correspondence between probability relations and the degrees of belief which they justify. This correspondence must clearly preserve the relations of greater and less, and so make the manifold of probability relations and that of degrees of belief similar in Mr Russell's sense. I think it is a pity that Mr Keynes did not see this clearly, because the exactitude of this correspondence would have provided quite as worthy material scepticism as did the numerical measurement of probability relations. Indeed some of his arguments against their numerical measurement appear to apply quite equally well against their exact correspondence with degrees of belief; for instance, he argues that if rates of insurance correspond to subjective, i.e. actual, degrees of

[1] J.M. Keynes, *A Treatise on Probability* (1921).

belief, these are not rationally determined, and we cannot infer that probability relations can be similarly measured. It might be argued that the true conclusion in such a case was not that, as Mr Keynes thinks, to the non-numerical probability relation corresponds a non-numerical degree of rational belief, but that degrees of belief, which were always numerical, did not correspond one to one with the probability relations justifying them. For it is, I suppose, conceivable that degrees of belief could be measured by a psychogalvanometer or some such instrument, and Mr Keynes would hardly wish it to follow that probability relations could all be derivatively measured with the measures of the beliefs which they justify.

But let us now return to a more fundamental criticism of Mr Keynes' views, which is the obvious one that there really do not seem to be any such things as the probability relations he describes. He supposes that, at any rate in certain cases, they can be perceived; but speaking for myself I feel confident that this is not true. I do not perceive them, and if I am to be persuaded that they exist it must be by argument; moreover I shrewdly suspect that others do not perceive them either, because they are able to come to so very little agreement as to which of them relates any two given propositions.

All we appear to know about them are certain general propositions, the laws of addition and multiplication; it is as if everyone knew the laws of geometry but no one could tell whether any given object were round or square; and I find it hard to imagine how so large a body of general knowledge can be combined with so slender a stock of particular facts. It is true that about some particular cases there is agreement, but these somehow paradoxically are always immensely complicated; we all agree that the probability of a coin coming down heads is $\frac{1}{2}$, but we can none of us say exactly what is the evidence which forms the other term for the probability relation about which we are then judging. If, on the other hand, we take the simplest possible pairs of propositions such as 'This is red' and 'That is blue' or 'This is red' and 'That is red', whose logical relations should surely be easiest to see, no one, I think, pretends to be sure what is the probability relation which connects them. Or, perhaps, they may claim to see the relation but they will not be able to say anything about it with certainty, to state if it is more or less than $\frac{1}{3}$, or so on. They may, of course, say that it is incomparable with any numerical relation, but a relation about which so little can be truly said will be of little scientific use and it will be hard to convince a sceptic of its existence. Besides this view is really rather paradoxical; for any believer in induction must admit that between 'This is red' as conclusion and 'This is round', together with a billion propositions of the form '*a* is round and red' as evidence, there is a finite probability relation; and it is hard to suppose that as we accumulate instances there is suddenly a point, say after 233 instances, at which the probability relation becomes finite and so comparable with some numerical relations.

It seems to me that if we take the two propositions '*a* is red', '*b* is red', we cannot really discern more than four simple logical relations between them; namely identity of form, identity of predicate, diversity of subject, and logical independence of import. If anyone were to ask me what probability one gave to the other, I should not try to answer by contemplating the propositions and trying to discern a logical relation between them, I should, rather, try to imagine that one of them was all that I knew, and to guess what degree of confidence I should then have in the other. If I were able to do this, I might no doubt still not be content with it, but might say 'This is what I should think, but, of course, I am only a fool' and proceed to consider what a wise man would think and call that the degree of probability. This kind of self-criticism I shall discuss later when developing my own theory; all that I want to remark here is that no one estimating a degree of probability simply contemplates the two propositions supposed to be related by it; he always considers *inter alia* his own actual or hypothetical degree of belief. This remark seems to me to be borne out by observation of my own behaviour; and to be the only way of accounting for the fact that we can all give estimates of probability in cases taken from actual life, but are quite unable to do so in the logically simplest cases in which, were probability a logical relation, it would be easiest to discern.

Another argument against Mr Keynes' theory can, I think, be drawn from his inability to adhere to it consistently even in discussing first principles. There is a passage in his chapter on the measurement of probabilities which reads as follows: –

> Probability is, *vide* Chapter 11 (§12), relative in a sense to the principles of *human* reason. The degree of probability, which it is rational for *us* to entertain, does not presume perfect logical insight, and is relative in part to the secondary propositions which we in fact know; and it is not dependent upon whether more perfect logical insight is or is not conceivable. It is the degree of probability to which those logical processes lead, of which our minds are capable; or, in the language of Chapter II, which those secondary propositions justify, which we in fact know. If we do not take this view of probability, if we do not limit it in this way and make it, to this extent, relative to human powers, we are altogether adrift in the unknown; for we cannot ever know what degree of probability would be justified by the perception of logical relations which we are, and must always be, incapable of comprehending.[2]

This passage seems to me quite unreconcilable with the view which Mr Keynes adopts everywhere except in this and another similar passage. For he generally holds that the degree of belief which we are justified in placing in the conclusion of an argument is determined by what relation of probability unites that conclusion to our premisses, There is only one such relation and consequently only one relevant true secondary proposition, which, of course, we may or may not know, but which is necessarily independent of the human mind. If we do not know it, we do not know it and cannot tell how far we ought to believe the conclusion. But often, he supposes, we do know it; probability relations are not ones which we are incapable of comprehending. But on this view of the matter the passage quoted above has no meaning: the relations which justify probable beliefs are probability relations, and it is nonsense to speak of them being justified by logical relations which we are, and must always be, incapable of comprehending. The significance of the passage for our present purpose lies in the fact that it seems to presuppose a different view of probability, in which indefinable probability relations play no part, but in which the degree of rational belief depends on a variety of logical relations. For instance, there

[2]p. 32, his italics.

might be between the premiss and conclusion the relation that the premiss was the logical product of a thousand instances of a generalization of which the conclusion was one other instance, and this relation, which is not an indefinable probability relation but definable in terms of ordinary logic and so easily recognizable, might justify a certain degree of belief in the conclusion on the part of one who believed the premiss. We should thus have a variety of ordinary logical relations justifying the same or different degrees of belief. To say that the probability of a given h was such-and-such would mean that between a and h was some relation justifying such-and-such a degree of belief. And on this view it would be a real point that the relation in question must not be one which the human mind is incapable of comprehending.

This second view of probability as depending on logical relations but not itself a new logical relation seems to me more plausible than Mr Keynes' usual theory; but this does not mean that I feel at all inclined to agree with it. It requires the somewhat obscure idea of a logical relation justifying a degree of belief, which I should not like to accept as indefinable because it does not seem to be at all a clear or simple notion. Also it is hard to say what logical relations justify what degrees of belief, and why; any decision as to this would be arbitrary, and would lead to a logic of probability consisting of a host of so-called 'necessary' facts, like formal logic on Mr Chadwick's view of logical constants.[3] Whereas I think it far better to seek an explanation of this 'necessity' after the model of the work of Mr Wittgenstein, which enables us to see clearly in what precise sense and why logical propositions are necessary, and in a general way why the system of formal logic consists of the propositions it does consist of, and what is their common characteristic. Just as natural science tries to explain and account for the facts of nature, so philosophy should try, in a sense, to explain and account for the facts of logic; a task ignored by the philosophy which dismisses these facts as being unaccountably and in an indefinable sense 'necessary'.

Here I propose to conclude this criticism of Mr Keynes' theory, not because there are not other respects in which it seems open to objection, but because I hope that what I have already said is enough to show that it is not so completely satisfactory as to render futile any attempt to treat the subject from a rather different point of view.

Degrees of Belief

The subject of our inquiry is the logic of partial belief, and I do not think we can carry it far unless we have at least an approximate notion of what partial belief is, and how, if at all, it can be measured. It will not be very enlightening to be told that in such circumstances it would be rational to believe a proposition to the extent of $\frac{2}{3}$, unless we know what sort of a belief in it that means. We must therefore try to develop a purely psychological method of measuring belief. It is not enough to

[3] "Logical Constants", *Mind*, 1927.

measure probability; in order to apportion correctly our belief to the probability we must also be able to measure our belief.

It is a common view that belief and other psychological variables are not measurable, and if this is true our inquiry will be vain; and so will the whole theory of probability conceived as a logic of partial belief; for if the phrase 'a belief two-thirds of certainty' is meaningless, a calculus whose sole object is to enjoin such beliefs will be meaningless also. Therefore unless we are prepared to give up the whole thing as a bad job we are bound to hold that beliefs can to some extent be measured. If we were to follow the analogy of Mr Keynes' treatment of probabilities we should say that some beliefs were measurable and some not; but this does not seem to me likely to be a correct account of the matter: I do not see how we can sharply divide beliefs into those which have a position in the numerical scale and those which have not. But I think beliefs do differ in measurability in the following two ways. First, some beliefs can be measured more accurately than others; and, secondly, the measurement of beliefs is almost certainly an ambiguous process leading to a variable answer depending on how exactly the measurement is conducted. The degree of a belief is in this respect like the time interval between two events; before Einstein it was supposed that all the ordinary ways of measuring a time interval would lead to the same result if properly performed. Einstein showed that this was not the case; and time interval can no longer be regarded as an exact notion, but must be discarded in all precise investigations. Nevertheless, time interval and the Newtonian system are sufficiently accurate for many purposes and easier to apply.

I shall try to argue later that the degree of a belief is just like a time interval; it has no precise meaning unless we specify more exactly how it is to be measured. But for many purposes we can assume that the alternative ways of measuring it lead to the same result, although this is only approximately true. The resulting discrepancies are more glaring in connection with some beliefs than with others, and these therefore appear less measurable. Both these types of deficiency in measurability, due respectively to the difficulty in getting an exact enough measurement and to an important ambiguity in the definition of the measurement process, occur also in physics and so are not difficulties peculiar to our problem; what is peculiar is that it is difficult to form any idea of how the measurement is to be conducted, how a unit is to be obtained, and so on.

Let us then consider what is implied in the measurement of beliefs. A satisfactory system must in the first place assign to any belief a magnitude or degree having a definite position in an order of magnitudes; beliefs which are of the same degree as the same belief must be of the same degree as one another, and so on. Of course this cannot be accomplished without introducing a certain amount of hypothesis or fiction. Even in physics we cannot maintain that things that are equal to the same thing are equal to one another unless we take 'equal' not as meaning 'sensibly equal' but a fictitious or hypothetical relation. I do not want to discuss the metaphysics or epistemology of this process, but merely to remark that if it is allowable in physics it is allowable in psychology also. The logical simplicity characteristic of the relations dealt with in a science is never attained by nature alone without any admixture of fiction.

But to construct such an ordered series of degrees is not the whole of our task; we have also to assign numbers to these degrees in some intelligible manner. We can of course easily explain that we denote full belief by 1, full belief in the contradictory by 0, and equal beliefs in the proposition and its contradictory by $\frac{1}{2}$. But it is not so easy to say what is meant by a belief $\frac{2}{3}$ of certainty, or a belief in the proposition being twice as strong as that in its contradictory. This is the harder part of the task, but it is absolutely necessary; for we do calculate numerical probabilities, and if they are to correspond to degrees of belief we must discover some definite way of attaching numbers to degrees of belief. In physics we often attach numbers by discovering a physical process of addition[4]: the measure-numbers of lengths are not assigned arbitrarily subject only to the proviso that the greater length shall have the greater measure; we determine them further by deciding on a physical meaning for addition; the length got by putting together two given lengths must have for its measure the sum of their measures. A system of measurement in which there is nothing corresponding to this is immediately recognized as arbitrary, for instance Mohs' scale of hardness[5] in which 10 is arbitrarily assigned to diamond, the hardest known material, 9 to the next hardest, and so on. We have therefore to find a process of addition for degrees of belief, or some substitute for this which will be equally adequate to determine a numerical scale.

Such is our problem; how are we to solve it? There are, I think, two ways in which we can begin. We can, in the first place, suppose that the degree of a belief is something perceptible by its owner; for instance that beliefs differ in the intensity of a feeling by which they are accompanied, which might be called a belief-feeling or feeling of conviction, and that by the degree of belief we mean the intensity of this feeling. This view would be very inconvenient, for it is not easy to ascribe numbers to the intensities of feelings; but apart from this it seems to me observably false, for the beliefs which we hold most strongly are often accompanied by practically no feeling at all; no one feels strongly about things he takes for granted.

We are driven therefore to the second supposition that the degree of a belief is a causal property of it, which we can express vaguely as the extent to which we are prepared to act on it. This is a generalization of the well-known view, that the differentia of belief lies in its causal efficacy, which is discussed by Mr Russell in his *Analysis of Mind*. He there dismisses it for two reasons, one of which seems entirely to miss the point. He argues that in the course of trains of thought we believe many things which do not lead to action. This objection is however beside the mark, because it is not asserted that a belief is an idea which does actually lead to action, but one which would lead to action in suitable circumstances; just as a lump of arsenic is called poisonous not because it actually has killed or will kill anyone, but because it would kill anyone if he ate it. Mr Russell's second argument is, however, more formidable. He points out that it is not possible to suppose that beliefs differ from other ideas only in their effects, for if they were otherwise identical their effects

[4]See N. Campbell, *Physics The Elements* (1920), p.277.

[5]Ibid., p.271.

would be identical also. This is perfectly true, but it may still remain the case that the nature of the difference between the causes is entirely unknown or very vaguely known, and that what we want to talk about is the difference between the effects, which is readily observable and important.

As soon as we regard belief quantitatively, this seems to me the only view we can take of it. It could well be held that the difference between believing and not believing lies in the presence or absence of introspectible feelings. But when we seek to know what is the difference between believing more firmly and believing less firmly, we can no longer regard it as consisting in having more or less of certain observable feelings; at least I personally cannot recognize any such feelings. The difference seems to me to lie in how far we should act on these beliefs: this may depend on the degree of some feeling or feelings, but I do not know exactly what feelings and I do not see that it is indispensable that we should know. Just the same thing is found in physics; men found that a wire connecting plates of zinc and copper standing in acid deflected a magnetic needle in its neighbourhood. Accordingly as the needle was more or less deflected the wire was said to carry a larger or a smaller current. The nature of this 'current' could only be conjectured: what were observed and measured were simply its effects. It will no doubt be objected that we know how strongly we believe things, and that we can only know this if we can measure our belief by introspection. This does not seem to me necessarily true; in many cases, I think, our judgment about the strength of our belief is really about how we should act in hypothetical circumstances. It will be answered that we can only tell how we should act by observing the present belief-feeling which determines how we should act; but again I doubt the cogency of the argument. It is possible that what determines how we should act determines us also directly or indirectly to have a correct opinion as to how we should act, without its ever coming into consciousness.

Suppose, however, I am wrong about this and that we can decide by introspection the nature of belief, and measure its degree; still, I shall argue, the kind of measurement of belief with which probability is concerned is not this kind but is a measurement of belief *qua* basis of action. This can I think be shown in two ways. First, by considering the scale of probabilities between 0 and 1, and the sort of way we use it, we shall find that it is very appropriate to the measurement of belief as a basis of action, but in no way related to the measurement of an introspected feeling. For the units in terms of which such feelings or sensations are measured are always, I think, differences which are just perceptible: there is no other way of obtaining units. But I see no ground for supposing that the interval between a belief of degree $\frac{1}{3}$ and one of degree $\frac{1}{2}$ consists of as many just perceptible changes as does that between one of $\frac{2}{3}$ and one of $\frac{5}{6}$, or that a scale based on just perceptible differences would have any simple relation to the theory of probability. On the other hand the probability of $\frac{1}{3}$ is clearly related to the kind of belief which would lead to a bet of 2 to 1, and it will be shown below how to generalize this relation so as to apply to action in general. Secondly, the quantitative aspects of beliefs as the basis of action are evidently more important than the intensities of belief-feelings. The latter are no doubt interesting, but may be very variable from individual to individual, and their practical interest is entirely due to their position as the hypothetical causes of beliefs *qua* bases of action.

It is possible that some one will say that the extent to which we should act on a belief in suitable circumstances is a hypothetical thing, and therefore not capable of measurement. But to say this is merely to reveal ignorance of the physical sciences which constantly deal with and measure hypothetical quantities; for instance, the electric intensity at a given point is the force which would act on a unit charge if it were placed at the point.

Let us now try to find a method of measuring beliefs as bases of possible actions. It is clear that we are concerned with dispositional rather than with actualized beliefs; that is to say, not with beliefs at the moment when we are thinking of them, but with beliefs like my belief that the earth is round, which I rarely think of, but which would guide my action in any case to which it was relevant.

The old-established way of measuring a person's belief is to propose a bet, and see what are the lowest odds which he will accept. This method I regard as fundamentally sound; but it suffers from being insufficiently general, and from being necessarily inexact. It is inexact partly because of the diminishing marginal utility of money, partly because the person may have a special eagerness or reluctance to bet, because he either enjoys or dislikes excitement or for any other reason, e.g. to make a book. The difficulty is like that of separating two different co-operating forces. Besides, the proposal of a bet may inevitably alter his state of opinion; just as we could not always measure electric intensity by actually introducing a charge and seeing what force it was subject to, because the introduction of the charge would change the distribution to be measured.

In order therefore to construct a theory of quantities of belief which shall be both general and more exact, I propose to take as a basis a general psychological theory, which is now universally discarded, but nevertheless comes, I think, fairly close to the truth in the sort of cases with which we are most concerned. I mean the theory that we act in the way we think most likely to realize the objects of our desires, so that a person's actions are completely determined by his desires and opinions. This theory cannot be made adequate to all the facts, but it seems to me a useful approximation to the truth particularly in the case of our self-conscious or professional life, and it is presupposed in a great deal of our thought. It is a simple theory and one which many psychologists would obviously like to preserve by introducing unconscious desires and unconscious opinions in order to bring it more into harmony with the facts. How far such fictions can achieve the required result I do not attempt to judge: I only claim for what follows approximate truth, or truth in relation to this artificial system of psychology, which like Newtonian mechanics can, I think, still be profitably used even though it is known to be false.

It must be observed that this theory is not to be identified with the psychology of the Utilitarians, in which pleasure had a dominating position. The theory I propose to adopt is that we seek things which we want, which may be our own or other people's pleasure, or anything else whatever, and our actions are such as we think most likely to realize these goods. But this is not a precise statement, for a precise statement of the theory can only be made after we have introduced the notion of quantity of belief.

Let us call the things a person ultimately desires 'goods', and let us at first assume that they are numerically measurable and additive. That is to say that if he prefers for its own sake an hour's swimming to an hour's reading, he will prefer two hours' swimming to one hour's swimming and one hour's reading. This is of course absurd in the given case but this may only be because swimming and reading are not ultimate goods, and because we cannot imagine a second hour's swimming precisely similar to the first, owing to fatigue, etc.

Let us begin by supposing that our subject has no doubts about anything, but certain opinions about all propositions. Then we can say that he will always choose the course of action which will lead in his opinion to the greatest sum of good.

It should be emphasized that in this essay good and bad are never to be understood in any ethical sense but simply as denoting that to which a given person feels desire and aversion.

The question then arises how we are to modify this simple system to take account of varying degrees of certainty in his beliefs. I suggest that we introduce as a law of psychology that his behaviour is governed by what is called the mathematical expectation; that is to say that, if p is a proposition about which he is doubtful, any goods or bads for whose realization p is in his view a necessary and sufficient condition enter into his calculations multiplied by the same fraction, which is called the 'degree of his belief in p'. We thus define degree of belief in a way which presupposes the use of the mathematical expectation.

We can put this in a different way. Suppose his degree of belief in p is $\frac{m}{n}$; then his action is such as he would choose it to be if he had to repeat it exactly n times, in m of which p was true, and in the others false. [Here it may be necessary to suppose that in each of the n times he had no memory of the previous ones.]

This can also be taken as a definition of the degree of belief, and can easily be seen to be equivalent to the previous definition. Let us give an instance of the sort of case which might occur. I am at a cross-roads and do not know the way; but I rather think one of the two ways is right. I propose therefore to go that way but keep my eyes open for someone to ask; if now I see someone half a mile away over the fields, whether I turn aside to ask him will depend on the relative inconvenience of going out of my way to cross the fields or of continuing on the wrong road if it is the wrong road. But it will also depend on how confident I am that I am right; and clearly the more confident I am of this the less distance I should be willing to go from the road to check my opinion. I propose therefore to use the distance I would be prepared to go to ask, as a measure of the confidence of my opinion; and what I have said above explains how this is to be done. We can set it out as follows: suppose the disadvantage of going x yards to ask is $f(x)$, the advantage of arriving at the right destination is r, that of arriving at the wrong one w. Then if I should just be willing to go a distance d to ask, the degree of my belief that I am on the right road is given by

$$p = 1 - \frac{f(d)}{r - w}.$$

For such an action is one it would just pay me to take, if I had to act in the same way n times, in np of which I was on the right way but in the others not.

For the total good resulting from not asking each time

$$= npr + n\,(1 - p)\,w$$
$$= nw + np\,(r - w)$$

that resulting from asking at distance x each time

$$= nr - nf(x). \qquad \text{[I now always go right.]}$$

This is greater than the preceding expression, provided

$$f(x) < (r - w)\,(1 - p)\,,$$

∴ the critical distance d is connected with p, the degree of belief, by the relation $f(d) = (r - w)(1 - p)$

$$\text{or} \quad p = 1 - \frac{f(d)}{r - w} \quad \text{as asserted above.}$$

It is easy to see that this way of measuring belief gives results agreeing with ordinary ideas; at any rate to the extent that full belief is denoted by 1, full belief in the contradictory by 0, and equal belief in the two by $\frac{1}{2}$. Further, it allows validity to betting as means of measuring beliefs. By proposing a bet on p we give the subject a possible course of action from which so much extra good will result to him if p is true and so much extra bad if p is false. Supposing, the bet to be in goods and bads instead of in money, he will take a bet at any better odds than those corresponding to his state of belief; in fact his state of belief is measured by the odds he will just take; but this is vitiated, as already explained, by love or hatred of excitement, and by the fact that the bet is in money and not in goods and bads. Since it is universally agreed that money has a diminishing marginal utility, if money bets are to be used, it is evident that they should be for as small stakes as possible. But then again the measurement is spoiled by introducing the new factor of reluctance to bother about trifles.

Let us now discard the assumption that goods are additive and immediately measurable, and try to work out a system with as few assumptions as possible. To begin with we shall suppose, as before, that our subject has certain beliefs about everything; then he will act so that what he believes to be the total consequences of his action will be the best possible. If then we had the power of the Almighty, and could persuade our subject of our power, we could, by offering him options, discover how he placed in order of merit all possible courses of the world. In this way all possible worlds would be put in an order of value, but we should have no definite way of representing them by numbers. There would be no meaning in the assertion that the difference in value between α and β was equal to that between γ and δ. [Here and elsewhere we use Greek letters to represent the different possible totalities of events between which our subject chooses – the ultimate organic unities.]

Suppose next that the subject is capable of doubt; then we could test his degree of belief in different propositions by making him offers of the following kind. Would you rather have world α in any event; or world β if *p* is true, and world γ if *p* is false? If, then, he were certain that *p* was true, simply compare α and β and choose between them as if no conditions were attached; but if he were doubtful his choice would not be decided so simply. I propose to lay down axioms and definitions concerning the principles governing choices of this kind. This is, of course, a very schematic version of the situation in real life, but it is, I think, easier to consider it in this form.

There is first a difficulty which must be dealt with; the propositions like *p* in the above case which are used as conditions in the options offered may be such that their truth or falsity is an object of desire to the subject. This will be found to complicate the problem, and we have to assume that there are propositions for which this is not the case, which we shall call ethically neutral. More precisely an atomic proposition *p* is called ethically neutral if two possible worlds differing only in regard to the truth of *p* are always of equal value; and a non-atomic proposition *p* is called ethically neutral if all its atomic truth-arguments[6] are ethically neutral.

We begin by defining belief of degree $\frac{1}{2}$ in an ethically neutral proposition. The subject is said to have belief of degree $\frac{1}{2}$ in such a proposition *p* if he has no preference between the options (1) α if *p* is true, β if *p* is false, and (2) α if *p* is false, β if *p* is true, but has a preference between α and β simply. We suppose by an axiom that if this is true of any one pair α, β, it is true of all such pairs.[7] This comes roughly to defining belief of degree $\frac{1}{2}$ as such a degree of belief as leads to indifference between betting one way and betting the other for the same stakes.

Belief of degree $\frac{1}{2}$ as thus defined can be used to measure values numerically in the following way. We have to explain what is meant by the difference in value between α and β being equal to that between γ and δ; and we define this to mean that, if *p* is an ethically neutral proposition believed to degree $\frac{1}{2}$, the subject has no preference between the options (1) α if *p* is true, δ if *p* is false, and (2) β if *p* is true, γ if *p* is false.

This definition can form the basis of a system of measuring values in the following way:–

Let us call any set of all worlds equally preferable to a given world a value: we suppose that if world α is preferable to β any world with the same value as α is preferable to any world with the same value as β and shall say that the value of α is greater than that of β. This relation 'greater than' orders values in a series. We shall use α henceforth both for the world and its value.

---

[6] I assume here Wittgenstein's theory of propositions; it would probably be possible to give an equivalent definition in terms of any other theory.

[7] α and β must be supposed so far undefined as to be compatible with both *p* and not-*p*.

Axioms

(1) There is an ethically neutral proposition p believed to degree $\frac{1}{2}$.
(2) If p, q are such propositions and the option

α if p, δ if not-p is equivalent to β if p, γ if not-p
then α if q, δ if not-q is equivalent to β if q, γ if not-q.
Def. In the above case we say $\alpha\beta = \gamma\delta$.

Theorems If $\alpha\beta = \gamma\delta$,
then $\beta\alpha = \delta\gamma$, $\alpha\gamma = \beta\delta$, $\gamma\alpha = \delta\beta$.

(2a) If $\alpha\beta = \gamma\delta$, then $\alpha > \beta$ is equivalent to $\gamma > \delta$
and $\alpha = \beta$ is equivalent to $\gamma = \delta$.

(3) If option A is equivalent to option B and B to C, then A to C.

Theorem If $\alpha\beta = \gamma\delta$ and $\beta\eta = \zeta\gamma$,
then $\alpha\eta = \zeta\delta$.

(4) If $\alpha\beta = \gamma\delta$, $\gamma\delta = \eta\zeta$, then $\alpha\beta = \eta\zeta$.
(5) $(\alpha, \beta, \gamma).\ E!(\iota x)\ (\alpha x = \beta\gamma)$
(6) $(\alpha, \beta).\ E!(\iota x)\ (\alpha x = x\beta)$
(7) Axiom of continuity: – Any progression has a limit (ordinal).
(8) Axiom of Archimedes.

These axioms enable the values to be correlated one-one with real numbers so that if α^1 corresponds to α, etc.

$$\alpha\beta = \gamma\delta. \equiv .\alpha^1 - \beta^1 = \gamma^1 - \delta^1.$$

Henceforth we use α for the correlated real number α^1 also.

Having thus defined a way of measuring value we can now derive a way of measuring belief in general. If the option of α for certain is indifferent with that of β if p is true and γ if p is false,[8] we can define the subject's degree of belief in p as the ratio of the difference between α and γ to that between β and γ; which we must suppose the same for all α's, β's and γ's that satisfy the conditions. This amounts roughly to defining the degree of belief in p by the odds at which the subject would bet on p, the bet being conducted in terms of differences of value as defined. The definition only applies to partial belief and does not include certain beliefs; for belief of degree 1 in p, α for certain is indifferent with α if p and any β if not-p.

We are also able to define a very useful new idea – the 'degree of belief in p given q'. This does not mean the degree of belief in 'If p then q', or that in 'p entails q', or that which the subject would have in p if he knew q, or that which he ought to have.

[8]Here β must include the truth of p, γ its falsity; p need no longer be ethically neutral. But we have to assume that there is a wolrd with any assigned value in which p is true, and one in which p is false.

It roughly expresses the odds at which he would now bet on p, the bet only to be valid if q is true. Such conditional bets were often made in the eighteenth century.

The degree of belief in p given q is measured thus. Suppose the subject indifferent between the options (1) α if q true, β if q false, (2) γ if p true and q true, δ if p false and q true, β if q false. Then the degree of his belief in p given q is the ratio of the difference between α and δ to that between γ and δ, which we must suppose the same for any α, β, γ, δ which satisfy the given conditions. This is not the same as the degree to which he would believe p, if he believed q for certain; for knowledge of q might for psychological reasons profoundly alter his whole system of beliefs.

Each of our definitions has been accompanied by an axiom of consistency, and in so far as this is false, the notion of the corresponding degree of belief becomes invalid. This bears some analogy to the situation in regard to simultaneity discussed above.

I have not worked out the mathematical logic of this in detail, because this would, I think, be rather like working out to seven places of decimals a result only valid to two. My logic cannot be regarded as giving more than the sort of way it might work.

From these definitions and axioms it is possible to prove the fundamental laws of probable belief (degrees of belief lie between 0 and 1):

(1) Degree of belief in p + degree of belief in $\overline{p} = 1$
(2) Degree of belief in p given q + degree of belief in $\overline{p}$ given $q = 1$.
(3) Degree of belief in (p and q) = degree of belief in $p \times$ degree of belief in q given p.
(4) Degree of belief in (p and q) + degree of belief in (p and $\overline{q}$) = degree of belief in p.

The first two are immediate. (3) is proved as follows.
Let degree of belief in $p = x$, that in q given $p = y$.
Then ξ for certain $\equiv \xi + (1-x)t$ if p true, $\xi - xt$ if p false for any t.

$$\xi + (1-x)\,t \text{ if } p \text{ true} \equiv$$

$$\begin{cases} \xi + (1-x)\,t + (1-y)\,u \text{ if `}p \text{ and } q\text{' true,} \\ \xi + (1-x)\,t - yu \text{ if } p \text{ true } q \text{ false;} \quad \text{for any } u. \end{cases}$$

Choose u so that $\xi + (1-x)\,t - yu = \xi - xt$,

$$\text{i.e. let } u = t/y \; (y \neq 0)$$

Then ξ for certain $\equiv$

$$\begin{cases} \xi + (1-x)\,t + (1-y)\,t/y \text{ if } p \text{ and } q \text{ true} \\ \xi - xt \text{ otherwise,} \end{cases}$$

$\therefore$ degree of belief in 'p and q' $= \frac{xt}{t+(1-y)t/y} = xy$. $(t \neq 0)$
If $y = 0$, take $t = 0$.

$$\text{Then } \xi \text{ for certain} \equiv \xi \text{ if } p \text{ true, } \xi \text{ if } p \text{ false}$$
$$\equiv \xi + u \text{ if } p \text{ true}, q \text{ true}; \xi \text{ if } p \text{ false}, q \text{ false}, \xi \text{ if } p \text{ false}$$
$$\equiv \xi + u, pq \text{ true}; \xi, pq \text{ false}$$

$\therefore$ degree of belief in $pq = 0$.

(4) follows from (2), (3) as follows: –

Degree of belief in $pq =$ that in $p \times$ that in q given p, by (3). Similarly degree of belief in $p\overline{q} =$ that in $p \times$ that in q given p

$\therefore$ sum $=$ degree of belief in p, by (2).

These are the laws of probability, which we have proved to be necessarily true of any consistent set of degrees of belief. Any definite set of degrees of belief which broke them would be inconsistent in the sense that it violated the laws of preference between options, such as that preferability is a transitive asymmetrical relation, and that if α is preferable to β, β for certain cannot be preferable to α if p, β if not-p. If anyone's mental condition violated these laws, his choice would depend on the precise form in which the options were offered him, which would be absurd. He could have a book made against him by a cunning better and would then stand to lose in any event.

We find, therefore, that a precise account of the nature of partial belief reveals that the laws of probability are laws of consistency, an extension to partial beliefs of formal logic, the logic of consistency. They do not depend for their meaning on any degree of belief in a proposition being uniquely determined as the rational one; they merely distinguish those sets of beliefs which obey them as consistent ones.

Having any definite degree of belief implies a certain measure of consistency, namely willingness to bet on a given proposition at the same odds for any stake, the stakes being measured in terms of ultimate values. Having degrees of belief obeying the laws of probability implies a further measure of consistency, namely such a consistency between the odds acceptable on different propositions as shall prevent a book being made against you.

Some concluding remarks on this section may not be out of place. First, it is based fundamentally on betting, but this will not seem unreasonable when it is seen that all our lives we are in a sense betting. Whenever we go to the station we are betting that a train will really run, and if we had not a sufficient degree of belief in this we should decline the bet and stay at home. The options God gives us are always conditional on our guessing whether a certain proposition is true. Secondly, it is based throughout on the idea of mathematical expectation; the dissatisfaction often felt with this idea is due mainly to the inaccurate measurement of goods. Clearly mathematical expectations in terms of money are not proper guides to conduct. It should be remembered, in judging my system, that in it value is actually defined by means of mathematical expectation in the case of beliefs of degree $\frac{1}{2}$, and so may be expected to be scaled suitably for the valid application in the case of other degrees of belief also.

Thirdly, nothing has been said about degrees of belief when the number of alternatives is infinite. About this I have nothing useful to say, except that I doubt if the mind is capable of contemplating more than a finite number of alternatives. It can consider questions to which an infinite number of answers are possible, but in order to consider the answers it must lump them into a finite number of groups. The difficulty becomes practically relevant when discussing induction, but even then there seems to me no need to introduce it. We can discuss whether past experience gives a high probability to the sun's rising to-morrow without bothering about what probability it gives to the sun's rising each morning for evermore. For this reason I cannot but feel that Mr Ritchie's discussion of the problem[9] is unsatisfactory; it is true that we can agree that inductive generalizations need have no finite probability, but particular expectations entertained on inductive grounds undoubtedly do have a high numerical probability in the minds of all of us. We all are more certain that the sun will rise to-morrow than that I shall not throw 12 with two dice first time, i.e. we have a belief of higher degree than $\frac{35}{36}$ in it. If induction ever needs a logical justification it is in connection with the probability of an event like this.

The Logic of Consistency

We may agree that in some sense it is the business of logic to tell us what we ought to think; but the interpretation of this statement raises considerable difficulties. It may be said that we ought to think what is true, but in that sense we are told what to think by the whole of science and not merely by logic. Nor, in this sense, can any justification be found for partial belief; the ideally best thing is that we should have beliefs of degree 1 in all true propositions and beliefs of degree 0 in all false propositions. But this is too high a standard to expect of mortal men, and we must agree that some degree of doubt or even of error may be humanly speaking justified.

Many logicians, I suppose, would accept as an account of their science the opening words of Mr Keynes' *Treatise on Probability*: "Part of our knowledge we obtain direct; and part by argument. The Theory of Probability is concerned with that part which we obtain by argument, and it treats of the different degrees in which the results so obtained are conclusive or inconclusive." Where Mr Keynes says 'the Theory of Probability', others would say Logic. It is held, that is to say, that our opinions can be divided into those we hold immediately as a result of perception or

---

[9] A. D. Ritchie, "Induction and Probability." *Mind*, 1926. p. 318. 'The conclusion of the foregoing discussion may be simply put. If the problem of induction be stated to be "How can inductive generalizations acquire a large numerical probability?" then this is a pseudo-problem, because the answer is "They cannot". This answer is not, however, a denial of the validity of induction but is a direct consequence of the nature of probability. It still leaves untouched the real problem of induction which is "How can the probability of an induction be increased?" and it leaves standing the whole of Keynes' discussion on this point.'

memory, and those which we derive from the former by argument. It is the business of Logic to accept the former class and criticize merely the derivation of the second class from them.

Logic as the science of argument and inference is traditionally and rightly divided into deductive and inductive; but the difference and relation between these two divisions of the subject can be conceived in extremely different ways. According to Mr Keynes valid deductive and inductive arguments are fundamentally alike; both are justified by logical relations between premiss and conclusion which differ only in degree. This position, as I have already explained, I cannot accept. I do not see what these inconclusive logical relations can be or how they can justify partial beliefs. In the case of conclusive logical arguments I can accept the account of their validity which has been given by many authorities, and can be found substantially the same in Kant, De Morgan, Peirce and Wittgenstein. All these authors agree that the conclusion of a formally valid argument is contained in its premisses; that to deny the conclusion while accepting the premisses would be self-contradictory; that a formal deduction does not increase our knowledge, but only brings out clearly what we already know in another form; and that we are bound to accept its validity on pain of being inconsistent with ourselves. The logical relation which justifies the inference is that the sense or import of the conclusion is contained in that of the premisses.

But in the case of an inductive argument this does not happen in the least; it is impossible to represent it as resembling a deductive argument and merely weaker in degree; it is absurd to say that the sense of the conclusion is partially contained in that of the premisses. We could accept the premisses and utterly reject the conclusion without any sort of inconsistency or contradiction.

It seems to me, therefore, that we can divide arguments into two radically different kinds, which we can distinguish in the words of Peirce as (1) 'explicative, analytic, or deductive' and (2) 'amplifiative, synthetic, or (loosely speaking) inductive'.[10] Arguments of the second type are from an important point of view much closer to memories and perceptions than to deductive arguments. We can regard perception, memory and induction as the three fundamental ways of acquiring knowledge; deduction on the other hand is merely a method of arranging our knowledge and eliminating inconsistencies or contradictions.

Logic must then fall very definitely into two parts: (excluding analytic logic, the theory of terms and propositions) we have the lesser logic, which is the logic of consistency, or formal logic; and the larger logic, which is the logic of discovery, or inductive logic.

What we have now to observe is that this distinction in no way coincides with the distinction between certain and partial beliefs; we have seen that there is a theory of consistency in partial beliefs just as much as of consistency in certain beliefs, although for various reasons the former is not so important as the latter. The theory of probability is in fact a generalization of formal logic; but in the process

---

[10]C.S. Peirce *Change Love and Logic*, p. 92.

of generalization one of the most important aspects of formal logic is destroyed. If p and $\overline{q}$ are inconsistent so that q follows logically from p, that p implies q is what is called by Wittgenstein a 'tautology' and can be regarded as a degenerate case of a true proposition not involving the idea of consistency. This enables us to regard (not altogether correctly) formal logic including mathematics as an objective science consisting of objectively necessary propositions. It thus gives us not merely the ἀνάγκη λέγειν, that if we assert p we are bound in consistency to assert q also, but also the ἀνάγκη εἶναι, that is p is true, so must q be. But when we extend formal logic to include partial beliefs this direct objective interpretation is lost; if we believe pq to the extent of $\frac{1}{3}$, and $p\overline{q}$ to the extent of $\frac{1}{3}$, we are bound in consistency to believe $\overline{p}$ also to the extent of $\frac{1}{3}$. This is the ἀνάγκη λέγειν; but we cannot say that if pq is $\frac{1}{3}$ true and $p\overline{q}$ $\frac{1}{3}$ true, $\overline{p}$ also must be $\frac{1}{3}$ true, for such a statement would be sheer nonsense. There is no corresponding ἀνάγκη εἶναι. Hence, unlike the calculus of consistent full belief, the calculus of objective partial belief cannot be immediately interpreted as a body of objective tautology.

This is, however, possible in a roundabout way; we saw at the beginning of this essay that the calculus of probabilities could be interpreted in terms of class-ratios; we have now found that it can also be interpreted as a calculus of consistent partial belief. It is natural, therefore, that we should expect some intimate connection between these two interpretations, some explanation of the possibility of applying the same mathematical calculus to two such different sets of phenomena. Nor is an explanation difficult to find; there are many connections between partial beliefs and frequencies. For instance, experienced frequencies often lead to corresponding partial beliefs, and partial beliefs lead to the expectation of corresponding frequencies in accordance with Bernouilli's Theorem. But neither of these is exactly the connection we want; a partial belief cannot in general be connected uniquely with any actual frequency, for the connection is always made by taking the proposition in question as an instance of a propositional function. What propositional function we choose is to some extent arbitrary and the corresponding frequency will vary considerably with our choice. . The pretensions of some exponents of the frequency theory that partial belief means full belief in a frequency proposition cannot be sustained. But we found that the very idea of partial belief involves reference to a hypothetical or ideal frequency; supposing goods to be additive, belief of degree $\frac{m}{n}$ is the sort of belief which leads to the action which would be best if repeated n times in m of which the proposition is true; or we can say more briefly that it is the kind of belief most appropriate to a number of hypothetical occasions otherwise identical in a proportion $\frac{m}{n}$ of which the proposition in question is true. It is this connection between partial belief and frequency which enables us to use the calculus of frequencies as a calculus of consistent partial belief. And in a sense we may say that the two interpretations are the objective and subjective aspects of the same inner meaning, just as formal logic can be interpreted objectively as a body of tautology and subjectively as the laws of consistent thought.

We shall, I think, find that this view of the calculus of probability removes various difficulties that have hitherto been found perplexing. In the first place it gives us a clear justification for the axioms of the calculus, which on such a system as Mr Keynes' is entirely wanting. For now it is easily seen that if partial beliefs are consistent they will obey these axioms, but it is utterly obscure why Mr Keynes' mysterious logical relations should obey them.[11] We should be so curiously ignorant of the instances of these relations, and so curiously knowledgeable about their general laws.

Secondly, the Principle of Indifference can now be altogether dispensed with; we do not regard it as belonging to formal logic to say what should be a man's expectation of drawing a white or a black ball from an urn; his original expectations may within the limits of consistency be any he likes; all we have to point out is that if he has certain expectations he is bound in consistency to have certain others. This is simply bringing probability into line with ordinary formal logic, which does not criticize premisses but merely declares that certain conclusions are the only ones consistent with them. To be able to turn the Principle of Indifference out of formal logic is a great advantage; for it is fairly clearly impossible to lay down purely logical conditions for its validity, as is attempted by Mr Keynes. I do not want to discuss this question in detail, because it leads to hair-splitting and arbitrary distinctions which could be discussed for ever. But anyone who tries to decide by Mr Keynes' methods what are the proper alternatives to regard as equally probable in molecular mechanics, e.g. in Gibbs' phase-space, will soon be convinced that it is a matter of physics rather than pure logic. By using the multiplication formula, as it is used in inverse probability, we can on Mr Keynes' theory reduce all probabilities to quotients of *a priori* probabilities; it is therefore in regard to these latter that the Principle of Indifference is of primary importance; but here the question is obviously not one of formal logic. How can we on merely logical grounds divide the spectrum into equally probable bands?

A third difficulty which is removed by our theory is the one which is presented to Mr Keynes' theory by the following case. I think I perceive or remember something but am not sure; this would seem to give me some ground for believing it, contrary to Mr Keynes' theory, by which the degree belief in it which it would be rational for me to have is that given by the probability relation between the proposition in question and the things I know for certain. He cannot justify a probable belief founded not on argument but on direct inspection. In our view there would be nothing contrary to formal logic in such a belief; whether it would be reasonable would depend on what I have called the larger logic which will be the subject of the next section; we shall there see that there is no objection to such a possibility, with which Mr Keynes' method of justifying probable belief solely by relation to certain knowledge is quite unable to cope.

---

[11]It appears in Mr Keynes' system as if the principal axioms – the laws of addition and multiplication – were nothing but definitions. This is merely a logical mistake; his definitions are formally invalid unless corresponding axioms are presupposed. Thus his definition of multiplication presupposes the law that if the probability of a given bh *is* equal to that of c given dh, and the probability of b given h is equal to that of d given h, then will the probabilities of ab given h and of cd given h be equal.

The Logic of Truth

The validity of the distinction between the logic of consistency and the logic of truth has been often disputed; it has been contended on the one hand that logical consistency is only a kind of factual consistency; that if a belief in p is inconsistent with one in q, that simply means that p and q are not both true, and that this is a necessary or logical fact. I believe myself that this difficulty can be met by Wittgenstein's theory of tautology, according to which if a belief in p is inconsistent with one in q, that p and q are not both true is not a fact but a tautology. But I do not propose to discuss this question further here.

From the other side it is contended that formal logic or the logic of consistency is the whole of logic, and inductive logic either nonsense or part of natural science. This contention, which would I suppose be made by Wittgenstein, I feel more difficulty in meeting. But I think it would be a pity, out of deference to authority, to give up trying to say anything useful about induction.

Let us therefore go back to the general conception of logic as the science of rational thought. We found that the most generally accepted parts of logic, namely, formal logic, mathematics and the calculus of probabilities, are all concerned simply to ensure that our beliefs are not self-contradictory. We put before ourselves the standard of consistency and construct these elaborate rules to ensure its observance. But this is obviously not enough; we want our beliefs to be consistent not merely with one another but also with the facts[12]: nor is it even clear that consistency is always advantageous; it may well be better to be sometimes right than never right. Nor when we wish to be consistent are we always able to be: there are mathematical propositions whose truth or falsity cannot as yet be decided. Yet it may humanly speaking be right to entertain a certain degree of belief in them on inductive or other grounds: a logic which proposes to justify such a degree of belief must be prepared actually to go against formal logic; for to a formal truth formal logic can only assign a belief of degree 1. We could prove in Mr Keynes' system that its probability is 1 on any evidence. This point seems to me to show particularly clearly that human logic or the logic of truth, which tells men how they should think, is not merely independent of but sometimes actually incompatible with formal logic.

In spite of this nearly all philosophical thought about human logic and especially induction has tried to reduce it in some way to formal logic. Not that it is supposed, except by a very few, that consistency will of itself lead to truth; but consistency combined with observation and memory is frequently credited with this power.

Since an observation changes (in degree at least) my opinion about the fact observed, some of my degrees of belief after the observation are necessarily

[12]Cf. Kant: 'Denn obgleich eine Erkenntnis der logischen Form völlig gemäss sein möchte, dass ist sich selbst nicht widerspräche, so kann sie doch noch immer dem Gegenstande widersprechen.' *Kritik der reinen Vernunft*, First Edition. p. 59.

inconsistent with those I had before. We have therefore to explain how exactly the observation should modify my degrees of belief; obviously if p is the fact observed, my degree of belief in q after the observation should be equal to my degree of belief in q given p before, or by the multiplication law to the quotient of my degree of belief in pq by my degree of belief in p. When my degrees of belief change in this way we can say that they have been changed consistently by my observation.

By using this definition, or on Mr Keynes' system simply by using the multiplication law, we can take my present degrees of belief, and by considering the totality of my observations, discover from what initial degrees of belief my present ones would have arisen by this process of consistent change. My present degrees of belief can then be considered logically justified if the corresponding initial degrees of belief are logically justified. But to ask what initial degrees of belief are justified, or in Mr Keynes' system what are the absolutely *a priori* probabilities, seems to me a meaningless question; and even if it had a meaning I do not see how it could be answered.

If we actually applied this process to a human being, found out, that is to say, on what *a prior* probabilities his present opinions could be based, we should obviously find them to be ones determined by natural selection, with a general tendency to give a higher probability to the simpler alternatives. But, as I say, I cannot see what could be meant by asking whether these degrees of belief were logically justified. Obviously the best thing would be to know for certain in advance what was true and what false, and therefore if any one system of initial beliefs is to receive the philosopher's approbation it should be this one. But clearly this would not be accepted by thinkers of the school I am criticising. Another alternative is to apportion initial probabilities on the purely formal system expounded by Wittgenstein, but as this gives no justification for induction it cannot give us the human logic which we are looking for.

Let us therefore try to get an idea of a human logic which shall not attempt to be reducible to formal logic. Logic, we may agree, is concerned not with what men actually believe, but what they ought to believe, or what it would be reasonable to believe. What then, we must ask, is meant by saying that it is reasonable for a man to have such and such a degree of belief in a proposition? Let us consider possible alternatives.

First, it sometimes means something explicable in terms of formal logic: this possibility for reasons already explained we may dismiss. Secondly, it sometimes means simply that were I in his place (and not e.g. drunk) I should have such a degree of belief. Thirdly, it sometimes means that if his mind worked according to certain rules, which we may roughly call 'scientific method', he would have such a degree of belief. But fourthly it need mean none of these things for men have not always believed in scientific method, and just as we ask 'But am I necessarily reasonable,' we can also ask 'But is the scientist necessarily reasonable?' In this ultimate meaning it seems to me that we can identify reasonable opinion with the opinion of an ideal person in similar circumstances. What, however, would this ideal person's opinion be? As has previously been remarked, the highest ideal would be always to have a true opinion and be certain of it; but this ideal is more suited to God than to man.[13]

We have therefore to consider the human mind and what is the most we can ask of it.[14] The human mind works essentially according to general rules or habits; a process of thought not proceeding according to some rule would simply be a random sequence of ideas; whenever we infer *A* from *B* we do so in virtue of some relation between them. We can therefore state the problem of the ideal as "What habits in a general sense would it be best for the human mind to have?" This is a large and vague question which could hardly be answered unless the possibilities were first limited by a fairly definite conception of human nature. We could imagine some very useful habits unlike those possessed by any men. [It must be explained that I use habit in the most general possible sense to mean simply rule or law of behaviour, including instinct: I do not wish to distinguish acquired rules or habits in the narrow sense from innate rules or instincts, but propose to call them all habits alike.] A completely general criticism of the human mind is therefore bound to be vague and futile, but something useful can be said if we limit the subject in the following way.

Let us take a habit of forming opinion in a certain way; e.g. the habit of proceeding from the opinion that a toadstool is yellow to the opinion that it is unwholesome. Then we can accept the fact that the person has a habit of this sort, and ask merely what degree of opinion that the toadstool is unwholesome it would be best for him to entertain when he sees it; i.e. granting that he is going to think always in the same way about all yellow toadstools, we can ask what degree of confidence it would be best for him to have that they are unwholesome. And the answer is that it will in general be best for his degree of belief that a yellow toadstool is unwholesome to be equal to the proportion of yellow toadstools which are in fact

---

[13][Earlier draft of matter of preceding paragraph in some ways better. – F.P.R.

What is meant by saying that a degree of belief is reasonable? First and often that it is what I should entertain if I had the opinions of the person in question at the time but was otherwise as I am now, e.g. not drunk. But sometimes we go beyond this and ask: 'Am I reasonable?' This may mean, do I conform to certain enumerable standards which we call scientific method, and which we value on account of those who practise them and the success they achieve. In this sense to be reasonable means to think like a scientist, or to be guided only by ratiocination and induction or something of the sort (i.e. reasonable means reflective). Thirdly, we may go to the root of why we admire the scientist and criticize not primarily an individual opinion but a mental habit as being conducive or otherwise to the discovery of truth or to entertaining such degrees of belief as will be most useful. (To include habits of doubt or partial belief.) Then we can criticize an opinion according to the habit which produced it. This is clearly right because it all depends on this habit; it would not be reasonable to get the right conclusion to a syllogism by remembering vaguely that you leave out a term which is common to both premisses.

We use reasonable in sense 1 when we say of an argument of a scientist this does not seem to me reasonable; in sense 2 when we *contrast* reason and superstition or instinct; in sense 3 when we *estimate* the value of new methods of thought such as soothsaying.].

[14]What follows to the end of the section is almost entirely based on the writings of C. S. Peirce. [Especially his "Illustrations of the Logic of Science", *Popular Science Monthly*, 1877 and 1878, reprinted in *Chance Love and Logic* (1923).].

unwholesome. (This follows from the meaning of degree of belief.) This conclusion is necessarily vague in regard to the spatio-temporal range of toadstools which it includes, but hardly vaguer than the question which it answers. (Cf. density at a point of gas composed of molecules.)

Let us put it in another way: whenever I make an inference, I do so according to some rule or habit. An inference is not completely given when we are given the premiss and conclusion; we require also to be given the relation between them in virtue of which the inference is made. The mind works by general laws; therefore if it infers q from p, this will generally be because q is an instance of a function φx and p the corresponding instance of a function ψx such that the mind would always infer φx from ψx. When therefore we criticize not opinions but the processes by which they are formed, the rule of the inference determines for us a range to which the frequency theory can be applied. The rule of the inference may be narrow, as when seeing lightning I expect thunder, or wide, as when considering 99 instances of a generalization which I have observed to be true I conclude that the 100th is true also. In the first case the habit which determines the process is 'After lightning expect thunder'; the degree of expectation which it would be best for this habit to produce is equal to the proportion of cases of lightning which are actually followed by thunder. In the second case the habit is the more general one of inferring from 99 observed instances of a certain sort of generalization that the 100th instance is true also; the degree of belief it would be best for this habit to produce is equal to the proportion of all cases of 99 instances of a generalization being true, in which the 100th is true also.

Thus given a single opinion, we can only praise or blame it on the ground of truth or falsity: given a habit of a certain form, we can praise or blame it accordingly as the degree of belief it produces is near or far from the actual proportion in which the habit leads to truth. We can then praise or blame opinions derivatively from our praise or blame of the habits that produce them.

This account can be applied not only to habits of inference but also to habits of observation and memory; when we have a certain feeling in connection with an image we think the image represents something which actually happened to us, but we may not be sure about it; the degree of direct confidence in our memory varies. If we ask what is the best degree of confidence to place in a certain specific memory feeling, the answer must depend on how often when that feeling occurs the event whose image it attaches to has actually taken place.

Among the habits of the human mind a position of peculiar importance is occupied by induction. Since the time of Hume a great deal has been written about the justification for inductive inference. Hume showed that it could not be reduced to deductive inference or justified by formal logic. So far as it goes his demonstration seems to me final; and the suggestion of Mr Keynes that it can be got round by regarding induction as a form of probable inference cannot in my view be maintained. But to suppose that the situation which results from this is a scandal to philosophy is, I think, a mistake.

We are all convinced by inductive arguments, and our conviction is reasonable because the world is so constituted that inductive arguments lead on the whole to true opinions. We are not, therefore, able to help trusting induction, nor if we could help it do we see any reason why we should, because we believe it to be a reliable process. It is true that if any one has not the habit of induction, we cannot prove to him that he is wrong; but there is nothing peculiar in that. If a man doubts his memory or his perception we cannot prove to him that they are trustworthy; to ask for such a thing to be proved is to cry for the moon, and the same is true of induction. It is one of the ultimate sources of knowledge just as memory is: no one regards it as a scandal to philosophy that there is no proof that the world did not begin two minutes ago and that all our memories are not illusory.

We all agree that a man who did not make inductions would be unreasonable: the question is only what this means. In my view it does not mean that the man would in any way sin against formal logic or formal probability; but that he had not got a very useful habit, without which he would be very much worse off, in the sense of being much less likely[15] to have true opinions.

This is a kind of pragmatism: we judge mental habits by whether they work, i.e. whether the opinions they lead to are for the most part true, or more often true than those which alternative habits would lead to.

Induction is such a useful habit, and so to adopt it is reasonable. All that philosophy can do is to analyse it, determine the degree of its utility, and find on what characteristics of nature this depends. An indispensable means for investigating these problems is induction itself, without which we should be helpless. In this circle lies nothing vicious. It is only through memory that we can determine the degree of accuracy of memory; for if we make experiments to determine this effect, they will be useless unless we remember them.

Let us consider in the light of the preceding discussion what sort of subject is inductive or human logic – the logic of truth. Its business is to consider methods of thought, and discover what degree of confidence should be placed in them, i.e. in what proportion of cases they lead to truth. In this investigation it can only be distinguished from the natural sciences by the greater generality of its problems. It has to consider the relative validity of different types of scientific procedure, such as the search for a causal law by Mill's Methods, and the modern mathematical methods like the *a priori* arguments used in discovering the Theory of Relativity. The proper plan of such a subject is to be found in Mill[16]; I do not mean the details of his Methods or even his use of the Law of Causality. But his way of treating the subject as a body of inductions about inductions, the Law of Causality governing lesser laws and being itself proved by induction by simple enumeration. The different scientific methods that can be used are in the last resort judged by induction by simple enumeration; we choose the simplest law that fits the facts, but unless we found that laws so obtained also fitted facts other than those they were made to fit, we should discard this procedure for some other.

[15] 'Likely' here simply means that I am not sure of this, but only have a certain degree of belief in it.

[16] Cf. also the account of 'general rules' in the Chapter 'Of Unphilosophical Probability' in Hume's *Treatise*.

Chapter 4
Probable Knowledge

Richard C. Jeffrey

The central problem of epistemology is often taken to be that of explaining how we can know what we do, but the content of this problem changes from age to age with the scope of what we take ourselves to know; and philosophers who are impressed with this flux sometimes set themselves the problem of explaining how we can get along, knowing as little as we do. For knowledge is sure, and there seems to be little we can be sure of outside logic and mathematics and truths related immediately to experience. It is as if there were some propositions – that this paper is white, that two and two are four – on which we have a firm grip, while the rest, including most of the theses of science, are slippery or insubstantial or somehow inaccessible to us. Outside the realm of what we are sure of lies the puzzling region of probable knowledge – puzzling in part because the sense of the noun seems to be cancelled by that of the adjective. The obvious move is to deny that the notion of knowledge has the importance generally attributed to it, and to try to make the concept of belief do the work that philosophers have generally assigned the grander concept. I shall argue that this is the right move.

A Pragmatic Analysis of Belief

To begin, we must get clear about the relevant sense of 'belief'. Here I follow Ramsey: 'the kind of measurement of belief with which probability is concerned is ... a measurement of belief *qua* basis of action'.[1]

[1]Frank P. Ramsey, 'Truth and probability', in *The Foundations of Mathematics and Other Logical Essays*, R. B. Braithwaite, ed., London and New York, 1931, p. 171.

R.C. Jeffrey (deceased)
Princeton University, Boston, MA, USA

H. Arló-Costa et al. (eds.), *Readings in Formal Epistemology*, Springer Graduate Texts in Philosophy 1, DOI 10.1007/978-3-319-20451-2_4

Fig. 4.1

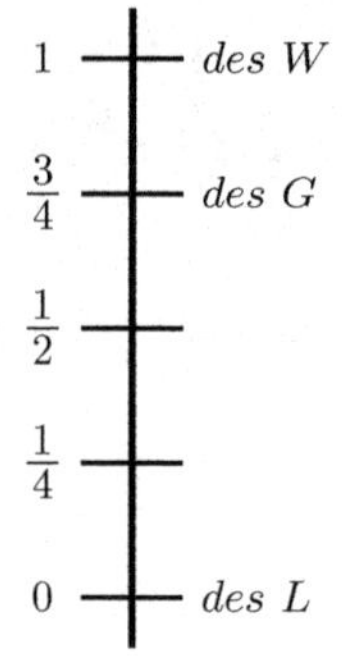

Ramsey's basic idea was that the desirability of a gamble G is a weighted average of the desirabilities of winning and of losing in which the weights are the probabilities of winning and of losing. If the proposition gambled upon is A, if the prize for winning is the truth of a proposition W, and if the penalty for losing is the truth of a proposition L, we then have

$$prob\ A = \frac{des\ G - des\ L}{des\ W - des\ L}. \tag{4.1}$$

Thus, if the desirabilities of losing and of winning happen to be 0 and 1, we have $prob\ A = des\ G$, as illustrated in Fig. 4.1, for the case in which the probability of winning is thought to be $\frac{3}{4}$.

On this basis, Ramsey[2] is able to give rules for deriving the gambler's subjective probability *and* desirability functions from his preference ranking of gambles, provided the preference ranking satisfies certain conditions of consistency. The probability function obtained in this way is a probability measure in the technical sense that, given any finite set of pairwise incompatible propositions which together exhaust all possibilities, their probabilities are non-negative real numbers that add up to 1. And in an obvious sense, probability so construed is a measure of the subject's willingness to act on his beliefs in propositions: it is a measure of degree of belief.

I propose to use what I take to be an improvement of Ramsey's scheme, in which the work that Ramsey does with the operation of forming gambles is done with the usual truth-functional operations on propositions.[3] The basic move is to restrict attention to certain 'natural' gambles, in which the prize for winning is the truth of the proposition gambled upon, and the penalty for losing is the falsity of that proposition. In general, the situation in which the gambler takes himself to be gambling on A with prize W and loss L is one in which he believes the proposition

[2] 'Truth and probability', F. P. Ramsey, *op. cit.*

[3] See Richard C. Jeffrey, *The Logic of Decision*, McGraw-Hill, 1965, the mathematical basis for which can be found in Ethan Bolker, *Functions Resembling Quotients of Measures*, Ph. D. Dissertation, Harvard University, 1965, and *Trans. Am. Math. Soc.*, 124, 1966, pp. 293–312.

$$G = AW \vee \overline{A}L.$$

If G is a natural gamble we have $W = A$ and $L = \overline{A}$, so that G is the necessary proposition, $T = A \vee \overline{A}$:

$$G = A\,A \vee \overline{A}\,\overline{A} = T.$$

Now if A is a proposition which the subject thinks good (or bad) in the sense that he places it above T (or below T) in his preference ranking, we have

$$prob\ A = \frac{des\ T - des\ \overline{A}}{des\ A - des\ \overline{A}}, \tag{4.2}$$

corresponding to Ramsey's formula (4.1).

Here the basic idea is that if $A_1, A_2, \ldots, A_n$ are an exhaustive set of incompatible ways in which the proposition A can come true, the desirability of A must be a weighted average of the desirabilities of the ways in which it can come true:

$$des\ A = w_1\ des\ A_1 + w_2\ des\ A_2 + \cdots + w_n\ des\ A_n, \tag{4.3}$$

where the weights are the conditional probabilities,

$$w_i = prob \quad A_i / prob \quad A. \tag{4.4}$$

Let us call a function *des* which attributes real numbers to propositions a *Bayesian desirability function* if there is a probability measure *prob* relative to which (4.3) holds for all suitable $A, A_1, A_2, \ldots, A_n$. And let us call a preference ranking of propositions *coherent* if there is a Bayesian desirability function which ranks those propositions in order of magnitude exactly as they are ranked in order of preference. One can show[4] that if certain weak conditions are met by a coherent preference ranking, the underlying desirability function is determined up to a fractional linear transformation, i.e., if *des* and *DES* both rank propositions in order of magnitude exactly as they are ranked in order of preference, there must be real numbers *a, b, c, d* such that for any proposition A in the ranking we have

$$DES\ \ A = \frac{a\ des\ A + b}{c\ des\ A + d}. \tag{4.5}$$

The probability measure *prob* is then determined by (4.2) up to a certain quantization. In particular, if *des* is Bayesian relative to *prob*, then *DES* will be Bayesian relative to *PROB*, where

[4] Jeffrey, *op. cit.*, chs. 6, 8.

$$PROB\,A = prob\,A\ \ (c\ des\,A + d)\,. \tag{4.6}$$

Under further plausible conditions, (4.5) and (4.6) are given either exactly (as in Ramsey's theory) or approximately by

$$DES\,A = a\ des\,A + b, \tag{4.7}$$

$$PROB\,A = prob\,A. \tag{4.8}$$

I take the principal advantage of the present theory over Ramsey's to be that here we work with the subject's actual beliefs, whereas Ramsey needs to know what the subject's preference ranking of relevant propositions would be if his views of what the world is were to be changed by virtue of his having come to believe that various arbitrary and sometimes bizarre causal relationships had been established via gambles.[5]

To see more directly how preferences may reflect beliefs in the present system, observe that by (4.2) we must have *prob A* > *prob B* if the relevant portion of the preference ranking is

$$\begin{array}{c} A,\quad B \\ T \\ \overline{B} \\ \overline{A} \end{array}$$

In particular, suppose that A and B are the propositions that the subject will get job 1 and that he will get job 2, respectively. Pay, working conditions, etc., are the same, so that he ranks A and B together. Now if he thinks himself more likely to get job 1 than job 2, he will prefer a guarantee of ($\overline{B}$) not getting job 2 to a guarantee of ($\overline{A}$) not getting job 1; for he thinks that an assurance of not getting job 2 leaves him more likely to get one or the other of the equally liked jobs than would an assurance of not getting job 1.

Probabilistic Acts and Observations

We might call a proposition *observational* for a certain person at a certain time if at that time he can make an observation of which the *direct* effect will be that his degree of belief in the proposition will change to 0 or to 1. Similarly, we might call a proposition *actual* for a certain person at a certain time if at that time he can perform an act of which the *direct* effect will be that his degree of belief in the proposition

[5]Jeffrey, *op. cit.*, pp. 145–150.

will change to 0 or to 1. Under ordinary circumstances, the proposition that the sun is shining is observational and the proposition that the agent blows his nose is actual. Performance of an act may give the agent what Anscombe calls[6] 'knowledge without observation' of the truth of an appropriate actual proposition. Apparently, a proposition can be actual or observational without the agent's knowing that it is; and the agent can be mistaken in thinking a proposition actual or observational.

The point and meaning of the requirement that the effect be 'direct', in the definitions of 'actual' and 'observational', can be illustrated by considering the case of a sleeper who awakens and sees that the sun is shining. Then one might take the observation to have shown him, directly that the sun is shining, and to have shown him indirectly that it is daytime. In general, an observation will cause numerous changes in the observer's belief function, but many of these can be construed as consequences of others. If there is a proposition E such that the *direct* effect of the observation is to change the observer's degree of belief in E to 1, then for any proposition A in the observer's preference ranking, his degree of belief in A after the observation will be the conditional probability

$$prob_E\, A = prob\ (A/E) = prob\, AE/prob\, E, \tag{4.9}$$

where *prob* is the observer's belief function before the observation. And conversely, if the observer's belief function after the observation is $prob_E$ and $prob_E$ is not identical with *prob*, then the *direct* effect of the observation will be to change the observer's degree of belief in E to 1. This completes a definition of *direct*.

But from a certain strict point of view, it is rarely or never that there is a proposition for which the direct effect of an observation is to change the observer's degree of belief in that proposition to 1; and from that point of view, the classes of propositions that count as observational or actual in the senses defined above are either empty or as good as empty for practical purposes. For if we care seriously to distinguish between 0.999 999 and 1.000 000 as degrees of belief, we may find that, after looking out the window, the observer's degree of belief in the proposition that the sun is shining is not quite 1, perhaps because he thinks there is one chance in a million that he is deluded or deceived in some way; and similarly for acts where we can generally take ourselves to be at best *trying* (perhaps with very high probability of success) to make a certain proposition true.

One way in which philosophers have tried to resolve this difficulty is to postulate a phenomenalistic language in which an appropriate proposition E can always be expressed, as a report on the immediate content of experience; but for excellent reasons, this move is now in low repute.[7] The crucial point is not that 0.999 999 is so close to 1.000 000 as to make no odds, practically speaking, for situations abound in which the gap is more like one half than one millionth. Thus, in examining a piece of cloth by candlelight one might come to attribute probabilities 0.6 and 0.4

[6]G. E. M. Anscombe, *Intention,* § 8, Oxford, 1957; 2nd ed., Ithaca, N.Y., 1963.

[7]See, e.g., J. L. Austin, *Sense and Sensibilia*, Oxford, 1962.

to the propositions G that the cloth is green and B that it is blue, without there being any proposition E for which the direct effect of the observation is anything near changing the observer's degree of belief in E to 1. One might think of some such proposition as that (E) *the cloth looks green or possibly blue*, but this is far too vague to yield *prob* $\left(\frac{G}{E}\right) = 0.6$ and *prob* $\left(\frac{B}{E}\right) = 0.4$. Certainly, there is *something* about what the observer sees that leads him to have the indicated degrees of belief in G and in B, but there is no reason to think the observer can express this something by a statement in his language. And physicalistically, there is some perfectly definite pattern of stimulation of the rods and cones of the observer's retina which prompts his belief, but there is no reason to expect him to be able to describe that pattern or to recognize a true description of it, should it be suggested.

As Austin[8] points out, the crucial mistake is to speak seriously of the *evidence* of the senses. Indeed the relevant experiences have perfectly definite characteristics by virtue of which the observer comes to believe as he does, and by virtue of which in our example he comes to have degree of belief 0.6 in G. But it does not follow that there is a proposition E of which the observer is certain after the observation and for which we have *prob* $\left(\frac{G}{E}\right) = 0.6$, *prob* $\left(\frac{B}{E}\right) = 0.4$, etc.

In part, the quest for such phenomenological certainty seems to have been prompted by an inability to see how uncertain evidence can be used. Thus C. I. Lewis:

> If anything is to be probable, then something must be certain. The data which themselves support a genuine probability, must themselves be certainties. We do have such absolute certainties, in the sense data initiating belief and in those passages of experience which later may confirm it. But neither such initial data nor such later verifying passages of experience can be phrased in the language of objective statement – because what can be so phrased is never more than probable. Our sense certainties can only be formulated by the expressive use of language, in which what is signified is a content of experience and what is asserted is the givenness of this content.[9]

But this motive for the quest is easily disposed of.[10] Thus, in the example of observation by candlelight, we may take the direct result of the observation (in a modified sense of 'direct') to be, that the observer's degrees of belief in G and B change to 0.6 and 0.4. Then his degree of belief in any proposition A in his preference ranking will change from *prob* A to

$$PROB\, A = 0.6\, prob\, (A/G) + 0.4\, prob\, (A/B)\,.$$

In general, suppose that there are propositions $E_1, E_2, \ldots, E_n$, in which the observer's degrees of belief after the observation are $p_1, p_2, \ldots, p_n$; where the E's are pairwise incompatible and collectively exhaustive; where for each i, *prob* E_i is

[8] Austin, *op. cit.*, ch. 10

[9] C. I. Lewis, *An Analysis of Knowledge and Valuation*, La Salle, Illinois, 1946, p. 186.

[10] Jeffrey, *op. cit.*, ch. 11.

neither 0 nor 1; and where for each proposition A in the preference ranking and for each i the conditional probability of A on E_i is unaffected by the observation:

$$PROB\ (A/E_i) = prob\ (A/E_i)\,. \tag{4.10}$$

Then the belief function after the observation may be taken to be *PROB*, where

$$PROB\,A = p_1\,prob\ (A/E_1) + p_2\,prob\ (A/E_2) + \cdots + p_n\,prob\,(A/E_n)\,, \tag{4.11}$$

if the observer's preference rankings before and after the observation are both coherent. Where these conditions are met, the propositions $E_1, E_2, \ldots, E_n$, may be said to form a *basis* for the observation; and the notion of a basis will play the role vacated by the notion of *directness*.

The situation is similar in the case of acts. A marksman may have a fairly definite idea of his chances of hitting a distant target, e.g. he may have degree of belief 0.3 in the proposition H that he will hit it. The basis for this belief may be his impressions of wind conditions, quality of the rifle, etc.; but there need be no reason to suppose that the marksman can express the relevant data; nor need there be any proposition E in his preference ranking in which the marksman's degree of belief changes to 1 upon deciding to fire at the target, and for which we have $prob\ \left(\frac{H}{E}\right) = 0.3$. But the pair H, $\overline{H}$ may constitute a *basis* for the act, in the sense that for any proposition A in the marksman's preference ranking, his degree of belief after his decision is

$$PROB\,A = 0.3\,prob\ (A/H) + 0.7\,prob\ \left(A/\overline{H}\right).$$

It is correct to describe the marksman as *trying* to hit the target; but the proposition that he is trying to hit the target can not play the role of E above. Similarly, it was correct to describe the cloth as *looking* green or possibly blue; but the proposition that the cloth looks green or possibly blue does not satisfy the conditions for directness.

The notion of directness is useful as well for the resolution of unphilosophical posers about probabilities, in which the puzzlıng element sometimes consists in failure to think of an appropriate proposition E such that the direct effect of an observation is to change degree of belief in E to 1, e.g. in the following problem reported by Mosteller.[11]

> Three prisoners, *a*, *b*, and *c*, with apparently equally good records have applied for parole. The parole board has decided to release two of the three, and the prisoners know this but not which two. A warder friend of prisoner *a* knows who are to be released. Prisoner *a* realizes that it would be unethical to ask the warder if he, *a*, is to be released, but thinks of asking for the name of *one* prisoner *other than himself* who is to be released. He thinks that before he asks, his chances of release are $\frac{2}{3}$. He thinks that if the warder says '*b* will be released,' his own chances have now gone down to $\frac{1}{2}$, because either *a* and *b* or *b* and *c* are to be released. And so *a* decides not to reduce his chances by asking. However, *a* is mistaken in his calculations. Explain.

[11]Problem 13 of Frederick Mosteller, *Fifty Challenging Problems in Probability,* Reading, Mass., Palo Alto, and London, 1965.

Here indeed the possible cases (in a self-explanatory notation) are

$$AB,\ AC,\ BC,$$

and these are viewed by a as equiprobable. Then *prob* A is $\frac{2}{3}$ but *prob* $\left(\frac{A}{B}\right)$ = *prob* $\left(\frac{A}{B}\right) = \frac{1}{2}$, and, since the warder must answer either 'b' or 'c' to a's question, it looks as if the direct result of the 'observation' will be that a comes to attribute probability 1 either to the proposition B that b will be released, or to the proposition C that c will be released. But this is incorrect. The relevant evidence-proposition would be more like the proposition *that the warder says*, 'b', or *that the warder says*, 'c', even though neither of these will quite do. For it is only in cases AB and AC that the warder's reply is dictated by the facts: in case BC, where b and c are both to be released, the warder must somehow choose *one* of the two true answers. If a expects the warder to make the choice by some such random device as tossing a coin, then we have *prob* (A/the warder says, 'b') = *prob* (A/the warder says, 'c') = *prob* $A = \frac{2}{3}$; while if a is sure that the warder will say 'b' if he can, we have *prob* (A/the warder says 'b') $= \frac{1}{2}$ but *prob* (A/the warder says 'c') $= 1$.

3. *Belief: reasons vs. causes*. Indeed it is desirable, where possible, to incorporate the results of observation into the structure of one's beliefs via a basis of form $E, \bar{E}$ where the probability of E after the observation is nearly 1. For practical purposes, E then satisfies the conditions of directness, and the 'direct' effect of the observation can be described as informing the observer of the truth of E. Where this is possible, the relevant passage of sense experience *causes* the observer to believe E; and if *prob* $\left(\frac{A}{E}\right)$ is high, his belief in E may be a *reason* for his believing A, and E may be spoken of as (inconclusive) *evidence* for A. But the sense experience is evidence neither for E nor for A. Nor does the situation change when we speak physicalistically in terms of patterns of irritation of our sensory surfaces, instead of in terms of sense experience: such patterns of irritation *cause* us to believe various propositions to various degrees; and sometimes the situation can be helpfully analyzed into one in which we are caused to believe $E_1, E_2, \ldots, E_n$, to degrees $p_1, p_2, \ldots, p_n$, whereupon those beliefs provide *reasons* for believing other propositions to other degrees. But patterns of irritation of our sensory surfaces are not reasons or evidence for any of our beliefs, any more than irritation of the mucous membrane of the nose is a *reason* for sneezing.

When I stand blinking in bright sunlight, I can no more believe that the hour is midnight than I can fly. My degree of belief in the proposition that the sun is shining has two distinct characteristics, (a) It is 1, as close as makes no odds. (b) It is compulsory. Here I want to emphasize the second characteristic, which is most often found in conjunction with the first, but not always. Thus, if I examine a normal coin at great length, and experiment with it at length, my degree of belief in the proposition that the next toss will yield a head will have two characteristics, (a) It is $\frac{1}{2}$. (b) It is compulsory. In the case of the coin as in the case of the sun, I cannot decide to have a different degree of belief in the proposition, any more than I can decide to walk on air.

In my scientific and practical undertakings I must make use of such compulsory beliefs. In attempting to understand or to affect the world, I cannot escape the fact that I am part of it: I must rather make use of that fact as best I can. Now where epistemologists have spoken of observation as a source of *knowledge*, I want to speak of observation as a source of compulsory *belief* to one or another degree. I do not propose to identify a very high degree of belief with knowledge, any more than I propose to identify the property of being near 1 with the property of being compulsory.

Nor do I postulate any *general* positive or negative connection between the characteristic of being compulsory and the characteristic of being sound or appropriate in the light of the believer's experience. Nor, finally, do I take a compulsory belief to be necessarily a permanent one: new experience or new reflection (perhaps, prompted by the arguments of others) may loosen the bonds of compulsion, and may then establish new bonds; and the effect may be that the new state of belief is sounder than the old. or less sound.

Then why should we trust our beliefs? According to K. R. Popper,

> ... the decision to accept a basic statement, and to be satisfied with it, is causally connected with our experiences – especially with our *perceptual experiences*. But we do not attempt to *justify* basic statements by these experiences. Experiences can *motivate a decision*, and hence an acceptance or a rejection of a statement, but a basic statement cannot be *justified* by them – no more than by thumping the table.[12]

I take this objection to be defective, principally in attempting to deal with basic statements (observation reports) in terms of *decisions* to *accept* or to *reject* them. Here acceptance parallels belief, rejection parallels disbelief (belief in the denial), and tentativeness or reversibility of the decision parallels *degree* of belief. Because logical relations hold between statements, but not between events and statements, the relationship between a perceptual experience (an event of a certain sort) and a basic statement cannot be a logical one, and therefore, Popper believes, cannot be of a sort that would justify the statement:

> Basic statements are accepted as the result of a decision or agreement; and to that extent they are conventions.[13]

But in the absence of a positive account of the nature of acceptance and rejection, parallel to the account of partial belief given in section 1, it is impossible to evaluate this view. Acceptance and rejection are apparently acts undertaken as results of decisions; but somehow the decisions are conventional – perhaps only in the sense that they may be *motivated* by experience, but not *adequately* motivated, if adequacy entails justification.

---

[12]K. R. Popper, *The Logic of Scientific Discovery,* London, 1959, p. 105.

[13]Popper, *op. cit.,* p. 106.

To return to the question, 'Why should we trust our beliefs?' one must ask what would be involved in *not* trusting one's beliefs, if belief is analyzed as in section 1 in terms of one's preference structure. One way of mistrusting a belief is declining to act on it, but this appears to consist merely in lowering the degree of that belief: to mistrust a partial belief is then to alter its degree to a new, more suitable value.

A more hopeful analysis of such mistrust might introduce the notion of sensitivity to further evidence or experience. Thus, agents 1 and 2 might have the same degree of belief – $\frac{1}{2}$ – in the proposition H_1 that the first toss of a certain coin will yield a head, but agent 1 might have this degree of belief because he is convinced that the coin is normal, while agent 2 is convinced that it is either two-headed or two-tailed, he knows not which.[14] There is no question here of agent 2's expressing his mistrust of the figure $\frac{1}{2}$ by lowering or raising it, but he can express that mistrust quite handily by aspects of his belief function. Thus, if H_i is the proposition that the coin lands head up the *i*th time it is tossed, agent 2's beliefs about the coin are accurately expressed by the function $prob_2$ where

$$prob_2\ H_i = \frac{1}{2},\ prob_2\ \left(H_i/H_j\right) = 1,$$

while agent 1's beliefs are equally accurately expressed by the function $prob_1$ where

$$prob_1\ \left(H_{i_1},\ H_{i_2}, \ldots, H_{i_n}\right) = 2^{-n},$$

if $i_1 < i_2 < \ldots < i_n$. In an obvious sense, agent 1's beliefs are *firm* in the sense that he will not change them in the light of further evidence, since we have

$$prob_1\ (H_{n+1}/H_1,\ H_2, \ldots,\ H_n) = prob_1\ H_{n+1} = \frac{1}{2},$$

while agent 2's beliefs are quite tentative and in that sense, mistrusted by their holder. Still, $prob_1\ H_i = prob_2\ H_i = \frac{1}{2}$.

After these defensive remarks, let me say how and why I take compulsive belief to be sound, under appropriate circumstances. Bemused with syntax, the early logical positivists were chary of the notion of truth; and then, bemused with Tarski's account of truth, analytic philosophers neglected to inquire how we come to believe or disbelieve simple propositions. Quite simply put, the point is: coming to have suitable degrees of belief in response to experience is a matter of training – a *skill* which we begin acquiring in early childhood, and are never quite done polishing. The skill consists not only in coming to have appropriate degrees of belief in appropriate propositions under paradigmatically good conditions of observation, but also in coming to have appropriate degrees of belief between zero and one when conditions are less than ideal.

Thus, in learning to use English color words correctly, a child not only learns to acquire degree of belief 1 in the proposition that the cloth is blue, when in bright sunlight he observes a piece of cloth of uniform hue, the hue being squarely in

[14]This is a simplified version of 'the paradox of ideal evidence', Popper, *op. cit.*, pp. 407–409.

the middle of the blue interval of the color spectrum: he also learns to acquire appropriate degrees of belief between 0 and 1 in response to observation under bad lighting conditions, and when the hue is near one or the other end of the blue region. Furthermore, his understanding of the English color words will not be complete until he understands, in effect, that blue is between green and violet in the color spectrum: his understanding of this point or his lack of it will be evinced in the sorts of mistakes he does and does not make, e.g. in mistaking green for violet he may be evincing confusion between the meanings of 'blue' and of 'violet', in the sense that his mistake is linguistic, not perceptual.

Clearly, the borderline between factual and linguistic error becomes cloudy, here: but cloudy in a perfectly realistic way, corresponding to the intimate connection between the ways in which we experience the world and the ways in which we speak. It is for this sort of reason that having the right language can be as important as (and can be in part identical with) having the right theory.

Then learning to use a language properly is in large part like learning such skills as riding bicycles and flying aeroplanes. One must train oneself to have the right sorts of responses to various sorts of experiences, where the responses are degrees of belief in propositions. This may, but need not, show itself in willingness to utter or assent to corresponding sentences. Need not, because e.g. my cat is quite capable of showing that it thinks it is about to be fed, just as it is capable of showing what its preference ranking is, for hamburger, tuna fish, and oat meal, without saying or understanding a word. With people as with cats, evidence for belief and preference is behavioral; and speech is far from exhausting behavior.[15]

Our degrees of beliefs in various propositions are determined jointly by our training and our experience, in complicated ways that I cannot hope to describe. And similarly for conditional subjective probabilities, which are certain ratios of degrees of belief: to some extent, these are what they are because of our training – because we speak the languages we speak. And to this extent, conditional subjective probabilities reflect *meanings*. And in this sense, there can be a theory of degree of confirmation which is based on analysis of meanings of sentences. Confirmation theory is therefore semantical and, if you like, logical.[16]

Discussion

L. HURWICZ: *Richard Jeffrey on the Three Prisoners.*

I would like to make a comment which I think is along the lines of Professor Jeffrey's discussion of the three prisoners. I would like to make the situation a little

[15] Jeffrey, *op. cit.*, pp. 57–59.

[16] Support of U.S. Air Force Office of Scientific Research is acknowledged, under Grant AF–AFOSR–529–65.

more explicit than it was earlier, although I shall not contradict anything that has been said: I think this will help us to see to what extent, if any, there is anything surprising or paradoxical in the situation.

First of all let me say this: there were three possible decisions by the warden – *AB, BC* and *AC*; then, as against that, there was also the question of what the warden would say to *a* who asked the question who else was being freed, and clearly the warden could only answer '*b*' or '*c*'. What I'm going to put down here is simply the bivariate or two-way probability distribution, and it doesn't matter at all at this stage whether we interpret it as a frequency or as a subjective probability, because it's just a matter of applying the mechanics of the Bayes theorem.

One other remark I'd like to make is this: the case that was considered by Professor Jeffrey was one where the *a priori* probabilities of *AB, BC* and *AC* were each one-third. This actually does not at all affect the reasoning, and I will stick with it just because it is close to my limitations in arithmetic.

So the marginal frequencies or probabilities are all equal to one-third. If the decision had been *AB*, then of course the warden could only answer '*b*', and similarly if the decision had been *AC*, he could only answer '*c*'. So the joint frequency or probability of the following event is one-third: the people chosen for freeing are *a* and *b*, and when the warden is asked, 'Who is the person other than *a* who is about to be freed?', his answer is '*b*'. The joint probability is also one-third that the choice was *AC* and that the warden answered '*c*'.

We now come to the only case where the warden has a choice of what he will say, namely, the case where the decision was *BC*. The question was raised, quite properly, of how he goes about making this choice.

Let me here say the following. In a sense what I'm doing here is a sally into enemy territory, because I personally am not particularly Bayesian in my approach to decision theory, so I would not myself assert that the only method is to describe the warden's decision, the warden's principle of choice, as a probabilistic one. However, if it is not probabilistic, then of course the prisoner, our *a*, would have to be using some other principle of choice on his part in order to decide what to do. Being an unrepentant conservative on this, I might choose, or *A* might choose, the minimax principle. However, in order to follow the spirit of the discussion here, I will assume that the whole thing is being done in a completely Bayesian or probabilistic way; in this case, to compute the remaining joint distribution we must make some probabilistic assumption about how the warden will behave when asked the question.

So let the principle be this, that he has a certain random device such that if the people to be freed are *b* and *c*, his answer to the question will be '*b*' with probability β and '*c*' with of course probability $1-\beta$. All I will assume for the moment about β is that it is between zero and one, and that's probably one of the few uncontroversial points so far.

It is clear that the sum of the two joint probabilities (*BC* and '*b*', and *BC* and '*c*') will be one-third; so the first will be $\frac{1}{3}\beta$, and the second $\frac{1}{3}(1-\beta)$. The marginal (or absolute) probabilities of '*b*' and '*c*' will be $\frac{1}{3}(1+\beta)$ and $\frac{1}{3}(2-\beta)$ respectively.

Inf. → Dec. ↓	‘b’	‘c’	Marginal
AB	$\frac{1}{3}$	0	$\frac{1}{3}$
BC	$\beta/3$	$(1-\beta)/3$	$\frac{1}{3}$
AC	0	$\frac{1}{3}$	$\frac{1}{3}$
Marginal	$(1+\beta)/3$	$(2-\beta)/3$	1

Now what are the probabilities after the warden has given his answer? Suppose that the answer that the warden gave is ‘b’: the problem now is, what is the probability that a is to be freed, given that the warden said that b is to be freed? This probability, which I will denote by ‘π_b’, I obtain in the following way using what I hope is a self-explanatory notation:

$$\begin{aligned}\pi_b &= p\,(A/\text{‘}b\text{’}) \\ &= \frac{p\,(A\cdot\text{‘}b\text{’})}{p\,(\text{‘}b\text{’})} \\ &= \frac{p\,(AB\cdot\text{‘}b\text{’}) + p\,(AC\cdot\text{‘}b\text{’})}{p\,(AB\cdot\text{‘}b\text{’}) + p\,(AC\cdot\text{‘}b\text{’}) + p\,(BC\cdot\text{‘}b\text{’})} \\ &= \frac{\frac{1}{3}+0}{\frac{1}{3}+0+\beta/3} \\ &= 1/\,(1+\beta)\,.\end{aligned}$$

Similarly I get π_c (the probability that a is to be freed, given that the warden said that b is to be freed) as follows:

$$\begin{aligned}\pi_c &= p\,(A/\text{‘}c\text{’}) \\ &= \frac{p\,(A\cdot\text{‘}c\text{’})}{p\,(\text{‘}c\text{’})} \\ &= \frac{p\,(AB\cdot\text{‘}c\text{’}) + p\,(AC\cdot\text{‘}c\text{’})}{p\,(AB\cdot\text{‘}c\text{’}) + p\,(AC\cdot\text{‘}c\text{’}) + p\,(BC\cdot\text{‘}c\text{’})} \\ &= \frac{0+\frac{1}{3}}{0+\frac{1}{3}+(1-\beta)\,/3} \\ &= 1/\,(2-\beta)\,.\end{aligned}$$

Now the question which we now have to ask is this: are these conditional probabilities, π, different from the marginal (absolute) probability that a is to be freed, $p(a)$?

And the answer is that they are except when β happens to be equal to one-half, in which case the probability remains at its marginal value of two-thirds. But except in this special case the probabilities π_b and π_c can vary from one-half to one.[17]

As I indicated before, there is no quarrel between us, but I do want to explore just one step further, and that is this. You remember when we were told this anecdote there was a wave of laughter and I now want to see what it was that was so funny. It is that this prisoner became doubtful about asking for this extra information, because he thought his probability of being released would go down after getting it. So it seemed that having this extra information would make him less happy, even though he didn't have to pay for it. That really was the paradox, not the fact that the probabilities changed. Clearly, the change in probabilities is itself not at all surprising; for example, if the warden had told a the names of *two* people other than himself who would be freed, his optimism would have gone down very drastically.[18]

What is surprising is that a thought he would be less happy with the *prospect* of having the extra piece of information than without this prospect. What I want to show now is that a was just wrong to think this; in other words, if this information was free, he should have been prepared to hear it.

Suppose for instance β is different from one-half: I think it is implicit in this little anecdote that the probability of a's being released either before or after getting the information, in some sense corresponds to his level of satisfaction. If his chances are good he is happy; if his chances are bad he is miserable. So these π's, though they happen to have been obtained as probabilities, may at the same time be interpreted as utilities or what Professor Jeffrey called desirabilities. Good. Now if a proceeds in the Bayesian way he has to do the following: he has to look at all these numbers, because before he asks for the information he does not know whether the answer will be 'b' or 'c'. Then he must ask himself the following: How happy will I be if he says 'b'? How happy will I be if he says 'c'? And then in the Bayesian (or de Finetti or Ramsey) spirit he multiplies the utilities, say $u(`b')$ and $u(`c')$ associated with hearing the warden say 'b' or 'c' by the respective probabilities, say $p(`b')$ and $p(`c')$, of hearing these answers. He thus obtains an expression for his expected[19] utility associated with getting the extra information, say

$$Eu^* = p\left('b'\right) \cdot u\left('b'\right) + p\left('c'\right) \cdot u\left('c'\right).$$

[17]In the problem as reported by Mosteller, it might be reasonable to take $\beta = \frac{1}{2}$. In that case, let us note, $\pi_b = 1/\left(1+\frac{1}{2}\right) = \frac{2}{3}$ (not $\frac{1}{2}$ as suggested in the statement of the problem!) and also $\pi_c = 1/(2-\beta) = 1/\left(2-\frac{1}{2}\right) = \frac{2}{3}$. Hence (for $\beta = \frac{1}{2}$) a was wrong to expect the probabilities to change. But, on the other hand, the warden's reply would give him no additional information.

[18]Or suppose, that $\beta = 1$ (and a knows this). Then if a hears the warden tell him that c is one of the persons to be released, he will have good reason to feel happy. For when $\beta = 1$, the warden will tell a about having selected c only if the selected pair was AC. On the other hand, still with $\beta = 1$, if the warden says that b is one of the persons to be released, this means (with equal probabilities) that either AB or BC has been chosen, but *not* AC. Hence, with the latter piece of information, a will be justifiably less optimistic about his chances of release. (With β close to one, a similar situation prevails.)

[19]In the sense of the mathematical expectation of a random variable.

Now the required probabilities are the marginal probabilities at the bottom of the table, i.e.,

$$p\left('b'\right) = \frac{1+\beta}{3},\ p\left('c'\right) = \frac{2-\beta}{3}.$$

As for utilities, it is implicit in the argument that they are linear[20] functions of the probabilities that *a* will be released given the warden's answer. So

$$u\left('b'\right) = \pi_b = \frac{1}{1+\beta},\ u\left('c'\right) = \pi_c = \frac{1}{2-\beta}.$$

Hence the (expected) utility associated with getting the extra information from the warden is

$$Eu^* = \frac{1+\beta}{3}.\frac{1}{1+\beta} + \frac{2-\beta}{3}.\frac{1}{2-\beta} = \frac{2}{3}.$$

On the other hand, the expected utility Eu°, associated with *not* asking the warden for extra information is simply equal to the original probability *p*(*a*) that *a* will be released,

$$Eu^\circ = p(a) = \frac{2}{3}.$$

Hence it so happens that (for a utility function linear in probabilities of release)

$$Eu^* = Eu^\circ,$$

i.e., the expected utility with extra information (Eu^*) is the same as without extra information (Eu°). Thus *a* should be willing (but not eager) to ask for extra information (if it is free of charge). 'On the average'[21], it won't do him any harm; nor will it help him.[22]

P. SUPPES: *Rational Changes of Belief.*

I am generally very much in agreement with Professor Jeffrey's viewpoint on belief and knowledge as expressed in his paper. The focus of my brief remarks is to point out how central and difficult are the problems concerning *changes* of belief. Jeffrey remarks that the familiar method of changing probable beliefs by explicitly conditionalizing the relevant probability measure is not adequate in many

---

[20]See footnote 22 on the next page.

[21]'On the average' expresses the fact that the decision is made on the basis of mathematical expectation. It need not imply a frequency interpretation of probabilities.

[22]When utilities are *non-linear* with respect to probabilities of release, the prospect of additional information may be helpful or harmful.

situations – in fact, in all those situations that involve a change in the probability assigned to evidence, but a change that does not make the probability of possible evidence 0 or 1.

My point is that once we acknowledge this fact about the probable character of evidence we open Pandora's box of puzzles for any theory of rational behavior. I would like to mention three problems. These problems are not dealt with explicitly by Jeffrey, but the focus I place on them is certainly consistent with his own expressed views.

Attention and Selection

I begin with the problem of characterizing how a man attempting to behave rationally is to go about selecting and attending to what seems to be the right kind of evidence. Formulations of the problem of evidence within a logically well-defined language or sample space have already passed over the problem of evaluating their appropriateness. Any man has highly limited capacities for attention and observation. Given a characterization of his powers and limitations what is a rational way to behave? Consider the familiar coin-flipping example. When is it appropriate to stop concentrating on the outcomes of the flips and to start observing closely the behavior of the gambler doing the flipping? To take another sort of example, how do we characterize the behavior of detectives who act on subthreshold cues and conjectures that can scarcely even be verbalized, let alone supported by explicit evidence? Put more generally, what is the rational way to go about discovering facts as opposed to verifying them? It is easy to claim that for a wide variety of situations rational discovery is considerably more important than rational verification. In these cases of discovery we need to understand much better the information-processing capacities of human beings in order to be able to characterize in a serious way *rational* information-processing. Above all, it is certainly not clear to me what proportion of rational information-processing should be verbalized or is even verbalizable.

Finite Memory

The second problem concerns the rational use of our inevitably restricted memory capacities. A full-blown theory of rationality should furnish guidelines for the kinds of things that should be remembered and the kind that should not. Again a solution, but certainly not a solution whose optimality can be seriously defended, is at least partly given by the choice of a language or a sample space for dealing with a given set of phenomena. But the amount of information impinging on any man in a day or a week or a month is phenomenal and what is accessible if he chooses to make it so is even more so. What tiny fraction of this vast potential should be absorbed and stored for ready access and use? Within the highly limited context of mathematical statistics, certain answers have been proposed. For example, information about the outcome of an experiment can be stored efficiently and with little loss of information in the form of the likelihood function or some other sufficient statistic, but this approach is not of much use in most situations, although elements of the approach can perhaps be generalized to less restricted circumstances. Perhaps even more importantly, it is not clear what logical structure is the most rational to impose on memory. The attempts at constructing associative computer memories, as they are often called, show how little we are able as yet to characterize explicitly a memory with the power and flexibility of a normal man's, not to speak of the memory of a normal man who is using his powers with utmost efficiency. Perhaps one of the most important weaknesses of confirmation theory and the Ramsey-sort of theory developed by Jeffrey and others is that little is said about imposing a logical structure on evidence. Part of the reason for this is that the treatment of evidence is fundamentally static rather than dynamic and temporal. In real life, evidence accumulates over time and we tend to pay more attention to later than earlier data, but the appropriate logical mechanisms for storing, organizing and compressing temporally ordered data are as yet far from being understood.

Concept Formation

The most fundamental and the most far-reaching cognitive changes in belief undoubtedly take place when a new concept is introduced. The history of science and technology is replete with examples ranging from the wheel to the computer, and from arithmetic to quantum mechanics. Perhaps the deepest problem of rational behavior, at least from a cognitive or epistemological standpoint, is to characterize when a man should turn from using the concepts he has available to solve a given problem to the search not just for new evidence but for new concepts with which to analyze the evidence. Perhaps the best current example of the difficulty of characterizing the kinds of concepts we apply to the solution of problems is the floundering and vain searching as yet typical of the literature on artificial intelligence. We cannot program a computer to think conceptually because we do not understand how men think conceptually, and the problem seems too difficult to conceive of highly nonhuman approaches. For those of us interested in rational behavior the lesson to be learned from the tantalizing yet unsatisfactory literature on artificial intelligence is that we are a long way from being able to say what a rational set of concepts for dealing with a given body of experience should be like, for we do not have a clear idea of what conceptual apparatus we actually use in any real sense.

To the problems about rationality I have raised in these remarks there is the pat answer that these are not problems of the theory of rational behavior but only of the theory of actual behavior. This I deny. A theory of rationality that does not take account of the specific human powers and limitations of attention, memory and conceptualization may have interesting things to say but not about human rationality.

R. C. JEFFREY: *Reply*.

Suppes' and Hurwicz' comments are interesting and instructive, and I find I have little to add to them. But perhaps a brief postscript is in order, in response to Suppes' closing remark:

> A theory of rationality that does not take account of the specific human powers and limitations of attention may have interesting things to say, but not about human rationality.

It may be that there is no real issue between us here, but the emphasis makes me uncomfortable. In my view, the logic of partial belief is a branch of decision theory, and I take decision theory to have the same sort of relevance to human rationality that (say) quantification theory has: the relevance is there, even though neither theory is directly about human rationality, and neither theory takes any account of the specific powers and limitations of human beings.

For definiteness, consider the following preference ranking of four sentences s, s', t, t', where s and s' are logically inconsistent, as are t and t'.

$$\begin{array}{c} s \\ s' \\ t \\ t' \end{array}$$

This ranking is *incoherent*: it violates at least one of the following two requirements, (a) Logically equivalent sentences are ranked together. (b) The disjunction of two logically incompatible sentences is ranked somewhere in the interval between them, endpoints included. Requirements (a) and (b) are part of (or anyway, implied by) a definition of 'incoherent'. To see that the given ranking is incoherent, notice that (a) implies that the disjunction of the sentences s, s' is ranked with the disjunction of the sentences t, t', while (b) implies that in the given ranking, the first disjunction is higher than the second. In my view, the point of classifying this ranking as incoherent is much like the point of classifying the pair s, s' as logically inconsistent: the two classifications have the same sort of relevance to human rationality. In the two cases, a rational man who made the classification would therefore decline to own the incoherent preference ranking or to believe both of the inconsistent sentences. (For simplicity I speak of belief here as an all-or-none affair.)

True enough: since there is no effective decision procedure for quantificational consistency there is no routine procedure a man can use – be he ever so rational – to correctly classify arbitrary rankings of sentences as incoherent or arbitrary sets of sentences as inconsistent. The relevance of incoherence and inconsistency to human rationality is rather that a rational man, once he comes to see that his preferences are incoherent or that his beliefs are inconsistent, will proceed to revise them. In carrying out the revision he may use decision theory or quantification theory as an aid; but neither theory fully determines how the revision shall go.

In fine, I take Bayesian decision theory to comprise a sort of *logic* of decision: the notion of coherence has much the same sort of relationship to human ideals of rationality that the notion of consistency has. But this is not to deny Suppes' point. The Bayesian theory is rather like a book of rules for chess which tells the reader what constitutes winning: there remains the question of ways and means.

Chapter 5
Fine-Grained Opinion, Probability, and the Logic of Full Belief

Bas C. van Fraassen

Personal or subjective probability entered epistemology as a cure for certain perceived inadequacies in the traditional notion of belief. But there are severe strains in the relationship between probability and belief. They seem too intimately related to exist as separate but equal; yet if either is taken as the more basic, the other may suffer.

After explaining the difficulties in some detail I will propose a single unified account which takes conditional personal probability as basic. Full belief is therefore a defined, derivative notion. Yet it is easiest to explain the resulting picture of opinion as follows: my subjective probability is only a grading of the possibilities left open by my beliefs. My conditional probabilities generally derive from the strongest belief I can maintain when admitting the relevant condition. Appendices will survey the literature.

Full Belief and Personal Probability

The most forceful answer I can give if asked for my opinion, is to say what I fully believe. The point of having beliefs is to construct a *single* (though in general incomplete) picture of what things are like. One obvious model of this part of my opinion is a set of propositions.[1] Their intersection is the proposition which captures

[1]This was essentially the model provided in Hintikka's book *Knowledge and Belief*. By propositions I mean the semantic content of statements; the same proposition can be expressed by many statements. I am not addressing how opinion is stored or communicated.

B.C. van Fraassen (✉)
San Francisco State University, San Francisco, CA, USA

H. Arló-Costa et al. (eds.), *Readings in Formal Epistemology*, Springer Graduate Texts in Philosophy 1, DOI 10.1007/978-3-319-20451-2_5

exactly that single picture of the world which has my full assent. Clearly a person's full beliefs leave open many alternatives. Alternatives left open by belief are then also represented by (sets of) propositions, namely ones that imply my beliefs. But these alternatives do not all have the same status for me, though they are all "possible for all I [know or] believe." Some seem more or less likely than others: enter personal (subjective) probability, as a grading of the possibilities left open by one's beliefs.

I will take for granted that the probability of a proposition is a real number in the interval [0, 1], with the empty proposition Λ (self-contradiction) receiving 0 and the universal proposition U (tautology) receiving 1. The assignment is a *measure*, that is, it is additive and continuous (equivalently, countably additive). It follows from this that the assignment of probabilities respects the ordering by logical implication:

$$\text{If } A \subseteq B \text{ then } P(A) \leq P(B)$$

though we must be careful in any extrapolation from propositions to sets of propositions unless they are countable. That is essentially because at most countably many disjoint propositions can receive finite positive probability. (*Reason*: at most one can receive probability greater than 1/2, at most two can receive more than 1/3, ... etc. The question of infinitesimal positive probability will be taken up in an Appendix.)

The so-called lottery paradox shows that we cannot equate belief with probability $\geq p$, if $p < 1$. For example, suppose $p = 0.99$ and a lottery which I believe to be fair has 1000 tickets, then my probability that the kth ticket will not win the (single) prize equals 0.999. Hence for each $k = 1, \ldots, 1000$, I would believe that the kth ticket will not win. My beliefs would then entail that all tickets will fail to win, which conflicts with my original belief that the lottery is fair. This argument is more important for what it presupposes than for what it shows. It is clearly based on the assumed role of full belief: to form a single, unequivocally endorsed picture of what things are like.[2]

In fact, the thesis that probabilities grade exactly the alternatives left open by full belief guarantees that all full beliefs have maximal personal probability.

So what if we simply set $p = 1$, that is, *identify* our full beliefs with the propositions which are maximally likely to us? The first problem with this is that we seem to be treating full beliefs as on a par with tautologies. Are there no distinctions to be drawn among the maximally likely propositions? There is a second problem for this proposal as well. In science we deal with continuous quantities. Therefore, in general, if we let science guide our opinion, the maximally likely propositions will not form a single picture – they will just give us a family of rival maximally likely pictures.

Example 5.1 Consider the mass of the moon reckoned in kilograms, and suppose I am sure that it is a number in the interval $[a, b]$. If my probability follows Lebesgue

[2]This has been denied, e.g. by Henry Kyburg, and doubted, e.g. by Richard Foley (1993, Ch. 4).

measure then my probability is *zero* that the number equals x, for $a \leq x \leq b$. Hence my probability equals 100 % that the number lies in the set $[a, b] - \{x\}$, for each such number x. Yet no real number belongs to all these sets – their intersection is empty. Probability measures of this sort (deriving from continuous probability densities) are ubiquitous in science, and informed opinion must be allowed to let itself be guided by them. We have here a transfinite lottery paradox, and we can't get out of it in the way that worked for the finite case (see Maher 1990).

Supposition and Two-Place Probability

There is a third aspect of opinion, besides belief and subjective grading, namely *supposition*. Much of our opinion can be elicited only by asking us to suppose something, which we may or may not believe. The respondent imaginatively puts himself in the position of someone for whom the supposition has some privileged epistemic status. But if his answer is to express his present opinion – which is surely what is requested – then this "momentary" shift in status must be guided by what his present opinion is. How does this guidance work?

One suggestion is that the respondent moves to a state of opinion derived from his own in two steps: (1) discarding beliefs so that the supposition receives more than minimal likelihood; (2) then (without further change in beliefs) regrading the alternatives left open so as to give the supposition maximal likelihood. This makes sense only if both steps are unambiguous. We can imagine a simple case. Suppose Peter has as "primary" beliefs A and B, and believes exactly what they jointly entail; he is asked to entertain the supposition $C - A$. In response he imaginatively moves into the epistemic position in which (1) B is the only primary belief, and (2) he assigns 0 to all alternatives left open by B which conflict with $(C - A)$ and then regrades the others in the same proportions as they had but with the maximum assigned to $(B \cap C - A)$.

This simple case already hinges on a certain hierarchical structure in Peter's opinion. Moreover it presupposes that those alternatives which were left open by B, but which conflict with his initial equally primary belief that A, had been graded proportionately as well. Even more structure must be present to guide the two steps in less simple cases. What if the beliefs had been, say, A, B, and D, and their joint consequences, and the supposition was compatible with each but not with the conjunction of any two? The discarding process can then be guided only if some hierarchy among the beliefs determines the selection.

Let us consider conditional personal probability as a possible means for describing structure of this sort. The intuitive Example 5.1 above about the mass of the moon is the sort often given to argue for the irreducibility of conditional probability. I could continue the example with: the mass of the moon seems to me to equally likely to be x as $(x + b)/2$, on the supposition that it is one of those two numbers. The two possibilities at issue here are represented by the degenerate intervals $[x]$, $[(x + b)/2]$, so both they and the supposition that one or other is the case (represented by set $\{x, (x+b)/2\}$ their union) receive probability 0. The usual calculation of conditional probability, which would set $P(B \mid A)$ equal to $P(B \cap C|A)$ divided by $P(C|A \cap C)$, can therefore not be carried out. The suggestion that conditional probability is irreducible means that *two-place probability* $P(\,|\,)$ – probability of one thing *given* (*on supposition of*) another – is autonomous and cannot be defined in terms of the usual one-place ("absolute") probability. Rather the reverse: we should define $P(\,) = P(\,|\,U)$, probability conditional on the tautology.

There is a good deal of literature on two-place ("irreducible conditional") probability (see Appendix). Despite many individual differences, general agreement concerning two-place probability extends to:

I. If P is a 2-place probability function then $P(- \mid A)$ is "normally" a (1-place) probability function with $P(A \mid A) = 1$.
II. These derivative 1-place probability functions [described in I.] are related at least by the *Multiplication Axiom:*

$$P(B \cap C|A) = P(B|A)\,P(C|A \cap C)$$

where A, B, C, ... are assumed to be in the domain and co-domain of the function. The "normally" restriction (eliminating at least $A = \Lambda$) is to be discussed below.

New non-trivial relationships between propositions are now definable. De Finetti suggested relations of local comparison of the following type:

$$A \text{ is } \textit{superior} \text{ to } B \text{ iff } P(A|A+B) = 1$$

where '+' marks exclusive disjunction: $A + B = (A - B) \cup (B - A)$.[3]

Example 5.2 Given any probability measure P it is easy to produce a 2-place function that has that character:

$$\begin{aligned} \text{define } P(A|B) &= P(A \cap B)/P(B) \text{ if } P(B) > 0 \\ &= 1 \text{ if } P(B) = 0 \end{aligned}$$

That is a trivial 2-place function since it is definable from a 1-place function.

[3]De Finetti (1936). I want to thank John M. Vickers for bringing this to my attention; De Finetti's idea is developed considerably further, with special reference to zero relative frequency, in Vickers (1988), Sections 3.6 and 5.4. The relation here defined is slightly different from the so-named one in my (1979) – to which the name was somewhat better suited – for convenience in some of the proofs.

Example 5.3 Let U be the set of natural number $\{0, 1, 2, \ldots\}$. For index $n = 0, 1, 2, \ldots$ let p_n be the probability measure defined on all subsets of U by the condition that it assigns 0.1 to $\{x\}$ if x is in the set $\{10n, \ldots, 10n+9\}$, and 0 otherwise. Define:

$$P(A|B) = p_n(A \cap B)/p_n(B) \text{ for the first index } n \text{ such that } p_n(B) > 0, \text{ if there is any such index}; = 1 \text{ otherwise.}$$

To verify the Multiplication Axiom note for instance that if $A \cap C$ is not empty, and $P(A \mid C) > 0$ then the first index n for which $p_n(A \cap C) > 0$ is the same as the first index m such that $p_m(C) > 0$. The "otherwise" clause will apply here only if $B = \Lambda$.

These examples are instances of the "lexicographic" probability models which I will discuss at some length below. We make the ideas of one- and two-place probability precise as follows.

A *space* is a couple $S = \langle U, F\rangle$ with U a non-empty set (the *worlds*) and F (the family of *propositions*) a sigma-field on U, that is:

(a) $U \in F$
(b) if $A, B \in F$ then $A - B \in F$
(c) if $\{A_i : i = 1, 2, \ldots\} \subseteq F$ then $\cup \{A_i\} \in F$

A (1-place) *probability measure* P on space $S = \langle U, F\rangle$ is a function mapping F into the real numbers, subject to

1. $0 = P(\Lambda) \leq P(A) \leq P(U) = 1$
2. $P(A \cup B) + P(A \cap B) = P(A) + P(B)$ (*finite additivity*)
3. If $E_1 \subseteq E_2 \subseteq \ldots \subseteq E_n \subseteq \ldots$ has union E, then $P(E) = \sup\{P(E_n): n = 1, 2, \ldots\}$ (*continuity*)

Property 3 is in this context equivalent to *countable additivity*:

4. If $\{E_n: n = 1, 2, \ldots\}$ are disjoint, with union E, then $P(E) = \Sigma\{P(E_n): n = 1, 2, \ldots\}$

and also to the dual continuity condition for countable intersection. The general class of two-place probability measures to be defined now will below be seen to contain a rich variety of non-trivial examples.

A *2-place probability measure* $P(-\mid-)$ on space $S = \langle U, F\rangle$ is a map of $F \times F$ into real numbers such that

I. (Reduction Axiom) The function $P(- \mid A)$ is either a probability measure on S or else has constant value $= 1$.
II. (Multiplication Axiom)

$$P(B \cap C|A) = P(B|A)\,P(C|B \cap A)$$

for all A, B, C, in F.

If $P(| A)$ is a (1-place) probability measure, I shall call A *normal* (for P), and otherwise *abnormal*. ("Absurd" might have been a better name; it is clearly a notion allied to self-contradiction.) The definition of 2-place probability allows for the totally abnormal state of opinion ($P(A \mid B) = 1$ for all A and B). It should not be excluded formally, but I shall tacitly exclude it during informal discussion. Here are some initial consequences of the definition. The variables range of course over propositions (members of family F in space $S = \langle U, F \rangle$).

[T2.1] $P(X|A) = P(X \cap A|A)$
[T2.2] If A is normal, so are its supersets
[T2.3] If A is abnormal, so are its subsets
[T2.4] B is abnormal iff $P(-B \mid B) = 1$; iff $P(B \mid A) = 0$ for all normal A.

Let us call the case in which only Λ is abnormal the "Very Fine" case; that there are Very Fine 2-place probability measures on infinite fields will follow from results below.

We add one consequence which is related to De Finetti's notion of conglomerability:

[T2.5] Condition Continuity: If $\{E_n\}$ is a countable increasing chain – i.e. E_n part of E_{n+1} – with union E, and $P(E_n \mid E) > 0$ for all n, then $P(X \mid E)$ is the limit of the numbers $\{P(X|E_n)\}$.

To prove this: $P(X|E_n) = P(X \cap E_n|E) / P(E_n \cap E)$, so since E is normal, this follows from the principle of condition continuity which can be demonstrated for (one-place) probability measures.[4]

The Idea of the *A Priori*

In any conception of our epistemic state there will be propositions which are not epistemically distinguishable from the tautology U – let us say these are *a priori* for the person. This notion is the opposite of the idea of abnormality:

$$\text{A is } a\ \textit{priori for } P \text{ iff } P(A|X) = 1 \text{ for all } X, \text{iff } U - A \text{ is abnormal for } P.$$

What is *a priori* for a person is therefore exactly what is certain for him or her on any supposition whatsoever. This notion generalizes unconditional certainty, i.e. $P(A) = 1$. The strongest unconditional probability equivalence relation between A and B is that their symmetric difference $(A + B)$ has measure zero. We can

[4]See B. De Finetti (1972), Section 5.22.

generalize this similarly. As our strictest epistemic equivalence relation between two propositions we have *a priori* equivalence (their symmetric difference has probability 0 on all normal suppositions):

$$A \langle P \rangle B \text{ iff } A + B \text{ is abnormal.}$$[5]

The abnormal propositions are the ones *a priori* equivalent to the empty set (the self-contradiction) and the *a prioris* are the ones *a priori* equivalent to the tautology. (Of course these are subjective notions: we are speaking of what is *a priori* for the person with this state of opinion.)

Note now that $A\langle P\rangle B$ iff $P(A \mid A+B)=1$ and $P(B \mid A+B)=1$, since additivity would not allow that if $A+B$ were normal. We can divide this equivalence relation into its two conjuncts:

Definition A $P{>}B$ iff $P(A \mid A+B)=1$.

This is the relationship of "superiority" mentioned above.

[T3.1] If A logically implies B then $B\ P{>}A$.

It follows from the definition that $A\ P{>}A$. In a later section I shall also show that $P{>}$is transitive. Clearly if $A+B$ is normal, then $A\ P{>}B$ means that A is comparatively superior to B, in the sense that A is certainly true and B certainly false, on the supposition that one but not both are the case. But if $A+B$ is abnormal then the relationship $A\ P{>}B$ amounts to $A\langle P\rangle B$. The right reading for "$P >$" is therefore "is superior to or *a priori* equivalent to". To be brief, however, I'll just say "A is superior to B" for "$A\ P{>}B$", and ask you to keep the qualifications in mind.

Full Belief Revisited

The beliefs I hold so strongly that they are *a priori* for me are those whose contraries are all abnormal. There is a weaker condition a proposition K can satisfy: namely that any normal proposition which implies K is superior to any that are contrary to K. Consider the following conditions and definitions:

(A1)	Normality:	K is normal
(A2)	Superiority:	If A is a non-empty subset of K while B and K are disjoint, then $A\ P{>}B$
(A3)	Contingency:	The complement $U-K$ of K is normal.

We can restate the "Superiority" condition informally as follows:

[5]From this point on I shall drop the ubiquitous "for P" unless confusion threatens, and just write "a priori", "abnormal", etc. leaving the context to specify the relevant 2-place probability measure.

Superiority: the alternatives *K* leaves open are all superior to any alternative that *K* excludes.

We can deduce from these conditions something reminiscent of Carnap's "regularity" (or Shimony's "strict coherence"):

(A4) Finesse: all non-empty subsets of *K* are normal.

Definition *K* is a *belief core* (for *P*) iff *K* satisfies (A1)–(A3).

Note that the *a priori* propositions satisfy (A1) and (A2), though definitely not (A3), but rather its opposite. However, the following elementary results show that all the *a prioris* are among the propositions implied by belief cores.

[T4.1] If K is a belief core then $P(K \mid U) = 1$.
[T4.2] If K is a belief core then A $P>K$ iff K implies A.
[T4.3] If K is a belief core and A is *a priori* then *K* is a subset of A.

To characterize the full beliefs we need take into account the extreme possibility of there being no belief cores; in that case we still want the *a prioris* to be full beliefs of course. (This corresponds to what I have elsewhere called "Zen minds": states of opinion in which nothing is fully believed if it is subjectively possible to withhold belief.) In view of the above we have several equivalent candidates for this characterization, of which we can choose one as definition:

Definition A is a *full belief* (for *P*) iff (i) there is a belief core, and $A\ P>K$ for some belief core *K*; or (ii) there is no belief core, and *A* is *a priori*.

[T4.4] The following conditions are equivalent:
(a) *A* is a full belief (for *P*).
(b) Some proposition *J* which is either *a priori* or a full belief core is such that $A\ P>J$.
(c) A is implied either by an *a priori* or by a belief core (for *P*).

Very little in this discussion of full belief hinges on the peculiarities of probability. Indeed, (conditional) probability enters here only to give us intelligible, non-trivial resources for defining the notions of subjective superiority and a-prioricity (equivalently, [ab]normality). If those notions could be acceptably primitive, the account of full belief here would just take the following form:

A belief core is a proposition *K* such that:

(a) *K* and its complement are both normal; (b) *K* does not leave open any abnormal alternatives; (c) any alternatives left open by *K* are superior to any alternatives *K* excludes.

A belief is any proposition implied by a belief core (or *a priori*).

As before, a proposition is here called an alternative left open by *K* exactly if it is non-empty and implies *K*.

Identification of Full Beliefs

By the definition I gave of full belief, beliefs are clustered: each belongs to a family $\{A: A\ P > K\}$ for some belief core K (if there are any at all), which by [T4.2] sums up that cluster exactly:

$$K \text{ is the intersection of } \{A: A \;\; P > K\}$$
$$\{A: A\ P > K\} = \{A: K \text{ implies } A\}.$$

We can now prove that these clusters form a chain, linearly ordered by set inclusion (implication).

[T5.1] If K, K' are belief cores, then either K is a subset of K' or K' is a subset of K.

For the proof let $A = K - K'$ and $B = K' - K$. Each is normal if non-empty, by (A4). If B is not empty then, since A is disjoint from K', it follows by (A2) that $B\ P{>}A$. By parity of reasoning, if A is not empty, then $A\ P{>}B$. But we cannot have both unless $A + B$ is abnormal, and hence empty by [T2.2] and (A4). So we conclude that either A or B or both are empty; hence at least one of K and K' is a subset of the other.

This result is crucial for the characterization of full belief. It is therefore worthwhile to note that the only ingredients needed for the proof were the features of Superiority and Finesse of belief cores, plus the following characteristic of the superiority relationship: if $A\ P{>}B$ and $B\ P{>}A$ then $A + B$ is abnormal. Here is an illustrative example:

Example 5.4 For the center of mass of the remains of Noah's ark, Petra has subjective probability 1 for each of the following three propositions: that it lies in the Northern Hemisphere (which part of the earth includes the Equator), that it lies in the Western Hemisphere, and thirdly that it lies North of the Equator. But only the first two of these propositions are among her full beliefs; the third is not. On the supposition that one but only one of these beliefs is true, she gives 100 % probability to the first proposition, that it lies in the Northern Hemisphere.

Note that the last sentence implies that the first proposition is superior to the second, although both are full beliefs. I will give a formal reconstruction of this example later on.

Writing K^* for the intersection of all the belief cores, we conclude that if A is a full belief, then K^* implies A. But is K^* itself a belief core? Does it have 100 % probability? Is it even non-empty? This is the problem of transfinite consistency of full belief in our new setting.

[T5.2] The intersection of a non-empty countable family of belief cores is a belief core.

For proof, assume there is at least one belief core; call it K. Assume also that the belief cores are countable and form a chain (the latter by [T5.1]), and call the intersection K^*. Countable additivity of the ordinary probability measure $P() = P(| U)$ is equivalent to just the right continuity condition needed here: the probabilities of the members of a countable chain of sets converge to the probability of its intersection. Since in our case all those numbers equal 1, so does $P(K^*)$. Therefore also K^* is not empty, and thus normal because it is a subset of at least one belief core. Moreover, so are its non-empty subsets, so that they are normal too. Its complement $U - K^*$ includes $U - K$, and is therefore also normal.

We have now seen that K^* satisfies conditions (A1), (A3), and (A4), and need still to establish (A2). If A is a normal subset of K^*, and hence of all belief cores, and B is disjoint of K^*, we have $P(A \mid A + (B - K')) = 1$ for all belief cores K'. But the sets $B - K'$ form an increasing chain whose union is $B - K^* = B$. Hence also the sets $A + (B - K')$ here form such a chain with union $A + B$. To conclude now that $P(A \mid A + B) = 1$, we appeal to [T2.5], the principle of Condition Continuity. This ends the proof.

The significance of this result may be challenged by noting that the intersection of countably many sets of measure 1 also has measure 1. So how have we made progress with the transfinite lottery paradox? In four ways. The first is that in the representation of opinion we may have a "small" family of belief cores even if probability is continuous and there are uncountably many propositions with probability 1. The second is that no matter how large a chain is, its intersection is one of its members if it has a first (= "smallest") element. The third is that the following is a condition typically met in spaces on which probabilities are defined even in the most scientifically sophisticated applications:

(*) Any chain of propositions, linearly ordered by set inclusion, has a countable subchain with the same intersection.

[T5.3] If (*) holds and there is at least one belief core, then the intersection of all belief cores is also a belief core.

This is a corollary to [T5.2].

Fourthly, farther below I will also describe an especially nice class of models of fine-grained opinion for which we can prove that the intersection of the belief cores, if any, is always also a belief core ("lexicographic probability"). There are no countability restrictions there.

To What Extent Does Belief Guide Opinion?

It is not to be expected that every two-place probability function is admissible as a representation of (possible) opinion. If we want to use this theory in descriptive epistemology, it is necessary to look for kinds of probability functions that have interesting structure. There are models in which there are no belief cores at all. Combining our previous Examples 5.1 and 5.2, take Lebesgue measure m on the unit interval, and trivially extend it to a two-place function by $P(A \mid B) = m(A \cap B)/m(B)$ if defined and $P(A \mid B) = 1$ if not (though A, B in domain of m). Then every unit set $\{x\}$ is in the domain and is abnormal. Therefore there is no set all of whose subsets are normal, and hence no belief cores. (The absence of belief cores in our present example derives from its triviality, and not from the continuity.) Obviously then, if this represents someone's opinion, his opinions are not guided or constrained by his beliefs (which include only the *a priori*).[6]

At the other extreme from this example, there is the Very Fine case of a probability function P for which every non-empty set is normal.

Definition P is *belief covered* if the union of the belief cores equals U.

In that case, P is Very Fine. For let A be any non-empty proposition; there will be some belief core K such that $K \cap A$ is not empty, hence normal, thus making A normal.

Example 5.3 furnishes us with a relatively simple example of this sort. Recall that P is there constructed from the series $p_0, p_1, \ldots, p_n, \ldots$ where the whole probability mass of p_n is concentrated (and evenly distributed) on the natural numbers $\{10n, \ldots, 10n+9\}$. In this example, the belief cores are exactly the sets

$$K_0 = \{0, \ldots, 9\}, K_1 = \{0, \ldots, 19\},$$
$$K_2 = \{0, \ldots, 29\}, \ldots K_i = \{0, \ldots, 10i + 9\}$$

Clearly K_i is normal, since $P(-|K_i) = P(-|K_0) = P_0$. The complement of K_i is normal too, for

$$P(-|U - K_i) = P(-|\{10(i+1), \ldots\}) = p_{i+1}.$$

If A is a non-empty subset of K_i and B is disjoint from K_i, then A is superior to B. Specifically, the first n such that $p_n(A) > 0$ can be no higher than i in that case, while the first m such that $p_m(B) > 0$ can be no lower than $i + 1$. Therefore, the first k such that $p_k(A + B) > 0$ will assign a positive probability to A and *zero* to B.

These belief cores clearly cover U; P is belief covered and Very Fine. Indeed, the belief cores are well-ordered,

Define the *belief remnants*

$$R_0 = K_0$$
$$R_{j+1} = K_{j+1} - K_j \, (j = 0, 1, 2, \ldots).$$

[6]Could a person's opinion be devoid of belief cores? Our definitions allow this, and it seems to me this case is related to the idea of a "Zen mind" which I have explored elsewhere (van Fraassen 1988).

Clearly $p_i = P(| R_i)$; for example, $p_1 = P(| \{10, \ldots, 19\})$. Probabilities conditional on belief remnants (beliefs remaining upon retrenchment to a weaker core) determine all probabilities in this case:

$$P(-|A) = P(-|A \cap R_i) \text{ for the first } i \text{ such that } P(A|R_i) > 0.$$

This says quite clearly that (in this case) belief guides opinion, for probabilities conditional on belief remnants are, so to speak, all the conditional probabilities there are.

The Multiplication Axiom Visualized

In the basic theory of two-place probability, the Multiplication Axiom places the only constraint on how the one-place functions $P(| A)$, $P(| B)$, ... are related to each other. It entails that the proposition $A - B$ is irrelevant to the value of $P(B \mid A)$ – that this value is the same as $P(A \cap B \mid A)$ – and that the usual ratio-formula calculates the conditional probability when applicable. Indeed, the ratio formula applies in the generalized form summarized in the following:

If X is a subset of A which is a subset of B, then:

[T7.1] if $P(A \mid B) > 0$ then $P(X|A) = P(X|B) : P(A|B)$

[T7.2] if X is normal, then $P(X|B) \leq P(X|A)$.

There is another way to sum up how the Multiplication Axiom constrains the relation between $P(| A)$ and $P(| B)$ in general. When we consider the two conditional probabilities thus assigned to any proposition that implies both A and B, we find a proportionality factor, which is constant when defined.

[T7.3] If $P(A \mid B) > 0$ then there is a constant $k \geq 0$ such that for all subsets X of $A \cap B$, $P(X|A) = kP(X|B)$. The constant $k = k(A, B) = [P(B|A) / P(A|B)]$, defined provided $P(A \mid \mathrm{B}) > 0$.

Equivalence Relations on Propositions

In the main literature on two-place probability we find an equivalence relationship other than *a priori equivalence*, which I shall call *surface equivalence*:[7]

[7] In earlier treatments of two-place probability this relationship has appeared as a special axiom: If $P(A \mid B) = P(B \mid A) = 1$ then $P(\cdot \mid A) = P(\cdot \mid B)$.

[T8.1] The following conditions are equivalent:
(a) $P(\cdot|A) = P(\cdot|B)$
(b) $P(A + B|A \cup B) = 0$ or else both A, B are abnormal
(c) $P(A|B) = P(B|A) = 1$.

(I use the dot for function notation: (a) means that $P(X|A) = P(X|B)$ for all X.) It is easy to prove that (a) implies (b), for if (a) and either A or B is normal then both are normal. Secondly suppose (b). If A, B abnormal then (c) follows. Suppose then that one of A, B is normal, so $A \cup B$ is normal. Then either $P(A|A \cup B)$ or $P(B|A \cup B)$ is positive; suppose the first. Since $A \cap B \subseteq A \subseteq A \cup B$ it follows by [T7.1] that $P(B \mid A) = P(B \cap A \mid A \cup B) : P(A \mid A \cup B)$. But $P(A - B|A \cup B) = 0$ so $P(A \cap B|A \cup B) = P(A|A \cup B) > 0$; hence we conclude that $P(B \mid A) = 1$. Accordingly, B too is normal and $P(B|A \cup B)$ is positive; the same argument leads *mutatis mutandis* to $P(A \mid B) = 1$. Therefore (b) implies (c).

Finally suppose (c). If A is abnormal, then so is B and (a) follows at once. If A is normal, then B and $A \cup B$ are also normal. But then $P(A \cap B|A \cup B) = P(A|A \cup B) = P(B|A \cup B) = 1$, using (c) and the Multiplication Axioms. Hence $P(X \cap A \cap B|A \cup B) = P(X \cap A|A \cup B) = P(X|A)$ and by similar reasoning $P(X \cap A \cap B|A \cup B) = P(X|B)$. Therefore (c) implies (a).

The relationship of *a priori* equivalence is much "deeper". As prelude let us introduce another operation on probability functions which is something like "deep conditionalization". Instead of raising a proposition to the status of subjective certainty it raises it to subjective aprioricity. To prevent confusion, I shall call this "relativization".[8]

Definition The *relativization of P to A* is the function $P//A$ defined by $P//A(X \mid Y) = P(X \mid Y \cap A)$ for all X, Y.[9]

[T8.2] The following are equivalent:
(i) $P(A|\cdot) = P(B|\cdot)$
(ii) $P(-|\cdot \cap A) = P(-|\cdot \cap B)$ (i.c. $P//A = P//B$)
(iii) $P(A|A + B) = P(B|A + B) = 1$ (*a priori* equivalence)
(iv) $A + B$ is abnormal (i. e. $A\langle P\rangle B$)

(For proof that (ii) implies (iv), use the *Lemma*: if A, B are disjoint and abnormal then $A \cup B$ is abnormal.).

[T8.3] *A priori* equivalence implies surface equivalence.

[8] As I have argued elsewhere (van Fraassen 1981a) this construction provides us with the "right" clue to the treatment of quantification and of intuitionistic implication in so-called probabilistic (or generally, subjective) semantics.

[9] In my (1981a), $P//A$ was designated as P^A and called "P conditioned on A." I now think this terminology likely to result in confusion, and prefer "P relativized to A."

The converse does not hold, as can be seen from our Example 5.3 in Section 2.1, where U is the set of natural numbers, and is surface equivalent, but not *a priori* equivalent, to $\{0, 1, \ldots, 9\}$. For $P(\{0, 1, \ldots, 9\} \mid \{10, \ldots, 19\}) = 0$ there.

Implication Relations; Superiority Is Transitive

We are representing propositions by means of a field of sets, whose elements are thought of as alternative possible situations or worlds. Accordingly, "A implies B" can be equated only with "$A \subseteq B$." But when two propositions are *a priori* equivalent for P then they are not distinguishable as far as P is concerned. Therefore we can introduce a less sensitive partial ordering as a "coarser" implication relationship with respect to a given two-place probability measure.

[T9.1] The following are equivalent:
(a) *A P-implies B*: $P(A \mid \cdot) \leq P(B \mid \cdot)$
(b) $A - B$ is abnormal
(c) for all X, if $P(A \mid X) = 1$ then $P(B \mid X) = 1$.

The superiority relation is not a (converse) implication relationship, despite formal similarities. If A is superior to B, A may still have probability *zero* conditional on B, for example. It is just that the supposition that A has to be given up – in fact, denied – before B comes into play in our thinking. The hierarchy so indicated John Vickers (*loc. cit.*) calls the De Finetti hierarchy. As he points out, it is crucial to this role that we can describe the comparison in terms of a transitive relationship.

In fact, only one further point needs to be made concerning normality to show that the propositions form a partially ordered set under superiority (with the abnormal propositions forming an equivalence class at the bottom of this p.o.set).

[T9.2] If X is a subset of Y and Y a subset of normal set E, and $P(Y \mid E) > 0$, then $P(X \mid Y) = 0$ iff $P(X \mid E) = 0$.
[T9.3] $P>$ is transitive.

Let it be given that $A\ P{>}B$ and $B\ P{>}C$; we need to prove that $A\ P{>}C$. To visualize the proof, think of a Venn Diagram with the following labels for relevant propositions:

$$1 = (A - B - C)\,,\;\; 2 = (AB - C)\,, 3 = B - A - C,$$

$$4 = AC - B$$

$$5 = BC - A, 6 = C - B - A$$

$$E = (A + B) \cup (B + C) = (A + B) \cup (B + C) \cup (A + C)$$

I will denote unions by concatenation of the labels; thus $E = 123456$ and $A - C = 12$. We now consider all possible cases.

(1) $A + C$ is abnormal. Then $P(A \mid A + C) = 1$
(2) $A + C$ is normal; then also E is normal.

Hence $P(A + B \mid E)$ or $P(B + C \mid E)$ or both are positive; we proceed to the possible subcases:

(2.1) $P(A + B \mid E) > 0$. By the given and [T9.2] it follows that $P(B - A \mid E) = 0$, i.e. $P(3 \mid E) = P(5 \mid E) = 0$.
(2.11) Assume $P(B + C \mid E) > 0$. By [T9.2], and the given, also $P(C - B \mid E) = 0$, so $P(4 \mid E) = P(6 \mid E) = 0$. Altogether, $P(3456 \mid E) = 0$ hence $P(12 \mid E) = P(A - C \mid E) = 1$. It follows that $P(A + C \mid E) > 0$, so by [T9.2] $P(56 \mid A + C) = P(C \mid A + C) = 0$, and therefore $P(A \mid A + C) = 1$.
(2.12) Assume $P(B + C \mid E) = 0$. Then $P(2346 \mid E) = 0$, so altogether $P(23456 \mid E) = 0$. Hence $P(1 \mid E) = 1$. It follows that $P(A + C \mid E) > 0$, therefore by [T9.2] again $P(56 \mid A + C) = 0$. It follows then that $P(A \mid A + C) = 1$.
(2.2) $P(A + B \mid E) = 0$ and $P(B + C \mid E) > 0$. The former entails that $P(1345 \mid E) = 0$. The latter entails by [T9.2] that $P(C - B \mid E) = P(46 \mid E) = 0$. Altogether therefore $P(13456 \mid E) = 0$ and $P(2 \mid E) = 1$. Therefore $P(A + C \mid E) > 0$, and so $P(C - A \mid A + C) = 0$ by [T9.2]; therefore $P(A \mid A + C) = 1$. This ends the proof.

Adding this to the fact that $P>$ is reflexive, we conclude that $P>$ is a partial ordering of the field of propositions. The abnormal propositions form a $\langle P \rangle$ equivalence class at the very bottom of this partial ordering.

A Large Class of Models

I will define a class of models such that P satisfies principles I–II iff P can be represented by one of these models, in the way to be explained. The class will be chosen large; a special subclass ("lexicographic models") will yield nontrivial, easily constructed examples to be used in illustrations and refutations. (The term "lexicographic" is used similarly in decision theory literature; see Blume et al. 1991a, b.)

A model begins with a sample space $S = \langle U, F \rangle$, where U is a non-empty set (the universe of possibilities) and F a sigma-field of sets on U (the propositions). We define the subfields:

$$\text{if } A \text{ is in } F \text{ then } FA = \{E \cap A : E \text{ in } F\};$$

thus FA is a field on A. For each such field designate as PA the set of probability measures defined on FA. (When A is empty, $FA = \{A\}$ and PA is empty.) The restriction of a member p of PA to a subfield FB, with B a subset of A, will be designed $p \mid FB$. Finally let PS be the union of all the sets PA, A in F.

A model M will consist of a sample space S as above, and a function π defined on a subset of F, with range in PS. That is, π associates some probability measure on some subfield with certain propositions. (These will be the normal propositions.) I will abbreviate "$\pi(A)$" to "πA", and when p is in FB I will also designate B as Up (the universe of p). Thus $B = U\pi A$ means that π associates with A a probability measure defined on the measurable subsets of B, i.e. on the propositions which imply B, i.e. on FB. The function π is subject to the following conditions:

(M1) $\pi A(A)$ is defined and positive.
(M2) If $\pi B(A)$ is defined and positive, then πA is defined
(M3) If $\pi B(A)$ is defined and positive, then $\pi A \mid F(A \cap B)$ is proportional to $\pi B \mid F(A \cap B)$.

This does not entail that if $\pi B(A \cap B)>0$ then $\pi A(A \cap B)>0$, because the proportionality constant can be 0 (in which case πA gives 0 to all members of $F(A \cap B)$ – see further the discussion of [T7.3] which suggested this condition). It is easy to see what the constant of proportionality has to be:

[T10.1] If $\pi B(A)$ is defined and positive, then
$\pi A\,|F\,(A \cap B) : \pi B|\,F\,(A \cap B)$
$= \pi A\,(A \cap B) : \pi B\,(A \cap B)$.

Finally we define what it means for one of these functions to represent a two-place function:

Definition Model $M = \langle S,\ \pi\rangle$ with $S = \langle U,\ F\rangle$ represents binary function P iff the domain of P is F and for all A, B in F, $P\,(A|B) = \pi B\,(A \cap B)\,/\pi B(B)$ if defined, and $= 1$ otherwise.

It is easy to prove that:

[T10.2] If P is represented by a model, then P is a two-place probability measure.

Conversely, suppose that P is a two-place probability measure in the sense of satisfying I–II, defined on F in space $S = \langle U,\ F\rangle$. For all normal sets A of P define πA on FA by:

$$\pi A(B) = P\,(B|A)\,.$$

That (M1) and (M2) are satisfied by $M = \langle S, \pi\rangle$ follows at once from the facts about normal sets. Suppose now, equivalently to the antecedent of (M3) that A and B are normal sets with $\pi(B)>0$. To prove that (M3) holds, suppose that $\pi B(A)$ is defined and positive, so that B and A are normal sets, $P(A \mid B)>0$. Then according to [T10.3], for each subset X of $A \cap B$ we have $P\,(X|A) = [P\,(B|A)\,/P\,(A|B)]\,P\,(X|B)$. Therefore here $\pi A(X) = [P(B \mid A)/P(A \mid B)]\pi B(X)$. In conclusion:

[T10.3] If P is a two-place probability measure satisfying the principles I–II, then P is represented by a model.

Having established this representation result, we now look for easily constructed models, for illustration, refutation of conjectures, and exploration of examples.

Definition Model $M = \langle S, \pi \rangle$ with $S = \langle U, F \rangle$ is *lexicographic* iff there is a sequence (well-ordered class) SEQ of 1-place probability measures defined on the whole of F, such that $\pi B(A) = q(A \cap B)/q(B)$ for the first member q of the sequence SEQ such that $q(B) > 0$; πB is undefined when there is no such q.

The members of SEQ correspond to the probabilities conditional on *belief remnants* (see discussion in Section 6 of Example 5.3). We will say that πA comes before πB in SEQ exactly when the first q in SEQ such that $q(A) > 0$ comes before the first q in SEQ such that $q(B) > 0$. It is easily checked that $M = \langle S, \pi \rangle$ is a model. Specifically, if A is a subset of B then πB will not come after πA, since whatever measure assigns a positive value to A will then assign one to B. Neither can πA come after πB if $\pi B(A) > 0$; in that case $\pi A = \pi B$. Consequently condition (M3) is easily verified: the proportionality constant = 1.

It is now very easy to make up examples of 2-place probability measures. Just take two or three or indeed any number, finite or infinite, of ordinary probability measures and well-order them. A special example, whose existence depends on the axiom of choice is this: let SEQ contain all one-place probability measures defined on given domain F. In that case, the only abnormal proposition is the empty set (the self-contradiction). Also the only *a priori* is the tautology. Short of this, we could of course have a sequence which does not contain literally all the definable probability measures, but contains all those which give 1 to a given set A. In that case, all propositions other than Λ that imply A are normal. Let us call P *Very Fine on A* in such a case. (The case of P Very Fine on U was already called "Very Fine" above.) Note that one of the defining conditions of a belief core K was that P had to be Very Fine on K.

Belief in a Lexicographic Model

I will first show that in a lexicographic model, the intersection of all belief cores, if any, is always a belief core too. Since this does not depend on cardinality or the character of the sample space, the result adds significantly to the previous theorems. Then I will construct a lexicographic model to show that in general not all propositions with probability 1 are full beliefs. This model will be a reconstruction of Example 5.4 (Petra and Noah's Arc).

[T11.1] If P is represented by lexicographic model $M = \langle S, p \rangle$ defined by w.o. sequence SEQ, and A, B are disjoint normal sets for P, then the following are equivalent:
(i) $A\ P{>}B$ and not $B\ P{>}A$
(ii) πA comes before πB in SEQ.

[T11.2] If P is represented by lexicographic model $M = \langle S, \pi \rangle$ defined by w.o. sequence SEQ, and K, $K1$, $K2$ are belief cores with K a proper subset of $K1$ and $K1$ a proper subset of $K2$ then $\pi\ (K1 - K)$ comes before $\pi(K2 - K1)$ in SEQ

For proof note that $K1 - K$ and $K2 - K1$ are disjoint normal sets (*belief remnants*), so [T11.1] applies.

[T11.3] If P is represented by lexicographic model $M = \langle S, \pi \rangle$ defined by w.o. sequence SEQ, then the intersection of all its belief cores is also a belief core.

For convenience in the proof, call $\pi(K'' - K')$ a marker of K' when both are belief cores and K' is a proper subset of K''. This measure exists in the model since the set is normal. If no superset of K is a core, call $\pi(U - K)$ its marker. We now consider the whole class of markers; it must have a first member in SEQ. Let that be $p^* = \pi(K2 - K1)$. It remains to prove that K1 is the intersection of all the belief cores. Suppose to the contrary that core K is a proper subset of $K1$. Then by [T11.2] marker $p\ (K1 - K)$ comes before p^* in SEQ – contra the hypothesis. This ends the proof.

At this point we know enough about lexicographic models in general to exploit their illustrative uses. Recall Example 5.4. Petra has subjective probability 1 for N: the center of mass of Noah's Ark lies in the Northern Hemisphere and also for W: that it lies in the Western Hemisphere. I shall take it here that the Equator is part of the Northern Hemisphere; she has probability 1 that –EQ: it does not lie on the Equator. Let me add here that she also has probability 1 that –GR: it does not lie on the Greenwich Meridian (which I shall here take to be part of the Western Hemisphere). But she has probability 1 that it lies in Northern Hemisphere on the supposition that $N + W$: it lies either in the Northern or in he Western Hemisphere but not both (which supposition has 0 probability for her).

For the sample space, let U be the entire surface are of the Earth, and let F be the family of measurable subsets of U in the usual sense, so we can speak of area and length where appropriate. Let us first define some measures and classes of measures:

$$mA(X) = \text{area of } X \cap A : \text{area of } A$$
$$1A(X) = \text{length of } X \cap A : \text{length of } A$$

where XX is a subset of F

$$M(XX) = \text{the class of all measures on } F$$

which give 1 to a member of XX.

The sequence SEQ is now pieced together from some other well ordered sequences as follows:

$$\text{SEQ} = m\,(N \cap W)\,,l\,(\text{GR} \cap N)\,,l\,(\text{EQ} \cap W)\,,\text{S1},\text{S2},mN,mW,mU,\text{S3},\text{S4}$$

where the indicated subsequences are:

S1: a well-ordering of $\{1A$: A is a subset of $N \cap W$ with 0 area but positive length$\}$,
S2: a well-ordering of $M\{A$ in F: A is a non-empty subset of $N \cap W$ with 0 area and 0 length$\}$,
S3: a well-ordering of $\{1A$: A is has 0 area but positive length$\}$,
S4: a well-ordering of $M\{A$ in F: A is a non-empty and has 0 area and 0 length$\}$.

Let us call the so constructed lexicographic model PETRA, in her honor. The construction is a little redundant: we would be constructing a Very Fine measure already if we just took the tail end mU, S3, S4. Preceding them with the others has the effect of establishing desired superiority relationships. For example, $N + W$ first receives a positive value from mN, which gives 0 to $W - N$, so that $P(N \mid N + W) = 1$. I made GReenwich similarly superior to EQuator.

[T11.4] $N \cap W$ is a belief core in PETRA.
[T11.5] No proper subset of $N \cap W - \text{GR}$ is a belief core.

For let X be such a subset; to be a belief core it must have probability 1 tout court, so its area equals that of $N \cap W$. Let X' be one of its non-empty subsets with 0 area. Then X′ as well as X itself are disjoint of $N \cap W \cap \text{GR} = Y$. The first measure to assign positive value to $X' + Y$ is the second member of SEQ, namely $m(\text{GR} \cap N)$, which assigns 1 to Y (because GR is part of W) and 0 to X'. Therefore X' is not superior to Y, and so X is not a belief core.

[T11.6] In PETRA some propositions not among its full beliefs have probability 1.

In view of the preceding, it suffices to reflect that $N \cap W - \text{GR}$ has proper subsets with probability 1; for example $N \cap W - \text{GR} - \text{EQ}$.

Appendix

A1. Previous Literature

The basic theory of two-place probability functions is a common part of a number of theories. Such probability functions have been called *Popper functions* because Popper's axioms originally presented in his *The Logic of Scientific Discovery* (1959) were adopted by other writers (see Harper 1976; Field 1977; van Fraassen 1979, 1981a, b). Carnap used essentially the same axioms for his "c-functions", but concentrated his research on those which derive trivially from one-place probability functions ("m-functions"). Reichenbach's probability was also irreducibly two-place. I have mentioned De Finetti's paper (1936) which introduced the idea of local comparisons (like my "superior"; Vickers' "thinner"); see also Section 4.18 in his *Theory of Probability*, vol. 1. The most extensive work on two-place probability theory is by Renyi (1955, 1970a,b). The theory of two-place probability here presented is essentially as explored in my (1979), but with considerable improvement in the characterization of the described classes of models. Finally, the discussion of supposition in section "Supposition and two-place probability" is related to work on belief revision, much of it indebted to ideas of Isaac Levi; see Gardenfors 1988 for a qualitative version.

A2. Transfinite Consistency

The ordering $P(A) \leq P(B)$ extends the partial ordering of logical implication: if $A \subseteq B$ then $P(A) \leq P(B)$. Unfortunately, the ordering $P(A) < P(B)$ does not extend in general the partial ordering of proper implication: $P(A) = P(B)$ is possible even when $A \neq B$. Indeed, this is inevitable if there are more than countably many disjoint propositions. As a corollary, the intersection of all propositions of maximal probability may itself even be empty. Kolmogoroff himself reacted to this problem by suggesting that we focus on probability algebras: algebras of propositions reduced by the relation of equivalence *modulo* differences of measure *zero*: $P(A + B) = 0$. (See Birkhoff (1967), XI, 5 and Kappos (1969), II, 4 and III, 3.)

The difficulty with this approach is that a probability algebra does not have the structure usually demanded of an algebra of propositions. For the latter, the notion of truth is relevant, so it should be possible to map the algebra homomorphically into $\{0, 1\}$. As example take the unit interval with Lebesgue measure, reduced by the above equivalence relation. This is a probability (sigma-)algebra. Let T be the class of elements designated as true, i.e. mapped into 1, and let A with measure x be in T. Then A is the join of two disjoint elements of measure $x/2$ each. Since the mapping is a homomorphism, one of these is in T. We conclude that T contains a countable downward chain $A_1, A_2, \ldots$ with the measures converging to *zero*. Therefore its meet is the *zero* element of the algebra. The meet should be in T because it is the countable meet of a family of "true" propositions; but it can't be in T, since the *zero* element is mapped into 0.

This "transfinite inconsistency" of the propositions which have probability *one*, was forcefully advanced by Patrick Maher (1990) as a difficulty for the integration of subjective probability and belief. My conclusion, contrary to Maher's, is that the role of subjective probability is to grade the alternatives left open by full belief. That automatically bestows maximal probability on the full beliefs, but allows for other propositions to also be maximally probable. The question became then: how are the two classes of maximally probable propositions to be distinguished?

A3. Rejection of the Bayesian Paradigm?

While I hold to the *probabilist* conviction that our opinion is to be represented by means of probability models, I reject many features associated with so-called "Bayesian" views in epistemology. In the present context, a main difference concerns the status of probability *one*. Conditionalization of an absolute (one-place) probability function cannot lower probability from *one* nor raise it from *zero*. As a result, such changes have often been relegated to epistemological catastrophes or irrational shifts of opinion. This is definitely not so in all probabilist work in epistemology (Isaac Levi and William Harper provide notable exceptions). In my view, probability *one* is easily bestowed, and as easily retracted, especially when it is *only* maximal unconditional probability (conditional on the tautology).

Obviously, then, I reject the naive Pascalian equation that a bet on *A*, with any payoff whatsoever, is worth to me my probability for *A* times that payoff. I think that Pascal's equation holds under restricted circumstances, with relevant assumptions kept fixed and in place. I mean this roughly in the sense of the "constructivist" view of subjective probability suggested in various ways by Glenn Shafer and Dick Jeffrey (and possibly meant by Dan Garber when he talks about the Bayesian hand-held calculator). In a given context I have a number of full beliefs which delimit the presently contemplated range of possibilities; it is the latter which I grade with respect to their comparative likelihood. The context may be anchored to a problem or type of problem, for which I go to this trouble. Some of the beliefs will indeed be "deepseated", and to some I subscribe so strongly that they would survive most any change of context. They are part of what I fall back on especially if I try to specify the context in which I am presently operating – for this involves seeing myself in a "larger" perspective.

A4. Infinitesimals?[10]

There is another solution on offer for most problems which two-place probability solves. That is to stick with one-place probability, but introduce infinitesimals. Any non-self-contradictory proposition can then receive a non-zero probability, though often it is infinitesimal (greater than zero but smaller than any rational fraction).

The infinitesimals solution is to say that all the non-self-contradictory propositions (that are not contrary to my full beliefs) receive not zero probability but an infinitesimal number as probability [in a non-standard model]. There is an important result, due to Vann McGee (1994) which shows that every finitely additive 2-place probability function $P(A \mid B)$ is the standard part of $p(A \cap B)/p(B)$ for some non-standard 1-place probability function p (and conversely). Despite this I see advantages to the present approach to conditional probability which eschews infinitesimals. First of all, there is really no such thing as "the" infinitesimals

[10]For related critiques of the 'infinitesimals' gambit, see Skyrms (1983), Hajek (1992).

solution. If you go to non-standard models, there will in principle be many ways to generalize the old or ordinary concept of measure so as to keep agreement with measure in the embedded standard models. You don't have a specific solution till you specify exactly which reconstruction you prefer, and what its properties are.

In addition, in the present approach it is very easy to see how you can have "layering" of the following sort. Task: produce a two-place probability measure P such that for given A, B, C, the following is the case:

$$P(A) = 1, P(B|A) = 0, P(C|A) = 0$$
A, B, C are normal
If $P(X|C) = P(X|B) = P(X|A) = 0$ then X is abnormal

It is easy to construct a small lexicographic model in which this is the case. Let C be a subset of B and B a subset of A; let p_1 give 1 to A but 0 to B and to C; p_2 give 1 to B but 0 to C; and p_3 give 1 to C. If these are all the measures in the sequence, then subsets of C which receive probability 0 conditional on C are all abnormal. Intuitively it would seem that in the infinitesimal approach this would require the construction in which there are exactly two layers L and M of infinitesimals: x in L is infinitesimal in comparison to standard numbers, Y in M is infinitesimal in comparison (even) to any number in L, and no numbers at all are infinitesimal in comparison to numbers in M. I leave this as an exercise for the reader.

As to the problem of belief, I wonder if the nonstandard reconstruction would have the desirable features for which we naturally turn to infinitesimals. Suppose for example that I choose a model in which each non-empty set has a positive (possibly infinitesimal) probability. Then my full beliefs are not just those which have probability 1, since that includes the tautology only. On the other hand, I can't make it a requirement that my full beliefs have probability $> 1 - d$, for any infinitesimal d one could choose. For the intersection of the sets with probability $\geq 1 - d$ will generally have a lower probability. Hence the lottery paradox comes back to haunt us. We would again face the trilemma of either restricting full beliefs to the tautology, or specifying them in terms of some factor foreign to the degrees-of-belief framework, or banishing them from epistemology altogether.

References

Birkhoff, G. (1967). *Lattice theory* (3rd ed.). Providence: American Mathematical Society.
Blume, L., Brandenburger, A., & Deckel, E. (1991a). Lexicographic probabilities and choice under uncertainty. *Econometrica, 59*, 61–79.
Blume, L., Brandenburger, A., & Deckel, E. (1991b). Lexicographic probabilities and equilibrium refinements. *Econometrica, 59*, 81–98.
De Finetti, B. (1936). Les probabilités nulles. *Bulletin des sciences mathématiques, 60*, 275–288.
De Finetti, B. (1972). *Theory of probability* (2 vols.). New York: Wiley.
Field, H. (1977). Logic, meaning and conceptual role. *Journal of Philosophy, 74*, 374–409.
Foley, R. (1993). *Working without a net*. Oxford: Oxford University Press.

Gardenfors, P. (1988). *Knowledge in flux: Modeling the dynamics of epistemic states*. Cambridge: MIT Press.

Hajek, A. (1992). *The conditional construal of conditional probability*. Ph.D. dissertation, Princeton University.

Harper, W. L. (1976). Rational belief change, Popper functions, and counter-factuals. In C. Hooker, & W. Harper (Eds.), *Foundations of probability theory* (Vol. 1, pp. 73–112). Dordrecht: Reidel Publishing Company.

Kappos, D. A. (1969). *Probability algebras and stochastic spaces*. New York: Academic Press.

Maher, P. (1990). Acceptance without belief. *PSA, 1*, 381–392.

McGee, V. (1994). Learning the impossible, In Eells, E. & Skyrms, B. (Eds.). *Probabilities and conditionals: Belief revision and rational decision* (pp. 179–199). Cambridge: Cambridge University Press.

Popper, K. (1959). *The logic of scientific discovery*. New York: Basic Books.

Renyi, A. (1955). On a new axiomatic theory of probability. *Acta Mathematica Hungarica, 6*, 285–333.

Renyi, A. (1970a). *Foundation of probability*. San Francisco: Holden-Day.

Renyi, A. (1970b). *Probability theory*. Amsterdam: North-Holland.

Skyrms, B. (1983). Three ways to give a probability function a memory. In J. Earman (Ed.), *Testing scientific theories* (Minnesota studies in the philosophy of science, Vol. X, pp. 157–161). Minneapolis: University of Minnesota Press.

van Fraassen, B. C. (1979). Foundations of probability: A modal frequency interpretation. In G. Toraldo di Francia (Ed.), *Problems in the foundations of physics* (pp. 344–387). Amsterdam/New York: North-Holland Publishing Company.

van Fraassen, B. C. (1981a). Probabilistic semantics objectified: I. Postulates and logics. *Journal of Philosophical Logic, 10*, 371–394.

van Fraassen, B. C. (1981b). Probabilistic semantics objectified: II. Implications in probabilistic model sets. *Journal of Philosophical Logic, 10*, 495–510.

van Fraassen, B. C. (1988). Identity in intensional logic: Subjective semantics. In U. Eco et al. (Eds.), *Meaning and mental representation* (pp. 201–219). Bloomington: Indiana University Press.

Vickers, J. M. (1988). *Chance and structure*. Oxford: Oxford University Press.

Chapter 6
A Theory of Higher Order Probabilities

Haim Gaifman

Introduction

The assignment of probabilities is the most established way of measuring uncertainties on a quantitative scale. In the framework of subjective probability, the probabilities are interpreted as someone's (the agent's) degrees of belief. Since justified belief amounts to knowledge, the assignment of probabilities, in as much as it can be justified, expresses knowledge. Indeed, knowledge of probabilities, appears to be the basic kind of knowledge that is provided by the experimental sciences today.

This is knowledge of a partial, or incomplete, nature, but not in the usual sense of "partial". Usually we mean by "partial knowledge" knowledge of some, but not all, of the facts in a certain domain. But knowing that a given coin is unbiased does not enable one to deduce any non-tautological Boolean combination of propositions which describe outcomes in the next, say 50 tosses. And yet it constitutes very valuable knowledge about these very same outcomes. What is the objective content of this knowledge? What kind of fact is the fact that the true probability of "heads" is 0.5, i.e., that the coin is unbiased? I shall not enter here into these classical problems[1]. I take it for granted that, among probability assignments, some are more successful, or better tuned to the actual world, than

A part of this paper has been included in a talk given in a NSF symposium on foundations of probability and causality, organized by W. Harper and B. Skyrms at UC Irvine, July 1985. I wish to thank the organizers for the opportunity to discuss and clarify some of these ideas.

[1]My Salzburg paper (1983) has been devoted to these questions. The upshot of the analysis there has been that even a "purely subjective" probability implies a kind of factual claim, for one can

H. Gaifman (✉)
Columbia University, New York, NY, USA

H. Arló-Costa et al. (eds.), *Readings in Formal Epistemology*, Springer Graduate Texts in Philosophy 1, DOI 10.1007/978-3-319-20451-2_6

others. Consequently probability assignments are themselves subject to judgement and evaluation. Having, for example, to estimate the possibility of rain I might give it, going by the sky's appearance, 70 %. But I shall be highly uncertain about my estimate and will adopt the different value given, five minutes later, in the weather forecast.

Thus we have two levels of uncertainty:

1. Uncertainty concerning the occurrence of a certain event – expressed through the assignment of probabilities.
2. Uncertainty concerning the probability values assigned in 1.

When this second level is itself expressed by assigning probabilities we get second order probabilities. An example of a second order probability is furnished by a cartoon in "The New Yorker" showing a forecaster making the following announcement:

> There is now 60 % chance of rain tomorrow, but, there is 70 % chance that later this evening the chance of rain tomorrow will be 80 %.

Just as we can iterate modal or epistemic operators, so in the system to be presented here we can iterate the probability-assignment operator to any depth. The goal of this paper is to present a general and adequate semantics for higher order probabilities and to obtain, via representation theorems, nice easily understood structures which give us a handle on the situation.

The basic structure to be defined here is a **HOP** (Higher Order Probability space). A *simple* HOP is based on a field of events, F, and on a binary operator $PR(,)$ which associates with every event A and every real closed interval Δ an event $PR(A, \Delta)$ in F. The intended meaning is that $PR(A, \Delta)$ is the event that A's true probability lies in the interval Δ.

"True probability" can be understood here as the probability assigned by an ideal expert or by someone fully informed. It is however up to us (or to the agent) to decide what in the given context constitutes an "ideal expert" or "someone fully informed". If "full information" means knowing *all* the facts then, of course, the true (unknown to us) probability has only two values 0 and 1; this will make the HOP trivial in a certain sense. In the other extreme, the agent may regard himself as being already fully informed and this leads to the "opposite" trivialization of the HOP. Generally, the agent will regard the expert as being more knowledgeable than himself, but not omniscient; e.g., the expert might know the true bias of a coin but not the outcomes of future tossings, or he might have statistical information for estimating the bias, which the agent lacks.

The agent himself at some future time can be cast in the role of being "fully informed". Thus, if P is the forecaster's present probability function and if PR

---

asses its success in the actual world. Rather than two different kinds, subjective and objective probabilities are better to be regarded as two extremes of a spectrum.

represents his state of knowledge later in the evening, then his announcement in "The New Yorker" cartoon can be summed up as follows, where $A =$ 'tomorrow it will rain':

$$P(A) = .6 \quad P(PR(A,\ [.8,.8])) = .7$$

In order to represent knowledge at different stages, we make PR into a 3-place operator: $PR(A, t, \Delta)$ is the event that the probability of A at stage t lies in Δ. The stages can be time-points, in which case t ranges over some ordered set. More generally, the set of stages is only partially ordered, where $s \leq t$ if the knowledge at stage t includes the knowledge at stage s. (Different agents may thus be represented in the structure.) This is how a HOP is defined in general. We shall first establish the properties of simple HOPs, then use them to derive those of the more general spaces.

We shall also define, in a separate section, a formal logical calculus, to be called *probability logic*, which is naturally associated with simple HOPs. Various modalities can be reconstructed within this calculus. The general HOPs give rise to stage-dependent modalities whose calculus will be outlined at the end of the paper.

The import of the subject for various branches of philosophy and for the foundations of probability is obvious. Also obvious should be its bearing upon applied probabilistic reasoning in distributed networks, or upon efforts to incorporate such reasoning in AI systems. Mathematically, most of this paper is rather easy. Our goal has not been to prove difficult theorems, but to clarify some basic concepts and to outline a general, conceptually "clean", framework within which one can use freely and to good effect statements such as: 'With probability 0.7 Adam will know at stage 3 Bob's probability for the event A, with error ≤ 0.01' (where Adam and Bob are either people or processors). Statements of this form express intuitive thinking which may underly involved technical proofs; to use them openly and precisely can help us as a guide for finding and organizing our arguments.

A theoretic framework for higher order probabilities may also yield insights into systems of reasoning which employ non-probabilistic certainty measures. For when probability is itself treated like a random variable, we can use various methods of "safe" estimation which do not necessarily yield a probability measure. For example, define the *certainty measure* of an event A to be the largest α such that, with probability 1, the probability of A is $\geq \alpha$. This is only one, apparently the most conservative, measure among various measures that can be used.

Higher order probabilities have been considered by De-Finetti, but rejected by him owing to his extreme subjectivist views. Savage considered the possibility but did not take it up, fearing that the higher order probabilities will reflect back on the ground level, leading to inconsistencies. Instances of higher order probabilities figure in works of Good (1965) and Jaynes (1958). More recent philosophical works are by Domotor (1981), Gardenfors (1975) (for qualitative probabilities), Miller (1966), Skyrms (1980a, b) – who did much to clarify matters, van-Frassen (1984), and others.

Due to limitations of space and deadline I have not entered into details of various proofs. Some of the material has been abridged; I have included some illustrative examples of simple HOPs, but not the more interesting ones of general HOPs (which arise naturally in distributed systems). Also the bibliography is far from complete.

Simple HOPs

Definition and Basic Properties

As in Kolmogoroff's framework (1933) we interpret propositions as subsets of some universal set, say *W*, and we refer to them as events. We can regard *W* as the set of all possible worlds. Thus we have X = set of all worlds in which *X* is true and we get the following standard correspondence:

$\vee$ (disjunction) $\mapsto$ $\cup$ (union)
$\wedge$ (conjunction) $\mapsto$ $\cap$ (intersection)
$\neg$ (negation) $\mapsto$ $-$ (complementation)

Terminology A Boolean Algebra (of sets) is a class of sets closed under finite unions (and intersections) and under complementation (with respect to some presupposed universal set, in our case – *W*). A field is a Boolean algebra closed under countable unions (what is known also as a σ-algebra). The field (Boolean algebra) generated by a class *S* of sets is the smallest field (Boolean algebra) which contains *S* as a subclass. Note that in generating a Boolean algebra we apply finitary operations only, whereas in generating a field infinitary countable operations are used. A field is countably generated if it has a countable set of generators. All probabilities are assumed here to be countably additive.

A **HOP** is a 4-tuple (W, F, P, PR), where *F* is a field of subsets of *W*, to be called events, *P* is a probability over *F* and *PR* is a mapping associating with every $A \in F$ and every real closed interval Δ an event $PR(A, \Delta)$,

$$PR : F \times set\ \ of\ \ closed\ \ intervals \rightarrow F$$

As explained in the introduction $PR(A,\ \Delta)$ is the event that the true (or the eventual, or the expert-assigned) probability of *A* lies in Δ. *P* is the agent's current subjective probability.

Among the closed intervals, we include also the empty interval, $\varnothing$. The minimal and maximal elements of *F* are, respectively, **0** and **1**; that is: $\mathbf{0}$ = empty subset of W = False, $1 = W$ = True.

In the explanations I shall use "probability" both for the agent's current subjective probability as well as for the true, or eventual one; the contexts indicate the intended reading.

The following axioms are postulated for a HOP:

(I) $PR(A, [0,1]) = \mathbf{1}$ (For every A, the event that A's probability lies in $[0,1]$ is W, i.e., true.)
(II) $PR[A, \varnothing] = \mathbf{0}$ (That A's probability lies in the empty interval is the empty event, i.e., false.)
(III) *If* $\Delta_1 \cup \Delta_2$ is an interval then $PR(A, \Delta_1 \cup \Delta_2) = PR(A, \Delta_1) \cup PR(A, \Delta_2)$ (A's probability lies in the interval $\Delta_1 \cup \Delta_2$ iff it lies either in Δ_1 or in Δ_2)

In the following two axioms "n" is a running index ranging over $\{1,2,\ldots\}$.

(IV) $\cap_n PR(A, \Delta_n) = PR(A, \cap_n \Delta_n)$ (A's probability lies in every Δ_n iff it lies in their intersecton).
(V) *If, for all* $n \neq m$, $A_n \cap A_m = \varnothing$, *then* $\cap_n PR(A, [\alpha_n, \beta_n]) \subset PR(\cup_n A_n, [\Sigma_n \alpha_n, \Sigma_n \beta_n])$ (For pairwise disjoint $A_n s$, if A_n*'s* probability lies in $[\alpha_n, \beta_n]$, $n = 1,2,\ldots$, then the probability of $\cup_n (A_n)$ lies in $[\Sigma_n \alpha_n, \Sigma_n \beta_n]$)).)

Note that axioms (I)–(V) involve only W, F and PR. The crucial axiom which connects PR with P will be stated later.

Theorem 6.1 *For every HOP, $H = (W, F, P, PR)$ there is a mapping p which associates with every x in W a probability, P_z, over F such that*

$$PR(A, \Delta) = \{x : p_x(A) \in \Delta\} \quad (6.1)$$

The mapping p is uniquely determined by (6.1) *and can be defined by:*

$$p_x(A) = \inf\{\alpha : x \in PR(A, [0, \alpha])\} \quad (6.2)$$

as well as by:

$$p_x(A) = \sup\{\alpha : x \in PR(A, [\alpha, 1])\}. \quad (6.2')$$

Vice versa, if, for every $x \in W$, P_x is a probability over F such that $\{x : p_x(A) \in \Delta\}$ is in F for all $A \in F$ and all real closed, Δ, and if we use (6.1) *as a definition of PR then Axioms* (I)–(V) *are satisfied.*

We call p the kernel of the HOP.

The proof of Theorem 6.1 is nothing more than a straight-forward derivation of all the required details from the axioms, using (6.2) as the definition of p_x. (The "vice versa" part is even more immediate than the first part.)

We can now extend PR and define $PR(A, \Xi)$, for arbitrary subsets Ξ of reals, as $\{x : p_z(A) \in \Xi\}$. If Ξ is a Borel set then $PR(A, \Xi)$ is in F.

The meaning of p_x is obvious: it is the probability which corresponds to the maximal state of knowledge in world x – the distribution chosen by the expert of that world.

Notation For $\alpha \in [0,1]$, $PR(A, \alpha) =_{\text{df}} PR(A, [\alpha, \alpha])$.

The picture is considerably simpler in the discrete case, where W is countable. Assuming with no loss of generality that $\{x\} \in F$ for every $x \in W$, the probability of some $A \subset W$ is simply the sum of the probabilities of the worlds in A. In that case, we can eliminate the closed intervals and consider only the special cases $PR(A,\alpha)$ where α ranges over $[0,1]$; also our 5 axioms can be replaced by 3 simpler ones. Discrete cases arise in many situations and are very useful as illustrative examples. But to consider only the discrete case is highly restrictive.

Notation For $x,y \in W$, $A \in F$, put: $p(x,A) =_{\text{df}} p_x(A)$ and (assuming $\{y\} \in F$) $p(x,y) =_{\text{df}} p(x,\{y\})$ and $P(y) =_{\text{df}} P(\{y\})$.

In the discrete case P is obviously determined by the values $P(x)$, $x \in W$. Thus, ordering W, we can represent P as a probability vector (a countable vector of non-negative entries which sum up to 1). Similarly the kernel p becomes a probability matrix (a countable square matrix in which every row is a probability vector). Examples (i) and (ii) in the Examples subsection can serve to illustrate the situation (the discussion there presupposes however the next subsection).

Mathematically, what we have got is a Markov process (with initial probability P and transition probabilities $p(x,\ )$, $x \in W$). But the interpretation is altogether different from the usual interpretation of such a structure. The connection between P and the kernel p is established in the sixth axiom.

Axiom (VI) and Its Consequences

Let $P(A|B)$ be the conditional probability of A, given B. It is defined in the case that $P(B) \neq 0$ as $P(A \cap B)/P(B)$. It is what the agent's probability for A should be had he known B.

Axiom ($\mathrm{VI_w}$) *If* $P(PR(A, [\alpha,\beta])) \neq 0$ *then* $\alpha \leq P(A \mid PR(A, [\alpha,\beta])) \leq \beta$.

Axiom ($\mathrm{VI_w}$) (the weak form of the forthcoming Axiom (VI)) is a generalization of Miller's Principle to the case of interval-based events. Rewritten in our notation, Miller's Principle is: $P(A \mid PR(A,\alpha)) = \alpha$. Axiom ($\mathrm{VI_w}$) appears to be the following rule: My probability for A should be no less than α and no more than β, were I to know that in a more informed state my probability for A will be within these bounds. Plausible as it sounds, the use of the hypothetical "were I to know that..." needs in this context some clarification. Now a well-known way of explicating conditional probabilities is through conditional bets. Using such bets van-Frassen (1984) gives a Dutch-book argument for the Principle: Its violation makes possible a system of bets (with odds in accordance with the agent's probabilities) in which the agent will incur a net loss in all circumstances. In this argument $PR(A,\alpha)$ is interpreted as the event that the agent's probability for A at a certain future time will be α, in which case he should accept at that time bets with odds α. The same kind of Dutch-book can be constructed if Axiom ($\mathrm{VI_w}$) is violated. (Here it is crucial that we use an interval, the argument fails if we replace $[\alpha,\beta]$ by a non-convex Borel set.)

Axiom (VI) is the interval-based form of the stronger version of Miller's Principle which was suggested by Skyrms (1980a).

Axiom (VI) *If C is a finite intersection of events of the form PR(B,Δ), and if* $P(C \cap PR(A,[\alpha,\beta])) \neq 0$, *then*

$$\alpha \leq P\left(A \middle| C \cap PR(A,[\alpha,\beta])\right) \leq \beta$$

The same intuition which prescribes (VI_w) prescribes (VI); also here the violation of the axiom makes possible a Dutch-book against the agent. What is essential is that events of the form $PR(B,\Delta)$ be, in principle, knowable to the agent, i.e., be known (if true) in the maximal states of knowledge as defined by our structure.[2]

In what follows integrating a function $f(t)$ with respect to a probability m is written as $\int f(t) \cdot m(dt)$.

Lemma 6.1 *Axiom* (VI_w) *implies that the following holds for all* $A \in F$:

$$P(A) = \int p(x,A) \cdot P(dx) \tag{6.3}$$

The proof consists in applying the formula $P(A) = \Sigma_i P(A \mid B_i) \cdot P(B_i)$, where the B_i's form a partition, passing to the limit and using the definition of an integral.

The implication (6.3) ⇒ (VI_w) is not true in general. Note that in the discrete case (6.3) becomes:

$$P(x) = \sum\nolimits_y p(x,y) \cdot P(y) \tag{6.3$_d$}$$

which means that the probability vector is an eigen-vector of the kernel.

Definition Call Two worlds $x,y \in W$ epistimically equivalent, (or, for short, equivalent) and denote it by $x \simeq y$, if $P_x = P_y$ • For S – a class of events, define $K[S]$ to be the field generated by all events of the form $PR(A,\Delta)$, $A \in S$, Δ – a real closed interval.

Epistemic equivalence means having the same maximal knowledge. Evidently $x \simeq y$ iff, for all A and all Δ, $x \in PR(A,\Delta) \Leftrightarrow y \in PR(A,\Delta)$. This is equivalent to: for all $C \in K[F]$, $x \in C \Leftrightarrow y \in C$. If $K[F]$ is generated by the countably many generators X_n, $n = 0, 1, \ldots$ then the equivalence classes are exactly all non-empty intersections $\cap_n X_n'$ where each X_n' is either X_n or its complement. Hence the equivalence classes are themselves in $K[F]$, they are exactly the atoms of this field. The next lemma

[2]It is important to restrict C in Axiom (VI) to an intersection of such events. The removal of this restriction will cause the p_x's to be two-valued functions, meaning that all facts are known in the maximal knowledge states.

shows that the condition that $K[F]$ be countably generated is rather mild, for it holds whenever F itself is countably generated (which is the common state of affairs):

Lemma 6.2 *If S is either countable or a countably generated field, then K[S] is countably generated.*

(As generators for $K[S]$ one can take all $PR(A,\Delta)$, $A\in S$, Δ – a rational closed interval; the second claim is proved by showing that if S' is a Boolean algebra that generates the field S then $K[S'] = K[S]$.)

Terminology A 0-set is a set of probability 0. Something is said to hold for almost all x if it holds for all x except for a 0-set. The probability in question is P, unless specified otherwise.

Theorem 6.2 *If F is countably generated then axiom* (VI) *is equivalent to each of the following conditions:*

(A) (6.3) *holds (for all A) and the following is true: Let C_x be the epistemic equivalence class to which x belongs, then*

$$p_x(C_x) = 1 \text{ for almost all } x.$$

(B) (6.3) *holds and, for almost all x, for all A:*

$$p_x(A) = \int p_y(A)\cdot p_x(dy) \tag{6.4}$$

The proof that axiom (VI) is equivalent to (A) and implies (B) uses only basic measure theory. The present proof of (B) ⇒ (A) relies on advanced ergodic theory[3] and I do not know if this can be avoided. Fortunately the rest of this paper does not rely on this implication (except the corresponding implication in Theorem 6.3). Note that in the discrete case (6.4) is equivalent to:

$$p(x,z) = \sum_y p(x,y)\cdot p(y,z) \tag{4_d}$$

(4_d) means that the kernel, as a matrix, is equal to its square.

Let $\{E_u : u \in U\}$ be the family of epistemic equivalence classes, with different indices attached to different classes. Let P_u be the common p_x for $x \in E_u$; let m be the probability, defined for all $V \subset U$ such that $\cup_{u\in U} E_u \in F$, by:

$$m(V) = P\left(\cup_{u\in V} E_u\right)$$

[3] I am thankful to my colleagues at the Hebrew University H. Furstenberg, I. Katzenelson and B. Weiss for their help in this item. Needless to say that errors, if any, are my sole responsibility.

Then (A) is equivalent to the following condition:

(C) *For all A in F*

$$P(A) = \int_U P_u(A) \cdot m(du)$$

and for almost all (with respect to m) u

$$P_u(E_u) = 1$$

The first equality in (C) is a recasting of (6.3); it can be equivalently described by saying that P is a mixture of the P_u's with weight function m. Altogether (C) means that we have here what is known as the disintegration of the probability space. It makes for a rather transparent structure.

For W – countable the disintegration means the following: After deleting from the kernel-matrix rows and columns which correspond to some set of probability 0, the rest decomposes into submatrices around the main diagonal in each of which all rows are equal, with 0's in all other places; P itself is a mixture of these rows. Such HOPs are exactly those that can be constructed as follows (hence this is the method for setting up higher order probabilities which avoid a Dutch book):

- *Chose a partition* $\{E_u : u \in U\}$ *of W into non-empty disjoint sets, with different u's marking different sets.*
- *Chose for each u in U a probability,* $P_{u'}$ *on W such that* $P_u(E_u) = 1$ *for all* $u \in U'$, *where* U' *is some non-empty susbset of U.*
- *Chose a probability, m, on U such that* $m(U') = 1$, *and let P be the mixture of the* P_u*'s with weight function m.*
- *For each* $u \in U$ *and each* $x \in E_x$ *put* $p_x = P_x$ *and define* $PR(A, \Delta)$ *to be* $\{x : p_x(A) \in \Delta\}$.

The construction is essentially the same for a general W (with a countably generated F); some additional stipulations of measurability should be included in order to make possible the formation of the mixture and to ensure that the $PR(A, \Delta)$*'s* are in F.

Definition Call P_u and its corresponding equivalence class, E_u, ontological if $P_u(E_u) = 1$, call it and its corresponding class coherent if P_u is a mixture of ontological P_u's. Call a world ontological (coherent) if it belongs to an ontological (coherent) equivalence class.

An ontological class is of course coherent. A coherent class which is not ontological must get the value 0 under its own P_u. It represents a state of knowledge in which the agent knows for sure that his eventual probability function will be different from his current one (and that it will be an ontological one).

The set of ontological worlds gets the value 1 under P and under each p_x where x is coherent. It is referred to as the ontological part of the HOP. Together with the structure induced by the original HOP it forms by itself a simple HOP. Similarly we define the coherent part of the HOP as the set of all coherent worlds (together with the induced structure). As far as calculating probabilities goes, only the ontological part matters. Coherent non-ontological worlds are useful as representatives of transitory states of knowledge.

Examples

Example 6.1 $W = \{w1, w2, w3\}$ $P = (1/3, 1/3, 1/3)$ and the kernel matrix is:

$$\begin{array}{ccc} .5 & .5 & 0 \\ 0 & .5 & .5 \\ .5 & 0 & .5 \end{array}$$

The agent's current probability assigns each world the value 1/3. Eventually, in world *w1* he will know that he is not in *w3* and he will assign each of the worlds *w1*, *w2* the value 0.5. This is the meaning of the first row. The other rows are similarly interpreted.

By direct checking one can verify that (VI_w) is satisfied. (The checking of all cases in this example is easy because $PR(A,\alpha) \neq \varnothing$ only for $\alpha = 0.5, 1$.) However the matrix is not equal to its square, hence Axiom (VI) is violated, as indeed the following case shows: Put $A = \{w1\}$, $C = PR(\{w2\}, 0.5)$. Then $C = \{x : p(x, w2) = 0.5\} = \{w1, w2\}$ and similarly $PR(A, 0.5) = \{w1, w3\}$. Hence $A = PR(A, 0.5) \cap C$ implying $P(A \mid PR(A, 0.5) \cap C) = 1 \neq 0.5$. This can be used to construct a Dutch book against the agent.

Note also that the epistemic equivalence classes are $\{w1\}$,$\{w2\}$ and $\{w3\}$ and that non is ontological; hence also there are no coherent worlds here.

Example 6.2 $W = \{w1, w2, \ldots, w8\}$, P is: (.1, .2, .2, .1, .4, 0, 0, 0} and the kernel matrix is:

$$\begin{array}{cccccccc} .2 & .4 & .4 & & & & & \\ .2 & .4 & .4 & & & & & \\ .2 & .4 & .4 & & & & & \\ & & & .2 & .8 & & & \\ & & & .2 & .8 & & & \\ & & & & & 1 & & \\ .05 & .2 & .2 & .05 & .2 & .5 & 0 & 0 \\ .2 & .1 & .1 & .2 & .1 & .1 & .1 & .1 \end{array}$$

where all undisplayed entries are 0. The sets $\{w1, w2, w3\}$, $\{w4, w5\}$ and $\{w6\}$ are equivalence classes which are ontological. P is a mixture of these 3 types of rows, with weights 0.5, 0.5, 0, respectively. Hence condition (C) is satisfied, therefore also Axiom (VI). *w7* is a coherent non-ontological world, because the 7th row is a mixture of the first three types (with weights .25, .25, .5) *w8* is not coherent. The ontological part consists of the upper left 6×6 matrix and the coherent part of the 7×7 one.

The example can be made more concrete by the following scenario. A number is to be chosen from $\{1,2,3\}$. For $i = 1,2,3$, the number chosen in *wi* is *i*, but in each of these 3 worlds the maximal knowledge consists in assigning probabilities 0.2, 0.4, 0.4 to the 3 possibilities. In *w4* the number chosen is *1* and in *w5* it is 2; in either of these worlds the maximal knowledge consists in assigning the probabilities 0.2, 0.8. In *w6* the number is 2 and it is also assigned probability 1. In the agent's current state he assigns probability 0 to finding himself eventually in the third state of maximal knowledge, and equal probabilities to the first and second states. World *w7* represent a similar situation but with different weights. We can imagine 3 lotteries for chosing the number; in each equivalence class the maximal knowledge is knowledge of the chosen lottery.

Example 6.3 Let H be the probability of "heads" of some given coin of unknown bias. Treat H as a random variable. The agent's knowledge is represented by a probability distribution for H. Say it is the uniform distribution over [0,1]. The expert does not know the value of H but he has some additional information. Say his additional information is the value of N – the number of "heads" in 50 independent tosses. Then our worlds can be regarded as pairs (h,n), such that in (h,n) the event $H = h \cap N = n$ is true; here h is a real number in [0,1] and n an integer between 0 and 50. The field F is generated by the sets $[\alpha, \beta] \times \{n\}, 0 \leq \alpha \leq \beta \leq 1, \; n = 0, \ldots, 50$.

Given $H = h$, we get the binomial distribution $b_{h,50}$ for N. This fact, together with the agent's uniform distribution for H, determines his probability P over F. The expert's probability in world (h,n) is obtained by conditioning on his information, it is $P(\;|N = n)$. There are 51 equivalence classes which correspond to the 51 possible values of N and all worlds are ontological.

As is well known, different values of N give rise to different conditional distributions of H. Therefore the events $N = n$ are in the field generated by the events[4] $PR(H \in [\alpha, \beta], \Delta)$. The whole field F is therefore generated by events which are either of the form $H \in [\alpha, \beta]$ or obtained from these by applying the operator *PR*. Consequently we can give an abstract description of this HOP which does not mention the fifty tosses. The only function of the tosses is to affect the distribution of H; in our framework such changes in distribution constitute themselves events which can be treated directly, without having to bring in their causes.

[4]Actually there are 51 real numbers α_n such that the event $N = n$ is the same as $PR(H \leq 1/2, \alpha_n)$.

The Case of a General Field

The restriction that F be countably generated is a mild one. The probability spaces which commonly appear in theory, or in applications, are essentially of this nature[5]. Usually we are interested in properties that involve only countably many generators. We will first show that for studying such properties we can always restrict ourselves to the case where the underlying field is countably generated.

Definition Given a simple HOP (W, F, P, PR) and given $S \subset F$, define $H[S]$ as the smallest field containing S and closed under PR (i.e., $A \in H[S] \Rightarrow PR(A, \Delta) \in H[S]$ for every real closed interval Δ).

$H[S]$, together with the restrictions of P and PR to it, forms a subHOP, where this notion is defined in the obvious way.

Lemma 6.3 *If S is a Boolean algebra and, for every A in S and every rational closed interval Δ, $PR(A, \Delta)$ is in S, then $H[S]$ is the field generated by S.*

This means that, once we have a Boolean algebra closed under $PR(, \Delta)$ for all Δ with rational endpoints, we get all the rest by countable Boolean operations without using PR.

Corollary *If S is either countable, or a countably generated field, then $H[S]$ is countably generated.*

Using this we can derive from Theorem 6.2 an analogous result for general fields:

Theorem 6.3 *Axiom* (VI) *is equivalent to each of the following conditions:*

(A′) (3) *holds and for every C in $K[F]$, for almost all x: $p_x(C) = 1$ if $x \in C$, $p_x(C) = 0$ otherwise.*
(B′) (3) *holds and for every A in F* (4) *is true for almost all x.*
(B′) differs from the analogous (B) of Theorem 6.2 in that the exceptional 0-set for (4) can depend on A

Say that A *is equal a.e. to* B if A-B and B-A are 0-sets. Say that two classes of sets are *equal modulo 0-sets* if every member of one is equal a.e. to some member of the other.

Assuming Axiom (VI) we get:

Corollary *If $S \subset F$, then:* (*i*) *The fields $K[S]$, $K[K[S]]$ and $K[H[S]]$ are equal modulo 0-sets.* (*ii*) *If S is a boolean algebra then $H[S]$ is equal modulo 0-sets to the field generated by $S \cup K[S]$.*

(To show, for example, that $K[S] = K[K[S]]$ modulo 0-sets, consider $C \in K[S]$; by Theorem 6.3, $\{x : p_x(C) \in \Delta\}$ is equal a.e. to C if $\Delta = [1,1]$, is equal a.e. to W-C if

[5]They are *seperable,* i.e., for some countably generated field every event in the space differs from a set in the field by a 0-set.

$\Delta = [0,0]$, and is a 0-set if $0,1 \notin \Delta$. Hence, for all Δ, $PR(C,\Delta)$ is equal a.e. to one of: C, W-C, W, $\varnothing$. Since $K[K[S]]$ is generated by such sets, the claim follows.)

Roughly speaking, (ii) means that, modulo 0-sets, nested applications of *PR* reduce to non-nested applications. A stronger, syntactical version of this is given in the next section.

Probability Logic

Let Ξ be a set of reals such that $0,1 \in \Xi$. Call an interval with end-points in Ξ a Ξ*-interval*.

Let PRL_Ξ be the calculus obtained by adjoining sentential operants, $PR(,\Delta)$, to the propositional calculus, where Δ ranges over all closed Ξ-intervals. Here, for the sake of convenience, I use '*PR*' for the syntactical operant, as well as for the operation in HOPs. Given some class $\{X_i: i \in I\}$ of sentential variables, the class of all wffs (well formed formulas) of PRL_Ξ is the smallest such that:

- Every sentential variable is a wff
- If ϕ and ψ are wffs, so are $\neg\phi$ and $\phi * \psi$ where * is any of the standard binary connectives.
- If ϕ is a wff and Δ is a closed Ξ-interval then $PR(\phi,\Delta)$ is a wff.

Let $\mathrm{H} = (W,F,P,PR)$ be a simple HOP and let r be a mapping which maps each sentential variable to a member of F. Then the value $|\phi|_{\mathrm{H},r}$ of the wff ϕ is defined by interpreting the sentential connectives as the corresponding Boolean operations and each syntactic operant $PR(,\Delta)$ as the operation $PR(,\Delta)$ of the HOP.

Definition A wff ϕ, is *p-valid*, to be denoted $\models_p \phi$, if, for every simple HOP H which satisfies Axiom (VI) and every r, the probability of $|\phi|_{\mathrm{H},\tau}$ is 1. Two wffs ϕ, ψ are *p-equivalent* if $\phi \leftrightarrow \psi$ is p-valid.

Call ϕ a *PC-formula* if it is a wff of the propositional calculus, i.e., does not contain any *PR*.

Theorem 6.4 *Every wff of* PRL_Ξ *is p-equivalent to a Boolean combination of PC-formulas and formulas of the form* $PR(\sigma,\Delta)$ *in which* σ *ranges over PC-formulas.*

This means that as far as probabilities are concerned (i.e., if we disregard 0-sets) we need not use nested *PR*'s.

Theorem 6.5 *Translate into* PRL_Ξ *the wffs of propositional modal logic with the necessity oprant N, by replacing each* $N(\psi)$ *by* $PR(\psi, [1,1])$. *Let* ϕ^* *be the translation of* ϕ. *Then*

$$S5 \vdash \phi \text{ iff } \models_p \phi*$$

Thus S5 becomes a fragment of PRL_Ξ. This relation becomes more explicit if we rewrite '$PR(\psi,\Delta)$' as '$N_\Delta(\psi)$'.

It can be shown that for Ξ = set of rationals the set of p-valid wffs is recursive. Also PRL_{Ξ} can be provided with a natural set of formal axioms so that, with modus ponens as derivation rule, p-validity coincides with provability.

Some Questions

Other validity notions can be considered (e.g., that $|\phi|_{H,\tau}$ always contains all coherent worlds in the HOP), as well as other interpretations of the necessity operant (e.g., as $\phi \bigwedge PR(\phi,[1,1])$). What modal logics are thereby obtained?

General HOPs

In general, a HOP is a structure of the form:

$$(W, F, P, T, PR)$$

where, as before, $(W\ F,\ P)$ is a probability space, $T = (T, <)$ is a partially ordered set and where

$$PR : F \times T \times \textit{set of closed intervals} \rightarrow F$$

$PR(A,\ t,\ \Delta)$ is the event that the probability of A at stage t lies in Δ. If the stages coincide with time points then the partial ordering of T is total. As before, P is the current subjective probability; here "current" is earlier (i.e., less than or equally informative) than the stages in T. Put:

$$PR_t\,(A, \Delta) =_{\text{df}} PR\,(A, t, \Delta)$$

The first five axioms **(I*)–(V*)** in this setting are the obvious generalizations of our previous axioms (I)–(V). Namely, we replace 'PR' by 'PR_t' and require that the condition hold for all t in T.

Theorem 6.1 generalizes in the obvious way and we get, for each $t \in T$ and each $x \in W$, a probability $P_{t,x}$ which determines PR_t; it represents the maximal state of knowledge at stage t in world x.

The "correct" generalization of Axiom (VI) is not as obvious, but is not difficult to find:

Axiom (VI*) *For each $t \in T$ the following holds: If C is a finite intersection of events of the form $PR_s(B, \Delta)$ where every s is $\leq t$, and $P\,(C \cap PR_t\,(A, [\alpha, \beta])) \neq 0$, then*

$$\alpha \leq P\left(A \middle| C \cap PR_t\,(A\,[\alpha,\ \ \beta])\right) \leq \beta$$

The argument for this axiom is the same as the argument for Axiom (VI). The essential point is that if $s \leq t$ then true events of the form $PR_t(\mathrm{B}, \Delta)$ are known at stage t. The same Dutch book argument works for Axiom (VI*).

As before, we consider fields generated by knowable events and define epistemic equivalence; but now these concepts depend on the stage parameter, to be displayed here as an additional subscript. Thus we put:

$$x \simeq_t y \Longleftrightarrow {}_{df} P_{t,x} = P_{t,y}$$

Then $x \simeq_t y$ iff $x \in A \Leftrightarrow y \in A$, for all $A \in K_t[F]$.

Theorem 6.6 *Assume F to be countably generated, then Axiom* (VI*) *is equivalent to the conjunction of:*

(D) *For each $t \in T$ the simple HOP* (W, F, P, PR_t) *satisfies Axiom* (VI).
and
(E) *For each $s \leq t$, $x \simeq_t y \Rightarrow x \simeq_s y$, for almost all x,y (i.e., for all $x,y \in W'$ where* $P(W') = 1$).

(E) means that, as we pass to more progressive stages, almost everywhere epistemic equivalence is the same or becomes stronger; the partition into equivalence classes can change only by becoming more refined.

Like Theorem 6.2 the last theorem has a version that applies to general fields but I shall not enter here into it. In the following theorem F is assumed to be countably generated.

Theorem 6.7 *Assume Axiom* (VI*). *Let $s \leq t$. Then, for almost all x, $p_{s,x}$ is a mixture of $p_{t,y}$'s (where y ranges over W). Consequently, for almost all x,* $(W, F, p_{s,x}, PR_t)$ *is a simple HOP satisfying Axiom* (VI).

Logic of HOPs and Stage Dependent Modalities

Fix a partially ordered set $T = (T, <)$. The logic $\mathrm{PRL}_{\Xi,T}$ (which corresponds to HOPs with set of stages T) is defined in the same way as PRL_{Ξ}, except that PR has an additional argument ranging over T. As before we employ a systematically ambiguous notation. Define ϕ to be p-valid if it gets probability 1 in all HOPs in which the set of stages is T.

Now consider a propositional modal language, M_T, in which we have, instead of a single necessity operant, an indexed family N_t, $t \in T$. $N_t\phi$ states that ϕ is necessary at stage t, i.e., necessary by virtue of the maximal knowledge available at that stage.

For $\phi \in \mathrm{M}_T$, let ϕ^* be the wff obtained by replacing each $N_t\psi$ by $PR_t(\psi, [1,1])$. It can be shown that the set of all ϕ in M_T such that ϕ^* is p-valid is exactly the set of wffs derivable, by modus ponens and the rule: if $\vdash \phi$ then $\vdash N_t\phi$, from the following axioms:

(i) All tautologies. (ii) For each $t \in T$, the axiom schemas of S5, with N replaced by N_t and

(iii) $N_s\phi \rightarrow N_t\phi$, for each $s \leq t$.

Note that (iii) accords well with the intended meaning of the N_t's: If something is necessary at stage s it is also necessary at later stages. On the other hand, something not necessary at stage s can be necessary later.

References[6]

Domotor, Z. (1981). *Higher order probabilities* (Manuscript).
Gaifman, H. (1983). *Toward a unified concept of probability.* Manuscript of invited lecture to the 1983 International Congress for Logic Philosophy and Methodology of science, salzburg. To appear in the proceedings of the congress, North Holland.
Gardenfors, P. (1975). Qualitative probability as intentional logic. *Journal of Philosophical Logic, 4*, 171–185.
Good, I. J. (1965). *The estimation of probabilities*. Cambridge: MIT Press.
Jaynes, E. T. (1958). *Probability theory in science and engineering* (Colloquium lectures in pure and applied science No 4). Dallas: Socony Mobil Oil Co Field Research Library.
Kolmogoroff, A. N. (1933). *Grundbegriffe der Wahrscheinlichkeit* (Ergebnisse der Mathematik 502 und ihrer Grenzgebitte, no 2). Berlin: Springer.
Miller, D. (1966). A paradox of information. *British Journal for the Philosophy of Science, 17*, 59–61.
Skyrms, B. (1980a). Higher order degree of belief. In D. H. Mellor (Ed.), *"Prospects for pragmatism" essays in honor of F.P. Ramsey*. Cambridge: Cambridge University Press.
Skyrms, B. (1980b). *Causal necessity* (*Appendix 2*). New-Haven: Yale.
van Frassen, B. (1984). Belief and the will. *Journal of Philosophy, 81*, 235–256.

[6]The list is far from being complete. Some important papers not mentioned in the abstract are to be found in Ifs, W. Harper ed. Boston Reidel, 1981. Material related in a less direct way: non-probabilistic measures of certainty (e.g., the Dempster-Shafer measure), expert systems involving reasoning with uncertainties, probabilistic protocols and distributed systems, has not been included, but should be included in a fuller exposition.

Chapter 7
On Indeterminate Probabilities

Isaac Levi

SOME men disclaim certainty about anything. I am certain that they deceive themselves. Be that as it may, only the arrogant and foolish maintain that they are certain about everything. It is appropriate, therefore, to consider how judgments of uncertainty discriminate between hypotheses with respect to grades of uncertainty, probability, belief, or credence. Discriminations of this sort are relevant to the conduct of deliberations aimed at making choices between rival policies not only in the context of games of chance, but in moral, political, economic, or scientific decision making. If agent X wishes to promote some aim or system of values, he will (*ceteris paribus*) favor a policy that guarantees him against failure over a policy that does not. Where no guarantee is to be obtained, he will (or should) favor a policy that reduces the probability of failure to the greatest degree feasible. At any rate, this is so when X is engaged in deliberate decision making (as opposed to habitual or routine choice).

Two problems suggest themselves, therefore, for philosophical consideration:

The Problem of Rational Credence: Suppose that an ideally rational agent X is committed at time t to adopting as certain a given system of sentences $K_{X,t}$ (in a suitably regimented L) and to assigning to sentences in L that are not in $K_{X,t}$ various degrees of (personal) probability, belief, or credence. The problem is to specify conditions that X's "corpus of knowledge" $K_{X,t}$ and his "credal state" $B_{X,t}$ (i.e., his system of judgments of probability or credence) should satisfy in order to be reasonable.

The Problem of Rational Choice: Given a corpus $K_{X,t}$ and a credal state $B_{X,t}$ at t, how should X make decisions between alternative policies from which he must choose one at t?

I. Levi (✉)
Columbia University, New York, NY, USA
e-mail: levi@columbia.edu

H. Arló-Costa et al. (eds.), *Readings in Formal Epistemology*, Springer Graduate Texts in Philosophy 1, DOI 10.1007/978-3-319-20451-2_7

Consideration of these two problems should lead to examination of a third. A rational agent X is entitled to count as certain at t not only logical, mathematical, and set-theoretical truths supplemented by suitably produced testimony of the senses, but theories, laws, and statistical claims as well. At the same time, the revisability of X's corpus at t should be recognized not only by others but by X himself. Moreover, just as X's judgments of certainty are liable to revision, so too are his judgments of probability or credence. Indeed, the two types of modification are apparently interdependent, and this interdependence itself deserves examination. The third problem, therefore, is as follows:

The Problem of Revision: Under what conditions should X modify his corpus $K_{X,t}$ or his credal state $B_{X,t}$, and, if he should do so, how should he choose between alternative ways of making revisions?

In this essay, I shall not attempt to solve the problem of revision. However, I shall indicate how a prima facie obstacle to offering anything other than a dogmatic or antirationalistic answer to the question can be eliminated.

The obstacle is a serious one; for it derives from a very attractive system of answers to the problem of rational credence and the problem of rational choice. I allude to what is called the "bayesian" view. Bayesians do not agree with one another in their answers to these questions in all respects. The views of Harold Jeffreys and the early views of Rudolf Carnap are not consonant in important ways with the ideas of Bruno de Finetti and Leonard J. Savage (or the later Carnap). Nonetheless, the answers these and a host of other authors offer to the first two questions share certain important ramifications for the problem of revision. One of these implications is the commitment to either dogmatism or antirationalism.

Of course, identifying an objectionable consequence of bayesianism, where the objection is grounded on a question of philosophical principle, is in itself unlikely to persuade devoted bayesians to abandon their position. Such authors will be tempted to modify philosophical principle so as to disarm the objection; and they will have good reasons for doing so. Bayesian doctrine does offer answers to the first two questions. These answers are derivable from a system of principles which are precise and simple. Even the disputes between bayesians can be formulated with considerable precision. Furthermore, the prescriptions bayesians recommend for making choices appear to conform to presystematic judgment at least in some contexts of decision. Rival attempts to answer the problems of rational credence and rational choice seem either eclectic or patently inadequate when compared with the bayesian approach.

Thus, it is not enough to complain of the defects of bayesianism. The serious challenge is to construct an alternative system of answers to the problems of rational credence and choice which preserves the virtues of bayesianism without its vices—in particular, the defects it exhibits relevant to the problem of revision.

In this paper, I shall outline just such a rival view.

I

X's corpus of knowledge $K_{X,t}$ at t identifies a set of options $A_1, A_2, \ldots, A_n$ as the options from which he will choose (at t' identical with or later than t) at least and at most one. In addition, $K_{X,t}$ implies that at least and at most one of the hypotheses h_1, $h_2, \ldots, h_m$ is true and that each of the h_j's is consistent with $K_{X,t}$. Finally, $K_{X,t}$ implies that, if X chooses A_i when h_j is true, the hypothesis o_{ij} asserting the occurrence of some "possible consequence" of A_i is true.

The problem of rational choice is to specify criteria for evaluating various choices of A_is from among those feasible for X according to what he knows at t. Such criteria may be construed as specifying conditions for "admissibility." Option A_i is admissible if and only if X is permitted as a rational agent to choose A_i from among the feasible options. If A_i is uniquely admissible, X is obliged, as a rational agent, to choose it. In general, however, unique admissibility cannot be guaranteed, and no theory of rational choice pretends to guarantee it.

Bayesians begin their answer to the problem of rational choice by assuming that X is an ideally rational agent in the following sense:

(i) X has a system of evaluations for the possible consequences (the o_{ij}s) representable by a real-valued "utility" function $u(o_{ij})$ unique up to a linear transformation (i.e., where utility assignments are nonarbitrary once a 0 point and a unit are chosen—as in the case of measuring temperature).

(ii) X has a system of assignments of degrees of credence to the o_{ij}s, given the choice of A_i representable by a real-valued function $Q(o_{ij}; A_i)$ conforming to the requirements of the calculus of probabilities. Often X will assign credence values to the "states of nature" $h_1, h_2, \ldots, h_n$ so that the h_js are probabilistically independent of the option chosen. When this is so, $Q(o_{ij}; A_i)$ equals the unconditional credence (given $K_{X,t}$) $Q(h_j)$. In the sequel, I shall suppose that we are dealing with situations of this kind.

Given such a utility function $u(o_{ij})$ and Q-function $Q(h_i)$, let $E(A_i) = \sum_{j=1}^{m} u(o_{ij})\, Q(h_j)$. $E(A_i)$ is the expected utility of the option A_i.

Bayesians adopt as their fundamental principle of rational choice the principle that an option is admissible only if it bears maximum expected utility among all the feasible options.

Very few serious writers on the topic of rational choice object to the principle of maximizing expected utility in those cases where X's values and credal state can be represented by a utility function unique up to a linear transformation and a unique probability function. The doubts typically registered concern the applicability of this principle. That is to say, critics doubt that ordinary men have the ability under normal circumstances to satisfy the conditions of ideal rationality stipulated by strict bayesians even to a modest degree of approximation.

The bayesian riposte to doubts about applicability is to insist that rational men should meet the requirements for applying the principle of maximizing expected utility and that, appearances to the contrary notwithstanding, men are quite capable of meeting these requirements and often do so.

I am not concerned to speculate on our capacities for meeting strict bayesian requirements for credal (and value) rationality. But even if men have, at least to a good degree of approximation, the abilities bayesians attribute to them, there are many situations where, in my opinion, rational men *ought not* to have precise utility functions and precise probability judgments. That is to say, on some occasions, we should avoid satisfying the conditions for applying the principle of maximizing expected utility even if we have the ability to satisfy them.

In this essay, reference to the question of utility will be made from time to time. I shall not, however, attempt to explain why I think it is sometimes (indeed, often) irrational to evaluate consequences by means of a utility function unique up to a linear transformation. My chief concern is to argue that rational men should sometimes avoid adopting numerically precise probability judgments.

The bayesian answer to the problem of rational choice presupposes at least part of an answer to the problem of rational credence. For a strict bayesian, a rational agent has a credal state representable by a numerically precise function on sentences (or pairs of sentences when conditional probability is considered) obeying the dictates of the calculus of probabilities.

There are, to be sure, serious disputes among bayesians concerning credal rationality. In his early writings, Carnap believed that principles of "inductive logic" could be formulated so that, given X's corpus $K_{X,t}$, X's credal state at t would be required by the principles of inductive logic to be represented by a specific Q-function that would be the same for anyone having that corpus.[1] Others (including the later Carnap[2]) despair of identifying such strong principles. Nonetheless, bayesian critics of the early Carnap's program for inductive logic continue to insist that ideally rational agents should assign precise probabilities to hypotheses.

II

X's corpus of knowledge $K_{X,t}$ shall be construed to be the set of sentences (in L) to whose certain truth X is committed at t. I am not suggesting that X is explicitly or consciously certain of the truth of every sentence in $K_{X,t}$, but only that he is committed to being certain. X might be certain at t of the truth of h and, hence, be committed to being certain of $h \vee g$, without actually being certain. Should it be brought to X's attention, however, that $h \vee g$ is a deductive consequence of h, he would be obliged as a rational agent either to cease being certain of h or to take $h \vee g$ to be certain. The latter alternative amounts to retaining his commitment; the former to abandoning it.

---

[1] *Logical Foundations of Probability* (Chicago: University Press, 2nd ed., 1962), pp. 219–241.

[2] "Inductive Logic and Rational Decisions," in Carnap and R. C. Jeffrey, eds., *Studies in Inductive Logic and Probability* (Berkeley: UCLA Press, 1971), p. 27.

In this sense, X's corpus of knowledge at t should be a deductively closed set of sentences. Insofar as we restrict our attention to changes in knowledge and credence which are changes in commitments, modifications of corpora of knowledge are shifts from deductively closed sets of sentences to other deductively closed sets of sentences. Such modifications come in three varieties:

1. *Expansions*, where X strengthens his corpus by adding new items. Some examples of expansion are acquiring new items via observation, from the testimony of others and through inductive or nondeductive inference leading to the "acceptance" of statistical claims, laws, or theories into the corpus.
2. *Contractions*, where X weakens his corpus by removing items. This can happen when X detects an inconsistency in his corpus due to his having added at some previous expansion step an observation report that contradicts assumptions already in his corpus, or when X finds himself in disagreement with Y (whose views he respects on the point at issue) and wishes to resolve the dispute without begging the question.
3. *Replacements*, where X shifts from a theory containing one assumption to another containing an assumption contradicting the first. This can happen when X substitutes one theory for another in his corpus.

No matter which kind of modification is made, I shall suppose that there is a "weakest" potential corpus UK (the "urcorpus") of sentences in L such that no rational agent should contract that corpus. UK is the deductively closed set of sentences in L such that every potential corpus in L is an expansion of UK (or is UK itself). I shall suppose that UK contains logical truths, set-theoretical truths, mathematical truths, and whatever else might be granted immunity from removal from the status of knowledge. (The items in UK are in this sense incorrigible.)

Replacement poses special problems for an account of the revision of knowledge. At t when X's corpus is $K_{X,t}$, why should he shift to a corpus K^* which is obtained by deleting items from $K_{X,t}$ and replacing them with other items inconsistent with the first? From X's point of view, at t, he is replacing a theory which he is certain is true by another which he is certain is false.

The puzzle can be avoided by regarding replacements for purposes of analysis as involving two steps: (a) contraction to a corpus relative to which no question is begged concerning the rival theories, and (b) subsequent expansion based on the information available in the contracted corpus, supplemented, perhaps, by the results of experiments conducted in the interim.

Those who insist on attempting to justify replacements without decomposing them into contractions followed by expansions confront the predicament that they cannot justify such shifts without begging questions. Such justification is no justification. The conclusion that beckons is that all replacements are forms of "conversion" to which men are subjected under revolutionary stress. This is the view which Thomas Kuhn has made so popular and which stands opposed to views that look to the formulation of objective criteria for the evaluation of proposed modifications of knowledge.

III

How does all this relate to bayesian views about the revision of credal states?

Consider X's corpus of knowledge $K_{X,t}$ at t. X's credal state $B_{X,t}$ at t is, according to strict bayesians, determined by $K_{X,t}$. Strict bayesians disagree among themselves concerning the appropriate way in which to formulate this determination. The following characterization captures the orthodox view in all its essentials.

Let K be any potential corpus (i.e., let it be UK or an expansion thereof). Let $C_{X,t}(K)$ be X's judgment at t as to what his credal state should be were he to adopt K as his corpus of knowledge. I shall suppose that X is committed to judgments of this sort for every feasible K in L. The resulting function from potential corpora of knowledge to potential credal states shall be called X's "confirmational commitment" at t.

According to strict bayesians, no matter what corpus K is (provided it is consistent), $C_{X,t}(K)$ is representable by a probability function where all sentences in K receive probability 1. In particular, $C_{X,t}(UK)$ is representable by a function $P(x;y)$—which I shall call a P-function, to contrast it with a Q-function representing $C_{X,t}(K)$ where K is an expansion of UK.

Strict bayesians adopt the following principle, which imposes restrictions upon confirmational commitments:

Confirmational Conditionalization: If K is obtained from UK by adding e (consistent with UK) to UK and forming the deductive closure, $P(x;y)$ represents $C_{x,t}(UK)$ and $Q(x;y)$ represents $C_{x,t}(K)$, $Q(h;f) = P(h;f\&e)$

In virtue of this principle, X's confirmational commitment is defined by specifying $C_{X,t}(UK) = C_{X,t}$ and employing confirmational conditionalization.[3] X's credal state at t, $B_{X,t}$, is then determined by $K_{X,t}$ and $C_{X,t}$ according to the following principle:

Total Knowledge: $C_{x,t}(K_{x,t}) = B_{x,t}$

Notice that the principle of confirmational conditionalization, even when taken together with the principle of total knowledge, does not prescribe how X should modify his credal state given a change in his corpus of knowledge.

To see this, suppose that at t_1 X's corpus is K_1 and that at t_2 his corpus K_2 is obtained from K_1 by adding e (consistent with K_1) and forming the deductive

[3]Confirmational commitments built on the principle of confirmational conditionalization are called "credibilities" by Carnap (*ibid.*, pp. 17–19). The analogy is not quite perfect. According to Carnap, a credibility function represents a permanent disposition of X to modify his credal states in the light of changes in his corpus of knowledge. When credibility is rational, it can be represented by a "confirmation function." Since I wish to allow for modifications of confirmational commitments as well as bodies of knowledge and credal states, I assign dates to confirmational commitments. Throughout I gloss over Carnap's distinction between credibility functions and confirmation functions (*ibid.*, pp. 24–27).

closure. From confirmational conditionalization and total knowledge, we can conclude that *if X does not alter his confirmational commitment in the interim from* t_1 *to* t_2, then, if Q_1 represents B_{X,t_1} and Q_2 represents B_{X,t_2}, $Q_2(h;f) = Q_1(h;f\&e)$. Should X renege at t_2 on the confirmational commitment he adopted at t_1, the change in knowledge just described need not and will not, in general, lead to a modification of credal state of the sort indicated.

Nonetheless, strict bayesians unanimously suppose that a rational agent will, save under unusual circumstances, modify his credal state in the fashion indicated. This mode of revising credal states is often called "conditionalization"; to distinguish it from confirmational conditionalization and other types of conditionalization, I shall call it "intertemporal credal conditionalization." I contend that the strict bayesian endorsement of intertemporal credal conditionalization presupposes commitment to the following principle:

Confirmational Tenacity: For every X, t, and t', $C_{x,t} = C_{x,t'}$

Thus, strict bayesians have an answer to the problem of revising credal states. X's confirmational commitment is to be held fixed over time. Given such a fixed commitment, the credal state he should adopt is determined for each possible modification of his corpus of knowledge which is a consistent expansion of *UK*. The problem of revising credal states reduces, therefore, to the problem of revising corpora of knowledge.

Is this answer to the problem of revision satisfactory? It would be, in my opinion, if the program for inductive logic envisaged by Carnap in his early writings on the subject could be realized. Inductive logic would then be strong enough to single out a standard *P*-function that all rational agents should adopt as their confirmational commitment. A fortiori, all such agents should hold that commitment fast at all times.

Few bayesians now think an inductive logic of the requisite power can be constructed. Their reasons (which, in my opinion, are sound) need not detain us. In response to this skepticism, most bayesians no longer require that all rational agents endorse a single standard confirmational commitment. They hold that rational X is perfectly free to pick any confirmational commitment consonant with the principles of inductive logic. Rational Y is quite free to pick a different commitment. However, bayesians tend to insist that, once X and Y have chosen their respective commitments, they should hold them fixed. To do this is to follow the probabilistic analogue of the method of tenacity so justly criticized by Peirce in "Fixation of Belief."

In the spirit of Peirce, it would have been far better to say that a rational X should not modify his confirmational commitment capriciously—i.e., without justification. To follow this approach, however, demands consideration of criteria for justified modifications of confirmational commitments. Bayesians not only fail to do this, but, as I shall now argue, they cannot do so without great difficulty. Given the bayesian answer to the problem of rational credence, no shift can be justified. If I am right, for bayesians, either tenacity should be obeyed, or, if not, justification is gratuitous. I think this implication of bayesian doctrine is to be deplored and should lead to scrutiny of other approaches.

IV

Modifying a confirmational commitment is not quite the same as modifying a corpus of knowledge. Yet, shifting from a confirmational commitment represented by a precise probability function to another confirmational commitment represented by a different precise probability function seems analogous to replacement in the following sense: The shift from confirmational commitment C_1 to confirmational commitment C_2 involves a shift to a confirmational commitment conflicting with C_1 in the sense that the P-function X uses to determine his credal state relative to his corpus when C_1 is adopted yields different precise subjective probability or credence assignments for hypotheses from those which X would make were he to adopt C_2 (and keep his corpus constant).

From X's vantage point at t when he endorses C_1, C_2 is illegitimate. He cannot justify shifting to C_2. At least, he cannot justify a direct shift. Can he do so indirectly by first performing a shift analogous to contraction from C_1 to C_3, which begs no questions concerning the merits of C_1 and C_2? Not from a strict bayesian point of view; for C_3 would, like C_1 and C_2, have to be representable by a precise P-function. The shift from C_1 to C_3 would be as problematic as the shift from C_1 to C_2.

Thus, from a bayesian point of view, no shift from one confirmational commitment to another can be justified. A rational man should conform to confirmational tenacity so that no justification is needed or else hold that some shifts are permitted without justification. Carnap sometimes seems to recognize shifts in confirmational commitments as a result of conceptual change.[4] Alternatively, one might allow shifts in confirmational commitment due to conversion under revolutionary stress. Except for the minimal requirement that the shift be to a commitment obeying requirements of inductive logic, no critical control is to be exercised. Bayesians are committed to being dogmatically tenacious or arbitrarily capricious.

The source of the difficulty should be apparent. Bayesians restrict the confirmational commitments a rational agent may adopt to those representable by numerically precise probability functions. This precludes shifting from a confirmational commitment C_1 to a confirmational commitment C_3 that begs no questions as to the merits of C_1 and another commitment C_2 that conflicts with C_1. My thesis is that not only are rational men allowed to make shifts to non-question-begging commitments but that on many occasions they ought to do so. That is to say, it is sometimes appropriate for a rational agent to adopt a confirmational commitment that is indeterminate in the sense that it cannot be represented by a numerically precise probability function. If we relax the stringent requirements imposed by bayesians on confirmational commitments and credal states so as to allow for such shifts, there is at least some hope that we can avoid endorsement of tenacity or capriciousness. Within the strict bayesian framework, we cannot expect to do so except by clinging desperately to Carnap's early program for constructing

[4]"A Basic System of Inductive Logic," in Carnap and Jeffrey, *op. cit.*, pp. 51–52.

an inductive logic so strong as to single out a standard P-function to represent the uniquely rational confirmational commitment (for a given language).

I propose to explore one way of relaxing strict bayesian requirements. The basic idea is to represent a credal state (confirmational commitment) by a *set* of Q-functions (P-functions). When the set is single-membered, the credal state (confirmational commitment) will be indistinguishable in all relevant respects from a strict bayesian credal state (confirmational commitment).

On this view, if X starts at t with the precise (i.e., single-membered) confirmational commitment C_1, he can then shift to a confirmational commitment that has as members all the P-functions in C_1 as well as the P-functions in some other confirmational commitment C_2. (As the technical formulation will indicate, other P-functions will be members of C_3 as well.)

C_3 will be "weaker" than C_1 or C_2 in that it will allow more P-functions to be "permissible" than either of the other two confirmational commitments alone does. It will allow as permissible all P-functions recognized as such according to C_1 and according to C_2. In this sense, the shift to C_3 will beg no questions as to the permissibility of the P-functions in the other two confirmational commitments.

Of course, the notion of a permissible P-function (and the correlative notion of a permissible Q-function according to a credal state) require elucidation. I shall offer only an indirect clarification. The account of rational credence (and confirmational commitment) based on the new proposal will be supplemented by criteria for rational choice which indicate how permissibility determines the admissibility of options. By indicating the connections between permissibility and rational choice, permissibility will have been characterized indirectly.

V

To simplify the technical details, I shall restrict the discussion to characterizing credal states and confirmational commitments for sentences in a given language L which belong to a set M generated as follows: Let $h_1, h_2, \ldots, h_n$ be a finite set of sentences in L all consistent with the urcorpus UK for L and such that UK logically implies the truth of at least and at most one h_i. M is the set of sentences in L which are equivalent, given UK, to a disjunction of zero or more distinct h_is. (A disjunction of zero h_is is, as usual, a sentence inconsistent with UK.)

With this understanding, X's credal state at t will be a set $B_{X,t}$ of functions $Q(x;y)$ where the sentences substituted for 'x' are in M and the sentences substituted for 'y' are in M and are consistent with $K_{X,t}$. When the sentence substituted for 'y' is a member of $K_{X,t}$, I shall write $Q(x) = Q(x;y)$.

The set $B_{X,t}$ must satisfy the following three conditions:

1. *Nonemptiness:* $B_{X,t}$ is nonempty.
2. *Convexity:* $B_{X,t}$ is a convex set—i.e., every weighted average of Q-functions in $B_{X,t}$ is in $B_{X,t}$.
3. *Coherence:* Every Q-function in $B_{X,t}$ is a probability measure where $Q(h;e) = 1$ if and only if h is deductively implied by e and $K_{X,t}$.

Every Q-function in $B_{X,t}$ is "permissible" according to $B_{X,t}$.

As before, X's confirmational commitment $C_{X,t}(K)$ is a function from feasible corpora of knowledge to potential credal states that X at t considers to be the credal states he should adopt were he to adopt K as his corpus of knowledge. The value of the function for given K, therefore, is a nonempty, convex set of Q-functions relative to K. $C_{X,t}(UK) = C_{X,t}$ is, therefore, a nonempty convex set of P-functions. The principle of confirmational conditionalization introduced previously must now be modified to conform to the new characterization of confirmational commitments and credal states:

Confirmational Conditionalization: Let K be obtained from UK by adding e (consistent with UK) to UK and forming the deductive closure. $C_{x,t}(K)$ is the set of Q-functions such that $Q(h;f) = P(h;f\&e)$ for some permissible P-function in $C_{x,t} = C_{x,t}(UK)$.

$B_{X,t}$ can be determined, as before, as follows:

Total Knowledge: $B_{x,t} = C_{x,t}(K_{x,t})$

Thus, X's confirmational commitment is defined by specifying the value of $C_{X,t}(UK)$.

A strict bayesian confirmational commitment, of course, allows a single P-function to be uniquely permissible. However, confirmational commitments are possible which contain more than one P-function. In general, I shall say that one confirmational commitment is stronger than another if the set of its P-functions is a subset of the set of P-functions in the other commitment.

On this view, the weakest confirmational commitment possible is that which contains all the P-functions that meet the requirements of inductive logic. I shall continue to follow Carnap in understanding inductive logic to be a system of principles that impose constraints on probability functions eligible for membership in confirmational commitments.

In contrast, the strongest confirmational commitment would be the empty one—which is inconsistent with our first requirement of nonemptiness. A strongest "consistent" confirmational commitment is single-membered.

We can, by the way, extend the notion of a confirmational commitment so as to define it for an inconsistent corpus. We can require that $C_{X,t}(K)$ where K is inconsistent, be empty. This means that our previous requirement that a credal state be nonempty is to be restricted to cases where K is consistent. Thus, X might adopt a consistent confirmational commitment (i.e., one that is nonempty). Yet, if he should, unfortunately, endorse an inconsistent K, his credal state should be empty.

As noted previously, strict bayesians have differed among themselves as to what constitutes a complete system of principles of inductive logic. These differences persist on the view I am now proposing. They may be viewed, however, in a new light. The disagreements over inductive logic turn out to be disagreements over what constitutes the "weakest" possible confirmational commitment—which I shall call "*CIL*(*UK*)."

"Coherentists" like de Finetti and Savage claim that the principle of coherence constitutes a complete inductive logic. On their view, $CIL(UK)$ is the set of all P-functions obeying the calculus of probabilities defined over M.

Some authors are prepared to add a further principle to the principle of coherence. This principle determines permissible Q-values for hypotheses about the outcome of a specific experiment on a chance device, given suitable knowledge about the experiment to be performed and the chances of possible outcomes of experiments of that type.

There is considerable controversy concerning the formulation of such a principle of "direct inference." In large measure, the controversy reflects disagreements over the interpretation of "chance" or "statistical probability," concerning the so-called "problem of the reference class" and random sampling. Indeed, the reason coherentists do not endorse a principle linking objective chance with credence is that they either deny the intelligibility of the notion of objective chance or argue in favor of dispensing with that notion.

Setting these controversies to one side, I shall call anyone who holds that a complete inductive logic consists of the coherence principle and an additional principle of direct inference from knowledge of chance to outcomes of random experiments an "objectivist."

There are many authors who are neither coherentists nor objectivists because they wish to supplement the principles of coherence and direct inference with additional principles. Some follow J. M. Keynes, Jeffreys, and Carnap in adding principles of symmetry of various kinds. Others, like I. Hacking,[5] introduce principles of irrelevance or other criteria which attempt to utilize knowledge about chances in a manner different from that employed in direct inference. Approaches of this sort stem by and large from the work of R. A. Fisher. I lack a good tag for this somewhat heterogeneous group of viewpoints. They all agree, however, in denying that objectivist inductive logic is a complete inductive logic.

Attempting to classify the views of historically given authors concerning inductive logic is fraught with risk. I shall not undertake a tedious and thankless task of textual analysis in the vain hope of convincing the reader that many eminent authors have been committed to an inductive logic whether they have said so or not. Yet much critical insight into controversies concerning probability, induction, and statistical inference can be obtained by reading the parties to the discussion as if they were committed to some form of inductive logic. If I am right, far from being a dead issue, inductive logic remains very much alive and debated (at least implicitly) not only by bayesians of the Keynes-Jeffreys-Carnap persuasion but by objectivists (to whose number I think J. Neyman, H. Reichenbach, and, with some qualifications, H. Kyburg belong) and the many authors, like Hacking, who are associated with the tradition of Fisher in various ways.

Assuming, for the sake of the argument, that the debate concerning what constitutes a complete set of principles of inductive logic is settled (I, for one, would

[5] *Logic and Statistical Inference* (New York: Cambridge, 1965), p. 135.

defend and will defend elsewhere adopting a variant of an objectivist inductive logic), there is yet another dimension to debates among students of probability, induction, and statistical inference.

Some authors seem to endorse the view that a rational agent should adopt the weakest confirmational commitment, *CIL*, consonant with inductive logic and hold it fast. They are, in effect, advocating confirmational tenacity. They do so, however, on the grounds that one should not venture to endorse a confirmational commitment stronger than the weakest allowed by inductive logic. (Their view is analogous to one that would require adopting the weakest corpus of knowledge *UK* and holding it fast.) I shall call advocates of such a view "necessitarians."

Again, classifying historically given authors is a risky business. However, Keynes, Jeffreys, and Carnap (in his early work) seem to be clear examples of necessitarians. What is more interesting is the implication that anyone is a necessitarian who insists that the only conditions under which a numerically precise probability can be assigned to a statement (other than a statement that is certainly true or false) are those derivable via direct inference from knowledge of chances. Such authors, on my view, are committed to saying that, when numerical probabilities are not assignable in this way, any numerical value is a permissible assignment provided that it is derived from Q-functions allowed by inductive logic.

To illustrate, suppose that X knows that a given coin has a .4 or a .6 chance of landing heads on a toss. Let h_1 be the first hypothesis that the chance is .4, and h_2 the second hypothesis. Let g be the hypothesis that the coin will land heads on the next toss. By direct inference, every permissible Q-function in X's credal state must be such that $Q(g;h_1) = .4$ and $Q(g;h_2) = .6$. By coherence, every Q-function in his credal state must be such that $Q(h_2) = 1 - Q(h_1)$, where $Q(h_1)$ is some real number between 0 and 1 and $Q(g) = Q(g;h_1)Q(h_1) + Q(g;h_2)Q(h_2) = .4Q(h_1) + .6(1 - Q(h_1))$.

According to the authors I have in mind, there is no unique numerical value that a rational X should adopt as uniquely permissible for $Q(h_1)$. As I am interpreting such authors as Kyburg, Neyman, Reichenbach, and Salmon, they mean to say that X's credal state should consist of all Q-functions meeting the conditions specified. The upshot is that the set of permissible Q-values for g should consist of all Q-values in the interval from .4 to .6. If I am reading them right, they endorse an objectivist logic and, at the same time, insist that X should adopt *CIL* as his confirmational commitment. They are "objectivist necessitarians."

The early Carnap, as noted previously, had hoped to identify an inductive logic that singled out a unique P-function as eligible for membership in confirmational commitments. Had his hope been realized, a rational agent would perforce have had to be a necessitarian. The weakest confirmational commitment would have been the strongest consistent one as well. Confirmational tenacity would have been necessitated by the principles of inductive logic.

But if Carnap's program is abandoned, necessitarianism is by no means the only response that one can make. Indeed, it seems to be of doubtful tenability, if for no other reason than that credal states formed on a necessitarian basis seem to be too weak for use in practical decision making or statistical inference. (Many objectivist necessitarians seem to deny this; but the matter is much too complicated to discuss here.)

Personalists, like de Finetti and Savage, abandon necessitarianism but continue to endorse confirmational tenacity—at least during normal periods free from revolutionary stress. It is this position that I contended earlier leads to dogmatism or capriciousness with respect to confirmational commitment.

The view I favor is *revisionism*. This view agrees with the personalist position in allowing rational men to adopt confirmational commitments stronger than *CIL*. It insists, however, that such commitments are open to revision. It sees as a fundamental epistemological problem the task of providing an account of the conditions under which such revision is appropriate and criteria for evaluating proposed changes in confirmational commitment on those occasions when such shifts are needed.

I shall not offer an account of the revision of confirmational commitments. The point I wish to emphasize here is that, once one abandons the strict bayesian approach to credal rationality and allows credal states to contain more than one permissible Q-function in the manner I am suggesting, the revisionist position can be seriously entertained. The strict bayesian view precludes it and leaves us with the dubious alternatives of necessitarianism and personalism. By relaxing the strict bayesian requirements on credal rationality, we can at least ask a question about revision which could not be asked before.

VI

According to the approach I am proposing, X's credal state at t is characterized by a set of Q-functions defined over sentences in a set M. Such a representation describes X's credal state globally. Nothing has been said thus far as to how individual sentences in M are to be assigned grades of credence or how the degrees of credence assigned to two or more sentences are to be compared with one another. The following definitions seem to qualify for this purpose:

Def. 1: $Cr_{x,t}(h;e)$ is the set of real numbers r such that there is a Q-function in $B_{x,t}$ according to which $Q(h;e) = r$.

Def. 2: $c_{x,t}(h;e)$ is the set of real numbers r such that there is a P-function in $C_{x,t}$ according to which $P(h;e) = r$.

In virtue of the convexity requirement, both the credence function $Cr_{x,t}(h;e)$ and the confirmation function $c_{x,t}(h;e)$ will take sets of values that are subintervals of the unit line—i.e., the interval from 0 to 1. The lower and upper bounds of such intervals have properties which have been investigated by I. J. Good,[6] C. A. B. Smith,[7] and A. P. Dempster.[8]

A partial ordering with respect to comparative credence or with respect to comparative confirmation can be defined as follows:

Def. 3: $(h; e) \overset{Cr_{x,t}}{\leq} (h'; e')$ if and only if, for every Q-function in $B_{x,t}$, $Q(h;e) \leq Q(h';e')$.

Def. 4: $(h; e) \overset{C_{x,t}}{\leq} (h'; e')$ if and only if, for every P-function in $C_{x,t}$, $P(h;e) \leq P(h';e')$.

The partial orderings induced by credal states and confirmational commitments conform to the requirements of B. O. Koopman's axioms for comparative probability.[9] Koopman pioneered in efforts to relax the stringent requirements imposed by bayesians on rational credence. Within the framework of his system, he was able not only to specify conditions of rational comparative probability judgment but to identify ways of generating interval-valued credence functions.

According to Koopman's approach, however, any two credal states (confirmational commitments) represented by the same partial ordering of the elements of M are indistinguishable. My proposal allows for important differences. Several distinct convex sets of probability distributions over the elements of M can induce the same partial ordering on the elements of M according to definitions 3 and 4.

Dempster, Good, Kyburg, Smith, and F. Schick, have all proposed modifying bayesian doctrine by allowing credal states and confirmational commitments to be represented by interval-valued probability functions.[10] Good, Smith, and Dempster have also explored the representation of credal states defined by interval-valued credence functions by means of sets of probability measures. Smith and Dempster explicitly consider convex sets of measures. Nonetheless, all these authors, including Dempster and Smith, seem to regard credal states (and confirmational commitments) represented by the same interval-valued function as indistinguishable. In contrast, my proposal recognizes credal states as different

[6]"Subjective Probability as the Measure of a Non-measurable Set," in P. Suppes, E. Nagel, and A. Tarski, *Logic, Methodology, and the Philosophy of Science* (Stanford: University Press, 1962), pp. 319–329.

[7]"Consistency in Statistical Inference and Decision" (with discussion), *Journal of the Royal Statistical Society*, series B, XXIII (1961): 1–25.

[8]"Upper and Lower Probabilities Induced by a mutivalued Mapping," *Annals of Mathematical Statistics*, XXXVIII (1967): 325–339.

[9]"The Bases of Probability," *Bulletin of the American Mathematical Society*, XLVI (1940): 763–774.

[10]Dempster, *op. cit.*; Good, *op. cit.*; Kyburg, *Probability and the Logic of Rational Belief* (Middletown, Conn.: Wesleyan Univ. Press, 1961): Smith, *op. cit.*; Schick, *Explication and Inductive Logic*, doctoral dissertation, Columbia University, 1958.

even though they generate the identical interval-valued function—provided they are different convex sets of Q-functions.[11]

Thus, the chief difference between my proposal and other efforts to come to grips with "indeterminate" probability judgments is that my proposal recognizes significant differences between credal states (confirmational commitments) where other proposals recognize none. Is this a virtue, or are the fine distinctions allowed by my proposal so much excess conceptual baggage?

I think that the distinctions between credal states recognized by the proposals introduced here are significant. Agents X and Y, who confront the same set of feasible options and evaluate the possible consequences in the same way may, nonetheless, be obliged as rational agents to choose different options if their credal states are different, even though their credal states define the same interval-valued credence function. That is to say, according to the decision theory that supplements the account of rational credence just introduced, differences in credal states recognized by my theory but not by Dempster's or Smith's, do warrant different choices in otherwise similar contexts of choice.

To explain this claim, we must turn to a consideration of rational choice. We would have to do so anyhow. One of the demands that can fairly be made of those who propose theories rival to bayesianism is that they furnish answers not only to the problems of rational credence and revision but to the questions about rational choice. Furthermore, the motivation for requiring credal states to be non-empty, convex sets of probability measures and the explanation of the notion of a permissible Q-function are best understood within the context of an account of rational choice. For all these reasons, therefore, it is time to discuss rational choice.

VII

Consider, once more, a situation where X faces a decision problem of the type described in section "I". No longer, however, will it be supposed that X's credal state for the "states of nature" $h_1, h_2, \ldots, h_n$ and for the possible consequences o_{i1} $o_{i2}, \ldots, o_{im}$ conditional on X choosing A_i are representable by a single Q-function. Instead, the credal state will be required only to be a nonempty convex set of Q-functions.[12]

[11]The difference between my approach and Smith's was drawn to my attention by Howard Stein. To all intents and purposes, both Dempster and Smith represent credal states by the largest convex sets that generate the interval-valued functions characterizing those credal states. Dempster (332/3) is actually more restrictive than Smith. Dempster, by the way, wrongly attributes to Smith the position I adopt. To my knowledge, Dempster is the first to consider this position in print—even if only to misattribute it to Smith.

[12]As in section "I", I am supposing that "states of nature" are "independent" of options in the sense that, for every permissible Q-function, $Q(h_j) = Q(o_{ij};A_i)$. I have done this to facilitate the exposition. No question of fundamental importance is, in my opinion, thereby seriously altered.

Although I have not focused attention here on the dubiety of requiring X's evaluations of the o_{ij}s to be representable by a utility function unique up to a linear transformation, I do believe that rational men can have indeterminate preferences and will, for the sake of generality, relax the bayesian requirement as follows: X's system of evaluations of the possible consequences of the feasible options is to be represented by a set G of "permissible" u-functions defined over the o_{ij}s which is (a) nonempty, (b) convex, and such that all linear transformations of u-functions in G are also in G. A bayesian G is, in effect, such that all u-functions in it are linear transformations of one another. It is this latter requirement that I am abandoning.

In those situations where X satisfies strict bayesian conditions so that his credal state contains only a single Q-function and G contains all and only those u-functions which are linear transformations of some specific u-function u_1, an admissible option A_i is, according to the principle of maximizing expected utility, an option that bears maximum expected utility $E(A_i) = \sum_{i=1}^{m} Q(h_i)\, u_1\left(o_{ij}\right)$. Notice that, if any linear transformation of u_1 is substituted for u_1 in the computation of expected utility, the ranking of options with respect to expected utility remains unaltered. Hence we can say that, according to strict bayesians, an option is admissible if it bears maximum expected utility relative to the uniquely permissible Q-function and to any of the permissible u-functions in G (all of which are linear transformations of u_1).

There is an obvious generalization of this idea applicable to situations where $B_{X,t}$ contains more than one permissible Q-function and G contains u-functions that are not linear transformations of one another. I shall say that A_i is *E-admissible* if and only if there is at least one Q-function in $B_{X,t}$ and one u-function in G such that $E(A_i)$ defined relative to that Q-function and u-function is a maximum among all the feasible options. The generalization I propose is the following:

E-admissibility: All admissible options are E-admissible.

The principle of E-admissibility is by no means novel. I. J. Good, for example, endorsed it at one time. Indeed, Good went further than this. He endorsed the converse principle that all E-admissible options are admissible as well.[13]

I disagree with Good's view on this. When X's credal state and goals select more than one option as E-admissible, there may be and sometimes are other considerations than E-admissibility which X, as a rational agent, should employ in choosing between them.

There are occasions where X identifies two or more options as E-admissible and where, in addition, he has the opportunity to defer decision between them. If that opportunity is itself E-admissible, he should as a rational agent "keep his options open." Notice that in making this claim I am not saying that the option of deferring choice between the other E-admissible options is "better" than the

[13] "Rational Decisions," *Journal of the Royal Statistical Society*, Ser. B, XIV (1952): 114.

other E-admissible options relative to X's credence and values and the assessments of expected utility based thereon. In general, E-admissible options will not be comparable with respect to expected utility (although sometimes they will be). The injunction to keep one's options open is a criterion of choice that is based not on appraisals of expected utility but on the "option-preserving" features of options. Deferring choice is better than the other E-admissible options in this respect, but not with respect to expected utility.

Thus, a *P-admissible* option is an option that is (a) E-admissible and (b) "best" with respect to E-admissible option preservation among all E-admissible options. I shall not attempt to provide an adequate explication of clause (b) here. In the subsequent discussion, I shall consider situations where there are no opportunities to defer choice. Nonetheless, it is important to notice that, given a suitably formulated surrogate for (b), the following principle holds:

P-admissibility: All admissible options are P-admissible.

My disagreement with Good goes still further than this; for I reject not only the converse of E-admissibility but that of P-admissibility as well.

To illustrate, consider a situation that satisfies strict bayesian requirements. X knows that a coin with a .5 chance of landing heads is to be tossed once. g is the hypothesis that the coin will land heads. Under the circumstances, we might say that X's credal state is such that all permissible Q-functions assign g the value $Q(g) = .5$. Suppose that X is offered a gamble on g where X gains a dollar if g is true and loses one if g is false. (I shall assume that X has neither a taste for nor an aversion to gambling and that, for such small sums, money is linear with utility). He has two options: to accept the gamble and to reject it. If he rejects it, he neither gains nor loses.

Under the circumstances described, the principle of maximizing expected utility may be invoked. It indicates that both options are optimal and, hence, in my terms E-admissible. Since there are no opportunities for delaying choice, both options (on a suitably formulated version of P-admissibility) become P-admissible.

Bayesians—and Good would agree with this—tend to hold that rational X is free to choose either way. Not only are both options E-admissible. They are both admissible. Yet, in my opinion, rational X should refuse the gamble. The reason is not that refusal is better in the sense that it has higher expected utility than accepting the gamble. The options come out equal on this kind of appraisal. Refusing the gamble is "better," however, with respect to the security against loss it furnishes X. If X refuses the gamble, he loses nothing. If he accepts the gamble, he might lose something. This appeal to security does not carry weight, in my opinion, when accepting the gamble bears higher expected utility than refusing it. However, in that absurdly hypothetical situation where they bear precisely the same expected utility, the question of security does become critical.

These considerations can be brought to bear on the more general situation where two or more options are E-admissible (even though they are not equal with respect to expected utility) and where the principle of P-admissibility does not weed out any options.

An S-admissible option (i.e., option admissible with respect to security) is an option that is P-admissible and such that there is a permissible u-function in G relative to which the minimum u-value assigned a possible consequence o_{ij} of option A_i is a maximum among all P-admissible options.[14]

S-admissibility: All admissible options are S-admissible.

I cannot think of additional criteria for admissibility which seem adequate. (But then I have no precise conditions of adequacy.) I think, perhaps, we should keep an open mind on this matter. Nonetheless, for the present, I shall tentatively assume that the converse of S-admissibility holds. This assumption will not alter the main course of the subsequent argument.

Even without detailed exploration of the ramifications of this decision theory, some of its main features are immediately apparent. It conforms to the strict bayesian injunction to maximize expected utility in those situations where X has a precise credal state and G contains u-functions that are all linear transformations of one another. In this sense, bayesian decision theory is a special case of mine.

Similarly, the proposed decision theory identifies situations where the well-known maximin criterion is applied legitimately. Customarily maximin is used to select that option from among all the *feasible* options which maximizes the minimum gain. This recommendation is legitimate, according to my theory, provided (1) G contains all and only u-functions that are linear transformations of one another, and (2) all feasible options are P-admissible. But even if condition (1) is satisfied, it can be the case that the maximin solution from among all the feasible options is not itself E-admissible and so cannot be considered to be S-admissible.

Finally, my proposal is able to discriminate between and cover a wider variety of situations where neither maximizing expected utility nor maximining can be invoked with much plausibility. Moreover, it does so with the aid of a unified system of criteria of rational credence and rational choice. Thus, it does offer answers to just those questions which Bayesian theory purports to solve. Move rover, it escapes the bayesian commitment to the dubious doctrines of necessitarianism or personalism.

[14]The possible consequences of a "mixed act" constructed by choosing between "pure options" A_i and A_j with the aid of a chance device with known chance probability of selecting one or the other option is the set of possible consequences of either A_i or A_j. Consequently, the security level of such a mixed option for a given u-function is the lowest of the security levels belonging to A_i and A_j. Thus, my conception of security levels for mixed acts differs from that employed by von Neumann and Morgenstern and by Wald in formulating maximin (or minimax) principles. For this reason, starting with a set of P-admissible pure options, one cannot increase the security level by forming mixtures of them. In any case, mixtures of E-admissible options are not always themselves E-admissible. I shall leave mixed options out of account in the subsequent discussion. See D. Luce and H. Raiffa, *Games and Decisions* (New York: Wiley, 1958), pp. 68–71, 74–76, 278–280.

VIII

Some elementary properties of credal states as nonempty convex sets will be illustrated and explained by applying the decision theory just outlined to simple gambling situations. Suppose X knows that a coin is to be tossed and has either a .4 or .6 chance of landing heads. g is the hypothesis that the coin will land heads. I shall suppose that X has neither a taste for nor an aversion to gambling and that X's values are such that G is a set of u-functions that are linear transformations of the monetary payoffs of the gambles to be considered.

Case 1: X is offered a gamble on a take-it-or-leave-it basis where he receives $S - P$ dollars if g is true and loses P dollars if g is false. (Both S and P are positive.)
Case 2: X is offered a gamble on a take-it-or-leave-it basis where he loses P dollars if g is true and receives $S - P$ dollars if g is false. (S and P have the same values as in case 1.)

h_1 is the hypothesis that the chance of heads is .4, and h_2 is the hypothesis that the chance of heads is .6. By the reasoning of page 455, every permissible Q-function in X's credal state should be such that $Q(g) = .4Q(h_1) + .6[1 - Q(h_1)]$.

According to strict bayesians, X should, therefore, adopt a credal state that selects a single such Q-function as permissible. This can be done by selecting a single value for $Q(h_1)$. If that value is r, $Q(g) = .4r + .6(1 - r) = .6 - .2r$.

Hence, the bayesian will find that accepting the case 1 gamble is uniquely admissible if and only if $Q(g) > P/S$, and will find accepting the case 2 gamble uniquely admissible if and only if $Q(g) > P/S$. (Otherwise rejecting the gamble for the appropriate case is uniquely admissible, assuming that ties in expected utility are settled in favor of rejection.) Hence, if P/S is less than .5, a bayesian must preclude the possibility of accepting the gamble being inadmissible both in case 1 and in case 2.

Suppose, however, that $Cr_{X,t}(h_1)$ takes a nondegenerate interval as a value. For simplicity, let that interval be [0, 1]. The set of permissible Q-values for g must be all values of $.6 - .2r$ where r takes any value from 0 to 1. Hence, $Cr_{X,t}(g) = [.4, .6]$.

Under these conditions, my proposal holds that, when P/S falls in the interval from .4 to .6, both options are E-admissible (and P-admissible) in case 1. The same is true in case 2. But in both case 1 and case 2, rejecting the gamble is uniquely S-admissible. Hence, in both cases, X should reject the gamble. *This is true even when P/S is less than .5*. In this case, my proposal allows a rational agent a system of choices that a strict bayesian would forbid. In adopting this position, I am following the analysis advocated by C. A. B. Smith for handling pairwise choices between accepting and rejecting gambles. Smith's procedure, in brief, is to characterize X's degree of credence for g by a pair of numbers (the "lower pignic probability" and the "upper pignic probability" for g) as follows: The lower pignic probability $\underline{s}$ represents the least upper bound of betting quotients P/S for which X is prepared to accept gambles on g for positive S. The upper pignic probability $\overline{s}$ for g is $1 - t$, where t is the least upper bound of betting quotients P/S for which X is prepared to accept gambles on $\sim g$ for positive S. Smith requires that $\underline{s} \leq \overline{s}$, but does not insist on equality as bayesians do. Given Smith's definitions of upper and lower pignic probabilities, it should be fairly clear that, in case 1 and case 2 where $Cr_{X,t}(g) = [.4, .6]$, Smith's analysis and mine coincide.[15]

Before leaving cases 1 and 2, it should be noted that, if X's credal state were empty, no option in case 1 would be admissible and no option in case 2 would be admissible either. If X is confronted with a case 1 predicament and an empty credal state, he would be constrained to act and yet as a rational agent enjoined not to act. The untenability of this result is to be blamed on adopting an empty credal state. Only when X's corpus is inconsistent, should a rational agent have an empty credal state. But, of course, if X finds his corpus inconsistent, he should contract to a consistent one.

Case 3: A_1 is accepting both the case 1 and the case 2 gamble jointly with a net payoff if g is true or false of $S - 2P$.

This is an example of decision making under certainty. Everyone agrees that if P is greater than $2S$ the gamble should be rejected; for it leads to certain loss. If P is less than $2S$ X should accept the gamble; for it leads to a certain gain. These results, by the way, are implied by the criteria proposed here as well as by the strict bayesian view.

Strict bayesians often defend requiring that Q-functions conform to the requirements of the calculus of probabilities by an appeal to the fact that, when credal states contain but a single Q-function, a necessary and sufficient condition for having credal states that do not license sure losses (Dutch books) is having a Q-function obeying the calculus of probabilities. The arguments also support the conclusion that, even when more than one Q-function is permissible according to a credal state, if all permissible Q-functions obey the coherence principle, no Dutch book can become E-admissible and, hence, admissible.

Case 4: B_1 is accepting the case 1 gamble, B_2 is accepting the case 2 gamble, and B_3 is rejecting both gambles.

[15]Smith, *op. cit*, pp. 3–5, 6–7. The agreement applies only to pairwise choices where one option is a gamble in which there are two possible payoffs and the other is refusing to gamble with 0 gain and 0 loss. In this kind of situation, it is clear that Smith endorses the principle of E-admissibility, but not its converse. However, in the later sections of his paper where Smith considers decision problems with three or more options or where the possible consequences of an option to be considered are greater than 2, Smith seems (but I am not clear about this) to endorse the converse of the principle of E-admissibility—counter to the analysis on the basis of which he defines lower and upper pignic probabilities. Thus, it seems to me that either Smith has contradicted himself or (as is more likely) he simply does not have a general theory of rational choice. The latter sections of the paper may then be read as interesting explorations of technical matters pertaining to the construction of such a theory, but not as actually advocating the converse of E-admissibility. At any rate, since it is the theory Smith propounds in the first part of his seminal essay which interests me, I shall interpret him in the subsequent discussion as having no general theory of rational choice beyond that governing the simple gambling situations just described.

Let the credal state be such that all values between 0 and 1 are permissible Q-values for h_1 and, hence, all values between .4 and .6 are permissible for g.

If P/S is greater than .6, B_3 is uniquely E-admissible and, hence, admissible. If P/S is less than .4, B_3 is E-inadmissible. The other two options are E-admissible and admissible.

If P/S is greater than or equal to .4 and less than .5, B_3 remains inadmissible and the other two admissible.

If P/S is greater than or equal to .5 and less than .6, all three options are E-admissible; but B_3 *is* uniquely S-admissible. Hence, B_3 should be chosen when P/S is greater than or equal to .5.

Three comments are worth making about these results.

(i) I am not sure what analysis Smith would propose of situations like case 4. At any rate, his theory does not seem to cover it (but see footnote 15).

(ii) When P/S is between .4 and .5, my theory recommends rejecting the gamble in case 1, rejecting the gamble in case 2, and yet recommends accepting one or the other of these gambles in case 4. This violates the so-called "principle of independence of irrelevant alternatives."[16]

(iii) If the convexity requirement for credal states were violated by removing as permissible values for g all values from $(S - P)/S$ to P/S, where P/S is greater than .5 and less than .6, but leaving all other values from .4 to .6, then—counter to the analysis given previously, B_3 would not be E-admissible in case 4. The peculiarity of that result is that B_1 is E-admissible because, for permissible Q-values from .6 down to P/S, it bears maximum expected utility, with B_3 a close second. B_2 is E-admissible because, for Q-values from .4 to $(S - P)/S$, B_2 is optimal, with B_3 again a close second. If the values between $(S-P)/S$ and P/S are also permissible, B_3 is E-admissible because it is optimal for those values. To eliminate such intermediate values and allow the surrounding values to retain their permissibility seems objectionable. Convexity guarantees against this.

Case 5: X is offered a gamble on a take-it-or-leave-it basis in which he wins 15 cents if f_1 is true, loses 30 cents if f_2 is true, and wins 40 cents if f_3 is true.

---

[16]See Luce and Raiffa, *op. cit.*, pp. 288/9. Because the analysis offered by Smith and me for cases 1 and 2 seems perfectly appropriate and the analysis for case 4 also appears impeccable, I conclude that there is something wrong with the principle of independence of irrelevant alternatives.

A hint as to the source of the trouble can be obtained by noting that if 'E-admissible' is substituted for 'optimal' in the various formulations of the principle cited by Luce and Raiffa, p. 289, the principle of independence of irrelevant alternatives stands. The principle fails because S-admissibility is used to supplement E-admissibility in weeding out options from the admissible set.

Mention should be made in passing that even when 'E-admissible' is substituted for 'optimal' in Axiom 9 of Luce and Raiffa, p. 292, the axiom is falsified. Thus, when $.5 \leq P/S \leq .6$ in case 4, all three options are E-admissible, yet some mixtures of B_1 and B_2 will not be.

Suppose X's corpus of knowledge contains the following information:

Situation a: X knows that the ratios of red, white, and blue balls in the urn are either (i) 1/8, 3/8, 4/8 respectively; (ii) 1/8, 4/8, 3/8; (iii) 2/8, 4/8, 2/8; or (iv) 4/8, 3/8, 1/8.

X's credal state for the f_is is determined by his credal state for the four hypotheses about the contents of the urn according to a more complex variant of the arguments used to obtain credence values for g in the first four cases. If we allow all Q-functions compatible with inductive logic of an objectivist kind to be permissible, X's credal state for the f_is is the convex set of all weighted averages of the four triples of ratios. $Cr_{X,t}(f_1) = (1/8, 4/8)$, $Cr_{X,t}(f_2) = (3/8, 4/8)$, and $Cr_{X,t}(f_3) = (1/8, 4/8)$. Both accepting and rejecting the gamble are E-admissible. Rejecting the gamble, however, is uniquely S-admissible. X should reject the gamble.

Situation b: X knows that the ratios of red, white, and blue balls is correctly described by (i), (ii), or (iv), but not by (iii). Calculation reveals that the interval-valued credence function is the same as in situation a. Yet it can be shown that accepting the gamble is uniquely E-admissible and, hence, admissible. X should accept the gamble.

Now we can imagine situations that are related as a and b are to one another except that the credal states do not reflect differences in statistical knowledge. Then, from the point of view of Dempster and Smith, the credal states would be indistinguishable. Because the set of permissible Q-distributions over the f_is would remain different for situations a and b, my view would recognize differences and recommend different choices. If the answer to the problem of rational choice proposed here is acceptable, the capacity of the account of credal rationality to make fine distinctions is a virtue rather than a gratuitous piece of pedantry.

The point has its ramifications for an account of the improvement of confirmational commitments; the variety of discriminations that can be made between confirmational commitments generates a variety of potential shifts in confirmational commitments subject to critical review. For intervalists, a shift from situation a to b is no shift at all. On the view proposed here, it is significant.

The examples used in this section may be used to illustrate one final point. The objective or statistical or chance probability distributions figuring in chance statements can be viewed as assumptions or hypotheses. Probabilities in this sense can be unknown. We can talk of a set of simple or precise chance distributions among which X suspends judgment. Such *possible* probability distributions represent hypotheses which are possibly true and which are themselves objects of appraisal with respect to credal probability. *Permissible* probability distributions which, in our examples, are defined over such *possible* probability distributions (like the hypotheses h_1 and h_2 of cases 1, 2, 3, and 4) are not themselves possibly true hypotheses. No probability distributions of a still higher type can be denned over them.[17]

[17]I mention this because I. J. Good, whose seminal ideas have been an important influence on the proposals made in this essay, confuses permissible with possible probabilities. As a consequence, he introduces a hierarchy of types of probability (Good, *op. cit.*, p. 327). For criticism of such

I have scratched the surface of some of the questions raised by the proposals made in this essay. Much more needs to be done. I do believe, however, that these proposals offer fertile soil for cultivation not only by statisticians and decision theorists but by philosophers interested in what, in my opinion, ought to be the main problem for epistemology—to wit, the improvement (and, hence, revision) of human knowledge and belief.

Acknowledgements Work on this essay was partially supported by N.S.F. grant GS 28992. Research was carried out while I was a Visiting Scholar at Leckhampton, Corpus Christi, Cambridge. I wish to thank the Fellows of Corpus Christi College and the Departments of Philosophy and History and Philosophy of Science, Cambridge University, for their kind hospitality. I am indebted to Howard Stein for his help in formulating and establishing some of the results reported here. Sidney Morgenbesser, Ernest Nagel, Teddy Seidenfeld, and Frederic Schick as well as Stein have made helpful suggestions.

---

views, see Savage, *The Foundations of Statistics* (New York: Wiley, 1954), p. 58. In fairness to Good, it should be mentioned that his possible credal probabilities are interpreted not as possibly true statistical hypotheses but as hypotheses entertained by X about his own unknown strictly bayesian credal state. Good is concerned with situations where strict bayesian agents having precise probability judgments cannot identify their credal states before decision and must make choices on the basis of partial information about themselves. [In *Decision and Value* (New York: Wiley, 1964), P. G. Fishburn devotes himself to the same question.] My proposals do not deal with this problem. I reject Good's and Fishburn's view that every rational agent is at bottom a strict bayesian limited only by his lack of self-knowledge, computational facility, and memory. To the contrary, I claim that, even without such limitations, rational agents should not have precise bayesian credal states. The difference in problem under consideration and presuppositions about rational agents has substantial technical ramifications which cannot be developed here.

Chapter 8
Why I am not a Bayesian

Clark Glymour

The aim of confirmation theory is to provide a true account of the principles that guide scientific argument in so far as that argument is not, and does not purport to be, of a deductive kind. A confirmation theory should serve as a critical and explanatory instrument quite as much as do theories of deductive inference. Any successful confirmation theory should, for example, reveal the structure and fallacies, if any, in Newton's argument for universal gravitation, in nineteenth-century arguments for and against the atomic theory, in Freud's arguments for psychoanalytic generalizations. Where scientific judgements are widely shared, and sociological factors cannot explain their ubiquity, and analysis through the lens provided by confirmation theory reveals no good explicit arguments for the judgements, confirmation theory ought at least sometimes to suggest some good arguments that may have been lurking misperceived. Theories of deductive inference do that much for scientific reasoning in so far as that reasoning is supposed to be demonstrative. We can apply quantification theory to assess the validity of scientific arguments, and although we must almost always treat such arguments as enthymematic, the premisses we interpolate are not arbitrary; in many cases, as when the same subject-matter is under discussion, there is a common set of suppressed premisses. Again, there may be differences about the correct logical form of scientific claims; differences of this kind result in (or from) different formalizations, for example, of classical mechanics. But such differences often make no difference for the assessment of validity in actual arguments. Confirmation theory should do as well in its own domain. If it fails, then it may still be of interest for many purposes, but not for the purpose of understanding scientific reasoning.

Who cares whether a pig-farmer is a Bayesian?—R. C. Jeffrey.

C. Glymour (✉)
Carnegie Mellon University, Pittsburgh, PA, USA
e-mail: cg09@andrew.cmu.edu

H. Arló-Costa et al. (eds.), *Readings in Formal Epistemology*, Springer Graduate Texts in Philosophy 1, DOI 10.1007/978-3-319-20451-2_8

The aim of confirmation theory ought not to be simply to provide precise replacements for informal methodological notions, that is, explications of them. It ought to do more; in particular, confirmation theory ought to *explain* both methodological truisms and particular judgements that have occurred within the history of science. By 'explain' I mean at least that confirmation theory ought to provide a rationale for methodological truisms, and ought to reveal some systematic connections among them and, further, ought, without arbitrary or question-begging assumptions, to reveal particular historical judgements as in conformity with its principles.

Almost everyone interested in confirmation theory today believes that confirmation relations ought to be analysed in terms of *probability* relations. Confirmation theory is the theory of probability plus introductions and appendices. Moreover, almost everyone believes that confirmation proceeds through the formation of conditional probabilities of hypotheses on evidence. The basic tasks facing confirmation theory are thus just those of explicating and showing how to determine the probabilities that confirmation involves, developing explications of such meta-scientific notions as 'confirmation', 'explanatory power', 'simplicity', and so on in terms of functions of probabilities and conditional probabilities, and showing that the canons and patterns of scientific inference result. It was not always so. Probabilistic accounts of confirmation really became dominant only after the publication of Carnap's *Logical Foundations of Probability* (1950), although of course many probabilistic accounts had preceded Carnap's. An eminent contemporary philosopher (Putnam 1967) has compared Carnap's achievement in inductive logic with Frege's in deductive logic: just as before Frege there was only a small and theoretically uninteresting collection of principles of deductive inference, but after him the foundation of a systematic and profound theory of demonstrative reasoning, so with Carnap and inductive reasoning. After Carnap's *Logical Foundations*, debates over confirmation theory seem to have focused chiefly on the interpretation of probability and on the appropriate probabilistic explications of various meta-scientific notions. The meta-scientific notions remain controversial, as does the interpretation of probability, although, increasingly, logical interpretations of probability are giving way to the doctrine that probability is degree of belief.[1] In very recent years a few philosophers have attempted to apply probabilistic analyses to derive and to explain particular methodological practices and precepts, and even to elucidate some historical cases.

I believe these efforts, ingenious and admirable as many of them are, are none the less misguided. For one thing, probabilistic analyses remain at too great a distance from the history of scientific practice to be really informative about that practice, and in part they do so exactly because they are probabilistic. Although considerations of probability have played an important part in the history of science, until very recently, explicit probabilistic arguments for the confirmation of various

[1]A third view, that probabilities are to be understood exclusively as frequencies, has been most ably defended by Wesley Salmon (1969).

theories, or probabilistic analyses of data, have been great rarities in the history of science. In the physical sciences at any rate, probabilistic arguments have rarely occurred. Copernicus, Newton, Kepler, none of them give probabilistic arguments for their theories; nor does Maxwell or Kelvin or Lavoisier or Dalton or Einstein or Schrödinger or... There are exceptions. Jon Dorling has discussed a seventeenth-century Ptolemaic astronomer who apparently made an extended comparison of Ptolemaic and Copernican theories in probabilistic terms; Laplace, of course, gave Bayesian arguments for astronomical theories. And there are people—Maxwell, for example—who scarcely give a probabilistic argument when making a case for or against scientific hypotheses but who discuss *methodology* in probabilistic terms. This is not to deny that there are many areas of contemporary physical science where probability figures large in confirmation; regression analysis is not uncommon in discussions of the origins of cosmic rays, correlation and analysis of variance in experimental searches for gravitational waves, and so on. It *is* to say that, explicitly, probability is a distinctly minor note in the history of scientific argument.

The rarity of probability considerations in the history of science is more an embarrassment for some accounts of probability than for others. Logical theories, whether Carnap's or those developed by Hintikka and his students, seem to lie at a great distance from the history of science. Still, some of the people working in this tradition have made interesting steps towards accounting for methodological truisms. My own inclination is to believe that the interest such investigations have stems more from the insights they obtain into syntactic versions of structural connections among evidence and hypotheses than to the probability measures they mesh with these insights. Frequency interpretations suppose that for each hypothesis to be assessed there is an appropriate reference class of hypotheses to which to assign it, and the prior probability of the hypothesis is the frequency of true hypotheses in this reference class. The same is true for statements of evidence, whether they be singular or general. The matter of how such reference classes are to be determined, and determined so that the frequencies involved do not come out to be zero, is a question that has only been touched upon by frequentist writers. More to the point, for many of the suggested features that might determine reference classes, we have no statistics, and cannot plausibly imagine those who figure in the history of our sciences to have had them. So conceived, the history of scientific argument must turn out to be largely a history of fanciful guesses. Further, some of the properties that seem natural candidates for determining reference classes for hypotheses—simplicity, for example—seem likely to give perverse results. We prefer hypotheses that posit simple relations among observed quantities, and so on a frequentist view should give them high prior probabilities. Yet simple hypotheses, although often very useful approximations, have most often turned out to be literally false.

At present, perhaps the most philosophically influential view of probability understands it to be degree of belief. The subjectivist Bayesian (hereafter, for brevity, simply Bayesian) view of probability has a growing number of advocates who understand it to provide a general framework for understanding scientific reasoning. They are singularly unembarrassed by the rarity of explicit probabilistic arguments in the history of science, for scientific reasoning need not be explicitly probabilistic in order to be probabilistic in the Bayesian sense. Indeed, a number of Bayesians have discussed historical cases within their framework. Because of its influence and its apparent applicability, in what follows it is to the subjective Bayesian account that I shall give my full attention.

My thesis is several-fold. First, there are a number of attempts to demonstrate a priori the rationality of the restrictions on belief and inference that Bayesians advocate. These arguments are altogether admirable, but ought, I shall maintain, to be unconvincing. My thesis in this instance is not a new one, and I think many Bayesians do regard these a priori arguments as insufficient. Second, there are a variety of methodological notions that an account of confirmation ought to explicate and methodological truisms involving these notions that a confirmation theory ought to explain: for example, variety of evidence and why we desire it, *ad hoc* hypotheses and why we eschew them, what separates a hypothesis integral to a theory from one 'tacked on' to the theory, simplicity and why it is so often admired, why 'de-Occamized' theories are so often disdained, what determines when a piece of evidence is relevant to a hypothesis, and what, if anything, makes the confirmation of one bit of theory by one bit of evidence stronger than the confirmation of another bit of theory (or possibly the same bit) by another (or possibly the same) bit of evidence. Although there are plausible Bayesian explications of some of these notions, there are not plausible Bayesian explications of others. Bayesian accounts of methodological truisms and of particular historical cases are of one of two kinds: either they depend on general principles restricting prior probabilities, or they don't. My claim is that many of the principles proposed by the first kind of Bayesian are either implausible or incoherent, and that, for want of such principles, the explanations the second kind of Bayesians provide for particular historical cases and for truisms of method are chimeras. Finally, I claim that there are elementary but perfectly common features of the relation of theory and evidence that the Bayesian scheme cannot capture at all without serious—and perhaps not very plausible—revision.

It is not that I think the Bayesian scheme or related probabilistic accounts capture nothing. On the contrary, they are clearly pertinent where the reasoning involved is explicitly statistical. Further, the accounts developed by Carnap, his predecessors, and his successors are impressive systematizations and generalizations, in a probabilistic framework, of certain principles of ordinary reasoning. But so far as understanding scientific reasoning goes, I think it is very wrong to consider our situation to be analogous to that of post-Fregean logicians, our subject-matter transformed from a hotchpotch of principles by a powerful theory whose outlines are clear. We flatter ourselves that we possess even the hotchpotch. My opinions are outlandish, I know; few of the arguments I shall present in their favour are new, and perhaps none of them is decisive. Even so, they seem sufficient to warrant taking seriously entirely different approaches to the analysis of scientific reasoning.

The theories I shall consider share the following framework, more or less. There is a class of sentences that express all hypotheses and all actual or possible evidence of interest; the class is closed under Boolean operations. For each ideally rational agent, there is a function defined on all sentences such that, under the relation of logical equivalence, the function is a probability measure on the collection of equivalence classes. The probability of any proposition represents the agent's degree of belief in that proposition. As new evidence accumulates, the probability of a proposition changes according to Bayes's rule: the posterior probability of a hypothesis on the new evidence is equal to the prior conditional probability of the hypothesis on the evidence. This is a scheme shared by diverse accounts of confirmation. I call such theories 'Bayesian', or sometimes 'personalist'.

We certainly have *grades* of belief. Some claims I more or less believe, some I find plausible and tend to believe, others I am agnostic about, some I find implausible and far-fetched, still others I regard as positively absurd. I think everyone admits some such gradations, although descriptions of them might be finer or cruder. The personalist school of probability theorists claim that we also have *degrees* of belief, degrees that can have any value between 0 and 1 and that ought, if we are rational, to be representable by a probability function. Presumably, the degrees of belief are to co-vary with everyday gradations of belief, so that one regards a proposition as preposterous and absurd just if his degree of belief in it is somewhere near zero, and he is agnostic just if his degree of belief is somewhere near a half, and so on. According to personalists, then, an ideally rational agent always has his degrees of belief distributed so as to satisfy the axioms of probability, and when he comes to accept a new belief, he also forms new *degrees* of belief by conditionalizing on the newly accepted belief. There are any number of refinements, of course; but that is the basic view.

Why should we think that we really do have *degrees* of belief? Personalists have an ingenious answer: people have them because we can measure the degrees of belief that people have. Assume that no one (rational) will accept a wager on which he expects a loss, but anyone (rational) will accept any wager on which he expects a gain. Then we can measure a person's degree of belief in proposition P by finding, for fixed amount v, the highest amount u such that the person will pay u in order to receive $u + v$ if P is true, but receive nothing if P is not true. If u is the greatest amount the agent is willing to pay for the wager, his expected gain on paying u must be zero. The agent's gain if P is the case is v; his gain if P is not the case is $-u$. Thus

$$v \cdot \text{prob}(P) + (-u) \cdot \text{prob}(\sim P) = 0.$$

Since prob $(\sim P) = 1 - \text{prob}(P)$, we have

$$\text{prob}(P) = u/\,(u + v)\,.$$

The reasoning is clear: any sensible person will act so as to maximize his expected gain; thus, presented with a decision whether or not to purchase a bet, he will make the purchase just if his expected gain is greater than zero. So the betting odds he will accept determine his degree of belief.[2]

I think that this device really does provide evidence that we have, or can produce, degrees of belief, in at least some propositions, but at the same time it is evident that betting odds are not an unobjectionable device for the measurement of degrees of belief. Betting odds could fail to measure degrees of belief for a variety of reasons: the subject may not believe that the bet will be paid off if he wins, or he may doubt that it is clear what constitutes winning, even though it is clear what constitutes losing. Things he values other than monetary gain (or whatever) may enter into his determination of the expected utility of purchasing the bet: for example, he may place either a positive or a negative value on risk itself. And the very fact that he is offered a wager on P may somehow change his degree of belief in P.

Let us suppose, then, that we do have degrees of belief in at least some propositions, and that in some cases they can be at least approximately measured on an interval from 0 to 1. There are two questions: why should we think that, for rationality, one's degrees of belief must satisfy the axioms of probability, and why should we think that, again for rationality, changes in degrees of belief ought to proceed by conditionalization? One question at a time. In using betting quotients to measure degrees of belief, it was assumed that the subject would act so as to maximize *expected* gain. The betting quotient determined the degree of belief by determining the coefficient by which the gain is multiplied in case that P is true in the expression for the expected gain. So the betting quotient determines a degree of belief, as it were, in the *role* of a probability. But why should the things, degrees of belief, that play this role be probabilities? Supposing that we do choose those actions that maximize the sum of the product of our degrees of belief in each possible outcome of the action and the gain (or loss) to us of that outcome. Why must the degrees of belief that enter into this sum be probabilities? Again, there is an ingenious argument: if one acts so as to maximize his expected gain using a degree-of-belief function that is not a probability function, and if for every proposition there were a possible wager (which, if it is offered, one believes will be paid off if it is accepted and won), then there is a circumstance, a combination of wagers, that one would enter into if they were offered, and in which one would suffer a net loss whatever the outcome. That is what the Dutch-book argument shows; what it counsels is prudence.

Some of the reasons why it is not clear that betting quotients are accurate measures of degrees of belief are also reasons why the Dutch-book argument is not conclusive: there are many cases of propositions in which we may have degrees of belief, but on which, we may be sure, no acceptable wager will be offered us;

[2]More detailed accounts of means for determining degrees of belief may be found in Jeffrey (1965). It is a curious fact that the procedures that Bayesians use for determining subjective degrees of belief empirically are an instance of the general strategy described in Glymour 1981, ch. 5. Indeed, the strategy typically used to determine whether or not actual people behave as rational Bayesians involves the bootstrap strategy described in that chapter.

again, we may have values other than the value we place on the stakes, and these other values may enter into our determination whether or not to gamble; and we may not have adopted the policy of acting so as to maximize our expected gain or our expected utility: that is, we may save ourselves from having book made against us by refusing to make certain wagers, or combinations of wagers, even though we judge the odds to be in our favour.

The Dutch-book argument does not succeed in showing that in order to avoid absurd commitments, or even the possibility of such commitments, one must have degrees of belief that are probabilities. But it does provide a kind of justification for the personalist viewpoint, for it shows that if one's degrees of belief are probabilities, then a certain kind of absurdity is avoided. There are other ways of avoiding that kind of absurdity, but at least the personalist way is one such.[3]

One of the common objections to Bayesian theory is that it fails to provide any connection between what is inferred and what is the case. The Bayesian reply is that the method guarantees that, in the long run, everyone will agree on the truth. Suppose that B_i are a set of mutually exclusive, jointly exhaustive hypotheses, each with probability $B(i)$. Let $\bar{x}_r$ be a sequence of random variables with a finite set of values and conditional distribution given by $P\left(\bar{x}_r = x_r \middle| B_i\right) = \varepsilon\left(x_r \middle| B_i\right)$; then we can think of the values x_r as the outcomes of experiments, each hypothesis determining a likelihood for each outcome. Suppose that no two hypotheses have the same likelihood distribution; that is, for $i \neq j$ it is not the case that for all values x_r of $\bar{x}_r$, $\varepsilon(x_r|B_i) = \varepsilon(x_r|B_j)$, where the ε's are defined as above. Let $\bar{x}$ denote the first n of these variables, where x is a value of $\bar{x}$. Now imagine an observation of these n random variables. In Savage's words:

Before the observation, the probability that the probability given x of whichever element of the partition actually obtains will be greater than α is

$$\sum_i B(i)P\left(P\left(B_i \middle| x\right) > \alpha \middle| B_i\right),$$

where summation is confined to those i's for which $B(i) \neq 0$. (1972: 49)

In the limit as n approaches infinity, the probability that the probability given x of whichever element of the partition actually obtains is greater than α is 1. That is the theorem. What is its significance? According to Savage, 'With the observation of an abundance of relevant data, the person is almost certain to become highly convinced of the truth, and it has also been shown that he himself knows this to be the case' (p. 50). That is a little misleading. The result involves second-order probabilities, but these too, according to personalists, are degrees of belief. So what has been shown seems to be this: in the limit as n approaches infinity, an ideally rational Bayesian has degree of belief 1 that an ideally rational Bayesian (with degrees of belief as in the theorem) has degree of belief, given x, greater than α

[3]For further criticisms of the Dutch-book argument see Kyburg 1978.

in whichever element of the partition actually obtains. The theorem does not tell us that in the limit any rational Bayesian will assign probability 1 to the true hypothesis and probability 0 to the rest; it only tells us that rational Bayesians are certain that he will. It may reassure those who are already Bayesians, but it is hardly grounds for conversion. Even the reassurance is slim. Mary Hesse points out (1974: 117–19), entirely correctly I believe, that the assumptions of the theorem do not seem to apply even approximately in actual scientific contexts. Finally, some of the assumptions of stable estimation theorems can be dispensed with if one assumes instead that all of the initial distributions considered must agree regarding which evidence is relevant to which hypotheses. But there is no evident a priori reason why there should be such agreement.

I think relatively few Bayesians are actually persuaded of the correctness of Bayesian doctrine by Dutch-book arguments, stable estimation theorems, or other a priori arguments. Their frailty is too palpable. I think that the appeal of Bayesian doctrine derives from two other features. First, with only very weak or very natural assumptions about prior probabilities, or none at all, the Bayesian scheme generates principles that seem to accord well with common sense. Thus, with minor restrictions, one obtains the principle that hypotheses are confirmed by positive instances of them; and, again, one obtains the result that if an event that actually occurs is, on some hypothesis, very unlikely to occur, then that occurrence renders the hypothesis less likely than it would otherwise have been. These principles, and others, can claim something like the authority of common sense, and Bayesian doctrine provides a systematic explication of them. Second, the restrictions placed a priori on rational degrees of belief are so mild, and the device of probability theory at once so precise and so flexible, that Bayesian philosophers of science may reasonably hope to explain the subtleties and vagaries of scientific reasoning and inference by applying their scheme together with plausible assumptions about the distribution of degrees of belief. This seems, for instance, to be Professor Hesse's line of argument. After admitting the insufficiency of the standard arguments for Bayesianism, she sets out to show that the view can account for a host of alleged features of scientific reasoning and inference. My own view is different: particular *inferences* can almost always be brought into accord with the Bayesian scheme by assigning degrees of belief more or less *ad hoc*, but we learn nothing from this agreement. What we want is an explanation of scientific argument; what the Bayesians give us is a theory of learning—indeed, a theory of personal learning. But arguments are more or less impersonal; I make an argument to persuade anyone informed of the premisses, and in doing so I am not reporting any bit of autobiography. To ascribe to me degrees of belief that make my slide from my premisses to my conclusion a plausible one fails to explain anything, not only because the ascription may be arbitrary, but also because, even if it is a correct assignment of my degrees of belief, it does not explain why what I am doing is *arguing*—why, that is, what I say should have the least influence on others, or why I might hope that it should. Now, Bayesians might bridge the gap between personal inference and argument in either of two ways. In the first place, one might give arguments in order to change others' beliefs because of the respect they have for his opinion. This is not very plausible; if that were the point of giving arguments, one would not bother with them, but would simply state one's opinion. Alternatively, and more hopefully, Bayesians may suggest that we give arguments exactly because there are general principles restricting belief, principles that are widely subscribed to, and in giving arguments we are attempting to show that, supposing our audience has certain beliefs, they must in view of these principles have other beliefs, those we are trying to establish. There is nothing controversial about this suggestion, and I endorse it. What is controversial is that the general principles required for argument can best be understood as conditions restricting prior probabilities in a Bayesian framework. Sometimes they can, perhaps; but I think that when arguments turn on relating evidence to theory, it is very difficult to explicate them in a plausible way within the Bayesian framework. At any rate, it is worth seeing in more detail what the difficulties may be.

There is very little Bayesian literature about the hotchpotch of claims and notions that are usually canonized as scientific method; very little seems to have been written, from a Bayesian point of view, about what makes a hypothesis *ad hoc*, about what makes one body of evidence more various than another body of evidence, and why we should prefer a variety of evidence, about why, in some circumstances, we should prefer simpler theories, and what it is that we are preferring when we do. And so on. There is little to nothing of this in Carnap, and more recent, and more personalist, statements of the Bayesian position are almost as disappointing. In a lengthy discussion of what he calls 'tempered personalism', Abner Shimony (1970) discusses only how his version of Bayesianism generalizes and qualifies hypothetico-deductive arguments. (Shimony does discuss simplicity, but only to argue that it is overvalued.) Mary Hesse devotes the later chapters of her book to an attempt to show that certain features of scientific method do result when the Bayesian scheme is supplemented with a postulate that restricts assignments of prior probabilities. Unfortunately, as we shall see, her restrictive principle is incoherent.[4]

One aspect of the demand for a variety of evidence arises when there is some definite set of alternative hypotheses between which we are trying to decide. In such cases we naturally prefer the body of evidence that will be most helpful in eliminating false competitors. This aspect of variety is an easy and natural one for Bayesians to take account of, and within an account such as Shimony's it is taken care of so directly as hardly to require comment. But there is more to variety. In some situations we have some reason to suspect that if a theory is false, its falsity will show up when evidence of certain kinds is obtained and compared. For example, given the tradition of Aristotelian distinctions, there was some reason to demand both terrestrial and celestial evidence for seventeenth-century theories of motion that subjected all matter to the same dynamical laws. Once again, I see no special reason why this kind of demand for a variety of evidence cannot be fitted into the Bayesian scheme. But there is still more. A complex theory may contain

[4]Moreover, I believe that much of her discussion of methodological principles has only the loosest relation to Bayesian principles.

a great many logically independent hypotheses, and particular bodies of evidence may provide grounds for some of those hypotheses but not for others. Surely part of the demand for a variety of evidence, and an important part, derives from a desire to see to it that the various independent parts of our theories are tested. Taking account of this aspect of the demand for a variety of evidence is just taking account of the relevance of evidence to pieces of theory. How Bayesians may do this we shall consider later.

Simplicity is another feature of scientific method for which some Bayesians have attempted to account. There is one aspect of the scientific preference for the simple that seems beyond Bayesian capacities, and that is the disdain for 'de-Occamized' hypotheses, for theories that postulate the operation of a number of properties, determinable only in combination, when a single property would do. Such theories can be generated by taking any ordinary theory and replacing some single quantity, wherever it occurs in the statement of the theory, by an algebraic combination of new quantities. If the original quantity was not one that occurs in the statement of some body of evidence for the theory, then the new, de-Occamized theory will have the same entailment relations with that body of evidence as did the original theory. If the old theory entailed the evidence, so will the new, de-Occamized one. Now, it follows from Bayesian principles that if two theories both entail e, then (provided the prior probability of each hypothesis is neither 1 nor 0), if e confirms one of them, it confirms the other. How then is the fact (for so I take it to be) that pieces of evidence just don't seem to *count* for de-Occamized theories to be explained? Not by supposing that de-Occamized theories have lower prior probabilities than un-de-Occamized theories, for being 'de-Occamized' is a feature that a theory has only with respect to a certain body of evidence, and it is not hard to imagine artificially restricted bodies of evidence with respect to which perfectly good theories might count as de-Occamized. Having extra wheels is a feature a theory has only in relation to a body of evidence; the only Bayesian relation that appears available and relevant to scientific preference is the likelihood of the evidence on the theory, and unfortunately the likelihood is the same for a theory and for its de-Occamized counterparts whenever the theory entails the evidence.

It is common practice in fitting curves to experimental data, in the absence of an established theory relating the quantities measured, to choose the 'simplest' curve that will fit the data. Thus linear relations are preferred to polynomial relations of higher degree, and exponential functions of measured quantities are preferred to exponential functions of algebraic combinations of measured quantities, and so on. The problem is to account for this preference. Harold Jeffreys, a Bayesian of sorts, offered an explanation (1979) along the following lines. Algebraic and differential equations may be ordered by simplicity; the simpler the hypothetical relation between two or more quantities, the greater is its prior probability. If measurement error has a known probability distribution, we can then compute the likelihood of any set of measurement results given an equation relating the measured quantities. It should be clear, then, that with these priors and likelihoods, ratios of posterior probabilities may be computed from measurement results. Jeffreys constructed a Bayesian significance test for the introduction of higher-degree terms in the equation relating the measured quantities. Roughly, if one's equation fits the data *too* well, then the equation has too many terms and too many arbitrary parameters; and if the equation does not fit the data well enough, then one has not included enough terms and parameters in the equation. The whole business depends, of course, entirely on the ordering of prior probabilities. In his *Theory of Probability* Jeffreys (1967) proposed that the prior probability of a hypothesis decreases as the number of arbitrary parameters increases, but hypotheses having the same number of arbitrary parameters have the same prior probability. This leads immediately to the conclusion that the prior probability of every hypothesis is zero. Earlier, Jeffreys proposed a slightly more complex assignment of priors that did not suffer from this difficulty. The problem is not really one of finding a way to assign finite probabilities to an infinite number of incompatible hypotheses, for there are plenty of ways to do that. The trouble is that it is just very implausible that scientists typically have their prior degrees of belief distributed according to any plausible simplicity ordering, and still less plausible that they would be rational to do so. I can think of very few simple relations between experimentally determined quantities that have with-stood continued investigation, and often simple relations are replaced by relations that are infinitely complex: consider the fate of Kepler's laws. Surely it would be naïve for anyone to suppose that a set of newly measured quantities will truly stand in a simple relation, especially in the absence of a well-confirmed theory of the matter. Jeffreys' strategy requires that we proceed in ignorance of our scientific experience, and that can hardly be a rational requirement (Jeffreys 1973).

Consider another Bayesian attempt, this one due to Mary Hesse. Hesse puts a 'clustering' constraint on prior probabilities: for any positive r, the conjunction of $r+1$ positive instances of a hypothesis is more probable than a conjunction of r positive instances with one negative instance. This postulate, she claims, will lead us to choose, *ceteris, paribus*, the most economical, the simplest, hypotheses compatible with the evidence. Here is the argument:

Consider first evidence consisting of individuals $a_1, a_2, \ldots, a_n$, all of which have properties P and Q. Now consider an individual a_{n+1} with property P. Does a_{n+1} have Q or not? If nothing else is known, the clustering postulate will direct us to predict $Q_{\alpha+1}$ since, *ceteris paribus*, the universe is to be postulated to be as homogeneous as possible consistently with the data... But this is also the prediction that would be made by taking the most economical general law which is both confirmed by the data and of sufficient content to make a prediction about the application of Q to a_{n+1}. For $h=$ 'All P are Q' is certainly more economical than the 'gruified' conflicting hypothesis of equal content h': 'All x up to a_n that are P are Q, and all other x that are P are Q.'

If follows in the [case] considered that if a rule is adopted to choose the prediction resulting from the most probable hypothesis on grounds of content, or, in case of a tie in content, the most economical hypothesis on those of equal content, this rule will yield the same predictions as the clustering postulate.

Here is the argument applied to curve-fitting:

Let f be the assertion that two data points $(x_1, y_1,), (x_2, y_2)$ are obtained from experiments... The two points are consistent with the hypothesis $y=a+bx$, and also of course with an indefinite number of other hypotheses of the form $y = a_0 + a_1 + \cdots + a_n x_1$, where the values of $a_0, \ldots, a_n$ are not determined by (x_1, y_1), (x_2, y_2). What is the most economical prediction of the y-value of a further point g, where the x-value of g is x_3? Clearly it is the prediction which uses only the information already contained in f, that is, the calculable values of a, b rather than a prediction which assigns arbitrary values to the parameters of a higher-order hypothesis. Hence the most economical prediction is about the point $g = (x_3, a + bx_3)$, which is also the prediction given by the 'simplest' hypothesis on almost all accounts of the simplicity of curves. Translated into probabilistic language, this is to say that to conform to intuitions about economy we should assign higher initial probability to the assertion that points $(x_1, a + bx_1)$, $(x_2, a + bx_2)$, $(x_3, a + bx_3)$ are satisfied by the experiment than to that in which the third point is inexpressible in terms of a and b alone. In this formulation economy is a function of finite descriptive lists of points rather than general hypotheses, and the relevant initial probability is that of a universe containing these particular points rather than that of a universe in which the corresponding general law is true... Description in terms of a minimum number of parameters may therefore be regarded as another aspect of homogeneity or clustering of the universe. (Hesse 1974: 230–2)

Hesse's clustering postulate applies directly to the curve-fitting case, for her clustering postulate then requires that if two paired values of x and y satisfy the predicate $y = ax + b$, then it is more probable than not that a third pair of values will satisfy the predicate. So the preference for the linear hypothesis in the next instance results from Hesse's clustering postulate and the probability axioms. Unfortunately, with trivial additional assumptions, everything results. For, surely, if $y = a + bx$ is a legitimate predicate, then so is $y = \alpha_1 + b_1 x^2$, for any definite values of a_1 and b_1. Now Hesse's first two data points can be equally well described by $(x_1, a_1 + b_1 x_1^2)$ and $(x_2, a_1 + b_1 x_2^2)$, where

$$b_1 = \frac{y_1 - y_2}{x_1^2 - x_2^2} \qquad a_1 = y_1 - x_1^2 \left(\frac{y_1 - y_2}{x_1^2 - x_2^2} \right),$$

Hence her first two data points satisfy both the predicate $y = a + bx$ and the predicate $y = a_1 + b_1 x^2$. So, by the clustering postulate, the probability that the third point satisfies the quadratic expression must be greater than one-half, and the probability that the third point satisfies the linear expression must also be greater than one-half, which is impossible.

Another Bayesian account of our preference for simple theories has recently been offered by Roger Rosencrantz (1976). Suppose that we have some criterion for 'goodness of fit' of a hypothesis to data—for example, confidence regions based on the χ^2 distribution for categorical data, or in curve-fitting perhaps that the average sum of squared deviations is less than some figure. Where the number of possible outcomes is finite, we can compare the number of such possible outcomes that meet the goodness-of-fit criterion with the number that do not. This ratio Rosencrantz calls the 'observed sample coverage' of the hypothesis. Where the possible outcomes are infinite, if the region of possible outcomes meeting the goodness-of-fit criterion is always bounded for all relevant hypotheses, we can compare the volumes of such regions for different hypotheses, and thus obtain a measure of comparative sample coverage.

It seems plausible enough that the smaller the observed sample coverage of a hypothesis, the more severely it is tested by observing outcomes. Rosencrantz's first proposal is this: the smaller the observed sample coverage, the simpler the hypothesis. But further, he proves the following for hypotheses about categorical data: if H_1 and H_2 are hypotheses with parameters, and H_1 is a special case of H_2 obtained by letting a free parameter in H_2 take its maximum likelihood value, then if we average the likelihood of getting evidence that fits each hypothesis well enough over all the possible parameter values, the average likelihood of H_1 will be greater than the average likelihood of H_2. The conclusion Rosencrantz suggests is that the simpler the theory, the greater the average likelihood of data that fit it sufficiently well. Hence, even if a simple theory has a lower prior probability than more complex theories, because the average likelihood is higher for the simple theory, its posterior probability will increase more rapidly than that of more complex theories. When sufficient evidence has accumulated, the simple theory will be preferred. Rosencrantz proposes to identify average likelihood with support.

Rosencrantz's approach has many virtues; I shall concentrate on its vices. First, observed sample coverage does not correlate neatly with simplicity. If H is a hypothesis, T another utterly irrelevant to H and to the phenomena about which H makes predictions, then H & T will have the same observed sample coverage as does H. Further, if H^* is a de-Occamization of H, then H^* and H will have the same observed sample coverage. Second, Rosencrantz's theorem does not establish nearly enough. It does not establish, for example, that in curve-fitting the average likelihood of a linear hypothesis is greater than the average likelihood of a quadratic or higher-degree hypothesis. We cannot explicate support in terms of average likelihood unless we are willing to allow that evidence supports a de-Occamized hypothesis as much as un-de-Occamized ones, and a hypothesis with tacked-on parts as much as one without such superfluous parts.

Finally, we come to the question of the relevance of evidence to theory. When does a piece of evidence confirm a hypothesis according to the Bayesian scheme of things? The natural answer is that it does so when the posterior probability of the hypothesis is greater than its prior probability, that is, if the conditional probability of the hypothesis on the evidence is greater than the probability of the hypothesis. That is what the condition of positive relevance requires, and that condition is the one most commonly advanced by philosophical Bayesians. The picture is a kinematic one: a Bayesian agent moves along in time having at each moment a coherent set of degrees of belief; at discrete intervals he learns new facts, and each time he learns a new fact, e, he revises his degrees of belief by conditionalizing on e. The discovery that e is the case has confirmed those hypotheses whose probability after the discovery is higher than their probability before. For several reasons, I think this account is unsatisfactory; moreover, I doubt that its difficulties are remediable without considerable changes in the theory.

The first difficulty is a familiar one. Let us suppose that we can divide the consequences of a theory into sentences consisting of reports of actual or possible observations, and simple generalizations of such observations, on the one hand; and on the other hand, sentences that are theoretical. Then the collection of 'observational' consequences of the theory will always be at least as probable as the theory itself; generally, the theory will be less probable than its observational consequences. A theory is never any better established than is the collection of its observational consequences. Why, then, should we entertain theories at all? On the probabilist view, it seems, they are a gratuitous risk. The natural answer is that theories have some special function that their collection of observational consequences cannot serve; the function most frequently suggested is explanation—theories explain; their collection of observational consequences do not. But however sage this suggestion may be, it only makes more vivid the difficulty of the Bayesian way of seeing things. For whatever explanatory power may be, we should certainly expect that goodness of explanation will go hand in hand with warrant for belief; yet, if theories explain, and their observational consequences do not, the Bayesian must deny the linkage. The difficulty has to do both with the assumption that rational degrees of belief are generated by probability measures and with the Bayesian account of evidential relevance. Making degrees of belief probability measures in the Bayesian way already guarantees that a theory can be no more credible than any collection of its consequences. The Bayesian account of confirmation makes it impossible for a piece of evidence to give us more total credence in a theory than in its observational consequences. The Bayesian way of setting things up is a natural one, but it is not inevitable, and wherever a distinction between theory and evidence is plausible, it leads to trouble.

A second difficulty has to do with how praise and blame are distributed among the hypotheses of a theory. Recall the case of Kepler's laws (discussed in Glymour 1981, ch. 2). It seems that observations of a single planet (and, of course, the sun) might provide evidence for or against Kepler's first law (all planets move on ellipses) and for or against Kepler's second law (all planets move according to the area rule), but no observations of a single planet would constitute evidence for or against Kepler's third law (for any two planets, the ratio of their periods equals the $^3/_2$ power of the ratio of their distances). Earlier [in Ch. 2 of Glymour's *Theory and Evidence*] we saw that hypothetico-deductive accounts of confirmation have great difficulty explaining this elementary judgement. Can the Bayesians do any better? One thing that Bayesians can say (and some have said) is that our degrees of belief are distributed—and historically were distributed—so that conditionalizing on evidence about one planet may change our degrees of belief in the first and second laws, but not our degree of belief in the third law.[5] I don't see that this is an explanation for our intuition at all; on the contrary, it seems merely to restate (with some additional claims) what it is that we want to be explained. Are there any reasons why people had their degrees of belief so distributed? If their beliefs had

[5]This is the account suggested by Horwich (1978).

been different, would it have been equally rational for them to view observations of Mars as a test of the third law, but not of the first? It seems to me that we never succeed in explaining a widely shared judgement about the relevance or irrelevance of some piece of evidence merely by asserting that degrees of belief happened to be so distributed as to generate those judgements according to the Bayesian scheme. Bayesians may instead try to explain the case by appeal to some structural difference among the hypotheses; the only gadget that appears to be available is the likelihood of the evidence about a single planet on various combinations of hypotheses. If it is supposed that the observations are such that Kepler's first and second laws entail their description, but Kepler's third law does not, then it follows that the likelihood of the evidence on the first and second laws—that is, the conditional probability of the evidence given those hypotheses—is unity, but the likelihood of the evidence on the third law may be less than unity. But any attempt to found an account of the case on these facts alone is simply an attempt at a hypothetico-deductive account. The problem is reduced to one already unsolved. What is needed to provide a genuine Bayesian explanation of the case in question (as well as of many others that could be adduced) is a *general* principle restricting conditional probabilities and having the effect that the distinctions about the bearing of evidence that have been noted here do result. Presumably, any such principles will have to make use of relations of content or structure between evidence and hypothesis. The case does nothing to establish that no such principles exist; it does, I believe, make it plain that without them the Bayesian scheme does not *explain* even very elementary features of the bearing of evidence on theory.

A third difficulty has to do with Bayesian kinematics. Scientists commonly argue for their theories from evidence known long before the theories were introduced. Copernicus argued for his theory using observations made over the course of millennia, not on the basis of any startling new predictions derived from the theory, and presumably it was on the basis of such arguments that he won the adherence of his early disciples. Newton argued for universal gravitation using Kepler's second and third laws, established before the *Principia* was published. The argument that Einstein gave in 1915 for his gravitational field equations was that they explained the anomalous advance of the perihelion of Mercury, established more than half a century earlier. Other physicists found the argument enormously forceful, and it is a fair conjecture that without it the British would not have mounted the famous eclipse expedition of 1919. Old evidence can in fact confirm new theory, but according to Bayesian kinematics, it cannot. For let us suppose that evidence e is known before theory T is introduced at time t. Because e is known at t, $\text{prob}_t(e) = 1$. Further, because $\text{prob}_t(e) = 1$, the likelihood of e given T, $\text{prob}_t(e, T)$, is also 1. We then have

$$\text{prob}_t(T, e) = \frac{\text{prob}_t(T) \times \text{prob}_t(e, T)}{\text{prob}_t(e)} = \text{prob}_t(T).$$

The conditional probability of T on e is therefore the same as the prior probability of T: e cannot constitute evidence for T in virtue of the positive relevance condition nor in virtue of the likelihood of *e* on *T*. None of the Bayesian mechanisms apply, and if we are strictly limited to them, we have the absurdity that old evidence cannot confirm new theory. The result is fairly stable. If the probability of *e* is very high but not unity, $prob_t(e, T)$ will still be unity if *T* entails *e*, and so $prob_r(T, e)$ will be very close to $prob_t(T)$. How might Bayesians deal with the old evidence/new theory problem?[6] Red herrings abound. The prior probability of the evidence, Bayesians may object, is not really unity; when the evidence is stated as measured or observed values, the theory does not really entail that those exact values obtain; an ideal Bayesian would never suffer the embarrassment of a novel theory. None of these replies will do: the acceptance of old evidence may make the degree of belief in it as close to unity as our degree of belief in some bit of evidence ever is; although the exact measured value (of, e.g., the perihelion advance) may not be entailed by the theory and known initial conditions, that the value of the measured quantity lies in a certain interval may very well be entailed, and that is what is believed anyway; and, finally, it is beside the point that an ideal Bayesian would never face a novel theory, for the idea of Bayesian confirmation theory is to explain scientific inference and argument by means of the assumption that good scientists are, about science at least, approximately ideal Bayesians, and we have before us a feature of scientific argument that seems incompatible with that assumption.

A natural line of defence lies through the introduction of counterfactual degrees of belief. When using Bayes's rule to determine the posterior probability of a new theory on old evidence, one ought not to use one's actual degree of belief in the old evidence, which is unity or nearly so; one ought instead to use the degree of belief one would have had in *e* if ... The problem is to fill in the blanks in such a way that it is both plausible that we have the needed counterfactual degrees of belief, and that they do serve to determine how old evidence bears on new theory. I tend to doubt that there is such a completion. We cannot merely throw *e* and whatever entails *e* out of the body of accepted beliefs; we need some rule for determining a counterfactual degree of belief in *e* and a counterfactual likelihood of *e* on *T*. To simplify, let us suppose that *T* does logically entail *e*, so that the likelihood is fixed.

If one flips a coin three times and it turns up heads twice and tails once, in using this evidence to confirm hypotheses (e.g. of the fairness of the coin), one does not take the probability of two heads and one tail to be what it is after the flipping—namely, unity—but what it was before the flipping. In this case there is an immediate and natural counterfactual degree of belief that is used in conditionalizing by Bayes's rule. The trouble with the scientific cases is that no such immediate and natural alternative distribution of degree of belief is available. Consider someone trying, in a Bayesian way, to determine in 1915 how much Einstein's derivation

[6] All of the defences sketched below were suggested to me by one or another philosopher sympathetic to the Bayesian view; I have not attributed the arguments to anyone for fear of misrepresenting them. None the less, I thank Jon Dorling, Paul Teller, Daniel Garber, Ian Hacking, Patrick Suppes, Richard Jeffrey, and Roger Rosencrantz for valuable discussions and correspondence on the point at issue.

of the perihelion advance confirmed general relativity. There is no single event, like the coin flipping, that makes the perihelion anomaly virtually certain. Rather, Leverrier first computed the anomaly in the middle of the nineteenth-century; Simon Newcomb calculated it again around 1890, using Leverrier's method but new values for planetary masses, and obtained a substantially higher value than had Leverrier. Both Newcomb and Leverrier had, in their calculations, approximated an infinite series by its first terms without any proof of convergence, thus leaving open the possibility that the entire anomaly was the result of a mathematical error. In 1912 Eric Doolittle calculated the anomaly by a wholly different method, free of any such assumption, and obtained virtually the same value as had Newcomb.[7] For actual historical cases, unlike the coin-flipping case, there is no single counterfactual degree of belief in the evidence ready to hand, for belief in the evidence sentence may have grown gradually—in some cases, it may have even waxed, waned, and waxed again. So the old evidence/new theory problem cannot be assimilated to coin flipping.

The suggestion that what is required is a counterfactual degree of belief is tempting, none the less; but there are other problems with it besides the absence of any unique historical degree of belief. A chief one is that various ways of manufacturing counterfactual degrees of belief in the evidence threaten us with incoherence. One suggestion, for example, is the following, used implicitly by some Bayesian writers. At about the time T is introduced, there will be a number of alternative competing theories available; call them $T_1, T_2, \ldots, T_k$, and suppose that they are mutually exclusive of T and of each other. Then $P(e)$ is equal to

$$P(T_1)\,P(e, T_1) + P(T_2)\,P(e, T_2) + \cdots + P(T_k)\,P(e, T_k) + P(\sim (T_1 \vee \cdots \vee T_k) \times P(e, \sim T_1 \vee \cdots \vee T_k))\,,$$

and we may try to use this formula to evaluate the counterfactual degree of belief in e. The problem is with the last term. Of course, one could suggest that this term just be ignored when evaluating $P(e)$, but it is difficult to see within a Bayesian framework any rationale at all for doing so. For if one does ignore this term, then the collection of prior probabilities used to evaluate the posterior probability of T will not be coherent unless either the likelihood of e on T is zero or the prior probability of T is zero. One could remedy this objection by replacing the last term by

$$P(T)P(e, T)\,,$$

but this will not do either, for if one's degree of belief in

$$P(T_1 \vee T_2 \vee \cdots \vee T_k \vee T)$$

[7]The actual history is still more complicated. Newcomb and Doolittle obtained values for the anomaly differing by about 2 s of are per century. Early in the 1920s. Grossmann discovered that Newcomb had made an error in calculation of about that magnitude.

is not unity, then the set of prior degrees of belief will still be incoherent. Moreover, not only will it be the case that if the actual degree of belief in e is replaced by a counterfactual degree of belief in e according to either of these proposals, then the resulting set of priors will be incoherent, it will further be the case that if we conditionalize on e the resulting conditional probabilities will be incoherent. For example, if we simply delete the last term, one readily calculates that

$$P(T_1 \vee \cdots \vee T_k,\ e) = \frac{P(T_1 \vee \cdots \vee T_k)\ P(e,\ T_1 \vee \cdots \vee T_k)}{P(e,\ T_1 \vee \cdots \vee T_k)\ P(T_1 \vee \cdots \vee T_k)} = 1,$$

and further that

$$P(T,\ e) = \frac{P(T)\ P(e,\ T)}{P(e,\ T_1 \vee \cdots \vee T_k)\ P(T_1 \vee \cdots \vee T_k)}.$$

But because T is supposed inconsistent with $T_1 \vee \cdots \vee T_k$ and $P(T, e)$ is not zero, this is incoherent.

Let us return to the proposal that when new theory confronts old evidence, we should look backwards to the time when the old evidence e had not yet been established and use for the prior probability of e whatever degree of belief we would have had at that time. We cannot just stick in such a counterfactual value for the prior probability of e and change nothing else without, as before, often making both prior and conditionalized probabilities incoherent. If we give all of our sentences the degree of belief they would have had in the relevant historical period (supposing we somehow know what period that is) and then conditionalize on e, incoherence presumably will not arise; but it is not at all clear how to combine the resulting completely counterfactual conditional probabilities with our actual degrees of belief. It does seem to me that the following rather elaborate procedure will work when a new theory is introduced. Starting with your actual degree of belief function P, consider the degree of belief you would have had in e in the relevant historical period, call it $H(e)$. Now change P by regarding $H(e)$ as an arbitrary change in degree of belief in e and using Richard Jeffrey's (1965) rule,

$$P'(S) = H(e)P(S, e) + (1 - H(e))\, P(S, \sim e)\,.$$

Jeffrey's rule guarantees that P' is a probability function. Finally, conditionalize on e:

$$P''(S) = P'(S, e)\,,$$

and let P'' be your new actual degree of belief function. (Alternatively, P'' can be formed by using Jeffrey's rule a second time.)

There remain a number of objections to the historical proposal. It is not obvious that there are, for each of us, degrees of belief we personally would have had in some historical period. It is not at all clear which historical period is the relevant one. Suppose, for example, that the gravitational deflection of sunlight had been determined experimentally around 1900, well before the introduction of general relativity.[8] In trying to assess the confirmation of general relativity, how far back in time should a twentieth-century physicist go under this supposition? If only to the nineteenth, then if he would have shared the theoretical prejudices of the period, gravitational deflection of light would have seemed quite probable. Where ought he to stop, and why? But laying aside these difficulties, it is implausible indeed that such a historical Bayesianism, however intriguing a proposal, is an accurate account of the principles by which scientific judgements of confirmation are made. For if it were, then we should have to condemn a great mass of scientific judgements on the grounds that those making them had not studied the history of science with sufficient closeness to make a judgement as to what their degrees of belief would have been in relevant historical periods. Combined with the delicacy that is required to make counterfactual degrees of belief fit coherently with actual ones, these considerations make me doubt that we should look to counterfactual degrees of belief for a plausible Bayesian account of how old evidence bears on new theory.

Finally, consider a quite different Bayesian response to the old evidence/new theory problem. Whereas the ideal Bayesian agent is a perfect logician, none of us are, and there are always consequences of our hypotheses that we do not know to be consequences. In the situation in which old evidence is taken to confirm a new theory, it may be argued that there is *something* new that is learned, and typically, what is learned is that the old evidence is entailed by the new theory. Some old anomalous result is lying about, and it is not this old result that confirms a new theory, but rather the new discovery that the new theory entails (and thus explains) the old anomaly. If we suppose that semi-rational agents have degrees of belief about the entailment relations among sentences in their language, and that

$$P\left(h \middle| - e\right) = 1 \textit{ implies } P(e, h) = 1,$$

this makes a certain amount of sense. We imagine the semi-rational Bayesian changing his degree of belief in hypothesis h in light of his new discovery that h entails e by moving from his prior degree of belief in h to his conditional degree of belief in h given that e, that $h \vdash e$, and whatever background beliefs there may be. Old evidence can, in this vicarious way, confirm a new theory, then, provided that

[8]Around 1900 is fanciful, before general relativity is not. In 1914 E. Freundlich mounted an expedition to Russia to photograph the eclipse of that year in order to determine the gravitational deflection of starlight. At that time, Einstein had predicted an angular deflection for light passing near the limb of the sun that was equal in value to that derived from Newtonian principles by Soldner in 1801. Einstein did not obtain the field equations that imply a value for the deflection equal to twice the Newtonian value until late in 1915. Freundlich was caught in Russia by the outbreak of World War I, and was interned there. Measurement of the deflection had to wait until 1919.

$$P\left(h, b\&e\&\left(h\middle| -e\right)\right) > P\,(h, b\&e)\,.$$

Now, in a sense, I believe this solution to the old evidence/new theory problem to be the correct one; what matters is the discovery of a certain logical or structural connection between a piece of evidence and a piece of theory, and it is in virtue of that connection that the evidence, if believed to be true, is thought to be evidence for the bit of theory. What I do not believe is that the relation that matters is simply the entailment relation between the theory, on the one hand, and the evidence, on the other. The reasons that the relation cannot be simply that of entailment are exactly the reasons why the hypothetico-deductive account (see Glymour 1981, ch. 2) is inaccurate; but the suggestion is at least correct in sensing that our judgement of the relevance of evidence to theory depends on the perception of a structural connection between the two, and that degree of belief is, at best, epiphenomenal. In the determination of the bearing of evidence on theory, there seem to be mechanisms and stratagems that have no apparent connection with degrees of belief, which are shared alike by people advocating different theories. Save for the most radical innovations, scientists seem to be in close agreement regarding what would or would not be evidence relevant to a novel theory; claims as to the relevance to some hypothesis of some observation or experiment are frequently buttressed by detailed calculations and arguments. All of these features of the determination of evidential relevance suggest that that relation depends somehow on structural, objective features connecting statements of evidence and statements of theory. But if that is correct, what is really important and really interesting is what these structural features may be. The condition of positive relevance, even if it were correct, would simply be the least interesting part of what makes evidence relevant to theory.

None of these arguments is decisive against the Bayesian scheme of things, nor should they be; for in important respects that scheme is undoubtedly correct. But taken together, I think they do at least strongly suggest that there must be relations between evidence and hypotheses that are important to scientific argument and to confirmation but to which the Bayesian scheme has not yet penetrated.

References

Carnap, R. (1950). *The logical foundations of probability*. Chicago: University of Chicago Press.
Glymour, C. (1981). *Theory and evidence*. Chicago: University of Chicago Press.
Hesse, M. (1974). *The structure of scientific inference*. Berkeley: University of California Press.
Horwich, P. (1978). An appraisal of Glymour's confirmation theory. *Journal of Philosophy, 75*, 98–113.
Jeffrey, R. (1965). *The logic of decision*. New York: McGraw-Hill.
Jeffreys, H. (1967). *Theory of probability*. Oxford: Clarendon.
Jeffreys, H. (1973). *Scientific inference*. Cambridge: Cambridge University Press.
Kyburg, H. (1978). Subjective probability: Criticisms, reflections and problems. *Journal of Philosophical Logic, 7*, 157–180.

Putnam, H. (1967). Probability and confirmation. In S. Morgenbesser (Ed.), *Philosophy of science today*. New York: Basic Books.
Rosencrantz, R. (1976). Simplicity. In W. Harper & C. Hooker (Eds.), *Foundations and philosophy of statistical inference*. Boston: Reidel.
Salmon, W. C. (1969). *Foundations of scientific inference*. Pittsburgh: University of Pittsburgh Press.
Savage, L. (1972). *The foundations of statistics*. New York: Dover.
Shimony, A. (1970). Scientific inference. In R. G. Colodny (Ed.), *The nature and function of scientific theories* (pp. 79–179). Pittsburgh: University of Pittsburgh Press.

Chapter 9
Discussion: A Mistake in Dynamic Coherence Arguments?

Brian Skyrms

Static Coherence of Degrees of Belief

The person whose degrees of belief are being tested for coherence acts as a bookie. She posts her fair prices for wagers corresponding to her degrees of belief. Her degrees of belief are *incoherent* if a cunning bettor can make a Dutch book against her with a finite system of wagers—that is, there is a finite set of wagers individually perceived as fair, whose net payoff is a loss in every possible future. Otherwise her degrees of belief are *coherent*. De Finetti ([1937] 1980) proved the following theorem: *Degrees of belief are coherent if and only if they are finitely additive probabilities*.

Obviously, if a Dutch book can be made with a finite number of fair transactions, it can be made with a finite number of uniformly favorable transactions. The bettor pays some small transaction premium ε to the bookie for each of the n transactions where $n\varepsilon$ is less than the guaranteed profit that the bettor gets under the Dutch book based on fair prices. Let us bear in mind that this point applies equally well in what follows.

Received April 1992; revised June 1992.
Send reprint requests to the author, Department of Philosophy, University of California, Irvine, CA 92717, USA.

B. Skyrms (✉)
Department of Philosophy, University of California, Irvine, CA, USA

H. Arló-Costa et al. (eds.), *Readings in Formal Epistemology*, Springer Graduate Texts in Philosophy 1, DOI 10.1007/978-3-319-20451-2_9

Dynamic Coherence for Updating Rules

The epistemologist acts as bookie. Her updating rule is public knowledge. Today she posts her fair prices, and does business. Tomorrow she makes an observation (with a finite number of possible outcomes each of which has positive prior probability) and updates her fair prices according to her updating rule. The updating rule is thus a *function* from possible observations to revised fair prices. The day after tomorrow she posts prices again, and does business. The pair consisting of the (1) her fair prices for today and (2) her updating function will be called the bookie's *epistemic strategy*.

The bookie's *epistemic strategy* is *coherent* if *there is no* possible bettor's strategy which makes a Dutch book against him (the bettor's strategy being a pair consisting of (1) a finite number of transactions today at the bookie's posted prices and (2) a function taking possible observations into a finite number of transactions the day after tomorrow at the prices that the bookie will post according to her epistemic strategy). Lewis (reported in Teller 1973) proves that *the epistemologist's strategy is coherent only if her degrees of belief today are finitely additive probabilities and her updating rule is Bayes's rule of conditioning.* The "only if" can be strengthened to "if and only if" (see section "The converse"). (For generalizations of this theorem see van Fraassen 1984 and Skyrms 1987, 1990.)

Notice that the relevant notions of coherence and incoherence here apply not just to the *pair* of degrees of belief for today and the day after tomorrow, but rather to an *epistemic strategy*, which is a more complicated object. A focus on the former notion leads understandably to skepticism regarding dynamic coherence, as in Hacking (1967), Kyburg (1978), and Christensen (1991).

The Dynamic Dutch Book

Coherence of degrees of belief today is the static case. It remains to show that for any non-Bayes updating rule, there is a bettor's strategy which makes a Dutch book. Let the conditional probability of A on e, that is $\Pr(A\ \&\ e)/\Pr(e)$, be symbolized as usual, as $\Pr(A|e)$, and let the probability that the updating rule gives A if e is observed be $\Pr_e(\text{A})$. If the predictor's rule disagrees with conditioning, then for some possible evidential result e and some A, $\Pr_e(A)$ is not $\Pr(A|e)$. Suppose that $\Pr(A|e) > \Pr_e(A)$. (The other case is similar.) Let the discrepancy be $\delta = \Pr(A|e) - \Pr_e(A)$. *Here is a bettor's strategy which makes a Dutch book:*

TODAY: Offer to sell the bookie at her fair price:

1: [\$1 if $A\ \&\ e$, 0 otherwise]
2: [\$$\Pr(A|e)$ if not-e, 0 otherwise]
3: [\$$\delta$ if e, 0 otherwise]

DAY AFTER TOMORROW:

If e was observed, offer to buy [$1 if A, 0 otherwise] for its current fair price, $\Pr_e(A) = \Pr(A|e) - \delta$.

Then in every possible situation, the bookie loses $\$\delta \Pr(e)$.

The Converse

If the bookie has the strategy of updating by Bayes's rule of conditioning, then every payoff that a bettor's strategy can achieve can be achieved by betting only today (see Skyrms 1987). This reduces our case to the static case. Thus, by de Finetti's result, if the epistemologist's prior degrees of belief are finitely additive probabilities and her updating rule is Bayes's rules of conditioning, then she is dynamically coherent.

Sequential Analysis 1: A Mistake in the Dynamic Coherence Argument?

Maher's (1992b) objection is that the bookie will see it coming and refuse to bet. This is made precise by modeling the bookie's situation as a sequential choice problem, as shown in Fig. 9.1. The bookie sees that if she bets today and e occurs, then at decision node 2, she will find the cunning bettor's offer fair according to her revised probability, $\Pr_e(A)$. Thus she sees that betting today leads to a sure loss. Since she prefers net gain of zero to a sure loss, she refuses to bet today—frustrating the cunning bettor who goes home unable to execute his plan.

The first thing that must be said about "Maher's objection" is that it is misleading to represent it as showing a "mistake" in the dynamic coherence theorem. Under the conditions of the theorem the bookie posts her fair prices for today and honors them. There is no provision for changing one's mind when approached by a cunning bettor who discloses his strategy, nor indeed any mention of a requirement that the cunning bettor disclose his strategy prior to the initial transaction. But Maher might be read as suggesting a different conception of dynamic coherence in this setting:

> The epistemologist acts as bookie. Her updating rule is public knowledge. Today she posts her tentative fair prices, but in fact does business only with bettors who disclose their strategies in advance, and does so on the basis of sequential decision analysis. Tomorrow she makes an observation (with a finite number of possible outcomes each of which has positive prior probability) and updates her probabilities according to her updating rule. The day after tomorrow she posts prices again, and does business according to those prices.
>
> She is *coherent* if *there is no* possible bettor's strategy which makes a Dutch book against her.

This is an interesting modification of the usual notion of dynamic coherence, and it merits investigation. Is it a better motivated conception of dynamic coherence? What differences does it make?

Sequential Analysis 2: A Mistake in the Mistake?

A natural reaction to Maher's line might be to say that the redefinition unfairly prejudices the case against dynamic coherence arguments. It is therefore of some interest to see that the dynamic Dutch book still goes through under the revised scenario.

There is a gratuitous assumption in the analysis presented in Fig. 9.1. Why is it assumed that the cunning bettor will just go home if the bookie refuses to bet today? The bettor's strategy which I presented says otherwise. The bettor will make an offer the day after tomorrow if *e* was observed. So the branch of the decision tree where the bookie refuses transactions today cannot simply be assumed to have payoff of zero, but requires further analysis. This is done in Fig. 9.2.

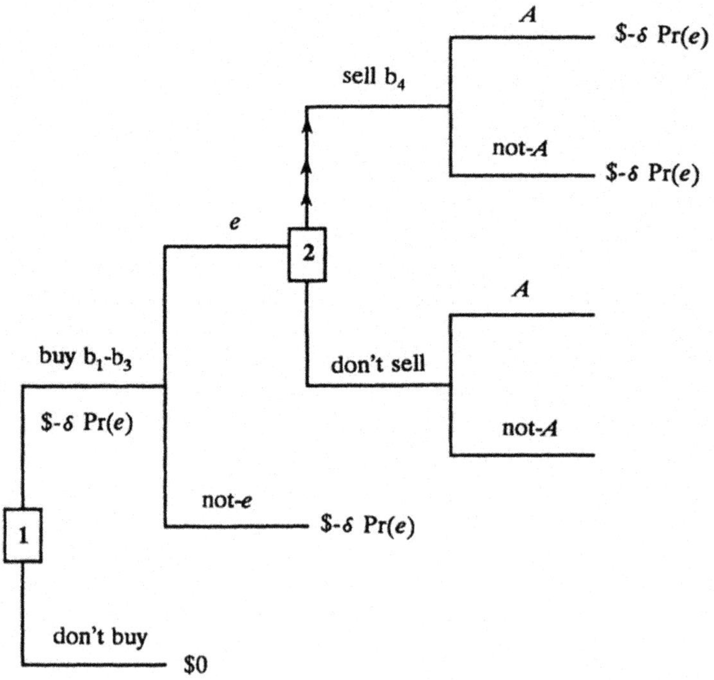

Fig. 9.1 Maher-Levi sequential analysis

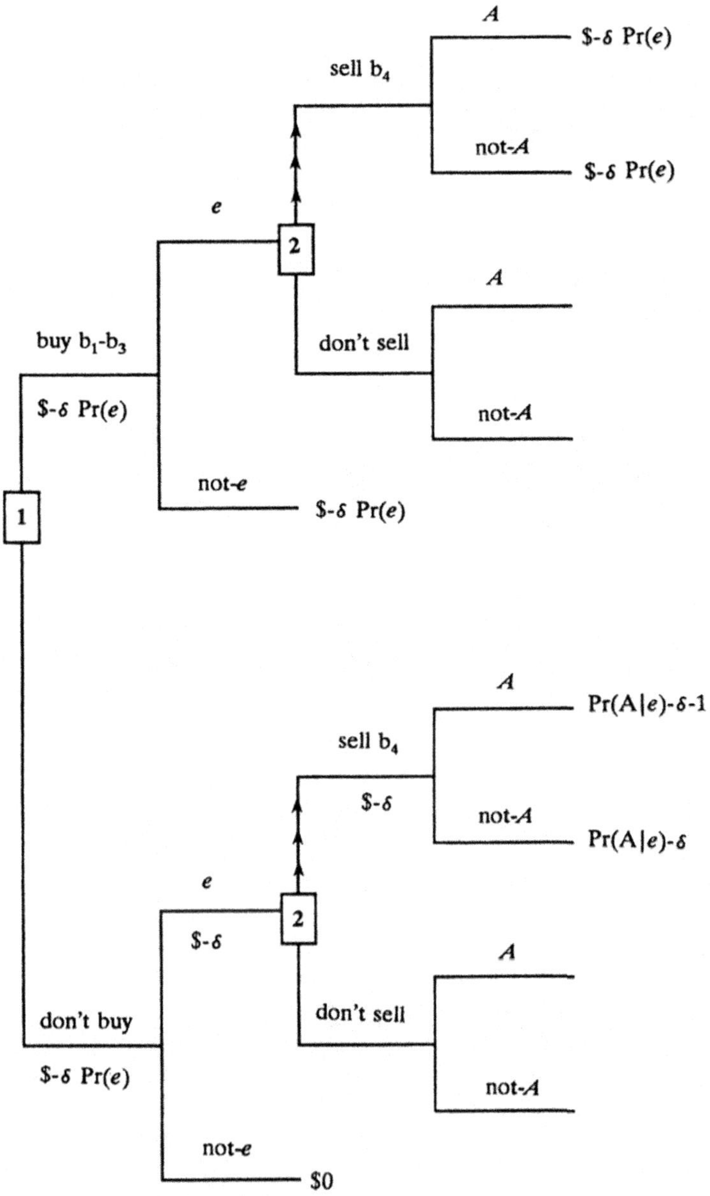

Fig. 9.2 Full sequential analysis

Note that the bookie knows that if e is observed, she will accept the offer the day after tomorrow for the same reason on the lower path as on the upper. Deciding now not to bet ever is not an option. If the offer the day after tomorrow is accepted but the offer today was not and e and A both happen, then the net payoff is the price the cunning bettor paid, $\$\Pr(A|e) - \delta$, less the lost bet, $\$-1$, as shown. If e occurs but A does not, the net payoff is just $\$\Pr(A|e) - \delta$. For the bookie's current analysis of this decision tree, to get the relevant expectation over A occurring or not we average using as weights her current conditional probabilities, $\Pr(A|e)$ and $\Pr(-A|e)$. Thus the value at the node where the bookie refused to bet today and where e is observed tomorrow is

$$\Pr\left(A|e\right)\$\left[\left\{\Pr\left(A|e\right)-\delta\right\}-1\right]+\left[1-\Pr\left(A|e\right)\right]\$\left[\Pr\left(A|e\right)-\delta\right]=\$-\delta.$$

Then the value at the node where the bookie refused to bet today is not 0 but rather $\$-\delta\Pr(e)$. This is just the same as the value at the node where the bookie agrees to bet today.

In fact, if we consider the version of the Dutch-book strategy where the bettor adds an ε premium for each transaction, the upper branch involves four transactions and the lower branch involves only one, so the upper branch has a higher payoff than the lower branch. *Even though the bookie sees it coming, she will prefer the sure loss of the upper branch because doing so looks strictly better to her than the alternative.*

Sequential Analysis 3: What Makes the Cunning Bettor Tick?

Why did the cunning bettor adopt a strategy of staying around if the bookie decided not to bet today? The official answer in sections "Dynamic coherence for updating rules" and "The dynamic Dutch book" is "Don't ask". *Any* bettor's strategy which makes a Dutch book will prove incoherence. But, as Levi (1991) points out, that sort of analysis proceeds in strategic normal form rather than in extensive form. Might it be that the cunning bettor's strategy described would have to be *sequentially irrational*? That is to say, might it not be that staying around and betting the day after tomorrow if the bookie decided not to bet today would not maximize expected utility for the cunning bettor in the belief state he would have in that case the day after tomorrow? If this could be shown, then the cunning bettor's strategy that I have described would have to rest on a noncredible threat, and the significance of the analysis of the previous section would be called into question. (For discussion of such noncredible threats in extensive form games and of sequential rationality, see Selten 1975 and Kreps and Wilson 1982.)

But such is not the case. Suppose that the bettor is a Bayesian; that he starts out with exactly the same degrees of belief as the bookie; and that he updates by conditioning. If e is observed tomorrow—whether or not the bookie accepted the bet today—he conditions on e and the day after tomorrow his fair price for $b4$ is \$ pr($A|e$). But his strategy only commits him to offering to pay the bookie's fair price, \$ pr($A|e$) $- \delta$, to buy back $b4$ for what he perceives as a net gain in expected utility of \$$\delta$. This bettor's threat to stick around and bet the day after tomorrow if e, even if the bookie declines to bet today, is perfectly credible and consistent with sequential rationality. If he is called upon to carry out the threat, he maximizes expected utility by doing so.

Strategic Rationality

Let us explicitly model the bookie's choice of an updating strategy. The bookie and the bettor start out with identical priors. The bettor updates by conditioning. First the bookie chooses an updating strategy. Then the bettor bets, the evidence comes in, the bookie updates according to her updating rule, and the bettor bets again. The bookie's initial strategy is either to choose updating by conditioning or not.

If the bookie chooses the strategy of updating by conditioning, then the fair prices of the bookie and bettor agree at all times. Thus either no transactions are made, or any transactions have net change in expected utility of 0 for both players. The bookie's expected utility of choosing the strategy of updating by conditioning is zero. If, however, the bookie chooses an updating strategy at variance with conditioning then, for the bettor, the expected utility of betting is greater than that of not betting (section "Sequential analysis 3: what makes the cunning bettor tick?") and the net expected utility for the bookie is negative (section "Sequential analysis 2: a mistake in the mistake?"). At the first choice point the bookie is strictly better off by choosing the rule of updating by conditioning.

Thus the strategy combination in which the bookie updates by conditioning and the bettor does not bet at all is an *equilibrium* in the sense that no player will perceive it in his or her interest at any decision node to deviate from that strategy. But no strategy combination in which the bookie chooses a strategy at variance with conditioning is such an equilibrium.

The Bottom Line

Two ways of strengthening the requirements for a dynamic Dutch book were suggested by the discussions of Levi (1987) and Maher: (1) We require the cunning bettor to disclose his strategy, and allow the bookie to use knowledge of that strategy in a sequential analysis when deciding whether to bet today or not, and (2) we require that the cunning bettor's strategy itself be sequentially rational. The somewhat surprising result is that the additional restrictions made no difference. The bookie whose epistemic strategy is at odds with conditioning is also subject to a Dutch book in this stronger sense. "Seeing it coming" does not help. It is at the very least a noteworthy property of the rule of conditioning that in this sort of epistemic situation, it alone is immune from a Dutch book under either the original or strengthened requirements.

Postscript: Conditioning, Coherence and Rationality

Many of the concerns of Levi and Maher have not been addressed in the foregoing. Levi is concerned to resist the doctrine of "confirmational tenacity", according to which the only legitimate way in which to update is by conditioning. Maher wishes to resist the doctrine that rationality requires dynamic coherence at all costs. Does the foregoing show that conditioning is the only coherent way to ever update one's probabilities? Does it show that rationality requires coherence at all costs?

I agree with Levi and Maher in answering "no" to both questions. With regard to the first, let me emphasize that the Lewis proof takes place within the structure of a very special epistemic model. In that context it shows that the rule of conditioning is the unique dynamically coherent updating rule. It does not show that one must have an updating rule. It does not apply to other epistemic situations which should be modeled differently. The modeling of a variety of epistemic situations and the investigation of varieties of dynamic coherence in such situations is an ongoing enterprise (in which I take it that both Levi and I are engaged; see Skyrms 1990 for further discussion).

Maher is concerned that an uncritical doctrine of "dynamic coherence at all costs" could lead one to crazy belief changes and disastrous actions. Should Ulysses have changed to 1 his prior probability of safe sailing conditional on hearing the Sirens' song so that subsequently his belief change would be in accordance with the rule of conditioning? Nothing in the foregoing implies that he should. In the first place, there is something a little odd in thinking that one achieves dynamic coherence by changing the original prior pr_1 to the revised prior pr_2 so that the change to pr_3 will agree with conditioning. What about the change from pr_1 to pr_2? But, more fundamentally, I would agree with Maher that rationality definitely does not require coherence *at all costs*. Where costs occur they need to be weighed against benefits. There are lucid discussions of this matter in Maher (1992a, b). These things said, it remains that in the Lewis epistemic model under the "sequentialized" notion of dynamic coherence, the unique coherent updating rule is the rule of conditioning.

Acknowledgement I would like to thank Brad Armendt, Ellery Eells, Isaac Levi, Patrick Maher and an anonymous referee for helpful comments on an earlier draft of this note. I believe that Maher, Levi and I are now in substantial agreement on the issues discussed here, although differences in emphasis and terminology may remain.

References

Christensen, D. (1991). Clever bookies and coherent beliefs. *The Philosophical Review, 100*, 229–247.

de Finetti, B. ([1937] 1980). Foresight: Its logical laws, its subjective sources, translated in H. E. Kyburg, Jr., & H. Smokler (Eds.), *Studies in subjective probability* (pp. 93–158). (Originally published as "La Prevision: ses lois logiques, ses sources subjectives", *Annales de l'Institut Henri Poincaré, 7*, 1–68.) Huntington: Kreiger.

Hacking, I. (1967). Slightly more realistic personal probability. *Philosophy of Science, 34*, 311–325.

Kreps, D., & Wilson, R. (1982). Sequential equilibria. *Econometrica, 50*, 863–894.

Kyburg, H. (1978). Subjective probability: Criticisms, reflections and problems. *The Journal of Philosophical Logic, 7*, 157–180.

Levi, I. (1987). The demons of decision. *The Monist, 70*, 193–211.

Levi, I. (1991). Consequentialism and sequential choice. In M. Bacharach & S. Hurley (Eds.), *Foundations of decision theory* (pp. 92–146). Oxford: Basil Blackwell.

Maher, P. (1992a). *Betting on theories*. Cambridge: Cambridge University Press.

Maher, P. (1992b). Diachronic rationality. *Philosophy of Science, 59*, 120–141.

Selten, R. (1975). Reexamination of the perfectness concept of equilibrium in extensive form games. *International Journal of Game Theory, 4*, 25–55.

Skyrms, B. (1987). Dynamic coherence and probability kinematics. *Philosophy of Science, 54*, 1–20.

Skyrms, B. (1990). *The dynamics of rational deliberation*. Cambridge: Harvard University Press.

Teller, P. (1973). Conditionalization and observation. *Synthese, 26*, 218–258.

van Fraassen, B. (1984). Belief and the will. *Journal of Philosophy, 81*, 235–256.

Chapter 10
Some Problems for Conditionalization and Reflection

Frank Arntzenius

I will present five puzzles which show that rational people can update their degrees of belief in manners that violate Bayesian Conditionalization and van Fraassen's Reflection Principle. I will then argue that these violations of Conditionalization and Reflection are due to the fact that there are two, as yet unrecognized, ways in which the degrees of belief of rational people can develop.

Two Roads to Shangri La

Every now and then the guardians to Shangri La will allow a mere mortal to enter that hallowed ground. You have been chosen because you are a fan of the Los Angeles Clippers. But there is an ancient law about entry into Shangri La: you are only allowed to enter, if, once you have entered, you no longer know by what path you entered. Together with the guardians you have devised a plan that satisfies this law. There are two paths to Shangri La, the Path by the Mountains, and the Path by the Sea. A fair coin will be tossed by the guardians to determine which path you will take: if heads you go by the Mountains, if tails you go by the Sea. If you go by the Mountains, nothing strange will happen: while traveling you will see the glorious Mountains, and even after you enter Shangri La you will forever retain your memories of that Magnificent Journey. If you go by the Sea, you will revel in the Beauty of the Misty Ocean. But, just as you enter Shangri La, your memory of this Beauteous Journey will be erased and be replaced by a memory of the Journey by the Mountains.

F. Arntzenius (✉)
University of Oxford, Oxford, UK
e-mail: frank.arntzenius@philosophy.ox.ac.uk

H. Arló-Costa et al. (eds.), *Readings in Formal Epistemology*, Springer Graduate Texts in Philosophy 1, DOI 10.1007/978-3-319-20451-2_10

Suppose that in fact you travel by the Mountains. How will your degrees of belief develop? Before you set out your degree of belief in heads will be ½. Then, as you travel along the Mountains and you gaze upon them, your degree of belief in heads will be 1. But then, once you have arrived, you will revert to having degree of belief ½ in heads. For you will know that you would have had the memories that you have either way, and hence you know that the only relevant information that you have is that the coin was fair.

This seems a bizarre development of degrees of belief. For as you are traveling along the Mountains, you know that your degree of belief in heads is going to go down from 1 to ½. You do not have the least inclination to trust those future degrees of belief. Those future degrees of belief will not arise because you will acquire any evidence, at least not in any straightforward sense of "acquiring evidence". Nonetheless you think you will behave in a fully rational manner when you acquire those future degrees of belief. Moreover, you know that the development of your memories will be completely normal. It is only because something strange would have happened to your memories had the coin landed tails, that you are compelled to change your degrees of belief to ½ when that counterfactual possibility would have occurred.

The Prisoner

You have just been returned to your cell on death row, after your last supper. You are to be executed tomorrow. You have made a last minute appeal to President George W. Bush for clemency. Since Dick Cheney is in hospital and can not be consulted, George W. will decide by flipping a coin: heads you die, tails you live. His decision will be made known to the prison staff before midnight. You are friends with the prison officer that will take over the guard of your cell at midnight. He is not allowed to talk to you, but he will tell you of Bush's decision by switching the light in your cell off at the stroke of midnight if it was heads. He will leave it on if it was tails. Unfortunately you don't have a clock or a watch. All you know is that it is now 6 pm since that is when prisoners are returned to their cells after supper. You start to reminisce and think fondly of your previous career as a Bayesian. You suddenly get excited when you notice that there is going to be something funny about the development of your degrees of belief. Like anybody else, you don't have a perfect internal clock. At the moment you are certain that it is 6 pm, but as time passes your degrees of belief are going to be spread out over a range of times. What rules should such developments satisfy?

Let us start on this problem by focusing on one particularly puzzling feature of such developments. When in fact it is just before midnight, say 11.59 pm, you are going to have a certain, non-zero, degree of belief that it is now later than midnight. Of course, at 11.59 pm the light in your cell is still going to be on. Given that at this time you will have a non-zero degree of belief that it is after midnight, and given that in fact you will see that the light is still on, you will presumably take it that the light provides some evidence that the outcome was tails. Indeed, it seems clear that as it gets closer to midnight, you will monotonically increase your degree of belief in tails. Moreover you know in advance that this will happen. This seems puzzling. Of course, after midnight, your degree of belief in tails will either keep on increasing, or it will flip to 0 at midnight and stay there after midnight. But that does not diminish the puzzlement about the predictable and inevitable increase in your degree of belief in tails prior to midnight. In fact, it seems that this increase is not merely puzzling, it seems patently irrational. For since this increase is entirely predictable, surely you could be made to lose money in a sequence of bets. At 6 pm you will be willing to accept a bet on heads at even odds, and at 11.59 pm you will, almost certainly, be willing to accept a bet on tails at worse than even odds. And that adds up to a sure loss. And surely that means you are irrational.

Now, one might think that this last argument shows that your degree of belief in tails in fact should not go up prior to midnight. One might indeed claim that since your degree of belief in heads should remain ½ until midnight, you should adjust your idea of what time it is when you see that the light is still on, rather than adjust your degree of belief in tails as time passes. But of course, this suggestion is impossible to carry out. Armed with an imperfect internal clock, you simply can not make sure that your degree of belief in heads stays ½ until midnight, while allowing it to go down after midnight. So how should they develop?

Let us start with a much simpler case. Let us suppose that there is no coin toss and no light switching (and that you know this). You go into you cell at 6 pm. As time goes by there will be some development of your degrees of belief as to what time it is. Let us suppose that your degrees of belief in possible times develop as pictured in the top half of Fig. 10.1.

Next, let us ask how your degrees of belief should develop were you to know with certainty that the guard will switch the light off at 12 pm. It should be clear that then at 11.59 pm your degree of belief distribution should be entirely confined

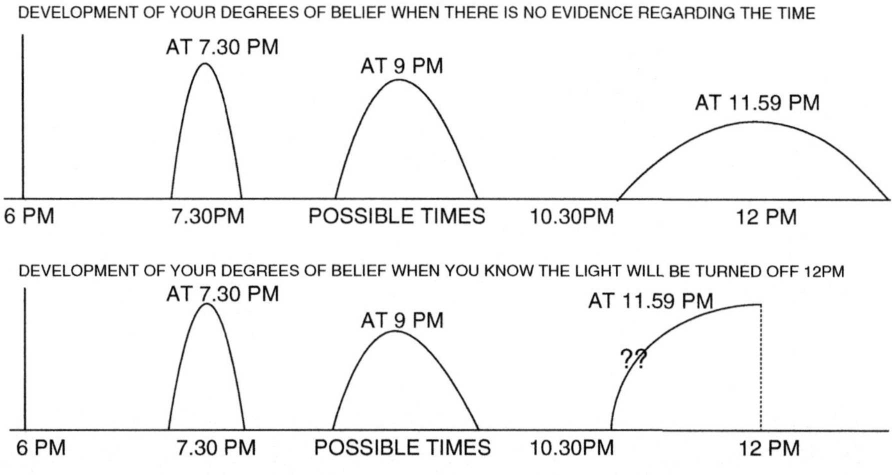

Fig. 10.1 Prisoner without evidence

to the left of 12 pm, as depicted in the bottom half of Fig. 10.1. For at 11.59 pm the light will still be on, so that you know that it must be before 12 pm. But other than that it should be entirely confined to the left of 12 pm, it is not immediately clear exactly what you degree of belief distribution should be at 11.59 pm. It is not even obvious that there should be a unique answer to this question. However, a very simple consideration leads to a unique answer.

Suppose that, even though the guard is going to switch the light off at 12 pm, you were not told that the guard is going to switch the light off at 12 pm. Then the development of your degrees of belief would be as pictured in the top half of Fig. 10.1. Next, suppose that at 11.59 pm your are told that the guard will switch the light off at 12 pm, but you are not told that it is now 11.59 pm. Obviously, since the light is still on you can infer that it is prior to 12 pm. Surely you should update your degrees of belief by conditionalization: you should erase that part of your degree of belief distribution that is to the right of 12 pm, and re-normalize the remaining part (increase the remaining part proportionally). Now it is clear that this is also the degree of belief distribution that you should have arrived at had you known all along that the guard would turn the light off at 12 pm. For either way you have accumulated exactly the same relevant information and experience by 11.59 pm. This uniquely determines how your degree of belief distribution should develop when you know all along that the guard will turn the light off at 12 pm. At any time this (constrained) distribution should be the distribution that you arrive at by conditionalizing the distribution that you have if you have no evidence regarding the time, on the fact that it is now before 12 pm. One can picture this development in the following way. One takes the development of the top part of Fig. 10.1. As this distribution starts to pass through the 12 pm boundary, the part that passes through this boundary gets erased, and, in a continuous manner, it gets proportionally added to the part that is to the left of the 12 pm boundary.

Now we are ready to solve the original puzzle. Your degrees of belief in that case can be pictured as being distributed over possible times in two possible worlds: see Fig. 10.2. The development is now such that when the bottom part of the degree of belief distribution hits midnight, it gets snuffed out to the right of midnight, and the rest of the degree of belief distribution is continuously re-normalized, i.e. the top part of the degree of belief distribution and the remaining bottom part are continuously proportionally increased as time passes. Note that Fig. 10.2 is essentially different from Fig. 10.1. In Fig. 10.2 the top distribution starts to increase its absolute size once the leading edge of the bottom distribution hits midnight. This does not happen in Fig. 10.1, since there the degree of belief distributions each were total degree of belief distributions in separate scenarios. Also, in Fig. 10.2 the bottom distribution starts to increase in size once its leading edge hits midnight, but it only increases half as much as it does in Fig. 10.1, since half of the "gains" is being diverted to the top degree of belief distribution.

Thus, at the very least until it actually is midnight, the top and the bottom degree of belief distribution will always be identical to each other, in terms of shape and size, to the left of midnight. Prior to midnight, your degrees of belief will be such that conditional upon it being prior to midnight, it is equally likely to

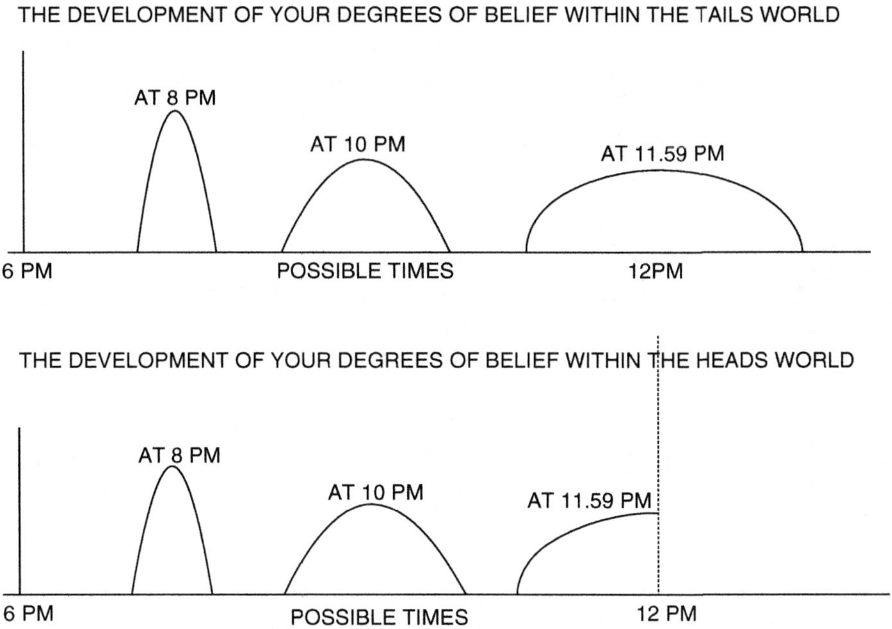

Fig. 10.2 Prisoner with evidence

be heads as tails. Your unconditional degree of belief in tails, however, will increase monotonically as you approach midnight.

After midnight there are two possible ways in which your degree of belief distribution can develop. If the light is switched off your degree of belief distribution collapses completely onto midnight and onto the heads world. If in fact it is not switched off your degree of belief distribution continues to move to the right in both worlds, and it continues to be snuffed out in the heads world to the right of midnight, and the remaining degrees of belief keep being proportionally increased.[1]

Now I can answer the questions that I started with. It is true, as I surmised, that your degree of belief in tails will have increased by 11.59 pm. You will take your internal sense of the passing of time, and combine it with the fact that the light is still on, and you will take this as providing some evidence that the outcome is tails. It is also true, as I surmised, that the light still being on will be taken by you as providing some evidence that it is not yet midnight. For at 11.59 pm your degree of belief distribution over possible times (averaged over the heads and tails worlds) will be further to the left that it would have been had you believed that the light would stay on no matter what. More generally, we have found a unique solution

[1]Thus, for instance, if the light is not switched off, there must be a moment (which could be before or after midnight) such that you have equal degree of belief in each of the 3 possibilities: heads and it is before midnight, tails and it is before midnight, tails and its after midnight.

to the puzzle of how a rational person's sense of time must interact with evidence, given how that person's sense of time works in the absence of evidence.

Rather surprisingly, this interaction can be such, as it is in my example, that you know in advance that at some specified later time you will, almost certainly, have increased your degree of belief in tails, and that you could not possibly have decreased your degree of belief in tails.[2] It is also interesting to note that nothing essential changes in this example if one assumes that the coin toss will take place exactly at midnight. Thus it can be the case that one knows in advance that one will increase one's degrees of belief that a coin toss, which is yet to occur, will land tails. Of course, at the time that one has this increased degree of belief one does not know that this coin toss is yet to occur. Nonetheless, such predictable increases in degrees of belief seem very strange.

John Collins' Prisoner

John Collins has come up with the following variation of the case of the prisoner that was described in the previous section. In Collins's variation the prisoner has 2 clocks in his cell, both of which run perfectly accurately. However, clock A initially reads 6 pm, clock B initially reads 7 pm. The prisoner knows that one of the clocks is set accurately, the other one is one hour off. The prisoner has no idea which one is set accurately; indeed he initially has degree of belief ½ that A is set accurately, and degree of belief ½ that B is set accurately. As in the original case, if the coin lands heads the light in his cell will be turned off at midnight, and it will stay on if it lands tails. So initially the prisoner has degree of belief 1/4 in each of the following 4 possible worlds:

W_1: Heads and clock A is correct
W_2: Heads and clock B is correct
W_3: Tails and clock A is correct
W_4: Tails and clock B is correct.

[2] One might wonder why I inserted the phrase "almost certainly" in this sentence. The reason for this is that there is a subtlety as to whether you know at 6 pm that you will have an increased degree of belief in tails at 11.59 pm. There is an incoherence in assuming that at 6 pm you know with certainty what your degree of belief distribution over possible times will be at 11.59 pm. For if you knew that you could simply wait until your degree of belief distribution is exactly like that. (You can presumably establish by introspection what your degree of belief distribution is.) And when you reach that distribution, you would know that it has to be 11.59 pm. So when that happens you should then collapse your degree of belief distribution completely on it being 11.59 pm. But this is incoherent. Thus, the fact that you do not have a perfect internal clock also implies that you can not know in advance what your degree of belief distribution is going to look like after it has developed (guided only by your internal clock). Thus you can not in advance be certain how your degree of belief distribution over possible times will develop. Nonetheless you can be certain at 6 pm that your degree of belief in tails will not decrease prior to midnight, and that it is extremely likely to have increased by 11.59 pm. At 6 pm your expectation for you degree of belief in tails at 11.59 pm will be substantially greater than 0.5.

When in fact it is 11.30 pm the light, for sure, will still be on. What will the prisoner's degrees of belief then be? Well, if the actual world is W_1, then, when it actually is 11.30 pm clock A will read 11.30 pm and clock B will read 12.30 am. In that case, since the prisoner sees that the light is still on, he will know that it can not be that the coin landed heads and clock B is correct. I.e. his degree of belief in W_2 will be 0, and his degrees of belief in the three remaining options will be 1/3 each. Similarly if the actual world is W_3 then at 11.30 pm prisoner will have degree of belief 0 in W_2 and degree of belief in 1/3 each of the remaining options. On the other hand if the actual world is W_2 or W_4, then when it is actually 11.30 pm, the clock readings will be 10.30 pm and 11.30 pm, and the prisoner will still have the degrees of belief that he started with, namely 1/4 in each of the 4 possibilities. The prisoner, moreover, knows all of this in advance.

This is rather bizarre, to say the least. For, in the first place, at 6 pm the prisoner knows that at 11.30 pm his degrees of belief in heads will be less or equal to what they now are, and can not be greater. So his current expectation of what his degrees of belief in heads will be at 11.30 pm, is less than his current degree of belief in heads. Secondly, there is a clear sense in which he does not trust his future degrees of belief, even though he does not think that he is, or will be, irrational, and even though he can acquire new evidence (the light being on or off). Let D_t denote the prisoner's degrees of belief at time t. Then, e.g., $D_{6.00}$(clock B is correct/$D_{11.30}$(clock B is correct) = 1/3) = 0. For $D_{11.30}$(clock B is correct) = 1/3 only occurs in worlds W_1 and W_3, and in each of those worlds clock B is not correct, and the prisoner knows this. Thus his current degrees of belief conditional upon his future degrees of belief do not equal those future degrees of belief. So he systematically distrusts his future degrees of belief. Strange indeed.

Sleeping Beauty

Some researchers are going to put Sleeping Beauty, SB, to sleep on Sunday night. During the two days that her sleep will last the researchers will wake her up either once, on Monday morning, or twice, on Monday morning and Tuesday morning. They will toss a fair coin Sunday night in order to determine whether she will be woken up once or twice: if it lands heads she will be woken upon Monday only, if it lands tails she will be woken up on Monday and Tuesday. After each waking, she will be asked what her degree of belief is that the outcome of the coin toss is heads. After she has given her answer she will be given a drug that erases her memory of the waking up, indeed it resets her mental state to the state that it was in on Sunday just before she was put to sleep. Then she is put to sleep again. The question now is: when she wakes up what should her degree of belief be that the outcome was heads?

Answer 1: Her degree of belief in heads should be 1/2. It was a fair coin and she learned nothing relevant by waking up.

Answer 2: Her degree of belief in heads should be 1/3. If this experiment is repeated many times, approximately 1/3 of the awakenings will be heads-awakenings, i.e. awakenings that happen on trials in which the coin landed heads.

Adam Elga[3] has argued for the second answer. I agree with him, and I agree with his argument. But let me amplify this view by giving a different argument for the same conclusion. Suppose that SB is a frequent and rational dreamer. Suppose in fact that every morning if SB is not woken up at 9 am, she dreams at 9 am that she is woken up at 9 am. Suppose that the dream and reality indistinguishable in terms of her experience, except that if SB pinched herself and she are dreaming, it does not hurt (and she doesn't wake up), while if she does this while she is awake it does hurt. And let us suppose that SB always remembers to pinch herself a few minutes after she experiences waking up (whether for real, or in a dream.) What should her degrees of belief when she experiences waking up? It seems obvious she should consider the 4 possibilities equally likely (the 4 possibilities being: Monday&Tails&Awake, Monday&Heads&Awake, Tuesday&Tails&Awake, Tuesday&Heads&Dreaming). If SB then pinches herself and finds herself to be awake, she should conditionalize and then have degree of belief 1/3 in each of the remaining 3 possibilities (Monday&Tails&Awake, Monday&Heads&Awake, Tuesday&Tails&Awake). Suppose now that at some point in her life SB loses the habit of dreaming. She no longer needs to pinch herself; directly upon waking she knows that she is not asleep. However, it seems clear that this lack of dreaming should make no difference as to her degrees of belief upon realizing that she is awake. The process now occurs immediately, without the need for a pinch, but the end result ought to be the same.

Here again the crucial assumption is commutativity: if the relevant evidence and experience collected is the same, then the order of collection should not matter for the final degrees of belief.[4] But there is clearly something very puzzling about such foreseeable changes in degrees of belief.

Duplication

Scenario 1 While you are at the beach, Vishnu tells you that, contrary to appearances, you have existed only for one month: Brahma created you one month ago, complete with all your memories, habits, bad back, and everything. What's more, says Vishnu, one month ago Brahma in fact created two human beings like you (you are one of them), in exactly the same environment, at two different ends of

---

[3]Elga, A. (2000): "Self-locating belief and the Sleeping Beauty problem", *Analysis* **60**, pp 143–147.

[4]Cian Dorr has independently arrived at the idea of using commutativity in order to argue for the degrees of belief that Elga advocates in the Sleeping Beauty case. See Dorr, C.: "Sleeping Beauty: in defence of Elga", forthcoming, *Analysis*.

the universe: one on earth, one on twin earth. Unfortunately, Vishnu has a further surprise for you: one month ago Shiva tossed a coin. If it landed heads Shiva will destroy the human being that is on twin earth one month from now. If it landed tails Shiva will do nothing. Vishnu does no tell you whether you are to be destroyed, but recommends that if you want to know, you should go check your mail at home. If there is a letter from president Bush for you, then you will be destroyed. Before running home, what degree of belief should you have in the 4 possibilities: Earth&Heads, Earth&Tails, Twin Earth&Heads, Twin Earth&Tails? It seems clear that you should have degree of belief 1/4 in each, or at the very least, that it is not irrational to have degree of belief 1/4 in each. You run home, and find no letter from Bush. What should your degrees of belief now be? Well, by conditionalization, they should now be 1/3 in each of the remaining possibilities (Earth&Tails, Twin Earth&Heads, Twin Earth&Tails). Consequently you should now have degree of belief 1/3 that the toss landed heads and 2/3 that it landed tails.
Scenario 2 same as scenario 1, except that Vishnu tells you that if the toss came heads, your identical twin was destroyed by Shiva one week ago. Since you were obviously not destroyed, you do not need to rush home to look for a letter from Bush. In essence you have learned the same as you learned in the previous scenario when you found you had no letter from Bush, and hence you should now have degree of belief 1/3 that the toss landed heads.
Scenario 3 same as scenario 2, except that Vishnu tells you that rather than that 2 beings were created one month ago by Brahma, one of them already existed and had the exactly the life you remember having had. This makes no relevant difference and you should now have degree of belief 1/3 that the coin landed heads.
Scenario 4 same as scenario 3, except that Vishnu tells you that if the die landed heads one month ago Shiva immediately prevented Brahma from creating the additional human being one month ago. The upshot is that only if the coin landed tails Brahma will have created the additional human being. Since the timing of the destruction/prevention makes no relevant difference you should again have degree of belief 1/3 that the coin landed heads.
Scenario 5[5] You are on earth, and you know it. Vishnu tells you that one month from now Brahma will toss a coin. If it lands tails Brahma will create, at the other end of the universe, another human being identical to you, in the same state as you will then be, and in an identical environment as you will then be. What do you now think that your degrees of belief should be in one month time? The answer is that they should be the same as they are in scenario 5, since in one month time you will be in exactly the epistemic situation that is described in scenario 5. Of course, it is plausible to claim that your future self will actually be on earth, since it is only your future continuation on earth that can plausibly be called "your future self". However, that does not mean that your future self can be sure that he is on earth. For your future self will know that he will have the same experiences and memories,

[5]This scenario is similar to the "Dr Evil scenario" in Elga, A. (manuscript): "Defeating Dr. Evil with self-locating belief".

whether or not he is on earth or on twin earth, and thus he will not know whether he can trust his memories. Thus you now have degree of belief ½ in heads, and yet you know that in one month time, you will have degree of belief 1/3. This is bizarre, to say the least.

Yet again, the crucial assumption in this reasoning is commutativity: your final degrees of belief should not depend on the order in which you receive all the relevant experience and evidence. You should end up with the same degrees of belief, namely degree of belief ½ in heads, whether you all along knew you were on Earth, or whether you only later found out that you were on Earth. But that can only be so if you had degree of belief 1/3 in heads prior to discovering that you were on Earth.

Diagnosis

Bas van Fraassen's Reflection Principle[6] says that one should trust one's future degrees belief in the sense that one's current degree of belief D_0 in any proposition X, given that one's future degree of belief D_t in X equals p, should be p: $D_0(X/D_t(X) = p) = p$. Given that one is sure that one will have precise degrees of belief at time t, the Reflection Principle entails that one's current degrees of belief equal the expectations of one's future degrees of belief: $D_0(X) = \sum pD_0(D_t(X) = p)$. The Reflection Principle is violated in each of the 5 puzzles that I have presented, for in each case there is a time at which one's expectation of one's future degree of belief in Heads differs from one's current degree of belief in Heads. This is presumably why we find these cases, prima facie, so worrying and strange.

The source of the problem, I claim, is that the degrees of belief of perfectly rational people, people who are not subject to memory loss or any other cognitive defect, can develop in ways that are as yet unrecognized, and indeed are not allowed according to standard Bayesian lore. Standard Bayesian lore has it that rational people satisfy the Principle of Conditionalization: rational people alter their degrees of belief only by strict conditionalization on the evidence that they acquire.[7] Strict conditionalization of one's degrees of belief upon proposition X can be pictured in the following manner. One's degrees of belief are a function on the set of possibilities that one entertains. Since this function satisfies the axioms of probability theory it is normalized: it integrates (over all possibilities) to 1.

---

[6]See van Fraassen (1995): "Belief and the problem of Ulysses and the sirens", *Philosophical Studies* 77: 7–37.

[7]Strict conditionalization: when one learns proposition X at t, one's new degrees of belief D_t equals one's old degrees of belief D_0 conditional upon X: $D_t(Y) = D_0(Y/X)$. One might also allow Jeffrey conditionalization. It matters not for our purposes.

Conditionalizing such a function on proposition X then amounts to the following: the function is set to 0 over those possibilities that are inconsistent with X, while the remaining non-zero part of the function is boosted (by the same factor) everywhere so that it integrates to 1 once again. Thus, without being too rigorous about it, it is clear that conditionalization can only serve to 'narrow down' one's degree of belief distribution (one really learns by conditionalization). In particular a degree of belief distribution that becomes more 'spread out' as time passes can not be developing by conditionalization, and a degree of belief distribution that exactly retains its shape, but is shifted as a whole over the space of possibilities, can not be developing by conditionalization. However, such spreading out and shifting is exactly what occurs in the 5 puzzles that I presented.

The reasons for such spreading and shifting are very simple. First let us consider shifting. Suppose that one knows exactly what the history of the world that one inhabits is like. And suppose that one is constantly looking at a clock one knows to be perfect. One's degrees of belief will then be entirely concentrated on one possible world, and at any given moment one's degrees of belief within that world will be entirely concentrated on one temporal location, namely the one that correspond to the clock reading that one is then seeing. And that of course means that the location that one's degree of belief distribution is concentrated at is constantly moving. That is to say, one's degree of belief distribution is constantly shifting, and such a constant shifting is simply not a case of Conditionalization. Self-locating beliefs will therefore genetically develop in ways that violate Conditionalization. John Collins's prisoner case involves exactly such a shifting of one's self-locating degrees of belief. The only difference is that in his case one additionally has an initial uncertainty as to which clock is accurate, i.e. one is initially uncertain whether one is in a world in which clock A is correct or one in which clock B is correct. It is somewhat surprising that this kind of violation of Conditionalization can be parlayed into a violation of Reflection. But Collins' prisoner case shows exactly how one can do this.

Next let us consider spreading. The simplest case of spreading is the case of the traveller that takes the path by the Mountains to Shangri La. His degrees of belief become more spread out when he arrives in Shangri La: at that time he goes from degrees of belief 1 in Heads and 0 in Tails, to degrees of belief ½ in Heads and ½ in Tails.[8] The reason why this happens is that there are two distinct possible

[8]Bas van Fraassen has, in conversation with me, has suggested that in such situations Conditionalization indeed should be violated, but Reflection should not. In particular he suggested that the degrees of belief of the traveler should become completely vague, upon arrival in Shangri La. This does not strike me as plausible. Surely upon arrival in Shangri La our traveler is effectively in the same epistemic situation as someone who simply knows that a fair coin has been tossed. One can make this vivid by considering two travelers, A and B. Traveler A never looks out of the window of the car, and hence maintains degree of belief ½ in heads all the way. (The memory replacement device does not operate on travelers who never look out of the window.) Traveler A, even by van Fraassen's lights, upon arrival in Shangri La, should still have degree of belief ½ in Heads. However, traveler B, does look out of the window during the trip. Upon arrival, by van Fraassen's lights, B's degrees of belief should become completely vague. But it seems odd to me that traveler B is epistemically penalized, i.e. is forced to acquire completely vague degrees of

experiential paths that end up in the same experiential state. That is to say the traveler's experiences earlier on determine whether possibility A is the case (Path by the Mountain), or whether possibility B is the case (Path by the Ocean). But because of the memory replacement that occurs if possibility B is the case, those different experiential paths merge into the same experience, so that that experience is not sufficient to tell which path was taken. Our traveler therefore has an unfortunate loss of information, due to the loss of the discriminating power of his experience. What is somewhat surprising is that this loss of discriminating power is not due to any loss of memory or any cognitive defect on his part: it is due to the fact that something strange would have happened to him had he taken the other path! This loss of discriminatory power of experience, and consequent spreading out of degrees of belief here does not involve self-locating degrees of belief. Suppose, e.g., that our traveler is the only person ever to travel along either paths. Then our traveler initially is unsure whether he is in a world in which path A is never taken or whether he is in a world in which path B is never taken. He then becomes sure that he is in a world in which path B is never taken. Even later, upon arrival, he again becomes unsure as to which world he is in. None of this has anything to do with self-locating beliefs.[9]

The source of the Sleeping Beauty and Duplication problems is exactly the same. In the case of Sleeping Beauty the possibility of memory erasure ensures that the self-locating degrees of belief of Sleeping Beauty, even on Monday when she has suffered no memory erasure, become spread out over two days. In the Duplication case, yet again, the possible duplication of experiences forces one to become uncertain as to where (or who) one is. The cause of the spreading of degrees of belief in both cases is "experience duplication", and has nothing to do with the self-locating nature of these beliefs.[10]

It is not very surprising that the spreading of degrees of belief can bring about a violation of Reflection. For instance, in the non-self locating case a predictable reduction from degree of belief 1 in some proposition X to anything less than 1 will

---

belief, just because he looked out of the window during the trip, when it seems clear that he ends up in exactly the same epistemic position as his companion, who did not look out of the window.

[9]It is obvious how to generalize this case to a case in which there are memory replacement devices at the end of both roads, where these memory replacement devices are indeterministic, i.e. when it is the case that for each possible path there are certain objective chances for certain memories upon arrival in Shangri La. For, given such chances (and the Principal Principle), one can easily calculate the degrees of belief that one should have (in heads and tails) given the memory state that one ends up with. And, generically, one will still violate Conditionalization and Reflection.

[10]Some people will balk at some of the degrees of belief that I have argued for in this paper, in particular in the self-locating cases. For instance, some people will insist that tomorrow one should still be certain that one is on Earth, even when one now knows (for sure) that a perfect duplicate of oneself will be created on Mars at midnight tonight. I beg to differ. However, even if in this case, and other cases, one disagrees with me as to which degrees of belief are rationally mandated, the main claim of this paper still stands. The main claim is that in such cases of possible experience duplication, it is at the very least rationally permissible that one's degrees of belief become more spread out as time progresses, and hence rational people can violate Conditionalization and Reflection.

immediately violate Reflection: now you know it, now you don't. The argument is slightly less straightforward in the self-locating case. Consider, e.g., a case in which one is on Earth and one knows that at midnight a duplicate of oneself will be created on Mars. One might claim that since one now is certain that one is on Earth, and at midnight one will be uncertain as to whether one is on Earth, that one has a clear violation of Reflection. However, this is too quick. To have a clear violation of Reflection it has to be the very same "content of belief" such that one's current degree of belief differs from one's expectation of one's future degree of belief. Depending on what one takes to be the contents of belief when it concerns self-locating beliefs (propositions? maps from locations to propositions? .....?), one might argue that the contents of belief are not the same at the two different times, and hence there is no violation of Reflection. However the arguments of sections IV and V show that one can in any case parlay such spreading of self-locating degrees of belief into violations of Reflection concerning such ordinary beliefs as to whether a coin lands Heads or Tails. So Reflection is suckered anyhow.

Finally, the original case of the prisoner involves both a spreading of degrees of belief and a shifting of degrees of belief. The shifting is due simply to the passage of time and the self-locating nature of the beliefs. The spreading is due to the fact that our prisoner does not have experiences that are discriminating enough to pick out a unique location in time.[11] The analysis of section II shows, yet again, that such a spreading and shifting of self-locating degrees of belief can be parlayed into a violation of Reflection concerning such ordinary beliefs as to whether a coin lands Heads or Tails.

Conclusions

The degrees of belief of rational people can undergo two as yet unrecognized types of development. Such degrees of belief can become more spread out due to the duplication of experiences, or more generally, due to the loss of discriminating power of experiences, and thereby violate Conditionalization. In addition self-locating degrees of belief will generically be shifted over the space of possible

[11]One might model the prisoner here as having unique distinct experiences at each distinct, external clock, time, and as initially having precise degrees of belief over the possible ways in which those experiences could correlate to the actual, external clock, time. If one were to do so then the prisoner would merely be initially uncertain as to which world he was in (where worlds are distinguished by how his experiences line up with the actual, external clock, time), but for each such possible world would be always certain as to where he was located in it. And, if one were to do so, then the original prisoner case would be essentially the same case as Collins's prisoner case: no uncertainty of location in any given world, merely an initial uncertainty as to which world one is in, and a subsequent shifting of the locally concentrated degrees of belief within each of the possible worlds. However, there is no need to represent the original prisoner case that way. Indeed it seems psychologically somewhat implausible to do so. More importantly, the arguments and conclusions of this paper do not depend on how one models this case.

locations, due to the passage of time, and thereby violate Conditionalization. Such violations of Conditionalization can be parlayed into violations of Reflection, and lead to a distrust of one's future degrees of belief. Strange, but not irrational.

Acknowledgements I would like to thank John Collins, Adam Elga, John Hawthorne, Isaac Levi, Barry Loewer, and Tim Maudlin for extensive and crucial comments and discussions on earlier versions of this paper.

Chapter 11
Stopping to Reflect

Mark J. Schervish, Teddy Seidenfeld, and Joseph B. Kadane

Our note is prompted by a recent article by Frank Arntzenius, "Some Problems for Conditionalization and Reflection".[1] Through a sequence of examples, that article purports to show limitations for a combination of two inductive principles that relate current and future rational degrees of belief: *Temporal Conditionalization* and *Reflection*:

(i) *Temporal Conditionalization* is the rule that, when a rational agent's corpus of knowledge changes between *now* and *later* solely by learning the (new) evidence, B, then coherent degrees of belief are updated using conditional probability according the formula, for each event A,

$$P_{later}(A) = P_{later}(A|B) = P_{now}(A|B)$$

(ii) *Reflection*[2] between *now* and *later* is the rule that current conditional degrees of belief defer to future ones according to the formula that, for each event A,[3]

$$P_{now}(A|P_{later}(A) = r) = r.$$

[1] *The Journal of Philosophy* Vol C, Number 7 (2003), 356–370.

[2] See B.van Frasseen's "Belief and the Will," this *Journal*, 81 (1984), 235–256. van Fraassen's *Reflection* has an antecedent in M.Goldstein's "Prevision of a Prevision," *JASA* 78 (1983): 817–819.

[3] Here and through the rest of this note 'r' is a standard designator for a real number – this in order to avoid Miller-styled problems. See, D.Miller's "A Paradox of Information," *Brit. J. Phil. Sci.* 17 (1966):144–147.

M.J. Schervish (✉) • T. Seidenfeld • J.B. Kadane
Carnegie Mellon University, Pittsburgh, PA, USA

H. Arló-Costa et al. (eds.), *Readings in Formal Epistemology*, Springer Graduate Texts in Philosophy 1, DOI 10.1007/978-3-319-20451-2_11

We will use the expression "*Reflection* holds with respect to the event A." to apply to this equality for a specific event A.

It is our view that neither of these principles is mandatory for a rational agent.[4] However, we do not agree with Arntzenius that, in the examples in his article, either of these two is subject to new restrictions or limitations beyond what is already assumed as familiar in problems of stochastic prediction.

To the extent that a rational person does not know *now* exactly what she or he will know in the future, anticipating one's future beliefs involves predicting the outcome of a stochastic process. The literature on stochastic prediction relies on two additional assumptions regarding states of information and the temporal variables that index them[5]:

(iii) When $t_2 > t_1$ are two fixed times, then the information the agent has at t_2 includes all the information that she or he had at time t_1.[6] This is expressed mathematically by requiring that the collection of information sets at all times through the future form what is called a *filtration*.

Second, since the agent may not know *now* the precise time at which some specific information may become known in the future, then future times are treated as *stopping times*. That is:

(iv) For each time T (random or otherwise) when a prediction is to be made, the truth or falsity of the event $\{T \leq t\}$ is known at time t, for all fixed t. Such (random) times T are called *stopping times*.

In this note, we apply the following three results[7] to the examples in Arntzenius' article. These results, we believe, help to explain why the examples at first appear puzzling and why they do not challenge either *Temporal Conditionalization* or *Reflection*. Result 11.1 covers the ordinary case, where *Reflection* holds. Results 11.2 and 11.3 establish that *Reflection* will fail when one or the other of the

---

[4]We have argued, for example, that when (subjective) probability is finitely but not countably additive, then there are simple problems where (*i*) is reasonable, but where (*i*) precludes (*ii*). See our "Reasoning to a Foregone Conclusion," *JASA* 91 (1996): 1228–1236. Also, Levi argues successfully, we think, that (*i*) is not an unconditional requirement for a rational agent. See his "The Demons of Decision, " *The Monist* 70 (1987): 193–211.

[5]See, for example, section 35 of P.Billingsley, *Probability and Measure* 3*rd* edition, J.Wiley, 1995.

[6]Here and through the rest of this note 't' is a standard designator for a real number for time. More precisely, we use the subscripted variable, e.g. 't_1' to denote a specific time as the agent of the problem is able to measure it. We presume that the agent has some real-valued "clock" that quantifies a transitive relation of "is later than." Subtleties about the differences between how time is so indexed for different observers is relevant to one of Arntzenius' puzzles, to wit, the Prisoner's Problem.

[7]Proofs for these three are given in the "Appendix". In this note, we assume that all probability is countably additive.

two additional assumptions, (*iii*) and (*iv*) fail. It is not hard to locate where these assumptions are violated in the examples that Arntzenius presents.

Result 11.1 When "*later*" is a *stopping time*, when the information sets of future times form a *filtration*, and assuming that degrees of belief are updated by *Temporal Conditionalization*, then *Reflection* between *now* and *later* follows.

Result 11.2 When the information known to the agent over time *fails* to form a filtration, not only is *Temporal Conditionalization* vacuously satisfied (as its antecedent fails), but then *Reflection* fails unless what is forgotten in the failure of filtration becomes practically certain (its probability becomes 0 or 1) in time for future predictions, *later*.

Result 11.3 However, if the information known to the agent over time forms a filtration and *Temporal Conditionalization* holds, but "*later*" is not a *stopping time*, then *Reflection* between *now* and *later* holds for the specific event A, i.e., $P_{now}(A|P_{later}(A) = r) = r$, subject to the necessary and sufficient condition, (11.1), below.

Let H_t be the event "$t = later$." When *later* is not a stopping time, the event H_t is news to the agent making the forecasts. The question at hand is whether this news is relevant to the forecasts expressed by *Reflection*. To answer that question, concerning such forecasts about the event A, define the quantity $y_t(A)$ by

$$y_t(A) = \frac{P_{now}(H_t|P_t(A) = r\&A)}{P_{now}(H_t|P_t(A) = r)}.$$

The quantity $y_t(A)$ is an index of the current conditional dependence between A and H_t, given that $P_t(A) = r$. For example, $y_t(A) = 1$ if and only if A and H_t are conditionally independent for the agent, *now*, given that $P_t(A) = r$. In other words, by symmetry of conditional independence, $y_t(A) = 1$ if and only if the agent's current conditional probability of A given that $P_{later}(A) = r$, is unchanged by the added information H_t.

Reflection holds for A between *now* and *later*, $P_{now}(A|P_{later}(A) = r) = r$ if and only if, given $P_{later}(A) = r$, the conditional expected value $y_T(A) = 1$. Specifically, if and only if

$$1 = \sum_t y_t(A) P_{now}(H_t|P_{later}(A) = r) \tag{11.1}$$

Thus, *Reflection* is satisfied between *now* and *later* if and only if (11.1) holds for each A.

Next, we illustrate the second and third results with examples that show how *Reflection* may fail.

Example 11.1 (Illustrating Result 11.2)

Suppose that the agent will observe a sequence of coin tosses, one at a time at a known rate, e.g. one toss per minute. Let $X_n = 1$ if the coin lands heads up on toss n, and let $X_n = 0$ otherwise. The agent does not know how the coin is loaded, but believes that it is *fair* (event A) with personal probability ½, and that with personal probability ½ it is biased with a chance of ¾ for landing tails (event A^c). Also, he believes that tosses are conditionally independent given the loading, i.e., given that the coin is *fair* or given that it is biased ¾ for tails.

Time is indexed for the agent by the number of the most recent coin toss. The time "*now*" occurs after the first toss ($t = n = 1$), and "*later*" denotes the time ($t = n = 2$) just after the second toss. Unfortunately, at each time t, the agent knows that he can remember only the most recent flip, X_t, though he knows which numbered toss it is because, e.g., he can see a clock. Suppose that the first toss lands heads up, which is the event $C = \{X_1 = 1\}$. The information that will be available to the forgetful agent later (at $t = 2$) will be only that either $B_1 = \{X_2 = 1\}$ or $B_0 = \{X_2 = 0\}$. He will not recall C because of his predictable memory lapse, and he knows all this. It is straightforward to compute:

$$P_{later}\,(A|B_1) = 2/3 \text{ and } P_{later}\,(A|B_0) = 2/5.$$

However, at $t = 1$, the agent's conditional probability for A, given event B_1 occurring at $t = 2$, satisfies $P_{now}(A \mid B_1) = 4/5$. Similarly, if *now* he conditions on event B_0 occurring at $t = 2$, his conditional probability will satisfy $P_{now}(A \mid B_0) = 4/7$.

Of course, *Temporal Conditionalization* holds vacuously at the *later* time, since the information sets available to the agent do not form a filtration. *Reflection* fails in this setting, as the agent does not remember at the *later* time what happened *now*, and he knows this all along. If B_1 occurs then $P_{later}(A) = P_{later}\,(A|B_1) = 2/3$, and if B_0 occurs then $P_{later}(A) = P_{later}\,(A|B_0) = 2/5$. Hence,

$$P_{now}\,(A|P_{later}(A) = 2/3) = 4/5$$

and

$$P_{now}\,(A|P_{later}(A) = 2/5) = 4/7. \qquad \square$$

Example 11.2 (Illustrating Result 11.3 when condition (11.1) fails and then *Reflection* fails too)

Modify Example 11.1 so that the agent has no memory failures and updates his degrees of belief by *Temporal Conditionalization*. Also, change the time "*now*" to denote the minute prior to the first toss, i.e., *now* is $t = n = 0$. Define the time "*later*" to be the *random time, T, just prior to the first toss that lands heads up*. From the point of view of the agent, the quantity T is not an observable random variable up to and including time T, and it is not a *stopping time* either. It is observable to the agent starting with time $T+1$, of course, as by then he will have seen when the first head occurs.

With probability 1 the possible values for T are $T = 0, 1, 2, \ldots$. It is straightforward to verify that: $P_{later}(A) = [1 + (3/2)^n]^{-1}$, when $T = n$, for $n = 0, 1, 2,$

.... Notice that $P_{later}(A) \leq ½$, no matter when T occurs, and $P_{later}(A) < ½$ for $T > 0$, since if $T > 0$, the initial sequence of tosses that the agent observes all land tails up. However, from the value of $P_{later}(A)$ and knowing it is this quantity, one may calculate T exactly and thus know the outcome of the $n + 1$st toss, which is a heads. But when the agent computes $P_{later}(A)$ at the time *later*, he does not then know that *later* has arrived. Thus, *later*, he is not in a position to use the extra information that he would get from knowing when T occurs to learn the outcome of the $n + 1$st toss. To repeat the central point, T is not a stopping variable.

It is evident that *Reflection* fails, $P_{now}(A \mid P_{later}(A) = r) \neq P_{later}(A)$. The extra information, namely that $P_{later}(A) = r$ rather than merely that $P_t(A) = r$ where t is the time on the agent's clock, is information that is relevant to his *current* probability of A, since it reveals the outcome of the next toss. Even *now*, prior to any coin tosses, when he computes $P_{now}(A \mid P_{later}(A) = r)$, the conditioning event reveals to him the value of T, since n is a function of r. In this case, the conditioning event entails the information of n and when the first heads occurs, namely, on the $n + 1$st toss. Then *Reflection* fails as

$$P_{now}\left(A|P_{later}(A) = [1 + (3/2)^n]^{-1}\right) = \left(1 + 3^n/2^{n+1}\right)^{-1}.$$

It remains only to see that (11.1) fails as well. Consider the quantity $y_t(A)$ used in condition (11.1). $y_t(A) = \frac{P_{now}(H_t|P_t(A)=r\&A)}{P_{now}(H_t|P_t(A)=r)}$. Given $P_t(A) = r$, the added information that A obtains is relevant to the agent's current probability when *later* occurs. Specifically, as $P_t(A) = [1 + (3/2)^n]^{-1}$ entails that $t = n$,

$$\begin{aligned} P_{now}\left(H_t|P_t(A) = [1 + (3/2)^n]^{-1}\right) &= P_{now}\left(X_{t+1} = 1|P_t(A) = [1 + (3/2)^n]^{-1}\right) \\ &= (1/2)\,[1 + (3/2)^n]^{-1} + (1/4)\,(3/2)^n[1 + (3/2)^n]^{-1} < \tfrac{1}{2} \\ &= P_{now}\left(X_{t+1} = 1|P_t(A) = [1 + (3/2)^n]^{-1}\&A\right) \\ &= P_{now}\left(H_t|P_t(A) = [1 + (3/2)^n]^{-1}\&A\right). \end{aligned}$$

Thus, $y_t > 1$.
Hence, $1 < \sum_t y_t(A) P_{now}\,(H_t|\, P_{later}(A) = r)$. □

Example 11.3 (Illustrating Result 11.3 when (11.1) obtains and *Reflection* holds even though *later* is not a *stopping time*)

In this example, consider a sequence of three times, $t = 0$, 1, and 2. *Now* is time $t = 0$. The available information increases with time, so that the information sets form a filtration, and the agent updates his degrees of belief by *Temporal Conditionalization*. Let the random time *later* be one of the two times $t = 1$, or $t = 2$, chosen at random, but which one is not revealed to the agent. Let the event H_i be that *later* $= i$, ($i = 1$, 2) and suppose that the occurrence of H_i (or its failure) while not known to the agent at any of the three times is independent of all else that the agent does know at all three times. In this case, for each event A (even for $A = H_i$) Eq. (11.1) is satisfied. That is, by the assumptions of the problem, either

$y_t(A) = \frac{P_{now}(H_t|y_t(A)=r\&A)}{P_{now}(H_t|P_t(A)=r)} = 1$, or if $A = H_i$ then $y_t(A) = \frac{P_{now}(H_t|P_{later}(A)=r)}{P_{now}(H_t|P_t(A)=r)} = 1$. Thus, $P_{now}(A \mid P_{later}(A) = r) = r$. That is, even though *later* is not a stopping time, *Reflection* holds in this case since, given that $P_{later}(A) = r$ no new (relevant) evidence about A is conveyed through knowing that *later* has arrived, H_i. □

We note that Result 11.2 applies to the *Sleeping Beauty*[8] *Shangri La*, and *Duplication* examples of Arntzenius' article, where known failures of memory are explicit to the puzzles. Result 11.3 applies to explain the failure of *Reflection* in the two versions of the "Prisoner" example where the *local time* in the story, as measured by an ordinary clock (e.g., "11:30 PM" in John Collins's example) is not a *stopping time* for the Prisoner.

It is our impression of Collins's *Prisoner* example that the reader is easily mistaken into thinking that the *local time*, as measured by an ordinary clock in the story, is a *stopping time* for all the characters in the story. Then *Reflection* holds for each of them, in accord with Result 11.1. In Collins' example, the *local time*, e.g., 11:30 PM, is a *stopping time* for the Jailor (and also for the reader), but *not* for the Prisoner. For the Prisoner, time is measured by real-valued increments over the starting point, denoted by "now." Increments of *local time* are *stopping times* for the Prisoner. This is because the Prisoner does not know at the start of the story which of two *local times* equals his time *now*. For subsequent times, he does know how much *local time* has elapsed since *now*. But that information is not equivalent to knowing the *local time*. That difference in what is a *stopping time* for different characters is what makes this a clever puzzle, we think.

Acknowledgements Our research was carried out under NSF Grant DMS 0139911. We thank Joseph Halpern for alerting one of us (*T.S.*) to the *Sleeping Beauty* problem, independent of Arntzenius' article.

[8]See also J.Y.Halpern's "Sleeping Beauty Reconsidered: Conditioning and Reflection in Asynchronous Systems," Dept. of Computer Science, Cornell University. September, 2003. We agree with Halpern that, in our words, coherence of a sequence of previsions does not require that they will be *well calibrated* – in a frequency sense of "well calibrated." That is, we think it is reasonable for Sleeping Beauty to give a prevision of ½ to the event that the known fair coin landed heads on the flip in question, each time she is woken up. What complicates the analysis is that the repeated trials in Sleeping Beauty's game do not form an independent sequence, and her mandated forgetfulness precludes any "feedback" about the outcome of past previsions. When repeated trials are dependent and there is no learning about past previsions, coherent previsions may be very badly calibrated in the frequency sense. For other examples and related discussion of this point see, e.g., Seidenfeld, T. (1985) "Calibration, Coherence, and Scoring Rules," *Philosophy of Science* 52: 274–294.

Appendix

Proof of Result 11.1[9] Assume that when X is a random variable and C is an event, the agent's expected value $E_P(X)$ and conditional expected value $E_P(X|\ C)$ exist with respect to the probability P. Let A be an event and let $X = P(A|Y)$ be a random variable, a function of the random variable Y. Then, as a consequence of the law of total probability, with C also a function of Y,

$$P\,(A|C) = E_P\,[X|C]\,. \tag{11.2}$$

Assume that the agent's degrees of belief *now* include his *later* degrees of belief as objects of uncertainty. That is, future events such as "$P_{later}(A) = r$" and "$P_{later}(A\ |\ C) = q$" are proper subjects, *now*, of the agent's current degrees of belief. Suppose that, *now*, the agent anticipates using (*i*) *Temporal Conditionalization* in responding to the new evidence $Y = y$ that he knows he will learn at the stopping time, *later*. For example, Y might be the result of a meter reading made at the *later* time, with a sample space of m many possible values $Y = \{y_1,\ \ldots,\ y_m\}$. Thus, by (*i*), for whichever value y of Y that results,

$$P_{later}(A) = P_{later}\,(A|Y = y) = P_{now}\,(A|Y = y)\,. \tag{11.3}$$

Then, by (*i*) and (11.2), for C also a function of Y, the agent *now* believes that

$$P_{now}\,(A|C) = E_{P_{now}}\,[P_{later}(A)|C]\,. \tag{11.4}$$

Let C be the event, "$P_{later}(A) = r$," which we presume is a possible value for $P_{later}(A)$ from the agent's current point of view. (This C is function of Y.) Then, because *later* is a stopping time,

$$P_{now}\,(A|P_{later}(A) = r) = E_{P_{now}}\,[P_{later}(A)|P_{later}(A) = r)]. \tag{11.5}$$

As

$$E_{P_{now}}\,[P_{later}(A)|P_{later}(A) = r)] = r, \tag{11.6}$$

[9] van Fraassen (1995) "Belief and the Problem of Ulysses and the Sirens," *Phil. Studies* 77: 7–37, argues (pp. 17–19) that *Temporal Conditionalization* implies *Reflection*. His argument (pp. 18–19) has an additional, tacit assumption that the time t at which conditioning applies for *Reflection* is a stopping time.

therefore

$$P_{now}\,(\mathrm{A}|P_{later}(A) = r) = r, \tag{11.7}$$

i.e., then *Reflection* holds as well. □

Proof of Result 11.2 To show that *Reflection* fails, consider two times $t_1 < t_2$. Call an event *forgotten* if its truth or falsity is known at time t_1 but not at time t_2. From the assumption that these times do not form a filtration, let E be forgotten between t_1 and t_2 and allow that at t_1 this is known to happen at t_2. Since $P_{t_1}(E) \in \{0, 1\}$, conditioning will not change this value, i.e.,

$$P_{t_1}(E) = P_{t_1}\,(E|P_{t_2}(E) = r) \tag{11.8}$$

for a set of r-values of probability 1 under P_{t_1}. But, since it is known at t_1 that E will be forgotten at t_2, $P_{t_1}\,(0 < P_{t_2}(E) < 1) = 1$. Hence *Reflection* fails as $0 < r < 1$ in (11.8). □

Proof of Result 11.3 Assume that the agent's information sets form a filtration over time and that *Temporal Conditionalization* holds between *now* and *later* but that *later* is not a stopping time for the agent. Let H_t be the event "*later* $= t$" for the specific time t. That is, assume that $0 < P_{later}(H_t) < 1$, when *later* occurs at t.

Later is the future time we will focus on in calculating whether *Reflection* holds, i.e. we will inquire whether for each event A, $P_{now}(\mathrm{A} \mid P_{later}(A) = r) = r$, or not. We calculate as follows.

$$P_{now}\,(A|P_{later}(A) = r)$$

$$= \sum\nolimits_t P_{now}\,(A\&H_t|P_{later}(A) = r)$$

by the law of total probability.

$$= \sum\nolimits_t P_{now}\,(A|P_{later}(A) = r\&H_t)\,P_{now}\,(H_t|P_{later}(A) = r)$$

by the multiplication theorem

$$= \sum\nolimits_t \frac{P_{now}\,(H_t|P_t(A) = r\&A)}{P_{now}\,(H_t|P_t(A) = r)} P_{now}\,(A|P_t(A) = r)\,P_{now}\,(H_t|P_{later}(A) = r)$$

by Bayes' theorem and the equivalence of

$$(P_{later}(A) = r\&H_t) \;\; and \;\; (P_t(A) = r\&H_t)$$

$$= r\sum_t \frac{P_{now}\,(H_t|P_t(A) = r\&A)}{P_{now}\,(H_t|P_t(A) = r)} P_{now}\,(H_t|P_{later}(A) = r)$$

as $P_{now}\,(A|P_t(A) = r) =$ r by Result 11.1.

$$= r\sum_t y_t(A)P_{now}\,(H_t|P_{later}(A) = r).$$

by the definition of $y_t(A)$

Hence, $P_{now}\,(A|P_{later}(A) = r) = r$ if and only if $\sum_t y_t(A)P_{now}(H_t|P_{later}(A) = r)$ $= 1$, which is condition (11.1). □

Part II
Belief Change

Chapter 12
Introduction

Horacio Arló-Costa, Vincent F. Hendricks, and Johan van Benthem

It is well known that the usual versions of probability kinematics have serious limitations. According to the classical notion of conditioning when one learns a piece of information A its probability raises to its maximum (one). Moreover no further instance of learning will be capable of defeating A. Once a piece of information is learned one should be maximally confident about it and this confidence should remain unaltered forever. It is clear that there are many instances of learning that cannot be accommodated in this Procrustean bed. There are various ways of amending this limited picture by enriching the Bayesian machinery. For example, one can appeal to a notion of primitive conditional probability capable of making sense of conditioning on zero measure events. But the detailed consideration of this alternative leads to similar limitations: the picture of learning that thus arises continues to be cumulative. There are many ways of overcoming these important limitations. Williamson considers one possible way of doing so in his essay reprinted in the section on Bayesian epistemology. One of the lessons that have been learned in recent years is that there is no apparent way of circumventing this rigidity of Bayesianism without introducing in some way a qualitative doxastic or epistemic notion as a primitive alongside probability. Here are two examples:

Horacio Arló-Costa was deceased at the time of publication.

H. Arló-Costa (deceased)
Carnegie Mellon University, Pittsburgh, PA, USA

V.F. Hendricks (✉)
Center for Information and Bubble Studies, University of Copenhagen, Copenhagen, Denmark
e-mail: vincent@hum.ku.dk

J. van Benthem
University of Amsterdam, Science Park 904, 1098 XH Amsterdam, The Netherlands

Stanford University, Stanford, United States
e-mail: johan@science.uva.nl

H. Arló-Costa et al. (eds.), *Readings in Formal Epistemology*, Springer Graduate Texts in Philosophy 1, DOI 10.1007/978-3-319-20451-2_12

Williamson proposes a model where knowledge is a primitive, while Levi appeals to a primitive notion of *full belief*.

The traditional work in belief revision has followed Levi in adopting some doxastic notion as a primitive (which does not need to be full belief, it could be Spohn's *plain belief*, etc). But if such a qualitative notion should be introduced primitively how should one go about representing it? And how one would go about characterizing changes of the represented notion? There are many possible options from sentential representations (that can be finite or not) to propositional representations that can, in turn, incorporate more fine-grained non-probabilistic structure than mere beliefs (entrenchment or some notion of plausibility, for example).

The traditional approach derived from the seminal work of Carlos Alchourrón, Peter Gärdenfors and David Makinson (AGM) opted for a sentential representation. They were ambivalent between using finite representations (usually called belief bases) and using more structured representations of belief (theories in logical parlance or belief sets). Isaac Levi proposed a philosophical justification for the use of theories: they are supposed to represent not belief but commitments to belief.

So, how should one go about contracting a sentence A from a theory K? Merely deleting A will not do, given that the sentence might be entailed by other sentences in K. The AGM solution to the problem is simple. Enter the *remainder set* of K with A (denoted $K \perp A$). This is the set of all maximal subsets of K that fail to entail A. One then makes a selection from the members of this set via the use of a selection function γ. This gives us $\gamma(K \perp A)$. Finally one takes the intersection of the resulting set. This is usually called a *partial meet* contraction. Of course there are many possible selection functions that can be used in the previous account. AGM would in turn require that the selection can be rationalized in the usual sense common in rational choice, i.e. by assuming that the selection function selects the *best* elements of $K \perp A$ (with respect to an underlying transitive preference relation). Then AGM manage to offer a set of rationality postulates that completely characterizes this operation. This is the main result presented in the article reprinted here. This article single handedly created an entire field of research by combining ideas from rational choice and classical proof techniques from philosophical logic.

If one reads the original AGM article carefully one sees that much of the inspiration from the use of selection functions in their model comes from the seminal work of Amartya Sen in rational choice. Exploiting results from the theory of choice Hans Rott (and previously Sten Lindström) systematically studies the relationship between functional constraints placed on selection functions and postulates of belief change. Among other things, Rott shows that certain functional constraints placed on propositional selection functions correspond in a one-to-one fashion to postulates of belief change. Rott's results forge a useful bridge between the mathematical theories of belief change and rational choice.

Still one might inquire why the feasible set from which one chooses rationally via a selection function should be limited to $K \perp A$. One common defense of this idea is in terms of minimizing information loss. The elements of $K \perp A$, usually called maxi-choice sets, do satisfy this requirement. But then if they are optimal until this point of view why not to use a selection function that picks singletons from K⊥*A*? AGM showed that this type of contraction is badly behaved. So is the opposite idea of taking directly the intersection of the entire K⊥*A*. So, partial meet appears as an Aristotelian middle ground that happens to satisfy a set of intuitive postulates. Or so argued AGM. Nevertheless the subsequent discussion focused on some controversial AGM postulates like *recovery* (requiring that if one contracts K with *A* and then adds *A* to the result of this contraction one returns to K). There are many putative counterexamples to recovery and this generated the interest in defining notions of contraction that fail to satisfy recovery. Isaac Levi is a well-known defender of this line of thought and in his article he characterizes a notion of contractions that does fail to satisfy recovery. The central idea he proposes is that what is minimized in contraction is not information loss but loss of *informational value*. The notion of information value is some sort of epistemic utility obeying basic structural postulates like:

(Weak Monotony) If $X \subset Y$, then $V(X) \leq V(Y)$.

This is an intuitive principle that makes permissible that two sets carry equal informational value even when one the sets carries more information than the other. The additional information might not be valuable at all and therefore the level of informational value of the larger set might remain equal to the informational value of the smaller set. What other constraints one should impose on information value? In the article reprinted here Levi presents a very specific form of information value that he uses to characterize a particular notion of withdrawal (some rational notion of contraction where recovery fails) that he calls *mild contraction*. Rott and Pagnucco offered an alternative model of the same notion that they call *severe withdrawal*. It is clear that when epistemic utility satisfies the constraints proposed by Levi this particular form of contraction obtains. What seems to be missing is a pre-systematic explanation of why epistemic utility should satisfy these constraints or a justification of some controversial properties of severe withdrawal (like the postulate of *antitony*). It is *A* true thought that the introduction of epistemic utility in models of belief change opens up an insightful research strategy that at the moment remains relatively unexplored.

Sven Ove Hansson offers another account of contraction that fails to obey recovery. Nevertheless he arrives at this conclusion in a completely different way. In fact, Hansson is one of the most prominent defenders of finite models of belief in terms of belief bases (finite sets of sentences that are one of the possible axiomatic bases of a given theory). It is easy to characterize a version of partial meet contraction for bases by using the obvious variant of the definition used for theories. Then one can proceed as follows: an operation of contraction on a belief set K is generated by a partial meet base contraction if and only if there is a belief base B for K and an operator $\sim$ of partial meet contraction for B such that the contraction of K with *A* yields the logical consequences of (B $\sim$ *A*) for all sentences *A* in the underlying language. Hansson shows that if an operation on a belief set is generated by some partial meet base contraction, then it satisfies the classical AGM postulates for contraction except recovery. In addition the operation satisfies other postulates encoding a specific notion of *conservativity*.

The article by Spohn articulates an important epistemological idea, namely that one should focus on changes of entire *epistemic states* endowed with more structure than mere belief. This approach, in a more general setting, is also independently pursued by Adnan Darwiche and Judea Pearl in Darwiche and Pearl (1996). Spohn focuses on a particular type of epistemic state that now is usually called a *ranking function*. Roughly a ranking function is a function from the set of propositions (= sets of possible worlds) to the set of natural, real, or ordinal numbers, similar to a probability measure. Epistemologically one can see such functions as numerical (but non-probabilistic) representations of a notion of *plausibility*. In the presence of a new input the current ranking is mapped to a new ranking incorporating the incoming information (in revision). This is an ideal setting to study the structure of iterated changes of view and as a matter of fact both articles offer the best existing articulation of principles regulating iterated change. This is an important area of research in this field that still remains relatively open.

Suggested Further Reading

There are a number of recent surveys and books that complement the reprinted papers appearing here. Regarding surveys the two most recent surveys are: Logic of Belief Revision, in *Stanford Encyclopedia of Philosophy*, 2006, by Sven Ove Hansson; and: Belief Revision in *The Continuum Companion to Philosophical Logic*, (eds.) L. Hornsten and R. Pettigrew, by Horacio Arlo-Costa and Paul Pedersen. These surveys contain references to previous surveys in the field. A classic book in this area that continues to be useful is Peter Gärdenfors's monograph: *Knowledge in Flux: Modeling the Dynamic of Epistemic States*, College Publications (June 2, 2008). A very useful textbook presentation of some of the main results in the theory of belief change is: *A Textbook of Belief Dynamics: Theory Change and Database Updating*, Springer 2010, by Sven Ove Hansson. The book focuses mainly on syntactic presentations of belief change and it contains a very detailed presentation of belief base updating. Some more recent topics like iterated belief change are not treated in detail though.

Decision theoretic foundations for belief change are provided in various books by Hans Rott and Isaac Levi (independently). A book-length argument articulating Rott's account (and extending the content of the article reprinted here) appears in: *Change, Choice and Inference: A Study of Belief Revision and Non-monotonic Reasoning*, Oxford Logic Guides, 2001. Some challenges to this type of foundational strategy are considered by Arlo-Costa and Pedersen in: "Social Norms, Rational Choice and Belief Change," in *Belief Revision Meets Philosophy of Science*, (eds.) E.J. Olsson and S. Enqvist, Springer, 2011. Isaac Levi has also published various essays where he presents decision theoretic foundations for belief change (but his account is rather different than Rott's). The most recent book presenting Levi's current views about belief change is: *Mild Contraction: Evaluating Loss of Information Due to Loss of Belief*, Oxford, 2004. Further references to his work can be found in this book.

The previous accounts tried to justify principles of belief change in the broader context of Bayesian or neo-Bayesian theory. An almost orthogonal view consists in deriving principles of belief change by taking some form of formal learning theory as an epistemological primitive. While all the previous accounts focused on justifying the next step of inquiry (or a finite and proximate sequence of steps) this second strategy focuses on selecting belief change methods capable of learning the truth in the long run. One important paper in this tradition is Kevin Kelly's: Iterated

Belief Revision, Reliability, and Inductive Amnesia, *Erkenntnis*, 50, 1998 pp. 11–58. Daniel Osherson and Eric Martin present a similarly motivated account that nevertheless is formally quite different from Kelly's theory in: *Elements of Scientific Inquiry*, MIT, 1998.

There are various attempts to extend the theory of belief revision to the multi-agent case and to present a theory of belief change as some form of *dynamic epistemic logic*. The idea in this case is to use traditional formal tools in epistemic logic to represent the process of belief change. Hans van Ditmarsch, Wiebe van der Hoek, and Barteld Kooi have recently published a textbook with some basic results in this area: *Dynamic Epistemic Logic*, Springer, 2011. Krister Segerberg has developed his own brand of dynamic doxastic logic in a series of articles since at least the mid 1990's. One recent paper including rather comprehensive results in this area is: "Some Completeness Theorems in the Dynamic Doxastic Logic of Iterated Belief Revision," *Review of Symbolic Logic*, 3(2):228–246, 2010.

The notion of *relevance* is quite central for a representation of belief and belief change. In a Bayesian setting there are standard ways of articulating relevance. But there is recent work that has used proof theoretic techniques to deal with relevance rather than probability theory. Rohit Parikh initiated this area of research with an article published in 1999: Beliefs, belief revision, and splitting languages, *Logic, language, and computation* (Stanford, California) (Lawrence Moss, Jonathan Ginzburg, and Maarten de Rijke, editors), vol. 2, CSLI Publications, pp. 266–278. Recently David Makinson has contributed as well an important article in collaboration with George Kourousias,: Parallel interpolation, splitting, and relevance in belief change, *Journal of Symbolic Logic* 72 September 2007 994-1002. This article contains a detailed bibliography of recent work in this area.

One recent paper including rather comprehensive results in this area is: "Some completeness theorems in the dynamic doxastic logic of iterated belief revision," *Review of Symbolic Logic, 3*, 02, 2010. For more on iterated belief revision please refer to: Darwiche and Pearl (Darwiche, A., & Pearl, J. (1996). On the logic of iterated belief revision. *Artificial Intelligence, 89*, 1–29) appears in: *Change, choice and inference: A study of belief revision and non-monotonic reasoning*, Oxford Logic Guides, 2001.

And there is also more to be found in Pagnucco and Rott (Pagnucco, M., & Rott, H. (1999). Severe withdrawal – and recovery. *Journal of Philosophical Logic, 28*, 501–547. See publisher's "Erratum" (2000), *Journal of Philosophical Logic, 29*, 121) and Lindström (Lindström, S. (1991). A semantic approach to nonmonotonic reasoning: Inference operations and choice. *Uppsala Prints and Preprints in Philosophy*, no. 6/1991, University of Uppsala).

Chapter 13
On the Logic of Theory Change: Partial Meet Contraction and Revision Functions

Carlos E. Alchourrón, Peter Gärdenfors, and David Makinson

1 Background

The simplest and best known form of theory change is *expansion*, where a new proposition (axiom), hopefully consistent with a given theory A, is set-theoretically added to A, and this expanded set is then closed under logical consequence. There are, however, other kinds of theory change, the logic of which is less well understood. One form is theory *contraction*, where a proposition x, which was earlier in a theory A, is rejected. When A is a code of norms, this process is known among legal theorists as the *derogation* of x from A. The central problem is to determine which propositions should be rejected along with x so that the contracted theory will be closed under logical consequence. Another kind of change is *revision*, where a proposition x, inconsistent with a given theory A, is added to A under the requirement that the revised theory be consistent and closed under logical consequence. In normative contexts this kind of change is also known as *amendment*.

Carlos E. Alchourrón died on 13 January 1996.

This paper was written while the third author was on leave from UNESCO. The contents are the responsibility of the authors and not of the institutions.

C.E. Alchourrón (deceased)
Universidad de Buenos Aires, Buenos Aires, Argentina

P. Gärdenfors
Lund University, Lund, Sweden

D. Makinson
London School of Economics, London, UK

H. Arló-Costa et al. (eds.), *Readings in Formal Epistemology*, Springer Graduate Texts in Philosophy 1, DOI 10.1007/978-3-319-20451-2_13

A basic formal problem for the processes of contraction and revision is to give a characterization of ideal forms of such change. In (Gärdenfors 1978) and (Gärdenfors 1982), Gärdenfors developed postulates of a more or less equational nature to capture the basic properties of these processes. It was also argued there that the process of revision can be reduced to that of contraction via the so-called Levi identity: if $A \dot{-} x$ denotes the contraction of A by x, then the revision of A by x, denoted $A \dot{+} x$, can be defined as $\mathrm{Cn}((A \dot{-} \neg x) \cup \{x\})$, where Cn is a given consequence operation.

In (Alchourron and Makinson 1982b), Alchourrón and Makinson tried to give a more explicit construction of the contraction process, and hence also of the revision process via the Levi identity. Their basic idea was to choose $A \dot{-} x$ as a *maximal* subset of A that fails to imply x. Contraction functions defined in this way were called "choice contractions" in (Alchourron and Makinson 1982b), but will here be more graphically referred to as "maxichoice contractions".

As was observed in (Alchourron and Makinson 1982b), the maxichoice functions have, however, some rather disconcerting properties. In particular, maxichoice revision $\dot{+}$, defined from maxichoice contraction as above, has the property that for every theory A, whether complete or not, $A \dot{+} x$ will be complete whenever x is a proposition inconsistent with A. Underlying this is the fact, also noted in (Alchourron and Makinson 1982b), that when A is a theory with $x \in A$, then for every proposition y, either $(x \vee y) \in A \dot{-} x$ or $(x \vee \neg y) \in A \dot{-} x$, where $\dot{-}$ is maxichoice contraction. The significance of these formal results is discussed briefly in (Alchourron and Makinson 1982b), and in more detail in Gärdenfors (1984) and Makinson (1985).

The "inflation properties" that ensue from applying the maxichoice operations bring out the interest of looking at other formal operations that yield smaller sets as values. In this paper, we will start out from the assumption that there is a selection function γ that picks out a class of the "most important" maximal subsets of A that fail to imply x. The contraction $A \dot{-} x$ is then defined as the intersection of all the maximal subsets selected by γ. Functions defined in this way will be called *partial meet contraction* functions, and their corresponding revision functions will be called *partial meet revision* functions. It will be shown that they satisfy Gärdenfors' postulates, and indeed provide a representation theorem for those postulates. When constrained in suitable ways, by relations or, more restrictedly, by transitive relations, they also satisfy his "supplementary postulates", and provide another representation theorem for the entire collection of "basic" plus "supplementary" postulates.

Acquaintance with (Makinson 1985) will help the reader with overall perspective, but it is not necessary for technical details.

Some background terminology and notation: By a *consequence operation* we mean, as is customary, an operation Cn that takes sets of propositions to sets of propositions, such that three conditions are satisfied, for any sets X and Y of propositions: $X \subseteq \mathrm{Cn}(X)$, $\mathrm{Cn}(X) = \mathrm{Cn}(\mathrm{Cn}(X))$, and $\mathrm{Cn}(X) \subseteq \mathrm{Cn}(Y)$ whenever $X \subseteq Y$. To simplify notation, we write $\mathrm{Cn}(x)$ for $\mathrm{Cn}(\{x\})$, where x is any individual proposition, and we also sometimes write $y \in \mathrm{Cn}(X)$ as $X \vdash y$. By a *theory*, we mean, as is customary, a set A of propositions that is closed under Cn; that is, such that $A = \mathrm{Cn}(A)$, or, equivalently, such that $A = \mathrm{Cn}(B)$ for some set B of propositions. As in (Alchourron and Makinson 1982b), we assume that Cn includes classical tautological implication, is compact (that is, $y \in \mathrm{Cn}(X')$ for some finite subset X' of X whenever $y \in \mathrm{Cn}(X)$), and satisfies the rule of "introduction of disjunctions in the premises" (that is, $y \in \mathrm{Cn}(X \cup \{x_1 \vee x_2\})$ whenever $y \in \mathrm{Cn}(X \cup \{x_1\})$ and $y \in \mathrm{Cn}(X \cup \{x_2\})$). We say that a set X of propositions is *consistent* (modulo Cn) iff for no proposition y do we have $y \,\&\neg y \in \mathrm{Cn}(X)$.

2 Partial Meet Contraction

Let Cn be any consequence operation over a language, satisfying the conditions mentioned at the end of the preceding section, and let A be any set of propositions. As in (Alchourron and Makinson 1982a) and (Alchourron and Makinson 1982b), we define $A \perp x$ to be the set of all maximal subsets B of A such that $B \nvdash x$. The maxichoice contraction functions $\dot{-}$ studied in (Alchourron and Makinson 1982b) put $A \dot{-} x$ to be an arbitrary element of $A \perp x$ whenever the latter is nonempty, and to be A itself in the limiting case that $A \perp x$ is empty. In the search for suitable functions with smaller values, it is tempting to try the operation $A \sim x$ defined as $\cap(A \perp x)$ when $A \perp x$ is nonempty, and as A itself in the limiting case that $A \perp x$ is empty. But as shown in Observation 2.1 of (Alchourron and Makinson 1982b), this set is in general far too *small*. In particular, when A is a theory with $x \in A$, then $A \sim x = A \cap \mathrm{Cn}(\neg x)$. In other words, the only propositions left in $A \sim x$ when A is a theory containing x are those which are already consequences of $\neg x$ considered alone. And thus, as noted in Observation 2.2 of (Alchourron and Makinson 1982b), if revision is introduced as usual via the Levi identity as $\mathrm{Cn}((A \sim \neg x) \cup \{x\})$, it reduces to $\mathrm{Cn}((A \cap \mathrm{Cn}(x)) \cup \{x\}) = \mathrm{Cn}(x)$, for any theory A and proposition x inconsistent with A. In other words, if we revise a theory A in *this* way to bring in a proposition x inconsistent with A, we get no more than the set of consequences of x considered alone—a set which is far too small in general to represent the result of an intuitive process of revision of A so as to bring in x.

Nevertheless, the operation of *meet contraction*, as we shall call $\sim$, is very useful as a point of reference. It serves as a natural *lower bound* on any reasonable contraction operation: any contraction operation $\dot{-}$ worthy of the name should surely have $A \sim x \subseteq A \dot{-} x$ for all A, x, and a function $\dot{-}$ satisfying this condition for a given A will be called *bounded over* A.

Following this lead, let A be any set of propositions and let γ be any function such that for every proposition x, $\gamma(A \perp x)$ is a nonempty subset of $A \perp x$, if the latter is nonempty, and $\gamma(A \perp x) = \{A\}$ in the limiting case that $A \perp x$ is empty. We call such a function a *selection function for* A. Then the operation $\dot{-}$ defined by putting $A \dot{-} x = \cap\gamma(A \perp x)$ for all x is called the *partial meet contraction over A determined by* γ. The intuitive idea is that the selection function γ picks out those elements in $A \perp x$ which are "most important" (for a discussion of this notion cf. Gärdenfors (1984)) and then the contraction $A \dot{-} x$ contains the propositions which are common to the selected elements of $A \perp x$. *Partial meet revision* is defined via the Levi identity as $A \dot{+} x = \mathrm{Cn}((A \dot{-} \neg x) \cup \{x\})$. Note that the identity of $A \dot{-} x$ and $A \dot{+} x$ depends on the choice function γ, as well, of course, as on the underlying consequence operation Cn. Note also that the concept of partial meet contraction includes, as special cases, those of maxichoice contraction and (full) meet contraction. The former is partial meet contraction with $\gamma(A \perp x)$ a singleton; the latter is partial meet contraction with $\gamma(A \perp x)$ the entire set $A \perp x$. We use the same symbols $\dot{-}$ and $\dot{+}$ here as for the maxichoice operations in (Alchourron and Makinson 1982b); this should not cause any confusion.

Our first task is to show that all partial meet contraction and revision functions satisfy Gärdenfors' postulates for contraction and revision. We recall (cf. (Alchourron and Makinson 1982b) and (Makinson 1985)) that these postulates may conveniently be formulated as follows:

($\dot{-}$1) $A \dot{-} x$ is a theory whenever A is a theory (closure).
($\dot{-}$2) $A \dot{-} x \subseteq A$ (inclusion).
($\dot{-}$3) If $x \notin \mathrm{Cn}(A)$, then $A \dot{-} x = A$ (vacuity).
($\dot{-}$4) If $x \notin \mathrm{Cn}(\varnothing)$, then $x \notin \mathrm{Cn}(A \dot{-} x)$ (success).
($\dot{-}$5) If $\mathrm{Cn}(x) = \mathrm{Cn}(y)$, then $A \dot{-} x = A \dot{-} y$ (preservation).
($\dot{-}$6) $A \subseteq \mathrm{Cn}((A \dot{-} x) \cup \{x\})$ whenever A is a theory (recovery).

The Gärdenfors postulates for revision may likewise be conveniently formulated as follows:

($\dot{+}$1) $A \dot{+} x$ is always a theory.
($\dot{+}$2) $x \in A + x$.
($\dot{+}$3) If $\neg x \notin \mathrm{Cn}(A)$, then $A \dot{+} x = \mathrm{Cn}\ (A \cup \{x\})$.
($\dot{+}$4) If $\neg x \notin \mathrm{Cn}(\varnothing)$, then $A \dot{+} x$ is consistent under Cn.
($\dot{+}$5) If $\mathrm{Cn}(x) = \mathrm{Cn}(y)$, then $A \dot{+} x = A \dot{+} y$.
($\dot{+}$6) $(A \dot{+} x) \cap A = A \dot{-} \neg x$, whenever A is a theory.

Our first lemma tells us that even the very weak operation of (full) meet contraction satisfies recovery.

Lemma 2.1 *Let A be any theory. Then* $A \subseteq \mathrm{Cn}((A \sim x) \cup \{x\})$.

Proof In the limiting case that $x \notin A$ we have $A \sim x = A$ and we are done. Suppose $x \in A$. Then, by Observation 2.1 of (Alchourron and Makinson 1982b), we have $A \sim x = A \cap \mathrm{Cn}(\neg x)$ so it will suffice to show $A \subseteq \mathrm{Cn}((A \cap \mathrm{Cn}(\neg x)) \cup \{x\})$. Let $a \in A$. Then since A is a theory, $\neg x \vee a \in A$. Also $\neg x \vee a \in \mathrm{Cn}(\neg x)$, so $\neg x \vee a \in A \cap \mathrm{Cn}(\neg x)$, so since Cn includes tautological implication, $a \in \mathrm{Cn}((A \cap \mathrm{Cn}(\neg x)) \cup \{x\})$. □

Corollary 2.2 *Let* $\dot{-}$ *be any function on pairs A, x. Let A be any theory. If* $\dot{-}$ *is bounded over A, then* $\dot{-}$ *satisfies recovery over A.*

Observation 2.3 *Every partial meet contraction function* $\dot{-}$ *satisfies the Gärdenfors postulates for contraction, and its associated partial meet revision function satisfies the Gärdenfors postulates for revision.*

Proof It is easy to show (cf. (Gärdenfors 1978) and (Gärdenfors 1982)) that the postulates for revision can all be derived from those for contraction via the Levi identity. So we need only verify the postulates for contraction. Closure holds, because when A is a theory, so too is each $B \in A \perp x$, and the intersection of theories is a theory; inclusion is immediate; vacuity holds because when $x \notin \text{Cn}(A)$ then $A \perp x = \{A\}$ so $\gamma(A \perp x) = \{A\}$; success holds because when $x \notin \text{Cn}(\varnothing)$ then by compactness, as noted in Observation 2.2 of (Alchourron and Makinson 1982a), $A \perp x$ is nonempty and so $A \dot{-} x = \cap\gamma(A \perp x) \nvdash x$; and preservation holds because the choice function is defined on families $A \perp x$ rather than simply on pairs A, x, so that when $\text{Cn}(x) = \text{Cn}(y)$ we have $A \perp x = A \perp y$, so that $\gamma\{A \perp x) = \gamma\ (A \perp y)$. Finally, partial meet contraction is clearly bounded over any set A, and so by Corollary 2.2 satisfies recovery. □

In fact, we can also prove a converse to Observation 2.3, and show that for theories, the Gärdenfors postulates for contraction *fully characterize* the class of partial meet contraction functions. To do this we first establish a useful general lemma related to 7.2 of (Alchourron and Makinson 1982b).

Lemma 2.4 *Let A be a theory and x a proposition. If $B \in A \perp x$, then $B \in A \perp y$ for all $y \in A$ such that $B \nvdash y$.*

Proof Suppose $B \in A \perp x$ and $B \nvdash y$, $y \in A$. To show that $B \in A \perp y$ it will suffice to show that whenever $B \subset B' \subseteq A$, then $B' \vdash y$. Let $B \subset B' \subseteq A$. Since $B \in A \perp x$ we have $B' \vdash x$. But also, since $B \in A \perp x$, $A \perp x$ is nonempty, so $A \sim x = \cap(A \perp x) \subseteq B$; so, using Lemma 2.1, $A \subseteq \text{Cn}(B \cup \{x\}) \subseteq \text{Cn}(B' \cup \{x\}) = \text{Cn}(B')$, so since $y \in A$ we have $B' \vdash$ y. □

Observation 2.5 *Let $\dot{-}$ be a function defined for sets A of propositions and propositions x. For every theory A, $\dot{-}$ is a partial meet contraction operation over A iff $\dot{-}$ satisfies the Gärdenfors postulates $(\dot{-}1) - (\dot{-}6)$ for contraction over A.*

Proof We have left to right by Observation 2.3. For the converse, suppose that $\dot{-}$ satisfies the Gärdenfors postulates over A. To show that $\dot{-}$ is a partial meet contraction operation, it will suffice to find a function such that:

(i) $\gamma(A \perp x) = \{A\}$ in the limiting case that $A \perp x$ is empty,
(ii) $\gamma(A \perp x)$ is a nonempty subset of $A \perp x$ when $A \perp x$ is nonempty, and
(iii) $A \dot{-} x = \bigcap\gamma(A \perp x)$.

Put $\gamma(A \perp x)$ to be $\{A\}$ when $A \perp x$ is empty, and to be $\{B \in A \perp x\text{: } A \dot{-} x \subseteq B\}$ otherwise. Then (i) holds immediately. When $A \perp x$ is nonempty, then $x \notin \text{Cn}(\varnothing)$ so by the postulate of success $A \dot{-} x \nvdash x$, so, using compactness, $\gamma(A \perp x)$ is nonempty, and clearly $\gamma(A \perp x) \subseteq A \perp x$, so (ii) also holds. For (iii) we have the inclusion $A \dot{-} x \subseteq \bigcap\gamma(A \perp x)$ immediately from the definition of γ. So it remains only to show that $\bigcap\gamma(A \perp x) \subseteq A \dot{-} x$.

In the case that $x \notin A$ we have by the postulate of vacuity that $A \dot{-} x = A$, so the desired conclusion holds trivially. Suppose then that $x \in A$, and suppose $a \notin A \dot{-} x$; we want to show that $a \notin \cap\gamma(A \perp x)$. In the case $a \notin A$, this holds trivially, so we suppose that $a \in A$. We need to find a $B \in A \perp x$ with $A \dot{-} x \subseteq B$ and $a \notin B$. Since $\dot{-}$ satisfies the postulate of recovery, and $a \in A$, we have $(A \dot{-} x) \cup \{x\} \vdash a$. But, by hypothesis, $a \notin A \dot{-} x = \mathrm{Cn}(A \dot{-} x)$ by the postulate of closure, so since Cn includes tautological implication and satisfies disjunction of premises, $(A \dot{-} x) \cup \{\neg x\} \nvdash a$, so $A \dot{-} x \nvdash x \vee a$. Hence by compactness there is a $B \in A \perp (x \vee a)$ with $A \dot{-} x \subseteq B$. Since $B \in A \perp (x \vee a)$ we have $B \nvdash x \vee$ a, so $a \notin B$. And also since $B \nvdash x \vee a$ we have $B \nvdash x$, so, by Lemma 2.4, and the hypothesis that $x \in A$, we have $B \in A \perp x$, and the proof is complete. □

A corollary of Observation 2.5 is that whenever $\dot{-}$ satisfies the Gärdenfors postulates for contraction over a theory A, then it is bounded over A. However, this consequence can also be obtained, under slightly weaker conditions, by a more direct argument. We first note the following partial converse of Lemma 2.1.

Lemma 2.6 *Let A be any theory. Then for every set B and every* $x \in A$, *if* $A \subseteq \mathrm{Cn}(B \cup \{x\})$, *then* $A \sim x \subseteq \mathrm{Cn}(B)$.

Proof Suppose $x \in A$, $A \subseteq \mathrm{Cn}(B \cup \{x\})$, and $a \in A \sim x$; we want to show that $a \in \mathrm{Cn}(B)$. Since A is a theory and $x \in A$ we have $A \sim x = \mathrm{Cn}(\neg x) \cap A$ by Observation 2.1 of (Alchourron and Makinson 1982b); so $\neg x \vdash a$, so $B \cup \{\neg x\} \vdash a$. But also since $a \in A \sim x \subseteq A \subseteq \mathrm{Cn}(B \cup \{x\})$ we have $B \cup \{x\} \vdash a$, so by disjunction of premises and the fact that Cn includes tautological implication, we have $a \in \mathrm{Cn}(B)$. □

Observation 2.7 *Let* $\dot{-}$ *be any function on pairs A, x. Let A be a theory. If* $\dot{-}$ *satisfies closure, vacuity and recovery over A, then* $\dot{-}$ *is bounded over A.*

Proof Suppose $\dot{-}$ satisfies closure, vacuity and recovery over A. Let x be any proposition; we need to show $A \sim x \subseteq A \dot{-} x$. In the case $x \in A$ we have trivially $A \sim x = A \dot{-} x$ by vacuity. In the case $x \in A$ we have $A \subseteq \mathrm{Cn}((A \dot{-} x) \cup \{x\})$ by recovery, so $A \sim x \subseteq \mathrm{Cn}(A \dot{-} x) = A \dot{-} x$ by Lemma 2.6 and closure. □

3 Supplementary Postulates for Contraction and Revision

Gärdenfors (1984) has suggested that revision should also satisfy two further "supplementary postulates", namely:

$(\dot{+}7)$ $A \dot{+} (x \,\&\, y) \subseteq \mathrm{Cn}((A \dot{+} x) \cup \{y\})$ for any theory A, and its conditional converse:
$(\dot{+}8)$ $\mathrm{Cn}((A \dot{+} x) \cup \{y\}) \subseteq A \dot{+} (x \,\&\, y)$ for any theory A, provided that $\neg y \notin A \dot{+} x$.

Given the presence of the postulates $(\dot{-}1)-(\dot{-}6)$ and $(\dot{+}1)-(\dot{+}6)$, these two supplementary postulates for $\dot{+}$ can be shown to be equivalent to various conditions on $\dot{-}$. Some such conditions are given in (Gärdenfors 1984); these can however be simplified, and one particularly simple pair, equivalent respectively to $(\dot{+}7)$ and $(\dot{+}8)$, are:

$(\dot{-}7)$ $(A \dot{-} x) \cap (A \dot{-} y) \subseteq A \dot{-} (x \,\&\, y)$ for any theory A.
$(\dot{-}8)$ $A \dot{-} (x \,\&\, y) \subseteq A \dot{-} x$ whenever $x \notin A \dot{-} (x \,\&\, y)$, for any theory A.

Observation 3.1 *Let $\dot{-}$ be any partial meet contraction operation over a theory A. Then it satisfies ($\dot{-}$7) iff it satisfies ($\dot{+}$7).*

Proof We recall that $\dot{+}$ is defined by the Levi identity $A \dot{+} x = \mathrm{Cn}((A \dot{-} \neg x) \cup \{x\})$. Let A be any theory and suppose that ($\dot{-}$7) holds for all x and y. We want to show that ($\dot{+}$7) holds for all x and y. Let

$$w \in A \dot{+} (x\&y) = \mathrm{Cn}\left((A \dot{-} \neg (x\&y)) \cup \{x\&y\}\right).$$

We need to show that

$$w \in \mathrm{Cn}\left((A \dot{+} x) \cup \{y\}\right) = \mathrm{Cn}\left(\mathrm{Cn}\left((A \dot{-} \neg x) \cup \{x\}\right) \cup \{y\}\right)$$
$$= \mathrm{Cn}\left((A \dot{-} \neg x) \cup \{x\&y\}\right)$$

by general properties of consequence operations. Noting that

$$\mathrm{Cn}(\neg x) = \mathrm{Cn}\left(\neg (x\&y) \,\&\, (\neg x \vee y)\right),$$

it will suffice by condition ($\dot{-}$7) to show that

$$w \in \mathrm{Cn}\left((A \dot{-} \neg (x\&y)) \cup \{x\&y\}\right) \text{ and } w \in \mathrm{Cn}\left((A \dot{-} (\neg x \vee y)) \cup \{x\&y\}\right).$$

But the former is given by hypothesis, so we need only verify the latter. Now by the former, we have $w \in \mathrm{Cn}(A \cup \{x \,\&\, y\})$, so it will suffice to show that

$$A \cup \{x\&y\} \subseteq \mathrm{Cn}\left((A \dot{-} (\neg x \vee y)) \cup \{x\&y\}\right).$$

But clearly $x \,\&\, y \in$ RHS, and moreover since $x \,\&\, y \vdash y \vdash \neg x \vee y$ we have by recovery that $A \subseteq$ RHS, and we are done.

For the converse, suppose that ($\dot{+}$7) holds for all x, y. Let $a \in (A \dot{-} x) \cap (A \dot{-} y)$; we need to show that $a \in A \dot{-} (x \,\&\, y)$. Noting that

$$\mathrm{Cn}(x) = \mathrm{Cn}\left(\neg ((\neg x \vee \neg y) \,\&\neg x)\right),$$

we have

$$a \in A \dot{-} \neg ((\neg x \vee \neg y) \,\&\neg x) \subseteq A \dot{+} ((\neg x \vee \neg y) \,\&\neg x)$$
$$\subseteq \mathrm{Cn}\left((A \dot{+} (\neg x \vee \neg y)) \cup \{\neg x\}\right).$$

A similar reasoning gives us also $a \in \mathrm{Cn}((A \dot{+} (\neg x \vee \neg y)) \cup \{\neg y\})$. So applying disjunction of premises and the fact that Cn includes tautological implication, we have

$$a \in \mathrm{Cn}\left(A \dot{+} (\neg x \vee \neg y)\right) = A \dot{+} (\neg x \vee \neg y) = Cn\left((A \dot{-} (x\&y)) \cup \{\neg (x\&y)\}\right).$$

But by recovery we also have $a \in \mathrm{Cn}((A \dot{-} (x \,\&\, y)) \cup \{x \,\&\, y\})$, so, again using disjunction of premises,

$$a \in \mathrm{Cn}\,(A \dot{-} (x\&y)) = A \dot{-} (x\&y)$$

by closure, and we are done. □

Observation 3.2 *Let* $\dot{-}$ *be any partial meet contraction function over a theory A. Then it satisfies* ($\dot{-}$8) *iff it satisfies* ($\dot{+}$8).

Proof Let A be a theory and suppose that ($\dot{-}$8) holds for all x and y. We want to show that ($\dot{+}$8) holds for all x and y. Noting that $\mathrm{Cn}(\neg x) = \mathrm{Cn}((\neg x \vee \neg y) \,\&\, \neg x)$ we have $A \dot{-} \neg x = A \dot{-} ((\neg x \vee \neg y) \,\&\, \neg x)$. But also, supposing for ($\dot{+}$8) that $\neg y \notin A \dot{+} x = \mathrm{Cn}((A \dot{-} \neg x) \cup \{x\})$, we have $\neg x \vee \neg y \notin A \dot{-} \neg x$. We may thus apply ($\dot{-}$8) to get

$$A \dot{-} \neg x = A \dot{-} ((\neg x \vee \neg y) \& \neg x) \subseteq A \dot{-} (\neg x \vee \neg y) = A \dot{-} \neg\,(x\&y)\,.$$

This inclusion justifies the inclusion in the following chain, whose other steps are trivial:

$$\begin{aligned}\mathrm{Cn}\left((A \dot{+} x) \cup \{y\}\right) &= \mathrm{Cn}\,(\mathrm{Cn}\,((A \dot{-} \neg x) \cup \{x\}) \cup \{y\}) \\ &= \mathrm{Cn}\,((A \dot{-} \neg x) \cup \{x\&y\}) \subseteq \mathrm{Cn}\,((A \dot{-} \neg\,(x\&y)) \cup \{x\&y\}) \\ &= A \dot{+} (x\&y)\,.\end{aligned}$$

For the converse, suppose ($\dot{+}$8) holds for all x and y, and suppose $x \notin A \dot{-} (x \,\&\, y)$. Then clearly

$$x \notin \mathrm{Cn}\,(A \dot{-} (x\&y) \cup \{\neg x \vee \neg y\}) = A \dot{+} \neg\,(x\&y)\,,$$

so we may apply ($\dot{+}$8) to get

$$\begin{aligned}\mathrm{Cn}\left((A \dot{+} \neg\,(x\&y)) \cup \{\neg x\}\right) \subseteq A \dot{+} (\neg\,(x\&y) \& \neg x) &= A \dot{+} \neg x \\ &= \mathrm{Cn}\,((A \dot{-} x) \cup \{\neg x\})\,.\end{aligned}$$

Thus, since $A \dot{-} (x \,\&\, y)$ is included in the leftmost term of this series, we have

$$A \dot{-} (x\&y) \subseteq \mathrm{Cn}\,((A \dot{-} x) \cup \{\neg x\})\,.$$

But using recovery we also have $A \dot{-} (x \,\&\, y) \subseteq A \subseteq \mathrm{Cn}((A \dot{-} x) \cup \{x\})$, so by disjunction of premises and the fact that Cn includes tautological implication, we have $A \dot{-} (x \,\&\, y) \subseteq \mathrm{Cn}(A \dot{-} x) = A \dot{-} x$ by closure, as desired. □

We end this section with some further observations on the powers of ($\dot{-}$7) and ($\dot{-}$8). Now postulate ($\dot{-}$7) does not tell us that $A \dot{-} x$ and $A \dot{-} y$, considered separately, are included in $A \dot{-} (x \& y)$. But it goes close to it, for it does yield the following "partial antitony" property.

Observation 3.3 *Let* $\dot{-}$ *be any partial meet contraction function over a theory A. Then* $\dot{-}$ *satisfies* ($\dot{-}$7) *iff it satisfies the condition*

$$(\dot{-}\mathrm{P})\ (A \dot{-} x) \cap \mathrm{Cn}(x) \subseteq A \dot{-} (x \& y) \text{ for all } x \text{ and } y.$$

Proof Suppose ($\dot{-}$7) is satisfied. Suppose $w \in A \dot{-} x$ and $x \vdash w$; we want to show that $w \in A \dot{-} (x \& y)$. If $x \notin A$ or $y \notin A$, then trivially $A \dot{-} (x \& y) = A$, so $w \in A \dot{-} (x \& y)$. So suppose that $x \in A$ and $y \in A$. Now

$$A \dot{-} (x \& y) = A \dot{-} ((\neg x \vee y)\ \&\ x),$$

so by ($\dot{-}$7) it will suffice to show that $w \in A \dot{-} (\neg x \vee y)$ and $w \in A \dot{-} x$. We have the latter by supposition. As for the former, recovery gives us $A \dot{-} (\neg x \vee y) \cup \{\neg x \vee y\} \vdash x$, so $A \dot{-} (\neg x \vee y) \cup \{\neg x\} \vdash x$, so $A \dot{-} (\neg x \vee y) \vdash x \vdash w$, so $w \in A \dot{-} (\neg x \vee y)$.

For the converse, suppose ($\dot{-}$P) is satisfied, and suppose $w \in (A \dot{-} x) \cap (A \dot{-} y)$; we want to show that $w \in A \dot{-} (x \& y)$. Since $w \in A \dot{-} x$, we have $x \vee w \in A \dot{-} x$, and so since $x \vdash x \vee w$, ($\dot{-}$P) gives us $x \vee w \in A \dot{-} (x \& y)$. Similarly, $y \vee w \in A \dot{-} (x \& y)$. Hence $w \vee (x \& y) = (x \vee w) \& (y \vee w) \in A \dot{-} (x \& y)$. But by recovery, $A \dot{-} (x \& y) \cup \{x \& y\} \vdash w$, so $w \vee \neg(x \& y) \in A \dot{-} (x \& y)$. Putting these together gives us $w \in A \dot{-} (x \& y)$ as desired. □

Condition ($\dot{-}$8) is related to another condition, which we shall call the *covering condition*:

($\dot{-}$C) For any propositions x, y, $A \dot{-} (x \& y) \subseteq A \dot{-} x$ or $A \dot{-} (x \& y) \subseteq A \dot{-} y$.

Observation 3.4 *Let* $\dot{-}$ *be any partial meet contraction function over a theory A. If* $\dot{-}$ *satisfies* ($\dot{-}$8) *over A, then it satisfies the covering condition* ($\dot{-}$C) *over A.*

Proof Let x and y be propositions. In the case $x \& y \in \mathrm{Cn}(\varnothing)$ we have, say, $x \in \mathrm{Cn}(\varnothing)$; so $A \dot{-} (x \& y) = A = A \dot{-} x$ and we are done. In the case $x \& y \notin \mathrm{Cn}(\varnothing)$, then by success we have $x \& y \notin A \dot{-} (x \& y)$, so either $x \notin A \dot{-} (x \& y)$ or $y \notin A \dot{-} (x \& y)$, so by ($\dot{-}$8) either $A \dot{-} (x \& y) \subseteq A \dot{-} x$ or $A \dot{-} (x \& y) \subseteq A \dot{-} y$. □

However, the converse of Observation 3.4 fails. For as we shall show at the end of the next section, there is a theory, finite modulo Cn, with a partial meet contraction over A that satisfies the covering condition (and indeed also supplementary postulate ($\dot{-}$7)), but that does not satisfy ($\dot{-}$8). Using Observation 3.4, it is easy to show that when A is a theory and $\dot{-}$ satisfies postulates ($\dot{-}$1) – ($\dot{-}$6), then ($\dot{-}$8) can equivalently be formulated as $A \dot{-} (x \& y) \subseteq A \dot{-} x$ whenever $x \notin A \dot{-} y$.

In (Alchourron and Makinson 1982b), it was shown that whilst the *maxichoice* operations do not in general satisfy ($\dot{-}$7) and ($\dot{-}$8), they do so when constrained by a relational condition of "orderliness". Indeed, it was shown that for the maxichoice operations, the conditions ($\dot{+}$7), ($\dot{+}$8), and orderliness are mutually equivalent, and also equivalent to various other conditions. Now as we have just remarked, in the general context of partial meet contraction, ($\dot{-}$7) does not imply ($\dot{-}$8), and it can also be shown by an example (briefly described at the end of next section) that the converse implication likewise fails. The question nevertheless remains whether there are relational constraints on the partial meet operations that correspond, perfectly or in part, to the supplementary postulates ($\dot{-}7$) and ($\dot{-}8$). That is the principal theme of the next section.

4 Partial Meet Contraction with Relational Constraints

Let A be a set of propositions and γ a selection function for A. We say that γ is *relational* over A iff there is a relation $\leq$ over 2^A such that for all $x \notin \mathrm{Cn}(\varnothing)$, $\leq$ *marks off* $\gamma(A \perp x)$ in the sense that the following identity, which we call the *marking off* identity, holds:

$$\gamma(A \perp x) = \{B \in A \perp x : B' \leq B \text{ for all } B' \in A \perp x\}.$$

Roughly speaking, γ is relational over A iff there is some relation that marks off the elements of $\gamma(A \perp x)$ as the *best* elements of $A \perp x$, whenever the latter is nonempty. Note that in this definition, $\leq$ is required to be fixed for all choices of x; otherwise all partial meet contraction functions would be trivially relational. Note also that the definition does not require any special properties of $\leq$ apart from being a relation; if there is a transitive relation $\leq$ such that for all $x \notin \mathrm{Cn}(\varnothing)$ the marking off identity holds, then γ is said to be *transitively relational* over A. Finally, we say that a partial meet contraction function $\dot{-}$ is relational (transitively relational) over A iff it is determined by some selection function that is so. "Some", because a single partial meet contraction function may, in the infinite case, be determined by two distinct selection functions. In the finite case, however, this cannot happen, as we shall show in Observation 4.6.

Relationality is linked with supplementary postulate ($\dot{-}7$), and transitive relationality even more closely linked with the conjunction of ($\dot{-}7$) and ($\dot{-}8$). Indeed, we shall show, in the first group of results of this section, that a partial meet contraction function $\dot{-}$ is transitively relational iff ($\dot{-}7$) and ($\dot{-}8$) are both satisfied. In the later part of this section we shall describe the rather more complex relationship between relationality and ($\dot{-}7$) considered alone. It will be useful to consider various further conditions, and two that are of immediate assistance are:

($\gamma 7$) $\gamma(A \perp x \,\&\, y) \subseteq \gamma(A \perp x) \cup \gamma(A \perp y)$ for all x and y.
($\gamma 8$) $\gamma(A \perp x) \subseteq \gamma(A \perp x \,\&\, y)$ whenever $A \perp x \cap \gamma(A \perp x \,\&\, y) \neq \varnothing$.

As with ($\dot{-}8$), it is easy to show that when A is a theory and γ is a selection function over A, then ($\gamma 8$) can equivalently be formulated as

$$\gamma(A \perp x) \subseteq \gamma(A \perp x \,\&\, y) \text{ whenever } A \perp x \cap \gamma(A \perp y) \neq \varnothing.$$

The following lemma will also be needed throughout the section.

Lemma 4.1 *Let A be any theory and* $x, y \in A$. *Then* $A \perp (x \,\&\, y) = A \perp x \cup A \perp y$.

Proof We apply Lemma 2.4. When $B \in A \perp (x \,\&\, y)$, then $B \nvdash x \,\&\, y$ so $B \nvdash x$ or $B \nvdash y$, so by 2.4 either $B \in A \perp x$ or $B \in A \perp y$. Conversely, if $B \in A \perp x$ or $B \in A \perp y$, then $B \nvdash x \,\&\, y$ so, by 2.4 again, $B \in A \perp (x \,\&\, y)$. □

Observation 4.2 *Let A be a theory and* $\dot{-}$ *a partial meet contraction function over A determined by a selection function* γ. *If* γ *satisfies the condition* (γ7), *then* $\dot{-}$ *satisfies* ($\dot{-}$7), *and if it satisfies* (γ8), *then* $\dot{-}$ *satisfies* ($\dot{-}$8).

Proof Suppose (γ7) holds. Then we have:

$$\begin{aligned}(A\dot{-}x) \cap (A\dot{-}y) &= \bigcap \gamma\,(A \perp x) \cap \bigcap \gamma\,(A \perp y) \text{ since } \gamma \text{ determines } \dot{-}\\ &= \bigcap (\gamma\,(A \perp x) \cup \gamma\,(A \perp y)) \text{ by general set theory}\\ &\subseteq \bigcap \gamma\,(A \perp (x\&y)) \text{ using condition } (\gamma 7)\\ &= A\dot{-}(x\&y)\,.\end{aligned}$$

Suppose now that (γ8) holds, and suppose $x \notin A \dot{-} (x \,\&\, y)$; that is, $x \notin \bigcap\gamma(A \perp x \,\&\, y)$. We need to show that $A \dot{-} (x \,\&\, y) \subseteq A \dot{-} x$. In the case $x \notin A$ we have $A \dot{-} (x \,\&\, y) = A = A \dot{-} x$. So suppose $x \in A$. Since $x \notin \bigcap\gamma(A \perp x \,\&\, y)$ there is a $B \in \gamma(A \perp x \,\&\, y)$ with $B \nvdash x$, so, by Lemma 2.4, $B \in A \perp x$ and thus $B \in A \perp x \cap \gamma(A \perp x \,\&\, y)$. Applying ($\gamma$8) we have $\gamma(A \perp x) \subseteq \gamma(A \perp x \,\&\, y)$, so $A \dot{-} (x \,\&\, y) = \bigcap\gamma(A \perp x \,\&\, y) \subseteq \bigcap\gamma(A \perp x) = A \dot{-} x$ as desired. □

Observation 4.3 *Let A be any theory and* γ *a selection function for A. If* γ *is relational over A then* γ *satisfies the condition* (γ7), *and if* γ *is transitively relational over A, then* γ *satisfies the condition* (γ8).

Proof In the cases that $x \in \mathrm{Cn}(\varnothing)$, $y \in \mathrm{Cn}(\varnothing)$, $x \notin A$ and $y \notin A$, both (γ7) and (γ8) hold trivially, so we may suppose that $x \notin \mathrm{Cn}(\varnothing)$, $y \notin \mathrm{Cn}(\varnothing)$, $x \notin A$ and $y \in A$.

Suppose γ is relational over A, and suppose $B \in \gamma(A \perp x \,\&\, y)$. Now $\gamma(A \perp x \,\&\, y) \subseteq A \perp x \,\&\, y = A \perp x \cup A \perp y$, so $B \in A \perp x$ or $B \in A \perp y$; consider the former case, as the latter is similar. Let $B' \in A \perp x$. Then $B' \in A \perp x \cup A \perp y = A \perp x \,\&\, y$, and so $B' \leq B$ since $B \in \gamma\,(A \perp x \,\&\, y)$ and γ is relational over A; and thus, by relationality again, $B' \in (A \perp x) \subseteq \gamma(A \perp x) \cup \gamma(A \perp y)$ as desired.

Suppose now that γ is transitively relational over A, and suppose $A \perp x \cap \gamma(A \perp x \,\&\, y) \neq \varnothing$. Suppose for reductio ad absurdum that there is a $B \in \gamma\,(A \perp x)$ with $B \notin \gamma\,(A \perp x \,\&\, y)$. Since $B \in \gamma\,(A \perp x) \subseteq A \perp x \subseteq A \perp x \,\&\, y$ by Lemma 4.1, whilst $B \notin \gamma\,(A \perp x \,\&\, y)$, we have by relationality that there is a $B' \in A \perp x \,\&\, y$ with $B' \leq B$. Now by the hypothesis $A \perp x \cap y\,(A \perp x \,\&\, y) \neq \varnothing$, there is a $B'' \in A \perp x$ with $B'' \in \gamma(A \perp x \,\&\, y)$. Hence by relationality $B' \leq B''$ and also $B'' \leq B$. Transitivity gives us $B' \leq B$ and thus a contradiction. □

When A is a theory and γ is a selection function for A, we define γ^*, the *completion* of γ, by putting $\gamma^*(A \perp x) = \{B \in A \perp x\colon \bigcap\gamma(A \perp x) \subseteq B\}$ for all $x \notin \mathrm{Cn}(\varnothing)$, and $\gamma^*(A \perp x) = \gamma(A \perp x) = \{A\}$ in the limiting case that $x \in \mathrm{Cn}(\varnothing)$.

It is easily verified that γ^* is also a selection function for A, and determines the same partial meet contraction function as γ does. Moreover, we clearly have $\gamma(A \perp x) \subseteq \gamma^*(A \perp x) = \gamma^{**}(A \perp x)$ for all x. This notion is useful in the formulation of the following statement:

Observation 4.4 *Let A be any theory, and $\dot{-}$ a partial meet contraction function over A, determined by a selection function γ. If $\dot{-}$ satisfies the conditions* ($\dot{-}$7) *and* ($\dot{-}$8) *then γ^* is transitively relational over A.*

Proof Define the relation $\leq$ over 2^A as follows: for all $B, B' \in 2^A$, $B' \leq B$ iff either $B' = B = A$, or the following three all hold:

(i) $B' \in A \perp x$ for some $x \in A$.
(ii) $B \in A \perp x$ and $A \dot{-} x \subseteq B$ for some $x \in A$.
(iii) For all x, if $B', B \in A \perp x$ and $A \dot{-} x \subseteq B'$, then $A \dot{-} x \subseteq B$.

We need to show that the relation is transitive, and that it satisfies the marking off identity $\gamma^*(A \perp x) = \{B \in A \perp x\text{: } B' \leq B \text{ for all } B' \in A \perp x\}$ for all $x \notin \mathrm{Cn}(\varnothing)$.

For the identity, suppose first that $B \in \gamma^*(A \perp x) \subseteq A \perp x$ since $x \notin \mathrm{Cn}(\varnothing)$. Let $B' \in A \perp x$; we need to show that $B' < B$. If $x \notin A$ then $B' = B = A$ so $B' \leq B$. Suppose that $x \in A$. Then clearly conditions (i) and (ii) are satisfied. Let y be any proposition, and suppose $B', B \in A \perp y$ and $A \dot{-} y \subseteq B'$; we need to show that $A \dot{-} y \subseteq B$. Now by covering, which we have seen to follow from ($\dot{-}$8), either $A \dot{-} (x \,\&\, y) \subseteq A \dot{-} x$ or $A \dot{-} (x \,\&\, y) \subseteq A \dot{-} y$. And in the latter case $A \dot{-} (x \,\&\, y) \subseteq A \dot{-} y \subseteq B' \in A \perp x$ so $x \notin A \dot{-} (x \,\&\, y)$; so by ($\dot{-}$8) again $A \dot{-} (x \,\&\, y) \subseteq A \dot{-} x$. Thus in either case $A \dot{-} (x \,\&\, y) \subseteq A \dot{-} x$. Now suppose for reductio ad absurdum that there is a $w \in A \dot{-} y$ with $w \notin B$. Then $y \vee w \in A \dot{-} y$ and so since $y \vdash y \vee w$ we have by ($\dot{-}$7) using Observation 3.3 that $y \vee w \in A \dot{-} (x \,\&\, y) \subseteq A \dot{-} x = \bigcap \gamma^*(A \perp x) \subseteq B$; so $y \vee w \in B$. But also since $B \in A \perp y$ and $w \notin B$ and $w \in A$, we have $B \cup \{w\} \vdash y$, so $\neg w \vee y \in B$. Putting these together gives us $(y \vee w) \,\&\, (y \vee \neg w) \in B$, so $y \in B$, contradicting $B \in A \perp y$.

For the converse, suppose $B \notin \gamma^*(A \perp x)$ and $B \in A \perp x$; we need to find a $B' \in A \perp x$ with $B' \nleq B$. Clearly the supposition implies that $x \in A$, so $B \neq A$. Since $B \in A \perp x$, the latter is nonempty, so $\gamma^*(A \perp x)$ is nonempty; let B' be one of its elements. Noting that $B', B \in A \perp x$, $B' \in \gamma^*(A \perp x)$, but $B \notin \gamma^*(A \perp x)$, we see that condition (iii) fails, so that $B' \nleq B$, as desired.

Finally, we check out transitivity. Suppose $B'' \leq B'$ and $B' \leq B$; we want to show that $B'' \leq B$. In the case that $B = A$ then clearly since $B' \leq B$ we have $B' = B = A$, and thus since $B'' \leq B'$ we have $B'' = B' = A$, so $B'' = B = A$ and $B'' \leq B$. Suppose for the principal case that $B \neq A$. Then since $B' \leq B$, clearly $B' \neq A$. Since $B' \leq B$ we have $B \in A \perp w$ and $A \dot{-} w \subseteq B$ for some $w \in A$, so (ii) is satisfied. Since $B'' \leq B'$ we have $B'' \in A \perp w$ for some $w \in A$, so (i) is satisfied. It remains to verify (iii). Suppose $B'', B \in A \perp y$ and $A \dot{-} y \subseteq B''$; we need to show that $A \dot{-} y \subseteq B$. First, note that since $B \neq A$ by the condition of the case, we have $y \in A$. Also, since $B'' \leq B'$ and $B' \neq A$, there is an $x \in A$ with $B' \in A \perp x$ and $A \dot{-} x \subseteq B'$. Since $x, y \in A$ we have by Lemma 4.1 that $A \perp x \,\&\, y = A \perp x \cup A \perp y$, so $B'', B', B \in A \perp x \,\&\, y$. Now by covering, either $A \dot{-} (x \,\&\, y) \subseteq A \dot{-} y$ or $A \dot{-} (x \,\&\, y) \subseteq A \dot{-} x$. The former case gives us $A \dot{-} (x \,\&\, y) \subseteq B''$, so since $B'' \leq B'$ and $B' \neq A$ We have $A \dot{-} (x \,\&\, y) \subseteq B'$, so again since $B' \leq B$ and $B \neq A$ we have $A \dot{-} (x \,\&\, y) \subseteq B$. Likewise, the latter case gives us $A \dot{-} (x \,\&\, y) \subseteq B'$, so since $B' \leq B$ and $B \neq A$ we have $A \dot{-} (x \,\&\, y) \subseteq B$. Thus in either case, $A \dot{-} (x \,\&\, y) \subseteq B$. Now let $w \in A \dot{-} y$ we need to show that $w \in B$. Since $w \in A \dot{-} y$ we have $y \vee w \in A \dot{-} y$; so by ($\dot{-}$7) and Observation 3.3, since $y \vee w \in Cn(y)$, we have $y \vee w \in A \dot{-} (x \,\&\, y) \subseteq B$. Hence $B \cup \{\neg y\} \vdash w$. But since $B \in A \perp y$ and $w \in A$, we also have $B \cup \{y\} \vdash w$, so $B \vdash w$ and thus $w \in B$ as desired. □

Corollary 4.5 *Let A be any theory, and $\dot{-}$ a partial meet contraction function over A determined by a selection function γ. Then $\dot{-}$ is transitively relational over A iff $\dot{-}$ satisfies both* ($\dot{-}$7) *and* ($\dot{-}$8).

Proof If $\dot{-}$ satisfies ($\dot{-}$7) and ($\dot{-}$8) then, by 4.4, γ^* is transitively relational, so since γ^* determines $\dot{-}$, the latter is transitively relational. Conversely, if $\dot{-}$ is transitively relational, then γ' is transitively relational for some γ' that determines $\dot{-}$; so, by 4.3, γ' satisfies (γ7) and (γ8); so, by 4.2, $\dot{-}$ satisfies ($\dot{-}$7) and ($\dot{-}$8). □

This result is the promised representation theorem for the collection of "basic" plus "supplementary" postulates. Since this collection of postulates can be independently motivated (cf. Gärdenfors (1978)), there is strong reason to focus on transitively relational partial meet contraction functions as an ideal representation of the intuitive process of contraction.

Note that Observation 4.4 and its corollary give us a sufficient condition for the transitive relationality of γ^*, and thus of $\dot{-}$, rather than of γ itself. The question thus arises: when can we get the latter? We shall show that in the finite case the passage from γ to $\dot{-}$ is injective, so that $\gamma = \gamma^*$, where γ is any selection function that determines $\dot{-}$. By the *finite case*, we mean the case where A is finite modulo Cn; that is, where the equivalence relation defined by $Cn(x) = Cn(y)$ partitions A into finitely many cells.

Observation 4.6 *Let A be any theory finite modulo* Cn, *and let y and y′ be selection functions for A. For every proposition x, if $\gamma\ (A \perp x) \neq \gamma'\ (A \perp x)$, then $\bigcap\gamma(A \perp x) \neq \cap\gamma'(A \perp x)$.*

Sketch of Proof Suppose $B \in \gamma(A \perp x)$, but $B \notin \gamma'(A \perp x)$. Then clearly $x \in A$ and $x \notin Cn(\varnothing)$. Since A is finite (we identify A with its quotient structure), so is B; put b to be the conjunction of its elements. Then it is easy to check that $b \in B$ but $b \notin B'$ for all $B' \notin \gamma'\ (A \perp x)$. Put $c = \neg b \vee x$: then it is easy to check that $c \notin B \supseteq \bigcap\gamma(A \perp x)$, but $c \in B'$ for all $B' \in \gamma'(A \perp x)$; that is, $c \in \bigcap\gamma'(A \perp x)$. □

Corollary 4.7 *Let A be any theory finite modulo* Cn, *and $\dot{-}$ a partial meet contraction function over A determined by a selection function γ. If $\dot{-}$ satisfies conditions* ($\dot{-}$7) *and* ($\dot{-}$8), *then y is transitively relational over A.*

Proof Immediate from 4.4 and 4.6. □

We turn now to the question of the relation of condition ($\dot{-}$7), considered alone, to relationality; and here the situation is rather more complex and less satisfying. Now we have from Observation 4.2 that when $\dot{-}$ is determined by γ, then if γ satisfies

(γ7), then $\dot{-}$ satisfies ($\dot{-}$7), and it is not difficult to show, by an argument similar to that of 4.6, that:

Observation 4.8 *If A is a theory finite modulo* Cn, *and* $\dot{-}$ *a partial meet contraction function over A determined by a selection function* γ, *then* $\dot{-}$ *satisfies* ($\dot{-}$7) *iff* γ *satisfies* (γ7). *Also,* $\dot{-}$ *satisfies* ($\dot{-}$8) *iff* γ *satisfies* (γ8).

But on the other hand, even in the finite case, (γ7) does *not* imply the relationality of γ or of $\dot{-}$:

Observation 4.9 *There is a theory A, finite modulo* Cn, *with a partial meet contraction function* $\dot{-}$ *over A, determined by a selection function* γ, *such that* $\dot{-}$ *satisfies* (γ7), *but* $\dot{-}$ *is not relational over A.*

Sketch of Proof Take the sixteen-element Boolean algebra, take an atom a_0 of this algebra, and put A to be the principal filter determined by a_0. This will be an eight-element structure, lattice-isomorphic to the Boolean algebra of eight elements. We take Cn in the natural way, putting $\mathrm{Cn}(X) = \{x\colon \wedge X \leq x\}$. We label the eight elements of A as $a_0, \ldots, a_7$, where a_0 is already defined, a_1, a_2, a_3 are the three atoms of A (not of the entire Boolean algebra), a_4, a_5, a_6 are the three dual atoms of A, and a_7 is the greatest element of A (i.e. the unit of the Boolean algebra). For each $i \leq 7$, we write $!a_i$ for $\{a_j \in A\colon a_i \leq a_j\}$. We define γ by putting $\gamma(A \perp a_7) = \gamma(A \perp \mathrm{Cn}(\varnothing)) = \{A\} = !a_0$ as required in this limiting case, $\gamma(A \perp a_0) = A \perp a_j$ for all j with $1 \leq j < 7$, and $\gamma(A \perp a_0) = \{!a_1\}$. Then it is easy to verify that for all $a_i \notin \mathrm{Cn}(\varnothing)$, $\gamma(A \perp a_i)$ is a nonempty subset of $A \perp a_i$, so γ is a selection function for A. By considering cases we easily verify (γ7) (and thus also by 4.2 ($\dot{-}$7)); and by considering the role of $!a_2$ it is easy to verify that γ (and hence by 4.6, $\dot{-}$ itself) is not relational over A. □

The question thus arises whether there is a condition on $\dot{-}$ or on γ that is equivalent to the relationality of $\dot{-}$ or of γ respectively. We do not know of any such condition for $\dot{-}$, but there is one for γ, of an infinitistic nature. It is convenient, in this connection, to consider a descending series of conditions, as follows:

(γ7:∞) $A \perp x \cap \bigcap_{i \in I}\{\gamma(A \perp y_i)\} \subseteq \gamma(A \perp x)$, whenever $A \perp x \subseteq \bigcup_{i \in I}\{A \perp y_j\}$.
(γ7:N) $A \perp x \cap \gamma(A \perp y_1) \cap \cdots \cap \gamma(A \perp y_n) \subseteq \gamma(A \perp x)$, whenever $A \perp x \subseteq A \perp y_1 \cup \cdots \cup A \perp y_n$, for all $n \geq 1$.
(γ7:2) $A \perp x \cap \gamma(A \perp y_1) \cap \gamma(A \perp y_2) \subseteq \gamma(A \perp x)$, whenever $A \perp x \subseteq A \perp y_1 \cup A \perp y_2$.
(γ7:1) $A \perp x \cap \gamma(A \perp y) \subseteq \gamma(A \perp x)$, whenever $A \perp x \subseteq A \perp y$.

Observation 4.10 *Let A be any theory and* γ *a selection function over A. Then* γ *is relational over A iff* (γ7: ∞) *is satisfied. Moreover, we have* (γ7:∞) $\rightarrow$ (γ7:N) $\leftrightarrow$ (γ7:2) $\rightarrow$ (γ7:1) $\leftrightarrow$ (γ7). *On the other hand,* (γ7:1) *does not imply* (γ7:2), *even in the finite case; although in the finite case,* (γ7:N) *is equivalent to* (γ7: ∞).

Sketch of Proof Writing (γR) for "γ is relational over A", we show first that (γR) $\rightarrow$ (γ7:∞). Suppose (γR), and suppose $A \perp x \subseteq \bigcup_{i \in I}\{A \perp y_i\}$. Suppose $B \in A \perp x \cap \bigcap_{i \in I}\{\gamma(A \perp y_i)\}$. We need to show that $B \in \gamma(A \perp x)$. Since $B \in \gamma(A \perp y_i)$

for all $i \in I$, we have by relationality that $B' \leq B$ for all $B' \in A \perp y_i$, for all $i \in I$; so, by the supposition, $B' \leq B$ for all $B' \in A \perp x$. Hence, since $B \in A \perp x$, so that also $x \notin \text{Cn}(\varnothing)$, we have by relationality that $B \in \gamma(A \perp x)$. To show the converse $(\gamma 7{:}\ \infty) \rightarrow (\gamma\text{R})$, suppose $(\gamma 7{:}\ \infty)$ holds, and define $\leq$ over 2^A by putting $B' \leq B$ iff there is an x with $B \in \gamma(A \perp x)$ and $B' \in A \perp x$; we need to verify the marking off identity. The left to right inclusion of the marking off identity is immediate. For the right to left, suppose $B \in A \perp x$ and for all $B' \in A \perp x$, $B' \leq B$. Then by the definition of $\leq$, for all $B_i \in \{B_i\}_{i \in I} = A \perp x$ there is a y_i with $B \in \gamma(A \perp y_i)$ and $B_i \in A \perp y_i$. Since $B_i \in A \perp y_i$, for all $B_i \in A \perp x$, we have $A \perp x \subseteq \bigcup_{i \in I}\{A \perp y_i\}$, so we may apply $(\gamma 7{:}\ \infty)$. But clearly $B \in A \perp x \cap \bigcap_{i \in I}\{\gamma(A \perp y_i)\}$. Hence by $(\gamma 7{:}\infty)$ we have $B \in \gamma(A \perp x)$, as desired.

The implications $(\gamma 7{:}\infty) \rightarrow (\gamma 7{:}N) \rightarrow (\gamma 7{:}2) \rightarrow (\gamma 7{:}1)$ are trivial, as is the equivalence of $(\gamma 7{:}\infty)$ to $(\gamma 7{:}N)$ in the finite case. To show that $(\gamma 7{:}2)$ implies the more general $(\gamma 7{:}N)$, it suffices to show that for all $n \geq 2$, $(\gamma 7{:}n) \rightarrow (\gamma 7{:}n+1)$: this can be done using the fact that when $y_n, y_{n+1} \in A$, $A \perp y_n \cup A \perp y_{n+1} = A \perp (y_n \,\&\, y_{n+1})$ by Lemma 4.1.

To show that $(\gamma 7{:}1) \rightarrow (\gamma 7)$, recall from 4.1 that when $x, y \in A$, then $A \perp x \,\&\, y = A \perp x \cup A \perp y$; so $A \perp x \subseteq A \perp x \,\&\, y$, and so, by $(\gamma 7{:}1)$, $(A \perp x) \cap \gamma\,(A \perp x \,\&\, y) \subseteq \gamma(A \perp x)$. Similarly $(A \perp y) \cap \gamma(A \perp x \,\&\, y) \subseteq \gamma(A \perp y)$. Forming unions on left and right, distributing on the left, and applying 4.1 gives us $\gamma(A \perp x \,\&\, y) \subseteq \gamma(A \perp x) \cup \gamma(A \perp y)$ as desired.

To show conversely that $(\gamma 7) \rightarrow (\gamma 7{:}1)$, suppose $(\gamma 7)$ is satisfied, suppose $A \perp x \subseteq A \perp y$ and consider the principal case that $x, y \in A$. Then using compactness we have $y \vdash x$, so $\text{Cn}(y) = \text{Cn}(x \,\&\, (\neg x \vee y))$, so by $(\gamma 7)$

$$\gamma(A \perp y) \subseteq \gamma(A \perp x) \cup \gamma(A \perp \neg x \vee y)\,,$$

so $A \perp x \cap \gamma(A \perp y) \subseteq \gamma(A \perp x) \cup \gamma(A \perp \neg x \vee y)$. The verification is then completed by showing that $A \perp x$ is disjoint from $\gamma\ (A \perp \neg x \vee y)$.

Finally, to show that $(\gamma 7{:}1)$ does not imply $(\gamma 7{:}2)$, even in the finite case, consider the same example as in the proof of Observation 4.9. We know from that proof that this example satisfies $(\gamma 7)$ and thus also $(\gamma 7{:}1)$, but that γ is not relational over A, so by earlier parts of this proof, $(\gamma 7{:}\infty)$ fails, so by finiteness $(\gamma 7{:}N)$ fails, so $(\gamma 7{:}2)$ fails. Alternatively, a direct counterinstance to $(\gamma 7{:}2)$ in this example can be obtained by putting $x = a_0$, $y_1 = a_1$, and $y_2 = a_2$. □

5 Remarks on Connectivity

It is natural to ask what the consequences are of imposing connectivity as well as transitivity on the relation that determines a selection function. Perhaps surprisingly, it turns out that in the infinite case it adds very little, and in the finite case nothing at all. This is the subject of the present section.

Let A be a set of propositions and γ a selection function for A. We say that γ is *connectively relational* over A iff there is a relation that is connected over 2^A such that for all $x \notin \mathrm{Cn}(\varnothing)$, the marking off identity of Sect. 4 holds. And a partial meet contraction function is called connectively relational iff it is determined by some selection function that is so.

We note as a preliminary that it suffices to require connectivity over the much smaller set $U_A = \bigcup_{x \in A}\{A \perp x\}$. For suppose that $\leq$ is connected over U_A. Put $\leq_0$ to be the restriction of $\leq$ to U_A; then $\leq_0$ will still be connected over U_A. Then put $\leq_1$ to be $\leq_0 \cup ((2^A - U_A) \times 2^A)$. Clearly $\leq_1$ will be connected over 2^A. Moreover, if $\leq$ satisfies the marking off identity, so does $\leq_1$.

Indeed, when $\leq$ is transitive, it suffices to require connectivity on the even smaller set $U_\gamma = \bigcup\{\gamma(A \perp x){:}x \in A, x \notin \mathrm{Cn}(\varnothing)\}$. For here likewise we can define $\leq_0$ as the restriction of $\leq$ to U_γ, and then define $\leq_1$, to be $\leq_0 \cup ((2^A - U_\gamma) \times 2^A)$. Then clearly $\leq_1$ is connected over 2^A, and is transitive if $\leq$ is transitive; and we can easily check, using the transitivity of $\leq$, that if $\leq$ satisfies the marking off identity for γ, so does $\leq_1$.

Observation 5.1 *Let A be any theory and $\dot{-}$ a partial meet contraction function over A. Then $\dot{-}$ is transitively relational iff it is transitively and connectively relational.*

Proof Suppose that $\dot{-}$ is determined by the transitively relational selection function γ. Then by 4.2 and 4.3, $\dot{-}$ satisfies the conditions ($\dot{-}$7) and ($\dot{-}$8), so the conditions of Observation 4.4 hold and the relation $\leq$ defined in its proof is transitive and satisfies the marking off identity for γ^*. By the above remarks, to show that $\dot{-}$ is transitively and connectively relational it suffices to show that $\leq$ is connected over the set U_{γ^*}.

Let $B', B \in U_{\gamma^*}$ and suppose $B' \not\leq B$. Since $B', B \in U_{\gamma^*}$, conditions (i) and (ii) of the definition of $\leq$ in the proof of 4.4 are satisfied for both B' and B. Hence since $B' \leq B$ we have by (iii) that there is an x with $B', B \in A \perp x$, $A \dot{-} x \subseteq B'$ and $A \dot{-} x \nsubseteq B$. But since $A \dot{-} x \subseteq B' \in A \perp x$ we have by the definition of γ^* that $B' \in \gamma^*(A \perp x)$, so by the marking off identity for γ^*, $\leq$ as verified in the proof of 4.4, since $B \in A \perp x$ we have $B \leq B'$ as desired. □

In the case that A is finite modulo Cn, this result can be both broadened and sharpened: Broadened to apply to relationality in general rather than only to transitive relationality, and sharpened to guarantee connectivity over U_A of any given relation under which the selection function γ is relational, rather than merely connectivity, as above, of a specially constructed relation under which the closure γ^* of γ is relational.

Observation 5.2 *Let A be a theory finite modulo* Cn, *and let $\dot{-}$ be a partial meet contraction function over A, determined by a selection function γ. Suppose that γ is relational, with the relation $\leq$ satisfying the marking off identity. Then $\leq$ is connected over U_A.*

Proof Let $B', B \in U_A = \bigcup_{x \in A}\{A \perp x\}$ · Since A is finite modulo Cn, there are $b', b \in A$ with $A \perp b' = \{B'\}$ and $A \perp b = \{B\}$—for example, put b to be the disjunction of all (up to equivalence modulo Cn) the elements $a \in A$ such that $B \nvdash a$. Now $A \perp b' \& b = A \perp b' \cup A \perp b = \{B', B\}$ by Lemma 4.1, and so since γ is a selection function, $\gamma(A \perp b' \& b)$ is a nonempty subset of $\{B', B\}$, which implies that either B' or B is in $\gamma(A \perp b' \& b)$. In the former case we have $B \leq B'$, and in the latter case we have the converse. □

Corollary 5.3 *Let A be a theory finite modulo* Cn, *and let* $\dot{-}$ *be a partial meet contraction function over A. Then* $\dot{-}$ *is relational iff it is connectively relational.*

Proof Immediate from 5.2. □

6 Maxichoice Contraction Functions and Factoring Conditions on $A \dot{-} (x \& y)$

The first topic of this section will be a brief investigation of the consequences of the following rather strong *fullness* condition:

($\dot{-}$F) If $y \in A$ and $y \notin A \dot{-} x$, then $\neg y \vee x \in A \dot{-} x$, for any theory A.

From the results in Gärdenfors (1982), it follows that if $\dot{-}$ is a partial meet contraction function, then this condition (there called (−6)) is equivalent with the following condition (called (21) in Gärdenfors (1982)) on partial meet revision functions:

($\dot{+}$ F) If $y \in A$ and $y \notin A \dot{+} x$, then $\neg y \in A \dot{+} x$, for any theory A.

The strength of the condition ($\dot{-}$F) is shown by the following simple representation theorem:

Observation 6.1 *Let* $\dot{-}$ *be any partial meet contraction function over a theory A. Then* $\dot{-}$ *satisfies* ($\dot{-}$*F*) *iff* $\dot{-}$ *is a maxichoice contraction function.*

Proof Suppose $\dot{-}$ satisfies ($\dot{-}$F). Suppose $B, B' \in \gamma(A \perp x)$ and assume for contradiction that $B \neq B'$. There is then some $v \in B'$ such that $y \notin B$. Hence $y \notin A \dot{-} x$ and since y $\in A$ it follows from ($\dot{-}$F) that $\neg y \vee x \in A \dot{-} x$. Hence $\neg y \vee x \in B'$, but since $y \in B'$ it follows that $x \in B'$, which contradicts the assumption that $B' \in A \perp x$. We conclude that $B = B'$ and hence that $\dot{-}$ is a maxichoice contraction function.

For the converse, suppose that $\dot{-}$ is a maxichoice contraction function and suppose that $y \in A$ and $y \in A \dot{-} x$. Since $A \dot{-} x = B$ for some $B \in A \perp x$, it follows that $y \notin B$. So by the definition of $A \perp x$, $x \in \mathrm{Cn}(B \cup \{y\})$. By the properties of the consequence operation we conclude that $\neg y \vee x \in B = A \dot{-} x$, and thus ($\dot{-}$F) is satisfied.

In addition to this representation theorem for maxichoice contraction functions, we can also prove another one based on the following *primeness* condition.

($\dot{-}$Q) For all $y, z \in A$ and for all x, if $y \vee z \in A \dot{-} x$, then either $y \in A \dot{-} x$ or $z \in A \dot{-} x$.

Observation 6.2 *Let $\dot{-}$ be any partial meet contraction function over a theory A. Then $\dot{-}$ satisfies ($\dot{-}$Q) iff $\dot{-}$ is a maxichoice contraction function.*

Proof Suppose first that $\dot{-}$ is a maxichoice function and suppose $y, z \in A$, $y \notin A \dot{-} x$ and $z \notin A \dot{-} x$. Then by maximality, $A \dot{-} x \cup \{y\} \vdash x$ and $A \dot{-} - x \cup \{z\} \vdash$, so $A \dot{-} x \cup \{y \vee z\} \vdash x$. But since say $y \in A$ and $y \notin A \dot{-} x$, we have $x \notin Cn(\emptyset)$, so $x \notin A \dot{-} x$. Thus $y \vee z \notin A \dot{-} x$, which shows that ($\dot{-}$Q) is satisfied.

For the converse, suppose that ($\dot{-}$Q) is satisfied. By Observation 6.1, it suffices to derive ($\dot{-}$F). Suppose $y \in A$ and $y \notin A \dot{-} x$. We need to show that $\neg y \vee x \in A \dot{-} x$. Now $(y \vee \neg y) \vee x \notin Cn(\emptyset)$, and so $(y \vee \neg y) \vee x = y \vee (\neg y \vee x) \in A \dot{-} x$. Also by hypothesis $y \in A$, and since $y \notin A \dot{-} x$ we have $x \in A$, so $\neg y \vee x \in A$. We can now apply the primeness condition ($\dot{-}$Q) and get either $y \in A \dot{-} x$ or $\neg y \vee x \in A \dot{-} x$. By hypothesis, the former fails, so the latter holds and ($\dot{-}$F) is verified.

With the aid of these results we shall now look at three "factoring" conditions on the contraction of a conjunction from a theory A. They are

Decomposition ($\dot{-}$D). For all x and y, $A \dot{-} (x \,\&\, y) = A \dot{-} x$ or $A \dot{-} (x \,\&\, y) = A \dot{-} y$.
Intersection ($\dot{-}$I). For all x and y in A, $A \dot{-} (x \,\&\, y) = A \dot{-} x \cap A \dot{-} y$.
Ventilation ($\dot{-}$V). For all x and y, $A \dot{-} (x \,\&\, y) = A \dot{-} x$ or $A \dot{-} (x \,\&\, y) = A \dot{-} y$ or $A \dot{-} (x \,\&\, y) = A \dot{-} x \cap A \dot{-} y$.

These bear some analogy to the very processes of maxichoice, full meet, and partial meet contraction respectively, and the analogy is even more apparent if we express the factoring conditions in their equivalent n-ary forms:

$$A \dot{-} (x_1 \& \cdots \& x_n) = A \dot{-} x_i \quad \text{for some } i \leq n;$$
$$A \dot{-} (x_1 \& \cdots \& x_n) = \bigcap_{i \leq n} \{A \dot{-} x_i\} \quad \text{whenever } x_1, \ldots, x_n \in A;$$
$$A \dot{-} (x_1 \& \cdots \& x_n) = \bigcap_{i \in I} \{A \dot{-} x_i\} \quad \text{for some } I, \text{ where } \emptyset \neq I \subseteq \{1, \ldots n\}.$$

This analogy of formulation corresponds indeed to quite close relationships between the three kinds of contraction process, on the one hand, and the three kinds of factorization on the other. We shall state the essential relationships first, to give a clear overall picture, and group the proofs together afterwards.

First, the relationship between maxichoice contraction and decomposition. In (Alchourron and Makinson 1982b) it was shown that if A is a theory and $\dot{-}$ is a maxichoice contraction function over A, then decomposition is equivalent to each of ($\dot{-}$7) and ($\dot{-}$8). In the more general context of partial meet contraction functions these equivalences between the conditions break down, and it is decomposition ($\dot{-}$D) that emerges as the strongest among them:

Observation 6.3 *Let A be a theory and $\dot{-}$ a partial meet contraction function over A. Then the following conditions are equivalent:*

(a) *$\dot{-}$ satisfies* ($\dot{-}$D).

(b) $\dot{-}$ *is a maxichoice contraction function and satisfies at least one of* ($\dot{-}$7) *and* ($\dot{-}$8).
(c) $\dot{-}$ *is a maxichoice contraction function and satisfies both of* ($\dot{-}$7) *and* ($\dot{-}$8).
(d) $\dot{-}$ *satisfies* ($\dot{-}$WD).
(e) $\dot{-}$ *is a maxichoice contraction function and satisfies* ($\dot{-}$C).

Here ($\dot{-}$WD) is the *weak decomposition* condition: for all x and y, $A \dot{-} x \subseteq A \dot{-} x \,\&\, y$ or $A \dot{-} y \subseteq A \dot{-} x \,\&\, y$.

The relationship of full meet contraction to the intersection condition ($\dot{-}$I) is even more direct. This is essentially because a full meet contraction function, as defined at the beginning of Sect. 2, is always transitively relational, and so always satisfies ($\dot{-}$7) and ($\dot{-}$8). For since $\gamma(A \perp x) = A \perp x$ for all $x \notin \mathrm{Cn}(\emptyset)$, γ is determined via the marking off identity by the total relation over 2^A or over $\bigcup x\{A \perp x\}$.

Observation 6.4 *Let A be a theory and* $\dot{-}$ *a partial meet contraction function over A. Then the following conditions are equivalent:*

(a) $\dot{-}$ *satisfies* ($\dot{-}$I).
(b) $\dot{-}$ *satisfies* ($\dot{-}$M).
(c) $\dot{-}$ *is a full meet contraction function.*

Here ($\dot{-}$M) is the *monotony condition:* for all $x \in A$, if $x \vdash y$ then $A \dot{-} x \subseteq A \dot{-} y$. This result gives us a representation theorem for full meet contraction. Note, as a point of detail, that whereas decomposition and ventilation are formulated for arbitrary propositions x and y, the intersection and monotony conditions are formulated under the restriction that x and y (respectively, x) are in A. For if $x \notin A$, then $x \,\&\, y \notin A$, so $A \dot{-} (x \,\&\, y) = A$ whilst $A \dot{-} x \cap A \dot{-} y = A \cap A \dot{-} y = A \dot{-} y \neq A$ if $y \in A$ and $y \notin \mathrm{Cn}(\emptyset)$.

Of the three factoring conditions, ventilation ($\dot{-}$V) is clearly the most "general" and the weakest. But it is still strong enough to imply the "supplementary postulates" ($\dot{-}$7) and ($\dot{-}$8):

Observation 6.5 *Let A be a theory and* $\dot{-}$ *a partial meet contraction function over A. Then* $\dot{-}$ *satisfies* ($\dot{-}$V) *iff* $\dot{-}$ *satisfies both* ($\dot{-}$7) *and* ($\dot{-}$8).

Proof of Observation 6.3 We know by the chain of equivalences in §8 of Alchourron and Makinson (1982b) that if $\dot{-}$ is a maxichoice contraction function then the conditions ($\dot{-}$7), ($\dot{-}$8) and ($\dot{-}$D) are mutually equivalent. This already shows the equivalence of (b) and (c), and also shows that they imply (a). (d) is a trivial consequence of (a). To prove the equivalence of (a)–(d) it remains to show that (d) implies (b).

Suppose that $\dot{-}$ satisfies ($\dot{-}$WD). Clearly it then satisfies ($\dot{-}$7), so we need only verify that $\dot{-}$ is a maxichoice function, for which it suffices by Observation 6.1 to verify ($\dot{-}$F); that is, that whenever $y \in A$ and $y \notin A \dot{-} x$ then $\neg y \vee x \in A \dot{-} x$. Suppose for reductio ad absurdum that $y \in A$, $y \notin A \dot{-} x$ and $\neg y \vee x \notin A \dot{-} x$. Note that this implies that $x \in A$. Now $\mathrm{Cn}(x) = \mathrm{Cn}((x \vee y) \,\&\, (x \vee \neg y))$, so by ($\dot{-}$WD) we have $A \dot{-} (x \vee y) \subseteq A \dot{-} x$ or $A \dot{-} (x \vee \neg y) \subseteq A \dot{-} x$. In the former case, $\neg y \vee x \in A \dot{-} (x \vee y)$. But by recovery $A \dot{-} (x \vee y) \cup \{x \vee y\} \vdash x$, so $A \dot{-} (x \vee y) \cup \{y\} \vdash x$, so

$\neg y \vee x \in A \dot{-} (x \vee y)$, giving a contradiction. And in the latter case, $y \notin A \dot{-} (x \vee \neg y)$, whereas by recovery $A \dot{-} (x \vee \neg y) \cup \{x \vee \neg y\} \vdash y$, so $A \dot{-} (x \vee \neg y) \cup \{\neg y\} \vdash y$, so $y \in A \dot{-} (x \vee \neg y)$, again giving a contradiction.

Finally, it must be shown that (e) is equivalent with (a)–(d). First note that it follows immediately from Observation 3.4 that (c) entails (e). To complete the proof we show that (e) entails (b). In the light of Observation 6.1 it suffices to show that ($\dot{-}$F) and ($\dot{-}$C) together entail ($\dot{-}$8). To do this assume that $x \notin A \dot{-} x \,\&\, y$. We want to show that $A \dot{-} x \,\&\, y \subseteq A \dot{-} x$. In the case when $x \notin A$, this holds trivially; so suppose that $x \in A$. It then follows from ($\dot{-}$F) that $\neg x \vee (x \,\&\, y) \in A \dot{-} x \,\&\, y$, so $\neg x \vee y \in A \dot{-} (x \,\&\, y) = A \dot{-} (\neg x \vee y) \,\&\, x$. By ($\dot{-}$C), $A \dot{-} x \,\&\, y = A \dot{-} (\neg x \vee y) \,\&\, x \subseteq A \dot{-} x \vee y$ or $A \dot{-} x \,\&\, y = A \dot{-} (\neg x \vee y) \,\&\, x \subseteq A \dot{-} x$. Since the second case is the desired inclusion, it will suffice to show that the first case implies the second. Suppose $A \dot{-} x \,\&\, y \subseteq A \dot{-} \neg x \vee y$. Then, since $\neg x \vee y \in A \dot{-} x \,\&\, y$, we have $\neg x \vee y \in A \dot{-} \neg x \vee y$, so by ($\dot{-}$4) $\neg x \vee y \in Cn(\emptyset)$. But this means that $Cn(x \,\&\, y) = Cn(x)$, so $A \dot{-} x \,\&\, y = A \dot{-} x$ by ($\dot{-}$5), and we are done. □

The last part of this proof shows that for maxichoice contraction functions the converse of Observation 3.4 also holds.

Proof of Observation 6.4 Suppose first that $\dot{-}$ is a full meet contraction function. We show that ($\dot{-}$I) is satisfied. If $x \in Cn(\emptyset)$ or $y \in Cn(\emptyset)$ then the desired equation holds trivially. Suppose that $x, y \notin Cn(\emptyset)$, and suppose that $x, y \in A$. Then we may apply Observation 4.1 to get

$$A \dot{-} (x\&y) = \bigcap\{A \perp x\&y\} = \bigcap\{A \perp x \cup A \perp y\} = \bigcap\{A \perp x\} \cap \bigcap\{A \perp y\}$$
$$= A \dot{-} x \cap A \dot{-} y,$$

so that $\dot{-}$ satisfies the intersection condition.

Trivially, intersection implies monotony. Suppose that $\dot{-}$ satisfies monotony; to prove (c) we need to show that $A \dot{-} x = A \sim x$, for which it clearly suffices to show that $A \dot{-} x \subseteq A \sim x$ in the light of Observation 2.7. In the case $x \notin A$ this holds trivially. In the case $x \in A$ we have by Observation 2.1 of (Alchourron and Makinson 1982b) that $A \sim x = A \cap Cn(\neg x)$, so we need only show $A \dot{-} x \subseteq Cn(\neg x)$. Suppose $y \in A \dot{-} x$. Then by ($\dot{-}$M), since $x \in A$ and $x \vdash x \vee y$, we have $y \in A \dot{-} (x \vee y)$, so $x \vee y \in A \dot{-} (x \vee y)$; so, by the postulate ($\dot{-}$4), $x \vee y \in Cn(\emptyset)$, so that $y \in Cn(\neg x)$ as desired. □

Proof of Observation 6.5 For the left to right implication, suppose $\dot{-}$ satisfies ($\dot{-}$V). Then ($\dot{-}$7) holds immediately. For ($\dot{-}$8), let x and y be propositions and suppose $x \notin A \dot{-} (x \,\&\, y)$; we need to show that $A \dot{-} (x \,\&\, y) \subseteq A \dot{-} x$. In the case that $x \notin A$ this holds trivially, so we suppose $x \in A$. Now $Cn(x \,\&\, y) = Cn(x \,\&\, (\neg x \vee y))$, so by ($\dot{-}$V) $A \dot{-} (x \,\&\, y)$ is identical with one of $A \dot{-} x$, $A \dot{-} (\neg x \vee y;)$ or $(A \dot{-} x) \cap (A \dot{-} (\neg x \vee y))$. In the first and last cases we have the desired inclusion, so we need only show that the middle case is impossible. Now by recovery, $A \dot{-} (\neg x \vee y) \cup \{\neg x \vee y\} \vdash x$, so $A \dot{-} (\neg x \vee y) \cup \{\neg x\} \vdash x$, so $x \in A \dot{-} (\neg x \vee y)$. But by hypothesis, $x \notin A \dot{-} (x \,\&\, y)$, so $A \dot{-} (x \,\&\, y) \neq A \dot{-} (\neg x \vee y)$, as desired.

The converse can be proven via the representation theorem (Observation 4.4), but it can also be given a direct verification as follows. Suppose that $\dot{-}$ satisfies ($\dot{-}$7) and ($\dot{-}$8), and suppose that $A \dot{-} (x \,\&\, y) \neq A \dot{-} x$ and $A \dot{-} (x \,\&\, y) \neq A \dot{-} y$; we want to show that $A \dot{-} (x \,\&\, y) = A \dot{-} x \cap A \dot{-} y$. By ($\dot{-}$7) it suffices to show that $A \dot{-} (x \,\&\, y) \subseteq A \dot{-} x \cap A \dot{-} y$, so it suffices to show that $A \dot{-} (x \,\&\, y) \subseteq \dot{-}\, x$ and $A \dot{-} (x \,\&\, y) \subseteq A \dot{-} y$. By ($\dot{-}$C), which we know by 3.4 to be an immediate consequence of ($\dot{-}$8), we have at least one of these inclusions. So it remains to show that under our hypotheses either inclusion implies the other. We prove one; the other is similar.

Suppose for reductio ad absurdum that $A \dot{-} (x \,\&\, y) \subseteq \dot{-}\, x$ but $A \dot{-} (x \,\&\, y) \not\subseteq A \dot{-} y$. Since by hypothesis $A \dot{-} (x \,\&\, y) \neq A \dot{-} x$, we have $A \dot{-} x \not\subseteq A \dot{-} (x \,\&\, y)$, so there is an $a \in A \dot{-} x$ with $a \notin A \dot{-} (x \,\&\, y)$. Since $A \dot{-} (x \,\&\, y) \not\subseteq A \dot{-} y$, we have by ($\dot{-}$8) that $y \in A \dot{-} (x \,\&\, y)$. Hence since $a \notin A \dot{-} (x \,\&\, y)$ we have $\neg y \vee a \notin A \dot{-} (x \,\&\, y)$. Hence by ($\dot{-}$7), $\neg y \vee a \notin A \dot{-} x$ or $\neg y \vee a \notin A \dot{-} y$. But since $a \in A \dot{-} x$ the former alternative is impossible. And the second alternative is also impossible, since by recovery $A \dot{-} y \cup \{y\} \vdash a$, so that $\neg y \vee a \in A \dot{-} y$. □

7 A Diagram for the Implications

To end the paper, we summarize the "implication results" of Sects. 4 and 6 in a diagram. The conditions are as named in previous pages with in addition ($\dot{-}$R) and ($\dot{-}$TR), meaning that $\dot{-}$ is relational, respectively transitively relational, over A, and (γR) and (γTR), meaning that γ is. ($\dot{-}$C) is the covering condition of Observation 3.4; (γC) is its analogue $\gamma(A \perp x) \subseteq \gamma(A \perp x \,\&\, y)$ or $\gamma(A \perp y) \subseteq \gamma(A \perp x \,\&\, y)$. ($\dot{-}$ P) is the partial antitony condition of 3.3; and (γP) is its obvious analogue $\gamma(A \perp x \,\&\, y) \cap A \perp x \subseteq \gamma(A \perp x)$. Conditions are understood to be formulated for an arbitrary theory A, selection function γ for A, and partial meet contraction function $\dot{-}$ over A determined by γ. Arrows are of course for implications, and conditions grouped into the same box are mutually equivalent in the finite case. Conversely, conditions in separate boxes are known to be nonequivalent, even for the finite case. The diagram should be read as a map of an ordering, but *not* as a lattice: a "∨" alignment does not necessarily mean that the bottom condition is equivalent to the conjunction of the other two. In some cases, it is—for example ($\dot{-}$TR) = ($\dot{-}$7) & ($\dot{-}$8) = ($\dot{-}$V), as proven in Corollary 4.5 and Observation 6.5; and again ($\dot{-}$D) = ($\dot{-}$F) & ($\dot{-}$8), as shown in Observation 6.3. But ($\dot{-}$7) & ($\dot{-}$C) is known *not* to be equivalent to ($\dot{-}$TR), and (γR) & ($\dot{-}$TR) may perhaps not be equivalent to (γTR). Finally, implications and nonimplications that follow from others by transitivity have not been written into the diagram, but are left as understood. Implications concerning connectivity from Sect. 5 have been omitted from the diagram, to avoid overcluttering.

All the general implications (arrows) have been proven in the text, or are immediate. The finite case equivalences issue from the injection result of Observation 4.6, and several were noted in Observation 4.10. Of the finite case non-equivalences, a first example serving to separate (γ7) from ($\dot{-}$R) was given in Observation 4.9, from which it follows immediately that (γ7) does not in the finite case imply ($\dot{-}$TR). The other nonequivalences need other examples, which we briefly sketch.

For the second example, take A to be the eight-element theory of Observation 4.9, but define γ as follows: In the limiting case of a_7, we put $\gamma(A \perp a_7) = \{!a_0\}$ as required by the fact that $a_7 \in \text{Cn}(\varnothing)$; put $\gamma(A \perp a_j) = A \perp a_j$ for all j with $2 \leq j < 7$; put $\gamma(A \perp a_1) = \{!a_3\}$; and put $\gamma(A \perp a_0) = \{!a_1, !a_3\}$. Then it can be verified that the partial meet contraction function $\dot{-}$ determined by γ satisfies ($\dot{-}$C), and so by finiteness also (γC), but not ($\dot{-}$8) and so a fortiori not ($\dot{-}$TR).

For the third example, take A as before, and put $\gamma(A \perp a_7) = \{!a_0\}$ as always; put $\gamma(A \perp a_1) = \{!a_2\}$; and put $\gamma(A \perp a_i) = A \perp a_i$ for all other a_i It is then easy to check that this example satisfies ($\dot{-}$8) but not ($\dot{-}$7), and so a fortiori not ($\dot{-}$R) and not ($\dot{-}$TR).

For the fourth and last example, take A as before, and put $\leq$ to be the least reflexive relation over 2^A such that $!a_1 \leq !a_2$, $!a_2 \leq !a_3$, $!a_3 \leq !a_2$ and $!a_3 \leq !a_1$. Define γ from $\leq$ via the marking off identity, and put $A \dot{-} x = \bigcap \gamma(A \perp x)$. Then it is easy to check that γ is a selection function for A, so that ($\dot{-}$R) holds. But ($\dot{-}$C) fails; in particular when $x = a_1$ and $y = a_2$ we can easily verify that $A \dot{-} (x \,\&\, y) \not\subseteq A \dot{-} x$ and $A \dot{-} (x \,\&\, y) \not\subseteq A \dot{-} y$. Hence, a fortiori, ($\dot{-}$8) and ($\dot{-}$TR) also fail.

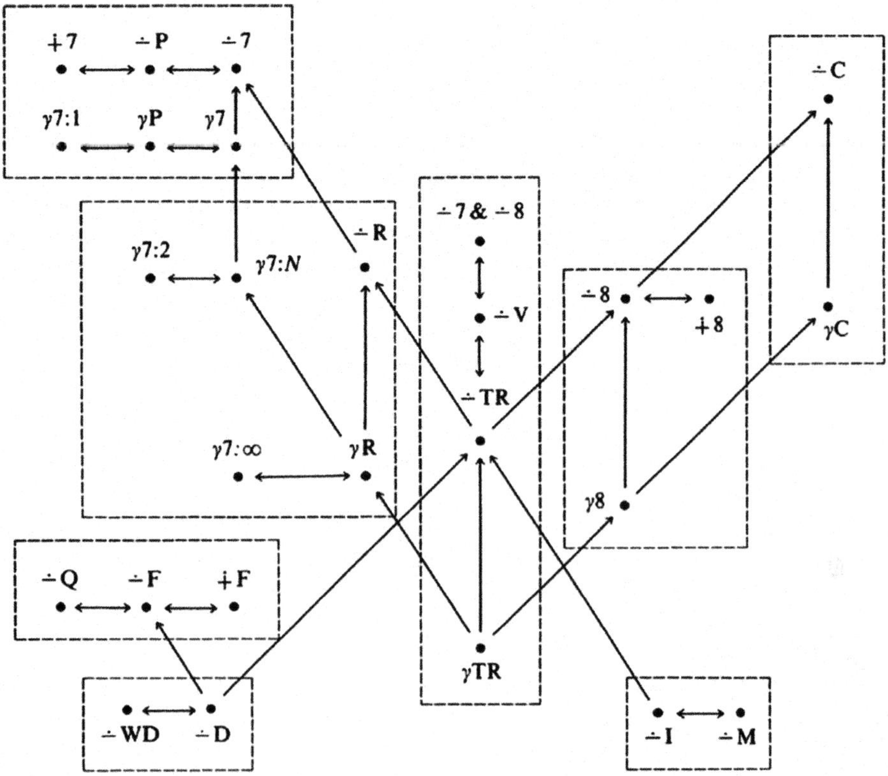

8 Added in Proof

The authors have obtained two refinements: the arrow $(\dot{-}\mathrm{D}) \rightarrow (\dot{-}\mathrm{TR})$ of the diagram on page 528 can be strengthened to $(\dot{-}\mathrm{D}) \rightarrow (\gamma\mathrm{TR})$; the implication $(\gamma 7{:}\infty) \rightarrow (\gamma 7{:}N)$ of Observation 4.10 can be strengthened to an equivalence. The former refinement is easily verified using the fact that any *maxichoice* contraction function over a theory is determined by a *unique* selection function over that theory. The latter refinement can be established by persistent use of the compactness of Cn.

Observation 4.10 so refined implies that for a theory A and selection function γ over A, γ is relational over A iff $(\gamma 7{:}2)$ holds. This raises an interesting *open question*, a positive answer to which would give a representation theorem for relational partial meet contraction, complementing Corollary 4.5: Can condition $(\gamma 7{:}2)$ be expressed as a condition on the contraction operation $\dot{-}$ determined by γ?

We note that a rather different approach to contraction has been developed by Alchourrón and Makinson in *On the logic of theory change: safe contraction*, to appear in ***Studia Logica***, vol. 44 (1985), the issue dedicated to Alfred Tarski; the relationship between the two approaches is studied by the same authors in *Maps between some different kinds of contraction function: the finite case*, also to appear in ***Studia Logica***, vol. 44 (1985).

References

Alchourron, C. E., & Makinson, D. (1982a). *Hierarchies of regulations and their logic, new studies in deontic logic* (pp. 125–148), R. Hilpinen (Ed.), Dordrecht: Reidel.

Alchourron, C. E., & Makinson, D. (1982b). On the logic of theory change: Contraction functions and their associated revision functions. *Theoria, 48*, 14–37.

Gärdenfors, P. (1978). Conditionals and changes of belief The logic and epistemology of scientific change. In I. Niiniluoto & R. Tuomela (Eds.), *Acta Philosophica Fennica* (Vol. 30, pp. 381–404).

Gärdenfors, P. (1982). *Rules for rational changes of belief 320311: Philosophical essays dedicated to Lennart Åqvist on his fiftieth birthday* (pp. 88–101), (T. Pauli, Ed., Philosophical studies no. 34). Uppsala: Department of Philosophy, University of Uppsala.

Gärdenfors, P. (1984). Epistemic importance and minimal changes of belief. *Australasian Journal of Philosophy, 62*, 136–157.

Makinson, D. (1985). How to give it up: A survey of some formal aspects of the logic of theory change. *Synthese, 62*, 347–363.

Chapter 14
Theory Contraction and Base Contraction Unified

Sven Ove Hansson

General Introduction

The AGM (Alchourrón-Gärdenfors-Makinson) model of belief change has acquired the status of a standard model. In that model, a belief state is represented by a set of sentences that is closed under logical consequence, the *belief set* or theory. (Alchourrón et al. 1985); (Gärdenfors 1988) Among the major rivals are models in which a belief state is represented by a *belief base*, a set that is not (except in a limiting case) closed under consequence. (Fuhrmann 1991); (Hansson 1989; 1991; 1992; 1993); (Nebel 1992) Obviously, the logical closure of a belief base is a belief set, and each non-trivial belief set can be represented by several different belief bases. It has been argued that different belief bases for one and the same belief set represent different ways of holding the same beliefs. Roughly, the elements of the belief base represent "basic", or independently grounded beliefs, in contrast to the "merely derived" beliefs that form the rest of the belief set. (Fuhrmann 1991, pp. 183–184); (Hansson 1989; 1992)

As has been accurately pointed out by Fuhrmann (1991), operations on a belief base generate operations on the corresponding belief set. In particular, if a belief base for a belief set **K** is contracted by a sentence α, then the logical closure of the contracted belief base is a belief set from which α has been contracted.

The purpose of this paper is to characterize the operators of contraction on a belief set **K** that can be obtained by assigning to it (1) a belief base, and (2) an operator of partial meet contraction for that belief base. The section

---

Sven Ove Hansson 1992-07-08.

S.O. Hansson (✉)
Division of Philosophy, KTH, Brinellvägen 32, 100 44 Stockholm, Sweden
e-mail: soh@kth.se

H. Arló-Costa et al. (eds.), *Readings in Formal Epistemology*, Springer Graduate Texts in Philosophy 1, DOI 10.1007/978-3-319-20451-2_14

"Partial meet contraction" contains introductory material. In the section "The new postulates", the postulates that will be used for the characterizations are introduced, and in the section "Axiomatic characterizations" axiomatic characterizations of various types of base-generated theory contractions are given. The section "Proofs" provides proofs of the results reported in the section "Axiomatic characterizations".

Partial Meet Contraction

We will assume a language L that is closed under truth-functional operations and a consequence operator Cn for L. Cn satisfies the standard conditions for a consequence operator, namely inclusion ($A \subseteq Cn(A)$), monotony (if $A \subseteq B$, then $Cn(A) \subseteq Cn(B)$), and iteration ($Cn(A) = Cn(Cn(A))$). Furthermore, it satisfies the properties of supraclassicality (if α follows by classical truth-functional logic from A, then $\alpha \in Cn(A)$), deduction (if $\beta \in Cn(A \cup \{\alpha\})$, then $(\alpha \to \beta) \in Cn(A)$) and compactness (if $\alpha \in Cn(A)$, then $\alpha \in Cn(A')$ for some finite set $A' \subseteq A$). $A \vdash \alpha$ will be used as an alternative notation for $\alpha \in Cn(A)$.

A *belief set* is a subset $\mathbf{K}$ of the language such that $\mathbf{K} = Cn(\mathbf{K})$. An operator of *contraction* for $\mathbf{K}$ is an operator $\div$ from $\mathcal{P}(L) \times L$ to $\mathcal{P}(L)$ such that for all sentences α, $\mathbf{K} \div \alpha \subseteq \mathbf{K}$, and if $\alpha \notin Cn(\varnothing)$, then $\alpha \notin Cn(\mathbf{K} \div \alpha)$. A particularly interesting type of contraction is *partial meet contraction*, which was introduced by Alchourrón, Gärdenfors, and Makinson (1985). (Gärdenfors 1984) It is defined by the identity:

$$\mathbf{K} \div \alpha = \bigcap \gamma(\mathbf{K} \perp \alpha),$$

where $\mathbf{K} \perp \alpha$ denotes the set of inclusion-maximal subsets of $\mathbf{K}$ that do not have α as a logical consequence. γ is a *selection function*, such that $\gamma(\mathbf{K} \perp \alpha)$ is a non-empty subset of $\mathbf{K} \perp \alpha$, unless the latter is empty, in which case $\gamma(\mathbf{K} \perp \alpha) = \{\mathbf{K}\}$.

Let γ be a selection function for $\mathbf{K}$. Then the *completion* of γ is the function γ^* such that $\gamma^*(\mathbf{K} \perp \alpha) = \{X \in \mathbf{K} \perp \alpha \mid \bigcap \gamma(\mathbf{K} \perp \alpha) \subseteq X\}$, unless $\mathbf{K} \perp \alpha$ is empty, in which case $\gamma^*(\mathbf{K} \perp \alpha) = \{\mathbf{K}\}$. γ^* is a selection function for $\mathbf{K}$, and gives rise to the same operator of partial meet contraction as γ. (Alchourrón et al. 1985, p. 519) It is in some contexts convenient to make the technical assumption that a selection function is *completed*, i.e. identical to its own completion.

Full meet contraction is the limiting case when $\gamma(\mathbf{K} \perp \alpha) = \mathbf{K} \perp \alpha$ for all non-empty $\mathbf{K} \perp \alpha$. In the other limiting case, when $\gamma(\mathbf{K} \perp \alpha)$ is a singleton for all α, $\div$ is an operator of *maxichoice* contraction. An operator of partial meet contraction is *relational* if and only if it is based on a selection function γ for which there is a relation $\preceq$ such that for all non-empty $\mathbf{K} \perp \alpha$ we have

$$\gamma(\mathbf{K} \perp \alpha) = \left\{\mathbf{K}' \in \mathbf{K} \perp \alpha \;\middle|\; \mathbf{K}'' \preceq \mathbf{K}' \text{ for all } \mathbf{K}'' \in \mathbf{K} \perp \alpha\right\}$$

If this condition holds for a transitive relation $\preceq$, then the operator is *transitively relational*.

Partial meet contraction derives much of its attractiveness from a representation theorem by Alchourrón et al. (1985): An operation $\div$ on a logically closed set $\mathbf{K}$ is a partial meet contraction if and only if it satisfies the following six postulates, the basic *Gärdenfors postulates*:

(G÷1) $\mathbf{K} \div \alpha$ is a theory if $\mathbf{K}$ is a theory (*closure*)
(G÷2) $\mathbf{K} \div \alpha \subseteq \mathbf{K}$ (*inclusion*)
(G÷3) If $\alpha \notin Cn(\mathbf{K})$ then $\mathbf{K} \div \alpha = \mathbf{K}$ (*vacuity*)
(G÷4) If $\alpha \notin Cn(\varnothing)$ then $\alpha \notin Cn(\mathbf{K} \div \alpha)$ (*success*)
(G÷5) If $Cn(\alpha) = Cn(\beta)$ then $\mathbf{K} \div \alpha = \mathbf{K} \div \beta$ (*preservation*)
(G÷6) $\mathbf{K} \subseteq Cn((\mathbf{K} \div \alpha) \cup \{\alpha\})$ whenever $\mathbf{K}$ is a theory (*recovery*).

Furthermore, an operator of partial meet contraction on a logically closed set is transitively relational if and only if it also satisfies:

(G÷7) $(\mathbf{K} \div \alpha) \cap (\mathbf{K} \div \beta) \subseteq \mathbf{K} \div (\alpha \& \beta)$ (*intersection*)
(G÷8) If $\alpha \notin \mathbf{K} \div (\alpha \& \beta)$ then $\mathbf{K} \div (\alpha \& \beta) \subseteq \mathbf{K} \div \alpha$ (*conjunction*).

A *belief base* for $\mathbf{K}$ is a set $\boldsymbol{B}$ such that $\mathbf{K} = Cn(\boldsymbol{B})$. Partial meet contraction for belief bases is defined in the same way as for belief sets, i.e., by the identity $\boldsymbol{B} \div \alpha = \bigcap \gamma(\boldsymbol{B} \perp \alpha)$. Full meet, maxichoice, relational, and transitively relational contraction is also defined in the same way as for belief sets. Furthermore, an operator of partial meet contraction is *transitively, maximizingly relational* (*TMR*) if and only if it is relational by a transitive relation $\underline{\ll}$ such that, for its strict counterpart $\ll$, if $A \subset B$, then $A \ll B$.[1]

Contractions on belief bases may be studied in their own right.[2] In this paper, however, they will be treated as means for contractions on belief sets.

Definition An operation $\div$ on a belief set $\mathbf{K}$ is *generated by* a partial meet base contraction if and only if there is a belief base $\boldsymbol{B}$ for $\mathbf{K}$ and an operator $\sim_\gamma$ of partial meet contraction for $\boldsymbol{B}$ such that $\mathbf{K} \div \alpha = Cn\left(\boldsymbol{B} \sim_\gamma \alpha\right)$ for all $\alpha \in L$.

We will see directly that if an operation on a belief set is generated by some partial meet base contraction, then it satisfies the first five of the basic Gärdenfors postulates, (G÷1)–(G÷5), but it does not, in general, satisfy recovery (G÷6).[3]

---

[1]The maximizing property may be interpreted as saying that all elements of the belief base have positive epistemic value. This property might at first hand seem superfluous. If $K' \subset K''$, then $K' \in K \perp \alpha$ and $K'' \in K \perp \alpha$ cannot both be true, so that K' and K'' cannot be candidates between which the selection function has to choose. However, when more than one contraction is taken into account, the property need not be superfluous. If $K_1 \subset K_2$, and K_3 is neither a subset nor a superset of either K_1 or K_2, then $\gamma(\{K_1,K_3\}) = \{K_1\}$ and $\gamma(\{K_2,K_3\}) = \{K_3\}$ may both hold for a transitively relational selection function γ, but not if it is in addition required to be maximizing.

[2]See (Hansson 1993) for some results, including an axiomatic characterization of the partial meet contractions on a belief base.

[3]In Makinson's (1987) terminology, an operation that satisfies (G÷1)-(G÷5) but not necessarily (G÷6) is a "withdrawal". However, I will use "contraction" in the wide sense indicated above, thus including Makinson's withdrawals.

Recovery is the most controversial of the six postulates, and has been seriously questioned by several authors. See (Hansson 1991); (Makinson 1987); (Niederée 1991).

The New Postulates

It has often been remarked that the only realistic belief sets are those that have a finite representation. (Gärdenfors and Makinson 1988) We are going to assume that both the original belief set and the belief sets that are obtained through contraction have finite representations (belief bases); thus for every α, there is some finite set A such that $\mathbf{K} \div \alpha = Cn(A)$. Furthermore, we will assume that although there may be infinitely many sentences by which the belief set can be contracted, there is only a finite number of belief sets that can be obtained through contraction, i.e. $\{\mathbf{K}' \mid (\exists\alpha)(\mathbf{K}' = \mathbf{K} \div \alpha)\}$ is finite. These two finiteness properties can be combined into the following:

There is a finite set A such that for every α, there is some $A' \subseteq A$ such that $\mathbf{K} \div \alpha = Cn(A')$. (*finitude*)

In the presence of vacuity (G$\div$3), *finitude* implies that there is some finite set A such that $\mathbf{K} = Cn(A)$.

If $\div$ is a contraction operator for $\mathbf{K}$, and α is not a logical theorem, then $\mathbf{K} \div \alpha$ does not contain α. However, to contract $\mathbf{K}$ by α is not the only way to exclude α from the belief set. Typically, there are several β distinct from α such that $\alpha \notin \mathbf{K} \div \beta$. This must be the case if α logically implies β, and it can be true in other cases as well. A contraction $\mathbf{K} \div \beta$ such that $\alpha \notin \mathbf{K} \div \beta$ will be called an α-*removal.*

Two different beliefs may have exactly the same justification(s). As an example, I believe that either Paris or Oslo is the capital of France (α). I also believe that either Paris or Stockholm is the capital of France (β). Both these beliefs are entirely based on my belief that Paris is the capital of France. Therefore, a contraction by some sentence δ removes α if and only if it removes β (namely if and only if it removes the common justification of these two beliefs). There is no contraction by which I can retract α without retracting β or vice versa. It is not unreasonable to require that if two beliefs in a belief set stand or fall together in this way, then their contractions are identical. In other words, *if all* α-*removals are* β-*removals and vice versa, then* $\mathbf{K} \div \alpha = \mathbf{K} \div \beta$. In the formal language:

If $\mathbf{K} \div \delta \vdash \alpha$ iff $\mathbf{K} \div \delta \vdash \beta$ for all δ, then $\mathbf{K} \div \alpha = \mathbf{K} \div \beta$. (*symmetry*)

An α-removal $\mathbf{K} \div \beta$ will be called a *preservative* α-*removal* if and only if $\mathbf{K} \div \alpha \subseteq \mathbf{K} \div \beta$, and a *strictly preservative* α-*removal* if and only if $\mathbf{K} \div \alpha \subset \mathbf{K} \div \beta$. A strictly preservative α-removal is an operation that removes α, and does this in a more economical way than what is done by the contraction by α.

Often, a belief can be removed more economically if more specified information is obtained. As an example, I believe that Albert Schweitzer was a German Missionary (α). Let α_1 denote that he was a German and α_2 that he was a Missionary, so that $\alpha \equiv \alpha_1 \& \alpha_2$. If I have to contract my belief set by α, then the contracted belief set will contain neither α_1 nor α_2. Admittedly it would be logically sufficient to withdraw one of them. However, they are both equally entrenched, so that I do not know which to choose in preference over the other. Therefore, both will have to go. On the other hand, if I have to contract my belief set by α_1, then I have no reason to let go of α_2.[4] To contract by α_1 is, given the structure of my belief state, a more specified way to remove α. Thus we may expect that $\mathbf{K} \div \alpha \subset \mathbf{K} \div \alpha_1$, so that $\mathbf{K} \div \alpha_1$ is a strictly preservative α-removal.

Let δ denote that Albert Schweitzer was a Swede, and let us consider the contraction of $\mathbf{K}$ by $\alpha_1 \vee \delta$, "Albert Schweitzer was a German or a Swede." Since I believe in $\alpha_1 \vee \delta$ only as a consequence of my belief in α_1, I can only retract $\alpha_1 \vee \delta$ by retracting α_1. Therefore, $\mathbf{K} \div (\alpha_1 \vee \delta)$ is not a proper superset of $\mathbf{K} \div \alpha_1$, i.e., it is not a more conservative α-withdrawal than $\mathbf{K} \div \alpha_1$. Indeed, the way my beliefs are structured, α_1 cannot be further subdivided in the way that α was subdivided into α_1 and α_2. There is no part of α_1 that stands on its own and can be retracted from $\mathbf{K}$ without the rest of α_1 being lost as well. In this sense, no α-removal can be more conservative than $\mathbf{K} \div \alpha_1$.

More generally, $\mathbf{K} \div \beta$ is a *maximally preservative α-removal* if and only if it is a preservative α-removal and there is no α-removal $\mathbf{K} \div \delta$ such that $\mathbf{K} \div \beta \subset \mathbf{K} \div \delta$. Intuitively, to perform a maximally preservative α-removal is to make the belief set not imply α, making use of information that is sufficiently specified to allow one to remove a part of α so small that no smaller part of it can be removed alone.

Contraction should be conservative in the sense that every element of $\mathbf{K}$ is retained in $\mathbf{K} \div \alpha$ unless there is some good reason to exclude it. As was noted in (Hansson 1991), (G÷1)–(G÷5) do not ensure this property, since they are satisfied by the operation $\div$ such that if $\alpha \notin Cn(\mathbf{K})$, then $\mathbf{K} \div \alpha = \mathbf{K}$, and otherwise $\mathbf{K} \div \alpha = \mathbf{K} \cap Cn(\varnothing)$. The same applies if *symmetry* is added to the list of postulates. We need some further postulate to prevent elements of $\mathbf{K}$ from being lost for no reason in operations of contraction.

One way to achieve this is to require that what is lost in the contraction by α must be incompatible with some reasonable way to remove α. In other words, if a unit of belief is lost in contraction by α, then there should be at least one preservative α-removal to which the lost unit cannot be added without α being implied. However, it would be too far-reaching to require this to hold for units of belief that cannot stand on their own (such as $\alpha_1 \vee \delta$ in our Albert Schweizer example). Such a unit of belief can be lost merely due to loss of the larger, self-sustained unit(s) of which it is a part. Thus, our principle of conservativity should refer only to units of belief that *can* stand on their own.

In order to formulate a conservativity principle, we therefore need a formal criterion that excludes non-self-sustained units of belief. In our example, there can be no sentence β such that $\mathbf{K} \div \beta$ consists exactly of $\alpha_1 \vee \delta$ and its logical consequences, since $\alpha_1 \vee \delta$ cannot stand on its own without α_1. Admittedly, this is a

[4]It is here assumed that $\alpha_1 \equiv \alpha_2$ is less entrenched than α_1 and α_2.

weak criterion, and it can be strengthened in various ways.[5] However, it turns out to be sufficient for our present purposes, and it will therefore be used in our postulate of conservativity:

If $\mathbf{K} \div \beta \nsubseteq \mathbf{K} \div \alpha$, then there is some δ such that $\mathbf{K} \div \alpha \subseteq \mathbf{K} \div \delta \nvdash \alpha$ and $(\mathbf{K} \div \beta) \cup (\mathbf{K} \div \delta) \vdash \alpha$. (*conservativity*)

An obvious way to strengthen *conservativity* is to require that if a unit of belief is lost in contraction by α, then α will be implied if it is added to $\mathbf{K} \div \alpha$ (and not merely if it is added to some preservative α-removal):

If $\mathbf{K} \div \beta \nsubseteq \mathbf{K} \div \alpha$ then $\mathbf{K} \div \alpha \nvdash \alpha$ and $(\mathbf{K} \div \beta) \cup (\mathbf{K} \div \alpha) \vdash \alpha$. (*strong conservativity*)

Strong conservativity is much less plausible than *conservativity*. This can be seen from our Albert Schweitzer example. In that example, it may reasonably be assumed that $\alpha_1 \notin \mathbf{K} \div (\alpha_1 \& \alpha_2)$, $\alpha_2 \notin \mathbf{K} \div (\alpha_1 \& \alpha_2)$, $\alpha_1 \in \mathbf{K} \div \alpha_2$, $\alpha_2 \in \mathbf{K} \div \alpha_1$, $\mathbf{K} \div (\alpha_1 \& \alpha_2) \subseteq \mathbf{K} \div \alpha_1$, and $\mathbf{K} \div (\alpha_1 \& \alpha_2) \subseteq \mathbf{K} \div \alpha_2$. However, this is incompatible with *strong conservativity*. Since $\mathbf{K} \div \alpha_1 \nsubseteq \mathbf{K} \div (\alpha_1 \& \alpha_2)$, this postulate requires that $\mathbf{K} \div \alpha_1 \cup \mathbf{K} \div (\alpha_1 \& \alpha_2) \vdash (\alpha_1 \& \alpha_2)$, contrary to our assumptions for this case. More generally, *strong conservativity* is implausible since it precludes the removal of two or more sentences (in this case α_1 and α_2), when it would have been logically sufficient to remove only one of them. Such epistemic behaviour is rational enough when the beliefs in question are equally entrenched, or have equal epistemic utility.

The concepts of epistemic entrenchment and epistemic utility refer to extra-logical reasons that one may have for preferring one way to remove a sentence α rather than another. It is conceivable for an epistemic agent to make no use of such extra-logical information. Such an agent is indecisive in the sense of not being able to make a choice among different ways to remove a belief, if these are on an equal footing from a logical point of view. Formally, this is close to a reversal of *conservativity*: If a (self-sustained) unit of belief conflicts with some way to remove α from $\mathbf{K}$, then it is not a part of $\mathbf{K} \div \alpha$:

If there is some δ such that $\mathbf{K} \div \delta \nvdash \alpha$ and $(\mathbf{K} \div \beta) \cup (\mathbf{K} \div \delta) \vdash \alpha$, then $\mathbf{K} \div \beta \nsubseteq \mathbf{K} \div \alpha$. (*indecisiveness*)

Non-logical considerations play an important role in actual human epistemic behaviour. Arguably, it would in many cases be irrational not to let them do so. Therefore, *indecisiveness* is not a plausible general property of rational belief change.

---

[5]Two such strengthened versions should be mentioned: (1) $\div$ can be extended to contraction by sets (multiple contraction), such that if $A \cap Cn(\varnothing) = \varnothing$, then $\mathbf{K} \div A$ is a logically closed subset of $\mathbf{K}$ that does not imply any element of A. It a unit of belief cannot stand on its own, then it should not be equal to $\mathbf{K} \div A$ for any set A. (2) Iterated contraction can be used for the same purpose: If a unit of belief cannot stand on its own, then it should not be equal to $\mathbf{K} \div \beta_1 \div \beta_2 \ldots \div \beta_n$ for any series $\beta_1 \ldots \beta_n$ of sentences.

The next postulate will again be concerned with maximally preservative removals, and thus with the smallest units of belief that can be removed from a belief set. In an orderly and coherent belief state, one would expect the identities and the relations of entrenchment of these units to be constant, i.e. they should be the same independently of what sentence we are contracting by.

As an illustration, let us extend the Albert Schweitzer example. Let α_1 denote that Schweitzer was a German, α_2 that he was a missionary and α_3 that he was a physician. Let us assume that $\mathbf{K} \div \alpha_1$ is a maximally preservative $\alpha_1 \& \alpha_2 \& \alpha_3$-removal, i.e. a maximally economical way to remove $\alpha_1 \& \alpha_2 \& \alpha_3$ from the belief set. Since $\mathbf{K} \div \alpha_1$ is also an $\alpha_1 \& \alpha_2$-removal, and since the $\alpha_1 \& \alpha_2$-removals are a subset of the $\alpha_1 \& \alpha_2 \& \alpha_3$-removals, it should also be maximally preservative among these. Furthermore, if $\mathbf{K} \div \alpha_2$ is equally economical as $\mathbf{K} \div \alpha_1$ in the context of removing $\alpha_1 \& \alpha_2$ (i.e., if it is also a maximaly preservative $\alpha_1 \& \alpha_2$-removal), then it should also be equally economical as $\mathbf{K} \div \alpha_1$ in the context of removing $\alpha_1 \& \alpha_2 \& \alpha_3$ (i.e., it should also be a maximally preservative $\alpha_1 \& \alpha_2 \& \alpha_3$-removal). In general:

If $\vdash \alpha \rightarrow \beta$ and the set of β-removals that are also maximally preservative α-removals is non-empty, then it coincides with the set of maximally preservative β-removals. (*regularity*)

It may be convenient to divide *regularity* into two parts[6]:

If $\vdash \alpha \rightarrow \beta$ and some maximally preservative α-removal is also a β-removal, then all maximally preservative β-removals are maximally preservative α-removals. (*regularity 1*)

If $\vdash \alpha \rightarrow \beta$ and $\mathbf{K} \div \delta$ is both a β-removal and a maximally preservative α-removal, then it is a maximally preservative β-removal. (*regularity 2*)

Clearly, *regularity* holds if and only if both *regularity 1* and *regularity 2* hold.

It does *not* follow from *regularity* that if, in our example, $\mathbf{K} \div (\alpha_1 \& \alpha_2 \& \alpha_3)$ is an $\alpha_1 \& \alpha_2$-removal, then $\mathbf{K} \div (\alpha_1 \& \alpha_2 \& \alpha_3) = \mathbf{K} \div (\alpha_1 \& \alpha_2)$. To see why this would not be plausible as a general principle, let us modify the example and assume that α_2 and α_3 are equally entrenched, whereas α_1 is more entrenched than both α_2 and α_3. Then we should expect $\alpha_3 \notin \mathbf{K} \div (\alpha_1 \& \alpha_2 \& \alpha_3)$ but $\alpha_3 \in \mathbf{K} \div (\alpha_1 \& \alpha_2)$. However, as a limiting case, we can imagine an epistemic agent who is never indifferent between different ways to remove a belief from her belief set. Such an agent will always, when contracting by α, remove one of the smallest removable parts of α. Thus, the contraction by α will itself be the only maximally preservative α-removal. In our example, when contracting by $\alpha_1 \& \alpha_2 \& \alpha_3$, this agent will remove exactly one of α_1, α_2, and α_3, and when contracting by $\alpha_1 \& \alpha_2$, she will remove exactly one of α_1 and α_2. Assuming that the relative entrenchment of her beliefs is context-independent, if she removes α_1 when contracting by $\alpha_1 \& \alpha_2 \& \alpha_3$, then α_1 is

[6] *Regularity 1* is closely related to Amartya Sen's β property for rational choice behaviour, and *regularity* 2 to his α property. (Sen 1970).

also the removed unit of belief when $\alpha_1 \& \alpha_2$ is contracted. In general, her epistemic behaviour should satisfy the following postulate:

If $\vdash \alpha \rightarrow \beta$ and $\mathbf{K} \div \alpha \nvdash \beta$, then $\mathbf{K} \div \beta = \mathbf{K} \div \alpha$. (*hyperregularity*)

Hyperregularity implies that for all α and β, $\mathbf{K} \div (\alpha\&\beta) = \mathbf{K} \div \alpha$ or $\mathbf{K} \div (\alpha\&\beta) = \mathbf{K} \div \beta$. This condition has also been called "decomposition" (Alchourrón et al. 1985, p. 525) As was noted by Gärdenfors (1988, p. 66), it is too strong a principle. Thus, *hyperregularity* is a limiting case of some interest, but not a plausible criterion of rational belief change. The same applies to *strong conservativity* and *indecisiveness*, whereas *symmetry*, *conservativity*, and *regularity* are proposed as reasonable postulates for rational belief change.

Axiomatic Characterizations

Symmetry and *conservativity* are sufficient to characterize, together with (G÷1)-(G÷5) and *finitude*, the contractions of belief sets that are generated by partial meet base contraction. Since these are all fairly plausible postulates, this result adds to the credibility of theory contraction through partial meet base contraction.

Theorem 14.1 An operation ÷ on a belief set **K** is generated by partial meet contraction of a finite base for **K** iff ÷ satisfies (G÷1), (G÷2), (G÷3), (G÷4), (G÷5), *finitude*, *symmetry* and *conservativity*.

If *conservativity* is strengthened to the (much less plausible) postulate of *strong conservativity*, then a characterization is obtained of operations that are generated by maxichoice contractions of finite bases.

Theorem 14.2 An operation ÷ on a belief set **K** is generated by maxichoice partial meet contraction of a finite base for **K** iff ÷ satisfies (G÷1), (G÷2), (G÷3), (G÷4), (G÷5), *finitude*, *symmetry*, and *strong conservativity*.

Indecisiveness, in combination with *conservativity*, is a characteristic postulate for operations based on full meet base contraction.

Theorem 14.3 An operation ÷ on a belief set **K** is generated by full meet contraction of a finite base for **K** iff ÷ satisfies (G÷1), (G÷2), (G÷3), (G÷4), (G÷5), *finitude*, *conservativity*, and *indecisiveness*.

Regularity ensures that the partial meet base contraction that generates ÷ is transitively, maximizingly relational (cf. the section "Partial meet contraction").

Theorem 14.4 An operation ÷ on a belief set **K** is generated by transitively, maximizingly relational partial meet contraction of a finite base for **K**, by a completed selection function,[7] iff ÷ satisfies (G÷1), (G÷2), (G÷3), (G÷4), (G÷5), *finitude*, *symmetry*, *conservativity*, and *regularity*.

---

[7]The completeness of γ is used in the proof. I do not know if it can be dispensed with.

For maxichoice operations that are transitively, maximizingly relational, the following axiomatic characterization has been obtained:

Theorem 14.5 An operation ÷ on a belief set **K** is generated by a transitively, maximizingly relational maxichoice contraction of a finite base for **K** iff ÷ satisfies (G÷1), (G÷2), (G÷3), (G÷4), (G÷5), *finitude*, *symmetry*, *strong conservativity*, and *hyperregularity*.

Some of the postulates used in Theorems 14.2, 14.3, and 14.5 were shown in the section "The new postulates" to be quite implausible. Indeed, maxichoice and full meet contraction are of interest only as limiting cases. In contrast, Theorems 14.1 and 14.4 only employ fairly plausible postulates of rational belief change. It is proposed that the classes of base-generated contractions of belief sets that are characterized in these theorems represent reasonable types of belief contraction.

Two further important properties have been obtained for the class of operations that were referred to in Theorem 14.4, namely contractions of belief sets that are generated by transitively, maximizingly relational partial meet base contractions:

Theorem 14.6 Let the operation ÷ on the belief set **K** be generated by some transitively, maximizingly relational partial meet contraction of a finite base for **K**. Then:

(1) If $\mathbf{K} \div \delta \subseteq (\mathbf{K} \div \alpha) \cap (\mathbf{K} \div \beta)$, then $\mathbf{K} \div \delta \subseteq \mathbf{K} \div (\alpha \& \beta)$. (*weak intersection*)
(2) If $\alpha \notin \mathbf{K} \div (\alpha \& \beta)$, then $\mathbf{K} \div (\alpha \& \beta) \subseteq \mathbf{K} \div \alpha$. (*conjunction*)

Weak intersection is a weaker form of Gärdenfors's (G÷7) postulate, namely that "the beliefs that are both in $\mathbf{K} \div \alpha$ and $\mathbf{K} \div \beta$ are also in $\mathbf{K} \div (\alpha \& \beta)$" (Gärdenfors 1988, p. 64).[8] It differs from Gärdenfors's original postulate in being restricted to beliefs that are self-sustained in the sense that was accounted for in section "The new postulates". To see that this is a reasonable restriction, let α and β be self-sustained beliefs that have the same degree of entrenchment, and such that $\alpha \vee \beta$ is believed only as a logical consequence of α and β. For a plausible practical example, let α denote that Algiers is a capital and β that Bern is a capital. I have both these beliefs, and they are equally entrenched. If I contract my belief set by α, then β is unperturbed, so that $\beta \in \mathbf{K} \div \alpha$, and as a consequence of that, $\alpha \vee \beta \in \mathbf{K} \div \alpha$. For symmetrical reasons, $\alpha \vee \beta \in \mathbf{K} \div \beta$. However, if I contract my belief set by $\alpha \& \beta$, then since α and β are equally entrenched I cannot choose between them, so that they must both go. Since neither α nor β is in $\mathbf{K} \div (\alpha \& \beta)$, and $\alpha \vee \beta$ was believed only as a consequence of (each of) these two beliefs, $\alpha \vee \beta$ will be lost as well. Thus, $\alpha \vee \beta \in (\mathbf{K} \div \alpha) \cap (\mathbf{K} \div \beta)$ but $\alpha \vee \beta \notin \mathbf{K} \div (\alpha \& \beta)$, contrary to (G÷7) but in accordance with *weak intersection*.

Conjunction is Gärdenfors's (G÷8) postulate. To motivate it, the Algiers and Bern example may again be used. In that case, to remove α is a way to remove a specified part of $\alpha \& \beta$. In general, the removal of a part of a certain belief is at least

[8]The formulas of the quotation have been adapted to the notational convention used here.

as economical (conservative, retentive of the original belief set) as the removal of that belief in in entirely. Therefore, if $\mathbf{K} \div (\alpha\&\beta)$ is an α-removal, then $\mathbf{K} \div \alpha$ should be at least as economical as $\mathbf{K} \div (\alpha\&\beta)$.

In conclusion, with base-generated theory contraction we can avoid the problematic *recovery* postulate (G÷6) of the AGM framework, but still have the plausible basic postulates (G÷1)-(G÷5) and the additional postulates (G÷7) (in a weakened but credibilized form) and (G÷8) (in its original form).[9]

Proofs

A set A is *finite-based* iff there is some finite set A' such that $Cn(A') = Cn(A)$. For any non-empty, finite-based set A, $\&A$ denotes the conjunction of all elements of some finite base of A. For any finite, non-empty set A, $\mathcal{V}(A)$ denotes the disjunction of the elements of A.

The following lemmas will be needed for the proofs:

Lemma 14.1 $\boldsymbol{B} \perp (\alpha\&\beta) \subseteq \boldsymbol{B} \perp \alpha \cup \boldsymbol{B} \perp \beta$.

Proof of Lemma 14.1 Let $W \in \boldsymbol{B} \perp (\alpha\&\beta)$. Then either $W \nvdash \alpha$ or $W \nvdash \beta$. It follows from $W \nvdash \alpha$ and $W \in \boldsymbol{B} \perp (\alpha\&\beta)$ that $W \in \boldsymbol{B} \perp \alpha$, and in the same way from $W \nvdash \beta$ and $W \in \boldsymbol{B} \perp (\alpha\&\beta)$ that $W \in \boldsymbol{B} \perp \beta$.

Lemma 14.2 If $X \not\subseteq Y \not\subseteq X$ for all $X \in \boldsymbol{B} \perp \alpha$ and $Y \in \boldsymbol{B} \perp \beta$, then $\boldsymbol{B} \perp (\alpha\&\beta) = \boldsymbol{B} \perp \alpha \cup \boldsymbol{B} \perp \beta$.

We will mostly use a special case of the lemma. Namely if $\{X\} = \boldsymbol{B} \perp \alpha$ and $\{Y\} = \boldsymbol{B} \perp \beta$, and $X \not\subseteq Y \not\subseteq X$, then $\{X, Y\} = \boldsymbol{B} \perp (\alpha\&\beta)$.

Proof of Lemma 14.2 One direction follows from *lemma* 14.1. For the other direction, let $X \in \boldsymbol{B} \perp \alpha$. Then $X \nvdash (\alpha\&\beta)$. In order to prove that $X \in \boldsymbol{B} \perp (\alpha\&\beta)$, suppose to the contrary that there is some W such that $X \subset W \subseteq \boldsymbol{B}$ and $W \nvdash (\alpha\&\beta)$. From $X \subset W$ it follows that $W \vdash \alpha$. With $W \nvdash (\alpha\&\beta)$ this yields $W \nvdash \beta$, from which it follows that $W \subseteq Y$ for some $Y \in \boldsymbol{B} \perp \beta$. We therefore have $X \subset W \subseteq Y$, contradicting the assumption that $X \not\subseteq Y \not\subseteq X$. We may conclude that $X \in \boldsymbol{B} \perp (\alpha\&\beta)$.

In the same way it follows that if $Y \in \boldsymbol{B} \perp \beta$ then $Y \in \boldsymbol{B} \perp (\alpha\&\beta)$.

Lemma 14.3 If $X \in \boldsymbol{B} \perp \alpha$ and $\boldsymbol{B}$ is finite, then there is some β such that $\vdash \alpha \rightarrow \beta$ and $\{X\} = \boldsymbol{B} \perp \beta$.

Proof of Lemma 14.3 If $X = \boldsymbol{B}$, then let $\beta = \alpha$. Otherwise, let $\boldsymbol{B} \backslash X = \{\xi_1, \ldots \xi_n\}$, and let β be $\alpha \vee \xi_1 \vee \cdots \vee \xi_n$. First suppose that $X \vdash \alpha \vee \xi_1 \vee \cdots \vee \xi_n$. It follows from $X \in \boldsymbol{B} \perp \alpha$ that that $X \vdash \xi_k \rightarrow \alpha$ for all $\xi_k \in \boldsymbol{B} \backslash X$. We therefore have $X \vdash$

[9]I do not know if *weak intersection* and *conjunction* can replace *regularity* in theorem 14.4.

α, contrary to the conditions. It may be concluded that $X \nvdash \alpha \vee \xi_1 \vee \cdots \vee \xi_n$. Since $X \cup \{\xi_k\} \vdash \alpha$, and thus $X \cup \{\xi_k\} \vdash \alpha \vee \xi_1 \vee \cdots \vee \xi_n$ holds for all $\xi_k \in \boldsymbol{B} \backslash X$, it follows that $X \in \boldsymbol{B} \perp (\alpha \vee \xi_1 \vee \cdots \vee \xi_n)$.

Next, let $Z \in \boldsymbol{B} \perp (\alpha \vee \xi_1 \vee \cdots \vee \xi_n)$. Since all elements of $\boldsymbol{B} \backslash X$ imply $\alpha \vee \xi_1 \vee \cdots \vee \xi_n$, we have $Z \subseteq X$. From this and $X \in \boldsymbol{B} \perp (\alpha \vee \xi_1 \vee \cdots \vee \xi_n)$ it follows that $\{X\} = \boldsymbol{B} \perp (\alpha \vee \xi_1 \vee \cdots \vee \xi_n)$.

Lemma 14.4 Let $\boldsymbol{B}$ be a finite set. If $\boldsymbol{B} \neq Z \in \boldsymbol{B} \perp \alpha$ for some α, then $\{Z\} = \boldsymbol{B} \perp \mathcal{V}(\boldsymbol{B} \backslash Z)$. ($\mathcal{V}(A)$ is the disjunction of all elements of A.)

Proof of Lemma 14.4 Let $\boldsymbol{B} \backslash Z = \{\xi_1, \ldots \xi_n\}$. It follows by $Z \in \boldsymbol{B} \perp \alpha$ that for each $\xi_k, Z \vdash \xi_k \rightarrow \alpha$. It follows from this and $Z \nvdash \alpha$ that $Z \nvdash (\xi_1 \vee \cdots \vee \xi_n)$ Since every Z' such that $Z \subset Z' \subseteq \boldsymbol{B}$ contains one of $\xi_1, \ldots \xi_n$, we can conclude that $Z \in \boldsymbol{B} \perp (\xi_1 \vee \cdots \vee \xi_n)$.

Next, suppose that $W \in \boldsymbol{B} \perp (\xi_1 \vee \cdots \vee \xi_n)$. Since all elements of $\boldsymbol{B} \backslash Z$ imply $\xi_1 \vee \cdots \vee \xi_n$, $W \cap (\boldsymbol{B} \backslash Z) = \varnothing$, i.e. $W \subseteq Z$. From this and $Z \in \boldsymbol{B} \perp (\xi_1 \vee \cdots \vee \xi_n)$ it follows that $W = Z$. We may conclude that $\{Z\} = \boldsymbol{B} \perp (\xi_1 \vee \cdots \vee \xi_n)$.

Lemma 14.5 Let $\boldsymbol{B}$ be a finite belief base and $\boldsymbol{B}''$ its closure under conjunction. ($\boldsymbol{B}''$ consists of the sentences that are elements of $\boldsymbol{B}$ or conjunctions of elements of $\boldsymbol{B}$.) If an operation $\div$ on $Cn(\boldsymbol{B})$ is generated by a partial meet contraction on $\boldsymbol{B}$, then it is generated by a partial meet contraction on $\boldsymbol{B}''$.

Definitions for the Proof of Lemma 14.5

1. Let A and $\boldsymbol{B}$ be sets of sentences. Then A is $\boldsymbol{B}$*-closed* iff $Cn(A) \cap \mathbf{B} \subseteq A$.
2. $\Re(\boldsymbol{B})$ is the set of $\boldsymbol{B}$-closed subsets of $\boldsymbol{B}$ (Hansson 1991).

Proof of Lemma 14.5 Let f be the function such that for each element A of $\Re(\boldsymbol{B})$, $f(A)$ is the closure under conjunction of A.

We first need to show that f is a one-to-one correspondence. Suppose that it is not. Then there are two elements A and A' of $\Re(\boldsymbol{B})$ such that $A \neq A'$ and $f(A) = f(A')$. We may, without loss of generality, assume that there is then some α such that $\alpha \in A$ and $\alpha \notin A'$. If follows from $\alpha \in A$ and $Cn(A) = Cn(f(A)) = Cn(f(A')) = Cn(A')$ that $\alpha \in Cn\,(A')$. Since A' is $\boldsymbol{B}$-closed, it follows from $\alpha \in Cn\,(A')$ and $\alpha \in \boldsymbol{B}$ that $\alpha \in A'$. We may conclude from this contradiction that f is a one-to-one correspondence.

In order to prove the lemma, suppose that the operation $\div$ on $Cn(\boldsymbol{B})$ is based on the partial meet contraction $\sim_\gamma$ on $\boldsymbol{B}$. Let γ'' be the function such that:

(1) If $\boldsymbol{B}'' \perp \alpha \neq \varnothing$, then $\gamma''\left(\boldsymbol{B}'' \perp \alpha\right) = \left\{X \in \boldsymbol{B}'' \perp \alpha \mid f^{-1}(X) \in \gamma(\boldsymbol{B} \perp \alpha)\right\}$
(2) If $\boldsymbol{B}'' \perp \alpha = \varnothing$, then $\gamma''\left(\boldsymbol{B}'' \perp \alpha\right) = \{\boldsymbol{B}''\}$.

We need to prove (1) that γ'' is a selection function for $\boldsymbol{B}''$, and (2) that for all α, $\mathbf{K} \div \alpha = Cn\left(\bigcap \gamma''\left(\boldsymbol{B}'' \perp \alpha\right)\right)$.

Part 1: In order to prove that γ'' is a selection function for $\boldsymbol{B}''$ we have to show that if $\boldsymbol{B}'' \perp \alpha \neq \varnothing$, then $\gamma''\left(\boldsymbol{B}'' \perp \alpha\right) \neq \varnothing$. Let $\boldsymbol{B}'' \perp \alpha \neq \varnothing$. Then $\boldsymbol{B} \perp \alpha \neq \varnothing$, from which follows that $\gamma(\boldsymbol{B} \perp \alpha)$ is nonempty. Let $X \in \gamma(\boldsymbol{B} \perp \alpha)$. It follows from $Cn(f(X)) = Cn(X)$ that $f(X) \nvdash \alpha$.

Suppose that there is some $Y \subseteq \boldsymbol{B''}$ such that $f(X) \subset Y \nvdash \alpha$. There is then a set Y with this property that is closed under conjunction. It follows that $X \subset f^{-1}(Y) \nvdash \alpha$ and $f^{-1}(Y) \subseteq \boldsymbol{B}$, contrary to $X \in \boldsymbol{B} \perp \alpha$. We may conclude from this contradiction that there is no $Y \subseteq \boldsymbol{B''}$ such that $f(X) \subset Y \nvdash \alpha$. Since $f(X) \nvdash \alpha$ it follows that $f(X) \in \boldsymbol{B''} \perp \alpha$. Since $f^{-1}(f(X)) = X \in \gamma(\boldsymbol{B} \perp \alpha)$, it follows from the construction of γ'' that $\gamma''(\boldsymbol{B''} \perp \alpha) \neq \varnothing$.

Part 2: Since, by the assumptions, $(Cn(\boldsymbol{B})) \div \alpha = Cn(\bigcap \gamma(\boldsymbol{B} \perp \alpha))$, it is sufficient to prove that $Cn(\bigcap \gamma''(\boldsymbol{B''} \perp \alpha)) = Cn(\bigcap \gamma(\boldsymbol{B} \perp \alpha))$. Since $\boldsymbol{B''} \perp \alpha \neq \varnothing$ if and only if $\boldsymbol{B} \perp \alpha \neq \varnothing$, and $Cn(\boldsymbol{B''}) = Cn(\boldsymbol{B})$, only the case when $\boldsymbol{B''} \perp \alpha \neq \varnothing$ requires further consideration.

For one direction, let $\delta \in \bigcap \gamma(\boldsymbol{B} \perp \alpha)$. Then $\delta \in Z$ for every $Z \in \gamma(\boldsymbol{B} \perp \alpha)$. By the construction of γ'', $\delta \in Z''$ for every $Z'' \in \gamma''(\boldsymbol{B''} \perp \alpha)$. It follows that $\delta \in \bigcap \gamma''(\boldsymbol{B''} \perp \alpha)$. Thus, $\bigcap \gamma(\boldsymbol{B} \perp \alpha) \subseteq \bigcap \gamma''(\boldsymbol{B''} \perp \alpha)$; from which $Cn(\bigcap \gamma(\boldsymbol{B} \perp \alpha)) \subseteq Cn(\bigcap \gamma''(\boldsymbol{B''} \perp \alpha))$ can be concluded.

For the other direction, suppose that $\varepsilon \in \bigcap \gamma''(\boldsymbol{B''} \perp \alpha)$. It follows from $\varepsilon \in \boldsymbol{B''}$ that there are elements $\varepsilon_1, \ldots \varepsilon_n$ of $\boldsymbol{B}$ such that $\varepsilon \equiv \varepsilon_1 \& \ldots \& \varepsilon_n$. Let $W \in \gamma(\boldsymbol{B} \perp \alpha)$. By the construction of γ'', $f(W) \in \gamma''(\boldsymbol{B''} \perp \alpha)$. It follows from $f(W) \in \boldsymbol{B''} \perp \alpha$ that f(W) is $\boldsymbol{B''}$-closed. We may conclude from $\{\varepsilon_1, \ldots \& \varepsilon_n\} \subseteq \boldsymbol{B''}$, $\varepsilon \in f(W)$ and the $\boldsymbol{B''}$-closure of f(W) that $\{\varepsilon_1, \ldots \& \varepsilon_n\} \subseteq f(W)$.

It follows from $W \in \boldsymbol{B} \perp \alpha$ that W is $\boldsymbol{B}$-closed. Since $Cn(W) = Cn(f(W))$ and $\{\varepsilon_1, \ldots \& \varepsilon_n\} \subseteq \boldsymbol{B}$, we may conclude from $\{\varepsilon_1, \ldots \& \varepsilon_n\} \subseteq f(W)$ that $\{\varepsilon_1, \ldots \& \varepsilon_n\} \subseteq W$. Since this holds for all $W \in \gamma(\boldsymbol{B} \perp \alpha)$, we have $\{\varepsilon_1, \ldots \& \varepsilon_n\} \subseteq \bigcap \gamma(\boldsymbol{B} \perp \alpha)$. Thus, $\varepsilon \in Cn(\bigcap \gamma(\boldsymbol{B} \perp \alpha))$. We have proved that $\bigcap \gamma''(\boldsymbol{B''} \perp \alpha) \subseteq Cn(\bigcap \gamma(\boldsymbol{B} \perp \alpha))$, from which $Cn(\bigcap \gamma''(\boldsymbol{B''} \perp \alpha)) \subseteq Cn(\bigcap \gamma(\boldsymbol{B} \perp \alpha))$ follows as desired.

Lemma 14.6 Let the operation $\div$ on the belief set **K** be generated by the partial meet contraction $\sim_\gamma$ on the finite base $\boldsymbol{B}$ for **K**. Then $\mathbf{K} \div \beta$ is a maximally preservative α-removal iff $\boldsymbol{B} \cap (\mathbf{K} \div \alpha) \subseteq \boldsymbol{B} \cap (\mathbf{K} \div \beta) \in \boldsymbol{B} \perp \alpha$.

Proof of Lemma 14.6 For the non-trivial direction, suppose that $\mathbf{K} \div \beta$ is a maximally preservative α-removal and that $\boldsymbol{B} \cap (\mathbf{K} \div \beta)$ is not an element of $\boldsymbol{B} \perp \alpha$. Then there must be some $X \subseteq \boldsymbol{B}$ such that $\boldsymbol{B} \cap (\mathbf{K} \div \beta) \subset X \in \boldsymbol{B} \perp \alpha$ and that there is no δ for which $Cn(X) = \mathbf{K} \div \delta$. However, this is impossible since by *lemma* 14.3 there is some δ such that $\{X\} = \boldsymbol{B} \perp \delta$, and by the definition of partial meet contraction $\boldsymbol{B} \sim_\gamma \delta = X$, so that $\mathbf{K} \div \delta = Cn(X)$.

Lemma 14.7 Let $\div$ be an operator on **K** that satisfies *closure* (G$\div$1) and *finitude*, and let $\boldsymbol{B} = \{\&X \mid (\exists \alpha)(X = \mathbf{K} \div \alpha)\}$. Then $Cn(\boldsymbol{B} \cap (\mathbf{K} \div \alpha)) = \mathbf{K} \div \alpha$.

Proof of Lemma 14.7 By the construction, $\&(\mathbf{K} \div \alpha) \in \boldsymbol{B}$. By *closure* (G$\div$1), $\&(\mathbf{K} \div \alpha) \in \mathbf{K} \div \alpha$. It follows that $\&(\mathbf{K} \div \alpha) \in (\boldsymbol{B} \cap (\mathbf{K} \div \alpha))$, so that

$$Cn(\{\&(\mathbf{K} \div \alpha)\}) \subseteq Cn(\boldsymbol{B} \cap (\mathbf{K} \div \alpha)).$$

We also have $\mathbf{K} \div \alpha \subseteq Cn(\{\&(\mathbf{K} \div \alpha)\})$, so that $\mathbf{K} \div \alpha \subseteq Cn(\boldsymbol{B} \cap (\mathbf{K} \div \alpha))$. The other direction follows directly from *closure* (G$\div$1).

Lemma 14.8 Let $\div$ be an operator on the belief set **K** that satisfies *closure* (G$\div$1), *success* (G$\div$4), *finitude* and *conservativity*, and let

$$\boldsymbol{B} = \{\&X \mid (\exists\alpha)(X = \mathbf{K} \div \alpha)\}.$$

Then:

$$\text{If } \{X\} = \boldsymbol{B} \perp \delta, \text{ then } X = \boldsymbol{B} \cap (\mathbf{K} \div \delta).$$

Proof of Lemma 14.8 Let $\{X\} = \boldsymbol{B} \perp \delta$. It follows by *success* (G$\div$4) that $\boldsymbol{B} \cap (\mathbf{K} \div \delta) \subseteq X$. Suppose that $X \not\subseteq \boldsymbol{B} \cap (\mathbf{K} \div \delta)$. Then, by the construction of $\boldsymbol{B}$, there is some ϕ such that $\&(\mathbf{K} \div \phi) \in X$ and $\&(\mathbf{K} \div \phi) \notin \boldsymbol{B} \cap (\mathbf{K} \div \delta)$. By *closure* (G$\div$1), $(\mathbf{K} \div \phi) \not\subseteq (\mathbf{K} \div \delta)$.

It follows by *conservativity* that there is some ψ such that $\mathbf{K} \div \delta \subseteq \mathbf{K} \div \psi \nvdash \delta$ and $(\mathbf{K} \div \phi) \cup (\mathbf{K} \div \psi) \vdash \delta$. However, it follows from $\mathbf{K} \div \psi \nvdash \delta$ and $\{X\} = \boldsymbol{B} \perp \delta$ that $\boldsymbol{B} \cap (\mathbf{K} \div \psi) \subseteq X$. Since both $\boldsymbol{B} \cap (\mathbf{K} \div \phi)$ and $\boldsymbol{B} \cap (\mathbf{K} \div \psi)$ are subsets of X, it follows by *lemma* 14.7 that $(\mathbf{K} \div \phi) \cup (\mathbf{K} \div \psi) \nvdash \delta$. We may conclude from this contradiction that $X = \boldsymbol{B} \cap (\mathbf{K} \div \delta)$.

Lemma 14.9 Let $\div$ be an operation on the belief set **K** that satisfies *closure* (G$\div$1), *success* (G$\div$4), *finitude, symmetry, conservativity* and *regularity 2*. Let $\boldsymbol{B} = \{\&X \mid (\exists\alpha)(X = \mathbf{K} \div \alpha)\}$. Then, if $\{X, Y\} \subseteq \boldsymbol{B} \perp \beta$ and $\mathbf{K} \div \beta \subseteq Cn(X)$, there is some β' such that $\{X, Y\} = \boldsymbol{B} \perp \beta'$ and $\mathbf{K} \div \beta' \subseteq Cn(X)$.

Proof of Lemma 14.9 Let $\boldsymbol{B} \perp \beta = \{X, Y, Z_1, \ldots Z_n\}$, and suppose that

$$\mathbf{K} \div \beta \subseteq Cn(X).$$

Then $\boldsymbol{B} \cap (\mathbf{K} \div \beta) \subseteq \boldsymbol{B} \cap Cn(X) = X$.

It follows from $X \in \boldsymbol{B} \perp \beta$, by *lemma* 14.3, that $\{X\} = \boldsymbol{B} \perp \delta$ for some δ such that $\vdash \beta \rightarrow \delta$. By *lemma* 14.8, $X = \boldsymbol{B} \cap (\mathbf{K} \div \delta)$, so that by *closure* (G$\div$1) and *lemma* 14.7, $\mathbf{K} \div \delta = Cn(X)$. Similarly, there are ε and $\zeta_1, \ldots \zeta_n$ such that $\{Y\} = \boldsymbol{B} \perp \varepsilon$ and $\mathbf{K} : \varepsilon = Cn(Y)$, and that $\{Z_k\} = \boldsymbol{B} \perp \zeta_k$ and $\mathbf{K} : \zeta_k = Cn(Z_k)$ for all k.

By repeated applications of *lemma* 14.2, $\boldsymbol{B} \perp \beta = \boldsymbol{B} \perp (\delta \& \varepsilon \& \zeta_1 \& \ldots \& \zeta_n)$. Thus, for all ϕ, $\boldsymbol{B} \cap (\mathbf{K} \div \phi) \vdash \beta$ iff $\boldsymbol{B} \cap (\mathbf{K} \div \phi) \vdash \delta \& \varepsilon \& \zeta_1 \& \ldots \& \zeta_n$. By *lemma* 14.7, $\mathbf{K} \div \phi \vdash \beta$ iff $\mathbf{K} \div \phi \vdash \delta \& \varepsilon \& \zeta_1 \& \ldots \& \zeta_n$. By *symmetry*, $\mathbf{K} \div \beta = \mathbf{K} \div (\delta \& \varepsilon \& \zeta_1 \& \ldots \& \zeta_n)$.

We have assumed that $\mathbf{K} \div \beta \subseteq Cn(X)$, i.e., $\mathbf{K} \div (\delta \& \varepsilon \& \zeta_1 \& \ldots \& \zeta_n) \subseteq \mathbf{K} \div \delta$. Suppose that $\mathbf{K} \div \delta$ is not a maximally preservative $\delta \& \varepsilon \& \zeta_1 \& \ldots \& \zeta_n$-removal. Then there is some ϕ such that $\mathbf{K} \div \delta \subset \mathbf{K} \div \phi \nvdash \delta \& \varepsilon \& \zeta_1 \& \ldots \& \zeta_n$. By *lemma* 14.7, $Cn(\boldsymbol{B} \cap (\mathbf{K} \div \delta)) \subset Cn(\boldsymbol{B} \cap (\mathbf{K} \div \phi))$, so that

$$\boldsymbol{B} \cap (\mathbf{K} \div \delta)) \subset \boldsymbol{B} \cap (\mathbf{K} \div \phi) \nvdash \delta \& \varepsilon \& \zeta_1 \& \ldots \& \zeta_n,$$

contrary to $\boldsymbol{B} \cap (\mathbf{K} \div \delta) \in \boldsymbol{B} \perp (\delta \& \varepsilon \& \zeta_1 \& \ldots \& \zeta_n)$. We may conclude that $\mathbf{K} \div \delta$ is a maximally preservative $\delta \& \varepsilon \& \zeta_1 \& \ldots \& \zeta_n$-removal. By *success* (G$\div$4), it is also a $\delta \& \varepsilon$-removal. It follows by *regularity 2* that $\mathbf{K} \div \delta$ is a maximally preservative

$\delta\&\varepsilon$-removal, so that $\mathbf{K} \div (\delta \div \varepsilon) \subseteq \mathbf{K} \div \delta$. By *lemma* 14.2, $\boldsymbol{B} \perp (\delta\&\varepsilon) = \{X, Y\}$. Since $Cn(X) = \mathbf{K} \div \delta$ we therefore have $\{X, Y\} = \boldsymbol{B} \perp (\delta\&\varepsilon)$ and $\mathbf{K} \div (\delta\&\varepsilon) \subseteq \mathbf{K} \div \delta = Cn(X)$ as desired.

Proof of Theorem 14.1, Left-to-Right Let $\div$ be an operation on $\mathbf{K}$ that is generated by an operator $\sim_\gamma$ of partial meet contraction on a finite belief base $\boldsymbol{B}$ for $\mathbf{K}$. It should be obvious that $\div$ satifies (G$\div$1)-(G$\div$5) and *finitude*.

Symmetry: We use the converse form of *symmetry*. Suppose that $\mathbf{K} \div \alpha \neq \mathbf{K} \div \beta$, i.e., that $Cn\,(\bigcap \gamma(\boldsymbol{B} \perp \alpha)) \neq Cn\,(\bigcap \gamma(\boldsymbol{B} \perp \beta))$. Then $\boldsymbol{B} \perp \alpha \neq \boldsymbol{B} \perp \beta$. Without loss of generality we may assume that there is some $X \in \boldsymbol{B} \perp \alpha$ such that $X \notin \boldsymbol{B} \perp \beta$. There are two cases:

Case 1, $X \vdash \beta$: By *lemma* 14.3 there is some δ such that $\{X\} = \boldsymbol{B} \perp \delta$. By the definition of partial meet contraction, $\boldsymbol{B}\sim_\gamma\delta = X$, so that $\mathbf{K} \div \delta = Cn(X)$. It follows that $\mathbf{K} \div \delta \nvdash \alpha$ and $\mathbf{K} \div \delta \vdash \beta$.

Case 2, $X \nvdash \beta$: Then there is some X' such that $X \subset X' \in \boldsymbol{B} \perp \beta$. *Lemma* 14.3 can be used in the same way as in case 1 to show that there is some δ such that $\mathbf{K} \div \delta = Cn\,(X')$. It follows that $\mathbf{K} \div \delta \nvdash \beta$ and $\mathbf{K} \div \delta \vdash \alpha$.

Conservativity: By *lemma* 14.5, we may assume that $\boldsymbol{B}$ is closed under conjunction.

Suppose that $\mathbf{K} \div \beta \nsubseteq \mathbf{K} \div \alpha$. Then $\&\,(\mathbf{K} \div \beta) \notin \boldsymbol{B}\sim_\gamma\alpha$ It follows by the definition of partial meet contraction that there is some $X \in \gamma(\boldsymbol{B} \perp \alpha)$ such that $\&\,(\mathbf{K} \div \beta) \notin X$. By *lemma* 14.3, there is some δ such that $\{X\} = \boldsymbol{B} \perp \delta$. By the definition of partial meet contraction, $X = \boldsymbol{B}\sim_\gamma\delta$. Since $\boldsymbol{B}$ is, by assumption, closed under conjunction, it follows from $\boldsymbol{B}\sim_\gamma\beta \subseteq \boldsymbol{B}$ that $\&\,(\boldsymbol{B}\sim_\gamma\beta) \in \boldsymbol{B}$. We also have $\&\,(\mathbf{K} \div \beta) = \&\,(\boldsymbol{B}\sim_\gamma\beta)$, and it therefore follows from $\&\,(\mathbf{K} \div \beta) \notin \boldsymbol{B}\sim_\gamma\delta \in \boldsymbol{B} \perp \alpha$ that $(\boldsymbol{B}\sim_\gamma\delta) \cup \{\&\,(\mathbf{K} \div \beta)\} \vdash \alpha$, from which $(\mathbf{K} \div \delta) \cup (\mathbf{K} \div \beta) \vdash \alpha$ can be concluded. It follows from $\boldsymbol{B}\sim_\gamma\delta \in \gamma\,(\boldsymbol{B} \perp \alpha)$ that $\boldsymbol{B}\sim_\gamma\alpha \subseteq \boldsymbol{B}\sim_\gamma\delta \nvdash \alpha$, from which we may conclude that $\mathbf{K} \div \alpha \subseteq \mathbf{K} \div \delta \nvdash \alpha$.

Proof of Theorem 14.1, Right-to-Left Let $\boldsymbol{B} = \{\&X |\, (\exists\alpha)(X = \mathbf{K} \div \alpha)\}$, and let γ be is defined as follows:

(1) If $\boldsymbol{B} \perp \alpha \neq \varnothing$, then $\gamma(\boldsymbol{B} \perp \alpha) = \{X \in \boldsymbol{B} \perp \alpha |\, (\mathbf{K} \div \alpha) \subseteq Cn(X)\}$.
(2) If $\boldsymbol{B} \perp \alpha = \varnothing$, then $\gamma(\boldsymbol{B} \perp \alpha) = \{\boldsymbol{B}\}$

We need to show (1) that $\boldsymbol{B}$ is a finite base for $\mathbf{K}$, (2) that γ is a function, (3) that γ is a selection function for $\boldsymbol{B}$, and (4) that for all α: $\mathbf{K} \div \alpha = Cn\,(\bigcap \gamma(\boldsymbol{B} \perp \alpha))$.

Part 1: It follows from *vacuity* (G$\div$3) and *finitude* that $\&\mathbf{K} \in \boldsymbol{B}$. Thus, $\mathbf{K} \subseteq Cn\,(\boldsymbol{B})$. It follows from *inclusion* (G$\div$2) that $\boldsymbol{B} \subseteq \mathbf{K}$. Thus, $Cn(\boldsymbol{B}) = \mathbf{K}$. By *finitude*, $\boldsymbol{B}$ is finite.

Part 2: Suppose that γ is not a function over the given domain. Then there are α and β such that $\boldsymbol{B} \perp \alpha = \boldsymbol{B} \perp \beta$ and

$$\{X \in \boldsymbol{B} \perp \alpha |\, (\mathbf{K} \div \alpha) \subseteq Cn(X)\} \neq \{X \in \boldsymbol{B} \perp \alpha |\, (\mathbf{K} \div \beta) \subseteq Cn(X)\}\,.$$

It follows that $\mathbf{K} \div \alpha \neq \mathbf{K} \div \beta$. However, from $\boldsymbol{B} \perp \alpha = \boldsymbol{B} \perp \beta$ it follows that for all δ, $\boldsymbol{B} \cap (\mathbf{K} \div \delta) \vdash \alpha$ iff $\boldsymbol{B} \cap (\mathbf{K} \div \delta) \vdash \beta$. By *lemma* 14.7, $Cn\,(\boldsymbol{B} \cap (\mathbf{K} \div \delta)) = \mathbf{K} \div \delta$.

Thus, $\mathbf{K} \div \delta \vdash \alpha$ iff $\mathbf{K} \div \delta \vdash \beta$. By *symmetry*, $\mathbf{K} \div \alpha = \mathbf{K} \div \beta$, contrary to what was just shown. This contradiction concludes part (2) of the proof.

Part 3: In order to prove that γ is a selection function for $\boldsymbol{B}$, it remains to be shown that if $\boldsymbol{B} \perp \alpha$ is non-empty, then so is $\gamma(\boldsymbol{B} \perp \alpha)$. If $\boldsymbol{B} \perp \alpha$ is non-empty, then α is not a logical truth. By *success* (G÷4), $\mathbf{K}{\div}\alpha \nvdash \alpha$. Thus $\boldsymbol{B} \cap (\mathbf{K} \div \alpha) \nvdash \alpha$, so that there is some X with $\boldsymbol{B} \cap (\mathbf{K} \div \alpha) \subseteq X \in \boldsymbol{B} \perp \alpha$. By *lemma* 14.7, $Cn(\boldsymbol{B} \cap (\mathbf{K} \div \alpha)) = \mathbf{K} \div \alpha$. It follows that $\mathbf{K} \div \alpha \subseteq Cn(X)$. Then by the definition of γ, $\gamma(\boldsymbol{B} \perp \alpha)$ is nonempty.

Part 4: If α is a logical theorem, then let $\beta \notin \mathbf{K}$. It follows by *vacuity* (G÷3) that $\mathbf{K}{\div}\beta = \mathbf{K}$. By *conservativity*, if $\mathbf{K} \div \beta \nsubseteq \mathbf{K} \div \alpha$, then there is some δ such that $\mathbf{K}{\div}\delta \nvdash \alpha$. By *closure* (G÷1), this is impossible. Thus $\mathbf{K}{\div}\beta \subseteq \mathbf{K}{\div}\alpha$, i.e. $\mathbf{K} \subseteq \mathbf{K} \div \alpha$. With *inclusion* (G÷2), this yields $\mathbf{K} = \mathbf{K}{\div}\alpha$. By the definition of partial meet contraction, $\bigcap \gamma(\boldsymbol{B} \perp \alpha) = \boldsymbol{B}$. Using the result of part 1 of the present proof, we obtain $Cn(\bigcap \gamma(\boldsymbol{B} \perp \alpha)) = Cn(\boldsymbol{B}) = \mathbf{K} = \mathbf{K} \div \alpha$.

If α is not a logical theorem, then we use the construction of $\boldsymbol{B}$ to obtain $\&(\mathbf{K}{\div}\alpha) \in \boldsymbol{B}$. By *closure* (G÷1), $\&(\mathbf{K} \div \alpha) \in \mathbf{K} \div \alpha$. The construction of γ yields $\&(\mathbf{K} \div \alpha) \in \bigcap \gamma(\boldsymbol{B} \perp \alpha)$. It follows that $\mathbf{K}{\div}\alpha \subseteq Cn(\bigcap \gamma(\boldsymbol{B} \perp \alpha))$.

For the other direction, suppose that $\varepsilon \notin \boldsymbol{B} \cap (\mathbf{K} \div \alpha)$. If there is no β such that $\varepsilon = \&(\mathbf{K}{\div}\beta)$, then $\varepsilon \notin \boldsymbol{B}$ so that $\varepsilon \notin \bigcap \gamma(\boldsymbol{B} \perp \alpha)$. If $\varepsilon = \&(\mathbf{K}{\div}\beta)$ for some β, then it follows from by *closure* (G÷1) from $\&(\mathbf{K} \div \beta) \notin \boldsymbol{B} \cap (\mathbf{K} \div \alpha)$ that $\mathbf{K} \div \beta \nsubseteq \mathbf{K} \div \alpha$. By *conservativity* there is some δ such that $\mathbf{K} \div \alpha \subseteq \mathbf{K} \div \delta \nvdash \alpha$ and $(\mathbf{K} \div \beta) \cup (\mathbf{K} \div \delta) \vdash \alpha$. By *lemma* 14.7, $Cn(\boldsymbol{B} \cap (\mathbf{K} \div \delta)) = \mathbf{K} \div \delta$, so that $(\boldsymbol{B} \cap (\mathbf{K} \div \delta)) \cup \{\&(\mathbf{K} \div \beta)\} \vdash \alpha$. It follows from this and $(\boldsymbol{B} \cap (\mathbf{K} \div \alpha)) \subseteq (\boldsymbol{B} \cap (\mathbf{K} \div \delta)) \nvdash \alpha$ that there is some Y such that $\boldsymbol{B} \cap (\mathbf{K} \div \alpha) \subseteq \boldsymbol{B} \cap (\mathbf{K} \div \delta) \subseteq Y \in \boldsymbol{B} \perp \alpha$ and $\&(\mathbf{K}{\div}\beta) \notin Y$. By *lemma* 14.7 and the definition of γ, $Y \in \gamma(\boldsymbol{B} \perp \alpha)$. Since $\&(\mathbf{K}{\div}\beta) \notin Y$, we have $\&(\mathbf{K} \div \beta) \notin \bigcap \gamma(\boldsymbol{B} \perp \alpha)$, i.e. $\varepsilon \notin \bigcap \gamma(\boldsymbol{B} \perp \alpha)$.

Thus if $\varepsilon \notin \boldsymbol{B} \cap (\mathbf{K} \div \alpha)$, then $\varepsilon \notin \bigcap \gamma(\boldsymbol{B} \perp \alpha)$. We may conclude that $\bigcap \gamma(\boldsymbol{B} \perp \alpha) \subseteq \boldsymbol{B} \cap (\mathbf{K} \div \alpha)$. It follows by *lemma* 14.7 that $Cn(\bigcap \gamma(\boldsymbol{B} \perp \alpha)) \subseteq \mathbf{K} \div \alpha$.

Proof of Theorem 14.2, Left-to-Right Let $\div$ be an operation on $\mathbf{K}$ that is generated by an operator $\sim_\gamma$ of maxichoice partial meet contraction on a finite belief base $\boldsymbol{B}$ for $\mathbf{K}$. We can make use of the corresponding part of the proof of Theorem 14.1, so that it only remains to prove that *strong conservativity* holds.

By *lemma* 14.5, we may assume that $\boldsymbol{B}$ is closed under conjunction. Suppose that $\mathbf{K}{\div}\beta \nsubseteq \mathbf{K}{\div}\alpha$. Then $\boldsymbol{B} \sim_\gamma \beta \nsubseteq \boldsymbol{B} \sim_\gamma \alpha$. It follows that α is not a logical theorem, so that $\boldsymbol{B} \sim_\gamma \alpha \nvdash \alpha$, i.e. $\mathbf{K}{\div}\alpha \nvdash \alpha$.

Since $\boldsymbol{B}$ is closed under conjunction, $\&(\boldsymbol{B} \sim_\gamma \beta) \in \boldsymbol{B}$. Since $\sim_\gamma$ is maxichoice, $\boldsymbol{B} \sim_\gamma \alpha \in \boldsymbol{B} \perp \alpha$. Since $\&\left(\boldsymbol{B} {\sim_\gamma} \beta\right) \notin \boldsymbol{B} {\sim_\gamma} \alpha$, we may conclude that $\boldsymbol{B} \sim_\gamma \alpha \cup \left\{\&\left(\boldsymbol{B} {\sim_\gamma} \beta\right)\right\} \vdash \alpha$, from which it follows that $\mathbf{K} \div \beta \cup \mathbf{K} \div \alpha \vdash \alpha$.

Proof of Theorem 14.2, Right-to-Left $\boldsymbol{B}$ and γ are constructed in the same way as in the proof of Theorem 14.1. We have to prove: (1) that $\boldsymbol{B}$ is a finite base for $\mathbf{K}$, (2) that γ is a function, (3) that γ is a selection function for $\boldsymbol{B}$, (4) that for all α: $\mathbf{K} \div \alpha = Cn(\bigcap \gamma(\boldsymbol{B} \perp \alpha))$, and (5) that γ is maxichoice. Parts 1–3 coincide with the corresponding parts of the proof of Theorem 14.1. Since *strong conservativity* implies *conservativity*, the proof of part 4 of Theorem 14.1 is also a proof of part 4 of the present proof.

Part 5: Let $\delta \in \boldsymbol{B}\backslash(\boldsymbol{B} \sim_{\gamma}\alpha)$. By the construction of $\boldsymbol{B}$, $\delta = \&(\mathbf{K}\div\beta)$ for some β. Since $\&(\mathbf{K}\div\beta) \in \boldsymbol{B}$ and $\boldsymbol{B} \sim_{\gamma}\alpha$ is $\boldsymbol{B}$-closed (cf. the definition for *lemma* 14.5), it follows from $\&(\mathbf{K} \div \beta) \notin \bigcap\gamma(\boldsymbol{B} \perp \alpha)$ that $\&(\mathbf{K} \div \beta) \notin Cn(\bigcap\gamma(\boldsymbol{B} \perp \alpha))$, thus by the result of part 4, $\&(\mathbf{K}\div\beta) \notin \mathbf{K}\div\alpha$. By *closure* (G÷1), $\mathbf{K}\div\beta \not\subseteq \mathbf{K}\div\alpha$. By *strong conservativity*, $\mathbf{K}\div\beta \cup \mathbf{K}\div\alpha \vdash \alpha$. Thus, $\boldsymbol{B} \sim_{\gamma}\alpha \cup \{\&(\mathbf{K}\div\beta)\} \vdash \alpha$, i.e., $\boldsymbol{B} \sim_{\gamma}\alpha \cup \{\delta\} \vdash \alpha$. Since this holds for all $\delta \in \boldsymbol{B}\backslash(\boldsymbol{B} \sim_{\gamma}\alpha)$, we can conclude that $\boldsymbol{B} \sim_{\gamma}\alpha \in \boldsymbol{B}\perp\alpha$.

Proof of Theorem 14.3, Left-to-Right Let $\div$ be the operation on $\mathbf{K}$ that is generated by the operator $\sim$ of full meet contraction on a finite belief base $\boldsymbol{B}$ for $\mathbf{K}$. Making use of the corresponding part of the proof of Theorem 14.1, it only remains for us to prove that *indecisiveness* holds. Just as in Theorem 14.1 we may, due to *lemma* 14.5, assume that $\boldsymbol{B}$ is closed under conjunction.

Suppose that $\mathbf{K}\div\delta \nvdash \alpha$ and $(\mathbf{K}\div\beta)\cup(\mathbf{K}\div\delta) \vdash \alpha$. Then $\boldsymbol{B}\sim\delta \nvdash \alpha$ and $(\boldsymbol{B}\sim\beta) \cup (\boldsymbol{B}\sim\delta) \vdash \alpha$. Since $\boldsymbol{B}\sim\beta$ and $\boldsymbol{B}\sim\delta$ are both subsets of $\boldsymbol{B}$ there is some subset X of $\boldsymbol{B}$ such that $\boldsymbol{B}\sim\delta \subseteq X \in \boldsymbol{B}\perp\alpha$ and $\boldsymbol{B}\sim\beta \not\subseteq X$. By the definition of full meet contraction, $\boldsymbol{B}\sim\alpha \subseteq X$. Suppose that $\boldsymbol{B}\sim\beta \subseteq \boldsymbol{B}\sim\alpha$. It would then follow from $(\boldsymbol{B}\sim\beta) \cup (\boldsymbol{B}\sim\delta) \vdash \alpha$, $\boldsymbol{B}\sim\delta \subseteq X$ and $\boldsymbol{B}\sim\alpha \subseteq X$ that $X \vdash \alpha$, contrary to $X \in \boldsymbol{B}\perp\alpha$. We may conclude that $\boldsymbol{B}\sim\beta \not\subseteq \boldsymbol{B}\sim\alpha$.

Since all elements of $\boldsymbol{B}\perp\alpha$ are $\boldsymbol{B}$-closed (cf. the definition for *lemma* 14.5), their intersection $\boldsymbol{B}\sim\alpha$ is also $\boldsymbol{B}$-closed. Similarly, $\boldsymbol{B}\sim\beta$ is $\boldsymbol{B}$-closed. It therefore follows from $\boldsymbol{B}\sim\beta \not\subseteq \boldsymbol{B}\sim\alpha$. that $Cn(\boldsymbol{B}\sim\beta) \not\subseteq Cn(\boldsymbol{B}\sim\alpha)$, i.e. $\mathbf{K}\div\beta \not\subseteq \mathbf{K}\div\alpha$.

Proof of Theorem 14.3, Right-to-Left Let $\boldsymbol{B} = \{\&X| (\exists\alpha)(X = \mathbf{K} \div \alpha)\}$. We need to show that (1) $\boldsymbol{B}$ is a finite base for $\mathbf{K}$, (2) for all α, $Cn(\bigcap(\boldsymbol{B} \perp \alpha)) \subseteq \mathbf{K} \div \alpha$, and (3) for all α, $\mathbf{K} \div \alpha \subseteq Cn(\bigcap(\boldsymbol{B} \perp \alpha))$. The proof of part 1 coincides with that of part 1 of Theorem 14.1.

Part 2: We are going to prove that $\bigcap(\boldsymbol{B} \perp \alpha) \subseteq \mathbf{K} \div \alpha$. Let $\zeta \notin \mathbf{K} \div \alpha$. If there is no β such that $\zeta = \mathbf{K} \div \beta$, then it follows by the construction of $\boldsymbol{B}$ that $\zeta \notin \bigcap(\boldsymbol{B} \perp \alpha)$. In the principal case, $\zeta = \&(\mathbf{K} \div \beta)$.

It follows from $\&(\mathbf{K} \div \beta) \notin \mathbf{K} \div \alpha$ and *closure* (G÷1) that $\mathbf{K}\div\beta \not\subseteq \mathbf{K}\div\alpha$. By *conservativity*, there is some δ such that $\mathbf{K} \div \alpha \subseteq \mathbf{K} \div \delta \nvdash \alpha$ and $(\mathbf{K} \div \beta) \cup (\mathbf{K} \div \delta) \vdash \alpha$. By *lemma* 14.7, $(\boldsymbol{B} \cap (\mathbf{K} \div \beta)) \cup (\boldsymbol{B} \cap (\mathbf{K} \div \delta)) \vdash \alpha$. Thus, there is some X such that $\&(\boldsymbol{B} \cap (\mathbf{K} \div \beta)) \notin X \in \boldsymbol{B} \perp \alpha$, i.e. $\&(\mathbf{K} \div \beta) \notin X \in \boldsymbol{B} \perp \alpha$. It follows that $\zeta = \&(\mathbf{K} \div \beta) \notin \bigcap(\boldsymbol{B} \perp \alpha)$.

Thus, if $\zeta \notin \mathbf{K} \div \alpha$, then $\zeta \notin \bigcap(\boldsymbol{B} \perp \alpha)$. i.e. $\bigcap(\boldsymbol{B} \perp \alpha) \subseteq \mathbf{K} \div \alpha$. By *closure* (G÷1), we can conclude that $Cn(\bigcap(\boldsymbol{B} \perp \alpha)) \subseteq \mathbf{K} \div \alpha$.

Part 3: We are going to show that $\boldsymbol{B}\cap(\mathbf{K} \div \alpha) \subseteq \bigcap(\boldsymbol{B} \perp \alpha)$. Let $\delta \notin \bigcap(\boldsymbol{B} \perp \alpha)$. If there is no β such that $\delta = \&(\mathbf{K} \div \beta)$, then clearly $\delta \notin \boldsymbol{B} \cap (\mathbf{K} \div \alpha)$. In the principal case, let $\delta = \&(\mathbf{K} \div \beta)$. It follows from $\&(\mathbf{K} \div \beta) \notin \bigcap(\boldsymbol{B} \perp \alpha)$ that there is some X such that $\&(\mathbf{K} \div \beta) \notin X \in \boldsymbol{B} \perp \alpha$.

By *lemma* 14.3, $\{X\} = \boldsymbol{B} \perp \phi$ for some ϕ. By *lemma* 14.8, $X = \boldsymbol{B} \cap (\mathbf{K} \div \phi)$. By *lemma* 14.7, $Cn(X) = \mathbf{K} \div \phi$. Therefore, it follows from $\&(\mathbf{K} \div \beta) \notin \boldsymbol{B} \cap (\mathbf{K} \div \phi) \in \boldsymbol{B} \perp \alpha$ that $\mathbf{K} \div \phi \nvdash \alpha$ and $(\mathbf{K} \div \beta) \cup (\mathbf{K} \div \phi) \vdash \alpha$. By *indecisiveness*,

$\mathbf{K}\div\beta \nsubseteq \mathbf{K}\div\alpha$. By *closure* (G÷1), $\&(\mathbf{K}\div\beta) \notin \mathbf{K}\div\alpha$, so that $\delta = \&(\mathbf{K}\div\beta) \notin \boldsymbol{B}\cap(\mathbf{K}\div\alpha)$.

We may conclude from this that $\boldsymbol{B}\cap(\mathbf{K}\div\alpha) \subseteq \bigcap(\boldsymbol{B}\perp\alpha)$. By *lemma* 14.7, $\mathbf{K}\div\alpha \subseteq Cn\left(\bigcap(\boldsymbol{B}\perp\alpha)\right)$.

Proof of Theorem 14.4, Left-to-Right Let γ be a complete selection function such that $\sim_\gamma$ is an operator of transitively, maximizingly, relational partial meet contraction on a finite belief base $\boldsymbol{B}$ for $\mathbf{K}$, and that $\sim_\gamma$ generates the operation $\div$ on $\mathbf{K}$. We can make use of the corresponding part of the proof of Theorem 14.1, so that it only remains to prove that *regularity 1* and *regularity* 2 hold.

Regularity 1: Suppose that $\vdash \alpha\rightarrow\beta$ and that there is some ζ such that $\mathbf{K}\div\zeta$ is both a β-removal and a maximally preservative α-removal. Let $Z = \boldsymbol{B}\cap(\mathbf{K}\div\zeta)$. By *lemma* 14.6, $\boldsymbol{B}\cap(\mathbf{K}\div\alpha) \subseteq Z \in \boldsymbol{B}\perp\alpha$.

We are first going to show that $\boldsymbol{B}\cap(\mathbf{K}\div a) = \boldsymbol{B}\sim_\gamma\alpha$. It follows from $\mathbf{K}\div\alpha = Cn\left(\boldsymbol{B}\sim_\gamma\alpha\right)$ that $\boldsymbol{B}\cap(\mathbf{K}\div\alpha) = \boldsymbol{B}\cap Cn\left(\boldsymbol{B}\sim_\gamma\alpha\right)$. Since each element of $\gamma(\boldsymbol{B}\perp\alpha)$ is $\boldsymbol{B}$-closed (cf. the definition for *lemma* 14.5), so is $\bigcap\gamma(\boldsymbol{B}\perp\alpha) = \boldsymbol{B}\sim_\gamma\alpha$, thus $\boldsymbol{B}\cap Cn\left(\boldsymbol{B}\sim_\gamma\alpha\right) = \boldsymbol{B}\sim_\gamma\alpha$, i.e. $\boldsymbol{B}\cap(\mathbf{K}\div\alpha) = \boldsymbol{B}\sim_\gamma\alpha$.

We now have $\boldsymbol{B}\sim_\gamma\alpha \subseteq Z \in \boldsymbol{B}\perp\alpha$ so that, by the completeness of γ, $Z \in \gamma(\boldsymbol{B}\perp\alpha)$. It follows from $Z \nvdash \beta$ and $\vdash \alpha\rightarrow\beta$ that $Z \in \boldsymbol{B}\perp\beta$.

Next, we are going to show that $\gamma(\boldsymbol{B}\perp\beta) \subseteq \gamma(\boldsymbol{B}\perp\alpha)$

Let $X \in \gamma(\boldsymbol{B}\perp\beta)$. Suppose that $X \notin \gamma(\boldsymbol{B}\perp\alpha)$. If $X \in \boldsymbol{B}\perp\alpha$, then $Z \in \gamma(\boldsymbol{B}\perp\alpha)$ yields $X \ll Z$. If $X \notin \boldsymbol{B}\perp\alpha$, then there is some X' such that $X \subset X' \in \boldsymbol{B}\perp\alpha$. It follows by the maximizing property that $X \ll X'$ and by $Z \in \gamma(\boldsymbol{B}\perp\alpha)$ that $X' \underline{\ll} Z$. Transitivity yields $X \ll Z$, in this case as well.

From $X \ll Z$ and $Z \in \boldsymbol{B}\perp\beta$ it follows that $X \notin \gamma(\boldsymbol{B}\perp\beta)$. From this contradiction we may conclude that if $X \in \gamma(\boldsymbol{B}\perp\beta)$ then $X \in \gamma(\boldsymbol{B}\perp\alpha)$, i.e., that $\gamma(\boldsymbol{B}\perp\beta) \subseteq \gamma(\boldsymbol{B}\perp\alpha)$.

Now let $\mathbf{K}\div\delta$ be a maximally preservative β-removal. By *lemma* 14.6,

$$\boldsymbol{B}\cap(\mathbf{K}\div\beta) \subseteq \boldsymbol{B}\cap(\mathbf{K}\div\delta) \in \boldsymbol{B}\perp\beta.$$

Since $\boldsymbol{B}\cap(\mathbf{K}\div\beta) = \boldsymbol{B}\sim_\gamma\beta$ (that follows in the same way as $\boldsymbol{B}\cap(\mathbf{K}\div\alpha) = \boldsymbol{B}\sim_\gamma\alpha$), we have $\boldsymbol{B}\sim_\gamma\beta \subseteq \boldsymbol{B}\cap(\mathbf{K}\div\delta) \in \boldsymbol{B}\perp\beta$ so that, by the completeness of γ, $\boldsymbol{B}\cap(\mathbf{K}\div\delta) \in \gamma(\boldsymbol{B}\perp\beta)$. We have just shown that $\gamma(\boldsymbol{B}\perp\beta) \subseteq \gamma(\boldsymbol{B}\perp\alpha)$, and may conclude that $\boldsymbol{B}\cap(\mathbf{K}\div\delta) \in \gamma(\boldsymbol{B}\perp\alpha)$.

It also follows from $\gamma(\boldsymbol{B}\perp\beta) \subseteq \gamma(\boldsymbol{B}\perp\alpha)$ that $\boldsymbol{B}\cap(\mathbf{K}\div\alpha) \subseteq \boldsymbol{B}\cap(\mathbf{K}\div\beta)$. From this and $\boldsymbol{B}\cap(\mathbf{K}\div\beta) \subseteq \boldsymbol{B}\cap(\mathbf{K}\div\delta)$ it follows that $\boldsymbol{B}\cap(\mathbf{K}\div\alpha) \subseteq \boldsymbol{B}\cap(\mathbf{K}\div\delta)$. We therefore have $\boldsymbol{B}\cap(\mathbf{K}\div\alpha) \subseteq \boldsymbol{B}\cap(\mathbf{K}\div\delta) \in \boldsymbol{B}\perp\alpha$. By *lemma* 14.6, $\mathbf{K}\div\delta$ is a maximally preservative α-removal.

Regularity 2: If β is a logical theorem, then $\mathbf{K}\div\alpha \vdash \beta$ so that *regularity 2* holds vacuously. For the principal case, suppose to the contrary that $\vdash \alpha\rightarrow\beta$, that $\mathbf{K}\div\delta$ is a maximally preservative α-removal, and that it is a β-removal but not a maximally preservative β-removal. It follows by *lemma* 14.6 that $\boldsymbol{B}\cap(\mathbf{K}\div\alpha) \subseteq \boldsymbol{B}\cap(\mathbf{K}\div\delta) \in \boldsymbol{B}\perp\alpha$. By $\boldsymbol{B}\sim_\gamma\alpha = \boldsymbol{B}\cap(\mathbf{K}\div\alpha)$ (cf. the proof for *regularity 1*) and the completeness of γ we have $\boldsymbol{B} \cap (\mathbf{K} \div \alpha) \subseteq \boldsymbol{B} \cap (\mathbf{K} \div \delta) \in \gamma\,(\boldsymbol{B} \perp \alpha)$. It follows from $\boldsymbol{B} \cap (\mathbf{K} \div \delta) \in \boldsymbol{B} \perp \alpha$, $\boldsymbol{B} \cap (\mathbf{K} \div \delta) \nvdash \beta$ and $\vdash \alpha \to \beta$ that $\boldsymbol{B} \cap (\mathbf{K} \div \delta) \in \boldsymbol{B} \perp \beta$.

Suppose that $\boldsymbol{B} \cap (\mathbf{K} \div \delta) \in \gamma\,(\boldsymbol{B} \perp \beta)$. Then $\boldsymbol{B} \sim_{\gamma} \beta \subseteq \boldsymbol{B} \cap (\mathbf{K} \div \delta)$, i.e. $\boldsymbol{B} \cap (\mathbf{K} \div \beta) \subseteq \boldsymbol{B} \cap (\mathbf{K} \div \delta)$. By *lemma* 14.6, $\boldsymbol{B} \cap (\mathbf{K} \div \beta) \subseteq \boldsymbol{B} \cap (\mathbf{K} \div \delta) \in \boldsymbol{B} \perp \beta$ contradicts the condition that $\mathbf{K} \div \delta$ is not a maximally preservative β-removal. We may conclude that $\boldsymbol{B} \cap (\mathbf{K} \div \delta) \notin \gamma\,(\boldsymbol{B} \perp \beta)$. It follows that there is some $Z \in \boldsymbol{B} \perp \beta$ such that $Z \underline{\ll} \boldsymbol{B} \cap (\mathbf{K} \div \delta)$ does not hold.

If $Z \in \boldsymbol{B} \perp \alpha$, then $\boldsymbol{B} \cap (\mathbf{K} \div \delta) \in \gamma\,(\boldsymbol{B} \perp \alpha)$ yields $Z \underline{\ll} \boldsymbol{B} \cap (\mathbf{K} \div \delta)$. If $Z \notin \boldsymbol{B} \perp \alpha$, then there is some Z' such that $Z \subset Z' \in \boldsymbol{B} \perp \alpha$. The maximizing property yields $Z \ll Z'$. It follows from $\boldsymbol{B} \cap (\mathbf{K} \div \delta) \in \gamma\,(\boldsymbol{B} \perp \alpha)$ that $Z' \underline{\ll} \boldsymbol{B} \cap (\mathbf{K} \div \delta)$. By transitivity, $Z \ll \boldsymbol{B} \cap (\mathbf{K} \div \delta)$, again contradicting what we have just shown. This contradiction concludes the proof.

Proof of Theorem 14.4, Right-to-Left Let $\boldsymbol{B} = \{\&X \mid (\exists\alpha)(X = \mathbf{K} \div \alpha)\}$. Let $\underline{\ll}$ be the relation such that $Y \underline{\ll} X$ iff either $Y \subset X$ or there is some β such that $\{X, Y\} \subset \boldsymbol{B} \perp \beta$ and $\mathbf{K} \div \beta \subseteq Cn(X)$. Furthermore, let the function γ be defined as follows:

(1) If $\boldsymbol{B} \perp \alpha \neq \varnothing$, then $\gamma\,(\boldsymbol{B} \perp \alpha) = \{\boldsymbol{B}' \in \boldsymbol{B} \perp \alpha \mid \boldsymbol{B}'' \underline{\ll} \boldsymbol{B}' \text{ for all } \boldsymbol{B}'' \in \boldsymbol{B} \perp \alpha\}$.
(2) Otherwise, $\gamma\,(\boldsymbol{B} \perp \alpha) = \{\boldsymbol{B}\}$.

We need to show that (1) $\boldsymbol{B}$ is a finite base for $\mathbf{K}$, (2) γ is a selection function for $\boldsymbol{B}$, (3) the partial meet contraction $\sim_{\gamma}$ on $\boldsymbol{B}$ generates the operation $\div$ on $\mathbf{K}$, (4) γ is a completed selection function, and (5) γ is transitively, maximizingly relational by $\underline{\ll}$. The proof of part 1 coincides with that of part 1 of Theorem 14.1.

Part 2: In order to prove that γ is a selection function for $\boldsymbol{B}$, we need to show that if $\boldsymbol{B} \perp \alpha \neq \varnothing$, then $\gamma(\boldsymbol{B} \perp \alpha)$ is non-empty.

It follows by *success* and *inclusion* that there is some $\boldsymbol{B}'$ such that $\boldsymbol{B} \cap (\mathbf{K} \div \alpha) \subseteq \boldsymbol{B}' \in \boldsymbol{B} \perp \alpha$. By *lemma* 14.3, there is some δ such that $\{\boldsymbol{B}'\} = \boldsymbol{B} \perp \delta$ and $\vdash \alpha \to \delta$. By *lemma* 14.8, $\boldsymbol{B}' = \boldsymbol{B} \cap (\mathbf{K} \div \delta)$.

Let $\boldsymbol{B}''$ be any element of $\boldsymbol{B} \perp \alpha$. By *lemma* 14.3, there is some ε such that $\{\boldsymbol{B}''\} = \boldsymbol{B} \perp \varepsilon$ and $\vdash \alpha \to \varepsilon$. By *lemma* 14.2, $\{\boldsymbol{B}', \boldsymbol{B}''\} = \boldsymbol{B} \perp (\delta \& \varepsilon)$.

By *lemma* 14.7, it follows from $\boldsymbol{B} \cap (\mathbf{K} \div \alpha) \subseteq \boldsymbol{B} \cap (\mathbf{K} \div \delta) \in \boldsymbol{B} \perp \alpha$ that $\mathbf{K} \div \delta$ is a maximally preservative α-removal. By *success* (G$\div$4), it is also a $\delta\&\varepsilon$-removal. Since $\vdash \alpha \to (\delta\&\varepsilon)$ it follows by *regularity 2* that $\mathbf{K} \div \delta$ is a maximally preservative $\delta\&\varepsilon$-removal. Thus $\mathbf{K} \div (\delta\&\varepsilon) \subseteq \mathbf{K} \div \delta$. We therefore have $\{\boldsymbol{B}', \boldsymbol{B}''\} = \boldsymbol{B} \perp (\delta\&\varepsilon)$ and (by *lemma* 14.7) $\mathbf{K} \div (\delta\&\varepsilon) \subseteq Cn\,(\boldsymbol{B} \cap (\mathbf{K} \div \delta)) = Cn\,(\boldsymbol{B}')$. By the definition of $\underline{\ll}$, $\boldsymbol{B}'' \underline{\ll} \boldsymbol{B}'$.

Since this holds for all $\boldsymbol{B}'' \in \boldsymbol{B} \perp \alpha$, we may conclude by the construction of γ that $\boldsymbol{B}' \in \gamma\,(\boldsymbol{B} \perp \alpha)$, so that $\gamma(\boldsymbol{B} \perp \alpha)$ is non-empty.

Part 3: By *lemma* 14.7, $Cn\,(\boldsymbol{B} \cap (\mathbf{K} \div \alpha)) = \mathbf{K} \div \alpha$. It is therefore sufficient to prove that $\boldsymbol{B} \cap (\mathbf{K} \div \alpha) = \bigcap \gamma\,(\boldsymbol{B} \perp \alpha)$.

If α is a logical theorem, then let $\beta \notin \mathbf{K}$. It follows by *vacuity* (G$\div$3) that $\mathbf{K} \div \beta = \mathbf{K}$. By *conservativity*, if $\mathbf{K} \div \beta \nsubseteq \mathbf{K} \div \alpha$, then there is some δ such that $\mathbf{K} \div \delta \nvdash \alpha$. By *closure* (G$\div$1), this is impossible. Thus $\mathbf{K} \div \beta \subseteq \mathbf{K} \div \alpha$, i.e. $\mathbf{K} \subseteq \mathbf{K} \div \alpha$. With *inclusion* (G$\div$2), this yields $\mathbf{K} = \mathbf{K} \div \alpha$. By the definition of partial meet contraction, $\bigcap\gamma(\boldsymbol{B} \perp \alpha) = \boldsymbol{B}$. Using the result of part 1 of the present proof, we obtain $Cn(\bigcap\gamma(\boldsymbol{B} \perp \alpha)) = Cn(\boldsymbol{B}) = \mathbf{K} = \mathbf{K} \div \alpha$.

It remains to prove the principal case, in which α is not a logical theorem.

Part 3a: We are going to show that $\boldsymbol{B} \cap (\mathbf{K} \div \alpha) \subseteq \bigcap\gamma\,(\boldsymbol{B} \perp \alpha)$, i.e., that if $X \in \gamma\,(\boldsymbol{B} \perp \alpha)$, then $\boldsymbol{B} \cap (\mathbf{K} \div a) \subseteq X$.

Let $X \in \gamma\,(\boldsymbol{B} \perp \alpha)$. By *lemma* 14.3, $\{X\} = B \perp \delta$ for some δ such that $\vdash \alpha \rightarrow \delta$. By *lemma* 14.8, $X = \boldsymbol{B} \cap (\mathbf{K} \div \delta)$ and by *lemma* 14.7, $Cn(X) = \mathbf{K} \div \delta$.

Let $Y_1, \ldots Y_n$ be the elements of $\boldsymbol{B} \perp \alpha$ apart from X. By *lemma* 14.3, for each Y_k there is some ε_k such that $\alpha \rightarrow \varepsilon_k$ and $\{Y_k\} = \boldsymbol{B} \perp \varepsilon_k$. By *lemma* 14.8, $Y_k = \boldsymbol{B} \cap (\mathbf{K} \div \varepsilon_k)$. By *lemma* 14.7, $Cn(Y_k) = \mathbf{K} \div \varepsilon_k$.

By repeated uses of *lemma* 14.2, $\boldsymbol{B} \perp \alpha = \boldsymbol{B} \perp (\delta \& \varepsilon_1 \& \ldots \& \varepsilon_n)$. Thus, for all ϕ, $\boldsymbol{B} \cap (\mathbf{K} \div \phi) \vdash \alpha$ iff $\boldsymbol{B} \cap (\mathbf{K} \div \phi) \vdash \delta \& \varepsilon_1 \& \ldots \& \varepsilon_n$. By *lemma* 14.7, $\mathbf{K} \div \phi \vdash \alpha$ iff $\mathbf{K} \div \phi \vdash \delta \& \varepsilon_1 \& \ldots \& \varepsilon_n$. By *symmetry*, $\mathbf{K} \div \alpha = \mathbf{K} \div (\delta \& \varepsilon_1 \& \ldots \& \varepsilon_n)$.

If $\mathbf{K} \div (\delta \& \varepsilon_1 \& \ldots \& \varepsilon_n) \nvdash \delta$, then it follows from $\{X\} = \boldsymbol{B} \perp \delta$ that $\boldsymbol{B} \cap (\mathbf{K} \div \alpha) = \boldsymbol{B} \cap (\mathbf{K} \div (\delta \& \varepsilon_1 \& \ldots \& \varepsilon_n)) \subseteq X$.

In the remaining case, when $\mathbf{K} \div (\delta \& \varepsilon_1 \& \ldots \& \varepsilon_n) \vdash \delta$, there is by *success* (G÷4) some ε_k such that $\mathbf{K} \div (\delta \& \varepsilon_1 \& \ldots \& \varepsilon_n) \nvdash \varepsilon_k$. Since $\{Yk\} = \boldsymbol{B} \perp \varepsilon_k$,

$$\boldsymbol{B} \cap (\mathbf{K} \div (\delta \& \varepsilon_1 \& \ldots \& \varepsilon_n)) \subseteq \boldsymbol{B} \cap (\mathbf{K} \div \varepsilon_k)\,,$$

thus by *lemma* 14.7, $\mathbf{K} \div (\delta \& \varepsilon_1 \& \ldots \& \varepsilon_n) \subseteq \mathbf{K} \div \varepsilon_k$.

We are now going to show that $\mathbf{K} \div \varepsilon_k$ is a maximally preservative $(\delta \& \varepsilon_1 \& \ldots \& \varepsilon_n)$-removal. Suppose that it is not. Then there is some ϕ such that $\mathbf{K} \div \varepsilon_k \subset \mathbf{K} \div \phi \nvdash (\delta \& \varepsilon_1 \& \ldots \& \varepsilon_n)$. Clearly, $\boldsymbol{B} \cap (\mathbf{K} \div \varepsilon_k) \subseteq \boldsymbol{B} \cap (\mathbf{K} \div \phi)$. Since $\boldsymbol{B} \cap (\mathbf{K} \div \varepsilon_k) = Y_k \in \boldsymbol{B} \perp (\delta \& \varepsilon_1 \& \ldots \& \varepsilon_n)$, it follows from

$$\mathbf{K} \div \phi \nvdash (\delta \& \varepsilon_1 \& \ldots \& \varepsilon_n)$$

that $\boldsymbol{B} \cap (\mathbf{K} \div \varepsilon_k) = \boldsymbol{B} \cap (\mathbf{K} \div \phi)$. On the other hand, it follows from the construction of $\boldsymbol{B}$ that $\&(\mathbf{K} \div \phi) \in \boldsymbol{B} \cap (\mathbf{K} \div \phi)$, whereas by $\mathbf{K} \div \varepsilon_k \subset \mathbf{K} \div \phi$ we have $\&(\mathbf{K} \div \phi) \notin \boldsymbol{B} \cap (\mathbf{K} \div \varepsilon_k)$, so that $\boldsymbol{B} \cap (\mathbf{K} \div \varepsilon_k) \neq \boldsymbol{B} \cap (\mathbf{K} \div \phi)$. By this contradiction, we can conclude that $\mathbf{K} \div \varepsilon_k$ is a maximally preservative $(\delta \& \varepsilon_1 \& \ldots \& \varepsilon_n)$-removal.

Since $\mathbf{K} \div \varepsilon_k$ is also a $(\delta \& \varepsilon_k)$-removal, it follows by *regularity 1* that all maximally preservative $\delta \& \varepsilon_k$-removals are maximally preservative $\delta \& \varepsilon_1 \& \ldots \& \varepsilon_n$-removals.

We are next going to show that $\mathbf{K} \div (\delta \& \varepsilon_1 \& \ldots \& \varepsilon_n) \subseteq \mathbf{K} \div (\delta \& \varepsilon_k)$. Suppose that this is not the case. Then by *conservativity* there is some ψ such that

$$\mathbf{K} \div (\delta \& \varepsilon_k) \subseteq \mathbf{K} \div \psi \nvdash \delta \& \varepsilon_k \text{ and } \mathbf{K} \div (\delta \& \varepsilon_1 \& \ldots \& \varepsilon_n) \cup \mathbf{K} \div \psi \vdash \delta \& \varepsilon_k.$$

Since $\mathbf{K} \div \psi$ is a preservative $\delta \& \varepsilon_k$-removal, there must, by *finitude*, be some maximally preservative $\delta \& \varepsilon_k$-removal $\mathbf{K} \div \psi'$ such that $\mathbf{K} \div \psi \subseteq \mathbf{K} \div \psi'$. Then clearly $\mathbf{K} \div (\delta \& \varepsilon_1 \& \ldots \& \varepsilon_n) \cup \mathbf{K} \div \psi' \vdash \delta \& \varepsilon_k$. However, we have just shown that all maximally preservative $\delta \& \varepsilon_k$-removals are maximally preservative $\delta \& \varepsilon_1 \& \ldots \& \varepsilon_n$-removals. Thus $\mathbf{K} \div (\delta \& \varepsilon_1 \& \ldots \& \varepsilon_n) \subseteq \mathbf{K} \div \psi'$, contradicting

$$\mathbf{K} \div (\delta\&\varepsilon_1\&\ldots\&\varepsilon_n) \cup \mathbf{K} \div \psi' \vdash \delta\&\varepsilon_k \text{ and } \mathbf{K} \div \psi' \nvdash \delta\&\varepsilon_k.$$

We can conclude that $\mathbf{K} \div (\delta\&\varepsilon_1\&\ldots\&\varepsilon_n) \subseteq \mathbf{K} \div (\delta\&\varepsilon_k)$.

From the definition of γ, it follows from $Y_k \in \boldsymbol{B} \perp \alpha$ and $X \in \gamma(\boldsymbol{B} \perp \alpha)$ that $Y_k \underline{\ll} X$ and $X \underline{\ll} X$. By the definition of $\underline{\ll}$ and *lemma* 14.9 there is some β such that $\{X, Y_k\} = \boldsymbol{B} \perp \beta$ and $\mathbf{K} \div \beta \subseteq Cn(X) = \mathbf{K} \div \delta$.

By *lemma* 14.2, $\{X, Y_k\} = \boldsymbol{B} \perp (\delta\&\varepsilon_k)$. It follows from $\boldsymbol{B} \perp \beta = \boldsymbol{B} \perp (\delta\&\varepsilon_k)$ that for all ζ, $\boldsymbol{B} \cap (\mathbf{K} \div \zeta) \vdash \beta$ iff $\boldsymbol{B} \cap (\mathbf{K} \div \zeta) \vdash \delta\&\varepsilon_k$. By *lemma* 14.7, $\mathbf{K} \div \zeta \vdash \beta$ iff $\mathbf{K} \div \zeta \vdash \delta\&\varepsilon_k$. By *symmetry*, $\mathbf{K} \div \beta = \mathbf{K} \div (\delta\&\varepsilon_k)$. Thus, $\mathbf{K} \div (\delta\&\varepsilon_k) \subseteq \mathbf{K} \div \delta$. We therefore have $\mathbf{K} \div \alpha = \mathbf{K} \div (\delta\&\varepsilon_1\&\ldots\&\varepsilon_n) \subseteq \mathbf{K} \div (\delta\&\varepsilon_k) \subseteq \mathbf{K} \div \delta = Cn(X)$, i.e., $\mathbf{K} \div \alpha \subseteq Cn(X)$ so that $\boldsymbol{B} \cap (\mathbf{K} \div \alpha) \subseteq \boldsymbol{B} \cap (Cn(X)) = X$. This holds for all $X \in \gamma(\boldsymbol{B} \perp \alpha)$, finishing this part of the proof.

Part 3b: In order to show that $\bigcap \gamma(\boldsymbol{B} \perp \alpha) \subseteq \boldsymbol{B} \cap (\mathbf{K} \div \alpha)$, we will assume that $\zeta \notin \boldsymbol{B} \cap (\mathbf{K} \div \alpha)$ and prove that there is some X such that $\zeta \notin X \in \gamma(\boldsymbol{B} \perp \alpha)$.

By the constructions of $\boldsymbol{B}$ and γ, this is trivially true unless $\zeta = \&(\mathbf{K} \div \beta)$ for some β. In that case, it follows by *closure* (G÷1) from $\&(\mathbf{K} \div \beta) \notin \boldsymbol{B} \cap (\mathbf{K} \div \alpha)$ that $\mathbf{K} \div \beta \nsubseteq \mathbf{K} \div \alpha$. By *conservativity*, there is some δ such that $\mathbf{K} \div \alpha \subseteq \mathbf{K} \div \delta \nvdash \alpha$ and $(\mathbf{K} \div \beta) \cup (\mathbf{K} \div \delta) \vdash \alpha$. It follows by *finitude* that there is some maximally preservative α-removal $\mathbf{K} \div \delta'$ such that $(\mathbf{K} \div \beta) \cup (\mathbf{K} \div \delta') \vdash \alpha$. It follows from $\mathbf{K} \div \delta' \nvdash \alpha$ and $(\mathbf{K} \div \beta) \cup (\mathbf{K} \div \delta') \vdash \alpha$ that $\mathbf{K} \div \beta \nsubseteq \mathbf{K} \div \delta'$.

Suppose that $\boldsymbol{B} \cap (\mathbf{K} \div \delta') \notin \boldsymbol{B} \perp \alpha$. There is then some W such that $\boldsymbol{B} \cap (\mathbf{K} \div \delta') \subset W \in \boldsymbol{B} \perp \alpha$. By *lemma* 14.3 there is some λ such that $\{W\} = \boldsymbol{B} \perp \lambda$. By *lemma* 14.8, $W = \boldsymbol{B} \cap (\mathbf{K} \div \lambda)$. By *closure* and *lemma* 14.7, $\mathbf{K} \div \delta' \subset \mathbf{K} \div \lambda \vdash \alpha$, contrary to the condition that $\mathbf{K} : \delta'$ is a maximally preservative α-removal. We may conclude that $\boldsymbol{B} \cap (\mathbf{K} \div \delta') \in \boldsymbol{B} \perp \alpha$.

By *lemma* 14.3, $\{\boldsymbol{B} \cap (\mathbf{K} \div \delta')\} = \boldsymbol{B} \perp \varepsilon$ for some ε such that $\vdash \alpha \rightarrow \varepsilon$, and by *lemma* 14.8 $\boldsymbol{B} \cap (\mathbf{K} \div \delta') = \boldsymbol{B} \cap (\mathbf{K} \div \varepsilon)$ so that by *lemma* 14.7 $\mathbf{K} \div \delta' = \mathbf{K} \div \varepsilon$.

Next, let $Y \in \boldsymbol{B} \perp \alpha$. By *lemmas* 14.3 and 14.8 there is some ϕ such that $\vdash \alpha \rightarrow \phi$, $Y = \boldsymbol{B} \cap (\mathbf{K} \div \phi)$, and $\{Y\} = \boldsymbol{B} \perp \phi$. By *lemma* 14.7, $Cn(Y) = \mathbf{K} \div \phi$. By *lemma* 14.2, $\{\boldsymbol{B} \cap (\mathbf{K} \div \varepsilon), \boldsymbol{B} \cap (\mathbf{K} \div \phi)\} = \boldsymbol{B} \perp (\varepsilon\&\phi)$. We have $\vdash \alpha \rightarrow (\varepsilon\&\phi)$, and $\mathbf{K} \div \varepsilon$ is both a maximally preservative α-removal and, by *success* (G÷4), an $\varepsilon\&\phi$-removal. It follows by *regularity 2* that $\mathbf{K} \div \varepsilon$ is a maximally preservative $\varepsilon\&\phi$-removal, so that $\mathbf{K} \div (\varepsilon\&\phi) \subseteq \mathbf{K} \div \varepsilon$. By the definition of $\underline{\ll}$, $(\boldsymbol{B} \cap (\mathbf{K} \div \phi)) \underline{\ll} (\boldsymbol{B} \cap (\mathbf{K} \div \varepsilon))$. Since this holds for all elements $Y = \boldsymbol{B} \cap (\mathbf{K} \div \phi) \in \boldsymbol{B} \perp \alpha$, it follows by the definition of γ that $(\boldsymbol{B} \cap (\mathbf{K} \div \varepsilon)) \in \gamma(\boldsymbol{B} \perp \alpha)$. We have already shown that $\mathbf{K} \div \beta \nsubseteq \mathbf{K} \div \delta' = \mathbf{K} \div \varepsilon$, so that by *closure* (G÷1), $\&(\mathbf{K} \div \beta) \notin (\boldsymbol{B} \cap (\mathbf{K} \div \varepsilon))$. This concludes this part of the proof.

Part 4: To prove that γ has the completion property, let $\boldsymbol{B} \sim_\gamma \alpha \subseteq X \in \boldsymbol{B} \perp \alpha$. Let $Y \in \boldsymbol{B} \perp \alpha$. Then $\{X, Y\} \subseteq \boldsymbol{B} \perp \alpha$, and by part 3 of the present proof, $\mathbf{K} \div \alpha = Cn(\boldsymbol{B} \sim_\gamma \alpha)$ so that $\mathbf{K} \div \alpha \subseteq Cn(X)$. It follows by the definition of $\underline{\ll}$ that $Y \underline{\ll} X$. Since this holds for all $Y \in \boldsymbol{B} \perp \alpha$, it follows by the definition of γ that $X \in \gamma(\boldsymbol{B} \perp \alpha)$.

Part 5: It follows directly by the construction that γ is relational by $\underline{\ll}$ and that $\underline{\ll}$ has the maximizing property. It remains to be shown that $\underline{\ll}$ is transitive.

We will use the symbol $\leq$ as follows:

$Y \leq X$ iff there is some β such that $\{X, Y\} = \boldsymbol{B} \perp \beta$ and $\mathbf{K} \div \beta \subseteq Cn(X)$.

By *lemma* 14.9, $X \ll Y$ iff either $X \subset Y$ or $X \leq Y$. The proof of transitivity can therefore be divided into the following four cases:

(1) If $X \subset Y$ and $Y \subset Z$, then either $X \subset Z$ or $X \leq Z$.
(2) If $X \leq Y$ and $Y \leq Z$, then either $X \subset Z$ or $X \leq Z$.
(3) If $X \subset Y$ and $Y \leq Z$, then either $X \subset Z$ or $X \leq Z$.
(4) If $X \leq Y$ and $Y \subset Z$, then either $X \subset Z$ or $X \leq Z$.

If $X = \boldsymbol{B}$, then $X \not\subset Y$, and $X \leq Y$ implies $X = Y$, so that $Y \ll Z$ implies $X \ll Z$. If $Y = \boldsymbol{B}$, then $Y \not\subset Z$, and $Y \leq Z$ implies $Y = Z$, so that $X \ll Y$ implies $X \ll Z$. If $Z = \boldsymbol{B}$, then either $X \subset Z$ or $X = Z$, in both cases yielding $X \ll Z$. Thus, in the proofs of the four cases we can assume that $X \neq \boldsymbol{B}$, $Y \neq \boldsymbol{B}$, and $Z \neq \boldsymbol{B}$.

Case 1: Trivial.

Case 2: Suppose that $X \leq Y$ and $Y \leq Z$. By *lemma* 14.4 there are sentences a, b, and c such that $\{X\} = \boldsymbol{B} \perp a$, $\{Y\} = \boldsymbol{B} \perp b$, and $\{Z\} = \boldsymbol{B} \perp c$. By *lemmas* 14.7 and 14.8, $\mathbf{K} \div a = Cn(X)$, $\mathbf{K} \div b = Cn(Y)$, and $\mathbf{K} \div c = Cn(Z)$. By *lemma* 14.2, $\{X, Y\} = \boldsymbol{B} \perp (a\&b)$, and $\{Y, Z\} = \boldsymbol{B} \perp (b\&c)$.

By $Y \leq Z$, there is some β such that $\{Y, Z\} = \boldsymbol{B} \perp \beta$ and $\mathbf{K} \div \beta \subseteq Cn(Z)$. By $\boldsymbol{B} \perp \beta = \boldsymbol{B} \perp (b\&c)$ it follows that for all ε, $\boldsymbol{B} \cap (\mathbf{K} \div \varepsilon) \vdash \beta$ iff $\boldsymbol{B} \cap (\mathbf{K} \div \varepsilon) \vdash (b\&c)$. By *lemma* 14.7, $\mathbf{K} \div \varepsilon \vdash \beta$ iff $\mathbf{K} \div \varepsilon \vdash (b\&c)$. By *symmetry*, $\mathbf{K} \div \beta = \mathbf{K} \div (b\&c)$. Thus we have $\mathbf{K} \div (b\&c) \subseteq Cn(Z)$. By a similar proof it follows that $\mathbf{K} \div (a\&b) \subseteq Cn(Y)$. From $\mathbf{K} \div (a\&b) \subseteq \mathbf{K} \div b$ and $\boldsymbol{B} \cap (\mathbf{K} \div b) = Y \in \boldsymbol{B} \perp (a\&b)$ it follows by *lemma* 14.6 that $\mathbf{K} \div b$ is a maximally preservative $a\&b$-removal.

We are now going to show that $Z \not\subset X$. Suppose to the contrary that $Z \subset X$. Then $\mathbf{K} \div c \subset \mathbf{K} \div a$. It follows from the above construction of a and c by *lemma* 14.4 that $\vdash a \rightarrow c$, so that $\vdash a\&b \rightarrow b\&c$. It follows from $\mathbf{K} \div (a\&b) \subseteq Cn(Y)$ and $Y \in \boldsymbol{B} \perp b$ that $\mathbf{K} \div (a\&b) \nvdash b\&c$. Furthermore, it follows by *lemma* 14.6 from $\mathbf{K} \div (b\&c) \subseteq \mathbf{K} \div c$ and $Z = \boldsymbol{B} \cap (\mathbf{K} \div c) \in \boldsymbol{B} \perp b\&c$ that $\mathbf{K} \div c$ is a maximally preservative $b\&c$-removal. We may then conclude from *regularity 1* (since $\mathbf{K} \div b$ is both a $b\&c$-removal and a maximally preservative $a\&b$-removal) that $\mathbf{K} \div c$ is a maximally preservative $a\&b$-removal, contrary to our assumption that $\mathbf{K} \div c \subset \mathbf{K} \div a \nvdash a\&b$. By this contradiction, we may conclude that $Z \not\subset X$.

Having excluded $Z \subset X$, we have three remaining subcases under case 2: $X \subset Z$, $X = Z$ and $X \not\subseteq Z \not\subseteq X$. If $X \subset Z$, then we are done. If $X = Z$, then we can use $\{X\} = \boldsymbol{B} \perp a$ and $\mathbf{K} \div a \subseteq Cn(X)$ to obtain $X \leq Z$ directly from the definition of $\leq$. The remaining subcase is $X \not\subseteq Z \not\subseteq X$.

When $X \not\subseteq Z \not\subseteq X$ it follows by *lemma* 14.2 that $\{X, Z\} = \boldsymbol{B} \perp (a\&c)$ and $\{X, Y, Z\} = \boldsymbol{B} \perp (a\&b\&c)$. We are first going to prove that $\mathbf{K} \div (a\&b\&c) \subseteq \mathbf{K} \div c$. By *success* (G÷4), $\mathbf{K} \div (a\&b\&c)$ does not imply all three of a, b, and c. If it does not imply a, then it follows from $\{X\} = \boldsymbol{B} \perp a$ that $\&(\mathbf{K} \div (a\&b\&c)) \in X$, so that $\mathbf{K} \div (a\&b\&c) \subseteq Cn(X) = \mathbf{K} \div a$. By similar reasoning for b and c it follows that $\mathbf{K} \div (a\&b\&c)$ is a subset of at least one of $\mathbf{K} \div a$, $\mathbf{K} \div b$, and $\mathbf{K} \div c$.

If $\mathbf{K} \div (a\&b\&c) \subseteq \mathbf{K} \div a$, then by *success* (G÷4), $\mathbf{K} \div (a\&b\&c) \nvdash a\&b$. We have already shown that $\mathbf{K} \div b$ is a maximally preservative $a\&b$-removal, and in the same way we can show that $\mathbf{K}\div a$ is a maximally preservative $a\&b\&c$-removal. Since it is also an $a\&b$-removal, it follows by *regularity 1* that all maximally preservative $a\&b$-removals are also maximally preservative $a\&b\&c$-removals. Thus, $\mathbf{K}\div b$ is a maximally preservative $a\&b\&c$-removal, so that

$$\mathbf{K} \div (a\&b\&c) \subseteq \mathbf{K} \div b.$$

Thus, if $\mathbf{K} \div (a\&b\&c) \subseteq \mathbf{K} \div a$ then $\mathbf{K} \div (a\&b\&c) \subseteq \mathbf{K} \div b$.

If $\mathbf{K} \div (a\&b\&c) \subseteq \mathbf{K} \div b$, then, by *success* (G÷4), $\mathbf{K}\div(a\&b\&c) \nvdash b\&c$. From $\mathbf{K} \div (b\&c) \subseteq \mathbf{K} \div c$ and $\boldsymbol{B} \cap (\mathbf{K} \div c) = Z \in \boldsymbol{B} \perp (b\&c)$ it follows by *lemma* 14.6 that $\mathbf{K}\div c$ is a maximally preservative $b\&c$-removal. In the same way, it follows that $\mathbf{K}\div b$ is a maximally preservative $a\&b\&c$-removal. Since it is also a $b\&c$-removal, it follows by *regularity 1* that all maximally preservative $b\&c$-removals are maximally preservative $a\&b\&c$-removals. Thus, $\mathbf{K}\div c$ is a maximally preservative $a\&b\&c$-removal, so that $\mathbf{K} \div (a\&b\&c) \subseteq \mathbf{K} \div c$.

Thus, $\mathbf{K}\div(a\&b\&c)$ is a subset of at least one of $\mathbf{K}\div a$, $\mathbf{K}\div b$, and $\mathbf{K}\div c$, and if it is a subset of $\mathbf{K}\div a$ then it is a subset of $\mathbf{K}\div b$ and if it is a subset of $\mathbf{K}\div b$ then it is a subset of $\mathbf{K}\div c$. We may conclude that $\mathbf{K} \div (a\&b\&c) \subseteq \mathbf{K} \div c$.

Since $\boldsymbol{B} \cap (\mathbf{K} \div c) \in \boldsymbol{B} \perp (a\&b\&c)$ it follows by *lemma* 14.6 that $\mathbf{K}\div c$ is a maximally preservative $a\&b\&c$-removal. It is also, by *success* (G÷4), an $a\&c$-removal. It follows by *regularity 2* that $\mathbf{K}\div c$ is a maximally preservative $a\&c$-removal, so that $\mathbf{K} \div (a\&c) \subseteq \mathbf{K} \div c$.

Since $\mathbf{K}\div c = Cn(Z)$ we therefore have $\{X, Z\} = \boldsymbol{B} \perp (a\&c)$ and $\mathbf{K} \div (a\&c) \subseteq Cn(Z)$, so that $X \leq Z$, concluding the proof of case 2.

Case 3: Suppose that $X \subset Y$ and $Y \leq Z$. By *lemma* 14.4 there are a, b, and c such that $\{X\} = \boldsymbol{B} \perp a$, $\{Y\} = \boldsymbol{B} \perp b$, $\{Z\} = \boldsymbol{B} \perp c$, and $\vdash b \rightarrow a$. By *lemmas* 14.7 and 14.8, $\mathbf{K}\div a = Cn(X)$, $\mathbf{K}\div b = Cn(Y)$, and $\mathbf{K}\div c = Cn(Z)$.

We are first going to show that $Z \not\subset X$. If $Z \subset X$, then $Z \subset Y$, contrary to $Y \leq Z$. Thus, $Z \not\subset X$. The subcases $X \subset Z$ and $X = Z$ are treated just as in case 2. It remains to treat the subcase when $X \nsubseteq Z \nsubseteq X$. In that case it follows by *lemma* 14.2 that $\{X, Z\} = \boldsymbol{B} \perp (a\&c)$ and $\{Y, Z\} = \boldsymbol{B} \perp (b\&c)$. In the same way as in case 2, it follows from $Y \leq Z$ that $\mathbf{K} \div (b\&c) \subseteq \mathbf{K} \div c$ and that $\mathbf{K}\div c$ is a maximally preservative $b\&c$-removal.

From $\vdash b \rightarrow a$ it follows that $\vdash b\&c \rightarrow a\&c$. By *success*, $\mathbf{K}\div c$ is an $a\&c$-removal. We can conclude by *regularity 2* that $\mathbf{K}\div c$ is a maximally preservative $a\&c$-removal, so that $\mathbf{K} \div (a\&c) \subseteq \mathbf{K} \div c$. Since $\mathbf{K}\div c = Cn(Z)$ we then have $\{X, Z\} = \boldsymbol{B} \perp (a\&c)$ and $\mathbf{K} \div (a\&c) \subseteq Cn(Z)$, so that $X \leq Z$. This concludes the proof of case 3.

Case 4: Suppose that $X \leq Y \subset Z$. By *lemma* 14.4 there are a, b, and c such that $\{X\} = \boldsymbol{B} \perp a$, $\{Y\} = \boldsymbol{B} \perp b$, $\{Z\} = \boldsymbol{B} \perp c$, and $\vdash c \rightarrow b$. By *lemmas* 14.7 and 14.8 we have $\mathbf{K}\div a = Cn(X)$, $\mathbf{K}\div b = Cn(Y)$, and $\mathbf{K}\div c = Cn(Z)$.

If $Z \subset X$, then $Y \subset X$, contrary to $X \leq Y$. Thus $Z \not\subset X$. The subcases $X = Z$ and $X \subset Z$ are treated as in case 2. In the remaining subcase, $X \nsubseteq Z \nsubseteq X$, it follows by *lemma* 14.2 that $\{X, Y\} = \boldsymbol{B} \perp (a\&b)$ and $\{X, Z\} = \boldsymbol{B} \perp (a\&c)$. By $X \leq Y$, we can show, just as in case 2, that $\mathbf{K} \div (a\&b) \subseteq \mathbf{K} \div b$ and that $\mathbf{K}\div b$ is a maximally preservative $a\&b$-removal.

Suppose that $\mathbf{K} \div (a\&c) \subseteq \mathbf{K} \div a$. Then, by *lemma* 14.6, $\mathbf{K}{\div}a$ is a maximally preservative $a\&c$-removal. By *success* (G÷4), $\mathbf{K}{\div}(a\&c) \nvdash a\&b$. From $\vdash c \rightarrow b$ it follows that $\vdash a\&c \rightarrow a\&b$. Since $\mathbf{K}{\div}a$ is both an $a\&b$-removal and a maximaly preservative $a\&c$-removal, it follows by *regularity 1* that all maximally preservative $a\&b$-removals are maximally preservative $a\&c$-removals. Thus, $\mathbf{K}{\div}b$ is a maximally preservative $a\&c$-removal. However, it follows from $Y \subset Z \nvdash c$ that $\mathbf{K} \div b \subset \mathbf{K} \div c \nvdash a\&c$, so that $\mathbf{K}{\div}b$ cannot be a maximally preservative $a\&c$-removal. From this contradiction we may conclude that $\mathbf{K}{\div}(a\&c) \not\subseteq \mathbf{K}{\div}a$. By *lemma* 14.7, $\boldsymbol{B} \cap (\mathbf{K} \div (a\&c)) \not\subseteq \boldsymbol{B} \cap (\mathbf{K} \div \alpha)$.

Since $\{\boldsymbol{B} \cap (\mathbf{K} \div a)\} = \boldsymbol{B} \perp a$, it follows from $\boldsymbol{B} \cap (\mathbf{K} \div (a\&c)) \not\subseteq \boldsymbol{B} \cap (\mathbf{K} \div \alpha)$ that $\boldsymbol{B} \cap (\mathbf{K} \div (a\&c)) \vdash a$, so that $\mathbf{K}{\div}(a\&c) \vdash a$. Therefore, by *success* (G÷4), $\mathbf{K}{\div}(a\&c) \nvdash c$. Since $\boldsymbol{B} \perp c = \{\boldsymbol{B} \cap (\mathbf{K} \div c)\}$, it follows that $\boldsymbol{B} \cap (\mathbf{K} \div (a\&c)) \subseteq \boldsymbol{B} \cap (\mathbf{K} \div c)$, and by *lemma* 14.7 that $\mathbf{K} \div (a\&c) \subseteq \mathbf{K} \div c$. Since $\mathbf{K}{\div}c = Cn(Z)$ we then have $\{X, Z\} = \boldsymbol{B} \perp (a\&c)$ and $\mathbf{K} \div (a\&c) \subseteq Cn(Z)$, so that $X \leq Z$. This concludes the proof.

Proof of Theorem 14.5, Left-to-Right Due to the corresponding part of the proof of Theorem 14.2, we only have to show that *hyperregularity* holds.

Suppose that $\vdash \alpha \rightarrow \beta$ and $\mathbf{K}{\div}\alpha \nvdash \beta$, i.e. $\boldsymbol{B} \sim_\gamma \alpha \nvdash \beta$. By the maxichoice property, $\boldsymbol{B}{\sim_\gamma}\alpha \in \boldsymbol{B} \perp a$. It follows from $\boldsymbol{B} \sim_\gamma \alpha \nvdash \beta$ that $\boldsymbol{B}{\sim_\gamma}\alpha \in \boldsymbol{B} \perp \beta$. Let $X \in \boldsymbol{B} \perp \beta$. It follows from $\vdash \alpha \rightarrow \beta$ that there is some X' such that $X \subseteq X' \in \boldsymbol{B} \perp \alpha$. It follows that $X \subseteq X' \underline{\ll} \boldsymbol{B}{\sim_\gamma}\alpha$. By the maximizing and transitive properties of $\underline{\ll}$, it follows that $X \underline{\ll} \boldsymbol{B} \sim_\gamma \alpha$. Since this holds for all $X \in \boldsymbol{B} \perp \beta$, and $\boldsymbol{B}{\sim_\gamma}\alpha \in \boldsymbol{B} \perp \beta$, $\boldsymbol{B}{\sim_\gamma}\alpha \in \gamma(\boldsymbol{B} \perp \beta)$. We may conclude by the maxichoice property that that $\boldsymbol{B} \sim_\gamma \beta = \boldsymbol{B} \sim_\gamma \alpha$, i.e. $\mathbf{K}{\div}\beta = \mathbf{K}{\div}\alpha$.

Proof of Theorem 14.5, Right-to-Left Let $\boldsymbol{B} = \{\&X | (\exists \alpha)(X = \mathbf{K} \div \alpha)\}$. Let $\underline{\ll}$ be the relation such that $Y \underline{\ll} X$ iff either $Y \subset X$ or there is some β such that $\{X, Y\} = \boldsymbol{B} \perp \beta$ and $\mathbf{K}{\div}\beta = Cn(X)$. Let γ be the function such that:

(1) If $\boldsymbol{B} \perp \alpha \neq \varnothing$, then $\gamma(\boldsymbol{B} \perp \alpha) = \{\boldsymbol{B}' \in \boldsymbol{B} \perp \alpha | \boldsymbol{B}'' \underline{\ll} \boldsymbol{B}'$ for all $\boldsymbol{B}'' \in \boldsymbol{B} \perp \alpha\}$
(2) Otherwise, $\gamma(\boldsymbol{B} \perp \alpha) = \{\boldsymbol{B}\}$

We need to show (1) that $\boldsymbol{B}$ is a finite base for $\mathbf{K}$, (2) that γ is a selection function for $\boldsymbol{B}$, (3) that γ is maxichoice, (4) that the partial meet contraction $\sim_\gamma$ on $\boldsymbol{B}$ generates the operation $\div$ on $\mathbf{K}$, and (5) that γ is transitively, maximizingly relational by $\underline{\ll}$. The proof of part 1 coincides with that of part 1 of Theorem 14.1.

Part 2: To prove that γ is a selection function, it is sufficient to show that if $\boldsymbol{B} \perp \alpha \neq \varnothing$, then $\gamma(\boldsymbol{B} \perp \alpha)$ is non-empty.

We are first going to show that $\boldsymbol{B} \cap (\mathbf{K} \div \alpha) \in \boldsymbol{B} \perp \alpha$. It follows by *inclusion* (G÷2) and *success* (G÷4) that that $\mathbf{K}{\div}\alpha$ is a subset of $\mathbf{K}$ that does not imply α. Therefore, $\boldsymbol{B} \cap (\mathbf{K} \div \alpha)$ is a subset of $\boldsymbol{B}$ that does not imply α. Suppose that it is not an element of $\boldsymbol{B} \perp \alpha$. Then there must be some $\phi \in \boldsymbol{B}$ such that $\phi \notin \boldsymbol{B} \cap (\mathbf{K} \div \alpha)$ and $(\boldsymbol{B} \cap (\mathbf{K} \div \alpha)) \cup \{\phi\} \nvdash \alpha$. By the construction of $\boldsymbol{B}$, $\phi = \&(\mathbf{K}{\div}\delta)$ for some δ. From $\& (\mathbf{K} \div \delta) \notin \boldsymbol{B} \cap (\mathbf{K} \div \alpha)$ it follows by *closure* (G÷1) that $\mathbf{K}{\div}\delta \not\subseteq \mathbf{K}{\div}\alpha$, and from $(\boldsymbol{B} \cap (\mathbf{K} \div \alpha)) \cup \{\& (\mathbf{K} \div \delta)\} \nvdash \alpha$ it follows that $(\mathbf{K} \div \alpha) \cup (\mathbf{K} \div \delta)\} \nvdash \alpha$.

This contradicts *strong conservativity*, and we may conclude that (for all α) $\boldsymbol{B} \cap (\mathbf{K} \div \alpha) \in \boldsymbol{B} \perp \alpha$.

By *lemma* 14.3 there is some ϕ such that $\vdash \alpha \rightarrow \phi$ and $\{\boldsymbol{B} \cap (\mathbf{K} \div \alpha)\} = \boldsymbol{B} \perp \phi$. By *lemma* 14.7, $\mathbf{K} \div \alpha \nvdash \phi$. Let $Y \in \boldsymbol{B} \perp \alpha$. By *lemma* 14.3 there is some ψ such that $\vdash \alpha \rightarrow \psi$ and $\{Y\} = \boldsymbol{B} \perp \psi$. By *lemma* 14.2, $\{\boldsymbol{B} \cap (\mathbf{K} \div \alpha), Y\} = \boldsymbol{B} \perp (\phi \& \psi)$. Since $\vdash \alpha \rightarrow (\phi \& \psi)$ and $\mathbf{K} \div \alpha \nvdash (\phi \& \psi)$ it follows by *hyperregularity* that $\mathbf{K} \div \alpha = \mathbf{K} \div (\phi \& \psi)$. By *lemma* 14.7 and our definition of $\underline{\ll}$, it follows that $Y \underline{\ll} \boldsymbol{B} \cap (\mathbf{K} \div \alpha)$. Since this holds for all $Y \in \boldsymbol{B} \perp \alpha$, it follows by the definition of γ that $\boldsymbol{B} \cap (\mathbf{K} \div \alpha) \in \gamma(\boldsymbol{B} \perp \alpha)$. Thus $\gamma(\boldsymbol{B} \perp \alpha)$ is non-empty whenever $\boldsymbol{B} \perp \alpha$ is non-empty.

Part 3: In order to prove that γ is maxichoice, suppose that it is not. Then there is some α such that there are distinct X and Y with $X, Y \in \gamma(\boldsymbol{B} \perp \alpha)$. It follows, by the definition of γ, that there are ϕ and ψ such that $\{X, Y\} = \boldsymbol{B} \perp \phi = \boldsymbol{B} \perp \psi$, $Cn(X) = \mathbf{K} \div \phi$ and $Cn(Y) = \mathbf{K} \div \psi$.

It follows from $\{X, Y\} = \boldsymbol{B} \perp \phi = \boldsymbol{B} \perp \psi$ that $\{X, Y\} = \boldsymbol{B} \perp (\phi \& \psi)$. By what was shown in part 2 of the present proof, $\boldsymbol{B} \cap (\mathbf{K} \div (\phi \& \psi)) \in \boldsymbol{B} \perp (\phi \& \psi)$. It follows from this and $\boldsymbol{B} \perp \phi = \boldsymbol{B} \perp (\phi \& \psi)$ that $\boldsymbol{B} \cap (\mathbf{K} \div (\phi \& \psi)) \nvdash \phi$. By *lemma* 14.7, $\mathbf{K} \div (\phi \& \psi) \nvdash \phi$. From this it follows by *hyperregularity* that $\mathbf{K} \div (\phi \& \psi) = \mathbf{K} \div \phi$. Similarly, $\mathbf{K} \div (\phi \& \psi) = \mathbf{K} \div \psi$, so that $\mathbf{K} \div \phi = \mathbf{K} \div \psi$. Thus $X = Y$, contrary to our conditions. We may conclude that γ is maxichoice.

Part 4: It was shown in part 2 of the present proof that for all α, $\boldsymbol{B} \cap (\mathbf{K} \div \alpha) \in \gamma(\boldsymbol{B} \perp \alpha)$. By part 3, $\{\boldsymbol{B} \cap (\mathbf{K} \div \alpha)\} = \gamma(\boldsymbol{B} \perp \alpha)$. It follows from this, by *lemma* 14.7, that $Cn(\boldsymbol{B} \sim_{\gamma} \alpha) = \mathbf{K} \div \alpha$.

Part 5: It follows directly by the construction that γ is relational by $\underline{\ll}$ and that the maximizing property holds. In the proof of transitivity, we will use the symbol $\leq$ as follows:

$Y \leq X$ iff there is some β such that $\{X, Y\} = \boldsymbol{B} \perp \beta$ and $\mathbf{K} \div \beta = Cn(X)$.

Thus, $X \underline{\ll} Y$ iff either $X \subset Y$ or $X \leq Y$. The proof of transitivity can therefore be divided into the following four cases:

(1) If $X \subset Y$ and $Y \subset Z$, then either $X \subset Z$ or $X \leq Z$.
(2) If $X \leq Y$ and $Y \leq Z$, then either $X \subset Z$ or $X \leq Z$.
(3) If $X \subset Y$ and $Y \leq Z$, then either $X \subset Z$ or $X \leq Z$.
(4) If $X \leq Y$ and $Y \subset Z$, then either $X \subset Z$ or $X \leq Z$.

If $X = \boldsymbol{B}$, then $X \not\subset Y$, and $X \leq Y$ implies $X = Y$, so that $Y \underline{\ll} Z$ implies $X \underline{\ll} Z$. If $Y = \boldsymbol{B}$, then $Y \not\subset Z$, and $Y \leq Z$ implies $Y = Z$, so that $X \underline{\ll} Y$ implies $X \underline{\ll} Z$. If $Z = \boldsymbol{B}$, then either $X \subset Z$ or $X = Z$, in both cases yielding $X \underline{\ll} Z$. Thus, in the proofs of the four cases we can assume that $X \neq \boldsymbol{B}$, $Y \neq \boldsymbol{B}$, and $Z \neq \boldsymbol{B}$.

Case 1: Trivial.

Case 2: Suppose that $X \leq Y$ and $Y \leq Z$. By *lemma* 14.4, there are sentences a, b, and c such that $\{X\} = \boldsymbol{B} \perp a$, $\{Y\} = \boldsymbol{B} \perp b$, and $\{Z\} = \boldsymbol{B} \perp c$. By *lemma* 14.2 we have $\{X, Y\} = \boldsymbol{B} \perp (a \& b)$ and $\{Y, Z\} = \boldsymbol{B} \perp (b \& c)$

By $X \leq Y$ there is some ϕ such that $\{X, Y\} = \boldsymbol{B} \perp \phi$ and $\mathbf{K}\div\phi = Cn(Y)$. It follows from $\boldsymbol{B} \perp \phi = \boldsymbol{B} \perp (a\&b)$ that for all ε, $\boldsymbol{B} \cap (\mathbf{K} \div \varepsilon) \vdash \phi$ iff $\boldsymbol{B} \cap (\mathbf{K} \div \varepsilon) \vdash (a\&b)$. By *lemma* 14.7, $\mathbf{K}\div\varepsilon \vdash \phi$ iff $\mathbf{K}\div\varepsilon \vdash (a\&b)$. By *symmetry*, $\mathbf{K}\div(a\&b) = \mathbf{K}\div\phi = Cn(Y)$. By a similar proof, $\mathbf{K}\div(b\&c) = Cn(Z)$.

Suppose that $Z \subset X$. By the construction from *lemma* 14.4 that was used above to obtain a, b, and c, $\vdash a \rightarrow c$. It follows that $\vdash a\&b \rightarrow b\&c$. Since $\mathbf{K}\div(a\&b) = Cn(Y)$ and $Y \nvdash b$, we have $\mathbf{K}\div(a\&b) \nvdash b\&c$. By *hyperregularity* we then have $\mathbf{K}\div(a\&b) = \mathbf{K}\div(b\&c)$, i.e., $Cn(Y) = Cn(Z)$, so that $Y = Z$. Then $X \leq Z$ follows directly from $X \leq Y$, and we are done.

If $X \subset Z$, then we are also done. If $X = Z$, then we can use $\{X\} = \boldsymbol{B} \perp \alpha$ and $\boldsymbol{B} \cap (\mathbf{K} \div \alpha) \in \boldsymbol{B} \perp \alpha$, that was obtained in part 2 of the present proof, in the definition of $\leq$, and obtain $X \leq X$.

Finally, we have the case when $Z \nsubseteq X \nsubseteq Z$. Then *lemma* 14.2 yields $\{X, Z\} = \boldsymbol{B} \perp (a\&c)$ and $\{X, Y, Z\} = \boldsymbol{B} \perp (a\&b\&c)$. By *hyperregularity* and $\mathbf{K}\div(a\&b) \nvdash b$ it follows that $\mathbf{K}\div(a\&b) = \mathbf{K}\div b$. Similarly, $\mathbf{K}\div(b\&c) = \mathbf{K}\div c$. By *success* (G$\div$4), $\mathbf{K}\div(a\&b\&c) \nvdash a$ or $\mathbf{K}\div(a\&b\&c) \nvdash b$ or $\mathbf{K}\div(a\&b\&c) \nvdash c$. By *hyperregularity*, $\mathbf{K}\div(a\&b\&c)$ is identical to one of $\mathbf{K}\div a$, $\mathbf{K}\div b$, and $\mathbf{K}\div c$. Similarly, $\mathbf{K}\div(a\&c)$ is identical to one of $\mathbf{K}\div a$ and $\mathbf{K}\div c$.

Suppose that $\mathbf{K}\div(a\&c) = \mathbf{K}\div a$. It then follows (since, by *hyperregularity*, for all κ and λ, either $\mathbf{K}\div(\kappa\&\lambda) = \mathbf{K}\div\kappa$ or $\mathbf{K}\div(\kappa\&\lambda) = \mathbf{K}\div\lambda$):

(A) Either $\mathbf{K}\div(a\&b\&c) = \mathbf{K}\div(a\&b) = \mathbf{K}\div b$ or $\mathbf{K}\div(a\&b\&c) = \mathbf{K}\div c$.
(B) Either $\mathbf{K}\div(a\&b\&c) = \mathbf{K}\div(a\&c) = \mathbf{K}\div a$ or $\mathbf{K}\div(a\&b\&c) = \mathbf{K}\div b$.
(C) Either $\mathbf{K}\div(a\&b\&c) = \mathbf{K}\div(b\&c) = \mathbf{K}\div c$ or $\mathbf{K}\div(a\&b\&c) = \mathbf{K}\div a$.

Since the three conditions are incompatible, we may conclude that $\mathbf{K}\div(a\&c) \neq \mathbf{K}\div a$. By parts 3 and 4 of the present proof, since $\boldsymbol{B} \perp (a\&c) = \{X, Z\}$, $\mathbf{K}\div(a\&c)$ is either $Cn(X)$ or $Cn(Z)$. In the same way it follows from $\boldsymbol{B} \perp a = \{X\}$ that $\mathbf{K}\div a = Cn(X)$ and from $\boldsymbol{B} \perp c = \{Z\}$ that $\mathbf{K}\div c = Cn(Z)$. Thus, $\mathbf{K}\div(a\&c) = Cn(Z)$. It follows that $X \leq Z$.

Case 3: Suppose that $X \subset Y \leq Z$. By *lemma* 14.4, there are sentences a, b, and c such that $\{X\} = \boldsymbol{B} \perp a$, $\{Y\} = \boldsymbol{B} \perp b$, $\{Z\} = \boldsymbol{B} \perp c$, and $\vdash b \rightarrow a$. $Z \subseteq X$ is impossible, because then $Z \subset Y$, which contradicts $Y \leq Z$. If $X \subset Z$, then we are done. In the remaining case, when $Z \nsubseteq X \nsubseteq Z$, *lemma* 14.2 yields $\{X, Z\} = \boldsymbol{B} \perp (a\&c)$ and $\{Y, Z\} = \boldsymbol{B} \perp (b\&c)$. In the same way as in case 2 we obtain $\mathbf{K}\div(b\&c) = Cn(Z)$. From $\vdash b \rightarrow a$ it follows that $\vdash b\&c \rightarrow a\&c$. Since $Z \nvdash c$ we have $\mathbf{K}\div(b\&c) \nvdash a\&c$. It follows by *hyperregularity* that $\mathbf{K}\div(b\&c) = \mathbf{K}\div(a\&c)$, so that $\mathbf{K}\div(a\&c) = Cn(Z)$. It follows that $X \leq Z$.

Case 4: Suppose that $X \leq Y \subset Z$. By *lemma* 14.4, there are sentences a, b, and c such that $\{X\} = \boldsymbol{B} \perp a, \{Y\} = \boldsymbol{B} \perp b, \{Z\} = \boldsymbol{B} \perp c$, and $\vdash c \rightarrow b$. $Z \subseteq X$ is impossible, because then $Y \subset X$, which contradicts $X \leq Y$. If $X \subset Z$, then we are done. In the remaining case, when $Z \nsubseteq X \nsubseteq Z$, *lemma* 14.2 yields $\{X, Y\} = \boldsymbol{B} \perp (a\&b)$ and $\{X, Z\} = \boldsymbol{B} \perp (a\&c)$. In the same way as in case 2, we obtain $\mathbf{K}\div(a\&b) = Cn(Y)$. We can also prove, in the same way as in case 2, that $\mathbf{K}\div(a\&c)$ is identical to either $\mathbf{K}\div a$ or $\mathbf{K}\div c$.

First, let $\mathbf{K}\div(a\&c) = \mathbf{K}\div a$. From $\vdash c \rightarrow b$ it follows that $\vdash a\&c \rightarrow a\&b$. Since $\mathbf{K}\div a \nvdash a$ we have $\mathbf{K}\div(a\&c) \nvdash a\&b$. It then follows by *hyperregularity* that

$\mathbf{K}\div(a\&c)=\mathbf{K}\div(a\&b)$, so that $\mathbf{K}\div(a\&b)=\mathbf{K}\div a$. By parts 3 and 4 of the present proof, $\mathbf{K}\div a=Cn(X)$. We therefore have $Cn(X)=Cn(Y)$, so that $X=Y$ and $X\subset Z$.

In the other case, when $\mathbf{K}\div(a\&c)=\mathbf{K}\div c$, we use parts 3 and 4 of the present theorem to obtain $\mathbf{K}\div c=Cn(Z)$. It follows from $\{X,Z\}=\boldsymbol{B}\perp(a\&c)$ and $\mathbf{K}\div(a\&c)=Cn(Z)$ that $X\underline{\ll}Z$.

Proof of Theorem 14.6 Let $\boldsymbol{B}$ be a base for $\mathbf{K}$ and γ a selection function for $\boldsymbol{B}$ that is transitively maximizingly relational by $\underline{\ll}$.

PART 1: We are first going to show that $(\boldsymbol{B}\sim_\gamma\alpha)\cap(\boldsymbol{B}\sim_\gamma\beta)\subseteq\boldsymbol{B}\sim_\gamma(\alpha\&\beta)$. This is trivial if $\vdash\alpha$ or $\vdash\beta$. For the principal case, in which $\nvdash\alpha$ and $\nvdash\beta$, suppose to the contrary that there is some ζ such that $\zeta\in\boldsymbol{B}\sim_\gamma\alpha$, $\zeta\in\boldsymbol{B}\sim_\gamma\beta$ and $\zeta\notin\boldsymbol{B}\sim_\gamma(\alpha\&\beta)$. It follows from $\zeta\notin\boldsymbol{B}\sim_\gamma(\alpha\&\beta)$ that there is some X such that $\zeta\notin X\in\gamma(\boldsymbol{B}\perp(\alpha\&\beta))$. By *lemma* 14.1, either $X\in\boldsymbol{B}\perp\alpha$ or $X\in\boldsymbol{B}\perp\beta$. Without loss of generality we may assume that $X\in\boldsymbol{B}\perp\alpha$.

Since $\zeta\in\boldsymbol{B}\sim_\gamma\alpha$, $X\notin\gamma(\boldsymbol{B}\perp\alpha)$. Let $Y\in\gamma(\boldsymbol{B}\perp\alpha)$. From $Y\in\boldsymbol{B}\perp\alpha$ it follows that there is some Y' such that $Y\subseteq Y'\in\boldsymbol{B}\perp(\alpha\&\beta)$. If $Y=Y'$ we have $X\ll Y'$ directly, and if $Y\subset Y'$ the maximizing property yields $Y\ll Y'$, which with $X\ll Y$ and transitivity yields $X\ll Y'$. Thus, in both cases $X\ll Y'$, $X\in\gamma(\boldsymbol{B}\perp(\alpha\&\beta))$ and $Y'\in\boldsymbol{B}\perp(\alpha\&\beta)$, contrary to our definitions. This contradiction concludes the proof that $(\boldsymbol{B}\sim_\gamma\alpha)\cap(\boldsymbol{B}\sim_\gamma\beta)\subseteq\boldsymbol{B}\sim_\gamma(\alpha\&\beta)$.

Now let $\mathbf{K}\div\delta\subseteq(\mathbf{K}\div\alpha)\cap(\mathbf{K}\div\beta)$. Then $\boldsymbol{B}\cap(\mathbf{K}\div\delta)\subseteq\mathbf{K}\div\alpha$, from which it follows that $\boldsymbol{B}\cap(\mathbf{K}\div\delta)\subseteq\boldsymbol{B}\sim_\gamma\alpha$. Similarly, $\boldsymbol{B}\cap(\mathbf{K}\div\delta)\subseteq\boldsymbol{B}\sim_\gamma\beta$. Since $(\boldsymbol{B}\sim_\gamma\alpha)\cap(\boldsymbol{B}\sim_\gamma\beta)\subseteq\boldsymbol{B}\sim_\gamma(\alpha\&\beta)$, we have $\boldsymbol{B}\cap(\mathbf{K}\div\delta)\subseteq\boldsymbol{B}\sim_\gamma(\alpha\&\beta)$, thus $Cn(\boldsymbol{B}\cap(\mathbf{K}\div\delta))\subseteq Cn(\boldsymbol{B}\sim_\gamma(\alpha\&\beta))$, i.e. $\mathbf{K}\div\delta\subseteq\mathbf{K}\div(\alpha\&\beta)$.

PART 2: This is trivial if $\vdash\alpha$, $\vdash\beta$, $\alpha\notin\mathbf{K}$ or $\beta\notin\mathbf{K}$. In the remaining case, suppose that $\alpha\notin\mathbf{K}\div(\alpha\&\beta)$. Then $\alpha\notin\bigcap\gamma(\boldsymbol{B}\perp(\alpha\&\beta))$, and there is some X such that $\alpha\notin X\in\gamma(\boldsymbol{B}\perp(\alpha\&\beta))$. Clearly, $X\in\boldsymbol{B}\perp\alpha$. We are going to prove that $\gamma(\boldsymbol{B}\perp\alpha)\subseteq\gamma(\boldsymbol{B}\perp(\alpha\&\beta))$. Let $Y\in\gamma(\boldsymbol{B}\perp\alpha)$.

We are first going to show that $Y\in\boldsymbol{B}\perp(\alpha\&\beta)$. Suppose to the contrary that $Y\notin\boldsymbol{B}\perp(\alpha\&\beta)$. Then there is some Y' such that $Y\subset Y'\in\boldsymbol{B}\perp(\alpha\&\beta)$. By the maximizing property of $\underline{\ll}$, $Y\ll Y'$. By $X\in\boldsymbol{B}\perp\alpha$ and $Y\in\gamma(\boldsymbol{B}\perp\alpha)$ it follows that $X\underline{\ll}Y$. By transitivity, $X\ll Y'$, which contradicts $X\in\gamma(\boldsymbol{B}\perp(\alpha\&\beta))$ and $Y'\in\boldsymbol{B}\perp(\alpha\&\beta)$. It follows that $Y\in\boldsymbol{B}\perp(\alpha\&\beta)$.

It follows by transitivity from $Y\in\boldsymbol{B}\perp(\alpha\&\beta)$, $X\in\gamma(\boldsymbol{B}\perp(\alpha\&\beta))$ and $X\underline{\ll}Y$ that $Y\in\gamma(\boldsymbol{B}\perp(\alpha\&\beta))$. Thus, $\gamma(\boldsymbol{B}\perp\alpha)\subseteq\gamma(\boldsymbol{B}\perp(\alpha\&\beta))$. This yields $\bigcap(\gamma(\boldsymbol{B}\perp(\alpha\&\beta))\subseteq\bigcap(\gamma(\boldsymbol{B}\perp\alpha))$ and $\mathbf{K}\div(\alpha\&\beta)\subseteq\mathbf{K}\div\alpha$.

Acknowledgement I would like to thank Peter Gärdenfors, Hans Rott, Wlodek Rabinowicz, and an anonymous referee for valuable comments on an earlier version.

References

Alchourrón, C. E., Gärdenfors, P., & Makinson, D. (1985). On the logic of theory change: Partial meet contraction and revision functions. *Journal of Symbolic Logic, 50*, 510–530.

Fuhrmann, A. (1991). Theory contraction through base contraction. *Journal of Philosophical Logic, 20*, 175–203.
Gärdenfors, P. (1984). Epistemic importance and minimal changes of belief. *Australasian Journal of Psychology, 62*, 136–157.
Gärdenfors, P. (1988). *Knowledge in flux. Modeling the dynamics of epistemic states*. Cambridge: MIT Press.
Gärdenfors, P., & Makinson, D. (1988). Revisions of knowledge systems using epistemic entrenchment. In M. Y. Vardi (Ed.), *Proceedings of the second conference on theoretical aspects of reasoning about knowledge* (pp. 83–95).
Hansson, S. O. (1989). New operators for theory change. *Theoria, 55*, 114–132.
Hansson, S. O. (1991). Belief contraction without recovery. *Studia Logica., 50*, 251–260.
Hansson, S. O. (1992). In defense of base contraction. *Synthese, 91*, 239–245.
Hansson, S. O. (1993). Reversing the Levi identity. *Journal of Philosophical Logic, 22*, 637–669.
Makinson, D. (1987). On the status of the postulate of recovery in the logic of theory change. *Journal of Philosophical Logic, 16*, 383–394.
Nebel, B. (1992). Syntax-based approaches to belief revision. In P. Gärdenfors (Ed.), *Belief Revision* (pp. 52–88). Cambridge: Cambridge University Press.
Niederée, R. (1991). Multiple contraction. A further case against Gärdenfors' principle of recovery. In A. Fuhrmann & M. Murreau (Eds.), *The logic of theory change* (Lecture notes in Artificial Intelligence 465, pp. 322–334). Berlin: Springer.
Sen, A. (1970). *Collective choice and social welfare*. San Francisco: Holden-Day, Inc.

Chapter 15
How Infallible but Corrigible Full Belief Is Possible

Isaac Levi

Justifying Changes in Full Belief

Inquirers ought to change beliefs for good reason (Levi 1980 ch. 1, 1991, ch. 1, 2004). What those good reasons are depend on the proximate goals of their inquiries. William James urged us to seek Truth and avoid Error in forming beliefs. He ought to have said: Seek Information and avoid Error. The common features of the proximate goals of scientific inquiries ought to be to answer questions of interest without error and in a manner that yields valuable information.

The beliefs inquirers seek to change are full beliefs. Agent X fully believes that *h* if and only if X is certain that *h* is true. That is to say, X rules out the logical possibility that *h* is false as a serious possibility, takes for granted that *h* is true and uses this information as evidence in efforts to increase the information available to X. Such evidence constitutes the basis for making assessments of credal probability used to evaluate risky choices.

Justifiably changing judgments of credal probability is important to inquiry only insofar as it contributes to the promotion of the goals of inquiry. One does not engage in inquiry in order to justify changes in credal probabilities. In inquiry new error free information is sought. Credal, belief or subjective probabilities are neither true nor false.[1] Changing credal probabilities neither succeeds in nor fails to avoid error. And changing credal probabilities fails to rule out logical possibilities as serious possibilities. So they do not add to information.

[1]As F.P. Ramsey (1990), B. de Finetti (1964) and L.J Savage (1954) rightly observed.

I. Levi (✉)
Columbia University, New York, NY, USA
e-mail: levi@columbia.edu

H. Arló-Costa et al. (eds.), *Readings in Formal Epistemology*, Springer Graduate Texts in Philosophy 1, DOI 10.1007/978-3-319-20451-2_15

The state of credal probability judgment to which X is committed is a function of X's state of full belief or "evidence and background knowledge" and a rule for deriving X's credal probability from X's evidence that I call X's *confirmational commitment* (Levi 1974; 1980). When X's credal state is derivable from X's state of full belief or evidence in accordance with a confirmational commitment, justifying a change in credal state is justifying a change either in X's state of full belief or justifying a change in X's confirmational commitment. As just stated, the proximate aim of a change in X's state of full belief is to avoid importing false belief while acquiring valuable information. Confirmational Commitments, like states of credal probability judgment lack truth values. The proximate aim of modifications of X's confirmational commitment do not involve avoidance of false confirmational commitments or acquiring information by ruling out confirmational commitments as serious possibilities. Changing confirmational commitments is relevant to inquiry aimed at acquiring new and valuable information only insofar as such change facilitate changes in full belief that result in new error free and informative states of full belief.

Philosophical reflection concerning the conditions under which states of credal probability judgment should be modified and how they should be modified is of first rate importance. Credal probability judgment is relevant to the assessment of risks both in practical and theoretical deliberation. But justifying changes in credal probability concerns either the justification of changes in states of full belief or changes in confirmational commitments. Changes in full belief and confirmational commitments alike are justified in terms of how well they promote the acquisition of belief states carrying new, valuable error free information.

Some authors think that in inquiry we seek to change degrees of belief (and disbelief) exhibiting a formal structure similar to Shackle's degrees of belief and surprise (disbelief) (1949; 1961). The goal is to change qualitative beliefs in a sense according to which X believes that *h* if and only if X's degree of belief reaches some suitable nonnegative threshold short of absolute certainty (Cohen 1970, 1977; Spohn 1988).

Degrees of belief in a satisficing sense and belief that *h* equivalent to a high satisficing degree of belief (Levi 1967a, 1997 ch. 2.4, 2002) that has reached a certain threshold are suitable as modes of appraisal in inquiry when the inquirer is considering whether to add an item of information to his or her stock of full beliefs. If the initial state of full belief supports a degree of belief that *h* to a sufficiently high degree relative to the initial state of full belief, the inquirer may be warranted in converting the high degree of belief to a full belief to be added to his or her full beliefs. However, to engage in inquiry or to justify changes in degrees of beliefs or mere beliefs loses its purpose except in a context where the changes in degrees of belief are instrumental to justifying a change in state of full belief.

Change in Doxastic Commitment

Studying changes in states of full belief requires some characterization of the structure of a system of potential states of full belief just as studying changes in the state of a mechanical system requires an understanding of a system of mechanical states. The concern here is, of course, inextricably normative concerned as it is with conditions under which changes from one potential state to another is legitimate and, indeed, justified. Neither classical, statistical nor quantum mechanics is normative in this respect. But the need to consider assumptions about the structure of a "space" of potential states remains the same in the case of changes in states of full belief, changes in classical mechanical states and changes in quantum mechanical states.

In this discussion, the space Ω of potential states of full beliefs is partially ordered by a consequence relation satisfying the conditions of a Boolean algebra. (The algebra is closed under meets and joins of whatever cardinality is required.) Inquirer X is in some state of full belief **K** located in this space. While in the state **K**, X is committed to full belief that each consequence of **K** in the algebra is true. **K** is thus a state of doxastic commitment to the truth of the members of the set of its consequences.

It is also a commitment to a standard for distinguishing the potential states in the algebra that are serious possibilities from those that are not. Because I doubt that there is an algebra that represents all potential states of full belief, I rest content with atomic algebras so that we may postulate a set $\boldsymbol{W}$ of atoms and every potential state in the algebra is the join of a subset of $\boldsymbol{W}$. The state of full belief **K** then distinguishes between the set $\boldsymbol{P}_{\mathbf{K}}$ of atoms that are serious possibilities according to **K** (which are all atoms that are consequences of **K**) and the impossibilities that are ruled out by **K**.

Insofar as a potential state **H** in Ω is representable by a set of sentences S in a regimented language L where the logical consequence relation in L preserves the consequence relation for Ω, **H** is also representable by the set $Cn(S)$ of sentences in L closed under the consequence relation for L. $\boldsymbol{W}$ is representable by a set W of maximally consistent sets of sentences in L and $\boldsymbol{P}_{\mathbf{K}}$ by the set P_{K} of maximally consistent extensions of K.

Thus, if we focus in inquiry on justifying changes in states of full belief we restrict attention to situations where both before and after the change the inquirer's belief state satisfies requirements for logically omniscience. This seems to be absurd. And it is absurd if the inquirer in belief state **K** is burdened with the obligation to recognize on demand all the consequences of his or her full beliefs. I have argued elsewhere for the view that the inquirer may be *committed* to fully believing that h without being able to fulfill the commitment on demand. The inquirer so committed ought to be prepared to acknowledge his or her failure when it is brought to his or her attention and should be prepared to improve his or her performance when opportunity and costs permit. But doxastic changes that implement such improvements are not changes in doxastic commitment. Even though they play an important role in inquiry, they are not the changes that are the target of inquiry.

The belief changes that are central to inquiry are changes in doxastic commitment. I have suggested for a long time that changes in doxastic commitment that are subject to justification are expansions and contractions.

Expansions are changes from weaker commitments to stronger ones. $\mathbf{K_N}$ is a consequence in the algebra of $\mathbf{K_I}$. The information contained in the initial commitment $\mathbf{K_I}$ is included in the information contained in the new commitment $\mathbf{K_N}$. Contractions are changes from strong commitments to weak one where the inquirer gives up information rather than acquiring it.

Expansions may be described in terms of the information added to an initial state **K**. The expansion of **K** by adding potential state of full belief **H** is $\mathbf{K} \wedge \mathbf{H}$ or $\mathbf{K}^{+}{}_{\mathbf{H}}$. To the extent that potential states are represented by deductively closed sets of sentences in L, the expansion is represented by $Cn(\mathrm{K} \cup \mathrm{H})$. $P_{\mathrm{Cn}(\{\mathrm{K} \cup \mathrm{H}\})} \subseteq P_{\mathrm{K}}$. When H is the set of consequences of a single sentence h, $Cn(\mathrm{K} \cup \{h\}) = \mathrm{K}^{+}{}_{\mathrm{h}}$ or expansion by adding h.

In contraction from **K**, some of the possibilities ruled out by **K** are added to the serious possibilities according to **K**. If **K** has as a consequence $\mathbf{H}^{c}$, contraction that removes this consequence has $\mathbf{K} \vee \mathbf{H}$ as a consequence. **H** is no longer impossible. But the contraction could be stronger. It could have $\mathbf{K} \vee \mathbf{X}$ as consequence where **X** is stronger than **H**. We shall explore this some more later on.

Many authors take another kind of change in doxastic commitment to be more central than expansion or contraction. Revisions are changes in doxastic commitment where the inquirer adds information to $\mathbf{K_I}$ whether or not the new information is consistent with the information in $\mathbf{K_I}$.[2]

Notice, however, that if **H** is a consequence of $\mathbf{K_I}$ and $\mathbf{K_N}$ has $\mathbf{H}^{c}$ as a consequence the inquirer starts with full belief that **H** is true and from the perspective to which he is then committed deliberately replaces this conviction with $\mathbf{H}^{c}$. From the inquirer's initial state of full belief $\mathbf{K_I}$, making the change from $\mathrm{K_I}$ to $\mathbf{K_N}$ is replacing true by false belief. Anyone seeking to avoid importing false belief in making a change in full belief should not deliberately engage in such replacement.

This consideration seems decisive as an objection against recognizing deliberate revision of $\mathbf{K_I}$ in the sense of Alchourrón et al. (1985) as a justifiable form of belief change. The objection is decisive provided the proximate goal of inquiry is to seek new, error free and valuable information at the stage where a change in belief is made. But alternative views of the goals of belief change are available.

(a) One can hold that avoidance of false belief is a desideratum in inquiry where truth and falsity are assessed according to an "external" standard severed from

[2] When the information added is implied by K, the revision is degenerate and identical with **K**. When the information added is consisitent with but not implied by **K**, the revision is a nondegenerate expansion. When the information is inconsistent with **K**, the revision is a *replacement* in the sense of Levi 1980.

any inquirer's point of view or as T.Nagel infamously put it, we consider truth according to the point of view from nowhere.[3] In that case, the inquirer should recognize that **H** in $\mathbf{K}_\mathrm{I}$ might be false as fallibilism requires even though $\mathbf{K}_\mathrm{I}$ is the inquirer's standard for serious possibility.

(b) One can follow John Dewey and Peter Gärdenfors in refusing to consider avoidance of false belief as a desideratum in inquiry.

(c) One can embrace the Messianic Realism of Charles Peirce and Karl Popper and regard convergence on the true complete story of the world as an ultimate goal of inquiry. This Messianic Realism may be supplemented by the thesis that inquirers ought to get closer to the truth as a proximate aim of inquiry as Niiniluoto continues to advocate.

(d) One can reject avoidance of error as a proximate aim of the next change in inquiry but insist that avoidance of error in some finite number of changes after the next change is such a goal.

(e) One can embrace Secular Realism as I and W.V. Quine do and maintain that inquirers ought to avoid error as judged according to the evolving doctrine (i.e., according to $\mathbf{K}_\mathrm{I}$) as a proximate aim at the next change in inquiry. (Niiniluoto 1984).

The secular realist response (e) that I favor argues against the justifiability of revising $\mathbf{K}_\mathrm{I}$ by adding information inconsistent with it in a single step. However, sometimes the net effect of such revision may be justified by justifying each step in a sequence of contractions and expansions.

The disagreement between secular realism and the alternatives is neither a metaphysical nor a semantic one. It is a question of values. Peirce and Popper thought that inquiry should be promoting progress toward the truth at the End of Days. Others might think that progress towards true answers is desirable but avoid the excesses of Messianic Realism. In any case, these approaches can spell out their conceptions of aiming at truth and allow truth and falsity to be judged from some agent's point of view. Cognitive values that should be pursued in inquiry are at issue. Truth and falsehood should be judged by the inquirer from the inquirer's initial state $\mathbf{K}_\mathrm{I}$ of full belief when assessing the options available for change in full belief. Retrospective assessment is from the inquirer's view in state $\mathbf{K}_\mathrm{N}$. And others may judge the truth or falsity of the inquirer's beliefs before and after change from their own belief states. Only assessment of the truth or falsity of the inquirer's beliefs from the point of view from nowhere is incoherent.[4]

In any case, according to Secular Realism (the position I endorsed under the epithet "myopic realism" in Levi 1980) the direct justifiability of replacements is decisively rejectable. Of course, changes of belief state by revision can be

[3]Donald Davidson criticized this view. He argued that inquirers cannot coherently aim at truth (1998). Aiming at truth is, indeed, incoherent if it is judged from the point of view from nowhere rather than as Quine put it from the "evolving doctrine".

[4]If someone insists on the coherence of that point of view, it remains obscure as to the relevance of seeking to avoid error as judged from that point of view.

represented as compositions of sequences of contractions and expansions. And such changes are justified, if each contraction and expansion in such a sequence is justified. I dissent, however, from the popular point of view according to which revision $\mathbf{K}^*_{\mathbf{H}}$ by adding **H** is a response to an input importing **H** in the sense that **K** is the initial state of full belief $\mathbf{K}_{\mathbf{I}}$ and $\mathbf{K}^*_{\mathbf{H}}$ is $\mathbf{K}_{\mathbf{N}}$.

As already noted, the central concern of a discussion of justifiable belief change ought to be changes in full belief rather than changes in degrees of belief understood as degrees of credal probability.[5] I have claimed for a long time that when X fully believes that **H**, X is committed to ruling out the *logical* possibility that **H** is false as a serious possibility. It is incoherent for X to fully believe (be certain) that **H** is true and at the same time to acknowledge that **H** might be false or that there is a small, perhaps infinitesimally small, probability that **H** is false.[6]

Many authors who preach the epistemological doctrine of fallibilism call into question the corrigibility of full beliefs i.e., the justifiable modification of full beliefs once endorsed.[7] Epistemological (or doxastic) infallibilism maintains that when X fully believes (is absolutely certain) that **H**, X rules out the logical possibility that **H** is false as a serious possibility. If there is no serious possibility according to X that *h* is false, it seems puzzling to many how X can justifiably come to doubt that **H**.

[5]When I first advanced this view, I used 'knows' rather than 'fully believes'. I continue to think that *according to X* at time t, X fully believes that *h* if and only if X knows that *h*. That is because I define 'knows' as 'truly believes'. This definition is an expression of the epistemic ideals I have borrowed from the pragmatists who do not require that X justify X's current beliefs but only changes in X's beliefs and the view that truth is judged relative to the evolving doctrine so that according to X everything X fully believes is true. Notice that agent Y can agree that X knows that *h* when Y also fully believes that *h* and will disagree other ways. X and Y will agree that X knows that *h* if and only if X truly believes that *h*.

[6]To fully believe, to be certain, or to know that **H** is not equivalent to judging that the probability that **H** is 1. Setting aside the issue of indeterminacy in probability judgment, subjective or credal probability judgment assigns numerical values in the closed unit interval to potential states of full belief (or to propositions) and their complements when both are judged seriously possible according to **K**. If **H** is fully believed and, hence, a consequence of **K**, it carries probability 1 and is seriously possible while its complement is assigned probability 0 and is ruled out as impossible.

[7]In a well-known example of the conflation of infallibility and incorrigibility, R.C. Jeffrey (1965) argued that reasonable agents should not assign credal probability 1 to propositions because one cannot coherently shift down from probability 1 in conformity with modifying credal probability by Conditionalization. This sort of change in credal state is derivable from the inquirer's state of full belief and the credal probability determined for that state of full belief by the inquirer's credibility function (Carnap 1960) or confirmational commitment provided that the confirmational commitment remains unchanged while the state of full belief is expanded by adding the proposition e and the confirmational commitment satisfies the principle of *confirmational conditionalization* (Levi 1980, 4.3). Confirmational Conditionalization is a synchronic constraint on confirmational commitments. Temporal credal conditionalization is a procedure for changing credal probabilities *if* the state of full belief is expanded by adding a new item *e* of information. That temporal principle does not forbid giving up e. Indeed, as long as the confirmational commitment remains fixed, a change from $\mathbf{K}^+_{\mathbf{e}}$ to K can be derived from confirmational conditionalization as long as the confirmational commitment remains fixed. Jeffrey took for granted that contraction of a state of full belief could not be justified. He did not offer a compelling case for this conclusion.

The puzzle is not premised on a contradiction between saying that X was certain that **H** but no longer is. There is no such contradiction. The issue is whether X can justifiably become uncertain (change from full belief that **H** to doubt as to the truth of **H**) if X is concerned to avoid false belief and maximize the value of information.

The justification should show how an inquirer X concerned to avoid importing false belief while seeking to increase the value of the information available could change views. The two desiderata of avoiding error and acquiring valuable information tend to be in conflict. The more probable X judges **H** to be, the lower the risk of error in coming to believe it and less the value of the information carried by *h* tends to be.

Suppose X initially fully believes that **H**, X may not change belief state, may replace full belief that **H** with full belief that $\mathbf{H}^c$ or move to a position of suspense between **H** and $\mathbf{H}^c$ by contracting from **K** by removing **H**.

From X's initial point of view, remaining with the status quo incurs no risk of false belief.

From the same point of view, replacing full belief that **H** with full belief that $\mathbf{H}^c$ imports false belief deliberately. Replacements are indefensible given the goal of avoiding false belief while increasing the value of information.

Finally moving to a position of suspense between **H** and $\mathbf{H}^c$ by contraction incurs no risk of importing false belief. But the inquirer X will be deliberately giving up information in doing this. And given the goals of inquiry, gratuitous surrender of information looks indefensible.

If not only replacement but contraction also is indefensible given the goals of inquiry, the only kind of change that does not deliberately import error or give up valuable information appears to be expansion. Expansion does to be sure incur a risk of importing false belief. But incurring the risk may be justifiable in a manner compatible with the concern to avoid error as long as the value of the information promised compensates for the risk incurred.

To the extent that expansion can be justified along these lines,[8] the inquirer can overcome the obstacles to giving up information in contraction. Contraction incurs a loss of valuable information. But subsequent inquiry stands some chance of enhancing the informational value of the state of full belief.

Changing from an initial state of full belief $\mathbf{K}_I$ to a contraction $\mathbf{K}_N$ implied by $\mathbf{K}_I$ is entertainably justifiable in at least two contexts:

A. $\mathbf{K}_I$ is inconsistent and contraction amounts to retreat from inconsistency.
B. $\mathbf{K}_I$ is consistent but has a consequence $\mathbf{H}^c$ where **H** is a conjecture that would contribute valuable information to X's store of information were it (counter to the verdict of X's current state of full belief $\mathbf{K}_N$) true. Contraction removing $\mathbf{H}^c$ from K_N is contemplated in order to give a hearing to **H**.

[8]In Levi 1983; 1980, and 1991 I provide an account of the justification of expansion along these lines based on ideas developed in Levi. 1967a and 1967b.

Contraction from Inconsistency

The first puzzle to consider regarding context A is that in order to retreat from inconsistency, the inquirer needs to be in a state of inconsistency. $\mathbf{K}_{\mathrm{I}} = \mathbf{K}_{\perp}$.

An inquirer's state of full belief can be inconsistent in one of two ways. X may be committed to a consistent state of full belief but his doxastic performances may be inconsistent. In that case, removing inconsistency involves either efforts at self-therapy or receiving help from others and their technologies. Indeed, not only must the agent extricate him or herself from inconsistency but must identify his or her consistent state of doxastic commitment. No one can humanly succeed in this endeavor completely. All flesh and blood inquirers are inconsistent in their performances. I have already indicated that I shall not be taking up the question of the considerable therapeutic and engineering tasks involved in realizing the local pockets of fulfillment of doxastic commitment that flesh and blood achieve. In this discussion, the kind of retreat from inconsistency involved in this achievement shall not be considered.

The sort of inconsistent state of full belief of concern here is an inconsistent state of doxastic commitment. The puzzle we need to address is how an inquirer who is rationally fulfilling his or her commitments could end up in such an inconsistent state without involvement in a performance failure? To deliberately expand into inconsistency is never justifiable if the common feature of the proximate goals of all efforts to change doxastic commitments is to obtain valuable new information without importing false beliefs. As long as X's state of full belief is consistent, X is committed to judging the inconsistent state $\mathbf{K}_{\perp}$ to be false. To deliberately expand into inconsistency is to deliberately import false belief into one's evolving doctrine.

The inquirer, however, may deliberately incur a risk of importing false belief, and do so rationally, if some appropriate benefit compensates for the risk. In inquiry where the proximate aim is the acquisition of new and valuable error-free information, making well designed observations is undertaken on the assumption that the beliefs formed in response to the observations made have a good chance of being true and informative. And the inquirer may sometimes be convinced that the testimony of witnesses and experts is reliable and substantive. Consulting external sources of information of both varieties incurs some risk of error. But often such consultation is the only available way of acquiring information relevant to a certain investigation. The value of the information acquired may compensate for the risk of error incurred.

The acquisition of new beliefs via observation and the testimony of experts and witnesses is "direct" in the familiar sense that it is not inference from premises. The consultation involves implementing a program for letting inputs (such as sensory excitations or verbal testimony) determine what information to add to a state of full belief. The sensory inputs and verbal testimonies, however, are not premises from which the inquirer infers a new belief. The addition of a new belief is an outcome of a process initiated by the input. Whatever the psychological details of the process might be, the occurrence of the inputs (the sensory stimulations or the testimony of the experts or witnesses) do not constitute a change in the inquirer's state of full belief in the doxastic commitment sense. If a change in doxastic commitment occurs, it is a response to that input in conformity with the program being implemented.

Pace Russell and generations of empiricists before and after him, the "data" acquired by such *routine expansion* is directly acquired without being immediately given. Even though the new beliefs are acquired *directly* (i.e., without inference), the acquisition presupposes background information relevant to assessing the reliability of the process of making observations or consulting experts.

Commitment to the program for routine expansion should be distinguished from commitment to implementing it. The inquirer may recognize someone as authoritative on some topic without consulting that agent. X may judge a procedure for making observations reliable without using it. Moreover, the inquirer may be committed in one or both respects without having undertaken the commitments deliberately. However, when undertaking commitment to a program for routine expansion is at issue or when implementing a program to which one is already committed is at issue, modeling the situation as a choice between programs for routine expansion is appropriate.

For purposes of analysis, it is useful to distinguish between the inquirer's point of view $\mathbf{K_{PI}}$ prior to deliberately undertaking commitment to a program for routine expansion and the inquirer's point of view $\mathbf{K_I}$ when deliberating as to whether to implement the program. We may say that at K_{PI}, the inquirer *precommits* to endorsing the deliverances of an implementation of the program should an implementation be undertaken. At $\mathbf{K_I}$, the inquirer *commits* to the deliverances of an implementation of the program to which the agent is already committed.

From the inquirer's point of view $\mathbf{K_{PI}}$ implementing a program for routine expansion being contemplated might lead to the formation of beliefs incompatible with $\mathbf{K_{PI}}$. And when the inquirer is considering implementation of a program already endorsed, implementation might lead to the formation of beliefs incompatible with the inquirer X's current full beliefs in $\mathbf{K_I}$. Routine expansion has the ability to be *conflict injecting*.

Yet, expansion into inconsistency does not deliberately add false belief to $\mathbf{K_I}$. The inquirer X has risked importing error for the sake of the information to be acquired and has lost the gamble. The outcome is that X has expanded into inconsistency $\mathbf{K_\perp}$ from $\mathbf{K_I}$.[9]

Inadvertent expansion into inconsistency is a byproduct of the need to use routine expansion to acquire new information that, perhaps, could not be obtained at the time by other means. This kind of inconsistency does not represent failure to fulfill

[9]Regardless of whether the implementation is the product of habitual or customary practice or is deliberate, the inquirer is committed to implementing the program prior to its implementation. The inquirer is precommitted. The inquirer is (pre) committed to the results of implementing the routine regardless of what they might be. When the result is expansion into inconsistency, the importation of false belief is inadvertent.

commitment. To the contrary, it fulfills a commitment to follow a program of routine expansion regardless of where it might lead.

$K_\perp$ fails as a standard for serious possibility and as a resource for specifying truth conditions for hypotheses in a sense of truth according to which an inquirer should be seeking to avoid false belief. It seems obvious that inquirers should retreat from such inconsistency – i.e., contract from $\mathbf{K}_\perp$.

Contraction from inconsistency cannot be justified as the choice of the best option among the (epistemic) options available. Such justification should be based on the inquirer's point of view prior to choice. Prior to choice, however, the inquirer's point of view is inconsistent. An inconsistent state of full belief cannot be used coherently as a standard for serious possibility, for judging truth or as evidence. In Levi 1980, footnote pp. 59–60, I worried about this question of coherence. I suggested treating the language in which the inconsistent state of full belief is expressed syntactically and rationalizing changes in a consistent metalanguage. Olsson (2003) rightly took me to task for not living up to my own commitments concerning the characterization of the state of full belief as a standard for serious possibility. Olsson's objection seems unanswerable according to the approach I favor. But according to that approach, his proposed cure is worse than the disease. Olsson maintained that there are admissible programs for routine expansion that avoid expansion into inconsistency and that programs for routine expansion should be restricted to these. In Levi 2003, I showed that routine expansion is inescapably conflict injecting – counter to what Olsson claimed. The only way to avoid routine expansion into inconsistency is to avoid routine expansion altogether. This remedy seemed to be as unacceptable to Olsson as it is to me. I concluded that we need to consider other ways to address the predicament of how to justify retreat from inconsistency.

In Levi 2003, I suggested that a program for routine expansion should be accompanied by a precommitment to a plan providing a response in case implementation of the program leads to expansion into inconsistency. From the point of view $\mathbf{K}_{PI}$ the inquirer can coherently identify a set of appropriate contractions from inconsistency and evaluate their merits.

Programs for routine expansion provide for expansions that yield many competing answers to a given question. As a consequence, expansion into inconsistency can take on a diversity of forms depending on the item of information h added to $\mathbf{K}_I$ that yields inconsistency. Given a specific h, the plan stipulates that one of the following contractions from $\mathbf{K}_\perp$ is implemented:

(i) Return to $\mathbf{K}_I$. The expansion prescribed by the routine is nullified.
(ii) Replacement of the consequence $\mathbf{H}^c$ of $\mathbf{K}_I$ by $\mathbf{H}$. The contraction from inconsistency recommended is representable as an expansion by adding $\mathbf{H}$ to a contraction of $\mathbf{K}_I$ by removing $\mathbf{H}^c$.[10]

[10]The equivalence of a replacement to an expansion of a contraction is the *Commensuration* Thesis. (Levi 1991, p.65) The Commensuration thesis is trivial as long the domain of potential states of full belief is partially ordered in a manner satisfying the requirements of a Boolean algebra.

(iii) Contraction of $\mathbf{K_I}$ by removing $\mathbf{H}^c$.

Each of the three plans recommends returning the state of full belief to a potential state that is formally representable as a transformation of $\mathbf{K_I}$. The important point to be emphasized is that, nonetheless, each if the three plans implements a potential contraction from inconsistency.

Exercising option (i) is contracting from $\mathbf{K}_\perp$ in a way that returns the state of full belief to the *status quo ex ante*.

Exercising option (ii) is also a move to a consistent potential state of full belief from $\mathbf{K}_\perp$. It is not a move from $\mathbf{K_I}$ to a replacement of $\sim h$ by h in $\mathbf{K_I}$. It is a move from $\mathbf{K}_\perp$ to the aforementioned replacement.

Similarly, option (iii) is a move from inconsistency to a contraction of $\mathbf{K_I}$ by removing $\mathbf{H}^c$.

We should look at (i), (ii), (iii) as three optional precommitment plans for retreating from inconsistency. The three alternatives may be formally representable as transformations of $\mathbf{K_I}$. But the challenge that provokes consideration of them is an expansion of $\mathbf{K_I}$ into inconsistency. The task to which all three are responsive is how to retreat from inconsistency.

Thus, even though retreating from $\mathbf{K}_\perp$ is an option as a precommitment plan, remaining with $\mathbf{K_I}$ is not an option. Similarly replacement of $\mathbf{H}^c$ in $\mathbf{K_I}$ by $\mathbf{H}$ is not an option even though shifting from inconsistency to the replacement is. Similar remarks apply to contraction of $\mathbf{K_I}$ by removing *h.*

Formally, replacement of $\mathbf{H}^c$ by $\mathbf{H}$ in $\mathbf{K_I}$ is equivalent to an expansion by adding $\mathbf{H}$ to a contraction by removing $\mathbf{H}^c$ from $\mathbf{K_I}$. But as is well known, there are many contractions of $\mathbf{K_I}$ that remove $\mathbf{H}^c$. We need some way of recommending which contractions removing $\mathbf{H}^c$ from $\mathbf{K_I}$ can serve as ingredients in contractions from $\mathbf{K_I}$ in precommitment plans of the three types mentioned above.

Deliberation aimed at adopting a precommitment plan at $\mathbf{K_{PI}}$ compares the admissible options of type (i), (ii) and (iii) and determines best or admissible precommitment plans from among these. Option of type (i) is formally the same as first contracting by removing $\mathbf{H}^c$ and then expanding by adding $\mathbf{H}^c$. The type (ii) option (replacement of $\mathbf{H}^c$ in $\mathbf{K_I}$ by $\mathbf{H}$) is formally equivalent to an expansion by adding $\mathbf{H}$ to a contraction of $\mathbf{K_I}$ by removing $\mathbf{H}^c$.

Both (i) and (ii), therefore, are representable formally as an expansion of the potential state of full belief that is the outcome of contraction type (iii) where the inquirer is in suspense with respect to the truth of $\mathbf{H}$. The upshot is that the third option yields a belief state that carries less information and, hence, no more informational value than the other two options.

Suppose that result of implementing (i) is a belief state carrying more (less) informational value than the belief state resulting from implementing (ii). On the assumption (that I favor) according to which the inquirer should seek to minimize loss of informational value, if the inquirer should have to retreat from inconsistency due to having expanded $\mathbf{K_I}$ by adding $\mathbf{H}^c$, the inquirer should follow option (i) or option (ii).

On the other hand, if options (i) and (ii) result in belief states carrying equal informational value or if they are noncomparable with respect to informational value, the inquirer should follow option (iii). This recommendation is based on the assumption that the informational value of the join of two potential states of full belief is the minimum informational value carried by the pair of states. This is the assumption I have used in evaluating loss of damped informational value in contraction in Levi 2004 and in Levi 1991 and 1997.

On this view, neither information acquired via observation nor from expert witnesses is categorically authoritative. If it were, option (ii) would be mandatory in all cases. All but the most rabid empiricists agree that the adoption of (ii) can be trumped when the information carried by the background beliefs to be given up is too valuable to surrender. Depending upon how valuable that information is, option (i) or (ii) is favored.

The expansion step used to define replacement of $\mathbf{H}^c$ by **H** in $\mathbf{K_I}$ is uniquely determined by the contraction of $\mathbf{K_I}$ by removing **H**. So both options (ii) and (iii) are determined once we can identify the contraction of $\mathbf{K_I}$ by removing $\mathbf{H^c}$. But such specification requires a choice from a roster of contraction strategies each of which removes $\mathbf{H}^c$ from $\mathbf{K_I}$.

Keep in mind that such a "choice" is not a deliberate decision to implement such a contraction. It is not a deliberate choice of a contraction from $\mathbf{K}_\perp$. Nor is it a deliberate choice of a contraction from $\mathbf{K_I}$. If there is any deliberate decision involved it is a deliberate choice of a precommitment at $\mathbf{K_{PI}}$ to a clause in the program for routine expansion precommitting the inquirer to an approach to contracting from inconsistency should routine expansion lead to inconsistency..

Even so, the decision concerning a contingency plan for contracting from inconsistency depends on answering the question: What should be a best or an admissible contraction removing **H** from K_I on the belief contravening supposition that a deliberate choice of a contraction removing $\mathbf{H}^c$ from $\mathbf{K_I}$ is required? The supposition of this question contravenes or should contravene the full beliefs of the inquiring agent. Even so, the question could be addressed as a problem of rational choice. Given a specification of the options and the payoffs, how should a decision be taken? In Levi 1991, I contended that that in such contexts of "coerced" contraction where the inquirer has no coherent option but to retreat from inconsistency, the inquirer still needs to decide "how to contract".

Deliberate Contraction

A decision concerning how to contract also arises in the context of deliberate contraction from consistent $\mathbf{K_I}$ in order to give a hearing to an informationally valuable conjecture.[11]

[11]In Levi 1991, ch. 4, I distinguished between coerced and uncoerced contraction and took note of the fact that both types of contraction require consideration of the problem of "How to Contract".

Sometimes inquirers can be justified in contracting from initial $\mathbf{K}_I$ even though $\mathbf{K}_I$ is consistent. $\mathbf{K}_I$ may be incompatible with **H** so that X is committed to being certain that **H** is false. Yet, X may regard **H** as a hypothesis that would, if it were true, explain propositions in some interesting and important domain.

Thus, the general theory of relativity was incompatible with received doctrine even though it could if true explain the perihelion of Mercury. The behavior of Mercury may have been consistent with $\mathbf{K}_I$ –i.e., with classical mechanics. However, the attempts to explain the perihelion of Mercury within the framework of $\mathbf{K}_I$ were not successful. The trajectory of Mercury represented a stubborn anomaly within the framework of classical mechanics.

Contracting $\mathbf{K}_I$ by removing $\mathbf{H}^c$ so as to recognize the general theory of relativity as a serious possibility is a retreat from anomaly but not from inconsistency. The contraction is deliberately chosen over the option of remaining with K_I in order to give a hearing for an important conjecture even though doing so incurs a loss of informational value.

The loss of informational value might be endured provided the inquirer has confidence that subsequent to contraction, inquiry to resolve the doubt (for example concerning the status of General Relativity) will be undertaken and will lead to removing doubts either by justifying return to $\mathbf{K}_I$ by rejecting General Relativity or expanding $\mathbf{K}_N$ by adding General Relativity. On that assumption, the inquirer can reason that either the informational value will be greater than that carried by $\mathbf{K}_I$ or there will be no net loss in informational value.

This reasoning assumes (1) that the informational value obtained by adding General Relativity Theory to the contraction of $\mathbf{K}_I$ by removing classical gravitational theory is greater than that $\mathbf{K}_I$ and (2) that contraction removing classical gravitational theory from $\mathbf{K}_I$ initiate inquiry warranting a resolution of the doubt. Otherwise reversion to $\mathbf{K}_I$ is justified.

There is another assumption that deserves mention here. It concerns the import of the thesis that the goal of inquiry seeking to modify full beliefs is to increase informational value while reducing risk of error.

According to secular realism the concern to avoid error or reduce risk of error is restricted to the change in belief state being contemplated. It does not concern changes subsequent to the next one. Nor does it concern convergence to the truth in the limit.

The quest for valuable information by way of contrast takes a longer view. In contraction we seek to minimize loss of informational value. But this concern can be trumped by the expectation of a gain in informational value afforded by the subsequent inquiry.

This is the question of determining what should constitute contraction removing *h* from **K**. I subsequently discussed this matter in Levi 1997. My final word on this topic (I think) is to be found in Levi 2004. I do not discuss this topic here but place emphasis on what I take to be the two important types of contraction.

There is no inconsistency in endorsing this asymmetry between the concern to avoid error and the aim of maximizing informational value. To be sure, I cannot demonstrate that the goals of inquiry should exhibit the asymmetry. There are many views of the aims of inquiry alternative to the one I am advocating. I think that the view of these goals being proposed gives us a sensible rationale for legitimate routine expansion, justified inductive expansion, coerced contraction and deliberate contraction superior to rival approaches that may be entertained.

Coerced contraction calls for a justification showing how to contract where it is determined already that some contraction from $\mathbf{K}_{\perp}$ is to be implemented, that the contraction to be implemented is a contraction removing some specific proposition **H** from $\mathbf{K}_{\mathbf{I}}$, but where it is unsettled as to which contraction meeting this requirement to implement as a contraction from $\mathbf{K}_{\perp}$.

In deliberate contraction, there is no issue concerning retreating from inconsistency. There is a question as to whether to contract by removing proposition **H**, proposition **G**, etc. from $\mathbf{K}_{\mathrm{I}}$ or not. And given that **H** is to be removed from $\mathbf{K}_{\mathrm{I}}$, there is a question as to which contraction removing **H** is to be implemented. The last issue raises a problem similar to the problem of how to contract that arises in connection to coerced contraction. Indeed, the similarity between the two problems concerning how to contract entitles us to explore them together.

How to Contract: The Available Options

In any decision problem, the rationality of the choice made depends on the set of options available to the decision maker – that is to say available to the decision maker *according to the decision maker's point of view.* In order for a proposition to represent an available option, the decision maker must be convinced of his or her ability to implement the option if he or she chooses to do so. The decision maker must also judge it a serious possibility that he or she will implement the option. If the decision maker fully believes that he or she will not make the choice, from the decision maker's point of view, deliberation directed at deciding whether to implement the option is pointless.

Deciding how to contract by removing **H** is, so I assume, a problem for choice where the inquirer seeks to implement the best option available that if implemented would remove **H** from $\mathbf{K}_{\mathrm{I}}$.

Given the nature of problem, the domain of options from which choice should be made should be some subset of the contractions removing **H** from $\mathbf{K}_{\mathbf{I}}$. Should the set of options include all contractions removing **H** from $\mathbf{K}_{\mathbf{I}}$?

This question is not well formed. A contraction removing **H** from $\mathbf{K}_{\mathrm{I}}$ is a potential state of full belief. We need to identify the Boolean algebra Ω of potential states of full belief and then ascertain the subset of Ω consisting of contractions from $\mathbf{K}_{\mathrm{I}}$ removing **H**. A customary view is to think of this algebra as the powerset of a set W of atoms conceptually accessible to the inquirer.

There are many reasons for resisting this idea:

(a) Given any set W of atoms, an inquiring agent is conceptually capable of refining elements of this set. This yields a new set of atoms whose powerset consists of elements conceptually accessible to the inquirer. There is no set W of categorically atomic potential states of full belief. There is no set of maximally consistent belief states. There are no possible worlds.
(b) Suppose, however, that we consider a very large domain W and the powerset generated by it. The inquirer may be interested in removing the proposition that G.W.Bush was properly elected President of the USA in 2000 from his or her state of full belief **K**. A potential contraction removing this item from K would be the join of **K** with a set T belonging to the powerset of W each element of which is inconsistent with **K** and at least one of which entails that Bush was not properly elected. But the elements of T might all specify information concerning famines in Africa, recessions in the USA and, indeed, detailed history of the world. This kind of information may be irrelevant according to the inquirer to topic of the inquirer's investigation. What we should be considering is a Boolean algebra of potential states of full belief that the inquirer is committed to judging relevant to this topic.

Instead of beginning with W and its powerset, consider a set U_{LK} of potential states of full belief where a potential state of full belief **LK** entails the truth of exactly one element of U_{LK} and all elements of U_{LK} are consistent with **LK**. Elements of U_{LK} represent the maximally specific potential states of belief that are deemed relevant by the inquirer to the topic under investigation. They generate an algebra of potential states of full belief and they should be conceptually accessible to the inquirer. However, members of U_{LK} need not constitute a logically exhaustive set so that **LK** need not be a logical truth or conceptual necessity. Similarly the members of U_{LK} may be refined if more specific judgments come to be recognized as relevant. The elements of U_{LK} serve as atoms in the Boolean algebra constituted by the powerset of U_{LK}. Elements of the algebra constitute the set of potential states of full belief to be used in trajectories formed by belief changes of relevance in the kind of inquiries under consideration. Perhaps some good reason will arise for refining U_{LK} or adopting a weaker minimal state of full belief. The domain of potential states extends well beyond any atomic algebra we may utilize in characterizing changes in states of full belief.

LK is the *minimal state of full belief* and U_{LK} is the *basic partition* for the purpose of characterizing changes in belief state that are relevant in the context of inquiry into a given budget of problems. The demands for information that are characterized by **LK** and U_{LK} may be modified in the ongoing process of inquiry and different inquiries and their inquirers have different demands that may require the use of different minimal states and basic partititions.

If the inquirer X is in state of full belief **K**, elements of U_{LK} consistent with **K** are elements of the *ultimate partition* U_{K} relative to **K**. Elements of U_{LK} inconsistent with **K** constitute the *dual ultimate partition* U^*_{K}.

Given **LK** and $U_{LK,}$ a potential contraction of **K** is representable by the join of **K** with the join of a subset T of U^*_K.[12] The contraction is a contraction removing **H** from **K** if and only if at least one element of T entails $\mathbf{H}^c$.

A maxichoice contraction removing **H** from **K** is a contraction removing **H** from **K** where T contains only one element and it entails $\mathbf{H}^c$.

A contraction removing **H** from **K** is saturatable if and only if exactly one element of T entails $\mathbf{H}^c$,

A partial meet contraction removing **H** from **K** is a contraction removing **H** from **K** where every element of T is a maxichoice contraction entailing $\mathbf{H}^c$.

The options available for choice to the agent X concerned to contract by removing **H** from **K** given **LK** and U_{LK} consist of all contractions removing **H** from **K** relative to **LK** and U_{LK} (or relative to **K** and U^*_K).

How to Contract: Minimizing Loss of Informational Value

According to a decision theoretic approach to contraction, one should begin with an explanation of how the goals of contraction determine a value structure for the set of available options (the potential contractions removing **H** from **K**). A value structure (Levi, 1986) is representable by a set of permissible utility functions. An option is V-admissible if and only if it comes out best among the available options according to at least one permissible utility function. In the special case where, the set of permissible utility functions is unique up to a positive affine transformation, thc V-admissible options coincide with the set of options that come out best according to all permissible utility functions.

I assume that when the inquirer is to contract in a manner that removes **H** from **K**, he should choose a contraction removing **H** from **K** that minimizes loss of informational value if such an option is available. Every contraction removing **H** from **K** yields a potential state relative to **LK** and U_{LK} that is the join of **K** with the join of a subset T of U_{LK} that contains at least one element that implies $\mathbf{H}^c$. Two factors determine the loss of informational value incurred:

(a) An evaluation of the informational values of elements of U_{LK}. I adopt a numerical measure $cont(\mathrm{x}) = 1 - M(\mathrm{x})$ as the measure of informational value of an element of U_{LK}. Here $M(\mathrm{x})$ is a finitely additive probability measure over the algebra generated by U_{LK}.
(b) An extension of the evaluation of elements of U_{LK} to the entire algebra generated by $U_{LK.}$.

[12] The contraction is also characterized as the intersection of the set of consequences of **K** and the set of consequences of T.

If the extension is the finitely additive probability mentioned in (a), the assessment of informational value is given b y $1 - M(x)$ for all x in the algebra. This is the *undamped* assessment of informational value.

There is an alternative assessment – the *damped assessment of informational value*. Given any finite subset S of the powerset of U_{LK}, the damped informational value of the join of S is the minimum value assigned to an element of S.[13]

In discussing contraction, we are interested in subsets of the dual ultimate partition U^*_K given **K**. The dual ultimate partition is, of course, a subset of the basic partition. The informational value of a subset T of U^*_K is the minimum of the informational values assigned elements of T and hence the maximum value of informational value determining probability M. This maximum exists as long as T is finite. We shall consider cases where U^*_K and, hence, T is finite.

In Levi 2004, I proposed that contraction from **K** removing **H** should be assessed by evaluating the damped informational value of every contraction from **K** removing **H** recognized by **K** and U^*_K and restricting choice to contractions that minimize loss of damped informational value.

I also proposed a rule for Ties that recommended choosing the weakest contraction minimizing loss of informational value.

This proposal is intended for cases where the inquirer is committed to evaluating damped informational value based on a numerically determinate probability distribution over U_{LK}.

In general, inquirers will not be committed to such numerically determinate assessments but rather to a set of permissible assessments. I contend that the set should be convex. The recommendation is to choose the weakest of the *V-admissible* contraction.

The result is that the appropriate contraction removing **H** from **K** should be what Rott and Pagnucco call "severe withdrawal" and I call "mild contraction" for situations where the evaluation of loss of informational value is numerically determinate or is representable the restriction of the set of probability distributions over U_{LK} to elements of U^*_K. that weakly order the elements of U^*_K and, hence, maxichoice contractions from **K** whether they remove **H** or not.

Moreover, the recommendation is based on appeal to an account of rational choice appropriate both for cases where the options are weakly ordered and where weal ordering fails.

If the domain of options were to be restricted to the domain of partial meet contractions removing **H** from **K**, and the assessment of informational remains numerically determinate or at least free from ordinal conflict, the recommendation would coincide with the recommendation of those who subscribe to the AGM account of contraction.

---

[13]In Levi 1991 and 1997, I proposed a variant on the damped informational value assessment. I abandoned it in Levi 2004. I called the earlier variant "version 1 damped informational value" and the later one version 2 damped informational value. In Rott (2006), Hans Rott showed that my characterization of version 1 was seriously defective. In Levi (2006), I showed how to repair the defect. But there were other reasons including comments by Hansson and Olsson (1995) that argue for abandoning version 1 damped informational value. I rehearse them in Levi (2004).

But the AGM account provides no decision theoretic basis for restricting the domain of options to just the partial meet contractions.

I do not mean to suggest that arguments have not been given in support of partial meet contraction. David Makinson has offered at least two. The first points out that every contraction removing **H** from **K** that is not the join of a subset T of U^*_K all of whose members entail $\mathbf{H}^c$ is the join of a subset T' that is the union of just such a subset with another whose members entail **H**. This "withdrawal incurs a greater loss of information than is incurred by using the partial meet contraction.

If we think of this argument decision theoretically, it presupposes that the goal of contraction should be to minimize loss of information rather than informational value. This is not a goal to which many, including Makinson, would subscribe; for if endorsed it favors restricting choice to maxichoice contractions removing **H** from **K**.[14]

[14]H. Rott (2006) declares that my recent proposal for assessing loss of informational value (damped informational value version 2) is strongly counterintuitive. The basis for his charge is that the recommended contraction could be the disjunction of a great number of maxichoice contractions each of which on its own incurs a great loss of informational value. But this basis presupposes what I deny – to wit, that the aim of contraction is to minimize loss of undamped informational value. Rott also insists that my proposal is sheer definition or stipulation. The only sense in which this is so is that damped informational value version 2 represents one among many utility functions that might be candidates for representing the value of information. I do not deny that rational agents could have different preferences. But the alternatives do not appear appetizing. Consider the interesting results reported in Rott (1993, 2001). Rott has explored choice functions over a domain of maxichoice functions and the preference relations thus generated. In cases where the choice function yields a set of two or more maxichoice contractions as optimal, he *stipulates without appeal to decision theoretic considerations* implementation of a corresponding partial meet contraction. Having done this he establishes a connection between choice consistency conditions for choice functions defined over the domain of maxichoice contractions and important axioms for contraction. Had he adopted a version 2 damped informational value utility function over the power set of the maxichoice contractions removing **H** the correspondence Rott identifies could have been rationalized decision theoretically rather than by stipulation. Even with this improvement, the approach still restricts the options to the set of partial meet contractions without any decision theoretic rationale for doing so whatsoever. I do not deny that Rott has established a correspondence between choice consistency over maxichoice contractions and axioms for contraction. Such a correspondence may satisfy philosophical logicians. But I do not see how it could satisfy anyone interested in a decision theoretic rationalization of contraction.

Pagnucco and Rott (1999) do consider the full range of contractions removing **H** as options. But, as Rott says, they think of the goals of contraction in such cases as being informational value and fairness (expressed by rule for Ties) which are conflicting primary desiderata.

I cannot prove anyone irrational who takes this position but I think it is obviously untenable. When two potential contractions removing **H** minimize loss of informational value, one may break ties by suspending judgment – which involves moving to a weaker contraction than either of the two – *provided that this contraction does not incur a greater loss in informational value*. If it does, invoking the rule for Ties is untenable. If it does not, then even if the rule for ties recommends favoring a contraction weaker than the two options that carry the same informational value that it does, it recommends minimizing loss of informational value. Why should Pagnucco and Rott deny this? I suspect that they think that suspending judgment between two options that minimize loss of information or informational value cannot coherently minimize loss of information or of informational value. This is so for information. But why do they insist that it is so for informational

The second argument alludes to the fact that every withdrawal is revision equivalent to a partial meet contraction. That is to say, expanding the partial meet contraction and expanding the revision equivalent withdrawal yield the same revision of K. Here the revision is an AGM revision.

Consider the following transformation of **K** by adding **K**. The Ramsey revision of **K** by adding **H** is the same as AGM revision in two cases: (i) **K** is inconsistent with **H**, (ii) **K** is consistent with both **H** and $\mathbf{H}^c$. In case (iii) where **K** is inconsistent with $\mathbf{H}^c$, AGM revision by adding **H** is an identity transformation. Ramsey revision requires contraction by removing **H** and then expanding by adding **H**.

If the Recovery postulate for contraction is applicable, AGM revision and Ramsey revision are equivalent. But Recovery should fail in many cases – most notably statistical examples. If X knows that coin has been tossed and landed heads and then contracts by removing the claim that the coin has been tossed, X should give up the claim that the coin landed heads (or, indeed, that it landed at all). Restoring the claim that the coin has been tossed will restore the claim that the coin landed on the surface but not that it landed heads. In that case recovery fails and AGM revision is no longer equivalent to Ramsey revision.

Although every withdrawal is AGM revision equivalent to a partial meet contraction, it is not Ramsey revision equivalent to a partial meet contraction. If the view of legitimate belief change I have been sketching is along right lines, neither form of revision plays a central role in justifying belief change. Revision comes into its own when an analysis of modal judgment on a supposition is on offer. I contend that such Ramsey revision is more adequate to this task than AGM revision.

Hans Rott (1993, 2001) has offered a representation of the preference among maxichoice contractions in terms of choice functions. Given a choice function over the domain of maxichoice contractions removing **H**, the value of the function is the set of optimal maxichoice contractions. removing **H**. The contraction determined by the choice function is the join of the set of optimal maxichoice contractions and is the partial meet contraction. This join is not, in general, a maxichoice contraction and, hence, is not a member of the value of the choice function examined by Rott. But it is join of the maxichoice contractions in the value of the choice function.

One could examine choice functions that take as arguments sets of contractions removing **H** whether they are maxichoice or not. Rott does not develop an account of preference over all contractions that shows that the meet contraction is best among all the options available to the decision maker. He does show that the most preferred maxichoice contractions removing **H** from **K** can be used to define a partial meet contraction. But given his restricted account of the domain of the choice functions, Rott cannot show that the recommended partial meet contraction is optimal.

It would have been easy for Rott to have obtained this definition of partial meet contraction in terms of choice functions (at least in the finite case) had he extended his choice functions from the domain of maxichoice contractions removing **H** from

value? If it were true, minimizing loss of informational value would lead straightforwardly to choosing maxichoice contractions a consequence he acknowledges is untenable.

K to the powerset of this domain (that is to say, the set of partial meet contractions). Given any set of maxichoice contractions, its value would be the best of the values of the maxichoice contractions.

But even if Rott had taken this step, he would not have provided a satisfactory decision theoretic account of contraction because he would not have explained decision theoretically why "withdrawals" are left out of account. Using damped informational value (that is to say the type 2 or second version), it is possible to account for withdrawals.

My aim here has been to sketch the rationalization I have offered for a decision theoretic approach to how to contract that completes the account of coerced contraction and deliberate contraction outlined in previous sections. Given that account, the corrigibility of the inquirer's point of view, from the inquirer's point of view, is justifiable even though the inquirer to be coherent is committed to ruling out the serious possibility that his or her current point of view is in error.

Curiosity and Corrigibilism

As I have done throughout most of my career, I have been maintaining that the belief states that are targets for justifiable change are states of full belief. Such states serve as standards for serious possibility. From the point of view of the inquirer who makes the judgment of serious possibility, there is no serious possibility that what the inquirer fully beliefs is false. The inquirer is committed to epistemological infallibilism.

The main philosophical tradition maintains that such a view is untenable because it implies epistemological incorrigibilism. I deny this. The inquiring agent is sometimes warranted in contracting his or her state of full belief.

My argument for this view is based on the legitimacy of both routine expansion and deliberate or inductive expansion.

Routine expansion is potentially conflict injecting. If we are to acknowledge the legitimacy of such expansion whether by appealing to the testimony of the senses or of competent witnesses and experts, programs for routine expansion must provide contingency plans for contraction in case inconsistency inadvertently arises.

Deliberate contraction to give informationally valuable propositions a hearing can be rationalized only on the assumption that subsequent to such contraction, expansion will be legitimate that affords a promise of removing the doubts that are raised. Sometimes routine expansion may be all that is required. But when the conjectures to be given a hearing are highly theoretical (as is the case with General Relativity Theory), deliberate or inductive expansion may be needed.

Expansion, so I claim, is legitimate as long as the quest for valuable information may be seen to compensate for the risk of importing error incurred in expansion. If we are to avoid the dogma that equates infallilibilism with incorrigibilism, we need to follow William James in rejecting W.K. Clifford's emphasis on the avoidance of error as the sole desideratum of inquiry.

References

Alchourrón, C., Gärdenfors, P., & Makinson, D. (1985). On the logic of theory change: Partial meet functions for contraction and revision. *Journal for Symbolic Logic, 50*, 510–530.

Carnap, R. (1960). The aim of inductive logic. In E. Nagel, P. Suppes, & A. Tarski (Eds.), *Logic, methodology and philosophy of science* (pp. 302–318). Stanford: Stanford University Press.

Cohen, L. J. (1970). *The implications of induction*. London: Methuen.

Cohen, L. J. (1977). *The probable and the provable*. Oxford: Oxford University Press.

Davidson, D. (1998). Truth rehabilitated, Unpublished Manuscript.

De Finetti, B. (1964). Foresight: Its logical laws, its subjective sources. In H. E. Kyburg & H. Smokler (Eds.), *Studies in subjective probability* (pp. 93–158). New York: Wiley.

Hansson, S. O., & Olsson, E. J. (1995). Levi contractions and AGM contractions: a comparison. *Notre Dame Journal of Formal Logic, 36*, 103–119.

Jeffrey, R. C. (1965). *The logic of decision*. New York: McGraw Hill.

Levi, I. (1967a). *Gambling with truth*. Cambridge: MIT Press, Paper, 1973.

Levi, I. (1967b). Information and inference. *Synthese, 17*, 369–391.

Levi, I. (1974). On indeterminate probabilities. *The Journal of Philosophy, 71*, 391–418.

Levi, I. (1980). *The enterprise of knowledge*. Cambridge: MIT Press, Paper, 1983.

Levi, I. (1983). Truth, fallibility and the growth of knowledge. In R. S. Cohen & M. W. Wartofsky (Eds.), *Language, logic and method* (pp. 153–174). Dordrecht: Reidel.

Levi, I. (1986). *Hard choices: Decision making under unresolved conflict*. New York: Cambridge University Press.

Levi, I. (1991). *The fixation of belief and its undoing*. Cambridge: Cambridge University Press, Paper, 2009.

Levi, I. (1997). *For the sake of the argument*. Cambridge: Cambridge University Press, Paper, 2007.

Levi, I. (2002). Maximizing and satisficing measures of evidential support. In M. David (Ed.), *Reading natural philosophy: Essays in the history and philosophy of science and mathematics* (pp. 315–333). Chicago: Open Court.

Levi, I. (2003). Contraction from epistemic hell is routine. *Synthese, 135*, 141–164.

Levi, I. (2004). *Mild contraction*. Oxford: Oxford University Press.

Levi, I. (2006). Informational value should be relevant and damped! Reply to Rott (2006).

Niiniluoto, I. (1984). *Is science progressive?* Dordrecht: D. Reidel.

Olsson, E. J. (2003). Avoiding epistemic hell, Levi on testimony and consistency. *Synthese, 135*, 119–140.

Pagnucco, M., & Rott, I. (1999). Severe withdrawal and recovery. *Journal of Philosophical Logic, 28*, 501–547. See 'Erratum' *Journal of Philosophical Logic* 29 (2000).

Ramsey, F. P., (1990). *Philosophial papers* (D. H. Mellor, Ed.). Cambridge: MIT Press.

Rott, H. (1993). Belief contraction in the context of the general theory of rational choice. *Journal of Symbolic Logic, 58*, 1426–1450.

Rott, H. (2001). *Change, choice and inference: A study of belief revision and nonmonotonic reasoning*. Oxford: Oxford University Press.

Rott, H. (2006). The value of truth and the value of information: On Isaac Levi's epistemology. In E. Olsson (Ed.), *Knowledge and inquiry* (pp. 179–200). Cambridge: Cambridge University Press.

Savage, L. J. (1954). *The foundations of statistics*. New York: Wiley.

Shackle, G. L. S. (1949, 1952). *Expectation in economics*, Cambridge: Cambridge University Press.

Shackle, G. L. S. (1961, 1969). *Decision, order and time*. Cambridge: Cambridge University Press.

Spohn, W. (1988). Ordinal conditional functions: A dynamic theory of epistemic states. In W. Harper & B. Skyrms (Eds.), *Causation in decision, belief change and statistics* (pp. 105–134). Dordrecht: Kluwer Academic Publishers.

Chapter 16
Belief Contraction in the Context of the General Theory of Rational Choice

Hans Rott

Introduction

The theory of partial meet contraction and revision was developed by Alchourrón, Gärdenfors and Makinson (henceforth, "AGM") in a paper published in the *Journal of Symbolic Logic* in 1985. That paper is the by now classic reference of the flourishing research programme of *theory change*, or as it is alternatively called, *belief revision* (Fuhrmann and Morreau 1991; Gärdenfors 1988, 1992; Katsuno and Mendelzon 1991). In particular, it has been shown that partial meet contraction is a powerful tool for the reconstruction of other kinds of theory change such as safe contraction, epistemic entrenchment contraction, and base contraction (Alchourrón and Makinson 1986; Nebel 1989, 1992; Rott 1991, 1992a).

The basic idea of partial meet contraction is as follows. In order to eliminate a proposition x from a theory A while obeying the constraint of deductive closure and minimizing the loss of information, it is plausible to look at the maximal subsets B of A that fail to imply x. In an earlier paper, Alchourrón and Makinson had proved that when $A = Cn(A)$ then taking one such B leaves an agent with too many propositions, while taking the intersection of all such B's leaves him with too few. In AGM (1985), AGM investigate the idea of taking the intersection of *a select set* of such B's. The choice which B's to take is made with the help of a *selection function*. A natural question is whether all these selections can be represented as the selections of *preferred* B's, where the preferences between maximally nonimplying subsets of A are independent of the proposition x to be deleted.

The purpose of the present paper is threefold. First, we put the theory of partial meet contraction in a broader perspective. We decompose it into two layers, each

H. Rott (✉)
Department of Philosophy, University of Regensburg, 93040 Regensburg, Germany
e-mail: hans.rott@ur.de

H. Arló-Costa et al. (eds.), *Readings in Formal Epistemology*, Springer Graduate Texts in Philosophy 1, DOI 10.1007/978-3-319-20451-2_16

of which can be cultivated with the help of methods developed in different research areas. On the one hand, in relating selection functions to preference relations we can draw on existing work in social choice theory (Chernoff 1954; Herzberger 1973; Sen 1982; Suzumura 1983). On the other hand, we shall elaborate on a remark of Grove (1988) and link maximally nonimplying subsets to "possible worlds" or models, thereby making it possible to compare partial meet contraction with semantical approaches to belief revision and nonmonotonic reasoning (Kraus et al. 1990; Katsuno and Mendelzon 1991; Lindström 1991; Lehmann and Magidor 1992). Exaggerating somewhat, we can say that the theory of partial meet contraction emerges from juxtaposing the general theory of rational choice and a tiny bit of model theory. After introducing abstract postulates for contraction functions, we reprove the two main representation theorems of AGM (1985, pp. 521, 530) concerning partial meet contraction and transitively relational partial meet contraction in a slightly more systematic fashion.

Second, we provide a partial solution to a problem left unanswered by AGM and still considered to be an interesting open question in Makinson and Gärdenfors (1991, p. 195). More precisely, we present two new results that lie strictly "between" those of AGM (1985), viz., representation theorems for *relational* and *negatively transitively relational* partial meet contraction. However, these results hold only under certain preconditions. If the theory to be contracted is logically finite, then all these conditions are met. Our decomposition allows for a uniform method of proof using so-called Samuelson preferences. It increases our understanding of the partial meet contraction mechanism by localizing exactly in which parts of the proofs the finiteness assumption is effective.

Third, as an application, we explore the logic of a variant of syntax-based belief change, namely *simple* and *prioritized base contractions* in the finite case. The basic idea here is that real life theories are generated from finite axiomatizations and that the axioms may carry different epistemic weight. Both the syntactical encoding of a theory and the prioritization are held to be relevant for theory change. We achieve a logical characterization of simple and prioritized base contraction by combining one of our representation theorems with observations originating with Lewis (1981) and Nebel (1989, 1992).

Independently from the research for this article, related work has been carried out by Katsuno and Mendelzon (1991) and Lindström (1991). Lindström proves numerous representation theorems, and like us, he heavily draws on results from the general theory of rational choice. However, there are some important differences. He adopts a semantic approach right from the outset, while our reconstruction starts by turning a syntactic approach into an essentially semantic one, with the notable feature that any two "worlds" in our sense are linguistically distinguishable (our models are "injective", to use a term of Michael Freund). Lindström's main concern is nonmonotonic inference relations and the related area of belief *revision*, whereas we shall focus on belief *contraction*. (Everything we might wish to say about the intimate connection between revisions and contractions is contained in AGM 1985 and Gärdenfors 1988). He applies *liberal* maximization where we apply *stringent* maximization. He uses different postulates for choice functions and a different notion of revealed preference. The main contrast between his approach and ours, however, is that Lindström permits revisions by—possibly infinite—*sets* of propositions whilst we stick to the original AGM model of changing theories by *single* propositions. Lindström's generalization allows him to revise by "worlds" and thus dispense with a finiteness assumption which will prove to be necessary at several places in the AGM framework. In fact, it is a major aim of our paper to make transparent the reasons where and why logical finiteness is needed in the AGM theory. Further important results on purely finitistic belief revision are found in Katsuno and Mendelzon (1991) who incorporate a revision operator into their object language. Both papers are highly recommended and should be read in conjunction with the present one.

Unfortunately, space limitations do not allow a presentation that makes for easy reading. Familiarity with earlier work either in the area of theory change—an excellent overview is Gärdenfors (1988)—or in the general theory of rational choice—of particular relevance is Herzberger (1973)— will greatly facilitate the reader's task. However, we will shortly repeat the basic definitions so that our presentation is in principle self-contained. It may be useful to inform the reader in advance about the fundamental entities that will show up as determinants of theory change. We shall meet contraction functions $\dot{-}$, selection functions γ, preference relations $\leq$ (and $\preccurlyeq$), maximally nonimplying sets M (and N), maximally consistent sets or "worlds" W, as well as simple and prioritized belief bases B and $\langle B_i \rangle$. We shall frequently jump to and fro between these kinds of entities. When we wish to generate a *c*ontraction function, *s*election function, *p*reference relation, *m*aximal nonimplying set, *w*orld, or belief *b*ase from another kind of entity, we shall use metalevel mappings denoted by $\mathscr{C}, \mathscr{S}, \mathscr{P}, \mathscr{M}, \mathscr{W}, \mathscr{B}$ respectively.

Selection Functions and Preference Relations

General Selection Functions

Let X be a set and $\mathscr{X}$ be a nonempty subset of $2^X - \{\emptyset\}$. A *selection function* (or *choice function*) over $\mathscr{X}$ is a function $\gamma : \mathscr{X} \to 2^X$ such that $\gamma(S)$ is a nonempty subset of S, for every $S \in \mathscr{X}$. Intuitively, selection functions are supposed to give us the best elements of each S in $\mathscr{X}$. The requirement that $\gamma(S)$ be nonempty means that a selection function is effective or "decisive": In every case it reaches a decision of which elements are best. As Lindström (1991) points out, the decisiveness of selection functions corresponds to a condition of consistency preservation in belief revision and nonmonotonic reasoning. Here consistency is judged by *some* monotonic background consequence relation, not necessarily by the classical one.

Domain conditions for selection functions are important. A set $\mathscr{X}$ of subsets of X is called *n-covering* ($n = 1, 2, 3, \ldots$) if it contains all subsets of X with exactly n elements, $\mathscr{X}$ is called *n_1n_2-covering* (or *$n_1n_2n_3$-covering*) if it is n_1-covering and

n_2-covering (and n_3-covering). $\mathscr{X}$ is called *ω-covering* if it is n-covering for all natural numbers $n = 1, 2, 3, \ldots$. The set $\mathscr{X}$ is called *additive* if it is closed under arbitrary unions, and it is called *finitely additive* if it is closed under finite unions; it is called *subtractive* if for every S and S' in $\mathscr{X}$ such that $S \not\subseteq S'$, $S - S'$ is also in $\mathscr{X}$, and it is called *full* if $\mathscr{X} = 2^X - \{\emptyset\}$. Of course, if $\mathscr{X}$ is 1-covering and finitely additive, then it is ω-covering. Finally, we say that $\mathscr{X}$ is *compact* if for every T and S_i, $i \in I$, from $\mathscr{X}$, if $T \subseteq \bigcup\{S_i : i \in I\}$ then $T \subseteq \bigcup\{S_i : i \in I_0\}$ for some finite $I_0 \subseteq I$. We say that a *selection function* γ with domain $\mathscr{X}$ is n-covering, additive, subtractive, etc., if its domain $\mathscr{X}$ is n-covering, additive, subtractive, etc.

Constraints on Selection Functions

We consider the following "coherence postulates" for selection functions. Their motivation, where it is not obvious, is given in Herzberger (1973), Sen (1982), and Suzumura (1983).

(I) For all $S, S' \in \mathscr{X}$, if $S \subseteq S'$ then $S \cap \gamma(S') \subseteq \gamma(S)$ ("Chernoff's axiom", Sen's Property α)

(II) For all $\{S_i : i \in I\} \subseteq \mathscr{X}$ such that $\bigcup\{S_i : i \in I\} \in \mathscr{X}$, $\bigcap\{\gamma(S_i) : i \in I\} \subseteq \gamma(\bigcup\{S_i : i \in I\})$ (Sen's Property γ)

(III) For all $S, S' \in \mathscr{X}$ such that $S \subseteq S'$, if $\gamma(S') \subseteq \gamma(S)$ then $\gamma(S) \subseteq \gamma(S')$ ("Superset Axiom", Sen's Property ε)

(IV) For all $S, S' \in \mathscr{X}$ such that $S \subseteq S'$, if $\gamma(S') \cap S \neq \emptyset$, then $\gamma(S) \subseteq \gamma(S')$ ("Arrow's axiom", Sen's Property $\beta+$)

Condition (III) is independent of condition (II), even for finite X and in the presence of condition (I). Example: Let $X = \{x, y, z\}$, $\mathscr{X}$ be the power set of X minus $\emptyset$, and consider γ_1 and γ_2 defined by $\gamma_1(\{x, y\}) = \gamma_2(\{x, y\}) = \{x, y\}$, $\gamma_1(\{x, z\}) = \{x\}$, $\gamma_1(\{y, z\}) = \{z\}$, $\gamma_1(\{x, y, z\}) = \{x\}$, $\gamma_2(\{x, z\}) = \{x, z\}$, $\gamma_2(\{y, z\}) = \{y, z\}$, and $\gamma_2(\{x, y, z\}) = \{x, y\}$. Evidently, γ_1 satisfies (I) and (II) and violates (III), while γ_2 satisfies (I) and (III) and violates (II). On the other hand, condition (IV) implies condition (II): Let $\{S_i : i \in I\} \subseteq \mathscr{X}$, $\bigcup\{S_i : i \in I\} \in \mathscr{X}$. Now $\emptyset \neq \gamma(\bigcup\{S_i : i \in I\}) \subseteq \bigcup\{S_i : i \in I\}$, so $\gamma(\bigcup\{S_i : i \in I\}) \cap S_{i_0} \neq \emptyset$ for some $i_0 \in I$. So by (IV), $\bigcap\{\gamma(S_i) : i \in I\} \subseteq \gamma(S_{i_0}) \subseteq \gamma(\bigcup\{S_i : i \in I\})$. It is trivial to see that condition (IV) also implies condition (III), since $\emptyset \neq \gamma(S') \subseteq \gamma(S)$ implies $\gamma(S') \cap S \neq \emptyset$.

Rational Choice

In the general theory of choice and preference we often find an idea which can be phrased in the slogan "*Rational choice is relational choice*". That is, rational choice is choice which can be construed as based on an underlying preference relation.

> The intended interpretation of the set $\gamma(S)$, called the *choice set* for S, is that its elements are regarded as equally adequate or satisfactory choices for an agent whose values are represented by the function γ, and who faces a decision problem represented by the set S. Following Chernoff (1954) ..., this relativistic concept of equiadequacy for a given decision problem bears sharp distinction from invariant concepts like preferential matching or indifference which for a given agent are not relativized to decision problems, and which may be subject to more stringent constraints, both for rational agents and for agents in general. (Herzberger 1973, p. 189, notation adapted)

Choice sets are taken to be sets of "best" elements. There are basically two ideas to make this precise. The first is based on a nonstrict (reflexive) preference relation $\leq$:

$$\gamma(S) = \{y \in S : y' \leq y \text{ for all } y' \in S\}$$

The second idea is based on a strict (asymmetric) preference relation $<$:

$$\gamma(S) = \{y \in S : y \not< y' \text{ for all } y' \in S\}$$

These suggestions are respectively referred to as *stringent* and *liberal maximization* in Herzberger (1973, p. 197), and *G-rationality* and *M-rationality* in Suzumura (1983, p. 21). (Stringent maximization is often attributed to Condorcet 1785.) If $<$ is the converse complement of $\leq$, then stringent and liberal maximization coincide. If $<$ is the asymmetric part of $\leq$, then every liberal maximizer with respect to $<$ is a stringent maximizer with respect to the *augmentation* $\leq^+$ of $\leq$ which is defined by $x \leq^+ y$ iff $x \leq y$ or $y \not\leq x$, i.e., iff not $y < x$. Clearly, $\leq^+$ is *connected*, i.e., it holds for every x and y that either $x \leq^+ y$ or $y \leq^+ x$. While $\leq$ allows us to keep apart indifferences (both $x \leq y$ and $y \leq x$) from incomparabilities (neither $x \leq y$ nor $y \leq x$), $\leq^+$ blurs just this distinction, for whenever we have neither $x \leq y$ nor $y \leq x$ we have both $x \leq^+ y$ and $y \leq^+ x$. Thus, we may not expect $\leq^+$ to be transitive. If, on the other hand, $\leq$ is already connected then $\leq^+ = \leq$, and stringent maximization with respect to the asymmetric part $<$ of $\leq$ coincides with liberal maximization with respect to $\leq$. For all this, cf. Herzberger (1973, § 3).

In accordance with AGM (1985) as well as with the dominant approach in the theory of choice and preference, we shall focus on stringent maximization.[1] From now on, when we say that γ is *relational with respect to* $\leq$ *over* X, we mean that $\gamma(S) = \{y \in S : y' \leq y \text{ for all } y' \in S\}$ for every $S \in \mathcal{X}$. In this case we write

[1]Intuitively, however, I think that liberal maximization is preferable. Liberal maximization is based on strict relations which do not allow to distinguish between incomparabilities and indifferences. Nonstrict relations do make this distinction, but stringent maximization tends to require connected relations which often can be had only if incomparabilities are turned into indifferences—i.e., if augmentations are used. The interpretation of nonstrict relations as the converse complements of—more intuitive—strict relations explains the crucial role of *negative* transitivity and *negative* well-foundedness in the following. Also compare the recommendation in Rott (1992b) to regard the nonstrict epistemic entrenchment relation $\leq_E$ of Gärdenfors and Makinson (1988) as the converse complement of a more intuitive strict relation $<_E$.

$\gamma = \mathscr{S}(\leq)$. When we say that γ is *relational*, we mean that there is some relation $\leq$ over X such that $\gamma = \mathscr{S}(\leq)$.

For any nonstrict relation $\leq$, $<$ is to denote the *asymmetric part* of $\leq$, which is defined by $x < y$ iff $x \leq y$ and not $y \leq x$. $\leq$ is called *n-acyclic*, if no n objects $x_1, x_2, \ldots, x_n$ form a cycle under $<$, i.e., if a chain $x_1 < x_2 < \cdots < x_n < x_1$ does not occur. 1-acyclicity is irreflexivity, 2-acyclicity is asymmetry. $\leq$ is called *acyclic*, if it is n-acyclic for every $n = 1, 2, 3, \ldots$. $\leq$ is called *negatively acyclic* if there is no cycle under $\not\leq$, i.e., if never $x_1 \not\leq x_2 \not\leq \cdots \not\leq x_n \not\leq x_1$, and it is called *negatively well-founded* if there is no infinite descending chain under $\not\leq$, i.e., if never $\cdots \not\leq x_3 \not\leq x_2 \not\leq x_1$. For connected relations $\leq$, negative well-foundedness coincides with converse well-foundedness where infinite ascending chains $x_1 < x_2 < x_3 < \ldots$ do not occur. Obviously, if $\leq$ is conversely (or negatively) well-founded then it is acyclic (negatively acyclic). $\leq$ is *smooth* with respect to $\mathscr{X}$ if there are no infinite descending $\not\leq$-chains in S, for every $S \in \mathscr{X}$. Smoothness is a restricted form of negative well-foundedness. $\leq$ is called *negatively transitive* (*virtually connected, modular, ranked*) if $x \not\leq y$ and $y \not\leq z$ implies $x \not\leq z$. It is quickly verified that a connected relation $\leq$ is negatively transitive iff it is *quasi-transitive* in the sense that its asymmetric part $<$ is transitive. Quasi-transitivity is a much disputed requirement in social choice theory (Herzberger 1973; Sen 1982; Suzumura 1983). It should be noted that all transitive relations $\leq$ are both acyclic and quasi-transitive.

Two Kinds of Revealed Preferences

In many contexts one can hope to recover the underlying preferences of an agent from observed choice behavior. The two most commonly used types of "revealed preference" relations are the *Samuelson preferences* (Samuelson (1950))

$$\leq_\gamma = \{\langle x, x'\rangle \in X \times X : \exists S \in \mathscr{X} \text{ such that } \{x, x'\} \subseteq S \quad \text{and } x' \in \gamma(S)\} \tag{16.1}$$

$$= \bigcup \{S \times \gamma(S) : S \in \mathscr{X}\} \tag{16.2}$$

and the so-called *base preferences* (Uzawa (1956); Arrow (1959)):

$$\leq_{\gamma,2} = \{\langle x, x'\rangle \in X \times X : \{x, x'\} \in \mathscr{X} \text{ and } x' \in \gamma(\{x, x'\})\} \tag{16.3}$$

$$= \bigcup \{S \times \gamma(S) : S \in \mathscr{X} \text{ and } S \text{ has at most two elements}\} \tag{16.4}$$

The terminology is taken from Herzberger (1973), where many other possibilities of defining notions of revealed preference are discussed. Obviously, $x \leq_{\gamma,2} y$ implies $x \leq_\gamma y$. Notice that neither of these relations is guaranteed to be reflexive, unless γ is 1-covering. $\leq_{\gamma,2}$ is defined for arbitrary γ's, but the definition makes good sense only for 2-covering ones for which $\{x, x'\}$ is always in $\mathscr{X}$. In this case, base preferences and Samuelson preferences are connected. Given a selection function γ, we shall also denote $\leq_\gamma$ by $\mathscr{P}(\gamma)$ and $\leq_{\gamma,2}$ by $\mathscr{P}_2(\gamma)$.[2]

Some Basic Properties and Reformulations

The following lemmas list a number of important facts which are basically common knowledge in the general theory of rational choice (cf. Herzberger 1973; Sen 1982; Suzumura 1983). For the straightforward proofs, see Appendix 2.

Lemma 1. (a) *If γ is relational then it satisfies (I) and (II).*
(b) *If γ is 12-covering and satisfies (I) and (II), then $\gamma = \mathscr{S}(\mathscr{P}_2(\gamma))$.*
(c) *If γ is 12-covering and satisfies (I), then $\mathscr{P}(\gamma) = \mathscr{P}_2(\gamma)$.*
(d) *If γ is 12-covering and satisfies (I) and (II), then $\gamma = \mathscr{S}(\mathscr{P}(\gamma))$.*
(e) *If γ is 12-covering and relational, then $\gamma = \mathscr{S}(\mathscr{P}(\gamma)) = \mathscr{S}(\mathscr{P}_2(\gamma))$.*
(f) *If γ is additive and satisfies (I) and (II), then $\gamma = \mathscr{S}(\mathscr{P}(\gamma))$.*

Lemma 2. (a) *If γ is 12n-covering and satisfies (I), then $\mathscr{P}(\gamma) = \mathscr{P}_2(\gamma)$ is n-acyclic. If γ is ω-covering and satisfies (I), then $\mathscr{P}(\gamma)$ is acyclic.*
(b) *If γ is 123-covering and satisfies (I) and (III), then $\mathscr{P}(\gamma)$ is negatively transitive.*
(c) *If γ is finitely additive and satisfies (IV), then $\mathscr{P}(\gamma)$ is transitive.*

Lemma 3. (a) *If $\leq$ is smooth with respect to $\mathscr{X}$, then $\mathscr{S}(\leq)$ is a selection function over $\mathscr{X}$ which satisfies (I) and (II).*
(b) *If $\leq$ is negatively transitive and negatively well-founded (or: if $\leq$ is negatively transitive and smooth and $\mathscr{X}$ is subtractive), then $\mathscr{S}(\leq)$ satisfies (III).*
(c) *If $\leq$ is transitive, then $\mathscr{S}(\leq)$ satisfies (IV).*

Now we bring conditions (I)–(IV) into a form which is more suitable for our purposes.

(I′) For all $S, S' \in \mathscr{X}$ such that $S \cup S' \in \mathscr{X}$, $\gamma(S \cup S') \subseteq \gamma(S) \cup \gamma(S')$
(II′) For all $S, S' \in \mathscr{X}$ such that $S \cup S' \in \mathscr{X}$, $\gamma(S) \cap \gamma(S') \subseteq \gamma(S \cup S')$
(III′) For all $S \in \mathscr{X}$ and S' such that $S \cup S' \in \mathscr{X}$, if $\gamma(S \cup S') \cap S' = \emptyset$ then $\gamma(S) \subseteq \gamma(S \cup S')$
(IV′) For all $S \in \mathscr{X}$ and S' such that $S \cup S' \in \mathscr{X}$, if $\gamma(S \cup S') \cap S \neq \emptyset$, then $\gamma(S) \subseteq \gamma(S \cup S')$

---

[2]It is worth pointing out that the characteristic definition of a relation of epistemic entrenchment (see Gärdenfors and Makinson 1988; Rott 1992b) between propositions from an observed contraction behavior, viz.

$$x \leq_E y \textit{ iff } x \notin A\dot{-}(x \wedge y) \text{ or } y \in A\dot{-}(x \wedge y)$$

can also be interpreted as a base preference (Rott 1992b, p. 61). In that paper it is argued that the instruction "remove $x \wedge y$" should be regarded as an instruction to remove x *or* remove y, where the agent holding theory A has free choice which proposition(s) out of $\{x, y\}$ to remove. [Note added in 2015: This sketch of an idea was turned into a theory in Rott (2003).]

(I&II) For all $S, S_i \in \mathscr{X}, i \in I$, if $S \subseteq \bigcup\{S_i : i \in I\}$ then $S \cap \bigcap\{\gamma(S_i) : i \in I\} \subseteq \gamma(S)$
(I&II′) For all $S, S', S'' \in \mathscr{X}$, if $S \subseteq S' \cup S''$ then $S \cap \gamma(S') \cap \gamma(S'') \subseteq \gamma(S)$

Note that condition (II′) is just a restriction of (II) to index sets with at most two elements, and similarly for (I&II) and (I&II′).

Lemma 4. (a) *If γ is subtractive, then condition (I) is equivalent to (I′).*
(b) *If γ is finitely additive, compact, and satisfies (I) and (II′), then $\gamma = \mathscr{S}(\mathscr{P}(\gamma))$.*
(c) *If γ satisfies (I), then (III) is equivalent to (III′).*
(d) *(IV) is equivalent to (IV′).*
(e) *If γ is additive, then the conjunction of (I) and (II) is equivalent to (I&II).*
(f) *If γ is finitely additive, then the conjunction of (I) and (II′) is equivalent to (I&II′).*

Lemma 4(b) is a substitute for Lemma 1(f) in the absence of infinite additivity. The conditions (I&II) and (I&II′) are generalized forms of the conditions (γ7:∞) and (γ7:2) in AGM (1985). AGM (1985, Observation 4.10) noticed that (I&II′) is the key to relationality in partial meet contraction (because the domain $\mathscr{X}$ encountered there is compact), but as they failed to decompose that condition into the simpler ones, (I) and (II′), it remained undigestible for them. As we shall see, they provided a postulate for contraction functions which corresponds to (I), but nothing in AGM (1985) corresponds to (II′).

The results of the theory of rational choice are quite nice, and we may be optimistic about their applicability to theory change as determined by partial meet contraction s. Still there are at least two problems which should not be underestimated. First, we do not know very much about selection functions which are neither 123-covering nor additive.[3] And second, in choice-and-preference theory we are faced with a bewildering pluralism of revealed preference relations (see Herzberger 1973). It is hard to decide in advance which notion of revealed preference is "the" right notion for a given purpose. It will turn out, happily, that the relation $\mathscr{P}(\gamma)$ defined above, and, to a smaller degree, also the relation $\mathscr{P}_2(\gamma)$ which is equivalent in the finite case are suitable for partial meet contraction s.

[3]This is no problem for Lindström (1991) whose selection functions are always ω-covering. Consequently, Lindström's constructions can always make use of the base preferences $\mathscr{P}_2(\gamma)$.

Selection Functions and Preference Relations in Partial Meet Contraction

The General Case

We presuppose a propositional language L, with the usual connectives $\neg, \wedge, \vee, \rightarrow$ and $\leftrightarrow$. From now on, small roman letters $a, b, c, \ldots, x, y, z, \ldots$ denote propositions of L, and capital roman letters $A, B, A', B', \ldots, M, N, M', N' \ldots, W, W', \ldots$ denote sets of propositions. Notational conventions: Capital A is to denote a theory, B an arbitrary set of propositions (in general, a nontheory), M and N maximally nonimplying subsets (of some A or B, respectively), and W a maximally consistent set of L. All these metavariables may occur with primes or subscripts.

As in AGM (1985), L is assumed to be governed by some reflexive, monotonic and idempotent logic (consequence operation) Cn which is supraclassical and compact and satisfies the deduction theorem: $B \subseteq Cn(B)$; if $B \subseteq B'$ then $Cn(B) \subseteq Cn(B')$; $Cn(Cn(B)) = Cn(B)$; if Cn_0 is classical tautological implication then $Cn_0(B) \subseteq Cn(B)$; if $y \in Cn(B)$ then $y \in Cn(B_0)$ for some finite subset B_0 of B; and finally, $y \in Cn(B \cup \{x\})$ iff $x \rightarrow y \in Cn(B)$. That is, Cn is a classical logic in the sense of Lindström (1991). Note that in the present context, idempotence is equivalent to the cut rule, and the deduction theorem is equivalent to disjunction in the antecedents.

We also write $B \vdash x$ for $x \in Cn(B)$. A is a called a *theory* if $A = Cn(A)$. By $[\![x]\!]$ and $[\![B]\!]$ we denote the set of all "possible worlds"(valuations, models) in which x (every x in B) is true. Here and throughout this paper, we identify a "world" satisfying x or B with a maximally Cn-consistent set of propositions which contains x (respectively, B). The set of all "worlds" is denoted $\mathbf{W}$.

Let $B \bot x = \{M \subseteq B : x \notin Cn(M)$ and $x \in Cn(N)$ for all N with $M \subset N \subseteq B\}$. Note that the elements of $B \bot x$ are theories if B is a theory. If $\not\vdash x$, then $B \bot x$ is nonempty, by the compactness of Cn.[4] We define $B\bot = \{B \bot x : x \in Cn(B) - Cn(\emptyset)\}$ and $U_B = \bigcup(B\bot) = \bigcup\{B \bot x : x \in Cn(B) - Cn(\emptyset)\}$. (Here we slightly deviate from the definitions in AGM (1985) which also include x's from $Cn(\emptyset)$.) If $Cn(B) \neq Cn(\emptyset)$, then $B\bot$ is a nonempty subset of $2^{2^B} - \{\emptyset\}$. The case $Cn(B) = Cn(\emptyset)$ will be handled separately in our reconstruction of partial meet contraction.

Lemma 5. *Let A be a theory and $x \in A$. Then $M \in A \bot x$ iff there is a maximally consistent set ("world") W such that $\neg x \in W$ and $M = A \cap W$.*

For the proof, compare § 4 of Grove (1988).

---

[4]This marks a difference with Lewis (1981) who identifies propositions with sets of extra-linguistic possible worlds and logical consequence with set-theoretic inclusion. Lacking compactness, Lewis has to ponder the impact of a "Limit Assumption" for premise semantics.

Now let A again be a theory. For $M \in U_A$, put $\mathscr{W}(M) = Cn(M \cup \{\neg x\})$ where x is any proposition in $A - M$, and put $\mathscr{M}(W) = W \cap A$, for any maximally consistent set W such that $A \not\subseteq W$. The reader is invited to verify: $\mathscr{W}$ is well-defined(!), $\mathscr{W}$ is a bijection from U_A to $V_A = \{W \in \mathbf{W} : A \not\subseteq W\}$, and $\mathscr{M}$ is the converse of $\mathscr{W}$, i.e., $\mathscr{M}(\mathscr{W}(M)) = M$ and $\mathscr{W}(\mathscr{M}(W)) = W$.

Given this representation of the elements of U_A, it is clear that they satisfy the following *fullness* and *primeness conditions*: If $x \in A - M$ and $y \in A$ then $x \to y \in M$, and if $x, y \in A - M$ then $x \vee y \in A - M$ (cf. Observations 6.1 and 6.2 in AGM 1985). U_A is just the set of all maximal proper subtheories of the theory A. Moreover, we immediately get

Corollary 1. *Let A be a theory and x, y, y_i be in $A - Cn(\emptyset)$.*

(i) *For $M \in U_A$, $x \notin M$ iff $M \in A \bot x$.*
(ii) $A\bot(x \wedge y) = A\bot x \cup A\bot y$.
(iii) $A\bot(x \vee y) = A\bot x \cap A\bot y$.
(iv) $A\bot(x \vee \neg y) = A\bot x - A\bot y$.
(v) *If $A\bot x \subseteq \bigcup\{A\bot y_i : i \in I\}$, then $A\bot x \subseteq \bigcup\{A\bot y_i : i \in I_0\}$ for some finite $I_0 \subseteq I$.*

Facts (i) and (ii) are contained in AGM (1985, Lemma 2.4 and Lemma 4.1). We see that $A\bot$, the special domain $\mathscr{X}$ of the selection functions which will figure in partial meet contraction s, is closed under finite unions, finite intersections, and differences. We give a direct proof of the compactness property (v). Let $A\bot x \subseteq \bigcup\{A\bot y_i : i \in I\}$. Then $\{y_i : i \in I\} \vdash x$. For otherwise, by Lindenbaum's Lemma and Lemma 5, there would be an $M \in A\bot x$ such that $\{y_i : i \in I\} \subseteq M$, so $M \notin A\bot y_i$ for every i, contradicting our hypothesis. Compactness of Cn gives us $\{y_i : i \in I_0\} \vdash x$ for some finite $I_0 \subseteq I$. Thus, there is, for every $M \in A\bot x$, an $i \in I_0$ such that $y_i \notin M$. Hence, by (i), there is, for every $M \in A\bot x$, an $i \in I_0$ such that $M \in A\bot y_i$. Thus $A\bot x \subseteq \bigcup\{A\bot y_i : i \in I_0\}$, as desired.

Now we can introduce selection functions for belief revision. Let A be a theory. A selection function $\gamma : A\bot \to 2^{2^A}$ is called a *selection function for* A. It follows from Corollary 1 that every selection function γ for A is finitely additive, subtractive and compact.

Let γ be a selection function for A. The *completion* γ^* of γ is defined by $\gamma^*(A\bot x) = \{M \in A\bot x : \bigcap \gamma(A\bot x) \subseteq M\}$, for all $x \in A - Cn(\emptyset)$. Following AGM (1985), we call a selection function γ *complete* if $\gamma = \gamma^*$. If γ is complete, then $M \in \gamma(A\bot x)$ whenever $\bigcap \gamma(A\bot x) \subseteq M \in U_A$ (Proof: Since $\{x \to y : y \in A\} \in \bigcap \gamma(A\bot x)$, $\bigcap \gamma(A\bot x) \subseteq M$ implies $M \nvdash x$, so $M \in A\bot x$ by Corollary 1(i).)

A *contraction function* over a theory A is a function $\dot{-}_A : L \to 2^L$. We write $A \dot{-} x$ for $\dot{-}_A(x)$, and as there will be no danger of confusion, we shall often write just $\dot{-}$ for $\dot{-}_A$. The term '$A \dot{-} x$' should be read as the 'the result of rationally removing x from A'. The idea of a contraction function dictates that it should satisfy at least the postulates $A \dot{-} x \subseteq A$ and $x \notin A \dot{-} x$ (unless $\vdash x$). More postulates will be discussed in section "Postulates for Contraction Functions". Intuitively, $A \dot{-} x$ is the minimal, most economical or rational, change of A needed to discard x.

A function $\dot{-}$ is the *partial meet contraction function over A determined by the selection function γ for A* if and only if

$$A\dot{-}x = \begin{cases} \bigcap \gamma(A\bot x) & \text{for every } x \in A - Cn(\emptyset) \\ A & \text{for every } x \notin A \text{ and every } x \in Cn(\emptyset) \end{cases}$$

Given a selection function γ for A, the partial meet contraction it determines will be denoted by $\mathscr{C}(\gamma)$. There is a slight deviation from AGM (1985) in order to avoid the application of γ to $\emptyset$ and to preserve the perfect correspondence between U_A and V_A. The function $\dot{-}$ is called a *partial meet contraction over A* if there is a selection function γ for A such that $\dot{-} = \mathscr{C}(\gamma)$.

A selection function γ for A is called *(transitively, quasi-transitively, connectively, acyclicly) relational over A* if there is a (transitive, quasi-transitive, connected, acyclic) preference relation $\leq$ over U_A (over 2^A in AGM 1985) such that for every $x \in A - Cn(\emptyset)$:

$$\gamma(A\bot x) = \{M \in A\bot x : M' \leq M \text{ for all } M' \in A\bot x\}.$$

This is an application of stringent maximization as discussed in section "Rational Choice", and it is called the *marking off identity* in AGM (1985). Recall that we write $\gamma = \mathscr{S}(\leq)$ if γ is determined by $\leq$ in that way.

A contraction function $\dot{-}$ is *(transitively, quasi-transitively, connectively, acyclicly) relational* if there is a (transitive, quasi-transitive, connective, acyclic) preference relation $\leq$ over U_A such that $\dot{-} = \mathscr{C}(\mathscr{S}(\leq))$.

Using the above functions $\mathscr{W}$ and $\mathscr{M}$, we find that the selection γ of sets in $A\bot x$ can equivalently be viewed as a selection γ_W of maximally consistent sets, or "worlds", in $[\![\neg x]\!]$. If we define

$$\gamma_W([\![\neg x]\!]) = \{\mathscr{W}(M) : M \in \gamma(A\bot x)\}$$

then clearly,

$$\gamma(A\bot x) = \{\mathscr{M}(W) : W \in \gamma_W([\![\neg x]\!])\}$$

In the principal case $x \in A - Cn(\emptyset)$, $A\dot{-}x = A \cap \bigcap(\gamma_W([\![\neg x]\!]))$. Conditions for the partial meet mechanism may thus be viewed as *model theoretic* or *semantical* conditions. The "possible worlds" view facilitates a visualization of the conditions on preference relations and selection functions and relates the work on partial meet contraction directly to the minimal models approach as adopted in Lewis (1973, 1981), Kraus et al. (1990), Katsuno and Mendelzon (1991), Lindström (1991), and Lehmann and Magidor (1992).

A set W' of maximal consistent sets of propositions is called *elementary* (Δ-*elementary*) if there is a proposition y (a set of propositions B) such that W' is exactly the set of all maximally consistent sets of propositions which contain y (respectively, which include B), in symbols $W' = [\![y]\!]$ ($W' = [\![B]\!]$). We call γ *elementary* (Δ-*elementary*) if $\gamma_W([\![\neg x]\!])$ is elementary (Δ-elementary) for every $x \in A - Cn(\emptyset)$. If γ is elementary (Δ-elementary) and $\gamma_W([\![\neg x]\!]) = [\![y]\!]$ ($\gamma_W([\![\neg x]\!]) = [\![B]\!]$), then $A \dot{-} x = Cn(\{z \vee y : z \in A\}) = A \cap Cn(y)$ (respectively, $A \dot{-} x = Cn(\{z \vee z' : z \in A$ and $z' \in B\}) = A \cap Cn(B)$).

Remark 1. γ is complete iff it is Δ-elementary.

Proof. From left to right. We show that $[\![(\bigcap \gamma(A \bot x)) \cup \{\neg x\}]\!] = \gamma_W([\![\neg x]\!])$, for every $x \in A - Cn(\emptyset)$. Clearly, $(\bigcap \gamma(A \bot x)) \cup \{\neg x\} \subseteq W$ for every $W \in \gamma_W([\![\neg x]\!])$. To show the converse, suppose for reductio that $(\bigcap \gamma(A \bot x)) \cup \{\neg x\} \subseteq W$ and $W \notin \gamma_W([\![\neg x]\!])$. Hence, by the latter, $\mathcal{M}(W) \notin \gamma(A \bot x)$, so by the completeness of γ, $\bigcap \gamma(A \bot x) \nsubseteq \mathcal{M}(W)$. But also $(\bigcap \gamma(A \bot x)) \cup \{\neg x\} \subseteq W$, and $\bigcap \gamma(A \bot x) \subseteq A$, so $\bigcap \gamma(A \bot x) \subseteq \mathcal{M}(W)$, and we have a contradiction.

From right to left. Suppose for reductio that there are $x \in A - Cn(\emptyset)$ and $M \in U_A$ such that $\bigcap \gamma(A \bot x) \subseteq M$ and $M \notin \gamma(A \bot x)$. Hence $\mathcal{W}(M) \notin \gamma_W([\![\neg x]\!])$. Since γ is Δ-elementary, there is a set B of propositions such that $\gamma_W([\![\neg x]\!]) = [\![B]\!]$. So $B \nsubseteq \mathcal{W}(M)$. Take some $y \in B - \mathcal{W}(M)$. Since $x \vee y \notin \mathcal{W}(M)$, we get, by the definition of $\mathcal{W}$, $x \vee y \notin M$. So by $\bigcap \gamma(A \bot x) \subseteq M$, $x \vee y \notin \bigcap \gamma(A \bot x)$. But since $x \in A$ and $y \in \bigcap \gamma_W([\![\neg x]\!])$, we get $x \vee y \in A \cap \bigcap \gamma_W([\![\neg x]\!]) = \bigcap \gamma(A \bot x)$, and we have again a contradiction. Q.E.D.

As the domain $A\bot$ of selection functions for A is finitely additive, subtractive and compact, all domain requirements mentioned in Lemma 4(a)–(d) are satisfied. We can further adapt the conditions (I′)–(IV′) to our needs. We now know that for all $x, y \in A - Cn(\emptyset)$, $A \bot x \cup A \bot y = A \bot (x \wedge y)$, and $\gamma(A \bot x) \cap A \bot y = \emptyset$ iff $y \in \bigcap \gamma(A \bot x)$. With $\mathcal{X} = A\bot$, the general conditions (I′)–(IV′) can therefore be transformed into the following conditions on selection functions for theories:

(I″) For all $x, y \in A - Cn(\emptyset)$, $\gamma(A \bot (x \wedge y)) \subseteq \gamma(A \bot x) \cup \gamma(A \bot y)$
(II″) For all $x, y \in A - Cn(\emptyset)$, $\gamma(A \bot x) \cap \gamma(A \bot y) \subseteq \gamma(A \bot (x \wedge y))$
(III″) For all $x, y \in A - Cn(\emptyset)$, if $y \in \bigcap \gamma(A \bot (x \wedge y))$, then $\gamma(A \bot x) \subseteq \gamma(A \bot (x \wedge y))$
(IV″) For all $x, y \in A - Cn(\emptyset)$, if $x \notin \bigcap \gamma(A \bot (x \wedge y))$, then $\gamma(A \bot x) \subseteq \gamma(A \bot (x \wedge y))$

From these conditions, it will be a rather short step to some interesting postulates for rational ("economical") theory contraction.

The Finite Case

A set B of propositions will be called *logically finite* (or *finite modulo Cn*, or simply *finite*) if Cn partitions B into finitely many cells. The finite case is much easier to handle than the general one. This is due to the fact that every selection function over a logically finite theory is ω-covering and even full.

Notice that for a theory A to be finite modulo a common logic Cn, it will be necessary that the underlying language has only finitely many atoms. For suppose there are infinitely many atoms $p_1, p_2, p_3, \ldots$ in our language. Then for every proposition x from A there are infinitely many atoms p_i not occurring in x. Thus the infinitely many $(x \vee p_i)$'s which are all contained in A will be mutually nonequivalent, so A is not finite. Conversely, if only a finite number of nonequivalent propositional operators of bounded arity is definable in Cn, as in classical propositional logic or in modal logics with finitely many modalities, then the finiteness of the number of propositional atoms is also sufficient for the logical finiteness of theories A phrased in the language in question.

Given a logically finite theory A, it is clear that each of the following sets is finite: $A \perp x$, for every x, $A \perp$, and U_A.

A *representative* of a logically finite set of propositions B is a conjunction of representatives of all the equivalence classes under Cn of propositions in B.

Henceforth, we shall use the following notational convention. For sets of propositions $A, B, \ldots, M, N, \ldots, W, \ldots$ which are finite modulo Cn, the lower case letters $a, b, \ldots, m, n, \ldots, w, \ldots$ denote their representatives. For any two sets A and B such that $B \subseteq A$, $\overline{b}_A$ denotes the disjunction of representatives of those equivalence classes under Cn, the elements of which are in $Cn(A) - Cn(B)$. When we are dealing with a fixed theory A, we simply write $\overline{b}$ instead of $\overline{b}_A$. If $B = A$, then $\overline{b}_A$ is defined to be the falsity, $\perp$. We may call $\overline{b}_A$ the *corepresentative* of B (relative to A).

It is easy to verify in the finite case that for $M \in U_A$ and $W = \mathscr{W}(M)$, w is equivalent with $\neg a \wedge m$, and that for $W \in V_A$ and $M = \mathscr{M}(W)$, m is equivalent with $a \vee w$. This helps us to get the following useful

Lemma 6. *Let A be a finite theory and $M \in U_A$.*

(i) *If w is the representative of $\mathscr{W}(M)$ and $\overline{w}_L$ is the corepresentative of $\mathscr{W}(M)$ relative to the set of all propositions in L, then the following four propositions are equivalent: $\overline{m}_A$, $m \rightarrow a$, $\overline{w}_L$ and $\neg w$.*
(ii) $A \perp \overline{m} = \{M\}$.
(iii) *For all $M_1, \ldots, M_n \in U_A$, if $m \vdash m_1 \vee \cdots \vee m_n$, then $M = M_i$ for some i.*

Lemma 6(ii), together with Corollary 1(ii), shows that in the finite case every subset $\mathbf{M}$ of U_A equals $A \perp x$ for some proposition x, viz., for $x = \bigwedge\{\overline{m} : M \in \mathbf{M}\}$. This in turn means that all selection functions for a finite theory A are ω-covering and in fact full. They are not only complete (Δ-elementary) but even elementary (cf. AGM 1985, Observation 4.6).

Postulates for Contraction Functions

We now turn to a set of rationality criteria which has gained some prominence in the literature on belief change. The *basic AGM postulates* are given by $(\dot{-}1)$ – $(\dot{-}6)$, and the two *supplementary* ones are $(\dot{-}7)$ and $(\dot{-}8)$. For their motivation, see Gärdenfors (1988).

$(\dot{-}1)$ $A \dot{-} x$ is a theory
$(\dot{-}2)$ $A \dot{-} x \subseteq A$

($\dot{-}$3) If $x \notin A$ then $A\dot{-}x = A$
($\dot{-}$4) If $x \in A\dot{-}x$ then $\vdash x$
($\dot{-}$5) $A \subseteq Cn((A\dot{-}x) \cup \{x\})$
($\dot{-}$6) If $Cn(x) = Cn(y)$ then $A\dot{-}x = A\dot{-}y$
($\dot{-}$7) $A\dot{-}x \cap A\dot{-}y \subseteq A\dot{-}(x \wedge y)$
($\dot{-}$8) If $x \notin A\dot{-}(x \wedge y)$ then $A\dot{-}(x \wedge y) \subseteq A\dot{-}x$.

These postulates, and all following postulates, are understood as quantified over all theories A and all propositions x and y. It follows from ($\dot{-}$1) and ($\dot{-}$5) that $A\dot{-}x = A$ for every $x \in Cn(\emptyset)$. We introduce two new conditions.

($\dot{-}$8r) $A\dot{-}(x \wedge y) \subseteq Cn(A\dot{-}x \cup A\dot{-}y)$
($\dot{-}$8c) If $y \in A\dot{-}(x \wedge y)$, then $A\dot{-}(x \wedge y) \subseteq A\dot{-}x$.

With two very recent exceptions,[5] I have never seen a condition like ($\dot{-}$8r) discussed in writings on the logic of theory change. Given ($\dot{-}$4), ($\dot{-}$8) implies the "covering condition"

$$A\dot{-}(x \wedge y) \subseteq A\dot{-}x \text{ or } A\dot{-}(x \wedge y) \subseteq A\dot{-}y$$

(AGM 1985, Observation 3.4) and hence ($\dot{-}$8r).[6] Postulate ($\dot{-}$8c) was found to be relevant in Rott (1992b), where it has the same name. The "r" in ($\dot{-}$8r) stands for "relational", and "c" stands for "cumulative". The first name will be explained by the present paper, the second one is explained in Rott (1992b). For contraction functions satisfying ($\dot{-}$4), ($\dot{-}$8c) is also a weakening of ($\dot{-}$8). However, there is no logical relationship between ($\dot{-}$8c) and ($\dot{-}$8r), not even in the finite case and in the presence of ($\dot{-}$1) – ($\dot{-}$7). To see this, consider the propositional language L over the two atoms p and q. In the following two figures, "$\binom{x}{y}$" is short for "$A\dot{-}x = Cn(y)$". It is easily verified that the contraction function $\dot{-}$ over $A = Cn(p \wedge q)$ defined in Fig. 16.1 satisfies ($\dot{-}$1) – ($\dot{-}$7) and ($\dot{-}$8r), but it does not satisfy ($\dot{-}$8c), because $q \in A\dot{-}(p \wedge q)$ but $Cn(q) = A\dot{-}(p \wedge q) \nsubseteq A\dot{-}p = Cn(\neg p \vee q)$. On the other hand, the contraction function $\dot{-}$ over the same theory A defined in Fig. 16.2 satisfies ($\dot{-}$1) – ($\dot{-}$7) and ($\dot{-}$8c), but it does not satisfy ($\dot{-}$8r), because $Cn(p \vee q) = A\dot{-}(p \wedge q) \nsubseteq Cn((A\dot{-}p) \cup (A\dot{-}q)) = Cn(\{\neg p \vee q, p \vee \neg q\}) = Cn(p \leftrightarrow q)$.

We now relate the abstract postulates for contraction functions to our previous requirements for selection functions in partial meet contraction.

[5]Both were discovered independently and concern belief revision rather than belief contraction. The first exception is condition (R8) in Katsuno and Mendelzon (1991). The second is the infinitary condition "Gamma" in Lindström (1991) which is labelled (BC7) in its variant for belief revision operations.

[6]Incidentally, it is proved in AGM (1985, Observation 6.5) that the conjunction of ($\dot{-}$7) and ($\dot{-}$8) is equivalent to the even stronger "ventilation condition"

$$A\dot{-}(x \wedge y) = A\dot{-}x \text{ or } A\dot{-}(x \wedge y) = A\dot{-}y \text{ or } A\dot{-}(x \wedge y) = A\dot{-}x \cap A\dot{-}y.$$

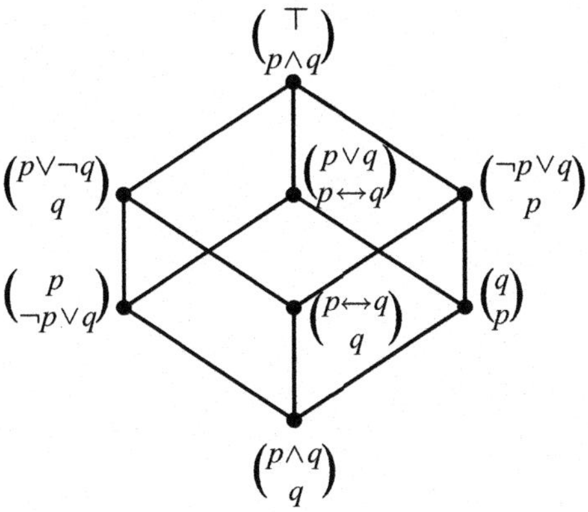

Fig. 16.1 Contraction function satisfying ($\dot{-}$1)–($\dot{-}$7) and ($\dot{-}$8r), but not ($\dot{-}$8c)

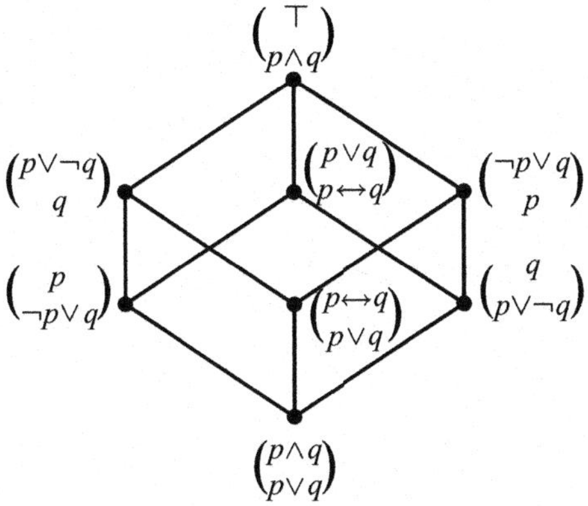

Fig. 16.2 Contraction function satisfying ($\dot{-}$1)–($\dot{-}$7) and ($\dot{-}$8c), but not ($\dot{-}$8r)

Lemma 7. *Let A be a theory. If* $\dot{-} = \mathscr{C}(\gamma)$ *for some selection function* γ *for A and* γ *satisfies*

$$\left\{\begin{array}{ll} (a) & - \\ (b) & (I'') \\ (c) & (II'') \text{ and } \gamma \text{ is complete} \\ (d) & (III'') \\ (e) & (IV'') \end{array}\right\}$$

then $\dot{-}$ *satisfies*

$$\left\{\begin{array}{ll} (a) & (\dot{-}1)-(\dot{-}6) \\ (b) & (\dot{-}1)-(\dot{-}7) \\ (c) & (\dot{-}1)-(\dot{-}6) \text{ and } (\dot{-}8r) \\ (d) & (\dot{-}1)-(\dot{-}6) \text{ and } (\dot{-}8c) \\ (e) & (\dot{-}1)-(\dot{-}6) \text{ and } (\dot{-}8) \end{array}\right\}.$$

Proof. Let $\dot{-}$ be a partial meet contraction function over A determined by γ.

(a) It is proved in AGM (1985, Observation 2.5), that $\dot{-}$ satisfies ($\dot{-}$1) – ($\dot{-}$6).

It is easy to verify that $(\dot{-}7)$ and $(\dot{-}8)$, and thus also $(\dot{-}8r)$ and $(\dot{-}8c)$, are satisfied whenever one of the limiting cases $x \notin A - Cn(\emptyset)$ or $y \notin A - Cn(\emptyset)$ holds. In the rest of this proof, we always presume the principal case $x, y \in A - Cn(\emptyset)$.

(b) Now let γ satisfy (I″), i.e., let $\gamma(A\bot(x \wedge y)) \subseteq \gamma(A\bot x) \cup \gamma(A\bot y)$. Hence $\bigcap(\gamma(A\bot x)) \cap \bigcap(\gamma(A\bot y)) = \bigcap(\gamma(A\bot x) \cup \gamma(A\bot y)) \subseteq \bigcap \gamma(A\bot(x \wedge y))$, i.e., $A\dot{-}x \cap A\dot{-}y \subseteq A\dot{-}(x \wedge y)$. That is, $\dot{-}$ satisfies $(\dot{-}7)$.

(c) Now let γ be complete and satisfy (II″). Let $z \in A\dot{-}(x \wedge y) = \bigcap \gamma(A\bot(x \wedge y))$. So by (II″), $z \in \bigcap(\gamma(A\bot x) \cap \gamma(A\bot y))$. Now suppose for reductio that $z \notin Cn(A\dot{-}x \cup A\dot{-}y)$. Then there is an $M \in A\bot z$ such that $(\bigcap \gamma(A\bot x)) \cup (\bigcap \gamma(A\bot y)) \subseteq M$. Since γ is complete, we get $M \in \gamma(A\bot x)$ and $M \in \gamma(A\bot y)$, so $M \in \gamma(A\bot x) \cap \gamma(A\bot y)$. But $z \notin M$, so $z \notin \bigcap(\gamma(A\bot x) \cap \gamma(A\bot y))$ which gives us a contradiction. So $\dot{-}$ satisfies $(\dot{-}8r)$.

(d) Now let γ satisfy (III″). Since the antecedents of (III″) and $(\dot{-}8c)$ are identical and $\gamma(A\bot x) \subseteq \gamma(A\bot(x \wedge y))$ entails $\bigcap \gamma(A\bot(x \wedge y)) \subseteq \bigcap \gamma(A\bot x)$, it is obvious that $\dot{-}$ satisfies $(\dot{-}8c)$.

(e) Now let γ satisfy (IV″). By exactly the same argument as in (d), $\dot{-}$ satisfies $(\dot{-}8)$. Q.E.D.

Given a contraction function $\dot{-}$ over A, we can derive from it a selection function $\gamma_{\dot{-}}$ for A. The idea is that an element of $A\bot x$ is a best element of $A\bot x$ if it includes everything which is included in the contraction of A with respect to x. So we define $\gamma_{\dot{-}}(A\bot x)$ as $\{M \in A\bot x : A\dot{-}x \subseteq M\}$, for every $x \in A - Cn(\emptyset)$. Instead of $\gamma_{\dot{-}}$, we shall also write $\mathscr{S}(\dot{-})$.

Lemma 8. *Let A be a theory. If $\dot{-}$ is a contraction function over A satisfying*

$$\left\{\begin{array}{ll} (a) & (\dot{-}1)-(\dot{-}6) \\ (b) & (\dot{-}1)-(\dot{-}7) \\ (c) & (\dot{-}1)-(\dot{-}6) \text{ and } (\dot{-}8r) \\ (d) & (\dot{-}1)-(\dot{-}6) \text{ and } (\dot{-}8c) \\ (e) & (\dot{-}1)-(\dot{-}6) \text{ and } (\dot{-}8) \end{array}\right\}$$

then $\dot{-}$ is a partial meet contraction function determined by the selection function $\mathscr{S}(\dot{-})$, i.e., $\dot{-} = \mathscr{C}(\mathscr{S}(\dot{-}))$, and $\mathscr{S}(\dot{-})$ is complete and satisfies

$$\left\{\begin{array}{ll} (a) & - \\ (b) & (I'') \\ (c) & (II'') \\ (d) & (III'') \\ (e) & (IV'') \end{array}\right\}.$$

Proof. Let $\dot{-}$ satisfy $(\dot{-}1)-(\dot{-}6)$.

(a) It is proved in (AGM 1985, Observation 2.5) that $\mathscr{S}(\dot{-})$ is a selection function for A and that $\dot{-}$ is determined by $\mathscr{S}(\dot{-})$. We add the check for well-definedness.

In the principal case $x, y \in A - Cn(\emptyset), A\bot x = A\bot y$ entails that $Cn(x) = Cn(y)$; for if it were the case that $x \nvdash y$, say, then there would be an element in $A\bot y$ containing x, contrary to $A\bot x = A\bot y$. But from $Cn(x) = Cn(y)$ and $(\dot{-}6)$ it follows by the definition of $\mathscr{S}(\dot{-})$ that $\gamma(A\bot x) = \gamma(A\bot y)$, as desired. By construction, $\mathscr{S}(\dot{-})$ is complete for every contraction function $\dot{-}$. It remains to prove the various additional properties of $\mathscr{S}(\dot{-})$.

(b) Let $\dot{-}$ in addition satisfy $(\dot{-}7)$, and let $M \in \gamma_{\dot{-}}(A\bot (x \wedge y))$. We want to show that $M \in \gamma_{\dot{-}}(A\bot x) \cup \gamma_{\dot{-}}(A\bot y)$. By $M \in \gamma_{\dot{-}}(A\bot (x \wedge y)), A\dot{-}(x \wedge y) \subseteq M$, so by $(\dot{-}7)$, $A\dot{-}x \cap A\dot{-}y \subseteq M$. Now suppose for reductio that $M \notin \gamma_{\dot{-}}(A\bot x) \cup \gamma_{\dot{-}}(A\bot y)$, i.e., neither $A\dot{-}x \subseteq M$ nor $A\dot{-}y \subseteq M$. Take $z \in (A\dot{-}x) - M$ and $z' \in (A\dot{-}y) - M$. By $z, z' \in A$, $z, z' \notin M$ and the primeness of all elements of U_A, $z \vee z' \notin M$. But on the other hand, since $z \vee z' \in A\dot{-}x \cap A\dot{-}y \subseteq M$, we get $z \vee z' \in M$, so we have a contradiction. So $\mathscr{S}(\dot{-})$ satisfies (I″).

(c) Now let $\dot{-}$ satisfy $(\dot{-}1)$ – $(\dot{-}6)$ and $(\dot{-}8r)$, and let $M \in \gamma_{\dot{-}}(A\bot x) \cap \gamma_{\dot{-}}(A\bot y)$. We want to show that $M \in \gamma_{\dot{-}}(A\bot (x \wedge y))$. But as our hypothesis means that $A\dot{-}x \cup A\dot{-}y \subseteq M$ and M is a theory, we have $Cn(A\dot{-}x \cup A\dot{-}y) \subseteq M$, so by $(\dot{-}8r)$ $A\dot{-}(x \wedge y) \subseteq M$, so $M \in \gamma_{\dot{-}}(A\bot (x \wedge y))$. So $\mathscr{S}(\dot{-})$ satisfies (II″).

(d) Now let $\dot{-}$ satisfy $(\dot{-}1)$ – $(\dot{-}6)$ and $(\dot{-}8c)$, and let $y \in \bigcap \gamma_{\dot{-}}(A\bot (x \wedge y)) = A\dot{-}(x \wedge y)$ and $M \in \gamma_{\dot{-}}(A\bot x)$. The latter condition means that $A\dot{-}x \subseteq M$, so by $(\dot{-}8c)$ $A\dot{-}(x \wedge y) \subseteq M$, so $M \in \gamma_{\dot{-}}(A\bot (x \wedge y))$, as desired. So $\mathscr{S}(\dot{-})$ satisfies (III″).

(e) Now let $\dot{-}$ satisfy $(\dot{-}1)$ – $(\dot{-}6)$ and $(\dot{-}8)$. By exactly the same argument as the last one, $\mathscr{S}(\dot{-})$ satisfies (IV″). Q.E.D.

Representation Theorems for Contraction Functions

We get four representation theorems for partial meet contraction as a result of conjoining previously established lemmas. Our Theorems 1 and 4 are proven in AGM (1985) although our construction for obtaining the latter appears to be quite different from the one used there. Theorems 2 and 3 are new; the "completeness" half of the former is generally valid, the remainder of the new results hold only under various additional conditions. All these conditions are satisfied if A is finite. It should be noted that our proofs of the completeness halves use one and the same method in that they all work with Samuelson revealed preferences.

Theorem 1. *A contraction function $\dot{-}$ over A satisfies $(\dot{-}1)$ – $(\dot{-}6)$ if and only if it is a partial meet contraction function.*

Proof. See Lemmas 7(a) and 8(a) above, according to which in the cases indicated $\mathscr{C}(\gamma)$ satisfies $(\dot{-}1)$ – $(\dot{-}6)$ and $\mathscr{S}(\dot{-})$ is a selection function for A such that $\mathscr{C}(\mathscr{S}(\dot{-})) = \dot{-}$. (Theorem 1 is Observation 2.5 of AGM 1985.) Q.E.D.

It is clear from Lemma 8 that in the proof of the completeness half of Theorem 1 the determining selection function is chosen complete, but we do not think it is necessary to state this in the theorem. The same comment applies to the following three representation theorems.

Theorem 2. *Every relational partial meet contraction function $\dot{-}$ over A satisfies $(\dot{-}1)$ – $(\dot{-}7)$, and if $\dot{-}$ is determined by a selection function that is both relational and complete (equivalently, Δ-elementary), then it satisfies $(\dot{-}8r)$. Conversely, every contraction function $\dot{-}$ over A satisfying $(\dot{-}1)$ – $(\dot{-}7)$ and $(\dot{-}8r)$ is a relational partial meet contraction function.*

Proof. For the first part, let $\dot{-}$ be a partial meet contraction function determined by a preference relation $\leq$. By Lemma 1(a), $\mathscr{S}(\leq)$ satisfies (I) and (II). Since $A\bot$ is subtractive, it also satisfies (I′) and (II′), by Lemma 2(a), and also (I″) and (II″), by reformulation. So by Lemma 7 (b), $\mathscr{C}(\mathscr{S}(\leq))$ satisfies $(\dot{-}1)$ – $(\dot{-}7)$, and by Lemma 7(c), it satisfies $(\dot{-}8r)$, if $\mathscr{S}(\leq)$ is complete, i.e., Δ-elementary.

For the converse, let $\dot{-}$ satisfy $(\dot{-}1)$ – $(\dot{-}7)$ and $(\dot{-}8r)$. By Lemma 8(b) and (c), $\mathscr{C}(\mathscr{S}(\dot{-})) = \dot{-}$ and $\mathscr{S}(\dot{-})$ satisfies (I″) and (II″), and also (I′) and (II′), by reformulation. Since $A\bot$ is subtractive, $\mathscr{S}(\dot{-})$ satisfies (I), by Lemma 4(a). Since $A\bot$ is also finitely additive and compact, $\mathscr{S}(\dot{-})$ is relational with respect to $\mathscr{P}(\mathscr{S}(\dot{-}))$, by Lemma 4(b). That is, $\mathscr{S}(\dot{-}) = \mathscr{S}(\mathscr{P}(\mathscr{S}(\dot{-})))$. Hence $\dot{-} = \mathscr{C}(\mathscr{S}(\mathscr{P}(\mathscr{S}(\dot{-}))))$, i.e., $\dot{-}$ is relational with respect to $\mathscr{P}(\mathscr{S}(\dot{-}))$. Q.E.D.

The completeness of γ is necessary in the "soundness" part of Theorem 2. The possible failure without completeness is due to the fact that for sets W' and W'' of maximally consistent sets of propositions, it does not necessarily hold that $\bigcap (W' \cap W'') \subseteq Cn\,(\bigcap W' \cup \bigcap W'')$.

Theorem 3. *Every relational partial meet contraction function $\dot{-}$ over A determined by a smooth and negatively transitive preference relation satisfies $(\dot{-}1)$ – $(\dot{-}7)$ and $(\dot{-}8c)$. In partial converse, every contraction function $\dot{-}$ over A satisfying $(\dot{-}1)$ – $(\dot{-}7)$, $(\dot{-}8r)$, and $(\dot{-}8c)$ is a relational partial meet contraction function, and if $A\bot$ is 123-covering, then the determining preference relation can be chosen negatively transitive.*

Proof. For the first part, let $\dot{-}$ be a partial meet contraction function determined by a smooth and negatively transitive preference relation $\leq$. We show in the same way as in the proof of Theorem 2 that $\mathscr{S}(\leq)$ satisfies (I′) and (II′), and we know from Lemmas 3(b) and 4(c) that it satisfies (III) and (III′). So by Lemma 7(b) and (d), $\mathscr{C}(\mathscr{S}(\leq))$ satisfies $(\dot{-}1)$ – $(\dot{-}7)$ and $(\dot{-}8c)$.

For the converse, let $\dot{-}$ satisfy $(\dot{-}1)$ – $(\dot{-}7)$, $(\dot{-}8r)$, and $(\dot{-}8c)$. As in the proof of Theorem 2, we get that $\dot{-}$ is relational with respect to $\mathscr{P}(\mathscr{S}(\dot{-}))$. Moreover, by Lemma 8(b), $\mathscr{S}(\dot{-})$ satisfies (I′), and by Lemma 8(d), $\mathscr{S}(\dot{-})$ satisfies (III′) and also (III), by Lemma 4(c). Hence if $A\bot$ is 123-covering, then $\mathscr{P}(\mathscr{S}(\dot{-}))$ is negatively transitive, by Lemma 2(b). Q.E.D.

Theorem 4. *A contraction function $\dot{-}$ over A satisfies $(\dot{-}1)$ – $(\dot{-}6)$, $(\dot{-}7)$, and $(\dot{-}8)$ if and only if it is a transitively relational partial meet contraction function.*

Proof. This is Observation 4.4 of AGM (1985), but we shall sketch our proof which is quite different from the construction offered there.

For the first part of the theorem, let $\dot{-}$ be a partial meet contraction function determined by a transitive preference relation $\leq$. We show in the same way as in the proof of Theorem 2 that $\mathscr{S}(\leq)$ satisfies (I′) and (II′), and we know from Lemmas 3(c) and 4(d) that it satisfies (IV) and (IV′). So by Lemma 7(b), $\mathscr{C}(\mathscr{S}(\leq))$ satisfies ($\dot{-}$1) – ($\dot{-}$7), and by Lemma 7(e), it satisfies ($\dot{-}$8).

For the converse, let $\dot{-}$ satisfy ($\dot{-}$1) – ($\dot{-}$7) and ($\dot{-}$8). As in the proof of Theorem 2, we get that $\dot{-}$ is relational with respect to $\mathscr{P}(\mathscr{S}(\dot{-}))$. Moreover, by Lemma 8(e), $\mathscr{S}(\dot{-})$ satisfies (IV′) and also (IV), by Lemma 4(d). Since $A\bot$ is finitely additive, $\mathscr{P}(\mathscr{S}(\dot{-}))$ is transitive, by Lemma 2(c). Q.E.D.

Corollary 2. *Let A be a logically finite theory. Then $\dot{-}$ is a*

$$\left\{\begin{array}{c} \text{—} \\ \textit{relational} \\ \textit{negatively transitively relational} \\ \textit{transitively relational} \end{array}\right\} \textit{partial meet contraction function iff it satisfies}$$

$$\left\{\begin{array}{c} (\dot{-}1) - (\dot{-}6) \\ (\dot{-}1) - (\dot{-}7) \textit{ and } (\dot{-}8r) \\ (\dot{-}1) - (\dot{-}7),\ (\dot{-}8r) \textit{ and } (\dot{-}8c) \\ (\dot{-}1) - (\dot{-}7) \textit{ and } (\dot{-}8) \end{array}\right\}.$$

Proof. By Theorems 1, 2, 3, and 4, because for finite theories A, γ over $A\bot$ is complete, every negatively transitive nonstrict preference relation $\leq$ over A is smooth, and $A\bot$ is 123-covering. □

It is easy to locate the impact of the finiteness assumption. In the case of relational partial meet contraction s (Theorem 2), it is a constraint (completeness) imposed by model theory which has to be met in order to make the "soundness" direction work. In the case of negatively transitively relational partial meet contraction s (Theorem 3), it is a constraint imposed by the theory of rational choice (that γ be 123-covering and smooth) which is satisfied without strains on intuitive plausibility only in the finite case.

It should be noted that although ($\dot{-}$8) in the context of the other postulates implies ($\dot{-}$8*r*) and ($\dot{-}$8*c*), transitive relationality does not imply negatively transitive relationality. However, transitivity in companion with connectivity implies negative transitivity. And it is known from AGM (1985, § 5) that a connectivity requirement on the underlying preference relation changes very little in the partial meet mechanism.[7]

We conclude this section with a direct representation of the Samuelson preferences over U_A revealed by contraction behavior. Let $\leq\, = \mathscr{P}(\mathscr{S}(\dot{-}))$. Then in the general case, $M \leq M'$ iff there is an x such that $A \dot{-} x \subseteq M'$ and $M \in A\bot x$, or there

[7]In order to reconcile this with your intuitions, cf. Footnote 1.

is no x such that $A\dot{-}x \subseteq M$. We recall, for contrast, the construction used in the proof of AGM (1985, Observation 4.4). There $M \leq M'$ is defined to hold between sets M and M' from U_A iff for all x such that $A\dot{-}x \subseteq M$ and $M' \in A\bot x$, it holds that $A\dot{-}x \subseteq M'$, and there is an $x \in A$ such that $A\dot{-}x \subseteq M'$.[8] Evidently, this definition is quite different from the one we used, and it is doubtful whether it is suitable for any case but the strongest one which is treated in Theorem 4. It is interesting, however, that AGM employ our construction in their proof of Observation 4.10 of AGM (1985). In the finite case, it is not difficult to find out that according to our definition, $M \leq M'$ iff $A\dot{-}\overline{m} \wedge \overline{m'} \subseteq M'$, or equivalently(!), iff $\overline{m'} \notin A\dot{-}\overline{m} \wedge \overline{m'}$.

The Logic of Prioritized Base Contractions in the Finite Case

In this section we present the concept of prioritized base contraction as studied in particular by Nebel (1989, 1992). On the face of it, these belief change operations are entirely different from the partial meet contractions discussed so far. As Nebel has shown, however, very close connections can be established between the two approaches. We shall further elaborate this theme and determine the logic of prioritized base contractions in terms of the above-mentioned postulates by applying one of our central representation theorems.

Let B be a finite set of propositions. B is called a *(simple) belief base* for the theory A if $Cn(B) = A$. B is called *prioritized* if there is a weak ordering, i.e., a transitive and connected relation $\preceq$ over B (Nebel's 1989, 1992 "epistemic relevance"). The relation $\sim = \preceq \cap \preceq^{-1}$ partitions B into finitely many equivalence classes $B_1, \ldots, B_n$. We label the classes in such a way that $x_i \preceq x_j$ for $x_i \in B_i$ and $x_j \in B_j$ with $i \leq j$. Generalizing slightly, we say that a *prioritized belief base* for A is a finite sequence $\langle B_1, \ldots, B_n \rangle$, or $\langle B_i \rangle$ for short, of finite sets of propositions such that $Cn(\bigcup B_i) = A$. Simple belief bases are now special cases of prioritized belief bases with $n = 1$. We do not require that the B_i's are disjoint or mutually consistent. Intuitively, the propositions in B_n are the most important (most relevant, most valuable, most certain, etc.) pieces of information, and the propositions in B_1 are the least important. One can think of the complement of $\bigcup B_i$ as the set B_0 of propositions for which there is no independent evidence. Although the finiteness of $B = \bigcup B_i$ does not ensure that the generated theory A is logically finite, our central result in this section, Theorem 7, will be restricted to this case.

Prioritized base contractions are contractions of theories A presented in the form of (generated by, associated with) a fixed prioritized belief base $\langle B_i \rangle$. The basic idea is to minimize the loss of propositions at every level of "priority".

For every $i \in \{1, \ldots, n\}$, let $B_{\geq i}$ be the set of all elements of $B = \bigcup B_i$ with a priority of at least i, i.e., $B_{\geq i} = \bigcup\{B_j : j \geq i\}$. Furthermore, let $C_i = C \cap B_i$ and

[8] This simplified rephrasing of the AGM construction makes use of the fact that for all $M \in U_A$ and contraction functions $\dot{-}$ satisfying the AGM postulates, $A\dot{-}x \subseteq M$ implies $M \in A\bot x$.

$C_{\geq i} = C \cap B_{\geq i}$ for any set of propositions C. Then, in the context of this section, C is a *preferred x-discarding* subset of B if $C \subseteq B$ and for every i, $C_{\geq i}$ is an inclusion maximal subset of $B_{\geq i}$ subject to the condition that x is not implied.

Two kinds of information are used in prioritized base contraction: the syntactical information encoded in the structure of the propositions in B, and the weighting of these propositions expressed by $\preceq$. If the belief base is simple, i.e., if $n = 1$, then we only exploit syntactical information. Prioritized base contractions and revisions are studied in an infinitistic framework by Nebel (1989, 1992).[9] He shows that prioritized base revisions can be represented by partial meet theory revisions. Then he proves that a certain variant of simple base contraction satisfies ($\dot{-}$1) – ($\dot{-}$7) but not ($\dot{-}$8), and a corresponding result for prioritized base revisions. However, Nebel leaves open the question which logic is *characterized* by prioritized base contractions. Building on Theorem 3 and its corollary, we shall answer this question for the finite case. Our concern will be the slightly more complicated case of prioritized base contractions, and we try to make our proof more transparent by introducing the concept of a base-oriented selection function.

Now consider the following strict preference relations between arbitrary sets of propositions. For every $i \in \{1, \ldots, n\}$ and any two sets C and C', we write $C \lll_i C'$ if and only if $C_i \subset C'_i$ and $C_j \subseteq C'_j$ for every $j > i$. We write $C \lll C''$ if there is an i such that $C \lll_i C'$. The relation $\lll$ is to reflect the fact that intuitively a set C' is better than another set C just in case it contains more important beliefs than C. In particular, $\lll$ satisfies a version of Hansson's (1992) maximizing property, because $C \cap B \subset C' \cap B$ implies $C \lll C'$. It is immediately verified that C is a preferred x-discarding subset of B if and only if it is a maximal element in $B \perp x$ under $\lll$. (Recall that $B \perp x$ is the set $\{N \subseteq B : x \notin Cn(N)$ and $x \in Cn(D)$ for all D with $N \subset D \subseteq B\}$.)

Following Nebel, we let $\underline{\lll}$ be the converse complement of $\lll$, i.e., $C \,\underline{\lll}\, C'$ iff not $C' \lll C$. We denote this preference relation over arbitrary sets of propositions by $\mathscr{P}(\langle B_i \rangle)$. Clearly, since $\lll$ is irreflexive, asymmetric, conversely well-founded and transitive, $\mathscr{P}(\langle B_i \rangle)$ is reflexive, connected, negatively well-founded and negatively transitive. But $\mathscr{P}(\langle B_i \rangle)$ is not transitive, not even in the special case of simple belief bases ($n = 1$) where $\underline{\lll}$ coincides with $\not\supset$.

As before, a selection function $\gamma = \mathscr{S}(\underline{\lll})$ can be defined by stringent maximization. But this time γ is a selection function for the base B as well as for the theory A, and its domain may in fact be construed as the class of all nonempty sets of sets propositions in L, that is $2^{2^L} - \{\emptyset\}$. In the following, it is understood that $\gamma(B \perp x) = \{N \in B \perp x : N' \underline{\lll} N$ for all $N' \in B \perp x\}$, while $\gamma(A \perp x) = \{M \in A \perp x : M' \underline{\lll} M$ for all $M' \in A \perp x\}$. Either way, γ is a selection function.

[9]Nebel's (1992) treatment of the fully general infinite case is not quite correct. Slips have crept into his claim that his $C \Downarrow \phi$ is nonempty, into his definition (9) of $\ll$, and into the proof of Proposition 8. As Nebel (personal communication) has suggested, they can be remedied by imposing a condition of converse well-foundedness on $\preceq$.

Let $\langle B_i \rangle$ be a prioritized belief base for A, and let $\gamma = \mathscr{S}(\mathscr{P}(\langle B_i \rangle))$ be the selection function over $A\bot$ determined by $\langle B_i \rangle$. A straightforward idea to get prioritized base contractions is the following:

$$A \dot{-} x = \begin{cases} \bigcap \{Cn(N) \colon N \in \gamma(B \bot x)\} & \text{for every } x \in A - Cn(\emptyset) \\ A & \text{for every } x \notin A \text{ and every } x \in Cn(\emptyset) \end{cases}$$

In the special case of simple belief bases ($n = 1$) we have $\gamma(B \bot x) = B \bot x$, so this definition boils down to the *full* meet $\bigcap \{B \bot x\}$.

Contraction functions obtained in this way satisfy most but not all of the postulates we would want them to satisfy.

Theorem 5. *If $\dot{-}$ is a prioritized base contraction function as determined by the above definition, then it satisfies ($\dot{-}$1) – ($\dot{-}$4), ($\dot{-}$6), ($\dot{-}$7), and ($\dot{-}$8c). However, even if the base is simple, $\dot{-}$ will in general fail to satisfy ($\dot{-}$5) and ($\dot{-}$8r).*

Proof. It is obvious that $\dot{-}$ satisfies ($\dot{-}$1) – ($\dot{-}$4) and ($\dot{-}$6).

For ($\dot{-}$7), assume that $N \vdash z$ for every $N \in \gamma(B \bot x) \cup \gamma(B \bot y)$. We need to show that $N' \vdash z$ for every $N' \in \gamma(B \bot (x \wedge y))$. Let $N' \in \gamma(B \bot (x \wedge y)) \subseteq B \bot (x \wedge y)$. First we note that every element of $B \bot (x \wedge y)$ is either in $B \bot x$ or in $B \bot y$. Without loss of generality, assume that $N' \in B \bot x$. We are ready if we can show that N' is in $\gamma(B \bot x)$. Suppose it is not. Then there is an $N_1 \in B \bot x$ such that $N' \lll N_1$. N_1 must not be in $B \bot (x \wedge y)$, since N' is in $\gamma(B \bot (x \wedge y))$. Because $N_1 \in B \bot x - B \bot (x \wedge y)$, we get that there is a proper superset N_2 of N_1 such that $N_2 \in B \bot y$. Since $N_1 \in B \bot x$, $N_2 \vdash x$. Since every proper superset of N_2 satisfies both x and y, N_2 is in $B \bot (x \wedge y)$. By the maximizing property, we have $N_1 \lll N_2$. By the transitivity of $\lll$, we get $N' \lll N_2$. This, however, contradicts $N' \in \gamma(B \bot (x \wedge y))$ and $N_2 \in B \bot (x \wedge y)$. Therefore our supposition must have been false.

For ($\dot{-}$8c), assume that, first, $N \vdash y$ and, second, $N \vdash z$ for every $N \in \gamma(B \bot (x \wedge y))$. We need to show that $N' \vdash z$ for every $N' \in \gamma(B \bot x)$. Let $N' \in \gamma(B \bot x)$ and suppose for reductio that $N' \nvdash z$. If N' is in $B \bot (x \wedge y)$ then there must be, by the second assumption, an N_1 in $B \bot (x \wedge y)$ such that $N' \lll N_1$. In fact, N_1 can be chosen from $\gamma(B \bot (x \wedge y))$, by the transitivity and converse well-foundedness of $\lll$. By the first assumption, $N_1 \in B \bot x$. But this contradicts $N' \lll N_1$ and $N' \in \gamma(B \bot x)$. Hence N' cannot be in $B \bot (x \wedge y)$. But because $N' \nvdash x \wedge y$, there must be a proper superset N_2 of N' in $B \bot (x \wedge y)$. We know that $N_2 \vdash x$, by $N' \in B \bot x$. By the first assumption then, N_2 cannot be in $\gamma(B \bot (x \wedge y))$. By the transitivity and converse well-foundedness of $\lll$, there must be an N_3 in $\gamma(B \bot (x \wedge y))$ such that $N_2 \lll N_3$. By our first assumption, N_3 is in $B \bot x$. By the maximizing property, we have $N' \lll N_2 \lll N_3$, so by transitivity $N' \lll N_3$. This, however, contradicts $N' \in \gamma(B \bot x)$ and $N_3 \in B \bot x$. Therefore our supposition must have been false.

The failure of ($\dot{-}$5) is obvious. Consider for instance $B = \{p \wedge q\}$, for which $A = Cn(p \wedge q) \nsubseteq Cn((A \dot{-} p) \cup \{p\}) = Cn(p)$.

The failure of ($\dot{-}$8r) is not so easy to see. The simplest examples involve three atoms p, q, r and look somewhat like the following. Let B consist of the following four propositions:

$$(p \wedge q) \vee (p \wedge r) \vee (q \wedge r), \quad r \to (p \leftrightarrow q),$$
$$q \wedge (p \leftrightarrow r), \quad p \wedge r.$$

Clearly $A = Cn(B) = Cn(p \wedge q \wedge r)$. Note that $q \wedge (p \leftrightarrow r)$ implies $r \to (p \leftrightarrow q)$ and that $p \wedge r$ implies $(p \wedge q) \vee (p \wedge r) \vee (q \wedge r)$. $(p \wedge q) \vee (p \wedge r) \vee (q \wedge r)$ and $r \to (p \leftrightarrow q)$ taken together imply $p \wedge q$. This gives us

$$B \perp p = \{\{(p \wedge q) \vee (p \wedge r) \vee (q \wedge r)\}, \{r \to (p \leftrightarrow q), q \wedge (p \leftrightarrow r)\}\}$$

and

$$B \perp q = \{\{r \to (p \leftrightarrow q)\}, \{(p \wedge q) \vee (p \wedge r) \vee (q \wedge r), p \wedge r\}\}.$$

Therefore, the above account yields $A \dot{-} p = Cn(q \vee (p \wedge r))$ and $A \dot{-} q = Cn(\neg(\neg p \wedge q \wedge r))$, so $Cn(A \dot{-} p \cup A \dot{-} q) = Cn((p \wedge r) \vee (q \wedge \neg r))$. On the other hand,

$$B \perp (p \wedge q) = \{\{(p \wedge q) \vee (p \wedge r) \vee (q \wedge r), p \wedge r\}, \{r \to (p \leftrightarrow q), q \wedge (p \leftrightarrow r)\}\},$$

yielding $A \dot{-} p \wedge q = Cn((p \wedge r) \vee (\neg p \wedge q \wedge \neg r))$. At last we realize that the union of $A \dot{-} p$ and $A \dot{-} q$ does not imply $(p \wedge r) \vee (\neg p \wedge q \wedge \neg r)$, because the former, but not the latter is compatible with $p \wedge q \wedge \neg r$. Q.E.D.

In so far as one considers the postulates ($\dot{-}5$) and ($\dot{-}8r$) to be plausible principles of theory change—which in fact we do—, Theorem 5 shows that the above definition is defective. For this reason we introduce a slightly more sophisticated concept, and our official definition reads as follows. Let a again be a representative of A.

The *prioritized base contraction* $\dot{-}$ over a logically finite theory A determined by $\langle B_i \rangle$ is then given by

$$A \dot{-} x = \begin{cases} \bigcap \{Cn(N \cup \{x \to a\}) : N \in \gamma(B \perp x)\}^{10} & \text{for every } x \in A - Cn(\emptyset) \\ A & \text{for every } x \notin A \text{ and } x \in Cn(\emptyset) \end{cases}$$

The singleton $\{x \to a\}$ is inserted in order to make sure that the recovery postulate ($\dot{-}5$) is satisfied. This trick is used by Nebel (1989).[11] We shall see that this modification also solves our problem with ($\dot{-}8r$). As a first indication

[10]Or equivalently, $Cn((\bigcap\{Cn(N) : N \in \gamma(B \perp x)\}) \cup \{x \to a\}) = Cn(A \dot{-}' x \cup \{x \to a\})$ where $A \dot{-}' x$ follows the first definition.

[11]Nebel's (1992) later paper deals with revisions where this problem does not arise. In fact, if revisions are construed as being generated by the so-called Levi-identity $A * x = Cn((A \dot{-} \neg x) \cup \{x\})$, then the modification made in our official definition does not have any effect on revisions.

of this, note that the official definition cures our above counterexample to ($\dot{-}8r$) by strengthening $A\dot{-}x$ to $Cn(y \wedge (x \rightarrow z))$ and $A\dot{-}y$ to $Cn(y \rightarrow (x \wedge z))$, so that $Cn(A\dot{-}x \cup A\dot{-}y) = Cn(x \wedge y \wedge z)$. Since this has been the full theory A, ($\dot{-}8r$) is clearly satisfied.

We denote the prioritized base contraction over A determined by the prioritized belief base $\langle B_i \rangle$ by $\mathscr{C}(\langle B_i \rangle)$. A contraction function $\dot{-}$ over A is a *simple* (or *prioritized*) *base contraction function* if there is a simple belief base B (a prioritized belief base $\langle B_i \rangle$) for A such that $\dot{-} = \mathscr{C}(B)$ (respectively, $\dot{-} = \mathscr{C}(\langle B_i \rangle)$). In conformity with the terminology of AGM (1985), simple base contractions could also be named *full meet* base contractions.

Lemma 9. *Let $B = \bigcup B_i$, $A = Cn(B)$, $M \in A \bot x$, and let $\gamma = \mathscr{S}(\mathscr{P}(\langle B_i \rangle))$. Then $M \in \gamma(A \bot x)$ iff $N \subseteq M$ for some $N \in \gamma(B \bot x)$.*

Let us call selection functions which harmonize choices in $A\bot$ with choices in $B\bot$ in the way exhibited in Lemma 9 base-oriented. The following result shows that the selection function γ induced by a prioritized belief base leads to the same result when applied directly to belief bases as when applied to the generated theory. Prioritized base contraction thus reduces to a special kind of partial meet theory contraction. A proof of almost the same fact has already been given by Nebel (1989, Theorem 14, 1992, Theorem 7). Our proof is somewhat more general in that it only turns on the fact that γ is base-oriented. Nothing hinges on the particular definition of $\underline{\lll}$.

Theorem 6. *Let $\langle B_i \rangle$ be a prioritized belief base for a logically finite theory A, and let $\gamma = \mathscr{S}(\mathscr{P}(\langle B_i \rangle))$. Then $\mathscr{C}(\langle B_i \rangle) = \mathscr{C}(\gamma)$.*

Proof. For the principal case $x \in A - Cn(\emptyset)$, we have to show that

$$\bigcap \{Cn(N \cup \{x \rightarrow a\}) : N \in \gamma(B \bot x)\} = \bigcap \gamma(A \bot x).$$

By Lemma 9, we know that γ is base-oriented.

"LHS $\subseteq$ RHS": Let $N \cup \{x \rightarrow a\} \vdash y$ for every $N \in \gamma(B \bot x)$, and let $M \in \gamma(A \bot x)$. Since γ is base-oriented, there is an $N \in \gamma(B \bot x)$ such that $N \subseteq M$. So $M \cup \{x \rightarrow a\} \vdash y$. But also, by the fullness of the elements in $A \bot x$, $x \rightarrow a \in M$. So $M \vdash y$, as desired.

"RHS $\subseteq$ LHS": Assume that there is an $N \in \gamma(B \bot x)$ such that $N \cup \{x \rightarrow a\} \nvdash y$. Hence, by the deduction theorem and supraclassicality, $N \cup \{y \rightarrow a\} \nvdash x$. Thus, there is an $M \in A \bot x$ such that $N \cup \{y \rightarrow a\} \subseteq M$. Since $N \subseteq M$ and γ is base-oriented, $M \in \gamma(A \bot x)$. Since $y \rightarrow a \in M$, $a \vdash x$ and $M \nvdash x$, we have $y \notin M$. So $y \notin \bigcap \gamma(A \bot x)$, as desired. Q.E.D.

We are now going to construct, for an arbitrary given preference relation $\leq$ over a logically finite theory A, an equivalent simple belief base B for A. This base will be denoted $\mathscr{B}(\leq)$. For $M \in U_A$, let $\hat{M}$ be the set $\{M' \in U_A : M' \nleq M\} \cup \{M\}$ of elements in U_A which are not covered by M, together with M itself. Let $\hat{m}$ be an abbreviation for the disjunction $\bigvee \{m' : M' \in \hat{M}\}$, with each m' being a representative of the respective M'. Then $B = \mathscr{B}(\leq)$ is defined as the set $\{\hat{m} : M \in U_A\}$. Since A is finite modulo Cn, everything else is. This construction and the proof idea for the direction "(iii)⇒(i)" in Theorem 7 is adapted from Lewis (1981).

Lemma 10. *Let A be a logically finite theory and $\leq$ be a reflexive and negatively transitive preference relation over U_A. Then $\mathscr{P}(\mathscr{B}(\leq)) = \leq$.*

Proof. As a preparatory consideration, let us consider what is contained in a set $M \cap B$, when $M \in U_A$ and $B = \mathscr{B}(\leq)$ for a given $\leq$. By definition, $M = Cn(m)$. Now any proposition from B is of the form $\widehat{m^*}$ for some $M^* \in U_A$, or more explicitly, of the form $\bigvee\{m^{**} : M^{**} \in U_A$, and $M^{**} \not\leq M^*$ or $M^{**} = M^*\}$. We know that a proposition $\widehat{m^*}$ from B is in M if and only if $m \vdash \widehat{m^*}$, i.e., $m \vdash \bigvee\{m^{**} : M^{**} \in U_A$, and $M^{**} \not\leq M^*$ or $M^{**} = M^*\}$. By Lemma 6(iii), this obtains just in case $M \not\leq M^*$ or $M = M^*$.

Now let $\leq^* = \mathscr{P}(\mathscr{B}(\leq))$. We want to show that $M \leq^* M'$ iff $M \leq M'$. By definition, $M \leq^* M'$ if and only if either $M \cap B \subseteq M' \cap B$ or $M' \cap B \not\subseteq M \cap B$. Applying our preparatory consideration, this means that either

(i) For every $M^* \in U_A$, if $M \not\leq M^*$ or $M = M^*$, then $M' \not\leq M^*$ or $M' = M^*$,
or
(ii) There is an $M^* \in U_A$ such that $M' \not\leq M^*$ or $M' = M^*$, and $M \leq M^*$ and $M \neq M^*$.

We need to show that this is equivalent with $M \leq M'$. That the latter implies (i) or (ii) is clear: If $M = M'$, then (i), and if $M \neq M'$, then (ii) is immediate by putting $M^* = M'$.

For the converse, suppose that $M \not\leq M'$. We verify that this is incompatible with each of (i) and (ii). From (i), we get by putting $M^* = M$ that $M' \not\leq M$ or $M' = M$. In the former case we get from $M \not\leq M'$ and negative transitivity that $M \not\leq M$, contradicting the reflexivity of $\leq$. The latter case doesn't fare better; it contradicts $M \not\leq M'$ as well. From (ii), we get that $M' \not\leq M^*$ or $M' = M^*$ for some M^*. In the former case, $M \not\leq M'$ and negative transitivity would give us $M \not\leq M^*$, contradicting (ii). In the latter case, $M \not\leq M'$ would equally well yield $M \not\leq M^*$, again contradicting (ii). Q.E.D.

Theorem 7. *Let $\dot{-}$ be a contraction function over a logically finite theory A. Then the following conditions are equivalent:*

(i) *$\dot{-}$ is a simple base contraction function;*
(ii) *$\dot{-}$ is a prioritized base contraction function;*
(iii) *$\dot{-}$ satisfies $(\dot{-}1)-(\dot{-}7)$, $(\dot{-}8r)$, and $(\dot{-}8c)$.*

Proof. That (i) entails (ii) is trivial.

To show that (ii) entails (iii), observe that by Theorem 6, $\mathscr{C}(\langle B_i \rangle) = \mathscr{C}(\mathscr{S}(\mathscr{P}(\langle B_i \rangle)))$ and that $\mathscr{P}(\langle B_i \rangle)$ is reflexive and negatively transitive. So by Corollary 2, $\mathscr{C}(\langle B_i \rangle)$ satisfies $(\dot{-}1)-(\dot{-}7)$, $(\dot{-}8r)$, and $(\dot{-}8c)$.

To show that (iii) entails (i), we conclude from (iii) with the help of Corollary 2 that there is a negatively transitive preference relation $\leq$ such that $\dot{-} = \mathscr{C}(\mathscr{S}(\leq))$.

We know from Lemma 10 that the latter is identical with $\mathscr{C}(\mathscr{S}(\mathscr{P}(\mathscr{B}(\leq))))$. So there is a simple belief base B, namely $B = \mathscr{B}(\leq)$, such that $\dot{-} = \mathscr{C}(\mathscr{S}(\mathscr{P}(B)))$. By Theorem 6 then, $\dot{-} = \mathscr{C}(B)$. Q.E.D.

Theorem 7 tells us that for every prioritized belief base $\langle B_i \rangle$ for A there is a simple belief base B for A such that $\mathscr{C}(B) = \mathscr{C}(\langle B_i \rangle)$. In the finite case, every prioritization can be encoded syntactically. Does this mean that prioritization is superfluous? In answering this question we first have to emphasize that our generation of B from $\langle B_i \rangle$ took a rather roundabout route: $B = \mathscr{B}(\mathscr{P}(\langle B_i \rangle))$. An interesting problem now is whether a more perspicuous construction of B from $\langle B_i \rangle$ is possible. This question, too, is put in informal terms, and as such it permits no absolutely precise answer. Still we think that the answer must be no. Even for the most straightforward prioritized belief bases $\langle B_i \rangle$ the generated simple base $B = \mathscr{B}(\mathscr{P}(\langle B_i \rangle))$ becomes grossly unintuitive, and there is no prospect of finding different solutions to the problem. Consider for example the base containing p and $p \to q$ with the alternative prioritizations $\langle \{p \to q\}, \{p\} \rangle$ and $\langle \{p\}, \{p \to q\} \rangle$. In the former case, $B = \{p \wedge q, p, p \vee q, q \to p\}$, while in the latter case, $B' = \{q, p \leftrightarrow q\}$ will lead to exactly the same results as the prioritized belief base. But in neither case is there anything like a transparent relation to the original base $\langle B_i \rangle$. It appears that prioritization is useful, notwithstanding its formal dispensability in the finite case.

Acknowledgements I gratefully acknowledge numerous comments and suggestions by David Makinson which have once again been extremely helpful.

Appendix 1: Relating Theory Change and Nonmonotonic Logic

In a recent paper, Makinson and Gärdenfors (1991) make considerable progress toward linking the areas of theory change and nonmonotonic reasoning. They close with the following problem which we quote in full:

> We end with an important open question. In their (1985), Alchourrón, Gärdenfors and Makinson established a representation theorem for theory revision operations $*$ satisfying conditions ($*$1)–($*$8), in terms of "transitively relational partial meet revisions". The proof went via a representation theorem for a contraction function $\dot{-}$ satisfying certain conditions ($\dot{-}$1) – ($\dot{-}$8). On the other hand, Kraus et al. (1990) have established a representation theorem for supraclassical, cumulative and distributive nonmonotonic inference relations $\mid\sim$ defined between individual propositions, in terms of classical stoppered preferential model structures. The former proof is relatively short and abstract; the latter seems more complex. Also, the latter has not been generalized to a representation theorem for supraclassical, cumulative and distributive inference operations $C : 2^L \to 2^L$ [...] Does the representation theorem for theory change hold the key for a solution to this problem of extending the Kraus/Lehmann/Magidor representation theorem to the infinite case—despite the failure of consistency preservation for preferential model structures? Or do we have two essentially different representation problems? (Makinson and Gärdenfors 1991, pp. 203–204, notation adapted)

This question does not have a simple answer. Three different points have to be taken into consideration.

First, the approaches of AGM and KLM are not as distinct as Makinson and Gärdenfors seemed to assume. AGM contract and revise by single propositions, and similarly KLM consider the nonmonotonic consequences of simple propositions. Makinson and Gärdenfors's equation $y \in A * x$ iff $x \mid\!\sim y$ (iff $y \in C(x)$) fully reflects this. A truly infinitistic stance towards both theory change and nonmonotonic logic is taken only by Lindström (1991). The question of whether the theory A is logically finite has no bearing on this issue. Here Makinson and Gärdenfors saw a difference which simply does not exist.

Secondly, Makinson and Gärdenfors tacitly passed over the fact that in KLM there is no counterpart to ($*8$) or ($\dot{-}8$). But this difference is indeed crucial, as is clear from a later paper of Lehmann and Magidor (1992). As regards the *preferential* logics dealt with by KLM, no "relatively short and abstract" proof seems to be possible. Thus it appears reasonable to construe Makinson and Gärdenfors's question as referring to the *rational* logics treated by Lehmann and Magidor.

But thirdly, Lehmann and Magidor's rational logics still differ from AGM-style theory revisions in that they are not required to satisfy a condition of consistency preservation which corresponds to postulate ($\dot{-}4$) for contractions. In this respect, Makinson and Gärdenfors show a keen sense of the intricacies of the situation. In unpublished notes, we have applied the techniques developed in this paper to the problem of providing rational logics with canonical models. We have found that it *is* possible to transfer our proof to the area of nonmonotonic logic, but that consistency preservation with respect to the underlying monotonic logic *Cn* is, in fact, indispensable for a perfect matching. The reason is that the compactness property presupposed for *Cn* runs idle if there are any "inaccessible worlds" (and this is just what a violation of consistency preservation amounts to). Since the results of our efforts bear considerable similarity with the venerable presentation in Lewis (1973)—except for the fact that the role of Lewis's extra-linguistic propositions (sets of "real" possible worlds) is played by the propositions of the object language—we had better refrain from expounding them here in more detail.

Appendix 2: Some Proofs and Examples[12]

Proof of Lemma 1. (a) Immediate from the definition of relationality. (If x is greatest in S' then it is so in all subsets of S' in which it is contained. If x is greatest in every S_i then it is so in $\bigcup S_i$.)

(b) Let γ be 12-covering and satisfy (I) and (II). We have to show that for every $S \in \mathscr{X}$, $\gamma(S) = \{x \in S :$ for all $y \in S$, $\{x,y\} \in \mathscr{X}$ and $x \in \gamma(\{x,y\})\}$. $LHS \subseteq RHS$: Let $x \in \gamma(S)$ and $y \in S$. As γ is 12-covering, $\{x,y\} \in \mathscr{X}$ and (I) gives us $x \in \gamma(\{x,y\})$.

[12]This appendix was not included in the original 1993 publication of this paper.

$RHS \subseteq LHS$: Let $x \in S$ and assume that for every $y \in S$, $\{x, y\} \in \mathscr{X}$ and $x \in \gamma(\{x, y\})$. Note that $\bigcup\{\{x, y\} : y \in S\} = S \in \mathscr{X}$ and $x \in \bigcup\{\{x, y\} : y \in S\}$. By (II), we get $x \in \gamma(\bigcup\{\{x, y\} : y \in S\}) = \gamma(S)$, as desired.

(c) Let γ be 12-covering and satisfy (I).

$\mathscr{P}_2(\gamma) \subseteq \mathscr{P}(\gamma)$: This direction is always valid.

$\mathscr{P}(\gamma) \subseteq \mathscr{P}_2(\gamma)$: Let $x \leq_\gamma y$. Then there is an $S \in \mathscr{X}$ such that $y \in \gamma(S) \subseteq S$ and $x \in S$. Because γ is 12-covering, $\{x, y\} \in \mathscr{X}$. But as $\{x, y\} \subseteq S$, (I) gives us $y \in \gamma(\{x, y\})$, so $x \leq_{\gamma,2} y$.

(d) Follows from (b) and (c).

(e) Follows from (a), (b), and (d).

(f) Let γ be additive and satisfy (I) and (II). We have to show that for every $S \in \mathscr{X}$, $\gamma(S) = \{x \in S :$ for all $y \in S$ there is a $T \in \mathscr{X}$ such that $x \in \gamma(T)$ and $y \in T\}$.

$LHS \subseteq RHS$: Take $T = S$.

$RHS \subseteq LHS$: Let $x \in S$ and assume that for all $y \in S$ there is a T^y such that $x \in \gamma(T^y)$ and $y \in T^y$. Thus $x \in \bigcap \gamma(T^y)$ and $S \subseteq \bigcup T^y$. From the former and the additivity of γ we get $x \in \gamma(\bigcup T^y)$, by (II), and now the latter and (I) yield $x \in \gamma(S)$, as desired. Q.E.D.

Proof of Lemma 2. (a) Let γ be $12n$-covering and satisfy (I). We show the claim for $\mathscr{P}_2(\gamma)$, which is identical with $\mathscr{P}(\gamma)$, by Lemma 1(c). Suppose for reductio that $x_1 <_\gamma^2 x_2 <_\gamma^2 \cdots <_\gamma^2 x_n <_\gamma^2 x_1$ for some $x_1, \ldots x_n \in X$. That is, $\gamma(\{x_i, x_{i+1}\}) = \{x_{i+1}\}$ with $+$ denoting addition modulo n. Now consider $\gamma(\{x_1, \ldots, x_n\}) \neq \emptyset$. Let $x_k \in \gamma(\{x_1, \ldots, x_n\})$. But $x_k \notin \gamma(\{x_k, x_k + 1\})$. This is in contradiction with (I). The rest of (a) is trivial.

(b) Let γ be 123-covering and satisfy (I) and (III). We show the claim for $\mathscr{P}_2(\gamma)$, which is identical with $\mathscr{P}(\gamma)$, by Lemma 1(c). Assume that $x \not\leq_{\gamma,2} y$ and $y \not\leq_{\gamma,2} z$. By definition of $\leq_{\gamma,2}$, this means that $y \notin \gamma(\{x, y\})$ and $z \notin \gamma(\{y, z\})$. Now consider $\gamma(\{x, y, z\})$. By (I), $y \notin \gamma(\{x, y, z\})$ and $z \notin \gamma(\{x, y, z\})$, so $\gamma(\{x, y, z\}) = \{x\}$ since $\gamma(\{x, y, z\})$ must be a non-empty subset of $\{x, y, z\}$. By (I), $x \in \gamma(\{x, z\})$, so $\gamma(\{x, y, z\}) \subseteq \gamma(\{x, z\})$. Hence, by (III), $\gamma(\{x, z\}) \subseteq \gamma(\{x, y, z\}) = \{x\}$, so $z \notin \gamma(\{x, z\})$, i.e. $x \not\leq_{\gamma,2} z$, as desired.

(c) Let γ be finitely additive and satisfy (IV), and let $x \leq_\gamma y$ and $y \leq_\gamma z$. This means that there is an $S \in \mathscr{X}$ such that $y \in \gamma(S)$ and $x \in S$ and a $T \in \mathscr{X}$ such that $z \in \gamma(T)$ and $y \in T$. Now consider $S \cup T$. Obviously, $x \in S \cup T$. In order to show that $x \leq_\gamma z$ it suffices to show that $z \in \gamma(S \cup T)$. By finite additivity, $S \cup T \in \mathscr{X}$. By $z \in \gamma(T)$ and (IV), it suffices to show that $\gamma(S \cup T) \cap T \neq \emptyset$. Suppose for reductio that $\gamma(S \cup T) \cap T = \emptyset$. Then, since $\emptyset \neq \gamma(S \cup T) \subseteq S \cup T$, $\gamma(S \cup T) \cap S \neq \emptyset$. So by (IV), $\gamma(S) \subseteq \gamma(S \cup T)$. So $y \in \gamma(S \cup T)$. But since also $y \in T$, $\gamma(S \cup T) \cap T \neq \emptyset$ after all, and we have a contradiction.

(Notice that an attempted proof of the transitivity of $\leq_{\gamma,2}$ would also need, apart from the 123-covering condition, (I) in order to come from $z \in \gamma(\{x, y, z\})$ to $z \in \gamma(\{x, z\})$, so we can rest content with (c).) Q.E.D.

Proof of Lemma 3. (a) Assume that $\mathscr{S}(\leq)$ is no selection function over $\mathscr{X}$. By the definition of $\mathscr{S}(\leq)$, this can only happen if some $S \in \mathscr{X}$ possesses no greatest element under $\leq$. Thus for every $x_i \in S$ there is an $x_{i+1} \in S$ such that $x_{i+1} \not\leq x_1$.

This, however, contradicts smoothness. That $\mathcal{S}(\leq)$ satisfies (I) and (II) follows from Lemma 1(a).

(b) Let γ be relational with respect to some negatively transitive and negatively well-founded relation $\leq$. (Or alternatively, let γ be subtractive and relational with respect to some negatively transitive and $\mathcal{X}$-smooth relation $\leq$.) Let $S, S' \in \mathcal{X}$, $S \subseteq S'$ and $\gamma(S') \subseteq \gamma(S)$. We want to show that $\gamma(S) \subseteq \gamma(S')$. Suppose for reductio that this is not the case, i.e., that there is some x which is in $\gamma(S)$ but not in $\gamma(S')$. The latter means that there is some y_1 in S' such that $y_1 \not\leq x$. As $x \in \gamma(S)$, $y_1 \in S' - S$. But because $\gamma(S') \subseteq \gamma(S) \subseteq S$, $y_1 \notin \gamma(S')$. So there is some $y_2 \in S'$ such that $y_2 \not\leq y_1$. By negative transitivity, $y_2 \not\leq x$. So by the same reasoning as before, $y_2 \in S' - S$ and $y_2 \notin \gamma(S')$. So there is some $y_3 \in S'$ such that $y_3 \not\leq y_2$. By negative transitivity again, $y_3 \not\leq x$, and the same reasoning can be continued again and again. What we get is an infinite chain $y_1, y_2, y_3, \ldots$ in $S' - S$ such that $\cdots \not\leq y_3 \not\leq y_2 \not\leq y_1$. But this contradicts the negative well-foundedness of $\leq$ (or the subtractivity of γ, which guarantees that $S' - S \in \mathcal{X}$, and smoothness of $\leq$).

(c) Let γ be relational with respect to some transitive relation $\leq$. Let $S, S' \in \mathcal{X}$, $S \subseteq S'$ and $x \in \gamma(S') \cap S$. By relationality, the latter conditions says that $y \leq x$ for all $y \in S'$. Let $z \in \gamma(S)$. We have to show that $z \in \gamma(S')$, i.e., by relationality, that $y \leq z$ for all $y \in S'$. But since $x \in S$ and $z \in \gamma(S)$, $x \leq z$, so since $y \leq x$ for all $y \in S'$, the desired conclusion follows from the transitivity of $\leq$. Q.E.D.

Proof of Lemma 4. (a) Let γ be subtractive.

(I) implies (I′): Let $S, S', S \cup S' \in \mathcal{X}$ and $x \in \gamma(S \cup S')$. As $S, S' \subseteq S \cup S'$ and x is either in S or in S', we get $x \in \gamma(S) \cup \gamma(S')$, by (I), as desired. *(I′) implies (I)*: Let $S, S' \in \mathcal{X}$ and $S \subseteq S'$. By subtractivity, $S' - S \in \mathcal{X}$ as well. Using $S' = S \cup (S' - S)$ and (I′), we get $\gamma(S') \subseteq \gamma(S) \cup \gamma(S' - S)$. Intersecting both sides with S, then, we get $\gamma(S') \cap S \subseteq (\gamma(S) \cap S) \cup (\gamma(S' - S) \cap S) = \gamma(S)$, since $\gamma(S) \subseteq S$

(b) First we note that for finitely additive γ, (II′) is equivalent to

(II$_{fin}$) For all $\{S_i : i \in I, I \text{ finite}\} \subseteq \mathcal{X}$ such that $\bigcup\{S_i : i \in I\} \in \mathcal{X}$, $\bigcap\{\gamma(S_i) : i \in I\} \subseteq \gamma(\bigcup\{S_i : i \in I\})$

Now we simply modify the proof of Lemma 1(f) by taking only finitely many of the T^y's into consideration. This is possible because compactness can be applied to the inclusion $S \subseteq \bigcup T^y$ in that proof.

(c) Let γ satisfy (I).

(III) implies (III′): Let $S, S \cup S' \in \mathcal{X}$ and $\gamma(S \cup S') \cap S' = \emptyset$. From the latter condition and $\gamma(S \cup S') \subseteq S \cup S'$ it follows that $\gamma(S \cup S') \subseteq S$. Hence, by (I), $\gamma(S \cup S') = S \cap \gamma(S \cup S') \subseteq \gamma(S)$. Thus, by (III), $\gamma(S) \subseteq \gamma(S \cup S')$, as desired. *(III′) implies (III)*: Let $S, S' \in \mathcal{X}$, $S \subseteq S'$ and $\gamma(S') \subseteq \gamma(S)$. Since $S' = S \cup (S' - S) \in \mathcal{X}$ and $\gamma(S') \subseteq \gamma(S) \subseteq S$, it holds that $\gamma(S \cup (S' - S)) \cap (S' - S) = \emptyset$. By (III′), we get $\gamma(S) \subseteq \gamma(S')$, as desired.

(d) Immediate.

(e) That (I&II) implies (I) is immediate if we put $I = \{i\}$ and $S_i = S'$, and that (I&II) implies (II) is immediate if we put $S = \bigcup S_i$ and observe that $\gamma(S_i) \subseteq S$.

In order to see that (I) and (II) taken together imply (I&II), we note that $\bigcup\{S_i : i \in I\} \in \mathscr{X}$, by additivity. So (II) gives us $S \cap \bigcap \gamma(S_i) \subseteq S \cap \gamma(\bigcup S_i)$, and (I) gives us $S \cap \gamma(\bigcup S_i) \subseteq \gamma(S)$, whenever $S \subseteq \bigcup S_i$. So in this case $S \cap \bigcap \gamma(S_i) \subseteq \gamma(S)$, as desired.

(f) Similar to (e). Q.E.D.

Re section "Representation Theorems for Contraction Functions": Example of a relational partial meet contraction function which does not satisfy ($\dot{-}$8r).

Consider a propositional language L with denumerably many atoms $p_1, p_2, p_3, \ldots$ and q and r. Let $A = Cn(p_1, p_2, p_3, \ldots, q, r)$. Let W_i, W_i^q, W_i^r, for $i = 1, 2, 3, 4, \ldots$, be the maximally consistent set of propositions defined by

$p_i \in W_i, W_i^q, W_i^r$, for every i,
$\neg p_j \in W_i, W_i^q, W_i^r$, for every i and every $j \neq i$,
$q \in W_i^q$ and $\neg q \in W_i, W_i^r$, for every i,
$r \in W_i^r$ and $\neg r \in W_i, W_i^q$, for every i.

Let $\mathbf{W}^{qr} = \bigcup\{\{W_i, W_i^q, W_i^r\} : i = 1, 2, 3, \ldots\}$. Remember that $V_A = \mathbf{W} - [\![A]\!]$. Let the relation $<_W$ over V_A be defined as follows:

$W <_W W'$ for every $W \in \mathbf{W}^{qr}$ and $W' \in V_A - \mathbf{W}^{qr}$
$W_j^q <_W W_{j+1}$, $W_j^q <_W W_{j+1}^q$, and $W_j^q <_W W_{j+1}^r$, for every $j \in \{0, 2, 4, 6, \ldots\}$
$W_k^r <_W W_{k+1}$, $W_k^r <_W W_{k+1}^q$, and $W_k^r <_W W_{k+1}^r$, for every $k \in \{1, 3, 5, \ldots\}$
These are all pairs standing in the relation $<_W$ (see Fig. 16.3).

Then define, for the purposes of this example, $M \leq M'$ iff $M' \not< M$ iff $\mathscr{W}(M) \not<_W \mathscr{W}(M')$. Then clearly, the $\leq$-greatest elements of $A \perp x$ correspond to the $<$-minimal elements of $[\![\neg x]\!]$. Note that $\leq$ is connected, acyclic and conversely well-founded (no infinite ascending chains), but not negatively transitive. Let $\dot{-} = \mathscr{C}(\mathscr{S}(\leq))$ be the relational partial meet contraction determined by $\leq$.

Now observe that the greatest elements of $A \perp (q \wedge r) = A \perp q \cup A \perp r$ are exactly $A \cap W_0$, $A \cap W_0^q$, and $A \cap W_0^r$. Since p_0 is an element of all these sets, we have $p_0 \in A \dot{-} (q \wedge r)$.

The greatest elements of $A \perp q$ are $A \cap W_k$ and $A \cap W_k^r$ for every $k \in \{0, 1, 3, 5, \ldots\}$. Obviously, $q \to x \in A \dot{-} q$ for every $x \in A$. Furthermore, $\neg p_n \vee \neg p_m \in A \dot{-} q$ for every $n \neq m$. Informally speaking, $A \dot{-} q$ contains the information that exactly one proposition p_k, for $k \in \{0, 1, 3, 5, \ldots\}$, is true. But this cannot be expressed in our first-order propositional language. Note in particular that $A \dot{-} q$ does not contain any finite disjunction of p_i's.

The greatest elements of $A \perp r$ are $A \cap W_j$ and $A \cap W_j^q$ for every $j \in \{0, 2, 4, 6, \ldots\}$. The information contained in $A \dot{-} r$ is analogous with that in $A \dot{-} q$, with the even positive integers playing the role of the odd ones in the previous case.

We find that $A \dot{-} q$ and $A \dot{-} r$ do not even jointly imply p_0. For consider the maximally consistent set W^* containing $\neg p_i$ for every $i \geq 0$, together with $\neg q$ and $\neg r$. Considering what we have just said about $A \dot{-} q$ and $A \dot{-} r$, it becomes clear that W^* contains both of these contracted theories, as well as $\neg p_0$.

Since $p_0 \in A \dot{-} (q \wedge r)$ but $p_0 \notin Cn(A \dot{-} q \cup A \dot{-} r)$, we have a violation of ($\dot{-}$8r).

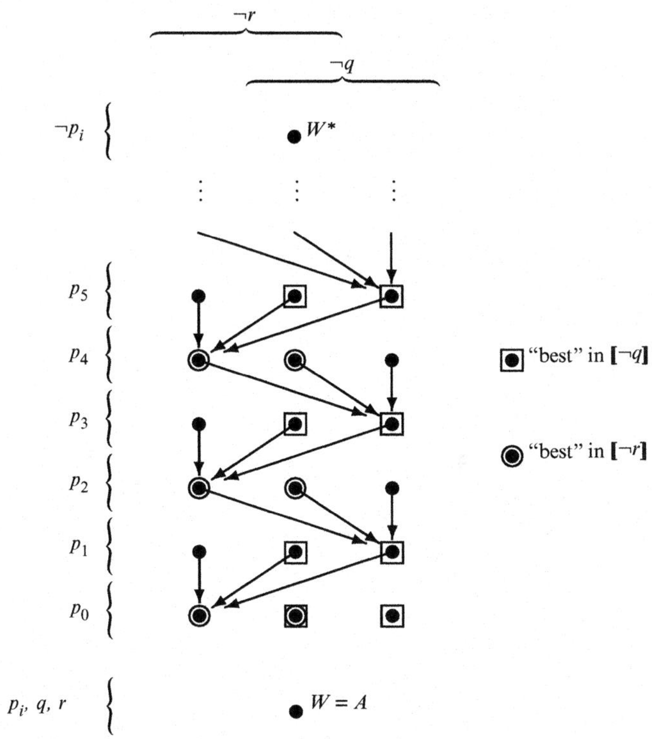

Fig. 16.3 Relational partial meet contraction not satisfying (∸8r)

Proof of Lemma 9. We show that for every $M \in A \perp x$,

(i) M is maximal in $A \perp x$ under $\lessdot$
  is equivalent to
(ii) $N \subseteq M$ for some N which is maximal in $B \perp x$ under $\lessdot$.

(i) implies (ii): (i) means that M is such that for no M' and i it holds that $M \lessdot_i M'$. Now take $N = M \cap B$. We have to show that N is maximal in $B \perp x$. First, it is clear that $N \nvdash x$. Next, we verify that N is a maximal subset of B which does not imply x. Suppose there is a $y \in B - N$ such that $N \cup \{y\} \nvdash x$, but then, with $y \in B_k$, $N_{\geq k} \cup \{y\} \nvdash x$. So there would be a superset M' of $N_{\geq k} \cup \{y\}$ in $A \perp x$ for which $M \lessdot_{\geq k} M'$, contradicting (i). Finally, we have to show that there is no $N' \in B \perp x$ which is better than N under $\lessdot$. But by the same argument as just used, it is impossible that there be some $y \in B_k$ such that $y \in N' - N$ and $N \lessdot_k N'$.

(ii) implies (i): Suppose for reductio that (ii) but not (i). By the latter, there is an M' such that $M \lessdot M'$. But then $M \cap B \lessdot M' \cap B$. Take, on the other hand, the N from (ii). Since $N \subseteq M \cap B$, we have $N \lessdot M' \cap B$. But since $M' \cap B \nvdash x$, N cannot be maximal in $B \perp x$, contradicting (ii). Q.E.D.

Re section "The Logic of Prioritized Base Contractions in the Finite Case": Illustration of the counterexample to ($\dot{-}$8r) in straightforward prioritized base contractions and its solution by the official definition. Let $B = \{(p \wedge q) \vee (p \wedge r) \vee (q \wedge r),\ r \to (p \leftrightarrow q),\ q \wedge (p \leftrightarrow r),\ p \wedge r\}, A = Cn(B), A \dot{-}' x = \bigcap \{Cn(N) : N \in \gamma(B \perp x)\}$ and $A \dot{-} x = \bigcap \{Cn(N \cup \{x \to a\}) : N \in \gamma(B \perp x)\}$. See Fig. 16.4

Re section "The Logic of Prioritized Base Contractions in the Finite Case": Example illustrating how prioritized belief bases can be replaced by simple ones.

Consider the propositional language L over the two atoms p and q. Let the belief base contain the propositions p and $p \to q$. We contrast the prioritizations $\langle \{p \to q\}, \{p\} \rangle$ and $\langle \{p\}, \{p \to q\} \rangle$. In the former case, $B = \mathscr{B}(\mathscr{P}(\langle \{p \to q\}, \{p\} \rangle)) = \{p \wedge q, p, p \vee q, q \to p\}$, while in the latter case, $B' = \mathscr{B}(\mathscr{P}(\langle \{p\}, \{p \to q\} \rangle)) = \{p \wedge q, q, p \leftrightarrow q, \top\}$.

In Figs. 16.5 and 16.6, "$\begin{pmatrix} x \\ y \\ z \end{pmatrix}$" should be read as "$A \dot{-} x = Cn(y)$ according to prioritized base contraction, while $A \dot{-} x = Cn(z)$ according to the corresponding simple base contraction". In the last two lines, the left argument of '$\wedge$' comes from straightforward base contraction and the right argument is the recovery-guaranteeing appendage used in our official definition of prioritized base contractions.

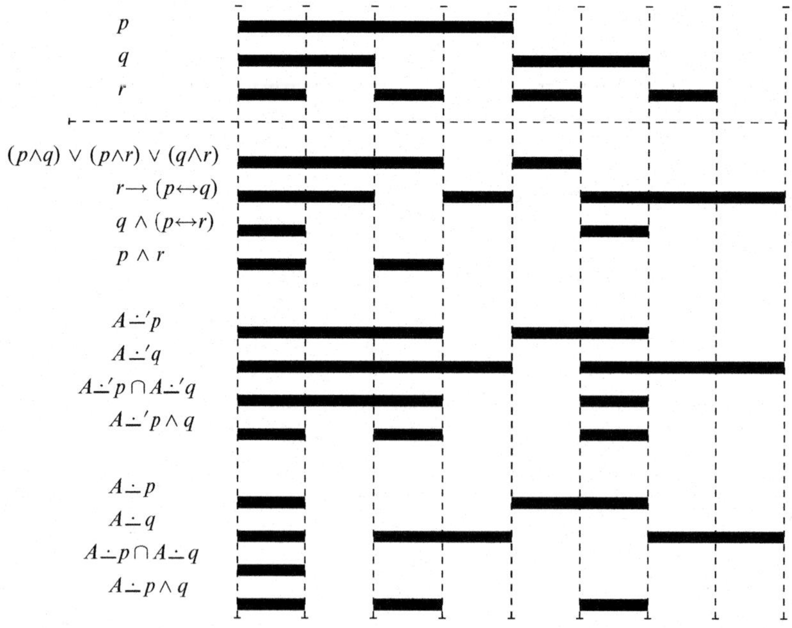

Fig. 16.4 Prioritized base contraction $\dot{-}'$ with simple base violating ($\dot{-}$8r), and strengthened contraction $\dot{-}$ satisfying ($\dot{-}$8r)

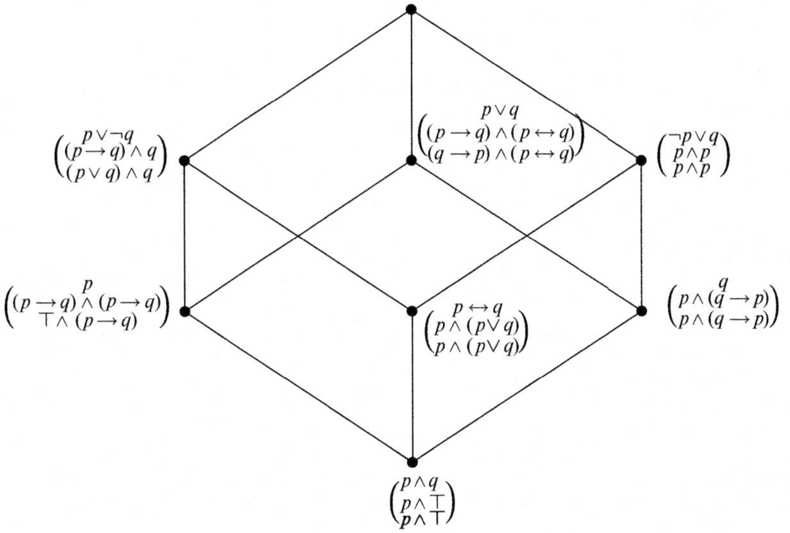

Fig. 16.5 Prioritized base contraction of a two-element base, with p having priority over $p \rightarrow q$

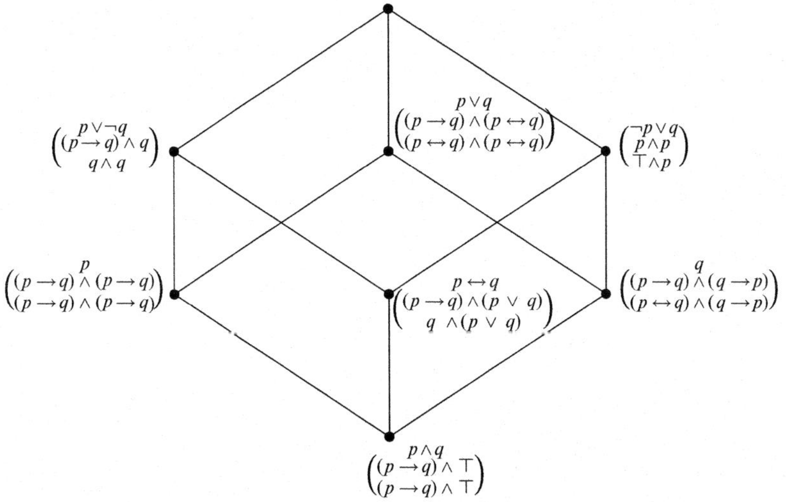

Fig. 16.6 Prioritized base contraction of a two-element base, with $p \rightarrow q$ having priority over p

Remarks. Prioritized and corresponding simple base contractions indeed lead to the same results—as they should, according to the proof of Theorem 7. But note that the recovery-guaranteeing appendage is essential in quite a few cases. As for the contraction function $\dot{-}$, the differences in prioritization are effective only in the case of $A\dot{-}(p \wedge q)$, $A\dot{-}q$ and $A\dot{-}(p \leftrightarrow q)$.

References

Alchourrón, C., Gärdenfors, P., & Makinson, D. (1985). On the logic of theory change: Partial meet contraction functions and their associated revision functions. *Journal of Symbolic Logic, 50*, 510–530.

Alchourrón, C., & Makinson, D. (1986). Maps between some different kinds of contraction function: The finite case. *Studia Logica, 45*, 187–198.

Arrow, K. J. (1959). Rational choice functions and orderings. *Economica*, N.S., *26*, 121–127.

Chernoff, H. (1954). Rational selection of decision functions. *Econometrica, 22*, 422–443.

de Condorcet, N. (1785). Essai sur l'application de l'analyse á la probabilité des décisions rendues á la pluralité des voix. Paris: Imprimerie Royale. Reprint Cambridge: Cambridge University Press (2014).

Fuhrmann, A., & Morreau, M. (Eds.) (1991). *The logic of theory change* (Lecture notes in computer science, Vol. 465). Berlin: Springer.

Gärdenfors, P. (1988). *Knowledge in flux: Modeling the dynamics of epistemic states*. Cambridge: Bradford Books/MIT.

Gärdenfors, P. (Ed.) (1992). *Belief revision*. Cambridge: Cambridge University Press.

Gärdenfors, P., & Makinson, D. (1988). Revisions of knowledge systems using epistemic entrenchment. In M. Vardi (Ed.), *Theoretical aspects of reasoning about knowledge* (pp. 83–95). Los Altos: Morgan Kaufmann.

Grove, A. (1988). Two modellings for theory change. *Journal of Philosophical Logic, 17*, 157–170.

Hansson, S. O. (1992). Similarity semantics and minimal changes of belief. *Erkenntnis, 37*, 401–429.

Herzberger, H. G. (1973). Ordinal preference and rational choice. *Econometrica, 41*, 187–237.

Katsuno, H., & Mendelzon, A. O. (1991). Propositional knowledge base revision and minimal change. *Artificial Intelligence, 52*, 263–294.

Kraus, S., Lehmann, D., & Magidor, M. (1990). Nonmonotonic reasoning, preferential models and cumulative logics. *Artificial Intelligence, 44*, 167–207.

Lehmann, D., & Magidor, M. (1992). What does a conditional knowledge base entail? *Artificial Intelligence, 55*, 1–60.

Lewis, D. (1973). *Counterfactuals*. Oxford: Blackwell.

Lewis, D. (1981). Ordering semantics and premise semantics for counterfactuals. *Journal of Philosophical Logic, 10*, 217–234.

Lindström, S. (1991). A semantic approach to nonmonotonic reasoning: Inference and choice. University of Uppsala, April 1991 (manuscript).

Makinson, D., & Gärdenfors, P. (1991). Relations between the logic of theory change and nonmonotonic logic. In Fuhrmann & Morreau (1991) (pp. 185–205).

Nebel, B. (1989). A knowledge level analysis of belief revision. In R. Brachman, H. Levesque, & R. Reiter (Eds.), *Proceedings of the 1st International Conference on Principles of Knowledge Representation and Reasoning* (pp. 301–311). San Mateo: Morgan Kaufmann.

Nebel, B. (1992). Syntax-based approaches to belief revision. In Gärdenfors (1992) (pp. 52–88).

Rott, H. (1991). Two methods of constructing contractions and revisions of knowledge systems. *Journal of Philosophical Logic, 20*, 149–173.

Rott, H. (1992a). On the logic of theory change: More maps between different kinds of contraction function. In Gärdenfors (1992) (pp. 122–141).

Rott, H. (1992b). Preferential belief change using generalized epistemic entrenchment. *Journal of Logic, Language and Information, 1*, 45–78.

Rott, H. (2003). Basic entrenchment. *Studia Logica, 73*, 257–280.

Samuelson, P. A. (1950). The problem of integrability in utility theory. *Economica*, N.S., *17*, 355–381.

Sen, A. K. (1982) *Choice, welfare and measurement*. Oxford: Blackwell.

Suzumura, K. (1983) *Rational choice, collective decisions, and social welfare*. Cambridge: Cambridge University Press.

Uzawa, H. (1956). Note on preference and axioms of choice. *Annals of the Institute of Statistical Mathematics, 8*, 35–40.

Chapter 17
A Survey of Ranking Theory

Wolfgang Spohn

Introduction

Epistemology is concerned with the fundamental laws of thought, belief, or judgment. It may inquire the fundamental relations among the objects or contents of thought and belief, i.e., among propositions or sentences. Then we enter the vast realm of formal logic. Or it may inquire the activity of judging or the attitude of believing itself. Often, we talk as if this would be a yes or no affair. From time immemorial, though, we know that judgment is firm or less than firm, that belief is a matter of degree. This insight opens another vast realm of formal epistemology.

Logic received firm foundations already in ancient philosophy. It took much longer, though, until the ideas concerning the forms of (degrees of) belief acquired more definite shape. Despite remarkable predecessors in Indian, Greek, Arabic, and medieval philosophy, the issue seemed to seriously enter the agenda of intellectual history only in the sixteenth century with the beginning of modern philosophy. Cohen (1980) introduced the handy, though somewhat tendentious opposition between Baconian and Pascalian probability. This suggests that the opposition was already perceivable with the work of Francis Bacon (1561–1626) and Blaise Pascal (1623–1662). In fact, philosophers were struggling to find the right mould. In that struggle, Pascalian probability, which *is* probability *simpliciter*, was the first to take a clear and definite shape, viz. in the middle of seventeenth century (cf. Hacking 1975), and since then it advanced triumphantly. The extent to which it interweaves with our cognitive enterprise has become nearly total (cf. the marvelous collection of Krüger et al. 1987). There certainly were alternative ideas. However, probability theory was always far ahead; indeed, the distance ever increased. The winner takes it all!

W. Spohn (✉)
Fachbereich Philosophie, Universität Konstanz, 78457 Konstanz, Germany
e-mail: wolfgang.spohn@uni-konstanz.de

H. Arló-Costa et al. (eds.), *Readings in Formal Epistemology*, Springer Graduate Texts in Philosophy 1, DOI 10.1007/978-3-319-20451-2_17

I use 'Baconian probability' as a collective term for the alternative ideas. This is legitimate since there are strong family resemblances among the alternatives. Cohen has chosen an apt term since it gives historical depth to ideas that can be traced back at least to Bacon (1620) and his powerful description of 'the method of lawful induction'. Jacob Bernoulli and Johann Heinrich Lambert struggled with a non-additive kind of probability. When Joseph Butler and David Hume spoke of probability, they often seemed to have something else or more general in mind than our precise explication. In contrast to the German Fries school British nineteenth century's philosophers like John Herschel, William Whewell, and John Stuart Mill elaborated non-probabilistic methods of inductive inference. And so forth.[1]

Still, one might call this an underground movement. The case of alternative forms of belief became a distinct hearing only in the second half of the twentieth century. On the one hand, there were scattered attempts like the 'functions of potential surprise' of Shackle (1949), heavily used and propagated in the epistemology of Isaac Levi since his (1967), Rescher's (1964) account of hypothetical reasoning, further developed in his (1976) into an account of plausible reasoning, or Cohen's (1970) account of induction which he developed in his (1977) under the label 'Non-Pascalian probability', later on called 'Baconian'. On the other hand, one should think that modern philosophy of science with its deep interest in theory confirmation and theory change produced alternatives as well. Indeed, Popper's hypothetical-deductive method proceeded non-probabilistically, and Hempel (1945) started a vigorous search for a qualitative confirmation theory. However, the former became popular rather among scientists than among philosophers, and the latter petered out after 25 years, at least temporarily.

I perceive all this rather as a prelude, preparing the grounds. The outburst came only in the mid 70s, with strong help from philosophers, but heavily driven by the needs of Artificial Intelligence. Not only deductive, but also inductive reasoning had to be implemented in the computer, probabilities appeared intractable[2], and thus a host of alternative models were invented: a plurality of default logics, non-monotonic logics and defeasible reasonings, fuzzy logic as developed by Zadeh (1975, 1978), possibility theory as initiated by Zadeh (1978) and developed by Dubois and Prade (1988), the Dempster-Shafer belief functions originating from Dempster (1967, 1968), but essentially generalized by Shafer (1976), AGM belief revision theory (cf. Gärdenfors 1988), a philosophical contribution with great success in the AI market, Pollock's theory of defeasible reasoning (summarized in Pollock 1995), and so forth. The field has become rich and complex. There are attempts of unification like Halpern (2003) and huge handbooks like Gabbay et al. (1994). One hardly sees the wood for trees. It seems that what had been forgotten for centuries had to be made good for within decades.

[1]This is not the place for a historical account. See, e.g., Cohen (1980) and Shafer (1978) for some details.

[2]Only Pearl (1988) showed how to systematically deal with probabilities without exponential computational explosion.

Ranking theory, first presented in Spohn (1983, 1988)[3], belongs to this field as well. Since its development, by me and others, is scattered in a number of papers, one goal of the present paper is to present an accessible survey of the present state of ranking theory.[4] This survey will emphasize the philosophical applications, thus reflecting my bias towards philosophy. My other goal is justificatory. Of course, I am not so blinded to claim that ranking theory would be *the* adequate account of Baconian probability. As I said, 'Baconian probability' stands for a collection of ideas united by family resemblances; and I shall note some of the central resemblances in the course of the paper. However, there is a multitude of epistemological purposes to serve, and it is entirely implausible that there is one account to serve all. Hence, postulating a reign of probability is silly, and postulating a duumvirate of probability and something else is so, too. Still, I am not disposed to see ranking theory as just one offer among many. On many scores, ranking theory seems to me to be superior to rival accounts, the central score being the notion of *conditional* ranks. I shall explain what these scores are, thus trying to establish ranking theory as one particularly useful account of the laws of thought.

The plan of the paper is simple. In the five subsections of section "The theory", pp. 305ff, I shall outline the main aspects of ranking theory. This central section will take some time. I expect the reader to get impatient meanwhile; you will get the strong impression that I am not presenting an alternative to (Pascalian) probability, as the label 'Baconian' suggests, but simply probability itself in a different disguise. This is indeed one way to view ranking theory, and a way, I think, to understand its virtues. However, the complex relation between probability and ranking theory, though suggested at many earlier points, will be systematically discussed only in section "Ranks and probabilities", pp. 328ff. The section "Further comparisons", pp. 335ff, will finally compare ranking theory to some other accounts of Baconian probability.

The Theory

Basics

We have to start with fixing the objects of the cognitive attitudes we are going to describe. This is a philosophically highly contested issue, but here we shall stay conventional without discussion. These objects are pure contents, i.e., propositions.

---

[3]There I called its objects ordinal conditional functions. Goldszmidt and Pearl (1996) started calling them ranking functions, a usage I happily adapted.

[4]In the meantime, my comprehensive book on ranking theory (Spohn 2012) has appeared. This paper may also serve as an introduction to this book. Reversely, various topics, which are only touched here and then referred back to older papers of mine, are developed in this book in a better and more comprehensive way.

To be a bit more explicit: We assume a non-empty set W of mutually exclusive and jointly exhaustive possible worlds or *possibilities*, as I prefer to say, for avoiding the grand associations of the term 'world' and for allowing to deal with *de se* attitudes and related phenomena (where doxastic alternatives are considered to be centered worlds rather than worlds). And we assume an algebra $\mathcal{A}$ of subsets of W, which we call *propositions*. All the functions we shall consider for representing doxastic attitudes will be functions defined on that algebra $\mathcal{A}$.

Thereby, we have made the philosophically consequential decision of treating doxastic attitudes as intensional. That is, when we consider sentences such as "a believes (with degree r) that p", then the clause p is substitutable salva veritate by any clause q expressing the same proposition and in particular by any logically equivalent clause q. This is so because by taking propositions as objects of belief we have decided that the truth value of such a belief sentence depends only on the proposition expressed by p and not on the particular way of expressing that proposition. The worries provoked by this decision are not our issue.

The basic notion of ranking theory is very simple:

Definition 17.1 Let $\mathcal{A}$ be an algebra over W. Then κ is a *negative ranking function*[5] for $\mathcal{A}$ iff κ is a function from $\mathcal{A}$ into $\mathbf{R}^* = \mathbf{R}^+ \cup \{\infty\}$ (i.e., into the set of non-negative reals plus infinity) such that for all $A, B \in \mathcal{A}$:

$$\kappa(W) = 0 \text{ and } \kappa(\varnothing) = \infty, \tag{17.1}$$

$$\kappa(A \cup B) = \min\{\kappa(A), \kappa(B)\} \quad [\textit{the law of disjunction (for negative ranks)}]. \tag{17.2}$$

$\kappa(A)$ is called the (*negative*) *rank* of A.

It immediately follows for each $A \in \mathcal{A}$:

$$\text{either } \kappa(A) = 0 \text{ or } \kappa(\overline{A}) = 0 \text{ or both} \quad [\textit{the law of negation}]. \tag{17.3}$$

A negative ranking function κ, this is the standard interpretation, expresses a *grading of disbelief* (and thus something negative, hence the qualification). If $\kappa(A) = 0$, A is not disbelieved at all; if $\kappa(A) > 0$, A is disbelieved to some positive degree. Belief in A is the same as disbelief in $\overline{A}$; hence, A is *believed* in κ iff $\kappa(\overline{A}) > 0$. This entails (via the law of negation), but is not equivalent to $\kappa(A) = 0$. The latter is compatible also with $\kappa(\overline{A}) = 0$, in which case κ is neutral or unopinionated concerning A. We shall soon see the advantage of explaining belief in this indirect way via disbelief.

A little example may be instructive. Let us look at Tweetie of which default logic is very fond. Tweetie has, or fails to have, each of the three properties: being a bird

[5]For systematic reasons I am slightly rearranging my terminology from earlier papers. I would be happy if the present terminology became the official one.

(B), being a penguin (P), and being able to fly (F). This makes for eight possibilities. Suppose you have no idea what Tweetie is, for all you know it might even be a car. Then your ranking function may be the following one, for instance:[6]

κ	$B\&\overline{P}$	$B \& P$	$\overline{B}\&\overline{P}$	$\overline{B}\&P$
F	0	4	0	11
$\overline{F}$	2	1	0	8

In this case, the strongest proposition you believe is that Tweetie is *either* no penguin and no bird $(\overline{B}\&\overline{P})$ *or* a flying bird and no penguin (F & B & $\overline{P}$). Hence, you neither believe that Tweetie is a bird (B) nor that it is not a bird ($\overline{B}$). You are also neutral concerning its ability to fly. But you believe, for instance: if Tweetie is a bird, it is not a penguin and can fly $(B \rightarrow \overline{P}\&F)$; and if Tweetie is not a bird, it is not a penguin $(\overline{B} \rightarrow \overline{P})$ – each if-then taken as material implication. In this sense you also believe: if Tweetie is a penguin, it can fly ($P \rightarrow F$); and if Tweetie is a penguin, it cannot fly $(P \rightarrow \overline{F})$ – but only because you believe that it is not a penguin in the first place; you simply do not reckon with its being a penguin. If we understand the if-then differently, as we shall do later on, the picture changes. The larger ranks in the last column indicate that you strongly disbelieve that penguins are not birds. And so we may discover even more features of this example.

What I have explained so far makes clear that we have already reached the first fundamental aim ranking functions are designed for: the *representation of belief*. Indeed, we may define $\mathcal{B}_\kappa = \{A \mid \kappa(\overline{A}) > 0\}$ to be the *belief set* associated with the ranking function κ. This belief set is finitely *consistent* in the sense that whenever $A_1, \ldots, A_n \in \mathcal{B}_\kappa$, then $A_1 \cap \ldots \cap A_n \neq \varnothing$; this is an immediate consequence of the law of negation. And it is finitely *deductively closed* in the sense that whenever $A_1, \ldots, A_n \in \mathcal{B}_\kappa$ and $A_1 \cap \ldots \cap A_n \subseteq B \in \mathcal{A}$, then $B \in \mathcal{B}_\kappa$; this is an immediate consequence of the law of disjunction. Thus, belief sets just have the properties they are normally assumed to have. (The finiteness qualification is a little cause for worry that will be addressed soon.)

There is a big argument about the rationality postulates of consistency and deductive closure; we should not enter it here. Let me only say that I am disappointed by all the attempts I have seen to weaken these postulates. And let me point out that the issue was essentially decided at the outset when we assumed belief to operate on propositions or truth-conditions or sets of possibilities. With these assumptions we ignore the relation between propositions and their sentential expressions or modes of presentation; and it is this relation where all the problems hide.

[6]I am choosing the ranks in an arbitrary, though intuitively plausible way (just as I would have to arbitrarily choose plausible subjective probabilities, if the example were a probabilistic one). The question how ranks may be measured will be taken up in section "The dynamics of belief and the measurement of belief", pp. 316ff.

When saying that ranking functions represent belief I do not want to further qualify this. One finds various notions in the literature, full beliefs, strong beliefs, weak beliefs, one finds a distinction of acceptance and belief, etc. In my view, these notions and distinctions do not respond to any settled intuitions; they are rather induced by various theoretical accounts. Intuitively, there is only one perhaps not very clear, but certainly not clearly divisible phenomenon which I exchangeably call believing, accepting, taking to be true, etc.

However, if the representation of belief were our only aim, belief sets or their logical counterparts as developed in doxastic logic (see already Hintikka 1962) would have been good enough. What then is the purpose of the ranks or degrees? Just to give another account of the intuitively felt fact that belief is graded? But what guides such accounts? Why should degrees of belief behave like ranks as defined? Intuitions by themselves are not clear enough to provide this guidance. Worse still, intuitions are usually tainted by theory; they do not constitute a neutral arbiter. Indeed, problems already start with the intuitive conflict between representing belief and representing degrees of belief. By talking of belief *simpliciter*, as I have just insisted, I seem to talk of *ungraded* belief.

The only principled guidance we can get is a theoretical one. The degrees must serve a clear theoretical purpose and this purpose must be shown to entail their behavior. For me, the theoretical purpose of ranks is unambiguous; this is why I invented them. It is the *representation of the dynamics of belief*; that is the second fundamental aim we pursue. How this aim is reached and why it can be reached in no other way will unfold in the course of this section. This point is essential; as we shall see, it distinguishes ranking theory from all similarly looking accounts, and it grounds its superiority.

For the moment, though, let us look at a number of variants of Definition 17.1. Above I mentioned the finiteness restriction of consistency and deductive closure. I have always rejected this restriction. An inconsistency is irrational and to be avoided, be it finitely or infinitely generated. Or, equivalently, if I take to be true a number of propositions, I take their conjunction to be true as well, even if the number is infinite. If we accept this, we arrive at a somewhat stronger notion:

Definition 17.2 Let $\mathcal{A}$ be a complete algebra over W (closed also under infinite Boolean operations). Then κ is a *complete negative ranking function* for $\mathcal{A}$ iff κ is a function from W into $\mathbf{N}^+ = \mathbf{N} \cup \{\infty\}$ (i.e., into the set of non-negative integers plus infinity) such that $\kappa^{-1}(0) \neq \varnothing$ and and $\kappa^{-1}(n) \in \mathcal{A}$ for each $n \in \mathbf{N}^+$. κ is extended to propositions by defining $\kappa(\varnothing) = \infty$ and $\kappa(A) = \min\{\kappa(w) \mid w \in A\}$ for each non-empty $A \in \mathcal{A}$.

Obviously, the propositional function satisfies the laws of negation and disjunction. Moreover, we have for any $\mathcal{B} \subseteq \mathcal{A}$:

$$\kappa(\cup\mathcal{B}) = \min\{\kappa(B) \mid B \in \mathcal{B}\} \qquad [\textit{the law of infinite disjunction}]. \qquad (17.4)$$

Due to completeness, we could start in Definition 17.2 with the point function and then define the set function as specified. Equivalently, we could have defined the set functions by the conditions (17.1) and (17.4) and then reduce the set function to a point function. Henceforth I shall not distinguish between the point and the set function. Note, though, that without completeness the existence of an underlying point function is not guaranteed; the relation between point and set function in this case is completely cleared up in Huber (2006).

Why are complete ranking functions confined to integers? The reason is condition (17.4). It entails that any infinite set of ranks has a minimum and hence that the range of a complete ranking function is well-ordered. Hence, the natural numbers are a natural choice. In my first publications (1983) and (1988) I allowed for more generality and assumed an arbitrary set of ordinal numbers as the range of a ranking function. However, since we want to calculate with ranks, this meant to engage into ordinal arithmetic, which is awkward. Therefore I later confined myself to complete ranking functions as defined above.

The issue about condition (17.4) was first raised by Lewis (1973, sect. 1.4) where he introduced the so-called Limit Assumption in relation to his semantics of counterfactuals. Endorsing (17.4), as I do, is tantamount to endorsing the Limit Assumption. Lewis finds reason against it, though it does not affect the *logic* of counterfactuals. From a semantic point of view, I do not understand his reason. He requests us to counterfactually suppose that a certain line is longer than an inch and asks how long it would or might be. He argues in effect that for each $\varepsilon > 0$ we should accept as true: "If the line would be longer than 1 inch, it would not be longer than $1 + \varepsilon$ inches." This strikes me as blatantly inconsistent, even if we cannot derive a contradiction in counterfactual logic (due to its ω–incompleteness). Therefore, I am accepting the Limit Assumption and, correspondingly, the law of infinite disjunction. This means in particular that in that law the minimum must not be weakened to the infimum.

Though I prefer complete ranking functions for the reasons given, the issue will have no further relevance here. In particular, if we assume the algebra of propositions to be finite, each ranking function is complete, and the issue does not arise. In the sequel, you can add or delete completeness as you wish.

Let me add another observation apparently of a technical nature. It is that we can mix ranking functions in order to form a new ranking function. This is the content of

Definition 17.3 Let Λ be a non-empty set of negative ranking functions for an algebra $\mathcal{A}$ of propositions, and let ρ be a complete negative ranking function over Λ. Then κ defined by

$$\kappa(A) = \min\{\lambda(A) + \rho(\lambda) \mid \lambda \in \Lambda\} \text{ for all } A \in \mathcal{A} \tag{17.5}$$

is obviously a negative ranking function for $\mathcal{A}$ as well and is called the *mixture of* Λ *by* ρ.

It is nice that such mixtures make formal sense. However, we shall see in the course of this paper that the point is more than a technical one; such mixtures will acquire deep philosophical importance later on.

So far, (degree of) disbelief was our basic notion. Was this necessary? Certainly not. We might just as well express things in positive terms:

Definition 17.4 Let $\mathcal{A}$ be an algebra over W. Then π is a *positive ranking function* for $\mathcal{A}$ iff π is a function from $\mathcal{A}$ into $\mathbf{R}^*$ such that for all $A, B \in \mathcal{A}$:

$$\pi(\varnothing) = 0 \text{ and } \pi(W) = \infty, \tag{17.6}$$

$$\pi(A \cap B) = \min\{\pi(A), \pi(B)\} \quad [\textit{the law of conjunction for positive ranks}]. \tag{17.7}$$

Positive ranks express *degrees of belief.* $\pi(A) > 0$ says that A is believed (to some positive degree), and $\pi(A) = 0$ says that A is not believed. Obviously, positive ranks are the dual to negative ranks; if $\pi(A) = \kappa(\overline{A})$ for all $A \in \mathcal{A}$, then π is a positive function iff κ is a negative ranking function.

Positive ranking functions seem distinctly more natural. Why do I still prefer the negative version? A superficial reason is that we have seen complete negative ranking functions to be reducible to point functions, whereas it would obviously be ill-conceived to try the same for the positive version. This, however, is only indicative of the main reason. Despite appearances, we shall soon see that negative ranks behave very much like probabilities. In fact, this parallel will serve as our compass for a host of exciting observations. (For instance, in the finite case probability measures can also be reduced to point functions.) If we were thinking in positive terms, this parallel would remain concealed.

There is a further notion that may appear even more natural:

Definition 17.5 Let $\mathcal{A}$ be an algebra over W. Then τ is a *two-sided ranking function*[7] for $\mathcal{A}$ iff τ is a function from $\mathcal{A}$ into $\mathbf{R} \cup \{-\infty, \infty\}$ such that there is a negative ranking function κ and its positive counterpart π for which for all $A \in \mathcal{A}$:

$$\tau(A) = \kappa(\overline{A}) - \kappa(A) = \pi(A) - \kappa(A).$$

Obviously, we have $\tau(A) > 0$, < 0, or $= 0$ according to whether A is believed, disbelieved, or neither. In this way, the belief values of all propositions are expressed in a single function. Moreover, we have the appealing law that $\tau(\overline{A}) = -\tau(A)$. For some purposes this is a useful notion that I shall readily employ. However, its formal behavior is awkward. Its direct axiomatic characterization would have been cumbersome, and its simplest definition consisted in its reduction to the other notions.

[7] In earlier papers I called this a belief function, obviously an unhappy term which has too many different uses. This is one reason fort the mild terminological reform proposed in this paper.

Still, this notion suggests an interpretational degree of freedom so far unnoticed.[8] We might ask: Why does the range of belief extend over all the positive reals in a two-sided ranking function and the range of disbelief over all the negative reals, whereas neutrality shrinks to rank 0? This looks unfair. Why may unopinionatedness not occupy a much broader range? Indeed, why not? We might just as well distinguish some positive rank or real z and define the closed interval $[-z, z]$ as the range of neutrality. Then $\tau(A) > z$ expresses belief in A and $\tau(A) < -z$ disbelief in A. This is a viable interpretation; in particular, consistency and deductive closure of belief sets would be preserved. However, 0 would still be a distinguished rank in this interpretation; it marks *central* neutrality, as it were, since it is the only rank x for which we may have $\tau(A) = \tau(\overline{A}) = x$.

The interpretational freedom appears quite natural. After all, the notion of belief is certainly vague and can be taken more or less strict. We can do justice to this vagueness with the help of the parameter z. The crucial point, though, is that we always get the formal structure of belief we want to get, however we fix that parameter. The principal lesson of this observation is, hence, that it is not the notion of belief which is of basic importance; it is rather the formal structure of ranks. The study of belief *is* the study of *that* structure. Still, it would be fatal to simply give up talking of belief in favor of ranks. Ranks express beliefs, even if there is interpretational freedom. Hence, it is of paramount importance to maintain the intuitive connection. In the sequel, I shall stick to my standard interpretation and equate belief in A with $\tau(A) > 0$, even though this is a matter of decision.

Let us pause for a moment and take a brief look back. What I have told so far probably sounds familiar. One has quite often seen all this, in this or a similar form – where the similar form may also be a comparative one: as long as only the ordering and not the numerical properties of the degrees of belief are relevant, a ranking function may also be interpreted as a weak ordering of propositions according to their plausibility, entrenchment, credibility, etc. Often things are cast in negative terms, as I primarily do, and often in positive terms. In particular, the law of negation securing consistency and the law of disjunction somehow generalizing deductive closure (we still have to look at the point more thoroughly) or their positive counterparts are pervasive. If one wants to distinguish a common core in that ill-defined family of Baconian probability, it is perhaps just these two laws.

So, why invent a new name, 'ranks', for familiar stuff? The reason lies in the second fundamental aim associated with ranking functions: to account for the dynamics of belief. This aim has been little pursued under the label of Baconian probability, but it is our central topic for the rest of this section. Indeed, everything stands and falls with our notion of conditional ranks; it is the distinctive mark of ranking theory. Here it is:

Definition 17.6 Let κ be a negative ranking function for $\mathcal{A}$ and $\kappa(A) < \infty$. Then the *conditional rank* of $B \in \mathcal{A}$ *given* A is defined as $\kappa(B \mid A) = \kappa(A \cap B) - \kappa(A)$.

[8] I am grateful to Matthias Hild for making this point clear to me.

The function $\kappa_A : B \mapsto \kappa(B \mid A))$ is obviously a negative ranking function in turn and called the *conditionalization of* κ *by* A.

We might rewrite this definition as a law:

$$\kappa(A \cap B) = \kappa(A) + \kappa(B \mid A) \quad [\textit{the law of conjunction (for negative ranks)}]. \tag{17.8}$$

This amounts to the highly intuitive assertion that one has to add the degree of disbelief in B given A to the degree of disbelief in A in order to get the degree of disbelief in A-and-B.

Moreover, it immediately follows for all $A, B \in \mathcal{A}$ with $\kappa(A) < \infty$:

$$\kappa(B \mid A) = 0 \ \text{ or } \kappa(\overline{B} \mid A) = 0 \quad [\textit{conditional law of negation}]. \tag{17.9}$$

This law says that even conditional belief must be consistent. If both, $\kappa(B \mid A)$ and $\kappa(\overline{B} \mid A)$, were > 0, both, B and $\overline{B}$, would be believed given A, and this ought to be excluded, as long as the condition A itself is considered possible.

Indeed, my favorite axiomatization of ranking theory runs reversely, it consists of the definition of conditional ranks and the conditional law of negation. The latter says that min $\{\kappa(A \mid A \cup B), \kappa(B \mid A \cup B)\} = 0$, and this and the definition of conditional ranks entail that min $\{\kappa(A), \kappa(B)\} = \kappa(A \cup B)$, i.e., the law of disjunction. Hence, the only substantial assumption written into ranking functions is conditional consistency, and it is interesting to see that this entails deductive closure as well. Huber (2007) has further improved upon this important idea and shown that ranking theory is indeed nothing but the assumption of dynamic consistency, i.e., the preservation of consistency under any dynamics of belief. (He parallels, in a way, the dynamic Dutch book argument for probabilities by replacing its assumption of no sure loss by the assumption of consistency under all circumstances.)

It is instructive to look at the positive counterpart of negative conditional ranks. If π is the positive ranking function corresponding to the negative ranking function κ, Definition 17.6 simply translates into: $\pi(B \mid A) = \pi(\overline{A} \cup B) - \pi(\overline{A})$. Defining $A \rightarrow B = \overline{A} \cup B$ as set-theoretical 'material implication', we may as well write:

$$\pi(A \rightarrow B) = \pi(B \mid A) + \pi(\overline{A}) \quad [\textit{the law of material implication}]. \tag{17.10}$$

Again, this is highly intuitive. It says that the degree of belief in the material implication $A \rightarrow B$ is added up from the degree of belief in its vacuous truth (i.e., in $\overline{A}$) and the conditional degree of belief of B given A.[9] However, again comparing the negative and the positive version, one can already sense the analogy between probability and ranking theory from (17.8), but hardly from (17.10). This analogy will play a great role in the following subsections.

[9]Thanks again to Matthias Hild for pointing this out to me.

Two-sided ranks have a conditional version as well; it is straightforward. If τ is the two-sided ranking function corresponding to the negative κ and the positive π, then we may simply define:

$$\tau(B \mid A) = \pi(B \mid A) - \kappa(B \mid A) = \kappa(\overline{B} \mid A) - \kappa(B \mid A). \tag{17.11}$$

It will sometimes be useful to refer to these two-sided conditional ranks.

For illustration of negative conditional ranks, let us briefly return to our example of Tweetie. Above, I already mentioned various examples of if-then sentences, some held vacuously true and some non-vacuously. Now we can see that precisely the if-then sentences non-vacuously held true correspond to conditional beliefs. According to the κ specified, you believe, e.g., that Tweetie can fly given it is a bird (since $\kappa(\overline{F} \mid B) = 1$)) and also given it is a bird, but not a penguin (since $\kappa(\overline{F} \mid B\&\overline{P}) = 2$)), that Tweetie cannot fly given it is a penguin (since $\kappa(F \mid P) = 3$) and even given it is a penguin, but not a bird (since $\kappa(F \mid \overline{B}\&P) = 3$)). You also believe that it is not a penguin given it is a bird (since $\kappa(P \mid B) = 1$) and that it is a bird given it is a penguin (since $\kappa(\overline{B} \mid P) = 7$)). And so forth.

Let us now unfold the power of conditional ranks and their relevance to the dynamics of belief in several steps.

Reasons and Their Balance

The first application of conditional ranks is in the theory of confirmation. Basically, Carnap (1950) told us, confirmation is positive relevance. This idea can be explored probabilistically, as Carnap did. But here the idea works just as well. A proposition A confirms or supports or speaks for a proposition B, or, as I prefer to say, A is a reason for B, if A strengthens the belief in B, i.e., if B is more strongly believed given A than given $\overline{A}$, i.e., iff A is positively relevant for B. This is easily translated into ranking terms:

Definition 17.7 Let κ be a negative ranking function for $\mathcal{A}$ and τ the associated two-sided ranking function. Then $A \in \mathcal{A}$ is a *reason for* $B \in \mathcal{A}$ w.r.t. κ iff $\tau(B \mid A) > \tau(B \mid \overline{A})$, i.e., iff $\kappa(\overline{B} \mid A) > \kappa(\overline{B} \mid \overline{A})$ or $\kappa(B \mid A) < \kappa(B \mid \overline{A})$.

If P is a standard probability measure on $\mathcal{A}$, then probabilistic positive relevance can be expressed by $P(B \mid A) > P(B)$ or by $P(B \mid A) > P(B \mid \overline{A})$. As long as all three terms involved are defined, the two inequalities are equivalent. Usually, then, the first inequality is preferred because its terms may be defined while not all terms of the second inequality are defined. If P is a Popper measure, this argument does not hold, and then it is easily seen that the second inequality is more adequate, just as in the case of ranking functions.[10]

[10]A case in point is the so-called problem of old evidence, which has a simple solution in terms of Popper measures and the second inequality; cf. Joyce (1999, pp. 203ff.).

Confirmation or support may take four different forms relative to ranking functions, which are unfolded in

Definition 17.8 Let κ be a negative ranking function for $\mathcal{A}$, τ the associated two-sided ranking function, and $A, B \in \mathcal{A}$. Then

$$A \text{ is a } \begin{Bmatrix} \textit{additional} \\ \textit{sufficient} \\ \textit{necessary} \\ \textit{insufficient} \end{Bmatrix} \textit{reason for } B \text{ w.r.t. } \kappa \text{ iff } \begin{Bmatrix} \tau(B \mid A) > \tau(B \mid \overline{A}) > 0 \\ \tau(B \mid A) > 0 \geq \tau(B \mid \overline{A}) \\ \tau(B \mid A) \geq 0 > \tau(B \mid \overline{A}) \\ 0 > \tau(B \mid A) > \tau(B \mid \overline{A}) \end{Bmatrix}.$$

If A is a reason for B, it must obviously take one of these four forms; and the only way to have two forms at once is by being a necessary and sufficient reason.[11]

Talking of reasons here is, I find, natural, but it stirs a nest of vipers. There is a host of philosophical literature pondering about reasons, justifications, etc. Of course, this is a field where multifarious philosophical conceptions clash, and it is not easy to gain an overview over the fighting parties. Here is not the place for starting a philosophical argument[12], but by using the term 'reason' I want at least to submit the claim that the topic may gain enormously by giving a central place to the above explication of reasons.

To elaborate only a little bit: When philosophers feel forced to make precise their notion of a (theoretical, not practical) reason, they usually refer to the notion of a *deductive* reason, as fully investigated in deductive logic. The deductive reason relation is reflexive, transitive, and not symmetric. By contrast, Definition 17.7 captures the notion of a *deductive or inductive* reason. The relation embraces the deductive relation, but it is reflexive, symmetric, and not transitive. Moreover, the fact that reasons may be additional or insufficient reasons according to Definition 17.8 has been neglected by the relevant discussion, which was rather occupied with necessary and/or sufficient reasons. Pursue, though, the use of the latter terms throughout the history of philosophy. Their deductive explication is standard and almost always fits. Often, it is clear that the novel inductive explication given by Definition 17.8 would be inappropriate. Very often, however, the texts are open to that inductive explication as well, and systematically trying to reinterpret these old texts would yield a highly interesting research program in my view.

The topic is obviously inexhaustible. Let me take up only one further aspect. Intuitively, we weigh reasons. This is a most important activity of our mind. We do not only weigh practical reasons in order to find out what to do, we also weigh theoretical reasons. We are wondering whether or not we should believe B, we are searching for reasons speaking in favor or against B, we are weighing these reasons, and we hopefully reach a conclusion. I am certainly not denying the phenomenon of

[11] In earlier publications I spoke of weak instead of insufficient reasons. Thanks to Arthur Merin who suggested the more appropriate term to me.

[12] I attempted to give a partial overview and argument in Spohn (2001a).

inference that is also important, but what is represented as an inference often rather takes the form of such a weighing procedure. ‘Reflective equilibrium’ is a familiar and somewhat more pompous metaphor for the same thing.

If the balance of reasons is such a central phenomenon the question arises: how can epistemological theories account for it? The question is less well addressed than one should think. However, the fact that there is a perfectly natural Bayesian answer is a very strong and more or less explicit argument in favor of Bayesianism. Let us take a brief look at how that answer goes:

Let P be a (subjective) probability measure over $\mathcal{A}$ and let B be the focal proposition. Let us look at the simplest case, consisting of one reason A for B and the automatic counter-reason $\overline{A}$ against B. Thus, in analogy to Definition 17.7, $P(B \mid A) > P(B \mid \overline{A})$. How does P balance these reasons and thus fit in B? The answer is simple, we have:

$$P(B) = P(B \mid A) \cdot P(A) + P(B \mid \overline{A}) \cdot P(\overline{A}) . \tag{17.12}$$

This means that the probabilistic balance of reason is a *beam balance* in the literal sense. The length of the lever is $P(B \mid A) - P(B \mid \overline{A})$; the two ends of the lever are loaded with the *weights* $P(A)$ and $P(\overline{A})$ of the reasons; $P(B)$ divides the lever into two parts of length $P(B \mid A) - P(B)$ and $P(B) - P(B \mid \overline{A})$ representing the *strength* of the reasons; and then $P(B)$ must be chosen so that the beam is in balance. Thus interpreted (17.12) is nothing but the law of levers.

Ranking theory has an answer, too, and I am wondering who else has. According to ranking theory, the balance of reasons works like a *spring balance*. Let κ be a negative ranking function for $\mathcal{A}$, τ the corresponding two-sided ranking function, B the focal proposition, and A a reason for B. So, $\tau(B \mid A) > \tau(B \mid \overline{A})$. Again, it easily proved that always $\tau(B \mid A) \geq \tau(B) \geq \tau(B \mid \overline{A})$. But where in between is $\tau(B)$ located? A little calculation shows the following specification to be correct:

$$\begin{aligned}
&\text{Let } x = \kappa(B \mid \overline{A}) - \kappa(B \mid A) \text{ and } y = \kappa(\overline{B} \mid A) - \kappa(\overline{B} \mid \overline{A}). \textit{ Then}\\
&\text{(a) } x, y \geq 0 \text{ and } \tau(B \mid A) - \tau(B \mid \overline{A}) = x + y,\\
&\text{(b) } \tau(B) = \tau(B \mid \overline{A}), \text{ if } \tau(A) \leq -x,\\
&\text{(c) } \tau(B) = \tau(B \mid A), \text{ if } \tau(A) \geq y,\\
&\text{(d) } \tau(B) = \tau(A) + \tau(B \mid \overline{A}) + x, \text{ if } -x < \tau(A) < y.
\end{aligned} \tag{17.13}$$

This does not look as straightforward as the probabilistic beam balance. Still, it is not so complicated to interpret (17.13) as a spring balance. The idea is that you hook in the spring at a certain point, that you extend it by the force of reasons, and that $\tau(B)$ is where the spring extends. Consider first the case where $x, y > 0$. Then you hook in the spring at point 0 $(= \tau(B \mid \overline{A}) + x)$ and exert the force $\tau(A)$ on the spring. Either, this force transcends the lower stopping point $-x$ or the upper stopping point y. Then the spring extends exactly till the stopping point, as (17.13b + c) say. Or, the force $\tau(A)$ is less. Then the spring extends exactly by $\tau(A)$, according to (17.13d).

The second case is that $x = 0$ and $y > 0$. Then you fix the spring at $\tau(B \mid \overline{A})$, the lower point of the interval in which $\tau(B)$ can move. The spring cannot extend below that point, says (17.13b). But according to (17.13c + d) it can extend above, by the force $\tau(A)$, but not beyond the upper stopping point. For the third case $x > 0$ and $y = 0$ just reverse the second picture. In this way, the force of the reason A, represented by its two-sided rank $\tau(A)$, pulls the two-sided rank of the focal proposition B to its proper place within the interval $[\tau(B \mid \overline{A}), \tau(B \mid A)]$ fixed by the relevant conditional ranks.

I do not want to assess these findings in detail. You might prefer the probabilistic balance of reasons, a preference I would understand. You might be happy to have at least one alternative model, an attitude I recommend. Or you may search for further models of the weighing of reasons; in this case, I wish you good luck. What you may not do is ignoring the issue; your epistemology is incomplete if it does not take a stance. And one must be clear about what is required for taking a stance. As long as one considers positive relevance to be the basic characteristic of reasons, one must provide some notion of conditional degrees of belief, conditional probabilities, conditional ranks, or whatever. Without some well-behaved conditionalization one cannot succeed.

The Dynamics of Belief and the Measurement of Belief

Our next point will be to define a reasonable dynamics for ranking functions that entails a dynamic for belief. There are many causes which affect our beliefs, forgetfulness as a necessary evil, drugs as an unnecessary evil, and so on. From a rational point of view, it is scarcely possible to say anything about such changes.[13] The rational changes are due to experience or information. Thus, it seems we have already solved our task: if κ is my present doxastic state and I get informed about the proposition A, then I move to the conditionalization κ_A of κ by A. This, however, would be a bad idea. Recall that we have $\kappa_A(\overline{A}) = \infty$, i.e., A is believed with absolute certainty in κ_A; no future evidence could cast any doubt on the information. This may sometimes happen; but usually information does not come so firmly. Information may turn out wrong, evidence may be misleading, perception may be misinterpreted; we should provide for flexibility. How?

One point of our first attempt was correct; if my information consists solely in the proposition A, this cannot affect my beliefs conditional on A. Likewise it cannot affect my beliefs conditional on $\overline{A}$. Thus, it directly affects only how firmly I believe A itself. So, how firmly should I believe A? There is no general answer. I propose to turn this into a parameter of the information process itself; somehow the way I get informed about A entrenches A in my belief state with a certain firmness x.

[13] Although there is a (by far not trivial) decision rule telling that costless memory is never bad, just as costless information; cf. Spohn (1976/78, sect. 4.4).

The point is that as soon as the parameter is fixed and the constancy of the relevant conditional beliefs is accepted, my posterior belief state is fully determined. This is the content of

Definition 17.9 Let κ be a negative ranking function for $\mathcal{A}$, $A \in \mathcal{A}$ such that $\kappa(A)$, $\kappa(\overline{A}) < \infty$, and $x \in \mathbf{R}^*$. Then the $A \to x$-conditionalization $\kappa_{A \to x}$ of κ is defined by $\kappa_{A \to x}(B) = \begin{cases} \kappa(B \mid A) \text{ for } B \subseteq A, \\ \kappa(B \mid \overline{A}) + x \text{ for } B \subseteq \overline{A} \end{cases}$. From this $\kappa_{A \to x}(B)$ may be inferred for all other $B \in \mathcal{A}$ with the law of disjunction.

Hence, the effect of the $A \to x$-conditionalization is to shift the possibilities in A (to lower ranks) so that $\kappa_{A \to x}(A) = 0$ and the possibilities in $\overline{A}$ (to higher ranks) so that $\kappa_{A \to x}(\overline{A}) = x$. If one is attached to the idea that evidence consists in nothing but a proposition, the additional parameter is a mystery. The processing of evidence may indeed be so automatic that one hardly becomes aware of this parameter. Still, I find it entirely natural that evidence comes more or less firmly. Consider, for instance, the proposition: "There are tigers in the Amazon jungle", and consider six scenarios: (a) I read a somewhat sensationalist coverage in the yellow press claiming this, (b) I read a serious article in a serious newspaper claiming this, (c) I hear the Brazilian government officially announcing that tigers have been discovered in the Amazon area, (d) I see a documentary in TV claiming to show tigers in the Amazon jungle, (e) I read an article in *Nature* by a famous zoologist reporting of tigers there, (f) I travel there by myself and see the tigers. In all six cases I receive the information that there are tigers in the Amazon jungle, but with varying and, I find, increasing certainty.

One might object that the evidence and thus the proposition received is clearly a different one in each of the scenarios. The crucial point, though, is that we are dealing here with a fixed algebra $\mathcal{A}$ of propositions and that we have nowhere presupposed that this algebra consists of all propositions whatsoever; indeed, that would be a doubtful presupposition. Hence $\mathcal{A}$ may be course-grained and unable to represent the propositional differences between the scenarios; the proposition in $\mathcal{A}$ which is directly affected in the various scenarios may be just the proposition that there are tigers in the Amazon jungle. Still the scenarios may be distinguished by the firmness parameter.

So, the dynamics of ranking functions I propose is simply this: Suppose κ is your prior doxastic state. Now you receive some information A with firmness x. Then your posterior state is $\kappa_{A \to x}$. Your beliefs change accordingly; they are what they are according to $\kappa_{A \to x}$. Note that the procedure is iterable. Next, you receive the information B with firmness y, and so you move to $(\kappa_{A \to x})_{B \to y}$. And so on. This point will acquire great importance later on.

I should mention, though, that this iterability need not work in full generality. Let us call a negative ranking function κ *regular* iff $\kappa(A) < \infty$ for all $A \neq \varnothing$. Then we obviously have that $\kappa_{A \to x}$ is regular if κ is regular and $x < \infty$. Within the realm of regular ranking functions iteration of changes works without restriction. Outside this realm you may get problems with the rank ∞.

There is an important generalization of Definition 17.9. I just made a point of the fact that the algebra $\mathcal{A}$ may be too coarse-grained to propositionally represent all possible evidence. Why assume then that it is just one proposition A in the algebra that is directly affected by the evidence? Well, we need not assume this. We may more generally assume that the evidence affects some evidential partition $\mathcal{E}$ of W and assigns some new ranks to the members of the partition, which we may sum up in a complete ranking function λ on $\mathcal{E}$. Then we may define the $\mathcal{E}\rightarrow\lambda$*-conditionalization* $\kappa_{\mathcal{E}\rightarrow\lambda}$ of the prior κ by $\kappa_{\mathcal{E}\rightarrow\lambda}(B) = \kappa(B \mid E_i) + \lambda(E_i)$ for $B \subseteq E_i$ $(i = 1, \ldots, n)$ and infer $\kappa_{\mathcal{E}\rightarrow\lambda}(B)$ for all other B by the law of disjunction. This is the most general law of doxastic change in terms of ranking functions I can conceive of. Note that we may describe the $\mathcal{E}\rightarrow\lambda$-conditionalization of κ as the mixture of all κ_{E_i} $(i = 1, \ldots, n)$. So, this is a first useful application of mixtures of ranking functions.

Here, at last, the reader will have noticed the great similarity of my conditionalization rules with Jeffrey's probabilistic conditionalization first presented in Jeffrey (1965, ch. 11). Indeed, I have completely borrowed my rules from Jeffrey. Still, let us further defer the comparison of ranking with probability theory. The fact that many things run similarly does not mean that one can dispense with the one in favor of the other, as I shall make clear in section "Ranks and probabilities", pp. 328ff.

There is an important variant of Definition 17.9. Shenoy (1991), and several authors after him, pointed out that the parameter x as conceived in Definition 17.9 does not characterize the evidence as such, but rather the result of the interaction between the prior doxastic state and the evidence. Shenoy proposed a reformulation with a parameter exclusively pertaining to the evidence:

Definition 17.10 Let κ be a negative ranking function for $\mathcal{A}$, $A \in \mathcal{A}$ such that $\kappa(A)$, $\kappa(\overline{A}) < \infty$, and $x \in \mathbf{R}^*$. Then the $A{\uparrow}x$*-conditionalization* $\kappa_{A\uparrow x}$ of κ is defined by $\kappa_{A\uparrow x}(B) = \begin{cases} \kappa(B) - y \text{ for } B \subseteq A, \\ \kappa(B) + x - y \text{ for } B \subseteq \overline{A}, \end{cases}$ where $y = \min\{\kappa(A), x\}$. Again, $\kappa_{A\uparrow x}(B)$ may be inferred for all other $B \in \mathcal{A}$ by the law of disjunction.

The effect of this conditionalization is easily stated. It is, whatever the prior ranks of A and $\overline{A}$ are, that the possibilities within A improve by exactly x ranks in comparison to the possibilities within $\overline{A}$. In other words, we always have $\tau_{A\uparrow x}(A) - \tau(A) = x$ (in terms of the prior and the posterior two-sided ranking function).

It is thus appropriate to say that in $A{\uparrow}x$-conditionalization the parameter x exclusively characterizes the evidential impact. We may characterize the $A{\rightarrow}x$-conditionalization of Definition 17.9 as *result-oriented* and the $A{\uparrow}x$-conditionalization of Definition 17.10 as *evidence-oriented*. Of course, the two variants are easily interdefinable. We always have $\kappa_{A\rightarrow x} = \kappa_{A\uparrow y}$, where $y = x - \tau(A) = x + \tau(\overline{A})$. Still, it is sometimes useful to change perspective from one variant to the other.[14]

[14]Generalized probabilistic conditionalization as originally proposed by Jeffrey was result-oriented as well. However, Garber (1980) observed that there is also an evidence-oriented version of generalized probabilistic conditionalization. The relation, though, is not quite as elegant.

For instance, the evidence-oriented version helps to some nice observations. We may note that conditionalization is reversible: $(\kappa_{A\uparrow x})_{\overline{A}\uparrow x} = \kappa$. So, there is always a possible second change undoing the first. Moreover, changes always commute: $(\kappa_{A\uparrow x})_{B\uparrow y} = (\kappa_{B\uparrow y})_{A\uparrow x}$. In terms of result-oriented conditionalization this law would look more awkward. Commutativity does not mean, however, that one could comprise the two changes into a single change. Rather, the joint effect of two conditionalizations according to Definition 17.9 or 17.10 can in general only be summarized as one step of generalized $\mathcal{E} \rightarrow \lambda$-conditionalization. I think that reversibility and commutativity are intuitively desirable.

Change through conditionalization is driven by information, evidence, or perception. This is how I have explained it. However, we may also draw a more philosophical picture, we may also say that belief change according to Definition 17.9 or 17.10 is driven by reasons. Propositions for which the information received is irrelevant do not change their ranks, but propositions for which that information is positively or negatively relevant do change their ranks. The evidential force pulls at the springs and they must find a new rest position for all the propositions for or against which the evidence speaks, just in the way I have described in the previous subsection.

This is a strong picture captivating many philosophers. However, I have implemented it in a slightly unusual way. The usual way would have been to attempt to give some substantial account of what reasons are on which an account of belief dynamics is thereafter based. I have reversed the order. I have first defined conditionalization in Definition 17.6 and the more sophisticated form in Definitions 17.9 and 17.10. With the help of conditionalization, i.e., from this account of belief dynamics, I could define the reason relation in a way sustaining this picture. At the same time this procedure entails dispensing with a more objective notion of a reason. Rather, what is a reason for what is entirely determined by the subjective doxastic state as represented by the ranking function at hand. Ultimately, this move is urged by inductive skepticism as enforced by David Hume and reinforced by Nelson Goodman. But it does not mean surrender to skepticism. On the contrary, we are about to unfold a positive theory of rational belief and rational belief change, and we shall have to see how far it carries us.[15]

If one looks at the huge literature on belief change, one finds discussed predominantly three kinds of changes: expansions, revisions, and contractions. Opinions widely diverge concerning these three kinds. For Levi, for instance, revisions are whatever results form concatenating contractions and expansions according to the so-called Levi identity, and so he investigates the latter (see his most recent account in Levi 2004). The AGM approach characterizes both, revisions and contractions, and claims nice correspondences back and forth by help of the Levi and the Harper identity (cf., e.g., Gärdenfors 1988, chs. 3 and 4). Or one might object to the characterization of contraction, but accept that of revision, and hence reject these identities. And so forth.

[15]Here it does not carry us far beyond the beginnings. In Spohn (1991, 1999) I have argued for some stronger rationality requirements and their consequences.

I do not really want to discuss the issue. I only want to point out that we have already taken a stance insofar as expansions, revisions, and contractions are all special cases of our $A \rightarrow x$–conditionalization. This is more easily explained in terms of result-oriented conditionalization:

If $\kappa(A) = 0$, i.e., if A is not disbelieved, then $\kappa_{A\rightarrow x}$ represents an *expansion* by A for any $x > 0$. If $\kappa(\overline{A}) = 0$, the expansion is genuine, if $\kappa(\overline{A}) > 0$, i.e., if A is already believed in κ, the expansion is vacuous. Are there many different expansions? Yes and no. Of course, for each $x > 0$ a different $\kappa_{A\rightarrow x}$ results. On the other hand, one and the same belief set is associated with all these expansions. Hence, the expanded belief set is uniquely determined.

Similarly for revision. If $\kappa(A) > 0$, i.e., if A is disbelieved, then $\kappa_{A\rightarrow x}$ represents a genuine *revision* by A for any $x > 0$. In this case, the belief in $\overline{A}$ must be given up and along with it many other beliefs; instead, A must be adopted together with many other beliefs. Again, there are many different revisions, but all of them result in the same revised belief set.

Finally, if $\kappa(A) = 0$, i.e., if A is not disbelieved, then $\kappa_{A\rightarrow 0}$ represents contraction by A. If $\kappa(\overline{A}) > 0$, i.e., if A is even believed, the contraction is genuine; then belief in A is given up after contraction and no new belief adopted. If $\kappa(\overline{A}) = 0$, the contraction is vacuous; there was nothing to contract in the first place. If $\kappa(A) > 0$, i.e., if $\overline{A}$ is believed, then $\kappa_{A\rightarrow 0} = \kappa_{\overline{A}\rightarrow 0}$ rather represents contraction by $\overline{A}$.[16]

As observed in Spohn (1988, footnote 20) and more fully explained in Gärdenfors (1988, pp. 73f.), it is easily checked that expansions, revisions, and contractions thus defined satisfy all of the original AGM postulates (K*1-8) and (K^{-}1-8) (cf. Gärdenfors 1988, pp. 54–56 and 61–64) (when they are translated from AGM's sentential framework into our propositional or set-theoretical one). For those like me who accept the AGM postulates this is a welcome result.

For the moment, though, it may seem that we have simply reformulated AGM belief revision theory. This is not so; $A \rightarrow x$-conditionalization is much more general than the three AGM changes. This is clear from the fact that there are many different expansions and revisions that cannot be distinguished by the AGM account. It is perhaps clearest in the case of vacuous expansion that is no change at all in the AGM framework, but may well be a genuine change in the ranking framework, a redistribution of ranks which does not affect the surface of beliefs. Another way to state the same point is that insufficient and additional reasons also drive doxastic changes, which, however, are inexpressible in the AGM framework. For instance, if A is still disbelieved in the *$A\uparrow x$-conditionalization* $\kappa_{A\uparrow x}$ of κ (since $\kappa(A) > x$), one has obviously received only an insufficient reason for A, and the *$A\uparrow x$-conditionalization* might thus be taken to represent what is called non-prioritized belief revision in the AGM literature (cf. Hansson 1997).

[16] If we accept the idea in section "Basics" (p. 311) of taking the interval $[-z, z]$ of two-sided ranks as the range of neutrality, contraction seems to become ambiguous as well. However, the contraction just defined would still be distinguishable as a *central* contraction since it gives the contracted proposition central neutrality.

This is not the core of the matter, though. The core of the matter is *iterated belief change*, which I have put into the center of my considerations in Spohn (1983, sect. 5.3, 1988). As I have argued there, AGM belief revision theory is essentially unable to account for iterated belief change. I take 20 years of multifarious, but in my view unsatisfactory attempts to deal with that problem (see the overview in Rott 2008) as confirming my early assessment. By contrast, changes of the type $A \rightarrow x$-conditionalization are obviously indefinitely iterable.

In fact, my argument in Spohn (1988) was stronger. It was that if AGM belief revision theory is to be improved so as to adequately deal with the problem of iterated belief change, ranking theory is the only way to do it. I always considered this to be a conclusive argument in favor of ranking theory.

This may be so. Still, AGM theorists, and others as well, remained skeptical. "What exactly is the meaning of numerical ranks?" they asked. One may well acknowledge that the ranking apparatus works in a smooth and elegant way, has a lot of explanatory power, etc. But all this does not answer this question. Bayesians have met this challenge. They have told stories about the operational meaning of subjective probabilities in terms of betting behavior, they have proposed an ingenious variety of procedures for measuring this kind of degrees of belief. One would like to see a comparative achievement for ranking theory.

It exists and is finally presented in Hild and Spohn (2008). There is no space here to fully develop the argument. However, the basic point can easily be indicated so as to make the full argument at least plausible. The point is that ranks do not only account for iterated belief change, but can reversely be measured thereby. This may at first sound unhelpful. $A \rightarrow x$-conditionalization refers to the number x; so even if ranks can somehow be measured with the help of such conditionalizations, we do not seem to provide a fundamental measurement of ranks. Recall, however, that (central) contraction by A (or $\overline{A}$) is just $A \rightarrow 0$-conditionalization and is thus free of a hidden reference to numerical ranks; it only refers to rank 0 which has a clear operational or surface interpretation in terms of belief. Hence, the idea is to measure ranks by means of iterated contractions; if that works, it really provides a fundamental measurement of ranks that is based only on the beliefs one now has and one would have after various iterated contractions.

How does the idea work? Recall our observation above that the positive rank of a material implication $A \rightarrow B$ is the sum of the degree of belief in B given A and the degree of belief in the vacuous truth of $A \rightarrow B$, i.e., of $\overline{A}$. Hence, after contraction by $\overline{A}$, belief in the material implication $A \rightarrow B$ is equivalent to belief in B given A, i.e., to the positive relevance of A to B. This is how the reason relation, i.e., positive relevance, manifests itself in beliefs surviving contractions. Similarly for negative relevance and irrelevance.

Next observe that positive relevance can be expressed by certain inequalities for ranks that compare certain differences between ranks (similarly for negative relevance and irrelevance). This calls for applying the theory of difference measurement, as paradigmatically presented by Krantz et al. (1971, ch. 4).

Let us illustrate how this might work in our Tweetie example, pp. 306f. There we had specified a ranking function κ for the eight propositional atoms, entailing ranks for all 256 propositions involved. Focusing on the atoms, we are thus dealing with a realm $X = \{x_1, \ldots x_8\}$ (where $x_1 = B \;\&\; \overline{P} \;\&\; F$, etc.) and a numerical function f such that

$$f(x_1) = 0, \; f(x_2) = 4, \; f(x_3) = 0, \; f(x_4) = 11,$$
$$f(x_5) = 2, \; f(x_6) = 1, \; f(x_7) = 0, \; f(x_8) = 8.$$

This induces a lot of difference comparisons. For instance, we have, $f(x_6) - f(x_5) < f(x_2) - f(x_1)$. It is easily checked that this inequality says that, given B (being a bird), P (being a penguin) is positively relevant to $\overline{F}$ (not being able to fly) and that this in turn is equivalent with $P \rightarrow \overline{F}$ or $\overline{P} \rightarrow F$ still being believed after iterated contraction first by $\overline{B}$ and then by P and $\overline{P}$ (only one of the latter is a genuine contraction). Or we have $f(x_2) - f(x_6) = f(x_4) - f(x_8)$. Now, this is an equality saying that, given P, B (and $\overline{B}$) is irrelevant to F (and $\overline{F}$), and this in turn is equivalent with none of the four material implications from B or $\overline{B}$ to F or $\overline{F}$ being believed after iterated contraction first by $\overline{P}$ and then by B and $\overline{B}$ (again, only one of the latter is a genuine contraction).

Do these comparisons help to determine f? Yes, the example was so constructed: First, we have $f(x_1) - f(x_3) = f(x_3) - f(x_1) = f(x_1) - f(x_7)$. This entails $f(x_1) = f(x_3) = f(x_7)$. Let us choose this as the zero point of our ranking scale; i.e., $f(x_1) = 0$. Next, we have $f(x_5) - f(x_6) = f(x_6) - f(x_1)$. If we choose $f(x_6) = 1$ as our ranking unit, this entails $f(x_5) = 2$. Then, we have $f(x_2) - f(x_5) = f(x_5) - f(x_1)$, entailing $f(x_2) = 4$, and $f(x_8) - f(x_2) = f(x_2) - f(x_1)$, entailing $f(x_8) = 8$. Finally, we have $f(x_4) - f(x_8) = f(x_2) - f(x_6)$, the equation I had already explained, so that $f(x_4) = 11$. In this way, the difference comparisons entailed by our specification of f determine f uniquely up to a unit and a zero point.

The theory of difference measurement tells us how this procedure works in full generality. The resulting theorem says the following: Iterated contractions behave thus and thus if and only if differences between ranks behave thus and thus; and if differences between ranks behave thus and thus, then there is a ranking function measured on a ratio scale, i.e., unique up to a multiplicative constant, which exactly represents these differences. (See theorems 4.12 and 6.21 in Hild and Spohn (2008) for what "thus and thus" precisely means.)

On the one hand, this provides for an axiomatization of iterated contraction going beyond Darwiche and Pearl (1997), who presented generally accepted postulates of iterated revision and contraction and partially agreeing and partially disagreeing with further postulates proposed.[17] This axiomatization is assessible on intuitive and other grounds. On the other hand, one knows that if one accepts this axiomatization of iterated contraction one is bound to accept ranks as I have proposed them. Ranks do not fall from the sky, then; on the contrary, they uniquely represent contraction behavior.

[17] For an overview over such proposals see Rott (2008). For somewhat more detailed comparative remarks see Hild and Spohn (2008, sect. 5).

Conditional Independence and Bayesian Nets

It is worthwhile looking a bit more at the details of belief formation and revision. For this purpose we should give more structure to propositions. They have a Boolean structure so far, but we cannot yet compose them from basic propositions as we intuitively do. A common formal way to do this is to generate propositions from (random) variables. I identify a variable with the set of its possible values. I intend variables to be specific ones. E.g., the temperature at March 15, 2005, in Konstanz (not understood as the actual temperature, but as whatever it may be, say, between -100 and $+ 100$ °C) is such a variable. Or, to elaborate, if we consider each of the six general variables temperature, air pressure, wind, humidity, precipitation, cloudiness at each of the 500 weather stations in Germany twice a day at each of the 366 days of 2004, we get a collection of $6 \times 500 \times 732$ specific variables with which we can draw a detailed picture of the weather in Germany in 2004.

So, let V be the set of specific variables considered, where each $v \in V$ is just at least a binary set. A possible course of events or a possibility, for short, is just a selection function w for V, i.e., a function w on V such that $w(v) \in v$ for all $v \in V$. Hence, each such function specifies a way how the variables in V may realize. The set of all possibilities then simply is $W = \times V$. As before, propositions are subsets of W. Now, however, we can say that propositions are *about* certain variables. Let $X \subseteq V$. Then we say that $w, w' \in W$ *agree on* X iff $w(v) = w'(v)$ for all $v \in X$. And we define that a proposition A is *about* $X \subseteq V$ iff, for each w in A, all w' agreeing with w on X are in A as well. Let $\mathcal{A}(X)$ be the set of propositions about X. Clearly, $\mathcal{A}(X) \subseteq \mathcal{A}(Y)$ for $X \subseteq Y$, and $\mathcal{A} = \mathcal{A}(V)$. In this way, propositions are endowed with more structure. We may conceive of propositions about single variables as *basic* propositions; the whole algebra $\mathcal{A}$ is obviously generated by such basic propositions (at least if V is finite). So much as preparation for the next substantial step.

This step consists in more closely attending to (doxastic) dependence and independence in ranking terms. In a way, we have already addressed this issue: dependence is just positive or negative relevance, and independence is irrelevance. Still, let me state

Definition 17.11 Let κ be a negative ranking function for $\mathcal{A}$ and $A, B, C \in \mathcal{A}$. Then A and B are *independent* w.r.t. κ, i.e., $A \perp B$, iff $\tau(B \mid A) = \tau(B \mid \overline{A})$, i.e., iff for all $A' \in \{A, \overline{A}\}$ and $B' \in \{B, \overline{B}\}$ $\kappa(A' \cap B') = \kappa(A') + \kappa(B')$. And A and B are *independent given* C w.r.t. κ, i.e., $A \perp B / C$, iff A and B are independent w.r.t. κ_C.

(Conditional) independence is symmetric. If A is independent from B, $\overline{A}$ is so as well. If A is independent from B and A' disjoint from A, then A' is independent from B iff $A \cup A'$ is. $\varnothing$ and W are independent from all propositions. And so on.

The more interesting notion, however, is dependence and independence among variables. Look at probability theory where research traditionally and overwhelmingly focused on independent series of random variables and on Markov processes that are characterized by the assumption that past and future variables are independent given the present variable. We have already prepared for explaining this notion in ranking terms as well.

Definition 17.12 Let κ be a ranking function for $\mathcal{A} = \mathcal{A}(V)$, and let $X, Y, Z \subseteq V$ be sets of variables. Then X and Y are *independent* w.r.t. κ, i.e., $X \perp Y$, iff $A \perp B$ for all $A \in \mathcal{A}(X)$ and all $B \in \mathcal{A}(Y)$. Let moreover $\mathcal{Z}(Z)$ be the set of atoms of $\mathcal{A}(Z)$, i.e., the set of the logically strongest, non-empty proposition in $\mathcal{A}(Z)$. Then X and Y are *independent given* Z w.r.t. κ, i.e., $X \perp Y / Z$, iff $A \perp B/C$ for all $A \in \mathcal{A}(X)$, $B \in \mathcal{A}(Y)$, and $C \in \mathcal{Z}(Z)$.

In other words, $X \perp Y/Z$ iff all propositions about X are independent from all propositions about Y given any full specification of the variables in Z. Conditional independence among sets of variables obey the following laws:

Let κ be a negative ranking function for $\mathcal{A}(V)$. Then for any mutually disjoint $X, Y, Z, U \subseteq V$:

(a) if $X \perp Y/Z$, then $Y \perp X/Z$ [Symmetry],

(b) if $X \perp Y \cup U/Z$, then $X \perp Y/Z$ and $X \perp U/Z$ [Decomposition],

(c) $X \perp Y \cup U/Z$, then $X \perp Y/Z \cup U$ [Weak Union],

(d) $X \perp Y/Z$ and $X \perp U/Z \cup Y$, then $X \perp Y \cup U/Z$ [Contraction],

(e) if κ is regular and if $X \perp Y/Z \cup U$ and $X \perp U/Z \cup Y$, then $X \perp Y \cup U/Z$ [Intersection]

(17.14)

These are nothing but what Pearl (1988, p. 88) calls the *graphoid* axioms; the labels are his (cf. p. 84). (Note that law (d), contraction, has nothing to do with contraction in belief revision theory.) That probabilistic conditional independence satisfies these laws was first proved in Spohn (1976/78, sect. 3.2) and Dawid (1979). The ranking Theorem (17.14) was proved in Spohn (1983, sect. 5.3, 1988, sect. 6). I conjectured in 1976, and Pearl conjectured, too, that the graphoid axioms give a complete characterization of conditional independence. We were disproved, however, by Studeny (1989) w.r.t. probability measures, but the proof carries over to ranking functions (cf. Spohn 1994a). Under special conditions, though, the graphoid axioms *are* complete, as was proved by Geiger and Pearl (1990) for probability measures and by Hunter (1991) for ranking functions (cf. again, Spohn 1994a).

I am emphasizing all this, because the main purport of Pearl's path-breaking book (1988) is to develop what he calls the theory of Bayesian nets, a theory that has acquired great importance and is presented in many text books (see, e.g., Neapolitan 1990 or Jensen 2001). Pearl makes very clear that the basis of this theory consists in the graphoid axioms; these allow representing conditional dependence and independence among sets of variables by Bayesian nets, i.e., by directed acyclic graphs, the nodes of which are variables. A vertex $u \rightarrow v$ of the graph then represents the fact that v is dependent on u given all the variables preceding v in some given order, for instance, temporally preceding v. A major point of this theory is that it can describe in detail how probabilistic change triggered at some node in the net propagates throughout the net. All this is not merely mathematics, it is intuitively sensible and philosophically highly significant; for instance, inference acquires a novel and fruitful meaning in the theory of Bayesian nets.

Of course, my point now is that all these virtues carry over to ranking theory with the help of observation (17.14). The point is obvious, but hardly elaborated; that should be done.[18] It will thus turn out that ranks and hence beliefs can also be represented and computationally managed in that kind of structure.

This is not yet the end of the story. Spirtes et al. (1993) (see also Pearl 2000) have made amply clear that probabilistic Bayesian nets have a most natural causal interpretation; a vertex $u \rightarrow v$ then represents that the variable v directly causally depends on the variable u. Spirtes et al. back up this interpretation, i.e., this connection of probability and causality, by their three basic axioms: the causal Markov condition, the minimality condition, and, less importantly, the faithfulness condition (cf. Spirtes et al. 1993, sect. 3.4). And they go on to develop a really impressive account of causation and causal inference on the basis of these axioms and thus upon the theory of Bayesian nets.

Again, all this carries over to ranking theory. Indeed, this is what ranks were designed for in the first place. In Spohn (1983) I gave an explication of probabilistic causation that entails the causal Markov condition and the minimality condition, and also Reichenbach's principle of the common cause, as I observed later in Spohn (1994b).[19] And I was convinced of the idea that, if the theory of causation is bound to bifurcate into a deterministic and a probabilistic branch, these two branches must at least be developed in perfect parallel. Hence, I proposed ranking theory in Spohn (1983) in order to realize this idea.[20] Of course, one has to discuss how adequate that theory of deterministic causation is, just as the adequacy of the causal interpretation of Bayesian nets is open to discussion. Here, my point is only that this deep philosophical perspective lies within reach of ranking theory; it is what originally drove that theory.

Objective Ranks?

Now, a fundamental problem of ranking theory is coming into sight. I have emphasized that ranking functions represent rational beliefs and their rational dynamics and are thus entirely subject-bound. You have your ranking function and I have mine. We may or may not harmonize. In any case, they remain our subjective property.

---

[18]It has been done in the meantime. See Hohenadel (2013).

[19]I have analyzed the relation between Spirtes' et al. axiomatic approach to causation and my definitional approach a bit more thoroughly in Spohn (2001b).

[20]For a recent presentation of the account of deterministic causation in terms of ranking functions and its comparison in particular with David Lewis' counterfactual approach see Spohn (2006).

I have also emphasized the analogy to probability theory. There, however, we find subjective *and* objective probabilities. There are radicals who deny the one or the other kind of probability; and the nature of objective probabilities may still be ill understood. So, we certainly enter mined area here. Still, the predominant opinion is that both, the subjective and the objective notion, are somehow meaningful.

We therefore face a tension. It increases with our remarks about causation. I said I have provided an analysis of causation in ranking terms. If this analysis were to go through, the consequence would be that causal relations obtain relative to a ranking function, i.e., relative to the doxastic state of a subject. David Hume endorsed and denied this consequence at the same time; he was peculiarly ambiguous. This ambiguity must, however, be seen as his great achievement with which all philosophers after him had and still have to struggle. In any case, it will not do to turn causation simply into a subjective notion, as I seem to propose. If my strategy is to work at all, then the actually existing causal relations have to be those obtaining relative to the objectively correct ranking function. Is there any way to make sense of this phrase? (It is not even a notion yet.)

Yes, partially. The beginning is easy. Propositions are objectively true or false, and so are beliefs. Hence, a ranking function may be called objectively true or false as well, according to the beliefs it embodies. However, this is a very small step. Ranking functions can agree in their belief sets or in the propositions receiving rank 0, and yet widely diverge in the other ranks and thus in inductive and dynamic behavior. So, the suggested beginning is a very small step, indeed.

Taking a bigger step is more difficult. In my (1993) I have made a precise and detailed proposal that I still take to be sound; there is no space to repeat it here. Let me only briefly explain the basic idea. It is simply this: If propositions and beliefs are objectively true or false, then other features of ranking functions can be objectified to the extent to which these features are uniquely reflected in the associated belief sets. One constructive task is then to precisely define the content of the phrase 'uniquely reflected' and the required presuppositions or restrictions. The other constructive task is to inquire which specific features can in this sense be objectified to which specific extent.

Very roughly, the results in my (1993) are this: First, positive relevance, i.e., the reason relation, is *not* objectifiable in this sense, even if restricted to necessary and/or sufficient reasons. Second, whenever A is a sufficient or necessary direct cause of B w.r.t. κ, there is an associated material implication of the form "if the relevant circumstances obtain, then if A, then B, or, respectively, if $\overline{A}$, then $\overline{B}$". I call the conjunction of all these material implications the *causal law* associated with κ. The causal law is a proposition, an objective truth-condition. The point now is that there is a rich class of ranking functions which, under certain presuppositions, can uniquely be reconstructed from their causal laws and which may thus be called causal laws as well. In this sense and to this extent, causal relations obtaining relative to a subjective raking function can be objectified and thus do hold objectively.

A special case treated in Spohn (2002, 2005a) is the case of strict or deterministic laws. A strict law is, by all means, a regularity, an invariable obtaining of a certain type of state of affairs. But not any regularity is a law. What I have proposed in

Spohn (2002) is that a law is an independent and identically distributed (infinite) repetition of the type of state in question or, rather, in order for that phrase to make sense, an independent and identically distributed repetition of a certain ranking assessment of that type of state. Hence, a law is a certain kind of ranking function. This sounds weird, because a law thus turns into a kind of doxastic attitude. The literature on lawlikeness shows, however, that this is not so absurd a direction; if, besides explanatory power or support of counterfactuals, projectibility or inductive behavior are made defining features of laws, they are characterized by their epistemic role and thus get somehow entangled with our subjective states (see also Lange 2000, ch. 7, on the root commitment associated with laws). The main point, though, is that the ranking functions expressing deterministic laws are again of the objectifiable kind. So, there is a way of maintaining even within this account that laws obtain mind-independently.

In fact, according to what I have sketched, a deterministic law is the precise ranking analogue of a statistical law. De Finetti (1937) has proposed an ingenious way of eliminating objective probabilities and statistical laws by showing, in his famous representation theorem, that beliefs (i.e., subjective probabilities) about statistical laws (describing an infinite sequence of independent and identically distributed trials) are strictly equivalent to symmetric or exchangeable subjective probabilities for these trials and that experience makes these symmetric probabilities converge to the true statistical law. The eliminativist intention of the story is mostly dismissed today; rather, objective probabilities are taken seriously. Still, de Finetti's account has remained a paradigm story about the relation between subjective and objective probabilities.

I am mentioning all this because this paradigm story can be directly transferred to ranking theory. Let κ be any ranking function for an infinite sequence of trials (= variables) which is regular and symmetric and according to which the outcome of a certain trial is not negatively relevant to the same outcome in the next trial. Then κ is a unique mixture of deterministic laws for that sequence of trials in the above-mentioned sense, and experience makes κ converge to the true deterministic law. (Cf. Spohn 2005a for all this, where I have treated only the simplest case of the infinite repetition of a binary variable or a trial having only two possible outcomes. With an additional condition, however, the results generalize to all variables taking finitely many values).

This may suffice as an overview over the basics of ranking theory and its elaboration into various directions; it got long enough. In a way, my overall argument in section "Further comparisons", pp. 335ff, when I shall make a bit more detailed comparative remarks about other members of the Baconian probability family, should be clear by now: Bayesian epistemology has enormous powers and virtues and rich details and ramifications. Small wonder that Pascal by far outstripped Bacon. In a nutshell, I have explained that many essential virtues can be duplicated in ranking theory; indeed, the duplications can stand on their own, having an independent significance. Bacon can catch up with Pascal. Of course, my rhetorical question will then be: Which other version of Baconian probability is able to come up with similar results?

Still, one might suspect that I can claim these successes only by turning Bacon into a fake Pascal. I have never left the Bayesian home, it may seem. Hence, one might even suspect that ranking theory is superfluous and may be reduced to the traditional Bayesian point of view. In other words, it is high time to study more closely the relation between probability and ranking theory. This will be our task in the next section.

Ranks and Probabilities

The relation between probabilities and ranks is surprisingly complex and fascinating. I first turn to the more formal aspects of the comparison before discussing the philosophical aspects.

Formal Aspects

The reader will have observed since long why ranks behave so much like probabilities. There is obviously a simple translation of probability into ranking theory: translate the sum of probabilities into the minimum of ranks, the product of probabilities into the sum of ranks, and the quotient of probabilities into the difference of ranks. Thereby, the probabilistic law of additivity turns into the law of disjunction, the probabilistic law of multiplication into the law of conjunction (for negative ranks), and the definition of conditional probabilities into the definition of conditional ranks. If the basic axioms and definitions are thus translated, then it is small wonder that the translation generalizes; take any probabilistic theorem, apply the above translation to it, and you are almost guaranteed to get a ranking theorem. This translation is obviously committed to negative ranks; therefore I always favored negative over positive ranks. However, the translation is not fool-proof; see, e.g., Spohn (1994a) for slight failures concerning conditional independence (between sets of variables) or Spohn (2005a) for slight differences concerning positive and non-negative instantial relevance. The issue is not completely cleared up.

Is there a deeper reason why this translation works so well? Yes, of course. The translation of products and quotients of probabilities suggests that negative ranks simply are the logarithm of probabilities (with respect to some base < 1). This does not seem to fit with the translation of sums of probabilities. But it does fit when the logarithmic base is taken to be some infinitesimal i (since for two positive reals $x \leq y$ $i^x + i^y = i^{x-j}$ for some infinitesimal j). That is, we may understand ranks as real orders of magnitude of non-standard probabilities. This is the basic reason for the pervasive analogy.

Does this mean that ranking epistemology simply reduces to non-standard Bayesianism? This may be one way to view the matter. However, I do not particularly like this perspective. Bayesian epistemology in terms of non-standard reals is really non-standard. Even its great proponent, David Lewis, mentions the possibility only in passing (for the first time in 1980, p. 268). It is well known that both, non-standard analysis and its continuation as hyperfinite probability theory, have their intricacies of their own, and it is highly questionable from an epistemological point of view whether one should buy these intricacies. Moreover, even though this understanding of ranks is in principle feasible, it is nowhere worked out in detail. Such an elaboration should also explain the slight failures of the above translation. Hence, even formally the relation between ranks and non-standard probabilities is not fully clear. Finally, there are algebraic incoherencies. As long as the probabilistic law of additivity and the ranking law of disjunction are finitely restricted, there is no problem. However, it is very natural to conceive probability measures as σ-additive (although there is an argument about this point), whereas it is very natural to conceive of ranking functions as complete (as I have argued). This is a further disanalogy, which is not resolved by the suggested understanding of ranks.

All in all, I prefer to stick to the realm of standard reals. Ranking theory is a standard theory, and it should be compared to other standard theories. So, let us put the issue of hyperfinite probability theory to one side.

Let us instead pursue another line of thought. I have heavily emphasized that the fundamental point of ranking theory is to represent the statics and the dynamics of belief or of taking-to-be-true; it *is* the theory of belief. So, instead of inquiring the relation between ranks and probabilities we might as well ask the more familiar question about the relation between belief and probability.

This relation is well known to be problematic. One naive idea is that belief vaguely marks some threshold in probability, i.e., that A is believed iff its subjective probability is greater than $1 - \varepsilon$ for some small ε. But this will not do, as is hightlighted by the famous lottery paradox (see Kyburg 1961, p. 197, and Hempel (1962, pp. 163–166). According to this idea you may believe A and believe B, but fail to believe A & B. However, this amounts to saying that you do not know the truth table of conjunction, i.e., that you have not grasped conjunction at all. So, this idea is a bad one, as almost all commentators to the lottery paradox agree. One might think then about more complicated relations between belief and probability, but I confess not to have seen any convincing one.

The simplest escape from the lottery paradox is, of course, to equate belief with probability 1. This proposal faces two further problems, though. First, it seems intuitively inadequate to equate belief with maximal certainty in probabilistic terms; beliefs need not be absolutely certain. Secondly, but this is only a theoretical version of the intuitive objection, only belief expansion makes sense according to this proposal, but no genuine belief revision. Once you assign probability 1 to a proposition, you can never get rid of it according to all rules of probabilistic change. This is obviously inadequate; of course, we can give up previous beliefs and easily do so all the time.

Jeffrey's radical probabilism (1991) is a radical way out. According to Jeffrey, all subjective probabilities are regular, and his generalized conditionalization provides a dynamics moving within regular probabilities. However, Jeffrey's picture and the proposal of equating belief with probability 1 do not combine; then we would believe in nothing but the tautology. Jeffrey did not deny beliefs, but he indeed denied their relevance for epistemology; this is what the adjective 'radical' in effect signifies. He did not believe in any positive relation between belief and probability, and probability is all you need – a viable conclusion from the lottery paradox perhaps, though only as a last resort.

The point that probability theory cannot account for belief revision may apparently be dealt with by an expansion of the probabilistic point of view, namely by resorting to Popper measures. These take conditional probability as the basic notion, and thus probabilities conditional on propositions having absolute probability 0 may be well defined. That is, you may initially believe A, i.e., assign probability 1 to A, and still learn that $\overline{A}$, i.e., conditionalize w.r.t. $\overline{A}$, and thus move to posterior probabilities and even beliefs denying A. In this way, one can stick to the equation of belief with probability 1 and escape the above objection. Have we thus reached a stable position?

No, we have not. One point of Spohn (1986) was to rigorously show that AGM belief revision is just the qualitative counterpart of Popper measures. Conversely, this entails that the inability of AGM belief revision theory to model iterated belief revision, which I criticized in my (1988), holds for Popper measures as well. In fact, Harper (1976) was the first to note this problem vis à vis Popper measures, and thus I became aware of the problem and noticed the parallel.

Harper proposed quite a complicated solution to the problem that is, as far as I know, not well received; but it may be worth revisiting. My conclusion was a different one. If AGM belief revision theory is incomplete and has to be evolved into ranking theory, the probabilistic point of view needs likewise to get further expanded. We need something like probabilified ranks or ranked probabilities; it is only in terms of them that we can unrestrictedly explain iterated probabilistic change.

A ranking function associates with each rank a set of propositions having that rank. A ranked probability measure associates with each rank an ordinary probability measure. The precise definition is straightforward. Hence, I confined myself to mentioning the idea in my (1988, sect. 7); only in my (2005b) I took the trouble to explicitly introduce it. One should note, though, that as soon as one assumes the probability measures involved to be σ-additive, one again forces the ranks to be well-ordered (cf. Spohn 1986); this is why in my (2005b) only the probabilification of complete ranking functions is defined.

One may say that ranking theory thus ultimately reduces to probability theory. I find this misleading, however. What I have just sketched is rather a unification of probability and ranking theory; after all, we have employed genuine ranking ideas in order to complete the probabilistic point of view. The unification is indeed a powerful one; all the virtues of standard Bayesianism which I have shown to carry over to ranking theory hold for this unification as well. It provides a unified account of confirmation, of lawlikeness, even of causation. It appears to be a surprising, but most desirable wedding of Baconian and Pascalian probability. I shall continue on the topic in the next subsection.

The previous paragraphs again urge the issue of hyperfinite probability; ranked probabilities look even more like probabilities in terms of non-standard reals. However, I cannot say more than I already did; I recommend the issue for further investigation.[21] I should use the occasion for clarifying a possible confusion, though. McGee (1994, pp. 181ff.) showed that Popper measures correspond to non-standard probability measures in a specific way. Now, I have suggested that ranked probabilities do so as well. However, my (1986, 1988) together entail that ranked probabilities are more general than Popper measures. These three assertions do not fit together. Yet, the apparent conflict is easily dissolved. The correspondence proved by McGee is not a unique one. Different non-standard probability measures may correspond to the same Popper measure, just as different ranked probabilities may. Hence, if McGee says that the two approaches, Popper's and the non-standard one, "amount to the same thing" (p. 181), this is true only for the respects McGee is considering, i.e., w.r.t. conditional probabilities. It is not true for the wider perspective I am advocating here, i.e., w.r.t. probability dynamics.

Philosophical Aspects

The relation between belief and probability is not only a formal issue, it is philosophically deeply puzzling. It would be disturbing if there should be two (or more) unrelated ways of characterizing our doxastic states. We must somehow come to grips with their relation.

The nicest option would be *reductionism*, i.e., reducing one notion to the other. This can only mean reducing belief to probability. As we have seen, however, this option seems barred by the lottery paradox. Another option is *eliminativism* as most ably defended in Jeffrey's radical probabilism also mentioned above. This option is certainly viable and most elegant. Still, I find it deeply unsatisfactory; it is unacceptable that our talk of belief should merely be an excusable error ultimately to be eliminated. Thus, both versions of *monism* seem excluded.

Hence, we have to turn to *dualism*, and then *interactionism* may seem the most sensible position. Of course, everything depends on the precise form of interaction between belief and probability. In Spohn (2005b) I had an argument with Isaac Levi whom I there described as the champion of interactionism. My general experience, though, is that belief and probability are like oil and water; they do not mix easily. Quite a different type of interactionism is represented by Hild (t.a.) who has many interesting things to say about how ranking and probability theory mesh, indeed how heavily ranking ideas are implicitly used in statistical methodology. I do not have space to assess this type of interactionism.

[21]For quite a different way of relating probabilities and ranks appealing neither to infinitesimals nor to Popperian conditional probabilities see Giang and Shenoy (1999).

When the fate of interactionism is unclear one might hope to return to reductionism and thus to monism, not in the form of reducing belief to probability, but in the form of *reducing both to something third*. This may be hyperfinite probability, or it may be ranked probabilities as suggested above. However, as already indicated, I consider this to be at best a formal possibility with admittedly great formal power of unification. Philosophically, I am not convinced. It is intuitively simply inadequate to equate belief with (almost) maximal probabilistic certainty, i.e., with probability 1 (minus an infinitesimal), even if this does not amount to unrevisability within these unifications. This intuition has systematic counterparts. For centuries, the behavioral connection of subjective probabilities to gambling and betting has been taken to be fundamental; many hold that this connection provides the only explanation of subjective probabilities. This fundamental connection does not survive these unifications. According to them, I would have to be prepared to bet my life on my beliefs; but this is true only of very few of my many beliefs. So, there are grave frictions that should not be plastered by formal means.

In view of all this, I have always preferred *separatism*, at least *methodologically*. If monism and interactionism are problematic, then belief and probability should be studied as two separate fields of interest. I sense the harshness of this position; this is why I am recommending it so far only as a methodological one and remain unsure about its ultimate status. However, the harshness is softened by the formal parallel which I have extensively exploited and which allows formal unification. Thus, separatism in effect amounts to *parallelism*, at least if belief is studied in ranking terms. Indeed, the effectiveness of the parallel sometimes strikes me as a pre-established harmony.

Thus, another moral to be drawn may perhaps be *structuralism*, i.e., the search for common structures. This is a strategy I find most clearly displayed in Halpern (2003). He starts with a very weak structure of degrees of belief that he calls plausibility measures and then discusses various conditions on those degrees that allow useful strengthenings of that structure such as a theory of conditioning, a theory of independence, a theory of expectation and integration, and so forth. Both, ranking and probability theory, but not only they are specializations of that structure and its various strengthenings. Without doubt, this is a most instructive procedure. Structuralism would moreover suggest that it is only those structures and not their specific realizations that matter. Halpern does not explicitly endorse this, and I think one should withstand it. For instance, one would thereby miss the essential purpose for which ranking theory was designed, namely the theory of belief. For this purpose, no less and no more than the ranking structure is required.

Hence, let me further pursue, in the spirit of methodological separatism, the philosophical comparison between ranks and standard probabilities. I have already emphasized the areas in which the formal parallel also makes substantial sense: inductive inference, confirmation, causation, etc. Let us now focus on three actual or apparent substantial dissimilarities, which in one or the other way concern the issue what our doxastic states have to do with reality.

The first aspect of this issue is the *truth connection*; ranks are related to truth in a way in which probabilities are not. This is the old point all over again.

Ranks represent beliefs that are true or false, whereas subjective probabilities do not represent beliefs and may be assessed in various ways, as well-informed, as reasonable, but never as true or false. Degrees of belief may perhaps conform to degrees of truthlikeness; however, it is not clear in the first place whether degrees of truthlikeness behave like probabilities (cf. Oddie 2001). Or degrees of belief may conform to what Joyce (1998) calls the norm of gradational accuracy from which he proceeds with an interesting argument to the effect that degrees of belief then have to behave like probabilities.[22] Such ideas are at best a weak substitute, however; they never yield an application of truth in probability theory as we have it in ranking theory.

This is a clear point in favor of ranking theory. And it is rich of consequences. It means that ranking theory, in contrast to probability theory, is able to connect up with traditional epistemology. For instance, Plantinga (1993, chs. 6 and 7) despairs of finding insights in Bayesianism he can use and dismisses it, too swiftly I find. This would have been different with ranking theory. The reason why ranking theory is connectible is obvious. Traditional epistemology is interested in knowledge, a category entirely foreign to probability theory; knowledge, roughly, is justified true belief and thus analyzed by notions within the domain of ranking theory. Moreover, the notion of justification has become particularly contested in traditional epistemology; one focal issue was then to give an account of the truth-conduciveness of reasons, again notions within the domain of ranking theory.

I am not claiming actual epistemological progress here. But I do claim an advantage of ranking over probability theory, I do claim that traditional epistemology finds in ranking theory adequate formal means for discussing its issues, and using such means is something I generally recommend as a formal philosopher.

The second aspect is the *behavioral connection*. Our doxastic states make some actions rational and others irrational, and our theories have to say which. Here, probability theory seems to have a clear advantage. The associated behavioral theory is, of course, decision theory with its fundamental principle of maximizing conditional expected utility. The power of this theory need not be emphasized here. Is there anything comparable on offer for ranking theory?

This appears excluded, for the formal reason that there is a theory of integration and thus of expectation in probabilistic, but none in ranking terms; this is at least what I had thought all along. However, the issue has developed. There are various remarkable attempts of stating a decision theory in terms of non-probabilistic or non-additive representations of degrees of belief employing the more general Choquet theory of integration.[23] Indeed, there is also one especially for ranking theory. Giang and Shenoy (2000) translate the axiomatic treatment of utility as it is

[22]Cf., however, Maher's (2002) criticism of Joyce's argument.

[23]Economists inquired the issue; see, e.g., Gilboa (1987), Schmeidler (1989), Jaffray (1989), Sarin and Wakker (1992) for early contributions, and Wakker (2005) for a recent one. The AI side concurs; see, e.g., Dubois and Prade (1995), Brafman and Tennenholtz (2000), and Giang and Shenoy (2005).

given by Luce and Raiffa (1957, sect. 2.5) in terms of simple and compound lotteries directly into the ranking framework, thus developing a notion of utility fitting to this framework. These attempts doubtlessly deserve further scrutiny (cf. also Halpern 2003, ch. 5).

Let me raise, though, another point relating to this behavioral aspect. Linguistic behavior is unique to humans and a very special kind of behavior. Still, one may hope to cover it by decision theoretic means, too. Grice's intentional semantics employs a rudimentary decision theoretic analysis, and Lewis' (1969) theory of conventions uses game (and thus decision) theoretic methods in a very sophisticated way. However, even Lewis' account of coordination equilibria may be reduced to a qualitative theory (in Lewis (1975) he explicitly uses only qualitative terminology). In fact, the most primitive linguistic behavioral law is the disquotation principle: if a seriously and sincerely utters "p", then a believes that p.[24] The point is that these linguistic behavioral laws and in particular the disquotation principle is stated in terms of belief. There is no probabilistic version of the disquotation principle, and it is unclear what it could be. The close relation between belief and meaning is obvious and undoubted, though perhaps not fully understood in the philosophy of language. I am not suggesting that there is a linguistic pragmatics in terms of ranking functions; there is hardly anything.[25] I only want to point out that the standing of ranking theory concerning this behavioral aspect is at least promising.

There is a third and final aspect, again apparently speaking in favor of probability theory. We do not only make decisions with the help of our subjective probabilities, we also do *statistics*. That is, we find a lot of *relative frequencies* in the world, and they are closely related to probabilities. We need not discuss here the exact nature of this relation. Concerning objective probabilities, it is extensively discussed in the debate about frequentism, and concerning subjective probabilities it is presumably best captured in Reichenbach's principle postulating that our subjective probabilities should rationally converge to the observed relative frequencies. What is clear, in any case, is that in some way or other relative frequencies provide a strong anchoring of probabilities in reality from which the powerful and pervasive application of statistical methods derives. Subjective probabilities are not simply free-floating in our minds.

For many years I thought that this is another important aspect in which ranking theory is inferior to probability theory. Recently, though, I have become more optimistic. Not that there would be any statistics in ranking terms[26]; I do not see ranks related to relative frequencies. However, a corresponding role is played by the notion of *exception* and thus by absolute frequencies. In section "Objective ranks?",

---

[24]If a speaks a foreign language, the principle takes a more complicated, but obvious form. There is also a disquotation principle for the hearer, which, however, requires a careful exchange of the hearer's and the speaker's role.

[25]See in particular Merin (2006, appendix B) and (2008) whose relevance-based pragmatics yields interesting results in probabilistic as well as in ranking-theoretic terms.

[26]However, I had already mentioned that Hild (t.a.) finds a much closer connection of probabilities and ranks within statistical methodology.

I left the precise account of objectifiable ranking functions in the dark. If one studies that account more closely, though, one finds that these objectifiable ranking functions, or indeed the laws as I have indicated them in section "Objective ranks?", are exception or fault counting functions. The rank assigned to some possible world by such a ranking function is just the number of exceptions from the law embodied in this function that occur in this world.

This is a dim remark so far, and here is not the place to elaborate on it. Still, I find the opposition of exceptions and relative frequencies appealing. Often, we take a type of phenomenon as more or less frequent, and then we apply our sophisticated statistical methodology to it. Equally often, we try to cover a type of phenomenon by a deterministic law, we find exceptions, we try to improve our law, we take recourse to a usually implicit ceteris paribus condition, etc. As far as I know, the methodology of the latter perspective is less sophisticated. Indeed, there is little theory. Mill's method of relevant variables, e.g., is certainly an old and famous attempt to such a theory (cf. its reconstruction in Cohen 1977, ch. 13). Still, both perspectives, the statistical and the deterministic one, are very familiar to us. What I am suggesting is that the deterministic perspective can be thoroughly described in terms of ranking theory.[27]

It would moreover be most interesting to attend to the vague borderline. Somewhere, we switch from one to the other perspective, from exceptions to small relative frequencies or the other way around. I am not aware of any study of this borderline, but I am sure it is worth getting inquired. It may have the potential of also illuminating the relation of belief and probability, the deterministic and the statistical attitude.

All these broad implications are involved in a comparison of ranks and probabilities. I would find it rather confusing to artificially combine them in some unified theory, be it hyperfinite or ranked probabilities. It is more illuminating to keep them separate. Also, I did not want to argue for any preference. I wanted to present the rich field of comparison in which both theories can show their great, though partially diverging virtues. There should be no doubt, however, that the driving force behind all these considerations is the formal *parallelism* which I have extensively used in section "The theory" (pp. 305ff) and explained in section "Formal aspects" (pp. 328ff).

Further Comparisons

Let me close the paper with a number of brief comparative remarks about alternative accounts subsumable under the vague label 'Baconian probability'. I have already

[27] I attempted to substantiate this suggestion with my account of strict and ceteris paribus laws in Spohn (2002) and with my translation of de Finetti's representation theorem into ranking theory in Spohn (2005a). (New addendum: For the most recent ranking-theoretic account of ceteris paribus laws see Spohn (2014).)

made a lot of such remarks *en passant*, but it may be useful to have them collected. I shall distinguish between the earlier and usually more philosophical contributions on the one hand and the more recent, often more technical contributions from the computer science side on the other hand. The borderline is certainly fuzzy, and I certainly do not want to erect boundaries. Still, the centuries old tendency of specialization and of transferring problems from philosophy to special fields may be clearly observed here as well.

Earlier and Philosophical Literature

It is perhaps appropriate to start with L. Jonathan Cohen, the inventor of the label. In particular his (1977) is an impressive document of dualism, indeed separatism concerning degrees of provability and degrees of probability or inductive (Baconian) and Pascalian probability. His work is, as far as I know, the first explicit and powerful articulation of the attitude I have taken here as well.[28]

However, his functions of inductive support are rather a preform of my ranking functions. His inductive supports correspond to my positive ranks. Cohen clearly endorsed the law of conjunction for positive ranks; see his (1970, pp. 21f. and p. 63). He also endorsed the law of negation; but he noticed its importance only in his (1977, pp. 177ff.), whereas in his (1970) it is well hidden as theorem 306 on p. 226. His presentation is a bit imperspicuous, though, since he is somehow attached to the idea that $\Box^i$, i.e., having an inductive support $\geq i$, behaves like iterable S4-necessity and since he even brings in first-order predicate calculus.

Moreover, Cohen is explicit on the relationality of inductive support; it is a two-place function relating evidence and hypothesis. Hence, one might expect to find a true account of conditionality. This, however, is not so. His conditionals behave like strict implication[29], a feature Lewis (1973, sect. 1.2–3) has already warned against. Moreover, Cohen discusses only laws of support with fixed evidence – with one exception, the consequence principle, as he calls it (1970, p. 62). Translated into my notation it says for a positive ranking function π that

$$\pi(C \mid A) \geq \pi(C \mid B) \text{ if } A \subseteq B, \qquad (17.15)$$

which is clearly not a theorem of ranking theory. These remarks sufficiently indicate that the aspect so crucial for ranking functions is scarcely and wrongly developed in Cohen's work.

The first clear articulation of the basic Baconian structure is found, however, not in Cohen's work, but in Shackle (1949, 1969). His functions of potential surprise

[28] I must confess, though, that I had not yet noticed his work when I basically fixed my ideas on ranking functions in 1983.

[29] This is particularly obvious from Cohen (1970, p. 219, def. 5).

clearly correspond to my negative ranking functions; axiom (17.9) in (1969, p. 81) is the law of negation, and axiom (17.4) and/or (17.6) in (1969, p. 90) express the law of disjunction. At least informally, Shackle also recognizes the duality of positive and negative ranks. He is explicit that potential surprise expresses certainty of wrongness, i.e., disbelief, and that there is conversely certainty of rightness (1969, p. 74).

His general attitude, however, is not so decidedly dualistic as that of Cohen. His concern is rather a general account of uncertainty, and he insists that probability does not exhaust uncertainty. Probability is an appropriate uncertainty measure only if uncertainty is 'distributional', whereas potential surprise accounts for 'non-distributional' uncertainty. So, he also ends up with an antagonistic structure; but the intention was to develop two special cases of a general theory.

It is most interesting to see how hard Shackle struggles with an appropriate law of conjunction for negative ranks. The first version of his axiom 7 (1969, p. 80) claims, in our terminology, that

$$\kappa(A \cap B) = \max\{\kappa(A), \kappa(B)\}. \tag{17.16}$$

He accepts the criticism this axiom has met, and changes it into a second version (1969, p. 83), which I find must be translated into

$$\kappa(B) = \max\{\kappa(A), \kappa(B \mid A)\} \tag{17.17}$$

(and is hence no law of conjunction at all). He continues that it would be fallacious to infer that

$$\kappa(A \cap B) = \min\left[\max\{\kappa(A), \kappa(B \mid A)\}, \max\{\kappa(B), \kappa(A \mid B)\}\right]. \tag{17.18}$$

In (1969, ch. 24) he is remarkably modern in discussing "expectation of change of own expectations". I interpret his formula (i) on p. 199 as slightly deviating from the second version of his axiom 7 in claiming that

$$\kappa(A \cap B) = \max\{\kappa(A), \kappa(B \mid A)\}. \tag{17.19}$$

And on pp. 204f. he even considers, and rejects (for no convincing reason), the equation

$$\kappa(A \cap B) = \kappa(A) + \kappa(B \mid A), \tag{17.20}$$

i.e., our law of conjunction for negative ranks. In all these discussions, conditional degrees of potential surprise appear to be an unexplained primitive notion. So, Shackle may have been here on the verge of getting things right. On the whole, though, it seems fair to say that his struggle has not led to a clear result.

Isaac Levi has always pointed to this pioneering achievement of Shackle, and he has made his own use of it. In a way he did not develop Shackle's functions of potential surprise; he just stuck to the laws of negation and of disjunction for negative ranks. In particular, there is no hint of any notion of conditionalization. This is not to say that his epistemology is poorer than the one I have. Rather, he finds a place for Shackle's functions in his elaborated doxastic decision theory, more precisely, in his account of belief expansion. He adds a separate account of belief contraction, and with the help of what is called Levi's identity he can thus deal with every kind of belief change. He may even claim to come to grips with iterated change.[30] One may thus sense that his edifice is at cross-purposes with mine.

A fair comparison is hence a larger affair. I have tried to give it in Spohn (2005b). Let me only mention one divergence specifically related to ranking functions. Since Levi considers ranking functions as basically identical with Shackle's functions of potential surprise and since he sees the latter's role in expansion, he continuously brings ranking functions into the same restricted perspective. I find this inadequate. I rather see the very same structure at work at expansions as well as at contractions, namely the structure of ranks. Insofar I do not see any need of giving the two kinds of belief change an entirely different treatment.

This brings me to the next comparison, with AGM belief revision theory (cf. e.g., Gärdenfors 1988). I have already explained that I came to think of ranking theory as a direct response to the challenge of iterated belief revision for AGM belief revision theory, and I have explained how $A \rightarrow x$-conditionalization for ranks unifies and generalizes AGM expansion, revision, and contraction. One may wonder how that challenge was taken up within the AGM discussion. With a plethora of proposals (see Rott 2008), that partially ventilated ideas that I thought to have effectively criticized already in Spohn (1988) and that do not find agreement, as far as I see, with the exception of Darwiche and Pearl (1997). As mentioned, Hild and Spohn (2008) gives a complete axiomatization of iterated contraction. Whether it finds wider acceptance remains to be seen.

By no means, though, one should underestimate the richness of the AGM discussion, of which, e.g., Rott (2001) or Hanson (1999) give a good impression. A pertinent point is that ranking theory generalizes and thus simply sides with the standard postulates for revision and contraction (i.e., (K*1-8) and (K^{-}1-8) in Gärdenfors 1988, pp. 54–56 and 61–64). The ensuing discussion has shown that these postulates are not beyond criticism and that many alternatives are worth discussing (cf., e.g., Rott 2001, pp. 103ff., who lists three alternatives of K*7, nine of K*8, six of K^{-}7, and ten of K^{-}8). I confess I would not know how to modify ranking theory in order to do justice to such alternatives. Hence, a fuller comparison with AGM belief revision theory would have to advance a defense of the standard postulates against the criticisms related with the alternatives.

The point is, of course, relevant in the debate with Levi, too. He prefers what he calls mild contraction to standard AGM contraction that can be represented in

[30]Many aspects of his epistemology are already found in Levi (1967). The most recent statement is given in Levi (2004), where one also gets a good idea of the development of his thought.

ranking theory only as a form of iterated contraction. Again, one would have to discuss whether this representation is acceptable.

It is worth mentioning that the origins of AGM belief revision theory clearly lie in conditional logic. Gärdenfors' (1978) epistemic semantics for conditionals was a response to the somewhat unearthly similarity spheres semantics for counterfactuals in Lewis (1973), and via the so-called Ramsey test Gärdenfors' interest more and more shifted from belief in conditionals to conditional beliefs and thus to the dynamics of belief. Hence, one finds a great similarity in the formal structures of conditional logic and belief revision theory. In particular, Lewis' similarity spheres correspond to Gärdenfors' entrenchment relations (1988, ch. 4). In a nutshell, then, the progress of ranking theory over Lewis' counterfactual logic lies in proceeding from an ordering of counterfactuality (as represented by Lewis' nested similarity spheres) to a cardinal grading of disbelief (as embodied in negative ranking functions).[31]

Indeed, the origins reach back farther. Conditional logic also has a history, the earlier one being somewhat indeterminate. However, the idea of having an ordering of levels of counterfactuality or of far-fetchedness of hypotheses is explicitly found already in Rescher (1964). If π is a positive ranking function taking only finitely many values $0, x_1, \ldots, x_m, \infty$, then $\pi^{-1}(\infty), \pi^{-1}(x_m), \ldots, \pi^{-1}(x_1), \pi^{-1}(0)$ is just a family of modal categories $M_0, \ldots, M_n$ ($n = m + 2$), as Rescher (1964, pp. 47–50) describes it. His procedure on pp. 49 f. for generating modal categories makes them closed under conjunction; this is our law of conjunction for positive ranks. And he observes on p. 47 that all the negations of sentences in modal categories up to M_{n-1} must be in $M_n = \pi^{-1}(0)$; this is our law of negation.

To resume, I cannot find an equivalent to the ranking account of conditionalization in all this literature. However, the philosophical fruits I have depicted in section "The theory", pp. 305ff., and also in section "Philosophical aspects", pp. 331ff., sprang from this account. Therefore, I am wondering to which extent this literature can offer similar fruits, and for all I know the answer tends to be negative.

More Recent Computer Science Literature

In view of the exploding computer science literature on uncertainty since the 80s even the brief remarks in the previous subsection on the earlier times were disproportionate. However, it is important, I think, not to forget about the origins. My comparative remarks concerning the more recent literature must hence be even more cursory. This is no neglect, though, since Halpern (2003), in book length, provides comprehensive comparisons of the various approaches with an emphasis on those aspects (conditionalization, independence, etc.) that I take to be important, too. Some rather general remarks must do instead and may nevertheless be illuminating.

[31] For my ideas how to treat conditionals in ranking-theoretic terms see Spohn (2015).

In the computer science literature, ranking theory is usually subsumed under the heading "uncertainty" and "degrees of belief". This is not wrong. After all, ranks are degrees, and if (absolute) certainty is equated with unrevisability, revisable beliefs are uncertain beliefs. Still, the subsumption is also misleading. My concern was *not* to represent uncertainty and to ventilate alternative models of doing so. Thus stated, this would have been an enterprise with too little guidance. My concern was exclusively to statically and dynamically represent *ungraded* belief, and my observation was that this necessarily leads to the ranking structure. If this is so, then, as I have emphasized, all the philosophical benefits of having a successful representation of ungraded belief are conferred to ranking theory. By contrast, if one starts modeling degrees of uncertainty, it is always an issue (raised, for instance, by the lottery paradox vis à vis probability) to which extent such a model adequately captures belief and its dynamics. So, this is a principled feature that sets ranking theory apart from the entire uncertainty literature.

The revisability of beliefs was directly studied in computer science under headings like "default logic" or "nonmonotonic reasoning". This is another large and natural field of comparison for ranking theory. However, let me cut things short. The relation between belief revision theory and nonmonotonic reasoning is meticulously investigated by Rott (2001). He proved far-reaching equivalences between many variants on both sides. This is highly illuminating. At the same time, however, it is a general indication that the concerns that led me to develop AGM belief revision theory into ranking theory are not well addressed in these areas of AI. Of course, such lump-sum statements must be taken with caution.

The uncertainty literature has observed many times that the field of nonmonotonic reasoning is within its reach. Among many others, Pearl (1988, ch. 10) has investigated the point from the probabilistic side, and Halpern (2003, ch. 8) has summarized it from his more comprehensive perspective. This direction of inquiry is obviously feasible, but the reverse line of thought of deriving kinds of uncertainty degrees from kinds of nonmonotonic reasoning is less clear (though the results in Hild and Spohn (2008) about the measurement of ranks with via iterated contractions may be a step in the reverse direction).

So, let me return to accounts of uncertainty in a bit more detail, and let me take up *possibility theory* first. It originates from Zadeh (1978), i.e. from fuzzy set theory and hence from a theory of vagueness. Its elaboration in the book by Dubois and Prade (1988) and many further papers shows its wide applicability, but never denies its origin. So, it should at least be mentioned that philosophical accounts of vagueness (cf., e.g., Williamson 1994) have nothing much to do with fuzzy logic. If one abstracts from this interpretation, though, possibility theory is formally very similar to ranking theory. If *Poss* is a possibility measure, then the basic laws are:

$$Poss\,(\emptyset) = 0, Poss(W) = 1, \text{ and } Poss\,(A \cup B) = \max\{Poss(A), Poss(B)\}\,. \tag{17.21}$$

So far, the difference is merely one of scale. Full possibility 1 is negative rank 0, (im)possibility 0 is negative rank ∞, and translating the scales translates the characteristic axiom of possibility theory into the law of disjunction for negative ranks. Indeed, Dubois and Prade often describe their degrees of possibility in such a way that this translation fits not only formally, but also materially.

Hence, the key issue is again how conditionalization is treated within possibility theory. There is some uncertainty. First, there is the motive that also dominated Shackle's account of the functions of potential surprise, namely to keep possibility theory as an ordinal theory where degrees of possibility have no arithmetical meaning. Then the idea is to stipulate that

$$Poss\,(A \cap B) = \min\{Poss(A), Poss\,(B \mid A)\} = \min\{Poss(B), Poss\,(A \mid B)\}\,. \tag{17.22}$$

This is just Shackle's proposal (17.19). Hisdal (1978) proposed to go beyond (17.19) just by turning (17.22) into a definition of conditional possibility by additionally assuming that conditionally things should be as possible as possible, i.e., by defining $Poss(B \mid A)$ as the maximal degree of possibility that makes (17.22) true:

$$Poss\,(B \mid A) = \left\{\begin{array}{l} P\,(A \cap B)\,,\ \text{if}\ Poss\,(A \cap B) < Poss(A) \\ 1,\ \text{if}\ Poss\,(A \cap B) = Poss(A) \end{array}\right\}. \tag{17.23}$$

Halpern (2003, Proposition 3.9.2, Theorem 4.4.5, and Corollary 4.5.8) entails that Bayesian net theory works also in terms of conditional possibility thus defined. Many things, though, do not work well. It is plausible that $Poss(B \mid A)$ is between the extremes 1 and $Poss(A \cap B)$. However, (17.23) implies that it can take only those extremes. This is unintelligible. (17.22) implies that, if neither $Poss(B \mid A)$ nor $Poss(A \mid B)$ is 1, they are equal, a strange symmetry. And so on. Such unacceptable consequences spread through the entire architecture.

However, there is a second way to introduce conditional possibilities (cf., e.g., Dubois and Prade 1998, p. 206), namely by taking numerical degrees of possibility seriously and defining

$$Poss\,(B \,||\, A) = Poss\,(A \cap B)\,/Poss(A). \tag{17.24}$$

This looks much better. Indeed, if we define $\kappa(A) = \log Poss(A)$, the logarithm taken w.r.t. some positive base < 1, then κ is a negative ranking function such that also $\kappa(B \mid A) = \log Poss(B \,||\, A)$. Hence, (17.24) renders possibility and ranking theory isomorphic, and all the philosophical benefits may be gained in either terms. Still, there remain interpretational differences. If we are really up to degrees of belief and disbelief, then the ranking scale is certainly more natural; this is particularly clear when we look at the possibilistic analogue to two-sided ranking functions. My remarks about objectifiable ranking functions as fault counting functions would make no sense for a possibilistic scale. And so on. Finally, one must be aware that the philosophical benefits resulted from adequately representing *belief*. Hence, it is doubtful whether the formal structure suffices to maintain the benefits for alternative interpretations of possibility theory.

Let me turn to some remarks about *(Dempster-Shafer) DS belief functions*. Shafer (1976) built on Dempster's ideas for developing a general theory of evidence. He saw clearly that his theory covered all known conceptions of degrees of belief. This, and its computational manageability, explains its enormous impact. However, before entering any formal comparisons the first argument that should be settled is a philosophical one about the nature of evidence. There is the DS theory of evidence, and there is a large philosophical literature on observation and confirmation, Bayesianism being its dominant formal expression. I have explained why ranking theory and its account of reasons is a member of this family, too. Of course, this argument cannot even be started here. My impression, though, is that it is still insufficiently fought out, certainly hampered by disciplinary boundaries.

In any case, it is to be expected that DS belief functions and ranking functions are interpretationally at cross-purposes. This is particularly clear from the fact that negative ranking functions, like possibility measures or Shackle's functions of potential surprise, are formally a special case of DS belief functions; they are *consonant* belief functions as introduced in Shafer (1976, ch. 10). There, p. 219, Shafer says that consonant belief functions "are distinguished by their failure to betray even a hint of conflict in the evidence"; they "can be described as 'pointing in a single direction'." From the perspective of Shafer's theory of evidence this may be an adequate characterization. As a description of ranking functions, however, it does not make any sense whatsoever. This emphasizes that the intended interpretations diverge completely.

Even formally things do not fit together. We saw that the virtues of ranking theory depend on the specific behavior of conditional ranks. This does not generalize to DS belief functions. There is again an uncertainty how to conditionalize DS belief functions; there are two main variants (cf. Halpern 2003, p. 103 and p. 132, which I use as my reference book in the sequel). The central tool of Shafer's theory of evidence is the rule of combination proposed by Dempster (1967); it is supposed to drive the dynamics of DS belief functions. Combination with certain evidence is identical with one of the two variants of conditionalization (cf. Halpern 2003, p. 94). According to Shafer, other uncertain evidence is also to be processed by this rule. One might think, though, instead to handle it with Jeffrey's generalized conditionalization, which is indeed definable for both kinds of conditional belief functions (cf. Halpern 2003, p. 107). However, both kinds of Jeffrey conditionalization diverge from the rule of combination (cf. Halpern 2003, p. 107 and p. 114).

Indeed, this was my argument in Spohn (1990, p. 156) against formally equating ranking functions with consonant belief functions: Ranking dynamics is driven by a ranking analogue to Jeffrey conditionalization, but it cannot be copied by the rule of combination since the corresponding combinations move outside the realm of consonant belief functions. And, as I may add now, it does not help to let the dynamics of DS belief functions be driven by Jeffrey conditionalization instead of the rule of combination: Consonant belief functions are not closed under Jeffrey conditionalization as well, whereas ranking functions are thus closed.[32] I conclude that there is no formal subsumption of ranking functions under DS belief functions. Hence, their interpretations do not only actually diverge, they are bound to do so.

Smets' transferable belief model (cf., e.g., Smets 1998) proposes a still more general model for changing DS belief functions in terms of his so-called specializations. One should check whether it offers means for formally subsuming ranking functions under his model. Even if this would be possible, however, the interpretational concerns remain. Smets' specializations are so much wedded to Shafer's conception of evidence that any subsumption would appear artificial and accidental. The philosophical argument about the nature of evidence is even more pressing here.

A final remark: There is a bulk of literature treating doxastic uncertainty not in terms of a specific probability measure, but in terms of convex sets of probability measures. The basic idea behind this is that one's uncertainty is so deep that one is not even able to fix one's subjective probability. In this case, doxastic states may be described as sets of measures or in terms of probability intervals or in terms of lower and upper probabilities. Again, the multiple ways of elaborating this idea and their relations are well investigated (see again Halpern 2003). Indeed, DS belief functions, which provide a very general structure, emerges as generalizations of lower probabilities. Even they, though, do not necessarily transcend the probabilistic point of view, as Halpern (2003, p. 279) argues; DS belief functions are in a way tantamount to so-called inner measures. May we say, hence, that the alternative formal structures mentioned ultimately reduce to probabilism (liberalized in the way explained)? We may leave the issue open, though it is obvious that the liberal idea of uncertainty conceived as sets of subjective probabilities is, in substance, a further step away from the ideas determining ranking theory. Even if probabilism were successful in this way, as far as ranking theory is concerned we would only be thrown back to our comparative remarks in section "Ranks and probabilities", pp. 328ff.

We may therefore conclude that ranking theory is a strong independent pillar in that confusingly rich variety of theories found in the uncertainty literature. This conclusion is the only point of my sketchy comparative remarks. Of course, it is not to deny that the other theories serve other purposes well. It is obvious that we are still far from an all-purpose account of uncertainty or degrees of belief.

---

[32]Does this contradict the fact that ranking functions are equivalent to possibility measures (with their second kind of conditionalization), that possibility measures may be conceived as a special case of DS belief (or rather: plausibility) functions, and that Jeffrey conditionalization works for possibility measures as defined by Halpern (2003, p. 107)? No. The reason is that Jeffrey conditionalization for possibility measures is not a special case of Jeffrey conditionalization for DS belief functions in general. Cf. Halpern (2003, p. 107).

References

Bacon, F. (1620), *Novum Organum.*

Brafman, R. I., & Tennenholtz, M. (2000). An axiomatic treatment of three qualitative decision criteria. *Journal of the Association of Computing Machinery, 47*, 452–482.

Carnap, R. (1950). *Logical foundations of probability*. Chicago: Chicago University Press.

Cohen, L. J. (1970). *The implications of induction*. London: Methuen.

Cohen, L. J. (1977). *The probable and the provable*. Oxford: Oxford University Press.

Cohen, L. J. (1980). Some historical remarks on the Baconian conception of probability. *Journal of the History of Ideas, 41*, 219–231.

Darwiche, A., & Pearl, J. (1997). On the logic of iterated belief revision. *Artificial Intelligence, 89*, 1–29.

Dawid, A. P. (1979). Conditional independence in statistical theory. *Journal of the Royal Statistical Society B, 41*, 1–31.

de Finetti, B. (1937). La Prévision: Ses Lois Logiques, Ses Sources Subjectives. *Annales de l'Institut Henri Poincaré, 7*; engl. translation: (1964) Foresight: Its logical laws, its subjective sources. In: H. E. Kyburg, Jr., H. E., & Smokler (Eds.), *Studies in subjective probability* (pp. 93–158). New York: Wiley.

Dempster, A. P. (1967). Upper and lower probabilities induced by a multivalued mapping. *Annals of Mathematical Statistics, 38*, 325–339.

Dempster, A. P. (1968). A generalization of Bayesian inference. *Journal of the Royal Statistical Society, Series B, 30*, 205–247.

Dubois, D., & Prade, H. (1988). *Possibility theory: An approach to computerized processing of uncertainty*. New York: Plenum Press.

Dubois, D., & Prade, H. (1995). *Possibility theory as basis for qualitative decision theory*. In: Proceedings of the 14th International Joint Conference on Artificial Iintelligence (IJCAI'95), Montreal, pp. 1925–1930.

Dubois, D., & Prade, H. (1998). Possibility theory: Qualitative and quantitative aspects. In D.M. Gabbay & P. Smets (Eds.), *Handbook of defeasible reasoning and uncertainty management systems* (Vol. 1) (pp. 169–226). Dordrecht: Kluwer.

Gabbay, D. M., et al. (Eds.). (1994). *Handbook of logic in artificial intelligence and logic programming, vol. 3, nonmonotonic reasoning and uncertainty reasoning*. Oxford: Oxford University Press.

Garber, D. (1980). Field and Jeffrey conditionalization. *Philosophy of Science, 47*, 142–145.

Gärdenfors, P. (1978). Conditionals and changes of belief. In I. Niiniluoto & R. Tuomela (Eds.), *The logic and epistemology of scientific change* (pp. 381–404). Amsterdam: North-Holland.

Gärdenfors, P. (1988). *Knowledge in flux*. Cambridge: MIT Press.

Geiger, D., & Pearl, J. (1990). On the logic of causal models. In R. D. Shachter, T. S. Levitt, J. Lemmer, & L. N. Kanal (Eds.), *Uncertainty in artificial intelligence 4* (pp. 3–14). Amsterdam: Elsevier.

Giang, P. G., & Shenoy, P. P. (1999). On transformations between probability and spohnian disbelief functions. In K. B. Laskey & H. Prade (Eds.), *Uncertainty in artificial intelligence* (Vol. 15, pp. 236–244). San Francisco: Morgan Kaufmann.

Giang, P. G., & Shenoy, P. P. (2000). A qualitative linear utility theory for Spohn's theory of epistemic beliefs. In C. Boutilier & M. Goldszmidt (Eds.), *Uncertainity in artificial intelligence* (Vol. 16, pp. 220–229). San Francisco: Morgan Kaufmann.

Giang, P. G., & Shenoy, P. P. (2005). Two axiomatic approaches to decision making using possibility theory. *European Journal of Operational Research, 162*, 450–467.

Gilboa, I. (1987). Expected utility with purely subjective non-additive probabilities. *Journal of Mathematical Economics, 16*, 65–88.

Goldszmidt, M., & Pearl, J. (1996). Qualitative probabilities for default reasoning, belief revision, and causal modeling. *Artificial Intelligence, 84*, 57–112.

Hacking, I. (1975). *The emergence of probability*. Cambridge: Cambridge University Press.

Halpern, J. Y. (2003). *Reasoning about uncertainty*. Cambridge: MIT Press.
Hansson, S.O. (ed.) (1997), *Special Issue on Non-Prioritized Belief Revision. Theoria* 63, 1–134.
Hansson, S. O. (1999). *A textbook of belief dynamics. Theory change and database updating*. Dordrecht: Kluwer.
Harper, W. L. (1976). Rational belief change, popper functions and counterfactuals. In W. L. Harper & C. A. Hooker (Eds.), *Foundations of probability theory, statistical inference, and statistical theories of science* (Vol. I, pp. 73–115). Dordrecht: Reidel.
Hempel, C. G. (1945). Studies in the logic of confirmation. *Mind, 54*, 1–26 + 97–121.
Hempel, C. G. (1962). Deductive-nomological vs. Statistical explanation. In H. Feigl & G. Maxwell (Eds.), *Minnesota studies in the philosophy of science, vol. III, scientific explanation, space, and time* (pp. 98–169). Minneapolis: University of Minnesota Press.
Hild, M. (t.a.). *Introduction to induction: On the first principles of reasoning*, Manuscript.
Hild, M., & Spohn, W. (2008). The measurement of ranks and the laws of iterated contraction. *Artificial Intelligence, 172*, 1195–1218.
Hintikka, J. (1962). *Knowledge and belief*. Ithaca: Cornell University Press.
Hisdal, E. (1978). Conditional possibilities – independence and noninteractivity. *Fuzzy Sets and Systems, 1*, 283–297.
Hohenadel, S. (2013). *Efficient epistemic updates in rank-based belief networks*. Dissertation, University of Konstanz. See http://nbn-resolving.de/urn:nbn:de:bsz:352-250406
Huber, F. (2006). Ranking functions and rankings on languages. *Artificial Intelligence, 170*, 462–471.
Huber, F. (2007). The consistency argument for ranking functions. *Studia Logica, 86*, 299–329.
Hunter, D. (1991). Graphoids, semi-graphoids, and ordinal conditional functions. *International Journal of Approximate Reasoning, 5*, 489–504.
Jaffray, J.-Y. (1989). Linear utility theory for belief functions. *Operations Research Letters, 8*, 107–112.
Jeffrey, R. C. (1965). *The logic of decision*. Chicago: University of Chicago Press, 2nd ed. 1983.
Jeffrey, R. C. (1991). *Probability and the art of judgment*. Cambridge: Cambridge University Press.
Jensen, F. V. (2001). *Bayesian networks and decision graphs*. Berlin: Springer.
Joyce, J. (1998). A nonpragmatic vindication of probabilism. *Philosophy of Science, 65*, 575–603.
Joyce, J. (1999). *The foundations of causal decision theory*. Cambridge: Cambridge University Press.
Krantz, D. H., Luce, R. D., Suppes, P., & Tversky, A. (1971). *Foundations of measurement* (Vol. I). New York: Academic.
Krüger, L., et al. (1987). *The probabilistic revolution. Vol. 1: ideas in history, Vol. 2: ideas in the sciences*. Cambridge: MIT Press.
Kyburg, H. E., Jr. (1961). *Probability and the logic of rational belief*. Middletown: Wesleyan University Press.
Lange, M. (2000). *Natural laws in scientific practice*. Oxford: Oxford University Press.
Levi, I. (1967). *Gambling with truth*. New York: A. A. Knopf.
Levi, I. (2004). *Mild contraction: Evaluating loss of information due to loss of belief*. Oxford: Oxford University Press.
Lewis, D. (1969). *Convention: A philosophical study*. Cambridge: Harvard University Press.
Lewis, D. (1973). *Counterfactuals*. Oxford: Blackwell.
Lewis, D. (1975). Languages and language. In K. Gunderson (Ed.), *Minnesota studies in the philosophy of science* (Vol. VII, pp. 3–35). Minneapolis: University of Minnesota Press.
Lewis, D. (1980). A subjectivist's guide to objective chance. In R. C. Jeffrey (Ed.), *Studies in inductive logic and probability* (Vol. II, pp. 263–293). Berkeley: University of California Press.
Luce, R. D., & Raiffa, H. (1957). *Games and decisions*. New York: Wiley.
Maher, P. (2002). Joyce's argument for probabilism. *Philosophy of Science, 69*, 73–81.
McGee, V. (1994). Learning the impossible. In E. Eells & B. Skyrms (Eds.), *Probability and conditionals. Belief revision and rational decision* (pp. 179–199). Cambridge: Cambridge University Press.
Merin, A. (2006). *Decision theory of rhetoric*, book manuscript, to appear.

Merin, A. (2008). Relevance and reasons in probability and epistemic ranking theory. A study in cognitive economy. In: Forschungsberichte der DFG-Forschergruppe *Logik in der Philosophie* (Nr. 130). University of Konstanz.

Neapolitan, R. E. (1990). *Probabilistic reasoning in expert systems: Theory and algorithms*. New York: Wiley.

Oddie, G. (2001). Truthlikeness. In E. N. Zalta (Ed.), *The Stanford encyclopedia of philosophy* (Fall 2001 Edition). http://plato.stanford.edu/archives/fall2001/entries/truthlikeness

Pearl, J. (1988). *Probabilistic reasoning in intelligent systems: Networks of plausible inference*. San Mateo: Morgan Kaufman.

Pearl, J. (2000). *Causality. Models, reasoning, and inference*. Cambridge: Cambridge University Press.

Plantinga, A. (1993). *Warrant: The current debate*. Oxford: Oxford University Press.

Pollock, J. L. (1995). *Cognitive carpentry*. Cambridge: MIT Press.

Rescher, N. (1964). *Hypothetical reasoning*. Amsterdam: North-Holland.

Rescher, N. (1976). *Plausible reasoning*. Assen; Van Gorcum.

Rott, H. (2001). *Change, choice and inference: A study of belief revision and nonmonotonic reasoning*. Oxford: Oxford University Press.

Rott, H. (2008). Shifting priorities: Simple representations for twenty seven iterated theory change operators. In D. Makinson, J. Malinowski, & H. Wansing (Eds.), *Towards mathematical philosophy*. Dordrecht: Springer.

Sarin, R., & Wakker, P. P. (1992). A simple axiomatization of nonadditive expected utility. *Econometrica, 60*, 1255–1272.

Schmeidler, D. (1989). Subjective probability and expected utility without additivity. *Econometrica, 57*, 571–587.

Shackle, G. L. S. (1949). *Expectation in economics*. Cambridge: Cambridge University Press.

Shackle, G. L. S. (1969). *Decision, order and time in human affairs* (2nd ed.). Cambridge: Cambridge University Press.

Shafer, G. (1976). *A mathematical theory of evidence*. Princeton: Princeton University Press.

Shafer, G. (1978). Non-additive probabilities in the work of Bernoulli and Lambert. *Archive for History of Exact Sciences, 19*, 309–370.

Shenoy, P. P. (1991). On Spohn's rule for revision of beliefs. *International Journal of Approximate Reasoning, 5*, 149–181.

Smets, P. (1998). The Transferable Belief Model for Quantified Belief Representation. In D.M. Gabbay & P. Smets (Eds.), *Handbook of defeasible reasoning and uncertainty management systems* (Vol. 1) (pp. 267–301). Dordrecht: Kluwer.

Spirtes, P., Glymour, C., & Scheines, R. (1993). *Causation, prediction, and search*. Berlin: Springer, 2nd ed. 2000.

Spohn, W. (1976/78). *Grundlagen der Entscheidungstheorie*. Ph.D. thesis, University of Munich 1976, published: Kronberg/Ts.: Scriptor 1978, out of print, pdf-version at: http://www.uni-konstanz.de/FuF/Philo/Philosophie/philosophie/files/ge.buch.gesamt.pdf

Spohn, W. (1983). *Eine Theorie der Kausalität*, unpublished Habilitationsschrift, Universität München, pdf-version at: http://www.uni-konstanz.de/FuF/Philo/Philosophie/philosophie/files/habilitation.pdf

Spohn, W. (1986). The representation of Popper measures. *Topoi, 5*, 69–74.

Spohn, W. (1988). Ordinal conditional functions. A dynamic theory of epistemic states. In W. L. Harper & B. Skyrms (Eds.), *Causation in decision, belief change, and statistics* (Vol. II, pp. 105–134). Dordrecht: Kluwer.

Spohn, W. (1990). A general non-probabilistic theory of inductive reasoning. In R. D. Shachter, T. S. Levitt, J. Lemmer, & L. N. Kanal (Eds.), *Uncertainty in artificial intelligence* (Vol. 4, pp. 149–158). Amsterdam: Elsevier.

Spohn, W. (1991). A reason for explanation: Explanations provide stable reasons. In W. Spohn, B. C. van Fraassen, & B. Skyrms (Eds.), *Existence and explanation* (pp. 165–196). Dordrecht: Kluwer.

Spohn, W. (1993). Causal laws are objectifications of inductive schemes. In J. Dubucs (Ed.), *Philosophy of probability* (pp. 223–252). Dordrecht: Kluwer.

Spohn, W. (1994a). On the properties of conditional independence. In P. Humphreys (Ed.), *Patrick suppes: Scientific philosopher. Vol. 1: Probability and probabilistic causality* (pp. 173–194). Dordrecht: Kluwer.

Spohn, W. (1994b). On Reichenbach's principle of the common cause. In W. C. Salmon & G. Wolters (Eds.), *Logic, language, and the structure of scientific theories* (pp. 215–239). Pittsburgh: Pittsburgh University Press.

Spohn, W. (1999). Two coherence principles. *Erkenntnis, 50*, 155–175.

Spohn, W. (2001a). Vier Begründungsbegriffe. In T. Grundmann (Ed.), *Erkenntnistheorie. Positionen zwischen Tradition und Gegenwart* (pp. 33–52). Paderborn: Mentis.

Spohn, W. (2001b). Bayesian nets are all there is to causal dependence. In M. C. Galavotti, P. Suppes, & D. Costantini (Eds.), *Stochastic dependence and causality* (pp. 157–172). Stanford: CSLI Publications.

Spohn, W. (2002). Laws, ceteris paribus conditions, and the dynamics of belief. *Erkenntnis, 57*, 373–394; also in: Earman, J., Glymour, C., Mitchell, S. (Eds.). (2002). *Ceteris paribus laws* (pp. 97–118). Dordrecht: Kluwer.

Spohn, W. (2005a). Enumerative induction and lawlikeness. *Philosophy of Science, 72*, 164–187.

Spohn, W. (2005b). Isaac Levi's potentially surprising epistemological picture. In E. Olsson (Ed.), *Knowledge and inquiry: Essays on the pragmatism of Isaac Levi*. Cambridge: Cambridge University Press.

Spohn, W. (2006). Causation: An alternative. *British Journal for the Philosophy of Science, 57*, 93–119.

Spohn, W. (2012). *The laws of belief. Ranking theory and its philosophical applications*. Oxford: Oxford University Press.

Spohn, W. (2014). The epistemic account of ceteris paribus conditions. *European Journal for the Philosophy of Science, 4*(2014), 385–408.

Spohn, W. (2015). Conditionals: A unified ranking-theoretic perspective. *Philosophers' Imprint 15*(1)1–30; see: http://quod.lib.umich.edu/p/phimp/3521354.0015.001/

Studeny, M. (1989). Multiinformation and the problem of characterization of conditional independence relations. *Problems of Control and Information Theory, 18*, 3–16.

Wakker, P. P. (2005). Decision-foundations for properties of nonadditive measures: General state spaces or general outcome spaces. *Games and Economic Behavior, 50*, 107–125.

Williamson, T. (1994). *Vagueness*. London: Routledge.

Zadeh, L. A. (1975). Fuzzy logics and approximate reasoning. *Synthese, 30*, 407–428.

Zadeh, L. A. (1978). Fuzzy sets as a basis for a theory of possibility. *Fuzzy Sets and Systems, 1*, 3–28.

Part III
Decision Theory

Chapter 18
Introduction

Horacio Arló-Costa, Vincent F. Hendricks, and Johan van Benthem

The classical account of decision-making derives from seminal work done by Frank P. Ramsey (1926) and later on by Von Neumann and Morgenstern (1947). This work culminates later on with the influential account of Leonard Savage (1954), Ascombe and Aumann (1963) and de Finetti (1974). We can recapitulate here in a compact form the classical presentation by Von Neumann and Morgenstern. Define a lottery as follows: If $A_1, \ldots, A_m$ is a partition of the possible outcomes of an experiment with $\alpha_j = Pr(A_j)$ for each j, then the lottery $(\alpha_1, \ldots, \alpha_m)$ awards prize z_j if A_j occur. We can assume that the choice of the partition events does not affect the lottery. We can then introduce some central axioms for preferences among lotteries.

Axiom 18.1 (Weak Order) There is a weak order, $\geq$, among lotteries such that $L_1 \geq L_2$ iff L_1 is not strictly preferred to L_2.

Then we have a second crucial axiom:

Axiom 18.2 (Independence) For each L, L_1, L_2, and $0 < a < 1$, $L_1 \geq L_2$ iff $aL_1 + (1 - a)L \geq aL_2 + (1 - a)L$.

Horacio Arló-Costa was deceased at the time of publication.

H. Arló-Costa (deceased)
Carnegie Mellon University, Pittsburgh, PA, USA

V.F. Hendricks (✉)
Center for Information and Bubble Studies, University of Copenhagen, Copenhagen, Denmark
e-mail: vincent@hum.ku.dk

J. van Benthem
University of Amsterdam, Science Park 904, 1098 XH Amsterdam, The Netherlands

Stanford University, Stanford, United States
e-mail: johan@science.uva.nl

H. Arló-Costa et al. (eds.), *Readings in Formal Epistemology*, Springer Graduate Texts in Philosophy 1, DOI 10.1007/978-3-319-20451-2_18

A third Archimedean axiom is often introduced to guarantee that utilities are real valued. These axioms suffice to prove that there exists a utility over prizes U such that $(\alpha_1, \ldots, \alpha_m) > (\beta_1, \ldots, \beta_m)$ iff $\sum_{I=1,m} \alpha_i U(z_i) \leq \sum_{I=1,m} \beta_i U(z_i)$ This utility is unique up to positive affine transformation. Anscombe and Aumann (1963) introduced a fourth axiom designed to explicitly say that preferences among prizes did not vary with the state, or, as it is usually put that utilities are state-independent. Savage (1954) provided an alternative set of axioms for preferences among acts (functions from states to consequences) that do not rely on an auxiliary randomization. He then shows that these axioms lead to a unique probability and state-independent utility such that acts are ranked according to their expected utilities. Savage's postulates are consistent with VonNeumann and Morgenstern's three central axioms. The corresponding theory offers what we can consider as the standard received view on models of preference by maximizing expected utility.

Savage's account of decision-making has been challenged by a series of paradoxes and counterexamples. Two of these paradoxes occupy a central role in recent theorizing. The first was offered by Maurice Allais (1953), the second by Daniel Ellsberg (1961). We will discuss here as well two additional paradoxes: one initially suggested by the physicist William Newcomb (presented explicitly in a published form by Robert Nozick (1969)) and a paradox due to Teddy Seidenfeld which appears in the article reprinted here that he coauthored with Mark J. Schervish and Joseph B. Kadane.

We can focus first on Allais's conundrum. Consider three rewards, $r_1 = \$0$, $r_2 = \$1$ million, $r_3 = \$5$ million. Now consider the following lotteries: $L_1 = 1$ million for certain; $L_2 = (.01, .89, .10)$ with prizes $z_i = r_i$; $L_3 = (.90, .10)$ with prizes $z_1 = r_1$ and $z_2 = r_3$; $L_4 = (.89, .11)$ with prizes $z_1 = r_1$ and $z_2 = r_2$. Most people choose L_1 over L_2, and L_3 over L_4. If we assume that choices reveal the underlying preference it is easy to see that this violates the classical theory of expected utility. There are, nevertheless, many cognitive explanations for this behavior. For example, Ariel Rubinstein has suggested (Similarity and Decision Making under Risk (Is there a utility theory resolution to the Allais Paradox?) *Journal of Economic Theory*, 46, 145–153, 1988) that in the first choice subjects simply go for the sure thing and in the second choice the probabilities .11 and .10 are sufficiently similar but r_3 clearly dominates r_2. Although the use of the corresponding heuristic usually leads to choices that can be justified by the theory of expected utility in this case the use of the heuristics clashes with expected utility (see Seidenfeld's article for a detailed explanation). Many psychologists concluded that in situations of this type agents reliably commit certain cognitive errors and tried to construct a theory capable of predicting this type of behavior. Daniel Kahnemann and Amos Tversky proposed a theory of this sort in a paper initially published in 1979 in the journal *Econometrica* (see the reference in the article reprinted here). The theory in question is usually called Prospect Theory (PT). The axiomatic presentation of PT abandons the corresponding version of Independence. Philosophers had different types of reactions to Allais but in general they accepted that there are at least some versions of the paradox that constitute examples of systematic errors caused by the bias induced by the use of certain heuristics.

The paradox proposed by Daniel Ellsberg is quite different. We can present here the simplest version of the paradox. Urn A contains exactly 100 balls. 50 of these balls are solid black and the remaining 50 are solid white. Urn B contains exactly 100 balls. Each of these balls is either solid black or solid white, although the ratio of black balls to white balls is unknown. Consider now the following questions: How much would you be willing to pay for a ticket that pays $25 ($0) if the next random selection from Urn A results in black (white) ball? Repeat then the same question for Urn B. It is well known that subjects tend to offer higher maximum buying prices for urn A than for urn B. This indicates that subjects do not have identical probabilities for both urns (.5 for each color) as Savage's theory predicts. It is considerably less clear that this behavior has to be interpreted as some sort of error. Ellsberg himself saw this behavior as an indication that Savage's theory has to be amended to deal with situations where uncertainty and vague or imprecise probabilities are involved. One can perfectly think, for example, that probabilities remain indeterminate in the case of Urn B. There is a vast literature dealing with decisions under ambiguity that is reviewed in the article by Gilboa and Marinacci reprinted here. As Seidenfeld's article indicates there are two main choices: either embracing a theory that abandons Axiom 18.1 (Ordering) or alternatively embracing a theory that abandons Axiom 18.2 (Independence). Seidenfeld argues that abandoning Independence (a solution that is rather popular and that Ellsberg himself supported) has a costly price: it leads to a form of sequential incoherence. Seidenfeld's argument requires the use of axioms for sequential decision making that many have found controversial. Seidenfeld's article remains mainly concerned with normative solutions to the paradoxes. The article by Tversky and Kahnemann reprinted here intends to extend the initial version of prospect theory to the case of uncertainty as well. So, they think that the common choices elicited by Ellsberg constitute also an error. This implies having a conservative attitude regarding the normative status of standard decision theory that clearly clashes with the motivation and some of the central theoretical ideas that motivated Ellsberg's work.

Mark J. Schervish, Teddy Seidenfeld and Josheph B. Kadane question in their paper another central tenet of the standard theories of decision making: the assumption that utility has to be state-independent. They show via an ingenious example that the uniqueness of probability in standard representations is relative to the choice of what counts as a constant outcome. Moreover they prove an important result showing how to elicit a unique state-dependent utility. The result does not assume that there are prizes with constant value by introducing a new kind of hypothetical kind of act in which both the prize and the state of nature are determined by an auxiliary experiment.

Our final paradox is the one proposed by the physicist William Newcomb. Consider an opaque box and a transparent box. An agent may choose one or the other taking into account the following: The transparent box contains one thousand dollars that the agent plainly sees. The opaque box contains either nothing or one million dollars, depending on a prediction already made. The prediction was about the agent's choice. If the prediction was that the agent will take both boxes, then the opaque box is empty. On the other hand, if the prediction was that the agent will take just the opaque box, then the opaque box contains a million dollars. The prediction is reliable. The agent knows all these features of his decision problem. So, we can depict the agent's options as follows:

	Prediction of one-boxing	Prediction of two-boxing
Take only one box	SM	\$0
Take two boxes	$SM + \$T$	$\$T$

It is clear that two-boxing dominates one-boxing (the prizes of two-boxing are better than the prizes of one-boxing in each state of nature). So, two-boxing is the adequate choice according to dominance. Given the hypothesis of reliability of prediction, a prediction of one-boxing has a high probability given one-boxing. Similarly, a prediction of two-boxing has a high probability given two-boxing. Therefore, one-boxing's expected utility exceeds two-boxing's expected utility. One-boxing is the rational choice according to the principle of expected-utility maximization. Should one be a one-boxer or a two-boxer?

The formula used to calculate expected utility in the second case is: $U(A) = \sum_{I=1,n} \rho(S_i \mid A)\, u(A, S_i)$, where A is an act, and S_i are relevant states of nature. Joyce and Gibbard argue that two-boxing can be rationalized if one appeals to a different way of calculating expected utility:

$$UC(A) = \sum_{I=1,n} \rho(A > S_i)\, u(A, S_i) = \sum_{I=1,n} \rho(S_i \setminus A)\, u(A, S_i),$$

where the connective ">" is a counterfactual conditional and $\rho(S_i \setminus A)$ indicates a deviant type of conditional probability (called *imaging*) proposed by David Lewis. This type of conditional probability can be articulated in a paradox-free manner such that the probability of a counterfactual conditional coincides with the corresponding conditional probability (that is that $\rho(A > S_i) = \rho(S_i \setminus A)$). Classical conditional probability cannot satisfy this equation on pain on triviality (this was shown also by Lewis in a seminal paper that appeared in 1976: Probabilities of Conditionals and Conditional Probabilities, *Philosophical Review* 85, 297–315). The bibliographical notes below contain various useful pointers to recent papers debating the tenability of the corresponding notion of causal decision theory.

Suggested Further Reading

A classical and still rather useful book presenting the received view in decision theory is Savage's influential and seminal book: *The Foundations of Statistics*, Dover Publications; 2 Revised edition (June 1, 1972). A slightly more accessible but pretty thorough textbook presentation of Savage's account and beyond is the monograph by David Kreps: *Notes on the Theory of Choice*, Westview Press (May 12, 1988).

The classical essay by Daniel Ellsberg introducing his now famous paradox continues to be a very important source in this area: "Risk, Ambiguity and the Savage Axioms," *Quarterly Journal of Economics*, 75: 643-669, 1961. Isaac Levi presented a unified normative view of both Allais and Ellsberg in: "The Paradoxes of Allais and Ellsberg," *Economics and Philosophy*, 2: 23-53, 1986.

This solution abandons ordering rather than independence unlike the solution proposed by Ellsberg himself. Solutions abandoning independence have been in general more popular. Some of the classical papers in this tradition appear in the bibliography of the paper by Gilboa and Marinacci reprinted here. Many of the responses to Allais have been descriptive rather than normative. Prospect theory is a classical type of response along these lines. We reprint here an article that intends to present a unified descriptive response to both Allais and Ellsberg. The reader can find an excellent, thorough and mathematically mature presentation of the contemporary state of the art in Prospect theory in a recent book published by Peter Wakker: *Prospect Theory for Risk and Ambiguity*, Cambridge University Press, Cambridge, 2011.

The debate about the foundations of causal decision theory is in a way still open. An excellent presentation of causal decision theory can be found in an important book by Jim Joyce: *The Foundations of Causal Decision Theory*, Cambridge Studies in Probability, Induction and Decision Theory, Cambridge, 2008. Joyce has also written an interesting piece answering challenges to causal decision theory: "Regret and Instability in Causal Decision Theory," forthcoming in the second volume of a special issue of *Synthese* devoted to the foundations of the decision sciences (eds.) Horacio Arlo-Costa and Jeffrey Helzner. This special issue contains as well an essay by Wolfgang Spohn that intends to articulate the main ideas of causal decision theory by appealing to techniques used in Bayesian networks: "Reversing 30 Years of Discussion: Why Causal Decision Theorists should be One-Box." This is a promising line of investigation that has also been considered preliminary in an insightful article by Christopher Meek and Clark Glymour: "Conditioning and Intervening", *British Journal for the Philosophy of Science* 45, 1001-1021, 1994. With regard to issues related to actual causation a useful collection is the book: *Causation and Counterfactuals*, edited by J. Collins, N. Hall and L.A. Paul, MIT Press, 2004.

Finally a paper by Joseph Halpern and Judea Pearl offers a definition of actual causes using structural equations to model counterfactuals, Halpern, J. Y. and Pearl, J. (2005) "Causes and explanations: a structural-model approach. Part I: Causes", *British Journal for Philosophy of Science* 56:4, 843-887. This paper articulates ideas about causation based on recent work on Bayesian networks and related formalisms. Current work in this area seems to point to an unification of causal decision theory and an account of causation based on Bayesian networks.

Chapter 19
Allais's Paradox

Leonard Savage

Introspection about certain hypothetical decision situations suggests that the sure-thing principle and, with it, the theory of utility are normatively unsatisfactory. Consider an example based on two decision situations each involving two gambles.[1]

Situation 1.	Choose between
Gamble 1.	\$500,000 with probability 1; and
Gamble 2.	\$2,500,000 with probability 0.1, \$500,000 with probability 0.89, status quo with probability 0.01.
Situation 2.	Choose between
Gamble 3.	\$500,000 with probability 0.11, status quo with probability 0.89; and
Gamble 4.	\$2,500,000 with probability 0.1, status quo with probability 0.9.

Many people prefer Gamble 1 to Gamble 2, because, speaking qualitatively, they do not find the chance of winning a *very* large fortune in place of receiving a large fortune outright adequate compensation for even a small risk of being left in the status quo. Many of the same people prefer Gamble 4 to Gamble 3; because, speaking qualitatively, the chance of winning is nearly the same in both gambles, so the one with the much larger prize seems preferable. But the intuitively acceptable pair of preferences, Gamble 1 preferred to Gamble 2 and Gamble 4 to Gamble 3, is not compatible with the utility concept or, equivalently, the sure-thing

Leonard Savage was deceased at the time of publication.

[1]This particular example is due to Allais (1953). Another interesting example was presented somewhat earlier by Georges Morlat (1954).

L. Savage (deceased)
Princeton University, New York, NY, USA

H. Arló-Costa et al. (eds.), *Readings in Formal Epistemology*, Springer Graduate Texts in Philosophy 1, DOI 10.1007/978-3-319-20451-2_19

principle. Indeed that pair of preferences implies the following inequalities for any hypothetical utility function.

$$U\ (\$500{,}000) > 0.1U\ (\$2{,}500{,}000) + 0.89U\ (\$500{,}000) + 0.1U\ (\$0),$$
(3)
$$0.1U\ (\$2{,}500{,}000) + 0.9U\ (\$0) > 0.11U\ (\$500{,}000) + 0.89U\ (\$0);$$

and these are obviously incompatible.

Examples[2] like the one cited do have a strong intuitive appeal; even if you do not personally feel a tendency to prefer Gamble 1 to Gamble 2 and simultaneously Gamble 4 to Gamble 3, I think that a few trials with other prizes and probabilities will provide you with an example appropriate to yourself.

If, after thorough deliberation, anyone maintains a pair of distinct preferences that are in conflict with the sure-thing principle, he must abandon, or modify, the principle; for that kind of discrepancy seems intolerable in a normative theory. Analogous circumstances forced D. Bernoulli to abandon the theory of mathematical expectation for that of utility (Bernoulli 1738). In general, a person who has tentatively accepted a normative theory must conscientiously study situations in which the theory seems to lead him astray; he must decide for each by reflection—deduction will typically be of little relevance—whether to retain his initial impression of the situation or to accept the implications of the theory for it.

To illustrate, let me record my own reactions to the example with which this heading was introduced. When the two situations were first presented, I immediately expressed preference for Gamble 1 as opposed to Gamble 2 and for Gamble 4 as opposed to Gamble 3, and I still feel an intuitive attraction to those preferences. But I have since accepted the following way of looking at the two situations, which amounts to repeated use of the sure-thing principle.

One way in which Gambles 1–4 could be realized is by a lottery with a hundred numbered tickets and with prizes according to the schedule shown in Table 19.1.

Table 19.1 Prizes in units of \$100,000 in a lottery realizing gambles 1–4

		Ticket number		
		1	2–11	12–100
Situation 1	Gamble 1	5	5	5
	Gamble 2	0	25	5
Situation 2	Gamble 3	5	5	0
	Gamble 4	0	25	0

[2]Allais has announced (but not yet published) an empirical investigation of the responses of prudent, educated people to such examples (Allais 1953).

Now, if one of the tickets numbered from 12 through 100 is drawn, it will not matter, in either situation, which gamble I choose. I therefore focus on the possibility that one of the tickets numbered from 1 through 11 will be drawn, in which case Situations 1 and 2 are exactly parallel. The subsidiary decision depends in both situations on whether I would sell an outright gift of $500,000 for a 10-to-1 chance to win $2,500,000—a conclusion that I think has a claim to universality, or objectivity. Finally, consulting my purely personal taste, I find that I would prefer the gift of $500,000 and, accordingly, that I prefer Gamble 1 to Gamble 2 and (contrary to my initial reaction) Gamble 3 to Gamble 4.

It seems to me that in reversing my preference between Gambles 3 and 4 I have corrected an error. There is, of course, an important sense in which preferences, being entirely subjective, cannot be in error; but in a different, more subtle sense they can be. Let me illustrate by a simple example containing no reference to uncertainty. A man buying a car for $2,134.56 is tempted to order it with a radio installed, which will bring the total price to $2,228.41, feeling that the difference is trifling. But, when he reflects that, if he already had the car, he certainly would not spend $93.85 for a radio for it, he realizes that he has made an error.

One thing that should be mentioned before this chapter is closed is that the law of diminishing marginal utility plays no fundamental role in the von Neumann-Morgenstern theory of utility, viewed either empirically or normatively. Therefore the possibility is left open that utility as a function of wealth may not be concave, at least in some intervals of wealth. Some economic-theoretical consequences of recognition of the possibility of non-concave segments of the utility function have been worked out by Friedman and myself (1948), and by Friedman alone (1953). The work of Friedman and myself on this point is criticized by Markowitz (1952).[3]

References

Allais, M. (1953). Le comportement de l'homme rationnel devant le risque: Critique des postulats et axioms de l'école Americaine. *Econometrica, 21*, 503–546.

Archibald, G. C. (1959). Utility, risk, and linearity. *Journal of Political Economy, 67*, 437–450.

Bernoulli, D. (1738). Specimen theoriae novae de mensura sortis. *Commentarii academiae scientiarum imperialis Petropolitanae (for 1730 and 1731)*, 5, 175–192.

Centre National de Recherche Scientifique Fondements et applications de la théorie du risque en économetrie, Paris, Centre National de la Recherche Scientifique. (1954) *Report of an international econometric colloquium on risk, in which there was much discussion of utility*, held in Paris, May 12–17, 1952.

Friedman, M. (1953). Choice, chance, and personal distribution of income. *Journal of Political Economy, 61*, 277–290.

Friedman, M., & Savage, L. J. (1948). The utility analysis of choices involving risk. *Journal of Political Economy, 56*, 279–304, Reprinted, with a correction, in Stigler and Boulding (1952).

[3]See also Archibald (1959) and Hakansson (1970).

Hakansson, N. H. (1970). Friedman-Savage utility functions consistent with risk aversion. *Quarterly Journal of Economics, 84,* 472–487.

Markowitz, H. (1952). The utility of wealth. *Journal of Political Economy, 60,* 151–158, Marshall, Alfred.

Stigler, G. J., & Boulding, K. E. (Eds.). (1952). *Readings in price theory*. Chicago: Richard D. Irwin.

Chapter 20
Decision Theory Without "Independence" or Without "Ordering"

What Is the Difference?

Teddy Seidenfeld

Introduction

It is a familiar argument that advocates accommodating the so-called paradoxes of decision theory by abandoning the "independence" postulate. After all, if we grant that choice reveals preference, the anomalous choice patterns of the Allais and Ellsberg problems (reviewed in section "Review of the Allais and Ellsberg "Paradoxes") violate postulate P2 ("sure thing") of Savage's (1954) system. The strategy of making room for new preference patterns by relaxing independence is adopted in each of the following works: Samuelson (1950), Kahneman and Tversky's "Prospect Theory" (1979), Allais (1979), Fishburn (1981), Chew and MacCrimmon (1979), McClennen (1983), and in closely argued essays by Machina (1982, 1983 [see the latter for an extensive bibliography]).

There is, however, a persistent underground movement that challenges instead the normative status of the "ordering" postulate for preference. Those whose theories evidence some misgivings about ordering include: Good (1952), C. A. B. Smith (1961), Levi (1974, 1980), Suppes (1974), Walley and Fine (1979), Wolfenson and Fine (1982), and Schick (1984). And abandoning ordering is a strategy that has been used to resolve group decision problems. For this see Savage (1954, section 7.2) and Kadane and Sedransk (1980), and see Kadane (1986) for an application to clinical trials. "Regret" models also involve a failure of ordering since choice-with-regret does not satisfy Sen's (1977) principle of "independence of irrelevant alternatives": Savage (1954, section 13.5), Bell and Raiffa (1979) and Bell (1982), Loomes and Sugden (1982), and Fishburn (1983) discuss regret.

T. Seidenfeld (✉)
Departments of Philosophy and Statistics, Carnegie Mellon University, Pittsburgh, PA 15213, USA
e-mail: teddy@stat.cmu.edu

H. Arló-Costa et al. (eds.), *Readings in Formal Epistemology*, Springer Graduate Texts in Philosophy 1, DOI 10.1007/978-3-319-20451-2_20

Expected Utility for Simple Lotteries – A Review

For ease of exposition, let us adopt an axiomatization similar to the von Neumann and Morgenstern (1947) theory, as condensed by Jensen (1967). Let R be a set of β-many rewards (or payoffs), $R = (r_\alpha\colon \alpha \leq \beta)$. In the spirit of Debreu's (1959, chapter 4) presentation, we can think of R as including (infinitely divisible) monetary rewards. A (simple) lottery over R is a probability measure P on R with the added requirement that $P(X) = 1$ for some *finite* subset of rewards.

Lotteries are individuated according to the following (0th) *reduction postulate*: Let L_1, L_2 be two lotteries with probabilities P_1, P_2 and let $R_n = (r_1, \ldots, r_n)$ be the finite set of the union of the payoffs under these two lotteries. A convex combination, $\alpha L_1 + (1-\alpha)L_2$ $(0 \leq \alpha \leq 1)$, of the two lotteries is again a lottery with probability measure $\alpha P_1 + (1-\alpha)P_2$ over R_n. Thus, the set of lotteries is a mixture set $M(R)$ in the sense of Herstein and Milnor (1953).

Three postulates comprise expected utility theory:

(1) An ordering requirement: preference, $\lesssim$, a relation over $M \times M$, is a weak-order. That is, $\lesssim$ is reflexive, transitive, and all pairs of lotteries are comparable under $\lesssim$. (Strict preference, $<$, and indifference, $\sim$, are defined relations.)

(2) An Archimedean requirement: If $L_1 < L_2$ and $L_2 < L_3$, there is a nontrivial convex combination of L_1 and L_3 strictly preferred (and another combination strictly dispreferred) to L_2. That is, there exist

$$0 < \alpha, \beta < 1 \text{ with } \alpha L_1 + (1-\alpha)\, L_3 < L_2 \text{ and } L_2 < \beta L_1 + (1-\beta)\, L_3.$$

The point in assuming that R includes (infinitely divisible) monetary payoffs is made clear by the additional stipulation that each lottery in M carries a sure-dollar equivalent (under $\lesssim$):

$$\forall L \in \mathrm{M}\ \exists \$x \in \mathrm{R}\ \ (L \sim L_{\$x}), \tag{*}$$

where $L_{\$x}$ is a degenerate lottery having only one prize, $\$x$. Then principle (2) deserves its title for, with (*) and the added stipulation that more is (strictly) better when it comes to money, (1) and (2) entail a real-valued utility representation for $\lesssim$ (continuous in \$).[1] To simplify still further, let us restrict attention to lotteries with none but monetary payoffs.

[1]By assuming (*), we fix it that $M/\sim$ (no longer assumed to be a mixture set) has a countable dense subset in the $<$—order on $M/\sim$, e.g., the rational-valued sure-dollar equivalents. Then our first two postulates ensure a real-valued utility u on M with the property that $L_1 < L_2$ if and only if $u(L_1) < u(L_2)$. The point is that, without "independence," the usual Archimedean axiom is neither necessary nor sufficient for a real valued utility. See Fishburn (1970, Section 3.1) for details, or Debreu (1959, Chapter 4), who discusses conditions for u to be continuous. Debreu uses a "continuity" postulate in place of (2) that, in our setting, requires that if the sequence $[L_i]$ converges (in distribution) to the lottery L_i, and $L_i < L_k$, then all but finitely many of the $L_i < L_k$. If we extend $\lesssim$ to general distributions over R, Debreu's continuity postulate entails countably

(3) The "independence" principle: For all L_i, L_j, and L_k, and for all α $(0<\alpha\leq 1)$, $L_i \precsim L_j \iff \alpha L_i + (1-\alpha)L_k \precsim \alpha L_i + (1-\alpha)L_k$.

Let us examine these postulates for the special case of lotteries on three rewards: $R=(r_1<r_2<r_3)$, where the reward r_i is identified with the degenerate lottery having point-mass $P(r_i)=1$ $(i=1,2,3)$. Following the excellent presentation by Machina (1982), we arrive at a simple geometric account of what is permitted by expected-utility theory. Figure 20.1 depicts the consequences of postulates (1)–(3).

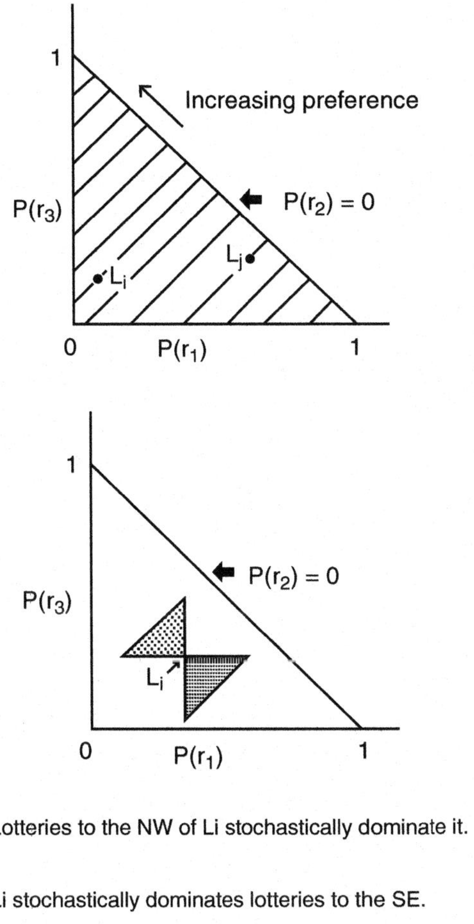

Fig. 20.1 Geometry of cardinal utility with three rewards

additive probability. See Seidenfeld and Schervish (1983) for some discussion of the decision-theoretic features of finitely additive probability.

Fig. 20.2 Geometry of the Allais paradox

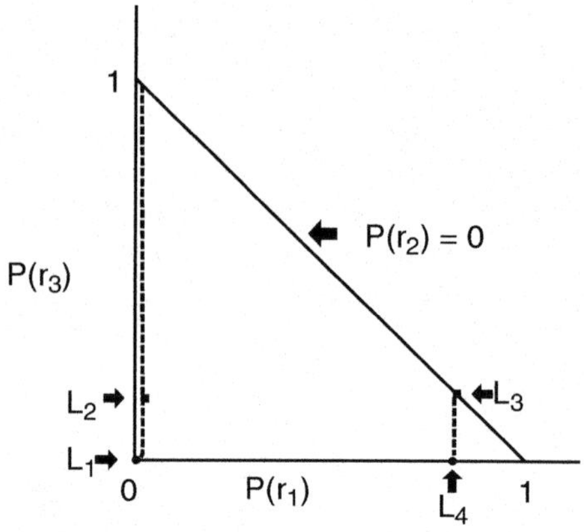

According to the postulates (1)–(3), indifference curves (∼) over lotteries are parallel, straight lines of (finite) positive slope. L_i is (strictly) preferred to L_j, $L_j < L_i$, is just in case the indifference curve for L_i is to the left of the indifference curve for L_j.

Consider a lottery L_i, as in Fig. 20.2. Stochastic dominance provides a (strict) preference for lotteries to the NW of L_i, whereas L_i is (strictly) preferred to lotteries to its SE.[2] Thus, the indifference lines must have positive slope. Hence, in this setting with lotteries over three rewards, expected-utility theory permits one degree of freedom for preferences, corresponding to the choice of a slope for the lines of indifference.

In a collaborated effort, Seidenfeld et al. (1987, Section 1), we apply this analysis to the selection of "sizes" (α-levels) for statistical tests of a simple null hypothesis against a simple rival hypothesis. The conclusion we derive is the surprising "incoherence" (conflict with expected-utility theory) of the familiar convention to choose statistical tests with a size, e.g., $\alpha = .01$ or $\alpha = .05$, independent of the sample size. This reasoning generalizes that of Lindley (1972, p. 14, where he gives his argument for the special case of "0–1" losses). In a purely "inferential" (nondecision-theoretic) Bayesian treatment for the testing of a simple hypothesis versus a composite alternative, Jeffreys (1971, p. 248) argues for the same caveat about constant α-levels.

[2]Recall, lottery L_2 (first order) stochastically dominates lottery L_1 if L_2 can be obtained from L_1 by shifting probability mass from less to more desirable payoffs. More precisely, L_2 stochastically dominates L_1 if, as a function of increasingly preferred rewards, the cumulative probability distribution for L_2 is everywhere less than (or equal to) the cumulative probability distribution for L_1. Of course, whenever L_2 stochastically dominates L_1, there is a scheme for payoffs, in accord with the two probability measures, where L_2 weakly dominates L_1.

Review of the Allais and Ellsberg "Paradoxes"

3.1

Allais (1953) poses the following question. For the three rewards, $r_1 = \$0$, $r_2 = \$1$ million, and $r_3 = \$5$ million (so $r_1 < r_2 < r_3$), what are your preferences, in the choice between lotteries L_1 and L_2, and in the choice between lotteries L_3 and L_4, where:

$$\begin{aligned} &L_1 - \text{with } P(r_2) = 1 \quad (\$1 \text{ million for certain}), \\ &L_2 - \text{with } P(r_1) = .01,\ P(r_2) = .89,\ \text{and } P(r_3) = .10, \\ &L_3 - \text{with } P(r_1) = .90 \text{ and } P(r_3) = .10, \text{ and} \\ &L_4 - \text{with } P(r_1) = .89 \text{ and } P(r_2) = .11? \end{aligned}$$

The common response, choose L_1 over L_2, and L_3 over L_4, violates EU theory (under the assumption that the choices reveal <). This is made evident by an application, Fig. 20.2, of (Machina's) figure 1.

The lines connecting the pairs of lotteries in the two choices are parallel. Thus, regardless of the slope of the parallel, straight-line indifference curves (from Fig. 20.1) imposed on lotteries over the three rewards, either L_2 and L_3 are preferred to their rivals, or else L_1 and L_4 are preferred. EU precludes the common answer to Allais' question.

3.2

Ellsberg's (1961) paradox of preference for lotteries with known risk ($\in M$) over uncertain lotteries ($\notin M$), bearing unknown risk, does not fit the simple mixture-set model, M. We can accommodate Ellsberg-styled problems by generalizing our concept of acts so that an act is a function f from states, a (finite, exhaustive) partition, to distributions on the reward set R. These more general acts are called "horse lotteries" by Anscombe and Aumann (1963). Denote by M' ($\supset M$) the generalized mixture-set for the class of horse lotteries.[3] Then lotteries of known risk belong to this enlarged (mixture) set M' as a special case: they are the "constant" acts. That is, the acts of known risk are those for which f^{-1} is a determinate probability measure.

Let us see how an Ellsberg-styled paradoxical choice violates postulate 3, supposing (1) and (2) obtain, when the postulates are applied to M'. Imagine I

[3]To define the generalized mixture-set M', it suffices to define the operation of convex-combination of two (generalized) lotteries. This is done exactly as in Anscombe and Aumann's (1963) treatment of "horse lotteries." Horse lotteries, the generalized postulates (1)–(3) for horse lotteries, and, with the addition of two minor assumptions (precluding a preference-interaction between payoffs and states), the subjective expected-utility theory that results, are discussed by Fishburn (1970, Chapter 13) and briefly in section Sequential coherence of Levi's decision theory here.

have placed \$10 in one of two pockets, which are otherwise empty. Consider the following three lotteries:

L_{left} – take the contents of my left pocket,
L_{right} – take the contents of my right pocket, and
L_{mix} – take the contents of my left pocket if a "fair" coin lands tails up, and take the contents of my right pocket if the fair coin lands head up.

Lotteries L_{left} and L_{right} are uncertain prospects. Suppose you are indifferent ($\sim$) between these two, which you evaluate as having a sure-dollar equivalent of \$2.50. However, the third option, L_{mix}, is (under the "reduction" postulate) a lottery of known risk. That is, L_{mix} is a lottery with an equal (.5; .5) probability distribution on the two payoffs (\$0, \$10). In the spirit of the Ellsberg paradoxical choice, suppose you evaluate the fair gamble on these two payoffs as having, say, a sure-dollar equivalent of \$4.00. You (strictly) prefer the lottery of known risk, L_{mix}, to either of the two uncertain lotteries. Finally, as the coin flip gives you no relevant information about the location of the \$10, your conditional preferences over the two uncertain lotteries (and their \$2.50 equivalent) are unaffected by the outcome of the coin flip. Then, as L_{mix} is (under reduction) equivalent to the ($\alpha = .5$) convex combination of L_{left} and L_{right}, preference for "risk" over "uncertainty" violates the independence postulate 3, assuming (1) and (2) hold.[4]

In fact, given (1) and (2), this version of the Ellsberg paradox conflicts with a principle (4), (strictly) weaker than principle (3).

(4) Mixture dominance ("betweenness"): Of lotteries L_1 and L_2, if each is (weakly or strictly) preferred (or dispreferred) to a lottery L_3; so, too, each convex combination of L_1 and L_2 is (weakly or strictly) preferred (or dispreferred) to L_3.

[4]These preferences are in conflict with Savage's (1954) "sure-thing" postulate P2. P2 is inconsistent with the following two preferences:

(i) $L_{\text{right}} < L_{\text{mix}}$.
(ii) $L_{\text{left}} \sim L_{\text{right}}$, given the coin lands heads up.

Consider the four-event partition generated by whether the coin lands heads (H) or tails (T), and whether the \$10 is in the left (L) or right (R) pocket. Then, by (i), the first row (below) is preferred to the second. Savage's theory uses "called-off" acts to capture conditional preference. Thus, by (ii), the agent is indifferent between the third and fourth rows.

	HL	HR	TL	TR
L_{mix}	\$10	\$0	\$0	\$10
L_{right}	\$0	\$10	\$0	\$10
$L_{\text{left}} \mid H$	\$10	\$0	\$0	\$0
$L_{\text{right}} \mid H$	\$0	\$10	\$0	\$0

In terms of (Machine's) figure 1, mixture dominance entails linear indifference curves. (This follows directly with (4), as then indifference is preserved under convex combinations. In Fig. 20.1, the set of convex combinations of two lotteries graphs as a straight line.) But the conjunction of (1), (2), and (4) does not entail (3). Samuelson's (1950) "Ysidro" ranking, and the "weighted utility" theory of Chew (1981) satisfy (1), (2) and (4) but fail (3).[5] Chew (1983) shows that the Allais paradoxical choices are admitted by his theory. What we find here is that Ellsberg-styled preference for risk over uncertainty cannot be so easily absorbed. In order to admit the Ellsberg-styled paradoxical choices, mixture dominance, (4), too, must fail.

Objections to the Denial of "Independence"

On Failures of "Stochastic Dominance"

Kahneman and Tversky's (1979) intriguing alternative to EU, "Prospect Theory," gives a reconstruction of Allais' paradoxical choice behavior at the expense of the independence postulate. Call a simple lottery *regular* provided not all its payoffs are (strictly) preferred to "status quo." Recall, the ranking of a lottery by expected utility uses the formula:

$$\sum_i P(r_i)\, u(r_i)\,.$$

For regular lotteries, the ranking of a lottery by prospect theory uses the formula:

$$\sum_i \pi\,[P(r_i)]\, \upsilon\,(r_i)\,,$$

where υ is a value-function for rewards (akin to the utility u), and π is some monotone-increasing function with $\pi(0)=0$ and $\pi(1)=1$. Again, let us consider (regular) lotteries on three rewards $r_1 < r_2 < r_3$, where we may take r_1 as status quo. If (and only if) π is linear do we have agreement between prospect theory and EU (for then $\pi(x)=x$ with the scalar constant absorbed into the utility u, defined up to

[5] Let u be a utility on payoffs and assume u is positive. Denote by $E_u(L_1)$ the expected utility of lottery L_1 under utility function u. Denote by $L_1{}^{-1}$ the lottery that has payoffs with (multiplicative) inverse utility to L_1. Samuelson's (1950) "Ysidro" ranking, $\lesssim_Y$, on lotteries is given by the function

$$Y(L_1) \;=\; \left[E_u(L_1)\,/E_u\left(L_1{}^{-1}\right)\right]^5.$$

Not only does $\lesssim_Y$ satisfy the ordering, Archimedean, and mixture dominance postulates while failing independence but in addition $\lesssim_Y$ respects stochastic dominance!

positive linear transformations). If π is not linear, so that prospect theory violates independence, stochastic dominance fails too.[6]

I do not know whether this aspect of prospect theory has been subjected to test for its descriptive accuracy. (I find it hard to believe that subjects would prefer a stochastically dominated lottery when the comparison involves just three rewards and the two lotteries involved assign identical probability to the status quo reward r_1.) In any event, it is normatively unacceptable to mandate a violation of stochastic dominance. What, after all, is left of the concern to avoid unnecessary losses (with respect to payoffs) when a theory *requires* a strict preference for an option dominated (on a set of positive probability)?

Thus, we shall examine only those violations of independence that induce preferences consistent with the partial order imposed by stochastic dominance. To that end, with an eye on the anticipated exchange between theories that abandon ordering versus those that abandon independence, I elevate respect for stochastic dominance to the status of a coherence condition.

Definition *A decision rule is* coherent *if (i) admissible choices under the rule are stochastically undominated, and (ii) admissibility is preserved under substitution (at choice points) of "indifferent" options.*

In section "Sequential coherence of Levi's decision theory", I summarize choice-based generalizations of "preference over rewards" (to explicate stochastic dominance) and "indifference over options" without assuming that choice induces a weak-order. However, in terminal decisions, when a choice rule induces an ordering $\lesssim$, condition (ii) adds nothing to (i). (See Sen's (1977) excellent discussion relating properties of choice rules to ordering.) That is, suppose lottery L_2 stochastically dominates L_1. By (i), L_1 is inadmissible when L_2 is available for choice, and $L_1 < L_2$. If, moreover, $L_1 \sim L_3$ and $L_2 \sim L_4$ then $L_3 < L_4$ by properties of $\lesssim$; so that inadmissibility of dominated options is preserved under substitution of indifferents, (ii). Therefore, in nonsequential decisions, and depending upon how indifference is defined without ordering, clause ii only serves as an added restriction on the coherence of choice rules that relax ordering. In sequential decisions the situation is rather different. As shown in section "Sequential incoherence – an example when mixture dominance fails, even though a choice rule induces a weak-order and respects stochastic dominance in nonsequential decisions, it may fail to be sequentially coherent.

The point of clause (ii) is to help identify a standard for evaluating decision rules predicated on the supposition that the agent's values for rewards (and for lotteries over those rewards) are stable over time. That is, this standard of coherence is

[6]The result is elementary and has been noted by many, including Kahneman and Tversky (1979, p. 283–284). Suppose π is not linear so that $\pi(p+q) > \pi(p) + \pi(q)$. Then by letting the value $\upsilon(r_2)$ approach the value $\upsilon(r_1)$, the agent is required (strictly) to prefer L_1: $P_1(r_1) = (1 - [p+q])$, $P_1(r_2) = p+q$, and $P_1(r_1) = 0$ – over L_2: $P_2(r_1) = P_1(r_1)$, $P_2(r_2). = p$, $P_2(r_3) = q$, even though L_2 stochastically dominates L_1. The argument for the other case is similar: $\pi(p+q) < \pi(p) + \pi(q)$.

offered for assessing the performance of a choice rule in sequential decisions when basic values are unchanging. Clause (ii) is not cogent, I would argue, when basic values are subject to revision over time. Then there may be a current preference between two rewards that are (to be) judged indifferent relative to the future, changed values. Thus there is no reason to demand that substitution of "future" indifferents preserves the inadmissibility of what is, by current values, a dominated option.

Of course, the agent's knowledge of events, chance occurrences, and preceding choices inevitably changes in the course of a sequential decision. In fact, these changes in evidence are what makes valuable adaptive experimental designs.

Sequential Incoherence – An Example When Mixture Dominance Fails

Respect for stochastic dominance provides a safeguard that choice over lotteries attends to sure-gains in payoffs. Single stage (nonsequential) decisions are thereby protected from violations of weak dominance over rewards. That is not the case, however, when we attend to sequential decisions. Specifically, coherence in nonsequential decisions, in choices over lotteries, does not entail the sequential version of coherence in choices over plans. This is illustrated by an example.

Consider what happens when mixture dominance (4) fails: *Example*: Let lotteries L_1 and L_2 be indifferent with a sure-dollar equivalent of \$5.00. Suppose, contrary to (4), that an equal ($\alpha = .5$) convex combination of them is strictly preferred with a sure-dollar equivalent of, e.g., \$6.00. Denote this by $L_3 = .5\,L_1 + .5\,L_2 \sim \$6.00$. Then, by continuity of preference for monetary payoffs, there is some fee, $\$\in$, that can be attached to the *payoffs* of L_1 and L_2 (resulting in the lotteries denoted by "$L_1 - \in$" and "$L_2 - \in$") satisfying

$$L_4 = (L_3 - \in) = .5\,(L_1 - \in) + .5\,(L_2 - \in) \sim \$5.75.$$

Also, we can find some dollar prize strictly dispreferred to both of the $\$\in$ – modifications of L_1 and L_2, e.g., suppose

$$\$4.00 < (L_1 - \in)\,,\,(L_2 - \in)\,.$$

Thus, we have the sequence:

$$\$4.00 < (L_1 - \in)\,,\,(L_2 - \in) < L_1 \sim L_2 \sim \$5.00 < L_4 \sim \$5.75 < L_3 \sim \$6.00$$

and, by assumption, these preferences respect stochastic dominance in dollar payoffs. So nonsequential choices among these options, according to these preferences, result in no incoherence.

Imagine, however, that an agent with these same preferences faces the following, two-stage sequential decision. Initially (at choice point A), the agent has two (sequential) alternatives, plans 1 and 2. Under sequential plan 1, a fair coin is flipped; (a) if it lands heads up, the agent chooses between L_1 and a dollar prize of \$5.50; and (b) if it lands tails up, the choice is between L_2 and the dollar prize of \$5.50. Under sequential option 2, the fair coin is flipped; (c) if it lands heads up, the agent chooses between $L_1 - \epsilon$ and \$4.00; and (d) if it lands tails up, the choice is between $L_2 - \epsilon$ and \$4.00. (The problem is depicted by Fig. 20.3, where $L_1 \sim L_2 \sim \$5.00 < .5\,L_1 + .5\,L_2 \sim \$6.00$, and where one finds the $\$\epsilon$ fee satisfying: $.5(L_1 - \epsilon) + .5(L_2 - \epsilon) \sim \$5.75$.)

How is the agent to choose between plans 1 and 2? It is clear, I think, that he should face up to what he knows his preferences are at choice nodes B, the choices he faces after the coin is flipped.[7] That is, the agent should assess the two (sequential) plans 1 and 2 in light of what he knows they lead to.

Under (1), if the coin lands heads up (a) he will choose \$5.50 over lottery L_1.[8] And if the coin lands tails up (b) again, he will choose the \$5.50 (over L_2). Thus, from the standpoint of (A), choosing plan 1 leads to a sure payoff of \$5.50.

Under (2), if the coin lands heads up (c) he will choose the lottery $L_1 - \epsilon$ over the dollar reward of \$4.00. And if the coin lands tails up (d), the lottery $L_2 - \epsilon$ is preferred to a sure \$4.00. Hence, from the standpoint of (A), choosing plan 2 leads

---

[7] Hammond's (1976, Section 3.3) felicitous phrase is that the agent uses "sophisticated" versus "myopic" choice.

[8] McClennen (1986, 1988, forthcoming) sketches a program of "resolute" choice to govern sequential decisions when independence fails. I am not very sure how resolute choice works. Part of my ignorance stems from my inability to find a satisfactory answer to several questions.

As I understand McClennen's notion of resolute choice, the agent's preferences for basic lotteries change across nodes in a sequential decision tree. (Then, the premises of the argument in Section 4 do not obtain.) In terms of the problem depicted in Fig. 4, at node A the agent resolves that he will choose L_1 at (a) of node B, and by so resolving increases its value at (a) of node B above the \$5.50 alternative.

There are several difficulties I find with this proposal. I suspect that the new value of L_1 at (a) will be fixed at \$6.00, and likewise for L_2 at (b). (The details of resolute choice are lacking on this point, but this suspicion is based on the observation that a minor variation in the sequential incoherence argument applies unless these two lotteries change their value from node A to node B as indicated. Just modify the construction so that the rejected cash alternative at B is \$6.00 – δ.) Then the assessed value of \$6.00 for L_3 (a mixture of the lotteries L_1 and L_2, now valued at \$6.00 each) is in accord with postulate (2). Such resolutions mandate that changes in preferences agree, sequentially, with the independence postulate. In terms of *sequential* decisions, is it not the case that resolute choice requires changes in values to agree with the independence postulate?

A second problem with resolute choice directs attention at the reasonableness of these mandatory changes in values. For example, consider the Ellsberg-styled choice problem described in Section 3.2. Cast in a sequential form, under this interpretation of resolute choice, if L_{mix} is most preferred, then the agent is required to increase the value for the option "take the contents of the right pocket," given that the coin lands heads up, over the value it has prior to the coin flip.

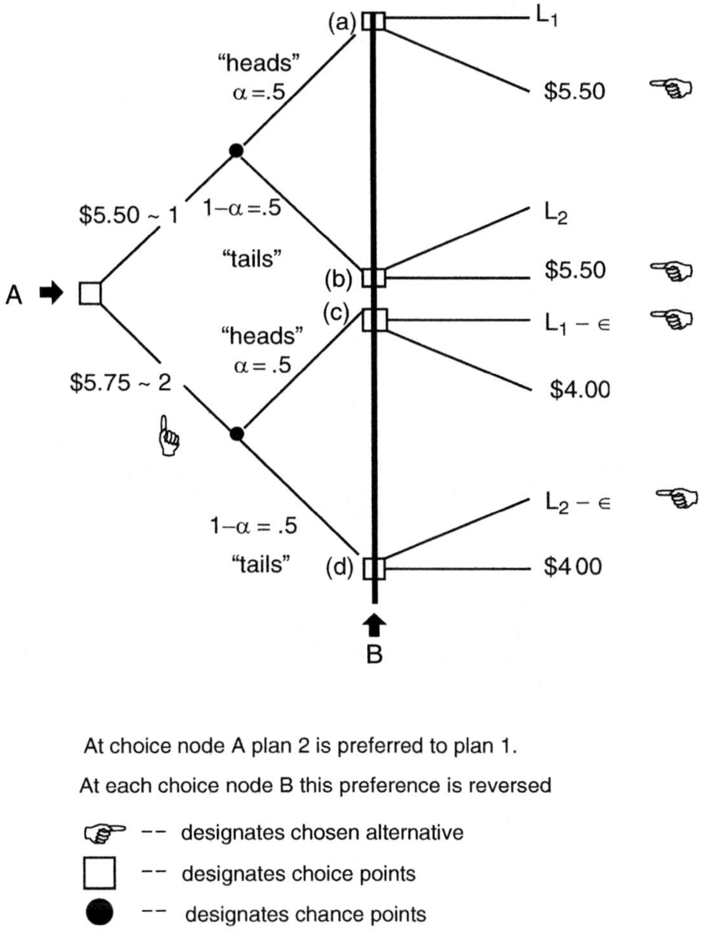

Fig. 20.3 An illustration of sequential incoherence for a failure of mixture of dominance ("betweenness")

to an equal ($\alpha = .5$) convex combination of the two lotteries $L_1 - \epsilon$ and $L_2 - \epsilon$. That is, from the perspective of choice node (A), sequential option 2 yields the lottery L_4, which is valued at \$5.75.

We make this reasoning precise with the following principle.

But the coin flip is irrelevant to a judgment of where the money is. However uncertain the agent is prior to the coin flip, is he not just as uncertain afterwards? Concern with uncertainty in the location of the money is the alleged justification for a failure of independence when comparing the three terminal options: L_{left}, L_{right}, L_{mix}, and declaring L_{mix} (strictly) better than the other two. What justifies the preference shift, given the outcome of the coin flip, when L_{right} becomes equivalued with an even-odds lottery over \$10 and \$0 despite the same state of uncertainty about the location of the money before and after the coin flip?

Dynamic Feasibility (DF)
To assess plan p *at a choice node* n_i, *anticipate how you will choose at its (potential) "future" choice nodes* n_j *and declare infeasible all future alternatives under* p *which are inadmissible at* n_j.

By this account, according to the principle DF, at (*A*) the agent prefers plan 2 over the rival plan 1. At (*A*) plan 2 is worth \$5.75 where plan 1 is worth only \$5.50. However, there is an embarrassment to these preferences. At choice nodes *B*, regardless of the fall of the coin, the agent prefers the choice he makes under plan 1 to what he chooses under plan 2.

If the coin lands heads up, the choice under (1), at (a), \$5.50 is preferred to the choice under (2), at (c), the lottery $L_1 - \in$. Likewise, if the coin lands tails up, the choice under (1), at (b), \$5.50 is preferred to the choice under (2), at (d), $L_2 - \in$. Therefore, though the agent prefers plan 2 to plan 1 initially [at (*A*)], he knows that this preference is reversed at (*B*), regardless of how the coin lands.

Under the indifferences of the ordering postulate and the preferences induced by stochastic dominance, at nodes *B*, the upshot is a contradiction in assessment of the sequential decision problem. The contradiction obtains as follows.

By postulate (1), the agent has a weak ordering of options at choice nodes *B*. There are some sure-dollar equivalents for the choices $L_1 - \in$ and $L_2 - \in$. By simple dominance, the sure-dollar equivalent for $L_1 - \in$ (or for $L_2 - \in$) is less than that for L_1. That is, each of $L_1 - \in$ and $L_2 - \in$ is worth less than \$5.00 and more than \$4.00. Since admissibility at choice nodes (such as nodes *B*) respects indifference, one's anticipation and knowledge [at (*A*)] of one's choices is only up to the level of indifferents. That is, the ordering postulate fixes choices only up to the equivalences of indifference, $\sim$. Thus, substitution of $\sim$–indifferents at choice nodes *B* leaves unchanged choices at those nodes. But plan 2 is dominated by plan 1 when, as dictated by postulate (1), the choices at nodes (c) and (d) [of (*B*)] are switched for their (indifferent) sure-dollar equivalents.

Of course, given the replacements of $L_1 - \in$ and $L_2 - \in$ by their sure-dollar equivalents at (*A*) plan 1 is preferred to plan 2 by dominance. (See Fig. 20.4, where, as in Fig. 20.3, $L_1 \sim L_2 \sim \$5.00 < .5\,L_1 + .5\,L_2 \sim \$6.00$, and where one finds the \$$\in$ fee satisfying: $.5(L_1 - \in) + .5(L_2 - \in) \sim \$5.75$.) Hence, subject to DF, applications of postulate (1) – ordering – at nodes *B* lead to contradictory conclusions with decisions made at node *A*.

Let us call this contradictory assessment an episode of *sequential incoherence*. That is, subject to Dynamic Feasibility, respect for stochastic dominance is not

---

Third, by what standards can the agent reassess the merits of a resolution made at an earlier time? In other words, how is the agent to determine whether or not to ignore an earlier judgment: the judgment to commit himself to a resolute future choice and thereby to change his future values. Once the future has arrived, why not instead frame the decision with the current choice node as the initial node without altering basic values? Unless this issue is addressed, the question of how to ratify a resolution is left unanswered, and the problem remains of how to make sense of the earlier resolution once the moment of choice is at hand.

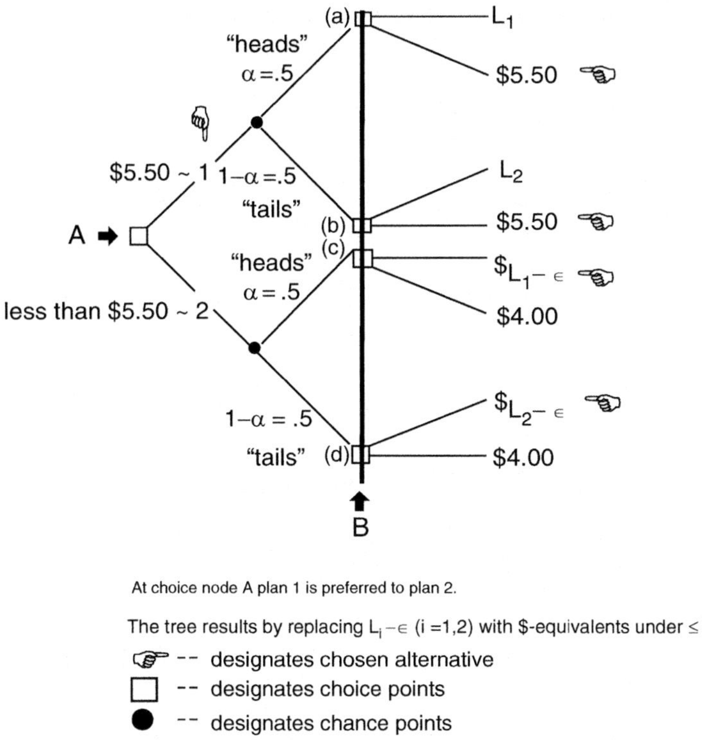

Fig. 20.4 An illustration of sequential incoherence for a failure of mixture of dominance ("betweenness")

preserved under substitution of indifferent alternatives at choice nodes. That is, clause ii of coherence fails with this choice rule.

It is important to understand that, at (*A*) (in the problem depicted in Fig. 20.3), the agent has *no* "terminal" options, no choices of lotteries. In particular, at choice node *A*, the agent does not have the terminal option L_3. Nor does he have any of the terminal options corresponding to the other seven lotteries that arise from an equal ($\alpha = .5$) convex combination of the options available to him at nodes *B*. Thus, at (*A*), he does not have the choice of $5.50 outright. What he does have as a choice at (*A*) is plan 1, which, by DF, he equates with a certain $5.50. But plan 1 calls for decisions at nodes *B*, depending upon how the coin lands, and these subsequent choices are not to be ignored at (*A*). The principle of Dynamic Feasibility achieves a limited reduction of plans to terminal options.

In the language of game theory (Luce and Raiffa 1957, chapter 3), the sequential decision problem (above) is in *extensive* form. What we learn from this problem is that, when mixture dominance fails, even with DF, sequential decisions in extensive form are not equivalent to the normal form one-stage (nonsequential) decisions that result by ignoring subsequent choice nodes like (*B*). In a normal form version of this sequential problem each of the two plans is represented by a set of four lotteries. In normal form, the decision is among the eight lotteries:

$$[L_3, (.5 \cdot \$5.50 + .5L_2), (.5L_1 + .5 \cdot \$5.50), \$5.50, \ldots, \$4.00].$$

Of these L_3 is most preferred, say. In normal form, then, plan 1 is chosen and is valued at \$6.00 ($\sim L_3$). However, the argument offered in this section (establishing sequential incoherence) does not presume the equivalence of extensive and normal forms.[9]

In the sequential problem, at node A, the agent knows L_3 is not available to him under plan 1. This is because he knows that (at nodes B) the dollar prize (\$5.50) is preferred to each of the lotteries L_1 and L_2. Under these preferences, the choice of plan 1 at (A) in the hope that L_1 will be chosen if heads and L_2 if tails is a pipe dream – mere wishful thinking that is brought up short by Dynamic Feasibility.[10]

Sequential Incoherence from Failures of Independence

What is the relation between failures of independence and episodes of sequential incoherence? An answer is given by the central critical result of this essay:

Theorem *If $\lesssim$ is a weak-order (1) of simple lotteries satisfying the Archimedean postulate (2) with sure-dollar equivalents for lotteries, and if $\lesssim$ respects stochastic dominance in payoffs ("a greater chance at more is better"), then a failure of independence, (3), entails an episode of sequential incoherence.*

---

[9]By contrast, Raiffa's (1968, pp. 83–85) classic objection to the failure of independence in the Allais paradox depends upon a reduction of extensive to normal forms. Also, in his interesting discussion, Hammond (1984) requires the equivalence of extensive and normal forms through his postulate of "consequentialism" in decision trees. These authors defend a strict expected-utility theory, in which the equivalence obtains. Likewise, LaValle and Wapman (1986) argue for the independence postulate with the aid of the assumption that extensive and normal forms are equivalent.

The analysis offered here does not presume this equivalence, nor does avoidance of sequential incoherence entail this equivalence, since, e.g., it is not satisfied in Levi's theory either – though his theory avoids such incoherence. Hence, for the purpose of separating decision theories without independence from those without ordering, it is critical to avoid equating extensive and normal forms of decision problems.

[10]One may propose that, by force of will, an agent can introduce new terminal options at an initial choice node, corresponding to the "normal" form version of a sequential decision given in "extensive" form. Thus, for the problem depicted in Fig. 20.3, the assumption is that the agent can create the terminal option L_3 at node A by opting for plan 1 at A and then choosing L_1 at (a) and L_2 at (b).

Whatever the merit of this proposal, it does not apply to the sequential decisions discussed here, since, by stipulation, the agent cannot avoid reconsideration at nodes B. There may be some problems in which agents can create new terminal options at will, but that is not a luxury we freely enjoy. Sometimes we have desirable terminal options and sometimes we can only plan. (See Levi's [1980, Chapter 17] interesting account of "using data as input" for more on this subject.)

Proof The proof is given in two cases. (The argument for the second case uses the full assumption of stochastic dominance rather than the weaker assumption [used in Case 20.1] that < respects simple dominance in dollar payoffs.)

Case 20.1 Let $L_1 \lesssim L_2$, yet for some L_3 and α, $\alpha L_2 + (1-\alpha)L_3 < \alpha L_1 + (1-\alpha)L_3$. Let $\$X \sim \alpha L_2 + (1-\alpha)L_3$, $\$Z \sim \alpha L_1 + (1-\alpha)L_3$, and $\$\mathcal{U} \sim L_3$, with $X < Z$. By our assumptions of a weak order for preference, continuity in dollar payoffs, and the strict preference for more (over less) money, there is some \$2∈ fee and amount \$$Y$ for which:

$$\$X < \alpha\,(L_1 - 2\in) + (1-\alpha)\,(L_3 - 2\in) \sim \$Y < \$Z.$$

Next, choose fees \$γ and \$δ so that:

$$L_2 - \gamma < L_1 - 2\in \text{ and } \$\,(\mathcal{U} - 2\delta) < L_3 - 2\in.$$

If we consider the sequential decision problem whose "tree" is depicted in Fig. 20.5 for Case 20.1, we discover by the same reasoning we used in the example above:

Fig. 20.5 Sequential incoherence for failures of "independence": Case 20.1

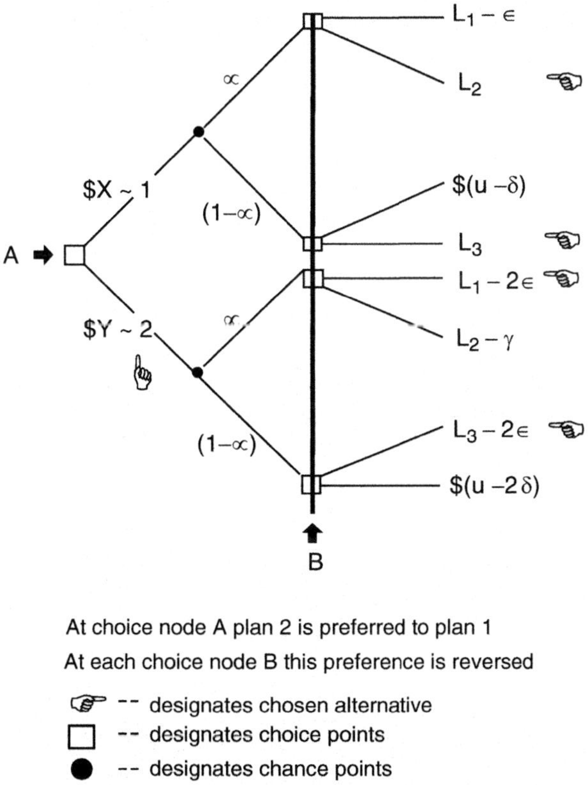

At node A, plan 1 is valued at \$$X$, whereas plan 2 is valued at \$$Y$. Thus, at node A, plan 2 is the preferred choice.

But at nodes B, regardless of which "chance event" occurs, the favored option under plan 1 is preferred to the favored option under plan 2. Thus, the preferences leading to a failure of independence in Case 20.1 succumb to sequential incoherence. An application of indifferences (from the ordering postulate 1 at nodes B leads to an inconsistent evaluation, at (A), of the sequential plans 1 and 2.

Case 20.2 $L_1 < L_2$, yet there are L_3 and $\alpha > 0$ with $\alpha L_1 + (1-\alpha)L_3 \sim \alpha L_2 + (1-\alpha) L_3$. Let \$$\mathcal{U} \sim L_3$ and \$$Z \sim \alpha L_1 + (1-\alpha)L_3$. Then choose an \$$\in$ fee so that $L_1 < L_2 - \in$. Let \$$X \sim \alpha(L_2 - \in) + (1-\alpha)L_3$. Then by stochastic dominance, $X < Z$. Next, by continuity, choose a \$$\delta$ fee to satisfy \$$Y \sim \alpha(L_1 - \delta) + (1-\alpha)(L_3 - \delta)$, where $X < Y < Z$. Choose an integer n so that $L_2 - n\in < L_1 - \delta$. Finally, find any \$$\gamma$ fee and choose an integer m so that \$$(u - m\gamma) < L_3 - \delta$.

Consider a sequential decision problem whose "tree" is depicted in Fig. 20.6 for Case 20.2. Once again we find an episode of sequential incoherence as:

Fig. 20.6 Sequential incoherence for failures of "independence": Case 20.2

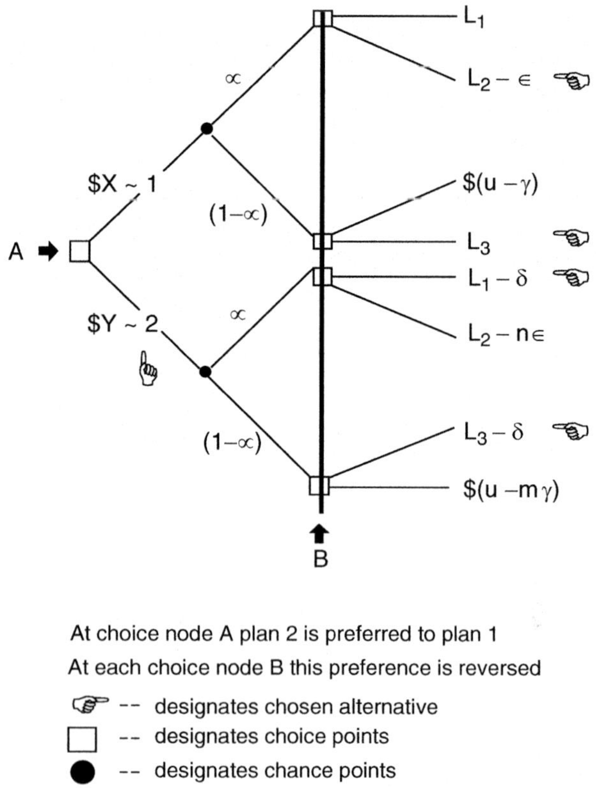

At (*A*), plan 1 is valued at $\$X$, whereas plan 2 is valued at $\$Y$. Thus, at node *A* plan 2 is the preferred choice.
At node *B*, regardless of which chance event occurs, the favored option under plan 1 is preferred to the favored option under plan 2. Thus, the preferences leading to a failure of independence in Case 20.2 succumb to sequential incoherence.

Concluding Remark

Can familiar "Dutch Book" arguments (de Finetti 1974; Shimony 1955) be used to duplicate the results obtained here? Do the considerations of book establish sequential incoherence when independence fails? I think they do not.

The book arguments require an assumption that the concatenation (conjunction) of favorable or indifferent or unfavorable gambles is, again, a favorable or indifferent or unfavorable gamble. That is, the book arguments presume payoffs with a simple, additive, utility-like structure. The existence of such commodities does not follow from a (subjective) expected-utility theory, like Savage's. And rivals to EU, such as Samuelson's (1950) "Ysidro" ranking, can fail to satisfy this assumption though they respect stochastic dominance in lotteries. Thus, in light of this assumption about combinations of bets, the Dutch Book argument is not neutral to the dispute over coherence of preference when independence fails. (Of course, that debate is not what Dutch Book arguments are designed for.)[11]

This objection to the use of a book argument does not apply to the analysis presented here. The argument for sequential incoherence is not predicated on the Dutch Book premise about concatenations of favorable gambles. That assumption is replaced by a weaker one, to wit: $\lesssim$ respects stochastic dominance in \$-rewards. There is no mystery why the weakening is possible. Here, we avoid the central question addressed by the Dutch Book argument: When are betting odds subjective probabilities? The book arguments pursue the representation of coherent belief as probabilities, given a particular valuation for combinations of payoffs. Instead, the spotlight here is placed on the notion of coherent sequential preference, given a preference (a weak ordering) of lotteries with canonical probabilities.

[11]See Frederic Schick's "Dutch Bookies and Money Pumps" (1986) for discussion of the import of this concatenation assumption in the Dutch Book argument. Its abuse in certain "intertemporal" versions of Dutch Book is discussed in Levi (1987).

Summary

Attempts to generalize EU by denying independence, while retaining the ordering and Archimedean postulates, fail the test of coherence in simple sequential choices over lotteries with dollar rewards.

Sequential Coherence of Levi's Decision Theory

A detailed analysis of Levi's Decision Theory (LDT), a theory without the ordering postulate, is beyond the scope of this essay. Here, instead, I shall merely report the central results that establish coherence of LDT in sequential choices over horse lotteries, a setting where both values (utility/security) and beliefs (probability) may be indeterminate. (I give proofs of these results in a technical report Seidenfeld (1987)).

To begin with, consider the following choice-based generalizations of the concepts: indifference, preference, and stochastic dominance. These generalizations are intended to apply in the domain of horse-lottery options, regardless of whether or not a decision rule induces a weak-ordering of acts.

The notational abbreviations I use are these. An option is denoted by o_i and, since acts are functions from states to outcomes, also by the function on states $o_i(s)$. Sets of feasible options are denoted by O, and the admissible options (according to a decision rule) from a feasible set O are denoted by the function C[O].

Call two options $\approx$ *–indifferent*, if and only if, whenever both are available either both are admissible or neither is.

Definition

$$\mathrm{o}_1 \approx \mathrm{o}_2 \iff \forall(\mathrm{O})\ (\{\mathrm{o}_1, \mathrm{o}_2\} \subset \mathrm{O} \Rightarrow (\mathrm{o}_1 \in \mathrm{C}[\mathrm{O}] \iff \mathrm{o}_2 \in \mathrm{C}[\mathrm{O}])).$$

When a choice rule induces a weak-order, denoted by $\lesssim$, then $\approx$ is just the symmetrized $\sim$ relation: $(\mathrm{o}_1 \sim \mathrm{o}_2) \iff (\mathrm{o}_1 \lesssim \mathrm{o}_2)$ *and* $(\mathrm{o}_2 \lesssim \mathrm{o}_1)$.

Next, define a choice-based relation of categorical preference over rewards using a restricted version of Anscombe and Aumann's (1963, p. 201) "Assumption 1," modified to avoid the ordering postulate. [This assumption is part of the value-neutrality of states. See Drèze's (1985) monograph for a helpful discussion of related issues.] Let o_1 and o_2 be two horse lotteries that differ only in that o_2 awards reward r_2 on states where o_1 awards reward r_1. Then reward r_2 is categorically preferred to reward r_1 just in case o_1 is inadmissible whenever o_2 is available. In symbols,

Definition *Reward* r_2 *is* categorically preferred *to reward* $\mathrm{r}_1 \iff \forall(\mathrm{O})\ \forall(\mathrm{o}_1 \neq \mathrm{o}_2)\ (\forall(\mathrm{s})[\mathrm{o}_1(\mathrm{s}) = \mathrm{o}_2(\mathrm{s}) \vee (\mathrm{o}_1(\mathrm{s}) = \mathrm{r}_1\ \&\ \mathrm{o}_2(\mathrm{s}) = \mathrm{r}_2)]\ \&\ \mathrm{o}_2 \in O \Rightarrow \mathrm{o}_1 \notin \mathrm{C}[\mathrm{O}])$.

Third, we need to make precise a suitable generalization of stochastic dominance among lotteries. Recall, when there is an ordering ($\lesssim$) of rewards, lottery L_2 stochastically dominates lottery L_1 if L_2 can be obtained from L_1 by shifting some distribution mass from less to more preferred rewards. Thus, L_2 stochastically dominates L_1 just in case $\alpha L_2 + (1-\alpha)L_3$ stochastically dominates $\alpha L_1 + (1-\alpha)L_3$ $(0 < \alpha \leq 1)$. Rely on this biconditional to formulate the following $\prec$ *–dominance* relation over horse lotteries, defined for an arbitrary choice function C.

Definition $o_1 \prec o_2 \iff \forall(O)\forall(o)\forall(\alpha > 0)\ ((\alpha o_2 + (1-\alpha)o) \in O \Rightarrow (\alpha o_1 + (1-\alpha)o) \notin C[O])$. Trivially, if $o_2 \prec$ *–dominates* o_1, *then (let* $\alpha = 1$*)* o_1 *is inadmissible whenever* o_2 *is available.*

Sequential coherence for a decision rule requires, then, that

(i) shifting distributional mass from categorically less to categorically more preferred rewards produces an $\prec$ –dominating option, and
(ii) the inadmissibility of $\prec$ –dominated options is preserved under substitution (at choice nodes) of $\approx$ –indifferents.

To see why LDT is sequentially coherent, recall that admissibility in Levi's (1974, 1980) decision theory is determined by a two-tiered lexicographic rule. The first tier is "*E*-admissibility." An option is *E*-admissible in the set O of feasible options, provided it maximizes expected utility (over O) for some pair $(P, \mathcal{U})$ – where P is a personal probability in the (convex) set **P** that represents the agent's beliefs about the states, and where $\mathcal{U}$ is a (cardinal) utility in the (convex) set **U** that represents one aspect of the agent's values over the rewards.

Definition o *is* E *-admissible* $\iff \exists(P,U)\ \forall(o' \in O)\ E_{PU}[o] \geq E_{PU}[o']$.[12]

The second tier in admissibility requires that an option be "S-admissible." This condition demands of an option that it be *E*-admissible and maximize a "security" index among the *E*-admissible options. The security of an option reflects yet another aspect of the agent's value structure. For purposes of this section, the notion of security is illustrated with three (real-valued) varieties:

$\text{Sec}_o[o] = 0$ – a vacuous standard, where all options have equal security;
$\text{sec}_1[o] = \text{infemum}_{U, R_o}\ \mathcal{U}[r]$, where R_o is the set of possible outcomes (rewards) for option o and $r \in R_o$. Thus, sec_1 is security indexed by the "worst" possible reward, a maximin consideration;

[12]Levi (1980, Section 5.6) offers a novel rule, here called rule', for determining expectation-inequalities when the partition of states, π, is finite but when events may have subjective probability 0. The motive for this emendation is to extend the applicability of "called-off" bets (Shimony 1955) to include a definition of conditional probability given an event of (unconditional) probability 0. Also, it extends the range of cases where a weak-dominance relation determines a strict preference.

Given a probability/utility pair $(P, \mathcal{U})$, maximizing †-expected utility (with rule') includes a weak-order that satisfies the independence axiom, though, †-expectations may fail to admit a real-valued representation, i.e., the "Archimedean" axiom is not then valid. Under rule†, given a pair $(P, \mathcal{U})$, †-expectations are represented by a lexicographic ordering of a vector-valued quantity.

$\sec_2[o] = \text{infemum}_{\mathbf{P}\times\mathbf{U}}\, E_{P,U}[o]$ – security indexed by the least expected utility, also called the "Γ-minimax" level for option o. (I avoid the minor details of defining $\sec_2$ when events have probability O and expectations are determined by rule†, as reported in note 12.)

Thus, an option is admissible in LDT just in case it is both E-admissible and maximizes security among those that likewise are E-admissible. As a special case, when security is vacuous or is indexed by $\sec_2$, and when both **P** and **U** are unit sets, C[O] is the subset of options that maximize subjective expected utility: the strict Bayesian framework. Then admissibility satisfies the ordering and independence postulates (and the Archimedean axiom, too, provided events have positive probability or rule† is not used).

The next few results, stated without proof, provide the core of the argument for sequential coherence of LDT. Condition i of coherence in nonsequential decisions, and more, is shown by the following.

Theorem 20.1 *If* o_2 *can be obtained from* o_1 *by shifting distribution masses from categorically less to more preferred rewards, then* $o_1 < o_2$, *and thus* o_1 *is inadmissible whenever* o_2 *is available.*

The theorem follows directly from a useful lemma about $<$ and categorical preference in LDT.

Lemma $o_1 \prec o_2 \iff \forall(P, U) E_{PU}(o_1) < E_{PU}(o_2)$. *Thus,* r_2 *is categorically preferred to* $r_1 \iff \forall(U)\, U(r_1) < U(r_2)$.

Thus, both $\prec$-dominance and categorical preference are strict partial orders, being irreflexive and transitive. In addition, following obviously from its definition, $\prec$-dominance satisfies the independence axiom. [Note the term "categorical preference" is borrowed from Levi (1986b, p. 91). The lemma provides the collateral for this conceptual loan.]

Condition ii of coherence for LDT in nonsequential decisions is shown by a result that two options are $\approx$ –related exactly when they have the same security index and are indiscernible by expectations:

Theorem 20.2 $o_1 \approx o_2 \iff (sec[o_1] = sec[o_2]\ \&\ \forall(P,U)(E_{PU}[o_1] = E_{PU}[o_2])$. *Thus,* $\approx$ *is an equivalence relation and, in nonsequential decisions, admissibility is preserved under substitution of* $\approx$ *–indifferent options.*

In order to extend the analysis to sequential decisions, the notion of $\approx$ –indifference is generalized to include conditional assessments, conditional upon the occurrence of either "chance" or "event" outcomes. The next corollary is elementary and indicates how conditional $\approx$ –indifference applies when for instance, choice nodes follow chance nodes.

Corollary 20.1 $\approx$ *is preserved under chance mixtures,*

$$\forall\,(i, j, k \text{ and } \alpha)\; o_{\mathrm{i}} \approx o_{\mathrm{i}} \Rightarrow (\alpha o_{\mathrm{i}} + (1-\alpha)\, o_{\mathrm{k}} \approx \alpha o_i + (1-\alpha)\, o_{\mathrm{k}}).$$

Under the principle of Dynamic Feasibility (which provides a limited reduction of sequential plans to nonsequential options), these findings combine to support the desired conclusion.

Sequential coherence for LDT

(i) Admissible options are $\prec$ –undominated among the dynamically feasible alternatives, and
(ii) Provided the agent updates his beliefs by Bayes' rule and does not change his value structure, admissibility is preserved under the substitution of $\approx$ –indifferent alternatives at choice nodes.

Conclusions About Relaxing Independence or Ordering

I have argued that decision theories that relax only the independence postulate succumb to sequential incoherence. That is, such programs face the embarrassment of choosing stochastically dominated options when, in simple two-stage sequential decisions, dollar equivalents are substituted for their indifferent options at terminal choice nodes. Moreover, the criticism does *not* presume an equivalence between sequential decisions (in extensive form) and their normal form reductions; instead, all decisions are subject to a principle of Dynamic Feasibility.

In section "Sequential coherence of Levi's decision theory", I generalize sequential coherence to choice rules that may not induce an ordering of options by preference. Also, I outline reasons for the claim that Levi's Decision Theory is sequentially coherent (in a setting where both belief and value are subject to indeterminacy). Since Levi's theory is one that fails the ordering postulate, the combined results establish a demarcation between these two strategies for relaxing traditional (subjective) expected-utility theory. The difference is that only one of the two approaches is sequentially coherent.

Acknowledgments I have benefitted from discussions with each of the following about the problems addressed in this essay: W. Harper. J. Kadane, M. Machina, P. Maher, E. F. McClennen, M. Schervish; and I am especially grateful to I. Levi.

Discussions

Editors' Note

Subjective expected-utility theory provides simple and powerful guidance concerning how to make rational decisions in circumstances involving risk. Yet actual decision making often fails, as has been well known for decades, to conform to the theory's recommendations. If subjective expected-utility theory represents the ideal of rational behavior, these failures may simply show that people often behave irrationally. Yet if the gap between ideal and actual behavior is too wide, or if behavior that on the best analysis we can make is rational but not consistent with subjective expected-utility theory, then we may come to doubt some of the axioms of the theory. Two main lines of revision have been suggested: either weakening the "ordering" axiom that requires preferences to be complete or surrendering the so-called independence principle. Although the issues are highly abstract and somewhat technical, the stakes are high; subjective expected-utility theory is critical to contemporary economic thought concerning rational conduct in public as well as private affairs.

In the preceding article, "Decision Theory without 'Independence' or without 'Ordering': What Is the Difference?" Teddy Seidenfeld argued for the sacrifice of ordering rather than independence by attempting to show that abandoning the latter leads to a kind of sequential incoherence in decision making that will not result from one specific proposal (Isaac Levi's) for abandoning ordering. In their comments in this section, Edward McClennen, who supports surrendering the independence postulate rather than ordering, and Peter Hammond, who argues against any weakening of subjective expected-utility theory, discuss Seidenfeld's argument from their quite different theoretical perspectives.

Economics and Philosophy, 4, 1988, 292–297. Printed in the United States of America.

References

Allais, M. (1953). Le comportement de l'homme rationnel devant le risque: Critique des postulats et axiomes de l'école americaine. *Econometrica, 21*, 503–546.

Allais, M. (1979). The so-called Allais Paradox and rational decisions under uncertainty. In M. Allais & O. Hagen (Eds.), *Expected utility hypotheses and the Allais Paradox* (pp. 437–681). Dordrecht: Reidel.

Anscombe, F. J., & Aumann, R. J. (1963). A definition of subjective probability. *Annals of Mathematical Statistics, 34*, 199–205.

Bell, D. (1982). Regret in decision making under uncertainty. *Operations Research, 20*, 961–981.

Bell, D., & Raiffa, H. (1979). *Decision regret: A component of risk aversion.* Unpublished manuscript, Harvard University.

Chew, S. H. (1981). *A mixture set axiomatization of weighted utility theory* (4th revision). Tuscon: Department of Economics, University of Arizona.

Chew, S. H. (1983). *A generalization of the quasilinear mean with applications to the measurement of income inequality and decision theory resolving the Allais Paradox.* Tuscon: Department of Economics, University of Arizona.

Chew, S. H., & MacCrimmon, K. R. (1979). *Alpha-Nu choice theory: A generalization of expected utility theory* (University of British Columbia Working Paper).

de Finetti, B. (1974). *Theory of probability* (2 vols.). New York: Wiley.

Debreu, G. (1959). *Theory of value.* New Haven: Yale University Press.

Drèze, J. H. (1985). *Decision theory with moral hazard and state-dependent preferences* (CORE discussion paper #8545). Belgium: Center for Operations Research and Econometrics, Université Catholique de Louvain.

Ellsberg, D. (1961). Risk, ambiguity, and the savage axioms. *Quarterly Journal of Economics, 75*, 643–699.

Fishburn, P. C. (1970). *Utility theory for decision making*. New York: Kriefer Publishing Company.

Fishburn, P. C. (1981). An axiomatic characterization of skew-symmetric bilinear functionals, with applications to utility theory. *Economic Letters, 8*, 311–313.

Fishburn, P. C. (1983). Nontransitive measurable utility. *Journal of Mathematical Psychology, 26*, 31–67.

Good, I. J. (1952). Rational decisions. *Journal of the Royal Statistical Society B, 14*, 107–114.

Hammond, P. J. (1976). Changing tastes and coherent dynamic choice. *Review of Economic Studies, 43*, 159–173.

Hammond, P. J. (1984). *Consequentialist behaviour in decision trees is Bayesian rational*. Stanford University.

Herstein, I. N., & Milnor, J. (1953). An axiomatic approach to measurable utility. *Econometrica, 21*, 291–297.

Jeffreys, H. (1971). *Theory of probability* (3rd ed.). Oxford: Oxford University Press.

Jensen, N. E. (1967). An introduction to Bernoullian utility theory: I. Utility functions. *Swedish Journal of Economics, 69*, 163–183.

Kadane, J. (1986). Toward a more ethical clinical trial. *Journal of Medicine and Philosophy, 11*, 385–404.

Kadane, J., & Sedransk, N. (1980). Toward a more ethical clinical trial. In J. Bernardo, M. H. DeGroot, D. V. Lindley, & A. F. M. Smith (Eds.), *Bayesian statistics* (pp. 329–338). Valencia: University Press.

Kahneman, D., & Tversky, A. (1979). Prospect theory: An analysis of decision under risk. *Econometrica, 47*, 263–291.

LaValle, I. H., & Wapman, K. R. (1986). Rolling back decision trees requires the independence axiom? *Management Science, 32*(3), 382–385.

Levi, I. (1974). On indeterminate probabilities. *Journal of Philosophy, 71*, 391–418.

Levi, I. (1980). *The enterprise of knowledge*. Cambridge: MIT Press.

Levi, I. (1986a). The paradoxes of Allais and Ellsberg. *Economics and Philosophy, 2*, 23–53.

Levi, I. (1986b). *Hard choices: Decision making under unresolved conflict*. Cambridge: Cambridge University Press.

Levi, I. (1987). The demons of decision. *The Monist, 70*, 193–211.

Lindley, D. V. (1972). *Bayesian statistics: A review*. Philadelphia: SIAM.

Loomes, G., & Sugden, R. (1982). Regret theory: An alternative theory of rational choice under uncertainty. *Economic Journal, 92*, 805–824.

Luce, D., & Raiffa, H. (1957). *Games and decisions*. New York: Wiley.

Machina, M. (1982). Expected utility analysis without the independence axiom. *Econometrica, 50*, 277–323.

Machina, M. (1983, October). *The economic theory of individual behavior toward risk: Theory, evidence and new directions* (Technical report #433). San Diego: Department of Economics, University of California.

McClennen, E. F. (1983). Sure-thing doubts. In B. Stigum & F. Wenstop (Eds.), *Foundations of utility and risk theory with applications* (pp. 117–136). Dordrecht: Reidel.

McClennen, E. F. (1986). Prisoner's dilemma and resolute choice. In R. Campbell & L. Sowden (Eds.), *Paradoxes of rationality and cooperation* (pp. 94–104). Vancouver: University of British Columbia Press.

McClennen, E. F. (1988). Dynamic choice and rationality. In B. R. Munier (Ed.), *Risk, decision, and rationality* (pp. 517–536). Dordrecht: Reidel.

McClennen, E. F. (forthcoming). *Rationality and dynamic choice: Foundational explorations*. Cambridge: Cambridge University Press.

Raiffa, H. (1968). *Decision analysis: Introductory lectures on choices under uncertainly*. Reading: Addison-Wesley.

Samuelson, P. (1950). Probability and the attempts to measure utility. *Economic Review, 1*, 167–173.

Savage, L. J. (1954). *The foundations of statistics*. New York: Wiley.
Schick, F. (1984). *Having reasons*. Princeton: Princeton University Press.
Schick, F. (1986). Dutch bookies and money pumps. *Journal of Philosophy, 83*, 112–119.
Seidenfeld, T. (1987). *Sequential coherence without the 'ordering' postulate: Levi's decision theory* (Technical report). Pittsburgh: Department of Philosophy, Carnegie-Mellon University.
Seidenfeld, T., & Schervish, M. (1983). A conflict between finite additivity and avoiding Dutch book. *Philosophy of Science, 50*, 398–412.
Seidenfeld, T., Schervish, M., & Kadane, J. (1987). Decisions without ordering. In W. Sieg (Ed.), *Acting and reflecting*. Dordrecht: Reidel (Forthcoming).
Sen, A. K. (1977). Social choice theory: A re-examination. *Econometrica, 45*, 53–89.
Shimony, A. (1955). Coherence and the axioms of confirmation. *Journal of Symbolic Logic, 20*, 1–28.
Smith, C. A. B. (1961). Consistency in statistical inference and decision. *Journal of the Royal Statistical Society B, 23*, 1–25.
Suppes, P. (1974). The measurement of belief. *Journal of the Royal Statistical Society B, 36*, 160–175.
von Neumann, J., & Morgenstern, O. (1947). *Theory of games and economic behavior* (2nd ed.). Princeton: Princeton University Press.
Walley, P., & Fine, T. L. (1979). Varieties of modal (classificatory) and comparative probability. *Synthese, 41*, 321–374.
Wolfenson, M., & Fine, T. (1982). Bayes-like decision making with upper and lower probabilities. *Journal of the American Statistical Association, 77*, 80–88.

Chapter 21
Ambiguity and the Bayesian Paradigm

Itzhak Gilboa and Massimo Marinacci

Introduction

Varying Probability Estimates

John and Lisa are offered additional insurance against the risk of a heart disease. They would like to know the probability of developing such a disease over the next 10 years. The happy couple shares some key medical parameters: they are 70 years old, smoke, and never had a blood pressure problem. A few tests show that both have a total cholesterol level of 310 mg/dL, with HDL-C (good cholesterol) of 45 mg/dL, and that their systolic blood pressure is 130. Googling "heart disease risk calculator", they find several sites that allow them to calculate their risk. The results (May 2010) are:

	John (%)	Lisa (%)
Mayo Clinic	25	11
National Cholesterol Education Program	27	21
American Heart Association	25	11
Medical College of Wisconsin	53	27
University of Maryland Heart Center	50	27

I. Gilboa (✉)
HEC, Paris, France

Tel-Aviv University, Tel Aviv, Israel
e-mail: tzachigilboa@gmail.com

M. Marinacci
Università Bocconi, Milano, Italy
e-mail: massimo.marinacci@unibocconi.it

H. Arló-Costa et al. (eds.), *Readings in Formal Epistemology*, Springer Graduate Texts in Philosophy 1, DOI 10.1007/978-3-319-20451-2_21

As we can see from the table, the estimates vary substantially: the highest for John is 100 % higher than the lowest, whereas for Lisa the ratio is 5:2. Opinion diverge in these examples, even though there are based on many causally independent observations that allow the use of statistical techniques such as logistic regression. However, in many important economic questions, such as the extent of global warming, there are very few past events to rely on. Further, many events, such as revolutions and financial crises, cannot be assumed independent of past observations. Thus, it appears that for many events of interest one cannot define an objective, agreed-upon probability.

Does Rationality Necessitate Probability?

Since the mid-twentieth century, economic theory is dominated by the Bayesian paradigm, which holds that any source of uncertainty can and should be quantified probabilistically.[1] According to this view, John and Lisa should have well-defined probabilities that they will develop a heart disease within the next 10 years, as should Mary for the temperature distribution anywhere on the globe 5 years hence. But where should John, Lisa, or Mary get these probabilities from? If they are to consult experts, they will typically obtain different estimates. Which experts are they to believe? Should they compute an average of the experts' estimates, and, if so, how much weight should each expert have in this average?

The standard line of reasoning of the Bayesian approach is that, in the absence of objective probabilities, the decision maker (DM, for short) should have her own, *subjective* probabilities, and that these probabilities should guide her decisions. Moreover, the remarkable axiomatic derivations of the Bayesian approach (culminating in Savage (1954)), show that axioms that appear very compelling necessitate that the DM behave as if she maximized expected utility relative to a certain probability measure, which is interpreted as her subjective probability. Thus, the axiomatic foundations basically say, "Even if you don't know what the probabilities are, you should better adopt some probabilities and make decisions in accordance with them, as this is the only way to satisfy the axioms."

There is a heated debate regarding the claim that rationality necessitates Bayesian beliefs. Knight (1921) and Keynes (1921, 1937) argued that not all sources of uncertainty can be probabilistically quantified. Knight suggested to distinguish between "risk", referring to situations described by known or calculable probabilities, and "uncertainty", where probabilities are neither given nor computable. Keynes (1937) wrote,

> By 'uncertain' knowledge, let me explain, I do not mean merely to distinguish what is known for certain from what is only probable. The game of roulette is not subject, in this

[1] As Cyert and DeGroot (1974) write on p. 524 "To the Bayesian, all uncertainty can be represented by probability distributions."

> sense, to uncertainty ... The sense in which I am using the term is that in which the prospect of a European war is uncertain, or the price of copper and the rate of interest twenty years hence ... About these matters there is no scientific basis on which to form any calculable probability whatever. We simply do not know.

Gilboa et al. (2008, 2009, 2012) argue that the axiomatic foundations of the Bayesian approach are not as compelling as they seem, and that it may be irrational to follow this approach. In a nutshell, their argument is that the Bayesian approach is limited because of its inability to express ignorance: it requires that the agent express beliefs whenever asked, without being allowed to say "I don't know" . Such an agent may provide arbitrary answers, which are likely to violate the axioms, or adopt a single probability and provide answers based on it. But such a choice would be arbitrary, and therefore a poor candidate for a rational mode of behavior.

Axiomatic derivations such as Savage's may convince the DM that she ought to have a probability, but they do not tell her which probability it makes sense to adopt. If there are no additional guiding principles, an agent who picks a probability measure arbitrarily should ask herself, is it so rational to make weighty decisions based on my arbitrarily-chosen beliefs? If there are good reasons to support my beliefs, others should agree with me, and then the probabilities would be objective. If, however, the probabilities are subjective, and others have different probabilities, what makes me so committed to mine? Wouldn't it be more rational to admit that these beliefs were arbitrarily chosen, and that, in fact, I don't know the probabilities in question?

Outline

The rest of this paper is organized as follows. Section "History and Background" discusses the history and background of the Bayesian approach. It highlights the fact that this approach has probably never been adopted with such religious zeal as it has within economic theory over the past 60 years. Section "Alternative Models" describes several alternatives to the standard Bayesian model. It surveys only a few of these, attempting to show that much of the foundations and machinery of the standard model need not be discarded in order to deal with uncertainty. Section "Ambiguity Aversion" surveys the notion of ambiguity aversion. The updating of non-Bayesian beliefs is discussed in section "Updating Beliefs". Section "Applications" briefly describes some applications of non-Bayesian models. The applications mentioned here are but a few examples of a growing literature. They serve to illustrate how non-Bayesian models may lead to different qualitative predictions than Bayesian ones. A few general comments are provided in section "Conclusion".

History and Background

Early Pioneers

Decision theory was born as a twin brother of probability theory through the works of a few scholars in the sixteenth and seventeenth century, originally motivated by the study of games of chance. Among them the works of Christiaan Huygens (1629–1695) and Blaise Pascal (1623–1662) are particularly relevant. We begin with Pascal, whose footsteps Huygens followed.

Pascal (1670) Since its very early days, probability had two different interpretations: first, it captures the notion of *chance*, referring to relative frequencies of occurrences in experiments that are repeated under the same conditions. This includes the various games of chance that provided the motivation for the early development of the theory. Second, probability can capture the notion of *degree of belief*, even when no randomness is assumed, and when nothing remotely similar to a repeated experiment can be imagined.

It is this second interpretation that, over time, evolved into the Bayesian approach in both decision and probability theory. In this regard, Pascal is perhaps the most important pioneer of probability theory. Though he made early key contributions to the probabilistic modeling of games of chance, it is his famous wager that is mostly relevant here. Roughly at the same time that Descartes and Leibniz were attempting to prove that God existed, Pascal changed the question from the proof of existence to the argument that it is worthwhile to believe in God, an option that he identified with the choice of a pious form of life based on the precepts of the Christian religion.[2] In so doing, he applied the mathematical machinery developed for objective probabilities in games of chance to the subjective question of God's existence, where no repeated experiment interpretation is possible. This led him to informally introduce several major ideas of modern decision theory, including the decision matrix, the notion of dominant strategies, subjective probability, expected utility maximization, and non-unique probability.[3]

Thus, the subjective interpretation of probabilities and their application as a tool to quantify beliefs showed up on the scene more or less as soon as did the objective interpretation and the application to games of chance. Further, as soon as the notion of subjective probability came on stage, it was accompanied by the possibility that this probability might not be known (see Shafer (1986), for related remarks on Bernoulli (1713), who introduced the law of large numbers).

---

[2]According to Pascal a pious life would ultimately induce faith. Importantly, Pascal did not assume that one can simply choose one's beliefs.

[3]Pascal did not finish his *Pensées*, which appeared in print in 1670, 8 years after his death. The text that was left is notoriously hard to read since he only sketches his thoughts (here we use the 1910 English edition of W. F. Trotter). Our rendering of his argument crucially relies on Hacking (1975)'s interpretation (see Hacking 1975, pp. 63–72; Gilboa 2009, pp. 38–40).

Huygens (1657) In the wake of the early probability discussions of Fermat and Pascal, Huygens (1657) first clearly proposed expected values to evaluate games of fortune.[4] Unlike Pascal's grand theological stand, Huygens only dealt with games of fortune ("cards, dice, wagers, lotteries, etc." as reported in the 1714 English version). Nevertheless, he was well aware of the intellectual depth of his subject.

Huygens' arguments are a bit obscure (at least for the modern reader; see Daston (1995)). His essay has, however, a few remarkable features from our perspective. First, he does not present the expected value criterion as an axiom; rather, he justifies its relevance by starting from more basic principles. For this reason his essay is articulated in a sequence of mathematical propositions that establish the expected value criterion for more and more complicated games. Huygens's propositions can be thus viewed as the very first decision-theoretic representation theorems, in which the relevance of a decision criterion is not viewed as self-evident, but needs to be justified through logical arguments based on first principles.

A second remarkable feature of Huygens' essay is the basic principle, his "postulat" in the English version, which he based his analysis upon. We may call it the principle of equivalent games, in which he assumes that the values of games of chances should be derived through the value of equivalent fair games. Ramsey's assumption of the existence of bets with equally likely outcomes (that he calls "ethically neutral") is an instance of this principle, as well as de Finetti's assumption of the existence of partitions of equally likely events. More recently, the central role that certainty equivalents play in many axiomatic derivations can be viewed as a later instance of Huygens' comparative principle of studying uncertain alternatives by means of benchmark alternatives with suitably simple structures.

Hacking (1975) makes some further observations on the relevance of Huygens' book for the history of subjective probability. We refer the interested reader to his book, with a warning on the difficulty of interpreting some of Huygens' arguments.

Subjective Probabilities and the Axiomatic Approach

Modern decision theory, and in particular the way it models uncertainty, is the result of the pioneering contributions of a truly impressive array of scholars. Some of the finest minds of the first half of last century contributed to the formal modeling of human behavior. Among them, especially remarkable are the works of Frank Plumpton Ramsey (1903–1930) with his early insights on the relations between utilities and subjective probabilities, John von Neumann (1901–1957) and Oskar Morgenstern (1902–1977) with their classic axiomatization of expected utility presented in the 1947 edition of their famous game theory book, Bruno de Finetti (1906–1985) with his seminal contributions to subjective probability, and Leonard J. Savage (1917–1971), who – in an unparalleled conceptual and mathematical tour de

[4]See Ore (1960).

force – integrated von Neumann-Morgenstern's derivation of expected utility with de Finetti's subjective probability.

For our purposes the contributions of de Finetti, Ramsey, and Savage are especially relevant since they shaped modern Bayesian thought and, through it, the modeling of uncertainty in economics. Next we briefly review their landmark contributions.

Ramsey (1926a) A main motivation of Ramsey (1926a) was Keynes (1921)'s logical approach to probability theory, in which the degrees of beliefs in different proposition were connected by necessary/objective relations, called probability relations. Skeptical regarding the existence of such relations, Ramsey argued that degrees of belief should be viewed and studied as subjective entities. To this end, he promoted the behavioral definition of subjective probability as willingness to bet, and claimed that if subjective probabilities, so defined, do not follow standard probability calculus, the individual will make incoherent decisions. These are two central ideas in the methodology of decision theory, which in about the same years were also advocated by Bruno de Finetti.

Specifically, the first tenet of Ramsey's approach is that the only sensible way to measure degrees of beliefs is not through introspection, but by considering them as a basis of action. The second main tenet is that the rules of standard probability calculus correspond to consistent betting behavior. By consistent betting behavior he meant behavior that was not subject to so-called "Dutch books" . Both ideas also appear in de Finetti (1931), and they are in line with the preaching of the logical positivist, culminating in the Received View, first stated by Rudolf Carnap in the 1920s (see Carnap 1923; Suppe 1977). de Finetti explicitly adopted the doctrine of Operationalism (see, e.g., the last chapter of his 1937 article), and saw the elicitation of subjective probabilities through betting behavior as methodologically akin to Vilfredo Pareto's ordinal utility theory based on the elicitation of indifference curves rather than on some psychological entities that could not measured (when data are assumed to be only a weak order over alternatives). Ramsey was motivated by similar methodological concerns, in a Pragmatist perspective,[5] and viewed this approach as akin to what was done in the physical sciences (see, section "Alternative Models" of his article).

de Finetti (1931) Bruno de Finetti, one of the greatest probabilists of the twentieth century, was a key figure in the development of the Bayesian approach. To the best of our knowledge, he was the first to promote the Bayesian approach as an all-encompassing method of reasoning about uncertainty, and he did so with a religious zeal. (See Cifarelli and Regazzini 1996). His two main papers in this regard are probably de Finetti (1931, 1937). In both papers he forcefully emphasized the two key ideas on subjective probabilities that we just discussed in relation with Ramsey's work.

---

[5]Operationalism started with Bridgman (1927), after Ramsey's articles of 1926a and 1926b.

In his 1931 article in Fundamenta Mathematicae (pp. 320–324), de Finetti first introduced the notion of a *qualitative probability*, which is a binary relation over events interpreted as "at least as likely as" . He viewed this relation as a primitive and did not explicitly relate it to betting behavior. (The connection was made explicit later on by Savage (1954).)

The novelty of de Finetti (1931) was both methodological and scientific. Methodologically, it is one of very first articles that adopted the axiomatic method based on a binary relation $\succsim$ and its numerical representation derived from suitable axioms on $\succsim$.[6] Scientifically, he provided the first result that axiomatized subjective probability, thereby establishing one of the two pillars which Savage's great synthesis relied upon.[7]

Savage (1954) de Finetti's derivation of subjective probability was conceptually complementary with von Neumann and Morgenstern's (vNM 1947) derivation of expected utility maximization under risk, which assumed known numerical probability measures. The integration of de Finetti's subjective probability with vNM's expected utility was achieved by Savage's (1954) book, which derived subjective expected utility maximization when neither probabilities nor utilities were given. For a description and interpretation of Savage's result, the reader is referred to Fishburn (1970), Kreps (1988), Gilboa (2009), Wakker (2010), and others.

Ellsberg Paradox

The classic Bayesian theory culminating in Savage's opus represents beliefs probabilistically, but it does not capture the degree of confidence that DMs have in their own probabilistic assessments, a degree that depends on the quality of the information that DMs use in forming these assessments. The classic theory focused on how to measure beliefs, without providing a way to assess the quality of such measurements.

Ellsberg (1961) provided two stark thought experiments that showed how this limitation may lead many people to violate Savage's otherwise extremely compelling axioms, and to express preferences that are incompatible with any (single, additive) probability measure. Ellsberg argued that a situation in which probabilities are not known, which he referred to as *ambiguity*,[8] induces different decisions than situations of risk, namely, uncertainty with known probabilities. Specifically, one of Ellsberg's experiments involves two urns, I and II, with 100 balls in each. The DM is told that

---

[6]Frisch (1926) was the first article we are aware of that adopted a similar approach in economic theory.

[7]See, e.g., chapter 8 of Kreps (1988).

[8]Today, the terms "ambiguity", "uncertainty" (as opposed to "risk"), and "Knightian uncertainty" are used interchangeably to describe the case of unknown probabilities.

(i) in both urns balls are either white or black;
(ii) in urn I there are 50 black and 50 white balls.

No information is given on the proportion of white and black balls in urn II. The DM has to choose both an urn and a color. A ball will be drawn at random from the urn that the DM named, and she will receive a prize if it is of the color stated.

The vast majority of decision makers are indifferent between betting on either color within each urn. However, not all are indifferent between the two urns. Specifically, many prefer either of the bets on the known urn (I) to either of the bets on the unknown urn (II). Clearly, no probability measure can justify such betting behavior.

Ellsberg's experiments revealed the phenomenon of *uncertainty aversion*, or *ambiguity aversion*: people tend to prefer situations with known probabilities to unknown ones, to the extent that these can be compared. Clearly, one can have the opposite phenomenon, of uncertainty/ambiguity liking, when people exhibit the opposite preferences. While gambling is an important exception, it is commonly assumed that people who are not uncertainty neutral tend to be uncertainty averse, in a way that parallels the common assumptions about attitudes toward risk.

Ellsberg's experiments are extremely elegant and they pinpoint precisely which of Savage's axioms is violated by DMs who are not indifferent between betting on the two urns.[9] But the elegance of these experiments is also misleading. Since they deal with balls and urns, and the information about the colors is completely symmetric, it is very tempting to adopt a probabilistic belief that would reflect this symmetry. Specifically, one may reason about the urn with unknown composition, "The number of red balls in it can be any number between 0 and 100. My information is completely symmetric, and there is no reason to believe that there are more red balls than black balls or vice versa. Hence, if I were to adopt a prior probability over the composition of the urn, from [0:100] to [100:0], I should choose a symmetric prior. That is, the probability that there are 3 red balls should be equal to the probability that there are 97 red balls, and so forth. In this case, the probability that a red ball is drawn out of the urn is precisely 50 %, and I should no longer express preferences for the known probabilities." Relatedly, one may also use the unknown urn to generate a bet with objective probability of 50 %: use an external chance device, which is known to be fair, and decide between betting on red or on black based on this device. If the DM has symmetric beliefs about the composition of the urn, she can thereby generate a bet that is equivalent to the bet on the urn with the known composition.

Based on such arguments, theorists often feel that there is no problem with subjective probabilities, at least as far as normative theories of choice are concerned. But this conclusion is wrong. In most real life examples there are no symmetries that allow the generation of risky bets. For example, suppose that Mary does not know what is the probability of the globe warming up by 4 degrees within the

[9] See Gilboa (2009) and Wakker (2010) for the analysis.

next 10 years. She cannot assume that this probability is 50 %, based on Laplace's Principle of Indifference (or "Principle of Insufficient Reason", Laplace 1814). The two eventualities, "average temperature increases by 4 degrees or more" and "average temperature does not increase by 4 degrees" are not symmetric. Moreover, if Mary replaces 4 degrees by 5 degrees, she will obtain two similar events, but she cannot generally assign a 50–50 % probability to any pair of complementary events. Nor will a uniform distribution over the temperature scale be a rational method of assigning probabilities.[10] The fundamental difficulty is that in most real life problems there is too much information to apply the Principle of Indifference, yet too little information to single out a unique probability measure.[11] Global warming and stock market crashes, wars and elections, business ventures and career paths face us with uncertainty that is neither readily quantified nor easily dismissed by symmetry considerations.

Other Disciplines

The Bayesian approach has proved useful in statistics, machine learning, philosophy of science, and other fields. In none of these fellow disciplines has it achieved the status of orthodoxy that it enjoys within economic theory. It is a respectable approach, providing fundamental insights and relishing conceptual coherence. It is worth pointing out, however, that in these disciplines the Bayesian approach is one among many. More importantly, in all of these disciplines the Bayesian approach is applied to a restricted state space, such as a space of parameters, whereas in economics it is often expected to apply also to a *grand state space*, whose elements describe anything that can possibly be of interest.

Consider statistics first. The statistical inference problem is defined by a set of distributions, or data generating processes, out of which a subset of distributions has to be chosen. In parametric problems, the set of distributions is assumed to be known up to the specification of finitely many parameters. Classical statistics does not allow the specification of prior beliefs over these parameters. By contrast, Bayesian

[10]Bertrand's (1907) early critique of the principle of indifference was made in the context of a continuous space. See also Gilboa (2009) and Gilboa et al. (2009).

[11]It is not entirely clear how one can justify the Principle of Indifference even in cases of ignorance. For example, Kass and Wasserman (1996) p. 1347 discuss the partition paradox and lack of parametric invariance, two closely related issues that arise with Laplace's Principle. Similar remarks from a Macroeconomics perspective can be found in Kocherlakota (2007) p. 357.

Based on a result by Henri Poincaré, Machina (2004) suggests a justification of the Laplace's Principle using a sequence of fine partitions of the state stace. This type of reasoning seems to underlie most convincing examples of random devices, such as tossing coins, spinning roulette wheels, and the like. It is tempting to suggest that this is the only compelling justification of the Principle of Indifference, and that this principle should not be invoked unless such a justification exists.

statistics demands that such beliefs be specified. Thus the Bayesian approach offers a richer language, within which the statistician can represent prior knowledge and intuition. Further, the Bayesian prior, updated to a posterior based on sampling, behaves in a much more coherent way than the techniques of classical statistics. (See, for example, Welch (1939), also described in DeGroot (1975), pp. 400–401.)

The main disadvantage of the Bayesian approach to statistics is its subjectivity: since the prior beliefs of the parameters is up to the statistician to choose, they will differ from one statistician to another. Admittedly, classical statistics cannot claim to be fully objective either, because the very formulation of the problem as well as the choice of statistics, tests, and significance levels leave room for the statistician's discretion. Yet, these are typically considered necessary evils, with objectivity remaining an accepted goal, whereas the Bayesian approach embraces subjective inputs unabashedly.[12] On the bright side, if a Bayesian statistician selects a sufficiently "diffused" or "uninformative" prior, she hopes not to rule out the true parameters a priori, and thereby to allow learning of objective truths in the long run, despite the initial reliance on subjective judgments.[13]

The Bayesian approach has a similar status in the related fields of computer science and machine learning.[14] On the one hand, it appears to be the most conceptually coherent model of inference. On the other, its conclusions depend on a priori biases. For example, the analysis of algorithms' complexity is typically conducted based on their worst case. The Bayesian alternative is often dismissed because of its dependence on the assumptions about the underlying distribution.

It is important to emphasize that in statistics and in computer science the state space, which is the subject of prior and posterior beliefs, tends to be a restricted space that does not grow with the data. For example, it can comprise of all combinations of values of finitely many parameters, which are held fixed throughout the sampling procedure. By contrast, the standard approach in economic theory suggests that the state of the world resolves all uncertainty, and thus describes everything that might be of relevance to the problem at hand, from the beginning of time until eternity. As a result, the state space that is often assumed in economics is much larger than in other disciplines. Importantly, it increases with the size of the data.

When one considers a restricted set of parameters, one may argue that the prior probability over this set is derived from past observations of similar problems, each with its own parameters, taken out of the same set. But when the grand state space is considered, and all past repetitions of the problem are already included in the

---

[12] See Lewis (1980) and chapter 4 of van Frassen (1989) (and the references therein) for a discussion of the relations between "objectivity" and subjective probabilities from a philosophical standpoint.

[13] Kass and Wasserman (1996), Bayarri and Berger (2004), and Berger (2004) discuss uninformative priors and related "objective" issues in Bayesian statistics (according to Efron (1986), some of these issues explain the relatively limited use of Bayesian methods in applied statistics).

[14] See Pearl (1986) and the ensuing literature on Bayesian networks.

description of each state, the prior probability should be specified on a rather large state space before any data were observed. With no observations at all, and a very large state space, the selection of a prior probability seems highly arbitrary.

In applications of the Bayesian approach in statistics, computer science, and machine learning, it is typically assumed that the basic structure of the process is known, and only a bounded number of parameters need to be learnt. Many non-parametric methods allow an infinitely dimensional parameter space, but one that does not grow with the number of observations. This approach is sufficient for many statistical inference and learning problems in which independent repetitions are allowed. But economics is often interested in events that do not repeat. Applying the Bayesian approach to these is harder to justify.

We are not fully aware of the origins of the application of the Bayesian approach to the grand state space. It is well known that de Finetti was a devout Bayesian. Savage, who followed his footsteps, was apparently much less religious in his Bayesian beliefs. Yet, he argued that a state of the world should "resolve all uncertainty" and, with a healthy degree of self-criticism, urged the reader to imagine that she had but one decision to be taken in her lifetime, and this is her choice of her strategy before being born. Harsanyi (1967, 1968) made a fundamental contribution to economics by showing how players' types should be viewed as part of the state of the world, and assumed that all unborn players start with a common prior over the grand state space that is thus generated. Aumann (1974, 1976, 1987) pushed this line further by assuming that all acts and all beliefs are fully specified in each and every state, while retaining the assumption that all players have a prior, and moreover, the same prior over the resulting state space.

Somewhere along recent history, with path-breaking contributions by de Finetti, Savage, Harsanyi, and Aumann, economic theory found itself with a state space that is much larger than anything that statisticians or computer scientists have in mind when they generate a prior probability. Surprisingly, the economic theory approach is even more idealized than the Bayesian approach in the philosophy of science. There is nothing wrong in formulating the grand state space as a canonical model within which claims can be embedded. But the assumption that one can have a prior probability over this space, or that this is the only rational way to think about it is questionable.

Summary

Since the mid-twentieth century economic theory has adopted a rather unique commitment to the Bayesian approach. By and large, the Bayesian approach is assumed to be the only rational way to describe knowledge and beliefs, and this holds irrespective of the state space under consideration. Importantly, economic theory clings to Bayesianism also when dealing with problems of unique nature, where nothing is known about the structure of the data generating process. Research in recent decades plainly shows that the Bayesian approach can be extremely fruitful even when applied to such unique problems. But it is also possible that the commitment to the Bayesian approach beclouds interesting findings and new insights.

The preceding discussion highlights our view that there is nothing irrational about violating the Bayesian doctrine in certain problems. As opposed to models of bounded rationality, psychological biases, or behavioral economics, the focus of this survey are models in which DMs may sometimes admit that they do not know what the probabilities they face are. Being able to admit ignorance is not a mistake. It is, we claim, more rational than to pretend that one knows what cannot be known.

Bounded rationality and behavioral economics models often focus on descriptive interpretations. At times, they would take a conditionally-normative approach, asking normative questions given certain constraints on the rationality of some individuals. Such models are important and useful. However, the models discussed here are different in that they are fully compatible with normative interpretations. When central bank executives consider monetary policies, and when leaders of a country make decisions about military actions, they will not make a mistake if they do not form Bayesian probabilities. On the contrary, they will be well advised to take into account those uncertainties that cannot be quantified.

Alternative Models

The Anscombe-Aumann Setup

Anscombe and Aumann (1963) developed a version of the subjective expected utility model of Savage that turned out to be especially well suited for subsequent extensions of the basic Bayesian decision model. For this reason, in this sub-section we present this important setup.

The basic feature of the Anscombe-Aumann (AA, for short) model is that acts map states into lotteries, that is, acts' consequences involve exogenous probabilities a la von Neumann-Morgenstern. This feature is important both conceptually and mathematically. We now turn to introduce the setting formally, in the version presented by Fishburn (1970).

The set of simple probabilities $\Delta(X)$ on some underlying space X of alternatives is the space of consequences considered by the AA model.[15] There is a space of states of the world S endowed with an event algebra Σ. The objects of choice are acts, which map states into lotteries. We denote by $\mathcal{F}$ the collection of all simple acts $f : S \to \Delta(X)$, that is, acts that are finitely valued and Σ-measurable.[16]

[15]Throughout the section we use interchangeably the terms lotteries and simple probabilities.

[16]Simple acts have the form $f = \sum_{i=1}^{n} p_i 1_{E_i}$, where $\{E_i\}_{i=1}^{n} \subseteq \Sigma$ is a partition of S and $\{p_i\}_{i=1}^{n} \subseteq \Delta(X)$ is a collection of lotteries.

A key feature of $\Delta(X)$ is its convexity, which makes it possible to combine acts. Specifically, given any $\alpha \in [0, 1]$, set

$$(\alpha f + (1-\alpha) g)(s) = \alpha f(s) + (1-\alpha) g(s), \qquad \forall s \in S. \tag{21.1}$$

The mixed act $\alpha f + (1-\alpha) g$ delivers in each state s the compound lottery $\alpha f(s) + (1-\alpha) g(s)$. In other words, *ex post*, after the realization of state s, the DM obtains a risky outcome governed by the lottery $\alpha f(s) + (1-\alpha) g(s)$.[17]

The possibility of mixing acts is a key dividend of the assumption that $\Delta(X)$ is the consequence space, which gives the AA setting a vector structure that the Savage setting did not have. The derivation of the subjective expected utility representation in the AA setting is based on this vector structure.

Risk preference The DM has a primitive preference $\succsim$ on $\mathcal{F}$. In turn, this preference induces a preference $\succsim_\Delta$ on lotteries by setting, for all $p, q \in \Delta(X)$,

$$p \succsim_\Delta q \Leftrightarrow f \succsim g,$$

where f and g are the constant acts such that $f(s) = p$ and $g(s) = q$ for all $s \in S$.

Constant acts are not affected by state uncertainty, only by the risk due to the lotteries' exogenous probabilities. For this reason, $\succsim_\Delta$ can be seen as the risk preference of the DM. This is an important conceptual implication of having $\Delta(X)$ as the consequence space. This richer consequence space mathematically delivers a most useful vector structure, while from a decision theoretic standpoint it enriches the setting with a risk preference that allows to consider the DMs' risk behavior separately. Differently put, the AA consequence space can be viewed as derived from an underlying consequence space X a la Savage, enriched by a lottery structure that allows to calibrate risk preferences.

Alternatively, one may view AA's model as an improved version of de Finetti's (1931, 1937) axiomatic derivation of expected value maximization with subjective probabilities. de Finetti assumed additivity or linearity in payoffs. This is a problematic assumption if payoffs are monetary, but it is more palatable if payoffs are probabilities of receiving a fixed desirable outcome. Replacing the payoffs in de Finetti's model by probabilities of outcomes, one obtains a model akin to AA's.

In a sense, the AA model is a hybrid between vNM's and Savage's. Mathematically it is akin to the former, as it starts with a vNM theorem on a particular mixture space, and imposes additional axioms to derive subjective probabilities. Conceptually, it is closer to Savage's model, as it derives probabilities from preferences. Many view this derivation as conceptually less satisfactory than Savage's, because the latter does not assume probabilities, or any numbers for that matter, to be part of the data. Anscombe and Aumann, however, viewed the use of objective probabilities as a merit, because they believed that people think in terms of subjective probabilities

[17]For this reason, mixing acts in this way is sometimes called "ex post randomization." For recent models with ex ante randomization, see Epstein et al. (2007), Ergin and Sarver (2009), Seo (2009), and Saito (2015).

after they have internalized the concept of objective probability. Be that as it may, there is no doubt that the AA model has become the main testbed for new models of decision under uncertainty.[18]

Axioms We now make a few assumptions on the primitive preference $\succsim$. The first one is a standard weak order axiom.

AA.1 WEAK ORDER: $\succsim$ on $\mathcal{F}$ is complete and transitive.

The next axiom is a monotonicity assumption: if state by state an act f delivers a weakly better (risky) consequence than an act g, then f should be weakly preferred to g. It is a basic rationality axiom.

AA.2 MONOTONICITY: for any $f, g \in \mathcal{F}$, if $f(s) \succsim_\Delta g(s)$ for each $s \in S$, then $f \succsim g$.

Next we have an independence axiom, which is peculiar to the AA setting since it relies on its vector structure.

AA.3 INDEPENDENCE: for any three acts $f, g, h \in \mathcal{F}$ and any $0 < \alpha < 1$, we have

$$f \succ g \Rightarrow \alpha f + (1-\alpha)h \succ \alpha g + (1-\alpha)h. \tag{21.2}$$

According to this axiom, the DM's preference over two acts f and g is not affected by mixing them with a common act h. In the special case when all these acts are constant, axiom AA.3 reduces to von Neumann-Morgenstern's original independence axiom on lotteries.

We close with standard Archimedean and nontriviality assumptions.[19]

AA.4 ARCHIMEDEAN: let f, g, and h be any three acts in $\mathcal{F}$ such that $f \succ g \succ h$. Then, there are $\alpha, \beta \in (0,1)$ such that $\alpha f + (1-\alpha)h \succ g \succ \beta f + (1-\beta)h$.

AA.5 NONDEGENERACY: there are $f, g \in \mathcal{F}$ such that $f \succ g$.

We can now state the AA subjective expected utility theorem.

Theorem 1. *Let $\succsim$ be a preference defined on $\mathcal{F}$. The following conditions are equivalent:*

(i) $\succsim$ satisfies axioms AA.1–AA.5;

(ii) there exists a non-constant function $u : X \to \mathbb{R}$ and a probability measure $P : \Sigma \to [0,1]$ such that, for all $f, g \in \mathcal{F}$, $f \succsim g$ if and only if

$$\int_S \left(\sum_{x \in \operatorname{supp} f(s)} u(x) f(s) \right) dP(s) \geq \int_S \left(\sum_{x \in \operatorname{supp} g(s)} u(x) g(s) \right) dP(s). \tag{21.3}$$

---

[18]See Ghirardato et al. (2003) for a subjective underpinning of the AA setup.

[19]See Gilboa (2009) for some more details on them.

Moreover, P is unique and u is cardinally unique.[20]

The preference functional $V : \mathcal{F} \to \mathbb{R}$ in (21.3) has the form

$$V(f) = \int_S \left(\sum_{x \in \text{supp} f(s)} u(x) f(s) \right) dP(s) \tag{21.4}$$

and consists of two parts. The inner part

$$\sum_{x \in \text{supp} f(s)} u(x) f(s) \tag{21.5}$$

is the expected utility of the lottery $f(s)$ that act f delivers when state s obtains. It is easy to see that this expected utility represents the DM's risk preference $\succsim_\Delta$. The outer part

$$\int_S \left(\sum_{x \in \text{supp} f(s)} u(x) f(s) \right) dP(s)$$

averages all expected utilities (21.5) according to the probability P, which quantifies the DM's beliefs over the state space.

The classical models of Savage and Anscombe-Aumann were considered the gold standard of decision under uncertainty, despite the challenge posed by Ellsberg's experiments. In the 1980s, however, several alternatives were proposed, most notably models based on probabilities that are not necessarily additive, or on sets of probabilities. We now turn to review these contributions and some of the current research in the area.

Choquet Expected Utility

The first general-purpose, axiomatically-based non-Bayesian decision model was the Choquet Expected Utility (CEU) model proposed by David Schmeidler in 1982, which appeared as Schmeidler (1989). Schmeidler's starting point was that the Bayesian model is a straightjacket that does not allow the DM to express her own degree of confidence in her beliefs. Schmeidler gave the example of two coins, one that has been tested extensively and is known to be fair, and the other about which nothing is known. He noted that a Bayesian would probably have 50–50 % beliefs regarding the result of the toss of either coin, but that these beliefs differ:

[20]Throughout the paper, cardinally unique means unique up to positive affine transformations.

in one case, the DM practically knows that each side of the coin has probability of 50 % of coming up. In the other case, the numbers 50–50 % are obtained with a shrug of one's shoulders, relying on symmetry of ignorance rather than symmetry of information.[21] Observe that Schmeidler's two-coin example is very close to Ellsberg's two-urn experiment. However, Schmeidler was not motivated by the desire to explain Ellsberg's results; rather, he considered the standard theory and found it counter-intuitive.

Schmeidler (1989) suggested to model probabilities by set functions that are not necessarily additive. For example, if H (T) designates the event "the unknown coin falls with H (T) up", and ν is the measure of credence, we may have

$$\nu(H) + \nu(T) < \nu(H \cup T)$$

Thus, the "probability" of events, as measured by our willingness to bet on them, may not satisfy the standard axioms of probability theory. Schmeidler referred to them as *non-additive probabilities*, and required that they be positive and monotone with respect to set inclusion. Such mathematical entities are also known by the term *capacities*. Formally, given an event algebra Σ of state space S, a set function $\nu : \Sigma \to [0,1]$ is a capacity if

(i) $\nu(\emptyset) = 0$ and $\nu(S) = 1$;
(ii) $E \subseteq E'$ implies $\nu(E) \leq \nu(E')$.

Dempster (1967) and Shafer (1976) also suggested a theory of belief in which the degree of belief in an event did not obey additivity. They focused on the representation of uncertainty by *belief functions*. There is a vast literature that followed, often referred to as "imprecise probabilities" (see Walley 1991). Most of this literature, however, does not address the question of decision making. By contrast, Schmeidler had decision theory in mind, and he sought a notion of integration that would generalize standard expectation when the capacity ν happens to be additive. Such a notion of integration was suggested by Choquet (1953).

To understand the gist of the Choquet integral,[22] suppose that Σ is a σ-algebra (e.g., the power set 2^S) and consider a positive and bounded Σ-measurable function $\phi : S \to \mathbb{R}$. The *Choquet integral* of ϕ with respect to a capacity ν is given by:

$$\int \phi \, d\nu = \int_0^\infty \nu\left(\{s \in S : \phi(s) \geq t\}\right) dt, \tag{21.6}$$

where on the right-hand side we have a Riemann integral. To see why the Riemann integral is well defined, observe that the sets $E_t = \{s \in S : \phi(s) \geq t\}$ define a

[21] See Fischhoff and Bruine De Bruin (1999) for experimental evidence on how people use 50–50 % statements in this sense.

[22] We refer the interested reader to Denneberg (1994) and to Marinacci and Montrucchio (2004) for detailed expositions of Choquet integration.

chain that is decreasing in t (in the sense of set inclusion), and, since a capacity, is monotone, $\nu(E_t)$ is a decreasing function of t. For more detailed explanation of the Choquet integral the reader is referred to Gilboa (2009).

Schmeidler (1989) axiomatized Choquet expected utility in the AA setup. The key innovation relative to the AA axioms AA.1–AA.4 was to restrict the Independence axiom AA.3 to *comonotonic* acts, that is, acts $f, g \in \mathcal{F}$ for which it is never the case that both $f(s) \succ f(s')$ and $g(s) \prec g(s')$ for some states of the world s and s'. This is the preference version of comonotonicity.

S.3 COMONOTONIC INDEPENDENCE: for any pairwise comonotonic acts $f, g, h \in \mathcal{F}$ and any $0 < \alpha < 1$,

$$f \succ g \Rightarrow \alpha f + (1-\alpha)h \succ \alpha g + (1-\alpha)h. \tag{21.7}$$

According to this axiom, the DM's preference between two comonotonic acts f and g is not affected by mixing them with another act h that is comonotonic with both. The intuition behind this axiom can best be explained by observing that the classical independence axiom may not be very compelling in the presence of uncertainty. For example, assume that there are two states of the world, and two vNM lotteries $P \succ Q$. Let $f = (P, Q)$ and $g = (Q, P)$. Suppose that, due to ignorance about the state of the world, the DM is driven to express indifference, $f \sim g$. By AA's independence, for every h we will observe

$$\frac{1}{2}f + \frac{1}{2}h \sim \frac{1}{2}g + \frac{1}{2}h$$

However, for $h = g$ this implies that $\frac{1}{2}f + \frac{1}{2}g \sim g$, despite the fact that the act $\frac{1}{2}f + \frac{1}{2}g$ is risky while g is uncertain.

In this example, g can serve as a hedge against the uncertainty inherent in f, but it clearly cannot hedge against itself. The standard independence axiom is too demanding, because it does not distinguish between mixing operations $\alpha f + (1-\alpha)h$ that reduce uncertainty (via hedging) and mixing operations that do not. Restricting the independence axiom to pairwise comonotonic acts neutralizes this asymmetric effect of hedging.

Using the Comonotonic Independence axiom S.3, Schmeidler (1989) was able to prove the following representation theorem, which generalizes the subjective expected utility representation established by Theorem 1 by allowing for possibly non-additive probabilities. The proof of the result is based on some results on Choquet integration established in Schmeidler (1986).

Theorem 2 ((ii)). *Let $\succsim$ be a preference defined on $\mathcal{F}$. The following conditions are equivalent:*

(i) $\succsim$ satisfies axioms AA.1, AA.2, S.3 (Comonotonic Independence), AA.4, and AA.5;

(ii) there exists a non-constant function $u : X \to \mathbb{R}$ *and a capacity* $\nu : \Sigma \to [0,1]$ *such that, for all* $f, g \in \mathcal{F}$, $f \succsim g$ *if and only if*

$$\int_S \left(\sum_{x \in \operatorname{supp} f(s)} u(x) f(s) \right) d\nu(s) \geq \int_S \left(\sum_{x \in \operatorname{supp} g(s)} u(x) g(s) \right) d\nu(s). \quad (21.8)$$

Moreover, ν *is unique and* u *is cardinally unique.*

Gilboa (1987), Wakker (1989a,b), and Nakamura (1990) established purely subjective versions of Schmeidler's representation result.[23] Sarin and Wakker (1992) showed that the existence of a suitable rich collection of unambiguous events substantially streamlines the derivation of Schmeidler's representation through a simple cumulative dominance condition.

Maxmin Expected Utility

Schmeidler's model is a generalization of Anscombe-Aumann's in a way that allows us to cope with uncertainty, or ambiguity. The capacity in the model can be interpreted as a lower bound on probabilities. Specifically, let $\Delta(\Sigma)$ be the collection of all finitely additive probability measures $P : \Sigma \to [0,1]$ and define the *core* of ν to be, as in cooperative game theory,

$$core(\nu) = \{P \in \Delta(\Sigma) : P(E) \geq \nu(E) \text{ for all } E \in \Sigma\}.$$

If $core(\nu) \neq \emptyset$, we may think of $\nu(E)$ as the lower bound on $P(E)$, and then ν is a concise way to represent a set of probabilities, presumably those that are considered possible. The lower envelope of a set of probabilities is also the common interpretation of belief functions (Dempster 1967; Shafer 1976).

Schmeidler (1986) has shown that if ν is *convex* in the sense that

$$\nu(E) + \nu(E') \leq \nu(E \cup E') + \nu(E \cap E'), \qquad \forall E, E' \in \Sigma,$$

(Shapley 1972) then

$$\int_S \phi \, d\nu = \min_{P \in core(\nu)} \int_S \phi \, dP \quad (21.9)$$

for every Σ-measurable bounded function $\phi : S \to \mathbb{R}$ (see also Rosenmueller 1971, 1972). Thus, when the capacity ν happens to be convex (e.g., a belief function a la

[23] Nakamura and Wakker's papers use versions of the so-called *tradeoff method* (see Kobberling and Wakker (2003), for a detailed study of this method and its use in the establishment of axiomatic foundations for choice models).

Dempster-Shafer), Choquet integration has a simple and intuitive interpretation: a DM who evaluated an act f by the Choquet integral of its utility profile $u \circ f$ can be viewed as if she entertained a *set* of possible probabilities, $core(\nu)$, and evaluated each act by its minimal expected utility, over all probabilities in the set.

There is a simple behavioral condition that characterizes CEU preferences with convex ν. To introduce it, denote by $B_0(\Sigma)$ the vector space of all simple functions $\phi : S \to \mathbb{R}$ and consider the Choquet functional $I : B_0(\Sigma) \to \mathbb{R}$ given by $I(\phi) = \int \phi \, d\nu$. This functional is easily seen to be concave when (21.9) holds. Actually, according to a classic result of Choquet (1953), I is concave if and only if its capacity ν is convex.[24] This concavity property suggests the following convexity axiom, due to Schmeidler (1989), which models a negative attitude toward ambiguity.

S.6 UNCERTAINTY AVERSION: for any $f, g \in \mathcal{F}$ and any $0 < \alpha < 1$, we have

$$f \sim g \Rightarrow \alpha f + (1-\alpha) g \succsim f.$$

Thus, uncertainty aversion states that mixing, through randomization, between equivalent acts can only make the DM better off. For example, in Ellsberg's example it is natural to expect that DMs prefer to hedge against ambiguity by mixing acts *IIB* and *IIW*, that is,

$$\alpha IIB + (1-\alpha) IIW \succsim IIB \sim IIW, \qquad \forall \alpha \in [0,1].$$

This mixing can be thus viewed as a form of hedging against ambiguity that the DM can choose.[25]

Theorem 3. *In Theorem 2, $\succsim$ satisfies axiom S.6 if and only if the capacity ν in (21.8) is convex.*

Theorem 3 (Schmeidler 1989) shows that convex capacities characterize ambiguity averse Choquet expected utility DMs (in the sense of axiom S.6). Since most DMs are arguably ambiguity averse, this is an important result in Choquet expected utility theory. Moreover, relating this theory to maximization of the worst-case expected utility over a set of probabilities has several advantages. First, it obviates the need to understand the unfamiliar concept of Choquet integration. Second, it provides a rather intuitive, if extreme, cognitive account of the decision process: as in classical statistics, the DM entertains several probability measures as potential beliefs. Each such "belief" induces an expected utility index for each act. Thus, each act has many expected utility values. In the absence of second-order beliefs, the cautious DM chooses the worst-case expected utility as summarizing the act's desirability. Wakker (1990, 1991) established several important behavioral properties and characterizations of concave/convex capacities in the CEU model.

---

[24]See Marinacci and Montrucchio (2004) p. 73. They show on p. 78 that (21.9) can be derived from this result of Choquet through a suitable application of the Hahn-Banach Theorem.

[25]Klibanoff (2001a,b) studied in detail the relations between randomization and ambiguity aversion.

Gilboa and Schmeidler (1989) This account of Choquet expected utility maximization also relates to the maxmin criterion of Wald (1950; see also Milnor 1954). However, there are many natural sets of probabilities that are not the core of any capacity. Assume, for example, that there are three states of the world, $S = \{1, 2, 3\}$. Assume that the DM is told that, if state 1 is not the case, then the (conditional) probability of state 2 is at least $2/3$. If this is all the information available to her, she knows only that state 2 is at least twice as likely than state 3. Hence the set of probability vectors $P = (p_1, p_2, p_3)$ that reflects the DM's knowledge consists of all vectors such that

$$p_2 \geq 2p_3$$

It is easy to verify that this set is not the core of a capacity. Similarly, one may consider a DM who has a certain probability measure P in mind, but allows for the possibility of error in its specification. Such a DM may consider a set of probabilities

$$C = \{Q \in \Delta(\Sigma) : \|P - Q\| < \varepsilon\}$$

for some norm $\|\cdot\|$ and $\varepsilon > 0$, and this set is not the core of any capacity (such sets were used in Nishimura and Ozaki (2007)).

It therefore makes sense to generalize Choquet expected utility with convex capacities to the maxmin rule, where the minimum is taken over general sets of probabilities. Decision rules of this type have been suggested first by Hurwicz (1951), under the name of Generalized Bayes-minimax principle, and then by Smith (1961), Levi (1974, 1980), and Gärdenfors and Sahlin (1982). Recently, related ideas appeared in mathematical finance (see Artzner et al. 1997, 1999).

Gilboa and Schmeidler (1989) provided an axiomatic model of maxmin expected utility maximization ("MMEU", also referred to as "MEU"). This model is also formulated in the AA framework and, like the Choquet expected utility model, is based on a suitable weakening of the Independence axiom AA.3. Schmeidler's Comonotonic Independence axiom restricted AA.3 to the case that all acts are pairwise comonotonic. This rules out obvious cases of hedging, but it may allow for more subtle ways in which expected utility can be "smoothed out" across states of the world.[26] A more modest requirement restricts the independence condition to the case in which the act h is constant:

---

[26]For example, assume that there are three states of the world, and two acts offer the following expected utility profiles: $f = (0, 10, 20)$ and $g = (4, 10, 14)$. Assume that the DM is indifferent between f and g, that is, that she is willing to give up 1 unit of expected utility in state 3 in order to transfer 5 units from state 3 to state 1. Comonotonic independence would imply that the DM should also be indifferent between f and g when they are mixed with any other act comonotonic with both, such as f itself. However, while f clearly doesn't offer a hedge against itself, mixing f with g can be viewed as reducing the volatility of the latter, resulting in a mix that is strictly better than f and g.

GS.3 C-INDEPENDENCE: for all acts $f, g \in \mathcal{F}$ and all lottery acts p,

$$f \succsim g \Rightarrow \alpha f + (1-\alpha)p \succsim \alpha g + (1-\alpha)p, \qquad \forall \alpha \in [0,1].$$

C-Independence is essentially weaker than Comonotonic Independence S.3 because lottery (constant) acts are comonotonic with all other acts.[27] The axiom is arguably easier to accept because the mixture with a lottery act can be viewed as a change of the unit of measurement. Indeed, this axiom may be viewed as the preference version of the following property of real-valued functionals: a functional $I : B_0(\Sigma) \to \mathbb{R}$ is said to be *translation invariant* if

$$I(\alpha\phi + k) = \alpha I(\phi) + I(k), \qquad \forall \alpha \in \mathbb{R},$$

given any $\phi \in B_0(\Sigma)$ and any constant function k.[28]

Gilboa and Schmeidler thus used a weaker version of the independence axiom, but they also imposed the uncertainty aversion axiom S.6. Both axioms GS.3 (C-Independence) and S.6 (Uncertainty Aversion) follow from the Independence axiom AA.3. Thus, the following representation result, due to Gilboa and Schmeidler (1989), generalizes Theorem 1 by allowing for possibly nonsingleton sets of probabilities.

Theorem 4. *Let $\succsim$ be a preference defined on $\mathcal{F}$. The following conditions are equivalent:*

(i) $\succsim$ satisfies axioms AA.1, AA.2, GS.3 (C-Independence), AA.4, AA.5, and S.6 (Uncertainty Aversion);
(ii) there exists a non-constant function $u : X \to \mathbb{R}$ and a convex and compact set $C \subseteq \Delta(\Sigma)$ of probability measures such that, for all $f, g \in \mathcal{F}$,

$$f \succsim g \Leftrightarrow \min_{P \in C} \int_S \left(\sum_{x \in \text{supp} f(s)} u(x) f(s) \right) dP(s)$$
$$\geq \min_{P \in C} \int_S \left(\sum_{x \in \text{supp} f(s)} u(x) g(s) \right) dP(s), \tag{21.10}$$

Moreover, C is unique and u is cardinally unique.

---

[27]Schmeidler required that all three acts be pairwise comonotonic, whereas C-Independence does not restrict attention to comonotonic pairs (f, g). Thus, C-Independence is not, strictly speaking, weaker than Comonotonic Independence. However, in the presence of Schmeidler's other axioms, Comonotonic Independence is equivalent to the version in which f and g are not required to be comonotonic.

[28]See Ghirardato et al. (1998) for details.

The set C is a singleton if and only if $\succsim$ satisfies the Independence axiom AA.3. A slightly more interesting result actually holds, which shows that maxmin expected utility DMs reduce to subjective expected utility ones when their choices do not involve any hedging against ambiguity.[29]

Proposition 5. *In Theorem 4, C is a singleton if and only if, for all* $f, g \in \mathcal{F}$,

$$f \sim g \Rightarrow \frac{1}{2}f + \frac{1}{2}g \sim g.$$

When C is not a singleton, the model can express more complex states of knowledge, reflected by various sets C of probabilities. For applications in economic theory, the richness of the maxmin model seems to be important. In particular, one may consider any model in economic theory and enrich it by adding some uncertainty about several of its parameters. By contrast, in order to formulate Choquet expected utility, one needs to explicitly consider the state space and the capacity defined on it. Often, this exercise may be intractable.

By contrast, for some practical applications such as in medical decision making, the richness of the maxmin model may prove a hindrance. Wakker (2010) presents the theory of decision making under risk and under ambiguity geared for such applications. He focuses on capacities as a way to capture ambiguity, rather than on sets of probabilities.[30]

The maxmin model allows for more degrees of freedom than the CEU model, but it does not generalize it. In fact, the overlap of the two models is described in Theorem 3 and occurs when the uncertainty averse axiom S.6 holds. But, whereas uncertainty aversion – through axiom S.6 – is built into the decision rule of the maxmin model, Choquet expected utility can express attitudes of uncertainty liking. This observation in part motivated the search by Ghirardato et al. (2004) of a class of preferences that may not satisfy S.6 and is able to encompass both CEU and MMEU preferences. We review this contribution below.

Finally, Casadesus-Masanell et al. (2000), Ghirardato et al. (2003), and Alon and Schmeidler (2014) established purely subjective versions of Gilboa and Schmeidler's representation result.[31]

Countably additive priors Theorem 4 considers the set $\Delta(\Sigma)$ of all finitely additive probabilities. In applications, however, it is often important to consider

[29]See Ghirardato et al. (2004) for details.

[30]Wakker (2010) also introduces the gain-loss asymmetry that is one of the hallmarks of Prospect Theory (Kahneman and Tversky 1979). The combination of gain-loss asymmtry with rank-dependent expected utility (Quiggin 1982; Yaari 1987) resulted in Cumulative Prospect Theory (CPT, Tversky and Kahneman 1992). When CPT is interpreted as dealing with ambiguity, it is equivalent to Choquet expected utility with the additional refinement of distinguishing gains from losses.

[31]For a critical review of the maxmin and other non-Bayesian models, see Al-Najjar and Weinstein (2009) (see Mukerji 2009; Siniscalchi 2009b, for a discussion).

countably additive probabilities, which have very convenient analytical properties that many important results in probability theory crucially rely upon.

The behavioral condition that underlies countably additive priors is Monotone Continuity, introduced by Arrow (1970) to characterize countable additivity of the subjective probability P in Savage's model.

MC MONOTONE CONTINUITY: If $f, g \in \mathcal{F}$, $x \in X$, $\{E_n\}_{n\geq 1} \in \Sigma$ with $E_1 \supseteq E_2 \supseteq \ldots$ and $\bigcap_{n\geq 1} E_n = \emptyset$, then $f \succ g$ implies that there exists $n_0 \geq 1$ such that $xE_{n_0}f \succ g$.

Marinacci (2002a) and Chateauneuf et al. (2005) showed that this condition keeps characterizing countable additivity in the MMEU model. Next we state a version of their results, a countably additive counterpart of Theorem 4. Here $Q \ll P$ means that Q is absolutely continuous with respect to P, i.e., $P(E) = 0$ implies $Q(E) = 0$ for all $E \in \Sigma$.

Theorem 6. *In Theorem 4, $\succsim$ satisfies Axiom MC if and only if all probabilities in C are countably additive. In this case, there exists $P \in C$ such that $Q \ll P$ for all $Q \in C$.*

Besides the countable additivity of priors, axiom MC also delivers the existence of a "control" prior $P \in C$ relative to which all other priors $Q \in C$ are absolutely continuous.[32]

In decision theory the use of countably additive priors has been often debated, most forcefully by de Finetti and Savage themselves, who argued that it is a purely technical property that, if anything, actually impairs the analysis (e.g., over countable state spaces it is not possible to define uniform priors that are countably additive). However, Arrow's characterization of countably additive priors in Savage's model through Monotone Continuity and its MMEU version in Theorem 6 show that behaviorally this technically most useful property requires a relatively small extra baggage compared to the basic axioms of the finitely additive case.[33]

Equivalent priors A minimal consistency requirement among priors in C is that they agree on what is possible or impossible. Formally, this is the case if any two priors P and P' in C are *equivalent*, i.e., if they are mutually absolutely continuous

---

[32] As Chateauneuf et al. (2005) show, this control prior exists because, under Axiom MC, the set C is weakly compact, a stronger compactness condition than the weak*-compactness that C features in Theorem 4. Their results have been generalized to variational preferences by Maccheroni et al. (2006a).

[33] In this regard, Arrow (1970) wrote that "the assumption of Monotone Continuity seems, I believe correctly, to be the harmless simplification almost inevitable in the formalization of any real-life problem." See Kopylov (2010) for a recent version of Savage's model under Monotone Continuity.

In many applications, countable additivity of the measure(s) necessitates the restriction of the algebra of events to be a proper subset of 2^S. Ignoring many events as "non-measurable" may appear as sweeping the continuity problem under the measurability rug. However, this approach may be more natural if one does not start with the state space S as primitive, but derives it as the semantic model of a syntactic system, where propositions are primitive.

($P(E) = 0$ if and only if $P'(E) = 0$ for all $E \in \Sigma$). Epstein and Marinacci (2007) provide a behavioral condition that ensures this minimal consistency among priors, which is especially important in dynamic problems that involve priors' updating.

Interestingly, this condition turns out to be a translation in a choice under uncertainty setup of a classic axiom introduced by Kreps (1979) in his seminal work on menu choices. Given any two consequences x and y, let

$$x \vee y = \begin{cases} x \text{ if x} \succsim \text{y} \\ y \text{ otherwise} \end{cases}$$

and given any two acts f and g, define the act $f \vee g$ by $(f \vee g)(s) = f(s) \vee g(s)$ for each $s \in S$.

GK GENERALIZED KREPS: For all $f, f', g \in \mathcal{F}, f \sim f \vee f' \Rightarrow f \vee g \sim (f \vee g) \vee f'$.

In every state, the act $f \vee f'$ gives the better of the two outcomes associated with f and f'. Thus we say that $f \vee f'$ weakly improves f in 'the direction' f'. GK requires that if an improvement of f in direction f' has no value, then the same must be true for an improvement in direction f' of any act (here $f \vee g$) that improves f. The next result of Epstein and Marinacci (2007) shows that for maxmin preferences this seemingly innocuous axiom is equivalent to the mutual absolute continuity of priors.

Theorem 7. *In Theorem 4, $\succsim$ satisfies Axiom GK if and only if the probabilities in C are equivalent.*

Unanimity Preferences

Another way to deal with ambiguity is to relax the completeness of preferences. Indeed, because of the poor information that underlies ambiguity, the DM may not be able to rank some pairs of acts. If so, one of the most basic assumptions in decision theory, namely, that preferences are complete, may be relaxed because of ambiguity.

This is the approach proposed by Truman Bewley. Incomplete preferences were already studied by Aumann (1962), interpreted as a DM's inability to decide between some pairs of alternatives. Building on Aumann's work, Bewley presented in 1986 a model of incomplete preferences in the context of uncertainty, which appeared as Bewley (2002). In his model the Weak Order Axiom AA.1 is replaced by two weaker assumptions.

B.1a PARTIAL ORDER: $\succsim$ on $\mathcal{F}$ is reflexive and transitive.

Hence, $\succsim$ is no longer required to be complete. The DM, however, knows her tastes: the only reason for incompleteness is ignorance about probabilities. For this reason, Bewley assumes the next weak form of completeness, which only applies to lottery acts.

B.1b C-COMPLETENESS: for every lottery acts $p, q \in \Delta(X)$, $p \succsim q$ or $q \succsim p$.

In other words, B.1 requires the risk preference $\succsim_\Delta$ to be complete. Using these two axioms, Gilboa et al. (2010) established the following general form of Bewley's representation theorem.[34]

Theorem 8. *Let $\succsim$ be a preference defined on $\mathcal{F}$. The following conditions are equivalent:*

(i) $\succsim$ satisfies axioms B.1, and AA.2–AA.5;
(ii) there exists a non-constant function $u : X \to \mathbb{R}$ and a convex and compact set $C \subseteq \Delta(\Sigma)$ of probability measures such that, for all $f, g \in \mathcal{F}$,

$$f \succsim g \Leftrightarrow \int_S \left(\sum_{x \in \text{supp} f(s)} u(x) f(s) \right) dP(s)$$
$$\geq \int_S \left(\sum_{x \in \text{supp} f(s)} u(x) g(s) \right) dP(s), \qquad \forall P \in C. \tag{21.11}$$

Moreover, C is unique and u is cardinally unique.

In this representation a set of probability measures C arises, interpreted as the probabilistic models that are compatible with the DM's information. Two acts f and g are comparable only when their expected utilities with respect to the probabilities in C unanimously rank one act over the other. If this is not the case – that is, if the probabilities in C do not agree in ranking of the two acts– the DM is unable to rank the two acts.

When preferences are incomplete, the model does not always specify what the DM will do. In particular, acts are not evaluated by a numerical index V that represents preferences and that makes it possible to formulate the optimization problems that most economic applications feature. To complete the model, one needs to add some assumptions about choices in case preferences do not have a maximum. One possibility is to assume that there exists a status quo, namely, an alternative that remains the default choice unless it is dethroned by another alternative that is unanimously better. This might be a rather reasonable descriptive model, especially of organizations, but it is considered by many to be less than rational. Recently, Ortoleva (2010) reconsidered Bewley's inertia insight from a different angle by showing, within a full-fledged axiomatic model, how status quo biases may lead to incomplete preferences.

Another approach suggests to complete preferences based on the same set of probabilities C. Gilboa et al. (2010) offer a model involving two preference

[34]A caveat: the unanimity rule (21.11) is slightly different from Bewley's, who represents strict preference by unanimity of strict inequalities. This is generally not equivalent to representation of weak preference by unanimity of weak inequalities.

relations, and show that certain axioms, stated on each relation separately as well as relating the two, are equivalent to a joint representation of the two relations by the same set of probabilities C: one by the unanimity rule, and the other – by the maxmin rule. Their results provide a bridge between the two classic representations (21.10) and (21.11), as well as a possible account by which maxmin behavior might emerge from incomplete preferences.

Unanimity, Scenarios, and Uncertainty Aversion

Ghirardato et al. (GMM, 2004) used some insights from Bewley's unanimity representation to remove the Uncertainty Aversion axiom S.6 in the derivation of Gilboa and Schmeidler (1989) and, in this way, to propose a class of preferences that encompasses both Choquet and maxmin preferences. To this end, they consider the following definition.

Definition 9. A preference $\succsim$ on $\mathcal{F}$ is said to be invariant biseparable if it satisfies axioms AA.1, AA.2, GS.3 (C-Independence), AA.4, and AA.5.

Invariant biseparable (IB) preferences thus satisfy all AA axioms, except for the Independence axiom AA.3, which is replaced by the C-Independence axiom GS.3 of Gilboa and Schmeidler (1989).[35] Thanks to this key weakening, invariant biseparable preferences include as special cases both CEU and MMEU preferences: the former constitute the special case when the Comonotonic Independence Axiom S.3 holds, while the latter – when the Uncertainty Aversion axiom S.6 holds.

The main tool that GMM use to study IB preferences is an auxiliary relation $\succsim^*$ on $\mathcal{F}$. Specifically, given any two acts $f, g \in \mathcal{F}$, act f is said to be *unambiguously (weakly) preferred* to g, written $f \succsim^* g$, if

$$\alpha f + (1-\alpha)\, h \succsim \alpha g + (1-\alpha)\, h$$

for all $\alpha \in [0, 1]$ and all $h \in \mathcal{F}$. In words, $f \succsim^* g$ holds when the DM does not find any possibility of hedging against or speculating on the ambiguity that she may perceive in comparing f and g. GMM argue that this DM's choice pattern reveals that ambiguity does not affect her preference between f and g, and this motivates the "unambiguously preferred" terminology.

The unambiguous preference relation is, in general, incomplete. This incompleteness is due to ambiguity

Lemma 10. *The following statements hold:*

(i) If $f \succsim^* g$, then $f \succsim g$.
(ii) $\succsim^*$ satisfies axioms B.1, AA.2, and AA.3

[35]The name biseparable originates in Ghirardato and Marinacci (2001, 2002), which we will discuss later.

(iii) $\succsim^*$ *is the maximal restriction of* $\succsim$ *satisfying the Independence axiom AA.3.*[36]

By (i) and (ii), the unambiguous preference $\succsim^*$ is a restriction of the primitive preference relation $\succsim$ that satisfies reflexivity, transitivity, monotonicity, and independence. By (iii), it is the maximal such restriction that satisfies independence.[37]

The next result proves, along the lines of the Bewley-type representation (21.11), that the unambiguous preference can be represented by a set of priors.

Proposition 11. *Let* $\succsim$ *be an IB preference on* $\mathcal{F}$. *Then, there exists a function* $u : X \to \mathbb{R}$ *and a convex and compact set* $C \subseteq \Delta(\Sigma)$ *of probability measures such that, for all* $f, g \in \mathcal{F}$,

$$f \succsim^* g \Leftrightarrow \int_S u(f)\, dP(s) \geq \int_S u(g)(s)\, dP(s), \qquad \forall P \in C. \tag{21.12}$$

In words, f is unambiguously weakly preferred to g if and only if every probability $P \in C$ assigns a weakly higher expected utility to f. It is natural to interpret each prior $P \in C$ as a "possible scenario" that the DM envisions, so that unambiguous preference corresponds to preference in every scenario. GMM thus argue that C represents the (subjective) perception of ambiguity of the DM, and that the DM perceives ambiguity in a decision problem if C is not a singleton.

The relation $\succsim^*$ thus makes it possible to elicit a set of priors C for a general IB preference $\succsim$. When $\succsim$ is a MMEU preference, C is the set of priors of the maxmin representation (21.10). When $\succsim$ is a CEU preference that satisfies axiom S.6, C is the core of the representing capacity ν.[38]

More generally, GMM prove a representation theorem for IB preferences based on the set C, which generalizes Theorems 2 and 4. To this end, given any act f consider its expected utility profile $\left\{\int_S u(f)\, dP(s) : P \in C\right\}$ under C. Write $f \asymp g$ if two acts f and g feature isotonic profiles, that is,

$$\int_S u(f(s))\, dP'(s) > \int_S u(f(s))\, dP''(s) \Leftrightarrow \int_S u(g(s))\, dP'(s) \geq \int_S u(g(s))\, dP''(s), \qquad \forall P', P'' \in C.$$

Intuitively, in this case the DM perceives a similar ambiguity in both acts. For example, $p \asymp q$ for all lottery acts, which are unambiguous.

It is easy to see that $\asymp$ is an equivalence relation. Denote by $[f]$ the relative equivalence class determined by an act f, and by $\mathcal{F}_{|\asymp}$ the quotient space of $\mathcal{F}$ that consists of these equivalence classes.

---

[36]That is, if $\succsim' \subseteq \succsim$ and $\succsim'$ satisfies independence, then $\succsim' \subseteq \succsim^*$.

[37]This latter feature of $\succsim^*$ relates this notion to an earlier one by Nehring (2001), as GMM discuss.

[38]GMM also show the form that C takes for some CEU preferences that do not satisfy S.6.

Theorem 12. *Let $\succsim$ be an IB preference on $\mathcal{F}$. Then, there exists a function $u : X \to \mathbb{R}$, a convex and compact set $C \subseteq \Delta(\Sigma)$ of probability measures, and a function $a : \mathcal{F}_{|\asymp} \to [0,1]$ such that $\succsim$ is represented by the preference functional $V : \mathcal{F} \to \mathbb{R}$ given by*

$$V(f) = a([f]) \min_{P \in C} \int_S u(f(s))\, dP(s) + (1 - a([f])) \max_{P \in C} \int_S u(f(s))\, dP(s), \tag{21.13}$$

where u and C represent $\succsim^*$ in the sense of (21.12).

Moreover, C is unique, u is cardinally unique, and a is unique on $\mathcal{F}_{|\asymp}$ (with the exclusion of the equivalence class $[p]$ of lottery acts).

In this representation, the revealed perception of ambiguity, embodied by the set C, is separated from the DM's reaction to it, modelled by the function a. Both C and a are derived endogenously within the model. When a is constant equal to 1, we get back to the maxmin representation. Otherwise, we have a more general choice criterion that may well exhibit ambiguity loving (the polar case is, clearly, when a is a constant equal to 0).

Giraud (2005) and Amarante (2009) studied invariant biseparable preferences, with novel important insights. Amarante established an alternative characterization of IB preferences through the two stage form

$$V(f) = \int_\Delta \left(\int_S u(f(s))\, dP(s) \right) d\nu(P)$$

where ν is a capacity over the set of measures $\Delta = \Delta(\Sigma)$ on S. In a statistical decision theory vein, the capacity ν quantifies DM's beliefs over the possible models P. Giraud thoroughly studies a similar representation, motivated by the desire to incorporate probabilistic information in a choice under ambiguity framework.

Finally, Siniscalchi (2006a) investigates an interesting class of invariant biseparable preferences that satisfy a local no-hedging condition that gives preferences a piecewise structure that makes them SEU on each component (see Castagnoli et al. 2003, for a related representation).

***a*-MEU Preferences** In the special case when the function a is constant the representation (21.13) reduces to

$$V(f) = a \min_{P \in C} \int_S u(f(s))\, dP(s) + (1 - a) \max_{P \in C} \int_S u(f(s))\, dP(s). \tag{21.14}$$

This is the a-MEU criterion that Jaffray (1989) suggested to combine Hurwicz (1951)'s criterion (see also Arrow and Hurwicz 1972) with a maxmin approach. Intuitively, $a \in [0,1]$ measures the degree of the individual's pessimism, where $a = 1$ yields the maxmin expected utility model, and $a = 0$ – its dual, the maxmax expected utility model. However, this apparently natural idea turned out to be surprisingly tricky to formally pin down. GMM provided a specific axiom that reduces the IB representation to (21.14), where C represent $\succsim^*$ in the sense of (21.12). Because of this latter clause, when $a \in (0, 1)$ it is not possible to take any pair u and C as a given and assume that the DMs' preferences are represented by the corresponding a-MEU criterion (21.14). In a nutshell, the issue is the uniqueness properties of C in (21.14), which are problematic when $a \in (0, 1)$. We refer the reader to GMM and to Eichberger et al. (2008, 2011) for more on this issue. (The latter paper shows that for finite state spaces the a-MEU axiomatized as a very special case of (21.13) by GMM only allows for $\alpha = 0$ or $\alpha = 1$).

Smooth Preferences

The MMEU model discussed above is often viewed as rather extreme: if, indeed, a set of probability measures C is stipulated, and each act f is mapped to a range of expected utility values, $\{\int_S u(f)dp \mid p \in C\}$, why should such an f be evaluated by the minimal value in this interval? This worst-case scenario approach seems almost paranoid: why should the DM assume that nature[39] will choose a probability as if to spite the DM? Isn't it more plausible to allow for other ways that summarize the interval by a single number?

The extreme nature of the maxmin model is not evident from the axiomatic derivation of the model. Indeed, this model is derived from Anscombe-Aumann's by relaxing their independence axiom in two ways: first, by restricting it to mixing with a constant act (h above) and, second, by assuming uncertainty aversion. These weaker axioms do not seem to reflect the apparently-paranoid attitude of the maxmin principle. A question then arises, how do these axioms give rise to such extreme uncertainty attitude?

In this context it is important to recall that the axiomatic derivation mentioned above is in the revealed preferences tradition, characterizing behavior that could be represented in a certain mathematical formula. An individual who satisfies the axioms can be thought of *as if* she entertained a set C of priors and maximized the minimal expected utility with respect to this set. Yet, this set of priors need not necessarily reflect the individual's knowledge. Rather, information and personal taste jointly determine the set C. Smaller sets may reflect both better information and a less averse uncertainty attitude. For example, an individual who bets on a flip of a coin and follows the expected utility axioms with respect to a probability $p = 0.5$ of "Head" may actually know that the probability p is 0.5, or she may have no clue about p but chooses the model $p = 0.5$ because she is insensitive to her ignorance about the true data generating process. Thus, information and attitude to uncertainty are inextricably intertwined in the set C. More generally, it is possible that the individual has objective information that the probability is in a set D, but

[39]Relations between ambiguity and games against nature are discussed in Hart et al. (1994), Maccheroni et al. (2006a,b), and Ozdenoren and Peck (2008).

behaves according to the maxmin expected utility rule with respect to a set $C \subset D$, reflecting her uncertainty attitude. This intuition has motivated the model of Gajdos et al. (2008) that axiomatically established the inclusion $C \subset D$ (some related ideas can be found in Wang (2003a) and Giraud (2005)).

If, however, the set of priors C is interpreted cognitively a la Wald, that is, as the set of probabilities that are consistent with objectively available information, one may consider alternatives to the maxmin rule that, under this Waldean interpretation, has an extreme nature. One approach to address this issue is to assume that the DM has a prior probability over the possible probability distributions in C. Thus, if $\Delta(\Sigma)$ is the space of all "first order" probability distributions (viewed as data generating processes), and μ is a "second order" prior probability over them, one can use μ to have an averaging of sorts over all expected utility values of an act f.

Clearly, the expectation of expectations is an expectation. Thus, if one uses μ to compute the expectation of the expected utility, there will exist a probability $\hat{p}$ on S, given by

$$\hat{p} = \int_{\Delta(\Sigma)} p d\mu$$

such that for every act f (and every utility function u)

$$\int_{\Delta(\Sigma)} \left(\int_S u(f) dp \right) d\mu = \int_S u(f) d\hat{p}$$

In this case, the new model cannot explain any new phenomena, as it reduces to the standard Bayesian model. However, if the DM uses a non-linear function to evaluate expected utility values, one may explain non-neutral attitudes to uncertainty. Specifically, assume that

$$\varphi : \mathbb{R} \to \mathbb{R}$$

is an increasing function, and an act f is evaluated by

$$V(f) = \int_{\Delta(\Sigma)} \varphi \left(\int_S u(f) dp \right) d\mu.$$

In this representation, μ is read as representing information (about the probability model p), whereas φ reflects attitude towards ambiguity, with a concave φ corresponding to ambiguity aversion, similarly to the way that concave utility represents risk aversion in the classical model of expected utility under risk. In this way we have a separation between ambiguity perception, an information feature modelled by μ and its support, and ambiguity attitude, a taste trait modelled by φ and its shape.

This decision rule has been axiomatized by Klibanoff et al. (2005). It has become to be known as the *smooth model* of ambiguity because, under mild assumptions,

V is a smooth functional, whereas the Choquet expected utility and the maxmin expected utility functionals are typically not everywhere differentiable (over the space of acts).

The notion of second order probabilities is rather old and deserves a separate survey.[40] This idea is at the heart of Bayesian statistics, where Bayes's rule is retained and a probability over probabilities over a state space is equivalent to a probability over the same space. Within decision theory, Segal (1987) already suggested that Ellsberg's paradox can be explained by second-order probabilities, provided that we allow the decision maker to violate the principle of reduction of compound lotteries. Specifically, Segal's model assumed that the second-order probabilities are used to aggregate first-order expectations via Quiggin's (1982) anticipated utility. Other related models have been proposed by Nau (2001, 2006, 2011), Chew and Sagi (2008), Ergin and Gul (2009), and Seo (2009). Halevy and Feltkamp (2005) proposed another approach according to which the decision maker does not err in the computation of probabilities, but uses a mis-specified model, treating a one-shot choice as if it were repeated.

As compared to Choquet expected utility maximization, the smooth preferences model, like the maxmin model, has the advantage of having a simple and intelligible cognitive interpretation. As opposed to both Choquet and maxmin expected utility models, smooth preferences have the disadvantage of imposing non-trivial epistemological demands on the DM: the smooth model requires the specification of a prior over probability models, that is, of a probability μ over a much larger space, $\Delta(\Sigma)$, something that may be informationally and observationally demanding.

That said, beyond the above mentioned separation, the smooth preferences model enjoys an additional advantage of tractability. If S is finite, one may choose μ to be a uniform prior over $\Delta(\Sigma)$ and specify a simple functional form for φ, to get a simple model in which uncertainty/ambiguity attitudes can be analyzed in a way that parallels the treatment of risk attitudes in the classical literature. Specifically, assume that

$$\varphi(x) = -\frac{1}{\alpha}e^{-\alpha x}$$

for $\alpha > 0$. In this case, the DM can be said to have a constant ambiguity aversion α; when $\alpha \to 0$, the DM's preferences converge to Bayesian preferences with prior $\hat{p}$, whereas when $\alpha \to \infty$, preferences converge to MMEU preferences relative to the support of μ. (See Klibanoff et al. 2005, for details.) Thus, the smooth ambiguity aversion model can be viewed as an extension of the maxmin model, in its Waldean interpretation.

[40]Bayes (1763) himself writes in his Proposition 10 that "the chance that the probability of the event lies somewhere between ..." (at the beginning of his essay, in Definition 6 Bayes says that "By chance I mean the same as probability").

Variational Preferences

Maccheroni et al. (MMR, 2006a) suggested and axiomatized an extension of the maxmin model in order to better understand the theoretical foundations of the works of Hansen and Sargent on model uncertainty in macroeconomics (see the surveys Hansen 2007; Hansen and Sargent 2008). These works consider agents who take into account the possibility that their (probabilistic) model Q may not be the correct one, but only an approximation thereof. For this reason, they rank acts f according to the following choice criterion

$$V(f) = \min_{P\in\Delta(\Sigma)} \left\{ \int_S u(f(s))\, dP(s) + \theta R(P\|Q) \right\}, \tag{21.15}$$

where $\theta > 0$, and $R(\cdot\|Q) : \Delta(\Sigma) \to [0,\infty]$ is the relative entropy with respect to Q.

Preferences $\succsim$ on $\mathcal{F}$ represented by criterion (21.15) are called *multiplier preferences* by Hansen and Sargent. The relative entropy $R(P\|Q)$ measures the relative likelihood of the alternative models P with respect to the reference model Q. The positive parameter θ reflects the weight that agents are giving to the possibility that Q might not be the correct model (as θ becomes larger, agents focus more on Q as the correct model, giving less importance to the alternatives P).

Model uncertainty, which motivated the study of multiplier preferences, is clearly akin to the problem of ambiguity, underlying maxmin preferences. Yet, neither class of preferences is nested in the other. A priori, it was not clear what are the commonalities between these models and how they can be theoretically justified. To address this issue, MMR introduced and axiomatized a novel class of preferences that includes both multiplier and maxmin preferences as special cases.

Specifically, observe that the maxmin criterion (21.10) can be written as

$$V(f) = \min_{P\in\Delta(\Sigma)} \left\{ \int_S u(f(s))\, dP(s) + \delta_C(P) \right\}, \tag{21.16}$$

where $\delta_C : \Delta \to [0,\infty]$ is the indicator function of C given by

$$\delta_C(P) = \begin{cases} 0 & \text{if} P \in C, \\ \infty & \text{otherwise.} \end{cases}$$

Like the relative entropy, the indicator function is a convex function defined on the simplex $\Delta(\Sigma)$. This suggests the following general representation

$$V(f) = \min_{P\in\Delta(\Sigma)} \left\{ \int u(f(s))\, dP(s) + c(P) \right\}, \tag{21.17}$$

where $c : \Delta(\Sigma) \to [0, \infty]$ is a convex function on the simplex. MMR call *variational* the preferences $\succsim$ on $\mathcal{F}$ represented by (21.17). Multiplier and maxmin preferences are the special cases of variational preferences where c is, respectively, the relative entropy $\theta R(\cdot \| Q)$ and the indicator function δ_C.

MMR establish a behavioral foundation for the representation (21.17), which in turn offers a common behavioral foundation for multiplier and maxmin preferences. Their axiomatization is based on a relaxation of the C-Independence GS.3 of Gilboa and Schmeidler. To understand it, consider the following equivalent form of GS.3 (C-Independence).

Lemma 13. *A binary relation $\succsim$ on $\mathcal{F}$ satisfies axiom GS.3 (C-Independence) if and only if, for all $f, g \in \mathcal{F}$, $p, q \in \Delta(X)$, and $\alpha, \beta \in (0, 1]$, we have:*

$$\alpha f + (1-\alpha)p \succsim \alpha g + (1-\alpha)p \Rightarrow \beta f + (1-\beta)q \succsim \beta g + (1-\beta)q.$$

Lemma 13 (MMR p. 1454) shows that axiom GS.3 actually involves two types of independence: independence relative to mixing with constants and independence relative to the weights used in such mixing. The next axiom, due to MMR, retains the first form of independence, but not the second one.

MMR.3 WEAK C-INDEPENDENCE: If $f, g \in \mathcal{F}$, $p, q \in \Delta(X)$, and $\alpha \in (0, 1)$,

$$\alpha f + (1-\alpha)p \succsim \alpha g + (1-\alpha)p \Rightarrow \alpha f + (1-\alpha)q \succsim \alpha g + (1-\alpha)q.$$

Axiom MMR.3 is therefore the special case of axiom GS.3 (C-Independence) in which the mixing coefficients α and β are required to be equal. In other words, axiom MMR.3 is a very weak independence axiom that requires independence only with respect to mixing with lottery acts, provided the mixing weights are kept constant.

This is a significant weakening of axiom GS.3 (C-Independence). One might wonder, why would the DM follow MMR.3 but not GS.3 in its full strength. To see this, consider the re-statement of axiom GS.3 in Lemma 13 in the case that the weights α and β are very different, say α is close to 1 and β is close to 0. Intuitively, acts $\alpha f + (1-\alpha)p$ and $\alpha g + (1-\alpha)p$ can then involve far more uncertainty than acts $\beta f + (1-\beta)q$ and $\beta g + (1-\beta)q$, which are almost constant acts. As a result, we expect that, at least in some situations, the ranking between the genuinely uncertain acts $\alpha f + (1-\alpha)p$ and $\alpha g + (1-\alpha)p$ can well differ from that between the almost constant acts $\beta f + (1-\beta)q$ and $\beta g + (1-\beta)q$. By contrast, Axiom MMR.3 is not susceptible to this critique: since only the same coefficient α is used in both sides of the implication, the axiom does not involve acts that differ in their overall uncertainty, as it were.

The representation result of MMR is especially sharp when the utility function u is unbounded (above or below), that is, when its image $u(X) = \{u(x) : x \in X\}$ is an unbounded set. In an AA setup this follows from the following assumption (see Kopylov 2001).

AA.7 UNBOUDEDNESS: There exist $x \succ y$ in X such that for all $\alpha \in (0,1)$ there exists $z \in X$ satisfying either $y \succ \alpha z + (1-\alpha)x$ or $\alpha z + (1-\alpha)y \succ x$.

We can now state the representation result of MMR, which generalizes Theorem 4 by allowing for general functions $c : \Delta(\Sigma) \to [0,\infty]$. Here x_f denotes the certainty equivalent of act f; i.e., $f \sim x_f$.

Theorem 14. *Let $\succsim$ be a binary relation on $\mathcal{F}$. The following conditions are equivalent:*

(i) $\succsim$ satisfies conditions AA.1, AA.2, MMR.3, AA.4, AA.5, S.6, and AA.7;
(ii) there exists an affine function $u : X \to \mathbb{R}$, with $u(X)$ unbounded, and a grounded,[41] convex, and lower semicontinuous function $c : \Delta(\Sigma) \to [0,\infty]$ such that, for all $f, g \in \mathcal{F}$

$$f \succsim g \Leftrightarrow \min_{P\in\Delta(\Sigma)} \left(\int_S u(f(s))\,dP(s) + c(P) \right) \geq \min_{P\in\Delta(\Sigma)} \left(\int u(g(s))\,dP(s) + c(P) \right). \tag{21.18}$$

For each u there is a unique $c : \Delta(\Sigma) \to [0,\infty]$ satisfying (21.18), given by

$$c(P) = \sup_{f\in\mathcal{F}} \left(u(x_f) - \int_S u(f(s))\,dP(s) \right). \tag{21.19}$$

MMR show how the function c can be viewed as an index of ambiguity aversion, as we will discuss later in section "Ambiguity Aversion". Alternatively, they observe that the function c can be interpreted as the cost of an adversarial opponent of selecting the prior P. In any case, formula (21.19) allows to determine the index c from behavioral (e.g., experimental) data in that it only requires to elicit u and the certainty equivalents x_f.

Behaviorally, maxmin preferences are the special class of variational preferences that satisfy the C-Independence axiom GS.3. For multiplier preferences, however, MMR did not provide the behavioral assumption that characterize them among variational preferences. This question left open by MMR was answered by Strzalecki (2010), who found the sought-after behavioral conditions. They turned out to be closely related to some of Savage's axioms. Strzalecki's findings thus completed the integration of multiplier preferences within the framework of choice under ambiguity.

The weakening of C-Independence in MMR.3 has a natural variation in which independence is restricted to a particular lottery act, but not to a particular weight α. Specifically, one may require that, for the worst possible outcome x_* (if such exists),

[41]The function $c : \Delta(\Sigma) \to [0,\infty]$ is grounded if its infimum value is zero.

$$\alpha f + (1-\alpha)x_* \succsim \alpha g + (1-\alpha)x_* \Leftrightarrow \beta f + (1-\beta)x_* \succsim \beta g + (1-\beta)x_*$$

for every two acts $f, g \in \mathcal{F}$ and every $\alpha, \beta \in (0, 1]$,

This condition has been used by Chateauneuf and Faro (2009), alongside other conditions, to derive the following representation: there exists a so-called *confidence* function φ on $\Delta(\Sigma)$, and a confidence threshold α, such that acts are evaluated according to

$$V(f) = \min_{\{P \in \Delta(\Sigma) | \varphi(P) \geq \alpha\}} \left[\frac{1}{\varphi(P)} \int_S u(f(s))dP(s) \right]$$

This decision rule suggests that the DM has a degree of confidence $\varphi(P)$ in each possible prior P. The expected utility associated with a prior P is multiplied by the inverse of the confidence in P, so that a low confidence level is less likely to determine the minimum confidence-weighted expected utility of f.

The intersection of the classes of variational preferences with confidence preferences is the maxmin model, satisfying C-Independence in its full force.[42] See also Ghirardato et al. (2005) for other characterizations of C-Independence.

Beyond Independence: Uncertainty Averse Preferences

All the choice models that we reviewed so far feature some violation of the Independence axiom AA.3, which is the main behavioral assumption questioned in the literature on choice under ambiguity in a AA setup. In order to better understand this class of models, Cerreia-Vioglio et al. (2011) recently established a common representation that unifies and classifies them. Since a notion of minimal independence among uncertain acts is, at best, elusive both at a theoretical and empirical level, this common representation does not use any independence condition on uncertain acts, however weak it may appear.

Cerreia-Vioglio et al. (2011) thus studied uncertainty averse preferences, that is, complete and transitive preferences that are monotone and convex, without any independence requirement on uncertain acts. This general class of preferences includes as special cases variational preferences, confidence preferences, as well as smooth preferences with a concave φ.

Though no independence assumption is made on uncertain acts, to calibrate risk preferences Cerreia-Vioglio et al. assumed standard independence on lottery acts.

CMMM.3 RISK INDEPENDENCE: If $p, q, r \in \Delta(X)$ and $\alpha \in (0, 1)$, $p \sim q \Rightarrow \alpha p + (1-\alpha) r \sim \alpha q + (1-\alpha) r$.

[42]This is so because one axiom relates preferences between mixtures with different coefficients α, β and the other – between mixtures with different constant acts x_*, p.

Along with the other axioms, CMMM.3 implies that the risk preference $\succsim_\Delta$ satisfies the von Neumann-Morgenstern axioms. In the representation result of Cerreia-Vioglio et al. (2011) functions of the form $G : \mathbb{R}\times\Delta(\Sigma) \to (-\infty, \infty]$ play a key role. Denote by $\mathcal{G}(\mathbb{R}\times\Delta(\Sigma))$ the class of these functions such that:

(i) G is quasiconvex on $\mathbb{R}\times\Delta(\Sigma)$,
(ii) $G(\cdot, P)$ is increasing for all $P \in \Delta(\Sigma)$,
(iii) $\inf_{P\in\Delta(\Sigma)} G(t, P) = t$ for all $t \in T$.

We can now state a version of their main representation theorem.

Theorem 15. *Let $\succsim$ be a binary relation on $\mathcal{F}$. The following conditions are equivalent:*

(i) $\succsim$ satisfies axioms AA.1, AA.2, CMMM.3, AA.4, AA.5, S.6, AA.7;
(ii) there exists a non-constant affine $u : X \to \mathbb{R}$, with $u(X) = \mathbb{R}$, and a lower semicontinuous $G : \mathbb{R}\times\Delta(\Sigma) \to (-\infty, \infty]$ that belongs to $\mathcal{G}(\mathbb{R}\times\Delta(\Sigma))$ such that, for all f and g in $\mathcal{F}$,

$$f \succsim g \Leftrightarrow \min_{P\in\Delta(\Sigma)} G\left(\int u(f)\, dP, P\right) \geq \min_{P\in\Delta(\Sigma)} G\left(\int u(g)\, dP, P\right). \quad (21.20)$$

The function u is cardinally unique and, given u, the function G in (21.20) is given by

$$G(t, P) = \sup_{f\in\mathcal{F}} \left\{ u(x_f) : \int u(f)\, dP \leq t \right\} \qquad \forall (t, p) \in \mathbb{R} \times \Delta(\Sigma). \quad (21.21)$$

In this representation DMs can be viewed as if they considered, through the term $G\left(\int u(f)\, dP, P\right)$, all possible probabilities P and the associated expected utilities $\int u(f)\, dP$ of act f. They then behave as if they summarized all these evaluations by taking their minimum. The quasiconvexity of G and the cautious attitude reflected by the minimum in (21.20) derive from the convexity of preferences. Their monotonicity, instead, is reflected by the monotonicity of G in its first argument.

The representation (21.20) features both probabilities and expected utilities, even though no independence assumption whatsoever is made on uncertain acts. In other words, this representation establishes a general connection between the language of preferences and the language of probabilities and utilities, in keeping with the tradition of the representation theorems in choice under uncertainty.

Cerreia-Vioglio et al. (2011) show that G can be interpreted as index of uncertainty aversion, in the sense of section "Ambiguity Aversion" below. Moreover, (21.21) shows that this index can be elicited from choice behavior.

Variational preferences correspond to additively separable functions G, i.e., these preferences are characterized by

$$G(t, P) = t + c(P)$$

where $c : \Delta(\Sigma) \to [0, \infty]$ is a convex function. In this case (21.20) reduces to the variational representation (21.18).

Smooth preferences with concave ϕ correspond to the uncertainty aversion index given by

$$G(t, P) = t + \min_{\nu \in \Gamma(P)} I_t(\nu \parallel \mu) \tag{21.22}$$

where $I_t(\cdot \parallel \mu)$ is a suitable statistical distance function that generalizes the classic relative entropy, and $\Gamma(P)$ is the set of all second-order probabilities ν that are absolutely continuous with respect to μ and that have P as their reduced, first-order, probability measure on S.

Other Classes of Preferences

The scope of this paper does not allow us to do justice to the variety of decision models that have been suggested in the literature to deal with uncertainty in a non-probabilistic way, let alone the otherwise growing literature in decision theory.[43] Here we only mention a few additional approaches to the problem of ambiguity.

As mentioned above, Segal (1987, 1990) suggested a risk-based approach to uncertainty, founded on the idea that people do not reduce compound lotteries. Recently, Halevy (2007) provided some experimental evidence on the link between lack of reduction of compound lotteries and ambiguity, and Seo (2009) carried out an in depth theoretical analysis of this issue. Since failure to reduce compound lotteries is often regarded as a mistake, this source of ambiguity has a stronger positive flavor than the absence of information, which is our main focus.

Stinchcombe (2003), Olszewski (2007), and Ahn (2008) model ambiguity through sets of lotteries, capturing exogenous or objective ambiguity. (See also Jaffray (1988), who suggested related ideas). Preferences are defined over these sets, with singleton and nonsingleton ones modelling risky and ambiguous alternatives, respectively. For example, these sets can be ranked either according to the criterion $V(A) = \left(\int_A \phi \circ u d\mu\right) / \mu(A)$ where ϕ and μ model ambiguity attitudes (Ahn 2008) or the criterion $V(A) = \alpha \min_{l \in A} U(l) + (1 - \alpha) \max_{l \in A} U(l)$ where α models ambiguity attitudes (Olszewski 2007). Viero (2009) combines this approach with the Anscombe-Aumann model.

Chateauneuf et al. (2007) axiomatize *neo-additive Choquet expected utility*, a tractable CEU criterion of the "Hurwicz" form $V(f) = \alpha \int u(f(s)) dP(s) + \beta \max_s u(f(s)) + (1 - \alpha - \beta) \min_s u(f(s))$. Through the values of the weights α and β, the preference functional V captures in a simple way different degrees of optimism and pessimism, whose extreme forms are given by the min and max of $u(f(s))$.

[43]Other sub-fields include choices from menus, decision under risk, minmax regret approaches, and others. On the first of these, see Limpan and Pesendorfer (2013).

Gajdos et al. (2008) axiomatize, as discussed before, a model with objective information. Preferences are defined over pairs (f, C) of acts and sets of probabilities (that represent objective information). Such pairs are ranked through the functional $V(f, C) = \min_{p \in \varphi(C)} \int u(f(s))\, dP(s)$, where $\varphi(C) \subseteq C$ is the subset of C that we denoted in the earlier discussion as D. When $\varphi(C) = C$, we get back to the MMEU model.

Gul and Pesendorfer (2008) suggested *subjective expected uncertain utility theory*, according to which acts can be reduced to bilotteries, each specifying probabilities for ranges of outcome values, where these probabilities need not be allocated to sub-ranges. Arlo-Costa and Helzner (2010a) propose to deal with the comparative ignorance hypothesis of Tversky and Fox (1995), and present experimental findings that challenge the explanation provided by the latter. (See also Arlo-Costa and Helzner 2010b).

Siniscalchi (2009a) axiomatizes *vector expected utility*, in which Savage's acts are assessed according to $V(f) = \int u(f(s))\, dP(s) + A\left(\left(\int \xi_i \cdot u(f(s))\, dP(s)\right)_{i=1,\dots,n}\right)$ where the first term on the right hand side is a baseline expected-utility evaluation and the second term is an adjustment that reflects DMs' perception of ambiguity and their attitudes toward it. In particular, ξ_i are random variables with zero mean that model different sources of ambiguity (see Siniscalchi 2009a, p. 803).

Given the variety of the models of decision making that allow for non-neutral approaches to ambiguity, one is led to ask, how should we select a model to work with? There are at least three possible approaches to this problem. First, one may follow the classical empirical tradition and compare the different models by a "horse-race". The model that best explains observed phenomena should be used for prediction, with the usual trade-offs between the model's goodness of fit and its simplicity and generality. The degree to which models fit the data should be measured both for their assumptions and for their conclusions. (Indeed, the assumptions are also, in a trivial sense, conclusions.) Thus, this approach calls both for experimental tests of particular axioms and of entire models, as well as for empirical tests of theories based on these models. Importantly, when engaging in such an endeavor, one should be prepared to find that a model may be the most appropriate for analyzing certain phenomena but not for others. Thus, for example, it is possible that smooth preferences are the best model for the behavior of organizations, whereas variational preferences are a better description for the behavior of individuals. Or that labor search models are best explained by the maxmin model, while financial investments call for the Hurwicz-Jaffray model, and so forth.

For qualitative analysis, one may adopt a second approach, which does not commit to a particular model of decision under uncertainty, but uses representatives of these models in order to gain robust insights. Adopting this approach, a researcher may start with a benchmark Bayesian model, and add the uncertainty ingredient using any of the models mentioned above, as a sensitivity analysis of the Bayesian model. In this approach, theoretical convenience may be an important guideline. However, it will be advisable to trust only the qualitative conclusions that emerge from more than one model. That is, sensitivity analysis itself should not be too sensitive.

Finally, in light of the variety of models and the theoretical difficulties in selecting a single one, one may choose a third approach, which attempts to obtain general conclusions within a formal model, without committing to a particular theory of decision making. This approach has been suggested in the context of risk by Machina (1982). In this celebrated paper, facing a variety of decision models under risk, Machina attempted to show that much of economic analysis of choice under risk can be carried through without specifying a particular model. More concretely, Machina stipulated a functional on lotteries (with given probabilities) that was smooth enough to allow local approximations by linear functions. The gradient of the functional was considered to be a *local* utility function. Machina has shown that some results in economic theory could be derived by allowing the local utility function to vary, as long as it satisfied the relevant assumptions. Machina's approach was therefore not about decision theory per se. It was about the degree to which decision theory mattered: it showed that, for some applications, economists need not worry about how people really make decisions, since a wide range of models were compatible with particular qualitative conclusions.

A similar approach has been suggested for decisions under uncertainty. An early example of this approach is the notion of biseparable preferences, suggested by Ghirardato and Marinacci (2001), and mentioned above. *Biseparable preferences* are any monotone and continuous preferences over general acts that, when restricted to acts f with only two outcomes, say, x and y, can be described by the maximization of

$$J(f) = u(x)\nu(A) + (u(x) - u(y))(1 - \nu(A))$$

where ν is a capacity and

$$f(s) = \begin{cases} x & s \in A \\ y & s \notin A \end{cases}$$

with $x \succ y$. Biseparable preferences include both CEU and MMEU. Ghirardato and Marinacci (2001) provide a definition of uncertainty aversion that does not depend on the specific model of decision making and applies to all biseparable preferences.

More recently, Machina (2005) suggested a general approach to preferences under uncertainty which, similarly to Machina (1982), assumes mostly smoothness and monotonicity of preferences, but remains silent regarding the actual structure of preferences, thereby offering a highly flexible model.

Ambiguity Aversion

Schmeidler's axiom S.6 provided a first important characterization of ambiguity aversion, modelled through a preference for hedging/randomization. Epstein (1999) and Ghirardato and Marinacci (2002) studied this issue from a different perspective, inspired by Yaari (1969)'s analysis of comparative risk attitudes.

Here we present the approach of Ghirardato and Marinacci because of its sharper model implications. This approach relies on two key ingredients:

(i) A comparative notion of ambiguity aversion that, given any two preferences $\succsim_1$ and $\succsim_2$ on $\mathcal{F}$, says when $\succsim_1$ is more ambiguity averse than $\succsim_2$.
(ii) A benchmark for neutrality to ambiguity; that is, a class of preferences $\succsim$ on $\mathcal{F}$ that are viewed as neutral to ambiguity.

The choice of these ingredients in turn determines the absolute notion of ambiguity aversion, because a preference $\succsim$ on $\mathcal{F}$ is classified as ambiguity averse provided it is more ambiguity averse than an ambiguity neutral one.

The comparative notion (i) is based on comparisons of acts with lottery acts that deliver a lottery p at all states. We consider them here because they are the most obvious example of unambiguous acts, that is, acts whose outcomes are not affected by the unknown probabilities.

Consider DM_1 and DM_2, whose preferences on $\mathcal{F}$ are $\succsim_1$ and $\succsim_2$, respectively. Suppose that

$$f \succsim_1 p,$$

that is, DM_1 prefers the possibly ambiguous act f to the unambiguous lottery act p. If DM_1 is more ambiguity averse than DM_2 it is natural to expect that DM_2 will also exhibit such preferences:

$$f \succsim_2 p.$$

For, if DM_1 is bold enough to have $f \succsim_1 p$, then DM_2 – who dislikes ambiguity no more than DM_1 – must be at least equally bold.

We take this as the behavioral characterization of the comparative notion of ambiguity aversion.

Definition 16. Given two preferences $\succsim_1$ and $\succsim_2$ on $\mathcal{F}$, $\succsim_1$ is *more ambiguity averse than* $\succsim_2$ if, for all $f \in \mathcal{F}$ and $p \in \Delta(X)$,

$$f \succsim_1 p \Rightarrow f \succsim_2 p. \tag{21.23}$$

As benchmark for neutrality to ambiguity we consider subjective expected utility (SEU) preferences on $\mathcal{F}$. These preferences intuitively embody ambiguity neutrality. They might not be the only preference embodying ambiguity neutrality, but they seem to be the most obvious ones.[44]

Methodologically, like the choice of lottery acts as the unambiguous acts in the comparison (21.23), also the neutrality benchmark is chosen by making the weakest

[44] Epstein (1999) takes the standard for ambiguity neutrality to be preferences that are probabilistically sophisticated in the sense of Machina and Schmeidler (1992). In his approach Theorem 18 below does not hold.

prejudgment on which preferences qualify for this role. Sharp model implications will follow, nevertheless, as we will see momentarily.

Having thus prepared the ground, we can define ambiguity aversion

Definition 17. A preference relation $\succsim$ on $\mathcal{F}$ is *ambiguity averse* if it is more ambiguity averse than some SEU preference on $\mathcal{F}$.

The next result, due to Ghirardato and Marinacci (2002), applies these notions to the maxmin expected utility (MEU) model. Here $u_1 \approx u_2$ means that there exist $\alpha > 0$ and $\beta \in \mathbb{R}$ such that $u_1 = \alpha u_2 + \beta$.

Theorem 18. *Given any two MMEU preferences $\succsim_1$ and $\succsim_2$ on $\mathcal{F}$, the following conditions are equivalent:*

(i) $\succsim_1$ is more ambiguity averse than $\succsim_2$,
(ii) $u_1 \approx u_2$ and $C_1 \subseteq C_2$ (provided $u_1 = u_2$).

Given that $u_1 \approx u_2$, the assumption $u_1 = u_2$ is just a common normalization of the two utility indices. Therefore, Theorem 18 says that more ambiguity averse MMEU preferences are characterized, up to a normalization, by smaller sets of priors C. Therefore, the set C can be interpreted as an *index of ambiguity aversion*.

This result thus provides a behavioral foundation for the comparative statics exercises in ambiguity through the size of the sets of priors C that play a key role in the economic applications of the MMEU model. In fact, a central question in these applications is how changes in ambiguity attitudes affect the relevant economic variables.

An immediate consequence of Theorem 18 is that, not surprisingly, MMEU preferences are always ambiguity averse. That is, they automatically embody a negative attitude toward ambiguity, an attitude inherited from axiom S.6.

The condition $u_1 \approx u_2$ ensures that risk attitudes are factored out in comparing the MMEU preferences $\succsim_1$ and $\succsim_2$. This is a dividend of the risk calibration provided by the AA setup via the risk preference $\succsim_\Delta$ discussed in section "The Anscombe-Aumann Setup". In a Savage setup, where this risk calibration is no longer available, Definition 16 has to be enriched in order to properly factor out risk attitudes, so that they do not interfere with the comparison of ambiguity attitudes (see Ghirardato and Marinacci (2002), for details on this delicate conceptual issue).

Maccheroni et al. (2006a) generalize Theorem 18 to variational preferences by showing that the condition $C_1 \subseteq C_2$ takes in this case the more general form $c_1 \leq c_2$. The function c can thus be viewed as an index of ambiguity aversion that generalizes the sets of priors C. Variational preferences are always ambiguity averse, a fact that comes as no surprise since they satisfy axiom S.6.

For CEU preferences, Ghirardato and Marinacci (2002) show that more ambiguity averse CEU preferences are characterized, up to a common normalization of utility indexes, by smaller capacities ν. More interestingly, they show that CEU preferences are ambiguity averse when the cores of the associated capacities are nonempty. Since convex capacities have nonempty cores, CEU preferences that satisfy axiom S.6 are thus ambiguity averse. The converse, however, is not true since there are capacities with nonempty cores that are not convex. Hence, there exist ambiguity averse CEU preferences that do not satisfy S.6, which is thus a sufficient but not necessary condition for the ambiguity aversion of CEU preferences. Ghirardato and Marinacci (2002) discuss at length this feature of CEU preferences, and we refer the interested reader to that paper for details (see also Chateauneuf and Tallon (2002), who present a notion of weak ambiguity aversion for CEU preferences, as well as Montesano and Giovannone (1996), who investigate how CEU preferences may reflect aversion to increasing ambiguity).

Unambiguous events Unambiguous events should be events over which decision makers do not perceive any ambiguity. Intuitively, in terms of functional forms an event E is unambiguous for a preference $\succsim$ if:

(i) $\nu(E) + \nu(E^c) = 1$ when $\succsim$ is CEU;
(ii) $P(E) = P'(E)$ for all $P, P' \in C$ when $\succsim$ is MMEU and, more generally, for all $P, P' \in \operatorname{dom} c$ when $\succsim$ is variational;[45]
(iii) $p(E) = k$ μ-a.e. for some $k \in [0, 1]$ when $\succsim$ is smooth.

A few behavioral underpinnings of these notions of unambiguous event have been proposed by Nehring (1999), Epstein and Zhang (2001), Ghirardato and Marinacci (2002), Zhang (2002), Ghirardato et al. (2004), Klibanoff et al. (2005), and Amarante and Feliz (2007) (who also provide a discussion of some of the earlier notions which we refer the interested to).

Updating Beliefs

How should one update one's beliefs when new information is obtained? In the case of probabilistic beliefs there is an almost complete unanimity that Bayes's rule is the only sensible way to update beliefs. Does it have an equivalent rule for the alternative models discussed above? The answer naturally depends on the particular non-Bayesian model one adopts. At the risk of over-generalizing from a small sample, we suggest that Bayes's rule can typically be extended to non-Bayesian beliefs in more than one way. Since the focus of this survey is on static preferences, we mention only a few examples, which by no means exhaust the richness of dynamic models.

For instance, if one's beliefs are given by a capacity ν, and one learns that an event B has obtained, one may assign to an event A the weight corresponding to the straightforward adaptations of Bayes's formula:

$$\nu(A|B) = \frac{\nu(A \cap B)}{\nu(B)}$$

[45] $\operatorname{dom} c$ is the effective domain of the function c; i.e., $\operatorname{dom} c = \{P \in \Delta(S) : c(p) < +\infty\}$.

However, another formula has been suggested by Dempster (1967, see also Shafer 1976) as a special case of his notion of merging of belief functions:

$$\nu(A|B) = \frac{\nu((A \cap B) \cup B^c) - \nu(B^c)}{1 - \nu(B^c)}$$

Clearly, this formula also boils down to standard Bayesian updating in case ν is additive. Yet, the two formulae are typically not equivalent if the capacity ν fails to be additive. Each of these formulae extends some, but not all, of the interpretations of Bayesian updating from the additive to the non-additive case.

If beliefs are given by a set of priors C, and event B is known to have occurred, a natural candidate for the set of priors on B is simply the same set C, where each probability is updated according to Bayes's rule. This results in *full Bayesian updating (FBU)*, defining the set of priors (on B)

$$C_B = \{p(\cdot|B) \mid p \in C\}$$

FBU allows standard learning given each possible prior, but does not reflect any learning about the set of priors that should indeed be taken into consideration. It captures Bayesian learning (conditional on a prior) but not the statistical inference typical of classical statistics, namely, the selection of subsets of distributions from an a priori given set of distributions. If we were to think of each prior p in C as an expert, who expresses her probabilistic beliefs, FBU can be interpreted as if each expert were learning from the evidence B, while the DM does not use the evidence to decide which experts' advice to heed.[46]

Following this line of reasoning, and in accordance with statistical principles, one may wish to select probabilities from the set C based on the given event B. One, admittedly extreme way of doing so is to adopt the maximum likelihood principle. This suggests that only the priors that a priori used to assign the highest probability to the event B should be retained among the relevant ones. Thus, *maximum likelihood updating (MLU)* is given by

$$C_B^M = \left\{ p(\cdot|B) \,\middle|\, p \in \arg\max_{q \in C} q(B) \right\}$$

If one's beliefs are given by a convex capacity, or, equivalently, by a set C which is the core of a convex capacity, MLU is equivalent to Dempster-Shafer's updating. This rule has been axiomatized by Gilboa and Schmeidler (1993), whereas FBU, suggested by Jean-Yves Jaffray, has been axiomatized by Pires (2002).

FBU and MLU are both extreme. Using the experts metaphor, FBU retains all experts, and gives as much weight to those who were right as to those who were

[46]See Seidenfeld and Wasserman (1993) who study counter-intuitive updating phenomena in this context.

practically proven wrong in their past assessments. By contrast, MLU completely ignores any expert who was not among the maximizers of the likelihood function. It therefore makes sense to consider intermediate methods, though, to the best of our knowledge, none has been axiomatized to date.

The tension between FBU and MLU disappears if the set of priors C is *rectangular* (Epstein and Miao 2003) in the sense that it can be decomposed into a set of current-period beliefs, coupled with next-period conditional beliefs, in such a way that any combination of the former and the latter is in C. Intuitively, rectangularity can be viewed as independence of sorts: it suggests that whatever happens in the present period does not teach us which prior (or expert) is to be trusted more in the next period. Formally, the set of conditional probabilities on the given event B using all priors and the set obtained using only the maximum likelihood ones coincide. Related arguments, in particular how rectangular sets of priors would lead to consistent dynamic MMEU behavior, were made by Sarin and Wakker (1998) (see in particular their Theorem 2.1). See also Epstein and Schneider (2007), who consider updating in a more explicit model, distinguishing between the set of parameters and the likelihood functions they induce.

Epstein and Miao (2003) consider preferences over consumption processes, and axiomatize a decision rule that extends MMEU to the dynamic set-up recursively. Their axioms also guarantee that the set of priors C is rectangular. The recursive structure means that the maxmin expected utility at a given period for the entire future can also be written as maxmin expected utility over the present period and the discounted continuation (MMEU) value starting in the following period. Wang (2003b) proposed a related recursive approach.

This recursive approach extends beyond the MMEU model. It has similarly been applied to extend smooth preferences (see Hayashi and Miao 2011; Klibanoff et al. 2009) and variational preferences to dynamic set-ups (see Maccheroni et al. 2006b). Equipped with a variety of models of behavior with ambiguous beliefs, which are adapted to deal with dynamic problems recursively, the stage is set to analyze economic problems in not-necessarily Bayesian ways.

Another approach to updating was proposed by Hanany and Klibanoff (2007, 2009). They retain dynamic consistency by allowing the update rule to depend not only on original beliefs and new information, but also on the choice problem. In the case of the MMEU model, their approach consists of selecting a subset of priors, and updating them according to Bayes rule, while the relevant subset of priors generally depends on the act chosen before the arrival of new information.

A different route was pursued by Siniscalchi (2006b), who investigated choices over decision trees rather than over temporal acts. This modification allows him to consider sophisticated choices, characterized through a natural notion of consistent planning, under ambiguity.

An important problem relating to updating is the long-run behavior of beliefs. Suppose that a non-Bayesian decision maker faces a process that is, in a well-defined sense, repeated under the same conditions. Will she learn the true process? Will the set of probabilities converge in the limit to the true one? A partial answer was given in the context of capacities, where laws of large numbers have been proved by

Marinacci (1999, 2002b) and Maccheroni and Marinacci (2005). The behavior of the set of probabilities in the context of the maxmin model was analyzed in Epstein and Schneider (2007).

Applications

There are many economic models that lead to different qualitative conclusions when analyzed in a Bayesian way as compared to the alternative, non-Bayesian theories. The past two decades have witnessed a variety of studies that re-visited classical results and showed that they need to be qualified when one takes ambiguity into account. The scope of this paper allows us to mention but a fraction of them. The following is a very sketchy description of a few studies, designed only to give a general idea of the scope of theoretical results that need to be re-examined in light of the limitations of the Bayesian approach.[47]

Dow and Werlang (1992) analyzed a simple asset pricing model. They showed that, if an economic agent is ambiguity averse as in the CEU or MMEU model, then there will be a *range* of prices at which she will wish neither to buy nor to sell a financial asset. This range will be of non-zero length even if one ignores transaction costs. To see the basic logic of this result, consider two states of the world, where the probability of the first state, p, is only known to lie in the interval $[0.4, 0.6]$. (This will also be the core of a convex capacity). Assume that a financial asset X yields 1 in the first state and -1 in the second. The MMEU model values both X and $-X$ at -0.2. In a Bayesian model, p would be known, and the agent would switch, at a certain price π, from demanding X to offering it. This is no longer the case when p is not known. In this case, assuming ambiguity aversion, there will be an interval of prices π at which neither X nor $-X$ will seem attractive to the agent. This may explain why people refrain from trading in certain markets. It can also explain why at times of greater volatility one may find lower volumes of trade: with a larger set of probabilities that are considered possible, there will be more DMs who prefer neither to buy nor to sell.[48] The question of trade among uncertainty averse agents has been also studied in Billot et al. (2000), Kajii and Ui (2006, 2009), and Rigotti et al. (2008).

Epstein and Miao (2003) use uncertainty aversion to explain the home bias phenomenon in international finance, namely, the observation that people prefer to trade stocks of their own country rather than foreign ones. The intuition is that agents know the firms and the stock market in their own country better than in foreign ones. Thus, there is more ambiguity about foreign equities than about domestic

---

[47] Mukerji and Tallon (2004) survey early works in this area.

[48] This argument assumes that the decision maker starts with a risk-free portfolio. A trader who already holds an uncertain position may be satisfied with it with a small set of probabilities, but wish to trade in order to reduce uncertainty if the set of probabilities is larger.

ones. A Bayesian analysis makes it more difficult to explain this phenomenon: when a Bayesian DM does not know the distribution of the value of a foreign equity, she should have beliefs over it, reducing uncertainty to risk. Thus, a Bayesian would behave in qualitatively similar ways when confronting known and unknown distributions. By contrast, the notion that agents are ambiguity averse may more readily explain why they prefer to trade when the value distribution is closer to being known than when there is a great deal of ambiguity about it.

There are many other applications of ambiguity aversion to models of asset pricing. For example, Epstein and Schneider (2008) show that models involving ambiguity can better capture market reaction to the quality of information than can Bayesian models (see also Epstein and Schneider 2010), while Gollier (2011) shows that ambiguity aversion may not reinforce risk aversion and investigates how this may affect asset prices. Other recent asset pricing applications include Garlappi et al. (2007), Caskey (2009), Miao (2009), Ju and Miao (2012), and Miao and Wang (2011) (see also Guidolin and Rinaldi 2013).

The MMEU model has also been employed in a job search model by Nishimura and Ozaki (2004). They ask how an unemployed agent will react to increasing uncertainty in the labor market. In a Bayesian model, greater uncertainty might be captured by higher variance of the job offers that the agent receives. Other things being equal, an increase in variance should make the agent less willing to accept a given offer, knowing that he has a chance to get better ones later on. This conclusion is a result of the assumption that all uncertainty is quantifiable by a probability measure. Nishimura and Ozaki (2004) show that for an ambiguity averse agent, using the MMEU model, the conclusion might be reversed: in the presence of greater uncertainty, modeled as a larger set of possible priors, the agent will be more willing to take a given job offer rather than bet on waiting for better ones in the future.

Hansen et al. (1999, 2002) compare savings behavior under expected utility maximization with savings behavior of a *robust DM* who behaves in accordance with the multiple prior model. They show that the behavior of a robust DM puts the market price of risk much closer to empirical estimates than does the behavior of the classical expected utility maximizer, and, in particular, can help account for the equity premium. Hansen and Sargent (2001, 2008) apply multiplier preferences to macroeconomic questions starting from the viewpoint that, whatever the probability model a policy maker might have, it cannot be known with certainty. They ask how robust economic policy would be to variations in the underlying probability, and find conclusions that differ qualitatively from classical results. See also Miao (2004), who studies the consumption-savings decision in a different set-up.

Other (published) applications of ambiguity averse preferences include Epstein and Wang (1994, 1995), who explain financial crashes and booms, Mukerji (1998), who explains incompleteness of contracts, Chateauneuf et al. (2000), who study optimal risk-sharing rules with ambiguity averse agents, Greenberg (2000), who finds that in a strategic set-up a player may find it beneficial to generate ambiguity about her strategy choice, Mukerji and Tallon (2001), who show how incompleteness of financial markets my arise because of ambiguity aversion, Rigotti and Shannon (2005), who characterize equilibria and optima and study how they depend on the degree of ambiguity, Bose et al. (2006), who study auctions under ambiguity, Nishimura and Ozaki (2007), who show that an increase in ambiguity changes the value of an investment opportunity differently than does an increase in risk, Easley and O'Hara (2009, 2010), who study how ambiguity affects market participation, and Treich (2010), who studies when the value of a statistical life increases under ambiguity aversion.

As mentioned above, this list is but a sample of applications and has no claim even to be a representative sample.

Conclusion

Uncertainty is present in practically every field of economic enquiry. Problems in growth and finance, labor and development, political economy and industrial organization lead to questions of uncertainty and require its modeling.

For the most part, economic theory has strived to have a unifying approach to decision making in general, and to decision under uncertainty in particular. It is always desirable to have simple, unifying principles, especially if, as is the case with expected utility theory, these principles are elegant and tractable.

At the same time, expected utility theory appears to be too simple for some applications. Despite its considerable generality, there are phenomena that are hard to accommodate with the classical theory. Worse still, using the classical theory alone may lead to wrong qualitative conclusions, and may make it hard to perceive certain patterns of economic behavior that may be readily perceived given the right language.

At this point it is not clear whether a single paradigm of decision making under uncertainty will ever be able to replace the Bayesian one. It is possible that different models will prove useful to varying degrees in different types of problems. But even if a single paradigm will eventually emerge, it is probably too soon to tell which one it will be.

For the time being, it appears that economic theory may benefit from having more than a single theory of decision under uncertainty in its toolbox. The Bayesian model is surely a great candidate to remain the benchmark. Moreover, often it is quite obvious that the insights learned from the Bayesian analysis suffice. For example, Akerlof's (1970) lemons model need not be generalized to incorporate ambiguity. Its insight is simple and clear, and it will survive in any reasonable model. But there are other models in which the Bayesian analysis might be misleadingly simple. In some cases, adding a touch of ambiguity to the model, often in whatever model of ambiguity one fancies, suffices to change the qualitative conclusions. Hence it seems advisable to have models of ambiguous beliefs in our toolbox, and to test each result, obtained under the Bayesian assumptions, for robustness relative to the presence of ambiguity.

Acknowledgements We thank Giulia Brancaccio, Simone Cerreia-Vioglio, Fabio Maccheroni, Andrew Postlewaite, Xiangyu Qu, and David Schmeidler for comments on earlier drafts of this survey. We are also grateful to many members of the "decision theory forum" for additional comments and references. Finally, we are indebted to Eddie Dekel for many comments and suggestions. Gilboa gratefully acknowledges the financial support of the Israel Science Foundation (grant 396/10) and of the European Reseach Council (advanced grant 269754), and Marinacci that of the European Reseach Council (advanced grant BRSCDP-TEA).

References

Ahn, D. (2008). Ambiguity without a state space. *Review of Economic Studies, 75*, 3–28.

Akerlof, G. A. (1970). The market for 'Lemons': Quality uncertainty and the market mechanism. *The Quarterly Journal of Economics, 84*, 488–500.

Al-Najjar, N., & Weinstein, J. L. (2009). The ambiguity aversion literature: A critical assessment. *Economics and Philosophy, 25*, 249–284.

Alon, S., & Schmeidler, D. (2014). Purely subjective maxmin expected utility. *Journal of Economic Theory, 152*, 382–412

Amarante, M. (2009). Foundations of Neo-Bayesian statistics. *Journal of Economic Theory, 144*, 2146–2173.

Amarante, M., & Feliz, E. (2007). Ambiguous events and maxmin expected utility. *Journal of Economic Theory, 134*, 1–33.

Anscombe, F. J., & Aumann, R. J. (1963). A definition of subjective probability. *Annals of Mathematics and Statistics, 34*, 199–205.

Arlo-Costa, H., & Helzner, J. (2010a). Ambiguity aversion: The explanatory power of indeterminate probabilities. *Synthese, 172*, 37–55.

Arlo-Costa, H., & Helzner, J. (2010b). Ellsberg choices: Behavioral anomalies or new normative insights? *Philosophy of Science, 3*, 230–253.

Arrow, K. J. (1970). *Essays in the theory of risk-bearing*. Amsterdam: North-Holland.

Arrow, K. J., & Hurwicz, L. (1972). An optimality criterion for decision making under ignorance. In C. F. Carter & J. L. Ford (Eds.), *Uncertainty and expectations in economics*. Oxford: Basil Blackwell.

Artzner, P., Delbaen, F., Eber, J. M., & Heath, D. (1997). Thinking coherently. *Risk, 10*, 68–71.

Artzner, P., Delbaen, F., Eber, J. M., & Heath, D. (1999). Coherent measures of risk. *Mathematical Finance, 9*, 203–228.

Aumann, R. J. (1962). Utility theory without the completeness axiom. *Econometrica, 30*, 445–462.

Aumann, R. J. (1974). Subjectivity and correlation in randomized strategies. *Journal of Mathematical Economics, 1*, 67–96.

Aumann, R. J. (1976). Agreeing to disagree. *Annals of Statistics, 4*, 1236–1239.

Aumann, R. J. (1987). Correlated equilibrium as an expression of bayesian rationality. *Econometrica, 55*, 1–18.

Bayarri, M. J., & Berger, J. O. (2004). The interplay of bayesian and frequentist analysis. *Statistical Science, 19*, 58–80.

Bayes, T. (1763). An essay towards solving a problem in the doctrine of chances. *Philosophical Transactions of the Royal Society of London, 53*, 370–418.

Berger, J. (2004). The case for objective bayesian analysis. *Bayesian Analysis, 1*, 1–17.

Bernoulli, J. (1713). *Ars Conjectandi*. Basel: Thurneysen Brothers (trans. E. D. Sylla, *The art of conjecturing*. Johns Hopkins University Press, 2005).

Bertrand, J. (1907). *Calcul de probabilite* (2nd ed.). Paris: Gauthiers Villars.

Bewley, T. (2002). Knightian decision theory: Part I. *Decisions in Economics and Finance, 25*, 79–110. (Working paper, 1986).

Billot, A., Chateauneuf, A., Gilboa, I., & Tallon, J.-M. (2000). Sharing beliefs: Between agreeing and disagreeing. *Econometrica, 68*, 685–694.

Bose, S., Ozdenoren, E., & Pape, A. (2006). Optimal auctions with ambiguity. *Theoretical Economics, 1*, 411–438.

Bridgman, P. W. (1927). *The logic of modern physics*. New York: Macmillan.

Carnap, R. (1923). Uber die Aufgabe der Physik und die Andwednung des Grundsatze der Einfachstheit. *Kant-Studien, 28*, 90–107.

Casadesus-Masanell, R., Klibanoff, P., & Ozdenoren, E. (2000). Maxmin expected utility over savage acts with a set of priors. *Journal of Economic Theory, 92*, 35–65.

Caskey, J. (2009). Information in equity markets with ambiguity-averse investors. *Review of Financial Studies, 22*, 3595–3627.

Castagnoli, E., Maccheroni, F., Marinacci, M. (2003). Expected utility with multiple priors. In *Proceedings of ISIPTA 2003*, Lugano.

Cerreia-Vioglio, S., Maccheroni, F., Marinacci, M., & Montrucchio, L. (2011). Uncertainty averse preferences. *Journal of Economic Theory, 146*(4), 1275–1330.

Chateauneuf, A., Dana, R.-A., & Tallon, J.-M. (2000). Optimal risk-sharing rules and equilibria with Choquet expected utility. *Journal of Mathematical Economics, 34*, 191–214.

Chateauneuf, A., Eichberger, J., & Grant, S. (2007). Choice under uncertainty with the best and worst in mind: Neo-additive capacities. *Journal of Economic Theory, 137*, 538–567.

Chateauneuf, A., Maccheroni, F., Marinacci, M., & Tallon, J.-M. (2005). Monotone continuous multiple priors. *Economic Theory, 26*, 973–982.

Chateauneuf, A., & Faro, J. H. (2009). Ambiguity through confidence functions. *Journal of Mathematical Economics, 45*, 535–558.

Chateauneuf, A., & Tallon, J.-M. (2002). Diversification, convex preferences, and non-empty core in the Choquet expected utility model. *Economic Theory, 19*, 509–523.

Chew, H. S., & Sagi, J. (2008). Small worlds: Modeling attitudes toward sources of uncertainty. *Journal of Economic Theory, 139*, 1–24.

Choquet, G. (1953). Theory of capacities. *Annales de l'Institut Fourier, 5*, 131–295.

Cifarelli, D. M., & Regazzini, E. (1996). de Finetti's contribution to probability and statistics. *Statistical Science, 11*, 253–282.

Cyert, R. M., & DeGroot, M. H. (1974). Rational expectations and bayesian analysis. *Journal of Political Economy, 82*, 521–536.

Daston, L. (1995). *Classical probability in the enlightenment*. Princeton: Princeton University Press.

de Finetti, B. (1931). Sul Significato Soggettivo della Probabilità. *Fundamenta Mathematicae, 17*, 298–329.

de Finetti, B. (1937). La Prevision: ses Lois Logiques, ses Sources Subjectives. *Annales de l'Institut Henri Poincare, 7*, 1–68. (trans. H. E. Kyburg & H. E. Smokler (Eds.) *Studies in subjective probability*. Wiley, 1963).

DeGroot, M. H. (1975). *Probability and statistics*. Reading: Addison-Wesley.

Dempster, A. P. (1967). Upper and lower probabilities induced by a multivalued mapping. *Annals of Mathematical Statistics, 38*, 325–339.

Denneberg, D. (1994). *Non-additive measure and integral*. Dordrecht: Kluwer.

Dow, J., & Werlang, S. R. C. (1992). Uncertainty aversion, risk aversion, and the optimal choice of portfolio. *Econometrica, 60*, 197–204.

Easley, D., & O'Hara, M. (2009). Ambiguity and nonparticipation: The role of regulation. *Review of Financial Studies, 22*, 1817–1843.

Easley, D., & O'Hara, M. (2010). Microstructure and ambiguity. *Journal of Finance, 65*, 1817–1846.

Eichberger, J., Grant, S., & Kelsey, D. (2008). Differentiating ambiguity: An expository note. *Economic Theory, 36*, 327–336.

Eichberger, J., Grant, S., Kelsey, D., & Koshevoy, G. A. (2011). The α-MEU model: A comment. *Journal of Economic Theory, 146*(4), 1684–1698.

Efron, B. (1986). Why Isn't everyone a Bayesian? *The American Statistician, 40*, 1–11. With discussion.
Ellsberg, D. (1961). Risk, ambiguity and the savage axioms. *Quarterly Journal of Economics, 75*, 643–669.
Epstein, L. (1999). A definition of uncertainty aversion. *Review of Economic Studies, 66*, 579–608.
Epstein, L. G., & Marinacci, M. (2007). Mutual absolute continuity of multiple priors. *Journal of Economic Theory, 137*, 716–720.
Epstein, L. G., Marinacci, M., & Seo, K. (2007). Coarse contingencies and ambiguity. *Theoretical Economics, 2*, 355–394.
Epstein, L. G., & Miao, J. (2003). A two-person dynamic equilibrium under ambiguity. *Journal of Economic Dynamics and Control, 27*, 1253–1288.
Epstein, L. G., & Schneider, M. (2007). Learning under ambiguity. *Review of Economic Studies, 74*, 1275–1303.
Epstein, L. G., & Schneider, M. (2008). Ambiguity, information quality and asset pricing. *Journal of Finance, 63*, 197–228.
Epstein, L. G., & Schneider, M. (2010). Ambiguity and asset markets. *Annual Review of Financial Economics, 2*, 315–346.
Epstein, L. G., & Wang, T. (1994). Intertemporal asset pricing under knightian uncertainty. *Econometrica, 62*, 283–322.
Epstein, L. G., & Wang, T. (1995). Uncertainty, risk-neutral measures and security price booms and crashes. *Journal of Economic Theory, 67*, 40–82.
Epstein, L. G., & Zhang, J. (2001). Subjective probabilities on subjectively unambiguous events. *Econometrica, 69*, 265–306.
Ergin, H., & Gul, F. (2009). A theory of subjective compound lotteries. *Journal of Economic Theory, 144*, 899–929.
Ergin, H., & Sarver, T. (2009). *A subjective model of temporal preferences*. Northwestern and WUSTL, Working paper.
Fischhoff, B., & Bruine De Bruin, W. (1999). Fifty–Fifty=50%? *Journal of Behavioral Decision Making, 12*, 149–163.
Fishburn, P. C. (1970). *Utility theory for decision making*. New York: Wiley.
Frisch, R. (1926). Sur un problème d'économie pure. *Norsk Matematisk Forenings Skrifter, 1*, 1–40.
Gajdos, T., Hayashi, T., Tallon, J.-M., & Vergnaud, J.-C. (2008). Attitude toward Imprecise Information. *Journal of Economic Theory, 140*, 27–65.
Gärdenfors, P., & Sahlin, N.-E. (1982). Unreliable probabilities, risk taking, and decision making. *Synthese, 53*, 361–386.
Garlappi, L., Uppal, R., & Wang, T. (2007). Portfolio selection with parameter and model uncertainty: a multi-prior approach. *Review of Financial Studies, 20*, 41–81.
Ghirardato, P. (2002). Revisiting savage in a conditional world. *Economic Theory, 20*, 83–92.
Ghirardato, P., Klibanoff, P., & Marinacci, M. (1998). Additivity with multiple priors. *Journal of Mathematical Economics, 30*, 405–420.
Ghirardato, P., & Marinacci, M. (2001). Risk, ambiguity, and the separation of utility and beliefs. *Mathematics of Operations Research, 26*, 864–890.
Ghirardato, P., & Marinacci, M. (2002). Ambiguity made precise: A comparative foundation. *Journal of Economic Theory, 102*, 251–289.
Ghirardato, P., Maccheroni, F., & Marinacci, M. (2004). Differentiating ambiguity and ambiguity attitude. *Journal of Economic Theory, 118*, 133–173.
Ghirardato, P., Maccheroni, F., & Marinacci, M. (2005). Certainty independence and the separation of utility and beliefs. *Journal of Economic Theory, 120*, 129–136.
Ghirardato, P., Maccheroni, F., Marinacci, M., & Siniscalchi, M. (2003). Subjective foundations for objective randomization: A new spin on roulette wheels. *Econometrica, 71*, 1897–1908.
Gilboa, I. (1987). Expected utility with purely subjective non-additive probabilities. *Journal of Mathematical Economics, 16*, 65–88.
Gilboa, I. (2009). *Theory of decision under uncertainty*. Cambridge: Cambridge University Press.

Gilboa, I., Maccheroni, F., Marinacci, M., & Schmeidler, D. (2010). Objective and subjective rationality in a multiple prior model. *Econometrica, 78*, 755–770.

Gilboa, I., Postlewaite, A., & Schmeidler, D. (2008). Probabilities in economic modeling. *Journal of Economic Perspectives, 22*, 173–188.

Gilboa, I., Postlewaite, A., & Schmeidler, D. (2009). Is it always rational to satisfy Savage's axioms? *Economics and Philosophy, 25*(03), 285–296.

Gilboa, I., Postlewaite, A., & Schmeidler, D. (2012). Rationality of belief or: Why Savage's axioms are neither necessary nor sufficient for rationality. *Synthese, 187*(1), 11–31.

Gilboa, I., & Schmeidler, D. (1989). Maxmin expected utility with a non-unique prior. *Journal of Mathematical Economics, 18*, 141–153. (Working paper, 1986).

Gilboa, I., & Schmeidler, D. (1993). Updating ambiguous beliefs. *Journal of Economic Theory, 59*, 33–49.

Giraud, R. (2005). Objective imprecise probabilistic information, second order beliefs and ambiguity aversion: An axiomatization. In *Proceedings of ISIPTA 2005*, Pittsburgh.

Gollier, C. (2011, forthcoming). Does ambiguity aversion reinforce risk aversion? Applications to portfolio choices and asset pricing. *Review of Economic Studies*.

Greenberg, J. (2000). The right to remain silent. *Theory and Decisions, 48*, 193–204.

Guidolin, M., & Rinaldi, F. (2013). Ambiguity in asset pricing and portfolio choice: A review of the literature. *Theory and Decision, 74*(2), 183–217.

Gul, F., & Pesendorfer, W. (2008). *Measurable ambiguity*. Princeton, Working paper.

Hacking, I. (1975). *The emergence of probability*. Cambridge: Cambridge University Press.

Halevy, Y. (2007). Ellsberg revisited: An experimental study. *Econometrica, 75*, 503–536.

Halevy, Y., & Feltkamp, V. (2005). A bayesian approach to uncertainty aversion. *Review of Economic Studies, 72*, 449–466.

Hanany, E., & Klibanoff, P. (2007). Updating preferences with multiple priors. *Theoretical Economics, 2*, 261–298.

Hanany, E., & Klibanoff, P. (2009). Updating ambiguity averse preferences. *The B.E. Journal of Theoretical Economics, 9*(Advances), Article 37.

Hansen, L. P. (2007). Beliefs, doubts, and learning: Valuing macroeconomic risk. *American Economic Review, 97*, 1–30.

Hansen, L. P., & Sargent, T. J. (2001). Robust control and model uncertainty. *American Economic Review, 91*, 60–66.

Hansen, L. P., & Sargent, T. J. (2008). *Robustness*. Princeton: Princeton University Press.

Hansen, L. P., Sargent, T. J., & Tallarini, T. D. (1999). Robust permanent income and pricing. *Review of Economic Studies, 66*(4), 873–907.

Hansen, L. P., Sargent, T. J., & Wang, N. E. (2002). Robust permanent income and pricing with filtering. *Macroeconomic Dynamics, 6*(01), 40–84.

Hart, S., Modica, S., & Schmeidler, D. (1994). A Neo^2 Bayesian foundation of the maxmin value for two-person zero-SUM games. *International Journal of Game Theory, 23*, 347–358.

Harsanyi, J. C. (1967). Games with incomplete information played by "Bayesian" players, I–III Part I. The basic model. *Management Science, INFORMS, 14*(3), 159–182.

Harsanyi, J. C. (1968). Games with incomplete information played by "Bayesian" players Part II. Bayesian equilibrium points. *Management Science, INFORMS, 14*(5), 320–334.

Hayashi, T., & Miao, J. (2011). Intertemporal substitution and recursive smooth ambiguity preferences. *Theoretical Economics, 6*(3), 423–472.

Hurwicz, L. (1951). Some specification problems and application to econometric models. *Econometrica, 19*, 343–344.

Huygens, C. (1657). *De Ratiociniis in Ludo Aleae*. Amsterdam: van Schooten (trans.: E. D. Sylla, *The art of conjecturing*. Johns Hopkins University Press, 2005).

Jaffray, J.-Y. (1988). Application of linear utility theory to belief functions. In *Uncertainty and intelligent systems* (pp. 1–8). Berlin: Springer.

Jaffray, J. Y. (1989). Coherent bets under partially resolving uncertainty and belief functions. *Theory and Decision, 26*(2), 99–105.

Ju, N., & Miao, J. (2012). Ambiguity, learning, and asset returns. *Econometrica, 80*(2), 559–591.

Kahneman, D., & Tversky, A. (1979). Prospect theory: An analysis of decision under risk. *Econometrica, 47*, 263–291.

Kajii, A., & Ui, T. (2006). Agreeable bets with multiple priors. *Journal of Economic Theory, 128*, 299–305.

Kajii, A., & Ui, T. (2009). Interim efficient allocations under uncertainty. *Journal of Economic Theory, 144*, 337–353.

Kass, R. E., & Wasserman, L. (1996). The selection of prior distributions by formal rules. *Journal of the American Statistical Association, 91*, 1343–1370.

Keynes, J. M. (1921). *A treatise on probability*. London: MacMillan.

Keynes, J. M. (1937). *The Quarterly Journal of Economics*. From *The Collected Writings of John Maynard Keynes* (Vol. XIV, pp. 109–123).

Klibanoff, P. (2001a). Stochastically independent randomization and uncertainty aversion. *Economic Theory, 18*, 605–620.

Klibanoff, P. (2001b). Characterizing uncertainty aversion through preference for mixtures. *Social Choice and Welfare, 18*, 289–301.

Klibanoff, P., Marinacci, M., & Mukerji, S. (2005). A smooth model of decision making under ambiguity. *Econometrica, 73*, 1849–1892.

Klibanoff, P., Marinacci, M., & Mukerji, S. (2009). Recursive smooth ambiguity preferences. *Journal of Economic Theory, 144*, 930–976.

Knight, F. H. (1921). *Risk, uncertainty, and profit*. Boston/New York: Houghton Mifflin.

Kobberling, V., & Wakker, P. P. (2003). Preference foundations for nonexpected utility: A generalized and simplified technique. *Mathematics of Operations Research, 28*, 395–423.

Kocherlakota, N. R. (2007). Model fit and model selection. *Federal Reserve Bank of St. Louis Review, 89*, 349–360.

Kopylov, I. (2001). *Procedural rationality in the multiple prior model*. Rochester, Working paper.

Kopylov, I. (2010). *Simple axioms for countably additive subjective probability*. UC Irvine, Working paper.

Kreps, D. M. (1979). A representation theorem for "preference for flexibility". *Econometrica: Journal of the Econometric Society, 47*(3), 565–577.

Kreps, D. (1988). *Notes on the theory of choice* (Underground classics in economics). Boulder: Westview Press.

Laplace, P. S. (1814). *Essai Philophique sur les Probabilites*. Paris: Gauthier-Villars (English ed., 1951, A philosophical essay on probabilities. New York: Dover).

Levi, I. (1974). On indeterminate probabilities. *Journal of Philosophy, 71*, 391–418.

Levi, I. (1980). *The enterprise of knowledge*. Cambridge: MIT.

Lewis, D. (1980). A subjectivist's guide to objective chance. In R. C. Jeffrey (Ed.), *Studies in inductive logic and probability*. Berkeley/Los Angeles: University of California Press.

Lipman, B., & Pesendorfer, W. (2013). Temptation. In D. Acemoglu, M. Arellano, & E. Dekel (Eds.), *Advances in economics and econometrics: Theory and applications*. Cambridge: Cambridge University Press.

Maccheroni, F., & Marinacci, M. (2005). A strong law of large numbers for capacities. *The Annals of Probability, 33*, 1171–1178.

Maccheroni, F., Marinacci, M., & Rustichini, A. (2006a). Ambiguity aversion, robustness, and the variational representation of preferences. *Econometrica, 74*, 1447–1498.

Maccheroni, F., Marinacci, M., & Rustichini, A. (2006b). Dynamic variational preference. *Journal of Economic Theory, 128*, 4–44.

Machina, M. J. (1982). 'Expected Utility' analysis without the independence axiom. *Econometrica, 50*, 277–323.

Machina, M. J. (2004). Almost-objective uncertainty. *Economic Theory, 24*, 1–54.

Machina, M. J. (2005). 'Expected Utility/Subjective Probability' analysis without the sure-thing principle or probabilistic sophistication. *Economic Theory, 26*, 1–62.

Machina, M. J., & Schmeidler, D. (1992). A more robust definition of subjective probability. *Econometrica, 60*, 745–780.

Marinacci, M. (1999). Limit laws for non-additive probabilities and their frequentist interpretation. *Journal of Economic Theory, 84*, 145–195.
Marinacci, M. (2002a). Probabilistic sophistication and multiple priors. *Econometrica, 70*, 755–764.
Marinacci, M. (2002b). Learning from ambiguous urns. *Statistical Papers, 43*, 143–151.
Marinacci, M., & Montrucchio, L. (2004). Introduction to the mathematics of ambiguity. In I. Gilboa (Ed.), *Uncertainty in economic theory*. New York: Routledge.
Miao, J. (2004). A note on consumption and savings under knightian uncertainty. *Annals of Economics and Finance, 5*, 299–311.
Miao, J. (2009). Ambiguity, risk and portfolio choice under incomplete information. *Annals of Economics and Finance, 10*, 257–279.
Miao, J., & Wang, N. (2011). Risk, uncertainty, and option exercise. *Journal of Economic Dynamics and Control, 35*(4), 442–461.
Milnor, J. (1954). Games against nature. In R. M. Thrall, C. H. Coombs, & R. L. Davis (Eds.), *Decision processes*. New York: Wiley.
Montesano, A., & Giovannone, F. (1996). Uncertainty aversion and aversion to increasing uncertainty. *Theory and Decision, 41*, 133–148.
Mukerji, S. (1998). Ambiguity aversion and the incompleteness of contractual form. *American Economic Review, 88*, 1207–1232.
Mukerji, S. (2009). Foundations of ambiguity and economic modelling. *Economics and Philosophy, 25*, 297–302.
Mukerji, S., & Tallon, J.-M. (2001). Ambiguity aversion and incompleteness of financial markets. *Review of Economic Studies, 68*, 883–904.
Mukerji, S., & Tallon, J.-M. (2004). An overview of economic applications of David Schmeidler's models of decision making under uncertainty. In I. Gilboa (Ed.), *Uncertainty in economic theory*. New York: Routledge.
Nakamura, Y. (1990). Subjective expected utility with non-additive probabilities on finite state spaces. *Journal of Economic Theory, 51*, 346–366.
Nau, R. F. (2001, 2006). Uncertainty aversion with second-order utilities and probabilities. *Management Science, 52*, 136–145. (see also *Proceedings of ISIPTA 2001*).
Nau, R. (2011). Risk, ambiguity, and state-preference theory. *Economic Theory, 48*(2–3), 437–467.
Nehring, K. (1999). Capacities and probabilistic beliefs: A precarious coexistence. *Mathematical Social Sciences, 38*, 197–213.
Nehring, K. (2001). Common priors under incomplete information: A unification. *Economic Theory, 18*(3), 535–553.
Nishimura, K., & Ozaki, H. (2004). Search and knightian uncertainty. *Journal of Economic Theory, 119*, 299–333.
Nishimura, K., & Ozaki, H. (2007). Irreversible investment and knightian uncertainty. *Journal of Economic Theory, 136*, 668–694.
Olszewski, W. B. (2007). Preferences over sets of lotteries. *Review of Economic Studies, 74*, 567–595.
Ore, O. (1960). Pascal and the invention of probability theory. *American Mathematical Monthly, 67*, 409–419.
Ortoleva, P. (2010). Status quo bias, multiple priors and uncertainty aversion. *Games and Economic Behavior, 69*, 411–424.
Ozdenoren, E., & Peck, J. (2008). Ambiguity aversion, games against nature, and dynamic consistency. *Games and Economic Behavior, 62*, 106–115.
Pascal, B. (1670). *Pensées sur la Religion et sur Quelques Autres Sujets*.
Pearl, J. (1986). Fusion, propagation, and structuring in belief networks. *Artificial Intelligence, 29*, 241–288.
Pires, C. P. (2002). A rule for updating ambiguous beliefs. *Theory and Decision, 33*, 137–152.
Quiggin, J. (1982). A theory of anticipated utility. *Journal of Economic Behaviorand Organization, 3*, 225–243.

Ramsey, F. P. (1926a). Truth and probability. In R. Braithwaite (Ed.), *The foundation of mathematics and other logical essays*, (1931). London: Routledge and Kegan.
Ramsey, F. P. (1926b). Mathematical logic. *Mathematical Gazette, 13*, 185–194.
Rigotti, L., & Shannon, C. (2005). Uncertainty and risk in financial markets. *Econometrica, 73*, 203–243.
Rigotti, L., Shannon, C., Strzalecki, T. (2008). Subjective beliefs and ex ante trade. *Econometrica, 76*, 1167–1190.
Rosenmueller, J. (1971). On core and value. *Methods of Operations Research, 9*, 84–104.
Rosenmueller, J. (1972). Some properties of convex set functions, Part II. *Methods of Operations Research, 17*, 287–307.
Saito, K. (2015). Preferences for flexibility and randomization under uncertainty. *The American Economic Review, 105*(3), 1246–1271.
Sarin, R., & Wakker, P. P. (1992). A simple axiomatization of nonadditive expected utility. *Econometrica, 60*, 1255–1272.
Sarin, R., & Wakker, P. P. (1998). Dynamic choice and nonexpected utility. *Journal of Risk and Uncertainty, 17*, 87–119.
Savage, L. J. (1954). *The foundations of statistics*. New York: Wiley. (2nd ed. in 1972, Dover)
Schmeidler, D. (1986). Integral representation without additivity. *Proceedings of the American Mathematical Society, 97*, 255–261.
Schmeidler, D. (1989). Subjective probability and expected utility without additivity. *Econometrica, 57*, 571–587. (Working paper, 1982).
Seo, K. (2009). Ambiguity and second-order belief. *Econometrica, 77*, 1575–1605.
Segal, U. (1987). The ellsberg paradox and risk aversion: An anticipated utility approach. *International Economic Review, 28*, 175–202.
Segal, U. (1990). Two-stage lotteries without the reduction axiom. *Econometrica, 58*, 349–377.
Seidenfeld, T., & Wasserman, L. (1993). Dilation for sets of probabilities. *The Annals of Statistics, 21*, 1139–1154.
Shafer, G. (1976). *A mathematical theory of evidence*. Princeton: Princeton University Press.
Shafer, G. (1986). Savage revisited. *Statistical Science, 1*, 463–486.
Shapley, L. S. (1972). Cores of convex games. *International Journal of Game Theory, 1*, 11–26. (Working paper, 1965).
Siniscalchi, M. (2006a). A behavioral characterization of plausible priors. *Journal of Economic Theory, 128*, 91–135.
Siniscalchi, M. (2006b). Dynamic choice under ambiguity. *Theoretical Economics, 6*(3). September 2011.
Siniscalchi, M. (2009a). Vector expected utility and attitudes toward variation. *Econometrica, 77*, 801–855.
Siniscalchi, M. (2009b). Two out of three ain't bad: A comment on 'The ambiguity aversion literature: A critical assessment'. *Economics and Philosophy, 25*, 335–356.
Smith, C. A. B. (1961). Consistency in statistical inference and decision. *Journal of the Royal Statistical Society, Series B, 23*, 1–25.
Stinchcombe, M. (2003). *Choice and games with ambiguity as sets of probabilities*. UT Austin, Working paper.
Strzalecki, T. (2010, forthcoming). Axiomatic foundations of multiplier preferences. *Econometrica*.
Suppe, F. (1977). *The structure of scientific theories*. Champaign: University of Illinois Press.
Treich, N. (2010). The value of a statistical life under ambiguity aversion. *Journal of Environmental Economics and Management, 59*, 15–26.
Tversky, A., & Fox, C. (1995). Weighing risk and uncertainty. *Psychological Review, 102*, 269–283.
Tversky, A., & Kahneman, D. (1992). Advances in prospect theory: Cumulative representation of uncertainty. *Journal of Risk and Uncertainty, 5*, 297–323.
van Fraassen, B. C. (1989). *Laws and symmetry*. Oxford: Oxford University Press.

Viero, M.-L. (2009) Exactly what happens after the Anscombe-Aumann race? Representing preferences in vague environments. *Economic Theory, 41*, 175–212.

von Neumann, J., & Morgenstern, O. (1947). *Theory of games and economic behavior* (2nd ed.). Princeton: Princeton University Press.

Wakker, P. P. (1989a). Continuous subjective expected utility with nonadditive probabilities. *Journal of Mathematical Economics, 18*, 1–27.

Wakker, P. P. (1989b). *Additive representations of preferences: A new foundation of decision analysis*. Dordrecht: Kluwer.

Wakker, P. P. (1990). Characterizing optimism and pessimism directly through comonotonicity. *Journal of Economic Theory, 52*, 453–463.

Wakker, P. P. (1991). Testing and characterizing properties of nonadditive measures through violations of the sure-thing principle. *Econometrica, 69*, 1039–1059.

Wakker, P. P. (2010). *Prospect theory*. Cambridge: Cambridge University Press.

Wald, A. (1950). *Statistical decision functions*. New York: Wiley.

Walley, P. (1991). *Statistical reasoning with imprecise probabilities*. London: Chapman and Hall.

Wang, T. (2003a). *A class of multi-prior preferences*. UBC, Working paper.

Wang, T. (2003b). Conditional preferences and updating. *Journal of Economic Theory, 108*, 286–321.

Welch, B. L. (1939). On confidence limits and sufficiency, and particular reference to parameters of location. *Annals of Mathematical Statistics, 10*, 58–69.

Yaari, M. E. (1969). Some remarks on measures of risk aversion and on their uses. *Journal of Economic Theory, 1*, 315–329.

Yaari, M. E. (1987). The dual theory of choice under risk. *Econometrica, 55*, 95–115.

Zhang, J. (2002). Subjective ambiguity, expected utility, and choquet expected utility. *Economic Theory, 20*, 159–181.

Chapter 22
State-Dependent Utilities

Mark J. Schervish, Teddy Seidenfeld, and Joseph B. Kadane

Introduction

Expected utility theory is founded upon at least one of several axiomatic derivations of probabilities and utilities from expressed preferences over acts, Savage (1954), deFinetti (1974), Anscombe and Aumann (1963), and Ramsey (1926). These theories provide for the simultaneous existence of a unique personal probability over the states of nature and a unique (up to positive affine transformations) utility function over the prizes such that the ranking of acts is by expected utility. For example, suppose that there are n states of nature which form the set $S = \{s_1, \ldots, s_n\}$ and m prizes in the set $Z = \{z_1, \ldots, z_m\}$. An example of an act is a function f mapping S to Z. That is, if $f(s_i) = z_j$, then we receive prize z_j if state s_i occurs. (We will consider more complicated acts than this later.) Now, suppose that

This research was reported, in part, at the Indo-United States Workshop on Bayesian Analysis in Statistics and Econometrics. The research was supported by National Science Foundation grants DMS-8805676 and DMS-8705646, and Office of Naval Research contract N00014-88-K0013. The authors would like to thank Morris DeGroot, Bruce Hill, Irving LaValle, Isaac Levi, and Herman Rubin for helpful comments during the preparation of this paper. We especially thank the associate editor for the patience and care that was given to this submission.

M.J. Schervish (✉)
Department of Statistics, Carnegie Mellon University, Pittsburgh, PA 15213, USA
e-mail: mark@cmu.edu

T. Seidenfeld
Departments of Philosophy and Statistics, Carnegie Mellon University, Pittsburgh, PA 15213, USA

J.B. Kadane
Departments of Statistics and Social and Decision Sciences, Carnegie Mellon University, Pittsburgh, PA 15213, USA

H. Arló-Costa et al. (eds.), *Readings in Formal Epistemology*, Springer Graduate Texts in Philosophy 1, DOI 10.1007/978-3-319-20451-2_22

there is a probability over the states such that $p_i = \Pr(s_i)$ and that there is a utility U over prizes. To say that acts are ranked by expected utility means that we strictly prefer act g to act f if and only if

$$\sum_{i=1}^{n} p_i U(f(s_i)) < \sum_{i=1}^{n} p_i U(g(s_i)). \tag{22.1}$$

If we allow the utilities of prizes to vary conditionally on which state of nature occurs, we can rewrite (22.1) as

$$\sum_{i=1}^{n} p_i U_i(f(s_i)) < \sum_{i=1}^{n} p_i U_i(g(s_i)), \tag{22.2}$$

where $U_i(z_j)$ is the utility of prize z_j given that state s_i occurs. However, without restrictions on the degree to which U_i can differ from $U_{i'}$ for $i \neq i'$, the uniqueness of the personal probability no longer holds. For example, let $q_1, \ldots, q_n$ be another probability over the states such that $p_i > 0$ if and only if $q_i > 0$. Then, for an arbitrary act f, $\sum_{i=1}^{n} q_i V_i(f(s_i)) = \sum_{i=1}^{n} p_i U_i(f(s_i))$, where $V_i(\cdot) = p_i U_i(\cdot)/q_i$ when $q_i > 0$ (V_i can be arbitrary when $q_i = 0$.) In this case, it is impossible to determine an agent's personal probability by studying the agent's preferences for acts. Rubin (1987) notes this fact and develops an axiom system that does not lead to a separation of probability and utility. Arrow (1974) considers the problem for insurance. A footnote in Arrow (1974) credits Herman Rubin with raising this same issue in an unpublished 1964 lecture.

DeGroot (1970) begins his derivation of expected utility theory by assuming that the concept of "at least as likely as" is an undefined primitive. This allows a construction of probability without reference to preferences. However, DeGroot also needs to introduce preferences among acts in order to derive a utility function. In section "State-Independent Utility", we will examine the axiomatization of VonNeumann and Morgenstern (1947) together with the extension of Anscombe and Aumann (1963) to see how it attempts to avoid the non-uniqueness problem just described. In section "Savage's Postulates", we look at the system of Savage (1954) with the same goal in mind. In section "An Example", we give an example that illustrates that the problem can still arise despite the best efforts of those who have derived the theories. This example leads to a critical examination of the theory of deFinetti (1974) in section "deFinetti's Gambling Approach". While reviewing an example from Savage (1954) in section "Savage's 'Small Worlds' Example", we see how close Savage was to discovering the non-uniqueness problem in connection with his own theory. In section "How to Elicit Unique Probabilities and Utilities Simultaneously", we describe a method for obtaining a unique personal probability and state-dependent utility based on a proposal of Karni et al. (1983).

State-Independent Utility

Following VonNeumann and Morgenstern (1947), we generalize the concept of act introduced in section "Introduction" by allowing randomization. That is, suppose that the agent is comfortable declaring probabilities for an auxiliary experiment the results of which he/she believes would, in no way, alter his/her preferences among acts. Furthermore, assume that this auxiliary experiment has events with arbitrary probabilities (for example, it may produce a random variable with continuous distribution). Define a *lottery* as follows. If $A_1, \ldots, A_m$ is a partition of the possible outcomes of the auxiliary experiment with $\alpha_j = \Pr(A_j)$ for each j, then the lottery $(\alpha_1, \ldots, \alpha_m)$ awards prize z_j if A_j occurs. We assume that the choice of the partition events $A_1, \ldots, A_m$ does not affect the lottery. That is, if $B_1, \ldots, B_m$ is another partition such that $\Pr(B_j) = \alpha_j$ for each j also, then the lottery that awards prize z_j when B_j occurs is, to the agent, *the same lottery* as the one described above in terms of the A_j. In this way, a lottery is just a simple probability distribution over prizes that is independent of the state of nature. Any two lotteries that award the prizes with the same probabilities are considered the same lottery. If L_1 and L_2 are two lotteries $(\alpha_1, \ldots, \alpha_m)$ and $(\beta_1, \ldots, \beta_m)$ respectively, then for $0 \leq \lambda \leq 1$, we denote by $\lambda L_1 + (1-\lambda)L_2$ the lottery $(\lambda\alpha_1 + (1-\lambda)\beta_1, \ldots, \lambda\alpha_m + (1-\lambda)\beta_m)$.

If we consider only lotteries, we can introduce some axioms for preferences among lotteries. For convenience, we will henceforth assume that there exist two lotteries such that the agent has a strict preference for one over the other. Otherwise, the preference relation is trivial and no interesting results are obtained.

Axiom 1 (Weak Order). There is a weak order $\preceq$ among lotteries such that $L_1 \preceq L_2$ if and only if L_1 is not strictly preferred to L_2.

This axiom requires that weak preference among lotteries be transitive, reflexive, and connected. If we define *equivalence* to mean "no strict preference in either direction", then equivalence is transitive also.

Definition 1. *Assuming Axiom 1, we say that* L_1 is equivalent to L_2 *(denoted* $L_1 \sim L_2$*) if* $L_1 \preceq L_2$ *and* $L_2 \preceq L_1$*. We say* L_2 *is* strictly preferred to L_1 *(denoted* $L_1 \prec L_2$*) if* $L_1 \preceq L_2$ *but not* $L_2 \preceq L_1$.

The axiom that does most of the work is one that entails stochastic dominance.

Axiom 2 (Independence). For each L, L_1, L_2, and $0 < \alpha < 1$, $L_1 \preceq L_2$ if and only if $\alpha L_1 + (1-\alpha)L \preceq \alpha L_2 + (1-\alpha)L$.

A third axiom is often introduced to guarantee that utilities are real valued.

Axiom 3 (Archimedean). If $L_1 \prec L_2 \prec L_3$, then there exists $0 < \alpha < 1$ such that $L_2 \sim (1-\alpha)L_1 + \alpha L_3$.

Axiom 3 prevents L_3 from being infinitely better than L_2 and it prevents L_1 from being infinitely worse than L_2.

With axioms equivalent to the three above, VonNeumann and Morgenstern (1947) prove that there exists a utility over prizes U such that $(\alpha_1, \ldots, \alpha_m) \preceq (\beta_1, \ldots, \beta_m)$ if and only if $\sum_{i=1}^{m} \alpha_i U(z_i) \leq \sum_{i=1}^{m} \beta_i U(z_i)$. This utility is unique up to positive affine transformation. In fact, it is quite easy (and useful for the example in section "An Example") to construct the utility function from the stated preferences. Pick an arbitrary pair of lotteries L_0 and L_1 such that $L_0 \prec L_1$. Assign these lotteries $U(L_0) = 0$ and $U(L_1) = 1$. For all other lotteries L, the utilities are assigned as follows. If $L_0 \preceq L \preceq L_1$, $U(L)$ is that α such that $(1-\alpha)L_0 + \alpha L_1 \sim L$. If $L \prec L_0$, then $U(L) = -\alpha/(1-\alpha)$, where $(1-\alpha)L + \alpha L_1 \sim L_0$ (hence $\alpha \neq 1$.) If $L_1 \prec L$, then $U(L) = 1/\alpha$, where $(1-\alpha)L_0 + \alpha L \sim L_1$. The existence of these α values is guaranteed by Axiom 3 and their uniqueness follows from Axiom 2.

To be able to handle acts in which the prizes vary with the state of nature, Anscombe and Aumann (1963) introduced a fourth axiom, which was designed to say that the preferences among prizes did not vary with the state. Before stating this axiom, we introduce a more general act, known as a *horse lottery*.

Definition 2. *A function mapping states of nature to lotteries is called a* horse lottery.

That is, if H is a horse lottery such that $H(s_i) = L_i$ for each i, then, if state s_i occurs, the prize awarded is the prize that lottery L_i awards. The L_i can be all different or some (or all) of them can be the same. If $H(s_i) = L$ for all i, then we say $H = L$. If H_1 and H_2 are two horse lotteries such that $H_j(s_i) = L_i^{(j)}$ for each i and j, and $0 \leq \alpha \leq 1$, then we denote by $\alpha H_1 + (1-\alpha)H_2$ the horse lottery H such that $H(s_i) = \alpha L_1^{(i)} + (1-\alpha)L_2^{(i)}$ for each i. Axioms 1 and 2, when applied to preferences among horse lotteries, imply that the choice of an act has no effect on the probabilities of the state of nature.

Definition 3. *A state of nature s_i is called* null *if, for each pair of horse lotteries H_1 and H_2 satisfying $H_1(s_j) = H_2(s_j)$ for all $j \neq i$, $H_1 \sim H_2$. A state is called* non-null *if it is not null.*

Axiom 4 (State-independence). For each non-null state s_i, each pair of lotteries (L_1, L_2), and each pair of horse lotteries H_1 and H_2 satisfying $H_1(s_j) = H_2(s_j)$ for $j \neq i$, $H_1(s_i) = L_1$, and $H_2(s_i) = L_2$, we have $L_1 \prec L_2$ if and only if $H_1 \prec H_2$.

Axiom 4 says that a strict preference between two lotteries is reproduced for every pair of horse lotteries that differ only in some non-null state, and their difference in that state is that each of them equals one of the two lotteries. With this setup, Anscombe and Aumann (1963) prove the following theorem.

Theorem 1 (Anscombe and Aumann). *Under Axioms 1, 2, 3, and 4, there exist a unique probability P over the states and utility U over prizes (unique up to positive affine transformation) such that $H_1 \preceq H_2$ if and only if $\sum_{i=1}^{n} P(s_i)U(H_1(s_i)) \leq \sum_{i=1}^{n} P(s_i)U(H_2(s_i))$, where, for each lottery $L = (\alpha_1, \ldots, \alpha_m)$, $U(L)$ stands for $\sum_{j=1}^{m} \alpha_j U(z_j)$.*

Even when the four axioms above hold, there is no requirement that the utility function U be the same conditional on each state of nature. As we did when we constructed Eq. (22.2), we could allow $U_i(z_j) = a_i U(z_j) + b_i$ where each $a_i > 0$. Then we could let $Q(s_i) = a_i P(s_i) / \sum_{k=1}^{n} a_k P(s_k)$. It would now be true that $H_1 \preceq H_2$ if and only if $\sum_{i=1}^{n} Q(s_i) U_i(H_1(s_i)) \leq \sum_{i=1}^{n} Q(s_i) U_i(H_2(s_i))$. The uniqueness of the probability in Theorem 1 depends on the use of a state-independent utility U. Hence, one cannot determine an agent's probability from the agent's stated preferences unless one assumes that the agent's utility is state-independent. This may not seem like a serious difficulty when Axiom 4 holds. However, as we will see in section "An Example", the problem is more complicated.

Savage's Postulates

Savage (1954) gives a set of postulates that does not rely on an auxiliary randomization in order to extract probabilities and utilities from preferences. Rather, the postulates rely on the use of prizes that can be considered as "constant" across states. Savage's most general acts are functions from states to prizes. Because he does not introduce an auxiliary randomization, he requires that there be infinitely many states. The important features of Savage's theory, for this discussion, are the first three postulates and a few definitions. Some of the axioms and definitions are stated in terms of *events*, which are sets of states. The postulates of Savage are consistent with the axioms of section "State-Independent Utility" in that they both provide models for preference by maximizing expected utility.

The first postulate of Savage is the same as Axiom 1. The second postulate requires a definition of conditional preference.

Definition 4. *Let B be an event. We say that $f \preceq g$* given *B if and only if*

- *$f' \preceq g'$ for each pair f' and g' such that $f'(s) = f(s)$ for all $s \in B$, $g'(s) = g(s)$ for all $s \in B$, and $f'(s) = g'(s)$ for all $s \notin B$.*
- *and $f' \preceq g'$ for every such pair or for none.*

The second postulate is an analog of Axiom 2. (See Fisburn 1970, p. 193.)

Postulate 2. *For each pair of acts f and g and each event B, either $f \preceq g$ given B or $g \preceq f$ given B.*

Savage has a concept of *null event* that is similar to the concept of null state from Definition 3.

Definition 5. *An event B is* null *if, for every pair of acts f and g, $f \preceq g$ given B. An event B is* non-null *if it is not null.*

The third postulate of Savage concerns acts that are constant such as $f(s) = z$ for all s, where z is a single prize. For convenience, we will call such an act f by the name z also.

Postulate 3. *For each non-null event B and each pair of prizes z_1 and z_2 (considered as constant acts), $z_1 \preceq z_2$ if and only if $z_1 \preceq z_2$ given B.*

Savage's definition of probability relies on Postulate 3.

Definition 6. *Suppose that A and B are events. We say that A* is at least as likely as *B if, for each pair of prizes z and w with $z \prec w$ we have $f_B \preceq f_A$, where $f_A(s) = w$ if $s \in A$, $f_A(s) = z$ if $s \notin A$, $f_B(s) = w$ if $s \in B$, and $f_B(s) = z$ if $s \notin B$.*

Postulate 2 guarantees that, with f_A and f_B as defined in Definition 6, either $f_B \preceq f_A$ no matter which pair of prizes z and w one chooses (so long as $z \prec w$) or $f_A \preceq f_B$ no matter which pair of prizes one chooses.

Postulate 3 says that the *relative* values of prizes cannot change between states. Savage (1954, p. 25) suggests that problems in locating prizes which satisfy this postulate might be solved by a clever redescription. For example, rather than describing prizes as "receiving a bathing suit" and "receiving a tennis racket" (whose relative values change depending on which of the two states "picnic at the beach" or "picnic in the park" occurs), Savage suggests that the prizes might be "a refreshing swim with friends," "sitting alone on the beach with a tennis racket," etc. However, we do not see how to carry out such redescriptions while satisfying Savage's structural assumption that each prize is available as an outcome under each state. (What does it mean to receive the prize "sitting alone on the beach with a tennis racket" when the state "picnic in the park" occurs?)

Our problem, however, is deeper than this. Definition 6 assumes that the *absolute* values of prizes do not change from state to state. For example, suppose that A and B are disjoint and the value of z is 1 for the states in A and 2 for the states in B. Similarly, suppose that the value of w is 2 for the states in A and 4 for the states in B. Then, even if A is more likely than B, but is not twice as likely, we would get $f_A \prec f_B$, and we would conclude, by Definition 6, that B is more likely than A. The example in section "An Example" (using just one of the currencies), as well as our interpretation of Savage's "small worlds" problem (in section "Savage's 'Small Worlds' Example") suggest that it might be very difficult to find prizes with the property that their "absolute" values do not change from state to state even though their "relative" values remain the same from state to state.

deFinetti's Gambling Approach

deFinetti (1974) assumes that there is a set of prizes with numerical values such that utility is linear in the numerical value. That is, a prize numbered 4 is worth twice as much as a prize numbered 2. More specifically, to say that utility is linear in the numerical values of prizes, we mean the following. For each pair of prizes, (z_1, z_2) with $z_1 < z_2$, and each $0 \le \alpha \le 1$, the lottery that pays z_1 with probability $1 - \alpha$ and pays z_2 with probability α (using the auxiliary randomization of section "State-Independent Utility") is equivalent to the lottery that pays prize

$(1 - \alpha)z_1 + \alpha z_2$ for sure. Using such a set of prizes, deFinetti supposes that an agent will accept certain gambles that pay these prizes. If f is an act, to gamble on f means to accept a contract that pays the agent the prize $c(f(s) - x)$ when state s occurs, where c and x are some values. A negative outcome means that the agent has to pay out, while a positive outcome means that the agent gains some amount.

Definition 7. *The* prevision *of an act f is the number x that one would choose so that all gambles of the form $c(f - x)$ would be accepted, for all small values of c, both positive and negative.*

If an agent is willing to gamble on each of several acts, then it is assumed that the agent will also gamble on them simultaneously. (For a critical discussion of this point, see Kadane and Winkler (1988) and Schick (1986).)

Definition 8. *A collection of previsions for acts is* coherent *if, for each finite set of the acts, say $f_1, \ldots, f_n$ with previsions $x_1, \ldots, x_n$ respectively, and each set of numbers $c_1, \ldots, c_n$, we have $\sup_{all\, s} \sum_{i=1}^{n} c_i(f_i(s) - x_i) \geq 0$. Otherwise, the previsions are* incoherent.

deFinetti (1974) proves that a collection of previsions of bounded acts is coherent if and only if there exists a finitely additive probability such that the prevision of each act is its expected value. This provides a method of eliciting probabilities by asking an agent to specify previsions for acts such as $f(s) = 1$ if $s \in A$ and $f(s) = 0$ if $s \notin A$. The prevision of such an act f would be its probability if the previsions are coherent. As plausible as this sounds, the example below casts doubt on the ability of deFinetti's program to elicit probabilities accurately.

An Example

Let the set of available prizes be various amounts of Dollars. We suppose that there are three states of nature, which we will describe in more detail later, and we suppose that the agent expresses preferences that satisfy the axioms of section "State-Independent Utility" and the postulates of Savage (1954). Furthermore, suppose that the agent's utility for money is linear. That is, for each state i, $U_i(\$cx) = cU_i(\$x)$. In particular, $U_i(\$0) = 0$. Now, we offer the agent three horse lotteries H_1, H_2, and H_3 whose outcomes are

	State of Nature		
	s_1	s_2	s_3
H_1	\$1	\$0	\$0
H_2	\$0	\$1	\$0
H_3	\$0	\$0	\$1

Suppose that the agent claims that these three horse lotteries are equivalent. If we assume that the agent has a state-independent utility, the expected utility of H_i is $U(\$1)P(s_i)$. It follows from the fact that the three horse lotteries are equivalent, that $P(s_i) = 1/3$ for each i.

Next, we alter the set of prizes to be various Yen amounts (the Japanese currency). Suppose that we offer the agent three Yen horse lotteries H_4, H_5, and H_6 whose outcomes are

State of Nature

	s_1	s_2	s_3
H_4	100Y	0Y	0Y
H_5	0Y	125Y	0Y
H_6	0Y	0Y	150Y

If the agent were to claim that these three horse lotteries were equivalent, and if we assumed that the agent used a state-independent utility for Yen prizes, then $P(s_1)U(100\text{Y}) = P(s_2)U(125\text{Y}) = P(s_3)U(150\text{Y})$. Supposing that the agent's utility is linear in Yen, as it was in dollars, we conclude that $P(s_1) = 1.25P(s_2) = 1.5P(s_3)$. It follows that $P(s_1) = .4054$, $P(s_2) = .3243$, and $P(s_3) = .2703$. It would seem incoherent for the agent to express both sets of equivalences since it appears that the agent is now committed to two different probability distributions over the three states. This is not correct, however, as we now see.

Suppose that the three states of nature represent three different exchange rates between Dollars and Yen. $s_1 = \{\$1 \text{ is worth } 100\text{Y}\}$, $s_2 = \{\$1 \text{ is worth } 125\text{Y}\}$, and $s_3 = \{\$1 \text{ is worth } 150\text{Y}\}$. Suppose further that the agent can change monetary units at the prevailing rate of exchange without any penalty. As far as this agent is concerned, H_i and H_{3+i} are worth exactly the same for $i = 1, 2, 3$ since, in each state the prizes they award are worth the same amount. The problem that arises in this example is that the two probability distributions were constructed under incompatible assumptions. The discrete uniform probability was constructed under the assumption that $U(\$1)$ is the same in all three states, while the other probability was constructed under the assumption that $U(100\text{Y})$ was the same in all three states. Clearly these cannot both be true given the nature of the states. What saves both Theorem 1 and Savage's theory is that preference can be represented by expected utility *no matter which of the two assumptions one makes*. Unfortunately, this same fact makes the uniqueness of the probability relative to the choice of which prizes count as constants in terms of utility. There are two different representations of the agent's preferences by probability and state-independent utility. But what is state-independent in one representation is state-dependent in the other.

If we allow both types of prizes at once, we can calculate the marginal exchange rate for the agent. That is, we can ask, "For what value x will the agent claim that \$1 and xY are equivalent?" This question can be answered using either of the two probability-utility representations and the answers will be the same. First, with

Dollars having constant value, the expected utility of a horse lottery paying \$1 in all three states is $U(\$1)$. The expected value of the horse lottery paying xY in all three states is

$$\frac{U_1(x\text{Y}) + U_2(x\text{Y}) + U_3(x\text{Y})}{3} = \frac{1}{3}\left(\frac{x}{100}U(\$1) + \frac{x}{125}U(\$1) + \frac{x}{150}U(\$1)\right)$$
$$= .008222xU(\$1),$$

using the linearity of utility and the state-specific exchange rates. By setting this expression equal to $U(\$1)$, we obtain that $x = 121.62$. Equivalently, we can calculate the exchange rate assuming that Yen have constant value over states. The act paying xY in all states has expected utility $U(x\text{Y}) = .01xU(100\text{Y})$. The act paying \$1 in all states has expected utility

$$.4054U_1(\$1) + .3243U_2(\$1) + .2703U_3(\$1)$$
$$= .4054U(100\text{Y}) + .3243U(125\text{Y}) + .2703U(150\text{Y})$$
$$= U(100\text{Y})[.4054 + .3243 \times 1.25 + .2703 \times 1.5]$$
$$= 1.2162U(100\text{Y}).$$

Setting this equal to $.01xU(100\text{Y})$ yields $x = 121.62$, which is the same exchange rate as calculated earlier.

The implications of this example for elicitation are staggering. Suppose we attempt to elicit the agent's probabilities over the three states by offering acts in Dollar amounts using deFinetti's gambling approach from section "deFinetti's Gambling Approach". The agent has utility that is linear in both Dollars and Yen without reference to the states, hence deFinetti's program will apply. To see this, select two prizes, such as \$0 and \$1, to have utilities 0 and 1 respectively. Then, for $0 < x < 1$, $U(\$x)$ must be the value c that makes the following two lotteries equivalent: $L_1 = \$x$, for certain and $L_2 = \$1$ with probability c and \$0 with probability $1 - c$. Assuming that Dollars have constant utility, it is obvious that $c = x$. Assuming that Yen have constant utility, the expected utility of L_1 is $1.2162xU(100\text{Y})$ and the expected utility of L_2 is $cU(121.62Y)$. These two are the same if and only if $x = c$. A similar argument works for x not between 0 and 1, and a similar argument works when the two prizes with utilities 0 and 1 are Yen prizes. Now, suppose that the agent actually uses the state-independent utility for Dollars and the discrete uniform distribution to rank acts, but the eliciter does not know this. The eliciter will try to elicit the agent's probabilities for the states by offering gambles in Yen (linear in utility). For example, the agent claims that the gamble $c(f - 40.54)$ would be accepted for all small values of c, where $f(s) = 150\text{Y}$ if $s = s_3$ and equals 0Y otherwise. The reason for this is that, since 150Y equals \$1 when s_3 occurs, the winnings are \$1 when s_3 occurs, which has probability 1/3. The marginal exchange rate is 121.62Y for \$1, so the appropriate amount to pay (no matter which state occurs), in order to win \$1 when s_3 occurs, is \$1/3, which equals $121.62\text{Y}/3 = 40.54\text{Y}$. Realizing that utility is linear in Yen, the eliciter now decides that $\Pr(s_3)$ must equal $40.54/150 = .2703$. Hence, the eliciter elicits the wrong probability, even though the agent is coherent!

The expressed preferences satisfy the four axioms of section "State-Independent Utility", all of Savage's postulates, and deFinetti's linearity condition, but we are still unable to determine the probabilities of the states based only on preferences. The problem becomes clearer if we allow both Dollar and Yen prizes at the same time. Now, it is impossible for a single utility to be state-independent for all prizes. That is, Axiom 4 and Postulate 3 would no longer hold. Things are more confusing in deFinetti's framework, because there is no room for state-dependent utilities. The agent would appear to have two different probabilities for the same event even though there would be no incoherency.

Savage's 'Small Worlds' Example

In Section 5.5 of Savage (1954), the topic of *small worlds* is discussed. An anomaly occurs in this discussion, and Savage seems to imply that it is an effect of the construction of the small world. In this section, we briefly introduce small worlds and then explain why we believe that the anomaly discovered by Savage is actually another example of the non-uniqueness illustrated in section "An Example". The fact that it arose in the discussion of small worlds is a mere coincidence. We show how precisely the same effect arises without any mention of small worlds.

A small world can be thought of as a description of the states of nature in which each state can actually be partitioned into several smaller states, but we don't actually do the partitioning when making comparisons between acts. For a mathematical example, Savage mentions the following case. Consider the unit square $S = \{(x, y) : 0 \leq x, y \leq 1\}$ as the finest possible partition of the states of nature. Suppose, however, that we consider as states the subsets $\overline{x} = \{(x, y) : 0 \leq y \leq 1\}$ for each $x \in [0, 1]$. The problem that Savage discovers in this example is the following. It is possible to define small world prizes in a natural way and for preferences among small world acts to satisfy all of his axioms and, at the same time, consistently define prizes in the "grand world" consisting of the whole square S. However, it is possible for the preferences among small world acts to be consistent with the preferences among grand world acts in such a way that the probability measure determined from the small world preferences is not the marginal probability measure over the sets $\overline{x}$ induced from the grand world probability. As we will see, the problem that Savage discovers is due to using different prizes as constants in the two problems. It is not due to the small world but actually will appear in the grand world as well.

Any grand world act can be considered a small world prize. In fact, the very reason for introducing small worlds is to deal with the case in which what we count as a prize turns out to actually be worth different amounts depending on which of the subdivisions of the small world state of nature occurs. So, suppose we let the grand world prizes be non-negative numbers and the grand world acts be all bounded measurable functions on S. The grand world probability is uniform over the square and the grand world utility is the numerical value of the prize. In order to guarantee that Savage's axioms hold in the small world, choose the small world prizes to be 0 and positive multiples of a single function h. Assuming that $U(h) = 1$, the small world probability of a set $\overline{B} = \{\overline{x} : x \in B\}$ is (from p. 89 of Savage 1954) $Q(\overline{B}) = \int_{\overline{B}} q(x)dx$, where

$$q(x) = \frac{\int_0^1 h(x, y)dy}{\int_0^1 \int_0^1 h(x, y)dydx}. \tag{22.3}$$

Unless $\int_0^1 h(x, y)dy$ is constant as a function of x, Q will not be the marginal distribution induced from the uniform distribution over S. However, even if $\int_0^1 h(x, y)dy$ is not constant, the ranking of small world acts is consistent with the ranking of grand world acts. Let $ch(\cdot, \cdot)$, considered as a small world prize, be denoted $\overline{c}$. Let $U(\overline{c}) = c$ denote the small world utility of small world prize $\overline{c}$. If $\overline{f}$ is a small world act, then for each $\overline{x}$, $\overline{f}(\overline{x}) = \overline{c}$ for some c. The expected small world utility of $\overline{f}$ is $\int_0^1 U(\overline{f}(\overline{x}))q(x)dx$. Let the grand world act f corresponding to $\overline{f}$ be defined by $f(x, y) = \overline{f}(\overline{x})h(x, y)$. It follows from (22.3) that $U(\overline{f}(\overline{x}))q(x) = \int_0^1 f(x, y)dy / \int_0^1 \int_0^1 h(x, y)dydx$. Hence, the expected small world utility of $\overline{f}$ is

$$\int_0^1 \frac{\int_0^1 f(x, y)dy}{\int_0^1 \int_0^1 h(x, y)dydx} dx,$$

which is just a constant times the grand world expected utility of f. Hence, small world acts are ranked in precisely the same order as their grand world counterparts, even though the small world probability is not consistent with the grand world probability.

We claimed that the inconsistency of the two probabilities is due to the choice of "constants" and not to the small worlds. To see this, let the grand world constants be 0 and the positive multiples of h. Then an act f in the original problem becomes an act f^* with $f^*(x, y) = f(x, y)/h(x, y)$. That is, the prize that f^* assigns to (x, y) is the number of multiples of $h(x, y)$ that $f(x, y)$ is. We define the new probability, for B a two-dimensional Borel set, $R(B) = \int_B h(x, y)dydx / \int_S h(x, y)dydx$. The expected utility of f^* is now $\int_S f^*(x, y)h(x, y)dydx / \int_S h(x, y)dydx = \int_S f(x, y)dydx / \int_S h(x, y)dydx$. This is just a constant times the original expected utility. Hence, acts are ranked in the same order by both probability-utility representations. Both representations are state-independent, but each one is relative to a different choice of constants. The constants in one representation have different utilities in different states in the other representation. Both representations satisfy Savage's axioms, however. (Note that the small world probability constructed earlier is the marginal probability associated with the grand world probability R, so that Savage's small world problem evaporates when the definition of constant is allowed to change.) The point to remember is that the uniqueness of the probability-utility representation for a collection of preferences is relative to what counts as a constant. To use Savage's notation in the example of section "An Example", suppose that we use Yen gambles to elicit probabilities. However, instead of treating multiples of 1Y as constants, we treat multiples of gamble $f(s_1) = 100\text{Y}, f(s_2) = 125\text{Y}, f(s_3) = 150\text{Y}$ as constants. Then we will elicit the discrete uniform probability rather than the non-uniform probability.

How to Elicit Unique Probabilities and Utilities Simultaneously

There is one obvious way to avoid the confusion of the previous examples. That would be to elicit a unique probability without reference to preferences. This is the approach taken by DeGroot (1970). This approach requires that the agent have an understanding of the primitive concept "at least as likely as" in addition to the more widely understood primitive "is preferred to". Some decision theorists prefer to develop the theory solely from preference without reference to the more statistical primitive "at least as likely as". It is these latter decision theorists who need an alternative to the existing theories in order to separate probability from utility.

Karni et al. (1983) (see also Karni 1985) propose a scheme for simultaneously eliciting probability and state-dependent utility. Their scheme is essentially as follows. In addition to preferences among horse lotteries, an agent is asked to state preferences among horse lotteries under the assumption that the agent holds a particular probability distribution over the states (explicitly, they say on p. 1024, "...contingent upon a strictly positive probability distribution p' on S".) And they require the agent to compare acts with different "contingent" probabilities as well. Karni (1985) describes these (in a slightly more general setting) as *prize-state lotteries* which are functions $\hat{f}$ from $Z \times S$ to $\Re^+$ such that $\sum_{\text{all }(z,s)} \hat{f}(z, s) = 1$, and such that the probability $\hat{f}(z, s)$ for each z and s is to be understood in the same sense as the probabilities involved in the lotteries of section "State-Independent Utility". That is, the results of a prize-state lottery are determined by an auxiliary randomization. The agent is asked to imagine that the state of nature could be chosen by the randomization scheme rather than by the forces of nature. This is intended to remove the uncertainty associated with how the state of nature is determined so that a pure utility can be extracted using Axioms 1, 2, and 3 applied to a preference relation among prize-state lotteries.

For example, suppose that the agent in section "An Example" expresses a strict preference for the prize-state lottery that awards \$1 in state 2 with probability 1 ($\hat{f}(\$1, s_2) = 1$) over $\hat{g}(\$1, s_1) = 1$. This preference would not be consistent with a state-independent utility for dollar prizes, however it would be consistent with a state-independent utility in Yen prizes.

The pure utility elicited in this fashion is a function of both prizes and states, so that it is actually a state-dependent utility. So long as the preferences among prize-state lotteries are consistent with the preferences among horse lotteries, the elicited state-dependent utility can then be assumed to be the agent's utility. There will then be a unique probability such that $H_1 \preceq H_2$ if and only if the expected utility of H_1 is at most as large as the expected utility of H_2. The type of consistency that (Karni et al. 1983) require between the two sets of preferences is rather more complicated than it needs to be. The following simple consistency axiom will suffice.

Axiom 5 (Consistency). For each non-null state s and each pair $(\hat{f}_1, \hat{f}_2)$ of prize-state lotteries satisfying $\sum_{\text{all } z} \hat{f}_i(z, s) = 1$, and some pair of horse lotteries H_1 and H_2 satisfying $H_1(s_i) = H_2(s_i)$ for all $s_i \neq s$ and $H_1(s) = f_1$ and $H_2(s) = f_2$, we have $H_1 \preceq H_2$ if and only if $\hat{f}_1 \preceq \hat{f}_2$, where f_1 and f_2 are lotteries that correspond to $\hat{f}_1$ and $\hat{f}_2$ as follows: $f_i = (\hat{f}_i(z_1, s), \ldots, \hat{f}_i(z_m, s))$, $i = 1, 2$, in the notation of section "State-Independent Utility".

All that Axiom 5 says is that preferences among prize-state lotteries with all of their probabilities on the same state must be reproduced as preferences between horse-lotteries which differ only in that common state.

Theorem 2. *Suppose that there are n states of nature and m prizes. Assume that preferences among horse lotteries satisfy Axioms 1, 2, and 3. Also assume that preferences among prize-state lotteries satisfy Axioms 1, 2, and 3. Finally, assume that Axiom 5 holds. Then there exists a unique probability P over the states and a utility* $U : Z \times S \to \Re$, *unique up to positive affine transformation, satisfying*

1. $H_1 \preceq H_2$ *if and only if* $\sum_{i=1}^{n} P(s_i) U(H_1(s_i), s_i) \leq \sum_{i=1}^{n} P(s_i) U(H_2(s_i), s_i)$, *where, for each lottery* $L = (\alpha_1, \ldots, \alpha_m)$, $U(L, s_i)$ *stands for* $\sum_{j=1}^{m} \alpha_j U(z_j, s_i)$,
2. $\hat{f} \preceq \hat{g}$ *if and only if* $\sum_{i=1}^{n} \sum_{j=1}^{m} \hat{f}(z_j, s_i) U(z_j, s_i) \leq \sum_{i=1}^{n} \sum_{j=1}^{m} \hat{g}(z_j, s_i) U(z_j, s_i)$.

The proof of Theorem 2 makes use of the following theorem from Fisburn (1970, p. 176):

Theorem 3 (Fishburn). *Under Axioms 1, 2, and 3, there exist real-valued functions* $W_1, \ldots, W_n$ *such that* $H_1 \preceq H_2$ *if and only if*

$$\sum_{i=1}^{n} W_i(H_1(s_i)) \leq \sum_{i=1}^{n} W_i(H_2(s_i)), \tag{22.4}$$

and the W_i *that satisfy (22.4) are unique up to a similar positive linear transformations, with* W_i *constant if and only if* s_i *is null.*

We provide only a sketch of the proof of Theorem 2. Let $(W_1, \ldots, W_n)$ be the state-dependent utility for horse lotteries guaranteed by Theorem 3, and let $\hat{V}$ be the utility for prize-state lotteries guaranteed by the theorem of VonNeumann and Morgenstern (1947). All we need to show is that there exist $c_1, \ldots, c_n$ and positive $a_1, \ldots, a_n$ such that for each $i = 1, \ldots, n$,

$$W_i(z) = a_i \hat{V}(z, s_i) + c_i, \text{ for all } z. \tag{22.5}$$

If (22.5) were true, then it follows directly from (22.4) that $U = \hat{V}$ would serve as the state-dependent utility and $P(s_i) = a_i / \sum_{k=1}^{n} a_k$ would be the probability. The uniqueness follows from the uniqueness of the W_i and of $\hat{V}$. To prove (22.5), let $s = s_j$ for some j and suppose that H_1, H_2, $\hat{f}_1$, $\hat{f}_2$, f_1, and f_2 are as in the statement of Axiom 5. Now, consider the set $\mathcal{H}_j$ of all horse lotteries H such that $H(s_i) = H_1(s_i)$ for all $i \neq j$. The stated preferences among this set of horse lotteries satisfies Axioms 1, 2, and 3. Hence there is a utility V_j for this set, and V_j is unique up to positive affine transformation. Clearly, W_j is such a utility, hence we will assume that $V_j = W_j$. Next, consider the set $\hat{\mathcal{H}}_j$ of all prize-state lotteries $\hat{f}$ that satisfy $\sum_{k=1}^{m} \hat{f}(z_k, s_j) = 1$. The stated preferences among elements of $\hat{\mathcal{H}}_j$ also satisfy Axioms 1, 2, and 3. Hence there is a utility $\hat{V}_j$, which is unique up to positive affine transformation. Clearly $\hat{V}$, with domain restricted to $\hat{\mathcal{H}}_j$, is such a utility, hence we will assume that $\hat{V}_j = \hat{V}$. The mapping $T_j : \mathcal{H}_j \to \hat{\mathcal{H}}_j$ defined by $T_j(H)(z, s) = 0$ for all (z, s) with $s \neq s_j$ and $T_j(H) = \alpha_i$ for $z = z_i$ and $s = s_j$, where $H(s_j) = (\alpha_1, \ldots, \alpha_m)$, is one-to-one and T_j preserves convex combination. It then follows from Axiom 5 that, for $H_1, H_2 \in \mathcal{H}_j$, $W_j(H_1) \leq W_j(H_2)$ if and only if $\hat{V}(T_j(H_1)) \leq \hat{V}(T_j(H_2))$. Since both $V_j = W_j$ and $\hat{V}_j = \hat{V}$ are unique up to positive affine transformation, we have $W_j = a_j \hat{V} + b_j$ for some positive a_j. This proves (22.5).

Discussion

The need for state-dependent utilities arises out of the possibility that what may appear to be a constant prize may not actually have the same value to an agent in all states of nature. Much of probability theory and statistical theory deals solely with probabilities and not with utilities. If probabilities are unique only relative to a specified utility, then the meaning of much of this theory is in doubt. Much of statistical decision theory makes use of utility functions of the form $U(\theta, d)$, where θ is a state of nature and d is a possible decision. The prize awarded when decision d is chosen and the state of nature is θ is not explicitly mentioned. Rather, the utility of the prize is specified without reference to the prize. Although it would appear that $U(\theta, d)$ is a state-dependent utility (as well it might be), one has swept comparisons between states "under the rug," For example, if $U(\theta, d) = -(\theta - d)^2$, one might ask how it was determined that an error of 1 when $\theta = a$ has the same utility as an error of 1 when $\theta = b$.

DeGroot (1970) avoids these problems by assuming that the concept of one event being "at least as likely as" another is understood without definition. He then proceeds to state axioms that imply the existence of a unique subjective probability distribution over states of nature. (For a discussion of attempts to derive quantitative probability from qualitative probability, see Narens 1980.) Further axioms could then be introduced that govern preference. These would then lead to a state-dependent utility function. Axioms, such as those of Savage (1954), Von-Neumann and Morgenstern (1947), Anscombe and Aumann (1963), and deFinetti (1974), which concern only preference among acts like horse lotteries, are not sufficient to guarantee a representation of preference by a unique state-dependent utility and probability. Direct comparisons need to be made between lotteries "in a specified state of nature" and other lotteries in another specified state of nature. These are the "prize-state" lotteries introduced by Karni (1985). Assuming that preferences among prize-state lotteries are consistent with preferences among horse lotteries, a unique state-dependent utility and probability can be recovered from the preferences.

References

Anscombe, F. J., & Aumann, R. J. (1963). A definition of subjective probability. *Annals of Mathematical Statistics, 34*, 199–205.

Arrow, K. J. (1974). Optimal insurance and generalized deductibles. *Scandinavian Actuarial Journal, 1*, 1–42.

deFinetti, B. (1974). *Theory of probability* (Vol. 2). New York: Wiley.

DeGroot, M. H. (1970). *Optimal statistical decisions*. New York: Wiley.

Fisburn, P. (1970). *Utility theory for decision making*. New York: Wiley.

Kadane, J. B., & Winkler, R. L. (1988). Separating probability elicitation from utilities. *Journal of the American Statistical Association, 83*, 357–363.

Karni, E. (1985). *Decision making under uncertainty*. Cambridge: Harvard University Press.

Karni, E., Schmeidler, D., & Vind, K. (1983). On state dependent preferences and subjective probabilities. *Econometrica, 51*, 1021–1031.

Narens, L. (1980). On qualitative axiomatizations for probability theory. *Journal of Philosophical Logic, 9*, 143–151.

Ramsey, F. P. (1926). Truth and probability. In H. E. Kyburg & H. E. Smokler (Eds.), *Studies in subjective probability* (pp. 23–52). Huntington: Krieger.

Rubin, H. (1987). A weak system of axioms for "rational" behavior and the non-separability of utility from prior. *Statistics and Decisions, 5*, 47–58.

Savage, L. J. (1954). *Foundations of statistics*. New York: Wiley.

Schick, F. (1986). Dutch book and money pumps. *Journal of Philosophy, 83*, 112–119.

VonNeumann, J., Morgenstern, O. (1947). *Theory of games and economic behavior* (2nd ed.). Princeton: Princeton University Press.

Chapter 23
Causal Decision Theory

James M. Joyce and Allan Gibbard

- EXAMPLE: *Prisoner's Dilemma with Twin (PDT).* You are caught in a standard, one-shot prisoner's dilemma (diagram next page), and the other player is your twin. You don't know for sure what Twin will do, but you know that Twin is amazingly like you psychologically. What you do, he or she too will likely do: news that you were going to rat would be good indication that Twin will rat, and news that you were going to keep mum would be a good sign that Twin will keep mum. Your sole goal is to minimize your own time in jail: Family feelings affect you not, and you care not a whit about loyalty, returning good for good, or how long Twin spends in jail. What course of action is rational for you in pursuit of your goals?

Many will find the answer easy—though they may disagree with each other on which the answer is. A standard line on the prisoner's dilemma rests on *dominance:* What you do won't affect what Twin does. Twin may rat or keep mum, but in either

---

In the nearly 20 years since this article was written there has been a revolution in the understanding of causal and counterfactual reasoning. This revolution had its roots in early work by Rubin (1974), Holland (1986) and Robbins (1986), which gave rise to the so-called "potential outcomes" framework. At roughly the same time the closely related "structural equations/causal graphs" approach was being developed and used to great effect by Spirtes et al. (1993), and Pearl (2000). In both treatments counterfactual reasoning plays a leading role in causal inference, just as in causal decision theory. While the core claims of this article remain true, and the basic structure of causal decision theory remains intact, these new models of provide us with far more sophisticated ways of representing and identifying causal relationships than were available and widely known when we wrote. As a result, some of our remarks about "the need for new advances in understanding of localization in relation to rational belief" have been rendered moot. Readers are encouraged to investigate these new developments, which we see as great advances.

J.M. Joyce (✉) • A. Gibbard
University of Michigan, Ann Arbor, MI, USA
e-mail: jjoyce@umich.edu; gibbard@umich.edu

H. Arló-Costa et al. (eds.), *Readings in Formal Epistemology*, Springer Graduate Texts in Philosophy 1, DOI 10.1007/978-3-319-20451-2_23

case, you yourself will do better to rat. Whichever Twin is doing, you would spend less time in jail if you were to rat than if you were to keep mum. Therefore the rational way to minimize your own time in jail is to rat.

	Mum	Rat
Mum	−1, −1	−10, 0
Rat	0, −10	−9, −9

Prisoner's Dilemma

Another line of argument leads to the opposite conclusion. Assess each act by its *auspiciousness*, by how welcome the news would be that you were about to perform it. News that you're about to rat would indicate that Twin is likewise about to rat. That's bad news; it means a long time in jail, for you as well as for Twin. News that you're about to keep mum, on the other hand, would be good news: It indicates that Twin is likewise about to keep mum, and your both keeping mum will mean a short time in jail. Keeping mum, then, is the *auspicious* act, and so—in terms of your selfish goals—you achieve best prospects by keeping mum.[1]

The two lines of reasoning, then, lead to opposite conclusions. One or the other, to be sure, may strike a reader as obviously wrong. Still, if one of them is cogent and the other not, decision theory should tell us why. Standard theories haven't spoken, though, with one voice on this matter. Savage himself (1972) was mostly silent on issues that would decide between the two lines: his system could be read in more than one way, and the few pertinent remarks he left us point in opposing directions. Various other decision-theoretic systems do have implications for this matter. Some imply that the argument from auspiciousness is correct: the principle of dominance, these systems entail, doesn't properly apply to a case like PDT. Taking the other side, a group called—perhaps somewhat misleadingly—*causal* decision theorists have formulated systems according to which the principle of dominance does apply to this case, and the rational thing to do is to rat.

"Causal" theorists maintain that decision theory requires a notion of causal dependency, explicit or implicit. Otherwise, they say, the theory will yield the wrong prescription for cases like PDT. We touch below on how causal notions might be made explicit for decision theorists' purposes, and how causality might be vindicated as empirically respectable. Auspiciousness theorists—or *evidential*

[1] An interesting model of Prisoner's Dilemma with a twin can be found in Howard (1988). Howard, who endorses a version of the auspiciousness argument, shows how to write a Basic program for playing the game which is capable of recognizing and cooperating with programs that are copies of itself.

decision theorists, as they are called in the literature—have no need for causal terms in their theory: they manage everything with standard subjective probabilities and conditional probabilities. Some evidential theorists deny that their theory, properly construed or developed, really does say not to rat in PDT. They deny, then, that causal notions must be introduced into decision theory, even if causal theorists are right about what to do in this case. We touch below on debates between "causal" theorists and this camp of "evidential" theorists, but mostly stick with the "causal" theory, explaining it and examining its potentialities.[2]

Cases with the structure of PDT can't be rare. The prisoner's dilemma itself is a parable, but economics, politics, war, and the like will be full of cases where one's own acts suggest how others are acting. Consider, for instance, a sophisticated speculator playing a market. Mustn't he reasonably take himself to model other sophisticated players? Why should he be unique? A rational agent interacting with others must escape the hubris of thinking that only he is smart and insightful—but then he'll have to take himself as a likely model for the schemings and reasonings of others. In such cases, different versions of decision theory may prescribe incompatible actions.

Dominance and Expected Utility: Two Versions

Savage (1972) encoded decision problems as matrices, with columns indicating "states of the world". For a Savage matrix with states $S_1, \ldots, S_n$, the expected utility of an act A is calculated by the Savage formula

$$\mathcal{V}(A) = \sum_{i=1}^{n} \rho(S_i)u(A, S_i), \tag{23.1}$$

where $u(A, S_i)$ is the utility of act A for state S_i and $\rho(S_i)$ is the subjective probability of S_i. From (23.1) follows a principle which we'll call the *Unqualified Principle of Dominance*, or *UPD:*

- UPD: If for each S_i, $u(A, S_i) > u(B, S_i)$, then $\mathcal{V}(A) > \mathcal{V}(B)$.

Which Savage matrix correctly represents a problem, though, must be decided with care, as is shown by a spoof due to Jeffrey (1967), p. 8:

[2]Nozick (1969) introduced PDT and other cases of this kind, focusing his discussion on Newcomb's problem, which he credits to physicist William Newcomb. He makes many of the points that causal theorists have come to accept, but recognizes only one kind of expected utility, the one we are calling auspiciousness. Stalnaker originated causal expected utility in a 1972 letter published only much later (Stalnaker 1981). Gibbard and Harper (1978) proselytize Stalnaker's proposal, and Lewis (1981) gives an alternative formulation which we discuss below. Gibbard and Harper (1978) and Lewis (1979b) also discuss PDT along with Newcomb's problem.

- EXAMPLE: *Better Red Than Dead (BRD)*. I'm an old-time American cold warrior with scruples, deciding whether or not my country is to disarm unilaterally. I construct a matrix as follows: My two possible states are that the Soviets invade and that they don't. In case they invade, better red than dead; in case they don't, better rich than poor. In either case, unilateral disarmament beats armament, and so by dominance, I conclude, it is rational to disarm.

Now whether or not unilateral disarmament would be rational all told, this argument can't be right. As the proponent of deterrence will point out, unilateral disarmament may decrease the likelihood of the Soviets' invading, and a scenario in which they don't invade is better than one in which they do. The argument from "dominance" treats these considerations as irrelevant—even if the Soviets are sure to invade if we disarm and to hold back if we arm.

Savage's states, then, must be act-independent: They must obtain or not independently of what the agent does. How, then, shall we construe this requirement? The first answer was developed, independently, by Jeffrey (1967) and by Luce and Krantz (1971): For dominance correctly to apply, they say, the states must be stochastically (or probabilistically) independent of acts. Where acts $A_1 \ldots A_m$ are open to the agent and $\rho(S/A_j)$ is the standard conditional probability of S given A_j,[3] S is stochastically act-independent iff

$$\rho\left(S/A_1\right) = \rho\left(S/A_2\right) = \cdots = \rho\left(S/A_m\right) = \rho(S) \tag{23.2}$$

The probabilities in question are subjective, or at any rate epistemic. Formula (23.2), then, means roughly that learning what one's going to do won't affect one's credence in S. One's act won't be evidence for whether S obtains. Requirement (23.2), then, is that any state S be *evidentially* act-independent. Theories that entail that (23.2) is what's needed for the principle of dominance to hold good we can thus call *evidential* theories of decision.

As applied to the prisoner's dilemma with your twin, evidential decision theory must treat the dominance argument as bad. Twin's act fails to be independent of yours evidentially. Let proposition C_t be that Twin cooperates, let C_y be that you cooperate, and let D_y be that you defect. C_t is not evidentially independent of what you do: $\rho(C_t/C_y)$ is high whereas $\rho(D_t/C_y)$ is low, since your cooperating is good evidence that she is cooperating, whereas your defecting is good evidence that she is defecting. Condition (23.2), then, won't hold for this case.

Evidential decision theory holds that the Savage formula (23.1) applies when condition (23.2) holds—that is, when the states of the matrix are evidentially act-independent. Requirements of act-independence, though, could be dropped if we changed formula (23.1) for computing expected utility. Use as weights not the probability of each state, as in (23.1), but its conditional probability—its probability conditional on the act's being performed. In this more general formulation, we have

[3]More precisely, A_i is the *proposition* that one performs a particular one of the alternative acts open to one in one's circumstances. We reserve the notation $\rho(S|A_j)$ for a more general use later in this chapter.

$$\mathcal{V}(A) = \sum_{i=1}^{n} \rho(S_i/A)u(A, S_i). \tag{23.3}$$

The Savage formula (23.1) is then a special case of (23.3), for conditions of evidential act-independence.[4] Since UPD follows from (23.1), it follows from (23.3) plus condition (23.2) of evidential act-independence. Evidential decision theory, then, has (23.3) as its general formula for expected utility. Its version of the principle of dominance is UPD qualified by condition (23.2) of evidential act-independence. In general, it recommends using "auspiciousness" to guide choices: a rational agent should select an act whose performance would be best news for him—roughly, an act that provides best evidence for thinking desirable outcomes will obtain.

Evidential theory has the advantage of avoiding philosophically suspect talk of causality: Its general formula (23.3) sticks to mathematical operations on conditional probabilities, and likewise, its requirement (23.2) of evidential act-independence—its condition, that is, for the Savage formula (23.1) to apply to a matrix—is couched in terms of conditional probabilities.

The causal theorist, in contrast, maintains that to apply a principle of dominance correctly, one can't avoid judgments of causality. One must form degrees of belief as to the causal structure of the world; one must have views on what is causally independent of what. Belief in Twin's causal isolation is a case in point. Dominance applies to PDT, their contention is, because you and your twin are causally isolated—and you know it. What you do, you know, will in no way affect what twin does. The argument, then, invokes a causal notion: the notion of what will and what won't causally *affect* what else. Causal decision theory then recommends using causal efficacy to guide choices: It holds, roughly, that a rational agent should select an act whose performance would be likely to bring about desirable results.

The causal decision theorist's requirement on a state S of a Savage matrix, then, is the following: The agent must accept that nothing he can do would causally affect whether or not S obtains. For each act A open to him, he must be certain that S's obtaining would not be causally affected by his doing A.

Can the causal theorist find a formula for expected utility that dispenses with this requirement of believed causal act-independence? A way to do so was proposed by Stalnaker (1968); see also Gibbard and Harper (1978). It requires a special conditional connective, which we'll render '$\Box\!\!\rightarrow$'. Read '$A \Box\!\!\rightarrow B$' as saying, "If A obtained then B would." In other words, either A's obtaining would cause B to obtain, or B obtains independently (causally) of whether or not A obtains. Then to say that S is causally independent of which act $A_1 \ldots A_n$ one performs is to say this: Either S would hold whatever one did, or whatever one did S would fail to hold. In other words, for every act A_i, we have $A_i \Box\!\!\rightarrow S$ iff S. We can now generalize the Savage formula (23.1) for the causal theorist's kind of expected utility. Use as weights, now, the probabilities of conditionals $\rho(A \Box\!\!\rightarrow Si)$, as follows:

[4]Formula (11.1) is introduced by Jeffrey (1967) and by Luce and Krantz (1971).

$$\mathcal{U}(A) = \sum_{i=1}^{n} \rho(A \,\square\!\!\rightarrow S_i) u(A, S_i) \tag{23.4}$$

Call this $\mathcal{U}(A)$ the *instrumental expected utility* of act A. The Savage formula

$$\mathcal{U}(A) = \sum_{i=1}^{n} \rho(S_i) u(A, S_i), \tag{23.5}$$

is then (23.4) for the special case where the following condition holds:

$$\text{For each } S_i, \rho\,(A \,\square\!\!\rightarrow S_i) = \rho\,(S_i)\,. \tag{23.6}$$

A sufficient condition for this to hold is that, with probability one, S_i is causally independent of A—in other words, that

$$\text{For each } S_i, \rho\,([A \,\square\!\!\rightarrow S_i] \leftrightarrow S_i) = 1. \tag{23.7}$$

Note that for the prisoner's dilemma with twin, condition (23.7) does hold. Twin, you know, is causally isolated from you. You know, then, that whether Twin would defect if you were to defect is just a matter of whether twin is going to defect anyway. In other words, you know that $D_y \,\square\!\!\rightarrow D_t$ holds iff D_t holds, and so for you, $\rho([D_y \,\square\!\!\rightarrow D_t] \leftrightarrow D_t) = 1$. This is an instance of (23.7), and similar informal arguments establish the other needed instances of (23.7) for the case.

In short, then, causal decision theory can be formulated taking formula (23.4) for instrumental expected utility as basic. It is instrumental expected utility as given by (23.4), the causal theorist claims, that is to guide choice. The Savage formula is then a special case of (23.4), for conditions of known causal act-independence—where (23.7) holds, so that for each state S_i, $\rho(A \,\square\!\!\rightarrow S_i) = \rho(S_i)$. The Unqualified Principle of Dominance for $\mathcal{U}$ is

- UPD: If for each S_i, $u(A, S_i) > u(B, S_i)$, then $\mathcal{U}(A) > \mathcal{U}(B)$.

Causal decision theory, in this formulation, has (23.4) as its general formula for the instrumental expected utility that is to guide choice, and its own version of the principle of dominance: UPD qualified by condition (23.7) of known causal act-independence.

Evidential and causal decision theorists, in short, accept different general formulas for the expected utility that is to guide choice, and consequently, they accept different conditions for the Savage formula to apply, and different principles of dominance. Causal theory—in the formulation we've been expounding—adopts (23.4) as its formula for expected utility, whereas evidential theory adopts (23.3). Causal theory, in other words, weighs the values of outcomes by the probabilities of the relevant conditionals, $\rho(A \,\square\!\!\rightarrow S_i)$, whereas evidential theory weighs them by the relevant conditional probabilities $\rho(S_i/A)$. Different conditions, then, suffice, according to the two theories, for the Savage formula correctly to apply to a matrix, and consequently for UPD to apply. That makes for distinct principles of dominance: For the causal theorist, UPD qualified by condition (23.7) of known causal act-independence, and for the evidential theorist, UPD qualified by condition (23.2) of evidential act-independence.

Conditionals and Their Probabilities

What, then, is the contrast on which all this hinges: the contrast between the probability $\rho(A \square\!\!\rightarrow S)$ of a conditional $A \square\!\!\rightarrow S$ and the corresponding conditional probability $\rho(S/A)$? Where probability measure ρ gives your *credences*—your degrees of belief—the conditional probability $\rho(S/A)$ is the degree to which you'd believe S if you learned A and nothing else. In the prisoner's dilemma with your twin, then, $\rho(D_t/D_y)$ measures how much you'd expect twin to rat on learning that you yourself were about to rat. If $\rho(D_t/D_y) \neq \rho(D_t/C_y)$, that doesn't mean that D_t is in any way causally dependent on whether D_y or C_y obtains. It just means that your act is somehow *diagnostic* of Twin's. Correlation is not causation. Probability $\rho(D_y \square\!\!\rightarrow D_t)$, on the other hand, is the degree to which you believe that if you were to defect, then Twin would. There are two circumstances in which this would obtain: Either Twin is about to defect whatever you do, or your defecting would cause Twin to defect. To the degree to which $\rho(D_y \square\!\!\rightarrow D_t) > \rho(D_y \square\!\!\rightarrow C_t)$, you give some credence to the proposition $[D_y \square\!\!\rightarrow D_t] \& \neg [D_y \square\!\!\rightarrow C_t]$, that twin would defect if you did, but not if you cooperated. This is credence in the proposition that your act will make a causal difference.

In daily life, we guide ourselves by judgments that seem to be conditional: What would happen if we did one thing, or did another? What would be the effects of the various alternatives we contemplate? We make judgments on these matters and cope with our uncertainties. Classic formulations of decision theory did not explicitly formalize such notions: notions of causal effects or dependency, or of "what would happen if". In Ramsey's and Savage's versions, causal dependency may be implicit in the representational apparatus, but this formal apparatus is open to interpretation. Other theorists had hoped that whatever causal or "would" beliefs are involved in rational decisions could be captured in the structure of an agent's conditional probabilities for non-causal propositions or events.

This last maneuver might have great advantages if it worked, but causal decision theorists argue that it doesn't. Causal or "would" notions must somehow be introduced into decision theory, they claim, if the structure of decision is to be elucidated by the theory. The introduction can be explicit, as in the general forumula (23.4) for $\mathcal{U}$ above, or it can be in the glosses we give—say, in interpretations of the Savage formula (23.1) or (23.5). If causal theorists are right, then, the theory of subjective conditional probability won't give us all we need for describing the beliefs relevant to decision. We'll need some way of displaying such beliefs and theorizing about them.

Causal theorists have differed, though, on how causal beliefs are best represented. So far, we've spoken in Stalnaker's terms, but we need to say more on what his treatment consists in, and what some of the alternatives might be for representing causal decision theory.

First, some terminology. Savage spoke of "states" and "events", and distinguished these from "acts" or "strategies". The philosophers who developed causal decision theory often spoke of "propositions", and include as propositions not only Savage's "events" and "states", but also acts and strategies. That is to say, propositions can characterize not only what happens independently of the agent, but also what the agent does—or even what he *would* do in various eventualities. A proposition can say that I perform act a or adopt strategy s. Such propositions can be objects of belief and of desire, and so can be assigned credences (subjective probabilities) and utilities.

Let A, then, be the proposition that I perform act a. Stalnaker constructs a conditional proposition $A \,\Box\!\!\rightarrow B$, which we read as "If I did A, then B would obtain." How does such a conditional proposition work? Much as Savage treats an event as a set of states, so Stalnaker treats a proposition as a set of *possible worlds* or maximally specific ways things might have been. Abstractly, the connective '$\Box\!\!\rightarrow$' is a two-place propositional function: To each pair of propositions it assigns a proposition.

Stalnaker hoped originally that this conditional function $\Box\!\!\rightarrow$ could be defined so that the probability of a conditional is always the corresponding conditional probability: So that whenever $\rho(C/A)$ is defined, $\rho(A \,\Box\!\!\rightarrow C) = \rho(C/A)$. Lewis (1976) proved that—with trivial exceptions—no such equality will survive conditionalization. Read ρ_A, in what follows, as probability measure ρ conditioned on A, so that by definition, $\rho_A(C) = \rho(C/A)$. What Lewis showed impossible is this: that for all propositions A, C and B for which $\rho(A\&B) > 0$, one has $\rho_B(A \,\Box\!\!\rightarrow C) = \rho_B(C/A)$. For if this did obtain, then one would have both $\rho_C(A \,\Box\!\!\rightarrow C) = \rho_C(C/A)$ and $\rho_{\neg C}$ (A W C) $= \rho_{\neg C}(C/A)$. But then

$$\begin{aligned}\rho\,(A \,\Box\!\!\rightarrow C) &= \rho_C\,(A \,\Box\!\!\rightarrow C)\,\rho(C) + \rho_{\neg C}\,(A \,\Box\!\!\rightarrow C)\,\rho\,(\neg C)\\ &= \rho_C\,(C/A)\,\rho(C) + \rho_{\neg C}\,(C/A)\,\rho\,(\neg C)\\ &= 1 \cdot \rho(C) + 0 \cdot \rho\,(\neg C)\\ &= \rho(C)\end{aligned}$$

We'd have $\rho(A \,\Box\!\!\rightarrow C) = \rho(C/A)$, then, at most when $\rho(C) = \rho(C/A)$. No such equality can survive conditionalization on an arbitrary proposition.

How, then, should we interpret the probability $\rho(A \,\Box\!\!\rightarrow C)$ of a conditional proposition $A \,\Box\!\!\rightarrow C$, if it is not in general the conditional probability $\rho(C/A)$. Many languages contrast two forms of conditionals, with pairs like this one[5]:

[5] Adams (1975) examines pairs like this.

If Shakespeare didn′t write *Hamlet*, someone else did. (23.8)

If Shakespeare hadn′t written *Hamlet*, someone else would have. (23.9)

Conditionals like (23.8) are often called *indicative*, and conditionals like (23.9) *subjunctive* or *counterfactual*. Now indicative conditional (23.8) seems epistemic: To evaluate it, you might take on, hypothetically, news that Shakespeare didn't write Hamlet. Don't change anything you now firmly accept, except as you would if this news were now to arrive. See, then, if given this news, you think that someone else did write Hamlet. You will, because you are so firmly convinced that Hamlet was written by someone, whether or not the writer was Shakespeare. The rule for this can be put in terms of a thinker's subjecive probabilities—or her *credences*, as we shall say: indicative conditional (23.8) is acceptable to anyone with a sufficiently high conditional credence $\rho(E/D)$ that someone else wrote Hamlet given that Shakespeare didn't. Subjunctive conditional (23.9) works differently: If you believe that Shakespeare did write Hamlet, you will find (23.9) incredible. You'll accept (23.8), but have near zero credence in (23.9). Your conditional credence $\rho(E/D)$ in someone else's having written Hamlet given that Shakespeare didn't will be high, but your credence $\rho(D \Box\!\!\rightarrow E)$ in the subjunctive conditional proposition (23.9) will be near zero. Here, then, is a case where one's credence in a conditional proposition (23.9) diverges from one's corresponding conditional credence. Speaking in terms of the subjective "probabilities" that we have been calling credences, we can put the matter like this: the probability $\rho(D \Box\!\!\rightarrow E)$ of a subjunctive conditional may differ from the corresponding conditional probability $\rho(E/D)$.

The reason for this difference lies in the meaning the $\Box\!\!\rightarrow$ operator. Stalnaker (1968) puts his account of conditionals in terms of alternative "possible worlds". A world is much like a "state" in Savage's framework (except that it will include a strategy one might adopt and its consequences). Think of a possible world as a maximally specific way things might have been, or a maximally specific consistent proposition that fully describes a way things might have been. Now to say what the proposition $A \Box\!\!\rightarrow C$ is, we have to say what conditions must obtain for it to be true. There is no difficulty when the antecedent A obtains, for then, clearly, $A \Box\!\!\rightarrow C$ holds true if and only if C obtains. The puzzle is for cases where A is false. In those situations, Stalnaker proposes that we imagine the possible world w^A in which A is true, and that otherwise is most *similar* to our actual world in relevant respects. $A \Box\!\!\rightarrow C$, then, holds true iff C holds in this world w^A. Stalnaker and Thomason offered a rigorous semantics and representation theorem for this explication. Stalnaker's distinctive axioms are these[6]:

- *Intermediate strength:* If A necessitates B, then $A \Box\!\!\rightarrow B$, and if $A \Box\!\!\rightarrow B$, then $\neg(A \& \neg B)$
- *Conditional non-contradiction:* For possible A, $\neg[(A \Box\!\!\rightarrow B) \& (A \Box\!\!\rightarrow \neg B)]$.

[6]Stalnaker (1968) p. 106, Stalnaker and Thomason (1970), slightly modified.

- *Distribution:* If $A \Box\!\!\rightarrow (B \vee C)$, *then* $(A \Box\!\!\rightarrow B) \vee (A \Box\!\!\rightarrow C)$.
- *Suppositional equivalence:* If $(A \Box\!\!\rightarrow B)$ and $(B \Box\!\!\rightarrow A)$, then $(A \Box\!\!\rightarrow C)$ iff $(B \Box\!\!\rightarrow C)$.

All this leaves mysterious the notion of *relevant similarity* invoked by Stalnaker's account. Formal axioms are easy: Stalnaker speaks of a *selection function f* which assigns a world $f(A, w) = w^A$ to each proposition A that has the possibility of being true. A compelling logic for $\Box\!\!\rightarrow$ can be developed on this basis.

How, though, do we interpret the notion of "relevant similarity" when applying this formal apparatus to real-life decision problems? Intuitive overall likeness of worlds won't do. Nixon, imagine, had in the Oval Office a red button to launch the missiles.[7] In his despair he considered pushing it, but drew back. We can say, "If Nixon had pushed the button, nuclear holocaust would have ensued." This is true, we would judge, if the apparatus was in working order and without safeguards. Of the worlds in which Nixon pushes the button, though, the one most similar overall to the actual world would be not one in which all was destroyed, but one in which the apparatus malfunctioned—a wire became temporarily nonconducting, say. After all, a world in which the missiles were launched would surely have a future radically different from the actual world. Little in any city would look the same.

Clearly, then, we cannot look to overall similarity of worlds to cash out the type of "relevant" similarity needed in the evaluation of subjunctive conditionals. Rather, we want to know what would have ensued from initial conditions that were much like those that actually obtained, but differed in some slight respect in Nixon's decision-making apparatus—differed in such a way that by natural laws, the outgrowth of those modified initial conditions would have been nuclear holocaust. Thus, the rough idea is that one evaluates it $A \Box\!\!\rightarrow C$ by considering a world in which A obtains that is as much like the actual world as possible both with regard to particular facts about the past as well as general facts about what might follow causally from what in the future. Lewis (1979a) attempts a general account of the kind of "relevant similarity" that fits our causal judgments, and derives from it an account of the regularities that give time its direction. As decision theorists, though, we need not be concerned whether such lofty philosophical ambitions can be fulfilled. We need only understand that where w_0 is the actual world, the value $f(A, w_0)$ of the Stalnaker selection function is the world as it would be if A obtained. It is that world, with all its ensuing history. In many situations, it will be clear that agents do have subjective probabilities for what that world would be like, and so the application of Stalnaker's apparatus to an agent's decision situation will be clear enough.[8]

Stalnaker's framework allows us to be more precise about how probabilities of subjunctive conditionals differ from ordinary conditionals probabilities. It is useful

[7] Fine (1975) gives this example to make roughly this point.

[8] Shin (1991a), for instance, devises a metric that seems suitable for simple games such as "Chicken".

to think of both $\rho(A \;\Box\!\!\rightarrow C)$ and $\rho(C/A)$ as capturing a sense of *minimal* belief revision. $\rho(C/A)$, as we have seen, is the credence that an agent with prior ρ should assign C if she gets pure and certain news of A's truth. Thus, the function $\rho_A = \rho(\bullet /A)$ describes the outcome of a belief revision process in which an agent learns that A. This revision is minimal in the sense that it changes ρ so as to make A certain without thereby altering any ratios of the form $\rho(X\&A) : \rho(Y\&A)$. In terms of possible worlds, ρ_A is obtained from ρ by setting the credence of $\neg A$ equal to zero, and spreading the residual probability uniformly over the worlds in which A obtains—thus leaving undisturbed any evidential relationships that might obtain among propositions that entail A. The function $\rho^A = \rho(A \;\Box\!\!\rightarrow \bullet)$ defines a rather different sort of minimal belief revision, *imaging* (Lewis 1976, 1981) or better, *imagining*. Instead of distributing the probability of $\neg A$ uniformly over the A-worlds, spread it with an eye to relevant similarities among worlds. Transfer the probability of each world w in which A is false to w^A, the A-world most similar to w, adding this probability to the probability w^A has already.

This whole treatment, though, rests on an assumption that is open to doubt: that there always exists a unique "most similar" world w^A. Perhaps there's no one definite way the world would be were I, say, now to flip this coin. The world might be indeterministic, or the supposition that I now flip the coin might be indeterminate—in that the exact force and manner in which I'd be flipping the coin isn't specified and isn't even under my control. It may then be the case neither that definitely were I to flip the coin it would land heads, nor that definitely were I to flip the coin it would not land heads. The law of *conditional excluded middle* will be violated:

$$(F \;\Box\!\!\rightarrow H) \vee (F \;\Box\!\!\rightarrow \neg H) \tag{23.10}$$

Conditional excluded middle obtains in Stalnaker's model. There are models and logics for conditionals in which it does not obtain.[9] In a case like this, however, the right weight to use for decision is clearly not the probability $\rho(F \;\Box\!\!\rightarrow H)$ of a conditional proposition $F \;\Box\!\!\rightarrow H$. If I'm convinced that neither $F \;\Box\!\!\rightarrow H$ nor $F \;\Box\!\!\rightarrow \neg H$ obtains, then my subjective probability for each is zero. The weight to use if I'm betting on the coin, though, should normally be one-half.

A number of ways have been proposed to handle cases like these. One is to think that with coins and the like, there's a kind of conditional chance that isn't merely subjective: the chance with which the coin would land heads *were* I to flip it. Write this $\pi_F(H)$. Then use as one's decision weight the following: one's *subjectively expected value for* this *objective conditional chance*. Suppose you are convinced that the coin is loaded, but don't know which way: You think that the coin is loaded either .6 toward heads or .6 toward tails, and have subjective probability of .5 for each of these possibilities:

[9]Lewis (1973) constructs a system in which worlds may tie for most similar, or it may be that for every A-world, there is an A-world that is more similar. He thus denies Conditional Excluded Middle: It fails, for instance, when two A-worlds tie for most similar to the actual world, one a C-world and the other a $\neg C$-world.

$$\rho\,(\pi_F(H) = .6) = .5 \qquad \text{and} \qquad \rho\,(\pi_F(H) = .4) = .5. \tag{23.11}$$

Your subjectively expected value for $\pi_F(H)$, then, will be the average of .6 and .4. Call this appropriate decision weight $\varepsilon_F(H)$. We can express this weighted averaging in measure theoretic terms, so that in general,

$$\varepsilon_A(C) = \int_0^1 x \cdot \rho\,(\pi_A(C) \in dx)\,. \tag{23.12}$$

$\varepsilon_A(C)$ thus measures one's subjective expectation of C's obtaining were A to occur: the sum of (i) the degree to which A's obtaining would tend to bring it about that C obtained, plus (ii) the degree to which C would tend to hold whether or not A obtained.

We can now write formulas for $\mathcal{U}$ using $\varepsilon_A(C)$ where we had previously used $\rho(A \;\Box\!\!\rightarrow C)$. Formula (23.4) above for instrumental expected utility now becomes

$$\mathcal{U}(A) = \sum_{i=1}^{n} \varepsilon_A(S_i)u(A, S_i) \tag{23.13}$$

Lewis (1981) gives an alternative formulation of causal decision theory, which is equivalent to the Stalnaker formulation we've been presenting whenever the Stalnaker framework applies as intended. He speaks of *dependency hypotheses:* complete hypotheses concerning what depends on what and how. Which dependency hypothesis obtains is causally independent of what the agent does, and so a Lewis dependency hypothesis can serve as a "state" in the Savage framework. He allows dependency hypotheses to contain an element of objective chance.[10]

Can an empiricist believe in such things as objective conditional chance or objective dependencies? This is a hotly debated topic, mostly beyond the scope of this article. Some philosophers think that objective dependency can be defined or characterized, somehow, in purely non-causal terms. Others doubt that any such direct characterization is possible, but think that a more indirect strategy may be available: Characterize a thinker's *beliefs* in causal dependencies—or his *degrees* of belief, his subjective probabilities, or as we have been calling them, his *credences*. His credences in causal propositions, this contention is, can be cashed out fully in terms of complex features of his credences in non-causal propositions—propositions that don't involve causal notions.

We ourselves would argue that causal propositions are genuine propositions of their own kind, basic to thinking about the world. They can't be fully explained in other terms, but they can be vindicated. We can be just as much empiricists about

[10]Skyrms (1980) offers another formulation, invoking a distinction between factors that are within the agent's control and factors that aren't. Lewis (1981) discusses both Skyrms and unpublished work of Jordan Howard Sobel, and Skyrms (Skyrms 1984), 105–6, compares his formulation with those of Lewis (1981) and Stalnaker (1981).

causes as we can about other features of the layout of the world. A rational thinker forms his credences in causal propositions in much the same Bayesian way he does for any other matter: He updates his subjective probabilities by conditionalizing on new experience. He starts with reasonable prior credences, and updates them. Subjective probability theorists like de Finetti long ago explained how, for non-causal propositions, updating produces convergence. The story depends on surprisingly weak conditions placed on the thinker's prior credence measure. The same kind of story, we suspect, could be told for credences in objective chance and objective dependence.

Lewis (1980) has told a story of this kind for credence in objective chance. His story rests on what he labels the "Principal Principle", a condition which characterizes reasonable credences in objective chances. Take a reasonable credence measure ρ, and a proposition about something that hasn't yet eventuated—say, that the coin I'm about to flip will land heads. Let me conditionalize his credences on the proposition that as of now, the objective probability of this coin's landing heads is .6. Then his resulting conditional credence in the coin's landing heads, the principle says, will likewise be .6. Many features of reasonable credence in objective chance follow from from this principle. From a condition on reasonable prior credences in objective chance follows an account of how one can learn about them from experience.

A like project for objective dependency hasn't been carried through, so far as we know, but the same broad approach would seem promising.[11] In the meantime, there is much lore as to how experience can lead us to causal conclusions—and even render any denial of a causal depencency wildly implausible. The dependence of cancer on smoking is a case in point. Correlation is not causation, we all know, and a conditional probability is not a degree of causal dependence. (In the notation introduced above, the point is that $\rho(C/A)$ need not be $\varepsilon_A(C)$, the subjectively expected value of the objective chance of C were one to do a.) Still, correlations, examined with sophistication, can *evidence* causation. A chief way of checking is to "screen off" likely common causes: A correlation between smoking and cancer might arise, say, because the social pressures that lead to smoking tend also to lead to drinking, and drinking tends to cause cancer. A statistician will check this possibility by separating out the correlation between smoking and cancer among drinkers on the one hand, and among non-drinkers on the other. More generally, the technique is this: A correlation between A and C, imagine, is suspected of being spurious—suspected *not* to arise from a causal influence of A on C or *vice versa*. Let F be a suspected common cause of A and C that might account for their correlation. Then see if the correlation disappears with F held constant. Econometricians elaborate such devices to uncover causal influences in an economy. The methodological literature on gleaning causal conclusions from experience includes classic articles by Herbert Simon (see Simon (1957), chs. 1–3).

[11]The work of Spirtes et al. (1993) and Pearl (2000) goes a long way toward realizing this goal.

Screening off is not a sure test of causality. A correlation might disappear with another factor held constant, not because neither factor depends causally on the other, but because the causal dependency is exactly counterbalanced by a contrary influence by a third factor.[12] Such a non-correlation might be robust, holding reliably with large sample sizes. But it will also be a coincidence: opposing tendencies may happen to cancel out, but we can expect such cases to be rare. Lack of correlation after screening off is *evidence* of lack of causal influence, but doesn't *constitute* lack of causal influence.

When controlled experiments can be done, in contrast, reasonable credences in a degree of objective dependency can be brought to converge without limit as sample size increases. Subjects are assigned to conditions in a way that we all agree has no influence on the outcome: by means of a chance device, say, or a table of pseudo-random numbers. Observed correlations then evidence causal dependence to whatever degree we can be confident that the correlation is no statistical fluke. With the *right* kind of partition, then, screening off does yield a reliable test of causality. But what makes a partition suitable for this purpose, we would claim, must be specified in terms that somehow invoke causality—in terms, for instance, of known causal independence.

How we can use evidence to support causal conclusions needs study. Standard statistical literature is strangely silent on questions of causation, however much the goals of statistical techniques may be to test and support causal findings. If we are right, then one class of treatments of causality will fail: namely, attempts to characterize causal beliefs in terms of the subjective probabilities and the like of non-causal propositions. Correct treatments must take causality as somehow basic. A constellation of relations—cause, chance, dependence, influence, laws of nature, what would happen if, what might likely happen if—are interrelated and may be intercharacterizable, but they resist being characterized purely from outside the constellation. Our hope should be that we can show how the right kind of evidence lets us proceed systematically from causal truisms to non-obvious causal conclusions. Fortunately, much of decision and game theory is already formulated in terms that are causal, implicitly at least, or that can be read or interpreted as causal. (When games are presented in normal form, for instance, it may be understood that no player's choice of strategies depends causally on the choice of any other.) A chief aim of causal decision theory is to make the role of causal beliefs in decision and game theory explicit.

Ratificationism

While some proponents of auspiciousness maximization have taken the heroic course and tried to argue that cooperating in Prisoner's Dilemma with Twin is rational,[13] most now concede that only defection makes sense. Nevertheless,

[12]Gibbard and Harper (1978), 140–2, construct an example of such a case.

[13]See, for example, Horgan (1981).

the promise of an analysis of rational choice free from causal or counterfactual entanglements remains appealing. A number of writers have therefore sought to modify the evidential theory so that it endorses the non-cooperative solution in PDT and yet does not appeal any unreconstructed causal judgements. The hope is that one will be able to find statistical techniques of the sort discussed toward the end of the last section to distinguish causal relationships from suprious correlations, and then employ these techniques to formulate a decision theory that can be sensitive to causal considerations without making explicit use of causal or subjunctive notions. The most influential of these attempts are found in Eells (1982) and Jeffrey (1983). Since the two approaches are similar, we focus on Jeffrey.

As we have seen, statisticians sometimes use "screening off" techniques to detect the effects of a common cause. Jeffrey's strategy is based on the insight that an agent's ability to anticipate her own decisions typically screens off any purely evidential import that her actions might possess. Prior to performing an act she will generally come to realize that she has decided on it. The act itself then ceases to be a piece of evidence for her since she has already discounted it. Letting Δ^A denote the decision to do a, we can put Jeffrey's claim like this:

- *Screening*. The decision to perform A screens off any purely evidential correlations between acts and states of the world, in the sense that $\rho(S/B\&\Delta^A) = \rho(S/\Delta^A)$ for all acts B and states S.

To ensure that these conditional credences are well-defined, Jeffrey must assume that the agent always assigns some positive probability to the prospect that she will fail to carry out a decision—due to a "trembling hand," a lack of nerve, or other factors beyond her control—so that $\rho(B\&\Delta^A)$ is non-zero for all acts B.

To see what happens when screening is introduced into PDT, imagine that during the course of your deliberations but prior to performing any act, you become certain that you will decide to cooperate, so that your credence in Δ_y^C moves to one. Since you are likely to carry out whatever decision you make, your probability for C_y also moves close to one, which gives you strong grounds for thinking your twin will cooperate. Indeed, if Screening obtains, you will have strong evidence for thinking that Twin is about to cooperate, no matter what news you get as to what you yourself are about do, because $\rho\left(C_t/C_y\&\Delta_y^C\right) = \rho\left(C_t/D_y\&\Delta_y^C\right) = \rho\left(C_t/\Delta_y^C\right) \approx 1$. Condition (23.2) is thus satisfied. You can then correctly apply the evidential dominance principle to conclude that defection is your most auspicious option. In this way, Screening ensures that if you are certain that you will eventually decide to cooperate, then you will assign defecting a higher auspiciousness than cooperating. On the other hand, if you are certain that you will decide to defect, you then have strong reason to suspect that twin is about to defect, whatever you learn that you yourself are about to do—and again, defecting turns out to be the more auspicious option. Therefore, no matter how auspicious cooperating might seem *before* you make up your mind about what to do, defecting is sure to look more auspicious *afterwards*—and this will be true no matter what decision you have made. Jeffrey proposes to uses this basic asymmetry—between what one decides and what one does—to argue that defection in PDT is the only rational course of action for an auspiciousness maximizer.

The case has already been made for an agent who is already certain about what she will decide. But what about agents who have yet to made up their minds? Here things get dicey. If you have not yet decided what to do, then the probabilities you assign to Δ_y^C and Δ_y^D will be far from one. This puts the auspiciousness values of C_y and D_y near those of $(C_y\&\Delta_y^C)$ and $(D_y\&\Delta_y^D)$ respectively, and since $\mathcal{V}(C\&\Delta_y^C) > \mathcal{V}(D\&\Delta_y^D)$, evidential decision theory tells you to choose cooperation. However, as soon as you make this choice, you will assign Δ_y^D a credence close to one, and as we have seen, you will then favor defection. Thus, the pursuit of good news forces you to make choices that you are certain to rue from the moment you make them—clearly something to avoid.

Jeffrey hopes to circumvent this difficulty by denying that evidential decision theory requires one to maximize auspiciousness as one *currently* estimates it. If you are savvy, he argues, you will realize that any choice you make will change some of your beliefs, thereby altering your estimates of auspiciousness. Thus, given that you want to make decisions that leave you better off for having made them, you should aim to maximize auspiciousness not as you currently estimate it, but as you *will* estimate it once your decision is made. You ought to, "choose for the person you expect to be when you have chosen,"[14] by maximizing expected utility computed relative to the personal probabilities you will have *after* having come to a firm decision about what to do. This is only possible if your choices conform to the maxim

- *Evidential Ratifiability*. An agent cannot rationally choose to perform A unless A is *ratifiable*, in the sense that $\mathcal{V}(A\&\Delta^A) \geq \mathcal{V}(B\&\Delta^A)$ for every act B under consideration.

This principle advises you to ignore your current views about the evidentiary merits of cooperating versus defecting, and to focus on maximizing future auspiciousness by making choices that you will regard as propitious from the epistemic perspective you will have once you have made them. Since in the presence of Screening, defection is the only such choice, the maxim of Evidential Ratifiability seems to provide an appropriately "evidentialist" rationale for defecting in PDT.

Unfortunately, though, the Screening condition need not always obtain. There are versions of PDT in which the actual performance of an act provides better evidence for some desired state than does the mere decision to perform it. This would happen, for example, if you and twin are bumblers who tend to have a similar problems carrying out your decisions. The fact that you were able to carry out a decision would then be evidentially correlated with you twin's act, and this correlation would not be screened off by the decision itself. In such cases the evidential ratifiability principle sanctions cooperation.[15] Therefore, the Jeffrey/Eells

[14] Jeffrey (1983) p. 16.

[15] Jeffrey, in his original published treatment of ratificationism (Jeffrey 1983), 20, gives this counterexample and credits it to Bas van Fraassen. Shin (1991b) treats cases in which the respective players' "trembles" are independent of each other.

strategy does not always provide a satisfactory evidentialist rationale for defecting in PDT. We regard this failure as reinforcing our contention that any adequate account of rational choice must recognize that decision makers have beliefs about causal or counterfactual relationships, beliefs that cannot be cashed out in terms of ordinary subjective conditional probabilities—in terms of conditional credences in non-causal propositions.

Ratificationism and Causality in the Theory of Games

Despite its failure, the Jeffrey/Eells strategy leaves the theory of rational choice an important legacy in the form of the Maxim of Ratifiability. We will see in this section that it is possible to assimilate Jeffrey's basic insight into casual decision theory and that so understood, it codifies a type of reasoning commonly employed in game theory. Indeed, the idea that rational players always play their part in a Nash equilibrium is a special case of ratificationism.

The notion of a ratifiable act makes sense within any decision theory with the resources for defining the expected utility of one act given the news that another will be chosen or performed. In causal decision theory the definition would be this:

$$\mathcal{U}(B/A) = \sum_{i=1}^{n} \rho((B \,\square\!\!\rightarrow S_i)/A)u(B, S_i)$$

Jeffrey's insight then naturally turns into the maxim,

- *Causal Ratifiability*. An agent cannot rationally choose to perform act A unless $\mathcal{U}(A/A) \geq \mathcal{U}(B/A)$ for every act B under consideration.

This says that a person should never choose A unless once she does, her expection of A's efficacity in bringing about desirable results is at least as great as that for any alternative to A. Notice that one no longer needs a "trembling hand" requirement to define the utility of one act conditional on another, since the conditional credence $\rho(B \,\square\!\!\rightarrow S/A)$ is well-defined even when A and B are incompatible. This means that an agent who is certain that she will perform A can still coherently wonder about how things *would* have gone *had* she done B—even though she can no longer wonder about how things are set to go if she's *going* to do B. This ability to assign utilities to actions that definitely will not be performed can be used to used to great advantage in game theory, for instance, when considering subgames and subgame perfection.

To see how an act might fail to be causally ratifiable, imagine yourself playing Matching Pennies with an opponent who you think can predict your move and will make a best response to it (see table). Neither pure act then turns out to be ratifiable. Suppose that you are strongly inclined to play [HEADS], so that $\rho(H_y)$ is close to one. Since your conditional probability for Twin playing heads given that you do is high, it also follows that your subjective probability for H_t will be high. Thus by recognizing that you plan to play [HEADS], you give yourself evidence for thinking that Twin will also play [HEADS]. Note, however, this does nothing to alter the fact that you still judge Twin's actions to be causally independent of your own; your subjective probabilities still obey

	HEADS	TAILS
HEADS	−1, 1	1, −1
TAILS	1, −1	−1, 1

Matching Pennies

$$\rho\left(\left(H_y\Box\!\!\rightarrow H_t\right)/H_y\right) \approx \rho\left(\left(T_u\Box\!\!\rightarrow H_t\right)/H_y\right).$$

Since $\rho(Ht) \approx 1$, your overall position is this: because you are fairly sure that you will play heads, you are fairly sure that Twin *will* play heads too; but you remain convinced that she *would* play heads even if (contrary to what you expect) you were to play tails. Under these conditions, the conditional expected utility associated with [TAILS] larger than that associated with [HEADS] on the supposition that [HEADS] is played. That is to say, [HEADS] is unratifiable. Similar reasoning shows that tails is also unratifiable. In fact, the only ratifiable act this game is the mixture ½ [HEADS] + ½ [TAILS].[16]

It is no coincidence that this mixture also turns out to be the game's unique Nash equilibrium; there is a deep connection between ratifiable acts and game-theoretic equilibria. Take any two-person game, such as Matching Pennies, for which the unique Nash equilibrium consists of mixed strategies (which the players in fact adopt). If a player has predicted the other's strategy correctly and with certainty, then by playing the strategy she does, she maximizes her expected utility. But this isn't the only strategy that, given her credences, would maximize her expected utility; any other probability mixture of the same pure strategies would do so too. The

[16] Piccione and Rubinstein (1997) present another kind of case in which considerations of ratifiability may be invoked: the case of the "absent-minded driver" who can never remember which of two intersections he is at. One solution concept they consider (but reject) is that of being "modified multi-selves consistent". In our terms, this amounts to treating oneself on other occasions as a twin, selecting a strategy that is ratifiable on the following assumption: that one's present strategy is fully predictive of one's strategy in any other situation that is subjectively just like it. This turns out to coincide with the "optimal" strategy, the strategy one would adopt if one could choose in advance how to handle all such situations.

strategy she adopts is unique, though, in this way: it is the only strategy that could be ratifiable, given the assumption that her opponent has predicted her strategy and is playing a best response to it. It should be clear that this argument extends straightforwardly to the n-person case. In any Nash equilibrium, all players perform causally ratifiable acts.

	C_1	C_2
R_1	1, 1	0, 4
R_2	4, 0	−15, −15

Chicken

In fact, the least restrictive of all equilibrium solution concepts—Aumann's correlated equilibrium—can be understood as a straightforward application of the maxim of ratifiability.[17] Aumann sought to generalize the Nash equilibrium concept by relaxing the assumption—implicit in most previous game-theoretic thinking—that players in normal-form games believe that everyone acts independently of everyone else. Take a game of Chicken. The table shown might give the utilities of drivers traveling along different roads into a blind intersection, who must decide whether to stop and lose time (R_1 and C_1) or to drive straight through and risk an accident (R_2 and C_2). Let the players assume that they will choose, respectively, strategies $p \cdot R_1 + (1-p) \cdot R_2$ and $q \cdot C_1 + (1-q) \cdot C_2$. What credences will they give to the various joint pure actions they may end up performing?

Game theorists have traditionally assumed that the players will treat the chance devices that implement their mixed strategies as evidentially independent, ascribing the credences in Table 23.1. Aumann imagines that they might ascribe the credences of Table 23.2: To see how the correlations in Table 23.2 might arise, imagine a "managed" game of Chicken in which, to minimize the chances of a disastrous outcome, things have been arranged so that an impartial arbitrator will throw a fair die and illuminate a traffic light at the intersection according to the "Arbitrator's Scheme" shown in the table. If each player intends to heed the signal, stopping on red and driving through on green, and believes his opponent intends to do so as well, then both will have the correlated priors of Table 23.2. Aumann showed how to define an equilibrium solution concept even when players' acts are correlated in this way. An *adapted* strategy is a rule s which dictates a unique pure act $s(A)$ for

[17]More precisely, correlated equilibrium is the weakest equilibrium solution concept which assumes that all players have common beliefs. When this assumption is relaxed one obtains a subjectively correlated equilibrium. For details see Aumann (1974, 1987).

Table 23.1 Chicken with Independence

	C_1	C_2
R_1	pq	$p(1-q)$
R_2	$(1-p)q$	$(1-p)(1-q)$

Table 23.2 Chicken with Correlation

	C_1	C_2
R_1	$p+q-1$	$1-q$
R_2	$1-p$	0

each act A signalled by the arbitrator: for instance, "Stop on red and go on green" or "Run all lights." A correlated equilibrium is simply a pair of adapted strategies r and c such that, for all alternatives r^* and c^*, the following condition holds (call it *CE*):

	one	two	three	four	five	six
ROW	Red	Red	Red	Red	Green	Green
COL	Red	Red	Green	Green	Red	Red

Arbitrator's Scheme

$$\sum_{ij} \rho(R_i \& C_j) \cdot \mathcal{U}_{\text{row}}(r(R_i), C_j) \geq \sum_{ij} \rho(R_i \& C_j) \cdot \mathcal{U}_{\text{row}}(r^*(R_i), C_j)$$

$$\sum_{ij} \rho(R_i \& C_j) \cdot \mathcal{U}_{\text{col}}((R_i), c(C_j)) \geq \sum_{ij} \rho(R_i \& C_j) \cdot \mathcal{U}_{\text{col}}(R_i c^*(C_j))$$

Table 23.3 Equilibrium for Chicken with Independence

	C_1	C_2
R_1	⅓	⅓
R_2	⅓	0

Thus, r and c constitute a correlated equilibrium iff no player has reason to deviate from her adapted strategy given her beliefs and the signal she receives. The reader may verify that, when things are as described in Table 23.3, both players' resolving to obey the light is a correlated equilibrium, whereas there is no correlated equilibrium in which either player decides to run red lights.

Definition CE makes it appear as if the existence of correlated equilibria depends on the availability of external signaling mechanisms. This was the view presented by Aumann in (1974), but in (1987), he shows that acts themselves can serve as appropriate signals. Specifically, Aumann established that a common probability distribution[18] ρ defined over strategy combinations comprises a correlated equilibrium iff given any pair of acts *R* and *C* assigned positive probabilities, one has *CE*:*

$$\sum_i \rho(C_j/R) \cdot \mathcal{U}_{\text{row}}(R, C_j) \geq \sum_i \rho(C_j/R) \cdot \mathcal{U}_{\text{row}}(R^*, C_j)$$

$$\sum_i \rho(R_j/C) \cdot \mathcal{U}_{\text{col}}(R_i, C) \geq \sum_i \rho(R_j/C) \cdot \mathcal{U}_{\text{col}}(R_i, C^*)$$

for all alternatives $R*$ and $C*$. CE* requires, then, that agents assign zero prior probability to any act, either of their own or of their adversaries', that does not maximize expected utility on the condition that it will be performed. Aumann regards this condition as "an expression of Bayesian rationality," since players satisfy it by maximizing expected utility at the time they act.

As a number of authors have noted, CE* is an application of the maxim of ratifiabilty.[19] It requires players to give zero credence to non-ratifiable acts. Hence for a group of players to end up in a correlated equilibrium, it is necessary and sufficient that all choose ratifiabile acts and expect others to do so as well. Aumann's "Bayesian rationality" thus coincides with the notion of rationality found in Jeffrey's ratificationism. More specifically, we contend that Aumann is requiring rational players to choose *causally* ratifiable actions.

[18]Note that such a distribution determines a unique mixed act for each player. Thus, it makes no difference whether one talks about the players' acts or the players' beliefs being in equilibrium.

[19]Shin (1991b), Skyrms (1990).

While Aumann has rather little to say about the matter, he clearly does not mean the statistical correlations among acts in correlated equilibrium to reflect causal connections. This is particularly obvious when external signaling devices are involved, for in such cases each player believes that his opponents would heed the arbitor's signal whether or not he himself were to heed it. Each player uses his knowledge of the arbitor's signal to him to make inferences about the signals given to others, to form beliefs about what his opponents expect him to do, and ultimately to justify his own policy of following the arbitor's signal. What he does not do is suppose that the correlations so discovered *would* continue to hold no matter what he decided to do. For example, if [ROW]'s credences are given in Table 23.3, it would be a mistake for him to run a red light in hopes of making it certain that [COLUMN] will stop; [COLUMN's action, after all, is determined by the signal she receives, not by what [ROW] does. Notice, however, that a straightforward application of evidential decision theory recommends running red lights in this circumstance—further proof that the view is untenable. In cases, then, where correlations are generated via external signaling, a causal interpretation of CE* clearly is called for. To make this explicit, we can rewrite CE* as the following condition *CE***:

$$\sum_j \rho((R\,\Box\!\!\rightarrow C_j)/R)\cdot\mathcal{U}_{\text{row}}(R,C_j) \geq \sum_j \rho((R^*\,\Box\!\!\rightarrow C_j)/R)\cdot\mathcal{U}_{\text{row}}(R^*,C_j)$$

$$\sum_j \rho((C\,\Box\!\!\rightarrow R_i)/C)\cdot\mathcal{U}_{\text{col}}(R_i,C) \geq \sum_i \rho((C^*\,\Box\!\!\rightarrow R_i)/C)\cdot\mathcal{U}_{\text{col}}(R_i,C^*)$$

This reduces to CE* on the assumption that [ROW] and [COLUMN] cannot influence each other's acts, so that both $\rho\left(\left(R*\;\Box\!\!\rightarrow C_j\right)/R\right) = \rho\left(C_j/R\right)$ and $\rho\left(\left(C*\;\Box\!\!\rightarrow R_i\right)/C\right) = \rho\left(R_i/C\right)$ hold for all R^* and C^*.

The situation is not appreciably different in cases where the correlation arises without any signaling device. Imagine playing [ROW] in the coordination game in Table 23.4, and consider the correlated equilibrium described by Table 23.5: These correlations need not have arisen through signaling. You might find the (R_1, C_1) equilibrium salient because it offers the vastly higher payoff, and, knowing that

Table 23.4 Correlation without Signaling

	C_1	C_2
R_1	25, 1	0, 0
R_2	0, 0	1, 2

Table 23.5 Equilibrium for Correlation without Signaling

	C_1	C_2
R_1	0.7	0.01
R_2	0.01	0.28

your opponent can appreciate its salience for you, you might conclude that she expects you to play it. That makes C_1 the better play for her, which reinforces your intention to play R_1 and your belief that she will play C_1, and so on. It would not be unreasonable under these conditions for the two of you to end up in the correlated equilibrium of Table 23.5. Still, there is no suggestion here that your initial inclination to play R_1 is somehow responsible causally for your opponent's credences. Her credences are what they are because she suspects that you are inclined to play R_1, but neither your decision to play R_1 nor your actually playing of R_1 is the cause of these suspicions – she develops them solely on the basis of her knowledge of the game's structure. As in the signaling case, the acts in a correlated equilibrium are evidentially correlated but causally independent.

This explicitly causal reading of Aumann's discussion helps to clear up a perplexing feature of correlated equilibria. CE only makes sense as a rationality constraint if agents are able to treat their own actions as bits of information about the world, for it is only then that the expressions appearing to the right of the "$\geq$" signs can be understood as giving utilities for the starred acts. As Aumann notes, his model (like that of Jeffrey before him)

> does away with the dichotomy usually perceived between uncertainty about acts of nature and of personal players… . In traditional Bayesian decision theory, each decision maker is permitted to make whatever decision he wishes, after getting whatever information he gets. In our model this appears not to be the case, since the decision taken by each decision maker is part of the description of the state of the world. This sounds like a restriction on the decision maker's freedom of action. (Aumann 1987), 8.

The problem here is that the utility comparisons in CE seem to portray acts as things that happen to agents rather than things they do. Moreover, it is not clear why an agent who learns that he is surely going to perform R should need to compare it with other acts that he is sure he will not perform. Aumann tries to smooth things over by suggesting that CE describes the perspective of agents, not as choosers, but as "outside observers." He writes,

> The "outside observer" perspective is common to all differential information models in economics… . In such models, each player gets some information or "signal"; he hears only the signal sent to him, not that of others. In analyzing his situation, [the] player must

> first look at the whole picture as if he were an outside observer; he cannot ignore the possibility of his having gotten a signal other than he actually got, even though he knows that he actually did not get such a signal. This is because the other players do not know what signal he got. [He] must take the ignorance of the other players into account when deciding on his own course of action, and he cannot do this if he does not explicitly include in the model signals other than the one he actually got. (Aumann 1987), 8.

The problem with this response is that it does not tell us how a player is supposed to use the knowledge gained from looking at things "externally" (i.e., as if he were not choosing) to help him with his "internal" decision (where he must choose how to act). The point is significant, since what's missing from the outside observer standpoint is the very thing that makes something a decision problem—the fact that a choice has to be made.

In our view, talk about outside observers here is misleading. Aumann's third-person, "outside observer" perspective is the first-person *subjunctive* perspective: a view on what *would* happen if *I were to do* something different. It is an advantage of causal decision theory that it allows one to assess the rationality of acts that certainly will not be performed in the same way as one assesses the rationality of any other option. One simply appeals to whatever facts about objective chances, laws of nature, or causal relations are required to imagine the "nearest" possible world in which the act is performed, the one most similar, in relevant respects, to the actual world in which one won't perform the act. One then sees what would ensue under those circumstances. Even, for instance, if one assigns zero credence to the proposition that one will intentionally leap off the bridge one is crossing, it still makes sense on the causal theory to speak of the utility of jumping. Indeed the abysmally low utility of this action is the principal reason why one is certain not perform it. When we interpret correlated equilibria causally, then, along the lines of CE**, there is no need for an "external" perspective in decision making. All decisions are made from the first-person subjunctive perspective of "What would happen if I performed that act?"

A number of other aspects of game-theoretic reasoning can likewise be analyzed in terms of causal ratifiability.[20] Some of the most interesting work in this area is due to Harper, who has investigated the ways in which causal decision theory and ratifiability can be used to understand extensive-form games (Harper 1986, 1991). This is a natural place for issues about causation to arise, since players' choices at early stages in extensive-form games can affect other players' beliefs—and thus their acts—at later stages.

Harper proposes using the maxim of ratifiability as the first step in an analysis of game-theoretic reasoning that is "eductive" or "procedural", an analysis that seeks to supply a list of rules and deliberative procedures that agents in states of indecision can use in order to arrive at an intuitively correct equilibrium choice.[21] Rational

[20] See Shin (1991b), for instance, for an interesting ratificationist gloss on Selten's notion of a "perfect" equilibrium.

[21] On the need for such "eductive" procedures see Binmore (1987, 1988).

players, he suggests, should choose actions that maximize their unconditional causal expected utility *from among the ratifiable alternatives*. (Harper regards cases where no ratifiable strategies exist as genuinely pathological.) The idea is not simply to choose acts that maximize unconditional expected utility, since these need not be ratifiable. Nor is it to choose acts with maximal expected utility on the condition that they are performed, since these may have low unconditional utility. Rather, one first eliminates unratifiable options, and then maximizes unconditional expected utility with what is left. In carrying out the second step of this process, each player imagines her adversaries choosing among their ratifiable options by assigning zero probability to any option that wouldn't be ratifiable.

Harper shows that both in normal and in extensive-form games, players who follow these prescriptions end up choosing the intuitively "right" act in a wide range of cases. In extensive-form games, his method produces choices that are in sequential equilibrium—and perhaps most interesting, the method seems to promote a strong form of "forward induction." To illustrate the latter point, and to get a sense of how Harper's procedure works in practice, consider the game of Harsanyi and Selten (1988) shown here. [ROW]'s act *A* yields fixed payoffs, whereas [ROW]'s act *B* leads to a strategic subgame with [COLUMN]. Harsanyi and Selten argue that *(C, e)* is the unique rational solution to the subgame, and they use backwards induction to argue that *(AC, e)* is the only rational solution in the full game. Their thought is that at the initial choice point [ROW] will know that *(C, e)* would be played if the subgame were reached, so he actually faces the "truncated game" as shown below, which makes *A* the only rational choice.

Kohlberg and Mertens (1986) have objected to this reasoning on the grounds that [ROW] can perform *B* as a way of signaling [COLUMN] that he has chosen *BD* (since it would be crazy to pass up *A* for *C*), and can thus force [COLUMN] to play *f* rather than *e*. Harsanyi and Selten respond by claiming that [COLUMN] would have to regard [ROW]'s playing *B* as a mistake since, "before deciding whether [ROW] can effectively signal his strategic intentions, we must first decide what strategies are rational for the two players in the subgame, and accordingly what strategy is the rational strategy for [ROW] in the truncation game" (Harsanyi and Selten 1988), 353 (see table). Thus, we have a dispute over what sorts of effects the playing of *B* would have on [COLUMN]'s beliefs at the second choice point, and thereby on her decision. This is just the sort of case where causal decision theory can be helpful.

Harper's procedure endorses Kohlberg's and Mertens' contention. Harsanyi and Selten's preferred act for [ROW] will be unratifiable, so long as each player knows that the other chooses only ratifiable options. For *A* to be ratifiable, it would at least have to be the case that

$$U(A/A) = 4 \geq \mathcal{U}(BD/A) = \rho((BD \,\Box\!\!\rightarrow e)/A) \cdot 0 + \rho((BD \,\Box\!\!\rightarrow f)/A) \cdot 10. \tag{23.14}$$

However, since [COLUMN] must choose at the third choice point knowing that [ROW] has played *B*, but without knowing whether he has played *C* or *D*, it follows that [ROW]'s credence for $BC \,\Box\!\!\rightarrow f$ must be $\rho(f/B)$. Now at the final choice point, [COLUMN] would know that [ROW] had chosen either *BC* or *BD* or some mixture of the two, and she would have to assume that the option chosen was ratifiable (if such an option is available). *BC* clearly cannot be ratifiable, since it is dominated by *A*. Harper also shows, using a somewhat involved argument, that no mixture of *BC* and *BD* can be ratifiable.[22] *BD*, however, is ratifiable, provided that $\rho\,(BD\ \Box\!\!\rightarrow f/BD) = \rho\,(f/B) \geq {}^{4}\!/_{10}$. Thus, since only one of [ROW]'s B-acts can be ratifiable, [COLUMN] would have to assign it a probability of one if she were to find herself at the second choice point. [ROW], knowing all this and knowing that f is [COLUMN]'s only ratifiable response to *BD*, will indeed assign a high value to $\rho(f/B)$, viz. $\rho(f/B) = 1$. This in turn ensures that $\mathcal{U}(BD/A) > \mathcal{U}(A/A)$, making *A* unratifiable. *BD* is thus the only ratifiable solution to the Harsanyi/Selten game. Hence, if Harper is correct, it seems that Kohlberg and Mertens were right to reject the backwards induction argument and to think that [ROW] can use *B* to warn [COLUMN] of his intention to play *D*.

We hope this example gives the reader something of the flavor of Harper's approach. Clearly his proposal needs to be elaborated more fully before we will be able to make an informed judgment on its merits. We are confident, however, that any adequate understanding of game-theoretic reasoning will rely heavily on causal decision theory and the maxim of ratifiability.

Foundational Questions

Before any theory of expected utility can be taken seriously, it must be supplemented with a representation theorem that shows precisely how its requirements are reflected in rational preference. To prove such a theorem one isolates a small set of axiomatic constraints on preference, argues that they are requirements of rationality, and shows that anyone who satisfies them will automatically act in accordance with the theory's principle of expected utility maximization. The best known result of this type is found in Savage (1972), where it was shown that any agent whose preferences conform to the well-known Savage axioms will maximize expected utility, as defined by Eq. (23.1), relative to a (unique) probability ρ and a (unique up to positive linear transformation) utility u. Unfortunately, this result does not provide a fully satisfactory foundation for either CDT or ETD, because, as we have seen, Savage's notion of expected utility is ambiguous between a causal and an evidential interpretation. This leaves some important unfinished business for evidential and causal decision theorists, since each camp has an obligation to present a representation theorem that unambiguously captures its version of utility theory.

Evidential decision theorists were first to respond to the challenge. The key mathematical result was proved in Bolker (1966), and applied to decision theory in Jeffrey (1983). The Jeffrey/Bolker approach differs from Savage's in two significant

[22] Harper (1991), 293.

ways: First, preferences are defined over a σ-algebra of propositions that describe not only states of the world, but consequences and actions as well. Second, Savage's "Sure-thing" Principle (his postulate P3) is replaced by the weaker

- *Impartiality Axiom*: Let X, Y and Z be non-null propositions such that (a) Z is incompatible with both X and Y, (b) X and Y are indifferent in the agent's preference ordering, and (c) $(X \vee Z)$ is indifferent with $(Y \vee Z)$ but not with Z. Then, $(X \vee Z^*)$ must be indifferent with $(Y \vee Z^*)$ for any proposition Z^* incompatible with both X and Y.

In the presence of the other Bolker/Jeffrey axioms, which do not differ substantially from those used by Savage, Impartiality guarantees that the agent's preferences can be represented by a function that satisfies Eq. (23.3).[23] It also ensures that the representation will be *partition independent* in the sense that, for any partitions $\{X_i\}$ and $\{Y_j\}$ and any act A, one will always have $\mathcal{V}(A) = \sum \rho(X_i/A)\mathcal{V}(A\&X_i) = \sum \rho(Y_j/A)V(A\&Y_j)$. Thus, in EDT it does not matter how one chooses the state partition relative to which expected utilities are computed.

This contrasts sharply with Savage's theory. Formula (23.1) shows how to compute expected utilities with respect to a single partition of states, but it gives no guarantee that different choices of partitions yield the same value for $\mathcal{V}(A)$. As a consequence, Savage had to place restrictions on the state partitions that could be legitimately employed in well-posed decision problems. Strictly speaking, he said, his axioms only apply to *grand-world* decisions whose act and state partitions slice things finely enough to ensure that each act/state pair produces a consequence that is sufficiently detailed to decide every question the agent cares about. Thus, on the official view, we can only be confident that (23.1) will yield the correct value for $\mathcal{V}(A)$ when applied to state partitions in "grand-world" decisions. Savage recognized this as a significant restriction on his theory, since owing to the extreme complexity of grand-world decisions, no actual human being can ever contemplate making one. He tried to make this restriction palatable by suggesting that his axioms might be usefully applied to certain "small-world" decisions, and expressing the hope that there would be only a "remote possibility" of obtaining values for $\mathcal{V}(A)$ inconsistent with those obtained in the grand-world case. This hope, however, was never backed-up by any rigorous proof. We take it as a mark in favor of EDT that it is can solve this "problem of the small-world" by giving such a proof.

The easiest way to prove a representation result for CDT is to co-opt Savage's theorem by stipulating that well-posed decision problems must be based on partitions of states that are certain to be causally independent of acts, and then imposing Savage's axioms on such problems. A number of causal decision

[23]This representation will not be unique (except in the rare case where $\mathcal{V}$ is unbounded), for, as a simple calculation shows, if the function $\mathcal{V}(A) = \sum \rho(S_i/A)u(A, S_i)$ represents a preference ordering, and if k is such that $1 + k\mathcal{V}(X) > 0$ for all propositions X in the algebra over which $\mathcal{V}_k(A) = \sum \rho k(S_i/A)u_k(A, S_i)$ is defined, then $V_k(A) = \sum \rho k\,(S_i/A)\,u_k\,(A, S_i)$ will also represent the ordering, when $\rho k(X) = \rho(X)(1 + k\mathcal{V}(X))$ and $\mathcal{V}_k(X) = [\mathcal{V}(X)(1 + k)]/(1 + k\mathcal{V}(X))$.

theorists have endorsed the view that there is a "right" partition of states to be used for computing instrumental expected utility.[24] The Lewis formulation of CDT in terms of dependency hypotheses mentioned in the section "Conditionals and their probabilities" above is an example of this strategy (Lewis 1981). Minor intramural squabbles aside, the idea is that each element in the privileged partition, often denoted $\mathbf{K} = \{K_j\}$ following Skyrms (1980), should provide a maximally specific description of one of the ways in which things that the agent cares about might depend on what she does.[25] It is characteristic of such partitions that they will be related to the agent's subjective probability by[26]

- *Definiteness of Outcome*: For any proposition O that the agent cares about (in the sense of not being indifferent between O and $\neg O$), any action A, and any $K_j \in \mathbf{K}$, either $\rho((A \Box\!\!\rightarrow O)/K_j) = 1$ or $\rho((A \Box\!\!\rightarrow \neg O)/K_j) = 1$.
- *Instrumental Act Independence*: $\rho([A \Box\!\!\rightarrow K_j] \leftrightarrow K_j) = 1$ for all acts A and states K_j.

The first of these ensures that $u\,(A,\ S_i)$ has the same value for each Savage-state S_i in K_j, and thus that

$$\mathcal{U}(A) = \sum_i \rho(A \Box\!\!\rightarrow S_i) u(A, S_i) = \sum_j \rho(A \Box\!\!\rightarrow K_j) \mathcal{U}(A, K_j)$$

The second condition then guarantees that $\mathcal{U}(A) = \sum_j \rho(K_j)\mathcal{U}(A, K_j)$. Since this equation has the form (23.5), it follows that if there exists a partition $\mathbf{K}$ that meets these two requirements, then one can appropriately apply the Savage axioms to actions whose outcomes are specified in terms of it. Fortunately, the required partition is certain to exist, because it is always possible to find a $\mathbf{K} = \{K_i\}$ such that (i) K_j entails either $(A \Box\!\!\rightarrow O)$ or $(A \Box\!\!\rightarrow \neg O)$ for every O the agent cares about, and (ii) $([A \Box\!\!\rightarrow K_j] \leftrightarrow K_j)$ is a truth of logic for all A and K_j.[27] The general existence of such partitions lets us use Savage's representation theorem as a foundation for CDT, subject to the proviso that the Savage axioms should only be applied to decisions framed in terms of a $\mathbf{K}$-partition.

The trouble with this strategy is that the partition dependence of Savage's theory is carried over into CDT. The need for a partition-independent formulation of CDT

[24] See, for example, Skyrms (1980), Lewis (1981), Armendt (1986).

[25] Notice that states are being viewed here as functions from acts to outcomes, whereas acts are taken as unanalyzed objects of choice (that is, as propositions the agent can make true or false as she pleases). This contrasts with Savage's well-known formalization in which acts are protrayed as functions from states to outcomes, and states are left as unanalyzed objects of belief. Less hangs on this distinction than one might think. When one adopts the perspective of Jeffrey (1967, 1983) and interprets both states and actions as propositions, and views outcomes as conjunctions of these propositions, the two analyses become interchangeable.

[26] Here we are following Gibbard (1986).

[27] An explicit construction of $\mathbf{K}$ can be found in Gibbard and Harper (1978).

has been argued by Sobel (1989), and Eells (1982) has suggested that EDT should be preferred to CDT on this basis alone. The main difficulty is the problem of "small-worlds", which threatens to make the theory inapplicable, in the strictest sense, to all the decision problems people actually consider. Gibbard (1986) goes a certain distance toward alleviating this difficulty by, in effect, showing how to find the smallest **K**-partition for a given decision problem, but his partition is still rather "grand", and a fully partition-invariant version of CDT would still be desirable.

Armendt (1986) proves a representation result that does provide a formulation of CDT that is partition-independent, even though it does not do away with the notion of a **K**-partition. He takes the conditional decision theory of Fishburn (1974) as his starting point. The basic concept here is that of the utility of a prospect X on the hypothesis that some condition C obtains. If $\{C_1, C_2, \ldots, C_n\}$ is a partition of C, these conditional utilities are governed by the (partition independent) equation:

$$\mathcal{U}(X/C) = \sum \rho(C_i/C)\mathcal{U}(X/C_i) \tag{23.15}$$

which shows how X's utility given C depends on its utilities given the various ways in which C might be true. Notice that (23.15) allows for a distinction between an act A's unconditional utility and its utility conditional on its own performance. These are given respectively by

$$\mathcal{U}(A) = \mathcal{U}(A/A \vee \neg A) = \sum \rho(S_i)\mathcal{U}(A/S_i)$$

$$\mathcal{U}(A/A) = \sum \rho(S_i/A)\mathcal{U}(A/A \,\&\, S_i)$$

where the state partition may be chosen arbitrarily. In a suggestion that bears some similarities to Jeffrey's ratificationist proposal, Armendt argues that decision problems in which an agent's unconditional preference for A differs from her preference for A conditional on itself are just the kinds of cases in which A's auspiciousness diverges from its instrumental expected utility. This suggests a way of characterizing **K**-partitions directly in terms of the agent's preferences. Armendt's thought is that the elements of an appropriate **K**-partition should "screen-off" differences in value between unconditional A and A-conditional-on-A, so that the agent is indifferent between A given K_i and A given $(A\&K_i)$ for every i. When this is so, we will have $\sum \rho(K_i)\mathcal{U}(A/K_i) = \sum \rho(K_i/A)\mathcal{U}(A/K_i)$, and Armendt shows that the conditional utilities $\mathcal{U}(A/K_i)$ can be eliminated in favor of the unconditional news values $\mathcal{V}(A\&K_i)$ as long as there exists at least one partition of "consequences" $\mathbf{O} = \{O_j\}$ such that the agent's unconditional utility for $(A\&O_j\&K_i)$ is equal to her utility for A conditional on $(A\&O_j\&K_i)$. When such a **K** and **O** exist, we have $\mathcal{U}(A) = \sum \rho(K_i)\mathcal{V}(A \,\&\, K_i)$, which is precisely the condition in which CDT and EDT coincide. What Armendt shows, then, is that an appropriate representation theorem for CDT can be obtained by supplementing Fishburn's conditional decision theory with the assumption that every act A can be associated with a **K**-partition such that $A/K_i \approx A/(A\&K_i)$ for all i, and a partition of consequences **O** (dependent on A and **K**) such that $(A\&O_j\&K_i) \approx A/(A\&O_j\&K_i)$.

This is a nice result. Since Fishburn's theory is partition independent, it follows that CDT will be as well, provided that at least one pair of partitions **K, O** exist for each act A. The crucial questions are whether such partitions do exist in general, and whether we should think that the condition that defines **K** really does guarantee that $\mathcal{U}(A)$ and $\mathcal{V}(A)$ coincide. On this latter point we have our doubts, but even if it is granted, it seems unlikely to us, in the absence of further argument, that the appropriate **K**-partitions will exist in all cases where we would want to apply CDT. Indeed, it would be useful to have a representation theorem for CDT that does not need to assume the existence of any special partition of states or any canonical form for a decision problem to take.

A representation theorem with these desirable features is proven in Joyce (1999). Joyce sees both EDT and CDT as instances of an abstract conditional expected utility whose basic concept is that of the utility of an act A on the supposition that some condition C obtains. Joyce begins by characterizing supposition, or provisional belief revision, in terms sufficiently general to subsume Bayesian conditioning and Lewis's imaging as special cases. Given a subjective probability ρ defined over a σ-algebra Ω, and a distinguished subset $\mathcal{C}$ of *conditions* in Ω,[28] a *supposition for ρ relative to $\mathcal{C}$* is a function $\rho(\bullet \mid \bullet)$ from $\Omega \times \mathcal{C}$ into the real numbers that satisfies

(a) $\rho(\bullet \mid C)$ is a countably additive probability on Ω for every $C \in \mathcal{C}$.
(b) $\rho(C \mid C) = 1$ for all $C \in \mathcal{C}$.
(c) $\rho(X \mid C \vee \neg C) = \rho(X)$ for all $X \in \Omega$.
(d) $\rho(X \mid B\&C) \geq \rho(X\&B \mid C)$ for all $X \in \Omega$ whenever $(B\&C) \in \mathcal{C}$.
(e) Let B and C be mutually incompatible conditions in $\mathcal{C}$. Then if one has $\rho(X \mid B) \geq \rho(X \mid C)$, then one has $\rho(X \mid B \vee C) \geq \rho(X \mid C)$, with equality if either $\rho(X \mid B) = \rho(X \mid C)$ or $\rho(B \mid B \vee C) = 0$.

The reader can verify that the ordinary conditional probability $\rho(X/C)$—that is, $\rho(X\&C)/\rho(C)$—is a supposition for p relative to $\mathcal{C} = \{C \in \Omega : \rho(C) > 0\}$. In fact, for any set of conditions $\mathcal{C}$ (even those containing conditions with zero prior probability), one can show that any map $\rho(\bullet \mid \bullet)$ defined on $\Omega \times \mathcal{C}$ that satisfies (a) – (c) plus

- *Bayes's Law*: $\rho(B \mid C)\rho(X \mid B\&C) = \rho(X \mid C)\rho(B \mid X\&C)$ whenever we have $(B\&C), (X\&C) \in \mathcal{C}$.

is a supposition.[29] The imaging function $p^C = \rho(C \;\Box\!\!\rightarrow \bullet)$ associated with a similarity relation among possible worlds is also a supposition, but not typically

[28]The set $\mathcal{C}$ always takes the form $\mathcal{C} = \Omega \sim I$, where the *ideal* I is a collection of Ω-propositions that contains the contradictory event $(X\&\neg X)$, is closed under countable disjunctions, and which contains $(X\&Y)$ whenever $X \in I$ and $Y \in \Omega$.

[29]These Bayesian suppositions were defined in Renyi (1955), and have come to be called "Popper measures" in the philosophical literature, after Popper (1934). Interested readers may consult van Fraassen (1995), Hammond (1994), and McGee (1994) for informative discussions of Popper measures.

one that satisfies Bayes's Law. There are also suppositions that are neither Bayesian nor instances of imaging.

Joyce, impressed by the need for partition-independent formulations of utility theories, uses the notion of a supposition to define an abstract conditional expected utility to be thought of as "utility under a supposition". Its (partition independent) basic equation is

$$\begin{aligned} \mathcal{V}(A|C) &= \sum_i \frac{\rho(S_i \,\&\, A|C)}{\rho(A|C)} u(A, S_i) \\ &= \sum_i \frac{\rho(X_i \,\&\, A|C)}{\rho(A|C)} \mathcal{V}(A \,\&\, X_i|C) \text{ for any partition } \{X_i\}, \end{aligned} \tag{23.16}$$

where $\rho(\bullet \mid \bullet)$ might be any supposition function for ρ defined relative to a set of conditions C that contains propositions describing all an agent's actions (as well as other things). Since (23.16) is just EDT's (23.3) with $\rho(\bullet \mid C)$ substituted in for $\rho(\bullet)$, $\mathcal{V}(A|C)$ gives A's auspiciousness on the supposition that condition C obtains, where this supposition may be any provisional belief revision that satisfies (a) – (e).

As with Fishburn's theory, there is no guarantee that A's unconditional utility, which is now just $\mathcal{V}(A)$, will coincide with its utility conditional on itself,

$$\mathcal{V}(A|A) = \sum_i \rho(S_i|A) u(A, S_i)$$

The sole exception occurs when $\rho(\bullet \mid \bullet)$ is Bayesian, for in that case we have $\mathcal{V}(A) = \mathcal{V}(A \mid A)$, because $\rho(Si \mid A) = \rho(S_i \& A)/\rho(A)$ for all i. Given that $\mathcal{V}(A)$ and $\mathcal{V}(A \mid A)$ can differ in general, it becomes a live question whether a decision maker should choose acts that maximize her unconditional expected utility or choose acts that maximize expected utility conditional on the supposition that they are performed. Joyce argues, on grounds having nothing to do with the conflict between CDT and EDT, that a choiceworthy action is always one that maximizes expected utility on the condition that it is performed. The rational decision maker's objective, in other words, should always be to choose an A such that $\mathcal{V}(A \mid A) \geq \mathcal{V}(B \mid B)$ for all alternatives B. Neither evidential nor causal decision theorists will dispute this point, since the former endorse the prescription to maximize $\mathcal{V}(A \mid A)$ when the supposition function is $\rho(A \mid C) = \rho_c(A)$, which makes $\mathcal{V}\,(A|A)$ equal A's auspiciousness, and the latter endorse it when $\rho(A \mid C) = \rho(C \,\Box\!\!\rightarrow A)$, which makes $\mathcal{V}\,(A \mid A)$ equal A's instrumental expected utility. Thus, EDT and CDT are both instances of abstract conditional utility theory. The difference between them has to do not with the basic form of the utility function or with the connection between expected utility and choiceworthiness, but with the correct type of supposition to use in decision making contexts.

Once we recognize this, it becomes clear that the problem of proving a representation theorem for CDT can be subsumed under the more general problem of proving a representation theorem for an abstract conditional utility theory. And, since the function $\mathcal{V}$ (• | C) obeys Eq. (23.3) relative to any fixed condition C, this latter problem can be solved by extending the Bolker/Jeffrey axioms for unconditional preferences to conditional preferences, and showing that anyone who satisfies the axioms is sure to have conditional preferences that can be represented by some function $\mathcal{V}(A \mid C)$ of form (23.16) that is defined relative to a supposition function $\rho(\bullet \mid \bullet)$ for her subjective probability ρ.

Joyce was able to accomplish this. We refer to reader to (1999) for the details, which turn out to be rather complicated, but the basic idea is straightforward. One starts by imagining an agent with a system of conditional preferences of the form: X on the supposition that B is weakly preferable to Y on the supposition that C, written $X \mid B \geq Y \mid C$. One assumes that this ranking obeys the usual axioms: transitivity, connectedness, a continuity principle, an Archimedean axiom, and so on. One also requires each *section* of the ranking $X \mid C \geq Y \mid C$, for C fixed, to satisfy the Bolker/Jeffrey axioms. Bolker's theorem then ensures that each section will be associated with a family r_C of ($\mathcal{V}$, ρ_C) pairs that satisfy Eq. (23.3) and that represent $X \mid C \geq Y \mid C$. Different r_C-pairs will be related by the equations $\rho_C{}^*(X) = \rho_C(X)[1 + k\mathcal{V}_C(X)]$ and $\mathcal{V}_C{}^*(X) = \mathcal{V}_C(X)[(k+1)/(1 + k\mathcal{V}_C(X)]$, where k is any real number such that $[1 + k\mathcal{V}(X)] > 0$ for all propositions X such that $\mathcal{V}(X)$ is defined. The trick to proving a representation theorem for conditional decision theory is to find further constraints on conditional preferences under which it is possible to select a unique ($\mathcal{V}$, P_C) pair from each r_C in such a way that $\mathcal{V}(X) \geq \mathcal{V}$ (X) is guaranteed to hold whenever $X \mid B \geq Y \mid C$. The main axiom that is needed is the following generalization of Impartiality:

- Let X_1, X_2, and X_3, and Y_1, Y_2, and Y_3, be mutually incompatible, and suppose that

$$X_1 \,|B \approx Y_1|\, C > (X_1 \vee X_2)\, |B \approx (Y_1 \vee Y_2)|\, C > X_2\, |B \approx Y_2|\, C \qquad (23.17)$$

holds for some conditions B and C. Then, if

$$X_1|B \approx Y_1|C \not\approx X_3|B \approx Y_3|C \not\approx (X_1 \vee X_3)|B \approx (Y_1 \vee Y_3)|C \qquad (23.18)$$

then

$$(X_1 \vee X_2 \vee X_3)\, |B \approx (Y_1 \vee Y_2 \vee Y_3)|\, C. \qquad (23.19)$$

This is not as complicated as it looks. If clause (23.17) holds, the only sort of conditional utility that will represent $X \mid C \geq Y \mid B$ will be one in which $\rho(X_1 \mid B)/\rho(X_1 \vee X_2 \mid B) = \rho(Y_1 \mid C)/\rho(Y_1 \vee Y_2 \mid C)$. Likewise, if (23.18) holds, then the representation must be one in which $\rho(X_1 \mid B)/\rho(X_1 \vee X_3 \mid B) = \rho(Y_1 \mid C)/\rho(Y_1 \vee Y_3 \mid C)$. Together these two equalities entail that $\rho(X_1 \mid B)/\rho(X_1 \vee X_2 \vee X_3 \mid B) = \rho(Y_1 \mid C)/\rho(Y_1 \vee Y_2 \vee Y_3 \mid C)$, and this is just what (23.19) guarantees.

Using this axiom as the main formal tool, Joyce is able to construct a full conditional expected utility representation for the ranking $X \mid C \geq Y \mid B$. By adding further conditions, one can ensure either that the representation's supposition function will be Bayesian or that it will arise via imaging from some similarity relation among possible worlds. In this way, both EDT and CDT are seen to have a common foundation in the abstract theory of conditional expected utility.

Conclusion

While the classical theory of Ramsey, de Finetti and Savage remains our best account of rational choice, its development has yet to be completed. An adequate theory should explain the role in decision making of causal thinking. True, a decision theory that did without causal propositions would have been nice: Cause and effect have long puzzled and divided philosophers and scientists, and theoretical discussion of causal methodology remains underdeveloped.[30] In decision making, though, we are stuck with causal thinking. Rational choice always involves judgements of how likely an option is to have various desirable consequences—and such judgements, we have argued, require the decision maker to have views, explicit or implicit, about causal or counterfactual relationships. Nothing else can substitute for these causal beliefs. The conditional credences employed by evidential decision theory cannot, because they are unable to distinguish causation from correlation. More refined "screening" techniques, while better at capturing causal connections, fail to apply in a important class of cases. To specify the kind of case to which they do apply, we must, one way or another, invoke causal relations.

We should not find this need for causal notions distressing. We draw causal conclusions all the time, after all, and scientists are able to glean causal tendencies from experiment and statistical data, using methods of high sophistication. Still, no one theory of causal notions has the precision and orthodox status of, say, the standard theory of subjective probability. Thus, an adequate decision theory, if we are right, must depend on new advances in our understanding of causation and its relation to rational belief. We might have wished that theoretical life had turned out easier, but as matters stand, important work on the foundations of utility theory remains to be done.

References

Adams, E. W. (1975). *The logic of conditionals*. Dordrecht: Reidel.

Armendt, B. (1986). A foundation for causal decision theory. *Topoi, 5*, 3–19.

Aumann, R. (1974). Subjectivity and correlation in randomized strategies. *Journal of Mathematical Economics, 1*, 67–96.

[30]Again, giant steps have been taken in this area since this article first appeared, especially by Spirtes et al. (1993) and Pearl (2000).

Aumann, R. (1987). Correlated equilibrium as an expression of Bayesian rationality. *Econometrica, 55*, 1–18.
Binmore, K. (1987). Modeling rational players I. *Economics and Philosophy, 3*, 179–212 (Reprinted, Binmore 1990).
Binmore, K. (1988). Modeling rational players II. *Economics and Philosophy, 4*, 9–55 (Reprinted Binmore 1990).
Binmore, K. (1990). *Essays on the foundations of game theory*. Oxford: Blackwell.
Bolker, E. (1966). Functions resembling quotients of measures. *Transactions of the American Mathematical Society, 124*, 292–312.
Eells, E. (1982). *Rational decision and causality*. Cambridge: Cambridge University Press.
Fine, K. (1975). Review of Lewis [24]. *Mind, 84*, 451–458.
Fishburn, P. (1974). A mixture-set axiomatization of conditional subjective expected utility. *Econometrica, 41*, 1–25.
Gibbard, A. (1986). A characterization of decision matrices that yield instrumental expected utility. In L. Daboni, A. Montesano, & M. Lines (Eds.), *Recent developments in the foundations of utility and risk theory* (pp. 139–148). Dordrecht: Reidel.
Gibbard, A., & Harper, W. L. (1978). Counterfactuals and two kinds of expected utility. In C. A. Hooker, J. J. Leach, & E. F. McClennen (Eds.), *Foundations and applications of decision theory* (Vol. I). Dordrecht: Reidel.
Hammond, P. (1994). Elementary non-Archimedean representations of probability for decision theory and games. In P. Humphries (Ed.), *Patrick Suppes: Scientific philosopher* (Vol. 1, pp. 25–61). Dordrecht: Kluwer Academic Publishers.
Harper, W. (1986). Mixed strategies and ratifiability in causal decision theory. *Erkenntnis, 24*, 25–26.
Harper, W. (1991). Ratifiability and refinements in two-person noncooperative games. In M. Bacharach & S. Hurley (Eds.), *Foundations of decision theory* (pp. 263–293). Oxford: Basil Blackwell.
Harsanyi, J., & Selten, R. (1988). *A general theory of equilibrium selection in games*. Cambridge: MIT Press.
Holland, P. W. (1986). Statistics and causal inference. *Journal of the American Statistical Association, 81*, 945–960.
Horgan, T. (1981). Counterfactuals and Newcomb's problem. *Journal of Philosophy, 68*, 331–356.
Howard, J. V. (1988). Cooperation in the prisoner's dilemma. *Theory and Decision, 24*, 203–213.
Jeffrey, R. (1967). *The logic of decision*. New York: McGraw Hill.
Jeffrey, R. (1983). *The logic of decision* (2nd ed.). Chicago: University of Chicago Press.
Joyce. (1999). *The foundations of causal decision theory*. Cambridge: Cambridge University Press.
Kohlberg, E., & Mertens, J. (1986). On the strategic stability of equilibria. *Econometrica, 54*, 1003–1037.
Lewis, D. K. (1973). *Counterfactuals*. Cambridge: Harvard University Press.
Lewis, D. K. (1976). Probabilities of conditionals and conditional probabilities. *Philosophical Review, 85*, 297–315 (Reprinted, Lewis 1986, 133–152).
Lewis, D. K. (1979). Counterfactual dependence and time's arrow. *Noûs, 13*, 455–476 (Reprinted, Lewis 1986, 32–52).
Lewis, D. K. (1979). Prisoner's dilemma is a Newcomb problem. *Philosophy and Public Affairs, 8*, 235–240 (Reprinted, Lewis 1986, 299–304).
Lewis, D. K. (1980). A subjectivist's guide to objective chance. In Richard C. Jeffrey (Ed.), *Studies in inductive logic and probability* (Vol. 2). Berkeley: University of California Press. Reprinted (Lewis 1986), 83–113.
Lewis, D. K. (1981). Causal decision theory. *Australasian Journal of Philosophy, 59*, 5–30 (Reprinted, Lewis 1986, 305–337).
Lewis, D. K. (1986). *Philosophical papers* (Vol. 2). Oxford: Oxford University Press.
Luce, R. D., & Krantz, D. H. (1971). Conditional expected utility. *Econometrica, 39*, 253–271.
McGee, V. (1994). Learning the impossible. In E. Eells & B. Skyrms (Eds.), *Probability and conditionals* (pp. 179–197). Cambridge: Cambridge University Press.

Nozick, R. (1969). Newcomb's problem and two principles of choice. In N. Rescher (Ed.), *Essays in honor of Carl G. Hempel*. Dordrecht-Holland: Reidel.
Piccione, M., & Rubinstein, A. (1997). On the interpretation of decision problems with imperfect recall. *Games and Economic Behavior, 20*, 3–24.
Popper, K. (1934). *Logik der Forschung*. Vienna: Springer. Translated as *The logic of scientific discovery* (London: Hutchinson, 1959).
Pearl, J. (2000). *Causality: Models, reasoning, and inference*. New York: Cambridge University Press.
Renyi, A. (1955). On a new axiomatic theory of probability. *Acta Mathematica Academiae Scientiarium Hungaricae, 6*, 285–335.
Robins, J. (1986). A new approach to causal inference in mortality studies with a sustained exposure period – applications to control of the healthy workers survivor effect. *Mathematical Modeling, 7*, 1393–1512.
Rubin, D. (1974). Estimating causal effects of treatments in randomized and nonrandomized studies. *Journal of Educational Psychology, 66*, 688–701.
Savage, L. J. (1972). *The foundations of statistics*. New York: Dover (First edition 1954).
Shin, H. S. (1991a). A reconstruction of Jeffrey's notion of ratifiability in terms of counterfactual beliefs. *Theory and Decision, 31*, 21–47.
Shin, H. S. (1991b). Two notions of ratifiability and equilibrium in games. In M. Bacharach & S. Hurley (Eds.), *Foundations of decision theory* (pp. 242–262). Oxford: Basil Blackwell.
Simon, H. A. (1957). *Models of man*. New York: Wiley.
Skyrms, B. (1980). *Causal necessity*. New Haven: Yale University Press.
Skyrms, B. (1984). *Pragmatics and empiricism*. New Haven: Yale University Press.
Skyrms, B. (1990). Ratifiability and the logic of decision. *Midwest Studies in Philosophy, 15*, 44–56.
Sobel, J. H. (1989). Partition theorems for causal decision theories. *Philosophy of Science, 56*, 70–95.
Spirtes, P., Glymour, C., & Scheines, R. (1993). *Causation, prediction, and search* (Lecture Notes in Statistics, Vol. 81). New York: Springer. ISBN: 978-1-4612-7650-0 (Print) 978-1-4612-2748-9 (Online).
Stalnaker, R. (1968). A theory of conditionals. In *Studies in logical theory* (American philosophical quarterly monograph series 2). Oxford: Blackwell.
Stalnaker, R. (1972). Letter to David Lewis. In W. Harper, R. Stalnaker, & G. Pearce (Eds.), *Ifs: Conditionals, belief, decision, chance, and time* (pp. 151–152). Dordrecht: Reidel.
Stalnaker, R. (1981). Letter to David Lewis of May 21, 1972. In W. L. Harper, R. Stalnaker, & G. Pearce (Eds.), *Ifs: Conditionals, belief, decision, chance, and time*. Dordrecht-Holland: Reidel.
Stalnaker, R., & Thomason, R. (1970). A semantic analysis of conditional logic. *Theoria, 36*, 23–42.
Van Fraassen, B. (1995). Fine-grained opinion, probability, and the logic of belief. *Journal of Philosophical Logic, 24*, 349–377.

Chapter 24
Advances in Prospect Theory: Cumulative Representation of Uncertainty

Amos Tversky and Daniel Kahneman

Expected utility theory reigned for several decades as the dominant normative and descriptive model of decision making under uncertainty, but it has come under serious question in recent years. There is now general agreement that the theory does not provide an adequate description of individual choice: a substantial body of evidence shows that decision makers systematically violate its basic tenets. Many alternative models have been proposed in response to this empirical challenge (for reviews, see Camerer 1989; Fishburn 1988; Machina 1987). Some time ago we presented a model of choice, called prospect theory, which explained the major violations of expected utility theory in choices between risky prospects with a small number of outcomes (Kahneman and Tversky 1979; Tversky and Kahneman 1986). The key elements of this theory are (1) a value function that is concave for gains, convex for losses, and steeper for losses than for gains, and (2) a nonlinear transformation of the probability scale, which overweights small probabilities and underweights moderate and high probabilities. In an important later development, several authors (Quiggin 1982; Schmeidler 1989; Yaari 1987; Weymark 1981) have advanced a new representation, called the rank-dependent or the cumulative functional, that transforms cumulative rather than individual probabilities. This article presents a new version of prospect theory that incorporates the cumulative functional and extends the theory to uncertain as well to risky prospects with any number of outcomes. The resulting model, called cumulative prospect theory,

Amos Tversky was deceased at the time of publication.

A. Tversky (deceased)
Stanford University, Stanford, CA, USA

D. Kahneman (✉)
Princeton University, Princeton, NJ, USA
e-mail: kahneman@princeton.edu

H. Arló-Costa et al. (eds.), *Readings in Formal Epistemology*, Springer Graduate Texts in Philosophy 1, DOI 10.1007/978-3-319-20451-2_24

combines some of the attractive features of both developments (see also Luce and Fishburn 1991). It gives rise to different evaluations of gains and losses, which are not distinguished in the standard cumulative model, and it provides a unified treatment of both risk and uncertainty.

To set the stage for the present development, we first list five major phenomena of choice, which violate the standard model and set a minimal challenge that must be met by any adequate descriptive theory of choice. All these findings have been confirmed in a number of experiments, with both real and hypothetical payoffs.

Framing Effects The rational theory of choice assumes description invariance: equivalent formulations of a choice problem should give rise to the same preference order (Arrow 1982). Contrary to this assumption, there is much evidence that variations in the framing of options (e.g., in terms of gains or losses) yield systematically different preferences (Tversky and Kahneman 1986).

Nonlinear Preferences According to the expectation principle, the utility of a risky prospect is linear in outcome probabilities. Allais's (1953) famous example challenged this principle by showing that the difference between probabilities of .99 and 1.00 has more impact on preferences than the difference between 0.10 and 0.11. More recent studies observed nonlinear preferences in choices that do not involve sure things (Camerer and Ho 1991).

Source Dependence People's willingness to bet on an uncertain event depends not only on the degree of uncertainty but also on its source. Ellsberg (1961) observed that people prefer to bet on an urn containing equal numbers of red and green balls, rather than on an urn that contains red and green balls in unknown proportions. More recent evidence indicates that people often prefer a bet on an event in their area of competence over a bet on a matched chance event, although the former probability is vague and the latter is clear (Heath and Tversky 1991).

Risk Seeking Risk aversion is generally assumed in economic analyses of decision under uncertainty. However, risk-seeking choices are consistently observed in two classes of decision problems. First, people often prefer a small probability of winning a large prize over the expected value of that prospect. Second, risk seeking is prevalent when people must choose between a sure loss and a substantial probability of a larger loss.

Loss Aversion One of the basic phenomena of choice under both risk and uncertainty is that losses loom larger than gains (Kahneman and Tversky 1984; Tversky and Kahneman 1991). The observed asymmetry between gains and losses is far too extreme to be explained by income effects or by decreasing risk aversion.

The present development explains loss aversion, risk seeking, and nonlinear preferences in terms of the value and the weighting functions. It incorporates a framing process, and it can accommodate source preferences. Additional phenomena that lie beyond the scope of the theory—and of its alternatives—are discussed later.

The present article is organized as follows. Section "Cumulative prospect theory" introduces the (two-part) cumulative functional; section "Relation to previous work" discusses relations to previous work; and section "Values and weights" describes the qualitative properties of the value and the weighting functions. These properties are tested in an extensive study of individual choice, described in section "Experiment", which also addresses the question of monetary incentives. Implications and limitations of the theory are discussed in section "Discussion". An axiomatic analysis of cumulative prospect theory is presented in the appendix.

Theory

Prospect theory distinguishes two phases in the choice process: framing and valuation. In the framing phase, the decision maker constructs a representation of the acts, contingencies, and outcomes that are relevant to the decision. In the valuation phase, the decision maker assesses the value of each prospect and chooses accordingly. Although no formal theory of framing is available, we have learned a fair amount about the rules that govern the representation of acts, outcomes, and contingencies (Tversky and Kahneman 1986). The valuation process discussed in subsequent sections is applied to framed prospects.

Cumulative Prospect Theory

In the classical theory, the utility of an uncertain prospect is the sum of the utilities of the outcomes, each weighted by its probability. The empirical evidence reviewed above suggests two major modifications of this theory: (1) the carriers of value are gains and losses, not final assets; and (2) the value of each outcome is multiplied by a decision weight, not by an additive probability. The weighting scheme used in the original version of prospect theory and in other models is a monotonie transformation of outcome probabilities. This scheme encounters two problems. First, it does not always satisfy stochastic dominance, an assumption that many theorists are reluctant to give up. Second, it is not readily extended to prospects with a large number of outcomes. These problems can be handled by assuming that transparently dominated prospects are eliminated in the editing phase, and by normalizing the weights so that they add to unity. Alternatively, both problems can be solved by the rank-dependent or cumulative functional, first proposed by Quiggin (1982) for decision under risk and by Schmeidler (1989) for decision under uncertainty. Instead of transforming each probability separately, this model transforms the entire cumulative distribution function. The present theory applies the cumulative functional separately to gains and to losses. This development extends prospect theory to uncertain as well as to risky prospects with any number of outcomes while preserving most of its essential features. The differences between the cumulative and the original versions of the theory are discussed in section "Relation to previous work".

Let S be a finite set of states of nature; subsets of S are called events. It is assumed that exactly one state obtains, which is unknown to the decision maker. Let X be a set of consequences, also called outcomes. For simplicity, we confine the present discussion to monetary outcomes. We assume that X includes a neutral outcome, denoted 0, and we interpret all other elements of X as gains or losses, denoted by positive or negative numbers, respectively.

An uncertain prospect f is a function from S into X that assigns to each state $s \in S$ a consequence $f(s) = x$ in X. To define the cumulative functional, we arrange the outcomes of each prospect in increasing order. A prospect f is then represented as a sequence of pairs (x_i, A_i), which yields x_i if A_i occurs, where $x_i > x_j$ iff $i > j$, and (A_i) is a partition of S. We use positive subscripts to denote positive outcomes, negative subscripts to denote negative outcomes, and the zero subscript to index the neutral outcome. A prospect is called strictly positive or positive, respectively, if its outcomes are all positive or nonnegative. Strictly negative and negative prospects are defined similarly; all other prospects are called mixed. The positive part of f denoted f^+, is obtained by letting $f^+(s) = f(s)$ if $f(s) > 0$, and $f^+(s) = 0$ if $f(s) \leq 0$. The negative part of f, denoted f^-, is defined similarly.

As in expected utility theory, we assign to each prospect f a number $V(f)$ such that f is preferred to or indifferent to g iff $V(f) \geq V(g)$. The following representation is defined in terms of the concept of *capacity* (Choquet 1955), a nonadditive set function that generalizes the standard notion of probability. A capacity W is a function that assigns to each $A \subset S$ a number $W(A)$ satisfying $W(\emptyset) = 0$, $W(S) = 1$, and $W(A) \geq W(B)$ whenever $A \supset B$.

Cumulative prospect theory asserts that there exist a strictly increasing value function v: $X \rightarrow \text{Re}$, satisfying $v(x_0) = v(0) = 0$, and capacities W^+ and W^-, such that for $f = (x_i, A_i)$, $-m \leq i \leq n$,

$$V(f) = V\left(f^+\right) + V\left(f^-\right),$$

$$V\left(f^+\right) = \sum_{i=0}^{n} \pi_i^+ v\left(x_i\right), \quad V\left(f^-\right) = \sum_{i=-m}^{0} \pi_i^- v\left(x_i\right), \tag{24.1}$$

where the decision weights $\pi^+\left(f^+\right) = \left(\pi_0^+, \ldots, \pi_n^+\right)$ and $\pi^-\left(f^-\right) = \left(\pi_{-m}^-, \ldots, \pi_0^-\right)$ are defined by:

$$\begin{aligned}
\pi_n^+ &= W^+\left(A_n\right), \pi_{-m}^- = W^-\left(A_{-m}\right), \\
\pi_i^+ &= W^+\left(A_i \cup \cdots \cup A_n\right) - W^+\left(A_{i+1} \cup \cdots \cup A_n\right), 0 \leq i \leq n-1, \\
\pi_i^- &= W^-\left(A_{-m} \cup \cdots \cup A_i\right) - W^-\left(A_{-m} \cup \cdots \cup A_{i-1}\right), 1-m \leq i \leq 0.
\end{aligned}$$

Letting $\pi_i = \pi_i^+$ if $i \geq 0$ and $\pi_i = \pi_i^-$ if $i < 0$, Eq. (24.1) reduces to

$$V(f) = \sum_{i=-m}^{n} \pi_i v\left(x_i\right). \tag{24.2}$$

The decision weight π_i^+, associated with a positive outcome, is the difference between the capacities of the events "the outcome is at least as good as x_i" and "the outcome is strictly better then x_i." The decision weight π_i^-, associated with a negative outcome, is the difference between the capacities of the events "the outcome is at least as bad as x_i" and "the outcome is strictly worse than x_i." Thus, the decision weight associated with an outcome can be interpreted as the marginal contribution of the respective event,[1] defined in terms of the capacities W^+ and W^-. If each W is additive, and hence a probability measure, then π_i is simply the probability of A_i. It follows readily from the definitions of π and W that for both positive and negative prospects, the decision weights add to 1. For mixed prospects, however, the sum can be either smaller or greater than 1, because the decision weights for gains and for losses are defined by separate capacities.

If the prospect $f=(x_i, A_i)$ is given by a probability distribution $p(A_i)=p_i$, it can be viewed as a probabilistic or risky prospect (x_i, p_i). In this case, decision weights are defined by:

$$\begin{aligned}
\pi_n^+ &= w^+(p_n), \pi_{-m}^- = w^-(p_{-m}), \\
\pi_i^+ &= w^+(p_i + \cdots + p_n) - w^+(p_{i+1} + \cdots + p_n), 0 \le i \le n-1, \\
\pi_i^- &= w^-(p_{-m} + \cdots + p_i) - w^-(p_{-m} + \cdots + p_{i-1}), 1-m \le i \le 0.
\end{aligned}$$

where w^+ and w^- are strictly increasing functions from the unit interval into itself satisfying $w^+(0)=w^-(0)=0$, and $w^+(1)=w^-(1)=1$.

To illustrate the model, consider the following game of chance. You roll a die once and observe the result $x = 1, \ldots, 6$. If x is even, you receive \$$x$; if x is odd, you pay \$$x$. Viewed as a probabilistic prospect with equiprobable outcomes, f yields the consequences (−5, −3, −1, 2, 4, 6), each with probability 1/6. Thus, $f^+=(0, 1/2; 2, 1/6; 4, 1/6; 6, 1/6)$, and $f^-=(-5, 1/6; -3, 1/6; -1, 1/6; 0, 1/2)$. By Eq. (24.1), therefore,

$$\begin{aligned}
V(f) &= V(f^+) + V(f^-) \\
&= v(2)\left[w^+(1/2) - w^+(1/3)\right] + v(4)\left[w^+(1/3) - w^+(1/6)\right] \\
&\quad + v(6)\left[w^+(1/6) - w^+(0)\right] \\
&\quad + v(-5)\left[w^-(1/6) \quad w^-(0)\right] + v(-3)\left[w^-(1/3) - w^-(1/6)\right] \\
&\quad + v(-1)\left[w^-(1/2) - w^-(1/3)\right].
\end{aligned}$$

Relation to Previous Work

Luce and Fishburn (1991) derived essentially the same representation from a more elaborate theory involving an operation ○ of joint receipt or multiple play. Thus, $f \bigcirc g$ is the composite prospect obtained by playing both f and g, separately. The key feature of their theory is that the utility function U is additive with respect to ○ that is, $U(f \bigcirc g) = U(f) + U(g)$ provided one prospect is acceptable (i.e.,

[1] In keeping with the spirit of prospect theory, we use the decumulative form for gains and the cumulative form for losses. This notation is vindicated by the experimental findings described in section "Experiment".

preferred to the status quo) and the other is not. This condition seems too restrictive both normatively and descriptively. As noted by the authors, it implies that the utility of money is a linear function of money if for all sums of money x, y, $U(x \bigcirc y) = U(x + y)$. This assumption appears to us inescapable because the joint receipt of x and y is tantamount to receiving their sum. Thus, we expect the decision maker to be indifferent between receiving a \$10 bill or receiving a \$20 bill and returning \$10 in change. The Luce-Fishburn theory, therefore, differs from ours in two essential respects. First, it extends to composite prospects that are not treated in the present theory. Second, it practically forces utility to be proportional to money.

The present representation encompasses several previous theories that employ the same decision weights for all outcomes. Starmer and Sugden (1989) considered a model in which $w^-(p) = w^+(p)$, as in the original version of prospect theory. In contrast, the rank-dependent models assume $w^-(p) = 1 - w^+(1 - p)$ or $W^-(A) = 1 - W^+(S - A)$. If we apply the latter condition to choice between uncertain assets, we obtain the choice model established by Schmeidler (1989), which is based on the Choquet integral.[2] Other axiomatizations of this model were developed by Gilboa (1987), Nakamura (1990), and Wakker (1989a, b). For probabilistic (rather than uncertain) prospects, this model was first established by Quiggin (1982) and Yaari (1987), and was further analyzed by Chew (1989), Segal (1989), and Wakker (1990). An earlier axiomatization of this model in the context of income inequality was presented by Weymark (1981). Note that in the present theory, the overall value $V(f)$ of a mixed prospect is not a Choquet integral but rather a sum $V(f^+) + V(f^-)$ of two such integrals.

The present treatment extends the original version of prospect theory in several respects. First, it applies to any finite prospect and it can be extended to continuous distributions. Second, it applies to both probabilistic and uncertain prospects and can, therefore, accommodate some form of source dependence. Third, the present theory allows different decision weights for gains and losses, thereby generalizing the original version that assumes $w^+ = w^-$. Under this assumption, the present theory coincides with the original version for all two-outcome prospects and for all mixed three-outcome prospects. It is noteworthy that for prospects of the form $(x,p;y, 1 - p)$, where either $x > y > 0$ or $x < y < 0$, the original theory is in fact rank dependent. Although the two models yield similar predictions in general, the cumulative version—unlike the original one—satisfies stochastic dominance. Thus, it is no longer necessary to assume that transparently dominated prospects are eliminated in the editing phase—an assumption that was criticized by some authors. On the other hand, the present version can no longer explain violations of stochastic dominance in nontransparent contexts (e.g., Tversky and Kahneman 1986). An axiomatic analysis of the present theory and its relation to cumulative utility theory and to expected utility theory are discussed in the appendix; a more comprehensive treatment is presented in Wakker and Tversky (1991).

---

[2]This model appears under different names. We use *cumulative utility theory* to describe the application of a Choquet integral to a standard utility function, and *cumulative prospect theory* to describe the application of two separate Choquet integrals to the value of gains and losses.

Values and Weights

In expected utility theory, risk aversion and risk seeking are determined solely by the utility function. In the present theory, as in other cumulative models, risk aversion and risk seeking are determined jointly by the value function and by the capacities, which in the present context are called cumulative weighting functions, or weighting functions for short. As in the original version of prospect theory, we assume that v is concave above the reference point ($v''(x) \leq 0, x \geq 0$) and convex below the reference point ($v''(x) \geq 0, x \leq 0$). We also assume that v is steeper for losses than for gains $v'(x) < v'(-x)$ for $x \geq 0$. The first two conditions reflect the principle of diminishing sensitivity: the impact of a change diminishes with the distance from the reference point. The last condition is implied by the principle of loss aversion according to which losses loom larger than corresponding gains (Tversky and Kahneman 1991).

The principle of diminishing sensitivity applies to the weighting functions as well. In the evaluation of outcomes, the reference point serves as a boundary that distinguishes gains from losses. In the evaluation of uncertainty, there are two natural boundaries—certainty and impossibility—that correspond to the endpoints of the certainty scale. Diminishing sensitivity entails that the impact of a given change in probability diminishes with its distance from the boundary. For example, an increase of .1 in the probability of winning a given prize has more impact when it changes the probability of winning from .9 to 1.0 or from 0 to .1, than when it changes the probability of winning from .3 to .4 or from .6 to .7. Diminishing sensitivity, therefore, gives rise to a weighting function that is concave near 0 and convex near 1. For uncertain prospects, this principle yields subadditivity for very unlikely events and superadditivity near certainty. However, the function is not well-behaved near the endpoints, and very small probabilities can be either greatly overweighted or neglected altogether.

Before we turn to the main experiment, we wish to relate the observed nonlinearity of preferences to the shape of the weighting function. For this purpose, we devised a new demonstration of the common consequence effect in decisions involving uncertainty rather than risk. Table 24.1 displays a pair of decision problems (I and II) presented in that order to a group of 156 money managers during a workshop. The participants chose between prospects whose outcomes were contingent on the difference d between the closing values of the Dow-Jones today and tomorrow. For example, f' pays \$25,000 if d exceeds 30 and nothing otherwise. The percentage of respondents who chose each prospect is given in brackets. The independence axiom of expected utility theory implies that f is preferred to g iff f' is preferred to g'. Table 24.1 shows that the modal choice was f in problem I and g' in problem II. This pattern, which violates independence, was chosen by 53 % of the respondents.

Essentially the same pattern was observed in a second study following the same design. A group of 98 Stanford students chose between prospects whose outcomes were contingent on the point-spread d in the forthcoming Stanford-Berkeley football game. Table 24.2 presents the prospects in question. For example, g pays \$10 if

Table 24.1 A test of independence (Dow-Jones)

		A	B	C	
		If $d < 30$	If $30 \leq d \leq 35$	If $35 < d$	
Problem I:	f	\$25,000	\$25,000	\$25,000	[68]
	g	\$25,000	0	\$75,000	[32]
Problem II:	f'	0	\$25,000	\$25,000	[23]
	g'	0	0	\$75,000	[77]

Note: Outcomes are contingent on the difference d between the closing values of the Dow-Jones today and tomorrow. The percentage of respondents ($N = 156$) who selected each prospect is given in brackets

Table 24.2 A test of independence (Stanford-Berkeley football game)

		A	B	C	
		If $d < 0$	If $0 \leq d \leq 10$	If $10 < d$	
Problem I:	f	\$10	\$10	\$10	[64]
	g	\$10	\$30	0	[36]
Problem II:	f'	0	\$10	\$10	[34]
	g'	0	\$30	0	[66]

Note: Outcomes are contingent on the point-spread d in a Stanford-Berkeley football game. The percentage of respondents ($N = 98$) who selected each prospect is given in brackets

Stanford does not win, \$30 if it wins by 10 points or less, and nothing if it wins by more than 10 points. Ten percent of the participants, selected at random, were actually paid according to one of their choices. The modal choice, selected by 46 % of the subjects, was f and g', again in direct violation of the independence axiom.

To explore the constraints imposed by this pattern, let us apply the present theory to the modal choices in Table 24.1, using \$1,000 as a unit. Since f is preferred to g in problem I,

$$v(25) > v(75)W^+(C) + v(25)\left[W^+(A \cup C) - W^+(C)\right]$$

or

$$v(25)\left[1 - W^+(A \cup C) + W^+(C)\right] > v(75)W^+(C).$$

The preference for g' over f' in problem II, however, implies

$$v(75)W^+(C) > v(25)W^+(C \cup B);$$

hence,

$$W^+(S) - W^+(S - B) > W^+(C \cup B) - W^+(C). \tag{24.3}$$

Thus, "subtracting" B from certainty has more impact than "subtracting" B from $C \cup B$. Let $W_+(D) = 1 - W^+(S - D)$, and $w_+(p) = 1 - w^+(1 - p)$. It follows readily that Eq. (24.3) is equivalent to the subadditivity of W_+, that is, $W_+(B) + W_+(D) \geq W + (B \cup D)$. For probabilistic prospects, Eq. (24.3) reduces to

$$1 - w^+ (1 - q) > w^+ (p + q) - w^+(p),$$

or

$$w_+(q) + w_+(r) \geq w_+ (q + r) , q + r < 1.$$

Allais's example corresponds to the case where $p(C) = .10$, $p(B) = .89$, and $p(A) = .01$.

It is noteworthy that the violations of independence reported in Tables 24.1 and 24.2 are also inconsistent with regret theory, advanced by Loomes and Sugden (1982a, b), and with Fishburn's (1988) SSA model. Regret theory explains Allais's example by assuming that the decision maker evaluates the consequences as if the two prospects in each choice are statistically independent. When the prospects in question are defined by the same set of events, as in Tables 24.1 and 24.2, regret theory (like Fishburn's SSA model) implies independence, since it is additive over states. The finding that the common consequence effect is very much in evidence in the present problems undermines the interpretation of Allais's example in terms of regret theory.

The common consequence effect implies the subadditivity of W_+ and of w_+. Other violations of expected utility theory imply the subadditivity of W^+ and of w^+ for small and moderate probabilities. For example, Prelec (1990) observed that most respondents prefer 2% to win \$20,000 over 1% to win \$30,000; they also prefer 1% to win \$30,000 and 32% to win \$20,000 over 34% to win \$20,000. In terms of the present theory, these data imply that $w^+(.02) - w^+(.01) \geq w^+(.34) - w^+(.33)$. More generally, we hypothesize

$$w^+(p + q) - w^+(q) \geq w^+(p + q + r) - w^+(q + r), \tag{24.4}$$

provided $p + q + r$ is sufficiently small. Equation (24.4) states that w^+ is concave near the origin; and the conjunction of the above inequalities implies that, in accord with diminishing sensitivity, w^+ has an inverted S-shape: it is steepest near the endpoints and shallower in the middle of the range. For other treatments of decision weights, see Hogarth and Einhorn (1990), Prelec (1989), Viscusi (1989), and Wakker (1990). Experimental evidence is presented in the next section.

Experiment

An experiment was carried out to obtain detailed information about the value and weighting functions. We made a special effort to obtain high-quality data. To this end, we recruited 25 graduate students from Berkeley and Stanford (12 men and

13 women) with no special training in decision theory. Each subject participated in three separate one-hour sessions that were several days apart. Each subject was paid $25 for participation.

Procedure

The experiment was conducted on a computer. On a typical trial, the computer displayed a prospect (e.g., 25 % chance to win $150 and 75 % chance to win $50) and its expected value. The display also included a descending series of seven sure outcomes (gains or losses) logarithmically spaced between the extreme outcomes of the prospect. The subject indicated a preference between each of the seven sure outcomes and the risky prospect. To obtain a more refined estimate of the certainty equivalent, a new set of seven sure outcomes was then shown, linearly spaced between a value 25 % higher than the lowest amount accepted in the first set and a value 25 % lower than the highest amount rejected. The certainty equivalent of a prospect was estimated by the midpoint between the lowest accepted value and the highest rejected value in the second set of choices. We wish to emphasize that although the analysis is based on certainty equivalents, the data consisted of a series of choices between a given prospect and several sure outcomes. Thus, the cash equivalent of a prospect was derived from observed choices, rather than assessed by the subject. The computer monitored the internal consistency of the responses to each prospect and rejected errors, such as the acceptance of a cash amount lower than one previously rejected. Errors caused the original statement of the problem to reappear on the screen.[3]

The present analysis focuses on a set of two-outcome prospects with monetary outcomes and numerical probabilities. Other data involving more complicated prospects, including prospects defined by uncertain events, will be reported elsewhere. There were 28 positive and 28 negative prospects. Six of the prospects (three nonnegative and three nonpositive) were repeated on different sessions to obtain the estimate of the consistency of choice. Table 24.3 displays the prospects and the median cash equivalents of the 25 subjects.

A modified procedure was used in eight additional problems. In four of these problems, the subjects made choices regarding the acceptability of a set of mixed prospects (e.g., 50 % chance to lose $100 and 50 % chance to win x) in which x was systematically varied. In four other problems, the subjects compared a fixed prospect (e.g., 50 % chance to lose $20 and 50 % chance to win $50) to a set of prospects (e.g., 50 % chance to lose $50 and 50 % chance to win x) in which x was systematically varied. (These prospects are presented in Table 24.6.)

---

[3] An IBM disk containing the exact instructions, the format, and the complete experimental procedure can be obtained from the authors.

Table 24.3 Median cash equivalents (in dollars) for all nonmixed prospects

	Probability								
Outcomes	.01	.05	.10	.25	.50	.75	.90	.95	.99
(0, 50)			9		21		37		
(0, –50)			−8		−21		−39		
(0, 100)		14		25	36	52		78	
(0, –100)		−8		−23.5	−42	−63		−84	
(0, 200)	10		20		76		131		188
(0, –200)	−3		−23		−89		−155		−190
(0, 400)	12								377
(0, –400)	−14								−380
(50, 100)			59		71		83		
(–50, –100)			−59		−71		−85		
(50, 150)		64		72.5	86	102		128	
(–50, –150)		−60		−71	−92	−113		−132	
(100, 200)		118		130	141	162		178	
(–100, –200)		−112		−121	−142	−158		−179	

Note: The two outcomes of each prospect are given in the left-hand side of each row; the probability of the second (i.e., more extreme) outcome is given by the corresponding column. For example, the value of $9 in the upper left corner is the median cash equivalent of the prospect (0, .9; $50, .1)

Results

The most distinctive implication of prospect theory is the fourfold pattern of risk attitudes. For the nonmixed prospects used in the present study, the shapes of the value and the weighting functions imply risk-averse and risk-seeking preferences, respectively, for gains and for losses of moderate or high probability. Furthermore, the shape of the weighting functions favors risk seeking for small probabilities of gains and risk aversion for small probabilities of loss, provided the outcomes are not extreme. Note, however, that prospect theory does not imply perfect reflection in the sense that the preference between any two positive prospects is reversed when gains are replaced by losses. Table 24.4 presents, for each subject, the percentage of risk-seeking choices (where the certainty equivalent exceeded expected value) for gains and for losses with low ($p \leq .1$) and with high ($p \geq .5$) probabilities. Table 24.4 shows that for $p \geq .5$, all 25 subjects are predominantly risk averse for positive prospects and risk seeking for negative ones. Moreover, the entire fourfold pattern is observed for 22 of the 25 subjects, with some variability at the level of individual choices.

Although the overall pattern of preferences is clear, the individual data, of course, reveal both noise and individual differences. The correlations, across subjects, between the cash equivalents for the same prospects on successive sessions averaged .55 over six different prospects. Table 24.5 presents means (after transformation to Fisher's z) of the correlations between the different types of prospects. For example, there were 19 and 17 prospects, respectively, with high probability of gain and high

Table 24.4 Percentage of risk-seeking choices

	Gain		Loss	
Subject	$p \leq .1$	$p \geq .5$	$p \leq .1$	$p \geq .5$
1	100	38	30	100
2	85	33	20	75
3	100	10	0	93
4	71	0	30	58
5	83	0	20	100
6	100	5	0	100
7	100	10	30	86
8	87	0	10	100
9	16	0	80	100
10	83	0	0	93
11	100	26	0	100
12	100	16	10	100
13	87	0	10	94
14	100	21	30	100
15	66	0	30	100
16	60	5	10	100
17	100	15	20	100
18	100	22	10	93
19	60	10	60	63
20	100	5	0	81
21	100	0	0	100
22	100	0	0	92
23	100	31	0	100
24	71	0	80	100
25	100	0	10	87
Risk seeking	78[a]	10	20	87[a]
Risk neutral	12	2	0	7
Risk averse	10	88[a]	80[a]	6

Note: The percentage of risk-seeking choices is given for low ($p \leq .1$) and high ($p \geq .5$) probabilities of gain and loss for each subject (risk-neutral choices were excluded). The overall percentage of risk-seeking, risk-neutral, and risk-averse choices for each type of prospect appear at the bottom of the table

[a]Values that correspond to the fourfold pattern

probability of loss. The value of .06 in Table 24.5 is the mean of the $17 \times 19 = 323$ correlations between the cash equivalents of these prospects.

The correlations between responses within each of the four types of prospects average .41, slightly lower than the correlations between separate responses to the same problems. The two negative values in Table 24.5 indicate that those subjects who were more risk averse in one domain tended to be more risk seeking in the other. Although the individual correlations are fairly low, the trend is consistent: 78 % of the 403 correlations in these two cells are negative. There is also a tendency for

Table 24.5 Average correlations between certainty equivalents in four types of prospects

	L^+	H^+	L^-	H^-
L^+	.41	.17	−.23	.05
H^+		.39	.05	−.18
L^-			.40	.06
H^-				.44

Note: Low probability of gain = L^+; high probability of gain = H^+; low probability of loss = L^-; high probability of loss = H^-

subjects who are more risk averse for high-probability gains to be less risk seeking for gains of low probability. This trend, which is absent in the negative domain, could reflect individual differences either in the elevation of the weighting function or in the curvature of the value function for gains. The very low correlations in the two remaining cells of Table 24.5, averaging .05, indicate that there is no general trait of risk aversion or risk seeking. Because individual choices are quite noisy, aggregation of problems is necessary for the analysis of individual differences.

The fourfold pattern of risk attitudes emerges as a major empirical generalization about choice under risk. It has been observed in several experiments (see, e.g., Cohen et al. 1987), including a study of experienced oil executives involving significant, albeit hypothetical, gains and losses (Wehrung 1989). It should be noted that prospect theory implies the pattern demonstrated in Table 24.4 within the data of individual subjects, but it does not imply high correlations across subjects because the values of gains and of losses can vary independently. The failure to appreciate this point and the limited reliability of individual responses has led some previous authors (e.g., Hershey and Schoemaker 1980) to underestimate the robustness of the fourfold pattern.

Scaling

Having established the fourfold pattern in ordinal and correlational analyses, we now turn to a quantitative description of the data. For each prospect of the form $(x, p; 0, 1-p)$, let c/x be the ratio of the certainty equivalent of the prospect to the nonzero outcome x. Figures 24.1 and 24.2 plot the median value of c/x as a function of p, for positive and for negative prospects, respectively. We denote c/x by a circle if $|x| < 200$, and by a triangle if $|x| \geq 200$. The only exceptions are the two extreme probabilities (.01 and .99) where a circle is used for $|x| = 200$. To interpret Figs. 24.1 and 24.2, note that if subjects are risk neutral, the points will lie on the diagonal; if subjects are risk averse, all points will lie below the diagonal in Fig. 24.1 and above the diagonal in Fig. 24.2. Finally, the triangles and the circles will lie on top of each other if preferences are homogeneous, so that multiplying the outcomes of a prospect f by a constant $k > 0$ multiplies its cash equivalent $c(kf)$ by the same

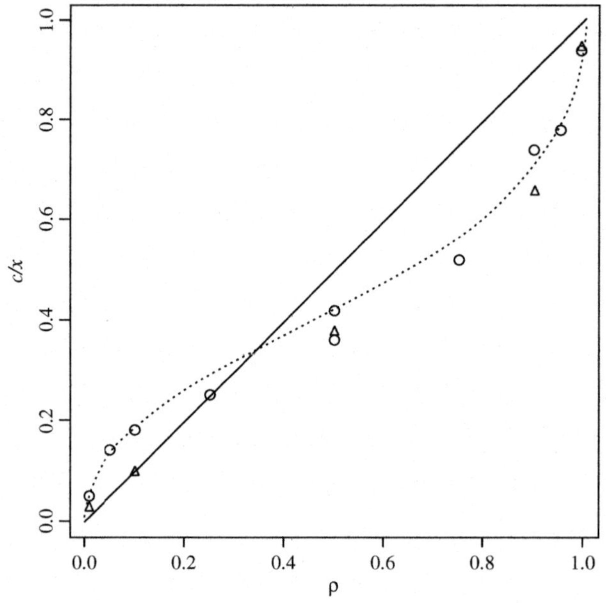

Fig. 24.1 Median *c/x* for all positive prospects of the form (*x*, *p*; 0,1 – *p*). Triangles and circles, respectively, correspond to values of *x* that lie above or below 200

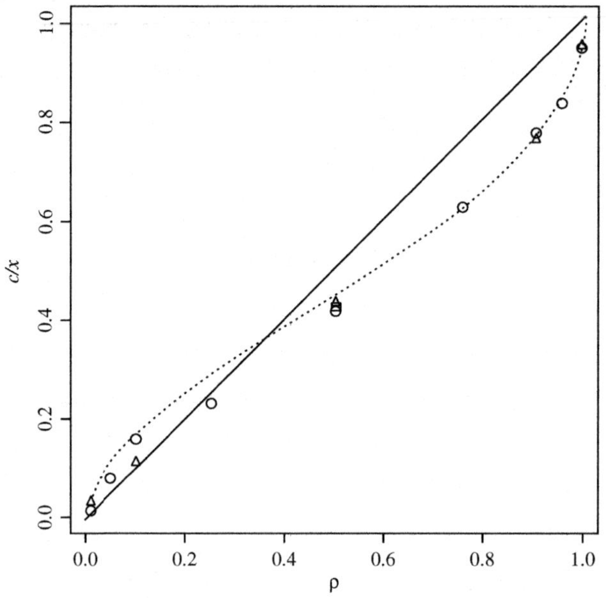

Fig. 24.2 Median *c/x* for all negative prospects of the form (*x*, *p*; 0,1 – *p*). Triangles and circles, respectively, correspond to values of *x* that lie below or above – 200

constant, that is, $c(kf) = kc(f)$. In expected utility theory, preference homogeneity gives rise to constant relative risk aversion. Under the present theory, assuming $X = \mathrm{Re}$, preference homogeneity is both necessary and sufficient to represent v as a two-part power function of the form

$$v(x) = \begin{cases} x^{\alpha}\text{if } x \geq 0 \\ -\lambda(-x)^{\beta}\text{if } x < 0. \end{cases} \tag{24.5}$$

Figures 24.1 and 24.2 exhibit the characteristic pattern of risk aversion and risk seeking observed in Table 24.4. They also indicate that preference homogeneity holds as a good approximation. The slight departures from homogeneity in Fig. 24.1 suggest that the cash equivalents of positive prospects increase more slowly than the stakes (triangles tend to lie below the circles), but no such tendency is evident in Fig. 24.2. Overall, it appears that the present data can be approximated by a two-part power function. The smooth curves in Figs. 24.1 and 24.2 can be interpreted as weighting functions, assuming a linear value function. They were fitted using the following functional form:

$$w^{+}(p) = \frac{p^{\gamma}}{(p^{\gamma} + (1-p)^{\gamma})^{1/\gamma}}, \quad w^{-}(p) = \frac{p^{\delta}}{(p^{\delta} + (1-p)^{\delta})^{1/\delta}}. \tag{24.6}$$

This form has several useful features: it has only one parameter; it encompasses weighting functions with both concave and convex regions; it does not require $w(.5) = .5$; and most important, it provides a reasonably good approximation to both the aggregate and the individual data for probabilities in the range between .05 and .95.

Further information about the properties of the value function can be derived from the data presented in Table 24.6. The adjustments of mixed prospects to acceptability (problems 1–4) indicate that, for even chances to win and lose, a prospect will only be acceptable if the gain is at least twice as large as the loss. This observation is compatible with a value function that changes slope abruptly at zero, with a loss-aversion coefficient of about 2 (Tversky and Kahneman 1991). The median matches in problems 5 and 6 are also consistent with this estimate: when the possible loss is increased by k the compensating gain must be increased by about $2\,k$. Problems 7 and 8 are obtained from problems 5 and 6, respectively, by positive translations that turn mixed prospects into strictly positive ones. In contrast to the large values of θ observed in problems 1–6, the responses in problems 7 and 8 indicate that the curvature of the value function for gains is slight. A decrease in the smallest gain of a strictly positive prospect is fully compensated by a slightly larger increase in the largest gain. The standard rank-dependent model, which lacks the notion of a reference point, cannot account for the dramatic effects of small translations of prospects illustrated in Table 24.6.

The estimation of a complex choice model, such as cumulative prospect theory, is problematic. If the functions associated with the theory are not constrained, the

Table 24.6 A test of loss aversion

Problem	a	b	c	x	θ
1	0	0	−25	61	2.44
2	0	0	−50	101	2.02
3	0	0	−100	202	2.02
4	0	0	−150	280	1.87
5	−20	50	−50	112	2.07
6	−50	150	−125	301	2.01
7	50	120	20	149	0.97
8	100	300	25	401	1.35

Note: In each problem, subjects determined the value of x that makes the prospect ($\$a$, ½, $\$b$, ½) as attractive as ($\$c$, ½, $\$x$, ½). The median values of x are presented for all problems along with the fixed values a, b, c. The statistic $\theta = (x - b)/(c - a)$ is the ratio of the "slopes" at a higher and a lower region of the value function

number of estimated parameters for each subject is too large. To reduce this number, it is common to assume a parametric form (e.g., a power utility function), but this approach confounds the general test of the theory with that of the specific parametric form. For this reason, we focused here on the qualitative properties of the data rather than on parameter estimates and measures of fit. However, in order to obtain a parsimonious description of the present data, we used a nonlinear regression procedure to estimate the parameters of Eqs. (24.5) and (24.6), separately for each subject. The median exponent of the value function was 0.88 for both gains and losses, in accord with diminishing sensitivity. The median λ was 2.25, indicating pronounced loss aversion, and the median values of γ and δ, respectively, were 0.61 and 0.69, in agreement with Eqs. (24.3) and (24.4) above.[4] The parameters estimated from the median data were essentially the same. Figure 24.3 plots w^+ and w^- using the median estimates of γ and δ.

Figure 24.3 shows that, for both positive and negative prospects, people overweight low probabilities and underweight moderate and high probabilities. As a consequence, people are relatively insensitive to probability difference in the middle of the range. Figure 24.3 also shows that the weighting functions for gains and for losses are quite close, although the former is slightly more curved than the latter (i.e., $\gamma < \delta$). Accordingly, risk aversion for gains is more pronounced than risk seeking for losses, for moderate and high probabilities (see Table 24.3). It is noteworthy that the condition $w^+(p) = w^-(p)$, assumed in the original version of prospect theory, accounts for the present data better than the assumption $w^+(p) = 1 - w^-(1 - p)$,

[4]Camerer and Ho (1991) applied Eq. (24.6) to several studies of risky choice and estimated γ from aggregate choice probabilities using a logistic distribution function. Their mean estimate (.56) was quite close to ours.

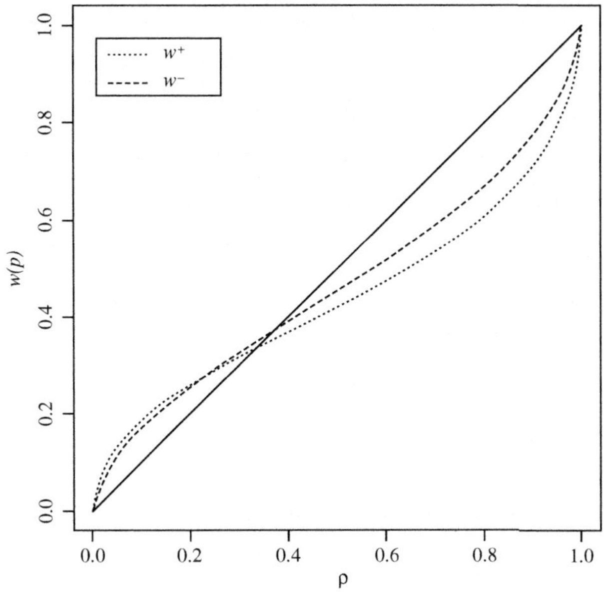

Fig. 24.3 Weighting functions for gains (w^+) and for losses (w^-) based on median estimates of γ and δ in Eq. (24.6)

implied by the standard rank-dependent or cumulative functional. For example, our estimates of w^+ and w^- show that all 25 subjects satisfied the conditions w^+ $(.5) < .5$ and w^- $(.5) < .5$, implied by the former model, and no one satisfied the condition w^+ $(.5) < .5$ iff w^- $(.5) > .5$, implied by the latter model.

Much research on choice between risky prospects has utilized the triangle diagram (Marschak 1950; Machina 1987) that represents the set of all prospects of the form $(x_1, p_1; x_2, p_2; x_3, p_3)$, with fixed outcomes $x_1 < x_2 < x_3$. Each point in the triangle represents a prospect that yields the lowest outcome (x_1) with probability p_1, the highest outcome (x_3) with probability p_3, and the intermediate outcome (x_2) with probability $p_2 = 1 - p_1 - p_3$. An indifference curve is a set of prospects (i.e., points) that the decision maker finds equally attractive. Alternative choice theories are characterized by the shapes of their indifference curves. In particular, the indifference curves of expected utility theory are parallel straight lines. Figures 24.4a, b illustrate the indifference curves of cumulative prospect theory for nonnegative and nonpositive prospects, respectively. The shapes of the curves are determined by the weighting functions of Fig. 24.3; the values of the outcomes (x_1, x_2, x_3) merely control the slope.

Figures 24.4a, b are in general agreement with the main empirical generalizations that have emerged from the studies of the triangle diagram; see Camerer (1992), and Camerer and Ho (1991) for reviews. First, departures from linearity, which violate expected utility theory, are most pronounced near the edges of the triangle. Second, the indifference curves exhibit both fanning in and fanning out. Third, the curves are concave in the upper part of the triangle and convex in the lower right.

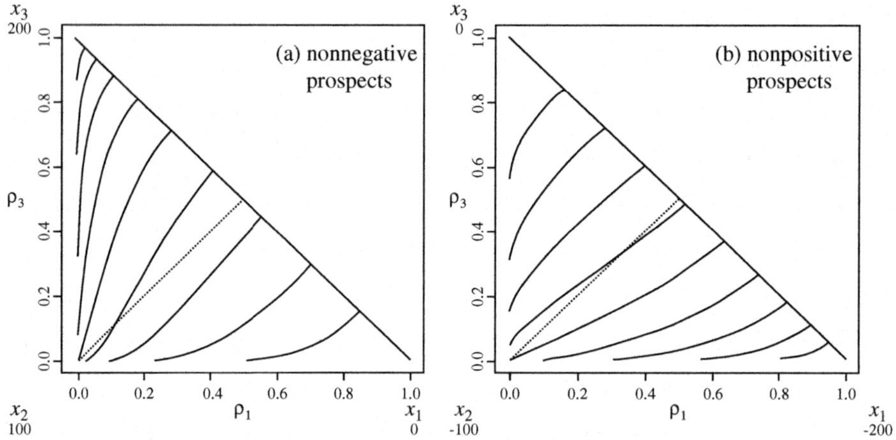

Fig. 24.4 Indifference curves of cumulative prospect theory (**a**) nonnegative prospects ($x_1 = 0$, $x_2 = 100$, $x_3 = 200$), and (**b**) nonpositive prospects ($x_1 = -200$, $x_2 = -100$, $x_3 = 0$). The curves are based on the respective weighting functions of Fig. 24.3, ($\gamma = .61$, $\delta = .69$) and on the median estimates of the exponents of the value function ($\alpha = \beta = .88$). The broken line through the origin represents the prospects whose expected value is x_2

Finally, the indifference curves for nonpositive prospects resemble the curves for nonnegative prospects reflected around the 45° line, which represents risk neutrality. For example, a sure gain of \$100 is equally as attractive as a 71 % chance to win \$200 or nothing (see Fig. 24.4a), and a sure loss of \$100 is equally as aversive as a 64 % chance to lose \$200 or nothing (see Fig. 24.4b). The approximate reflection of the curves is of special interest because it distinguishes the present theory from the standard rank-dependent model in which the two sets of curves are essentially the same.

Incentives

We conclude this section with a brief discussion of the role of monetary incentives. In the present study we did not pay subjects on the basis of their choices because in our experience with choice between prospects of the type used in the present study, we did not find much difference between subjects who were paid a flat fee and subjects whose payoffs were contingent on their decisions. The same conclusion was obtained by Camerer (1989), who investigated the effects of incentives using several hundred subjects. He found that subjects who actually played the gamble gave essentially the same responses as subjects who did not play; he also found no differences in reliability and roughly the same decision time. Although some studies found differences between paid and unpaid subjects in choice between simple prospects, these differences were not large enough to change any significant qualitative conclusions. Indeed, all major violations of expected utility theory (e.g. the common consequence effect, the common ratio effect, source dependence, loss aversion, and preference reversals) were obtained both with and without monetary incentives.

As noted by several authors, however, the financial incentives provided in choice experiments are generally small relative to people's incomes. What happens when the stakes correspond to three- or four-digit rather than one- or two-digit figures? To answer this question, Kachelmeier and Shehata (1991) conducted a series of experiments using Masters students at Beijing University, most of whom had taken at least one course in economics or business. Due to the economic conditions in China, the investigators were able to offer subjects very large rewards. In the high payoff condition, subjects earned about three times their normal monthly income in the course of one experimental session! On each trial, subjects were presented with a simple bet that offered a specified probability to win a given prize, and nothing otherwise. Subjects were instructed to state their cash equivalent for each bet. An incentive compatible procedure (the BDM scheme) was used to determine, on each trial, whether the subject would play the bet or receive the "official" selling price. If departures from the standard theory are due to the mental cost associated with decision making and the absence of proper incentives, as suggested by Smith and Walker (1992), then the highly paid Chinese subjects should not exhibit the characteristic nonlineariry observed in hypothetical choices, or in choices with small payoffs.

However, the main finding of Kachelmeier and Shehata (1991) is massive risk seeking for small probabilities. Risk seeking was slightly more pronounced for lower payoffs, but even in the highest payoff condition, the cash equivalent for a 5 % bet (their lowest probability level) was, on average, three times larger than its expected value. Note that in the present study the median cash equivalent of a 5 % chance to win $100 (see Table 24.3) was $14, almost three times the expected value of the bet. In general, the cash equivalents obtained by Kachelmeier and Shehata were higher than those observed in the present study. This is consistent with the finding that minimal selling prices are generally higher than certainty equivalents derived from choice (see, e.g., Tversky et al. 1990). As a consequence, they found little risk aversion for moderate and high probability of winning. This was true for the Chinese subjects, at both high and low payoffs, as well as for Canadian subjects, who either played for low stakes or did not receive any payoff. The most striking result in all groups was the marked overweighting of small probabilities, in accord with the present analysis.

Evidently, high incentives do not always dominate noneconomic considerations, and the observed departures from expected utility theory cannot be rationalized in terms of the cost of thinking. We agree with Smith and Walker (1992) that monetary incentives could improve performance under certain conditions by eliminating careless errors. However, we maintain that monetary incentives are neither necessary nor sufficient to ensure subjects' cooperativeness, thoughtfulness, or truthfulness. The similarity between the results obtained with and without monetary incentives in choice between simple prospects provides no special reason for skepticism about experiments without contingent payment.

Discussion

Theories of choice under uncertainty commonly specify (1) the objects of choice, (2) a valuation rule, and (3) the characteristics of the functions that map uncertain events and possible outcomes into their subjective counterparts. In standard applications of expected utility theory, the objects of choice are probability distributions over wealth, the valuation rule is expected utility, and utility is a concave function of wealth. The empirical evidence reported here and elsewhere requires major revisions of all three elements. We have proposed an alternative descriptive theory in which (1) the objects of choice are prospects framed in terms of gains and losses, (2) the valuation rule is a two-part cumulative functional, and (3) the value function is S-shaped and the weighting functions are inverse S-shaped. The experimental findings confirmed the qualitative properties of these scales, which can be approximated by a (two-part) power value function and by identical weighting functions for gains and losses.

The curvature of the weighting function explains the characteristic reflection pattern of attitudes to risky prospects. Overweighting of small probabilities contributes to the popularity of both lotteries and insurance. Underweighting of high probabilities contributes both to the prevalence of risk aversion in choices between probable gains and sure things, and to the prevalence of risk seeking in choices between probable and sure losses. Risk aversion for gains and risk seeking for losses are further enhanced by the curvature of the value function in the two domains. The pronounced asymmetry of the value function, which we have labeled loss aversion, explains the extreme reluctance to accept mixed prospects. The shape of the weighting function explains the certainty effect and violations of quasi-convexity. It also explains why these phenomena are most readily observed at the two ends of the probability scale, where the curvature of the weighting function is most pronounced (Camerer 1992).

The new demonstrations of the common consequence effect, described in Tables 24.1 and 24.2, show that choice under uncertainty exhibits some of the main characteristics observed in choice under risk. On the other hand, there are indications that the decision weights associated with uncertain and with risky prospects differ in important ways. First, there is abundant evidence that subjective judgments of probability do not conform to the rules of probability theory (Kahneman et al. 1982). Second, Ellsberg's example and more recent studies of choice under uncertainty indicate that people prefer some sources of uncertainty over others. For example, Heath and Tversky (1991) found that individuals consistently preferred bets on uncertain events in their area of expertise over matched bets on chance devices, although the former are ambiguous and the latter are not. The presence of systematic preferences for some sources of uncertainty calls for different weighting functions for different domains, and suggests that some of these functions lie entirely above others. The investigation of decision weights for uncertain events emerges as a promising domain for future research.

The present theory retains the major features of the original version of prospect theory and introduces a (two-part) cumulative functional, which provides a convenient mathematical representation of decision weights. It also relaxes some descriptively inappropriate constraints of expected utility theory. Despite its greater generality, the cumulative functional is unlikely to be accurate in detail. We suspect that decision weights may be sensitive to the formulation of the prospects, as well as to the number, the spacing and the level of outcomes. In particular, there is some evidence to suggest that the curvature of the weighting function is more pronounced when the outcomes are widely spaced (Camerer 1992). The present theory can be generalized to accommodate such effects, but it is questionable whether the gain in descriptive validity, achieved by giving up the separability of values and weights, would justify the loss of predictive power and the cost of increased complexity.

Theories of choice are at best approximate and incomplete. One reason for this pessimistic assessment is that choice is a constructive and contingent process. When faced with a complex problem, people employ a variety of heuristic procedures in order to simplify the representation and the evaluation of prospects. These procedures include computational shortcuts and editing operations, such as eliminating common components and discarding nonessential differences (Tversky 1969). The heuristics of choice do not readily lend themselves to formal analysis because their application depends on the formulation of the problem, the method of elicitation, and the context of choice.

Prospect theory departs from the tradition that assumes the rationality of economic agents; it is proposed as a descriptive, not a normative, theory. The idealized assumption of rationality in economic theory is commonly justified on two grounds: the conviction that only rational behavior can survive in a competitive environment, and the fear that any treatment that abandons rationality will be chaotic and intractable. Both arguments are questionable. First, the evidence indicates that people can spend a lifetime in a competitive environment without acquiring a general ability to avoid framing effects or to apply linear decision weights. Second, and perhaps more important, the evidence indicates that human choices are orderly, although not always rational in the traditional sense of this word.

Appendix: Axiomatic Analysis

Let $F = \{f : S \to X\}$ be the set of all prospects under study, and let F^+ and F^- denote the positive and the negative prospects, respectively. Let $\lesssim$ be a binary preference relation on F, and let $\approx$ and $>$ denote its symmetric and asymmetric parts, respectively. We assume that $\lesssim$ is complete, transitive, and strictly monotonic, that is, if $f \neq g$ and $f(s) \geq g(s)$ for all $s \in S$, then $f > g$.

For any f, $g \in F$ and $A \subset S$, define $h = fAg$ by: $h(s) = f(s)$ if $s \in A$, and $h(s) = g(s)$ if $s \in S - A$. Thus, fAg coincides with f on A and with g on $S - A$. A preference relation $\lesssim$ on F satisfies *independence* if for all f, g, f', $g' \in F$ and $A \subset S$, $fAg \lesssim fAg'$ iff $f'\ Ag \lesssim f'\ Ag'$. This axiom, also called the sure thing principle (Savage 1954), is one of the basic qualitative properties underlying expected utility theory, and it is violated by Allais's common consequence effect. Indeed, the attempt to accommodate Allais's example has motivated the development of numerous models, including cumulative utility theory. The key concept in the axiomatic analysis of that theory is the relation of comonotonicity, due to Schmeidler (1989). A pair of prospects $f, g \in F$ are *comonotonic* if there are no $s, t \in S$ such that $f(s) > f(t)$ and $g(t) > g(s)$. Note that a constant prospect that yields the same outcome in every state is comonotonic with all prospects. Obviously, comonotonicity is symmetric but not transitive.

Cumulative utility theory does not satisfy independence in general, but it implies independence whenever the prospects $fAg, fAg', f'Ag$, and $f'Ag'$ above are pairwise comonotonic. This property is called *comonotonic independence*.[5] It also holds in cumulative prospect theory, and it plays an important role in the characterization of this theory, as will be shown below. Cumulative prospect theory satisfies an additional property, called *double matching:* for all $f, g \in F$, if $f^+ \approx g^+$ and $f^- \approx g^-$, then $f \approx g$.

To characterize the present theory, we assume the following structural conditions: S is finite and includes at least three states; $X = \mathrm{Re}$; and the preference order is continuous in the product topology on Re^k, that is, $\{f \in F : f \geq g\}$ and $\{f \in F : g \geq f\}$ are closed for any $g \in F$. The latter assumptions can be replaced by restricted solvability and a comonotonic Archimedean axiom (Wakker 1991).

Theorem 24.1 Suppose $(F^+, \precsim)$ and $(F^-, \precsim)$ can each be represented by a cumulative functional. Then $(F, \precsim)$ satisfies cumulative prospect theory iff it satisfies double matching and comonotonic independence.

The proof of the theorem is given at the end of the appendix. It is based on a theorem of Wakker (1992) regarding the additive representation of lower-diagonal structures. Theorem 24.1 provides a generic procedure for characterizing cumulative prospect theory. Take any axiom system that is sufficient to establish an essentially unique cumulative (i.e., rank-dependent) representation. Apply it separately to the preferences between positive prospects and to the preferences between negative prospects, and construct the value function and the decision weights separately for F^+ and for F^-. Theorem 24.1 shows that comonotonic independence and double matching ensure that, under the proper rescaling, the sum $V(f^+) + V(f^-)$ preserves the preference order between mixed prospects. In order to distinguish more sharply between the conditions that give rise to a one-part or a two-part representation, we need to focus on a particular axiomatization of the Choquet functional. We chose Wakker's (1989a, b) because of its generality and compactness.

For $x \in X$, $f \in F$, and $r \in S$, let $x\{r\}f$ be the prospect that yields x in state r and coincides with f in all other states. Following Wakker (1989a), we say that a preference relation satisfies *tradeoff consistency*[6] (TC) if for all $x, x', y, y' \in X, f, f', g, g' \in F$, and $s, t \in S$.

---

[5]Wakker (1989b) called this axiom *comonotonic coordinate independence*. Schmeidler (1989) used *comonotonic independence* for the mixture space version of this axiom: $f \precsim g$ iff. $\alpha f + (1-\alpha)h \precsim \alpha g + (1-\alpha)h$.

[6]Wakker (1989a, b) called this property *cardinal coordinate independence*. He also introduced an equivalent condition, called the absence of *contradictory tradeoffs*.

$$x\{s\}f \lesssim y\{s\}g, x'\{s\}f \gtrsim y'\{s\}g \text{ and } x\{t\}f' \gtrsim y\{t\}g' \text{ imply } x'\{t\}f' \gtrsim y'\{t\}g'.$$

To appreciate the import of this condition, suppose its premises hold but the conclusion is reversed, that is, $y'\{t\}g' > x'\{t\}f'$. It is easy to verify that under expected utility theory, the first two inequalities, involving $\{s\}$, imply $u(y) - u(y') \geq u(x) - u(x')$, whereas the other two inequalities, involving $\{t\}$, imply the opposite conclusion. Tradeoff consistency, therefore, is needed to ensure that "utility intervals" can be consistently ordered. Essentially the same condition was used by Tversky et al. (1988) in the analysis of preference reversal, and by Tversky and Kahneman (1991) in the characterization of constant loss aversion.

A preference relation satisfies *comonotonic tradeoff consistency* (CTC) if TC holds whenever the prospects $x\{s\}f$, $y\{s\}g$, $x'\{s\}f$, and $y'\{s\}g$ are pairwise comonotonic, as are the prospects $x\{f\}f'$, $y\{t\}g'$, $x'\{t\}f'$, and $y'\{t\}g'$ (Wakker 1989a). Finally, a preference relation satisfies *sign-comonotonic tradeoff consistency* (SCTC) if CTC holds whenever the consequences x, x', y, y' are either all nonnegative or all nonpositive. Clearly, TC is stronger than CTC, which is stronger than SCTC. Indeed, it is not difficult to show that (1) expected utility theory implies TC, (2) cumulative utility theory implies CTC but not TC, and (3) cumulative prospect theory implies SCTC but not CTC. The following theorem shows that, given our other assumptions, these properties are not only necessary but also sufficient to characterize the respective theories.

Theorem 24.2 Assume the structural conditions described above.

(a) (Wakker 1989a) Expected utility theory holds iff $\lesssim$ satisfies TC.
(b) (Wakker 1989b) Cumulative utility theory holds iff $\lesssim$ satisfies CTC.
(c) Cumulative prospect theory holds iff $\lesssim$ satisfies double matching and SCTC.

A proof of part c of the theorem is given at the end of this section. It shows that, in the presence of our structural assumptions and double matching, the restriction of tradeoff consistency to sign-comonotonic prospects yields a representation with a reference-dependent value function and different decision weights for gains and for losses.

Proof of Theorem 24.1 The necessity of comonotonic independence and double matching is straightforward. To establish sufficiency, recall that, by assumption, there exist functions π^+, π^-, v^+, v^-, such that $V^+ = \sum \pi^+ v^+$ and $V^- = \sum \pi^- v^-$ preserve $\lesssim$ on F^+ and on F^-, respectively. Furthermore, by the structural assumptions, π^+ and π^- are unique, whereas v^+ and v^- are continuous ratio scales. Hence, we can set $v^+(1) = 1$ and $v^-(-1) = \theta < 0$, independently of each other.

Let Q be the set of prospects such that for any $q \in Q$, $q(s) \neq q(t)$ for any distinct $s, t \in S$. Let F_g denote the set of all prospects in F that are comonotonic with G. By comonotonic independence and our structural conditions, it follows readily from a theorem of Wakker (1992) on additive representations for lower-triangular subsets of Re^k that, given any $q \in Q$, there exist intervals scales $\{U_{qi}\}$, with a common unit, such that $U_q = \sum_i U_{qi}$ preserves $\lesssim$ on F_q. With no loss of generality we can set $U_{qi}(0)=0$ for all i and $U_q(1)=1$. Since V^+ and V^- above are additive representations of $\lesssim$ on F_q^+ and F_q^-, respectively, it follows by uniqueness that there exist aq, $bq>0$ such that for all i, U_{qi} equals $a_q\pi_i^+v^+$ on Re^+, and U_{qi} equals $b_q\pi_i^-v^-$ on Re^-.

So far the representations were required to preserve the order only within each F_q. Thus, we can choose scales so that $b_q=1$ for all q. To relate the different representations, select a prospect $h\neq q$. Since V^+ should preserve the order on F^+, and U_q should preserve the order within each F_q, we can multiply V^+ by a_h, and replace each a_q by a_q/a_h. In other words, we may set $a_h=1$. For any $q\in Q$, select $f\in F_q$, $g\in F_h$ such that $f^+\approx g^+>0$, $f^-\approx g^->0$, and $g\approx 0$. By double matching, then, $f\approx g\approx 0$. Thus, $a_qV^+(f^+)+V^-(f^-)=0$, since this form preserves the order on F_q. But $V^+(f^+)=V^+(g^+)$ and $V^-(f^-)=V^-(g^-)$, so $V^+(g^+)+V^-(g^-)=0$ implies $V^+(f^+)+V^-(f^-)=0$. Hence, $a_q=1$, and $V(f)=V^+(f^+)+V^-(f^-)$ preserves the order within each F_q.

To show that V preserves the order on the entire set, consider any f, $g\in F$ and suppose $f\lesssim g$. By transitivity, $c(f)\lesssim c(g)$ where $c(f)$ is the certainty equivalent of f. Because $c(f)$ and $c(g)$ are comonotonic, $V(f)=V(c(f))\geq V(c(g))=V(g)$. Analogously, $f>g$ implies $V(f)>V(g)$, which complete the proof of theorem 24.1.

Proof of Theorem 24.2 (part c) To establish the necessity of SCTC, apply cumulative prospect theory to the hypotheses of SCTC to obtain the following inequalities:

$$\begin{aligned}
V(x\{s\}f) &= \pi_s v(x) + \sum_{r\in S-s} \pi_r v(f(r)) \\
&\leq \pi'_s v(y) + \sum_{r\in S-s} \pi'_r v(g(r)) = V(y\{s\}g) \\
V(x'\{s\}f) &= \pi_s v(x') + \sum_{r\in S-s} \pi_r v(f(r)) \\
&\geq \pi'_s v(y') + \sum_{r\in S-s} \pi'_r v(g(r)) = V(y'\{s\}g).
\end{aligned}$$

The decision weights above are derived, assuming SCTC, in accord with Eqs. (24.1) and (24.2). We use primes to distinguish the decision weights associated with g from those associated with f. However, all the above prospects belong to the same comonotonic set. Hence, two outcomes that have the same sign and are associated with the same state have the same decision weight. In particular, the weights associated with $x\{s\}f$ and $x'\{s\}f$ are identical, as are the weights associated with $y\{s\}g$ and with $y'\{s\}g$. These assumptions are implicit in the present notation. It follows that

$$\pi_s v(x) - \pi'_s v(y) \leq \pi_s v(x') - \pi'_s v(y').$$

Because x, y, x', y' have the same sign, all the decision weights associated with state s are identical, that is, $\pi_s = \pi'_s$. Cancelling this common factor and rearranging terms yields $v(y)-v(y')\geq v(x)-v(x')$.

Suppose SCTC is not valid, that is, $x\{t\}f \lesssim y\{t\}g'$ but $x'\{t\}f' < y'\{t\}g'$. Applying cumulative prospect theory, we obtain

$$\begin{aligned} V\left(x\{t\}f'\right) &= \pi_t v(x) + \sum_{r\in S-t} \pi_r v\left(f'(r)\right) \\ &\geq \pi_t v(y) + \sum_{r\in S-t} \pi_r v\left(g'(r)\right) = V\left(y\{t\}g'\right) \\ V\left(x'\{t\}f'\right) &= \pi_t v\left(x'\right) + \sum_{r\in S-t} \pi_r v\left(f'(r)\right) \\ &< \pi_t v\left(y'\right) + \sum_{r\in S-t} \pi_t v\left(g'(r)\right) = V\left(y'\{t\}g'\right). \end{aligned}$$

Adding these inequalities yields $v(x) - v(x') > v(y) - v(y')$ contrary to the previous conclusion, which establishes the necessity of SCTC. The necessity of double matching is immediate.

To prove sufficiency, note that SCTC implies comonotonic independence. Letting $x = y$, $x' = y'$, and $f = g$ in TC yields $x\{t\}f' \lesssim x\{t\}g'$ implies $x'\{t\}f' \lesssim x'\{t\}g'$, provided all the above prospects are pairwise comonotonic. This condition readily entails comonotonic independence (see Wakker 1989b).

To complete the proof, note that SCTC coincides with CTC on $(F^+, \lesssim)$ and on $(F^-, \lesssim)$. By part b of this theorem, the cumulative functional holds, separately, in the nonnegative and in the nonpositive domains. Hence, by double matching and comonotonic independence, cumulative prospect theory follows from Theorem 24.1.

References

Allais, M. (1953). Le comportement de l'homme rationel devant le risque, critique des postulates et axiomes de l'ecole americaine. *Econometrica, 21*, 503–546.

Arrow, K. J. (1982). Risk perception in psychology and economies. *Economic Inquiry, 20*, 1–9.

Camerer, C. F. (1989). An experimental test of several generalized utility theories. *Journal of Risk and Uncertainty, 2*, 61–104.

Camerer, C. F. (1992). Recent tests of generalizations of expected utility theory. In W. Edwards (Ed.), *Utility: Theories, measurement and applications*. Boston: Kluwer Academic Publishers.

Camerer, C. F., & Ho, T.-H., (1991). Nonlinear weighting of probabilities and violations of the betweenness axiom. Unpublished manuscript, The Wharton School, University of Pennsylvania.

Chew, S.-H. (1989). An axiomatic generalization of the quasilinear mean and the gini mean with application to decision theory, Unpublished manuscript, Department of Economics, University of California at Irvine.

Choquet, G. (1955). Theory of capacities. *Annales de L'Institut Fourier, 5*, 131–295.

Cohen, M., Jaffray, J.-Y., & Said, T. (1987). Experimental comparison of individual behavior under risk and under uncertainty for gains and for losses. *Organizational Behavior and Human Decision Processes, 39*, 1–22.

Ellsberg, D. (1961). Risk, ambiguity, and the savage axioms. *Quarterly Journal of Economics, 75*, 643–669.

Fishburn, P. C. (1988). *Nonlinear preference and utility theory*. Baltimore: The Johns Hopkins University Press.

Gilboa, I. (1987). Expected utility with purely subjective non-additive probabilities. *Journal of Mathematical Economics, 16*, 65–88.

Heath, C., & Tversky, A. (1991). Preference and belief: Ambiguity and competence in choice under uncertainty. *Journal of Risk and Uncertainty, 4*, 5–28.

Hershey, J. C., & Schoemaker, P. J. H. (1980). Prospect theory's reflection hypothesis: A critical examination. *Organizational Behavior and Human Performance, 25*, 395–418.

Hogarth, R., & Einhorn, H. (1990). Venture theory: A model of decision weights. *Management Science, 36*, 780–803.

Kachelmeier, S. J., & Shehata, M. (1992). Examining risk preferences under high monetary incentives: Experimental evidence from the People's Republic of China. *American Economic Review, 82*(5), 1120–1141.

Kahneman, D., & Tversky, A. (1979). Prospect theory: An analysis of decision under risk. *Econometrica, 47*, 263–291.

Kahneman, D., & Tversky, A. (1984). Choices, values and frames. *American Psychologist, 39*, 341–350.

Kahneman, D., Slovic, P., & Tversky, A. (Eds.). (1982). *Judgment under uncertainty: Heuristics and biases*. New York: Cambridge University Press.

Loomes, G., & Sugden, R. (1982a). Regret theory: An alternative theory of rational choice under uncertainty. *The Economic Journal, 92*, 805–824.

Loomes, G., & Sugden, R. (1982b). Some implications of a more general form of regret theory. *Journal of Economic Theory, 41*, 270–287.

Luce, R. D., & Fishburn, P. C. (1991). Rank- and sign-dependent linear utility models for finite first-order gambles. *Journal of Risk and Uncertainty, 4*, 29–59.

Machina, M. J. (1987). Choice under uncertainty: Problems solved and unsolved. *Economic Perspectives, 1*(1), 121–154.

Marschak, J. (1950). Rational behavior, uncertain prospects, and measurable utility. *Econometrica, 18*, 111–114.

Nakamura, Y. (1990). Subjective expected utility with non-additive probabilities on finite state space. *Journal of Economic Theory, 51*, 346–366.

Prelec, D. (1989). On the shape of the decision weight function. Unpublished manuscript, Harvard Graduate School of Business Administration.

Prelec, D. (1990). A 'Pseudo-endowment' effect, and its implications for some recent non-expected utility models. *Journal of Risk and Uncertainty, 3*, 247–259.

Quiggin, J. (1982). A theory of anticipated utility. *Journal of Economic Behavior and Organization, 3*, 323–343.

Savage, L. J. (1954). *The foundations of statistics*. New York: Wiley.

Schmeidler, D. (1989). Subjective probability and expected utility without additivity. *Econometrica, 57*, 571–587.

Segal, U. (1989). Axiomatic representation of expected utility with rank-dependent probabilities. *Annals of Operations Research, 19*, 359–373.

Smith, V. L., & Walker, J. M. (1992). Monetary rewards and decision cost in experimental economics. Unpublished manuscript, Economic Science Lab, University of Arizona.

Starmer, C., & Sugden, R. (1989). Violations of the independence axiom in common ratio problems: An experimental test of some competing hypotheses. *Annals of Operations Research, 19*, 79–102.

Tversky, A. (1969). The intransitivity of preferences. *Psychology Review, 76*, 31–48.

Tversky, A., Kahneman, D. (1986). Rational choice and the framing of decisions, *The Journal of Business 59*(4), part 2, S251–S278.

Tversky, A., & Kahneman, D. (1991). Loss aversion in riskless choice: A reference dependent model. *Quarterly Journal of Economics, 107*(4), 1039–1061.

Tversky, A., Sattath, S., & Slovic, P. (1988). Contingent weighting in judgment and choice. *Psychological Review, 95*(3), 371–384.

Tversky, A., Slovic, P., & Kahneman, D. (1990). The causes of preference reversal. *The American Economic Review, 80*(1), 204–217.

Viscusi, K. W. (1989). Prospective reference theory: Toward an explanation of the paradoxes. *Journal of Risk and Uncertainty, 2*, 235–264.

Wakker, P. P. (1989a). *Additive representations of preferences: A new foundation in decision analysis*. Dordrecht: Kluwer Academic Publishers.

Wakker, P. P. (1989b). Continuous subjective expected utility with nonadditive probabilities. *Journal of Mathematical Economics, 18*, 1–27.

Wakker, P. P. (1990). Separating marginal utility and risk aversion. Unpublished manuscript, University of Nijmegen, The Netherlands.

Wakker, P. P. (1991). Additive representations of preferences, a new foundation of decision analysis; the algebraic approach. In J. D. Doignon & J. C. Falmagne (Eds.), *Mathematical psychology: Current developments* (pp. 71–87). Berlin: Springer.

Wakker, P. (1993). Additive representations on rank-ordered sets: II. The topological approach. *Journal of Mathematical Economics, 22*(1), 1–26.

Wakker, P. P., & Tversky, A. (1991). An axiomatization of cumulative prospect theory. Unpublished manuscript. University of Nijmegan, the Netherlands.

Wehrung, D. A. (1989). Risk taking over gains and losses: A study of oil executives. *Annals of Operations Research, 19*, 115–139.

Weymark, J. A. (1981). Generalized gini inequality indices. *Mathematical Social Sciences, 1*, 409–430.

Yaari, M. E. (1987). The dual theory of choice under risk. *Econometrica, 55*, 95–115.

Part IV
Logics of Knowledge and Belief

Chapter 25
Introduction

Horacio Arló-Costa, Vincent F. Hendricks, and Johan van Benthem

This part, like all others in this book, consists of a mix of classic papers that have defined the area and modern ones illustrating important current issues. These texts provide rich fare, and they defy simple labels summarizing their content. Moreover, a look at the list of authors reveals a mixture of different academic cultures, from philosophy to computer science. One might also add that the texts do not all agree: they span a landscape with many positions and perspectives.

Epistemic logic as the systematic study of reasoning with knowledge and belief started with Jaakko Hintikka's classic book Knowledge and Belief: An Introduction to Logic of the Two Notions, which set the agenda for many subsequent lines of research and debate. It interpreted knowledge as what is true in some current range of epistemically accessible worlds, while doing something similar for belief and doxastic accessibility. Thus, general methods from modal logic became available for studying knowledge, and the resulting axiomatic systems have shaped many philosophical discussions for or against principles of 'omniscience', 'closure', and positive and negative 'introspection'. One general interest behind such specific issues has been the search for satisfactory definitions of knowledge, an interest with a long epistemological history running from Plato's "justified true belief" to post-Gettier strengthenings involving forms of robustness of true beliefs under new information, under new relevant considerations, or across counterfactual variations

H. Arló-Costa (deceased)
Carnegie Mellon University, Pittsburgh, PA, USA

V.F. Hendricks (✉)
Center for Information and Bubble Studies, University of Copenhagen, Copenhagen, Denmark
e-mail: vincent@hum.ku.dk

J. van Benthem
University of Amsterdam, Science Park 904, 1098 XH Amsterdam, The Netherlands

Stanford University, Stanford, United States
e-mail: johan@science.uva.nl

H. Arló-Costa et al. (eds.), *Readings in Formal Epistemology*, Springer Graduate Texts in Philosophy 1, DOI 10.1007/978-3-319-20451-2_25

of the actual world – and in a related spirit, in various forms of faithful tracking during a history of investigation. The past half-century of formal studies has even produced further perspectives, such as the availability of proof or evidence in some precise sense, or convergence in the limit in some process of inquiry.

The paper by Dretske makes the original Hintikka semantics a more dynamic process, showing how knowledge claims are always based on some current range of relevant worlds, which can change under the pressure of legitimate new considerations. In a related vein, Lewis provides systematic principles guiding this process of selecting relevant worlds. Following a further intuition, Nozick proposes a counterfactual idea of knowledge as a true belief that would stay attuned to the facts in non-actual worlds close to ours. While these approaches are semantic, Artemov's 'justification logic' brings explicit proof and evidence into epistemic logic, allowing us to syntactically manipulate reasons for our beliefs. Finally, Kelly discusses the learning-theoretic view of temporal convergence for knowledge of complete histories of the world.

Since this part is about logics of knowledge and belief, many readers will be interested not just in formal languages and semantics, but also in complete calculi for reasoning capturing the operational proof-theoretic aspects of reasoning with knowledge or belief. Stalnaker's paper presents a broad logical view of possible modal systems and defensible epistemic and doxastic principles, and in another perspective, so does Artemov. Interestingly, not all notions of knowledge proposed in the 'robustness' tradition have been studied in this systematic manner, and many questions remain open, though actively pursued by some young philosophical logicians.

Another running theme is the issue of which epistemic attitudes form a natural family that requires scrutiny in its entirety. Knowledge and belief need not be enough, and for instance, Stalnaker's survey of doxastic and epistemic logics proposes a new notion of 'safe belief' that will survive true new information, as an intermediate between logic and belief simpliciter. Parikh even suggests an algebraic framework that ties together a wide abstract range of knowledge, belief, and action in general. Specializing general action again to informational action, we arrive at what has been Hintikka's guiding interest throughout: the combination of knowledge as based on our current semantic information with acts that systematically change that information, such as questions, and games of inquiry over time.

It is only a short step then to a dynamic focus on learning rather than the mere statics of knowledge. Kelly's article looks at this dynamics from the viewpoint of learning theory, and investigates which truths can be acquired in the limit, i.e., which processes of inquiry will reliably converge to stable true belief about the answer to the main question at stake. In a somewhat related mode, Williamson focuses on the scope of purely operational definitions of knowledge, and shows that knowledge-based epistemology remains indispensable. And these are just two dynamic or computational aspects of knowledge and belief. There is more to information dynamics when one begins to study effects of specific local acts of knowledge update or belief revision as an agent navigates the world. Many of these topics will be addressed in the next section on interactive epistemology, since much information flow that is crucial to humans involves more than one party: be it a group of agents, or just one agent interacting with Nature.

There are many further themes to ponder when reading these articles. Does the semantics describe internal first-person views of epistemic agents, or the theorist's external view of their situation? Do different epistemic attitudes correlate with different sorts of information? How does knowledge of propositions tie in with knowledge of objects, "that" or "whether" versus "which", and why not then also discuss knowledge "how" and "why"? And finally, what is the status of all these logical theories? Are they normative prescriptions, of do they represent some existing cognitive practice, if only idealized and at higher levels of abstraction? Reading the papers in this section will not necessarily answer all these questions, but it will make readers much better equipped to pursue these issues for themselves.

Suggested Further Reading

Starting with a classical trailblazer, J. Hintikka, Knowledge and Belief: An Introduction to the Logic of the Two Notions, Cornell University Press 1962 and King's College Publications 2005, set the whole subject on its course. A series of later books broadened the paradigm to a general view of information and inquiry, as represented in J. Hintikka, Logic, Language-Games and Information: Kantian Themes in the Philosophy of Logic, Clarendon Press Oxford, 1973. Putting inquiry at center stage in epistemology has also been the persistent theme of Robert Stalnaker's work in the field, with Inquiry, The MIT Press, 1987, as a classic source. Meanwhile, a richer view of possible epistemic and doxastic attitudes suggested by natural language was investigated in W. Lenzen, "Recent Work in Epistemic Logic", Acta Philosophica Fennica 30 (1978): 1–219, which also discusses links with probability. Also still in the 1970s, epistemic and doxastic logic were rediscovered in the foundations of game theory, but references for this will be found in another part of these readings. But arguably the major development invigorating epistemic logic has been its crossing into computer science, in the study of information-driven agency. Two major books demonstrating the resulting outburst of new research are R. Fagin, J. Y. Halpern, Y Moses & M. Vardi, Reasoning about Knowledge, The MIT Press, 1995, and W. van der Hoek & J-J Meijer, Epistemic Logic for AI and Computer Science, Cambridge University Press, 1995. An even more radically computational algorithmic understanding has been that of formal learning theory, inspired by learning methods for infinite structures like human languages or the process of scientific inquiry. The classic source for this is K. Kelly, The Logic of Reliable Inquiry, Oxford University Press, 1996. We conclude with two other angles on knowledge that bring in further mathematical paradigms. One is the verificationist perspective on knowledge through proof and evidence, for which a classic text is M. Dummett, Truth and Other Enigmas, Harvard University Press, 1978. Finally, while these publications are concerned with knowledge and belief, another broad stream has taken information flow through Shannon-type channels to be the basic underlying notion, following Dretske's classic Knowledge and the Flow of Information, The MIT Press, 1981. An innovative logical framework taking this road much further is J. Barwise & J. Seligman, Information Flow, Cambridge University Press, 1995.

Chapter 26
Epistemology Without Knowledge and Without Belief

Jaakko Hintikka

Knowledge and Decision-Making

Epistemology seems to enjoy an unexpectedly glamorous reputation in these days. A few years ago, William Safire (1995) wrote a popular novel called *The Sleeper Spy*. It depicts a distinctly post-Cold War world in which it is no longer easy to tell the good guys—including the good spies—from the bad ones. To emphasize this sea change, Safire tells us that his Russian protagonist has not been trained in the military or in the police, as he would have been in the old days, but as an epistemologist.

But is this with-it image deserved? Would the theory of knowledge that contemporary academic epistemologists cultivate be of any help to a sleeper spy? This question prompts a critical survey of the state of the art or, rather, the state of the theory of knowledge. I submit that the up-to-date image is not accurate and that most of the current epistemological literature deals with unproductive and antiquated questions. This failure is reflected in the concepts that are employed by contemporary epistemologists.

What are those concepts? It is usually thought and said that the most central concepts of epistemology are knowledge and belief. The prominence of these two notions is reflected in the existing literature on epistemology. A large chunk of it consists in discussions of how the concept of knowledge is to be defined or is not to be defined. Are those discussions on the target? An adequate analysis of such concepts as knowledge and belief, whether it is calculated to lead us to a formal

Jaakko Hintikka was deceased at the time of publication.

J. Hintikka (deceased)
Boston University, Helsinki, Finland

H. Arló-Costa et al. (eds.), *Readings in Formal Epistemology*, Springer Graduate Texts in Philosophy 1, DOI 10.1007/978-3-319-20451-2_26

definition or not, should start from the role that they play in real life. Now in real life we are both producers and consumers of knowledge. We acquire knowledge in whatever ways we do so, and we then put it to use in our actions and decision-making. I will here start from the latter role, which takes us to the question: What is the role that the notion of knowledge plays in that decision-making?

To take a simple example, let us suppose that I am getting ready to face a new day in the morning. How, then, does it affect my actions if I know that it will not rain today? You will not be surprised if I say that what it means is that I am entitled to behave as if it will not rain—for instance to leave my umbrella home. However, you may be surprised if I claim that most of the important features of the logical behavior of the notion of knowledge can be teased out of such simple examples. Yet this is the case. My modest example can be generalized. The role of knowledge in decision-making is to rule out certain possibilities. In order to use my knowledge, I must know which possibilities it rules out. In other words, any one scenario must therefore be either incompatible or compatible with what I know, for I am either entitled or not entitled to disregard it. Thus the totality of incompatible scenarios determines what I know and what I do not know, and vice versa. In principle, all that there is to logic of knowledge is this dichotomy between epistemically impossible and epistemically possible scenarios.

It is also clear how this dichotomy serves the purposes of decision-making, just as it does in my mini-example of deciding whether or not to take an umbrella with me. But the connection with overt behavior is indirect, for what the dichotomy merely demarcates are the limits of what I am *entitled to* disregard. And being entitled to do something does not always mean that I do it. It does not always show up in the overt ways one actually or even potentially acts. For other considerations may very well enter into my decision-making. Maybe I just want to sport an umbrella even though I know that it need not serve its function of shielding myself from rain. Maybe I am an epistemological *akrates* and act against what I know. The connection is nevertheless real, even though it is a subtle one. There is a link between my knowledge and my decisions, but it is, so to speak, a *de jure* connection and not a *de facto* connection. I think that this is a part of what John Austin (1961a) was getting at when he compared "I know" with "I promise." To know something does not mean simply to have evidence of a superior degree for it, nor does it mean to have a superior kind of confidence in it. If my first names were George Edward, I might use the open-question argument to defend these distinctions. By saying "I promise," I entitle you to expect that I fulfill my promise. By saying "I know," I claim that I am entitled to disregard those possibilities that do not agree with what I know. There is an evaluative element involved in the concept of knowledge that does not reduce to the observable facts of the case. Hence, it is already seen to be unlikely that you could define what it means to know by reference to matters of fact, such as the evidence that the putative knower possesses or the state of the knower's mind.

This evaluative element is due to the role of knowledge in guiding our life in that it plays a role in the justification of our decisions. This role determines in the last analysis the logic and in some sense the meaning of knowledge. A Wittgensteinean might put this point by saying that decision-making is one of the language-games that constitute the logical home of the concept of knowledge. You can remove knowledge from the contexts of decision-making, but you cannot remove a relation to decision-making from the concept of knowledge. For this reason, it is among other things misguided in a fundamental way to try to separate epistemic possibility from actual (natural) possibility. Of course, the two are different notions, but the notion of epistemic possibility has conceptual links to the kind of possibility that we have to heed in our decision-making. For one thing, the set of scenarios involved in the two notions must be the same.

But the main point here is not that there is an evaluative component to the notion of knowledge. The basic insight is that there is a link between the concept of knowledge and human action. The evaluative element is merely a complicating factor in the equation. The existence of a link between the two is not peculiar to the notion of knowledge. There is a link, albeit of a different kind, also in the case of belief. In fact, the conceptual connection is even more obvious in the case of belief. Behavioral scientists have studied extensively decision principles where belief constitutes one component, as, for instance, in the principle of maximizing expected utility. It usually comes in the form of degrees of belief. (They are often identified with probabilities.) Typically, utilities constitute another component. Whether or not such explicit decision principles capture the precise links between belief and behavior, they illustrate the existence of the link and yield clues to its nature.

Indeed, from a systematic point of view, the relative roles assigned to knowledge and to belief in recent epistemology and recent decision theory cannot but appear paradoxical. Belief is in such studies generally thought of as a direct determinant of our decisions, whereas knowledge is related to action only indirectly, if at all. Yet common sense tells us that one of the main reasons for looking for more knowledge is to put us in a better position in our decision-making, whereas philosophers often consider belief—especially when it is contrasted with knowledge—as being initially undetermined by our factual information and therefore being a much worse guide to decision-making. Probability is sometimes said to be a guide to life, but surely knowledge is a better one. Or, if we cannot use black-or-white concepts here, shouldn't rational decision-making be guided by degrees of knowledge rather than degrees of mere belief?

The same point can perhaps be made by noting that in many studies of decision-making, a rational agent is supposed to base his or her decisions on the agent's beliefs (plus, of course, utilities) and then by asking: Would it not be even more rational for the agent to base his or her decisions on what the agent *knows*?

In order for a rational agent to act on his or her belief, this belief clearly must be backed up by some evidence. Otherwise, current decision theory makes little sense. The difference is that the criteria of what entities are to act are different in the case of belief from what they are in the case of knowledge. If I act on a belief, that belief must satisfy my personal requirements for that role. They may vary from person to person. In contrast, the criteria of knowing are impersonal and not dependent on the agent in question. In order to define knowledge as distinguished from beliefs, we would have to spell out those impersonal criteria. This is obviously an extremely difficult task at best.

Another fact that complicates the connection between knowledge and behavior—that is, between what I know and what I do—is that in principle, this link is holistic. What matters to my decisions in the last analysis is the connection between the totality of my knowledge. There is not always any hard-and-fast connection between particular items of knowledge and my behavior. In principle, the connection is via my entire store of knowledge. This is reflected by the fact emphasized earlier that the dichotomy that determines the logic of knowledge is a distinction between scenarios that are ruled out by the *totality* of what I know and scenarios that are compatible with the *totality* of my knowledge and that I therefore must be prepared for. The same feature of the concept of knowledge also shows up in the requirement of total evidence that is needed in Bayesian inference and which has prompted discussion and criticism there. (See, e.g., Earman 1992.)

To spell out the criteria of the justification involved in the applications of the concept of knowledge is to define what knowledge is as distinguished from other propositional attitudes. Characterizing these conditions is obviously a complicated task. I will return to these criteria later in this chapter.

The Logic of Knowledge and Information

Meanwhile, another dimension of the concept of knowledge is brought out by homely examples of the kind I am indulging in. By this time it should be clear—I hope—that it is extremely hard to specify the kind of entitlement or justification that knowing something amounts to. This difficulty is perhaps sufficiently attested to by the inconclusiveness of the extensive discussions about how to define knowledge that one can find in the literature. (See, e.g., Shope 1983.) But another aspect of this notion is in principle as clear as anything one can hope to find in philosophical analysis (or synthesis). It may be difficult to tell whether a certain propositional attitude amounts to knowledge, belief, opinion or whatnot, but there is typically no difficulty in spelling out the *content* of any one of these attitudes on some particular occasion. Here, the lesson drawn from my rain-and-umbrella example is applicable. It was seen that what someone knows specifies, and is specified by, the class of possible scenarios that are compatible with what he or she knows. And such classes of scenarios or of "possible worlds" can be captured linguistically as the classes of scenarios (alias possible worlds) in which a certain sentence is true. Indeed, for Montague (1974, p. 153) such classes of possible worlds (or, strictly speaking, the characteristic functions of these classes, in the sense of functions from possible worlds to truth-values) *are* propositions. In this way, the content of a propositional attitude can normally be captured verbally. For another instance, for Husserl (1983, sec. 124), the task would be to capture the noematic *Sinn* of an act, which he says can in principle always be accomplished linguistically—that is, in Husserl's terminology, through *Bedeutungen*.

Let us now call the members of the class of scenarios admitted by someone's knowledge that someone's epistemic alternatives. That I know that it will not rain today means that none of the scenarios under which the wet stuff falls down are among my epistemic alternatives, and likewise for all *knowing that* statements. What the concept of knowledge involves in a purely logical perspective is thus a dichotomy of the space of all possible scenarios into those that are compatible with what I know and those that are incompatible with my knowledge. What was just seen is that this dichotomy is directly conditioned by the role of the notion of knowledge in real life. Now this very dichotomy is virtually all we need in developing an explicit logic of knowledge, better known as epistemic logic. This conceptual parentage is reflected by the usual notation of epistemic logic. In it, the epistemic operator K_a ("a knows that") receives its meaning from the dichotomy between excluded and admitted scenarios, while the sentence within its scope specifies the content of the item of knowledge in question.

Basing epistemic logic on such a dichotomy has been the guiding idea of my work in epistemic logic right from the beginning. I have seen this idea being credited to David Lewis, but I have not seen any uses of it that predate my work.

But here we seem to run into a serious problem in interpreting epistemic logic from the vantage point of a dichotomy of excluded and admitted scenarios. Such an interpretation might seem to exclude "quantifying in"—that is to say, to exclude applications of the knowledge operator to open formulas for them, it would not make any sense to speak of scenarios in which the content of one's knowledge is true or false. Such "quantifying in" is apparently indispensable for the purpose of analyzing the all-important *wh*-constructions with *knows*. For instance, "John *knows* who murdered Roger Ackroyd" apparently must be expressed by

$$(\exists x)\, K_{John}\, (x \text{ murdered Roger Ackroyd}) \tag{26.1}$$

as distinguished from

$$K_{John}\, (\exists x)\, (x \text{ murdered Roger Ackroyd}) \tag{26.2}$$

which says that John knows that someone murdered the victim and hence can serve as the presupposition of the question, "Who murdered Roger Ackroyd?"

But in (26.1), the notion of knowledge apparently cannot be interpreted by reference to a distinction between admitted and excluded scenarios. The reason is that the knowledge operator in (26.1) is prefixed to an open formula. Such an open formula cannot be said to be true or false in a given scenario, for its truth depends on the value of the variable x. Hence it cannot implement the required dichotomy.

In order for our epistemic discourse to express the *wh*-constructions, the knowledge operator must apparently be allowed to occur also internally, prefixed to open formulas rather than sentences (formulas without free variables). This prompts a serious interpretational problem. Indeed we can see here the reason for the deep theoretical interest of the problem of "quantifying in," which otherwise might strike one as being merely the logicians' technical problem. Fortunately, this apparent problem can be solved by means of suitable analysis of the relations between different logical operators (see section "Information acquisition as a questioning procedure").

An epistemic logic of this kind can obviously be developed within the framework of possible worlds semantics. (For a sketch of how this can be done, see Hintikka 2003b.) In fact, the truth condition for *knows that* is little more than a translation of what was just said: "b knows that S" is true in a world W if and only if S is true in all the epistemic b-alternatives to W. These alternatives are all the scenarios or "worlds" compatible with everything b knows in W. In certain important ways, this truth condition for knowledge statements is clearer than its counterpart in the ordinary (alethic) modal semantics, in that in epistemic logic the interpretation of the alternativeness relation (alias accessibility relation) is much clearer than in the logic of physical or metaphysical modalities.

Here we have already reached a major conclusion. Epistemic logic presupposes essentially only the dichotomy between epistemically possible and epistemically excluded scenarios. How this dichotomy is drawn is a question pertaining to the definition of knowledge. However, we do not need to know this definition in doing epistemic logic. Thus the logic and the semantics of knowledge can be understood independently of any explicit definition of knowledge. Hence it should not be surprising to see that a similar semantics and a similar logic can be developed for other epistemic notions—for instance, belief, information, memory, and even perception. This is an instance of a general law holding for propositional attitudes. This law says that the content of a propositional attitude can be specified independently of differences between different attitudes. This law has been widely recognized, even if it has not always been formulated as a separate assumption. For instance, in Husserl (1983, e.g., sec. 133) it takes the form of separating the noematic *Sinn* from the thetic component of a noema. As a consequence, the respective logics of different epistemic notions do not differ much from each other. In particular, they do not differ at all in those aspects of their logic that depend merely on the dichotomical character of their semantics. These aspects include prominently the laws that hold for quantifiers and identity, especially the modifications that are needed in epistemic contexts in the laws of the substitutivity of identity and existential generalization.

The fact that different epistemic notions, such as knowledge, belief, and information, share the same dichotomic logic should not be surprising in the light of what has been said. The reason is that they can all serve the same purpose of guiding our decisions, albeit in different ways. Hence the same line of thought can be applied to them as was applied earlier to the concept of knowledge, ending up with the conclusion that their logic is a dichotomic logic not unlike the logic that governs the notion of knowledge. The common ingredient in all these different logics is then the true epistemic logic. But it turns out to be a logic of information rather than a logic of knowledge.

This distinction between what pertains to the mere dichotomy between admitted and excluded scenarios and what pertains to the criteria relied on in this dichotomy is not a novelty. It is at bottom only a restatement in structural terms of familiar contrast, which in the hands of different thinkers has received apparently different formulations. The dichotomy defines the content of a propositional attitude, while the criteria of drawing it determine which propositional attitude we are dealing with.

Hence we are naturally led to the project of developing a generic logic of contents of attitudes, independent of the differences between different attitudes.

This generic logic of epistemology can be thought of as the logic of information. Indeed, what the content of a propositional attitude amounts to can be thought of as a certain item of information. In attributing different attitudes to agents, different things are said about this information—for instance, that it is known, believed, remembered, and so on. This fits in well with the fact that the same content can be known by one person, believed by another, remembered by a third one, and so on. This idea that one and the same objective content may be the target of different people's different attitudes is part of what Frege (see, e.g., 1984) was highlighting by his notion of *the thought*. Thus it might even be happier to talk about the logic of information than about epistemic logic. John Austin (1961b) once excused his use of the term "performative" by saying that even though it is a foreign word and an ugly word that perhaps does not mean very much, it has one good thing about it: It is not a deep word. It seems to me that epistemology would be in much better shape if instead of the deep word "knowledge," philosophers cultivated more the ugly foreign word "information," even though it perhaps does not capture philosophers' profound sense of knowing. In any case, in the generic logic of epistemology here envisaged, philosophers' strong sense of knowledge plays no role.

Information Acquisition as a Questioning Procedure

But what about the other context in which we encounter knowledge in real life—the context of knowledge acquisition? As was noted, what the concept of knowledge amounts to is revealed by two questions: What is it that we are searching for in the process of knowledge acquisition? What purpose can the product of such an inquiry serve? The second question has now been discussed. It remains to examine the crucial first question. Surely the first order of business of any genuine theory of knowledge—the most important task both theoretically and practically—is how new acquired, not merely how previously obtained information can be evaluated. A theory of information (knowledge) acquisition is both philosophically and humanly much more important than a theory of whether or not already achieved information amounts to knowledge. Discovery is more important than the defense of what you already know. In epistemology, as in warfare, offense frequently is the best defense.

This point can be illustrated in a variety of ways. For instance, a thinker who does not acquire any information cannot even be a skeptic, for he or she would not have anything to be skeptical about. And a skeptic's doubts must be grounded on some grasp as to how that information is obtained, unless these doubts are totally irrational. Epistemology cannot start from the experience of wonder or doubt. It should start from recognition of where the item of information that we are wondering about or doubting came from in the first place. Any rational justification or rational distinction of such wonder or doubt must be based on its ancestry.

Fortunately we now have available to us a framework in which to discuss the logic and epistemology of knowledge acquisition or, rather, if I have the terminological courage of my epistemological convictions, information acquisition. The framework is what is referred to as the interrogative model of inquiry or interrogative approach to inquiry. (See Hintikka 1999.) Its basic idea is the same as that of the oldest explicit form of reasoning in philosophy, the Socratic method of questioning or *elenchus*. In it, all new information enters into an argument or a line of reasoning in the form of answers to questions that the inquirer addresses to a suitable source of information.

It might at first seem implausible that this approach might yield a viable theory of ampliative reasoning in general, for several different reasons. Fortunately all these objections can be overcome. First, it might not seem likely that this model can be developed into a form explicit and detailed enough to allow for precise conclusions. This objection would have been eminently appropriate as recently as a decade or two ago. For it is only in the last several years that there has existed a general and explicit logical theory of all the relevant kinds of questions. This logic of questions and answers is the backbone of the interrogative model. This theory has not yet been presented in a monographic or textbook form, but its basic ideas are explained in recent and forthcoming papers of mine. (See, e.g., Hintikka 2003a.) This logic of questions and answers is an extension and application of epistemic logic (logic of knowledge). It has been made possible by a quiet revolution in epistemic logic. One of the main problems in representing questions is to specify which ingredients of the aimed-at information are the questioned elements—that is to say, are supposed to be made known by the answer. It turns out that their specification can sometimes be accomplished only by means of the independence indicators whose logic is only now being explored, even though it cannot be done in the earlier "first-generation" epistemic logic. The details of the new "second-generation" epistemic logic that makes use of the notion of independence need not concern us here. It may nevertheless be noted that this new logic solves the problem of "quantifying in" in that in it, the epistemic operator K always occurs sentence-initially. There is no problem of quantifying in, one might say here, only quantifying (binding variables) independently of an epistemic operator.

Another main requirement that can be addressed to the interrogative approach—and indeed to the theory of any goal-directed activity—is that it must do justice to the strategic aspects of inquiry. Among other things, it ought to be possible to distinguish the definitory rules of the activity in question from its strategic rules. The former spell out what is possible at each stage of the process. The latter express what actions are better and worse for the purpose of reaching the goals of the activity. This requirement can be handled most naturally by doing what Plato already did to the Socratic *elenchus* and by construing knowledge-seeking by questioning as a game that pits the questioner against the answerer. Then the study of the strategies of knowledge acquisition becomes another application of the mathematical theory of games, which perhaps ought to be called "strategy theory" rather than "game theory" in the first place. The distinction between the definitory rules—usually called simply the rules of the game—and strategic principles is built right into the structure of such games.

The greatest obstacle to generality might seem to be the apparently restricted range of applicability of the interrogative model. Some of the resistance to this approach, which I have referred to as the idea of "inquiry as inquiry," can be dispelled by pointing out that questions and answers can be understood in a wide sense, and have to be so understood if the generality claim is to be acceptable. Sources of answers to explicit or implicit questions have to include not only human witnesses and other informants or databases in a computer, but observation and experimentation as well as memory and tacit knowledge. One of the leading ideas of the interrogative approach is that all information used in an argument must be brought in as an answer to a question. In claiming such generality for the interrogative model, I can appeal to such precedents as Collingwood's (1940) and Gadamer's (1975) "logic of questions and answers," even though what they called logic really was not. My claims of generality on behalf of the interrogative approach are not even as sweeping as Collingwood's thesis that every proposition may be considered as an answer to a question. Likewise in construing experiments as questions to nature, I can cite Kant (Kant 1787) and Bacon.

Interrogation and Justification

But the context of knowledge acquisition is vital even if the aim of your game is justification and not discovery. Suppose that a scientist has a reason to think that one of his or her conclusions is not beyond doubt. What is he or she to do? Will the scientist try to mine his or her data so as to extract from them grounds for a decision? Sometimes, perhaps, but in an overwhelming majority of actual scientific situations, the scientist will ask what further information one should in such circumstances try to obtain in order to confirm or disconfirm the suspect proposition—for instance, what experiments it would be advisable to perform or what kinds of observation one should try to make in order to throw light on the subject matter. Unfortunately such contexts—or should I say, such language-games—of verification by means of new information have not received much attention from recent philosophers. They have been preoccupied with the justification of already acquired knowledge rather than with the strategies of reaching new knowledge.

Thus we must extend the scope of the interrogative model in such a way that it enables us to cope with justification and not just pure discovery. What we need is a rule or rules that authorize the rejection—which is tentative and may be only temporary—of some of the answers that an inquirer receives. The *terminus technicus* for such rejection is *bracketing*. The possibility of bracketing widens the scope of epistemological and logical methods tremendously. After this generalization has been carried out, the logic of interrogative inquiry can serve many of the same purposes as the different variants of non-monotonic reasoning, and serve them without the tacit assumptions that often make nonmonotonic reasoning epistemologically restricted or even philosophically dubious. A telling example is offered by what is known as circumscriptive reasoning. (See McCarthy 1990.)

It relies on the assumption that the premises present the reasoner with all the relevant information, so that the reasoner can assume that they are made true in the intended models in the simplest possible way. This is an assumption that in fact can often be made, but it is not always available on other occasions. As every puzzle fan knows, often a key to the clever reasoning needed to solve a puzzle lies precisely in being able to imagine circumstances in which the normal expectations evoked by the specification of the puzzle are not realized. Suppose a puzzle goes as follows: "Evelyn survived George by more than 80 years, even though she was born many decades before him. How come?" The explanation is easy if you disregard the presumption that "George" is a man's name and "Evelyn" a woman's. Evelyn Waugh in fact survived George Eliot by 86 years. Here the solution of the puzzle depends entirely on going beyond the *prima facie* information provided by the putative—in other words, on violating the presuppositions of a circumscriptive inference. Reasoning by circumscription is enthymemic reasoning. It involves tacit premises that may be false.

Thus by introducing the idea of bracketing, we can dispense with all modes of ampliative reasoning. The only rules besides rules of logical inference are the rules for questioning and the rule allowing bracketing. This may at first look like a cheap trick serving merely to sweep all the difficulties of epistemic justification under the rug of bracketing. In reality, what is involved is an important insight. What is involved is not a denial of the difficulties of justification, but an insight into their nature as problems. Once a distinction is made between strategic and definitory rules, it is realized that the definitory rules can only be permissive, telling what one may do in order to reach knowledge and to justify it. The problem of justification is a strategic problem. It pertains to what one ought to do in order to make sure that the results of one's inquiry are secure. This is to be done by the double process of disregarding dubious results and confirming the survivors through further inquiry. The only new permissive rule needed for the purpose is the rule that allows bracketing.

Thus the question as to which answers to bracket is always at bottom a strategic problem. It is therefore futile in principle to try to capture the justificatory process by means of definitory rules of this or that kind. To attempt to do so is a fallacy that in the last analysis vitiates all the usual "logics" of ampliative reasoning. This mistake is committed not only by non-monotonic logics but also by inductive logic and by the current theories of belief revision. Ampliative logics can be of considerable practical interest and value, but in the ultimate epistemological perspective, they are but types of enthymemic reasoning, relying on tacit premises quite as much as circumscriptive reasoning. An epistemologist's primary task here is not to study the technicalities of such modes of reasoning, fascinating though they are in their own right. It is to uncover the tacit premises on which such euthymemic reasoning is in reality predicated.

Allowing bracketing is among other things important because it makes it possible to conceive of interrogative inquiry as a model also of the confirmation of hypotheses and other propositions in the teeth of evidence. The interrogative model can thus also serve as a general model of the justification of hypotheses. It should in fact be obvious that the processes of discovery and justification cannot be sharply separated from each other in the practice or in the theory of science. Normally, a new discovery in science is justified by the very same process—for instance, by the same experiments—by means of which it was made, or could have been made And this double duty service of questioning is not due only to the practical exigencies of "normal science." It has a firm conceptual basis. This basis is the fact that information (unlike many Federal appropriations) does not come to an inquirer earmarked for a special purpose—for instance, for the purpose of discovery rather than justification. The inquirer may ask a question for this or that proximate purpose in mind, but there is nothing in the answer that rules out its being used for other purposes as well.

And such an answer can only be evaluated in terms of its service for both causes. This is because from game theory we know that in the last analysis, game-like goal-directed processes can be evaluated only in terms of their strategies, not in terms of what one can say of particular moves—for instance, what kinds of "warrants" they might have. As a sports-minded logician might explain the point, evaluating a player's skills in a strategic game is in principle like judging a figure-skating performance rather than keeping score in a football game. In less playful terms, one can in generally associate utilities (payoffs) only with strategies, not with particular moves. But since discovery and justification are aspects of the same process, they have to be evaluated in terms of the different possible strategies that are calculated to serve both purposes.

When we realize this strategic inseparability of the two processes, we can in fact gain a better understanding of certain otherwise puzzling features of epistemic enterprise. For instance, we can now see why it sometimes is appropriate to jump to a conclusion on the basis of relatively thin evidence. The reason is that finding what the truth is can help us mightily in our next order of business of finding evidence for that very truth. Sherlock Holmes has abductively "inferred" that the stablemaster has stolen the famous racing horse "Silver Blaze" (see the Conan Doyle story with this title) in order to lame it partially. He still has to confirm this conclusion, however, and in that process he is guided by the very content of that abductive conclusion—for instance, in directing his attention to the possibility that the stablemaster had practiced his laming operation on the innocent sheep grazing nearby. He puts a question to the shepherd as to whether anything had been amiss with them of late. "Well, sir, not of much account, but three of them have gone lame, sir." Without having already hit on the truth, Holmes could not have thought of asking this particular question.

If you disregard the strategic angle, the frequent practice of such "jumps to a conclusion" by scientists may easily lead one to believe that scientific discovery is not subject to epistemological rules. The result will then be the hypothetico-deductive model of scientific reasoning, which is hence seen to rest on a fallacious dismissal of the strategic angle.

Thus we reach a result that is neatly contrary to what were once prevalent views. It used to be held that discovery cannot be subject to explicit epistemological theory, whereas justification can. We have found out that not only can discovery be approached epistemologically, but that justification cannot in the long run be done justice to by a theory that does not also cover discovery.

A critical reader might initially have been wondering why contexts of verification and of other forms of justification do not constitute a third logical home of the notion of knowledge, besides the contexts of decision-making and information-acquisition. The answer is that processes of justification can only be considered as aspects of processes of information-acquisition.

The Generality of the Interrogative Model

The most general argument for the generality of the interrogative approach relies only on the assumption that the inquirer's line of thought can be rationally evaluated. What is needed for such an evaluation? If no new information is introduced into an argument by a certain step, then the outcome of that step is a logical consequence of earlier statements reached in the argument. Hence we are dealing with a logical inference step that has to be evaluated by the criteria of logical validity. It follows that interrogative steps are the ones in which new information enters into the argument. In order to evaluate the step, we must know what the source of this information is, for the reliability of the information may depend on its source. We must also know what else might have resulted from the inquirer's approaching this particular source in this particular way and with what probabilities. If so, what the inquirer did can be thought of as a question addressed to that source of information. Likewise, we must know what other sources of information the inquirer could have consulted and what the different results might have been. This amounts to knowing what other sources of answers the inquirer might have consulted. But if all of this is known, we might as well consider what the inquirer did as a step in interrogative inquiry.

In an earlier work (Hintikka 1998), I have likened such tacit interrogative steps to Peircean abductions, which Peirce insists are inferences even though they have interrogative and conjectural aspects.

The interrogative model can be thought of as having also another kind of generality—namely, generality with respect to the different kinds of questions. Earlier epistemic logic was incapable of handling questions more complicated than simple *wh*-questions. In particular, it could not specify the logical form of questions in which the questioned ingredient was apparently within the scope of a universal quantifier, which in turn was in the scope of a *knows that* operator. This defect was eliminated by means of the independence indicator (slash) /. (See Hintikka 2003b.) What characterizes the questioned ingredient is its independence of the epistemic operator, and such independence is perfectly compatible with its being dependent on a universal quantifier, which is in turn dependent on the universal quantifier. In symbols we can now write, for instance, $K\ (\forall x)\ (\exists y/K)$ without having to face the impossible task of capturing the threefold dependence structure by means of scopes—that is, by ordering K, $(\forall x)$, and $(\exists y)$ linearly so as to capture their dependence relations.

In this way, we can treat all *wh*-questions and all prepositional questions (involving questions where the two kinds of question ingredients are intermingled). The question ingredient of prepositional questions turns out to be of the form $(\vee/K)$ and the question ingredient of *wh*-questions of the form $(\exists x/K)$. We can also close a major gap in our argument so far. The connection between knowledge and decision-making discussed in section "Knowledge and decision-making" is apparently subject to the serious objection mentioned in section "The logic of knowledge and information". It helps to understand a knowledge operator K only when it occurs clause-initially, prefixed to a closed sentence. For it is only such sentences, not all and sundry formulas, that express a proposition that can serve as a justification of an action. Occurrences of K inside a sentence prefixed to an open formula cannot be interpreted in the same way. Now we can restrict K to a sentence-initial position, which eliminates this objection. This also helps to fulfill the promise made in section "The logic of knowledge and information" of constructing a general logic for the epistemic operator. Here we are witnessing a major triumph of second-generation epistemic logic, which relies on the notion of independence. It solves once and for all the problem of "quantifying in." It turns out that we do not at bottom *quantify into* a context governed by the epistemic operator K. What we in effect do is to *quantify independently* of this operator.

Why-questions and *how*-questions require a special treatment, which nevertheless is not hard to do. (See, e.g., Hintikka and Halonen 1995.)

The most persuasive argument for the interrogative model nevertheless comes from the applications of the interrogative viewpoint to different problems in epistemology. An important role in such applications is played by the presuppositions of questions and by the presuppositions of answers, better known as their conclusiveness conditions. Examples of such application are offered in Chaps. 4 and 5 of this volume.

The Place of Knowledge in Inquiry

It would take me too far afield here to essay a full-fledged description of the interrogative model. It is nevertheless easy to make an inventory of the concepts that are employed in it. In an explicit model, question-answer steps are interspersed with logical inference steps. Hence the concepts of ordinary deductive logic are needed. As long as the inquirer can trust all the answers, the concepts that are needed are the presuppositions of a question, the conclusiveness condition of an answer (which might be called the "presupposition" of the answer), and the notion of information. To describe an interrogative argument with uncertain answers (responses), we need the notion of tentative rejection of an answer, also known as *bracketing*, and hence also the converse operation of unbracketing, plus ultimately also the notion of probability needed to judge the conditions of bracketing and unbracketing.

What is remarkable about this inventory is that it does not include the concept of knowledge. One can construct a full epistemological theory of inquiry as inquiry without ever using the k-word. This observation is made especially significant by the generality of the interrogative model. As was indicated, not only is it by means of an interrogative argument that all new information can be thought of as having been discovered, it is by the same questioning method that its credibility must be established in principle.

What this means is that by constructing a theory of interrogative inquiry we apparently can build up a complete theory of epistemology without using the concept of knowledge. We do not need the notion of knowledge in our theory of knowledge—or so it seems. We do not need it either in the theory of discovery or in the theory of justification.

This conclusion might seem to be too strange to be halfway plausible. It is not, but it needs explanations to be seen in the right perspective.

It might perhaps seem that the concept of knowledge is smuggled into interrogative argumentation by the epistemic logic that has to be used in it. This objection is in fact a shrewd one. I said earlier that the logic of questions and answers, which is the backbone of the interrogative model, is part of the logic of knowledge. And this need to resort to epistemic notions is grounded deeply in the facts of the case. It might at first seem that in an interrogative inquiry, no epistemic notions are needed. The presuppositions of questions, questions themselves, and replies to them can apparently be formulated without using epistemic notions.

However, this first impression turns out to be misleading. The structure of and the rules governing it cannot be specified without using some suitable epistemic logic. For one thing, many of the properties of questions and answers are best explained by reference to what is known as the desideratum of a question. This desideratum specifies the epistemic state that the questioner wants to be brought about (in the normal use of questions). For instance, the desideratum of "Who murdered Roger Ackroyd?" is "I know who murdered Roger Ackroyd." But the desideratum with its *prima facie* knowledge operator is not only a part of a theory of question-answer sequences, it is a vital ingredient of the very interrogative process.

In particular, it is needed to solve Meno's problem (Plato 1924) applied to interrogative inquiry. In the initial formulation of the rules for interrogative inquiry, it is apparently required that we must know not only the initial premises of inquiry but also their ultimate conclusion. This seems to mean that we can use interrogative inquiry only to explain conclusions we have already reached but not to solve problems—in other words, answer questions by means of questions. But in trying to answer a question by means of interrogative inquiry, we apparently do not know what the ultimate conclusion is. We are instead looking for it. How, then, can we use interrogative inquiry for the purpose of answering questions? The answer is that we must formulate the logic of inquiry in terms of what the inquirer knows (in the sense of being informed about) at each stage. Then we can solve Meno's problem merely by using the desideratum of the overall question as the ultimate conclusion. But then we seem to need the notion of knowledge with vengeance.

What is true is that a viable theory of questions and answers will inevitably involve an intensional operator, and in particular an epistemic operator in a wide sense of the word. However, the epistemic attitude this operator expresses is not knowledge in any reasonable sense of the word, not just not in the philosopher's solemn sense. Here, the results reached in section "The logic of knowledge and information" are applicable. Before an interrogative inquiry has reached its aim—that is, knowledge—we are dealing with information that has not yet hardened into knowledge. It was seen earlier that the logic of such unfinished epistemological business is indeed a kind of epistemic logic, but a logic of information rather than of knowledge.

This point is worth elaborating. Indeed the real refutation of the accusation of having smuggled the concept of knowledge into interrogative inquiry in the form of the epistemic operator used in questions and answers lies in pointing out the behavior of this operator in epistemic inquiry. It may sound natural to say that after having received what is known as a conclusive answer to a question, the inquirer now knows it. But the notion of knowledge employed here is a far cry from the notion of knowledge that philosophers have tried to define. It looks much more like the ugly foreign notion of information. It does not even carry the implication of truth, for the answer might very well have to be bracketed later in the same inquiry. By the same token, it does not even presuppose any kind of stable belief in what is "known." Instead of saying that after having received a conclusive answer, the inquirer knows it, it would be more accurate to say that he or she has been informed about it. Here the advantages of the less deep notion of information are amply in evidence. Unlike knowledge, information need not be true. If an item of information offered to me turns out to be false, I can borrow a line from *Casablanca* and ruefully say, "I was misinformed." The epistemic operator needed in the logic of questions and answers is therefore not a knowledge operator in the usual sense of the term. My emphasis on this point is a penance, for I now realize that my statements in the past might have conveyed to my readers a different impression. What is involved in the semantics of questions and answers is the logic of information, not the logic of knowledge. This role of the notion of information in interrogative inquiry is indeed crucial, but it does not involve epistemologists' usual concept of knowledge at all.

This point is so important as to be worth spelling out even more fully. Each answer presents the inquirer with a certain item of information, and the distinction between question-answer steps and logical inferences steps hinges on the question of whether this information must be old or whether it can be new information. But it is important to realize that such information does not amount to knowledge. In an ongoing interrogative inquiry, there are no propositions concerning which question is ever raised, whether they are known or not. There may be a provisional presumption that, barring further evidence, the answers that an inquirer receives are true, but there is not even a whiff of a presumption that they are known. Conversely, when an answer is bracketed, it does not mean that it is definitively declared not to be known, for further answers may lead the inquirer to unbracket it. In sum, it is true in the strictest possible sense that the concept of knowledge in anything like philosophers' sense is not used in the course of interrogative inquiry.

These observations show the place of knowledge in the world of actual inquiry, and it also shows the only context in which questions about the definition of knowledge can legitimately be asked. The notion of knowledge may or may not be a discussion-stopper, but it is certainly an inquiry-stopper.

It might be suspected that this is due to the particular way the interrogative model is set up. Such a suspicion is unfounded, however. The absence of the concept of knowledge from ampliative inquiry is grounded in the very nature of the concept of knowledge. Questions of knowledge do not play any role in the questioning process itself, only in evaluating its results. For what role was it seen to play in human life? It was seen as what justifies us to act in a certain way. The concept of knowledge is therefore related to interrogative inquiry by asking: When has an interrogative inquiry reached far enough to justify the inquirer's acting on the basis of the conclusions it has so far reached? Or, to align this question with the locutions used earlier, when has the inquiry entitled the inquirer to dismiss the scenarios that are incompatible with the propositions accepted in the inquiry at the time? This is a genuine question, and it might seem to bring the concept of knowledge to the center of the theory of interrogative inquiry.

In a sense it does that. But this sense does not bring the notion of knowledge back as a concept that can possibly figure in the definitory rules of inquiry. It brings knowledge back to the sphere of strategic aspects of inquiry. The question as to whether a conclusion of inquiry has been justified strongly enough for it to qualify as knowledge is on a par with the question as to whether or not a step in an inquiry (typically an answer to a question) should perhaps be bracketed (however tentatively). Both are strategic questions. It is hopeless to try to model knowledge acquisition in a way that turns these decisions into questions of definitory correctness.

Any context-free definition of knowledge would amount to a definitory rule in the game of inquiry—namely, a definitory rule for stopping an inquiry. And once one realizes that this is what a definition of knowledge would have to do in the light of the conception of inquiry as inquiry, one realizes that the pursuit of such a definition is a wild goose chase.

It is important to realize that this conclusion does not only apply to attempted definitions of knowledge that refer only to the epistemic situation that has been reached at the putative end stage of the "game" of inquiry. In other words, it does not apply only to the state of an inquirer's evidence at the end of an inquiry. It also applies to definitions in which the entire history of inquiry so far is taken into account.

This conclusion is worth spelling out more fully. What the conclusion says is that no matter how we measure the credence of the output of interrogative inquiry, there is no reason to believe that an answer to the question as to when an inquirer is justified to act on his or her presumed knowledge depends only on the process of inquiry through which the inquirer's information has been obtained independently of the subject matter of the inquiry. In an old terminology, the criteria of justification cannot be purely *ad argumentum*, but must also be *ad hoc*. Neither the amount of information nor the amount of justification that authorizes an agent to stop his or her inquiry and act on its results can always be specified independently of the subject matter—for instance, independently of the seriousness of the consequences of being wrong about the particular question at hand. And if the justification depends on the subject matter, then so does the concept of knowledge, because of the roots of our concept of knowledge in action.

But since the notion of knowledge was seen to be tied to the justification of acting on the basis of what one knows, the concept of knowledge depends on the subject matter and not only on the epistemological situation. Accordingly, no general definition of knowledge in purely epistemological terms is possible.

This point is not a relativistic one as far as the possibility of *a priori* epistemology is concerned. If anything, the divorce of knowledge from inquiry underlines the objectivity of inquiry and its independence of the value aspects of the subject matter. The fashionable recent emphasis on the alleged value-ladenness of science is misleading in that it is typically predicated on forgetting or overlooking that the question as to when the results of scientific inquiry authorize acting on them is different from questions concerning the methodology of scientific inquiry itself. The dependence of the criteria of knowledge on subject matter ought to be a platitude. It is one thing for Einstein to claim that he knew that the special theory of relativity was true notwithstanding *prima facie* contrary experimental evidence, and another thing for a medical researcher to be in a position to claim to know that a new vaccine is safe enough to be administered to sixty million people. But some relativists mistakenly take this platitude to be a deep truth about scientific methodology and its dependence on subject matter. This is a mistake in the light of the fact that the allegedly value-laden concept of knowledge does not play any role in the actual process of inquiry.

Here, a comparison with such decision principles as the maximization of expected utility is instructive. What an inquiry can provide is only the expectations (probabilities). But they do not alone determine the decision, which depends also on the decider's utilities. Hence the criteria of knowing cannot be defined by any topic-neutral general epistemology alone. But this dependence does not mean that the probabilities used—misleadingly called "subjective" probabilities—should in rational decision-making depend on one's utilities. Decision-making based on such probability estimates would be paradigmatically irrational.

The influence of subject matter on the notion of knowledge does not imply that the interrogative process through which putative knowledge has been obtained is irrelevant for the evaluation of its status. Here lies, in fact, a promising field of work for applied epistemologists. Material for such work is available in, among many other places, different kinds of studies of risk-taking. Even though considerations of strategies do not help us to formulate a topic-neutral definition of knowledge, in such a topic-sensitive epistemology they are bound to play a crucial role. This is a consequence of the general fact that in game-like processes, only strategies, not individual moves, can in the last analysis be evaluated.

Comparisons with Other Epistemologists

Relativizing our humanly relevant concept of knowledge to some particular subject matter also provides a strategy of answering a philosophical skeptic. If knowledge claims depend for their very meaning on the criteria governing some particular walk of human action, then so also must reasonable doubts. It is only unspecific "philosophical" doubts that do not have built into their own logic standards that show how they can be surmounted.

One philosopher who would have agreed with my thesis concerning the dependence of the criteria of knowledge on the subject matter, and who in fact supplied reasons for it, is Ludwig Wittgenstein. In Hintikka forthcoming, I have shown that according to Wittgenstein's mature views, the concept of knowledge cannot be used in what I have called "primary language-games." These language-games are for Wittgenstein the direct links between language and reality. In them, we cannot, in Wittgenstein's metaphor, drive a wedge between language and what it expresses. Such a primary language-game does not operate by means of criteria, but by means of spontaneous responses. If I try to say in such a primary language-game "I know that I am in pain," all that I can express is the same as "I am in pain." And in a primary language-game, to utter "I am in pain" is but a form of pain-behavior.

In Wittgenstein's view, epistemic concepts can be used only in what I have called "secondary language-games." These secondary language-games presuppose primary ones. They do not operate through spontaneous responses, verbal or behavioral, and hence they must involve criteria. For this reason, epistemic vocabulary can be used in them. But those criteria are different in different secondary games. Hence the force of epistemic terms depends on the particular secondary game in which they are being used. Saying this is very nearly nothing but Ludwigspeak for saying that the criteria of knowing depend on the subject matter.

Other epistemologists have not been unaware, either, of connections between the justifiability of knowledge claims and the subject matter involved. (See, e.g., DeRose 1995; Cohen 1998; Williams 2001, ch. 14; Bonjour 2002, pp. 267–271.) They seem to have ascribed the dependence in question to the context of inquiry rather than to its subject matter, however. Unless and until the notion of context used here is clarified, I remain doubtful of such claims of context-dependence. For instance, criteria of knowing that a vaccine is safe depend on the life-or-death character of the subject matter, but they presumably should not depend on the context, which may be an administrative decision to initiate compulsory vaccination or a pharmaceutical company's promise to produce the requisite vaccine. However, if the notion of context is interpreted in such a way that it includes first and foremost the subject matter of inquiry, contextualist epistemology might very well converge with the views expressed here. In this work, contextualism is not examined further, however.

Moreover, contextual epistemologists seem to have assimilated the insight into the context-dependence of knowledge to another insight—namely, to the insight that every epistemological inquiry concerns some particular model, a "system" as physicists would call it, which typically is not an entire world. (See here Hintikka 2003a.) All epistemological inquiry is therefore contextual in this sense of being relative to a model (scenario or "possible world"). But this does not make epistemology itself contextual or relative as a scientific theory is made contextual or relative by the fact that it is inevitably applied to reality system by system. Hence the impact of the line of thought pursued here is diametrically opposed to the most common form of contextualism. This form of contextualism aims at the rejection of global epistemological questions. (See Bonjour 2002, p. 267). For us, global epistemological questions concern in the first place the nature of interrogative inquiry, and they are in no sense context-dependent or even dependent on the subject matter.

The Folly of Trying to Define Knowledge

The concept of knowledge thus belongs to applied epistemology, not to general epistemology. The criteria of knowledge concern the conditions on which the results of epistemological inquiry can be relied as a basis of action. It follows that it is an exercise in futility to try to define knowledge in any general epistemological theory. Such a definition could never help Satire's sleeper spy. But my point is not only about what is not useful in practice. The extensive discussions about how to define knowledge are not only useless for applications, they are theoretically misguided. Here the true relations of the concepts knowledge and truth to definability are almost precisely opposite to what they have been taken to be recently. Tarski (1956) proved certain results concerning the undefinability of truth. Philosophers and other thinkers have taken Tarski's results at their apparent face value, without realizing how restrictive the assumptions are on which these impossibility results are predicated. (See Hintikka 2002.) They have even let Tarski's results discourage them to the extent of giving up attempts to define truth. Tarski notwithstanding, a truth predicate can be formulated for sufficiently rich languages in a philosophically relevant sense in the same language. In contrast, no major philosopher has to the best of my knowledge openly maintained it to be a folly to try to define knowledge. Yet if someone has done so, that philosopher would have been much closer to truth than a philosopher who argues that it is foolish to try to define truth. (See Davidson 1996.)

Belief as a Product of Inquiry

The notion of knowledge belongs to applied epistemology because it is connected conceptually with the notions of acting and decision-making. The particular connection is not crucial. But if it does not matter, similar conclusions must hold also for those other epistemic concepts that are connected conceptually with behavior, especially with decision-making. The concept of belief is a case in point. And conclusions similar to the ones that have been reached here concerning the notion of knowledge can in fact be drawn concerning the notion of belief. If you are inspired by this line of thought to review the structure of interrogative inquiry with a view to finding a role for the notion of belief there, you will not find such a role. Receiving an answer and incorporating it into one's interrogative inquiry is not the same as adopting a new belief. Acceptance is not the same thing as belief-formation. (For a discussion of their relationship, see Cohen 1992.) For one thing, at no stage of an interrogative inquiry are there any indications whether or not the inquirer is prepared to act on the truth of the propositions that the inquirer has at that stage accepted (and not bracketed). Hence the entire theory of knowledge acquisition can—and must—be developed without using the notion of belief. This notion does not play any role in an interrogative inquiry, only in the evaluation of its putative end-point. If one thinks about it, the notion of belief does not play much of a role in the methodology of science. What I am suggesting is that is should not play any more of a role in general epistemology either.

There is thus a certain partial epistemological parallelism between belief and knowledge. This parallelism has not been appreciated by epistemologists. Ever since Plato, the two notions are habitually contrasted to each other. This contrast is nevertheless seriously misleading, as far the epistemology of belief is concerned.

It seems to me that the same point is unwittingly attested to by all the decision theorists who are using beliefs as an ingredient in rational decision-making. Such a use would be pointless unless there were some previous reasons to think that the beliefs in question can rationally be expected to be true. And such reasons must somehow come from the inquirer's previous experience, if one is a good empiricist.

Belief, too, is connected with criteria as to when I am ready to act on a certain item of information I have received. But whereas the criteria of knowing are impersonal (even though they can be relative to the subject matter), the criteria of belief can be personal and dependent on an even wider selection of the aspects of the subject matter. In claiming to know, I am making a commitment to others, but in forming a belief, I am usually responsible only to myself.

There are also intermediate cases. For instance, a scientist's beliefs *qua* scientist are subject to the standards of acceptance in his or her scientific community. The crucial point is that those beliefs are, in the case of a scientist, formed as a result of an inquiry, rather than, so to speak, as a response to the question, "What do you think about it?" One may very well catch a physicist asking whether he or she should believe a certain hypothesis in the light of available evidence. But one is even likelier to find a scientific inquirer asking what new information he or she should try to acquire—for instance, what experiments to carry out—in order to be in a position to entertain a certain belief.

In general, the same things can thus be said of belief and its standards as were said earlier of knowledge. Belief statements, like knowledge statements, express entitlement of a certain sort. In neither case does an agent have to avail himself or herself of such entitlement. Beliefs need not manifest themselves in overt behavior any more than knowledge. Hence, decision theorists' frequent assumption that an agent's beliefs (or degrees of belief) together with utilities determine his, her, or its behavior is in need of scrutiny. Above all, beliefs, too, must be thought of as being formed by means of inquiry.

What I take to be a related point has been expressed by Timothy Williamson by pointing out that a "reason is needed for thinking that beliefs tend to be true." (Quoted from the abstract of his contribution to the conference on "Modalism and Mentalism in Modern Epistemology," Copenhagen, January 29–31, 2004.) The relationship is mediated by the fact that, if I am right, interrogative inquiry is, in the last analysis, the only way of arriving at true beliefs.

The conclusions reached here have repercussions for the entire research strategies that should be pursued in epistemology. For instance, there is a major school of thought that conceives of inquiry as a series of belief revisions. But is this at all realistic as a description of what good reasoners actually do? Georges Simenon's Inspector Maigret is sometimes asked what he believes about the case he is investigating. His typical answer is: "I don't believe anything." And this does not mean, contrary to what one might first suspect, that Maigret wants only to know and not to believe and that he has not yet reached that state of knowledge. No—in one story he says, "The moment for believing or not believing hasn't come yet." (Georges Simenon, *Maigret and the Pickpocket*, Harcourt Brace Jovanovich, San Diego, 1985.) It is not that Maigret has not carried his investigation far enough to be in a position to know something. He has not reached for enough to form a belief. (The mere possibility of using the locution "belief formation" is instructive.) In serious inquiry, belief too is a matter whether an inquiry has reached far enough.

Belief, too, concerns the question of when to stop an inquiry. That is the place of this concept in the framework of the interrogative approach. The difference between belief and knowledge does not lie merely in the degree of justification the believer has reached. It does not mean that there is an evaluative component in knowledge but not in belief. The difference lies in the kind of evaluation involved. It is much more like the difference between satisfying an agent's own freely chosen standards of epistemic confidence and satisfying certain impersonal standards that are appropriate to the subject matter.

In linguists' terminology, knowing is an achievement verb. In a way, although not in a literal sense, believing is in the context of interrogative inquiry likewise an achievement notion. What should be studied in epistemology is belief-formation and not only belief change. The notion of belief cannot serve the role as a determinant of human action that is assigned to it in decision theory if it is not influenced by what the agent knows. But such influence is usually not studied in decision theory.

One corollary to the results we have reached concerns philosophers' research strategies. What we can see now is that the interrogative model is not only a rational reconstruction of knowledge acquisition, it can also be used as a model of belief formation. The insight that belief, too, is typically a product of inquiry lends some renewed interest to the "true belief" type of attempted definitions of knowledge. What they perhaps succeed in capturing is admittedly not philosophers' strong sense of knowledge. But there may be other uses (senses?) of the words *knowledge* and *knowing* that can be approached by means of such characterizations.

Philosophers tend to downplay the role of certainty, especially of experienced certainty, in explicating the notion of knowledge. There is nevertheless a third-person use of knowledge attributions in which the meaning of knowing is very close to true conviction reached through inquiry. In such cases, the inquirer has convinced himself or herself by means of inquiry of the truth of some proposition or other even when, by some standards, the inquirer has not yet reached sufficient justification.

A typical context is when an investigator has reached a correct conclusion—for instance, identified the perpetrator—through inquiry and has become convinced of this conclusion even though his or her reasons would not satisfy the standards of evidence in a court of law. It is interesting to note that in such a usage, the true conclusion must have been reached through a viable strategy. A mere guess would not amount to knowledge even in such cases. (Notice that one could not attribute knowledge in this sense to an automaton or to a database.) This observation may be related to Frank Ramsey's (1978) attempt to characterize knowledge as true belief obtained through a reliable method. This sense of knowing seems to be much closer to colloquial usage than the one philosophers have in vain been trying to define.

Repercussions for Other Approaches

From the point of view we have reached, we can also see some serious problems about the Bayesian approach to inquiry. (See, e.g., Earman 1992.) This approach deals with belief change rather than belief-formation. Insofar as we can find any slot for belief-formation within the Bayesian framework from the point of view of any simple application of, it is pushed back to the selection of priors. In other words, it is made entirely *a priori*, at least locally. This is by itself difficult to implement in the case of theory-formation (belief-formation) in science. Is it, for instance, realistic to assume that a scientist can associate an *a priori* probability with each and every possible law of nature? And these doubts are reinforced by general conceptual considerations. Assignments of priors amount to assumptions concerning the world. What is more, prior probabilities pertain to the entire system (model, "world") that the inquirer is investigating bit by bit. How can the inquirer choose such priors on the basis of his or her limited knowledge of the world? These difficulties might not be crucial if there existed a Bayesian theory of belief-change that included a study of changes of priors. Even though such changes have been studied, it seems to me that their theory has not been developed far enough in the Bayesian framework to cover all possibilities.

All sorts of difficult questions face us here. For instance, in order to use Bayesian inference, we need to know the prior probabilities. It seems to be thought generally that this does not amount to asking very much. This may be true in situations in which the primary data is reasonably reliable, as in typical scientific contexts. However, if our evidence is likely to be relatively unreliable, the situation may be different—for instance, when we are dealing with testimony as our basic form of evidence. I may easily end up asking: Do I really have enough information to make the guess concerning the world that was seem to be involved in the choice of the priors?

For one thing, even though the matter is highly controversial, fascinating evidence to this effect comes from the theory of so-called cognitive fallacies studied by mathematical psychologists such as Amos Tversky and Daniel Kahneman. (See, e.g., Kahneman et al. 1982; Piatelli-Palmarini 1994.) These alleged fallacies include the conjunction fallacy and the base-rate fallacy. As I have suggested in Chap. 9 of this volume (and in Hintikka 2004), at least in certain "crucial experiment" cases, the alleged mistakes are not fallacious at all, but rather point to certain subtle but very real ways in which one's prior probabilities can (and must) be changed in the light of new evidence. They do not show that certain fallacious ways of thinking are hardwired into human beings. Rather, what they show is that Bayesians have so far failed to master certain subtle modes of ampliative reasoning. Tversky's and Kahneman's Nobel Prize notwithstanding, epistemologists should take a long critical look at the entire theory of cognitive fallacies.

Here I can only give indications of how to view the cognitive fallacies conundrum. Very briefly, in the kind of situation that is at issue in the alleged conjunctive fallacy, the prior probabilities that one in effect relies on include the degrees of probability (credibility) assigned to the reports one receives. But that credibility can not only be affected by suitable new evidence, it can be affected by the very report itself. If the report shows that the reporter is likely to know more about the subject matter than another one, it is not fallacious to assign a higher prior probability to his or her report, even though it is a conjunction of a less credible report and further information.

In the case of an alleged base-rate fallacy, there is no conceivable mistake present if the intended sample space consists simply of the different possible courses of events concerning the crucial event—for example, a traffic accident. Base rates enter into the picture only when a wider class of courses of events is considered—for example, all possible courses of events that might have led to the accident. This means considering a larger sample space. Either sample space can of course be considered entirely consistently, depending on one's purposes. A fallacy would inevitably be committed only if the only legitimate application of our language and our epistemological methods was to the entire world—in this case, the larger sample space. But such an exclusive preference of the larger sample space is but an instance of the one-world assumption, which I have criticized elsewhere. (See Hintikka 2003a.)

Whither Epistemology?

The moral of the insights we have thus reached is not merely to avoid certain words in our epistemological theorizing. It calls for rethinking our overall research strategies in epistemology. And the spirit in which we should do so is perhaps illustrated by the first epistemologist in the Western philosophical tradition. Socrates did not claim that he knew anything. In the manner of a practitioner of my interrogative method, what he did was to ask questions. I suspect that it is only in Plato's dialogues that he was looking for a definition of knowledge. And Plato put this question (and other questions of definition) into Socrates's mouth because Plato shared the widespread Greek assumption that the definition of X gives us the "blueprint" that enables us to bring about X. (See Hintikka 1974, ch. 1–2.) This applies both to the generic search for knowledge and to the quest of particular items of knowledge. Thus, insofar as Plato contemplated knowledge-seeking (information-seeking) by questioning in our sense, he would have had to say that we must know what we are looking for there and that it is this knowledge alone that can guide our search. (No wonder he was worried about Meno's problem.) By the same token, all search for knowledge would have had to be guided by our knowledge of what knowledge is.

Hence it is seen that Plato had in one important respect the same focus as we: the quest for knowledge rather than the justification of beliefs. The definition of knowledge was thought of by Plato as a means for this quest. If so, the pursuit of the definition of knowledge would indeed have been the alpha and omega of epistemology. But we do not think in that way. The training that Satire's spymaster is supposed to have received did not aim exclusively at learning the definition of knowledge. For us, the fact that knowledge can be considered the end product of inquiry shows on the contrary that it cannot play any role in the process of inquiry. Hence the wild goose chase of the definition of knowledge only shows that too many contemporary epistemologists are still bewitched by Plato's assumptions. This is one of the reasons why at the beginning of this chapter, I called contemporary academic epistemology antiquated. Maybe it is time for its practitioners to take up some more up-to-date problems.

References

Austin, J. (1961a) (original 1946). Other minds. In *Philosophical papers*. Oxford: Clarendon Press, especially pp. 67–68.

Austin, J. (1961b). Performative utterances. In *Philosophical papers*. Oxford: Clarendon Press, ch. 10, especially p. 220.

Bonjour, L. (2002). *Epistemology: Classical problems and contemporary responses*. Lanham: Rowman & Littlefield.

Cohen, L. J. (1992). *Essay on belief and acceptance*. Oxford: Clarendon.

Cohen, S. (1998). Contextual solutions to epistemological problems. *Australasian Journal of Philosophy, 76*, 289–306.

Collingwood, R. G. (1940). *An essay on metaphysics*. Oxford: Clarendon Press, especially ch. I, sec. 5.

Davidson, D. (1996). The folly of trying to define truth. *Journal of Philosophy, 93*, 263–278.

DeRose, K. (1995). Solving the skeptical problem. *Philosophical Review, 104*, 1–52.

Earman, J. (1992). *Bayes or bust? A critical examination of Bayesian confirmation theory*. Cambridge: MIT Press.

Frege, G. (1984) (original 1918–19). Thoughts. In *Collected papers*. Oxford: Basil Blackwell.
Gadamer, H. -G. (1975) (original 1960). *Truth and method*. Continuum, New York, especially the section "Logic of Questions and Answers," pp. 333–341.
Hintikka, J. (1974). *Knowledge and the known*. Dordrecht: Reidel.
Hintikka, J. (1998). What is abduction? The fundamental problem of contemporary epistemology. *Transactions of the Charles Peirce Society, 34*, 503–533. A revised version, "Abduction—Inference, Conjecture, or an Answer to a Question," appears as Chapter 2 in this volume.
Hintikka, J. (1999). *Inquiry as inquiry: A logic of scientific discovery*. Dordrecht: Kluwer.
Hintikka, J. (2002). Post-tarskian truth. *Synthese, 126*, 17–36.
Hintikka, J. (2003a), A distinction too few or too many: A vindication of the snalytic vs. synthetic distinction. In Carol C. Gould (Ed.), *Constructivism and practice: Toward a historical epistemology* (pp. 47–74). Lanham: Rowman & Littlefield.
Hintikka, J. (2003b). A second-generation epistemic logic and its general significance. In Vincent F. Hendricks, et al. (Eds.), *Knowledge contributors* (pp. 33–56). Dordrecht: Kluwer Academic. And as Chap. 3 in this volume.
Hintikka, J. (2004). A fallacious fallacy? *Synthese, 140*, 25–35. And as Chap. 9 in this volume.
Hintikka, J. (forthcoming). *Wittgenstein on knowledge and skepticism* (Working paper).
Hintikka, J., & Halonen, I. (1995). Semantics and pragmatics for why-questions. *Journal of Philosophy, 92*, 636–657.
Husserl, E. (1983) (original 1913). *Ideas pertaining to a pure phenomenology. First book*. The Hague: Martinus Nijhoff.
Kahneman, D., Slovic, P., & Tversky, A. (Eds.). (1982). *Judgment under uncertainty: Heuristics and biases*. Cambridge: Cambridge University Press.
Kant, I. (1787). *Kritik der reinen Vernunft*, 2nd ed. (see Preface, p. xiii).
McCarthy, J. (1990). Circumscription—A form of non-monotonic reasoning. *Artificial Intelligence, 13*, 27–39 and 171–172.
Montague, R. (1974). *Formal philosophy*. New Haven: Yale University Press.
Piatelli-Palmarini, M. (1994). *Inevitable illusions*. New York: Wiley.
Plato. (1924). *Meno, Plato: With an English translation* (Loeb classical library, vol. IV). Cambridge: Harvard University Press.
Ramsey, F. (1978) (original 1929). Knowledge. In *Foundations: Essays in philosophy, logic, mathematics, and economics* (pp. 126–127). London: Routledge & Kegan Paul.
Safire, W. (1995). *The sleeper spy*. New York: Random House.
Shope, R. K. (1983). *The analysis of knowing: A decade of research*. Princeton: Princeton University Press.
Tarski, A. (1956) (original 1935). The concept of truth in formalized languages. In *Logic, semantics, metamathematics* (pp. 152–278). Oxford: Clarendon Press.
Williams, M. (2001). *Problems of knowledge: A critical introduction to epistemology*. Oxford: Oxford University Press.

Chapter 27
Epistemic Operators

Fred I. Dretske

Suppose *Q* is a necessary consequence of *P*. Given only this much, it is, of course, quite trivial that if it is true that *P*, then it must also be true that *Q*. If it is a fact that *P*, then it must also be a fact that *Q*. If it is necessary that *P*, then it is necessary that *Q*; and if it is possible that *P*, then it must also be possible that *Q*.

I have just mentioned four prefixes: ‘it is true that’ ‘it is a fact that’, ‘it is necessary that’, and ‘it is possible that’. In this paper I shall refer to such affixes as *sentential operators* or simply *operators;* when affixed to a sentence or statement, they operate on it to generate another sentence or statement. The distinctive thing about the four operators I have just mentioned is that, if *Q* is a necessary consequence of *P*, then the statement we get by operating on *Q* with one of these four operators is a necessary consequence of the statement we get by operating on *P* with the same operator. This may be put more succinctly if we let ‘O’ stand for the operator in question and ‘O(*P*)’ for the statement we get by affixing the operator ‘O’ to the statement ‘*P*’. We can now say that the above four operators share the following property: if *P* entails *Q*, then O(*P*) entails O(*Q*). I shall call any operator having this property a *penetrating operator* (Or, when emphasis is required, a fully *penetrating* operator). In operating on *P* these operators penetrate to every necessary consequence of *P*.

Versions of this paper were read to the philosophy departments of several universities in the United States and Canada during the year 1969/1970. I profited greatly from these discussions. I wish especially to thank Paul Dietl who helped me to see a number of points more clearly (perhaps still not clearly enough in his opinion). Finally, my exchanges with Mr. Don Affeldt were extremely useful; I am much indebted to him in connection with some of the points made in the latter portions of the paper.

F.I. Dretske (deceased)
University of Wisconsin, Madison, WI, USA

H. Arló-Costa et al. (eds.), *Readings in Formal Epistemology*, Springer Graduate Texts in Philosophy 1, DOI 10.1007/978-3-319-20451-2_27

We are now in a position to ask ourselves a preliminary question. The answer to this question is easy enough, but it will set the stage for more difficult questions. Are all sentential operators fully penetrating operators? Are all operators such that if *P* entails *Q*, then O(*P*) entails O(*Q*)? If *all* operators are penetrating operators, then each of the following statements must be true (when *P* entails *Q*):

1. You cannot have a reason to believe that *P* unless you have a reason to believe that *Q*.
2. You cannot know that *P* unless you know that *Q*.
3. You cannot explain why *P* is the case unless you can explain why *Q* is the case.
4. If you assert that *P*, then you assert that *Q*.
5. If you hope that *P*, then you hope that *Q*.
6. If it is strange (or accidental) that *P*, then it must be strange (or accidental) that Q.
7. If it was a mistake that *P*, then it was a mistake that *Q*.

This list begins with two epistemic operators, I 'reason to believe that' and 'know that'. Since I shall be concerned with these later in the paper, let me skip over them now and look at those appearing near the end of the list. They will suffice to answer our opening question, and their status is much less problematic than that of some of the other operators.

'She lost' entails 'Someone lost'. Yet, it may be strange that she lost, not at all strange that someone lost. 'Bill and Susan married each other' entails that Susan got married; yet, it may be quite odd that (strange that, incredible that) Bill and Susan married each other but quite unremarkable, not at all odd that, Susan got married. It may have been a mistake that they married each other, not a mistake that Susan got married. Or finally, 'I hit the bull's-eye' entails that I either hit the bull's-eye or the side of the barn; and though I admit that it was lucky that (accidental that) I hit the bull's-eye, I will deny that it was lucky, an accident, that I hit either the bull's-eye or the side of the barn.

Such examples show that not all operators are fully penetrating. Indeed, such operators as 'it is strange that', 'it is accidental that' and 'it is a mistake that' fail to penetrate to some of the most elementary logical consequences of a proposition. Consider the entailment between "*P* and *Q*" and '*Q*'. Clearly, it may be strange that *P* and *Q*, not at all strange that *P*, and not at all strange that *Q*. *A* concatenation of factors, no one of which is strange or accidental may itself be strange or accidental. Taken by itself, there is nothing odd or suspicious about Frank's holding a winning ticket in the first race. The same could be said about any of the other races: there is nothing odd or suspicious about Frank's holding a winning ticket in the nth race. Nonetheless, there is something very odd, very suspicious, in Frank's having a winning ticket in n races.

Therefore, not only are these operators not fully penetrating, they lie, as it were, on the other end of the spectrum. They fail to penetrate to some of the most elementary consequences of a proposition. I shall refer to this class of operators as nonpenetrating operators. I do not wish to suggest by this label that such operators are totally impotent in this respect (or that they are all uniform in their degree of penetration). I mean it, rather, in a rough, comparative, sense: their *degree of penetration* is less than that of any of the other operators I shall have occasion to discuss.

We have, then, two ends of the spectrum with examples from both ends. Anything that falls between these two extremes I shall call a *semi-penetrating* operator. And with this definition I am, finally, in a position to express my main point, the point I wish to defend in the rest of this paper. It is, simply, that all epistemic operators are semi-penetrating operators. There is both a trivial and a significant side to this claim. Let me first deal briefly with the trivial aspect.

The epistemic operators I mean to be speaking about when I say that all epistemic operators are semi-penetrating include the following:

(a) *S* knows that ...
(b) *S* sees (or can see) that ...
(c) *S* has reason (or a reason) to believe that ...
(d) There is evidence to suggest that ...
(e) *S* can prove that ...
(f) *S* learned (discovered, found out) that ...
(g) In relation to our evidence it is probable that ...

Part of what needs to be established in showing that these are all semi-penetrating operators is that they all possess a degree of penetration greater than that of the nonpenetrating operators. This is the trivial side of my thesis. I say it is trivial because it seems to me fairly obvious that if someone knows that *P* and *Q*, has a reason to believe that *P* and *Q*, or can prove that *P* and *Q*, he thereby knows that *Q*, has a reason to believe that *Q*, or can prove (in the appropriate epistemic sense of this term) that *Q*. Similarly, if *S* knows that Bill and Susan married each other, he (must) know that Susan got married (married someone). If he knows that *P* is the case, he knows that *P* or *Q* is the case (where the 'or' is understood in a sense which makes '*P* or *Q*' a necessary consequence of '*P*'). This is not a claim about what it would be appropriate to say, what the person himself thinks he knows or would say he knows. It is a question, simply, of what he knows. It may not be appropriate to *say* to Jim's wife that you know it was either her husband, Jim, or Harold who sent the neighbor lady an expensive gift *when you know it was Harold*. For, although you do know this, it is misleading to say you know it – especially to jim's wife.

Let me accept, therefore, without further argument that the epistemic operators are not, unlike 'lucky that', 'strange that', 'a mistake that' and 'accidental that', nonpenetrating operators. I would like to turn, then, to the more significant side of my thesis. Before I do, however, I must make one point clear lest it convert my entire thesis into something as trivial as the first half of it. When we are dealing with the epistemic operators, it becomes crucial to specify whether the agent in question knows that *P* entails *Q*. That is to say, *P* may entail *Q*, and *S* may know that *P*, but he may not know that *Q because,* and perhaps *only* because, he fails to appreciate the fact that *P* entails *Q*. When *Q* is a simple logical consequence of *P* we do not expect this to happen, but when the propositions become very complex, or the relationship between them very complex, this might easily occur. Let *P* be a set of axioms, *Q* a *theorem*. *S*'s knowing *P* does not entail *S*'s knowing *Q* just because

P entails *Q*; for, of course, *S* may not know that *P* entails *Q*, may not know that *Q* is a theorem. Hence, our epistemic operators will turn out not to be penetrating because, and perhaps only because, the agents in question are not fully cognizant of all the implications of what they know to be the case, can see to be the case, have a reason to believe is the case, and so on. Were we all ideally astute logicians, were we all fully apprised of all the necessary consequences (supposing this to be a well defined class) of every proposition, perhaps then the epistemic operators would turn into fully penetrating operators. That is, assuming that if *P* entails *Q*, we know that *P* entails *Q*, then every epistemic operator is a penetrating operator: the epistemic operators penetrate to all the known consequences of a proposition.

It is this latter, slightly modified, claim that I mean to reject. Therefore, I shall assume throughout the discussion that when *Q* is a necessary consequence of *P*, every relevant agent *knows that it is.* I shall be dealing with only the *known consequences* (in most cases because they are immediate and obvious consequences). What I wish to show is that, even under this special restriction, the epistemic operators are *only* semi-penetrating.

I think many philosophers would disagree with this contention. The conviction is that the epistemic worth of a proposition is hereditary under entailment, that whatever the epistemic worth of *P*, *at least* the same value must be accorded the known consequences of *P*. This conviction finds expression in a variety of ways. Epistemic logic: if *S* knows that *P*, and knows that *P* entails *Q*, then *S* knows that *Q*. Probability theory: if *A* is probable, and *B* is a logical consequence of *A*, then *B* is probable (relative to the same evidence, of course). Confirmation theory: if evidence *e* tends to confirm hypothesis *h*, then *e* indirectly confirms all the logical consequences of *h*. But perhaps the best evidence in favor of supposing that most philosophers have taken the epistemic operators to be fully penetrating is the way they have argued and the obvious assumptions that structure their arguments. Anyone who has argued in the following way seems to me to be assuming the thesis of penetrability (as I shall call it): if you do not know whether *Q* is true or not, and *P* cannot be true unless *Q* is true, then you (obviously) do not know whether *P* is true or not. *A* slightly more elaborate form of the same argument goes like this: If *S* does not know whether or not *Q* is true, then for all he knows it might be false. If *Q* is false, however, then *P* must also be false. Hence, for all *S* knows, *P* may be false. Therefore, *S* does not know that *P* is true. This pattern of argument is sprinkled throughout the epistemological literature. Almost all skeptical objections trade on it. *S* claims to know that this is a tomato. *A* necessary consequence of its being a tomato is that it is not a clever imitation which only looks and feels (and, if you will, tastes) like a tomato. But *S* does not know that it is not a clever imitation that only looks and feels (and tastes) like a tomato. (I assume here that no one is prepared to argue that anything that looks, feels, and tastes like a tomato to *S must be* a tomato.) Therefore, *S* does not know that this is a tomato. We can, of course, reply with G. E. Moore that we certainly do know it is a tomato (after such an examination) and since tomatoes are not imitations we know that this is not an imitation. It is interesting to note that this reply presupposes the same principle as does the skeptical objection: they both assume that if *S* knows that this is a *P*, and knows that every *P* is a *Q*, then *S* knows that this is a Q. The only difference is that the skeptic performs a modus tollens, Moore a modus ponens. Neither questions the principle itself.

Whether it be a question of dreams or demons, illusions or fakes, the same pattern of argument emerges. If you know this is a chair, you must know that you are not dreaming (or being deceived by a cunning demon), since its being a (real) chair entails that it is not simply a figment of your own imagination. Such arguments assume that the epistemic operators, and in particular the operator 'to know', penetrate to all the known consequences of a proposition. If these operators were not penetrating, many of these objections might be irrelevant. Consider the following exchange:

S: How strange. There are tomatoes growing in my apple tree.
K: That isn't strange at all. Tomatoes, after all, are physical objects and what is so strange about physical objects growing in your apple tree?

What makes K's reply so silly is that he is treating the operator 'strange that' as a fully penetrating operator: it cannot be strange that there are tomatoes growing in the apple tree unless the consequences of this (e.g., there are objects growing in your apple tree) are also strange. Similarly, it may not be at all relevant to object to someone who claims to know that there are tomatoes in the apple tree that he does not know, cannot be absolutely sure, that there are really any material objects. Whether or not this is a relevant objection will depend on whether or not this particular consequence of there being tomatoes in the apple tree is one of the consequences to which the epistemic operators penetrate. What I wish to argue in the remainder of this paper is that the traditional skeptical arguments exploit precisely those consequences of a proposition to which the epistemic operators do not penetrate, precisely those consequences which distinguish the epistemic operators from the fully penetrating operators.

In support of this claim let me begin with some examples which are, I think, fairly intuitive and then turn to some more problematic cases. I shall begin with the operator 'reason to believe that' although what I have to say could be said as well with any of them. This particular operator has the added advantage that if it can be shown to be only semi-penetrating, then many accounts of knowledge, those which interpret it as a form of justified true belief, would also be committed to treating 'knowing that' as a semi-penetrating operator. For, presumably, 'knowing that' would not penetrate any deeper than one's 'reasons for believing that'.

Suppose you have a reason to believe that the church is empty. *Must* you have a reason to believe that it is a church? I am not asking whether you generally have such a reason. I am asking whether one can have a reason to believe the church empty without having a reason to believe that it is a church which is empty. Certainly your reason for believing that the church is empty is not *itself* a reason to believe it is a church; or it *need not* be. Your reason for believing the church to be empty may be that you just made a thorough inspection of it without finding anyone. That is a good reason to believe the church empty. Just as clearly, however, it is not a reason, much less a good reason, to believe that what is empty is a church. The fact is, or so it seems to me, I do not have to have any reason to believe it is a church. Of course,

I would never *say* the church was empty, or that I had a reason to believe that the church was empty, unless I believed, and presumably had a reason for so believing, that *it was* a church which was empty, but this is a presumed condition of my *saying* something, not of my having a reason to believe something. Suppose I had simply assumed (correctly as it turns out) that the building was a church. Would this show that I had no reason to believe that the church was empty?

Suppose I am describing to you the "adventures" of my brother Harold. Harold is visiting New York for the first time, and he decides to take a bus tour. He boards a crowded bus and immediately takes the last remaining scat. The little old lady he shouldered aside in reaching his seat stands over him glowering. Minutes pass. Finally, realizing that my brother is not going to move, she sighs and moves resignedly to the back of the bus. Not much of an adventure, but enough, I hope, to make my point. I said that the little old lady realized that my brother would not move. Does this imply that she realized that, or knew that, *it was my brother* who refused to move? Clearly not. We can say that *S* knows that *X is Y* without implying that *S* knows that *it is X* which is *Y*. We do not *have* to describe our little old lady as knowing that *the* man or *the person* would not move. We can say that she realized that, or knew that, my brother would not move (minus, of course, this pattern of emphasis), and we can say this because saying this does not entail that the little old lady knew that, or realized that, it was my brother who refused to move. She knew that my brother would not move, and she knew this despite the fact that she did not know something that was necessarily implied by what she did know–viz., that the person who refused to move was my brother.

I have argued elsewhere that to see that *A is B*, that the roses are wilted for example, is not to see, not even to be able to see, that they are roses which are wilted.[1] To see that the widow is limping is not to see that it is a widow who is limping. I am now arguing that this same feature holds for all epistemic operators. I can know that the roses are wilting without knowing that they are roses, know that the water is boiling without knowing that it is water, and prove that the square root of 2 is smaller than the square root of 3 and, yet, be unable to prove what is entailed by this–viz., that the number 2 *has* a square root.

The general point may be put this way: there are certain presuppositions associated with a statement. These presuppositions, although their truth is entailed by the truth of the statement, are not part of what is *operated* on when we operate on the statement with one of our epistemic operators. The epistemic operators do not *penetrate to* these presuppositions. For example, in saying that the coffee is boiling I assert that the coffee is boiling, but in asserting this I do not assert that *it is* coffee which is boiling. Rather, this is taken for granted, assumed, presupposed, or what have you. Hence, when I say that I have a reason to believe that the coffee is boiling, I am not saying that this reason applies to the fact that it is coffee which is boiling. This is *still* presupposed. I may have such a reason, of course, and chances are good

[1] *Seeing and Knowing* (Chicago: University Press, 1969), pp. 93–112, and also "Reasons and Consequences," *Analysis* (April 1968).

that I do have such a reason or I would not have referred to what I believe to be boiling as *coffee,* but to have a reason to believe the coffee is boiling is not, thereby, to have a reason to believe it is coffee which is boiling.

One would expect that if this is true of the semi-penetrating operators, then it should also be true of the nonpenetrating operators. They also should fail to reach the presuppositions. This is exactly what we find. It may be accidental that the two trucks collided, but not at all accidental that it was two trucks that collided. Trucks were the only vehicles allowed on the road that day, and so it was not at all accidental or a matter of chance that the accident took place between two trucks. Still, it was an accident that the two trucks collided. Or suppose Mrs. Murphy mistakenly gives her cat some dog food. It need not be a mistake that she gave the food to *her* cat, or *some* food to a cat. This was intentional. What was a mistake was that it was dog food that she gave to her cat.

Hence, the first class of consequences that differentiate the epistemic operators from the fully penetrating operators is the class of consequences associated with the presuppositions of a proposition. The fact that the epistemic operators do not penetrate to these presuppositions is what helps to make them semi-penetrating. And this is an extremely important fact. For it would appear that if this is true, then to know that the flowers are wilted I do not have to know that they are flowers (which are wilted) and, therefore, do not have to know all those consequences which follow from the fact that they are flowers, real flowers, which I know to be wilted.

Rather than pursue this line, however, I would like to turn to what I consider to be a more significant set of consequences– "more significant" because they are the consequences that are directly involved in most skeptical arguments. Suppose we assert that x is A. Consider some predicate, 'B', which is incompatible with A, such that nothing can be both A and B. It then follows from the fact that x *is* A that x is not B. Furthermore, if we conjoin B with any other predicate, Q, it follows from the fact that x *is* A that x is not(B and Q). I shall call this type of consequence a *contrast consequence,* and I am interested in a particular subset of these; for I believe the most telling skeptical objections to our ordinary knowledge claims exploit a particular set of these contrast consequences. The exploitation proceeds as follows: someone purports to know that x *is* A, that the wall is red, say. The skeptic now finds a predicate 'B' that is incompatible with 'A'. In this particular example we may let 'B' stand for the predicate 'is white'. Since 'x is red' entails 'x is not white' it also entails that x is not-(white and Q) where 'Q' is any predicate we care to select. Therefore, the skeptic selects a 'Q' that gives expression to a condition or circumstance under which a white wall would appear exactly the same as a red wall. For simplicity we may let 'Q' stand for: 'cleverly illuminated to look red'. We now have this chain of implications: 'x is red' entails 'x is not white' entails 'x is not white cleverly illuminated to look red'. If 'knowing that' is a penetrating operator, then if anyone knows that the wall is red he must know that it is not white cleverly illuminated to look red. (I assume here that the relevant parties know that if x is red, it cannot be white made to look red.) He must know that this particular contrast consequence is true. The question is: do we, generally speaking, know anything of the sort? Normally we never take the trouble to check the lighting.

We seldom acquire any *special* reasons for believing the lighting normal although we can talk vaguely about there being no reason to think it unusual. The fact is that we habitually take such matters for granted, and although we normally have good reasons for making such routine assumptions, I do not think these reasons are sufficiently good, not without special precautionary checks in the particular case, to say of the particular situation we are in that we know conditions are normal. To illustrate, let me give you another example–a silly one, but no more silly than a great number of skeptical arguments with which we are all familiar. You take your son to the zoo, see several zebras, and, when questioned by your son, tell him they are zebras. Do you know they are zebras? Well, most of us would have little hesitation in saying that we did know this. We know what zebras look like, and, besides, this is the city zoo and the animals are in a pen clearly marked "Zebras." Yet, something's being a zebra implies that it is not a mule and, in particular, not a mule cleverly disguised by the zoo authorities to look like a zebra. Do you know that these animals are not mules cleverly disguised by the zoo authorities to look like zebras? If you are tempted to say "Yes" to this question, think a moment about what reasons you have, what evidence you can produce in favor of this claim. The evidence you *had* for thinking them zebras has been effectively neutralized, since it does not count toward their not being mules cleverly disguised to look like zebras. Have you checked with the zoo authorities? Did you examine the animals closely enough to detect such a fraud? You might do this, of course, but in most cases you do nothing of the kind. You have some general uniformities on which you rely, regularities to which you give expression by such remarks as, "That isn't very likely" or "Why should the zoo authorities do that?" Granted, the hypothesis (if we may call it that) is not very plausible, given what we know about people and zoos. But the question here is not whether this alternative is plausible, not whether it is more or less plausible than that there are real zebras in the pen, but whether *you know* that this alternative hypothesis is false. I don't think you do. In this I agree with the skeptic. I part company with the skeptic only when he concludes from this that, therefore, you do not know that the animals in the pen are zebras. I part with him because I reject the principle he uses in reaching this conclusion–the principle that if you do not know that Q is true, when it is known that P entails Q, then you do not know that P is true.

What I am suggesting is that we simply admit that we do not know that some of these contrasting "skeptical alternatives" are not the case, but refuse to admit that we do not know what we originally said we knew. My knowing that the wall is red certainly entails that the wall is red; it also entails that the wall is not white and, in particular, it entails that the wall is not white cleverly illuminated to look red. But it does not follow from the fact that I know that the wall is red that I *know* that it is not white cleverly illuminated to look red. Nor does it follow from the fact that I know that those animals are zebras that I know that they are not mules cleverly disguised to look like zebras. These are some of the contrast consequences to which the epistemic operators do not penetrate.

Aside from asserting this, what arguments can be produced to support it? I could proceed by multiplying examples, but I do not think that examples alone will support the full weight of this view. The thesis itself is sufficiently counterintuitive to render controversial most of the crucial examples. Anyone who is already convinced that skepticism is wrong and who is yet troubled by the sorts of skeptical arguments I have mentioned will, no doubt, take this itself as an argument in favor of my claim that the epistemic operators are only semi-penetrating. This, however, hardly constitutes an argument against skepticism. For this we need *independent* grounds for thinking that the epistemic operators do not penetrate to the contrast consequences. So I shall proceed in a more systematic manner. I shall offer an analogy with three other operators and conclude by making some general remarks about what I think can be learned from this analogy. The first operator is 'explains why' or, more suggestively (for the purposes of this analogy):

(A) *R* is the reason (explanatory reason) that (or why) ...

For example, the reason why *S* quit smoking was that he was afraid of getting cancer. The second operator has to do with reasons again, but in this case it is a reason which tends to *justify* one in doing something:

(B) *R* is a reason for ... (*S* to do *Y*).[2]

For example, the fact that they are selling the very same (type of) car here much more cheaply than elsewhere is a reason to buy it here rather than elsewhere. The status of this as a reason will, of course, depend on a variety of circumstances, but situations can easily be imagined in which this would be a reason for someone to buy the car here. Finally, there is a particular modal relationship which may be construed as a sentential operator:

(C) *R* would not be the case unless ...

For example, he would not have bid seven no-trump unless he had all four aces. I shall abbreviate this operator as '$R \Rightarrow \ldots$'; hence, our example could be written 'he bid seven no-trump $\Rightarrow$ he had all four aces'.

Each of these operators has features similar to those of our epistemic operators. If one retraces the ground we have already covered, one will find, I think, that these operators all penetrate deeper than the typical nonpenetrating operator. If *R* explains why (or is the reason that) *P* and *Q* are the case, then it explains why (is the reason that) *Q* is the case.[3] If I can explain why Bill and Harold are always invited to every party, I can explain why Harold is always invited to every party. From the fact that it was a mistake for me to quit my job it does not follow that it was a mistake for

[2]Unlike our other operators, this one does not have a propositional operand. Despite the rather obvious differences between this case and the others, I still think it useful to call attention to its analogous features.

[3]One must be careful not to confuse sentential conjunction with similar sounding expressions involving a relationship between two things. For example, to say Bill and Susan got married (if it is intended to mean that they married *each other),* although it entails that Susan got married, does not do so by sim*plification.* 'Reason why' penetrates through logical simplification, not through the type of entailment represented by these two propositions. That is, the reason they got married is that they loved each other; that they loved each other is not the reason Susan got married.

me to do something, but if I had a reason to quit my job, it does follow that I had a reason to do something. And if the grass would not be green unless it had plenty of sunshine and water, it follows that it would not be green unless it had water.

Furthermore, the similarities persist when one considers the presuppositional consequences. I argued that the epistemic operators fail to penetrate to the presuppositions; the above three operators display the same feature. In explaining why he takes his lunch to work, I do not (or need not) explain why he goes to work or why he works at all. The explanation may be obvious in some cases, of course, but the fact is I need not be able to explain why he works (he is so wealthy) to explain why he takes his lunch to work (the cafeteria food is *so* bad). The reason why the elms on Main Street are dying is not the reason there are elms on Main Street. I have a reason to feed my cat, no reason (not, at least, the same reason) to have a cat. And although it is quite true that he would not have known about our plans if the secretary had not told him, it does not follow that he would not have known about our plans if *someone other than the secretary* had told him. That is, (He knew about our plans) $\Rightarrow$ (The secretary told him) even though it is not true that (He knew about our plans) $\Rightarrow$ (It was the secretary who told him). Yet, the fact that *it was the secretary* who told him is (I take it) a presuppositional consequence of the fact that *the secretary told* him. Similarly, if George is out to set fire to the first empty building he finds, it may be true to say that George would not have set fire to the church unless it (the church) was empty, yet false to say that George would not have set fire to the church unless *it was a church.*

I now wish to argue that these three operators do not penetrate to a certain set of contrast consequences. To the extent that the epistemic operators are similar to these operators, we may then infer, by analogy, that they also fail to penetrate to certain contrast consequences. This is, admittedly, a weak form of argument, depending as it does on the grounds there are for thinking that the above three operators and the epistemic operators share the same logic in this respect. Nonetheless, the analogy is revealing. Some may even find it persuasive.[4]

(A) The pink walls in my living room clash with my old green couch. Recognizing this, I proceed to paint the walls a compatible shade of green. This is the reason I have, and give, for painting the walls green. Now, in having this explanation for why I painted the walls green, I do not think I have an explanation for two other things, both of which are entailed by what I do have an explanation for. I have not explained why I did not, *instead* of painting the walls green, buy a new couch or cover the old one with a suitable slipcover. Nor have I explained why,

[4]I think that those who are inclined to give a causal account of knowledge should be particularly interested in the operator '$R \Rightarrow \ldots$' since, presumably, it will be involved in many instances of knowledge ("many" not "all," since one might wish to except some form of immediate knowledge-knowledge of one's own psychological state-from the causal account). If this operator is only semipenetrating, then any account of knowledge that relies on the relationship expressed by this operator (as I believe causal accounts must) will be very close to giving a "semi-penetrating" account of 'knowing that'.

instead of painting the walls green, I did not paint them white and illuminate them with green light. The same effect would have been achieved, the same purpose would have been served, albeit at much greater expense.

I expect someone to object as follows: although the explanation given for painting the walls green does not, by itself, explain why the couch was not changed instead, it nonetheless succeeds as an explanation for why the walls were painted green only in so far as there is an explanation for why the couch was not changed instead. If there is no explanation for why I did not change the couch instead, there has been no real, no complete, examination for why the walls were painted green.

I think this objection wrong. I may, of course, have an explanation for why I did not buy a new couch: I love the old one or it has sentimental value. But then again I may not. It just never occurred to me to change the couch; or (if someone thinks that its not occurring to me *is* an explanation of why I did not change the couch) I may have thought of it but decided, for what reasons (if any) I cannot remember, to keep the couch and paint the walls. That is to say, I cannot explain why I did not change the couch. I thought of it but I did not do it. I do not know why. Still, I *can* tell you why I painted the walls green. They clashed with the couch.

(B) The fact that they are selling Xs so much more cheaply here than elsewhere may be a reason to buy your Xs here, but it certainly need not be a reason to do what is a necessary consequence of buying your Xs here–viz., not *stealing your* Xs here.

(C) Let us suppose that S is operating in perfectly normal circumstances, a set of circumstances in which it is true to say that the wall he sees would not (now) look green to him unless it was green (if it were any other color it would look different to him). Although we can easily imagine situations in which this is true, it does not follow that the wall would not (now) look green to S if it were white cleverly illuminated to look green. That is,

(i) The wall looks green (to S) $\Rightarrow$ the wall is green.
(ii) The wall is green *entails* the wall is not white cleverly illuminated to look green (to S).

are both true; yet, it is *not true* that

(iii) The wall looks green (to S) $\Rightarrow$ the wall is not white cleverly illuminated to look green (to S).

There are dozens of examples that illustrate the relative impenetrability of this operator. We can truly say that A and B would not have collided if B had not swerved at the last moment and yet concede that they would have collided without any swerve on the part of B *if* the direction in which A was moving had been suitably altered in the beginning.[5]

[5]The explanation for why the modal relationship between R and P ($R \Rightarrow P$) fails to carry over (penetrate) to the logical consequences of P (*i.e.*, $R \Rightarrow Q$ where Q is a logical consequence of P)

The structure of these cases is virtually identical with that which appeared in the case of the epistemic operators, and I think by looking just a little more closely at this structure we can learn something very fundamental about our class of epistemic operators and, in particular, about what it means to know something. If I may put it this way, within the context of these operators no fact is an island. If we are simply rehearsing the facts, then we can say that it is a fact that Brenda did not take any dessert (though it was included in the meal). We can say this without a thought about what sort of person Brenda is or what she might have done had she ordered dessert. However, if we put this fact into, say, an explanatory context, if we try to explain this fact, it suddenly appears within a network of related facts, a network of possible alternatives which serve to define *what it is that is being explained.* What is being explained is a function of two things-not only the fact (Brenda did not order any dessert), but also the range of relevant alternatives. *A* relevant alternative is an alternative that might have been realized in the existing circumstances if the actual state of affairs had not materialized.[6] When I explain why Brenda did not order any dessert by saying that she was full (was on a diet, did not like anything on the dessert menu), I explain why she did not order any dessert rather *than, as opposed to, or instead of* ordering some dessert and eating *it.* It is this competing possibility which helps to define what it is that I am explaining when I explain why Brenda did not order any dessert. Change this contrast, introduce a different set of relevant alternatives, and you change what it is that is being explained and, therefore, what counts as an explanation, even though (as it were) the same fact is being explained. Consider the following contrasts: ordering some dessert and throwing it at the waiter; ordering some dessert and taking it home to a sick friend.

---

is to be found in the set of circumstances that are taken as *given, or held fixed,* in subjunctive conditionals. There are certain logical consequences of *P* which, by bringing in a reference to circumstances tacitly held fixed in the original subjunctive ($R \Rightarrow P$), introduce a possible variation in these circumstances and, hence, lead to a *different* framework of fixed conditions under which to assess the truth of *R* @ *Q*. For instance, in the last example in the text, when it is said that *A* and *B* would not have collided if *B* had not swerved at the last moment, the truth of this conditional clearly takes it *as given that A* and *B* possessed the prior trajectories they in fact had on the occasion in question. Given certain facts, including the fact that they were traveling in the direction they were, they would not have collided if *B* had not swerved. Some of the logical consequences of the statement that *B* swerved do not, however, leave these conditions unaltered–e.g., *B* did not move in a perfectly straight line in a direction $2'$ counterclockwise to the direction it actually moved. This consequence "tinkers" with the circumstances originally taken *as given* (held fixed), and a failure of penetration will usually arise when this occurs. It *need not be* true that *A* and *B* would not have collided if *B* had moved in a perfectly straight line in a direction $2'$ counterclockwise to the direction it actually moved.

[6]I am aware that this characterization of "a relevant alternative" is not, as it stands, very illuminating. I am not sure I can make it more precise. What I am after can be expressed this way: if Brenda had ordered dessert, she *would not* have thrown it at the waiter, stuffed it in her shoes, or taken it home to a sick friend (she has no sick friend). These are not alternatives that might have been realized in the existing circumstances if the actual state of affairs had not materialized. Hence, they are not relevant alternatives. In other words, the 'might have been' in my characterization of a relevant alternative will have to be unpacked in terms of counterfactuals.

With these contrasts none of the above explanations are any longer explanations of why Brenda did not order dessert. Anyone who really wants to know why Brenda did not order dessert and throw it at the waiter will not be helped by being told that she was full or on a diet. This is only to say that, within the context of explanation and within the context of our other operators, the proposition on which we operate must be understood as embedded within a matrix of relevant alternatives. We explain why *P*, but we do so within a framework of competing alternatives *A, B,* and C. Moreover, if the possibility D is not within this contrasting set, not within this network of relevant alternatives, then even though not-D follows necessarily from the fact, *P*, which we do explain, we do not explain why not-D. Though the fact that Brenda did not order dessert and throw it at the waiter follows necessarily from the fact that she did not order dessert (the fact that is explained), this necessary consequence is not explained by the explanation given. The only contrast consequences to which this operator penetrates are those which figured in the original explanation as relevant alternatives.

So it is with our epistemic operators. To know that *x* is *A* is to know that *x is A* within a framework of relevant alternatives, *B, C,* and D. This set of contrasts, together with the fact that *x is A*, serve to define what it is that is known when one knows that *x is A*. One cannot change this set of contrasts without changing what a person is said to know when he is said to know that *x is A*. We have subtle ways of shifting these contrasts and, hence, changing what a person is said to know *without changing the sentence that we use to express what he knows.* Take the fact that Lefty killed Otto. By changing the emphasis pattern we can invoke a different set of contrasts and, hence, alter what it is that *S* is said to know when he is said to know that Lefty killed Otto. We can say, for instance, that *S* knows that *Lefty* killed Otto. In this case (and I think this is the way we usually hear the sentence when there is no *special* emphasis) we are being told that *S* knows the identity of Otto's killer, that *it was Lefty* who killed Otto. Hence, we expect *S*'s reasons for believing that Lefty killed Otto to consist in facts that single out Lefty as the assailant *rather than* George, Mike, or someone else. On the other hand, we can say that *S* knows that Lefty *killed* Otto. In this case we are being told that *S* knows *what Lefty did to Otto;* he killed him *rather than* merely injuring him, killed him *rather than* merely threatening him, etc. *A* good reason for believing that Lefty *killed* Otto (rather than merely injuring him) is that Otto is dead, but this is not much of a reason, if it is a reason at all, for believing that *Lefty* killed Otto. Changing the set of contrasts (from 'Lefty rather than George or Mike' to 'killed rather than injured or threatened') by shifting the emphasis pattern changes what it is that one is alleged to know when one is said to know that Lefty killed Otto.[7] The same point can be made here as we made in the case of explanation: the operator will penetrate only to those contrast consequences which form part of the network of relevant alternatives structuring

[7]The same example works nicely with the operator '*R*⇒ . . . '. **It** may be true to say that Otto would not be dead unless Lefty killed him (unless what Lefty did **to** him was kill him) without its being true that **Otto** would not be dead unless *Lefty* killed him (unless it was Lefty who killed him).

the original context in which a knowledge claim was advanced. just as we have not explained why Brenda did not order some dessert and throw it at the waiter when we explained why she did not order some dessert (although what we have explained-her not ordering any dessert-entails this), so also in knowing that Lefty *killed* Otto (knowing that what Lefty did to Otto was kill him) we do not nec*essarily* (although we may) know that *Lefty* killed Otto (know that *it was Lefty* who killed Otto). Recall the example of the little old lady who knew that my brother would not move without knowing that it was my brother who would not move.

The conclusions to be drawn are the same as those in the case of explanation. Just as we can say that within the original setting, within the original framework of alternatives that defined what we were trying to explain, we *did explain* why Brenda did not order any dessert, so also within the original setting, within the set of contrasts that defined what it was we were claiming to know, we did know that the wall was red and *did know* that it was a zebra in the pen.

To introduce a novel and enlarged set of alternatives, as the skeptic is inclined to do with our epistemic claims, is to exhibit consequences of what we know, or have reason to believe, which we may not know, may not have a reason to believe; but it does not show that we did not know, did not have a reason to believe, whatever it is that has these consequences. To argue in this way is, I submit, as much a mistake as arguing that we have not explained why Brenda did not order dessert (within the original, normal, setting) because we did not explain why she did not order some and throw it at the waiter.

Chapter 28
Elusive Knowledge

David Lewis

We know a lot. I know what food penguins eat. I know that phones used to ring, but nowadays squeal, when someone calls up. I know that Essendon won the 1993 Grand Final. I know that here is a hand, and here is another.

We have all sorts of everyday knowledge, and we have it in abundance. To doubt that would be absurd. At any rate, to doubt it in any serious and lasting way would be absurd; and even philosophical and temporary doubt, under the influence of argument, is more than a little peculiar. It is a Moorean fact that we know a lot. It is one of those things that we know better than we know the premises of any philosophical argument to the contrary.

Besides knowing a lot that is everyday and trite, I myself think that we know a lot that is interesting and esoteric and controversial. We know a lot about things unseen: tiny particles and pervasive fields, not to mention one another's underwear. Sometimes we even know what an author meant by his writings. But on these questions, let us agree to disagree peacefully with the champions of "post-knowledgeism." The most trite and ordinary parts of our knowledge will be problem enough.

For no sooner do we engage in epistemology – the systematic philosophical examination of knowledge – than we meet a compelling argument that we know next to nothing. The sceptical argument is nothing new or fancy. It is just this: it seems as if knowledge must be by definition infallible. If you claim that S knows that P, and yet you grant that S cannot eliminate a certain possibility in which not-P, it certainly seems as if you have granted that S does not after all know that P.

David Lewis was deceased at the time of publication.

D. Lewis (deceased)
Princeton University, Princeton, NJ, USA

H. Arló-Costa et al. (eds.), *Readings in Formal Epistemology*, Springer Graduate Texts in Philosophy 1, DOI 10.1007/978-3-319-20451-2_28

To speak of fallible knowledge, of knowledge despite uneliminated possibilities of error, just sounds contradictory.

Blind Freddy can see where this will lead. Let your paranoid fantasies rip – CIA plots, hallucinogens in the tap water, conspiracies to deceive, old Nick himself – and soon you find that uneliminated possibilities of error are everywhere. Those possibilities of error are farfetched, of course, but possibilities all the same. They bite into even our most everyday knowledge. We never have infallible knowledge.

Never – well, hardly ever. Some say we have infallible knowledge of a few simple, axiomatic necessary truths; and of our own present experience. They say that I simply cannot be wrong that a part of a part of something is itself a part of that thing; or that it seems to me now (as I sit here at the keyboard) exactly as if I am hearing clicking noises on top of a steady whirring. Some say so. Others deny it. No matter; let it be granted, at least for the sake of the argument. It is not nearly enough. If we have only that much infallible knowledge, yet knowledge is by definition infallible, then we have very little knowledge indeed – not the abundant everyday knowledge we thought we had. That is still absurd.

So we know a lot; knowledge must be infallible; yet we have fallible knowledge or none (or next to none). We are caught between the rock of fallibilism and the whirlpool of scepticism. Both are mad!

Yet fallibilism is the less intrusive madness. It demands less frequent corrections of what we want to say. So, if forced to choose, I choose fallibilism. (And so say all of us.) We can get used to it, and some of us have done. No joy there – we know that people can get used to the most crazy philosophical sayings imaginable. If you are a contented fallibilist, I implore you to be honest, be naive, hear it afresh. "He knows, yet he has not eliminated all possibilities of error." Even if you've numbed your ears, doesn't this overt, explicit fallibilism still sound wrong?

Better fallibilism than scepticism; but it would be better still to dodge the choice. I think we can. We will be alarmingly close to the rock, and also alarmingly close to the whirlpool, but if we steer with care, we can – just barely – escape them both. Maybe epistemology is the culprit. Maybe this extraordinary pastime robs us of our knowledge. Maybe we do know a lot in daily life; but maybe when we look hard at our knowledge, it goes away. But only when we look at it harder than the sane ever do in daily life; only when we let our paranoid fantasies rip. That is when we are forced to admit that there always are uneliminated possibilities of error, so that we have fallible knowledge or none. Much that we say is context-dependent, in simple ways or subtle ways. Simple: "it's evening" is truly said when, and only when, it is said in the evening. Subtle: it could well be true, and not just by luck, that Essendon played rottenly, the Easybeats played brilliantly, yet Essendon won. Different contexts evoke different standards of evaluation. Talking about the Easybeats we apply lax standards, else we could scarcely distinguish their better days from their worse ones. In talking about Essendon, no such laxity is required. Essendon won because play that is rotten by demanding standards suffices to beat play that is brilliant by lax standards.

Maybe ascriptions of knowledge are subtly context-dependent, and maybe epistemology is a context that makes them go false. Then epistemology would be an investigation that destroys its own subject matter. If so, the sceptical argument might be flawless, when we engage in epistemology-and only then![1]

If you start from the ancient idea that justification is the mark that distinguishes knowledge from mere opinion (even true opinion), then you well might conclude that ascriptions of knowledge are context-dependent because standards for adequate justification are context-dependent. As follows: opinion, even if true, deserves the name of knowledge only if it is adequately supported by reasons; to deserve that name in the especially demanding context of epistemology, the arguments from supporting reasons must be especially watertight; but the special standards of justification that this special context demands never can be met (well, hardly ever). In the strict context of epistemology we know nothing, yet in laxer contexts we know a lot.

But I myself cannot subscribe to this account of the context-dependence of knowledge, because I question its starting point. I don't agree that the mark of knowledge is justification.[2] First, because justification is not sufficient: your true opinion that you will lose the lottery isn't knowledge, whatever the odds. Suppose you know that it is a fair lottery with one winning ticket and many losing tickets, and you know how many losing tickets there are. The greater the number of losing tickets, the better is your justification for believing you will lose. Yet there is no number great enough to transform your fallible opinion into knowledge – after all, you just might win. No justification is good enough – or none short of a watertight deductive argument, and all but the sceptics will agree that this is too much to demand.[3]

Second, because justification is not always necessary. What (non-circular) argument supports our reliance on perception, on memory, and on testimony?[4] And yet we do gain knowledge by these means. And sometimes, far from having supporting arguments, we don't even know how we know. We once had evidence, drew conclusions, and thereby gained knowledge; now we have forgotten our

---

[1]The suggestion that ascriptions of knowledge go false in the context of epistemology is to be found in Barry Stroud, "Understanding Human Knowledge in General" in Marjorie Clay and Keith Lehrer (eds.), Knowledge and Skepticism (Boulder: Westview Press, 1989); and in Stephen Hetherington, "Lacking Knowledge and lustification by Theorising About Them" (lecture at the University of New South Wales, August 1992). Neither of them tells the story just as I do, however it may be that their versions do not conflict with mine.

[2]Unless, like some, we simply define "justification" as "whatever it takes to turn true opinion into knowledge" regardless of whether what it takes turns out to involve argument from supporting reasons.

[3]The problem of the lottery was introduced in Henry Kyburg, Probability and the Logic of Rational Belief (Middletown, CT: Wesleyan University Press, 1961), and in Carl Hempel, "Deductive-Nomological vs. Statistical Explanation" in Herbert Feigl and Grover Maxwell (eds.), Minnesota Studies in the Philosophy of Science, Vol. II (Minneapolis: University of Minnesota Press, (962). It has been much discussed since, as a problem both about knowledge and about our everyday, non-quantitative concept of belief.

[4]The case of testimony is less discussed than the others; but see C. A. J. Coady, Testimony: A Philosophical Study (Oxford: Clarendon Press, 1992) pp. 79–129.

reasons, yet still we retain our knowledge. Or we know the name that goes with the face, or the sex of the chicken, by relying on subtle visual cues, without knowing what those cues may be.

The link between knowledge and justification must be broken. But if we break that link, then it is not – or not entirely, or not exactly – by raising the standards of justification that epistemology destroys knowledge. I need some different story.

To that end, I propose to take the infallibility of knowledge as my starting point.[5] Must infallibilist epistemology end in scepticism? Not quite. Wait and see. Anyway, here is the definition. Subject S knows proposition P iff (that is, if and only if) P holds in every possibility left uneliminated by S's evidence; equivalently, iff S's evidence eliminates every possibility in which not-P.

The definition is short, the commentary upon it is longer. In the first place, there is the proposition, P. What I choose to call "propositions" are individuated coarsely, by necessary equivalence. For instance, here is only one necessary proposition. It holds in every possibility; hence in every possibility left uneliminated by S's evidence, no matter who S may be and no matter what his evidence may be. So the necessary proposition is known always and everywhere. Yet this known proposition may go unrecognised when presented in impenetrable linguistic disguise, say as the proposition that every even number is the sum of two primes. Likewise, the known proposition that I have two hands may go unrecognised when presented as the proposition that the number of my ands is the least number n such that every even number is the sum of n primes. (Or if you doubt the necessary existence of numbers, switch to an example involving equivalence by logic alone.) These problems of disguise shall not concern us here. Our topic is modal, not hyperintensional, epistemology.[6]

Next, there are the possibilities. We needn't enter here into the question whether these are concreta, abstract constructions, or abstract simples. Further, we needn't decide whether they must always be maximally specific possibilities, or whether they need only be specific enough for the purpose at hand. A possibility will be specific enough if it cannot be split into subcases in such a way that anything we have said about possibilities, or anything we are going to say before we are done, applies to some subcases and not to others. For instance, it should never happen that proposition P holds in some but not all subcases; or that some but not all sub-cases are eliminated by S's evidence.

But we do need to stipulate that they are not just possibilities as to how the whole world is; they also include possibilities as to which part of the world is oneself, and as to when it now is. We need these possibilities de se et nunc because the

---

[5]I follow Peter Unger, Ignorance: A Case for Skepticism (New York: Oxford University Press, 1975). But I shall not let him lead me into scepticism.

[6]See Robert Stalnaker. Inquiry (Cambridge, MA: MIT Press, 1984) pp. 59–99.

propositions that may be known include propositions de se et nunc.[7] Not only do I know that there are hands in this world somewhere and somewhen. I know that I have hands. or anyway I have them now. Such propositions aren't just made true or made false by the whole world once and for all. They are true for some of us and not for others, or true at some times and not others, or both.

Further, we cannot limit ourselves to "real" possibilities that conform to the actual laws of nature, and maybe also to actual past history. For propositions about laws and history are contingent, and mayor may not be known.

Neither can we limit ourselves to "epistemic" possibilities for S – possibilities that S does not know not to obtain. That would drain our definition of content. Assume only that knowledge is closed under strict implication. . (We shall consider the merits of this assumption later.) Remember that we are not distinguishing between equivalent propositions. Then knowledge of a conjunction is equivalent to knowledge of every conjunct. P is the conjunction of all propositions not- W, where W is a possibility in which not-P. That suffices to yield an equivalence: S knows that P iff, for every possibility W in which not-P, S knows that not-W. Contraposing and cancelling a double negation: iff every possibility which S does not know not to obtain is one in which P. For short: iff P holds throughout S's epistemic possibilities. Yet to get this far, we need no substantive definition of knowledge at all! To turn this into a substantive definition, in fact the very definition we gave before, we need to say one more thing: S's epistemic possibilities are just those possibilities that are uneliminated by S's evidence.

So, next, we need to say what it means for a possibility to be eliminated or not. Here I say that the uneliminated possibilities are those in which the subject's entire perceptual experience and memory are just as they actually are. There is one possibility that actually obtains (for the subject and at the time in question); call it actuality. Then a possibility W is uneliminated iff the subject's perceptual experience and memory in W exactly match his perceptual experience and memory in actuality. (If you want to include other alleged forms of basic evidence, such as the evidence of our extrasensory faculties, or an innate disposition to believe in God, be my guest. If they exist, they should be included. If not, no harm done if we have included them conditionally.)

Note well that we do not need the "pure sense-datum language" and the "incorrigible protocol statements" that for so long bedevilled foundationalist epistemology. It matters not at all whether there are words to capture the subject's perceptual and memory evidence, nothing more and nothing less. If there are such words, it matters not at all whether the subject can hit upon them. The given does not consist of basic axioms to serve as premises in subsequent arguments. Rather, it consists of a match between possibilities.

[7]See my 'Attitudes De Dicta and De Se', The Philosophical Review 88 (1979) pp. 513–543; and R. M. Chisholm, "The Indirect Reflexive" in C. Diamond and J. Teichman (eds.), Intention and Intentionality: Essays in Honour of G. E. M. Anscomhe (Brighton: Harvester, 1979).

When perceptual experience E (or memory) eliminates a possibility W. that is not because the propositional content of the experience conflicts with W. (Not even if it is the narrow content.) The propositional content of our experience could, after all, be false. Rather, it is the existence of the experience that conflicts with W: W is a possibility in which the subject is not having experience E. Else we would need to tell some fishy story of how the experience has some sort of infallible, ineffable, purely phenomenal propositional content. . . . Who needs that? Let E have propositional content P. Suppose even – something I take to be an open question – that E is, in some sense, fully characterized by P. Then I say that E eliminates W iff W is a possibility in which the subject's experience or memory has content different from P. I do not say that E eliminates W iff W is a possibility in which P is false.

Maybe not every kind of sense perception yields experience; maybe, for instance, the kinaesthetic sense yields not its own distinctive sort of sense experience but only spontaneous judgements about the position of one's limbs. If this is true, then the thing to say is that kinaesthetic evidence eliminates all possibilities except those that exactly resemble actuality with respect to the subject's spontaneous kinaesthetic judgements. In saying this, we would treat kinaesthetic evidence more on the model of memory than on the model of more typical senses.

Finally, we must attend to the word "every." What does it mean to say that every possibility in which not-P is eliminated? An idiom of quantification, like "every," is normally restricted to some limited domain. If I say that every glass is empty, so it's time for another round, doubtless I and my audience are ignoring most of all the glasses there are in the whole wide world throughout all of time. They are outside the domain. They are irrelevant to the truth of what was said.

Likewise, if I say that every uneliminated possibility is one in which P, or words to that effect, I am doubtless ignoring some of all the uneliminated alternative possibilities that there are. They are outside the domain, they are irrelevant to the truth of what was said.

But, of course, I am not entitled to ignore just any possibility I please. Else true ascriptions of knowledge, whether to myself or to others, would be cheap indeed. I may properly ignore some uneliminated possibilities; I may not properly ignore others. Our definition of knowledge requires a sotto voce proviso. S knows that P iff S's evidence eliminates every possibility in which not-P – Psst! – except for those possibilities that we are properly ignoring.

Unger suggests an instructive parallel.[8] Just as P is known iff there are no uneliminated possibilities of error, so likewise a surface is flat iff there are no bumps on it. We must add the proviso: Psst! – except for those bumps that we are properly ignoring. Else we will conclude, absurdly, that nothing is flat. (Simplify by ignoring departures from flatness that consist of gentle curvature.) We can restate the definition. Say that we presuppose proposition Q iff we ignore all possibilities

[8]Peter Unger, Ignorance, chapter II. I discuss the case, and briefly foreshadow the present paper, in my "Scorekeeping in a Language Game," Journal of Philosophical Logic 8 (1979) pp. 339–359, esp. pp. 353–355.

in which not-Q. To close the circle: we ignore just those possibilities that falsify our presuppositions. Proper presupposition corresponds, of course, to proper ignoring. Then S knows that P iff S's evidence eliminates every possibility in which not-P – Psst! – except for those possibilities that conflict with our proper presuppositions.[9]

The rest of (modal) epistemology examines the sotto voce proviso. It asks: what may we properly presuppose in our ascriptions of knowledge? Which of all the uneliminated alternative possibilities may not properly be ignored? Which ones are the ""relevant alternatives"? – relevant, that is, to what the subject does and doesn't know?[10] In reply, we can list several rules.[11] We begin with three prohibitions: rules to tell us what possibilities we may not properly ignore.

First, there is the Rule of Actuality. The possibility that actually obtains is never properly ignored; actuality is always a relevant alternative; nothing false may properly be presupposed. It follows that only what is true is known, wherefore we did not have to include truth in our definition of knowledge. The rule is "externalist" – the subject himself may not be able to tell what is properly ignored. In judging which of his ignorings are proper, hence what he knows, we judge his success in knowing-not how well he tried.

When the Rule of Actuality tells us that actuality may never be properly ignored, we can ask: whose actuality? Ours, when we ascribe knowledge or ignorance to others? Or the subject's? In simple cases, the question is silly. (In fact, it sounds like the sort of pernicious nonsense we would expect from someone who mixes up what is true with what is believed.) There is just one actual world, we the ascribers live in that world, the subject lives there too, so the subject's actuality is the same as ours.

But there are other cases, less simple, in which the question makes perfect sense and needs an answer. Someone mayor may not know who he is; someone may or may not know what time it is. Therefore I insisted that the propositions that may be known must include propositions de se et nunc; and likewise that the possibilities that may be eliminated or ignored must include possibilities de se et nunc. Now we have a good sense in which the subject's actuality may be different from ours. I ask today what Fred knew yesterday. In particular, did he then know who he was? Did he know what day it was? Fred's actuality is the possibility de se et nunc of being

---

[9]See Robert Stalnaker, "Presuppositions," Journal of Philosophical Logic 2 (1973) pp. 447–457; and "Pragmatic Presuppositions" in Milton Munitz and Peter Unger (eds.), Semantics and Philosophy (New York: New York University Press. 1974). See also my "Score keeping in a Language Game." The definition restated in terms of presupposition resembles the treatment of knowledge in Kenneth S. Ferguson, Philosophical Scepticism (Cornell University doctoral dissertation, 1980).

[10]See Fred Dretske, "Epistemic Operators," The Journal of Philosophy 67 (1970) pp. 1007–1022, and "The Pragmatic Dimension of Knowledge," Philosophical Studies 40 (1981) pp. 363–378; Alvin Goldman, "Discrimination and Perceptual Knowledge:' The Journal of Philosophy 73 (1976) pp. 771–791; G. C. Stine, "Skepticism. Relevant Alternatives, and Deductive Closure," Philosophical Studies 29 (1976) pp. 249–261: and Stewart Cohen. "How to be A Fallibilist," Philosophical Perspectives 2 (1988) pp. 91–123.

[11]Some of them, but only some, taken from the authors just cited.

Fred on September 19th at such-and-such possible world; whereas my actuality is the possibility de se et nunc of being David on September 20th at such-and-such world. So far as the world goes, there is no difference: Fred and I are worldmates, his actual world is the same as mine. But when we build subject and time into the possibilities de se et nunc, then his actuality yesterday does indeed differ from mine today.

What is more, we sometimes have occasion to ascribe knowledge to those who are off at other possible worlds. I didn't read the newspaper yesterday. What would I have known if I had read it? More than I do in fact know. (More and less: I do in fact know that I left the newspaper unread, but if I had read it, I would not have known that I had left it unread.) I-who-did-not-read-the-newspaper am here at this world, ascribing knowledge and ignorance. The subject to whom I am ascribing that knowledge and ignorance, namely I-as-I-would-have-been-if-I-had-read-the-newspaper, is at a different world. The worlds differ in respect at least of a reading of the newspaper. Thus the ascriber's actual world is not the same as the subject's. (I myself think that the ascriber and the subject are two different people: the subject is the ascriber's otherworldly counterpart. But even if you think the subject and the ascriber are the same identical person, you must still grant that this person's actuality qua subject differs from his actuality qua ascriber.)

Or suppose we ask modal questions about the subject: what must he have known, what might he have known? Again we are considering the subject as he is not here, but off at other possible worlds. Likewise if we ask questions about knowledge of knowledge: what does he (or what do we) know that he knows? So the question "whose actuality?" is not a silly question after all. And when the question matters, as it does in the cases just considered, the right answer is that it is the subject's actuality, not the ascriber's, that never can be properly ignored.

Next, there is the Rule of Belief. A possibility that the subject believes to obtain is not properly ignored, whether or not he is right to so believe. Neither is one that he ought to believe to obtain – one that evidence and arguments justify him in believing – whether or not he does so believe.

That is rough. Since belief admits of degree, and since some possibilities are more specific than others, we ought to reformulate the rule in terms of degree of belief, compared to a standard set by the unspecificity of the possibility in question. A possibility may not be properly ignored if the subject gives it, or ought to give it, a degree of belief that is sufficiently high, and high not just because the possibility in question is unspecific.

How high is "sufficiently high"? That may depend on how much is at stake. When error would be especially disastrous, few possibilities may be properly ignored. Then even quite a low degree of belief may be "sufficiently high" to bring the Rule of Belief into play. The jurors know that the accused is guilty only if his guilt has been proved beyond reasonable doubt.[12]

[12]Instead of complicating the Rules of Belief as I have just done. 1 might equivalently have introduced a separate Rule of High Stakes saying that when error would be especially disastrous. few possibilities are properly ignored.

Yet even when the stakes are high, some possibilities still may be properly ignored. Disastrous though it would be to convict an innocent man, still the jurors may properly ignore the possibility that it was the dog, marvellously well-trained, that fired the fatal shot. And, unless they are ignoring other alternatives more relevant than that, they may rightly be said to know that the accused is guilty as charged. Yet if there had been reason to give the dog hypothesis a slightly less negligible degree of belief – if the world's greatest dog-trainer had been the victim's mortal enemy – then the alternative would be relevant after all.

This is the only place where belief and justification enter my story. As already noted, I allow justified true belief without knowledge, as in the case of your belief that you will lose the lottery. I allow knowledge without justification, in the cases of face recognition and chicken sexing. I even allow knowledge without belief, as in the case of the timid student who knows the answer but has no confidence that he has it right, and so does not believe what he knows.[13] Therefore any proposed converse to the Rule of Belief should be rejected. A possibility that the subject does not believe to a sufficient degree, and ought not to believe to a sufficient degree, may nevertheless be a relevant alternative and not properly ignored.

Next, there is the Rule of Resemblance. Suppose one possibility saliently resembles another. Then if one of them may not be properly ignored, neither may the other. (Or rather, we should say that if one of them may not properly be ignored in virtue of rules other than this rule, then neither may the other. Else nothing could be properly ignored; because enough little steps of resemblance can take us from anywhere to anywhere.) Or suppose one possibility saliently resembles two or more others, one in one respect and another in another, and suppose that each of these may not properly be ignored (in virtue of rules other than this rule). Then these resemblances may have an additive effect, doing more together than any one of them would separately.

We must apply the Rule of Resemblance with care. Actuality is a possibility uneliminated by the subject's evidence. Any other possibility W that is likewise uneliminated by the subject's evidence thereby resembles actuality in one salient respect: namely, in respect of the subject's evidence. That will be so even if W is in other respects very dissimilar to actuality – even if, for instance, it is a possibility in which the subject is radically deceived by a demon. Plainly, we dare not apply the Rules of Actuality and Resemblance to conclude that any such W is a relevant alternative – that would be capitulation to scepticism. The Rule of Resemblance was never meant to apply to this resemblance! We seem to have an ad hoc exception to the Rule, though one that makes good sense in view of the function of attributions of knowledge. What would be better, though, would be to find a way to reformulate the Rule so as to get the needed exception without ad hocery. I do not know how to do this.

[13] A. D. Woozley, "Knowing and Not Knowing:' Proceedings of the Aristotelian Society 53 (1953) pp. 151–172: Colin Radford. "Knowledge – by Examples," Analysis 27 (1966) pp. 1–11.

It is the Rule of Resemblance that explains why you do not know that you will lose the lottery, no matter what the odds are against you and no matter how sure you should therefore be that you will lose. For every ticket, there is the possibility that it will win. These possibilities are saliently similar to one another: so either everyone of them may be properly ignored, or else none may. But one of them may not properly be ignored: the one that actually obtains.

The Rule of Resemblance also is the rule that solves the Gettier problems: other cases of justified true belief that are not knowledge.[14]

(1) I think that Nogot owns a Ford, because I have seen him driving one; but unbeknownst to me he does not own the Ford he drives, or any other Ford. Unbeknownst to me, Havit does own a Ford, though I have no reason to think so because he never drives it, and in fact I have often seen him taking the tram. My justified true belief is that one of the two owns a Ford. But I do not know it; I am right by accident. Diagnosis: I do not know, because I have not eliminated the possibility that Nogot drives a Ford he does not own whereas Havit neither drives nor owns a car. This possibility may not properly be ignored. Because, first. actuality may not properly be ignored; and, second. this possibility saliently resembles actuality. It resembles actuality perfectly so far as Nogot is concerned; and it resembles actuality well so far as Havit is concerned, since it matches actuality both with respect to Havit's carless habits and with respect to the general correlation between carless habits and carlessness. In addition, this possibility saliently resembles a third possibility: one in which Nogot drives a Ford he owns while Havit neither drives nor owns a car. This third possibility may not properly be ignored, because of the degree to which it is believed. This time, the resemblance is perfect so far as Havit is concerned, rather good so far as Nogot is concerned.

(2) The stopped clock is right twice a day. It says 4:39. as it has done for weeks. I look at it at 4:39; by luck I pick up a true belief. I have ignored the uneliminated possibility that I looked at it at 4:22 while it was stopped saying 4:39. That possibility was not properly ignored. It resembles actuality perfectly so far as the stopped clock goes.

(3) Unbeknownst to me, I am travelling in the land of the bogus barns; but my eye falls on one of the few real ones. I don't know that I am seeing a barn, because I may not properly ignore the possibility that I am seeing yet another of the

[14]See Edmund Gettier, "Is lustified True Belief Knowledge?," Analysis 23 (1963) pp. 121–123. Diagnoses have varied widely. The four examples below come from: (1) Keith Lehrer and Thomas Paxson Jr., "Knowledge: Undefeated True Belief," The Journal of Philosophy 66 (1969) pp. 225–237; (2) Bertrand Russell, Human Knowledge: Its Scope and Limits (London: Allen and Unwin, 1948) p. 154; (3) Alvin Goldman, "Discrimination and Perceptual Knowledge," op. cit.; (4) Gilbert Harman, Thought (Princeton. N1: Princeton University Press, 1973) p. 143.

Though the lottery problem is another case of justified true belief without knowledge, it is not normally counted among the Gettier problems. It is interesting to find that it yields to the same remedy.

abundant bogus barns. This possibility saliently resembles actuality in respect of the abundance of bogus barns, and the scarcity of real ones, hereabouts.

(4) Donald is in San Francisco, just as I have every reason to think he is. But, bent on deception, he is writing me letters and having them posted to me by his accomplice in Italy. If I had seen the phoney letters, with their Italian stamps and postmarks, I would have concluded that Donald was in Italy. Luckily, I have not yet seen any of them. I ignore the uneliminated possibility that Donald has gone to Italy and is sending me letters from there. But this possibility is not properly ignored, because it resembles actuality both with respect to the fact that the letters are coming to me from Italy and with respect to the fact that those letters come, ultimately, from Donald. So I don't know that Donald is in San Francisco.

Next, there is the Rule of Reliability. This time, we have a presumptive rule about what may be properly ignored; and it is by means of this rule that we capture what is right about causal or reliabilist theories of knowing. Consider processes whereby information is transmitted to us: perception, memory, and testimony. These processes are fairly reliable.[15] Within limits, we are entitled to take them for granted. We may properly presuppose that they work without a glitch in the case under consideration. Defeasibly – very defeasibly! – a possibility in which they fail may properly be ignored.

My visual experience, for instance, depends causally on the scene before my eyes, and what I believe about the scene before my eyes depends in turn on my visual experience. Each dependence covers a wide and varied range of alternatives.[16] Of course, it is possible to hallucinate – even to hallucinate in such a way that all my perceptual experience and memory would be just as they actually are. That possibility never can be eliminated. But it can be ignored. And if it is properly ignored – as it mostly is – then vision gives me knowledge. Sometimes, though, the possibility of hallucination is n properly ignored; for sometimes we really do hallucinate. The Rule of Reliability may be defeated by the Rule of Actuality. Or it may be defeated by the Rules of Actuality and of Resemblance working together, in a Gettier problem: if I am not hallucinating, but unbeknownst to me I live in a world where people mostly do hallucinate and I myself have only narrowly escaped, then the uneliminated possibility of hallucination is too close to actuality to be properly ignored.

We do not, of course, presuppose that nowhere ever is there a failure of, say, vision. The general presupposition that vision is reliable consists, rather, of a

[15]See Alvin Goldman, "A Causal Theory of Knowing," The Journal of Philosophy 64 (1967) pp. 357–372; D. M. Armstrong, Belief, Truth and Knowledge (Cambridge: Cambridge University Press, 1973).

[16]See my "Veridical Hallucination and Prosthetic Vision," Australasian Journal of Philosophy 58 (1980) pp. 239–249. John Bigelow has proposed to model knowledge-delivering processes generally on those found in vision.

standing disposition to presuppose, concerning whatever particular case may be under consideration, that we have no failure in that case.

In similar fashion, we have two permissive Rules of Method. We are entitled to presuppose – again, very defeasibly – that a sample is representative; and that the best explanation of our evidence is the true explanation. That is, we are entitled properly to ignore possible failures in these two standard methods of nondeductive inference. Again, the general rule consists of a standing disposition to presuppose reliability in whatever particular case may come before us.

Yet another permissive rule is the Rule of Conservatism. Suppose that those around us normally do ignore certain possibilities, and it is common knowledge that they do. (They do, they expect each other to, they expect each other to expect each other to, . . .) Then – again, very defeasibly! – these generally ignored possibilities may properly be ignored. We are permitted, defeasibly, to adopt the usual and mutually expected presuppositions of those around us.

(It is unclear whether we need all four of these permissive rules. Some might be subsumed under others. Perhaps our habits of treating samples as representative, and of inferring to the best explanation, might count as normally reliable processes of transmission of information. Or perhaps we might subsume the Rule of Reliability under the Rule of Conservatism, on the ground that the reliable processes whereby we gain knowledge are familiar, are generally relied upon, and so are generally presupposed to be normally reliable. Then the only extra work done by the Rule of Reliability would be to cover less familiar – and merely hypothetical? – reliable processes, such as processes that relied on extrasensory faculties. Likewise, mutatis mutandis. we might subsume the Rules of Method under the Rule of Conservatism. Or we might instead think to subsume the Rule of Conservatism under the Rule of Reliability, on the ground that what is generally presupposed tends for the most part to be true, and the reliable processes whereby this is so are covered already by the Rule of Reliability. Better redundancy than incompleteness, though. So. leaving the question of redundancy open, I list all four rules.)

Our final rule is the Rule of Attention. But it is more a triviality than a rule. When we say that a possibility is properly ignored, we mean exactly that; we do not mean that it could have been properly ignored. Accordingly, a possibility not ignored at all is ipso facto not properly ignored. What is and what is not being ignored is a feature of the particular conversational context. No matter how far-fetched a certain possibility may be, no matter how properly we might have ignored it in some other context, if in this context we are not in fact ignoring it but attending to it, then for us now it is a relevant alternative. It is in the contextually determined domain. If it is an uneliminated possibility in which not-P, then it will do as a counter-example to the claim that P holds in every possibility left uneliminated by S's evidence. That is, it will do as a counter-example to the claim that S knows that P.

Do some epistemology. Let your fantasies rip. Find uneliminated possibilities of error everywhere. Now that you are attending to them. just as I told you to, you are no longer ignoring them, properly or otherwise. So you have landed in a context with an enormously rich domain of potential counter-examples to ascriptions of knowledge. In such an extraordinary context, with such a rich domain, it never can happen (well, hardly ever) that an ascription of knowledge is true. Not an ascription of knowledge to yourself (either to your present self or to your earlier self, untainted by epistemology); and not an ascription of knowledge to others. That is how epistemology destroys knowledge. But it does so only temporarily.

The pastime of epistemology does not plunge us forevermore into its special context. We can still do a lot of proper ignoring, a lot of knowing, and a lot of true ascribing of knowledge to ourselves and others, the rest of the time.

What is epistemology all about? The epistemology we've just been doing, at any rate, soon became an investigation of the ignoring of possibilities. But to investigate the ignoring of them was ipso facto not to ignore them. Unless this investigation of ours was an altogether atypical sample of epistemology, it will be inevitable that epistemology must destroy knowledge. That is how knowledge is elusive. Examine it, and straightway it vanishes.

Is resistance useless? If you bring some hitherto ignored possibility to our attention, then straightway we are not ignoring it at all, so a fortiori we are not properly ignoring it. How can this alteration of our conversational state be undone? If you are persistent, perhaps it cannot be undone – at least not so long as you are around. Even if we go off and play backgammon, and afterward start our conversation afresh, you might turn up and call our attention to it all over again.

But maybe you called attention to the hitherto ignored possibility by mistake. You only suggested that we ought to suspect the butler because you mistakenly thought him to have a criminal record. Now that you know he does not – that was the previous butler – you wish you had not mentioned him at all. You know as well as we do that continued attention to the possibility you brought up impedes our shared conversational purposes. Indeed, it may be common knowledge between you and us that we would all prefer it if this possibility could be dismissed from our attention. In that case we might quickly strike a tacit agreement to speak just as if we were ignoring it; and after just a little of that, doubtless it really would be ignored.

Sometimes our conversational purposes are not altogether shared, and it is a matter of conflict whether attention to some far-fetched possibility would advance them or impede them. What if some farfetched possibility is called to our attention not by a sceptical philosopher, but by counsel for the defence? We of the jury may wish to ignore it, and wish it had not been mentioned. If we ignored it now, we would bend the rules of cooperative conversation; but we may have good reason to do exactly that. (After all, what matters most to us as jurors is not whether we can truly be said to know; what really matters is what we should believe to what degree, and whether or not we should vote to convict.) We would ignore the far-fetched possibility if we could – but can we? Perhaps at first our attempted ignoring would be make-believe ignoring, or self-deceptive ignoring; later, perhaps, it might ripen into genuine ignoring. But in the meantime, do we know? There may be no definite answer. We are bending the rules, and our practices of context-dependent attributions of knowledge were made for contexts with the rules unbent.

If you are still a contented fallibilist, despite my plea to hear the sceptical argument afresh, you will probably be discontented with the Rule of Attention. You will begrudge the sceptic even his very temporary victory. You will claim the right to resist his argument not only in everyday contexts, but even in those peculiar contexts in which he (or some other epistemologist) busily calls your attention to farfetched possibilities of error. Further, you will claim the right to resist without having to bend any rules of cooperative conversation. I said that the Rule of Attention was a triviality: that which is not ignored at all is not properly ignored. But the Rule was trivial only because of how I had already chosen to state the sotto voce proviso. So you, the contented fallibilist, will think it ought to have been stated differently. Thus, perhaps: "Psst! – except for those possibilities we could properly have ignored". And then you will insist that those far-fetched possibilities of error that we attend to at the behest of the sceptic are nevertheless possibilities we could properly have ignored. You will say that no amount of attention can, by itself, turn them into relevant alternatives.

If you say this, we have reached a standoff. I started with a puzzle: how can it be, when his conclusion is so silly, that the sceptic's argument is so irresistible? My Rule of Attention, and the version of the proviso that made that Rule trivial, were built to explain how the sceptic manages to sway us – why his argument seems irresistible, however temporarily. If you continue to find it eminently resistible in all contexts, you have no need of any such explanation. We just disagree about the explanandum phenomenon.

I say S knows that P iff P holds in every possibility left uneliminated by S's evidence – Psst! – except for those possibilities that we are properly ignoring. "We" means: the speaker and hearers of a given context; that is, those of us who are discussing S's knowledge together. It is our ignorings, not S's own ignorings, that matter to what we can truly say about S's knowledge. When we are talking about our own knowledge or ignorance, as epistemologists so often do, this is a distinction without a difference. But what if we are talking about someone else?

Suppose we are detectives; the crucial question for our solution of the crime is whether S already knew, when he bought the gun, that he was vulnerable to blackmail. We conclude that he did. We ignore various far-fetched possibilities, as hard-headed detectives should. But S does not ignore them. S is by profession a sceptical epistemologist. He never ignores much of anything. If it is our own ignorings that matter to the truth of our conclusion, we may well be right that S already knew. But if it is S's ignorings that matter, then we are wrong, because S never knew much of anything. I say we may well be right; so it is our own ignorings that matter, not S's.

But suppose instead that we are epistemologists considering what S knows. If we are well-informed about S (or if we are considering a well-enough specified hypothetical case), then if S attends to a certain possibility, we attend to S's attending to it. But to attend to 5's attending to it is ipso facto to attend to it ourselves. In that case, unlike the case of the detectives. the possibilities we are properly ignoring must be among the possibilities that S himself ignores. We may ignore fewer possibilities than S does, but not more.

Even if S himself is neither sceptical nor an epistemologist, he may yet be clever at thinking up farfetched possibilities that are uneliminated by his evidence. Then again, we well-informed epistemologists who ask what S knows will have to attend to the possibilities that S thinks up. Even if S's idle cleverness does not lead S himself to draw sceptical conclusions, it nevertheless limits the knowledge that we can truly ascribe to him when attentive to his state of mind. More simply: his cleverness limits his knowledge. He would have known more, had he been less imaginative.[17]

Do I claim you can know P just by presupposing it?! Do I claim you can know that a possibility W does not obtain just by ignoring it? Is that not what my analysis implies, provided that the presupposing and the ignoring are proper? Well, yes. And yet I do not claim it. Or rather, I do not claim it for any specified P or W. I have to grant, in general, that knowledge just by presupposing and ignoring is knowledge; but it is an especially elusive sort of knowledge, and consequently it is an unclaimable sort of knowledge. You do not even have to practise epistemology to make it vanish. Simply mentioning any particular case of this knowledge, aloud or even in silent thought, is a way to attend to the hitherto ignored possibility. and thereby render it no longer ignored, and thereby create a context in which it is no longer true to ascribe the knowledge in question to yourself or others. So. just as we should think, presuppositions alone are not a basis on which to claim knowledge.

In general, when S knows that P some of the possibilities in which not-P are eliminated by S's evidence and others of them are properly ignored. There are some that can be eliminated, but cannot properly be ignored. For instance, when I look around the study without seeing Possum the cat, I thereby eliminate various possibilities in which Possum is in the study; but had those possibilities not been eliminated, they could not properly have been ignored. And there are other possibilities that never can be eliminated, but can properly be ignored. For instance, the possibility that Possum is on the desk but has been made invisible by a deceiving demon falls normally into this class (though not when I attend to it in the special context of epistemology).

There is a third class: not-P possibilities that might either be eliminated or ignored. Take the farfetched possibility that Possum has somehow managed to get into a closed drawer of the desk-maybe he jumped in when it was open, then I closed it without noticing him. That possibility could be eliminated by opening the drawer and making a thorough examination. But if uneliminated, it may nevertheless be ignored, and in many contexts that ignoring would be proper. If I look all around the study, but without checking the closed drawers of the desk, I may truly be said to know that Possum is not in the study – Dr at any rate, there are many contexts in which that may truly be said. But if I did check all the closed drawers, then I would know better that Possum is not in the study. My knowledge would be better in the

[17]See Catherine Elgin, "The Epistemic Efficacy of Stupidity," Synthese 74 (1988) pp. 297–311. The "efficacy" takes many forms; some to do with knowledge (under various rival analyses), some to do with justified belief. See also Michael Williams, Unnatural Doubts: Epistemological Realism and the Basis of Scepticism (Oxford: Blackwell, 1991) pp. 352–355, on the instability of knowledge under reflection.

second case because it would rest more on the elimination of not-P possibilities, less on the ignoring of them.[18,19]

Better knowledge is more stable knowledge: it stands more chance of surviving a shift of attention in which we begin to attend to some of the possibilities formerly ignored. If, in our new shifted context, we ask what knowledge we may truly ascribe to our earlier selves, we may find that only the better knowledge of our earlier selves still deserves the name. And yet, if our former ignorings were proper at the time, even the worse knowledge of our earlier selves could truly have been called knowledge in the former context.

Never – well, hardly ever – does our knowledge rest entirely on elimination and not at all on ignoring. So hardly ever is it quite as good as we might wish. To that extent. the lesson of scepticism is right – and right permanently, not just in the temporary and special context of epistemology.[20]

What is it all for? Why have a notion of knowledge that works in the way I described? (Not a compulsory question. Enough to observe that short-cuts – like satisficing, like having indeterminate degrees of belief – that we resort to because we are not smart enough to live up to really high, perfectly Bayesian, standards of rationality. You cannot maintain a record of exactly which possibilities you have eliminated so far, much as you might like to. It is easier to keep track of which possibilities you have eliminated if you – Psst! – ignore many of all the possibilities there are. And besides, it is easier to list some of the propositions that are true in all the uneliminated, unignored possibilities than it is to find propositions that are true in all and only the uneliminated, unignored possibilities.

If you doubt that the word "know" bears any real load in science or in metaphysics, I partly agree. The serious business of science has to do not with knowledge per se; but rather, with the elimination of possibilities through the evidence of perception, memory, etc., and with the changes that one's belief system would (or might or should) undergo under the impact of such eliminations. Ascriptions of

[18]Mixed cases are possible: Fred properly ignores the possibility W1 which Ted eliminates; however, Ted properly ignores the possibility W2 which Fred eliminates. Ted has looked in all the desk drawers but not the file drawers, whereas Fred has checked the file drawers but not the desk. Fred's knowledge that Possum is not in the study is better in one way, Ted's is better in another.

[19]To say truly that X is known, I must be properly ignoring any uneliminated possibilities in which not-X; whereas to say truly that Y is better known than X, I must be attending to some such possibilities. So I cannot say both in a single context. If I say "X is known, but Y is better known," the context changes in mid-sentence: some previously ignored possibilities must stop being ignored. That can happen easily. Saying it the other way around – "Y is better known than X, but even X is known" – is harder, because we must suddenly start to ignore previously unignored possibilities. That cannot be done, really; but we could bend the rules and make believe we had done it, and no doubt we would be understood well enough. Saying "X is flat, but Y is flatter" (that is, "X has no bumps at all, but Y has even fewer or smaller bumps") is a parallel case. And again, "Y is flatter, but even X is flat" sounds clearly worse – but not altogether hopeless.

[20]Thanks here to Stephen Hetherington. While his own views about better and worse knowledge are situated within an analysis of knowledge quite unlike mine, they withstand transplantation.

knowledge to yourself or others are a very sloppy way of conveying very incomplete information about the elimination of possibilities. It is as if you had said:

> The possibilities eliminated, whatever else they may also include, at least include all the not-P possibilities; or anyway, all of those except for some we are presumably prepared to ignore just at the moment.

The only excuse for giving information about what really matters in such a sloppy way is that at least it is easy and quick! But it is easy and quick; whereas giving full and precise information about which possibilities have been eliminated seems to be extremely difficult, as witness the futile search for a "pure observation language." If I am right about how ascriptions of knowledge work, they are a handy but humble approximation. They may yet be indispensable in practice, in the same way that other handy and humble approximations are.

If we analyse knowledge as a modality, as we have done, we cannot escape the conclusion that knowledge is closed under (strict) implication.[21] Dretske has denied that knowledge is closed under implication; further, he has diagnosed closure as the fallacy that drives arguments for scepticism. As follows: the proposition that I have hands implies that I am not a handless being, and a fortiori that I am not a handless being deceived by a demon into thinking that I have hands. So, by the closure principle, the proposition that I know I have hands implies that I know that I am not handless and deceived. But I don't know that I am not handless and deceived – for how can I eliminate that possibility? So, by modus tollens, I don't know that I have hands. Dretske's advice is to resist scepticism by denying closure. He says that although having hands does imply not being handless and deceived, yet knowing that I have hands does not imply knowing that I am not handless and deceived. I do know the former, I do not know the latter.[22]

What Dretske says is close to right, but not quite. Knowledge is closed under implication. Knowing that I have hands does imply knowing that I am not handless and deceived. Implication preserves truth – that is, it preserves truth in any given, fixed context. But if we switch contexts midway, all bets are off. I say (1) pigs fly; (2) what I just said had fewer than three syllables (true); (3) what I just said had fewer than four syllables (false). So "less than three" does not imply "less than four"? No! The context switched midway, the semantic value of the context-dependent

[21] A proof-theoretic version of this closure principle is common to all "normal" modal logics: if the logic validates an inference from zero or more premises to a conclusion, then also it validates the inference obtained by prefixing the necessity operator to each premise and to the conclusion. Further, this rule is all we need to take us from classical sentential logic to the least normal modal logic. See Brian Chellas, Modal Logic: An Introduction (Cambridge: Cambridge University Press, 1980) p. 114.

[22] Dretske, "Epistemic Operators." My reply follows the lead of Stine, "Skepticism, Relevant Alternatives, and Deductive Closure," op. cit.; and (more closely) Cohen, "How to be a Fallibilist," op. cit.

phrase "what I just said" switched with it. Likewise in the sceptical argument the context switched midway, and the semantic value of the context-dependent word "know" switched with it. The premise "I know that I have hands" was true in its everyday context, where the possibility of deceiving demons was properly ignored. The mention of that very possibility switched the context midway. The conclusion "I know that I am not handless and deceived" was false in its context, because that was a context in which the possibility of deceiving demons was being mentioned, hence was not being ignored, hence was not being properly ignored. Dretske gets the phenomenon right, and I think he gets the diagnosis of scepticism right; it is just that he misclassifies what he sees. He thinks it is a phenomenon of logic, when really it is a phenomenon of pragmatics. Closure, rightly understood, survives the test. If we evaluate the conclusion for truth not with respect to the context in which it was uttered, but instead with respect to the different context in which the premise was uttered, then truth is preserved. And if, per impossible, the conclusion could have been said in the same unchanged context as the premise, truth would have been preserved.

A problem due to Saul Kripke turns upon the closure of knowledge under implication. P implies that any evidence against P is misleading. So, by closure, whenever you know that P. you know that any evidence against P is misleading. And if you know that evidence is misleading, you should pay it no heed. Whenever we know – and we know a lot, remember – we should not heed any evidence tending to suggest that we are wrong, But that is absurd. Shall we dodge the conclusion by denying closure? I think not. Again, I diagnose a change of context. At first, it was stipulated that S knew, whence it followed that S was properly ignoring all possibilities of error. But as the story continues, it turns out that there is evidence on offer that points to some particular possibility of error. Then, by the Rule of Attention, that possibility is no longer properly ignored, either by S himself or by we who are telling the story of S. The advent of that evidence destroys S's knowledge, and thereby destroys S's licence to ignore the evidence lest he be misled.

There is another reason, different from Dretske's. why we might doubt closure. Suppose two or more premises jointly imply a conclusion. Might not someone who is compartmentalized in his thinking – as we all are – know each of the premises but fail to bring them together in a single compartment? Then might he not fail to know the conclusion? Yes; and I would not like to plead idealization-of-rationality as an excuse for ignoring such cases. But I suggest that we might take not the whole compartmentalized thinker, but rather each of his several overlapping compartments, as our "subjects." That would be the obvious remedy if his compartmentalization amounted to a case of multiple personality disorder; but maybe it is right for milder cases as well.[23]

A compartmentalized thinker who indulges in epistemology can destroy his knowledge, yet retain it as well. Imagine two epistemologists on a bushwalk. As they walk, they talk. They mention all manner of far-fetched possibilities of error.

---

[23]See Stalnaker, Inquiry, pp. 79–99.

By attending to these normally ignored possibilities they destroy the knowledge they normally possess. Yet all the while they know where they are and where they are going! How so? The compartment in charge of philosophical talk attends to far-fetched possibilities of error. The compartment in charge of navigation does not. One compartment loses its knowledge, the other retains its knowledge. And what does the entire compartmentalized thinker know? Not an altogether felicitous question. But if we need an answer, I suppose the best thing to say is that S knows that P iff anyone of S's compartments knows that P. Then we can say what we would offhand want to say: yes, our philosophical bushwalkers still know their whereabouts.

Context-dependence is not limited to the ignoring and non-ignoring of farfetched possibilities. Here is another case. Pity poor Bill! He squanders all his spare cash on the pokies, the races, and the lottery. He will be a wage slave all his days. We know he will never be rich. But if he wins the lottery (if he wins big), then he will be rich. Contrapositively: his never being rich, plus other things we know, imply that he will lose. So, by closure, if we know that he will never be rich, we know that he will lose. But when we discussed the case before, we concluded that we cannot know that he will lose. All the possibilities in which Bill loses and someone else wins saliently resemble the possibility in which Bill wins and the others lose; one of those possibilities is actual; so by the Rules of Actuality and of Resemblance, we may not properly ignore the possibility that Bill wins. But there is a loophole: the resemblance was required to be salient. Salience, as well as ignoring, may vary between contexts. Before, when I was explaining how the Rule of Resemblance applied to lotteries, I saw to it that the resemblance between the many possibilities associated with the many tickets was sufficiently salient. But this time, when we were busy pitying poor Bill for his habits and not for his luck, the resemblance of the many possibilities was not so salient. At that point, the possibility of Bill's winning was properly ignored; so then it was true to say that we knew he would never be rich. Afterward I switched the context. I mentioned the possibility that Bill might win, wherefore that possibility was no longer properly ignored. (Maybe there were two separate reasons why it was no longer properly ignored, because maybe I also made the resemblance between the many possibilities more salient.) It was true at first that we knew that Bill would never be rich. And at that point it was also true that we knew he would lose – but that was only true so long as it remained unsaid! (And maybe unthought as well.) Later, after the change in context, it was no longer true that we knew he would lose. At that point, it was also no longer true that we knew he would never be rich.

But wait. Don't you smell a rat? Haven't I, by my own lights, been saying what cannot be said? (Or whistled either.) If the story I told was true, how have I managed to tell it? In trendyspeak, is there not a problem of reflexivity? Does not my story deconstruct itself?

I said: S knows that P iff S's evidence eliminates every possibility in which not-P – Psst! – except for those possibilities that we are properly ignoring. That "psst" marks an attempt to do the impossible – to mention that which remains unmentioned. I am sure you managed to make believe that I had succeeded. But I could not have done.

And I said that when we do epistemology, and we attend to the proper ignoring of possibilities, we make knowledge vanish. First we do know, then we do not. But I had been doing epistemology when I said that. The uneliminated possibilities were Hot being ignored – not just then. So by what right did I say even that we used to know?[24]

In trying to thread a course between the rock of fallibilism and the whirlpool of scepticism, it may well seem as if I have fallen victim to both at once. For do I not say that there are all those uneliminated possibilities of error? Yet do I not claim that we know a lot? Yet do I not claim that knowledge is, by definition, infallible knowledge?

I did claim all three things. But not all at once! Or if I did claim them all at once, that was an expository shortcut, to be taken with a pinch of salt. To get my message across, I bent the rules. If I tried to whistle what cannot be said, what of it? I relied on the cardinal principle of pragmatics, which overrides every one of the rules I mentioned: interpret the message to make it make sense – to make it consistent, and sensible to say.

When you have context-dependence, ineffability can be trite and unmysterious. Hush! [moment of silence] I might have liked to say, just then, "All of us are silent." It was true. But I could not have said it truly, or whistled it either. For by saying it aloud, or by whistling, I would have rendered it false.

I could have said my say fair and square, bending no rules. It would have been tiresome, but it could have been done. The secret would have been to resort to "semantic ascent." I could have taken great care to distinguish between (1) the language I use when I talk about knowledge, or whatever, and (2) the second language that I use to talk about the semantic and pragmatic workings of the first language. If you want to hear my story told that way, you probably know enough to do the job for yourself. If you can, then my informal presentation has been good enough.

[24]Worse still: by what right can I even say that we used to be in a position to say truly that we knew? Then, we were in a context where we properly ignored certain uneliminated possibilities of error. Now, we are in a context where we no longer ignore them. If now I comment retrospectively upon the truth of what was said then, which context governs: the context now or the context then? I doubt there is any general answer, apart from the usual principle that we should interpret what is said so as to make the message make sense.

Chapter 29
Knowledge and Scepticism

Robert Nozick

You think you are seeing these words, but could you not be hallucinating or dreaming or having your brain stimulated to give you the experience of seeing these marks on paper although no such thing is before you? More extremely, could you not be floating in a tank while super-psychologists stimulate your brain electrochemically to produce exactly the same experiences as you are now having, or even to produce the whole sequence of experiences you have had in your lifetime thus far? If one of these other things was happening, your experience would be exactly the same as it now is. So how can you know none of them is happening? Yet if you do not know these possibilities don't hold, how can you know you are reading this book now? If you do not know you haven't always been floating in the tank at the mercy of the psychologists, how can you know anything-what your name is, who your parents were, where you come from?

The sceptic argues that we do not know what we think we do. Even when he leaves us unconverted, he leaves us confused. Granting that we do know, how *can* we? Given these other possibilities he poses, how is knowledge possible? In answering this question, we do not seek to convince the sceptic, but rather to formulate hypotheses about knowledge and our connection to facts that show how knowledge can exist even given the sceptic's possibilities. These hypotheses must reconcile our belief that we know things with our belief that the sceptical possibilities are logical possibilities.

The sceptical possibilities, and the threats they pose to our knowledge, depend upon our knowing things (if we do) mediately, through or by way of something else. Our thinking or believing that some fact p holds is connected somehow to the

---

Robert Nozick was deceased at the time of publication.

R. Nozick (deceased)
Harvard University, Boston, MA, USA

H. Arló-Costa et al. (eds.), *Readings in Formal Epistemology*, Springer Graduate Texts in Philosophy 1, DOI 10.1007/978-3-319-20451-2_29

fact that *p*, but is not itself identical with that fact. Intermediate links establish the connection. This leaves room for the possibility of these intermediate stages holding and producing our belief that *p*, without the fact that *p* being at the other end. The intermediate stages arise in a completely different manner, one not involving the fact that *p* although giving rise to the appearance that *p* holds true.

Are the sceptic's possibilities indeed logically possible? Imagine reading a science fiction story in which someone is raised from birth floating in a tank with psychologists stimulating his brain. The story could go on to tell of the person's reactions when he is brought out of the tank, of how the psychologists convince him of what had been happening to him, or how they fail to do so. This story is coherent, there is nothing self-contradictory or otherwise impossible about it. Nor is there anything incoherent in imagining that you are now in this situation, at a time before being taken out of the tank. To ease the transition out, to prepare the way, perhaps the psychologists will give the person in the tank thoughts of whether floating in the tank is possible, or the experience of reading a book that discusses this possibility, even one that discusses their easing his transition. (Free will presents no insuperable problem for this possibility. Perhaps the psychologists caused all your experiences of choice, including the feeling of freely choosing; or perhaps you do freely choose to act while they, cutting the effector circuit, continue the scenario from there.)

Some philosophers have attempted to demonstrate there is no such coherent possibility of this sort. However, for any reasoning that purports to show this sceptical possibility cannot occur, we can imagine the psychologists of our science fiction story feeding it to their tank-subject, along with the (inaccurate) feeling that the reasoning is cogent. So how much trust can be placed in the apparent cogency of an argument to show the sceptical possibility isn't coherent? The sceptic's possibility is a logically coherent one, in tension with the existence of (almost all) knowledge; so we seek a hypothesis to explain how, even given the sceptic's possibilities, knowledge is possible. We may worry that such explanatory hypotheses are ad hoc, but this worry will lessen if they yield other facts as well, fit in with other things we believe, and so forth. Indeed, the theory of knowledge that follows was not developed in order to explain how knowledge is possible. Rather, the motivation was external to epistemology; only after the account of knowledge was developed for another purpose did I notice its consequences for scepticism, for understanding how knowledge is possible. So whatever other defects the explanation might have, it can hardly be called ad hoc.

Knowledge

Our task is to formulate further conditions to go alongside

(1) *p* is true
(2) S believes that *p*.

We would like each condition to be necessary for knowledge, so any case that fails to satisfy it will not be an instance of knowledge. Furthermore, we would like the conditions to be jointly sufficient for knowledge, so any case that satisfies all of them will be an instance of knowledge. We first shall formulate conditions that seem to handle ordinary cases correctly, classifying as knowledge cases which are knowledge, and as non-knowledge cases which are not; then we shall check to see how these conditions handle some difficult cases discussed in the literature.

One plausible suggestion is causal, something like: the fact that *p* (partially) causes S to believe that *p*, that is, (2) because (1). But this provides an inhospitable environment for mathematical and ethical knowledge; also there are well-known difficulties in specifying the type of causal connection. If someone floating in a tank oblivious to everything around him is given (by direct electrical and chemical stimulation of the brain) the belief that he is floating in a tank with his brain being stimulated, then even though that fact is part of the cause of his belief, still he does not know that it is true. Let us consider a different third condition:

(3) If *p* were not true, S would not believe that *p*.

Throughout this work, let us write the subjunctive "if-then" by an arrow, and the negation of a sentence by prefacing "not-" to it. The above condition thus is rewritten as:

(3) not-$p \rightarrow$ not-(S believes that *p*).

This subjunctive condition is not unrelated to the causal condition. Often when the fact that *p* (partially) causes someone to believe that *p*, the fact also will be causally necessary for his having the belief without the cause, the effect would not occur. In that case, the subjunctive condition (3) also will be satisfied. Yet this condition is not equivalent to the causal condition. For the causal condition will be satisfied in cases of causal overdetermination, where either two sufficient causes of the effect actually operate, or a back-up cause (of the same effect) would operate if the first one didn't; whereas the subjunctive condition need not hold for these cases.[1,2] When the two conditions do agree, causality indicates knowledge because it acts in a manner that makes the subjunctive (3) true.

The subjunctive condition (3) serves to exclude cases of the sort first described by Edward Gettier, such as the following. Two other people are in my office and I am justified on the basis of much evidence in believing the first owns a Ford car; though he (now) does not, the second person (a stranger to me) owns one. I believe truly and justifiably that someone (or other) in my office owns a Ford car, but I do not know someone does. Concluded Gettier, knowledge is not simply justified true belief.

[1]See Hilary Putnam, Reason, Truth and History (Cambridge, 1981), ch. I.

[2]I should note here that I assume bivalence throughout this chapter, and consider only statements that are true if and only if their negations are false.

The following subjunctive, which specifies condition (3) for this Gettier case, is not satisfied: if no one in my office owned a Ford car, I wouldn't believe that someone did. The situation that would obtain if no one in my office owned a Ford is one where the stranger does not (or where he is not in the office); and in that situation I still would believe, as before, that someone in my office does own a Ford, namely, the first person. So the subjunctive condition (3) excludes this Gettier case as a case of knowledge.

The subjunctive condition is powerful and intuitive, not so easy to satisfy, yet not so powerful as to rule out everything as an instance of knowledge. A subjunctive conditional "if p were true, q would be true," $p \rightarrow q$, does not say that p entails q or that it is logically impossible that p yet not-q. It says that in the situation that would obtain if p were true, q also would be true. This point is brought out especially clearly in recent "possible-worlds" accounts of subjunctives: the subjunctive is true when (roughly) in all those worlds in which p holds true that are closest to the actual world, q also is true. (Examine those worlds in which p holds true closest to the actual world, and see if q holds true in all these.) Whether or not q is true in p worlds that are still farther away from the actual world is irrelevant to the truth of the subjunctive. I do not mean to endorse any particular possible-worlds account of subjunctives, nor am I committed to this type of account.[3] I sometimes shall use it, though, when it illustrates points in an especially clear way.

The subjunctive condition (3) also handles nicely cases that cause difficulties for the view that you know that p when you can rule out the relevant alternatives to p in the context. For, as Gail Stine writes, "what makes an alternative relevant in one context and not another? ... if on the basis of visual appearances obtained under optimum conditions while driving through the countryside Henry identifies an object as a barn, normally we say that Henry knows that it is a barn. Let us suppose, however, that unknown to Henry, the region is full of expertly made papier-mache facsimiles of barns. In that case, we would not say that Henry knows that the object is a barn, unless he has evidence against it being a papier-mache facsimile, which is now a relevant alternative. So much is clear, but what if no such facsimiles exist in Henry's surroundings, although they once did? Are either of these circumstances sufficient to make the hypothesis (that it's a papier-mache object) relevant? Probably not, but the situation is not so clear:"[4] Let p be the statement that the object in the field is a (real) barn, and q the one that the object in the field is a papier-mache barn. When papier-mache barns are scattered through the area, if p were false, q would be true or might be. Since in this case (we are supposing) the person still would believe p, the subjunctive

[3]See Robert Stalnaker, "A Theory of Conditionals," in N. Rescher, ed., *Studies in Logical Theory* (Oxford 1968); David Lewis, Counteifactuals (Cambridge 1973); and Jonathan Bennett's critical review of Lewis, "Counterfactuals and Possible Worlds," *Canadian Journal of Philosophy*, 4/2 (Dec. 1974), 381–402. Our purposes require, for the most part, no more than an intuitive understanding of subjunctives.

[4]G. C. Stine, "Skepticism, Relevant Alternatives and Deductive Closure," *Philosophical Studies*, 29 (1976), 252, who attributes the example to Carl Ginet.

(3) not-$p \rightarrow$ not-(S believes that p)

is not satisfied, and so he doesn't know that p. However, when papier-mache barns are or were scattered around another country, even if p were false q wouldn't be true, and so (for all we have been told) the person may well know that p. A hypothesis q contrary to p clearly is relevant when if p weren't true, q would be true; when not-$p \rightarrow q$. It clearly is irrelevant when if p weren't true, q also would not be true; when not-$p \rightarrow$ not-q. The remaining possibility is that neither of these opposed subjunctives holds; q might (or might not) be true if p weren't true. In this case, q also will be relevant, according to an account of knowledge incorporating condition (3) and treating subjunctives along the lines sketched above. Thus, condition (3) handles cases that befuddle the "relevant alternatives" account; though that account can adopt the above subjunctive criterion for when an alternative is relevant, it then becomes merely an alternate and longer way of stating condition (3).

Despite the power and intuitive force of the condition that if p weren't true the person would not believe it, this condition does not (in conjunction with the first two conditions) rule out every problem case. There remains, for example, the case of the person in the tank who is brought to believe, by direct electrical and chemical stimulation of his brain, that he is in the tank and is being brought to believe things in this way; he does not know this is true. However, the subjunctive condition is satisfied: if he weren't floating in the tank, he wouldn't believe he was.

The person in the tank does not know he is there, because his belief is not sensitive to the truth. Although it is caused by the fact that is its content, it is not sensitive to that fact. The operators of the tank could have produced any belief, including the false belief that he wasn't in the tank; if they had, he would have believed that. Perfect sensitivity would involve beliefs and facts varying together. We already have one portion of that variation, subjunctively at least: if p were false he wouldn't believe it. This sensitivity as specified by a subjunctive does not have the belief vary with the truth or falsity of p in all possible situations, merely in the ones that would or might obtain if p were false. The subjunctive condition

(3) not-$p \rightarrow$ not-(S believes that p)

tells us only half the story about how his belief is sensitive to the truth-value of p. It tells us how his belief state is sensitive to p's falsity, but not how it is sensitive to p's truth; it tells us what his belief state would be if p were false, but not what it would be if p were true. To be sure, conditions (1) and (2) tell us that p is true and he does believe it, but it does not follow that his believing p is sensitive to p's being true. This additional sensitivity is given to us by a further subjunctive: if p were true, he would believe it.

(4) $p \rightarrow$ S believes that p.

Not only is p true and S believes it, but if it were true he would believe it. Compare: not only was the photon emitted and did it go to the left, but (it was then true that): if it were emitted it would go to the left. The truth of antecedent and consequent is not alone sufficient for the truth of a subjunctive; (4) says more than

(1) and (2). Thus, we presuppose some (or another) suitable account of subjunctives. According to the suggestion tentatively made above, (4) holds true if not only does he actually truly believe *p*, but in the 'close' worlds where *p* is true, he also believes it. He believes that *p* for some distance out in the *p*. neighbourhood of the actual world; similarly, condition (3) speaks not of the whole not-*p* neighbourhood of the actual world, but only of the first portion of it. (If, as is likely, these explanations do not help, please use your own intuitive understanding of the subjunctives (3) and (4).)

The person in the tank does not satisfy the subjunctive condition (4). Imagine as actual a world in which he is in the tank and is stimulated to believe he is, and consider what subjunctives are true in that world. It is not true of him there that if he were in the tank he would believe it; for in the close world (or situation) to his own where he is in the tank but they don't give him the belief that he is (much less instill the belief that he isn't) he doesn't believe he is in the tank. Of the person actually in the tank and believing it, it is not true to make the further statement that if he were in the tank he would believe it – so he does not know he is in the tank.

The subjunctive condition (4) also handles a case presented by Gilbert Harman.8 The dictator of a country is killed; in their first edition, newspapers print the story, but later all the country's newspapers and other media deny the story, falsely. Everyone who encounters the denial believes it (or does not know what to believe and so suspends judgement). Only one person in the country fails to hear any denial and he continues to believe the truth. He satisfies conditions (1) – (3) (and the causal condition about belief) yet we are reluctant to say he knows the truth. The reason is that if he had heard the denials, he too would have believed them, just like everyone else. His belief is not sensitively tuned to the truth, he doesn't satisfy the condition that if it were true he would believe it. Condition (4) is not satisfied.

There is a pleasing symmetry about how this account of knowledge relates conditions (3) and (4), and connects them to the first two conditions. The account has the following form.

(1)
(2)
(3) not-1 $\rightarrow$ not-2
(4) 1 $\rightarrow$ 2

I am not inclined, however, to make too much of this symmetry, for I found also that with other conditions experimented with as a possible fourth condition there was some way to construe the resulting third and fourth conditions as symmetrical answers to some symmetrical looking questions, so that they appeared to arise in parallel fashion from similar questions about the components of true belief.

Symmetry, it seems, is a feature of a mode of presentation, not of the contents presented. A uniform transformation of symmetrical statements can leave the results non-symmetrical. But if symmetry attaches to mode of presentation, how can it possibly be a deep feature of, for instance, laws of nature that they exhibit symmetry? (One of my favourite examples of symmetry is due to Groucho Marx. On his radio programme he spoofed a commercial, and ended. "And if you are not completely satisfied, return the unused portion of our product and we will return the unused portion of your money.") Still, to present our subject symmetrically makes the connection of knowledge to true belief especially perspicuous. It seems to me that a symmetrical formulation is a sign of our understanding, rather than a mark of truth. If we cannot understand an asymmetry as arising from an underlying symmetry through the operation of a particular factor, we will not understand why that asymmetry exists in that direction. (But do we also need to understand why the underlying asymmetrical factor holds instead of its opposite?)

A person knows that *p* when he not only does truly believe it, but also would truly believe it and wouldn't falsely believe it. He not only actually has a true belief, he subjunctively has one. It is true that *p* and he believes it; if it weren't true he wouldn't believe it, and if it were true he would believe it. To know that *p* is to be someone who would believe it if it were true, and who wouldn't believe it if it were false.

It will be useful to have a term for this situation when a person's belief is thus subjunctively connected to the fact. Let us say of a person who believes that *p*, which is true, that when (3) and (4) hold, his belief *tracks* the truth that *p*. To know is to have a belief that tracks the truth. Knowledge is a particular way of being connected to the world, having a specific real factual connection to the world: tracking it.

Scepticism

The sceptic about knowledge argues that we know very little or nothing of what we think we know, or at any rate that this position is no less reasonable than the belief in knowledge. The history of philosophy exhibits a number of different attempts to refute the sceptic: to prove him wrong or show that in arguing against knowledge he presupposes there is some and so refutes himself. Others attempt to show that accepting scepticism is unreasonable, since it is more likely that the sceptic's extreme conclusion is false than that all of his premisses are true, or simply because reasonableness of belief just means proceeding in an anti-sceptical way. Even when these counter-arguments satisfy their inventors, they fail to satisfy others, as is shown by the persistent attempts against scepticism. The continuing felt need to refute scepticism, and the difficulty in doing so, attests to the power of the sceptic's position, the depth of his worries.

An account of knowledge should illuminate sceptical arguments and show wherein lies their force. If the account leads us to reject these arguments, this had better not happen too easily or too glibly. To think the sceptic overlooks something obvious, to attribute to him a simple mistake or confusion or fallacy, is to refuse to acknowledge the power of his position and the grip it can have upon us. We thereby cheat ourselves of the opportunity to reap his insights and to gain self-knowledge in understanding why his arguments lure us so. Moreover, in fact, we cannot lay the spectre of scepticism to rest without first hearing what it shall unfold.

Our goal is not, however, to refute scepticism, to prove it is wrong or even to argue that it is wrong. We have elsewhere distinguished between philosophy that attempts to prove, and philosophy that attempts to explain how something is possible. Our task here is to explain how knowledge is possible, given what the sceptic says that we do accept (for example, that it is logically possible that we are dreaming or are floating in the tank). In doing this, we need not convince the sceptic, and we may introduce explanatory hypotheses that he would reject. What is important for our task of explanation and understanding is that we find those hypotheses acceptable or plausible, and that they show us how the existence of knowledge fits together with the logical possibilities the sceptic points to, so that these are reconciled within our own belief system. These hypotheses are to explain to ourselves how knowledge is possible, not to prove to someone else that knowledge is possible.[5,6]

Sceptical Possibilities

The sceptic often refers to possibilities in which a person would believe something even though it was false: really, the person is cleverly deceived by others, perhaps by an evil demon, or the person is dreaming, or he is floating in a tank near Alpha Centauri with his brain being stimulated. In each case, the *p* he believes is false, and he believes it even though it is false.

How do these possibilities adduced by the sceptic show that someone does not know that *p*? Suppose that someone is you; how do these possibilities count against your knowing that *p*? One way might be the following. (I shall consider other ways later.) If there is a possible situation where *p* is false yet you believe that *p*, then in that situation you believe that *p* even though it is false. So it appears you do not satisfy condition (3) for knowledge.

(3) If *p* were false, S wouldn't believe that *p*.

For a situation has been described in which you do believe that *p* even though *p* is false. How then can it also be true that if *p* were false, you wouldn't believe it? If the sceptic's possible situation shows that (3) is false, and if (3) is a necessary condition for knowledge, then the sceptic's possible situation shows that there isn't knowledge.

So construed, the sceptic's argument plays on condition (3); it aims to show that condition (3) is not satisfied. The sceptic may seem to be putting forth

[5]Gilbert Harman, *Thought* (Princeton; 1973), ch. 9, 142–54.

[6]From the perspective of explanation rather than proof, the extensive philosophical discussion, deriving from Charles S. Peirce, of whether the sceptic's doubts are real is beside the point. The problem of explaining how knowledge is possible would remain the same, even if no one ever claimed to doubt that there was knowledge.

R: Even if *p* were false, S still would believe *p*.

This conditional, with the same antecedent as (3) and the contradictory consequent, is incompatible with the truth of (3). If (3) is true, then R is not. However, R is stronger than the sceptic needs in order to show (3) is false. For (3) is false when if *p* were false, S might believe that *p*. This last conditional is weaker than R, and is merely (3)'s denial:

T: not-[not-*p* → not-(S believes that *p*)]

Whereas R does not simply deny (3), it asserts an opposing subjunctive of its own. Perhaps the possibility the sceptic adduces is not enough to show that R is true, but it appears at least to establish the weaker T; since this T denies (3), the sceptic's possibility appears to show that (3) is false. However, the truth of (3) is not incompatible with the existence of a possible situation where the person believes *p* though it is false. The subjunctive

(3) not-*p* → not-(S believes *p*)

does not talk of all possible situations in which *p* is false (in which not-*p* is true). It does not say that in all possible situations where not-*p* holds, S doesn't believe *p*. To say there is no possible situation in which not-*p* yet S believes *p*, would be to say that not-*p* entails not-(S believes *p*), or logically implies it. But subjunctive conditionals differ from entailments; the subjunctive (3) is not a statement of entailment. So the existence of a possible situation in which *p* is false yet S believes *p* does not show that (3) is false; (3) can be true even though there is a possible situation where not-*p* and S believes that *p*.

What the subjunctive (3) speaks of is the situation that would hold if *p* were false. Not every possible situation in which *p* is false is the situation that would hold if *p* were false. To fall into possible worlds talk, the subjunctive (3) speaks of the not-*p* world that is closest to the actual world, or of those not-*p* worlds that are closest to the actual world. And it is of this or these not-*p* worlds that it says (in them) S does not believe that *p*. What happens in yet other more distant not-*p* worlds is no concern of the subjunctive (3). The sceptic's possibilities (let us refer to them as SK), of the person's being deceived by a demon or dreaming or floating in a tank, count against the subjunctive

(3) if *p* were false then S wouldn't believe that *p*

only if (one of) these possibilities would or might obtain if *p* were false. Condition (3) says: if *p* were false, S still would not believe *p*. And this can hold even though there is some situation SK described by the sceptic in which *p* is false and S believes *p*. If *p* were false S still would not believe *p*, even though there is a situation SK in which *p* is false and S does believe *p*, provided that this situation SK wouldn't obtain if *p* were false. If the sceptic describes a situation SK which would not hold even if *p* were false then this situation SK doesn't show that (3) is false and so does not (in this way at least) undercut knowledge. Condition C acts to rule out sceptical hypotheses.

C: not-$p \rightarrow$ SK does not obtain.

Any sceptical situation SK which satisfies condition C is ruled out. For a sceptical situation SK to show that we don't know that p, it must fail to satisfy C which excludes it; instead it must be a situation that might obtain if p did not, and so satisfy C's denial:

not-(not-$p \rightarrow$ SK does not obtain)

Although the sceptic's imagined situations appear to show that (3) is false, they do not; they satisfy condition C and so are excluded.

The sceptic might go on to ask whether we know that his imagined situations SK are excluded by condition C, whether we know that if p were false SK would not obtain. However, typically he asks something stronger: do we know that his imagined situation SK does not actually obtain? Do we know that we are not being deceived by a demon, dreaming, or floating in a tank? And if we do not know this, how can we know that p? Thus we are led to the second way his imagined situations might show that we do not know that p.

Sceptical Results

According to our account of knowledge, S knows that the sceptic's situation SK doesn't hold if and only if

(1) SK doesn't hold
(2) S believes that SK doesn't hold
(3) If SK were to hold, S would not believe that SK doesn't hold
(4) If SK were not to hold, S would believe it does not.

Let us focus on the third of these conditions, The sceptic has carefully chosen his situations SK so that if they held we (still) would believe they did not. We would believe we weren't dreaming, weren't being deceived, and so on, even if we were. He has chosen situations SK such that if SK were to hold, S would (still) believe that SK doesn't hold – and this is incompatible with the truth of (3).

Since condition (3) is a necessary condition for knowledge, it follows that we do not know that SK doesn't hold. If it were true that an evil demon was deceiving us, if we were having a particular dream, if we were floating in a tank with our brains stimulated in a specified way, we would still believe we were not. So, we do not know we're not being deceived by an evil demon, we do not know we're not in that tank, and we do not know we're not having that dream. So says the sceptic, and so says our account. And also so we say – don't we? For how could we know we are not being deceived that way, dreaming that dream? If those things were happening to us, everything would seem the same to us. There is no way we can know it is not happening for there is no way we could tell if it were happening; and if it were happening we would believe exactly what we do now – in particular, we still would believe that it was not. For this reason, we feel, and correctly, that we don't know –

how could we? – that it is not happening to us. It is a virtue of our account that it yields, and explains, this result.

The sceptic asserts we do not know his possibilities don't obtain, and he is right. Attempts to avoid scepticism by claiming we do know these things are bound to fail. The sceptic's possibilities make us uneasy because, as we deeply realize, we do not know they don't obtain; it is not surprising that attempts to show we do know these things leave us suspicious, strike us even as bad faith. Nor has the sceptic merely pointed out something obvious and trivial. It comes as a surprise to realize that we do not know his possibilities don't obtain. It is startling, shocking. For we would have thought, before the sceptic got us to focus on it, that we did know those things, that we did know we were not being deceived by a demon, or dreaming that dream, or stimulated that way in that tank. The sceptic has pointed out that we do not know things we would have confidently said we knew. And if we don't know these things, what can we know? So much for the supposed obviousness of what the sceptic tells us.

Let us say that a situation (or world) is doxically identical for S to the actual situation when if S were in that situation, he would have exactly the beliefs (*doxa*) he actually does have. More generally, two situations are doxically identical for S if and only if he would have exactly the same beliefs in them. It might be merely a curiosity to be told there are non-actual situations doxically identical to the actual one. The sceptic, however, describes worlds doxically identical to the actual world in which almost everything believed is false.[7]

Such worlds are possible because we know mediately, not directly. This leaves room for a divergence between our beliefs and the truth. It is as though we possessed only two-dimensional plane projections of three-dimensional objects. Different three-dimensional objects, oriented appropriately, have the same two-dimensional plane projection. Similarly, different situations or worlds will lead to our having the very same beliefs. What is surprising is how very different the doxically identical world can be – different enough for almost everything believed in it to be false. Whether or not the mere fact that knowledge is mediated always makes room for such a very different doxically identical world, it does so in our case, as the sceptic's possibilities show. To be shown this is non-trivial, especially when we recall that we do not know the sceptic's possibility doesn't obtain: we do not know that we are not living in a doxically identical world wherein almost everything we believe is false.

What more could the sceptic ask for or hope to show? Even readers who sympathized with my desire not to dismiss the sceptic too quickly may feel this has gone too far, that we have not merely acknowledged the force of the sceptic's position but have succumbed to it.

---

[7]I say almost everything, because there still could be some true beliefs such as "I exist." More limited sceptical possibilities present worlds doxically identical to the actual world in which almost every belief of a' certain sort is false, for example, about the past, or about other people's mental states.

The sceptic maintains that we know almost none of what we think we know. He has shown, much to our initial surprise, that we do not know his (nontrivial) possibility SK doesn't obtain. Thus, he has shown of one thing we thought we knew, that we didn't and don't. To the conclusion that we know almost nothing, it appears but a short step. For if we do not know we are not dreaming or being deceived by a demon or floating in a tank, then how can I know, for example, that I am sitting before a page writing with a pen, and how can you know that you are reading a page of a book?

However, although our account of knowledge agrees with the sceptic in saying that we do not know that not-SK, it places no formidable barriers before my knowing that I am writing on a page with a pen. It is true that I am, I believe I am, if! weren't I wouldn't believe I was, and if I were, I would believe it. Also, it is true that you are reading a page (please, don't stop now!), you believe you are, if you weren't reading a page you wouldn't believe you were, and if you were reading a page you would believe you were. So according to the account, I do know that I am writing on a page with a pen, and you do know that you are reading a page. The account does not lead to any general scepticism.

Yet we must grant that it appears that if the sceptic is right that we don't know we are not dreaming or being deceived or floating in the tank, then it cannot be that I know I am writing with a pen or that you know you are reading a page. So we must scrutinize with special care the sceptic's "short step" to the conclusion that we don't know these things, for either this step cannot be taken or our account of knowledge is incoherent.

Nonclosure

In taking the "short step," the sceptic assumes that if S knows that *p* and he knows that "*p* entails *q*" then he also knows that *q*. In the terminology of the logicians, the sceptic assumes that knowledge is closed under known logical implication; that the operation of moving from something known to something else known to be entailed by it does not take us outside of the (closed) area of knowledge. He intends, of course, to work things backwards, arguing that since the person does not know that *q*, assuming (at least for the purposes of argument) that he does know that *p* entails *q*. it follows that he does not know that *p*. For if he did know that *p*, he would also know that *q*, which he doesn't.

The details of different sceptical arguments vary in their structure, but each one will assume some variant of the principle that knowledge is closed under known logical implication. If we abbreviate "knowledge that *p*" by "K*p*" and abbreviate "entails" by the fishhook sign"⥽," we can write this principle of closure as the subjunctive principle

p : K(*p*⥽*q*) & K*p* → K*q*.

If a person were to know that p entails q and he were to know that p then he would know that q. The statement that q follows by modus ponens from the other two stated as known in the antecedent of the subjunctive principle p; this principle counts on the person to draw the inference to q. You know that your being in a tank on Alpha Centauri entails your not being in place X where you are. (I assume here a limited readership.) And you know also the contrapositive, that your being at place X entails that you are not then in a tank on Alpha Centauri. If you knew you were at X you would know you're not in a tank (of a specified sort) at Alpha Centauri. But you do not know this last fact (the sceptic has argued and we have agreed) and so (he argues) you don't know the first. Another intuitive way of putting the sceptic's argument is as follows. If you know that two statements are incompatible and you know the first is true then you know the denial of the second.

You know that your being at X and your being in a tank on Alpha Centauri are incompatible; so if you knew you were at X you would know you were not in the (specified) tank on Alpha Centauri. Since you do not know the second, you don't know the first.

No doubt, it is possible to argue over the details of principle p, to point out it is incorrect as it stands. Perhaps, though Kp, the person does not know that he knows that p (that is, not-KKp) and so does not draw the inference to q. Or perhaps he doesn't draw the inference because not-KK($p ⥽ q$). Other similar principles face their own difficulties: for example, the principle that $K(p \to q) \to (Kp \to Kq)$ fails if Kp stops $p \to q$ from being true, that is, if $Kp \to$ not-$(p \to q)$; the principle that $K(p ⥽ q) \to K(Kp \to Kq)$ faces difficulties if Kp makes the person forget that ($p ⥽ q$) and so he fails to draw the inference to q. We seem forced to pile K upon K until we reach something like $KK(p ⥽ q)$ & $KKp \to Kq$; this involves strengthening considerably the antecedent of p and so is not useful for the sceptic's argument that p is not known. (From a principle altered thus it would follow at best that it is not known that p is known.)

We would be ill-advised, however, to quibble over the details of p. Although these details are difficult to get straight, it will continue to appear that something like p is correct. If S knows that "p entails q," and he knows that p and knows that "(p and p entails q) entails q" and he does draw the inference to q from all this and believes q via the process of drawing this inference, then will he not know that q? And what is wrong with simplifying this mass of detail by writing merely principle p, provided we apply it only to cases where the mass of detail holds, as it surely does in the sceptical cases under consideration? For example, I do realize that my being in the Van Leer Foundation Building in Jerusalem entails that I am not in a tank on Alpha Centauri; I am capable of drawing inferences now; I do believe I am not in a tank on Alpha Centauri (though not solely via this inference, surely); and so forth. Won't this satisfy the correctly detailed principle, and shouldn't it follow that I know I am not (in that tank) on Alpha Centauri? The sceptic agrees it should follow; so he concludes from the fact that I don't know I am not floating in the tank on Alpha Centauri that I don't know I am in Jerusalem. Uncovering difficulties in the details of particular formulations of p will not weaken the principle's intuitive appeal; such quibbling will seem at best like a wasp attacking a steamroller, at worst like an effort in bad faith to avoid being pulled along by the sceptic's argument.

Principle *p* is wrong, however, and not merely in detail. Knowledge is not closed under known logical implication. S knows that *p* when S has a true belief that *p*, and S wouldn't have a false belief that *p* (condition (3)) and S would have a true belief that *p* (condition (4)). Neither of these latter two conditions is closed under known logical implication. Let us begin with condition

(3) if *p* were false, S wouldn't believe that *p*.

When S knows that *p*, his belief that *p* is contingent on the truth of *p*. contingent in the way the subjunctive condition (3) describes. Now it might be that *p* entails *q* (and S knows this), that S's belief that *p* is subjunctively contingent on the truth of *p*, that S believes *q*, yet his belief that *q* is not subjunctively dependent on the truth of *q*. in that it (or he) does not satisfy:

(3′) if *q* were false, S wouldn't believe that *q*.

For (3′) talks of what S would believe if *q* were false, and this may be a very different situation from the one that would hold if *p* were false, even though *p* entails *q*. That you were born in a certain city entails that you were born on earth.[8] Yet contemplating what (actually) would be the situation if you were not born in that city is very different from contemplating what situation would hold if you weren't born on earth. Just as those possibilities are very different, so what is believed in them may be very different. When *p* entails *q* (and not the other way around) *p* will be a stronger statement than *q*, and so not-*q* (which is the antecedent of (3′)) will be a stronger statement than not-*p* (which is the antecedent of (3)). There is no reason to assume you will have the same beliefs in these two cases, under these suppositions of differing strengths.

There is no reason to assume the (closest) not-*p* world and the (closest) not-*q* world are doxically identical for you, and no reason to assume, even though *p* entails *q*, that your beliefs in one of these worlds would be a (proper) subset of your beliefs in the other.

Consider now the two statements:

$p =$ I am awake and sitting on a chair in Jerusalem;
$q =$ I am not floating in a tank on Alpha Centauri being stimulated by electrochemical means to believe that *p*.

The first one entails the second: *p* entails *q*. Also, I know that *p* entails *q*; and I know that *p*. If *p* were false, I would be standing or lying down in the same city, or perhaps sleeping there, or perhaps in a neighbouring city or town. If *q* were false, I would be floating in a tank on Alpha Centauri. Clearly these are very different situations, leading to great differences in what I then would believe. If *p* were false,

[8] Here again I assume a limited readership, and ignore possibilities such as those described in James Blish, *Cities in Flight* (New York, 1982).

if 1 weren't awake and sitting on a chair in Jerusalem, 1 would not believe that *p*. Yet if *q* were false, if I was floating in a tank on Alpha Centauri, 1 would believe that *q*, that I was not in the tank, and indeed, in that case, I would still believe that *p*. According to our account of knowledge, I know that *p* yet I do not know that *q*, even though (1 know) *p* entails *q*.

This failure of knowledge to be closed under known logical implication stems from the fact that condition (3) is not closed under known logical implication; condition (3) can hold of one statement believed while not of another known to be entailed by the first. It is clear that any account that includes as a necessary condition for knowledge the subjunctive condition (3), not-$p \rightarrow$ not-(S believes that *p*), will have the consequence that knowledge is not closed under known logical implication. When *p* entails *q* and you believe each of them, if you do not have a false belief that *p* (since *p* is true) then you do not have a false belief that *q*. However, if you are to know something not only don't you have a false belief about it, but also you wouldn't have a false belief about it. Yet, we have seen how it may be that *p* entails *q* and you believe each and you wouldn't have a false belief that *p* yet you might have a false belief that *q* (that is, it is not the case that you wouldn't have one). Knowledge is not closed under the known logical implication because 'wouldn't have a false belief that' is not closed under known logical implication.

If knowledge were the same as (simply) true belief then it would be closed under known logical implication (provided the implied statements were believed). Knowledge is not simply true belief, however; additional conditions are needed. These further conditions will make knowledge open under known logical implication, even when the entailed statement is believed, when at least one of the further conditions itself is open. Knowledge stays closed (only) if all of the additional conditions are closed. I lack a general non-trivial characterization of those conditions that are closed under known logical implication: possessing such an illuminating characterization, one might attempt to prove that no additional conditions of that sort could provide an adequate analysis of knowledge.

Still, we can say the following. A belief that *p* is knowledge that *p* only if it somehow varies with the truth of *p*. The causal condition for knowledge specified that the belief was "produced by" the fact, but that condition did not provide the right sort of varying with the fact. The subjunctive conditions (3) and (4) are our attempt to specify that varying. But however an account spells this out, it will hold that whether a belief that *p* is knowledge partly depends on what goes on with the belief in some situations when *p* is false. An account that says nothing about what is believed in any situation when *p* is false cannot give us any mode of varying with the fact.

Because what is preserved under logical implication is truth, any condition that is preserved under known logical implication is most likely to speak only of what happens when *p*, and *q*, are true, without speaking at all of what happens when either one is false. Such a condition is incapable of providing "varies with"; so adding only such conditions to true belief cannot yield an adequate account of knowledge.

A belief's somehow varying with the truth of what is believed is not closed under known logical implication. Since knowledge that *p* involves such variation, knowledge also is not closed under known logical implication. The sceptic cannot easily deny that knowledge involves such variation, for his argument that we don't know that we're not floating in that tank, for example, uses the fact that knowledge does involve variation. ("If you were floating in the tank you would still think you weren't, so you don't know that you're not.") Yet, though one part of his argument uses that fact that knowledge involves such variation, another part of his argument presupposes that knowledge does not involve any such variation. This latter is the part that depends upon knowledge being closed under known logical implication. as when the sceptic argues that since you don't know that not-SK, you don't know you are not floating in the tank, then you also don't know, for example, that you are now reading a book. That closure can hold only if the variation does not. The sceptic cannot be right both times. According to our view he is right when he holds that knowledge involves such variation and so concludes that we don't know, for example, that we are not floating in that tank: but he is wrong when he assumes knowledge is closed under known logical implication and concludes that we know hardly anything.[9]

Knowledge is a real factual relation, subjunctively specifiable, whose structure admits our standing in this relation, tracking, to p without standing in it to some q which we know p to entail. Any relation embodying some variation of belief with this fact. with the truth (value), will exhibit this structural feature. The sceptic is right that we don't track Some particular truths – the ones stating that his sceptical possibilities SK don't hold – but wrong that we don't stand in the real knowledge-relation of tracking to many other truths, including ones that entail these first mentioned truths we believe but don't know.

The literature on scepticism contains writers who endorse these sceptical arguments (or similar narrower ones), but confess their inability to maintain their

[9]Reading an earlier draft of this chapter, friends pointed out to me that Fred Dretske already had defended the view that knowledge (as one among many epistemic concepts) is not closed under known logical implication. (See his "Epistemic Operators," *Journal of Philosophy*, 67, (1970). 1007–23.) Furthermore, Dretske presented a subjunctive condition for knowledge (in his "Conclusive Reasons," *Australasian Journal of Philosophy*, 49, (1971), 1–22), holding that S knows that p on the basis of reasons R only if: R would not be the case unless p were the case. Here Dretske ties the evidence subjunctively to the fact, and the belief based on the evidence subjunctively to the fact through the evidence. The independent statement and delineation of the position here I hope will make clear its many merits.

After Goldman's paper on a causal theory of knowledge, in *Journal of Philosophy*, 64, (1967), an idea then already "in the air," it required no great leap to consider subjunctive conditions. Some 2 months after the first version of this chapter was written, Goldman himself published a paper on knowledge utilizing counterfactuals ("Discrimination and Perceptual Knowledge," Essay II in this collection), also talking of relevant possibilities (without using the counterfactuals to identify which possibilities are relevant); and R. Shope has called my attention to a paper of L. S. Carrier ("An Analysis of Empirical Knowledge," *Southern Journal of Philosophy*, 9, (1971), 3–11) that also used subjunctive conditions including our condition (3). Armstrong's reliability view of knowledge (Belief, Truth and Knowledge, Cambridge, 1973, pp. 166, 169) involved a lawlike connection between the belief that p and the state of affairs that makes it true. Clearly, the idea is one whose time has come.

sceptical beliefs at times when they are not focusing explicitly on the reasoning that led them to sceptical conclusions. The most notable example of this is Hume:

> I am ready to reject all belief and reasoning, and can look upon no opinion even as more probable or likely than another... Most fortunately it happens that since reason is incapable of dispelling these clouds, nature herself suffices to that purpose, and cures me of this philosophical melancholy and delirium, either by relaxing this bent of mind, or by some avocation, and lively impression of my senses, which obliterate all these chimeras. I dine, I play a game of backgammon, I converse, and am merry with my friends; and when after three or four hours' amusement, I would return to these speculations, they appear so cold, and strained, and ridiculous, that I cannot find in my heart to enter into them any farther. (A Treatise of Human Nature, Book I, Part IV, section VII.)
>
> The great subverter of Pyrrhonism or the excessive principles of skepticism is action, and employment, and the occupations of common life. These principles may flourish and triumph in the schools; where it is, indeed, difficult, if not impossible, to refute them. But as soon as they leave the shade, and by the presence of the real objects, which actuate our passions and sentiments, are put in opposition to the more powerful principles of our nature, they vanish like smoke, and leave the most determined skeptic in the same condition as other mortals... And though a Pyrrhonian may throw himself or others into a momentary amazement and confusion by his profound reasonings; the first and most trivial event in life will put to flight all his doubts and scruples, and leave him the same, in every point of action and speculation, with the philosophers of every other sect, or with those who never concerned themselves in any philosophical researches. When he awakes from his dream, he will be the first to join in the laugh against himself, and to confess that all his objections are mere amusement. (An Enquiry Concerning Human Understanding, Section XII, Part II.)

The theory of knowledge we have presented explains why sceptics of various sorts have had such difficulties in sticking to their far-reaching sceptical conclusions "outside the study," or even inside it when they are not thinking specifically about sceptical arguments and possibilities SK.

The sceptic's arguments do show (but show only) that we don't know the sceptic's possibilities SK do not hold; and he is right that we don't track the fact that SK does not hold. (If it were to hold, we would still think it didn't.) However, the sceptic's arguments don't show we do not know other facts (including facts that entail not-SK) for we do track these other facts (and knowledge is not closed under known logical entailment). Since we do track these other facts – you, for example, the fact that you are reading a book; I, the fact that I am writing on a page – and the sceptic tracks such facts too, it is not surprising that when he focuses on them, on his relationship to such facts, the sceptic finds it hard to remember or maintain his view that he does not know those facts. Only by shifting his attention back to his relationship to the (different) fact that not-SK, which relationship is not tracking, can he revive his sceptical belief and make it salient. However, this sceptical triumph is evanescent, it vanishes when his attention turns to other facts. Only by fixating on the sceptical possibilities SK can he maintain his sceptical virtue; otherwise, unsurprisingly, he is forced to confess to sins of credulity.

Chapter 30
On Logics of Knowledge and Belief

Robert Stalnaker

Introduction

Formal epistemology, or at least the approach to formal epistemology that develops a logic and formal semantics of knowledge and belief in the possible worlds framework, began with Jaakko Hintikka's book *Knowledge and Belief*, published in 1962. Hintikka's project sparked some discussion of issues about iterated knowledge (does knowing imply knowing that one knows?) and about "knowing who", and quantifying into knowledge attributions. Much later, this kind of theory was taken up and applied by theoretical computer scientists and game theorists.[1] The formal semantic project gained new interest when it was seen that it could be applied to contexts with multiple knowers, and used to clarify the relation between epistemic and other modal concepts.

Edmund Gettier's classic refutation of the Justified True Belief analysis of knowledge (Gettier 1963) was published at about the same time as Hintikka's book, and it immediately spawned an epistemological industry—a project of attempting to revise the refuted analysis by adding further conditions to meet the counterexamples. Revised analyses were met with further counterexamples, followed by further refinements. This kind of project flourished for some years, but eventually became an internally driven game that was thought to have lost contact with the fundamental epistemological questions that originally motivated it. This way of approaching epistemological questions now seems hopelessly out of date,

[1]See Fagin et al. (1995) and Battigalli and Bonanno (1999) for excellent surveys of the application of logics of knowledge and belief in theoretical computer science and game theory.

R. Stalnaker (✉)
Department of Linguistics and Philosophy, MIT, Cambridge, MA, USA
e-mail: stal@mit.edu

H. Arló-Costa et al. (eds.), *Readings in Formal Epistemology*, Springer Graduate Texts in Philosophy 1, DOI 10.1007/978-3-319-20451-2_30

but I think there may still be some insights to be gained by looking back, if not at the details of the analyses, at some of the general strategies of analysis that were deployed.

There was little contact between these two very different epistemological projects. The first had little to say about substantive questions about the relation between knowledge, belief, and justification or epistemic entitlement, or about traditional epistemological issues, such as skepticism. The second project ignored questions about the abstract structure of epistemic and doxastic states. But I think some of the abstract questions about the logic of knowledge connect with traditional questions in epistemology, and with the issues that motivated the attempt to find a definition of knowledge. The formal semantic framework provides the resources to construct models that may help to clarify the abstract relationship between the concept of knowledge and some of the other concepts (belief and belief revision, causation and counterfactuals) that were involved in the post-Gettier project of defining knowledge. And some of the examples that were originally used in the post-Gettier literature to refute a proposed analysis can be used in a different way in the context of formal semantic theories: to bring out contrasting features of some alternative conceptions of knowledge, conceptions that may not provide plausible analyses of knowledge generally, but that may provide interesting models of knowledge that are appropriate for particular applications, and that may illuminate, in an idealized way, one or another of the dimensions of the complex epistemological terrain.

My aim in this paper will be to bring out some of the connections between issues that arise in the development and application of formal semantics for knowledge and belief and more traditional substantive issues in epistemology. The paper will be programmatic; pointing to some highly idealized theoretical models, some alternative assumptions that might be made about the logic and semantics of knowledge, and some of the ways in which they might connect with traditional issues in epistemology, and with applications of the concept of knowledge. I will bring together and review some old results, and make some suggestions about possible future developments. After a brief sketch of Hintikka's basic logic of knowledge, I will discuss, in section "Partition models", the S5 epistemic models that were developed and applied by theoretical computer scientists and game theorists, models that, I will argue, conflate knowledge and belief. In section "Belief and knowledge", I will discuss a basic theory that distinguishes knowledge from belief and that remains relatively noncommittal about substantive questions about knowledge, but that provides a definition of belief in terms of knowledge. This theory validates a logic of knowledge, S4.2, that is stronger than S4, but weaker than S5. In the remaining four sections, I will consider some alternative ways of adding constraints on the relation between knowledge and belief that go beyond the basic theory: in section "Partition models and the basic theory", I will consider the S5 partition models as a special case of the basic theory; in section "Minimal and maximal extensions", I will discuss the upper and lower bounds to an extension of the semantics of belief to a semantics for knowledge; in section "Belief revision and the defeasibility analysis",

I will discuss a version of the defeasibility analysis of knowledge, and in section "The causal dimension", a simplified version of a causal theory.

The basic idea that Hintikka developed, and that has since become familiar, was to treat knowledge as a modal operator with a semantics that parallels the possible worlds semantics for necessity. Just as necessity is truth in all possible worlds, so knowledge is truth in all *epistemically* possible worlds. The assumption is that to have knowledge is to have a capacity to locate the actual world in logical space, to exclude certain possibilities from the candidates for actuality. The epistemic possibilities are those that remain after the exclusion, those that the knower cannot distinguish from actuality. To represent knowledge in this way is of course not to provide any kind of reductive analysis of knowledge, since the abstract theory gives no substantive account of the criteria for determining epistemic possibility. The epistemic possibilities are defined by a binary accessibility relation between possible worlds that is a primitive component of an epistemic model. (Where x and y are possible worlds, and 'R' is the accessibility relation, 'xRy' says that y is epistemically possible for the agent in world x.). The idea was to give a precise representation of the structure of an epistemic state that was more or less neutral about more substantive questions about what constitutes knowledge, but that sharpened questions about the logic of knowledge. This form of representation was, however, far from innocent, since it required, from the start, an extreme idealization: Even in its most neutral form, the framework required the assumption that knowers know all logical truths, and all of the consequences of their knowledge, since no matter how the epistemically possible worlds are selected, all logical truths will be true in all of them, and for any set of propositions true in all of them, all of their logical consequences will also be true in all of them. There are different ways of understanding the character of this idealization: on the one hand, one might say that the concept of knowledge that is being modeling is knowledge in the ordinary sense, but that the theory is intended to apply only to idealized knowers—those with superhuman logical capacities. Alternatively, one might say that the theory is intended to model an idealized sense of knowledge—the information that is implicit in one's knowledge—that literally applies to ordinary knowers. However the idealization is explained, there remain the questions whether it is fruitful to develop a theory that requires this kind of deviation from reality, and if so why.[2] But I think these questions are best answered by looking at the details of the way such theories have been, and can be developed.

The most basic task in developing a semantics for knowledge in the possible worlds framework is to decide on the properties of the epistemic accessibility relation. It is clear that the relation should be reflexive, which is necessary to validate the principle that knowledge implies truth, an assumption that is just about the only principle of a logic of knowledge that is uncontroversial. Hintikka argued

[2] I explore the problem of logical omniscience in two papers, Stalnaker (1991, 1999b) both included in Stalnaker (1999a). I don't attempt to solve the problem in either paper, but only to clarify it, and to argue that it is a genuine problem, and not an artifact of a particular theoretical framework.

that we should also assume that the relation is transitive, validating the much more controversial principle that knowing implies knowing that one knows. Knowing and knowing that one knows are, Hintikka claimed, "virtually equivalent." Hintikka's reasons for this conclusion were not completely clear. He did not want to base it on a capacity for introspection: he emphasized that his reasons were logical rather than psychological. His proof of the KK principle rests on the following principle: If $\{K\varphi, \sim K \sim \psi\}$ is consistent, then $\{K\varphi, \psi\}$ is consistent, and it is clear that if one grants this principle, the KK principle immediately follows.[3] The reason for accepting this principle seems to be something like this: Knowledge requires conclusive reasons for belief, reasons that would not be defeated by any information compatible with what is known. So if one knows that φ while ψ is compatible with what one knows, then the truth of ψ could not defeat one's claim to know that φ. This argument, and other considerations for and against the KK principle deserve more careful scrutiny. There is a tangle of important and interesting issues underlying the question whether one should accept the KK principle, and the corresponding semantics, and some challenging arguments that need to be answered if one does.[4] I think the principle can be defended (in the context of the idealizations we are making), but I will not address this issue here, provisionally following Hintikka in accepting the KK principle, and a semantics that validates it.

The S4 principles (Knowledge implies truth, and knowing implies knowing that one knows) were as far as Hintikka was willing to go. He unequivocally rejects the characteristic S5 principle that if one lacks knowledge, then one knows that one lacks it.("unless you happen to be as sagacious as Socrates"[5]), and here his reasons seem to be clear and decisive[6]:

> The consequences of this principle, however, are obviously wrong. By its means (together with certain intuitively acceptable principles) we could, for example, show that the following sentence is self sustaining:
>
> (13) $p \supset K_a P_a p$. [In Hintikka's notation, 'P_a' is the dual of the knowledge operator, 'K_a': '$\sim K_a \sim$'. I will use 'M' for $\sim K \sim$]

The reason that (13) is clearly unacceptable, as Hintikka goes on to say, is that it implies that one could come to know by reflection alone, of any truth, that it was compatible with one's knowledge. But it seems that a consistent knower might believe, and be justified in believing, that she knew something that was in fact false. That is, it might be, for some proposition φ that $\sim\varphi$, and $BK\varphi$. In such a case, if the subject's beliefs are consistent, then she does not believe, and so does not know, that $\sim\varphi$ is compatible with her knowledge. That is, $\sim K \sim K\varphi$, along with $\sim\varphi$, will be true, falsifying (13).

---

[3]Substituting '$\sim K\varphi$' for ψ, and eliminating a double negation, the principle says that if $\{K\varphi, \sim KK\varphi\}$ is consistent, then $\{K\varphi, \sim K\varphi\}$ is consistent.

[4]See especially, Williamson (2000) for some reasons to reject the KK principle.

[5]Hintikka (1962, 106).

[6]Ibid, 54.

Partition Models

Despite Hintikka's apparently decisive argument against the S5 principle, later theorists applying epistemic logic and semantics, both in theories of distributive computer systems and in game theory assumed that S5 was the right logic for (an idealized concept of) knowledge, and they developed semantic models that seem to support that decision. But while such models, properly interpreted, have their place, I will argue that they have conflated knowledge and belief in a way that has led to some conceptual confusion, and that they have abstracted away from some interesting problems within their intended domains of application that more general models might help to clarify. But before getting to this issue, let me first take note of another way that more recent theorists have modified, or generalized, Hintikka's original theory.

Hintikka's early models were models of the knowledge of a single knower, but much of the later interest in formal epistemic models derives from a concern with situations in which there are multiple knowers who may know or be ignorant about the knowledge and ignorance of the others. While Hintikka's early work did not give explicit attention to the interaction of different knowers, the potential to do so is implicit in his theory. Both the logic and the semantics of the knowledge of a single knower generalize in a straightforward way to a model for multiple knowers. One needs only a separate knowledge operator for each knower, and in the semantics, a separate relation of epistemic accessibility for each knower that interprets the operator. One can also introduce, for any group of knowers, an operator for the *common* knowledge shared by the member of the group, where a group has common knowledge that φ if and only if all know that φ, all know that all know that φ, all know that all know that all know, etc. all the way up. The semantics for the common knowledge operator is interpreted in terms of an accessibility relation that is definable in terms of the accessibility relations for the individual knowers: the common-knowledge accessibility relation for a group G is the transitive closure of the set of epistemic accessibility relations for the members of that group.[7] If R^G is this relation, then the knowers who are members of G have common knowledge that φ (in possible world x) iff φ is true in all possible worlds that are R^G related to world x. The generalization to multiple knowers and to common knowledge, works the same way, whatever assumptions one makes about the accessibility relation, and one can define notions of common belief in an exactly analogous way. The properties of the accessibility relations for common knowledge and common belief will derive from the properties of the individual accessibility relations, but they won't necessarily be the same as the properties of the individual

[7]More precisely, if R^i is the accessibility relation for knower i, then the common-knowledge accessibility relation for a group G is defined as follows; xR^Gy iff there is a sequence of worlds, $z_1, \ldots z_n$ such that $z_1 = x$ and $z_n = y$ and for all j between 1 and n−1, there is a knower $i \in G$, such that $z_jR^i\, z_{j+1}$.

accessibility relations. (Though if the logic of knowledge is S4 or S5, then the logic of common knowledge will also be S4 or S5, respectively).

Theoretical computer scientists have used the logic and semantics for knowledge to give abstract descriptions of distributed computer systems (such as office networks or email systems) that represent the distribution and flow of information among the components of the system. For the purpose of understanding how such systems work and how to design protocols that permit them to accomplish the purposes for which they are designed, it is useful to think of them as communities of interacting rational agents who use what information they have about the system as a whole to serve their own interests, or to play their part in a joint project. And it is useful in turn for those interested in understanding the epistemic states of rational agents to think of them in terms of the kind of simplified models that theoretical computer scientists have constructed.

A distributed system consists of a set of interconnected components, each capable of being in a range of local states. The way the components are connected, and the rules by which the whole system works, constrain the configurations of states of the individual components that are possible. One might specify such a system by positing a set of n components and possible local states for each. One might also include a component labeled "nature" whose local states represent information from outside the system proper. *Global* states will be n-tuples of local states, one for each component, and the model will also specify the set of global states that are *admissible*. Admissible global states are those that are compatible with the rules governing the way the components of the system interact. The admissible global states are the possible worlds of the model. This kind of specification will determine, for each local state that any component might be in, a set of global states (possible worlds) that are compatible with the component being in that local state. This set will be the set of epistemically possible worlds that determines what the component in that state knows about the system as a whole.[8] Specifically, if 'a' and 'b' denote admissible global states, and 'a_i' and 'b_i' denote the ith elements of a and b, respectively (the local states of component i.), then global world-state b is epistemically accessible (for i) to global world-state b if and only if $a_i = b_i$. So, applying the standard semantic rule for the knowledge operator, component (or knower) i will know that φ, in possible world a, if and only if φ is true

[8] A more complex kind of model would specify a set of admissible *initial* global states, and a set of transition rules taking global states to global states. The possible worlds in this kind of model are the admissible global *histories*—the possible ways that the system might evolve. In this kind of model, one can represent the distribution of information, not only about the current state of the system, but also about how it evolved, and where it is going. In the more general model, knowledge states are time-dependent, and the components may have or lack information not only about which possible world is actual, but also about where (temporally) it is in a given world. The dynamic dimension, and the parallels with issues about indexical knowledge and belief, are part of the interest of the distributed systems models, but I will ignore these issues here.

in all possible worlds in which i has the same local state that it has in world-state a. One knows that φ if one's local state carries the information that φ.[9]

Now it is obvious that this epistemic accessibility relation is an equivalence relation, and so the logic for knowledge in a model of this kind is S5. Each of the epistemic accessibility relations partitions the space of possible worlds, and the cross-cutting partitions give rise to a simple and elegant model of common knowledge, also with an S5 logic. Game theorists independently developed this kind of partition model of knowledge and have used such models to bring out the consequences of assumptions about common knowledge. For example, it can be shown that, in certain games, players will always make certain strategy choices when they have common knowledge that all players are rational. But as we have seen, Hintikka gave reasons for rejecting the S5 logic for knowledge, and the reasons seemed to be decisive. It seems clear that a consistent and epistemically responsible agent might take herself to know that φ in a situation in which φ was in fact false. Because knowledge implies truth, it would be false, in such a case, that the agent knew that φ, but the agent could not know that she did not know that φ without having inconsistent beliefs. If such a case is possible, then there will be counterexamples to the S5 principle, $\sim K\varphi \rightarrow K \sim K\varphi$. That is, the S5 principles require that rational agents be immune to error. It is hard to see how any theory that abstracts away from the possibility of error could be relevant to epistemology, an enterprise that begins with skeptical arguments using scenarios in which agents are systematically mistaken and that seeks to explain the relation between knowledge and belief, presupposing that these notions do not coincide.

Different theorists have different purposes, and it is not immediately obvious that the models of knowledge that are appropriate to the concerns of theoretical computer scientists and game theorists need be relevant to issues in epistemology. But I think that the possibility of error, and the differences between knowledge and belief are relevant to the intended domains of application of those models, and that some of the puzzles and problems that characterize epistemology are reflected in problems that may arise in applying those theories.

As we all know too well, computer systems sometimes break down or fail to behave as they were designed to behave. In such cases, the components of a distributed system will be subject to something analogous to error and illusion. Just as the epistemologist wants to explain how and when an agent knows some things even when he is in error about others, and is interested in methods of detecting and avoiding error, so the theoretical computer scientist is interested in the way that the components of a system can avoid and detect faults, and can continue to

---

[9] Possible worlds, on this way of formulating the theory, are not primitive points, as they are in the usual abstract semantics, but complex objects—sequences of local states. But an equivalent formulation might begin with a given set of primitive (global) states, together with a set of equivalence relations, one for each knower, and one for "nature". The local states could then be defined as the equivalence classes.

function appropriately even when conditions are not completely normal. To clarify such problems, it is useful to distinguish knowledge from something like belief.

The game theorist, or any theorist concerned with rational action, has a special reason to take account of the possibility of false belief, even under the idealizing assumption that in the actual course of events, everyone's beliefs are correct. The reason is that decision theorists and game theorists need to be concerned with causal or counterfactual possibilities, and to distinguish them from epistemic possibilities. When I deliberate, or when I reason about why it is rational to do what I know that I am going to do, I need to consider possible situations in which I make alternative choices. I know, for example, that it would be irrational to cooperate in a one-shot prisoners' dilemma because I know that in the counterfactual situation in which I cooperate, my payoff is less than it would be if I defected. And while I have the capacity to influence my payoff (negatively) by making this alternative choice, I could not, by making this choice, influence your prior beliefs about what I will do; that is, your prior beliefs will be the same, in the counterfactual situation in which I make the alternative choice, as they are in the actual situation Since you take yourself (correctly, in the actual situation) to know that I am rational, and so that I will not cooperate, you therefore also take yourself to know, in the counterfactual situation I am considering, that I am rational, and so will not cooperate. But in that counterfactual situation, you are wrong—you have a false belief that you take to be knowledge. There has been a certain amount of confusion in the literature about the relation between counterfactual and epistemic possibilities, and this confusion is fed, in part, by a failure to make room in the theory for false belief.[10]

Even in a context in which one abstracts away from error, it is important to be clear about the nature of the idealization, and there are different ways of understanding it that are sometimes confused. But before considering the alternative ways of making the S5 idealization, let me develop the contrast between knowledge and belief, and the relation between them, in a more general setting.

Belief and Knowledge

Set aside the S5 partition models for the moment, and consider, from a more neutral perspective, the logical properties of belief, and the relation between belief and knowledge. It seems reasonable to assume, at least in the kind of idealized context we are in, that agents have introspective access to their beliefs: if they believe that φ, then they know that they do, and if they do not, then they know that they do not. (The S5, "negative introspection" principle, $\sim K\varphi \rightarrow K \sim K\varphi$, was problematic for knowledge because it is in tension with the fact that knowledge implies truth, but the corresponding principle for belief does not face this problem.) It also seems reasonable to assume that knowledge implies belief. Given the fact that our idealized

[10]These issues are discussed in Stalnaker (1996).

believers are logically omniscient, we can assume, in addition, that their beliefs will be consistent. Finally, to capture the fact that our intended concept of belief is a strong one—subjective certainty—we assume that believing implies believing that one knows. So our logic of knowledge and belief should include the following principles in addition to those of the logic S4:

(PI)	$\vdash B\varphi \rightarrow KB\varphi$	positive introspection
(NI)	$\vdash \sim B\varphi \rightarrow K \sim B\varphi$	negative introspection
(KB)	$\vdash K\varphi \rightarrow B\varphi$	knowledge implies belief
(CB)	$\vdash B\varphi \rightarrow \sim B \sim \varphi$	consistency of belief
(SB)	$\vdash B\varphi \rightarrow BK\varphi$	strong belief

The resulting combined logic for knowledge and belief yields a pure belief logic, KD45, which is validated by a doxastic accessibility relation that is serial, transitive and euclidean.[11] More interestingly, one can prove the following equivalence theorem: $\vdash B\varphi \leftrightarrow MK\varphi$ (using 'M' as the epistemic possibility operator, '∼K∼'). This equivalence permits a more economical formulation of the combined belief-knowledge logic in which the belief operator is defined in terms of the knowledge operator. If we substitute 'MK' for 'B' in our principle (CB), we get $MK\varphi \rightarrow KM\varphi$, which, if added to S4 yields the logic of knowledge, S4.2. All of the other principles listed above (with 'MK' substituted for 'B') are theorems of S4.2, so this logic of knowledge by itself yields a combined logic of knowledge and belief with the appropriate properties.[12]

The assumptions that are sufficient to show the equivalence of belief with the epistemic possibility of knowledge (one believes that φ, in the strong sense, if and only if it is compatible with one's knowledge that one knows that φ) might also be made for a concept of *justified* belief, although the corresponding assumptions will be more controversial. Suppose (1) one assumes that justified belief is a necessary condition for knowledge, and (2) one adopts an *internalist* conception of justification that supports the positive and negative introspection conditions (if one has justified belief that φ, one knows that one does, and if one does not, one knows that one does not), and (3) one assumes that since the relevant concept of belief is a strong one, one is justified in believing that φ if and only if one is justified in believing that one knows that φ. Given these assumptions, justified belief will also

[11]KD45 adds to the basic modal system K the axioms (D), which is our CB, (4) $B\varphi \rightarrow BB\varphi$, which follows immediately from our (PI) and (KB), and (5) $\sim B\varphi \rightarrow B \sim B\varphi$, which follows immediately from (NI) and (KB). The necessitation rule for B (If $\vdash \varphi$, then $\vdash B\varphi$) and the distribution principle ($B(\varphi \rightarrow \psi) \rightarrow (B\varphi \rightarrow B\psi)$) can both be derived from our principles.

[12]The definability of belief in terms of knowledge, and the point that the assumptions about the relation between knowledge and belief imply that the logic of knowledge should be S4.2, rather than S4, were first shown by Wolfgang Lenzen. See his classic monograph, *Recent Work in Epistemic Logic. Acta Philosophica Fennica* **30** (1978). North-Holland, Amsterdam.

coincide with the epistemic possibility that one knows, and so belief and justified belief will coincide. The upshot is that for an internalist, a divergence between belief (in the strong sense) and justified belief would be a kind of internal inconsistency. If one is not fully justified in believing φ, one knows this, and so one knows that a necessary condition for knowledge that φ is lacking. But if one believes that φ, in the strong sense, then one believes that one knows it. So one both knows that one lacks knowledge that φ, and believes that one has knowledge that φ.

The usual constraint on the accessibility relation that validates S4.2 is the following convergence principle (added to the transitivity and reflexivity conditions): if xRy and xRz, then there is a w such that yRw and zRw. But S4.2 is also sound and complete relative to the following stronger convergence principle: for all x, there is a y such that for all z, if xRz, then zRy. The weak convergence principle (added to reflexivity and transitivity) implies that for any *finite* set of worlds accessible to x, there is a single world accessible with respect to all of them. The strong convergence principle implies that there is a world that is accessible to *all* worlds that are accessible to x. The semantics for our logic of knowledge requires the stronger convergence principle.[13]

Just as, within the logic, one can define belief in terms of knowledge, so within the semantics, one can define a doxastic accessibility relation for the derived belief operator in terms of the epistemic accessibility relation. If 'R' denotes the epistemic accessibility relation and 'D denotes the doxastic relation, then the definition is as follows: $xDy =_{df} (z)\ (xRz \rightarrow zRy)$. Assuming that R is transitive, reflexive and strongly convergent, it can be shown that D will be serial, transitive and euclidean—the constraints on the accessibility relation that characterize the logic KD45.

One can also define, in terms of D, and so in terms of R, a third binary relation on possible worlds that is relevant to describing the epistemic situation of our ideal knower: Say that two possible worlds x and y are *epistemically indistinguishable* to an agent (xEy) if and only if she has exactly the same beliefs in world x as she has in world y. That is, $xEy =_{df} (z)\ (xDz \leftrightarrow yDz)$. E is obviously an equivalence relation, and so any modal operator interpreted in the usual way in terms of E would be an S5 operator But while this relation is definable in the semantics in terms of the epistemic accessibility relation, we cannot define, in the object language with just the knowledge operator, a modal operator whose semantics is given by this accessibility relation.

So the picture that our semantic theory paints is something like this: For any given knower i and possible world x, there is, first, a set of possible worlds that are subjectively indistinguishable from x, to i (those worlds that are E-related to x); second, there is a subset of that set that includes just the possible worlds compatible with what i *knows* in x (those worlds that are R-related to x); third, there is a subset of

[13]The difference between strong and weak convergence does not affect the *propositional* modal logic, but it will make a difference to the quantified modal logic. The following is an example of a sentence that is valid in models satisfying strong convergence (along with transitivity and reflexivity) but not valid in all models satisfying weak convergence: $MK\ ((x)\ (MK\varphi \rightarrow \varphi))$.

that set that includes just the possible worlds that are compatible with what i *believes* in x (those worlds that are D-related to x). The world x itself will necessarily be a member of the outer set and of the R-subset, but will not necessarily be a member of the inner D-subset. But if x is itself a member of the inner D-set (if world x is itself compatible with what i believes in x), then the D-set will coincide with the R-set.

Here is one way of seeing this more general theory as a generalization of the distributive systems models, in which possible world-states are sequences of local states: one might allow *all* sequences of local states (one for each agent) to count as possible world-states, but specify, for each agent, a subset of them that are *normal*—the set in which the way that agent interacts with the system as a whole conforms to the constraints that the system conforms to when it is functioning as it is supposed to function. In such models, two worlds, x and y, will be subjectively indistinguishable, for agent i ($xE_i\ y$), whenever $x_i = y_i$ (so the relation that was the epistemic accessibility relation in the unreconstructed S5 distributed systems model is the subjective indistinguishability relation in the more general models). Two worlds are related by the doxastic accessibility relation ($xD_i\ y$) if and only if $x_i = y_i$, *and in addition, y is a normal world, with respect to agent i.*[14] This will impose the right structure on the D and E relations, and while it imposes some constraints on the epistemic accessibility relation, it leaves it underdetermined. We might ask whether R can be defined in a plausible way in terms of the components of the model we have specified, or whether one might add some independently motivated components to the definition of a model that would permit an appropriate definition of R. This question is a kind of analogue of the question asked in the more traditional epistemological enterprise—the project of giving a definition of knowledge in terms of belief, truth, justification, and whatever other normative and causal concepts might be thought to be relevant. Transposed into the model theoretic framework, the traditional problem of adding to true belief further conditions that together are necessary and sufficient for knowledge is the problem of extending the doxastic accessibility relation to a reflexive relation that is the right relation (at least in the idealized context) for the interpretation of a knowledge operator. In the remainder of this paper, I will consider several ways that this might be done, and at the logics of knowledge that they validate.

[14]We observed in note 7 that an equivalent formulation of the S5 distributed systems models would take the global world-states as primitive, specifying an equivalence relation for each agent, and defining local states as equivalence classes of global states. In an equivalent formulation of this kind of the more general theory, the assumption that every sequence of local states is a possible world will be expressed by a recombination condition: that for every sequence of equivalence classes (one for each agent) there is a possible world that is a member of their intersection. I have suggested that a recombination condition of this kind should be imposed on game theoretic models (where the equivalence classes are types, represented by probability functions), defending it as a representation of the conceptual independence of the belief states of different agents.

Partition Models and the Basic Theory

One extreme way of defining the epistemic accessibility relation in terms of the resources of our models is to identify it with the relation of subjective indistinguishability, and this is one way that the S5 partition models have implicitly been interpreted. If one simply assumes that the epistemic accessibility relation is an equivalence relation, this will suffice for a collapse of our three relations into one. Subjective indistinguishability, knowledge, and belief will all coincide. This move imposes a substantive condition on knowledge, and so on belief, when it is understood in the strong sense as belief that one knows, a condition that is appropriate for the skeptic who thinks that we are in a position to have genuine knowledge only about our own internal states—states about which we cannot coherently be mistaken. On this conception of knowledge, one can have a false belief (in the strong sense) only if one is internally inconsistent, and so this conception implies a bullet-biting response to the kind of argument that Hintikka gave against the S5 logic for knowledge. Hintikka's argument was roughly this: S5 validates the principle that any proposition that is in fact false, is known by any agent to be compatible with his knowledge, and this is obviously wrong: The response suggested by the conception of knowledge that identifies knowledge with subjective indistinguishability is that if we assume that all we can know is how things seem to us, and also assume that we are infallible judges of the way things seem to us, then it will be reasonable to conclude that we are in a position to know, of anything that is in fact false, that we do not know it.

There is a less radical way to reconcile our basic theory of knowledge and belief with the S5 logic and the partition models. Rather than making more restrictive assumptions about the concept of knowledge, or about the basic structure of the model, one may simply restrict the intended domain of application of the theory to cases in which the agent in question has, in fact, only true beliefs. On this way of understanding the S5 models, the model theory does not further restrict the relations between the three accessibility relations, but instead assumes that the *actual* world of the model is a member of the inner D-set.[15] This move does not provide us with

[15]In most formulations of a possible-worlds semantics for propositional modal logic, a *frame* consists simply of a set of worlds and an accessibility relation. A model on a frame determines the truth values of sentences, relative to each possible world. On this conception of a model, one cannot talk of the truth of a sentence in a model, but only of truth *at a world* in a model. Sentence validity is defined, in formulations of this kind, as truth in all worlds in all models. But in some formulations, including in Kripke's original formal work, a frame (or model structure, as Kripke called it at the time) included, in addition to a set of possible worlds and an accessibility relation, a designated possible world—the actual world of the model. A sentence is true in a model if it is true in the designated actual world, and valid if true in all models. This difference in formulation was a minor detail in semantic theories for most of the normal modal logics, since any possible world of a model might be the designated actual world without changing anything else. So the two ways of defining sentence validity will coincide. But the finer-grained definition of a frame allows for theories in which the constraints on R, and the semantic rules for operators, make reference to the

a way to define the epistemic accessibility relation in terms of the other resources of the model; but what it does is to stipulate that the actual world of the model is one for which the epistemic accessibility relation is determined by the other components. (That is, the set of worlds y that are epistemically accessible to the actual world is determined) Since the assumptions of the general theory imply that all worlds outside the D-sets are epistemically inaccessible to worlds within the D-sets, and that all worlds within a given D-set are epistemically accessible to each other, the assumption that the actual world of the model is in a D-set will determine the R-set for the actual world, and will validate the logic S5.

So long as the object language that is being interpreted contains just one modal operator, an operator representing the knowledge of a single agent, the underdetermination of epistemic accessibility will not be reflected in the truth values in a model of any expressible proposition. Since all possible worlds outside of any D-set will be invisible to worlds within it, one could drop them from the model (taking the set of all possible worlds to be those R-related to the actual world) without affecting the truth values (at the actual world) of any sentence. This generated submodel will be a simple S5 model, with a universal accessibility relation. But as soon as one enriches the language with other modal and epistemic operators, the situation changes. In the theory with two or more agents, even if one assumes that all agents have only true beliefs, the full S5 logic will not be preserved. The idealizing assumption will imply that Alice's beliefs coincide with her knowledge (in the actual world), and that Bob's do as well, but it will not follow that Bob knows (in the actual world) that Alice's beliefs coincide with her knowledge. To validate the full S5 logic, in the multiple agent theory, we need to assume that it is not just true, but common knowledge that everyone has only true beliefs. This stronger idealization is needed to reconcile the partition models, used in both game theory and in distributed systems theory, with the general theory that allows for a distinction between knowledge and belief. But even in a context in which one makes the strong assumption that it is common knowledge that no one is in error about anything, the possible divergence of knowledge and belief, and the failure of the S5 principles to be *necessarily* true will show itself when the language of knowledge and common knowledge is enriched with non-epistemic modal operators, or in semantic models that represent the interaction of epistemic and non-epistemic concepts. In game theory, for example, an adequate model of the playing of a game must represent not just the epistemic possibilities for each of the players, but also the capacities of players to make each of the choices that are open to that player, even when it is known that the player will not make some of those choices. One might assume that it is common knowledge that Alice will act rationally in a certain game, and it might be that it is known that Alice would be acting irrationally if she chose option X. Nevertheless, it would distort the representation of the game to deny that Alice has the option of choosing action

actual world of the model. In such theories, truth in all worlds in all models may diverge from truth in all models, allowing for semantic models of logics that fail to validate the rule of necessitation.

X, and the counterfactual possibility in which she exercises that option may play a role in the deliberations of both Alice and the other players, whose knowledge that Alice will not choose option X is based on their knowledge of what she knows would happen if she did. So even if one makes the idealizing assumption that all agents have only true beliefs, or that it is common belief that everyone's beliefs are true, one should recognize the more general structure that distinguishes belief from knowledge, and that distinguishes both of these concepts from subjective indistinguishability. In the more general structure that recognizes these distinctions, the epistemic accessibility relation is underdetermined by the other relations.

Minimal and Maximal Extensions

So our task is to say more about how to extend the relation D of doxastic accessibility to a relation R of epistemic accessibility. We know, from the assumption that knowledge implies belief, that in any model meeting our basic conditions on the relation between knowledge and belief, R will be an extension of D (for all x and y, if xDy, then xRy), and we know from the assumption that knowledge implies truth that the extension will be to a reflexive relation. We know by the assumption that belief is strong belief (belief that one knows) that R coincides with D, within the D-set (for all x and y, if xDx, then xRy if and only if xDy). What remains to be said is what determines, for a possible world x that is *outside* of a D-set, which other possible worlds outside that D-set are epistemically accessible to x. If some of my beliefs about what I know are false, what can be said about other propositions that I think that I know?

The assumptions of the neutral theory put clear upper and lower bounds on the answer to this question, and two ways to specify R in terms of the other resources of the model are to make the minimal or maximal extensions. The *minimal* extension of D would be the reflexive closure of D. On this account, the set of epistemically possible worlds for a knower in world x will be the set of doxastically accessible worlds, plus x. To make this minimal extension is to adopt the true belief analysis of knowledge, or in case one is making the internalist assumptions about justified belief, it would be to adopt the justified true belief analysis. The logic of true belief, S4.4, is stronger than S4.2, but weaker than S5.[16] The true belief analysis has its defenders, but most will want to impose stronger conditions on knowledge, which in our setting means that we need to go beyond the minimal extension of R.

It follows from the positive and negative introspection conditions for belief that for any possible world x, all worlds epistemically accessible to x will be subjectively indistinguishable from x (for all x and y, if xRy, then xEy) and this sets the upper bound on the extension of D to R. To identify R with the *maximal* admissible

[16]See the appendix for a summary of all the logics of knowledge discussed, their semantics, and the relationships between them.

extension is to define it as follows: $xRy =_{df}$ either (xDx and xDy) or (not xDx and xEy). This account of knowledge allows one to know things that go beyond one's internal states only when *all* of one's beliefs are correct. The logic of this concept of knowledge, S4F, is stronger than S4.2, but weaker than the logic of the minimal extension, S4.4. The maximal extension would not provide a plausible account of knowledge in general, but it might be the appropriate idealization for a certain limited context. Suppose one's information all comes from a single source (an oracle), who you presume, justifiably, to be reliable. Assuming that all of its pronouncements are true, they give you knowledge, but in possible worlds in which any one of its pronouncements is false, it is an unreliable oracle, and so nothing it says should be trusted. This logic, S4F, has been used as the underlying logic of knowledge in some theoretical accounts of a nonmonotonic logic. Those accounts don't provide an intuitive motivation for using this logic, but I think a dynamic model, with changes in knowledge induced by a single oracle who is presumed to be reliable, can provide a framework that makes intuitive sense of these nonmonotonic theories.[17]

Belief Revision and the Defeasibility Analysis

Any attempt to give an account of the accessibility relation for knowledge that falls between the minimal and maximal admissible extensions of the accessibility relation for belief will have to enrich the resources of the theory. One way to do this, a way that fits with one of the familiar strategies for responding to the Gettier counterexamples to the justified true belief analysis, is to add to the semantics for belief a theory of belief revision, and then to define knowledge as belief (or justified belief) that is stable under any potential revision by a piece of information that is in fact true. This is the defeasibility strategy followed by many of those who responded to Gettier's challenge: the idea was that the fourth condition (to be added to justified true belief) should be a requirement that there be no "defeater"—no true proposition that, if the knower learned that it was true, would lead her to give up the belief, or to be no longer justified in holding it.[18] There was much discussion in the post-Gettier literature, about exactly how defeasibility should be defined, but in the context of our idealized semantic models, supplemented by a semantic version of the standard belief revision theory, a formulation of a defeasibility analysis of knowledge is straightforward. First, let me sketch the outlines of the so-called AGM theory of belief revision,[19] and then give the defeasibility analysis.

[17] See Schwarz and Truszczyński (1992).

[18] See Lehrer and Paxson (1969) and Swain (1974) for two examples.

[19] See Gärdenfors (1988) for a survey of the basic ideas of the AGM belief revision theory, and Grove (1988) for a semantic formulation of the theory.

The belief revision project is to define, for each belief state (the prior belief state), a function taking a proposition (the potential new evidence) to a posterior belief state (the state that would be induced in one in the prior state by receiving that information as one's total new evidence). If belief states are represented by sets of possible worlds (the doxastically accessible worlds), and if propositions are also represented by sets of possible worlds, then the function will map one set of worlds (the prior belief set) to another (the posterior belief set), as a function of a proposition. Let **B** be the set representing the prior belief state, φ the potential new information, and $\mathbf{B}(\varphi)$ the set representing the posterior state. Let **E** be a superset of **B** that represents the set of all possible worlds that are potential candidates to be compatible with some posterior belief state. The formal constraints on this function are then as follows: (1) $\mathbf{B}(\varphi) \rightarrow \varphi$ (the new information is believed in the posterior belief state induced by that information). (2) If $\varphi \cap \mathbf{B}$ is nonempty, then $\mathbf{B}(\varphi) = \varphi \cap \mathbf{B}$ (If the new information is compatible with the prior beliefs, then nothing is given up—the new information is simply added to the prior beliefs.). (3) $\mathbf{B}(\varphi)$ is nonempty if and only if $\varphi \cap \mathbf{E}$ is non-empty (the new information induces a consistent belief state whenever that information is compatible with the knower being in the prior belief state. and only then). (4) If $\mathbf{B}(\varphi) \cap \psi$ is nonempty, then $\mathbf{B}(\varphi \cap \psi) = \mathbf{B}(\varphi) \cap \psi$. The fourth conditions is the only one that is not straightforward. What it says is that if ψ is compatible, not with Alice's *prior* beliefs, but with the *posterior* beliefs that she would have if she learned φ, then what Alice should believe upon learning the *conjunction* of φ and ψ should be the same as what she would believe if she first learned φ, and then learned ψ. This condition can be seen as a generalization of condition (2), which is a modest principle of methodological conservativism (Don't give up any beliefs if your new information is compatible with everything you believe). It is also a kind of path independence principle. The order in which Alice receives two compatible pieces of information should not matter to the ultimate belief state.[20]

To incorporate the standard belief revision theory into our models, add, for each possible world x, and for each agent i, a function that, for each proposition φ, takes i's belief state in x, $B_{x,i} = \{y\text{: } xD_iy\}$, to a potential posterior belief state, $B_{x,i}(\varphi)$. Assume that each of these functions meets the stated conditions, where the set **E**, for the function $B_{x,i}$ is the set of possible worlds that are subjectively indistinguishable from x to agent i. We will also assume that if x and y are subjectively indistinguishable to i, then i's belief revision function will be the same in x as it is in y. This is to extend the positive and negative introspection assumptions

[20]The third principle is the least secure of the principles; there are counterexamples that suggest that it should be given up. See Stalnaker (1994) for a discussion of one. The defeasibility analysis of knowledge can be given with either the full AGM belief revision theory, or with the more neutral one that gives up the fourth condition.

to the agent's belief revision policies. Just as she knows what she believes, so she knows how she would revise her beliefs in response to unexpected information.[21]

We have added some structure to the models, but not yet used it to interpret anything in the object language that our models are interpreting. Suppose our language has just belief operators (and not knowledge operators) for our agents, and only a doxastic accessibility relation, together with the belief revision structure, in the semantics The defeasibility analysis suggests that we might add, for knower i, a knowledge operator with the following semantic rule: $K_i\varphi$ is true in world x iff $B_i\ \varphi$ is true in x, and for any proposition ψ that is true in x, $\mathbf{B}_{x,i}(\psi) \rightarrow \varphi$. Alice knows that φ if and only if, for any ψ that is true, she would still believe that φ after learning that ψ. Equivalently, we might define an epistemic accessibility relation in terms of the belief revision structure, and use it to interpret the knowledge operator in the standard way. Let us say that $xR_i\ y$ if and only if there exists a proposition φ such that $\{x, y\} \subseteq \varphi$, and $y \notin \mathbf{B}_{x,i}(\varphi)$. The constraints imposed on the function $\mathbf{B}_{x,i}$ imply that this relation will extend the doxastic accessibility relation D_i, and that it will fall between our minimal and maximal constraints on this extension. The relation will be transitive, reflexive, and strongly convergent, and so meet all the conditions of our basic theory. It will also meet an additional condition: it will be weakly connected (if xRy and xRy, then either yRz, or zRy). This defeasibility semantics will validate a logic of knowledge, S4.3, that is stronger than S4.2, but weaker than either S4F or S4.4.[22]

So a nice, well behaved version of our standard semantics for knowledge falls out of the defeasibility analysis, yielding a determinate account, in terms of the belief revision structure, of the way that epistemic accessibility extends doxastic accessibility. But I doubt that this is a plausible account of knowledge in general, even in our idealized setting. The analysis is not so demanding as the S4F theory, but like that theory, it threatens to let any false belief defeat too much of our knowledge, even knowledge of facts that seem unrelated. Consider the following kind of example: Alice take herself to know that the butler didn't do it, since she saw him in the drawing room, miles away from the scene of the crime, at the time

---

[21]It should be noted that even with the addition of the belief revision structure to the epistemic models I have been discussing, they remain static models. A model of this kind represents only the agents' beliefs at a fixed time, together with the policies or dispositions to revise her beliefs that she has at that time. The model does not represent any actual revisions that are made when new information is actually received. The models can be enriched by adding a temporal dimension to represent the dynamics, but doing so requires that the knowledge and belief operators be time indexed, and that one be careful not to confuse belief changes that are changes of mind with belief changes that result from a change in the facts. (I may stop believing that the cat is on the mat because I learn that what I thought was the cat was the dog, or I may stop believing it because the cat gets up and leaves, and the differences between the two kinds of belief change are important).

[22]In game theoretic models, the strength of the assumption that there is common knowledge of rationality depends on what account one gives of knowledge (as well as on how one explains rationality). Some backward induction arguments, purporting to show that common knowledge of rationality suffices to determine a particular course of play (in the centipede game, or the iterated prisoners' dilemma, for example) can be shown to work with a defeasibility account of knowledge, even if they fail on a more neutral account. See Stalnaker (1996).

of the murder (or so she thinks). She also takes herself to know there is zucchini planted in the garden, since the gardener always plants zucchini, and she saw the characteristic zucchini blossoms on the vines in the garden (or so she thinks). As it happens, the gardener, quite uncharacteristically, failed to plant the zucchini this year, and coincidentally, a rare weed with blossoms that resemble zucchini blossoms have sprung up in its place. But it really was the butler that Alice saw in the drawing room, just as she thought. Does the fact that her justified belief about the zucchini is false take away her knowledge about the butler? It is a fact that either it wasn't really the butler in the drawing room, or the gardener failed to plant zucchini. Were Alice to learn just this disjunctive fact, she would have no basis for deciding which of her two independent knowledge claims was the one that was wrong. So it seems that, on the simple defeasibility account, the disjunctive fact is a defeater. The fact that she is wrong about one of her knowledge claims seems to infect other, seemingly unrelated claims. Now it may be right that if Alice was in fact reliably informed that one of her two knowledge claims was false, without being given any information about which, she would *then* no longer know that it was the butler that she saw. But if the mere fact that the disjunction is true were enough to rob her of her knowledge about the butler, then it would seem that almost all of Alice's knowledge claims will be threatened. The defeasibility account is closer than one might have thought to the maximally demanding S4F analysis, according to which we know nothing except how things seem to us unless we are right about everything we believe.

I think that one might plausibly defend the claim that the defeasibility analysis provides a *sufficient* condition for knowledge (in our idealized setting), and so the belief revision structure might further constrain the ways in which the doxastic accessibility relation can be extended to an epistemic accessibility relation. But it does not seem to be a plausible *necessary* and sufficient condition for knowledge. In a concluding section, I will speculate about some other features of the relation between a knower and the world that may be relevant to determining which of his true beliefs count as knowledge.

The Causal Dimension

What seems to be driving the kind of counterexample to the defeasibility analysis that I have considered is the fact that, on this analysis, a belief with a normal and unproblematic causal source could be defeated by the fact that some different source had delivered misinformation about some independent and irrelevant matter. Conditions were normal with respect to the explanation of Alice's beliefs about the butler's presence in the drawing room. There were no anomalous circumstances, either in her perceptual system, or in the conditions in the environment, to interfere with the normal formation of that belief. This was not the case with respect to the explanation of her belief about what was planted in the garden, but that does not seem, intuitively, to be relevant to whether her belief about the butler constituted knowledge. Perhaps the explanation of epistemic accessibility, in the case where conditions are not fully normal, and not all of the agent's beliefs are true, should focus more on the causal sources of beliefs, rather than on how agents would respond to information that they do not in fact receive. This, of course, is a strategy that played a central role in many of the responses to the Gettier challenge. I will describe a very simple model of this kind, and then mention some of the problems that arise in making the simple model even slightly more realistic.

Recall that we can formulate the basic theory of belief this way: a relation of subjective indistinguishability, for each agent, partitions the space of possibilities, and there will be a nonempty subset of each partition cell which is the set of worlds compatible with what the agent believes in the worlds in that cell. We labeled those worlds the normal one, since they are the worlds in which everything determining the agent's beliefs is functioning normally, all of the beliefs are true in those worlds, and belief and knowledge coincide. The problem was to say what the agent knows in the worlds that lie outside of the normal set. One idea is to give a more detailed account of the normal conditions in terms of the way the agent interacts with the world he knows about; we start with a crude and simple model of how this might be done. Suppose our agent receives his information from a fixed set of independent sources—different informants who send messages on which the agent's knowledge is based. The "informants" might be any kind of input channel. The agent might or might not be in a position to identify or distinguish different informants. But we assume that the informants are, in fact, independent in the sense that there may be a fault or corruption that leads one informant to send misinformation (or more generally, to be malfunctioning) while others are functioning normally. So we might index normal conditions to the informant, as well as to the agent. For example, if there are two informants, there will be a set of worlds that is normal with respect to the input channel for informant one, and an overlapping set that is normal for informant two. Possible worlds in which conditions are fully normal will be those in which all the input channels are functioning normally—the worlds in the intersection of the two sets.[23] This intersection will be the set compatible with the agent's beliefs, the set where belief and knowledge coincide. If conditions are abnormal with respect to informant one (if that information channel is corrupted) then while that informant may influence the agent's beliefs, it won't provide any knowledge. But if the other channel is uncorrupted, the beliefs that have it as their sole source will be knowledge. The formal model suggested by this picture is a simple and straightforward generalization of the S4F model, the maximal admissible extension of the doxastic accessibility relation. Here is a definition of the epistemic accessibility relation for the S4F semantics, where **E**(x) is the set of worlds subjectively indistinguishable from x (to the agent in question) and **N**(x) is the subset of that set where conditions are normal (the worlds compatible with what the agent believes in world x): xRy if and only if x $\in$ **N**(x) and y $\in$ **N**(x), or x $\notin$ **N**(x) and y $\in$ **E**(x). In the generalization, there is a finite set of normal-conditions properties, $\mathbf{N}^j$, one for each informant j, that each determines a subset of **E**(x), $\mathbf{N}^j$(x), where conditions are functioning normally in the relation between that informant and the agent. The definition of R will say that the analogue of the

[23]It will be required that the intersection of all the normal-conditions sets be nonempty.

S4F condition holds for each $\mathbf{N}^j$. The resulting logic (assuming that the number of independent information channels or informants is unspecified) will be the same as the basic theory: S4.2.

Everything goes smoothly if we assume that information comes from discrete sources, even if the agent does not identify or distinguish the sources. Even when the agent makes inferences from beliefs derived from multiple sources, some of which may be corrupt and other not, the model will determine which of his true beliefs count as knowledge, and which do not. But in even a slightly more realistic model, the causal explanations for our beliefs will be more complex, with different sources not wholly independent, and deviations from normal conditions hard to isolate. Beliefs may have multiple interacting sources—there will be cases of overdetermination and preemption There will be problems about how to treat cases where a defect in the system results, not in the reception of misinformation, but from the failure to receive a message. (It might be that had the system been functioning normally, I would have received information that would have led me to give up a true belief.) And along with complicating the causal story, one might combine this kind of model with a belief revision structure, allowing one to explore the relation between beliefs about causal structure and policies for belief revision, and to clarify the relation between the defeasibility analysis and an account based on the causal strategy. The abstract problems that arise when one tries to capture a more complex structure will reflect, and perhaps help to clarify, some of the patterns in the counterexamples that arose in the post-Gettier literature. Our simple model avoids most of these problems, but it is a start that may help to provide a context for addressing them.

Appendix

To give a very concise summary of all the logics of knowledge I have discussed, and their corresponding semantics, I will list, first the alternative constraints on the accessibility relation, and then the alternative axioms. Then I will distinguish the different logics, and the semantic conditions that are appropriate to them in terms of the items on the lists.

Conditions on R:

(Ref)	$(x)xRx$
(Tr)	$(x)(y)(z)((xRy \,\&\, yRz) \rightarrow xRz)$
(Cv)	$(x)(y)(z)((xRy \,\&\, yRz) \rightarrow (\exists w)(yRw \,\&\, zRw))$
(SCv)	$(x)(\exists z)(y)(xRy \rightarrow yRz)$
(WCt)	$(x)(y)(z)((xRy \,\&\, xRz) \rightarrow (yRz \vee zRy))$
(F)	$(x)(y)\big(xRy \rightarrow ((z)(xRz \rightarrow yRz) \vee (z)(xRz \rightarrow zRy))$
(TB)	$(x)(y)(xRy \,\&\, x \neq y)(z)(xRz \rightarrow zRy)\big)$
(E)	$(x)(y)(z)((xRy \,\&\, xRz) \rightarrow yRz)$

Axioms:

(T)	$K\varphi \rightarrow \varphi$
(4)	$K\varphi \rightarrow KK\varphi$
(4.2)	$MK\varphi \rightarrow KM\varphi$
(4.3)	$(K(\varphi \rightarrow M\psi) \vee K(\psi \rightarrow M\varphi))$
(f)	$((M\varphi \& MK\psi) \rightarrow K(M\varphi \vee \psi))$
(4.4)	$((\varphi \& MK\psi) \rightarrow K(\varphi \vee \psi))$
(5)	$M\varphi \rightarrow KM\varphi$

The logics for knowledge we have considered, and semantic constraints on R relative to which they are sound and complete, are as follows: The logics are of increasing order of strength, the theorems of each including those of the previous logics on the list.

S4	K + T + 4	Ref + Tr		
S4.2	S4 + 4.2	Ref + Tr + SCv	OR	Ref + Tr + Cv
S4.3	S4 + 4.3	Ref + Tr + SCv + WCt	OR	Ref + Tr + WCt
S4F	S4 + f	Ref + Tr + F		
S4.4	S4 + 4.4	Ref + Tr + TB		
S5	S4 + 5	Ref + Tr + E		

In each of the logics of knowledge we have considered, from S4.2 to S4.4, the derived logic of belief, with belief defined by the complex operator MK, will be KD45. (In S4, belief is not definable, since in that logic, the complex operator MK does not satisfy the K axiom, and so is not a normal modal operator. In S5, belief and knowledge coincide, so the logic of belief is S5.) KD45 is K + D + 4 + 5, where D is $(K\varphi \rightarrow M\varphi)$. The semantic constraints are Tr + E + the requirement that the accessibility relation be *serial*: $(x)(\exists y)xRy$.

In a semantic model with multiple knowers, we can add a common knowledge operator, defined in terms of the transitive closure of the epistemic accessibility relations for the different knowers. For any of the logics, from S4 to S4.4, with the corresponding semantic conditions, the logic of *common* knowledge will be S4, and the accessibility relation will be transitive and reflexive, but will not necessarily have any of the stronger properties. If the logic of knowledge is S5, then the logic of common knowledge will also be S5, and the accessibility relation will be an equivalence relation.

References

Battigalli, P., & Bonanno, G. (1999). Recent results on belief, knowledge and the epistemic foundations of game theory. *Research in Economics, 53*, 149–225.

Fagin, R., Halpern, J. Y., Moses, Y., & Vardi, M. (1995). *Reasoning about knowledge*. Cambridge, MA: MIT Press.

Gärdenfors, P. (1988). *Knowledge in flux: Modeling the dynamics of epistemic states*. Cambridge, MA: MIT Press.

Gettier, E. (1963). Is justified true belief knowledge? *Analysis, 6*, 121–123.

Grove, A. (1988). Two modeling for theory change. *Journal of Philosophical Logic, 17*, 157–170.

Hintikka, J. (1962). *Knowledge and belief*. Ithaca: Cornell University Press.

Lehrer, K., & Paxson, T. (1969). Knowledge: Undefeated justified true belief. *The Journal of Philosophy, 66*, 225–237.

Lenzen, W. (1978). *Recent work in epistemic logic* (Acta philosophica Fennica, Vol. 30). Amsterdam: North-Holland.

Schwarz, G., & Truszczyński, M. (1992). Modal logic S4F and the minimal knowledge paradigm. In *Proceedings of the fourth conference on theoretical aspects of reasoning about knowledge* (pp. 184–198). Sam Mateo: Morgan Kaufmann Publishers, Inc.

Stalnaker, R. (1991). The problem of logical omniscience, I. *Synthese, 89*, 425–440. (Reprinted in Stalnaker (1999a), 240–254).

Stalnaker, R. (1994). What is a non-monotonic consequence relation? *Fundamenta Informaticae, 21*, 7–21.

Stalnaker, R. (1996). Knowledge, belief and counterfactual reasoning in games. *Economics and Philosophy, 12*, 133–162.

Stalnaker, R. (1999a). *Context and content: Essays on intentionality in speech and thought*. Oxford: Oxford University Press.

Stalnaker, R. (1999b). "The Problem of Logical Omniscience II" in Stalnaker (1999a), 255–273.

Swain, M. (1974). Epistemic defeasibility. *The American Philosophical Quarterly, 11*, 15–25.

Williamson, T. (2000). *Knowledge and its limits*. Oxford: Oxford University Press.

Chapter 31
Sentences, Belief and Logical Omniscience, or What Does Deduction Tell Us?

Rohit Parikh

Introduction

In deductive reasoning, if ϕ is deduced from some set Γ, then ϕ is already implicit in Γ. But then how do we learn anything from deduction? That we do not learn anything is the (unsatisfying) answer suggested by Socrates in Plato's *Meno*. This problem is echoed in the problem of logical omniscience prominent in epistemic logic according to which an agent knows all the consequences of his/her knowledge. An absurd consequence is that someone who knows the axioms of Peano Arithmetic knows all its theorems.

Since knowledge presumes belief, the lack of closure of (actual) beliefs under deduction remains an important issue.

The post-Gettier (1963) literature has concentrated on the gap between justified true belief and knowledge, but has not concerned itself with what *belief* is. This question, or at least an analysis of sentences of the form *Jack believes that frogs have ears*, **has** been prominent in the philosophy of language. But even there, less attention has been paid to *how* we know what someone believes. This will turn out to be an important factor.

The question "*How we know what someone believes*" has, however, been addressed by Ramsey and de Finetti as well as by Savage in the context of decision theory and the foundations of subjective probability. Beliefs are revealed by the choices we make, the bets we accept and the bets we refuse. And among these choices are the choices of what to say and what to assent to. Of course the second kind of choice can only be made by humans, or at least by creatures possessing

R. Parikh (✉)
City University of New York, 365 Fifth Avenue, New York, NY 10016, USA
e-mail: rparikh@gc.cuny.edu

H. Arló-Costa et al. (eds.), *Readings in Formal Epistemology*, Springer Graduate Texts in Philosophy 1, DOI 10.1007/978-3-319-20451-2_31

language. But the first kind of choice is perfectly open to animals and to pre-lingual children.[1]

We argue that this way of thinking about beliefs not only allows us to address the issue of logical omniscience and to offer formal models of "inconsistent" beliefs, it also allows us to say something useful about Frege's problem, whether it is sentences or something else that we believe, and whether *believes* can be a binary relation at all.[2]

We start with a dialogue between Socrates, Meno, and Meno's slave boy from Plato's *Meno*. In this dialogue, Socrates carries on a conversation with the boy and asks him the area (space) of a square whose side is two. The boy correctly answers that the area is four. And then Socrates raises the question of *doubling* the area to eight. He wants to know what the side should be. The boy makes two conjectures, first that the side of the larger square should also be double (i.e., four) – but that yields an area of sixteen, twice what is wanted. The boy's second guess is that the side should be three – but that yields an area of nine, still too large.

> **Socrates:** Do you see, Meno, what advances he has made in his power of recollection? He did not know at first, and he does not know now, what is the side of a figure of eight feet: but then he thought that he knew, and answered confidently as if he knew, and had no difficulty; now he has a difficulty, and neither knows nor fancies that he knows.
> **Meno:** True.
> **Socrates:** Is he not better off in knowing his ignorance?
> **Meno:** I think that he is.
> **Socrates:** If we have made him doubt, and given him the "torpedo's shock," have we done him any harm?
> **Meno:** I think not.
> **Socrates:** We have certainly, as would seem, assisted him in some degree to the discovery of the truth; and now he will wish to remedy his ignorance, but then he would have been ready to tell all the world again and again that the double space should have a double side.

Now Socrates suggests that the diagonal of the smaller square would work as the side of the larger square and the boy agrees to this. We continue with the quotation:

> **Socrates:** And that is the line which the learned call the diagonal. And if this is the proper name, then you, Meno's slave, are prepared to affirm that the double space is the square of the diagonal?
> **Boy:** Certainly, Socrates.

[1]In Ruth Marcus (1990, 1995) describes a man and his dog in a desert, deprived of water and thirsty. When they both react the same way to a mirage, it is hard to deny that they both have the same false belief that they are seeing water. The fact that animals can experience mirages would seem to be substantiated by the fact that the Sanskrit word for a mirage is *mrigajal* which literally means 'deer water,' faux water which deer pursue to their deaths. Frans de Waal makes a much more detailed case for animal intentionality in Waal (2005).

[2]In Levi (1997), Isaac Levi considers the doxastic commitment we have to *try to achieve* logical closure of our beliefs, even when, as he admits, we cannot actually achieve such logical closure. I am sympathetic to Levi's requirement, but in this paper, my concern is to develop a theory of actual beliefs rather than of doxastic commitments. The latter are less problematic from a purely logical point of view. If the agent's full beliefs are consistent, then his doxastic commitments will satisfy a suitable modal logic, perhaps the logic KD4.

Socrates: What do you say of him, Meno? Were not all these answers given out of his own head?
Meno: Yes, they were all his own.
Socrates: And yet, as we were just now saying, he did not know?
Meno: True.
Socrates: But still he had in him those notions of his – had he not?
Meno: Yes.
Socrates: Then he who does not know may still have true notions of that which he does not know?
Meno: He has.

Here Socrates appears to be arguing that the beliefs which Meno's slave could be brought to have via a discussion were beliefs which he *already* had.[3] There is of course a tension in this line of argument, for if this is so, then deduction would appear to be dispensable, and yet, leading the boy to make deductions was Socrates' own method in bringing him to a new state of mind.

A conclusion similar to that of Socrates also follows from the Kripke Semantics which has become a popular tool for formalizing logics of knowledge. The semantics for the logic of knowledge uses Kripke structures with an accessibility relation R, typically assumed to be reflexive, symmetric, and transitive. If we are talking about belief rather than knowledge, then R would be serial, transitive, and Euclidean. Then some formula ϕ is said to be believed (or known) at state s iff ϕ is true at all states R-accessible from s. Formally,

$$s \models B(\phi) \text{ iff } (\forall t)(sRt \rightarrow t \models \phi)$$

If follows immediately that if a formula is logically valid then it is true at all states and hence it is both known and believed. Moreover, if ϕ and $\phi \rightarrow \psi$ are believed at s then both are true at all t such that sRt, and hence ψ is true at all such t. Thus ψ is also believed at s. Moreover, a logically inconsistent formula can be neither known nor believed. Aumann's (1976) semantics uses partitions rather than Kripke structures, but is known to be equivalent (Fagin et al. 1995) and suffers from the same difficulty.

There have been suggestions, (Fagin et al. 1995; Moses 1988) that these 'problems' can be dealt with by using tools like *awareness*, *impossible possible worlds*, or by referring to *computational complexity*. In our view, these methods do not address the fundamental problem, namely *what is it that we are trying to model?* In order to have a theory of knowledge, we need to have some criterion for "measuring" what someone knows. Just to start with a logic which yields logical

---

[3]This impression is surely strengthened by the remarks which Socrates makes elsewhere in *Meno* to the effect that if knowledge was always present then the soul must be eternal. *The soul, then, as being immortal, and having been born again many times, and having seen all things that exist, whether in this world or in the world below, has knowledge of them all; and it is no wonder that she should be able to call to remembrance all that she ever knew about virtue, and about everything.*

omniscience, and then fiddling with it by various methods, leaves aside the question of what our goal is. Unless we have a clear sight of the goal we are not likely to reach it.

For instance, we do not know if the Goldbach conjecture is true, but if it is true, then it is necessarily true. Surely we cannot argue that we do not know which only because we are not *aware* of it. That is surely not the problem. Nor can computational complexity help us, because *if* it is true, then there is a sound formal system which has it as an axiom, and then a proof of the Goldbach conjecture is going to be very fast in that system. If it is false, then of course the falsity is provable in arithmetic as it only requires us to take notice of a particular even number which is not the sum of two (smaller) primes.

The issue of computational complexity can only make sense for an infinite family of questions, whose answers may be undecidable or at least not in polytime. But for *individual* mathematical questions whose answers we do not know, the appeal to computational complexity misses the issue.

In sum, our goal is to first answer the question, *How do we know what people believe?* and then try to develop a theory of what people do believe.

Inconsistent Beliefs

Moreover, beliefs are not always consistent, so one could say, *thank heaven for a lack of logical omniscience!* A person, some of whose beliefs conflict with others, and who believes all logical consequences of her belief, would end up believing *everything!*

The following two examples are from Daniel Kahneman's Nobel lecture, (2002)

> *A bat and a ball cost \$1.10 in total. The bat costs \$1 more than the ball. How much does the ball cost?" Almost everyone reports an initial tendency to answer "10 cents" because the sum \$1.10 separates naturally into \$1 and 10 cents, and 10 cents is about the right magnitude. Frederick found that many intelligent people yield to this immediate impulse: 50 % (47/93) of Princeton students, and 56 % (164/293) of students at the University of Michigan gave the wrong answer. Clearly, these respondents offered a response without checking it.*

> *Linda is 31 years old, single, outspoken and very bright. She majored in philosophy. As a student she was deeply concerned with issues of discrimination and social justice and also participated in antinuclear demonstrations.*
> *#6* Linda is a bank teller
> *#8* Linda is a bank teller and active in the feminist movement
> *89 % of respondents rated item #8 higher in probability than item #6.*

But the set of bank tellers who are active in the feminist movement is a proper subset (perhaps even a rather small subset) of the set of all bank tellers, so #8 *cannot* have higher probability[4] than #6.

---

[4]The "conjunction fallacy" is committed when someone assigns higher probability to a conjunction than to one of the conjuncts. Gigerenzer (1996), Levi (2004) and Hintikka (2004) all dispute that the 89 % of people who responded the way indicated were actually committing the conjunction

Clearly human states of belief are not usually consistent. Nor are they usually closed under logical inference. Various researchers including ourselves have looked into this issue (Fagin et al. 1995; Gaifman 2004; Parikh 1987, 1995; Stalnaker 1999). There is also more recent work by Artemov and Nogina (2005), and by Fitting (2004), aimed towards the logic of proofs.

And yet we do operate in the world without getting into trouble, and when we discover an inconsistency or incompleteness in our thinking, we remove it, either by adding some beliefs which we did not have before, or by deleting some which we had. What we need is a formal account of this activity and a more general representation of beliefs than is afforded by the current Kripke semantics. In particular, logically omniscient belief states need to be represented as a *proper subset* of all belief states.

The proposal we make here draws on the work of Ramsey (1931), de Finetti (1937), Hayek (1948), and Savage (1954) with echoes from Marcus (1990), Marcus (1995), Millikan (2006), and Whyte (1990).[5] According to this view, an agent's beliefs are revealed by the *choices* which an agent makes, and while 'incoherent' choices are unwise, they are not inconsistent, for if they were, they would be *impossible*. For an agent to make such unwise, incoherent, choices is perfectly possible and is done all the time.

Imagine that Carol assigns probabilities of .3, .3, and .8 respectively to events $X, Y, X \cup Y$. One could say that these probabilities are *inconsistent*. But in fact nothing prevents Carol from *accepting* bets based on these probabilities. What makes them incoherent is that we can make Dutch book against Carol – i.e., place bets in such a way that no matter what happens, she will end up losing money.[6] Thus incoherent beliefs, on this account, are *possible*, (and hence consistent) but unwise. It will be important to keep in mind this distinction between inconsistency and incoherence.

We now look at some previous work which contains strong indications of the direction in which to proceed.

Hayek (1948) considers an isolated person acting over a period according to a preconceived plan. The plan

> *may, of course, be based on wrong assumptions concerning external facts and on this account may have to be changed. But there will always be a conceivable set of external events which would make it possible to execute the plan as originally conceived.*

---

fallacy. However, I assume that the dispute is about the interpretation of this particular experiment, and that these three writers would not dispute the *more general point* that people do sometimes reason incorrectly.

[5]However, unlike Ramsey et al., we shall not try to explain probabilistic belief.

[6]For instance we can bet \$3 on X, \$3 on Y, and \$2 against $X \cup Y$. If either X or Y happens, we earn \$7 (at least), and lose (at most) \$5. If neither happens, we gain \$8 and lose \$6, so that we again make a profit – and Carol makes a loss.

Beliefs are related here to an agent's plans, and it is implicit in Hayek's view that the agent believes that the world is such that his plans will work out (or have a good chance of doing so).

But note that the belief states which are implicit in plans are more general than the belief states which correspond to Kripke structures, and the first may be incoherent and/or logically incomplete. An agent may plan to buy high and sell low, and expect to make money. It is not possible for the agent to actually do so, but it is perfectly possible for the agent to have such a plan.

Animal Beliefs

Now, for a creature to have a plan it is not necessary that the plan be formulated explicitly in language, or even that the planner *has* a language. It is perfectly possible for an animal to have a plan and to a smaller extent, it is also possible for a pre-lingual child to engage in deliberate behaviour which is plan-like.

This is an important point made explicitly by Marcus (1995), Searle (1994), and also fairly clear in Millikan (2006), that we ought not to limit states like belief and desire to language using creatures, i.e., to adult humans and older children. Frans de Waal is even more emphatic on this point.[7] We should also attribute some kinds of intentional states to higher animals and to pre-lingual children. Going back further, Hume (1988) is quite explicit on this point:

> *Next to the ridicule of denying an evident truth is that of taking much pains to defend it; and no truth appears to me more evident, than that beasts are endow'd with thought and reason as well as men.*

Ramsey (1990) has a related comment.[8]

> *It is for instance possible to say that a chicken believes a caterpillar of a certain sort to be poisonous,and mean by that merely that it abstains from eating such caterpillars on account of unpleasant experiences connected with them.*

And so does Millikan (2006).

> *Reasoning is just trial and error in thought. Dennett (1996) calls animals capable of trial and error in thought "Popperian." The reference is to Popper's remark that* it is better to let one's hypotheses die in one's stead. *The Popperian animal is capable of thinking*

---

[7]"It wasn't until an ape saved a member of our own species that there was public awakening to the possibility of nonhuman kindness. This happened on August 16, 1996 when an eight-year old female gorilla named Binti Jua helped a three-year-old boy who had fallen eighteen feet into the primate exhibit at Chicago's Brookfield Zoo. Reacting immediately, Binti scooped up the boy and carried him to safety." De Waal is quite disdainful of Katherine Hepburn's remark in *The African Queen*: "Nature, Mr. Allnut, is what we are put in this world to rise above."

[8]Of course we need not and should not attribute to the chicken the specific belief that such caterpillars are *poisonous*. Davidson (1982) is right on this particular point. But we *can* attribute to it the belief that eating them will lead to bad consequences.

hypothetically, of considering possibilities without yet fully believing or intending them. The Popperian animal discovers means by which to fulfill its purposes by trial and error with inner representations. It tries things out in its head, which is, of course, quicker and safer than trying them out in the world. It is quicker and safer than either operant conditioning or natural selection. One of many reasonable interpretations of what it is to be rational is that being rational is being a Popperian animal. The question whether any non-human animals are rational would then be the question whether any of them are Popperian.

Finally Whyte (1990) suggests that we can even define truth in this way. He appeals to Ramsey's principle (R):

(R) *A belief's truth condition is that which guarantees the fulfilment of any desire by the action which that belief and desire would combine to cause.*

Defining Belief

However, we need not address the issue of truth here. Brandom (1994) subjects Whyte to some criticism, but it only has to do with whether the criterion of truth is adequate. We are only interested here in a *representation* of belief. Such a representation must accommodate all the following groups: language possessing, logically omniscient humans[9]; language possessing but fallible humans; and non-lingual creatures like animals and very young children.[10]

However, if beliefs are not (always) expressed by sentences, or coincide with propositions expressed by sentences, then we need another way of representing belief states, and then relate language-oriented belief states to some specific species of such belief states.

We shall consider two kinds of beliefs. Non-linguistic beliefs which may also be possessed by animals, and linguistic beliefs which can only be possessed by

[9]Of course I do not personally know any logically omniscient humans, but in a limited context it is possible for a human to show full logical competence. Suppose that p stands for *Pandas live in Washington DC*, q stands for *Quine was born in Ohio*, and r stands for *Rabbits are called* gavagai *at Harvard.* Suppose that Jill believes that p is true and that q and r have the same truth values. Then she is allowing two truth valuations, $v = (t, t, t)$, and $v' = (t, f, f)$. Given a formula ϕ on p, q, r in disjunctive normal form, she can evaluate $v(\phi)$ and $v'(\phi)$. If both are t she can say that she believes ϕ. If both are f, she disbelieves ϕ, and otherwise she is suspending judgment. Then Jill *will* be logically omniscient in this domain. But note that she will actually have to *make the calculations* rather than just sit back and say, "Now *do* I believe ϕ?" In fact if it so happens that ϕ is a complex formula logically equivalent to p, then ϕ represents the same proposition as p, and is therefore believed by Jill. And yet, Jill will not agree to ϕ *because* it is the same 'proposition' as p, but rather that she will agree to the *formula* ϕ whose truth value she has calculated. See also, Dennett (1985) p. 11.

[10]It is plausible that when a vervet monkey utters a leopard call, then it is saying, '*Ware leopard!*', but surely nothing in the behaviour of such monkeys would justify us to think that they might utter *If there were no leopards here, then this would be a wonderful spot to picnic.*

humans; adults and older children. Of course the last two groups will also have non-linguistic beliefs which must be somehow correlated with their linguistic beliefs.

Let $\mathcal{B}$ be the space (so far unspecified) of belief states of some agent. Then the elements of $\mathcal{B}$ will be correlated with the choices which the agent makes. Roughly speaking, if I believe that it is raining, I will take my umbrella, and if I believe that it is not raining, then I won't. What I believe is revealed by what I do.

But clearly the choice of whether to take my umbrella or not is correlated with my belief only if I don't want to get wet. So my preferences enter *in addition* to my beliefs.

Conventional theories like that of Savage (1954) take the (actual and potential) choices of an agent, assume that these choices satisfy certain axioms, and simultaneously derive both the agent's beliefs (her subjective probability) and the agent's preferences (expressed by a utility function). But it is known that agents which are rational in Savage's sense, i.e., obey his axioms, are not very common and so we would like to retain the *rough framework* without the inappropriate precision of Savage's theory.

So we will just assume that the agent has *some* space $\mathcal{P}$ of preferences, and that the choices are governed by the beliefs as well as the preferences. We will use $\mathcal{S}$ for the set of choice situations, and C for the set of choices. Thus the set $\{U, \neg U\}$ could be a choice situation, or an element of $\mathcal{S}$, (with U standing for *take the umbrella*); and both U and $\neg U$ are elements of C. We could say very roughly that an agent who prefers not to get wet will choose U from $\{U, \neg U\}$ iff she believes that it is raining.

An agent who does have language can also be subjected to a purely linguistic choice. If asked *Do you think it is raining?* the agent may choose from the set *{Yes, No, Not sure}*. And it is going to be *usually* the case that the agent will choose U in the situation, $\{U, \neg U\}$ iff she chooses *Yes* in the situation where she hears *Do you think that it is raining?* But this is not a *logical* requirement, only a pragmatic one.

We will say that an agent *endorses* (agrees with) a sentence ϕ iff she chooses *Yes* when asked, *Do you think* ϕ*?*, and *denies* (or disagrees with) ϕ iff she chooses *No*. She may also choose *Not sure*, in which case of course she neither endorses nor denies.

Note that nothing prevents an agent from endorsing ϕ as well as $\neg\phi$, but few agents are that 'out of it'. But of course an agent may endorse ϕ, $\phi \rightarrow \psi$, and either deny ψ or at least fail to endorse ψ. We will say in the first case that the agent is logically incoherent, and in the second that the agent is an incomplete reasoner – or simply incomplete.

Given the fact that $(\phi \wedge \neg\phi) \rightarrow \psi$ for arbitrary ψ is a tautology, it is obvious that an agent who is logically incoherent, but also logically complete, and who endorses both ϕ and $\neg\phi$, will end up endorsing *everything*. Fortunately, most of us, though we *are* logically incoherent, tend also to be incomplete. If we endorse the statements that all men are equal, that Gandhi and Hitler are men, and that Gandhi and Hitler are not equal, we are still not likely to agree that pigs fly – which we would if we were also *complete.*

As we have made clear, elements of $\mathcal{B}$ cannot be identified with propositions, for an agent may agree to one sentence expressing a proposition and disagree (or not agree with) another sentence expressing the same proposition. An agent in some state $b \in \mathcal{B}$ may agree with *Hesperus is bright this evening*, while disagreeing with *Phosphorus is bright this evening*. But surely we *may* think of an agent speaking some language $\mathcal{L}$ as agreeing with or endorsing a particular *sentence* ϕ.

Definition 1. A belief state b is *incomplete* when there are sentences $\phi_1, \ldots, \phi_n$ which are endorsed in b, ψ follows logically from $\phi_1, \ldots, \phi_n$, and ψ is not endorsed in b.

A belief state b is *incoherent* when there are sentences $\phi_1, \ldots, \phi_n$ which are endorsed in b, ψ follows logically from $\phi_1, \ldots, \phi_n$, and ψ is denied in b.

There is of course an entirely different sort of incoherence which arises when an agent's linguistic behaviour does not comport with his choices. Suppose an agent who prefers not to get wet, *says* that it is raining, but does not take her umbrella, then she may well be quite coherent in her linguistic behaviour, but her linguistic behaviour and her non-linguistic choices have failed to match. Whether we then conclude that the agent is irrational, is being deceptive, or does not know the language, will depend on the weight of the evidence.

We can easily see now how deduction changes things. Meno's slave was initially both incoherent and incomplete. He believed that a square of side four had an area of eight. Since he knew some arithmetic and some geometry, his belief state was not actually coherent. At the end of the conversation with Socrates, he came to endorse the sentence, *The square whose side is the diagonal of a square of side two, has an area of eight.* It is not quite clear what practical application this information had for him in this case, but surely carpenters carrying out deductions of the same kind will manage to make furniture which does not collapse and bears the load of people and objects.

Beliefs can change not only as a result of deductions, as those Meno's slave did, they can also change as a result of experience, e.g., raindrops falling on your head, or as the result of hearing something, like *It is raining.*

A Philosophical Aside

In this paper I shall avoid taking sides in philosophical disputes, but one thing does need to be mentioned. The representational account of belief seems simply wrong to me. We *can* certainly think of beliefs as being stored in the brain in some form and called forth as needed, but when we think of the details, we can soon see that the stored belief model is too meager to serve. I will offer an analogy.

Suppose Jennifer owns a bookstore. If someone calls her and asks if she has Shakespeare's *Hamlet*, she will look on her shelves, and answer yes if *Hamlet* is in her *store*. But suppose now that someone asks her if she has Shakespeare's *Tragedies*. These include *Hamlet* of course, but also *Macbeth*, *Romeo and Juliet* and *King Lear*. If they are stored alphabetically by title, then they will be in different locations and they won't be in her store as *one item*. But she can *create* the set and ship it out as one item to the customer. Did she have the set when the customer called? Surely yes. It would sound silly to say to the customer, "I do have *Hamlet, Macbeth, Romeo and Juliet* and *King Lear*, but unfortunately I don't have Shakespeare's *Tragedies*".

Let us go one step further. Suppose she actually only has a CD containing the files for the four plays, a printer, and a binder. She can still create the set using what she has, even though at the time of the call she had no physical object corresponding to the noun phrase *Shakespeare's Tragedies*.

It is what she *has*, namely the CD, the printer, and the binder, and what she *can do*, namely print and bind, which together allow her to fulfill the order. There are elements of pure storage, and elements which are algorithmic which together produce the item in question. These two kinds of elements may not always be easy to separate. It is wiser just to concentrate on what she can supply to her customer.

It is the same, in my view, with beliefs. No doubt there are certain beliefs which are stored in some way, but there may be other equally valid beliefs which can be produced *on the spot* so to say, without having been there to start with. And note that if Jennifer has a large number of actual physical books, but also CDs for some, it may be easier for her to produce a book from a CD than to find a book which exists physically in the store.

If someone asks me if I believe that 232345456 is even, I can answer yes at once (since the number ends in a 6), even though that particular belief had never been stored. But if someone asks me the name of some classmate from high school, I might take a long time to remember. Retrieving from storage is *one* way to exhibit a belief, but not the only one, and often, not even the best one.

In this paper, I shall not assume any particular representation of beliefs, but deal with them purely in terms of how a person acts, and how the potential for acting in some way is affected by various kinds of update.

Some Technical Details

We assume given a space $\mathcal{B}$ for some agent whose beliefs we are considering. The elements of $\mathcal{B}$ are the belief states of that agent, and these are not assumed to be sentences in Mentalese although for some restricted purposes they could be. There are three important update operations on $\mathcal{B}$ coming about as a result of (i) events observed, (ii) sentences heard, and (iii) deductions made. Elements of $\mathcal{B}$ are also used to make *choices*. Thus in certain states of belief an agent may make the choice to take his umbrella and we could then say that the agent believes it is raining. Many human agents are also likely to make the choice to *say*, "I think it is raining and so I am taking my umbrella" but clearly only if the agent is English speaking. Thus two agents speaking different languages, both of whom are taking their umbrellas, but making different noises, have the same belief in one sense but a different one in another. Both of these will matter. Later on we shall look into the connection.

Deduction is an update which does not require an input from the outside. But it *can* result in a change in $\mathcal{B}$. Suppose for instance that Jane thinks it is clear, and is about to leave the building without an umbrella. She might say, "Wait, didn't I just see Jack coming in with a wet umbrella? It must be raining." The sight of Jack with a wet umbrella might not have caused her to believe that it was raining, perhaps she was busy with a phone call. But the memory of that wet umbrella may later cause a deduction to take place of the fact that it is raining.

Thus our three update operations are:

$$\mathcal{B} \times \mathcal{E} \rightarrow_e \mathcal{B}$$

A belief state gets revised by witnessing an event.

$$\mathcal{B} \times \mathcal{L} \rightarrow_s \mathcal{B}$$

A belief state gets revised through hearing a sentence.[11,12]

And hearing two logically equivalent sentences s, s' need not result in the same change occurring, although they may, in case the agent knows they are equivalent.[13]

$$\mathcal{B} \rightarrow_d \mathcal{B}$$

A deduction causes a change in the belief state (which we may sometimes represent as an *addition*).[14]

Here E is the set of events which an agent may witness and $\mathcal{L}$ is some language which an agent understands (i.e., uses successfully). The last map $\rightarrow_d$ for deduction is non-deterministic as an agent in the same state of belief may make one deduction or another.

Finally, we also have a space $\mathcal{S}$ of *choice sets* where an agent makes a particular choice among various alternatives. This gives us the map

$$\mathcal{B} \times \mathcal{S} \rightarrow_{ch} \mathcal{B} \times \mathcal{C}$$

---

[11]Of course, hearing a sentence is also an event, but its effect on speakers of the language goes beyond just the event. It is this second part which falls under $\rightarrow_s$.

[12]Even a dog may revise its state of belief on hearing *Sit!*, see for instance Parikh and Ramanujam (2003). Note also that if the sentence heard is inconsistent with one's current beliefs *and* one notices the inconsistency, then some theory like that in Alchourron et al. (1985) may need to be deployed.

[13]Clearly Lois Lane will react differently to the sentences *Superman flew over the Empire State Building*, and *Clark Kent flew over the Empire State Building*. Similarly, Kripke's Pierre (1979) will react differently to the questions, *Would you like a free trip to Londra?* and *Would you like a free trip to London?* Indeed in the second case he might *offer to pay* in order not to go to London!

[14]See, however, Alchourron et al. (1985) where more complex kinds of reactions to sentences heard are described.

An agent in a certain belief state makes a choice among various alternatives, and may arrive at a different state of belief *after* making that choice.

If we want to explicitly include preferences, we could write,

$$\mathcal{B} \times \mathcal{P} \times \mathcal{S} \rightarrow_{ch} \mathcal{B} \times C$$

While $\mathcal{S}$ is the family of choice sets, C is the set of possible choices and $\mathcal{P}$ is *some* representation of the agent's preferences. Thus *{take umbrella, don't take umbrella}* is a choice set and an element of $\mathcal{S}$ but *take umbrella* is a choice and an element of C.

Example. Suppose that Vikram believes it is not raining. In that case he will be in a belief state b such that $ch(b, \{U, \neg U\}) = (b', \neg U)$. Given a choice between taking an umbrella or not, he chooses not to take the umbrella, and goes into state b'.

Suppose, however, that he looks out the window and sees drops of rain falling. Let r be the event of rain falling. Then the update operation $\rightarrow_e$ causes him to go into state c such that in state c he chooses U from $\{U, \neg U\}$. Thus $ch(c, \{U, \neg U\}) = (c', U)$

Note that state c will also have other properties beside choosing to take an umbrella. It may also cause him to say to others, "You know, it is raining," or to complain, "Gosh, does it always rain in this city?" There is no such thing as *the state* of believing that it is raining. Every belief state has many properties, most of them unrelated to rain.

Among the choices that agents make are choices to assent to or dissent from sentences. But there is no *logical* reason why an agent who assents to "It is raining" must take an umbrella or a raincoat. It is just the more pragmatic choice to take the umbrella when one says that it is raining, because otherwise either one gets wet, or one is suspected of insincerity. We could say that the agent believes the *sentence* "It is raining", and dis-believes the *proposition* that it is raining. But we would feel uncomfortable with such an account and might prefer to say that the agent is either lying or confused.

Of course an agent may prefer to get wet and in that case, saying "It is raining," and not taking an umbrella are perfectly compatible choices. This shows that an agent's preferences need to be taken into account when correlating the actions which an agent takes and what an agent believes. But we usually do not want to get wet and to make such choices, and usually we do not say what we do not believe. It does not work for us.

Thus our theory of an agent presupposes such a belief set $\mathcal{B}$, and appropriate functions $\rightarrow_e, \rightarrow_s, \rightarrow_d, \rightarrow_{ch}$. We can understand an agent (with some caveats) if what we *see* as the effects of these maps conforms to some theory of what an agent wants and what the agent thinks. And we succeed pretty well. *Contra* Wittgenstein (1958), we not only have a theory of what a lion wants, and what it means when it growls, we even have theories for bees and bats. Naturally these theories do not have the map $\rightarrow_s$ except with creatures like dogs or cats and some parrots (who not only "parrot" our words but understand them to some extent (Pepperberg 2004)).

Partial Similarity of Belief States

A familiar notion of similarity used in mathematics is that of *isomorphism*. If two structures $< X, R >$, $< Y, R' >$ are isomorphic, then it means that there is a 1–1 function f from X onto Y such that for all $a, b \in X$, $R(a, b)$ iff $R'(f(a), f(b))$. Isomorphic structures have the same logical properties.

The notion which will be relevant for us, however, will be the notion of *bi-simulation* (Johan van Benthem 1976; Milner 1989; Park 1981).

In the same context as above, we will say that $< X, R >$ and $< Y, R' >$ are *bi-similar* if there is a subset B (the bi-simulation) of $X \times Y$ such that X is the domain of B, Y is the co-domain of B, and we have the following conditions.

- If $B(a, b)$ and there is c such that $R(a, c)$, then there is a d such that $B(c, d)$ and $R'(b, d)$.
- If $B(a, b)$ and there is d such that $R'(b, d)$, then there is a c such that $B(c, d)$ and $R(a, c)$.

Intuitively we can say that if a, b are related by the bi-simulation, then for anything c which is related by R to a, there is something d which is related by R' to b, and vice versa. Moreover, c, d are also related by B.

Thus the structures $< X, R >$ and $< Y, R' >$ can imitate each other without having to be isomorphic. Isomorphism implies bi-similarity but not vice versa.

If we have richer structures, $< X, R, S >$ and $< Y, R', S' >$ then it may well happen that $< X, R >$ and $< Y, R' >$ are bi-similar, but $< X, R, S >$ and $< Y, R', S' >$ are not. Then the imitation will be partial.

It can thus happen that two belief structures $\mathcal{B}$ and $\mathcal{B}'$ can be bi-similar in *some* ways, without being bi-similar in every way. Two students in the same school may show similar behavior in the cafetaria, and be bi-similar in their cafetaria personae, but not be bi-similar in, say, their dating behavior. One could say then, somewhat loosely, that they have the same beliefs about food but not about girls. (I am assuming here that their preferences are the same.)

Similarly, Jack and Jill can have the same beliefs and preferences about ice cream, and exhibit similar behavior in ice cream choosing settings, but may have different beliefs about politics.

So far we have only talked about belief states as being indicated by choices. But of course many beliefs are expressed (or so we think) by sentences. We may suggest, with Stalnaker, that what we believe are propositions. Though I do not think such accounts can work, the fact that we find them appealing requires some explanation.

We shall now introduce two notions of belief, e-belief (which is choice-based) and i-belief (which is sentence-based and only appropriate for language users). These two notions will be used to sort out various belief puzzles.

The Setting

In our setting we imagine an observer o who is pondering on what some agent i believes. We assume (for convenience) that o thinks of a proposition expressed by a sentence as a set of possible worlds where that sentence is true, but that the observee i need not even have a language or a notion of truth. However, it is assumed that i does have some plans. Even if i is just a dog digging for a bone, o understands that i has a plan and roughly what that plan is. And we shall use this plan to make it possible for o to attribute beliefs to i.

We also assume that there is a context C which is the set of *relevant* possible worlds, and that worlds outside C, even though they are there, are not considered in deliberating about i's belief or beliefs. It is not assumed that i knows that there are worlds outside C; in some sense i lives inside C, but we will assume that o does know that there are possible worlds outside C. The purpose of the context is to avoid considering cases which are possible but strange, like the laws of physics failing, or Bush suddenly joining the Green party. A plan is assumed to work in *normal* circumstances, and an agent i is only required to have the sort of belief which would be enough for the plan to be carried out in normal circumstances.

For instance, chickens, when given a choice between drinking water and drinking mercury tend to prefer the latter, presumably because it looks more like water (is shinier) than water itself does. We certainly do not want to attribute to the chickens a belief that mercury is good to drink, merely that if you are thirsty, you should go for a shiny liquid.

So let P be i's plan at the moment, and let $\pi(P)$ be the set of worlds w in C such that the plan is *possible* at w.

There are two senses in which the plan may be possible. One is that the plan can actually be carried out. For instance for the dog digging for a bone, that possibility means that the ground is soft enough. The other sense is that the plan actually yields i the benefit which i wants; in this case, the actual presence of a bone.

So formally,

$$\pi(P) = \{w|\ w \in C \wedge w\, enables\, P\}.$$

Let ϕ be a sentence. Then $||\phi|| = \{w|w \models \phi\}$, the set of worlds where ϕ is true, is the *proposition* corresponding to the sentence ϕ. Note that if ϕ and ψ are logically equivalent, then $||\phi|| = ||\psi||$.

Definition 2. We will say that i *e-believes* ϕ, $B^i_e(\phi)$ if $\pi(P) \subseteq ||\phi||$. We will suppress the superscript i when it is clear from context.

It is obvious in terms of the semantics which we just gave that the statement "The dog e-believes that there is a bone where he is digging" will be true from *our* point of view. A Martian observing the same dog, but not having the notion of a bone (I assume there are no vertebrates on Mars) will obviously not assign the same belief to the dog; but the Martian may still have his own theory of the dog's belief state which will allow him to predict the dog's behavior. In other words, the Marian will assign a space $\mathcal{B}'$ to the dog which will be bi-similar in relevant ways to the space $\mathcal{B}$ which *we* would assign to the dog.

It is easy to see that if an agent e-believes ϕ and ψ then the agent also e-believes $\phi \wedge \psi$ and that if the agent e-believes ϕ and $\phi \rightarrow \psi$ then the agent e-believes ψ.[15] Oddly enough, creatures which do not use language do not suffer from a lack of logical omniscience!

A lot of logic goes along with e-belief, but only within the context of a single plan. For instance, one may drive a colleague to the airport for a two week vacation in Europe and then forget, and arrange a meeting three days later at which this colleague's presence is essential. But within the context of a single (short) plan, consistency and logical omniscience will tend to hold. The situation is more complex with multiple plans. And there is nothing to prevent an agent from having one e-belief in one plan and another contradicting e-belief in another plan. It is pragmatic considerations – the logical counterpart of avoiding Dutch Book – which will encourage the agent i to be consistent and to use logical closure.

Suppose someone has a plan P consisting of, "If ϕ then do α, else do β" and another plan P' consisting of "If ϕ then do γ, else do δ". Now we find him doing α and also doing δ (we are assuming that the truth value of ϕ has not changed). We could accuse him of being illogical, but there is no need to appeal to logic. For he is doing Dutch book against himself.

Presumably he assumed[16] that $u(\alpha|\phi) > u(\beta|\phi)$ but $u(\alpha|\neg\phi) < u(\beta|\neg\phi)$. Thus given ϕ, α was better than β but with $\neg\phi$ it was the other way around. Similarly, $u(\gamma|\phi) > u(\delta|\phi)$, but $u(\gamma|\neg\phi) < u(\delta|\neg\phi)$. And that is why he had these plans. But then his choice of α, δ results in a loss of utility whether ϕ is true or not. If ϕ is true then he lost out doing δ and if ϕ is false, then he lost out doing α.

For a concrete example of this, suppose that on going out I advise you to take your umbrella, but fail to take mine. If it is raining, there will be a loss of utility for I will get wet. If it is not raining, there will be a loss of utility because you will be annoyed at having to carry an umbrella for no good reason. My choice that I advise you to take your umbrella, but fail to take mine, is not *logically* impossible. It just makes no pragmatic sense. A similar argument will apply if someone endorses ϕ, endorses $\phi \rightarrow \psi$ and denies ψ. If such a person makes plans comporting with these three conditions, then he will make choices which do not maximise his utility. Of course such arguments go back to Ramsey (1931) and Savage (1954).

---

[15]To see this, if $\pi(P) \subseteq ||\phi||$, and $\pi(P) \subseteq ||\psi||$ then clearly $\pi(P) \subseteq ||\phi|| \cap ||\psi|| = ||\phi \wedge \psi||$. The proof for the other case is similar using the fact that $||\phi \rightarrow \psi|| = (C - ||\phi||) \cup ||\psi||$. Since $\pi(P)$ is contained in $||\phi||$, it is disjoint from $C - ||\phi||$. Hence it can be contained in $(C - ||\phi||) \cup ||\psi||$ if and only if it contained in $||\psi||$.

[16]The use of the letter u for utility is not meant to suggest that we have a formal notion of utility in mind; only a rough one.

If we assume that over time, people learn to maximise their utility (they do not always but often do), then they will 'learn' a certain amount of logic and they will make certain obvious logical inferences.[17]

A little girl who was in kindergarten was in the habit of playing with her older sister in the garden every day when she came home. Once, her older sister was sick. So the little girl went to visit her sick sibling in her bedroom, and then, as usual, went out into the garden to play with her. Clearly the little girl had not yet learned to maximise her utility!

A Second Notion of Belief: Language Enters

We now define a second notion of belief which does *not* imply logical omniscience. This is a more self-conscious, language-dependent notion of belief.

For agents i who do have a language (assumed to be English from now on), their plan may contain linguistic elements. At any moment of time they have a finite stock of currently believed sentences. This stock may be revised as time passes. These agents may perform atomic actions from time to time, and also make observations which may result in a revision in their stock of believed sentences.[18]

Thus Lois seeing Superman in front of her will add the sentence "Superman is in front of me", to her stock, but, since she does not know that Clark Kent is Superman, she will *not* add the sentence "Clark Kent is in front of me". Someone else may add the sentence "I see the Evening Star", but not the sentence "I see the Morning Star" at 8 PM on a summer night. A person who knows that $ES = MS$, may add the sentence, "Venus is particularly bright tonight." In any case, this stock consists of sentences and not of propositions.

The basic objects in the agents' *plans* are atomic actions and observations which may be active (one looks for something) or passive (one happens to see something). These are supplemented by the operations of concatenation (sequencing), *if then else*, and *while do*, where the tests in the *if then else* and *while do* are on *sentences*. There may also be recursive calls to the procedure: *find out if the sentence ϕ or its negation is derivable within the limits of my current resources, from my current stock of beliefs*. Thus if i's plan has currently a test on ϕ, then, to be sure, the stock of sentences will be consulted to see if ϕ or its negation is in the stock. But there may also be a recursive call to a procedure for deciding ϕ. If someone asks "Do you

[17]In Bicchieri (1997) suggests that co-operation also comes about as a result of such a learning process. Such suggestions have of course also been made by many others. Since we are only considering the one-agent case here, we shall not go into this issue any further. See, however, our Parikh (1991).

[18]It may seem to the reader as if I am endorsing a representational theory after all, but not so. First, the stock may not literally exist, but may simply refer to those sentences which the agent assents to *quickly*. Secondly, the agent's beliefs need not be restricted to this stock – just as earlier, the bookseller Jill was not restricted to selling sets of books which were in her store *as a set*.

know the time?", we do not usually say, "I don't", but look at our watches. Thus consulting our stock of sentences is typically only the first step in deciding if some sentence or its negation can be derived with the resources we have.

This difference between sentences and propositions matters as we now show.[19]

It has been suggested in this context (e.g. by Stalnaker 1999) that such issues can be addressed by using the notion of fine grain. By this account, if I understand it correctly, logical equivalence is too coarse a grain and that a finer grain may be needed. So if two sentences are in the same fine grain and an agent knows or believes one, then the agent will also know or believe the other. But if we try to flesh out this metaphor then we can see that it is not going to work.

For instance if $G(\phi, \psi)$ means that ϕ and ψ are in the same grain, then G will be an equivalence relation. But surely we cannot have transitivity in reality, because we could have a sequence $\phi_1,\ldots,\phi_n$ of sentences, any two successive ones of which are easily seen to be equivalent, whereas it is quite hard to see the equivalence of ϕ_1 and ϕ_n.

Moreover, "being in the same grain" sounds interpersonal. If two molecules are in the same rice grain for you, then they are also in the same fine grain for me – it is just a fact of the matter. But in reality people differ greatly in their ability to perceive logical equivalences.

Thus suppose some set theorist thinks of some new axiom ϕ and wonders if ϕ implies the continuum hypothesis, call it ψ. The set theorist may find it quite easy to decide this question even if we see no resemblance between ϕ and ψ. And if he does not find it easy, he may look in the literature or ask another set theorist, processes which cannot easily be built into a formal theory. And they should not be! For if they were easy to build into a formal theory, then the representation is almost certain to be wrong.

Or a chess champion may be able to see 20 moves (40 half-moves) ahead in the end game, but you and I cannot. And he too usually cannot do this during the middle game. Thus context, habit and expertise matter a lot.

Suppose for instance that Lois Lane has invited Clark Kent to dinner but he has not said yes or no. So she forms the plan, *While I do not have a definite answer one way or another,* if *I see Clark Kent, I will ask him if he is coming to dinner.* Here *seeing Clark Kent* is understood to consist of an observation followed by the addition of the sentence "I am seeing Clark Kent" to her stock.

Suppose now that she sees Superman standing on her balcony. She will *not* ask him if he is coming to dinner as the sentence "I am seeing Clark Kent" will not be in

[19]We might compare entertaining a proposition as a bit like entering a building. If Ann and Bob enter the same building through different doors, they need not be in the same spot, and indeed they might never meet. But what makes it the same building is that they *could* meet without going out into the street. Thus if Ann and Bob are apt to assent to sentences s, s' respectively where *we* know that s, s' are equivalent; then it need not follow that there is a sentence they share. But they *could* through a purely deductive process, and without appealing to any additional facts, be brought to a situation where Ann assents to s' and Bob to s (unless one of them withdraws a belief, which may also happen).

her stock of sentences. And this is the sense in which she does *not* know that when she is seeing Superman, she is also seeing Clark Kent. If she *suspects* that Clark Kent is Superman, then it may happen that her recursive call to the procedure "decide if I am seeing Clark Kent" will take the form of the question, "Are you by any chance Clark Kent, and if so, are you coming to dinner?" addressed to Superman.

Have we given up too much by using sentences and not propositions as the objects of i-belief? Suppose her plan is, "If I see Jack and Jill, I will ask them to dinner" but she sees Jill first and then Jack so that she adds the sentence "Jill and Jack are in front of me" to her stock. Will this create a problem? Not so, because if a sentence ϕ is in her stock, ϕ easily implies ψ, and she needs to know the value of ψ so she can choose, then the program *find out about* ψ which she calls will probably find the sentence she needs. If the program terminates without yielding an answer she may well have a default action which she deems safest.

So here we make use of the fact that Lois does have a reasonable amount of intelligence. Even if she does not explicitly add some sentence ψ to her stock, when she comes to a point where the choice of action depends on ψ, she *will* ask if ψ or its negation is derivable from her present stock of sentences, possibly supplemented by some actions which add to this stock.

Definition 3. If an agent a comes to a point in her plan where her appropriate action is *If* ϕ *then do* α *else do* β, and she does α, then we will say that she *i-believes* ϕ. If, moreover, ϕ is true, and we believe that in a similar context she would judge it to be true only if it *is* true, then (within the context of this plan) we will say that she *i-knows* ϕ.

A common example of such a plan is the plan to answer a question correctly. Thus if an agent is asked "Is ϕ true?", the agent will typically call the procedure "decide if ϕ is true", and then answer "yes", "no", or "I don't know" in the appropriate cases.

Now note that if an agent means to deceive, then the same procedure will be called, but the answers given will be the opposite of the ones indicated by the procedure. But if we ourselves know that the agent's plan is to deceive, then we can clearly take the statement "ϕ is false" to indicate that the agent believes ϕ.

We no longer have the law that if the agent i-knows ϕ and ϕ implies ψ then the agent of necessity i-knows ψ. But if the agent has the resources to decide ϕ and the proof of ψ from ϕ is easy, then she might well also know ψ. But her notion of "easy" may be different from ours, and how much effort she devotes to this task will depend on her mood, how much energy she has, etc. Dennett (1985), who makes a related point, does not seem to make the distinction between linguistic and non-linguistic belief which I am making here.

For instance if a customer sits down on a bar stool and the bartender sees him, we do not need to ask, "Does the bartender know he wants a drink?" Of course he knows. But suppose a woman suffering from a persistent cough calls her husband and says, "I got an appointment with the doctor for 2:30 PM", she may later think, "I wonder if he realized that I cannot pick up our daughter at 3". When i knows ϕ and ψ is deducible from ϕ, then whether we assume that i knows ψ will depend less on some objective distance between ϕ and ψ than on what we know or can assume about i's habits.

Note now that we often tend to assign knowledge and beliefs to agents even when they are not in the midst of carrying out a plan. Even when we are brushing our teeth we are regarded as knowing that the earth goes around the sun. This can be explained by a continuity assumption. Normally when we *are asked* if the earth goes around the sun or vice versa, we say that the former is the case. An agent is then justified in assuming that even in between different occasions of being so asked, if we *were* asked, we would give the same answer. This assumption, which is usually valid, is what accounts for such attributions of belief.

Conclusion

We have tried to give an account of belief which starts with behavior rather than with some ad hoc logical system. There are many things to be worked out. For instance, why do we perform deductions at all? They must benefit us in some way, but a detailed logical account of how systems can evolve towards better logical acumen is yet to be developed. Yet another question to ask is about partial beliefs. In some circumstances these partial beliefs will be correlated with a probability function, but again, a concordance with the Kolmogorov axioms will be a norm which is not always observed in practice.

Acknowledgements We thank Sergei Artemov, Can Başkent, Samir Chopra, Horacio Arló Costa, Juliet Floyd, Haim Gaifman, Isaac Levi, Mike Levin, Larry Moss, Eric Pacuit, Catherine Wilson, and Andreas Witzel for comments. The information about chess came from Danny Kopec. This research was supported by a grant from the PSC-CUNY faculty research assistance program. Earlier versions of this paper were given at *TARK-05*, *ESSLLI-2006*, at the *Jean Nicod Institute*, at a seminar in the philosophy department at Bristol University, and at the Philosophy Colloquium at the City University Graduate Center. Some of the research for this paper was done when the author was visiting Boston University and the Netherlands Institute for Advanced Study. A very preliminary version of some of the ideas was presented at Amsterdam, and published as Parikh (2001). This research was partially supported by grants from the PSC-CUNY program at the City university of New York.

References

Alchourron, C., Gärdenfors, P., & Makinson, D. (1985). On the logic of theory change: Partial meet contraction and revision functions. *The Journal of Symbolic Logic, 50*, 510–530.

Artemov, S., & Nogina, E. (2005). On epistemic logics with justifications. In R. Meyden (Ed.), *Theoretical aspecits of rationality and knowledge* (pp. 279–294). Singapore: University of Singapore press.

Aumann, R. (1976). Agreeing to disagree. *Annals of Statistics, 4*, 1236–1239.

van Benthem, J. (1976). *Modal correspondence theory*. Doctoral dissertation, University of Amsterdam.

Bicchieri, C. (1997). Learning to co-operate. In C. Biccheri, R. C. Jeffrey, & B. Skyrms (Eds.), *The dynamics of norms* (pp. 17–46). Cambridge: Cambridge University Press.

Brandom, R. (1994). Unsuccessful seminatics. *Analysis, 54*, 175–178.

Davidson, D. (1982). Rational animals. *Dialectica, 36*, 318–327.

Dennett, D. (1985). *Brainstorms* (2nd ed.). MIT, cf. page 11. Singapore.

Dennett, D. (1996). *The Intentional Stance (6th printing)*, Cambridge, Massachusetts: The MIT Press.

Fagin, R., Halpern, J. Y., Moses, Y., & Vardi, M. (1995). *Reasoning about knowledge*. Cambridge: MIT.

de Finetti, B. (1937). Foresight: Its logical laws, its subjective sources. *Annales de l'Institut Henri Poincare, 7*. (Trans.: Kyburg, H. (1980). *Studies in subjective probability*, Kyburg and Smokler (Eds.), (pp. 53–118). Krieger Publishing.)

Fitting, M. (2004). A logic of explicit knowledge. *Logica Yearbook*, 11–22.

Gaifman, H. (2004). Reasoning with limited resources and assigning probabilities to arithmetical statements. *Synthese, 140*, 97–119.

Gettier, E. (1963). Is justified true belief knowledge? *Analysis, 23*, 121–123.

Gigerenzer, G. (1996). On narrow norms and vague heuristics: A reply to Kahneman and Tversky (1996). *Psychological Review, 103*, 592–596.

Hayek, F. A. (1948). *Individualism and economic order* (See especially chapters II and IV). Chicago: University of Chicago Press.

Hintikka, J. (2004). A fallacious fallacy? *Synthese, 140*, 25–35.

Hume, D. (1988). *A treatise of human nature*, (L. A. Selby-Brigge (Ed.), pp. 176–179). New York: Oxford University Press.

Kahneman, D. (2002). *Maps of Bounded Rationality*, Nobel prize lecture.

Kripke, S. (1979). A puzzle about belief. In A. Margalit (Ed.), *Meaning and use*. Dordrecht/Boston: Reidel

Levi, I. (1997). Rationality and commitment. In *The covenant of reason*. Cambridge/New York: Cambridge University Press.

Levi, I. (2004). Jaakko Hintikka. *Synthese, 140*, 37–41.

Marcus, R. (1990). Some revisionary proposals about belief and believing. *Philosophy and Phenomenological Research, 50*, 133–153.

Marcus, R. (1995). The anti-naturalism of some language centere accounts of belief. *Dialectica, 49*, 112–129.

Millikan, R. (2006). Styles of rationality. In M. Nudds & S. Hurley (Eds.), *Rationality in animals* (pp. 117–126). Oxford: Oxford University Press.

Milner, R. (1989). *Communication and concurrency*. New York: Prentice Hall.

Moses, Y. (1988). Resource bounded knowledge. In M. Vardi (Ed.), *Proceedings of the 2nd Conference on Theorectical Aspects of Reasoning About Knowledge* (pp. 261–276). San Francisco: Morgan Kaufmann.

Parikh, R. (1987). Knowledge and the problem of logical omniscience. In *International symposium on methodology for intelligent systems (ISMIS- 87)*, Charlotte (pp. 432–439). North Holland.

Parikh, R. (1991). Finite and infinite dialogues. In Moschovakis (Ed.) *Proceedings of a Workshop on Logic from Computer Science* (pp. 481–498). Berlin: Springer/MSRI Publications.

Parikh, R. (2001). Propositions, propositional attitudes and belief revision. In K. Segerberg, M. Zakharyaschev, M. de Rijke, & H. Wansing (Eds.), *Advances in modal logic* (Vol. 2). Stanford: CSLI Publications.

Parikh, R. (1995). Logical omniscience. In Leivant (Ed.), *Logic and computational complexity* (Lecture notes in computer science, Vol. 960, pp. 22–29). Berlin/New York: Springer.

Parikh, R., & Ramanujam, R. (2003). A knowledge based semantics of messages. *Journal of Logic, Language and Information, 12*, 453–467.

Park, D. (1981). Concurrency and automata on infinite sequences. In P. Deussen (Ed.), *Proceedings of the 5th GI-Conference Karlsruhe*. Berlin/Heidelberg: Springer.

Pepperberg, I. (2004). Talking with Alex: Logic and speech in parrots; exploring intelligence. *Scientific American Mind*, August 2004.

Plato. *Meno*. (380 BC). (trans. Benjamin Jowett). Available online at http://classics.mit.edu//Plato/meno.html

Ramsey, F. P. (1990). Facts and propositions. In D. H. Mellor (Ed.), *Philosophical papers* (pp. 34–51). Cambridge/New York: Cambridge University Press.
Ramsey, F. P. (1931). Truth and probability. In *The foundations of mathematics* (pp. 156–198). London: Routledge and Kegan Paul.
Savage, L. J. (1954). *The foundations of statistics*. New York: Wiley.
Schwitzgebel, E. (2002). A phenomenal, dispositional account of belief. *Nous, 36:2*, 249–275.
Schwitzgebel, E. (2006, Fall). Belief. In E. N. Zalta (Ed.), *The stanford encyclopedia of philosophy* (Fall 2006 ed.). http://plato.stanford.edu/archives/fall2006/entries/belief/.
Searle, J. (1994). Animal minds. In P. French & H. Wettstein (Eds.), *Philosophical naturalism, midwest studies in philosophy* (XIX, pp. 206–219). Notre Dame: University of Notre Dame Press.
Stalnaker, R. (1999). *Context and content*. Oxford/New York: Oxford University Press.
de Waal, F. (2005). *Our inner ape*. Penguin. Singapore.
Whyte, J. T. (1990). Success semantics. *Analysis, 50*, 149–157.
Wittgenstein, L. (1958). *Philosophical investigations*. New York: MacMillan.

Chapter 32
The Logic of Justification

Sergei Artemov

Introduction

The celebrated account of Knowledge as Justified True Belief commonly attributed to Plato (cf. Gettier 1963; Hendricks 2003) was widely accepted until 1963 when a paper by Edmund Gettier (1963) opened the door to a broad philosophical discussion of the subject (cf. Dretske 1971; Goldman 1967; Lehrer and Paxson 1969; Nozick 1981; Stalnaker 1996 and many others).

Meanwhile, commencing from seminal works (Hintikka 1962; von Wright 1951), the notions of Knowledge and Belief have acquired formalization by means of modal logic with atoms $\mathbf{K}F$ (*F is known*) and $\mathbf{B}F$ (*F is believed*). Within this approach, the following analysis was adopted: for a given agent,

$$F \text{ is known} \quad \sim \quad F \text{ holds in all epistemically possible situations.} \tag{32.1}$$

The resulting *Epistemic Logic* has been remarkably successful in terms of developing a rich mathematical theory and applications (cf. Fagin et al. 1995; Meyer and van der Hoek 1995, and other sources). However, the notion of justification, which has been an essential component of epistemic studies, was conspicuously absent in the mathematical models of knowledge within the epistemic logic framework. This deficiency is displayed most prominently, in the *Logical Omniscience* defect of the modal logic of knowledge (cf. Fagin and Halpern 1985, 1988; Hintikka 1975;

This work has been supported by NSF grant 0830450, CUNY Collaborative Incentive Research Grant CIRG1424, and PSC CUNY Research Grant PSCREG-39-721.

S. Artemov (✉)
Graduate Center CUNY, 365 Fifth Avenue, New York, NY 10016, USA
e-mail: sartemov@gc.cuny.edu

H. Arló-Costa et al. (eds.), *Readings in Formal Epistemology*, Springer Graduate Texts in Philosophy 1, DOI 10.1007/978-3-319-20451-2_32

Moses 1988; Parikh 1987). In the provability domain, the absence of an adequate description of the logic of justifications (here mathematical proofs) remained an impediment to both formalizing the Brouwer-Heyting-Kolmogorov semantics of proofs and providing a long-anticipated exact provability semantics for Gödel's provability logic S4 and intuitionistic logic (Artemov 1999, 2001, 2007; van Dalen 1986). This lack of a justification component has, perhaps, contributed to a certain gap between epistemic logic and mainstream epistemology (Hendricks 2003, 2005). We would like to think that Justification Logic is a step towards filling this void.

The contribution of this paper to epistemology can be briefly summarized as follows.

> *We describe basic logical principles for justifications and relate them to both mainstream and formal epistemology. The result is a long-anticipated mathematical notion of justification, making epistemic logic more expressive. We now have the capacity to reason about justifications, simple and compound. We can compare different pieces of evidence pertaining to the same fact. We can measure the complexity of justifications, which leads to a coherent theory of logical omniscience. Justification Logic provides a novel, evidence-based mechanism of truth-tracking which seems to be a key ingredient of the analysis of knowledge. Finally, Justification Logic furnishes a new, evidence-based foundation for the logic of knowledge, according to which*

$$F \textit{ is known} \quad \sim \quad F \textit{ has an adequate justification.} \tag{32.2}$$

There are several natural interpretations of Justification Logic. Justification assertions of the format $t : F$ read generically as

$$t \textit{ is a justification of } F. \tag{32.3}$$

There is also a more strict 'justificationist' reading in which $t : F$ is understood as

$$t \textit{ is accepted by agent as a justification of } F. \tag{32.4}$$

The language and tools of Justification Logic accommodate both readings of $t : F$. Moreover, Justification Logic is general enough to incorporate other semantics that are not necessarily terminologically related to justifications or proofs. For example, $t : F$ can be read as

$$t \textit{ is a sufficient resource for } F. \tag{32.5}$$

Tudor Protopopescu suggests that $t : F$ could also be assigned an externalist, non-justificationist reading, something like

$$F \textit{ satisfies conditions } t. \tag{32.6}$$

In this setting, t would be something like a set of causes or counterfactuals. Such a reading would still maintain the distinction between partial and factive justifications, since t may not be all that is required for *belief that* F to count as *knowledge that* F.

Within Justification Logic, we do not directly analyze what it means for t to justify F beyond the format $t : F$, but rather attempt to characterize this relation axiomatically. This is similar to the way Boolean logic treats its connectives, say, disjunction: it does not analyze the formula $p \vee q$ but rather assumes certain logical axioms and truth tables about this formula.

There are several design decisions made for this installment of Justification Logic.

1. We decide to limit our attention, at this stage, to *propositional* and *quantifier-free* systems of Justification Logic, and leave quantified systems for further study.
2. We build our systems on the simplest base: *classical Boolean logic*, though we are completely aware that there are much more elaborate logical models, e.g., intuitionistic and substructural logics, conditionals, relevance logics, and logics of counterfactual reasoning, just to name a few. There are several good reasons for choosing the Boolean logic base here. At this stage, we are concerned first with *justifications*, which provide a sufficiently serious challenge on even the simplest Boolean base. Once this case is sorted out in a satisfactory way, we can move on to incorporating justifications into other logics. Second, the paradigmatic examples which we will consider (e.g., Goldman-Kripke and Gettier), can be handled with Boolean Justification Logic. Third, the core of Epistemic Logic consists of modal systems with a classical Boolean base (K, T, K4, S4, K45, KD45, S5, etc.). We provide each of them with a corresponding Justification Logic companion based on Boolean logic.
3. Within the Justification Logic framework, we treat both partial and factive justifications. This helps to capture the essence of discussion on these matters in epistemology, where justifications are not generally assumed to be factive.
4. In this paper, we consider the case of one agent only, although several multi-agent Justification Logic systems have already been developed (Artemov 2006; Artemov and Nogina 2005; Yavorskaya (Sidon) 2006).

Formal logical methods do not directly solve philosophical problems, but rather provide a tool for analyzing assumptions and to ensure that we draw correct conclusions. Our hope is that Justification Logic will do just that.

Preliminary Analysis of Principles Involved

In this section, we will survey the Logic of Proofs, Gettier's examples (Gettier 1963), and examine some classical post-Gettier sources to determine what logical principles in the given Justification Logic format (propositional Boolean logic with justification assertions $t : F$) may be extracted. As is usual with converting informally stated principles into formal ones, a certain amount of good will is required. This does not at all mean that the considerations adduced in Dretske (1971), Goldman (1967), Lehrer and Paxson (1969), Nozick (1981), and Stalnaker (1996) may be readily formulated in the Boolean Justification Logic. The aforementioned papers are written in natural language, which is richer than any formal one; a more sophisticated formal language could probably provide a better account here, which we leave to future studies.

The Logic of Proofs

The Logic of Proofs LP was suggested by Gödel in (1995) and developed in full in Artemov (1995, 2001). LP gives a complete axiomatization of the notion of mathematical proof with natural operations 'application,' 'sum,' and 'proof checker.' We discuss these operations below in a more general epistemic setting.

In LP, justifications are represented by *proof polynomials*, which are terms built from *proof variables* $x, y, z, \ldots$ and *proof constants* $a, b, c, \ldots$ by means of two binary operations: *application* '$\cdot$' and *sum* (*union, choice*) '$+$,' and one unary operation *proof checker* '!'. The formulas of LP are those of propositional classical logic augmented by the formation rule: *if t is a proof polynomial and F a formula, then $t : F$ is again a formula.*

The Logic of Proofs LP contains the postulates of classical propositional logic and the rule of *Modus Ponens* along with

$s : (F \rightarrow G) \rightarrow (t : F \rightarrow (s \cdot t) : G)$ (*Application*)
$s : F \rightarrow (s+t) : F, \quad t : F \rightarrow (s+t) : F$ (*Sum*)
$t : F \rightarrow !t : (t : F)$ (*Proof Checker*)
$t : F \rightarrow F$ (*Reflection*).

Proof constants in LP represent 'atomic' proofs of axioms which are not analyzed any further. In addition to the usual logical properties, such as being closed under substitution and respecting the Deduction Theorem, LP enjoys the Internalization property:

If $\vdash F$, then there is a proof polynomial p such that $\vdash p : F$.

Gettier Examples

Gettier in Gettier (1963) described two situations, Case I and Case II, that were supposed to provide examples of justified true beliefs which should not be considered knowledge. In this paper we will focus on formalizing Case I, which proved to be more challenging. Case II can be easily formalized in a similar fashion.

Here is a shortened exposition of Case I from Gettier (1963).

Suppose that Smith and Jones have applied for a certain job. And suppose that Smith has strong evidence for the following conjunctive proposition:

(*d*) *Jones is the man who will get the job, and Jones has ten coins in his pocket. Proposition (d) entails:*

(e) The man who will get the job has ten coins in his pocket.

Let us suppose that Smith sees the entailment from (d) to (e), and accepts (e) on the grounds of (d), for which he has strong evidence. In this case, Smith is clearly justified in believing that (e) is true. But imagine, further, that unknown to Smith, he himself, not Jones, will get the job. And, also, unknown to Smith, he himself has ten coins in his pocket. Then, all of the following are true:

(1) (e) is true,
(2) Smith believes that (e) is true, and
(3) Smith is justified in believing that (e) is true.

But it is equally clear that Smith does not know that (e) is true. . ..

Gettier uses a version of the epistemic closure principle, closure of justification under logical consequence:

> *... if Smith is justified in believing P, ... and Smith deduces Q from P ..., then Smith is justified in believing Q.*

Here is its natural formalization:

> *Smith is justified in believing P* can be formalized as "for some t, $t : P$";
> *Smith deduces Q from P*—"there is a deduction of $P \to Q$ (available to Smith)";
> *Smith is justified in believing Q*—"$t : Q$ for some t."

Such a rule holds for the Logic of Proofs, as well as for all other Justification Logic systems considered in this paper. It is a combination of the Internalization Rule:

$$\textit{if} \vdash F, \textit{ then} \vdash s : F \textit{ for some } s \tag{32.7}$$

and the Application Axiom:

$$s : (P \to Q) \to (t : P \to (s \cdot t) : Q). \tag{32.8}$$

Indeed, suppose $t : P$ and there is a deduction of $P \to Q$. By the Internalization Rule, $s : (P \to Q)$ for some s. From the Application Axiom, by *Modus Ponens* twice, we get $(s \cdot t) : Q$.

Goldman's Reliabilism

Goldman in Goldman (1967) offered the 'fourth condition' to be added to the Justified True Belief definition of knowledge. According to Goldman (1967),

> *a subject's belief is justified only if the truth of a belief has caused the subject to have that belief (in the appropriate way), and for a justified true belief to count as knowledge, the subject must also be able to 'correctly reconstruct' (mentally) that causal chain.*

Goldman's principle makes it clear that a justified belief (in our language, a situation *t justifies F* for some *t*) for an agent occurs **only if** *F* is true, which provides the Factivity Axiom for 'knowledge-producing' justifications

$$t : F \rightarrow F \qquad \textit{(Factivity Axiom).} \tag{32.9}$$

The Factivity Axiom is assumed for *factive justifications* (systems JT, LP, JT45 below) but not for general justification systems J, J4, J45, JD45.

With a certain amount of good will, we can assume that *the 'causal chain' leading from the truth of F to a justified belief that F* manifests itself in the Principle of Internalization which holds for many Justification Logic systems:

If F is valid, then one could construct a justification p such that p : F is valid. (32.10)

Internalization is usually represented in an equivalent form (in the presence of the Completeness Theorem) as a meta-rule (32.7). The algorithm which builds a justified belief $p : F$ from a strong evidence (proof) of the validity of F seems to be an instance of Goldman's 'causal chain.'

Lehrer and Paxson's Indefeasibility Condition

Lehrer and Paxson in Lehrer and Paxson (1969) offered the following 'indefeasibility condition':

> *there is no further truth which, had the subject known it, would have defeated [subject's] present justification for the belief.*

The 'further truth' here could refer to a possible update of the subject's database, or some possible-worlds situation, etc.: these readings lie outside the scope of our language of Boolean Justification Logic. A natural reading of 'further truth' in our setting could be 'other postulate or assumption of the system,' which means a simple consistency property which vacuously holds for all Justification Logic systems considered here. Another plausible reading of 'further truth' could be 'further evidence,' and we assume this particular reading here. Since there is no temporal or update component in our language yet, 'any further evidence' could be understood for now as 'any other justification,' or just 'any justification.'

Furthermore, Lehrer and Paxson's condition seems to involve a negation of an existential quantifier over justifications 'there is no further truth . . . ,' or

there is no justification. . . .

However, within the classical logic tradition, we can read this as a universal quantifier over justifications followed by a negation

for any further evidence, it is not the case. . . .

Denoting 'present justification for the belief' as the assertion $s : F$, we reformulate Lehrer–Paxson's condition as

given $s : F$, *for any evidence* t, *it is not the case that* t *would have defeated* $s : F$.

The next step is to formalize 't does not defeat $s : F$.' This informal statement seems to suggest an implication

> *if* $s : F$ *holds, then the joint evidence of* s *and* t, *which we denote here as* $s + t$, *is also an evidence for* F, *i.e.*, $(s + t) : F$ *holds.*

Here is the resulting formal version of Lehrer–Paxson's condition: for any proposition F and any justifications s and t, the following holds

$$s : F \rightarrow (s + t) : F \qquad (\textit{Monotonicity Axiom}). \tag{32.11}$$

Further Assumptions

In order to build a formal account of justification, we will make some basic structural assumptions: *justifications are abstract objects which have structure, operations on justifications are potentially executable, agents do not lose or forget justifications, agents apply the laws of classical logic and accept their conclusions, etc.*

In the following, we consider both: **justifications**, which do not necessarily yield the truth of a belief, and **factive justifications**, which yield the truth of the belief.

Basic Principles and Systems

Application

The *Application* operation takes justifications s and t and produces a justification $s \cdot t$ such that if $s : (F \rightarrow G)$ and $t : F$, then $(s \cdot t) : G$. Symbolically,

$$s : (F \rightarrow G) \rightarrow (t : F \rightarrow (s \cdot t) : G). \tag{32.12}$$

This is a basic property of justifications assumed in combinatory logic and λ-calculi (cf. Troelstra and Schwichtenberg 1996), BHK-semantics (Troelstra and van Dalen 1988), Kleene realizability (Kleene 1945), the Logic of Proofs LP (Artemov 2001), etc. Application Principle (32.12) is related to the epistemological closure principle (cf., for example, Dretske 2005; Luper 2005) that one knows everything that one knows to be implied by what one knows. However, (32.12) does not rely on this closure principle, since (32.12) deals with a broader spectrum of justifications, not necessarily linked to knowledge.

Note that the epistemological closure principle which could be formalized using the knowledge modality **K** as

$$\mathbf{K}(F \rightarrow G) \rightarrow (\mathbf{K}F \rightarrow \mathbf{K}G), \tag{32.13}$$

smuggles the *logical omniscience* defect into modal epistemic logic. The latter does not have the capacity to measure how hard it is to attain knowledge (Fagin and Halpern 1985, 1988; Hintikka 1975; Moses 1988; Parikh 1987). Justification Logic provides natural means of escaping logical omniscience by keeping track of the size of justification terms Artemov and Kuznets (2006).

Monotonicity of Justification

The *Monotonicity* property of justification has been expressed by the operation *sum* '+,' which can be read from (32.11). If $s : F$, then whichever evidence t occurs, the combined evidence $s + t$ remains a justification for F. Operation '+' takes justifications s and t and produces $s + t$, which is a justification for everything justified by s or by t.

$$s : F \rightarrow (s + t) : F \quad \text{and} \quad s : F \rightarrow (t + s) : F.$$

A similar operation '+' is present in the Logic of Proofs LP, where the sum '$s + t$' can be interpreted as a concatenation of proofs s and t.

Correspondence Theorem 10 uses Monotonicity to connect Justification Logic with epistemic modal logic. However, it is an intriguing challenge to develop a theory of non-monotonic justifications which prompt belief revision. Some Justification Logic systems without Monotonicity have been studied in Artemov and Strassen (1993) and Krupski (2001, 2006).

Basic Justification Logic $\mathbf{J}_0$

Justification terms (polynomials) are built from justification variables $x, y, z, \ldots$ and justification constants $a, b, c, \ldots$ (with indices $i = 1, 2, 3 \ldots$ which we will be omitting whenever it is safe) by means of the operations **application** '$\cdot$' and **sum** '+'.[1] Constants denote atomic justifications which the system no longer analyzes; variables denote unspecified justifications.

Basic Logic of Justifications J_0:

[1]More elaborate models considered below in this paper also use additional operations on justifications, e.g., verifier '!' and negative verifier '?'.

A1. *Classical propositional axioms and rule Modus Ponens,*
A2. *Application Axiom* $s : (F \to G) \to (t : F \to (s \cdot t) : G)$,
A3. *Monotonicity Axiom* $s : F \to (s + t) : F$, $s : F \to (t + s) : F$,

J_0 is the logic of general (not necessarily factive) justifications for an absolutely skeptical agent for whom no formula is provably justified, i.e., J_0 does not derive $t : F$ for any t and F. Such an agent is, however, capable of making *relative justification conclusions* of the form

$$\textit{if } x : A,\ y : B,\ \ldots,\ z : C \textit{ hold, then } t : F.$$

J_0 is able, with this capacity, to adequately emulate other Justification Logic systems in its language.

Logical Awareness and Constant Specifications

The *Logical Awareness principle* states that logical axioms are justified ex officio: an agent accepts logical axioms (including the ones concerning justifications) as justified. As stated here, Logical Awareness is too restrictive and Justification Logic offers a flexible mechanism of Constant Specifications to represent all shades of logical awareness.

Justification Logic distinguishes between an assumption and a justified assumption. Constants are used to denote justifications of assumptions in situations when we don't analyze these justifications any further. Suppose we want to postulate that an axiom A is justified for a given agent. The way to say it in Justification Logic is to postulate

$$e_1 : A$$

for some evidence constant e_1 with index 1. Furthermore, if we want to postulate that this new principle $e_1 : A$ is also justified, we can postulate

$$e_2 : (e_1 : A)$$

for the similar constant e_2 with index 2, etc. Keeping track of indices is not necessary, but it is easy and helps in decision procedures (cf. Kuznets 2008). The set of all assumptions of this kind for a given logic is called a *Constant Specification*. Here is a formal definition.

A **Constant Specification** CS for a given logic $\mathcal{L}$ is a set of formulas

$$e_n : e_{n-1} : \ldots : e_1 : A \quad (n \geq 1),$$

where A is an axiom of $\mathcal{L}$, and $e_1, e_2, \ldots, e_n$ are similar constants with indices $1, 2, \ldots, n$. We also assume that CS contains all intermediate specifications, i.e., whenever $e_n : e_{n-1} : \ldots : e_1 : A$ is in CS, then $e_{n-1} : \ldots : e_1 : A$ is in CS too. In this paper, we will distinguish the following types of constant specifications:

- *empty*: $CS = \emptyset$. This corresponds to an absolutely skeptical agent (cf. a comment after axioms of J_0).
- *finite*: CS is a finite set of formulas. This is a representative case, since any specific derivation in Justification Logic concerns only finite sets of constants and constant specifications.
- *axiomatically appropriate*: for each axiom A there is a constant e_1 such that $e_1 : A$ is in CS, and if

  $$e_n : e_{n-1} : \ldots : e_1 : A \in CS,$$

  then

  $$e_{n+1} : e_n : e_{n-1} : \ldots : e_1 : A \in CS.$$

  Axiomatically appropriate CS's are necessary for ensuring the Internalization property.
- *total*: for each axiom A and **any** constants $e_1, e_2, \ldots, e_n$,

  $$e_n : e_{n-1} : \ldots : e_1 : A \in CS.$$

  We are reserving the name TCS for the total constant specification (for a given logic). Naturally, the total constant specification is axiomatically appropriate.

Logic of Justifications with given Constant Specification

$$\mathsf{J}_{CS} = \mathsf{J}_0 + CS.$$

Logic of Justifications

$$\mathsf{J} = \mathsf{J}_0 + R4,$$

where R4 is the **Axiom Internalization Rule**:

For each axiom A and any constants $e_1, e_2, \ldots, e_n$, *infer* $e_n : e_{n-1} : \ldots : e_1 : A$.

Note that J_0 is $\mathsf{J}_\emptyset$, and J coincides with J_{TCS}. The latter reflects the idea of the unrestricted Logical Awareness for J. A similar principle appeared in the Logic of Proofs LP; it has also been anticipated in Goldman's Goldman (1967). Note that any **specific** derivation in J may be regarded as a derivation in J_{CS} for a corresponding **finite** constant specification CS, hence finite CS's constitute an important representative class of constant specifications.

Logical Awareness expressed by axiomatically appropriate constant specifications is an explicit incarnation of the Necessitation Rule in modal epistemic logic:

$$\vdash F \quad \Rightarrow \vdash \mathbf{K}F \tag{32.14}$$

applied to axioms.

Let us consider some basic examples of derivations in J. In Examples 1 and 2, only constants of level 1 have been used; in such situations we skip indices completely.

Example 1. This example shows how to build a justification of a conjunction from justifications of the conjuncts. In the traditional modal language, this principle is formalized as

$$\Box A \wedge \Box B \rightarrow \Box (A \wedge B).$$

In J we express this idea in a more precise justification language.

1. $A \rightarrow (B \rightarrow (A \wedge B))$, *a propositional axiom*;
2. $c : [A \rightarrow (B \rightarrow (A \wedge B))]$, *from 1, by R4*;
3. $x : A \rightarrow (c \cdot x) : (B \rightarrow (A \wedge B))$, *from 2, by A2 and Modus Ponens*;
4. $x : A \rightarrow (y : B \rightarrow ((c \cdot x) \cdot y) : (A \wedge B))$, *from 3, by A2 and some propositional reasoning*;
5. $x : A \wedge y : B \rightarrow ((c \cdot x) \cdot y) : (A \wedge B)$, *from 5, by propositional reasoning*.

The derived formula 5 contains constant c, which was introduced in line 2, and the complete reading of the result of this derivation is

$$x : A \wedge y : B \rightarrow ((c \cdot x) \cdot y) : (A \wedge B), \textit{ given } c : [A \rightarrow (B \rightarrow (A \wedge B))].$$

Example 2. This example shows how to build a justification of a disjunction from justifications of either of the disjuncts. In the usual modal language this is represented by

$$\Box A \vee \Box B \rightarrow \Box (A \vee B).$$

Let us see how this would look in J.

1. $A \rightarrow (A \vee B)$, *by A1*;
2. $a : [A \rightarrow (A \vee B)]$, *from 1, by R4*;
3. $x : A \rightarrow (a \cdot x) : (A \vee B)$, *from 2, by A2 and Modus Ponens*;
4. $B \rightarrow (A \vee B)$, *by A1*;
5. $b : [B \rightarrow (A \vee B)]$, *from 4, by R4*;
6. $y : B \rightarrow (b \cdot y) : (A \vee B)$ *from 5, by A2 and Modus Ponens*;
7. $(a \cdot x) : (A \vee B) \rightarrow (a \cdot x + b \cdot y) : (A \vee B)$, *by A3*;
8. $(b \cdot y) : (A \vee B) \rightarrow (a \cdot x + b \cdot y) : (A \vee B)$, *by A3*;
9. $(x : A \vee y : B) \rightarrow (a \cdot x + b \cdot y) : (A \vee B)$ *from 3, 6, 7, 8, by propositional reasoning*.

The complete reading of the result of this derivation is

$$(x : A \vee y : B) \rightarrow (a \cdot x + b \cdot y) : (A \vee B), \textit{ given } a : [A \rightarrow (A \vee B)] \textit{ and } b : [B \rightarrow (A \vee B)].$$

Explicit mention of Constant Specifications of Justification Logic systems is normally used when semantic issues are concerned: e.g., arithmetical, symbolic, and epistemic semantics. To define the truth value of a formula under a given interpretation, one should be given a specification of constants involved.

For each constant specification CS, J_{CS} enjoys the Deduction Theorem, because J_0 contains propositional axioms and *Modus Ponens* as the only rule of inference.

Theorem 3. *For each axiomatically appropriate constant specification CS, J_{CS} enjoys Internalization:*

$$\textit{If} \vdash F, \textit{ then} \vdash p : F \textit{ for some justification term } p.$$

Proof. Induction on derivation length. Suppose $\vdash F$. If F is an axiom, then, since CS is axiomatically appropriate, there is a constant e such that $e : F$ is in CS, hence an axiom of J_{CS}. If F is in CS, then, since CS is axiomatically appropriate, $e : F$ is in CS for some constant e. If F is obtained by *Modus Ponens* from $X \rightarrow F$ and X, then, by the Induction Hypothesis, $\vdash s :(X \rightarrow F)$ and $\vdash t :X$ for some s, t. By the Application Axiom, $\vdash (s \cdot t) :F$. Note that Internalization can require a growth of constant specification sets; if $\vdash F$ with a Constant Specification CS, then the proof of $p :F$ may need some Constant Specification CS' which is different from CS.

Q.E.D.

Red Barn Example and Tracking Justifications

We begin illustrating new capabilities of Justification Logic with a paradigmatic Red Barn Example which Kripke developed in 1980 in objection to Nozick's account of knowledge (cf. article *The Epistemic Closure Principle* in Stanford Encyclopedia of Philosophy (Luper 2005), from which we borrow the formulation, with some editing for brevity).

> *Suppose I am driving through a neighborhood in which, unbeknownst to me, papier-mâché barns are scattered, and I see that the object in front of me is a barn. Because I have barn-before-me percepts, I believe that the object in front of me is a barn. Our intuitions suggest that I fail to know barn. But now suppose that the neighborhood has no fake red barns, and I also notice that the object in front of me is red, so I know a red barn is there. This juxtaposition, being a red barn, which I know, entails there being a barn, which I do not, "is an embarrassment".*[2]

[2]Dretske (2005).

We proceed in the spirit of the Red Barn Example and consider it a general test for theories that explain knowledge. What we want is a way to represent what is going on here which maintains epistemic closure,

one knows everything that one knows to be implied by what one knows, (32.15)

but also preserves the problems the example was intended to illustrate.

We present plausible formal analysis of the Red Barn Example in epistemic modal logic (sections "Red Barn in Modal Logic of Belief" and "Red Barn in Modal Logic of Knowledge") and in Justification Logic (sections "Red Barn in Justification Logic of Belief" and "Red Barn in Justification Logic of Knowledge"). We will see that epistemic modal logic is capable only of telling us that there is a problem, whereas Justification Logic helps to analyse what has gone wrong. We see that closure holds as it is supposed to, and we see that if we keep track of justifications we can analyse why we had a problem.

Red Barn in Modal Logic of Belief

In our first formalization, the logical derivation will be made in epistemic modal logic with 'my belief' modality $\Box$. We then interpret some of the occurrences of $\Box$ as 'knowledge' according to the problem's description. We will not try to capture the whole scenario formally; to make our point, it suffices to formalize and verify its "entailment" part. Let

- B be 'the object in front of me is a barn,'
- R be 'the object in front of me is red,'
- $\Box$ be 'my belief' modality.

The formulation considers observations 'I see a barn' and 'I see a red barn,' and claims logical dependencies between them. The following is a natural formalization of these assumptions in the epistemic modal logic of belief:

1. $\Box B$, 'I believe that the object in front of me is a barn';
2. $\Box(B \wedge R)$, 'I believe that the object in front of me is a red barn.'

At the metalevel, we assume that 2 is knowledge, whereas 1 is not knowledge by the problem's description. So, we could add factivity of 2, $\Box(B \wedge R) \rightarrow (B \wedge R)$, to the formal description, but this would not matter for our conclusions. We note that indeed 1 logically follows from 2 in the modal logic of belief K:

3. $(B \wedge R) \rightarrow B$, *logical axiom*;
4. $\Box[(B \wedge R) \rightarrow B]$, *from 3, by Necessitation. As a logical truth, this is a case of knowledge too*;
5. $\Box(B \wedge R) \rightarrow \Box B$, *from 4, by modal logic.*

Within this formalization, it appears that Closure Principle (32.15) is violated: $\Box(B\wedge R)$ is knowledge by the problem's description, $\Box[(B\wedge R)\rightarrow B]$ is knowledge as a simple logical axiom, whereas $\Box B$ is not knowledge.

Red Barn in Modal Logic of Knowledge

Now we will use epistemic modal logic with 'my knowledge' modality **K**. Here is a straightforward formalization of Red Barn Example assumptions:

1. $\neg\mathbf{K}B$, 'I do not know that the object in front of me is a barn';
2. $\mathbf{K}(B\wedge R)$, 'I know that the object in front of me is a red barn.'

It is easy to see that these assumptions are inconsistent in the modal logic of knowledge. Indeed,

3. $\mathbf{K}(B\wedge R)\rightarrow(\mathbf{K}B\wedge\mathbf{K}R)$, *by normal modal logic*;
4. $\mathbf{K}B\wedge\mathbf{K}R$, *from 2 and 3, by Modus Ponens*;
5. $\mathbf{K}B$, *from 4, by propositional logic.*

Lines 1 and 5 formally contradict each other.

Modal logic of knowledge does not seem to apply here.

Red Barn in Justification Logic of Belief

Justification Logic seems to provide a more fine-grained analysis of the Red Barn Example. We naturally refine assumptions by introducing individual justifications u for belief that B, and v for belief that $B\wedge R$. The set of assumptions in the Justification Logic is

1. $u : B$, 'u is the reason to believe that the object in front of me is a barn';
2. $v : (B\wedge R)$, 'v is the reason to believe that the object in front of me is a red barn.' *On the metalevel, the description states that this is a case of knowledge, not merely a belief.*

Again, we can add the factivity condition for 2, $v : (B\wedge R)\rightarrow(B\wedge R)$, but this does not change the analysis here. Let us try to reconstruct the reasoning of the agent in J:

3. $(B\wedge R)\rightarrow B$, *logical axiom*;
4. $a : [(B\wedge R)\rightarrow B]$, *from 3, by Axiom Internalization. This is also knowledge, as before*;
5. $v : (B\wedge R)\rightarrow(a\cdot v) : B$, *from 4, by Application and Modus Ponens*;
6. $(a\cdot v) : B$, *from 2 and 5, by Modus Ponens.*

Closure holds! Instead of deriving 1 from 2 as in section "Red Barn in Modal Logic of Belief", we have obtained a correct conclusion that $(a \cdot v) : B$, i.e., 'I know B for reason $a \cdot v$,' which seems to be **different** from u: the latter is the result of a perceptual observation, whereas the former is the result of logical reasoning. In particular, we cannot conclude that 2, $v : (B \wedge R)$, entails 1, $u : B$; moreover, with some basic model theory of J in section "Basic Epistemic Semantics", we can show that 2 **does not entail** 1. Hence, after observing a red façade, I indeed know B, but this knowledge does not come from 1, which remains a case of belief rather than of knowledge.

Red Barn in Justification Logic of Knowledge

Within this formalization, $t : F$ is interpreted as

'I know F for reason t.'

As in section "Red Barn in Modal Logic of Knowledge", we assume

1. $\neg u : B$, 'u is not a sufficient reason to know that the object is a barn';
2. $v : (B \wedge R)$, 'v is a sufficient reason to know that the object is a red barn.'

This is a perfectly consistent set of assumptions in the logic of factive justifications

$$\mathsf{J} + \textit{Factivity Principle}\ (t : F \rightarrow F).$$

As in section "Red Barn in Justification Logic of Belief", we can derive $(a \cdot v) : B$ where $a : [(B \wedge R) \rightarrow B]$, but this does not lead to a contradiction. Claims $\neg u : B$ and $(a \cdot v)$: B naturally co-exist. They refer to different justifications u and $a \cdot v$ of the same fact B; one of them insufficient and the other quite sufficient for my knowledge that B.

It appears that in sections "Red Barn in Justification Logic of Belief" and "Red Barn in Justification Logic of Knowledge", Justification Logic represents the structure of the argument made by Kripke in his Red Barn Example, and which was not captured by traditional epistemic modal tools. The Justification Logic formalization represents what seems to be happening in such a case; we can maintain closure of knowledge under logical entailment, even though 'barn' is not perceptually known.

In all fairness to modal tools, we could imagine a formalization of the Red Barn example in a sort of bi-modal language with distinct modalities for knowledge and belief. However, it seems that such a resolution will, intellectually, involve repeating Justification Logic arguments in a way that obscures, rather than reveals, the truth. Such a bi-modal formalization would distinguish $u : B$ from $(a \cdot v) : B$ not because they have different reasons (which reflects the true epistemic structure of the problem), but rather because the former is labelled 'belief' and the latter 'knowledge.' But what if we need to keep track of a larger number of different unrelated reasons? By introducing a number of distinct modalities and then imposing various assumptions governing the inter-relationships between these modalities, one would essentially end up with a reformulation of the language of Justification Logic itself (with distinct terms replaced by distinct modalities). This suggests that there may not really be a 'halfway point' between the modal language and the language of Justification Logic, at least inasmuch as one tries to capture the essential structure of examples involving the deductive failure of knowledge (e.g., Kripke's Red Barn example). Accordingly, one is either stuck with modal logic and its inferior account of these examples or else moves to Justification Logic and its superior account of these examples. This move can either come about by taking a multi-modal language and imposing inter-dependencies on different modals–ending up with something essentially equivalent to the language of Justification Logic–or else one can use the language of Justification Logic from the start. Either way, all there is to move to is Justification Logic.

Basic Epistemic Semantics

The standard epistemic semantics for J has been provided by the proper adaptation of Kripke-Fitting models Fitting (2005) and Mkrtychev models Mkrtychev (1997).

A Kripke-Fitting **J-model** $\mathcal{M} = (W, R, \mathcal{E}, \Vdash)$ is a Kripke model $(W, R, \Vdash)$ enriched with an **admissible evidence function** $\mathcal{E}$ such that $\mathcal{E}(t, F) \subseteq W$ for any justification t and formula F. Informally, $\mathcal{E}(t, F)$ specifies the set of possible worlds where t is considered admissible evidence for F. The intended use of $\mathcal{E}$ is in the truth definition for justification assertions:

$u \Vdash t : F$ *if and only if*

1. *F holds for all possible situations, that is, $v \Vdash F$ for all v such that uRv;*
2. *t is an admissible evidence for F at u, that is, $u \in \mathcal{E}(t, F)$.*

An admissible evidence function $\mathcal{E}$ must satisfy the closure conditions with respect to operations '$\cdot$' and '$+$':

- *Application*: $\mathcal{E}(s, F \to G) \cap \mathcal{E}(t, F) \subseteq \mathcal{E}(s \cdot t, G)$. This condition states that whenever s is an admissible evidence for $F \to G$ and t is an admissible evidence for F, their 'product,' $s \cdot t$, is an admissible evidence for G.
- *Sum*: $\mathcal{E}(s, F) \cup \mathcal{E}(t, F) \subseteq \mathcal{E}(s+t, F)$. This condition guarantees that $s+t$ is an admissible evidence for F whenever either s is admissible for F or t is admissible for F.

These are natural conditions to place on $\mathcal{E}$ because they are necessary for making basic axioms of Application and Monotonicity valid.

We say that $\mathcal{E}(t, F)$ *holds at a given world u* if $u \in \mathcal{E}(t, F)$.

Given a model $\mathcal{M} = (W, R, \mathcal{E}, \Vdash)$, the forcing relation $\Vdash$ is extended from sentence variables to all formulas as follows: for each $u \in W$,

1. $\Vdash$ respects Boolean connectives at each world ($u \Vdash F \wedge G$ iff $u \Vdash F$ and $u \Vdash G$; $u \Vdash \neg F$ iff $u \not\Vdash F$, etc.);
2. $u \Vdash t : F$ iff $u \in \mathcal{E}(t, F)$ and $v \Vdash F$ for every $v \in W$ with uRv.

Note that an admissible evidence function $\mathcal{E}$ may be regarded as a Fagin-Halpern awareness function (Fagin et al. 1995) equipped with the structure of justifications.

A model $\mathcal{M} = (W, R, \mathcal{E}, \Vdash)$ *respects a Constant Specification CS at* $u \in W$ if $u \in \mathcal{E}(c, A)$ for all formulas $c : A$ from *CS*. Furthermore, $\mathcal{M} = (W, R, \mathcal{E}, \Vdash)$ *respects a Constant Specification CS* if $\mathcal{M}$ respects *CS* at each $u \in W$.

Theorem 4. *For any Constant Specification CS,* J_{CS} *is sound and complete for the class of all Kripke-Fitting models respecting CS.*

Proof. Fix a Constant Specification *CS* and consider J_{CS}.

Soundness is straightforward. Induction on derivations in J_{CS}. Let us check the axioms.

Application. Suppose $u \Vdash s : (F \rightarrow G)$ and $u \Vdash t : F$. Then, by the definition of forcing, $u \in \mathcal{E}(s, F \rightarrow G)$ and $u \in \mathcal{E}(t, F)$, hence, by the closure condition for $\mathcal{E}$, $u \in \mathcal{E}(s \cdot t, G)$. Moreover, for each v such that uRv, $v \Vdash F \rightarrow G$ and $v \Vdash F$, hence $v \Vdash G$. Thus $u \Vdash (s \cdot t) : G$ and $u \Vdash s : (F \rightarrow G) \rightarrow (t : F \rightarrow (s \cdot t) : G)$.
Sum. Suppose $u \Vdash t : F$. Then $u \in \mathcal{E}(t, F)$, hence, by the closure condition for $\mathcal{E}$, $u \in \mathcal{E}(s + t, F)$. In addition, $v \Vdash F$ for each v such that uRv, hence $u \Vdash (s + t) : F$. Thus $u \Vdash t : F \rightarrow (s + t) : F$.

Axioms from *CS* hold at each world, since the models respect *CS*. The Induction Step corresponds to the use of *Modus Ponens*, which is clearly a sound rule here.

To establish completeness, we use standard canonical model construction. The canonical model $\mathcal{M} = (W, R, \mathcal{E}, \Vdash)$ for J_{CS} is defined as follows:

- W is the set of all maximal consistent sets in J_{CS}. Following an established tradition, we denote elements of W as Γ, Δ, etc.;
- $\Gamma R \Delta$ iff $\Gamma^{\sharp} \subseteq \Delta$, where $\Gamma^{\sharp} = \{F \mid t : F \in \Gamma \text{ for some } t\}$;
- $\mathcal{E}(s, F) = \{\Gamma \in W \mid s : F \in \Gamma\}$;
- $\Gamma \Vdash p$ iff $p \in \Gamma$.

The Truth Lemma claims that for all F's,

$$\Gamma \Vdash F \quad \textit{if and only if} \quad F \in \Gamma.$$

This is established by standard induction on the complexity of F. The atomic cases are covered by the definition of '$\Vdash$.' The Boolean induction steps are standard. Consider the case when F is $t : G$ for some t and G.

If $t : G \in \Gamma$, then $G \in \Delta$ for all Δ such that $\Gamma R \Delta$ by the definition of R. By the Induction Hypothesis, $\Delta \Vdash G$. In addition, $\Gamma \in \mathcal{E}(t, G)$ by the definition of $\mathcal{E}$. Hence $\Gamma \Vdash t : G$, i.e., $\Gamma \Vdash F$.

If $t : G \notin \Gamma$, then $\Gamma \notin \mathcal{E}(t, G)$, i.e., $\Gamma \not\Vdash t : G$ and $\Gamma \not\Vdash F$.

Furthermore, $\mathcal{M}$ respects *CS* at each node. Indeed, by the construction of $\mathcal{M}$, $CS \subseteq \Gamma$ for each $\Gamma \in W$. By the Truth Lemma, $\Gamma \Vdash c : A$ for each $c : A \in CS$.

The conclusion of the proof of Theorem 4 is standard. Let F be not derivable in J_{CS}. Then the set $\{\neg F\}$ is consistent. Using the standard saturation construction (Fagin et al. 1995; Meyer and van der Hoek 1995), extend $\{\neg F\}$ to a maximal consistent set Γ. By consistency, $F \notin \Gamma$. By the Truth Lemma, $\Gamma \not\Vdash F$. Q.E.D.

There are several features of the canonical model which could be included into the formulation of the Completeness Theorem to make it stronger.

Strong Evidence. We can show that the canonical model considered in this proof satisfies the Strong Evidence property

$$\Gamma \in \mathcal{E}(t, F) \quad \textit{implies} \quad \Gamma \Vdash t : F.$$

Indeed, let $\Gamma \in \mathcal{E}(t, F)$. By the definition of $\mathcal{E}$, $t : F \in \Gamma$, hence $F \in \Gamma^\sharp$ and $F \in \Delta$ for each Δ such that $\Gamma R \Delta$. By the Truth Lemma, $\Delta \Vdash F$, hence $\Gamma \Vdash t : F$. In a model with the Strong Evidence property there are no void or irrelevant justifications; if t is an admissible evidence for F, then t is a 'real evidence' for F, i.e., F holds at all possible worlds.

Fully Explanatory property for axiomatically appropriate Constant Specifications:

$$\textit{If } \Delta \Vdash F \textit{ for all } \Delta \textit{ such that } \Gamma R \Delta, \textit{ then } \Gamma \Vdash t : F \textit{ for some } t.$$

Note that for axiomatically appropriate constant specifications CS, the Internalization property holds: if G is provable in J_{CS}, then $t : G$ is also provable there for some term t. Here is the proof of the Fully Explanatory property for canonical models.[3] Suppose $\Gamma \not\Vdash t : F$ for any justification term t. Then the set $\Gamma^\sharp \cup \{\neg F\}$ is consistent. Indeed, otherwise for some $t_1 : X_1, t_2 : X_2, \ldots, t_n : X_n \in \Gamma$, $X_1 \to (X_2 \to \ldots \to (X_n \to F)\ldots)$ is provable. By Internalization, there is a justification s such that $s : (X_1 \to (X_2 \to \ldots \to (X_n \to F)\ldots))$ is also provable. By Application, $t_1 : X_1 \to (t_2 : X_2 \to \ldots \to (t_n : X_n \to (s \cdot t_1 \cdot t_2 \cdot \ldots \cdot t_n) : F)\ldots)$ is provable, hence $\Gamma \vdash t : F$ for $t = s \cdot t_1 \cdot t_2 \cdot \ldots \cdot t_n$. Therefore, $\Gamma \Vdash t : F$—a contradiction. Let Δ be a maximal consistent set extending $\Gamma^\sharp \cup \{\neg F\}$. By the definition of R, $\Gamma R \Delta$, by the Truth Lemma, $\Delta \not\Vdash F$, which contradicts the assumptions.

Mkrtychev semantics is a predecessor of Kripke-Fitting semantics (Mkrtychev 1997). *Mrktychev models* are Kripke-Fitting models with a single world, and the proof of Theorem 4 can be easily modified to establish completeness of J_{CS} with respect to Mkrtychev models.

Theorem 5. *For any Constant Specification CS, J_{CS} is sound and complete for the class of Mrktychev models respecting CS.*

Proof. Soundness follows immediately from Theorem 4. For completeness, define the canonical model as in Theorem 4 except for R, which should be taken empty.

[3]This proof for LP was offered by Fitting in Fitting (2005).

This assumption makes the condition '$\Delta \Vdash F$ for all Δ such that $\Gamma R \Delta$' vacuously true, and the forcing condition for justification assertions $\Gamma \Vdash t : F$ becomes equivalent to $\Gamma \in \mathcal{E}(t, F)$, i.e., $t : F \in \Gamma$. This simplification immediately verifies the Truth Lemma.

The conclusion of the proof of Theorem 5 is standard. Let F be not derivable in J_{CS}. Then the set $\{\neg F\}$ is consistent. Using the standard saturation construction, extend it to a maximal consistent set Γ containing $\neg F$. By consistency, $F \notin \Gamma$. By the Truth Lemma, $\Gamma \nVdash F$. The Mkrtychev model consisting of this particular Γ is the desired counter-model for F. The rest of the canonical model is irrelevant.

Q.E.D.

Note that Mkrtychev models built in Theorem 5 are not reflexive, and possess the Strong Evidence property. On the other hand, Mkrtychev models cannot be Fully Explanatory, since '$\Delta \Vdash F$ for all Δ such that $\Gamma R \Delta$' is vacuously true, but $\Gamma \Vdash t : F$ is not.

Theorem 5 shows that the information about Kripke structure in Kripke-Fitting models can be completely encoded by the admissible evidence function. Mkrtychev models play an important theoretical role in Justification Logic (Artemov 2008; Brezhnev and Kuznets 2006; Krupski 2006; Kuznets 2000; Milnikel 2007). On the other hand, as we will see in section "Formalization of Gettier Examples", Kripke-Fitting models can be useful as counter-models with desired properties since they take into account both epistemic Kripke structure and evidence structure. Speaking metaphorically, Kripke-Fitting models naturally reflect two reasons why a certain fact F can be unknown to an agent: F fails at some possible world or an agent does not have a sufficient evidence of F.

Another application area of Kripke-Fitting style models is Justification Logic with both epistemic modalities and justification assertions (cf. Artemov 2006; Artemov and Nogina 2005).

Corollary 6 (Model existence). *For any constant specification CS,* J_{CS} *is consistent and has a model.*

Proof. J_{CS} is consistent. Indeed, suppose J_{CS} proves $\bot$, and erase all justification terms (with ':'s) in each of its formulas. What remains is a chain of propositional formulas provable in classical logic (an easy induction on the length of the original proof) ending with $\bot$—contradiction.

To build a model for J_{CS}, use the Completeness Theorem (Theorem 4). Since J_{CS} does not prove $\bot$, by Completeness, there is a J_{CS}-model (where $\bot$ is false, of course). Q.E.D.

Factivity

Unlike Application and Monotonicity, *Factivity* of justifications is not required in basic Justification Logic systems, which makes the latter capable of representing both partial and factive justifications.

Factivity states that justifications of F are factive, i.e., sufficient for an agent to conclude that F is true. This yields the Factivity Axiom

$$t : F \rightarrow F, \tag{32.16}$$

which has a similar motivation to the Truth Axiom in epistemic modal logic

$$\mathbf{K}F \rightarrow F, \tag{32.17}$$

widely accepted as a basic property of knowledge (Plato, Wittgenstein, Hintikka, etc.).

The Factivity Axiom (32.16) first appeared in the Logic of Proofs LP as a principal feature of mathematical proofs. Indeed, in this setting (32.16) is valid: if there is a mathematical proof t of F, then F must be true.

We adopt the Factivity Axiom (32.16) for justifications that lead to knowledge. However, factivity alone does not warrant knowledge, which has been demonstrated by Russell and Gettier examples (Russell 1912; Gettier 1963)

Logic of Factive Justifications:

$$\mathsf{JT}_0 = \mathsf{J}_0 + \mathrm{A4},$$

$$\mathsf{JT} = \mathsf{J} + \mathrm{A4},$$

with

A4. *Factivity Axiom* $t : F \rightarrow F$.

Systems JT_{CS} corresponding to Constant Specifications CS are defined as in section "Logical Awareness and Constant Specifications".

JT-models are J-models with reflexive accessibility relations R. The reflexivity condition makes each possible world accessible from itself which exactly corresponds to the Factivity Axiom. The direct analogue of Theorem 3 holds for JT_{CS} as well.

Theorem 7. *For any Constant Specification CS, each of the logics* JT_{CS} *is sound and complete with respect to the class of* JT*-models respecting CS.*

Proof. We now proceed as in the proof of Theorem 4. The only addition to soundness is establishing that the Factivity Axiom holds in reflexive models. Let R be reflexive. Suppose $u \Vdash t : F$. Then $v \Vdash F$ for all v such that uRv. By reflexivity of R, uRu, hence $u \Vdash F$ as well.

For completeness, it suffices to check that R in the canonical model is reflexive. Indeed, if $s : F \in \Gamma$, then, by the properties of the maximal consistent sets, $F \in \Gamma$ as well, since JT derives $s : F \rightarrow F$ (with any CS). Hence $\Gamma^\sharp \subseteq \Gamma$ and $\Gamma R \Gamma$. Q.E.D.

Mkrtychev JT-models are singleton JT-models, i.e., JT-models with singleton W's.

Theorem 8. *For any Constant Specification CS, each of the logics* JT_{CS} *is sound and complete with respect to the class of Mkrtychev* JT*-models respecting CS.*

Proof. Soundness follows from Theorem 7. For completeness, we follow the footprints of Theorems 4 and 5, but define the accessibility relation R as

$$\Gamma R \Delta \ \textit{iff} \ \Gamma = \Delta.$$

Q.E.D.

Russell's Example: Induced Factivity

Here is Russell's well-known example from Russell (1912) of an epistemic scenario which can be meaningfully analyzed in Justification Logic.

> *If a man believes that the late Prime Minister's last name began with a 'B,' he believes what is true, since the late Prime Minister was Sir Henry Campbell Bannerman.*[4] *But if he believes that Mr. Balfour was the late Prime Minister, he will still believe that the late Prime Minister's last name began with a 'B,' yet this belief, though true, would not be thought to constitute knowledge.*

As in the Red Barn Example (section "Red Barn Example and Tracking Justifications"), we have to handle a wrong reason for a true justified fact. Again, the tools at Justification Logic seem to be useful and adequate here.

Let B stand for

the late Prime Minister's last name began with a 'B.'

Furthermore, let w be a wrong reason for B and r the right (hence factive) reason for B. Then, Russell's example yields the following assumptions:

$$\{w : B,\ r : B,\ r : B \rightarrow B\}. \tag{32.18}$$

In the original setting (32.18), we do not claim that w is a factive justification for B; moreover, such factivity is not completely consistent with our intuition. Paradoxically, however, in the basic Justification Logic J, we can logically deduce factivity of w from (32.18):

1. $r : B$—*an assumption*;
2. $r : B \rightarrow B$—*an assumption*;
3. B—*from 1 and 2, by Modus Ponens*;
4. $B \rightarrow (w : B \rightarrow B)$—*a propositional axiom*;
5. $w : B \rightarrow B$—*from 3 and 4, by Modus Ponens.*

---

[4]Which was common knowledge back in 1912.

However, this derivation utilizes the fact that r is a factive justification for B to conclude $w : B \rightarrow B$, which constitutes the case of 'induced factivity' of $w : B$. The question is, how can we distinguish the 'real' factivity of $r : B$ from an 'induced factivity' of $w : B$? Again, some sort of truth-tracking is needed here, and Justification Logic seems to do the job. The natural approach would be to consider the set of assumptions (32.18) **without** $r : B$, i.e.,

$$\{w : B,\ r : B \rightarrow B\}, \tag{32.19}$$

and establish that factivity of w, i.e., $w : B \rightarrow B$ is not derivable from (32.19). Here is a J-model $\mathcal{M} = (W, R, \mathcal{E}, \Vdash)$ in which (32.19) holds but $w : B \rightarrow B$ does not.

$W = \{\mathbf{0}\}$, $R = \emptyset$, $\mathbf{0} \nVdash B$, and $\mathcal{E}(t, F)$ holds for all pairs (t, F) except (r, B). It is easy to see that the closure conditions *Application* and *Sum* on $\mathcal{E}$ are fulfilled. At $\mathbf{0}$, $w : B$ holds, that is,

$$\mathbf{0} \Vdash w : B,$$

since w is an admissible evidence for B at $\mathbf{0}$ and there are no possible worlds accessible from $\mathbf{0}$. Furthermore,

$$\mathbf{0} \nVdash r : B,$$

since, according to $\mathcal{E}$, r is not an admissible evidence for B at $\mathbf{0}$. Hence

$$\mathbf{0} \Vdash r : B \rightarrow B.$$

On the other hand,

$$\mathbf{0} \nVdash w : B \rightarrow B$$

since B does not hold at $\mathbf{0}$.

Additional Principles and Systems

In this section, we discuss other principles and operations which may or may not be added to the core Justification Logic systems.

Positive Introspection

One of the common principles of knowledge is identifying *knowing* and *knowing that one knows*. In the formal modal setting, this corresponds to

$$\mathbf{K}F \rightarrow \mathbf{KK}F.$$

This principle has an adequate explicit counterpart: the fact that the agent accepts t as a sufficient evidence of F serves as a sufficient evidence that $t : F$. Often, such meta-evidence has a physical form, e.g., a referee report certifying that a proof of a paper is correct, a computer verification output given a formal proof t of F as an input, a formal proof that t is a proof of F, etc. *Positive Introspection* assumes that given t, the agent produces a justification $!t$ of $t : F$ such that

$$t : F \rightarrow !t : (t : F).$$

Positive Introspection in this operational form first appeared in the Logic of Proofs LP (Artemov 1995, 2001). A similar suggestion was made by Gödel (1995).

We define

$$\mathsf{J4} = \mathsf{J} + \mathrm{A5}$$

and

$$\mathsf{LP} = \mathsf{JT} + \mathrm{A5},$$ [5]

with

A5. *Positive Introspection Axiom* $t : F \rightarrow !t : (t : F)$.

We also define $\mathsf{J4}_0$, $\mathsf{J4}_{CS}$, LP_0, and LP_{CS} in the natural way (cf. section "Logical Awareness and Constant Specifications"). The direct analogue of Theorem 3 holds for $\mathsf{J4}_{CS}$ and LP_{CS} as well.

Note that in the presence of the Positive Introspection Axiom, one could limit the scope of the Axiom Internalization Rule R4 to internalizing axioms which are not yet of the form $e : A$. This is how it has been done in LP: the Axiom Internalization can then be emulated by using $!!e : (!e : (e : A))$ instead of $e_3 : (e_2 : (e_1 : A))$, etc. The notion of Constant Specification could also be simplified accordingly.

Such modifications are minor and they do not affect the main theorems and applications of Justification Logic.

Negative Introspection

Pacuit and Rubtsova considered in Pacuit (2005, 2006) and Rubtsova (2005, 2006) the *Negative Introspection* operation '?' which verifies that a given justification assertion is false. A possible motivation for considering such an operation could be that the positive introspection operation '!' may well be regarded as capable

[5]In our notation, LP can be assigned the name JT4. However, in virtue of a fundamental role played by LP for Justification Logic, we suggest keeping the name LP for this system.

of providing conclusive verification judgments about the validity of justification assertions $t : F$. So, when t is not a justification for F, such a '!' should conclude that $\neg t : F$. This is normally the case for computer proof verifiers, proof checkers in formal theories, etc. This motivation is, however, nuanced: the examples of proof verifiers and proof checkers work with both t and F as inputs, whereas the Pacuit-Rubtsova format $?t$ suggests that the only input for '?' is a justification t, and the result $?t$ is supposed to justify propositions $\neg t : F$ uniformly for all F's for which $t : F$ does not hold. Such an operation '?' does not exist for formal mathematical proofs since $?t$ should be a single proof of infinitely many propositions $\neg t : F$, which is impossible.[6] For what it's worth, we include Negative Introspection in the list of additional justification principles, and leave the decision of whether to accept it or not to the user.

A6. *Negative Introspection Axiom* $\neg t : F \rightarrow ?t : (\neg t : F)$.

We define systems

$$\mathsf{J45} = \mathsf{J4} + \mathrm{A6},$$

$$\mathsf{JD45} = \mathsf{J45} + \neg t : \bot,$$

$$\mathsf{JT45} = \mathsf{J45} + \mathrm{A4},$$

and naturally extend these definitions to $\mathsf{J45}_{CS}$, $\mathsf{JD45}_{CS}$, and $\mathsf{JT45}_{CS}$.

The direct analogue of Theorem 3 holds for $\mathsf{J45}_{CS}$, $\mathsf{JD45}_{CS}$, and $\mathsf{JT45}_{CS}$.

More Epistemic Models

We now define epistemic models for other Justification Logic systems.

- J4-models are J-models with *transitive* R and two additional conditions:
  Monotonicity with respect to R, i.e., $u \in \mathcal{E}(t, F)$ and uRv yield $v \in \mathcal{E}(t, F)$,
  Introspection closure: $\mathcal{E}(t, F) \subseteq \mathcal{E}(!t, t : F)$;
- LP-models are J4-models with *reflexive* R (these are the original Kripke-Fitting models);
- J45-models are J4-models satisfying conditions:
  Negative Introspection closure: $[\mathcal{E}(t, F)]^c \subseteq \mathcal{E}(?t, \neg t : F)$ (Here $[X]^c$ denotes the complement of X.)
  Strong Evidence: $u \Vdash t : F$ for all $u \in \mathcal{E}(t, F)$ (i.e., only 'actual' evidence is admissible).

[6] A proof-compliant way to represent negative introspection in Justification Logic was suggested in Artemov et al. (1999), but we will not consider it here.

Note that J45-models satisfy the *Stability* property: uRv yields '$u \in \mathcal{E}(t, F)$ iff $v \in \mathcal{E}(t, F)$.' In other words, $\mathcal{E}$ is monotone with respect to R^{-1} as well. Indeed, the direction '$u \in \mathcal{E}(t, F)$ yields $v \in \mathcal{E}(t, F)$' is due to Monotonicity. Suppose $u \notin \mathcal{E}(t, F)$. By Negative Introspection closure, $u \in \mathcal{E}(?t, \neg t : F)$. By Strong Evidence, $u \Vdash ?t : (\neg t : F)$. By the definition of forcing, $v \Vdash \neg t : F$, i.e., $v \nVdash t : F$. By Strong Evidence, $v \notin \mathcal{E}(t, F)$.

Note also that the Euclidean property of the accessibility relation R is not required for J45-models and is not needed to establish the soundness of J45 with respect to J45-models. However, the canonical model for J45 is Euclidean, hence both soundness and completeness claims trivially survive an additional requirement that R is Euclidean.

- JD45-models are J45-models with the *Serial* condition on the accessibility relation R: for each u there is v such that uRv holds.
- JT45-models are J45-models with reflexive R. Again, the Euclidean property (or, equivalently, symmetry) of R is not needed for soundness. However, these properties hold for the canonical JT45-model, hence they could be included into the formulation of the Completeness Theorem.

Theorem 9. *Each of the logics* $\mathsf{J4}_{CS}$*,* LP_{CS}*,* $\mathsf{J45}_{CS}$*,* $\mathsf{JT45}_{CS}$ *for any Constant Specification is sound and complete with respect to the corresponding class of epistemic models.* $\mathsf{JD45}_{CS}$ *is complete w.r.t. its epistemic models for axiomatically appropriate CS.*

Proof. We will follow the footprints of the proof of Theorem 4.

1. J4. For soundness, it now suffices to check the validity of the Positive Introspection Axiom at each node of any J4-model. Suppose $u \Vdash t : F$. Then $u \in \mathcal{E}(t, F)$ and $v \Vdash F$ for each v such that uRv. By the closure condition, $u \in \mathcal{E}(!t, t : F)$, and it remains to check that $v \Vdash t : F$. By monotonicity of $\mathcal{E}$, $v \in \mathcal{E}(t, F)$. Now, take any w such that vRw. By transitivity of R, uRw as well, hence $w \Vdash F$. Thus $v \Vdash t : F$, $u \Vdash\ !t : t : F$, and $u \Vdash t : F \rightarrow\ !t : t : F$.

   Completeness is again established as in Theorem 4. It only remains to check that the accessibility relation R is transitive, the admissible evidence function $\mathcal{E}$ is monotone, and the additional closure condition on $\mathcal{E}$ holds.

   Monotonicity. Suppose $\Gamma R \Delta$ and $\Gamma \in \mathcal{E}(t, F)$, i.e., $t : F \in \Gamma$. By maximality of Γ, $!t : t : F \in \Gamma$ as well, since $\mathsf{J4} \vdash t : F \rightarrow !t : t : F$. By definition, $t : F \in \Delta$, i.e., $\Delta \in \mathcal{E}(t, F)$.
   Transitivity. Suppose $\Gamma R \Delta$, $\Delta R \Sigma$, and $t : F \in \Gamma$. Then, by monotonicity, $t : F \in \Delta$. By the definition of R, $F \in \Sigma$, hence $\Gamma R \Sigma$.
   Closure. Suppose $\Gamma \in \mathcal{E}(t, F)$, i.e., $t : F \in \Gamma$. Then as above, $!t : t : F \in \Gamma$, hence $\Gamma \in \mathcal{E}(!t, t : F)$.

2. LP. This is the well-studied case of the Logic of Proofs, (cf. Fitting 2005).
3. J45. Soundness. We have to check the Negative Introspection Axiom. Let $u \Vdash \neg t : F$, i.e., $u \nVdash t : F$. By the Strong Evidence condition, $u \notin \mathcal{E}(t, F)$. By Negative Introspection closure, $u \in \mathcal{E}(?t, \neg t : F)$. By Strong Evidence, $u \Vdash ?t : (\neg t : F)$.

Completeness. We follow the same canonical model construction as in J and J4. The only addition is checking **Negative Introspection closure**. Let $\Gamma \not\in \mathcal{E}(t, F)$. Then $t : F \not\in \Gamma$. By maximality, $\neg t : F \in \Gamma$. By the Negative Introspection Axiom, $?t : (\neg t : F) \in \Gamma$, hence $\Gamma \in \mathcal{E}(?t, \neg t : F)$.

Here is an additional feature of the canonical model that can be included in the formulation of the Completeness Theorem to make it more specific.

R is Euclidean. Let $\Gamma R \Delta$ and $\Gamma R \Delta'$. It suffices to show that $\Delta^\sharp \subseteq \Delta'$. Let $F \in \Delta^\sharp$. Then for some t, $t : F \in \Delta$, i.e., $\Delta \in \mathcal{E}(t, F)$. By Stability, $\Gamma \in \mathcal{E}(t, F)$, hence $t : F \in \Gamma$ and $F \in \Gamma^\sharp$. By the definition of R, $F \in \Delta'$.

4. JD45. The proof can be found in Kuznets (2008).
5. JT45. For soundness, it suffices to check the Factivity Axiom, which easily follows from the reflexivity of R. For completeness, follow the footprints of 3 and note that R is reflexive. Indeed, $\Gamma^\sharp \subseteq \Gamma$ for reflexive theories.

   The additional features of the canonical model are as follows: *R is an equivalence relation, the admissible evidence function does not distinguish equivalent worlds*. This follows easily from 5.

Q.E.D.

Historical survey. The first Justification Logic system LP was introduced in 1995 in Artemov (1995) (cf. also Artemov 2001). Such basic properties of Justification Logic as internalization, realization, arithmetical semantics Artemov (1995, 2001), symbolic models and complexity estimates (Brezhnev and Kuznets 2006; Kuznets 2000; Milnikel 2007; Mkrtychev 1997), and epistemic semantics and completeness (Fitting 2003, 2005) were first established for LP.

A fair amount of work has already been done on Justification Logics other than LP. Systems J, J4, and JT were first considered in Brezhnev (2000) under different names and in a slightly different setting.[7] JT45 appeared independently in Pacuit (2005, 2006) and Rubtsova (2005, 2006), and JD45 in Pacuit (2005, 2006). J45 has, perhaps, first been considered in this work. Systems combining epistemic modalities and justifications were studied in Artemov (2006) and Artemov and Nogina (2004, 2005).

Mkrtychev semantics for J, JT, and J4 with Completeness Theorem were found in Kuznets (2000). Complexity bounds for LP and J4 were found in Kuznets (2000); Milnikel (2007). A comprehensive overview of all decidability and complexity results can be found in Kuznets (2008).

Forgetful Projection and the Correspondence Theorem

An intuitive connection between justification assertions and the justified belief modality $\Box$ involves the informal **existential quantifier**: $\Box F$ is read as

[7]Brezhnev (2000) also considered variants of Justification Logic systems which, in our notations, would be called "JD" and "JD4."

$$\textit{for some } x,\ x : F.$$

The language of Justification Logic does not have quantifiers over justifications, but instead has a sufficiently rich system of operations (polynomials) on justifications. We can use Skolem's idea of replacing quantifiers by functions and view Justification Logic systems as Skolemized logics of knowledge/belief. Naturally, to convert a Justification Logic sentence to the corresponding Epistemic Modal Logic sentence, one can use the **forgetful projection** '$\rightsquigarrow$' that replaces each occurrence of $t : F$ by $\Box F$.

Example: the sentence

$$x : P \rightarrow f(x) : Q$$

can be regarded as a Skolem-style version of

$$\exists x(x : P) \rightarrow \exists y(y : Q),$$

which can be read as

$$\Box P \rightarrow \Box Q,$$

which is the forgetful projection of the original sentence $x : P \rightarrow f(x) : Q$ (here, P, Q are assumed to be atomic sentences for simplicity's sake).

Examples (P, Q are atomic propositions):

$$\begin{aligned} t : P \rightarrow P \quad &\rightsquigarrow \quad \Box P \rightarrow P, \\ t : P \rightarrow !t : (t : P) \quad &\rightsquigarrow \quad \Box P \rightarrow \Box\Box P, \\ s : (P \rightarrow Q) \rightarrow (t : P \rightarrow (s \cdot t) : Q) \quad &\rightsquigarrow \quad \Box(P \rightarrow Q) \rightarrow (\Box P \rightarrow \Box Q). \end{aligned}$$

Forgetful projection sometimes forgets too much, e.g., a logical triviality $x : P \rightarrow x : P$, a meaningful principle $x : P \rightarrow (x + y) : P$, and a non-valid formula $x : P \rightarrow y : P$ have the same forgetful projection $\Box P \rightarrow \Box P$. However, '$\rightsquigarrow$' always maps valid formulas of Justification Logic to valid formulas of Epistemic Logic. The converse also holds: any valid formula of Epistemic Logic is a forgetful projection of some valid formula of Justification Logic. This follows from Correspondence Theorem 10. We assume that '$\rightsquigarrow$' is naturally extended from sentences to logics.

Theorem 10 (Consolidated Correspondence Theorem).

1. J $\rightsquigarrow$ K
2. JT $\rightsquigarrow$ T
3. J4 $\rightsquigarrow$ K4
4. LP $\rightsquigarrow$ S4
5. J45 $\rightsquigarrow$ K45
6. JD45 $\rightsquigarrow$ KD45
7. JT45 $\rightsquigarrow$ S5

Proof. It is straightforward that the forgetful projection of each of the Justification Logic systems J, JT, J4, LP, J45, JD45, JT45 is derivable in the corresponding epistemic modal logics K, T, K4, S4, K45, KD45, S5, respectively.

The core of Theorem 10 is the Realization Theorem:

> *One can recover justification terms for all modal operators in valid principles of epistemic modal logics* K, T, K4, S4, K45, KD45, *and* S5 *such that the resulting formula is derivable in the corresponding Justification Logic system* J, JT, J4, LP, J45, JD45, *and* JT45.

The important feature of the Realization Theorem is that it recovers realizing functions according to the **existential reading of the modality**, i.e., negative occurrences of the modality are realized by (distinct) free variables, and the positive occurrences by justification polynomials, depending on these variables. For example, $\Box F \rightarrow \Box G$ will be realized by $x : F' \rightarrow f(x) : G'$ where F', G' are realizations of F and G, respectively.

The Realization Theorem was first established for S4/LP (case 4) in Artemov (1995, 2001), cases 1–3 are covered in Brezhnev (2000). The Realization Theorem for 7 is established in Rubtsova (2006) using a very potent method from Fitting (2005), and the proof for 5 and 6 is very similar to Fitting (2005) and Rubtsova (2006) and can be safely omitted here. Q.E.D.

The Correspondence Theorem shows that the major epistemic modal logics K, K4, K45, KD45 (for belief) and T, S4, S5 (for knowledge) have exact Justification Logic counterparts J, J4, J45, JD45 (for partial justifications) and JT, LP, JT45 (for factive justifications).

Foundational Consequences of the Correspondence Theorem

Is there anything new that we have learned from the Correspondence Theorem about epistemic modal logics?

First of all, this theorem provides a new semantics for major modal logics. In addition to the traditional Kripke-style 'universal' reading of $\Box F$ as

F holds in all possible situations,

there is now a rigorous 'existential' semantics for $\Box F$ that reads as

there is a witness (proof, justification) for F.

Perhaps the justification semantics plays a similar role in modal logic to that played by Kleene realizability in intuitionistic logic. In both cases, the intended semantics was **existential**: the Brouwer-Heyting-Kolmogorov interpretation of intuitionistic logic (Heyting 1934; Troelstra and van Dalen 1988; van Dalen 1986) and Gödel's provability reading of S4 (Gödel 1933; Gödel 1995). In both cases, a later possible-world semantics of **universal** character became a highly potent and dominant technical tool. However, in both cases, Kripke semantics did not solve the original semantical problems. It took Kleene realizability (Kleene 1945; Troelstra 1998) to reveal the computational semantics of intuitionistic logic and the Logic of Proofs (Artemov 1995, 2001) to provide exact BHK semantics of proofs for intuitionistic and modal logic.

In the epistemic context, Justification Logic and the Correspondence Theorem add a new 'justification' component to modal logics of knowledge and belief. Again, this new component was in fact an old and central notion which has been widely discussed by mainstream epistemologists but has remained out of the scope of formal logical methods. The Correspondence Theorem tells us that justifications are compatible with Hintikka-style systems and hence can be regarded as a foundation for epistemic modal logic.

Another comparison suggests itself here: Skolem functions for first-order logic which provide a functional reading of quantifiers. It might seem that Skolem functions do not add much, since they do not suggest altering first-order logic. However, Skolem functions proved to be very useful for foundations (e.g., Henkin and Herbrand models, etc.), as well as for applications (Resolution, Logic Programming, etc.).

Note that the Realization Theorem is not at all trivial. For cases 1–4, realization algorithms are known that use cut-free derivations in the corresponding modal logics (Artemov 1995, 2001; Brezhnev 2000; Brezhnev and Kuznets 2006). For 5–7, the Realization Theorem has been established by Fitting's method or its proper modifications (Fitting 2005; Rubtsova 2006). In principle, these results also produce realization procedures which are based on exhaustive search.

It would be a mistake to draw the conclusion that **any** modal logic has a reasonable Justification Logic counterpart. For example, the logic of formal provability GL (Artemov and Beklemishev 2005; Boolos 1993) contains the *Löb Principle*

$$\Box(\Box F \to F) \to \Box F, \tag{32.20}$$

which does not seem to have an epistemically acceptable explicit version. Let us consider, for example, a case when F is the propositional constant $\bot$ for *false*. A Skolem-style reading of (32.20) suggests that there are justification terms s and t such that

$$x : (s : \bot \to \bot) \to t : \bot. \tag{32.21}$$

This is intuitively false for factive justification, though. Indeed, $s : \bot \to \bot$ is the Factivity Axiom. Apply Axiom Internalization R4 to obtain $c : [s : \bot \to \bot]$ for some constant c. This choice of c makes the antecedent of (32.21) intuitively true and the conclusion of (32.21) false[8]. In particular, (32.20) is not valid for proof interpretation (cf. Goris 2007 for a total account of which principles of GL are realizable).

[8] To be precise, we have to substitute c for x everywhere in s and t.

Quantifier-Free First-Order Justification Logic

In this section, we extend J from the propositional language to the quantifier-free first-order language. To simplify formalities, we will regard here the first-order language without functional symbols, but with equality. Later, in section "Formalization of Gettier Examples", we will introduce definite descriptions in the form $\iota x F(x)$.

The language under consideration in this section is the first-order predicate language with individual variables and constants, predicate symbols of any arity and the equality symbol '=,' along with justification terms (including operations '$\cdot$' and '+') and the formula formation symbol ':' as in section "Basic Justification Logic J_0". Formulas are defined in the usual first-order way (without quantifiers) with an additional clause that if F is a formula and t is a justification polynomial, then $t : F$ is again a formula. The 'quantifier-free J' has all the axioms and rules of J, plus the equality axioms.

The formal system qfJ_0 has the following postulates:

A1. *Classical axioms of quantifier-free first-order logic with equality and Modus Ponens,*

A2. *Application Axiom* $s : (F \rightarrow G) \rightarrow (t : F \rightarrow (s \cdot t) : G)$,

A3. *Monotonicity Axiom* $s : F \rightarrow (s + t) : F,\ s : F \rightarrow (t + s) : F$,

E1. $g = g$ *for any individual term g (reflexivity of equality)*;

E2. $f = g \rightarrow (P[f/x] \rightarrow P[g/x])$ *(substitutivity of equality)*, where f and g are individual terms, P is any atomic formula, $P[f/x]$ and $P[g/x]$ are the results of replacing all the occurrences of a variable x in P by f and g respectively; we will use notations $P(f)$, $P(g)$ for that.

The system qfJ is qfJ_0 + R4, where

R4. *For each axiom A and any constants* $e_1, e_2, \ldots, e_n$, *infer* $e_n : e_{n-1} : \ldots : e_1 : A$.

As in section "Logical Awareness and Constant Specifications", we define Constant Specifications and systems qfJ_{CS}. In particular, $\mathsf{qfJ}_{\emptyset}$ is qfJ_0 and qfJ_{TCS} is qfJ.

The following proposition follows easily from the definitions.

Proposition 11. *Deduction Theorem holds for* qfJ_{CS} *for any constant specification CS. Internalization holds for* qfJ_{CS} *for an axiomatically appropriate constant specification CS.*

The following theorem provides a way to resolve the Frege puzzle Frege (1952) in an epistemic environment: equality of individual objects alone does not warrant substitutivity, but justified equality does.

Theorem 12 (Justified substitution). *For any individual terms f and g, justification variable u, and atomic formula* $P(x)$, *there is a justification term* $s(u)$ *such that* qfJ *proves*

$$u : (f = g) \rightarrow s(u) : [P(f) \leftrightarrow P(g)].$$

The same holds for any qfJ_{CS} *with an axiomatically appropriate constant specification CS.*

Proof. Taking into account Example 1, it suffices to establish that for some $t(u)$,

$$u : (f = g) \rightarrow t(u) : [P(f) \rightarrow P(g)].$$

From E2 it follows that qfJ proves

$$(f = g) \rightarrow [P(f) \rightarrow P(g)].$$

By R4, there is a justification constant c such that qfJ proves

$$c : \{(f = g) \rightarrow [P(f) \rightarrow P(g)]\}.$$

By A2, qfJ proves

$$c : \{(f = g) \rightarrow [P(f) \rightarrow P(g)]\} \rightarrow \{u : (f = g) \rightarrow (c \cdot u) : [P(f) \rightarrow P(g)]\}.$$

By Modus Ponens, qfJ proves

$$u : (f = g) \rightarrow (c \cdot u) : [P(f) \rightarrow P(g)].$$

It suffices now to pick $c \cdot u$ as $t(u)$. Q.E.D.

An unjustified substitution can fail in qfJ. Namely, for any individual variables x and y, a predicate symbol P, and justification term s, the formula

$$(x = y) \rightarrow s : [P(x) \leftrightarrow P(y)] \tag{32.22}$$

is not valid. To establish this, one needs some model theory for qfJ.

We define *qfJ-models* as the usual first-order Kripke models[9] equipped with admissible evidence functions. A model is $(W, \{D_w\}, R, \mathcal{E}, \Vdash)$ such that the following properties hold.

- W is an nonempty set of worlds.
- $\{D_w\}$ is the collection of nonempty domains D_w for each $w \in W$.
- R is the binary (accessibility) relation on W.
- $\mathcal{E}$ is the admissible evidence function which for each justification term t and formula F, returns the set of worlds $\mathcal{E}(t, F) \subseteq W$. Informally, these are the worlds

[9] Equality is interpreted as identity in the model.

Fig. 32.1 Kripke-Fitting counter-model for unjustified substitution

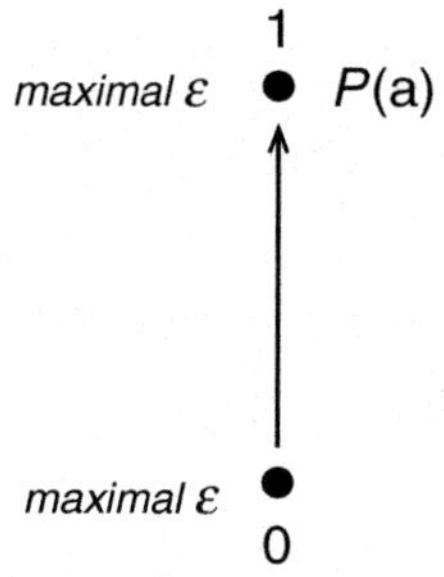

where t is admissible evidence for F. We also assume that $\mathcal{E}$ satisfies the usual closure properties *Application* and *Sum* (section "Basic Epistemic Semantics").

- $\Vdash$ is the forcing (truth) relation such that

  $\Vdash$ assigns elements of D_w to individual variables and constants for each $w \in W$,
  for each n-ary predicate symbol P, and any $a_1, a_2, \ldots, a_n \in D_w$, it is specified whether $P(a_1, a_2, \ldots, a_n)$ holds in D_w,
  $\Vdash$ is extended to all the formulas by stipulating that
  $w \Vdash s = t$ iff '$\Vdash$' maps s and t to the same element of D_w,
  $w \Vdash P(t_1, t_2, \ldots, t_n)$ iff '$\Vdash$' maps t_i's to a_i's and $P(a_1, a_2, \ldots, a_n)$ holds in D_w,
  $w \Vdash F \wedge G$ iff $w \Vdash F$ and $w \Vdash G$,
  $w \Vdash \neg F$ iff $w \not\Vdash F$,
  $w \Vdash t : F$ iff $v \Vdash F$ for all v such that wRv, and $w \in \mathcal{E}(t, F)$.

The notion of a model respecting given constant specification is directly transferred from section "Basic Epistemic Semantics".

The following Theorem is established in the same manner as the soundness part of Theorem 4.

Theorem 13. *For any Constant Specification CS,* qfJ_{CS} *is sound with respect to the corresponding class of epistemic models.*

We are now ready to show that instances of unjustified substitution can fail in qfJ. To do this, it now suffices to build a qfJ-counter-model for (32.22) with the total constant specification. Obviously, the *maximal* $\mathcal{E}$ (i.e., $\mathcal{E}(t, F)$ contains each world for any t and F) respects any constant specification.

The Kripke-Fitting counter-model in Fig. 32.1 exploits the traditional modal approach to refute a belief assertion by presenting a possible world where the object of this belief does not hold. In the picture, only true atomic formulas are shown next to possible worlds.

- $W = \{\mathbf{0}, \mathbf{1}\}$; $R = \{(\mathbf{0}, \mathbf{1})\}$; $D_0 = D_1 = \{a, b\}$;
- $\mathbf{1} \Vdash P(a)$ and $\mathbf{1} \not\Vdash P(b)$; the truth value of P at $\mathbf{0}$ does not matter;
- x and y are interpreted as a at $\mathbf{0}$; x is interpreted as a and y as b at $\mathbf{1}$;
- $\mathcal{E}$ is maximal at $\mathbf{0}$ and $\mathbf{1}$.

Obviously, $\mathbf{0} \Vdash x = y$. Since $\mathbf{1} \nVdash P(x) \leftrightarrow P(y)$, for any justification term s, $\mathbf{0} \nVdash s:[P(x) \leftrightarrow P(y)]$. Hence

$$\mathbf{0} \nVdash x = y \rightarrow s:[P(x) \leftrightarrow P(y)].$$

Formalization of Gettier Examples

We consider Gettier's Case I in detail; Case II is much simpler logically and can be given similar treatment. We will present a complete formalization of Case I in qfJ with a definite description operation. Let

- $J(x)$ be the predicate *x gets the job*;
- $C(x)$ be the predicate *x has (ten) coins (in his pocket)*;
- `Jones` and `Smith` be individual constants denoting Jones and Smith, respectively[10];
- u be a justification variable.

Natural Model for Case I

Gettier's assumptions (d) and (e) contain a definite description

the man who will get the job. (32.23)

In this section, we will formalize Case I using a definite description ι-operation such that $\iota x P(x)$ is intended to denote

the x such that $P(x)$.

We interpret $\iota x P(x)$ in a given world of a qfJ-model as the element a such that $P(a)$ if there exists a unique a satisfying $P(a)$. Otherwise, $\iota x P(x)$ is undefined and any atomic formula where $\iota x P(x)$ actually occurs is taken to be false. Definite description terms are non-rigid designators: $\iota x P(x)$ may be given different interpretations in different worlds of the same qfJ-model (cf. Fitting 2007). The use of a definite description

Jones is the man who will get the job

as a justified belief by Smith hints that Smith has strong evidence for the fact that at most one person will get the job. This is implicit in Gettier's assumption.

---

[10]Assuming that there are people seeking the job other than Jones and Smith does not change the analysis.

We now present a Fitting model $\mathcal{M}$ which may be regarded as an exact epistemic formulation of Case I.

1. At the actual world **0**, $J(\mathtt{Smith})$, $C(\mathtt{Smith})$, and $C(\mathtt{Jones})$[11] hold and $J(\mathtt{Jones})$ does not hold.
2. There is a possible belief world **1** for Smith at which $J(\mathtt{Jones})$ and $\neg J(\mathtt{Smith})$ hold. These conditions follow from proposition (d)

   Jones is the man who will get the job, and Jones has coins

   or, in logic form,

   $$(\mathtt{Jones} = \iota x J(x)) \;\wedge\; C(\mathtt{Jones})$$

   for which Smith has a strong evidence. In addition, Smith has no knowledge of 'Smith has coins' and there should be a possible world at which $C(\mathtt{Smith})$ is false; we use **1** to represent this possibility.
3. World **1** is accessible from **0**.
4. Smith has a *strong evidence of (d)*, which we will represent by introducing a justification variable u such that

   $$u : [(\mathtt{Jones} = \iota x J(x)) \wedge C(\mathtt{Jones})] \tag{32.24}$$

   holds at the actual world **0**. We further assume that the admissible evidence function $\mathcal{E}$ respects the justification assertion (32.24), which yields

   $$\mathbf{0} \in \mathcal{E}(u, (\mathtt{Jones} = \iota x J(x)) \wedge C(\mathtt{Jones})).$$

   To keep things simple, we can assume that $\mathcal{E}$ is the maximal admissible evidence function, i.e., $\mathcal{E}(t, F) = \{\mathbf{0}, \mathbf{1}\}$ for each t, F.

   These observations lead to the following model $\mathcal{M}$ on Fig. 32.2.

Fig. 32.2 Natural Kripke-Fitting model for Gettier Case I

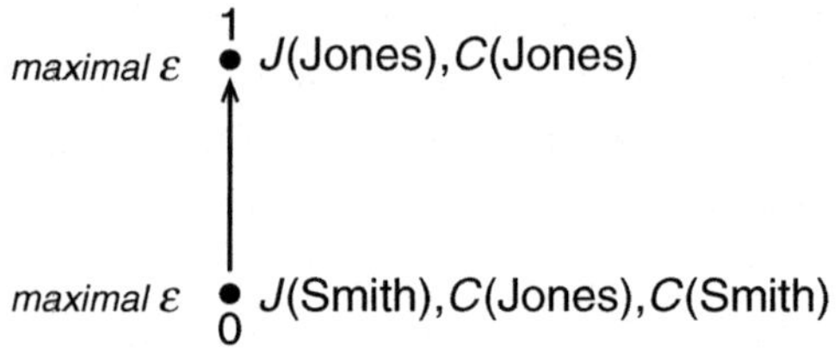

[11] Strictly speaking, Case I explicitly states only that Smith has a strong evidence that $C(\mathtt{Jones})$, which is not sufficient to conclude that $C(\mathtt{Jones})$, since Smith's justifications are not necessarily factive. However, since the actual truth value of $C(\mathtt{Jones})$ does not matter in Case I, we assume that in this instance, Smith's belief that $C(\mathtt{Jones})$ was true.

- $W = \{\mathbf{0}, \mathbf{1}\}$; $R = \{(\mathbf{0}, \mathbf{1})\}$;
- $D_{0,1} = \{\text{Jones}, \text{Smith}\}$, `Jones` is interpreted as 'Jones' and `Smith` as 'Smith';
- $\mathbf{0} \Vdash J(\texttt{Smith}),\ C(\texttt{Jones}),\ C(\texttt{Smith}), \neg J(\texttt{Jones})$;
- $\mathbf{1} \Vdash J(\texttt{Jones}),\ C(\texttt{Jones}),\ \neg J(\texttt{Smith}), \neg C(\texttt{Smith})$;
- $\iota x J(x)$ at $\mathbf{0}$ is interpreted as `Smith` and at $\mathbf{1}$ as `Jones`;
- $\mathcal{E}$ is maximal at both $\mathbf{0}$ and $\mathbf{1}$.

It is interesting to compare this model with the axiomatic description of Case I. Here is the list of explicit assumptions:

$$J(\texttt{Smith}),\ C(\texttt{Smith}),\ C(\texttt{Jones}),\ \neg J(\texttt{Jones}),\ u : [(\texttt{Jones} = \iota x J(x)) \wedge C(\texttt{Jones})]. \tag{32.25}$$

It follows from the Soundness Theorem 13 that assumptions (32.25) provide a sound description of the actual world:

Proposition 14. qfJ + *(32.25)* $\vdash F$ *entails* $\mathbf{0} \Vdash F$.

Example 15. The description of a model by (32.25) is not complete. For example, conditions (32.25) do not specifically indicate whether $t : C(\texttt{Smith})$ holds at the actual world for some t, whereas it is clear from the model that $\mathbf{0} \nVdash t : C(\texttt{Smith})$ for any t since $\mathbf{1} \nVdash C(\texttt{Smith})$ and $\mathbf{1}$ is accessible from $\mathbf{0}$. Model $\mathcal{M}$ extends the set of assumptions (32.25) to a possible **complete** specification: every ground proposition F in the language of this example is either true or false at the 'actual' world $\mathbf{0}$ of the model.

Formalizing Gettier's Reasoning

Gettier's conclusion in Case I states that *Smith is justified in believing that 'The man who will get the job has ten coins in his pocket.'* In our formal language, this amounts to a statement that for some justification term t,

$$t : C(\iota x J(x)) \tag{32.26}$$

is derivable in qfJ from assumptions of Case I.

Theorem 16. *Gettier's conclusion* $t : C(\iota x J(x))$ *is derivable in* qfJ *from assumptions (32.25) of Case I. Furthermore,* $t : C(\iota x J(x))$ *holds at the 'actual world'* $\mathbf{0}$ *of the natural model* $\mathcal{M}$ *of Case I.*

Proof. In order to find t we may mimic Gettier's informal reasoning. First, we formally derive (e) (i.e., $C(\iota x J(x))$) from (d) (i.e., $\texttt{Jones} = \iota x J(x) \wedge C(\texttt{Jones})$) and then use the fact that (d) is justified (i.e., $u : [\texttt{Jones} = \iota x J(x) \wedge C(\texttt{Jones})]$). We will now show that this argument can be formalized in qfJ. Note that in qfJ, we may reason as follows:

1. $\texttt{Jones} = \iota x J(x) \rightarrow [C(\texttt{Jones}) \rightarrow C(\iota x J(x))]$, *an axiom of* qfJ;
2. $[\texttt{Jones} = \iota x J(x) \wedge C(\texttt{Jones})] \rightarrow C(\iota x J(x))$, *by propositional reasoning, from 1;*
3. $s : \{[\texttt{Jones} = \iota x J(x) \wedge C(\texttt{Jones})] \rightarrow C(\iota x J(x))\}$, *by Internalization, from 2;*
4. $u : [\texttt{Jones} = \iota x J(x) \wedge C(\texttt{Jones})] \rightarrow (s \cdot u) : C(\iota x J(x))$, *by Axiom A2 and Modus Ponens, from 3;*
5. $u : [\texttt{Jones} = \iota x J(x) \wedge C(\texttt{Jones})]$, *an assumption from (32.25);*
6. $(s \cdot u) : C(\iota x J(x))$, *by Modus Ponens, from 4 and 5.*

Now we can pick t to be $s \cdot u$. So,

$$\mathsf{qfJ} + (32.25) \vdash (s \cdot u) : C(\iota x J(x))$$

and, by Proposition 14,

$$\mathbf{0} \Vdash (s \cdot u) : C(\iota x J(x)).$$

Q.E.D.

Eliminating Definite Descriptions, Russell-Style

We can eliminate definite descriptions from Case I using, e.g., Russell's translation (cf. Fitting and Mendelsohn 1998; Neale 1990; Russell 1905, 1919) of definite descriptions. According to Russell, $C(\iota x J(x))$ contains a hidden *uniqueness assumption* and reads as

$$\exists x[J(x) \wedge \forall y(J(y) \rightarrow y = x) \wedge C(x)], \tag{32.27}$$

and $\texttt{Jones} = \iota x J(x)$ as

$$J(\texttt{Jones}) \wedge \forall y(J(y) \rightarrow y = \texttt{Jones}). \tag{32.28}$$

In addition, in the universe of Case I consisting of two objects *Jones, Smith*, a universally quantified sentence $\forall y F(y)$ reads as

$$F(\texttt{Jones}) \wedge F(\texttt{Smith}),$$

and an existentially quantified statement $\exists x G(x)$ reads as

$$G(\texttt{Jones}) \vee G(\texttt{Smith}).$$

Taking into account all of these simplifying observations, we may assume that for Smith (and the reader), $\forall y(J(y) \rightarrow y = \mathtt{Jones})$ reads as

$$[J(\mathtt{Jones}) \rightarrow (\mathtt{Jones} = \mathtt{Jones})] \wedge [J(\mathtt{Smith}) \rightarrow (\mathtt{Smith} = \mathtt{Jones})],$$

which is equivalent[12] to

$$\neg J(\mathtt{Smith}).$$

Now, (32.28) is equivalent to

$$J(\mathtt{Jones}) \wedge \neg J(\mathtt{Smith}),$$

and the whole Gettier proposition (d) collapses to

$$J(\mathtt{Jones}) \wedge \neg J(\mathtt{Smith}) \wedge C(\mathtt{Jones}). \tag{32.29}$$

The assumption that (d) is justified for Smith can now be represented by

$$v : [J(\mathtt{Jones}) \wedge \neg J(\mathtt{Smith}) \wedge C(\mathtt{Jones})], \tag{32.30}$$

for some justification variable v.

Smith's justified belief

'the man who will get the job has coins,' (32.31)

according to Russell, should read as

$$\exists x[J(x) \wedge \forall y(J(y) \rightarrow y = x) \wedge C(x)]. \tag{32.32}$$

The same considerations as above show that

$$\forall y[J(y) \rightarrow (y = \mathtt{Jones})]$$

is equivalent to

$$\neg J(\mathtt{Smith}),$$

and

$$\forall y[J(y) \rightarrow (y = \mathtt{Smith})]$$

---

[12] We assume that everybody is aware that $\mathtt{Smith} \neq \mathtt{Jones}$.

is equivalent to

$$\neg J(\texttt{Jones}).$$

Since an existentially quantified formula $\exists x G(x)$ is logically equivalent to a disjunction $G(\texttt{Jones}) \vee G(\texttt{Smith})$, formula (32.32) is equivalent to

$$[J(\texttt{Jones}) \wedge \neg J(\texttt{Smith}) \wedge C(\texttt{Jones})] \vee [J(\texttt{Smith}) \wedge \neg J(\texttt{Jones}) \wedge C(\texttt{Smith})]. \tag{32.33}$$

Finally, the formalization of (32.31) in our language amounts to stating that for some justification term p,

$$p : \{[J(\texttt{Jones}) \wedge \neg J(\texttt{Smith}) \wedge C(\texttt{Jones})] \vee [J(\texttt{Smith}) \wedge \neg J(\texttt{Jones}) \wedge C(\texttt{Smith})]\}. \tag{32.34}$$

Theorem 17. *Gettier's claim (32.34) is derivable in* qfJ *from the assumption (32.30) of Case I, and holds in the 'actual world'* **0** *of the natural model* $\mathcal{M}$ *of Case I.*

Proof. After all the preliminary work and assumptions, there is not much left to do. We just note that (32.29) is a disjunct of (32.33). A derivation of (32.34) from (32.30) in qfJ reduces now to repeating steps of Example 2, which shows how to derive a justified disjunction from its justified disjunct. Q.E.D.

Comment 18. *One can see clearly the essence of Gettier's example. In (32.33), one of two disjuncts is justified but false, whereas the other disjunct is unjustified but true. The resulting disjunction (32.33) is both justified and true, but not really known to Smith.*

Hidden Uniqueness Assumption Is Necessary

In this subsection, we study what happens if we deviate from Russell's reading of definite descriptions, in particular if we skip the uniqueness of the defined object. For example, let us read Gettier's proposition (d) as

Jones will get the job, and Jones has ten coins in his pocket, (32.35)

and proposition (e) as

A man who will get the job has ten coins in his pocket. (32.36)

Then a fair formalization of (32.35) would be

$$J(\texttt{Jones}) \wedge C(\texttt{Jones}), \tag{32.37}$$

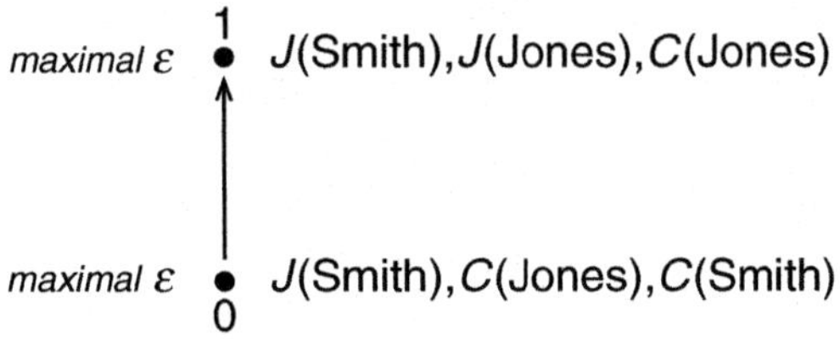

Fig. 32.3 Counter-model for Case I without uniqueness

and the assumption that (32.35) is justified for Smith is formalized as

$$u : [J(\texttt{Jones}) \wedge C(\texttt{Jones})]. \tag{32.38}$$

In this case, the set of explicitly made non-logical assumptions is

1. $u : [J(\texttt{Jones}) \wedge C(\texttt{Jones})]$, *assumption (32.38)*;
2. $\neg J(\texttt{Jones})$ *(Jones does not get the job)*;
3. $J(\texttt{Smith})$ *(Smith gets the job)*;
4. $C(\texttt{Smith})$ *(Smith has coins).*

Condition (32.36) naturally formalizes as

$$[J(\texttt{Jones}) \to C(\texttt{Jones})] \wedge [J(\texttt{Smith}) \to C(\texttt{Smith})]. \tag{32.39}$$

The claim that (32.39) is justified for Smith is formalized as

$$t : \{[J(\texttt{Jones}) \to C(\texttt{Jones})] \wedge [J(\texttt{Smith}) \to C(\texttt{Smith})]\} \tag{32.40}$$

for some justification term t.

We show that the assumptions 1–4 above do not suffice for proving (32.40).

Proposition 19. *For any justification term t, formula (32.40) is not derivable in* qfJ *from assumptions 1–4.*

Proof. Suppose (32.40) is derivable in qfJ from assumptions 1–4. Then, by the Deduction Theorem, qfJ would derive

$$\text{'Conjunction of 1–4'} \to (32.40). \tag{32.41}$$

It now suffices to build a Fitting qfJ-model (Fig. 32.3) where (32.41) does not hold at a certain world.

At **0**, all assumptions 1–4 hold, but (32.40) is false at **0** for all t's. Indeed, (32.39) is false at **1**, since its conjunct

$$J(\texttt{Smith}) \to C(\texttt{Smith})$$

is false at **1**, and **1** is accessible from **0**. Q.E.D.

Streamlined Case I: No Coins/Pockets Are Needed

In this subsection, we show that references to coins and pockets, as well as definite descriptions, are redundant for making the point in Gettier example Case I. Here is a simpler, streamlined case based on the same material.

Smith has strong evidence for the proposition:

(d) Jones will get the job.
Proposition (d) entails:
(e) Either Jones or Smith will get the job.

Let us suppose that Smith sees the entailment from (d) to (e), and accepts (e) on the grounds of (d), for which he has strong evidence. In this case, Smith is clearly justified in believing that (e) is true. But imagine further that unknown to Smith, he himself, not Jones, will get the job. Then

(1) (e) is true,
(2) Smith believes that (e) is true, and
(3) Smith is justified in believing that (e) is true.

But it is equally clear that Smith does not know that (e) is true....

In this version, the main assumption is

$$\textit{Smith has a strong evidence that Jones gets the job.} \tag{32.42}$$

Its straightforward formalization is

$$v : J(\texttt{Jones}). \tag{32.43}$$

The claim is that

$$\textit{Smith is justified in believing that either Jones or Smith will get the job.} \tag{32.44}$$

The natural formalization of the claim

$$t : [J(\texttt{Jones}) \vee J(\texttt{Smith})]. \tag{32.45}$$

The set of formal assumptions is

$$v : J(\texttt{Jones}),\ \ J(\texttt{Smith}),\ \ \neg J(\texttt{Jones}).$$

It is easy now to derive (32.45) in qfJ from assumption (32.43).

1. $v : J(\texttt{Jones})$, *assumption (32.43)*;
2. $J(\texttt{Jones}) \to J(\texttt{Jones}) \vee J(\texttt{Smith})$, *propositional axiom*;
3. $c : [J(\texttt{Jones}) \to J(\texttt{Jones}) \vee J(\texttt{Smith})]$, *from 2, by Axiom Internalization R4*;

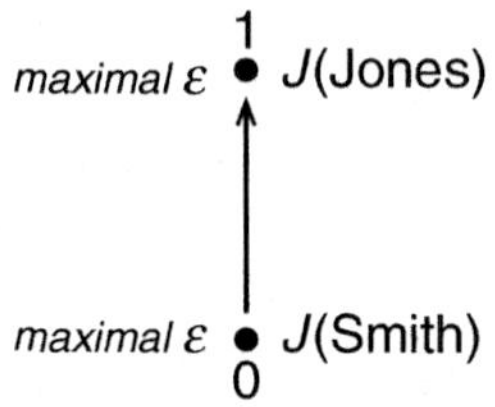

Fig. 32.4 Natural Kripke-Fitting model for the streamlined Case I

4. $c : [J(\texttt{Jones}) \rightarrow J(\texttt{Jones}) \vee J(\texttt{Smith})] \rightarrow [v : J(\texttt{Jones}) \rightarrow (c \cdot v) : (J(\texttt{Jones}) \vee J(\texttt{Smith}))]$, *Axiom A2*;
5. $(c \cdot v) : [J(\texttt{Jones}) \vee J(\texttt{Smith})]$, *from 4, 3, and 1, by Modus Ponens twice.*

At the actual world **0**, both hold:

$$J(\texttt{Jones}) \vee J(\texttt{Smith}) \text{ (meaning } \textit{(e) is true}\text{)}$$

and

$$(c \cdot v) : [J(\texttt{Jones}) \vee J(\texttt{Smith})] \text{(meaning } \textit{(e) is justified}\text{)}.$$

The desired Gettier-style point is made on the same material but without the unnecessary use of quantifiers, definite descriptions, coins, and pockets (Fig. 32.4).

It is fair to note, however, that Gettier example Case II in Gettier (1963) does not have these kinds of redundancies and is logically similar to the streamlined version of Case I presented above.

Gettier Example and Factivity

Theorem 20. *Gettier assumptions (32.25) in Case I are inconsistent in Justification Logic systems with factive justifications.*

Proof. Here is an obvious derivation of a contradiction in qfJT from (32.25):

$u : [(\texttt{Jones} = \iota x J(x)) \wedge C(\texttt{Jones})]$, *by (32.24)*;
$\texttt{Jones} = \iota x J(x)$, *by the Factivity Axiom and some propositional logic*;
$(\texttt{Jones} = \iota x J(x)) \rightarrow J(\texttt{Jones})$, *an assumed natural property of definite descriptions*;
$J(\texttt{Jones})$, *by Modus Ponens. This contradicts the condition* $\neg J(\texttt{Jones})$ *from (32.25).*

Q.E.D.

The question is, what we have learned about Justification, Belief, Knowledge, and other epistemic matters?

Within the domain of formal epistemology, we now have a basic logic machinery to study justifications and their connections with Belief and Knowledge. Formalizing Gettier is a case study that demonstrates the method.

We show that Gettier reasoning was formally correct, with some hidden assumptions related to definite descriptions. Gettier examples belong to the area of Justification Logic dealing with partial justifications and are inconsistent within Justification Logic systems of factive justifications and knowledge. All this, perhaps, does not come as a surprise to epistemologists. However, these observations show that models provided by Justification Logic behave in a reasonable manner.

For epistemology, these developments are furthering the study of justification, e.g., the search for the 'fourth condition' of the JTB definition of knowledge. Justification Logic provides systematic examples of epistemological principles such as Application, Monotonicity, Logical Awareness, and their combinations, which look plausible, at least, within the propositional domain. Further discussion on these and other Justification Logic principles could be an interesting contribution to this area.

Conclusions

Justification Logic extends the logic of knowledge by the formal theory of justification. Justification Logic has roots in mainstream epistemology, mathematical logic, computer science, and artificial intelligence. It is capable of formalizing a significant portion of reasoning about justifications. In particular, we have seen how to formalize Kripke, Russell, and Gettier examples in Justification Logic. This formalization has been used for the resolution of paradoxes, verification, hidden assumption analysis, and eliminating redundancies.

Among other known applications of Justification Logic, so far there are

- intended provability semantics for Gödel's provability logic S4 with the Completeness Theorem (Artemov 1995, 2001);
- formalization of Brouwer-Heyting-Kolmogorov semantics for intuitionistic propositional logic with the Completeness Theorem (Artemov 1995, 2001);
- a general definition of the Logical Omniscience property, rigorous theorems that evidence assertions in Justification Logic are not logically omniscient (Artemov and Kuznets 2006). This provides a general framework for treating the problem of logical omniscience;
- an evidence-based approach to Common Knowledge (so-called Justified Common Knowledge) which provides a rigorous semantics to McCarthy's 'any fool knows' systems (Antonakos 2007; Artemov 2006; McCarthy et al. 1978). Justified Common Knowledge offers formal systems which are less restrictive than the usual epistemic logics with Common Knowledge (Artemov 2006).
- analysis of Knower and Knowability paradoxes (Dean and Kurokawa 2007, 2010).

It remains to be seen to what extent Justification Logic can be useful for analysis of empirical, perceptual, and *a priori* types of knowledge. From the perspective of Justification Logic, such knowledge may be considered as justified by constants (i.e., atomic justifications). Apparently, further discussion is needed here.

Acknowledgements The author is very grateful to Walter Dean, Mel Fitting, Vladimir Krupski, Roman Kuznets, Elena Nogina, Tudor Protopopescu, and Ruili Ye, whose advice helped with this paper. Many thanks to Karen Kletter for editing this text. Thanks to audiences at the CUNY Graduate Center, Bern University, the Collegium Logicum in Vienna, and the 2nd International Workshop on Analytic Proof Systems for comments on earlier versions of this paper. This work has been supported by NSF grant 0830450, CUNY Collaborative Incentive Research Grant CIRG1424, and PSC CUNY Research Grant PSCREG-39-721.

References

Antonakos, E. (2007). Justified and common knowledge: Limited conservativity. In S. Artemov & A. Nerode (Eds.), *Logical Foundations of Computer Science. International Symposium, LFCS 2007, Proceedings*, New York, June 2007 (Lecture notes in computer science, Vol. 4514, pp. 1–11). Springer.

Artemov, S. (1995). Operational modal logic. Technical report MSI 95-29, Cornell University.

Artemov, S. (1999). Understanding constructive semantics. In *Spinoza Lecture for European Association for Logic, Language and Information*, Utrecht, Aug 1999.

Artemov, S. (2001). Explicit provability and constructive semantics. *Bulletin of Symbolic Logic, 7*(1), 1–36.

Artemov, S. (2006). Justified common knowledge. *Theoretical Computer Science, 357*(1–3), 4–22.

Artemov, S. (2007). On two models of provability. In D. M. Gabbay, M. Zakharyaschev, & S. S. Goncharov (Eds.), *Mathematical problems from applied logic II* (pp. 1–52). New York: Springer.

Artemov, S. (2008). Symmetric logic of proofs. In A. Avron, N. Dershowitz, & A. Rabinovich (Eds.), *Pillars of computer science, essays dedicated to Boris (Boaz) Trakhtenbrot on the occasion of his 85th birthday* (Lecture notes in computer science, Vol. 4800, pp. 58–71). Berlin/Heidelberg: Springer.

Artemov, S., & Beklemishev, L. (2005). Provability logic. In D. Gabbay & F. Guenthner (Eds.), *Handbook of philosophical logic* (2nd ed., Vol. 13, pp. 189–360). Dordrecht: Springer.

Artemov, S., Kazakov, E., & Shapiro, D. (1999). Epistemic logic with justifications. Technical report CFIS 99-12, Cornell University.

Artemov, S., & Kuznets, R. (2006). Logical omniscience via proof complexity. In *Computer Science Logic 2006*, Szeged (Lecture notes in computer science, Vol. 4207, pp. 135–149).

Artemov, S., & Nogina, E. (2004). Logic of knowledge with justifications from the provability perspective. Technical report TR-2004011, CUNY Ph.D. Program in Computer Science.

Artemov, S., & Nogina, E. (2005). Introducing justification into epistemic logic. *Journal of Logic and Computation, 15*(6), 1059–1073.

Artemov, S., & Strassen, T. (1993). Functionality in the basic logic of proofs. Technical report IAM 93-004, Department of Computer Science, University of Bern, Switzerland.

Boolos, G. (1993). *The logic of provability*. Cambridge: Cambridge University Press.

Brezhnev, V. (2000). On explicit counterparts of modal logics. Technical report CFIS 2000-05, Cornell University.

Brezhnev, V., & Kuznets, R. (2006). Making knowledge explicit: How hard it is. *Theoretical Computer Science, 357*(1–3), 23–34.

Dean, W., & Kurokawa, H. (2007). From the knowability paradox to the existence of proofs. Manuscript (submitted to *Synthese*).
Dean, W., & Kurokawa, H. (2010). The knower paradox and the quantified logic of proofs. In A. Hieke (Ed.), *Austrian Ludwig Wittgenstein society. Synthese, 176*(2), 177–225.
Dretske, F. (1971). Conclusive reasons. *Australasian Journal of Philosophy, 49*, 1–22.
Dretske, F. (2005). Is knowledge closed under known entailment? The case against closure. In M. Steup & E. Sosa (Eds.), *Contemporary Debates in Epistemology* (pp. 13–26). Malden: Blackwell.
Fagin, R., & Halpern, J. Y. (1985). Belief, awareness, and limited reasoning: Preliminary report. In *Proceedings of the Ninth International Joint Conference on Artificial Intelligence (IJCAI-85)*, (pp. 491–501). Los Altos, CA: Morgan Kaufman.
Fagin, R., & Halpern, J. Y. (1988). Belief, awareness, and limited reasoning. *Artificial Intelligence, 34*(1), 39–76.
Fagin, R., Halpern, J. Y., Moses, Y., & Vardi, M. (1995). *Reasoning about knowledge*. Cambridge: MIT Press.
Fitting, M. (2003). A semantics for the logic of proofs. Technical report TR-2003012, CUNY Ph.D. Program in Computer Science.
Fitting, M. (2005). The logic of proofs, semantically. *Annals of Pure and Applied Logic, 132*(1), 1–25.
Fitting, M. (2007). Intensional logic. Stanford Encyclopedia of Philosophy (http://plato.stanford.edu), Feb 2007.
Fitting, M., & Mendelsohn, R. L. (1998). *First-order modal logic*. Dordrecht: Kluwer Academic.
Frege, G. (1952). On sense and reference. In P. Geach & M. Black (Eds.), *Translations of the philosophical writings of Gottlob Frege*. Oxford: Blackwell.
Gettier, E. (1963). Is justified true belief knowledge? *Analysis, 23*, 121–123.
Gödel, K. (1933). Eine Interpretation des intuitionistischen Aussagenkalkuls. *Ergebnisse Math. Kolloq., 4*, 39–40. English translation in: S. Feferman et al., (Eds.) (1986). *Kurt Gödel collected works* (Vol. 1, pp. 301–303). Oxford: Oxford University Press/New York: Clarendon Press.
Gödel, K. (1995) Vortrag bei Zilsel/Lecture at Zilsel's (*1938a). In S. Feferman, J. W. Dawson Jr., W. Goldfarb, C. Parsons, & R. M. Solovay (Eds.), *Unpublished essays and lectures* (Kurt Gödel collected works, Vol. III, pp. 86–113). Oxford University Press.
Goldman, A. (1967). A causal theory of knowing. *The Journal of Philosophy, 64*, 335–372.
Goris, E. (2007). Explicit proofs in formal provability logic. In S. Artemov & A. Nerode (Eds.), *Logical Foundations of Computer Science. International Symposium, LFCS 2007, Proceedings*, New York, June 2007 (Lecture notes in computer science, Vol. 4514, pp. 241–253). Springer.
Hendricks, V. F. (2003). Active agents. *Journal of Logic, Language and Information, 12*(4), 469–495.
Hendricks, V. F. (2005). *Mainstream and formal epistemology*. New York: Cambridge University Press.
Heyting, A. (1934). *Mathematische Grundlagenforschung. Intuitionismus. Beweistheorie*. Berlin: Springer.
Hintikka, J. (1962). *Knowledge and belief*. Ithaca: Cornell University Press.
Hintikka, J. (1975). Impossible possible worlds vindicated. *Journal of Philosophical Logic, 4*, 475–484.
Kleene, S. (1945). On the interpretation of intuitionistic number theory. *The Journal of Symbolic Logic, 10*(4), 109–124.
Krupski, N. V. (2006). On the complexity of the reflected logic of proofs. *Theoretical Computer Science, 357*(1), 136–142.
Krupski, V. N. (2001). The single-conclusion proof logic and inference rules specification. *Annals of Pure and Applied Logic, 113*(1–3), 181–206.
Krupski, V. N. (2006). Referential logic of proofs. *Theoretical Computer Science, 357*(1), 143–166.
Kuznets, R. (2000). On the complexity of explicit modal logics. In *Computer Science Logic 2000* (Lecture notes in computer science, Vol. 1862, pp. 371–383). Berlin/Heidelberg: Springer.

Kuznets, R. (2008). *Complexity issues in justification logic*. PhD thesis, CUNY Graduate Center. http://kuznets.googlepages.com/PhD.pdf.

Lehrer, K., & Paxson, T. (1969). Knowledge: Undefeated justified true belief. *The Journal of Philosophy, 66*, 1–22.

Luper, S. (2005). The epistemic closure principle. Stanford Encyclopedia of Philosophy.

McCarthy, J., Sato, M., Hayashi, T., & Igarishi, S. (1978). On the model theory of knowledge. Technical report STAN-CS-78-667, Stanford University.

Meyer, J. -J. Ch., & van der Hoek, W. (1995). *Epistemic logic for AI and computer science*. Cambridge: Cambridge University Press.

Milnikel, R. (2007). Derivability in certain subsystems of the logic of proofs is Π_2^p-complete. *Annals of Pure and Applied Logic, 145*(3), 223–239.

Mkrtychev, A. (1997). Models for the logic of proofs. In S. Adian & A. Nerode (Eds.), *Logical Foundations of Computer Science '97, Yaroslavl'* (Lecture notes in computer science, Vol. 1234, pp. 266–275). Springer.

Moses, Y. (1988). Resource-bounded knowledge. In M. Vardi (Ed.), *Proceedings of the Second Conference on Theoretical Aspects of Reasoning About Knowledge*, Pacific Grove, March 7–9, 1988 (pp. 261–276). Morgan Kaufmann Publishers.

Neale, S. (1990). *Descriptions*. Cambridge: MIT.

Nozick, R. (1981). *Philosophical explanations*. Cambridge: Harvard University Press.

Pacuit, E. (2005). A note on some explicit modal logics. In *5th Panhellenic Logic Symposium*, Athens, July 2005.

Pacuit, E. (2006). A note on some explicit modal logics. Technical report PP-2006-29, University of Amsterdam. ILLC Publications.

Parikh, R. (1987). Knowledge and the problem of logical omniscience. In Z. Ras & M. Zemankova (Eds.), *ISMIS-87 International Symposium on Methodology for Intellectual Systems* (pp. 432–439). North-Holland.

Rubtsova, N. (2005). Evidence- for based knowledge S5. In *2005 Summer Meeting of the Association for Symbolic Logic, Logic Colloquium '05*, Athens, 28 July–3 August 2005. Abstract. Association for Symbolic Logic. (2006, June). *Bulletin of Symbolic Logic, 12*(2), 344–345. doi:10.2178/bsl/1146620064.

Rubtsova, N. (2006). Evidence reconstruction of epistemic modal logic S5. In *Computer Science – Theory and Applications. CSR 2006* (Lecture notes in computer science, Vol. 3967, pp. 313–321). Springer.

Russell, B. (1905). On denoting. *Mind, 14*, 479–493.

Russell, B. (1912). *The problems of philosophy*. London: Williams and Norgate/New York: Henry Holt and Company.

Russell, B. (1919). *Introduction to mathematical philosophy*. London: George Allen and Unwin.

Stalnaker, R. C. (1996). Knowledge, belief and counterfactual reasoning in games. *Economics and Philosophy, 12*, 133–163.

Troelstra, A. S. (1998). Realizability. In S. Buss (Ed.), *Handbook of proof theory* (pp. 407–474). Amsterdam: Elsevier.

Troelstra, A. S., & Schwichtenberg, H. (1996). *Basic proof theory*. Amsterdam: Cambridge University Press.

Troelstra, A. S., & van Dalen, D. (1988). *Constructivism in mathematics* (Vols. 1 & 2). Amsterdam: North–Holland.

van Dalen, D. (1986). Intuitionistic logic. In D. Gabbay & F. Guenther (Eds.), *Handbook of philosophical logic* (Vol. 3, pp. 225–340). Dordrecht: Reidel.

von Wright, G. H. (1951). *An essay in modal logic*. Amsterdam: North-Holland.

Yavorskaya (Sidon), T. (2006). Multi-agent explicit knowledge. In D. Grigoriev, J. Harrison, & E. A. Hirsch (Eds.), *Computer Science – Theory and Applications. CSR 2006* (Lecture notes in computer science, Vol. 3967, pp. 369–380). Springer.

Chapter 33
Learning Theory and Epistemology

Kevin T. Kelly

Introduction

Learning is the acquisition of new knowledge and skills. It spans a range of processes from practice and rote memorization to the invention of entirely novel abilities and scientific theories that extend past experience. Learning is not restricted to humans: machines and animals can learn, social organizations can learn, and a genetic population can learn through natural selection. In that broad sense, learning is adaptive change, whether in behavior or in belief.

Learning can occur through the receipt of unexpected information, as when a detective learns where the suspect resides from an anonymous informant. But it can also be a process whose arrival at a correct result is in some sense guaranteed before the new knowledge is acquired. Such a learning process may be said to be *reliable* at the time it is adopted. *Formal Learning Theory* is an a priori, mathematical investigation of that strategic conception of reliability. It does not examine how people learn or whether people actually know, but rather, how reliable any system, human or otherwise, could possibly be. Thus, learning theory is related to traditional psychological and epistemological issues, but retains its own, distinct emphasis and character.

Reliability is a notoriously vague concept, suggesting a disposition to acquire new knowledge or skill over a broad range of relevantly possible environments. Learning theory deals with the vagueness not by insisting on a single, sharp "explication" of reliability, but by studying a range of possible explications, no one of which is insisted upon. That approach subtly shifts the focus from intractable

K.T. Kelly (✉)
Carnegie Mellon University, Pittsburgh, PA, USA
e-mail: kk3n@andrew.cmu.edu

H. Arló-Costa et al. (eds.), *Readings in Formal Epistemology*, Springer Graduate Texts in Philosophy 1, DOI 10.1007/978-3-319-20451-2_33

debates about what reliability *is* to the more objective task of determining which precise sens*es* of reliability are achievable in a given, precisely specified learning *problem*.

A learning problem specifies (1) what is to learned, (2) a range of relevantly possible environments in which the learner must succeed, (3) the kinds of inputs those environments provide to the learner, (4) what it means to learn over a range of relevantly possible environments, and (5) the sorts of learning strategies that will be entertained as solutions. A learning strategy *solves* a learning problem just in case it is admitted as a potential solution by the problem and succeeds in the specified sense over the relevant possibilities. A problem is *solvable* just in case some admissible strategy solves it.

Solvability is the basic question addressed by formal learning theory. To establish a positive solvability result, one must construct an admissible learning strategy and prove that the strategy succeeds in the relevant sense. A negative result requires a general proof that every allowable learning strategy fails. Thus, the positive results appear "methodological" whereas the negative results look "skeptical". Negative results and positive results lock together to form a whole that is more interesting than the sum of its parts. For example, a learning method may appear unimaginative and pedestrian until it is shown that no method could do better (i.e., no harder problem is solvable). And a notion of success may sound too weak until it is discovered that some natural problem is solvable in that sense but not in the more ambitious senses one would prefer.

There are so many different parameters in a learning problem that it is common to hold some of them fixed (e.g., the notion of success) and to allow others to vary (e.g., the set of relevantly possible environments). A partial specification of the problem parameters is called a learning *paradigm* and any problem agreeing with these specifications is an *instance* of the paradigm.

The notion of a paradigm raises more general questions. After several solvability and unsolvability results have been established in a paradigm, a pattern begins to emerge and one would like to know what it is about the combinatorial structure of the solvable problems that makes them solvable. A rigorous answer to that question is called a *characterization theorem*.

Many learning theoretic results concern the relative difficulty of two paradigms. Suppose we change a parameter (e.g., success) in one paradigm to produce another paradigm. There will usually remain an obvious correspondence between problems in the two paradigms (e.g., identical sets of serious possibilities). A *reduction* of paradigm P to another paradigm P' transforms a solution to a problem in P' into a solution to the corresponding problem in P. Then one may say that P is *no harder* than P'. Inter-reducible paradigms are *equivalent*. Equivalent paradigms may employ intuitively different standards of success, but the equivalence in difficulty shows that the quality of information provided by the diverse criteria is essentially the same. Paradigm equivalence results may therefore be viewed as epistemic analogues of the conservation principles of physics, closing the door on the temptation to get something (more reliability) for nothing by fiddling with the notion of success.

Learning in Epistemology

Epistemology begins with the irritating stimulus of unlearnability arguments. For example, Sextus Empiricus records the classical problem of inductive justification as follows:

> [Dogmatists] claim that the universal is established from the particulars by means of induction. If this is so, they will effect it by reviewing either all the particulars or some of them. But if they review only some, their induction will be unreliable, since it is possible that some of the particulars omitted in the induction may contradict the universal. If, on the other hand, their review is to include all the particulars, theirs will be an impossible task, because particulars are infinite and indefinite (Sextus 1985): 105.

That argument may be modelled in the following *input stream paradigm*. An *input stream* is just an infinite sequence e of natural numbers encoding discrete "observations". By stage n of inquiry the learner has seen observations $e(0)$, $e(1)$, ... $e(n-1)$. An *empirical proposition* is a proposition whose truth or falsity depends only on the input stream, and hence may be identified with a set of input streams. A learning strategy *decides* a given empirical proposition *with certainty* just in case in each relevantly possible input stream, it eventually halts and returns the truth value of the proposition.

Let the hypothesis to be assessed be "zeros will be observed forever", which corresponds to the empirical proposition whose only element is the everywhere zero input stream. Let every Boolean-valued input stream be a relevant alternative. To show that no possible learning strategy decides the hypothesis with certainty over these alternatives, construct a "demonic strategy" for presenting input in response to the successive outputs of an arbitrary learning strategy in such a way that the learner fails to halt with the right answer on the input stream presented. The demon presents the learner with the everywhere zero sequence until the learner halts and returns "true". If that never happens, the learner fails on the everywhere zero input stream. If the learner halts with "true", there is another relevantly possible input stream that agrees with the everywhere zero input stream up to the present and that presents only ones thereafter. The demon then proceeds to present this alternative input stream, on which the learner has already halted with the wrong answer. So whatever the learner's strategy does, it fails on some relevantly possible input stream and hence does not decide the hypothesis with certainty. That is the simplest example of a negative learning theoretic argument.

The argument actually shows something stronger. *Verification* with certainty requires, asymmetrically, that the learner's strategy halt with the output "true" if the hypothesis under assessment is true and that the strategy always say "false" otherwise, possibly without ever halting. The preceding argument shows that the "zeros forever" hypothesis is not verifiable with certainty.

Karl Popper's falsificationist epistemology was originally based on the observation that although universal hypotheses cannot be verified with certainty, they can be *refuted* with certainty, meaning that a method exists that halts with "false" if the hypothesis is false and that always says "true" otherwise. In the "zeros forever" example, the refutation method simply returns "true" until a nonzero value is observed and then halts inquiry with "false".

When reliability demands verification with certainty, there is no tension between the static concept of conclusive justification and the dynamical concept of reliable success, since convergence to the truth occurs precisely when conclusive justification is received. Refutation with certainty severs that tie: the learner reliably stabilizes to the truth value of h, but when h is true there is no time at which that guess is certainly justified. The separation of reliablity from complete justification was hailed as a major epistemological innovation by the American Pragmatists.[1] In light of that innovation, one may either try to invent some notion of *partial* empirical justification (e.g., a theory of *confirmation*), or one may, like Popper, side entirely with reliability.[2] Learning theory has nothing to say about whether partial epistemic justification exists or what it might be. Insofar as such notions are entertained at all, they are assessed either as components of reliable learning strategies or as extraneous constraints on admissible strategies that may make reliability more difficult or even impossible to achieve. Methodological principles with the latter property are said to be *restrictive*.[3]

"Hypothetico-deductivism" is sometimes viewed as a theory of partial inductive support (Glymour 1980), but it can also been understood as a strategy for *reducing* scientific discovery to hypothesis assessment (Popper 1968; Kemeny 1953; Putnam 1963). Suppose that the relevant possibilities are covered by a countable family of hypotheses, each of which is refutable with certainty and informative enough to be interesting. A *discovery* method produces empirical hypotheses in response to its successive observations. A discovery method *identifies* these hypotheses *in the limit* just in case, on each relevantly possible input stream, the method eventually stabilizes to some true hypothesis in the family. Suppose that there is an assessment method that refutes each hypothesis with certainty. The corresponding hypothetico-deductive method is constructed as follows. It enumerates the hypotheses (by "boldness", "abduction", "plausibility", "simplicity", or the order by which they are produced by "creative intuition") and outputs the first hypothesis in the enumeration that is not rejected by the given refutation method. That reduction has occurred to just about everyone who has ever thought about inductive methodology. But things needn't be quite so easy. What if the hypotheses aren't even refutable with certainty? Could enumerating the right hypotheses occasion computational difficulties? Those

---

[1] "We may talk of the *empiricist* and the *absolutist* way of believing the truth. The absolutists in this matter say that we not only can attain to knowing truth, but we can know when we have attained to knowing it; while the empiricists think that although we may attain it, we cannot infallibly know when." (James 1948: 95–96).

[2] "Of course theories which we claim to be no more than conjectures or hypotheses need no justification (and least of all a justification by a nonexistent 'method of induction', of which nobody has ever given a sensible description)." (Popper 1982: 79).

[3] Cf. section "A foolish consistency" below.

are just the sorts of questions of principle that are amenable to learning theoretic analysis, as will be seen below.

Another example of learning theoretic thinking in the philosophy of science is Hans Reichenbach's "pragmatic vindication" of the "straight rule" of induction (Reichenbach 1938). Reichenbach endorsed Richard Von Mises' frequentist interpretation of probability. The relative frequency of an outcome in an input stream at position n is the number of occurrences of the outcome up to position n divided by n. The *probability* of an outcome in an input stream is the limit of the relative frequencies as n goes to infinity. Thus, a probabilistic statement determines an empirical proposition: the set of all input streams in which the outcome in question has the specified limiting relative frequency.

To discover limiting relative frequencies, Reichenbach recommended using the *straight rule*, whose guess at the probability of an outcome is the currently observed relative frequency of that outcome. It is immediate by definition that if the relevant possibilities include only input streams in which the limiting relative frequency of an event type is defined, then following the straight rule *gradually identifies* the true probability value, in the sense that on each relevantly possible input stream, for each nonzero distance from the probability, the conjectures of the rule eventually stay within that distance.

If the straight rule is altered to output an open interval of probabilities of fixed width centered on the observed relative frequency, then the modified method evidently identifies a true interval in the limit (given that a probability exists). That is the same property that hypothetico-deductive inquiry has over countable collections of refutable hypotheses.

So are probability intervals refutable with certainty? Evidently not, for each finite input sequence is consistent with each limiting relative frequency: simply extend the finite sequence with an infinite input sequence in which the probability claim is true. Is there any interesting sense in which open probability intervals can be reliably assessed? Say that a learner *decides* a hypothesis *in the limit* just in case in each relevantly possible environment, the learner eventually stabilizes to "true" if the hypothesis is true and to "false" if the hypothesis is false. According to that notion of success, the learner is guaranteed to end up with the correct truth value, even though no relevantly possible environment affords certain verification or refutation. But even assuming that some limiting relative frequency exists, open probability intervals are not decidable even in that weak, limiting sense (Kelly 1996).[4] A learner

[4] This footnote has been added in 2015. The results that follow are valid for the Reichenbach-Von Mises view that probability is limiting relative frequency. But that is not the most natural way to relate learning theoretic results to statistical inference. A better idea is to view chance as an unanalyzed probability distribution over outcomes. Learners receive sequences of random samples as inputs. Hypotheses are sets of possible chance distributions. Let $0 < \alpha \leq 1$. A learner α-verifies hypothesis H iff, in each possible chance distribution p, hypothesis H is true in p exactly when there exists a sample size at which the learner outputs "true" with chance $1 - \alpha$. Then open intervals of probability are α-verifiable, for $0 < \alpha \leq 1$. All of the learning theoretic criteria of success discussed below can be re-interpreted probabilistically in a similar way.

verifies a hypothesis *in the limit* just in case, on each relevantly possible input stream, she converges to "true" if the hypothesis is true and fails to converge to "true" otherwise. That even weaker notion of success is "one sided", for when the hypothesis is true, it is only guaranteed that "false" is produced infinitely often (possibly at ever longer intervals).[5] Analogously, *refutation in the limit* requires convergence to "false" when the hypothesis is false and anything but convergence to "false" otherwise. It turns out that open probability intervals are verifiable but not decidable in the limit given that some probability (limiting relative frequency) exists.[6]

Identification in the limit is possible even when the possible hypotheses are merely verifiable in the limit. Indeed, identification in the limit is in general reducible to limiting verification, but the requisite reduction is a bit more complicated than the familiar hypothetico-deductive construction. Suppose we have a countable family of hypotheses covering all the relevant possibilities and a limiting verifier for each of these hypotheses. Enumerate the hypotheses so that each hypothesis occurs infinitely often in the enumeration. At a given stage of inquiry, find the first remaining hypothesis whose limiting verifier currently returns "true". If there is no such, output the first hypothesis and go to the next stage of inquiry. If there is one, output it and delete all hypotheses occurring prior to it from the hypothesis enumeration. It is an exercise to check that this method identifies a true hypothesis in the limit. So although limiting verification is an unsatisfying sense of reliable assessment, it sufficees for limiting identification. If the hypotheses form a partition, limiting verifiability of each cell is also necessary for limiting identification (Kelly 1996). So limiting verification is more important than it might first have appeared.

Neyman and Pearson justified their theory of statistical testing in terms of the frequentist interpretation of probability:

> It may often be proved that if we behave according to such a rule, then in the long run we shall reject h when it is true not more, say, than once in a hundred times, and in addition we may have evidence that we shall reject h sufficiently often when it is false (Neyman and Pearson 1933: 142).

The significance level of a test is a fixed upper bound on the limiting relative frequency of false rejection of the hypothesis under test over all possible input streams. A test is "useless" if the limiting frequency of mistaken acceptances exceeds one minus the significance, for then one could have done better at reducing the limiting relative frequency of error by ignoring the inputs and flipping a coin biased according to the significance level. "Useful" testability can be viewed as

---

[5]If there were any schedule governing the rate at which the outputs "false" spread apart through time, that schedule could be used to produce a method that decides the hypothesis in the limit: the new rule outputs "false" until the simulated rule produces more "true" outputs than the schedule allows for. Thus, the potential for ever rarer "false" outputs when the hypothesis is false is crucial to the extra lenience of this criterion.

[6]Conjecturing "true" while the observed frequency is in the interval and "false" otherwise does suffice unless we exclude possible input streams in which the limiting relative frequency approaches its limit from one side, for all but finitely many stages along the input stream. A reliable method is presented in Kelly (1996).

a learning paradigm over input streams. How does it relate to the "qualitative" paradigms just discussed? It turns out that the existence of a useful test for a hypothesis is equivalent to the hypothesis being either verifiable or refutable in the limit (Kelly 1996). That is an example of a paradigm equivalence theorem, showing that useful statistical tests provide essentially no more "information" than limiting verification or refutation procedures, assuming the frequentist interpretation of probability.

It is standard to assume in statistical studies that the relevant probabilities exist, but is there a sense in which that claim could be reliably assessed? Demonic arguments reveal the existence of a limiting relative frequency to be neither verifiable in the limit nor refutable in the limit over arbitrary input streams. But that hypothesis is *gradually verifiable* in the sense that there is a method that outputs numbers in the unit interval such that the numbers approach one just if the hypothesis is true (Kelly 1996). A demonic argument shows that the existence of a limiting relative frequency is not *gradually refutable,* in the sense of producing a sequence of numbers approaching zero just in case that hypothesis is false.

Gradual decidability requires that the learner's outputs gradually converge to the truth value of the hypothesis whatever the truth value happens to be. Unlike gradual verification and refutation, which we have just seen to be weaker than their limiting analogues, gradual decision is inter-reducible with limiting decision: simply choose a cutoff value (e.g. 0.5) and output "true" if the current output is less than 0.5 and "false" otherwise. Gradual decision is familiar as the sense of success invoked in Bayesian convergence arguments. Since Bayesian updating by conditionalization can never retract a zero or a one on input of nonzero probability, those outputs indicate certainty (inquiry may as well be halted), so limiting decision may only be accomplished gradually.

This short discussion illustrates how familiar epistemological issues as diverse as the problem of induction, Popper's falsificationism, Reichenbach's vindication of the straight rule, statistical testability, and Bayesian convergence all fit within a single, graduated system of learnability concepts.

Computable Learning

The preceding discussion frames traditional epistemological topics in learning theoretic terms. But unlike traditional epistemology, formal learning theory develops systematic connections between inductive inference and computation.

One of the earliest examples of a computationally driven unlearnability argument was presented by Hilary Putnam in 1963 in an article criticizing Rudolph Carnap's (1950) approach to inductive logic. Following suggestions by Wittgenstein, Carnap viewed inductive logic as a theory of "partial entailment", in which the conditional probability of the hypothesis given the inputs is interpreted as the proportion of logical possibilities satisfying the "premise" that also satisfy the intended "conclusion".

An inductive logic determines a *prediction* function: given the inputs encountered so far, output the most probable guess at the next datum to be seen. Interpret a tie as a refusal to select a prediction and count it as a failure at that round. Since the relevant probabilities are computable in Carnap's inductive logic, so is the induced prediction function.

In the *extrapolation paradigm*, the goal in each relevantly possible input stream is to eventually produce only correct predictions. Putnam showed that no computable prediction function can extrapolate the set of all total computable input streams, from which it follows that Carnap's inductive logic cannot extrapolate the computable input streams. Let an arbitrary, computable prediction strategy be given. At each stage, the demon calculates the computable prediction strategy's next prediction in light of the input already presented. If the prediction is one or greater, the demon presens a zero. If the prediction is zero, the demon presents a one. Evidently, every prediction made by the computable extrapolator along the resulting input stream is wrong. Since both the demon's strategy and the learner's strategy are both computable. The input stream presented by the demon is computable and, hence, relevantly possible.[7]

On the other hand, the problem is solved by an obvious, but uncomputable, hypothetico-deductive method. Enumerate a set of computer programs that compute all and only the total computable functions (i.e., no programs that go into infinite loops are included). Each such program is computably refutable with certainty, by calculating its prediction for the current stage of inquiry and rejecting it if the prediction does not agree with what is observed. That method identifies a correct program in the limit. To turn it into a reliable extrapolator, just compute what the currently output hypothesis says will happen at the next stage (another example of a paradigm reduction).

The only part of that procedure that is not computable is enumerating a collection of programs covering exactly the total computable functions. Since the prediction problem is computably unsolvable, it follows immediately that no such program enumeration is computable. So computable predictors fail on this problem "because" they cannot enumerate the right collection of hypotheses.[8]

The *computable function identification* paradigm poses the closely related problem of identifying in the limit a computer program correctly predicting each position in the input stream. The preceding hypothetico-deductive method uncomputably identifies the computable input streams in that sense, but in a seminal paper, E. M. Gold (1965) showed that the problem is not computably solvable. The computable demonic construction employed in the proof of Gold's result is more subtle than in

---

[7]Putnam's actual argument was more complicated than the version presented here.

[8]Putnam concluded that a scientific method should always be equipped with an extra input slot into which hypotheses that occur to us during the course of inquiry can be inserted. But such an "open minded" method must hope that the external hypothesis source (e.g., "creative intuition") does not suggest any programs that go into infinite loops, since the inability to distinguish such programs from "good" ones is what restricted the reliability of computable predictors to begin with!

the extrapolation case, because it is a nontrivial matter for a computable demon to figure out what the computable learner's current hypothesis predicts the next datum to be. For all the computable demon knows, the prediction may be undefined (i.e., the hypothesis may go into an infinite loop).

The demon proceeds in stages as follows[9]: At stage n, the learner has received a sequence of n inputs. The demon employs a fixed, computable enumeration of all ordered pairs of natural numbers. He then seeks the first pair (i, j) in the enumeration such that, after reading the current input followed by i zeros, the learner outputs a program that halts in j steps of computation with a prediction of zero for the next datum. If the search terminates with some such pair (i, j), then the demon adds i zeros to the input presented so far, and then presents a one (falsifying the hypothesis output by the learner after seeing the last zero). Otherwise, the demon continues searching forever and never proceeds to the next stage.

Suppose that the demon's construction runs through infinitely many stages. Then the search for a pair always terminates, so the resulting input stream falsifies the learner's conjecture infinitely often. The input stream is computable because it is produced by the interaction of two computable strategies. Suppose, then, that the demon's construction eventually gets stuck at a given stage. Then the demon's search for a pair fails. So on the input stream consisting of the input presented so far followed by all zeros, the learner never produces a hypothesis that correctly predicts the next zero. Thats input stream is also computable: use a finite lookup table to handle the input presented so far and output zero thereafter. So in either case, the demon never identifies a correct program along some relevantly possible input stream.

Since the demon makes the learner's conjecture false infinitely often, his strategy wins even if we weaken the criterion of success to *unstable* identification in the limit, according to which the learner must eventually output only true hypotheses, but need not stabilize to a particular hypothesis.[10]

Each total computer program is computably refutable with certainty (compute its successive predictions and compare them to the input), so we now know that computable refutability with certainty reduces neither computable extrapolation nor computable limiting identification. Does computable identification in the limit reduce computable extrapolation? One might suppose so: just compute the prediction of the limiting identifier's current conjecture, which must eventually be right since the identifier's conjectures are eventually correct. But although the limiting identifier eventually produces programs without infinite loops, nothing prevents it from producing defective programs in the short run. If a computer attempts to derive

[9]This construction (Case and Smith 1983) is a bit stronger than Gold's. It produces an input stream on which infinitely many outputs of the learner are wrong. Gold's construction merely forces the learner to vacillate forever (possibly among correct conjectures).

[10]Cf. the preceding footnote. In the learning theoretic literature, unstable identification is called BC identification for "behaviorally correct", whereas stable identification is called EX identification for "explanatory". Osherson and Weinstein (1986) call stable identification "intensional" and unstable identification "extensional".

predictions from those conjectures in the manner just described, it may get caught in an infinite loop and hang for eternity.

Blum and Blum (1975) constructed a learning problem that is computably identifiable in the limit but not computably extrapolable for just that reason. Consider a problem in which an unknown Turing machine without infinite loops is hidden in a box and the successive input are the (finite) runtimes of that program on successive inputs. The learner's job is to guess some computer program whose runtimes match the observed runtimes for each input (a task suggestive of fitting a computational model to psychological reaction time input). In that problem, every program is computably refutable with certainty: simulate it and see if it halts precisely when the input say it should. Infinite loops are no problem, for one will observe in finite time that the program doesn't halt when it should have. Since the set of all programs is computably enumerable (we needn't restrict the enumeration to *total* programs this time), a computable implementation of the hypothetico-deductive strategy identifies a correct hypothesis in the limit. Nonetheless, computable extrapolation of runtimes is not possible. Let a computable extrapolator be given. The demon is a procedure that wastes computational cycles in response to the computable predictor's last prediction. So at a given stage, the demonic program simulates the learner's program on the successive runtimes of the demonic program on earlier inputs. Whatever the learner's prediction is, the demon goes into a wasteful subroutine that uses at least one more step of computation than the predictor expected.

Another question raised by the preceding discussion is whether stable identification is equivalent to or harder than unstable identification for computable learners in the computable function identification paradigm. That question is answered affirmatively by Case and Smith (1983). To see why the answer might be positive, consider the function identification problem in which the relevant possibilities are the "almost self-describing input streams". A unit variant of an input stream is a partial computable function that is just like the input stream except that it may disagree or be undefined in at most one position. An input stream is *almost self-describing* just in case it is a unit variant of the function computed by the program whose index (according to a fixed, effective encoding of Turing programs into natural numbers) occurs in the input stream's first position. In other words, an "almost self-describing" input stream "gives away" a nearly correct hypothesis, but it doesn't say where the possible mismatch might be. An unstable learner can succeed by continually patching the "given away" program with ever larger lookup tables specifying what has been seen so far, since eventually the lookup table corrects the mistake in the "given away" program. But a stable learner would have to know *when* to stop patching, and that information was not given away.

In the problem just described, it is trivial to stably identify an almost correct program (just output the first datum) whereas no computable learner can stably identify an exactly correct program. Indeed, for each finite number of allowed errors there is a learning problem that is computably solvable under that error error fewer but not with one error fewer (Case and Smith 83). That result, known as the *anomaly hierarchy theorem,* can be established by means of functions that are self-describing up to *n* possible errors.

There are many more sophisticated results of the kind just presented, all of which share the following points in common. (1) Uncomputability is taken just as seriously as the problem of induction from the very outset of the analysis. That is different from the approach of traditional epistemology, in which idealized logics of justification are proposed and passed along to experts in computation for advice on how to satisfy them (e.g., Levi 1991). (2) When computability is taken seriously, the *halting problem* (the *formal* problem of determining whether a computer program is in an infinite loop on a given input) is very similar to the classical problem of induction: for as soon as one is sure that a computation will never end, it might, for all the simulator knows *a priori,* halt at the next stage. (3) Thus, computable learners fail when ideal ones succeed because computable solvability requires the learner to solve an *internalized* problem of induction (Kelly and Schulte 1997).

Some Other Paradigms

E. M. Gold's *language learnability* paradigm (1967) was intended to model child language acquition. In that setting, a *language* is just a computably enumerable set and a hypothesis is a code number (index) of a procedure that *accepts* all and only the members of the set.[11] Different kinds of relevantly possible environments are considered. An *informant* for a language is an enumeration of all possible strings labelled as positive or negative examples of the language. A *text* for a language is an enumeration of the elements of the language, and hence provides only positive information about membership.

Gold showed a number of results that attracted attention in cognitive science. The results for informants are similar to those for computable function identification. For example, (1) the obvious hypothetico-deductive method (non-computably) identifies all languages and (2) even the set of all computably decidable languages is not computably identifiable in the limit (the proof is similar to the one showing that the total computable functions are not identifiable in the limit). But the results for text are much weaker. For example, no collection of languages containing one infinite language and all finite subsets of that language is identifiable in the limit, even by non-computable learners.[12] Since children seem to learn language with few negative examples or corrections (Brown and Hanlon 1970), there have been attempts to obtain stronger positive results. For example, Wexler and Culicover (1980) modelled the environment as a presentation of context-utterance pairs, exchanging language learning from positive examples for the easier problem of computable function

[11] I.e., the procedure halts on members of the set (indicating acceptance) and not on any other inputs.

[12] The demon presents a text for the infinite language until the learner outputs a grammar for it, then keeps repeating the preceding datum until the learner produces a grammar for the input presented so far, then starts presenting the text from where he left off last, etc.

identification. Many other variations of the language learnability paradigm have been examined.[13]

The special difficulty with learning from text is "over-generalization", or leaping to a language that properly extends the actual language, for then no further input will correct the error. If there is no way to avoid positioning a language prior to one of its proper subsets (e.g., an infinite language must occur prior to all but finitely many of its finite subsets), hypothetico-deductivism must fail, since it will converge to the large language when one of its subsets is true. What is required is a way to use evidence to avoid overgeneralizing. That can be accomplished if (†) each possible language has a finite, characteristic sample such that once that sample is seen, the language can be produced without risk of overgeneralization. Then one may proceed by enumerating the relevantly possible grammars and conjecturing the first in the enumeration that is consistent with the input and whose characteristic sample has been observed. If no such grammar exists, stick with the preceding conjecture. Condition (†) is both necessary and sufficient for a collection of languages to be identifiable in the limit from text (Angluin 1980; Osherson and Weinstein 1986), providing our first example of a learning theoretic *characterization theorem*. Computable identification from text is characterized by the existence of a procedure that enumerates the characteristic sample for a language when provided with the index of a formal verification program for that language.

The *logical paradigm* (Shapiro 1981; Osherson and Weinstein 1986, 1989a, b; Kelly and Glymour 1989, 1990), situates learning theoretic ideas in a more traditional epistemological setting. In that paradigm, there is a first-order language in which to frame hypotheses and the underlying world is a countable relational structure interpreting that language. An environment consists of such a structure together with a variable assignment onto the domain of the structure and an enumeration of the set of all quantifier-free formulas true under that assignment.[14] The relevant possibilities are all the environments presenting models of some theory representing the learner's background knowledge.

An hypothesis assessment method tries to guess the truth value of a particular sentence or theory in light of the increasing information provided by the environment, and successful assessment can be interpreted in any of the senses introduced above. So for example, the dense order postulate (each pair of points has a point between them) is refutable but not verifiable in the limit given as background the theory of total orders with endpoints (Osherson and Weinstein 1989a, b).

The characterization theorem for this paradigm explains the grain of truth in the positivist's program of linking "cognitive significance" to logical form. An hypothesis is refutable (respectively, verifiable) with certainty given background theory K just in case the hypothesis is equivalent in K to a sentence in prenex

[13]A systematic compendium of results on language learnability is Osherson and Weinstein (1986).

[14]The "onto" assumption can be dropped if empirical adequacy rather than truth is the goal Lauth (1993).

normal form[15] with a purely universal (respectively, existential) quantifier prefix. Similarly, an hypothesis is refutable (respectively, verifiable) in the limit given K just in case it is equivalent in K to a prenex sentence with a prefix of form $\forall\exists$ (respectively, $\exists\forall$) (Osherson and Weinstein 1989a, b; Kelly and Glymour 1990). As one might expect, decision with certainty is possible just in case the hypothesis is equivalent to a quantifier-free sentence in K and decision in the limit (and hence gradual decision) is possible just in case the hypothesis is equivalent in K to a finite Boolean combination of purely universal and existential sentences.

A discovery method outputs theories in response to the information provided. As the goal of discovery, one can require that the method converge to the complete true theory in some fragment of the language (e.g., the purely universal sentences). *Uniform* theory identification requires that after some time the outputs of the method are true and entail the complete theory of the required fragment. For example, the complete truth is uniformly identifiable in the limit in a language with only unary predicates, but if there is a binary predicate or a unary predicate and a function symbol in the language, then neither the purely universal nor the purely existential fragment of the complete truth is identifiable in the limit (Kelly and Glymour 1989; Kelly 1996). *Nonuniform* or *pointwise* theory identification requires only that each true sentence in the specified fragment is eventually always entailed by the scientist's successive conjectures and that each false sentence is eventually never entailed. The theory of all true Boolean combinations of universal and existential sentences is identifiable in the limit in that sense. Thus, nonuniform theory identification provides a logical conception of scientific progress that, unlike Popper's "deductivist" epistemology, treats verifiable and refutable hypotheses symmetrically.

Nonuniform theory identification bears on another Popperian difficulty. Popper held that hypothetico-deductivism leads us ever closer to the truth in the limit. David Miller (1974) argued that "closeness" to the truth is not a semantic notion since it is not preserved under translation. Thomas Mormann (1988) traced the difficulty to mathematics: translation is a type of topological equivalence, but topological equivalence permits "stretching" and hence does not preserve distance (e.g., verisimilitude). Nonuniform identification is a topological rather than a metrical notion, and hence is preserved under translation, thereby avoiding Miller-style objections. Nonetheless it constitutes a nontrivial account of scientific progress toward the complete truth that does not imply that any future theory produced by science will be literally true.

[15] I.e., the sentence has the form of a quantifier-free sentence preceded by a sequence of quantifiers.

Reliability and Complexity

Learnability is a matter of how the possible futures making different hypotheses correct branch off from one another through time. The more complex the temporal entanglement of the futures satisfying incompatible hypotheses, the more difficult learning will be. Learnability is governed by the *topological* complexity of the possible hypotheses, and computable learnability depends on their *computational* complexity.[16]

Input streams can be topologized in an epistemologically relevant manner as follows. A *fan* of input streams is the set of all input streams extending some finite input sequence, which we may call the *handle* of the fan. A fan with a given handle is just the empirical proposition asserting that the handle has occurred in the input. An empirical proposition is *open* just in case it is a union of fans and is *closed* just in case its complement is open.[17] Then we have the following characterization: an empirical proposition is verifiable with certainty just in case it is open, is refutable with certainty just in case it is closed, and is decidable with certainty just in case it is both closed and open. For suppose that a hypothesis is open. To verify it with certainty, just wait until the observed input sequence is the handle of a fan contained in the hypothesis and halt inquiry with "true". Conversely, if a given method verifies a hypothesis with certainty, the hypothesis can be expressed as the union of all fans whose handles are finite input sequences on which the method halts with "true".

To characterize limiting and gradual success, topological generalizations of the open and closed propositions are required. Call the open and closed propositions the Σ_1 and Π_1 propositions, respectively. For each n, the Σ_{n+1} propositions are countable unions (disjunctions) of Π_n propositions and the Π_{n+1} propositions are countable intersections of Σ_n propositions. At each level n, a proposition is Δ_n just in case it is both Π_n and Σ_n. These are known as the *finite Borel* complexity classes, which have been familiar in functional analysis since early in this century (Hinman 1978). Then it can be shown that limiting verifiability, refutability, and decidability are characterized by Σ_2, Π_2, and Δ_2, respectively and that gradual verifiability, refutability, and decidability are characterized by Π_3, Σ_3 and Δ_2, respectively. It can also be shown that when the hypotheses are mutually incompatible, stable identification in the limit is characterized by each hypothesis being Σ_2.[18]

In computable inquiry, attaching hypotheses to propositions is a nontrivial matter, so instead of bounding the complexity of empirical propositions, one must

[16]The computational versions of these ideas are in Gold (1965), Putnam (1965), and Kugel (1977). The topological space is introduced in Osherson and Weinstein (1986) and the characterizations are developed in Kelly (1992, 1996) Logical versions of the characterizations are developed in Osherson and Weinstein (1991) and Kelly and Glymour (1990).

[17]These are, in fact, the open sets of an extensively studied topological space known as the *Baire space* (Hinman 1978).

[18]Necessity of the condition fails if the hypotheses are mutually compatible or if we drop the stability requirement.

consider the overall *correctness relation* $C(e, h)$, which expresses that hypothesis h is correct in environment e. In computable function identification, for example, correctness requires that h be the index of a computer program that computes e. In language learning from text, h must be the index of a positive test procedure for the range of e. By suitable coding conventions, language learning from informant and logical learning can also modeled with correctness relations in the input stream paradigm. Computational analogs of the Borel complexity classes can be defined for correctness relations, in which case analogous characterization theorems hold for computable inquiry (Kelly 1996).

The moral of this discussion is that the problem of induction, or empirical underdeterination, comes in degrees corresponding to standard topological and computational complexity classes, which determine the objective sense in which reliable inquiry is possible.

A Foolish Consistency

A *consistent* learner never produces an output that is incorrect of every relevantly possible input stream extending the current input sequence. For non-computable learners, consistency makes a great deal of sense: why should someone who aims to find the truth say what has to be wrong? On the other hand, we have seen that formal relations can pose an "internal" problem of induction for computable learners. Since we do not require omniscience on the empirical side, why should we do so on the formal side when the underlying structure of the problem of induction is the same on both sides?

That raises an interesting question. Could insistence on computationally achievable consistency *preclude* computationally achievable empirical reliability? The answer is striking. One can construct an empirical proposition with the following properties. (1) The proposition is computably refutable with certainty. (2) Some computable, consistent method exists for the proposition (the method that always says "false" suffices since the proposition is never verified). But (3) no consistent, computable method of even a highly idealized, uncomputable kind[19] can even gradually decide the hypothesis. Thus, where traditional epistemology sees consistency as a *means* for finding the truth sooner, enforcing *achievable* consistency may prevent computable learners from finding truths they could otherwise have reliably found. So if the aim of inquiry is to find the truth, inconsistency may be an epistemic *obligation* (rather than a merely forgivable lapse) for computable agents. Such results exemplify the sharp difference in emphasis between computational learning theory and traditional, justificationist epistemology.[20]

[19]i.e., hyperarithmetically definable.

[20]Osherson and Weinstein (1986) contains many restrictiveness results carrying a similar moral. Also, see Osherson and Weinstein (1988).

Gambling with Success

Suppose that each learning problem comes equipped with an assignment of probabilities to empirical propositions. More precisely, suppose that the probability assignment is defined on the set of all *Borel propositions* (i.e., the least set that contains all the open (Σ_1) propositions and that is closed under countable union and complementation). A *probability assignment* on the Borel propositions is a function taking values in the unit interval that assigns unity to the vacuous proposition and that is *finitely additive* in the sense that the probability of a finite union of mutually incompatible Borel propositions is the sum of the probabilities of the propositions the union is taken over. *Countable additivity* extends finite additivity to countable, disjoint unions. While Kolmogorov's familiar mathematical theory of probability assumes countable additivity as a postulate, limiting relative frequencies do not satisfy it, and the usual foundations of Bayesian probability theory do not entail it (e.g., DeFinetti 1990; Savage 1972).

Say that an hypothesis is gradually decidable *with probability* r just in case there exists some empirical proposition of probability r over which the hypothesis is gradually decidable in the usual sense, and similarly for the other assessment criteria. Probabilistic success can be much easier to achieve than success in each relevant possibility. If the probability assignment is countably additive, then, remarkably, every Borel hypothesis is (1) decidable in the limit with unit probability and (2) decidable with certainty with arbitrarily high but non-unit probability. (1) can be improved to the result that the method of updating the given probability measure by conditionalization gradually decides the hypothesis with unit prior probability (e.g., Halmos 1974). That is a very general version of the familiar Bayesian claim that prior probabilities are eventually "swamped" by the input.

Compared with the purely topological analysis of section "Reliability and complexity", these probabilistic results sound too good to be true, since of error Borel propositions can be infinitely more complex than Δ_2 propositions (Hinman 1978). What accounts for the dramatic difference? Suppose we want to decide the "zeros forever" hypothesis with a small, nonzero probability of error r. The negation of that hypothesis is the countable, disjoint union of the hypotheses h_i = "the first nonzero occurs at position i". So by countable additivity, the probability that the "zeros forever" hypothesis is false is the sum of the probabilities of the propositions h_i. Since the infinite sum converges to a finite value, there is some position n such that the sum of the probabilities of $h_n, h_{n+1}, \ldots$ is less than r. So our probability of failure is less than r if we halt with "true" at stage n if no nonzero datum has been seen by position n and halt with "false" as soon as a nonzero datum is seen. In other words, countable additivity asserts that when a high prior probability of successful learning suffices, only finitely many of the demon's opportunities to make the hypothesis false matter.

Without countable additivity, it is possible that the probability that the hypothesis is false exceeds the mass distributed over the h_n, say by a value of r. Since that "residual" probability mass is not distributed over the propositions h_i, the learner never "gets past" it, so whenever the learner halts inquiry with "true", the probability that the conclusion was in error remains at least as high as r. The residual probability reflects the demon's *inexhaustible* opportunities to falsify the hypothesis in the infinite future, providing a probabilistic model of Sextus' demonic argument. In fact, both (1) and (2) can fail when countable additivity is dropped (Kelly 1996), highlighting the pivotal epistemological significance of that seemingly "technical" asumption.

Concept Learning and the PAC Paradigm

In the *Meno*, Plato outlined what has come to be known as the *concept learning paradigm*, which has captured the imagination of philosophers, psychologists, and artificial intelligence researchers ever since. A concept learning problem specifies a domain of *examples* described as vectors of *values* (e.g., blue, five kilos) of a corresponding set of *attributes* (e.g., color, weight), together with a set of possible *target concepts*, which are sets of examples. The learner is somehow presented with examples labelled either as positive or as negative examples of the concept to be learned, and the learner's task is to converge in some specified sense to a correct definition. In contemporary artificial intelligence and cognitive science, the "concepts" to be learned are defined by neural networks, logic circuits, and finite state automata, but the underlying paradigm would still be familiar to Socrates.

Socrates ridiculed interlocutors who proposed disjunctive concept definitions, which suggests that he would countenance only conjunctively definable concepts as relevant possibilities. Socrates' notorious solution to the concept learning problem was to have the environment "give away" the answer in a mystical flash of insight. But J. S. Mill's (i.e., Francis Bacon's) well-known inductive methods need no mystical help to identify conjunctive concepts with certainty: the first conjecture is the first positive example sampled. On each successive positive example in the sample, delete from the current conjecture each conjunct that disagrees with the corresponding attribute value of the example (the "method of difference"). On each successive negative example that agrees with the current conjecture everywhere except on one attribute, underline the value of that attribute in the current conjecture (the "method of similarity"). When all conjuncts in the current conjecture are underlined, halt inquiry.

Boolean concepts are also identifiable with certainty over a finite set of attribute values: wait for all possible examples to come in and then disjoin the positive ones. Bacon's methods sound plausible in the conjunctive case, whereas that "gerrymandering" procedure for learning Boolean concepts sounds hopeless (it is, in fact, just what Socrates ridiculed). Yet both procedures identify the truth with certainty, since the set of examples is finite. The PAC (Probably Approximately Correct) paradigm distinguishes such "small" problems in terms of *tractable* rather than merely *computable* inquiry.[21]

[21] An excellent source presenting all of the results mentioned here is Kearns and Vazirani (1994), which provides detailed descriptions and bibliographic notes for all the results mentioned below.

In the PAC paradigm, examples are sampled with replacement from an urn in which the probability of selecting an example is unknown. There is a collection of relevantly possible concepts and also a collection of hypotheses specifying the possible forms in which the learner is permitted to define a relevantly possible concept. Say that a hypothesis is ϵ-accurate just in case the sampling probability that a single sampled individual is a counterexample is less than ϵ. The learner is given a *confidence* parameter δ and an *error* parameter ϵ. From these parameters, the learner specifies a sample size and upon inspecting the resulting sample, she outputs a hypothesis. A learning strategy is *probably approximately correct* (PAC) just in case for each probability distribution on the urn and for each ϵ, δ exceeding zero, the strategy has a probability of at least $1-\epsilon$ of producing an ϵ-accurate hypothesis.

It remains to specify what it means for a PAC learning strategy to be *efficient*. Computational complexity is usually analyzed in terms of asymptotic growth *rate* over an infinite sequence of "similar" but "ever larger" examples of the problem. Tractability is understood as resource consumption bounded almost everywhere by some polynomial function of problem size. The size of a concept learning problem is determined by (1) the number of attributes (2) the size of the smallest definition of the target concept, (3) the reciprocal of the confidence parameter, and (4) the reciprocal of the error parameter (higher accuracy and reliability requirements make for a "bigger" inference problem). A *input efficient* PAC learner takes a sample in each problem whose size is bounded by a polynomial in these four arguments. Otherwise, the learner requires samples that are exponentially large in the same parameters, which is considered to be intractable.

There is an elegant combinatorial characterization of how large the sample required for PAC learning should be. Say that a concept class *shatters* a set S of examples just in case each subset of S is the intersection of S with some concept in the class. The *Vapnik-Chervonenkis* (VC) dimension of the concept class is the cardinality of the largest set of instances shattered by the class. There exists a fixed constant c such that if the VC dimension of the concept class is d, it suffices for PAC learnability that a sample of size s be taken, where:

$$s \geq c\left(\frac{1}{\epsilon}\log\frac{1}{\delta} + \frac{d}{\epsilon}\log\frac{1}{\epsilon}\right)$$

For example, the VC dimension of the conjunctive concepts over n Boolean attributes is $2n$ (and in fact is just n if $n > 1$) so the problem is input-efficiently solvable by setting the sample size according to the above formula and then using any method producing conjectures consistent with the input (e.g., Bacon's method of similarity). Calculating the VC dimension of the concepts decidable by neural networks reveals that they are also input-efficiently learnable.

On the negative side, it can be shown that if the VC dimension of a concept class is d, then on some concept and in some sampling distribution, a sample size of at least d/ϵ is required. Since the VC dimension of the Boolean concepts over n Boolean attributes is 2^n, exponentially large samples will sometimes be required. Thus, any algorithm that takes a sample whose size depends only on the problem and not the size of the (unknown) target concept itself will be input-inefficient (since the sample size grows non-polynomially when concept size is held fixed at the minimum value).

A *computationally efficient* PAC learner is a PAC learner whose runtime is bounded by a polynomial of the sort described in the definition of input efficiency. Since scanning a sampled instance takes time, computational efficiency implies input efficiency. Since Bacon's method is computationally trivial and requires small samples, it is a computationally efficient PAC learner. Bacon's method can be generalized to efficiently PAC learn k-CNF concepts (i.e., conjunctions of k-ary disjunctions of atomic or negated atomic sentences), for fixed k.

Sometimes computational difficulties arise entirely because it is hard for the learner to frame her conjecture in the required hypothesis language. It is known, for example, that the k-term DNF concepts (i.e., disjunctions of k purely conjunctive concepts) are not efficiently PAC learnable using k-term DNF hypotheses (when $k \geq 2$),[22] whereas they are efficiently PAC learnable using k-CNF hypotheses.

For some time it was not known whether there exist efficiently solvable PAC problems that are unsolvable neither due to sample-size complexity nor due to output representation. It turns out (Kearns and Valiant 1994) that under a standard cryptographic hypothesis,[23] the Boolean concepts of length polynomial in the number of attributes have that property, as does the neural network training problem.

An alternative way to obtain more refined results in a non-probabilistic context is to permit the learner to ask questions. A *membership oracle* accepts an example from the learner and returns "in" or "out" to indicate whether it is a positive or a negative example. A *Socratic oracle* responds to an input conjecture with a counterexample, if there is one.[24] One such result is that Socratic and membership queries suffice for identification of finite state automata with certainty in polynomial time (Angluin 1987).

Learning Theory and Epistemology

To coherentists, learning theory looks like a naive form of foundationalism, in which incorrigible beliefs are the fulcrum driving inquiry to the truth. But foundationalists are also disappointed because positive learning theoretic results depend on substantial, contingent assumptions such as the nature of the signals from the environment, the structure of time, and the range of relevant possibilities. Externalists would prefer to investigate *our* reliability directly, instead of taking a mathematical detour into possible methods and problems. And contextualists will object to the fixity

[22] This negative result holds only under the familiar complexity-theoretic hypothesis that $P \neq NP$.

[23] I.e, that computing discrete cube roots is intractable even for random algorithms.

[24] In the learning theoretic literature, Socratic queries are referred to as "equivalence" queries.

of truth through time, ignoring the possibility of meaning shifts due to conceptual change.

But on a more careful examination, learning theory amplifies some recent epistemological trends. The search for incorrigible foundations for knowledge is no longer considered a serious option, so the fact that reliability depends on contingent assumptions is hardly a penetrating objection. Indeed, it can be shown by learning theoretic means that if some background knowledge is necessary for reliability, that knowledge cannot be reliably assessed according to the same standard, blocking any attempt at an entirely reliability-based foundationalism.

Externalist epistemologies sidestep the foundational demand that the conditions for reliability be known by requiring only that we be reliable, without necessarily being aware of that fact. Knowledge attributions are then empirical hypotheses that can be studied by ordinary empirical means. But mature empirical investigations are always focused by general mathematical constraints on what is possible. Accordingly, learning theoretic results constrain naturalistic epistemology by specifying how reliable an arbitrary system, whether computable or otherwise, could possibly be in various learning situations.

Externalism has encountered the objection (Lehrer 1990) that reliability is insufficient for knowledge if one is not justified in believing that one is reliable (e.g., someone has a thermometer implanted in her brain that suddenly begins to produce true beliefs about the local temperature). The intended point of such objections is that reliable belief-forming processes should be embedded in a coherent belief system incorporating beliefs about the agent's own situation and reliability therein. Learning theory may then be viewed as defining the crucial relation of *methodological coherence* between epistemic situations, ambitions, and means. Unlearnability arguments isolate methodological incoherence and positive arguments suggest methods, background assumptions, or compromised ambitions which, if adopted, could bring a system of beliefs into methodological coherence.

Incorporating learning theoretic structure into the concept of coherence addresses what some coherentists take to be the chief objection to their position.

> ... [A]lthough any adequate epistemological theory must confront the task of bridging the gap between justification and truth, the adoption of a nonstandard conception of truth, such as a coherence theory of truth, will do no good unless that conception is independently motivated. Therefore, it seems that a coherence theory of justification has no acceptable way of establishing the essential connection with truth (Bonjour 1985): 110.

Whether a methodological principle guarantees or prevents reliable convergence to the truth is, of course, the unshakable focus of learning theoretic analysis. Where coherence is at issue, one must consider a multitude of possible interpretations of reliability and of one's epistemic situation, backing and filling until the analysis seems apt and fits with the rest of one's beliefs. That pluralistic attitude is reflected in the wide variety of success criteria, paradigms and problems considered in the learning theoretic literature.

Contextualists may also find some value in learning theoretic results. The first moral of the subject is that reliability is highly sensitive to the finest details of the input presentation, the range of possible alternatives, the kinds of hypotheses or skills at issue, the learner's cognitive powers and resources, and the methodological principles to which she is committed. Reliable methodology is unavoidably piecemeal, contextual methodology, optimized to the special features of the problem at hand.

A remaining contextualist objection is that learning theory presupposes a fixed "conceptual scheme" in which truth is a fixed target, whereas in light of conceptual revolutions, meaning and hence truth changes as the beliefs of the learner change through time. That objection does apply to the usual learning theoretic paradigms, but the concept of reliability is flexible enough to accommodate it. If truth feints as inquiry lunges, then success can be defined as a methodological *fixed point* in which the beliefs of the learner are eventually true *with respect to themselves* (Kelly 1996; Kelly and Glymour 1992). Unlike norms of justification, which may change through time, convergence to the *relative* truth provides a strategic aim that plausibly survives successive changes in the underlying scientific tradition.

Bibliography

Angluin, D. (1980). Inductive inference of formal languages from positive data. *Information and Control, 45*(2), 117–135.

Angluin, D. (1987). Learning regular sets from queries and counterexamples. *Information and Computation, 75*, 87–106.

Blum, M., & Blum, L. (1975). Toward a mathematical theory of inductive inference. *Information and Control, 28*, 125–155.

Bonjour, L. (1985). *The structure of empirical knowledge*. Cambridge: Harvard University Press.

Brown, R., & Hanlon, C. (1970). Derivational complexity and the order of acquisition of child speech. In J. Hayes (Ed.), *Cognition and the development of language*. New York: Wiley.

Carnap, R. (1950). *The logical foundations of probability*. Chicago: University of Chicago Press.

Case, J., & Smith, C. (1983). Comparison of identification criteria for machine inductive inference. *Theoretical Computer Science, 24*, 193–220.

DeFinetti. (1990). *The theory of probability*. New York: Wiley.

Glymour, C. (1980). *Theory and evidence*. Cambridge: M.I.T. Press.

Gold, E. M. (1965). Limiting recursion. *Journal of Symbolic Logic, 30*, 27–48.

Gold, E. M. (1967). Language identification in the limit. *Information and Control, 10*, 447–474.

Halmos, P. (1974). *Measure theory*. New York: Springer.

Hinman, P. (1978). *Recursion theoretic hierarchies*. New York: Springer.

James, W. (1948). The will to believe. In A. Castell (Ed.), *Essays in pragmatism*. New York: Collier Macmillan.

Kearns, M., & Valiant, L. (1994). Cryptographic limitations on learning boolean formulae and finite automata. *Journal of the ACM, 41*, 57–95.

Kearns, M., & Vazirani, U. (1994). *An introduction to computational learning theory*. Cambridge: M.I.T. Press.

Kelly, K. (1992). Learning theory and descriptive set theory. *Logic and Computation, 3*, 27–45.

Kelly, K. (1996). *The logic of reliable inquiry*. New York: Oxford University Press.

Kelly, K., & Glymour, C. (1989). Convergence to the truth and nothing but the truth. *Philosophy of Science, 56*, 185–220.

Kelly, K., & Glymour, C. (1990). Theory discovery from data with mixed quantifiers. *Journal of Philosophical Logic, 19*, 1–33.

Kelly, K., & Glymour, C. (1992). Inductive inference from theory-laden data. *Journal of Philosophical Logic, 21*, 391–444.

Kelly, K., & Schulte, O. (1995). The computable testability of theories making uncomputable predictions. *Erkenntnis, 43*, 29–66.

Kelly, K., & Schulte, O. (1997). Church's thesis and Hume's problem. In M. L. Dalla Chiara et al. (Eds.), *Logic and scientific methods*. Dordrecht: Kluwer.

Kemeny, J. (1953). The use of simplicity in induction. *Philosophical Review, 62*, 391–408.

Kugel, P. (1977). Induction pure and simple. *Information and Control, 33*, 236–336.

Lauth, B. (1993). Inductive inference in the limit for first-order sentences. *Studia Logica, 52*, 491–517.

Lehrer, K. (1990). *Theory of knowledge*. San Francisco: Westview.

Levi, I. (1991). *The fixation of belief and its undoing*. Cambridge: Cambridge University Press.

Miller, D. (1974). On Popper's definitions of verisimilitude. *British Journal of the Philosophy of Science, 25*, 155–188.

Mormann, T. (1988). Are all false theories equally false? *British Journal for the Philosophy of Science, 39*, 505–519.

Neyman, J., & Pearson, E. (1933). On the problem of the most efficient tests of statistical hypotheses. *Philosophical Transactions of the Royal Society, 231*(A), 289–337.

Osherson, D., & Weinstein, S. (1986). *Systems that learn*. Cambridge: M.I.T. Press.

Osherson, D., & Weinstein, S. (1988). Mechanical learners pay a price for Bayesianism. *Journal of Symbolic Logic, 56*, 661–672.

Osherson, D., & Weinstein, S. (1989a). Paradigms of truth detection. *Journal of Philosophical Logic, 18*, 1–41.

Osherson, D., & Weinstein, S. (1989b). Identification in the limit of first order structures. *Journal of Philosophical Logic, 15*, 55–81.

Osherson, D., & Weinstein, S. (1991). A universal inductive inference machine. *Journal of Symbolic Logic, 56*, 661–672.

Popper, K. (1968). *The logic of scientific discovery*. New York: Harper.

Popper, K. (1982). *Unended quest: An intellectual autobiography*. LaSalle: Open Court.

Putnam, H. (1963). Degree of confirmation' and inductive logic. In A. Schilpp (Ed.), *The philosophy of Rudolph Carnap*. LaSalle: Open Court.

Putnam, H. (1965). Trial and error predicates and a solution to a problem of Mostowski. *Journal of Symbolic Logic, 30*, 49–57.

Reichenbach, H. (1938). *Experience and prediction*. Chicago: University of Chicago Press.

Savage, L. (1972). *The foundations of statistics*. New York: Dover.

Sextus Empiricus. (1985). *Selections from the major writings on scepticism, man and god* (P. Hallie, Ed., S. Etheridge, Trans.). Indianapolis: Hackett.

Shapiro, E. (1981). *Inductive inference of theories from facts*. Report YLU 192. New Haven: Department of Computer Science, Yale University.

Wexler, K., & Culicover, P. (1980). *Formal principles of language acquisition*. Cambridge: M.I.T. Press.

Chapter 34
Some Computational Constraints in Epistemic Logic

Timothy Williamson

Introduction

This paper concerns limits that some epistemic logics impose on the complexity of an epistemic agent's reasoning, rather than limits on the complexity of the epistemic logic itself.

As an epistemic agent, one theorizes about a world which contains the theorizing of epistemic agents, including oneself. Epistemic logicians theorize about the abstract structure of epistemic agents' theorizing. This paper concerns the comparatively simple special case of epistemic logic in which only one agent is considered. Such an epistemic agent theorizes about a world which contains that agent's theorizing. One has knowledge about one's own knowledge, or beliefs about one's own beliefs. The considerations of this paper can be generalized to multi-agent epistemic logic, but that will not be done here. Formally, single-agent epistemic logic is just standard monomodal logic; we call it 'epistemic' in view of the envisaged applications.

In epistemic logic, we typically abstract away from some practical computational limitations of all real epistemic agents. For example, we are not concerned with their failure to infer from a proposition q the disjunction $q \vee r$ for every unrelated proposition r. What matters is that if some propositions do in fact follow from the agent's theory (from what the agent knows, or believes), then so too do all their logical consequences. For ease of exposition, we may idealize epistemic agents and describe them as knowing whatever follows from what they know, or as believing whatever follows from what they believe, but we could equally well redescribe the matter in less contentious terms by substituting 'p follows from what one knows'

T. Williamson (✉)
University of Oxford, Oxford, UK
e-mail: timothy.williamson@philosophy.ox.ac.uk

H. Arló-Costa et al. (eds.), *Readings in Formal Epistemology*, Springer Graduate Texts in Philosophy 1, DOI 10.1007/978-3-319-20451-2_34

for 'one knows *p*' or '*p* follows from what one believes' for 'one believes *p*' throughout the informal renderings of formulas, at the cost only of some clumsiness. Thus, if we so wish, we can make what looks like the notorious assumption of logical omniscience true by definition of the relevant epistemic operators. On suitable readings, it is a triviality rather than an idealization. It does not follow that no computational constraints are of any concern to epistemic logic. For if one's knowledge is logically closed by definition, that makes it computationally all the harder to know that one does *not* know something: in the standard jargon, logical omniscience poses a new threat to negative introspection. That threat is one of the phenomena to be investigated in this paper.

In a recursively axiomatizable epistemic logic, logical omniscience amounts to closure under a recursively axiomatizable system of inferences. Thus all the inferences in question can in principle be carried out by a single Turing machine, an idealized computer. Epistemic logicians do not usually want to make assumptions which would require an epistemic agent to exceed every Turing machine in computational power. In particular, such a requirement would presumably defeat the purpose of the many current applications of epistemic logic in computer science. By extension, epistemic logicians might prefer not to make assumptions which would permit an epistemic agent not to exceed every Turing machine in computational power only under highly restrictive conditions. Of course, such assumptions might be perfectly appropriate in special applications of epistemic logic to cases in which those restrictive conditions may be treated as met. But they would not be appropriate in more general theoretical uses of epistemic logic.

As an example, let us consider the so-called axiom of *negative introspection* alluded to above. It may be read as the claim that if one does not know *p* then one knows that one does not know *p*, or that if one does not believe *p* then one believes that one does not believe *p*. In terms of theories: if one's theory does not entail *p*, then one's theory entails that one's theory does not entail *p*. That assumption is acceptable in special cases for special values of '*p*'. However, for a theory to be consistent is in effect for there to be some *p* which it does not entail. On this reading, negative introspection implies that if one's theory is consistent then it entails its own consistency. But, by Gödel's second incompleteness theorem, if one's theory is recursively axiomatizable and includes Peano arithmetic, then it entails its own consistency only if it is inconsistent. Thus, combined with the incompleteness theorem, negative introspection implies that if one's theory is recursively axiomatizable then it includes Peano arithmetic only if it is inconsistent. Yet, in a wide range of interesting cases, the output of a Turing machine, or the theory of an epistemic agent of equal computational power, is a consistent recursively axiomatizable theory which includes Peano arithmetic. Thus, except in special circumstances, the negative introspection axiom imposes an unwarranted constraint on the computational power of epistemic agents.

Naturally, such an argument must be made more rigorous before we can place much confidence in it. That will be done below. The problem for the negative introspection axiom turns out to be rather general: it arises not just for extensions of Peano arithmetic but for any undecidable recursively axiomatizable theory, that is, for any theory which is the output of some Turing machine while its complement is not. It is very natural to consider epistemic agents whose theories are of that kind.

The aim of this paper is not primarily to criticize the negative introspection axiom. Rather, it is to generalize the problem to which that axiom gives rise, to formulate precisely the conditions which a system of epistemic logic must satisfy in order not to be susceptible to such problems, and to investigate which systems satisfy those conditions. The conditions in question will be called *r.e. conservativeness* and *r.e. quasi-conservativeness*. Very roughly indeed, a system satisfies these conditions if it has a wide enough variety of models in which the epistemic agent is computationally constrained. Such models appear to be among the intended models on various applications of epistemic logic. As already noted, systems of epistemic logic which do not satisfy the conditions may be appropriate for other applications. But it is time to be more precise.

Elementary Epistemic Logic

Let L be the language consisting of countably many propositional variables p_0, p_1, p_2, ... (p and q represent arbitrary distinct variables), the falsity constant $\bot$ and the material conditional $\supset$. Other operators are treated as metalinguistic abbreviations in the usual way. We expand L to the language $L_\Box$ of propositional modal logic by adding the operator $\Box$. $\Diamond\alpha$ abbreviates $\neg\Box\neg\alpha$. Unless otherwise specified, the metalinguistic variables α, β, γ, ... range over all formulas of $L_\Box$. We use the necessity symbol $\Box$ from modal logic to make various formulas and formal systems look familiar, without prejudice to its interpretation. We reinterpret $\Box$ as something like 'I know that' or 'I believe that'. To generalize over reinterpretations, we use the neutral verb 'cognize' for $\Box$ in informal renditions of formulas.

A *theory* in $L_\Box$ is a subset of $L_\Box$ containing all truth-functional tautologies and closed under modus ponens for $\supset$ (MP). A *model* M of $L_\Box$ induces a function $M() : L_\Box \to \{0, 1\}$ where $M(\bot) = 0$ and $M(\alpha \supset \beta) = 1$ if and only if $M(\alpha) \leq M(\beta)$. Intuitively, $M(\alpha) = 1$ if and only if α is true in M; $M(\alpha) = 0$ if and only if α is false in M. An *application* of epistemic logic determines a class of its intended models. The logic of the application is the set of formulas α such that $M(\alpha) = 1$ for every intended model M; thus the logic is a theory in $L_\Box$. Of course, we can also define a relation of logical consequence on the models, but for present purposes it is simpler to identify a logic with the set of its theorems.

Since atomic sentences are treated simply as propositional variables, we may substitute complex formulas for them. More precisely, we assume that for each intended model M and uniform substitution σ there is an intended model M^σ such that for every α $M^\sigma(\alpha) = M(\sigma\alpha)$. Thus the logic of the application is closed under uniform substitution (US).

A *modal logic* is a theory in $L_\Box$ closed under US. The logic of an application is a modal logic. The smallest modal logic is PC, the set of all truth-functional tautologies. If Σ is a modal logic, we write $\vdash_\Sigma \alpha$ when $\alpha \in \Sigma$. For any $X \subseteq L_\Box$, we define $X \vdash_\Sigma \alpha$ if and only if $\vdash_\Sigma \wedge X_0 \supset \alpha$ for some finite $X_0 \subseteq X$ ($\wedge X_0$ and $\vee X_0$ are the conjunction and disjunction respectively of X_0 on a fixed ordering of the language). X is Σ-consistent unless $X \vdash_\Sigma \bot$. A maximal Σ-consistent set is a Σ-consistent set not properly included in any Σ-consistent set.

If M is a model, let $\Box^{-1}M = \{\alpha : M(\Box\alpha) = 1\}$. Thus $\Box^{-1}M$ expresses what the agent cognizes in M. If Σ is the logic of an application on which $\Box^{-1}M$ is a theory in $L_\Box$ for every intended model M, then for all formulas α and β, $\vdash_\Sigma \Box(\alpha \supset \beta) \supset (\Box\alpha \supset \Box\beta)$ (axiom schema K) and if $\vdash_{PC} \alpha$ then $\vdash_\Sigma \Box\alpha$ (rule RN_{PC}). A modal logic satisfying RN_{PC} and K is *prenormal.* If cognizing is knowing or believing, then prenormality is an extreme idealization, a form of logical omniscience. But if cognizing is the closure of knowing or believing under at least truth-functional consequence, then prenormality is innocuous. The rule RN_{PC} differs from the stronger and better-known rule RN (necessitation or epistemization): if $\vdash_\Sigma \alpha$ then $\vdash_\Sigma \Box\alpha$. A modal logic Σ satisfying RN and K is *normal.* Unlike RN_{PC}, RN requires the agent to cognize all theorems of the epistemic logic, not just all truth-functional tautologies. For instance, $\Box\top$ is a theorem of every prenormal logic by RN_{PC}, but since it is not a theorem of PC we cannot iterate the rule; $\Box\Box\top$ is not a theorem of the smallest prenormal logic. By contrast, we can iterate RN, and $\Box\Box\top$ is a theorem of every normal modal logic. Prenormality does not imply that agents cognize their own cognizing. It merely implies that they can formulate propositions about cognizing, for since $\Box\alpha \supset \Box\alpha$ is a truth-functional tautology, $\Box(\Box\alpha \supset \Box\alpha)$ is a theorem of every prenormal logic. Since normality entails prenormality, results about all prenormal logics apply to all normal modal logics. Every logic extending a prenormal logic is prenormal; by contrast, some nonnormal logics extend normal logics, although any extension of a normal logic is at least prenormal.

Any normal logic Σ has a possible worlds semantics where $\Box\alpha$ is true at a world w in a model M if and only if α is true at every world in M to which w has the accessibility relation of M. Intuitively, a world x is accessible from w if and only if what the agent cognizes at w is true at x. In other words, one world is accessible from another if and only if for all one cognizes in the latter one is in the former. The formulas α such that $\Box\alpha$ is true at w express what the agent cognizes at w. For every normal logic Σ there is a class C of models such that Σ consists of exactly the formulas true at every world in every model in C.

Many authors require the accessibility relation to be an equivalence relation (reflexive, symmetric and transitive) for every intended model of their application. A common attitude is expressed by the authors of a standard text, who write that the postulate 'seems reasonable for many applications we have in mind', but 'we can certainly imagine other possibilities' (Fagin et al. 1995, 33). For example, if x is accessible from w if and only if appearances to the agent are identical in x and w, then accessibility is an equivalence relation because identity in any given respect is an equivalence relation. The logic of the class of all possible worlds models in which accessibility is an equivalence relation is the modal system known as S5 : $\vdash_{S5} \alpha$ if and only if α is true in every model for which accessibility is an equivalence relation. Since equivalence relations correspond to partitions of the set of worlds, S5 is also known as the logic of the *partitional* conception of knowledge. S5 is the smallest normal modal logic with the theorem schemas T($\Box\alpha \supset \alpha$) and E ($\neg\Box\alpha \supset \Box\neg\Box\alpha$). T (*truthfulness*) says that the agent cognizes only truths; it is appropriate for applications on which one cognizes only what follows from what one knows. T corresponds to the condition that accessibility be reflexive. For applications on which one cognizes what follows from what one *believes*, T would have to be dropped, perhaps replaced by the weaker principle D ($\Box\alpha \supset \Diamond\alpha$). D requires cognition to be consistent in the sense that an agent who cognizes something does not also cognize its negation. D corresponds to the condition that accessibility be serial (from every world some world is accessible). E is the principle of *negative introspection*: cognition records its omissions in the sense that agents who do not cognize something cognize that they do not cognize it. E corresponds to the condition that accessibility be euclidean (worlds accessible from a given world are accessible from each other). In S5 we can derive the principle of *positive introspection* 4 ($\Box\alpha \supset \Box\Box\alpha$), that cognition records its contents in the sense that agents who cognize something cognize that they cognize it. 4 corresponds to the condition that accessibility be transitive. If T is dropped or weakened to D then 4 is no longer derivable from E, so 4 might be added as an independent schema. Accessibility is reflexive (T) and euclidean (E) if and only if it is an equivalence relation.

Computational Constraints

To formulate computational constraints, we generalize concepts from recursion theory to $L_\Box$ using a standard intuitively computable coding procedure. A model M is r.e. if and only if $\Box^{-1}$M (which expresses what the agent cognizes in M) is an r.e. (recursively enumerable) theory in $L_\Box$. In that sense, the agent's cognition in an r.e. model does not exceed the computational capacity of a sufficiently powerful Turing machine.

Consider the restriction of $\Box^{-1}$M to the $\Box$-free sublanguage L, L $\cap$ $\Box^{-1}$M. Let $\Box^{-1}$M be an r.e. theory in $L_\Box$. Thus L $\cap$ $\Box^{-1}$M is an r.e. theory in L. It is the part of the agent's overall theory in M which is not specifically epistemic. From the standpoint of general epistemic logic, can we reasonably impose any further constraints on L $\cap$ $\Box^{-1}$M beyond recursive enumerability?

If $\Box^{-1}$M is required to be consistent, L $\cap$ $\Box^{-1}$M is consistent too. Can we limit the possible values of L $\cap$ $\Box^{-1}$M still further? For many applications we cannot. L $\cap$ $\Box^{-1}$M simply expresses what the agent cognizes in M about some aspect of reality. The agent can store any r.e. theory in L as a recursive axiomatization (Craig 1953). If the agent might cognize that aspect of reality simply by having learned a theory about it on the testimony of a teacher, any (consistent) r.e. theory in L is possible. In particular, we can interpret the propositional variables as mutually independent. For example, given a black box which may or may not flash a light on input of a symbol for a natural number, we can read p_i as 'The light flashes on input *i*'. Then any (consistent) r.e. theory in L could exhaust everything expressible in L which the agent (with only the computational power of a Turing machine) has learned about the black box. Such situations seem quite reasonable. If we want an epistemic logic to have a generality beyond some local application, it should apply to them: such situations should correspond to intended models. Now any application which has all those intended models thereby satisfies (*) or ($*_{con}$), depending on whether the epistemic agent's theory is required to be consistent:

(*) For every r.e. theory R in L, $L \cap \Box^{-1}M = R$ for some r.e. intended model M.

($*_{con}$) For every consistent r.e. theory R in L, $L \cap \Box^{-1}M = R$ for some r.e. intended model M.

(*) is appropriate for readings of $\Box$ like 'It follows from what I believe that ...', if the agent is not required to be consistent. For readings of $\Box$ like 'It follows from what I know that ...', only ($*_{con}$) is appropriate, for one can know only truths and any set of truths is consistent. We can define corresponding constraints on a modal logic Σ without reference to models:

Σ is *r.e. conservative* if and only if for every r.e. theory R in L, there is a maximal Σ-consistent set X such that $\Box^{-1}X$ is r.e. and $L \cap \Box^{-1}X = R$.

Σ is *r.e. quasi-conservative* if and only if for every consistent r.e. theory R in L, there is a maximal Σ-consistent set X such that $\Box^{-1}X$ is r.e. and $L \cap \Box^{-1}X = R$.

Here $\Box^{-1}X = \{\alpha \in L_\Box : \Box\alpha \in X\}$. Roughly, if Σ is r.e. (quasi-)conservative then every (consistent) r.e. theory in the language without $\Box$ is conservatively extended by an r.e. theory in the language with $\Box$ such that it is consistent in Σ for R to be exactly what the agent cognizes in the language without $\Box$ while what the agent cognizes in the language with $\Box$ constitutes an r.e. theory. If an application satisfies (*), its logic is r.e. conservative, for X can be the set of formulas true in M. Conversely, any r.e. conservative logic is the logic of some application which satisfies (*), for some appropriate kind of intended model. The same relationships hold between ($*_{con}$) and r.e. quasi-conservativeness. For many applications of epistemic logic, the class of intended models is quite restricted and even ($*_{con}$) does not hold. But if the application interprets $\Box$ as something like 'It follows from what I believe/know that', without special restrictions on the epistemic subject, then situations of the kind described above will correspond to intended models and the logic of the application will be r.e. [quasi-] conservative. In this paper we do not attempt to determine which informally presented applications of epistemic logic satisfy (*) or ($*_{con}$). We simply investigate which logics are r.e. [quasi-] conservative.

Trivially, every r.e. conservative modal logic is r.e. quasi-conservative. Examples will be given below of r.e. quasi-conservative normal modal logics which are not r.e. conservative. For prenormal modal logics, r.e. conservativeness can be characterized in terms of r.e. quasi-conservativeness in a simple way which allows us to transfer results about one to the other:

Proposition 34.1 Let Σ be a prenormal modal logic. Then Σ is r.e. conservative if and only if Σ is r.e. quasi-conservative and not $\vdash_\Sigma \Diamond\top$.

Proof Let $\Box$ L = {$\Box\alpha$: $\alpha \in$ L}. ($\Rightarrow$) Trivially, Σ is r.e. quasi-conservative if r.e. conservative. Suppose that $\vdash_\Sigma \Diamond\top$. Since $\Box$ L $\vdash_\Sigma \Box\neg\top$, $\Box$ L is Σ-inconsistent. Thus L $\cap\ \Box^{-1}$X = L for no Σ-consistent set X. Since L is an r.e. theory in L, Σ is not r.e. conservative. ($\Leftarrow$) Suppose that Σ is r.e. quasi-conservative but not r.e. conservative. Since L is the only inconsistent theory in L, there is no maximal Σ-consistent set X such that $\Box^{-1}$X is r.e. and L $\cap\ \Box^{-1}$X = L. If $\Box$ L is Σ-consistent, then some maximal Σ-consistent set X extends $\Box$ L, so L $\cap\ \Box^{-1}$X = L; but for α $\in$ L $\Box \vdash_\Sigma \bot \supset \alpha$, so $\vdash_\Sigma \Box\bot \supset \Box\alpha$ by prenormality, so $\Box^{-1}$X = $L_\Box$ because $\Box\bot \in X$, so $\Box^{-1}$X is r.e. Thus $\Box$ L is Σ-inconsistent, i.e., for some $\alpha_0, \ldots, \alpha_m \in$ L, $\vdash_\Sigma \neg\wedge\{\Box\alpha_i : i \leq m\}$. But for $i \leq m$, $\vdash_\Sigma \Box\neg\top \supset \Box\alpha_i$ by prenormality, so $\vdash_\Sigma \neg\Box\neg\top$.

Examination of the proof shows that the prenormality condition can be weakened to this: if $\vdash_{PC} \alpha \supset \beta$ then $\vdash_\Sigma \Box\alpha \supset \Box\beta$. An example of a reading of $\Box$ which verifies this weaker condition but falsifies prenormality is 'There is a subjective probability of at least x that', where $0 < x < 1$, for prenormality implies that $\vdash_\Sigma (\Box p \wedge \Box q) \supset \Box(p \wedge q)$, whereas this reading invalidates that formula. Prenormality can be weakened in similar ways for subsequent propositions.

R.e. conservativeness and r.e. quasi-conservativeness do not state upper or lower bounds on the epistemic agent's computational capacity. Rather, they state upper bounds on the strength of the epistemic logic itself; evidently a modal logic with an r.e. [quasi-] conservative extension is itself r.e. [quasi-] conservative. But too strong a logic can impose unwarranted restrictions on the agent's theory of the world given an upper bound on the agent's computational capacity.

Some Non-R.e. Quasi-Conservative Logics

Which modal logics are not r.e. [quasi-] conservative? Obviously, since $\vdash_{S5} \Diamond\top$, the logic S5 is not r.e. conservative. Since S5 is decidable, this does not result from non-recursiveness in S5 itself. More significantly:

Proposition 34.2 S5 is not r.e. quasi-conservative.

Proof (Skyrms 1978, 377 and Shin and Williamson 1994, Proposition 34.1 have similar proofs of related facts about S5): Let R be a non-recursive r.e. theory in L; R is consistent. Suppose that $\Box^{-1}$X is r.e. and L $\cap\ \Box^{-1}$X = R for some maximal S5-consistent set X. Now L $-$ R = {α : $\Box\neg\Box\alpha \in$ X} $\cap$ L. For if $\alpha \in$ L $-$ R then $\Box\alpha \notin$ X, so $\neg\Box\alpha \in$ X; but $\vdash_{S5} \neg\Box\alpha \supset \Box\neg\Box\alpha$, so $\Box\neg\Box\alpha \in$ X since X is maximal S5-consistent. Conversely, if $\Box\neg\Box\alpha \in$ X then $\neg\Box\alpha \in$ X since $\vdash_{S5} \Box\neg\Box\alpha \supset \neg\Box\alpha$, so $\Box\alpha \notin$ X, so $\alpha \notin$ R since L $\cap\ \Box^{-1}$X = R. Since $\Box^{-1}$X is r.e., so is {α : $\Box\neg\Box\alpha \in$ X) $\cap$ L, i.e. L $-$ R. Contradiction.

Thus the partitional conception of knowledge prevents a subject with the computational capacity of a Turing machine from having as the restriction of its theory to the $\Box$-free language any non-recursive r.e. theory (for other problems with the S5 schema in epistemic logic and further references see Williamson (2000, 23–24, 166–167, 226–228, 316–317)). Thus S5 is unsuitable as a general epistemic logic for Turing machines.

The proof of Proposition 34.2 depends on the existence of an r.e. set whose complement is not r.e. By contrast, the complement of any recursive set is itself recursive; decidability, unlike semi-decidability, is symmetric between positive and negative answers. The analogue of Proposition 34.2 for a notion like r.e. quasi-conservativeness but defined in terms of recursiveness rather than recursive enumerability would be false. For it is not hard to show that if R is a consistent recursive theory in L, then there is a maximal S5-consistent set X in $L_\Box$ such that $\Box^{-1}X$ is recursive and $L \cap \Box^{-1}X = R$. Thus S5 imposes computational constraints not on very clever agents (whose theories need not be r.e.) or on very stupid agents (whose theories must be recursive) but on half-clever agents (whose theories must be r.e. but need not be recursive).

Proposition 34.2 is the rigorous version of the argument sketched in the introduction. Can we generalize it? The next result provides a rather unintuitive necessary condition for r.e. quasi-conservativeness which nevertheless has many applications.

Theorem 34.3 Let Σ be a modal logic such that for some formulas $\alpha_0, \ldots, \alpha_n \in L_\Box$ and $\beta_0, \ldots, \beta_n \in L$, $\vdash_\Sigma \bigvee\{\Box\alpha_i : i \leq n\}$ and, for each $i \leq n$, $\vdash_\Sigma (\Box\alpha_i \wedge \Box\beta_i) \supset \Box\bot$ and not $\vdash_{PC} \neg\beta_i$. Then Σ is not r.e. quasi-conservative.

Proof There are pairwise disjoint r.e. subsets $I_0, I_1, I_2, \ldots$ of the natural numbers $\mathbb{N}$ such that for every total recursive function f, $i \in I_{f(i)}$ for some $i \in \mathbb{N}$. For let $f[0]$, $f[1], f[2], \ldots$ be a recursive enumeration of all partial and total recursive functions on $\mathbb{N}$ and set $I_i = \{j : f[j](j) \text{ is defined and} = i\}$; then $j \in I_{f[j](i)}$ whenever $f[j]$ is total, I_i is r.e. and $I_i \cap I_j = \{\}$ whenever $i \neq j$. Now suppose that (i) $\vdash_\Sigma \bigvee\{\Box\alpha_i : i \leq n\}$; (ii) $\vdash_\Sigma (\Box\alpha_i \wedge \Box\beta_i) \supset \Box\bot$ for each $i \leq n$; (iii) $\vdash_{PC} \neg\beta_i$ for no $i \leq n$. Let m be the highest subscript on any propositional variable occurring in $\beta_0, \ldots, \beta_n$. For all $i \in \mathbb{N}$, let σ_i and τ_i be substitutions such that $\sigma_i\, p_j = p_{i(m+1)+j}$ and $\tau_i\, p_{i(m+1)+j} = p_j$ for all $j \in \mathbb{N}$. Set $U = \{\sigma_i\beta_j : i \in I_j\}$. Since the σ_i are recursive and the I_j are r.e., U is r.e. Now $\vdash_{PC} \neg\sigma_i\beta_j$ for no i, j, otherwise $\vdash_{PC} \neg\tau_i\sigma_i\beta_j$, i.e., $\vdash_{PC} \neg\beta_j$, contrary to (iii). Moreover, if $h \neq i$ then $\sigma_h\beta_j$ and $\sigma_i\beta_k$ have no propositional variable in common. Thus if $h \in I_j$ and $i \in I_k$ and $\sigma_h\beta_j$ has a variable in common with $\sigma_i\beta_k$, then $h = i$, so $j = k$ because the I_j are pairwise disjoint. Hence no two members of U have a propositional variable in common. Thus U is consistent. Let R be the smallest theory in L containing U; R is consistent and r.e. Suppose that for some maximal Σ-consistent set X, $\Box^{-1}X$ is r.e. and $L \cap \Box^{-1}X = R$. Let the total recursive function g enumerate $\Box^{-1}X$. Fix $j \in \mathbb{N}$. By (i), $\vdash_\Sigma \bigvee\{\Box\sigma_j\alpha_i : i \leq n.\}$ since Σ is closed under US, so $\Box\sigma_j\alpha_i \in Y$ for some $i \leq n$ since Y is maximal Σ-consistent. Thus $g(k) = \sigma_j\alpha_i$ for some k; let $k(j)$ be the least k such that $g(k) \in \{\sigma_j\alpha_i : i \leq n\}$. Let $f(j)$ be the least $i \leq n$ such that $g(k(j)) = \sigma_j\alpha_i$. Since g enumerates $\Box^{-1}X$, $\Box\sigma_j\alpha_{f(j)} \in X$. Since g and σ_j are total recursive, k is total recursive, so f is total recursive. Thus $j \in I_{f(j)}$ for some $j \in \mathbb{N}$, so $\sigma_j\beta_{f(j)} \in U \subseteq R$ since $f(j) \leq n$. Since $L \cap \Box^{-1}X = R$, $\Box\sigma_j\beta_{f(j)} \in X$. By (ii), $\vdash_\Sigma (\Box\alpha_{f(j)} \wedge \Box\beta_{f(j)}) \supset \Box\bot$, so $\vdash_\Sigma (\Box\sigma_j\alpha_{f(j)} \wedge \Box\sigma_j\beta_{f(j)}) \supset \Box\bot$; since X is maximal Σ-consistent, $\Box\bot \in X$. Thus $\bot \in R$, contradicting the consistency of R. Thus no such set as X can exist, so Σ is not r.e. quasi-conservative.

In other words, a necessary condition for Σ to be r.e. quasi-conservative is that for all formulas $\alpha_0, \ldots, a_n \in L_\Box$ and $\beta_0, \ldots, \beta_n \in L$, if $\vdash_\Sigma \vee(\Box\alpha_j : i \leq n)$ and, for each $i \leq n$, $\vdash_\Sigma(\Box\alpha_i \wedge \Box\beta_i) \supset \Box\bot$ then, for some $i \leq n$, $\vdash_{PC} \neg\beta_i$. Of course, if Σ is prenormal and contains the D axiom (requiring the agent to be consistent) then the condition that $\vdash_\Sigma(\Box\alpha_i \wedge \Box\beta_i) \supset \Box\bot$ can be simplified to the condition that $\vdash_\Sigma \neg\Box(\alpha_i \wedge \beta_i)$.

Open Problem Is the necessary condition for r.e. quasi-conservativeness in Theorem 34.3 (or some natural generalization of it) also sufficient?

Observation The proof of Theorem 34.3 uses significantly more recursion theory than does the proof of Proposition 34.2, which relies only on the existence of an r.e. set whose complement is not r.e. Samson Abramsky observed (informal communication) that the proof of Proposition 34.2 would generalize to a setting in which r.e. sets were replaced by open sets in a suitable topology (in which not all open sets have open complements). It would be interesting to see whether a generalization along such lines yielded a smoother theory. One might then seek an intuitive interpretation of the topology.

To see that Proposition 34.2 is a special case of Theorem 34.3, put $n = 1$, $\alpha_0 = \Diamond\neg p$, $\alpha_1 = \Diamond p$, $\beta_0 = p$ and $\beta_1 = \neg p$. Now; $\vdash_{S5} \Box\Diamond\neg p \vee \Box\Box p$; $\vdash_{S5} (\Box\Diamond\neg p \wedge \Box p) \supset \Box\bot$ because $\vdash_{S5} \Box p \supset \Box\Box p$ and S5 is normal; likewise $\vdash_{S5} (\Box\Diamond p \wedge \Box\neg p) \supset \Box\bot$; finally, neither $\vdash_{S5} p$ nor $\vdash_{S5} \neg p$. These features of S5 follow easily from the fact that it is a consistent normal extension of $K4G_1$, the smallest normal logic Σ including both 4 and G_1 ($\Diamond\Box\alpha \supset \Diamond\Box\alpha$). Since the inconsistent logic is certainly not r.e. quasi-conservative, we have this generalization of Proposition 34.2:

Corollary 34.4 No normal extension of KG_14 is r.e. quasi-conservative.

We can use Corollary 34.1 to show several familiar weakenings of S5 not to be r.e. quasi-conservative. G_1 corresponds to the condition that accessibility be convergent, in the sense that if x and y are both accessible from w, then some world z is accessible from both x and y. Informally, G_1 says that agents either cognize that they do not cognize α or cognize that they do not cognize $\neg\alpha$. Any normal logic satisfying E also satisfies G_1, so Corollary 34.4 implies in particular the failure of r.e. quasi-conservativeness for the logics K4E and KD4E. Those two logics are the natural analogues for belief of S5 as a logic for knowledge, since they retain positive and negative introspection while dropping truthfulness altogether (K4E) or weakening it to consistency (KD4E). Thus they are often used as logics of belief. But positive and negative introspection together violate the computational constraint in a normal logic even in the absence of truthfulness. Thus, in a generalized context, K4E or KD4E impose unacceptably strong computational constraints as logics of belief, just as S5 does as a logic of knowledge.

For more examples, consider the schemas

$$B(\alpha \supset \Box\Diamond\alpha) \text{ and } D_1(\Box(\Box\alpha \supset \beta) \vee \Box(\Box\beta \supset \alpha)).$$

B corresponds to the condition that accessibility be symmetric, D_1 to the condition that accessibility be connected, in the sense that if x and y are both accessible from w, then either x is accessible from y or y is accessible from x. Any normal logic satisfying B or D_1 also satisfies G_1, so KB4 and KD_14 are not r.e. quasi-conservative. *A fortiori*, the same holds if one requires the agent to be consistent or truthful by adding D or T respectively. Thus KD4E, KDG14, KTG_14 (= S4.2), KDD_14 and KTD_14 (= S4.3) are also not r.e. quasi-conservative. All these are sublogics of S5; we shall need to weaken S5 considerably to find an r.e. quasi-conservative logic.

Theorem 34.3 is also applicable to logics without positive introspection. We can use T rather than 4 to derive $(\Box\Diamond\neg p \wedge \Box p) \supset \Box\bot$, so:

Corollary 34.5 No normal extension of KTG_1 is r.e. quasi-conservative.

Again, consider Alt_n $(\vee\{\Box(\wedge\{p_j : j<i\} \supset p_i) : i \leq n\})$, e.g., Alt_2 is $\Box p_0 \vee \Box(p_0 \supset p_1) \vee \Box((p_0 \wedge p_1) \supset p_2)$. Alt_n corresponds to the condition that from each world at most n worlds be accessible; informally, the agent rules out all but n specific possibilities. Setting $\alpha_i = \wedge\{p_j : j<i\} \supset p_i$ and $\beta_i = \neg\alpha_i$ in Theorem 34.3 gives:

Corollary 34.6 For any n, no r.e. quasi-conservative prenormal modal logic contains Alt_n.

An epistemic logic which imposes an upper bound on how many possibilities the agent can countenance thereby excludes the agent from having some consistent r.e. theories about the black box.

Some R.e. Conservative Logics

Since every modal logic with an r.e. [quasi-] conservative extension is itself r.e. [quasi-] conservative, an efficient strategy is to seek very strong r.e. [quasi-] conservative logics, even if they are implausibly strong for most epistemic applications, because we can note that the weaker and perhaps more plausible logics which they extend will also be r.e. [quasi-] conservative.

A large class of r.e. conservative logics arises as follows. Let Σ be any epistemic logic. The agent might cognize each theorem of Σ. Moreover, an epistemic logic Σ^* may imply this, in that $\vdash_{\Sigma^*} \Box\alpha$ whenever $\vdash_\Sigma \alpha$. Σ and Σ^* may be distinct, even incompatible. For example, let Ver be the smallest normal modal logic containing $\Box\bot$. Interpreted epistemically, Ver implies that the agent is inconsistent; but Ver itself is consistent. An epistemic theory consisting just of Ver falsely but consistently self-attributes inconsistency, and an epistemic logic may report that the agent self-attributes inconsistency without itself attributing inconsistency to the agent. Thus Ver* may contain $\Box\Box\bot$ without $\Box\bot$. Similarly, let Triv be the smallest normal modal logic containing all theorems of the form $\alpha \equiv \Box\alpha$. Interpreted epistemically, Triv implies that the agent cognizes that his beliefs contain all and only truths; but Triv itself does not contain all and only truths (neither $\vdash_{\mathrm{Triv}} p$ nor $\vdash_{\mathrm{Triv}} \neg p$). Thus

Triv* may contain $\Box(p \equiv \Box p)$ without $p \equiv \Box p$. To be more precise, for any modal logics Λ and Σ let $\Lambda\Box\Sigma$ be the smallest normal extension of Λ containing $\{\Box\alpha : \vdash_\Sigma \alpha\}$. We will prove that if Σ is consistent and normal then $K\Box\Sigma$ is r.e. conservative. $K\Box\Sigma$ is an epistemic logic for theorizing about theories that incorporate the epistemic logic Σ. R.e. conservativeness implies no constraint on what epistemic logic the agent uses beyond consistency (if Σ is inconsistent, then $K\Box\Sigma$ contains Alt_0 and so is not even r.e. quasi-conservative). In particular, the smallest normal logic K itself is r.e. conservative. Moreover, if Σ is consistent and normal, then $K4\Box\Sigma$ is r.e. conservative; that is, we can add positive introspection. In particular, K4 itself is r.e. conservative. We prove this by proving that K□Ver and K□Triv are r.e. conservative. Since K□Ver and K□Triv contain $\Box\Box\bot$ and $\Box(p \equiv \Box p)$ respectively, they are too strong to be useful epistemic logics themselves, but equally they are strong enough to contain many other logics of epistemic interest, all of which must also be r.e. conservative. By contrast, Ver and Triv are not themselves even r.e. quasi-conservative, for $\vdash_{Ver} Alt_0$ and $\vdash_{Triv} Alt_1$.

For future reference, call a mapping ϕ from $L_\Box$ into $L_\Box$ *respectful* if and only if $\phi p = p$ for all propositional variables p, $\phi\bot = \bot$ and $\phi(\alpha \supset \beta) = \phi\alpha \supset \phi\beta$ for all formulas α and β.

Lemma 34.7 K□Triv is r.e. conservative.

Proof Let R be an r.e. theory in L. Let δ and κ be respectful mappings from $L_\Box$ to L such that $\delta\Box\alpha = \delta\alpha$; $\kappa\Box\alpha = \top$ if R $\vdash_{PC} \delta\alpha$ and $\kappa\Box\alpha = \bot$ otherwise for all formulas α. (i) Axiomatize Triv with all truth-functional tautologies and formulas of the form $\alpha \equiv \Box\alpha$ as the axioms and MP as the only rule of inference (schema K and rule RN are easily derivable). By an easy induction on the length of proofs, $\vdash_{Triv} \alpha$ only if $\vdash_{PC} \delta\alpha$. (ii) Axiomatize K□Triv with all truth-functional tautologies and formulas of the forms $\Box(\alpha \supset \beta) \supset (\Box\alpha \supset \Box\beta)$ and $\Box\gamma$ whenever $\vdash_{Triv} \gamma$ as the axioms and MP as the only rule of inference (RN is a derived rule; its conclusion is always an axiom because the logic so defined is a sublogic of Triv). We show by induction on the length of proofs that $\vdash_{K\Box Triv} \alpha$ only if $\vdash_{PC} \kappa\alpha$. Basis: If $\vdash_{PC} \alpha$, $\vdash_{PC} \kappa\alpha$. If $\kappa\Box(\alpha \supset \beta) = \top$ and $\kappa\Box\alpha = \top$ then R $\vdash_{PC}\delta\alpha \supset \delta\beta$ and R $\vdash_{PC}$ $\delta\alpha$, so R $\vdash_{PC}\delta\beta$, so $\kappa\Box\beta = \top$, so R $\vdash_{PC}$ $\kappa(\Box(\alpha \supset \beta) \supset (\Box\alpha \supset \Box\beta))$; otherwise $\kappa\Box(\alpha \supset \beta) = \bot$ or $\kappa\Box\alpha = \bot$ and again R $\vdash_{PC}$ $\kappa(\Box(\alpha \supset \beta) \supset (\Box\alpha \supset \Box\beta))$. If $\vdash_{Triv}$ γ then $\vdash_{PC} \delta\gamma$ by (i), so R $\vdash_{PC} \delta\gamma$, so $\kappa\Box\gamma = \top$, so $\vdash_{PC} \kappa\Box\gamma$. Induction step: trivial. (iii) Put $Y = \{\Box\alpha \in L_\Box$: R $\vdash_{PC} \delta\gamma\} \cup \{\neg\Box\alpha \in L_\Box$: not R $\vdash_{PC} \delta\alpha\}$. Y is K□Triv-consistent, for if $Y_0 \subseteq Y$ is finite and $\vdash_{K\Box Triv} \wedge Y_0 \supset \bot$ then $\vdash_{PC} \kappa\ (\wedge Y_0 \supset \top)$ by (ii), i.e. $\vdash_{PC} \wedge\{\kappa\alpha : \alpha \in Y_0\} \supset \bot$, which is impossible since $\{\kappa\alpha : \alpha \in Y\} \subseteq \{\top, \neg\bot\}$. Let X be a maximal K□Triv-consistent extension of Y. By definition of Y, $\Box^{-1}X = \{\alpha : R \vdash_{PC} \delta\alpha\}$, which is r.e. because R is r.e. and δ is recursive (although κ need not be). If $\alpha \in L$, $\delta\alpha = \alpha$, so $\Box\alpha \in X$ if and only if R $\vdash_{PC} \alpha$, i.e., if and only if $\alpha \in R$ because R is a theory; thus $L \cap \Box^{-1}X = R$. Hence K□Triv is r.e. conservative.

Lemma 34.8 K□Ver is r.e. conservative.

Proof Like Lemma 34.7, but in place of δ use a respectful mapping λ such that $\lambda\Box\alpha = \top$.

A notable sublogic of K□Ver is GL, the smallest normal modal logic including $\Box(\Box\alpha \supset \alpha) \supset \Box\alpha$. Thus a corollary of Lemma 34.8 is that GL is r.e. conservative. GL is in a precise sense the logic of what is provable in Peano arithmetic (PA) about provability in PA (Boolos 1993 has exposition and references). More generally, if R is an ω-consistent r.e. extension of PA, then GL is the logic of what is provable in R about provability in R. Since a Turing machine's theory of arithmetic is presumably at best an ω-consistent r.e. extension of PA, GL is therefore a salient epistemic logic for Turing machines, and its r.e. conservativeness is not surprising.

Caution We must be careful in our informal renderings of results about provability logic. A provability operator creates an intensional context within which the substitution of coextensive but not provably coextensive descriptions can alter the truth-value of the whole sentence; this point applies in particular to descriptions of agents or their theories. On a provability interpretation of □, occurrences of □ within the scope of other occurrences of □ in effect involve just such occurrences of descriptions of agents or their theories in an intensional context, so which logic is validated can depend on the manner in which a given agent or theory is described. The validity of GL as an epistemic logic is relative to a special kind of descriptive self-presentation of the theory T in the interpretation of □, by a coding of its axioms and rules of inference. GL is not valid relative to some extensionally equivalent but intensionally distinct interpretations of □, e.g. the indexical reading 'I can prove that' as uttered by an epistemic subject with the computational capacity of a Turing machine (Shin and Williamson 1994; Williamson 1996, 1998).

Proposition 34.9 If Σ is a consistent normal modal logic, $K\Box\Sigma$ and $K4\Box\Sigma$ are r.e. conservative.

Proof By Makinson (1971), either $\Sigma \subseteq \text{Triv}$ or $\Sigma \subseteq \text{Ver}$. Hence either $K\Box\Sigma \subseteq K\Box\text{Triv}$ or $K\Box\Sigma \subseteq K\Box\text{Ver}$. But Schema 4 is easily derivable in both K□Triv and K□Ver, so $K4\Box\Sigma \subseteq K\Box\text{Triv}$ or $K4\Box\Sigma \subseteq K\Box\text{Ver}$. By Lemmas 34.7 and 34.8, K□Triv and K□Ver are r.e. conservative, so $K4\Box\Sigma$ is.

All the logics salient in this paper are decidable, and therefore r.e., but we should note that an epistemic logic need not be r.e. to be r.e. conservative:

Corollary 34.10 Not all r.e. conservative normal modal logics are r.e.

Proof (i) We show that for any normal modal logic Σ, $\vdash_\Sigma \alpha$ if and only if $\vdash_{K\Box\Sigma} \Box\alpha$. Only the $\Leftarrow$ direction needs proving. Axiomatize $K\Box\Sigma$ with all truth-functional tautologies and formulas of the forms $\Box(\alpha \supset \beta) \supset (\Box\alpha \supset \Box\beta)$ and $\Box\gamma$ whenever $\vdash_\Sigma \gamma$ as the axioms and MP as the only rule of inference (RN is a derived rule; its conclusion is always an axiom because the logic so defined is a sublogic of Σ). Let η be a respectful mapping from $L_\Box$ to $L_\Box$ such that $\eta\Box\alpha = \alpha$ for all formulas α (η is distinct from δ in the proof of Lemma 34.7 since $\eta\Box\Box p = \Box p$ whereas $\delta\Box\Box p = p$). By induction on the length of proofs, $\vdash_{K\Box\Sigma} \alpha$ only if $\vdash_\Sigma \eta\alpha$. Hence $\vdash_{K\Box\Sigma} \Box\alpha$ only if $\vdash_\Sigma \alpha$. (ii) By (i), for any normal modal logics Σ_1 and Σ_2, $K\Box\Sigma_1 = K\Box\Sigma_2$ if and only if $\Sigma_1 = \Sigma_2$. But there are continuum many consistent normal modal logics (Blok 1980 has much more on these lines). Hence there are continuum many corresponding logics of the form K□Σ; all are r.e. conservative by Proposition 34.9. Since only countably many modal logics are r.e., some of them are not r.e.

One limitation of Proposition 34.9 is that K□Σ and K4□Σ never contain the consistency schema D. In a sense this limitation is easily repaired. For any modal logic Σ, let Σ [*D*] be the smallest extension of Σ containing D; thus $\vdash_{\Sigma[D]} \alpha$ just in case $\vdash_{\Sigma} \Diamond\top \supset \alpha$.

Proposition 34.11 For any r.e. conservative modal logic Σ, Σ[D] is r.e. quasi-conservative.

Proof For any consistent theory R, any maximal Σ-consistent set X such that $L \cap \Box^{-1}X = R$ is Σ[D]-consistent because $\Diamond\top \in X$.

Corollary 34.12 If Σ is a consistent normal modal logic, (K□Σ)[D] and (K4□Σ)[D] are r.e. quasi-conservative.

Proof By Propositions 34.9 and 34.11.

Although Σ[D] is always prenormal, it may not be normal, even if Σ is normal; sometimes not $\vdash_{\Sigma[D]} \Box\Diamond\top$. But we can also consider epistemic interpretations of normal logics with the D schema, e.g., KD and KD4. Such logics contain $\Box\Diamond\top$; they require agents to cognize their own consistency. By Gödel's second incompleteness theorem, this condition cannot be met relative to a Gödelian manner of representing the theory in itself; no consistent normal extension of the provability logic GL contains D. But $\Box\Diamond\top$ is true on other epistemic interpretations; for example, we know that our knowledge (as opposed to our beliefs) does not imply a contradiction. Since GL ⊆ K□Ver, Proposition 34.9 does not generalize to the r.e. quasi-conservativeness of KD□Σ. But we can generalize Lemma 34.7 thus:

Proposition 34.13 If Σ ⊆ Triv then KD□Σ and KD4□Σ are r.e. quasi-conservative.

Proof It suffices to prove that KD□Triv (=KD4□Triv) is r.e. quasi-conservative. Argue as for Lemma 34.1, adding $\Diamond\top$ as an axiom for KD□Triv and noting that if R is consistent then $\kappa\Box\neg\top = \bot$, so $\vdash_{PC} \kappa\Diamond\top$.

In particular, KD and KD4 are themselves r.e. quasi-conservative; they are our first examples of r.e. quasi-conservative logics which are not r.e. conservative.

We now return to systems with the T schema. Since T implies D, only r.e. quasi-conservativeness is at issue. That constraint was motivated by the idea that any consistent r.e. theory in the non-modal language might be exactly the restriction of the agent's total r.e. theory to the non-modal language. On many epistemic interpretations, it is in the spirit of this idea that the agent's total theory might be true in the envisaged situation (for example, the agent's theory about the black box might be true, having been derived from a reliable witness). To require an epistemic logic Σ to leave open these possibilities is to require that Σ[T] be r.e. quasi-conservative, where Σ[T] is the smallest extension of Σ containing all instances of T. As with Σ[D], Σ[T] need not be normal even when Σ is; sometimes not

$\vdash_{\Sigma[T]} \Box(\Box\alpha \supset \alpha)$ (Williamson 1998, 113–116 discusses logics of the form Σ[T]). Agents may not cognize that they cognize only truths. Nevertheless, particularly when □ is interpreted in terms of knowledge, one might want an epistemic logic such as KT containing $\Box(\Box\alpha \supset \alpha)$.

Proposition 34.11 and Corollary 34.12 have no analogues for T in place of D. For any modal logic Σ, if $\vdash_{\Sigma} \alpha$ then $\vdash_{(K\Box\Sigma)[T]} \Box\alpha$, but $\vdash_{(K\Box\Sigma)[T]} \Box\alpha \supset \alpha$, so $\vdash_{(K\Box\Sigma)[T]} \alpha$; thus (K□Σ)[T] extends Σ and is r.e. quasi-conservative only if Σ is. Similarly, Proposition 34.12 would be false with T in place of D (counterexample: $\Sigma = S5$). Therefore, needing a different approach, we start with the system GL[T]. GL[T] has intrinsic interest, for it is the provability logic GLS introduced by Solovay and shown by him to be the logic of what is true (rather than provable) about provability in PA; more generally, it is the logic of what is true about provability in an ω-consistent r.e. extension of PA. GLS is therefore a salient epistemic logic for Turing machines, and its r.e. quasi-conservativeness is not surprising. Although GLS is not normal and has no consistent normal extension, we can use its r.e. quasi-conservativeness to establish that of normal logics containing T.

Proposition 34.14 GLS is r.e. quasi-conservative.

Proof Let R be an consistent r.e. theory in L. Axiomatize a theory R+ in $L_\Box$ with all members of R, truth-functional tautologies and formulas of the forms $\Box(\alpha \supset \beta) \supset (\Box\alpha \supset \Box\beta)$ and $\Box(\Box\alpha \supset \alpha) \supset \Box\alpha$ as the axioms and MP and RN as the rules of inference. Since R is r.e., so is R+. Let λ be the respectful mapping such that $\lambda\Box\alpha = \top$ for all formulas α. By an easy induction on the length of proofs, if $\vdash_{R+} \alpha$ then $R \vdash_{PC} \lambda\alpha$. But if $\alpha \in L$ then $\lambda\alpha = \alpha$, so $\vdash_{R+} \alpha$ only if $R \vdash_{PC} \alpha$, i.e., $\alpha \in R$; conversely, if $\alpha \in R$ then $\vdash_{R+} \alpha$; thus $L \cap R+ = R$. Let $Y \subseteq L$ be a maximal consistent extension of R. Define a set $X \subseteq L_\Box$ inductively: $p_i \in X \iff p_i \in Y$; $\bot \notin X$; $\alpha \supset \beta \in X \iff \alpha \notin X$ or $\beta \in X$; $\Box\alpha \in X \iff \vdash_{R+} \alpha$. For $\alpha \in L_\Box$, either $\alpha \in X$ or $\neg\alpha \in X$. We show by induction on the length of proofs that if $\vdash_{R+} \alpha$ then $\alpha \in X$. Basis: If $\alpha \in R$ then $\alpha \in Y \subseteq X$. If $\Box(\alpha \supset \beta) \in X$ and $\Box\alpha \in X$ then $\vdash_{R+} \alpha \supset \beta$ and $\vdash_{R+} \alpha$, so $\vdash_{R+} \beta$, so $\Box\beta \in X$; thus $\Box(\alpha \supset \beta) \supset (\Box\alpha \supset \Box\beta) \in X$. If $\Box(\Box\alpha \supset \alpha) \in X$ then $\vdash_{R+} \Box\alpha \supset \alpha$, so $\vdash_{R+} \Box(\Box\alpha \supset \alpha)$ because R+ is closed under RN; but $\vdash_{R+} \Box(\Box\alpha \supset \alpha) \supset \Box\alpha$, so $\vdash_{R+} \Box\alpha$, so $\vdash_{R+} \alpha$, so $\Box\alpha \in X$; thus $\Box(\Box\alpha \supset \alpha) \supset \Box\alpha \in X$. Induction step: Trivial. Now axiomatize GLS with all theorems of GL and formulas of the form $\Box\alpha \supset \alpha$ as the axioms and MP as the only rule of inference. We show by induction on the length of proofs that, for all formulas α, if $\vdash_{GLS} \alpha$ then $\alpha \in X$. Basis: If $\vdash_{GL} \alpha$ then $\vdash_{R+} \alpha$ because $GL \subseteq R+$, so $\alpha \in X$ by the previous induction. If $\Box\alpha \in X$ then $\vdash_{R+} \alpha$, so again $\alpha \in X$; thus $\Box\alpha \supset \alpha \in X$. Induction step: Trivial. Hence $GLS \subseteq X$, so X is maximal GLS-consistent. Now $L \cap \Box^{-1}X = L \cap R+ = R$ and $\Box^{-1}X = R+$ is r.e. Thus GLS is r.e. quasi-conservative.

We can extend Proposition 34.14 to another system of interest in relation to provability logic. Grz is the smallest normal modal logic containing all formulas of the form $\Box(\Box(\alpha \supset \Box\alpha) \supset \alpha) \supset \alpha$. Grz turns out to be in a precise sense the logic of what is both provable and true in PA (Boolos 1993, 155–161 has all the facts about Grz used here). Grz is intimately related to GLS in a way which allows us to extend the r.e. quasi-conservativeness of GLS to Grz:

Proposition 34.15 Grz is r.e. quasi-conservative.

Proof Let R be a consistent theory in L. By Proposition 34.14, for some maximal GLS-consistent X, $L \cap \Box^{-1}X = R$ and $\Box^{-1}X$ is r.e. Let τ be the respectful mapping from $L_\Box$ to $L_\Box$ such that $\tau\Box\alpha = \Box\tau\alpha \wedge \tau\alpha$ for all formulas α. Put $\tau^{-1}X = \{\alpha : \tau\alpha \in X\}$. Now $Grz \subseteq \tau^{-1}X$, for $\vdash_{Grz} \alpha$ if and only if $\vdash_{GLS} \tau\alpha$ (Boolos 1993, 156), so $\tau\alpha \in X$ since X is maximal GLS-consistent, so $\alpha \in \tau^{-1}X$. Since X is maximal GLS-consistent, $\tau^{-1}X$ is maximal Grz-consistent. Suppose $\alpha \in L$, so $\tau\alpha = \alpha$, so $\tau\Box\alpha = \Box\alpha \wedge \alpha$; if $\Box\alpha \in X$ then $\alpha \in X$ because $\vdash_{GLS} \Box\alpha \supset \alpha$, so $\tau\Box\alpha \in X$, so $\Box\alpha \in \tau^{-1}X$; conversely, if $\Box\alpha \in \tau^{-1}X$ then $\tau\Box\alpha \in X$, so $\Box\alpha \in X$. Thus $L \cap \Box^{-1}\ \tau^{-1}X = L \cap \Box^{-1}\ X = R$. Moreover, $\Box^{-1}\ \tau^{-1}X$ is r.e. because X is r.e. and τ is recursive. Thus Grz is r.e. quasi-conservative.

Grz is not plausible as the logic of other epistemic applications. It is not a sublogic of S5 and $\vdash_{Grz} \neg\Box(\Box p \wedge \Diamond\neg p)$, which in effect forbids agents to cognize that they do not cognize whether p is true. Yet you can know that what you know neither entails that the coin came down heads nor entails that it did not. However, since Grz extends the epistemically more plausible S4, the smallest normal modal logic including both the T and 4 schemas, its r.e. quasi-conservativeness entails that of S4. Truthfulness and positive introspection are together consistent with r.e. quasi-conservativeness.

Corollary 34.16 [Compare Shin and Williamson 1994 Proposition 4.] S4 is r.e. quasi-conservative.

Since S4 is r.e. quasi-conservative while S5, its extension by E, is not, and K4 is r.e. conservative while K4E is not, one might be tempted to blame E for the failure to satisfy the constraints, and to suppose that no normal logics with E is r.e. quasi-conservative. That would be a mistake; the next two propositions show that E is harmless when not combined with 4.

Proposition 34.17 KDE is r.e. quasi-conservative.

Proof Let R be a consistent r.e. theory in L. Let μ and θ be respectful mappings from $L_\Box$ to L such that for all formulas α, $\theta\Box\alpha = \top$ if $\vdash_{PC} \theta\alpha$ and $\theta\Box\alpha = \bot$ otherwise; $\mu\Box\alpha = \top$ if $R \vdash_{PC} \theta\alpha$ and $\mu\Box\alpha = \bot$ otherwise. Axiomatize KDE with all truth-functional tautologies and formulas of the forms $\Box(\alpha \supset \beta) \supset (\Box\alpha \supset \Box\beta)$, $\neg\Box\bot$ and $\neg\Box\alpha \supset \Box\neg\Box\alpha$ as the axioms and MP and RN as the rules of inference. We show by induction on the length of proofs that for all formulas α, $\vdash_{KDE} \alpha$ only if $\vdash_{PC} \theta\alpha$ and $\vdash_{PC} \mu\alpha$. Basis: If $R \vdash_{PC} \theta\alpha$, then $\mu\Box\alpha = \top$, so $\mu(\neg\Box\alpha \supset \Box\neg\Box\alpha) = \neg\top \supset \mu\Box\neg\Box\alpha$; if not, then not $\vdash_{PC} \theta\alpha$, so $\theta\Box\alpha = \bot$, so $\theta\Box\alpha = \neg\bot$, so $R \vdash_{PC} \theta\ \neg\Box\alpha$, so $\mu\Box\neg\Box\alpha = \top$, so $\mu(\neg\Box\alpha \supset \Box\neg\Box\alpha) = \neg\bot \supset \top$; either way, $\vdash_{PC} \mu(\neg\Box\alpha \supset \Box\neg\Box\alpha)$. The rest of the induction is by now routine. The rest of the proof is like that of Lemma 34.7, with θ and μ in place of δ and κ respectively.

Corollary 34.18 KE is r.e. conservative.

Proof KE is r.e. quasi-conservative by Proposition 34.17. Since not $\vdash_{KE} \Diamond\top$, KE is r.e. conservative by Proposition 34.1.

Although both positive and negative introspection are individually consistent with r.e. [quasi-] conservativeness, their conjunction is not. Part of the explanation is this: without positive introspection, an r.e. but non-recursive theory R can count as satisfying negative introspection by falsely equating the agent's theory with a recursive subtheory of R; the idea behind the clause for $\theta\Box\alpha$ in the proof of Proposition 34.17 is to use PC as such a subtheory. That R satisfies negative introspection by making false equations is crucial, for KE[T] is S5 itself. Although both negative introspection and truthfulness are individually consistent with r.e. [quasi-] conservativeness, their conjunction is not.

Related Non-computational Constraints

Although r.e. conservativeness and r.e. quasi-conservativeness are defined in computational terms, something remains when the computational element is eliminated. For given any [consistent] theory R in L, r.e. or not, we might require an epistemic logic to leave open the possibility that R is exactly the restriction of the agent's theory to L. On this view, an epistemic logic should impose no constraint beyond consistency on the agent's non-epistemic theorizing. Thus we define a modal logic Σ to be *conservative* if and only if for every theory R in L, $L \cap \Box^{-1}X = R$ for some maximal Σ-consistent set X. Σ is *quasi-conservative* if and only if for every consistent theory R in L, $L \cap \Box^{-1}X = R$ for some maximal Σ-consistent set X. Equivalently, Σ is [quasi-] conservative if and only if for every [consistent] theory R in L, $\{\Box\alpha : \alpha \in R\} \cup \{\neg\Box\alpha : \alpha \in L - R\}$ is Σ-consistent. We can assess how far r.e. conservativeness and r.e. quasi-conservativeness are specifically computational constraints by comparing them with conservativeness and quasi-conservativeness respectively.

Theorem 34.19 A prenormal modal logic Σ is quasi-conservative if and only if for no $n \vdash_\Sigma \mathrm{Alt}_n$.

Proof ($\Rightarrow$) Suppose that $\vdash_\Sigma \mathrm{Alt}_n$. Put $X = \{\Box\alpha : \alpha \in PC\} \cup \{\neg\Box\alpha : \alpha \in L - PC\}$. For all $i \leq n$, not $\vdash_{PC} \wedge\{p_j : j < i\} \supset p_i$, so $\neg\Box(\wedge\{p_j : j < i\} \supset p_i) \in X$. Hence $X \vdash_\Sigma \neg\mathrm{Alt}_n$, so $X \vdash_\Sigma$ is $\perp$. Since PC is a theory in L, Σ is not quasi-conservative. ($\Leftarrow$) Suppose that R is a consistent theory in L and $\{\Box\alpha : \alpha \in R\} \cup \{\neg\Box\alpha : \alpha \in L - R\}$ is not $\Sigma -$ consistent. Thus for some $\alpha_0, \ldots, \alpha_m \in R$ and $\beta_0, \ldots, \beta_n \in L - R$ (such β_i exist because R is consistent), $\vdash_\Sigma \wedge\{\Box\alpha_i : i \leq m\} \supset \vee\{\Box\beta_i : i \leq n\}$. Let $i \leq n$; since $\alpha_0, \ldots, \alpha_m \in R$, $\beta_i \in L - R$ and R is a theory, it follows that for some valuation υ_i of L onto $\{0, 1\}$ (where $\upsilon_i(\perp) = 0$ and $\upsilon_i(\gamma_1 \supset \gamma_2) = 1$ just in case $\upsilon_i(\gamma_1) \leq \upsilon_i(\gamma_2)$), $\upsilon_i(\alpha_j) = 1$ for all $j \leq m$ and $\upsilon_i(\beta_i) = 0$. Put $\upsilon_{n+1} = \upsilon_0$. Set $\delta_i = \wedge\{p_j : j < i\} \wedge \neg p_i$ for $i \leq n$ and $\delta_{n+1} = \wedge\{p_j : j \leq n\}$. Let σ be the substitution such that for all j, $\sigma\, p_j = \vee\{\delta_i : \upsilon_i(p_j) = 1, i \leq n+1\}$. Since Σ is closed under US, $\vdash_\Sigma \wedge\{\Box\sigma\alpha_i : i \leq m\} \supset \vee\{\Box\sigma\beta_i : i \leq n\}$. We can prove by induction on the complexity of γ that for all $\gamma \in L$ and $i \leq n+1$, if $\upsilon_i(\gamma) = 1$ then $\vdash_{PC} \delta_i \supset \sigma\gamma$ and if $\upsilon_i(\gamma) = 0$ then $\vdash_{PC} \delta_i \supset \neg\sigma\gamma$. Basis: Immediate by definition of σ, for $\vdash_{PC} \delta_i \supset \neg\delta_k$ whenever $i \neq k$. Induction step:

Routine. Now for $i \leq n+1$ and $j \leq m$, $\upsilon_i(\alpha_j) = 1$, so $\vdash_{PC} \delta_i \supset \sigma\alpha_j$; since $\vdash_{PC} \vee\{\delta_i : i \leq n+1\}$, $\vdash_{PC} \sigma\alpha_j$, so $\vdash_{PC} \top \supset \sigma\alpha_j$. Hence by prenormality $\vdash_\Sigma \Box\top \supset \Box\sigma\alpha_j$ and so $\vdash_\Sigma \Box\sigma\alpha_j$. Thus $\vdash_\Sigma \vee\{\Box\sigma\beta_i : i \leq n\}$. Moreover, for each $i \leq n$, $\upsilon_i(\beta_i) = 0$, so $\vdash_{PC} \delta_i \supset \neg\sigma\beta_i$, so $\vdash_{PC} \sigma\beta_i \supset (\wedge\{p_j : j < i\} \supset p_i)$, so $\vdash_\Sigma \Box\sigma\beta_i \supset \Box\ (\wedge\{p_j : j < i\} \supset p_i)$. Thus $\vdash_\Sigma \text{Alt}_n$.

Proposition 34.20 A prenormal modal logic Σ is conservative if and only if Σ is quasi-conservative and not $\vdash_\Sigma \Diamond\top$.

Proof Like Proposition 34.1 with 'r.e.' omitted.

Thus S5 is a quasi-conservative normal modal logic which is not r.e. quasi-conservative; K4E is a conservative normal modal logic which is not r.e. conservative. Most of the examples given above of logics which are not r.e. [quasi-] conservative are [quasi-] conservative. It is the distinctively computational requirements of r.e. quasi-conservativeness and r.e. conservativeness which those logics fail to meet.

Corollary 34.21 Every r.e. quasi-conservative prenormal modal logic is quasi-conservative; every r.e. conservative prenormal modal logic is conservative.

Proof From Proposition 34.1, Corollary 34.6, Theorem 34.19 and Proposition 34.20.

Although quasi-conservativeness exceeds r.e. quasi-conservativeness in requiring an epistemic logic to leave open the possibility that the restriction of the subject's theory to the language L is any given non-r.e. theory in L, this requirement is met by any epistemic logic which leaves open the corresponding possibility for every consistent r.e. theory in L.

Conclusion

Our investigation has uncovered part of a complex picture. The line between those modal logics weak enough to be r.e. conservative or r.e. quasi-conservative and those that are too strong appears not to coincide with any more familiar distinction between classes of modal logics, although a solution to the problem left open in the section "Some non-r.e. quasi-conservative logics" about the converse of Theorem 34.3 might bring clarification. What we have seen is that some decidable modal logics in general use as logics of knowledge (such as S5) or belief (such as KD45 and K45) when applied in generalized settings impose constraints on epistemic agents that require them to exceed every Turing machine in computational power. For many interpretations of epistemic logic, such a constraint is unacceptably strong.

The problem is not the same as the issue of logical omniscience, since many epistemic logics (such as S4 and various provability logics) do not impose the unacceptably strong constraints, although they do impose logical omniscience. Interpretations that finesse logical omniscience by building it into the definition of the propositional attitude that interprets the symbol □ do not thereby finesse the computational issue that we have been investigating. Nevertheless, the two questions are related, because the deductive closure of a recursively axiomatised theory is what makes its theorems computationally hard to survey. In particular, it can be computationally hard to check for *non*-theoremhood, which is what negative introspection and similar axioms require. In fact, negative introspection by itself turned out not to impose unacceptable computational requirements (Corollary 34.18), but its combination with independently more plausible axioms does so. Perhaps the issues raised in this paper will provide a more fruitful context in which to discuss some of the questions raised by the debate on logical omniscience and bounded rationality.

The results proved in the paper also suggest that more consideration should be given to the epistemic use of weaker modal logics that are r.e. conservative or quasi-conservative. The plausibility of correspondingly weaker axioms must be evaluated under suitable epistemic interpretations. Weaker epistemic logics present a more complex picture of the knowing subject, but also a more nuanced one, because they make distinctions that stronger logics erase. We have seen that the more nuanced picture is needed to express the limits in general cognition of creatures whose powers do not exceed those of every Turing machine.

Acknowledgements Material based on this paper was presented to colloquia of the British Society for the Philosophy of Science and the Computer Science Laboratory at Oxford. I thank participants in both for useful comments.

References

Blok, W. J. (1980). The lattice of modal logics: An algebraic investigation. *Journal of Symbolic Logic, 45*, 221–236.

Boolos, G. (1993). *The logic of provability*. Cambridge: Cambridge University Press.

Craig, W. (1953). On axiomatizability within a system. *Journal of Symbolic Logic, 18*, 30–32.

Fagin, R., Halpern, J. Y., Moses, Y., & Vardi, M. (1995). *Reasoning about knowledge*. Cambridge, MA: MIT Press.

Makinson, D. C. (1971). Some embedding theorems in modal logic. *Notre Dame Journal of Formal Logic, 12*, 252–254.

Shin, H. S., & Williamson, T. (1994). Representing the knowledge of Turing machines. *Theory and Decision, 37*, 125–146.

Skyrms, B. (1978). An immaculate conception of modality. *The Journal of Philosophy, 75*, 368–387.

Williamson, T. (1996). Self-knowledge and embedded operators. *Analysis, 56*, 202–209.

Williamson, T. (1998). Iterated attitudes. In T. Smiley (Ed.), *Philosophical logic* (pp. 85–133). Oxford: Oxford University Press for the British Academy.

Williamson, T. (2000). *Knowledge and its limits*. Oxford: Oxford University Press.

Part V
Interactive Epistemology

Chapter 35
Introduction

Horacio Arló-Costa, Vincent F. Hendricks, and Johan van Benthem

The last few decades have witnessed the growing importance of multi-agent perspectives in epistemic matters. While traditional epistemology has largely centered on what single agents know, barring the occasional encounter with a skeptic, there has been a growing focus on interaction in many disciplines, turning from single-reasoner to many-reasoners problems, the way physicists turned to many-body constellations as the essence of nature. This trend may be seen in social epistemology, speaker-hearer views of meaning, dialogical foundations of logic, or multi-agent systems instead of single computing devices in computer science. While an inference or an observation may be the basic informational act for a single agent, think of a question plus answer as the unit of social communication. This agent exchange involves knowledge about facts and about others, and the information that flows and thus changes the current epistemic state of both agents in systematic ways. Existing epistemic and doxastic logics can describe part of this setting, since they allow for iteration for different agents, expressing thinks like "agent 1 believes that

---

Horacio Arló-Costa was deceased at the time of publication.

H. Arló-Costa (deceased)
Carnegie Mellon University, Pittsburgh, PA, USA

V.F. Hendricks (✉)
Center for Information and Bubble Studies, University of Copenhagen, Copenhagen, Denmark
e-mail: vincent@hum.ku.dk

J. van Benthem
University of Amsterdam, Science Park 904, 1098 XH Amsterdam, The Netherlands

Stanford University, Stanford, United States
e-mail: johan@science.uva.nl

H. Arló-Costa et al. (eds.), *Readings in Formal Epistemology*, Springer Graduate Texts in Philosophy 1, DOI 10.1007/978-3-319-20451-2_35

agent 2 knows whether the heater is on". But the next level of social interaction involves the formation of groups with their own forms of knowledge, based perhaps on shared information.

One crucial notion concerning groups is 'common knowledge', which has come up in several disciplines independently. The article by Lewis gives an early philosophical motivation and development for common knowledge as a foundation for coordinated behavior, while the piece by Barwise points out logical subtleties in defining this notion that still have not all been disentangled decades later. Another crucial phenomenon, as we said, is the nature of the dynamic actions and events that drive the flow of information and interaction. Such actions have been brought inside the scope of logic by combining ideas from epistemic logic and logics of programs in the seminal paper by Baltag, Solecki and Moss. Baltag and Smets generalize this social dynamics to include belief revision and related phenomena. In particular, these papers show how a wide variety of actions of knowledge update and belief change can be described in 'dynamic-epistemic logics' allowing for groups with individual differences in observational power and habits of revision.

All these things and more come together in the compass of games, where information-driven agents pursue goals based on their evaluation of outcomes, through an interaction over time involving individual strategies responding to what others do. Indeed, the foundations of game theory have been a 'philosophical lab' since the mid 1970s, and the classic paper by Aumann on agreeing to disagree shows how group knowledge, investigative procedure, and eventually communication of disagreement can be subject to surprising logical laws. Perhaps the most well-studied notions in game theory are epistemic conditions such as common knowledge or common belief in rationality guaranteeing, under broad conditions, that players follow the backward induction solution, or other important solution concepts in games. The paper by Aumann & Brandenburger provides key examples of this style of analysis, as a major representative of what is by now a large body of literature. The philosophical community has started picking up on these developments, and related ones in the study of multi-agent systems: partly due to their intrinsic interest, and partly because games form a concrete microcosm where about all issues that have occupied philosophical logicians occur together. The article by Stalnaker shows how this encounter of logic and game theory can be of clear mutual benefit, and the perceptive paper by Halpern analyzes the ensuing debate across communities about the status of rationality.

Suggested Further Reading

Epistemic foundations of game theory as initiated by Aumann is a very rich area. Here is a selection of just a few papers that will set the reader thinking: J. Geanakoplos & H. Polemarchakis, 'We Can't Disagree Forever', Journal of Economic Theory 28, 1982, 192–200; P. Battigalli & G. Bonanno, 'Recent Results on Belief, Knowledge and the Epistemic Foundations of Game Theory', Research in Economics 53(2), 1999, 149–225; A. Brandenburger, 'The Power of Paradox: some Recent Developments in Interactive Epistemology', International Journal of Game Theory 35(4): 465–492. D. Samet & P. Jehiel, 'Learning to Play Games in Extensive Form by Valuation', Journal of Economic Theory 124, 2005, 129–148.

Epistemic considerations on agency in computer science are an equally rich area. We have given some textbook references in connection with logics of knowledge and belief, but here is an influential classical paper: J. Y. Halpern & Y. Moses, 'Knowledge and Common Knowledge in a Distributed Environment', Journal of the ACM 37:3, 1990, 549–587. An excellent monograph tying together logics of agency with game theory and much more is Y. Shoham & K. Leyton-Brown, Multiagent Systems, Cambridge University Press, 2009. Zooming in on 'dynamic-epistemic' foundations of information-driven agency, a comprehensive treatment is in J. van Benthem, Logical Dynamics of Information and Interaction, Cambridge University Press, 2011. Additional specifics of the interface of logic and games are found in W. van der Hoek & M. Pauly, 'Modal Logic and Games', in P. Blackburn, J. van Benthem & F. Wolter, eds. Handbook of Modal Logic, Elsevier, 2006, 1077–1148. A comprehensive philosophical analysis is found in B. de Bruin, Explaining Games: The Epistemic Programme in Game Theory, Springer, 2010. Finally, there is the growing stream of social epistemology, an area with many philosophical motivations beyond game theory, with books such as A. Goldman, Knowledge in a Social World, Oxford University Press, 1999, and Ch. List & Ph. Petit, Group Agency, Oxford University Press, 2011. Another important stream has been the influence of evolutionary game theory as a paradigm for philosophical studies, with B. Skyrms' The Stag Hunt and the Evolution of Social Structure, Cambridge University Press, 2004, as an outstanding example of the insights emerging in this way.

Chapter 36
Convention (An Excerpt on Coordination and Higher-Order Expectations)

David Lewis

Sample Coordination Problems

Use of language belongs to a class of situations with a conspicuous common character: situations I shall call *coordination problems.* I postpone a definition until we have seen a few examples. We begin with situations that might arise between two people—call them "you" and "I."

(1) Suppose you and I both want to meet each other. We will meet if and only if we go to the same place. It matters little to either of us where (within limits) he goes if he meets the other there; and it matters little to either of us where he goes if he fails to meet the other there. We must each choose where to go. The best place for me to go is the place where you will go, so I try to figure out where you will go and to go there myself. You do the same. Each chooses according to his expectation of the other's choice. If either succeeds, so does the other; the outcome is one we both desired.

(4) Suppose several of us are driving on the same winding two-lane roads. It matters little to anyone whether he drives in the left or the right lane, provided the others do likewise. But if some drive in the left lane and some in the right, everyone is in danger of collision. So each must choose whether to drive in the left lane or in the right, according to his expectations about the others: to drive in the left lane if most or all of the others do, to drive in the right lane if most or all of the others do (and to drive where he pleases if the others are more or less equally divided).

This chapter is an excerpt of David Lewis' PhD thesis *Convention*. It contains the primary content of Chapter 1: *Coordination and Convention*, where Lewis presents the idea of conventions as a mutually expected regularity in coordination in a recurrent situation.

D. Lewis (deceased)
Princeton University, Princeton, NJ, USA

H. Arló-Costa et al. (eds.), *Readings in Formal Epistemology*, Springer Graduate Texts in Philosophy 1, DOI 10.1007/978-3-319-20451-2_36

(7) Suppose we are contented oligopolists. As the price of our raw material varies, we must each set new prices. It is to no one's advantage to set his prices higher than the others set theirs, since if he does he tends to lose his share of the market. Nor is it to anyone's advantage to set his prices lower than the others set theirs, since if he does he menaces his competitors and incurs their retaliation. So each must set his prices within the range of prices he expects the others to set.

Analysis of Coordination Problems

With these examples, let us see how to describe the common character of coordination problems.

Two or more agents must each choose one of several alternative actions. Often all the agents have the same set of alternative actions, but that is not necessary. The outcomes the agents want to produce or prevent are determined jointly by the actions of all the agents. So the outcome of any action an agent might choose depends on the actions of the other agents. That is why—as we have seen in every example—each must choose what to do according to his expectations about what the others will do.

To exclude trivial cases, a coordination problem must have more than one coordination equilibrium. But that requirement is not quite strong enough. Figure 36.1 shows two matrices in which, sure enough, there are multiple coordination equilibria (two on the left, four on the right). Yet there is still no need for either agent to base his choice on his expectation about the other's choice. There is no need for them to try for the same equilibrium—no need for coordination—since if they try for different equilibria, some equilibrium will nevertheless be reached. These cases exhibit another kind of triviality, akin to the triviality of a case with a unique coordination equilibrium.

A combination is an equilibrium if each agent likes it *at least as well as* any other combination he could have reached, given the others' choices. Let us call it a *proper* equilibrium if each agent likes it *better than* any other combination he could have

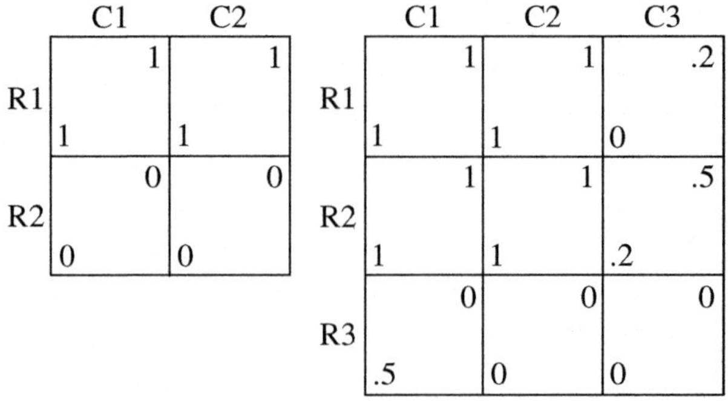

Fig. 36.1

	C1	C2	C3
R1	2 2	0 0	0 0
R2	0 0	2 2	0 0
R3	0 0	1 1	1 1

Fig. 36.2

reached, given the others' choices. In a two-person matrix, for instance, a proper equilibrium is preferred by Row-chooser to all other combinations in its column, and by Column-chooser to all other combinations in its row. In the matrices in Fig. 36.1, there are multiple coordination equilibria, but all of them are improper.

There is no need to stipulate that all equilibria in a coordination problem must be proper; it seems that the matrix in Fig. 36.2 ought to be counted as essentially similar to our clear examples of coordination problems, despite the impropriety of its equilibrium ⟨*R3*, *C3*⟩. The two proper coordination equilibria—⟨*R1*, *C1*⟩ and ⟨*R2*, *C2*⟩—are sufficient to keep the problem nontrivial. I stipulate instead that a coordination problem must contain at least two proper coordination equilibria.

This is only one—the strongest—of several defensible restrictions. We might prefer a weaker restriction that would not rule out matrices like those in Fig. 36.3. But a satisfactory restriction would be complicated and would entail too many qualifications later. And situations like those of Fig. 36.3 can be rescued even under the strong restriction we have adopted. Let *R2′* be the disjunction of *R2* and *R3*, and *C2′* the disjunction of *C2* and *C3* in the left-hand matrix. Then the same situation can be represented by the new matrix in Fig. 36.4, which does have two proper coordination equilibria. The right-hand matrix can be consolidated in a similar way. But matrices like the one in Fig. 36.5, which are ruled out by the strong restriction, and ought to be ruled out, cannot be rescued by any such consolidation.

To sum up: Coordination problems—situations that resemble my 11 examples in the important respects in which they resemble one another[1]—are situations of interdependent decision by two or more agents in which coincidence of interest predominates and in which there are two or more proper coordination equilibria. We could also say—though less informatively than one might think—that they are situations in which, relative to *some* classification of actions, the agents have a common interest in all doing the same one of several alternative actions.

---

[1]See Michael Slote, "The Theory of Important Criteria," *Journal of Philosophy*, 63 (1966), pp. 211–224. Slote shows that we commonly introduce a class by means of examples and take the defining features of the class to be those distinctive features of our examples which seem important for an understanding of their character. That is what I take myself to be doing here and elsewhere.

	C1	C2	C3
R1	1, 1	0, 0	0, 0
R2	0, 0	1, 1	1, 1
R3	0, 0	1, 1	1, 1

	C1	C2	C3	C4
R1	1, 1	1, 1	0, 0	0, 0
R2	1, 1	1, 1	0, 0	0, 0
R3	0, 0	0, 0	1, 1	1, 1
R4	0, 0	0, 0	1, 1	1, 1

Fig. 36.3

	C1	C2′
R1	1, 1	0, 0
R2′	0, 0	1, 1

Fig. 36.4

	C1	C2	C3
R1	1, 1	1, 1	0, 0
R2	0, 0	1, 1	1, 1
R3	0, 0	0, 0	1, 1

Fig. 36.5

Solving Coordination Problems

Agents confronted by a coordination problem may or may not succeed in each acting so that they reach one of the possible coordination equilibria. They might succeed just by luck, although some of them choose without regard to the others' expected actions (doing so perhaps because they cannot guess what the others will do, perhaps because the chance of coordination seems so small as to be negligible).

But they are more likely to succeed—if they do—through the agency of a system of suitably concordant mutual expectations. Thus in example (1) I may go to a certain place because I expect you to go there, while you go there because you expect me to; in example (2) I may call back because I expect you not to, while you do not because you expect me to; in example (4) each of us may drive on the right because he expects the rest to do so; and so on. In general, each may do his part of one of the possible coordination equilibria because he expects the others to do theirs, thereby reaching that equilibrium.

If an agent were completely confident in his expectation that the others would do their parts of a certain proper coordination equilibrium, he would have a decisive reason to do his own part. But if—as in any real case—his confidence is less than complete, he must balance his preference for doing his part if the others do theirs against his preferences for acting otherwise if they do not. He has a decisive reason to do his own part if he is *sufficiently* confident in his expectation that the others will do theirs. The degree of confidence which is sufficient depends on all his payoffs and sometimes on the comparative probabilities he assigns to the different *ways* the others might not all do their parts, in case not all of them do. For instance, in the coordination problem shown in Fig. 36.6, Row-chooser should do his part of the coordination equilibrium ⟨*R1*, *C1*⟩ by choosing *R1* if he has more than .5 confidence that Column-chooser will do his part by choosing *C1*. But in the coordination problems shown in Fig. 36.7, Row-chooser should choose R1 only if he has more than .9 confidence that Column-chooser will choose *C1*. If he has, say, .8 confidence that Column-chooser will choose *C1*, he would do better to choose *R2*, sacrificing his chance to achieve coordination at ⟨*R1*, *C1*⟩ in order to hedge against the possibility that his expectation was wrong. And in the coordination problem shown in Fig. 36.8, Row-chooser might be sure that if Column-chooser fails to do

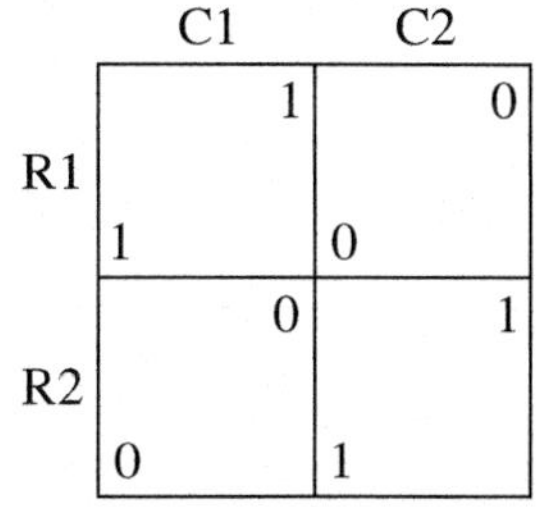

Fig. 36.6

	C1	C2
R1	1, 1	−8, −8
R2	0, 0	1, 1

	C1	C2
R1	1, 1	0, 0
R2	0, 0	9, 9

	C1	C2
R1	3, 3	−26, −26
R2	0, 0	1, 1

Fig. 36.7

	C1	C2	C3
R1	1, 1	0, 0	−8, −8
R2	0, 0	1, 1	9, 9

Fig. 36.8

his part of ⟨*R1*, *C1*⟩, at least he will choose *C2*, not *C3*; if so, Row-chooser should choose *R1* if he has more than .5 confidence that Column-chooser will choose *C1*. Or Row-chooser might think that if Column-chooser fails to choose *R1*, he is just as likely to choose *C3* as to choose *C2*; if so, Row-chooser should choose *R1* only if he has more than .9 confidence that Column-chooser will choose *C1*. Or Row-chooser might be sure that if Column-chooser does not choose *C1*, he will choose *C3* instead; if so, Row-chooser's minimum sufficient degree of confidence is about .95. The strength of concordant expectation needed to produce coordination at a certain equilibrium is a measure of the difficulty of achieving coordination there, since however the concordant expectations are produced, weaker expectations will be produced more easily than stronger ones. (We can imagine cases in which so much mutual confidence is required to achieve coordination at an equilibrium that success is impossible. Imagine that a millionaire offers to distribute his fortune equally among a thousand men if each sends him $10; if even one does not, the millionaire will keep whatever he is sent. I take it that no matter what the thousand do to increase their mutual confidence, it is a practical certainty that the millionaire will not have to pay up. So if I am one of the thousand, I will keep my $10.)

We may achieve coordination by acting on our concordant expectations about each other's actions. And we may acquire those expectations, or correct or corroborate whatever expectations we already have, by putting ourselves in the other fellow's shoes, to the best of our ability. If I know what you believe about the matters of fact that determine the likely effects of your alternative actions, and if I know your preferences among possible outcomes and I know that you possess a modicum of practical rationality, then I can replicate your practical reasoning to figure out what you will probably do, so that I can act appropriately.

In the case of a coordination problem, or any other problem of interdependent decision, one of the matters of fact that goes into determining the likely effects of your alternative actions is my own action. In order to figure out what you will do by replicating your practical reasoning, I need to figure out what *you* expect *me* to do.

I know that, just as I am trying to figure out what you will do by replicating your reasoning, so you may be trying to figure out what I will do by replicating my reasoning. This, like anything else you might do to figure out what I will do, is itself part of your reasoning. So to replicate your reasoning, I may have to replicate your attempt to replicate my reasoning.

This is not the end. I may reasonably expect *you* to realize that, unless I already know what you expect me to do, I may have to try to replicate your attempt to replicate my reasoning. So I may expect you to try to replicate my attempt to replicate your attempt to replicate my reasoning. So my own reasoning may have to include an attempt to replicate your attempt to replicate my attempt to replicate your attempt to replicate my reasoning. And so on.

Before things get out of hand, it will prove useful to introduce the concept of *higher-order expectations*, defined by recursion thus:

> A first-order expectation about something is an ordinary expectation about it.
> An $(n+1)$th-order expectation about something $(n \geq 1)$ is an ordinary expectation about someone else's nth-order expectation about it.

For instance, if I expect you to expect that it will thunder, then I have a second-order expectation that it will thunder.

Whenever I replicate a piece of your practical reasoning, my second-order expectations about matters of fact, together with my first-order expectations about your preferences and your rationality, justify me in forming a first-order expectation about your action. In the case of problems of interdependent decision—for instance, coordination problems—some of the requisite second order expectations must be about my own action.

Consider our first sample coordination problem: a situation in which you and I want to meet by going to the same place. Suppose that after deliberation I decide to come to a certain place. The fundamental practical reasoning which leads me to that choice is shown in Fig. 36.9. (In all diagrams of this kind, heavy arrows represent implications; light arrows represent causal connections between the mental states or actions of a rational agent.) And if my premise for this reasoning—my expectation that you will go there—was obtained by replicating your reasoning, my replication is shown in Fig. 36.10. And if my premise for this replication—my expectation that you will expect me to go there—was obtained by replicating your replication of my reasoning, my replication of your replication is shown in Fig. 36.11. And so on. The whole of my reasoning (simplified by disregarding the rationality premises) may be represented as in Fig. 36.12 for whatever finite number of stages it may take for

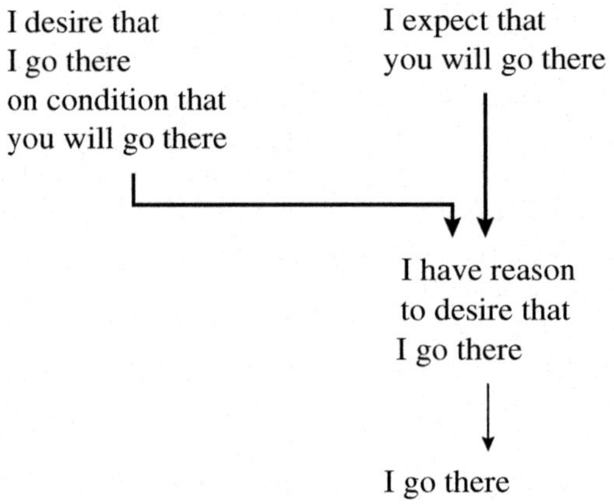

Fig. 36.9

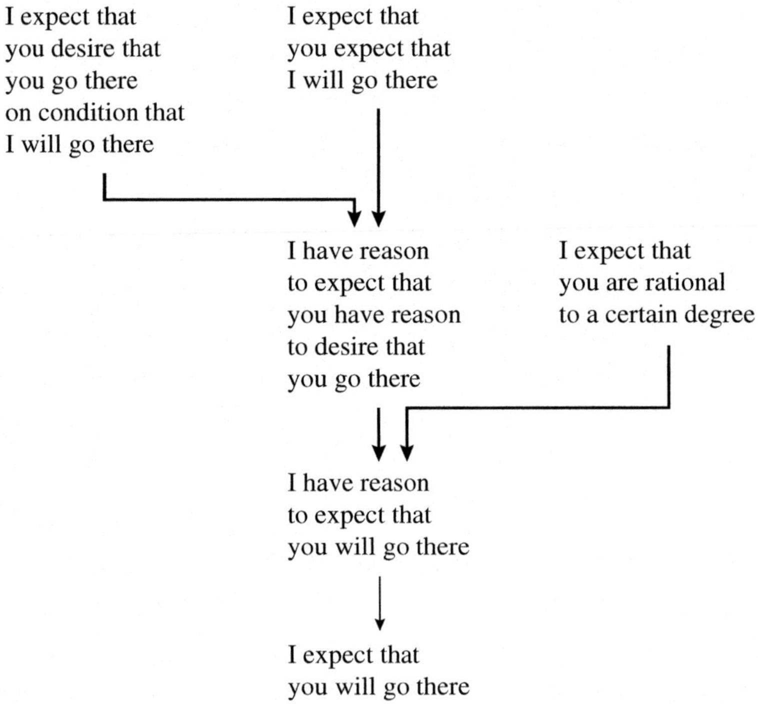

Fig. 36.10

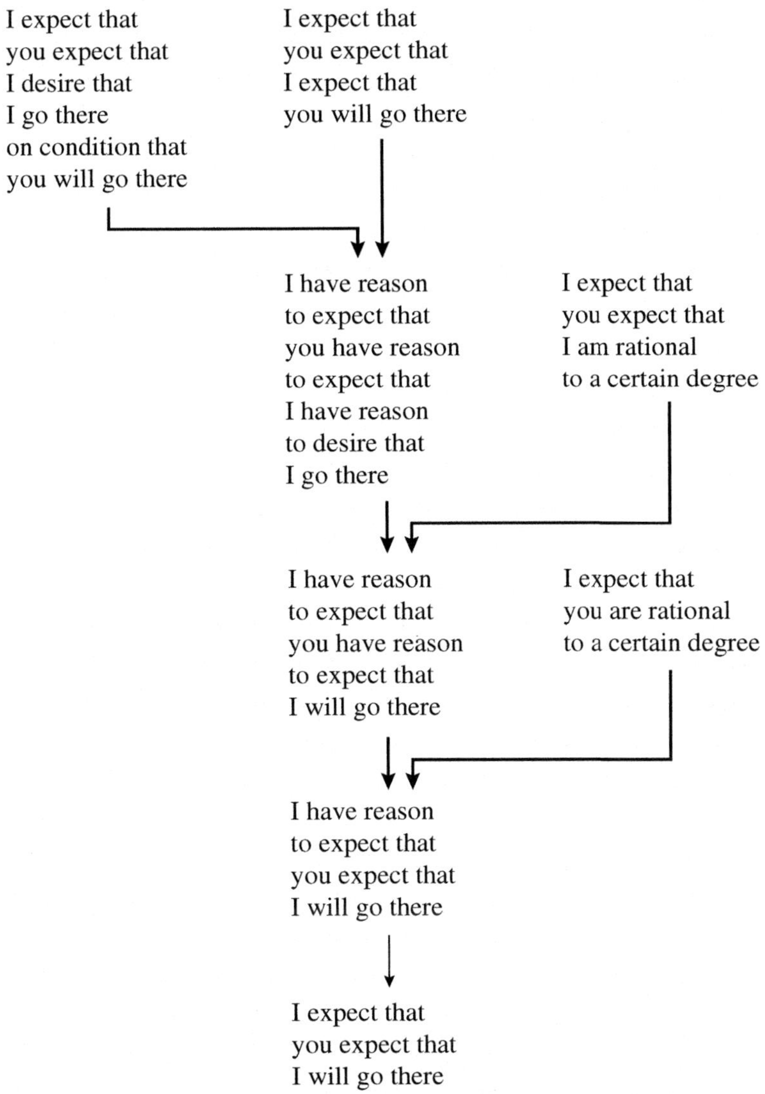

Fig. 36.11

me to use whatever higher-order expectations may be available to me regarding our actions and our conditional preferences. Replications are nested to some finite depth: my reasoning (outer boundary) contains a replication of yours (next boundary), which contains a replication of your replication of mine (next boundary), and so on.

So if I somehow happen to have an *n*th-order expectation about action in this two-person coordination problem, I may work outward through the nested replications to

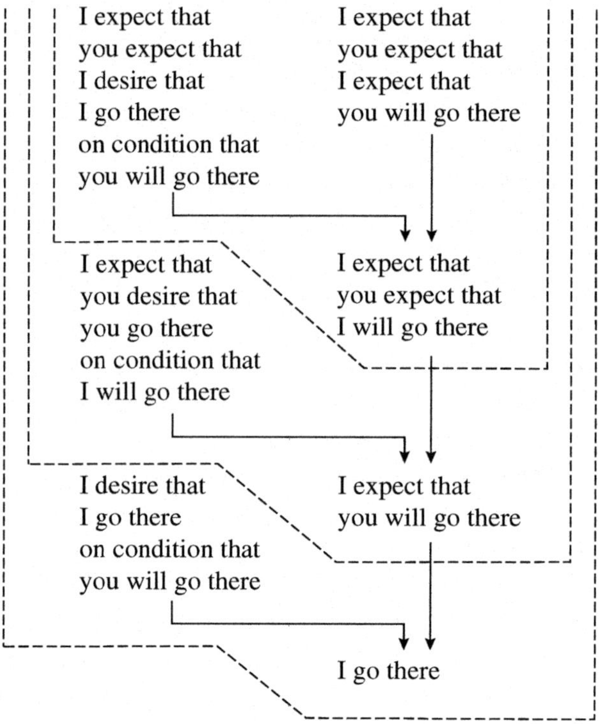

Fig. 36.12

lower- and lower-order expectations about action. Provided I go on long enough, and provided all the needed higher-order expectations about preferences and rationality are available, I eventually come out with a first-order expectation about your action—which is what I need in order to know how I should act.

Clearly a similar process of replication is possible in coordination problems among more than two agents. In general, my higher-order expectations about something are my expectations about x_1's expectations about x_2's expectations ... about it. (The sequence x_1, x_2 ... may repeat, but x_1 cannot be myself and no one can occur twice in immediate succession.) So when m agents are involved, I can have as many as $(m-1)^n$ different nth-order expectations about anything, corresponding to the $(m-1)^n$ different admissible sequences of length n. Replication in general is ramified: it is built from stages in which $m-1$ of my various $(n+1)$th-order expectations about action, plus ancillary premises, yield one of my nth-order expectations about action. I suppressed the ramification by setting $m=2$, but the general case is the same in principle.

Note that replication is *not* an interaction back and forth between people. It is a process in which *one* person works out the consequences of his beliefs about the world—a world he believes to include other people who are working out the consequences of their beliefs, including their belief in other people who ... By our interaction in the world we acquire various high-order expectations that can serve us as premises. In our subsequent reasoning we are windowless monads doing our best to mirror each other, mirror each other mirroring each other, and so on.

Of course I do not imagine that anyone will solve a coordination problem by first acquiring a seventeenth-order expectation from somewhere and then sitting down to do his replications. For one thing, we rarely do have expectations of higher order than, say, fourth. For another thing, any ordinary situation that could justify a high-order expectation would also justify low-order expectations directly, without recourse to nested replications.

All the same, given the needed ancillary premises, an expectation of arbitrarily high order about action does give an agent *one* good reason for a choice of action. The one may, and normally will, be one reason among the many which jointly suffice to justify his choice. Suppose the agent is originally justified somehow in having expectations of several orders about his own and his partners' actions. And suppose the ancillary premises are available. Then each of his original expectations independently gives him a reason to act one way or another. If he is lucky, all these independent reasons will be reasons for the same action.[2] Then that action is strongly, because redundantly, justified; he has more reason to do it than could have been provided by any one of his original expectations by itself.

I said earlier that coordination might be rationally achieved with the aid of concordant mutual expectations about action. We have seen that these may be derived from first- and higher-order expectations about action, preferences, and rationality. So we generalize: coordination may be rationally achieved with the aid of a system of concordant mutual expectations, of first or higher orders, about the agents' actions, preferences, and rationality.

The more orders of expectation about action contribute to an agent's decision, the more independent justifications the agent will have; and insofar as he is aware of those justifications, the more firmly his choice will be determined. Circumstances that will help to solve a coordination problem, therefore, are circumstances in which the agents become justified in forming mutual expectations belonging to a concordant system. And the more orders, the better.

---

[2]Michael Scriven, in "An Essential Unpredictability in Human Behavior," *Scientific Psychology: Principles and Approaches,* ed. B. B. Wolman (New York: Basic Books, 1965), has discussed mutual replication of practical reasoning between agents in a game of conflict who want *not* to conform to each other's expectations. There is a cyclic alternation: from my $(n+4)$th-order expectation that I will go to Minsk to my $(n+3)$th-order expectation that you will go to Pinsk to my $(n+2)$th-order expectation that I will go to Pinsk to my $(n+1)$th-order expectation that you will go to Minsk to my nth-order expectation that I will go to Minsk ... Scriven notices that we cannot both act on complete and accurate replications of each other's reasoning. He takes this to prove human unpredictability. But perhaps it simply proves that the agents cannot both have enough time to finish their replications, since the time either needs increases with the time the other uses. See David Lewis and Jane Richardson, "Scriven on Human Unpredictability," *Philosophical Studies*, 17 (1966), pp. 69–74.

In considering how to solve coordination problems, I have postponed the answer that first comes to mind: by agreement. If the agents can communicate (without excessive cost), they can ensure a common understanding of their problem by discussing it. They can choose a coordination equilibrium—an arbitrary one, or one especially good for some or all of them, or one they can reach without too much mutual confidence. And each can assure the rest that he will do his part of the chosen equilibrium. Coordination by means of an agreement is not, of course, an alternative to coordination by means of concordant mutual expectations. Rather, agreement is one means of producing those expectations. It is an especially effective means, since it produces strong concordant expectations of several orders.

Suppose you and I want to meet tomorrow; today we happen to meet, and we make an appointment. Each thereby gives evidence of his interest in going where the other goes and of his intention to go to a certain place. By observing this evidence, we form concordant first-order expectations about each other's preferences and action. By observing each other observing it, we may also form concordant second-order expectations. By observing each other observing each other observing it, we may even form concordant third-order expectations. And so on; not forever, of course, but limited by the amount of reasoning we do and the amount we ascribe to each other—perhaps one or two steps more. The result is a system of concordant mutual expectations of several orders, conducive to coordination by means of replication.

The agents' agreement might be an exchange of formal or tacit promises. But it need not be. Even a man whose word is his bond can remove the promissory force by explicit disavowal, if not otherwise. An exchange of declarations of present intention will be good enough, even if each explicitly retains his right to change his plans later. No one need bind himself to act against his own interest. Rather, it will be in the interest of each to do just what he has led the others to expect him to do, since that action will be best for him if the others act on their expectations.

If one does consider himself bound by a promise, he has a second, independent incentive. His payoffs are modified, since he has attached the onus of promise breaking to all but one choice. Indeed, he may modify his payoffs so much by promising that the situation is no longer a coordination problem at all. For instance, the agent's promised action might become his dominant choice: he might wish to keep his promise no matter what, coordination or no coordination.

If such a strong promise is made publicly, the others will know that they must go along with the one who has promised, for they know what he will do. Such forceful promising is a way of getting rid of coordination problems, not a way of solving them.

Explicit agreement is an especially good and common means to coordination—so much so that we are tempted to speak of coordination otherwise produced as *tacit* agreement. But agreement (literally understood) is not the only source of concordant expectations to help us solve our coordination problems. We do without agreement by choice if we find ourselves already satisfied with the content and strength of our mutual expectations. We do without it by necessity if we have no way to communicate, or if we can communicate only at a cost that outweighs our improved chance of coordination (say, if we are conspirators being shadowed).

Schelling has experimented with coordination problems in which the agents cannot communicate. His subjects know only that they share a common understanding of their problem—for instance, they may get instructions describing their problem and stating that everyone gets the same instructions. It turns out that sophisticated subjects in an experimental setting can often do very well—much better than chance—at solving novel coordination problems without communicating. They try for a coordination equilibrium that is somehow *salient:* one that stands out from the rest by its uniqueness in some conspicuous respect. It does not have to be uniquely *good*; indeed, it could be uniquely bad. It merely has to be unique in some way the subjects will notice, expect each other to notice, and so on. If different coordination equilibria are unique in different conspicuous ways, the subjects will need to be alike in the relative importance they attach to different respects of comparison; but often they are enough alike to solve the problem.

How can we explain coordination by salience? The subjects might all tend to pick the salient as a last resort, when they have no stronger ground for choice. Or they might expect each other to have that tendency, and act accordingly; or they might expect each other to expect each other to have that tendency and act accordingly, and act accordingly; and so on. Or—more likely—there might be a mixture of these. Their first- and higher-order expectations of a tendency to pick the salient as a last resort would be a system of concordant expectations capable of producing coordination at the salient equilibrium.

If their expectations did produce coordination, it would not matter whether anyone really would have picked the salient as a last resort. For each would have had a good reason for his choice, so his choice would not have been a last resort.

Thus even in a novel coordination problem—which is an extreme case—the agents can sometimes obtain the concordant expectations they need without communicating. An easier, and more common, case is that of a *familiar* coordination problem without communication. Here the agents' source of mutual expectations is precedent: acquaintance with past solved instances of their present coordination problem.

Convention

Let us start with the simplest case of coordination by precedent and generalize in various ways. In this way we shall meet the phenomenon I call *convention,* the subject of this book.

Suppose we have been given a coordination problem, and we have reached some fairly good coordination equilibrium. Given exactly the same problem again, perhaps each of us will repeat what he did before. If so, we will reach the same solution. If you and I met yesterday—by luck, by agreement, by salience, or however—and today we find we must meet again, we might both go back to yesterday's meeting place, each hoping to find the other there. If we were cut off on the telephone and you happened to call back as I waited, then if we are cut off again in the same call, I will wait again.

We can explain the force of precedent just as we explained the force of salience. Indeed, precedent is merely the source of one important kind of salience: conspicuous uniqueness of an equilibrium because we reached it last time. We may tend to repeat the action that succeeded before if we have no strong reason to do otherwise. Whether or not any of us really has this tendency, we may somewhat expect each other to have it, or expect each other to expect each other to have it, and so on—that is, we may each have first- and higher-order expectations that the others will do their parts of the old coordination equilibrium, unless they have reason to act otherwise. Each one's expectation that the others will do their parts, strengthened perhaps by replication using his higher-order expectations, gives him some reason to do his own part. And if his original expectations of some order or other were strong enough, he will have a decisive reason to do his part. So he will do it.

I have been supposing that we are given a coordination problem, and then given the same problem again. But, of course, we could never be given exactly the same problem twice. There must be this difference at least: the second time, we can draw on our experience with the first. More generally, the two problems will differ in several independent respects. We cannot do exactly what we did before. Nothing we could do this time is exactly like what we did before—like it in every respect—because the situations are not exactly alike.

So suppose not that we are given the original problem again, but rather that we are given a new coordination problem analogous somehow to the original one. Guided by whatever analogy we notice, we tend to follow precedent by trying for a coordination equilibrium in the new problem which uniquely corresponds to the one we reached before.

There might be alternative analogies. If so, there is room for ambiguity about what would be following precedent and doing what we did before. Suppose that yesterday I called you on the telephone and I called back when we were cut off. Today you call me and we are cut off. We have a precedent in which I called back and a precedent—the same one—in which the original caller called back. But this time you are the original caller. No matter what I do this time, I do something analogous to what we did before. Our ambiguous precedent does not help us.

In fact, there are always innumerable alternative analogies. Were it not that we happen uniformly to notice some analogies and ignore others—those we call "natural" or "artificial," respectively—precedents would always be completely ambiguous and worthless. *Every* coordination equilibrium in our new problem (every other combination, too) corresponds uniquely to what we did before under *some* analogy, shares *some* distinctive description with it alone. Fortunately, most of the analogies are artificial. We ignore them; we do not tend to let them guide our choice, nor do we expect each other to have any such tendency, nor do we expect each other to expect each other to, and so on. And fortunately we have learned that all of us will mostly notice the same analogies. That is why precedents can be unambiguous in practice, and often are. If we notice only one of the analogies between our problem and the precedent, or if one of those we notice seems far more conspicuous than the others, or even if several are conspicuous but they all happen to agree in indicating the same choice, then the other analogies do not matter. We are not in trouble unless conflicting analogies force themselves on our attention.

The more respects of similarity between the new problem and the precedent, the more likely it is that different analogies will turn out to agree, the less room there will be for ambiguity, and the easier it will be to follow precedent. A precedent in which I, the original caller, called back is ambiguous given a new problem in which you are the original caller—but not given a new problem in which I am again the original caller. That is why I began by pretending that the new problem was like the precedent in all respects.

Salience in general is uniqueness of a coordination equilibrium in a preeminently conspicuous respect. The salience due to precedent is no exception: it is uniqueness of a coordination equilibrium in virtue of its preeminently conspicuous analogy to what was done successfully before.

So far I have been supposing that the agents who set the precedent are the ones who follow it. This made sure that the agents given the second problem were acquainted with the circumstances and outcome of the first, and expected each other to be, expected each other to expect each other to be, and so on. But it is not an infallible way and not the only way. For instance, if yesterday I told you a story about people who got separated in the subway and happened to meet again at Charles Street, and today we get separated in the same way, we might independently decide to go and wait at Charles Street. It makes no difference whether the story I told you was true, or whether you thought it was, or whether I thought it was, or even whether I claimed it was. A fictive precedent would be as effective as an actual one in suggesting a course of action for us, and therefore as good a source of concordant mutual expectations enabling us to meet. So let us just stipulate that somehow the agents in the new problem are acquainted with the precedent, expect each other to be acquainted with it, and so on.

So far I have been supposing that we have a single precedent to follow. But we might have several. We might all be acquainted with a class of previous coordination problems, naturally analogous to our present problem and to each other, in which analogous coordination equilibria were reached. This is to say that the agents' actions conformed to some noticeable regularity. Since our present problem is suitably analogous to the precedents, we can reach a coordination equilibrium by all conforming to this same regularity. Each of us wants to conform to it if the others do; he has a *conditional preference* for conformity. If we do conform, the explanation has the familiar pattern: we tend to follow precedent, given no particular reason to do anything else; we expect that tendency in each other; we expect each other to expect it; and so on. We have our concordant first- and higher-order expectations, and they enable us to reach a coordination equilibrium.

It does not matter *why* coordination was achieved at analogous equilibria in the previous cases. Even if it had happened by luck, we could still follow the precedent set. One likely course of events would be this; the first case, or the first few, acted as precedent for the next, those for the next, and so on. Similarly, no matter how our precedents came about, by following them this time we add this case to the stock of precedents available henceforth.

Several precedents are better than one, not only because we learn by repetition but also because differences between the precedents help to resolve ambiguity. Even if our present situation bears conflicting natural analogies to any one precedent, maybe only one of these analogies will hold between the precedents; so we will pay attention only to that one. Suppose we know of many cases in which a cut-off telephone call was restored, and in every case it was the original caller who called back. In some cases I was the original caller, in some you were, in some neither of us was. Now we are cut off and I was the original caller. For you to call back would be to do something analogous—under one analogy—to what succeeded in some of the previous cases. But we can ignore that analogy, for under it the precedents disagree.

Once there are many precedents available, without substantial disagreement or ambiguity, it is no longer necessary for all of us to be acquainted with precisely the same ones. It is enough if each of us is acquainted with some agreeing precedents, each expects everyone else to be acquainted with some that agree with his, each expects everyone else to expect everyone else to be acquainted with some precedents that agree with his, etc. It is easy to see how that might happen: if one has often encountered cases in which coordination was achieved in a certain problem by conforming to a certain regularity, and rarely or never encountered cases in which it was not, he is entitled to expect his neighbors to have had much the same experience. If I have driven all around the United States and seen many people driving on the right and never one on the left, I may reasonably infer that almost everyone in the United States drives on the right, and hence that this man driving toward me also has mostly seen people driving on the right—even if he and I have not seen any of the *same* people driving on the right.

Our acquaintance with a precedent need not be very detailed. It is enough to know that one has learned of many cases in which coordination was achieved in a certain problem by conforming to a certain regularity. There is no need to be able to specify the time and place, the agents involved, or any other particulars; no need to be able to recall the cases one by one. I cannot cite precedents one by one in which people drove on the right in the United States; I am not sure I can cite even one case; nonetheless, I know very well that I have often seen cars driven in the United States, and almost always they were on the right. And since I have no reason to think I encountered an abnormal sample, I infer that drivers in the United States do almost always drive on the right; so anyone I meet driving in the United States will believe this just as I do, will expect me to believe it, and so on.

Coordination by precedent, at its simplest, is this: achievement of coordination by means of shared acquaintance with the achievement of coordination in a single past case exactly like our present coordination problem. By removing inessential restrictions, we have come to this: achievement of coordination by means of shared acquaintance with a *regularity* governing the achievement of coordination in a class of past cases which bear some conspicuous analogy to one another and to our present coordination problem. Our acquaintance with this regularity comes from our experience with some of its instances, not necessarily the same ones for everybody.

Given a regularity in past cases, we may reasonably extrapolate it into the (near) future. For we are entitled to expect that when agents acquainted with the past regularity are confronted by an analogous new coordination problem, they will succeed in achieving coordination by following precedent and continuing to conform to the same regularity. We come to expect conforming actions not only in past cases but in future ones as well. We acquire a general belief, unrestricted as to time, that members of a certain population conform to a certain regularity in a certain kind of recurring coordination problem for the sake of coordination.

Each new action in conformity to the regularity adds to our experience of general conformity. Our experience of general conformity in the past leads us, by force of precedent, to expect a like conformity in the future. And our expectation of future conformity is a reason to go on conforming, since to conform if others do is to achieve a coordination equilibrium and to satisfy one's own preferences. And so it goes—we're here because we're here because we're here because we're here. Once the process gets started, we have a metastable self-perpetuating system of preferences, expectations, and actions capable of persisting indefinitely. As long as uniform conformity is a coordination equilibrium, so that each wants to conform conditionally upon conformity by the others, conforming action produces expectation of conforming action and expectation of conforming action produces conforming action.

This is the phenomenon I call convention. Our first, rough, definition is:

A regularity R in the behavior of members of a population P when they are agents in a recurrent situation S is a *convention* if and only if, in any instance of S among members of P,

(1) everyone conforms to R;
(2) everyone expects everyone else to conform to R;
(3) everyone prefers to conform to R on condition that the others do, since S is a coordination problem and uniform conformity to R is a proper coordination equilibrium in S.

Sample Conventions

Chapter II will be devoted to improving the definition. But before we hide the concept beneath its refinements, let us see how it applies to examples. Consider some conventions to solve our sample coordination problems.

(1) If you and I must meet every week, perhaps at first we will make a new appointment every time. But after we have met at the same time and place for a few weeks running, one of us will say, "See you here next week," at the end of every meeting. Later still we will not say anything (unless our usual arrangement is going to be unsatisfactory next week). We will just both go regularly to a certain place at a certain time every week, each going there to meet the other and confident that he will show up. This regularity that has gradually developed in our behavior is a convention.

Chapter 37
Three Views of Common Knowledge

Jon Barwise

Introduction

As the pioneering work of Dretske[1] has shown, knowing, believing, and having information are closely related and are profitably studied together. Thus while the title of this paper mentions common knowledge, I really have in mind the family of related notions including common knowledge, mutual belief and shared information. Even though I discuss common knowledge in this introduction, the discussion is really intended to apply to all three notions.

Common knowledge and its relatives have been written about from a wide variety of perspectives, including psychology, economics, game theory, computer science, the theory of convention, deterrence theory, the study of human-machine interaction, and the famous Conway paradox, just to mention a few. There are literally hundreds of papers that touch on the topic. However, while common knowledge is widely recognized to be an important phenomenon, there is no agreement as to just what it amounts to. Or rather, as we will see, what agreement there is presupposes a set of simplifying assumptions that are completely unrealistic. This paper offers a comparison of three competing views in a context which does not presuppose them to be equivalent, and explores their relationships in this context.[2]

Jon Barwise was deceased at the time of publication.

[1] F. Dretske, *Knowledge and the Flow of Information* (Cambridge, Mass.: Bradford Books/MIT Press, 1981).

[2] Obviously I have not read all, or even most, of the papers on common knowledge, so it could be that some or all of the points made in this paper are made elsewhere. If so, I would appreciate

J. Barwise (deceased)
Stanford University, Stanford, CA, USA

H. Arló-Costa et al. (eds.), *Readings in Formal Epistemology*, Springer Graduate Texts in Philosophy 1, DOI 10.1007/978-3-319-20451-2_37

I take it that these accounts are after characterizations of common knowledge in terms of ordinary knowledge, of mutual belief in terms of belief, and of having shared information in terms of having information. Such accounts should be compatible with, but presumably distinct from, an account that shows how it is that common knowledge comes about. They should also be compatible with some explanation of how common knowledge is used.

We are going to compare the following approaches to common knowledge: (1) the *iterate* approach, (2) the *fixed-point* approach, and (3) the *shared-environment* approach. In order to review these three accounts, let's consider a special case where there are just two agents, say p and q, with common knowledge of some fact σ. Let τ be this additional fact, of the common knowledge of σ. We are looking for a characterization of τ in terms of p, q, σ and ordinary (private) knowledge.

By far the most common view of common knowledge is that τ is to be understood in terms of *iterated* knowledge of σ: p knows σ, q knows σ, p knows q knows σ, q knows p knows, p knows q knows p knows σ, and so forth. On this account, for p and q to have common knowledge of σ is for all members of this infinite collection of other facts to obtain. This is the approach taken in David Lewis' influential book[3] on convention, for example. It is, without doubt, the orthodox account, at least in the field of logic. It is, for example, the one that is the basis of the mathematical modeling of common knowledge in the logic of distributed systems.[4]

The two other accounts we want to investigate replace this infinite hierarchy with some sort of circularity. One such account was explicitly proposed by Harman.[5] Harman's proposal is that the correct analysis of τ is as:

$$p \text{ and } q \text{ know } (\sigma \text{ and } \tau)$$

Notice that on this fixed-point account, τ is in some sense a proper constituent of itself. Harman seems to suggest that this is nothing but a succinct representation of the first infinite hierarchy.

This fixed point approach is also the view of common knowledge that is implicit in Aumann's pioneering paper modeling common knowledge in game theory, as

---

learning about it. But even if this is so, I am reasonably sure that the particular model I develop below is original, depending as it does on recent work in set theory by Peter Aczel.

[3]David Lewis, *Convention, A Philosophical Study* (Cambridge, Mass.: Harvard University Press, 1969).

[4]See, for example, the paper by Halpern and Moses, "Knowledge and common knowledge in distributed environments," Proc. 3rd ACM Symp. on Principles of Distributed Computing (1984), 50–61, and the paper by Fagin, Halpern and Vardi, "A model-theoretic analysis of knowledge: preliminary report," Proc. 25th IEEE Symposium on Foundations of C.S., 268–278.

[5]See Gilbert Harman's review of *Linguistic Behavior* by Jonathan Bennett, *Language* 53 (1977): 417–24.

was pointed out by Tommy Tan and Sergio Ribeiro da Costa Werlang.[6] Aumann suggests that this approach is equivalent to the iterate approach. Tan and Ribeiro da Costa Werlang develop a mathematical model of the iterate approach and show that it is equivalent to Aumann's fixed point model. Similarly, one sees from the work of Halpern and Moses, that while they start with the iterate approach, in their set-up, this is equivalent to a fixed point. One of the aims of this paper is to develop a mathematical model where both iterate and fixed point accounts fit naturally, but where they are *not* equivalent. Only in such a framework can we explicitly isolate the assumptions that are needed to show them equivalent. We will see that these assumptions are simply false (in the case of knowledge), so that the issue as to which of the two, if either, is the "right" analysis of the notion is a live one.

The final approach we wish to discuss, the shared-environment approach, was proposed by Clark and Marshall,[7] in response to the enormous processing problems associated with the iterate account. On their account, p and q have common knowledge of σ just in case there is a situation s such that:

- $s \models \sigma$,
- $s \models p_1$ knows s,
- $s \models p_2$ knows s.

Here $s \models \theta$ is a notation for: θ is a fact of s. The intuitive idea is that common knowledge amounts to perception or other awareness of some situation, part of which includes the fact in question, but another part of which includes the very awarenesses of the situation by both agents. Again we note the circular nature of the characterization.

What Are We Modeling: Knowing or Having Information?

It is these three characterizations of common knowledge, and their relatives for the other notions of mutual belief and shared information, that we wish to compare. Among common knowledge, mutual belief, and shared information, we focus primarily on the case of having information, secondarily on the case of knowledge. Part of the claim of the paper is that these two notions are often conflated, and that it is this conflation that lends some credibility to the assumptions under which the first two approaches to common knowledge are equivalent. So I need to make clear

[6]R. J. Aumann, "Agreeing to disagree," *Annals of Statistics*, 4 (1976), 1236–1239, and the working paper "On Aumann's Notion of Common Knowledge – An alternative approach," Tan and Ribeiro da Costa Werlang. University of Chicago Graduate School of Business, 1986.

[7]H. Clark and C. Marshall, "Definite reference and mutual knowledge," in *Elements of Discourse Understanding*, ed. A. Joshi, B. Webber, and I. Sag (Cambridge: Cambridge University Press, 1981), 10–63.

what I take to be the difference between an agent p knowing some fact σ, and the agent simply having the information σ.

Here I am in agreement with Dretske.[8] Knowing σ is stronger than having the information σ. An agent knows σ if he not only has the information σ, but moreover, the information is "had" in a way that is tied up with the agent's abilities to act. When might this not be the case? The most notorious example (and by no means the only one) is when I know one fact σ, and another fact σ' logically follows from σ, but I disbelieve the latter because I don't know that the one follows from the other. Obviously there is a clear sense in which I have the information σ', but I certainly don't know it in the ordinary sense of the word. Another arises with certain forms of perceptual information. If I see the tallest spy hide a letter under a rock, then there is a clear sense in which I have the information the tallest spy has hidden the letter. However, if I don't know that he is a spy, say, then I don't know that the tallest spy has hidden a letter. Information travels at the speed of logic, genuine knowledge only travels at the speed of cognition and inference.

Much of the work in logic which seems to be about knowledge is best understood in terms of having information. And for good reason. For example, in dealing with computers, there is a good reason for our interest in the latter notion. We often use computers as information processors, after all, for our own ends. We are often interested less in what the computer does with the information it has, than in just what information it has and what we can do with it. Or, in the design of a robot, we may be aiming at getting the robot to behave in a way that is appropriate given the information it has. One might say, we are trying to make it know the information it has.

So, as noted earlier, this paper focuses primarily on the case of having information. The model I am going to develop originated with an analysis of shared perceptual information,[9] but it also works quite well for primary epistemic perceptional information[10] and the relation of having information.

Let me say all this in another way, since it seems to be a confusing point. In the section that follows, I could interpret the model as a model of knowledge if I were to make the same idealization that is made in most of the literature on common knowledge. However, part of what I want to do here is make very explicit just what the role of this idealization is in the modeling of common knowledge. Thus, I am forced to work in a context where we do not make it. Once we are clear about its role, we can then decide if we want to make it.

---

[8] Op. Cit.

[9] See ch. 2 of Fred Dretske, *Seeing and Knowing* (Chicago: University of Chicago Press, 1969); or J. Barwise, "Scenes and other Situations", *Journal of Philosophical Logic* 78 (1981): 369–97; or ch. 8 of J. Barwise and J. Perry, *Situations and Attitudes* (Cambridge, Mass.: Bradford Books/MIT Press, 1983).

[10] See ch. 3 of *Seeing and Knowing* or ch. 9 of *Situations and Attitudes.*

Summary of Results

Our results suggest that the fixed point approach gives the right theoretical analysis of the pretheortic notion of common knowledge. On the other hand, the shared-environment approach is the right way to understand how common knowledge usually arises and is maintained over an extended interaction. It does not offer an adequate characterization of the pretheoretic notion, though, since a given piece of common knowledge may arise from many different kinds of shared environments. The fixed point gets at just what is in common to the various ways a given piece of common knowledge can arise.

What about the iterate approach? We will show that for the relation of having information, the fixed-point approach is equivalent to the iterate approach, provided we restrict ourselves to finite situations. Without this assumption, though, the iterate approach, with only countably many iterations, is far too weak. In general, we must iterate on indefinitely into the transfinite.

Not only is the iterate approach too weak. When we move from having information to knowing, then even two iterations are unjustified. In general, the iterate approach is incomparable and really seems to miss the mark. We will see just what assumptions *are* needed to guarantee that the iterate account is equivalent to the fixed-point account.

Modeling Shared Information

In developing our model, we will follow the general line used in *The Liar*[11] in three ways. First, we take our metatheory to be ZF/AFA, a theory of sets that admits of circularity. We do this because ZF/AFA offers the most elegant mathematical setting we know for modeling circularity. Space does not permit us to give an introduction to this elegant set theory. We refer the reader to Chap. 3 of this book, or to Aczel's lectures[12] for an introduction.

Second, we follow the approach taken in *The Liar* in paying special attention to "situations," or "partial possible worlds." As far as this paper goes, the reader can think of a situation as simply representing an arbitrary set of basic facts, where a fact is simply some objects standing in some relation. Actually, in this paper, situations play a dual role. On the one hand they represent parts of the world. On the other hand they represent information about parts of the world. Thus, for example, we will define what it means for one situation s_0 to support another situation s_1, in the sense that s_0 contains enough facts to support all the facts in s_1.

[11] J. Barwise and J. Etchemendy, *The Liar: An Essay on Truth and Circularity* (New York: Oxford University Press, 1987).

[12] P. Aczel, *Non-well-founded Sets* (CSLI Lecture Notes (Chicago: University of Chicago Press, 1987 (to appear)).

Finally, on the trivial side, we also follow *The Liar* in considering a domain of card players as our domain to be modeled. We use this domain because it is simple, and because the existence of common knowledge is absolutely transparent to anyone who has ever played stud poker. And while the example is simple, there is enough complexity to illustrate many of the general points that need making. However, there is nothing about the results that depend on this assumption. You could replace the relation of having a given card with any relation whatsoever, and the results would still obtain.

Example 37.1 Simply by way of illustration, we have a running example, a game of stud poker. To make it very simple, we will use two card stud poker,[13] with two players, Claire and Max. We will assume that the players have the following cards:

Player	Down card	Up card
Claire	$A♠$	$3♣$
Max	$3♠$	$3◇$

Except for the rules and the idiosyncrasies of the other players, all the information available to the players is represented in this table. Note that based on what he sees, Max knows that he has the winning hand, or at least a tie, but Claire thinks she has a good chance of having the winning hand. The question before us is how best to model the informational difference between up cards and down cards.

Notice how different this situation would be from draw poker, where all cards are down, even if each player had cheated and learned the value of the second card. Anyone who has played poker will realize the vast difference. The reason is that in the standard case, the values of all the up cards is common knowledge, but in the second it isn't. Our aim, then, is to use tools from logic to model the three approaches to the common knowledge and shared information present in such a situation.

We reiterate that we use this simple card domain simply by way of making things concrete. We could equally well treat the more general case, if space permitted. We use S for the relation of seeing (or more generally of having information), H for the relation of having a card, and appropriate tuples to represent facts involving these relations. Thus, the fact that Max has the $3♠$ will be represented by the triple $\langle H, \text{Max}, 3♠\rangle$. The fact that Claire sees this will be represented by $\langle S, \text{Claire}, \{\langle H,$

[13]For the reader unfamiliar with two card stud poker, here is all you need to know to follow the example. First each player is dealt one card which only he is allowed to see, and there is a round of betting. Then each player is dealt one card face up on the table and there is another round of betting. Hands are ranked and players bet if they think their hand is best. But they can also drop out of the round at any point. After both rounds of betting are over, the hands are displayed, so that all players can see who won. As far as the ranking, all that matters is that a hand with a matching pair is better than a hand with no pairs. But among hands with no pairs, a hand with an ace is better than a hand with no ace.

Max, 3♠⟩}⟩. The question is how to adequately represent the common knowledge, or public information, of the up cards, like the fact ⟨H, Max, 3◇⟩ that Max has the 3◇. Thus for our formal development we have primitives: players $p_1, \ldots, p_n$, cards A♠, K♠, . . . , 2♣, and relations H for the relation of having some card and S for the relation of seeing or otherwise having the information contained in some situation.

Comparing the Iterate and Fixed Point Accounts

Definition 37.1

1. The (models of) *situations* and *facts*[14] form the largest classes *SIT*, *FACT* such that:
   - $\sigma \in FACT$ iff σ is a triple, either of the form $\langle H, p, c\rangle$, where p is a player and c is a card, or of the form $\langle S, p, s\rangle$, where p is a player and $s \in SIT$.
   - A set s is in *SIT* iff $s \subseteq FACT$.
2. The *wellfounded situations* and *wellfounded facts* form the smallest classes *Wf-SIT* and *Wf-FACT* satisfying the above conditions.

Routine monotonicity considerations suffice to show that there are indeed largest and smallest such collections. If our working metatheory were ordinary ZF set theory, then these two definitions would collapse into a single one. However, working in ZF/AFA, there are many nonwellfounded situations and facts. A fact $\sigma = \langle R, a, b\rangle$ being in some situation s represents the fact of the relation R holding of the pair a, b in s, and is said to be a *fact of s*.

Example 37.1, Cont'd The basic situation s_0 about which player has which cards is represented by the following situation: $s_0 =$

$$\{\langle H, \text{Claire}, A♠\rangle, \langle H, \text{Max}, 3\diamondsuit\rangle, \langle H, \text{Claire}, 3♣\rangle, \langle H, \text{Max}, 3♠\rangle\}$$

Abbreviations We sometimes write (p_i Hc) for the fact ⟨H, p_i, c⟩, and similarly (p_i Ss) for the fact ⟨S, p_i, s⟩. We write (p_i $S\sigma$) for (p_i Ss) where $s = \{\sigma\}$. All of our facts are atomic facts. However, our situations are like conjunctive facts. Hence we sometimes write $\sigma \wedge \tau$ for the situation $s = \{\sigma, \tau\}$, and so we can write (p_i S ($\sigma \wedge \tau$)) for p_i Ss, where $s = \{\sigma, \tau\}$. Similarly when there are more conjuncts.

Example 37.1, Cont'd With these tools and abbreviations, we can discuss the first two approaches to the public information about the up cards in our example. Toward this end, let $s_u =$

$$\{\langle H, \text{Claire}, 3♣\rangle, \langle H, \text{Max}, 3\diamondsuit\rangle\}$$

which represents situation concerning the up cards.

[14]In order to keep this paper within bounds, I am restricting attention only to positive, nondisjunctive facts.

Iterates On this account, the fact that s_u is public information would be represented by an infinite number of distinct wellfounded facts: (Claire Ss_u), (Max Ss_u), Claire S(Claire Ss_u), (Max S(Claire Ss_u)), etc., in other words, by a wellfounded though infinite situation.

Fixed-Point On this account, the fact that s_u is publicly perceived by our players can be represented by the following public situation s_p:

$$s_p = \{\text{Claire } S\,(s_u \cup s_p)\,, (\text{Max } S\,(s_u \cup s_p)\}$$

By contrast with the iterate approach, this situation contains just two facts. However, it is circular and so not wellfounded. The Solution Lemma of ZF/AFA guarantees that the sets used to represent the situation s_p exists.

It will be useful for later purposes to have a notation for some of the situations that play a role in our example. First, let the situations s_1, s_2 represent the visual situations, as seen by each of Claire and Max, respectively, including both the up cards and what each sees about what the others see. Consider also the larger situation s_w that represents the whole. Let $s_w = s_0$ (from above) union the set of the following facts:

$$\langle S, \text{Claire}, (\text{Claire } HA\spadesuit)\rangle, \langle S, \text{Max}, (\text{Max } H3\spadesuit)\rangle,$$
$$\langle S, \text{Claire}, s_1\rangle, \langle S, \text{Max}, s_2\rangle$$

where the first two facts represent what each player sees about his own down cards, and, e.g., s_1 is everything relevant seen by Claire, with facts s_u (= the "up" cards, as above) plus the fact (S, Max, s_2). Notice that s_1 is a constituent of s_2, and vice versa, so that s_w is a circular, nonwellfounded situation.

The next task is to define what it means for a fact a to hold in a situation s, which we write $s \models \sigma$, so that we can show that the situation s_w does satisfy the fixed point fact situation s_p defined above, as well as the above iterates.

Definition 37.2 The relation $\models$ is the largest subclass of $SIT \times FACT$ satisfying the following conditions:

- s $\models (pHc)$ iff $\langle H, p, c\rangle \in s$
- $s \models (pSs_0)$ iff there is an s_1 such that $\langle S, p, s_1\rangle \in s$, and for each $\sigma \in s_0$, $s_1 \models \sigma$.

The motivation for the second clause should be fairly obvious. If, in s, a player p sees (or otherwise has the information) s_1, and if s_1 satisfies each $\sigma \in s_0$, then in s that same player p sees (or otherwise has the information) s_0. This would not be a reasonable assumption about the usual notion of knowledge, since knowledge is not closed under logical entailment.

There is a difference with the possible worlds approach that sometimes seems puzzling to someone familiar with the traditional modal approach to knowledge. In p.w. semantics, partial situations are represented by the set of all possible worlds compatible with them. As a result, whereas we can use an *existential* quantifier in clause (2) of our definition over situations about which p has information, the p.w. approach is forced to use a universal quantifier over possible worlds.

The reader can verify that all of the facts of s_p and the hierarchy of iterates of our running example indeed hold in the situation s_w. We also note that it follows from the definition that for all facts σ, if $\sigma \in s$ then $s \models \sigma$. However, the converse does not hold.

We extend our notation a bit and write $s_1 \models s_2$ provided $s_1 \models \sigma$ for each $\sigma \in s_2$.

As a companion of this notion of holding in, there is a notion of hereditary subsituation.[15] Intuitively, s_1 is a hereditary subsituation of s_2, written $s_1 \sqsubseteq s_2$, if all the information present in s_1 is present in s_2.

Definition 37.3 The hereditary subsituation relation $\sqsubseteq$ is the largest relation on $SIT \times SIT$ satisfying: $s_1 \sqsubseteq s_2$ iff:

- If $\langle H, p, c\rangle \in s_1$, then $\langle H, p, c\rangle \in s_2$;
- If $\langle S, p, s_0\rangle \in s_1$, then there is an s such that $s_0 \sqsubseteq s$ and $\langle S, p, s\rangle \in s_2$.

Proposition 37.1

1. *If* $s_1 \models \sigma$ *and* $s_1 \sqsubseteq s_2$, *then* $s_2 \models \sigma$.
2. *For all situations* s_0 *and* s_1, *the following are equivalent:*

   (a) $s_1 \sqsubseteq s_2$
   (b) $s_2 \models \sigma$ *for each* $\sigma \in s_1$.

Proof Limitations of space in this volume prevent us from doing more than hint at the proofs of the results in this paper. In this case we note that (1) is a simple consequence of the maximality of the $\models$ relation. Likewise, the implication from (2a) to (2b) is a consequence of the maximality of the $\models$ relation. The converse is a simple consequence of the maximality of the $\sqsubseteq$ relation. □

We say that situations s_0, s_1 are *informationally equivalent*, $s_0 \equiv s_1$, if the same facts hold in them. This is clearly an equivalence relation on situations. By the above lemma, $s_0 \equiv s_1$ if and only if each is a hereditary subsituation of the other. Distinct situations are often informationally equivalent. For example, suppose s_0' is a proper subset of the set s_1' of facts. Consider the situation $s_1 = \{\langle S, \text{Max}, s_1\rangle\}$, where Max has the information s_1, with the situation s_0 where there are two facts, that Max has the information s_0' and that he has the information s_1'. Using the fact just mentioned it is clear that $s_0 \equiv s_1$.

To compare the iterate and the fixed point approaches, we will show how an arbitrary fact θ (or situation s) gives rise to a transfinite sequence of wellfounded facts θ^α (or wellfounded situations s^α), for arbitrary ordinal α, finite or infinite. We use *Tr* for the conjunction of the empty situation, a fact that holds in every situation.

[15] In more recent joint work with Aczel, a generalization of this relation takes center stage.

Definition 37.4 The transfinite sequence $\langle \theta^\alpha \mid \alpha \in \mathit{Ordinals} \rangle$ of wellfounded facts associated with an arbitrary fact θ is defined by induction on ordinals as follows: for any θ, $\theta^0 = Tr$, and for $\alpha > 0$ we have:

$$(pHc)^\alpha = (pHc)$$
$$(pSs)^\alpha = (pS\ s^{<\alpha})$$

where

$$s^{<\alpha} = \left\{ \sigma^\beta \middle| \sigma \in s, \beta < \alpha \right\}$$

Similarly, for any situation s we define the transfinite sequence $\langle s^\alpha \mid \alpha \in \mathit{Ordinals} \rangle$ by letting $s^\alpha = \{\sigma^{\ \alpha} | \sigma \in s\}$.

The reader should verify that if we apply this definition to the fixed point fact in our example, we generate the iterates for all the finite ordinals, but then we go on beyond them into the transfinite.

We say that a fact σ *entails* a fact τ, written $\sigma \Rightarrow \tau$, if for every situation s, if $s \models \sigma$ then $s \models \tau$.

Theorem 37.2 Let θ be some fact.

1. *For all* α, $\theta \Rightarrow \theta^\alpha$.
2. *If each approximation fad* θ^α *holds in a situation* s, *then so does* θ.
3. *Assume that* κ *is a regular cardinal, and that* s *is a situation of size less than* κ. *If each approximation* θ^α, *for* $\alpha < k$, *holds in* s, *then so does* θ.

Proof The first is proved by means of a routine induction on α. The second is a consequence of the maximality of $\models$ and is not too difficult to prove. The third is a strengthening of the second involving routine cardinality considerations. □

Corollary 37.3 *Let* θ *be any fact, and let* s_w *be the set of all finite approximations of* θ. *Then, for any finite situation* s, $s \models \theta$ *iff* $s \models s_w$.

Refinement (3) of (2) of Theorem 37.2, and so the above corollary, were not present in the original working paper referred to above. They were discovered later in joint work with Peter Aczel. This result shows that the finite approximations of a circular fact will be equivalent to it, with respect to *finite* situations. This is a bit unsatisfactory, since the iterates themselves form an infinite situation. Still, it is the best we can hope for. However, in general, when we drop this restriction to finite models, one must look at the whole transfinite sequence of approximations. No initial segment is enough, as simple examples show. In this sense, the usual iterate approach is actually weaker than the simpler fixed-point approach.

When we move from having shared information to knowing, additional considerations must be brought to bear, as we will see below.

Comparing the Fixed Point and Shared Environment Approaches

To compare the shared environment approach with the fixed point approach, we introduce a simple second-order language which allows us to make existential claims about situations of just the kind made in the shared environment approach. We call the statements of this language ∃-statements. Before giving the definition, let's give an example. The following ∃-statement

$$\exists e\,[e = ((\text{Claire } H3\clubsuit) \wedge (\text{Claire } S\ \ e) \wedge (\text{Max } S\ \ e))]$$

is one shared environment analysis of the fact that Claire and Max share the information that Claire has the 3♣. Notice that what we have here is a simple, finite, wellfounded statement, but one that could only hold of nonwellfounded situations. Similarly, there is a fairly simple ∃-statement explicitly describing the situation s_w in our running example.

To define our language, we introduce variables $e_1, e_2, \ldots$ ranging over situations, in addition to constants for the cards and players. In fact, we do not bother to distinguish between a card or player and the constant used to denote it in statements. For atomic statements we have those of the form $(p_i Hc)$ (where P_i is a player and c is a card) and $(p_i S e_j)$. The set of ∃-statements forms the smallest set containing these atomic statements and closed under conjunction ($\wedge$), existential quantification over situations ($\exists e_j$) and the rule: if Φ is a statement so is $(e_j \models \Phi)$. We are thus using $\models$ both for a relation symbol of our little language, as well as a symbol in our metalanguage. No more confusion should result from this than from the similar use of constants for cards and people. Finally, given any function f which assigns situations to variables, we define what it means for a statement Φ to hold in a situation s relative to f, written $s \models \Phi[f]$, in the expected way.

Definition 37.5

1. If Φ is an atomic statement, then $s \models \Phi[f]$ iff the appropriate fact is an element of s. In particular, if Φ is $(p_i S e_j)$, then $s \models \Phi[f]$ iff $\langle S, p_i, f(e_j)\rangle \in s$.
2. If Φ is $\Phi_1 \wedge \Phi_2$ then $s \models \Phi[f]$ iff $s \models \Phi_1[f]$ and $s \models \Phi_2[f]$
3. If Φ is $\exists e_j \Phi_0$ then $s \models \Phi[f]$ iff there is a situation s_j so that $s \models \Phi_0[f(e_j/s_j)]$
4. If Φ is $(e_j \models \Phi_0)$ then $s \models \Phi[f]$ iff the situation $s_j = f(e_j)$ satisfies $s_j \models \Phi_0[f]$.

A *closed* ∃-statement is one with no free variables, as usual. If Φ is closed, we write $s \models \Phi$ if some (equivalently, every) assignment f satisfies $s \models \Phi[f]$.

Notice that the ∃-statements are all finite and wellfounded. (The results that follow would hold equally well if we allowed infinite conjunctions and infinite strings of quantifiers, except for the word "finite" in Theorem 37.5 below.) Nevertheless, some of them can only hold of nonwellfounded situations, as the above example shows.

We want to show that any ∃-statement can be approximated in a certain sense by a fixed point situation. In particular, if we take as our ∃-statement one that expresses a shared environment approach to shared information, the resulting situation will be the one that characterizes the fixed point approach. Then, using the transfinite wellfounded iterates approximating the fixed point approach, we obtain a transfinite sequence of wellfounded facts approximating any ∃-statement.

Let us say that a situation s_Φ almost characterizes the ∃-statement Φ if $s_\Phi \models \Phi$ and for every situation $s \models \Phi$, we have $s \models s_\Phi$. For example, if we take our above example of an ∃-statement, then the following situation can easily be seen to almost characterize it:

$$s = \{\langle H, Claire, 3\clubsuit\rangle, \langle S, Claire, s\rangle, \langle S, \text{Max}, s\rangle\}$$

Clearly our statement is true in this model. It is also easy to see that s is a hereditary subsituation of any situation which is a model of our statement, so by Proposition 37.1, s almost characterizes the statement. This definition is justified by the following result, which is an easy consequence of Proposition 37.1.

Proposition 37.4 *Suppose that the situation s almost characterizes the ∃-statement Φ. Then for any fact σ, the following are equivalent:*

1. *σ is entailed by Φ, i.e., σ holds in all models of Φ*
2. $s \models \sigma$

The following is the main result of this paper. It shows the extent to which the shared environment approach can be approximated by the fixed point and iterate approaches.

Theorem 37.5 *Every ∃-statement Φ is almost characterized by some finite situation s_Φ.*

Proof First one establishes a normal form lemma for ∃-statements, where all the quantifiers are pulled out front. One then uses the Solution Lemma of AFA to define the desired situation. The proof that it almost characterizes the statement uses Proposition 37.1. □

However, there is a distinct sense in which ∃-statements are more discriminating than the situations that almost characterize them. For example, compare our above example of an ∃-statement with the following:

$$\exists e_1, e_2\, [e_1 \models ((\text{Claire } H 3\clubsuit) \wedge (\text{Claire } S\ \ e_2)) \wedge e_2 \models ((\text{Claire } H 3\clubsuit)\, J \wedge (\text{Max } S\ \ e_1))]$$

Clearly any model of our first statement is a model of our second. However, it is easy to see that there are models of our second that are not models of our first. (Think of a case where the card is not an up card, but is down, but where there are suitably placed mirrors.) On the other hand, these two statements are almost characterized by exactly the same situations. Or, in view of Proposition 37.4, the two statements entail the same facts, both wellfounded and circular.

Intuitively, what is going on here is that both of these statements represent ways in which Max and Claire might share the information that Claire has the 3♣. The first would be the one predicted by a literal reading of the Clark and Marshall account, but the second is clear in the spirit of that account. However, this means that since they are not equivalent, neither one can be the right characterization of the shared information. Rather, what they represent are two distinct ways, among many, that Max and Claire might have come to have the shared information. We leave it to the reader to work out analogous inequivalent ∃-statements that also give rise to the shared information in our running example.

We conclude this section by observing that the results can be extended to the case where we allow disjunctions to occur in ∃-statements, if one also allows disjunctive facts.

Conclusions

In thinking about shared information and common knowledge, it is important to keep three questions separate: (i) What is the correct analysis of common knowledge? (ii) Where does it come from? (iii) How is it used?

It would be neat if these three questions got their answers from the three different approaches in the literature. The results discussed above prompt us to propose that the fixed-point approach is the right analysis of the notion, and that it typically arises through some sort of shared environment.

However, by definition, the epistemically neutral case we have been studying is divorced from questions of use. To think about how shared information gets used, we turn to the epistemic case. Let us suppose that the fixed-point approach, or something like it, characterizes common knowledge, and the shared-environment approach characterizes the way in which common knowledge commonly arises. Does it follow that the iterate approach approximates common knowledge, or perhaps how it is used?

It seems that it can't. A clear difference between having information and knowing arises in the respective relationships between the fixed-point facts and its approximations. In the nonepistemic case, it is a matter of logical entailment. However, in the latter case, the fixed-point fact will simply not entail the analogous approximations. To see why, let's consider an example.

Example 37.2 Consider the following situation s, where we use K for the relation of knowing of a situation:

$$\langle H, \text{Max}, 3\diamondsuit \rangle\ ,\ \langle K, \text{Claire}, s \rangle\ ,\ \langle K, \text{Dana}, s \rangle\ ,\ \langle K, \text{Max}, s \rangle$$

It seems clear that the fact

$$\theta = (\text{Max}\ H 3\diamondsuit) \wedge (\text{Claire}\ K\theta) \wedge (\text{Dana}\ K\theta) \wedge (\text{Max}\ K\theta)$$

holds in this situation. However, is it a fact in this situation that, say, Max knows that Dana knows that Claire knows that he, Max, has the 3$\diamond$? And even more iterations?

It seems clear that it will not in general be true. After all, some sort of inference is required to get each iteration, and the players might not make the inference. They are, after all, only 3 years old. And even if Claire makes her inference, Dana may have legitimate doubts about whether Claire has made her inference. But once one player has the least doubt about some other player's making the relevant inference, the iterated knowledge facts breaks down. That is, once the making of an inference is implausible, or even just in doubt, the next fact in the hierarchy is not really a *fact* at all.

It is usually said that the iterate account assumes that all the agents are perfectly rational, that is, that they are perfect reasoners. This example also shows that it in fact assumes more: it assumes that it is *common knowledge* among the agents that they are all perfectly rational. It is only by making this radical idealization, plus restricting attention to finite situations, that the iterate account is equivalent to the fixed-point account. And the idealization requires the very notion that one is trying to understand in the first place.

We began this section by asking three questions. We have proposed answers to the last two of them, and suggested that the third question, about how common knowledge is used, is not answered by the iterate approach. But then how *do* people make use of common knowledge in ordinary situations?

My own guess is that common knowledge per se, the notion captured by the fixed-point analysis, is not actually all that useful. It is a necessary but not a sufficient condition for action. What suffices in order for common knowledge to be useful is that it arise in some fairly straightforward shared situation. The reason this is useful is that such shared situations provide a basis for perceivable situated action, action that then produces further shared situations. That is, what makes a shared environment work is not just that it gives rise to common knowledge, but also that it provides a stage for maintaining common knowledge through the maintainence of a shared environment. This seems to me to be part of the moral of the exciting work of Parikh, applying ideas of game theory to the study of communication.[16]

It seems to me that the consequences of this view of common knowledge are startling, if applied to real world examples, things like deterrence (mutual assured destruction, say). Indeed, it suggests a strategy of openness that is the antithesis of the one actually employed. But that goes well beyond the scope of this conference.

Finally, let me note that the results here do not lend themselves to an immediate comparison with other mathematical models of common knowledge, especially the approaches in game theory. It would be interesting to see a similar analysis there, one that pinpoints the finiteness or compactness assumption that must be lurking behind the Tan and Ribeiro da Cost Werlang result.

[16]Prashant Parikh, "Language and strategic inference," Ph.D. Dissertation, Stanford University, 1987.

Chapter 38
The Logic of Public Announcements, Common Knowledge, and Private Suspicions

Alexandru Baltag, Lawrence S. Moss, and Sławomir Solecki

Introduction: Example Scenarios and Their Representations

We introduce the issues in this paper by presenting a few *epistemic scenarios*. These are all based on the Muddy Children scenario, well-known from the literature on knowledge. The intention is to expose the problems that we wish to address. These problems are first of all to get models which are faithful to our intuitions, and then to build and study logical systems which capture some of what is going on in the scenarios.

The cast of characters consists of three children: A, B, and C. So that we can use pronouns for them in the sequel, we assume that A is male, and B and C are female. Furthermore, A and B are dirty, and C is clean. Each of the children can see all and only the others. It is known to all (say, as a result of a shout from one of the parents) that at least one child is dirty. Furthermore, each child must try to figure out his or her state only by stating "I know whether I'm dirty or not" or "I don't know whether I'm dirty or not." They must tell the truth, and they are perfect reasoners in the sense that they know all of the semantic consequences of their knowledge. The opening situation and these rules are all assumed to be common knowledge.

Scenario 1. After reflection, A and B announce to everyone that at that point they do not know whether they are dirty or not. (The reason we are having A and B make this announcement rather than all three children is that it fits in better with our scenarios to follow.) Let α denote this announcement.

A. Baltag (✉)
ILLC, University of Amsterdam, The Netherlands
e-mail: thealexandrubaltag@gmail.com

L.S. Moss • S. Solecki
Mathematics Department, Indiana University, Bloomington, IN 47401, USA
e-mail: lsm@cs.indiana.edu; ssolecki@indiana.edu

H. Arló-Costa et al. (eds.), *Readings in Formal Epistemology*, Springer Graduate Texts in Philosophy 1, DOI 10.1007/978-3-319-20451-2_38

As in the classical Muddy Children, there are intuitions about knowledge before and after α. Here are some of those intuitions. Before α, nobody should know that he or she is dirty. However, A should think that it is possible that B knows. (For if A were clean, B would infer that she must the dirty one.) After α, A and B should each know that they are dirty, and hence they know whether they are dirty or not. On the other hand, C should not know whether she is dirty or not.

Scenario 1.5. This scenario begins after α. At this point, A and B announce to all three that they *do* know whether or not they are dirty. We'll call this event α'. Our intuition is that after α', C should know that she is not dirty. Moreover, A and B should know that C knows this. Actually, the dirty-or-not states of all the children should be *common knowledge* to all three.

Scenario 2. As an alternative to the first scenario, let's assume that C falls asleep for a minute. During this time, A and B got together and told each other that they didn't know whether they were dirty or not. Let β denote this event. After β, C wakes up. Part of what we mean by β is that C does not even consider it possible that β occurred, and that it's common knowledge to A and B that this is the case. Then our intuitions are that after β, C should "know" (actually: believe) that A does not know whether he is dirty (and similarly for B); and this fact about C is common knowledge for all three children. Of course, it should also be common knowledge to A and B that they are dirty.

Scenario 2.5. Following Scenario 2, we again have α': A and B announce that they do know whether they are dirty or not. Our intuitions are not entirely clear at this point. Surely C should suspect some kind of cheating or miscalculation on the part of the others. However, we will not have much to say about the workings of this kind of real-world sensibility. Our goal will be more in the direction of modeling different alternatives.

Scenario 3. Now we vary Scenario 2. C merely feigned sleep and thought she heard both A and B whispering. C cannot be sure of this, however, and also entertains the possibility that nothing was communicated. (In reality, A and B did communicate.) A and B for their part, still believe that C was sleeping. We call this event γ.

One might at first glance think that A and B's "knowledge" of C's epistemic state is unchanged by γ. After all, the communication was not about C. However, we work with a semantic notion of knowledge, and after γ, A and B know that they are dirty, hence then know that C knows that they are dirty. A and B did not know this at the outset.

So we need to revise the initial intuition. What is correct is that if C knows some fact φ before γ, then after γ, A and B know (or rather, believe) that C knows φ. This is because after γ, A and B not only know the clean-or-dirty state of everyone, they (therefore) also know exactly which possibilities everyone is aware of, which they discard as impossible, etc. So each of them can reconstruct C's entire epistemic state. They believe that their reconstruction is current, but of course, what they reconstruct is C's original one, before γ.

Conversely, if after γ, A and B "know" that C knows φ, then before γ, C really did know φ. That is, the reconstruction is accurate. For example, after γ, A believes that C should not consider it possible that A knows that he is dirty. However, C thinks it is possible that A knows he is dirty.

There is a stronger statement that is true: C knows φ before γ iff after γ, it is common knowledge to A and B that each of them knows that C knows φ. Intuitively, this hold because each of A and B knows that both of them are able to carry out the reconstruction of C's state.

Our final intuition is that after γ, C should know that if A were to subsequently announce that he knows that he is dirty, then C would know that B knows that she is dirty.

Scenario 3.5. Again, continue Scenario 3 by α'. At this point, C should know that her suspicions were confirmed, and hence that she is not dirty. For their part, A and B should think that C is confused by α': they should think that C is as she was following Scenario 2.5.

Scenario 4. A and B are on one side of the table and C is on the other, dozing. C wakes up at what looks to her like the middle of a joint confession by A and B. The two sides stare each other down. In fact, A and B have already communicated. We call this action δ. So C suspects that δ is what happened, but can't tell if it was δ or nothing. For their part, A and B see that C suspects but does not know that δ happened.

The basic intuition is that after δ, it should be common knowledge to all three that C suspects that the communication happened. Even if C thinks that A and B did not communicate, C should not think that she is sure of this.

One related intuition is that after δ, it should be common knowledge that C suspects that A *knows that he is dirty.* As it happens, this intuition is wrong. Here is a detailed analysis: C thinks it possible that everyone is dirty at the outset, and if this were the case then the announcement of B's ignorance would not help A to learn that he is dirty; from A's point of view, he still could be clean and B would not know that she is dirty. C's view on this does not change as a result of δ, so afterwards, C still thinks that it could be the case that A says, "It's possible that B and C are the dirty one and I am clean, Hence C would see my clean face and not suspect that I know that I am dirty." So it certainly should not be common knowledge that C suspects that A knows he is dirty.

Notice also that C would say after δ: "I think it is possible that no announcement occurred, and yet A thinks it possible that B is the only dirty one. In that case, what A would think that I suspect that A told B that he knows that he is *not* dirty. Of course, this is not what I actually suspect." The point is that C's reasoning about A and B's reasoning about her involves suspicion of a different announcement than we at first considered.

Scenario 4.5. Once again, we continue with α'. Our intuition is that this is tantamount to an admission of private communication by A and B. If we disregard this and only look at higher order knowledge concerning who is and is not dirty, we expect that the epistemic state after α' is the same for all three children as it is at the end of Scenario 1.5.

Models

Now that we have detailed a few scenarios and our intuitions about them, it is time to construct some Kripke models as representations for them.

The Models *U and V*. We begin with a representation of the situation before α. We take the Kripke model U whose worlds are $u_1, \ldots, u_7$ and whose atomic information and structure is given below:

$$u_1 : \mathsf{D}_C \quad u_2 : \mathsf{D}_B \quad u_3 : \mathsf{D}_B, \mathsf{D}_C$$
$$u_4 : \mathsf{D}_A \quad u_5 : \mathsf{D}_A, \mathsf{D}_C \quad u_6 : \mathsf{D}_A, \mathsf{D}_B \quad u_7 : \mathsf{D}_A, \mathsf{D}_B, \mathsf{D}_C$$

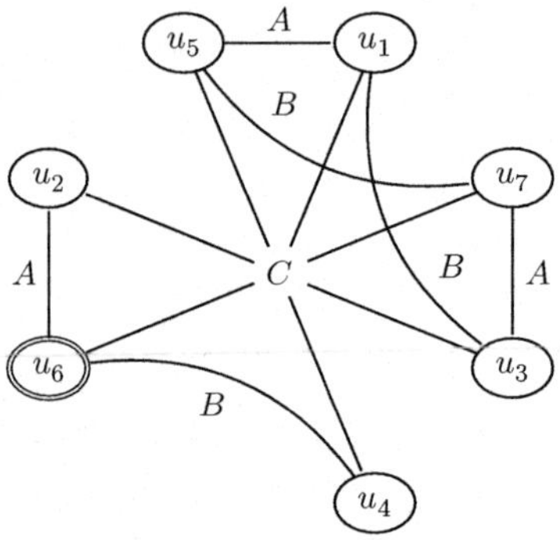

Note that we have not indicated any direction on the edges; they are all intended to be bidirectional. In addition, we have not shown self-loops for the agents. However, we intend in this model that all of the self-loops be present on all nodes for all agents.

As an example of reading the picture, in world u_3, A is clean, but B and C are dirty. Also, the worlds which A thinks are possible are u_3 and u_7. Thus, A sees that B and C are dirty, so A infers that the world is either u_3 or u_7. The lines for C are intended to go across the middle. The rest of the structure is explained similarly, except for the doubled ellipse around u_6. This specifies u_6 as the actual world in the model, the one which corresponds to our description of the model before α. Note that U incorporates some of the conventions stated in Scenario 1. For example, in each world, each child has a complete and correct assessment of which worlds are possible for all three reasoners.

Each of our intuitions about knowledge before α turns into a statement in the modal logic of knowledge. This logic has atomic sentences D_A, D_B, and D_C standing for "A is dirty", etc.; it has knowledge operators $\Box_A$, $\Box_B$, and $\Box_C$ along with the usual boolean connectives. We are going to use the standard Kripke semantics for multi-modal logic throughout this paper. So given a model-world pair, say $\langle A, a\rangle$, and some agent, say D, we'll write

$$\langle K, k\rangle \models \Box_D \varphi \quad \text{iff} \quad \text{whenever } k \xrightarrow{D} l \text{ in } K, \text{ we have } \langle K, l\rangle \models \varphi.$$

The boolean connectives will be interpreted classically. We can then check the following:

$$\langle U, u_6\rangle \models \neg\Box_A \mathsf{D}_A \wedge \neg\Box_A\neg\mathsf{D}_A \wedge \neg\Box_B\mathsf{D}_B \wedge \neg\Box_B\neg\mathsf{D}_B \wedge \neg\Box_C\mathsf{D}_C \wedge \neg\Box_C\neg\mathsf{D}_C$$
$$\langle U, u_6\rangle \models \Diamond_A\Box_B\mathsf{D}_B$$

The model after α is the Kripke model V, shown below:

$$v_1 : \mathsf{D}_C \;\; v_3 : \mathsf{D}_B, \mathsf{D}_C \;\; v_5 : \mathsf{D}_A, \mathsf{D}_C \;\; v_6 : \mathsf{D}_A, \mathsf{D}_B \;\; v_7 : \mathsf{D}_A, \mathsf{D}_B, \mathsf{D}_C$$

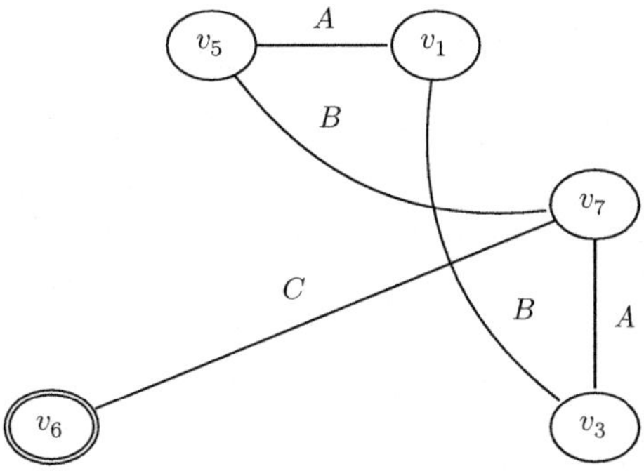

The way we got V from U was to discard the worlds u_2 and u_4 of U, since in U at each of those worlds, either A or B would know if they were dirty. We also changed the u's to v's to avoid confusion, and to stress the fact that we get a new model. Turning back to our intuitions, we can see that the following holds:

$$\langle V, v_6\rangle \models \Box_A\mathsf{D}_A \wedge \Box_B\mathsf{D}_B \wedge \neg(\Box_C\mathsf{D}_C \vee \Box\neg\mathsf{D}_C)$$

The Model *W*. Scenario 1.5 elaborates Scenario 1 by the event α'. So we discard the worlds where this is false in V, and we obtain a one-world model $W = \{w_6\}$. In w_6, all three agents are dirty. The picture is

(We have renamed v_6 to w_6. We continue to omit the three self-loops, but shortly our pictures will begin to incorporate those when they are appropriate.) This model reflects our intuition that at this point, C should know that she is not dirty.

The Model X. This corresponds to Scenario 2. We start with U and see the effect of the private announcement β. The resulting model X is too large to show in a small diagram, so instead we use a chart:

World	A	B	C	$\xrightarrow{A}$	$\xrightarrow{B}$	$\xrightarrow{C}$
$u_1, \ldots, u_7$						
x_1			•	x_1, x_5	x_1, x_3	u_1
x_3		•	•	x_3, x_7	x_1, x_3	u_2, u_3
x_5	•		•	x_1, x_5	x_5, x_7	u_4, u_5
$x_6 \surd$	•	•		x_6	x_6	u_6, u_7
x_7	•	•	•	x_3, x_7	x_5, x_7	u_6, u_7

The "real world" is x_6. Notice that the worlds $u_1, \ldots, u_7$ are also worlds in X. We did not put any information in the chart above for those worlds since it should be exactly the same as in U above. The reason for having these "old worlds" in X is that since C was asleep, the worlds that C considers possible after β should be just the ones that were possible before β. We can check that

$$\langle X, x_6 \rangle \models \neg\Box_C(\Box_A \mathsf{D}_A \vee \Box_A \neg \mathsf{D}_A).$$

Let φ be the sentence above. Then also, $\langle X, x_6 \rangle \models \Box^*_{\{A,B,C\}}\varphi$. This is our formal statement that it is common knowledge in the group of three children that φ holds. The semantics of this is that for all sequences $D_1, \ldots, D_m \in \{A, B, C\}^*$, $\langle X, x_6 \rangle \models \Box_{D_1} \cdots \Box_{D_m}\varphi$. Note that we have no way of saying in the modal language that C suspects that an announcement happened; the best we can do is (roughly) to say that C thinks that some sentence ψ is possible in the sense that ψ holds in some possible world. Of course, we have no way to say that A and B know that C was asleep, either.

Note as well that in X, we do not have $x_6 \xrightarrow{C} x_6$. In other words, the real world would not be possible for C. This is some indication that something strange is going on in this model. Further, we consider the model of what happens after A and B's announcement. Then in this model, *no worlds* would be accessible for C from the actual world. These anomalies should justify our interest in the more complicated scenarios and models involving suspicions of announcements.

The Model obtained by announcing α' in X. This would be the model with one world, say x_6^* where A and B are dirty, and whose structure is given by

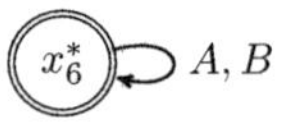

We have not only deleted the worlds where either A or B does not know that they are dirty in X, but we also discarded all worlds not reachable from the new version x_6^* of x_6. The anomaly here is that C thinks no worlds are possible.

World	A	B	C	$\xrightarrow{A}$	$\xrightarrow{B}$	$\xrightarrow{C}$
$u_1, \ldots, u_7$						
x_1, x_3, x_5, x_6, x_7						
y_1			•	x_1, x_5	x_1, x_3	y_1, y'_1
y_3		•	•	x_3, x_7	x_1, x_3	y_3, y'_2, y'_3
y_5	•		•	x_1, x_5	x_5, x_7	y_5, y'_4, y'_5
$y_6 \surd$	•	•		x_6	x_6	y_6, y_7, y'_6, y'_7
y_7	•	•	•	x_3, x_7	x_5, x_7	y_6, y_7, y'_6, y'_7
y'_1			•	u_1, u_5	u_1, u_3	y_1, y'_1
y'_2		•		u_2, u_6	u_2	y_3, y'_2, y'_3
y'_3		•	•	u_3, u_7	y_1, y_3	y_3, y'_2, y'_3
y'_4	•			u_4	u_4, u_6	y_5, y_4, y'_5
y'_5	•		•	u_1, u_5	u_5, u_7	y_5, y'_4, y'_5
y'_6	•	•		u_2, u_6	u_4, u_6	y_6, y_7, y'_6, y'_7
y'_7	•	•	•	u_3, u_7	u_5, u_7	y_6, y_7, y'_6, y'_7

Fig. 38.1 The model Y

The Model *Y*. We consider γ from Scenario 3, in which C thought she might have heard A and B, while A and B think that C is unaware of γ. We get the model Y displayed in Fig. 38.1 above. Y has 24 worlds, and so we won't justify all of them individually. We will give a more principled construction of Y from W and γ, once we have settle on a mathematical model of γ. For now, the ideas are that the y worlds are those where the announcement happened, and the y' worlds are those in which it did not. Note that some of the y worlds are missing, since the truthful announcement by A and B presupposes that they don't know whether they are dirty in U at the corresponding world. The x's and u's are from above, and they inherit the accessibility relations which we have seen.

Now our main intuition here is that $\langle U, u_6\rangle \models \Box_C\varphi$ iff $\langle Y, y_6\rangle \models \Box^+_{\{A,B\}}\Box_C\varphi$. (The sentence $\Box^+_{\{A,B\}}\chi$ means that A knows φ, A knows B knows χ, etc. It differs from $\Box^*_{\{A,B\}}\chi$ in that it does not entail that χ is true.) To see this, note that $u_6 \xrightarrow{C} u_6, u_7$ and no other worlds. And the only worlds reachable from y_6 using one or more $\xrightarrow{A}$ or $\xrightarrow{B}$ transitions followed by a $\xrightarrow{C}$ transition are again u_6 and u_7.

Another intuition is that in $\langle Y, y_6\rangle$, C should think that it is possible that A knows that he is dirty. This is justified since $y_6 \xrightarrow{C} x_6$, and $\langle Y, x_6\rangle \cong \langle X, x_6\rangle$ (that is, the submodels of X and Y generated by x_6 are isomorphic), and $\langle X, x_6\rangle \models \Box_A \mathsf{D}_A$.

Our final intuition is that in $\langle Y, y_6\rangle$, C should know that if A were to subsequently announce that he knows that he is dirty, then C would know that B knows that she is dirty. To check this, we need to modify Y by deleting the worlds where A does not know that he is dirty. These include y_7, y'_6 and y'_7. In the updated model, the only world accessible for C from (the new version of) y_6 is y_6 itself, and at y_6 in the new structure, B correctly knows she is dirty.

The Model obtained by announcing α' in Y. As when α' is announced in X, we only keep the worlds of Y worlds where both A or B do know they are dirty. So we drop y_7, y'_6, and y'_7.

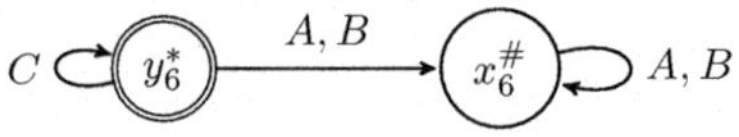

We also only keep the worlds accessible from y_6 (this change is harmless). C knows she is not dirty. Technically, A and B "know" this, but this is for the nonsensical reason that they "know" that C knows everything.

The Model Z. Corresponding to Scenario 4, we get the model Z shown below.

w	A	B	C	$\xrightarrow{A}$	$\xrightarrow{B}$	$\xrightarrow{C}$	w	A	B	C	$\xrightarrow{A}$	$\xrightarrow{B}$	$\xrightarrow{C}$
z_1			•	z_1, z_5	z_1, z_3	z_1, z'_1	z'_1			•	z'_1, z'_5	z'_1, z'_3	z_1, z'_1
z_2		•		z_2	z_2	z_2, z_3, z'_2, z'_3	z'_2		•		z'_2, z'_6	z_2, z'_2	z_2, z_3, z'_2, z'_3
z_3		•	•	z_3, z_7	z_1, z_3	z_2, z_3, z'_2, z'_3	z'_3		•	•	z'_3, z'_7	z'_1, z'_3	z_2, z_3, z'_2, z'_3
z_4	•			z_4	z_4	z_4, z_5, z'_4, z'_5	z'_4	•			z'_4, z'_6	z'_4, z'_6	z_4, z_5, z'_4, z'_5
z_5	•		•	z_1, z_5	z_5, z_7	z_4, z_5, z'_4, z'_5	z'_5	•		•	z'_1, z'_5	z'_5, z'_7	z_4, z_5, z'_4, z'_5
z_6 √	•	•		z_6	z_6	z_6, z_7, z'_6, z'_7	z'_6	•	•		z'_2, z'_6	z'_4, z'_6	z_6, z_7, z'_6, z'_7
z_7	•	•	•	z_3, z_7	z_5, z_7	z_6, z_7, z'_6, z'_7	z'_7	•	•	•	z'_3, z'_7	z'_5, z'_7	z_6, z_7, z'_6, z'_7

Recall our last point in Scenario 4, that we need to consider a few possible announcements for C to suspect. This is reflected in the fact that the z worlds are of three types. In z_2, B announced that she knows whether she is dirty, and A announced that he doesn't. Similar remarks apply to z_4. In all other z worlds, both announced that they do not know. The worlds accessible from each of these is based on the relevant announcement. For example, in z_2, neither A nor B thinks any other world is possible. (One might think that $z_2 \xrightarrow{A} z_6$. But in z_6, B could not announce that she knows she is dirty. So if the world were z_2 and the relevant announcement made, then A would not think z_6 is possible.) The z' worlds are those in which no announcement actually happened.

Our key intuition was that it is common knowledge that C suspects that δ happened. This will not correspond to anything in the formal language $\mathcal{L}([\alpha], \Box^*)$ introduced later in this paper. (However, it will be representable in an auxiliary language about actions; see Example 3.) Informally, the intuition is valid for Z because for every z_i (or z'_i) there is some z_j (unprimed) such that $z_i \xrightarrow{C} z_j$ (or $z'_i \xrightarrow{C} z_j$). In addition, in this particular model there is a sentence in our formal language which happens to hold only at the worlds where an announcement occurred. Here is one:

$$\chi \quad \equiv \quad \Box_A \mathsf{D}_A \ \vee\ \Box_A \neg \mathsf{D}_A \ \vee\ \Diamond^*_{\{A,B\}} \Diamond_C \Box_A \mathsf{D}_A$$

So $\langle Z, z_6\rangle \models \Box^*_{\{A,B,C\}} \Diamond_C\, \chi$.

The explanation of the mistaken intuition in Scenario 4 is that $z_6 \xrightarrow{C} z_7 \xrightarrow{A} z_3$, and $\Diamond_A \Box_A \mathsf{D}_A$ fails in z_2, z_3, z_2', and z_3'. Overall, $\langle Z, z_6 \rangle \models \neg\Box_C \Box_A \Diamond_A \Box_A \mathsf{D}_A$.

The point that C's suspicion varies corresponds to the fact that $\Diamond_C \Diamond_A \Diamond_C \Box_A \neg \mathsf{D}_A$ holds at $\langle Z, z_6 \rangle$. Indeed $z_6 \xrightarrow{C} z_6' \xrightarrow{A} z_2' \xrightarrow{C} z_2$, and $\langle Z, z_2 \rangle \models \Box_A \neg \mathsf{D}_A$.

A few more involved statements are true in Z. For example, $\Box_{\{A,B,C\}^*}(\Box_A \mathsf{D}_A \to \Diamond_C \Box_A \mathsf{D}_A)$. It is common knowledge to all three that *if* A knows he is dirty, then C thinks it possible that A knows this.

The Model obtained by announcing α' in Z. This model is W from above. (Actually, it is bisimilar to W; see section "The larger language $\mathcal{L}([\alpha], \Box^*)$".) This corresponds to the intuition that Scenarios 2.5 and 4.5 lead to the same model.

Epistemic Actions

We will formalize a language in section "A Logical Language with Epistemic Actions" along with the notions of (epistemic) *action structure* and *actions*. Before we do that, it makes sense to present the idea informally based on the examples which we have already dealt with.

α and α': announcements to everyone. We first consider α of Scenario 1. Let ψ be given by

$$\psi \quad := \quad \neg(\Box_A \mathsf{D}_A \vee \Box_A \neg \mathsf{D}_A) \wedge \neg(\Box_B \mathsf{D}_B \vee \Box_B \neg \mathsf{D}_B) \tag{38.1}$$

So ψ says that neither A nor B know whether or not they are dirty. This is the *precondition* of the announcement, but it is not the *structure*. The structure of this announcement is quite simple (so much so that the reader will need to read further to get an idea for what we mean by structure). It is the following Kripke structure K: we take one point, call it k, and we set $k \xrightarrow{D} k$ for all $D \in \{A, B, C\}$. We call $\langle K, k \rangle$ an *action structure*. Along with K, we also have a *precondition*; this will be ψ from (38.1). To deal with action structures with more than one point, the precondition will be a function PRE from worlds to sentences. In this case, the function PRE is just $\{\langle k, \psi \rangle\}$. The tuple $\langle K, k, \text{PRE} \rangle$ will be an example of what we call an *action*. This particular action is our model of the announcement α. Henceforth we use the symbol α to refer ambiguously to the pretheoretic notion of the announcement event and to our mathematical model of it.

Another example of an announcement to everyone is α'. Here we just change ψ from (38.1) to the sentence ψ' which says that both A and B know whether or not they are dirty. Yet another example is the *null announcement*. This models the announcement of a tautology *true* to everyone. We'll write this as τ.

β: a secure announcement to a set of agents. Next, suppose we have an announcement made to some possibly proper subset $\mathcal{B} \subseteq \mathcal{A}$ in the manner of Scenario 2. So there is some dispute as to what happened: the agents in $\mathcal{B}$ think that there was an announcement, while those out of $\mathcal{B}$ are sure that nothing happened. We formalize this with a Kripke structure of two points, l and t. We set $l \xrightarrow{D} l$ for all $D \in \mathcal{B}$, $l \xrightarrow{D} t$ for $D \notin \mathcal{B}$, and $t \xrightarrow{D} t$ for all D. The point is that l here is the actual announcement, and the agents in $\mathcal{B}$ know that this is the announcement. The agents not in $\mathcal{B}$ think that t is for sure the only possible action, and t in this model will behave just like the null announcement. The precondition function will be called PRE in all of our examples. Here PRE is given by $\text{PRE}(l) = \psi$ and $\text{PRE}(t) = \textit{true}$, where ψ is from (38.1). The action overall is $\langle L, l, \text{PRE}\rangle$, where $L = \{l, t\}$. We call this action β.

γ: an announcement with a suspicious outsider. This is based on Scenario 3. The associated structure has four points, as shown below:

$$\text{PRE}(m) = \psi \;\; \text{PRE}(n) = \textit{true} \;\; \text{PRE}(l) = \psi \;\; \text{PRE}(n) = \textit{true}$$

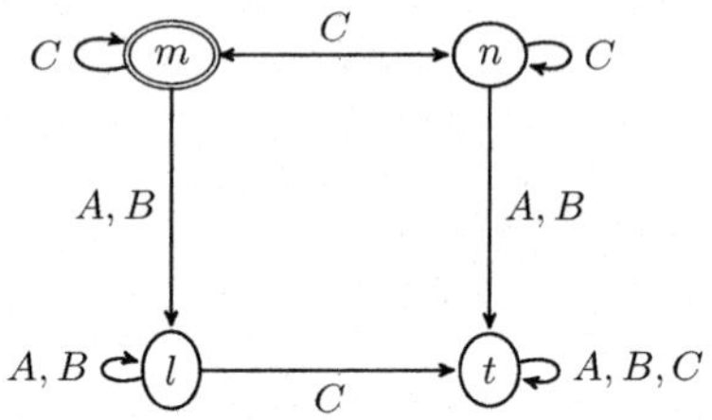

The idea is that m is the (private) announcement that C suspects, and n is other announcement that C thinks is possible (where nothing was communicated by A and B). Then if m happened, A and B were sure that what happened was l; similarly, if n happened, A and B would think that t was what happened. We call this action γ; technically it is $\langle\{m, n, l, t\}, m, \text{PRE}\rangle$. We get a different action, say γ' if we use the same model as above but change the designated ("real") world from m to n.

δ: an announcement with common knowledge of suspicion. Corresponding to Scenario 4, we have the following model. In it ψ_A denotes the sentence saying that A knows whether he is dirty but B does not, ψ_B the sentence saying that B knows whether she is dirty but A does not, and $\psi_\emptyset$ the sentence stating that neither knows.

$$\text{PRE}(o) = \psi \;\; \text{PRE}(p) = \psi_A \;\; \text{PRE}(q) = \psi_B \;\; \text{PRE}(r) = \psi_\emptyset \;\; \text{PRE}(s) = \textit{true}$$

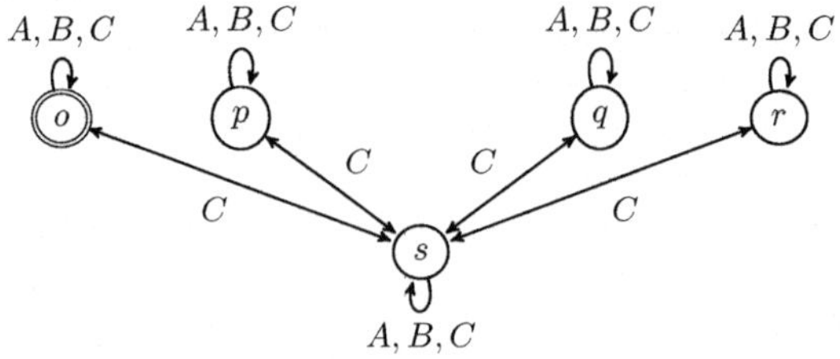

We call this action δ. There are five possible actions here, depending on whether it was ψ, ψ_A, ψ_B, $\psi_\emptyset$ or nothing which was announced. In each case, A and B are sure of what happened. Even if nothing actually happened (s), C would suspect one of the other four possibilities. In those, C still considers it possible that nothing happened.

Still to come. The reader is perhaps wondering what the actual connection is between the (formal) actions just introduced and the concrete models of the previous section. The connection is that there is a way of taking a model and an action and producing another model. When applied to the specific model U and the actions of this section, we get the models $V, \ldots, Z$. We delay this connection until section "Semantics" below, since it is high time that we introduce our language of epistemic actions and its semantics. The point is that there is a principled reason behind the models.

The question also arises as to whether there are any principles behind the particular actions which we put down in this section. As it happens, there is more which can be said on this matter. We postpone that discussion until section "More on Actions", after we have formally defined the syntax and semantics of our logical languages.

The Issues

The main issue we address in this paper is to formally represent *epistemic updates*, i.e., changes in the information states of agents in a distributed system. We think of these changes as being induced by specific information-updating actions, which can be of various types: (1) information-gathering and processing (e.g., realizing the possibility of other agents' hidden actions, and more generally, learning of any kind of new possibility via experiment, computation, or introspection); (2) information-exchange and communication (learning by sending/receiving messages, public announcements, secret interception of messages, etc.); (3) information-hiding (lying or other forms of deceiving actions, such as communication over secret channels, sending encrypted messages, holding secret suspicions); (4) information-loss and misinformation (being lied to, starting to have gratuitous suspicions, non-introspective learning, wrong computations or faulty observations, paranoia); (5) and more generally sequential or synchronous combinations of all of the above.

Special cases of our logic, dealing only with public or semi-public *announcements to mutually isolated groups*, have been considered in Plaza (1989), Gerbrandy (1999a,b), and Gerbrandy and Groeneveld (1997). These deal with actions such as α and β in our Introduction. Our examples γ and δ go beyond what is possible in the setting of these papers. But our overall setting is much more liberal setting, since it allows for all the above-mentioned types of actions. We feel it would be interesting to study further examples with an eye towards applications, but we leave this to other papers.

In our formal system, we capture only the *epistemic aspect* of these real actions, disregarding other (intentional) aspects. In particular, for simplicity reasons, we only deal with "'purely epistemic" actions; i.e., the ones that do not change the facts of the world, but affect only the agents' beliefs about the world. However, this is *not* an essential limitation, as our formal setting can be easily adapted to express fact-changing actions (see the end of section "More on Actions" and also section "Two Extensions").

On the semantical side, the main original technical contribution of our paper lies in our decision to represent not only the epistemic states, but also the epistemic actions, by Kripke structures. While for states, these structures represent in the usual way the uncertainty of each agent concerning the current state of the system, we similarly use action-structures to represent the uncertainty of each agent concerning the current action taking place. The intuition is that we are dealing with potentially "half-opaque/half-transparent" actions, about which the agents may be incompletely informed, or even completely misinformed. Besides the structure, actions have preconditions, defining their domain of applicability: not every action is possible in every state. We model the update of a state by an action as a partial *update* operation, given by a restricted product of the two structures: the uncertainties present in the given state and the given action are multiplied, while the "impossible" combinations of states and actions are eliminated (by testing the actions' preconditions on the state). The underlying intuition is that the agent's uncertainties concerning the state and the ones concerning the action are mutually independent, except for the consistency of the action with the state.

On the syntactical side, we use a mixture of dynamic and epistemic logic, with dynamic modalities associated to each action-structure, and with common-knowledge modalities for various groups of agents (in addition to the usual individual-knowledge operators). We give a complete and decidable axiomatization for this logic, and we prove various expressivity results. From a proof-theoretical point of view, the main originality of our system is the presence of our Action Rule, an inference rule capturing what might be called a notion of "epistemic (co)recursion". We understand this rule and our Knowledge-Action Axiom (a generalization of Ramsey's axiom to half-opaque actions) as expressing fundamental formal features of the interaction between action and knowledge in multi-agent systems, features that we think have not been formally expressed before.

Further Contents of This Paper

Section "A Logical Language with Epistemic Actions" gives our basic logic $\mathcal{L}([\alpha])$ of epistemic actions and knowledge. The idea is to define the logic together with the action structures which we have just looked at informally. So in $\mathcal{L}([\alpha])$ we finally will present the promised formal versions of the announcements of section "Epistemic Actions". In section "A Logic for $\mathcal{L}([\alpha])$" we present a sound and complete axiomatization of $\mathcal{L}([\alpha])$. We add the common knowledge operators to get $\mathcal{L}([\alpha], \Box^*)$ in section "A Logic for $\mathcal{L}([\alpha], \Box^*)$". Completeness for this logic is proved in section "Completeness for $\mathcal{L}([\alpha], \Box^*)$". Two results on the expressive power are presented in section "Results on Expressive Power". An Appendix contains some technical results which, while needed for our work, seem to interrupt the flow of the paper.

A Logical Language with Epistemic Actions

Syntax

We begin with a set AtSen of atomic sentences, and we define two sets simultaneously: the language $\mathcal{L}([\alpha])$, and a set of *actions* (*over* $\mathcal{L}([\alpha])$).

$\mathcal{L}([\alpha])$ is the smallest collection which includes AtSen and which is closed under $\neg$, $\wedge$, $\Box_A$ for $A \in \mathcal{A}$, and $[\alpha]\varphi$, where α is an action over $\mathcal{L}([\alpha])$, and $\varphi \in \mathcal{L}([\alpha])$.

An *action structure* (*over* $\mathcal{L}([\alpha])$) is a pair $\langle K, \text{PRE}\rangle$, where K is a finite Kripke frame over the set $\mathcal{A}$ of agents, and PRE is a map $\text{PRE} : K \to \mathcal{L}$. We will usually write K for the action structure $\langle K, \text{PRE}\rangle$. An *action (over* $\mathcal{L}([\alpha])$*)* is a tuple $\alpha = \langle K, k, \text{PRE}\rangle$, where $\langle K, \text{PRE}\rangle$ is an action structure over $\mathcal{L}([\alpha])$, and $k \in K$. Each action α thus is a finite set with relations $\xrightarrow{D}$ for $D \in \mathcal{A}$, together with a precondition function and a specified *actual* world.

The actions themselves constitute a Kripke frame Actions in the natural way, by setting

$$\langle K, k, \text{PRE}\rangle \xrightarrow{D} \langle L, l, \text{PRE}'\rangle \quad \text{iff} \quad K = L,\ \text{PRE} = \text{PRE}', \text{ and } k \xrightarrow{D} l \text{ in } K. \qquad (38.2)$$

When $\alpha = \langle K, k, \text{PRE}\rangle$, we set $\text{PRE}(\alpha) = \text{PRE}(k)$. That is, $\text{PRE}(\alpha)$ is the precondition associated to the distinguished world of the action. For this reason, we often write $\text{PRE}(\alpha)$ instead of $\text{PRE}(k)$.

Examples 1. All of the sentences mentioned in section "Models" are sentences of $\mathcal{L}([\alpha])$, except for the ones that use $\Box^*_{\{A,B,C\}}$. This construct gives us a more expressive language, as we shall see. The structures α, τ, β, γ, γ', δ, and δ' described informally in section "Epistemic Actions" are bona fide actions. As examples of the accessibility relation on the class of actions, we have the following facts: $\alpha \xrightarrow{D} \alpha$ and $\tau \xrightarrow{D} \tau$ for all $D \in \{A, B, C\}$; $\beta \xrightarrow{B} \beta$; $\beta \xrightarrow{C} \beta$; $\beta \xrightarrow{C} \tau$; $\gamma \xrightarrow{AB} \beta$; $\gamma, \gamma' \xrightarrow{C} \gamma, \gamma'$; $\gamma' \xrightarrow{AB} \tau$; $\delta \xrightarrow{AB} \delta$, $\delta' \xrightarrow{AB} \delta'$, and $\delta, \delta' \xrightarrow{C} \delta, \delta'$.

Many other types of examples are possible. We can represent misleading epistemic actions, e.g. lying, or more generally acting such that some people do not suspect that your action is possible. We can also represent gratuitous suspicion ("paranoia"): maybe no "real" action has taken place, except that some people start suspecting some action (e.g., some private communication) has taken place.

Semantics

As with the syntax, we define two things simultaneously: the semantic relation $\langle W, w\rangle \models \varphi$, and a partial operation $(\langle W, w\rangle, \alpha) \mapsto \langle W, w\rangle^\alpha$. Before this, we need another definition. Given a model W and an action structure K, we define the model W^K as follows:

1. The worlds of W^K are the pairs $(w, k) \in W \times K$ such that $\langle W, w\rangle \models \text{PRE}(k)$.
2. For such pairs,

$$(w, k) \xrightarrow{A} (w', k') \quad \text{iff} \quad w \xrightarrow{A} w' \text{ and } k \xrightarrow{A} k'. \tag{38.3}$$

3. We interpret the atomic sentences by setting $v_{W^K}((w, k)) = v_W(w)$. That is, p is true at (w, k) in W^K iff p is true at w in W.

Given an action $\alpha = \langle K, k\rangle$ and a model-world pair $\langle W, w\rangle$, we say that $\langle W, w\rangle^\alpha$ is defined iff $\langle W, w\rangle \models \text{PRE}(k)$, and in that case we set $\langle W, w\rangle^\alpha = \langle W, w\rangle^{\langle K,k\rangle} = \langle W^K, (w, k)\rangle$. One can now check that the following holds for these definitions.

$$\langle W, w\rangle^\alpha \xrightarrow{A} \langle W, x\rangle^\beta \quad \text{iff} \quad \langle W, w\rangle^\alpha \text{ and } \langle W, x\rangle^\beta \text{ are defined, } w \xrightarrow{A} x \text{ in } W, \text{ and } \alpha \xrightarrow{A} \beta.$$

The semantics is given by extending the usual clauses for modal logic by one for actions:

$$\langle W, w\rangle \models [\alpha]\varphi \quad \text{iff} \quad \langle W, w\rangle^\alpha \text{ is defined implies } \langle W, w\rangle^\alpha \models \varphi.$$

As is customary, we abbreviate $\neg[\alpha]\neg\varphi$ by $\langle\alpha\rangle\varphi$. Then we have

$$\langle W, w\rangle \models \langle\alpha\rangle\varphi \quad \text{iff} \quad \langle W, w\rangle^\alpha \text{is defined and } \langle W, w\rangle^\alpha \models \varphi.$$

We also abbreviate the boolean connectives classically, and we let *true* denote some tautology such as $p \vee \neg p$.

The larger language $\mathcal{L}([\alpha], \Box^*)$ We also consider a larger language $\mathcal{L}([\alpha], \Box^*)$. This is defined by adding operators $\Box^*_{\mathcal{B}}$ for all subsets $\mathcal{B} \subseteq \mathcal{A}$. (When we do this, of course we get more actions as well.) The semantics works by taking $\Box^*_{\mathcal{B}}\varphi$ to abbreviate the infinitary conjunction

$$\bigwedge_{\langle A_1,\ldots,A_n\rangle \in \mathcal{B}^*} \Box_{A_1} \cdots \Box_{A_n}\varphi.$$

Here $\mathcal{B}^*$ is the set of all sequences from $\mathcal{B}$. This includes the empty sequence, so $\Box^*_{\mathcal{B}}\varphi$ logically implies φ.

Bisimulation Given two models, say K and L, over the same set of $\mathcal{A}$ of agents, a *bisimulation* between K and L is a relation $R \subseteq K \times L$ such that if kRl and $A \in \mathcal{A}$, then:

1. For all atomic p, $\langle K, k\rangle \models p$ iff $\langle L, l\rangle \models p$.
2. For all $k \xrightarrow{A} k'$ there is some $l \xrightarrow{A} l'$ such that $k'Rl'$.
3. For all $l \xrightarrow{A} l'$ there is some $k \xrightarrow{A} k'$ such that $k'Rl'$.

Given two model-world pairs $\langle K, k\rangle$ and $\langle L, l\rangle$, we write $\langle K, k\rangle \equiv \langle L, l\rangle$ iff there is some bisimulation R such that kRl. It is a standard fact that if $\langle K, k\rangle \equiv \langle L, l\rangle$, then the two pairs agree on all sentences of standard modal logic. In our setting, we also can speak about actions being bisimilar: we change condition (1) above to refer to say that $\text{PRE}(k) = \text{PRE}(l)$. It is easy now to check two things simultaneously: (1) bisimilar pairs agree on all sentences of $\mathcal{L}([\alpha])$; and (2) if $\langle K, k\rangle \equiv \langle L, l\rangle$ and $\alpha \equiv \beta$, then $\langle K, k\rangle^{\alpha} \equiv \langle L, l\rangle^{\beta}$. Furthermore, these results extend to $\mathcal{L}([\alpha], \Box^*)$.

Examples 2. We look back at section "Models" for some examples. We use $\cong$ to denote the relation of isomorphism on model-world pairs. It is not hard to check the following: $\langle U, u_6\rangle^{\alpha} \cong \langle V, v_6\rangle$, $\langle U, u_6\rangle^{\beta} \cong \langle X, x_6\rangle$, $\langle U, u_6\rangle^{\gamma} \cong \langle Y, y_6\rangle$, and $\langle U, u_6\rangle^{\delta} \cong \langle Z, z_6\rangle$. For example, the isomorphism which shows that $\langle U, u_6\rangle^{\delta} \cong \langle Z, z_6\rangle$ is $(u_i, o) \mapsto z_i$ for $i \neq 2, 4$, $(u_2, q) \mapsto z_2$, $(u_4, p) \mapsto z_4$, and $(u_i, r) \mapsto z_i'$ for all i.

Let α' be the action of announcing to all agents that both A and B do know whether they are dirty. Then $\langle V, v_6\rangle^{\alpha'} \cong \langle X, x_6\rangle$. Moreover, $\langle Z, v_6\rangle^{\alpha'} \equiv \langle X, x_6\rangle$. Note that in this case we only have bisimilarity. However, we know that our languages will not discriminate between bisimilar pairs, so we can regard them as the same. This models our intuition that the epistemic states at the end of Scenarios 1.5 and 4.5 should be the same.

Finally, all of the semantic facts about the various models in section "Models" now turn into precise statements. For example, $\langle U, u_6\rangle \models [\alpha] \Diamond_A \Box_B \mathrm{D}_B$. Also, $\langle U, u_6\rangle \models [\alpha][\alpha']\Box^*_{A,B,C}\Box_C \mathrm{D}_C$. This formalizes our intuition that if we start with $\langle U, u_6\rangle$, first announce that each of A and B do not know their state, then second announce that they each do know it, then at that point it will be common knowledge to all three that C knows she is dirty.

More on Actions

In this section, we have a few remarks on actions. The point here is to clarify the relation between the scenarios of section "Introduction: Example Scenarios and Their Representations" and the intuitions concerning them, and the corresponding actions of section "Epistemic Actions".

First and foremost, here are the the conceptual points involved in our formalization. The idea is that epistemic actions present a lot of uncertainty. Indeed, what might be thought of as a single action (or event) is naturally interpreted by agents in different ways. The various agents might be unclear on what exactly happened, and again they might well have different interpretations on what is happening. Our formalization reflects this by making epistemic actions into Kripke models. So our use of possible-worlds modeling of actions is on a par with other uses of these models, and it inherits all of the features and bugs of those approaches.

Next, we want to spell out in words what our proposal amounts to. The basic problem is to decide how to represent what happens to a Kripke model W after an announcement α. (Of course, we are modeling α by an action in our formal sense.) Our solution begins by considering copies of W, one for each action token k of α in which $\text{PRE}(\alpha)$ holds. We can think of tagging the worlds of W with the worlds of α, and then we must give an account of the accessibility relation between them. The intuition is that the agents' relations to alternative worlds should be independent from their relations to other possibilities for α. So the accessibility relations of K and W should be combined *independently*. This is expressed formally in (38.3).

The auxiliary language $\hat{\mathcal{L}}$ has as atomic sentences all sentences φ of $\mathcal{L}([\alpha], \Box^*)$. It has all boolean connectives, standard modal operators $\Box_A$ for $A \in \mathcal{A}$, and also group knowledge operators $\Box^*_{\mathcal{B}}$ for $\mathcal{B} \subseteq \mathcal{A}$.

We interpret $\hat{\mathcal{L}}$ on actions using the standard clauses for the connectives and modal operators, and by interpreting the atomic sentences as follows $\langle K, k\rangle \models p$ iff $\text{PRE}(k) = p$.

Examples 3. The idea here is that the auxiliary language formalizes talk about what the different agents think is happening in our announcements. We refer back to the actions of section "Epistemic Actions". For example, $\alpha \models \Box^*_{\{A,B,C\}}\psi$. Intuitively, in α, it is common knowledge that ψ was announced. Another example: that

$$\delta \models \Box^*_{\{A,B,C\}} \Diamond_C (\psi \vee \psi_A \vee \psi_B).$$

That is, in δ, it is common knowledge that C thinks it is possible that some non-trivial announcement happened. Recall that this was one of our basic intuitions about δ, one which is not in general statable in our main language $\mathcal{L}([\alpha], \Box^*)$.

Definition. Let $\langle K, k\rangle$ be a model-world pair, and let φ be a sentence of $\hat{\mathcal{L}}$. Then χ *characterizes* $\langle K, k\rangle$ iff for all $\langle L, l\rangle$, $\langle L, l\rangle \models \chi$ iff $\langle L, l\rangle \equiv \langle K, k\rangle$.

Proposition 4. *Let $\langle K, k\rangle$ be a model-world pair with K finite. Then there is a sentence χ of $\hat{\mathcal{L}}$ which characterizes $\langle K, k\rangle$.*

Proof. By replacing $\langle K, k\rangle$ by its quotient under the largest auto-bisimulation, we may assume that if $l \neq m$, then $\langle K, l\rangle \not\equiv \langle K, m\rangle$. It is well-known that the relation of elementary equivalence in modal logic is a bisimulation on models in which each world has finitely many arrows coming in and out. It follows from this and the overall finiteness of K that we can find sentences φ_l for $l \in K$ with the property that for all l and m, $\langle K, m\rangle \models \varphi_l$ iff $m = l$. Let ψ be the following sentence

$$\psi \equiv \bigwedge_{l \in K, A \in \mathcal{A}} \left(\varphi_l \to \Box_A \bigvee_{l \xrightarrow{A} l'} \varphi_{l'} \wedge \bigwedge_{l \xrightarrow{A} l'} \Diamond_A \varphi_{l'} \right)$$

Going back to our original $\langle K, k \rangle$, let χ be $\varphi_k \wedge \Box^*_{\mathcal{A}} \psi$. It is easy to check that each $\langle K, l \rangle$ satisfies ψ; hence each satisfies $\Box^*_{\mathcal{A}} \psi$. Therefore $\langle K, k \rangle \models \chi$. We claim that χ characterizes $\langle K, k \rangle$. To see this, suppose that $\langle J, j \rangle \models \chi$. Consider the relation $R \subseteq K \times J$ given by

$$k' R j' \quad \text{iff} \quad \langle J, j' \rangle \models \varphi_{k'} \wedge \Box^*_{\mathcal{A}} \psi.$$

It is sufficient to see that R is a bisimulation. We'll verify half of this: suppose that $k' R j'$ and $j' \xrightarrow{A} j''$. By using ψ, we see that there is some k'' such that $k' \xrightarrow{A} k''$ and $j'' \models \varphi_{k''}$. And also, since $\models \Box^*_{\mathcal{A}} \psi \to \Box_A \Box^*_{\mathcal{A}} \psi$, we see that $\langle J, j'' \rangle \models \Box^*_{\mathcal{A}} \psi$. This completes the proof.

The connection of this result and our discussion of actions is that it is often difficult to go from an informal description of an an epistemic action to a formal one along our lines. (For example, our formulation of δ was the last of several versions.) Presumably, one way to get a formal action in our sense is to think carefully about which properties the action should have, express them in the auxiliary language, and then write a characterizing sentence such as ψ in the proof of Proposition 4. Then one can construct the finite model by standard methods. Although this would be a tedious process, it seems worthwhile to know that it is available.

Our formalization of actions reflects some choices which one might wish to modify. One of these choices is to take the range of the function PRE to be some language. Another option would be to have the range to be the *power set* of that language. This would make actions into Kripke models over the whole set of sentences. (And so what we have done is like considering modal logic with the restriction that at any world satisfies exactly one atomic sentence.) Taking this other option thus brings actions and models closer. This idea is pursued in Baltag (1999), a continuation of this work which develops a "calculus of epistemic actions." This replaces the "semantic" actions of this paper with *action expressions*. These expressions have nicer properties than the auxiliary language of this paper, but it would take us too far afield to discuss this further.

On a different matter, it makes sense to restrict attention from the full collection of actions as we have defined it to the smaller collection of *S5 actions*, where each accessibility $\xrightarrow{A}$ is an equivalence relation. This corresponds to the standard move of restricting attention to models with this property, and the reasons for doing this are similar. Intuitively, an S5 action is one in which every agent is introspective (with respect to their own suspicions about actions). Moreover, the introspection is accurate, and this fact is common knowledge.

A final modification which is quite natural is to allow actions which *change the world*. One would do this by adding to our notion of action a *sentential update u*. This would be a function defined on AtSen and written in terms of update equations such as $u(p) := p \wedge q$; $u(q) = \mathit{false}$, etc. We are confident that our logical systems can be modified to reflect this change, and we discuss this at certain points below. We decided not to make this change mostly in order to keep the basic notions as simple as possible.

With respect to both of the changes mentioned in the last two paragraphs, it is not hard to modify our logical work to get completeness results for the new systems. We discuss all of this in section "Two Extensions".

A Logic for $\mathcal{L}([\alpha])$

In Fig. 38.2 below we present a logic for $\mathcal{L}([\alpha], \Box^*)$ which we shall study later. In this section, we shall restrict the logic to the simpler language $\mathcal{L}([\alpha])$. We do so partly to break up the study of a system with many axioms and rules, and partly to emphasize the significance of adding the infinitary operators $\Box^*_{\mathcal{B}}$ to $\mathcal{L}([\alpha])$. To carry

Basic Axioms

All sentential validities

([α]-normality) $\vdash [\alpha](\varphi \to \psi) \to ([\alpha]\varphi \to [\alpha]\psi)$

($\Box_A$-normality) $\vdash \Box_A(\varphi \to \psi) \to (\Box_A\varphi \to \Box_A\psi)$

* ($\Box^*_{\mathcal{C}}$-normality) $\vdash \Box^*_{\mathcal{C}}(\varphi \to \psi) \to (\Box^*_{\mathcal{C}}\varphi \to \Box^*_{\mathcal{C}}\psi)$

Action Axioms

(Atomic Permanence) $\vdash [\alpha]p \leftrightarrow (\mathrm{PRE}(\alpha) \to p)$

(Partial Functionality) $\vdash [\alpha]\neg\chi \leftrightarrow (\mathrm{PRE}(\alpha) \to \neg[\alpha]\chi)$

(Action-Knowledge) $\vdash [\alpha]\Box_A\varphi \leftrightarrow (\mathrm{PRE}(\alpha) \to \bigwedge\{\Box_A[\beta]\varphi : \alpha \xrightarrow{A} \beta\})$

* **Mix Axiom** $\vdash \Box^*_{\mathcal{C}}\varphi \to \varphi \wedge \bigwedge\{\Box_A\Box^*_{\mathcal{C}}\varphi : A \in \mathcal{C}\}$

* **Composition Axiom** $\vdash [\alpha][\beta]\varphi \leftrightarrow [\alpha \circ \beta]\varphi$

Modal Rules

(Modus Ponens) From $\vdash \varphi$ and $\vdash \varphi \to \psi$, infer $\vdash \psi$

([α]-necessitation) From $\vdash \psi$, infer $\vdash [\alpha]\psi$

($\Box_A$-necessitation) From $\vdash \varphi$, infer $\vdash \Box_A\varphi$

* ($\Box^*_{\mathcal{C}}$-necessitation) From $\vdash \varphi$, infer $\vdash \Box^*_{\mathcal{C}}\varphi$

* **Action Rule**

Let ψ be a sentence, and let $\mathcal{C}$ be a set of agents. Let there be sentences χ_β for all β such that $\alpha \to^*_{\mathcal{C}} \beta$ (including α itself), and such that

1. $\vdash \chi_\beta \to [\beta]\psi$.
2. If $A \in \mathcal{C}$ and $\beta \xrightarrow{A} \gamma$, then $\vdash (\chi_\beta \wedge \mathrm{PRE}(\beta)) \to \Box_A\chi_\gamma$.

From these assumptions, infer $\vdash \chi_\alpha \to [\alpha]\Box^*_{\mathcal{C}}\psi$.

Fig. 38.2 The logical system for $\mathcal{L}([\alpha], \Box^*)$. For $\mathcal{L}([\alpha])$, we drop the $*$ axioms and rules out the restriction, we forget the axioms and rules of inference in Fig. 38.2 which are marked by a $*$. In particular $\alpha \circ \beta$ will be defined later (section "A Logic for $\mathcal{L}([\alpha], \Box^*)$").

The rules of the system are all quite standard from modal logic. The Action Axioms are the interesting new ones. In the Atomic Permanence axiom, p is an atomic sentence. The axiom then says that announcements do not change the brute fact of whether or not p holds. This axiom reflects the fact that our actions do not change any kind of local state. (We discuss an extension of our system in section "Two Extensions" where this axiom is not sound.) The Partial Functionality Axiom corresponds to the fact that the operation $\langle W, w \rangle \mapsto \langle W, w \rangle^\alpha$ is a partial function. The key axiom of the system is the Action-Knowledge Axiom, giving a criterion for knowledge after an announcement. We will check soundness of this axiom leaving checking soundness of other unstarred axioms and rules to the reader.

Proposition 1. *The Action-Knowledge Axiom*

$$[\alpha]\Box_A\varphi \leftrightarrow (\text{PRE}(\alpha) \to \bigwedge\{\Box_A[\beta]\varphi : \alpha \xrightarrow{A} \beta\})$$

is sound.

Proof. We remind the reader that the relevant definitions and notation are found in section "Semantics". Let α be the action $\langle K, k \rangle$. Fix a pair $\langle W, w \rangle$. If $\langle W, w \rangle \models \neg\text{PRE}(\alpha)$, then both sides of our biconditional hold. We therefore assume that $\langle W, w \rangle \models \text{PRE}(\alpha)$ in the rest of this proof. Assume that $\langle W, w \rangle^\alpha \models \Box_A\varphi$. Take some β such that $\alpha \xrightarrow{A} \beta$. This β is of the form $\langle K, k' \rangle$ for some k' such that $k \xrightarrow{A} k'$. Let $w \xrightarrow{A} w'$. We have two cases: $\langle W, w' \rangle \models \text{PRE}(k')$, and $\langle W, w' \rangle \models \neg\text{PRE}(k')$. In the latter case, $\langle W, w' \rangle \models [\beta]\varphi$ trivially. We'll show this in the former case, so assume $\langle W, w' \rangle \models \text{PRE}(k')$. Then (w', k') is a world of W^K, and indeed $(w, k) \xrightarrow{A} (w', k')$. Now our assumption that $\langle W, w \rangle^\alpha \models \Box_A\varphi$ implies that $\langle W^K, (w', k') \rangle \models \varphi$. This means that $\langle W, w' \rangle^\beta \models \varphi$. Hence $\langle W, w' \rangle \models [\beta]\varphi$. Since β and w' were arbitrary, $\langle W, w \rangle \models \bigwedge_\beta \Box_A[\beta]\varphi$.

The other direction is similar.

The rest of this section is devoted to the completeness result for $\mathcal{L}([\alpha])$. The reader not interested in this may omit the rest of this section, but at some points later we will refer back to the *term rewriting system* $\mathcal{R}$ which we shall describe shortly. Our completeness proof is based on a translation of $\mathcal{L}([\alpha])$ to ordinary modal logic $\mathcal{L}$. And this translation is based on a term rewriting system to be called $\mathcal{R}$.

The rewriting rules of $\mathcal{R}$ are:

$$\begin{array}{ll}
[\alpha]p & \rightsquigarrow \text{PRE}(\alpha) \to p \\
[\alpha]\neg\psi & \rightsquigarrow \text{PRE}(\alpha) \to \neg[\alpha]\psi \\
[\alpha](\psi \wedge \chi) & \rightsquigarrow [\alpha]\psi \wedge [\alpha]\chi \\
[\alpha]\Box_A\psi & \rightsquigarrow \text{PRE}(\alpha) \to \bigwedge\{\Box_A[\beta]\psi : \alpha \xrightarrow{A} \beta\}
\end{array}$$

As in all rewrite systems, we apply the rules of $\mathcal{R}$ at arbitrary subsentences of a given sentence. (For example, consider what happens with something like $[\alpha][\beta]\varphi$. We might rewrite $[\beta]\varphi$ using some rule, say to ψ. Then we might rewrite $[\alpha]\psi$ to something like $[\gamma]\psi$, etc.)

Lemma 2. *There is a relation $<$ on the sentences of $\mathcal{L}([\alpha])$ such that*

1. *$<$ is wellfounded.*
2. *For all rules $\varphi \rightsquigarrow \psi$ of $\mathcal{R}$, $\psi < \varphi$.*
3. *A sentence $\varphi \in \mathcal{L}([\alpha])$ is a normal form iff it is a modal sentence (that is, φ cannot be rewritten iff no actions occur in φ).*

This takes some work, and because the details are less important than the facts themselves, we have placed the entire matter in an Appendix to this paper. (The Appendix also discusses an extension of the rewrite system $\mathcal{R}$ to a system $\mathcal{R}^*$ for the larger language $\mathcal{L}([\alpha], \Box^*)$, so if you read it at this point you will need to keep this in mind.)

In the next result, we let $\mathcal{L}$ be ordinary modal logic over AtSen (where of course there are no actions).

Proposition 3. *There is a translation $t : \mathcal{L}([\alpha]) \to \mathcal{L}$ such that for all $\varphi \in \mathcal{L}([\alpha])$, φ is semantically equivalent to φ^t.*

Proof. Every sentence φ of $\mathcal{L}([\alpha])$ may be rewritten to a normal form. By Lemma 2, the normal forms of φ is a sentence in $\mathcal{L}$. We therefore set φ^t to be any normal form of φ, say the one obtained by carrying out leftmost reductions. The semantic equivalence follows from the fact that the rewrite rules themselves are sound, and from the fact that semantic equivalence is preserved by substitutions.

Lemma 4 (Substitution). *Let φ be any sentence, and let $\vdash \chi \leftrightarrow \chi'$. Suppose that $\varphi[p/\chi]$ comes from φ by replacing p by χ at some point, and $\varphi[p/\chi']$ comes similarly. Then $\vdash \varphi[p/\chi] \leftrightarrow \varphi[p/\chi']$.*

Proof. By induction on φ. The key point is that we have necessitation rules for each $[\alpha]$.

Theorem 5. *This logical system for $\mathcal{L}([\alpha])$ is strongly complete: $\Sigma \vdash \varphi$ iff $\Sigma \models \varphi$.*

Proof. The soundness half being easy, we only need to show that if $\Sigma \models \varphi$, then $\Sigma \vdash \varphi$. First, $\Sigma^t \models \varphi^t$. Since our system extends the standard complete proof system of modal logic, $\Sigma^t \vdash \varphi^t$. Now for each χ of $\mathcal{L}([\alpha])$, $\vdash \chi \leftrightarrow \chi^t$. (This is an easy induction on $<$ using Lemma 4.) As a result, $\Sigma \vdash \chi^t$ for all $\chi \in \Sigma$. So $\Sigma \vdash \varphi^t$. As we know $\vdash \varphi^t \leftrightarrow \varphi$. So we have our desired conclusion: $\Sigma \vdash \varphi$.

Strong completeness results of this kind may also be found in Plaza (1989) and in Gerbrandy and Groeneveld (1997). We discuss some of the history of the subject in section "Conclusions and Historical Remarks".

A Logic for $\mathcal{L}([\alpha], \Box^*)$

At this point, we turn to the completeness result for $\mathcal{L}([\alpha], \Box^*)$. It is easy to check that there is no hope of getting a *strong* completeness result (where one has arbitrary sets of hypotheses). The best one can hope for is weak completeness: $\vdash \varphi$ if and only if $\models \varphi$. Also, in contrast to our translations results for $\mathcal{L}([\alpha])$, the larger language $\mathcal{L}([\alpha], \Box^*)$ cannot be translated into $\mathcal{L}$ or even to $\mathcal{L}(\Box^*)$ (modal logic with extra modalities $\Box^*_{\mathcal{B}}$). We prove this in Theorem 2 below. So completeness results for $\mathcal{L}([\alpha], \Box^*)$ cannot simply be based on translation.

Our logical system is listed in Fig. 38.2 above. We discussed the fragment of the system which does not have the $*$ axioms and rules in section "A Logic for $\mathcal{L}([\alpha])$". The $\Box^*_{\mathcal{C}}$-normality Axiom and $\Box^*_{\mathcal{C}}$-necessitation Rule are standard, as is the Mix Axiom. We leave checking their soundness to the reader. The key features of the system are thus the Composition Axiom and the Action Rule. We begin with the Action Rule, restated below:

The Action Rule Let ψ be a sentence, and let $\mathcal{C}$ be a set of agents. Let there be sentences χ_β for all β such that $\alpha \to^*_{\mathcal{C}} \beta$ (including α itself), and such that

1. $\vdash \chi_\beta \to [\beta]\psi$.
2. If $A \in \mathcal{C}$ and $\beta \xrightarrow{A} \gamma$, then $\vdash (\chi_\beta \wedge \text{PRE}(\beta)) \to \Box_A \chi_\gamma$.

From these assumptions, infer $\vdash \chi_\alpha \to [\alpha]\Box^*_{\mathcal{C}}\psi$.

Remark. We use $\to^*_{\mathcal{C}}$ as an abbreviation for the reflexive and transitive closure of the relation $\bigcup_{A\in\mathcal{C}} \xrightarrow{A}$. Recall that there are only finitely many β such that $\alpha \to^*_{\mathcal{C}} \beta$, since each is determined by a world of the same finite Kripke frame that determines α. So even though the Action Rule might look like it takes infinitely many premises, it really only takes finitely many.

Another point: if one so desires, the Action Rule could be replaced by a (more complicated) axiom scheme which we will not state here.

Lemma 1. *$\langle W, w\rangle \models \langle\alpha\rangle \Diamond^*_{\mathcal{C}}\, \varphi$ iff there is a sequence of worlds from W*

$$w = w_0 \quad \to_{A_1} \quad w_1 \quad \to_{A_2} \quad \cdots \quad \to_{A_{k-1}} \quad w_{k-1} \quad \to_{A_k} \quad w_k$$

where $k \geq 0$, and also a sequence of actions of the same length k,

$$\alpha = \alpha_0 \quad \to_{A_1} \quad \alpha_1 \quad \to_{A_2} \quad \cdots \quad \to_{A_{k-1}} \quad \alpha_{k-1} \quad \to_{A_k} \quad \alpha_k$$

such that $A_i \in \mathcal{C}$ and $\langle W, w_i\rangle \models \text{PRE}(\alpha_i)$ for all $0 \leq i \leq k$, and $\langle W, w_k\rangle \models \langle\alpha_k\rangle\varphi$.

Remark. The case $k = 0$ just says that $\langle W, w\rangle \models \langle\alpha\rangle \Diamond^*_{\mathcal{C}}\, \varphi$ is implied by $\langle W, w\rangle \models \langle\alpha\rangle\varphi$.

Proof. $\langle W, w\rangle \models \langle\alpha\rangle \Diamond^*_{\mathcal{C}}\, \varphi$ iff $\langle W, w\rangle \models \text{PRE}(\alpha)$ and $\langle W^\alpha, (w, \alpha)\rangle \models \Diamond^*_{\mathcal{C}}\varphi$; iff $\langle W, w\rangle \models \text{PRE}(\alpha)$ and there is a sequence in W^α,

$$(w,\alpha) = v_0 \quad \rightarrow_{A_1} \quad v_1 \quad \rightarrow_{A_2} \quad \cdots \quad \rightarrow_{A_{k-1}} \quad v_{k-1} \quad \rightarrow_{A_k} \quad v_k$$

where $k \geq 0$ such that $A_i \in \mathcal{C}$ and $\langle W^\alpha, v_k\rangle \models \varphi$. Now suppose such sequences exist in W^α. Then we get a sequence of worlds w_i in W and actions α_i such that $v_i = (w_i, \alpha_i)$ and $\langle W, w_i\rangle \models \text{PRE}(\alpha_i)$. The condition that $\langle W^\alpha, v_k\rangle \models \varphi$ translates to $\langle W, w_k\rangle \models \langle\alpha_k\rangle\varphi$. Conversely, if we have a sequence in W with these properties, we get one in W^α by taking $v_i = (w_i, \alpha_i)$.

Proposition 2. *The Action Rule is sound.*

Proof. Assume that $\langle W, w\rangle \models \chi_\alpha$ but also $\langle W, w\rangle \models \langle\alpha\rangle \Diamond_{\mathcal{C}}^* \neg\psi$. According to Lemma 1, there is a labeled sequence of worlds from W

$$w = w_0 \quad \rightarrow_{A_1} \quad w_1 \quad \rightarrow_{A_2} \quad \cdots \quad \rightarrow_{A_{k-1}} \quad w_{k-1} \quad \rightarrow_{A_k} \quad w_k$$

where $k \geq 0$ and each $A_i \in \mathcal{C}$, and also a sequence of actions of length k, with the same labels,

$$\alpha = \alpha_0 \quad \rightarrow_{A_1} \quad \alpha_1 \quad \rightarrow_{A_2} \quad \cdots \quad \rightarrow_{A_{k-1}} \quad \alpha_{k-1} \quad \rightarrow_{A_k} \quad \alpha_k$$

such that $\langle W, w_i\rangle \models \text{PRE}(\alpha_i)$ for all $0 < i \leq k$, and $\langle W, w_k\rangle \models \langle\alpha_k\rangle\neg\psi$. If $k = 0$, we have $\langle W, w\rangle \models \langle\alpha\rangle\neg\psi$. But since $\vdash \chi_\alpha \rightarrow [\alpha]\psi$, we have $\langle W, w\rangle \models [\alpha]\psi$. This is a contradiction.

Now we argue the case $k > 0$. We show by induction on $1 \leq i \leq k$ that $\langle W, w_i\rangle \models \chi_{\alpha_i} \wedge [\alpha_i]\psi$. In particular, $\langle W, w_k\rangle \models [\alpha_k]\psi$. This is a contradiction.

We close with a discussion of the Composition Rule, beginning with a general definition.

Definition. Let $\alpha = \langle K, k\rangle$ and $\beta = \langle L, l\rangle$ be actions. Then the *action composition* $\alpha \circ \beta$ is the action defined as follows. Consider the product set $K \times L$. We turn this into a Kripke frame using the restriction of the product arrows. We get an action structure by setting

$$\text{PRE}((k', l')) \quad = \quad \text{PRE}(k') \wedge [\langle K, k'\rangle]\text{PRE}(l').$$

Finally, we set $\alpha \circ \beta = \langle K \times L, (k, l)\rangle$.

Proposition 3. *Concerning the composition operation:*

1. $(W^\alpha)^\beta \cong W^{\alpha\circ\beta}$ *via the restriction of* $((w,k'),l') \mapsto (w,(k',l'))$ *to* $(W^\alpha)^\beta$.
2. *The Composition Axiom is sound:* $[\alpha][\beta]\varphi \leftrightarrow [\alpha \circ \beta]\varphi$.
3. $\alpha \circ (\beta \circ \gamma) \cong (\alpha \circ \beta) \circ \gamma$.
4. $\alpha \circ \tau \cong \alpha \cong \tau \circ \alpha$, *where the null action* τ *is from section "Epistemic Actions".*

Proof. Let $\alpha = \langle K, k\rangle$ and $\beta = \langle L, l\rangle$. For (1), note that the worlds of $(W^\alpha)^\beta$ are of the form $((w,k'),l')$, where $(w,k') \in W^\alpha$ and $\langle W^\alpha, (w,k')\rangle \models \text{PRE}(l')$. For such

$((w,k'),l')$, $\langle W,w\rangle \models \text{PRE}(k')$ and $\langle W,w\rangle \models [\langle K,k'\rangle]\text{PRE}(l')$. That is, $(w,(k',l')) \in W^{\alpha\circ\beta}$. The converse is similar, and the rest of the isomorphism properties are easy.

Part (2) follows from (1). We use the obvious isomorphism $((k,l),m) \mapsto (k,(l,m))$ in part (3). We use the Composition and $[\alpha]$-necessitation axioms to show that this isomorphism preserves the PRE function up to logical equivalence. Part (4) is easy, using the fact that $\models [\tau]\varphi \leftrightarrow \varphi$.

Extending the rewriting system $\mathcal{R}$ to $\mathcal{L}([\alpha],\Box^*)$. We consider $\mathcal{L}([\alpha],\Box^*)$. The rewriting system $\mathcal{R}$ extends naturally to this larger language, taking new symbols for the operators $\Box^*_{\mathcal{B}}$. We also add a rule corresponding to the Composition Axiom: $[\alpha][\beta]\varphi \rightsquigarrow [\alpha\circ\beta]\varphi$. We call this rewriting system $\mathcal{R}^*$.

Lemma 4. *There is a relation $<$ on the sentences and actions of $\mathcal{L}([\alpha],\Box^*)$ such that*

1. *$<$ is wellfounded.*
2. *For all rules $\varphi \rightsquigarrow \psi$ of $\mathcal{R}^*$, $\psi < \varphi$.*
3. *If ψ is a proper subsentence of φ, then $\psi < \varphi$.*
4. *A sentence $\varphi \in \mathcal{L}([\alpha],\Box^*)$ is a normal form iff it is built from atomic sentences using $\neg$, $\wedge$, $\Box_A$, and $\Box^*_{\mathcal{B}}$, or if it is of the form $[\alpha]\Box^*_{\mathcal{B}}\psi$, where α is an action in normal form, and ψ too is in normal form.*
5. *An action α is a normal form if whenever $\alpha \xrightarrow{*} \beta$, PRE$(\beta)$ is a normal form sentence.*
6. *If $\alpha \xrightarrow{*} \beta$, then $[\alpha]\Box^*_{\mathcal{C}}\psi > [\beta]\psi$.*
7. *$nf(\varphi) \leq \varphi$.*

Once again, the details are in the Appendix.

In section "A Logic for $\mathcal{L}([\alpha])$", we saw a translation t from $\mathcal{L}([\alpha])$ to $\mathcal{L}$ can be extended to a translation from $\mathcal{L}([\alpha],\Box^*)$ to the infinitary language $\mathcal{L}_\infty$, where we have countable conjunctions and disjunctions. This extension is defined using Part (4) of Lemma 4. The additional clauses in the definition of t are

$$\begin{array}{ll}(\Box^*_{\mathcal{B}}\varphi)^t & = \bigwedge_{\langle A_1,\dots,A_n\rangle\in\mathcal{B}^*}(\Box_{A_1}\cdots\Box_{A_n}\varphi)^t \\ ([\alpha]\Box^*_{\mathcal{B}}\psi)^t & = \bigwedge_{\langle A_1,\dots,A_n\rangle\in\mathcal{B}^*}([\alpha]\Box_{A_1}\cdots\Box_{A_n}\psi)^t\end{array}$$

In this way, we see that $\mathcal{L}([\alpha],\Box^*)$ may be regarded as a fragment of infinitary modal logic.

Remark. It is possible to drop the Composition Axiom in favor of a more involved version of the Action Rule. The point is the Composition Axiom simplifies the normal forms of the $\mathcal{L}([\alpha],\Box^*)$: Without the Composition Axiom, the normal forms of sentences of $\mathcal{L}([\alpha],\Box^*)$ would be of the form $[\alpha_1][\alpha_2]\cdots[\alpha_r]\psi$, where each α_i is a normal form action and ψ is a normal form sentence. The Composition Axiom insures that the normal forms are of the form $[\alpha]\psi$. So if we were to drop the Composition Axiom, we would need a formulation of the Action Rule which involved *sequences* of actions. It is not terribly difficult to formulate such a rule, and completeness can be obtained by an elaboration of the work which we shall do. We did not present this work, mostly because adding the Composition Axiom leads to shorter proofs.

This completes the discussion of the axioms and rules of our logical system for $\mathcal{L}([\alpha], \Box^*)$.

Completeness for $\mathcal{L}([\alpha], \Box^*)$

In this section, we prove the completeness of the logical system for $\mathcal{L}([\alpha], \Box^*)$. Section "Some Syntactic Results" has some technical results which culminate in the Substitution Lemma 3. This is used in some of our work on normal forms in the Appendix, and that work figures in the completeness theorem of section "Completeness".

Some Syntactic Results

Lemma 1. *For all $A \in C$ and all β such that $\alpha \to_A \beta$,*

1. $\vdash [\alpha]\Box_C^*\psi \to [\alpha]\psi$.
2. $\vdash [\alpha]\Box_C^*\psi \wedge \text{PRE}(\alpha) \to \Box_A[\beta]\Box_C^*\psi$.

Proof. Part (1) follows easily from the Mix Axiom and modal reasoning. For part (2), we start with a consequence of the Mix Axiom: $\vdash \Box_C^*\psi \to \Box_A\Box_C^*\psi$. Then by modal reasoning, $\vdash [\alpha]\Box_C^*\psi \to [\alpha]\Box_A\Box_C^*\psi$. By the Action-Knowledge Axiom, we have $\vdash [\alpha]\Box_C^*\psi \wedge \text{PRE}(\alpha) \to \Box_A[\beta]\Box_C^*\psi$.

Definition. Let α and α' be actions. We write $\vdash \alpha \leftrightarrow \alpha'$ if α and α' are based on the same Kripke frame W and the same world w, and if for all $v \in W$, $\vdash \text{PRE}(v) \leftrightarrow \text{PRE}'(v)$, where PRE is the announcement function for α, and PRE' for α'.

We note the following bisimulation-like properties:

1. If $\vdash \alpha \leftrightarrow \alpha'$, then also $\vdash \text{PRE}(\alpha) \leftrightarrow \text{PRE}(\alpha')$.
2. Whenever β' is such that $\alpha' \to_C^* \beta'$, then there is some β such that $\vdash \beta \leftrightarrow \beta'$ and $\alpha \to_C^* \beta$.

These follow easily from the way we defined PRE on actions in terms of functions on frames.

Lemma 2. *If $\vdash \alpha \leftrightarrow \alpha'$, then for all ψ, $\vdash [\alpha]\psi \leftrightarrow [\alpha']\psi$.*

Proof. By induction on ψ. For ψ atomic, our result is easy. The induction steps for $\neg$ and $\wedge$ are trivial. The step for $\Box_A$ is not hard, and so we omit it. Assuming the result for ψ gives the result for $[\chi]\psi$, using the Composition Axiom and the induction hypothesis. This leaves the step for sentences of the form $\Box_B^*\psi$, assuming the result for ψ. We use the Action Rule to show that $\vdash [\alpha]\Box_C^*\psi \to [\alpha']\Box_C^*\psi$. For each β', we let $\chi_{\beta'}$ be $[\beta]\Box_C^*\psi$, where β is such that $\vdash \beta \leftrightarrow \beta'$. We need to show that for all relevant β' and γ',

(a) $\vdash [\beta]\Box_C^*\psi \to [\beta']\psi$; and
(b) If $\beta' \xrightarrow{A} \gamma'$, then $\vdash [\beta]\Box_C^*\psi \wedge \text{PRE}(\beta') \to \Box_A[\gamma]\Box_C^*\psi$.

For (a), we know from Lemma 1, part (1) that $\vdash [\beta]\Box_C^*\psi \to [\beta]\psi$. By induction hypothesis on ψ, $\vdash [\beta]\psi \leftrightarrow [\beta']\psi$. And this implies (a). For (b), Lemma 1, part (2) tells us that under the assumptions,

$$\vdash [\beta]\Box_C^*\psi \wedge \text{PRE}(\beta) \to \Box_A[\gamma]\Box_C^*\psi.$$

As we know, $\vdash \text{PRE}(\beta) \leftrightarrow \text{PRE}(\beta')$. This implies (b).

This completes the induction on ψ.

Lemma 3 (Substitution). *Let t be a sentence or action of $\mathcal{L}([\alpha], \Box^*)$, and let $\vdash \chi \leftrightarrow \chi'$. Suppose that $t[p/\chi]$ comes from t by replacing p by χ at some point, and $t[p/\chi']$ comes similarly. Then $\vdash t[p/\chi] \leftrightarrow t[p/\chi']$.*

Proof. By induction on t, using Lemma 2.

Lemma 4. *For every sentence $\varphi \in \mathcal{L}([\alpha], \Box^*)$ there is some normal form $nf(\varphi) \leq \varphi$ such that $\vdash \varphi \leftrightarrow nf(\varphi)$.*

Proof. Given φ, there is a finite sequence $\varphi_0 \rightsquigarrow \cdots \rightsquigarrow \varphi_n = \varphi'$ such that $\varphi_0 = \varphi$, and φ_n is in normal form. This is a consequence of the fact that $<$ is wellfounded and the rules of the system are reducing. By Lemma 3, we see that for all i, $\vdash \varphi_i \leftrightarrow \varphi_i'$.

Completeness

The proof of completeness and decidability is based on the filtration argument for completeness of PDL due to Kozen and Parikh (1981). We show that every consistent φ has a finite model, and that the size of the model is recursive in φ. We shall need to use some results concerning the rewriting system $\mathcal{R}^*$ from section "A Logic for $\mathcal{L}([\alpha], \Box^*)$".

Definition. Let $s(\varphi)$ be the set of subsentences of φ, including φ itself. This includes all sentences occurring in actions which occur in φ and their subsentences. For future use, we note that

$$s([\alpha]\Box_C^*\varphi) = \{[\alpha]\Box_C^*\varphi, \Box_C^*\varphi\} \cup s(\varphi) \cup \bigcup\{s(\text{PRE}(\beta)) : \alpha \to_C^* \beta\} \quad (38.4)$$

We define a function $f : \mathcal{L}([\alpha], \Box^*) \to \mathcal{P}(\mathcal{L}([\alpha], \Box^*))$ by recursion on the wellfounded relation $<$ as follows: For *normal forms*, f works as follows:

$$
\begin{array}{lcl}
f(p) & = & \{p\} \\
f(\neg\varphi) & = & f(\varphi) \cup \{\neg\varphi\} \\
f(\varphi \wedge \psi) & = & f(\varphi) \cup f(\psi) \cup \{\varphi \wedge \psi\} \\
f(\Box_A\varphi) & = & f(\varphi) \cup \{\Box_A\varphi\} \\
f(\Box^*_{\mathcal{B}}\varphi) & = & f(\varphi) \cup \{\Box^*_{\mathcal{B}}\varphi\} \cup \{\Box_A\Box^*_{\mathcal{B}}\varphi : A \in \mathcal{B}\} \\
f([\alpha]\Box^*_{\mathcal{C}}\varphi) & = & \{\Box_A[\beta]\Box^*_{\mathcal{C}}\varphi \ : \ \alpha \to^*_{\mathcal{C}} \beta \ \& \ A \in \mathcal{C}\} \\
& & \cup \{[\beta]\Box^*_{\mathcal{C}}\varphi \ : \ \alpha \to^*_{\mathcal{C}} \beta \ \& \ A \in \mathcal{C}\} \\
& & \cup \bigcup\{f(\chi) \ : \ (\exists\beta)\ \alpha \to^*_{\mathcal{C}} \beta \ \& \ \chi \in s(\text{PRE}(\beta))\} \\
& & \cup f(\Box^*_{\mathcal{C}}\varphi) \\
& & \cup \bigcup\{f([\beta]\varphi) \ : \ \alpha \to^*_{\mathcal{C}} \beta\}
\end{array}
$$

For φ not in normal form, let $f(\varphi) = f(nf(\varphi))$. (Note that we need to define f on sentences which are not normal forms, because $f([\beta]\psi)$ figures in $f([\alpha]\Box^*_{\mathcal{C}}\varphi)$. Also, the definition makes sense because the calls to f on the right-hand sides are all $<$ the arguments on the left-hand sides, and since $nf(\varphi) \leq \varphi$ for all φ; see Lemma 4.)

Lemma 5. *For all* φ*:*

1. $f(\varphi)$ *is a finite set of normal form sentences.*
2. $nf(\varphi) \in f(\varphi)$.
3. *If* $\psi \in f(\varphi)$*, then* $f(\psi) \subseteq f(\varphi)$.
4. *If* $\psi \in f(\varphi)$*, then* $s(\psi) \subseteq f(\varphi)$.
5. *If* $[\gamma]\Box^*_{\mathcal{C}}\chi \in f(\varphi)$, $\gamma \to^*_{\mathcal{C}} \delta$*, and* $A \in \mathcal{C}$*, then* $f(\varphi)$ *also contains* $\Box_A[\delta]\Box^*_{\mathcal{C}}\chi$, $[\delta]\Box^*_{\mathcal{C}}\chi$, PRE$(\delta)$*, and* $nf([\delta]\chi)$.

Proof. All of the parts are by induction on φ in the well-order $<$. For part (1), note that if $[\alpha]\Box^*_{\mathcal{C}}\psi$ is a normal form, then each sentence $\Box_A[\beta]\Box^*_{\mathcal{C}}\psi$ and all subsentences of this sentence are normal forms. For part (2), note that when φ is a normal form, $\varphi \in f(\varphi)$.

In part (3), we only need to consider φ in normal form. The result is immediate when φ is an atomic sentence p. The induction steps for $\neg$, $\wedge$, and $\Box_A$ are easy. For $\Box^*_{\mathcal{B}}\varphi$, note that since $\varphi < \Box^*_{\mathcal{B}}\varphi$, our induction hypothesis implies the result for φ; we verify it for $\Box^*_{\mathcal{B}}\varphi$. The only interesting case is when ψ is $\Box_A\Box^*_{\mathcal{B}}\varphi$ for some $A \in \mathcal{B}$. And in this case

$$
f(\psi) \quad = \quad f(\Box^*_{\mathcal{B}}\varphi) \cup \{\Box_A\Box^*_{\mathcal{B}}\varphi\} \quad \subseteq \quad f(\Box^*_{\mathcal{B}}\varphi).
$$

To complete part (3), we consider $[\alpha]\Box^*_{\mathcal{C}}\varphi$. If there is some $\chi < [\alpha]\Box^*_{\mathcal{C}}\varphi$ such that $\psi \in f(\chi)$ and $f(\chi) \subseteq f([\alpha]\Box^*_{\mathcal{C}}\varphi)$, then we are easily done by the induction hypothesis. This covers all of the cases except for $\psi = [\beta]\Box^*_{\mathcal{C}}\varphi$ and $\psi = \Box_A[\beta]\Box^*_{\mathcal{C}}\varphi$. For the first of these, we use the transitivity of $\to^*_{\mathcal{C}}$ to check that $f([\beta]\Box^*_{\mathcal{C}}\varphi) \subseteq f([\alpha]\Box^*_{\mathcal{C}}\varphi)$. And now the second case follows:

$$
f(\Box_A[\beta]\Box^*_{\mathcal{C}}\varphi) \quad = \quad f([\beta]\Box^*_{\mathcal{C}}\varphi) \cup \{\Box_A[\beta]\Box^*_{\mathcal{C}}\varphi\} \quad \subseteq \quad f([\alpha]\Box^*_{\mathcal{C}}\varphi).
$$

Part (4) is similar to part (3), using equation (38.4) at the beginning of this subsection.

For part (5), assume that $[\gamma]\Box^*_C\chi \in f(\varphi)$. By part (1), $[\gamma]\Box^*_C\chi$ is a normal form. We show that $\Box_A[\delta]\Box^*_C\chi$, $[\delta]\Box^*_C\chi$, $\mathrm{PRE}(\delta)$, and $nf([\delta]\chi)$ all belong to $f([\gamma]\Box^*_C\chi)$, and then use part (3). The first two of these sentences are immediate by the definition of f; the third one follows from part (4); and the last comes from part (2) since $nf([\delta]\chi) \in f([\delta]\chi) \subseteq f([\gamma]\Box^*_C\chi$.

The set $\Delta = \Delta(\varphi)$ Fix a sentence φ. We set $\Delta = f(\varphi)$ (i.e., we drop φ from the notation). This set Δ is the version for our logic of the Fischer-Ladner closure of φ. Let $\Delta = \{\psi_1, \ldots, \psi_n\}$. Given a maximal consistent set U of $\mathcal{L}([\alpha], \Box^*)$, let

$$[\![U]\!] \quad = \quad \pm\psi_1 \wedge \cdots \wedge \pm\psi_n,$$

where the signs are taken in accordance with membership in U. That is, if $\psi_i \in U$, then ψ is a conjunct of $[\![U]\!]$; but if $\psi_i \notin U$, then $\neg\psi_i$ is a conjunct.

Two (standard) observations are in order. Notice that if $[\![U]\!] \neq [\![V]\!]$, then $[\![U]\!] \wedge [\![V]\!]$ is inconsistent. Also, for all $\psi \in \Delta$,

$$\vdash \psi \leftrightarrow \bigvee\{[\![W]\!] : W \text{ is maximal consistent and } \psi \in W\}. \tag{38.5}$$

and

$$\vdash \neg\psi \leftrightarrow \bigvee\{[\![W]\!] : W \text{ is maximal consistent and } \neg\psi \in W\}. \tag{38.6}$$

(The reason is that ψ is equivalent to the disjunction of *all* complete conjunctions which contain it. However, some of those complete conjunctions are inconsistent and these can be dropped from the big disjunction. The others are consistent and hence can be extended to maximal consistent sets.)

Definition. The *filtration* $\mathcal{F}$ is the model whose worlds are the equivalence classes $[U]$, where U is a maximal consistent set in the logic for $\mathcal{L}([\alpha], \Box^*)$, and the equivalence relation is $U \equiv V$ iff $[\![U]\!] = [\![V]\!]$ (iff $U \cap \Delta = V \cap \Delta$). We set $\langle \mathcal{F}, [U]\rangle \models p$ iff $p \in U \cap \Delta$. Furthermore,

$$[U] \xrightarrow{A} [V] \text{ in } \mathcal{F} \quad \text{iff} \quad \text{whenever } \Box_A\psi \in U \cap \Delta, \text{ then also } \psi \in V\,. \tag{38.7}$$

This condition is independent of the choice of representatives: we use part (4) of Lemma 5 to see that if $\Box_A\chi \in \Delta$, then also $\chi \in \Delta$.

A *good path from* $[V_0]$ *for* $\langle\alpha\rangle \Diamond^*_C \psi$ is a path in $\mathcal{F}$

$$[V_0] \quad \to_{A_1} \quad [V_1] \quad \to_{A_2} \quad \cdots \quad \to_{A_{k-1}} \quad [V_{k-1}] \quad \to_{A_k} \quad [V_k]$$

such that $k \geq 0$, each $A_i \in \mathcal{C}$, and such that there exist actions

$$\alpha = \alpha_0 \quad \to_{A_1} \quad \alpha_1 \quad \to_{A_2} \quad \cdots \quad \to_{A_{k-1}} \quad \alpha_{k-1} \quad \to_{A_k} \quad \alpha_k$$

such that $\text{PRE}(\alpha_i) \in V_i$ for all $0 \leq i \leq k$, and $\langle \alpha_k \rangle \psi \in V_k$.

The idea behind a good path comes from considering Lemma 1 in $\mathcal{F}$. Of course, the special case of that result would require that $\langle \mathcal{F}, [V_i] \rangle \models \text{PRE}(\alpha_i)$ rather than $\text{PRE}(\alpha_i) \in V_i$, and similarly for $\langle \alpha_k \rangle \psi$ and V_k. The exact formulation above was made in order that the Truth Lemma will go through for sentences of the form $\langle \alpha \rangle \diamond^*_C \psi$ (see the final paragraphs of the proof of Lemma 8).

Lemma 6. *Let* $[\alpha]\Box^*_C \psi \in \Delta$. *If there is a good path from* $[V_0]$ *for* $\langle \alpha \rangle \diamond^*_C \neg\psi$, *then* $\langle \alpha \rangle \diamond^*_C \neg\psi \in V_0$.

Proof. By induction on the length k of the path. If $k = 0$, then $\langle \alpha \rangle \neg\psi \in V_0$. If $\langle \alpha \rangle \diamond^*_C \neg\psi \notin V_0$, then $[\alpha]\Box^*_C \psi \in V_0$. By Lemma 1, part (1), we have $[\alpha]\psi \in V_0$. This is a contradiction.

Assume the result for k, and suppose that there is a good path from $[V_0]$ for $\langle \alpha \rangle \diamond^*_C \neg\psi$ of length $k + 1$. Then there is a good path of length k from $[V_1]$ for $\langle \alpha_1 \rangle \diamond^*_C \neg\psi$. Also, $[\alpha_1]\Box^*_C \psi \in \Delta$, by Lemma 5, part (5). By induction hypothesis, $\langle \alpha_1 \rangle \diamond^*_C \neg\psi \in V_1$.

If $\langle \alpha \rangle \diamond^*_C \neg\psi \notin V_0$, then $[\alpha]\Box^*_C \psi \in V_0$. By Lemma 1, part (2), V_0 contains $[\alpha]\Box^*_C \psi \wedge \text{PRE}(\alpha) \to \Box_A[\alpha_1]\Box^*_C \psi$. So V_0 contains $\Box_A[\alpha_1]\Box^*_C \psi$. Again, this sentence belongs to Δ by Lemma 5, part (5). Now by definition of $\xrightarrow{A}$ in $\mathcal{F}$, we see that $[\alpha_1]\Box^*_C \psi \in V_1$. This is a contradiction.

Lemma 7. *If* $[\![V_0]\!] \wedge \langle \alpha \rangle \diamond^*_C \psi$ *is consistent, then there is a good path from* $[V_0]$ *for* $\langle \alpha \rangle \diamond^*_C \psi$.

Proof. For each β such that $\alpha \to^*_C \beta$, let S_β be the (finite) set of all $[W] \in \mathcal{F}$ such that there is *no* good path from $[W]$ for $\langle \beta \rangle \diamond^*_C \psi$. We need to see that $[V_0] \notin S_\alpha$; suppose toward a contradiction that $[V_0] \in S_\alpha$. Let

$$\chi_\beta \quad = \quad \bigvee \{[\![W]\!] : W \in S_\beta\}.$$

Note that $\neg\chi_\beta$ is logically equivalent to $\bigvee\{[\![W']\!] : [W'] \in \mathcal{F} \text{and} W' \notin S_\beta\}$. Since we assumed $V_0 \in S_\alpha$, we have $\vdash [\![V_0]\!] \to \chi_\alpha$.

We first claim that $\chi_\beta \wedge \langle \beta \rangle \psi$ is inconsistent. Otherwise, there would be $[W] \in S_\beta$ such that $\chi_\beta \wedge \langle \beta \rangle \psi \in W$. Note that by the Partial Functionality Axiom, $\vdash \langle \beta \rangle \psi \to \text{PRE}(\beta)$. But then the one-point path $[W]$ is a good path from $[W]$ for $\langle \beta \rangle \diamond^*_C \psi$. Thus $[W] \notin S_\beta$, and this is a contradiction. So indeed, $\chi_\beta \wedge \langle \beta \rangle \psi$ is inconsistent. Therefore, $\vdash \chi_\beta \to [\beta]\neg\psi$.

We will need the following standard claim, an argument for which can be found in Kozen and Parikh (1981). We will also use this claim in the proof of Lemma 8.

Claim. If $[\![U]\!] \wedge \diamond_A [\![V]\!]$ is consistent, then $[U] \to_A [V]$.

Proof of Claim. Assume $\Box_A\psi \in U \cap \Delta$. If $\psi \notin V$, then $\neg\psi \in V$, so since $\psi \in \Delta$, $\vdash [\![V]\!] \to \neg\psi$. Thus, $\vdash \Diamond_A[\![V]\!] \to \Diamond_A\neg\psi$, and so $\vdash [\![U]\!]\wedge\Diamond_A[\![V]\!] \to \Box_A\psi\wedge\Diamond_A\neg\psi$, whence $[\![U]\!] \wedge \Diamond_A[\![V]\!]$ is inconsistent. This contradiction establishes the claim.

We next show that for all $A \in \mathcal{C}$ and all β such that $\beta \to_A \gamma$, $\chi_\beta \wedge \text{PRE}(\beta) \wedge \Diamond_A\neg\chi_\gamma$ is inconsistent. Otherwise, there would be $[W] \in S_\beta$ with χ_β, $\text{PRE}(\beta)$, and $\Diamond_A\neg\chi_\gamma$ in it. Then $\bigvee\{\Diamond_A[\![W']\!] : W' \notin S_\gamma\}$, being equivalent to $\Diamond_A\neg\chi_\beta$, would belong to W. It follows that $\Diamond_A[\![W']\!] \in W$ for some $W' \notin S_\gamma$. By the claim, $[W] \to_A [W']$. Since $[W'] \notin S_\gamma$, there is a good path from $[W']$ for $\langle\gamma\rangle\Diamond^*_{\mathcal{C}}\psi$. But since $\beta \to_A \gamma$ and W contains $\text{PRE}(\beta)$, we also have a good path from $[W]$ for $\langle\beta\rangle\Diamond^*_{\mathcal{C}}\psi$. This again contradicts $[W] \in S_\beta$. As a result, for all relevant A, β, and γ, $\vdash \chi_\beta \wedge \text{PRE}(\beta) \to \Box_A\chi_\gamma$.

By the Action Rule, $\vdash \chi_\alpha \to [\alpha]\Box^*_{\mathcal{C}}\neg\psi$. Now $\vdash [\![V_0]\!] \to \chi_\alpha$. So $\vdash [\![V_0]\!] \to [\alpha]\Box^*_{\mathcal{C}}\neg\psi$. This contradicts the assumption with which we began this proof.

Lemma 8 (Truth Lemma). *Consider a sentence φ, and also the set $\Delta = f(\varphi)$. For all $\chi \in \Delta$ and $[U] \in \mathcal{F}$: $\chi \in U$ iff $\langle\mathcal{F}, [U]\rangle \models \chi$.*

Proof. We argue by induction on the wellfounded $<$ that if $\chi \in \Delta$, then: $\chi \in U$ iff $\langle\mathcal{F}, [U]\rangle \models \chi$. The case of χ atomic is trivial. Now assume this Truth Lemma for sentences $< \chi$. Note that by soundness, we may assume that χ is in normal form. We argue by cases on χ.

The cases that χ is either a negation or conjunction are trivial.

Suppose next that $\chi \equiv \Box_A\psi$. Suppose $\Box_A\psi \in U$; we show $\langle\mathcal{F}, [U]\rangle \models \Box_A\psi$. Let $[V]$ be such that $[U] \xrightarrow{A} [V]$. Then by definition of $\xrightarrow{A}$, $\psi \in V$. The induction hypothesis applies to ψ, since $\psi < \Box_A\psi$, and since $\psi \in \Delta$ by Lemma 5, part (4). So by induction hypothesis, $\langle\mathcal{F}, [V]\rangle \models \psi$. This gives half of our equivalence. Conversely, suppose that $\langle\mathcal{F}, [U]\rangle \models \Box_A\psi$. Suppose towards a contradiction that $\Diamond_A\neg\psi \in U$. So $[\![U]\!] \wedge \Diamond_A\neg\psi$ is consistent. We use Eq. (38.6) and the fact that $\Diamond_A$ distributes over disjunctions to see that $[\![U]\!] \wedge \Diamond_A\neg\psi$ is logically equivalent to $\bigvee([\![U]\!]\wedge\Diamond_A[\![V]\!])$, where the disjunction is taken over all V which contain $\neg\psi$. Since $[\![U]\!] \wedge \Diamond_A\neg\psi$ is consistent, one of the disjuncts $[\![U]\!] \wedge \Diamond_A[\![V]\!]$ must be consistent. The induction hypothesis again applies, and we use it to see that $\langle\mathcal{F}, [V]\rangle \models \neg\psi$. By the claim in the proof of Lemma 7, $[U] \xrightarrow{A} [V]$. We conclude that $\langle\mathcal{F}, [U]\rangle \models \Diamond_A\neg\psi$, and this is a contradiction.

For χ of the form $\Box^*_{\mathcal{C}}\psi$, we use the standard argument for PDL (see Kozen and Parikh 1981). This is based on lemmas that parallel Lemmas 6 and 7. The work is somewhat easier than what we do below for sentences of the form $[\alpha]\Box^*_{\mathcal{C}}\psi$, and so we omit these details.

We conclude with the case when χ is a normal form sentence of the form $[\alpha]\Box^*_{\mathcal{C}}\psi \in \Delta$. Assume that $[\alpha]\Box^*_{\mathcal{C}}\psi \in \Delta$. First, suppose that $[\alpha]\Box^*_{\mathcal{C}}\psi \notin U$. Then by Lemma 7, there is a good path from $[U]$ for $\langle\alpha\rangle\Diamond^*_{\mathcal{C}}\neg\psi$. We want to apply Lemma 1 in $\mathcal{F}$ to assert that $\langle\mathcal{F}, [U]\rangle \models \langle\alpha\rangle\Diamond^*_{\mathcal{C}}\neg\psi$. Let k be the length of the good path. For $i \leq k$, $\text{PRE}(\alpha_i) \in U_i$. Now each $\text{PRE}(\alpha_i)$ belongs to Δ by Lemma 5, part (5), and is $< [\alpha]\Box^*_{\mathcal{C}}\psi$. So by induction hypothesis, $\langle\mathcal{F}, [U_i]\rangle \models \text{PRE}(\alpha_i)$. We also need to check that $\langle\mathcal{F}, [U_k]\rangle \models \langle\alpha_k\rangle\neg\psi$. For this, recall from Lemma 5, part (5) that Δ contains

$nf([\alpha_k]\psi) \leq [\alpha_k]\psi$. By Lemma 4, $nf([\alpha_k]\psi) \leq [\alpha_k]\psi < [\alpha]\Box_C^*\psi$. Since the path is good, U_k contains $\langle\alpha_k\rangle\neg\psi$ and hence $\neg[\alpha_k]\psi$. It also must contain the normal form of this, by Lemma 4. So by induction hypothesis, $\langle\mathcal{F}, [U_k]\rangle \models nf(\neg[\alpha_k]\psi)$. By soundness, $\langle\mathcal{F}, [U_k]\rangle \models \langle\alpha_k\rangle\neg\psi$. Now it does follow from Lemma 1 that $\langle\mathcal{F}, [U]\rangle \models \langle\alpha\rangle \Diamond_C^* \neg\psi$.

Going the other way, suppose that $\langle\mathcal{F}, [U]\rangle \models \langle\alpha\rangle \Diamond_C^* \neg\psi$. By Lemma 1, we get a path in $\mathcal{F}$ witnessing this. The argument of the previous paragraph shows that this path is a good path from $[U]$ for $\langle\alpha\rangle \Diamond_C^* \neg\psi$. By Lemma 6, U contains $\langle\alpha\rangle \Diamond_C^* \neg\psi$. This completes the proof.

Theorem 9 (Completeness). *For all φ, $\vdash \varphi$ iff $\models \varphi$. Moreover, this relation is decidable.*

Proof. By Lemma 4, $\vdash \varphi \leftrightarrow nf(\varphi)$. Let φ be consistent. By the Truth Lemma, $nf(\varphi)$ holds at some world in the filtration $\mathcal{F}$. So $nf(\varphi)$ has a model; thus φ has one, too. This establishes completeness. For decidability, note that the size of the filtration is computable in the size of the original φ.

Two Extensions

We briefly mention two extensions of the Completeness Theorem 9. These extensions come from our discussion at the end of section "More on Actions".

First, consider the case of S5 actions. We change our logical system by restricting to these S5 actions, and we add the S5 axioms to our logical system. We interpret this new system on S5 models. It is easy to check that applying an S5 action to an S5 model gives another S5 model. Further, the S5 actions are closed under composition. Finally, if α is an S5 action and $\alpha \to_A \beta$, then β also is an S5 action. These easily imply the soundness of the new axioms. For completeness, we need only check that if we assume the S5 axioms, then the filtration $\mathcal{F}$ from the previous section has the property that each $\xrightarrow{A}$ is an equivalence relation. This is a standard exercise in modal logic (see, e.g., Fagin et al. (1995), Theorem 3.3.1).

Our second extension concerns the move from actions as we have been working them to actions which change the truth values of atomic sentences. If we make this move, then the axiom of Atomic Permanence is no longer sound. However, it is easy to formulate the relevant axioms. For example, if we have an action α which effects the change $p := p \wedge \neg q$, then we would take an axiom $[\alpha]p \leftrightarrow (\text{PRE}(\alpha) \to p \wedge \neg q)$. Having made these changes, all of the rest of the work we have done goes through. In this way, we get a completeness theorem for this logic.

Results on Expressive Power

In this section, we present two results which show that adding announcements to modal logic with $\Diamond^*$ adds expressive power as does adding private announcements to modal logic with $\Diamond^*$ and public announcements. To show these results it will be sufficient to take the set $\mathcal{A}$ of agents to be $\{A, B\}$ and consider only languages contained in a language built-up from the atomic sentences p and q, using $\Diamond_A$, $\Diamond_B$, $\Diamond^*_A$, $\Diamond^*_B$, and $\Diamond^*_{AB}$, and the actions $[\varphi]_A$, $[\varphi]_B$ of announcing φ to A or B privately, and $[\varphi]_{AB}$ the action of announcing φ to A and B publicly. Let $\mathcal{L}_{all}$ stand for this language. We use here the customary notation ($[\varphi]_A$, $[\varphi]_B$, $[\varphi]_{AB}$) for announcements, but $[\varphi]_A$ is simply the action with the Kripke structure $K = \{k\}$ with $\xrightarrow{A}$ from k to k and $\text{PRE}(k) = \varphi$. We think of $[\varphi]_B$ similarly. $[\varphi]_{AB}$ is the action with the Kripke structure $K = \{k\}$ with $\xrightarrow{A}$ and $\xrightarrow{B}$ going from k to k and $\text{PRE}(k) = \varphi$.

We need to define a *rank* $|\varphi|$ on sentences from $\mathcal{L}_{all}$. Let $|p| = 0$ for p atomic, $|\neg\varphi| = |\varphi|$, $|\varphi \wedge \psi| = \max(|\varphi|, |\psi|)$, $|\neg\varphi| = |\varphi|$, $|\Diamond_X \varphi| = 1 + |\varphi|$, for $X = A$ or $X = B$, $|\Diamond^*_X \varphi| = 1 + |\varphi|$ for $X = A$, $X = B$, or $X = AB$, and $|[\varphi]_X\psi| = \max(|\varphi|, |\psi|)$ for $X = A$, $X = B$, or $X = AB$.

First we present a lemma which allows us, in certain circumstances, to do the following: from the existence of a sentence in a language $\mathcal{L}_1$ which is not equivalent to any sentence in a language $\mathcal{L}_0$ infer that there exists a sentence in $\mathcal{L}_1$ not equivalent to any theory in $\mathcal{L}_0$.

Lemma 1. *Let $\mathcal{L}_0$ be a language included in $\mathcal{L}_{all}$, and let ψ be a sentence in $\mathcal{L}_{all}$. Assume that for each n we have models F_n and G_n with some worlds $f_n \in F_n$ and $g_n \in G_n$ such that $\langle F_n, f_n\rangle$ satisfies $\neg\psi$, $\langle G_n, g_n\rangle$ satisfies ψ, and $\langle F_n, f_n\rangle$ and $\langle G_n, g_n\rangle$ agree on all sentences in $\mathcal{L}_0$ of rank $\leq n$. Then $\Diamond_A\psi$ is not equivalent with any theory in $\mathcal{L}_0$.*

Proof. For a sequence of model-world pairs $\langle H_n, h_n\rangle$, $n \in D \subseteq \omega$, we let $\bigoplus_{n\in D}(H_n, h_n)$ be a model-world pair defined as follows. Let h be a new world. Take disjoint copies of the H_n's and add an A-arrow from h to each h_n. All other arrows are within the H_n's and stay the same as in H_n. No atomic sentences are true at h. Atomic sentences true in the worlds belonging to the copy of H_n in $\bigoplus_{n\in D}(H_n, h_n)$ are precisely those true at the corresponding worlds of H_n.

Let F be $\bigoplus_{n\in\omega}(F_n, f_n)$ with the new world denoted by f. Define also F^m, for $m \in \omega$, to be $\bigoplus_{n\in\omega}(H_n, h_n)$ with the new world f^m where $H_m = G_m$, $h_m = g_m$ and for all $n \neq m$, $H_n = F_n$ and $h_n = f_n$

Now assume towards a contradiction that $\Diamond_A\psi$ is equivalent with a theory Φ in $\mathcal{L}_0$. Clearly $\Diamond_A\psi$ fails in $\langle F, f\rangle$. Thus some sentence $\varphi \in \Phi$ fails in $\langle F, f\rangle$. On the other hand, each $\langle F^m, f^m\rangle$ satisfies $\Diamond_A\psi$, whence $\langle F^m, f^m\rangle$ satisfies φ. Let $m_0 = |\varphi|$. The following claim shows that both $\langle F, f\rangle$ and $\langle F^{m_0}, f^{m_0}\rangle$ make φ true or both of them make it false, which leads to a contradiction.

Claim. Let φ be a sentence in Φ of rank $\leq m$. Let H_n, K_n, $n \in D$, with $h_n \in H_n$ and $k_n \in K_n$ be models such that $\langle H_n, h_n\rangle$ and $\langle K_n, k_n\rangle$ agree on sentences in Φ of rank $\leq m$. Then $\langle\bigoplus_n(H_n, h_n), h\rangle$ and $\langle\bigoplus_n(K_n, k_n), k\rangle$ agree on φ.

This claim is proved by induction on complexity of φ. It is clear for atomic sentences. The induction steps for boolean connectives are trivial. A moment of thought gives the induction step for $\Diamond$ and $\Diamond^*$ with various subscripts. It remains to consider the case when $\varphi = [\varphi_1]_A\varphi_2$. (The cases when $\varphi = [\varphi_1]_B\varphi_2$ and $\varphi = [\varphi_1]_{AB}\varphi_2$ are similar.) Fix H_n, K_n, $h_n \in H_n$, $k_n \in K_n$, with $n \in D$, such that $\langle H_n, h_n\rangle$

and $\langle K_n, k_n\rangle$ agree on sentences in Φ of rank $\leq m$. Note that, for each $n \in D$, $\langle H_n, h_n\rangle \models \varphi_1$ if and only if $\langle K_n, k_n\rangle \models \varphi_1$. Let D_1 be the set of all $n \in D$ for which $\langle H_n, h_n\rangle \models \varphi_1$. Let H'_n and K'_n be models obtained by updating H_n and K_n by $[\varphi_1]_A$. By the definition of rank and the fact that $|\varphi_1| \leq m$, we have that $\langle H'_n, h_n\rangle$ and $\langle K'_n, k_n\rangle$ agree on sentences from Φ of rank $\leq m$. Therefore, by our inductive hypothesis

$$\langle \bigoplus_{n\in D_1} H'_n, h\rangle \models \varphi_2 \quad \text{iff} \quad \langle \bigoplus_{n\in D_1} K'_n, k\rangle \models \varphi_2.$$

However,

$$\langle \bigoplus_{n} K_n, k\rangle \models \varphi \quad \text{iff} \quad \langle \bigoplus_{n\in D_1} K'_n, k\rangle \models \varphi_2,$$

and

$$\langle \bigoplus_{n} H_n, h\rangle \models \varphi \quad \text{iff} \quad \langle \bigoplus_{n\in D_1} H'_n, h\rangle \models \varphi_2$$

and we are done.

Announcements Add Expressive Power to Modal Logic with $\Box^*$

In the result below, there will be only one agent A, and so we omit the letter A from the notation. We let $\mathcal{L}([\,], \Diamond^*)$ be modal logic with announcements (to this A) and $\Diamond^* = \Diamond^*_A$. We also let $\mathcal{L}(\Diamond^*)$ be the obvious sublanguage.

Theorem 2. *There is a sentence of $\mathcal{L}([\,], \Diamond^*)$ which cannot be expressed by any set of sentences of $\mathcal{L}(\Diamond^*)$.*

Proof. We show first that $[p]\,\Diamond^+ q = [p]\,\Diamond\,\Diamond^* q$ cannot be expressed by any single sentence of $\mathcal{L}(\Box^*)$. (Incidentally, the same holds for $[p]\,\Diamond^*\,q$.) Fix a natural number n. We define structures $\mathcal{A} = \mathcal{A}_n$ and $\mathcal{B} = \mathcal{B}_n$ as follows. First $\mathcal{B}$ has $2n + 3$ points arranged cyclically as

$$0 \to 1 \to \cdots \to n \to n+1 \to -n \to \cdots \to -1 \to 0.$$

For the atomic sentences, we set p true at all points except $n + 1$, and q true only at 0.

The structure $\mathcal{A}$ is a copy of $\mathcal{B}$ with n more points $\overline{1}, \ldots, \overline{n}$ arranged as

$$0 \to \overline{1} \to \cdots \to \overline{n} \to 0.$$

The shape of $\mathcal{A}$ is a figure-8. In both structures, every point is reachable from every point by the transitive closure of the $\to$ relation. At the points $\overline{i}$, p is true and q is false. Notice that $1 \models [p]\,\Diamond^+ q$ in $\mathcal{A}$, but $1 \not\models [p]\,\Diamond^+ q$ in $\mathcal{B}$.

The main technique in the proof is an adaptation of Fraisse-Ehrenfeucht games to the setting of modal logic. Here is a description of the relevant game $G_n(\langle U,u\rangle, \langle V,v\rangle)$. For $n = 0$, *II* immediately wins if the following holds: for all $p \in$ AtSen, $\langle U,u\rangle \models p$ iff $\langle V,v\rangle \models p$. And if u and v differ on some atomic sentence, i immediately wins. Continuing, here is how we define $G_{n+1}(\langle U,u\rangle, \langle V,v\rangle)$. As in the case of G_0, we first check if u and v differ on some atomic sentence. If they do, then i immediately wins. Otherwise, the play continues. Now i can make two types of moves.

1. $\Diamond$-move

i has a choice of playing from U or from V. If i chooses U, then i continues by choosing some u' such that $u \to u'$ in U. Then *II* replies with some $v' \in V$ such that $v \to v'$. Of course, if i had chosen in V, then *II* would have chosen in U. Either way, points u' and v' are determined, and the two players then play $G_n(\langle U,u'\rangle, \langle V,v'\rangle)$.

2. $\Diamond^*$-move

i plays by selecting U (or V, but we ignore this symmetric case below), and then playing some u' (say) reachable from u in the reflexive-transitive closure $\to^*$ of $\to$; *II* responds with a point in the other model, V, which is similarly related to v.

We write $\langle U,u\rangle \sim_n \langle V,v\rangle$ if *II* has a winning strategy in the game $G_n(\langle U,u\rangle, \langle V,v\rangle)$. It is easy to check that by induction on m that if $\langle U,u\rangle \sim_n \langle V,v\rangle$ and $m < n$, then $\langle U,u\rangle \sim_m \langle V,v\rangle$.

Claim 1. If $\langle U,u\rangle \sim_n \langle V,v\rangle$, then for all φ with $|\varphi| \leq n$, $\langle U,u\rangle \models \varphi$ iff $\langle V,v\rangle \models \varphi$.

The proof will be done by induction on φ. Let φ be atomic. Suppose $\langle U,u\rangle \sim_n \langle V,v\rangle$. Then since *II* has a winning strategy, the atomic sentences satisfied by u and v must be the same. So we are done in this case.

The induction steps for the boolean connectives are trivial. For $\Box\varphi$, suppose that $|\Box\varphi| \leq n$, $\langle U,u\rangle \sim_n \langle V,v\rangle$, and $\langle U,u\rangle \models \Box\varphi$. Suppose towards a contradiction that $\langle V,v\rangle \models \Diamond\neg\varphi$. Let v' be such that $v \to v'$ in V and $\langle V,v'\rangle \models \neg\varphi$. Let i begin a play of $G_{n-1}(\langle U,u\rangle, \langle V,v\rangle)$ by choosing to play $v' \in V$. Then *II*'s winning strategy responds with some u' such that $(U,u') \sim_{n-1} (V,v')$. Since $|\varphi| \leq n-1$, our induction hypothesis implies that $\langle U,u'\rangle \models \Diamond\neg\varphi$. This is a contradiction.

The argument for $\Box^*\varphi$ is similar and we leave it to the reader. The claim is proved.

We return to the models $\mathcal{A}$ and $\mathcal{B}$ described in the beginning of this proof. For $0 \leq i \leq n$, we let $S_i \subseteq \mathcal{A} \times \mathcal{B}$ be the following set

$$\begin{aligned} S_i \quad = \quad & \{(0,0),\ldots,(n,n),(n+1,n+1),(-n,-n),\ldots,(-1,-1)\} \\ & \cup \{(\overline{n},-1),(\overline{n-1},-2)\ldots,(\overline{2},-(n-1)),(\overline{1},-n)\} \\ & \cup \{(\overline{1},1),\ldots,(\overline{n-i},n-i)\} \end{aligned}$$

In the case of $i = n$, then the last disjunct is empty. Note that $S_0 \supset S_1 \supset \cdots \supset S_n$. Also, for $0 \leq i \leq n$, every point of one structure is related by S_i to some point of the other.

Claim 2. If $0 \leq i \leq n$ and $(a, b) \in S_i$, then $\langle \mathcal{A}, a \rangle \sim_i \langle \mathcal{B}, b \rangle$.

The proof is by induction on i. If $i = 0$, this is due to the fact that pairs in S_0 agree on the atomic formulas. Assume the statement for i, and that $i + 1 \leq n$. Let $(a, b) \in S_{i+1}$. We only need to show that *II* can respond to any play and have the resulting pair belong to S_i. Suppose first that i plays a $\Diamond$-move. Suppose also that $a = b$, so that (a, a) comes from the first subset of S_{i+1}. In this case, we only need to notice that $(a+1, a+1) \in S_i$ if $|a| \leq n$, $(-n, -n) \in S_i$ if $a = n + 1$, and $(\overline{1}, 1) \in S_i$ if $a = 0$, since $i < n$. The case of (a, b) from the second subset is similar. Finally, if $(\overline{a}, a)$ belongs to the third subset of S_{i+1}, then $a \leq n - (i + 1) = n - i - 1$. So $a + 1 \leq n - i$, and $(\overline{a+1}, a+1)$ belongs to the third subset of S_i. This tells *II* how to play.

We remarked above that each S_i is a total relation. Moreover, each world can be reached from any other one in $\mathcal{A}$ and in $\mathcal{B}$. This implies that if i makes a $\Diamond^*$-move, *II* can respond. This completes the proof of the claim.

It follows that $\langle \mathcal{A}_n, 1 \rangle \sim_n \langle \mathcal{B}_n, 1 \rangle$. So by Claim 1, for each sentence $\varphi \in \mathcal{L}(\Box^*)$ and all $n \geq |\varphi|$, $\langle \mathcal{A}_n, 1 \rangle \models \varphi$ iff $\langle \mathcal{B}_n, 1 \rangle \models \varphi$. This shows that $[p]\ \Diamond^+\ q$ cannot be expressed by a single sentence in $\mathcal{L}(\Diamond^*)$. To prove the stronger result as stated in Theorem 2, we only need to quote Lemma 1.

Private Announcements Add Expressive Power

In this section, $\mathcal{L}([\ ]_{AB}, \Diamond^*)$ denotes the set of sentences built from p using $[\varphi]_{AB}$, $\Diamond_A$, $\Diamond_B$, $\Diamond^*_A$, $\Diamond^*_B$, and $\Diamond^*_{AB}$. $\mathcal{L}([\ ]_A, \Diamond^*_A)$ denotes the set built from p using $[\varphi]_A$, $\Diamond^*_A$, and $\Diamond_B$.

Theorem 3. *There is a sentence of $\mathcal{L}([\ ]_A, \Diamond^*_A)$ which cannot be expressed by any set of sentences in $\mathcal{L}([\ ]_{AB}, \Diamond^*)$.*

Proof. We consider $\chi \equiv [p]_A\ \Diamond^*_A\ \Diamond_B \neg p$.

Let G_n be the following model. We begin with a cycle in $\xrightarrow{A}$:

$$a_1 \xrightarrow{A} a_\infty \xrightarrow{A} b \xrightarrow{A} a_n \xrightarrow{A} a_{n-1} \xrightarrow{A} \cdots \xrightarrow{A} a_2 \xrightarrow{A} a_1 \qquad (38.8)$$

We add edges $a_i \xrightarrow{A} b$ for all i (including $i = \infty$), and also $x \xrightarrow{A} a_\infty$ for all x (again including $x = a_\infty$). The only $\xrightarrow{B}$ edge is $a_1 \xrightarrow{B} b$. The atomic sentence p is true at all points except b.

The first thing to note is that after a *private* update of p to A, $\langle G_n, a_i \rangle \models \chi$ for all $i < \infty$. The relevant path is $a_i \xrightarrow{A} \cdots \xrightarrow{A} a_1 \xrightarrow{B} b$; the important point is that since the announcement was private, the edge $a_1 \xrightarrow{B} b$ survives the update. On the other hand, $\langle G_n, a_\infty \rangle \models \neg\chi$. This is because the only way to go from a_∞ to b is to go through b, and the edge $a_\infty \xrightarrow{A} b$ is lost in the update.

Suppose towards a contradiction that χ were equivalent to $\varphi \in \mathcal{L}([\ ]_{AB}, \Diamond^*)$. Let $i = |\varphi|$, and let $n = i + 1$. As we know from our discussion of χ, $\langle G_n, a_n \rangle \models \chi$ and $\langle G_n, a_\infty \rangle \models \neg\chi$. However, this contradicts the claim below.

Claim. Assume that $1 < j \leq n$, $\varphi \in \mathcal{L}([\ ]_{AB}, \diamond^*)$ and $|\varphi| < j$. Then $\langle G_n, a_j\rangle \models \varphi$ iff $\langle G_n, a_\infty\rangle \models \varphi$.

The proof is by induction on φ. For $\varphi = p$, the result is clear, as are the induction steps for $\neg$ and $\wedge$. For $\diamond_A\varphi$, suppose that $a_j \models \diamond_A\varphi$. Either $a_\infty \models \varphi$, in which case $a_\infty \models \diamond_A\varphi$, or else $a_{j-1} \models \varphi$. In the latter case, by induction hypothesis, $a_\infty \models \varphi$; whence $a_\infty \models \diamond_A\varphi$. The converse is similar.

The case of $\diamond_B\varphi$ is trivial: $a_j \models \neg \diamond_B \varphi$ and $a_\infty \models \neg \diamond_B \varphi$.

For $\diamond_A^*\varphi$, note that since we have a cycle (38.8) containing all points, the truth value of $\diamond_A^*\varphi$ does not depend on the point. The cases of $\diamond_B^*\varphi$ and $\diamond_{AB}^*\varphi$ are similar.

For $[\varphi]_{AB}\psi$, assume the result for φ and ψ, and let $|[\varphi]_{AB}\psi| < j$. Then also $|\varphi| < j$ and $|\psi| < j$. Let $H = \{x : x \models \varphi\}$ be the updated model, and recall that $\langle G_n, x\rangle \models [\varphi]_{AB}\psi$ iff $x \in H$ and $\langle H, x\rangle \models \psi$. We have two cases: First, $H = G_n$. Then $\langle G_n, x\rangle \models [\varphi]_{AB}\psi$ iff $\langle G_n, x\rangle \models \psi$. So we are done by the induction hypothesis.

The other case is when there is some $x \notin H$. If $a_k \notin H$ for some $k \geq j$ or for $k = \infty$, then *all* these a_k do not belong to H. In particular, neither a_j nor a_∞ belong. And so both a_j and a_∞ satisfy $[\varphi]_{AB}\psi$. If $b \notin H$, then H is bisimilar to a one-point model. This is because every $a_i \in H$ would have some $\xrightarrow{A}$-successor in H (e.g., a_∞), and there would be no $\xrightarrow{B}$ edges. So we assume $b \in H$. Thus $a_i \notin H$ for some $i < j$. Let k be least so that for $k \leq l \leq \infty$, $a_l \models \varphi$. Then $1 < k \leq j$. Let $A_{\geq k} = \{a_l : k \leq l \leq \infty\}$. The submodels generated by a_j and a_∞ contain the same worlds: all worlds in $A_{\geq k}$ and b. We claim that $(A_{\geq k} \times A_{\geq k}) \cup \{\langle b, b\rangle\}$ is a bisimulation on H. The verification here is easy.

So in H, a_j and a_∞ agree on all sentences in *any* language which is invariant for bisimulation. Now $\mathcal{L}([\ ]_{AB}, \diamond^*)$ has this property (as do all the languages which we study: they are translatable into infinitary modal logic). In particular, $\langle H, a_j\rangle \models \psi$ iff $\langle H, a_\infty\rangle \models \psi$. This concludes the claim.

We get Theorem 3 directly from the claim, the observation that $\langle G_n, a_n\rangle \models \chi$ and $\langle G_n, a_\infty\rangle \models \neg\chi$, and Lemma 1.

We feel that our two results on expressive power are just a sample of what could be done in this area. We did not investigate the next natural questions: Do announcements with suspicious outsiders extend the expressive power of modal logic with all secure private announcements and common knowledge operators? And then do announcements with common knowledge of suspicion add further expressive power?

Conclusions and Historical Remarks

The work of this paper builds on the long tradition of epistemic logic as well as technical results in other areas. In recent times, one very active arena for work on knowledge is distributed systems, and the main source of work in recent times on knowledge in distributed systems is the book **Reasoning About Knowledge** (Fagin et al. 1995) by Fagin, Halpern, Moses, and Vardi. We depart from Fagin et al. (1995) by introducing the new operators for epistemic actions, and by doing without temporal logic operators. In effect, our Kripke models are simpler, since they do not incorporate all of the runs of a system; the new operators can be viewed as a compensation for that. We have not made a detailed comparison of our work with the large body of work on knowledge on distributed systems, and such a comparison would require both technical and conceptual results. On the technical side, we suspect that neither framework is translatable into the other. One way to show this would be by expressivity results. Perhaps another way would use complexity results. In this direction, we note that Halpern and Vardi (1989) examines 96 logics of knowledge and time. Thirty-two of these contain common knowledge operators, and of these, all but twelve of these are undecidable. But overall, our logics are based on differing conceptual points and intended applications, and so we are confident that they differ.

As far as we know, the first paper to study the interaction of communication and knowledge in a formal setting is Plaza's paper "Logics of Public Communications" (Plaza 1989). As the title suggests, the epistemic actions studied are announcements to the whole group, as in our α and α'. Perhaps the main result of the paper is a completeness theorem for the logic of public announcements and knowledge. This result is closely related to a special case of our Theorem 5. The difference is that Plaza restricts attention to the case when all of the accessibility relations are equivalence relations. Incidentally, Plaza's proof involves a translation to multi-modal logic, just as ours does. In addition to this, Plaza (1989) contains a number of results special to the logic of announcements which we have not generalized, and it also studies an extension of the logic with non-rigid constants.

Other predecessors to this paper are the papers of Gerbrandy (1999a,b) and Gerbrandy and Groeneveld (1997). These study epistemic actions similar to our β, where an announcement is made to set of agents in a private way with no suspicions. They presented a logical system which included the common knowledge operators. An important result is that all of the reasoning in the original Muddy Children scenario can be carried out in their system. This shows that in order to get a formal treatment of the problem, one need not posit models which maintain histories. They did not obtain the completeness/decidability result for their system, but it would be the version of Theorem 9 restricted to actions which are compositions of private announcements. So it follows from our work that all of the reasoning in the Muddy Children can be carried out in a decidable system.

We should mention that the systems studied in Gerbrandy (1999a,b) and Gerbrandy and Groeneveld (1997) differ from ours in that they are variants of dynamic logic rather than propositional logic. That is, announcements are particular types of *programs* as opposed to modalities. This is a natural move, and although we have not followed it in this paper, we have carried out a study of expressive power issues of various fragments of a dynamic logic with announcement operators. We have shown, for example, that the dynamic logic formulations are more expressive than the purely propositional ones. Details on this will appear in a forthcoming paper.

Incidentally, the semantics in Gerbrandy (1999a,b) and Gerbrandy and Groeneveld (1997) use non-wellfounded sets. In other words, they work with models modulo bisimulation. The advantages of moving from these to arbitrary Kripke models are that the logic can be used by those who do not know about non-wellfounded sets, and also that completeness results are slightly stronger with a more general semantics. The relevant equivalence of the two semantics is the subject of the short note (Moss 1999).

The following are the new contributions of this paper:

1. We formulated a logical system with modalities corresponding to intuitive group-level epistemic actions. These actions include natural formalizations of announcements such as γ and δ, which allow various types of suspicion by outsiders. Our apparatus also permits us to study epistemic actions which apparently have not yet been considered in this line of work, such as actions in which nothing actually happens but one agent suspects that a secret communication took place.
2. We formulated a logical system with these modalities and with common knowledge operators for all groups. Building on the completeness of PDL and using a bit of term rewriting theory, we axiomatized the validities in our system.
3. We obtained some results on expressive power: in the presence of common knowledge operators, it is not possible to translate away public announcements, and in our framework, private announcements add expressive power to public ones.

Appendix: The Lexicographic Path Order

In this appendix, we give the details on the *lexicographic path ordering* (*LPO*), both in general and in connection with $\mathcal{L}([\alpha])$ and $\mathcal{L}([\alpha], \Box^*)$.

Fix some many-sorted signature Σ of terms. In order to define the LPO $<$ on the Σ-terms, we must first specify a well-order $<$ on the set of function symbols of Σ. The LPO determined by such choices is the smallest relation $<$ such that:

(LPO1) If $(t_1, \ldots, t_n) < (s_1, \ldots, s_n)$ in the lexicographic ordering on n-tuples, and if $t_j < f(s_1, \ldots, s_n)$ for $1 \leq j \leq n$, then $f(t_1, \ldots, t_n) < f(s_1, \ldots, s_n)$.
(LPO2) If $t \leq s_i$ for some i, then $t < f(s_1, \ldots, s_n)$.
(LPO3) If $g < f$ and $t_i < f(s_1, \ldots, s_n)$ for all $i \leq m$, then $g(t_1, \ldots, t_m) < f(s_1, \ldots, s_n)$.

Here is how this is applied in this paper. We shall take two sorts: *sentences* and *actions*. Our signature contains the usual sentence-forming operators p (for $p \in$ AtSen) $\neg$, $\wedge$, and $\Box_A$ for all $A \in \mathcal{A}$. Here each p is 0-ary, $\neg$ and $\Box_A$ are unary, and $\wedge$ is binary. We also have an operator *app* taking actions and sentences to sentences. We think of $app(\psi, \alpha)$ as merely a variation on $[\alpha]\psi$. (The order of arguments to *app* is significant.) We further have a binary operator $\circ$ on actions. (This is a departure from the treatment of this paper, since we used $\circ$ as a metalinguistic abbreviation instead of as a formal symbol. It will be convenient to make this change because this leads to a smoother treatment of the Composition Axiom.) Finally, for each finite Kripke frame K over $\mathcal{L}([\alpha])$ and each $1 \leq i \leq |K|$, we have a symbol F^i_K taking $|K|$ sentences and returning an action.

Each sentence φ has a formal version $\overline{\varphi}$ in this signature, and each action α also has a formal version $\overline{\alpha}$. These are defined by the recursion which is obvious except for the clauses

$$\begin{array}{ll}\overline{[\alpha]\varphi} & = app(\overline{\varphi}, \overline{\alpha}) \\ \overline{\alpha} & = \mathsf{F}^i_K(\overline{\text{PRE}(k_1)}, \ldots, \overline{\text{PRE}(k_n)})\end{array}$$

Here $\alpha = \langle K, k_i, \text{PRE}\rangle$ with $K = \{k_1, \ldots, k_n\}$ in some specified order. However, outside of the proof of Proposition 2 we shall not explicitly mention the formal versions at all, since they are harder to read than the standard notation.

We must also first fix a wellfounded relation $<$ on the function symbols. We set *app* to be greater than all other function symbols. In all other cases, distinct function symbols are unordered.

Theorem 1 (Kamin and Levy 1980; Dershowitz 1982). *Let $<$ be an LPO on Σ-terms.*

1. *$<$ is transitive.*
2. *$<$ has the* subterm property*: if t is a proper subterm of u, then $t < u$.*
3. *$<$ is* monotonic *(it has the* replacement property*): if $y < x_i$ for some i, then*

$$f(x_1, \ldots, y, \ldots x_n) < f(x_1, \ldots, x_i, \ldots x_n).$$

4. *$<$ is wellfounded.*
5. *Consider a term rewriting system every rule of which of the form $l \rightsquigarrow r$ with $r < l$. Then the system is* terminating*: there are no infinite sequences of rewritings.*

Proof. Here is a sketch for part (1): We check by induction on the construction of the least relation $<$ that if $s < t$, then for all u such that $t < u$, $s < u$. For this, we use induction on the term u. We omit the details. Further, (2) follows easily from (1) and (LPO2), and (3) from (LPO1), (1) and (2). Moreover, (5) follows easily from (4) and (3), since the latter implies that any replacement according to the rewrite system results in a smaller term in the order $<$.

Here is a proof of of the wellfoundedness property (4), taken from on Buchholz (1995). (We generalized it slightly from the one-sorted to the many-sorted setting and from the assumption that $<$ is a finite linear order on Σ to the assumption that $<$ is any wellfounded order.)

Let W be the set of terms t such that the order $<$ is wellfounded below t. W is then itself wellfounded under $<$. So for all n, W^n is wellfounded under the induced lexicographic order. We prove by induction on the given wellfounded relation on function symbols of Σ that for all n-ary f, $f[W^n] \subseteq W$. So assume that for $g < f$, say with arity m, $g[W^m] \subseteq W$. We check this for f by using induction on W^n. Fix $\vec{s} \in W^n$, and assume that whenever $\vec{u} < \vec{s}$ in W^n, that $f(\vec{u}) \in W$. We prove that $f(\vec{s}) \in W$ by checking that for all t such that $t < f(\vec{s})$, $t \in W$. And *this* is done by induction on the structure of t. If $t = f(\vec{u}) < f(\vec{s})$ via (LPO1), then $\vec{u} < \vec{s}$ lexicographically, and each

$u_i < f(\vec{s})$. This last point implies that $\vec{u} \in W^n$ by induction hypothesis on t, so $t \in W$ by induction hypothesis on W^n. If $t \leq s_i$ so that $t < f(\vec{s})$ via (LPO2), then $t \in W$ by definition of W. And if $t = g(u_1, \ldots, u_m) < f(\vec{s})$ via (LPO3), then $g < f$ and each $u_i < f(\vec{s})$. By induction hypothesis on t, each $u_i \in W$. So by induction hypothesis on f, $g(\vec{u}) \in W$.

Now that we know that each f takes tuples in W^n to elements of W, it follows by induction on terms that all terms belong to W.

For more on the LPO, its generalizations and extensions, see the surveys Dershowitz (1982) and Plaisted (1985).

Proposition 2. *Consider the LPO $<$ on $\mathcal{L}([\alpha], \Box^*)$ defined above.*

1. *If $\alpha \xrightarrow{*} \beta$, then* $\text{PRE}(\beta) < \alpha$.
2. *If $\alpha \xrightarrow{*} \beta$, then* $[\beta]\psi < [\alpha]\Box^*_C\psi$.
3. $\text{PRE}(\alpha) \to p < [\alpha]p$.
4. $\text{PRE}(\alpha) \to \neg[\alpha]\psi < [\alpha]\neg\psi$.
5. $[\alpha]\psi \wedge [\alpha]\chi < [\alpha](\psi \wedge \chi)$.
6. $\text{PRE}(\alpha) \to \bigwedge\{\Box_A[\beta]\psi : \alpha \xrightarrow{A} \beta\} < [\alpha]\Box_A\psi$.
7. $[\alpha \circ \beta]\varphi < [\alpha][\beta]\psi$.

In particular, for all rules $\varphi \rightsquigarrow \psi$ of the rewriting system $\mathcal{R}^*$, $\psi < \varphi$.

Proof. Part (1) holds because we regard α as a term $\alpha = \mathsf{F}^i_K(\overline{\gamma_1}, \ldots, \overline{\gamma_n})$, for some frame K and i. So whenever $\alpha \xrightarrow{*} \beta$, each $\text{PRE}(\beta)$ is a proper subterm of α.

Here is the argument for part (2): We need to see that $app(\psi, \overline{\beta}) < app(\Box^*_C\psi, \overline{\alpha})$. Now lexicographically, $(\psi, \overline{\beta}) < (\Box^*_C\psi, \overline{\alpha})$. So we only need to know that $\overline{\beta} < app(\Box^*_C\psi, \overline{\alpha})$. Let $\alpha = \mathsf{F}^i_K(\overline{\gamma_1}, \ldots, \overline{\gamma_n})$. Now according to Eq. (38.2) in section "Syntax", $\overline{\beta}$ is $\mathsf{F}^j_K(\overline{\gamma_1}, \ldots, \overline{\gamma_n})$, for the same K and $\gamma_1, \ldots, \gamma_n$ but perhaps for $j \neq i$. Then it is clear by (LPO2) that $\overline{\gamma_i} < app(\Box^*_C\psi, \overline{\alpha})$ for all i. So by (LPO3), $\overline{\beta} < app(\Box^*_C\psi, \overline{\alpha})$.

The remaining parts are similar.

A *normal form* in a rewriting system is a sentence which cannot be rewritten in the system. Of course, we are interested in the systems $\mathcal{R}$ and $\mathcal{R}^*$ from sections "A Logic for $\mathcal{L}([\alpha])$" and "A Logic for $\mathcal{L}([\alpha], \Box^*)$", respectively. It follows from the wellfoundedness of $<$ that for every φ there is a normal form $nf(\varphi) \leq \varphi$ obtained by rewriting φ in some arbitrary fashion until a normal form is reached.

Lemma 3. *A sentence $\varphi \in \mathcal{L}([\alpha])$ is a normal form of $\mathcal{R}^*$ iff φ is a modal sentence (i.e., iff φ contains no actions). Moreover, the rule $[\alpha][\beta]\varphi \rightsquigarrow [\alpha \circ \beta]\varphi$ is not needed to reduce φ to normal form. So for $\mathcal{L}([\alpha])$, $\mathcal{R}$ has the same normal forms as $\mathcal{R}^*$.*

A sentence $\varphi \in \mathcal{L}([\alpha], \Box^*)$ is a normal form of $\mathcal{R}^*$ iff φ is built from atomic sentences using $\neg$, $\wedge$, $\Box_A$, and $\Box^*_{\mathcal{B}}$, or if φ is of the form $[\alpha]\Box^*_{\mathcal{B}}\psi$, where α is a normal form action, and ψ is a normal form. An action α is a normal form if whenever $\alpha \xrightarrow{*} \beta$, then $\text{PRE}(\beta)$ *is a normal form.*

Proof. It is immediate that every modal sentence is a normal form in $\mathcal{L}([\alpha])$, that every $[\alpha]\Box_C^*\varphi$ is a normal form in $\mathcal{L}([\alpha], \Box^*)$, and that if each $\text{PRE}(\beta)$, with $\alpha \xrightarrow{*} \beta$, is a normal form, then α is a normal form action. Going the other way, we check that if $\varphi \in \mathcal{L}([\alpha])$, $[\alpha]\varphi$ is *not* a normal form. So we see by an easy induction that the normal forms of $\mathcal{L}([\alpha])$ are exactly the modal sentences. We also argue by induction for $\mathcal{L}([\alpha], \Box^*)$, and we note that every $[\alpha][\beta]\varphi$ is *not* a normal form, using the rule $[\alpha][\beta]\varphi \rightsquigarrow [\alpha \circ \beta]\varphi$.

One fine point concerning $\mathcal{R}$ and our work in section "A Logic for $\mathcal{L}([\alpha])$" is that to reduce sentences of $\mathcal{L}([\alpha])$ to normal form we may restrict ourselves to rewriting sentences which are not subterms of actions. This simplification accounts for the differences between parallel results of sections "A Logic for $\mathcal{L}([\alpha])$" and "A Logic for $\mathcal{L}([\alpha], \Box^*)$".

Acknowledgements We thank Jelle Gerbrandy and Rohit Parikh for useful conversations on this work. An earlier version of this paper was presented at the 1998 Conference on Theoretical Aspects of Rationality and Knowledge.

References

Baltag, A. (1999). *A logic of epistemic actions, ms.* Amsterdam: CWI.

Buchholz, W. (1995). Proof-theoretic analysis of termination proofs. *Annals of Pure and Applied Logic, 75*, 57–65.

Dershowitz, N. (1982). Orderings for term-rewriting systems. *Theoretical Computer Science, 17*, 279–301.

Fagin, R., Halpern, J. Y., Moses, Y., & Vardi, M. Y. (1995). *Reasoning about knowledge*. Cambridge: MIT.

Gerbrandy, J. (1999a). Dynamic epistemic logic. In L. S. Moss, et al. (Eds.), *Logic, language, and information* (Vol. 2). Stanford: CSLI Publications, Stanford University.

Gerbrandy, J. (1999b). *Bisimulations on planet Kripke*. Ph.D. dissertation, University of Amsterdam.

Gerbrandy, J., & Groeneveld, W. (1997). Reasoning about information change. *Journal of Logic, Language, and Information, 6*, 147–169.

Halpern, J. Y., & Vardi, M. Y. (1989). The complexity of reasoning about knowledge and time. I. Lower bounds. *Journal of Computer and System Sciences, 38*(1), 195–237.

Kamin, S., & Levy, J. J. (1980). *Two generalizations of the recursive path orderings*. Unpublished note, Department of Computer Science, University of Illinois, Urbana.

Kozen, D., & Parikh, R. (1981). An elementary proof of the completeness of PDL. *Theoretical Computer Science, 14*, 113–118.

Moss, L. S. (1999). From hypersets to Kripke models in logics of announcements. In J. Gerbrandy, et al. (Eds.), *JFAK. Essays dedicated to Johan van Benthem on the occasion of his 50th birthday*. Amsterdam: Vossiuspers, Amsterdam University Press.

Plaisted, D. A. (1985). Termination orderings. In D. Gabbay, et al. (Eds.), *Handbook of logic in artificial intelligence and logic programming* (Vol. I, pp. 273–364).

Plaza, J. (1989). Logics of public communications. In *Proceedings of the 4th international symposium on methodologies for intelligent systems*, Charlotte.

Chapter 39
A Qualitative Theory of Dynamic Interactive Belief Revision

Alexandru Baltag and Sonja Smets

Introduction

This paper contributes to the recent and on-going work in the logical community (Aucher 2003; Baltag and Sadrzadeh 2006; Baltag and Smets 2006a,b,c; van Benthem 2007; van Ditmarsch 2005) on dealing with mechanisms for belief revision and update within the Dynamic-Epistemic Logic (DEL) paradigm. DEL originates in the work of Gerbrandy and Groeneveld (1997) and Gerbrandy (1999), anticipated by Plaza in (1989), and further developed by numerous authors Baltag et al. (1998), Gerbrandy (1999), van Ditmarsch (2000, 2002), Baltag (2002), Kooi (2003), Baltag and Moss (2004), and van Benthem et al. (2006a,b) etc. In its standard incarnation, as presented e.g., in the recent textbook by van Ditmarsch et al. (2007), the DEL approach is particularly well fit to deal with *complex multi-agent learning actions* by which groups of interactive agents update their beliefs (including *higher-level beliefs* about the others' beliefs), *as long as the newly received information is consistent with the agents' prior beliefs*. On the other hand, the classical AGM theory and its more recent extensions have been very successful in dealing with the problem of *revising one-agent, first-level (factual) beliefs when they are contradicted by new information*. So it is natural to look for a way to combine these approaches.

A. Baltag (✉) • S. Smets
ILLC, University of Amsterdam, The Netherlands
e-mail: thealexandrubaltag@gmail.com; S.J.L.Smets@uva.nl

H. Arló-Costa et al. (eds.), *Readings in Formal Epistemology*, Springer Graduate Texts in Philosophy 1, DOI 10.1007/978-3-319-20451-2_39

We develop here a notion of *doxastic actions*,[1] general enough to cover most examples of multi-agent communication actions encountered in the literature, but also flexible enough to deal with *(both static and dynamic) belief revision*, and in particular to *implement various "belief-revision policies" in a unified setting*. Our approach can be seen as a natural extension of the work in Baltag and Moss (2004) and Baltag et al. (1998) on "epistemic actions", incorporating ideas from the AGM theory along the lines pioneered in Aucher (2003) and van Ditmarsch (2005), but using a *qualitative* approach based on *conditional beliefs*, in the line of Stalnaker (1968), Bonanno (2005), Board (2002), and van Benthem (2007).

Our paper assumes the general distinction, made in Baltag and Smets (2006a); van Benthem (2007); van Ditmarsch (2005), between *"dynamic" and "static" belief revision*. It is usually acknowledged that the classical AGM theory in Alchourrón et al. (1985) and Gärdenfors (and embodied in our setting by the *conditional belief* operators $B_a^P Q$) is indeed "static", in the sense that it captures *the agent's changing beliefs about an unchanging world*. But in fact, when we take into account all the higher-level beliefs, the "world" (that these higher-level beliefs are about) includes all agent's (real) beliefs.[2] Thus, such a world is *always changed by our changes of beliefs*! So we can better understand a belief conditional on P as capturing the agent's beliefs *after revising with P* about the state of the world *before the revision*: the statement $B_a^P Q$ says that, *if agent a would learn P, then she would come to believe that Q was the case (before the learning)*. In contrast, "dynamic" belief revision uses dynamic modalities to capture the agent's revised beliefs about the world *as it is after revision*: $[!P]B_a Q$ says that *after learning P, agent a would come to believe that Q is the case (in the world after the learning)*. The standard alternative (Katsuno and Mendelzon 1992) to the AGM theory calls this *belief update*, but like the AGM approach, it only deals with "first-level" (factual) beliefs from a non-modal perspective, neglecting any higher-order "beliefs about beliefs". As a result, *it completely misses the changes induced* (in our own or the other agents' epistemic-doxastic states) *by the learning actions themselves* (e.g., the learning of a Moore sentence, see the third section on ""Dynamic" Belief Revision"). This is reflected in the acceptance in Katsuno and Mendelzon (1992) of the AGM "Success Axiom": in dynamic notation, this is the axiom $[!P]B_a P$ (which cannot accommodate Moore sentences). Instead, Katsuno and Mendelzon (1992) exclusively concentrate on the possible changes of (ontic) facts that may have occurred during our learning (but *not due to our learning*). In contrast, our approach to belief update (following the DEL tradition) may be thought of as "dual" to the one in Katsuno and Mendelzon

[1] Or "doxastic events", in the terminology of van Benthem (2007).

[2] To verify that a higher-level belief about another belief is "true" we need to check the content of that higher-level belief (i.e., the existence of the second, lower-level belief) against the "real world". So the real world has to include the agent's beliefs.

(1992): we completely neglect here the ontic changes,[3] considering only the changes induced by *"purely doxastic" actions* (learning by observation, communication, etc.).

Our formalism for "static" revision can best be understood as a modal-logic implementation of the well-known view of belief revision in terms of *conditional reasoning* (Stalnaker 1968, 2006). In Baltag and Smets (2006a,c), we introduced two equivalent semantic settings for conditional beliefs in a multi-agent epistemic context (*conditional doxastic models* and *epistemic plausibility models*), taking the first setting as the basic one. Here, we adopt the second setting, which is closer to the standard semantic structures used in the literature on modeling belief revision (Board 2002; Friedmann and Halpern 1994; Grove 1988; Spohn 1988; Stalnaker 2006; van Benthem 2007, 2004). We use this setting to define notions of *knowledge* K_aP, *belief* B_aP and *conditional belief* B_a^QP. Our concept of "knowledge" is the standard S5-notion, partition-based and fully introspective, that is commonly used in Computer Science and Economics, and is sometimes known as "Aumann knowledge", as a reference to Aumann (1999). The conditional belief operator is a way to "internalize", in a sense, the "static" (AGM) belief revision within a modal framework: saying that, at state s, agent a believes P conditional on Q is a way of saying that Q belongs to a's revised "theory" (capturing her revised beliefs) after revision with P (of a's current theory/beliefs) at state s. Our conditional formulation of "static" belief revision is close to the one in Stalnaker (1968), Ryan and Schobbens (1997), Board (2002), Bonanno (2005), and Rott (1989). As in Board (2002), the preference relation is assumed to be well-preordered; as a result, the logic CDL of conditional beliefs is equivalent to the strongest system in Board (2002).

We also consider other modalities, capturing other "doxastic attitudes" than just knowledge and conditional belief. The most important such notion expresses a form of "weak (non-introspective) knowledge" $\Box_aP$, first introduced by Stalnaker in his modal formalization (Stalnaker 1968, 2006) of Lehrer's *defeasibility analysis of knowledge* (Lehrer 1990; Lehrer and Paxson 1969). We call this notion *safe belief*, to distinguish it from our (Aumann-type) concept of knowledge. Safe belief can be understood as belief that is *persistent under revision with any true information*. We use this notion to give a new solution to the so-called "Paradox of the Perfect Believer". We also solve the open problem posed in Board (2002), by providing a *complete axiomatization of the "static" logic $K\Box$ of conditional belief, knowledge and safe belief*. In a forthcoming paper, we apply the concept of safe belief to Game Theory, improving on Aumann's epistemic analysis of backwards induction in games of perfect information.

Moving thus on to *dynamic belief revision*, the first thing to note is that (unlike the case of "static" revision), *the doxastic features of the actual "triggering event"* that induced the belief change *are essential* for understanding this change (as a "dynamic

[3]But our approach can be easily modified to incorporate ontic changes, along the lines of van Benthem et al. (2006b).

revision", i.e., in terms of the revised beliefs about the state of the world after revision). For instance, our beliefs about *the current situation after* hearing a *public* announcement (say, of some *factual* information, denoted by an atomic sentence p) are different from our beliefs after receiving a *fully private* announcement with the same content p. Indeed, in the public case, we come to believe that p is now *common knowledge* (or at least *common belief*). While, in the private case, we come to believe that the content of the announcement forms now our *secret knowledge*. So the agent's *beliefs about the learning actions* in which she is currently engaged affect the way she updates her previous beliefs.

This distinction is irrelevant for "static" revision, since e.g., in both cases above (public as well as private announcement) we learn the same thing about the situation that existed *before the learning*: our beliefs about that past situation will change in the same way in both cases. More generally, our beliefs about the "triggering action" are irrelevant, as far as our "static" revision is concerned. This explains a fact observed in van Benthem (2007), namely that by and large, the standard literature on belief revision (or belief update) *does not usually make explicit the doxastic events that "trigger" the belief change* (dealing instead only with types of abstract operations on beliefs, such as update, revision and contraction etc). The reason for this lies in the "static" character of AGM revision, as well as its restriction (shared with the "updates" of Katsuno and Mendelzon 1992) to one-agent, first-level, factual beliefs.

A "truly dynamic" logic of belief revision has to be able to capture the *doxastic-epistemic features (e.g., publicity, complete privacy etc.) of specific "learning events"*. We need to be able to model the agents' "dynamic beliefs", i.e., their *beliefs about the learning action itself*: the *appearance* of this action (while it is happening) to each of the agents. In Baltag and Moss (2004), it was argued that a natural way to do this is to use *the same type of formalism that was used to model "static" beliefs*: *epistemic actions should be modeled in essentially the same way as epistemic states*; and this common setting was taken there to be given by *epistemic Kripke models*.

A similar move is made here in the context of our richer doxastic-plausibility structures, by introducing *plausibility pre-orders on actions* and developing a notion of "action plausibility models", that extends the "epistemic action models" from Baltag and Moss (2004), along similar lines to (but without the quantitative features of) the work in Aucher (2003) and van Ditmarsch (2005).

Extending to (pre)ordered models the corresponding notion from Baltag and Moss (2004), we introduce an operation of *product update* of such models, based on the *anti-lexicographic order* on the product of the state model with the action model. The simplest and most natural way to define a connected pre-order on a Cartesian product from connected pre-orders on each of the components is to use either the *lexicographic* or the *anti-lexicographic* order. Our choice is the second, which we regard as the *natural generalization of the AGM theory*, giving *priority to incoming information* (i.e., to "actions" in our sense). This can also be thought of as a generalization of the so-called *"maximal-Spohn" revision*. We call this type of update rule the *"Action-Priority" Update*. The intuition is that the beliefs encoded in the action model express *the "incoming" changes of belief*, while the state model only captures that *past beliefs*. One could say that the new "beliefs about actions" are *acting* on the prior "beliefs about states", producing the updated (posterior) beliefs. This is embedded in the Motto that we give in the paragraph on "Action Models" in the third section, the Motto is: *"beliefs about changes encode (and induce) changes of beliefs"*.

By abstracting away from the quantitative details of the plausibility maps when considering the associated *dynamic logic*, our approach to dynamic belief revision is in the spirit of the one in van Benthem (2007): instead of using "graded belief" operators as in e.g., Aucher (2003) and van Ditmarsch (2005), or probabilistic modal logic as in Kooi (2003), both our account and the one in van Benthem (2007) concentrate on the simple, qualitative language of *conditional beliefs, knowledge and action modalities* (to which we add here the *safe belief* operator). As a consequence, we obtain *simple, elegant, general logical laws of dynamic belief revision*, as natural generalizations of the ones in van Benthem (2007). These "reduction laws" give a *complete axiomatization of the logic of doxastic actions*, "reducing" it to the "static" logic $K\Box$. Compared both to our older axiomatization in Baltag and Smets (2006c) and to the system in Aucher (2003), one can easily see that the introduction of the safe belief operator leads to a major simplification of the reduction laws.

Our qualitative logical setting (in this paper and in Baltag and Smets 2006a,b,c), as well as the closely related setting in van Benthem (2007), are conceptually very different from the more "quantitative" approaches to dynamic belief revision taken in (Aucher 2003; van Ditmarsch 2005; van Ditmarsch and Labuschagne 2007), approaches based on "degrees of belief" given by ordinal plausibility functions. This is not just a matter of interpretation, but it makes a difference for the choice of dynamic revision operators. Indeed, the update mechanisms proposed in Spohn (1988), Aucher (2003), and van Ditmarsch (2005) are essentially quantitative, using various binary functions in transfinite ordinal arithmetic, in order to compute the degree of belief of the output-states in terms of the degrees of the input-states and the degrees of the actions. This leads to an increase in complexity, both in the computation of updates and in the corresponding logical systems. Moreover, there seems to be no canonical choice for the arithmetical formula for updates, various authors proposing various formulas. No clear intuitive justification is provided to any of these formulas, and we see no transparent reason to prefer one to the others. In contrast, classical (AGM) belief revision theory is a qualitative theory, based on natural, intuitive postulates, of great generality and simplicity.

Our approach retains this qualitative flavor of the AGM theory, and aims to build a theory of "dynamic" belief revision of equal simplicity and naturality as the classical "static" account. Moreover (unlike the AGM theory), it aims to provide a *"canonical"* choice for a dynamic revision operator, given by our "Action Priority" update. This notion is a *purely qualitative one*,[4] based on a *simple, natural relational*

[4]One could argue that our plausibility pre-order relation is equivalent to a quantitative notion (of ordinal degrees of plausibility, such as in Spohn (1988)), but unlike in Aucher (2003) and van

definition. From a *formal point of view*, one might see our choice of the anti-lexicographic order as *just one of the many possible options* for developing a belief-revision-friendly notion of update. As already mentioned, it is a generalization of the "maximal-Spohn" revision, already explored in van Ditmarsch (2005) and Aucher (2003), among many other possible formulas for combining the "degrees of belief" of actions and states. But here we justify our option, arguing that our *qualitative interpretation of the plausibility order makes this the only reasonable choice*.

It may seem that by making this choice, we have confined ourselves to *only one of the bewildering multitude of "belief revision policies"* proposed in the literature by Spohn (1988), Rott (1989), Segerberg (1998), Aucher (2003), van Ditmarsch (2005), van Benthem (2004), and van Benthem (2007). But, as argued below, *this apparent limitation is not so limiting after all*, but can instead be regarded as an *advantage*: the power of the "action model" approach is reflected in the fact that *many different belief revision policies* can be recovered as *instances of the same type of update operation*. In this sense, our approach can be seen as a *change of perspective*: the diversity of possible revision policies is replaced by the diversity of possible action models; the differences are now viewed as *differences in input, rather than having different "programs"*. For a computer scientist, this resembles "Currying" in lambda-calculus: if every "operation" is encoded as an input-term, then *one operation* (functional application) *can simulate all operations*.[5] In a sense, this is nothing but the idea of Turing's universal machine, which underlies universal computation.

The title of our paper is a paraphrase of Oliver Board's "Dynamic Interactive Epistemology" (Board 2002), itself a paraphrase of the title ("Interactive Epistemology") of a famous paper by Aumann (1999). We interpret the word "interactive" as referring to the *multiplicity of agents* and the *possibility of communication*. Observe that "interactive" does not necessarily imply "dynamic": indeed, Board and Stalnaker consider Aumann's notion to be "static" (since it doesn't accommodate any non-trivial belief revision). But even Board's logic, as well as Stalnaker's (2006), are "static" in our sense: they cannot directly capture the effect of learning *actions* (but can only express "static" conditional beliefs). In contrast, our DEL-based approach has all the "dynamic" features and advantages of DEL: in addition to "simulating" a range of individual belief-revision policies, it can deal with an even wider range of *complex types of multi-agent learning and communication actions*. We thus think it is realistic to expect that, *within its own natural limits*,[6] our Action-Priority Update Rule could play the role of a *"universal machine" for qualitative dynamic interactive belief-revision*.

Ditmarsch (2005) the way belief update is defined in our account does not make any use of the ordinal "arithmetic" of these degrees.

[5]Note that, as in untyped lambda-calculus, the input-term encoding the operation (i.e., our "action model") and the "static" input-term to be operated upon (i.e., the "state model") are essentially *of the same type*: epistemic plausibility models for the same language (and for the same set of agents).

[6]E.g., our update cannot deal with "forgetful" agents, since "perfect recall" is in-built. But finding out what exactly are the "natural limits" of our approach is for now an open problem.

"Static" Belief Revision

Using the terminology in van Benthem (2007) and Baltag and Smets (2006a,b,c, 2007a), "static" belief revision is about *pre-encoding potential belief revisions as conditional beliefs*. A conditional belief statement $B_a^P Q$ can be thought of as expressing a "doxastic predisposition" or a "plan of doxastic action": the agent is determined to believe that Q was the case, if he learnt that P was the case. The semantics for conditional beliefs is usually given in terms of plausibility models (or equivalent notions, e.g., "spheres", "onions", ordinal functions etc.) As we shall see, both *(Aumann, S5-like) knowledge* and *simple (unconditional) belief* can be defined in terms of conditional belief, which itself could be defined in terms of a *unary belief-revision operator*: $*_a P$ captures *all the revised beliefs* of agent a after revising (her current beliefs) with P.

In addition, we introduce a *safe belief* operator $\Box_a P$, meant to express a weak notion of "defeasible knowledge" (obeying the laws of the modal logic $S4.3$). This concept was defined in Stalnaker (2006) and Board (2002) using a higher-order semantics (quantifying over conditional beliefs). But this is in fact equivalent to a first-order definition, as the Kripke modality for the (converse) plausibility relation. This observation greatly simplifies the task of completely axiomatizing the logic of safe belief and conditional beliefs: indeed, our proof system $K\Box$ below is a solution to the open problem posed in Board (2002).

Plausibility Models: The Single Agent Case

To warm up, we consider first the case of only *one agent*, a case which fits well with the standard models for belief revision.

A *single-agent plausibility frame* is a structure $(S, \leq)$, consisting of a set S of "states" and a "well preorder" $\leq$, i.e., a reflexive, transitive binary relation on S such that *every non-empty subset has minimal elements*. Using the notation

$$\text{Min}_{\leq} P := \{s \in P : s \leq s' \text{ for all } s' \in P\}$$

for the set of $\leq$-minimal elements of P, the last condition says that: For every set $P \subseteq S$, if $P \neq \emptyset$ then $\text{Min}_{\leq} P \neq \emptyset$.

The usual reading of $s \leq t$ is that "state s is *at least as plausible* as state t". We keep this reading for now, though we will later get back to it and clarify its meaning. The "minimal states" in $\text{Min}_{\leq} P$ are thus the "most plausible states" satisfying proposition P. As usual, we write $s < t$ iff $s \leq t$ but $t \not\leq s$, for the *"strict" plausibility relation* (s is *more plausible* than t). Similarly, we write $s \cong t$ iff both $s \leq t$ and $t \leq s$, for the *"equi-plausibility"* (or *indifference*) relation (s and t are *equally plausible*).

S-propositions and models. Given an epistemic plausibility frame S, an *S-proposition* is any subset $P \subseteq S$. Intuitively, we say that *a state s satisfies the proposition P* if $s \in P$. Observe that a plausibility frame is just a special case of a *relational frame* (or *Kripke frame*). So, as it is standard for Kripke frames in general, we can define a *plausibility model* to be a structure $\mathbf{S} = (S, \leq, \|\cdot\|)$, consisting of a plausibility frame $(S, \leq)$ together with a valuation map $\|\cdot\| : \Phi \to \mathcal{P}(S)$, mapping every element of a given set Φ of "atomic sentences" into S-propositions.

Interpretation. The elements of S will represent the *possible states* (or "possible worlds") of a system. The atomic sentences $p \in \Phi$ represent *"ontic" (non-doxastic) facts*, that might hold or not in a given state. The valuation tells us which facts hold at which worlds. Finally, the plausibility relations $\leq$ capture the agent's *(conditional) beliefs about the state* of the system; if e.g., the agent was given the information that the state of the system is either s or t, she would believe that the system was in the *most plausible* of the two. So, if $s < t$, the agent would believe the real state was s; if $t < s$, she would believe it was t; otherwise (if $s \cong t$), the agent would be indifferent between the two alternatives: she will not be able to decide to believe any one alternative rather than the other.

Propositional operators, Kripke modalities. For every model $\mathbf{S}$, we have the usual Boolean operations with S-propositions

$$P \wedge Q := P \cap Q, \quad P \vee Q := P \cup Q,$$

$$\neg P := S \setminus P, \quad P \to Q := \neg P \vee Q,$$

as well as Boolean constants $\top_S := S$ and $\bot_S := \emptyset$. Obviously, one may also introduce *infinitary* conjunctions and disjunctions. In addition, any binary relation $R \subseteq S \times S$ on S gives rise to a *Kripke modality* $[R] : \mathcal{P}(S) \to \mathcal{P}(S)$, defined by

$$[R]Q := \{s \in S : \forall t\,(sRt \Rightarrow t \in Q)\}.$$

Accessibility relations for belief, conditional belief and knowledge. To talk about beliefs, we introduce a *doxastic accessibility relation* $\to$, given by

$$s \to t \text{ iff } t \in \mathrm{Min}_{\leq} S\,.$$

We read this as saying that: when the actual state is s, the agent believes that *any* of the states t with $s \to t$ *may be* the actual state. This matches the above interpretation of the preorder: the states believed to be possible are the minimal (i.e., "most plausible") ones.

In order to talk about *conditional beliefs*, we can similarly define a *conditional doxastic accessibility relation* for each S-proposition $P \subseteq S$:

$$s \to^P t \text{ iff } t \in \mathrm{Min}_{\leq} P\,.$$

We read this as saying that: when the actual state is s, if the agent is given the information (that) P (is true at the actual state), then she believes that *any* of the states t with $s \to t$ *may be* the actual state.

Finally, to talk about knowledge, we introduce a relation of *epistemic possibility* (or "indistinguishability") $\sim$. Essentially, this is just the universal relation:

$$s \sim t \text{ iff } s, t \in S .$$

So, in single-agent models, *all* the states in S are assumed to be "epistemically possible": the only thing *known* with absolute certainty about the current state is that it belongs to S. This is natural, in the context of a single agent: the states known to be impossible are *irrelevant* from the point of doxastic-epistemic logic, so they can simply be excluded from our model S. (As seen below, this cannot be done in the case of multiple agents!)

Knowledge and (conditional) belief. We define *knowledge* and *(conditional) belief* as the Kripke modalities for the epistemic and (conditional) doxastic accessibility relations:

$$KP := [\sim]P ,$$

$$BP := [\rightarrow]P ,$$

$$B^Q P := [\rightarrow^Q]P .$$

We read KP as saying that the (implicit) agent *knows* P. This is "knowledge" in the strong Leibnizian sense of "truth in all possible worlds". We similarly read BP as "P is believed" and $B^Q P$ as "P is believed given (or conditional on) Q". As for *conditional belief* statements $s \in B^Q P$, we interpret them in the following way: if the actual state is s, then after coming to believe that Q is the case (at this actual state), the agent will believe that P *was* the case (at the same actual state, *before* his change of belief). In other words, conditional beliefs B^Q give descriptions of the agent's *plan* (or *commitments*) about what he will believe about the current state after receiving new (believable) information. To quote Johan van Benthem in (2007), conditional beliefs are *"static pre-encodings" of the agent's potential belief changes* in the face of new information.

Discussion on interpretation. Observe that our interpretation of the plausibility relations is *qualitative*, in terms of *conditional beliefs* rather than "degrees of belief": there is no scale of beliefs here, allowing for "intermediary" stages between believing and not believing. Instead, all these beliefs are equally "firm" (*though conditional*): given the condition, something is either believed or not. To repeat, writing $s < t$ is for us just a way to say that: if *given* the information that the state of the system is either s or t, the agent would believe it to be s. So plausibility relations are special cases of conditional belief. This interpretation is based on the following (easily verifiable) equivalence:

$$s < t \text{ iff } s \in B^{\{s,t\}}\{s\} \text{ iff } t \in B^{\{s,t\}}\{s\}.$$

There is nothing quantitative here, no need for us to refer in any way to the "strength" of this agent's belief: though she might have beliefs of unequal strengths, we are not interested here in modeling this quantitative aspect. Instead, we give the agent some information about a state of a virtual system (that it is either s or t) and we ask her a *yes-or-no question* ("Do you believe that virtual state to be s ?"); we write $s < t$ iff the agent's answer is "yes". This is a firm answer, so it expresses a firm belief. "Firm" does not imply "un-revisable" though: if later we reveal to the agent that the state in question was in fact t, she should be able to accept this new information; after all, the agent should be introspective enough to realize that her belief, however firm, was just a belief.

One possible objection against this qualitative interpretation is that our postulate that $\leq$ is a well-preorder (and so in particular a connected pre-order) introduces a hidden "quantitative" feature; indeed, any such preorder can be equivalently described using a plausibility map as in e.g., Spohn (1988), assigning ordinals to states. Our answer is that, first, the specific ordinals will not play any role in our definition of a dynamic belief update; and second, all our postulates can be given a justification in purely qualitative terms, using conditional beliefs. The transitivity condition for $\leq$ is just a *consistency* requirement imposed on a rational agent's conditional beliefs. And the existence of minimal elements in any non-empty subset is simply the natural extension of the above setting to *general* conditional beliefs, not only conditions involving two states: more specifically, for any possible condition $P \subseteq S$ about a system S, the S-proposition $\mathrm{Min}_{\leq} P$ is simply a way to encode everything that the agent would believe about the current state of the system, if she was given the information that the state satisfied condition P.

Note on other models in the literature. Our models are the same as Board's "belief revision structures" (Board 2002), i.e., nothing but "Spohn models" as in Spohn (1988), but with a purely relational description. Spohn models are usually described in terms of a map assigning ordinals to states. But giving such a map is equivalent to introducing a well pre-order $\leq$ on states, and it is easy to see that all the relevant information is captured by this order.

Our conditions on the preorder $\leq$ can also be seen as a *semantic analogue* of Grove's conditions for the (relational version of his) models in Grove (1988). The standard formulation of Grove models is in terms of a "system of spheres" (weakening Lewis' similar notion), but it is equivalent (as proved in Grove 1988) to a relational formulation. Grove's postulates are still *syntax-dependent*, e.g., existence of minimal elements is required only for subsets that are *definable* in his language: this is the so-called "smoothness" condition, which is weaker than our "well-preordered" condition. We prefer a purely semantic condition, independent of the choice of a language, both for reasons of elegance and simplicity and because we want to be able to consider more than one language for the same structure.[7]

[7] Imposing syntactic-dependent conditions in the very definition of a class of structures makes the definition meaningful only for one language; or else, the meaning of what, say, a plausibility model is won't be *robust*: it will change whenever one wants to extend the logic, by adding a few more

So, following Board (2002) and Stalnaker (2006) and others, we adopt the natural semantic analogue of Grove's condition, simply requiring that *every* subset has minimal elements: this will allow our conditional operators to be well-defined on sentences of *any* extension of our logical language.

Note that the minimality condition implies, by itself, that the relation $\leq$ is both *reflexive* (i.e., $s \leq s$ for all $s \in S$) and *connected*[8] (i.e., either $s \leq t$ or $t \leq s$, for all $s, t \in S$). In fact, a "well-preorder" is the same as a *connected, transitive, well-founded*[9] *relation*, which is the setting proposed in Board (2002) for a logic of conditional beliefs equivalent to our logic CDL below. Note also that, when the set S is *finite*, a well-preorder is nothing but a *connected preorder*. This shows that our notion of frame subsumes, not only Grove's setting, but also some of the other settings proposed for conditionalization.

Multi-agent Plausibility Models

In the multi-agent case, *we cannot exclude from the model the states that are known to be impossible* by some agent a: they may still be considered possible by a second agent b. Moreover, they might still be relevant for a's beliefs/knowledge about what b believes or knows. So, in order to define an agent's knowledge, we cannot simply quantify over *all* states, as we did above: instead, we need to consider, as usually done in the Kripke-model semantics of knowledge, only the "possible" states, i.e., the ones that are *indistinguishable* from the real state, as far as a given agent is concerned. It is thus natural, in the multi-agent context, to explicitly specify the agents' epistemic indistinguishability relations $\sim_a$ (labeled with the agents' names) as part of the basic structure, in addition to the plausibility relations $\leq_a$. Taking this natural step, we obtain *epistemic plausibility frames* $(S, \sim_a, \leq_a)$. As in the case of a single agent, specifying epistemic relations turns out to be *superfluous*: the relations $\sim_a$ can be recovered from the relations $\leq_a$. Hence, we will simplify the above structures, obtaining the equivalent setting of *multi-agent plausibility frames* $(S, \leq_a)$.

Before going on to define these notions, observe that it doesn't make sense anymore to require the plausibility relations $\leq_a$ to be connected (and even less sense to require them to be well-preordered): if two states s, t are *distinguishable* by an agent a, i.e., $s \not\sim_a t$, then a will never consider both of them as epistemically possible in the same time. If she was given the information that the real state is either s or t, agent a will immediately *know* which of the two: if the real state was s, she would be able to distinguish this state from t, and would thus know the state

operators. This is very undesirable, since then one cannot compare the expressivity of different logics on the same class of models.

[8]In the Economics literature, connectedness is called "completeness", see e.g., Board (2002).

[9]I.e., there exists no infinite descending chain $s_0 > s_1 > \cdots$.

was s; similarly, if the real state was t, she would know it to be t. Her beliefs will play no role in this, and it would be meaningless to ask her which of the two states is more plausible to her. So only the states in the same $\sim_a$-equivalence class could, and should, be $\leq_a$-comparable; i.e., *$s \leq_a t$ implies $s \sim_a t$*, and the restriction of $\leq_a$ to each $\sim_a$-equivalence class is connected. Extending the same argument to arbitrary conditional beliefs, we can see that *the restriction of $\leq_a$ to each $\sim_a$-equivalence class must be well-preordered.*

Epistemic plausibility frames. Let $\mathcal{A}$ be a finite set of labels, called *agents*. An *epistemic plausibility frame* over $\mathcal{A}$ (EPF, for short) is a structure $\mathbf{S} = (S, \sim_a, \leq_a)_{a\in\mathcal{A}}$, consisting of a set S of "states", endowed with a family of equivalence relations $\sim_a$, called *epistemic indistinguishability relations*, and a family of *plausibility relations* $\leq_a$, both labeled by "agents" and assumed to satisfy two conditions: (1) $\leq_a$-comparable states are $\sim_a$-indistinguishable (i.e., $s \leq_a t$ implies $s \sim_a t$); (2) the restriction of each plausibility relation $\leq_a$ to each $\sim_a$-equivalence class is a well-preorder. As before, we use the notation $\mathrm{Min}_{\leq_a} P$ for the set of $\leq_a$-minimal elements of P. We write $s <_a t$ iff $s \leq_a t$ but $t \not\leq_a s$ (the *"strict" plausibility relation*), and write $s \cong_a t$ iff both $s \leq_a t$ and $t \leq_a s$ (the *"equi-plausibility" relation*). The notion of *epistemic plausibility models* (EPM, for short) is defined in the same way as the plausibility models in the previous section.

Epistemic plausibility models. We define a *(multi-agent) epistemic plausibility model* (EPM, for short) as a multi-agent EPF together with a valuation over it (the same way that single-agent plausibility models were defined in the previous section).

It is easy to see that our definition of EPFs includes superfluous information: in an EPF, the knowledge relation $\sim_a$ can be recovered from the plausibility relation $\leq_a$, via the following rule:

$$s \sim_a t \text{ iff either } s \leq_a t \text{ or } t \leq_a s\,.$$

In other words, two states are indistinguishable for a iff they are *comparable* (with respect to $\leq_a$).

So, in fact, one could present epistemic plausibility frames simply as *multi-agent plausibility frames*. To give this alternative presentation, we use, for *any preorder relation* $\leq$, the notation $\sim$ for the associated *comparability relation*

$$\sim\; :=\; \leq \cup \geq$$

(where $\geq$ is the converse of $\leq$). A *comparability class* is a set of the form $\{t : s \leq t \text{ or } t \leq s\}$, for some state s. A relation $\leq$ is called *locally well-preordered* if it is a preorder such that its restriction to each comparability class is well-preordered. Note that, when the underlying set S is *finite*, a locally well-preordered relation is nothing but a *locally connected preorder*: a preorder whose restrictions to any comparability class are connected. More generally, *a locally well-preordered relation is the same as a locally connected and well-founded preorder.*

Multi-agent plausibility frames. A *multi-agent plausibility frame* (MPF, for short) is a structure $(S, \leq_a)_{a \in \mathcal{A}}$, consisting of a set of states S together with a family of locally well-preordered relations $\leq_a$, one for each agent $a \in \mathcal{A}$. Oliver Board (2002) calls multi-agent plausibility frames "belief revision structures". A *multi-agent plausibility model* (MPM, for short) is an MPF together with a valuation map.

Bijective correspondence between EPFs and MPFs. *Every MPF can be canonically mapped into an EPF*, obtained by defining epistemic indistinguishability via the above rule ($\sim_a \; := \; \leq_a \cup \geq_a$). Conversely, every EPF gives rise to an MPF, via the map that "forgets" the indistinguishability structure. It is easy to see that these two maps are the inverse of each other. Consequently, from now on we *identify MPFs and EPFs, and similarly identify MPMs and EPMs*; e.g., we can talk about "knowledge", "(conditional) belief" etc. in an MPM, defined in terms of the associated EPM.

So from now on we identify the two classes of models, via the above canonical bijection, and talk about "plausibility models" in general. One can also see how this approach relates to another widely adopted definition for conditional beliefs; in Board (2002), van Ditmarsch (2005), and van Benthem (2007), this definition involves the assumption of a *"local plausibility" relation at a given state* $s \leq_a^w t$, to be read as: "at state w, agent a considers state s at least as plausible as state t ". Given such a relation, the conditional belief operator is usually defined in terms that are equivalent to putting $s \to_a^P t$ iff $t \in \mathrm{Min}_{\leq_a^s} P$. One could easily restate our above definition in this form, by taking:

$$s \leq_a^w t \;\text{ iff }\; \text{ either } w \not\sim_a t \text{ or } s \leq_a t.$$

The converse problem is studied in Board (2002), where it is shown that, if full introspection is assumed, then one can recover "uniform" plausibility relations $\leq_a$ from the relations $\leq_a^w$.

Information cell. The equivalence relation $\sim_a$ induces a partition of the state space S, called *agent a's information partition*. We denote by $s(a)$ the *information cell* of s in a's partition, i.e., the $\sim_a$-equivalence class of s:

$$s(a) := \{t \in S : s \sim_a t\}.$$

The information cell $s(a)$ captures *all the knowledge possessed by the agent* at state s: when the actual state of the system is s, then agent a knows only the state's equivalence class $s(a)$.

Example 1. Alice and Bob play a game, in which an anonymous referee puts a coin on the table, lying face up but in such a way that the face is covered (so Alice and Bob cannot see it). Based on previous experience, (it is common knowledge that) Alice and Bob believe that the upper face is Heads (since e.g., they noticed that the referee had a strong preference for Heads). And in fact, they're right: the coin lies Heads up. Neglecting the anonymous referee, the EPM for this example is the following model **S**:

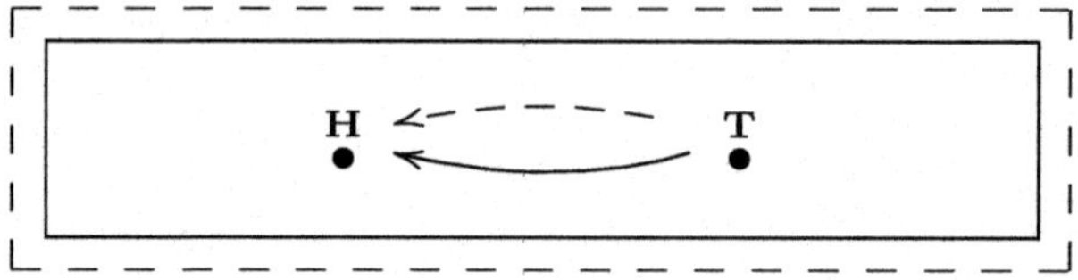

Here, the arrows represent *converse plausibility relations* $\geq$ *between distinct states only* (going from less plausible to more plausible states): since these are always reflexive, we choose to *skip all the loops* for convenience. The squares represent the *information cells* for the two agents. Instead of labels, we use *dashed arrows and squares for Alice*, while using *continuous arrows and squares for Bob*. In this picture, the actual state of the system is the state *s* on the left (in which **H** is true). Henceforth, in our other examples, we will refer to this particular plausibility model as **S**.

By deleting the squares, we obtain a representation of the corresponding MPM, also denoted by **S** (where we now use labels for agents instead of different types of lines):

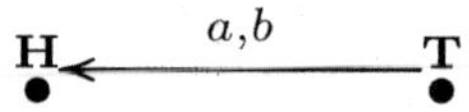

Example 2. In front of Alice, the referee shows the face of the coin to Bob, but Alice cannot see the face. The EPM is now the following model **W**:

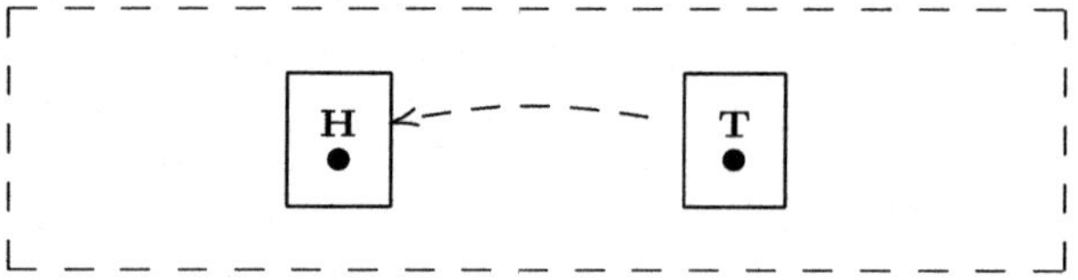

while the MPM is

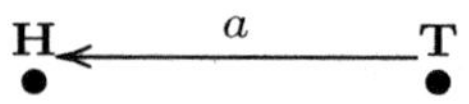

Since Bob now knows the state of the coin, his local plausibility relation consists only of loops, and hence we have no arrows for Bob in this diagrammatic representation.

(Conditional) doxastic appearance and (conditional) doxastic accessibility. As in the previous section, we can define a doxastic and epistemic accessibility relations, except that now we have to select, for each state s, the most plausible states in its information cell $s(a)$ (instead of the most plausible in S). For this, it is convenient to introduce some notation and terminology: the *doxastic appearance* of state s to agent a is the set

$$s_a := \mathrm{Min}_{\leq_a} s(a)$$

of the "most plausible" states that are consistent with the agent's knowledge at state s. The doxastic appearance of s captures *the way state s appears to the agent*, or (in the language of Belief Revision) *the agent's current "theory" about the world s*. We can extend this to capture *conditional beliefs* (in full generality), by associating to each S-proposition $P \subseteq S$ and each state $s \in S$ the *conditional doxastic appearance* s_a^P *of state s to agent a, given (information) P*. This can be defined as the S-proposition

$$s_a^P := \mathrm{Min}_{\leq_a} s(a) \cap P$$

given by the set of all $\leq_a$-minimal states of $s(a) \cap P$: these are the "most plausible" states satisfying P that are consistent with the agent's knowledge at state s. The conditional appearance S_a^P gives *the agent's revised theory (after learning P) about the world s*. We can put these in a relational form, by defining *doxastic accessibility relations* $\to_a$, $\to_a^P$, as follows:

$$s \to_a t \text{ iff } t \in s_a ,$$

$$s \to_a^P t \text{ iff } t \in s_a^P .$$

Knowledge and (conditional) belief. As before, we define the *knowledge* and *(conditional) belief* operators for an agent a as the Kripke modalities for a's epistemic and (conditional) doxastic accessibility relations:

$$K_a P := [\sim_a]P = \{s \in S : s(a) \subseteq P\} ,$$

$$B_a P := [\to_a]P = \{s \in S : s_a \subseteq P\} ,$$

$$B_a^Q P := [\to_a^Q]P = \{s \in S : s_a^Q \subseteq P\}.$$

We also need a notation for the *dual of the K modality* ("epistemic possibility"):

$$\tilde{K}_a P := \neg K_a \neg P.$$

Doxastic propositions. Until now, our notion of proposition is "local", being specific to a given model: we only have "S-propositions" for each model S. As long as the model is fixed, this notion is enough for interpreting sentences over the given model. But, since later we will proceed to study systematic *changes* of models (when dealing with *dynamic* belief revision), we need a notion of proposition that is not confined to one model, but makes sense on *all* models:

A *doxastic proposition* is a map $\mathbf{P}$ assigning to each plausibility model $\mathbf{S}$ some S-proposition $\mathbf{P}_{\mathbf{S}} \subseteq S$. We write $s \models_{\mathbf{S}} \mathbf{P}$, and say that the proposition $\mathbf{P}$ is true at $s \in \mathbf{S}$, iff $s \in (\mathbf{P})_{\mathbf{S}}$. We skip the subscript and write $s \models \mathbf{P}$ when the model is understood.

We denote by Prop the family of all doxastic propositions. All the Boolean operations on S-propositions as sets can be *lifted* pointwise to operations on Prop: in particular, we have the "always true" $\top$ and "always false" $\bot$ propositions, given by $(\bot)_{\mathbf{S}} := \emptyset, (\top)_{\mathbf{S}} := S$, negation $(\neg\mathbf{P})_{\mathbf{S}} := S \setminus \mathbf{P}_{\mathbf{S}}$, conjunction $(\mathbf{P} \wedge \mathbf{Q})_{\mathbf{S}} := \mathbf{P}_{\mathbf{S}} \cap \mathbf{Q}_{\mathbf{S}}$, disjunction $(\mathbf{P} \vee \mathbf{Q})_{\mathbf{S}} := \mathbf{P}_{\mathbf{S}} \cup \mathbf{Q}_{\mathbf{S}}$ and all the other standard Boolean operators, including *infinitary* conjunctions and disjunctions. Similarly, we can define pointwise the *epistemic and (conditional) doxastic modalities*: $(K_a\mathbf{P})_{\mathbf{S}} := K_a\mathbf{P}_{\mathbf{S}}$, $(B_a\mathbf{P})_{\mathbf{S}} := B_a\mathbf{P}_{\mathbf{S}}$, $(B_a^{\mathbf{Q}}\mathbf{P})_{\mathbf{S}} := B_a^{\mathbf{Q}_{\mathbf{S}}}\mathbf{P}_{\mathbf{S}}$. It is easy to check that we have: $B_a\mathbf{P} = B_a^{\top}\mathbf{P}$. Finally, the relation of *entailment* $\mathbf{P} \models \mathbf{Q}$ between doxastic propositions is given pointwise by inclusion: $\mathbf{P} \models \mathbf{Q}$ iff $\mathbf{P}_{\mathbf{S}} \subseteq \mathbf{Q}_{\mathbf{S}}$ for all $\mathbf{S}$.

Safe Belief and the Defeasibility Theory of Knowledge

Ever since Plato's *identification of knowledge with "true justified (or justifiable) belief"* was shattered by Gettier's celebrated counterexamples (Gettier 1963), philosophers have been looking for the "missing ingredient" in the Platonic equation. Various authors identify this missing ingredient as "robustness" (Hintikka 1962), "indefeasibility" (Klein 1971; Lehrer 1990; Lehrer and Paxson 1969; Stalnaker 2006) or "stability" (Rott 2004). According to this *defeasibility theory of knowledge* (or "stability theory", as formulated by Rott), a belief counts as "knowledge" if it is *stable under belief revision with any new evidence*: "if a person has knowledge, than that person's justification must be sufficiently strong that it is not capable of being defeated by evidence that he does not possess" (Pappas and Swain 1978).

One of the problems is interpreting what "evidence" means in this context. There are at least two natural interpretations, each giving us a concept of "knowledge". The first, and the most common,[10] interpretation is to take it as meaning "any *true* information". The resulting notion of "knowledge" was formalized by Stalnaker in (2006), and defined there as follows: "an agent knows that φ if and only if φ is true, she believes that φ, and she continues to believe φ if any *true* information is received". This concept differs from the usual notion of knowledge ("Aumann knowledge") in Computer Science and Economics, by the fact that it does not satisfy the laws of the modal system S5 (in fact, negative introspection fails); Stalnaker

[10]This interpretation is the one virtually adopted by all the proponents of the defeasibility theory, from Lehrer to Stalnaker.

shows that the complete modal logic of this modality is the modal system S4.3. As we'll see, this notion ("Stalnaker knowledge") corresponds to what we call "safe belief" $\Box P$. On the other hand, another natural interpretation, considered by at least one author Rott (2004), takes "evidence" to mean "*any* proposition", i.e., to include possible *misinformation*: "real knowledge" should be robust even in the face of false evidence. As shown below, this corresponds to our "knowledge" modality KP, which could be called "absolutely unrevisable belief". This is a partition-based concept of knowledge, identifiable with "Aumann knowledge" and satisfying all the laws of S5. In other words, this last interpretation provides a perfectly decent "defeasibility" defense of S5 and of negative introspection!

In this paper, we adopt the pragmatic point of view of the formal logician: instead of debating which of the two types of "knowledge" is the real one, we simply formalize both notions in a common setting, compare them, axiomatize the logic obtained by combining them and use their joint strength to express interesting properties. Indeed, as shown below, conditional beliefs can be *defined* in terms of knowledge *only* if we combine both the above-mentioned types of "knowledge".

Knowledge as unrevisable belief. Observe that, for all propositions **P**, we have

$$K_a\mathbf{Q} = \bigwedge_{\mathbf{P}} B_a^{\mathbf{P}}\mathbf{Q}$$

(where the conjunction ranges over *all* doxastic propositions), or equivalently, we have for every state s in every model **S**:

$$s \models K_a\mathbf{Q} \quad \text{iff} \quad s \models B_a^{\mathbf{P}}\mathbf{Q} \text{ for all } \mathbf{P}. \tag{39.1}$$

This gives a characterization of *knowledge as "absolute" belief, invariant under any belief revision*: a given belief is "known" iff it cannot be revised, i.e., it would be still believed in any condition.[11] Observe that this resembles the defeasibility analysis of knowledge, but only if we adopt the *second interpretation* mentioned above (taking "evidence" to include misinformation). Thus, our "knowledge" is more robust than Stalnaker's: it resists any belief revision, not capable of being defeated by *any* evidence (including false evidence). This is a very "strong" notion of knowledge (implying "absolute certainty" and full introspection), which seems to us to fit better with the standard usage of the term in Computer Science literature. Also, unlike the one in Stalnaker (2006), our notion of knowledge *is negatively introspective*.

[11]This of course assumes agents to be "rational" in a sense that excludes "fundamentalist" or "dogmatic" beliefs, i.e., beliefs in unknown propositions but refusing any revision, even when contradicted by facts. But this "rationality" assumption is already built in our plausibility models, which satisfy an epistemically friendly version of the standard AGM postulates of rational belief revision. See Baltag and Smets (2006a) for details.

Another identity[12] that can be easily checked is:

$$K_a\mathbf{Q} = B_a^{\neg\mathbf{Q}}\mathbf{Q} = B_a^{\neg\mathbf{Q}}\bot \tag{39.2}$$

(where $\bot$ is the "always false" proposition). This captures in a different way the "absolute un-revisability" of knowledge: something is "known" if it is believed even if conditionalizing our belief with its negation. In other words, this simply expresses the *impossibility* of accepting its negation as evidence (since such a revision would lead to an inconsistent belief).

Safe belief. To capture "Stalnaker knowledge", we introduce the Kripke modality $\Box_a$ associated to the converse $\geq_a$ of the plausibility relation, going from any state s to all the states that are "at least as plausible" as s. For S-propositions $P \subseteq S$ over any given model $\mathbf{S}$, we put

$$\Box_a P := [\geq_a]P = \{s \in S : t \in P \text{ for all } t \leq_a s\},$$

and this induces pointwise an operator $\Box_a\mathbf{P}$ on doxastic propositions. We read $s \models \Box_a\mathbf{P}$ as saying that: *at state s, agent a's belief in* $\mathbf{P}$ *is safe*; or at *state s, a safely believes that* $\mathbf{P}$. We will explain this reading below, but first observe that: $\Box_a$ is an $S4$-modality (since $\geq_a$ is reflexive and transitive), but not necessarily $S5$; i.e., *safe beliefs are truthful* ($\Box_a\mathbf{P} \models \mathbf{P}$) and *positively introspective* ($\Box_a\mathbf{P} \models \Box_a\Box_a\mathbf{P}$), but not necessarily negatively introspective: in general, $\neg\Box_a\mathbf{P} \not\models \Box_a\neg\Box_a\mathbf{P}$.

Relations between knowledge, safe belief and conditional belief. First, *knowledge entails safe belief*

$$K_a\mathbf{P} \models \Box_a\mathbf{P},$$

and *safe belief entails belief*

$$\Box_a\mathbf{P} \models B_a\mathbf{P}.$$

The last observation can be strengthened to characterize safe belief in a similar way to the above characterization (39.1) of knowledge (as belief invariant under any revision): *safe beliefs are precisely the beliefs which are persistent under revision with any true information.* Formally, this says that, for every state s in every model $\mathbf{S}$, we have

$$s \models \Box_a\mathbf{Q} \quad \text{iff:} \quad s \models B_a^{\mathbf{P}}\mathbf{Q} \text{ for every } \mathbf{P} \text{ such that } s \models \mathbf{P} \tag{39.3}$$

[12]This identity corresponds to the definition of "necessity" in Stalnaker (1968) in terms of doxastic conditionals.

We can thus see that *safe belief coincides indeed with Stalnaker's notion of "knowledge"*, given by the first interpretation ("evidence as true information") of the defeasibility theory. As mentioned above, we prefer to keep the name "knowledge" for the strong notion (which gives absolute certainty), and call this weaker notion "safe belief"; indeed, these are beliefs that are "safe" to hold, in the sense that no future learning of truthful information will force us to revise them.

Example 3 (Dangerous Knowledge). This starts with the situation in Example 1 (when none of the two agents has yet seen the face of the coin). Alice has to get out of the room for a minute, which creates an opportunity for Bob to quickly raise the cover in her absence and take a peek at the coin. He does that, and so he sees that the coin is Heads up. After Alice returns, she obviously doesn't know whether or not Bob took a peek at the coin, but she believes he didn't do it: taking a peek is against the rules of the game, and so she trusts Bob not to do that. The model is now rather complicated, so we only represent the MPM:

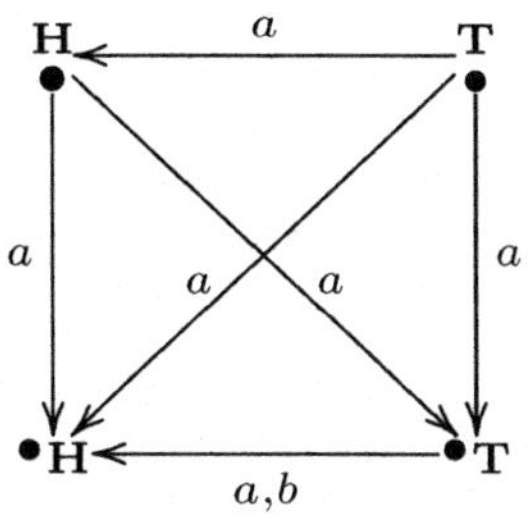

Let us call this model $\mathbf{S}'$. The actual state s'_1 is the one in the upper left corner, in which Bob took a peek and saw the coin Heads up, while the state t'_1 in the upper right corner represents the other possibility, in which Bob saw the coin lying Tails up. The two lower states s'_2 and t'_2 represent the case in which Bob *didn't take a peek*. Observe that the above drawing includes the (natural) assumption that Alice keeps her previous belief that the coin lies Heads up (since there is no reason for her to change her mind). Moreover, we assumed that she will keep this belief even if she'd be told that Bob took a peek: this is captured by the a-arrow from t'_1 to s'_1. This seems natural: Bob's taking a peek doesn't change the upper face of the coin, so it shouldn't affect Alice's prior belief about the coin.

In both Examples 1 and 3 above, Alice holds a *true belief* (at the real state) that the coin lies Heads up: the actual state satisfies $B_a\mathbf{H}$. In both cases, this true belief is *not knowledge* (since Alice doesn't know the upper face), but nevertheless in Example 1, this belief is *safe* (although it is *not known by the agent to be safe*): no additional truthful information (about the real state s) can force her to revise this belief. (To see this, observe that any *new* truthful information would reveal to Alice the real state s, thus confirming her belief that Heads is up.) So in the model $\mathbf{S}$ from Example 1, we have $s \models \Box_a \mathbf{H}$ (where s is the actual state). In contrast, in Example 2, Alice's belief (that the coin is Heads up), though true, is *not safe*. There is some piece of correct information (about the real state s_1') which, if learned by Alice, would make her change this belief: we can represent this piece of correct information as the doxastic proposition $\mathbf{H} \to \mathbf{K_b H}$. It is easy to see that the actual state s_1' of the model $\mathbf{S}'$ satisfies the proposition $B_a^{\mathbf{H} \to \mathbf{K_b H}}\mathbf{T}$ (since $(\mathbf{H} \to K_b\mathbf{H})_{\mathbf{S}'} = \{s_1', t_1', t_2'\}$ and the minimal state in the set $s_1'(a) \cap \{s_1', s_1', t_2'\} = \{s_1', t_1', t_2'\}$ is t_2', which satisfies $\mathbf{T}$.) So, if given this information, Alice would come to wrongly believe that the coin is Tails up! This is an example of a *dangerous truth*: a true information whose learning can lead to wrong beliefs.

Observe that *an agent's belief can be safe without him necessarily knowing this* (in the "strong" sense of knowledge given by K): "safety" (similarly to "truth") is an *external* property of the agent's beliefs, that can be ascertained only by comparing his belief-revision system with reality. Indeed, *the only way* for an agent to *know a belief to be safe* is to actually *know it to be truthful*, i.e., to have actual *knowledge* (not just a belief) of its truth. This is captured by the valid identity

$$K_a \Box_a \mathbf{P} = K_a \mathbf{P}. \tag{39.4}$$

In other words: *knowing that something is safe to believe is the same as just knowing it to be true*. In fact, *all beliefs held by an agent "appear safe" to him*: in order to believe them, he has to believe that they are safe. This is expressed by the valid identity

$$B_a \Box_a \mathbf{P} = B_a \mathbf{P} \tag{39.5}$$

saying that: *believing that something is safe to believe is the same as just believing it*. Contrast this with the situation concerning "knowledge": in our logic (as in most standard doxastic-epistemic logics), we have the identity

$$B_a K_a \mathbf{P} = K_a \mathbf{P}. \tag{39.6}$$

So *believing that something is known is the same as knowing it*!

The Puzzle of the Perfect Believer. The last identity is well-known and has been considered "paradoxical" by many authors. In fact, the so-called "Paradox of the Perfect Believer" in Gochet and Gribomont (2006), Voorbraak (1993), Hoek (1993), Meyer and Hoek (1995), Williamson (2001), and Friedmann and Halpern (1994) is based on it. For a "strong" notion of belief as the one we have here ("belief" = belief with certainty), it seems reasonable to assume the following "axiom":

$$B_a \varphi \to B_a K_a \varphi. \qquad (?)$$

Putting this together with (39.6) above, we get a paradoxical conclusion:

$$B_a \varphi \to K_a \varphi. \qquad (?!)$$

So this leads to a triviality result: *knowledge and belief collapse to the same thing*, and *all beliefs are always true*! One solution to the "paradox" is to reject (?), as an (intuitive but) *wrong* "axiom". In contrast, various authors Friedmann and Halpern (1994); Hoek (1993); Voorbraak (1993); Williamson (2001) accept (?) and propose other solutions, e.g., giving up the principle of "negative introspection" for knowledge.

Our solution to the paradox, as embodied in the contrasting identities (39.5) and (39.6), combines the advantages of both solutions above: the "axiom" (?) is *correct if we interpret "knowledge" as safe belief* $\Box_a$, since then (?) becomes equivalent to identity (39.5) above; but then *negative introspection fails for this interpretation*! On the other hand, if we interpret "knowledge" as our K_a-modality then negative introspection holds; but then *the above "axiom" (?) fails*, and on the contrary we have the identity (39.6).

So, in our view, *the paradox of the perfect believer arises from the conflation of two different notions of "knowledge":* "Aumann" (partition-based) knowledge and "Stalnaker" knowledge (i.e., safe belief).

(Conditional) beliefs in terms of "knowledge" notions. An important observation is that *one can characterize/define (conditional) beliefs only in terms of our two "knowledge" concepts (K and* $\Box$*)*: For simple beliefs, we have

$$B_a\mathbf{P} \;=\; \tilde{K}_a\Box_a\mathbf{P} \;=\; \Diamond_a\Box_a\mathbf{P}\,,$$

recalling that $\tilde{K}_a\mathbf{P} = \neg K_a\neg\mathbf{P}$ is the Diamond modality for K_a, and $\Diamond_a\mathbf{P} = \neg\Box_a\neg\mathbf{P}$ is the Diamond for $\Box_a$.

The equivalence $B_a\mathbf{P} = \Diamond_a\Box_a\mathbf{P}$ has recently been observed by Stalnaker in (2006), who took it as the basis of a philosophical analysis of "belief" in terms of "defeasible knowledge" (i.e., safe belief). Unfortunately, this analysis does not apply to conditional belief: one can easily see that *conditional belief cannot be defined in terms of safe belief only*! However, one *can generalize* the identity $B_a\mathbf{P} = \tilde{K}_a\Box_a\mathbf{P}$ above, defining conditional belief in terms of *both our "knowledge" concepts*:

$$B_a^{\mathbf{P}}\mathbf{Q} \;=\; \tilde{K}_a\mathbf{P} \to \tilde{K}_a(\mathbf{P} \wedge \Box_a(\mathbf{P} \to \mathbf{Q}))\,. \tag{39.7}$$

Other Modalities and Doxastic Attitudes

From a modal logic perspective, it is natural to introduce the Kripke modalities $[>_a]$ and $[\cong_a]$ for the other important relations (strict plausibility and equiplausibility): For S-propositions $P \subseteq S$ over a given model $\mathbf{S}$, we put

$$[>_a]P \;:=\; \{s \in S : t \in P \text{ for all } t <_a s\},$$
$$[\cong_a]P \;:=\; \{s \in S : t \in P \text{ for all } t \cong_a s\},$$

and as before these pointwise induce corresponding operators on Prop. The intuitive meaning of these operators is not very clear, but they can be used to define other interesting modalities, capturing various "doxastic attitudes".

Weakly safe belief. We can define a *weakly safe belief* operator $\Box^{\text{weak}}\mathbf{P}$ in terms of the strict order by putting:

$$\Box_a^{\text{weak}}\mathbf{P} \;=\; \mathbf{P} \wedge [>_a]\mathbf{P}\,.$$

Clearly, this gives us the following truth clause:

$$s \models \Box_a^{\text{weak}}\ \mathbf{P} \text{ iff: } s \models \mathbf{P} \text{ and } t \models \mathbf{P} \text{ for all } t < s\,.$$

But a more useful characterization is the following:

$$s \models \Box_a^{\text{weak}}\mathbf{Q} \quad \text{iff:} \quad s \models \neg B_a^{\mathbf{P}}\neg\mathbf{Q} \text{ for every } \mathbf{P} \text{ such that } s \models \mathbf{P}\,.$$

So "weakly safe beliefs" are *beliefs which (might be lost but) are never reversed (into believing the opposite) when revising with any true information.*

The unary revision operator. Using the strict plausibility modality, we can also define a unary "belief revision" modality $*_a$, which in some sense *internalizes the standard (binary) belief revision operator*, by putting:

$$*_a\mathbf{P} \;=\; \mathbf{P} \wedge [>_a]\neg\mathbf{P}\,.$$

This gives us the following truth clause:

$$s \models *_a\mathbf{P} \text{ iff } s \in s_a^{\mathbf{P}}\,.$$

It is easy to see that $*_a\mathbf{P}$ *selects from any given information cell* $s(a)$ *precisely those states that satisfy agent* a*'s revised theory* s_a^P:

$$*_aP \cap s(a) = s_a^P\,.$$

Recall that $s_a^P \;=\; \text{Min}_{\leq_a} s(a) \cap P$ is the conditional appearance of s to a given P, representing the agent's "revised theory" (after revision with P) about s. This explains our interpretation: the proposition $*_a\mathbf{P}$ is a *complete description of the agent's* P*-revised "theory" about the current state.*

Another interesting identity is the following:

$$B_a^{\mathbf{P}}\mathbf{Q} \;=\; K_a(*_a\mathbf{P} \to \mathbf{Q})\,. \tag{39.8}$$

In other words: $\mathbf{Q}$ is *a conditional belief (given a condition* $\mathbf{P}$*) iff it is a known consequence of the agent's revised theory (after revision with* $\mathbf{P}$*).*

Degrees of belief. Spohn's "degrees of belief" in Spohn (1988) were captured by Aucher (2003) and van Ditmarsch (2005) using logical operators $B^n_a\mathbf{P}$. Intuitively, 0-belief $B^0_a\mathbf{P}$ is the same as simple belief $B_a\mathbf{P}$; 1-belief $B^1_a\mathbf{P}$ means that $\mathbf{P}$ is believed conditional on learning that not all the 0-beliefs are true etc. Formally, this can be introduced e.g., by defining by induction a sequence of appearance maps s^n_a for all states s and natural numbers n:

$$s^0_a = \mathrm{Min}_{\leq_a} s(a) \ , \ s^n_a = \mathrm{Min}_{\leq_a}\Big(s(a) \setminus \bigcup_{i<n} s^i_a\Big)$$

and defining

$$s \models B^n_a\mathbf{P} \ \text{ iff } \ t \models \mathbf{P} \text{ for all } t \in s^n_a.$$

A state s has degree of belief n if we have $s \in s^n_a$. An interesting observation is that the *finite degrees of belief $B^n_a\mathbf{P}$ can be defined using the unary revision operator $*_a\mathbf{P}$ and the knowledge operator K_a* (and, as a consequence, they can be defined using the plausibility operator $[>_a]\mathbf{P}$ and the knowledge operator). To do this, first put inductively:

$$b^0_a := *_a\top \ , \ b^n_a := *_a\Big(\bigwedge_{m<n} \neg b^m_a\Big) \quad \text{for all } n \geq 1$$

and then put

$$B^n_a\mathbf{P} := \bigwedge_{m<n} \neg K_a(b^m_a \to \mathbf{P}) \wedge K_a(b^n_a \to \mathbf{P}) \, .$$

"Strong belief". Another important doxastic attitude can be defined in terms of knowledge and safe belief as:

$$Sb_a\mathbf{P} \ = \ B_a\mathbf{P} \wedge K_a(\mathbf{P} \to \Box_a\mathbf{P}) \, .$$

In terms of the plausibility order, it means that *all the **P**-states in the information cell $s(a)$ of s are bellow (more plausible than) all the non-**P** states in $s(a)$* (and that, moreover, *there are such **P**-states in $s(a)$*). This notion is called "strong belief" by Battigalli and Siniscalchi (2002), while Stalnaker (1996) calls it "robust belief". Another characterization of strong belief is the following

$s \models Sb_a\mathbf{Q}$ iff:

$$s \models B_a\mathbf{Q} \text{ and } s \models B^{\mathbf{P}}_a\mathbf{Q} \text{ for every } \mathbf{P} \text{ such that } s \models \neg K_a(\mathbf{P} \to \neg\mathbf{Q}).$$

In other words: *something is strong belief if it is believed and if this belief can only be defeated by evidence (truthful or not) that is known to contradict it.* An example is the "presumption of innocence" in a trial: requiring the members of the jury to hold the accused as "innocent until proven guilty" means asking them to start the trial with a "strong belief" in innocence.

The Logic of Conditional Beliefs

The logic CDL ("conditional doxastic logic") introduced in Baltag and Smets (2006a) is a logic of conditional beliefs, equivalent to the strongest logic considered in Board (2002). The *syntax* of CDL (without common knowledge and common belief operators[13]) is:

$$\varphi := p \mid \neg\varphi \mid \varphi \wedge \varphi \mid B_a^{\varphi}\varphi$$

while the *semantics* is given by an *interpretation map* associating to each sentence φ of CDL a doxastic proposition $\|\varphi\|$. The definition is by induction, in terms of the obvious compositional clauses (using the doxastic operators $B_a^{\mathbf{P}}\mathbf{Q}$ defined above).

In this logic, *knowledge and simple (unconditional) belief are derived operators*, defined as abbreviations by putting $K_a\varphi := B_a^{\neg\varphi}\varphi$, $B_a\varphi := B_a^{\top}\varphi$ (where $\top := \neg(p \wedge \neg p)$ is some tautological sentence).

Proof system. In addition to the rules and axioms of propositional logic, the *proof system* of CDL includes the following:

Necessitation Rule:	From $\vdash \varphi$ infer $\vdash B_a^{\psi}\varphi$.
Normality:	$\vdash B_a^{\theta}(\varphi \to \psi) \to (B_a^{\theta}\varphi \to B_a^{\theta}\psi)$
Truthfulness of Knowledge:	$\vdash K_a\varphi \to \varphi$
Persistence of Knowledge:	$\vdash K_a\varphi \to B_a^{\theta}\varphi$
Full Introspection:	$\vdash B_a^{\theta}\varphi \to K_a B_a^{\theta}\varphi$,
	$\vdash \neg B_a^{\theta}\varphi \to K_a \neg B_a^{\theta}\varphi$
Success of Belief Revision:	$\vdash B_a^{\varphi}\varphi$
Minimality of Revision:	$\vdash \neg B_a^{\varphi}\neg\psi \to (B_a^{\varphi\wedge\psi}\theta \leftrightarrow B_a^{\varphi}(\psi \to \theta))$

Proposition 4 (Completeness and Decidability). *The above system is complete for MPMs (and so also for EPMs). Moreover, it is decidable and has the finite model property.*

Proof. The proof is essentially the same as in Board (2002). It is easy to see that the proof system above is equivalent to Board's strongest logic in Board (2002) (the one that includes axiom for full introspection), and that our models are equivalent to the "full introspective" version of the semantics in Board (2002). Q.E.D.

[13]The logic in Baltag and Smets (2006a) has these operators, but for simplicity we decided to leave them aside in this presentation.

The Logic of Knowledge and Safe Belief

The problem of finding a complete axiomatization of the logic of "defeasible knowledge" (safe belief) and conditional belief was posed as an *open question* in Board (2002). We answer this question here, by extending the logic CDL above to a complete logic $K\Box$ of *knowledge and safe belief*. Since this logic can *define* conditional belief, it is in fact equivalent to the logic whose axiomatization was required in Board (2002). Solving the question posed there becomes in fact trivial, once we observe that the higher-order definition of "defeasible knowledge" in Stalnaker (2006) and Board (2002) (corresponding to our identity (39.3) above) is in fact equivalent to our simpler, first-order definition of "safe belief" as a Kripke modality.

Syntax and semantics. The *syntax* of the logic $K\Box$ is:

$$\varphi \ := \ p \mid \neg\varphi \mid \varphi \wedge \varphi \mid \Box_a\varphi \mid K_a\varphi$$

while the *semantics* over plausibility models is given as for CDL, by inductively defining an interpretation map from sentences to doxastic propositions, using the obvious compositional clauses. *Belief and conditional belief are derived operators here, defined as abbreviations*:

$$B_a^{\varphi}\psi \ := \ \tilde{K}_a\varphi \rightarrow \tilde{K}_a(\varphi \wedge \Box_a(\varphi \rightarrow \psi))\,,$$
$$B_a\varphi \ := \ B_a^{\top}\varphi\,,$$

where $\tilde{K}_a\varphi := \neg K_a\neg\varphi$ is the Diamond modality for K, and $\top = \neg(p \wedge \neg p)$ is some tautological sentence. So *the logic $K\Box$ is more expressive than CDL.*

Proof system. In addition to the rules and axioms of propositional logic, the *proof system* for the logic $K\Box$ includes the following:

- the Necessitation Rules for both K_a and $\Box_a$;
- the $S5$-axioms for K_a;
- the $S4$-axioms for $\Box_a$;
- $K_aP \rightarrow \Box_aP$;
- $K_a(P \vee \Box_aQ) \wedge K_a(Q \vee \Box_aP) \rightarrow K_aP \vee K_aQ$.

Theorem 5 (Completeness and Decidability). *The logic $K\Box$ is* (weakly) complete *with respect to MPMs (and so also with respect to EPMs). Moreover, it is* decidable *and has* the finite model property.

Proof. A *non-standard frame (model)* is a structure $(S, \geq_a, \sim_a)_a$ (together with a valuation, in the case of models) such that $\sim_a$ are equivalence relations, $\geq_a$ are preorders, $\geq_a \ \subseteq \ \sim_a$ and the restriction of $\geq_a$ to each $\sim_a$-equivalence class is connected. For a logic with two modalities, $\Box_a$ for $\geq_a$ and K_a for the relation $\sim_a$, we can use well-known results in Modal Correspondence Theory to see that each of these semantic conditions corresponds to one of our modal axioms above. By general classical results on canonicity and modal correspondence,[14] we immediately obtain *completeness for non-standard models. Finite model property for these non-standard models* follows from the same general results. But every *finite* strict preorder relation $>$ is well-founded, and an MPM is nothing but a non-standard model whose strict preorders $>_a$ are well-founded. So *completeness for ("standard") MPMs* immediately follows. Then we can use Proposition 4 above to obtain *completeness for EPMs*. Finally, *decidability* follows, in the usual way, from finite model property together with *completeness* (with respect to a *finitary* proof system) and with the *decidability of model-checking on finite models*. (This last property is obvious, given the semantics.) Q.E.D.

"Dynamic" Belief Revision

The revision captured by conditional beliefs is of a *static*, purely *hypothetical*, nature. We *cannot* interpret B_a^φ as referring to the agent's revised beliefs about the situation *after revision*; if we did, then the "Success" axiom

$$\vdash B_a^\varphi \varphi$$

would *fail for higher-level beliefs*. To see this, consider a "Moore sentence"

$$\varphi := p \wedge \neg B_a p\,,$$

saying that some fact p holds but that agent a doesn't believe it. The sentence φ is consistent, so it may very well happen to be true. But agent a's beliefs about the situation after learning that φ was true *cannot* possibly include the sentence φ itself: after learning this sentence, agent *a knows p*, and so he believes p, contrary to what φ asserts. Thus, after learning φ, agent *a knows that φ is false now* (after the learning). This directly contradicts the Success axiom: far from believing the sentence after learning it to be true, the agent (knows, and so he correctly) believes that it has become false. There is nothing paradoxical about this: sentences may obviously change their truth values, due to our actions. Since learning the truth of a sentence is itself an action, it is perfectly consistent to have a case in which learning changes the truth value of the very sentence that is being learnt. Indeed, this is always the case with Moore sentences. Though not paradoxical, the existence of Moore sentences shows that the "Success" axiom does not correctly describe a rational agent's (higher-level) beliefs about what is the case after a new truth is being learnt.

[14]See e.g., Blackburn et al. (2001) for the general theory of modal correspondence and canonicity.

The only way to understand the "Success" axiom in the context of higher-level beliefs is to insist on the above-mentioned "static" interpretation of conditional belief operators B_a^φ, as expressing the agent's *revised belief* about how the state of the world *was before the revision*.

In contrast, a *belief update* is a dynamic form of belief revision, meant to capture the actual change of beliefs induced by learning: the updated belief is about the state of the world as it is *after the update*. As noticed in Gerbrandy (1999), Baltag et al. (1998), and Baltag and Moss (2004), the original model does not usually include enough states to capture all the epistemic possibilities that arise in this way. While in the previous section the models were kept unchanged during the revision, all the possibilities being already there (so that both the unconditional and the conditional beliefs *referred to the same model*), we now have to allow for belief updates that *change the original model*.

In Baltag and Moss (2004), it was argued that *epistemic events should be modeled in essentially the same way as epistemic states*, and this common setting was taken to be given by *epistemic Kripke models*. Since in this paper we enriched our state models with doxastic plausibility relations to deal with (conditional) beliefs, it is natural to follow Baltag and Moss (2004) into extending the similarity between actions and states to this setting, thus obtaining *(epistemic) action plausibility models*. The idea of such an extension was first developed in Aucher (2003) (for a different notion of plausibility model and a different notion of update product), then generalized in van Ditmarsch (2005), where many types of action plausibility models and notions of update product, that extend the so-called *Baltag-Moss-Solecki (BMS) update product* from Baltag et al. (1998) and Baltag and Moss (2004), are explored. But both these works are based on a *quantitative* interpretation of plausibility ordinals (as "degrees of belief"), and thus they define the various types of products using complex formulas of transfinite ordinal arithmetic, for which no intuitive justification is provided.

In contrast, our notion of update product is a *purely qualitative one*, based on a *simple and intuitive relational definition*: the simplest way to define a total pre-order on a Cartesian product, given total pre-orders on each of the components, is to use either the *lexicographic* or the *anti-lexicographic* order. We choose the second option, as the closest in spirit to the classical AGM theory: it gives *priority to the new, incoming information* (i.e., to "actions" in our sense).[15] We justify this choice by interpreting the action plausibility model as representing the agent's *"incoming" belief*, i.e., the *belief-updating event*, which *"performs" the update, by "acting" on the "prior" beliefs* (as given in the state plausibility model).

[15]This choice can be seen as a generalization of the so-called *"maximal-Spohn" revision*.

Action Models

An *action plausibility model*[16] (APM, for short) is a plausibility frame $(\Sigma, \leq_a)_{a \in \mathcal{A}}$ together with a *precondition map* pre : $\Sigma \to \mathsf{Prop}$, associating to each element of Σ some doxastic proposition pre_σ. We call the elements of Σ *(basic) doxastic actions* (or "events"), and we call pre_σ the *precondition* of action σ. The basic actions $\sigma \in \Sigma$ are taken to represent *deterministic belief-revising actions* of a particularly simple nature. Intuitively, the precondition defines the *domain of applicability* of action σ: it can be executed on a state s iff s satisfies its precondition. The relations $\leq_a$ give the agents' beliefs about which actions are more plausible than others.

To model *non-determinism*, we introduce the notion of epistemic program. A *doxastic program over a given action model* $\mathbf{\Sigma}$ (or $\mathbf{\Sigma}$*-program*, for short) is simply a *set* $\Gamma \subseteq \Sigma$ of doxastic actions. We can think of doxastic programs as non-deterministic actions: each of the basic actions $\gamma \in \Gamma$ is a possible "deterministic resolution" of Γ. For simplicity, when $\Gamma = \{\gamma\}$ is a singleton, we ambiguously identify the program Γ with the action γ.

Observe that $\mathbf{\Sigma}$-programs $\Gamma \subseteq \Sigma$ are formally the "dynamic analogues" of $\mathbf{S}$-propositions $P \subseteq S$. So the dynamic analogue of the conditional doxastic appearance s_a^P (representing agent a's revised theory about state s, after revision with proposition P) is the set σ_a^Γ.

Interpretation: beliefs about changes encode changes of beliefs. The name "doxastic actions" might be a bit misleading, and from a philosophical perspective Johan van Benthem's term "doxastic events" seems more appropriate. The elements of a plausibility model do not carry information about agency or intentionality and cannot represent "real" actions in all their complexity, but only the *doxastic changes* induced by these actions: each of the nodes of the graph represents a *specific kind of change of beliefs (of all the agents)*. As in Baltag and Moss (2004), we only deal here with pure "belief changes", i.e., actions that do not change the "ontic" facts of the world, but only the agents' beliefs.[17] Moreover, we think of these as *deterministic* changes: there is at most one output of applying an action to a state.[18] Intuitively, the precondition defines the *domain of applicability* of σ: this action can be executed on a state s iff s satisfies its precondition. The plausibility pre-orderings $\leq_a$ give *the agents' conditional beliefs about the current action*. But this should be interpreted as *beliefs about changes*, that *encode changes of beliefs*. In this sense, we use such

[16] Van Benthem calls this an "event model".

[17] We stress this is a minor restriction, and it is very easy to extend this setting to "ontic" actions. The only reason we stick with this restriction is that it simplifies the definitions, and that it is general enough to apply to all the actions we are interested here, and in particular to all *communication actions*.

[18] As in Baltag and Moss (2004), we will be able to represent non-deterministic actions as sums (unions) of deterministic ones.

"beliefs about actions" as a way to represent doxastic changes: the information about how the agent changes her beliefs is captured by our action plausibility relations. So we read $\sigma <_a \sigma'$ as saying that: if agent a is informed that either σ or σ' is currently happening, then she cannot distinguish between the two, but she believes that σ is in fact happening. As already mentioned, doxastic programs $\Gamma \subseteq \Sigma$ represent *non-deterministic* changes of belief. Finally, for an action σ and a program Γ, the program σ_a^Γ represents *the agent's revised theory (belief) about the current action σ after "learning" that (one of the deterministic resolutions γ in) Γ is currently happening*.

Example 4 (Private "Fair-Game" Announcements). Let us consider the *action* that produced the situation represented in Example 2 above. In front of Alice, Bob looked at the coin, in such a way that (it was common knowledge that) only he saw the face. In the DEL literature, this is sometimes known as a "fair game" announcement: everybody is commonly aware that an insider (or a group of insiders) privately learns some information. It is "fair" since the outsiders are *not "deceived" in any way*: e.g., in our example, Alice knows that Bob looks at the coin (and he knows that she knows etc.). In other words, Bob's looking at the coin is not an "illegal" action, but one that obeys the (commonly agreed) "rules of the game". To make this precise, let us assume that this is happening in such a way that Alice has no strong beliefs about which of the two possible actions (Bob-seeing-Heads-up and Bob-seeing-Tails-up) is actually happening. Of course, we assumed that before this, she already believed that the coin lies Heads up, but apart from this we now assume that *the way the action (of "Bob looking") is happening gives her no indication of what face he is seeing*. We represent these actions using a two-node plausibility model $\mathbf{\Sigma}_2$ (where as in the case of state models we draw arrows for the converse plausibility relations $\geq_a$, disregarding all the loops):

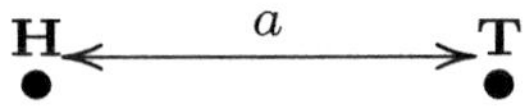

Example 5 (Fully Private Announcements). Let us consider the *action* that produced the situation represented in Example 3 above. This was the action of Bob taking a peek at the coin, while Alice was away. Recall that we assumed that Alice *believed that nothing was really happening* in her absence (since she assumed Bob was playing by the rules), though obviously she *didn't know* this (that nothing was happening). In the DEL literature, this action is usually called *a fully private announcement*: Bob learns which face is up, while the outsider Alice believes nothing of the kind is happening. To represent this, we consider an action model $\mathbf{\Sigma}_3$ consisting of three "actions": the actual action σ in which Bob takes a peek and sees the coin lying Heads up; the alternative possible action ρ is the one in which Bob sees the coin lying Tails up; finally, the action τ is the one in which "nothing is really happening" (as Alice believes). The plausibility model $\mathbf{\Sigma}_3$ for this action is:

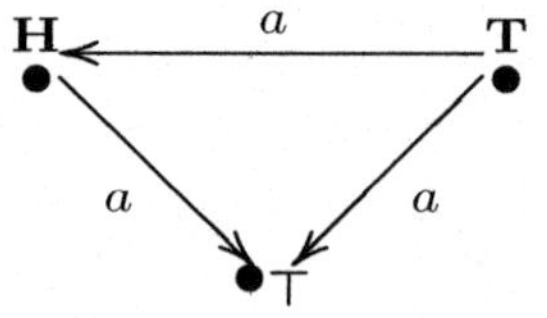

Here, the action σ is the one in the upper left corner, having precondition **H**: indeed, this can happen iff the coin is really lying Heads up; similarly, the action ρ in the upper right corner has precondition **T**, since it can only happen iff the coin is Tails up. Finally, the action τ is the lower one, having as precondition the "universally true" proposition $\top$: indeed, this action can always happen (since in it, nothing is really happening!). The plausibility relations reflect the agents' beliefs: in each case, both Bob and Charles know exactly what is happening, so their local plausibility relations are the identity (and thus we draw no arrows for them). Alice believes nothing is happening, so τ is the most plausible action for her (to which all her arrows are pointing); so she keeps her belief that **H** is the case, thus considering σ as more plausible than ρ.

Examples of doxastic programs. Consider the program $\Gamma = \{\sigma, \rho\} \subseteq \mathbf{\Sigma}_3$ over the action model $\mathbf{\Sigma}_3$ from Example 5. The program Γ represents *the action of "Bob taking a peek at the coin", without any specification of which face he is seeing*. Although expressed in a non-deterministic manner (as a collection of two possible actions, σ and ρ), this program corresponds is in fact *deterministic*, since in each possible state only one of the actions σ or ρ can happen: there is no state satisfying both **H** and **T**. The whole set Σ gives another doxastic program, one that is really non-deterministic: it represents the non-deterministic choice of Bob between taking a peek and not taking it.

Appearance of actions and their revision: Examples. As an example of an agent's "theory" about an action, consider the appearance of action ρ to Alice: $\rho_a = \{\tau\}$. Indeed, if ρ happens (Bob taking a peek and sees the coin is Tails up), Alice believes that τ (i.e., nothing) is happening: this is the "apparent action", as far as Alice is concerned. As an example of a "revised theory" about an action, consider the conditional appearance ρ_a^Γ of ρ to Alice given the program $\Gamma = \{\sigma, \rho\}$ introduced above. It is easy to see that we have $\rho_a^\Gamma = \{\sigma\}$. This captures our intuitions about Alice's revised theory: if, while ρ was happening, she were told that Bob took a peek (i.e., she'd revise with Γ), then she would believe that he saw the coin lying Heads up (i.e., that σ happened).

Example 6 (Successful Lying). Suppose now that, *after* the previous action, i.e., after we arrived in the situation described in Example 3, Bob sneakily announces: "I took a peek and saw the coin was lying *Tails up*". We formalize the content of this announcement as $K_b\mathbf{T}$, i.e., saying that "Bob knows the coin is lying Tails up". This is a *public announcement*, but *not a truthful one* (though it does convey some truthful information): it is a *lie*! We assume it is in fact a *successful lie*: it is common knowledge that, even after Bob admitted having taken a peek, Alice still believes him. This action is given by the *left node* in the following model $\mathbf{\Sigma}_4$:

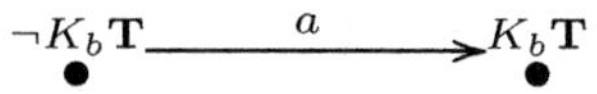

The Action-Priority Update

We are ready to define our *update operation*, representing the way an action from a (action) plausibility model $\mathbf{\Sigma} = (\Sigma, \leq_a, \mathrm{pre})_{a \in \mathcal{A}}$ "acts" on an input-state from a given (state) plausibility model $\mathbf{S} = (S, \leq_a, \|\cdot\|)_{a \in \mathcal{A}}$. We denote the updated state model by $\mathbf{S} \otimes \mathbf{\Sigma}$, and call it the *update product* of the two models. The construction is similar to a point to the one in Baltag et al. (1998) and Baltag and Moss (2004), and thus also somewhat similar to the ones in Aucher (2003) and van Ditmarsch (2005). In fact, the set of updated states, the updated valuation and the updated indistinguishability relation are *the same* in these constructions. The main difference lies in our definition of the *updated plausibility relation*, via the *Action Priority Rule*.

Updating Single-Agent Models: The Anti-lexicographic Order

To warm up, let us first define the update product for the single-agent case. Let $\mathbf{S} = (S, \leq, \|\cdot\|)$ be a single-agent plausibility state model and let $\mathbf{\Sigma} = (\Sigma, \leq, \mathrm{pre})$ be a single-agent plausibility action model.

We represent the *states of the updated model* $\mathbf{S} \otimes \mathbf{\Sigma}$ as pairs (s, σ) of input-states and actions, i.e., as elements of the Cartesian product $S \times \Sigma$. This reflects that the basic actions in our action models are assumed to be *deterministic*: For a given input-state and a given action, there can only be at most one output-state. More specifically, we select the pairs which are *consistent, in the sense that the input-state satisfies the precondition of the action*. This is natural: the precondition of an action is a specification of its domain of applicability. So the *set of states* of $\mathbf{S} \otimes \mathbf{\Sigma}$ is taken to be

$$S \otimes \Sigma := \{(s, \sigma) : s \models_{\mathbf{S}} \mathrm{pre}(\sigma)\}.$$

The *updated valuation* is essentially given by the *original valuation* from the input-state model: For all $(s, \sigma) \in S \otimes \Sigma$, we put $(s, \sigma) \models p$ iff $s \models p$. This "conservative" way to update the valuation expresses the fact that we only consider here actions that are *"purely doxastic"*, i.e., pure "belief changes", that do not affect the ontic "facts" of the world (captured here by atomic sentences).

We still need to define the updated plausibility relation. To motivate our definition, we first consider two examples:

Example 7 (A Sample Case). Suppose that we have two states $s, s' \in \mathbf{S}$ such that $s < s'$, $s \models \neg\mathbf{P}$, $s' \models \mathbf{P}$. This means that, if given the supplementary information that the real state is either s or s', the agent believes $\neg\mathbf{P}$:

Suppose then an event happens, in whose model there are two actions σ, σ' such that $\sigma > \sigma'$, $\text{pre}_\sigma = \neg\mathbf{P}$, $\text{pre}_{\sigma'} = \mathbf{P}$. In other words, if given the information that either σ or σ' is happening, the agent believes that σ' is happening, i.e., she believes that $\mathbf{P}$ is learnt. This part of the model behaves just like a *soft public announcement* of $\mathbf{P}$:

Naturally, we expect the agent to *change her belief* accordingly, i.e., her updated plausibility relation on states should now go the other way:

Example 8 (A Second Sample Case). Suppose the initial situation was the same as above, but now the two actions σ, σ' are assumed to be equi-plausible: $\sigma \cong \sigma'$. This is a *completely unreliable announcement of* $\mathbf{P}$, in which the veracity and the falsity of the announcement are equally plausible alternatives:

In the AGM paradigm, it is natural to expect the agents to *keep their original beliefs* unchanged after this event:

The anti-lexicographic order. Putting the above two sample cases together, we conclude that the updated plausibility relation should be the *anti-lexicographic preorder relation* induced on pairs $(s, \sigma) \in S \times \Sigma$ by the preorders on $\mathbf{S}$ and on $\mathbf{\Sigma}$, i.e.:

$$(s, \sigma) \leq (s', \sigma') \quad \text{iff:} \quad \text{either } \sigma < \sigma', \text{ or else } \sigma \cong \sigma' \text{ and } s \leq s'.$$

In other words, the updated plausibility order gives *"priority" to the action plausibility relation, and apart from this it keeps as much as possible the old order*. This reflects our commitment to an AGM-type of revision, in which the new information has priority over old beliefs. The "actions" represent here the "new information", although (unlike in AGM) this information comes in *dynamic form* (as action plausibility order), and so it is not fully reducible to its propositional content (the action's precondition). In fact, this is a generalization of one of the belief-revision policies encountered in the literature (the so-called "*maximal-Spohn revision*"). But, in the context of our qualitative (conditional) interpretation of plausibility models, we will argue below that this is essentially the only reasonable option.

Updating Multi-agent Models: The General Case

In the multi-agent case, the construction of *the updated state space and updated valuation is the same as above*. But for the updated plausibility relation we need to take into account *a third possibility*: the case when either the initial states or the actions are *distinguishable*, belonging to *different information cells*.

Example 9 (A Third Sample Case). Suppose that we have two states $s, s' \in \mathbf{S}$ such that $s \models \neg\mathbf{P}$, $s' \models \mathbf{P}$, but $s \not\sim_a s'$ are *distinguishable* (i.e., non-comparable):

This means that, if given the supplementary information that the real state is either s or s', the agent immediately *knows* which of the two is the real states, and thus *she knows whether* **P** *holds or not*. It is obvious that, after any of the actions considered in the previous two examples, a perfect-recall agent *will continue to know* whether **P** held or not, and so *the output-states after* σ *and* σ' *will still be distinguishable (non comparable)*.

The "Action-Priority" Rule. Putting this together with the other sample cases, we obtain our update rule, in full generality:

$$(s,\sigma) \leq_a (s',\sigma') \text{ iff either } \sigma <_a \sigma' \text{ and } s \sim_a s', \text{ or else } \sigma \cong_a \sigma' \text{ and } s \leq_a s'$$

We regard this construction as the most natural analogue in a belief-revision context of the similar notion in Baltag and Moss (2004) and Baltag et al. (1998). Following a suggestion of Johan van Benthem, we call this the Action-Priority Update Rule.

Sanity check: Examples 2 and 3 revisited. To check the correctness of our update operation, take first the update product $\mathbf{S} \otimes \mathbf{\Sigma}_2$ of the model **S** in Example 1 from the previous section with the action model $\mathbf{\Sigma}_2$ in Example 4 from the previous section. As predicted, the resulting state model is isomorphic to the model **W** from

Example 2. Similarly, if $\mathbf{\Sigma}_3$ is the action model from Example 5, then we can see that the product $\mathbf{S} \otimes \mathbf{\Sigma}_3$ is isomorphic to the state model $\mathbf{S}'$ from Example 3.

"In-sanity check": Successful lying. Applying the action model $\mathbf{\Sigma}_4$ in Example 6, representing the "successful lying" action, to the state model $\mathbf{S}'$ from Example 3, we obtain indeed the intuitively correct output of "successful lying", namely the following model $\mathbf{S}' \otimes \mathbf{\Sigma}_4$:

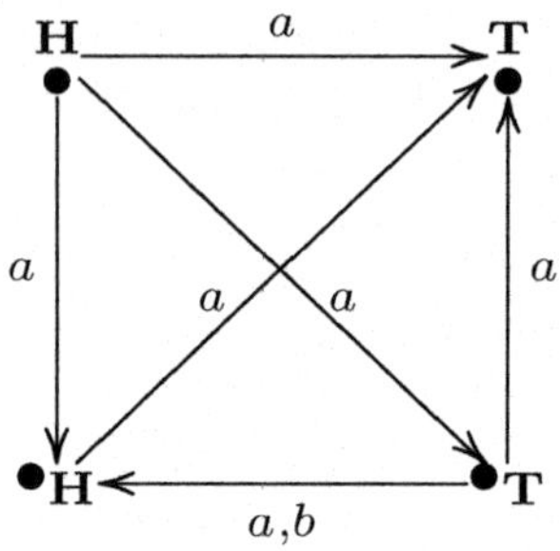

Interpretation. As its name makes explicit, the Action-Priority Rule gives "priority" to the *action* plausibility relation. This is not an arbitrary choice, but it is motivated by our specific interpretation of action models, as embodied in our Motto above: *beliefs about changes* (i.e., the action plausibility relations) *are nothing but ways to encode changes of belief* (i.e., reversals of the original plausibility order). So *the (strict) order on actions encodes changes of order on states*. The Action-Priority Rule is a consequence of this interpretation: it just says that a strong plausibility order $\sigma <_a \sigma'$ on actions corresponds indeed to a change of ordering, (from whatever the ordering was) between the original (indistinguishable) input-states $s \sim_a s'$, to the order $(s, \sigma) <_a (s', \sigma')$ between output-states; while equally plausible actions $\sigma \cong_a \sigma'$ will leave the initial ordering unchanged: $(s, \sigma) \leq_a (s', \sigma')$ iff $s \leq_a s'$. Giving priority to action plausibility does not in any way mean that the agent's belief in actions is stronger than her belief in states; it just captures the fact that, at the time of updating with a given action, *the belief about the action is what is actual, it is the current belief about what is going on, while the beliefs about the input-states are in the past.*[19]

In a nutshell: *the doxastic action is the one that changes the initial doxastic state, and not vice-versa.* The belief update induced by a given action is nothing but an update with the (presently) believed action. If the believed action σ requires the agent to revise some past beliefs, then so be it: this is the whole point of believing σ, namely to use it to revise one's past beliefs. For example, in a successful lying, the action plausibility relation makes the hearer believe that the speaker is telling

[19]Of course, *at a later moment*, the above-mentioned belief about action (*now* belonging to the past) might be itself revised. But this is another, *future update*.

the truth; so she'll accept this message (unless contradicted by her knowledge), and change her past beliefs appropriately: this is what makes the lying successful.

Action-priority update generalizes product update. Recall the definition of the epistemic indistinguishability relation $\sim_a$ in a plausibility model: $s \sim_a s'$ iff either $s \leq_a s'$ or $s' \leq_a s$. It is easy to see that the Action Priority Update implies the familiar update rule from Baltag et al. (1998) and Baltag and Moss (2004), known in Dynamic Epistemic Logic as the "product update":

$$(s,\sigma) \sim_a (s',\sigma') \text{ iff } s \sim_a s' \text{ and } \sigma \sim_a \sigma'.$$

Program transitions. For every state model $\mathbf{S}$, every program $\Gamma \subseteq \Sigma$ over an action model $\mathbf{\Sigma}$ induces a transition relation $\xrightarrow{\Gamma}_{\mathbf{S}} \subseteq \mathbf{S} \times (S \otimes \Sigma)$ from $\mathbf{S}$ to $\mathbf{S} \otimes \mathbf{\Sigma}$, given by:

$$s \xrightarrow{\Gamma}_{\mathbf{S}} (s',\gamma) \text{ iff } s = s',\ (s,\gamma) \in S \otimes \Sigma \text{ and } \gamma \in \Gamma .$$

Simulating Various Belief-Revision Policies

We give here three examples of *multi-agent belief-revision policies* that can be simulated by our product update: *truthful public announcements of "hard facts"*, *"lexicographic update"* and *"conservative upgrade"*. They were all introduced by van Benthem in (2007), as multi-agent versions of revision operators previously considered by Rott (1989) and others.

Public announcements of "hard facts". A *truthful public announcement* $!\mathbf{P}$ *of some "hard fact"* $\mathbf{P}$ is not really about belief revision, but about the learning of *certified true information*: it establishes *common knowledge* that $\mathbf{P}$ was the case. This is the action described in van Benthem (2007) as (public) "belief change under hard facts". As an operation on models, this is described in van Benthem (2007) as taking any state model $\mathbf{S}$ and *deleting all the non-*$\mathbf{P}$ *states, while keeping the same indistinguishability and plausibility relations between the surviving states.* In our setting, the corresponding action model consists of only one node, labeled with $\mathbf{P}$. It is easy to see that the above operation on models can be exactly "simulated" by taking the anti-lexicographic product update with this one-node action model.

Public announcements of "soft facts": The "lexicographic upgrade". To allow for "soft" belief revision, an operation $\Uparrow\mathbf{P}$ was introduced in van Benthem (2007), essentially adapting to public announcements the 'lexicographic' policy for belief revision described in Rott (1989). This operation, called "lexicographic update" consists of changing the current plausibility order on any given state model as follows: *every* $\mathbf{P}$*-world becomes "better" (more plausible) than all* $\neg\mathbf{P}$*-worlds in* *the same information cell, and within the two zones (**P** and ¬**P**), the old ordering remains.* In our setting, this action corresponds to the following local plausibility action model:

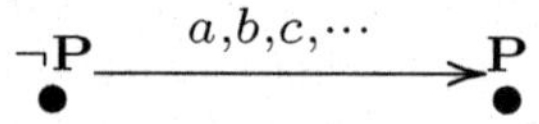

Taking the anti-lexicographic update product with this action will give an exact "simulation" of the lexicographic upgrade operation.

"Conservative upgrade". The operation $\uparrow\mathbf{P}$ of "conservative upgrade", also defined in van Benthem (2007), changes any model as follows: *in every information cell, the best* **P**-*worlds become better than all the worlds in that cell* (i.e., in every cell the most plausible **P**-states become the most plausible overall in that cell), *and apart from that, the old order remains*. In the case of a system *with only one agent*, it is easy to see that we have $\uparrow\mathbf{P} = \Uparrow(*_a\mathbf{P})$, where $*_a$ is the unary "revision modality" introduced in the previous section. In the case of a set $\mathcal{A} = \{1, \cdots, n\}$ with $n > 1$ agents, we can simulate $\uparrow\mathbf{P}$ using a model with 2^n actions $\{\uparrow_I\mathbf{P}\}_{I\subseteq\mathcal{A}}$, with

$$\mathrm{pre}_{\uparrow_I\mathbf{P}} = \bigwedge_{i\in I} *_i\mathbf{P} \wedge \bigwedge_{j\notin I} \neg *_j \mathbf{P},$$

$$\uparrow_I\mathbf{P} \leq_k \uparrow_J\mathbf{P} \quad \text{iff} \quad J \cap \{k\} \subseteq I.$$

Operations on Doxastic Programs

First, we introduce *dynamic modalities*, capturing *the "weakest precondition" of a program* Γ. These are the natural analogues of the PDL modalities for our program transition relations $\stackrel{\Gamma}{\rightarrow}$ between models.

Dynamic modalities. Let $\mathbf{\Sigma}$ be some action plausibility model and $\Gamma \subseteq \Sigma$ be a doxastic model over $\mathbf{\Sigma}$. For every doxastic proposition **P**, we define a doxastic proposition $[\Gamma]\mathbf{P}$ given by

$$([\Gamma]\mathbf{P})_{\mathbf{S}} := [\stackrel{\Gamma}{\rightarrow}_{\mathbf{S}}]\mathbf{P}_{\mathbf{S}} = \{s \in S : \forall t \in S \otimes \Sigma\, (s \stackrel{\Gamma}{\rightarrow}_{\mathbf{S}} t \Rightarrow t \models_{\mathbf{S}\otimes\mathbf{\Sigma}} \mathbf{P})\,\}.$$

For *basic doxastic actions* $\sigma \in \Sigma$, we define the dynamic modality $[\sigma]$ via the above-mentioned identification of actions σ with singleton programs $\{\sigma\}$:

$$([\sigma]\mathbf{P})_{\mathbf{S}} := ([\{\sigma\}]\mathbf{P})_{\mathbf{S}} = \{s \in S : \text{ if } (s, \sigma) \in \mathbf{S} \otimes \mathbf{\Sigma} \text{ then } (s, \sigma) \in \mathbf{P}_{\mathbf{S}\otimes\mathbf{\Sigma}}\}.$$

The dual (Diamond) modalities are defined as usually: $\langle\Gamma\rangle\mathbf{P} := \neg[\Gamma]\neg\mathbf{P}$.

We can now introduce operators on doxastic programs that are the analogues of the *regular operations* of PDL.

Sequential composition. The *sequential composition* $\mathbf{\Sigma};\mathbf{\Delta}$ of two action plausibility models $\mathbf{\Sigma}=(\Sigma,\leq_a,\text{pre})$, $\mathbf{\Delta}=(\Delta,\leq_a,\text{pre})$ is defined as follows:

- the set of basic actions is the Cartesian product $\Sigma\times\Delta$
- the preconditions are given by $\text{pre}_{(\sigma,\delta)} := \langle\sigma\rangle\,\text{pre}_\delta$
- the plausibility order is given by putting $(\sigma,\delta)\leq_a(\sigma',\delta')$ iff: either $\sigma <_a \sigma'$ and $\delta\sim_a\delta'$, or else $\sigma\cong_a\sigma'$ and $\delta\leq_a\delta'$.

We think of (σ,δ) as the action of *performing first* σ *then* δ, and thus use the notation

$$\sigma;\delta := (\sigma,\delta).$$

We can extend this notation to doxastic programs, by defining the *sequential composition of programs* $\Gamma\subseteq\Sigma$ and $\Lambda\subseteq\Delta$ to be a program $\Gamma;\Lambda\subseteq\Sigma;\Delta$ over the action model $\mathbf{\Sigma};\mathbf{\Delta}$, given by:

$$\Gamma;\Lambda := \{(\gamma,\lambda):\gamma\in\Gamma,\lambda\in\Lambda\}.$$

It is easy to see that this behaves indeed like a sequential composition:

Proposition 12. *For every state plausibility model* $\mathbf{S}$*, action plausibility models* $\mathbf{\Sigma}$ *and* $\mathbf{\Delta}$*, and programs* $\Gamma\subseteq\Sigma$*,* $\Lambda\subseteq\Delta$*, we have the following:*

1. *The state plausibility models* $(\mathbf{S}\otimes\mathbf{\Sigma})\otimes\mathbf{\Delta}$ *and* $\mathbf{S}\otimes(\mathbf{\Sigma};\mathbf{\Delta})$ *are isomorphic, via the canonical map* $F:(S\otimes\Sigma)\otimes\Delta\to S\otimes(\Sigma;\Delta)$ *given by*

$$F(\,((s,\sigma),\delta)\,) \;:=\; (s,(\sigma,\delta)).$$

2. *The transition relation for the program* $\Gamma;\Delta$ *is the relational composition of the transition relations for* Γ *and for* Δ *and of the isomorphism map* F*:*

$s\xrightarrow{\Gamma;\Delta}_{\mathbf{S}}s'$ iff there exist $w,t\in S\otimes\Sigma$ such that

$$s\xrightarrow{\Gamma}_{\mathbf{S}}w\xrightarrow{\Delta}_{\mathbf{S}\otimes\mathbf{\Sigma}}t \text{ and } F(t)=s'.$$

Union (non-deterministic choice). If $\mathbf{\Sigma}=(\Sigma,\leq_a,\text{pre})$ and $\mathbf{\Delta}=(\Delta,\leq'_a,\text{pre}')$ are two action plausibility models, their *disjoint union* $\mathbf{\Sigma}\sqcup\mathbf{\Delta}$ is simply given by taking as set of states the disjoint union $\Sigma\sqcup\Delta$ of the two sets of states, taking as plausibility order the disjoint union $\leq_a\sqcup\leq'_a$ and as precondition map the disjoint union $\text{pre}\sqcup\text{pre}'$ of the two precondition maps. If $\Gamma\subseteq\Sigma$ and $\Lambda\subseteq\Delta$ are doxastic programs over the two models, we define their *union* to be the program over the model $\mathbf{\Sigma}\sqcup\mathbf{\Delta}$ given by the disjoint union $\Gamma\sqcup\Lambda$ of the sets of actions of the two programs.

Again, it is easy to see that *this behaves indeed like a non-deterministic choice operator*:

Proposition 13. *Let $i_1 : \Sigma \to \Sigma \sqcup \Delta$ and $i_2 : \Delta \to \Sigma \sqcup \Delta$ be the two canonical injections. Then the following are equivalent:*

- $s \overset{\Gamma \sqcup \Delta}{\to}_{\mathbf{S}} s'$
- *there exists t such that:*

$$\text{either } s \overset{\Gamma}{\to}_{\mathbf{S}} t \text{ and } i_1(t) = s', \text{ or else } s \overset{\Delta}{\to}_{\mathbf{S}} t \text{ and } i_2(t) = s'.$$

Other operators. *Arbitrary unions* $\bigsqcup_i \Gamma_i$ can be similarly defined, and then one can define *iteration* $\Gamma^* := \bigsqcup_i \Gamma^i$ (where $\Gamma^0 = !\top$ and $\Gamma^{i+1} = \Gamma; \Gamma^i$).

The Laws of Dynamic Belief Revision

The "laws of dynamic belief revision" are the fundamental equations of Belief Dynamics, allowing us *to compute future doxastic attitudes from past ones*, given the doxastic events that happen in the meantime. In modal terms, these can be stated as "reduction laws" for inductively computing dynamic modalities $[\Gamma]\mathbf{P}$, by reducing them to modalities $[\Gamma']\mathbf{P}'$ in which either the propositions $\mathbf{P}'$ or the programs Γ' have *lower complexity*.

The following immediate consequence of the definition of $[\Gamma]\mathbf{P}$ allows us to reduce modalities for non-deterministic programs Γ to the ones for their deterministic resolutions $\gamma \in \Gamma$:

Deterministic Resolution Law. For every program $\Gamma \subseteq \Sigma$, we have

$$[\Gamma]\mathbf{P} = \bigwedge_{\gamma \in \Gamma} [\gamma]\mathbf{P}.$$

So, for our other laws, we can restrict ourselves to *basic actions* in Σ.

The Action-Knowledge Law. For every action $\sigma \in \Sigma$, we have:

$$[\sigma]K_a\mathbf{P} = \mathrm{pre}_\sigma \to \bigwedge_{\sigma' \sim_a \sigma} K_a[\sigma']\mathbf{P}.$$

This Action-Knowledge Law is essentially the same as in Baltag et al. (1998) and Baltag and Moss (2004): *a proposition* **P** *will be known after a doxastic event iff, whenever the event can take place, it is known that* **P** *will become true after all events that are indistinguishable from the given one*.

The Action-Safe-Belief Law. For every action $\sigma \in \Sigma$, we have:

$$[\sigma]\Box_a \mathbf{P} \;=\; \mathrm{pre}_\sigma \rightarrow \bigwedge_{\sigma' <_a \alpha} K_a[\sigma']\mathbf{P} \wedge \bigwedge_{\sigma'' \cong_a \sigma} \Box_a[\sigma'']\mathbf{P}\,.$$

This law embodies the essence of the Action-Priority Rule: *a proposition* **P** *will be safely believed after a doxastic event iff, whenever the event can take place, it is known that* **P** *will become true after all more plausible events and in the same time it is safely believed that* **P** *will become true after all equi-plausible events.*

Since we took knowledge and safe belief as the basis of our static logic $K\Box$, the above two laws are the "fundamental equations" of our theory of dynamic belief revision. But note that, as a consequence, one can obtain *derived laws for (conditional) belief* as well. Indeed, using the above-mentioned characterization of conditional belief in terms of K and $\Box$, we obtain the following:

The Derived Law of Action-Conditional-Belief. For every action $\sigma \in \Sigma$, we have:

$$[\sigma]B_a^{\mathbf{P}}\mathbf{Q} \;=\; \mathrm{pre}_\sigma \rightarrow \bigvee_{\Gamma \subseteq \Sigma} \Big(\bigwedge_{\gamma \in \Gamma} \tilde{K}_a\langle\gamma\rangle\mathbf{P} \wedge \bigwedge_{\gamma' \notin \Gamma} \neg\tilde{K}_a\langle\gamma'\rangle\mathbf{P} \wedge B_a^{\langle\sigma_a^\Gamma\rangle\mathbf{P}}[\sigma_a^\Gamma]\mathbf{Q}\Big)\,.$$

This derived law, a version of which was first introduced in Baltag and Smets (2006c) (where it was considered a fundamental law), allows us to predict future conditional beliefs from current conditional beliefs.

To explain the meaning of this law, we re-state it as follows: For every $s \in \mathbf{S}$ and $\sigma \in \mathbf{\Sigma}$, we have:

$$s \models [\sigma]B_a^{\mathbf{P}}\mathbf{Q} \;\text{ iff }\; s \models \mathrm{pre}_\sigma \rightarrow B_a^{\langle\sigma_a^\Gamma\rangle\mathbf{P}}[\sigma_a^\Gamma]\mathbf{Q}\,,$$
$$\text{where } \Gamma = \{\gamma \in \Sigma : s \models_{\mathbf{S}} \tilde{K}_a\langle\gamma\rangle\mathbf{P}\}\,.$$

It is easy to see that this "local" (state-dependent) version of the reduction law is equivalent to the previous (state-independent) one. The set Γ encodes the extra information about the current action that is given to the agent by the context s and by the post-condition **P**; while σ_a^Γ is the action's *post-conditional contextual appearance*, i.e., the way it appears to the agent in the view of this extra-information Γ. Indeed, a given action might "appear" differently in a given context (i.e., at a state s) than it does in general: the information possessed by the agent at the state s might imply the negation of certain actions, hence their impossibility; this information will then be used to revise the agent's beliefs about the actions, obtaining her contextual beliefs. Moreover, in the presence of further information (a "post-condition" **P**), this appearance might again be revised. The "post-conditional contextual appearance" is the result of this double revision: the agent's belief about action σ is revised with the information given to her by the context s and the post-condition **P**. This information is encoded in a set $\Gamma = \{\gamma \in \Sigma : s \models_{\mathbf{S}} \tilde{K}_a\langle\gamma\rangle\mathbf{P}\}$ of "admissible" actions: the actions for which the agent considers epistemically possible (at s) that they can be performed and they can achieve the post-condition **P**. The "post-conditional contextual appearance" σ_a^Γ of action σ captures the agent's revised theory about σ after revision with the relevant information Γ.

So the above law says that: *the agent's future conditional beliefs* $[\sigma]B_a^{\mathbf{P}}$ *can be predicted, given that action* σ *happens, by her current conditional beliefs* $B_a^{\langle\sigma_a^\Gamma\rangle\mathbf{P}}[\sigma_a^\Gamma]$ *about what will be true after the apparent action* σ_a^Γ (as it appears in the given context and in the view of the given post-condition **P**), *beliefs conditioned on the information* ($\langle\sigma_a^\Gamma\rangle\mathbf{P}$) *that the apparent action* σ_a^Γ *actually can lead to the fulfillment of the post-condition* **P**.

Special cases. As special cases of the Action-Conditional-Belief Law, we can derive *all the reduction laws* in van Benthem (2007) for (conditional) belief after the events $!\mathbf{P}$, $\Uparrow\mathbf{P}$ and $\uparrow\mathbf{P}$:

$$[!\mathbf{P}]B_a^{\mathbf{Q}}\mathbf{R} = \mathbf{P} \to B_a^{\mathbf{P}\wedge[!\mathbf{P}]\mathbf{Q}}[!\mathbf{P}]\mathbf{R},$$

$$[\Uparrow\mathbf{P}]B_a^{\mathbf{Q}}\mathbf{R} = (\tilde{K}_a^{\mathbf{P}}[\Uparrow\mathbf{P}]\mathbf{Q} \wedge B_a^{\mathbf{P}\wedge[\Uparrow\mathbf{P}]\mathbf{Q}}[\Uparrow\mathbf{P}]\mathbf{R}) \vee (\neg\tilde{K}_a^{\mathbf{P}}[\Uparrow\mathbf{P}]\mathbf{Q} \wedge B_a^{[\Uparrow\mathbf{P}]\mathbf{Q}}[\Uparrow\mathbf{P}]\mathbf{R}),$$

$$[\uparrow\mathbf{P}]B_a^{\mathbf{Q}}\mathbf{R} = (\tilde{B}_a^{\mathbf{P}}[\uparrow\mathbf{P}]\mathbf{Q} \wedge B_a^{\mathbf{P}\wedge[\uparrow\mathbf{P}]\mathbf{Q}}[\uparrow\mathbf{P}]\mathbf{R}) \vee (\neg\tilde{B}_a^{\mathbf{P}}[\uparrow\mathbf{P}]\mathbf{Q} \wedge B_a^{[\uparrow\mathbf{P}]\mathbf{Q}}[\uparrow\mathbf{P}]\mathbf{R}),$$

where

$$K_a^{\mathbf{P}}\mathbf{Q} := K_a(\mathbf{P} \to \mathbf{Q}), \qquad \tilde{K}_a^{\mathbf{P}}\mathbf{Q} := \neg K_a^{\mathbf{P}}\neg\mathbf{Q}, \qquad \tilde{B}_a^{\mathbf{P}}\mathbf{Q} := \neg B_a^{\mathbf{P}}\neg\mathbf{Q}.$$

Laws for other doxastic attitudes. The *equi-plausibility modality behaves dynamically "almost" like knowledge*, while *the strict plausibility modality behaves like safe belief*, as witnessed by the following laws:

$$[\sigma][\cong_a]\mathbf{P} = pre_\sigma \to \bigwedge_{\sigma'\cong_a\sigma} [\cong_a][\sigma']\mathbf{P},$$

$$[\sigma][>_a]\mathbf{P} = pre_\sigma \to \bigwedge_{\sigma'<_a\alpha} K_a[\sigma']\mathbf{P} \wedge \bigwedge_{\sigma''\cong_a\sigma} [>_a][\sigma'']\mathbf{P}.$$

From these, we can *derive laws for all the other doxastic attitudes* above.

The Logic of Doxastic Actions

The problem of finding a general syntax for action models has been tackled in various ways by different authors. Here we use the *action-signature approach* from Baltag and Moss (2004).

Signature. A doxastic *action signature* is a *finite plausibility frame* $\mathbf{\Sigma}$, together with an *ordered list without repetitions* $(\sigma_1, \ldots, \sigma_n)$ of some of the elements of $\mathbf{\Sigma}$. The elements of $\mathbf{\Sigma}$ are called *action types*. A type σ is called *trivial* if it is *not* in the above list.

Example 10. The *"hard" public announcement signature* HardPub is a singleton frame, consisting of one action type !, identity as the order relation, and the list (!).

The *"soft" public announcement signature* SoftPub is a two-point frame, consisting of types $\Uparrow$ and $\Downarrow$, with $\Downarrow <_a \Uparrow$ for all agents a, and the list $(\Uparrow, \Downarrow)$.

Similarly, one can define the signatures of *fully private announcements with n alternatives, private "fair-game" announcements, conservative upgrades* etc. As we will see below, *there is no signature of "successful (public) lying": public lying actions fall under the type of "soft" public announcements*, so they are generated by that signature.

Languages. For each action signature $(\mathbf{\Sigma}, (\sigma_1, \ldots, \sigma_n))$, the language $L(\mathbf{\Sigma})$ consists of a set of *sentences* φ and a set of *program terms* π, defined by simultaneous recursion:

$$\varphi ::= p \mid \neg\varphi \mid \varphi \wedge \varphi \mid K_a\varphi \mid \Box_a\varphi \mid [\pi]\varphi$$

$$\pi ::= \sigma\varphi_1 \ldots \varphi_n \mid \pi \sqcup \pi \mid \pi ; \pi$$

where $p \in \Phi$, $a \in \mathcal{A}$, $\sigma \in \Sigma$, and $\sigma\varphi_1 \ldots \varphi_n$ is an expression consisting of σ and a string of n sentences, where n is the length of the list $(\sigma_1, \ldots, \sigma_n)$.

Syntactic action model. The expressions of the form $\sigma\vec{\varphi}$ are called *basic programs*. The preorders on $\mathbf{\Sigma}$ induce in a natural way preorders on the basic programs in $L(\mathbf{\Sigma})$:

$$\sigma\vec{\varphi} \leq_a \sigma\vec{\psi} \text{ iff } \sigma \leq_a \sigma' \text{ and } \vec{\varphi} = \vec{\psi}.$$

The given listing can be used to assign syntactic preconditions for basic programs, by putting: $\text{pre}_{\sigma_i\vec{\varphi}} := \varphi_i$, and $\text{pre}_{\sigma\vec{\varphi}} := \top$ (the trivially true sentence) if σ is not in the listing. Thus, the basic programs of the form $\sigma\vec{\varphi}$ form a *"syntactic plausibility model"* $\mathbf{\Sigma}\vec{\varphi}$; i.e., every given interpretation $\|\cdot\| : L(\mathbf{\Sigma}) \to$ Prop of sentences as doxastic propositions will convert this syntactic model into a "real" (semantic) plausibility model, called $\mathbf{\Sigma}\,\|\vec{\varphi}\|$.

Action models induced by a signature. For a given signature $\mathbf{\Sigma}$, let $(\sigma_1, \ldots, \sigma_n)$ be its list of non-trivial types, and let $\vec{\mathbf{P}} = (\mathbf{P}_1, \ldots, \mathbf{P}_n)$ be a matching list of doxastic propositions. The *action model generated by the signature* $\mathbf{\Sigma}$ *and the list of propositions* $\vec{\mathbf{P}}$ is the model $\mathbf{\Sigma}\vec{\mathbf{P}}$, having $\mathbf{\Sigma}$ as its underlying action frame and having a precondition map given by: $\text{pre}_{\sigma_i} = \mathbf{P}_i$, for non-trivial types σ_i; and $\text{pre}_\sigma = \top$ (the trivially true proposition), for trivial types σ. When referring to σ as an *action* in $\mathbf{\Sigma}\vec{\mathbf{P}}$, we will denote it by $\sigma\vec{\mathbf{P}}$, to distinguish it from the action *type* $\sigma \in \Sigma$.

We can obviously extend this construction to *sets of action types*: given a signature $\mathbf{\Sigma}$ and a list $\vec{\mathbf{P}} = (\mathbf{P}_1, \dots, \mathbf{P}_n)$, every set $\Gamma \subseteq \Sigma$ gives rise to a doxastic program $\Gamma\vec{\mathbf{P}} := \{\sigma\vec{\mathbf{P}} : \sigma \in \Sigma\} \subseteq \Sigma\vec{\mathbf{P}}$.

Example 11. The action model of a hard public announcement $!\,\mathbf{P}$ is generated as $!(\mathbf{P})$ by the hard public announcement signature $\mathsf{HardPub} = \{!\}$ and the list $(\mathbf{P})$. Similarly, the action model $\mathsf{SoftPub}(\mathbf{P})$ generated by the *soft* public announcement signature $\mathsf{SoftPub}$ and a list $(\mathbf{P}, \mathbf{Q})$ of two propositions consists of two actions $\Uparrow(\mathbf{P}, \mathbf{Q})$ and $\Downarrow(\mathbf{P}, \mathbf{Q})$, with $\Uparrow(\mathbf{P}, \mathbf{Q}) <_a \Downarrow(\mathbf{P}, \mathbf{Q})$, $\mathrm{pre}_{\Uparrow(\mathbf{P},\mathbf{Q})} = \mathbf{P}$ and $\mathrm{pre}_{\Downarrow(\mathbf{P},\mathbf{Q})} = \mathbf{Q}$:

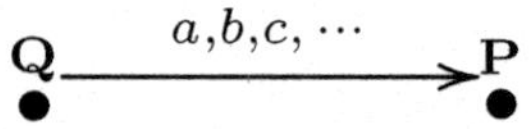

This represents an event during which all agents share a common belief that $\mathbf{P}$ is announced; but they might be wrong and maybe $\mathbf{Q}$ was announced instead. However, it is common knowledge that either $\mathbf{P}$ or $\mathbf{Q}$ was announced.

Successful (public) lying Lie $\mathbf{P}$ *(by an anonymous agent, falsely announcing* $\mathbf{P}$*)* can now be expressed as $\mathrm{Lie}\,\mathbf{P} := \Downarrow(\mathbf{P}, \neg\mathbf{P})$. The *truthful* soft announcement is $\mathrm{True}\,\mathbf{P} := \Uparrow(\mathbf{P}, \neg\mathbf{P})$. Finally, the soft public announcement (lexicographic update) $\Uparrow\mathbf{P}$, as previously defined, is given by the non-deterministic union $\Uparrow\mathbf{P} := \mathrm{True}\,\mathbf{P} \sqcup \mathrm{Lie}\,\mathbf{P}$.

Semantics. We define by simultaneous induction two *interpretation maps*, one taking sentences φ into doxastic propositions $\|\varphi\| \in \mathsf{Prop}$, the second taking program terms π into doxastic programs $\|\pi\|$ over some plausibility frames. The inductive definition uses the obvious semantic clauses. For programs: $\|\sigma\vec{\varphi}\|$ is the action $\sigma\,\|\vec{\varphi}\|$ (or, more exactly, the singleton program $\{\sigma\,\|\vec{\varphi}\|\}$ over the frame $\mathbf{\Sigma}\,\|\vec{\varphi}\|$), $\|\pi \sqcup \pi'\| := \|\pi\| \sqcup \|\pi'\|$, $\|\pi; \pi'\| := \|\pi\|; \|\pi'\|$. For sentences: $\|p\|$ is as given by the valuation, $\|\neg\varphi\| := \neg\|\varphi\|$, $\|\varphi \wedge \psi\| := \|\varphi\| \wedge \|\psi\|$, $\|K_a\varphi\| := K_a\|\varphi\|$, $\|\Box_a\varphi\| := \Box_a\|\varphi\|$, $\|[\pi]\varphi\| := [\,\|\pi\|\,]\,\|\varphi\|$.

Proof system. In addition to the axioms and rules of the logic $K\Box$, the logic $L(\mathbf{\Sigma})$ includes the following Reduction Axioms:

$$[\alpha]p \leftrightarrow \mathrm{pre}_\alpha \rightarrow p$$

$$[\alpha]\neg\varphi \leftrightarrow \mathrm{pre}_\alpha \rightarrow \neg[\alpha]\varphi$$

$$[\alpha](\varphi \wedge \psi) \leftrightarrow \mathrm{pre}_\alpha \rightarrow [\alpha]\varphi \wedge [\alpha]\psi$$

$$[\alpha]K_a\varphi \leftrightarrow \mathrm{pre}_\alpha \rightarrow \bigwedge_{\alpha' \sim_a \alpha} K_a[\alpha']\varphi$$

$$[\alpha]\Box_a\varphi \leftrightarrow \mathrm{pre}_\alpha \rightarrow \bigwedge_{\alpha' <_a \alpha} K_a[\alpha']\varphi \wedge \bigwedge_{\alpha'' \cong_a \alpha} \Box_a[\alpha'']\varphi$$

$$[\pi \sqcup \pi']\varphi \leftrightarrow [\pi]\varphi \wedge [\pi']\varphi$$

$$[\pi; \pi']\varphi \leftrightarrow [\pi]\,[\pi']\varphi$$

where p is any atomic sentence, π, π' are program terms, α is a *basic* program term in $L(\Sigma)$, pre is the syntactic precondition map defined above, and $\sim_a, <_a, \cong_a$ are respectively the (syntactic) epistemic indistinguishability, the strict plausibility order and the equi-plausibility relation on basic programs.

Theorem 16. *For every signature* Σ*, the above proof system for the dynamic logic* $L(\Sigma)$ *is* complete, decidable *and* has the finite model property. *In fact, this dynamic logic has the same expressive power as the "static" logic* $K\Box$ *of knowledge and safe belief.*

Proof (Sketch). The proof is similar to the ones in Baltag and Moss (2004), Baltag et al. (1998), and van Ditmarsch et al. (2007). We use the reduction laws to inductively simplify any formula until it is reduced to a formula of the $K\Box$-logic, then use the completeness of the $K\Box$ logic. Note that this is *not* an induction on subformulas, but (as in Baltag et al. 1998) on an appropriate notion of "complexity" ordering of formulas. Q.E.D.

Current and Future Work, Some Open Questions

In our papers Baltag and Smets (2007a,b), we present a *probabilistic version* of the theory developed here, based on *discrete (finite) Popper-Renyi conditional probability spaces* (allowing for conditionalization on events of non-zero probability, in order to cope with non-trivial belief revisions). We consider subjective probability to be the proper notion of "degree of belief", and we investigate its relationship with the qualitative concepts developed here. We develop a probabilistic generalization of the Action Priority Rule, and show that the logics presented above are *complete for the (discrete) conditional probabilistic semantics.*

We mention here a number of open questions: (1) Axiomatize the full (static) logic of doxastic attitudes introduced in this paper. It can be easily shown that they can all be reduced to the modalities K_a, $[>_a]$ and $[\cong_a]$. There are a number of obvious axioms for the resulting logic $K[>][\cong]$ (note in particular that $[>]$ satisfies the Gödel-Löb formula!), but the completeness problem is still open. (2) Axiomatize the logic of *common safe belief and common knowledge*, and their *dynamic versions*. More generally, explore the logics obtained by adding *fixed points*, or at least "epistemic regular (PDL-like) operations" as in van Benthem et al. (2006b), on top of our doxastic modalities. (3) Investigate the *expressive limits* of this approach *with respect to belief-revision policies*: what policies can be simulated by our update? (4) Extend the work in Baltag and Smets (2007a,b), by investigating and axiomatizing doxastic logics on *infinite* conditional probability models. (5) Extend the logics with *quantitative (probabilistic)* modal operators $B^P_{a,x}Q$ (or $\Box_{a,x}Q$) expressing that *the degree of conditional belief in* Q *given* P (or the *degree of safety* of the belief in Q) is at least x.

Acknowledgements Sonja Smets' contribution to this research was made possible by the postdoctoral fellowship awarded to her by the Flemish Fund for Scientific Research. We thank Johan van Benthem for his insights and help, and for the illuminating discussions we had with him on the topic of this paper. His pioneering work on dynamic belief revision acted as the "trigger" for our own. We also thank Larry Moss, Hans van Ditmarsch, Jan van Eijck and Hans Rott for their most valuable feedback. Finally, we thank the editors and the anonymous referees of the LOFT7-proceedings for their useful suggestions and comments.

During the republication of this paper in 2015, the research of Sonja Smets was funded by the European Research Council under the European Community's Seventh Framework Programme (FP7/2007-2013)/ERC Grant agreement no.283963.

References

Alchourrón, C. E., Gärdenfors, P., & Makinson, D. (1985). On the logic of theory change: partial meet contraction and revision functions. *Journal of Symbolic Logic, 50*(2), 510–530.

Aucher, G. (2003). *A combined system for update logic and belief revision*. Master's thesis, University of Amsterdam. ILLC Publications MoL-2003-03.

Aumann, R. J. (1999). Interactive epistemology I: Knowledge. *International Journal of Game Theory, 28*(3), 263–300.

Baltag, A. (2002). A logic for suspicious players: Epistemic actions and belief updates in games. *Bulletin of Economic Research, 54*(1), 1–46.

Baltag, A., & Moss, L. S. (2004). Logics for epistemic programs. *Synthese, 139*(2), 165–224.

Baltag, A., Moss, L. S., & Solecki, S. (1998). The logic of public announcements, common knowledge, and private suspicions. In I. Gilboa (Ed.), *Proceedings of the 7th Conference on Theoretical Aspects of Rationality and Knowledge (TARK 98)*, Morgan Kaufmann Publishers Inc. San Francisco, CA, USA, (pp. 43–56).

Baltag, A., & Sadrzadeh, M. (2006). The algebra of multi-agent dynamic belief revision. *Electronic Notes in Theoretical Computer Science, 157*(4), 37–56.

Baltag, A., & Smets, S. (2006). Conditional doxastic models: A qualitative approach to dynamic belief revision. *Electronic Notes in Theoretical Computer Science, 165*, 5–21.

Baltag, A., & Smets, S. (2006b) Dynamic belief revision over multi-agent plausibility models. In Bonanno et al. (2006) (pp. 11–24).

Baltag, A., & Smets, S. (2006c). The logic of conditional doxastic actions: A theory of dynamic multi-agent belief revision. In S. Artemov, & Parikh, R. (Eds.), *Proceedings of ESSLLI Workshop on Rationality and Knowledge*, (pp. 13–30). ESSLLI.

Baltag, A., & Smets, S. (2007a). From conditional probability to the logic of doxastic actions. In D. Samet (Ed.), *Proceedings of the 11th Conference on Theoretical Aspects of Rationality and Knowledge (TARK)*, Brussels (pp. 52–61). UCL Presses Universitaires de Louvain.

Baltag, A., & Smets, S. (2007b). Probabilistic dynamic belief revision. In J. F. A. K. van Benthem, S. Ju, & F. Veltman (Eds.), *A Meeting of the Minds: Proceedings of the Workshop on Logic, Rationality and Interaction*, Beijing, 2007 (Texts in computer science, Vol. 8). London: College Publications.

Battigalli, P., & Siniscalchi, M. (2002). Strong belief and forward induction reasoning. *Journal of Economic Theory, 105*(2), 356–391.

Blackburn, P., de Rijke, M., & Venema, Y. (2001). *Modal logic* (Cambridge tracts in theoretical computer science, Vol. 53). Cambridge: Cambridge University Press.

Board, O. (2002). Dynamic interactive epistemology. *Games and Economic Behaviour, 49*(1), 49–80.

Bonanno, G. (2005). A simple modal logic for belief revision. *Synthese, 147*(2), 193–228.

Bonanno, G., van der Hoek, W., & Wooldridge, M. (Eds.). (2006). *Proceedings of the 7th Conference on Logic and the Foundations of Game and Decision Theory (LOFT7)*, University of Liverpool UK.

Friedmann, N., & Halpern, J. Y. (1994). Conditional logics of belief revision. In *Proceedings of the of 12th National Conference on Artificial Intelligence (AAAI-94)*, Seattle, 31 July–4 Aug 1994 (pp. 915–921). Menlo Park: AAAI.

Gärdenfors, P. *Knowledge in flux: Modelling the dynamics of epistemic states*. Gardenfors. 1988, MIT Press, Cambridge/London.

Gerbrandy, J. (1999). Dynamic epistemic logic. In L. S. Moss, J. Ginzburg, & M. de Rijke (Eds.), *Logic, language and information* (Vol. 2, p. 67–84). Stanford: CSLI Publications/Stanford University.

Gerbrandy, J., & Groeneveld, W. (1997). Reasoning about information change. *Journal of Logic, Language and Information, 6*(2), 147–169.

Gerbrandy, J. D. (1999). *Bisimulations on planet Kripke*. PhD thesis, University of Amsterdam. ILLC Publications, DS-1999-01.

Gettier, E. (1963). Is justified true belief knowledge? *Analysis, 23*(6), 121–123.

Gochet, P., & Gribomont, P. (2006). Epistemic logic. In D. M. Gabbay & J. Woods (Eds.), *Handbook of the history of logic* (Vol. 7, p. 99–195). Oxford: Elsevier.

Grove, A. (1988). Two modellings for theory change. *Journal of Philosophical Logic, 17*(2), 157–170.

Hintikka, J. (1962). *Knowledge and belief*. Ithaca: Cornell University Press.

Katsuno, H., & Mendelzon, A. O. (1992). On the difference between updating a knowledge base and revising it. In P. Gärdenfors (Ed.), *Belief revision* (Cambridge tracts in theoretical computer science, pp. 183–203). Cambridge/New York: Cambridge University Press.

Klein, P. (1971). A proposed definition of propositional knowledge. *Journal of Philosophy, 68*(16), 471–482.

Kooi, B. P. (2003). Probabilistic dynamic epistemic logic. *Journal of Logic, Language and Information, 12*(4), 381–408.

Lehrer, K. (1990). *Theory of knowledge*. London: Routledge.

Lehrer, K., & Paxson, T. Jr. (1969). Knowledge: Undefeated justified true belief. *Journal of Philosophy, 66*(8), 225–237.

Meyer, J.-J. Ch. & van der Hoek, W. (1995). *Epistemic logic for AI and computer science* (Cambridge tracts in theoretical computer science, Vol. 41). Cambridge: Cambridge University Press.

Pappas, G., & Swain, M. (Eds.). (1978). *Essays on knowledge and justification*. Ithaca: Cornell University Press.

Plaza, J. A. (1989). Logics of public communications. In M. L. Emrich, M. S. Pfeifer, M. Hadzikadic, & Z. W. Ras (Eds.), *Proceedings of the 4th International Symposium on Methodologies for Intelligent Systems Poster Session Program* (pp. 201–216). Oak Ridge National Laboratory, ORNL/DSRD-24.

Rott, H. (1989). Conditionals and theory change: Revisions, expansions, and additions. *Synthese, 81*(1), 91–113.

Rott, H. (2004). Stability, strength and sensitivity: Converting belief into knowledge. *Erkenntnis, 61*(2–3), 469–493.

Ryan, M., & Schobbens, P.-Y. (1997). Counterfactuals and updates as inverse modalities. *Journal of Logic, Language and Information, 6*(2), 123–146.

Segerberg, K. (1998). Irrevocable belief revision in dynamic doxastic logic. *Notre Dame Journal of Formal Logic, 39*(3), 287–306.

Spohn, W. (1988). Ordinal conditional functions: A dynamic theory of epistemic states. In W. L. Harper & B. Skyrms (Eds.), *Causation in decision, belief change, and statistics* (Vol. II, pp. 105–134). Dordrecht/Boston: Kluwer Academic

Stalnaker, R. (1968). A theory of conditionals. In N. Rescher (Ed.), *Studies in logical theory* (APQ monograph series, Vol. 2). Oxford: Blackwell.

Stalnaker, R. (1996). Knowledge, belief and counterfactual reasoning in games. *Economics and Philosophy, 12*, 133–163.
Stalnaker, R. (2006). On logics of knowledge and belief. *Philosophical Studies, 128*(1), 169–199.
van Benthem, J. F. A. K. (2007). Dynamic logic for belief revision. *Journal of Applied Non-classical Logics, 17*(2), 129–155.
van Benthem, J. F. A. K., Gerbrandy, J., & Kooi, B. (2006a) Dynamic update with probabilities. In Bonanno et al. (2006) (pp. 237–246).
van Benthem, J. F. A. K., & Liu, F. (2004). Dynamic logic of preference upgrade. Technical report, University of Amsterdam. ILLC Publications, PP-2005-29.
van Benthem, J. F. A. K., van Eijck, J., & Kooi, B. P. (2006b). Logics of communication and change. *Information and Computation, 204*(11), 1620–1662.
van der Hoek, W. (1993). Systems for knowledge and beliefs. *Journal of Logic and Computation, 3*(2), 173–195.
van Ditmarsch, H. P. (2000). *Knowledge games*. PhD thesis, University of Groningen. ILLC Pubications, DS-2000-06.
van Ditmarsch, H. P. (2002). Descriptions of game actions. *Journal of Logic, Language and Information, 11*(3), 349–365.
van Ditmarsch, H. P. (2005) Prolegomena to dynamic logic for belief revision. *Synthese, 147*(2), 229–275.
van Ditmarsch, H. P., & Labuschagne, W. (2007). My beliefs about your beliefs: A case study in theory of mind and epistemic logic. *Synthese, 155*(2), 191–209.
van Ditmarsch, H. P., van der Hoek, W., & Kooi, B. P. (2007). *Dynamic epistemic logic* (Synthese library, Vol. 337). Dordrecht: Springer.
Voorbraak, F. P. J. M. (1993). *As far as I know*. PhD thesis, Utrecht University, Utrecht (Quaestiones infinitae, Vol. VII).
Williamson, T. (2001). Some philosophical aspects of reasoning about knowledge. In J. van Benthem (Ed.), *Proceedings of the 8th Conference on Theoretical Aspects of Rationality and Knowledge (TARK'01)* (p. 97). San Francisco: Morgan Kaufmann.

Chapter 40
Agreeing to Disagree

Robert J. Aumann

If two people have the same priors, and their posteriors for a given event A are common knowledge, then these posteriors must be equal. This is so even though they may base their posteriors on quite different information. In brief, people with the same priors *cannot agree to disagree.*

We publish this observation with some diffidence, since once one has the appropriate framework, it is mathematically trivial. Intuitively, though, it is not quite obvious; and it is of some interest in areas in which people's beliefs about each other's beliefs are of importance, such as game theory[1] and the economics of information.[2] A "concrete" illustration that may clarify matters (and that may be read at this point) is found at the end of the paper.

The key notion is that of "common knowledge." Call the two people 1 and 2. When we say that an event is "common knowledge," we mean more than just that both 1 and 2 know it; we require also that 1 knows that 2 knows it, 2 knows that 1 knows it, 1 knows that 2 knows that 1 knows it, and so on. For example, if 1 and 2 are both present when the event happens and see each other there, then the event

AMS 1970 *subject classifications.* Primary 62A15, 62C05; Secondary 90A05, 90D35.

Key words and phrases. Information, subjective probability, posterior, statistics, game theory, revising probabilities, concensus, Harsanyi doctrine.

[1]Cf. Harsanyi (1967–1968); also Aumann (1974), especially Section 9j (page 92), in which the question answered here was originally raised.

[2]Cf., e.g., Radner (1968, 1972); also the review by Grossman and Stiglitz (1976) and the papers quoted there.

R.J. Aumann (✉)
Federmann Center for the Study of Rationality, The Hebrew University of Jerusalem, Jerusalem, Israel
e-mail: raumann@math.huji.ac.il

H. Arló-Costa et al. (eds.), *Readings in Formal Epistemology*, Springer Graduate Texts in Philosophy 1, DOI 10.1007/978-3-319-20451-2_40

becomes common knowledge. In our case, if 1 and 2 tell each other their posteriors and trust each other, then the posteriors are common knowledge. The result is not true if we merely assume that the persons know each other's posteriors.

Formally, let $(\Omega, \mathscr{B}, p)$ be a probability space, $\mathscr{P}_1$ and $\mathscr{P}_2$ partitions of Ω whose join[3] $\mathscr{P}_1 \vee \mathscr{P}_2$ consists of nonnull events.[4] In the interpretation, $(\Omega, \mathscr{B})$ is the space of states of the world, p the common prior of 1 and 2, and $\mathscr{P}_i$ the information partition of i; that is, if the true state of the world is ω, then i is informed of that element $\mathbf{P}_i(\omega)$ of $\mathscr{P}_i$ that contains ω. Given ω in Ω, an event E is called *common knowledge at* ω if E includes that member of the meet[5] $\mathscr{P}_1 \wedge \mathscr{P}_2$ that contains ω. We will show below that this definition is equivalent to the informal description given above.

Let A be an event, and let $\mathbf{q}_i$ denote the posterior probability $p(A \mid \mathscr{P}_i)$ of A given i's information; i.e., if $\omega \in \Omega$, then $\mathbf{q}_i(\omega) = p(A \cap \mathbf{P}_i(\omega))/p(\mathbf{P}_i(\omega))$.

Proposition *Let* $\omega \in \Omega$, *and let* q_1 *and* q_2 *be numbers. If it is common knowledge at* ω *that* $\mathbf{q}_1 = q_1$ *and* $\mathbf{q}_2 = q_2$, *then* $q_1 = q_2$.

Proof Let P be the member of $\mathscr{P}_1 \wedge \mathscr{P}_2$ that contains ω. Write $P = \cup_j P^j$, where the P^j are disjoint members of $\mathscr{P}_1$. Since $\mathbf{q}_1 = q_1$ throughout P, we have $p(A \cap P^j)/p(P^j) = q_1$ for all j; hence $p(A \cap P^j) = q_1 p(P^j)$, and so by summing over j we get $p(A \cap P) = q_1 p(P)$. Similarly $p(A \cap P) = q_2 p(P)$, and so $q_1 = q_2$. This completes the proof.

To see that the formal definition of "common knowledge" is equivalent to the informal description, let $\omega \in \Omega$, and call a member ω' of Ω *reachable from* ω if there is a sequence $P^1, P^2, \ldots, P^k$ such that $\omega \in P^1$, $\omega' \in P^k$, and consecutive P^j intersect and belong alternatively to $\mathscr{P}_1$ and $\mathscr{P}_2$. Suppose now that ω is the true state of the world, $P^1 = \mathbf{P}_1(\omega)$, and E is an event. To say that 1 "knows" E means that E includes P^1. To say that 1 knows that 2 knows E means that E includes all P^2 in $\mathscr{P}_2$ that intersect P^1. To say that 1 knows that 2 knows that 1 knows E means that E includes all P^3 in $\mathscr{P}_1$ that intersect P^2 in $\mathscr{P}_2$ that intersect P^1. And so on. Thus all sentences of the form "i knows that i' knows that i knows ... E" (where $i' = 3 - i$) are true if and only if E contains all ω' reachable from ω. But the set of all ω' reachable from ω is a member of $\mathscr{P}_i \wedge \mathscr{P}_2$; so the desired equivalence is established.

The result fails when people merely know each other's posteriors. Suppose Ω has 4 elements $\alpha, \beta, \gamma, \delta$ of equal (prior) probability, $\mathscr{P}_1 = \{\alpha\beta, \gamma\delta\}$, $\mathscr{P}_2 = \{\alpha\beta\gamma, \delta\}$, $A = \alpha\delta$, and $\omega = \alpha$. Then 1 knows that $\mathbf{q}_2$ is $\frac{1}{3}$, and 2 knows that $\mathbf{q}_1$ is $\frac{1}{2}$; but 2 thinks that 1 may not know what $\mathbf{q}_2$ is ($\frac{1}{3}$ or 1).

Worthy of note is the implicit assumption that the information partitions $\mathscr{P}_1$ and $\mathscr{P}_2$ are themselves common knowledge. Actually, this constitutes no loss of

[3]Coarsest common refinement of $\mathscr{P}_1$ and $\mathscr{P}_2$.

[4]Events whose (prior) probability does not vanish.

[5]Finest common coarsening of $\mathscr{P}_1$ and $\mathscr{P}_2$.

generality. Included in the full description of a state ω of the world is the manner in which information is imparted to the two persons. This implies that the information sets $\mathbf{P}_1(\omega)$ and $\mathbf{P}_2(\omega)$ are indeed defined unambiguously as functions of ω, and that these functions are known to both players.

Consider next the assumption of equal priors for different people. John Harsanyi (1968) has argued eloquently that differences in subjective probabilities should be traced exclusively to differences in information—that there is no rational basis for people who have always been fed precisely the same information to maintain different subjective probabilities. This, of course, is equivalent to the assumption of equal priors. The result of this paper might be considered evidence against this view, as there are in fact people who respect each other's opinions and nevertheless disagree heartily about subjective probabilities. But this evidence is not conclusive: even people who respect each other's acumen may ascribe to each other errors in calculating posteriors. Of course we do not mean simple arithmetical mistakes, but rather systematic biases such as those discussed by Tversky and Kahneman (1974). In private conversation, Tversky has suggested that people may also be biased because of psychological factors that may make them disregard information that is unpleasant or does not conform to previously formed notions.

There is a considerable literature about reaching agreement on subjective probabilities; a recent paper is DeGroot (1974), where a bibliography on the subject may be found. A "practical" method is the Delphi technique (see, e.g., Dalkey (1972)). It seems to me that the Harsanyi doctrine is implicit in much of this literature; reconciling subjective probabilities makes sense if it is a question of implicitly exchanging information, but not if we are talking about "innate" differences in priors. The result of this paper might be considered a theoretical foundation for the reconciliation of subjective probabilities.

As an illustration, suppose 1 and 2 have a uniform prior on the parameter of a coin, and let A be the event that the coin will come up H (heads) on the next toss. Suppose that each person is permitted to make one previous toss, and that these tosses come up H and T (tails) respectively. If each one's information consists precisely of the outcome of his toss, then the posteriors for A will be $\frac{2}{3}$ and $\frac{1}{3}$ respectively. If each one then informs the other one of his posterior, then they will both conclude that the previous tosses came up once H and once T, so that both posteriors will be revised to $\frac{1}{2}$.

Suppose now that each person is permitted to make several previous tosses, but that neither one knows how many tosses are allowed to the other one. For example, perhaps both make 4 tosses, which come up *HHHT* for 1, and *HTTT* for 2. They then inform each other that their posteriors are $\frac{2}{3}$ and $\frac{1}{3}$ respectively. Now these posteriors may result from a single observation, from 4 observations, or from more. Since neither one knows on what observations the other's posterior is based, he may be inclined to give more weight to his own observations. Some revision of posteriors would certainly be called for even in such a case; but it does not seem clear that it would necessarily lead to equal posteriors.

Presumably, such a revision would take into account each person's prior on the number of tosses available to him and to the other person. By assumption these two priors are the same, but each person gets additional private information—namely, the actual number of tosses he is allotted. By use of the prior and the information that the posteriors are, respectively, $\frac{2}{3}$ and $\frac{1}{3}$, new posteriors may be calculated. If the players inform each other of these new posteriors, further revision may be called for. Our result implies that the process of exchanging information on the posteriors for A will continue until these posteriors are equal.

Acknowledgments This work was supported by National Science Foundation Grant SOC74-11446 at the Institute for Mathematical Studies in the Social Sciences, Stanford University.

References

Aumann, R. J. (1974). Subjectivity and correlation in randomized strategies. *Journal of Mathematical Economics, 1*, 67–96.

Dalkey, N. C. (1972). *Studies in the quality of life*. Lexington: Lexington Books.

DeGroot, M. H. (1974). Reaching a consensus. *Journal of the American Statistical Association, 69*, 118–121.

Grossman, S. J., & Stiglitz, J. E. (1976). Information and competitive price systems. *The American Economic Review, 46*, 246–253.

Harsanyi, J. (1967–1968). Games of incomplete information played by Bayesian players, Parts I–III. *Management Science, 14*, 159–182, 320–334, 486–502.

Radner, R. (1968). Competitive equilibrium under uncertainty. *Econometrica, 36*, 21–58.

Radner, R. (1972). Existence of equilibrium plans, prices, and price expectations in a sequence of markets. *Econometrica, 40*, 289–304.

Tversky, A., & Kahneman, D. (1974). Judgment under uncertainty: Heuristics and biases. *Science, 185*, 1124–1131.

Chapter 41
Epistemic Conditions for Nash Equilibrium

Robert J. Aumann and Adam Brandenburger

Introduction

Game theoretic reasoning has been widely applied in economics in recent years. Undoubtedly, the most commonly used tool has been the strategic equilibrium of Nash (1951), or one or another of its so-called "refinements." Though much effort[1] has gone into developing these refinements, relatively little attention has been paid to a more basic question: Why consider Nash equilibrium in the first place?

A Nash equilibrium is defined as a way of playing the game—an n-tuple of strategies—in which each player's strategy is optimal for him, given the strategies of the others. The definition seems beautifully simple and compelling; but when considered carefully, it lacks a clear motivation. What would make the players play such an equilibrium? How, exactly, would it come about?

Over the years, much has been written about the connection between Nash equilibrium and common knowledge.[2] According to conventional wisdom, Nash

[1] Selten (1965), (1975), Myerson (1978), Kreps and Wilson (1982), Kohlberg and Mertens (1986), and many others.

[2] An event is called *common knowledge* if all players know it, all know that all know it, and so on ad infinitum (Lewis 1969).

R.J. Aumann (✉)
Federmann Center for the Study of Rationality, The Hebrew University of Jerusalem, Jerusalem, Israel
e-mail: raumann@math.huji.ac.il

A. Brandenburger
Stern School of Business, Tandon School of Engineering, NYU Shanghai, New York University, New York, NY 10012, USA
e-mail: adam.brandenburger@stern.nyu.edu; adambrandenburger.com

H. Arló-Costa et al. (eds.), *Readings in Formal Epistemology*, Springer Graduate Texts in Philosophy 1, DOI 10.1007/978-3-319-20451-2_41

equilibrium is "based on" common knowledge of (a) the structure of the game (i.e., the payoff functions[3]), (b) the rationality[4] of the players, and (c) the strategies actually played. These ideas sound appealing; the circularity of Nash equilibrium—each player chooses his strategy only because the others choose theirs—does seem related to the infinite hierarchy of beliefs inherent in common knowledge. But a formalization has proved elusive. What, precisely, does "based on" mean? Can the above wisdom be turned into a theorem?

It is our purpose here to clarify these issues in a formal framework. Specifically, we seek *epistemic* conditions for a given strategy profile to be a Nash equilibrium: conditions involving what the players know or believe about one another—in particular, about payoff functions, strategy choices, decision procedures, and beliefs about these matters.[5]

Surprisingly, we will find that common knowledge of the payoff functions and of rationality are *never* needed; much weaker epistemic conditions suffice. Common knowledge of the strategies actually being played is also irrelevant. What *does* turn out to be relevant is common knowledge of the beliefs that the players hold about the strategies of the others; but here, too, common knowledge is relevant only when there are at least three players.

The main results are informally described in section "Description of the results." The background for a formal presentation is given in section "Interactive belief systems"; the underlying tool is that of an *interactive belief system*, which provides a framework for formulating epistemic conditions. Illustrations of such systems are given in section "Illustrations." The formal statements and proofs of the main results, in section "Formal statements and proofs of the theorems," are preceded by some lemmas in section "Properties of belief systems." Sections "The main counterexamples" and "Additional counterexamples" contain a series of counterexamples showing the results to be sharp; the examples—particularly those of section "The main counterexamples"—provide insight into the role played by the various epistemic conditions. Section "General (infinite) belief systems" shows that the results apply to infinite as well as finite belief systems. Section "Discussion" is devoted to a discussion of conceptual aspects and of the related literature. An appendix treats extensions and converses of the main results.

The reader wishing to understand just the main ideas should read sections "Description of the results" and "The main counterexamples," and skim sections "Interactive belief systems" and "Illustrations."

---

[3]See section "Structure of the game" for a discussion of why the payoff functions can be identified with the "structure of the game."

[4]I.e., that the players are optimizers; that given the opportunity, they will choose a higher payoff. A formal definition is given below.

[5]Other epistemic conditions for Nash equilibrium have been obtained by Armbruster and Boege (1979) and Tan and Werlang (1988).

Description of the Results

An event is called *mutual knowledge* if all players simply know it (to be distinguished from common knowledge, which also requires higher knowledge levels—knowledge about knowledge, and so on). Our first and simplest result is Theorem 41.1: *Suppose that each player is rational and knows his own payoff function, and that the strategy choices of the players are mutually known. Then these choices constitute a Nash equilibrium in the game being played.*

The proof is immediate: Since each player knows the choices of the others, and is rational, his choice must be optimal given theirs; so by definition, we are at a Nash equilibrium.

Note that neither the players' rationality, nor their payoff functions, nor their strategy choices are assumed common knowledge. For strategies, only mutual knowledge is assumed. For rationality and the structure of the game, not even mutual knowledge is assumed; only that the players are in fact rational, and know their own payoff functions.[6]

Theorem 41.1 applies to all pure strategy profiles. It applies also to mixed strategy profiles, under the traditional view of mixed strategies as conscious randomizations; in that case, of course, it is the mixture that must be mutually known, not just their pure realizations.

In recent years, a different view of mixed strategies has emerged.[7] In this view, players do not randomize; each player chooses some definite pure strategy. But the other players need not know which one, and the mixture represents their uncertainty, their probability assessment of his choice. This is the view adopted in the sequel; it fits in well with the Bayesian approach to game theory, in which uncertainty about strategic choices of others is, like any other uncertainty, subject to probability assessment by each player.

For brevity, let us refer to pure strategies as *actions*. Define the *conjecture* of a player as his probability assessment of the actions of the other players. Call a player *rational* if his action maximizes his expected payoff given his conjecture.

When there are two players (and only then), the conjecture of each player is a mixed strategy of the other player. Because of this, the results in the two-person and n-person cases are quite different. For *two*-person games, we have the following (Theorem 41.2): *Suppose that the game being played (i.e., both payoff functions), the rationality of the players, and their conjectures are all mutually known. Then the conjectures constitute a Nash equilibrium.*

Theorem 41.2 differs from Theorem 41.1 in two ways. First, in both the conclusion and the hypothesis, strategy choices are replaced by conjectures; thus we get "conjectural equilibrium"—an equilibrium in conjectures, not in strategies

[6]When a game is presented in strategic form, as here, knowledge of one's own payoff function may be considered tautologous. See section "Interactive belief systems."

[7]Harsanyi (1973), Armbruster and Boege (1979), Aumann (1987), Tan and Werlang (1988), Brandenburger and Dekel (1989), among others.

actually played. Second, the hypothesis calls not just for the fact of rationality, but for mutual knowledge of this fact, and for mutual knowledge of the payoff functions. But common knowledge still does not enter the picture.

Since we are now viewing mixed strategies as conjectures, it is natural that conjectures replace choices in the result. So with n players, too, one might expect a theorem roughly analogous to Theorem 41.2; i.e., that mutual knowledge of the conjectures (when combined with appropriate assumptions about rationality and the payoff functions) is sufficient for them to be in equilibrium. But here we are in for a surprise: when $n > 2$, the conditions for a conjectural equilibrium become much more stringent.

To understand the situation, note that the conjecture of each player i is a probability mixture of $(n-1)$-tuples of pure strategies of the other players. So when $n > 2$, it is not itself a mixed strategy; however, it induces a mixed strategy[8] for each player j other than i, called *i's conjecture about j*. One difficulty is that different players other than j may have different conjectures about j, in which case it is not clear how to define j's component of the conjectural equilibrium we seek to construct.

To present Theorem 41.3, the n-person "conjecture theorem", one more concept is needed. We say that the players have a *common prior*[9] if all differences between their probability assessments are due only to differences in their information; more precisely, if there is an outside observer O with no private information,[10] such that for all players i, if O were given i's information, his probability assessments would be the same as i's.

Theorem 41.3 is now as follows: *In an n-player game, suppose that the players have a common prior, that their payoff functions and their rationality are mutually known, and that their conjectures are commonly known. Then for each player j, all the players i agree on the same conjecture σ_j about j, and the resulting profile $(\sigma_1, \ldots, \sigma_n)$ of mixed strategies is a Nash equilibrium.*

The above three theorems give sufficient epistemic conditions for Nash equilibrium. The conditions are not necessary; it is always possible for the players to blunder into a Nash equilibrium "by accident," so to speak, without anybody knowing much of anything. Nevertheless, all three theorems are "sharp," in the sense that they cannot be improved upon; none of the conditions can be dispensed with, or, so far as we can see, significantly weakened.

The presentation in this section, while correct, has been informal. For a formal presentation, one needs a framework for describing "epistemic" situations in game contexts; in which, for example, one can describe a situation where each player maximizes against the choices of the others, all know this, but not all know that all know this. Such frameworks are available in the differential information literature; a particular adaptation is presented in the next section.

---

[8] The marginal on j's strategy space of i's overall conjecture.

[9] Aumann (1987); for a formal definition, see section "Interactive belief systems." Harsanyi (1967–1968) uses the term "consistency" to describe this situation.

[10] That is, what O knows is common knowledge among the players; each player knows everything that O knows.

Interactive Belief Systems

Let us be given a strategic *game form*; that is, a finite set $\{1, \ldots, n\}$ (the *players*), together with an action set A_i for each player i. Set $A := A_1 \times \cdots \times A_n$. An *interactive belief system* (or simply *belief system*) for this game form is defined to consist of:

(1) for each player i, a set S_i (i's *types*),
and for each type s_i of i,
(2) a probability distribution on the set S^{-i} of $(n-1)$-tuples of types of the other players (s_i's *theory*),
(3) an action a_i for i (s_i's *action*),
and
(4) a function $g_i : A \to \mathbb{R}$ (s_i's *payoff function*).

The action sets A_i are assumed finite. One may also think of the spaces S_i as finite; the ideas are then more transparent. For a general definition, where the S_i are measurable spaces and the theories are probability measures,[11] see section "General (infinite) belief systems."

A belief system is a formal description of the players' beliefs—about each other's actions and payoff functions, about these beliefs, and so on. Specifically, the theory of a type s_i represents the probabilities that s_i ascribes to the types of the other players, and so to their actions, their payoff functions, and their theories. See section "Belief systems" for further discussion.

Set $S := S_1 \times \cdots \times S_n$. Call the members $s = (s_1, \ldots, s_n)$ of S *states of the world*, or simply *states*. An *event* is a subset E of S. Denote by $p(\cdot; s_i)$ the probability distribution on S induced by s_i's theory; formally, if E is an event, then $p(E; s_i)$ is the probability assigned by s_i to $\{s^{-i} \in S^{-i} : (s_i, s^{-i}) \in E\}$.

A function $g : A \to \mathbb{R}^n$ (an n-tuple of payoff functions) is called a *game*.

Set $A^{-i} := A_1 \times \cdots \times A_{i-1} \times A_{i+1} \times \cdots \times A_n$; for a in A set $a^{-i} := (a_1, \ldots, a_{i-1}, a_{i+1}, \ldots, a_n)$. When referring to player i, the phrase "at s" means "at s_i". Thus, "i's action at s" means s_i's action (see (3)); we denote it $\mathbf{a}_i(s)$, and write $\mathbf{a}(s)$ for the n-tuple $(\mathbf{a}_1(s), \ldots, \mathbf{a}_n(s))$ of actions at s. Similarly, "i's payoff function at s" means s_i's payoff function (see (4)); we denote it $\mathbf{g}_i(s)$, and write $\mathbf{g}(s)$ for the n-tuple $(\mathbf{g}_1(s), \ldots, \mathbf{g}_n(s))$ of payoff functions[12] at s. Viewed as a function of a, we call $\mathbf{g}(s)$ "the game being played at s," or simply "the game at s."

[11]Readers unfamiliar with measure theory may think of the type spaces S_i as finite throughout the paper. All the examples involve finite S_i only. The results, too, are stated and proved without reference to measure theory, and may be understood completely in terms of finite S_i. On the other hand, we do not *require* finite S_i; the definitions, theorems, and proofs are all worded so that when interpreted as in section "General (infinite) belief systems," they apply without change to the general case. One can also dispense with finiteness of the action spaces A_i; but that is both more involved and less important, and we will not do it here.

[12]Thus i's actual payoff at the state s is $\mathbf{g}_i(s)(\mathbf{a}(s))$.

Functions defined on S (like $\mathbf{a}_i(s)$, $\mathbf{a}(s)$, $\mathbf{g}_i(s)$, and $\mathbf{g}(s)$) may be viewed like random variables in probability theory. Thus if $\mathbf{x}$ is such a function and x is one of its values, then $[\mathbf{x} = x]$, or simply $[x]$, denotes the event $\{s \in S : \mathbf{x}\,(s) = x\}$. For example, $[a_i]$ denotes the event that i chooses the action a_i; and $[g]$ denotes the event that the game g is being played.

A *conjecture* φ^i of i is a probability distribution on A^{-i}. For $j \neq i$, the marginal of φ^i on A_j is called the *conjecture of i about j induced by* φ^i. The theory of i at a state s yields a conjecture $\boldsymbol{\varphi}^i(s)$, called *i's conjecture at s*, given by $\boldsymbol{\varphi}^i(s)\,(a^{-i}) := p\,([a^{-i}]; s_i)$. We denote the n-tuple $(\boldsymbol{\varphi}^1(s), \ldots, \boldsymbol{\varphi}^n(s))$ of conjectures at s by $\boldsymbol{\varphi}(s)$.

Player i is called *rational at s* if his action at s maximizes his expected payoff given his information (i.e., his type s_i); formally, letting $h_i := \mathbf{g}_i(s)$ and $b_i, := \mathbf{a}_i(s)$, this means that Exp $(h_i\,(b_i, \mathbf{a}^{-i})\,|s_i) \geq$ Exp $(h_i\,(a_i, \mathbf{a}^{-i})\,|s_i)$ for all a^i in A^i. Another way of saying this is that i's actual choice b_i maximizes the expectation of his actual payoff h_i when the other players' actions are distributed according to his actual conjecture $\boldsymbol{\varphi}^i(s)$.

Player i is said to *know* an event E *at s* if at s, he ascribes probability 1 to E. Define K_iE as the set of all those s at which i knows E. Set $K^1E := K_1E \cap \cdots \cap K_nE$; thus K^1E is the event that all players know E. If $s \in K^1E$, call E *mutually known at s*. Set $CKE := K^1E \cap K^1K^1E \cap K^1K^1K^1E \cap \cdots$; if $s \in CKE$, call E *commonly known at s*.

A probability distribution P on S is called a *common prior* if for all players i and all of their types s_i, the conditional distribution of P given s_i is $p(\cdot; s_i)$; this implies that for all i, all events E and F, and all numbers π,

(5) if $p(E; s_i) = \pi\, p(F; s_i)$ for all $s_i \in S_i$, then $P(E) = \pi P(F)$.

In words, (5) says that for each player i, if two events have proportional probabilities given any s_i, then they have proportional prior probabilities.

Another regularity condition that is sometimes used in the differential information literature is "mutual absolute continuity." We do not define this here because we have no use for it.

Belief systems provide a formal language for stating epistemic conditions. When we say that a player knows some event E, or is rational, or has a certain conjecture φ^i or payoff function g_i, we mean that that is the case at some specific state s of the world. Thus at s, Rowena may know E, but not know that Colin knows E; or at s, it may be that Colin is rational, and that Rowena knows this, but that Colin does not know that Rowena knows it. Some illustrations of these ideas are given in the next section.

Illustrations

Example 41.1 We start with a belief system in which all types of each player i have the same payoff function g_i, namely, that depicted in Fig. 41.1. Thus the game being played is commonly known. Call the row and column players (Players 1 and 2) "Rowena" and "Colin" respectively.

Fig. 41.1

	c	d
C	2, 2	0, 0
D	0, 0	1, 1

Fig. 41.2

	c_1	d_1	d_2
C_1	2/5, 2/3	3/5, 3/5	0, 0
D_1	1/2, 1/3	0, 0	1/2, 1/2
D_2	0, 0	2/3, 2/5	1/3, 1/2

The theories are depicted in Fig. 41.2; here C_1 denotes a type of Rowena whose action is C, whereas D_1 and D_2 denote two different types of Rowena whose actions are D. Similarly for Colin. Each square denotes a state, i.e., a pair of types. The two entries in each square denote the probabilities that the corresponding types of Rowena and Colin ascribe to that state. For example, Colin's type d_2 attributes $1/2 - 1/2$ probabilities to Rowena's type being D_1 or D_2. So at state (D_2, d_2), he knows that Rowena will choose the action D. Similarly, Rowena knows at (D_2, d_2) that Colin will choose d. Since d and D are optimal against each other, both players are rational at (D_2, d_2), and (D, d) is a Nash equilibrium.

We have here a typical instance of Theorem 41.1 (see the beginning of section "Description of the results"), which also shows that the folk wisdom cited in the introduction is misleading. At (D_2, d_2), there is mutual knowledge of the actions D and d, and both players are in fact rational. But the actions are not common knowledge. Thus, though Colin knows that Rowena will play D, she doesn't know that he knows this; indeed, she attributes probability 2/3 to his attributing probability 3/5 to her playing C. Moreover, though both players are rational at (D_2, d_2), there isn't even mutual knowledge of rationality there. For example, Colin's type d_1 chooses d, with an expected payoff of 2/5, rather than c, with an expected payoff of 6/5; thus this type is "irrational." At (D_2, d_2), Rowena attributes probability 2/3 to Colin being of this irrational type.

Note that the players have a common prior (Fig. 41.3). But Theorem 41.1 has nothing to do with common priors. If, for example, we change the theory of Rowena's type D_2 from $\frac{2}{3}d_1 + \frac{1}{3}d_2$ to $\frac{1}{2}d_1 + \frac{1}{2}d_2$, then there is no longer a common prior; but (D, d) is still Nash equilibrium, for the same reasons as above. (As usual,

Fig. 41.3

	c_1	d_1	d_2
C_1	0.2	0.3	0
D_1	0.1	0	0.1
D_2	0	0.2	0.1

Fig. 41.4

	h	t
H	1, 0	0, 1
T	0, 1	1, 0

Fig. 41.5

	h_1	t_1	t_2
H_1	1/2, 1/2	1/2, 1/2	0, 0
T_1	1/2, 1/2	0, 0	1/2, 1
T_2	0, 0	1, 1/2	0, 0

$\frac{2}{3}d_1 + \frac{1}{3}d_2$ stands for the $\frac{2}{3} - \frac{1}{3}$ probability combination of d_1 and d_2. Similar notation will be used throughout the sequel.)

Example 41.2 As in the previous example, the game being played here—"matching pennies" (Fig. 41.4)—is commonly known. The theories are depicted in Fig. 41.5. At the state (H_1, h_1), the conjectures of Rowena and Colin are $\frac{1}{2}h + \frac{1}{2}t$ and $\frac{1}{2}H + \frac{1}{2}T$ respectively, and these conjectures are mutually known (i.e., each knows that these are the conjectures). Moreover, the rationality of both players is mutually known. Thus Theorem 41.2 (section "Description of the results") implies that $(\frac{1}{2}H + \frac{1}{2}T, \frac{1}{2}h + \frac{1}{2}t)$ is a Nash equilibrium, which indeed it is.

Note that neither the conjectures of the players nor their rationality are commonly known. Indeed, at (T_1, t_2), Colin knows that Rowena plays T, so that his conjecture is not $\frac{1}{2}H + \frac{1}{2}T$ but T; so it is irrational for him to play t, which yields him 0, rather than 1 that he could get by playing h. At the state (H_1, h_1), Colin attributes probability 1/2 to Rowena attributing probability 1/2 to the state (T_1, t_2); so at (H_1, h_1), there is common knowledge neither of Colin's conjecture $\frac{1}{2}H + \frac{1}{2}T$, nor of his rationality.

Fig. 41.6

	h_1	t_1	t_2
H_1	0.2	0.2	0
T_1	0.2	0	0.2
T_2	0	0.2	0

Note, too, that like in the previous example, this belief system has a common prior (Fig. 41.6). But also like there, this is not essential; the discussion would not be affected if, say, we changed the theory of Rowena's type T_2 from t_1 to $\frac{1}{2}t_1 + \frac{1}{2}t_2$.

Properties of Belief Systems

In this section we formally establish some basic properties of belief systems, which are needed in the sequel. These properties are intuitively fairly obvious, and some are well known in various formalizations of interactive knowledge theory; so this section can be omitted at a first reading.

Lemma 41.1 *Player i knows that he attributes probability π to an event E if and only if he indeed attributes probability π to E.*

Proof If: Let F be the event that i attributes probability π to E; that is, $F := \{t \in S : p(E; t_i) = \pi\}$. If $s \in F$, then $p\ (E; s_i) = \pi$, so all states u with $u_i = s_i$ are in F. Therefore $p\ (F; s_i) = 1$; that is, i knows F at s, so $s \in K_iF$.

Only if: Suppose that i attributes probability $\rho \neq \pi$ to E. By the "if" part of the proof, he must know this, contrary to his knowing that he attributes probability π to E. ■

Corollary 41.1 *Let φ be an n-tuple of conjectures. Suppose that at some state s, it is mutually known that $\boldsymbol{\varphi} = \varphi$. Then $\boldsymbol{\varphi}(s) = \varphi$. (In words: if it is mutually known that the conjectures are φ, then they are indeed φ.)*

Corollary 41.2 *A player is rational if and only if he knows that he is rational.*

Corollary 41.3 *$K_iK_iE = K_iE$ (a player knows something if and only if he knows that he knows it), and $K_i\neg K_iE = \neg K_iE$ (a player doesn't know something if and only if he knows that he doesn't know it).*

Lemma 41.2 *$K_i\ (E_1 \cap E_2 \cap \cdots) = K_iE_1 \cap K_iE_2 \cap \cdots$ (a player knows each of several events if and only if he knows that they all obtain).*

Proof At s, player i ascribes probability 1 to $E_1 \cap E_2 \cap \cdots$ if and only if he ascribes probability 1 to each $E_1, E_2, \ldots$. ■

Lemma 41.3 *$CKE \subset K_i CKE$ (if something is commonly known, then each player knows that it is commonly known).*

Proof Since $K_i K^1 F \supset K^1 K^1 F$ for all F, Lemma 41.2 yields $K_i CKE = K_i(K^1 E \cap K^1 K^1 E \cap \cdots) = K_i K^1 E \cap K_i K^1 K^1 E \cap \cdots \supset K^1 K^1 E \cap K^1 K^1 K^1 E \cap \cdots \supset CKE$. ■

Lemma 41.4 *Suppose P is a common prior, $K_i H \supset H$, and $p\,(E; s_i) = \pi$ for all $s \in H$. Then $P(E \cap H) = \pi P(H)$.*

Proof Let H_i be the projection of H on S_i. From $K_i H \supset H$ it follows that $p\,(H; s_i) = 1$ or 0 according to whether s_i is or is not[13] in H_i. So when $s_i \in H_i$, then $p(E \cap H; s_i) = p\,(E; s_i) = \pi = \pi p\,(H; s_i)$; and when s_i, $\notin H_i$, then $p(E \cap H; s_i) = 0 = \pi p(H; s_i)$. The lemma now follows from (5). ■

The following lemma is not needed for the proofs of the theorems, but relates to the examples in section "The main counterexamples." Set $K^2 E := K^1 K^1 E$, $K^3 E := K^1 K^2 E$, and so on. If $s \in K^m E$, call E *mutually known of order m* at s.

Lemma 41.5 *$K^m E \subset K^{m-1} E$ for $m > 1$ (mutual knowledge of order m implies mutual knowledge of orders 1, ..., m − 1).*

Proof By Lemma 41.2 and Corollary 41.3, $K^2 E = K^1 K^1 E = \cap_{i=1}^n K_i K^1 E \subset K_i K^1 E = K_i \cap_{j=1}^n K_j E \subset K_i K_i E$. Since this is so for all i, we get $K^2 E \subset \cap_{i=1}^n K_i E = K^1 E$. The result for $m > 2$ follows from this by substituting $K^{m-2}\,E$ for E. ■

Formal Statements and Proofs of the Theorems

We now formally state and prove Theorems 41.1, 41.2, and 41.3. For more transparent paraphrases (using the same terminology), see section "Description of the results."

Theorem 41.1 *Let a be an n-tuple of actions. Suppose that at some state s, all players are rational, and all know that $\mathbf{a} = a$. Then a is a Nash equilibrium.*

Proof Immediate (see beginning of section "Description of the results"). ■

Theorem 41.2 *With $n = 2$ (two players), let g be a game, φ a pair of conjectures. Suppose that at some state, it is mutually known that $\mathbf{g} = g$, that the players are rational, and that $\boldsymbol{\varphi} = \varphi$. Then (φ^2, φ^1) is a Nash equilibrium of g.*

The proof uses a lemma; we state it for the n-person case, since it is needed again in the proof of Theorem 41.3.

Lemma 41.6 *Let g be a game, φ an n-tuple of conjectures. Suppose that at some state s, it is mutually known that $\mathbf{g} = g$, that the players are rational, and that $\boldsymbol{\varphi} = \varphi$.*

[13] In particular, i always knows whether or not H obtains.

Let a_j be an action of a player j to which the conjecture φ^i of some other player i assigns positive probability. Then a_j maximizes g_j against[14]φ^j.

Proof By Corollary 41.1, the conjecture of i at s is φ^i. So i attributes positive probability at s to $[a_j]$. Also, i attributes probability 1 at s to each of the three events [j is rational], $[\varphi^j]$, and $[g_j]$. When one of four events has positive probability, and the other three each have probability 1, then their intersection is non-empty. So there is a state t at which all four events obtain: j is rational, he chooses a_j, his conjecture is φ^j, and his payoff function is g_j. So a_j maximizes g_j against φ^j. ■

Proof of Theorem 41.2 By Lemma 41.6, every action a_1 with positive probability in φ^2 is optimal against φ^1 in g, and every action a_2 with positive probability in φ^1 is optimal against φ^2 in g. This implies that (φ^2, φ^1) is a Nash equilibrium of g. ■

Theorem 41.3 *Let g be a game, φ an n-tuple of conjectures. Suppose that the players have a common prior, which assigns positive probability to it being mutually known that $\mathbf{g} = g$, mutually known that all players are rational, and commonly known that $\boldsymbol{\varphi} = \varphi$. Then for each j, all the conjectures φ^i of players i other than j induce the same conjecture σ_j for j, and $(\sigma_1, \ldots, \sigma_n)$ is a Nash equilibrium of g.*

The proof requires a lemma.

Lemma 41.7 *Let Q be a probability distribution on A with*[15] $Q(a) = Q(a_i)\, Q(a^{-i})$ *for all a in A and all i. Then $Q(a) = Q(a_1) \cdots Q(a_n)$ for all a.*

Proof By induction. For $n = 1$ and 2 the result is immediate. Suppose it true for $n-1$. From $Q(a) = Q(a_1)Q(a^{-1})$ we obtain, by summing over a_n, that $Q(a^{-n}) = Q(a_1)\, Q(a_2, \ldots, a_{n-1})$. Similarly $Q(a^{-n}) = Q(a_1)\, Q(a_i, \ldots, a_{i-1}, a_{i+1}, \ldots, a_{n-1})$ whenever $i < n$. So the induction hypothesis yields $Q(a^{-n}) = Q(a_1)Q(a_2) \cdots Q(a_{n-1})$. Hence $Q(a) = Q(a^{-n})\, Q(a_n) = Q(a_1)\, Q(a_2) \cdots Q(a_n)$. ■

Proof of Theorem 41.3 Set $F := CK[\varphi]$, and let P be the common prior. By assumption, $P(F) > 0$. Set $Q(a) := P([a] \mid F)$. We show that for all a and i,

$$Q(a) = Q(a_i)\, Q\left(a^{\ i}\right). \tag{41.1}$$

Set $H := [a_i] \cap F$. By Lemmas 41.2 and 41.3, $K_i H \supset H$, since i knows his own action. If $s \in H$, it is commonly, and so mutually, known at s that $\boldsymbol{\varphi} = \varphi$; so by Corollary 41.1, $\boldsymbol{\varphi}(s) = \varphi$; that is, $p([a^{-i}]; s_i) = \varphi^i(a^{-i})$. So Lemma 41.4 (with $E = [a^{-i}]$) yields $P([a] \mid F) = P([a^{-i}] \cap H) = \varphi^i(a^{-i})\, P(H) = \varphi^i(a^{-i})\, P([a_i] \mid F)$. Dividing by $P(F)$ yields $Q(a) = \varphi^i(a^{-i})\, Q(a_i)$; then summing over a_i, we get

$$Q\left(a^{-i}\right) = \varphi_i\left(a^{-i}\right). \tag{41.2}$$

Thus $Q(a) = Q(a^{-i})\, Q(a_i)$, which is (41.1).

---

[14]That is, $\text{Exp } g_j(a_j, a^{-j}) \geq \text{Exp } g_j(b_j, a^{-j})$ for all b_j in A_j, when a^{-j} is distributed according to φ^j.

[15]We denote $Q(a^{-i}) := Q(A_i \times \{a^{-i}\})$, $Q(a_i) := Q(A^{-i} \times \{a_i\})$, and so on.

For each j, define a probability distribution σ_j on A_j by $\sigma_j(a_j) := Q(a_j)$. Then (41.2) yields $\varphi^i(a_j) = Q(a_j) = \sigma_j(a_j)$ for $j \neq i$. Thus for all i, the conjecture for j induced by φ^i is σ_j, which does not depend on i. Lemma 41.7, (41.1), and (41.2) then yield

$$\varphi^i\left(a^{-i}\right) = \sigma_1\left(a_1\right)\cdots\sigma_{i-1}\left(a_{i-1}\right)\sigma_{i+1}\left(a_{i+1}\right)\cdots\sigma_n\left(a_n\right); \qquad (41.3)$$

that is, the distribution φ^i is the product of the distributions σ_j with $j \neq i$.

Since common knowledge implies mutual knowledge, the hypothesis of the theorem implies that there is a state at which it is mutually known that $\mathbf{g} = g$, that players are rational, and that $\boldsymbol{\varphi} = \varphi$. So by Lemma 41.6, each action a_j with $\varphi^i(a_j) > 0$ for some $i \neq j$ maximizes g_j against φ^j. By (41.3), these a_j are precisely the ones that appear with positive probability in σ_j. Again using (41.3), we conclude that each action appearing with positive probability in σ_j maximizes g_j against the product of the distributions σ_k with $k \neq j$. This implies that $(\sigma_1, \ldots, \sigma_n)$ is a Nash equilibrium of g.■

The Main Counterexamples

This section explores possible variations on Theorem 41.3 (the result giving sufficient epistemic conditions, when $n \geq 3$, for the players' conjectures to yield a Nash equilibrium). For simplicity, let $n = 3$. Each player's "overall" conjecture is then a distribution on pairs of actions of the other two players; so the three conjectures form a triple of probability mixtures of action *pairs*. On the other hand, an equilibrium is a triple of mixed actions. Our discussion hinges on the relation between these two kinds of objects.

First, since our real concern is with mixtures of *actions* rather than of action *pairs*, could we not formulate conditions that deal directly with each player's "individual" conjectures—his conjectures about each of the other players—rather than with his overall conjecture? For example, one might hope that it would be sufficient to assume common knowledge of each player's individual conjectures.

Example 41.3 shows that this hope is vain, even when the priors are common and rationality is commonly known. Overall conjectures do play an essential role.

Nevertheless, *common* knowledge of the overall conjectures seems a rather strong assumption. Couldn't we get away with less—say, with mutual knowledge of the overall conjectures, or with mutual knowledge of a high order?

Again, the answer is no. In Example 41.4, there is mutual knowledge of overall conjectures (which may be of an arbitrarily high order), common knowledge of rationality, and common prior, but the individual conjectures do not constitute a Nash equilibrium.

What drives this example is that different players have different individual conjectures about some particular player j, so there isn't even a clear *candidate* for a Nash equilibrium.[16] This raises the question of whether (sufficiently high order) mutual knowledge of the overall conjectures implies Nash equilibrium of the individual conjectures when the players do happen to agree, in *addition* to assuming (sufficiently high order) mutual knowledge of the overall conjectures. Do we get Nash equilibrium?

Again, the answer is no; this is shown in Example 41.5.

Finally, Example 41.6 shows that the common prior assumption is really needed; it exhibits a situation with common knowledge of the overall conjectures and of rationality, where the individual conjectures agree; but there is no common prior, and the agreed-upon individual conjectures do *not* form a Nash equilibrium.

Summing up, one must consider the overall conjectures, and nothing less than common knowledge of these conjectures, together with common priors, will do.

Except in Example 41.6, the belief systems in this section have common priors, and these are used to describe them. In all the examples, the game being played is (like in section "Illustrations") fixed throughout the belief system, and so is commonly known. Each example has three players, Rowena, Colin, and Matt, who choose the row, column, and matrix (west or east) respectively. As in section "Illustrations," each type is denoted by the same letter as its action, and a subscript is added.

Example 41.3 Here the individual conjectures are commonly known and agreed upon, rationality is commonly known, and there is a common prior, and yet we don't get Nash equilibrium. Consider the game of Fig. 41.7, with theories induced by the common prior in Fig. 41.8.

At each state, Colin and Matt agree on the conjecture $\frac{1}{2}U + \frac{1}{2}D$ about Rowena, and this is commonly known. Similarly, it is commonly known that Rowena and Matt agree on the conjecture $\frac{1}{2}L + \frac{1}{2}R$ about Colin, and Rowena and Colin agree on $\frac{1}{2}W + \frac{1}{2}E$ about Matt. All players are rational at all states, so rationality is common knowledge at all states. But $(\frac{1}{2}U + \frac{1}{2}D, \frac{1}{2}L + \frac{1}{2}R, \frac{1}{2}W + \frac{1}{2}E)$ is not a Nash equilibrium, because if these were independent mixed strategies, Rowena could gain by moving to D.

Fig. 41.7

	L	R
U	1, 1, 1	0, 0, 0
D	1, 0, 0	1, 1, 1

W

	L	R
U	0, 0, 0	1, 1, 1
D	1, 1, 1	0, 0, 0

E

[16]It is *not* what drives Example 41.3; since the individual conjectures are commonly known there, they must agree (Aumann 1976).

Fig. 41.8

W_1:

	L_1	R_1
U_1	1/4	0
D_1	0	1/4

E_1:

	L_1	R_1
U	0	1/4
D	1/4	0

Fig. 41.9

	L_1	L_2	L_3
U_1	0.4 W_1	0.2 E_1	
U_2		0.2 W_2	0.1 E_2
U_3			0.1 W_3

Note that the overall conjectures are not commonly (nor even mutually) known at any state. For example, at (U_1, L_1, W_1), Rowena's conjecture is $(\frac{1}{2}LW + \frac{1}{2}RE)$, but nobody else knows that that is her conjecture.

Example 41.4 Here we have mutual knowledge of the overall conjectures, common knowledge of rationality, and common priors; yet individual conjectures don't agree, so one can't even identify a candidate for a Nash equilibrium.

Consider a three-person game in which Rowena and Colin each have just one action—say U for Rowena and L for Colin—and Matt has two actions, W and E. The payoffs are unimportant in this example,[17] since we are only interested in showing that the individual conjectures do not agree. Let the belief system be as in Fig. 41.9. As usual, Rowena's types are depicted by rows, Colin's by columns. Matt's types are indicated in the individual boxes in the diagram; note that in this case, he knows the true state.

Consider the state (U_2, L_2, W_2), or simply W_2 for short. At this state, Rowena's conjecture is $\frac{2}{3}LW + \frac{1}{3}LE$, Colin's is $\frac{1}{2}UW + \frac{1}{2}UE$, and Matt's is UL. Rowena knows Colin's and Matt's conjectures, as they are the same at the only other state (E_2) that she considers possible.[18] Similarly, Colin knows Rowena's and Matt's conjectures, as they are the same at the only other state (E_1) that he considers possible. Matt knows Rowena's and Colin's conjectures, since he knows that the true state is W_2.

[17] They can easily be chosen so that rationality is common knowledge.

[18] I.e., to which she assigns positive probability.

Fig. 41.10

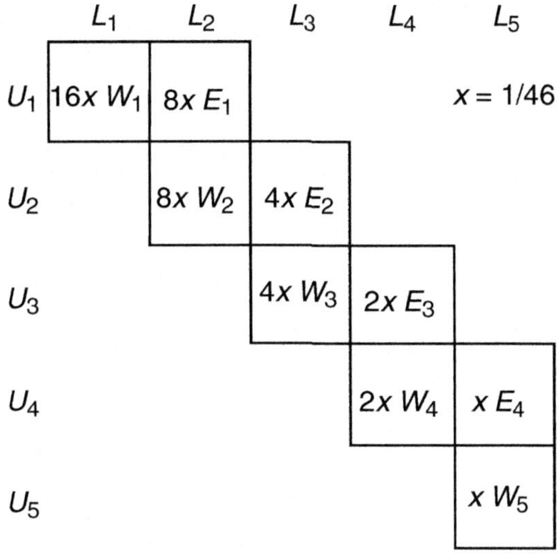

So the conjectures are mutually known. Yet Rowena's conjecture for Matt, $\frac{2}{3}W+\frac{1}{3}E$, is different from Colin's, $\frac{1}{2}W + \frac{1}{2}E$.

The same idea yields examples of this kind with higher order mutual knowledge of the conjectures. See Fig. 41.10. At W_3, there is third order mutual knowledge of the conjectures. By lengthening the staircase[19] and choosing a state in its middle, one can get mutual knowledge of arbitrarily high order that Rowena's conjecture for Matt is $\frac{2}{3}W + \frac{1}{3}E$, while Colin's is $\frac{1}{2}W + \frac{1}{2}E$.

The perspicacious reader will have realized that this example is not intrinsically game-theoretic. It really boils down to a question about "agreeing to disagree": Suppose that two people with the same prior get different information, and that given this information, their posterior probabilities for some event A are mutual knowledge of some order. Are the posterior probabilities then equal? (Here the individuals are Rowena and Colin, and the event A is that Matt chooses W; but actually that's just window dressing.) An example in Aumann (1976) provides a negative answer to the question in the case of first order mutual knowledge. Geanakoplos and Polemarchakis (1982) showed that the answer is negative also for higher order mutual knowledge. The ingenious example presented here is due to John Geanakoplos.[20]

Example 41.5 Here we have mutual knowledge of the overall conjectures, agreement of individual conjectures, common knowledge of rationality, and a common

[19] Alternatively, one can use a single probability system with countably many states, represented by a staircase anchored at the top left and extending infinitely downwards to the right (Jacob's ladder?). By choosing a state sufficiently far from the top, one can get mutual knowledge of any given order.

[20] Private communication. The essential difference between the 1982 example of Geanakoplos and Polemarchakis and the above example of Geanakoplos is that in the former, Rowena's and Colin's

Fig. 41.11

	h	t
H	1, 0, 3	0, 1, 0
T	0, 1, 0	1, 0, 3

W

	h	t
H	1, 0, 2	0, 1, 2
T	0, 1, 2	1, 0, 2

E

Fig. 41.12

$x = 1/52$

	h_1	t_1	h_2	t_2	h_3	t_3
H_1	$9x\ W_1$	$9x\ E_1$				
T_1	$9x\ E_1$	$9x\ W_1$				
H_2			$3x\ W_2$	$3x\ W_1$		
T_2			$3x\ W_1$	$3x\ W_2$		
H_3					$x\ W_3$	$x\ W_2$
T_3					$x\ W_2$	$x\ W_3$

prior, and yet the individual conjectures do not form a Nash equilibrium. Consider the game of Fig. 41.11. For Rowena and Colin, this is simply "matching pennies" (Fig. 41.4); their payoffs are not affected by Matt's choice. So at a Nash equilibrium, they must play $\frac{1}{2}H + \frac{1}{2}T$ and $\frac{1}{2}h + \frac{1}{2}t$ respectively. Thus Matt's expected payoff is 3/2 for W, and 2 for E; so he must play E. Hence $(\frac{1}{2}H + \frac{1}{2}T, \frac{1}{2}h + \frac{1}{2}t, E)$ is the unique Nash equilibrium of this game.

Consider now the theories induced by the common prior in Fig. 41.12. Rowena and Colin know which of the three "boxes" contains the true state, and in fact this is commonly known between the two of them. In each box, Rowena and Colin "play matching pennies optimally"; their conjectures about each other are $\frac{1}{2}H + \frac{1}{2}T$ and $\frac{1}{2}h + \frac{1}{2}t$. Since these conjectures obtain at each state, they are commonly known (among all three players); so it is also commonly known that Rowena and Colin are rational.

---

probabilities for A approach each other as the other m of mutual knowledge approaches ∞, whereas in the latter, they remain at 2/3 and 1/2 no matter how large m is.

As for Matt, suppose first that he is of type W_1 or W_2. Each of these types intersects two adjacent boxes in Fig. 41.12; it consists of the diagonal states in the left box and the off-diagonal ones in the right box. The diagonal states on the left have equal probability, as do the off-diagonal ones on the right; but on the left, the probability is three times that on the right. So Matt assigns the diagonal states three times the probability of the off-diagonal states; i.e., his conjecture is $\frac{3}{8}Hh + \frac{3}{8}Tt + \frac{1}{8}Th + \frac{1}{8}Ht$. Therefore his expected payoff from choosing W is $\frac{3}{8}3 + \frac{3}{8}3 + \frac{1}{8}0 + \frac{1}{8}0 = 2\frac{1}{4}$, whereas from E it is only 2 (as all his payoffs in the eastern matrix are 2). So W is indeed the optimal action of these types; so they are rational. It may be checked that also E_1 and W_3 are rational. Thus the rationality of all players is commonly known at all states.

Consider now the state $s := (H_2, h_2, W_2)$ (the top left state in the middle box). Rowena and Colin know at s that they are in the middle box, so they know that Matt's type is W_1 or W_2. We have just seen that these two types have the same conjecture, so it follows that Matt's conjecture is mutually known at s. Also Rowena's and Colin's conjectures are mutually known at s (Rowena's is $\frac{1}{2}hW + \frac{1}{2}tW$, Colin's is $\frac{1}{2}HW + \frac{1}{2}TW$).

Finally, the individual conjectures derived from Matt's overall conjecture $\frac{3}{8}Hh + \frac{3}{8}Tt + \frac{1}{8}Th + \frac{1}{8}Ht$ are $\frac{1}{2}H + \frac{1}{2}T$ for Rowena and $\frac{1}{2}h + \frac{1}{2}t$ for Colin. These are the same as Rowena's and Colin's conjectures for each other. Also, since Matt plays W throughout the middle box, both Rowena and Colin conjecture W for Colin there. Thus throughout the middle box, individual conjectures are agreed upon.

To sum up: There is a common prior; at all states, the game is commonly known and all players are commonly known to be rational. At the top left state in the middle box, the overall conjectures of all players are mutually known, and the individual conjectures are agreed: $\sigma_R = \frac{1}{2}H + \frac{1}{2}T$, $\sigma_C = \frac{1}{2}h + \frac{1}{2}t$, $\sigma_M = W$. But $(\sigma_R, \sigma_C, \sigma_M)$ is not a Nash equilibrium.

To understand the construction, note that in the east matrix, Matt gets 2 no matter what the others do; in the west matrix, he gets 3 if the others play on the (main) diagonal, 0 if they play off. Thus he is motivated to play W when his conjecture assigns a high probability to their playing on the diagonal. The common prior in Fig. 41.12 generates such a conjecture that is mutually known, in spite of the overall probability of the diagonal being only 1/2. The reason for using the diagonal[21] is to avoid correlation with Rowena's or Colin's information, so as to assure that Matt's conjectures about them will agree with their conjectures about each other.

As in Example 41.4, one can construct a similar example for which the mutual knowledge of the conjectures is of an arbitrarily high order, simply by using more boxes; the result follows as before.

[21] Rather than the upper row, say.

Fig. 41.13

	h	t
H	3/8	1/8
T	1/8	3/8

W

	h	t
H	0	0
T	0	0

E

Fig. 41.14

	h_1	t_1
H_1	1/2, 1/2, 3/8	1/2, 1/2, 1/8
T_1	1/2, 1/2, 1/8	1/2, 1/2, 3/8

W_1

Agreement between individual conjectures may be viewed as a kind of "consistency" between the overall conjectures φ^i. A stronger consistency condition,[22] suggested by Kenneth Arrow, is that there be a single probability distribution ϕ on $A = \times_{i=1}^{n} A_i$ whose marginals on the spaces A^{-i} are the φ^i. One may ask whether with this stronger condition, one can replace common by mutual knowledge in Theorem 41.3. The answer is still no, even with mutual knowledge of arbitrarily high order. The above example satisfies Arrow's condition, with $\phi = \frac{3}{8}HhW + \frac{3}{8}TtW + \frac{1}{8}ThW + \frac{1}{8}HtW$ (Fig. 41.13).

Example 41.6 Here we show that one cannot dispense with common priors in Theorem 41.3. Consider again the game of Fig. 41.11, with the theories depicted in Fig. 41.14 (presented in the style of Figs. 41.2 and 41.5; note that Matt has no type[23] whose action is E). At each state there is common knowledge of rationality, of overall conjectures (which are the same as in the previous example), and of the game. As before, Arrow's condition is satisfied, and it follows that the individual conjectures are in agreement. And as before, the individual conjectures $(\frac{1}{2}H + \frac{1}{2}T, \frac{1}{2}h + \frac{1}{2}t)$ do not constitute a Nash equilibrium.

[22] Though similar in form, this condition neither implies nor is implied by common priors. We saw in Example 41.4 that common priors do not even imply agreement between individual forecasts; a fortiori, they do not imply Arrow's condition. In the opposite direction, Example 41.6 satisfies Arrow's condition, but has no common prior.

[23] If one wishes, one can introduce a type E_1 of Matt to which Rowena's and Colin's types ascribe probability 0, and whose theory is, say, $\frac{1}{4}Hh + \frac{1}{4}Tt + \frac{1}{4}Th + \frac{1}{4}Ht$.

Fig. 41.15

	h_1	t_1
H_1	1/6	1/6
T_1	1/3	1/3

Fig. 41.16

Game T

	S
U	1, 1
D	0, 0

Game B

	S
U	0, 1
D	1, 0

Additional Counterexamples

In the previous section we saw that the assumption of a common prior and of common knowledge of the conjectures are essential in Theorem 41.3. This section explores the assumption of mutual knowledge of rationality and of the game being played (in both Theorems 41.2 and 41.3), and shows that they, too, cannot be substantially weakened.

Example 41.7 Here we show that in Theorems 41.2 and 41.3, mutual knowledge of rationality cannot be replaced by the simple fact of rationality (as in Theorem 41.1). Consider again the game of "matching pennies" (Fig. 41.4), this time with theories induced by the common prior in Fig. 41.15.

At the state (H_1, h_1), Colin's and Rowena's conjectures are commonly known to be $\frac{1}{3}H + \frac{2}{3}T$ and $\frac{1}{2}h + \frac{1}{2}t$ respectively, and both are in fact rational (indeed Rowena's rationality is commonly known); but $(\frac{1}{3}H + \frac{2}{3}T, \frac{1}{2}h + \frac{1}{2}t)$ is not a Nash equilibrium, since the only equilibrium of "matching pennies" is $(\frac{1}{2}H + \frac{1}{2}T, \frac{1}{2}h + \frac{1}{2}t)$. Note that at the state (H_1, h_1), Rowena does not know that Colin is rational.

Example 41.8 Here we show that in Theorems 41.2 and 41.3, knowing one's own payoff function does not suffice; one needs mutual knowledge of all the payoff functions. Consider a two-person belief system where one of two games, T (top) or B (bottom), is being played (Fig. 41.16).

Fig. 41.17

	S_1
TU_1	1/2
BD_1	1/2

Fig. 41.18

Game T

	S
U	0, 1
D	0, 0

Game B

	S
U	1, 1
D	0, 0

The theories are given by the common prior in Fig. 41.17. Thus Rowena knows Colin's payoff function, but Colin does not know Rowena's. Rowena's type TU_1 has the *Top* payoff function and plays *Up*, whereas BD_1 has the *Bottom* payoff function and plays *Down*. Colin has just a single type, S_1. At both states, both players are rational: Rowena, who knows the game, always plays an action that is strictly dominant in the true game; Colin has no choice, so what he does is rational. So there is common knowledge of rationality. At both states, Colin's conjecture about Rowena is $\frac{1}{2}U + \frac{1}{2}D$; Rowena's conjecture about Colin is S. But $(\frac{1}{2}U + \frac{1}{2}D, S)$ is not a Nash equilibrium in either of the games; Rowena prefers U in the top game, D in the bottom game.

Example 41.9 Here we show that the hypotheses of Theorem 41.3 do not imply that rationality is commonly known (as is the case when the game g is commonly known—see Proposition 41.A1). Consider a two-person belief system where one of the two games of Fig. 41.18 is being played. The theories are given by the common prior in Fig. 41.19. Of Rowena's three types, TU_1 and TD_1 are rational, whereas BD_1 (who plays *Down* in Game B) is irrational. Colin's two types, S_1 and S_2, differ only in their theories, and both are rational. The conjectures $\frac{1}{2}U + \frac{1}{2}D$ and S are common knowledge at all states. At the state (TU_1, S_1), it is mutually known that T

Fig. 41.19

	S_1	S_2
TU_1	1/4	1/4
TD_1	1/4	0
BD_1	0	1/4

is the game being played, and that both players are rational; but neither rationality nor the game being played are commonly known.

General (Infinite) Belief Systems

For a general definition of a belief system, we specify that the type spaces S_i be measurable spaces. As before, a *theory* is a probability measure on $S^{-i} = \times_{j \neq i} S_j$, which is now endowed with the standard product structure.[24] The state space $S = \times_j S_j$, too, is endowed with the product structure. An *event* is now a measurable subset of S. The "action functions" $\mathbf{a}_i$ ((3)) are assumed measurable; so are the payoff functions $\mathbf{g}_i$ ((4)), as functions of s_i, for each action n-tuple a separately. Also the "theory functions" ((2)) are assumed measurable, in the sense that for each event E and player i, the probability $p(E; s_i)$ is measurable as a function of the type s_i. It follows that also the conjectures $\boldsymbol{\varphi}^i$ are measurable functions of s_i.

With these definitions, the statements of the results make sense, and the proofs remain correct, without any change.

Discussion

(a) Belief Systems

An interactive belief system is not a prescriptive model; it does not suggest actions to the players. Rather, it is a formal framework—a language—for *talking* about actions, payoffs, and beliefs. For example, it enables us to say whether a given player is behaving rationally at a given state, whether this is known to another player, and

[24]The σ-field of measurable sets is the smallest σ-field containing all the "rectangles" $\times_{j \neq i} T_j$, where T_j is measurable in S_j.

so on. But it does not prescribe or even suggest rationality; the players do whatever they do. Like the disk operating system of a personal computer, the belief system simply organizes things, so that we can coherently discuss what they do.

Though entirely apt, use of the term "state of the world" to include the actions of the players has perhaps caused confusion. In Savage (1954), the decision maker cannot affect the state; he can only react to it. While convenient in Savage's one-person context, this is not appropriate in the interactive, many-person world under study here. Since each player must take into account the actions of the others, the actions should be included in the description of the state. Also the plain, everyday meaning of the term "state of the world" includes one's actions: Our world is shaped by what we do.

It has been objected that prescribing what a player must do at a state takes away his freedom. This is nonsensical; the player may do what he wants. It is simply that whatever he does is part of the description of the state. If he wishes to do something else, he is heartily welcome to do it, but he thereby changes the state.

Historically, belief systems were introduced by John Harsanyi (1967–1968), to enable a coherent formulation of games in which the players need not know each other's payoff functions. To analyze such games, it is not enough to specify each player's beliefs about (i.e., probability distributions on) the payoff functions of the others; one must also specify the beliefs of the players about the beliefs of the players about the payoff functions, the beliefs of the players about *these* beliefs, and so on ad infinitum. This complicated infinite regress seemed to make useful analysis very difficult.

Harsanyi's ingenious solution was to think of each player as being one of several possible "types," where a type determines both a player's own payoff function and a belief about the types of the others. The belief of a player about the types of the others induces a belief about their payoff functions; it also induces a belief about their beliefs about the types, and so a belief about the beliefs about the payoff functions. The reasoning continues indefinitely. Thus from an *I-game* ("I" for incomplete information), as Harsanyi calls his type-model, one can read off the entire infinite regress of beliefs.

Belief systems as defined in section "Interactive belief systems" are formally just like Harsanyi's I-games, except that in belief systems, a player's type determines his action as well as his payoff function and his belief about other players' types.[25] As above, it follows that the player's type determines his entire *belief hierarchy*—i.e., the entire infinite regress of his beliefs about actions, beliefs about beliefs about actions, and so on—in addition to the infinite regress of beliefs about payoff functions, and how these two kinds of beliefs affect each other.

Traditionally, payoff functions have been treated as exogenous, actions as endogenous. It was thought that unlike payoff functions, actions should be "predicted" by the theory. Belief systems wipe out this distinction; they treat uncertainty

---

[25]For related ideas, see Armbuster and Boege (1979), Boege and Eisele (1979), and Tan and Werlang (1988).

about actions just like uncertainty about payoff functions. Indeed, in this paper the focus is on actions;[26] uncertainty about payoff functions was included as an afterthought, because we realized that more comprehensive results can be obtained at almost no cost in the complexity of the proofs.

(b) Common Knowledge of the Belief System

Is the belief system itself common knowledge among players? If so, how does it get to be common knowledge? If not, how do we take into account the players' uncertainty about it?

A related question is whether the belief system is exogenous, like a game or a market model. If not, where does it come from?

The key to these issues was provided[27] in a fundamental paper of Mertens and Zamir (1985). They treat the Harsanyi case, in which the "underlying variables"[28] are the payoff functions only (see (a) above); but the result applies without change to our situation, where actions as well as payoff functions are underlying variables.

At (a) above, we explained how each type in a belief system determines a belief hierarchy. Mertens and Zamir reverse this procedure: They start with the belief hierarchies, and construct belief systems from them. Specifically, they define the *universal belief space* as a belief system in which the type space of each player is simply the set of *all* his belief hierarchies that satisfy certain minimal consistency conditions. Thus the universal belief space is not exogenously given, like a game or market; rather, it is an analytic tool, like the payoff matrix of a game originally given in extensive form. It follows that the universal belief space may be considered common knowledge.

Though the universal belief space is infinite, it is the disjoint union of infinitely many "common knowledge components,"[29] many of which are finite. Mertens and Zamir call any union of such components a *subspace* of the universal belief space. It follows that when the belief system is a subspace, then *the belief system itself is common knowledge.*

It may be shown that *any* belief system B for $A_1 \times \cdots\cdots \times A_n$—including, of course, a finite one—is "isomorphic" to a subspace of the universal belief space.[30] From all this we conclude that the belief system itself may *always* be considered common knowledge.

---

[26]As is apparent from the examples in sections "Illustrations," "The main counterexamples," and "Additional counterexamples," in most of which the game g is common knowledge.

[27]See also Armbuster and Boege (1979) and Boege and Eisele (1979).

[28]The variables about which beliefs—of all orders—are held.

[29]At each state s of such a component S, the identity of that component is commonly known (i.e., it is commonly known at s that the true state is in S, though it need not be commonly known that the true state is s).

[30]This is because each type in B determines a belief hierarchy (see (a)). The set of n-tuples of all these belief hierarchies is the subspace of the universal belief space that is isomorphic to B.

(c) Knowledge and Belief

In this paper, "know" means "ascribe probability 1 to." This is sometimes called "believe," while "know" is reserved for absolute certainty, with no possibility at all for error. In the formalism of belief systems[31] (section "Interactive belief systems"), absolute certainty has little concrete meaning; a player can be absolutely certain only on his own action, his own payoff function, and his own theory, not of anything pertaining to anybody else.

Choosing between the terms "know" and "believe" caused us many sleepless nights. In the end, we decided on "know" because it enables simpler, less convoluted language, and because we were glad to de-emphasize the relatively minor conceptual difference between probability 1 and absolute certainty.[32]

Note that since our conditions are sufficient, our results are stronger with probability 1 than with absolute certainty. If probability 1 knowledge of certain events implies that σ is a Nash equilibrium, then a fortiori, so does absolute certainty of those events.

(d) Structure of the Game

In section "Introduction," we identified the "structure of the game" with an n-tuple g of payoff functions. As pointed out by John Harsanyi (1967–1968), this involves no conceptual loss of generality, as one can always pick the action sets A_j sufficiently large to justify this.

(e) Alternative Formalisms

Instead of using belief systems to formulate and prove our results and examples, one may use knowledge partitions (e.g., Aumann 1976). The advantage of the partition formalism is that with it one can represent absolute certainty, not just probability 1 knowledge (see (c)). Also some of the proofs may become marginally simpler.[33] On the other hand, the partition formalism—especially when combined with probabilities—is itself more complex than that of belief systems, and it is desirable to use as simple, transparent, and unified an approach as possible.

[31]Unlike that of partitions (see (e) below).

[32]Another reason is that "belief" often refers to a general probability distribution, which does not go well with using "know" to mean "ascribe probability 1 to."

[33]This is natural, as the results are slightly weaker (see (c)).

(f) Independent Conjectures

Theorem 41.3 deduces independence of the individual conjectures from common knowledge of the overall conjectures. An alternative approach is embodied in the following:

Remark 41.1 *Let σ be an n-tuple of mixed strategies. Suppose that at some state, it is mutually known that the players are rational, that the game g is being played, that the conjecture of each player i about each other player j is σ_j, and that it is independent of i's conjecture about all other players. Then σ is a Nash equilibrium in g.*

The proof is as in the last part of Theorem 41.3's proof, after (41.3) has been established.

Here we assume mutual rather than common knowledge of conjectures and do not assume common priors. On the other hand, we assume outright that the individual conjectures are agreed upon, and that each player's conjectures about the others are independent. Because of the strength of these assumptions, the result is of limited interest.

Mutual independence of the conjectures of a player i about different players j seems a particularly untenable assumption. It may well be that Colin and Matt "choose their actions independently" in the sense of not consulting each other or even communicating, but that by no means implies that Rowena's conjecture about Colin is stochastically independent of her conjecture about Matt. It is possible to have considerable stochastic dependence even when there is no possibility of communication or correlation between Colin and Matt—even when they are placed in separate rooms and then informed of the game they are to play. The probability is in Rowena's head, it is a question of *her* beliefs, *her* ignorance or knowledge, it is not directly determined by some kind of physical process such as "independently" choosing or being in separate rooms or anything like that. Suppose, for example, that Colin and Matt could each act either "boldly" or "carefully." Even if Rowena knows "nothing at all" about Colin and Matt (whatever that might mean; perhaps she believes they came from unspecified different directions in outer space), her probability that Matt acts boldly given that Colin does might well be higher than her unconditional probability. After all, it is a question of *her* outlook on life, how *she* thinks people react to things; if she is at all sophisticated, she might easily say to herself, "Well, I'm not sure of human nature, I don't know what people do in this kind of situation; but if Colin plays boldly, that tells me something about people in general, and it raises my probability that Matt will also play boldly." In fact, it is quite unlikely that the conjectures would be stochastically independent, one would have great difficulty in constructing a set of circumstances that would justify such a conclusion.

When teaching probability, we are at pains to point out that a sequence of coin tosses is *not* in general represented by an i.i.d. sequence of random variables. This would be the case only if the parameter of the coin were known with certainty before the first toss, a highly unlikely proposition. And this in spite of the fact that the tosses themselves are of course "physically" independent (whatever that may mean; we're not sure it means anything). The relation between "physical" and stochastic independence is murky at best.

Independence of individual conjectures is an appropriate assumption when one thinks of mixed strategies in the old way, in terms of explicit randomizations. With that kind of interpretation, "acting independently" is of course closely associated with stochastic independence. But not when the probabilities represent other players' ignorance.

To be sure, common knowledge of the conjectures is also a rather strong assumption. Moreover, we do *conclude* that the individual conjectures are independent. But there is a difference between an assumption that is merely strong and one that is groundless. More to the point, there is a qualitative difference between assuming common knowledge of the conjectures and assuming independence. Common knowledge of the conjectures describes what might almost be called a "physical" state of affairs. It might, for example, be the outgrowth of experience or learning, like common knowledge of a language. Independence, on the other hand, is in the mind of the decision maker. It isn't reasonable to make such an assumption when one can't describe some clear set of circumstances under which it would obtain—and that is very difficult to do. By contrast, common knowledge of the conjectures may itself be considered a "clear set of circumstances."

(g) Epistemic "Equivalence" Conditions

This paper is devoted to sufficient epistemic conditions for Nash equilibrium. Another kind of epistemic result is illustrated by the following:

Remark 41.2 *Let a be an n-tuple of actions. Suppose that at some state s, it is commonly known that* $\mathbf{g} = g$ *and* $\mathbf{a} = a$. *Then rationality is commonly known at s iff a is a Nash equilibrium in g.*

For another result of this kind, which relates to Theorem 41.2 as the above relates to Theorem 41.1, see Brandenburger and Dekel (1989). These results provide sufficient conditions for the *equivalence* between Nash equilibrium and common knowledge of rationality, rather than directly for Nash equilibrium.

(h) Global Epistemic Conditions

The epistemic conditions discussed in this paper are *local*; they say when the players' actions or conjectures at a specific state constitute an equilibrium. By contrast, one can treat *global* conditions, which refer to the belief system as a whole.

These are usually quite different from the local conditions treated here. For example, the condition for correlated equilibrium in Aumann (1987) is global.

A global epistemic condition for Nash equilibrium[34] is as follows:

Remark 41.3 *Suppose that the game g is fixed,*[35] *that there is a common prior P, and that all players are rational at all states. Then the distribution Q of action n-tuples is a Nash equilibrium in g if and only if for each player i, the expectation of i's conjecture given one of his actions* a_i *is the same for all* a_i *that are assigned positive probability by Q.*

Roughly speaking, the condition says that if the players use only information that is relevant to their choice of an action, then their conjectures are always the same, independently of their information.

(i) Irrationality, Probability Assessments, and Payoff Functions

There is a difficulty when explicitly allowing for irrationality, as we do here. The problem rears its ugly head on several levels. Most fundamentally, a player is by definition rational if his choice maximizes his utility function. Again by definition, the utility function represents his preferences, and yet again by definition, the player prefers x to y iff he chooses x when given the choice between x and y. So "rationality" appears tautologous: utility is always maximized; it is *defined* as a function that the choice maximizes.

One may attempt to resolve this by noting that whereas rationality requires the choice x to maximize the utility function, that does not mean that any x maximizes some utility function. Indeed there may not be any utility function at all. Rationality must be understood in terms of a richer structure, one involving preferences (choices between pairs or from other subsets of the available alternatives). If these preferences do not satisfy basic requirements (such as transitivity), then we may speak of irrationality.

But the difficulty reappears at a different level. In a belief system, all players, rational or not, have theories—i.e., probability assessments over the types of others, and so over their actions and so on. In the foundations of decision theory, e.g., à la Savage, probabilities and utilities are derived from axioms that apply to rational players only. How, then, can an irrational player entertain probabilities—indeed, how can he have a payoff function, which is presumably derived from his utility function?

There are several ways to address this problem. One is to change the definition of a belief system, by specifying for each type whether or not it is "rational," and then specifying theories and payoff functions only for types that are specified

[34]Stated, but not proved, in Aumann (1987).

[35]Constant throughout the belief system.

as rational.[36] This would rescue Theorems 41.1 and 41.2, since in those results, probability assessments and payoff functions of irrational types play no role.

It would, however, not rescue Theorem 41.3, since there one needs common knowledge of the conjectures, but only mutual knowledge of rationality. Thus it appears that irrational players must entertain conjectures; for this they should be rational. To resolve this problem, one may distinguish between *subjective* and *objective* theories—and so also between subjective and objective knowledge, conjectures, and rationality. Thus, think of the common prior in Theorem 41.3 as referring to the "objective" assessment of a fixed outside observer Otto. Call type s_i *objectively* rational if its choice maximizes its expected payoff when calculated according to the objective conjecture (the one that Otto would hold if given s_i's information), and according to its own payoff function. This differs from *subjective* rationality of s_i (the notion of rationality hitherto used), which requires that s_i's choice be maximal according to s_i's own conjecture, not Otto's. Similarly, say that a type s_i knows some event E *objectively* if given s_i's information, Otto would ascribe probability 1 to E. Theorem 41.3 can then be understood as asking for mutual objective knowledge of the players' payoff functions and of objective rationality, and for common objective knowledge of objective conjectures. These assumptions do not demand that irrational types entertain conjectures, but they may. If they do, they are now to be thought of as subjectively—but not necessarily objectively—rational.

Needless to say, our results remain true—and almost as interesting—when rationality *is* commonly known. Thus readers who are dissatisfied with the above interpretations may simply assume common knowledge of rationality.

(j) Knowledge of Equilibrium

The conclusions of our theorems state that a specified (mixed) strategy n-tuple σ is an equilibrium; they do not state that the players know it to be an equilibrium, or that this is commonly known. In the case of Theorems 41.2 and 41.3, though, it is in fact mutual knowledge of order 1—but not necessarily of any higher order—that σ is a Nash equilibrium. In the case of Theorem 41.1, it need not even be mutual knowledge of order 1 that σ is a Nash equilibrium; but this does follow if, in addition to the stated assumptions, one assumes mutual knowledge of the payoff functions.

[36]It would be desirable to see whether and how one can derive this kind of modified belief system from a Mertens–Zamir-type construction.

(k) Conclusions

Where does all this leave us? Are the "foundations" of Nash equilibrium more secure or less secure than previously imagined? Do our results strengthen or weaken the case for Nash equilibrium?

First, in assessing the validity of a game-theoretic solution concept, one should not place undue emphasis on its a priori rationale. At least as important is the question of how successful the concept is in providing insight into applications, how tractable it is, and, relatedly, even the extent of its aesthetic appeal. On all these counts, Nash equilibrium has proved its worth.

This said, the present results do indicate that the a priori case for Nash equilibrium is a little stronger than conventional wisdom has granted. Common knowledge turns out to play a more limited role—at least in games with two players—than previously thought. The reader may object that even mutual knowledge of, say, payoff functions is implausible; but indisputably, common knowledge of payoff functions is more so. It is true that our epistemic conditions for Nash equilibrium in games with more than two players involve common knowledge (of conjectures); indeed, it was surprising to discover that the conditions for equilibrium in the n-person case are stronger than in the two-person case. Perhaps Nash equilibrium rests on firmer foundations in two-person than in n-person games.

It should be remembered that the conditions for Nash equilibrium described here are sufficient, but not necessary. Perhaps there are other ways of looking at Nash equilibrium epistemically; if so, their nature is as yet unclear.

There also are non-epistemic ways of looking at Nash equilibrium, such as the evolutionary approach (e.g., Maynard Smith 1982). Related to this is the idea that a Nash equilibrium represents a societal norm. In the end, these viewpoints will perhaps provide a more compelling basis for Nash equilibrium than those involving games played by consciously maximizing, rational players.

Finally, the apparatus of this paper—belief systems, conjectures, knowledge, mutual knowledge, common knowledge, and the like—has an appeal that extends beyond our immediate purpose of providing epistemic conditions for Nash equilibrium. The apparatus offers a way of analyzing strategic situations that corresponds nicely to the concerns that we have all experienced in practice—what is the other person thinking, what is his true motivation, does he see the world as I do, and so on.

Appendix: Extensions and Converses

We start with some remarks on Theorem 41.3 and its proof. First, the conclusions of Theorem 41.3 continue to hold under the slightly weaker assumption that the common prior assigns positive probability to $\boldsymbol{\varphi} = \varphi$ being commonly known, and there is a state at which $\boldsymbol{\varphi} = \varphi$ is commonly known and $\mathbf{g} = g$ and the rationality of the players are mutually known.

Second, note that the rationality assumption is not used until the end of Theorem 41.3's proof, after (41.3) is established. Thus if we assume only that there is a common prior that assigns positive probability to the conjectures φ^i being commonly known, we may conclude that all players i have the same conjecture σ_j for other players j, and that each φ^i is the product of the σ_j with $j \neq i$; that is, the $n-1$ conjectures of each player about the other players are independent.

Third, if in Theorem 41.3 we assume that the game being played is commonly (not just mutually) known, then we can conclude that also the rationality of the players is commonly known.[37] That is, we have

Proposition 41.A1 *Suppose that at some state s, the game g and the conjectures φ^i are commonly known and rationality is mutually known. Then at s, rationality is commonly known. (Note that common priors are not assumed here.)*

Proof Set $G := [g]$, $F := [\varphi]$, $R_j := [j \text{ is rational}]$, and $R :=$ [all players are rational] $= R_1 \cap \cdots \cap R_n$. In these terms, the proposition says that $CK(G \cap F) \cap K^1 R \subset CKR$. We assert that it is sufficient for this to prove

$$K^2(G \cap F) \cap K^1 R \subset K^2 R. \tag{41.A1}$$

Indeed, if we have (A1), an inductive argument using Lemma 41.5 and that $E \subset E'$ implies $K^1(E) \subset K^1(E')$ (which follows from Lemma 41.2) yields $K^m(G \cap F) \cap K^1 R \subset K^m R$ for any m; so taking intersections, $CK\,(G \cap F) \cap K^1 R \subset CKR$ follows.

Let j be a player, B_j the set of actions a_j of j to which the conjecture φ^i of some other player i assigns positive probability. Let $E_j := [\mathbf{a}_j \in B_j]$ (the event that the action chosen by j is in B_j). Since the game g and the conjectures φ^i are commonly known at s, they are a fortiori mutually known there; so by Lemma 41.6, each action in B_j maximizes g_j against φ^j. Hence $E_j \cap G \cap F \subset R_j$. At each state in F, each player other than j knows that j's action is in B_j; that is, $F \subset \cap_{i \neq j} K_i E_j$. So $G \cap F \subset (\cap_{i \neq j} K_i E_j) \cap (G \cap F)$. So Lemmas 41.2 and 41.5 yield

$$\begin{aligned} &K^2(G \cap F) \subset K^2\left(\cap_{i\neq j} K_i E_j\right) \cap K^2(G \cap F) \subset K^1\left(\cap_{i\neq j} K_i E_j\right) \cap K^2(G \cap F) \subset \\ &K^1\left(\cap_{i\neq j} K_i E_j\right) \cap K^1\left(\cap_{i\neq j} K_i (G \cap F)\right) = K^1\left(\cap_{i\neq j} K_i \left(E_j \cap G \cap F\right)\right) \subset \\ &\qquad \times K^1\left(\cap_{i\neq j} K_i R_j\right). \end{aligned}$$

Hence using $R \subset R_j$ and $R_j = K_j R_j$ (Corollary 41.2), we obtain

$$\begin{aligned} &K^2(G \cap F) \cap K^1 R \subset K^1\left(\cap_{i\neq j} K_i R_j\right) \cap K^1 R \subset K^1\left(\cap_{i\neq j} K_i R_j\right) \cap K^1 R_j = \\ &K^1\left(\cap_{i\neq j} K_i R_j\right) \cap K^1 K_j R_j = K^2 R_j. \end{aligned}$$

Since this holds for all j, Lemma 41.2 yields[38] (A.2). ■

---

[37]This observation, for which we are indebted to Ben Polak, is of particular interest because in many applied contexts there is only one game under consideration, so it is of necessity commonly known.

[38]The proof would be simpler with a formalism in which known events are true ($K_j E \subset E$). See section "Alternative formalisms."

Our fourth remark is that in both Theorems 41.2 and 41.3, mutual knowledge of rationality may be replaced by the assumption that each player knows the others to be rational; in fact, all players may themselves be irrational at the state in question. (Recall that "know" means "ascribe probability 1"; thus a player may be irrational even though another player knows that he is rational.)

We come next to the matter of converses to our theorems. We have already mentioned (at the end of section "Description of the results") that the conditions are not necessary, in the sense that it is quite possible to have a Nash equilibrium even when they are not fulfilled. In Theorem 41.1, the action n-tuple $\mathbf{a}(s)$ at a state s may well be a Nash equilibrium even when $\mathbf{a}(s)$ is not mutually known, whether or not the players are rational. (But if the actions are mutually known at s and $\mathbf{a}(s)$ is a Nash equilibrium, then the players *are* rational at s; cf. Remark 41.2.) In Theorem 41.2, the conjectures at a state s in a two-person game may constitute a Nash equilibrium even when, at s, they are not mutually known and/or rationality is not mutually known. Similarly for Theorem 41.3.

Nevertheless, there is a sense in which the converses hold: Given a Nash equilibrium in a game g, one can construct a belief system in which the conditions are fulfilled. For Theorem 41.1, this is immediate: Choose a belief system where each player i has just one type, whose action is i's component of the equilibrium and whose payoff function is g_i. For Theorems 41.2 and 41.3, we may suppose that as in the traditional interpretation of mixed strategies, each player chooses an action by an independent conscious randomization according to his component σ_i of the given equilibrium σ. The types of each player correspond to the different possible outcomes of the randomization; each type chooses a different action. All types of player i have the same theory, namely, the product of the mixed strategies of the other $n-1$ players appearing in σ, and the same payoff function, namely g_i. It may then be verified that the conditions of Theorems 41.2 and 41.3 are met.

These "converses" show that the sufficient conditions for Nash equilibrium in our theorems are not too strong, in the sense that they do not imply more than Nash equilibrium; every Nash equilibrium is attainable with these conditions. Another sense in which they are not too strong—that the conditions cannot be dispensed with or even appreciably weakened—was discussed in sections "The main counterexamples" and "Additional counterexamples."

References

Armbruster, W., & Boege, W. (1979). Bayesian game theory. In O. Moeschlin & D. Pallaschke (Eds.), *Game theory and related topics*. Amsterdam: North-Holland.

Aumann, R. (1976). Agreeing to disagree. *Annals of Statistics, 4*, 1236–1239.

Aumann, R. (1987). Correlated equilibrium as an expression of Bayesian rationality. *Econometrica, 55*, 1–18.

Boege, W., & Eisele, T. (1979). On solutions of Bayesian games. *International Journal of Game Theory, 8*, 193–215.

Brandenburger, A., & Dekel, E. (1989). The role of common knowledge assumptions in game theory. In F. Hahn (Ed.), *The economics of missing markets, information, and games*. Oxford: Oxford University Press.
Geanakoplos, J., & Polemarchakis, H. (1982). We can't disagree forever. *Journal of Economic Theory, 28*, 192–200.
Harsanyi, J. (1967–1968). Games of incomplete information played by 'Bayesian' players, I-III. *Management Science, 14*, 159–182, 320–334, 486–502.
Harsanyi, J. (1973). Games with randomly disturbed payoffs: A new rationale for mixed strategy equilibrium points. *International Journal of Game Theory, 2*, 1–23.
Kohlberg, E., & Mertens, J.-F. (1986). On the strategic stability of equilibria. *Econometrica, 54*, 1003–1037.
Kreps, D., & Wilson, R. (1982). Sequential equilibria. *Econometrica, 50*, 863–894.
Lewis, D. (1969). *Conventions: A philosophical study*. Cambridge: Harvard University Press.
Maynard Smith, J. (1982). *Evolution and the theory of games*. Cambridge: Cambridge University Press.
Mertens, J.-F., & Zamir, S. (1985). Formulation of Bayesian analysis for games with incomplete information. *International Journal of Game Theory, 14*, 1–29.
Myerson, R. (1978). Refinements of the Nash equilibrium concept. *International Journal of Game Theory, 7*, 73–80.
Nash, J. (1951). Non-cooperative games. *Annals of Mathematics, 54*, 286–295.
Savage, L. (1954). *The foundations of statistics*. New York: Wiley.
Selten, R. (1965). Spieltheoretische Behandlung eines Oligopolmodels mit Nachfragetragheit. *Zietschrift fur die gesante Staatswissenschaft, 121*, 301–324.
Selten, R. (1975). Reexamination of the perfectness concept for equilibrium points in extensive games. *International Journal of Game Theory, 4*, 25–55.
Tan, T., & Werlang, S. (1988). The Bayesian foundations of solution concepts of games. *Journal of Economic Theory, 45*, 370–391.

Chapter 42
Knowledge, Belief and Counterfactual Reasoning in Games

Robert Stalnaker

Introduction

Deliberation about what to do in any context requires reasoning about what will or would happen in various alternative situations, including situations that the agent knows will never in fact be realized. In contexts that involve two or more agents who have to take account of each others' deliberation, the counterfactual reasoning may become quite complex. When I deliberate, I have to consider not only what the causal effects would be of alternative choices that I might make, but also what other agents might believe about the potential effects of my choices, and how their alternative possible actions might affect my beliefs. Counterfactual possibilities are implicit in the models that game theorists and decision theorists have developed – in the alternative branches in the trees that model extensive form games and the different cells of the matrices of strategic form representations – but much of the reasoning about those possibilities remains in the informal commentary on and motivation for the models developed. Puzzlement is sometimes expressed by game theorists about the relevance of what happens in a game 'off the equilibrium path': of what would happen if what is (according to the theory) both true and known by the players to be true were instead false. My aim in this paper is to make some suggestions for clarifying some of the concepts involved in counterfactual reasoning in strategic contexts, both the reasoning of the rational agents being modeled, and the reasoning of the theorist who is doing the modeling, and to bring together some ideas and technical tools developed by philosophers and logicians that I think might be relevant to the analysis of strategic reasoning, and more generally to the conceptual foundations of game theory.

R. Stalnaker (✉)
Department of Linguistics and Philosophy, MIT, Cambridge, MA, USA
e-mail: stal@mit.edu

H. Arló-Costa et al. (eds.), *Readings in Formal Epistemology*, Springer Graduate Texts in Philosophy 1, DOI 10.1007/978-3-319-20451-2_42

There are two different kinds of counterfactual possibilities – causal and epistemic possibilities – that need to be distinguished. They play different but interacting roles in a rational agent's reasoning about what he and others will and should do, and I think equivocation between them is responsible for some of the puzzlement about counterfactual reasoning. In deliberation, I reason both about how the world might have been different if I or others did different things than we are going to do, and also about how my beliefs, or others' beliefs, might change if I or they learned things that we expect not to learn. To take an often cited example from the philosophical literature to illustrate the contrast between these two kinds of counterfactual suppositions, compare: *if Shakespeare didn't write Hamlet, someone else did,* with *if Shakespeare hadn't written Hamlet, someone else would have.*[1] The first expresses a quite reasonable disposition to hold onto the belief that someone wrote Hamlet should one receive the unexpected information that Shakespeare did not; the second expresses a causal belief, a belief about objective dependencies, that would be reasonable only if one held a bizarre theory according to which authors are incidental instruments in the production of works that are destined to be written. The content of what is supposed in the antecedents of these contrasting conditionals is the same, and both suppositions are or may be counterfactual in the sense that the person entertaining them believes with probability one that what is being supposed is false. But it is clear that the way it is being supposed is quite different in the two cases.

This contrast is obviously relevant to strategic reasoning. Beliefs about what it is rational to do depend on causal beliefs, including beliefs about what the causal consequences would be of actions that are alternatives to the one I am going to choose. But what is rational depends on what is believed, and I also reason about the way my beliefs and those of others would change if we received unexpected information. The two kinds of reasoning interact, since one of the causal effects of a possible action open to me might be to give unexpected information to another rational agent.[2]

It is obvious that a possible course of events may be causally impossible even if it is epistemically open, as when you have already committed yourself, but I have not yet learned of your decision. It also may happen that a course of events is causally open even when it is epistemically closed in the sense that someone believes, with probability one, that it will not happen. But can it be true of a causally open course of events that someone not only believes, but also *knows* that it will not occur? This is less clear; it depends on how we understand the concept of knowledge. It does not seem incoherent to suppose that you know that I am rational, even though irrational choices are still causally possible for me. In fact, the concept of rationality seems applicable to actions only when there are options open to an agent. If we are to make sense of assumptions of knowledge and common knowledge of rationality, we need

---

[1]Ernest Adams (1970) first pointed to the contrast illustrated by this pair of conditionals. The particular example is Jonathan Bennett's.

[2]The relation between causal and evidential reasoning is the central concern in the development of causal decision theory. See Gibbard and Harper (1981), Skyrms (1982) and Lewis (1980).

to allow for the possibility that an agent may know what he or another agent is going to do, even when it remains true that the agent could have done otherwise.

To clarify the causal and epistemic concepts that interact in strategic reasoning, it is useful to break them down into their component parts. If, for example, there is a problem about exactly what it means to assume that there is common knowledge of rationality, it ought to be analyzed into problems about exactly what rationality is, or about what knowledge is, or about how common knowledge is defined in terms of knowledge. The framework I will use to represent these concepts is one that is designed to help reveal the compositional structure of such complex concepts: it is a formal semantic or model theoretic framework – specifically, the Kripkean 'possible worlds' framework for theorizing about modal, causal and epistemic concepts. I will start by sketching a simple conception of a model, in the model theorist's sense, of a strategic form game. Second, I will add to the simple conception of a model the resources to account for one kind of counterfactual reasoning, reasoning about belief revision. In these models we can represent concepts of rationality, belief and common belief, and so can define the complex concept of common belief in rationality, and some related complex concepts, in terms of their component parts. The next step is to consider the concept of knowledge, and the relation between knowledge and belief. I will look at some different assumptions about knowledge, and at the consequences of these different assumptions for the concepts of common knowledge and common knowledge of rationality. Then to illustrate the way some of the notions I discuss might be applied to clarify some counterfactual reasoning about games, I will discuss some familiar problems about backward induction arguments, using the model theory to sharpen the assumptions of those arguments, and to state and prove some theorems about the consequences of assumptions about common belief and knowledge.

Model Theory for Games

Before sketching the conception of a model of a game that I will be using, I will set out some assumptions that motivate it, assumptions that i think will be shared by most though not all, game theorists. First, I assume that a game is a *partial* description of a set or sequence of interdependent Bayesian decision problems. The description is partial in that while it specifies all the relevant utilities motivating the agents, it does not give their degrees of belief. Instead, qualitative constraints are put on what the agents are assumed to believe about the actions of other agents; but these constraints will not normally be enough to determine what the agents believe about each other, or to determine what solutions are prescribed to the decision problems. Second, I assume that all of the decision problems in the game are problems of individual decision making. There is no special concept of rationality for decision making in a situation where the outcomes depend on the actions of more than one agent. The acts of other agents are, like chance events, natural disasters and acts of God, just facts about an uncertain world that agents have beliefs and degrees of belief about. The utilities of other agents are relevant to an agent only as information that, together with beliefs about the rationality of those agents, helps to predict their actions. Third, I assume that in cases where degrees of belief are undetermined, or only partially determined, by the description of a decision problem, then no action is prescribed by the theory unless there is an action that would be rational for every system of degrees of belief compatible with what is specified. There are no special rules of rationality telling one what to do in the absence of degrees of belief, except this: decide what you believe, and then maximize expected utility.

A model for a game is intended to represent a completion of the partial specification of the set or sequence of Bayesian decision problems that is given by the definition of the game, as well as a representation of a particular play of the game. The class of all models for a game will include all ways of filling in the relevant details that are compatible with the conditions imposed by the definition of the game. Although a model is intended to represent one particular playing of the game, a single model will contain many possible worlds, since we need a representation, not only of what actually happens in the situation being modeled, but also what might or would happen in alternative situations that are compatible with the capacities and beliefs of one or another of the agents. Along with a set of possible worlds, models will contain various relations and measures on the set that are intended to determine all the facts about the possible worlds that may be relevant to the actions of any of the agents playing the game in a particular concrete context.

The models considered in this paper are models for finite games in normal or strategic form. I assume, as usual, that the game Γ itself consists of a structure $\langle N, \langle C_i, u_i \rangle_{i \in N} \rangle$, where N is a finite set of players, C_i is a finite set of alternative strategies for player i, and u_i is player i′s utility function taking a strategy profile (a specification of a strategy for each player) into a utility value for the outcome that would result from that sequence of strategies. A model for a game will consist of a set of possible worlds (a state space), one of which is designated as the actual world of the model. In each possible world in the model, each player has certain beliefs and partial beliefs, and each player makes a certain strategy choice. The possible worlds themselves are simple, primitive elements of the model; the information about them – what the players believe and do in each possible world – is represented by several functions and relations given by a specification of the particular model. Specifically, a model for a game Γ will consist of a structure $\langle W, a, \langle S_i, R_i, P_i \rangle_{i \in N} \rangle$, where W is a nonempty set (the possible worlds), a is a member of W (the actual world), each S_i is a function taking possible worlds into strategy choices for player i, each R_i is a binary relation on W, and each P_i is an additive measure function on subsets of W.

The R relations represent the qualitative structure of the players' beliefs in the different possible worlds in the following way: the set of possible worlds that are compatible with what player i believes in world w is the set $\{x : wR_ix\}$. It is assumed that the R relations are *serial, transitive,* and *euclidean.*[3] The first assumption is simply the requirement that in any possible world there must be at least one possible world compatible with what any player believes in that world. The other two constraints encode the assumption that players know their own minds: they are necessary and sufficient to ensure that players have introspective access to their beliefs: if they believe something, they believe that they believe it, and if they do not, they believe that they do not.

The S functions encode the facts about what the players do – what strategies they choose – in each possible world. It is assumed that if xR_iy, then $S_i(x) = S_i(y)$. Intuitively, this requirement is the assumption that players know, at the moment of choice, what they are doing – what choice they are making. Like the constraints on the structure of the R relations, this constraint is motivated by the assumption that players have introspective access to their own states of mind.

The measure function P_i, encodes the information about the player's *partial* beliefs in each possible world in the following way: player i's belief function in possible world w is the relativization of P_i to the set $\{x{:}wR_ix\}$. That is, for any proposition ϕ, $P_{i,w}(\phi) = P_i(\phi \cap \{x{:}wR_ix\})/P_i(\{x{:}wR_ix\})$. The assumptions we are making about R_i and P_i will ensure that $P_i(\{x{:}wR_ix\})$ is nonzero for all w, so that this probability will always be defined.

The use of a single measure function for each player, defined on the whole space of possible worlds, to encode the information required to define the player's degrees of belief is just a technical convenience – an economical way to specify the many different belief functions that represent that player's beliefs in different possible worlds. No additional assumptions about the players' beliefs are implicit in this form of representation, since our introspection assumptions already imply that any two different belief states for a single player are disjoint, and any set of probability measures on disjoint sets can be represented by a single measure on the union of all the sets. This single measure will contain some extraneous information that has no representational significance – different total measures will determine the same set of belief functions – but this artifact of the model is harmless.[4]

---

[3]That is, for all players i $(x)(\exists y)xR_iy$, $(x)(y)(z)((xR_iy \ \& \ yR_iz) \rightarrow xR_iz)$, and $(x)(y)(z)((xR_iy \ \& \ xR_iz) \rightarrow yRiz)$.

[4]It has been suggested that there is a substantive, and implausible, assumption built into the way that degrees of belief are modeled: namely, that any two worlds in which a player has the same *full* beliefs he also has the same *partial* beliefs. But this assumption is a tautological consequence of the introspection assumption, which implies that a player fully believes that he himself has the partial beliefs that he in fact has. It does follow from the introspection assumptions that player j cannot be uncertain about player i's partial beliefs while being certain about all of i's full beliefs. But that is just because the totality of i's full beliefs includes his beliefs about his own partial beliefs, and by the introspection assumption, i's beliefs about his own partial beliefs are complete and correct. Nothing, however, prevents there being a model in which there are different worlds in which player i has full beliefs about objective facts that are exactly the same, even though the degrees of belief about such facts are different. This situation will be modeled by disjoint but isomorphic sets of possible worlds. In such a case, another player j might be certain about player i's full beliefs about everything except i's own partial beliefs, while being uncertain about i's partial beliefs.

In order to avoid complications that are not relevant to the conceptual issues I am interested in, I will be assuming throughout this discussion that our models are finite, and that the measure functions all assign nonzero probability to every nonempty subset of possible worlds.

We need to impose one additional constraint on our models, a constraint that is motivated by our concern with counterfactual reasoning. A specification of a game puts constraints on the causal consequences of the actions that may be chosen in the playing of the game, and we want these constraints to be represented in the models. Specifically, in a strategic form game, the assumption is that the strategies are chosen independently, which means that the choices made by one player cannot influence the beliefs or the actions of the other players. One could express the assumption by saying that certain counterfactual statements must be true in the possible worlds in the model: if a player had chosen a different strategy from the one he in fact chose, the other players would still have chosen the same strategies, and would have had the same beliefs, that they in fact had. The constraint we need to add is a closure condition on the set of possible worlds – a requirement that there be enough possible worlds of the right kind to represent these counterfactual possibilities.

For any world w and strategy s for player i, there is a world f(w,s) meeting the following four conditions:

1. for all $j \neq i$, if wR_jx, then $f(w,s)R_jx$.
2. if wR_ix, then $f(w,s)R_if(x,s)$.
3. $S_i(f(w,s)) = s$
4. $P_i(f(w,s)) = P_i(w)$.

Intuitively, f(w,s) represents the counterfactual possible world that, in w, is the world that would have been realized if player i, believing exactly what he believes in w about the other players, had chosen strategy s.

Any of the (finite) models constructed for the arguments given in this paper can be extended to (finite) models satisfying this closure condition. One simply adds, for each $w \in W$ and each strategy profile c, a world corresponding to the pair (w,c), and extending the R's, P's, and S's in a way that conforms to the four conditions.[5]

Because of our concern to represent counterfactual reasoning, it is essential that we allow for the possibility that players have false beliefs in some possible worlds, which means that a world in which they have certain beliefs need not itself be compatible with those beliefs. Because the epistemic structures we have defined allow for false belief, they are more general than the partition structures that will be more familiar to game theorists. An equivalence relation meets the three conditions we have imposed on our R relations, but in addition must be reflexive. To impose this

[5]More precisely, for any given model $M = \langle W, a, \langle S_i, R_i, P_i \rangle_{i \in N} \rangle$, not necessarily meeting the closure condition, define a new model M′ as follows: $W' = W \times C$; $a' = \langle a, S(a) \rangle$; for all $w \in W$ and $c \in C$, $S'(\langle w,c \rangle) = c$; for all $x,y \in W$ and $c,d \in C$, $\langle x,c \rangle R'_i \langle y,d \rangle$ if the following three conditions are met: (i) xR_iy, (ii) $c_i = d_i$, and (iii) for all $j \neq i$, $S_j(y) = d_j$; $P'_i(\langle x,c \rangle) = P_i(x)$. This model will be finite if the original one was, and will satisfy the closure condition.

additional condition would be to assume that all players necessarily have only true beliefs. But even if an agent in fact has only true beliefs, counterfactual reasoning requires an agent to consider possible situations in which some beliefs are false. First, we want to consider belief contravening, or epistemic counterfactuals: how players would revise their beliefs were they to learn they were mistaken. Second, we want to consider deliberation which involves causal counterfactuals: a player considers what the consequences would be of his doing something he is not in fact going to do. In both cases, a player must consider possible situations in which either she or another player has a false belief.

Even though the R relations are not, in general, equivalence relations, there is a relation definable in terms of R that does determine a partition structure: say that two worlds x and y are *subjectively indistinguishable* for player i ($x \approx_i y$) if player i's belief state in x is the same as it is in y. That is, $x \approx_i y$ if and only if $\{z{:}xR_iz\} = \{z{:}yR_iz\}$. Each equivalence class determined by a subjective indistinguishability relation will be divided into two parts: the worlds compatible with what the player believes, and the worlds that are not. In the regular partition models, all worlds are compatible with what the player believes in the world, and the two relations, R_i and $\approx_i$, will coincide.

To represent counterfactual reasoning, we must also allow for possible worlds in which players act irrationally. Even if I am resolved to act rationally, I may consider in deliberation what the consequences would be of acting in ways that are not. And even if I am certain that you will act rationally, I may consider how I would revise my beliefs if I learned that I was wrong about this. Even models satisfying some strong condition, such as common belief or knowledge that everyone is rational, will still be models that contain counterfactual possible worlds in which players have false beliefs, and worlds in which they fail to maximize expected utility.

The aim of this model theory is generality: to make, in the definition of a model, as few substantive assumptions as possible about the epistemic states and behavior of players of a game in order that substantive assumptions can be made explicit as conditions that distinguish some models from others. But of course the definition inevitably includes a range of idealizing and simplifying assumptions, made for a variety of reasons. Let me just mention a few of the assumptions that have been built into the conception of a model, and the reasons for doing so.

First, while we allow for irrational action and false belief, we do assume (as is usual) that players all have coherent beliefs that can be represented by a probability function on some nonempty space of possibilities. So in effect, we make the outrageously unrealistic assumption that players are logically omniscient. This assumption is made only because it is still unclear, either conceptually or technically, how to understand or represent the epistemic situations of agents that are not ideal in this sense. This is a serious problem, but not one I will try to address here.

Second, as I have said, it is assumed that players have introspective access to their beliefs. This assumption could be relaxed by imposing weaker conditions on the R relations, although doing so would raise both technical and conceptual problems. It is not clear how one acts on one's beliefs if one does not have introspective access to them. Some may object to the introspective assumption on the ground that a person may have unconscious or inarticulate beliefs, but the assumption is not incompatible with this: if beliefs can be unconscious, so can beliefs about beliefs. It is not assumed that one knows how to say what one believes.

Third, some have questioned the assumption that players know what they do. This assumption might be relaxed with little effect; what is its motivation? The idea is simply that in a static model for a strategic form game, we are modeling the situation at the moment of choice, and it seems reasonable to assume that at that moment, the agent knows what choice is being made.

Fourth, it is assumed that players know the structure of the game – the options available and the utility values of outcomes for all of the players. This assumption is just a simplifying assumption made to avoid trying to do too much at once. It could easily be relaxed with minimal effect on the structure of the models, and without raising conceptual problems. That is, one could consider models in which different games were being played in different possible worlds, and in which players might be uncertain or mistaken about what the game was.

Finally, as noted we assume that models are finite. This is again just a simplifying assumption. Relaxing it would require some small modifications and add some mathematical complications, but would not change the basic story.

In any possible worlds model, one can identify *propositions* with subsets of the set of possible worlds, with what economists and statisticians call 'events'. The idea is to identify the content of what someone may think or say with, its truth conditions – that is, with the set of possible worlds that would realize the conditions that make what is said or thought true. For any proposition ϕ and player i, we can define the proposition that i fully believes that ϕ as the set $\{x \in W:\{y \in W:xR_iy\} \subseteq \phi\}$, and the proposition that i believes that ϕ to at least degree r as the set $\{x \in W:P_{i,x}(\phi) \geq r\}$. So we have the resources to interpret unlimited iterations of belief in any proposition, and the infinitely iterated concept of *common belief* (all players believe that ϕ, and all believe that all believe that ϕ, and all believe that all believe that all believe that ϕ, and ... etc.) can be defined as the intersection of all the propositions in this infinite conjunction. Equivalently, we can represent common belief in terms of the *transitive closure* R^*, of the set all the R relations. For any proposition ϕ, it is, in possible world x, common belief among the players that ϕ if and only if ϕ is true in all possible worlds compatible with common belief, which is to say if and only if $\{y:xR^*y\} \subseteq \phi$.

If rationality is identified with maximizing expected utility, then we can define, in any model, the propositions that some particular player is rational, that all players are rational, that all players believe that all players are rational, and of course that it is common belief among the players that all players are rational. Here is a sequence of definitions, leading to a specification of the proposition that there is common belief that all players are rational[6]: first, the *expected utility* of an action (a strategy choice) s for a player i in a world x is defined in the familiar way:

[6]In these and other definitions, a variable for a strategy or profile, enclosed in brackets, denotes the proposition that the strategy or profile is realized. So, for example, if $e \in C_{-i}$ (if e is a strategy profile for players other than player i) then $[e] = \{x \in W:S_j(x) = e_j \text{ for all } j \neq i\}$.

$$eu_{i,x}(s) = \sum_{e \in C_{-i}} P_{i,x}([e]) \times u_i((s, e))$$

Second, we define the set of strategies that maximize expected utility for player i in world x:

$$r_{i,x} = \left\{ s \in C_i \ : \ eu_{i,x}(s) \geq eu_{i,x}(s') \text{ for all } s' \in C_i \right\}$$

Third, the proposition *that player i is rational* is the set of possible worlds in which the strategy chosen maximizes expected utility in that world:

$$A_i = \{x \in W \ : \ S_i(x) \in r_{i,x}\}$$

Fourth, the proposition *everyone is rational* is the intersection of the A_i's:

$$A = \cap_{i \in N} A_i$$

Fifth, the proposition *there is common belief that everyone is rational* is defined as follows:

$$Z = \{x \in W \ : \ \{y \in W : xR^*y\} \subseteq A\}.$$

Any specification that determines a proposition relative to a model can also be used to pick out a class of models – all the models in which the proposition is true in that model's actual world. So for any given game, we can pick out the class of models of that game that satisfy some intuitive condition, for example, the class of models in which the proposition Z, that there is common belief in rationality, is true (in the actual world of the model). A class of models defined this way in turn determines a set of strategy profiles for the game: a profile is a member of the set if and only if it is realized in the actual world of one of the models in the class of models. This fact gives us a way that is both precise and intuitively motivated of defining a solution concept for games, or of giving a proof of adequacy for a solution concept already defined. The solution concept that has the most transparent semantic motivation of this kind is rationalizability: we can define rationalizability semantically as the set of strategies of a game that are realized in (the actual world of) some model in which there is common belief in rationality.[7] Or, we can give

[7]This model theoretic definition of rationalizability coincides with the standard concept defined by Bernheim (1984) and Pearce (1984) only in two person games. In the general case, it coincides with the weaker concept, correlated rationalizability. Model theoretic conditions appropriate for the stronger definition would require that players' beliefs about each other satisfy a constraint that (in games with more than two players) goes beyond coherence: specifically, it is required that no player can believe that any information about another player's strategy choices would be evidentially relevant to the choices of a different player. I think this constraint could be motivated, in general, only if one confused causal with evidential reasoning. The structure of the game ensures

a direct nonsemantic definition of the set of strategies – the set of strategies that survive the iterated elimination of strictly dominated strategies – and then prove that this set is *characterized* by the class of models in which there is common belief in rationality: a set of strategies is characterized by a class of models if the set includes exactly the strategies that are realized in some model in the class.[8]

Belief Revision

There are many ways to modify and extend this simple conception of a model of a game. I will consider here just one embellishment, one that is relevant to our concern with counterfactual reasoning. This is the addition of some structure to model the players' policies for revising their beliefs in response to new information. We assume, as is usual, that rational players are disposed to revise their beliefs by conditionalization, but there is nothing in the models we have defined to say how players would revise their beliefs if they learned something that had a prior probability of 0 – something incompatible with the initial state of belief. A belief revision policy is a way of determining the sets of possible worlds that define the posterior belief states that would be induced by such information. The problem is not to generate such belief revision policies out of the models we already have – that is impossible. Rather, it is to say what new structure needs to be added to the model in order to represent belief revision policies, and what formal constraints the policies must obey.

Since we are modeling strategic form games, our models are static, and so there is no representation of any actual change in what is believed. But even in a static situation, one might ask how an agent's beliefs are disposed to change were he to learn that he was mistaken about something he believed with probability one, and the answer to this question may be relevant to his decisions. These dispositions to change beliefs, in contrast to the potential changes that would display the dispositions, are a part of the agent's prior subjective state – the only state represented in the worlds of our models.

I said at the start that one aim in constructing this model theory was to clarify, in isolation, the separate concepts that interact with each other in strategic contexts, and that are the component parts of the complex concepts used to describe those contexts. In keeping with this motivation, I will first look at a pure and simple abstract version of belief revision theory, for a single agent in a single possible

---

that players' strategy choices are made independently: if player one had chosen differently, it could not have influenced the choice of player two. But this assumption of causal independence has no consequences about the evidential relevance of information about player one's choice for the beliefs that a third party might rationally have about player two. (Brian Skyrms (1992, pp. 147–8) makes this point.)

[8]This characterization theorem is proved in Stalnaker (1994).

world, ignoring degrees of belief, and assuming nothing about the subject matter of the beliefs. After getting clear about the basic structure, I will say how to incorporate it into our models, with many agents, many possible worlds, and probability measures on both the prior and posterior belief states. The simple theory that I will sketch is a standard one that has been formulated in a number of essentially equivalent ways by different theorists.[9] Sometimes the theory is formulated syntactically, with prior and posterior belief states represented by sets of sentences of some formal language, but I will focus on a purely model theoretic formulation of the theory in which the agent's belief revision policy is represented by a set of possible worlds – the prior belief state – and a function taking each piece of potential new information into the conditional belief state that corresponds to the state that would be induced by receiving that information. Let B be the set representing the prior state, and let B′ be the set of all the possible worlds that are compatible with any new information that the agent could possibly receive. Then if ϕ is any proposition which is a subset of B′, B(ϕ) will be the set that represents the posterior belief state induced by information ϕ.

There are just four constraints that the standard belief revision theory imposes on this belief revision function:

1. For any ϕ, $B(\phi) \subseteq \phi$
2. If ϕ is nonempty, then B(ϕ) is nonempty
3. If $B \cap \phi$ is nonempty, then $B(\phi) = B \cap \phi$
4. If $B(\phi) \cap \psi$ is nonempty, then $B(\phi \& \psi) = B(\phi) \cap \psi$

The first condition is simply the requirement that the new information received is believed in the conditional state. The second is the requirement that consistent information results in a consistent conditional state. The third condition requires that belief change be conservative in the sense that one should not give up any beliefs unless the new information forces one to give something up: if ϕ is compatible with the prior beliefs, the conditional belief state will simply add ϕ to the prior beliefs. The fourth condition is a generalization of the conservative condition. Its effect is to require that if two pieces of information are received in succession, the second being compatible with the posterior state induced by the first, then the resulting change should be the same as if both pieces of information were received together.

Any belief revision function meeting these four conditions can be represented by an ordering of all the possible worlds, and any ordering of a set of possible worlds will determine a function meeting the four conditions. Let Q be any binary transitive and connected relation on a set B′. Then we can define B as the set of highest ranking members of B′, and for any subset ϕ of B′, we can define B(ϕ) as the set of highest ranking members of ϕ:

---

[9]The earliest formulation, so far as I know, of what has come to be called the AGM belief revision theory was given by William Harper (1975). For a general survey of the belief revision theory, see Gärdenfors (1988). Other important papers include Alchourón and Makinson (1982), Alchourón et al. (1985), Grove (1988), Makinson (1985) and Spohn (1987).

$$B(\phi) = \{x \in \phi \ : \ yQx \ \text{ for all } y \in \phi\}$$

It is easy to show that this function will satisfy the four conditions. On the other hand, given any revision function meeting the four conditions, we can define a binary relation Q in terms of it as follows:

$$xQy \ \text{ if } y \in B(\{x, y\}).$$

It is easy to show, using the four conditions, that Q, defined this way, is transitive and connected, and that $B(\phi) = \{x \in \phi$: yQx for all $y \in \phi\}$. So the specification of such a Q relation is just an alternative formulation of the same revision theory.

Now to incorporate this belief revision theory into our models, we need to give each player such a belief revision policy in each possible world. This will be accomplished if we add to the model a binary relation Q for each player. We need just one such relation for each player, if we take our assumption that players know their own states of mind to apply to belief revision policies as well as to beliefs themselves. Since the belief revision policy is a feature of the agent's subjective state, it is reasonable to assume that in all possible worlds that are subjectively indistinguishable for a player, he has the same belief revision policies.

Subjective indistinguishability (which we defined as follows: $x \approx_i y$ if and only if $\{z{:}xR_iz\} = \{z{:}yR_iz\}$) is an equivalence relation that partitions the space of all possible worlds for each player, and the player's belief revision function will be the same for each world in the equivalence class. (The equivalence class plays the role of B' in the simple belief revision structure.) What we need to add to the game model is a relation Q_i for each player that orders all the worlds within each equivalence class with respect to epistemic plausibility, with worlds compatible with what the player believes in the worlds in that class having maximum plausibility. So Q_i must meet the following three conditions:

(q1) $x \approx_i y$, if and only if xQ_iy or yQ_ix.
(q2) Q_i is transitive.
(q3) xR_iy if and only if wQ_iy for all w such that $w \approx_i x$.

For any proposition ϕ, we can define the conditional belief state for player i in world x, $B_{i,x}(\phi)$ (the posterior belief state that would be induced by learning ϕ),[10] in terms of Q_i as follows:

$$B_{i,x}(\phi) = \{w \in \phi : \text{for all } y \in \phi \cap \{z : z{\approx_i}x\}, yQ_iw\}.$$

[10]There is this difference between the conditional belief state $B_{i,x}(\phi)$ and the posterior belief state that would actually result if the agent were in fact to learn that ϕ: if he were to learn that ϕ, he would believe that he *then* believed that ϕ, whereas in our static models, there is no representation of what the agent comes to believe in the different possible worlds at some later time. But the potential posterior belief states and the conditional belief states as defined do not differ with respect to any information represented in the model. In particular, the conditional and posterior belief states do not differ with respect to the agent's beliefs about his *prior* beliefs.

Once we have added to our models a relation Q_i for each player that meets these three conditions, the R relations become redundant, since they are definable in terms of Q.[11] For a more economical formulation of the theory, we drop the R_i's when we add the Q_i's, taking condition (q1) as above the new definition of subjective indistinguishability, and condition (q3) as the definition of R_i. Formulated this way, the models are now defined as follows:

A model is a structure $<W,a,<S_i,Q_i,P_i>_{i \in N}>$. W, a, P_i, and S_i are as before; Each Q_i is a binary reflexive transitive relation on W meeting in addition the following condition: any two worlds that are Q_i related (in either direction) to a third world are Q_i related (in at least one direction) to each other. One can then prove that each R_i, defined as above, is serial, transitive, and euclidean. So our new models incorporate and refine models of the simpler kind.

To summarize, the new structure we have added to our models expresses exactly the following two assumptions:

1. In each possible world each player has a belief revision policy that conforms to the conditions of the simple AGM belief revision theory sketched above, where (for player i and world x) the set B is $\{y{:}xR_iy\}$, and the set B′ is $\{y{:}\ y \approx_i x\}$
2. In each world, each player has a correct belief about what his own belief revision policy is.

Each player's belief revision structure determines a ranking of all possible worlds with respect to the player's degree of epistemic success or failure in that world. In some worlds, the player has only true beliefs; in others, he makes an error, but not as serious an error as he makes in still other possible worlds. Suppose I am fifth on a standby waiting list for a seat on a plane. I learn that there is only one unclaimed seat, and as a result I feel certain that I will not get on the plane. I believe that the person at the top of the list will certainly take the seat, and if she does not, then I am certain that the second in line will take it, and so on. Now suppose in fact that my beliefs are mistaken: the person at the top of the list turns the seat down, and the next person takes it. Then my initial beliefs were in error, but not as seriously as they would be if I were to get the seat. If number two gets the seat, then I was making a simple first degree error, while if I get the seat, I was making a fourth degree error.

It will be useful to define, recursively, a sequence of propositions that distinguish the possible worlds in which a player's beliefs are in error to different degrees:

---

[11]The work done by Q is to rank the worlds incompatible with prior beliefs; it does not distinguish between worlds compatible with prior beliefs – they are ranked together at the top of the ordering determined by Q. So Q encodes the information about what the prior beliefs are – that is why R becomes redundant. A model with both Q and R relations would specify the prior belief sets in two ways. Condition (q3) is the requirement that the two specifications yield the same results.

Here is a simple abstract example, just to illustrate the structure: suppose there are just three possible worlds, x y and z, that are subjectively indistinguishable in those worlds to player i. Suppose {x} is the set of worlds compatible with i′s beliefs in x, y, and z, which is to say that the R relation is the following set: $\{<x,x>,<y,x>,<z,x>\}$. Suppose further that y has priority over z, which is to say if i were to learn the proposition {y,z}, his posterior or conditional belief state would be {y}. In other words, the Q relation is the following set: $\{<x,x>,<y,x>,<z,x>,<y,y>,<z,y>,<z,z>\}$.

$E^1{}_i$ is the proposition that player i has at least some false belief – makes at least a simple first degree error.

$E_i^1 = \{x \in W$: for some y such that $y \approx_i x$, not $yQ_ix\}(= \{x \in W$: not $xR_ix\})$

$E^{k+1}{}_i$ is the proposition that player i makes at least a $k+1$ degree error:

$$E_i^{k+1} = \left\{x \in E_i^k \ : \text{ for some } y \in E_i^k \text{ such that } y \approx_i x, \text{ not } yQ_ix\right\}.$$

The belief revision structure provides for epistemic distinctions between propositions that are all believed with probability one. Even though each of two propositions has maximum degree of belief, one may be believed *more robustly* than the other in the sense that the agent is more disposed to continue believing it in response to new information. Suppose, to take a fanciful example, there are three presidential candidates, George, a Republican from Texas, Bill, a Democrat from Arkansas, and Ross, an independent from Texas. Suppose an agent believes, with probability one, that George will win. She also believes, with probability one, that a Texan will win and that a major party candidate will win, since these follow, given her other beliefs, from the proposition that George will win. But one of these two weaker beliefs may be more robust than the other. Suppose the agent is disposed, on learning that George lost, to conclude that Bill must then be the winner. In this case, the belief that a major party candidate will win is more robust than the belief that a Texan will win.

The belief revision structure is purely qualitative, but the measure functions that were already a part of the models provide a measure of the partial beliefs for conditional as well as for prior belief states. The Q relations, like the R relations, deliver the sets of possible worlds relative to which degrees of belief are defined. The partial beliefs for conditional belief state, like those for the prior states, are given by relativizing the measure function to the relevant set of possible worlds. Just as player i's partial beliefs in possible world x are given by relativizing the measure to the set $B_{i,x} = \{y{:}xR_iy\}$, so the partial beliefs in the conditional belief state for player i, world x and condition ϕ is given by relativizing the measure to the set $B_{i,x}(\phi) = \{y \in \phi$: for all $z \in \phi$ such that $z \approx_i x$, $zQ_iy)$.

So with the help of the belief revision function we can define *conditional* probability functions for each player in each world:

$$P_{i,x}(\phi/\psi) = P_i(\phi \cap B_{i,x}(\psi))/P_i(B_{i,x}(\psi))$$

In the case where the condition ψ is compatible with i's prior beliefs – where $P_{i,x}(\psi) > 0$ – this will coincide with conditional probability as ordinarily defined. (This is ensured by the conservative condition on the belief revision function.) But this definition extends the conditional probability functions for player x in world i to any condition compatible with the set of worlds that are subjectively indistinguishable for x in i.[12]

[12]These extended probability functions are equivalent to *lexicographic probability systems*. See Blume et al. (1991a, b) for an axiomatic treatment of lexicographic probability in the context of decision theory and game theory. These papers discuss a concept equivalent to the one defined below that I am calling perfect rationality.

The belief revision theory, and the extended probability functions give us the resources to introduce a refinement of the concept of rationality. Say that an action is *perfectly rational* if it not only maximizes expected utility, but also satisfies a tie-breaking procedure that requires that certain *conditional* expected utilities be maximized as well. The idea is that in cases where two or more actions maximize expected utility, the agent should consider, in choosing between them, how he should act if he learned he was in error about something. And if two actions are still tied, the tie-breaking procedure is iterated – the agent considers how he should act if he learned that he were making an error of a higher degree. Here is a sequence of definitions leading to a definition of perfect rationality.

Given the extended conditional probability functions, the definition of conditional expected utility is straightforward:

$$\mathrm{eu}_{i,x}(s/\phi) = \sum_{e \in C_{-i}} P_{i,x}([e]/\phi) \times u_i((s,e))$$

Second, we define, recursively, a sequence of sets of strategies that maximize expected utility, and also satisfy the succession of tie-breaking rules:

$$\begin{aligned} r^0_{i,x} &= r_{i,x} \left(\text{that is, } \{s \in C_i : \mathrm{eu}_{i,x}(s) \geq \mathrm{eu}_{i,x}(s') \text{ for all } s' \in C_i\}\right) \\ r^{k+1}_{i,x} &= \left\{s \in r^k_{i,x} : \mathrm{eu}_{i,x}\left(s/E^{k+1}\right) \geq \mathrm{eu}_{i,x}\left(s'/E^{k+1}\right) \text{ for all } s' \in r^k_{i,x}\right\}. \\ r^+_{i,x} &= \cap r^k_{i,x} \text{ for all k such that } E^k \cap \{y : x \approx_i y\} \text{ is nonempty.} \end{aligned}$$

The set $r^+_{i,x}$ is the set of strategies that are perfectly rational for player i in world x. So the proposition that player i is perfectly rational is defined as follows:

$$A^+_i = \left\{x \in W : S_i(x) \in r^+_{i,x}\right\}.$$

I want to emphasize that this refinement is defined wholly within individual decision theory. The belief revision theory that we have imported into our models is a general, abstract structure, as appropriate for a single agent facing a decision problem to which the actions of other agents are irrelevant as it is for a situation in which there are multiple agents. It is sometimes said that while states with probability 0 are

I don't want to suggest that this is the only way of combining the AGM belief revision structure with probabilities. For a very different kind of theory, see Mongin (1994). In this construction, probabilities are nonadditive, and are used to represent the belief revision structure, rather than to supplement it as in the models I have defined. I don't think the central result in Mongin (1994) (that the same belief revision structure that I am using is in a sense equivalent to a nonadditive, and so non-Bayesian, probability conception of prior belief) conflicts with, or presents a problem for, the way I have defined extended probability functions: the probability numbers just mean different things in the two constructions.

relevant in game theory, they are irrelevant to individual decision making,[13] but I see no reason to make this distinction. There is as much or as little reason to take account, in one's deliberation, of the possibility that nature may surprise one as there is to take account of the possibility that one may be fooled by one's fellow creatures.

Perfect rationality is a concept of individual decision theory, but in the game model context this concept may be used to give a model theoretic definition of a refinement of rationalizability. Say that a strategy of a game Γ is *perfectly rationalizable* if and only if the strategy is played in some model of Γ in which the players have common belief that they all are perfectly rational. As with ordinary correlated rationalizability, one can use a simple algorithm to pick out the relevant class of strategies, and prove a characterization theorem that states that the model theoretic and algorithmic definitions determine the same class of strategies. Here is the theorem:

Strategies that survive the elimination of all weakly dominated strategies followed by the iterated elimination of strictly dominated strategies are all and only those that are realized in a model in which players have common belief that all are perfectly rational.[14]

Before going on to discuss knowledge, let me give two examples of games to illustrate the concepts of perfect rationality and perfect rationalizability.

First, consider the following very simple game: Alice can take a dollar for herself alone, ending the game, or instead leave the decision up to Bert, who can either decide whether the two players get a dollar each, or whether neither gets anything. Figure 42.1 represents the strategic form of this game.

Both strategies for both players are rationalizable, but only Tt is perfectly rationalizable. If Alice is certain that Bert will play t, then either of her strategies would maximize expected utility. But only choice T will ensure that utility is maximized also on the condition that her belief about Bert's choice is mistaken. Similarly, Bert may be certain that Alice won't give him the chance to choose, but if he has to commit himself to a strategy in advance, then if he is perfectly rational, he will opt for the choice that would maximize expected utility if he did get a chance to choose.

---

[13]For example, Fudenberg and Tirole (1992) make the following remark about the relation between game theory and decision theory: 'Games and decisions differ in one key respect: probability-0 events are both exogenous and irrelevant in decision problems, whereas what *would* happen if a player played differently in a game is both important and endogenously determined'.

To the extent that this is true, it seems to me an accident of the way the contrasting theories are formulated, and to have no basis in any difference in the phenomena that the theories are about.

[14]The proof of this theorem, and others stated without proof in this paper, are available from the author. The argument is a variation of the proof of the characterization theorem for simple (correlated) rationalizability given in Stalnaker (1994). See Dekel and Fudenberg (1990) for justification of the same solution concept in terms of different conditions that involve perturbations of the payoffs.

I originally thought that the set of strategies picked out by this concept of perfect rationalizability coincided, in the case of two person games, with perfect rationalizability as defined by Bernheim (1984), but Pierpaolo Battigalli pointed out to me that Bernheim's concept is stronger.

Fig. 42.1

		BERT t	BERT l
ALICE	T	1, 0	1, 0
ALICE	L	1, 1	0, 0

Fig. 42.2

		BERT t	BERT l
ALICE	T	2	2
ALICE	LT	1	3
ALICE	LL	1	0

Second, consider the following pure common interest game, where the only problem is one of coordination. It is also a perfect information game. One might think that coordination is no problem in a perfect information game, but this example shows that this is not necessarily true.

Alice can decide that each player gets two dollars, ending the game, or can leave the decision to Bert, who may decide that each player get one dollar, or may give the decision back to Alice. This time, Alice must decide whether each player gets three dollars, or neither gets anything. Figure 42.2 represents the strategic form of this game.

Now suppose Bert believes, with probability one, that Alice will choose T; what should he do? This depends on what he thinks Alice would do on the hypothesis that his belief about her is mistaken. Suppose that, if he were to be surprised by Alice choosing L on the first move, he would conclude that, contrary to what he previously believed, she is irrational, and is more likely to choose L on her second choice as well. Given these belief revision policies, only choice t is perfectly rational for him. But why should Alice choose T? Suppose she is sure that Bert will choose t, which as we have just seen, is the only perfectly rational choice for him to make if his beliefs about Alice are as we have described. Then Alice's only rational choice is T. So it might be that Alice and Bert both know each others' beliefs about each other, and are both perfectly rational, but they still fail to coordinate on the optimal outcome for both. Of course nothing in the game requires that Bert and Alice should have these beliefs and belief revision policies, but the game is compatible with them, and with the assumption that both Bert and Alice are perfectly rational.

Now one might be inclined to question whether Bert really believes that Alice is fully rational, since he believes she would choose L on her second move, if she got a second move, and this choice, being strictly dominated, would be irrational. Perhaps if Bert believed that Alice was actually disposed to choose L on her second move, then he wouldn't believe she was fully rational, but it is not suggested that he believes this. Suppose we divide Alice's strategy T into two strategies, TT and TL, that differ only in Alice's counterfactual dispositions: the two strategies are 'T, and I *would* choose T again on the second move if I were faced with that choice', and 'T, but I *would* choose L on the second move if I were faced with that choice'. One might argue that only TT, of these two, could be fully rational, but we may suppose that Bert believes, with probability one, that Alice will choose TT, and not TL. But were he to learn that he is wrong – that she did not choose TT (since she did not choose T on the first move) he would conclude that she instead chooses LL. To think there is something incoherent about this combination of beliefs and belief revision policy is to confuse epistemic with causal counterfactuals – it would be like thinking that because I believe that if Shakespeare hadn't written Hamlet, it would have never been written by anyone, I must therefore be disposed to conclude that Hamlet was never written, were I to learn that Shakespeare was in fact not its author.

Knowledge

As has often been noted, rationalizability is a very weak constraint on strategy choice, and perfect rationalizability is only slightly more restrictive. Would it make any difference if we assumed, not just common *belief* in rationality, or perfect rationality, but common *knowledge* as well? Whether it makes a difference, and what difference it makes, will depend on how knowledge is analyzed, and on what is assumed about the relation between knowledge and belief. I will consider a certain analysis of knowledge with roots in the philosophical literature about the definition of knowledge, an analysis that can be made precise with the resources of the belief revision structure that we have built into our models. But before getting to that analysis, I want to make some general remarks about the relation between knowledge and belief.

Whatever the details of one's analysis of knowledge and belief, it is clear that the central difference between the two concepts is that the first, unlike the second, can apply only when the agent is in fact correct in what he believes: the claim that i knows that ϕ, in contrast with the claim that i believes that ϕ, entails that ϕ, is true. Everyone knows that knowledge is different from belief – even from the extreme of belief, probability one – in this way, but sometimes it is suggested that this difference does not matter for the purposes of decision theory, since the rationality of a decision is independent of whether the beliefs on which it is based are in fact correct. It is *expected* utility, not the value of the actual payoff that I receive in the end, that is relevant to the explanation and evaluation of my actions, and expected utility cannot be influenced by facts about the actual world that do not affect my beliefs. But as soon as we start looking at one person's beliefs and knowledge about another's beliefs and knowledge, the difference between the two notions begins to matter. The assumption that Alice believes (with probability one) that Bert believes (with probability one) that the cat ate the canary tells us nothing about what Alice believes about the cat and the canary themselves. But if we assume instead that Alice *knows* that Bert *knows* that the cat ate the canary, it follows, not only that the cat in fact ate the canary, but that Alice knows it, and therefore believes it as well.

Since knowledge and belief have different properties, a concept that conflates them will have properties that are appropriate for neither of the two concepts taken separately. Because belief is a subjective concept, it is reasonable to assume, as we have, that agents have introspective access to what they believe, and to what they do not believe. But if we switch from belief to knowledge, an external condition on the cognitive state is imposed, and because of this the assumption of introspective access is no longer tenable, even for logically omniscient perfect reasoners whose mental states are accessible to them. Suppose Alice believes, with complete conviction and with good reason that the cat ate the canary, but is, through no fault of her own, factually mistaken. She *believes,* let us suppose, that she knows that the cat ate the canary, but her belief that she knows it cannot be correct. Obviously, no amount of introspection into the state of her own mind will reveal to her the fact that she lacks this knowledge. If we conflate knowledge and belief, assuming in general that i knows that ϕ if and only if i's degree of belief for ϕ is one, then we get a concept that combines the introspective properties appropriate only to the internal, subjective concept of belief with the success properties appropriate only to an external concept that makes claims about the objective world. The result is a concept of knowledge that rests on equivocation.

The result of this equivocation is a concept of knowledge with the familiar partition structure, the structure often assumed in discussions by economists and theoretical computer scientists about common knowledge, and this simple and elegant structure has led to many interesting results.[15] But the assumption that knowledge and common knowledge have this structure is the assumption that there can be no such thing as false belief, that while ignorance is possible, error is not. And since there is no false belief, there can be no disagreement, no surprises, and no coherent counterfactual reasoning.[16]

---

[15]Most notably, Robert Aumann's important and influential result on the impossibility of agreeing to disagree, and subsequent variations on it all depend on the partition structure, which requires the identification of knowledge with belief. See Aumann (1976) and Bacharach (1985). The initial result is striking, but perhaps slightly less striking when one recognizes that the assumption that there is no disagreement is implicitly a premise of the argument.

[16]If one were to add to the models we have defined the assumption that the R relation is reflexive, and so (given the other assumptions) is an equivalence relation, the result would be that the three relations, R_i, Q_i, and $\approx_i$, would all collapse into one. There would be no room for belief revision,

It is sometimes suggested that if one were to analyze knowledge simply as true belief, then the result would be a concept of knowledge with this partition structure, but this is not correct. The conjunctive concept, true belief, will *never* determine a partition structure unless it is assumed that it is necessary that *all* beliefs are true, in which case the conjunctive concept would be redundant. For suppose there might be a false belief – that it might be that some person i believed that ϕ, but was mistaken. Then it is false that i truly believes that ϕ, and so if true belief satisfied the conditions of the partition structure, it would follows that i truly believes that he does not truly believe that ϕ, from which (since he believes ϕ) he could infer that ϕ is false. The point is that to assume negative introspection for true belief is to assume that a believer can distinguish, introspecrively, her true beliefs from her false beliefs, which implies (at least if she is consistent) that she won't have any false beliefs.

While it can never be reasonable to equate knowledge and belief in general, we can specify the contingent conditions under which knowledge and belief will coincide. If we assume about a particular situation that *as a matter of fact,* a person has no false beliefs, then (and only then) can we conclude that in that situation, knowledge and belief coincide. To get this conclusion, we need to make no assumptions about knowledge beyond the minimal one that knowledge implies true belief. The assumption we need to make is that full belief is a state that is subjectively indistinguishable from knowledge: that fully believing that ϕ is the same as fully believing that one knows that ϕ.

If we make the idealizing assumption about a particular game situation being modeled that no one has any false beliefs, and that it is common belief that no one has any false beliefs, then we can have the benefits of the identification of knowledge and belief without the pernicious consequences that come from equivocating between them. What we cannot and need not assume is that it is a *necessary* truth – true in all possible worlds in the model – that no one has any false beliefs. Even if players *actually* have only true beliefs, there will inevitably be *counterfactual* possible worlds in the model in which players have false beliefs. These counterfactual possible worlds must be there to represent the causal possibilities that define the structure of the game, and to represent the belief revision policies of the players. If we assumed that it was a necessary truth that there was no false belief, then it would be impossible for one player to believe that a second player was rational in any model for any game in which irrational options are available to the second player.

In terms of this idealizing assumption about knowledge and belief, we can define a refinement of rationalizability, which I have called *strong rationalizability*. Here

since it would be assumed that no one had a belief that could be revised. Intuitively, the assumption would be that it is a necessary truth that all players are Cartesian skeptics: they have no probability-one beliefs about anything except necessary truths and facts about their own states of mind. This assumption is not compatible with belief that another player is rational, unless it is assumed that it is a necessary truth that the player is rational.

is the model theoretic definition: for any game Γ a strategy profile is strongly rationalizable if and only if it is realized in a model in which there is no error, common belief that all players are rational, and common belief that there is no error. The set of strategy profiles characterized by this condition can also be given an algorithmic definition, using an iterated elimination procedure intermediate between the elimination of strictly dominated and of weakly dominated strategies.[17] We can also define a further refinement, *strong perfect rationalizability*: just substitute 'perfect rationality' for 'rationality' in the condition defining strong rationalizability. A minor variation of the algorithm will pick out the set of strategy profiles characterized by these conditions.

Knowledge and belief coincide on this demanding idealization, but suppose we want to consider the more general case in which a person may know some things about the world, even while being mistaken about others. How should knowledge be analyzed? The conception of knowledge that I will propose for consideration is a simple version of what has been called, in the philosophical literature about the analysis of knowledge, the *defeasibility analysis*. The intuitive idea behind this account is that 'if a person has knowledge, then that person's justification must be sufficiently strong that it is not capable of being *defeated* by evidence that he does not possess' (Pappas and Swain 1978). According to this idea, if evidence that is unavailable to you would give you reason to give up a belief that you have, then your belief rests in part on your ignorance of that evidence, and so even if that belief is true, it will not count as knowledge.

We can make this idea precise by exploiting the belief revision structure sketched above, and the notion of robustness that allowed us to make epistemic distinctions between propositions believed with probability one. The analysis is simple: i knows that ϕ if and only if i believes that ϕ (with probability one), *and that belief is robust with respect to the truth*. That is, i knows that ϕ in a possible world x if and only if ϕ receives probability one from i in x, and also receives probability one in every conditional belief state for which the condition is true in x. More precisely, the proposition that i knows that ϕ is the set $\{x \in W\text{: for all } \psi \text{ such that } x \in \psi,\ B_{i,x}(\psi) \subseteq \phi\}$.

Let me illustrate the idea with the example discussed above of the presidential candidates. Recall that there are three candidates, George, Bill and Ross, and that the subject believes, with probability one, that George will win. As a result she also believes with probability one that a Texan will win, and that a major party candidate will win. But the belief that a major party candidate will win is more robust than the belief that a Texan will win, since our subject is disposed, should she learn that George did not win, to infer that the winner was Bill. Now suppose, to everyone's surprise, Ross wins. Then even though our subject's belief that a Texan would win turned out to be true, it does not seem reasonable to say that she *knew* that a Texan would win, since she was right only by luck. Had she known more (that George

[17]The algorithm, which eliminates iteratively profiles rather than strategies, is given in Stalnaker (1994), and it is also proved there that the set of strategies picked out by this algorithm is characterized by the class of models meeting the model theoretic condition.

would lose), then that information would have undercut her belief. On the other hand, if Bill turns out to be the winner, then it would not be unreasonable to say that she knew that a major party candidate would win, since in this case her belief did not depend on her belief that it was George rather than Bill that would win.

The defeasibility conception of knowledge can be given a much simpler definition in terms of the belief revision structure. It can be shown that the definition given above is equivalent to the following: the proposition *i knows that* ϕ is the set $\{x: \{y:xQ_iy\} \subseteq \phi\}$. This exactly parallels the definition of the proposition that i believes that ϕ: $\{x: \{y:xR_iy\} \subseteq \phi\}$. On the defeasibility analysis, the relations that define the belief revision structure are exactly the same as the relations of epistemic accessibility in the standard semantics for epistemic logic.[18] And common knowledge (the infinite conjunction, everyone knows that ϕ, everyone knows that everyone knows that ϕ, ...) exactly parallels common belief: the proposition *there is common knowledge that* ϕ is $\{x:\{y:xQ^*y\} \subseteq \phi\}$, where Q^* is the transitive closure of the Q_i relations.

The defeasibility analysis provides us with two new model theoretic conditions that can be used to define solution concepts: first, the condition that there is common knowledge of rationality; second, the condition that there is common knowledge of perfect rationality. The conditions are stronger (respectively) than the conditions we have used to characterize rationalizability and perfect rationalizability, but weaker than the conditions that characterize the concepts I have called strong rationalizability and strong perfect rationalizability. That is, the class of models in which there is common belief in (perfect) rationality properly includes the class in which there is common knowledge, in the defeasibility sense, of (perfect) rationality, which in turn properly includes the class in which there is no error, common belief that there is no error, and common belief in (perfect) rationality. So the defeasibility analysis gives us two distinctive model theoretic solution concepts, but surprisingly, the sets of strategy profiles characterized by these new model theoretic conditions are the same as those characterized, in one case, by the weaker condition, and in the other case by the stronger condition. That is, the following two claims are theorems:

1. *Any strategy realized in a model in which there is common belief in (simple) rationality is also realized in a model in which there is common knowledge (in the defeasibility sense) of rationality.*
2. *Any strategy profile realized in a model in which there is common knowledge of perfect rationality is also realized in a model meeting in addition the stronger condition that there is common belief that no one has a false belief.*[19]

[18]The modal logic for the knowledge operators in a language that was interpreted relative to this semantic structure would be S4.3. This is the logic characterized by the class of Kripke models in which the accessibility relation is transitive, reflexive, and weakly connected (if xQ_iy and xQ_iz, then either yQ_iz or zQ_iy). The logic of common knowledge would be S4.

[19]Each theorem claims that any strategy that is realized in a model of one kind is also realized in a model that meets more restrictive conditions. In each case the proof is given by showing how to modify a model meeting the weaker conditions so that it also meets the more restrictive conditions.

Backward Induction

To illustrate how some of this apparatus might be deployed to help clarify the role in strategic arguments of assumptions about knowledge, belief and counterfactual reasoning, I will conclude by looking at a puzzle about backward induction reasoning, focusing on one notorious example: the finite iterated prisoners' dilemma. The backward induction argument purports to show that if there is common belief, or perhaps common knowledge, that both players are rational, then both players will defect every time, from the beginning. Obviously rational players will defect on the last move, and since they know this on the next to last move, they will defect then as well, and so on back through the game. This kind of argument is widely thought to be paradoxical, but there is little agreement about what the paradox consists in. Some say that the argument is fallacious, others that it shows an incoherence in the assumption of common knowledge of rationality, and still others that it reveals a self-referential paradox akin to semantic paradoxes such as the liar. The model theoretic apparatus we have been discussing gives us the resources to make precise the theses that alternative versions of the argument purport to prove, and to assess the validity of the arguments. Some versions are clearly fallacious, but others, as I will show, are valid.

The intuitive backward induction argument applies directly to games in extensive form, whereas our game models are models of static strategic form games.[20] But any extensive form game has a unique strategic form, and proofs based on the idea of the intuitive backward induction argument can be used to establish claims about the strategic form of the game. A backward induction argument is best seen as an argument by mathematical induction about a class of games that is closed with respect to the subgame relation – in the case at hand, the class of iterated prisoners' dilemmas of length n for any natural number n.

The conclusions of the backward induction arguments are conditional theses: if certain conditions obtain, then players will choose strategies that result in defection every time. The conditions assumed will correspond to the constraints on models that we have used to characterize various solution concepts, so the theses in question will be claims that only strategy profiles that result in defection every time will satisfy the conditions defining some solution concept. If, for example, the conditions are that there is common belief in rationality, then the thesis would be that only strategies that result in defection every time are rationalizable. It is clear that a backward induction argument for this thesis must be fallacious since many

[20]Although in this paper we have considered only static games, it is a straightforward matter to enrich the models by adding a temporal dimension to the possible worlds, assuming that players have belief states and perform actions at different times, actually revising their beliefs in the course of the playing of the game in accordance with a belief revision policy of the kind we have supposed. Questions about the relationship between the normal and extensive forms of games, and about the relations between different extensive-form games with the same normal form can be made precise in the model theory, and answered.

cooperative strategies are rationalizable. Pettit and Sugden (1989) have given a nice diagnosis of the fallacy in this version of the argument. But what if we make the stronger assumption that there is common *knowledge* of rationality, or of perfect rationality? Suppose, first, that we make the idealizing assumption necessary for identifying knowledge with belief: that there is no error and common belief that there is no error, and common belief that both players are rational. Are all *strongly* rationalizable strategy pairs in the iterated prisoners' dilemma pairs that result in defection every time? In this case the answer is positive, and the theorem that states this conclusion is proved by a backward induction argument.

To prove this backward induction theorem, we must first prove a lemma that is a general claim about multi-stage games – a class of games that includes iterated games. First, some notation and terminology: let Γ be any game that can be represented as a multi-stage game with observed action (a game that can be divided into stages where at each stage all players move simultaneously, and all players know the result of all previous moves). Let $\Gamma^{\#}$ be any subgame – any game that begins at the start of some later stage of Γ. For any strategy profile c of Γ that determines a path through the subgame $\Gamma^{\#}$, let $c^{\#}$ be the profile for $\Gamma^{\#}$ that is determined by c, and let $C^{\#}$ be the set of all strategy profiles of Γ that determine a path through $\Gamma^{\#}$. By 'an SR model', I will mean a model in which there is (in the actual world of the model) no error, common belief that there is no error, and common belief that all players are rational. Now we can state the multi-stage game lemma:

If profile c is strongly rationalizable in Γ, and if c determines a path through $\Gamma^{\#}$, then $c^{\#}$ is strongly rationalizable in $\Gamma^{\#}$.

This is proved by constructing a model for $\Gamma^{\#}$ in terms of a model for Γ, and showing that if the original model is an SR model, so is the new one. Let M be any SR model for Γ in which c is played in the actual world of the model. Let $\Gamma^{\#}$ be any subgame that contains the path determined by c. We define a model $M^{\#}$ for $\Gamma^{\#}$ in terms of M as follows: $W^{\#} = \{x \in W{:}S(x) \in C^{\#}\}$. The $Q_i{}^{\#}$'s and $P_i{}^{\#}$'s are simply the restrictions of the Q_i's and P_i's to $W^{\#}$. The $S_i{}^{\#}$'s are defined so that for each $x \in W^{\#}$, $S^{\#}(x)$ is the profile for the game $\Gamma^{\#}$ that is determined by the profile S(x). (That is, if $S(x) = e$, then $S^{\#}(x) = e^{\#}$.)

To see that $M^{\#}$ is an SR model for $\Gamma^{\#}$, note first that if there is no error and common belief that there is no error in the original model, then this will also hold for the model of the subgame: if $\{x{:}aR^{*}x\} \subseteq \{x{:}xR_ix \text{ for all } i\}$, then $\{x{:}aR^{\#*}x\} \subseteq \{x{:}xR_i{}^{\#}x \text{ for all } i\}$. This is clear, since $\{x{:}aR^{\#*}x\} \subseteq \{x{:}aR^{*}x) \cap W^{\#}$, and $(x{:}xR_i{}^{\#}x \text{ for all } i) = \{x{:}xR_ix \text{ for all } i\} \cap W^{\#}$. Second, because of the fact that players know all previous moves at the beginning of each stage, they can make their strategy choices conditional on whether a subgame is reached. (More precisely, for any player i and pair of strategies s and s′ for i, that are compatible with $\Gamma^{\#}$ being reached, there is a strategy equivalent to this: s if $\Gamma^{\#}$ is reached, s′ if not.) This implies that for any world w, player i and subgame such that it is compatible with i′s beliefs that that subgame be reached, a strategy will be rational for i only if the strategy determined for the subgame is rational, conditional on the hypothesis that the subgame is reached. This ensures that rationality is preserved in all worlds when the model is modified. So $c^{\#}$ is strongly rationalizable in $\Gamma^{\#}$.

An analogous result about strong *perfect* rationalizability can be shown by essentially the same argument.

One further observation before turning to the backward induction theorem itself: for any game Γ, if profile c is compatible with common belief in (the actual world of) an SR model for Γ, then c itself is strongly rationalizable. It is obvious that if $S(x) = c$ and aR^*x, then the same model, with x rather than a as the actual world will be an SR model if the original model was.

Now the backward induction theorem:

Any strongly rationalizable strategy profile in a finite iterated prisoners' dilemma is one in which both players defect every time.

The proof is by induction on the size of the game. For the base case – the one shot PD – it is obvious that the theorem holds, since only defection is rational. Now assume that the theorem holds for games of length k. Let Γ be a game of length $k+1$, and Γ^- be the corresponding iterated PD of length k. Let M be any SR model of Γ, and let c be any strategy profile that is compatible with common belief (that is, c is any profile for which there exists an x such that $S(x) = c$, and aR^*x). By the observation just made, c is strongly rationalizable, so by the multi-stage game lemma, c^- (the profile for Γ^- determined by c) is strongly rationalizable in Γ^-. But then by hypothesis of induction, c^- is a profile in which both players defect every time. So c (in game Γ) is a profile in which both players defect every time after the first move. But c is any profile compatible with common belief in the actual world of the model, so it follows that in the model M, it is common belief that both players will choose strategies that result in defection every time after the first move. Given these beliefs, any strategy for either player that begins with the cooperative move is strictly dominated, relative to that player's beliefs. So since the players are both rational, it follows that they choose a strategy that begins with defection, and so one that results in defection on every move.

Our theorem could obviously be generalized to cover some other games that have been prominent in discussions of backward induction such as the centipede game and (for strong *perfect* rationalizability) the chain store game. But it is not true, even in perfect information games, that the strong or strong and perfect rationalizability conditions are always sufficient to support backward induction reasoning. Recall the perfect information, pure coordination game discussed above in which Alice and Bert failed to coordinate on the backward induction equilibrium, even though the conditions for strong perfect rationalizability were satisfied. In that example, the strategy profile played was a perfect, but not subgame perfect, equilibrium. One can show in general that in perfect information games, all and only Nash equilibrium strategy profiles are strongly rationalizable (see Stalnaker (1994) for the proof).

As I noted at the end of the last section, it can be shown that the set of strongly and perfectly rationalizable strategy profiles is characterized also by the class of models in which there is common knowledge (in the defeasibility sense) of perfect rationality. So we can drop the strong idealizing assumption that there is no error, and still get the conclusion that if there is common knowledge (in the defeasibility sense) of perfect rationality, then players will choose strategies that result in defection every time.

Pettit and Sugden, in their discussion of the paradox of backward induction, grant that the argument is valid when it is common knowledge rather than common belief that is assumed (though they don't say why they think this, or what they are assuming about knowledge). But they suggest that there is nothing surprising or paradoxical about this, since the assumption of common *knowledge* of rationality is incompatible with the possibility of rational deliberation, and so is too strong to be interesting. Since knowledge logically implies truth, they argue, the argument shows that 'as a matter of logical necessity, both players *must* defect and presumably therefore that they know they must defect' (Pettit and Sugden 1989). But I think this remark rests on a confusion of epistemic with causal possibilities. There is no reason why I cannot both *know* that something is true, and also entertain the counterfactual possibility that it is false. It is of course inconsistent to suppose, counterfactually or otherwise, the conjunction of the claim that ϕ is false with the claim that I know that ϕ is true, but it is not inconsistent for me, knowing (in the actual world) that ϕ is true, to suppose, counterfactually, that ϕ is false. As Pettit and Sugden say, the connection between knowledge and truth is a matter of logical necessity, but that does not mean that if I know that I will defect, I therefore *must* defect, 'as a matter of logical necessity'. One might as well argue that lifelong bachelors are powerless to marry, since it is a matter of logical necessity that lifelong bachelors never marry.

The semantic connection between knowledge and truth is not, in any case, what is doing the work in this version of the backward induction argument: it is rather the assumption that the players believe in common that neither of them is in error about anything. We could drop the assumption that the players beliefs are all actually true, assuming not common knowledge of rationality, but only common belief in rationality and common belief that no one is in error about anything. This will suffice to validate the induction argument.

Notice that the common belief that there will not, in fact, be any surprises, does not imply the belief that there couldn't be any surprises. Alice might think as follows: 'Bert expects me to defect, and I will defect, but I could cooperate, and if I did, he would be surprised. Furthermore, I expect him to defect, but he could cooperate, and if he did, I would be surprised'. If these 'could's were epistemic or subjective, expressing uncertainty, then this soliloquy would make no sense, but it is unproblematic if they are counterfactual 'could's used to express Alice's beliefs about her and Bert's capacities. A rational person may know that she will not exercise certain of her options, since she may believe that it is not in her interest to do so.

It is neither legitimate nor required for the success of the backward induction argument to draw conclusions about what the players would believe or do under counterfactual conditions. In fact, consider the following 'tat for tit' strategy: defect on the first move, then on all subsequent moves, do what the other player did on the previous move, until the last move; defect unconditionally on the last move. Our backward induction argument does not exclude the possibility that the players should each adopt, in the actual world, this strategy, since this pair of strategies results in defection every time. This pair is indeed compatible with the conditions for strong and perfect rationalizability. Of course unless each player assigned a very low probability to the hypothesis that this was the other player's strategy, it would not be rational for him to adopt it, but he need not rule it out. Thus Pettit and Sugden are wrong when they say that the backward induction argument can work only if it is assumed that each player would maintain the beliefs necessary for common belief in rationality 'regardless of what the other does' (Pettit and Sugden 1989, p. 178). All that is required is the belief that the beliefs necessary for common knowledge of rationality will, in fact, be maintained, given what the players in fact plan to do. And this requirement need not be *assumed*: it is a consequence of what is assumed.

Conclusion

The aim in constructing this model theory was to get a framework in which to sharpen and clarify the concepts used both by rational agents in their deliberative and strategic reasoning and by theorists in their attempts to describe, predict and explain the behavior of such agents. The intention was, first, to get a framework that is rich in expressive resources, but weak in the claims that are presupposed or implicit in the theory, so that various hypotheses about the epistemic states and behavior of agents can be stated clearly and compared. Second, the intention was to have a framework in which concepts can be analyzed into their basic components, which can then be considered and clarified in isolation before being combined with each other. We want to be able to consider, for example, the logic of belief, individual utility maximization, belief revision, and causal-counterfactual structure separately, and then put them together to see how the separate components interact. The framework is designed to be extended, both by considering further specific substantive assumptions, for example, about the beliefs and belief revision policies of players, and by adding to the descriptive resources of the model theory additional structure that might be relevant to strategic reasoning or its evaluation, for example temporal structure for the representation of dynamic games, and resources for more explicit representation of counterfactual propositions. To illustrate some of the fruits of this approach we have stated some theorems that provide model theoretic characterizations of some solution concepts, and have looked closely at one familiar form of reasoning – backward induction – and at some conditions that are sufficient to validate this form of reasoning in certain games, and at conditions that are not sufficient. The focus has been on the concepts involved in two kinds of counterfactual reasoning whose interaction is essential to deliberation in strategic contexts, and to the evaluation of the decisions that result from such deliberation: reasoning about what the consequences would be of actions that are alternatives to the action chosen, and reasoning about how one would revise one's beliefs if one were to receive information that one expects not to receive. We can get clear about why people do what they do, and about what they ought to do, only by getting clear about the relevance of what they could have done, and might have learned, but did not.[21]

References

Adams, E. (1970). Subjunctive and indicative conditionals. *Foundations of Language, 6*, 89–94.

Alchourón, C., & Makinson, D. (1982). The logic of theory change: Contraction functions and their associated revision functions. *Theoria, 48*, 14–37.

Alchourón, C., Gärdenfors, P., & Makinson, D. (1985). On the logic of theory change: Partial meet functions for contraction and revision. *Journal of Symbolic Logic, 50*, 510–530.

Aumann, R. (1976). Agreeing to disagree. *Annals of Statistics, 4*, 1236–1239.

Bacharach, M. (1985). Some extensions of a claim of Aumann in an axiomatic model of knowledge. *Journal of Economic Theory, 37*, 167–190.

Bernheim, B. (1984). Rationalizable strategic behavior. *Econometrica, 52*, 1007–1028.

Blume, L., Brandenburger, A., & Dekel, E. (1991a). Lexicographic probabilities and choice under uncertainty. *Econometrica, 59*, 61–79.

Blume, L., Brandenburger, A., & Dekel, E. (1991b). Lexicographic probabilities and equilibrium refinements. *Econometrica, 59*, 81–98.

Dekel, E., & Fudenberg, D. (1990). Rational behavior with payoff uncertainty. *Journal of Economic Theory, 52*, 243–267.

Fudenberg, D., & Tirole, J. (1992). *Game theory*. Cambridge, MA: MIT Press.

Gärdenfors, P. (1988). *Knowledge in flux: Modeling the dynamics of epistemic states*. Cambridge, MA: MIT Press.

Gibbard, A., & Harper, W. (1981). Counterfactuals and two kinds of expected utility. In C. Hooker et al. (Eds.), *Foundations and applications of decision theory*. Dordrecht/Boston: Reidel.

Grove, A. (1988). Two modelings for theory change. *Journal of Philosophical Logic, 17*, 157–170.

Harper, W. (1975). Rational belief change, popper functions and counterfactuals. *Synthese, 30*, 221–262.

Lewis, D. (1980). Causal decision theory. *Australasian Journal of Philosophy, 59*, 5–30.

Makinson, D. (1985). How to give it up: A survey of some formal aspects of the logic of theory change. *Synthese, 62*, 347–363.

Mongin, P. (1994). The logic of belief change and nonadditive probability. In D. Prawitz & D. Westerstahl (Eds.), *Logic and philosophy of science in Uppsala*. Dordrecht: Kluwer.

Pappas, G., & Swain, M. (1978). *Essays on knowledge and justification*. Ithaca: Cornell University Press.

Pearce, G. (1984). Rationalizable strategic behavior and the problem of perfection. *Econometrica, 52*, 1029–1050.

Pettit, P., & Sugden, R. (1989). The backward induction paradox. *Journal of Philosophy, 86*, 169–182.

Skyrms, B. (1982). Causal decision theory. *Journal of Philosophy, 79*, 695–711.

Skyrms, B. (1992). *The dynamics of rational deliberation*. Cambridge, MA: Harvard University Press.

Spohn, W. (1987). Ordinal conditional functions: A dynamic theory of epistemic states. In W. Harper & B. Skyrms (Eds.), *Causation in decision, belief change and statistics* (Vol. 2, pp. 105–134). Dordrecht: Reidel.

Stalnaker, R. (1994). On the evaluation of solution concepts. *Theory and Decision, 37*, 49–73.

[21] I would like to thank Pierpaolo Battigalli, Yannis Delmas, Drew Fudenberg, Philippe Mongin, Hyun Song Shin, Brian Skyrms, and an anonymous referee for helpful comments on several earlier versions of this paper.

Chapter 43
Substantive Rationality and Backward Induction

Joseph Y. Halpern

Starting with the work of Bicchieri (1988, 1989), Binmore (1987), and Reny (1992), there has been intense scrutiny of the assumption of common knowledge of rationality, the use of counterfactual reasoning in games, and the role of common knowledge and counterfactuals in the arguments for backward induction in games of perfect information. Startlingly different conclusions were reached by different authors.

These differences were clearly brought out during a 2.5 h round table discussion on "Common knowledge of rationality and the backward induction solution for games of perfect information" at the 1998 TARK (Theoretical Aspects of Rationality and Knowledge) conference. During the discussion, Robert Aumann and Robert Stalnaker stated the following theorems:

Aumann's Theorem (Informal version). *Common knowledge of substantive rationality implies the backwards induction solution in games of perfect information.*

Stalnaker's Theorem (Informal version). *Common knowledge of substantive rationality does not imply the backwards induction solution in games of perfect information.*

The discussion during the round table was lively, but focused on more philosophical, high-level issues. My goal in this short note is to explain the technical differences between the framework of Aumann (1995) and Stalnaker (1996) that lead to the different results, and to show what changes need to be made to Aumann's

---

Supported in part by NSF under grant IRI-96-25901.

J.Y. Halpern (✉)
Computer Science Department, Cornell University, Ithaca, NY 14853, USA
e-mail: halpern@cs.cornell.edu

H. Arló-Costa et al. (eds.), *Readings in Formal Epistemology*, Springer Graduate Texts in Philosophy 1, DOI 10.1007/978-3-319-20451-2_43

framework to get Stalnaker's result.[1] I believe that the points that I make here are well known to some (and certainly were made informally during the discussion). Indeed, many of the key conceptual points I make already appear in Stalnaker's discussion of Aumann's result in Stalnaker (1998, pp. 45–50). However, since Stalnaker uses belief rather than knowledge and must deal with the complications of having probability, it is not so easy to directly compare his results in Stalnaker (1998) with Aumann's. I hope that the simpler model I present here will facilitate a careful comparison of the differences between Aumann's and Stalnaker's results and thus clarify a few issues, putting the debate on a more rational footing.

There are three terms in the theorems that need clarification:

- (common) knowledge
- rationality
- *substantive* rationality

I claim that Stalnaker's result can be obtained using exactly the same definition of (common) knowledge and rationality as the one Aumann (1995) used. The definition of knowledge is the standard one, given in terms of partitions. (I stress this point because Stalnaker (1996) has argued that probability-1 belief is more appropriate than knowledge when considering games.) The definition of rationality is that a player who uses strategy s is rational at vertex v if there is no other strategy that he knows will give him a better payoff, conditional on being at vertex v. Both Aumann and Stalnaker give substantive rationality the same reading: "rationality at all vertices v in the game tree". They further agree that this involves a counterfactual statement: "for all vertices v, if the player were to reach vertex v, then the player would be rational at vertex v". The key difference between Aumann and Stalnaker lies in how they interpret this counterfactual. In the rest of this note, I try to make this difference more precise.

The Details

I start by considering Aumann's model. Fix a game Γ of perfect information for n players. As usual, we think of Γ as a tree. Because Γ is a game of perfect information, the players always know which vertex in the tree describes the current situation in the game. The nonleaf vertices in Γ are partitioned into n sets, $G_1, \ldots, G_n$, one for each player. The vertices in G_i are said to belong to i; these are the ones where player i must move. A *model of* Γ is a tuple $(\Omega, \mathcal{K}_1, \ldots, \mathcal{K}_n, \mathbf{s})$, where Ω is a set of states of the world, $\mathcal{K}_1, \ldots, \mathcal{K}_n$ are partitions, one for each player $i = 1, \ldots, n$ ($\mathcal{K}_i$ is i's information partition), and $\mathbf{s}$ maps each world $\omega \in \Omega$

[1]The model that I use to prove Stalnaker's result is a variant of the model that Stalnaker (1996) used, designed to be as similar as possible to Aumann's model so as to bring out the key differences. This, I believe, is essentially the model that Stalnaker had in mind at the round table.

to a strategy profile $\mathbf{s}(\omega) = (s_1, \ldots, s_n)$; s_i is i's strategy in game Γ at state ω. As usual, a strategy for i in Γ is just a mapping from i's vertices in Γ to actions. I write $\mathbf{s}_i(\omega)$ for s_i.

Let $\mathcal{K}_i(\omega)$ denote the cell in partition $\mathcal{K}_i$ that includes ω. Define the operator K_i on events as usual:

$$K_i(E) = \{\omega : \mathcal{K}_i(\omega) \subseteq E\}.$$

$K_i(E)$ is the event that i knows E. Let $A(E) = K_1(E) \cap \ldots \cap K_n(E)$. $A(E)$ is the event that everyone (*all* the players) know E. Let

$$CK(E) = A(E) \cap A(A(E)) \cap A(A(A(E))) \cap \ldots$$

$CK(E)$ is the event that E is common knowledge.

Aumann and Stalnaker (and everyone else who has written on this subject that I am aware of) assume that the players know their strategies. Formally, that means that if $\omega' \in \mathcal{K}_i(\omega)$, then $\mathbf{s}_i(\omega) = \mathbf{s}_i(\omega')$; that is, i uses the same strategy at all the states in a cell of $\mathcal{K}_i$.

Next we need to define rationality. Note that a strategy profile s and vertex v uniquely determine a path in Γ that would be followed if s were played starting at v. Let $h_i^v(s)$ denote i's payoff if this path is followed. Informally, i is rational at vertex v if there is no strategy that i could have used that i knows would net him a higher payoff than the strategy he actually uses. More precisely, *i is rational at vertex v in ω* if, for all strategies $s^i \neq \mathbf{s}_i(\omega)$, $h_i^v(\mathbf{s}(\omega')) \geq h_i^v(\mathbf{s}_{-i}(\omega'), s^i))$ for some $\omega' \in \mathcal{K}_i(\omega)$. That is, i cannot do better by using s^i than $\mathbf{s}_i(\omega)$ against all the strategy profiles of the other players that he considers possible at ω. This is a weak notion of rationality (which is certainly satisfied by expected utility maximization). By taking such a weak notion, Aumann's Theorem becomes stronger. As will be clear from the example, Stalnaker's Theorem holds even if we strengthen the requirements of rationality (to require strict inequality, for example).

Aumann then defines *substantive rationality* to mean rationality at all vertices in the game tree. That is, i is substantively rational in state ω if i is rational at vertex v in ω for every vertex $v \in G_i$. For future reference, I call Aumann's notion of substantive rationality A-rationality.

Using these definitions, Aumann can and does prove his theorem (using a straightforward backward induction argument).

Stalnaker's definition of substantive rationality is different from Aumann's although, as I indicated above, he is trying to capture the same intuition. His definition tries to enforce the intuition that, for every vertex $v \in G_i$, *if* i were to actually reach v, then what he would do in that case would be rational. The key point is that, according to Stalnaker's definition, in order to evaluate, at state ω, whether i is being rational at vertex v by performing the action dictated by his strategy at ω, we must consider i's beliefs in the state "closest" to ω according to i where v is actually reached.

To formalize this, we must add one more component to Aumann's model: for each player i, we must have a selection function f mapping states and i's vertices to states. An *extended model of* Γ is a tuple $(\Omega, \mathscr{K}_1, \ldots, \mathscr{K}_n, \mathbf{s}, f)$, where $(\Omega, \mathscr{K}_1, \ldots, \mathscr{K}_n, \mathbf{s})$ is a model of Γ, $f : \Omega \times G \rightarrow \Omega$, and $G = \cup_{i=1}^{n} G_i$. Intuitively, if $f(\omega, v) = \omega'$, then state ω' is the state closest to ω where vertex v is reached.[2]

Given this intuition, we may want to assume that f satisfies some constraints such as the following:

F1. v is reached in $f(\omega, v)$ (that is, v is on the path determined by $\mathbf{s}(f(\omega, v))$).
F2. If v is reached in ω, then $f(\omega, v) = \omega$.
F3. $\mathbf{s}(f(\omega, v))$ and $\mathbf{s}(\omega)$ agree on the subtree of Γ below v.

F1 guarantees that v is actually reached in $f(\omega, v)$, while F2 says that if v is actually reached in ω, then ω is the closest state to itself where v is reached. F3 is intended to capture the intuitive meaning of a strategy. If, according to $\mathbf{s}_i(\omega)$, player i performs action a at vertex v, it seems reasonable to expect that at the closest world where v is actually reached, player i does in fact perform a. This follows from F3. However, F3 says more than this. It says that at all the vertices below v, all the players also perform the actions dictated by $\mathbf{s}(\omega)$. This extra requirement arguably makes F3 too strong. However, as I shall show, Stalnaker's Theorem continues to hold even with this strong assumption.[3]

According to Stalnaker, i is substantively rational in state ω if i is rational at vertex v in $f(\omega, v)$ for every vertex $v \in G_i$. Let us call this notion *S-rationality*. Thus, the crucial (and, in fact, only) difference between Aumann's approach and Stalnaker's approach is that A-rationality requires i to be rational at vertex v in ω and S-rationality requires i to be rational at vertex v in $f(\omega, v)$. Notice that, by F3, if it is i's move at vertex v, then i makes the same move at both ω and $f(\omega, v)$. However, this move may be rational at $f(\omega, v)$ and irrational at ω, since, in general, i's beliefs are different at ω and $f(\omega, v)$.

The difference between A-rationality and S-rationality can perhaps be best understood by considering the game described in Fig. 43.1, which is also considered by Stalnaker (1998) and is a variant of a game introduced by Aumann (1995).

Consider the following five strategy profiles:

- s^1 is the strategy profile (da, d) (i.e., Ann goes down at v_1 and across at v_3 and Bob goes down at v_2);
- s^2 is the strategy profile (aa, d);

[2] Again, I should stress that this is not exactly the model that Stalnaker uses in (1996), but it suffices for my purposes. I remark that in Halpern (1999), I use selection functions indexed by the players, so that player 1 may have a different selection function than player 2. I do not need this greater generality here, so I consider the simpler model where all players use the same selection function.

[3] There are certainly other reasonable properties we could require of the selection function. For example, we might want to require that if v is reached in some state in $\mathscr{K}_i(\omega)$, then $f(\omega, v) \in \mathscr{K}_i(\omega)$. I believe that it is worth trying to characterize the properties we expect the selection function should have, but this issue would take us too far afield here. (See Stalnaker RC, 1999, Counterfactual propositions in games, Unpublished manuscript, for further discussion of this point.) Note that F1–F3 are properties that seem reasonable for arbitrary games, not just games of perfect information.

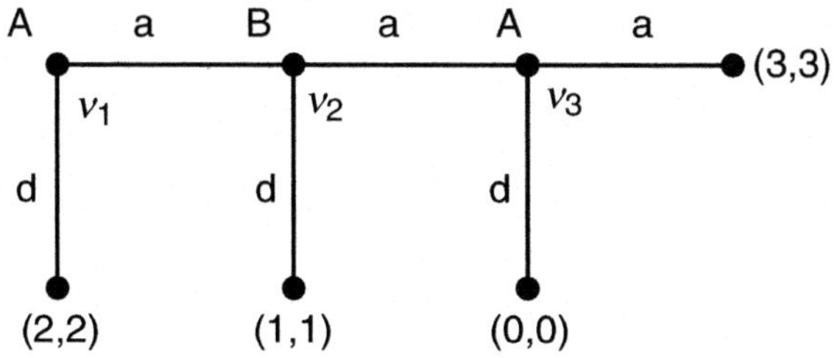

Fig. 43.1 A variant of Aumann's Ann-Bob game

- s^3 is the strategy profile (ad, d);
- s^4 is the strategy profile (aa, a);
- s^5 is the strategy profile (ad, a).

Note that s^4 is the backward induction solution.

Now consider the extended model $(\omega, \mathcal{K}_{Ann}, K_{Bob}, \mathbf{s}, f)$ of this game, where

- $\Omega = \{\omega_1, \omega_2, \omega_3, \omega_4, \omega_5\}$;
- $\mathcal{K}_{Ann}(\omega_i) = \{\omega_i\}$, for $i = 1, \ldots, 5$;
- $\mathcal{K}_{Bob}(\omega_i) = \{\omega_i\}$ for $i = 1, 4, 5$; $\mathcal{K}_{Bob}(\omega_2) = \mathcal{K}_{Bob}(\omega_3) = \{\omega_2, \omega_3\}$;
- $\mathbf{s}(\omega_i) = s^i$, for $i = 1, \ldots, 5$;
- f is the unique selection function satisfying F1–F3 (so that $f(\omega_i, v_1) = \omega_i$ for all i; $f(\omega_1, v_2) = \omega_2$ and $f(\omega_i, v_2) = \omega_i$ for $i = 2, \ldots, 5$; $f(\omega_1, v_3) = f(\omega_2, v_2) = \omega_4$, $f(\omega_3, v_3) = \omega_5$, and $f(\omega_i, v_3) = \omega_i$ for $i = 4, 5$).

It is easy to check that, at ω_1, it is common knowledge that strategy profile s^1 is being used. It is also common knowledge at ω_1 that, if vertex v_2 were reached, Bob would play down.

In this extended model, clearly Bob is not rational at vertex v_2 in ω_1, since he plays down. This means that we do not have A-rationality at ω_1 (and, a fortiori, we do not have common knowledge of A-rationality at ω_1). On the other hand, Bob is rational at vertex v_2 in ω_2, since Bob considers it possible that Ann may go down at v_3 (since $\mathcal{K}_{Bob}(\omega_2) = \{\omega_2, \omega_3\}$). Similarly, Ann is rational at vertex v_3 in ω_4. Since $f(\omega_1, v_2) = \omega_2$ and $f(\omega_1, v_3) = \omega_4$, it follows that we have S-rationality at ω_1, and hence common knowledge of S-rationality at ω_1.

This example is an instance of Stalnaker's Theorem: we have common knowledge of substantive rationality in the sense of S-rationality at ω_1, yet the backward induction solution is not played at ω_1. Nevertheless, it does not contradict Aumann's Theorem, since we do not have common knowledge of A-rationality.

With this machinery, we can now state Aumann's Theorem and Stalnaker's Theorem more formally. Let *S-RAT* consist of all states where all the players are S-rational; let *A-RAT* consist of all states where all the players are A-rational; let *BI* consist of all states where the backward induction solution is played.

Aumann's Theorem. *If Γ is a nondegenerate*[4] *game of perfect information, then in all models of Γ, we have $CK(A\text{-}RAT) \subseteq BI$.*

Stalnaker's Theorem. *There exists a nondegenerate game Γ of perfect information and an extended model of Γ in which the selection function satisfies F1–F3 such that $CK(S\text{-}RAT) \not\subseteq BI$.*

Note that, in an extended model of the Ann-Bob game, it is consistent for Ann to say "Although it is common knowledge that I would play across if v_3 were reached, if I were to play across at v_1, Bob would consider it possible that I would play down at v_3." This is not possible in Aumann's framework because, without selection functions, Aumann has no way of allowing the players to revise their beliefs. (This point is essentially made by Samet (1996).) In the definition of A-rationality, for any vertex v, player i's beliefs in state ω about the possible strategies player j may be using if vertex v is reached are the same (and are determined by $\mathcal{K}_i(\omega)$). It is crucial for Aumann's result (and, I believe, a weakness in his model) that players do not (and cannot) revise their beliefs about other players' strategies when doing such hypothetical reasoning.

It is not hard to place a condition on selection functions that guarantees that players' beliefs about other players' strategies do not change when they are doing hypothetical reasoning.

F4. For all players i and vertices v, if $\omega' \in \mathcal{K}_i(f(\omega, v))$ then there exists a state $\omega'' \in \mathcal{K}_i(\omega)$ such that $\mathbf{s}(\omega')$ and $\mathbf{s}(\omega'')$ agree on the subtree of Γ below v.[5]

Combined with F1–F3, F4 this gives us the properties we want. In fact, we have the following result, similar in spirit to a result proved by Stalnaker (1998, p. 43).

Theorem 1. *If Γ is a nondegenerate game of perfect information, then for every extended model of Γ in which the selection function satisfies F1–F4, we have $CK(S\text{-}RAT) \subseteq BI$. Moreover, there is an extended model of Γ in which the selection function satisfies F1–F4.*

Proof. For the first half of the theorem, suppose $(\Omega, \mathcal{K}_1, \ldots, \mathcal{K}_n, \mathbf{s}, f)$ is an extended model of Γ in which the selection function satisfies F1–F4. Further suppose that $\omega \in CK(S\text{-}RAT)$. We want to show $\omega \in BI$. The proof basically mimics Aumann's proof of his theorem, since F1–F4 essentially gives us his framework.

I first recall a standard characterization of common knowledge. Define the notion of ω'' *being reachable from* ω' *in* k *steps* inductively: ω'' *is reachable from* ω' *in 1 step* iff $\omega'' \in \mathcal{K}_i(\omega')$ for some $i \in \{1, \ldots, n\}$; ω'' *is reachable from* ω' *in* $k + 1$

[4] A game is nondegenerate if the payoffs are different at all the leaves.

[5] Actually, F4 says that player i considers at least as many strategies possible at ω as at $f(\omega, v)$. To capture the fact that player i's beliefs about other players' possible strategies do not change, we would need the opposite direction of F4 as well: if $\omega' \in \mathcal{K}_i(\omega)$ then there exists a state $\omega'' \in \mathcal{K}_i(f(\omega, v))$ such that $\mathbf{s}(\omega')$ and $\mathbf{s}(\omega'')$ agree on the subtree of Γ below v. I do not impose this requirement here simply because it turns out to be unnecessary for Aumann's Theorem.

steps iff there exists a state ω''' that is reachable from ω' in 1 step such that ω'' is reachable from ω''' in k steps. We say that ω'' *is reachable from* ω' if ω'' is reachable from ω' in k steps for some k. It is well known (Aumann 1976) that $\omega' \in CK(E)$ iff $\omega'' \in E$ for all ω'' reachable from ω'.

I show by induction on k that for all states ω' reachable from ω, if v is a vertex which is at height k in the game tree (i.e., k moves away from a leaf), the move dictated by the backward induction solution (for the subgame of Γ rooted at v) is played at v in state ω'.

For the base case, suppose v is at height 1 and ω' is reachable from ω. Since $\omega \in CK(S\text{-}RAT)$, we must have $\omega' \in S\text{-}RAT$. Suppose player i moves at ω'. Since $\omega' \in S\text{-}RAT$, player i must make the move dictated by the backwards induction solution at $f(\omega', v)$. By F3, he must do so at ω' as well.

For the inductive step, suppose that v is at height $k+1$, player i moves at v, and ω' is reachable from ω. Suppose, by way of contradiction, that a is the action indicated by the backward induction solution at v but $\mathbf{s}_i(\omega')(v) = a' \neq a$. Note that by the induction hypothesis, at every vertex below v, all the players play according to the backward induction solution in state ω'. Since $\omega' \in S\text{-}RAT$, we must have that i is rational at v in $f(\omega', v)$. By F3, it follows that i plays a' at vertex v in $f(\omega', v)$ and at every vertex below v, the players play according to the backward induction solution. Thus, there must be a state $\omega'' \in \mathcal{K}_i(f(\omega', v))$ such that by using $\mathbf{s}_i(f(\omega', v))$, player i does at least as well in ω'' as by using the backward induction strategy starting from v. By F4, there must exist some $\omega''' \in \mathcal{K}_i(\omega')$ such that $\mathbf{s}(\omega'')$ and $\mathbf{s}(\omega''')$ agree on the subtree of Γ below v. Since ω''' is reachable from ω, by the induction hypothesis, all players use the backward induction solution at vertices below v. By F3, this is true at ω'' as well. However, this means that player i does better at ω'' playing a at v than a', giving us the desired contradiction.

For the second half, given a nondegenerate game Γ of perfect information, let s be the strategy where, at each vertex v, the players play the move dictated the backward induction solution in the game defined by the subtree below v. For each vertex v, let s_v be the strategy where the players play the actions required to reach vertex v, and then below v, they play according to s. Note that if v is reached by s, then $s_v = s$. In particular, if r is the root of the tree, then $s_r = s$. Consider the extended model $(\Omega, \mathcal{K}_1, \ldots, \mathcal{K}_n, \mathbf{s}, f)$ where $\Omega = \{\omega_v : v \text{ is a vertex of } \Gamma\}$, $\mathcal{K}_i(\omega_v) = \{\omega_v\}$, $\mathbf{s}(\omega_v) = s_v$, and $f(\omega, v)$ is ω if v is reached by $\mathbf{s}(\omega)$ and ω_v otherwise. I leave it to check that this gives us an extended model where the selection function satisfies F1–4. ■

Remarks. 1. As mentioned earlier, Theorem 1 is similar in spirit to the theorem proved by Stalnaker (1998, p. 43). To distinguish this latter theorem from what I have been calling "Stalnaker's Theorem" in this paper, I call the latter theorem "Stalnaker's probabilistic theorem", since it applies to his formal model with probabilities. To state Stalnaker's probabilistic theorem precisely, I need to explain some differences between my model and his. First, as I mentioned earlier, Stalnaker considers belief rather than knowledge, where belief is essentially "with probability 1". He defines a probabilistic analogue of extended models and a notion of *perfect rationality* in such models which is an analogue of substantive rationality. In addition, Stalnaker considers an epistemic independence condition on the players' beliefs. This condition says that a player treats information about disjoint sets of players as epistemically independent. This means, in particular, that if v was the result of player i's move, then a player's belief about the rationality of players other than i will be the same in ω and $f(\omega, v)$.

F4 plays essentially the same role as this independence condition, although formally it is somewhat different. F4 says that the set of strategies a player considers possible at $f(\omega, v)$ is a subset of that he considers possible in ω. The stronger version (discussed in Footnote 5) says that he considers possible the same set of strategies in ω and $f(\omega, v)$. The stronger version can be viewed as saying that a player's beliefs about the strategies that will be used below v (including his own) is epistemically independent of the move made.

In any case, with this background, I can state Stalnaker's probabilistic theorem. It also gives a condition sufficient to ensure that the backward induction solution is played. It says that if Γ is a nondegenerate game of perfect information in which each player moves at most once and M is an extended model (in Stalnaker's sense) of Γ where it is common belief that all players are perfectly rational and players treat information about disjoint sets of players as epistemically independent, then the backwards induction solution is played in M.

The restriction in Stalnaker's Theorem to games where each player moves at most once is not that serious. Given an arbitrary game Γ, we can consider the *agent-normal form* Γ' of Γ (Selten 1975), by associating a different "self" of player i with each vertex of Γ where player i moves. We view these different selves as different agents. Given an extended model M of Γ, let M' be the corresponding extended model of Γ' where, to each world ω in M, there is a corresponding world ω' in M' such that each "self" of player i is endowed with the same beliefs at world ω' as player i has at ω. M' satisfies Stalnaker's epistemic independence assumption; since in Γ' each player makes at most one move, Stalnaker's probabilistic theorem applies. Note that the epistemic independence assumption now says player i's early moves do not affect other players' beliefs about whether player i will player rationally at later in the game (since the later vertices are under the control of different agents—different selves of agent i). Thus, it should be clear that epistemic independence plays essentially the same role in Stalnaker's result as F4 does in Theorem 1.

2. Theorem 1 and the earlier discussion suggests that one possible culprit for the confusion in the literature regarding what is required to force the backwards induction solution in games of perfect information is the notion of a *strategy*. Exactly what should it mean to say that Ann's strategy at a state ω is s? For example, consider the game in Fig. 43.1. According to strategy s_A^1, Ann plays across at vertex v_3. But v_3 is a vertex that cannot be reached if Ann uses s_A^1, since according to this strategy, Ann plays down at v_1. The standard reading of $s_A^1(v_3) = a$ is that "*if* v_3 is reached, then Ann plays across". But this reading leaves a number of questions unanswered. How Ann plays (if she is rational) depends on her beliefs. Should we read this as "no matter what Ann's beliefs are,

if v_3 is reached, Ann will play a"? Or perhaps it should be "given her current beliefs (regarding, for example, what move Bob will make), if v_3 is reached, Ann will play a". Or perhaps it should be "in the state 'closest' to the current state where v_3 is actually reached, Ann plays a".

I have taken the last reading here (where 'closest' is defined by the selection function); assumption F3 essentially forces it to be equivalent to the second reading.

However, without F4, this equivalence is not maintained with regard to Bob's beliefs. That is, consider the following two statements:

- Bob currently believes that, given Ann's current beliefs, Ann will play a if v_3 is reached;
- in the state closest to the current state where v_3 is reached, Bob believes that Ann plays a at v_3.

The first statement considers Bob's beliefs at the current state; the second considers Bob's beliefs at a different state. Without F4, these beliefs might be quite different. It is this possible difference that leads to Stalnaker's Theorem.

3. Strategies themselves clearly involve counterfactual reasoning. If we take strategies as primitive objects (as both Aumann and Stalnaker do, and as I have done for consistency), we have two sources of counterfactuals in extended models: selection functions and strategies. Stalnaker (1996, p. 135) has argued that "To clarify the causal and epistemic concepts that interact in strategic reasoning, it is useful to break them down into their component parts." This suggests that it would be useful to have a model where strategy is *not* a primitive, but rather is defined in terms of counterfactuals. This is precisely what Samet (1996) does.[6]

   Not surprisingly, in Samet's framework, Aumann's Theorem does not hold without further assumptions. Samet shows that what he calls a *common hypothesis* of rationality implies the backward induction solution in nondegenerate games of perfect information. Although there are a number of technical differences in the setup, this result is very much in the spirit of Theorem 1.

Acknowledgements I'd like to thank Robert Stalnaker for his many useful comments and criticisms of this paper.

References

Aumann, R. J. (1976). Agreeing to disagree. *Annals of Statistics, 4*(6), 1236–1239.

Aumann, R. J. (1995). Backwards induction and common knowledge of rationality. *Games and Economic Behavior, 8*, 6–19.

[6]Samet does not use selection functions to capture counterfactual reasoning, but *hypothesis transformations*, which map cells (in the information partition) to cells. However, as I have shown (Halpern 1999), we can capture what Samet is trying to do by using selection functions.

Bicchieri, C. (1988). Strategic behavior and counterfactuals. *Synthese, 76*, 135–169.

Bicchieri, C. (1989). Self refuting theories of strategic interaction: A paradox of common knowledge. *Erkenntnis, 30*, 69–85.

Binmore, K. (1987). Modeling rational players I. *Economics and Philosophy, 3*, 179–214. Part II appeared ibid., *4*, 9–55.

Halpern, J. Y. (1999). Hypothetical knowledge and counterfactual reasoning. *International Journal of Game Theory, 28*(3), 315–330.

Reny, P. (1992). Rationality in extensive form games. *Journal of Economic Perspectives, 6*, 103–118.

Samet, D. (1996). Hypothetical knowledge and games with perfect information. *Games and Economic Behavior, 17*, 230–251.

Selten, R. (1975). Reexamination of the perfectness concept for equilibrium points in extensive games. *International Journal of Game Theory, 4*, 25–55.

Stalnaker, R. (1998). Belief revision in games: Forward and backward induction. *Mathematical Social Sciences, 36*, 31–56.

Stalnaker, R. C. (1996). Knowledge, belief and counterfactual reasoning in games. *Economics and Philosophy, 12*, 133–163.

Index

H. Arló-Costa et al. (eds.), *Readings in Formal Epistemology*, Springer Graduate Texts in Philosophy 1, DOI 10.1007/978-3-319-20451-2

Printed by Printforce, the Netherlands